CHILTON'S TRUCK & VAN REPAIR MANUAL 1971-1978

Sr. Vice President	Ronald A. Hoxter
Publisher and Editor-In-Chief	Kerry A. Freeman, S.A.E.
Managing Editors	Peter M. Conti, Jr. □ W. Calvin Settle, Jr., S.A.E.
Assistant Managing Editor	Nick D'Andrea
Senior Editors	Richard J. Rivele, S.A.E. □ Ron Webb
Director of Manufacturing	Mike D'Imperio
Manager of Manufacturing	John F. Butler

CHILTON BOOK COMPANY

ONE OF THE *DIVERSIFIED PUBLISHING COMPANIES,*
A PART OF *CAPITAL CITIES/ABC, INC.*

Manufac‌
© 1971
Chilton
ISBN 0-

13 14 1

D1295627

CONTENTS

TRUCK SERVICE SECTIONS

GENERAL REPAIR SECTIONS

TRANSMISSION SECTIONS

Index

TUNE-UP SPECIFICATIONS

CU. IN. DISPLACE-MENT (cu. in.)	YEAR	SPARK PLUG GAP (in.)	DISTRIBUTOR POINT DWELL (deg.)	POINT GAP (in.)	IGNITION TIMING (DEGREES)	CRANKCASE COMP. PRESSURE	VALVE CLEARANCE Int. Exh.	NO LOAD GOV. RPM (rpm)	PUMP FUEL PRESS (psi)	IDLE SPEED* (rpm) STD.	AUTO.
				SIX	**CYLINDER**						
250	1971	0.035	31-34	0.019	4B	130	Hyd.		3.5-4.5	550	500
	1972	0.035	31-34	0.019	4B②	130	Hyd.		3.5-4.5	700	600
	1973	0.035	31-34	0.019	6B③	130	Hyd.		3.5-4.5	700	600④
	1974	0.035	31-34	0.019	8B⑩	130	Hyd.		3.5-4.5	850⑤	600
	1975-78	0.035	Elec.	Elec.	10B	130	Hyd.		3.5-4.5	900N	550DR
292	1971	0.035	31-34	0.019	4B	130	Hyd.		3.5-4.5	550	500
	1972	0.035	31-34	0.019	4B	130	Hyd.		3.5-4.5	700	700
	1973	0.035	31-34	0.019	4B⑧	130	Hyd.		3.5-4.5	700⑨	700⑨
	1974	0.035	31-34	0.019	8B	130	Hyd.		3.5-4.5	700	700
	1975-78	0.035	Elec.	Elec.	8B	130	Hyd.		3.5-4.5	600	600
305C (V6)	1973	0.040	31-34	0.019	7-½B	125	0.012-0.018	3600	5.0-7.0	550	550
	1974	0.035	31-34	0.019	7-½B	125	0.012-0.018	3600	5.0-7.0	600	600
379 (V6)	1973	0.035	31-34	0.019	8B	125	0.012-0.018	4000	5.0-7.0	550	550
	1974	0.035	31-34	0.019	6B	125	0.012-0.018	4000	5.0-7.0	575	575
432 (V6)	1973	0.035	31-34	0.019	8B	125					
	1974	0.035	31-34	0.019	6B	125	0.012-0.018	3200	5.0-7.0	525	525
				EIGHT	**CYLINDER**						
307	1971 (200hp)	0.035	28-32	0.019	4B⑫	150	Hyd.		5.0-6.5	600	550
	1971 (215hp)	0.035	28-32	0.019	4B	150	Hyd.		5.0-6.5	550	500
	1972	0.035	28-32	0.019	4B⑬	150	Hyd.		5.0-6.5	900⑭	600
	1973	0.035	29-31	0.019	4B⑮	150	Hyd.		5.0-6.5	900⑯	600
350	1971	0.035	28-32	0.019	4B⑳	150	Hyd.		5.0-6.5	600	550
	1972	0.035	29-31	0.019	4B⑳	150	Hyd.		5.0-6.5	750	600
	1973	0.035	29-31	0.019	㉑	150	Hyd.		5.0-6.5	900㉒	600㉒
	1974	0.035	29-31	0.019	㉓	150	Hyd.		5.0-6.5	900㉔	600㉔
	1975 (2bbl)	0.060	Elec.	Elec.	6B	150	Hyd.		5.0-6.5	——	600
	1975	0.060	Elec.	Elec.	6B	150	Hyd.		5.0-6.5	700	600
	1975-78 (Calif.)	0.060	Elec.	Elec.	2B	150	Hyd.		5.0-6.5	700	600
366	1971	0.035	28-32	0.019	8B	150	Hyd.		5.0-6.5	500	500
	1972	0.035	28-32	0.019	8B	150	Hyd.		5.0-6.5	550	550
	1973	0.035	28-32	0.019	8B	150	Hyd.		5.0-6.5	550㉓	550㉓
	1974	0.035	28-32	0.019	8B	150	Hyd.		5.0-6.5	600	600
	1975 (Fed.)	0.035	28-32	0.019	8B	150	Hyd.		5.0-6.5	700	700
	1975 (w/AT-475)	0.035	28-32	0.019	8B	150	Hyd.	3600	5.0-6.5	700	700
	1975-76 (Calif.)	0.060	Elec.	Elec.	8B	150	Hyd.		5.0-6.5	700	700
400	1975-78 (fed.)	0.045	Elec.	Elec.	4B	150	Hyd.		5.0-6.5	700	700
	1975-78 (Calif.)	0.045	Elec.	Elec.	2B	150	Hyd.		5.0-6.5	700	700
402	1971	0.035	28-32	0.019	8B	150	Hyd.		7.0-8.5	600	600
	1972	0.035	28-32	0.019	8B	150	Hyd.		7.0-8.5	750	600
427	1971	0.035	28-32	0.019	8B	150	Hyd.		7.0-8.5	500	500
	1972	0.035	28-32	0.019	8B	150	Hyd.		7.0-8.5	550	550
	1973	0.035	28-32	0.019	8B	150	Hyd.		7.0-8.5	550㉓	550㉓
	1974	0.035	28-32	0.019	8B	150	Hyd.		7.0-8.5	600	600
	1975-76 (Calif.)	0.060	Elec.	Elec.	8B	150	Hyd.		7.0-8.5	700	700
	1975-76 (Fed.)	0.035	28-32	0.019	8B	150	Hyd.		7.0-8.5	700	700

TUNE-UP SPECIFICATIONS

CU. IN. DISPLACE-MENT (cu. in.)	YEAR	SPARK PLUG GAP (in.)	DISTRIBUTOR POINT DWELL (deg.)	POINT GAP (in.)	IGNITION TIMING (DEGREES)	CRANKCASE COMP. PRESSURE	VALVE CLEARANCE Int. Exh.	NO LOAD GOV. RPM (rpm)	PUMP FUEL PRESS (psi)	IDLE SPEED* (rpm) STD.	AUTO.
EIGHT CYLINDER											
454	1973 (Fed.)	0.035	28-32	0.019	10B	150	Hyd.		7.0-8.5	900④	600④
	1973 (Calif.)	0.035	28-32	0.019	㉘	150	Hyd.		7.0-8.5	900④	600④
	1974	0.035	29-31	0.019	10B㉙	150	Hyd.		7.0-8.5	800④	600④
	1975-76 (Fed.)	0.045	Elec.	Elec.	16B㉙	150	Hyd.		7.0-8.5	——	650④
	1975-78 (Calif.)	0.045	Elec.	Elec.	8B	150	Hyd.		7.0-8.5	700	700

—— Not Applicable
Elec.—Electronic Ignition
Hyd.—Hydraulic valve lifters
N—Neutral
DR—Drive
B—BTDC

① Not used
② TDC on K-20 Suburban California only
③ 4B All C-K 20 except Suburban; all C-P 30 series and all G-30 except Sportvan
④ 700 rpm All C-K 20 except Suburban; all C-P 30 series and all G-30 except Sportvan
⑤ 600 rpm All C-K 20 except Suburban; all C-P 30 series and all G-30 except Sportvan
⑥ Not used
⑦ Not used
⑧ 8B California
⑨ 600 rpm California
⑩ 68 All C-K 20 series except Suburban; all C-P 30 series and all G-30 except Sportvan
⑪ Not used
⑫ 8B w/automatic transmission
⑬ 8B 10 series w/automatic transmission only
⑭ 950 rpm California
⑮ 8B All 10 series, C-K 20 Suburban, G-20 & 30 Sportvans w/auto transmission TDC All others
⑯ 600 rpm All C-K 20 except Suburban; all C-P 30 series and all G-30 except Sportvan
⑰ Not used
⑱ Not used
⑲ Not used
⑳ 8B w/automatic transmission
㉑ C-20 Suburban 2B
All 10 series, K-20 Suburban, G-20, and G-30 Sportvan:
  w/manual transmission 8B
  w/automatic transmission 12B
All others 4B
㉒ All C-K 20 except Suburban; all C-P 30 series and all G-20 & 30 except Sportvan 700 rpm
㉓ Federal except C-K 10 & 20 Suburban and G-20 & 30 Sportvan 8B
Federal C-K 10 & 20 Suburban and G-20 & 30 Sportvan:
  w/automatic transmission 12B
  w/automatic transmission (except Suburban) 8B
  Suburban (w/manual transmission) 6B
California:
  w/automatic transmission 8B
  Suburban (w/manual transmission) 6B
  w/manual transmission 4B
㉔ All C-K 20 except Suburban, all C-P 30 series and all G-20 & 30 except Sportvan 600 rpm
㉕ 750 rpm California
㉖ Not used
㉗ Not used
㉘ All 10 series, C-K 20 Suburban, G-20 and G-30 Sportvan 10 B
All C-K 20 except Suburban, all C-P 30 Series and all G-30 except Sportvan:
  w/manual transmission 5B
  w/automatic transmission 8B
㉙ All C-K 20 except Suburban, all C-P 30 series and all G-30 except Sportvan 8B

FIRING ORDER AND ROTATION

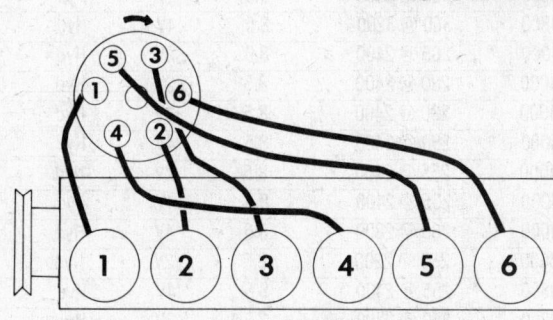

250 & 292 six cylinder—firing order 1-5-3-6-2-4

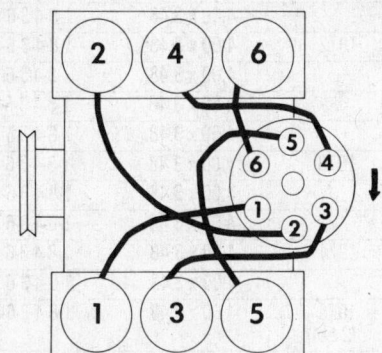

305C, 379, 432 V6 engines—firing order 1-6-5-4-3-2

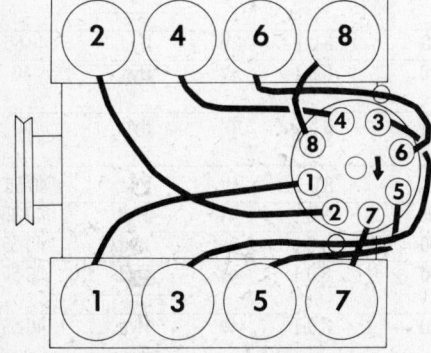

307, 350, 366, 402, 427, 454 engines
—firing order 1-8-4-3-6-5-7-2

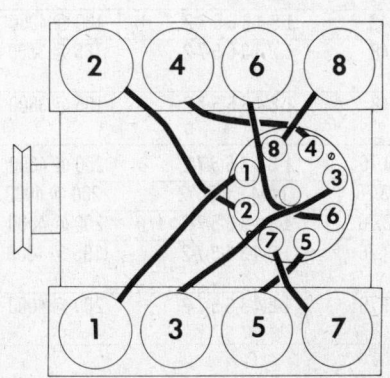

305, 350, 366, 400, 454 engines
—firing order 1-8-4-3-6-5-7-2

Chevrolet Trucks · Vans · Blazer

GENERAL ENGINE SPECIFICATIONS

CU. IN. DISPLACE-MENT	YEAR	BORE X STROKE	FIRING ORDER	HORSEPOWER @ R.P.M.	TORQUE @ R.P.M.	COMPRESSION RATIO	CARBURETOR	VALVE LIFTER TYPE	NORMAL OIL PRESSURE
SIX CYLINDER									
250	1971	3.875 x 3.53	1-5-3-6-2-4	145 @ 4200	230 @ 1600	8.5	1V	Hyd.	40-60
	1972	3.875 x 3.53	1-5-3-6-2-4	110 @ 3800	185 @ 1600	8.5	1V	Hyd.	40-60
	1973-74	3.875 x 3.53	1-5-3-6-2-4	100 @ 3600	175 @ 2000	8.25	1V	Hyd.	40-60
	1975	3.875 x 3.53	1-5-3-6-2-4	105 @ 3800	185 @ 1200	8.25:1	1V	Hyd.	40-60
	1976 (L.D.)	3.875 x 3.53	1-5-3-6-2-4	105 @ 3800	185 @ 1200	8.25:1	1V	Hyd.	40-60
	1976 (H.D.)	3.875 x 3.53	1-5-3-6-2-4	100 @ 3600	175 @ 1800	8:25:1	1V	Hyd.	40-60
	1977-78 (L.D.)	3.87 x 3.53	1-5-3-6-2-4	110 @ 3800	195 @ 1600	8.3:1	1V	Hyd.	40-60
	1977-78 (H.D.)	3.87 x 3.53	1-5-3-6-2-4	100 @ 3600	175 @ 1800	8.0	1V	Hyd.	40-60
292	1971	3.875 x 4.125	1-5-3-6-2-4	165 @ 4000	270 @ 1600	8.0	1V	Hyd.	40-60
	1972	3.875 x 4.125	1-5-3-6-2-4	125 @ 3600	225 @ 2400	8.0	1V	Hyd.	40-60
	1973-74	3.875 x 4.125	1-5-3-6-2-4	120 @ 3600	225 @ 2000	8.0	1V	Hyd.	40-60
	1975	3.87 x 4.12	1-5-3-6-2-4	120 @ 3600	215 @ 2000	8.0:1	1V	Hyd.	40-60
	1976-78	3.87 x 4.12	1-5-3-6-2-4	120 @ 3600	215 @ 2000	8.0:1	1V	Hyd.	40-60
305C (V6)	1973-74	4.250 x 3.58	1-6-5-4-3-2	148 @ 4000	238 @ 1600	7.5	2V	Mech.	57
379 (V6)	1973-74	3.56 x 3.86	1-6-5-4-3-2	170 @ 3600	280 @ 1600	7.5	2V	Mech.	57
432 (V6)	1973-74	4.875 x 3.86	1-6-5-4-3-2	190 @ 3200	331 @ 1600	7.5	2V	Mech.	60
EIGHT CYLINDER									
307	1971	3.875 x 3.25	1-8-4-3-6-5-7-2	200 @ 4600	300 @ 2400	8.5	2V	Hyd.	30
	1971	3.875 x 3.25	1-8-4-3-6-5-7-2	215 @ 4800	305 @ 2800	8.5	2V	Hyd.	30
	1972	3.875 x 3.25	1-8-4-3-6-5-7-2	130 @ 4000	230 @ 2400	8.5	2V	Hyd.	30
	1972	3.875 x 3.25	1-8-4-3-6-5-7-2	135 @ 4000	230 @ 2400	8.5	2V	Hyd.	30
	1973	3.875 x 3.25	1-8-4-3-6-5-7-2	115 @ 3600	205 @ 2000	8.5	2V	Hyd.	30
	1973	3.875 x 3.25	1-8-4-3-6-5-7-2	130 @ 4000	220 @ 2200	8.5	2V	Hyd.	30
350	1971	4.00 x 3.48	1-8-4-3-6-5-7-2	215 @ 4000	335 @ 2800	8.0	2V	Hyd.	50-65
		4.00 x 3.48	1-8-4-3-6-5-7-2	245 @ 4800	350 @ 2800	8.5	2V	Hyd.	50-65
		4.00 x 3.48	1-8-4-3-6-5-7-2	250 @ 4600	350 @ 3000	8.5	4V	Hyd.	50-65
		4.00 x 3.48	1-8-4-3-6-5-7-2	255 @ 4600	355 @ 3000	9.0	4V	Hyd.	50-65
		4.00 x 3.48	1-8-4-3-6-5-7-2	270 @ 4800	360 @ 3200	8.5	4V	Hyd.	50-65
	1972	4.00 x 3.48	1-8-4-3-6-5-7-2	155 @ 4000	265 @ 2400	8.0	2V	Hyd.	50-65
		4.00 x 3.48	1-8-4-3-6-5-7-2	165 @ 4000	280 @ 2400	8.5	2V	Hyd.	50-65
		4.00 x 3.48	1-8-4-3-6-5-7-2	175 @ 4000	280 @ 2400	8.5	4V	Hyd.	50-65
		4.00 x 3.48	1-8-4-3-6-5-7-2	175 @ 4000	290 @ 2400	8.5	4V	Hyd.	50-65
	1973	4.00 x 3.48	1-8-4-3-6-5-7-2	145 @ 4000	255 @ 2400	8.5	2V	Hyd.	40
		4.00 x 3.48	1-8-4-3-6-5-7-2	155 @ 4000	255 @ 2400	8.5	4V	Hyd.	40
		4.00 x 3.48	1-8-4-3-6-5-7-2	175 @ 4000	260 @ 2800	8.5	4V	Hyd.	40
	1974	4.00 x 3.48	1-8-4-3-6-5-7-2	145 @ 3600	250 @ 2200	8.5	2V	Hyd.	40
		4.00 x 3.48	1-8-4-3-6-5-7-2	160 @ 3800	255 @ 2400	8.5	4V	Hyd.	40
	1975 (2 bbl)	4.00 x 3.48	1-8-4-3-6-5-7-2	145 @ 3800	250 @ 2200	8.5:1	2V	Hyd.	40-60
	1975 (4 bbl)	4.00 x 3.48	1-8-4-3-6-5-7-2	160 @ 3800	250 @ 2400	8.5:1	4V	Hyd.	40-60
	1976	4.00 x 3.48	1-8-4-3-6-5-7-2	160 @ 4000	265 @ 2400	8.0:1	2V	Hyd.	40-60
	1977-78 (L.D.)	4.00 x 3.48	1-8-4-3-6-5-7-2	165 @ 3800	260 @ 2400	8.5:1	4V	Hyd.	40
	1977-78 (H.D.)	4.00 x 3.48	1-8-4-3-6-5-7-2	165 @ 3800	260 @ 2400	8.5:1	4V	Hyd.	40
366	1971-72	3.937 x 3.76	1-8-4-3-6-5-7-2	200 @ 4000	295 @ 3200	8.0	4V	Hyd.	40-55
	1973-74	3.937 x 3.76	1-8-4-3-6-5-7-2	200 @ 4000	310 @ 2800	8.0	4V	Hyd.	40-55
	1975	3.937 x 3.76	1-8-4-3-6-5-7-2	200 @ 4000	305 @ 2800	8.0:1	4V	Hyd.	40-55
	1976-78 (sing. exh.)	3.937 x 3.76	1-8-4-3-6-5-7-2	195 @ 4000	290 @ 2800	8.0:1	4V	Hyd.	40-55
	1976-78 (dual exh.)	3.937 x 3.76	1-8-4-3-6-5-7-2	200 @ 4000	305 @ 2800	8.0:1	4V	Hyd.	40-55

GENERAL ENGINE SPECIFICATIONS, continued

CU. IN. DISPLACEMENT	YEAR	BORE X STROKE	FIRING ORDER	HORSEPOWER @ R.P.M.	TORQUE @ R.P.M.	COMPRESSION RATIO	CARBURETOR	VALVE LIFTER TYPE	NORMAL OIL PRESSURE
400	1975	4.125 x 3.75	1-8-4-3-6-5-7-2	175 @ 3600	290 @ 2800	8.5:1	4V	Hyd.	40-60
	1976-78	4.125 x 3.75	1-8-4-3-6-5-7-2	175 @ 3600	290 @ 2800	8.5:1	4V	Hyd.	40-60
402 240 H.P. 1971		4.126 x 3.76	1-8-4-3-6-5-7-2	240 @ 4400	340 @ 3200	8.5	4V	Hyd.	40
260 H.P. 1971		4.126 x 3.76	1-8-4-3-6-5-7-2	260 @ 4400	345 @ 3200	8.5	4V	Hyd.	40
210 H.P. 1972		4.126 x 3.76	1-8-4-3-6-5-7-2	210 @ 4000	320 @ 2800	8.5	4V	Hyd.	45-55
427	1971-72	4.25 x 3.76	1-8-4-3-6-5-7-2	230 @ 4000	360 @ 2400	8.0	4V	Hyd.	40
	1973-74	4.25 x 3.76	1-8-4-3-6-5-7-2	230 @ 4000	360 @ 2800	8.0	4V	Hyd.	40-55
	1975	4.25 x 3.76	1-8-4-3-6-5-7-2	220 @ 4000	360 @ 2400	8.0:1	4V	Hyd.	40-55
	1976-78	4.25 x 3.76	1-8-4-3-6-5-7-2	220 @ 4000	360 @ 2400	8.0:1	4V	Hyd.	40-55
454	1971	4.251 x 4.00	1-8-4-3-6-5-7-2	285 @ 4000	390 @ 3200	8.5	4V	Hyd.	40
		4.251 x 4.00	1-8-4-3-6-5-7-2	325 @ 5600	390 @ 3600	9.0	4V	Hyd.	40
	1972	4.251 x 4.00	1-8-4-3-6-5-7-2	270 @ 4000	390 @ 3200	8.5	4V	Hyd.	40
	1973	4.251 x 4.00	1-8-4-3-6-5-7-2	240 @ 4000	355 @ 2800	8.25	4V	Hyd.	40
		4.251 x 4.00	1-8-4-3-6-5-7-2	245 @ 4000	375 @ 2800	8.25	4V	Hyd.	40
		4.251 x 4.00	1-8-4-3-6-5-7-2	250 @ 4000	365 @ 2800	8.25	4V	Hyd.	40
	1974	4.251 x 4.00	1-8-4-3-6-5-7-2	230 @ 4000	350 @ 2800	8.25	4V	Hyd.	40
		4.251 x 4.00	1-8-4-3-6-5-7-2	245 @ 4000	365 @ 2800	8.25	4V	Hyd.	40
	1975 (L.D.)	4.251 x 4.00	1-8-4-3-6-5-7-2	215 @ 4000	350 @ 2400	8.15:1	4V	Hyd.	40
	1975 (H.D.)	4.251 x 4.00	1-8-4-3-6-5-7-2	245 @ 4000	355 @ 3000	8.15:1	4V	Hyd.	40
	1976-78 (L.D.)	4.251 x 4.00	1-8-4-3-6-5-7-2	245 @ 3800	365 @ 2800	8.15:1 ①	4V	Hyd.	40
	1976-78 (H.D., Fed.)	4.251 x 4.00	1-8-4-3-6-5-7-2	240 @ 3800	370 @ 2800	8.15:1	4V	Hyd.	40
	1976-78 (H.D., Calif.)	4.251 x 4.00	1-8-4-3-6-5-7-2	250 @ 3800	385 @ 2800	8.15:1	4V	Hyd.	40

Fed.—Federal (all states except California)
Calif.—California only

L.D.—Light duty emissions (under 6,000 lb GVW)
H.D.—Heavy duty emissions (over 6,000 lb GVW)
① 8:25:1 on C-10 w/H.D. chassis and C-10, 20 Suburban

CRANKSHAFT BEARING JOURNAL SPECIFICATIONS

CU. IN. DISPLACEMENT	YEAR	MAIN BEARING JOURNALS				CONNECTING ROD BEARING JOURNALS		
		JOURNAL DIAMETER	OIL CLEARANCE	SHAFT END PLAY	THRUST ON NO.	JOURNAL DIAMETER	OIL CLEARANCE	END PLAY
SIX CYLINDER								
250	1971-72	2.2983-2.2993	.0003-.0029	.002-.006	Rear	1.999-2.000	.0007-.0027	.0009-.0014
	1973-78	2.2983-2.2993	.0003-.0029	.002-006	Rear	1.999-2.000	.0007-.0027	.0006-.0017
292	1971-72	2.2983-2.2993	.0008-.0034	.002-.006	5	2.099-2.100	.0007-.0028	.0009-.0014
	1973-78	2.2983-2.2993	.0008-.0034	.002-.006	5	2.099-2.100	.0007-.0027	.0006-.0017
305C (V6)	1973-74	3.1247-3.1237[1]	.0013-.0039	.003-.008	—	2.8112-2.8122	.001-.003	.006-.011
379 (V6)	1973-74	3.1247-3.1237[1]	.0023-.0039[2]	.003-.008	—	2.8112-2.8122	.0015-.0035	.006-.011
432 (V6)	1973-74	3.1247-3.1237[1]	.0023-.0039[2]	.003-.008	—	2.8112-2.8122	.0015-.0035	.006-.011
EIGHT CYLINDER								
307	1971-73	2.4484-2.4493[3]	[5]	.002-.006	5	2.1990-2.200	.0013-.0035	.008-.014
350	1971-78	2.4484-2.4493[3]	[5]	.002-.006	5	2.199-2.200	.0013-.0035	.008-.014
366	1971-76	2.7481-2.7490[6]	.0013-.0025[7]	.006-.010	5	2.1985-2.1995	.0014-.0030	.019-.025
400	1975-76	2.4484-2.4493[3]	[5]	.002-.006	5	2.199-2.200	.0013-.0035	.008-.014
402	1971-72	2.7487-2.7496[10]	.0007-.0019[11]	.002-.006	5	2.1985-2.1995	.0009-.0025	.013-.023
427	1971-75	2.7481-2.7490[6]	.0013-.0025[7]	.006-.010	5	2.199-2.1998	.0014-.0030	.019-.025
454	1971-78	2.7485-2.7494[12]	.0013-.0025[13]	.006-.010	5	2.1985-2.1995	.0009-.0025	.013-.023

1. Rear—3.1229-3.1239 in.
2. No. 4—.0031-.0047 in.
3. Rear—2.4479-2.4488 in.
4. Not used
5. No. 1—.0008-.0020; No. 2, 3, 4—.0011-.0023; No. 5—.0017-.0033
6. Rear—2.7473-2.7483 in.

7. Rear—.0029-.0045 in.
8. Not used
9. Not used
10. No. 3-4—2.7841-2.7490; No. 5—2.7473-2.7483
11. No. 2 3-4—.0013-.0025; No. 5—.0019-.0033
12. No. 2 3-4—2.7481-2.7490; No. 5—2.7478-2.7488
13. No. 5—.0024-.0040

RING SIDE CLEARANCE (IN.)

ENGINE (cu in.)	YEAR	TOP COMPRESSION	BOTTOM COMPRESSION	OIL CONTROL
250	1971-76	.0012-.0027	.0012-.0032	.0000-.0050
292	1971-76	.0020-.0040	.0020-.0040	.0005-.0055
305 (V6)	1973-74	.0030-.0050	①	.0010-.0040
379 (V6)	1973-74	.0030-.0045	②	.0025-.0040
432 (V6)	1973-74	.0030-.0045	②	.0025-.0040
307	1971-73	.0012-.0027	.0012-.0032	.0000-.0050
350	1971-76	.0012-.0032	.0012-.0032	.0000-.0050
366	1971-76	.0018-.0032	.0018-.0032	.0020-.0035
400	1975-78	.0012-.0032	.0012-.0032	.0000-.0050
402	1971-72	.0017-.0032	.0017-.0032	.0005-.0065
454	1973-78	.0017-.0032	.0017-.0032	.0005-.0065

① —2nd compression—.0030-.0045
3rd compression—.0025-.0040
② —2nd compression—.0025-.0040
3rd compression—.0025-.0040

RING GAP SPECIFICATIONS (IN.)

ENGINE	YEAR	TOP COMPRESSION	BOTTOM COMPRESSION	OIL CONTROL
250	1971-76	.010-.020	.010-.020	.015-.055
292	1971-76	.010-.020	.010-.020	.015-.055
305 (V6)	1973-74	.017-.027	①	No Gap
379 (V6)	1973-74	.022-.032	②	No Gap
432 (V6)	1973-74	.024-.034	③	No Gap
307	1971-73	.010-.020	.010-.020	.015-.055
350	1971-76	.010-.020	.013-.025	.015-.055
366	1971-76	.010-.020	.010-.020	.010-.023
400	1975-76	.010-.020	.010-.020	.010-.035
402	1971-72	.010-.020	.010-.020	.015-.055
427	1971-76	.010-.020	.010-.020	.010-.023
454	1973-78	.010-.020	.010-.020	.010-.030

① —2nd compression—.015-.025
3rd compression—.015-.025
② —2nd compression—.022-.032
3rd compression—.015-.025
③ —2nd compression—.024-.034
3rd compression—.015-.025

VALVE SPECIFICATIONS

CU. DISPLACE-MENT	YEAR	LASH (HOT) (INCHES) INT.	EXH.	ANGLE (DEGREES) FACE	SEAT	STEM DIA. (INCHES) INT.	EXH.	STEM CLEARANCE INTAKE	EXHAUST	CAM LOBE LIFT (INCHES)	VALVE SPRING TENSION (LBS @ INCHES) OPEN	CLOSED	FREE LENGTH (INCHES)
SIX CYLINDER													
250	1971	0+1	Turn	45	46	.341	.341	.001-.003	.002-.004	.390	186 @ 1.27	60 @ 1.66	2.08
	1972-75	0+1	Turn	45	46	—	—	.0010-.0027	.0015-.0032	.221	185 @ 1.27	60 @ 1.66	1.90
292	1971-76	0+1	Turn	45	46	—	—	.0010-.0027	.0015-.0032	.2315	185 @ 1.27	90 @ 1.69	1.90
	1977-78	0+1	Turn	45	46	—	—	.0010-.0027	.0015-.0032	.2315	180 @ 1.30	90 @ 1.69	1.94
305C (V6)	1973-74	.012	.018	—	30	.341	.340	.0015-.003	.002-.0035	.454²	203 @ 1.50	80 @ 1.92	2.27
379 (V6)	1973-74	.012	.018	—	30	.373	.438	.0015-.003	.0019-.0036	.454²	203 @ 1.50	80 @ 1.92	2.27
432 (V6)	1973-74	.012	.018	—	30	.373	.438	.0015-.003	.0019-.0036	.454²	203 @ 1.50	80 @ 1.92	2.27
EIGHT CYLINDER													
307	1971-73	0+1	Turn	45	46	.341	.341	.001-.003	.001-.003	³	189 @ 1.20	80 @ 1.61	1.91
350	1971-78	0+1	Turn	45	46	.341	.341	.001-.003	.001-.003	³	200 @ 1.25	80 @ 1.70	2.03
366	1971-76	0+1	Turn	45	46	.341	.341	.001-.003	.001-.003	.234	220 @ 1.40	90 @ 1.80	2.05
400	1975-78	0+1	Turn	45	46	—	—	.001-.0027	.001-.0027	³	⁵	⁶	2.03
402	1971-72	0+1	Turn	45	46	—	—	.001-.0027	.0012-.0029	.234	245 @ 1.38	75 @ 1.88	2.06
427	1971-78	0+1	Turn	45	46	.341	.341	.001-.003	.001-.002	⁴	220 @ 1.40	90 @ 1.80	2.05
454	1971-78	0+1	Turn	45	46	—	—	.001-.0027	.001-.0029	.234	290 @ 1.38	80 @ 1.88	2.12

1. .450-Exhaust
2. .464-Exhaust
3. .260-Intake; .273-Exhaust
4. .234-Intake; .253-Exhaust
5. 200 @ 1.25 intake, 200 @ 1.16 exhaust
6. 80 @ 1.70 intake, 80 @ 1.61 exhaust

TORQUE SPECIFICATIONS

CU. IN. DISPLACE-MENT	YEAR	CYLINDER HEAD BOLTS (FT. LBS.)	ROD BEARING BOLTS (FT. LBS.)	MAIN BEARING BOLTS (FT. LBS.)	CRANKSHAFT BALANCER BOLT (FT. LBS.)	FLYWHEEL TO CRANKSHAFT BOLTS (FT. LBS.)	MANIFOLDS INTAKE (FT. LBS.)	MANIFOLDS EXHAUST (FT. LBS.)
				SIX CYLINDER				
250	1971-78	95	35-45	65	Pressed on	60	[1]	[1]
292	1971-78	95	40	65	Pressed on	55-65	[1]	[1]
305C (V6)	1973-74	60-65	55-65	170-180[2]	180-200	100-110	30-35	15-20
379 (V6)	1973-74	90-100	55-65	170-180[2]	240-260	100-110	30-35	15-20
432 (V6)	1973-74	90-100	55-65	170-180[2]	240-260	100-110	30-35	15-20
				EIGHT CYLINDER				
307	1971-73	65	45	70	Pressed on	60	30	20[3]
350	1971-78	65	45	70	60	60	30	20[3]
366	1971-78	80	55	100[4]	85	60	30	20
402	1971-72	80	50	110	85	65	30	20
427	1971-78	80	50	110	85	60	30	20
454	1971-74	80	50	110	85	65	30	20

1. Outer—20 ft. lbs.; others—30 ft. lbs.
2. Rear—90-100 ft lbs.
3. Inside bolts—30 ft lbs.
4. Outer—95 ft. lbs.

WHEEL ALIGNMENT SPECIFICATIONS

MODEL	YEAR	CASTER (deg)	CAMBER (deg)	TOE-IN (in.)	KINGPIN INCLINATION (in.)
		10-20-30 Series (Light Duty)			
P-10, 20, 30	1971	①	①	1/8-1/4	——
	1972-78	②	1/4P	1/8-1/4	——
K-10, 20	1971	1 1/2P*	4P*	1/16-5/16	8 1/2
	1972-78	1 1/2P*	4P*	1/8-1/4	8 1/2
G-10, 20, 30	1971	1/4N-1/4P	0-1/2P	1/8-1/4	——
	1972	1/4N-1/4P②	0-1/2P	1/8-1/4	——
	1973-78	②	0-1/2P	1/8-1/4	——
P-10, 20, 30	1971	①	①	1/8-1/4	——
	1972-78	②	1/4P	1/8-1/4	——
		40 thru 65 Series (Medium Duty)			
C-40, 50, 60	1971-72	2P-3P	1/2P-1 1/2P③	3/32-3/16④	7 1/4⑤
C-50	1973	2P-3P	1/2P-1 1/2P	1/8-3/16	7 1/4
	1974-78	2P-3P	1/2P-1 1/2P	1/16-1/8	7 1/4
C-60, 65	1973	2P-3P	1/2P-1 1/2P	3/32-3/16	7 1/4
	1974-78	2P-3P	1/2P-1 1/2P③	1/16-1/8④	7 1/4⑤
P-40, 45	1974-78	2P-3P	1/2P-1 1/2P	1/16-1/8	7 1/4
M-60, 65	1971-73	2P-3P	1/2P-1 1/2P③	3/32-3/16④	7 1/4⑤
M-60, 65	1974-78	2F-3P	1/2P-1 1/2P③	1/16-1/8⑥	7 1/4⑤
S-50, 60 (exc long w.b.)	1971-72	2P-3P	1/2P-1 1/2P	3/32-3/16	7 1/4
S-60	1973	2P-3P	1/2P-1 1/2P	3/32-3/16	7 1/4
	1974-78	2P-3P	1/2P-1 1/2P	1/16-1/8	7 1/4
T-50, 60	1970	2P-3P	1/2-1 1/2P⑦	3/32-3/16⑧	7 1/4⑩
T-50, 60, 65	1971-78	3/4P-1 3/4P	1P-2P⑨	1/8-1/4	7⑩

P—Positive
N—Negative
L.H.—Left hand
R.H.—Right hand
*—No provision for adjustment
—— Not Applicable

FOOTNOTE 1

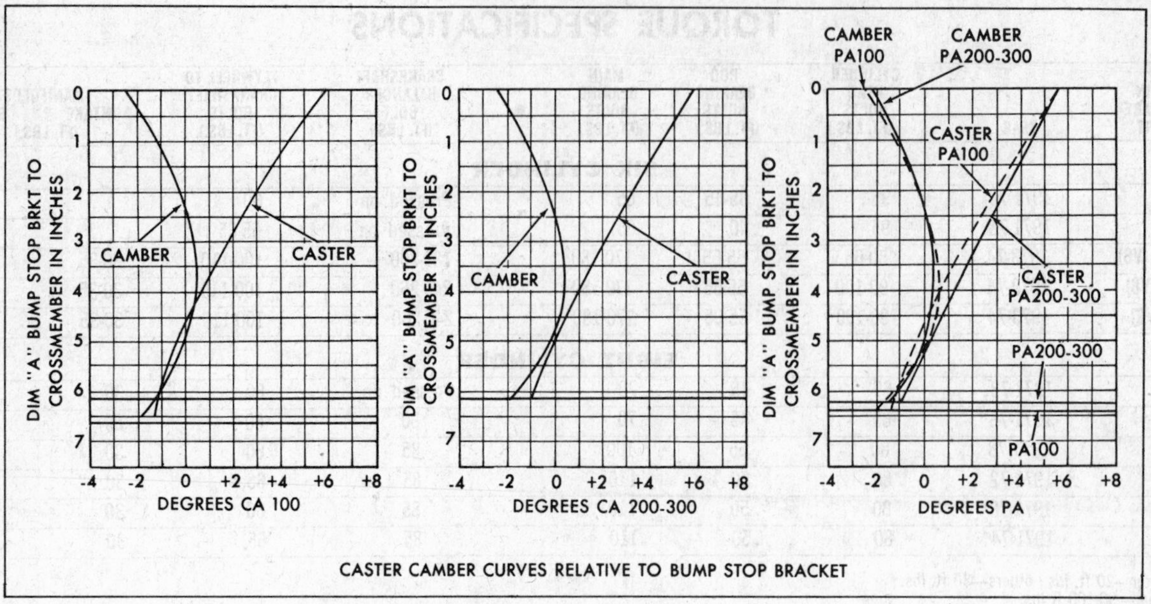
CAMBER
CASTER
DIM "A" BUMP STOP BRKT TO CROSSMEMBER IN INCHES
DEGREES CA 100
DEGREES CA 200-300
CAMBER PA100
CAMBER PA200-300
CASTER PA100
CASTER PA200-300
PA200-300
PA100
DEGREES PA
CASTER CAMBER CURVES RELATIVE TO BUMP STOP BRACKET

1971 C-10, 20, 30 and F-10, 20, 30 (© Chevrolet Div., G.M. Corp.)

FOOTNOTE 2

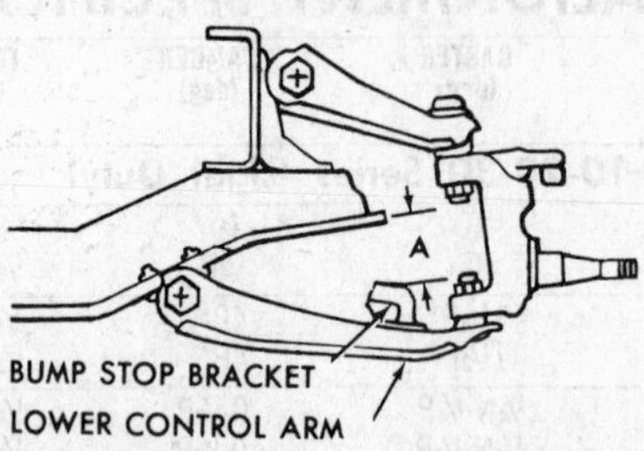
A
BUMP STOP BRACKET
LOWER CONTROL ARM

Measure the distance from the bump stop bracket to the frame and consult the appropriate chart. (© Chevrolet Div., G.M. Corp.)

		CASTER										
Dimension "A" in inches		2½"	2¾"	3"	3¼"	3½"	3¾"	4"	4¼"	4½"	4¾"	5"
1972	**C-10**	—	—	+2°	+1¾°	+1½°	+1¼°	+1°	+¾°	+½°	+¼°	0°
1972	**C-20,30**	+2¾°	+2½°	+2°	+1¾°	+1½°	+1¼°	+1°	+¾°	+½°	+¼°	0°
1973-78	**C-10**	—	—	+2°	+1½°	+1¼°	+1°	+¾°	+½°	+¼°	0°	−½°
1973	**C-20, 30**	+2°	+1½°	+1¼°	+1°	+¾°	+½°	+¼°	0°	−¼°	−½°	−¾°
1974-78	**C-20, 30**	+1½°	+1¼°	+1°	+¾°	+½°	+¼°	0°	−¼°	−½°	−¾°	−1°
1972-78	**G-10, 20, 30**	+2¼°	+2°	+1½°	+1¼°	+1°	+¾°	+½°	+¼°	0°	−¼°	−½°
1972	**P-10**	—	—	+2°	+1¾°	+1½°	+1¼°	+1°	+¾°	+½°	+¼°	0°
1972	**P-20, 30**	+2¾°	+2½°	+2°	+1¾°	+1½	+1¼°	+1°	+¾°	+½°	+¼°	0°
1973-78	**P-10**	—	—	+2°	+1½°	+1¼°	+1°	+¾°	+½°	+¼°	0°	−½°
1973	**P-20, 30**	+2°	+1½°	+1¼°	+1°	+¾°	+½°	+¼°	0°	−¼°	−½°	−¾°
1974-78	**P-20, 30**	+1½°	+1¼°	+1°	+¾°	+½°	+¼°	0°	−¼°	−½°	−¾°	−1°

③ C-M 60, 65 w/9000 and 12000 lb axles—¼P (L.H.); ¼N (R.H.)
④ C-M 60, 65 w/9000 and 12000 lb axles—⅛-7/32 in.
⑤ C-M 60, 65 (1971-73) w/9000 and 12000 lb axle—5¾° (L.H.); 6¼° (R.H.)
⑥ C-M 65 (1974-76 w/9000 and 12000 lb axle—⅛-¼in.
⑦—Not used
⑧—Not used
⑨ T-50, 60, 65 w/9000 and 12000 lb axles—¼N-¾P (L.H.); ¾N-¼P (R.H.)
⑩ T-50, 60 w/9000 and 11000 lb axles and T-50, 60, 65 w/9000 and 12000 lb axles—5¾° (L.H.); 6¼° (R.H.)

TRUCK MODELS AND ENGINE APPLICATION

Identification Plate: The vehicle identification plate containing the VIN is located on the left door pillar of all conventional models. On all later Forward Control models, on the dash and toe panel. On cowl models, the plate is attached to the engine side of the cowl.

Engines (cu in.)	Year	Models Available	Engine Make
250	1971	C-10, 20, 30, 40 K-10, 20 P-10, 20, 30, 40 G-10, 20, 30 S-40	Own
	1972	C-10, 20, 30, 40 K-10, 20 P-10, 20, 30 G-10, 20, 30	Own
	1973-74	C-10, 20, 30, 50 K-10, 20 P-10, 20, 30 G-10, 20, 30	Own
	1975-78	C-10, K-10, G-10	Own
292	1971	C-10, 20, 30, 40, ③ 50 K-10, 20③ P-20, 30, 40 S-40, 50 T-50	Own
	1972	C-10, 20, 30, 40, ③ 50 K-10, 20③ P-20, 30 S-50	Own
	1973	C-20, 30, 50, 60 S-60	Own
	1974	C-20, 30, 50, 60 K-20 P-20, 30, 40 S-60	Own
	1975-78	C-20, 30, 50, 60 K-20 P-10, 20, 30 G-20, 30 S-60	Own
305 (V6)	1973	C-60, S-60, T-60	Own
	1974	S-60, T-60	Own
379 (V6)	1973-74	C-60, 65 S-60 T-60, 65	Own
432 (V6)	1973	C-65, T-65	Own
	1974	C-65, M-65, T-65	
307	1971-73	C-10, 20, 30 K-10, 20 P-20, 30 G-10	Own
305	1974-78	C-10, 20, 30 K-10, 20 P-20, 30 G-10	Own
350	1971-72	C-10, 20, 30, 40, 50 K-10, 20 P-20, 30 G-10, 20, 30 M-50, S-50, T-50	Own
	1973-74	C-10, 20, 30, 50, 60 K-10, 20 P-20, 30 G-10, 20, 30 S-60, T-60	Own
	1975-78	C-10, 20, 30, 50, 60 K-10, 20 P-20, 30 G-20, 30 S-60, T-60	Own
366	1971-72	C-50, 60 M-60, S-50 T-50, 60	Own
	1973-76	C-60,65 S-60 T-60, 65 M-65	Own
400	1975-78	K-10, 20 G-20, 30	Own
402	1971-72	C-10, 20, 30 ③	Own
427	1971-72	C-60, M-60, T-60	Own
	1973-74	C-65, M-65, T-65	Own
	1975-76	C-65, M-65, T-65	Own
454	1973-78	C-10, 20,30 ③ P-30	Own

①—Not used
②—Not used
③—Except Blazer

VEHICLE IDENTIFICATION NUMBER

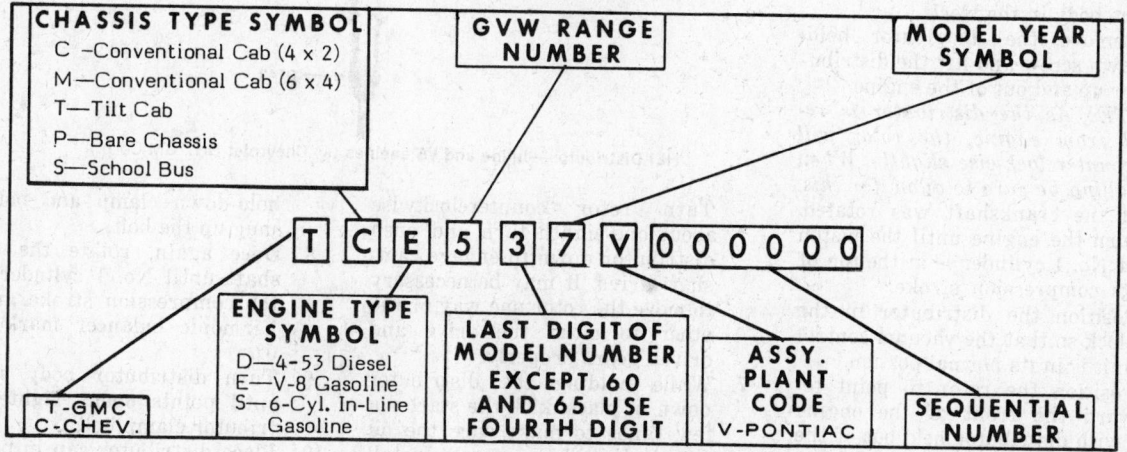

CHASSIS TYPE SYMBOL
C —Conventional Cab (4 x 2)
M—Conventional Cab (6 x 4)
T—Tilt Cab
P—Bare Chassis
S—School Bus

GVW RANGE NUMBER

MODEL YEAR SYMBOL

ENGINE TYPE SYMBOL
D—4-53 Diesel
E—V-8 Gasoline
S—6-Cyl. In-Line Gasoline

LAST DIGIT OF MODEL NUMBER EXCEPT 60 AND 65 USE FOURTH DIGIT

ASSY. PLANT CODE
V-PONTIAC

SEQUENTIAL NUMBER

T-GMC
C-CHEV.

T C E 5 3 7 V 000000

Distributor

Detailed information on direction of distributor rotation, cylinder numbering, firing order, point gap, cam dwell, spark plugs, and idle speed will be found in the Specifications tables.

Engine diagnosis is found in the General Repair Section.

H.E.I. System—1975-78

1975-78 light duty Chevrolet truck engines use a High Energy Ignition system. Two types are used. V8 and 1978 and later six-cylinder distributors combine all ignition components in one unit. The coil is in the distributor cap and connects directly to the rotor. The 6 cylinder distributor trrough 1977 has an externally mounted coil. Both units operate in basically the same manner. The module and pick-up coil replace the conventional breaker points. The module automatically controls the dwell, stretching it with increased engine speed. The system also features a longer spark duration due to the greater amount of energy stored in the primary coil.

The centrifugal and vacuum advance mechanisms are basically the same type of unit as on a conventional ignition distributor.

The electronic module is serviced by complete replacement.

WARNING: Do not remove the spark plug wires with the engine running. Severe shock could result.

Distributor Removal and Installation

Inline Engines

1. Remove distributor cap, primary wire and vacuum line.
2. Scribe a mark on the distributor body, locating the position of the rotor. Scribe another mark on the distributor body and engine block, showing the position of the body in the block.
3. Remove the distributor hold-down screw and lift the distributor up and out of the engine.

NOTE: As the distributor is removed from engine, the rotor will turn counterclockwise slightly. When reinstalling be sure to allow for this.

4. If the crankshaft was rotated, turn the engine until the piston of No. 1 cylinder is at the top of its compression stroke.
5. Position the distributor in the block so that the vacuum control unit is in its normal postion.
6. Position the rotor to point toward the front of the engine (with distributor held out of the block, but in installed position).

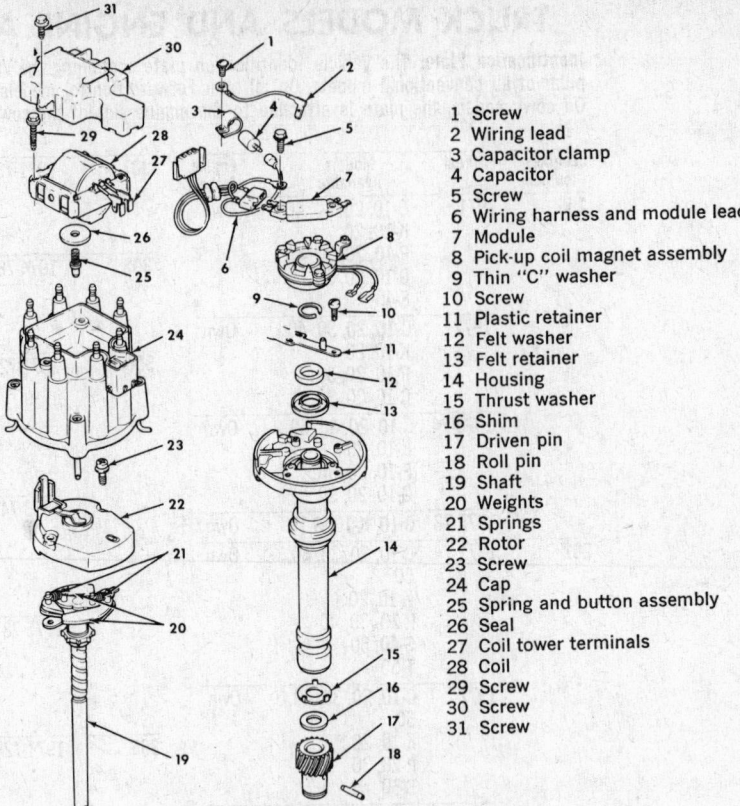

1. Screw
2. Wiring lead
3. Capacitor clamp
4. Capacitor
5. Screw
6. Wiring harness and module leads
7. Module
8. Pick-up coil magnet assembly
9. Thin "C" washer
10. Screw
11. Plastic retainer
12. Felt washer
13. Felt retainer
14. Housing
15. Thrust washer
16. Shim
17. Driven pin
18. Roll pin
19. Shaft
20. Weights
21. Springs
22. Rotor
23. Screw
24. Cap
25. Spring and button assembly
26. Seal
27. Coil tower terminals
28. Coil
29. Screw
30. Screw
31. Screw

HEI Distributor—V8 engine—exploded view (© G.M.C.)

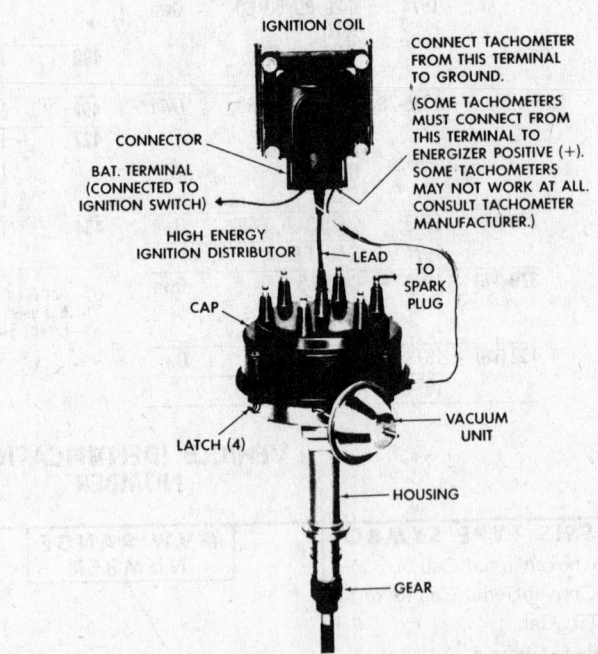

HEI Distributor—Inline and V6 engines (© Chevrolet Div., GM Corp.)

Turn rotor counterclockwise about one-eighth turn and push distributor down to engage camshaft drive. It may be necessary to move the rotor one way or the other to mesh the drive and driven gears properly.

7. While holding the distributor down in place, kick the starter a few times to make sure the oil pump shaft is engaged. Install

hold-down clamp and bolt and snug up the bolt.

8. Once again, rotate the crankshaft until No. 1 cylinder is on the compression stroke and the harmonic balancer mark is on 0°.
9. Turn distributor body slightly until points open. Tighten distributor clamp bolt.
10. Place distributor cap in position

and see that the rotor lines up with the terminal for the No. 1 spark plug.

11. Install cap, distributor primary wire, and double check plug wires in the cap towers.

12. Start engine and set timing according to the Tune-up chart.

13. Reconnect vacuum hose to vacuum control assembly.

CAUTION: When using an auxiliary starter switch for bumping the engine into position for timing or compression test, the primary distributor lead must be disconnected from the negative post of the ignition coil and the ignition switch must be on. Failure to do this may cause damage to the grounding circuit in the ignition switch. This will also prevent the sudden starting of engine and possible serious injury.

V8 Engines

If it becomes necessary to remove the distributor, carefully mark the position of the rotor so that, if the engine is not turned after the distributor is taken out, the rotor can be returned to the position from which it was removed without difficulty.

1. To remove the distributor, take off the carburetor air cleaner, disconnect the coil primary wire and the vacuum line, remove the distributor cap, take out the single hold-down bolt located under the distributor body. With a pencil, mark the position of the body relative to the block, and then work the distributor up out of the block.

NOTE: If necessary, remove secondary leads from cap after first marking cap tower for No. 1 lead.

2. Remove No. 1 spark plug and, with finger on plug hole, crank the engine until compression is felt in No. 1 cylinder. Continue cranking until pointer lines up with the timing mark on the crankshaft pulley.

3. Position distributor in opening of the block in normal installed

attitude; have rotor pointing to front of engine.

4. Turn the rotor counterclockwise about one-eighth of a turn (from straight front toward the left cylinder bank). Push the distributor down to engage the camshaft and while holding, turn the engine with the starter so that distributor shaft engages the oil pump shaft.

5. Return engine to compression stroke of No. 1 piston with timing mark on pulley aligned with the pointer. Adjust the distributor so that the points are opening. Install the cap being sure the rotor points to the contact for No. 1 spark plug. Connect the timing light and check that spark occurs as timing mark and pointer are aligned.

CAUTION: On the V8, the distributor body is involved in the engine lubricating system. The lubricating circuit can be interrupted to the right bank valve train by misalignment of the distributor body. This can cause serious trouble and may be hard to diagnose. See Firing Order and Timing illustrations.

Contact Point Replacement

Inline Engines

1. Release distributor cap hold-down screws, remove cap.

2. Remove rotor.

3. Pull primary and condenser lead wires from contact point quick disconnect terminal.

4. Remove condenser hold-down screw and replace condenser.

5. Remove contact set attaching screw, lift contact point set from breaker plate.

6. Clean breaker plate of oil and dirt.

7. Place new contact point assembly in position on breaker plate, install attaching screw.

CAUTION: Carefully wipe protective film from point set prior to installation.

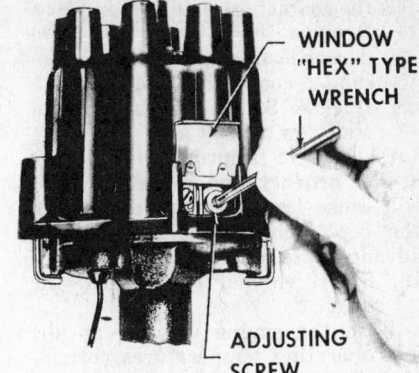

Thru cap point adjustment
(© Chevrolet Div., G.M. Corp.)

NOTE: Pilot on contact set must engage matching hole in breaker plate.

8. Connect primary and condenser lead wires to quick disconnect terminal on contact point set.

9. Check and adjust points for proper alignment and breaker arm spring tension. Use an aligning tool to bend stationary contact support if points need alignment.

10. Set point gap.

11. Reinstall rotor and distributor cap.

12. Start engine and check dwell.

V8 Engines

1. The contact point set is replaced as one complete assembly and only dwell angle requires adjustment after replacement. Breaker lever spring tension and point alignment are factory set.

2. Remove the distributor cap by placing a screw driver in the slot head of the latch, press down and turn ¼ turn in either direction. Remove two attaching screws, which hold rotor to weight base, and remove rotor.

3. (On '73-'74 models, remove the two screws which hold R.F.I. shield in place, and remove shield.) Remove the two attaching screws which hold the base of

ADJUST DWELL ANGLE SETTING OR POINT OPENING

Adjusting contact points (© Chevrolet Div., G.M. Corp.)

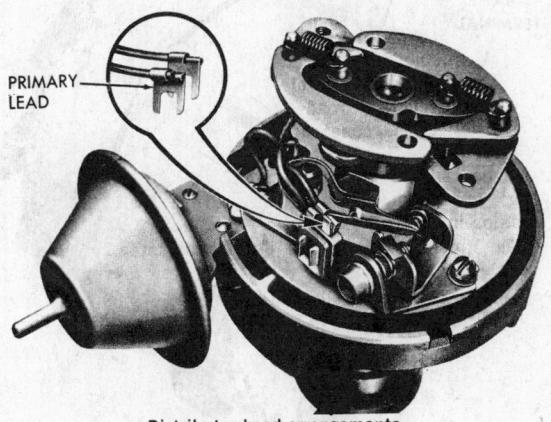

PRIMARY LEAD

Distributor head arrangements
(© Chevrolet Div., G.M. Corp.)

the contact set assembly in place.

4. Remove the primary and condenser leads from their nylon insulated connection in contact set.

5. Reverse Steps 2, 3 and 4 to install new contact set.

CAUTION: **Improper installation of the primary and condenser leads will cause lead interference between the cap, weight base and breaker advance plate.**

6. Start engine and check the dwell.

With the engine running at idle and operating temperatures normalized, the dwell is adjusted by raising the window provided in the cap and inserting a "Hex" type wrench into the adjusting screw head.

Turn the adjusting screw until the specified dwell angle is obtained.

Dwell Angle—H.E.I. System (1975 and later)

The dwell angle is fixed and is not adjustable. No attempt should be made to adjust the unit.

Timing Light Connections— H.E.I. System

Timing light connections should be made in parallel using an adapter at the distributor No. 1 terminal.

Tachometer Connections— H.E.I. System

There is a "tach" terminal on the distributor cap or on the remote-mounted coil. Connect the tachometer to this terminal and ground. Follow the tachometer manufacturer's instructions. CAUTION: **Grounding the tach terminal could damage the H.E.I. ignition module.**

Ignition Timing

On conventional ignition engines,

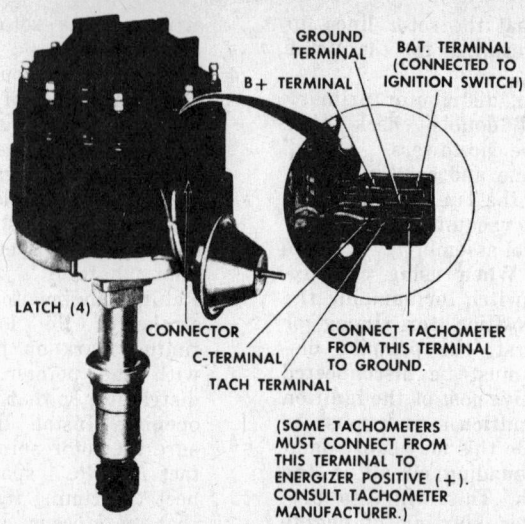

V8 HEI distributor tachometer connection point (© Chevrolet Div., GM Corp.)

remove the spark plug wire from No. 1 plug and attach a timing light between the wire and the plug. On H.E.I. systems, connect the timing light in parallel at the No. 1 tower on the distributor cap. Disconnect the distributor spark advance hose and plug the vacuum opening. Start the engine and run it at idle speed. Aim the timing light at the degree scale just over the harmonic balancer. The markings on the scale are in 2° increments with the greatest number of markings on the *before* side of the 0. Adjust the timing by loosening the securing clamp and rotating the distributor until the desired ignition advance is achieved, then tighten the clamp. To advance the timing, rotate the distributor opposite to the normal direction of rotor rotation. Retard the timing by rotating the distributor in the normal direction of rotor rotation.

NOTE: On conventional ignition engines, if engine miss or rough idle

occurs, connect the dwell meter and accelerate the engine to 1700 rpm. If the dwell reading varies more than 3°, check the distributor for worn distributor shaft, worn bushings or a loose distributor breaker plate.

Alternator

Delcotron, the alternator by Delco-Remy is used on Chevrolet trucks. These units are furnished in two types with companion voltage regulators, which will be a two-unit external regulator or transistorized internal regulator.

Repair and test details on the alternator and its regulators are covered in the General Repair Section.

Alternator Removal and Installation

1. Disconnect the battery ground strap at battery to prevent dam-

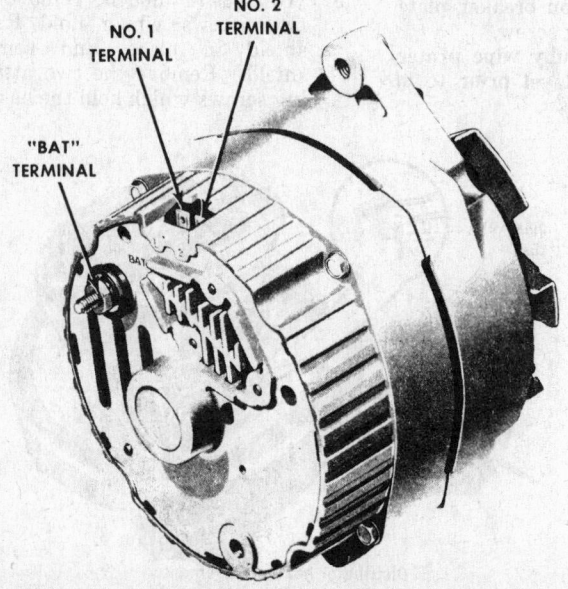

Alternator wiring terminals (© Chevrolet Div., G.M. Corp.)

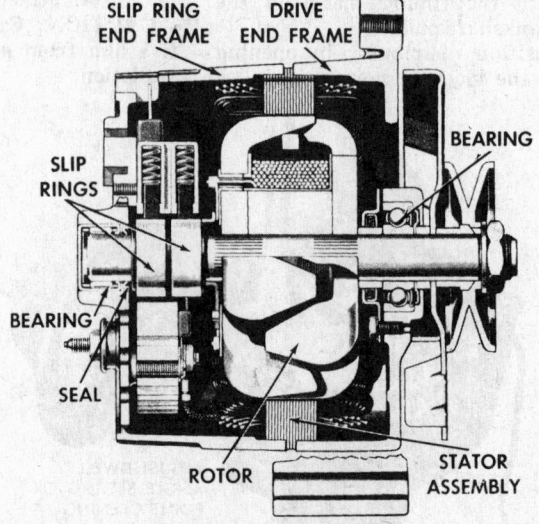

Cross section of alternator (© Chevrolet Div., G.M. Corp.)

aging diodes or wiring harness (also prevents accidentally reversing polarity).

2. Disconnect wiring leads at Delcotron.

3. Remove the alternator brace bolt, (if power steering equipped, loosen pump brace and mount nuts) then detach drive belt(s).

4. Support the generator and remove alternator mount bolt (6.2″ Delcotron uses 2 mounting bolts) or bolts and remove from vehicle.

5. Reverse the removal procedure to install, then adjust drive belt(s).

6. If no belt tension tool is available, force alternator away from the engine until fan belt has 5/16 in. deflection when forced downward from normal position with light pressure applied between the alternator and the fan.

CAUTION: Since the Delcotron and regulator are designed for use on a single polarity system, the following precautions must be observed:

1. The polarity of the battery, generator, and regulator must be matched and considered before making any electrical connections in the system.

2. When connecting a booster battery, be sure to connect the negative battery terminals with one another, and the positive battery terminals with one another.

3. When connecting a charger to the battery, connect the charger positive lead to the battery positive terminal. Connect the charger negative lead to the battry negative terminal.

4. Never operate the Delcotron on uncontrolled open circuit. Be sure that all connections in the circuit are clean and tight.

5. Do not short across or ground any of the terminals on the Delcotron regulator.

6. Do not attempt to polarize the Delcotron.

7. Do not use test lamps of more than 12 volts for checking diode continuity.

8. Avoid long soldering times when replacing diodes or transistors. Prolonged heat is damaging to these units.

9. Disconnect the battery ground terminal when servicing any AC system. This will prevent the possibility of accidentally reversing polarity.

External Voltage Regulator

Removal and Installation

1. Disconnect the ground cable from the battery.

2. Disconnect the wiring harness from the regulator.

3. Remove the mounting screws and remove the regulator.

4. Make sure that the regulator base gasket is in place before installation.

5. Clean the attaching area for proper grounding.

6. Install the regulator. Do not overtighten the mounting screws, as this will cancel the cushioning effect of the rubber grommets.

Voltage Adjustment

10-30 Series (1971-72)

1. Insert a ¼ ohm-25 watt fixed resistor into the charging circuit at the horn relay junction block, between both leads and the terminal.

2. Install a voltmeter as shown in the figure.

3. Warm the engine by running it for several minutes at 1,500 rpm or more.

4. Cycle the voltage regulator by disconnecting and reconnecting the regulator connector.

5. Read the voltage on the voltmeter. If it is between 13.5 and 15.2, the regulator does not need adjustment or replacement. If the voltage is not within these limits, leave the engine running at 1,500 rpm.

6. Disconnect the four-terminal connector and remove the regulator cover (except on transistorized regulators). Reconnect the four-terminal connector and adjust the voltage to between 14.2 and 14.6 volts by turning the adjusting screw while observing the voltmeter.

7. Disconnect the terminal, install the cover, and then reconnect the terminal.

8. Continue running the engine at 1,500 rpm to re-establish the regulator internal temperature.

9. Cycle the regulator by disconnecting/reconnecting the regulator connector. Check the voltage. If the voltage is between 13.5 and 15.2, the regulator is good.

CAUTION: Always disconect the regulator before removing or installing the cover in order to prevent damage by short-circuiting.

Starter

Starter Motor Removal and Installation

The following procedure is a general guide for all vehicles and will vary slightly depending on the truck series and model. (On Forward Control Vans, vehicle must be raised and supported.)

1. Disconnect battery ground cable at the battery.

2. Disconnect engine wiring harness and battery leads at solenoid terminals.

3. Remove starter mounting bolts and retaining nuts and disengage starter assembly from the flywheel housing. Light duty gasoline powered engines use conventional nose housing or pad mounting. Intermediate and heavy duty use conventional flange. On these, scribe mark on flange and flywheel housing as nose housing can be mounted in several positions.

4. Position starter motor assembly to the flywheel housing and install the mounting bolts and retaining nuts. Torque the mounting bolts 25 - 35 ft. lbs.

5. Connect all wiring leads at the solenoid terminals.

6. Connect the battery ground cable and check operation of the unit.

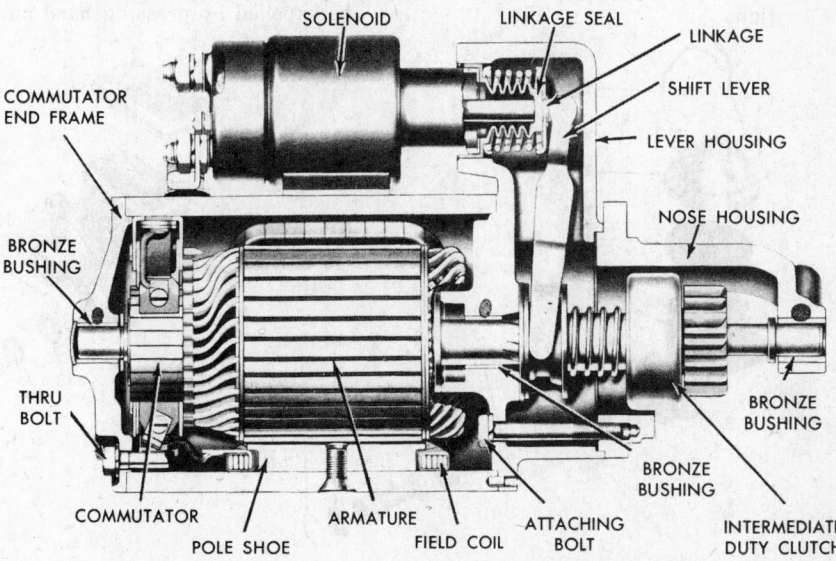

Overrunning clutch type starter (© Chevrolet Div., G.M. Corp.)

Brakes

Specific information will be found in General Brake Section on adjustments, bleeding, master cylinder and wheel cylinder overhaul procedures and trouble shooting.

Refer to Power Brake Section for details concerning power brakes.

Master Cylinders

Twin (Brake & Clutch) Type Master Cylinder

Removal

1. Disconnect clutch and brake pedal return springs.
2. Detach push rod boots from cylinders.
3. Remove clutch and brake hydraulic lines.
4. Remove three bolts holding cylinder to dash and slide cylinder off push rod.

NOTE: Wipe hydraulic fittings clean and place dry cloth under lines to absorb any fluid spillage. Cover fittings and lines to prevent any foreign matter from entering system.

Installation

1. Place new gaskets and push rod boots over cylinder tubes.
2. Hold cylinder next to dash and insert push rods, making sure they are centered.
3. Bolt assembly loosely to dash. This freedom of assembly will allow hydraulic lines to be started in cylinder without stripping fittings.
4. Tighten assembly and hydraulic lines securely.
5. Replace pedal return springs.
6. Fill reservoir and bleed both clutch and brake cylinders.
7. Check pedal free play and operation.

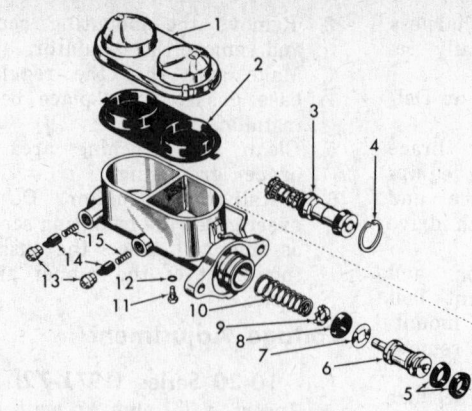

1. Cover
2. Diaphragm
3. Rear piston assembly
4. Snap ring
5. Secondary cups
6. Front piston
7. Cup protector
8. Primary cup
9. Cup retainer
10. Front piston return spring
11. Piston top screw
12. "O" ring
13. Tube seat inserts
14. Check valves
15. Check valve springs

Dual type master cylinder   (© Chevrolet Div., G.M. Corp.)

Standard Conventional Single Master Cylinder

Removal

1. Clean area at fitting and place dry cloth under line to absorb leakage.
2. Disconnect hydraulic line at cylinder and cover ends with clean cloth to prevent any foreign matter from entering system.
3. Disconnect push rod from pedal.
4. Remove the two nuts and washers holding cylinder to firewall. Remove Cylinder.

Installation

1. Position cylinder at dash, align push rod through boot and secure loosely. This freedom of assembly will allow hydraulic line and push rod to be installed with minimum of effort.
2. Tighten nuts and line. Check free play.
3. Fill cylinder and bleed. Bleeding can be accomplished by slowly pressing down on brake pedal and at same time tightening the hydraulic line fitting. Any air still trapped in line at fitting can be expelled by pressing hard on

brake pedal to a point of just below free play. By having a slight pressure on pedal, the piston is held forward enough to clear ports in reservoir and check valve is held off its seat, so air can be released.
4. Final check of fluid and brake operation.

Standard Dual Master Cylinder

Removal

1. Clean area at fittings and place dry cloth under lines to absorb leakage.
2. Disconnect both lines at cylinder and cover to prevent foreign matter from entering system.
3. Disconnect any stop light or brake warning light wires.
4. Unbolt cylinder and remove, allowing push rod to fall loose.

Installation

1. Install new boot on push rod.
2. Position cylinder making certain push rod and boot are in proper position and fasten loosely to firewall. This freedom of assembly will allow both hydraulic lines to be started easily.

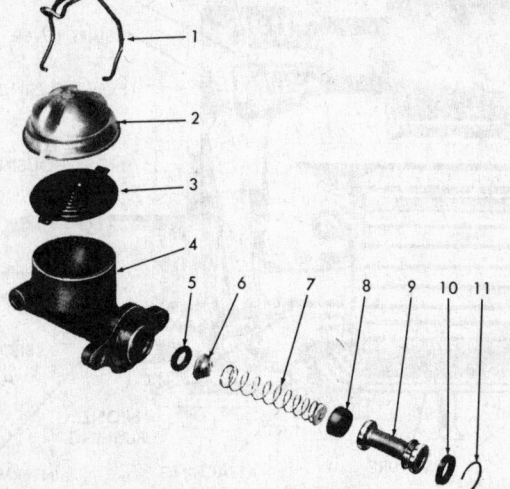

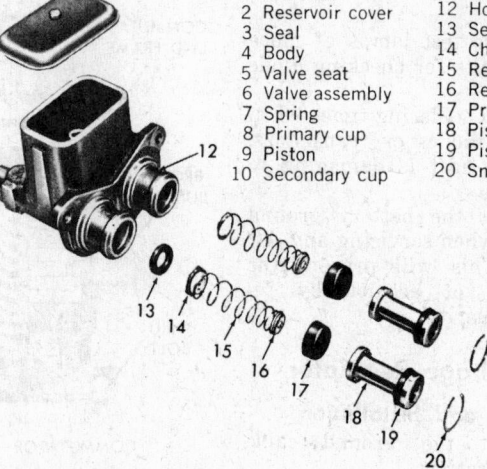

1. Bail wire
2. Reservoir cover
3. Seal
4. Body
5. Valve seat
6. Valve assembly
7. Spring
8. Primary cup
9. Piston
10. Secondary cup
11. Lock ring
12. Housing
13. Seal
14. Check valve
15. Return spring
16. Retainer
17. Primary cup
18. Piston
19. Piston seal
20. Snap ring

Single & clutch assist type cylinders (© Chevrolet Div., G.M. Corp.)

3. Tighten mounting nuts and lines. Check free play.
4. Connect any stop light or brake warning light wires.
5. Fill reservoir and bleed. Test brakes before moving truck.

Wheel Cylinders

Four types of cylinders are used, identified by type of brake system used.

Duo servo—One double-end cylinder mounted at toe ends of shoes.

Twinplex—Two double-end cylinders mounted between shoes at toe and heel.

Wagner F (FA)—Two single end cylinders (single piston, single direction) mounted so as to be an anchor for one and powering other.

Wagner FR3 (FR3A)—Two double-end cylinders mounted between shoes.

Wheel Cylinder

Removal

1. Jack up axle and support. Remove wheel and drum.
NOTE: To remove drum it may be necessary to back off brake adjustment, also if rear drum, release hand brake cable if so equipped.
CAUTION: To gain access to adjusting starwheel, a knockout lanced area is located in web of drum. After knocking out metal be sure to clean all metal particles from brake compartment. A new cover plug must be installed.
2. Release shoe return springs and spread shoes to clear wheel cylinder links. Make sure any lubricant or brake fluid does not get on facings by covering same.
3. At front—Disconnect metal line from flexible hose and remove hose if accessible or remove hose later after cylinder is removed. At rear—Disconnect metal line from cylinder.
4. Remove shield over cylinder and connecting line between cylinders, if so equipped.
5. Remove cap screws and washers holding cylinder to backing plate. Remove cylinder being careful of any fluid spillage.

Installation

1. Clean mounting surface and reverse above procedures.
2. Bleed and readjust brakes. Check pedal before moving vehicle.
NOTE: Twinplex—Upper and lower cylinders are not interchangeable due to position of connector tube openings. Upper cylinder has threaded bleeder valve opening drilled at outer edge of bore.
Wagner F & FA—Two wheel cylinder (upper and lower) are identical, however cylinders on right and left brakes have opposite castings.
Wagner FR3 & FR3A—Upper and lower cylinders on both right and left brakes are interchangeable.

Disc Brakes

Disc Brakes are used on the light trucks, light vans, and on four wheel drive units.

Specific information will be found in the Disc Brake Sections on replacement and overhaul.

Power Brakes

Power Hydraulic

Specific information will be found in Power Brake Section on adjustments, bleeding, overhaul and trouble shooting.

Various vacuum-hydraulic or air-hydraulic brake systems are used, although sizes and shapes differ the basic function is the same.

Master-Vac

The Master-Vac, a self contained hydraulic and vacuum unit is used on light duty trucks. This hydrovac is of the diaphragm type. The multi-vac unit was designed for use with the low input-high output system, while the newer hydrovac is used in equal displacement system. In equal displacement hydraulic system the fluid displaced by the master cylinder is equal to the fluid displaced by the power cylinder. Vacuum powered cylinders on 40-50-60 series, with hy-

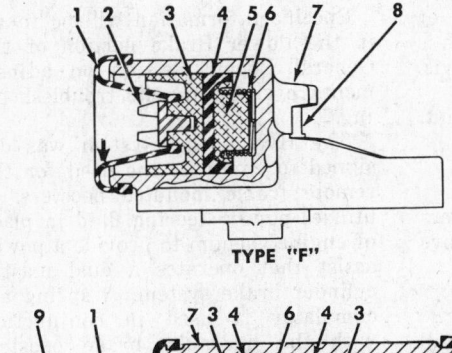

TYPE "F"

TYPE "FR-3"

Duo servo and Twin Plex cylinders
(© Chevrolet Div., G.M. Corp.)

1 Boot
2 Brake shoe guide
3 Piston
4 Piston cup
5 Cup filler
6 Piston spring
7 Cylinder
8 Brake shoe anchor slot
9 Push rod

DUO-SERVO

TWINPLEX

Wheel cylinder used with "F", "FA", "FR3" & "FR3A" brakes
(© Chevrolet Div., G.M. Corp.)

1 Push rod
2 Boot
3 Piston
4 Piston cup
5 Spring
6 Housing

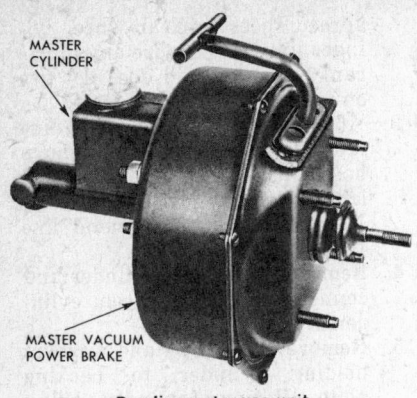

MASTER
CYLINDER

MASTER VACUUM
POWER BRAKE

Bendix master vac unit
(© Chevrolet Div., G.M. Corp.)

draulic brakes, are either single or tandem diaphragms. The tandem diaphragm unit is used with single master cylinder. Single diaphragm units, two used, are on dual cylinder units. One for each system.

Air-Pak

Air-pak is a self contained hydraulic-air pressure power unit for use on trucks with an air compressor, reservoir, regulator and check valve. It consists of three elements.

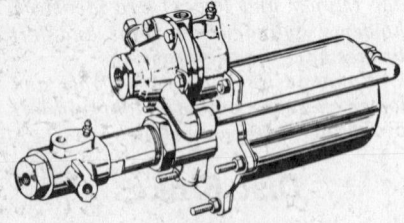

Air-Pak unit (© Chevrolet Div., G.M. Corp.)

1. An air cylinder, consisting of a cylinder, a piston and a push rod that connects power piston to hydraulic piston.
2. A hydraulic cylinder, consists of a hydraulic piston with built in check valve, a residual check valve and a compensating check valve.
3. An air pressure control valve, which controls the power output of the air pressure cylinder in relation to the hydraulic pressure in the master cylinder.

The operation of the air-pak is similar to the multi-pak unit. The difference between units is that the air-pak is designed to operate in equal displacement hydraulic systems. The multi-pak is made to work in the low-input, high-output systems. In the equal displacement system, the fluid output of the master cylinder is equal to the fluid output of the power cylinder.

NOTE: The correct master cylinder MUST be used with the proper power unit. Air-pak and multi-pak units are NOT interchangeable.

Power-Vacuum

Removal
NOTE: Wipe hydraulic fittings clean, place dry cloth under lines to absorb any fluid leakage, cover lines to keep system clean.
1. Disconnect push rod clevis at pedal, if clearance hole in dash is not large enough, remove clevis. (mark position)
2. Remove vacuum hose from unit. (check valve)
3. Disconnect hydraulic lines.
4. Remove any stop light wires.
5. Remove 4 nuts and washers holding unit to firewall, remove unit (and bracket)

Installation
1. Mount unit in place and install loosely (freedom of assembly will allow easy starting of fittings), secure push rod to pedal and check free-play.
2. Tighten mounting nuts and hydraulic lines.
3. Install vacuum line.
4. Connect any stop light wire.
5. Bleed brakes, (bench-bleed unit before installing, units with 2 bleeder valves. Bleed valve nearest to shell first).
CAUTION: Pressure bleeding must be done with engine off (no vacuum). In manual bleeding, use engine (start engine, allow vacuum to build up).
6. Check brakes and stoplight before moving vehicle.
7. Units requiring lubrication, remove 1/8 inch pipe plug in

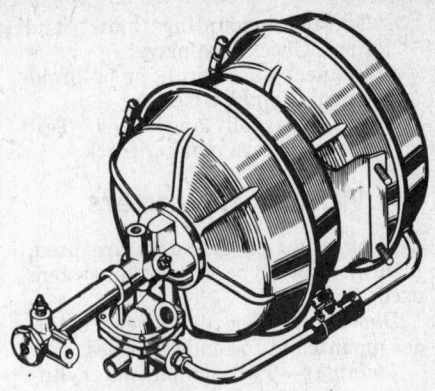

Dual vacuum rooster unit
(© Chevrolet Div., G.M. Corp.)

front end of shell (engine off). Fill with vacuum cylinder oil until oil runs out filler hole.

Hydro-Boost
Specific information will be found in the Power Brake portion of the General Repair Section on adjustments, overhauling and troubleshooting.

The Hydro-boost system was designed to eliminate the need for the remote frame mounted boosters. It utilizes power steering fluid in place of engine vacuum to provide a power assist that operates a dual master cylinder brake system. A spring accumulator is used in conjunction with the hydraulic brake booster. The accumulator is a sealed hydraulic cylinder with a port at each end. On 1974 models the accumulator and booster are mounted separately. In 1975 all "C" model vehicles equipped with the hydro-boost incorporate an accumulator which is integral with the booster.

CAUTION: The accumulator contains a spring compressed under high pressure. Any attempt to disassemble or cut could cause personal injury.

Hydraulic Brake Booster Removal and Installation

Motor Home Chassis
1. Make sure all pressure is discharged from the accumulator

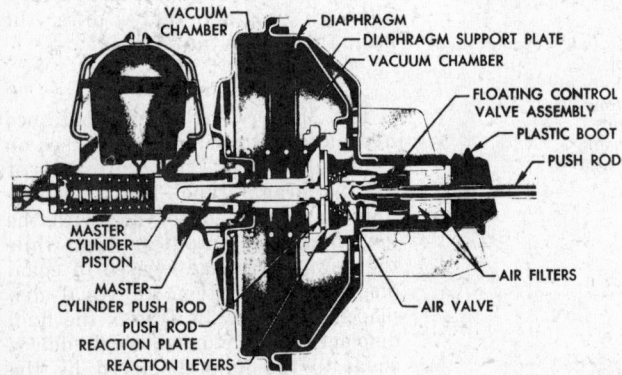

VACUUM CHAMBER — DIAPHRAGM — DIAPHRAGM SUPPORT PLATE — VACUUM CHAMBER — FLOATING CONTROL VALVE ASSEMBLY — PLASTIC BOOT — PUSH ROD — AIR FILTERS — AIR VALVE

MASTER CYLINDER PISTON — MASTER CYLINDER PUSH ROD — PUSH ROD REACTION PLATE — REACTION LEVERS

Moraine unit (© Chevrolet Div., G.M. Corp.)

Hy-Power vacuum booster unit
(© Chevrolet Div., G.M. Corp.)

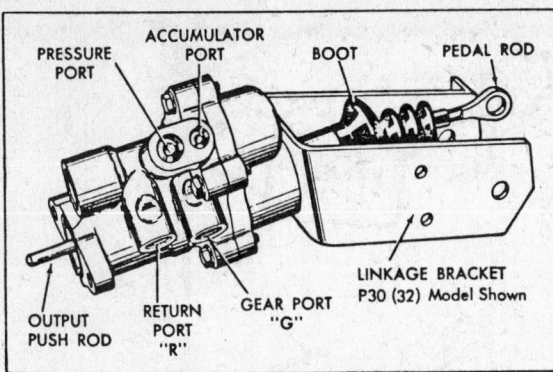

Hydro-Boost—1974 (© Chevrolet Div., G.M. Corp.)

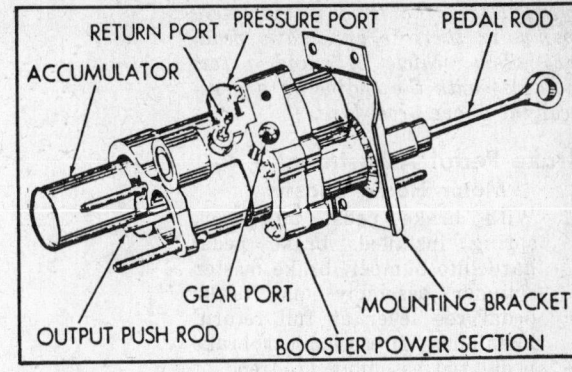

Hydro-Boost—1975 Cabs, Suburban (© Chevrolet Div., G.M. Corp.)

by depressing and releasing the brake pedal several times.
2. Raise the vehicle on a hoist.
3. Clean all the dirt from the booster at the hydraulic line connections and master cylinder.
4. Remove the nuts and lockwashers that secure the master cylinder to the booster and support bracket. Support the master cylinder leaving the hydraulic lines attached to the master cylinder.
5. Disconnect and plug the hydraulic lines from the booster ports.
6. Remove the cotter pin, nut, bolt and washers that secure the operating lever to the vertical brake rod.
7. Remove the six nuts, lockwashers and bolts that secure the booster linkage bracket to the front and rear support brackets, then slide the booster off the rear support studs and remove the booster from the vehicle.
8. Remove the cotter pin, nut, washer and bolt that secures the operating lever to the pedal rod.
9. Remove the brake pedal rod lever nut and bolt and then remove the lever, sleeve and bushings.
10. To install reverse the removal procedures. Bleed the booster-power steering hydraulic system and check the brake pedal and stoplamp switch adjustment.

Conventional Cab & Suburban
1. Make sure all pressure is discharged from the accumulator

by depressing and releasing the brake pedal several times.
2. Remove the nuts and lockwashers that secure the master cylinder to the booster and support bracket. Support the master cylinder leaving the hydraulic lines attached to the master cylinder.
3. Remove the booster pedal push rod cotter pin, and washer and disconnect the push rod from the brake pedal.
4. Remove the booster support bracket.
5. Remove the booster bracket to dash panel or support bracket nuts and remove the booster assembly.
6. To install reverse the removal procedure. Bleed the booster-power steering hydraulic system and check brake pedal and stop-lamp switch adjustment.

Forward Control Chassis—(Vans)
1. Make sure all pressure is discharged from the accumulator by depressing and releasing the brake pedal several times.
2. Clean all the dirt from the booster at the hydraulic line connections and master cylinder.
3. Remove the nuts and lockwashers that secure the master cylinder to the booster and the support bracket. Support the master cylinder leaving the hydraulic lines attached to the master cylinder.
4. Remove the booster pedal push rod cotter pin and washer and

disconnect the push rod from the booster bracket pivot lever.
5. Remove the booster supper braces.
6. Remove the booster bracket nuts and remove the booster.
7. To install reverse the removal procedure. Bleed the booster-power steering hydraulic system and check the brake pedal and stop lamp switch adjustment.

Bleeding Hydro-Boost System
1. Fill the power steering pump to the proper level.
2. Start the engine for approximately two seconds. Check the fluid and add if necessary.
3. Repeat step 2 until the fluid level remains constant.
4. Raise the front end of the vehicle so that the tires are clear of the ground.
5. Start the engine and run approximately 1500 rpm. Depress and release the brake pedal several times then turn the steering wheel right and left, lightly contacting the wheel stops.
6. Turn off the engine and check the fluid level in the reservoir and add fluid if necessary.
7. Lower the vehicle, start the engine and run at approximately 1500 rpm. Depress and release the brake pedal several times and turn the steering wheel to full right and left.
8. Turn the engine off and check the fluid level in the reservoir. Add fluid if necessary.

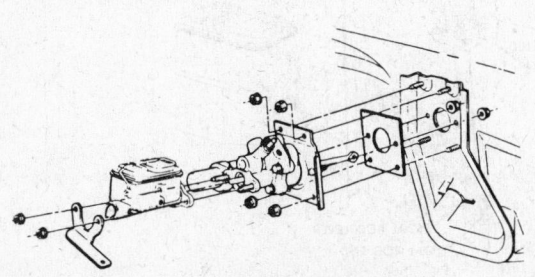

Booster Installation—Cab & Suburban
(© Chevrolet Div., G.M. Corp.)

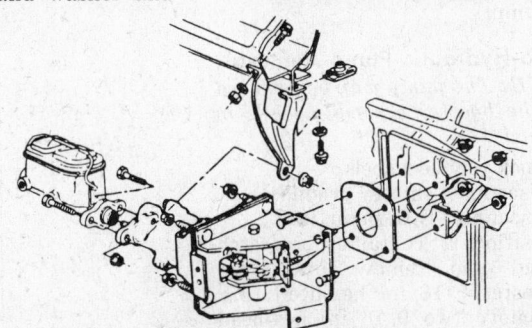

Booster Installation—Vans
(© Chevrolet Div., G.M. Corp.)

NOTE: If the fluid is extremely foamy, or there is an erratic pedal feel allow vehicle to stand a few moments with the engine off and repeat the above procedure.

Brake Pedal Adjustment
Motor Home Chassis

1. With brake pedal pull back spring installed, brake pedal hard into bumper, brake master cylinder assembly and brake pedal rod lever at full return, install the preassembled brake pedal rod assembly (rod end at boot) and adjust to 31.75".
2. Turn the brake pedal rod end and adjust the free pedal travel to .06" to .36".
3. Fasten the boot to the floor pan assembly and compress the boot to 2.54" installed height.

Forward Control Chassis—(Vans)

1. Adjust the pedal rod to 9.90" and install in the vehicle.

NOTE: The brake pedal push rod is not adjustable on cab and Suburban models.

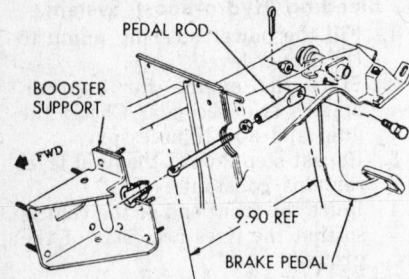

Brake pedal adjustment—forward control chassis (© Chevrolet Div., G.M. Corp.)

Electro-Hydraulic

The booster unit is hydraulically operated with an electrical pump attached as a back-up unit in the event of primary pump failure. The booster system features a dual braking system, increased pressure output, excellent pedal feel, separate hydraulic pump, and an electrical powered back-up system.

The hydraulic booster is powered by a standard power steering vane type pump.

Electro-Hydraulic Pump Removal

NOTE: The pump may be removed from the booster assembly while in the vehicle.

1. Block vehicle wheels.
2. Disconnect battery ground.
3. Disconnect E.H. pump lead.
4. Position a container to catch fluid and remove bracket-to-booster 9/16 in. hex-head bolt.
5. Remove two 9/16 in. hex-head mounting bolts. Remove pump and two O-rings.

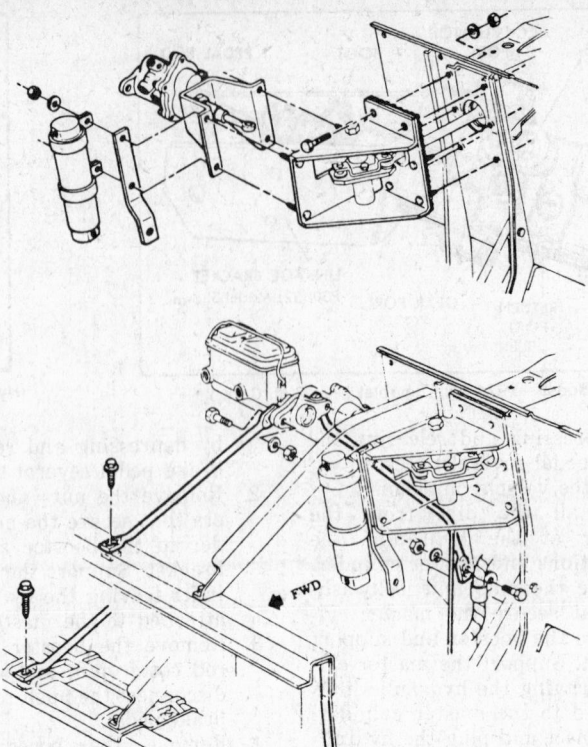

Booster Installation—Forward Control chassis exc. Light duty Vans (© Chevrolet Div., G.M. Corp.)

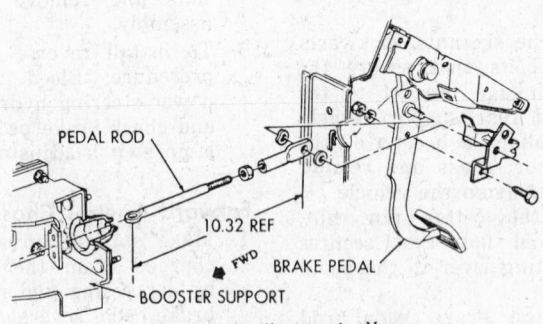

Brake pedal adjustment—Vans
(© Chevrolet Div., G.M. Corp.)

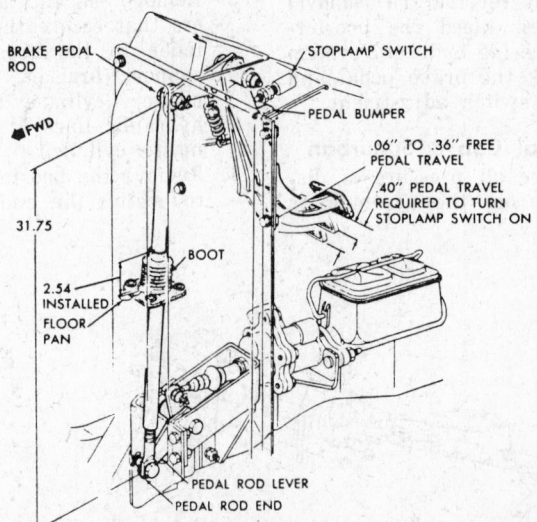

Brake pedal and stop lamp adjustments—Motor Home Chassis (© Chevrolet Div., G.M. Corp.)

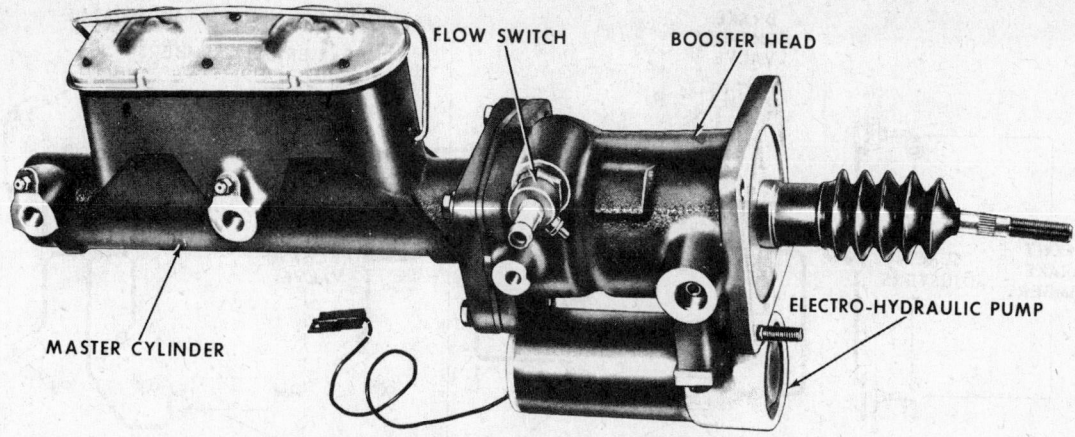

Electro-Hydraulic brake booster and master cylinder (© G.M.C.)

Installation

1. To install reverse the removal procedure. Use new O-rings and torque the two ⅜-16 x 1⅛ in mounting bolts to 16-30 ft. lbs.

Booster Assembly Removal
Conventional Cab

1. Block vehicle wheels.
2. Disconnect battery ground cable.
3. Disconnect electrical leads to E.H. pump and flow switch.

NOTE: Booster head may be removed from vehicle without removing master cylinder and disconnecting brake lines. Remove two nuts retaining brake line clips to line support. Remove bolt and clips in booster head. Remove four 9/16 in. hex head master cylinder to booster head mounting bolts and move master cylinder forward from booster. Secure in an upright position. Complete booster assembly removal procedure following:

4. Disconnect hydraulic lines from booster. Use a container to catch fluid. DO NOT reuse fluid.
5. Remove cotter key from push rod pin. Remove nut and bolt from pedal push rod eye. Thread push rod and nut from booster push rod.
6. Remove line and hose supports from booster head.

NOTE: Use care when proceeding to next two steps. This is a heavy unit, approximately 50 pounds including master cylinder, and should be handled as such. After removing from vehicle, use care in handling so that flow switch or E.H. pump is not damaged and bail or cover on master cylinder is not damaged.

7. Remove two ⅜-24 thread hex-head nuts and washers from inside cab at dash panel.
8. Remove upper two mounting bolts and washers from booster head at dash. Remove booster assembly.
9. Remove four 9/16 in. hex-head bolts which attach brake master cylinder to booster head. Remove

master cylinder, gasket, and brake line support.

Installation

1. To install reverse the removal procedure. Torque the four ⅜-16 x 1⅜ in. hex-head bolts attaching the master cylinder to the booster head to 16-30 ft lbs. The two hex-nuts on the ⅜-24 in. studs are torqued to 25-30 ft lbs. The two mounting bolts are torqued to 25-30 ft lbs.

Tilt Cab

1. Block vehicle wheels.
2. Disconnect battery ground cable.
3. Disconnect electrical lead to E.H. pump.
4. Disconnect hydraulic lines from booster and master cylinder. Use a suitable container to catch fluid. Do not reuse fluid.
5. Loosen nuts on each end of push rod extension. Turn extension until free of booster push rod. Bellcrank ball joint may have to be removed to facilitate turning extension.
6. Remove two 9/16 in. hex-head bolts and nuts retaining support to cab sill.
7. Remove two 9/16 in. hex-head bolts and nuts retaining bracket to sill.
8. Remove booster assembly. Move push rod end forward and down while revolving unit to clear E.H. pump, support, and bracket.
9. Remove four 9/16 in. hex-head bolts which attach brake master cylinder to booster head and remove master cylinder and gasket.

Flow Switch Removal

1. Position container to catch fluid from disconnected hose at flow switch.
2. Loosen outer hose clamp at switch and remove hydraulic line from hose. Drain fluid into previously placed container. DO NOT reuse fluid.

3. Loosen inner hose clamp at switch and remove hose.
4. Disconnect electrical lead to switch.
5. Remove switch using a one-inch thin blade wrench.

Installation

1. Use a new O-ring seal and torque to 20-30 ft lbs.

Stop Lamp Switch Adjustment

1. Release the brake pedal to its normal position.
2. Loosen the switch locknut and rotate the switch in its bracket. Electrical contact should be made when pedal travel is ⅜"-⅝".

Air Brakes

Specific information will be found in Air Brake Section on adjustments, overhauling and troubleshooting.

Full air brakes completely replace ALL hydraulic parts with more durable components, capable of producing and using greater braking energy.

Brakes are applied by pushing on the pedal, which controls the application valve. Varying amounts of pressurized air will fill the brake chambers depending on brake pedal travel. Cam type shoe actuators (wedge type on stopmaster) are connected to push rods attached to diaphragms in the brake chamber. When air pressure is passed to brake chambers, the diaphragm then converts air pressure energy to mechanical force, the pressured diaphragms move the cam type actuators (wedge type on stopmaster) spreading the brake shoes and thereby applying brakes.

When the brake pedal (application valve) is released, a rapid discharge of air pressure from brake chambers is necessary to speed brake shoe release. A front and rear quick release valve aids in this function.

Many safety devices are used. A

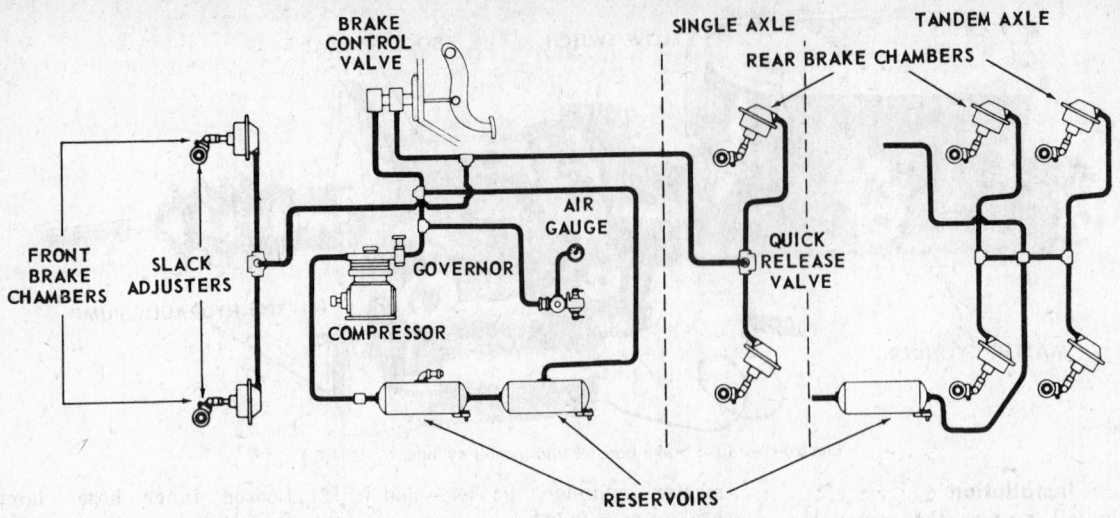

Full air brake system (© Chevrolet Div., G.M. Corp.)

low air pressure warning buzzer sounds when pressure falls below a safe level. An air pressure gauge on dash shows the air pressure in system. Normal air pressure, for brake application, is at least 70 lbs. "Wet" and "dry" reserve tanks (also called primary and secondary) serve to remove moisture from air and also to provide a reserve of braking power. Drain cocks in tanks are provided to drain condensation. A pressure relief valve on the "wet" tank will release pressures over 150 lbs. A check valve located ahead of "wet" tank will retain air pressure in event of compressor failure or leaks.

Components

Compressor

Belt driven on gas engines and usually gear driven on diesel.

The air compressor serves only to supply and maintain sufficient pressure for brakes and air operated accessories. When pressure in system reaches top of normal range, an un-

loading valve opens and nullifies compressor action.

The average compressor is a single stage reciprocating piston type, usually one cylinder. Larger units are two cylinders. Compressors are lubricated by the engine system.

NOTE: Water cooled compressors, in event of freezing weather, must be drained as well as engine block.

Governor

Controls load and unload mechanism to automatically maintain maximum and minimum air pressures in reservoirs. Pressure ranges or settings are adjustable. The governor, by regulating the load and unloading mechanism, establishes an intermittent compressor pumping cycle.

Brake Control (Application) Valve (Foot Operated)

Provides quick and sensitive control of air pressure (FORCE) from reservoir to brake chambers. The amount of force applied to brakes is

proportional to the amount of pedal depression.

Reservoir(s)

"Wet and Dry" tanks serve to remove moisture and provide a sufficient reserve of air under pressure for several brake applications (safety factor). Drain cocks are provided to drain condensed moisture. A dash mounted gauge will show amount of reservoir pressure.

Safety Valve

Usually mounted on reservoir, allows air to escape when air pressure exceeds a predetermined setting (adjustable).

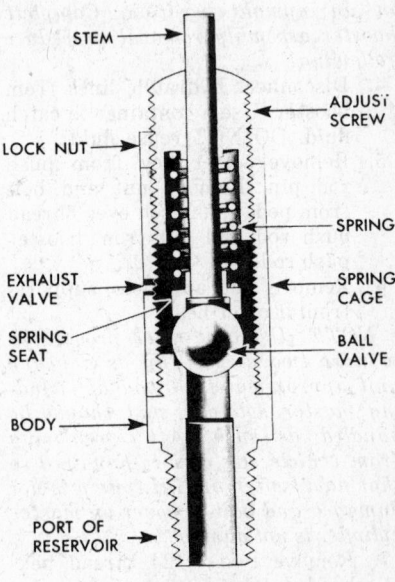

Safety valve
(© Chevrolet Div., G.M. Corp.)

Check Valve

Between "Wet" tank and compressor to retain air pressure in the event of compression (compressor or lines) failure.

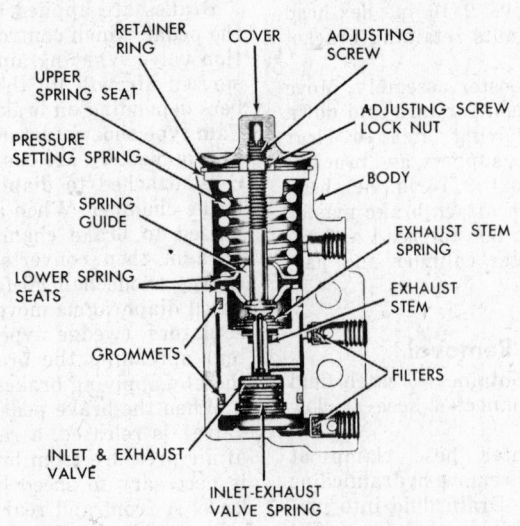

Governor assembly (© Chevrolet Div., G.M. Corp.)

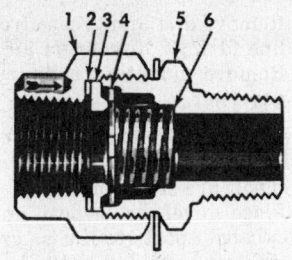

1 Valve body 4 Valve disc
2 Washer 5 Screw cap
3 Valve seat 6 Valve spring

Air tank check valve
(© Chevrolet Div., G.M. Corp.)

Low Pressure Signal

A safety device (buzzer) sounds when air pressure is absent or low.

Air Gauge

Located on instrument panel, shows air pressure in system, works in conjunction with low pressure switch to warn of low pressure.

Pressure Protection Valve

Mounted in the delivery port of application valve. Its function is to close all air lines to auxiliary systems in event of loss of air pressure.

Relay Valve

A relay station to speed the application and release of brakes because of long air lines and volume of air necessary. A relay valve is usually at rear wheels. It is connected to application valve and meters air directly to rear brake chambers from an auxiliary reservoir.

Quick Release Valve

When brake pedal is released, a rapid discharge of air is necessary to speed the return of brake shoes. Two valves are used, one with front brake chambers and one with rear chambers.

Moisture Ejector Valve

Mounted on bracket on cab step support close to wet air tank. Valve operates when brakes are applied and released to evacuate moisture from system.

Brake Chambers

Converts energy of compressed air into mechanical force required for brake application.

Cam Type

Air, admitted by control valve, en-

ters brake chamber and pressurizes diaphragm with attached push rod. Push rod rotates lever arm of slack adjuster exerting a turning force on camshaft with an "S" design on end. This "S" cam operates between rollers on free ends of brake shoes and serves to expand shoes. Adjustment is manual at slack adjusters.

Wedge Type

Features two brake chambers per wheel. Wedge type actuators, operating between roller assemblies, force *each* shoe evenly against drum. Stop-master brakes have automatic adjusters and do not need slack adjusters. *Fail-Safe* and *Super Fail-Safe* can be operated by either air or spring pressure, with additional features such as spring applied parking brake and is a safety factor in event of air brake failure.

NOTE: These units have a manual release bolt, in center of chamber cap, to permit safe handling for service. See note and caution under R&R.

Slack Adjusters

Used with cam type brakes to provide convenient means of adjustment for brake lining wear. With brakes

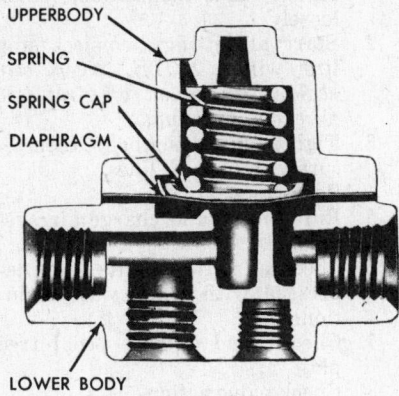

Pressure protection valve
(© Chevrolet Div., G.M. Corp.)

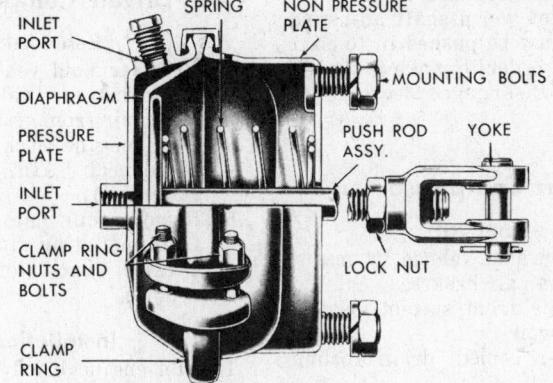

Air brake chamber (© Chevrolet Div., G.M. Corp.)

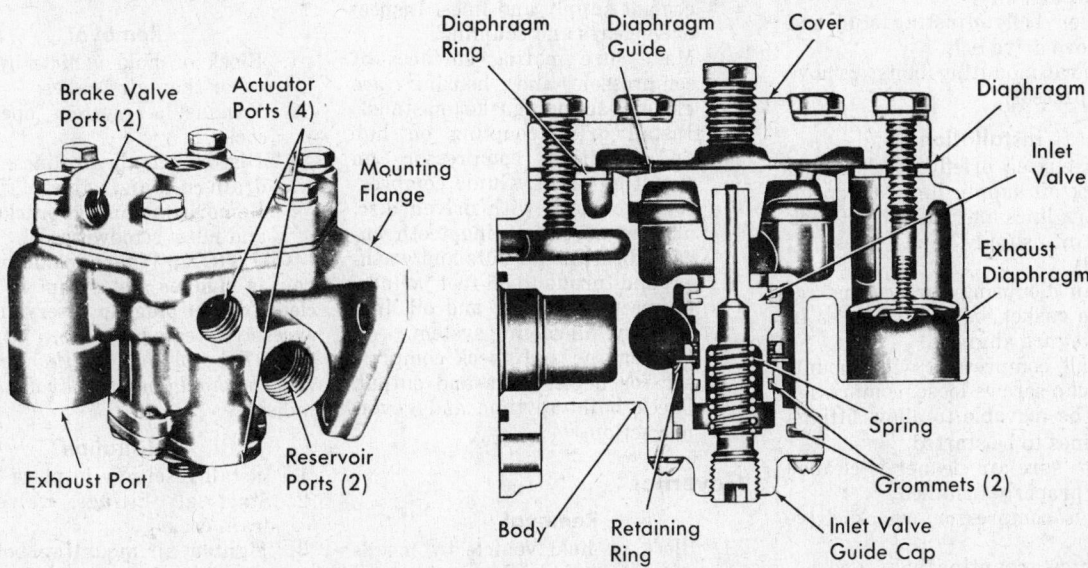

Relay valve components (© Chevrolet Div., G.M. Corp.)

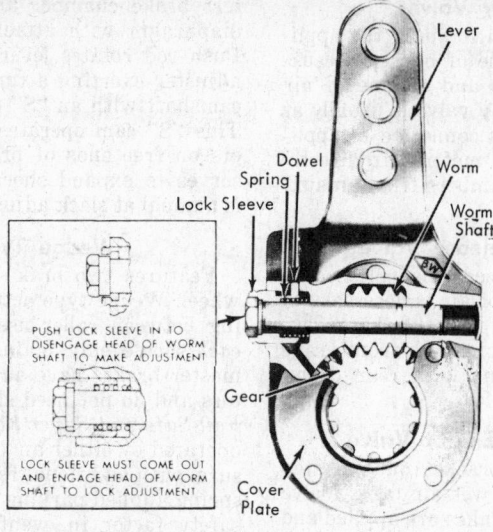

Lever

Dowel

Spring

Lock Sleeve

Worm

Worm Shaft

PUSH LOCK SLEEVE IN TO DISENGAGE HEAD OF WORM SHAFT TO MAKE ADJUSTMENT

Gear

LOCK SLEEVE MUST COME OUT AND ENGAGE HEAD OF WORM SHAFT TO LOCK ADJUSTMENT

Cover Plate

Slack adjusters (© Chevrolet Div., G.M. Corp.)

APPLIED the angle formed by slack adjuster lever and brake chamber push rod should be approximately 90 degrees, and all adjusters to be about the same angle. Excessive travel of push rod shortens the life of chamber diaphragms and also results in slow braking response. Some slack adjusters have a locking sleeve, which engages head of worm shaft adjusting bolt, that must be pushed in to clear bolt head in order to make brake adjustments. Re-engage sleeve when finished.

Belt Driven Compressor

Removal

1. Block or hold vehicle by means other than air brakes.
2. Drain air from system, usually at reservoirs.
3. If water cooled, drain cooling system.
4. Disconnect ALL lines. (air, water and oil).
5. Loosen belt adjusting stud and remove drive belt.
6. Remove mounting bolts, remove compressor.

Installation

1. Run engine briefly to clear and check oil supply lines. Clean oil return lines and passages. Check coolant supply and lines (if used).
2. Clean mounting surface and replace gasket, be sure oil holes in gasket are aligned.
3. Install compressor with mounting cap screws loose, compressor will be movable to allow fittings on lines to be started.
4. Make sure air cleaner is cleaned and properly installed.
5. Align compressor, check drive belt.
6. Tighten mounting bolts and adjust belt tension.

7. Tighten all lines, fill cooling system if drained.
8. Run engine and check compressor for noises, leaks and output. Soapy water will help pin-point any air leaks. Check build-up time and governor.

Gear Driven Compressor

Removal

1. Block or hold vehicle by means other than air brakes.
2. Drain air from system.
3. Drain engine block.
4. Disconnect ALL lines (air, water, oil).
5. Remove four nuts and washers from mounting studs, pull compressor back off studs and remove.

Installation

1. Run engine briefly to clear and check oil supply lines. Check oil return lines and passages. Check coolant supply and lines. Inspect drive gears and coupling.
2. Make sure mating surfaces of compressor and housing are clean. Place new gasket on studs.
3. Install drive coupling on hub and position compressor on mounting studs. Guide compressor into mesh with driven disc, making sure coupling teeth engage disc. Install nuts and washers and torque to 65 foot pounds.
4. Connect water, air and oil lines securely, fill cooling system.
5. Run engine and check compressor for noise, leaks and output. Check build-up time and governor action.

Governor

Removal

1. Block or hold vehicle by means other than air brakes.
2. Drain air system.

3. Remove dirt and grease from air line fittings, disconnect air lines.
4. Remove mounting bolts, remove governor.

Installation

1. Make sure both air lines, to governor, are clean and open.
2. Place governor in position with exhaust port towards ground, tighten bolts finger-tight allowing unit to move. Start fittings.
3. After fittings are started, tighten governor mounting bolts then securely tighten fittings.
4. Test governor and check for leaks.

Application Control Valve

Removal

1. Block or hold vehicle by means other than air brakes.
2. Exhaust air system.
3. Remove all lines and wires.
4. Remove pedal clevis pin and pedal.
5. Remove mounting bolts, remove valve.

Installation

1. Mount valve in position, fasten loosely.
2. Start all fittings, connect stop lite wires. *NOTE. Keep any sealant compound off first two threads of fittings.*
3. Tighten mounting bolts, then securely tighten all lines.
4. Replace pedal.
5. Run engine and charge air system.
6. Check all fittings with pedal depressed with soapy water solution.
7. Check (and adjust) pedal free play.
8. Check valve action.

Reservoir

Removal

1. Block or hold vehicle by means other than air brakes.
2. Exhaust air system, open drain cocks.
3. Disconnect all air lines, remove drain cock and valves.
4. Remove mounting bracket bolts and nuts, remove tank.

CAUTION: Where inside of reservoir is sludged and steam is used to clean, do not plug up reservoir or use excessive steam pressure.

NOTE—In cold weather, more attention should be given to draining of moisture.

Installation

1. Install reservoir in place loosely.
2. Start all fittings, valves and drain cock.
3. Tighten all mounting bolts and nuts to make sure reservoir does not vibrate in service. (Vibra-

tion causes premature line failures).

4. Securely tighten all lines, drain cock and valves.
5. Run engine, charge air system.
6. Check for leaks.

Valves, Signals, and Gauges

Removal

1. Block or hold vehicle by means other than air brakes.
2. Exhaust air system, make certain ignition is off.
3. Clean work area, remove any wires, air lines and brackets. Remove unit.

Installation

1. Position replacement unit in system, making certain of any markings showing air flow direction, and secure.
2. Connect any wires and air lines.
3. Run engine and charge air system.
4. Test for leaks and operation of unit.

Cam Type Chamber (Uses Slack Adjusters, one per wheel)

Removal

1. Disconnect air hose from chamber.
2. Remove clevis pin from yoke.
3. Remove nuts and washers from mounting studs.
4. Remove chamber.

Installation

1. Place chamber on mounting bracket and secure with stud nuts and lock washers.
2. Connect air hose.
3. Install slack adjuster yoke clevis pin, after adjusting for minimum travel. (Angle made by push rod and slack adjuster lever should *not* be less than 90°, brakes applied). Lock yoke with locking nut. Push rod travel should be as short as possible without brakes dragging.
4. Check for leaks, and possible brake shoe adjustment.

Wedge Type Chamber (Two per Wheel)

Removal

1. Block or hold vehicle by means other than air brakes.
2. Disconnect air hose from chamber.
3. Remove lock washer tangs from notches in spanner nut and spider housing.
4. Loosen spanner nut and unscrew air chamber from housing.

Installation

1. Screw air chamber in spider housing until it "bottoms", then

back off (no more than one turn) until chamber air port aligns with air hose. The plastic guide will assure proper position of wedge. Lock brake chamber in position with spanner nut and lock washer.
2. Start engine, charge air system, check for leaks. Pump brake pedal to allow automatic adjusters to adjust brakes.

NOTE: When brakes are equipped with "fail-safe" brake chambers, cage power spring before starting any disassembly or removal of wheels or drums to avoid possible injury. When a vehicle is disabled, due to low or lost air pressure, cage power spring before attempting to move the vehicle. Cage the power spring by rotating the release bolt approximately 18 to 21 turns clockwise. Caging and uncaging can be made easier by applying air pressure, 65 lbs needed (if possible). This takes spring load off release bolt.

CAUTION: **Before removing or caging brake chamber, block wheels since parking brake will not be applied.**

Slack Adjuster

Removal

1. Remove clevis pin at lever.
2. Remove lock ring and washer on splined camshaft (some front slack adjusters are held on by a retaining screw).
3. Slide adjuster off splined shaft.

Installation

1. Make sure spacer washers are in place (if used).
2. Slide slack adjuster on splined camshaft, lock in place with snap ring and washer. If held on by retaining screw, allow .010"

end play. Stake edge of screw to lock.
3. Connect yoke clevis pin. Angle made by push rod and slack adjuster lever should NOT be less than 90° brakes applied. Push rod travel should be as short as possible without brakes dragging.
4. Lubricate adjuster.

Parking Brake Adjustment

NOTE: Except in case of an emergency, set parking brake only after vehicle is brought to a complete stop. Parking brakes are not designed to take the place of service brakes.

Drive Shaft—Band Type

This type, using external band and drum mounted on rear of transmission, provides a hand controlled brake independent of service brakes. Band, with lining, is centered on transmission bracket (anchor) and both sides are arranged to contract equally. Brake is actuated by lever and rod, or cable, to a cam and "J" bolt.

1. Block wheels and release hand brake.
2. Remove locking wire from anchor bolt (19) and adjust to obtain .010" clearance between lining and drum—rewire anchor.
3. Loosen lock nut (6) on locating bolt and tighten adjusting nut (7) until there is a clearance of .020" between lining and drum. Measure clearance about 3 inches from end of lining. Tighten lock nut.
4. Loosen lock nut (12) on adjusting "J" bolt (17) and tighten adjusting nut (11) to obtain a

1	Brake band
2	Cams
3	Links
4	Clevis pins
5	Cam shoe
6	Lock nut
7	Adjusting nut
8	Locating bolt
9	Tension spring
10	Washer
11	Adjusting nut
12	Lock nut
13	Lock washer
14	Brake lining
15	Release spring
16	Brake drum
17	Adjusting bolt
18	Anchor bar
19	Anchor screw
20	Lock wire

Contracting bank (external) parking brake
(© Chevrolet Div., G.M. Corp.)

1 Brake cable
2 Adjustable clevis
3 Connecting lever
4 Return spring
5 Link
6 Lock nut
7 Operating lever
8 Inner shoe
9 Outer shoe
10 Brake drum
11 Adjusting nut
12 Lock nut
13 Adjusting bolt
14 return spring
15 Adjuster bracket

Two shoe (Duo Grip) parking brake
(© Chevrolet Div., G.M. Corp.)

1 Return springs
2 Anchor pin link
3 Camshaft
4 Control lever
5 Link
6 Relay lever
7 Brake shoe
8 Plate bolt
9 Adjusting screw spring
10 Adjusting screw
11 Support plate

Internal expanding type brake—installed
(© Chevrolet Div., G.M. Corp.)

clearance of .020″ approximately 3 inches from end of lining. Tighten lock nut.

5. Check hand brake operation by applying lever. If more than ½ the number of notches on sector are needed to lock the propeller shaft, make final adjustment on lever rod, at clevis end. Remove blocks.

Propeller Shaft—Shoe Type

Hand brake drum is mounted on rear of transmission with internal and external shoes (duo-grip) or with two internal expanding shoes. Shoes are forced against drum by lever and cam actuated by cable. Propeller shaft brakes lock the driveline for parking.

Duo-Grip

1. Block wheels and release hand brake.
2. Loosen lock nut (12) and tighten adjusting bolt (13) to obtain .010″ clearance between outer shoe lining and drum at this point. Hold bolt and lock nut securely.
3. Loosen lock nut (6) and tighten adjusting nut (11) to obtain .010″ between inner shoe lining and drum, at center of shoe. Hold nut (11) and tighten nut (6).

NOTE: Some vehicles, such as tilt cab, use a compound hand brake lever with flexible cable that features a cable adjustment. This allows the operator to adjust the "over-center" position of lever required to lock parking brake.

Two Shoe Internal Expanding

1. Block wheels and release hand brake, it may be necessary to remove clevis pin to assure full release of shoes.
2. Align slot in drum with star wheel adjuster. Knock out lanced area in drum, if necessary to gain access to star wheel. Be sure to remove all metal from brake compartment.

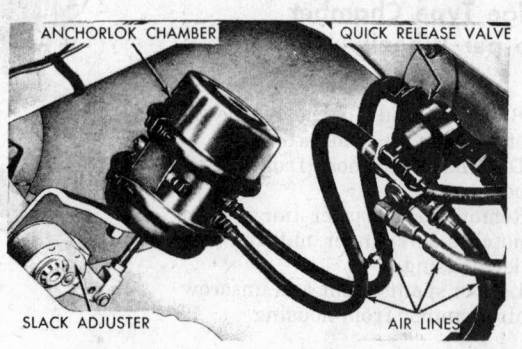

Internal expanding type adjustment
(© Chevrolet Div., G.M. Corp.)

"Anchor Lock" brake chamber—installed
(© Chevrolet Div., G.M. Corp.)

3. Engage star wheel with brake spoon and rotate star wheel to obtain a .010" clearance between lining and drum. Moving handle end of spoon down expands shoes. Check with .010" feeler gauge rotating drum. Shoes can be expanded so as to lock drum then backed off 5 notches similar to a wheel adjustment. Check drum, jack up one rear wheel, for free rotation.

4. Replace clevis, install drum hole cover, remove jack and blocks.

Rear Wheel—Cable Type

Hand lever or foot pedal operated, cable actuated, rear wheel service brakes are used for parking brakes.

1. Block front wheels and apply hand brake two notches from released position.
2. Jack up both rear wheels.
3. All cables and linkage connected, pull back spring in place, loosen cable adjusting lock nut and tighten adjusting nut until a slight drag is felt when rotating rear wheels. Tighten lock nut.
4. Fully release parking brake and check both rear wheels. No drag should be present.
5. Remove jack and blocks.

NOTE: Adjustment of cable should only be made when service brakes are in full adjustment. Cables should be lubricated when rear drums are off.

Rear Wheel—Air Brake

NOTE: For stopmaster fail-safe and super fail-safe information see brake chamber section. (automatic adjusters)

Anchorlok

Anchorlok chamber is mounted "piggy back" on service chamber. It contains a diaphragm under air pressure (60 pounds or more) to contain a powerful spring in compression in

STUD, NUT, AND WASHER STOWED AT SIDE OF CHAMBER

STUD, NUT, AND WASHER INSTALLED (SPRING COMPRESSED)

Spring compressor assembly
(© Chevrolet Div., G.M. Corp.)

1 Brake Support Bracket
2 Parallel Adjusting Screws
3 Front Lever Arm Pin
4 Pin Retaining Screw
5 Brake Cable Clevis
6 Brake Lever
7 Brake Shoe Pin Retainer
8 Brake Shoe Pin
9 Front Brake Shoe
10 Rear Brake Shoe
11 Front Lever Arm
12 Brake Disc
13 Tension Spring
14 Spring
15 Rear Lever Arm
16 Adjusting Nut
17 Tie Rod

Tru-stop type parking brake (© Chevrolet Div., G.M. Corp.)

normal opration. When air pressure drops below 60 pounds, the coiled spring starts to move out. If air pressure continues to drop, the spring will keep expanding to apply brakes until at approximately 30 pounds the spring will have applied the brakes sufficiently to bring the vehicle to a safe, even stop.

To permit moving vehicle, when air pressure is not available to compress spring, a caging tool consisting of a stud and nut will be found stored on service chamber housing. Remove rubber plug in center of anchorlok and insert stud ¼ turn. Turning nut on stud will cage spring.

CAUTION: Cage spring before removing or servicing anchorlock.

TRU-STOP (Disc Type)

This type brake is used only on models equipped with an auxilliary transmission. It uses a ventilated brake disc which is mounted between the propeller shaft flange and the auxiliary transmission shaft companion flange. The brake shoes are mounted in the opposed positions with the brake disc between. When the brake is applied the shoes are forced against the disc.

1. Disconnect the brake cable or rod clevis from the brake lever.
2. Tighten the adjusting nut until the spring exerts enough pressure to bring the lever against the front lever arm.
3. Insert a 1/32" shim between the rear shoe lining and brake disc.
4. Tighten the adjusting nut until the front shoe lining is firmly against the disc yet still allowing removal of the shim.
5. Make sure that the tensions spring is in place. Make sure that both linings are parallel with the disc by adjusting the parallel

adjusting screws. This provides 1/64" clearance between the front and rear shoelinings and brake disc at all points.

6. Check to see that brake lever in the cab is in the full released position. Adjust the clevis on the brake cable as necessary to permit installation of the clevis pin through the clevis and brake lever without changing position of the lever. Install the clevis pin and cotter pin.
7. Make sure the lock nuts on the brake cable and adjusting screws are firmly tightened.

Fuel System

This section contains brief information on removal, installation and minor external adjustments. For more detailed information see carburetor general section.

Carburetor

Various carburetors are designed to meet requirements of engine, transmission and vehicle; therefore carburetors may look alike but are not always interchangeable. All carburetors have conventional float, idle, low speed, power or high speed and accelerating circuits with either manual or automatic chokes.

NOTE: Some symptoms indicate carburetor trouble but in reality are ignition. Before any extensive repairs on carburetor, check first—heat riser, intake manifold and ignition.

Removal
1. Remove air cleaner.
2. Disconnect fuel, vacuum, spark control and governor lines.

3. Disconnect choke and hand throttle controls.
4. Disconnect throttle and automatic transmission linkage at carburetor. Remove pull back spring.
5. Remove mounting nuts and washers.
6. Lift carburetor off manifold and drain. Discard gasket.

Installation

1. Clean carburetor mounting surface.
2. Install new gasket on manifold. Be sure vacuum port and gasket slots are aligned.
3. Place carburetor on manifold, reconnect finger-tight all lines before (evenly) tightening carburetor mounting nuts.
4. Tighten all lines.
5. Reconnect choke and hand throttle controls, replace throttle and automatic transmission linkage, reconnect all vacuum and electrical lines. Connect pull back spring. Check choke and throttle operation.
6. Install air cleaner, check element.
7. Start engine and warm to operating temperature. During warm-up time, torque intake manifold.
8. Adjust idle mixture and idle speed screws, be sure choke is fully open.

CAUTION: Do not force idle mixture screw against seat, this will damage needle. On transmission controlled spark engines (TCS), fast idle adjustment must be set with electrical leads disconnected at solenoid and transmission in neutral.

NOTE: If possible (safely), fill carburetor bowl before installing, will save time and battery drain as well as reducing possible backfiring. Dirt is greatest troublemaker for carburetors, check all filters. Use starter briefly to clear fuel lines (before reconnecting) of any metal flakes that are always present when fuel lines (metal) are disturbed.

NOTE: For Idle Speed and Mixture adjustments (see Tune-Up decal) in the engine compartment. For all other adjustments refer to the appropriate carburetor in the General Repair Section.

Fuel Pump

Two types of pumps are used, mechanical and electric. Mechanical pump is diaphragm type, consisting of a single fuel chamber or a combination fuel and vacuum chambers. In-line engines the pump is actuated by an eccentric lobe on camshaft, V6 eccentric is attached to front of camshaft, V8 engines use a push rod between eccentric lobe on camshaft and pump rocker arm. The single pump is non-serviceable while the combination pump is rebuildable. Electrical pumps are used with step fuel tanks or as stand-by emergency units.

Removal

(Diaphragm Type)

1. Disconnect fuel lines at pump, also vacuum lines on combination unit. Be ready to cap gas feed line should it be necessary (trucks without shut-off valve).
2. Remove two cap screws and washers holding pump to block.

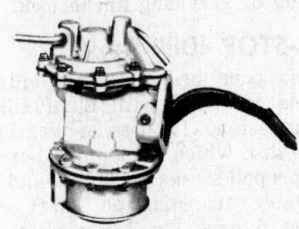

Mechanical fuel pump
Single
Combination
(© Chevrolet Div., G.M. Corp.)

Remove pump and mounting gasket. Be careful of push rod on V8's.

Installation

1. Crank engine to position camshaft lobe on lowest point.
2. Check feed line for restrictions.
3. On V8s—place heavy grease on one end of push rod and slide into position.
4. Hold fuel pump, with new gasket, in position and start two cap screws finger tight. CAUTION: Be sure rocker arm contacts eccentric in correct position.
5. Start fuel line fittings, using flexibility of pump to insure against crossing threads, also start vacuum lines if dual pump.
6. Tighten mounting bolts and lines. Start engine and check for leaks. (open shut-off valve if closed)

CAUTION: When engine is cranked at starter with jumper cable, remove distributor lead from negative post on coil and ignition switch must be on. Failure to do this can result in damage to the ground circuit of ignition switch.

(Electric)

1. Disconnect battery ground cable.
2. Disconnect wiring harness from pump connector.
3. Remove fuel line from pump.
4. Remove bolts and washers holding pump to tank, rotate pump 90 degrees counterclockwise and remove from tank.

Electric fuel pump
(© Chevrolet Div., G.M. Corp.)

Installation

1. Carefully position pump into tank opening. Reconnect fuel outlet line to fitting.

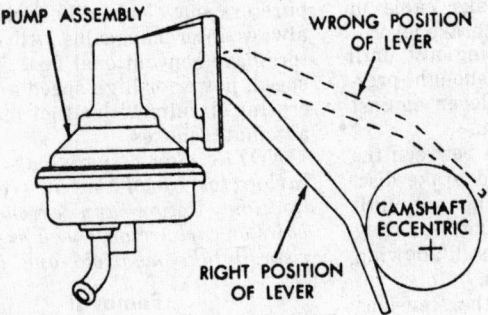

PUMP ASSEMBLY
WRONG POSITION OF LEVER
CAMSHAFT ECCENTRIC
RIGHT POSITION OF LEVER

Fuel pump installation
(© Chevrolet Div., G.M. Corp.)

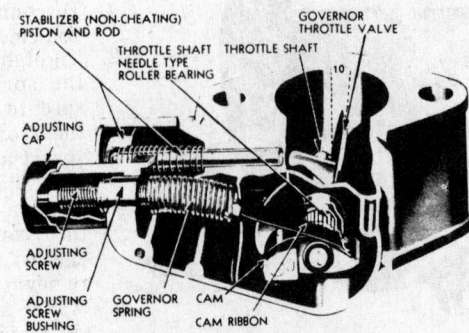

STABILIZER (NON-CHEATING) PISTON AND ROD
GOVERNOR THROTTLE VALVE
THROTTLE SHAFT   THROTTLE SHAFT
NEEDLE TYPE ROLLER BEARING
ADJUSTING CAP
ADJUSTING SCREW
ADJUSTING SCREW BUSHING
GOVERNOR SPRING
CAM
CAM RIBBON

Diagrammatic view of governor
(© Chevrolet Div., G.M. Corp.)

2. Install attaching bolts and torque to 4-6 ft.-lbs.
3. Reconnect pump wiring harness and battery ground cable.
4. Check operation of pumps by using pump selector switch.

Fuel Tank

Following procedure is intended only as a guide. It will vary according to truck model and tank-type.

Removal and Installation

1. On trucks with dual tanks check shut-off valve position.
2. Be sure ignition switch is off or battery ground is disconnected.
3. Remove seat back rest if cab mounted.
4. Drain tank. If tank does not have a drain plug, disconnect gas line and use opening to drain tank. If not accessible, siphon fuel. Do not siphon by mouth, use equipment for that purpose or make a siphon hose as shown below and use air pressure if available.

5. Remove filler neck, cap and vent hose.
6. Disconnect tank gage wire and any ground lead.
7. Remove tank support straps or mounting bolts. Remove tank.
8. Clean all lines, check filters. (Blow clean only after disconnecting other end of line).
9. To install, reverse removal procedures.
CAUTION: Do not use drop cord in area. A bulb breakage could have disastrous effects. Use only safety cans for fuel storage. Do not overtighten lines, as this could distort or twist and lead to leaks.

Governors

An electronic governor is built into the HEI system used on the 1977- and later medium duty 350,366, and 427 V8 engines. The unit is set at 4000 rpm at the factory and is not adjustable.

Velocity Type

Governor is mounted between carburetor and intake manifold and automatically governs the maximum speed of engine which in turn limits maximum speed of vehicle.

Operation

Governor is operated by vacuum intake manifold opposing a calibrated (adjusting) spring, which in turn is connected to a throttle shaft and valve. Velocity of gas mixture from carburetor tends to close valve but this action is opposed by governor spring tension. The calibration of velocity versus spring tension is very sensitive.

Adjustment

Adjustment of spring pull is accomplished by number of spring coils operating (active). Turn adjusting cap counter-clockwise for higher speeds. If too sensitive use special hollow wrench to turn adjusting screw nut. Caps are usually wire locked and sealed.

Vacuum Spinner Type

Governor consists of two compo-

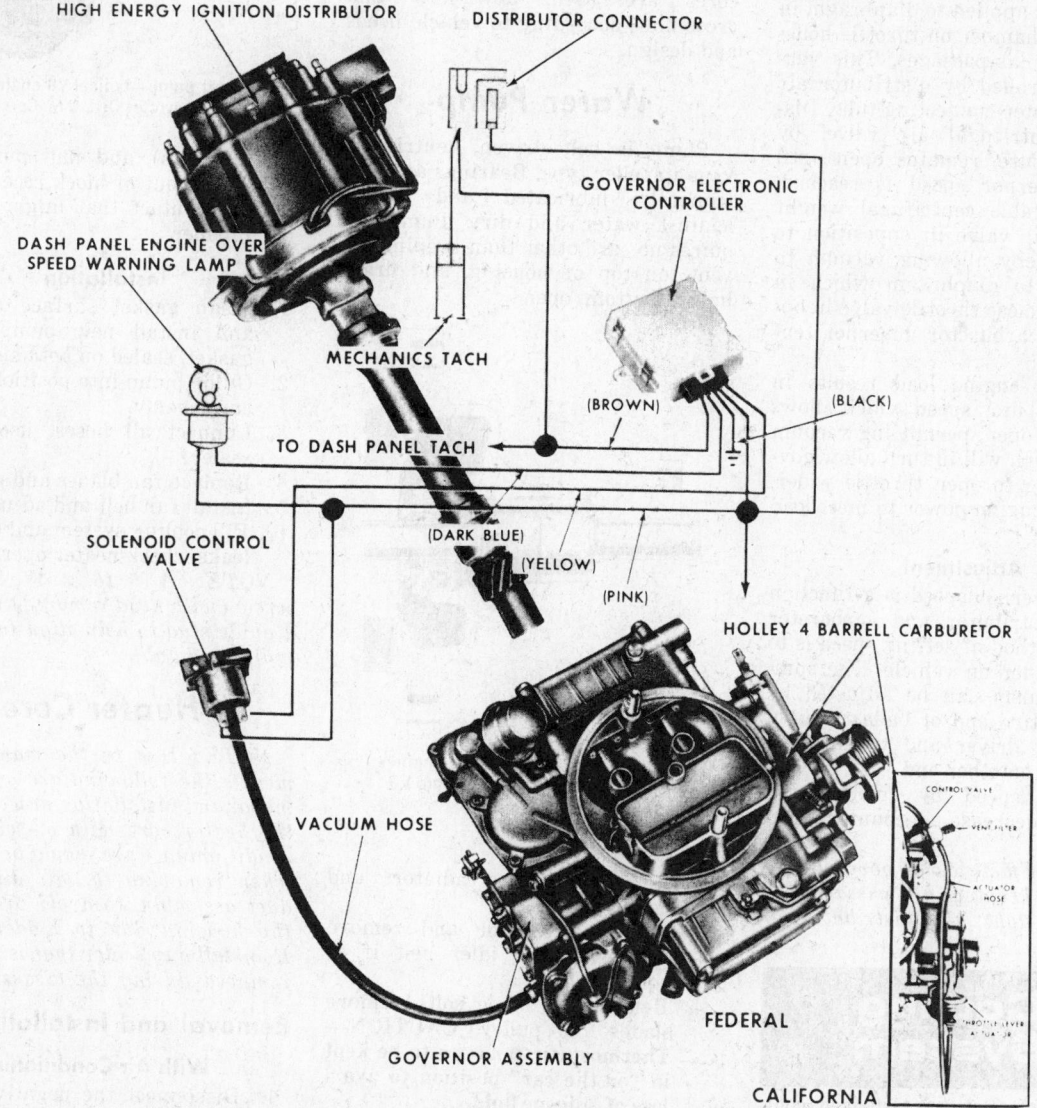

Holley governor control valve and electronic controller with HEI system (© G.M.C.)

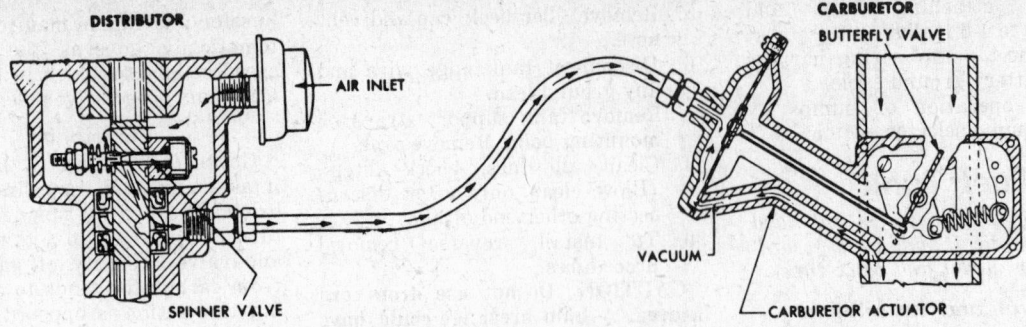

Spinner type governor installation (© Chevrolet Div., G.M. Corp.)

nents connected by tubing. One is in distributor housing and other on carburetor throttle body. This type permits full horsepower usage without excessive engine speeds. An overspeed warning device is usually incorporated in system. A system of calibration of vacuum versus spring tension.

Operation

To limit engine speed while permitting greater throttle openings when additional power is required.

Vacuum is applied to diaphragm in carburetor chamber on throttle housing by internal passages. This vacuum is controlled by distributor air valve thru inter-connecting tube. Distributor centrifugal air valve, by spring pressure, remains open until desired governor speed is reached. Then adjustable centrifugal weight will close air valve in opposition to spring, thereby allowing vacuum to be applied to diaphragm which in turn acts to close throttle valve in opposition to carburetor governor tension.

Increased engine load results in lower distributor speed which allows air valve to open, permitting vacuum to drop, which will in turn allow governor spring to open throttle wider, increasing engine power to meet load increase.

Adjustment

Since governor speed is a function of both distributor and carburetor the best method of setting speed is to adjust spinner on vehicle. The spinner mechanism can be adjusted by holding slotted end of ballast weight with screw driver and turning adjusting nut on other end.

Increase speed by turning nut clockwise, decrease — counter-clockwise.

NOTE: Efficiency of governor depends on keeping all passageways clean and tight, also distributor air filter clean.

Cooling System

All models have pressurized cooling system, thermostatically controlled bypass, sealed by radiator cap. System is designed to operate with coolant's boiling point raised which increases efficiency of radiator. Pressure cap contains both pressure and vacuum relief valves, spring loaded. Pressure valve allows excess pressure out overflow and vacuum valve relieves when system cools off. Thermostat is pellet or poppet construction designed to open and close at predetermined temperatures, and incorporates a by-pass. Two types of cores are used, down-flow and cross-flow, according to vehicle needs and design.

Water Pump

Pump is belt driven, centrifugal vane impeller type. Bearings are permanently lubricated and sealed against water and dirt. Pump requires no care other than keeping air vent, on top of housing, and drain hole in bottom, open.

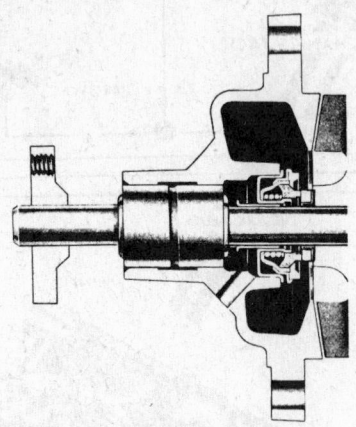

Water pump—typical inline engines
(© Chevrolet Div., G.M. Corp.)

Removal

1. Drain system.
2. Remove lower radiator and heater hoses.
3. Loosen alternator and remove fan belt (and idler belt if so equipped).
4. Remove fan blade bolts, remove blades and pulley. **CAUTION— Thermostatic fans are to be kept in "on the car" position to avoid loss of silicone fluid.**
5. Remove pump attaching bolts

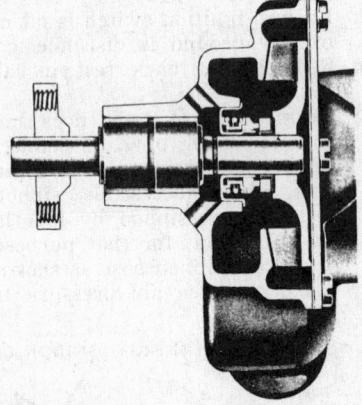

Water pump—typical V8 engines
(© Chevrolet Div., G.M. Corp.)

(or nuts) and pull pump cautiously out of block recess, avoid any contact that might damage impeller.

Installation

1. Clean gasket surface on block and install new pump-to-block gasket, sealed on *both* sides.
2. Guide pump into position and secure evenly.
3. Connect all hoses, use a good sealer.
4. Replace fan blades and pulley.
5. Install fan belt and adjust.
6. Fill cooling system and check for leaks, check heater operation.

NOTE: A 5/16 x 1" SAE cap screw (with head removed) will make a guide stud to help align fan blades, pulley and hub.

Heater Core

NOTE: Due to the many vehicle models the following are general removal and installation procedures for the heater core with or without air conditioning. Care should be exercised when removing the air distribution duct assembly, to avoid breakage of the housing due to hidden screws. Reinstall any sealer that is broken or removed during the disassembly.

Removal and Installation

With Air Conditioning

1. Disconnect the negative battery cable and drain the cooling sys-

tem. Plug the heater core outlets to avoid leakage.

2. Remove the glove box and door assembly.
3. Remove the center duct to selector duct and instrument panel screws. Remove the center lower and upper ducts.
4. Disconnect the Bowden cable at the temperature door.
5. Remove the nuts from the three selector duct studs, projecting into the engine compartment.
6. Remove the selector duct to dash panel screw, (inside the vehicle).
7. Pull the duct assembly rearward to clear the dash panel by the heater core tubes.
8. Lower the unit to gain access to the electrical and vacuum harnesses and disconnect them.
9. Remove the duct assembly from the vehicle. Remove the core straps and remove the heater core.
10. To install, reverse the procedure. Refill the cooling system, start the engine, and check the heater system for proper operation.

Without Air Conditioning

1. Disconnect the negative battery cable, drain the coolant and plug the core tubes.
2. Remove the nuts from the distribution duct studs projecting into the engine compartment.
3. Remove the glove box and door assemblies and disconnect the temperature and air-defrost cable.
4. Remove the floor outlet, the defroster duct to air distribution duct screws and the distribution duct to dash panel screws.
5. Pull the assembly to the rear and remove the wiring harness.
6. Remove the heater distribution unit from the vehicle and remove the core retaining screws.
7. Remove the heater core from the distribution unit.
8. To install, reverse the procedure. Refill the cooling system, start the engine, and check the heater system for proper operation.

Engine

Emission Control

Emission control systems are designed to control the emissions of Hydrocarbons (HC), Carbon Monoxide (CO), and Oxides of Nitrogen (NO_x), at the levels specified by the federal and state governments. Emission control systems vary in their usage, in relation to the engine, transmission, and series application.

The units are covered in the Emission Control Section of the General Repair Section.

Evaporative Emission System

This system was designed to reduce fuel vapor emissions that are normally vented into the atmosphere from the gas tank and the carburetor fuel bowl, through the use of a carbon canister and liquid-vapor separator.

Transmission Controlled Spark System

This system controls emissions by eliminating the vacuum advance in the low gears. The control is accomplished by the use of a solenoid vacuum switch which regulates the vacuum to the distributor. This system also incorporates a thermal override switch which provides full vacuum in all gears when engine is cold.

Exhaust Gas Recirculation System

This system helps reduce nitrogen oxides emitted by the engine exhaust. This is accomplished by releasing small amounts of exhaust gas into the cylinders by means of the E.G.R. valve. This lowers the peak combustion temperatures, reducing the amounts of oxides produced.

Air Injection Reactor

The A.I.R. system injects compressed air into the exhaust system, close enough to the exhaust valves to continue the burning of the normally unburned segment of the exhaust gases. To do this it employs an air injection pump and a system of hoses, valves, tubes, etc., necessary to carry the compressed air from the pump to the exhaust manifolds. Carburetors and distributors for A.I.R. engines have specific modifications to adapt them to the air injection system; these components should not be interchanged with those intended for use on engines that do not have the system.

A diverter valve is used to prevent backfiring. The valve senses sudden increases in manifold vacuum and ceases the injection of air during fuel-rich periods. During coasting, this valve diverts the entire air flow through the muffler and during high engines speeds, expels it through a relief valve. Check valves in the system prevent exhaust gases from entering the pump.

Early Fuel Evaporation System

1975 and later models are equipped with this system to reduce engine warm-up time, improve driveability, and reduce emissions. On start-up, a vacuum motor acts to close a heat valve in the exhaust manifold which causes exhaust gases to enter the intake manifold heat riser passages. Incoming fuel mixture is then heated and more complete fuel exaporation is provided during warm-up.

Catalytic Converters

The converters are used to oxidize hydrocarbons (HC) and carbon monoxide (CO). They are necessary because of even stricter emission standards for the 1975 and later models.

The catalysts are made of noble metals (platinum and palladium) which are bonded to either a monolithic (one-piece) element or to individual pellets. The catalyst causes the HC and CO to break down without taking part in the reaction; hence, a catalyst life of 50,000 miles is expected.

Some engines equipped with the converters require an air injection pump to supply air for the reaction; others will not.

For more detailed information refer to the General Section.

Engine R & R

NOTE: Due to the varied engine-transmission combinations used in the Chevrolet truck line, the procedures for removal and installation of the engines are given as a general outline. It may be necessary to alter the procedures somewhat to compensate.

Conventional Cab, Pickups, Panel and 4 Wheel Drive

Removal

1. Disconnect and remove battery.
2. Drain cooling system.
3. Drain engine oil.
4. Remove air cleaner and ducts.
5. Remove hood and radiator (also shroud if equipped) on larger models (1½ ton and up) it will be necessary to remove front end sheet metal. *NOTE: Scribe alignment marks on hood hinges.*
6. Disconnect wires at:
   Starter solenoid.
   Delcotron.
   Temperature switch.
   Oil Pressure switch.
   Transmission controlled spark solenoid.
   Coil.
   Neutral safety switch.
7. Disconnect:
   Accelerator linkage.
   Choke cable at carburetor (if so equipped).
   Fuel line to fuel pump. (cap line).
   Heater hoses.
   Transmission dipstick and tube, plug hole (if so equipped).
   Oil pressure line to gauge. (if so equipped).

Oil fill tube (and dipstick).
Vacuum or air lines.
Parking brake cable (if used).
Power steering lines (if so equipped).
Engine ground straps.
Exhaust pipe (support if necessary).

8. Loosen fan belt, remove fan blades and pulley.

Exploded view—V8 engine typical (© Chevrolet Div., G.M. Corp.)

9. Remove clutch cross-shaft or disconnect clutch slave cylinder (if so equipped).
10. Remove road draft tube (if so equipped).
11. Remove rocker arm cover (s) and attach engine lifting tool. Take engine weight off mounting bolts.
12. Remove propeller shaft and plug end of transmission housing on light trucks. On 1½ ton and up, support transmission and disconnect from engine.
13. On light duty trucks, disconnect speedometer cable at transmission, shift linkage and any clutch linkage as required.
14. Disconnect transmission cooler lines (if so equipped).
15. On light trucks equipped with automatic or 4 speed transmission, remove rear engine mount bolts and rear crossmember.
16. Remove front engine mount bolts. **CAUTION: Make final check that all necessary disconnects have been made.**
17. On light trucks, raise engine and transmission and pull forward until removed. Larger trucks remove rear mounting bolts,

raise engine and pull forward until disconnected from transmission continue to raise engine until removed from vehicle. Be careful not to damage clutch disc.

Installation

Install in reverse order of removal, fill cooling system and crankcase (check transmission). Start engine and check for leaks and operation. Adjust carburetor and road test.

Van (Panel and Sport)
Removal

Because of its location, the engine is removed from below and both engine and transmission removed as a unit.

1. Disconnect battery ground cable.
2. Drain cooling system.
3. Remove engine cover.
4. Disconnect:
   Air cleaner, extensions and heat tubes.
   Accelerator linkage.
   Choke cable at carburetor (if used).
   Transmission controlled spark solenoid (if used)
   Primary wire at coil.
   Oil fill tube (and dipstick).
   Starter solenoid.
   Temperature sender at unit.
   Delcotron.
   Oil pressure line or sender switch.
   Transmission dipstick and tube, plug hole (if so equipped).
   Evaporative Control System lines at air cleaner and carburetor.
5. Remove upper radiator hose and heater hoses.
6. Loosen fan belt, remove blades, pulley and shroud (if used).
7. Remove distributor cap and spark plug wiring harness.
   NOTE: Disconnect any wiring harness retaining clips.
8. Raise vehicle on hoist, drain crankcase and disconnect:
   Engine splash shields (if used).
   Engine ground straps.
   Lower radiator hose.
   Neutral safety switch, at transmission (if equipped).
   Fuel line to fuel pump. (capline).
   Transmission controlled spark switch, at transmission (if used).
   Transmission cooler lines (if used).
   Speedometer cable at transmission (and plug).

Engine cradle under engine
(© Chevrolet Div., G.M. Corp.)

Power steering hoses (if equipped).
Parking brake cable (if used).
Steering idler and pitman arms.
Positive battery cable at starter.

9. Disconnect exhaust pipe, support if necessary or block out of way.

10. Remove propeller shaft at transmission and plug extension housing to prevent oil leakage.

11. On manual transmissions, disconnect shift levers and clutch linkage; on automatics, shift linkage.

12. Place jack, with engine cradle attached, under engine and take weight off mounting bolts. CAUTION: Place cradle under engine as far as possible, (to the fifth oil pan cap screw from flywheel at least) and secure with chain over flywheel housing.

13. Remove engine front mounting bolts (and brackets for additional room).

14. Remove bolts at rear mount to crossmember.

15. Remove transmission support crossmember bolts and remove crossmember. CAUTION: Make final check that all necessary disconnects have been made.

16. Lower engine and transmission assembly slowly, pulling to rear to clear front crossmember, and move unit out from under vehicle.

NOTE: On 1974 and later Chevy-Van and Sportvan, provisions have been made so that the grille, filler panels and tie bars can be removed easier than on previous models. This allows removal of the engine from the front of the vehicle instead of from underneath, which makes engine R & R quicker and easier.

Installation

Install in reverse order of removal, fill cooling system and crankcase (check transmission). Start engine and check for leaks and operation. Adjust carburetor and road test.

Step Vans

Removal

1. Disconnect battery and remove.
2. Drain cooling system, drain engine oil.
3. Remove engine box, drivers seat, floor panels at stepwells, floor panel around steering column and pedals, and inspection plate on firewall above engine box.
4. Remove air cleaner (and any ducts).
5. Remove radiator and shroud.
6. Loosen fan belt and remove fan blades.
7. Remove engine splash pans (if used).

8. Disconnect neutral safety wire at converter (if used).
9. Remove upper and lower radiator hoses.
10. Disconnect wires at:
Starter solenoid.
Delcotron.
Transmission controlled spark solenoid (if used).
Temperature switch.
Oil pressure switch.
Coil.
11. Disconnect:
Accelerator Linkage.
Choke cable at carburetor (if so equipped).
Fuel line to pump.
Heater hoses.
Oil pressure gauge line (if so equipped).
Parking brake cable (if used).
Vacuum or airlines.
Power steering lines (if so equipped).
Engine ground straps.
Exhaust pipe (support if necessary).
12. Remove clutch cross-shaft or disconnect clutch slave cylinder (if so equipped).
13. Remove oil fill tube and dipstick also transmission dipstick and tube, plug holes (if used)
14. Remove rocker arm cover (s) and attach engine lift tool.
15. Push arm of engine lift crane in right side door opening. Attach to lifting device or sling and take engine weight off mounts.
CAUTION: Make final check that all necessary disconnects have been made.
16. Support transmission and remove shift controls (cover opening), speedometer cable (plug hole), transmission spark control switch wire and oil cooler lines (if used).
17. Remove propeller shaft at transmission. Plug extension housing to prevent leakage.
18. Remove 2 top transmission to clutch housing cap screws and insert 2 guide pins. NOTE: Take 2 bolts, the same diameter and threads as just removed, but at least 4" long and cut heads of bolts off. Using a hack saw, slot ends just cut for screwdriver and use for guides.
19. Remove 2 lower transmission to clutch housing cap screws and slide transmission back until clear of clutch disc. When transmission is free from engine, lower and remove from under vehicle.
20. Remove engine mounting bolts, front and rear.
21. Raise engine slightly and push forward to clear crossmember, then lift up and remove engine through door opening.

Installation

A careful check of clutch components should be made. Install in reverse order of removal, fill cooling system and crankcase (check transmission). Start engine and check for leaks and operation. Adjust carburetor and road test.

Tilt Cab

Removal

1. Tilt cab to expose engine area and secure.
2. Block wheels and exhaust air (if so equipped).
3. Disconnect battery ground cable.
4. Drain cooling system.
5. Drain crankcase.
6. Remove air cleaner and any ducts.
7. Remove radiator and heater hoses.
8. Remove radiator support and shroud assembly.
9. Disconnect hoses and remove surge tank.
10. Disconnect choke cable at carburetor.
11. Disconnect hand throttle at carburetor.
12. Disconnect shift linkage at control island.
13. Remove control island bolts and swing island out of way.
14. Remove both right and left island supports.
15. Disconnect cab safety lock and remote cab support.
16. Disconnect emergency brake cable.
17. Remove any engine splash shields.
18. Disconnect wires at:
Starter solenoid.
Delcotron.
Temperature sender.
Oil pressure sender.
Coil.
Governor speed warning.
Transmission controlled spark solenoid at carburetor.
19. Remove:
Accelerator linkage.
Fuel line to pump.
Lines or wires to dash guages.
Vacuum or air lines at engine.
Power steering lines at pump (if equipped).
Engine ground straps.
Exhaust pipe or crossover pipe (support if necessary).
20. Remove fan blades, pulley and support bracket assembly.
21. Remove clutch cross-shaft or disconnect clutch slave cylinder.
22. Remove rocker arm covers and install lift tool or sling.
23. Hoist engine and take engine weight off motor mounts.
NOTE: According to work to be performed select either of the following.

Removal of Engine and Transmission as a Unit

1. Disconnect speedometer (and plug).
2. Disconnect shift linkage at transmission (cover opening).
3. Disconnect any clutch linkage not yet removed.
4. Disconnect and drop propeller shaft at transmission, also any power take-off or auxiliary transmission couplings (cover or plug all openings).
5. Disconnect oil cooler lines or transmission spark control switch (if so equipped).
6. On Roadrangers remove air lines and parking brake drum (if used).
7. Remove all engine mounting bolts and raise slightly to support transmission weight. CAUTION: Make final check that all necessary disconnects have been made.
8. Remove any transmission to support bolts.
9. Raise engine and transmission assembly out of chassis as a unit.

Removal of Engine Only

1. Support transmission, remove flywheel under-pan, and disconnect from engine. NOTE: If possible, install transmission guide pins in top (transmission to clutch housing) holes to allow engine to slide forward to clear clutch disc splines. This will prevent bending clutch hub.
2. Remove all engine mounting bolts. CAUTION: Make final check that all necessary disconnects have been made.
3. Raise slightly and pull forward until clear of transmission.
4. Continue to raise engine until high enough to clear chassis.

Installation

Install in reverse order of removal, fill cooling system and crankcase (check transmission). Start engine and check for leaks and operation. Make any minor adjustments and road test.

Engine Manifolds

NOTE: The use of a good chemical solvent on exhaust manifold bolts and nuts will facilitate the operation.

CAUTION: Pay particular attention to heat risers, their failure increases warm-up time and failure to open can cause lean mixtures at higher speeds.

Inline Engines (Combination Manifold)

Removal

1. Remove air cleaner (and any ducts).
2. Disconnect throttle controls, rods, linkage and return spring.
3. Disconnect fuel and vacuum lines at carburetor, also choke cable or control (if used).
4. Disconnect crankcase ventilation valve, vacuum brake or transmission spark control hoses (if used).
5. Remove carburetor.
6. Remove oil filter support bracket and swing filter to one side (if so equipped).
7. Disconnect exhaust pipe at flange and support if necessary (discard gasket or packing).
8. Remove manifold attaching bolts and clamps, remove manifold assembly (be careful of locating rings). Discard gaskets.

Installation

Reverse removal procedures after cleaning all gasket surfaces, checking for cracks, check heat riser and alignment. Lay straight edge on manifold to head surface to check. If intake and exhaust are not in line, loosen center bolts where they are joined and do not tighten until assembly is bolted to head, then retighten). Install finger-tight, check that pilot or locating rings and gaskets are in place, then torque to specifications in proper sequence. Torque on end bolts is 15 to 20 ft. lbs., all others are 20-30 ft. lbs.

Warm engine, adjust idle and check for leaks.

V Engines—Intake Manifold

Removal

1. Drain radiator, remove top hose at thermostat housing also any bypass or heater hoses.

2. Remove battery ground cable.
3. Remove carburetor air cleaner and any ducts.
4. Remove oil fill tube and cap.
5. Disconnect gas line, all vacuum hoses, throttle linkage and return spring, choke cable and crankcase ventilation valve.
6. Disconnect wires to temperature sender, coil and transmission spark controlled solenoid (if so equipped).
7. Remove carburetor.
8. Remove distributor cap and mark position of rotor. Remove distributor.
9. Exhaust air and remove air compressor, disconnect oil drain line at manifold (if so equipped and interferes).
10. Remove manifold attaching bolts, remove manifold.

NOTE: If manifold is not to be replaced, some components can be left on such as carburetor, oil fill tube, thermostat housing and temperature sender.

Installation

Clean gasket and seal surfaces on manifold, block and heads. Install new gaskets and seals, coated with a good sealer particularly at water passages. Position manifold, use guide pins to prevent gaskets moving, and check mating angle at heads. (Angle could be incorrect due to excessive cylinder head resurfacing)

Install bolts finger tight, then torque to specifications in proper sequence. Reverse the removal procedures fill radiator, warm engine, adjust timing and carburetor idle if necessary and check for leaks.

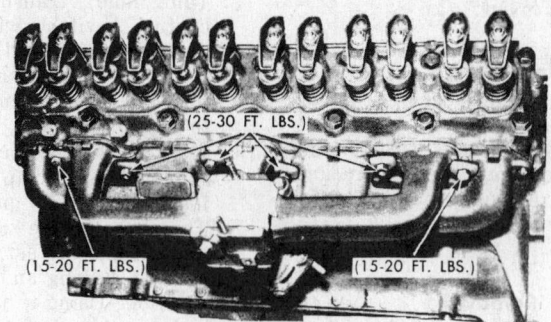

Manifold torque (typical)
(© Chevrolet Div., G.M. Corp.)

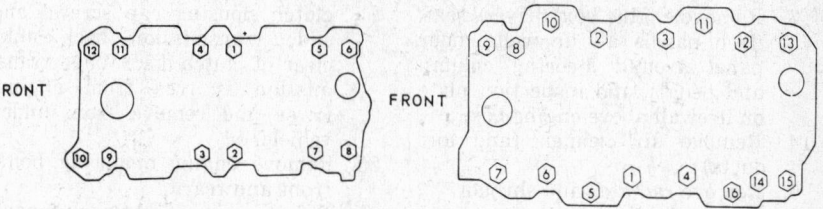

"SMALL V8"   "MARK IV V8"

Intake manifold bolt torque sequence (© Chevrolet Div., G.M. Corp.)

V Engines—Exhaust Manifold

Removal

1. Use a good liquid penetrant freely on attaching bolts and nuts.
2. Remove the carburetor heater if so equipped.
3. Remove exhaust pipe (support if necessary).
4. Open french locks and remove manifold attaching bolts. *NOTE: A 916", thin wall, 6 point socket sharpened at leading edges, placed over head of bolt then tapped with hammer will speed the opening of french locks.*
5. Remove Delcotron (on left side), remove manifold.
NOTE: On 366 and 427 engs. spark plugs must also be removed before removing manifold.

Installation

Clean mating surfaces, install new gaskets where used, check heat riser and check for cracks. Reverse removal procedures. Start engine and check for leaks.

Cylinder Head

Inline Engines

Removal

1. Drain cooling system and remove battery ground strap at head.
2. Remove air cleaner and any ducts.
3. Remove choke cable, accelerator rod and return spring, fuel and vacuum lines at carburetor.
4. Remove manifold to head bolts and clamps, pull manifold and carburetor assembly clear of head and support.
5. Remove fuel and vacuum lines from retaining clip at thermostat housing.
6. Disconnect temperature sender wire, remove wiring harness from rocker cover clip. Remove coil wires and coil.
7. Remove top radiator hose, at thermostat housing.
8. Remove spark plug wires and distributor cap.
9. Remove rocker arm cover. **CAUTION: Never pry rocker arm cover loose—bump cover rearward in a gasket shearing manner.**
10. Engines with rocker arm shafts, back off adjusting nuts, rotate rocker arm to clear push rod and remove push rods. Engines using pedestal rocker arms, remove rocker arm ball nuts, arms and push rods.
11. Remove push rod cover.
12. Remove cylinder head bolts, cylinder head and discard gasket.

NOTE: Place rocker arm mechanism and cylinder head bolts in a rack so they can be re-installed in same locations. (mated)
Check cylinder head for warpage with a straight edge. Inspect for cracks and burnt valves.

Installation

Reverse removal procedures and adjust valves after cleaning gasket surfaces. Engines using a steel (shim) gasket, coat both sides with a good sealer, bead side up. Do not reuse gaskets. Cylinder head bolt threads in block and threads on bolts must be clean. Coat threads on bolts with sealer before installing. Tighten each cylinder head bolt a little at a time, in correct sequence, until specified torque is reached. Engines using composition (steel asbestos) gaskets must have heads retorqued after warm-up. (retightening heads effects valve lash) Refer to specifications at beginning of this section for nut tightening sequence.

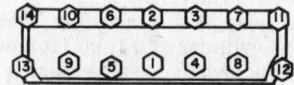

Six 250 and 292 engines

V Engines

Removal

1. Remove intake manifold (with carburetor) and exhaust manifold—see manifolds R & R. *NOTE: If only one head is to be removed, remove inlet manifold bolts on that side only and leave inlet manifold assembly in place.*
2. Loosen belt and remove power steering pump (if so equipped).
3. Remove rocker arm covers. **CAUTION: Never pry rocker arm cover loose—bump cover rearward in a gasket shearing manner.**
4. Loosen rocker arm adjusting nut, turn rocker arm to clear push rods and remove push rods. Exhaust push rods are longer than intake push rods in some engines so to be sure—place in sequence so that they can be installed in same location (mated).
5. Remove cylinder head bolts, cylinder head and discard gasket.
6. Check cylinder head for warpage with a straight edge. Inspect for cracks and burnt valves.

Installation

Reverse removal procedures and adjust valves, after cleaning gasket surfaces. If heads are to be resurfaced check alignment at intake manifold. Cylinder head bolt threads in block and threads on bolts must be clean, coat threads with sealer.

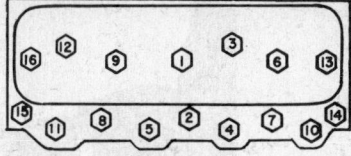

V-8 366 and 427 engine

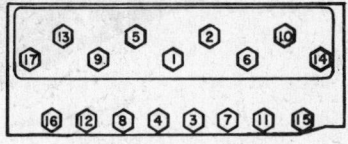

V-8 307 and 350 engine

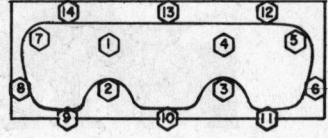

V6—401 & 478 engines

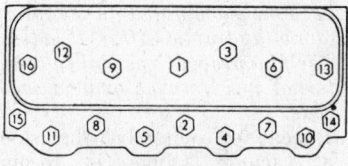

V-8 400, 454 engines
(© Chevrolet Div., G.M. Corp.)

Tighten each cylinder head bolt a little at a time, in correct sequence, until specified torque is reached. Engines using steel (shim) gasket, coat both sides with a good sealer, bead side up. Engines using composition (steel asbestos) gaskets must be retorqued after warm-up (retightening heads effects valve lash).

Refer to specifications at beginning of this section for nut tightening sequence.

Valve System

Adjustments—Hydraulic lifters

Engine Running

1. Run engine to normalize (stabilize oil temperature), remove rocker cover (s) bump off—do not pry off). Leave old gasket on head to aid against oil overflow or use oil deflector clips.
2. Reduce engine idle as low as possible, tighten cylinder head bolts (and rocker supports if used). Check camshaft lobe lift.
3. Back off rocker arm adjusting nut until rocker arm starts to clatter, then turn nut down slowly until clatter stops. This is zero lash position.
4. Turn adjusting nut down 1/4 turn and pause 10 seconds until engine runs smoothly. Repeat operation 3 more times until 1 full turn down from zero lash position is reached. *NOTE: This 1 turn pre-load adjustment must*

Valve adjustment (© Chevrolet Div., G.M. Corp.)

be done slowly and in stages to allow hydraulic lifter to adjust itself to prevent possibility of internal interference or bent push rods.

5. Repeat for each valve, the use of a vacuum gauge is recommended.
6. Install new rocker cover gaskets and torque rocker covers. Reset engine idle.

Valve Arrangement

Front
E I I E E I I E E I I E
6 cylinder

Front
E I I E I E
E I E I I E
V6 engine

Front
E I I E E I I E
E I I E E I I E
Small block V8s

Front
I E I E I E I E
E I E I E I E I
Big block V8s

Engine Not Running

1. Remove rocker cover (s) (bump off—do not pry off).
2. Tighten cylinder head bolts (and rocker arm supports if used).
3. Disconnect primary wire at negative terminal of coil.
4. Mark distributor housing with chalk at each spark plug tower (double mark no. 1 cylinder).
5. Remove distributor cap and crank engine until rotor points to No. 1 chalk mark. (No. 1 cylinder is approximately at TDC and both valves can be adjusted). Valve adjustment is made by backing off rocker arm nut until push rod can be rotated and then slowly tightening until

push rod does not turn. This is zero lash position. Turn adjusting nut down 1 full turn to complete adjustment.
6. Adjust the remaining valves, one cylinder at a time (following firing order) in same manner.
7. Install distributor cap.
8. Install new rocker cover gasket (s) and torque rocker cover (s).

Noisy Lifters

Locate a noisy lifter by using a mechanic's stethoscope or hose, can also be detected by placing finger on valve spring retainer (a distinct shock will be felt each time valve returns to seat). Forcing push rod down will cause lifter check valve to unload and remain open. A noisy or defective push rod is usually indicated by a free-spinning push rod.

Rocker Arms

Rocker arms are trough shaped, pressed steel levers that transfer lifter motion to valves. Rocker arms are supported on individual pedestals and have an oval hole in center to fit over stud and pivot on ball seats. Oil is fed to rocker arms by means of hollow push rods. Whenever arms or ball seats are being installed, coat all bearing surfaces with engine oil.

Rocker arm studs are pressed in cylinder head and are available in oversize for replacement.

CAUTION: Do not try to press oversized stud in head without reaming stud hole first.

Pressed in studs can be replaced by a stud threaded on both ends. Head is to be threaded in stud hole to accept threaded stud.

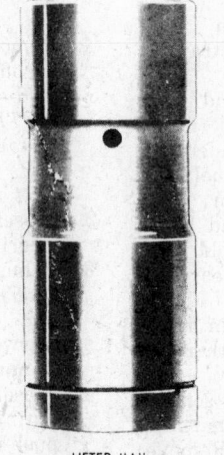
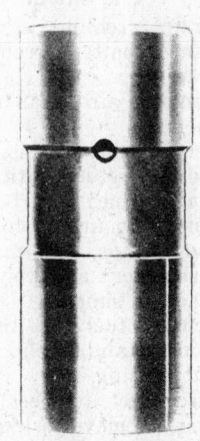

LIFTER "A"   LIFTER "B"

1 Lifter body
2 Push rod seat
3 Metering valve (Lifter A)
  inertia valve (Lifter B)
4 Check ball
5 Check ball retainer
6 Push rod seat retainer
7 Plunger
8 Check ball spring
9 Plunger spring

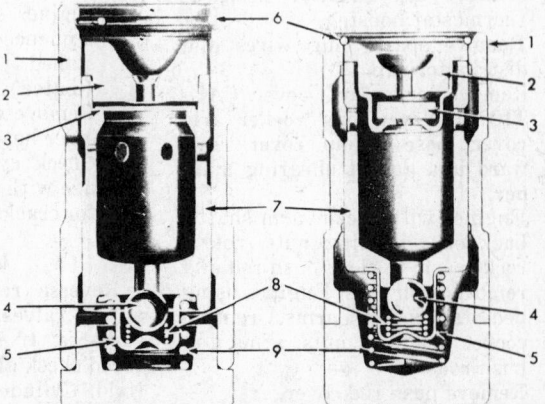

Hydraulic valve lifters (© Chevrolet Div., G.M. Corp.)

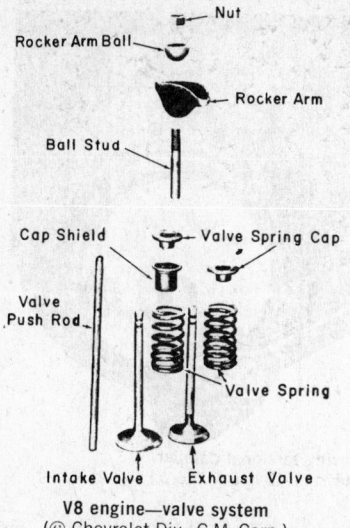

V8 engine—valve system
(© Chevrolet Div., G.M. Corp.)

NOT USED ON 194,230 & 250 CU. IN. L6

Valve spring installation
(inline & small V8 engines)
(© Chevrolet Div., G.M. Corp.)

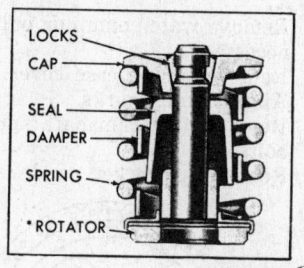

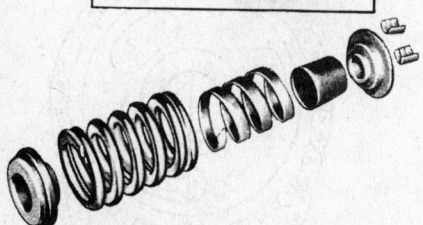

Exhaust valve spring installation
(Mark IV V8 truck, 366 and 427 engines)
(© Chevrolet Div., G.M. Corp.)

Checking Engine Valve Timing

6 cyl. (Inline Engine)

1. Remove the valve rocker arm cover and the push rod front cover.

2. Loosen the nut and the no. 2 intake valve rocker arm, swing the rocker arm away from the push rod and remove the push rod and valve lifter.
3. Temporarily install a flat face mechanical lifter in place of the hydraulic lifter.
4. Turn the crankshaft until the no. 2 exhaust valve opens and the notch on the pulley dampner is aligned with the "O" mark on the timing pointer.
5. Position a deal indicator to measure lifter movement and set the indicator at zero. Turn the crankshaft to 360 degrees and read the indicator. On engines that are correctly timed the indicator will read as follows:

   250 engines . . . . 0.014"
   292 engines . . . . 0.016"

   If the reading is not as above, re-set the indcator at zero and turn the crankshaft 360 degrees, then read the indicator again. If the read is now correct, the engine is timed properly. The following chart shows indicator readings result from improperly indexed gears.
6. If after checking an out-of-time condition exist, remove the engine front cover and check for proper indexing of time marks on gears.

V6 Engine

1. Remove the left-hand rocker cover.
2. Facing the front of the engine turn the engine clockwise to the top dead center No. 1 mark at the crankshaft pulley on the compression stroke. Both the intake and the exhaust valve on the No. 1 cylinder will then be closed.
3. On all V-6 engines adjust the clearance to exactly 0.099 inches at the No. 1 exhaust valve (front valve).
4. Turn the engine clockwise until the No. 1 exhaust valve opens and begins to close, then with fingers, try turning the push rod of the No. 1 exhaust valve as the engine is cranked slowly. When the push rod rotates with finger pressure, the 5-degree (before top dead center mark) on the pulley should be at the pointer. This will be about one revolution from the starting point. If the pushrod can be rotated at any point between the 10-degree mark and the top dead center No. 1 mark, the valve timing is correct. After making the timing

check be sure to adjust the exhaust valve clearance to 0.018".
NOTE: If the timing chain has been installed improperly, there will be a 15 degrees out-of-time condition for each mismatched tooth on the sprocket.

V8 Engine

1. Remove the rocker arm covers.
2. Turn the crankshaft so that the timing mark is at the "O" mark on the pointer and the No. 1 cylinder is ready to fire.
3. Move the crankshaft, and the exhaust valve in the No. 6 cylinder will just close and the intake valve will just begin to open.
4. If the exhaust valve or the intake valve is open when the pointer is at the "O" timing mark the camshaft is out of time
5. When the No. 6 cylinder is ready to fire and the pointer is at the "O" timing mark the No. 1 cylinder exhaust valve will have just closed and the intake valve will begin to open.

Timing Gears and Chain

Crankshaft Pulley or Damper

Removal

1. Drain radiator; remove hoses.
2. Remove radiator—See Radiator R & R
NOTE: On V-8 engines if additional operations such as camshaft removal are not being performed, radiator removal will not be necessary.
3. Loosen fan belt and remove any accessory drive.
4. Use puller to remove pulley or damper.
NOTE: On some early models, pulley is bolted to Hub. Remove pulley first then, using pulley holes, remove hub. More recent models have the hub bonded to inertia weight. Care must be exercised when using a puller.

Installation

1. Clean area, inspect oil seal in cover. (Now would be good time to install new seal even if just for preventative maintenance.)
2. Coat timing case cover oil seal with light oil, inspect hub for seal (grooved) wear.
3. Position damper (or hub) on

Engine (Cu. In.)	Camshaft Part No. & Valve Lift	Gears Properly Indexed	One Tooth Adv.	One Tooth Ret.
250	3864896-.388	.014" ± .004"	.0351"	.0055"
292	384800-.405	.016" ±.004"	.0379"	.0068"

Installing oil seal (cover installed)
(© Chevrolet Div., G.M. Corp.)

Removing torsional damper
(© Chevrolet Div., G.M. Corp.)

crankshaft, aligning keyway, and tap lightly into position.
4. Pull damper (or hub) into position with damper retaining bolt and washer. (Make sure damper retaining bolt has good thread engagement before applying force). Torque to specifications.
5. Reverse balance of removal steps.

Timing Cover Oil Seal Replacement

1. Remove torsional damper—see damper R & R.
2. Pry old seal out, be careful not to mar crankshaft or bend cover.
3. Install new seal (lip of seal toward block), tap lightly into position. Coat seal with oil.
4. Reverse removal procedures.

Timing Gear Replacement

Inline Engines

NOTE: When necessary to install a new camshaft gear the camshaft should be removed. However gear can be removed from camshaft without removing camshaft from engine. Cam gear can be split and hub section pulled off but extreme care must be taken not to allow any impact, either removing or installing gear, on the shaft. Camshaft must be totally blocked so as not to allow any movement to disturb oil sealing welsh plug at camshaft rear bearing.

Removal

1. Drain and remove radiator.
2. Remove front end sheet metal or grille.
3. Remove damper or pulley.
4. Remove oil pan (See Oil Pan R & R).
5. Remove timing case cover (2 bolts inside oil pan at front main bearing cap).
6. Remove rocker arm cover(s) and remove lifters.
7. Remove fuel pump and distributor (mark position of rotor).

8. Align timing gear marks (check rotor mark) then remove 2 thrust plate screws by reaching through 2 holes in camgear.
9. Remove camshaft and rear assembly by pulling and turning shaft carefully so lobes will not mar bearings and will clear lifters.
10. Use press to remove cam gear using caution not to damage thrust plate by woodruff key.
11. Clean all gasket surfaces and inspect.
12. Check camshaft alignment and lobes for wear. Check camshaft bearings in block, check crank gear and oil spray nozzle.

Installation

1. Support camshaft at back end of front bearing in a press, place thrust plate over end of shaft, install woodruff key in key way, align camgear with key and press gear on shaft until clearance at thrust plate (and front end of front bearing) is .001-.003".
2. Install camshaft and gear as-

sembly in block, turning shaft carefully so lobes clear lifters and bearings, until it almost bottoms. *NOTE: Coat lobes and bearing surfaces with engine oil.*
3. Turn camshaft and mesh timing marks, tighten thrust plate.
4. Check cam gear for runout (should not exceed .005").
5. Check backlash at gears. (.004-.006").
6. Reverse remaining removal procedures.
7. Fill radiator, and oil and start engine. Adjust valves, check timing and inspect for leaks.

Timing Chain Removal

V Engines

1. Drain and remove radiator.
2. Remove torsional damper.
3. Remove oil pan (see oil pan R & R).
4. Remove water pump or pulley if necessary.
5. Remove timing case cover.
6. Align timing marks.
7. Remove (3) camshaft sprocket bolts.
8. Remove sprocket and chain.

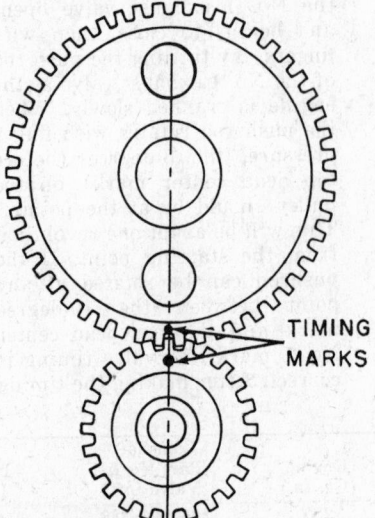

Timing mark alignment inline engines

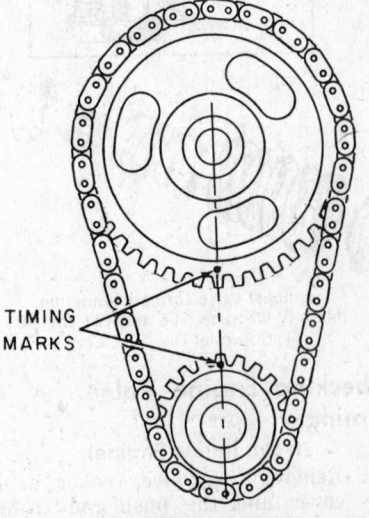

Valve timing—V8 engines

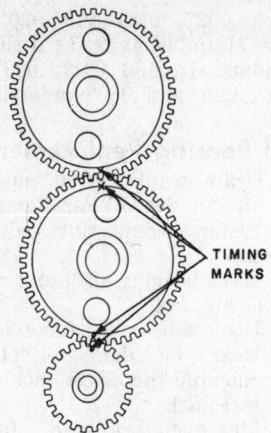

Timing mark alignment V6 401 & 478 engines

NOTE: Sprocket is a light press fit on camshaft, especially at dowel pin. If sprocket is tight, tap lightly with plastic hammer on lower edge of sprocket.

9. Clean all gasket and seal areas and inspect.
10. Check teeth on both sprockets for wear.

Installation

1. Suspend chain on camshaft sprocket with timing mark in approximate position.
2. Place chain over crankshaft sprocket and position camshaft sprocket on dowel. Recheck timing marks.
3. Draw camshaft sprocket in place using the three mounting bolts. Torque to specifications. **CAUTION: do not drive sprocket on camshaft as welsh plug at rear of camshaft can be dislodged.**
4. Lubricate chain with engine oil, make sure oil slinger is in place.
5. Reverse removal procedures.
6. Fill radiator, add engine oil and start engine. Check for leaks.

Check for Worn Chain

1. Check fan belt tension, adjust if too loose.

2. Remove distributor cap, loosen spark plugs.
3. Move fan blades until rotor moves. Mark distributor housing and balancer at pointer.
4. Move fan in opposite direction until rotor moves, remark both units.
5. Marks in excess of 4° apart, 2 graduations (usually) on balancer, indicate excess wear.

Pistons and Connecting Rods

Pistons

Pistons are of various designs, flat, cup, hump and dome. Heads are usually notched to indicate front or pistons are marked on pin boss.

Connecting Rods

Rod forging and cap have mating numbers and must be on same side when installed or cap is on "backwards." Rod number on in-line engines go to camshaft side. Numbers on "V" engines go to outside of block. The oil spurt (cylinder wall oiling) hole goes toward camshaft. On in-line engines the spurt hole and number are both on camshaft side. On "V" engines the spurt hole is in center, toward camshaft, and numbers to outside.

Removal

1. Drain cooling system and remove cylinder head(s). See cylinder head removal.
2. Drain crankcase oil and remove oil pan. See Oil Pan Removal.
3. Remove any ridge and/or deposits from upper end of cylinder bores with a ridge reamer. *NOTE: Move piston to bottom of its travel and place a cloth on top of piston to collect cuttings. After ridge and/or deposits are removed turn crankshaft until piston is at top of its stroke and carefully remove cloth with its cuttings.*

4. Check connecting rods and pistons for cylinder number identification and if necessary, mark them.
5. Remove conecting rod nuts and caps. Push rods away from crankshaft and install caps and nuts loosely to their respective rods.
6. Push piston and rod assemblies away from crankshaft and out of cylinders. See Engine Rebuilding—General Section.

Installation

1. Lightly coat pistons, rings and cylinder walls with light engine oil, making sure everything is clean and free of dirt and foreign material.

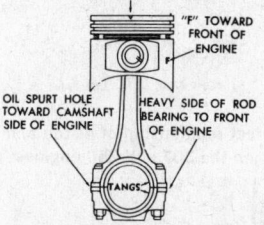

Correct relationship of piston and rod on 250 engine

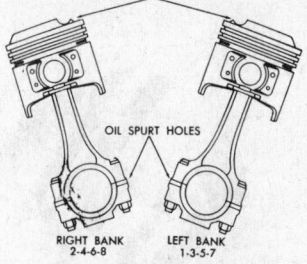

Correct relation of piston and rod 366, 400, 427, 454 engines
(© Chevrolet Div., G.M. Corp.)

2. With bearing caps removed, and ring compressor tool installed, install each piston in its respective bore.

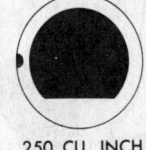

250 CU. INCH

INSERT PISTONS WITH NOTCHES TOWARD FRONT OF ENGINES

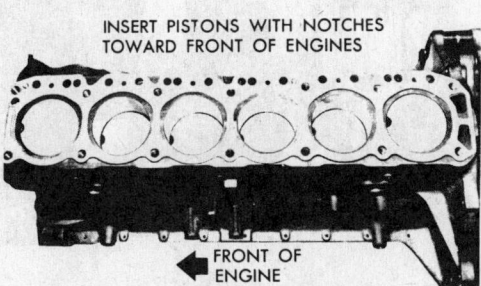

Pistons installed (inline engines)
(© Chevrolet Div., G.M. Corp.)

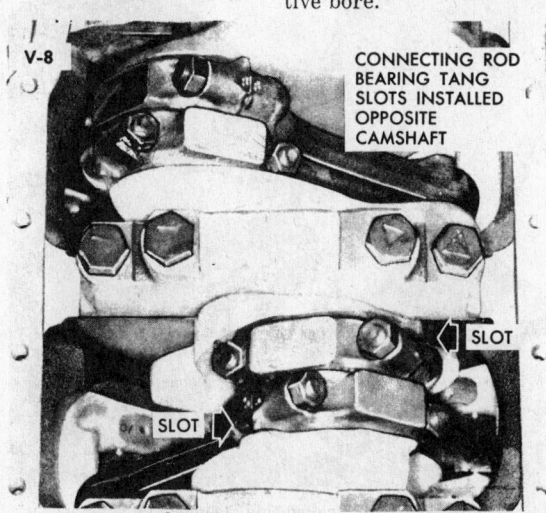

Connecting rods installed V8 engines

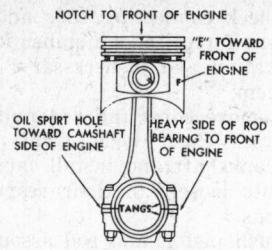

Correct relation of piston & rod
6—292 engine

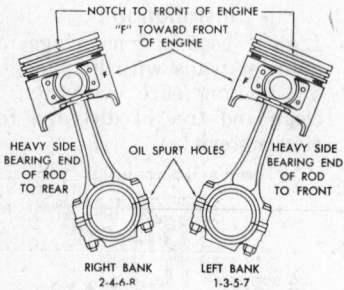

Correct relationship of piston and rod
on the 307 and 350 engines

Install piston & rod assemblies
(© Chevrolet Div., G.M. Corp.)

3. Install bearing caps and check bearing clearance. See Engine Rebuilding—General Section.
4. Install oil pan gaskets, seals and oil pan. See Oil Pan Installation.
5. Install cylinder head gasket(s) and head(s). See Cylinder Head Installation.
6. Refill crankcase and cooling system and check for leaks.

Piston Rings

Piston rings are available in standard size as well as .020″, .030″, and .040″ oversizes.

Removal

1. With pistons removed from cylinders, remove piston rings by expanding them and sliding them off piston.
2. Clean piston ring grooves by removing all particles of carbon. Check for burrs or nicks that might cause rings to hang up.

Installation

See Engine Rebuilding—General Section.

Main and Rod Bearings

Main bearings and connecting rod bearings are replaceable inserts, precision fit and held in place by locking tangs. Excessive bearing clearances reduce oil pressure. Never replace the lower half of any bearing without replacing the upper half. Do not file any bearing cap. Make sure, on main bearings, the upper half oil hole is aligned. Be certain oil passages in the crankshaft are open. Mark rod caps and upper forgings, also main caps and block, in numerical order to aid in reassembly. Rod bearings are available in standard size as well as

.001″, .002″, .010″ and .020″ undersizes. Main bearings are furnished in standard size and .001″, .002″, .009″, .010″, .020″ and .030″ undersizes.

Rod Bearing Replacement

1. Drain crankcase oil and remove oil pan. See Oil Pan Removal.
2. Remove connecting rod bearing cap.
3. Wipe bearing shell and crankpin clean of oil.
4. Inspect bearings for evidence of wear or damage. (Bearings showing the above should not be installed).
5. Measure crankpin for out-of-round or taper with a micrometer. If within specifications measure bearing clearance with plastigage or its equivalent. See Engine Rebuilding—General Section.
6. Install bearing in connecting rod and cap.
7. Coat bearing surface with oil, install rod cap and torque nuts to specifications.
8. Rotate crankshaft after bearing adjustment to be sure bearings are not too tight. Check side clearance.
9. Install oil pan gaskets, seals and oil pan. See Oil Pan Installation.

Main Bearing Replacement

NOTE: Main bearings may be replaced with or without removing crankshaft.

(Engine in Vehicle)

1. Drain crankcase oil and remove oil pan. See Oil Pan Removal.
2. Remove oil pump.
3. Loosen or remove spark plugs for easier crankshaft rotation.
4. Starting with rear main bearing, remove bearing cap and wipe oil from journal and cap.

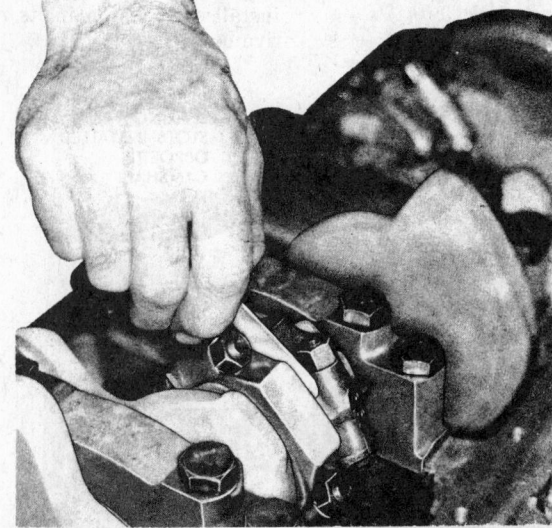

Measuring connecting rod side clearance V8 engines
(© Chevrolet Div., G.M. Corp.)

Rear main bearing removal (upper half)
(© Chevrolet Div., G.M. Corp.)

5. Inspect bearings for evidence of wear or damage.
6. Measure bearing clearance with plastigage or its equivalent. The crankshaft should be supported at damper and flywheel to remove clearance from upper bearing. Total clearance can then be measured between lower bearing and journal. See Engine Rebuilding—General Section.
7. Remove bearing shell from cap.
8. On in-line engine crankshaft, rear main bearing has no oil hole. Replace rear main bearing upper half as follows:

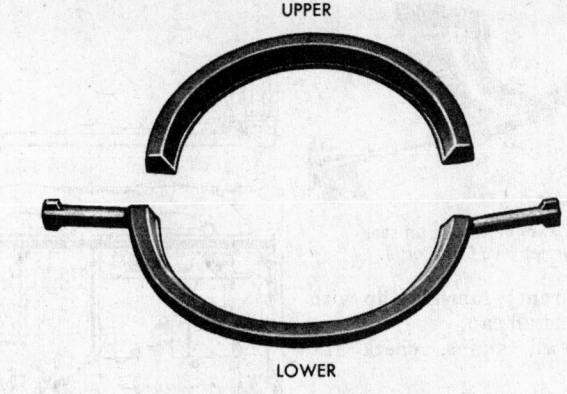

Rear main oil seal 6 cyl. engines
(© Chevrolet Div., G.M. Corp.)

Wire attached on oil seal
(© Chevrolet Div., G.M. Corp.)

Rear main oil seal installation
(© Chevrolet Div., G.M. Corp.)

a. Use a small drift punch and hammer to start upper bearing half rotating out of block.
b. Use a pair of pliers (with taped jaws) to hold bearing thrust surface to oil slinger and rotate crankshaft to remove bearing.
c. Oil new selected size upper bearing and insert plate (unnotched) and between crankshaft and indented or notched side of block.
d. Use pliers as in removing to rotate bearing into place. The last ¼ movement may be done by holding just the slinger with pliers or tap in place with a drift punch.
9. All other crankshaft journals (in-line and "V" models) have oil holes. Replace main bearing upper half as follows:

a. Install a main bearing removing and installing tool in oil hole in crankshaft journal. *NOTE: If such a tool is not available, a cotter pin (with head flattened) may be used.*
b. Rotate crankshaft clockwise as viewed from front of engine. This will roll upper bearing out of block.
c. Oil new selected size upper bearing and insert plain (unnotched) end between crankshaft and indented or notched side of block. Rotate bearing into place and remove tool from oil hole in crankshaft journal.
10. Oil new lower bearing and install in bearing cap.
11. Install main bearing cap according to markings. Torque bearing cap bolts to specifications.
12. Install oil pump and oil pan. Refill crankcase. Install and tighten spark plugs.

Rear Main Bearing Oil Seal Replacement (Engine in Vehicle)

1. Raise vehicle and drain oil.
2. Remove oil pan and oil pump. See oil pan and oil pump removal.
3. Remove rear main cap and discard seal.
4. Loosen all mains (except no. 1) and block crankshaft down for maximum clearance at rear main.
5. Use wooden dowel, so as not to mar journal, tap lightly until seal can be gripped and removed. Rotating shaft may help.
6. Wick type seal: Use "Chinese Finger" type of tool, or insert a piece of soft wire through seal approximately ¼ from end, then wrap end with wire. Insert wire through seal opening in crankcase and around crankshaft. Start upper half of seal in place, lubricated with light sealer. Pull seal into place (rotating shaft may help) until centered. Cut each end of seal so that ¼" protrudes. Install new lower half of oil seal in cap and roll or pack into position. Cut ends flush with cap parting face. Replace cap. Molded type seal: Insert new upper half, lubricated with light sealer, into channel and apply firm pressure with hammer handle until seal is centered. (Lip of seal facing toward front of engine.) Install lower half of oil seal in cap (lip

Rear main oil seal removal
(© Chevrolet Div., G.M. Corp.)

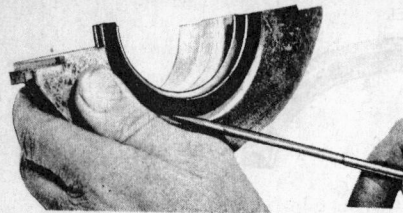

Removing lower half—rear oil seal
(© Chevrolet Div., G.M. Corp.)

toward front), lubricate lip with oil and install cap.

7. Torque all mains, check for drag.
8. Replace oil pump, oil pan and add oil.
9. Start engine and check for leaks.

Engine Lubrication

Oil Pan

Removal and Installation

Vans

1. Support on stands.
2. Disconnect battery ground cable, loosen fan belt and remove radiator shroud top belt.
3. Remove radiator fan and pulley.
4. Raise vehicle, clean road dirt from oil pan.
5. Drain oil, replace drain plug gasket.
6. Remove engine splash shields if equipped.
7. Remove starter, leave electrical connections intact, swing out of way.
8. Automatic transmisson models—remove oil cooler lines and converter pan.
9. Remove front motor mount bolts.
10. Remove accessory drive pulley (if used).
11. Engines with radiator shroud—drain radiator, remove lower radiator hose, remove lower shroud bolts and lower shroud out of way.
12. Using a jack, raise and support front of engine.
13. Remove crossmember to frame bolts, remove crossmember.
14. Remove oil pan bolts, remove oil pan. **CAUTION: If any prolonged operations are planned (with pan off), it would be safer to re-install crossmember and lower engine on mounts.**
15. Reverse removal steps to install after cleaning gasket and seal surfaces.
16. Lower vehicle, fill crankcase and radiator, start engine and check for leaks. *NOTE: Use gasket sealer as a retainer to hold side gaskets in place on block. Bolts*

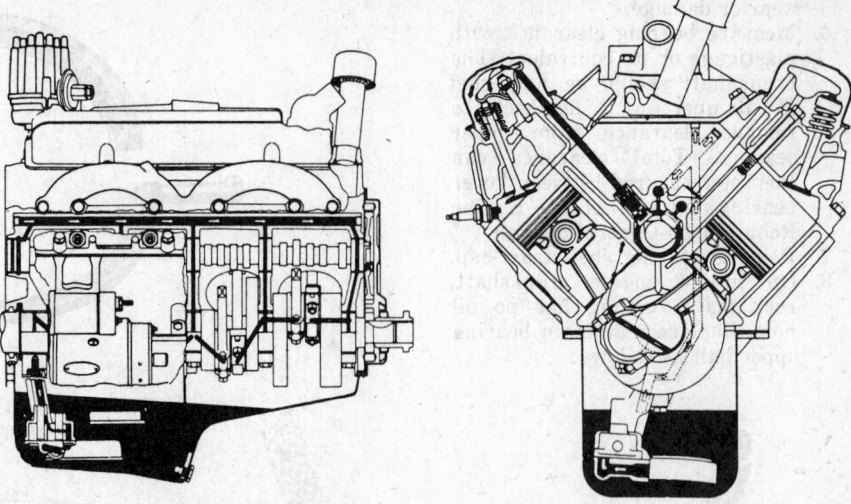

Engine lubrication on the 307 and 350 engines (© G.M.C.)

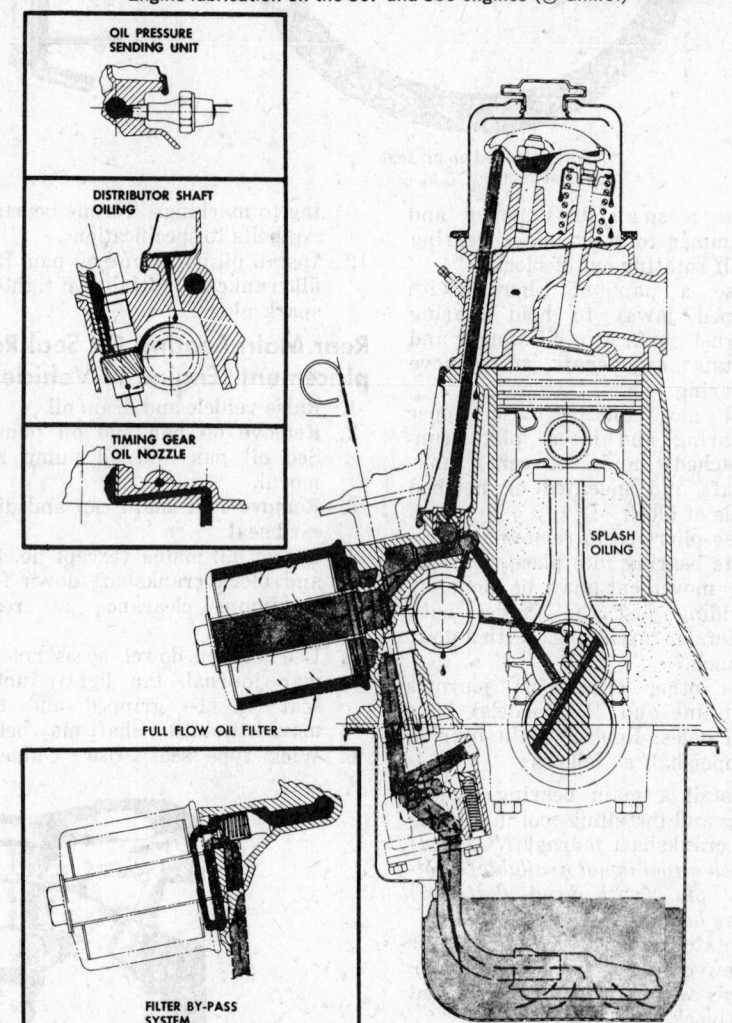

OIL PRESSURE SENDING UNIT

DISTRIBUTOR SHAFT OILING

TIMING GEAR OIL NOZZLE

FULL FLOW OIL FILTER

FILTER BY-PASS SYSTEM

SPLASH OILING

Engine lubrication on the 250 and 292 engines (© G.M.C.)

in front cover should be installed last. They are installed at an angle and holes line up after rest of pan bolts are tightened.

(Except Vans)

1. Raise front of vehicle and support on stands.
2. Remove road dirt from pan and drain plug.
3. Drain oil, replace drain plug gasket.
4. Remove converter pan on automatic transmission.
5. Remove battery ground cable, remove starter (leave electrical

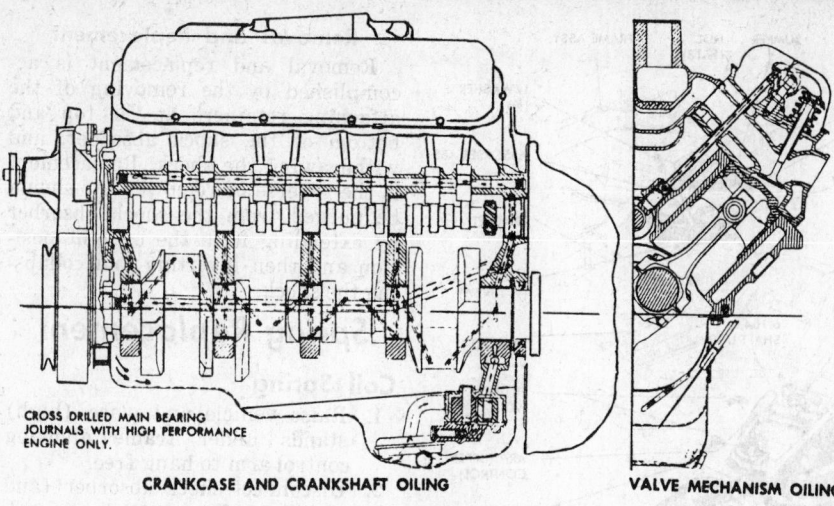

CROSS DRILLED MAIN BEARING JOURNALS WITH HIGH PERFORMANCE ENGINE ONLY.

CRANKCASE AND CRANKSHAFT OILING

VALVE MECHANISM OILING

Engine lubrication on the 366 engine (© G.M.C.)

connections intact and swing aside).

6. V8—Remove dipstick and tube, remove exhaust crossover pipe.

7. C10-30—Remove front motor mount bolts, using a block of wood under oil pan, raise engine with a jack high enough to insert blocks at motor mounts.

8. Remove oil pan and discard gaskets and seals.

9. Reverse removal steps to install after cleaning all gasket and seal surfaces.

10. Lower vehicle, fill crankcase to level, start engine and check for leaks.

NOTE: Use gasket sealer as a retainer to hold side gaskets in place on block.

Oil Filter

Oil filters are related to oil change periods, which in turn are related to quality of oil used, type of service and operating conditions. Severe conditions such as frequent and prolonged idle periods may warrant more changes. Heavy sludge in pan or filter indicates oil change intervals should be shortened. Filter replacements should be made at time of oil change. It is recommended that filter be replaced with initial oil change

and every second oil change thereafter.

Filters on in-line engines are located at right front of engine, "V" engines at left rear.

Replacement
Element Type

1. Remove drain plug (some models) in shell and drain, replace drain plug (if used).

2. Remove center stud, withdraw shell and empty (if no drain plug).

3. Lift out element and clean inside of shell thoroughly.

4. Clean base and replace shell gasket, check bypass valve.

5. Install new element (with element gasket) in shell.

6. Install shell on base, drain plug away from engine, and tighten retaining bolt to engine and check for leaks, check oil level.

Disposable
(Spin-on) type

1. Turn filter from mounting base and discard.

2. Clean base and inspect gasket area.

3. Apply oil film to gasket and turn filter on stud until gasket seats then tighten ½ turn more.

4. Run engine, check for leaks, check oil level.

NOTE: On some vans the right hand engine splash shield will have to be removed.

Oil Pump

The pumps used are distributor driven gear type. Oil pump consists of two spur gears and a relief valve in a two piece housing. On in-line engines, pump is mounted on cylinder block while on "V" engines pump is mounted on rear main cap inside oil pan. Pump gears and body are not serviced separately, replace pump as a unit. A baffle, incorporated on the pickup screen, eliminates oil pressure loss due to surging.

Removal and Installation

1. Remove oil pan—see oil pan removal.

2. In-line engines—Remove oil suction pipe at housing. CAUTION: Donot disturb screen on pick up pipe.

3. Remove two flange mounting bolts, remove pump.

4. Reverse removal steps to install, using new pan gasket, watch slot alignment with distributor tang. *NOTEs Pump should slide easily into place, if not remove and relocate slot.*

5. Refill with oil, start engine and check for leaks.

Front Suspension

General instructions covering the front suspension and how to repair and adjust it are given in the General Repair Section.

Figures covering the caster, camber, toe-in, kingpin inclination, and turning radius can be found in the Wheel Alignment table of this section.

Coil Spring Type
This suspension consists of upper

THROW-AWAY TYPE **REPLACEABLE ELEMENT TYPE**

Engine oil filters (© Chevrolet Div., G.M. Corp.)

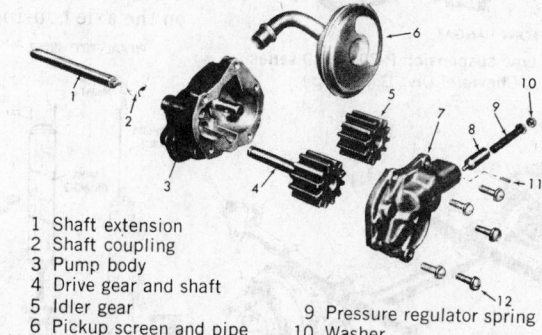

1 Shaft extension
2 Shaft coupling
3 Pump body
4 Drive gear and shaft
5 Idler gear
6 Pickup screen and pipe
7 Pump cover
8 Pressure regulator valve
9 Pressure regulator spring
10 Washer
11 Retaining pin
12 Screws

Oil pump V8 engines (© Chevrolet Div., G.M. Corp.)

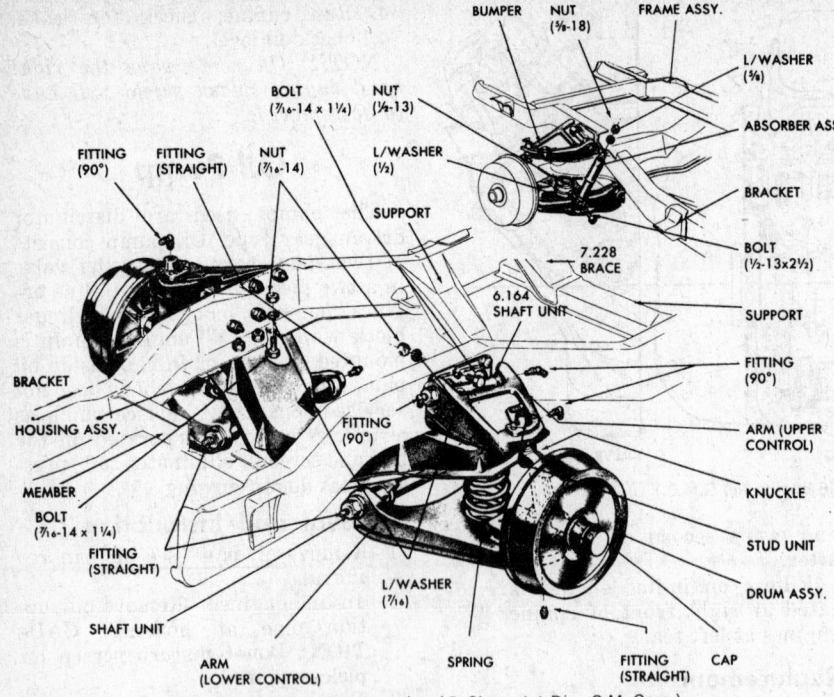

Coil spring type suspension (© Chevrolet Div., G.M. Corp.)

and lower control arms, pivoting on steel threaded bushings on upper and lower control arm inner shafts which are attached to the crossmember. Control arms are connected to the steering knuckle by ball joints. A coil spring is seated between the upper and lower control arms, thus the lower control arm is the load carrying member.

I Beam Type

Front axle is the conventional reverse Elliott type with king pins that use full floating bushings up to approximately 7,000 lb. Front leaf spring is semi-elliptical, steel bushed at front eye, shackled at rear and seated on top of axle.

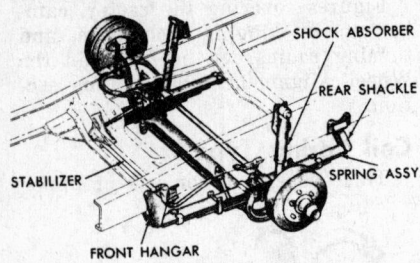

I beam type suspension P-20—P-20 series (© Chevrolet Div., G.M. Corp.)

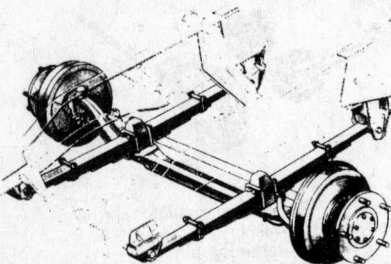

I beam type suspension 40, 50 and 60 series (© Chevrolet Div., G.M. Corp.)

Shock Absorbers

Shock absorbers are used to dampen the rebound of the two types of springs used coil and leaf.

Leaf Type

The top of the shock absorbers are mounted to the frame and the bottom is usually mounted to the U bolt bracket at the axle area or to a bracket welded to the axle housing.

Coil Spring

Front

The shock absorbers are usually attached to the lower control arm at the bottom and to the frame rail at the top. On some models, the shock absorbers may be mounted through the coil spring.

Rear

The top of the shock absorber is mounted to the body or to a crossmember with the bottom mounted to a stud or bracket welded or mounted on the axle housing.

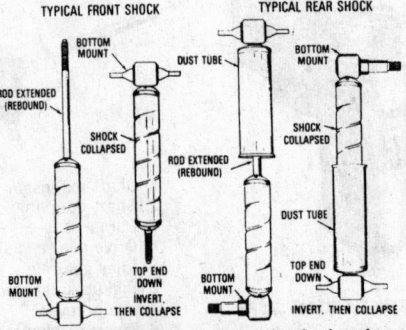

Positions for purging air from shock absorbers (© Chevrolet Div., GM Corp.)

Removal and Replacement

Removal and replacement is accomplished by the removing of the attaching retainers at the top and bottom of the shock absorber, and withdrawing the shock. Replacement is the reverse of removal. Air should be purged from the shock absorber by extending it in the upright position and then inverting and collapsing the shock.

Spring Replacement

Coil Spring

1. Raise vehicle and place (high) stands under frame allowing control arm to hang free.
2. Disconnect shock absorber (and stabilizer if used) at lower end.
3. Using a floor jack, under center of lower control arm inner shaft, raise and remove tension from shaft. CAUTION: Install a safety chain through spring.
4. Remove both clamps or "U" bolts securing inner shaft to crossmember.
5. Release jack very cautiously, slowly lowering arm with spring until spring is free. Remove safety chain, remove spring.
6. Inspect front end especially at ball joints and both upper and lower control arm inner shaft bushings.
7. Reverse removal steps to install, use a long tapered drift to align holes of inner control arm shaft and crossmember while slowly jacking arm into place.

Leaf Spring

1. Wire brush all road dirt from threaded areas on U bolts, shock absorbers and stabilizer links, apply a good penetrant on threads.
2. Disconnect shock and stabilizer link at lower bracket.
3. Loosen both spring U bolts.
4. Using jack under I-beam, raise front of vehicle and support at frame side rail with stand. Finish removing U bolts and rebound bumper. Lower jack until spring clears I-beam or tire rests on ground. Remove any caster wedges (shims) and set aside for installation. NOTE: Thick end of shim goes to rear of vehicle for increased caster.
5. Remove front spring eye bolt.
6. Remove rear shackle (and hanger cam if equipped).
7. Remove spring, inspect hangers and spring seat (center bolt index).
8. Reverse removal procedures to install after placing spring on axle with center bolt head indexed in seat and caster shims in place. Torque all nuts and lubricate.

Coil Spring Suspension

Ball Joint Replacement

Upper

1. Place jack under lower control arm, at coil spring, and raise vehicle until tire clears floor.
2. Remove tire, wheel and drum assembly.
3. Remove upper ball stud nut and break the stud taper from the steering knuckle by rapping sides of knuckle flats at stud. Separate stud from knuckle.
4. Remove rivets and bolt in new ball joint. Rivets can be chiseled off, ground off or drilled out.

AXIAL MOVEMENT

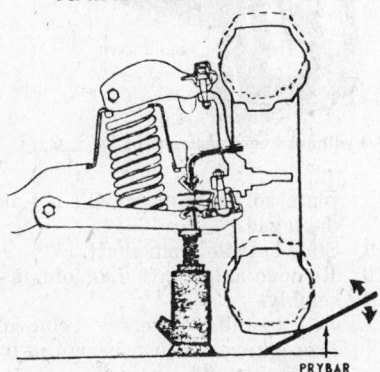

Ball joint check—unloaded
(© Chevrolet Div., G.M. Corp.)

TIRE SIDEWALL MOVEMENT

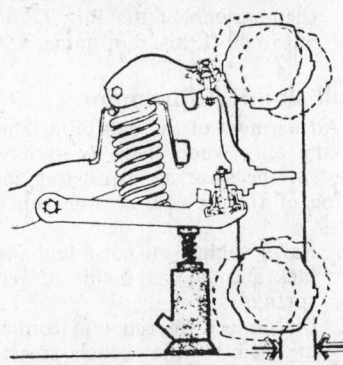

Ball joint check—unloaded
(© Chevrolet Div., G.M. Corp.)

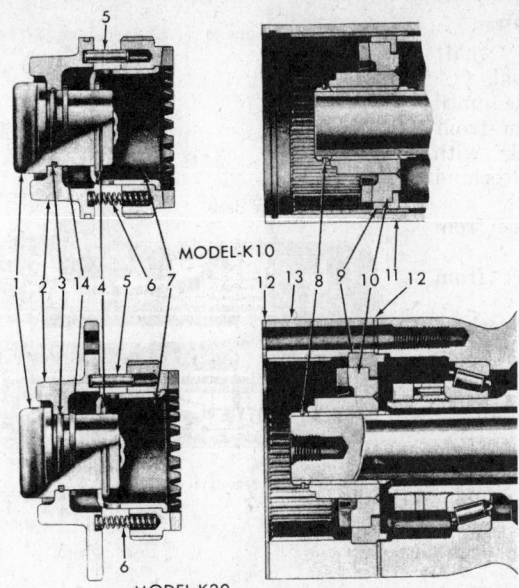

MODEL-K10

MODEL-K20

Hub assembly (© Chevrolet Div., G.M. Corp.)

1 Key knob
2 Retainer cap
3 Key knob seal
4 Key knob retainer pin
5 Cap aligning pin
6 Spring
7 Outer clutch gear assembly
8 Axle shaft lock ring
9 Collar (bushing) internal clutch gear
10 Internal clutch gear
11 Wheel hub housing
12 Extension housing gaskets
13 Extension housing
14 Snap ring

CAUTION: Use special hardened bolts only when installing joint (furnished with joint).

5. Reverse remaining removal steps to complete installation.

Lower

1. Jack vehicle at lower control arm spring seat and remove tire, wheel and drum assembly.
2. Remove coil spring, see coil spring R&R.
3. Remove ball stud nut at knuckle, use stud jack or rap stud loose from knuckle.
4. Remove lower control arm assembly.
5. Press out ball joint. Press new ball joint into arm. Make sure ball joint assembly is fully seated and square with arm. Check inner shaft bushings.
6. Reverse remaining removal steps to complete installation.

Wheel Bearing Adjustment

1. Check the bearing for a tight or loose fit by gripping the wheel at the top and bottom and moving the wheel in and out on the spindle. The end play should be .001 to .005 inch.
2. If adjustment is needed, remove the cotter pin and tighten the spindle nut to 12 ft. lbs. to fully seat the bearings.
3. Loosen the nut until either hole in the spindle lines up with the slot in the nut.
4. Install the cotter pin and bend the ends against the nut. The end play should be between .001 and .005 inch.
5. Install the dust cover and wheel cover, if equipped.

Front Drive Axle Assembly

The front drive axle is a hypoid type gear axle unit equipped with steering knuckles. Axle assembly number and production dates are stamped on the left axle tube. Conventional truck brakes are provided on all four wheel drive units.

Refer to the GMC truck section for the method of removal and installation of the ball joints used in the front drive axle.

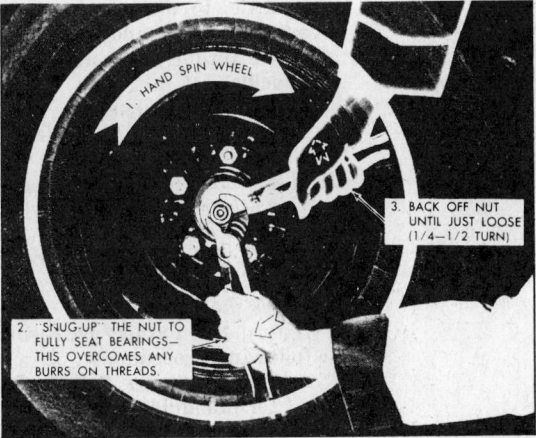

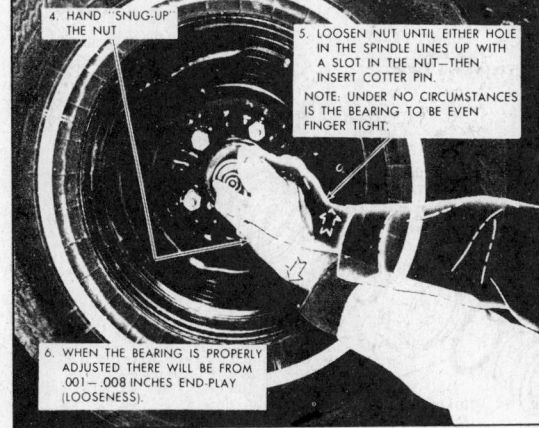

Front wheel bearing adjustment—typical (© G.M.C.)

Removal and Installation

1. Disassemble propeller shaft from front axle differential.
2. Raise front of vehicle until weight is removed from front springs. Support truck with stands on frame just behind front springs.
3. Disconnect connecting rod from steering arm.
4. Disconnect brake hoses from frame fittings.
5. Disconnect shock absorbers from axle brackets.
6. Dismount "U" bolts from axle to separate axle from truck springs.
7. Raise truck enough to clear axle and roll axle out from underneath.
8. To install axle, reverse removal procedures.

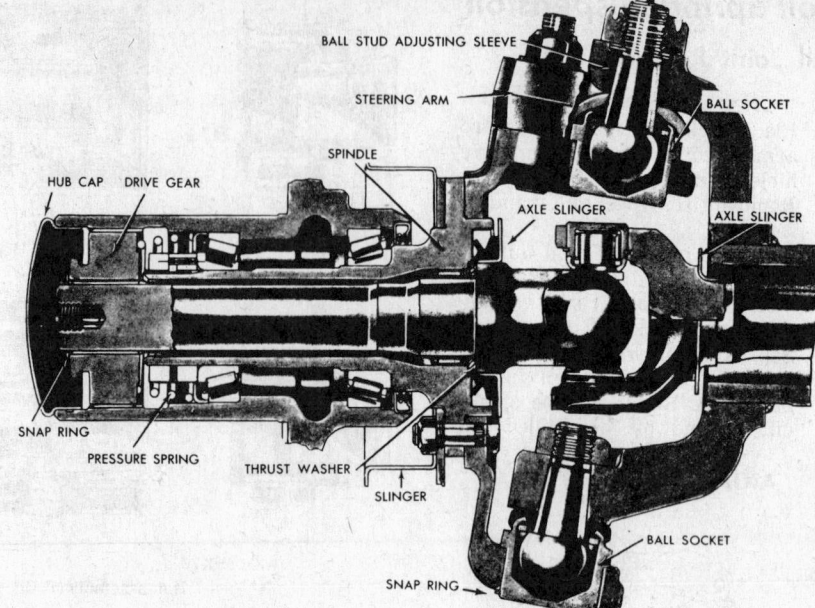

Steering knuckle and hub assembly—4 wheel drive without free-wheeling hubs (© G.M.C.)

Front Axle Shaft

Removal and Installation

1. Support vehicle with front wheel free.
2. Place drain pan under wheel and hub grease cap.
3. Remove snap ring from axle shaft.
4. Remove drive flange as follows:
   a. On K10 and K14 models, pull splined drive flange from shaft and hub and remove spacer from hub.
   b. On K20 and K25 models remove stud nuts and washers, remove drive flange and gasket.
5. Remove bearing lock nut and lock ring and remove adjusting nut.

6. Remove wheel, hub and drum as an assembly.
7. Remove bolts and lockwashers holding brake flange plate and spindle to steering knuckle.
8. Remove flange plate from steering knuckle and support the plate so brake hose will not be damaged.
9. Slide spindle from shaft.
10. Remove axle shaft and joint assembly.
11. To install, reverse removal procedures. Adjust bearings by tightening nut snug while rotating drum. Loosen nut 1/8 turn and assemble lock-washer by loosening nut to nearest hole, then assemble outer nut. Tighten nut to 40 ft. lbs. minimum.

Ball Joint Adjustment

Adjustment of the ball joint is necessary only when there is excessive play in the steering, persistent loosening of the tie rod, or wear on the tires.

1. Raise vehicle on hoist and place jack stands just inside of front springs.
2. Disconnect tie rod and connecting rod to allow each steering knuckle to move independently.
3. Apply torque wrench, at top of knuckle, to one of the steering arms attaching and stud nuts and check torque necesstry to turn knuckle.
   Maximum torque is:
   15 ft.-lbs.   Axle 44-5B
   20 ft.-lbs.   Axle 44- B
   NOTE: Knuckle should turn smoothly through turning arc and have no vertical end play.
   IMPORTANT: Temperature will affect turning torque considerably so allowances should be made.
4. If torque is too low perform the following procedures:
   a. Remove upper ball stud cotter pin and lock nut.
   b. Torque stud adjusting sleeve to 50 ft.-lbs.

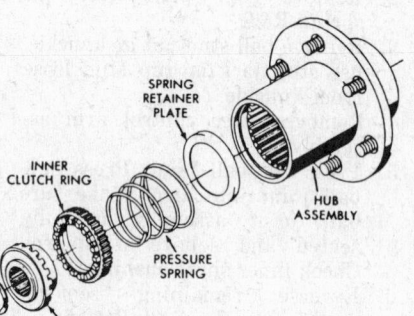

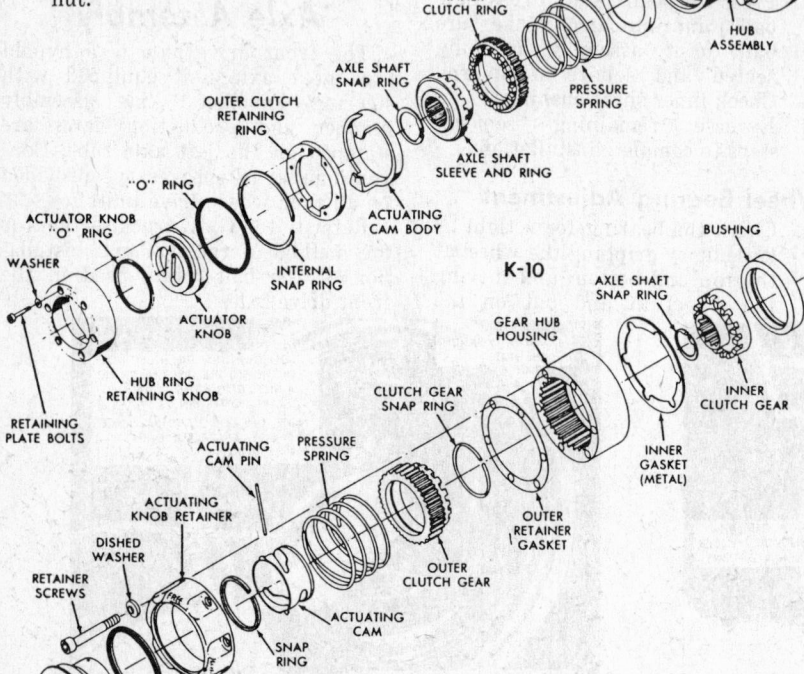

Exploded view of free-wheeling hub assemblies (© G.M.C.)

c. Install lock nut and new cotter pin.
d. Recheck torque necessary to turn steering knuckle. If torque is still too low retorque sleeve to 60-ft.-lbs. and recheck.

NOTE: *If torque specifications for turning of knuckle can not be reached, replace required parts.*

Free Wheeling Hub

Free-wheeling hubs are available for the front wheels of four wheel drive vehicles. The purpose of these hubs is to reduce friction and wear by disengaging the front axle shafts, differential and drive line from the front wheels when the vehicle is operated under conditions where front wheel drive is not needed.

The engagement and disengagement of free-wheeling hubs is a manual operation which must be performed at each front wheel. The transfer case control lever must be in 2-wheel drive position when locking or unlocking hubs. Both hubs must be in the fully locked or fully unlocked position. They must not be in the free-wheeling position when low all wheel drive is used as the additional torque output in this position can subject the rear axle to severe strain and rear axle failure may result.

K20 Series

Removal

1. Turn hub key to set hub to *"free"* or *"disengage"* position.
2. Remove outer clutch ring, cap and cam assemblies as follows: Remove internal snap ring from hub and pull assembly out of hub.
   a. Remove bolts retaining hub cap assembly to wheel hub, pull assembly and extension housing away from hub and remove gasket.
3. Remove snap ring from end of axle shaft.
4. Pull bushing and inner clutch ring assembly shaft.

Installation

1. Place bushing and inner clutch ring over axle spline with clutch teeth outward. Slide on until bushing bottoms in wheel hub and install snap ring in axle shaft.
2. Install outer clutch ring, cap and cam assembly as follows: Place gasket on hub, align mounting holes in clutch ring and cap with holes in hub and install bolts and lockwashers.
3. Turn knob to check for proper locking.

K10 Series

Removal

1. Turn actuator lever to set hub to "lock" position and raise vehicle.
2. Remove retaining plate bolts and remove retaining plate actuating knob and "O" ring.
3. Remove internal snap ring, outer clutch retaining ring and actuating cam body.
4. Remove axle shaft snap ring.
5. Remove axle shaft sleeve and clutch ring assmbly and inner clutch ring and bushing assembly.
6. Remove pressure spring and spring retainer plate.

Installation

1. Lubricate all parts. Install spring retainer plate (flange side facing bearing) over spindle nuts and seat retainer against bearing outer cup.
2. Install pressure spring into position. Large O.D. seats against spring retaining plate. NOTE: *When spring is seated, spring extends past the spindle nuts by approximately 7/8".*
3. Place inner clutch ring and bushing assembly into axle shaft sleeve and clutch ring assembly

and install as an assembly into the axle shaft. Press in on assembly and install axle shaft snap ring. NOTE: *Install 7/16x20 bolt in axle shaft end and pull outward on axle shaft to aid in installing snap rings.*
4. Install activating cam body (cams facing outward), outer clutch retaining ring and internal snap ring.
5. Install "O" ring on retaining plate and install activating knob and retaining plate. NOTE: *Install activating knob with knob in "lock" position, grooves in knob must fit into activator cam body.*
6. Install cover bolts and seals.
7. Turn knob to *"free"* position to check for proper operation.
8. Lower vehicle.

Steering Gear

Manual Steering Gear

Instructions covering the overhaul of the steering gear will be found in the General Repair Section.

Power Steering Gears

Troubleshooting and repair instructions covering power steering gears are given in the General Repair Section.

Steering Gear Replacement

Except Van and Forward Control

1. Set wheels in straight ahead position. Mark alignment of upper and lower steering shafts. Loosen champ bolt at coupling (coupling is splined), or separate at joint. On some models to remove coupling, either lower steering gear or raise mast jacket assembly.
2. Mark pitman arm and sector shaft, for reassembly, and using

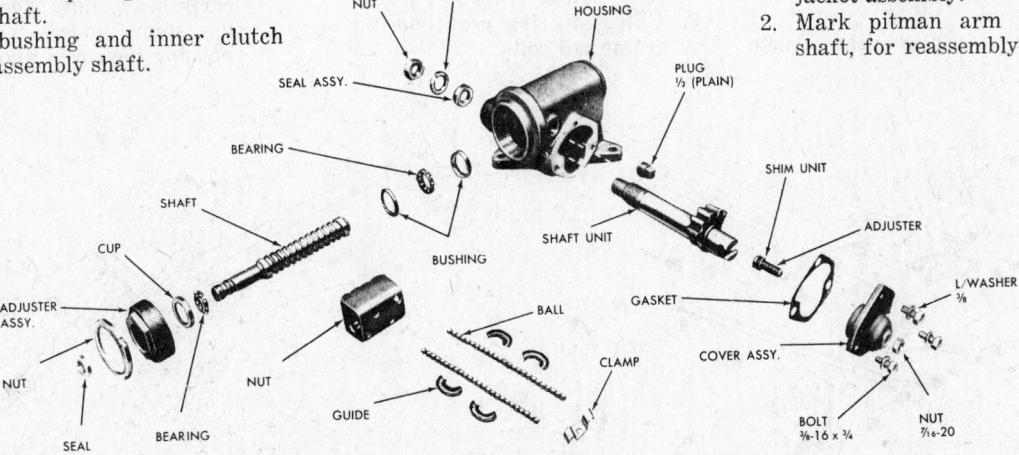

NUT · SEAL ASSY. · L/WASHER 7/8 · HOUSING · PLUG 1/2 (PLAIN) · SHIM UNIT · ADJUSTER · L/WASHER 3/8 · BEARING · SHAFT · CUP · BUSHING · SHAFT UNIT · GASKET · COVER ASSY. · NUT 7/16-20 · ADJUSTER ASSY. · NUT · NUT · GUIDE · BALL · CLAMP · BOLT 3/8-16 x 3/4 · SEAL · BEARING

Manual steering gear (© Chevrolet Div., G.M. Corp.)

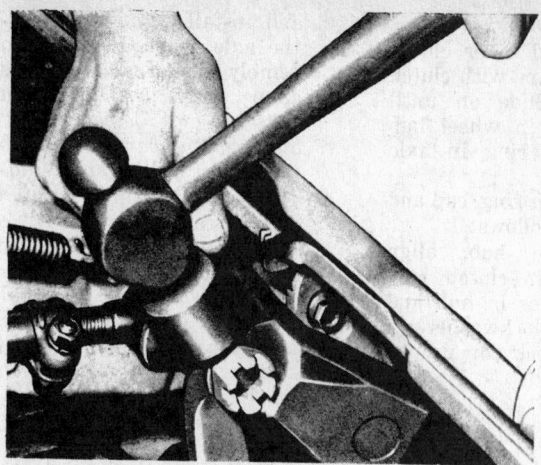

Ball stud removal (typical)
(© Chevrolet Div., G.M. Corp.)

K-10-20 steering linkage
(© Chevrolet Div., G.M. Corp.)

puller remove pitman arm, or remove drag link from pitman arm if sufficient room between drag link stud and frame rail. If clamp type pitman arm is used, spread just enough to remove.

3. Remove power steering hoses from gear assembly, cap lines and plug openings (if equipped). from vehicle.
4. Remove gear housing to frame rail bolts (watch shims), remove from vehicle.
5. Reverse removal steps to install. *NOTE: Do not hammer pitman arm on sector shaft, use nut to pull into position. On power steering vehicles, raise front when bleeding.*

Van and Forward Control

1. Disconnect battery ground cable.
2. Set wheels in straight ahead position.
3. Disconnect steering shaft flange bolts at flexible coupling.
4. Mark position of pitman arm on shaft. Remove pitman shaft nut and remove arm using a puller. *NOTE: If clamp type pitman arm is used, spread arm, using a wedge, just enough to remove pitman from shaft.*
5. Remove steering gear frame

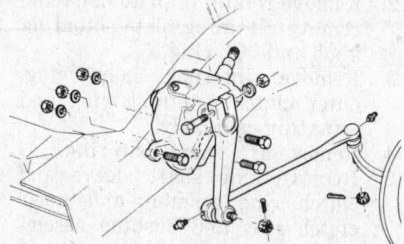

20-40-50-60 ser. steering linkage (typical)
(© Chevrolet Div., G.M. Corp.)

mounting bolts and remove steering gear assembly.

6. Reverse removal steps to install. *NOTE: Do not hammer pitman arm onto sector shaft, use nut to pull into position.*

Adjustments (on Vehicle)

Manual Steering

1. Check tire pressure and inspect steering linkage.
2. Check gear housing for lubricant, tighten cover side plate.
3. Check steering gear housing to frame rail bolts.

4. Set wheels in straight ahead position.
5. Disconnect drag link from pitman arm, remove pitman arm on Vans.
6. Loosen sector (cross) shaft lock nut and back off lash adjuster ¼ turn. This lessens steering worm bearing load by reducing tooth mesh contact.
7. Check load on steering gear by measuring pull on steering wheel with scale. Pull is measured at rim of wheel with scale tangent to rim of wheel (½ to 3 lb. pull according to size of truck).
8. If pull is not within limits, adjust worm bearings. Loosen worm bearing lock nut and turn adjusting nut in until there is no perceptible end play. Tighten lock nut. Using an inch-pound torque wrench and socket on steering wheel nut (remove horn wire and button), measure torque (3 inch pounds to 14 inch pounds according to size of truck). A rough feeling when rotating steering wheel indicates defective worm bearings. Some early heavy duty steering gears use shims under steering gear housing top cover, and also use a

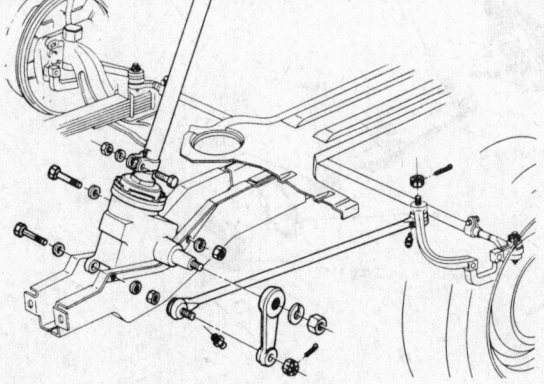

(© Chevrolet Div., G.M. Corp.)
Chevy Van steering gear mounting

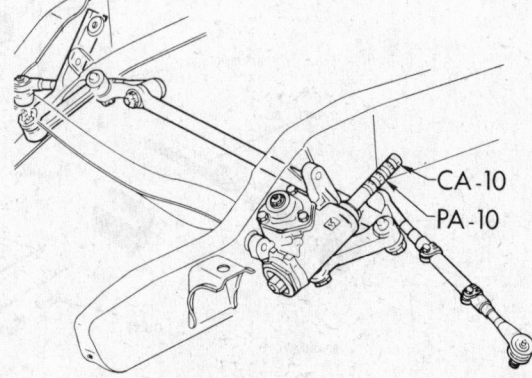

C-10-20-30 steering linkage
(© Chevrolet Div., G.M. Corp.)

CA-10
PA-10

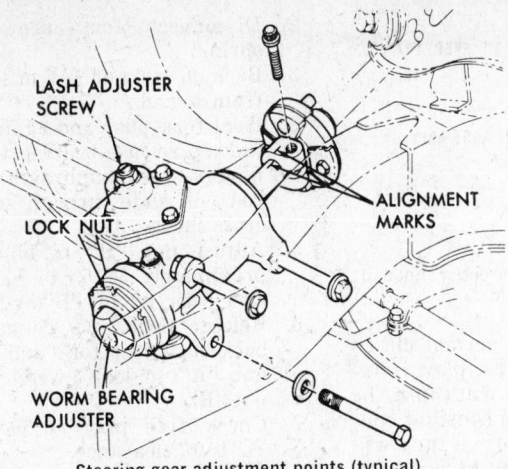

Steering gear adjustment points (typical)
(© Chevrolet Div., G.M. Corp.)

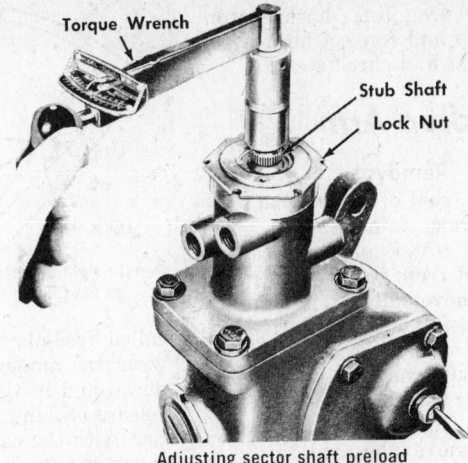

Adjusting sector shaft preload
(© Chevrolet Div., G.M. Corp.)

back-up adjuster. Removal of shims decreases worm bearing end play. Back-up adjuster setting is made last. After adjusting worm bearings and lash, then tighten back-up adjuster until adjuster bottoms against ball nut return guide clamp, then back-off adjuster ⅛ to ¼ turn and secure with lock nut.

9. After proper adjustment of worm bearings is obtained, center steering wheel by turning wheel gently from one stop to the other, counting the number of turns. Turn wheel back exactly half way to center position (high point). Mark steering wheel rim at top or bottom center with tape. Turn lash (slotted) adjuster clockwise to take out all lash in gear teeth and tighten lock nut. Check steering free play. *NOTE: If maximum adjustment is exceeded, turn lash adjuster screw back (counter-clockwise) and then come in (clockwise) slowy on adjustment.* **CAUTION: Do not bounce steering wheel hard against stops with drag link removed, worm ball guide damage can result.**

10. Connect drag link or replace pitman arm, road test and check steering.

Power Steering

Power steering gear is adjusted in the same manner as manual steering gear however, sector lash adjustment is the only power steering gear adjustment that can be made on the vehicle. In order to make this adjustment, it is necessary to check the combined valve drag and worm bearing preload.

1. Check power steering fluid level, check belt tension and hose for leaks or kinks.
2. Remove drag link from pitman arm.
3. Disconnect horn wire, remove horn button assembly.
4. Center steering wheel—turn through its full travel then locate wheel at center of its travel.
5. Loosen sector lash adjusting screw locknut and back off (slotted) adjusting screw to the limit of its travel.
6. Check combined valve drag and worm bearing preload with inch-pound torque wrench and socket on steering shaft nut, by rotating wheel approximately

(© Chevrolet Div., G.M. Corp.)

40-50-60 series power steering gear and central valve (typical)

20° in each direction. Note highest reading.

7. Tighten sector lash adjusting screw until torque at steering wheel meets specifications (4 to 18 inch pounds according to truck size). Secure lock nut. *NOTE: If maximum adjustment is exceeded, turn lash adjuster screw back (counter-clockwise) and then come in (clockwise) slowly on adjustment.* **CAUTION: Do not bounce steering wheel hard against stops with drag link removed, worm ball guide damage can result.**

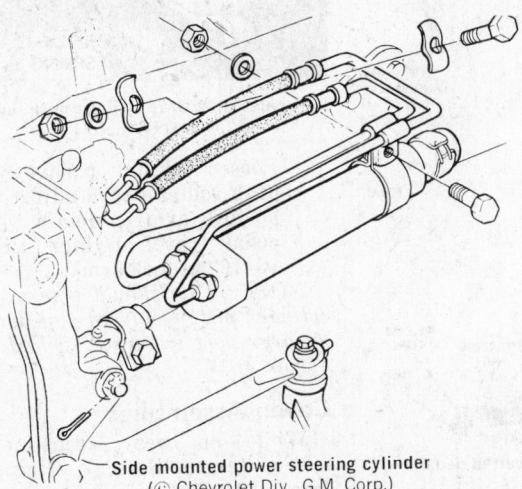

Side mounted power steering cylinder
(© Chevrolet Div., G.M. Corp.)

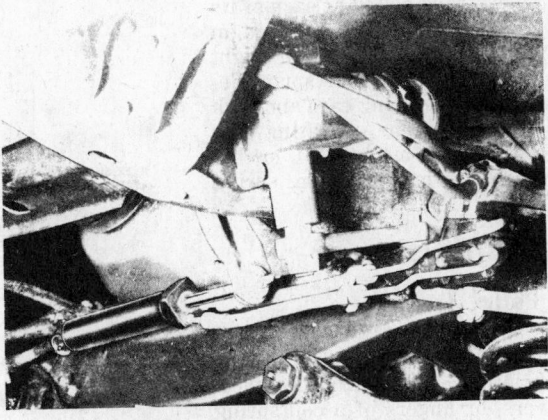

C-10-20-30 series control valve and power cyl. installed
(© Chevrolet Div., G.M. Corp.)

8. Replace drag link, horn button assembly and connect horn wire.
9. Road test and check steering.

Idler Arm

Removal

1. Jack up front of vehicle and support or raise on hoist.
2. Remove cotter pin and castellated nut from ball stud at relay rod. Remove ball stud from relay rod by using a tapered driver or tapping on relay rod boss. Support relay rod so impact will center on ball stud.
3. Remove idler arm to frame bolts, remove idler arm.

Installation

1. Position idler arm on frame and install mounting bolts. (special plain washers under bolt heads). Torque nuts.
2. Make sure threads on ball stud and in ball stud nut are good—ball stud may turn in seat if threads are damaged. Check ball and seal.
3. Install idler arm ball stud in relay rod (seal in place) and install castellated nut and cotter pin. Torque nut to specifications.
4. Lower vehicle or remove from hoist.

Clutch

Clutches used on inline engines are single (driven) disc with either of two types of pressure (drive) plates, diaphragm or coil spring. "V" engines use a single disc with coil spring pressure plate. Heavy duty clutches use two discs. The two plate clutch consists of three basic assemblies, the cover with rear pressure plate assembly, front pressure plate and two discs. The front pressure plate, located between the two driven discs, has two friction surfaces and is coupled to rear pressure plate through steel drive straps bolted at each of its four driving bosses. Diaphragm spring covers operate with light pedal pressure while coil spring levers combine operating ease and high torque capacity. The operating controls are either mechanical or hydraulic. Discs have torsion spring centers.

Adjustment

Free Pedal Travel

This adjustment is for the amount of pedal travel (measured at pedal) before clutch release bearing contacts the levers, or fingers of a coil spring cover or the diaphragm spring of a diaphragm clutch cover. This is

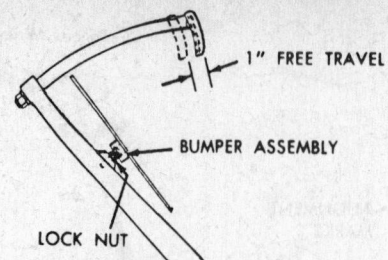

Pedal height adjustment (conventional pedal)
(© Chevrolet Div., G.M. Corp.)

called free-play. With normal clutch wear the amount of free-play is reduced and in time this will cause the release bearing to be in constant contact with the cover. This in turn will cause clutch disc slippage resulting in premature failure of disc and release bearing. It is necessary to maintain sufficient free pedal travel for clutch efficiency and long life.

Mechanical Linkage

1. Check linkage for excessive wear (will create false free play).

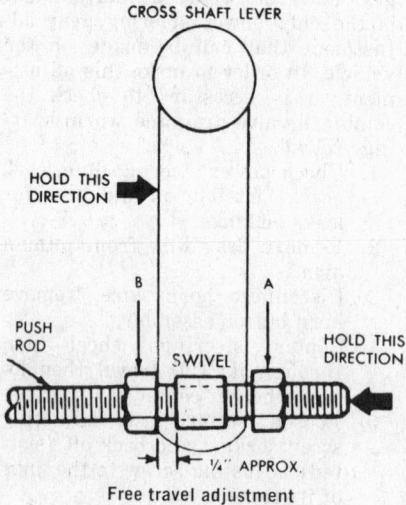

Free travel adjustment
(© Chevrolet Div., G.M. Corp.)

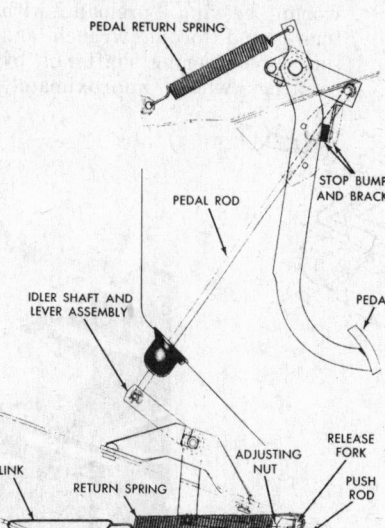

Mechanical linkage (inverted pedal)
(© Chevrolet Div., G.M. Corp.)

2. Disconnect fork arm return spring.
3. Back off locknut "A" at least ½" from swivel.
4. Hold fork push rod against fork to keep release bearing touching fingers (or diaphragm). Push rod will slide through swivel at cross shaft.
5. Adjust nut "B" to obtain approximately 3/16" to ¼" clearance between nut "B" and swivel.
6. Release push rod, connect pull back spring at fork and tighten nut "A" to lock swivel against nut "B".
7. Check free play at pedal for ¾" to 1" clearance.

Hydraulic Controls

1. Check master cylinder, cylinder must be filled and no air in system.

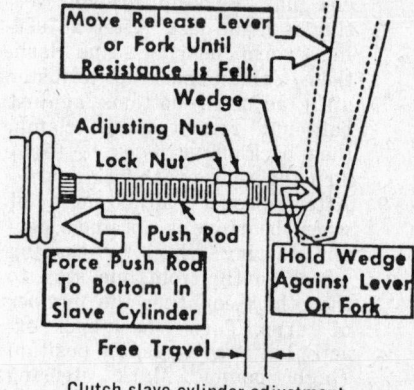

Clutch slave cylinder adjustment
(© Chevrolet Div., G.M. Corp.)

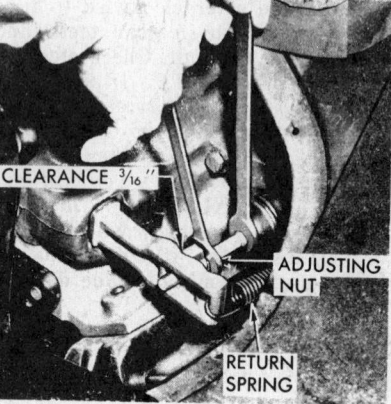

Adjusting slave cylinder push rod
(© Chevrolet Div., G.M. Corp.)

2. Loosen lock nut on slave cylinder push rod and turn adjusting nut at fork arm to obtain ¼" free pedal travel, tighten lock nut (3/16" at fork arm).
NOTE: This should insure master cylinder piston moving sufficiently to uncover port in master cylinder reservoir.

Cover Assemblies

While no wear adjustment is needed (except Spicer), original settings at time of manufacture must be retained for good clutch operation. If a diaphragm clutch cover fails to release properly, after first checking pedal travel and pedal lash and linkage for looseness, replace diaphragm retracting springs. This can be done in the vehicle. If trouble still persists the diaphragm is probably overstressed.

Clutch Replacement

1. Remove transmission (see transmission R & R).
2. Block clutch fingers down on coil spring covers for additional clearance. On heavy duty clutches use blocks between release bearing and spring plate hub. On Lipe-Rollaway use bolts and washer in three holes provided
3. Disconnect pedal linkage and fork arm pull back springs.
4. Disconnect slave cylinder from fork arm (if equipped).
5. Remove release bearing from fork arm or disconnect grease hose and remove bearing assembly from yoke.
6. Remove fork arm from ball stud or remove yoke and shaft.

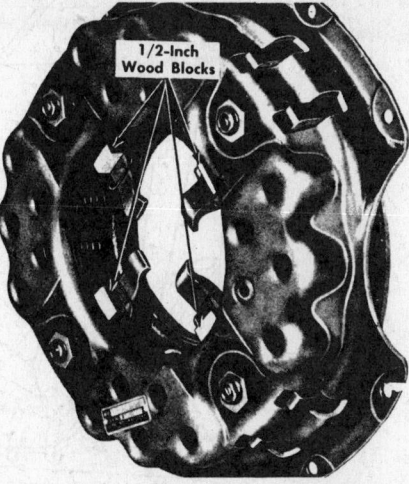

Use of wood blocks between release levers and cover (© Chevrolet Div., G.M. Corp.)

Use of blocks between release bearing and spring plate hub (© Chevrolet Div., G.M. Corp.)

Disc(s) can be removed and marked for positions. Remove remaining cover bolts and slide cover assembly off stud. Stud will also aid in installation.
10. Clean flywheel and pressure plates, check for scores and heat cracks. Excessive bluing indicates abnormally high operating temperatures. Torque flywheel bolts.
11. Check pilot bearing or bushing (lubricate sparingly), check splines on clutch shaft, check re-

lease bearing (do not wash bearing).
12. Reverse removal steps to install. Use pilot tool or dummy shaft to align disc(s). Be certain of disc's position in relation to flywheel. Tighten cover bolts evenly. Do not try to pull clutch into place with impact wrench. This procedure can crack or break pilot shoulders on bolts. Lubricate with a light coat of grease, fork arm ball seat and inside of release bearing. Make

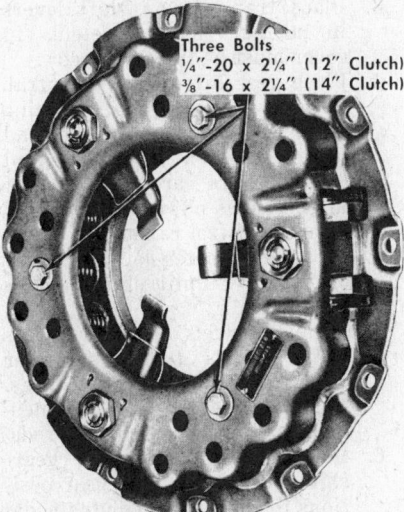

Lipe clutch hold-down bolts installed (typical) (© Chevrolet Div., G.M. Corp.)

7. Fork arm ball stud can be removed if necessary.
8. Punch mark cover and flywheel for alignment if cover assembly is to be re-used.
9. Loosen cover bolts a turn or two at a time to prevent cover distortion. Support cover and disc with pilot tool (or sling) to prevent damage to clutch when last cover bolt is removed. A good practice on heavy units is to remove one cover bolt and install a support stud. Rotate flywheel and locate stud at top, loosen all other bolts evenly until pressure is released from disc(s).

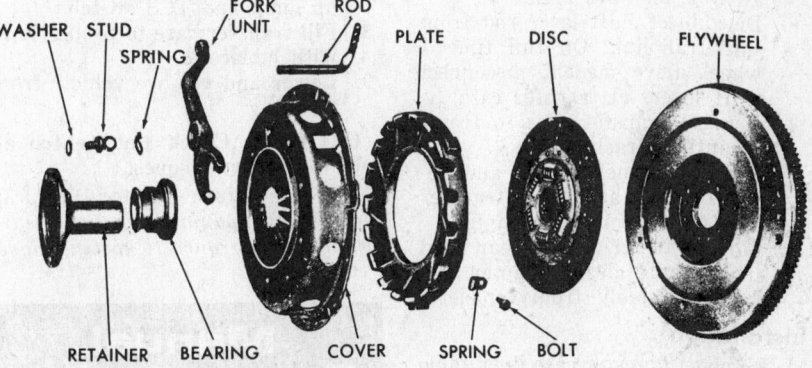

101 Diaphragm spring type clutch parts (typical) (© Chevrolet Div., G.M. Corp.)

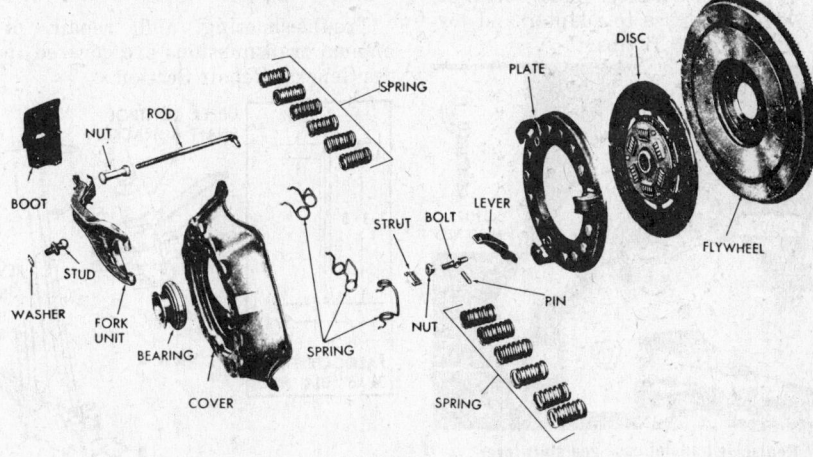

Coil spring clutch parts (typical) (© Chevrolet Div., G.M. Corp.)

sure spring retainer (if used) for release fork ball stud is in correct position. Install retainer with high side up, away from bottom of ball socket and with open end of retainer horizontal. After clutch and transmission are installed and clutch pedal free travel adjusted, check disc(s) for release. This can be done by putting transmission in gear and pulling down slowly on clutch pedal while applying torque to transmission propeller shaft flange. If release can be felt, then complete installation.

Transfer Case

Information on repair and overhaul of the transfer case can be found in the General Repair Section.

Removal

1. Raise and support vehicle on hoist. Drain transfer case.
2. Disconnect speedometer cable, back-up lamp and TCS switch.
3. Remove skid plate and crossmember supports as necessary.
4. Disconnect rear prop shaft from transfer case and tie up away from work area.
5. Disconnect front prop shaft from transfer case and tie up shaft away from work area.
6. Disconnect shift lever rod from shift rail link. On full time 4 wheel drive models, disconnect shift levers at transfer case.
7. Remove transfer case to frame mounting bracket bolts.
8. Support transfer case and remove bolts attaching transfer case to transmission adapter.
9. Move transfer case to rear until input shaft clears adapter and lower assembly from vehicle.

Installation

1. Support transfer case in suitable stand and position case to transmission adapter. Install bolts attaching case to adapter and torque to 45 ft. lbs.

Replacing transfer case gearshift lever
(© Chevrolet Div., G.M. Corp.)

2. Remove stand as required and install bolts attaching transfer case to frame rail. Bend lock tabs after assembly.
3. Install connecting rod to shift rail link or connect shift levers to transfer case, as applicable.
4. Connect front prop shaft to transfer case front output shaft.
5. Connect rear prop shaft to transfer case rear output shaft.
6. Install crossmember support and skid plate, If removed.
7. Connect speedometer cable, back-up lamp and TCS switch.
8. Fill transfer case to proper level with lubricant.
9. Lower and remove vehicle from hoist.

CAUTION: Check and tighten all bolts to specified torques.

NOTE: Before connecting prop shafts to companion flanges be sure locknuts are torqued to specifications.

Manual Transmission

Troubleshooting and repair of manual transmissions are covered in the General Repair Section.

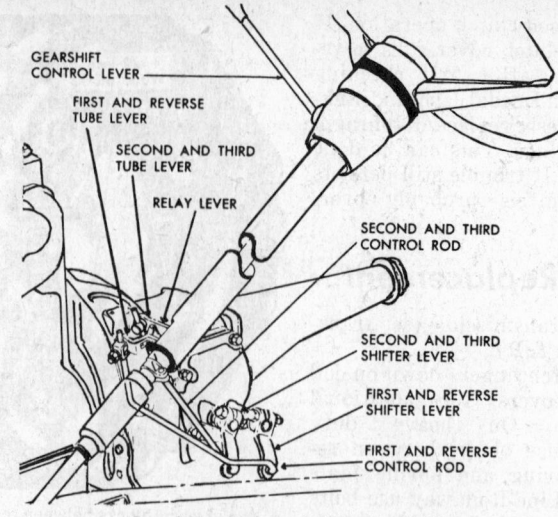

Steering column gearshift—3 speed late
(© Chevrolet Div., G.M. Corp.)

Shift Linkage Adjustment

3 Speed Column Shift

1. Raise vehicle and support on stands.
2. Disconnect control rods at transmission levers.
3. Place transmission shift levers in neutral (neutral detents in cover must be fully engaged).
4. Place gearshift lever in neutral, on early models remove housing cover at base of mast jacket and make sure shifter gates and inner levers are aligned. If alignment is off, loosen first and reverse control rod swivel clamp at housing outer lever and adjust swivel until shifter gates are aligned.
5. Adjust swivels on control rods until swivels (or rods) enter transmission shift lever holes. Make sure levers remain in neutral position. Lock control rods.
6. Lower vehicle and move gearshift lever through all gear positions to check (keep clutch pedal depressed to aid shifting).

4 Speed Column Shift

1. Place gearshift lever in neutral. Raise vehicle on hoist.

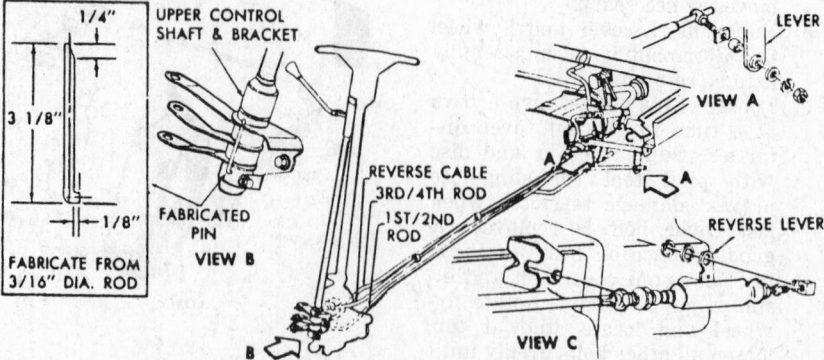

Steering column gearshift—4 speed (© Chevrolet Div., G.M. Corp.)

2. Disconnect first and second shift rod from cross shaft lever. Disconnect third and fourth shift rod from transmission lever. Disconnect reverse cable from reverse lever by removing "C" clip. Manually shift all transmission controls into neutral, including reverse lever.
3. Remove engine splash shield. Install a fabricated pin (see illustration for details) through upper control shaft bracket into cutouts in shift levers and into hole provided at base of control shaft as shown.
4. Adjust swivel on end of first and second rod to freely enter cross shaft lever hole. Reconnect rod to lever.
5. Adjust swivel on end of third and fourth rod to freely enter transmission lever. Reconnect rod to lever.
6. Adjust swivel on end of reverse cable to freely enter reverse lever hole. If more adjustment is needed at swivel, adjust cable assembly by using cable to bracket attaching nuts. Install washer and "C" clip. Tighten swivel lock nut.
7. Remove fabricated pin, replace splash shield, lower vehicle and move gearshift lever through all gear positions to check. Depressing clutch pedal will aid in shifting.

Van Column Shift

1. Raise vehicle and support on stands.
2. Remove control rods at transmission levers.
3. Move both transmission levers until transmission is in neutral. Neutral detents must be fully engaged.
4. Move gearshift lever into neutral position, align shifter relay levers on mast jacket, install pin

in holes of levers to hold levers in alignment and in neutral position.
5. Adjust swivel on end of low and reverse control rod until swivel enters transmission lever freely, lock with retaining ring, tighten swivel locknut.
6. Similarly, install second and third control rod, be sure levers remain in neutral.
7. Lower vehicle and move gearshift lever through all gear positions to check. Keep clutch pedal depressed to aid in shifting.

Floor Shift (and Related Controls)

Removal of gearshift lever, remote control and control island and any linkage adjustments are carried in transmission removal and installation section.

3 and 4 Speed Light Duty Transmission

3 Speed Except Van and Blazer with 4 Wheel Drive

Removal

1. Raise vehicle and support on jack stands.
2. Drain transmission.
3. Disconnect speedometer cable, TCS switch and back-up lamp wire at transmission.
4. Disconnect shift control levers from transmission.
5. Disconnect parking brake lever and controls (if used).
6. Remove drive shaft after marking position of shaft to flange.
7. Position jack under transmission to support weight of transmission.
8. Remove crossmember. Visually inspect to see if other equipment, brackets or lines, must be removed to permit removal of

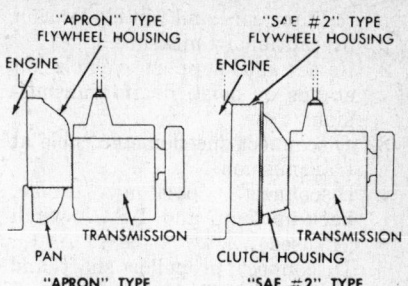

transmission. *NOTE: Mark position of crossmember when removing to prevent incorrect installation.*
9. Remove flywheel housing underpan.
10. Remove the top two transmission to housing bolts and insert two guide pins. *NOTE: The use of guide pins will not only support the transmission but will prevent damage to the clutch disc. Guide pins can be made by taking two bolts, the same as those just removed only longer, and cutting off the heads. (Slot for screwdriver)*
11. Remove two remaining bolts and slide transmission straight back from engine. Use care to keep the transmission drive gear straight in line with clutch disc hub. NOTE: Be sure to support release bearing when removing transmission to avoid having bearing fall into flywheel housing.
12. When transmission is free from engine, move from under vehicle.

Installation

1. Place transmission on guide pins, slide forward starting main drive gear into clutch disc's splines. *NOTE: Place transmission in gear and rotate transmission flange or output yoke to aid entry of main drive gear into disc's splines. Make sure clutch release bearing is in position.*
2. Install two lower transmission mounting bolts, and flywheel lower pan (if equipped).
3. Remove guide pins and install upper mounting bolts.
4. Install propeller shaft, watch align marks.
5. Connect parking brake, back-up lamp and T.S.C. switch (if used).
6. Connect shift levers, see section on adjustment if needed.
7. Connect speedometer cable, refill transmission.
8. Lower vehicle and road test.

Vans

Removal

1. Place heavy cardboard between

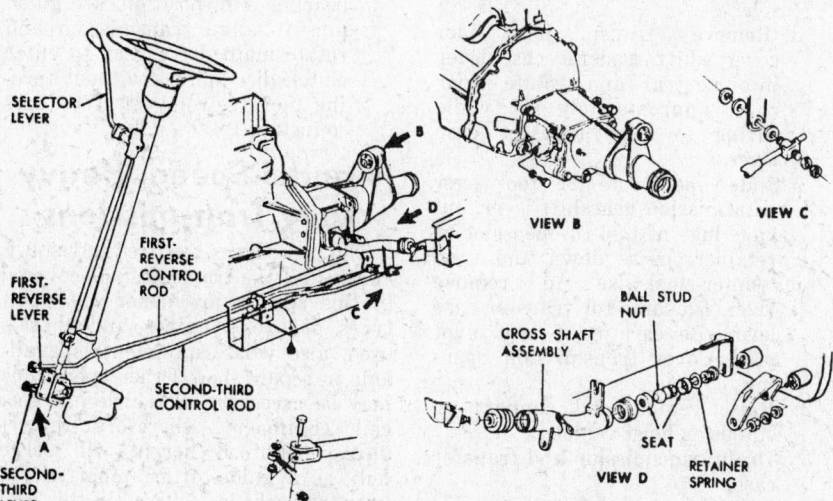

Column shift linkage—vans (© Chevrolet Div., G.M. Corp.)

radiator core and fan blades as a precautionary measure.

2. Raise and support vehicle on stands or hoist, drain transmission.
3. Disconnect speedometer cable at transmission.
4. Disconnect parking brake, back-up lamp and T.S.C. switch (if used).
5. Disconnect propeller shaft and power take off (if equipped).
6. 3 Speed (Column Shift)
   a. Remove shift controls from transmission.
   4 Speed (Floor Shift)
   a. Remove floor mat.
   b. Remove floor pan.
   c. Place transmission in neutral and remove gearshift lever by sliding open side of tool over lever, engage lugs of tool in the open slot of retainer, press down on tool and turn to left to disengage the lugs on retainer. Cover transmission opening. Be careful of pivot pin.
   d. Remove reverse lever cable and bracket at transmission.
   4 Speed (Column Shift)
   a. Remove shift controls from levers.
   b. Remove reverse lever cable and bracket at transmission.
7. Remove clutch shaft frame bolts and accelerator linkage at manifold bellcrank.
8. Place jack under bell housing and raise enough to relieve weight at transmission rear support.
9. Remove transmission rear support crossmember or on early models, remove support bolt and lower engine carefully to allow transmission rear mount to clear support bracket.
10. Position jack under transmission and adjust to carry weight of transmission (if 4 speed).
11. Remove 2 top transmission to bell housing bolts and install 2 guide pins to prevent damaging clutch disc.
12. Remove two lower transmission mounting bolts.
13. Visually inspect to determine if other equipment or lines need to be removed.
14. Slide transmission back on guide pins, four speed units aided by jack (support release bearing), until transmission clears engine. Remove transmission from under vehicle.
CAUTION: If other work is to be performed support engine more securely after transmission is removed.

Installation

1. Clean bell housing and transmission mating surfaces, lightly lubricate main drive gear bearing retainer and clutch pilot bushing or bearing. Make sure release bearing is in position.
2. Move transmission into position on guide pins, shift transmission into any gear.
3. Slide transmission forward rotating transmission flange or yoke to aid entry of main drive gear into clutch disc splines.
4. Install two lower transmission mounting bolts. Remove guide pins and install two upper mounting bolts. Remove transmission jack.
5. Carefully raise engine and transmission to normal position and install transmission rear mounting bolt or crossmember. Remove jack from under bell housing. Remove cardboard from radiator.
6. Connect speedometer cable, parking brake, back-up light and T.S.C. switch (if equipped).
7. Connect propeller shaft and power take off (if equipped).
8. Connect clutch shaft and accelerator linkage.
9. Reinstall shift controls on transmission or gearshift lever on four speed. Install reverse cable and bracket. See section on shift linkage adjustment if needed.
10. Replace floor mat and floor pans on four speed units.
11. Refill transmission, lower vehicle and road test.

4 Wheel Drive (Including Blazer)

Removal and Installation

1. Floor shift models.
   a. Remove shift lever boots and retainers on both transfer case and transmission.
   b. Remove floor mat or carpet, seat and accelerator pedal.
   c. Center console models: remove center outlet from heater distributor duct and remove console.
   d. Remove transmission floor cover, shift transfer case lever into neutral and rotate floor cover approximately 90° while lifting to clear transfer case lever.
   e. Slide open side of tool over transmission gearshift lever, engage lugs of tool in open slot of retainer, press down and turn counter-clockwise to remove lever. Do same for transfer case lever. Be careful of any pivot pins. Cover transmission openings.
2. Raise vehicle and support on stands or hoist vehicle.
3. Drain transmission and transfer case.
4. Disconnect back-up light and T.S.C. switches (if equipped).
5. Disconnect parking brake.

6. Disconnect speedometer cable (at transfer case on some models).
7. Disconnect front and rear auxiliary drive shafts at transfer case and tie up out of work area.
8. Remove bolts attaching transfer case to adapter (remove side access cover to reach two bolts).
9. Support rear of engine with jack.
10. Support transfer case on cradle or dolly. Remove two transmission adapter mounting bolts.
11. Remove transfer case mounting bolts, remove transfer case (all except Blazer).
12. Column Shift Models:
    a. Disconnect shift control rods from levers at transmission.
    b. On 4 speed, remove reverse cable and bracket at transmission.
13. Remove two upper transmission mounting bolts and install two guide pins (longer bolts with heads cut off and slotted). Use of guide pins will prevent damage to clutch disc.
14. Remove flywheel under pan and remove two lower transmission mounting bolts.
15. On V-8 engines, remove exhaust crossover pipe.
16. On Blazer, remove transmission frame crossmember bolts. Remove crossmember (rotating to clear frame rails).
17. Visually inspect to determine if other equipment or lines need to be removed.
18. Slide transmission and adapter (transmission with transfer case on Blazer) back on guide pins, aided by transmission dolly, until main drive gear clears clutch (watch clutch release bearing), remove from under vehicle.
19. Lubricate pilot bushing or bearing, make sure clutch release bearing is in position, use guide pins to align transmission and rotate main drive gear to enter clutch disc splines without forcing. Reverse removal procedures to install.

4 and 5 Speed, Heavy Duty Transmissions

The procedures required to remove and install the transmissions covered in this section are dependent upon types of cabs, engines and chassis used, also what equipment is available in repair shop. Other operations may be necessary if vehicle has special equipment, therefore, procedures contained herein will serve only as a guide. It is important to note that vehicles covered in this section will have either an "apron" or "S.A.E. 2" type of flywheel housing.

The apron type is identified by sheet metal pan, also note it is a one piece housing. The "S.A.E. 2" type completely surrounds the flywheel. A separate clutch housing is used in addition to the flywheel housing. Transmission replacement procedures are different for each type of flywheel housing used.

Removal

1. Drain transmission.
2. On transmission equipped with remote controls.
   a. Disconnect control rods from shift levers at transmission.
3. On transmissions with a conventional floor gearshift lever.
   a. Remove steering jacket grommet from floor and slide grommet up mast jacket out of way (if used).
   b. Remove floor mat and accelerator pedal.
   c. Disconnect and remove parking brake lever.
   d. Remove transmission floor pan (s), place gearshift lever in neutral.
   e. Remove gearshift lever (and control tower on some models).

NOTE: On models with New Process Transmissions remove lever by sliding open side of tool over lever. Engage lugs of tool in open slot of retainer, press down on tool and turn to left to disengage lugs on retainer. Lift lever out of cover, be careful of pivot pin. Cover opening in transmission.

On Spicer models, press down on shift lever cup and drive locking pin out of lever. Lift off cup, spring, cap

Removing gearshift lever
(© Chevrolet Div., G.M. Corp.)

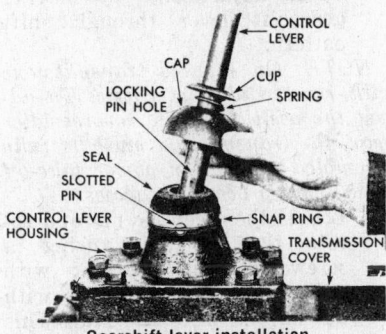

Gearshift lever installation
(© Chevrolet Div., G.M. Corp.)

Reverse idler gear and adjustment
(© Chevrolet Div., G.M. Corp.)

and seal. Remove snap ring from groove on lever housing and tap out slotted pin. Lift lever out of housing and cover opening.

On Clark models remove shift lever housing cover to transmission bolts. Lift lever and control tower from transmission. Cover transmission opening.

On Air Control Shift models, bleed air tanks, remove range shift lines at air valve on transmission. Remove gearshift lever and control tower assembly from transmission. Cover transmission opening and tape or plug air valves.

4. Disconnect and drop propeller shaft at transmission.
5. If unit is equipped with power take-off, disconnect drive shaft and controls.

6. Remove reverse shift control cable and bracket on 4 speed units.
7. Disconnect any clutch control linkage on transmission.
8. Disconnect speedometer cable at transmission.
9. Remove engine ground strap, back up lamp switch and "T.S.C." switch (if used).
10. Place transmission jack into position and adjust to carry weight of transmission. Use locking chain to secure transmission to jack. *NOTE: On vehicles which have rear engine mountings attached to the clutch housing (except "apron" type flywheel housing models), position a jack under flywheel housing and adjust to carry the weight of the engine. Remove rear engine mounts.*
11. Remove bolts attaching transmission to rear crossmember support brackets (if used).
12. Remove clutch housing-to-flywheel housing bolts. (except on "apron" type flywheel housing models) *NOTE: On models with "apron" type flywheel housing, remove flywheel housing under pan (also access*

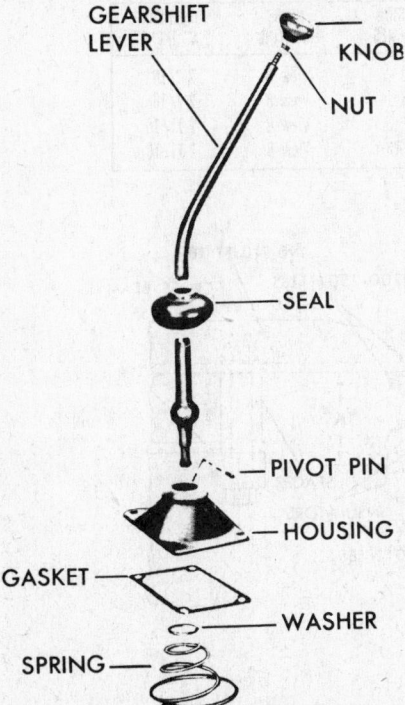

Gearshift lever and housing
(© Chevrolet Div., G.M. Corp.)

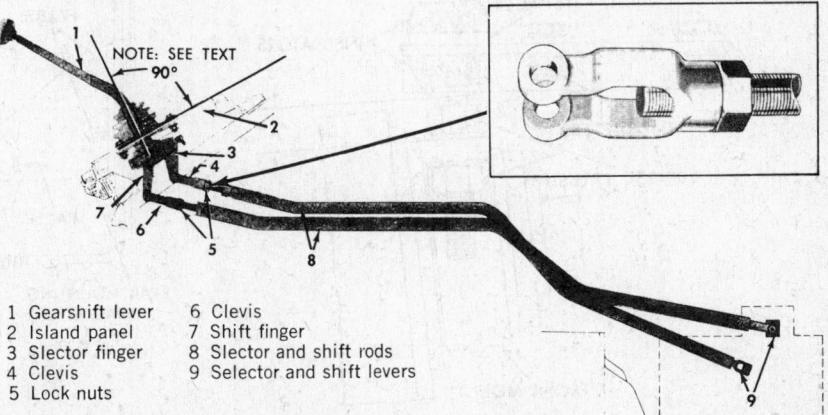

1 Gearshift lever
2 Island panel
3 Slector finger
4 Clevis
5 Lock nuts
6 Clevis
7 Shift finger
8 Slector and shift rods
9 Selector and shift levers

Tilt cab transmission control linkage (© Chevrolet Div., G.M. Corp.)

panel on Spicer) and transmission to flywheel housing bolts. The use of guidepins in two top holes of "apron" type flywheel housing or in two top side holes on "S.A.E." type will maintain alignment during both removal and installation of transmission.

13. Visually inspect to determine if other equipment or lines must be removed.
14. Move transmission straight back, using guide pins to keep transmission main drive gear in alignment with clutch disc, until free from engine. Be sure to support clutch release bearing during removal of transmission. Lower transmission and move from under the vehicle.

Installation

1. Clean transmission mating surfaces and apply a light film of grease to main drive gear bearing retainer and clutch pivot bearing.
2. Place transmission on jack and move into position.
3. On "apron" type, place clutch release bearing and support assembly inside flywheel housing. Be sure clutch fork engages bearing. On "S.A.E. 2" type, make sure clutch release bearing is in position.
4. Using guide pins to align transmission main drive gear with clutch disc, move transmission forward rotating main drive gear so gear can enter clutch disc splines without forcing.

5. Reverse removal procedures to install.
NOTE: On transmissions having remote controls make the following control island shift mechanism adjustments if necessary.
1. Place transmission selector and shift levers in neutral.
2. Adjust selector and shift rods to provide 90° angle at the lower end of the gearshift lever to the control island panel. Adjustment is made by rotating the adjustable clevis at either the control island or transmission end of the selector and shift rods. Tighten lock nuts.
3. Check adjustments by moving gearshift lever through shift pattern.
NOTE: On 4 speed transmissions with reverse idler eccentric. To adjust the position of the reverse idler gear, the transmission must be fully assembled except for power take-off cover. Then proceed as follows:
1. Loosen the eccentric nut and rotate the eccentric, using a screwdriver in the slot with end of electric, until slot with dot on end is to the rear. This places the reverse idler in its extreme rear position and will provide for maximum engagement when the transmission is shifted into reverse.
2. Shift transmission into second. Check for interference between reverse idler and first and reverse gear. If interference exists, rotate eccentric in a counter-clockwise direction to obtain approximately 1/32" clearance

this clearance can be checked through the power take-off opening.
3. Shift transmission into reverse and check clearance between reverse idler gear and transmission case. If necessary, rotate the eccentric an additional amount, in counter-clockwise direction, to obtain running clearance at this point.
4. Tighten eccentric nut and lock.
5. Install power take-off cover and new gasket.

Auxiliary Transmissions

The spicer auxiliary transmission is supported at the front by a support bracket attached to the frame side rails and at the rear by a support beam attached to frame brackets. The gears are shifted by a lever in the cab, which is interconnected to the auxiliary transmission with control rods. The hand brake and speedometer drive gear are located at the rear of the transmission.

Removal

1. Drain the lubricant.
2. Disconnect and support the propeller shafts from the input and output ends of the transmission.
3. Disconnect the shift control rods from the front of the transmission.
4. Disconnect the speedometer cable from the adapter at the rear of the transmission.
5. Disconnect the parking brake linkage if applicable.
6. Remove all connections to the

TRUCK MODELS	ENGINE	MAIN TRANSMISSION	AUXILIARY TRANSMISSION	FRONT MOUNTING	DIMENSION "A" INCHES	REAR MOUNTING	DIMENSION "A" INCHES
ME65	366	NP542	SP6041	View A	1-1/8	View B	2-7/16
ME65	366	CL285V	SP6041	View A	1-1/8	View B	2-7/16
ME65	427	CL325V	SP7041	View B	2-11/16	View B	2-11/16
ME65	427	SP5652B	SP7041	View B	2-11/16	View B	2-11/16

Alignment data chart (© Chevrolet Div., G.M. Corp.)

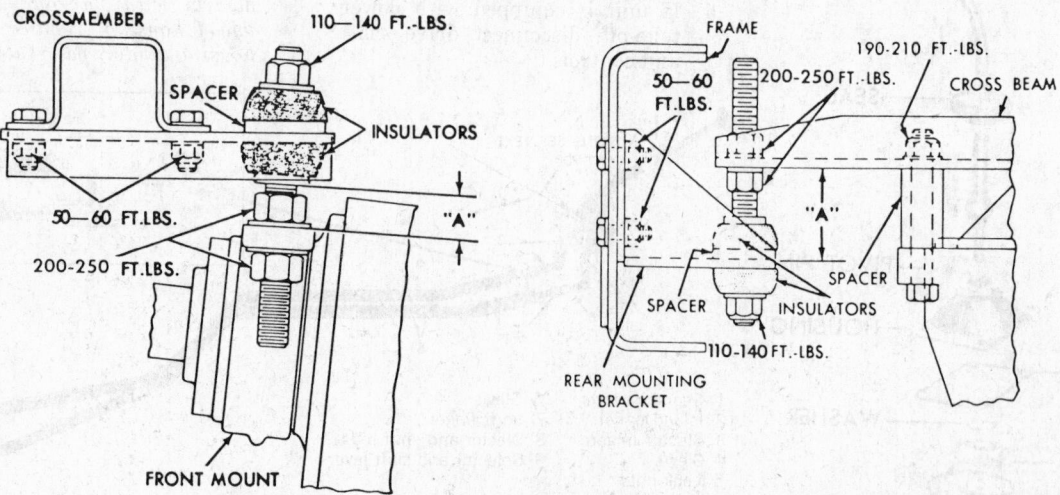

VIEW A NOTE: See Chart for Dimension "A" **VIEW B**

Auxiliary transmission mounting (© Chevrolet Div., G.M. Corp.)

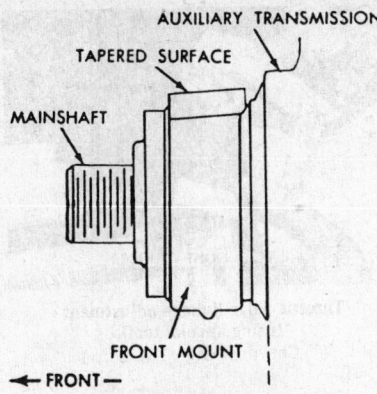

Front mount installed
(© Chevrolet Div., G.M. Corp.)

auxiliary transmission power take-off.

7. Place a suitable dolly or jack under the transmission and adjust its position so it can safely carry the weight.
8. Disconnect the front and rear mountings and lower the transmission away from the chassis.

Installation

1. Make sure the tapered surface of the front mount face the front of the vehicle as shown in the illustration.
2. Move the transmission into position under the vehicle and adjust the front and rear height. (See alignment Data Chart)
3. Torque the attaching parts to the proper specifications (See
4. Reconnect the propeller shafts to the input and output ends of the transmission.
NOTE: It is important that all angles of the driveline be checked with a bevel protractor. Also the auxiliary transmission must be the same as the engine and main transmission. Adjustments may be made by raising or lowering the front or rear of the auxiliary transmission or by adding plates, washers spacers etc. (See Drive shaft alignment in U-joints-Drive Line Section.
5. Connect the power take-off if applicable.
6. Connect the parking brake linkage.
7. Connect the speedometer cable to the adapter at the rear of the transmission.
8. Connect the shift control rods and adjust if necessary.
9. Refill the transmission with the recommended lubricant.

Linkage Adjustment

1. Disconnect the control rods from the shift control tower under the cab.
2. Place the auxiliary transmission gearshift lever and shift rods in the "Neutral" position.
3. Adjust the length of each control rod by rotating its adjust-

able clevis to provide a free clevis pin fit.
4. Reconnect the control rods to the control tower and shift the transmission through its entire shift pattern.
5. Replace any worn or damaged cotter pins, tighten the locknuts firmly and lubricate the control linkage.

Automatic Transmission

See specific chapter in General Repair Section for overhaul procedures for each make.

Draining and Refilling Automatic Transmission

Powerglide and Turbo Hydra-Matic

1. Raise the vehicle and support safely.
2. Place a fluid receptacle under the transmission pan. Remove the pan attaching bolts from the front and side of the pan.
3. Loosen the rear pan attaching bolts approximately four turns and pry the pan loose to allow the fluid to drain.
4. Remove the remaining pan screws and remove the pan and gasket. Discard the gasket.
5. Remove the strainer to valve body screws and remove the strainer (filter) and gasket and discard.
NOTE: On the 400/475 transmissions, remove the filter retaining bolt, filter, and intake pipe O ring. When installing, replace the filter and the intake pipe O ring. Tighten the retaining bolt to 10 ft. lbs.
6. Install the new strainer and gasket and install the strainer to valve body screws and tighten.
7. Install a new gasket on the oil pan and install the oil pan. Tighten the pan bolts to 12 ft. lbs. torque. Connect and tighten the filler tube.
8. Lower the vehicle and install 2.5 quarts of transmission fluid into the transmission and start the engine.
9. Move the selector lever through the detents for each range. Add fluid to bring the level to ¼ inch below the ADD mark on the dipstick.

Powerglide Transmission

Shift Linkage Adjustment

1. Place the gearshift lever in

Drive (D) position, as determined by detent.
2. Loosen adjustment swivel at mast jacket lever and rotate transmission lever so that it contacts the Drive stop in the steering column.
3. Tighten the swivel and recheck the adjustment.

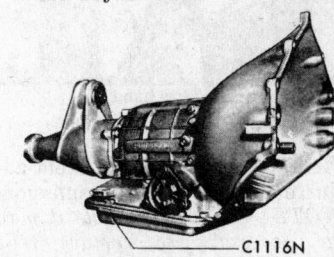

Power Glide and Turbo Hydramatic 350 identification number located on right rear vertical surface of oil pan
(© Chevrolet Div., G.M. Corp.)

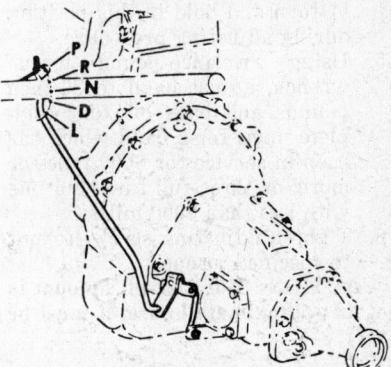

Power glide shift linkage adjustment
(© Chevrolet Div., G.M. Corp.)

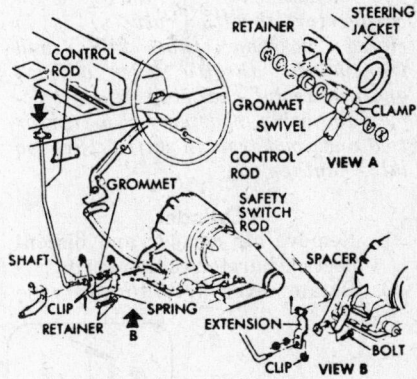

Powerglide shift linkage
(© Chevrolet Div., G.M. Corp.)

4. Readjust indicator pointer, if necessary, to agree with transmission detent positions.
5. Readjust neutral safety switch, if necessary.

Low Band Adjustment

Adjustments should be made at periodical intervals or sooner if necessary, determined by operating performances.
1. Raise vehicle and support on stands.
2. Place gearshift lever in Neutral.

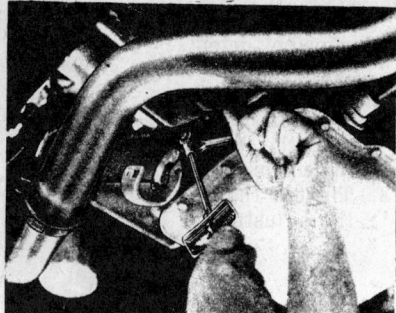

Adjusting low band
(© Chevrolet Div., G.M. Corp.)

SIX CYLINDER

Throttle valve linkage adjustment
(using special tool)
(© Chevrolet Div., G.M. Corp.)

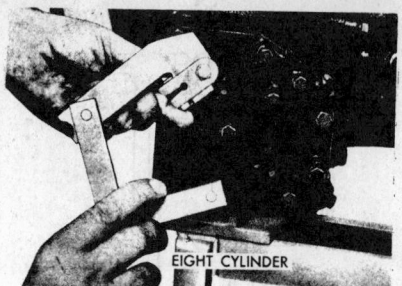

EIGHT CYLINDER

Throttle valve linkage adjustment
(using special tool)
(© Chevrolet Div., G.M. Corp.)

3. Remove protective cap from adjustment screw on transmission. *NOTE: On some models it may be necessary to remove rear mounting bolts from crossmember and move transmission slightly to passenger side for additional clearance.*

4. Loosen adjusting screw locknut ¼ turn and hold in this position during adjusting procedure.

5. Using an inch-pound torque wrench, adjust band to 70 inch pounds and back off four complete turns for a band which has been in service for 6000 miles or more, or three full turns for one with less than 6000 miles.

6. Tighten adjusting screw locknut to specified torque.

CAUTION: The back-off amount is not an approximate figure, it must be exact.

Throttle Linkage Adjustment

NOTE: Powerglide throttle linkage adjustments are made when carburetor throttle valve(s) is in closed position (hot idle) and transmission throttle lever is back against its internal stop. If shift occurs too early, shorten rod a turn or two and road test. If shift occurs too late, lengthen rod.

6 Cylinder

1. Remove air cleaner and disconnect carburetor return spring.

2. Rotate lever (A) to the wide

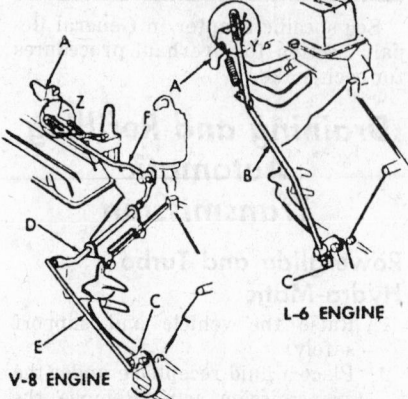

V-8 ENGINE

L-6 ENGINE

Powerglide throttle valve linkage adjustment
(© Chevrolet Div., G.M. Corp.)

open throttle position and T.V. lever through detent. (Depress accelerator fully).

3. Hold slot on rod (B) against pin on lever (A) as shown in circle Y—adjust swivel on rod so that it freely enters hole on lever (C).

4. Hold rod (B) perpendicular to pin on lever (A) and then tighten swivel nut. Rod must be 1/64″-1/16″ off lever stop with transmission lever against internal stop. Connect carburetor return spring.

5. Check for throttle linkage freedom and proper operation.

V8

1. Remove air cleaner.

2. Disconnect accelerator linkage at carburetor.

3. Disconnect accelerator return and throttle valve rod return springs.

4. With right hand, pull T.V. upper rod (F) forward until transmission lever (C) is through detent. With left hand, rotate carburetor lever (A) to wide open throttle position. Carburetor must reach wide open throttle position at the same time pin or ball stud on lever (A) contacts end of slot in upper T.V. rod (F)—see circle Z.

5. Adjust swivel on upper end of T.V. rod (F) to obtain setting described in step 4 (approximately 1/32″).

6. Connect and adjust accelerator linkage. Connect carburetor return springs.

7. Check for throttle linkage freedom and proper operation.

Transmission Removal

1. Raise vehicle on hoist or support on stands.

2. Drain transmission oil if equipped with drain plug (or remove oil pan to drain, then replace pan using several bolts). Oil can be drained after transmission removal, if desired.

3. On models with long oil fill pipe, remove dipstick and pipe.

4. Disconnect oil cooler lines (if

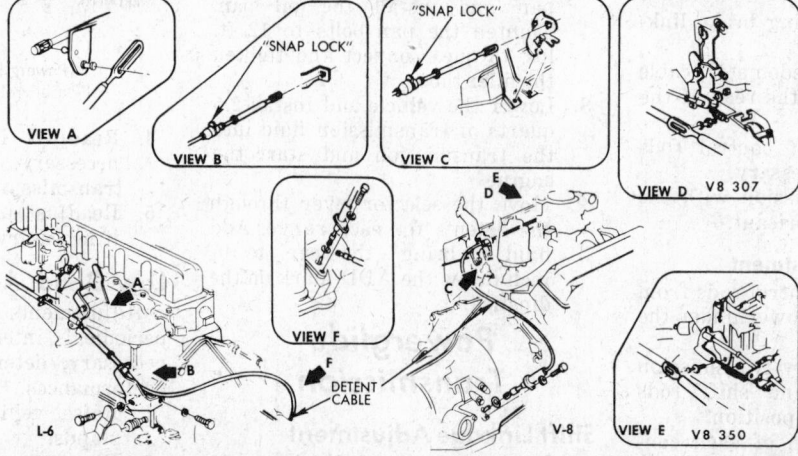

VIEW A

"SNAP LOCK"

VIEW B

"SNAP LOCK"

VIEW C

VIEW D   V8 307

L-6

VIEW F

DETENT CABLE

V-8

VIEW E   V8 350

Throttle valve linkage adjustment—6 cyl.—V8 engines (© Chevrolet Div., G.M. Corp.)

equipped), vacuum modulator hose and speedometer cable at transmission. Support lines out of work area.

5. Disconnect ground cable, T.S.C. switch, neutral switch and back-up lamp switch (if used).

6. Remove shift and throttle valve controls from transmission.

7. Disconnect propeller shaft from rear of transmission.

8. Position transmission jack and secure transmission to jack with safety chain.

9. Raise jack slightly to relieve engine rear mount on transmission extension and remove transmission support crossmember (Vans—remove thru-bolt). *NOTE: Use care to remove any shims (if used) at crossmember. It is vital that exactly the same number be reinstalled.*

10. Remove flywheel cover or converter under pan, scribe flywheel-converter marks for reassembly, remove flywheel to converter attaching bolts.

11. Support engine at oil pan rail with a jack to support engine weight when transmission is removed.

12. Vans—Remove linkage cross shaft frame bolts and accelerator linkage at bellcrank before lowering engine. Lower engine to allow rear mount housing to clear support bracket.

13. Lower rear of transmission slightly so that the upper transmission to engine bolts can be reached using a universal socket and a long extension. Remove upper bolts. **CAUTION: On V8 engines do not lower too far because of distributor—firewall interference.**

14. Remove remaining transmission mounting bolts.

15. Remove transmission by moving it slightly to the rear and downward, remove from under vehicle. *NOTE: Watch converter when moving transmission back. If it does not move with transmission pry it free of flywheel before proceeding.* **CAUTION: Keep front of transmission up to prevent the converter from falling out. Install a converter hold tool or improvise.** For Overhaul Procedures—See General Repair Section

Installation

1. Reverse removal procedures. *NOTE: If reusing converter align scribed marks, if not through flywheel cover opening align as closely as possible the white stripe painted on engine side of flywheel outer rim (heavy side) with the blue stripe painted on end of converter*

(light side) to maintain balance.

2. Refill transmission and check for proper operation and leaks.

Turbo-Hydra-Matic 250 Automatic Transmission

Removal and Installation

1. Disconnect the negative battery cable and raise the vehicle and support safely.

2. Disconnect the detent downshift cable at the carburetor and release the parking brake.

3. Remove the drive shaft, speedometer cable, modulator vacuum line and the cooler lines at the transmission.

4. Disconnect the shift linkage at the transmission and support the transmission with a jack.

5. Disconnect the rear mount from the crossmember and remove the crossmember from the frame.

6. Remove the converter housing under pan.

7. Remove the converter to flywheel bolts.

8. Lower the engine-transmission assembly and remove the transmission to engine bolts.

9. Remove the filler tube from the transmission.

10. Support the rear of the engine and slide the transmission rearward, down and out of the vehicle. *NOTE: Loss of the converter can occur if the transmission front is lower than the rear.*

11. The installation is in the reverse of the removal procedure.

12. Fill the transmission with 2.5 quarts of transmission fluid, start the engine and move the selector lever through all gear positions, recheck the level, fill to ¼ inch below the ADD mark on the dipstick.

Shift Linkage Adjustment

1. Position the transmission selector lever in the D (Drive) position.

2. Remove the control rod from the transmission selector lever, and place the column selector lever in the D (Drive) position and assure that the trunnion on the control rod has a free fit in the transmission lever.

3. Check to assure that the column selector lever must be lifted to place the transmission in the Reverse and Low ranges. *NOTE: Do not use the indicator pointer to position the shift selector lever.*

4. Install the control rod retainer.

5. Adjust the indicator needle by loosening and repositioning the cable end mounted to the steering housing shift tube.

6. Adjust the neutral start switch to allow the engine to start in Park and Neutral only, by loosening the attaching screws and moving the switch either right or left. Tighten the attaching screws.

Intermediate Band Adjustment

1. Place the selector lever in Neutral and raise the vehicle and support safely.

2. Loosen the adjusting screw lock nut, tighten the adjusting screw to 30 inch pounds torque, and back off exactly three turns.

3. Holding the adjusting screw in position, tighten the lock nut.

Downshift Cable Detent Adjustment

With the snap lock disengaged, position the carburetor to the wide open throttle position and push the snap lock downward until the top is flush with the cable, thereby adjusting the cable length.

Turbo-Hydra-Matic 350, 400, 475

Shift Linkage Adjustment

Cabs, Suburbans, 4-wheel drive 1974-78 Forward Control exc. Vans

1. Place gearshift lever in Drive (D), as determined by transmission detent. Obtain Drive position by rotating transmission lever counterclockwise to low detent, then clockwise two detent positions to Drive.

2. Loosen adjustment swivel at mast jacket lever and rotate transmission lever so that it contacts the Drive stop in the steering column.

3. Tighten swivel and recheck adjustment.

4. Readjust indicator pointer, if necessary, to agree with transmission detent positions.

5. Readjust Neutral safety switch if necessary.

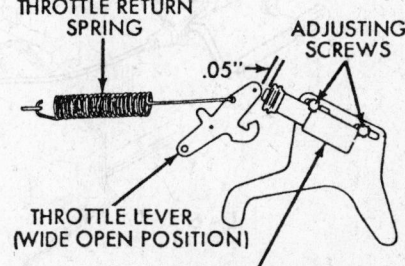

THROTTLE RETURN SPRING

ADJUSTING SCREWS

.05"

THROTTLE LEVER (WIDE OPEN POSITION)

DOWNSHIFT SWITCH (PLUNGER FULLY DEPRESSED) Downshift Linkage—AT-475 Trans. (© Chevrolet Div., G.M. Corp.)

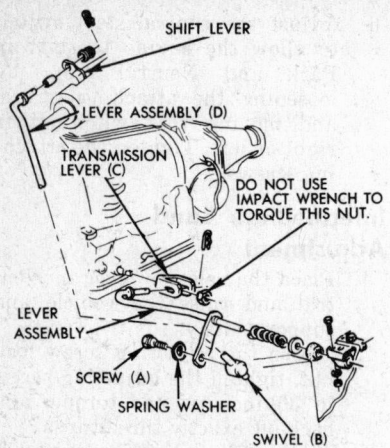

Turbo-Hydramatic control rod linkage—all models except Vans (© G.M.C.)

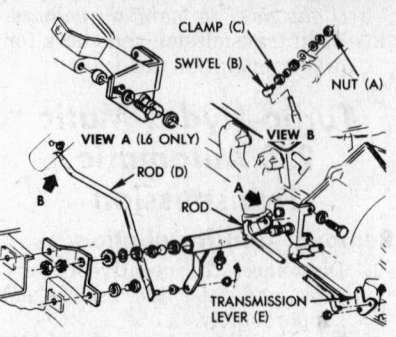

Turbo Hydra Matic Control Rod Linkage—
(© Chevrolet Div., G.M. Corp.)

1971-73—Forward Control
1971-74—Vans

1. Set transmission lever in Drive position. Obtain Drive position by rotating transmission lever counter-clockwise to Low detent, then clockwise two detent positions to Drive.
2. Attach control rod to lever and shaft assembly with retainers.
3. Assemble ring, washers, grommet, swivel, retainer and nut loosely on shaft.
4. Insert control rod in swivel and retainer, attach opposite end to tube and lever assembly.
5. Set tube lever assembly in Drive position and tighten nut.

NOTE: When above procedure is done, the following conditions must be met by manual operation of the gearshift lever. From Reverse to Drive position travel, the transmission detent feel must be noted and related to indicated position on dial. When in Drive and Reverse position, pull lever toward steering wheel and

then release. It must drop back into position with no restriction.

1975-78 Vans

1. The shift tube and lever assembly must be free in the mast jacket.
2. Set transmission lever (C) in "neutral" position by one of the following optional methods.

NOTE: Obtain "neutral" position by moving transmission lever (C) counter-clockwise to "LI" detent, then clockwise three detent positions to "neutral" or obtain "neutral" position by moving transmission lever (C) clockwise to the "park" detent then counter-clockwise two detents to "neutral".

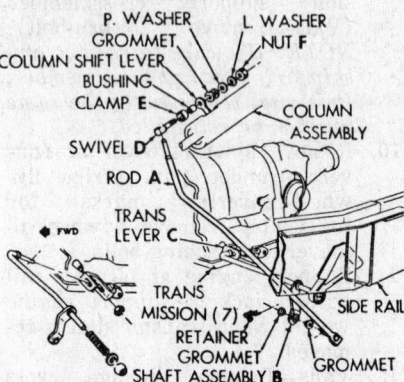

Turbo Hydramatic control rod linkage—1975 and later Vans (© G.M.C.)

3. Set the column shift lever in "neutral" position. This is obtained by rotating shift lever until it locks into mechanical stop in the column assembly.

NOTE: Do not use indicator pointer as a reference to position the shift lever.

4. Attach rod (A) to shaft assembly (B) as shown.
5. Slide swivel (D) and clamp (E) onto rod (A) align the column shift lever and loosely attach as shown.
6. Hold column lever against "neutral" stop "park" position side.
7. Tighten nut (F) to 18 foot pounds.
8. Readjust indicator needle if necessary to agree with the transmission detent positions.
9. Readjust neutral start switch if necessary to provide the correct relationship to the transmission detent positions.

CAUTION: Any inaccuracies in the above adjustments may result in premature failure of the transmission due to operation without controls in full detent position. Such operation results in reduced oil pressure and partial engagement of clutches.

Kickdown or Detent Switch Adjustment

Turbo-Hydramatic 350

1. Disengage the snap lock on the detent cable.

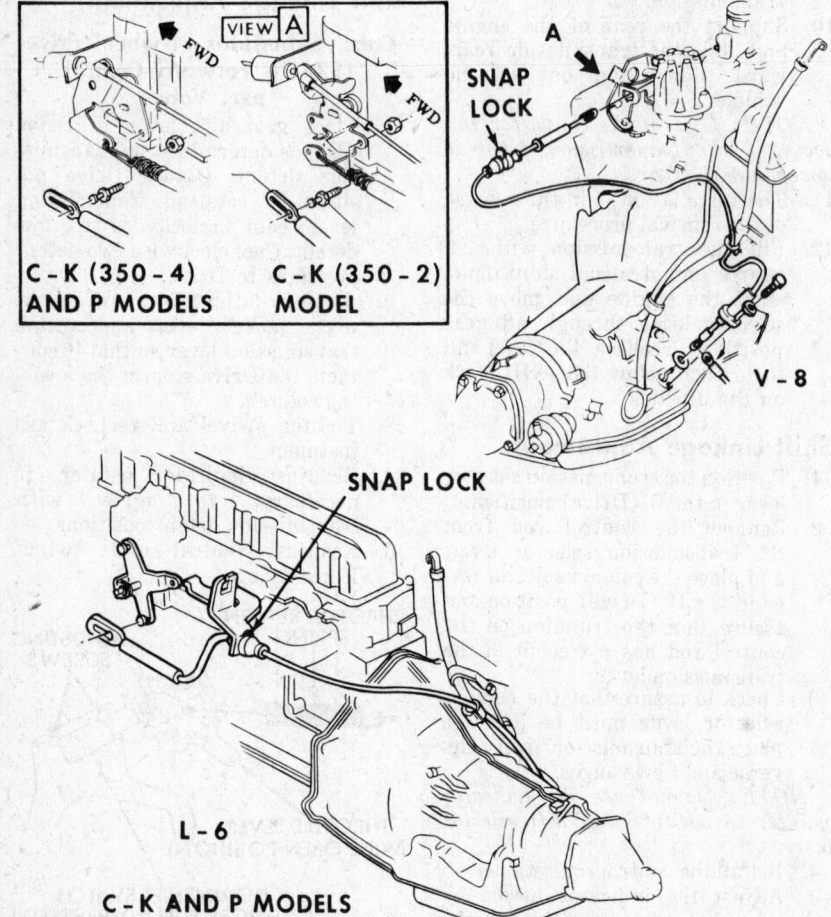

Throttle valve linkage adjustments (© Chevrolet Div., G.M. Corp.)

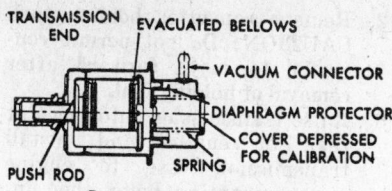

Detent cable adjustment
(© Chevrolet Div., G.M. Corp.)

2. Place the carburetor in the wide open position (W.O.P.).
3. Holding the carburetor in the wide open position, push the snap lock on the detent cable downward until the top is flush with the cable.

Turbo-Hydramatic 400/475

A detent solenoid, activated by an electrical switch on carburetor, controls downshifts.

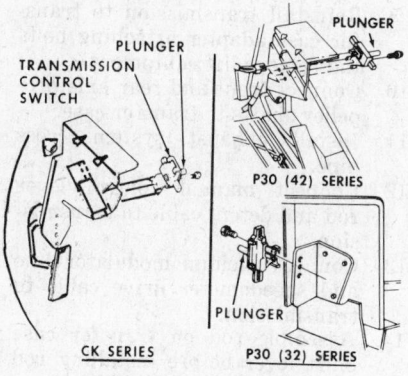

Detent switch adjustment
(© Chevrolet Div., G.M. Corp.)

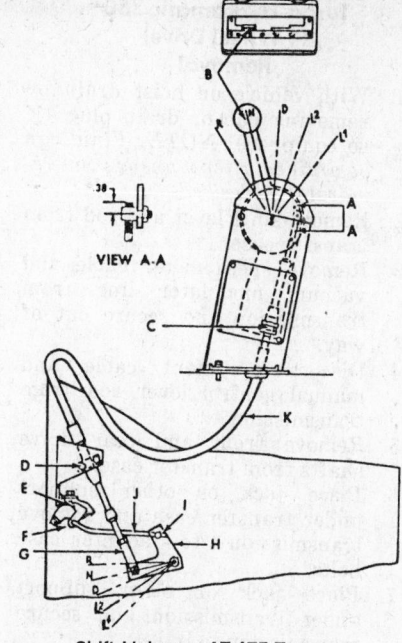

VIEW A-A

Shift Linkage—AT-475 Trans.
(© Chevrolet Div., G.M. Corp.)

Neutral Safety Switch Adjustment

Column Mounted

1. Place gearshift lever in Neutral (N).
2. Loosen retainer screws holding switch, install 3/32" drill (or pin) through hole in lower switch arm and bracket. Adjust position of switch until engine turns over (with ignition switch in start).

WITH ACCELERATOR CONTROLS PROPERLY ADJUSTED, ROTATE LEVER TO OBTAIN WIDE OPEN THROTTLE. POSITION SWITCH TO OBTAIN DIM. A SHOWN IN TABLE BETWEEN PLUNGER AND LEVER. TIGHTEN SWITCH ATTACHING SCREWS.

	DIM. A
ALL EXCEPT 350 V8	.05
350 V8	.20

6 CYLINDER ENGINE

307 V-8 ENGINE

350 ENGINE

Detent switch adjustment (© Chevrolet Div., G.M. Corp.)

Transmission Mounted

1. Place gearshift lever in Neutral (N), loosen transmission lever extension bolt.
2. Pin switch lever in Neutral position with 3/32" drill or pin.
3. Install rod into switch lever, adjust swivel on rod to allow free entry of rod into lever.
4. Secure rod with retainer, tighten transmission lever extension bolt.
5. Check adjustment by testing for cranking in both Neutral and Park.

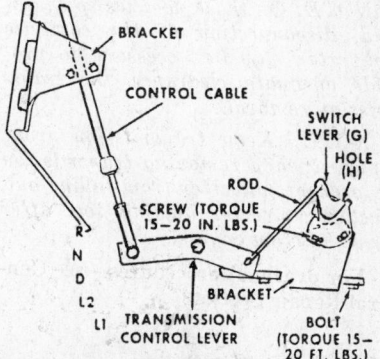

Neutral safety and back-up lamp switch 1975-76–AT-475 Trans. (© Chevrolet Div., G.M. Corp.)

Transmission R & R

Turbo-Hydramatic 350 (Except 4 Wheel Drive)
Removal

1. With vehicle on hoist drain by removing pan or drain plug (if so equipped). *NOTE: Fluid can be drained after transmission is removed if so desired.*
2. Remove the vacuum modulator line and speedometer cable from the transmission and secure out of way.
3. Remove detent cable and manual control lever from the transmission.
4. Disconnect the drive shaft and remove.
5. Place suitable jack or other support under transmission and secure transmission to it.
6. At transmission extension disconnect rear engine mount and then remove support crossmember.
7. Remove converter underpan. Place marks on the flywheel and converter to insure proper installation and then remove the flywheel-to-converter bolts.
8. Support engine at oil pan rail with jack capable of supporting the weight of the engine when transmission is removed.
9. Lower rear of the transmission slightly so that upper, housing-to-engine, transmission bolts can be reached with a long extension and universal socket. Remove upper bolts. *NOTE: Have an as-*

sistant watch upper engine parts to make sure everything clears when rear of transmission is being lowered.

10. Remove remaining transmission-to-housing bolts.
11. Remove transmission by moving it slightly to the rear and downward. Remove from under vehicle.

CAUTION: Watch converter when removing transmission to be sure that it moves with transmission. If it does not move pry it free from flywheel before proceeding any further.

NOTE: On those vehicles so equipped, disconnection of the catalytic converter may be necessary to provide adequate clearance for transmission removal.

NOTE: Keep transmission front upward when removing transmission to prevent converter from falling out. Install converter holding tool after removal from engine.

For overhaul procedures—see General Repair Section.

Installation

1. Mount transmission on transmission lifting equipment installed on jack or other lifting device.
2. Remove converter holding tool.

CAUTION: Do not permit converter to move forward after removal of holding tool.

3. Raise transmission into place at rear of engine and install transmission case to engine upper mounting bolts, then install remainder of the mounting bolts.
4. Remove support from beneath engine, then raise rear of transmission to final position.
5. If scribed during removal, align scribe marks on flywheel and converter cover. Install converter to flywheel attaching nuts and bolts.
6. Install converter underpan.
7. Reinstall transmission support crossmember to transmission and frame.
8. Remove transmission lift equipment.
9. Connect propeller shaft to transmission.
10. Connect manual control lever rod and detent cable to transmission.
11. Connect vacuum modulator line, and speedometer drive cable to transmission.
12. Lower vehicle.
13. Refill transmission.
14. Check transmission for proper operation and for leakage. Check and, if necessary, adjust linkage.
15. Remove vehicle from hoist.

Turbo-Hydramatic 350 (4 Wheel Drive)
Removal

1. With vehicle on hoist drain by removing pan or drain plug (if so equipped). *NOTE: Fluid can be drained after transmission removal if so desired.*
2. Remove shift lever and rod from transfer case.
3. Remove speedometer cable and vacuum modulator line from transmission and secure out of way.
4. Disconnect detent cable and manual control lever rod from transmission.
5. Remove front and rear drive shafts from transfer case.
6. Place jack or other support under transfer case and remove transmission-to-adapter case bolts.
7. Place jack or other support under transmission and secure transmission to it.
8. Remove transfer case-to-frame bracket bolts and remove transfer case.
9. On V8 engines remove exhaust crossover pipe.
10. Disconnect and remove rear transmission crossmember.
11. Remove converter underpan. Place marks on the flywheel and converter to insure proper installation and then remove the flywheel-to-converter bolts.
12. Support engine at oil pan rail with jack capable of supporting engine weight when transmission is removed.
13. Lower rear of transmission slightly so that upper housing-to-engine transmission bolts can be removed with a long extension and universal socket. Remove upper bolts. *NOTE: Have an assistant watch upper engine parts to make sure everything clears when transmission is lowered.*
14. Remove the remaining transmission-to-housing bolts.
15. Remove the transmission by moving it slightly to the rear and downward. Remove from under vehicle.

CAUTION: Watch converter when removing transmission to make sure converter moves with transmission. If it does not, pry it loose from flywheel before proceeding any further.

NOTE: Keep transmission front upwards when removing from engine to prevent converter holding tool after removal from engine. See General Repair Section for overhaul procedures.

Installation

1. Mount transmission on transmission lifting equipment installed on jack or other lifting device.

2. Remove converter holding tool. **CAUTION: Do not permit converter to move forward after removal of holding tool.**
3. Raise transmission into place at rear of engine and install transmission case to engine upper mounting bolts, then install remainder of the mounting bolts.
4. Remove support from beneath engine, then raise rear of transmission to final position.
5. If scribed during removal, align scribe marks on flywheel and converter cover. Install converter to flywheel attaching bolts.
6. Install flywheel cover.
7. Place transfer case and adapter assembly at rear of transmission on suitable lift equipment and install transfer case to frame bracket attaching bolts.
8. Reinstall transmission to transfer case adapter attaching bolts and remove lift equipment.
10. Connect front and rear axle propeller shafts to transfer case.
11. Install exhaust system cross pipe.
12. Connect manual control lever rod and detent cable to transmission.
13. Connect vacuum modulator line and speedometer drive cable to transmission.
14. Assemble rod on transfer case shift lever before installing rod to transfer case shift linkage.
15. Lower vehicle.
16. Refill transmission.
17. Check transmission for proper operation and for leakage. Check, and if necessary, adjust linkage.
18. Remove from hoist.

Turbo-Hydramatic 400/475 Removal and Installation

Before raising the truck, disconnect the battery and release the parking brake.

1. Raise truck on hoist.
2. Remove propeller shaft.
3. Disconnect speedometer cable, electrical lead to case connector, vacuum line at modulator, and oil cooler pipes.
4. Disconnect shift control linkage.
5. Support transmission with transmission jack.
6. Disconnect rear mount from frame crossmember.
7. Remove two bolts at each end of frame crossmember and remove crossmember.
8. Remove converter under pan.
9. Remove converter to flywheel bolts.
10. Loosen exhaust pipe to manifold bolts approximately 1/4 inch, and lower transmission until jack is barely supporting it.

11. Remove transmission to engine mounting bolts and remove oil filler tube at transmission.

12. Raise transmission to its normal position, support engine with jack and slide transmission rearward from engine and lower it away from vehicle.

CAUTION: Use converter holding tool when lowering transmission or keep rear of transmission lower than front so as not to lose converter.

The installation of the transmission is the reverse of the removal with the following additional steps.

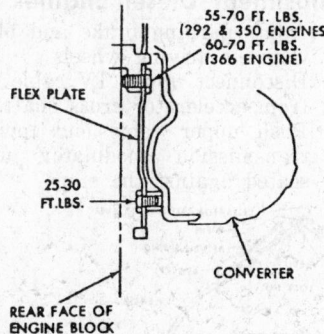

Flex plate installed (inline and V-8 engines)—AT-475 (© Chevrolet Div., G.M. Corp.)

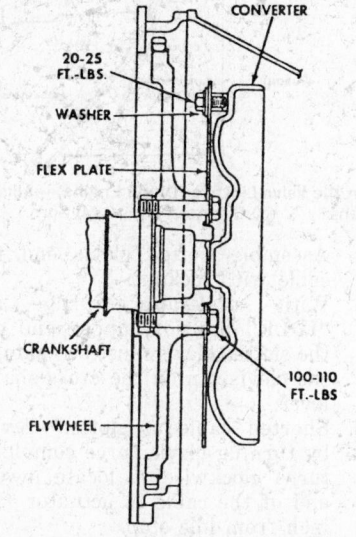

Flex plate installed (V-6 engine)—AT-475 (© Chevrolet Div., G.M. Corp.)

Before installing the flex plate to converter bolts, make certain that the weld nuts on the converter are flush with the flex plate and the converter rotates freely by hand in this position. Then, hand start all three bolts and tighten finger tight before torquing to specification. This will insure proper converter alignment.

NOTE: After installation of transmission check linkage for proper adjustment and check for leaks.

Allison Transmissions
Fluid and Filter Changes

AT and MT Models

1. Have the transmission at normal operating temperature (160 to 220 degrees F), and the transmission in neutral.

2. Remove the fill tube from the pan or the drain plug from the right side of the transmission pan. Allow the fluid to drain into a large container.
NOTE: Do not allow the fluid to spill and splash. Burns can result.

3. Remove the pan bolts, loosen and remove the pan and gasket. Discard the gasket.

4. Remove the one screw that retains the filter, remove the filter and discard.
NOTE: Later models will have a suction tube that separates from the filter. Retain the tube for use with the new filter, the tubes will have a sealring. Replace with a new sealring upon installation.

5. To clean or replace the governor feed screen, the valve body must be removed on the AT Models, and the screen taken from the governor feed bore. Refer to the valve body removal and installation procedures outlined in the General Repair Section. The MT Models have discontinued the use of the primary governor screen, and if one is found in the governor feed tube area of the valve body, discard it.

6. The MT Models have the main governor screen located in the rear cover. Early models will have a screen to be cleaned and reinstalled while the later models will have a replaceable cartridge type filter.
NOTE: The screen or filter is inserted into the rear cover open end first.

7. Retain the screen or filter with the plug or cap.

8. Install the new filter, suction tube and seal ring in the sump area and secure with the retaining bolt. Torque to 10 to 15 ft. lbs. for the MT models and 10 to 13 ft. lbs. for the AT models.

9. Install the pan with a new gasket. Torque the pan bolts to 10 to 15 ft. lbs. for the MT models and 10 to 13 ft. lbs. for the AT models.

10. Install the filler tube or drain plug into the transmission or pan.

11. Install 10 quarts of transmission fluid into the AT models, and 15 quarts into the MT models. Start the engine, check for leaks, move the selector through the gear positions, and recheck the fluid level of the transmission and refill to the full mark on the dipstick.

Shift Linkage Adjustment

AT540, MT640, MT650 Transmissions

Gas and Diesel Engines

1. Disconnect clevis from transmission shift lever by removing cotter pin and clevis pin.

2. Disconnect control cable from anchor point at bracket on transmission and remove cable retainer clips from underside of cab.

3. Remove four cross recess screws retaining range selector cover to tower or bracket, depending on

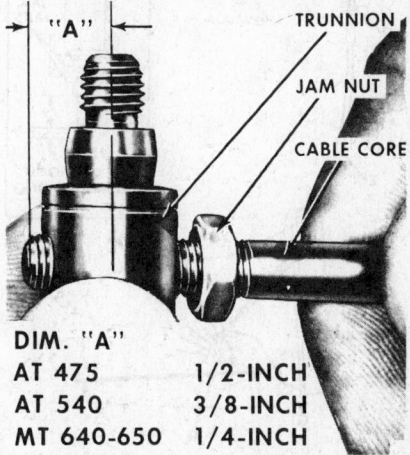

DIM. "A"	
AT 475	1/2-INCH
AT 540	3/8-INCH
MT 640-650	1/4-INCH

Trunion adjustment—Allison Trans. (© Chevrolet Div., G.M. Corp.)

Engine rear mounting—AT-475 (© Chevrolet Div., G.M. Corp.)

truck model. Lift range selector assembly out to inspect cable attachment at range selector lever and hanger.

4. Check cable at trunnion for dimension "A" as shown. This dimension is necessary to allow for proper cable length at clevis. If adjustment is necessary, remove cable from range selector lever and hanger assembly as follows:

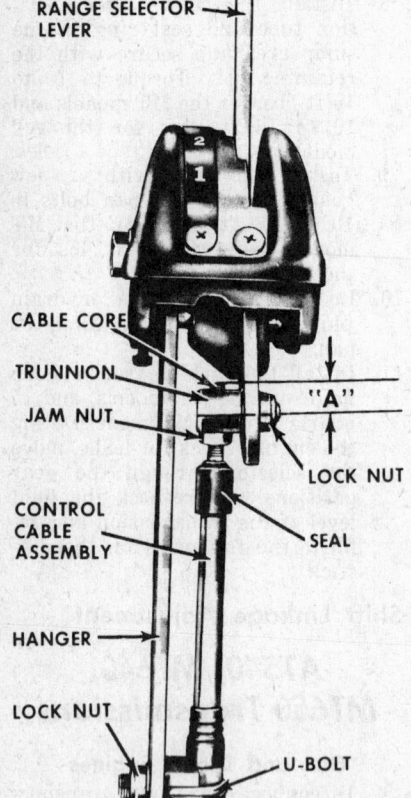

RANGE SELECTOR LEVER

CABLE CORE

TRUNNION

JAM NUT

"A"

LOCK NUT

CONTROL CABLE ASSEMBLY

SEAL

HANGER

LOCK NUT

U-BOLT

Trunion and Cable Core Assembly—Allison Trans. (© Chevrolet Div., G.M. Corp.)

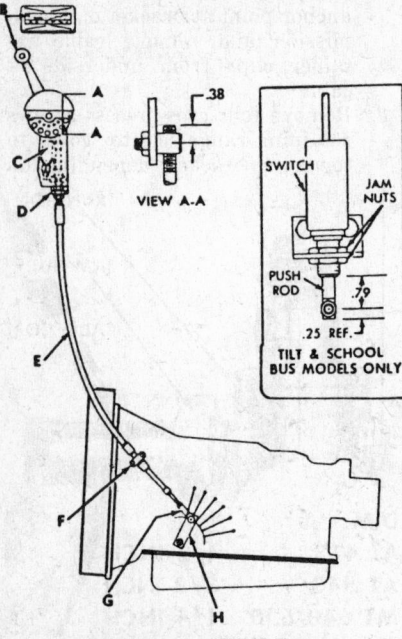

.38

VIEW A-A

SWITCH

JAM NUTS

PUSH ROD

.79

.25 REF.

TILT & SCHOOL BUS MODELS ONLY

Shift Linkage—AT-540 Trans. (© Chevrolet Div., G.M. Corp.)

a. Remove ½-inch locknut from trunnion.
b. Disconnect cable from anchor point on hanger by removing two locknuts and U-bolt.

NOTE: Locknuts can be removed and reinstalled up to six times before their replacement becomes necessary.

c. Mark location of trunnion. Remove trunnion with attached cable core from lever.

CAUTION: The trunnion has been placed in the proper hole at the factory. Any change in the location could cause vehicle operation to be dangerous.

5. With control cable removed, loosen jam nut at trunnion.

NOTE: Do not use pliers on cable core when loosening jam nut or when adjusting trunnion. Cable core finish may be damaged resulting in cable core seal damage.

6. Turn trunnion clockwise, or counterclockwise to attain dimension "A". Tighten jam nut against trunnion.
7. Install trunnion in its original hole and tighten locknut securely.
8. Anchor cable to hanger with U-bolt, washers and locknuts. Tighten locknuts 3-5 foot-pounds torque.
9. Install shift control cover in tower or bracket depending on truck model. Install four cross recess screws. Tighten screws 3-5 foot-pounds torque. Install cable clips to underside of cab.
10. Anchor cable to bracket at transmission. Tighten two cross recess screws securely.

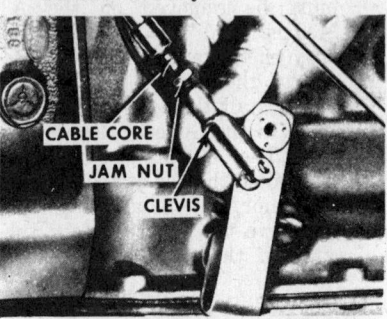

CABLE CORE

JAM NUT

CLEVIS

Clevis adjustment—Allison Trans. (© Chevrolet Div., G.M. Corp.)

11. Locate transmission shift lever in "R" (Reverse) position.

NOTE: "R" position if full counterclockwise (except on AT 475 "full clockwise") movement of lever.

12. Locate range selector lever against stop in "R" (Reverse) position.
13. Loosen jam nut at clevis. Turn clevis clockwise or counterclockwise on threaded cable core until holes in clevis align with hole in shift lever. Install clevis pin, note that pin should enter freely, if it does not, adjust clevis slightly by ½ turn in each direc-

tion until pin does enter freely.
14. Turn clevis one full turn clockwise to allow for cable backlash.
15. Connect clevis to shift lever with clevis pin and a new cotter pin, but do not spread cotter pin at this time.
16. Move the range selector lever through all drive ranges. The transmission detents should fully engage just before the range selector lever hits the stops incorporated in the shift control cover.

Throttle Valve Linkage Adjustment Diesel Engines

1. Apply parking brake and block vehicle's driving wheels.
2. Disconnect upper TV cable end from accelerator cross shaft.
3. Push upper cable end toward transmission modulator until seated against the stop.

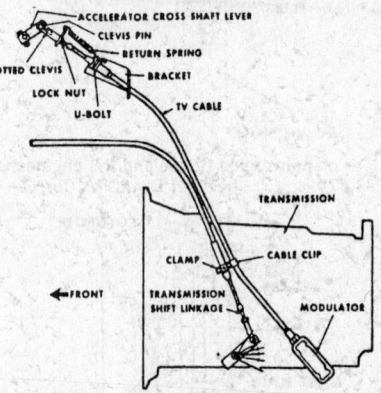

ACCELERATOR CROSS SHAFT LEVER

CLEVIS PIN

RETURN SPRING

SLOTTED CLEVIS

BRACKET

LOCK NUT

TV CABLE

U-BOLT

TRANSMISSION

CLAMP

CABLE CLIP

FRONT

TRANSMISSION SHIFT LINKAGE

MODULATOR

Throttle Valve Linkage (Diesel Engine)—Allison Trans. (© Chevrolet Div., G.M. Corp.)

4. Assemble slotted clevis end to cable with locknut.
5. With accelerator linkage in "IDLE" position, upper end of the slotted clevis must be against the clevis pin on the cross shaft lever.
6. Shorten cable at slotted clevis by turning clevis three complete turns clockwise to locate lower end of the cable in actuator ⅛-inch from idle stop.
7. Tighten locknut securely.
8. Assemble slotted clevis to cross shaft lever with clevis pin and a new cotter pin.
9. Attach TV return spring to bracket.

Neutral Safety and Backup Lamp Switch Adjustment

Allison—(AT540) Transmission Tilt & Schoolbus Models

NOTE: "Shift Linkage Adjustment" should be performed as described previously, prior to adjustment of the neutral safety and backup lamp switch.

1. Block driving wheels, apply parking brake, and perform the following to prevent the vehicle from accidentally starting.

NOTE: Pull the secondary wire out of center socket in the distributor cap and ground wire to prevent possible damage to coil.

2. Move selector lever (B) to "N" (Neutral) position. Then, loosen jam nuts and adjust length of push rod to dimension shown.
3. With switch push rod properly adjusted, tighten jam nuts securely.
4. Check each range position of shift linkage to make sure the starter does not operate with the selector lever in any position other than "N." Have assistant check for proper operation of back-up lights with selector lever in "R." If necessary readjust switch.
5. Reconnect secondary wire to distributor cap.

NOTE: The neutral safety and back-up lamp switches are non-adjustable on the Allison-MT-640, MT-650 trans. and all diesel engines with automatic transmissions.

Transmission R & R

AT540 Removal

NOTE: It may be necessary to remove the air tanks, fuel tanks, special equipment, etc., on some vehicles to provide clearance before the transmission is removed.

1. Block vehicle so that it cannot move. Disconnect ground strap from battery negative (—) post. Remove the spark plugs so the engine can be turned over manually.
2. Remove the level gauge (dipstick). Drain transmission by disconnecting filler tube at right side of transmission pan. Remove bracket holding filler tube to transmission and remove filler tube from vehicle. Replace dipstick in tube and cover the pan opening to prevent entry of foreign material.
3. Disconnect cooler lines from fittings on right side of transmission case. Plug line ends and case openings with lint-free material.
4. Disconnect the range selector cable from shift lever at left-side of transmission.

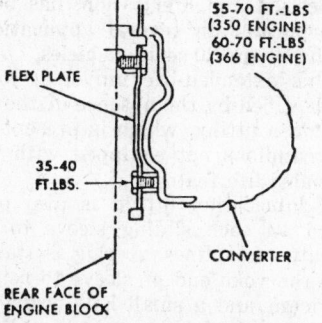

55-70 FT.-LBS. (350 ENGINE)
60-70 FT.-LBS. (366 ENGINE)
FLEX PLATE
35-40 FT.-LBS.
CONVERTER
REAR FACE OF ENGINE BLOCK

Flex plate installation (V-8 engine)—AT-540
(© Chevrolet Div., G.M. Corp.)

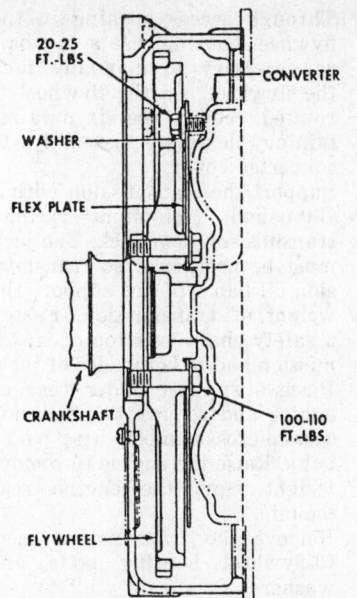

20-25 FT.-LBS.
CONVERTER
WASHER
FLEX PLATE
CRANKSHAFT
100-110 FT.-LBS.
FLYWHEEL

Flex plate installation (V-6 engine)—AT-540
(© Chevrolet Div., G.M. Corp.)

5. Disconnect vacuum modulator line from modulator. Also, on conventional cab models, disconnect wiring from neutral safety and back-up lamp switches.
6. Disconnect the speedometer shaft fitting from adapter at rear of transmission.
7. Disconnect the propeller shaft from transmission.
8. Disconnect the mechanical parking brake linkage at the right side of the transmission (if used).
9. Through the opening in the flywheel housing, use a pry-bar, as necessary to manually turn the flywheel. As the flywheel is rotated, remove the six bolts retaining flywheel flex plate assembly to converter cover.
10. Support the transmission with a 500-pound (minimum) transmission floor jack. The jack must be positioned so transmission oil pan will not support the weight of transmission. Fasten a safety chain over top of transmission and to both sides of jack.
11. Place a support under rear of engine and remove transmission case-to-crossmember support bolts. Raise the engine to remove weight from the engine rear mounts.
12. Remove the transmission case-to-flywheel housing bolts and washers.
13. Carefully inspect transmission and surrounding area to be sure no lines, hoses, or wires will interfere with transmission removal.

NOTE: When removing transmission, keep rear of transmission lower than the front so as not to lose converter.

14. Move transmission assembly from the engine, lower the assembly carefully and move it out from the vehicle.

Installation

1. Raise vehicle sufficiently to allow installation of transmission. With transmission assembly mounted on transmission jack move transmission into position aligning converter with flywheel. Check for and clean away any foreign material in flywheel pilot hole, flywheel flex-plate assembly, and front face of transmission case. Rotate flywheel as necessary so that the six bolt holes in flex-plate are aligned with bolt holes in converter cover. Carefully move transmission assembly toward engine so flex-plate-to-converter cover bolts can be loosely installed and so that pilot on transmission converter enters pilot hole in center of flywheel.
2. Install bolts and washers that attach transmission case to flywheel housing. Tighten bolts to 25-30 foot-pounds torque.
3. Tighten the six flex-plate-to-converter cover bolts to 35-40 foot-pounds torque.
4. Carefully lower engine and transmission assembly onto engine rear mounts. Tighten engine rear mounting bolts to 60-70 foot-pounds torque. Then bend lock tabs down over head of each bolt. Remove lifting equipment from beneath vehicle.
5. Remove plugs from oil cooler lines and transmission case fittings. Be sure fittings are clean and lint-free, then connect oil cooler lines to transmission.
6. Install oil filler tube and bracket on right side of transmission. Install oil level gauge (dipstick).
7. Connect the speedometer shaft fitting to adapter at rear of transmission.
8. Connect propeller shaft to transmission.
9. Connect parking brake linkage (if used) at side of transmission.
10. Connect the range selector cable to shift lever at left side of transmission.
11. Connect the vacuum modulator line to modulator. Also, on conventional cab models, connect wiring to neutral safety and back-up lamp switches.

NOTE: Make sure the ignition switch is in the off position before proceeding to the next step.

12. Install spark plugs and connect battery ground strap, previously disconnected.
13. Connect any other lines, hoses, or wires which were discon-

nected to aid in transmission removal.

14. Adjust the shift linkage. (see "Shift Linkage Adjustment")
15. Refill the transmission with the proper lubricant.

MT640, MT650
Removal

NOTE: It may be necessary to remove the air tanks, fuel tanks, special equipment, etc., on some vehicles to provide clearance before the transmission is removed.

1. Block vehicle so that it cannot move. Disconnect ground strap from battery negative (—) post. Remove the spark plugs so the engine can be turned over manually.
2. Remove the level gauge (dipstick). Drain transmission by disconnecting filler tube at right side of transmission pan. Remove bracket holding filler tube to transmission and remove filler tube from vehicle. Replace dipstick in tube and cover the pan opening to prevent entry of foreign materials.
3. Disconnect cooler lines from fittings on right side of transmission case. Plug line ends and case openings with lint-free material.
4. Disconnect shift cable from shift lever at left side of transmission.
5. Disconnect vacuum modulator line from modulator. Also, disconnect wiring from back-up lamp switch (right side of transmission) and neutral safety switch (left side of transmission).
6. Disconnect the speedometer shaft fitting from adapter at rear of transmission.
7. Disconnect the propeller shaft from transmission.
8. Disconnect the mechanical parking brake linkage at the right side of the transmission (if used).

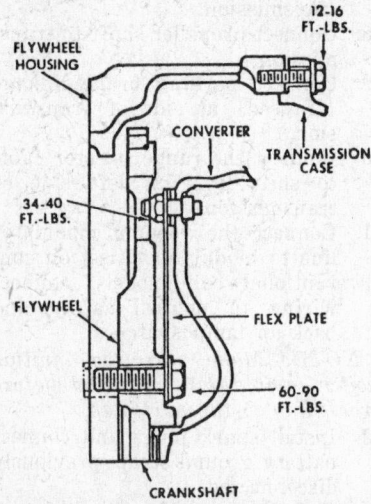

Flex plate installation—MT-640, MT-650
(© Chevrolet Div., G.M. Corp.)

9. Through access opening in the flywheel housing, use a pry bar, as necessary to manually turn the flywheel. As the flywheel is rotated, remove the six nuts retaining flex-plate assembly to converter cover.
10. Support the transmission with a 750-pound (minimum rating) transmission floor jack. The jack must be positioned so transmission oil pan will not support the weight of transmission. Fasten a safety chain over top of transmission and to both sides of jack.
11. Place a support under rear of engine and remove transmission case-to-crossmember support bolts. Raise the engine to remove weight from the engine rear mounts.
12. Remove the transmission case-to-flywheel housing bolts and washers.
13. Carefully inspect transmission and surrounding area to be sure no lines, hoses, or wires will interfere with transmission removal.

NOTE: When removing transmission keep rear of transmission lower than the front so as not to lose converter.

14. Move transmission assembly from the engine, lower the assembly carefully and move it out from the vehicle.

Installation

1. Raise vehicle sufficiently to allow installation of transmission. With transmission assembly mounted on transmission jack move transmission into position aligning converter with flywheel. Check for and clean away any foreign material in flywheel pilot hole, flex-plate assembly, and front face of transmission case. Rotate flywheel as necessary so that the six studs in converter cover are aligned with holes in flex plate. Carefully move transmission assembly toward engine so flex-plate-to-converter cover nuts can be loosely installed and so that pilot on transmission converter enters pilot hole in center of flywheel.
2. Install bolts and washers that attach transmission case-to-flywheel housing. Tighten bolts to 12-16 foot-pounds torque.
3. Tighten the six flex-plate-to-converter cover nuts to 34-40 foot-pounds torque.
4. Carefully lower engine and transmission assembly onto engine rear mounts. Tighten engine rear mounting nuts to 190-210 foot-pounds torque. Remove lifting equipment from beneath vehicle.
5. Remove plugs from oil cooler

lines and transmission case fittings. Be sure fittings are clean and lint-free, then connect oil cooler lines-to-transmission.
6. Install oil filler tube and bracket on right side of transmission. Install oil level gauge (dipstick).
7. Connect the speedometer shaft fitting to adapter at rear of transmission.
8. Connect propeller shaft to transmission.
9. Connect parking brake linkage (if used) at side of transmission.
10. Connect shift cable to shift lever at left side of transmission.
11. Connect the vacuum modulator line to modulator. Also, connect wiring to neutral safety switch (left side of transmission) and back-up lamp switch (right side of transmission).

NOTE: Make sure the ignition switch is in the off position before proceeding to the next step.

12. Install spark plugs and connect battery ground strap, previously disconnected.
13. Connect any other lines, hoses, or wires which were disconnected to aid in transmission removal.
14. Adjust the shift linkage. (see "Shift Linkage Adjustment")
15. Refill the transmission with the proper lubricant.

Drive Shaft

Tubular type drive shafts and needle bearing type universal joints are used on all model trucks. An internally splined sleeve which compensates for variation in distance between rear axle and transmission is located at the forward end of single or rear shafts.

The number of shafts used is dependent upon the wheel base of the vehicle. On vehicles which use two or more shafts, each shaft (except the rear) is supported near its splined end in a rubber cushioned ball bearing which is mounted in a bracket attached to a frame cross member. The ball bearing is a permanently sealed and lubricated type.

An extended-life universal joint, which does not require periodic inspection and lubrication, has been incorporated in several applications on the 10 and 20 series vehicles. This extended-life universal joint is identified by the absence of the lubrication fitting, which is present on all trunnions not equipped with the extended-life feature.

A lubrication fitting is also provided on each sliding sleeve to lubricate the splines. A plug is staked into the yoke end of sleeve to retain lubricant and a small hole is drilled in the end of this plug to relieve trapped air. The opposite end of the

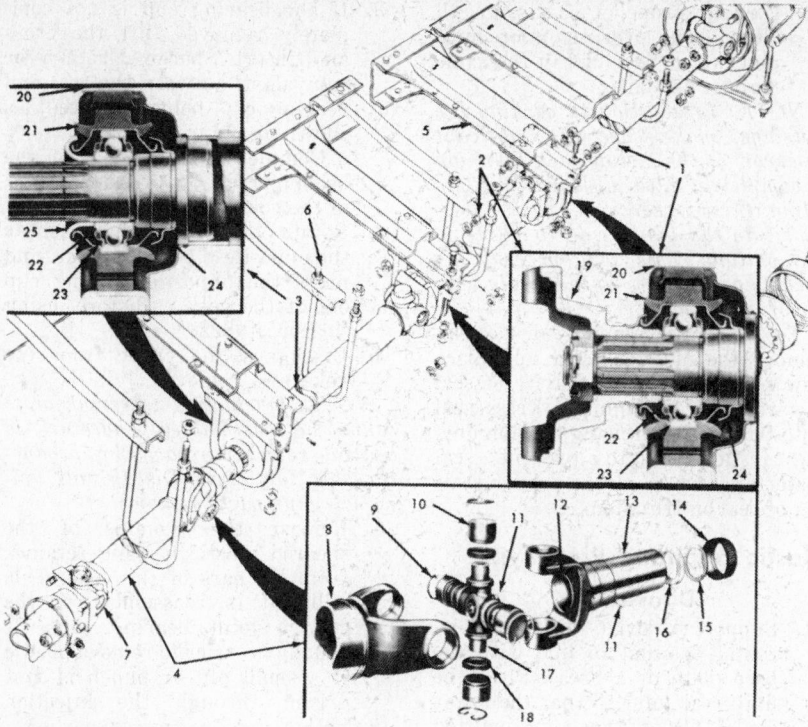

1 Front propeller shaft and bearing suuport assy.
2 Front intermediate propeller shaft and bearing support assy.
3 Rear intermediate shaft and bearing support assy.
4 Rear propeller shaft and sleeve assy.
5 Frame crossmember
6 Guard
7 "U" clamp
8 Rear propeller shaft
9 Lock ring
10 Bearing assembly
11 Lubrication fitting
12 Plug
13 Sleeve
14 Retainer
15 Washer
16 Cork packing
17 Trunnion
18 Seal ring
19 Flange and deflector assy.
20 Brakcet
21 Cushion
22 Slinger
23 Grease retainer
24 Inner deflector
25 Dust shield

Propeller shafts, universal joints and bearing supports (© Chevrolet Div., G.M. Corp.)

sleeve is sealed by means of a cork packing in a retainer which screws on the end of the sleeve.

Shaft Removal

Single or Rear

Remove rear trunnion "U" clamps, lower the rear of shaft and pull back to disengage the sleeve at front of shaft. Remove shaft from under vehicle.

Front

Remove four front flange nuts at transmission and, if equipped with intermediate shaft, remove the rear trunnion "U" clamps. Remove nuts and lock washers attaching bearing support to frame crossmember and pull shaft assembly from vehicle.

Intermediate or Rear Intermediate Shaft

Remove the front trunnion "U" clamps and the bearing support mounting bolt nuts and lock washers. Lower the front of the shaft and pull forward to disengage splines at rear of shaft. Remove shaft and bearing support assembly from under vehicle.

Front Intermediate Shaft

Remove the front and rear trun-nion "U" clamps and the bearing support mounting bolt nuts and lock washers. Lower shaft and bearing support assembly from vehicle.

Inspection

Wash ends of propeller shaft in cleaning solvent, inspect for damage and excessive wear on splines, trunnions and bearings. Examine sleeve seal, washer and retainer for damage or deterioration.

CAUTION: When trunnion bearing "U" clamps are removed to remove the propeller shaft, tape the bearings to keep them clean and from becoming damaged. Propeller shaft guards may be removed, if necessary, by removing nut at each end of the guard.

Shaft Installation

Drive shafts may be installed by reversing the procedure used in removal when the following notes are observed.

NOTE 1: Before installing a rear shaft and sleeve assembly, slide seal retainer, steel washer and cork seal on spline of mating shaft. Assemble these parts to sleeve by turning retainer onto sleeve after rear propeller shaft is installed.

NOTE 2: Over torquing "U" clamp nuts will result in bearing cap distortion which will reduce roller bearing life.

NOTE 3: To prevent excessive driveline vibration on some models, the rear propeller shaft must be installed so that centerline of sleeve yoke is positioned from vertical to 7 splines clockwise from vertical. The centerline of either yoke at the transmission end is perpendicular to the ground.

NOTE 4: The shaft to pinion flange fastener is an important attaching part in that it could affect the performance of vital components and systems and/or could result in major repair expense. It must be replaced with one of the same part number or with an equivalent part if replacement becomes necessary. Do not use a replacement part of lesser quality or substitute design. Torque values must be used as specified during reassembly to assure proper retention of this part.

Universal Joints

Snap Ring Type

Disassembly

1. Remove trunnion bearings from propeller shaft yoke as follows:
   a. Remove lock rings from yoke and lubrication fitting from trunnion.
   b. Support yoke in a bench vise.

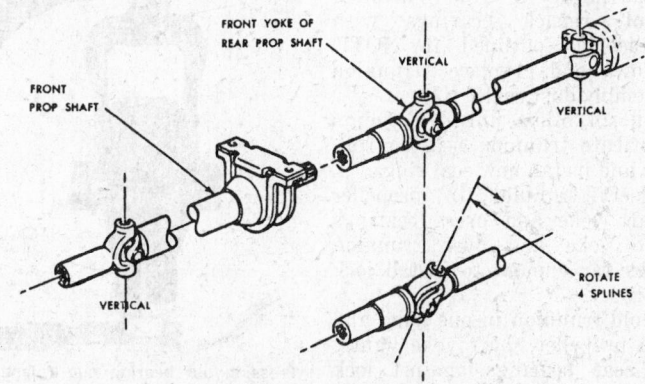

Aligning universal joints (© Chevrolet Div., G.M. Corp.)

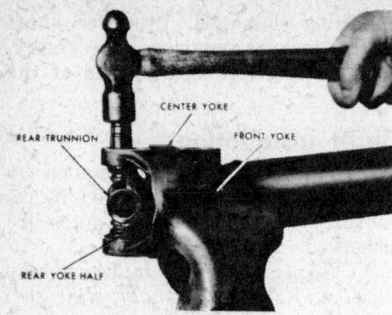

Driving out U-joint bearing cups (snap ring type) (© Chevrolet Div., G.M. Corp.)

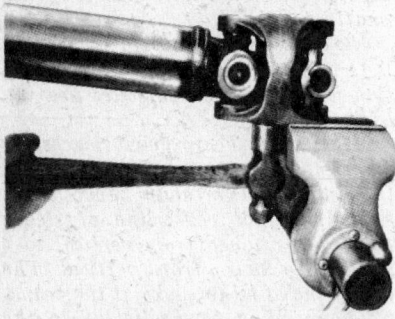

Removing U-joint bearing cups (snap ring type) (© Chevrolet Div., G.M. Corp.)

c. Using soft drift and hammer, drive on one trunnion bearing to drive opposite bearing from yoke.

NOTE: The bearing cap cannot be driven completely out.

d. Grasp cap in vise and work out.

e. Support other side of yoke and drive other bearing cap from yoke and remove as in step d.

f. Remove trunnion from propeller shaft yoke.

2. If equipped with sliding sleeve, remove trunnion bearings from sleeve yoke in the same manner as above. Remove seal retainer from end of sleeve and pull seal and washer from retainer.

Assembly

1. Assemble trunnion bearings to propeller shaft as follows:

a. On extended life universal joints when performing service operations that require disassembly of the universal joint, repack bearings with grease as outlined in NOTE below and replace trunnion assembly dust seals.

b. On all other universal joints lubricate trunnion bearing rollers and install new seal rings.

c. Insert trunnion in propeller shaft yoke and press bearings into yoke and over trunnion hubs far enough to install lock rings.

d. Hold trunnion in one hand and tap propeller shaft yoke lightly to seat bearings against lock rings.

2. On rear propeller shafts, install sleeve yoke over trunnion hubs and install bearings in the same manner as above.

NOTE: In addition to packing the bearings, make sure the lubricant reservoir at the end of each trunnion is completely filled with lubricant. In filling these reservoirs, pack lubricant into the hole so as to fill from the bottom. This will prevent air pockets and ensure an adequate supply of lubricant

To replace trunnion dust seal, remove the old dust seal and place new seal on trunnion—cavity of seal toward end of trunnion—Press seal onto trunnion exercising caution during installation to prevent seal distortion and to assure proper seating of seal on trunnion.

Plastic Retaining Ring Type

Disassembly

1. Support the drive shaft in a horizontal position in line with the base plate of a press. Place the universal joint so that the lower ear of the shaft yoke is supported on a 1⅛″ socket. Place the cross press, J-9522-3 or equivalent, on the open horizontal bearing cups, and press the lower bearing cup out of the yoke ear as shown in the illustration. This will shear the plastic retaining the lower bearing cup.

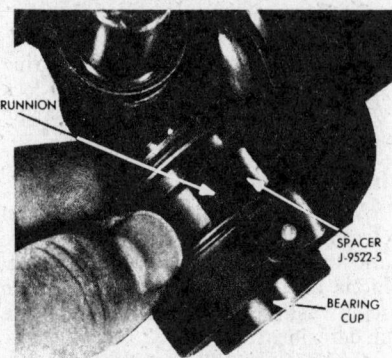

Using spacer to remove bearing cup (plastic retaining ring type) (© Chevrolet Div., G.M. Corp)

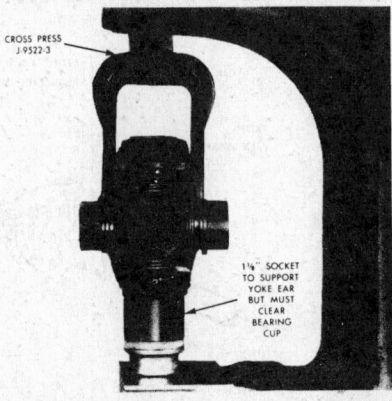

Pressing out bearing cup (plastic retaining ring type) (© Chevrolet Div., G. M. Corp)

2. If the bearing cup is not completely removed, lift the cross and insert Spacer J-9522-5 or equivalent, between the seal and bearing cup being removed, as shown in figure 2.

Complete the removal of the bearing cup, by pressing it out of the yoke.

3. Rotate the drive shaft, shear the opposite plastic retainer, and press the opposite bearing cup out of the yoke as before, using Spacer J-9522.

4. Disengage the cross from the yoke and remove.

NOTE: Production universal joints cannot be reassembled. There are no bearing retainer grooves in production bearing cups. Discard all universal joint parts removed.

5. Remove the remains of the sheared plastic bearing retainer from the ears of the yoke. This will aid in reassembly of the service joint bearing cups. It usually is easier to remove plastic if a small pin or punch is first driven through the injection holes.

6. If the front universal joint is being serviced, remove the pair of bearing cups from the slip yoke in the same manner.

Reassembly

1. A universal joint service kit is used when reassembling this joint. This kit includes one pre-greased cross assembly, four service bearing cup assemblies with seals, needle rollers, washers, grease and four bearing retainers.

2. Make sure that the seals are in place on the service bearing cups to hold the needle rollers in place for handling.

3. Remove all of the remains of the sheared plastic bearing retainers from the grooves in the yokes. The sheared plastic may prevent the bearing cups from being pressed into place, and this

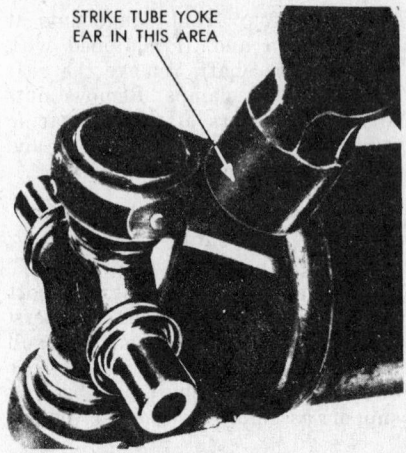

Seating plastic snap rings (© Chevrolet Div., G.M. Corp.)

Installing trunion into yoke (plastic retaining ring type) (© Chevrolet Div., G.M. Corp.)

prevents the bearing retainers from being properly sealed.

4. Install one bearing cup part way into one side of the yoke, and turn this yoke ear to the bottom.
5. Insert cross into yoke so that the trunnion seats freely into bearing cup as shown in illustration.
6. Install opposite bearing cup part way. Make sure that both trunnions are started straight and true into both bearing cups.
7. Press against opposite bearing cups, working the cross all of the time to check for free movement of the trunnions in the bearings. If there isn't, stop pressing and recheck needle rollers to determine if one or more of them has been tipped under the end of the trunnion.

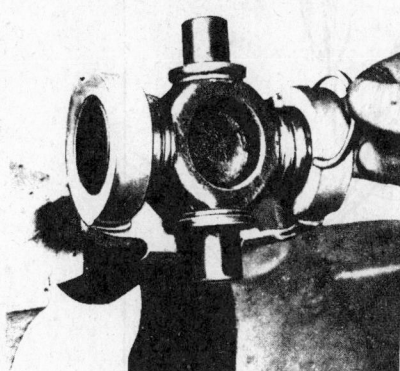

Install plastic snap rings
(© Chevrolet Div., G.M. Corp.)

8. As soon as one bearing retainer groove clears the inside of the yoke, stop pressing and snap the bearing retainer into place as shown in the illustration.
9. Continue to press until the opposite bearing retainer can be snapped into place. If difficulty is encountered, strike the yoke firmly with a hammer to aid in seating bearing retainers. This springs the yoke ears slightly.
10. Assemble the other half of the universal joint in the same manner.
11. Check the freedom of rotation of both sets of trunnions of the cross. If too tight, again rap the

yoke ears as described above. This will loosen the bearings and help seat the bearing retainers.

Constant Velocity Joint

Disassembly

1. Remove front propeller shaft from vehicle.
2. Remove rear trunnion snap rings from center yoke. Remove grease fitting.
3. Place propeller shaft in vise and drive one rear trunnion bearing cap from center yoke until it protrudes approximately $3/8''$. *NOTE: Keep rear portion of propeller shaft up to avoid interference of rear yoke half with center yoke.*
4. Once the bearing cap protrudes $3/8''$, release vise. Grasp protruding portion of cap in vise and drive on center yoke until cap is removed. Remove cap seal by prying off with a thin screwdriver.
5. Repeat steps 3 and 4 for remaining bearing caps.
6. Once the center yoke caps have been removed remove rear yoke half bearing caps. Remove rear trunnion.
7. Gently pull rear yoke half from propeller shaft. Remove all loose needle bearings. Remove spring seal.
8. Remove front trunnion from center and front yoke in same manner as described in Steps 2, 3 & 4.
   NOTE: Before front trunnion can be removed all four (4) bearing caps must be removed.

Assembly

1. Clean and inspect all needle bearings, caps, seals, fittings, trunnions and yokes. Assemble all needle bearings in caps (27 per cap) ; assemble needle bearings in front yoke (28 total). Retain bearings with a heavy grease. Assemble seals to bearing caps.
2. Place front trunnion in drive shaft. Place center yoke on front trunnion. Install one bearing cap and seal assembly in front yoke. Drive in to a depth that the snap ring can be installed. Install snap ring. Install remaining cap and seal in front yoke. Install snap ring.
3. Install front trunnion bearing caps in center yoke in same manner.
4. With front trunnion completely installed, install seal on propeller shaft (large face first). Gently slip rear yoke half on propeller shaft using care not to upset rollers. Insert rear trunnion in center yoke. Install rear

yoke half bearing caps on rear trunnion. Install one rear trunnion bearing cap in center yoke and press into yoke until snap ring can be installed. Install remaining cap and snap ring.
5. Before assembly is reinstalled in vehicle, grease universal at all three (3) fittings (2 conventional type and one, in rear yoke half), that requires a needle nose grease gun adapter.

Bearing Support

Removal

1. Remove dust shield, or, if equipped with flange, remove cotter pin and nut and pull flange and deflector assembly from shaft.
2. Pull support bracket from rubber cushion and pull cushion from bearing.
3. Pull bearing assembly from shaft. Remove grease retainers and slingers (if used) from bearing.
4. Remove inner deflector from shaft if replacement is necessary.
   NOTE: The ball bearing is a permanently sealed and lubricated type.

Installation

1. Install inner deflector on propeller shaft, if removed, and prick punch deflector at two opposite points to make sure it is tight on shaft.
2. Pack retainers with grease. Insert a slinger (if used) inside one retainer and press this retainer over bearing outer race.
3. Start bearing and slinger assembly straight on shaft journal. Support propeller shaft and, using suitable length of pipe over splined end of shaft, press bearing and inner slinger against shoulder on shaft.
4. Install second slinger on shaft and press second retainer over bearing outer race.
5. Install dust shield over shaft, small diameter first and press into position against outer slinger or, if equipped with flange, install flange and deflector assembly as follows:
   a. Install deflector on flange, if removed, and prick punch at two opposite points to make sure it is tight on flange.
   b. Align centerline of flange yoke with centerline of propeller shaft yoke and start flange straight on splines of shaft with end of flange against slinger.
   c. Install retaining nut and tighten to 160-180 ft. lbs. torque. Install cotter pin.
6. Force rubber cushion onto bear-

ing and coat outside diameter of cushion with brake fluid.

7. Force bracket onto cushion.

Drive Shaft Alignment

Correct drive line angles are necessary to prevent excessive torsional vibrations, especially tandem rear axle models. On some vehicles adjustable auxiliary transmission mountings are provided for adjusting the angle of the various drive line components. On vehicles not having adjustable auxiliary transmission mountings and adjustable torque rods at rear axles, proper adjustment of the angle of the drive line compo-nents must be accomplished by the use of spacers or shims at the frame crossmember or hangers. All angles must be checked with a maximum amount of exactness. The use of a bubble level is not sufficient, a bevel protractor must be used. The vehicle should be checked on a reasonably flat surface.

Clean machined surface at rear of (main) tranmission to check engine (and transmission) angle. This is the key angle and auxiliary transmission (if equipped) and rear axle pinion must be set to this angle. Make sure all drive line components from (main) transmission to rear axle are properly centered. Clean dirt and paint off machined surface of propeller shaft yoke, make sure surface is free of nicks or burrs. Set bevel protractor to zero, place protractor on yoke surface at right angle to propeller shaft and rotate shaft until bubble is centered in glass. Reposition protractor on yoke, in-line with propeller shaft, and note shaft angle.

Shaft angle must be held within a maximum of 1° less than engine. Check rear axles on machined surfaces on differential carrier housing, at right angle to pinion shaft. Make sure protractor is held straight up to get correct angle. On rear axles

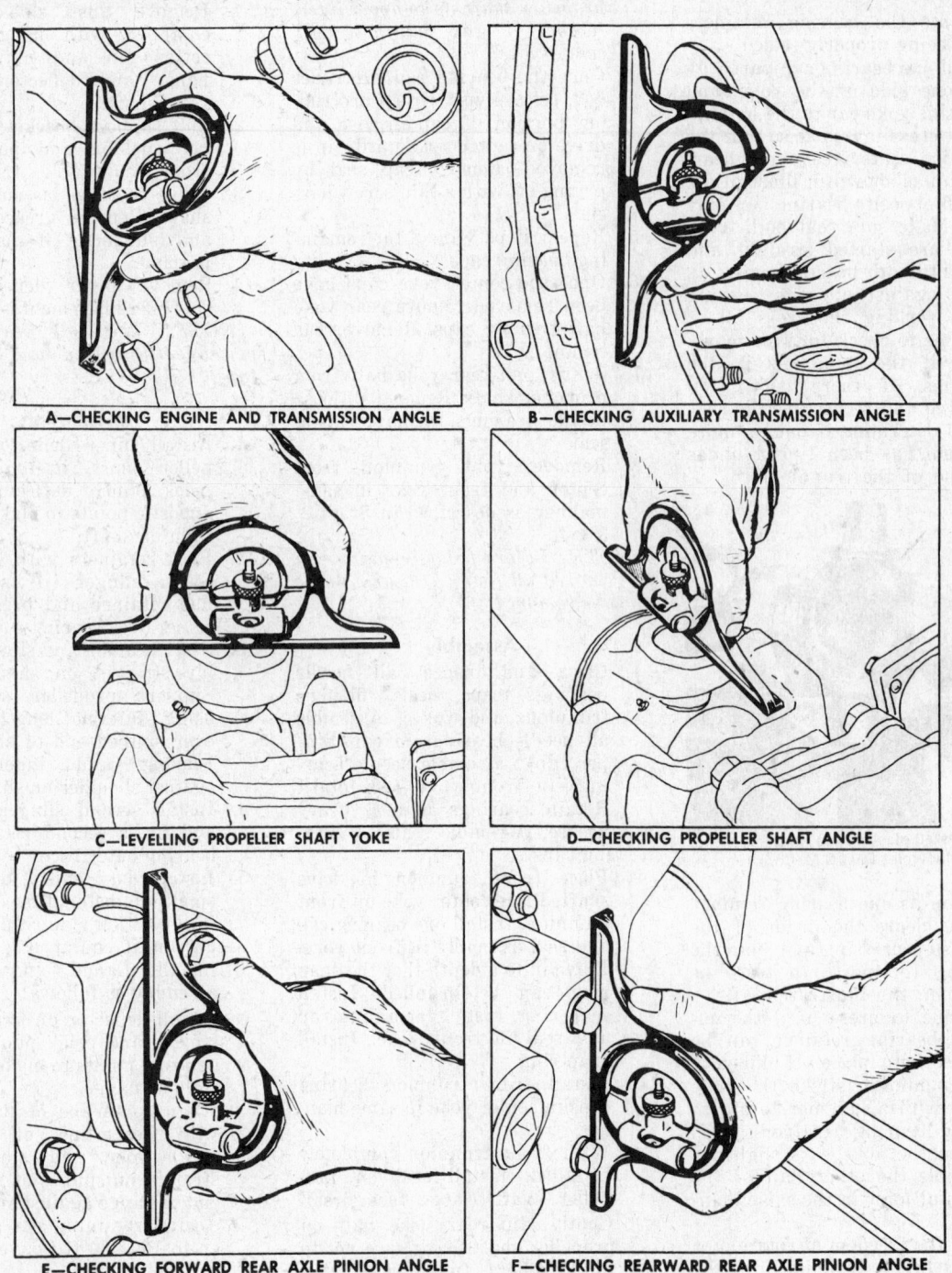

A—CHECKING ENGINE AND TRANSMISSION ANGLE

B—CHECKING AUXILIARY TRANSMISSION ANGLE

C—LEVELLING PROPELLER SHAFT YOKE

D—CHECKING PROPELLER SHAFT ANGLE

E—CHECKING FORWARD REAR AXLE PINION ANGLE

F—CHECKING REARWARD REAR AXLE PINION ANGLE

Methods of checking drive line angles—typical (© G.M.C.)

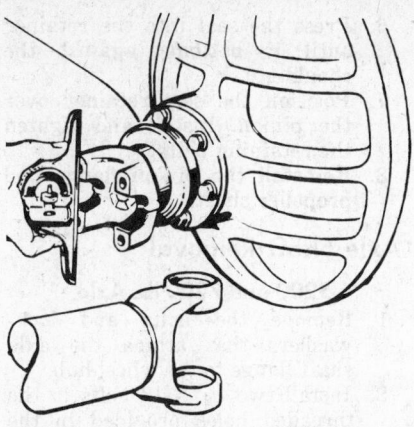

Method of checking pinion angle, single speed rear—typical (© G.M.C.)

that do not have a machined surface it will be necessary to remove propeller shaft. Rotate the pinion yoke into a vertical position, clean the four machined ends of yoke of dirt, paint, nicks and burrs. Place the protractor across ends of yoke, on either side, and in as close as possible to a vertical position. Rear axle angle should be same as engine.

Rear Axle

Chevrolet Semi-Floating Single Speed

Two types of rear axles are used. The removable carrier type with Hotchkiss drive and the Salisbury type with an integral carrier.

The following applies to both rear axles except where noted.

The drive pinion is mounted on two preloaded taper roller bearings. The ring gear is bolted to the differential case which is mounted on preloaded taper roller bearings. There are two side gears and two differential pinion gears.

Axle Shaft

Removal
1. Remove the brake drum.
2. Drain lubricant from the differential and remove the housing cover.
3. Remove the differential pinion shaft lock screw, pinion shaft and axle shaft spacer.
4. Push the axle shaft in and remove the "C" washer from the inner end of the axle shaft.
5. Remove the axle shaft from the housing.

Installation
NOTE: If a new axle shaft is to be installed.
1. Position the axle shaft gasket to the axle shaft flange.

2. Apply heavy shellac or paint to both sides of the gasket and axle shaft oil deflector.
3. Install the axle shaft oil deflector over the gasket aligning the oil pocket with the notch in the flange.
4. Insert six special axle shaft bolts and force the heads down to the deflector.
5. Peen the end of the shoulder on the bolts into the countersink around the bolt holes in the flange.
6. Slide the axle shaft into place. CAUTION: Exercise care that the splines on the end of the shaft do not cut the axle shaft oil seal and that they engage with the splines of the differential side gears.
7. Install the "C" washer on the inner end of the shaft.
8. Pry the shafts apart so that the "C" washers are seated in the counterbore in the differential side gears and install the pinion gears.
9. Select the proper axle shaft spacer to give free fit to .014" maximum clearance between the end of the axle shaft and the spacer.
10. Install the spacer and pinion shaft, locking in place with the special screw.
11. Install the axle housing cover and gasket and refill the differential.
12. Install the drum and wheel.
13. Road test for leaks and noise.

Axle Shaft (Spline Drive Type)

Removal
Procedure for removal of axle shafts is same with assembly removed or installed in the vehicle.
1. Remove cap screws and hub cap from hub.
2. Install slide hammer adapter into tapped hole in axle flange.
3. Install slide hammer into adapter and remove axle shaft.

Installation
1. Dip small end of splined shaft in axle lubricant, and insert shaft into hub.
2. Turn shaft as necessary to index shaft splines with differential side gear splines. As shaft is pushed inward, rotate hub to align axle shaft flange splines to hub. Push shaft into place.
3. Install new gasket on hub cap and install hub cap to hub with cap screws. Torque cap screws 15 to 20 foot-pounds.

Axle Shaft Bearing or Oil Seal
Removal
1. Remove the wheel, drum and

axle shaft (see axle shaft removal).

Bearing and oil seal removal
(© Chevrolet Div., G.M. Corp.)

2. Using a slide hammer, remove the bearing, bearing retainer and oil seal.
3. Inspect the bore and dress out the old stake points.

Installation
1. Using the proper driver, place the oil seal, bearing and inside bearing retainer on the driver in that order.
2. Place a light coat of sealer on the outside of the seal to insure proper sealing of the seal in the housing bore.
3. Start the bearing into the axle housing and tap the tool with a hammer to seat the parts.
4. Remove the driver and stake the oil seal in place with a punch.
5. Assemble the axle shafts (see Axle Shaft Installation).

Chevrolet Full Floating

Single Speed
The rear axle is a full floating type with hypoid ring gear and pinion. The full floating construction enables removal of the axle shafts without removing the truck load or jacking up the rear axle. The drive pinion is straddle mounted being supported at the rear end on a roller bearing and at the front end on a double row bearing.

The ring gear is bolted to the differential case and some models are provided with a ring gear thrust pad to prevent distortion when starting under heavy loads.

Some models have a two pinion differential while others have a four pinion differential.

Two-speed
The Chevrolet two-speed axle is available in the 15,000 and 17,000 lb. capacity. In low gear, torque is transmitted to the differential case through the planetary pinions. The straddle mounted drive pinion and the ring gears operate to produce the high range reduction, the planet and sun gears being locked to revolve with the ring gear.

Early models have a two way vacuum system for axle shifts. Later models use an electric shift system.

Maintenance and adjustments for the two speed axle are performed the same as those outlined for the Chevrolet single-speed axles.

Differential Carrier

Removal

1. Drain the lubricant from the differential.
2. Remove the axle shafts. (See Axle Shaft Removal)
3. Disconnect the rear universal joint and swing the propeller shaft to one side.
4. On two speed axles, remove the electric or vacuum lines.
5. Remove the bolts and lockwashers which retain the carrier assembly to the axle housing.

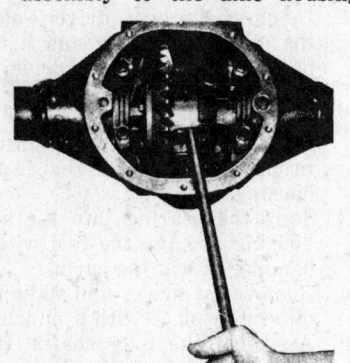

Differential case removal
(© Chevrolet Div., G.M. Corp.)

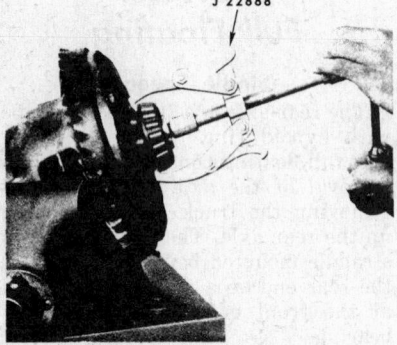

Differential bearing removal
(© Chevrolet Div., G.M. Corp.)

Support the differential housing with a floor jack and roll it from under the truck.
For overhaul—see General Section.

Installation

1. Clean the axle housing and differential housing gasket surfaces and place a new gasket over the axle housing.
2. Assembly the carrier to the axle housing, install the lockwashers and bolts and tighten securely.
3. Assemble the rear universal joint.
4. On two speed axles, connect the electric or vacuum lines.
5. Install the axle shafts. (See Axle Shaft Installation)
6. Fill with lubricant and road test for proper operation.

Installing differential shim
(© Chevrolet Div., G.M. Corp.)

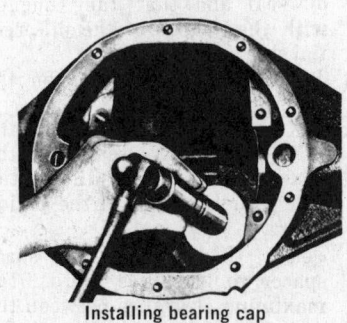

Installing bearing cap
(© Chevrolet Div., G.M. Corp.)

Drive Pinion Oil Seal

Replacement

1. Disconnect the propeller shaft and remove the pinion flange and deflector.
2. Remove the bolts retaining the oil seal retainer to the carrier, and withdraw the retainer from the pinion.
3. Pry the old seal from the bore.
4. Clean all foreign matter from the retainer.
5. Pack the cavity of the new seal with a high melting point bearing lubricant, position the seal on an installer.

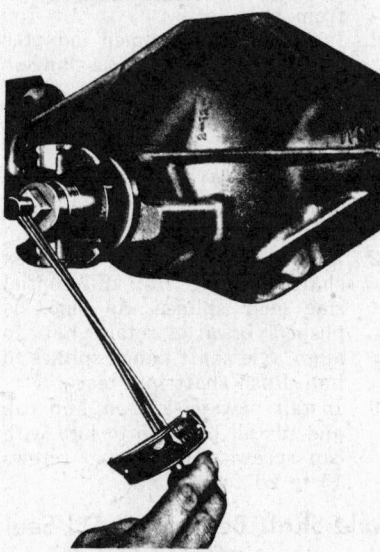

Measuring pinion bearing preload
(© Chevrolet Div., G.M. Corp.)

6. Press the seal into the retainer until it bottoms against the shoulder.
7. Position the seal retainer over the pinion. Install and tighten the retaining bolts.
8. Reinstall the pinion flange and propeller shaft.

Axle Shaft Removal

5200 and 7200 lb. Axle

1. Remove the bolts and lockwashers that attach the axle shaft flange to the wheel hub.
2. Install two ½" - 13 bolts in the threaded holes provided in the axle shaft flange. By turning these bolts alternately the axle shaft may be started and then removed from the housing.

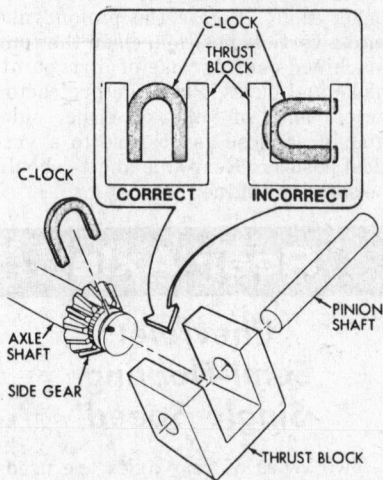

Correct C lock position (© G.M.C.)

Installation

1. Clean both the axle flange and the wheel hub.
2. Place a new gasket over the axle shaft and position the axle shaft in the housing so that the shaft splines enter the differential side gear.
3. Install the bolts and torque to 85-95 ft. lbs.

11,000-13,500 and 1500 lb. Axle

1. Remove the hub cap and install a slide hammer and adapter in the tapped hole on the shaft flange
2. Remove the axle shaft.

Installation

1. Clean the old gasket from the hub and hub cap. Clean the axle shaft flange and mating surfaces in the wheel hub.
2. Install the axle shaft so that the shaft splines index into the hub splines.
3. Tap the shaft into position. Install a new hub cap gasket, position the hub cap to the hub and install the attaching bolts. Torque the bolts to 11-18 ft. lbs.

Axle shaft removal—10½ ring gear axle (© G.M.C.)

Axle shaft removal—12¼ ring gear axle (© G.M.C.)

17,000 lb. Axle

1. Remove the axle shaft flange—to—hub nuts.
2. Strike the flange with a lead hammer to loosen the flange and dowels.
3. Remove the tapered dowels from the studs and pull the axle shaft from the housing.

Installation

1. Clean the old gasket from the wheel hub and axle shaft flange and install a new gasket over the hub studs.
2. Install the axle shaft so that the splines are aligned with the differential side gear and the flange holes index over the hub studs.
3. Install the tapered dowel over each hub stud. Install and tighten the stud nuts to 80-100 ft. lbs

Spline Drive Type

Procedure for removal of axle shafts is same with assembly removed or installed in the vehicle.

1. Remove cap screws and hub cap from hub.
2. Install slide hammer adapter into tapped hole in axle flange.
3. Install slide hammer into adapter and remove axle shaft.

Installation

1. Dip small end of splined shaft in axle lubricant, and insert shaft into hub.
2. Turn shaft as necessary to index shaft splines with differential side gear splines. As shaft is pushed inward, rotate hub to align axle shaft flange splines to hub. Push shaft into place.
3. Install new gasket on hub cap and install hub cap to hub with cap screws. Torque cap screws 15 to 20 foot-pounds.

Hub and Drum

Removal

1. Remove the wheel assembly and axle shaft. (See Axle Shaft Removal)
2. Disengage the tang of the nut lock from the slot or flat of the adjusting nut and remove the

nut lock. Using an appropriate tool, remove the adjusting nut. *NOTE: On 5200 through 15,000 lb. axles, remove the thrust washer from the housing tube.*

3. Pull the hub and drum assembly straight off the axle housing. *NOTE: On 11,000 through 17,000 lb. axles avoid dropping the outer bearing inner race and roller assembly.*

Bearing and Bearing Cup

Replacement

Replace the inner cup (all axles) and outer bearing cup for 17,000 lb. axle as follows:

1. Place an appropriate press-out tool behind the bearing cup, index the tool in provided notches, and press out the cup. *NOTE: The hub outer bearing (all axles except 17,000 lb. axle) cannot be replaced with the inner bearings in position; therefore, replace the outer bearings (if required) before proceeding.*
2. Position the cup in the hub, with the thick edge of the cup toward the shoulder of the hub. Using an applicable cup installer, press the cup into the hub until it seats on the hub shoulder.

Replace the outer bearing assembly (all axles except 17,000 lb.) as follows:

NOTE: The inner bearing assembly must be removed before attempting to replace the outer bearing.

1. Using a punch, tap the bearing outer race away from the bearing retaining ring. Then remove the retaining ring from the hub.

Removing bearing adjusting nut (typical)
(© Chevrolet Div., G.M. Corp.)

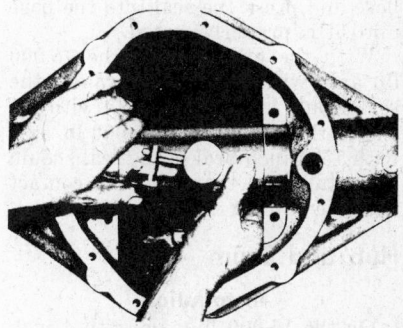

PINION SHIM IDENTIFICATION

Thickness	Identification Notches
.021	None
.024	1
.027	2
.030	3
.033	4

Measuring pinion shim requirement
(© Chevrolet Div., G.M. Corp.)

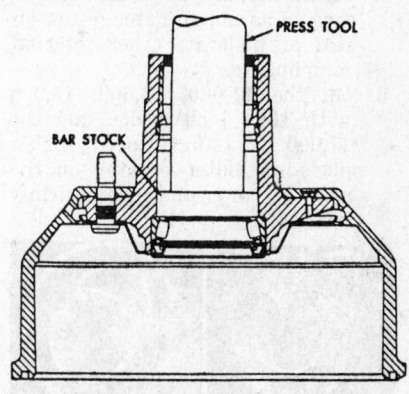

Removing hub inner bearing cup
(© Chevrolet Div., G.M. Corp.)

2. On 5,200 and 7,200 lb. axles, remove the outer bearing by using a brass drift. On 11,000, 13,500 and 15,000 lb. axles, remove the bearings by driving on the axle spacer, using the splined flange cut from an old axle.
3. On 11,000, 13,500 and 15,000 lb. axles place the axle shaft spacer in the hub first. Place the inner race and roller assembly in the hub, larger O.D. towards the outer end of the hub. Position the bearing cup in the hub, then end of the cup toward the outer end of the hub. Press the cup

into the hub, install the retainer ring, then press the cup into positive contact with the retainer ring.

NOTE: The bearing cup to retainer ring seating is essential to assure accurate wheel bearing adjustment.

Wheel Hub Oil Seal

Replacement

Pry out the old seal from the hub bore. Pack the cavity between the new seal lips with wheel bearing grease. Position the seal in the hub bore and press the seal into the bore until it is properly seated.

With the exception of the 15,000 lb. axle with 15x4 inch brakes, the seal should be installed flush with the end of the hub. On the 15,000 lb. axle with 15x4 inch brakes the seal should be installed so that it makes contact with the bearing race.

Hub and Drum

Installation

On the 15,000 lb. axle with 4 inch brakes, install the inner bearing oil seal in the inner bearing race and position the bearing race on the axle housing.

1. Using a smooth cup grease, pack the bearings and apply a light coat of grease to the inside of the bearing hub and the outside of the axle housing tube.
2. Install the hub and drum assembly on the axle housing, exercise care so as not to damage the oil seal or dislocate other internal components.
3. On the 11,000, 13,500, 15,000 with 15x4 inch brake, and the 17,000 lb. (single speed) axles, place the outer bearing on the axle housing and press firmly into the hub.

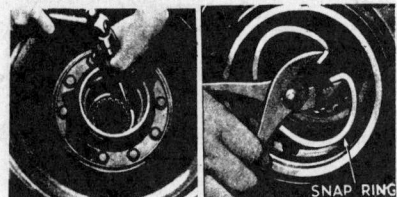

Removing bearing rotation ring 5200 and 7200 lb. axle   (© Chevrolet Div., G.M. Corp.)

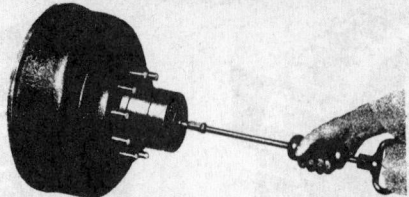

Removing axle shaft—11,000, 13,500 and 15,000 lb. axle   (© Chevrolet Div., G.M. Corp.)

Tightening wheel hub bearing nut
(© Chevrolet Div., G.M. Corp.)

4. On 5,200 through 15,000 lb. with 4 inch brake axles, install the thrust washer so that the tang is in the keyway on the axle housing.
5. Install the adjusting nut and adjust the bearings.

Bearing Adjustment

Before checking the bearing adjustment, make sure the brakes are fully released and do not drag. Check bearing play by grasping the tire at the top and pulling back and forth, or by using a pry bar under the tire. If the bearings are properly adjusted, movement of the brake drum in relation to the brake flange plate will be barely noticeable and the wheel will turn freely. If movement is excessive, adjust the bearings as follows:

1. Remove the axle shaft and raise the vehicle until the wheel is free to rotate. (See Axle Shaft Removal)
2. Disengage the nut lock from the lock nut and remove them from the axle housing tube.
3. Using an appropriate tool, tighten the adjusting nut to specifications, at the same time rotating the hub.
   5,200 and 7,200 lb. axles—55 ft. lbs.
   11,000 and 13,500 lb. axles—90 ft. lbs.

15,000 lb. axle
4 inch brakes—90 ft. lbs.
5 inch brakes—50 ft. lbs.
17,000 lb. axle—65 ft. lbs.
Then back the nut off ⅛ to ¼ turn to align the nearest slot with the short tang on the nut lock.
Install the nut lock.

4. Install the lock nut and tighten to specifications.
   5,200 and 7,200 lb. axles—175 ft. lbs.

Checking ring gear backlash
(© Chevrolet Div., G.M. Corp.)

Measuring differential shim requirement
(© Chevrolet Div., G.M. Corp.)

11,000 and 15,000 lb. axles with 15x4 inch brake—250 ft. lbs.
13,500 lb. axle—135 ft. lbs.
15,000 lb. axle with 15x5 inch brake—135 ft. lbs.
17,000 lb. axle—135 ft. lbs.

5. Bend the tang of the nut lock over the flat or slot of the lock nut.
   Final bearing check should show 0.001" to 0.007" end play.
6. Lower the vehicle and install the axle shaft. (See Axle Shaft Installation)

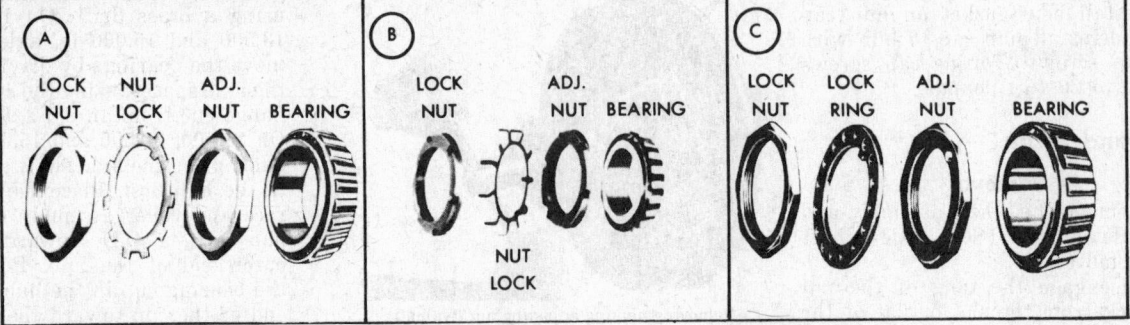

Typical lock types used on rear axle wheel bearing nuts—full floating axle systems   (© Chevrolet Div., GM Corp.)

Eaton Full Floating

Single Speed

This axle is equipped with a straddle mounted drive pinion. Pinion bearings are of the opposed tapered roller bearing type.

A straight roller type pilot bearing is pressed onto the inner end of the drive pinion and seats in the bore of the differential case.

The differential carrier assembly may be removed, while the axle is still installed in the truck, after the axle shafts have been removed.

The differential is a conventional four pinion type. On early models the ring gear is riveted to the differential case. On later models the ring gear is bolted to the differential case.

Some models have a thrust pad mounted on the end of an adjusting screw which is threaded into an opening in the differential carrier. This thrust pad limits the deflection of the drive gear under severe loads.

Axle shaft, oil seal, wheel bearings, pinion seal and differential removal and installtion are performed the same as those outlined for Chevrolet axles. Refer to Rear Axle (Chevrolet).

For overhaul—see General Section.

Two Speed

The differential and planetary assembly is installed in a two-piece support case. The ring gear is installed between the halves of the support case and retained in place by the same bolts which fasten the support case halves together.

The planetary assembly is composed of a high speed clutch plate, and four planetary pinions.

An electric power shifting arrangement is used to assist in making ratio changes. For schematic of Electric Shift see Chevrolet Two Speed Axle.

Axle shaft, oil seal, wheel bearings, pinion seal and differential removal and installation are performed the same as those outlined for Chevrolet Axles. Refer to Rear Axle (Chevrolet).

For overhaul—see General Section.

Tandem Axle

The Hendrickson type tandem axle suspension uses equalizing beams to tie the front to the rear axle and to permit independent vertical movement of each axle. The torque rods are used to maintain proper drive line alignment and to stabilize the driving and braking forces. Bolts are used on some models to hold the spring to the top saddle pad, while U-bolts are used on other models.

NOTE: When major overhaul is required, the complete tandem axle should be removed as a unit. The torque rods, springs, equalizing beam and other parts may be removed separately as required.

CAUTION: Block the vehicle securely before removal of the assembly to avoid rolling or pivoting at the equalizer beams when the torque rods are disconnected. The use of a helper is suggested, along with proper lifting tools so that personal injury does not occur.

Removal and Installation

1. Block all wheels and disconnect all applicable brake lines or hoses, differential lock lines, or electrical wiring from the rear axles.
2. Remove the rebound bolts from the rear spring brackets.
3. Remove all nuts and washers from the front spring brackets.
4. If equipped with ball stud torque rods, remove the stud nuts and tap each ball stud loose with a soft hammer. Remove the ball studs from the axle brackets.
5. If equipped with straddle mount torque rods, remove the mounting bolts from the rear axle bracket.
6. Support the rear axle differential with a floor jack, and disconnect the drive shaft from the forward rear axle.
7. Using a hoist, raise the rear of the frame high enough to clear the tandem axle assembly. Roll the assembly out from under the frame.
8. Installation is the reverse of removal.

Rear Suspension

Coil Spring Type

Control Arm

Removal

1. Remove the load from the spring by jacking at the frame.
2. Disconnect the parking brake cable from the control arm.
3. Remove the spring clamp bolt from the underside of the control arm.
4. Remove the "U" bolt nuts and separate the shock absorber bracket from the control arm. Separate the control arm from the "U" bolts and lower the rear of the arm.
5. Remove the pivot bolt and remove the arm from the vehicle.

Installation

1. Position the bushed end of control arm and insert the pivot bolt. Place the nut on the bolt finger tight.
2. Position the clamp inside the spring, raise the control arm, then pass the bolt, with the flat washer installed, up through the control arm and clamp. Install the lock washer and nut. Torque from 40 to 50 ft. lbs.
3. Place the arm adjacent to the axle. Pass the "U" bolt over the axle and through the holes in the arm.
4. Place the shock absorber bracket on the "U" bolt, install the nuts and torque from 200 to 225 ft. lbs.
5. Lower the vehicle to put full weight of the unloaded vehicle on the front and rear suspension, torque the control arm pivot bolt from 125 to 165 ft. lbs.

Coil Spring

Removal

1. Raise the vehicle and adjust the axle to frame height so that the spring is not under tension.
2. Remove the shock absorber bolt from the mounting bracket at the control arm.
3. Remove the upper and lower clamps from the spring.
4. Lower the control arm sufficiently to permit removal of the spring.

Installation

1. Place the spring lower clamp inside the spring. Position the clamp so that the end of the spring coil is within the notch. Locate the spring and clamp over the bolt hole in the control arm.
2. Pass the clamp bolt and washer up through the hole in the control arm and loosely install the nut.
3. Position the upper clamp inside the spring and install the bolt and washer. Torque to 45-55 ft. lbs.
4. Connect the shock absorber. Torque the nut to 110-150 ft. lbs.
5. Torque the spring lower clamp bolt to 40-50 ft. lbs.
6. Lower the vehicle.

Leaf Spring Type

Spring R & R

Light Duty Trucks with Spring Hanger and Shackle Pin Lubricating Fittings

1. Jack the vehicle at the frame to relieve tension on the spring.
2. Remove the lubrication fitting

from the spring eye and rear shackle pin.

3. Remove the lock bolts and nuts or pins from the shackle pin and spring eye pin.
4. Using appropriate tools, remove the spring eye pin and shackle pin.
5. Remove the "U" bolt nuts, withdraw the "U" bolts and remove the spring from the vehicle.

Installation

1. Position the spring assembly on the axle housing. Install the spacer assembly between the axle housing and spring, if so equipped, then install the "U" bolts.
2. Install the "U" bolt retaining nuts, torque the nuts alternately and evenly to properly seat the spring.
3. Jack as required to align the spring eyes; install the spring eye and shackle pins, lock bolts and nuts or lock pins and lubricating fittings. Lubricate the spring bushings and lower the vehicle.

Light Duty Trucks with Rubber Spring Hanger and Shackle Bushings

1. Jack the vehicle at the frame to relieve tension on the spring.
2. Remove the "U" bolt retaining nuts and withdraw the "U" bolts.
3. Loosen the shackle bolts and remove the lower bolt.
4. Remove the nut and bolt securing the spring to the front hanger.
5. Remove the spring from the vehicle.

Installation

1. Position the spring assembly, and spacers if so equipped, on the axle housing. *NOTE: On springs with metal encased pressed in type bushings the shackle assembly must be attached to the rear spring eye before installing the shackle to the rear hanger.*
2. Position the "U" bolts and loosely install the "U" bolt retaining nuts.
3. Jack as required to align the spring eyes with the front hanger and rear shackle, install the eye bolts.
4. Lower the vehicle.

REAR SUSPENSION TORQUE SPECIFICATIONS

PART	FT. LBS TORQUE
Spring U-Bolt Nuts (All "C/S" Series)	190-210
(All "T" Series)	190-210
Shock Absorber Nuts (when used) Upper Nut	85-90
Lower Nut	25-30
Rear Spring Radius Leaf Bolt Nut	175-225
(All T Series)	290-320
(All "C/S" Series with ⅝" x 18 Bolt)	60-80
(All "C/S" Series with 1" x 14 Bolt)	150-200
Rebound Pin Retainer Bolt (All Except "T" Series)	10-14
("T" Series)	
60	5-10
65	20-25
Rear Axle Bumper Bolt Nuts (All Except "T" Series)	6-8
("T" Series)	10-12

5. Tighten the "U" bolt retaining nuts alternately and evenly to properly seat the spring, and tighten the front hanger and rear shackle bolts.

Medium Duty Trucks

1. Raise vehicle frame to take weight off the spring. Make sure vehicle is supported safely.
2. Remove rear wheels to provide access to spring assembly.
3. Safely support axle on floor jack.
4. Install a C-clamp on radius leaf, to relieve load on radius leaf eye bolt on 45 Series vehicles.
5. On 45 Series at the front and rear hanger, remove rebound pin retainer bolt, then remove retainer. Install suitable puller into tapped hole at end of rebound pin, then remove pin.
6. Remove spring U-bolt nuts, shock absorber bracket (when used) U-bolt anchor plate and U-bolts and U-bolt spacer, then lower axle slightly.
7. Remove spring eye on radius bolt nut and washer, then remove spring eye bolt from spring eye or radius leaf.

NOTE: When tapered shim is used, the position of shim thin and thick edge should be noted so that shim can be installed properly at assembly.

Installation

1. Set spring assembly and tapered shim or spacer (if used) at axle pad.

NOTE: Tapered shim must be installed on axle in same position that was noted at removal.

2. Install U-bolt spacer over center bolt.
3. Seat U-bolts in spacer grooves, then secure spring to axle by installing anchor plates, shock absorber bracket (when used) and nuts on U-bolts.

4. Lower frame until ends of spring enter the hanger and touch the cam surface of hanger. Compress radius leaf with C-clamp until radius leaf eye and hanger holes are aligned and torque to specifications.
5. Remove C-clamp from radius leaf.
6. Install rebound pin at front and rear hangers. Install rebound pin retainer and secure with retainer bolt.
7. Install wheels.
8. Remove blocking and lower frame to place weight on springs. Check U-bolt nuts for proper torque.

Single or Tandem Axles Springs—Heavy Duty

1. Raise the rear of the vehicle, place floor jacks under the axle(s) and remove the dual wheels from the hubs to facilitate the removal of the spring eye pin and to expose the other nuts and bolts.
2. Remove the saddle cap stud nuts and/or spring U-bolts.
3. Remove the rebound pin locks or retainers, and then remove the rebound pins.
4. Remove the eye bolts or radius lead pin clamp bolts, then remove the lubrication fitting from the inner end of the pin, if equipped.
5. Remove the pins from the springs and lower the axle(s) or raise the frame until the spring will clear the brackets.
6. Remove the spring from the vehicle.
7. The installation is in the reverse of the removal procedure.
8. Torque the U-bolts or saddle cap stud nuts to specifications after the vehicle is lowered to the floor.

Index

TUNE-UP SPECIFICATIONS

CU. IN. DISPLACEMENT	YEAR	SPARK PLUG GAP	DISTRIBUTOR		IGNITION TIMING DEG. BTC (±2°)	CRANKING COMP. PRESS.	VALVE CLEARANCE		GOV. RPM NO LOAD	FUEL PUMP PRESS	IDLE SPEED	
			POINT DWELL	POINT GAP			INTAKE	EXHAUST			STD.	AUTO.
SIX CYLINDER												
198	1971	.035	40-45	.020	TDC	130	.012H	.024H	3500	3½-5	550	650N
225	1971	.035	42-47	.020	TDC	125	.012H	.024H	3500	3½-5	650	650N
225	1972	.035	41-46	.020	TDC	100	.012H	.024H	3500	3½-5	750	750N
225 Lt.Duty	1973	.035	Electronic		TDC	100	.012H	.024H	3600	3½-5	700 (800)	700N (800)
225 Heavy Duty	1973	.035	Electronic		TDC (2½A)	100	.012H	.024H	3600	3½-5	700 (800)	700N (800)
225 Lt. Duty	1974	.035	Electronic		TDC	100	.012H	.024H	3600	3½-5	800	750N
225 Heavy Duty	1974	.035	Electronic		2½A	100	.012H	.024H	3600	3½-5	800	750N
225 Lt. Duty	1975	.035①	Electronic		TDC	100	.012H	.024H	3600	3½-5	800	750N
225 Heavy Duty	1975	.035①	Electronic		TDC	100	.012H	.024H	3600	3½-5	700	700N
225	1976	.035①	Electronic		③	100	.012H	.024H	3600	3½-5	③	③
225 LD-1V	1977-78	.035①	Electronic		2	100	.012H	.020H	3600	3½-5	750	750
225 HD-2V	1977-78	.035①	Electronic		TDC	100	.012H	.020H	3600	3½-5	700	700
EIGHT CYLINDER												
318 S. Tr.	1971	.035	28-32	.017	5B	140	—	—	3900	5-7	500	—
318 A. Tr.	1971	.035	28-32	.017	10B	140	—	—	3900	5-7	—	500N
318 w/exh. em.	1971	.035	28-32	.017	5A	140	—	—	3800	5-7	650	600N
318-1	1971	.035	28-32	.017	5B	140	—	—	3800	5-7	500	500N
318-3	1971	.035	28-32	.017	12½B	110	—	—	3800	5-7	500	500N
318	1972	.035	32	.016	TDC	110	—	—	3800	5-7	750	700N
318-1 Lt. Duty	1973	.035	Electronic		2½B②	100	—	—	3900	5-7	750	700(750)N
318-1 Heavy Duty	1973	.035	Electronic		5B (TDC)	100	—	—	3900	5-7	750	700(750)N
318-3 Heavy Duty	1973	.035	Electronic		2½B	100	—	—	3700	5-7	700	700N
318-1 Lt. Duty	1974	.035	Electronic		TDC	100	—	—	3900	5-7	750	750N
318-1 Heavy Duty	1974	.035	Electronic		2½A	100	—	—	3700	5-7	750	750N
318-3 Heavy Duty	1974	.035	Electronic		TDC	100	—	—	3900	5-7	700	700N
318-1 Lt. Duty	1975	.035	Electronic		2B (TDC)	100	—	—	3900	5-7	750	750N
318-1 Heavy Duty	1975	.035	Electronic		2A (TDC)	100	—	—	3800	5-7	750 (700)	750N (700)
318-3 Heavy Duty	1975	.035	Electronic		TDC (2B)	100	—	—	3800	5-7	700	700N
318	1976	.035	Electronic		③	100	—	—	3800	5-7	③	③
318-3	1976	.035	Electronic		③	100	—	—	3800	5-7	③	③
318 LD	1977-78	.035	Electronic		2	100	—	—	3800	5-7	750	750
318-1 HD	1977-78	.035	Electronic		2ATDC	100	—	—	3800	5-7	750	750
318-3 HD	1977-78	.035	Electronic		TDC	100	—	—	3800	5-7	700	700
360	1971-72	.035	32	.016	TDC (2½B)	110	—	—	3900	5-7	750	700N
360 Lt. Duty	1973	.035	Electronic		TDC	100	—	—	3900	5-7	750	700N(750)N
360 Heavy Duty	1973	.035	Electronic		TDC	100	—	—	3900	5-7	750	700N(750)N
360 Lt. Duty	1974	.035	Electronic		2½B	100	—	—	3900	5-7	750	750N
360 Heavy Duty	1974	.035	Electronic		TDC	100	—	—	3900	5-7	750	750N
360 Lt. Duty	1975	.035	Electronic		TDC (4B)	100	—	—	3800	5-7	750 (700)	750N (700)
360 Heavy Duty	1975	.035	Electronic		TDC	100	—	—	3800	5-7	750 (700)	750N (700)
360	1976	.035	Electronic		③	100	—	—	3800	5-7	③	③
360-3	1976	.035	Electronic		③	100	—	—	3800	5-7	③	③
360 LD	1977-78	.035	Electronic		6	100	—	—	3800	5-7	700	700
360-1 HD	1977-78	.035	Electronic		TDC	100	—	—	3800	5-7	750	750
360-3 HD	1977-78	.035	Electronic		TDC	100	—	—	3800	5-7	750	750
361-3, 4	1971-72	.035	28-32	.016	5B	120	—	—	3600	3½-5	500	—
361-3, 4	1973	.035	Electronic		5B	100	—	—	3600	3½-5	600 (700)	700N
361-3, 4	1974-75	.035	Electronic		2½B	100	—	—	3600	6-7½	700	700N
361-3	1976	.035	Electronic		③	100	—	—	3600	6-7½	③	③
361-4 HD	1977-78	.035	Electronic		2.5	100	—	—	—	5-7	700	700

TUNE-UP SPECIFICATIONS

CU. IN. DISPLACE-MENT	YEAR	SPARK PLUG GAP	DISTRIBUTOR POINT DWELL	DISTRIBUTOR POINT GAP	IGNITION TIMING DEG. BTC (± 2°)	CRANKING COMP. PRESS.	VALVE CLEARANCE INTAKE	VALVE CLEARANCE EXHAUST	GOV. RPM NO LOAD	FUEL PUMP PRESS	IDLE SPEED STD.	IDLE SPEED AUTO.
EIGHT CYLINDER												
383 Std.	1971	.035	28-33	.017	TDC	140	—	—	—	3½-5	650	—
383 Auto.	1971	.035	28-33	.017	7½B	140	—	—	—	3½-5	—	600N
400	1971-72	.035	32	.016	7½B	—	—	—	—	—	700	700N
400 Lt. Duty	1973	.035	Electronic		10B	100	—	—	—	3½-5	700	700N
400 Heavy Duty	1973	.035	Electronic		2½B	100	—	—	—	3½-5	700	700N
400 Lt. Duty	1974	.035	Electronic		7½B	100	—	—	—	5-7	750	750N
400 Heavy Duty	1974	.035	Electronic		2½B	100	—	—	—	5-7	750	750N
400	1976	.035	Electronic		③	100	—	—	—	5-7	③	③
400-1 HD	1977-78	.035	Electronic		2④	100	—	—	—	5-7	700	700
400-1 HD	1977-78	.035	Electronic		8⑤	100	—	—	—	5-7	700	700
413	1971	.035	27-32	.017	5B	130	—	—	3600	3½-5	500	500N
413-1, 3	1972	.035	28-32	.015	5B	100	—	—	3600	3½-5	(700) 600	(700)N 600N
413-2	1972	.035	28½-32½	.018	5B	100	—	—	3600	3½-5	600 (700)	600N (700)
413-1	1973	.035	Electronic		5B	100	—	—	3600	3½-5	600 (700)	600N (700)N
413-2	1973	.035	Electronic		5B (2½B)	100	—	—	3600	3½-5	600 (700)	500N (700)N
413-3	1973	.035	Electronic		5B	100	—	—	3600	3½-5	600 (700)	600N (700)N
413-2	1974	.035	Electronic		2½B	100	—	—	3600	6-7½	700	700N
413-3 Heavy Duty	1975	.035	Electronic		5B	100	—	—	3600	6-7½	700	700N
413-3 Bus	1975	.035	Electronic		TDC	100	—	—	3600	6-7½	700	700N
413-3	1976	.035	Electronic		③	100	—	—	3600	5-7	③	③
413-3 HD	1977-78	.035	Electronic		5	100	—	—	—	5-7	700	700
440 Lt. Duty	1974	.035	Electronic		10B (5B)	100	—	—	—	5-7	700	700N
440 Heavy Duty	1974	.035	Electronic		7½B	100	—	—	—	5-7	700	700N
440	1975	.035	Electronic		8B	100	—	—	3800	5-7	700	700N
440	1976	.035	Electronic		③	100	—	—	—	5-7	③	③
440-3	1976	.035	Electronic		③	100	—	—	3800	5-7	③	③
440-1 HD	1977-78	.035	Electronic		8	100	—	—	—	8	700	700
478	1971-73	.035	28-32	.016	10B	140	—	—	3400	4-4¾	500	500N
549	1971-72	.035	28-32	.016	7B	140	—	—	3200	4-4¾	500	500N

① Uses taper seat plug without tube or gasket, torque to 10 ft. lbs.
② TDC with automatic.
③ See underhood specifications sticker.
④ 2 bbl Carb
⑤ 4 bbl Carb
LD—Light duty cycle
HD—Heavy duty cycle
Lt. Duty—Under 6,000 lbs. GVW.
Heavy Duty—Over 6,000 lbs. GVW.
Figures in parentheses are for California only.

Note: The underhood specifications sticker often reflects tune up specifications changes made in production. Sticker figures must be used if they disagree with those in this Chart.

GENERAL ENGINE SPECIFICATIONS

CU. IN. DISPLACE-MENT	YEAR	BORE & STROKE	FIRING ORDER	ESTIMATED H.P. @ RPM	ESTIMATED TORQUE @ RPM	COMPRESSION RATIO	CARBU-RETOR	VALVE LIFTER TYPE	NORMAL OIL PRESSURE
SIX CYLINDER									
198	1971	3.406 x 3.64	1-5-3-6-2-4	130 @ 4000	182 @ 1600	8.4	1V	Mech.	50
225	1971	3.40 x 4.125	1-5-3-6-2-4	140 @ 3900	215 @ 1600	8.4	1V	Mech.	50
225	1972-74	3.40 x 4.125	1-5-3-6-2-4	100 @ 3900	180 @ 1600	8.4	1V	Mech.	55
225	1975-76	3.40 x 4.125	1-5-3-6-2-4	90 @ 3600	170 @ 1600	8.4	1V	Mech.	55
225	1977-78	3.40 x 4.125	1-5-3-6-2-4	100 @ 3600	170 @ 1600	8.4 x 1	1V	Std.	35-65
225	1977-78	3.40 x 4.125	1-5-3-6-2-4	110 @ 3600	180 @ 2000	8.4 x 1	2V	Std.	35-65
EIGHT CYLINDER									
318-3	1971	3.91 x 3.312	1-8-4-3-6-5-7-2	202 @ 3900	288 @ 2400	7.5	2V	Hyd.	70
318	1971	3.91 x 3.312	1-8-4-3-6-5-7-2	210 @ 2800	318 @ 2800	8.5	2V	Hyd.	60

FIRING ORDER AND ROTATION

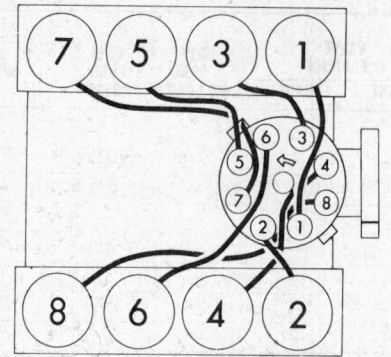

383, 400, 440, V8 engines

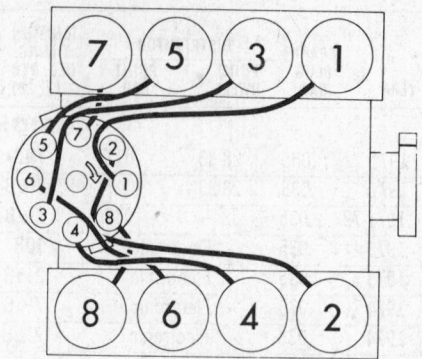

318, 340, 360, V8 engines

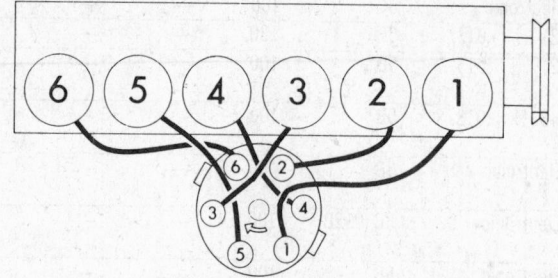

6 cylinder engine

GENERAL ENGINE SPECIFICATIONS

CU. IN. DISPLACE-MENT	YEAR	BORE & STROKE	FIRING ORDER	ESTIMATED H.P. @ RPM	ESTIMATED TORQUE @ RPM	COMPRESSION RATIO	CARBU-RETOR	VALVE LIFTER TYPE	NORMAL OIL PRESSURE
EIGHT CYLINDER									
318-2	1971	3.91 x 3.312	1-8-4-3-6-5-7-2	212 @ 4000	322 @ 2800	8.0	2V	Hyd.	70
318	1972-74	3.91 x 3.312	1-8-4-3-6-5-7-2	150 @ 3600	260 @ 2000	7.8/8.6	2V	Hyd.	30-80
318	1975	3.91 x 3.312	1-8-4-3-6-5-7-2	150 @ 4000	255 @ 1600	7.8/8.6	2V	Hyd.	30-80
318	1976	3.91 x 3.312	1-8-4-3-6-5-7-2	150 @ 4000	255 @ 1600	8.4	2V	Hyd.	30-80
318-3	1976	3.91 x 3.312	1-8-4-3-6-5-7-2	150 @ 4000	255 @ 1600	7.8	2V	Hyd.	30-80
318	1977-78	3.91 x 3.31	1-8-4-3-6-5-7-2	145 @ 4000	245 @ 1600	8.5 x 1	2V	Hyd.	35-65
318-3	1977-78	3.91 x 3.31	1-8-4-3-6-5-7-2	150 @ 4200	250 @ 1600	8.5 x 1	2V	Hyd.	35-65
360	1971-75	4.00 x 3.58	1-8-4-3-6-5-7-2	170 @ 4000	285 @ 2400	8.4	2V	Hyd.	30-80
360	1976	4.00 x 3.58	1-8-4-3-6-5-7-2	170 @ 4000	285 @ 2400	8.7	2V	Hyd.	30-80
360-3	1976	4.00 x 3.58	1-8-4-3-6-5-7-2	170 @ 4000	285 @ 2400	7.9	2V	Hyd.	30-80
360	1977-78	4.00 x 3.58	1-8-4-3-6-5-7-2	155 @ 3600	275 @ 2000	8.4 x 1	2V	Hyd.	35-65
360-3	1977-78	4.00 x 3.58	1-8-4-3-6-5-7-2	165 @ 3600	285 @ 2000	8.4 x 1	4V	Hyd.	35-65
361-2	1971-75	4.125 x 3.375	1-8-4-3-6-4-7-2	186 @ 4000	300 @ 2400	7.5	2V	Hyd.	70
361-3	1971-75	4.125 x 3.375	1-8-4-3-6-5-7-2	194 @ 3600	310 @ 2400	7.5	2V	Hyd.	70
361-4	1971-75	4.125 x 3.375	1-8-4-3-6-5-7-2	204 @ 3600	335 @ 2400	7.5	2V	Hyd.	70
361	1972-75	4.125 x 3.375	1-8-4-3-6-5-7-2	155 @ 3600	295 @ 2000	7.5	2V	Hyd.	70
361-4	1977-78	4.125 x 3.375	1-8-4-3-6-5-7-2	150 @ 3600	290 @ 2000	7.5 x 1	2V	Hyd.	35-75
383	1971	4.25 x 3.38	1-8-4-3-6-5-7-2	258 @ 4400	375 @ 2800	9.2	2V	Hyd.	45-65
400	1971-74	4.34 x 3.38	1-8-4-3-6-5-7-2	180 @ 3600	315 @ 2400	8.2	2V, 4V	Hyd.	30-80
400	1976	4.34 x 3.38	1-8-4-3-6-5-7-2	180 @ 3600	315 @ 2400	8.2	2V	Hyd.	30-80
400	1977-78	4.34 x 3.38	1-8-4-3-6-5-7-2	195 @ 3600	305 @ 3200	8.2 x 1	4V	Hyd.	50-75
413-2	1971	4.188 x 3.75	1-8-4-3-6-5-7-2	220 @ 3600	386 @ 2200	7.5	2V	Hyd.	70
413-3	1971	4.188 x 3.75	1-8-4-3-6-5-7-2	283 @ 3600	407 @ 2000	7.5	4V	Hyd.	70
413	1972-78	4.188 x 3.75	1-8-4-3-6-5-7-2	180 @ 3200	334 @ 2000	7.5	2V	Hyd.	70
440	1973-76	4.32 x 3.75	1-8-4-3-6-5-7-2	235 @ 4000	340 @ 2400	8.2	4V	Hyd.	30-80
440-3	1976	4.32 x 3.75	1-8-4-3-6-5-7-2	235 @ 4000	340 @ 2400	7.5	4V	Hyd.	30-80
440	1977-78	4.32 x 3.75	1-8-4-3-6-5-7-2	195 @ 3600	320 @ 2000	8.2 x 1	4V	Hyd.	50-75
478	1971-73	4.50 x 3.75	1-8-7-3-6-5-4-2	206 @ 3400	403 @ 1800	8.2	2V	Hyd.	50-70
549	1971-72	4.50 x 4.312	1-8-7-3-6-5-4-2	230 @ 2300	505 @ 2000	7.6	4V	Hyd.	50-70

CRANKSHAFT BEARING JOURNAL SPECIFICATIONS

CU. IN. DISPLACEMENT	YEAR	MAIN BEARING JOURNALS				CONNECTING ROD BEARING JOURNALS		
		JOURNAL DIAMETER	OIL CLEARANCE	SHAFT END PLAY	THRUST ON NO.	JOURNAL DIAMETER	OIL CLEARANCE	END PLAY
SIX CYLINDER								
198	1971	2.7495-2.7505	.0005-.0015	.004-.008	3	2.1865-2.1875	.0005-.0015	.006-.025
225	1971-78	2.7495-2.7505	.0005-.0020	.002-.009	3	2.1865-2.1875	.0005-.0020	.006-.025
EIGHT CYLINDER								
318	1971-78	2.4495-2.5005	.0005-.0020	.002-.009	3	2.124-2.125	.0005-.0020	.006-.014
360	1971-78	2.8095-2.8105	.0005-.0020	.002-.009	3	2.124-2.125	.0005-.0020	.006-.014
361	1971-78	2.6245-2.6255	.0015-.0025	.002-.009	3	2.374-2.375	.001-.002	.009-.017
383	1971	2.6245-2.6255	.0005-.0015	.002-.007	3	2.374-2.375	.0005-.0020	.009-.017
400	1971-78	2.6245-2.6255	.0005-.0020	.002-.009	3	2.374-2.375	.0005-.0020	.009-.017
413	1971-78	2.7495-2.7505	.002-.0022	.002-.009	3	2.374-2.375	.0005-.0025	.009-.017
440	1973-78	2.7495-2.7505	.002-.0022	.002-.009	3	2.374-2.375	.0005-.0020	.009-.017
478	1971-73	3.123-3.124	.0014-.0044	.004-.009	3	2.623-2.624	.0017-.0042	.010-.018
549	1971-72	3.123-3.124	.0014-.0044	.004-.009	3	2.623-2.624	.0017-.0042	.010-.018

VALVE SPECIFICATIONS

CU. IN. DISPLACEMENT	YEAR	LASH (HOT) INCHES		ANGLE DEGREE		STEM DIA. INCHES		STEM CLEARANCE		VALVE LIFT INCHES	VALVE SPRING LBS. @ INCHES		FREE LENGTH INCHES
		INT.	EXH.	FACE	SEAT	INT.	EXH.	INTAKE	EXHAUST		OPEN	CLOSED	
SIX CYLINDER													
198	1971	.012	.024	①	45	.372	.371	.001-.003	.002-.004	.395	144 @ 1⁵/₁₆	53 @ 1¹¹/₁₆	1.92
225	1971-72	.012	.024	①	45	.372	.371	.001-.003	.002-.004	③	144 @ 1⁵/₁₆	53 @ 1¹¹/₁₆	1.92
225	1973-74	.012	.024	①	45	.372	.371	.001-.003	.002-.004	⑰	144 @ 1⁵/₁₆	53 @ 1¹¹/₁₆	1.92
225	1975-76	.012	.024	①	45	.372	.371	.001-.003	.002-.004	⑱	144 @ 1⁵/₁₆	53 @ 1¹¹/₁₆	1.92
EIGHT CYLINDER													
318-1	1971-72	no adj.	no adj.	①	45	.372	.371	.001-.003	.002-.004	㉑	177 @ 1⁵/₁₆	83 @ 1¹¹/₁₆	2.00
318-3	1971-72	no adj.	no adj.	45	45	.372	.371	.001-.003	.002-.004	㉑	177 @ 1⁵/₁₆	83 @ 1¹¹/₁₆⑳	2.00㉒
318-1	1973-74	no adj.	no adj.	①	45	.372	.371	.001-.003	.002-.004	㉑	177 @ 1⁵/₁₆	83 @ 1¹¹/₁₆	2.00
318-3	1973-74	no adj.	no adj.	45	45	.372	.371	.001-.003	.002-.004	㉑	177 @ 1⁵/₁₆	83 @ 1¹¹/₁₆	2.00
318-1	1975-76	no adj.	no adj.	45	45	.372	.371	.001-.003	.002-.004	㉑	177 @ 1⁵/₁₆㉓	83 @ 1¹¹/₁₆㉔	2.00㉕
318-3	1975-78	no adj.	no adj.	45	45	.372	.371	.001-.003	.002-.004	㉑	185 @ 1¼㉖	93 @ 1²¹/₃₂	2.00
360	1971-72	no adj.	no adj.	①	45	.373	.372	.001-.003	.002-.004	⑭	177 @ 1⁵/₁₆	83 @ 1¹¹/₁₆	2.00
360	1973-78	no adj.	no adj.	①	45	.372	.371	.001-.003	.002-.004	㉗	177 @ 1⁵/₁₆㊱	83 @ 1¹¹/₁₆㉟	2.00
361	1971-72	no adj.	no adj.	45	45	.372	.433	.001-.003	.003-.005	.360	⑤	⑥	2.31⑫
361	1973-78	no adj.	no adj.	45	45	.372	.433	.001-.003	.003-.005	.360	180 @ 1¹⁵/₃₂㉘	80 @ 1⁵⁵/₆₄㉙	2.31⑩
383	1971	no adj.	no adj.	45	45	.372	.371	.001-.003	.002-.004	⑦	200 @ 1⁷/₁₆	125 @ 1⁵⁵/₆₄	2.63
400	1971-78	no adj.	no adj.	45	45	.372	.372	.002-.003	.002-.003	.430	200 @ 1⁷/₁₆	125 @ 1⁵⁵/₆₄	2.63
413	1971-72	no adj.	no adj.	45	45	.372	.433	.001-.003	.003-.005	.360	⑤	⑥	2.31⑫
413	1973-78	no adj.	no adj.	45	45	.372	.433	.001-.003	.003-.005	.360	180 @ 1¹⁵/₃₂㉘	80 @ 1⁵⁵/₆₄㉙	2.31⑩
440	1973-74	no adj.	no adj.	45	45	.372	.372	.002-.003	.002-.003	.464	105 @ 1⁵⁵/₆₄	105 @ 1⁵⁵/₆₄㉛	2.23
440	1975-78	no adj.	no adj.	45	45	.372	.372	.002-.003	.002-.003	.425⑦	200 @ 1⁷/₁₆㉜	125 @ 1⁷/₈㉜	2.63
478	1971-73	0 + 1turn		15⑧	14⑨	.434	.434	.0015-.004	.0025-.005	—	83 @ 1¹⁷/₃₂	⑩	2.28⑬
549	1971-72	0 + 1turn		15⑧	14⑨	.434	.434	.0015-.004	.0025-.004	—	83 @ 1¹⁷/₃₂	⑩	2.28⑬

① Intake 45, Exhaust 43
③ Intake .397, Exhaust .392
⑤ Intake 180 @ 1¹⁵/₃₂, Exhaust Outer 115 @ 1⁸/₈, Inner 49 @ 1¹¹/₆₄
⑦ Intake .425, Exhaust .435
⑧ Intake, Exhaust 45
⑨ Intake, Exhaust 44
⑩ Inner, Outer 116.5
⑫ Intake, Exhaust (Outer) 2.125, (Inner) 1.875
⑬ Inner, Outer 2.562
⑭ Intake .410, Exhaust .412
⑰ Intake .406, Exhaust .414
⑱ Intake .394, Exhaust .390
⑳ Exhaust 85 @ 1³⁵/₆₄
㉑ Intake, .372, Exhaust .400
㉒ Intake, Exhaust 1.88

㉓ 185 @ 1⁵/₆₄
㉔ Exhaust 85 @ 1³¹/₆₄
㉕ Intake, Exhaust 1.81
㉖ Exhaust 192 @ 1⁹/₃₂
㉗ Intake, .410, Exhaust .400
㉘ Intake, Exhaust 175 @ 1²¹/₆₄
㉙ Intake, Exhaust 85 @ 1¾
㉚ Intake, Exhaust 2.13
㉛ Intake, Exhaust 246 @ 1²³/₆₄
㉜ Intake, Exhaust 208 @ 1⁵/₁₆
㉞ 1977-78 184 @ 1⁵/₁₆
㉟ 1977-78 88 @ 1¹¹/₁₆
㊱ 1977-78 200 @ 1¹¹/₁₆
㊲ 1977-78 118 @ 1²¹/₃₂

TORQUE SPECIFICATIONS

CU. IN. DISPLACE-MENT	YEAR	CYLINDER HEAD BOLTS (FT. LBS.)	ROD BEARING BOLTS (FT. LBS.)	MAIN BEARING BOLTS (FT. LBS.)	CRANKSHAFT BALANCER BOLT (FT. LBS.)	FLYWHEEL TO CRANKSHAFT BOLTS (FT. LBS.)	MANIFOLD INTAKE (FT. LBS.)	EXHAUST
				SIX CYLINDER				
198	1971	65	45	85	Press	55	10	10
225	1971-78	65①	45	85	Press	55	10	10
				EIGHT CYLINDER				
318	1971	85	45	85	135	60	30	25
318	1972-78	95	45	85	100	55	35	20
360	1971-78	95	45	85	100	55	35	20
361	1971-78	70	45	85	135	55	45	30
383	1971	70	45	85	135	60	50	30
400	1971-78	70	45	85	135	55	40	30
413	1971-78	70	45	85	135	55	45	30
440	1973-78	70	45	85	135	55	45	30
478	1971-73	90-100	60-70	100-110	—	90-100	—	30
549	1971-72	90-100	60-70	100-110	—	90-100	—	30

① 70 starting 1973

WHEEL ALIGNMENT SPECIFICATIONS

YEAR	MODEL	CASTER (Deg.)*	CAMBER (Deg.)	TOE-IN (in.)	KING PIN INCLINATION (Deg.)
1971	D100	1½ to 2½	1½	0-⅛	4
1972-78	D100	0 to +½	0 to +½	1/16-⅛	—
1971-78	AW, PW, W100	3	1½	0-⅛	7½
1971-78	B, PB, CB, MB100, 200, 300	½⑤	+¼	1/16-⅛	—
1971	D200	1½	2	0-⅛	4
1971	D200 crew cab	1½	1½	0-⅛	7
1972-78	D200	0 to +½	0 to +½	1/16-⅛	—
1971-72	P200	1½	1½	0-⅛	4
1971-78	W200⑥	3	1½	0-⅛	7½
1971	D300	1½	2	0-⅛	7
1973-78	D300	0 to +½	0 to +½	1/16-⅛	—
1971-73	W300	1½	3½	0-⅛	8
1974-78	W300⑥	3	½	0-⅛	8½
1971-72	M375	1½ to 2½	2	0-⅛	7
1971	D400, P400	1½ to 2½	2	0-⅛	7
1972-73	P400	1④	2	0-⅛	7
1971-78	D500, S500②	½ to 2	2	0-⅛	7
1971	S500	0 to +½	1	0-⅛	5½
1971	W500	2½	¾	0-⅛	8
1971-72	C600②	½ to 2	2	0-⅛	7
1971-72	C600③	½ to 1	2	0-⅛	7
1971-78	D600②	½ to 2	2	0-⅛	7
1971-78	S600	0 to +1	1	0-⅛	5½
1972-78	W600	2½	¾	0-⅛	8
1971-72	C700	1④	1	0-⅛	5½
1971-78	D700②	½ to 2	2	0-⅛	7
1971-78	D700③	0 to +1	1	0-⅛	5½
1975-78	S700	0 to +1	1	0-⅛	5½
1971-78	800 All	0-1④	1	0-⅛	5½
1971-76	850 All	1③	1	0-⅛	5½
1971-76	900 All	0 to +1④	1	0-⅛	5½
1971-76	1000 All	0 to +1④	1	0-⅛	5½

* No load
① Not used
② 5000 lb. axle
③ 7000 lb. axle
④ With power steering 4-5 degrees
⑤ 2¼ with power steering
⑥ Use W300 figures for W200 with Spicer 60 or 70F front axle

Truck Model and Engine Application

<table>
<tr><td colspan="3">

GASOLINE ENGINES

</td></tr>
<tr><th>ENGINE</th><th>YEARS AVAILABLE</th><th>ENGINE MAKE</th></tr>
<tr><td>6-198</td><td>1971</td><td>Own</td></tr>
<tr><td>6-225</td><td>1971-78</td><td>Own</td></tr>
<tr><td>V8-318</td><td>1971-78</td><td>Own</td></tr>
<tr><td>V8-360</td><td>1971-78</td><td>Own</td></tr>
<tr><td>V8-361</td><td>1971-78</td><td>Own</td></tr>
<tr><td>V8-383</td><td>1971</td><td>Own</td></tr>
<tr><td>V8-400</td><td>1971-78</td><td>Own</td></tr>
<tr><td>V8-413</td><td>1971-78</td><td>Own</td></tr>
<tr><td>V8-440</td><td>1973-78</td><td>Own</td></tr>
<tr><td>V8-478</td><td>1971-73</td><td>IH</td></tr>
<tr><td>V8-549</td><td>1971-72</td><td>IH</td></tr>
</table>

<table>
<tr><td colspan="2">

DIESEL ENGINES

</td></tr>
<tr><th>ENGINE</th><th>ENGINE MAKE</th></tr>
<tr><td>6-243</td><td>Mitsubishi</td></tr>
<tr><td>6-71N</td><td>Detroit</td></tr>
<tr><td>8V-71N</td><td>Detroit</td></tr>
<tr><td>8V-71NE</td><td>Detroit</td></tr>
<tr><td>NH230</td><td>Cummins</td></tr>
<tr><td>NHC250</td><td>Cummins</td></tr>
<tr><td>NTC280</td><td>Cummins</td></tr>
<tr><td>NTC335</td><td>Cummins</td></tr>
<tr><td>V8-210</td><td>Cummins</td></tr>
<tr><td>V903</td><td>Cummins</td></tr>
</table>

NOTE: Ramcharger and Trailduster service procedures are the same as those for conventional trucks, except where specific and separate procedures are given.

Distributor

Distributor—All Types

Removal

1. Disconnect primary lead wire at coil. On electronic ignition, disconnect the distributor lead wire at the connector.
2. Disconnect vacuum hose at distributor.
3. On the Holley distributor, disconnect tachometer drive and governor inlet and outlet lines.
4. Unfasten distributor cap retaining clips and remove distributor cap.
5. Scribe a line on the distributor housing and engine block to indicate positioning of the rotor and housing.
6. Remove distributor hold-down clamp or arm screw.
7. Carefully lift out distributor assembly.

NOTE: Do not disturb engine position.

Installation

1. If the crankshaft has not been rotated, insert distributor into block with the rotor and body aligned to the previously scribed marks. Make sure O-ring seal is in groove of shank. *NOTE: Distributors on 6 cylinder engines have the drive gear mounted on the bottom of the distributor shaft and a slight rotation will occur when installing. Allow for this rotation when aligning rotor with scribed line on housing.*
2. If engine has been cranked while distributor was removed, it will be necessary to correctly time

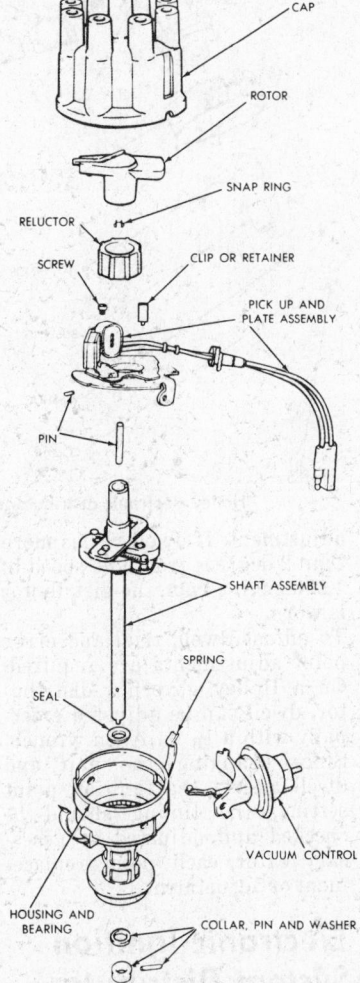

V8 electronic ignition distributor (© Chrysler Corp.)

the distributor with the camshaft. This is done by rotating the crankshaft until No. 1 piston is at top dead center of compression stroke. With rotor in No. 1 cylinder firing position with respect to the distributor cap, insert distributor into engine.

3. Connect primary lead or electronic ignition lead wire.
4. Install distributor cap and check that all high tension leads are se-

curely in position.
5. On the Holley distributor, connect governor lines and tachometer drive cable.
6. Set ignition timing.
7. Tighten distributor arm or clamp screw.
8. Connect vacuum advance line.

Breaker Points and Condenser Replacement, Dwell Angle Adjustment

1. Remove the distributor cap and rotor and inspect them for burned or corroded conditions.
2. Loosen the terminal screw nut and remove the primary and con-

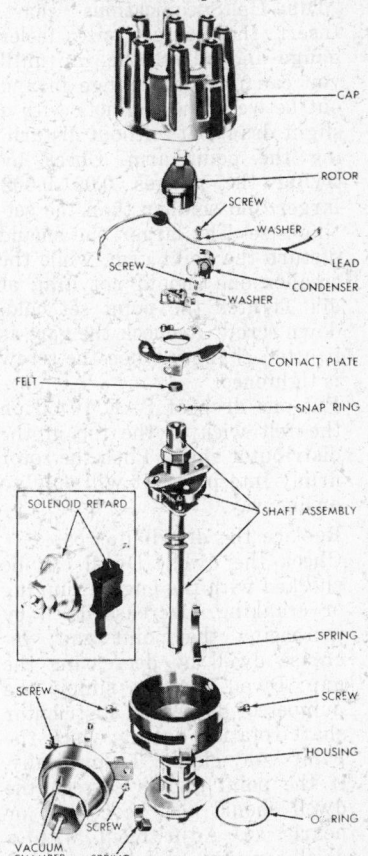

Typical point type distributor-V8 engines (© Chrysler Corp.)

denser lead wires. Remove the stationary contact lockscrew and remove the point set. Be very careful not to drop any of these screws inside the distributor. If the hold-down screw is lost, it must be replaced with one that is no longer than the original to avoid interference with the distributor advance mechanism workings. Remove the point set.

3. If the points are to be reused, clean them with a few strokes of a fine file.
4. Remove the condenser retaining screw and the condenser.
5. Rub a matchhead size dab of high melting point grease on the cam lobes, and install the new condenser.
6. Replace the point set and leave the screw slightly loose. Replace the two wire terminals, making sure that the wires don't interfere with anything.
7. Check that the contacts meet squarely. If they don't, bend the tab supporting the fixed contact.
8. Turn the engine until a high point on the cam which opens the points contacts the rubbing block on the point arm.
9. There is a screwdriver slot near the contacts. Insert a screwdriver and lever the points open or closed until they appear to be at about the gap specified in the "Tune-Up Specifications" chart.
10. Insert the correct size feeler gauge and adjust the gap until you can push the gauge in and out between the contacts with a slight drag, but without disturbing the point arm. Check by trying the gauges 0.001-0.002 larger and smaller than the setting size. The larger one should disturb the point arm, while the smaller one should not drag at all. Tighten the point set hold-down screw. Recheck the gap, as it often changes when the screw is tightened.
11. Put one drop of SAE 10 oil on the felt wick in the top of the distributor shaft. Push the rotor firmly into place. It will only go on one way.
12. Replace the distributor cap.
13. Check the dwell. Dwell can be checked with the engine running or cranking. Decrease dwell by increasing the point gap; increase dwell by decreasing the gap. Dwell angle is simply the number of degrees of distributor shaft rotation during which the points stay closed. Theoretically, if the point gap is correct, the dwell should also be correct or nearly so. Adjustment with a dwell meter produces more exact, consistent results than a feeler gauge since it is a dynamic

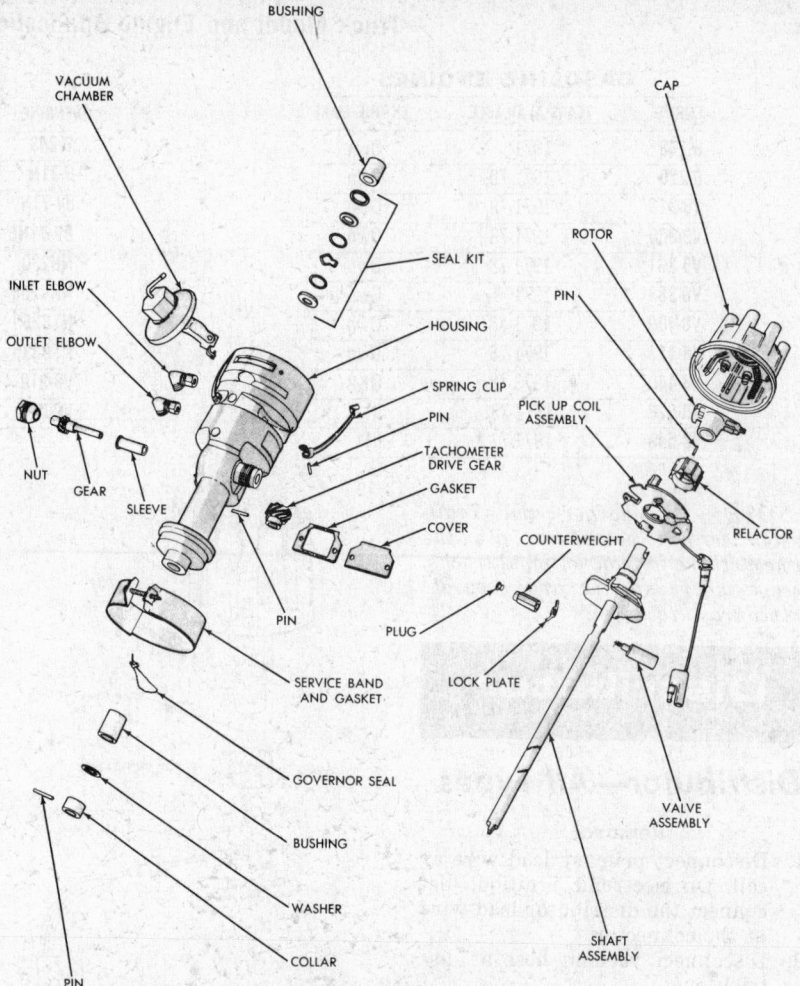

Holley electronic distributor-exploded view (© Chrysler Corp.)

adjustment. If dwell varies more than 2 degrees from idle speed to 1,500 engine rpm, the distributor is worn.

14. To adjust dwell, trial and error point adjustments are required. On a Holley governor distributor, dwell can be adjusted externally with a ⅛ in. Allen wrench.
15. Since changing the gap and dwell affects the ignition point setting, the timing should be checked and adjusted as necessary after each point replacement or adjustment.

Electronic Ignition System Distributor

No internal distributor maintenance is required with this system; it does not use contact points or a condenser. This system is easily identified by a double wire lead from the distributor and a control unit in the engine compartment. It is on all engines beginning 1973; there is also an electronic Holley governor distributor.

NOTE: The dwell reading that is obtained with a dwell meter is of no significance in servicing the ignition

system and since dwell is non-adjustable, no changes should be attempted.

Electronic Ignition Distributor Air Gap Adjustment

This adjustment is not required at regular intervals. It is not a normal tune-up service.

1. Release the spring clips and remove the distributor cap. Pull off the rotor.
2. Align one reluctor tooth with the pick-up coil tooth by turning the engine. The reluctor is the six or eight-toothed ring around the distributor shaft.
3. Insert an 0.008 in. *nonmagnetic* (brass) feeler gauge between the reluctor tooth and the pick-up coil tooth.
4. Loosen the hold-down screw and adjust the gap using the screwdriver slot in the mounting plate. Contact should be made between the reluctor tooth, the feeler gauge, and the pick-up coil tooth.
5. Tighten the hold-down screw.
6. Remove the feeler gauge. No force should be required.
7. Check the gap with a 0.010 in. *nonmagnetic* feeler gauge. It

should not fit; don't force it into the gap.

8. Turn the distributor shaft and apply vacuum to the vacuum advance unit. If it is adjusted properly and nothing is bent, the pick-up coil tooth will not hit the reluctor teeth.

Ignition Timing Marks

On all engines the timing plate is located on the timing case (front) cover and the timing mark is on the crankshaft pulley damper. On all models the ignition is timed to the No. 1 cylinder spark plug. Always remove and plug the vacuum advance line when setting ignition timing.

Holley Governor Distributor

Dwell setting may be adjusted externally with the use of a ⅛" Allen wrench. See Specifications for correct dwell angle. Disconnect and plug the vacuum advance line and connect dwell meter. Insert wrench while engine is running and turn adjusting screw until correct dwell angle reading is obtained.

Alternator

Reference
For voltage regulator circuit tests and for alternator off-the-vehicle service, see The Electrical Section.

Removal
1. Disconnect battery ground cable at the negative terminal.
2. Disconnect alternator output "BAT" and field "FLD" leads and disconnect ground wire.
3. Remove mounting bolts and remove alternator.

Installation
1. Install alternator and adjust drive belt.
2. Connect output "BAT" and field "FLD" leads and connect ground wire.
3. Connect battery ground cable.
4. Start engine, and observe alternator operation.
5. Test current output and adjust regulator voltage setting, if necessary. See Electrical Section.
   NOTE: Late models use a non-adjustable sealed electronic regulator.

Adjustable Voltage Regulator

Removal and Installation
1. Disconnect the battery ground cable at battery negative terminal.

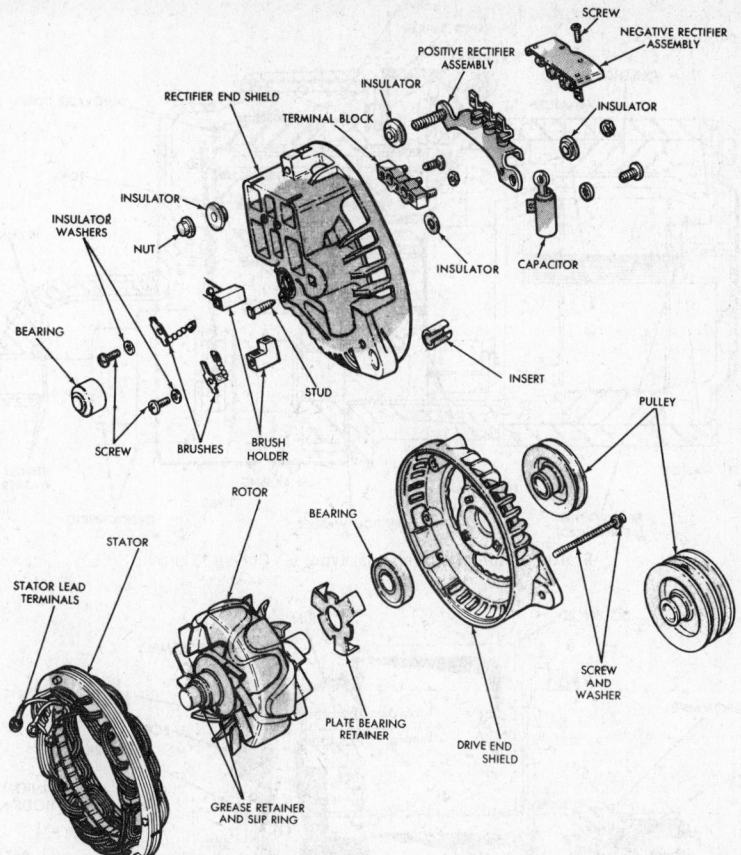

Exploded view of alternator (© Chrysler Corp.)

2. Disconnect the field lead from alternator and ignition lead from regulator terminals.
3. Remove three regulator assembly mounting screws and remove regulator.
4. Installation is the reverse of the above procedure.

Adjustment
1. Turn the adjustment screw in the cover to get the correct voltage. This is 13.5-14.5 Volts, measured at 70°F. There is a stop in the regulator cover to limit the voltage adjustment.
2. If Step 1 didn't produce the correct voltage, remove the cover and check the upper contact gap. It should be .010-.018 in. Adjust by bending the upper contact bracket. Make sure the contacts are in alignment.
3. If voltage still can't be adjusted correctly, adjust the air gap by bending the fixed contact bracket above the winding. If the voltage difference is more than .7 Volt, bend the bracket down. If it is less than .2 Volt, bend the bracket up.
4. When replacing the cover, make sure that the rubber grommet is centered over the adjusting screw.
5. Replace the regulator if voltage cannot be brought within specifications.

Electronic Voltage Regulator

Removal and Installation
1. Release the spring clips and pull off the regulator wiring plug.
2. Unbolt and remove the regulator.
3. Installation is the reverse of removal. Be sure that the spring clips engage the wiring plug.

Regulator Fusible Wire Replacement

Excessive current loads through the two fusible wires in the adjustable regulator will cause them to melt and break the circuit. Remove what is left of the old wire and solder new fusible wire to terminals, using only resin core solder. See the Electrical Section for service and testing of alternator systems.

Starter

Reference
For starter motor overhaul procedures see the Electrical Section.

Starter Removal and Installation
1. Disconnect the battery ground cable.
2. Remove the cable at the starter.
3. If the solenoid is mounted on the

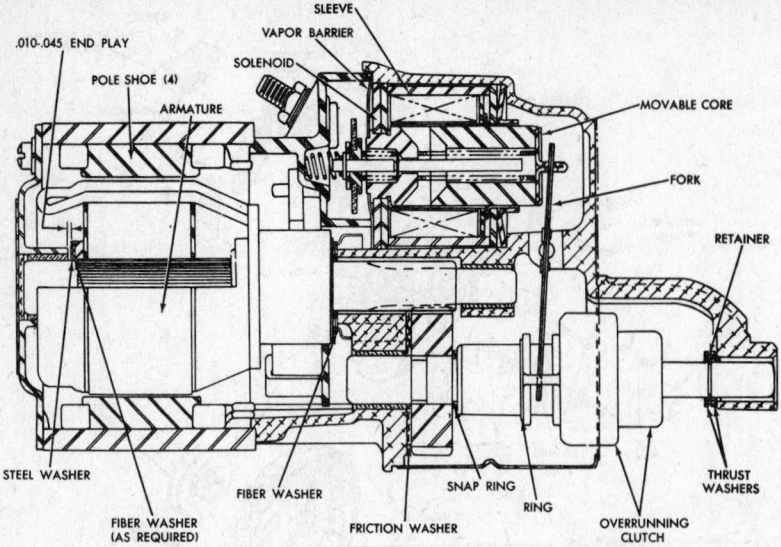

Starter motor (reduction gear type) (© Chrysler Corp.)

Labels: .010-.045 END PLAY, POLE SHOE (4), ARMATURE, SLEEVE, VAPOR BARRIER, SOLENOID, MOVABLE CORE, FORK, RETAINER, STEEL WASHER, FIBER WASHER, FIBER WASHER (AS REQUIRED), FRICTION WASHER, SNAP RING, RING, OVERRUNNING CLUTCH, THRUST WASHERS

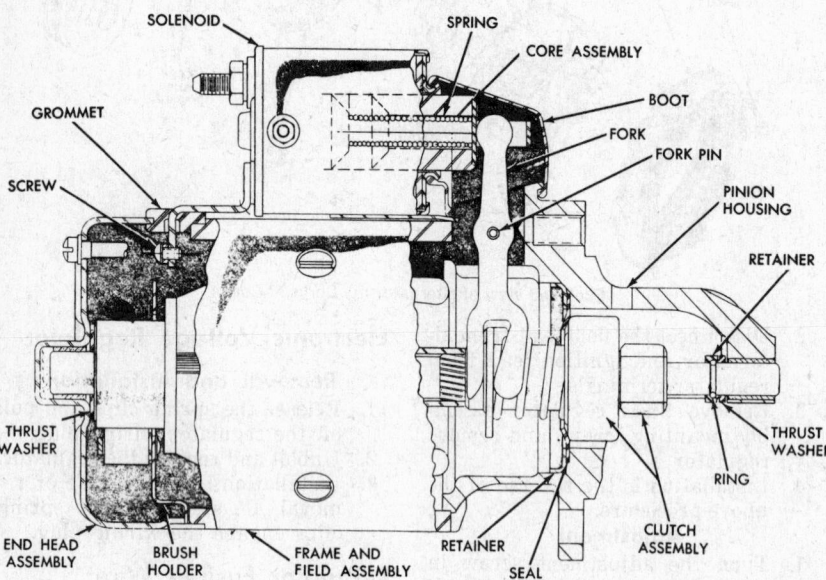

Starter motor (direct drive type) (© Chrysler Corp.)

Labels: SOLENOID, SPRING, CORE ASSEMBLY, BOOT, FORK, FORK PIN, PINION HOUSING, RETAINER, GROMMET, SCREW, THRUST WASHER, RING, THRUST WASHER, END HEAD ASSEMBLY, BRUSH HOLDER, FRAME AND FIELD ASSEMBLY, RETAINER, SEAL, CLUTCH ASSEMBLY

starter, disconnect the wires at the solenoid terminals. On Diesel engine starters, record the wiring harness color coding at the solenoid to insure proper installation.

4. Remove the starter to flywheel housing mounting bolts. Remove automatic transmission oil cooler tube bracket off the stud if necessary. Remove the starter and removable seal if so equipped.

5. Before installing the starter, make sure the starter and flywheel housing mounting surfaces are free of dirt and oil. These surfaces must be clean.

6. Install the starter to flywheel housing removable seal if so equipped.

7. Position the starter to the flywheel housing and, if necessary, install the automatic transmission oil cooler bracket. Install

mounting bolts. Tighten securely. *NOTE: When tightening the mounting bolt and nut on the starter hold the starter away from the engine for correct alignment.*

8. If the solenoid is mounted on the starter, connect the wires to the solenoid terminals.

9. Connect the cable to the starter terminal.

10. Connect the battery ground cable and test operation of the starter for proper engine cranking.

Brakes

Reference

For master, wheel, slave cylinder, and disc brake caliper overhaul, brake shoe and pad replacement and service procedures, bleeding of the hydraulic system, dual master cylin-

der and vacuum-hydraulic booster system, refer to the Hydraulic Brake Section in the General Repair Section.

Master Cylinder

Firewall Mounted R & R

1. From within the cab, disconnect push rod from brake pedal.
2. Disconnect hydraulic line from master cylinder.
3. Remove four nuts and washers from master cylinder mounting studs and remove master cylinder assembly.
4. Installation is the reverse of the above procedure.
5. Bleed hydraulic system.

Tandem Type Dual Cylinder R & R

The tandem master cylinder is of the compensating type with reservoirs cast integrally on the dash mounted type and with removable reservoirs on the booster mounted type. Both consist of a front and rear piston (in tandem) with two outlets, each containing a residual pressure valve and spring.

The dual hydraulic brake system is basically a safety system in which the front brakes are actuated by one master cylinder and the rear brakes by the other. A hydraulic safety switch closes a warning light circuit whenever unequal pressure exists between the two systems whenever a pressure failure occurs in either.

The procedure for removing and installing dual master cylinders is the same as that for single master cylinders as described above.

Booster Mounted Type

Removal

1. Disconnect both hydraulic lines at master cylinder outlets.
2. Remove capscrew holding lower support strap to cylinder.
3. Remove master cylinder to booster mounting nuts and lift cylinder from vehicle.

Installation

1. Position seal ring in groove on master cylinder body mounting flange, then install unit on power booster. Tighten mounting nuts to 18-20 ft. lbs. torque.
2. Install capscrew holding lower support strap to cylinder.
3. Connect and tighten both hydraulic brake lines at cylinder outlet.
4. Bleed entire brake system.

Frame Mounted Type

Removal and Installation

1. Remove bolts and screws that attach the airscoop to the frame.

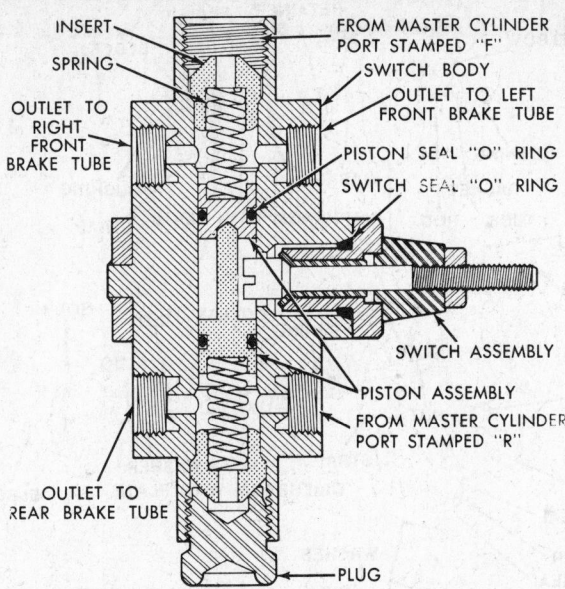

Dual brake system safety switch (© Chrysler Corp.)

INSERT
SPRING
OUTLET TO RIGHT FRONT BRAKE TUBE
FROM MASTER CYLINDER PORT STAMPED "F"
SWITCH BODY
OUTLET TO LEFT FRONT BRAKE TUBE
PISTON SEAL "O" RING
SWITCH SEAL "O" RING
SWITCH ASSEMBLY
PISTON ASSEMBLY
FROM MASTER CYLINDER PORT STAMPED "R"
OUTLET TO REAR BRAKE TUBE
PLUG

2. Slide scoop out (toward front) from under vehicle.
3. Disconnect brake tube at rear of master cylinder.
4. Disconnect stop light switch wires.
5. Remove bolts and nuts that attach master cylinder to frame bracket.
6. Slide unit toward rear to clear push rod and remove from under vehicle.
7. Installation is the reverse of the above procedure.

Vans and Wagons

Removal

1. Disconnect the front and rear tubes from the master cylinder.
2. Install plugs in the reservoir outlets.
3. Remove mounting nuts.
4. Disconnect the brake pedal push rod from pedal.
5. Slide the master cylinder straight out.

Installation

1. Put the cylinder in position, aligning the push rod with the cowl panel opening or power brake cylinder piston.
2. Slide the cylinder on over the mounting studs. Install the attaching nuts, tightening to 200 inch-pounds.
3. Connect and tighten brake tubes.
4. Bleed brake system.

Wheel Cylinder

Removal

1. Remove the wheel, drum, and brake shoes.
2. Disconnect the brake hose from the brake tube at the frame bracket for front wheels, and disconnect the brake tube from the wheel cylinder for the rear wheels.
3. Remove the wheel cylinder attaching bolts and slide the cylinder from the brake support plate.
4. Overhaul or replace the cylinder as required.

Installation

1. Position the wheel cylinder on the brake support plate and install the cylinder attaching bolts.
2. On the front wheel cylinders, tighten the brake hose to the cylinder before attaching the brake tube to the hose at the frame location. Tighten the attaching bolts.
3. With the rear wheel cylinder loose on the brake support plate, connect the brake tube to the cylinder and then, tighten the attaching bolts and the brake tube to the wheel cylinder.
4. Install the brake shoes, drum, and wheel.
5. Bleed the hydraulic system.

Disc Brakes

Disc brake removal, installation, and overhaul procedures will be found in the general Repair Section under Disc Brakes.

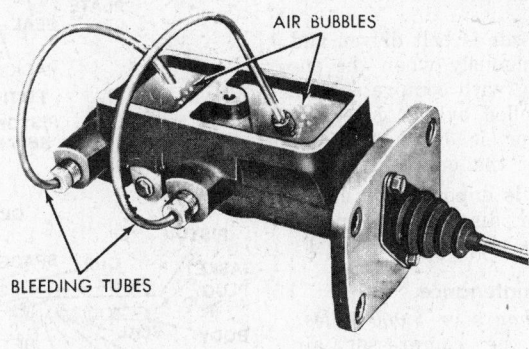

AIR BUBBLES
BLEEDING TUBES

Bleeding dual master cylinder (© Chrysler Corp.)

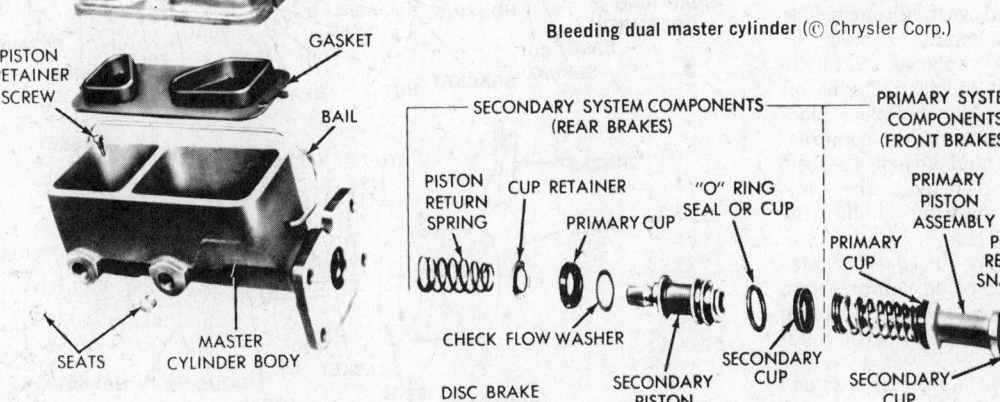

COVER
GASKET
PISTON RETAINER SCREW
BAIL
SEATS
MASTER CYLINDER BODY

SECONDARY SYSTEM COMPONENTS (REAR BRAKES)
PRIMARY SYSTEM COMPONENTS (FRONT BRAKES)

PISTON RETURN SPRING
CUP RETAINER
PRIMARY CUP
"O" RING SEAL OR CUP
PRIMARY PISTON ASSEMBLY
PRIMARY CUP
PISTON RETAINER SNAP RING
CHECK FLOW WASHER
DISC BRAKE
SECONDARY PISTON
SECONDARY CUP
SECONDARY CUP

Typical tandem master cylinder—exploded view (© Chrysler Corp.)

Power Brakes

Vacuum-Hydraulic Booster

Removal

1. Depress brake pedal several times to remove all vacuum from the system.
2. Disconnect all lines, hoses and wires from the unit.
3. Remove the brake booster mounting brackets, then remove booster.

For overhaul procedures see Brake Section in General Repair Section.

Installation

1. Position the assembly on the mounting brackets, and install the attaching bolts.
2. Connect all lines, hoses and wires to the unit. *NOTE: On Tandem Booster units, remove the lubricating plugs from the end and center the plates. Add vacuum cylinder oil to the level of the filler holes, then install the lubricating plugs.*
3. Bleed the hydraulic system.

Air Brakes

Compressor

The air compressor used on Dodge trucks is a two cylinder single stage reciprocating type using automatic inlet valves. The compressor for the single system is rated at 7¼ cubic feet per minute at 1250 rpm and the compressor for the dual system is rated at 12 cubic feet per minute at 1250 rpm.

The compressor is belt driven and operates continuously when the engine is running with compressed air delivery controlled by the governor. The compressor is lubricated and cooled through the engine systems.

Proper cooling is important in maintaining the air discharge temperatures below the maximum of 400°F.

Maintenance

- *Every 100 hours or 5,000 miles:* Remove the compressor air strainer and wash all parts. The strainer element should be cleaned or replaced. Saturate the element in clean engine oil and squeeze dry before replacing the strainer in the compressor. Check and adjust the belt tension, and inspect the compressor mounting bolts for tightness.
- *Every 350 hours or 10,000 miles:* On self-lubricated compressors, drain the compressor crankcase and flush and refill with clean engine oil.
- *Every 1,000 hours or 35,000 miles:* Remove the compressor discharge valve cap and nuts

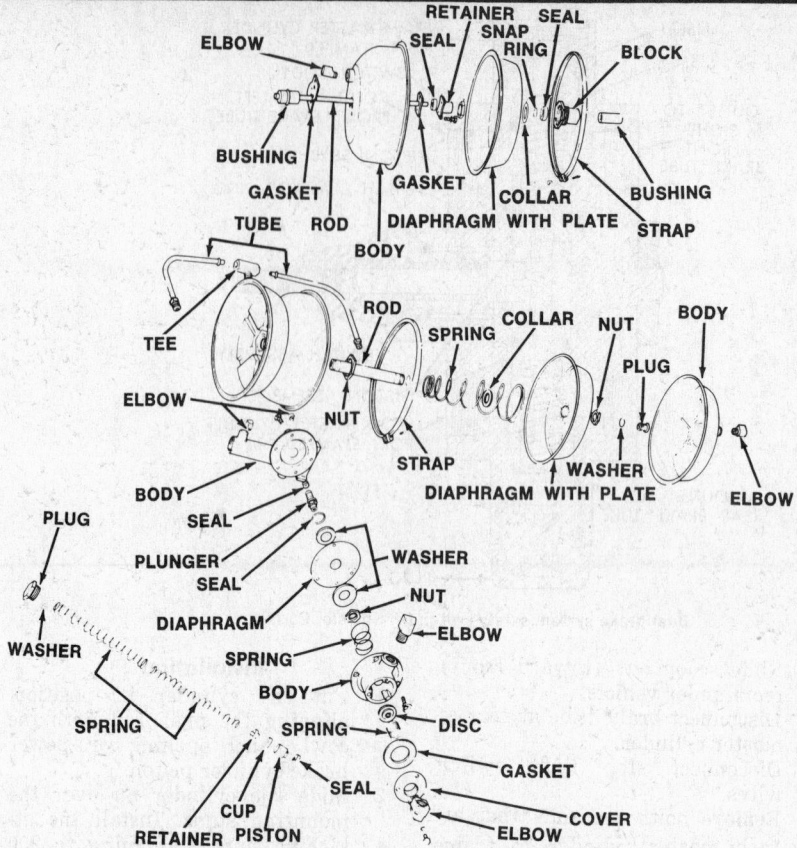

Double diaphragm power brake unit—exploded view (© Chrysler Corp.)

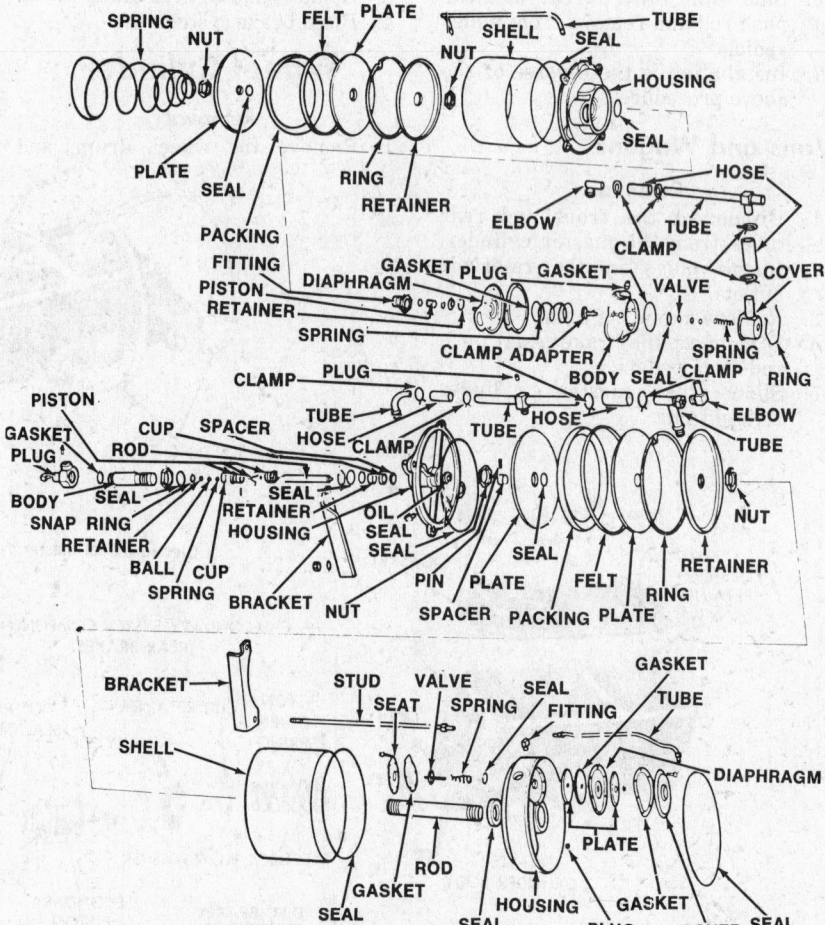

Tandem power brake unit-exploded view (© Chrysler Corp.)

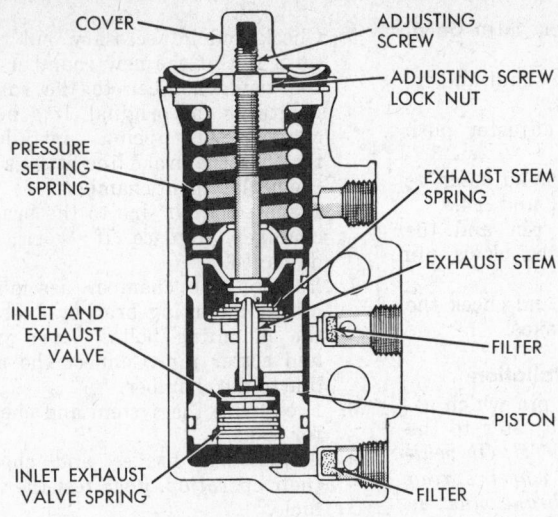

Governor assembly (© Chrysler Corp.)

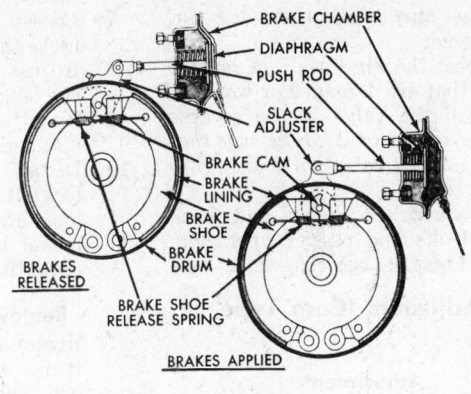

Cam type brake chamber installation (front)—typical
(© Chrysler Corp.)

and check for excessive carbon deposits. Check the discharge line for carbon. If excessive carbon is found in the cylinder head, or in the above checks, the cylinder head or discharge line should be replaced. If the compressor is a self-lubricating model, service the crankcase breather by washing in a suitable solvent.

• *Every 3,000 hours or 100,000 miles:* Disassemble the compressor, clean and inspect all parts thoroughly replacing any component that is worn or damaged.

Governor

The governor is mounted at the rear of the compressor to automati-

cally control the pressure in the air brake system. Pressure in the system reservoirs is maintained between the desired pressures of 80-95 psi for the single system and 95-120 psi for the dual system.

1. With the engine running, observe the registered pressure at which the governor cuts, stopping any further pressure build up in the system. The governor should cut out between 85-125 psi (depending on system desired pressure).
2. Make several brake applications and observe the gauge pressure at which the governor cuts-in. Air compression should begin between 80-125 psi.
3. Adjust the governor only if cut-out and cut-in pressures have

been checked with an accurate air pressure gauge to confirm the original dash gauge readings.

Air Brake System Valves

Removal and Installation

NOTE: Removal and installation of the different valves (application, relay, quick release, and so on) in the air brake system is very straight forward. The following procedure is a general outline to be followed for all valves.

1. Block and hold the vehicle's wheels.
2. *Exhaust the air pressure from the brake system.*
3. Disconnect the air lines at the valve.
4. Remove the mounting bolts and

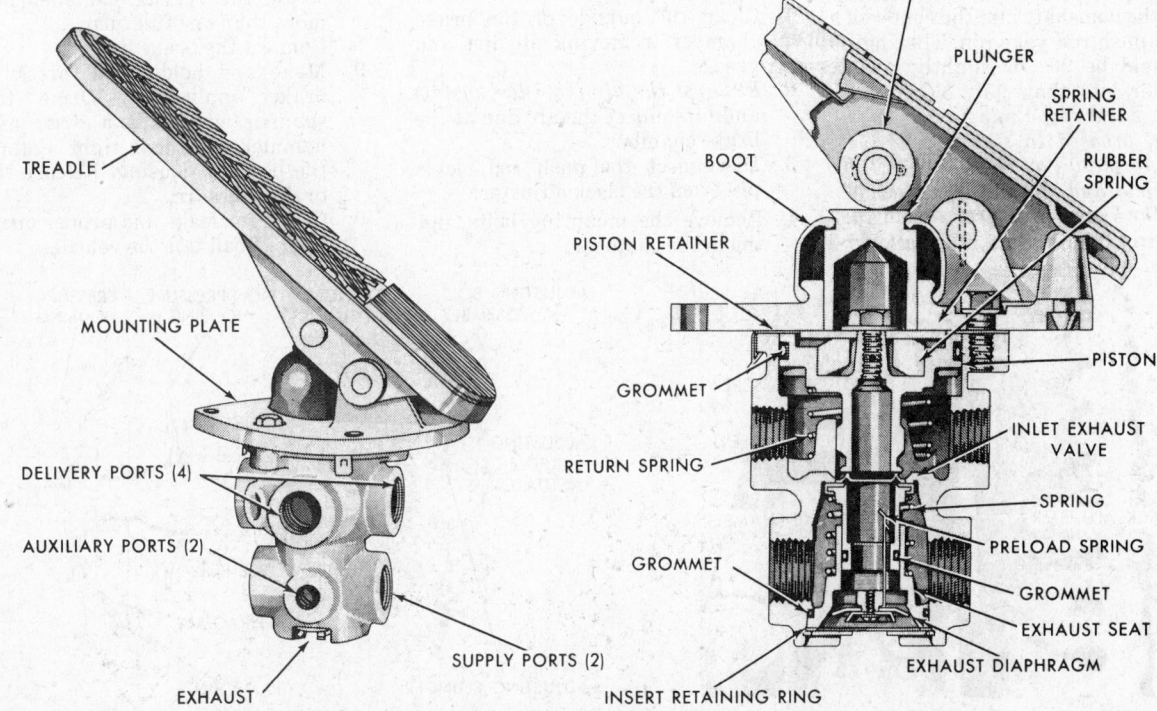

Single system air brake control valve (application valve)—typical (© Chrysler Corp.)

the valve.

5. Inspect the valve ports (in use) to be sure that they are not plugged.
6. Inspect the air lines and replace any that are damaged or worn.
7. Install the valve in the reverse order of removal. *Make sure that any unused valve ports are properly sealed.*
8. Pressurize the system and check for leaks and proper operation. Road test the vehicle.

Slack Adjusters (Cam Type Brakes)

Adjustment

1. Raise the wheels.
2. Turn the adjusting screw of the slack adjuster in the direction that will rotate the camshaft in the apply direction. *NOTE: The apply direction rotation may be clockwise or counterclockwise depending on the particular installation on the rear wheels. The front brakes will be clockwise. On rear wheels if the adjustment has a locking sleeve, it will be necessary to depress the sleeve against the spring tension with a wrench while turning the adjusting screw.*
3. Adjust the slack adjuster adjustment screw until the brake shoes are firmly against the brake drum.
4. Back off the adjusting screw just until the wheels will turn freely.
5. Bring the air pressure up to the operating limits and apply full pedal. With the brakes fully applied, check the angle of the push rod to a line through the center of the camshaft and the center of the push rod yoke pin. The angle should be 90° or slightly more, never less than 90°. *NOTE: If the push rod angle is less than 90°, proceed to step #6. If the angle is 90° or more, lower the vehicle and check the operation of the brakes.*
6. Release the brakes and back off

the slack adjuster adjustment screw.
7. Loosen the clevis locknut and remove the clevis pin.
8. Adjust the slack adjuster push rod length.
9. Re-adjust the slack adjuster as detailed in steps #3 and #4.
10. Install the clevis pin and the locknut. Tighten the clevis pin locknut.
11. Lower the vehicle and check the operation of the brakes.

Removal and Installation

1. Remove the clevis pin which attaches the slack adjuster to the brake chamber. *NOTE: On vehicles with a rear wheel spring brake, the spring brake must be released to remove the clevis pin.*
2. Remove the lock ring attaching the slack adjuster to the camshaft.
3. Mark the position of the slack adjuster on the camshaft, then slide the slack adjuster off the camshaft.
4. Place the slack adjuster on the camshaft, aligning the locating marks. If there is excessive end play, install an additional spacer at the cam end of the camshaft. Install the lock ring.
5. Connect the brake chamber push rod to the slack adjuster by installing the clevis pin in the upper hole, and install the cotter pin.
6. Lubricate the slack adjuster, and adjust the brakes.

Cam Type Brake Chamber

Removal and Installation

1. Clean the outside of the brake chamber, removing all dirt and grease.
2. *Exhaust the air from the system* and disconnect the air line at the brake chamber.
3. Disconnect the push rod clevis pin from the slack adjuster.
4. Remove the mounting bolts and the brake chamber.

5. Check and if necessary, cut the push rod of the new chamber to the new chamber to the same length as the original. If a new chamber is being installed, transfer the brake line fittings to the replacement chamber.
6. Inspect the air line to the brake chamber, replace if worn or damaged.
7. Position the chamber assembly on the mounting bracket. Install the mounting bolts, clevis pin, and cotter pin. Connect the air line to the chamber.
8. Pressurize the system and check for leaks.
9. Adjust the brakes and check their operation. Road test the vehicle.

Wedge Type Brake Chamber

Removal & Installation

1. *Exhaust the air pressure from the brake system.*
2. Disconnect the air lines.
3. Loosen the spanner nut with a drift and a light hammer.
4. Unscrew the brake chamber from the wedge housing.
5. Check the position of the wedge in the plunger housing to make sure that the wedge assembly is properly seated. Be certain that the automatic adjusting identification ring on the chamber tube is replaced.
6. Screw the brake chamber into the plunger housing until it bottoms (spanner nut loose).
7. Align the connection ports with the brake lines, if necessary, unscrew the service chamber not more than one full turn.
8. Connect the brake lines.
9. Make and hold a full pressure brake application. Drive the spanner nut with a drift and hammer until it is tight against the plunger housing. Release the brake pressure.
10. Check for leaks and proper operation. Road test the vehicle.

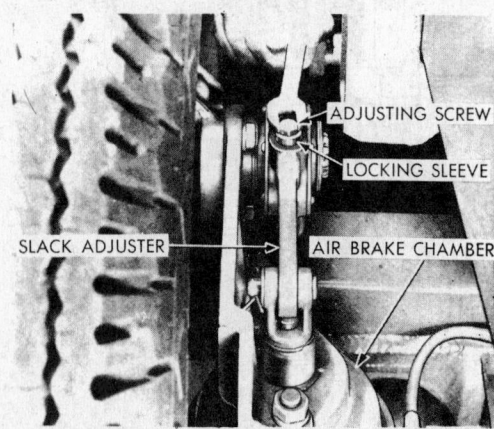

Slack adjuster adjustment (© Chrysler Corp.)

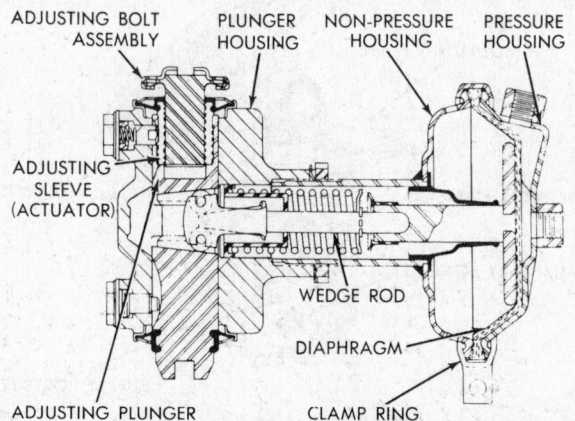

Wedge type brake chamber—typical (© Chrysler Corp.)

Parking Brake

External Contracting Driveshaft Type

Removal

1. Disconnect the brake cable.
2. Remove adjusting bolt nut.
3. Remove guide bolt adjusting lock nuts.
4. Remove anchor adjusting screw.
5. Pull band assembly away from transmission and off propeller shaft.

Installation

1. Position brake band and lining assembly over propeller shaft and on brake drum.
2. Install brake band anchor adjusting screw and adjusting guide bolt nut and lock nuts.
3. Connect brake cable.

Internal Expanding Driveshaft Type (Early)

NOTE: This procedure is for the early type brake which has an internal adjuster similar to the type used on wheel brakes. There is an adjusting screw cover plate on the bottom of the backing plate.

Removal

1. Disconnect propeller shaft at transmission.
2. Remove companion flange nut, lockwasher and flatwasher.
3. Install a puller tool on companion flange and remove flange and brake drum while using a holding tool to prevent rotation.
4. Disengage ball end of cable from operating lever.
5. Separate shoes at bottom, allowing brake shoe adjusting nut screws, and sleeve to drop out and release shoes.
6. Pry brake shoe return spring up and over to the right and brake shoe in.
7. Then, work spring out of assembly.
8. Pry out brake shoe retaining washer and remove outer guide.
9. Slide each shoe out from under guide spring (as shoes are removed, the operating lever strut will drop out of place).
10. Separate the operating lever from the right hand brake shoe by removing nut, lockwasher and bolt.

Installation

1. Assemble operating lever to right hand brake shoe.
2. Slide the right and left parking brake shoes under guide.
3. Spread shoes and insert operating lever.
4. Work return spring under guide spring and upward to engage retaining pin on left hand shoe.
5. Force the other end of return spring upward and over retaining pin on right hand shoe.
6. Install adjusting nut, screw and sleeve.
7. Place outer anchor guide over anchor and secure shoes with retaining washer.
8. Turn shoe adjusting nut until shoes are in a released position and install brake drum.
9. Adjust brake shoe and control cable.

CAUTION: If parking brake is adjusted incorrectly, the automatic shifting of the Load-Flite Transmission will be affected.

Internal Expanding Driveshaft Type (Late)

NOTE: This procedure is for the late type brake that has an external linkage clevis adjustment.

Removal

1. Disconnect the adjuster clevis at the operating lever.
2. Remove the universal joint trunnion bolts and support the driveshaft.
3. Remove the mainshaft locknut and washer. Remove the yoke and drum.
4. Unbolt the drum from the yoke.
5. Unbolt the brake assembly from the transmission mounting flange. Separate the shoe assembly from the actuating cam and lever.

Installation

1. Place the brake assembly, with the cam lugs between the ends of the shoes, on the mounting flange.
2. Align the brake on the flange, install the mounting bolts and lockwashers, torquing to 75 ft. lbs.
3. Bolt the drum to the yoke, torquing to 75 ft. lbs.
4. Install the drum to the mainshaft. Torque the nut to 120 ft. lbs.
5. You can check drum runout with a dial indicator. If it exceeds .015 in., loosen the drum to yoke bolts ¼ turn and tap the high side with a soft hammer. Tighten the bolts and recheck the runout. If runout persists, but doesn't exceed .018 in., make 6-8 moderate brake applications from 20 mph and recheck runout. If runout still exceeds .015 in., a new drum is required.

Air Spring Type
Reference

The air spring type parking brake is air actuated and is operated in conjunction with the rear wheel service brake air chamber. The spring is held in the retracted (or off) position by air pressure from a protected third

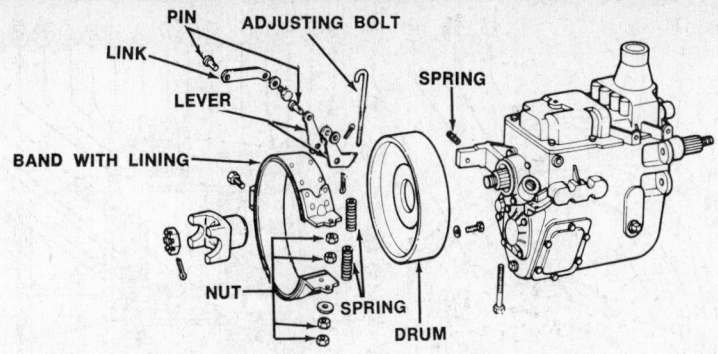

Typical external contracting type parking brake—exploded view (© Chrysler Corp.)

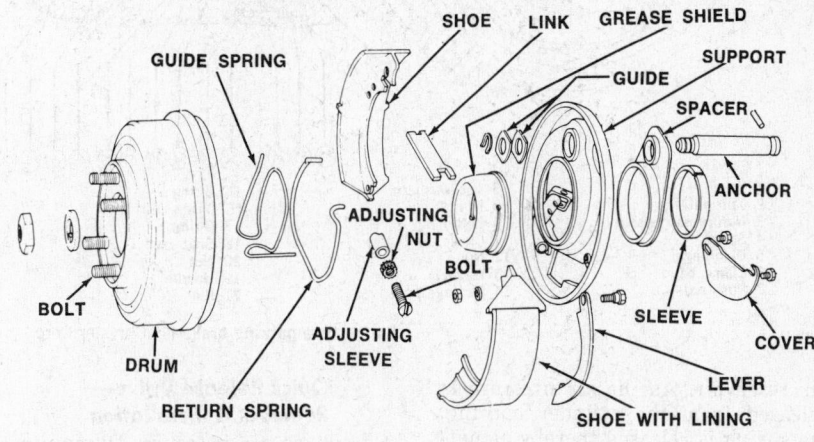

Typical internal expanding type parking brake-exploded view (© Chrysler Corp.)

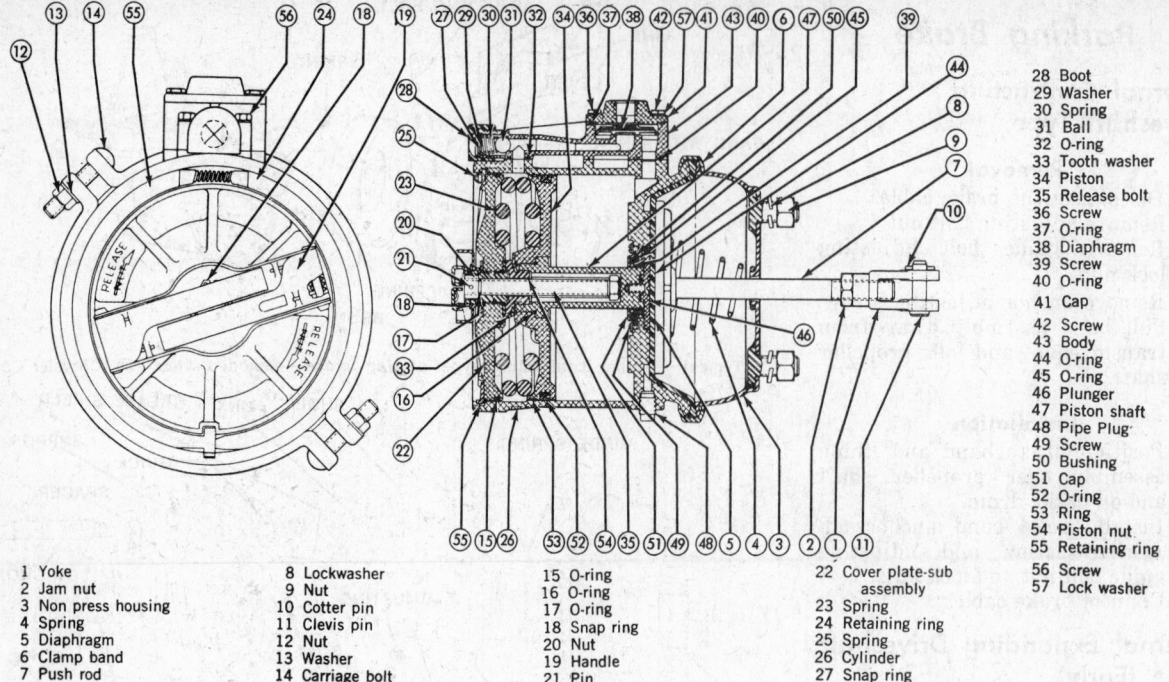

					28	Boot
1	Yoke	8	Lockwasher	15	O-ring	29	Washer
2	Jam nut	9	Nut	16	O-ring	30	Spring
3	Non press housing	10	Cotter pin	17	O-ring	31	Ball
4	Spring	11	Clevis pin	18	Snap ring	32	O-ring
5	Diaphragm	12	Nut	20	Nut	33	Tooth washer
6	Clamp band	13	Washer	19	Handle	34	Piston
7	Push rod	14	Carriage bolt	21	Pin	35	Release bolt

28 Boot
29 Washer
30 Spring
31 Ball
32 O-ring
33 Tooth washer
34 Piston
35 Release bolt
36 Screw
37 O-ring
38 Diaphragm
39 Screw
40 O-ring
41 Cap
42 Screw
43 Body
44 O-ring
45 O-ring
46 Plunger
47 Piston shaft
48 Pipe Plug
49 Screw
50 Bushing
51 Cap
52 O-ring
53 Ring
54 Piston nut
55 Retaining ring
56 Screw
57 Lock washer

1 Yoke
2 Jam nut
3 Non press housing
4 Spring
5 Diaphragm
6 Clamp band
7 Push rod
8 Lockwasher
9 Nut
10 Cotter pin
11 Clevis pin
12 Nut
13 Washer
14 Carriage bolt
15 O-ring
16 O-ring
17 O-ring
18 Snap ring
20 Nut
19 Handle
21 Pin
22 Cover plate-sub assembly
23 Spring
24 Retaining ring
25 Spring
26 Cylinder
27 Snap ring

Spring type parking brake (© Chrysler Corp.)

air reservoir. As the air pressure is released from the cylinder housing, the spring is released thereby actuating the service brake push rod and applying the brakes.

Air—Removal & Installation

1. Back off slack adjuster adjusting screw.
2. Disconnect brake chamber push rod by removing clevis pin.
3. Disconnect air lines.
4. Remove nuts and washers from mounting bolts, remove brake chamber and spring brake assembly.
5. To install reverse the above procedure, and adjust brakes.

Quick Release Valve— Removal & Installation

1. Relieve air pressure. Disconnect air line.
2. Remove four capscrews and remove cap.
3. Remove remaining two capscrews holding release valve to body.
4. Remove boot from small end of body, clean all parts with solvent and blow dry with compressed air. All "O" rings and rubber parts must be replaced.
5. To install, reverse the above procedure. Tighten all six capscrews to 15 ft. lb. Connect air

line pressurize system and check for leaks.

Fail-Safe

This unit is mounted "Piggy Back" on the air chamber non-pressure housing of the stopmaster brake. The power spring is held in a compressed position as long as the air pressure is 65 PSI or more. Anytime the air pressure drops below 65 PSI the power spring pushes the piston against the diaphragm plate, forcing the wedge head between the rollers, which spreads the plungers apart and applies the brake.

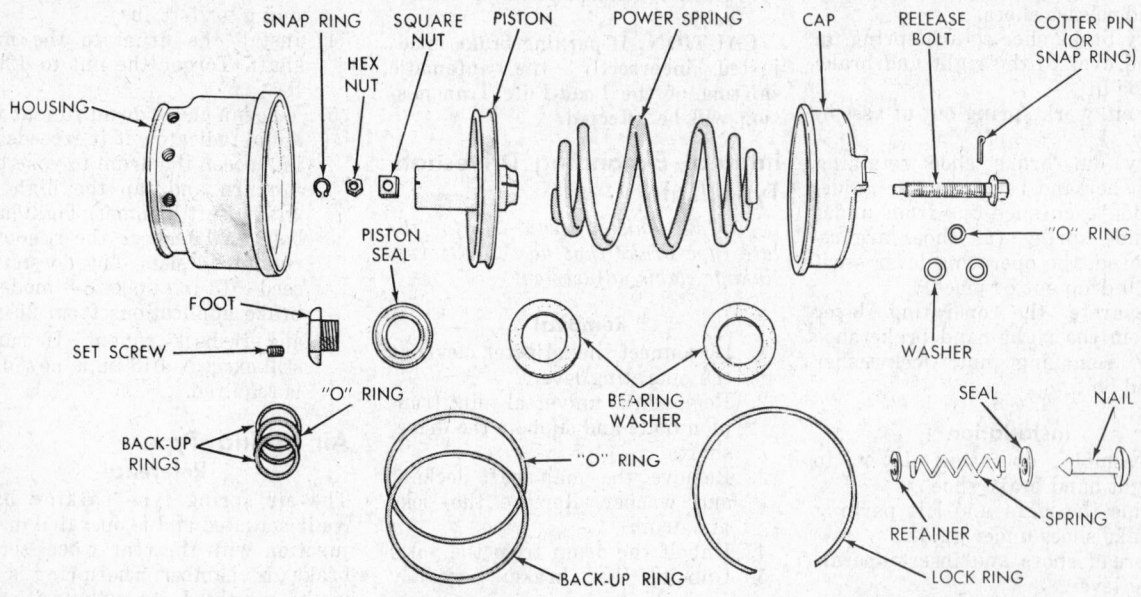

Fail safe unit (© Chrysler Corp.)

Removal

1. Cage power spring. Relieve air pressure and remove air lines from unit.
2. Mark non-pressure housing, pressure housing and cap with punch or chisel to aid in alignment at reassembly. Loosen or remove clamp ring and remove unit from non-pressure housing.

Installation

1. Align cap and housing with marks made during removal.
2. Position unit on non-pressure housing and install clamp ring and bolt.
3. Connect air lines and apply air pressure. Uncage power spring and check operation of system.

Shortstop Spring Brake

This spring type parking brake assembly is used in conjunction with the wedge type air brakes.

Removal

1. Bleed air from system. Disconnect air lines (mark lines for proper assembly).
2. Remove breather cap. Using a 9/16″ deep socket, unscrew release bolt completely.
3. Mark clamp ring, spring brake housing and service brake housing to aid in proper alignment at assembly. Remove bolt from clamp ring. Spring brake is now free for further work.

Installation

1. Align service brake housing, spring brake housing and clamp ring. Install and tighten clamp bolt.
2. Connect and tighten air lines using marks made during removal.
3. Charge system with air, screw release bolt down.
4. Check for air leaks, clean and replace breather cap.

Rear Wheel Cable Type

In this system, the rear wheel brakes also act as parking brakes. They are operated mechanically by a lever and strut connected to a steel cable. This cable is connected to the rear wheel brake cables via an equalizer.

Removal

1. Raise the vehicle, release the brakes, and remove the rear wheels.
2. Remove the brake drum from the rear axle.
3. Remove the brake shoe return spring and brake shoe retaining springs.
4. Remove the brake shoe strut and spring from the support plate.
5. Disconnect the cable from the operating arm.
6. Compress the retainers on the end of the cable housing and remove the cables from the support plate.
7. Remove the retaining bolt and nut from the cable bracket. Remove the clips at the frame bracket.
8. Disconnect the cable from the equalizer bar, and remove the assembly.

Installation

1. Lubricate the cable with short fiber grease at the point of contact.
2. Insert the cable and housing into the frame bracket and install the retaining clips.
3. Engage the end with the equalizer bar.
4. Insert the rear end of the cable and housing into the brake support plate. Make sure the housing retainers lock firmly in place.
5. Insert the end of the cable into the brake shoe operating lever and install the brake shoes.
6. Install the retaining springs, return springs, brake drum, and wheel.
7. Connect the brake cable bracket.
8. Adjust the brakes and cable.

Parking Brake Adjustment

Internal Expanding Driveshaft Type (Early)

Disconnect front end of propeller shaft to permit turning brake drum by hand.

1. Remove adjusting screw cover plate, loosen brake cable clamping bolt and back off the cable adjusting nut.
2. Turn adjusting nut to decrease shoe-to-drum clearances, until a slight drag is felt on drum.
3. Back off adjusting nut at least one full notch, using spanner wrench. CAUTION: Make sure the two raised shoulders on adjusting nut are seated in the grooves on the adjusting sleeve.
4. The cable length adjusting nut should be positioned against cable housing so there is at least .005 inch but no more than .010 inch clearance, between the operating lever and brake shoe cable. To lock adjustment, tighten cable housing clamp securely.
5. Tighten cable adjusting nut against the housing.
6. Check parking brake lever for travel.
7. Install adjusting screw cover plate and connect propeller shaft.
8. With rear wheels off the floor, start engine, apply brakes a few times, check for binding and proper adjustment.

Internal Expanding Driveshaft Type (Late)

1. Release the brake handle.
2. Adjustment is made by removing the pin and adjusting the length of the clevis. The brake should be free of drag when rotated. Shoe to drum clearance should be .020 in. on the 10 in. brake used on D800 and S600 models.

External Contracting Driveshaft Type

1. Remove the lockwire from the anchor adjusting screw on the left side.
2. Using a feeler gauge between drum and lining tighten or loosen adjusting screw to give .015 to .020 inch clearance. Install lock wire. CAUTION: The lockwire which retains anchor screw must not be drawn up tight. It will cause uneven wear and poor brake application.
3. Adjust the small diameter guide bolt on the right side to adjust the clearance of the lower part of the band.
4. Adjust the large adjusting bolt (the one with the springs) on the right side to adjust the clearance of the upper part of the band. Adjust until the drum is free with the brake released.
5. Adjust the cable as necessary.

Rear Wheel Cable Type

1. Inspect all components and correct any deficiencies such as rust, kinks, or bent parts.
2. Raise the vehicle on a lift and release the brake.
3. Loosen the adjustment until both cables have slack.
4. Tighten the cable adjuster until a slight drag is felt while rotating the wheel. Loosen the adjuster until there is no drag, then back off two turns.
5. Apply the brake several times, then recheck the adjustment.

Fuel System

Reference

Carburetor specifications, exploded views, and basic adjustments are found in the General Repair Section under Carburetor Repairs.

Fuel Tank

Vans and Wagons

Removal

1. Disconnect the battery ground cable.
2. Remove the fuel tank filler cap.
3. Raise the vehicle on a lift. Pump all fuel from the tank into an ap-

proved holding tank, and raise the vehicle.

4. Disconnect the fuel line and wire lead to the gauge unit. Remove the ground strap.
5. Remove the vent hose shield and the hose clamps from the hoses running to the vapor vent tube.
6. Remove the filler tube hose clamps and disconnect the hose from the tank.
7. Place a transmission jack under the center of the tank and apply sufficient pressure to support the tank.
8. Disconnect the two J-bolts and remove the retaining straps at the rear of the tank. Lower the tank from the vehicle. Feed the two vent tube hoses and filler tube vent hose through the grommets in the frame as the tank is being lowered. Remove the tank gauge unit.

Installation

1. Inspect the fuel filter, and if it is clogged or damaged, replace it.
2. Insert a new gasket in the recess of the fuel gauge opening and slide the gauge into the tank. Align the positioning tangs on the gauge with those on the tank. Install the lock ring, and tighten securely.
3. Position the tank on a transmission jack and hoist it into place, feeding the vent hoses through the grommets on the way up.
4. Connect the J-bolts and retaining straps, and tighten to 40 inch-lbs. Remove the jack.
5. Connect the filler tube and all vent hoses.
6. Connect the fuel supply line, ground strap, and gauge unit wire lead.
7. Refill the tank and inspect it for leaks. Connect the battery ground cable.

Conventional

Removal

1. Disconnect the battery ground cable. Remove the fuel tank filler cap.
2. Pump all fuel out of the tank and the auxiliary tank (if so equipped) into a safe holding tank.
3. Remove all seat mounting nuts from under the cab, and remove the seat from the cab.
4. Disconnect the gauge wire, vent lines, and fuel line and remove the grommet from the outer end of the filler tube. If the vehicle is equipped with an auxiliary fuel tank (through 1973), disconnect the two large hose clamps on the lower filler tube inside the cab and slide the hose down far enough to allow the top filler

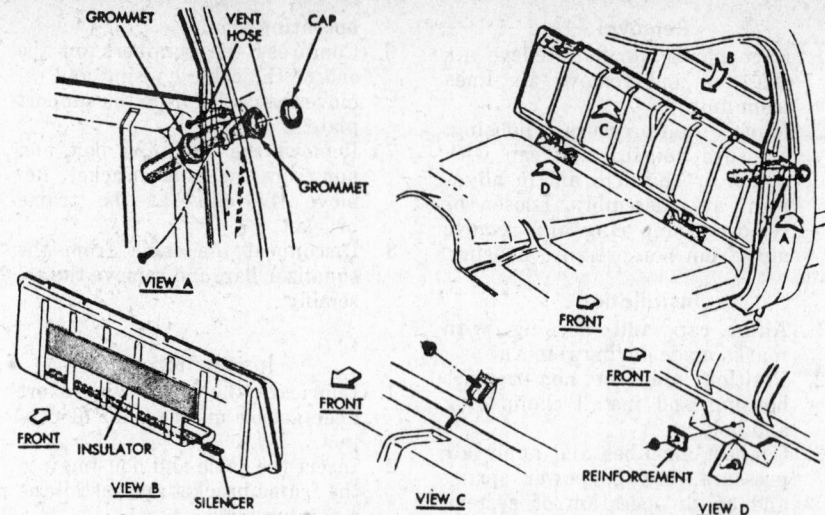

Conventional truck in-cab fuel tank, 1972 and later (© Chrysler Corp.)

tube to be removed with the tank. Disconnect the auxiliary fuel tank vent hose from the filler tube.
5. Remove the screw from the filler tube mounting bracket located on the face of the cab pillar.
6. Loosen the four nuts from the mounting studs along the top of the tank and remove two bolts from the bottom of the tank.
7. Tip the bottom of the tank forward and remove the tank from the vehicle, working through the passenger side door opening.
8. Remove the tank gauge unit with a special spanner wrench. Slide the gauge unit assembly from the tank and discard the gasket.

Installation

1. Insert a new gasket and the gauge assembly into the fuel gauge opening recess. Align the gauge positioning tangs with the tangs in the tank. Install the lock ring and secure it.
2. Put the tank into position in the cab and tighten the bolts and nuts securely.
3. Connect the gauge wire, vent tubes, and fuel line. If there is an auxiliary fuel tank, connect the lower hoses and tighten the hose clamps.
4. Install the filler tube grommet over the end tube and install the retaining screw into the filler tube mounting bracket.
5. Mount the seat in the vehicle and securely install the nuts on the seat mounting studs under the vehicle.
6. Refill the fuel tank, and inspect for leaks. If no leaks are present, install the battery and ground cable and gauge wire.

Conventional Chassis Mounted

Removal

1. Disconnect the battery ground.

Remove the fuel tank filler cap.
2. Pump all fuel from the tank into an approved holding tank.
3. Put the vehicle on a lift and disconnect the vent hoses and filler hose. Remove the vent hoses from the hose routing bracket.
4. Remove the nut from the outboard end of the center retaining strap, pull off the end of the strap, and allow it to hang free.
5. Place a transmission jack under the tank and apply sufficient pressure to support it.
6. Remove the two remaining outboard retaining strap nuts and lower the tank slightly to permit disconnecting the gauge wire and fuel line. Remove these, and then lower the tank further and remove it from the vehicle.

Installation

1. Place the tank on top of the transmission jack and raise it high enough to connect the fuel line and gauge wire.
2. Connect the fuel line and gauge wire. Raise the tank into position and connect the two retaining straps.
3. Remove the jack and connect the center strap. Tighten all three straps firmly, but cautiously, to avoid overstraining straps.
4. Route the vent hoses through the hose holding bracket and connect them to the adapters at the back of the cab.
5. Connect the filler hose to the adaptor at the back of the cab.
6. Refill the tank and inspect for leaks, verifying safety before reconnecting battery ground.

Ramcharger and Trail Duster

Removal

1. If there is a tank skid plate, remove it.
2. Disconnect the battery ground cable.

3. Remove the tank filler cap.
4. Pump or siphon the contents of the tank into a safe container.
CAUTION: Siphoning should not be started by mouth. Only fuel-safe pumps should be used.
5. Raise the vehicle on a hoist and disconnect the fuel line and tank sending unit wire. Remove the ground strap or wire.
6. Remove the hose clamps from the vent dome hose.
7. Remove the filler tube hose clamps. Detach the hoses from the tank.
8. Support the tank with a padded transmission jack.
9. Disconnect the two J-bolts and remove the straps at the rear of the tank.
10. Remove the tank gauge sending unit.

Installation

1. Use a new tank gauge sending unit gasket. Check the filter on the end of the fuel suction tube.
2. Use a new or undamaged tank to frame insulator. Raise the tank into positon.
3. Connect the J-bolts and retaining straps. Tighten the bolts until about .97 in. of threads protrude.
4. Connect the filler tube and all hoses. Tighten the clamps.
5. Connect the fuel line, ground strap or wire, and tank sending unit wire. Make sure that all fuel line heat shields are in place.
6. Reconnect the battery ground cable and replace the skid plate.

Carburetor

Reference

For idle speeds see the Specifications at the beginning of this section.

Removal and Installation

The following is a general removel procedure for all carburetors.
1. Disconnect the battery ground cable.
2. Remove the air cleaner.
3. Remove the fuel tank pressure-vacuum filler cap. The tank could be under a small amount of pressure.
4. Disconnect and plug the fuel lines. Use two wrenches to avoid twisting the fuel line. A container is also useful to catch any fuel which spills from the lines.
5. Disconnect the throttle and choke linkage.
6. Disconnect any vacuum lines.
7. Remove the mounting bolts.
8. Carefully remove the carburetor from the engine and carry it in a level position to a clean work place.
9. Installation is the reverse of re-moval. Adjust the curb idle speed.

Idle Speed and Mixture Adjustment

NOTE: Adjust with the air cleaner installed.
1. Run the engine at fast idle to stabilize engine temperature.
2. Make sure that the choke plate is fully released.
3. Attach a tachometer to the engine. With electronic ignition, connect the meter to the negative primary coil terminal and to a ground.
NOTE: Not all tachometers or dwell/tachometers will work with electronic ignition; some may be damaged. Check the manufacturer's instructions carefully.
4. Connect an exhaust analyzer to the engine and insert the probe as far into the tailpipe as possible. On vehicles with dual exhaust, insert the probe into the left tailpipe as this is the side without the heat riser valve.
5. Check ignition timing and adjust it as required.
6. If the truck has air conditioning, turn the air conditioner off. On six-cylinder engines, turn the headlights on high beam.
7. Place the manual transmission in neutral; put the automatic in Park. Make sure the hot idle compensator valve (if any) on the carburetor is fully seated in the closed position.
8. Turn the engine idle speed adjustment screw in or out to adjust idle speed to specification. If the carburetor has an electric solenoid, turn the solenoid adjusting screw in or out to obtain the specified rpm. Then, adjust the curb idle speed screw until it just touches the stop on the carburetor body. Now, back the curb idle speed adjusting screw out one full turn.
9. Turn each idle mixture adjustment screw 1/16 turn richer (counterclockwise). Wait 30 seconds and observe the reading on the exhaust gas analyzer. Continue this procedure until the meter indicates a definite increase in the richness of the mixture.
NOTE: This step is very important. A carburetor that is set too lean will cause the exhaust gas analyzer to give a false reading indicating a rich mixture. Because of this, the carburetor must first be known to have a rich mixture to verify the reading on the exhaust gas analyzer.
10. After verifying the reading obtained on the meter, adjust the mixture screws to get an air/fuel ratio of 14.2:1. Turn the mixture screws clockwise (leaner) to raise the meter reading or counterclockwise (richer) to lower the meter reading.
NOTE: On 1975 and later models, adjust to get the air/fuel ratio and percentage of CO indicated on the engine compartment sticker.

Idle Speed Solenoid Adjustment

This solenoid is energized whenever the ignition circuit is on. Its function is to allow the throttle plates to close farther when the ignition is switched off, thereby preventing engine over-running.
1. Bring the engine to operating temperature and attach a tachometer.
2. With the engine running, adjust the solenod screw to the proper rpm.
3. Adjust the slow curb idle screw until the screw end just contacts the stop on the carburetor body. Back the screw off one full turn.
4. Test the above procedure by disconnecting the solenoid wire at the connector. Be sure not to let the lead short to the engine. The solenoid should de-energize and idle speed should drop down below normal. Now reconnect the wire. After you reconnect the solenoid, move the throttle linkage by hand since the solenoid isn't strong enough to move it.

Chokes

Well Type Automatic Choke

This choke requires no servicing as the unit is set when manufactured. Move the choke rod up and down to check for free movement in the pivot. If the unit binds, a new choke unit must be installed. NOTE: Do not attempt to repair or change the setting. When installing the well type choke unit, be certain that the coil housing does not contact the sides of the well in the exhaust manifold.

Electric Assist Choke System— 1973 and later

Description

Two types of electric choke controls are used to shorten the choke duration during both winter and summer operation.
The single stage choke control operates a slower choke opening at temperatures of 58° or below, and a rapid choke opening at temperatures of 68° and higher.
The dual stage choke control provides partial power to the choke coil at temperatures of 58° and below, and full power to the choke coil at temperatures of 68° and above, and will stop the current to the choke coil at temperatures of approximately 130° and higher.

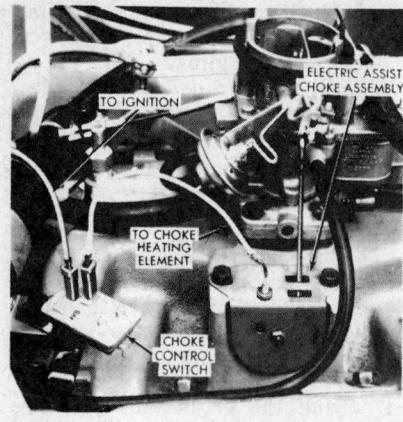

Typical electric choke system (© Chrysler Corp.)

SINGLE STAGE CONTROL

DUAL STAGE CONTROL

RESISTOR

Two types of electric choke controls
(© Chrysler Corp.)

Engines started in the winter will experience three levels of current to the choke coil from the dual choke control. LOW during the engine warm-up, HIGH after the engine warm-up, and NONE after the engine reaches normal operating temperatures. Engines started in summer weather will not have the LOW system in operation, nor will an engine that is restarted when hot, have a HIGH.

NOTE: All temperature readings are in Fahrenheit.

Single Stage Control Switch Test

1. Remove the "BAT" connector from the control unit.
2. Connect a test lamp to the small terminal of the control to ground.
3. Start the engine and warm it up to normal operating temperature.
4. Reconnect the "BAT" terminal wire to its post on the control and observe the test lamp.
5. The test lamp should light. It may remain on for a few seconds or for a longer duration, but must not remain on for over five minutes. If so, replace the control switch.

Dual Stage Control Switch Test

1. The test procedure is the same for the dual stage control switch as for the single stage switch, except the brightness or intensity of the test lamp should match that of battery current during the test. If the intensity is less,

or the light remains on for over five minutes, the control switch is defective and should be replaced.

Choke Heating Element Test—

1. Disconnect the electric heating element wire at the control switch.
2. Connect an ohmmeter lead to the crimped junction of the element wire at the choke end, avoiding connection with the heater casing. Ground the other lead of the ohmmeter.
3. Resistance of twelve ohms is acceptable. Replace the unit if resistance is outside this range.
4. Make sure choke linkage moves freely when hot and when cold.

Governors

Sandwich Velocity Type Governor

Speed Adjustment

1. See Specifications at the beginning of this section for engine no load governed rpm.
2. Remove seal.
3. Turn speed screw adjusting cap, one-half turn at a time, counterclockwise to increase speed and clockwise to decrease speed. Each ½ turn equals approximately 150 rpm. More than two complete counterclockwise turns from factory setting is not recommended.

Surge Adjustment

1. Remove adjusting screw cap assembly and locate special hollow wrench in calibrating nut.
2. Locate hex wrench through hollow wrench into adjusting screw.

3. Block throttle linkage to produce surge at governed speed.
4. Turn calibrating nut clockwise ¼ turn at a time while holding adjusting screw with hex wrench until surge is minimized.
5. Reinstall adjusting screw cap assembly and reset governor speed if necessary.
6. Install a new seal.

Slow Action Adjustment

If governor does not cut in quickly at maximum speed or does not open promptly at governed speed when load is applied, governor is said to be slow acting. This is corrected by the same procedure as that described for surge adjustments above, except that the calibrating nut is turned counterclockwise.

Holley Governors

Adjustment

1. Connect tachometer to engine. run it up to governed speed.
2. After the engine has warmed up, run it up to governed speed.
3. Stop the engine for adjustment.
4. Remove the cover band and seal from the distributor body. Turn the engine until the governor counterweight adjusting screw is accessible.
5. Remove the plug from the counterweight with a ⅛ in. Allen wrench.
6. Insert a slotted adjusting tool into the counterweight hole and engage the adjusting tangs. Turn clockwise to decrease speed and counterclockwise to increase. ¼ turn changes governed speed about 100 rpm.
7. Remove the adjusting tool and replace the plug and cover band

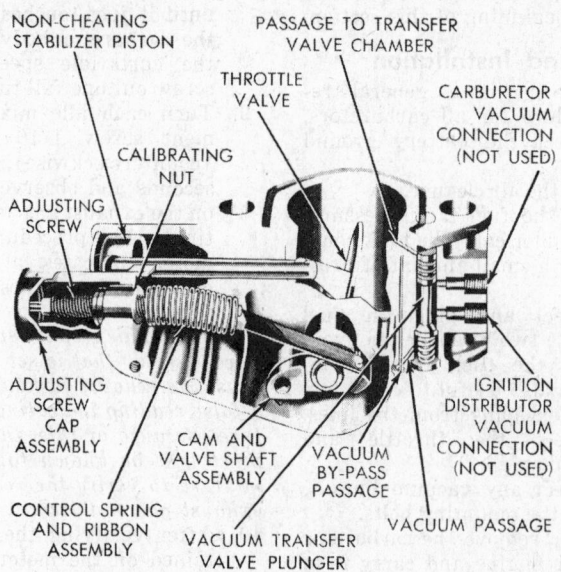

NON-CHEATING STABILIZER PISTON

PASSAGE TO TRANSFER VALVE CHAMBER

THROTTLE VALVE

CARBURETOR VACUUM CONNECTION (NOT USED)

CALIBRATING NUT

ADJUSTING SCREW

ADJUSTING SCREW CAP ASSEMBLY

CAM AND VALVE SHAFT ASSEMBLY

CONTROL SPRING AND RIBBON ASSEMBLY

VACUUM BY-PASS PASSAGE

VACUUM TRANSFER VALVE PLUNGER

IGNITION DISTRIBUTOR VACUUM CONNECTION (NOT USED)

VACUUM PASSAGE

Sandwich type velocity governor (© Chrysler Corp.)

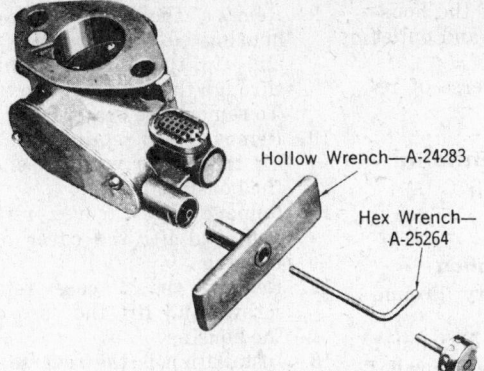

Sandwich governor surge adjustment
(© Chrysler Corp.)

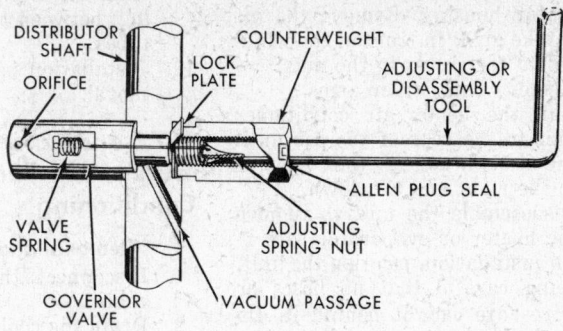

Holley governor distributor adjustment—361 and 413 engines
(© Chrysler Corp.)

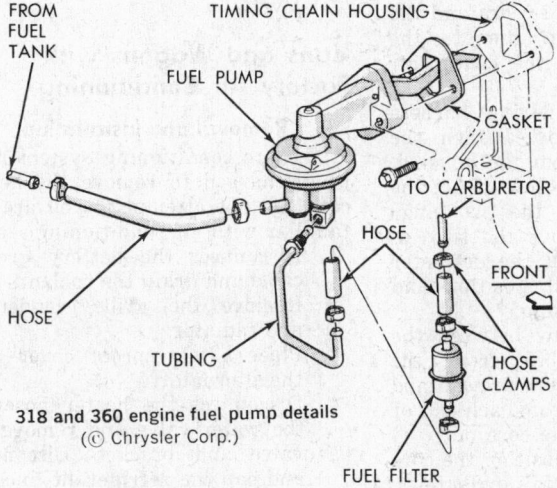

318 and 360 engine fuel pump details
(© Chrysler Corp.)

Slant six fuel pump (© Chrysler Corp.)

before starting the engine.

8. Install new band seal.

Fuel Pump

Removal and Installation

1. Disconnect the fuel lines from the inlet and output sides of the fuel pump.
2. Plug these lines to prevent gasoline from leaking out.
3. Unbolt the retaining bolts from the fuel pump and remove the fuel pump from the engine and pushrod (if equipped).
4. Remove the old gasket from the engine and/or fuel pump.
5. Clean all mounting surfaces.
6. Using a new gasket, install the fuel pump. Installation is the reverse of removal.

Cooling System

Heater Core

Conventional Trucks without Air Conditioning—1971

Removal and Installation

1. Drain the coolant.

2. Mark and disconnect the coolant hoses from the heater.
3. Disconnect the motor wiring and heater control cables.
4. Remove the three heater mounting bolts or nuts from inside the engine compartment.
5. Lower the heater unit to the floor and remove the defroster hoses and intake duct.
6. Disassemble the heater unit to remove the core.
7. On installation, position the heater unit and install the defroster hoses and intake duct.
8. Install the mounting nuts or bolts.
9. Connect the wiring and control cables. Connect the coolant hoses.
10. Refill the cooling system and let the engine warm up with the heater on. Check the coolant level.

Conventional Trucks with Factory Air Conditioning—1971

Removal and Installation

The air conditioning system must be discharged to remove the heater core. Do not attempt this if you are not familiar with air conditioning service.

1. Disconnect the battery ground

cable, drain the coolant, remove the air cleaner, disconnect the coolant hoses and plug the outlets to prevent spillage.
2. Detach the refrigerant lines from the evaporator core tubes. Cap all openings.
3. Disconnect the two vacuum hoses from the water valve and the one from the intake manifold. Push the hoses and the grommet through the firewall.
4. Remove the three mounting nuts from the studs.
5. Remove the glove box and disconnect the actuator vacuum hose cluster from the back of the control switch. Remove both cables from the bellcrank on top of the housing. Disconnect the wiring at the blower motor resistor block and evaporator temperature control switch.
6. Remove the ashtray and disconnect the left spot cooler and defroster ducts. Disconnect the left side support bracket.
7. Remove the three retaining screws from the face of the housing. Remove the appearance shield. Collapse the distribution duct toward the back of the instrument panel to allow the housing to be rolled out.
8. Remove the support bracket

bolts at the back of the blower motor housing. Remove the air intake grille in front of the windshield for access to the nuts.

9. Remove both wiper arms.
10. Pull the heater/air conditioner unit to the rear of the cab until the studs are clear, then move it to the right and tilt down.
11. Disassemble the unit to remove the heater or evaporator core.
12. On installation, position the unit, being careful that no hoses or wires are caught behind it. Install the stud nuts, center the studs in their holes, and tighten the nuts. Install the rear support bracket bolts.
13. Install the left and right support bracket.
14. Connect the vacuum hose cluster to the control switch. Connect the control cables to the bellcrank. Connect all electrical and vacuum connectors.
15. Install the left spot cooler, defroster, and distribution ducts.
16. Install the ashtray.
17. Connect the refrigerant lines, using new O-rings lubricated with refrigerant oil. Connect the heater hoses.
18. Replace the glove box.
19. Fill the cooling system, replace the air cleaner, and install the battery cable. Let the engine warm up with the heater on and check the coolant. Charge the air conditioner.

1972 and Later Conventional Trucks without Air Conditioning

Removal and Installation

1. Disconnect the battery ground cable.
2. Drain the radiator.
3. Disconnect the heater hose from the heater.
4. Disconnect the wiring from the resistor.
5. Remove the defroster ducts.
6. Disconnect the ground wire and the cooling tube from the engine side of the blower motor.
7. Remove the bracket from the righthand side of the instrument panel and pull the panel toward the rear of the truck.
8. Remove the seven retaining nuts from the engine side of the firewall. Remove one nut from inside the truck at the right-hand kick panel.
9. Roll the heater out from its mounting.
10. Separate the front and rear housings. Two of the retaining screws are located inside the unit at the right-hand end.
11. Remove the screws at each end of the core, holding the core to the housing. Remove the screws

from the front side of the housing, between the inlet and outlet tubes.
12. Installation is the reverse of removal.

1972 and Later Conventional Trucks With Factory Air Conditioning

Removal and Installation

1. Disconnect the battery ground cable.
2. Drain the coolant and disconnect the heater hoses at the firewall.
3. Remove the glove box and ashtray.
4. Remove the right and left ducts.
5. Remove the four screws and the distribution duct. Remove the two screws and the center air outlet and duct.
6. Disconnect the wiring harness from the resistor. Detach the vacuum lines from the housing.
7. Remove the 22 screws from the housing. Remove the two plugs in order to remove the two inside screws. Hold the defroster door to the heat position and separate the housing.
8. Remove the screw between the heater core tubes from the engine compartment firewall and the two screws from each end of the core. Slide the core out.
9. On installation, position the core and install the retaining screws.
10. Assemble the housing, holding the defroster door to the heat position and making sure that the shaft lines up in the hole.
11. Install the vacuum lines, wiring, ducts, glovebox, and ashtray.
12. Connect the heater hoses and the battery cable. Fill the cooling system.
13. Let the engine warm up with the heater on, then check the coolant level.

Vans and Wagons Without Air Conditioning

Removal and Installation

1. Disconnect the battery ground cable.
2. Drain the radiator.
3. Cover the alternator with a waterproof cover.
4. Disconnect the blower motor resistor and ground wires from the heater.
5. Disconnect and plug the heater core hoses.
6. Disconnect the control cables.
7. Remove the retaining screws from the water valve. Do not disconnect the hoses from the water valve; place the water valve with hoses attached to one side.
8. Remove the blower motor cooler tube.

9. Remove the nuts holding the housing to the mounting studs and tip the complete unit out through the hood opening. To remove the heater core:
10. Remove the 3 retaining nuts and lift the blower with wheel out of the housing.
11. Remove the 4 cover retaining nuts and lift the cover off the housing.
12. Remove the 4 core retaining screws and lift the core out of the housing.
13. Installation is the reverse of removal. Fill the cooling system.
14. Let the engine warm up with the heater on, then check the coolant level.

Vans and Wagons with Factory Air Conditioning

Removal and Installation

The air conditioning system must be discharged to remove the heater core. Do not attempt if you are not familiar with air conditioning servce.

1. Disconnect the battery ground cable and drain the coolant.
2. Remove the grille, condenser, and radiator.
3. Place a waterproof cover over the alternator.
4. Disconnect the heater hoses at the water valve and remove the valve and bracket. Disconnect and cap the refrigerant lines.
5. Remove the glovebox, spot cooler bezel, and appearance shield.
6. Working through the glovebox opening, remove the evaporator housing to firewall screws and nuts.
7. Remove the wiper motor. Detach all evaporator housing vacuum and electrical connections. Detach the blower motor cooling hose and the drain hoses. all evaporator housing vacuum
8. Remove the two 2¼ in. bolts from the crossbar and the four screws from the sealplate on the front of the housing. Separate the evaporator and blower motor housings, remove the evaporator housing.
9. Remove the receiver drier and cap all the openings. Carefully pry the heater core out, leaving the air seal at the front intact.
10. On installation, connect the hoses to the core. Position the evaporator housing on top of the blower housing. Install the mounting screws and nuts.
11. Position the crossbar under the lip on the blower housing opening and install the two 2¼ in. bolts. Install the four seal plate screws at the front of the housing.
12. Replace the wiper motor and connect the vacuum and electrical lines. Connect the blower

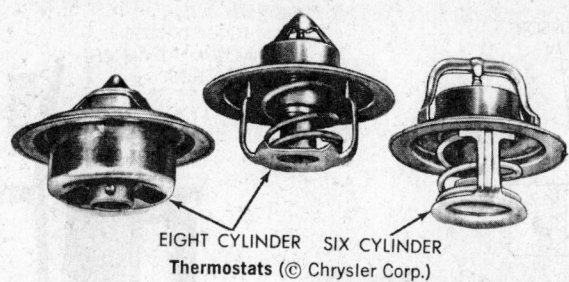

EIGHT CYLINDER   SIX CYLINDER
Thermostats (© Chrysler Corp.)

motor cooler hose and the drain hoses.

13. Connect the heater hoses to the water valve.

14. Install the receiver drier and connect the refrigerant lines.

15. Install the radiator, condenser, and grille.

16. Replace the glovebox, spot cooler bezel, and appearance shield.

17. Install the battery ground cable and fill the cooling system. Let the engine warm up with the heater on, then check the coolant. Charge the air conditioner.

Water Pump

NOTE: Only the 361/413 water pump is rebuildable.

Removal—All Engines

1. Drain cooling system and remove the fan belt and fan shroud. Remove the radiator on 1974 and later V8s.

2. Unscrew fan blade bolts and remove fan blade, spacers and bolts as an assembly. **CAUTION: Silicone drive fans must be kept in their normal attitude. If the shaft points down, silicone fluid will contaminate the fan drive bearing.**

3. Position the six-cylinder lower clamp in the center of the by-pass hose. Disconnect or remove heater and radiator hoses.

4. Remove water pump retaining bolts and remove water pump assembly.

NOTE: On air conditioned equipped vehicles with V-8 engines, the compressor clutch assembly and the front mounting brackets may have to be removed to allow for the removal of the water pump assembly.

Installation

1. Install a new by-pass hose if necessary, with clamps positioned in the center of the hose.

2. Use a new gasket and install water pump. Install and tighten pump retaining bolts to 30 foot pounds. Position by-pass hose clamps. Install the heater and radiator hoses.

3. Install fan blade, spacer and bolt assembly. Start all bolts, then tighten to 15-18 foot pounds.

4. Install fan belt and adjust belt tension. Fill cooling system.

Engine

Emission Controls

Description

Two basic approaches to emission control have been used on Dodge engines. The first is engine design modifications, which apply to some extent to all engines. The second is a group of specific emission control systems, the application of which varies with model and power train. Those models rated under 6000 lbs GVW are subject to stricter light-duty emission regulations, and therefore use more of these systems.

The design modifications and emission control systems are listed here; tests and adjustments for the systems will be found in the General Repair Section on Emission Controls.

Engine Design Modifications

Engine design modifications have been made from model year to model year, to aid in the reduction of harmful emissions.

The intake manifold has been modified to aid in the rapid vaporization of the fuel, and modification of the combustion chamber design allows better combustion of the fuel.

Compression ratios were reduced on most engines, to allow the use of a lower octane fuel, and the cam shaft design was changed to allow greater valve overlap to reduce engine emissions.

Carburetors have continually been modified to aid in fuel distribution and the Electronic Ignition System was introduced to reduce the need for continual adjustment of the ignition system and also provides for better and more precise spark.

Electric Assist Choke System

All engines are equipped with an electric assist choke to shorten the period of choke operation when control temperatures are above 60° F. This device consists of a heating element docated in the choke well and a control unit adjacent to the choke well. Heat applied to the thermostatic choke coil by the heating element causes the choke to open in a shorter time than previously.

NOTE: The heating element should not be exposed to or immersed in any fluid for any purpose. This especially applies to cleaning agents. A short in the wiring to the heater will result in a short circuit in the ignition.

Heated Air Intake System

The carburetor air preheater is a device which is part of the air cleaner

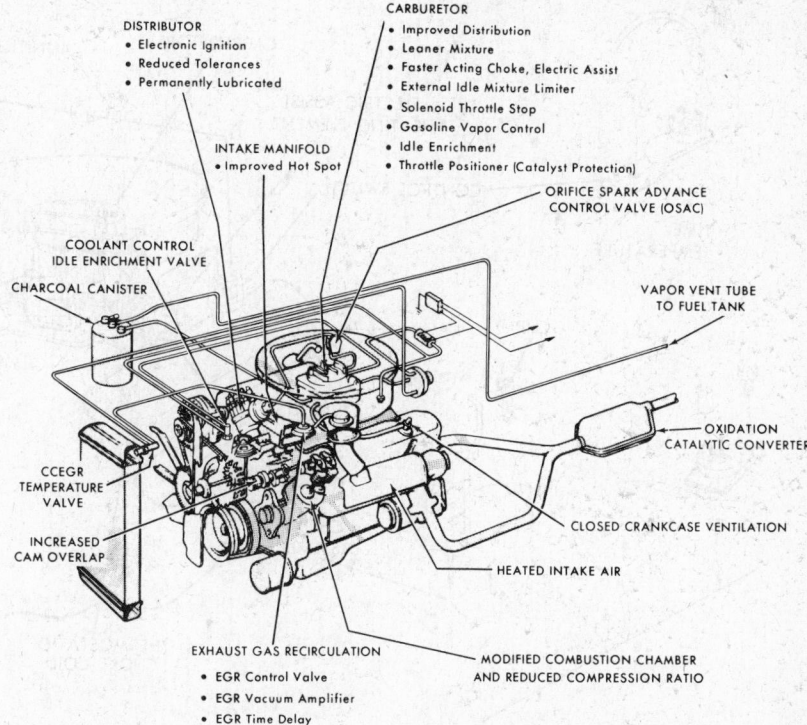

Light Duty (under 6,000 lbs. GVW) emission controls (© Chrysler Corp.)

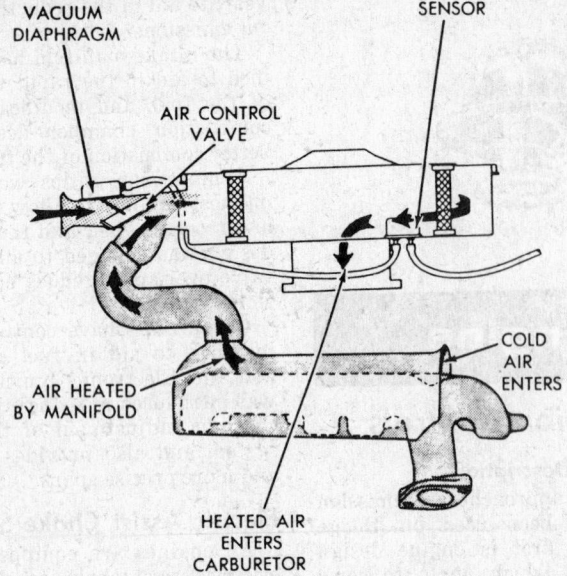

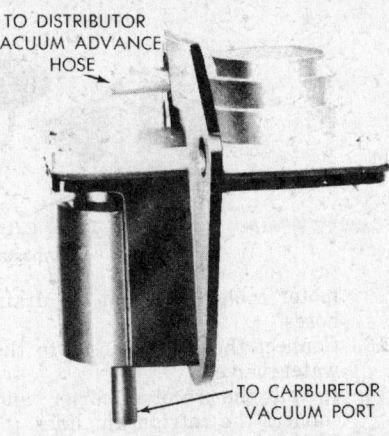

OSAC valve (© Chrysler Corp.)

Heated inlet air system (© Chrysler Corp.)

and which keeps the air entering the carburetor at about 100° F when underhood temperatures are less than 100° F. By using this device, the carburetor can be calibrated much leaner to improve engine warm-up characteristics.

The heated air intake system is basically a two circuit airflow system. When underhood temperatures are less than 100° F, the air will flow into the stove, through a flexible connector, into the adaptor on the bottom of the snorkel, and then into the carburetor. When the underhood temperature is above 100° F, the airflow will be through the snorkel.

Modulation of the induction air is performed through intake manifold vacuum, a temperature sensor, and a vacuum diaphragm which operates the heat control door in the snorkel.

Orific Spark Advance Control (OSAC)

The OSAC system is used to control NO_x. The system controls the amount of vacuum supplied to the vacuum advance mechanism of the distributor.

Exhaust Gas Recirculation (EGR)

EGR is used in conjunction with the vacuum spark advance control to limit peak flame temperatures and thus retard the formation of NO_x. Two alternate systems are used.

Ported Vacuum Control System

This system uses a slot type port in the carburetor throttle body which is exposed to an increasing percentage of manifold vacuum by the opening movement of the throttle plate. The throttle bore port is directly connected to the EGR valve through an external nipple. The flow rate of exhaust gases is determined by manifold vacuum, throttle position, and exhaust gas backpressure. Wide open throttle recycling of exhaust gases is prevented by calibrating the valve opening point above manifold vacuum available at wide open throttle, since port vacuum cannot exceed manifold vacuum.

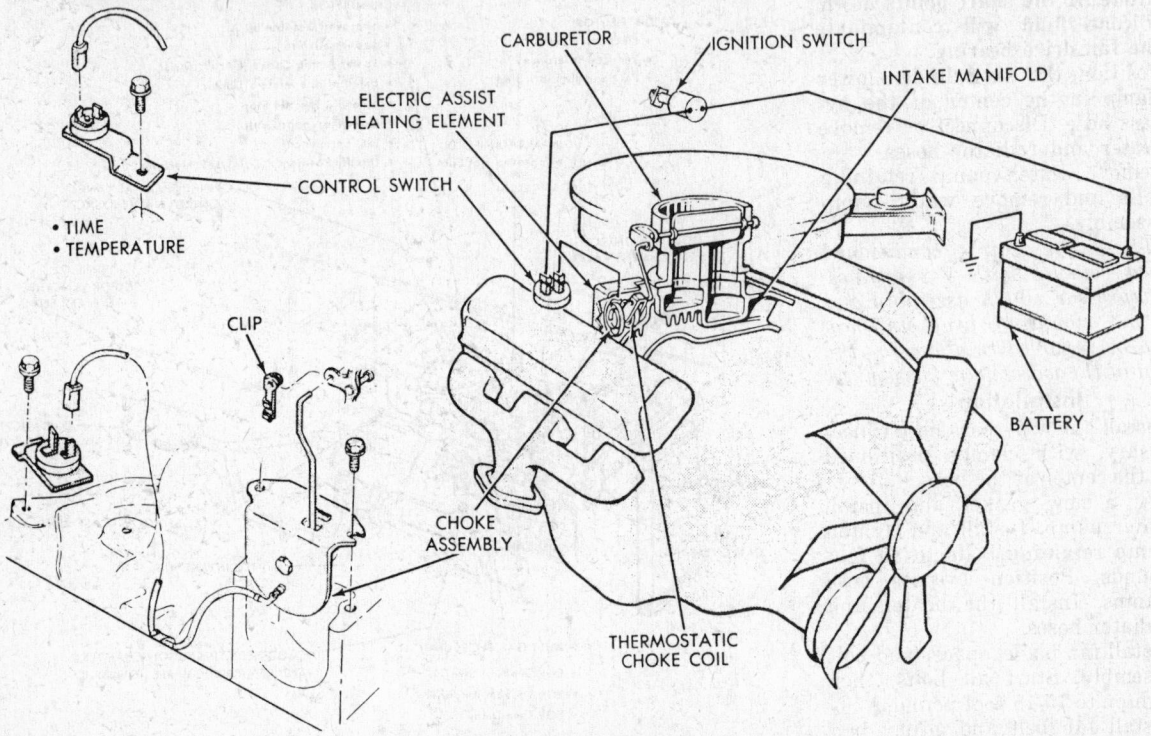

Electric assist choke system (© Chrysler Corp.)

Venturi Vacuum Control System

This system uses a vacuum tap at the throat of the carburetor venturi to provide a control signal. Because the signal is so low however, a vacuum amplifier is used to increase the strength of the signal. The amplifier uses stored manifold vacuum to provide the source for amplification. Elimination of EGR at wide open throttle is accomplished by a "dump" diaphragm which compares venturi and manifold vacuum to determine when wide open throttle is achieved. At wide open throttle, the internal reservoir is "dumped", limiting the output of the EGR valve to manifold vacuum. As with the ported vacuum control system, the valve opening point is set above the manifold vacuum available at wide open throttle, permitting the valve to be closed at wide open throttle.

Coolant Control Exhaust Gas Recirculation Valve

Trucks with EGR are equipped with a CCEGR valve located in the top of the radiator tank. When coolant in the top radiator tank reaches 65° F, the valve opens to apply vacuum to the EGR valve to recirculate exhaust gases.

EGR Delay System

Some trucks are equipped with an EGR delay system, which is an electrical timer on the dash to control an engine mounted solenoid. The timer prevents exhaust gas recirculation for about 35 seconds after the ignition is turned on.

Air Injection System

This system adds a controlled amount of air through special passages in the cylinder head, to exhaust gases in the exhaust ports, causing oxidation of the gases and thereby reducing carbon monoxide and hydrocarbon emissions to the required levels.

The air injection system consists of a belt-driven air pump, rubber hose, a check valve to protect the hoses and pump from hot gases, injection tubes, and a combination diverter/pressure relief valve assembly.

Evaporative Control System

The function of the evaporative control system is to prevent the emission of raw gasoline vapors from the fuel tank and carburetor into the atmosphere. When fuel evaporates in the tank or float bowl, the vapors pass through lines and into the charcoal canister where they are temporarily stored until they can be drawn into the intake manifold and burned. A vacuum port located in the base of the carburetor governs vapor flow from the canister to the engine.

Closed Crankcase Ventilation System

The closed PCV system operates as follows:

In place of a vented oil filler cap, an air intake line is installed between the carburetor air filter and a crankcase opening in the valve cover.

A sealed oil filler cap and dipstick are used.

A separate PCV air filter is used. The filter is located where the intake air line connects to the valve cover.

Under normal engine operation, air enters through the intake line from the air filter. Under heavy acceleration, any excess vapors back up through the air intake line and are forced to mix with incoming air into the carburetor and are burned in the combustion chamber. Back-up fumes cannot escape into the atmosphere, creating a closed system.

The PCV valve is used to control the rate at which crankcase vapors are returned to the intake manifold. The action of the valve plunger is controlled by intake manifold vacuum and the spring. During deceleration and idle, when manifold vacuum is high, it overcomes the tension of the valve spring and the plunger bottoms in the manifold end of the valve housing. Because of the valve construction, it reduces, but does not stop, the passage of vapors to the intake manifold. When the engine is lightly accelerated or operated at constant speed, spring tension matches intake manifold vacuum pull and the plunger takes a mid-position in the valve body, allowing more vapors to flow into the manifold.

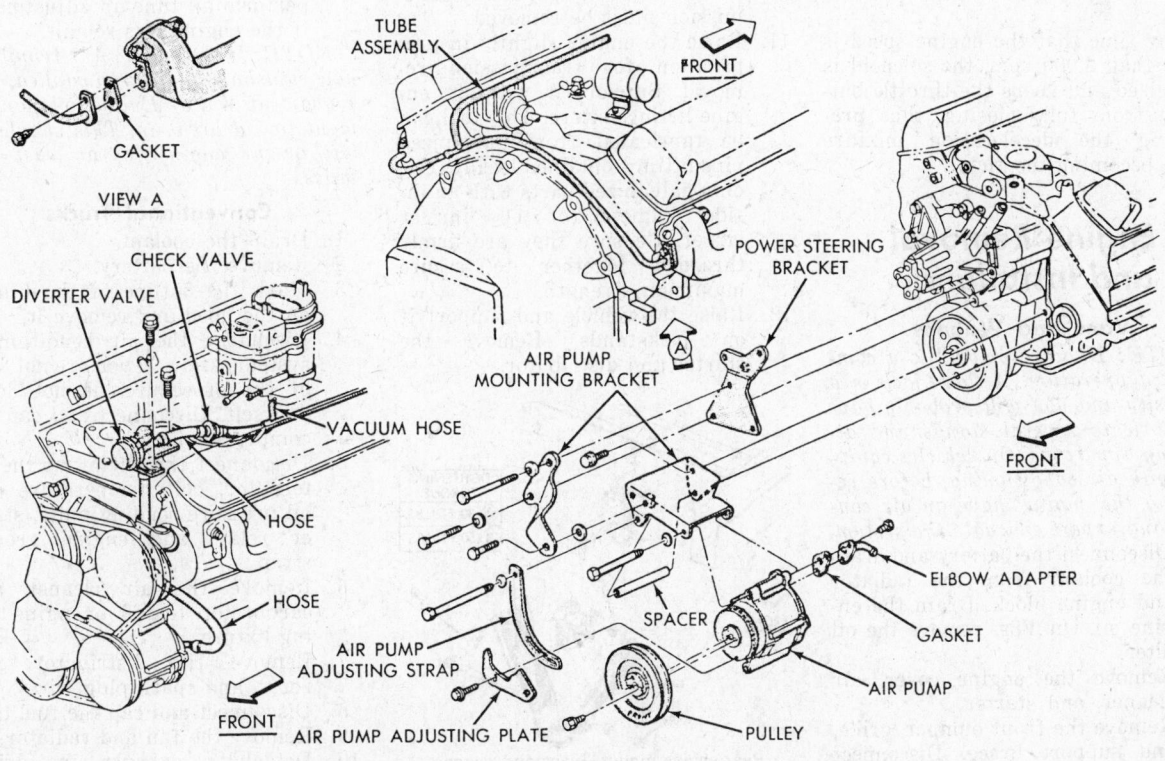

Air injection system installation for 360 engine (© Chrysler Corp.)

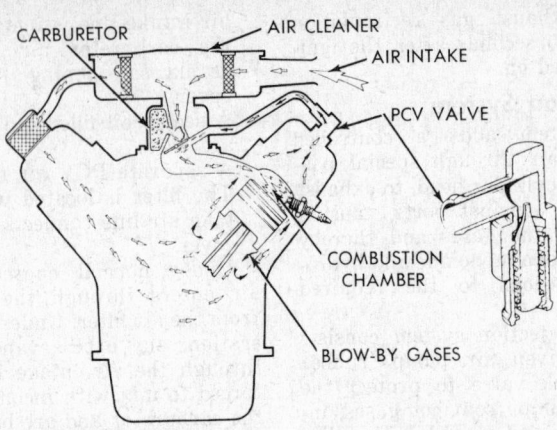

Closed crankcase ventilation system (© Chrysler Corp.)

Catalytic Converter

Most 1975 and later models less than 6,000 lbs GVW, are equipped with catalytic converters. These devices are used to oxidize excess carbon monoxide (CO) and hydrocarbons (HC) in the exhaust system before they can escape out the tailpipe and into the atmosphere. The converter is installed in front of the mufflers, underneath the truck, and protected by a heat shield.

The expected catalyst life is 50,000 miles, provided that the engine is kept in tune and unleaded fuel is used.

To keep the catalyst from being overheated by an overly rich mixture during deceleration, a catalyst protection system (CPS) is sometimes used. The system consists of a throttle positioner solenoid (not to be confused with the idle stop solenoid), a control box, and an engine rpm sensor.

Any time that the engine speed is more than 2,000 rpm, the solenoid is energized and keeps the throttle butterfly from fully closing, thus preventing the deceleration mixture from becoming too rich.

Engine Removal and Installation

Vans and Wagons

NOTE: Engine removal is a complicated operation. A floor jack is a necessity and you will probably have to fabricate several stands and attaching apparatus. On vehicles equipped with air conditioning, before removing the engine, have an air conditioning expert evacuate the system.

1. Disconnect the battery and drain the coolant from the radiator and engine block. Drain the engine oil. On V8s, remove the oil filter.
2. Remove the engine cover, air cleaner, and starter.
3. Remove the front bumper, grille, and support brace. Disconnect both radiator hoses and remove the radiator and support brace as a unit.
4. Remove the power steering and air pumps with the hoses attached and lay them aside.
5. Disconnect the throttle linkage, heater and vacuum hoses and all electrical connections to the ignition, alternator, and all other electrical connections.
6. Remove the alternator, fan, pulley, and drive belts.
7. Remove the heater blower motor.
8. Remove and plug the inlet line to the fuel pump.
9. Remove the oil dipstick tube. On V8s, remove the intake manifold and left exhaust manifold. If equipped with air conditioning, remove the right side valve cover.
10. To provide clearance for engine removal, the oil pan and transmission must be removed.
11. Raise the engine slightly in preparation for transmission removal. Support it with an engine lifting fixture. This tool can be fabricated from galvanized pipe fittings obtained locally. Use only galvanized parts with an inside diameter of 1½ in. or larger. Be sure they are firmly threaded together to assure maximum strength.
12. Raise the vehicle and support it on jackstands. Remove the starter and distributor.

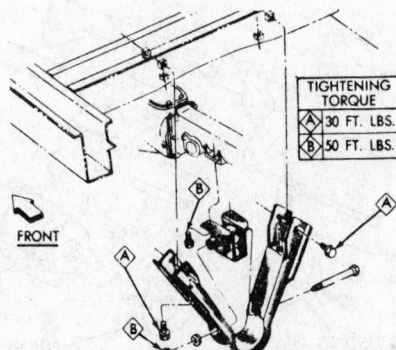

TIGHTENING TORQUE

Ⓐ 30 FT. LBS.
Ⓑ 50 FT. LBS.

FRONT

Rear engine mount—vans and wagons
(© Chrysler Corp.)

13. Remove the driveshaft and engine rear support. Remove the rear support by removing the rear mount through-bolt and the U-shaped bracket from the crossmember. Remove the insulator from the bottom face of the transmission housing.
14. If equipped with an automatic transmission, remove the transmission intact with the filler tube and the torque converter separated from the drive plate.
15. Raise the rear of the engine approximately 2 in. and remove the clutch or drive plate and the flywheel.
15a. On V8s, position the cut-out in the crankshaft flange at 3 o'clock. Remove the oil pan screws and lower the oil pan far enough to reach inside and turn the oil pump pick-up tube slightly to the right to clear the pan. Remove the oil pan.
16. Lower the vehicle.
17. Using a boom hoist attached to the engine with the shortest hook-up possible, take up all tension and support the engine. The boom hoist is the ideal tool to use. If one is not available, it may be possible to support the engine on a stationary hoist and roll the vehicle out from under the engine.
18. Remove the engine front mounts and insulators.
19. Carefully remove the engine from the vehicle.
20. Installation is the reverse of removal. Check all fluid levels and perform all tune-up adjustments if the engine was rebuilt.

NOTE: If the engine was rebuilt or new camshaft or lifters installed, add 1 quart of MoPar Engine oil Supplement to aid break-in. This should be left in the engine for at least 500 miles.

Conventional Trucks

1. Drain the coolant.
2. Remove the battery.
3. Mark the outline of the hinges on the hood and remove it.
4. Discharge the air conditioning system safely. If you are not sure of the procedure, do not do it yourself. Disconnect and cap the compressor lines.
5. Disconnect the wiring at the alternator, coil, temperature and oil pressure sending units, starter relay, and engine ground strap.
6. Remove the air cleaner and carburetor. Install an engine lifting fixture.
7. Remove the distributor cap, rotor, and spark plug wires.
8. Disconnect and cap the fuel line.
9. Remove the fan and radiator.
10. Detach the exhaust pipe, driveshaft, linkage, and oil cooler

lines.

11. Support the rear of the engine and remove the engine rear crossmember and transmission. On the 361 and 413 engines, remove the transmission only.

12. Unbolt the engine mounts and lift the engine out. On D500 and larger models, the grille and support assembly should be removed first.

Ramcharger and Trail Duster, W100, 200, 300

1. Drain the coolant from the radiator and cylinder block.

2. Disconnect the battery ground cable. Remove the battery on V8 models.

3. Scribe the outline of the hood hinges and remove the hood.

4. If equipped with air conditioning, remove the compressor with lines attached and lay it aside.

CAUTION: Do not disconnect any refrigerant lines. Bodily injury could result.

5. Disconnect the electrical connections at the alternator, ignition coil, temperature and oil pressure sending units, starter-to-solenoid, and engine/body ground.

6. Remove the air cleaner and carburetor. Install an engine lifting fixture.

7. Remove the distributor cap and rotor.

8. Disconnect and plug the fuel pump line.

9. Disconnect the radiator and heater hoses. Disconnect and plug the oil cooler lines.

10. Remove the fan, spacer, fluid drive, and radiator. Do not store the fan drive unit with the shaft pointing downward. Fluid will leak out.

11. Raise the truck and support the rear of the engine.

12. Disconnect the exhaust pipes at the manifolds.

13. Remove the starter on V8 models.

14. Remove the automatic transmission dust cover and attach a C-clamp to the front bottom of the torque converter housing to prevent it from failling out. Remove the drive plate bolts from the torque converter. On manual transmission models, remove the rear crossmember, transmission, transfer case and adapter. You can leave the transfer case in place on six-cylinder models.

15. Support the transmission and remove the transmission attaching bolts.

16. Lower the truck and attach a hoist to the engine.

17. Remove the front motor mount bolt stud nuts and washers.

18. Carefully remove the engine.

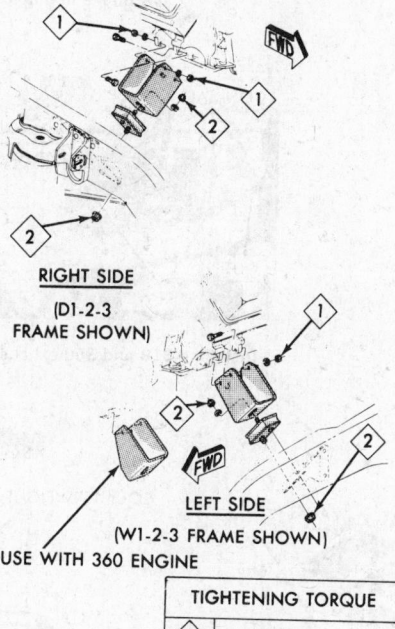

Rear engine mount—100-300 series

RIGHT SIDE
(D1-2-3 FRAME SHOWN)

LEFT SIDE
(W1-2-3 FRAME SHOWN)
USE WITH 360 ENGINE

TIGHTENING TORQUE	
1	55 FT.LBS.
2	75 FT.LBS.

V8 engine mounts—100-300 series, conventional trucks (© Chrysler Corp.)

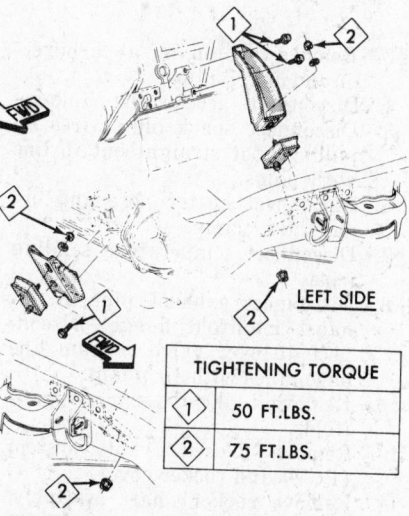

LEFT SIDE

TIGHTENING TORQUE	
1	50 FT.LBS.
2	75 FT.LBS.

RIGHT SIDE

Six cylinder, except 4WD, engine mounts—100-300 series, conventional trucks

19. Installation is the reverse of removal. Fill the engine with coolant and fresh oil. Adjust the transmission linkage, carburetor, and ignition timing.

Manifolds

Removal and Installation

6 Cylinder Combination Manifold

1. Remove the air cleaner, lines and tubes to the carburetor.

2. Disconnect all the linkages to the carburetor and remove the carburetor from the manifold.

3. Disconnect the exhaust pipe from the manifold, remove the manifold attaching washers and retaining nuts, and remove the manifold from the cylinder head.

4. Separate the exhaust manifold from the intake manifold, if necessary, and install a new gasket between the two upon reassembly. *NOTE: Do not tighten the three securing bolts until the manifold assembly has been installed on the cylinder head.*

5. Position the manifold on the cylinder head using a new gasket, and install the conical and triangular washers, the retaining nuts, and torque the retaining nuts and the three securing bolts to the specified torque.

6. Attach the exhaust pipe to the exhaust manifold flange.

7. Install the carburetor and attach all the lines, tubes, and linkages. Install the air cleaner assembly.

V8 Intake Manifold

1. Drain cooling system and disconnect battery.

2. Remove alternator, carburetor air cleaner, and fuel line.

3. Disconnect accelerator linkage.

4. Remove vacuum control between carburetor and distributor.

5. Remove distributor cap and wires.

6. Disconnect coil wires, temperature sending unit wire, heater hoses and bypass hose.

7. Remove intake manifold, ignition coil and carburetor as an assembly.

8. Installation is the reverse of the above procedure. Tighten the intake manifold to head bolts in the sequence illustrated, from center alternating out. Tighten manifold mounting bolts to torque listed in Specifications, tightening in 15 ft. lb. increments.

V8 Exhaust Manifold

1. Disconnect the exhaust manifold at the flange where it mates to the exhaust pipe.

2. If the vehicle is equipped with air injection and/or a carbure-

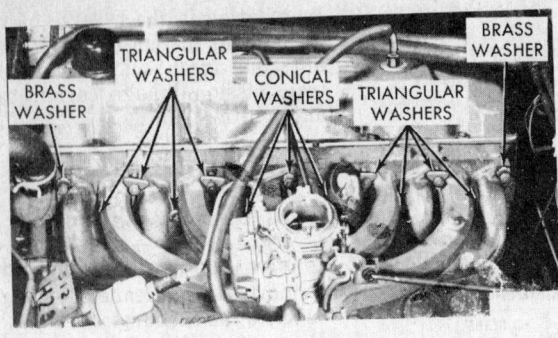

Intake and exhaust manifold installation,
slant six (© Chrysler Corp.)

V8 Intake manifold bolt tightening sequence (© Chrysler Corp.)

tor-heated air stove, remove them.

3. Remove the exhaust manifold by removing the securing bolts and washers. To reach these bolts, it may be necessary to jack the engine slightly off its front mounts. When the exhaust manifold is removed, sometimes the securing studs will screw out with the nuts. If this occurs, the studs must be replaced with the aid of sealing compound on the coarse thread ends. If this is not done, water leaks may develop at the studs. To install the exhaust manifold, reverse the removal procedure. On the center branch of the 318 and 360 manifold, no conical washers are used.

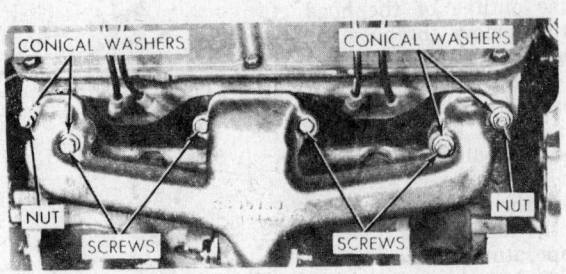

Installing 318 and 360 exhaust manifold (© Chrysler Corp.)

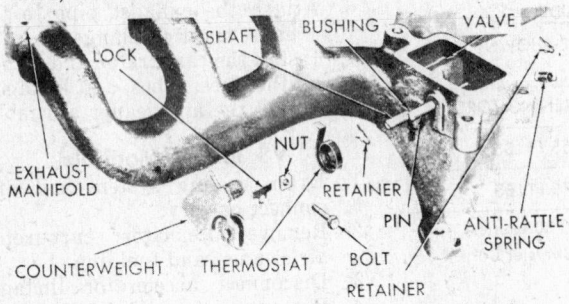

Slant six manifold heat control valve (© Chrysler Corp.)

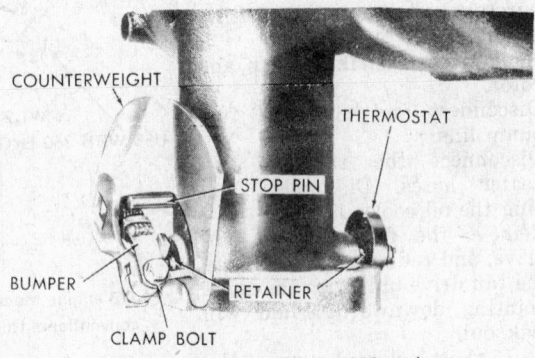

Manifold heat control valve 383 engine
(© Chrysler Corp.)

Manifold Heat Control Valve

Inspect operation of manifold heat control valve by racing engine momentarily to wide open throttle. The counterweight should respond by moving about ½". If it does not move it may be frozen with deposits or the thermostatic spring may be weak or broken. Use a manifold heat control valve solvent to loosen valve shaft at bushings. CAUTION: Be sure that manifold is cool when applying solvent and let it soak for a few minutes before attempting to free the valve.

Cylinder Head

Removal and Installation

6 Cyl. OHV
1. Drain cooling system.
2. Remove air cleaner and fuel line.
3. Remove vacuum line at carburetor and distributor.
4. Disconnect accelerator linkage.
5. Disconnect spark plug wires by pulling boot straight out in line with plugs.
6. Disconnect heater hose and by-pass hose clamp.
7. Disconnect temperature sending wire.
8. Disconnect exhaust pipe at exhaust manifold flange. Disconnect diverter valve vacuum line on engines with air pump.
9. Remove intake and exhaust manifolds.
10. Remove closed vent system (PCV) and rocker cover.
11. Remove rocker shaft assembly.
12. Remove pushrods in sequence and mark them in such a way that they may be put back into their original positions.
13. Remove head bolts.
14. Remove spark plugs and tubes (through 1974).
15. To install, clean all gasket surfaces of cylinder block and cylinder head and install spark plugs.
16. Check all surfaces with a straightedge if there is any reason to suspect leakage.
17. Install gasket, using sealer, and cylinder head.
18. Install cylinder head attaching bolts. Tighten all bolts in the sequence illustrated to 50 ft. lbs. torque, then repeat tightening sequence torquing to 65 ft. lbs.
19. Install rocker arms and shaft assembly with the flat (through 1973) or the oil hole on the end of the shaft on top and to the front. The special bolt goes to the rear. Install rocker shaft retainers between rocker arms so they

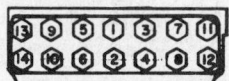

Slant six cylinder head bolt tightening sequence

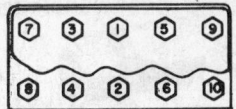

Cylinder head bolt tightening sequence,
360 and smaller V8

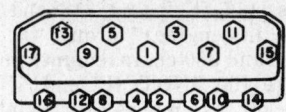

Cylinder head bolt tightening sequence,
361 and larger V8

seat on rocker shaft and not on extended bushing of rocker arm. Be sure to install long retainer in center position only. Tighten bolts to 25 ft. lbs. Tighten the special bolt to 200 in. lbs.

20. Loosen the 3 bolts holding intake manifold to exhaust manifold. This is necessary for proper alignment.

21. Install the manifolds as detailed under manifold removal and installation.

22. Connect heater hose and by-pass hose clamp.

23. Connect heat indicator sending unit wire, the accelerator linkage and spark plug wires.

24. Install the vacuum control tube from carburetor to distributor. Install the air tube assembly with a new gasket to the head, tightening it to 100 inch-pounds. Install the diverter valve vacuum line.

25. Connect exhaust pipe to exhaust manifold flange.

26. Install fuel line and carburetor air cleaner.

27. Fill cooling system, start and warm up engine, and adjust valve tappet clearance.

28. Install valve cover using new gasket, tightening retaining nuts to 3-4 ft. lbs.

29. Install crankcase ventilation system.

V8 Engines

1. Drain cooling system and disconnect battery. On vehicles with 361 or 413 cu. in. engines, remove battery.

2. Remove alternator, air cleaner and fuel line.

3. Disconnect accelerator linkage.

4. Remove vacuum control hose between carburetor and distributor.

5. Remove cooling and heater hoses

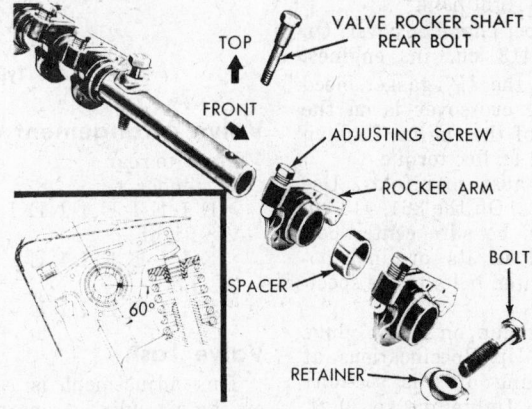

Slant six rocker shaft, 1974 and later models don't have the flat shown (© Chrysler Corp.)

from head and, if so equipped, remove air compressor.

6. Remove distributor cap and high tension leads as an assembly.

7. Remove heat indicator sending unit wire.

8. Remove crankcase ventilation system and valve covers.

9. Remove spark plugs and, on 361, 413 cu. in. engines, remove distributor and governor.

10. Remove intake manifold, carburetor and, if attached to manifold, ignition coil as an assembly.

11. Remove tappet chamber cover.

12. On 361, 413 cu. in. engines, remove the bolts that secure water pump housing to each cylinder head.

13. Remove exhaust manifolds. On the 361, 413 cu. in. engines, tag the center bolts for reinstallation in same hole.

14. Remove rocker arms and shaft assemblies.

15. Remove pushrods and *identify to insure installation in original location.*

16. Remove cylinder head bolts and cylinder heads.

17. Before installing cylinder heads, clean all gasket surfaces of block and heads.

18. Inspect all surfaces with a straightedge if there is any indication of leakage.

19. Coat new head gaskets with sealant and install on block.

20. Install cylinder head and head bolts. Tighten bolts in the sequence illustrated, first to 50 ft. lbs. torque, then in sequence again to the specified torque (see Specifications at the beginning of this section). CAUTION: Do not retighten cylinder head bolts after engine has been operated if embossed steel head gaskets are used.

21. Inspect pushrods for wear or bending and install in original positions. Use an aligning rod.

22. Install rocker shaft assembly, making sure long stamped steel retainers are in the No. 2 and No. 4 positions. Tighten rocker shaft mounting bolts to 25 ft.

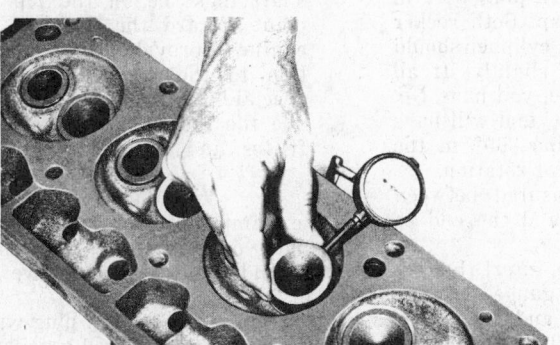

Measuring valve stem to guide clearance (© Chrysler Corp.)

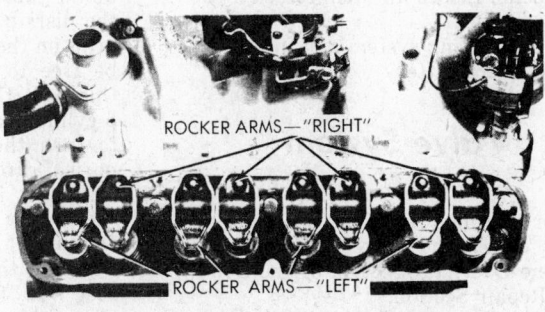

Proper rocker arm location on shaft (© Chrysler Corp.)

lbs. (17 ft. lbs on 318 and 360 cu. in. engine) torque. On the 318 and 360 cu. in. engines, make sure the "NOTCH" end of the rocker assemblies are pointing toward the centerline of the engine and toward the front on the left bank and toward the rear on the right bank.

23. Install tappet chamber cover. On the 361, 413 cu. in. engines, make sure the ⅛" gasket bleed hole at the crossover is on the right side of the engine. Tighten bolts to 7-8 ft. lbs. torque.

24. Install exhaust manifolds. Use new gaskets. On the 361, 413 cu. in. engines, be sure center bolt is inserted in its original position. Tighten bolts to the specified torque.

25. Set correct gap on spark plugs (see Tune-Up Specifications at the beginning of this section) and install, tightening to 30 ft. lbs. torque.

26. Install intake manifold and carburetor assembly, tightening bolts from center outward, first to 25 ft. lbs. torque, then again to correct torque (see Torque Specifications at the beginning of this section). On the 318 and 360 cu. in. engine, coat the gaskets with sealant and install with bead down.

27. If they were removed, install distributor and governor.

28. Install distributor cap and high tension leads.

29. Connect vacuum control hose between distributor and carburetor, throttle linkage, heat indicator sending unit wire, heater and coolant hoses, fuel line and, if so equipped, manual choke cable.

30. On the 361, 413 cu. in. engines, install bolts which secure water pump housing to each cylinder head.

31. Install rocker cover, using new gasket. Tighten bolts to 3-4 ft. lbs. torque.

32. Install crankcase ventilation system and air cleaner.

33. Install alternator and drive belts. Install air compressor if so equipped.

34. Fill cooling system and install or connect battery.

Valve System

Reference

For complete instructions for servicing valves and valve train see the Engine Overhaul Section of the General Repair Section.

For valve angles, stem dimensions, clearances and spring tensions see Valve Specifications at the beginning of this section.

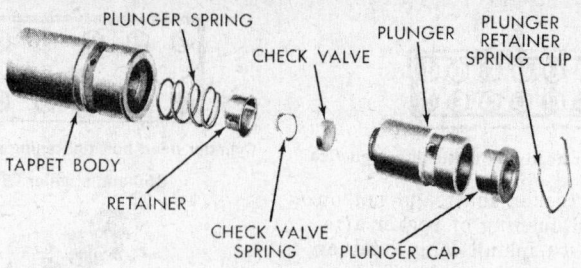

Typical hydraulic tappet (© Chrysler Corp.)

Valve Arrangement

Front to rear
6 Cylinder
E I E I E I I E I E I E
V-8 Right
E I I E E I I E
    Left
E I I E E I I E

Valve Lash

This adjustment is required only on the six-cylinder engine. It should be done at every tune-up. It should also be done whenever there is excessive noise from the valve mechanism.

No valve lash adjustment is necessary or possible on any other Chrysler-built engine. Hydraulic valve lifters automatically maintain zero clearance. After engine reassembly these lifters adjust themselves as soon as engine oil pressure builds up.

NOTE: Do not set the valve lash closer than specified in an attempt to quiet the valve mechanism. This will result in burned valves.

The manufacturer recommends that the valves be adjusted with the engine running, but the following procedure can also be used.

1. The engine must be at normal operating temperature. Mark the crankshaft pulley into three equal 120° segments, starting at the TDC mark.

2. Remove the valve (rocker) cover and the distributor cap.

3. Set the engine at TDC on the No. 1 cylinder by aligning the mark on the crankshaft pulley with the 0° mark on the timing cover pointer. The distributor rotor should point at the position of the No. 1 spark plug wire in the distributor cap. Both rocker arms on the No. 1 cylinder should be free to move slightly. If all this isn't the case, you have No. 6 cylinder at TDC and will have to turn the engine 360° in the normal direction of rotation.

4. The lash is measured between the rocker arm and the end of the valve.

5. To check the lash, insert the correct size feeler gauge between the rocker arm and the valve. Press down lightly on the other end of the rocker arm. If the gauge cannot be inserted, loosen

the self-locking adjustment nut on top of the rocker arm. Tighten the nut until the gauge can just be inserted and withdrawn without buckling.

6. After both valves for the No. 1 cylinder are adjusted, turn the engine so that the pulley turns 120° in the normal direction of rotation (clockwise). The distributor rotor will turn 60°, since it turns at half engine speed.

7. Check that the rocker arms are free and adjust the valves for the next cylinder in the firing order, 5. The firing order is 1-5-3-6-2-4.

8. Turn the engine 120° to adjust each of the remaining cylinders in the firing order. When you are done the engine will have made two complete revolutions (720°) and the rotor one complete revolution (360°).

9. Replace the rocker cover with a new gasket. Replace the distributor cap. Start the engine and check for leaks.

Rocker Shaft Removal and Installation

Six-Cylinder Engines

1. Remove the closed ventilation system.

2. Remove the evaporative control system (if so equipped).

3. Remove the valve cover with its gasket.

4. Take out the rocker arm and shaft assembly securing bolts and remove the rocker arm and shaft.

5. Reverse the above for installation. The flat (through 1973) or the oil hole on the end of the shaft must be on the top and point toward the front of the engine to provide proper lubrication to the rocker arms. The special bolt goes to the rear. Torque the rocker arm bolts to 25 ft. lbs. and adjust the valves.

V8 Engines

The stamped steel rocker arms are arranged on one rocker arm shaft per cylinder head. To remove the rocker arms and shaft:

1. Disconnect the spark plug wires.

2. Disconnect the closed ventilation and evaporative control system (if so equipped).

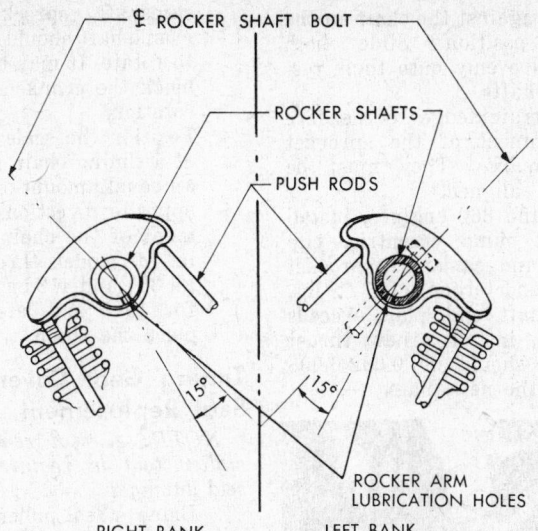

On 361 and larger V8s, the rocker arm lubrication holes should be aligned as shown (© Chrysler Corp.)

3. Remove the valve covers with their gaskets.
4. Remove the rocker shaft bolts and retainers, and lift off the rocker arm assembly.
5. Reverse the above procedure to install. The notch on the end of both rocker shafts on the 318 and 360 should point to the engine centerline and toward the front of the engine on the left cylinder head, or toward the rear on the right cylinder head. On the 361, 383, 400, 413, and 440, the rocker arm lubrication holes must point down and toward the valves. Torque the rocker shaft bolts to 17 ft. lbs. on the 318 and 316, and 25 ft. lbs. on the others.

Valve Stem Oil Seal Replacement

If valve stem oil seals are found to be the cause of excessive oil consumption, they may be replaced without removing the cylinder heads.
1. Remove the air cleaner.
2. Remove the rocker arm covers and spark plugs.
3. Detach the coil wire from the distributor.
4. Turn the engine so that No. 1

cylinder is at Top Dead Center on the compression stroke. Both valves for No. 1 cylinder should be fully closed and the crankshaft damper timing mark at TDC. The distributor rotor will point at the No. 1 spark plug wire location in the cap.
5. Remove the rocker shaft and install a dummy shaft.
6. Apply 90-100 psi air pressure to No. 1 cylinder, using a spark plug hole air hose adaptor.
7. Use a valve spring compressor to compress each No. 1 cylinder valve spring and remove the retainer locks and the spring. Remove the old seals.
8. Install a cup shield on the exhaust valve stem. Position it down against the valve guide.
9. Push the intake valve stem seal firmly and squarely over the valve guide.
10. Compress the valve spring only enough to install the lock.
11. Repeat the operation on each successive cylinder in the firing order, making sure that the crankshaft is exactly on TDC for each cylinder. See the Firing Order and Distributor Rotation illustrations in the Specifications

Section of this Chapter for cylinder numbering.
12. Replace the rocker arms, covers, spark plugs, and coil wire.

Timing Cover and Chain

NOTE: On models through 1974 it is normal to find particles of rubber collected between the seal retainer and the crankshaft oil slinger after the seal has been in service. Check the slack in the chain after installation.

Removal and Installation

Six-Cylinder Engines
1. Drain the cooling system and disconnect the battery.
2. Remove the radiator and fan.
3. With a puller, remove the vibration damper.
4. Loosen the oil pan bolts to allow clearance, and remove the timing case cover and gasket.
5. Slide the crankshaft oil slinger off the front of the crankshaft.
6. Remove the camshaft sprocket bolt.
7. Remove the timing chain with the camshaft sprocket.
8. On installation: Turn the crankshaft to line up the timing mark on the crankshaft sprocket with the centerline of the camshaft (without the chain).
9. Install the camshaft sprocket and chain. Align the timing marks.

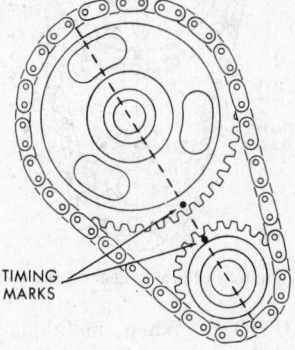

Alignment of timing gear marks—6 cylinder

10. Torque the camshaft sprocket bolt to 35 ft. lbs.
11. Replace the oil slinger.
12. Reinstall the timing case cover with a new gasket and torque the bolts to 17 ft. lbs. Retighten the engine oil pan to 17 ft. lbs.
13. Press the vibration damper back on.
14. Replace the radiator and hoses.
15. Refill the cooling system.

V8 Engines
1. Disconnect the battery and drain

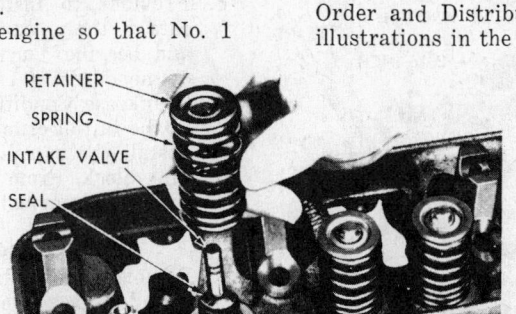

Valve stem seal installation

the cooling system. Remove the radiator.

2. Remove the vibration damper pulley. Unbolt and remove the vibration damper with a puller. On 318 and 360 engines, remove the fuel lines and fuel pump, then loosen the oil pan bolts and remove the front bolt on each side. On 361 and 413 engines, you will have to remove the front engine mount to get out the two front pan bolts.

3. Remove the timing gear cover and the crankshaft oil slinger.

4. On 318 and 360 engines, remove the camshaft sprocket lockbolt, securing cup washer, and fuel pump eccentric. Remove the timing chain with both sprockets. On 383, 361, 400, 413, and 440 engines, remove the camshaft sprocket lockbolt and remove the timing chain with the camshaft and crankshaft sprockets.

5. To begin the installation procedure, place the camshaft and crankshaft sprockets on a flat surface with the timing indicators on an imaginary centerline through both sprocket bores. Place the timing chain around both sprockets. Be sure that the timing marks are in alignment.

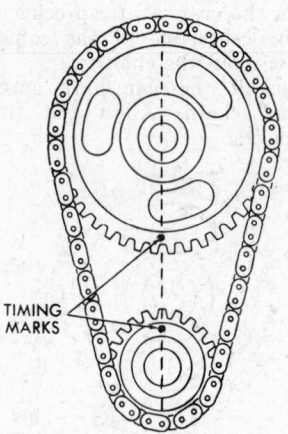

TIMING MARKS

Alignment of timing gear marks—V8

CAUTION: When installing the timing chain, have an assistant support the camshaft with a screwdriver to prevent it from contacting the plug in the rear of the engine block. Remove the distributor and the oil pump/distributor drive gear. Position the screwdriver against the rear side of the cam gear and be careful not to damage the cam lobes.

6. Turn the crankshaft and camshaft to align them with the keyway location in the crankshaft sprocket and the keyway or dowel hole in the camshaft sprocket.

7. Lift the sprockets and timing chain while keeping the sprock-

ets tight against the chain in the correct position. Slide both sprockets evenly onto their respective shafts.

8. Use a straightedge to measure the alignment of the sprocket timing marks. They must be perfectly aligned.

9. On 318 and 360 engines, install the fuel pump eccentric, cup washer, and camshaft sprocket lockbolt and torque to 35 ft. lbs. If camshaft end play exceeds 0.010 in., install a new thrust plate. It should be 0.002-0.006 in. with the new plate.

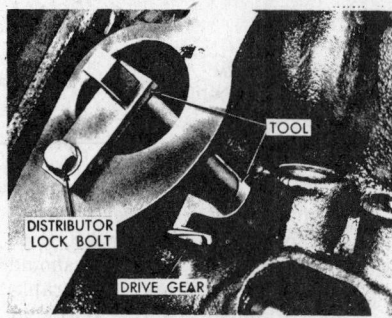

TOOL

DISTRIBUTOR LOCK BOLT

DRIVE GEAR

The V8 camshaft should be held forward while installing the chain and sprockets

On 383, 361, 400, 413, and 440 V8s, install the washer and camshaft sprocket lockbolt(s) and then torque the lockbolt to 35-40 ft. lbs. Check to make sure that the rear face of the camshaft sprocket is flush with the camshaft end.

Checking Timing Chain Slack

1. Position a scale next to the timing chain to detect any movement in the chain.

2. Place a torque wrench and socket on the camshaft sprocket attaching bolt. Apply either 30 ft. lbs. (if the cylinder heads are installed on the engine) or 15 ft. lbs. (cylinder heads removed) of force to the bolt and rotate the bolt in the direction of crankshaft rotation in order to remove all slack from the chain.

3. While applying torque to the

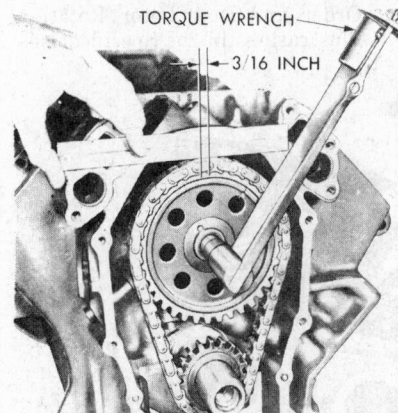

TORQUE WRENCH

3/16 INCH

Measuring timing chain stretch
(© Chrysler Corp.)

camshaft sprocket bolt, the crankshaft should not be allowed to rotate. It may be necessary to block the crankshaft to prevent rotation.

4. Position the scale over the edge of a timing chain link and apply an equal amount of torque in the opposite direction. If the movement of the chain exceeds 3/16 in. for models through 1972 and all 361 and 413 engines, or 1/8 in. for 1973 and later models, replace the chain.

Timing Gear Cover Seal Replacement

NOTE: A seal remover and installer tool is required to prevent seal damage.

1. Using a seal puller, separate the seal from the retainer.

2. Pull the seal from the case.

3. To install the seal place it face down in the case with the seal lips downward.

4. Seat the seal tightly against the cover face. There should be a maximum clearance of .0014 in. between the seal and the cover. Be careful not to over-compress the seal.

Camshaft

Removal and Installation

Six-Cylinder Engines

1. Remove the cylinder head, timing gear cover, camshaft sprocket, and timing chain.

2. Remove the valve tappets, keeping them in order to ensure installation in their original locations.

3. Remove the crankshaft sprocket.

4. Remove the distributor and oil pump.

5. Remove the fuel pump.

6. Install a long bolt into the front of the camshaft to facilitate its removal.

7. Remove the camshaft, being careful not to damage the cam bearings with the cam lobes.

8. Previous to installation, lubricate the camshaft lobes and bearing journals. It is recommended that 1 pt. of Chrysler Crankcase Conditioner be added to the initial crankcase oil fill.

9. Install the camshaft in the engine block. From this point, reverse the removal procedure.

V8 Engines

1. Remove the intake manifold, cylinder head covers, rocker arm assemblies, push rods, and valve tappets, keeping them in order to insure the installation in their original locations.

2. Remove the timing gear cover, the camshaft and crankshaft

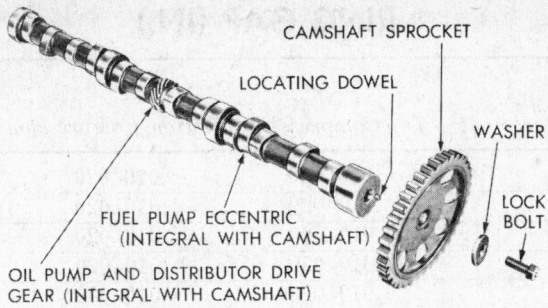

Camshaft OHV 6 cyl. engines (© Chrysler Corp.)

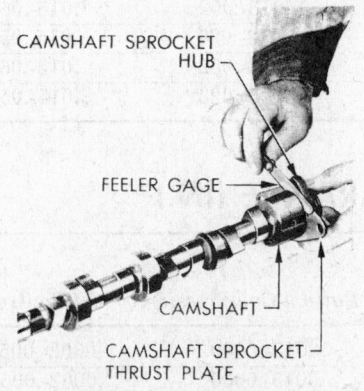

Measuring camshaft end play
(© Chrysler Corp.)

sprockets, and the timing chain.

3. Remove the distributor and lift out the oil pump and distributor driveshaft.

4. Remove the camshaft thrust plate (on 318 and 360).

5. Install a long bolt into the front of the camshaft and remove the camshaft, being careful not to damage the cam bearings with the cam lobes.

6. Previous to installation, lubricate the camshaft lobes and bearing journals. It is recommended that 1 pt. of Chrysler Crankcase Conditioner be added to the initial crankcase oil fill. Insert the camshaft into the engine block within 2 in. of its final position in the block.

7. Have an assistant support the camshaft with a screwdriver to prevent the camshaft from contacting the plug in the rear of the engine block. Position the screwdriver against the rear side of the cam gear and be careful not to damage the cam lobes.

8. Replace the camshaft thrust plate. If camshaft end play exceeds 0.010 in., install a new thrust plate. It should be 0.002-0.006 in. with the new plate.

9. Install the timing chain and sprockets, timing gear cover, and pulley.

10. Install the tappets, pushrods, rocker arms, and cylinder head covers.

11. Install the distributor and oil pump driveshaft. Install the distributor.

12. After starting the engine, adjust the ignition timing.

Pistons and Connecting Rods

The notch on the top of each piston must face the front of the engine.

To position the connecting rod correctly, the oil squirt hole should point to the right-side on all six-cylinder engines. On all V8 engines, the larger chamfer of the lower connecting rod bore must face to the rear on the right bank and to the front on the left bank.

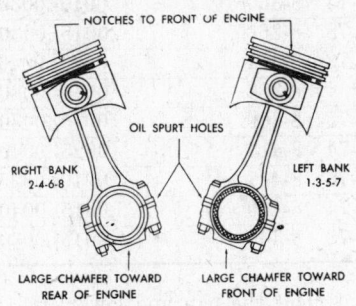

Relation of piston to rod V8 engines
(© Chrysler Corp.)

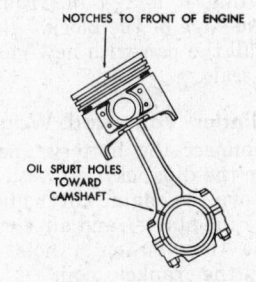

Relation of piston to rod OHV 6 engines
(© Chrysler Corp.)

Bearings

Reference

Detailed procedures for fitting main and rod bearings can be found in the Engine Rebuilding Section.

6 Cylinder Engine

The maximum allowable bearing clearance is .001″. No. 1, No. 2 and No. 4 lower inserts are interchangeable. No. 2 and No. 4 upper inserts are interchangeable. No. 1 upper insert has a chamfer on the tab side for timing chain oiling and is identified by the red mark on the edge of the insert. No. 3 upper and lower inserts are flanged. Bearing caps are not interchangeable and are numbered for correct installation. Maximum end play is .0085″. Replace No. 3 (thrust) bearing if end play exceeds that amount.

V8 Engine

A Maltese Cross stamped on the engine (except on the 318 and 360) numbering pad indicates that the engine is equipped with a crankshaft which has one or more connecting rods and/or main bearing journal finished .001″ undersize. The position of the undersize journal(s) is stamped on a machine surface of the No. 3 counterweight. The letter "R" or "M" signifies whether the undersize journal is a rod or main, and the number following the letter indicates which one it is. A Maltese Cross with an "X" indicates that all those journals are .010″ undersize. On the 318 and 360 engines, .001 in. undersize journals are indicated by marks on the No. 8 crankshaft counterweight. If the "R" or "M" is followed by "X", all those journals are .010 in. undersize.

Upper and lower bearing inserts are not interchangeable on any of the V8 engines due to oil hole and V-groove in the uppers. On the 318 and 360 cu. in. engine lower bearing halves No. 1, No. 2 and No. 4 are interchangeable; No. 1, No. 2 and No. 4 upper bearing halves are interchangeable. No. 3 bearing is the thrust bearing and No. 5 is the wider rear main bearing. On 361, 383, 400, 413, and 440 cu. in. engines the No. 1, No. 2, No. 4 and No. 5 lower bearing halves are interchangeable; No. 2, No. 4 and No. 5 upper bearing halves

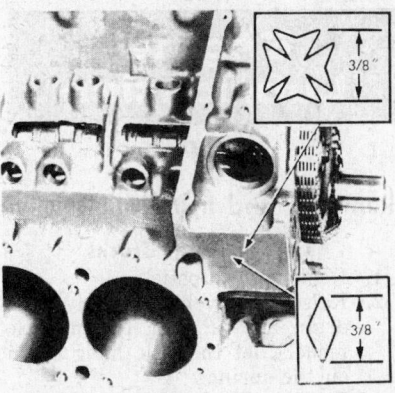

The engine numbering pad shows marks which indicate undersize crankshaft journals-except 318 and 360 engines
(© Chrysler Corp.)

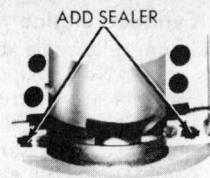

ADD SEALER

360 ENGINE BEARING CAP

SEALS

318 ENGINE BEARING CAP

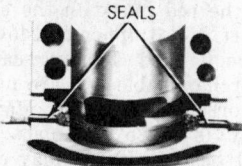

Typical rear main bearing caps (© Chrysler Corp.)

RING GAP (IN.)

Year	Engine No. Cyl- Displacement	Top Compression	Bottom Compression	Oil Control
1971	6-198	.010-.020	.010-.020	.015-.055
1971	6-225	.010-.020	.010-.020	.015-.025
1972-78	6-225	.010-.020	.010-.020	.015-.055
1971-78	8-318	.010-.020	.010-.020	.015-.055
1971-78	8-360	.010-.020	.010-.020	.015-.055
1971-78	8-361	.013-.025	.013-.025	.015-.055
1971	8-383	.013-.052	.013-.052	.015-.062
1971-78	8-400	.013-.052	.013-.052	.015-.062
1971-78	8-413	.013-.025	.013-.025	.015-.055
1973-74	8-440	.013-.052	.013-.052	.015-.062
1975-78	8-440	.013-.023	.013-.023	.015-.055

are interchangeable. No. 1 upper insert has a chamfer on the tab side for timing chain oiling and is identified by the red marking on the edge. No. 3 bearing is a thrust bearing and should be replaced if end play exceeds .007".

Remove main bearing caps one at a time and check clearance. Check number of cap for proper location.

On the 361, 383, 400, 413, and 440 cu. in. engines, the rear main bearing lower seal is held in place by a seal retainer. On the 318 and 360 cu. in. engine, the rear main bearing lower seal is held in place by the rear main bearing cap. Note that the oil pump is mounted on this cap and that there is a hollow dowel which must be in place when the cap is installed.

RING SIDE CLEARANCE (IN.)

Year	Engine No. Cyl- Displacement	Top Compression	Bottom Compression	Oil Control
1971	6-198	.0015-.0040	.0015-.0040	.0002-.0050
1971-78	6-225	.0015-.0030	.0015-.0030	.0002-.0050
1971	8-318	.0015-.0030	.0015-.0030	.0010-.0050
1973-74	8-318	.0015-.0030	.0015-.0030	.0002-.0050
1975-78	8-318	.0015-.0030	.0015-.0030	.0005-.0050
1971-74	8-360	.0015-.0030	.0015-.0030	.0002-.0050
1975-78	8-360	.0015-.0030	.0015-.0030	.0005-.0050
1971-78	8-361	.0025-.0040	.0025-.0040	.0010-.0030
1971	8-383	.0015-.0030	.0015-.0030	.0002-.0050
1971-78	8-400	.0015-.0030	.0015-.0030	.0002-.0050
1971-74	8-413	.0025-.0040	.0025-.0040	.0010-.0030
1975-78	8-413	.0010-.0025	.0010-.0025	.0010-.0030
1973-74	8-440	.0015-.0040	.0015-.0040	.0000-.0050
1975-78	8-440	.0015-.0030	.0015-.0030	.0000-.0050

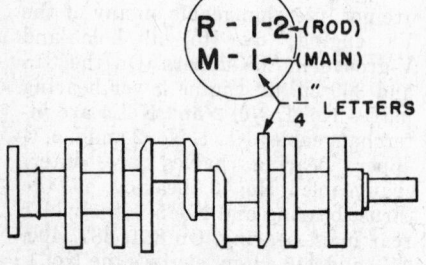

R - 1-2 (ROD)
M - 1 (MAIN)
¼" LETTERS

Location of undersize markings on counter weight; 318 and 360 engines are marked on the No. 8 counterweight (© Chrysler Corp.)

Engine Lubrication

Oil Pan

Removal and Installation

Conventional Trucks

1. Remove the dipstick.
2. Raise the vehicle safely and drain the oil. On I-beam axle models, let the axle hang down on the springs.
3. Remove the optional frame reinforcement.
4. Remove the left bellhousing brace.
5. Unbolt the pan and lower it.
6. On installation, make sure that the six cylinder oil pickup screen contacts the bottom of the pan and that it is 1⅛ in. from the inside edge of the block.
7. Install the pan with new gaskets and seals.

Six Cylinder Vans and Wagons

1. Disconnect the battery and remove the dipstick.
2. Remove the engine cover and remove the starter and air cleaner.
3. Raise the van on a hoist and drain the crankcase oil.
4. Install an engine support as described under "Engine Removal."
5. Disconnect and tie out of the way: driveshaft, transmission linkage, and exhaust pipe at the manifold.
6. Remove the clutch torque shaft (if equipped) and the oil cooler lines (if equipped).
7. Disconnect the speedometer cable and electrical connections to the transmission.
8. Remove the support bracket, inspection plate, and drive plate-to-converter attaching screws if equipped.
9. Remove the bolts which attach the transmission to the clutch

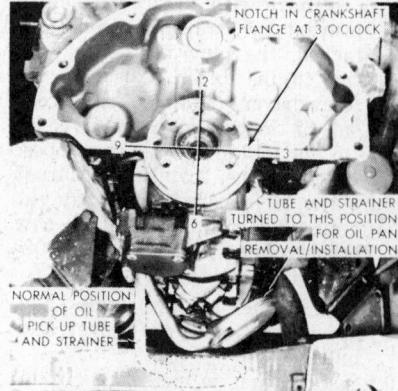

NOTCH IN CRANKSHAFT FLANGE AT 3 O'CLOCK

TUBE AND STRAINER TURNED TO THIS POSITION FOR OIL PAN REMOVAL/INSTALLATION

NORMAL POSITION OF OIL PICK UP TUBE AND STRAINER

The crankshaft and oil pickup tube positioned for oil pan removal or installation on six cylinder vans (© Chrysler Corp.)

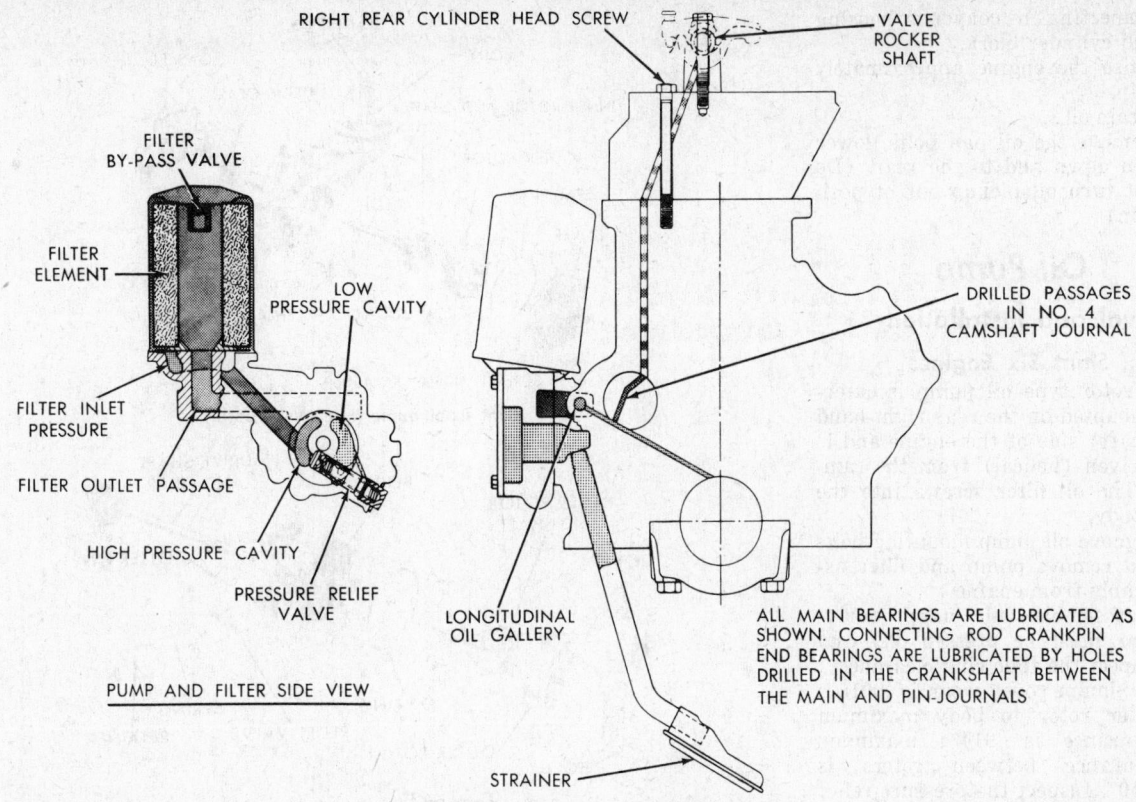

RIGHT REAR CYLINDER HEAD SCREW

VALVE ROCKER SHAFT

FILTER BY-PASS VALVE

FILTER ELEMENT

LOW PRESSURE CAVITY

FILTER INLET PRESSURE

DRILLED PASSAGES IN NO. 4 CAMSHAFT JOURNAL

FILTER OUTLET PASSAGE

HIGH PRESSURE CAVITY

PRESSURE RELIEF VALVE

LONGITUDINAL OIL GALLERY

ALL MAIN BEARINGS ARE LUBRICATED AS SHOWN. CONNECTING ROD CRANKPIN END BEARINGS ARE LUBRICATED BY HOLES DRILLED IN THE CRANKSHAFT BETWEEN THE MAIN AND PIN-JOURNALS

PUMP AND FILTER SIDE VIEW

STRAINER

Slant six engine oiling system (© Chrysler Corp.)

housing. Carefully work the transmission and converter rearward off the engine dowels and disengage the converter hub from the end of the crankshaft if so equipped. Remove the transmission.

10. Support the rear of the engine and raise it two inches.

11. Remove the oil pan attaching bolts. Positioning the crankshaft so that the counterweights will clear the pan, rotate the pan to the steering gear side and remove it. You may have to turn the pump pickup tube for clearance.

12. Installation is the reverse of removal. Make sure that the pick-up screen contacts the bottom of the pan. Fill the engine with oil and check for leaks.

1971 V8 Vans and Wagons

1. Drain the cooling system and disconnect the battery.
2. Remove the engine cover.
3. Disconnect the radiator hoses.
4. Disconnect the exhaust pipes. Remove the left exhaust manifold, oil dipstick, and tube.
5. Raise the vehicle and support it. Drain the engine oil.
6. Remove the clutch dust pan, inspection plate, and engine-to-torque converter left housing strut.
7. Remove the starter and clutch torque shaft or shifter torque shaft.

8. Disconnect and cap the inlet line to the fuel pump.
9. Remove the front engine mount insulator top nut and insulator.
10. Loosen the transmission mount through-bolt and jack up and support the engine.
11. Remove the oil pan attaching screws and position the crankshaft so that the pan will clear the counterweights. Remove the pan.
12. Installation is the reverse of removal. Check all fluid levels and be sure that there are no leaks.

1972 and Later 318 and 360 V8 Vans and Wagons

1. Disconnect the battery ground cable. Remove the dipstick and tube, engine cover, and air cleaner.
2. Disconnect the throttle linkage at the rear of the engine and the clutch or automatic transmission linkage.
3. Raise the engine slightly and support it with the device described under "Engine Removal".
4. Raise the vehicle and drain the oil. Remove the starter.
5. Remove the driveshaft and engine rear support.
6. Remove the transmission from the van. Remove the automatic transmission with the filler tube installed and the torque converter separated from the drive plate.

7. Remove the clutch assembly and flywheel (or driveplate) from the crankshaft.
8. Raise the engine about 2 in.
9. Rotate the crankshaft so that the counterweights will clear the oil pan. Maximum clearance is with the notch in the crankshaft flange at the 3 o'clock position. Remove the oil pan. It will be necessary to reach inside the oil pan and turn the oil pick-up tube and strainer slightly to the right to clear the pan.
10. Installation is the reverse of removal. Be sure to check all fluid levels and be sure that there are no leaks.

Two-wheel Drive Ramcharger and Trail Duster

1. Disconnect the battery cable and remove the dipstick.
2. Raise and support the truck.
3. Drain the oil.
4. Remove the torque converter or clutch housing brace.
5. If necessary, remove the exhaust pipe.
6. Remove the oil pan bolts and remove the pan.
7. Installation is the reverse of removal.

Four-Wheel Drive Ramcharger and Trail Duster, W100, 200, 300

1. Raise vehicle on a hoist.
2. Remove the two front engine mounting bolts.
3. Remove the left-side support,

Dodge and Plymouth Trucks ·

connecting the converter housing and cylinder block.
4. Raise the engine approximately 2 in.
5. Drain oil.
6. Remove the oil pan bolts, lower pan down and to the rear. (Do not turn oil pickup out of position)

Oil Pump

Removal and Installation

Slant Six Engines

The rotor type oil pump is externally mounted on the rear right-hand (camshaft) side of the engine and is gear driven (helical) from the camshaft. The oil filter screws into the pump body.
1. Remove oil pump mounting bolts and remove pump and filter assembly from engine.
2. Disassemble oil pump (drive gear must be pressed off) and inspect the following clearances: maximum cover wear is .0015"; outer rotor to body maximum clearance is .014"; maximum clearance between rotors is .010". Inspect the pressure relief valve for scoring and free operation. Relief valve spring should have a free length of 2-9/32 to 2-19/64 in. through 1971 and 2¼ in. beginning 1972.
3. Install new oil seal rings between cover and body, tightening cover attaching bolts to 95 in. lbs.
4. Install oil pump to engine block using a new gasket and tightening mounting bolts to 200 in. lbs.

Slant six oil pump pickup screen on conventional trucks must be positioned 1⅛ inch from the inside edge of the block (© Chrysler Corp.)

318, 360 Cu. In. V8 Engine

The rotor type oil pump is gear (helical) driven from the camshaft and is mounted on the rear main bearing cap.
1. Drain engine oil and remove oil pan.
2. Remove oil pump mounting bolts and remove pump from rear

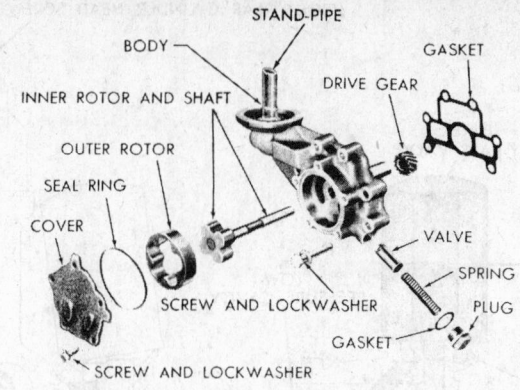

Slant six oil pump (© Chrysler Corp.)

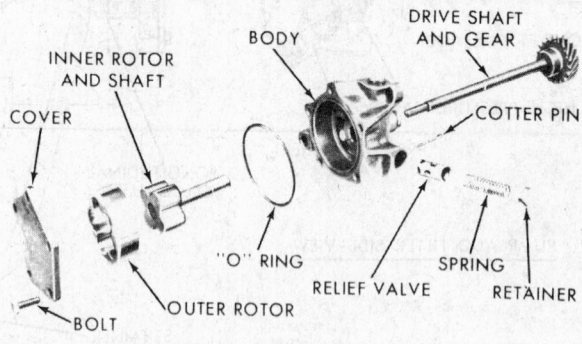

318 and 360 V8 oil pump (© Chrysler Corp.)

main bearing cap.
3. The pressure relief valve cap is pressed in and is destroyed when removed. Remove cotter pin first, then drill cap and use a self-tapping sheet metal screw to get a grip on the cap. Free length of spring is 2 1/32—2 3/64 in.
4. When assembling pump, use a new oil seal ring between cover and body. Prime pump before installing onto rear main bearing cap. The distributor drive gear slot should align with the front left intake manifold bolt with No. 1 cylinder on TDC. Tighten mounting bolts to 35 ft. lbs. torque. Use new gaskets when installing oil pan.

361, 383, 400, 413, 440 Cu. In. V8 Engines

The rotor type oil pump is externally mounted and gear driven from the camshaft. The oil filter screws into the pump body.
1. Drain engine oil.
2. Remove oil pump and filter assembly.
3. Disassemble and inspect pump components for wear. If 0.0015" feeler gauge can be inserted between cover and straight edge, replace cover. Install outer rotor in pump body and holding against one side of body measure clearance between rotor and body. If clearance is greater than 0.014", replace oil pump body.

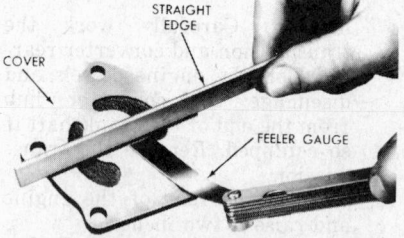

Measuring the oil pump cover wear-typical (© Chrysler Corp.)

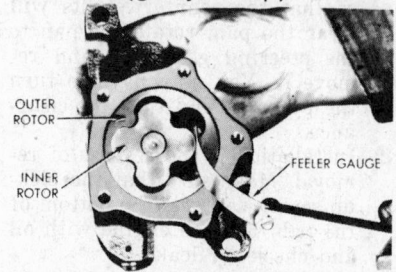

Measuring the clearance between the rotors-typical (© Chrysler Corp.)

Install inner rotor into pump body and place straight edge across pump body between bolt holes. If feeler gauge greater than 0.004" can be inserted between rotors and body, replace oil pump body. Measure clearance between tips of inner and outer rotor where they are opposed. If clearance exceeds 0.010", replace inner and outer rotors. Use new oil seal rings between filter base and body. Tighten bolts to 10 ft. lbs. Use a new "O" ring seal on pilot of

110

Measuring the clearance over the rotors-typical
(© Chrysler Corp.)

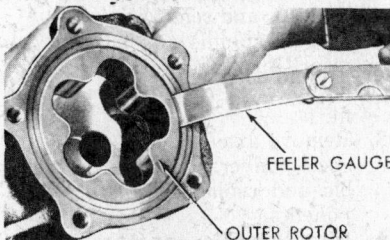

Measuring the outer rotor clearance—typical
(© Chrysler Corp.)

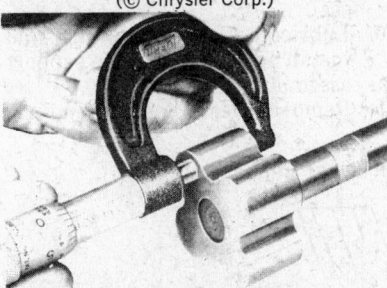

Measuring the inner rotor clearance—typical
(© Chrysler Corp.)

oil pump before attaching pump to engine block.
4. Install oil pump on engine using new gasket and tightening bolts to 30 ft. lbs. The distributor drive gear slot should parallel the crankshaft with no. 1 cylinder on TDC.
5. Install oil pan and fill crankcase with oil.

Rear Main Bearing Oil Seal

Replacement

Service replacement seals are of the split rubber type composition. This type of seal makes it possible to replace the upper rear seal without removing the crankshaft. The seal must be used as an upper and lower set and cannot be used with the rope type seal.

NOTE: Rope type seals are included in overhaul gasket sets, for use when the crankshaft has been removed, on all engines, except the 360 V-8, which uses only the composition seal.

The following procedure is for removing the rope type rear main seal and replacing it with the rubber type seal.

1. Remove the oil pan, and both the rear seal retainer and the rear main bearing cap, if separate.

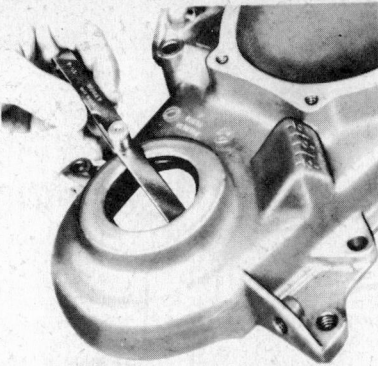

Inspection of oil seal for proper seating
(© Chrysler Corp.)

2. Remove the lower rope seal from the cap or retainer by prying the seal out of the groove.
3. With the use of suitable tools, either pull or push the seal from its seat, while rotating the crankshaft, being careful not to damage the surface of the journal. If necessary, loosen all the main bearing caps slightly, to lower the crankshaft, which will aid in the removal and replacement of the seal.
4. Clean and lubricate the crankshaft journal. Hold the seal tight against the crankshaft with the painted stripe to the rear, and install the seal into the block groove.
5. Rotate the crankshaft while pushing the seal into the groove. Be careful that the sharp edges of the block groove. *Do not cut or nick the rear of the seal.*
6. Install the lower half of the seal into the lower seal retainer or the main bearing cap, if separate, with the paint stripe facing to the rear.
7. Install the lower seal retainer and/or the rear main bearing cap. Torque all main bearing caps to specifications.
8. Install the oil pan, add oil and check for oil leaks.

Front Suspension

I-Beam Axle

I-Beam Axle King Pin and Bushing Removal and Installation

1. Raise the front of the vehicle and safely support it on stands. Remove the wheels and drums.
2. Remove the brake support plate attaching bolts, and remove the support plate from the steering knuckle. Secure the plate to the frame so that it does not hang by the brake hose. If equipped with

air brakes, disconnect the push rod and remove the air chamber.
3. Remove the steering arm from the steering knuckle.
4. Remove the pivot pin locking screw or pin from the knuckle. Some models may have two locking screws.
5. Remove the upper pivot pin oil seal plug from the knuckle and drive the pivot pin down, forcing the lower oil seal plug from its seat. *NOTE: On some models, a lock ring is used to hold the oil seal plug in place. Others have caps with hold-down screws.*
6. Remove the knuckle from the axle and if equipped with bronze bushings, press the old ones out and press the new ones in and line ream them to fit the new pin. If equipped with Delrin or Zytel type bushings, bronze bushings must be used as service replacements. Upon installation of the bushings, align the grease hole in the bushing to that of the grease fitting hole in the knuckle.
7. Install the knuckle on the axle. position the thrust bearing, and install the pivot pin through the knuckle and axle, securing it with the locking pins or screws.
8. Install the oil seal plugs and secure them by staking in four locations.
9. Complete the assembly by reversing steps one to three. Lubricate to assure grease channels are open.

Front Leaf Spring Removal and Installation

This procedure applies to all models with front leaf springs.

1. Raise truck until weight is removed from springs.
2. Install stands under side frame members as a safety precaution.
3. Disconnect the sway bar at the spring plate. Remove nuts, lockwashers and U-bolts securing spring to axle.
4. Remove spring shackle bolts, shackles and spring front eye bolt.
5. Remove spring.
6. To install, line up spring fixed eye with bolt hole in bracket and install spring bolt and nut.
7. Install shackle bolts, shackles and nuts. Tighten shackle bolt nuts and fixed eye bolt until slack is taken up.
8. Position spring on axle so spring center bolt enters locating hole in axle pad.
9. Install U-bolts, new lockwashers and nuts, tightening securely.
10. Remove stands from under frame, lower truck so weight is resting on wheels. Tighten U-bolt nuts, spring eye bolt nuts and shackle bolt nuts.

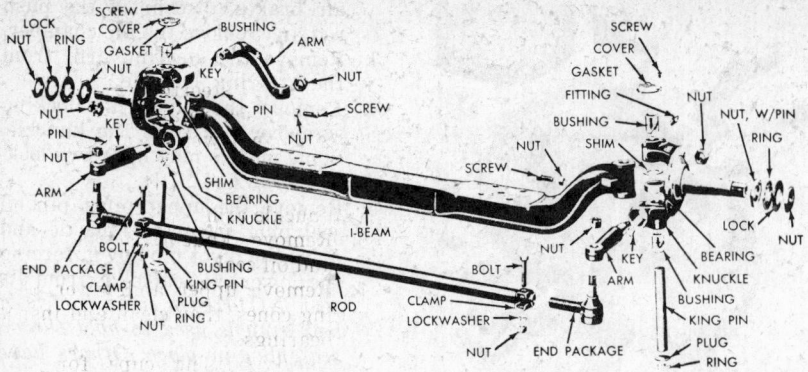

Typical "I" beam front axle (© Chrysler Corp.)

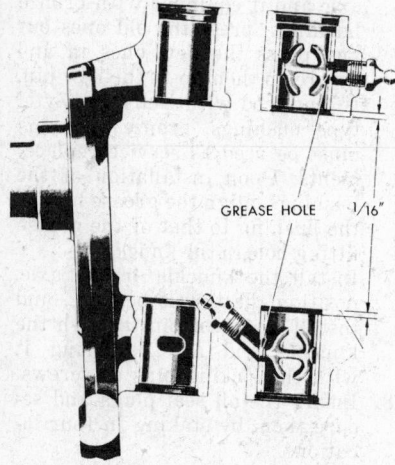

King pin bushing installation—D100,200
(© Chrysler Corp.)

11. Lubricate spring bolts and shackle bolts with chassis lubricant. *Do not lubricate rubber*

Front Drive Axle

Front Drive Axle Assembly Removal & Installation

1. Raise the truck and install stands under the frame rails, behind the front springs.
2. Disconnect front driveshaft at drive pinion yoke.
3. Disconnect steering linkage at drag link.
4. Disconnect front shock absorbers and brake line at frame. Disconnect the sway bar link assembly from the spring clip plate.
5. Remove nuts from the spring hold down bolts and remove axle assembly from under vehicle.
6. To install, place axle assembly under vehicle and line up spring center bolts with locating hole in axle housing pad.
7. Install spring clips or spring U-bolts, new lock washer and nuts.
8. Connect the shock absorbers, and the brake line at the frame.
9. Connect the steering linkage to the drag link, and the driveshaft to the pinion yoke. Check lubri-

cants and bleed the brakes.
10. Lower the vehicle and test the operation.

Front Drive Axle Shaft Removal and Installation

Enclosed U-Joint, Pivot Pin Type
1. After removing wheel assembly and locking hubs, remove grease cap from drive flange.

2. Remove drive flange bolts and snap ring. Then, using a suitable puller, remove drive flange.
3. Remove locknut, lockwasher and adjusting nut.
4. Remove wheel hub and brake drum assembly, being careful not to damage oil seal.
5. Remove brake assembly and wire to frame, leaving hydraulic line connected.
6. Tap end of spindle with a soft hammer and remove spindle.
7. Inspect spindle and bushing for wear. Bushing is pressed on and requires no reaming for proper fit.
8. Remove axle shaft assembly. Inspect universal joint. Disassemble and replace any worn components.
9. Install axle assembly, being careful not to damage inner oil seal.
10. Lubricate flanged bushing and install spindle and brake support assembly.
11. Remove, clean, inspect and re-

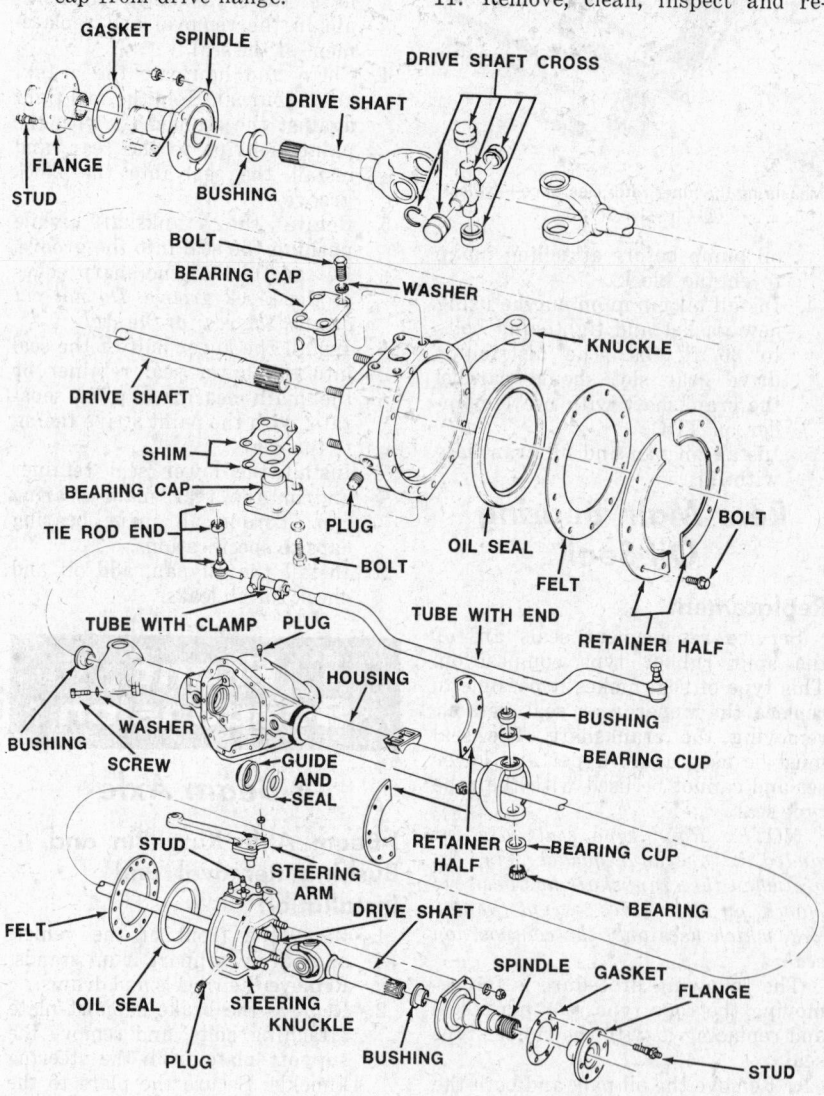

Axle housing (W500)—exploded view (© Chrysler Corp.)

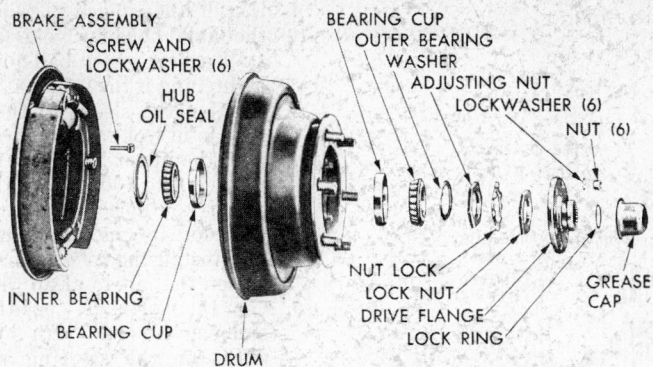

BRAKE ASSEMBLY
SCREW AND
LOCKWASHER (6)
HUB
OIL SEAL

BEARING CUP
OUTER BEARING
WASHER
ADJUSTING NUT
LOCKWASHER (6)
NUT (6)

INNER BEARING
BEARING CUP
DRUM

NUT LOCK
LOCK NUT
DRIVE FLANGE
LOCK RING

GREASE
CAP

Typical brake hub and drum assembly (© Chrysler Corp.)

pack inner and outer wheel bearings.

12. Install inner wheel bearing, position hub, drum and wheel assembly on spindle, then install outer bearing, washer and adjusting nut.

13. Adjust bearing by tightening adjusting nut to 50 ft. lbs. torque, then back off 1/4 to 1/3 turn. Tighten locknut and bend tab of lock-washer.

14. Install drive flange, retaining nuts, snap ring and grease cup.

Exposed U-Joint, Ball Joint Type

This procedure is for disc-braked vehicles, but may be adapted to those with drum brakes.

1. Remove the locking hubs. Remove the cotter key and loosen the axle shaft nut.
2. Block the brake pedal up. Remove the wheel.
3. Remove and hang the caliper out of the way. Remove the inner pad.
4. Working through the hole in the disc hub, remove the socket head capscrews.
5. Remove the axle shaft nut and use a hub puller to remove the disc and hub.
6. Remove the O-ring from the steering knuckle. Remove the disc brake adapter from the knuckle. Punch out the inner oil seal from the rear of the knuckle.
7. Slide the axle shaft from the housing.
8. On replacement, first slide the shaft into place, then drive a new seal into the steering knuckle.
9. Install the disc brake adapter and torque the mounting bolts to 85 ft. lbs.
10. Install a new O-ring on the steering knuckle.
11. Slide the disc and hub, retainer and bearing assembly over the shaft and start it into the housing. Install the axle shaft nut.
12. Install the capscrews holding the retainer to the steering knuckle flange. Tighten them to 30 ft. lbs. in a criss-cross pattern.
13. Torque the axle shaft nut to 100 ft. lbs. and tighten further until

the cotter key can be installed.

14. Locate the inner pad on the adapter with the shoe flanges in the adapter ways. Slide the caliper into position, being careful not to pull the dust boot from its grooves.
15. Install the anti-rattle springs, making sure that the inner one is on top of the retainer spring plate. Tighten the retaining clips to 200 in. lbs.
16. Install the wheel and lower the vehicle.

Front Drive Axle Steering Knuckle Service

Enclosed U-Joint, Pivot Pin Type

1. Remove, wheel, hub, brake drum assembly, brake support assembly, spindle and axle shaft.

2. Disconnect tie rod ends from steering knuckles and drag link from steering knuckle arm.
3. Remove steering knuckle felt and oil seal retainers.
4. Remove lower bearing cap and shims.
5. Remove upper bearing cap or knuckle arm and shims.
6. Remove knuckle housing, felt and oil seal.
7. Remove upper and lower bearing cones, then clean and inspect bearings.
8. Check bearing cups for wear and remove with drift if required.
9. To assemble, place felt and oil seal over end of axle housing.
10. Install steering knuckle arm or upper bearing cap and shim pack and tighten retaining bolts or nuts. If a new shim pack is used, use *only* one which is .060" thick.
11. Lubricate and install cone (press in) on upper bearing pivot, making sure serration or key and slot are properly located, if so equipped.
12. Seat bearing cups in yokes of axle housing. Slide knuckle assembly over yoke and enter bronze cone in its cup.
13. Lubricate and insert lower bearing cone, tilting knuckle to provide access. Install lower bearing cap and shim pack. If a new pack is required, use *only* .025"

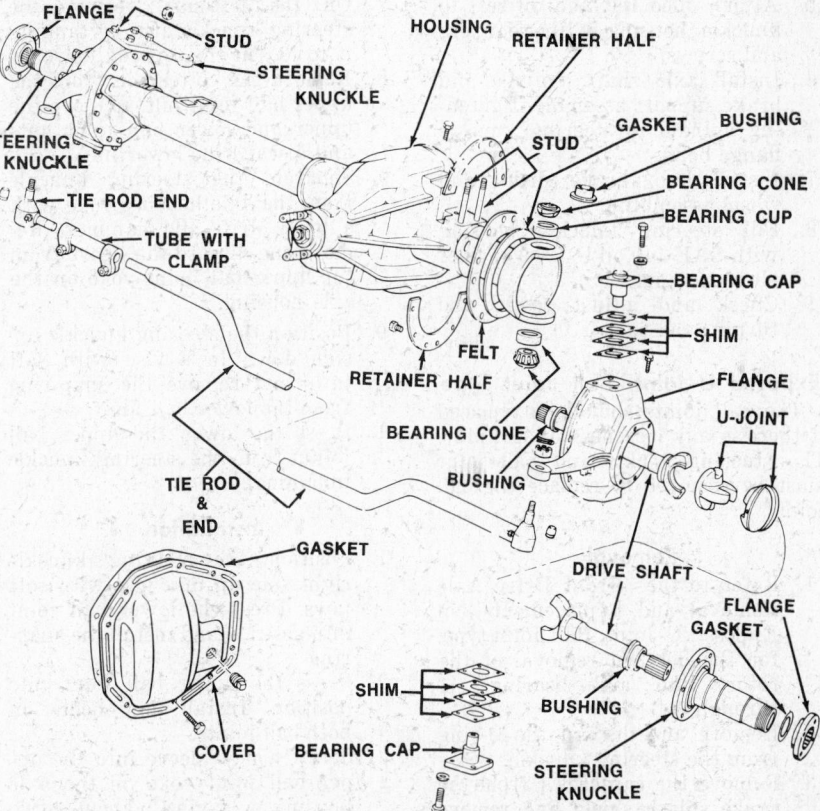

FLANGE
STUD
STEERING KNUCKLE
HOUSING
RETAINER HALF
STUD
GASKET
BUSHING
BEARING CONE
BEARING CUP
BEARING CAP
STEERING KNUCKLE
TIE ROD END
TUBE WITH CLAMP
SHIM
FLANGE
U-JOINT
RETAINER HALF
FELT
BEARING CONE
FLANGE
BUSHING
TIE ROD & END
GASKET
DRIVE SHAFT
FLANGE
GASKET
SHIM
BUSHING
COVER
BEARING CAP
STEERING KNUCKLE

70F axle and steering knuckle-exploded view (© Chrysler Corp.)

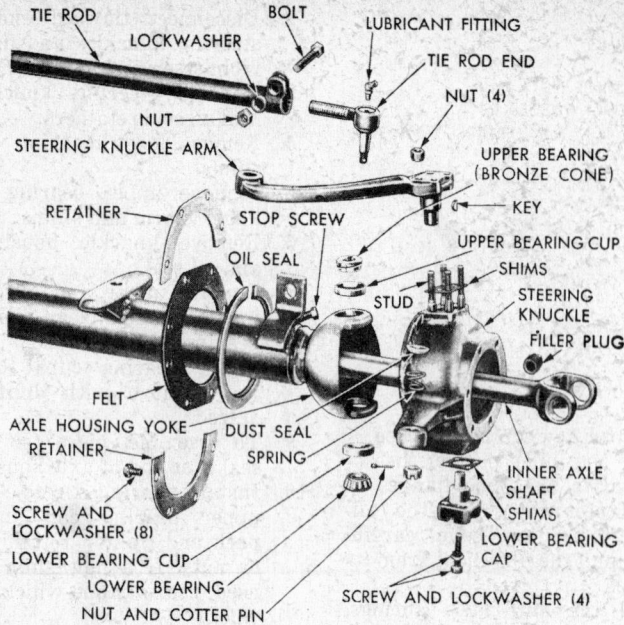

TIE ROD
LOCKWASHER
BOLT
LUBRICANT FITTING
TIE ROD END
NUT (4)
NUT
STEERING KNUCKLE ARM
UPPER BEARING (BRONZE CONE)
KEY
RETAINER
STOP SCREW
UPPER BEARING CUP
OIL SEAL
SHIMS
STUD
STEERING KNUCKLE
FILLER PLUG
FELT
AXLE HOUSING YOKE
RETAINER
DUST SEAL
SPRING
SCREW AND LOCKWASHER (8)
LOWER BEARING CUP
LOWER BEARING
NUT AND COTTER PIN
INNER AXLE SHAFT
SHIMS
LOWER BEARING CAP
SCREW AND LOCKWASHER (4)

Steering knuckle (W-100 & W-200) (© Chrysler Corp.)

thick pack for 44-3F axle or .055" thick pack for 44-3HF or 70F axle.

14. Using a torque wrench on a top outer knuckle bearing cap retaining bolt or nut, measure bearing preload as knuckle is turning. It must read 5-15 ft. lbs. for 44-3F axle and 15-35 ft. lbs. for the 44-3HDF or 70-F axles. Remove or add shims to adjust preload.

15. Attach oiled felt and oil seal to knuckle housing with retainers and screws.

16. Install axle shaft, spindle and brake support assembly, tightening spindle to steering knuckle flange bolts.

17. Install hub, brake drum and wheel assembly.

18. Fill steering knuckle housing with SAE 140 oil to level of filler plug opening.

19. Check and adjust toe-in and turning radius.

Exposed U-Joint, Ball Joint Type

The ball joints should be replaced if there is any looseness or end-play. The steering knckle and ball joint must be removed to replace the ball joint.

Removal

1. Refer to the "Front Drive Axle Removal and Replacement," on Exposed U-Joint, Ball Joint type, for the detailed removal of the rotor, hub and bearings, 1 through 6.

2. Remove and discard the O-ring from the steering knuckle.

3. Remove the capscrews from the brake splash shield and remove the splash shield. Remove the

brake disc adaptor from the steering knuckle.

4. Disconnect the tie-rod from the steering knuckle. On the left side, disconnect the drag link from the steering knuckle arm.

5. Using a punch and hammer, remove the inner oil seal from the rear of the steering knuckle.

6. Carefully, slide the outer and inner axle shaft complete with U-joint from the axle housing.

7. On the left-side, remove the steering knuckle arm by tapping it to loosen the tapered dowels.

8. Remove the cotter pin from the upper ball joint nut. Remove the upper and lower ball joint nuts and discard the lower nut.

9. Separate the steering knuckle from the axle housing yoke with a brass drift and hammer. Remove and discard the sleeve from the upper ball joint yoke on the axle housing.

10. Position the steering knuckle upside down in a vise with soft jaws and remove the snap-ring from the lower ball joint.

11. Press the lower and upper ball joints from the steering knuckle individually.

Installation

1. Position the steering knuckle right side up in a vise with soft jaws. Press the lower ball joint into position and install the snap-ring.

2. Press the upper ball joint into position. Install new boots on both ball joints.

3. Screw a new sleeve into the upper ball joint yoke on the axle housing, leaving about two threads showing at the top.

4. Install the steering knuckle on the axle housing yoke and install a new lower ball joint nut, tightening it to 80 ft. lbs.

5. Tighten the sleeve in the upper ball joint yoke to 40 ft. lbs. Install the upper ball joint nut and tighten it to 100 ft. lbs. Align the cotter key hole in the stud with the slot in the castellated nut and install the cotter pin. Do not loosen the nut to align the holes.

6. On the left-side, position the steering knuckle arm over the studs on the steering knuckle. Install the tapered dowels and nuts. Tighten the nuts to 90 ft. lbs. Install the drag link on the steering arm. Install the nut and tighten it to 60 ft. lbs. Install the cotter pin.

7. Install the tie-rod end on the steering knuckle. Tighten the nut to 45 ft. lbs. and install the cotter pin.

8. Install the axle shaft. Install the brake splash shield and tighten the screws to 13 ft. lbs. Install the brake disc adaptor and tighten the bolts to 85 ft. lbs.

9. Install a new O-ring in the steering knuckle.

10. Clean any rust from the axle shaft splines.

11. Carefully, slide the hub, rotor and retainer, and bearing onto the axle shaft and start it into the housing. Install the axle shaft nut.

12. Align the retainer with the steering knuckle flange. Install the retainer screws and tighten them in a criss-cross pattern to 30 ft. lbs.

13. Tighten the axle shaft nut to 100 ft. lbs. Tighten the nut until the next slot in the nut aligns with the hole in the axle shaft. Install the cotter pin.

14. Install the inboard brake shoe on the adaptor with the shoe flanges in the adaptor ways. Install the caliper in the adaptor and over

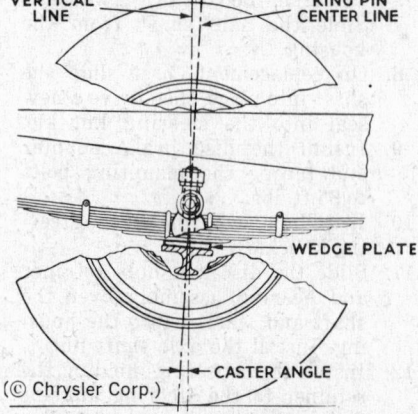

VERTICAL LINE
KING PIN CENTER LINE
WEDGE PLATE
CASTER ANGLE
(© Chrysler Corp.)

Caster angle can be adjusted by placing wedges between the spring and axle on both driving and I-beam front axles

the disc. Align the caliper on the machined ways of the adaptor. Be careful not to pull the dust boot from its grooves as the piston and boot slide over the inboard shoe.

15. Install the anti-rattle springs and retaining clips. Torque to 16-17 ft. lbs. The inboard shoe anti-rattle spring must always be installed on top of the retainer spring plate.
16. Install the wheel, tire, and locking hub and lower the truck. Lubricate all fittings.

Warn Front Locking Hubs

Removal and Disassembly
1. Straighten the lock tabs and remove the six hub mounting bolts.
2. Tap the hub gently with a mallet to remove.
3. Separate the clutch assembly from the body assembly.
4. Remove the snap ring from the rear of the body assembly, using snap-ring pliers. Slip the axle shaft hub out of the body from the front.
5. Remove the Allen screw from the inner side of the clutch, and remove the bronze dial assembly from the front side of the clutch housing assembly.
6. Remove the clutch assembly from the rear of the housing, complete with the twelve roller pins.

Assembly and Installation
1. Coat the moving parts with a water-proof grease.
2. Slide the axle shaft hub into the body from the front, and replace the snap-ring.
3. Replace the bronze dial assembly and the inner clutch. Tighten the Allen screw and stake the edge of the screw with a center punch to prevent loosening.
4. With the dial in the FREE position, rotate the outer clutch body into the inner assembly until it bottoms in the housing. Back it up to the nearest hole and install the roller pins.
5. Position the hub and clutch assembly together with a new gasket in between.
6. Position the hub assembly over the end of the axle and replace

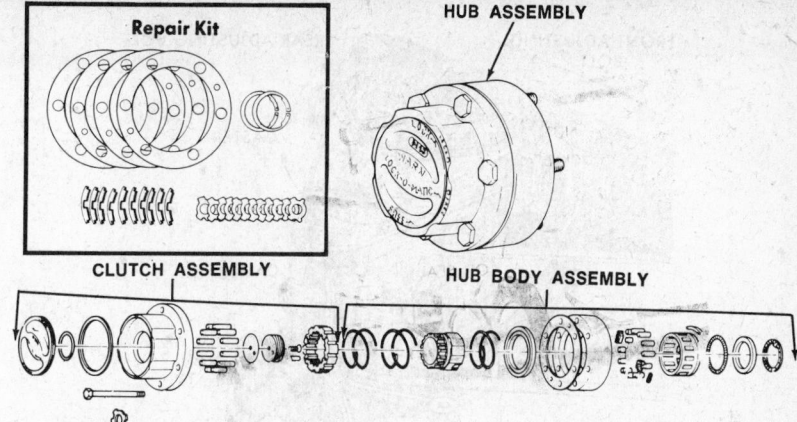

Warn locking hub-exploded view (© Chrysler Corp.)

the six hub mounting bolts and lock tabs.
7. Torque the bolts to 35 foot-pounds and bend the lock tabs to anchor the bolts.
8. Verify the operation by road testing.

Dana Front Locking Hubs

Removal and Disassembly
1. Place hub in lock position. Remove Allen head mounting bolts and washers.
2. Carefully remove retainer, O-ring seal and knob. Separate knob from retainer.
3. Remove large internal snap-ring. Slide retainer ring and cam from hub.
4. While pressing against sleeve and ring assembly, remove axle shaft snap-ring. Relieve pressure and remove sleeve and ring, ring and bushing, spring and plate.
5. Inspect all parts for wear, nicks and burrs. Replace all parts which appear questionable.

Assembly and Installation
1. Slide plate and spring (large coils first) into wheel hub housing.
2. Assemble ring and bushing, sleeve and bushing. Slide complete assembly into housing.
3. Compress spring and install axle shaft snap-ring.
4. Position cam and retainer in housing and install large internal snap-ring.
5. Place small O-ring seal on knob, lubricate with waterproof

grease and install in retainer at lock position.
6. Place large O-ring seal on retainer. Align retainer and retainer ring and install washers and Allen head mounting screws.
7. Check operation.

Independent Front Suspension

NOTE: These procedures apply to all vans, wagons, and conventional trucks with independent front suspension.

Coil Spring Removal and Installation
1. Raise the vehicle and support it with jackstands under the front ends of the frame rails.
2. Remove the wheel.
3. Remove the shock absorber and upper shock absorber bushing bushing and sleeve.
4. If equipped, remove the sway bar.
5. Remove the strut.
6. Install a spring compressor and tighten finger-tight.
7. Remove the cotter pins and ball joint nuts.
8. Install a ball joint breaker tool and turn the threaded portion of the tool to lock it against the lower stud.
9. Spread the tool to place the lower stud under pressure, then strike the steering knuckle sharply with a hammer to free the stud. Do not attempt to force the stud out of the steering knuckle with the tool.
10. Remove the tool. Slowly release the spring compressor until all tension is relieved from the spring.
11. Remove the spring compressor and spring.
12. Installation is the reverse of removal. Compress the spring until the ball joint can be properly positioned in the steering knuckle.

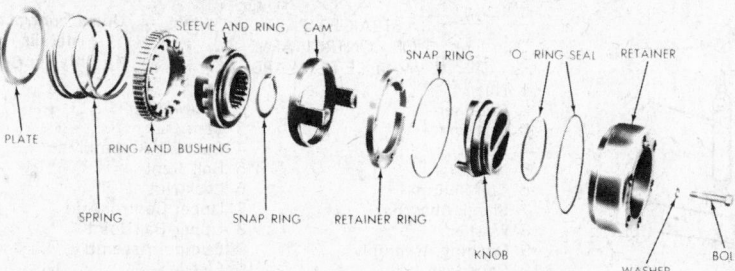

Dana locking hub—exploded view (© Chrysler Corp.)

Dodge and Plymouth Trucks ·

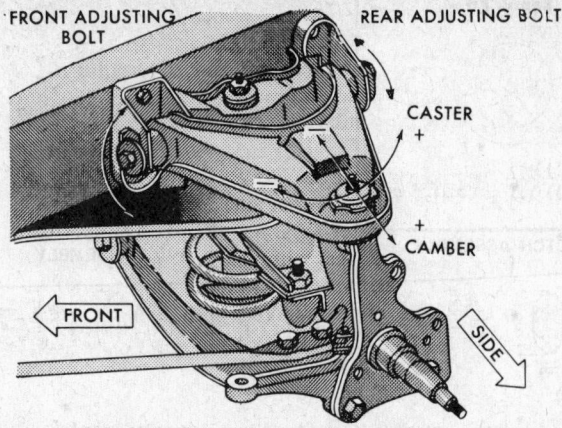

Independent front suspension alignment points (© Chrysler Corp.)

Shock Absorber Removal and Installation

1. Raise and support the vehicle with jackstands positioned at the extreme front ends of the frame rails.
2. Remove the wheel.
3. Remove the upper nut and retainer.
4. Remove the two lower mounting bolts.
5. Remove the shock absorber.
6. Installation is the reverse of removal.

Upper Control Arm Removal and Installation

NOTE: Any time the control arm is removed, it is necessary to align the front end.

1. Raise and support the vehicle with jackstands under the frame rails.
2. Remove the wheel.
3. Remove the shock absorber and shock absorber upper bushing and sleeve.
4. Install a spring compressor and tighten it finger-tight.
5. Remove the cotter pins and ball joint nuts.

6. Install a ball joint breaker and turn the threaded portion of the tool, locking it securely against the upper stud. Spread the tool enough to place the upper ball joint under pressure and strike the steering knuckle sharply to loosen the stud. Do not attempt to remove the stud from the steering knuckle with the tool.
7. Remove the tool.
8. Remove the eccentric pivot bolts, after making their relative positions in the control arm.
9. Remove the upper control arm.
10. Installation is the reverse of removal. Tighten the ball joint nuts to 135 ft. lbs. Tighten the eccentric pivot bolts to 70 ft. lbs.
11. Adjust the caster and camber.

Lower Control Arm Removal and Installation

1. Follow the procedure outlined under "Coil Spring Removal and Installation."
2. Remove the mounting bolt from the crossmember.
3. Remove the lower control arm from the vehicle.
4. Installation is the reverse of re-

moval. After the vehicle has been lowered to the ground, tighten the mounting bolt to 210 ft. lbs.

Lower Ball Joint Removal and Installation

1. Remove the lower control arm.
2. Remove the ball joint seal.
3. Using an arbor press and a sleeve, press the ball joint from the control arm.
4. Installation is the reverse of removal. Be sure that the ball joint is fully seated. Install a new ball joint seal.
5. Install the lower control arm. Be sure to install the ball joint cotter pins.

Upper Ball Joint Removal and Installation

1. Install a jack under the outer end of the lower control arm and raise the vehicle.
2. Remove the wheel.
3. Remove the ball joint nuts. Using a ball joint breaker, loosen the upper ball joint.
4. Unscrew the ball joint from the control arm.
5. Screw a new ball joint into the control arm and tighten 125 ft. lbs.
6. Install the new ball joint seal, using a 2 in. socket. Be sure that the seal is seated on the ball joint housing.
7. Insert the ball joint into the steering knuckle and install the ball joint nuts. Tighten the nuts to 135 ft. lbs. and install the cotter pins.
8. Install the wheel and lower the truck to the ground.

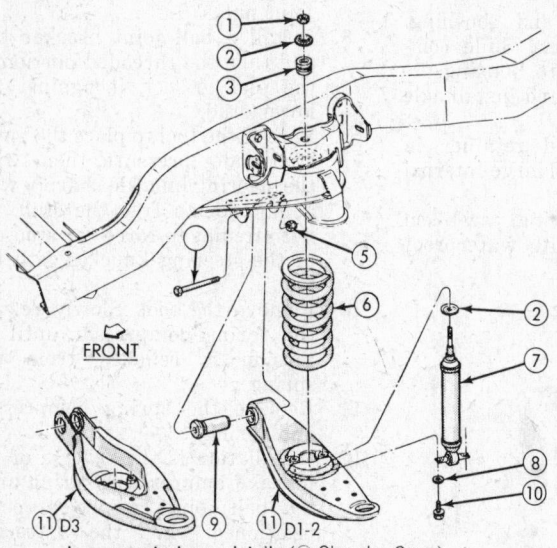

Lower control arm details (© Chrysler Corp.)

1 Nut
2 Retainer
3 Bushing
4 Bolt
5 Nut
6 Coil Spring
7 Shock Absorber
8 Washer
9 Bushing Assembly
10 Capscrew
11 Lower Control Arm

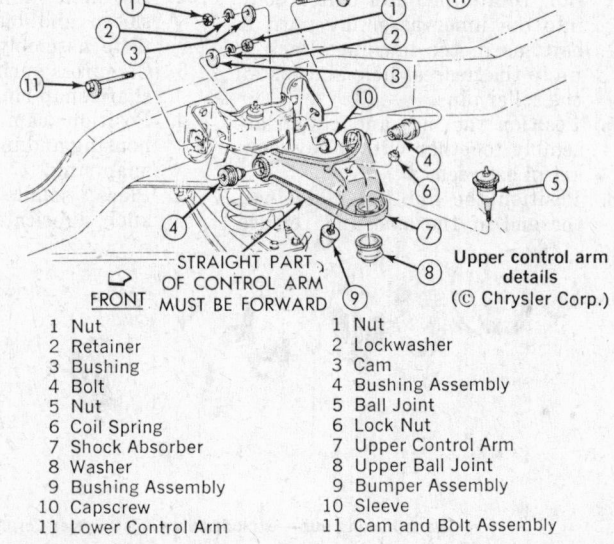

Upper control arm details (© Chrysler Corp.)

1 Nut
2 Lockwasher
3 Cam
4 Bushing Assembly
5 Ball Joint
6 Lock Nut
7 Upper Control Arm
8 Upper Ball Joint
9 Bumper Assembly
10 Sleeve
11 Cam and Bolt Assembly

Steering Gear

Reference

Refer to the Manual Steering chapter in the General Repair Section for the overhaul of the manual steering gears.

Manual Steering

NOTE: Before making any adjustments, inspect steering linkage for lash and wheel bearing adjustment. Ross and Saginaw steering gears, except the model 24J Ross, may be adjusted for cross shaft lash and worm bearing preload without removing the gear from the vehicle. On the 24J, worm bearing preload adjustment must be done with the unit removed.

Worm Bearing Preload Adjustment

All Except Ross Model 24J

1. On Ross gears, remove the horn button and spring.
2. Remove the steering arm from the cross (sector) shaft. Mark the relation of the steering arm to the shaft before removing.
3. On long worm shaft type gears, check steering column for alignment.
4. Loosen cross (sector) shaft adjusting screw and back out adjusting screw about two turns.
5. Turn steering wheel two complete turns from the straight-ahead position. **CAUTION: Do not turn steering wheel hard against the stops as this may damage gear.**
6. Using a torque wrench on the steering wheel nut, rotate steering wheel through "lash" area toward straight-ahead position. Torque required to keep wheel moving indicates worm bearing preload.
7. If preload is not within limits, remove or add shims beneath the lower worm bearing cover on Ross gears. Cover retaining bolts are tightened to 20 ft. lbs. torque. Preload on the Saginaw gear is made by turning the large adjuster. Adjuster locknut is tightened to 85 ft. lbs. torque.
8. Recheck preload, adjusting again if necessary.

Ross Model 24J

This gear must be removed to adjust the worm bearing preload. Torque required to keep the worm shaft moving through "lash" (center position) should be 5-10 in. lbs. Adjustment is made by removing or adding shims between the worm shaft cover plate and the gear body. Tighten

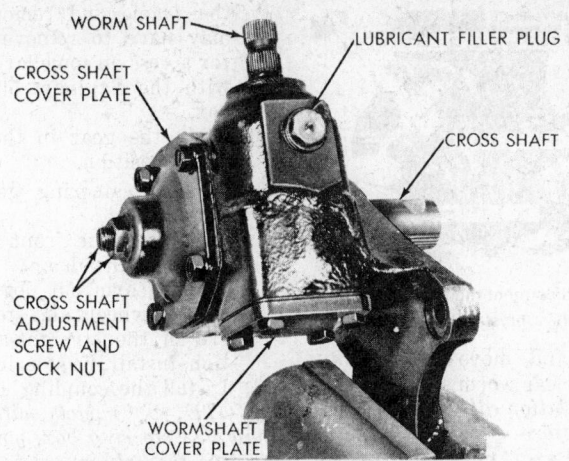

WORM SHAFT

CROSS SHAFT COVER PLATE

LUBRICANT FILLER PLUG

CROSS SHAFT

CROSS SHAFT ADJUSTMENT SCREW AND LOCK NUT

WORMSHAFT COVER PLATE

Model (24J) Ross steering gear (© Chrysler Corp.)

cover bolts to 18-20 ft. lbs. torque. Perform cross shaft adjustment after worm gear preload is adjusted.

Cross or Sector Gear Mesh Adjustment

This adjustment is made only after worm bearing preload is adjusted. Steering arm is still removed from cross (sector) shaft.

1. Center steering wheel (wormshaft) by counting turns from full right to full left and counting back exactly half way.
2. Loosen locknut on cross (sector) shaft adjusting screw and turn adjusting screw in until all lash is gone.
3. Check torque required to move steering wheel (torque wrench on steering wheel nut) through high-spot (center) position.
4. Readjust if necessary and retighten locknut.
5. Reinstall steering arm on cross (sector) shaft and tighten clamp bolt to 85 ft. lbs. torque.

Steering Gear	Worm Bearing Preload (in. lbs)	Cross or Sector Gear Mesh (in. lbs)
Ross (Gemmer) 24J	5-10	7-11 over worm preload
Ross (Gemmer) Y4D-335, 35J, 376	7-14	20-27 total
Ross (Gemmer) 6D-400	7-14	25-32 total
Saginaw 525	4-6	14 total
Saginaw 553	13-18	25-29 total
All Vans and Wagons, Saginaw	1½-4½	8¼-11¼ total

NOTE: D100,200,300 and W100, 200 use the Ross 24J (early) and Saginaw 525 (1972 and later). W300 uses Ross 35J or 376.

Manual Steering Gear Removal and Installation

Ross Models With Long Shaft

1. Disconnect battery negative cable at battery, horn wire at connec-

tor and any gear shift linkage or directional signal controls that are attached to steering column jacket.
2. Remove horn button.
3. Remove steering wheel and column clamp to instrument panel.
4. Disconnect drag link from steering arm. *Mark relation of arm to cross shaft before removal.*
5. Using a puller, remove steering arm from cross shaft.
6. If so equipped, remove transmission shift linkage.
7. Remove gear housing mounting bolts and lower jacket mounting bolts and remove gear and tube assembly from vehicle.
8. To install, insert gear and tube assembly from below frame.
9. Install mounting bolts.
10. Install clamp bolts and secure tube to instrument panel.
11. Tighten gear housing to frame bolts.
12. Loosen instrument panel to tube clamp bolts. Column tube will not change position if steering gear is in alignment. If column tube shifted position when clamp was loosened, relocate column tube bracket before tightening clamp.
13. Install spring, retainer, steering wheel, horn button spring, retainer and wheel nut, tightening nut to 24 ft. lbs. torque.
14. Attach directional signal switch.
15. Install spring, contact plate, horn wire spring and horn button.
16. Connect and adjust shift linkage (see "Manual Transmissions") if so equipped.
17. Install steering arm and drag link, tighten steering arm nut to 85 ft. lbs. torque.

Ross Models With Short Shaft

1. Remove bolt from column coupling clamp at upper end of steering worm shaft. Disconnect turn signal and horn wire con-

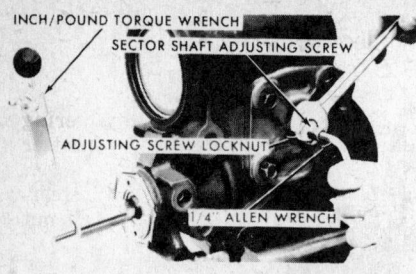

Sector shaft adjustment model 710 steering gear (© Chrysler Corp.)

nectors and move column upward to clear worm shaft splines.

2. *Mark relation of arm to gear shaft before removal.* Remove steering arm nut and washer from gear shaft and remove steering arm with a puller.

3. Remove steering gear housing mounting bolts. On 6D-400 models, remove two bolts at trunnion cap.

4. Remove steering gear from vehicle.

5. To install, center gear and install on the frame, aligning housing with frame. Tighten 7/16" and 1/2" mounting bolts to 70 ft. lbs. torque and 9/16" bolts to 100 ft. lbs. torque.

6. Install universal joint clamp to worm shaft and tighten to 30 ft. lbs. torque. On pot type joint, center the steering column shaft in the joint to insure adequate travel.

7. Set front wheels straight ahead.

8. Install steering arm (use marks made during removal to position arm in original location). Tighten bolt clamp to 170 ft. lbs. and cross shaft nut to 250 ft. lbs.

Saginaw Models

1. Unbolt the coupling clamp at the bottom of the column.

2. Remove the steering gear arm nut.

3. Use a puller to remove the steering gear arm.

4. Unbolt the steering gear from

the frame and remove it. You may have to remove the grille for access on some larger models, with the Saginaw 553 steering box.

5. Align the gear in the straight-ahead position.

6. Bolt the steering gear to the frame.

7. Make sure the front wheels are straight ahead and install the steering arm but not the nut.

8. If both wheels and steering gear are in the straight-ahead position, install the steering arm nut.

9. Install the coupling clamp.

NOTE: On models with the Saginaw 553 steering box, you will have to slide the column up to install the coupling.

Vans and Wagons

1. Remove the drag link from the steering arm.

2. Raise the hood and disconnect the battery. Disconnect the wires from the windshield washer motor and lift the reservoir up and out of the way.

3. Remove the steering column.

4. Remove the retaining bolts and the retaining nut which hold the gear onto the frame.

5. Remove the two bolts from the bottom of the splash shield going in under the left wheel house. Force the bottom of the shield out about 1/2 inch.

6. Lift the gear up off the mounting stud and out through the hood opening.

7. To replace the gear, position it to the frame in reverse of the above step, and install the mounting bolts, nut, and lockwasher. Tighten the bolts and nut to 100 ft-lbs.

8. Center the worm shaft and steering wheel.

9. Install the drag link to the steering arm.

10. Install the two splash shield bolts.

11. Install and adjust the steering column.

12. Connect the windshield washer motor wiring and battery. Mount the washer reservoir in its retaining bracket.

Power Steering

The linkage assist type power steering consists of a power cylinder attached to the front axle and tie rod, a control valve installed on the drag link between the steering knuckle arm and the steering arm and a belt-driven hydraulic pump mounted on the engine. Integral Power steering was introduced on light trucks in 1972.

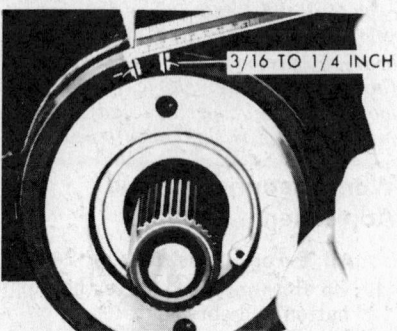

Mark housing in relationship with tool hole in the adjuster plug (© Chrysler Corp.)

Power Steering Pump Pressure Test

Model .94, 1.14

This vane type pump is used on D and W 100 through 300 series trucks. The 1.14 pump used on D500 and larger trucks is similar. The number .94 or 1.14 indicates the fluid displacement per revolution.

1. Fill fluid reservoir and check drive belt tension.

2. Disconnect the high pressure hydraulic line going to the control valve at the control valve and install a pressure gauge and valve, connecting the gauge end to the pump and the valve end to the control valve.

3. After warming up the fluid to 150-170° (let engine run and turn steering back and forth), note pressure reading while turning steering wheel to extreme in either direction as engine idles at 600-800 rpm. Maximum pressure should be at least 900 psi.

4. If maximum pressure is below 900 psi, momentarily close valve and note maximum pressure reading. If pressure was less than 900 psi, the pump is faulty. If pressure was at least 1000 psi, then either the control valve, power cylinder, or integral steering gear is faulty.

5. When removing test gauge and valve, be sure to reinstall high pressure line in its original position to avoid interference with

Model 553 Saginaw steering gear (© Chrysler Corp.)

Model 1.14 power steering pump
(© Chrysler Corp.)

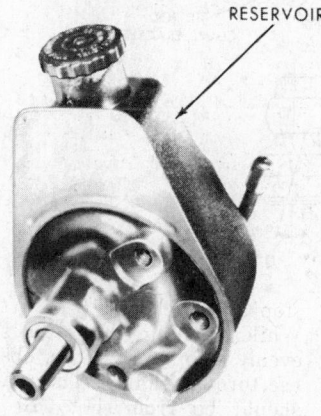

Second type .94 power steering pump
(© Chrysler Corp.)

engine or sheet metal.

6. For power steering pump overhaul procedures, see "Power Steering" in the General Repair Section.

Model 1.2

This slipper type pump is used on 1972 and later D500 and larger trucks.

The pressure test procedure is the same as that described for the model .94 above. In this test, however, the engine idles at 500 rpm. It is possible with this pump that the pressure relief valve may be faulty if the test indicated that the pump was faulty. Before removing pump and overhauling pump, retest with a new pressure relief valve.

Pump Removal and Installation

NOTE: See "Power Steering" in General Repair Section for power steering pump overhaul procedures.

1. Loosen pump lower mounting and locking bolts and remove the belt.
2. Place a container under pump and disconnect both pump hoses.
3. Remove mounting and locking bolts and remove pump and brackets.
4. To install, position pump on engine and install retaining and locking bolts.
5. Install drive belt and adjust. Tighten pump brackets to 30 ft. lbs. torque.
6. Connect pressure and return hoses, routing them in the same position they were in before removal.
7. Fill pump reservoir.
8. Start engine and turn steering wheel all the way left and right to bleed the system. Stop engine and recheck fluid level, refilling if necessary.

Linkage Assist Type

Control Valve Removal and Installation

NOTE: See "Power Steering" in the General Repair Section for control valve overhaul and service procedures.

1. Disconnect pressure and return hoses at valve assembly.
2. Cap hoses and connections at valve assembly. Fasten hoses so that ends are above fluid level in reservoir and tag them for reinstallation identification.
3. Disconnect ball ends at steering arm and steering knuckle arm and remove control valve and drag link assembly.
4. To install, connect control valve sliding sleeve end at steering arm and install nut and cotter pin.
5. Connect drag link end to steering knuckle arm and install nut and cotter pin.
6. Connect pressure and return hydraulic lines at valve assembly, tightening securely.
7. Refill pump reservoir.
8. Bleed system by turning steering wheel back and forth several times while the engine is idling.
9. Recheck fluid level and refill if necessary.

Power Cylinder Removal and Installation

1. Disconnect hydraulic lines at power cylinder and cap ends and connections at cylinder.
2. Remove cotter pin and nut at ball studs (and tie rod clamp U-bolts on lightweight trucks). Remove cylinder.
3. To install, insert ball stud into place on axle bracket and install nut and cotter pin, tighten nut securely.
4. Rotate steering wheel to extreme right turn position.
5. Install tie rod clamp loosely (on medium weight trucks loosen tie rod clamp and install ball stud and nut).
6. Position cylinder piston rod in fully retracted position and tighten U-bolt clamps on tie rod. On medium weight trucks the ball stud will have to be connected to tie rod clamp bracket before tightening clamp bolts. Clamp should be mounted on tie rod at a 20° angle to the vertical. Nuts on U-bolts are torqued to 40 ft. lbs.
7. Connect hydraulic lines to power cylinder in their original positions.
8. Bleed system by turning the steering wheel back and forth several times with the engine idling.

Integral Type
Removal and Installation

Conventional Trucks

1. Center the steering gear.
2. Pull off the steering arm.
3. Disconnect the pressure and return hoses.
4. Disconnect the steering shaft coupling.
5. Unbolt the gear from the frame and remove it.
6. Install the mounting bolts finger tight on installation.
7. Connect the coupling. Align the gear to the frame to prevent binding; tighten the bolts.

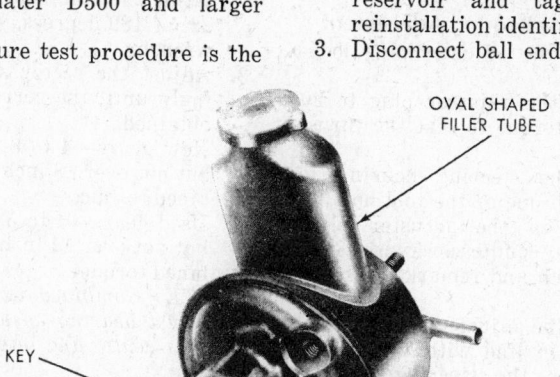

OVAL SHAPED
FILLER TUBE

KEY

Early model .94 power steering pump, later models have a pear shaped fluid reservoir (© Chrysler Corp.)

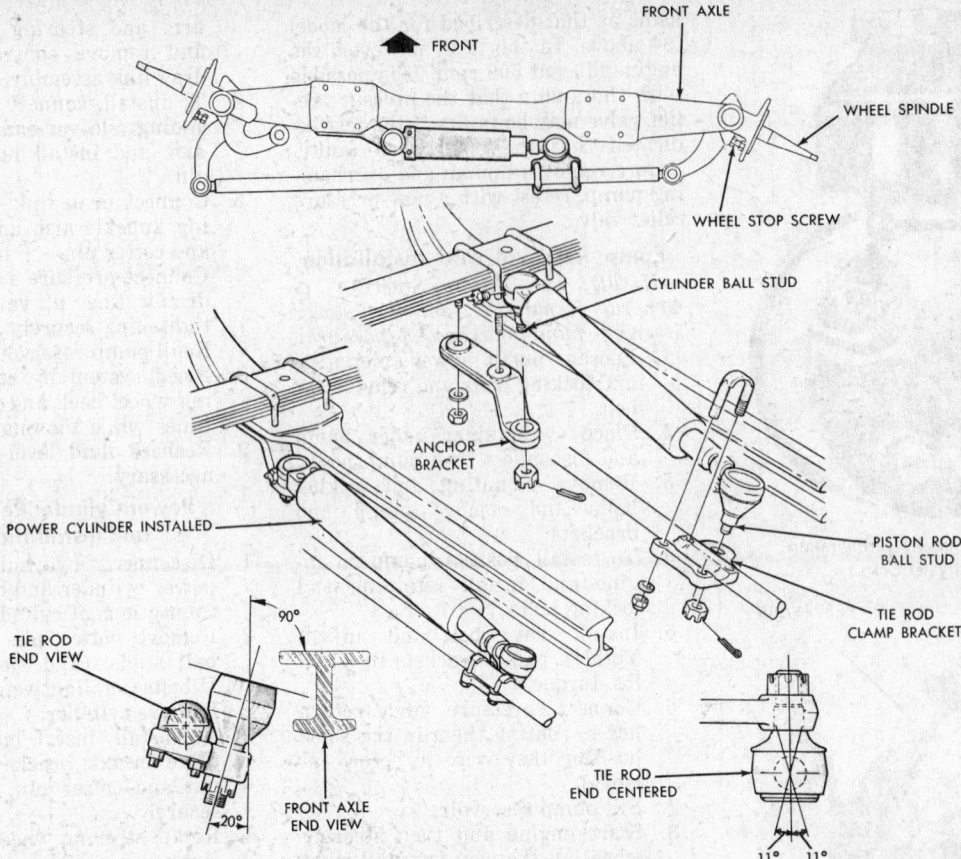

FRONT

FRONT AXLE

WHEEL SPINDLE

WHEEL STOP SCREW

CYLINDER BALL STUD

ANCHOR BRACKET

POWER CYLINDER INSTALLED

PISTON ROD BALL STUD

TIE ROD CLAMP BRACKET

TIE ROD END VIEW

90°

20°

FRONT AXLE END VIEW

TIE ROD END CENTERED

11° 11°

Power cylinder adjustment (© Chrysler Corp.)

8. Connect the pressure and return hoses.
9. Center the steering gear, make sure the wheels are straight ahead, replace the steering arm and nut.
10. Idle the engine and turn the steering wheel gently from stop to stop to bleed the system of air.

Vans and Wagons

1. Disconnect the battery and remove the windshield washer reservoir.
2. Remove the steering column.
3. Remove the optional steering arm shield.
4. Use a puller to remove the drag link from the steering arm.
5. Detach the hoses and raise their ends to prevent spillage.
6. Remove the two gear to frame bolts and the nut from the stud.
7. Remove the two bolts from the bottom of the splash shield (left side) and pull it out to lift the gear off the stud.
8. Turn the steering arm and lift the steering gear out through the hood opening.
9. On installation, move the parking brake cable away from the mounting stud. Position the gear. Install the mounting bolts and nut.
10. Install the drag link to the steering arm.
11. Replace the splash shield bolts.
12. Connect the hoses and check the

reservoir fluid level.
13. Replace the steering column.
14. Replace the windshield washer reservoir and connect the battery.
15. Idle the engine and turn the steering wheel gently from stop to stop to bleed the system of air.

Adjustment

The power steering gear used in conventional trucks should not be adjusted while in the chassis or filled with hydraulic fluid. The unit in vans and wagons may be adjusted in the vehicle.

Worm Thrust Bearing Adjustment

1. Remove the adjuster plug locknut.
2. Bottom the adjuster plug to 20 pounds torque to seat the thrust bearings.
3. Mark the steering housing in line with one of the tool hole locations on the adjuster plug. Measure counterclockwise 3/16 to ¼ inch and remark the housing.
4. Loosen the adjuster until the tool hole is in line with the second mark on the steering housing and install the lock nut and tighten while maintaining the alignment of the adjuster tool hole with the mark on the housing.
5. With the aid of a torque wrench, turn the stub shaft to the right

stop and back off ¼ of a turn. While turning the stub shaft evenly counterclockwise, observe the torque reading. The reading should be from 4 to 10 inch pounds.
6. Continue the adjustments as necessary to obtain the specified torque readings.

Cross Shaft Over-Center Adjustment

1. With the gear on center, loosen the adjusting nut and tighten the sector adjusting screw.
2. Tighten the lock nut and check the overcenter torque while rotating the stub shaft through an arc of 180 degrees, with a torque wrench.
3. Adjust the sector shaft accordingly until the correct torque is obtained.
   New gears—4 to 8 inch pounds, but not over 18 inch pounds combined torque.
   Used gears—4 to 5 inch pounds, but not over 14 inch pounds combined torque.
NOTE: Combined torque includes the thrust bearing adjustment reading, over-center and internal friction.

Sector Adjustment, Vans and Wagons

1. Disconnect the center link from the steering gear arm.
2. Start the engine and allow to run at normal idle speed.

3. Rotate the steering wheel from lock to lock. Carefully count the number of turns required, then rotate the wheel back, exactly to the midpoint of its travel.
4. Loosen the adjusting screw until backlash in the steering gear arm becomes apparent.

NOTE: Backlash is felt by holding the end of the steering gear arm lightly between your thumb and forefinger.

5. Tighten the adjusting screw just enough so that the backlash disappears. Continue to tighten the screw for another ⅜ to ½ turn from this point. Tighten the locknut to 28 ft. lbs.
6. Attach the center link to the steering gear arm.

Clutch

Mechanical Clutch Linkage Adjustment

The only adjustment required is pedal free-play. Adjust the clutch actuating fork rod by turning the self-locking adjusting nut to provide ⅛ in. (3/32 in. on vans and wagons) free movement at the end of the fork. This will provide the recommended 1½ in. (1 in. on Vans, Wagons, Ramcharger, and Trailduster) free-play at the pedal.

Hydraulic Clutch Adjustment, Medium Duty (D400 up) 1971 Models

1. Inside the cab, adjust clutch pedal free play to 3/16 in. by turning adjusting nut to allow .010″ free play between push rod and master cylinder piston.
2. At the slave cylinder, adjust clutch linkage free play at clutch fork pivot pin to 3/16″. Total pedal free play should then be approximately 1-½″. This figure should be used as a check and not for setting free play.

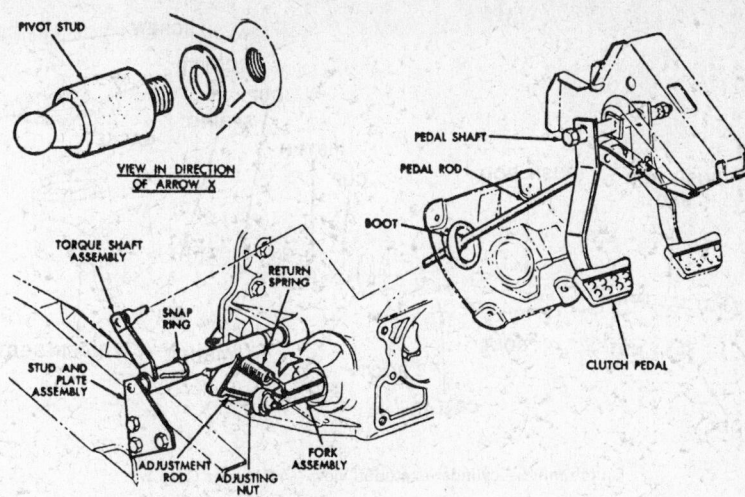

D and W 100, 200 clutch linkage (© Chrysler Corp.)

3. Inside the cab, check clutch pedal reserve. This is that portion of pedal travel remaining after free play and complete clutch disengagement. There should be a minimum of 1″ pedal reserve. If there is insufficient pedal reserve, bleed the hydraulic system as described below. Extremely low reserve indicates excessive wear in either of the cylinders.
4. Check for binding, incorrect parts or improperly installed parts and make corrections, repeating Steps 1 through 3 if necessary.
5. Over center spring tension (pedal return) is increased by backing off end nut on rod at top of clutch pedal.

Bleeding Hydraulic Clutch System

1. Clean away dirt from around master cylinder reservoir filler cap, remove cap and fill reservoir. Make sure to check reservoir frequently during bleeding process and fill when necessary.

2. Clean bleeder valve on slave cylinder and attach bleeder hose to valve.
3. Place other end of hose in jar half full of brake fluid.
4. Bleed intermittently as clutch pedal is being depressed by opening and closing bleeder valve. Continue bleeding until no air bubbles come from bleeder hose.

Clutch Master Cylinder Removal and Installation

1. Remove pedal return spring.
2. Disconnect push rod end at clutch pedal and hydraulic fluid line at master cylinder.
3. Unbolt master cylinder from firewall.
4. Installation is the reverse of the above procedure. Adjust master cylinder push rod to .010″ free play. Bleed hydraulic system.

Clutch Slave Cylinder Removal and Installation

Slave cylinder is located on the right side of the clutch housing. Disconnect hydraulic line, then remove cylinder mounting bolts. After installation, be sure to bleed hydraulic system and adjust as described above.

Clutch Removal & Installation

All Models

Removal

1. Support engine on a suitable jack, if necessary.
2. Remove crossmember, if necessary.
3. Remove transmission. Remove transfer case, if equipped.
4. Remove clutch housing pan if so equipped.
5. Remove clutch fork, clutch bearing and sleeve assembly if not

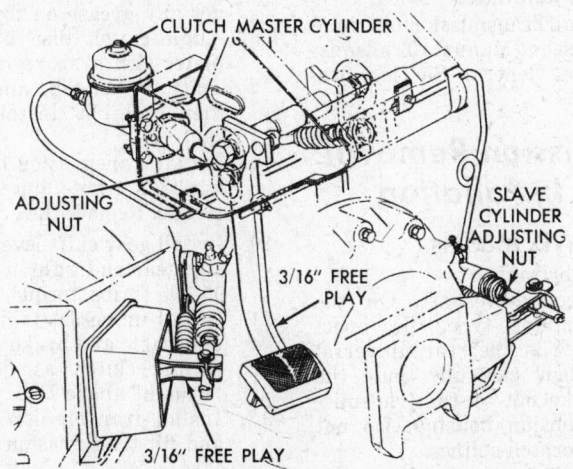

Hydraulic clutch adjustments (© Chrysler Corp.)

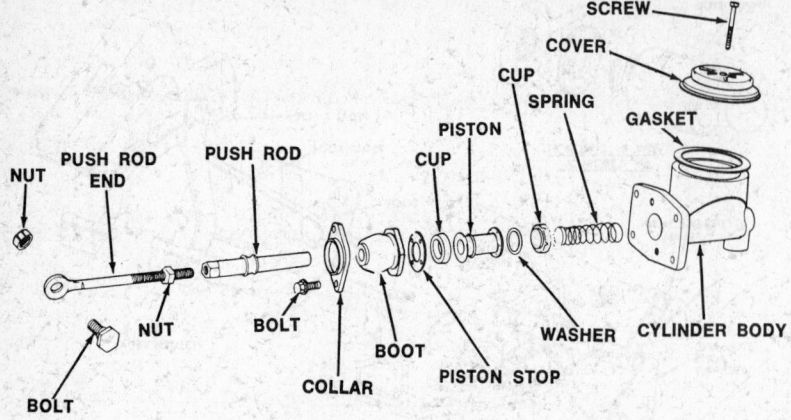

Clutch master cylinder-exploded view (© Chrysler Corp.)

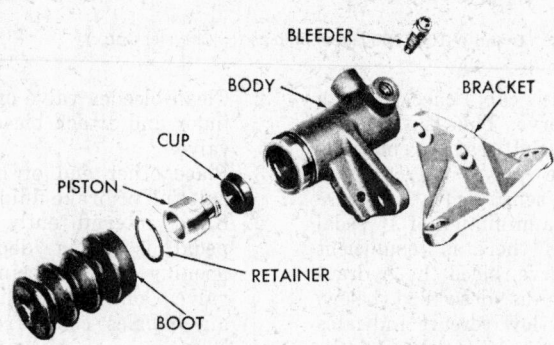

Clutch slave cylinder (© Chrysler Corp.)

removed with transmission.

6. Mark clutch cover and flywheel, with a prick punch, to assure correct reassembly.

7. Remove clutch cover retaining bolts, loosening them evenly so clutch cover will not be distorted.

8. Pull pressure plate assembly clear of flywheel and, while supporting pressure plate, slide clutch disc from between flywheel and pressure plate.

Installation

1. Thoroughly clean all working surfaces of flywheel and pressure plate.

2. Grease radius at back of bushing.

3. Rotate clutch cover and pressure plate assembly for maximum clearance between flywheel and frame crossmember if crossmember was not removed during clutch removal.

4. Tilt top edge of clutch cover and pressure plate assembly back and move it up into the clutch housing. Support clutch cover and pressure plate assembly and slide clutch disc into position.

5. Position clutch disc and plate against flywheel and insert spare transmission main drive gear shaft or clutch installing tool through clutch disc hub and into main drive pilot bearing.

6. Rotate clutch cover until the prick punch marks on cover and flywheel line up.

7. Bolt cover loosely to flywheel. Tighten cover bolts a few turns at a time, in progression, until tight. Then tighten bolts to 20 ft. lbs. torque.

8. Install transmission.

9. Install frame crossmembers and insulator, tighten all bolts.

Manual Transmission

Reference

For Manual Transmission overhaul procedures see "Manual Transmission" in the General Service Section.

Transmission Removal and Installation

3-Speed 2WD Models

1. Drain lubricant.

2. Disconnect driveshaft. On the sliding spline type, disconnect driveshaft at the rear universal joint, then carefully pull the shaft yoke out of the transmission extension housing. Do not nick or scratch splines.

3. Disconnect gearshift control rods and speedometer cable.

4. Remove backup light switch if so equipped.

5. Support engine.

6. Remove crossmember and rubber insulator on 1971 models and 1975 and later models with A-390 transmission. On all other models, unbolt the insulator or mount from the crossmember. Support the transmission with a jack.

7. Remove transmission to clutch housing bolts.

8. Slide transmission rearward until pinion shaft clears clutch completely, then lower transmission from vehicle.

9. Installation is the reverse order of the above procedure. Before inserting transmission drive shaft into clutch, make sure clutch housing bore, disc and face are aligned. Tighten clutch housing to transmission bolts to 50 ft. lbs. torque.

10. Fill with lubricant.

11. Adjust shift linkage.

12. Road test.

4-Speed 2 WD Models

1. Shift transmission into any gear.

2. Disconnect universal joint and loosen yoke retaining nut.

3. Disconnect parking brake (if so equipped) and speedometer cables at transmission.

4. Remove lever retainer by pressing down, rotating retainer counter-clockwise slightly, then releasing.

5. Remove lever and its springs and washers.

6. Support the rear of the engine and remove the crossmember. Remove transmission to clutch bell housing retaining bolts and pull transmission rearward until drive pinion clears clutch, then remove transmission.

7. To install, place ½ teaspoon of short fibre grease in pinion shaft pilot bushing, taking care not to get any grease on flywheel face.

8. Align clutch disc and backing plate with a spare drive pinion shaft or clutch aligning tool, then carefully install transmission.

9. Install transmission to bell housing bolts, tightening to 50 ft. lbs. torque. Replace the crossmember.

10. Install gear shift lever, shift into any gear and tighten yoke nut to 95-105 ft. lbs. torque.

11. Install universal joint, speedometer cable and brake cable.

12. Adjust clutch as described in "Clutch" above.

13. Install transmission drain plug and fill transmission with lubricant.

14. Road test.

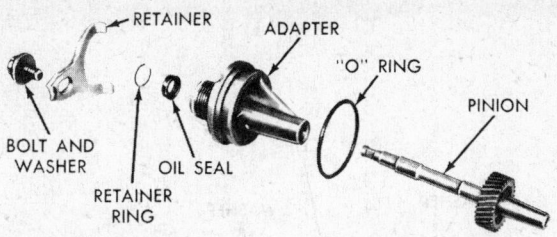

Speedometer drive gear assembly—exploded view (© Chrysler Corp.)

3 and 4-Speed-4WD Models

1. Raise and support the truck.
2. Remove the skid plate, if any.
3. Disconnect the speedometer cable.
4. Disconnect the front and rear driveshafts. Suspend each shaft from a convenient place; do not allow them to hang free.
5. Disconnect the shift rods at the transfer case. On 4-speed transmissions, remove the shift lever retainer by pressing down and turning it counterclockwise. Remove the shift lever springs and washers.
6. Remove the rear driveshaft. Matchmark the driveshaft and rear U-joints before removing the driveshaft.
7. Support the transfer case.
8. Remove the extension-to-transfer case mounting bolts.
9. Move the transfer case rearward to disengage the front input shaft spline.
10. Lower and remove the transfer case.
11. Disconnect the back-up light switch.
12. Support the engine.
13. Support* the transmission.
14. Remove the transmission crossmember.
15. Remove the transmission-to-clutch housing bolts.
16. Slide the transmission rearward until the mainshaft clears the clutch disc.
17. Lower and remove the transmission.
18. Installation is the reverse of removal. The transmission pilot bushing in the end of the crankshaft requires high-temperature grease. Multipurpose grease should be used. Do not lubricate the end of the mainshaft, clutch splines, or clutch release levers. Adjust the gearshift linkage on 3-speed transmissions.

5-Speed Transmission

1. Drain lubricant.
2. Remove gearshift lever, floor mat and floor cover over transmission.
3. Disconnect parking brake cable, speedometer cable and driveshaft.
4. Loosen flange nut and drop center bearing if so equipped.
5. Position a transmission jack under transmission and remove clutch housing to transmission bolts.
6. Slide transmission rearward until pinion shaft clears clutch (about 6"), then lower jack slightly. Move transmission to the left just enough for the main drive pinion to clear clutch housing, then lower transmission.
7. Installation is the reverse order of the above procedure. Align clutch backing plate and disc with a spare drive pinion shaft or clutch aligning tool before inserting transmission drive pinion and be sure splines on shaft align with clutch hub splines before inserting. Tighten bell housing to transmission bolts to 100 ft. lbs. torque and driveshaft flange nut to 125 ft. lbs. torque.
8. Fill with lubricant and adjust clutch.
9. Road test.

Heavy Duty Transmission —Five Speed, Six Speed and Up

1. a. *Conventional Gearshift Lever:* Remove the floor mat and transmission pan cover and place the gearshift lever in the neutral position. Remove the gearshift lever and control tower assembly from the transmission.
   b. *Range or Splitter Power Shift:* Bleed the air tanks and disconnect the air lines at the air valve on the transmission. Remove the gear shift lever and control tower assembly from the transmission. On Tilt cab models, disconnect the shift control linkage from the remote control assembly at the transmission or remove the cover assembly.
2. Place a clean lint free cloth over the opening in the transmission to keep dirt out.
3. Disconnect the wire from the back-up light switch mounted on the transmission (if equipped).
4. Drain the transmission lubricant.
5. Disconnect the speedometer cable at the transmission adapter.
6. Disconnect the clutch control linkage.
7. Disconnect the parking brake lever and controls (if equipped).
8. Disconnect the propeller shaft from the transmission.
9. Remove any exhaust system brackets mounted on the transmission.
10. If necessary, support the saddle tanks and remove the tank cross struts.
11. Remove the engine ground strap and the battery cable support clip if attached to the transmission or clutch housing.
12. Remove the power take off unit and controls if equipped.
13. On vehicles equipped with range or splitter air power shift, disconnect the air lines from the intake at the air filter, the air input, and shift control. Cover the air filter opening to keep dirt out.
14. Position a transmission jack or dolly under the transmission and adjust to carry the weight of the transmission.
15. Remove the parts that attach the transmission to the rear mount, if used.
16. Visually inspect to determine if other equipment, lines or brackets must be removed to permit removal of the transmission.
17. On vehicles which have a rear engine mounting attached to the clutch housing, it will be necessary to support the engine with a suitable jack positioned under the flywheel housing. Remove the engine rear mounting as necessary to free the transmission.
18. Remove the clutch housing-to-flywheel housing bolts or the transmission-to-flywheel housing bolts if equipped with an apron type flywheel housing.
19. Move the transmission assembly straight back from the engine until the main drive gear shaft is clear. Be careful to keep the transmission main shaft in alignment with the clutch disc.
20. Lower the transmission and remove from the vehicle. If additional clearance is needed, raise the rear wheels of the vehicle.
21. Inspect the clutch components and replace any worn or damaged parts.
22. Apply a light film of multi-purpose grease to the main drive gear retainer and the splined portion of the main shaft to assure smooth assembly. *NOTE: Do not apply an excessive amount of grease to the mainshaft components as under operating conditions the grease would be thrown onto the clutch facings causing clutch failure.*

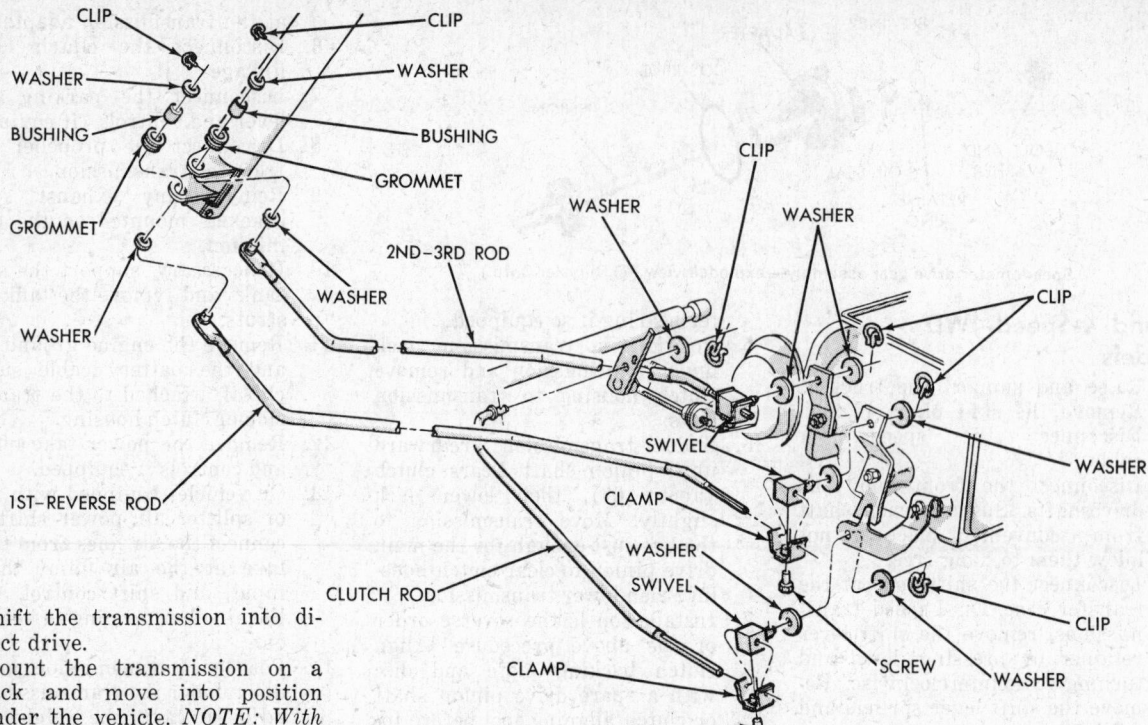

CLIP · CLIP · WASHER · WASHER · BUSHING · BUSHING · GROMMET · GROMMET · GROMMET · WASHER · WASHER · 2ND-3RD ROD · WASHER · CLIP · WASHER · WASHER · CLIP · 1ST-REVERSE ROD · SWIVEL · CLAMP · WASHER · CLUTCH ROD · WASHER · SWIVEL · CLIP · CLAMP · SCREW · WASHER · SCREW

A-250 transmission linkage with clutch interlock—van with six-cylinder (© Chrysler Corp.)

23. Shift the transmission into direct drive.

24. Mount the transmission on a jack and move into position under the vehicle. *NOTE: With the apron type flywheel housing, position the clutch release bearing and support assembly inside the flywheel housing. Be sure that the clutch fork properly engages the clutch release bearing.*

25. Align the transmission mainshaft with the clutch disc hub by rotating the companion flange. Move the transmission forward, guiding the mainshaft into the clutch disc splines. *NOTE: Avoid springing the clutch when the transmission is being installed on the engine. Do not force the transmission into the clutch disc splines. Do not let the transmission drop or hang unsupported in the splined hub or clutch release bearing.*

26. Install the clutch housing-to-engine flywheel housing mounting bolts and lock washers, tighten to specification.

27. Install the engine mountings if they were removed and tighten to specification.

28. Install the remaining parts and components in the reverse of removal.

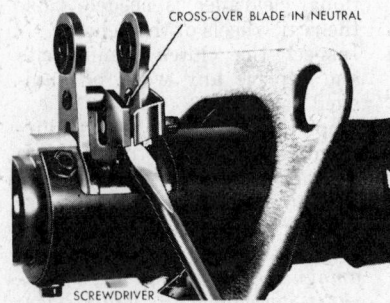

CROSS-OVER BLADE IN NEUTRAL

SCREWDRIVER

Holding the three speed column shift linkage in neutral position
(© Chrysler Corp.)

Clutch Interlock

Adjustment

A-250 3-Speed

This adjustment is required only on the A-250 3-speed transmission.

This is a top cover unit used only as base equipment on light duty six cylinder models. It has synchromesh only on second and third gears.

1. Disconnect the clutch rod swivel from the interlock pawl. Adjust the clutch pedal free play.

2. Shift the transmission to neutral. Loosen swivel clamp bolt and slide the swivel onto the rod untl the pawl is positioned fully within the slot in the first-reverse lever. Install the washers and clip.

3. Hold the interlock pawl forward and tighten the swivel clamp bolt. The clutch pedal must be in fully returned position during the adjustment. *NOTE: Do not pull the clutch rod rearward to engage the swivel in the pawl.*

4. Shift the transmission into first and reverse and release the clutch pedal while in either gear to check for normal clutch action. Then, shift halfway between neutral and either gear and release clutch. The interlock should hold it to within one or two inches of the floor.

Column Shift Linkage

Adjustment

Conventional Trucks

1. Remove both shift rod swivels from transmission shift levers. Make sure transmission shift levers are in neutral (middle detent) position.

2. Move shift lever to line up locating slots in bottom of steering column shift housing and bearing housing. Install suitable tool in slot, if any.

3. Place screwdriver or suitable tool between crossover blade and second-third lever at steering column so that both lever pins are engaged by cross-over blade.

4. Set first-reverse lever on transmission to reverse position (rotate clockwise).

5. Adjust first-reverse rod swivel by loosening clamp bolt and sliding swivel along rod so it will enter first-reverse lever at transmission. Install washers and clip. Tighten swivel bolt.

6. Remove gearshift housing locating tool and shift column lever into neutral position.

7. Adjust second-third rod swivel by loosening clamp bolt and sliding swivel along rod so it will enter second-third lever at transmission. Install washers and clip. Tighten swivel bolt.

8. Remove tool from crossover blade at steering column, and

shift through all gears to check adjustment and cross-over smoothness.

Vans and Wagons

1. Adjust the length of the 2-3 shift rod so the position of the shift lever on the steering column will be correct.
2. Assemble the 1st-reverse and 2-3 shift rods, and place each in its normal position, secured with a clip. Loosen both swivel clamp bolts.
3. Move the 2-3 shift lever into 3rd position (this means moving the forward lever forward). Move the steering column lever until it is about five degrees above the horizontal. Tighten the shift rod swivel clamp bolt.
4. Shift the transmission to neutral. Place a screwdriver between the crossover blade and the 2-3 lever at the steering column so that both lever pins are engaged by the cross over blade.
5. Set the 1st-reverse lever in neutral. Tighten the swivel clamp bolt.
6. Remove the tool from the cross over blade, and check all shifts for smoothness.

Transfer Case

Reference

See General Section for overhaul procedures.

Transfer Case Removal and Installation

W100, 200, 300 through 1974

This is the new process 205 two-speed transfer case.
1. Drain lubricant.
2. Disconnect speedometer cable.
3. Disconnect input and output shafts, securing driveshafts to frame.
4. Remove shift rod from shift rail link.
5. Using a suitable jack, support transfer case while removing frame crossmember.
6. Remove mounting bolts and lower transfer case from truck.
7. Installation is the reverse of the above procedure. Tighten mounting bolts to 35 ft. lbs. and driveshaft yoke bolts to 300-400 ft. lbs. torque.

Ramcharger and Trail Duster, 1975 and Later W100, 200, 300

This is the New Process 203 two-speed, full-time four-wheel drive transfer case. It incorporates a differential unit which allows the front and rear axles to remain in continuous drive in normal HI and LO positions. Both driveshafts transmit power simultaneously at the same rpm in Hi-Loc and Lo-Loc positions with the differential locked. The two-speed feature gives direct drive in High range and 2.01:1 reduction in Low range.

1. Raise and support the truck.
2. Remove the skid plate, if any.
3. Drain the transfer case by removing the bottom bolt from the front output rear cover.
4. Disconnect the speedometer cable.
5. Disconnect the front and rear output shafts. Suspend these from a convenient location; do not allow them to hang free.
6. Disconnect the shift rods at the transfer case.
7. Support the transfer case.
8. Remove the adaptor-to-transfer case mounting bolts and move the transfer case rearward to disengage the front input splines.
9. Lower and remove the transfer case.
10. Installation is the reverse of removal. Adjust the linkage.

W500, 600

This is the new process T-223 two speed, two lever, transfer case. On Canadian models through 1973, it is the new process T-201.
1. Drain lubricant.
2. Disconnect speedometer and parking brake cables.
3. Disconnect input and output shafts. *Do not allow shafts to hang.*
4. Remove clevis pins, disconnecting de-clutch and shift control rods from transfer case rails. Secure rods out of the way.
5. Using a jack, support transfer case while removing support bracket bolts.
6. Carefully lower transfer case from truck.
7. To install, raise transfer case with a jack and align threaded mounting bolt holes in transfer case with holes in support brackets. Install bolts and tighten securely.
8. Connect speedometer cable and parking brake cable, adjusting parking brake as described in "Parking Brake" above.
9. Install and adjust transfer case shift and de-clutch rods.
10. Connect driveshafts.

Transfer Case Linkage
Adjustment

W500, 600

1. Make sure that the front axle

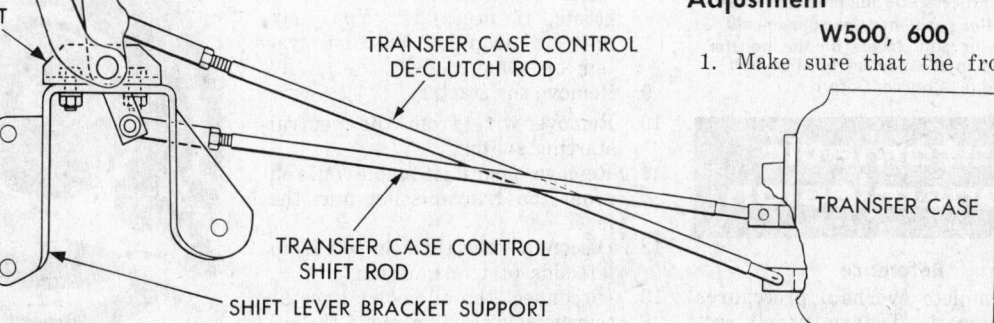

TRANSFER CASE CONTROL SHIFT LEVER

TRANSFER CASE CONTROL DE-CLUTCH LEVER

A

SHIFT LEVER BRACKET

TRANSFER CASE CONTROL DE-CLUTCH ROD

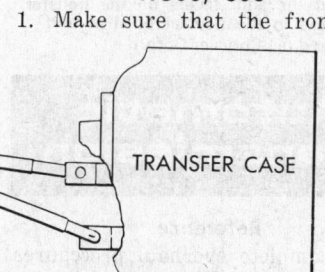

TRANSFER CASE

TRANSFER CASE CONTROL SHIFT ROD

SHIFT LEVER BRACKET SUPPORT

New Process T-233 and T-201 transfer case linkage adjustment, measurement A should be ¼ in. (© Chrysler Corp.)

drive is engaged (the de-clutch rail on the case must be fully in). Make sure that the transfer case is in low range (the shift rail on the case must be fully out).

2. Disconnect the de-clutch and shift rods at the case by removing the clevis pins.
3. Adjust the de-clutch rod length so that the lever clears the rear end of the cab floor slot by ½ in. Replace the clevis pin.
4. Adjust the shift rod length until the distance between the lever contact surfaces is ¼ in. Replace the clevis pin.

Ramcharger and Trailduster, 1975 and Later W100, 200, 300

1. Loosen the lockscrews in both swivel rod clamps at the shifter assembly. The rods must be free to slide in the swivels.
2. Place the selector lever in the cab in neutral and insert a 11/64 in. diameter rod through the alignment holes in the shifter housing.
3. Place the range shift lever (outboard lever) on the transfer case in the neutral position.
4. Place the lockout shift lever on the transfer case (the inboard lever) in the unlocked position.
5. Retighten the rod swivel screws.
6. Remove the alignment rod from the shifter housing.

W100, 200, 300 through 1974

There is no external linkage adjustment necessary or possible on this unit.

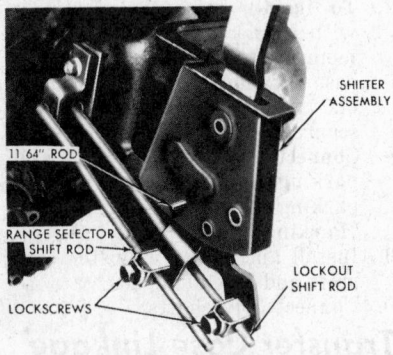

The New Process 203 full-time four wheel drive shifter positioned for adjustments, note that the shift levers on the transfer case are in the neutral position and unlocked (© Chrysler Corp.)

Automatic Transmission

Reference

For complete overhaul procedures see "Automatic Transmission" section of the General Repair Section. *NOTE: Internal components vary*

with model and drive train. Use the part number stamped on the left side of the fluid pan flange for positive identification.

Loadflite and New Process Automatic Transmissions

Transmission R&R

Removal

1. Remove the transmission and converter as an assembly; otherwise the converter drive plate, pump bushing, and oil seal will be damaged. The drive plate will not support a load. Therefore, none of the weight of transmission should be allowed to rest on the plate during removal. Remove the transfer case, as necessary.
2. Attach a remote control starter switch to the starter solenoid so the engine can be rotated from under the vehicle.
3. Disconnect high tension cable from the ignition coil.
4. Remove cover plate from in front of converter assembly to provide access to the converter drain plug and mounting bolts.
5. Rotate engine to bring drain plug to "6 o'clock" position. Drain the converter and transmission.
6. Mark converter and drive plate to aid in reassembly.
7. Rotate the engine with the remote control switch to locate two converter to drive plate bolts at "5 and 7 o'clock" positions.

   Remove the two bolts, rotate engine again and remove the other two bolts. **CAUTION: Do not rotate converter on drive plate by prying with a screwdriver or similar tool as drive plate might become distorted. Also the starter should never be engaged if drive plate is not attached to converter with at least one bolt or if transmission case to engine block bolts have been loosened.**
8. Disconnect battery ground cable. Remove engine to transmission struts, if necessary. You may have to drop the exhaust system on some models.
9. Remove the starter.
10. Remove wire from the neutral starting switch.
11. Remove gearshift cable or rod from the transmission and the lever.
12. Disconnect the throttle rod from left side of transmission.
13. Disconnect the oil cooler lines at transmission and remove the oil filler tube. Disconnect the speedometer cable.

14. Disconnect the driveshaft.
15. Install engine support fixture to hold up the rear of the engine.
16. Raise transmission slightly with jack to relieve load and remove support bracket or crossmember. Remove all bell housing bolts and carefully work transmission and converter rearward off engine dowels and disengage converter hub from end of crankshaft. **CAUTION: Attach a small "C" clamp to edge of bell housing to hold converter in place during transmission removal; otherwise the front pump bushing might be damaged.**

Installation

NOTE: Install transmission and converter as an assembly. The drive plate will not support a load. Do not allow weight of transmission to rest on the plate during installation.

1. Rotate pump rotors until the rotor lugs are vertical.
2. Carefully slide converter assembly over input shaft and reaction shaft. Make sure converter impeller shaft slots are also vertical and fully engage front pump inner rotor lugs.
3. Use a "C" clamp on edge of converter housing to hold converter in place during transmission installation.

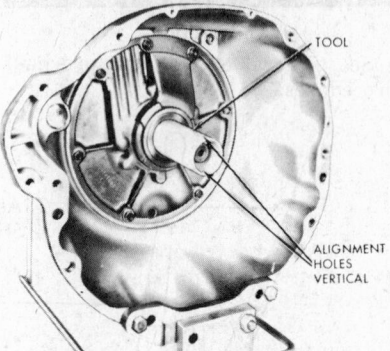

Aligning the pump rotors—typical
(© Chrysler Corp.)

Converter and drive plate markings—typical
(© Chrysler Corp.)

4. Converter drive plate should be free of distortion and drive plate to crankshaft bolts tightened to 55 ft. lbs. torque.
5. Using a jack, position transmission and converter assembly in alignment with engine.
6. Rotate converter so mark on converter (made during removal) will align with mark on drive plate. The offset holes in plate are located next to the 1/8" hole in inner circle of the plate. A stamped "V" mark identifies the offset hole in converter front cover. Carefully work transmission assembly forward over engine block dowels with converter hub entering the crankshaft opening.
7. Install converter housing to engine bolts and tighten to 28 ft. lbs.
8. Install the two lower drive plate to converter bolts and tighten to 270 in. lbs. torque.
9. Install engine to transmission struts, if required. Install starting motor and connect battery ground cable.
10. Rotate engine and install two remaining drive plate to converter bolts.
11. Install crossmember and tighten attaching bolts to 90 ft. lbs. torque. Lower transmission so that extension housing is aligned and rests on the rear mount. Install bolts and tighten to 40 ft. lbs. torque.
12. Remove transmission jack and engine support fixture, then install tie-bars under the transmission.
13. Replace the driveshaft.
14. Connect oil cooler lines, install oil filler tube and connect the speedometer cable.
15. Connect gearshift cable or rod and torqueshaft assembly to the transmission case and to the lever.
16. Connect throttle rod to the lever at left side of transmission bell housing.
17. Connect wire to back-up and neutral starting switch.
18. Install cover plate in front of the converter assembly.
19. Refill transmission with fluid.
20. Adjust throttle and shift linkage.

Chrysler Auto Transmissions— Adjustment

Loadflite Kickdown Throttle

1971 Conventional Trucks
1. Warm up engine until choke and fast idle cam are released.

2. On all except 318 cu. in. V8 engines, remove clip, spring, flat washer and transmission throttle rod from the carburetor pin. Holding transmission throttle lever forward against its stop, adjust rod end so rear of slot contacts carburetor pin. Reinstall washer, spring and clip.
3. On 318 cu. in. engine, remove cotter pin, washer and transmission throttle rod from carburetor pin. Unhook spring from adjustable rod end. Hold transmission throttle lever forward against its stop and adjust rod end so front of slot contacts carburetor pin. Replace spring, washer and cotter pin.
4. Road test vehicle. If transmission does not kickdown, adjust transmission throttle rod one more turn to move the lever at the transmission farther back.

1972 and Later Conventional Trucks, Six-Cylinder
1. Block the choke valve fully open and release the fast idle cam.
2. Hold the transmission lever forward firmly against its stop while adjusting the linkage. If there is a solenoid idle stop, the plunger must be fully extended.
3. Loosen the slotted link lock bolt and pull forward on the slotted adjuster link until it contacts the carburetor lever pin and all slack is removed.
4. Tighten the lock bolt.
5. Check the adjustment by moving the slotted adjuster link all the way back and releasing it slowly. It should return to the full forward position.

1972 and Later Conventional Trucks, V8
1. Set the idle speed to specifications.
2. Disconnect the choke at the carburetor or block the choke open in the wide open position.
3. Open the throttle slightly to release the fast idle cam and return the carburetor to curb idle.
4. It is important that the transmission lever remain firmly against the stop during the next two steps. Have an assistant hold the transmission throttle lever forward against the stop while the adjustment is made. On engines with solenoid idle stops, the solenoid plunger must be fully extended.
5. With a 3/16 in. diameter rod placed in the holes in the upper bellcrank and lever, adjust the length of the intermediate transmission rod by means of the threaded adjustment at the upper end. The ball socket must align with the ball end with a slight downward pressure on the rod.
6. Assemble the ball socket to the ball end and remove the 3/16 in. rod from the upper bellcrank and lever.
7. Disconnect the return spring, clip, and washer and adjust the length of the carburetor rod by pushing rearward on the rod and turning the threaded adjustment. The rear end of the slot should contact the carburetor lever pin with no backlash when the slotted adjuster link is in its normal operating position.
8. Assemble the slotted adjustment to the carburetor lever pin and install the washer and retainer

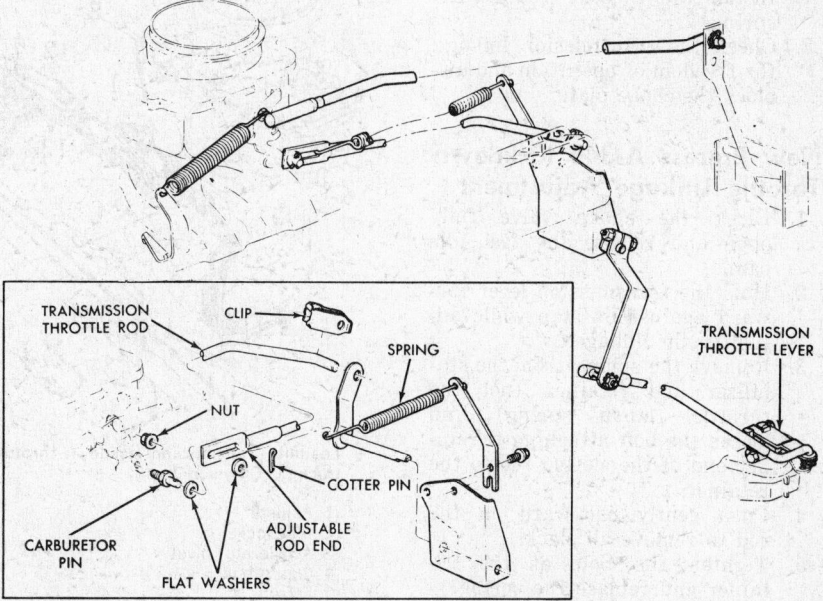

Loadflite transmission kickdown throttle linkage—conventional truck 318 V8 through 1971 © Chrysler Corp.

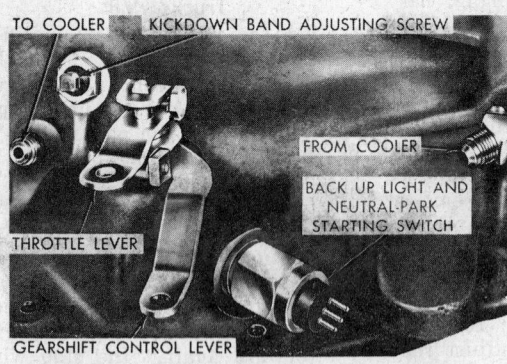

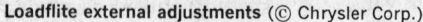

Loadflite external adjustments (© Chrysler Corp.)

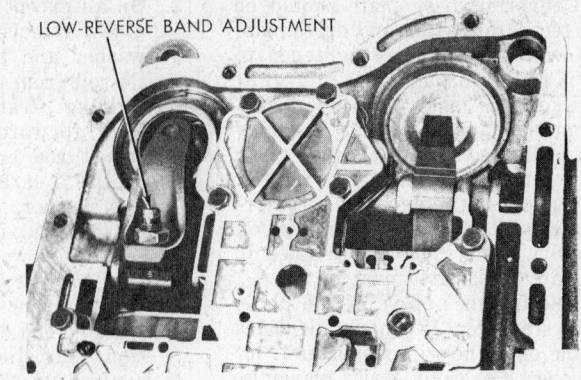

LOW-REVERSE BAND ADJUSTMENT

Loadflite low-reverse band adjustment—A-345 similar (© Chrysler Corp.)

clip. Install the transmission linkage return spring.

9. Check the freedom of operation by moving the slotted link at the carburetor to the full rearward position and allow it to return slowly. Be sure that it returns to the full rearward position.

10. Connect the choke rod or unblock the choke plate.

Vans and Wagons

1. Warm the engine to operating temperature.
2. Block the choke plate fully open.
3. Remove the throttle return spring from the carburetor.
4. Remove the clip, washer and slotted throttle rod from the carburetor pin.
5. Rotate the threaded end of the rod so that the rear edge of the slot in the rod contacts the carburetor pin when the transmission throttle lever is held forward against its stop.
6. Install the washer and clip to retain the throttle rod to the carburetor.
7. Install the throttle rod return spring.
8. Check the transmission linkage for freedom of operation and unblock the choke plate.

New Process A-345 Kickdown Throttle Linkage Adjustment

1. Block the choke valve fully open and release the fast idle cam.
2. Hold the transmission lever forward against its stop while adjusting the linkage.
3. Remove the spring from the stabilizer and retainer (not the throttle return spring) and loosen the bolt attaching the upper end of the slotted rod to the retainer.
4. Push gently backward on the rod to remove all slack.
5. Tighten the bolt at the retainer and replace the spring.
6. Check that the linkage works freely.

Shift Linkage Adjustment
Loadflite

1. Place the gearshift lever in the Park position.
2. Move the shift control lever on the transmission all the way to the rear (in the Park detent).
3. Set the adjustable rod to the proper length and install it with no load in either direction. Tighten the swivel bolt.
4. The shift linkage must be free of binding and be positive in all positions. Make sure that the engine can start only when the gearshift lever is in the Park or Neutral position. Be sure that the gearshift lever will not jump into an unwanted gear.

New Process A-345

1. Detach the clevis from the transmission lever by removing the pin.
2. Place the shift lever in Neutral.
3. Place the transmission lever in Neutral, the second detent from the rear.
4. Adjust and install the clevis.
5. Make sure that the engine can only be started in Neutral.

Kickdown Band

The kickdown band adjusting screw is located on the left-hand side of the transmission case near the throttle lever shaft.

1. Loosen the locknut and back it off about five turns. Be sure that the adjusting screw is free in the case.
2. Torque the adjusting screw to 72 in. lbs.
3. Back off the adjusting screw 2 turns. On 1973 and later six-cylinder engines, and 1974 and later conventional truck V8s back off 2½ turns. On 1974 and later 440 V8, V8 van and wagon, and A-345 four-speed back off 2 turns. Then, keep the screw from turning and torque the locknut to 35 ft. lbs. (29 ft. lbs. through 1972).

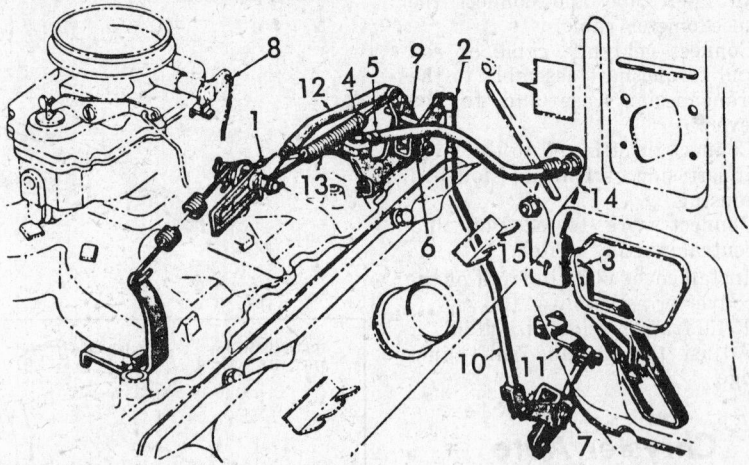

Loadflite transmission kickdown throttle linkage—1972 and later V8 conventional truck (© Chrysler Corp.)

1 Adjuster
2 Ball socket
3 Accelerator pivot
4 Clamp bolt
5 Ferrule
6 Upper bellcrank
7 Pivot pin
8 Choke
9 3/16 in. adjusting rod
10 Intermediate transmission rod
11 Transmission throttle lever
12 Carburetor rod
13 Transmission linkage return spring
14 Throttle base
15 Retaining pin

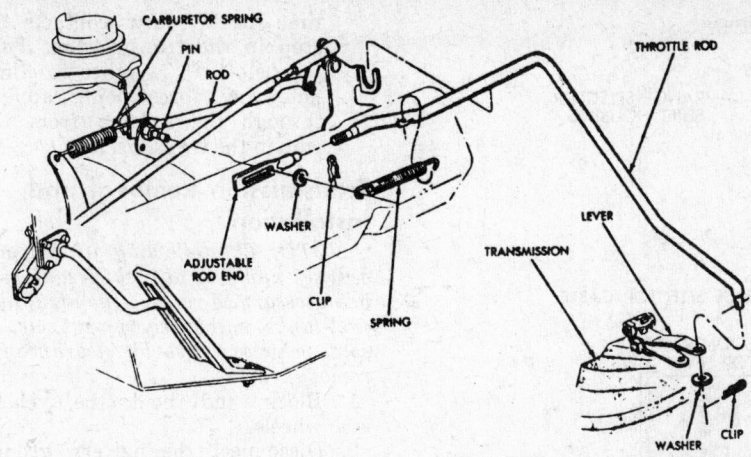

Loadflite transmission kickdown throttle linkage—V8 van (© Chrysler Corp.)

Low and Reverse Band

The pan must be removed from the transmission to gain access to the low and reverse band adjusting screw.

1. Remove the skidplate, if any. Drain the transmission and remove the pan.
2. Loosen the band adjusting screw locknut and back it off about five turns. Be sure that the adjusting screw turns freely in the lever.
3. Torque the adjusting screw to 72 in. lbs.
4. Back off the adjusting screw 2 turns. Keep the screw from turning and torque the locknut to 30 ft. lbs.
5. Using a new gasket, install the pan and torque the pan bolts to 150 in. lbs. Refill the transmission.

Neutral Safety/Backup Light Switch Replacement

The neutral safety switch is mounted in the transmission case. When the gearshift lever is placed in either the Park or Neutral position, a cam, which is attached to the transmission throttle lever inside the transmission, contacts the neutral safety switch and provides a ground to complete the starter solenoid circuit.

The back-up light switch is incorporated into the neutral safety switch. The center terminal is for the neutral safety switch and the two outer terminals are for the back-up lamps.

There is no adjustment for the switch. If a malfunction occurs, the switch must be removed and replaced.

To remove the switch, disconnect the electrical leads and unscrew the switch. Use a drain pan to catch the transmission fluid. Using a new seal, install the new switch and torque it to 24 ft. lbs. Refill the transmission.

Pan Removal and Installation, Filter Change

1. Operate the transmission until it is thoroughly warmed up.
2. Remove the skidplate, if any. Unbolt the pan. So be ready with a large container to drain the fluid into.

NOTE: If the fluid removed smells burnt, serious transmission troubles, probably due to overheating should be suspected.

3. Remove the access plate in front of the torque converter. Rotate the engine clockwise to bring the converter drain to the bottom. Position the container under the converter, remove the drain plug, and allow the fluid to drain. Drain the compounder housing on the A-345.
   Replace the converter drain plug and torque it to 110 in. lbs. for a 7/16 in. head plug and 90 in. lbs. for a 5/16 in. plug. Install the access plate.
4. Unscrew and discard the filter.
5. Install a new filter. The proper torque is 35 in. lbs.
6. Clean out the pan, being extremely careful not to leave any lint from rags inside.
7. Replace the pan with a new gasket. Tighten the bolts to 150 in. lbs. in a crisscross pattern.
8. Pour six quarts (eight on the A-345) of DEXRON® automatic transmission fluid through the dipstick tube.
9. Start the engine in Neutral and let it idle for two minutes or more.
10. Hold your foot on the Brake and shift through D,2,1, and R and back to N.
11. Add enough fluid to bring the level to the ADD ONE PINT mark.
12. Operate the truck until the transmission is thoroughly

warmed up, then check the level. It should be between FULL and ADD ONE PINT. Add fluid as necessary.

NOTE: The manufacturer recommends MOPAR automatic transmission sealer be added to reduce fluid leakage resulting from hardening or shrinking of the seals in high-mileage vehicles.

Allison Automatic Transmissions

Adjustment

Selector Lever Linkage

1. Place the selector lever in the R (reverse) position.
2. Remove the cotter pin from the connection at the transmission selector lever. Move the connector out of the, then shift the lever all the way forward (toward the front of the vehicle).
3. With both the selector lever in the cab and the lever on the transmission in the reverse ("R") position, the pin in the cable connector should be aligned with the hole in the transmission lever.
4. If alignment is not correct, loosen the jam nut at the connector and turn the connector in or out until alignment is obtained and the connector pin can be installed easily.
5. Secure the connector pin with a cotter pin and tighten the jam nut securely at the connector.
6. Test the selector operation in all positions.

Neutral Safety Switch

1. Check the starter circuit at all range selector lever positions by positioning the lever and turning the starter switch to the "Start" position.
2. The engine should start only with the selector lever in the neutral ("N") position. If it is necessary to make an adjustment, first loosen the neutral safety switch bracket screws and rotate the switch clockwise so the starter circuit is closed when the selector lever is placed in the neutral position. Be sure that the switch mounting screws are tightened securely.
3. If the engine fails to start with the selector lever in neutral, check the switch lead wires and switch with a test light.

Throttle Linkage

1. Remove the throttle control return spring and governor arm spring (if equipped).

CAUTION: Do not stretch or distort the governor spring in any way

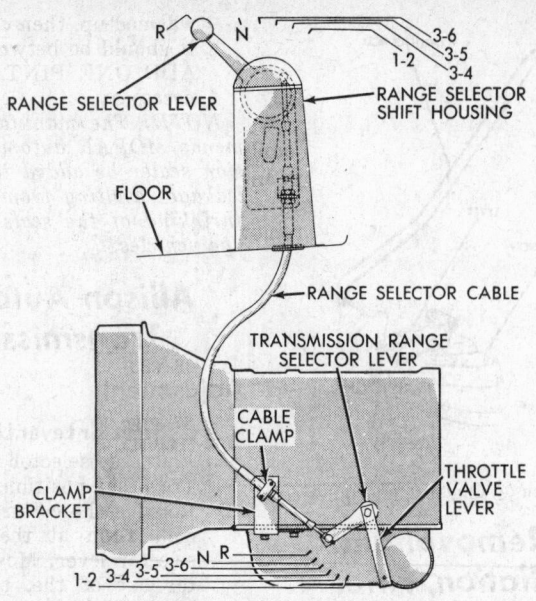

Selector lever adjustment—typical (© Chrysler Corp.)

as it will throw off the governor calibration.

2. At the transmission, remove the throttle control shift rod swivel cotter pin and withdraw the swivel from the TV lever on the transmission.
3. Remove the carburetor air cleaner.
4. Place the carburetor linkage in the idle position.
5. Adjust the throttle control rod so that the accelerator control bell crank stop is against the bracket.
6. Tighten the nuts on the throttle control rod.
7. Move the carburetor throttle lever to the wide open position.
8. The wide open throttle stop bolt must be adjusted so that the bolt head contacts the bell crank when the throttle is in the wide open position.
9. With the throttle control shaft pushed upward until the travel of the rod is halted by the wide open throttle stop, rotate the transmission throttle control lever forward until the TV lever is firm against the internal stop in the transmission. In this position the swivel pin should freely enter the lower hole in the transmission throttle valve (TV) control leved. Finger tighten the front and rear adjusting nuts against the swivel.
10. Loosen the swivel 3 turns to back off the valve. Connect the linkage and tighten the nuts.
11. On all models the through detent pedal stop, located on the throttle control bell crank, must be adjusted so that the pedal forces cannot be applied to the TV linkage. To do this, adjust the stop so that it contacts the bracket bolted to the engine at the same time the TV lever contacts the stop in the transmission. Turn the bolt 1/16″ further to eliminate any force being applied through the linkage from the stop in the transmission.

Transmission Removal and Installation

NOTE: The following general procedures apply to all vehicles. It may be necessary to remove the air tanks, fuel tanks, special equipment, etc. on some vehicles to provide clearance for removal.

1. Block and hold the vehicle wheels.
2. Disconnect the battery ground cable and remove the spark plugs so that the engine can be turned over manually.
3. Remove the dipstick. Drain the transmission by disconnecting the oil filler tube at the right side of the transmission oil pan. Remove the bracket holding the oil filler tube to the transmission and remove the filler tube. Cover the oil pan opening to prevent entry of foreign material.
4. Disconnect the oil cooler lines from the fittings on the right side of the transmission case. Plug the ends of the lines and the fittings in the transmission case.
5. Disconnect the range selector cable from the shift lever.
6. Disconnect the vacuum modulator line from the modulator.
7. Disconnect the speedometer shaft fitting from the adapter.
8. Disconnect the propeller shaft from the transmission.
9. Disconnect the mechanical parking brake linkage (if equipped).
10. Remove the bolts holding the flywheel flex plate to the converter by working through the opening in the flywheel housing and turning the flywheel manually.
11. Support the transmission with a 500 lb. (minimum) transmission floor jack. The jack should be positioned so that the transmission oil pan will not support the weight of the transmission. Fasten a safety chain over the transmission.
12. Place a support under the rear of the engine and remove the transmission case-to-crossmember support bolts. Raise the engine to remove the weight from the rear engine mounts.
13. Remove the transmission case-to-flywheel housing bolts and washers.
14. Carefully inspect the transmission and surrounding area to be sure that there are no lines, hoses, or wires which will interfere with the transmission removal.

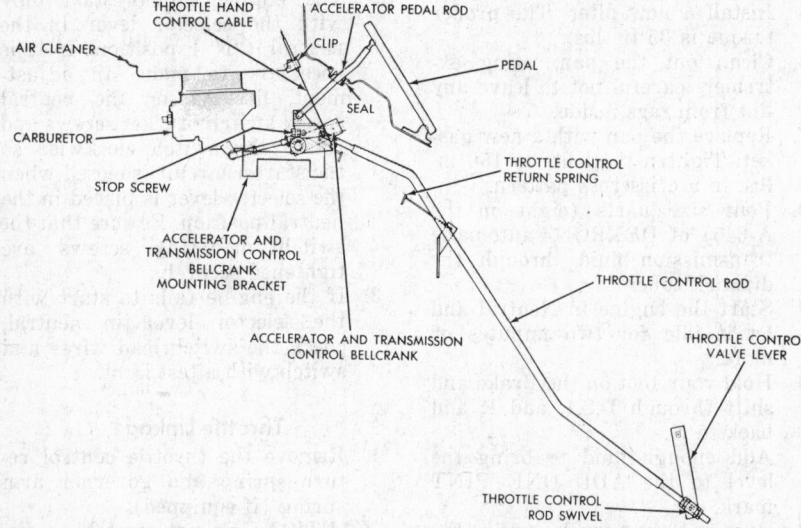

Throttle linkage assembly—typical (© Chrysler Corp.)

CAUTION: During removal of the transmission, keep the rear of the transmission lower than the front to prevent the loss of the converter.

NOTE: Position the jack or hoist sling relative to the transmission center of gravity. The torque converter is free to move forward when the transmission is disconnected from the engine. Be sure that the converter is not allowed to separate from the transmission while the assembly is being removed. Install a retainer strap to hold the converter in place as soon as the transmission is clear of its mountings.

15. Move the transmission assembly straight away from the engine and lower the assembly carefully. Remove the transmission from the vehicle.

16. Install the transmission in the reverse order of removal.

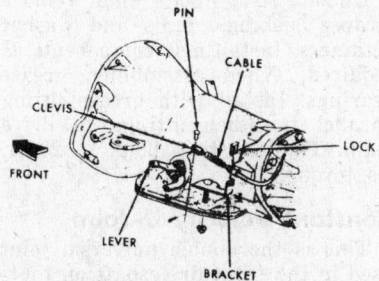

A-345 4-speed automatic transmission shift linkage (© Chrysler Corp.)

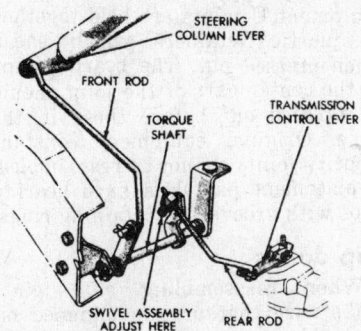

Typical Loadflite transmission shift linkage (© Chrysler Corp.)

Drive Shaft, U Joints

Driveshaft Removal and Installation

Single Section Type

This driveshaft has a universal joint at either end and no external supports.

1. Raise and support the truck with the rear higher.
2. Matchmark the shaft and pinion flange to assure proper balance at installation.

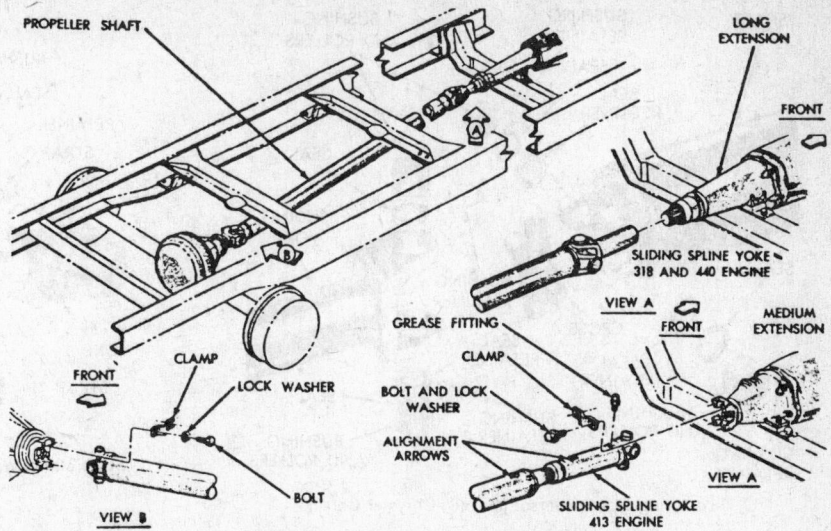

Single section driveshaft (© Chrysler Corp.)

3. Remove both rear U-joint roller and bushing clamps from the rear axle pinion flange. Do not disturb the retaining strap which holds the bushing assemblies on the U-joint cross.

NOTE: Do not allow the driveshaft to hang during removal. Suspend it from the frame with a piece of wire. Before removing the driveshaft, raise the rear end of the truck to prevent loss of transmission fluid.

4. Slide the driveshaft, with the front sliding yoke, off the transmission output shaft.
5. Installation is the reverse of removal. Align the matchmarks made during removal.

Two-Section Type

This driveshaft has a universal joint at either end, with a third universal joint and a support bearing at the center.

1. Matchmark the shaft and the rear axle pinion hub yoke. Matchmark the center bearing spline and slip yoke.

NOTE: Do not allow the driveshaft to hang down during removal. Suspend it from the frame. Raise the rear of the truck to prevent loss of transmission fluid.

2. Remove both rear U-joint roller and bushing assembly clamps from the rear axle pinion yoke. Do not disturb the retaining strap used to hold the bushing assemblies on the U-joint cross.
3. Slide the rear half of the shaft off the front shaft splines at the center bearing. Remove the rear half.
4. At the transmission end of the front half, remove the bushing retaining bolts and clamps, after matchmarking. If there is a driveshaft brake, there will be flange nuts.
5. Unbolt the center bearing mounting nuts and bolts and remove the front half of the shaft.
6. On installation, align the matchmarks at the transmission and start all the bolts and nuts at the front U-joint and the center sup-

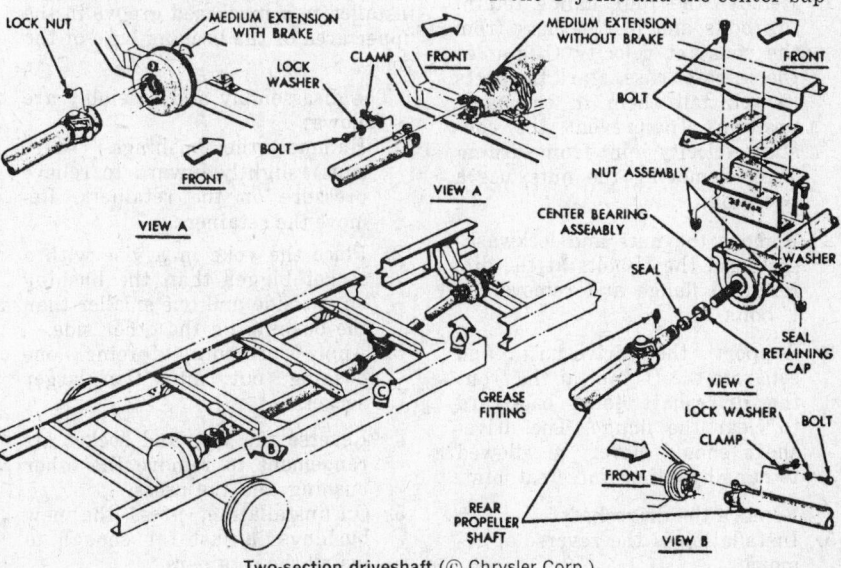

Two-section driveshaft (© Chrysler Corp.)

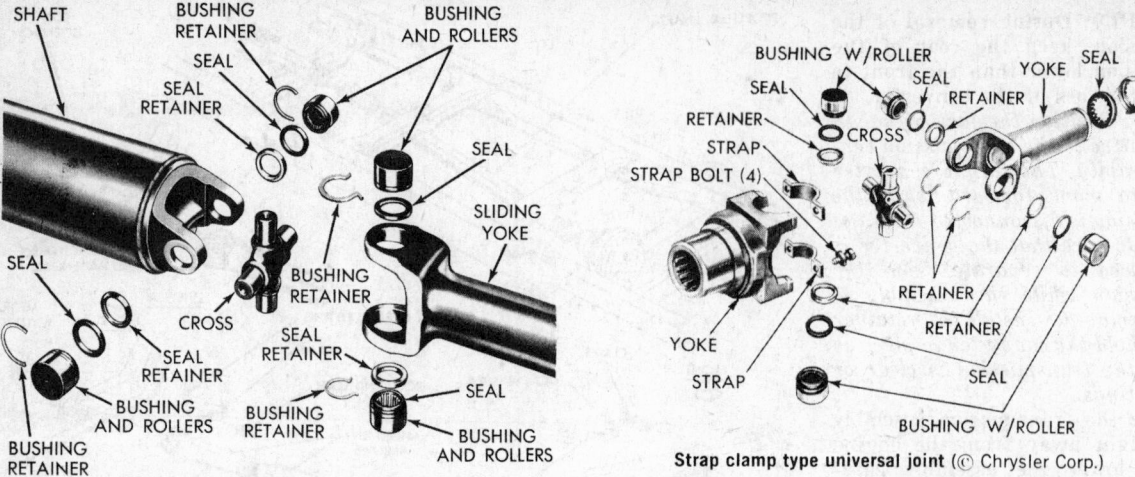

Lock ring type universal joint (© Chrysler Corp.)

Strap clamp type universal joint (© Chrysler Corp.)

port bearing.

7. Tighten ¼ in. clamp bolts to 170 in. lbs, and 5/16 in. bolts to 300 in. lbs. Tighten driveshaft brake flange nuts to 35 ft. lbs. Leave the center bearing bolts just snug.

8. Align the rear shaft matchmarks and slide the yoke onto the front shaft splines.

9. Align the rear U-joint matchmarks and install the bushing clamps and bolts. Tighten the bolts to the torque given in Step 7. Grease the joints and splines.

10. Jack up the rear wheels and let the engine drive the shaft. The center support bearing will align itself.

11. Tighten the center bearing bolts to 50 ft. lbs.

Four-Wheel Drive Front Driveshaft

This applies to Ramcharger and Trailduster, W100, 200, and 300.

1. Remove the four flange retaining bolts and lockwashers from the constant velocity U-joint at the transfer case. Mark the parts to reinstall them in the same position. To prevent the constant velocity joint from turning while removing the nuts, use a press bar.

2. Remove the nuts and lockwashers from the U-bolts at the differential flange and remove the U-bolts.

3. Support the driveshaft and separate the U-joint at the front the driveshaft joint backward to clear the flange. The driveshaft should never be allowed to hang by either universal joint.

4. Remove the driveshaft.

5. Installation is the reverse of removal.

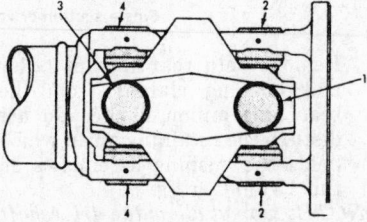

Constant velocity joint bearing cup removal sequence (© Chrysler Corp.)

Universal Joint Replacement

The Lock Ring and the Snap Ring Type

The lock-ring type and the snapring type universal joints are basically the same, except for the locations of the retainers. The lock-ring retainers hold the bearing cups in the yoke by being installed in a machined groove on the bearing cup, which is located on the inner side of the yoke when the joint is assembled.

The snap-ring type retainer holds the bearing cup in the yoke by being installed in a machined groove in the upper area of the bearing bore of the yoke.

The disassembly and assembly are as follows:

1. Hammer the bushings (roller cups) slightly inward to relieve pressure on the retainers. Remove the retainers.

2. Place the yoke in a vise with a socket bigger than the bushing on one side and one smaller than the bushing on the other side.

3. Apply pressure, forcing one bushing out into the larger socket.

4. Reverse the vise and socket arrangement to remove the other bushing and the cross.

5. On installation, press the new bushings in just far enough to install the retainers.

Strap Clamp Type (Rear Axle Yoke)

Unbolt strap bolts and remove straps, bushings, seals and washer retainers. Install new components as required. When assembling, grease bearings. Install with grease fitting parallel to other fittings in drive train. Tighten strap bolts to 20 ft. lbs. torque.

Constant Velocity U-Joint

This is the double universal joint used in the front driveshaft on four-wheel drive models. These are disassembled in the same way as the snap-ring type U-Joint. Original equipment U-joints are held together by plastic retainers which shear when pressed out. The bearing cups in the center part of the joint should be pressed out before those in the yoke. Original equipment constant velocity joints cannot be reassembled. Replacement part kits have bearing cups with grooves for retaining rings.

Slip Joints

When reassembling slip joints make sure that arrows stamped on each side are matched. This will assure proper universal joint alignment.

Rear Axle

Reference

Numerous rear axles are used on Dodge and Plymouth trucks. Those used through the 300 series can be identified by the chart.

The gear ratio is usually identified by a metal tag under one of the cover or housing bolts. The tag gives either the gear ratio or the number of teeth on the ring and pinion gears. Larger axles (up to 15,000 lb. capacity) are the Chrysler RA115 12½ in. axle, the F140D 12½ in. Rockwell, the F147 13¼ in. Rockwell, and the Eaton 13802 and 15201 two-speed units.

Maker	Ring Gear Size (in.)	Identification
Chrysler	8⅜	10 bolt cover, front filler plug
Chrysler	8¾	Welded cover, front filler plug
Chrysler	9¼	12 bolt cover, filler plug in cover
Spicer 60, 60HD	9¾	10 bolt cover, filler plug in cover
Spicer 70	10½	10 bolt cover, filler plug in cover

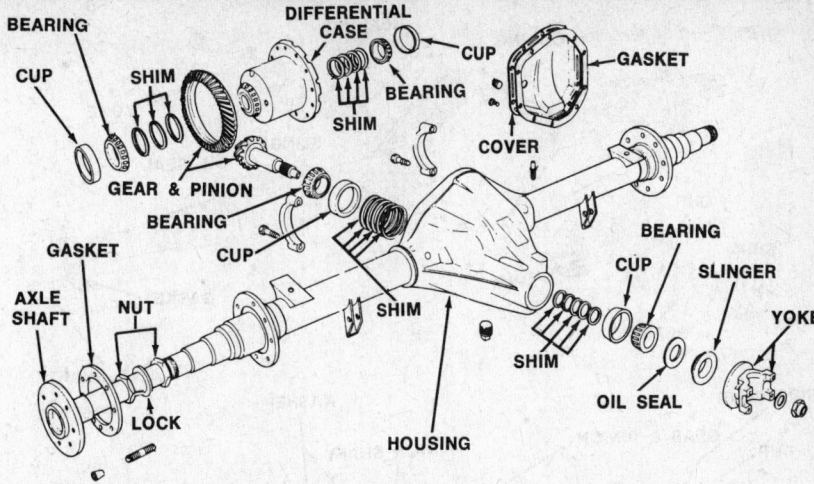

Spicer (Dana) 60, 60HD rear axle-exploded view (© Chrysler Corp.)

Axle Assembly

Removal and Installation

1. Raise vehicle and support at front of rear springs.
2. Block brake pedal in the up position.
3. Remove rear wheels.
4. Disconnect hydraulic brake hose at "T" fitting or at each wheel.
5. Disconnect parking brake cable.
   NOTE: to insure proper drive line balance when reassembling, make scribe marks on the driveshaft universal joint and differential pinion flange before removal.
6. Disconnect driveshaft at rear universal joint bearing clamps and secure with wire to prevent damage to front universal joint.
7. Disconnect shock absorbers and remove rear spring nuts and U-bolts.
8. Remove assembly from vehicle.
9. To install, position rear axle assembly spring pads over the spring center bolts.
10. Install U-bolts and tighten nuts securely.
11. Connect shock absorbers.
12. Connect parking brake cable.
13. Connect hydraulic brake lines. Install brake drums and adjust. Bleed hydraulic brake system.
14. Connect driveshaft universal joint in its original position, matching scribe marks made during removal. Tighten universal joint clamp bolts.

Differential Service

For service and overhaul procedures on differentials see "Drive Axles" in the General Repair Section.

Axle Shaft

Removal and Installation

8⅜ and 9¼ in Axles

NOTE: There is no provision for axle shaft end-play adjustment on this axle.

1. Raise the vehicle and remove the rear wheels.
2. Clean all dirt from the housing cover and remove the housing cover to drain the lubricant.
3. Remove the brake drum.
4. Rotate the differential case until the differential pinion shaft lockscrew can be removed. Remove the lockscrew and pinion shaft.
5. Push the axle shafts toward the center of the vehicle and remove the C-locks from the grooves on the axle shafts.
6. Pull the axle shafts from the housing, being careful not to damage the bearing which remains in the housing.
7. Inspect the axle shaft and bearings and replace any doubtful parts. Whenever the axle shaft is replaced, the bearings should also be replaced.
8. Remove the axle shaft seal from the bore in the housing.
9. Remove the axle shaft bearing from the housing.

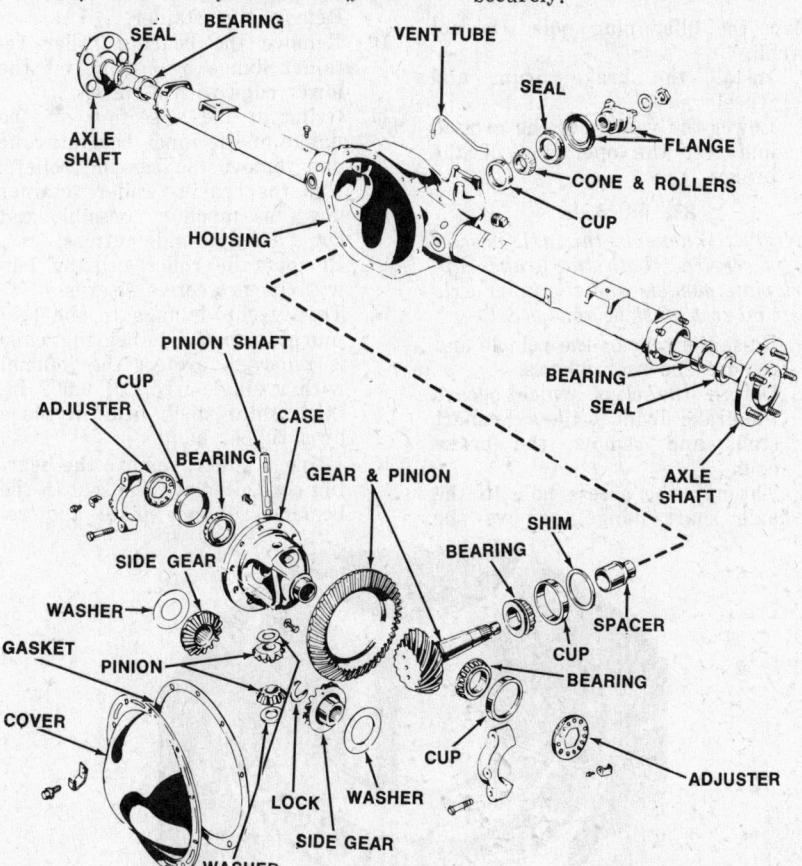

8⅜ inch rear axle, the 9¼ is similar (© Chrysler Corp.)

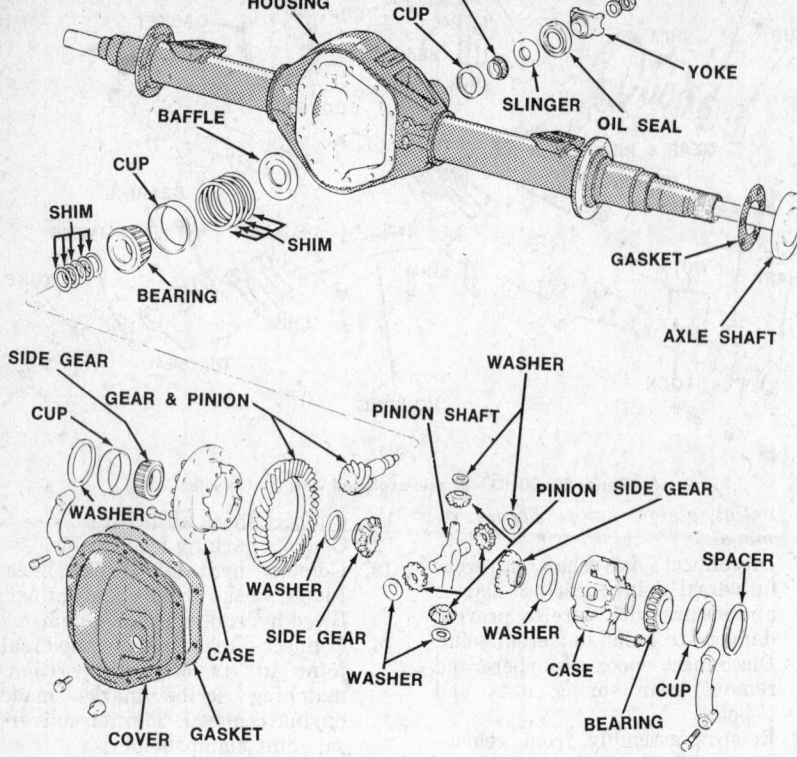

Spicer (Dana) 70 rear axle-exploded view (© Chrysler Corp.)

10. Check the bearing shoulder in the axle housing for imperfections, and should be corrected.
11. Clean the axle shaft bearing cavity.
12. Install the axle shaft bearing in the cavity. Be sure that the bearing is seated firmly against the shoulder.
13. Install the axle shaft bearing seal. It should be seated beyond the end of the flange face.
14. Insert the axle shaft, making sure that the splines do not damage the seal. Be sure that the splines are properly engaged with the differential side gear splines.
15. Install the C-locks in the grooves on the axle shafts. Pull the shafts outward so that the C-locks seat in the counterbore of the differential side gears.
16. Install the differential pinion shaft through the case and pinions. Install the lockscrew and secure it in position.
17. Install the cover and a new gasket.

NOTE: Replacement gaskets may not be available for differential covers. In this case, the use of a gel type nonsticking sealant is recommended.

Be sure that the rear axle ratio identification tag is replaced under one of the cover bolts. Refill the axle with the specified lubricant to ½ in.

below the filler plug hole. Do not overfill.
18. Install the brake drum and wheel.
19. Lower the vehicle to the ground and test the operation of the brakes.

8¾ in. Axle

NOTE: Whenever this axle assembly is serviced, both the brake support plate gaskets and the inner axle shaft oil seal must be renewed.

1. Raise the rear of the vehicle and remove the rear wheels.
2. Detach the clips which secure the brake drum to the axle shaft studs and remove the brake drum.
3. Through the access hole in the axle shaft flange, remove the

axle shaft retaining nuts. The right-side axle shaft has a threaded adjuster in the retainer plate and a lock under one of its studs which should be removed at this time.
4. Remove the parking brake strut.
5. Attach a puller to the axle shaft flange and remove the axle shaft.
6. Remove the brake assembly from the axle housing.
7. Remove the axle shaft oil seal from the axle housing.

CAUTION: It is advisable to position some sort of a protective sleeve over the axle shaft seal surface next to the bearing collar to protect the seal surface. Never use a torch or other heat source as an aid in removing any axle shaft components; this will result in serious damage to the axle assembly.

8. Wipe the axle housing seal bore clean. Install a new axle shaft oil seal.

NOTE: All 8¾ in. rear axle shaft bearings are packed with a special lubricant at the factory. If the roller bearing must be repacked, the factory lubricant must be washed out. The service lubricant is not compatible with the factory lubricant.

9. Place the axle shaft retainer retaining collar in a vise. With a chisel, cut deeply into the retaining collar at 90° intervals. Remove the retainer.
10. Remove the bearing roller retainer flange by cutting off the lower edge with a chisel.
11. Grind or file a section off the flange of the inner bearing cone and remove the bearing rollers.
12. Pull the bearing roller retainer down as much as possible and cut it off with side cutters.
13. Remove the roller bearing cup with its protective sleeves.
14. To prevent damage to the seal journal when the bearing cone is removed, protect the journal with a single wrap of 0.002 in. thick shim stock held in place by a rubber band.
15. Using a puller, remove the bearing cone. Remove the seal in the bearing retainer plate and re-

Grind the flange off the inner cone to remove the 8¾ in. axle bearing rollers (© Chrysler Corp.)

place it.

16. To assemble the axle, first install the retainer plate and seal assembly on the axle shaft.
17. Grease the wheel bearings and install them.
18. Install a new axle shaft bearing cup, cone, and collar on the shaft. Check the axle shaft seal journal for imperfections and if necessary, polish with No. 600 crocus cloth.
19. Thoroughly clean the axle housing flange face and brake support. Install a new rubber/asbestos gasket onto the axle housing studs. Next, install the brake support plate assembly on the left side of the axle housing.
20. Lightly grease the outside edge of the bearing cup. Install the bearing cup in the bearing bore.
21. Replace the foam gasket on the studs of the left-side axle housing and very carefully slide the axle shaft assembly through the oil seal and engage the splines of the differential side gear.
22. Using a non-metallic hammer, lightly tap the axle shaft bearing in the recess end of the axle shaft to position the axle housing. Install the retainer plate over the axle housing studs and, starting with the bottom securing nut, torque the nuts to 30-35 ft. lbs.
23. Repeat steps 19-22 for the right-side axle housing.
24. At the right side of the axle housing, back off the threaded adjuster until the inner face of the adjuster is flush with the inner face of the retainer plate. Very carefully slide the axle shaft assembly through the oil seal and engage the splines of the differential side gear. Then repeat step 22.
25. Mount a dial indicator on the left brake support. Turn the adjuster clockwise until both wheel bearings are seated and there is zero end-play in the axle shafts. Back off the adjuster about four notches to establish proper end-play (0.008-0.018).
26. Lightly tap the end of the left axle shaft with a non-metallic hammer. This will seat the right wheel bearing cup against the adjuster. Turn the axle shaft several times so that a true end-play reading is obtained.
27. Remove one retainer plate nut and install the adjuster lock. If the lock tab does not mate with the notch in the adjuster, turn the adjuster slightly until it does. Refit the nut and torque it to 30-35 ft. lbs.
28. Recheck the axle shaft end-play. If it is not within specifications, repeat the adjustment. When the

adjustment is complete, remove the dial indicator.
29. Install the parking brake strut. Replace the brake drum and retaining clips.
30. Install the rear wheels and lower the vehicle.

RA115, Spicer 60 and 70, F147, F140D, Eaton 2-Speed Axles

1. Remove axle shaft flange nuts and washers.
2. Rap axle shafts sharply in center of flange with hammer to free dowels.
3. Remove tapered dowels and axle shafts.
4. Clean gasket contact area with suitable solvent and install a new flange gasket.
5. Install axle shaft into axle housing.
6. On axles having an outer wheel bearing seal, install new gaskets on each side of seal mounting flange.
7. Install tapered dowels, lock washers and nuts. Torque the nuts to 30-35 ft. lbs. on the RA115, 40-70 ft. lbs. on Spicer axles with 7/16 in. thread size, and 65-105 on Spicer with ½ in.

Rear Suspension

Rear Spring Removal and Installation

Conventional Trucks, 100-300

1. Raise rear of truck until weight is removed from springs, wheels just touching the floor.
   NOTE: Truck must be lifted by jack or hoist under frame side rail at crossmember behind the axle being careful not to bend flange of side rail.
2. Place stands under side frame members as a safety precaution.
3. Remove nuts, lockwashers and U-bolts securing spring to axle.
4. Remove spring shackle bolts, shackle and spring front bolt, then remove spring.
5. To install, position spring on axle so spring center bolt enters locating hole in axle housing pad.
6. Line up spring front eye with bolt hole in bracket and install spring bolt and nut.
7. Install the rear shackle, bolts and nuts. Tighten shackle bolt nut until slack is taken up.
8. On headless type spring bolts install the bolts with lock bolt groove lined up with lock bolt hole in bracket. Install lock bolt and tighten lock bolt nut. Install lubrication fittings.
9. Install U-bolts, new lockwashers

and nuts, tightening until nuts push lockwashers against axle. Align auxiliary spring parallel with main spring.
10. Remove stands from under vehicle, lower truck so weight is resting on wheels. Tighten U-bolt nuts, spring eye nuts and shackle bolt nuts.
11. Lubricate spring bolts and shackle bolts with chassis lubricant when equipped with lubrication fittings.

Vans and Wagons

1. Raise the vehicle until springs are accessible. Place jackstands under the bumper brackets.
2. Remove U-bolt nuts, U-bolts, and plate.
3. Remove the front pivot bolt nut. Remove the bolt.
4. Remove the rear shackle bolt nuts. Remove the shackle plate.
5. Remove the outer shackle and bolt assembly from the hanger. Remove the spring. On vehicles equipped with one piece shackles, remove the nut, remove the shackle to spring bolt, and remove the spring.
6. To install, position the spring and shackle assembly in the rear hanger.
7. Install the shackle plates and nuts, tightening the nuts to 40 ft-lbs. On vehicles with the one piece shackle, first position the spring to the shackle, and then install the bolt and nut.
8. Position the spring in the front pivot hanger and install the bolt and nut.
9. Position the spring properly on the axle and install the U-bolt plate.
10. Install the U-bolts and nuts. Make sure the shackled end of the spring is above the shackle bracket pivot.
11. Lower the vehicle to the floor, and tighten all nuts.

Installing New Leaf

1. Clamp spring in a vise, remove center bolt and bend clamp type clips back from spring leaves.
2. Insert long drift in center bolt hole and release vise slowly.
3. Remove assembly from vise and replace broken leaf.
4. Place spring assembly in vise, slowly tightening vise while holding spring leaves in alignment with drift.
5. Remove drift and install new center bolt.
6. Install nut, tightening to 15 ft. lbs. torque.
7. Remove spring from vise.

Medium Duty Models, 400-800

1. Raise truck frame until weight is off rear springs with wheels

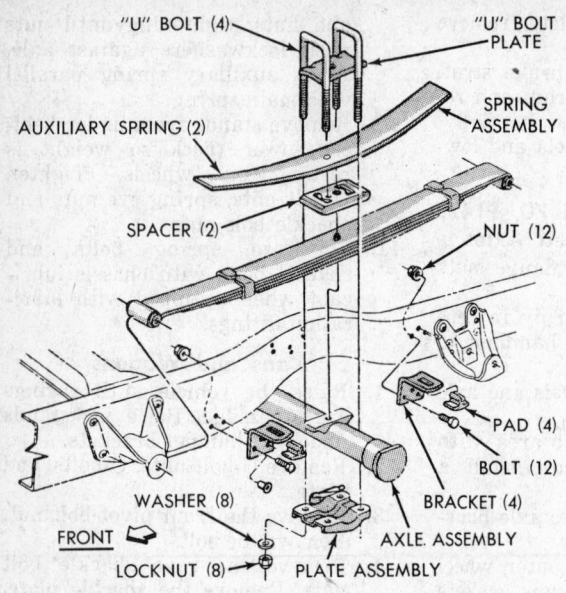

Rear suspension, D100, 200 shown (© Chrysler Corp.)

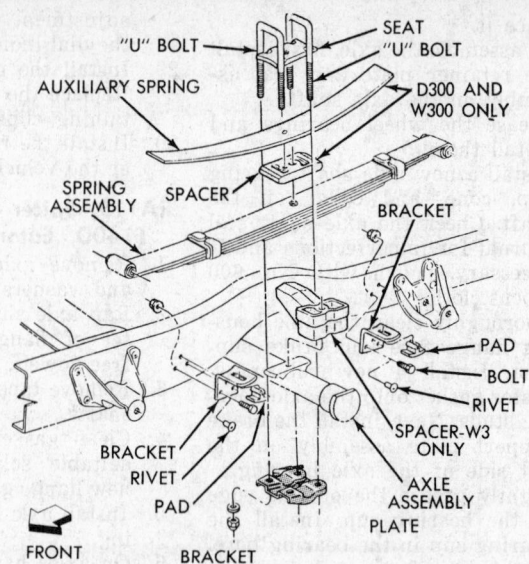

Rear suspension, W200, W300, and D300 shown (© Chrysler Corp.)

still touching the ground.

2. Remove nuts from spring clips and remove spring clip plate. Remove spring clips.

3. If truck is so equipped, remove the auxiliary spring and spacer.

4. Install a "C" clamp on radius leaf to relieve load on spring eye bolt. Remove front eye bolt retaining nut.

5. Remove spring eye bolt and slowly release tension on radius leaf.

6. Remove rebound pin retaining lock from front bracket. Remove spring rebound pin. Repeat for rear rebound pin.

7. Remove rear spring.

8. If necessary, replace radius leaf eye rubber bushing.

9. To install, position spring in front bracket and install rebound pin. Install retaining lock.

10. Position spring in rear bracket, and install rebound pin. Install retaining lock.

11. Align spring eye and install spring eye bolt and nut. It may be necessary to draw the radius leaf eye into position with a "C" clamp.

12. Install auxiliary spring assembly if so equipped.

13. Install spring U-bolt plate and U-bolt seat. Install spring clips and nuts and washers, tightening nuts until they are snug.

14. Lower truck to the floor and tighten spring U-bolts.

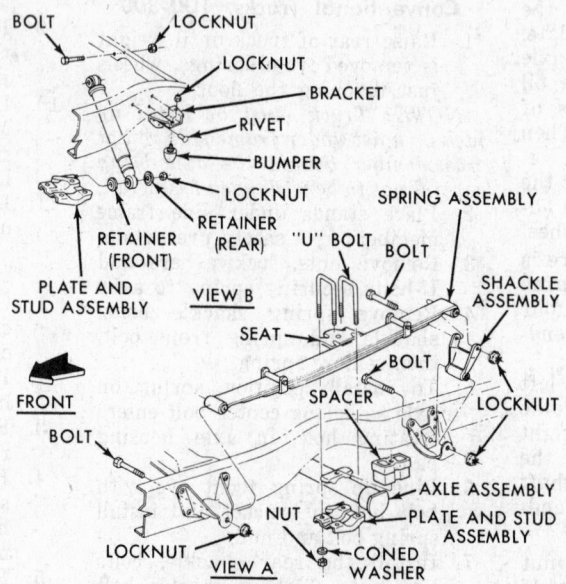

Rear suspension Ramcharger, Trail Duster (© Chrysler Corp.)

Index

TUNE-UP SPECIFICATIONS

CU. IN. DISPLACEMENT	YEAR	SPARK PLUG GAP	DISTRIBUTOR POINT DWELL	POINT GAP	IGNITION TIMING DEGREES	CRANKING COMP. PRESSURE	VALVE CLEARANCE INLET	EXHAUST	GOV. R.P.M. NO LOAD	FUEL PUMP PRESS.	IDLE SPEED STD.	AUTO.
SIX CYLINDER											
170	1971-72	.034	35-40	.027	6	175	.018H	.018H	——	4-6	525	—
200	1973-75	.034	37	.027	6B	175	Zero		——	4-6	500	650
240	1971-72	.034	33-38⑤	.027⑥	6B	175	Zero		4000	4-6	500	500
	1973	.034	33-39	.027	①	⑫	Zero		4000	4-6	①	①
	1974	①	①	①	①	⑫	Zero		①	4-6	①	①
300	1971-72	.034	33-38	.027⑥	6B	175	Zero		——	4-6	775	500⑦
	1973	.034	33-39	.027	①	⑫	Zero		——	4-6	①	①
	1974-78	①	①	①	①	⑫	Zero		——	4-6	⑨	⑨
EIGHT CYLINDER											
302	1971-72	.034	24-29	.017⑧	6B	150	Zero		——	5	500	500
	1973	.034	24-30	.017	①	150	Zero		——	5	①	①
	1974-78	①	①	①	①	⑫	Zero		——	5	①	①
330	1971-73	.030	24-30②	.017③	①	140	Zero		3900	4½-6½	①	①
	1974-78	①	①	①	①	⑫	Zero		3900	4½-6½	①	①
351W	1976-78	①	⑪	⑪	①	⑫	Zero		——	4½-6½	①	①
351M	1977-78	①	⑪	⑪	①	⑫	Zero		——	4½-6½	①	①
360/359	1971-72	.034	24-29⑨	.021⑩	6B	140	Zero		——	4½-6½	650	550
	1973	.034	24-30	.017	①	⑫	Zero		——	4½-6½	①	①
	1974-78	①	①	①	①	⑫	Zero		——	4½-6½	①	①
361	1974-78	①	①	①	①	⑫	Zero		3800	4½-6½	①	①
390	1971-72	.034	24-39⑨	.021	6	140	Zero		3800	4½-6½	650	550
	1973	.034	24-30	.017	①	⑫	Zero		3800	4½-6½	①	①
	1974-78	①	①	①	①	⑫	Zero		3800	4½-6½	①	①
391/389	1971-73	.030	26-31②	.017③	①	140	Zero		3800	4½-6½	550①	500①
	1974-78	①	①	①	①	⑫	Zero		3800	4½-6½	①	①
400	1977-78	①	⑪	⑪	①	⑫	Zero		——	4½-6½	①	①
401	1971-74	.030	26-31②	.017③	8①	150	.020	.020	3600	④	525①	500D①
	1975-78	①	①	①	①	⑫	.020	.020	①	⑧	①	①
460	1973-78	①	①	①	①	⑫	Zero		——	4½-6½	①	①
475	1977-78	①	①	①	①	⑫	.020	.020	①	⑧	①	①
477	1971-74	.030	26-31②	.017③	8①	150	.020	.020	3400	④	525①	500D①
	1975-78	①	①	①	①	⑫	.020	.020	①	⑧	①	①
534	1971-74	.030	22-24	.020	8①	150	.020	.020	3200	④	550①	500①
	1975-78	①	①	①	①	⑫	.020	.020	①	⑧	①	①

① Set to specifications shown on engine decal.
② Transistor Ignition 22-24.
③ Transistor Ignition .020.
④ Electric.
⑤ F-250 and 1972 Models—34-40.
⑥ 1971 F-250—.025.
⑦ Over 6000 GVW California—550.

⑧ .021 w/Dual Diaphragm Distributor.
⑨ 26-31 on F-250 and 1972 Models.
⑩ .017 on F-250 and 1972 Models.
⑪ Electronic ignition.
⑫ Take the highest reading and compare it to the lowest reading. The lower reading must be within 75% of the highest.

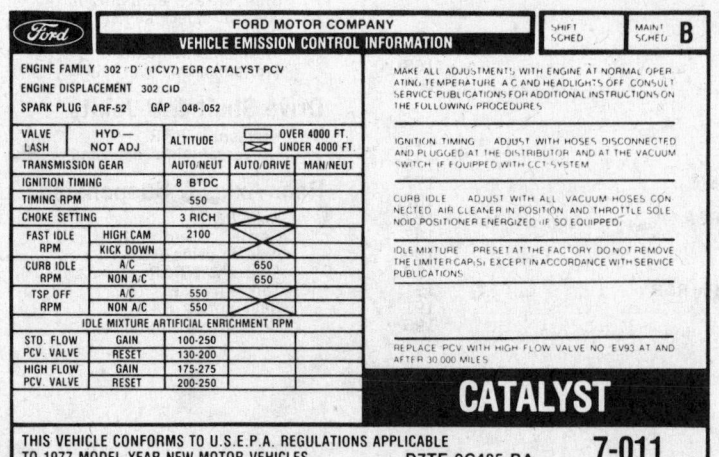

Typical emission certificate (tune-up specification) decal (© Ford Motor Co.)

FIRING ORDER AND ROTATION

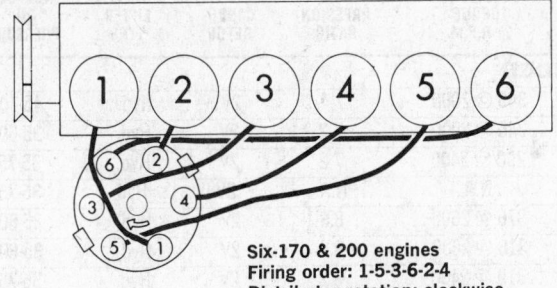

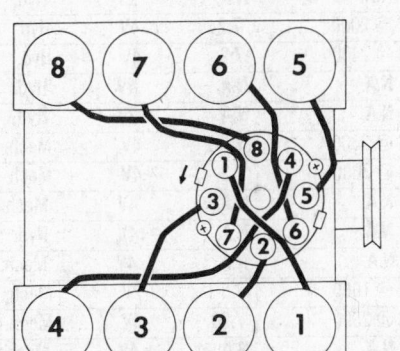

Six-170 & 200 engines
Firing order: 1-5-3-6-2-4
Distributor rotation: clockwise

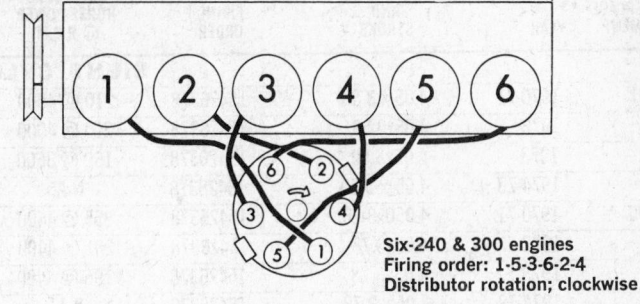

Six-240 & 300 engines
Firing order: 1-5-3-6-2-4
Distributor rotation; clockwise

V8-351, 351W & 400 engines
Firing order: 1-3-7-2-6-5-4-8
Distributor rotation: counterclockwise

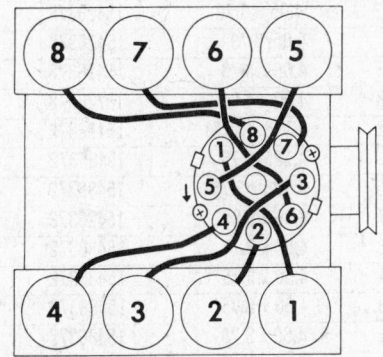

V8-302, 330, 359, 360, 361, 389, 390, 401, 460, 475, 477, & 534 engines
Firing order: 1-5-4-2-6-3-7-8
Distributor rotation: counterclockwise

GENERAL ENGINE SPECIFICATIONS

CU. IN. DISPLACE-MENT	YEAR	BORE AND STROKE	FIRING ORDER	DEVELOPED HORSE POWER @ R.P.M.	DEVELOPED TORQUE @ R.P.M.	COM-PRESSION RATIO	CARBU-RETOR	VALVE LIFTER TYPE	NORMAL OIL PRESSURE
SIX CYLINDER									
170	1971	3.50 x 2.94	153624	100 @ 4000	156 @ 2200	8.7	1V	Mech.	35-55
	1972	3.50 x 2.94	153624	97 @ 4400	137 @ 2600	8.3	1V	Mech.	35-60
200	1973-75	3.683 x 3.26	153624	84 @ 3600	151 @ 1800	8.3	2V	Hyd.	35.55
240	1970-71	4.00 x 3.18	153624	150 @ 4000	234 @ 2200	9.2-1	1V	Hyd.	35-60
	1972	4.00x3.18	153624	121 @ 4000	186 @ 2400	8.5	1V	Hyd.	35-60
	1973-74	4.00x3.18	153624	119 @ 3800	190 @ 2200	8.5	1V	Hyd.	40-60
300	1970-71	4.00 x 3.98	153624	165 @ 3600	284 @ 2000	8.8-1	1V	Hyd.	35-60
	1972	4.00 x 3.98	153624	132 @ 3600	240 @ 2000	8.4	1V	Hyd.	35-60
	1972	4.00 x 3.98	153624	165 @ 3600	294 @ 2000	7.9	1V	Hyd.	35-60
	1973	4.00 x 3.98	153624	128 @ 3600	234 @ 1600	7.9	1V	Hyd.	40-60
	1973	4.00 x 3.98	153624	132 @ 3600	241 @ 1800	7.9	1V	Hyd.	40-60
	1973-78	4.00 x 3.98	153624	N.A.	N.A.	N.A.	1V	Hyd.	40-60
EIGHT CYLINDER									
302	1970-71	4.00x3.00	15426378	205 @ 4600	300 @ 2600	8.6-1	2V	Hyd.	35-60
	1972	4.00x3.00	15426378	159 @ 4000	250 @ 2400	8.2	2V	Hyd.	35-60
	1973	4.00x3.00	15426378	157 @ 4000	249 @ 2600	8.2	2V	Hyd.	40-60
	1974-78	4.00 x 3.00	15426378	N.A.	N.A.	N.A.	2V	Hyd.	40-60
330	1970-72	3.875x3.50	15426378	190 @ 4000	306 @ 2000	7.4-1	2V	Hyd.	45-70
	1973	3.875x3.50	15426378	154 @ 3600	276 @ 2000	7.4	2V	Hyd.	35-60
	1973	3.875x3.50	15426378	155 @ 3600	262 @ 2600	7.4	2V	Hyd.	35-60
	1974-78	3.815 x 3.50	15426378	N.A.	N.A.	N.A.	2V	Hyd.	35-60
351W	1976-78	4.00 x 3.50	13726548	N.A.	N.A.	N.A.	2V	Hyd.	40-65
351M	1977-78	4.00 x 3.50	13726548	N.A.	N.A.	N.A.	2V	Hyd.	50-75
360/359	1970-71	4.05 x 3.50	15426378	215 @ 4400	327 @ 2600	8.4-1	2V	Hyd.	35-60
	1972	4.05x3.50	15426378	196 @ 4600	291 @ 2000	8.0	2V	Hyd.	35-60
	1973	4.05x3.50	15426378	189 @ 4600	287 @ 2400	8.0	2V	Hyd.	35-60
	1974-78	4.05 x 3.50	15426378	N.A.	N.A.	N.A.	2V	Hyd.	35-60

GENERAL ENGINE SPECIFICATIONS

CU. IN. DISPLACEMENT	YEAR	BORE AND STROKE	FIRING ORDER	DEVELOPED HORSE POWER @ R.P.M.	DEVELOPED TORQUE @ R.P.M.	COMPRESSION RATIO	CARBURETOR	VALVE LIFTER TYPE	NORMAL OIL PRESSURE
				EIGHT CYLINDER					
361	1970-71	4.05 x 3.50	15426378	210 @ 4000	345 @ 2000	7.4-1	2V	Hyd.	45-70
	1972	4.05x3.50	15426378	210 @ 4000	345 @ 2000	7.2	2V	Hyd.	35-60
	1973	4.05x3.50	15426378	169 @ 3600	290 @ 2400	7.2	2V	Hyd.	35-75
	1974-78	4.05 x 3.50	15426378	N.A.	N.A.	N.A.	2V	Hyd.	35-75
390	1970-71	4.050x3.78	15426378	255 @ 4400	376 @ 2600	8.6-1	2V	Hyd.	35-60
	1972	4.05x3.78	15426378	201 @ 4400	316 @ 2600	8.2	2V	Hyd.	35-60
	1973	4.05x3.78	15426378	195 @ 4400	319 @ 2400	8.2	2V	Hyd.	35-75
	1974-78	4.05 x 3.78	15426378	N.A.	N.A.	N.A.	2V	Hyd.	35-75
391/389	1970-72	4.05 x 3.79	15426378	235 @ 4000	372 @ 2000	7.2	4V	Hyd.	35-60
	1973	4.05x3.79	15426378	192 @ 3600	333 @ 2400	7.2	4V	Hyd.	35-75
	1974-78	4.05 x 3.79	15426378	N.A.	N.A.	N.A.	4V	Hyd.	35-75
400	1977-78	4.00 x 4.00	13726548	N.A.	N.A.	N.A.	2V	Hyd.	50-75
401	1970-72	4.125 x 3.75	15486372	226 @ 3600	343 @ 2300	7.5-1	4V	Mech.	35-65
	1973	4.125x3.75	15486372	185 @ 3400	293 @ 3000	7.3	4V	Mech.	35-60
	1974-76	4.125 x 3.75	15486372	N.A.	N.A.	N.A.	4V	Mech.	35-60
460	1973-77	4.36 x 3.85	15426378	N.A.	N.A.	N.A.	4V	Hyd.	40-65
475	1977-78	4.50 x 3.75	15486372	N.A.	N.A.	N.A.	4V	Mech.	35-60
477	1970-71	4.50 x 3.75	15486372	231 @ 3400	412 @ 1600	7.5-1	2V	Mech.	35-65
	1972-73	4.50 x 3.75	15486372	209 @ 3200	353 @ 2800	7.20	4V	Mech.	35-60
	1974-76	4.50 x 3.75	15486372	N.A.	N.A.	N.A.	4V	Mech.	35-60
534	1970-72	4.50 x 4.20	15486372	266 @ 3200	481 @ 1700	7.5-1	4V	Mech.	35-65
	1973-78	4.50 x 4.20	15486372	222 @ 3000	400 @ 1800	7.3	4V	Mech.	35-65

CRANKSHAFT BEARING JOURNAL SPECIFICATIONS

CU. IN. DISPLACEMENT	YEAR	MAIN BEARING JOURNALS — JOURNAL DIAMETER	OIL CLEARANCE	SHAFT END PLAY	THRUST ON NO.	CONNECTING ROD BEARING JOURNALS — JOURNAL DIAMETER	OIL CLEARANCE	END PLAY
				SIX CYLINDER				
170	1971-72	2.2482-2.2490	.0007-.0026	.004-.008	3	2.1232-2.1240	.0006-.0026	.0035-.0105
200	1973	2.2482-2.2490	.0005-.0026	.004-.008	5	2.1232-2.1240	.0002-.0024	.0035-.0105
240	1971-74	2.3982-2.3990	.0005-.0022	.004-.008	5	2.1228-2.1236	.0008-.0024	.006-.013
300	1971-78	2.3982-2.3990	.0009-.0028	.004-.008	5	2.1228-2.1236	.0009-.0027	.006-.013
				EIGHT CYLINDER				
302	1971-78	2.2482-2.2490	.0005-.0015	.004-.008	3	2.1228-2.1236	.0010-.0015	.010-.020
330 MD.	1971-78	2.7484-2.7492	.0011-.0031	.004-.008	3	2.4380-2.4388	.0006-.0023	.010-.020
HD.	1971-78	2.7479-2.7487	1	.004-.010	3	2.4377-2.4385	.0009-.0027	.010-.020
351W	1976-78	2.9994-3.002	3	.004-.008	3	2.3103-2.311	.0008-.0026	.010-.020
351M	1977-78	2.9994-3.0002	.0008-.0025	.004-.008	3	2.3103-2.3111	.0008-.0025	.010-.020
360/359	1971-78	2.7484-2.7492	.0005-.0024	.004-.010	3	2.4380-2.4388	.0008-.0026	.010-.020
361	1971-78	2.7479-2.7487	1	.004-.010	3	2.4377-2.4385	.0010-.0029	.010-.020
390	1971-78	2.7484-2.7492	.0005-.0024	.004-.010	3	2.4380-2.4385	.0008-.0026	.010-.020
391/389	1971-78	2.7479-2.7487	.0012-.0033	.004-.008	3	2.4377-2.4385	.0012-.0033	.006-.016
400	1977-78	2.9994-3.0002	.0008-.0025	.004-.008	3	2.3103-2.3111	.0008-.0025	.010-.020
401	1971-76	3.1246-3.1254	.0018-.0025	.004-.008	3	2.7092-2.7100	.0017-.0036	.006-.014
460	1973-78	2.9994-3.002	.0008-.0026 4	.004-.008	3	2.4992-2.500	.008-.0015	.010-.020
475	1977-78	3.1246-3.1254	.0018-.0039	.004-.008	3	2.7092-2.7100	.0017-.0036	.006-.014
477	1971-78	3.1246-3.1254	.0018-.0039	.004-.008	3	2.7092-2.7100	.0017-.0036	.006-.014
534	1971-78	3.1246-3.1254	.0018-.0039	.004-.008	3	2.7092-2.7100	.0017-.0036	.006-.014

MD—Medium Duty.
HD—Heavy Duty.

1—No. 1, 3 .0012-.0036, No. 2, 4, 5 .0010-.0033.
2—No. 1 .0007-.0031, others .0005-.0028.

3—N. 1—.0005-.0015; All others .0008-.0015.
4—No. 1 only—All others, .0008-.0015.

VALVE SPECIFICATIONS

CU. IN. DISPLACE-MENT	YEAR	LASH (HOT) INCHES INT.	EXH.	ANGLE DEGREE FACE	SEAT	STEM DIA. INCHES INT.	EXH.	STEM CLEARANCE INTAKE	EXHAUST	VALVE LIFT INCHES	VALVE SPRING LBS. @ INCHES OPEN	CLOSED	LENGTH INCH FREE
SIX CYLINDER													
170	1971-72	.018	.018	44	45	.3104	.3102	.0008-.0025	.0017-.0027	.2405[1]	117 @ 1.222	52 @ 1.585	2.00
200	1973-75	Zero		45	44	.3103	.3101	.0008-.0025	.0010-.0027	—	150 @ 1.20	—	—
240	1971-74	Zero		44	45	.3420	.3420	.0010-.0027	.0010-.0027	.249[5]	197 @ 1.30	80 @ 1.70	1.99
300	1971-73	Zero		44	45	.3420	.3420[4]	.0010-.0027	.0010-.0027	.249	192 @ 1.180[2]	80 @ 1.58[2]	1.87[3]
	1974-78	Zero		44	45	.3420	.3420	.0010-.0027	.0010-.0027	.249	197 @ 1.30[6]	80 @ 1.70[6]	1.87[7]
EIGHT CYLINDER													
302	1971-78	Zero		44	45	.3420	.3420	.0010-.0027	.0015-.0032	.244[8]	180 @ 1.230	75 @ 1.66	1.97
330 MD	1971-78	Zero		44	45	.3715	.3705	.0010-.0024	.0028-.0042	.2446[9]	189 @ 1.420	99 @ 1.82	2.26
HD.	1971-78	Zero		44	45	.3715	.4343	.0010-.0024	.0020-.0037	.2446[9]	185 @ 1.240	80 @ 1.67	2.00
351W	1975-78	Zero		44	45	.3420	.3415	.0010-.0027	.0015-.0032	.370[10]	200 @ 1.34	75 @ 1.79	2.06[11]
351M	1977-78	Zero		44	45	.3420	.3415	.0010-.0027	.0015-.0032	.235	225 @ 1.39	80 @ 1.82	2.06
360	1971-76	Zero		44	45	.3715	.3715	.0010-.0024	.0020-.0034	.247[12]	221 @ 1.30[13]	80 @ 1.67[13]	2.12[14]
359	1974-78	Zero		44	45	.3715	.4343	.0010-.0027	.0020-.0040	.2448	220 @ 1.38[15]	90 @ 1.82[15]	2.12[16]
361	1971-72	Zero		44	45	.3715	.4343	.0010-.0027	.0020-.0037	.2448	185 @ 1.240	80 @ 1.67	2.00
	1973-78	Zero		44	45	.3715	.4343	.0010-.0027	.0020-.0040	.2448	220 @ 1.38[15]	90 @ 1.82[15]	2.12[16]
390	1971-78	Zero		44	45	.3715	.3715	.0010-.0024	.0020-.0034	.248	220 @ 1.370	90 @ 1.67	2.12
391/389	1971-78	Zero		44	45	.3715	.4343	.0010-.0024	.0020-.0037	.2448	220 @ 1.38[17]	90 @ 1.82[17]	2.12
400	1977-78	Zero		44	45	.3420	.3415	.0010-.0027	.0015-.0032	.2474[18]	225 @ 1.39	80 @ 1.82	2.06
401	1971-76	.020	.020	44	45	.4354	.4340	.0010-.0026	.0024-.0040	.278	185 @ 1.28	77 @ 1.70	2.07
460	1973-78	Zero		44	45	.3420	.3420	.0010-.0027	.0010-.0027	.253[19]	252 @ 1.33	80 @ 1.81	2.07
475	1977-78	.020	.020	44	45	.4354	.4340	.0010-.0026	.0024-.0040	.278	185 @ 1.280	77 @ 1.70	2.07
477	1971-78	.020	.020	44	45	.4354	.4340	.0010-.0026	.0024-.0040	.278	185 @ 1.280	77 @ 1.70	2.07
534	1971-78	.020	.020	44	45	.4354	.4340	.0010-.0026	.0024-.0040	.278	185 @ 1.280	77 @ 1.70	2.07

MD—Medium Duty.
HD—Heavy Duty.
[1] Exhaust—.2395 inch.
[2] HD engine: intake—197 @ 1.30/80 @ 1.70; exhaust—192 @ 1.180/79 @ 1.580.
[3] HD engine: intake—1.99 inch; exhaust 1.87 inch.
[4] HD engine: exhaust—.3415 inch.
[5] 1971-72 exhaust valve lift—.233 inch.
[6] Exhaust: 1974—192 @ 1.180/79 @ 1.690; 1975—192 @ 1.180/79 @ 1.580.
[7] Intake valve spring free length—1.99 inch.
[8] Exhaust—.2490 inch.
[9] Exhaust—.2328 inch.
[10] Exhaust—.382 inch.
[11] Exhaust—.212 inch.
[12] Exhaust—.249 inch.
[13] F100: intake—220 @ 1.380/90 @ 1.82; Exhaust—185 @ 1.240/80:1.67.
[14] F100: 2.00 inch.
[15] Exhaust—190 @ 1.24/80 @ 1.67
[16] Exhaust—2.00 inch.
[17] Exhaust—177 @ 1.70/77 @ 1.28
[18] Exhaust—.250 inch.
[19] Exhaust—.278 inch.

RING SIDE CLEARANCE SPECIFICATIONS

ENGINE (CU IN.)	YEAR	TOP COMPRESSION (IN.)	BOTTOM COMPRESSION (IN.)	OIL CONTROL (IN.)
170-6	1971-72	.002-.004	.002-.004	Snug
200-6	1973-75	.0019-.0036	.0025-.0045	Snug
240-6	1971-74	.0019-.0036	.0025-.0045	Snug
300-6	1971-78	.0019-.0036	.0025-.0045	Snug
302-V8	1971-78	.002-.004	.002-.004	Snug
351W-V8	1976-78	.002-.004	.002-.004	Snug
351M-V8	1977-78	.0019-.0036	.002-.004	Snug
330-V8	1971-78	.0024-.0046	.003-.005	Snug[1]
360-V8	1971-78	.002-.004	.002-.004	Snug
361-V8	1971-72	.0029-.0046	.0029-.0046	.0036-.0039
361-V8 359-V8	1973-78	.002-.005	.002-.005	.003-.0046
390-V8	1971-78	.002-.004	.002-.004	Snug
391/389-V8[2]	1971-72	.0029-.0046	.0029-.0046	.0036-.0039
391-V8[2]	1973-78	.003-.005	.003-.005	.003-.0046
400-V8	1977-78	.0019-.0036	.002-.004	Snug
401-V8 477-V8 534-V8	1971-72	.0029-.0046	.0025-.0045[2]	.0014-.0045
401-V8 477-V8 534-V8	1973-76	.0029-.0046	.0029-.0046	.0014-.0031
475-V8 477-V8 534-V8	1977-78	.0029-.0046	.0025-.0045	.0014-.0031
460-V8	1976-78	.002-.004	.002-.004	Snug

[1] 330 Heavy Duty—.0036-.0039
[2] Intermediate compression same as others

RING GAP SPECIFICATIONS

ENGINE (CU IN.)	YEAR	TOP COMPRESSION (IN.)	BOTTOM COMPRESSION (IN.)	OIL CONTROL (IN.)
170-6	1971-72	.010-.020	.010-.020	.015-.055
200-6	1973-75	.010-.020	.010-.020	.015-.055
240-6	1971-74	.010-.020	.010-.020	.015-.055
300-6	1971-78	.010-.020	.010-.020	.015-.055
302-V8	1971-78	.010-.020	.010-.020	.015-.055
351W-V8	1976-78	.010-.020	.010-.020	.015-.055
351M-V8	1977-78	.010-.020	.010-.020	.010-.035
330-V8①	1971-78	.010-.015	.010-.015	.015-.055
360-V8①	1971-78	.015-.023	.010-.015	.015-.055
361-V8	1971-72	.015-.023	.010-.020	.015-.030
359-V8 361-V8	1973-78	.015-.023	.010-.020	.015-.025
390-V8①	1971-78	.015-.023	.010-.020	.015-.055
391-V8	1971-72	.015-.023	.010-.020	.015-.030
391-V8	1973-78	.015-.023	.010-.020	.015-.025

ENGINE (CU IN.)	YEAR	TOP COMPRESSION (IN.)	BOTTOM COMPRESSION (IN.)	OIL CONTROL (IN.)
400-V8	1977-78	.010-.020	.010-.020	.010-.035
401-V8 477-V8 534-V8	1970-72	.018-.039	.015-.036②	.013-.039
401-V8 477-V8 534-V8	1973	.018-.028	.018-.028②	.013-.028
401V8 477-V8 534-V8	1974-76	.018-.028	.015-.025②	.013-.028
475-V8 477-V8 534-V8	1977-78	.018-.028	.015-.025	.013-.028
460	1975-76	.010-.020	.010-.020	.015-.055

① If equipped with intermediate ring, gap is .015-.023 in.
② Intermediate ring same as bottom compression.

TORQUE SPECIFICATIONS

CU. IN. DISPLACEMENT	YEAR	CYLINDER HEAD BOLTS (FT. LBS.)	ROD BEARING BOLTS (FT. LBS.)	MAIN BEARING BOLTS (FT. LBS.)	CRANKSHAFT BALANCER BOLT (FT. LBS.)	FLYWHEEL TO CRANKSHAFT BOLTS (FT. LBS.)	MANIFOLD INTAKE (FT. LBS.)	MANIFOLD EXHAUST
SIX CYLINDER								
170	1971-72	70-75	19-24	60-70	85-100	75-85	—	13-18
200	1973-75	70-75	19-24	60-70	85-100	75-85	—	13-18
240	1971-74	70-75	40-45	60-70	130-150	75-85	23-28	23-28
300	1971-78	70-85②	40-45	60-70	130-150	75-85	22-32	28-33
EIGHT CYLINDER								
302	1971-78	65-72③	19-24	60-70	70-90	75-85	23-25	20-30
330 MD.	1971-78	80-90	40-45	95-105	70-90	75-85	32-35	23-28
HD.	1971-78	80-90	40-45	95-105	150-175	75-85	32-35	23-28
351W	1976-78	65-70	19-24	60-70	70-90	75-85	23-25	18-24
351 M	1977-78	95-105④	40-45	35-45	70-90	75-85	⑤	18-24
360/359	1971-78	80-90	40-45	95-105	70-90①	75-85	32-35	12-18
361	1971-78	80-90	40-45	95-105	150-175	75-85	32-35	18-24
390	1971-78		40-45	95-105	70-90①	75-85	32-35	12-18
391/389	1971-78	80-90	40-45	95-105	150-175	75-85	32-35	18-24
400	1977-78	95-105④	40-45	35-45	70-90	75-85	⑤	18-24
401	1971-76	170-180	60-65	130-150	130-175	100-110	23-28	23-28
460	1973-78	130-140	40-45	95-105	70-90	75-85	25-30	28-33
477/475	1971-78	170-180⑥	60-65	130-150	130-175	100-110	25-32	22-32
534	1971-78	170-180⑥	60-65	150-165	130-175	100-110	25-32	22-32

MD—Medium Duty.
HD—Heavy Duty.
① 1975-76—130-150.
② Torque in steps; first to 55 ft. lbs., then to 65, final 70-85.
③ Torque in steps; first to 55-65 ft. lbs., then to 65-72.
④ Torque in steps; first to 75, then to 95-105.
⑤ 3/8 in-22-32 ft. lbs.; 5/16"-17-25 ft. lbs.
⑥ Torque in steps; first to 140 ft. lbs., then to 160 ft. lbs., final 170-180 ft. lbs.

WHEEL ALIGNMENT SPECIFICATIONS — EXCEPT BRONCO

YEAR	MODEL	CASTER (deg)			CAMBER (deg)			TOE-IN (in.)
		MAXIMUM	MINIMUM	OPTIMUM	MAXIMUM	MINIMUM	OPTIMUM	
1971	E-100, 200	5½P	½P	—	3½P	½P	—	⅛
	E-300	7½P	2½P	—	3½P	½P	—	⅛
	F-100, 250	5½P	½P	—	2½P	½N	—	3/32
	F-350	5½P	½P	—	3P	0	—	3/32
	F-100 (4WD)	3¼P	1¾P	2½P	2P	1P	½P	⅛
	F-250 (4WD)	4½P	3½P	4P	2P	1P	1½P	⅛
	P-350 P-3500	5¼P	3¾P	4½P	⅞P	⅜P	⅝P	⅛
	P-400, 500 P-4000, 5000	4¼P	2¾P	3½P	1P	0	½P	⅛
1972	E-100, 200	8½P	3½P	—	3½P	½P	—	⅛
	E-300	7½P	2½P	—	3½P	½P	—	⅛
	F-100, 250	8½P	3½P	—	2½P	½N	—	3/32
	F-350	8½P	3½P	—	3P	0	—	3/32
	F-100 (4WD)	4¼P	2¾P	3½P	2P	1P	1½P	⅛
	F-250 (4WD)	4½P	3½P	2½P	2P	1P	1½P	⅛
	P-350 P-3500	5¼P	3¾P	4½P	⅞P	⅜P	⅝P	⅛
	P-400, 500 P-4000, 5000	4¼P	2¾P	3½P	1P	0	½P	⅛
1973	E-100, 200	8½P	3½P	—	3½P	½P	—	⅛
	E-300	7½P	3½P	—	3½P	½P	—	⅛
	F-150, 250, 350	8½P	3½P	—	3½P	½P	—	⅛
	F-100 (4WD)	4¼P	2¾P	3½P	2P	1P	½P	5/32
	F-250 (4WD)	4½P	3½P	2½P	2P	1P	½P	5/32
	P-350 P-3500	5¼P	3¾P	4½P	⅞P	⅜P	⅝P	⅛
	P-400, 4000 P-500, 5000	4¼P	2¾P	3½P	1P	0	½P	⅛
1974	E-100, 150, 200, 250	8½P	2½P	—	3½P	½P	—	⅛
	E-300, 350	7½P	2½P	—	3½P	½P	—	⅛
	F-150, 250	8½P	½P	—	2½P	½N	—	⅛
	F-350	8½P	½P	—	3P	0	—	⅛
	F-100 (4WD) F-150 (4WD)	4¼P	2¾P	3½P	2P	1P	1½P	5/32
	F-250 (4WD)	4½P	3½P	4P	2P	1P	1½P	5/32
	P-350 P-3500	5¼P	3¾P	4½P	⅞P	⅜P	⅜P	⅛
	P-400, 500 P-4000, 5000	4¼P	2¾P	3½P	1P	0	-½P	⅛
1975	E-100, 150, 250	8½P	½P	—	2½P	½N	—	⅛
	E-350	7½P	2½P	—	3½P	½P	—	⅛
	F-100, 250	8½P	½P	—	2½P	½N	—	⅛
	F-350	8½P	½P	—	3P	0	—	3/32
	F-100 (4WD)	4¼P	2¾P	3½	2P	1P	1½P	5/32
	F-250 (4WD)	4½P	3½P	4P	2P	1P	1½P	5/32
	P-350, 3500	5¼P	3¾P	4½	⅞P	⅜P	⅝P	⅛
	P-400, 500, 4000, 5000	4¼P	2¾P	3½	1P	0	½P	⅛
1976	E-100, 150, 260	8½P	½P	—	3½P	½P	—	⅛
	E-350	7½P	2½P	—	3½P	½P	—	⅛
	F-100, 250	8½P	½P	—	2½P	½N	—	⅛
	F-350	8½P	½P	—	3P	0	—	⅛
	F-100, 150 (4WD)	4¼P	2¾P	4½P	2P	1P	1½P	5/32
	F-250 (4WD)	4½P	3½P	4P	2P	1P	1½P	5/32
	P-350	5¼P	3¾P	4½P	⅞P	⅜P	⅝P	⅛
	P-400, 500	4¼P	2¾P	3½P	1P	0	½P	⅛

WHEEL ALIGNMENT SPECIFICATIONS — EXCEPT BRONCO

YEAR	MODEL	CASTER (deg)			CAMBER (deg)			TOE-IN (in.)
		MAXIMUM	MINIMUM	OPTIMUM	MAXIMUM	MINIMUM	OPTIMUM	
1977-78	E-100, 150	5¾P	2P	—	1¾P	¾N	—	⅛
	E-250, 350	8¼P	4P	—	2¼P	½N	—	⅛
	F-100, 150, 250 (6200-6900 GVW)	9P	4P	—	3P	1N	—	⅛
	F-250 (7800-8000 GVW) F-250 Super Cab	8½P	3¼P	—	3¼P	0	—	⅛
	F-250 (8100 GVW S.C.) F-350	9P	3¼P	—	3¾P	0	—	⅛
	F-150 (4WD)	4¼P	2¾P	3½P	2P	1P	1½P	5/32
	F-250 (4WD)	5P	3P	4P	1½P	½P	1P	5/32
	P-500	4¼P	2¾P	3½P	1P	0	½P	⅛

Note: Maximum variation between wheels—½°
P—Positive
N—Negative

CASTER SPECIFICATIONS
— MEDIUM AND HEAVY DUTY TRUCKS

VEHICLE	AXLE CAPACITY (lbs)	YEAR	CASTER (deg) ①	
			Minimum	Maximum
F-B 700, 750	All	1971-78	3P	4P
L-N 500, 750	All	1971-78	4P	5P
C-CT Series	6000, 7000, 9000, 12000 Conventional, 15000	1971-78	2½P	3½P
	12000 Center-Point	1971-78	0	1P
L-LT-LN-LNT 800-900, 9000	6000, 7000	1971-78	3¼P	4¼P
	9000,	1973-76	3½P	4½P
	9000, 12000 Conventional	1971-72	3½P	4½P
	12000 Steer-Ease	1972	3¼P	4¼P
	16000, 18000, 20000	1972-78	3P	4P
	12000 Conventional 12000 Steer-Ease	1973-78	1¹⁵/₁₆P	3¹/₁₆P
W-WT 9000	12000 Conventional 15000 Conventional	1971-73 1971-78	3P	4P
	12000 Center-Point	1971-73	0	1P
	12000 Conventional	1974-78	2½P	3½P
F-600 (4WD)	7500	1971-78	2½P	3½P

① Vehicle unladen
P—Positive
N—Negative

Note: Caster Specifications are for a level frame front to rear. If the frame is lower at the rear, subtract the frame angle from the angle on the checking equipment. If the frame is lower at the front, add the frame angle to the angle on the checking equipment.

CAMBER SPECIFICATIONS
— MEDIUM AND HEAVY DUTY TRUCKS

VEHICLE	AXLE CAPACITY (lbs)	YEAR	CAMBER (deg)①	
			Minimum	Maximum
All	5000, 5500, 6000, 7000	1971-78	¼P	1P
All Except L-LT-LTS-LN-LNT 800-9000	9000, 12000 Conventional	1971-78	D(L), ½N(R)	¾P(L), ¼P(R)
	12000 Center-Point	1971-78	⅛P(L), ⅜N(R)	⅞P(L), ⅜(PR)
	15000	1971-78	¼P	1P
L-LT-LTS-LN-LNT 800-9000	9000, 12000 Conventional, 12000 Steer-Ease, 16000, 18000, 20000	1971-78	¼P	1P
F-600 (4WD)	7500	1971-78	½P	1½P

① Vehicle Empty
P—Positive
N—Negative
L—Left
R—Right

TOE-IN SPECIFICATIONS
— MEDIUM AND HEAVY DUTY TRUCKS

VEHICLE	AXLE CAPACITY (lbs)	YEAR	TOE-IN (in.)① Minimum	Maximum
All Exc. w-wt w/Center Point	5000, 5500, 6000, 7000, 12000 Center-Point	1971-78	³⁄₁₆	⁵⁄₁₆
All Exc. L-w-wt Series conventional or Steer-Ease	9000, 12000 Conventional, 12000 Steer-Ease, 15000, 16000, 18000, 20000	1971-78	⁵⁄₁₆	⁷⁄₁₆
W-WT 9000 w/Center-Point	1200 Center-Point②	1971-78	³⁄₁₆	¹¹⁄₁₆
F-600 (4WD)	7500	1971-78	¹⁄₁₆	³⁄₁₆
L-w-wt Series w/12000 Conv. or 12000 Steer-Ease	12000 Conventional 12000 Steer-Ease	1973-78	³⁄₁₆	⁵⁄₁₆

① Vehicle Unladen
② No part of the tie-rod clamp can be in the area between 10 o'clock and 2 o'clock.

BRONCO WHEEL ALIGNMENT

YEAR	MODEL	CASTER (Deg.)	CAMBER (Deg.)	TOE-IN (In.)	KING PIN INCLINATION (Deg.)
1970		3½¹	1½¹	⅛	8½
1971		2½	1½	⅛	8½
1972		3½	1½	⅛	8½
1973-75		3½	1½	⁵⁄₃₂	8½
1976		4½	1½	⁵⁄₃₂	8½
1977-78		3½	1½	⁵⁄₃₂	8½

1—The Caster and Camber angles are designed into the front axle and can not be adjusted.

TRUCK MODEL AND ENGINE APPLICATION

IDENTIFICATION PLATE: On light and medium cowl and windshield vehicles, the rating plate is attached to the right side of the cowl top panel or the upper cowl panel under the hood. The rating plate is on the right side of the radiator support on Parcel Delivery trucks, and on the inside of the glove compartment door on Broncos. On all other models, the plate is attached to the rear face of the left front door.

ENGINE PLATE: An identification tag is attached to the engine, giving the cubic inch displacement, building date and year.

GASOLINE ENGINE— APPLICATIONS

CUBIC INCH DISPLACEMENT	YEAR	ENGINE MAKE
6-170	1971-72	Own
6-200	1973-75	Own
6-240	1971-74	Own
6-300	1971-78	Own
V-8-302	1971-78	Own
V8-330	1971-78	Own
V8-351W	1975-78	Own
V8-351M	1977-78	Own
V8-360 / V8-359	1971-78	Own
V8-361	1971-78	Own
V8-389	1975-78	Own
V8-390	1971-78	Own
V8-391	1971-78	Own
V8-400	1977-78	Own
V8-401	1971-76	Own
V8-460	1975-78	Own
V8-475	1977-78	Own
V8-477	1971-78	Own
V8-534	1971-78	Own

Distributor

Note: For the 1978 and later Bronco, use F-150 4 x 4 procedures.

General Information

All 1971-74 engines are equipped with Autolite Motorcraft dual advance distributors. Starting in 1975, all light trucks began using the Motorcraft electronic ignition system. The medium and heavy duty trucks with gas engines continue to have standard point type ignition systems.

Rotation is clockwise on the six cylinder engines and counterclockwise on the V8 engines.

Distributor Removal & Installation

All Types

1. a. On a conventional ignition system, disconnect the primary wire at the coil.
   b. On a transistor ignition system, disconnect the primary wire from quick disconnect terminal.
   c. On a electronic ignition, disconnect the primary wire from the wiring harness.
2. Disconnect the vacuum line(s) at the distributor.
3. Remove the distributor cap.
4. Scribe a mark on the distributor body, indicating the position of the rotor. Scribe another mark on the body and engine block, indicating the position of the body in the block. These marks will insure that the distributor will be correctly timed when it is reinstalled.
5. Remove hold down bolt and clamp or retaining bolt and lockwasher and lift the distributor out of the block. *NOTE: Do not rotate the crankshaft while the*

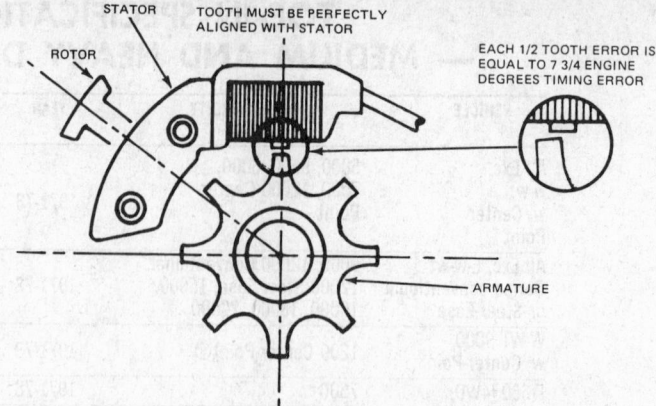

Electronic ignition static timing position (© Ford Motor Co.)

distributor is removed, or it will be necessary to time the engine.

6. To install, position the distributor in the block with the rotor aligned to the mark previously scribed on the distributor and the marks on the distributor and block aligned.
7. If the crankshaft has been rotated while the distributor was removed, the distributor must be timed with respect to the crankshaft.
   a. For Conventional and Transistor ignitions: Rotate the engine until the No. 1 piston is at TDC of the compression stroke. Position the distributor in the block with the rotor in the No. 1 firing position. Rotate the distributor until the points just about open. *NOTE: Make sure the oil pump intermediate shaft properly engages the distributor shaft. It may be necessary to crank the engine with the starter after the distributor gear is engaged in order to engage the oil pump intermediate shaft.*

   b. For Electronic ignition systems: Rotate the engine until the No. 1 piston is at TDC on the compression stroke. Align the correct initial timing mark with the pointer on the crankshaft damper. Position the distributor in the block with one of the armature segments aligned with the pickup on the stator and the rotor in the No. 1 firing position. The terminal housing is marked with an embossed "No.1" at the No. 1 terminal. *NOTE: Make sure the oil pump intermediate shaft properly engages the distributor shaft. It may be necessary to crank the engine with the starter after the distributor drive is partially engaged in order to engage the oil pump intermediate shaft. If the previous step was necessary, return the crankshaft to the initial timing alignment.*

8. Install retaining clamp and bolt or retaining bolt and lockwasher, but do not tighten. On electronic ignition distributors, rotate the distributor to advance the timing to the position where the armature tooth is properly aligned.
9. Connect primary wire and install distributor cap.
10. Check and reset ignition timing as described below.
11. Tighten hold down clamp bolt or retaining bolt.
12. Connect vacuum hose(s).

Setting Ignition Timing

Before checking and adjusting the ignition timing, on conventional systems inspect the breaker points for alignment and adjust if necessary. Rotate the distributor until breaker rubbing block rests on the peak of a cam lobe and check the breaker point gap, adjusting if necessary. See Specifications at the beginning of this section for correct gap setting.

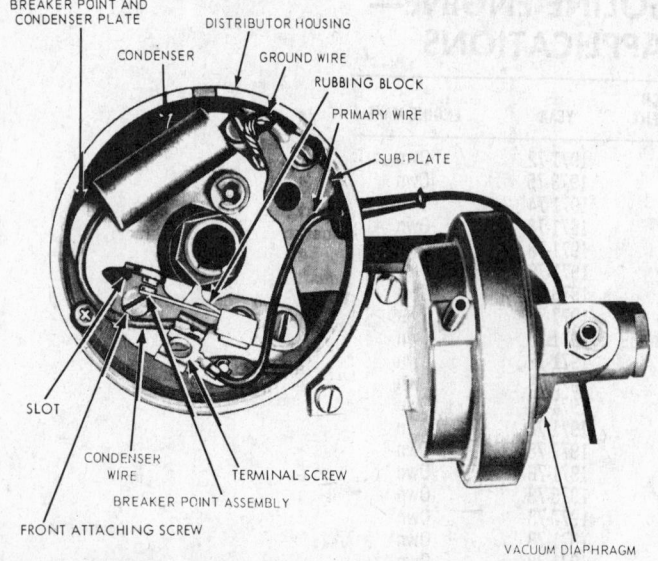

Autolite dual diaphragm distributor—6 cyl. (© Ford Motor Co.)

NOTE: The higher coil charging currents from the Dura Spark II ignition system can cause false triggering of timing lights with capacitive coupled pickups. The false triggering will result in apparent multiple sparks or an erratic timing indication. Timing lights with inductive pickup should be used for correct readings.

1. Clean and mark timing marks. All timing marks are on the crankshaft belt drive pulley, except on the 240 cu. in. engine in Econoline models and 240 and 300 cu. in. engines in P series models, where the marks are on the flywheel or flexplate (automatic transmission).
2. Disconnect vacuum line(s) and plug the disconnected line(s).
3. Connect a timing strobe light to the No. 1 spark plug lead and install a tachometer.
4. Start engine and set idle speed.
5. Loosen hold down clamp bolt or retainer bolt and rotate distributor housing until timing is correctly set.
6. Check the operation of the centrifugal advance by accelerating the engine to 2000 rpm. If the timing does not advance 9-14 degrees, the mechanical advance mechanism is faulty and must be removed for service.
7. Unplug vacuum line(s) and connect to distributor diaphragm. Momentarily accelerate engine to 2000 rpm and note timing advance. If it is not considerably more than the mechanical advance alone, the disphragm must be replaced.

8. Remove timing light and tachometer. Reset idle to correct speed.

Breaker Point Replacement

1. Remove the distributor cap and rotor.
2. Disconnect the primary and condenser wires from breaker point assembly.
3. Remove the breaker point assembly and condenser retainer screws and lift the breaker point assembly and condenser out of the vehicle.
4. To install, place breaker point assembly and condenser in position and install retaining screws. Be sure to place the ground wire under the breaker point assembly screw farthest from the breaker point contacts on V8 engine distributor or under the condenser retaining screw on

6-cylinder engine distributor.
5. Align and adjust the breaker points.
6. Connect primary and condenser wires to the breaker point assembly.
7. Install rotor and distributor cap.
8. Set ignition timing to specifications.

Transistor Ignition System

The tachometer block is used to connect a tachometer or other test equipment into the circuit.

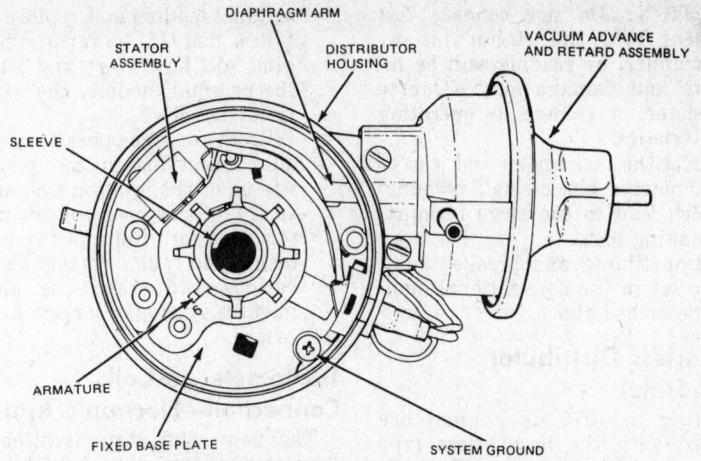

Electronic ignition distributor (© Ford Motor Co.)

Field relay transistor ignition (© Ford Motor Co.)

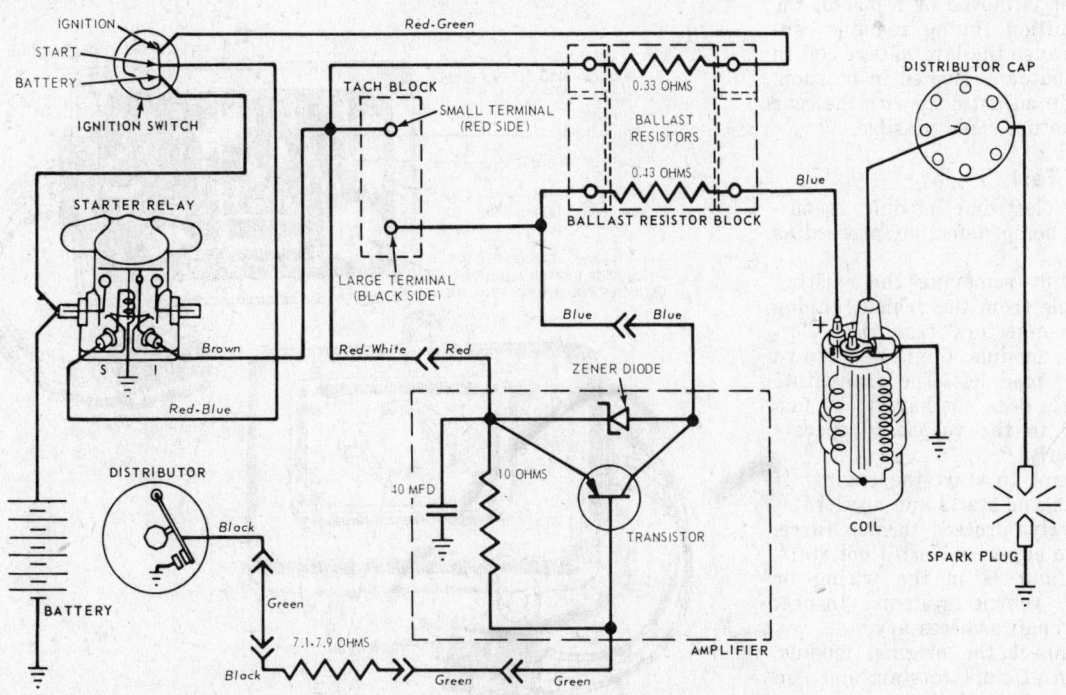

Transistor ignition system (© Ford Motor Co.)

CAUTION: Do not connect test equipment into the circuit in any other manner, or reading will be inaccurate and damage may occur to the resistor, or change its operating characteristics.

Connect the tachometer red lead to the tachometer block small terminal and black lead to the large terminal when making tests.

Ignition timing and breaker point gap are set in the conventional manner as described above.

Breakerless Distributor (Solid State)

Starting in 1975 all engines are equipped with the breakerless type electronic ignition system. The conventional contact breaker points and condenser in the distributor are replaced by a permanent magnet low-voltage generator. The generator consists of an armature with four or six gear-like teeth mounted on the top of the distributor shaft, and a permanent magnet inside a small coil. The coil is riveted in place to provide a preset air gap with the armature. The distributor base, cap, rotor and vacuum and centrifugal spark advance are about the same as the conventional system.

The distributor is wired to a solid state module in the engine compartment. Inside the module is an electronic circuit board which consists of inner connecting resistors, capacitors, transistors and diodes. The module senses a signal from the magnetic generator to perform the switching function of conventional points and it senses and controls dwell

Unless a malfunction occurs, or the distributor is moved or replaced, the initial ignition timing remains constant. Because the low voltage coil in the distributor is riveted in position. the air gap adjustment with the rear tooth armature is not possible.

Module Test

If the electronic module is suspected of being defective, proceed as follows:

1. Without removing the existing module from the vehicle, unplug the connectors from the electronic module. Connect a known good module. The substitute module does not have to be fastened to the vehicle to operate properly.
2. Attempt to start the engine. If the engine starts and accelerates properly, proceed to step three. If the engine will still not start, the fault is in the wiring or other vehicle systems. Inspect and repair as necessary.
3. Reconnect the original module. Again attempt to start and run the engine. If the engine once again will not start, remove the

original module and replace with a new one. If, however, the engine will now start and run on the original module, the module is not defective.

4. With the engine operating, check all connections in the primary wiring of the ignition system for such faults as poor wire crimp to terminal or improper engagement. The faulty connection will be observed when the engine misfires or stops. Correct as necessary.

Tachometer-to-Coil Connection—Electronic Ignition

The new solid state ignition coil connector allows a tachometer test lead with an alligator-type clip to be connected to the DEC (Distributor Electronic Control) terminal without removing the connector.

When engine rpm must be checked, install the tachometer alligator clip into the "TACH TEST" cavity as shown. If the coil connector must be removed, grasp the wires and pull horizontally until it disconnects from the terminals.

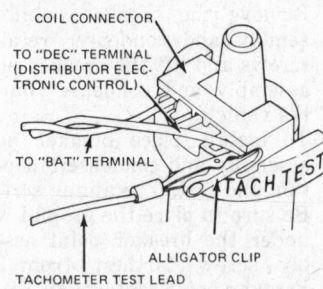

COIL CONNECTOR

TO "DEC" TERMINAL (DISTRIBUTOR ELECTRONIC CONTROL)

TO "BAT" TERMINAL

TACH TEST

ALLIGATOR CLIP

TACHOMETER TEST LEAD

Tachometer to coil connection——electronic ignition (© Ford Motor Co.)

Alternator

Reference

Procedures for diagnosis and repair of the charging system can be found in the "Electrical Section" in the General Repair portion of this manual.

Alternator Removal and Installation

1. Disconnect battery ground cable.
2. Loosen the alternator mounting bolts and remove the adjustment arm to alternator attaching bolt.
3. Remove the electrical connectors from the alternator. The stator and field connectors are of the push-on type and should be pulled straight off to prevent damage to the terminal studs.
4. Disengage the alternator belt.
5. Remove the alternator mounting bolt and alternator.
6. To install, first install the wiring harness, then position the alternator on engine, installing spacer (if used) and mounting bolt. Tighten bolt only finger tight.
7. Install the adjustment arm to alternator attaching bolt.
8. Adjust belt tension as described below.
9. Tighten adjusting arm bolts and mounting bolt to 22-32 ft. lbs. torque.

Alternator Belt Adjustment

1. Loosen the alternator mounting bolt to a snug position and loosen the adjusting arm bolt.

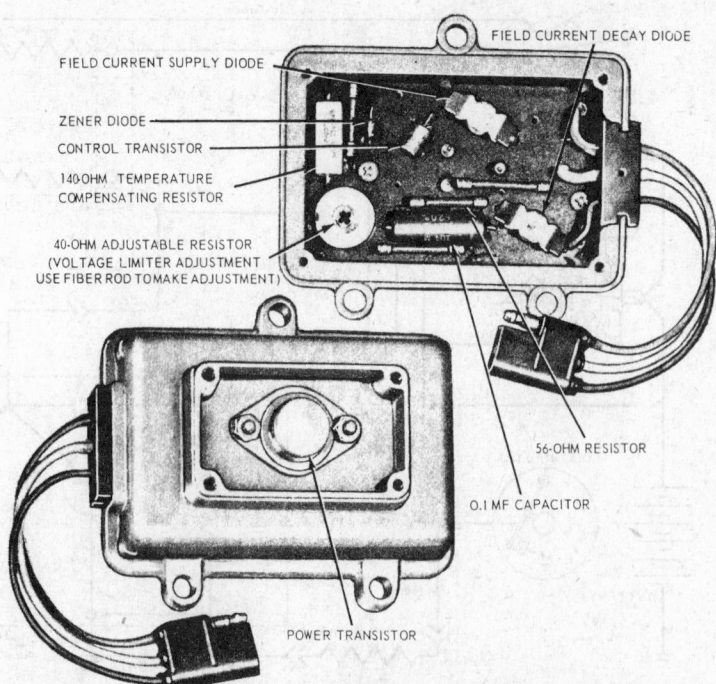

FIELD CURRENT DECAY DIODE

FIELD CURRENT SUPPLY DIODE

ZENER DIODE

CONTROL TRANSISTOR

140-OHM TEMPERATURE COMPENSATING RESISTOR

40-OHM ADJUSTABLE RESISTOR (VOLTAGE LIMITER ADJUSTMENT USE FIBER ROD TO MAKE ADJUSTMENT)

56-OHM RESISTOR

0.1 MF CAPACITOR

POWER TRANSISTOR

1971-73 Transistorized alternator regulator (© Ford Motor Co.)

2. *Apply pressure on the alternator front housing only* and tighten the adjusting arm to alternator bolt.

3. Check for correct tension. Belt tension is correctly adjusted when belt can be deflected by hand ½-¾". Readjust if necessary.

4. Tighten all mounting bolts to 22-32 ft. lbs. torque.

Regulator Adjustments

Motorcraft

The Motorcraft electro-mechanical regulator used on 38, 42, 44, 55 and 61 Ampere alternators is not adjustable and must be replaced if charging system tests indicate that it is defective.

Transistorized Regulator—1971-73

1. Remove regulator mounting screws and regulator.
2. Remove bottom cover from regulator.
3. With regulator at normal operating temperature, use a fiber rod as a screwdriver to turn 40-ohm adjustable resistor.
4. Install bottom cover and mount regulator.

Transistorized Regulator 1974-78

NOTE: The only adjustment that can be made on this regulator is the voltage limiter adjustment.

1. With the regulator at normal operating temperature, remove the cover screws and cover.
2. Using a fiber rod as a screwdriver, turn the voltage adjusting screw clockwise to increase voltage setting or counterclockwise to decrease the setting.
3. Reinstall the cover.

Regulator Removal and Installation

1. Remove battery ground cable.

2. Remove the regulator mounting screws.
3. Disconnect the regulator from the wiring harness.
4. To install, connect wiring harness.
5. Mount the regulator to the regulator mounting plate. The radio supression condenser (electro-mechanical regulator) mounts under one of the mounting screws. The ground lead mounts under the other mounting screw.
6. Connect battery ground cable and test the charging system for proper voltage regulation. See the "Electrical Diagnosis" in the General Repair Section for testing procedures.

Fuse Link

The fuse link is a short thin length of insulated wire integral with the engine compartment wiring harness. If there is a heavy reverse current flow (such as caused by an improperly connected booster battery or when a short occurs in the wiring harness), the fuse link burns out and thus protects the alternator from damage. Production fuse links are black and replacement are green or black. All have the words FUSE LINK printed on the insulation. Burn-out of a fuse link may be evidenced by disfigured or bubbled insulation or by bare wire ends protruding from the insulation.

To test continuity of the fuse link (Bronco and F-Series LD) check that the battery is OK, then check with a voltmeter for voltage at the BAT terminal of the alternator. No voltage indicates that the fuse link is probably burned out. On P-Series trucks, disconnect the battery ground cable and, using an ohmmeter or a self-powered test light, check for continuity between points A and B (see illustration). A good fuse link will light the test bulb or show zero resistance on the ohmmeter. If fuse link is

burned out, replace as described below.

Fuse Link Replacement

1. If the 5/16" eyelet terminal is not required, cut it off as close to the terminal as possible.
2. Disconnect battery ground cable.
3. Disconnect the fuse link terminal at the starter relay (alternator on P-Series).
4. Remove the complete fuse link at the splice, when applicable, remove the old terminal from the battery stud of the starter relay.
5. Cut out original splice(s), then splice and solder the new fuse link with the existing wires from the original splice(s). See illustration. Wrap splice completely with tape. If there were two wires connected to the fuse link eyelet, cut link from eyelet and position the second wire with the eyelet back on the starter relay terminal.
6. Install all other wires removed during service.
7. Connect battery ground cable.

Starter

Reference

For complete diagnostic and overhaul procedures for starter motor and drive mechanisms, see "Starters" in the General Repair Section.

Autolite & Motorcraft Positive Engagement Starter
Removal and Installation

1. Disconnect starter cable at the starter terminal.
2. Remove the starter mounting bolts.
3. Remove the starter assembly.
4. To install, insert the starter into the pilot hole in the flywheel housing, making sure that the starter housing pilot completely enters the pilot hole all the way around, and that the starter housing face is square and tight to the engine rear cover plate. Install mounting bolts.
5. Snug all bolts, torquing them to 23–28 ft. lbs.
6. Connect starter cable.

Motorcraft Solenoid Actuated Starter

Removal and Installation

1. Disconnect the battery ground cable and raise the vehicle.
2. Disconnect the cable and wires at the solenoid terminals.
3. Turn the front wheels all the way to the right and remove the

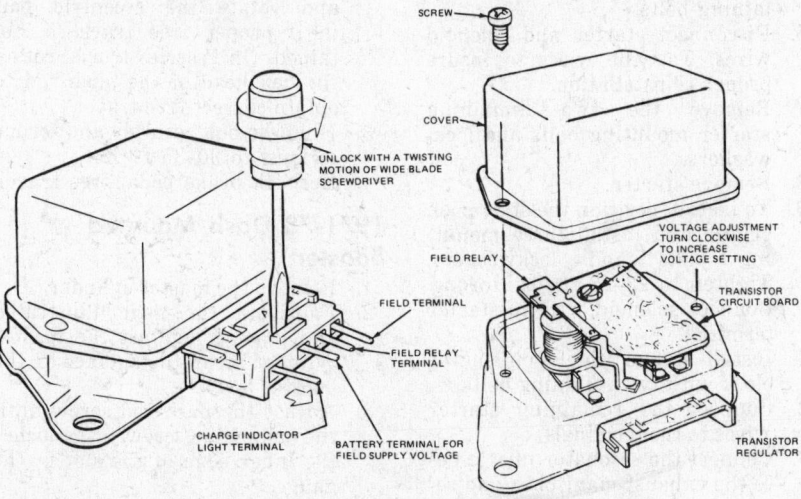

SCREW

COVER

UNLOCK WITH A TWISTING MOTION OF WIDE BLADE SCREWDRIVER

VOLTAGE ADJUSTMENT TURN CLOCKWISE TO INCREASE VOLTAGE SETTING

FIELD RELAY

FIELD TERMINAL

TRANSISTOR CIRCUIT BOARD

FIELD RELAY TERMINAL

CHARGE INDICATOR LIGHT TERMINAL

BATTERY TERMINAL FOR FIELD SUPPLY VOLTAGE

TRANSISTOR REGULATOR

1974-78 Transistorized alternator regulator (© Ford Motor Co.)

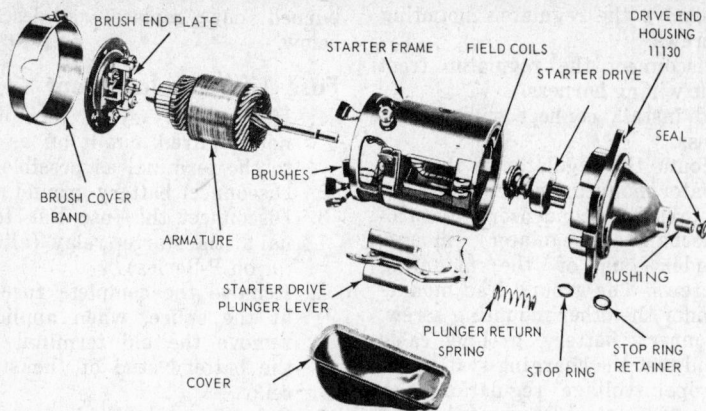

BRUSH END PLATE
STARTER FRAME
FIELD COILS
DRIVE END HOUSING 11130
STARTER DRIVE
SEAL
BRUSHES
BRUSH COVER BAND
ARMATURE
STARTER DRIVE PLUNGER LEVER
PLUNGER RETURN SPRING
BUSHING
STOP RING
STOP RING RETAINER
COVER

Autolite starter—disassembled (© Ford Motor Co.)

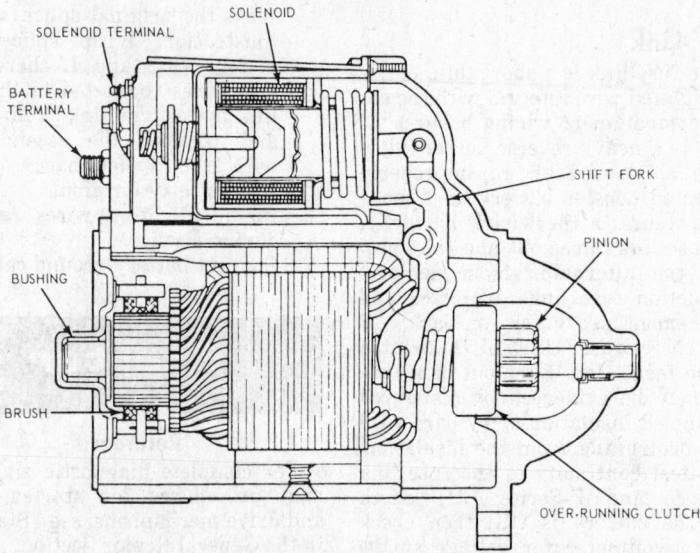

SOLENOID
SOLENOID TERMINAL
BATTERY TERMINAL
SHIFT FORK
PINION
BUSHING
BRUSH
OVER-RUNNING CLUTCH

Solenoid Actuated starter—cross section (© Ford Motor Co.)

bolts attaching the idler arm to the frame.
4. Remove the starter mounting bolts and the starter.
5. Position the starter on the mounting plate and start the bolts.
6. With the starter held firmly against the mounting surface and fully in the pilot hole, run the mounting bolts in until they are snug.
7. Tighten the mounting bolts to 15-20 ft-lbs.
8. Connect the cable and wires to the solenoid terminals. The battery cable should be tightened to 45-95 in-lbs.
9. Install the idler arm bracket on the frame and tighten the bolts to 28-35 ft-lbs.
10. Lower the vehicle and connect the battery ground cable.

Delco Remy Starter

Removal and Installation

1. Disconnect the battery cables.
2. Remove one bolt and lockwasher securing the battery cable retain-

er to the right rear cylinder head cover.
3. Remove the starter upper mounting bolt.
4. Remove the clamps securing the resonator inlet pipes to the exhaust manifolds, and position the inlet pipes out of the way.
5. Remove the starter shield retaining bolts.
6. Disconnect starter and solenoid wires. Tag the wires to insure proper reinstallation.
7. Remove the two remaining starter mounting bolts and lockwashers.
8. Remove starter.
9. To install, position the starter on engine and install three mounting bolts and lockwashers. Tighten to 23–28 ft. lbs. torque.
10. Connect solenoid wire to starter terminal.
11. Install starter shield on engine block with two retaining bolts.
12. Connect the remaining starter wires to the terminals.
13. Connect the resonator inlet pipes to the exhaust manifolds, and secure with clamps.

14. Connect the battery cables and close battery box.
15. Check operation of starter.

Prestolite Starter

Removal and Installation

1. Disconnect the starter cable at the starter terminals.
2. Remove the heat shield from the manifold.
3. Remove the starter mounting bolts.
4. Tilt the starter slightly so the starter drive clears the flywheel housing and remove the starter
5. Installation is the reverse of removal. On trucks with automatic transmission the dipstick tube bracket is mounted under the starter mounting bolt.

Brakes

Note: For the 1978 and later Bronco, use F-150 4 x 4 procedures.

Reference

Complete overhaul and service information for hydraulic brake components can be found in the "Hydraulic Brake", "Power Brake" and "Disc Brake" sections of the Unit Repair portion of this manual.

All service and overhaul operations for air brake systems are under the heading of "Air Brakes" in the Unit Repair portion of this manual.

Brake Pedal Adjustment

1971-72 Standard or Frame Mounted Booster

If the pedal free travel in a standard hydraulic brake system or frame mounted hydraulic booster system is not within 3/16-3/8", the pedal should be adjusted as follows:
1. Push the brake pedal down by hand pressure, and check the free travel.
2. Loosen locknut on eccentric bolt, and rotate the eccentric bolt until proper free travel is obtained. On P-series trucks rotate the hex head of the push rod to obtain correct free travel.
3. Hold the bolt securely and torque locknut to 30–35 ft. lbs.
4. Recheck brake pedal free travel.

1971-78 Dash Mounted Booster

1. Remove the master cylinder.
2. Fabricate the gage illustrated and place it against the master cylinder mounting surface on the booster body.
3. Adjust the push rod screw until the end of the screw just touches the inner edge of the slot in the gauge.
4. Install the master cylinder.

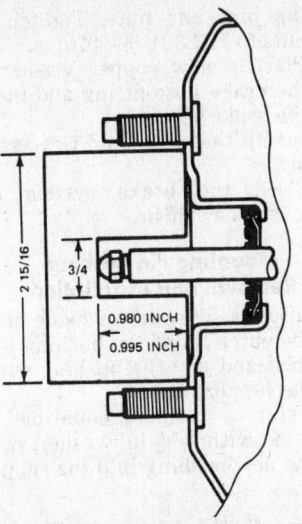

Pushrod (brake pedal) adjustment—Bendix dash mounted brake booster, Midland Ross similar (© Ford Motor Co.)

Bleeding System

Conventional Hydraulic Brake System

1. Fill master cylinder reservoir with fluid. Check level of fluid frequently during bleeding procedure.
2. If hydraulic system is equipped with a vacuum booster, bleed the booster before bleeding the rest of the system.
3. If vehicle is equipped with dual slave cylinders, bleed the upper one first. If there are two bleeder screws, bleed the one nearest the power chamber first.
4. Bleed the wheel cylinder with the longest hydraulic line first.
5. Attach bleeder tube to bleeder screw and place other end of tube in a container partially filled with fluid.
6. Loosen bleeder screw, then slowly depress the brake pedal by hand, allowing it to return slowly to the fully-released position. Repeat until all bubbles cease to flow fom bleeder tube.
7. Close bleeder screw and remove tube.
8. Repeat this procedure at each wheel until system is completely free of air bubbles.

Dual Hydraulic Brake Systems

The primary and secondary hydraulic brake system are individual systems and are bled separately. Bleed the longest line first on the individual system being serviced. Be sure to keep reservoir filled during bleeding operation

After bleeding it is necessary to centralize the pressure differential valve. On 500-900 series trucks, remove the brake warning light switch

from the pressure differential valve to prevent damage to the switch assembly.

1. Bleed master cylinder at the outlet port side of the system being serviced. If there are no bleed screws, loosen the hydraulic line nut. Do not use the secondary piston stop screw located on the bottom of the master cylinder—the stop screw or piston could easily be damaged.
2. Operate pedal slowly until fluid is free of air bubbles, then tighten bleed screw.
3. Follow Steps 1 through 8 of Conventional Bleeding procedure above.
4. Centralize the pressure differential valve. Turn the ignition switch to ACC or ON position. Loosen the pressure differential valve inlet tube nut on the system opposite the system which was bled last. This will result in unequal pressure in the other direction and allow the valve to center. Slowly depress brake pedal until light goes out. Tighten inlet tube nut. On 500-900 series trucks, disconnect wires from warning light switch on the differential valve, remove switch. Springs inside valve will center valve. Replace switch and wire.
5. Check fluid level in reservoir and fill to within ¼" of top.

Disc Brake Hydraulic System

The hydraulic system may be bled in the conventional manner, with the following additional steps.

1. First bleed master cylinder, then

rear brake cylinders (longest line first).

2. On front disc brakes, the bleeder button on the metering valve must be depressed to allow the brake fluid to reach the caliper assemblies. When the bleeding operation is complete, fill the reservoir to within ¼" of top.
3. Centralize the pressure differential valve as described above in "Bleeding Dual Hydraulic Brake Systems".

Master Cylinder

Removal and Installation

On models equipped with power boosters, depress brake pedal while engine is not running to expel vacuum or air from booster system.

1. If stoplight switch is mounted on the master cylinder, disconnect wires.
2. On dash-mounted master cylinders, disconnect the dust boot from the rear of the master cylinder at the dash panel. If the boot is connected to the master cylinder only, leave it in place.
3. Disconnect the hydraulic line(s) from the master cylinder and pump the brake pedal all the way several times to evacuate all fluid from the master cylinder into a suitable container.
4. Disconnect the pushrod from the brake pedal. On dash-mounted master cylinders the pushrod is connected to the brake pedal lever with a bolt, and there may be a stoplight switch mounted by that bolt. Remove the bolt and

BENDIX   GPD

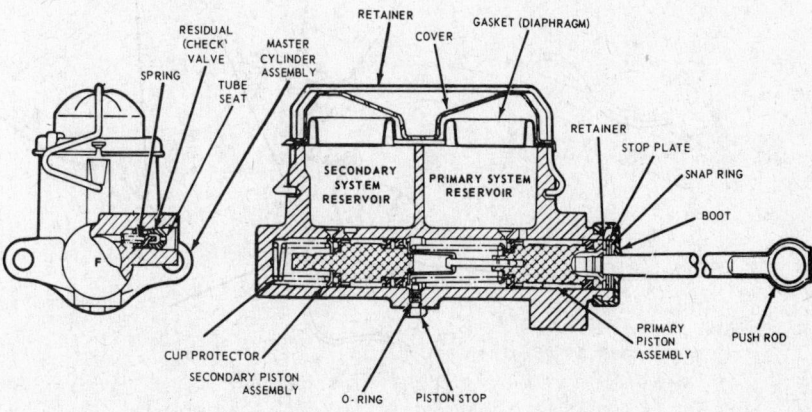

Typical dual master cylinder (© Ford Motor Co.)

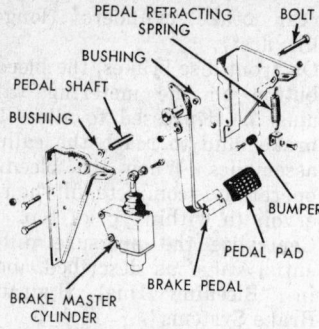

Brake cylinder installation—C series
(© Ford Motor Co.)

stoplight switch. If the master cylinder is under the floorboard, remove clevis pin from pushrod yoke. Mark or tag bushings and spacers for correct position.

5. On trucks equipped with conventional brakes, remove the master cylinder mounting bolts and master cylinder. If the truck is equipped with a dash-mounted booster, remove the retaining nuts and washers and pull master cylinder off mounting studs.
6. To install, position master cylinder so that mounting bolts or retainer nuts may be installed. Tighten mounting bolts or nuts securely.
7. Connect hydraulic lines(s) loosely to master cylinder fitting(s).
8. If rubber dust boot attaches to both cylinder and dash, make sure it is properly installed at this time.
9. Connect pushrod to brake pedal lever. If there is a bushing in the hole at the end of the pushrod, lubricate bushing. Install stoplight switch at this time if it was removed when disconnecting pushrod from brake pedal lever. Make sure all bushings and spacers are properly installed.
10. Connect stoplight switch wires.
11. Bleed hydraulic system as described above. Bleed master cylinder first.

Wheel Cylinder

Removal and Installation

1. Remove the wheel, drum, and brake shoes.
2. Remove the cylinder-to-shoe connecting links.
3. Disconnect the brake line from the wheel cylinder.
4. Remove the wheel cylinder retaining bolts and remove the cylinder from the brake backing plate.
5. Position the brake cylinder on the backing plate and install the retaining bolts and lockwashers.
6. Install a new gasket on the brake line (if equipped) and connect the line to the wheel cylinder.
7. Install the brake shoes and the connecting links between the shoes and cylinder. Install the brake drum and wheel.
8. Adjust the brakes and bleed the system (as outlined previously). Check pedal operation before moving the vehicle.

Disc Brakes

Floating Caliper

Caliper Removal and Installation

1. Raise and secure the vehicle.
2. Remove the wheel and tire assembly.
3. Disconnect the brake hose from the caliper.
4. Remove the pins and nuts holding the caliper to the anchor plate, remove the caliper.
5. Coat the mounting pins with a light film of chassis lube. Place the caliper assembly on the anchor plate and install the retain-

ing pins and nuts. Tighten the nuts to 17-23 ft-lbs. torque.
6. Place a new copper washer on the brake hose fitting and install the brake hose.
7. Install the wheel and tire assembly.
8. Bleed the brake system and lower the vehicle.

Mounting Pin Bushing Removal and Installation

1. Insert a screwdriver blade under the outer lip of the bushing steel shell and pry the bushing out of the support boss.
2. Using a caliper mounting pin fitted with a ½ in. washer, press the new bushing into the support boss.

Sliding Caliper

Removal and Installation

1. Siphon or dip part of the brake fluid out of the large section of the master cylinder to avoid overflow when the caliper piston is pressed into the cylinder bore.
2. Raise the vehicle and remove the tire and wheel assembly.
3. Position a 8 in. C-clamp on the caliper and tighten the clamp to bottom the piston in the caliper cylinder bore.
4. Remove the key retaining screw.
5. Drive the caliper support key and spring out with a brass rod and light hammer.
6. Disconnect the brake hose from the inlet port. Cap the hose and inlet port to prevent fluid leakage.
7. Remove the caliper from the spindle assembly by pushing it downward against the spindle and rotating the upper end up-

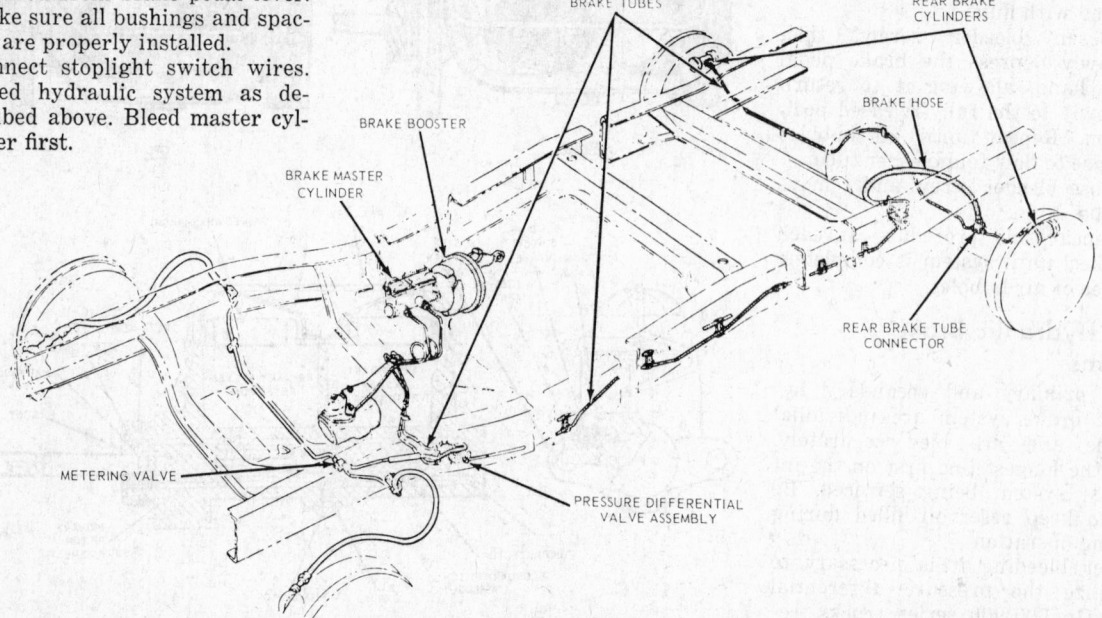

Disc brake system (© Ford Motor Co.)

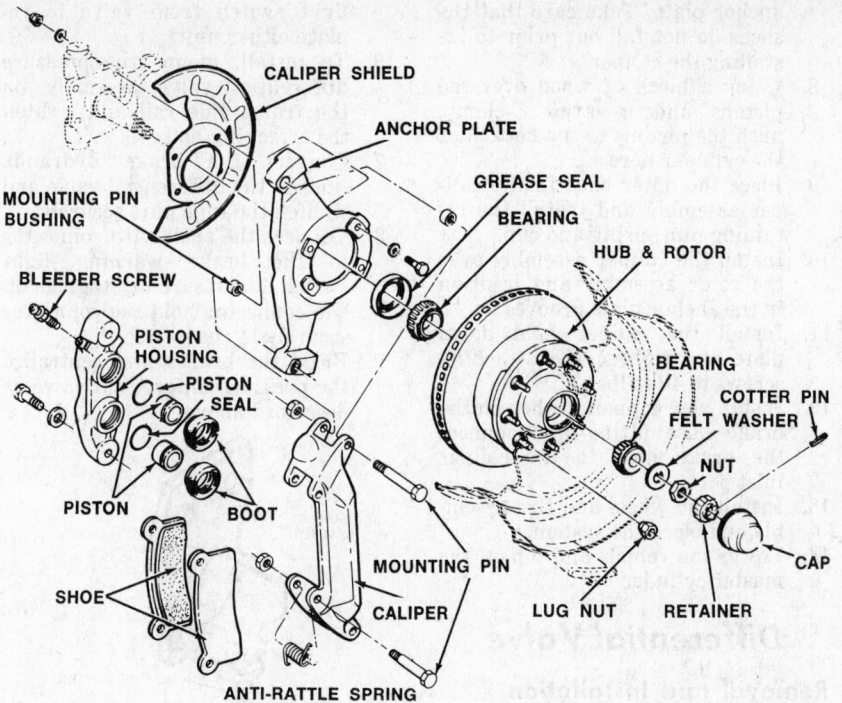

Exploded view of floating caliper assembly (© Ford Motor Co.)

ward out of the spindle assembly.

8. Remove the outer shoe and lining from the caliper. Remove the inner shoe and lining from the spindle assembly. Remove the shoe antirattle clip from the lower shoe abutment surface on the spindle assembly.

9. Thoroughly clean the areas of the caliper and spindle assembly that come in contact during the sliding action of the caliper.

10. Position a new anti-rattle clip in the lower shoe abutment in the spindle assembly. Make sure that the tabs on the clip are positioned properly and the loop-type spring is away from the rotor.

11. Place the lower end of the inner shoe and lining in the spindle assembly shoe abutment against the anti-rattle clip. Slide the Upper end of the shoe into position. Make certain that the clip is still in the proper position.

12. Check to be sure that the caliper piston is fully bottomed in caliper piston pore. If it is necessary, use a 8-in. C-clamp to bottom the piston.

13. Place the outer shoe and lining on the caliper and using your fingers, press the shoe tabs into place. If the shoe cannot be pressed into place by hand, use a C-clamp to press the shoe into position. Be careful not to damage the lining.

14. Place the caliper on the spindle assembly by pivoting the caliper around the spindle upper mounting surface. Be careful not to

tear or cut the boot as it slips over the inner shoe.

15. Using a brake adjusting tool or screwdriver, hold the upper machined surface of the caliper against the surface of the spindle assembly. Install a new caliper support spring and a new caliper

support key. Drive the key and spring assembly into position with a soft faced mallet. Install the key retaining screw. Tighten the screw to 12-20 ft-lbs torque.

16. Install new copper washers on the brake hose fittings. Connect the brake hose to the caliper inlet port.

17. Bleed the brake system, install the wheel and tire assembly, and lower the vehicle.

Rail Slider Two Piston Sliding Caliper

Removal and Installation

1. Raise and secure the vehicle. Remove the wheel and tire assembly.

2. Disconnect the brake hose from the caliper and plug the hose and ilnet port.

3. Remove the key retaining screw.

4. Using a brass rod and light hammer drive out the key and spring.

5. Remove the caliper from its support assembly by rotating the key and spring end out and away from the rotor. Slide the opposite end of the caliper clear of the slide in the support end of the rotor.

6. Clean the areas of the caliper and support that come in to contact during the sliding action of the caliper.

7. Clean any brake fluid, grease or

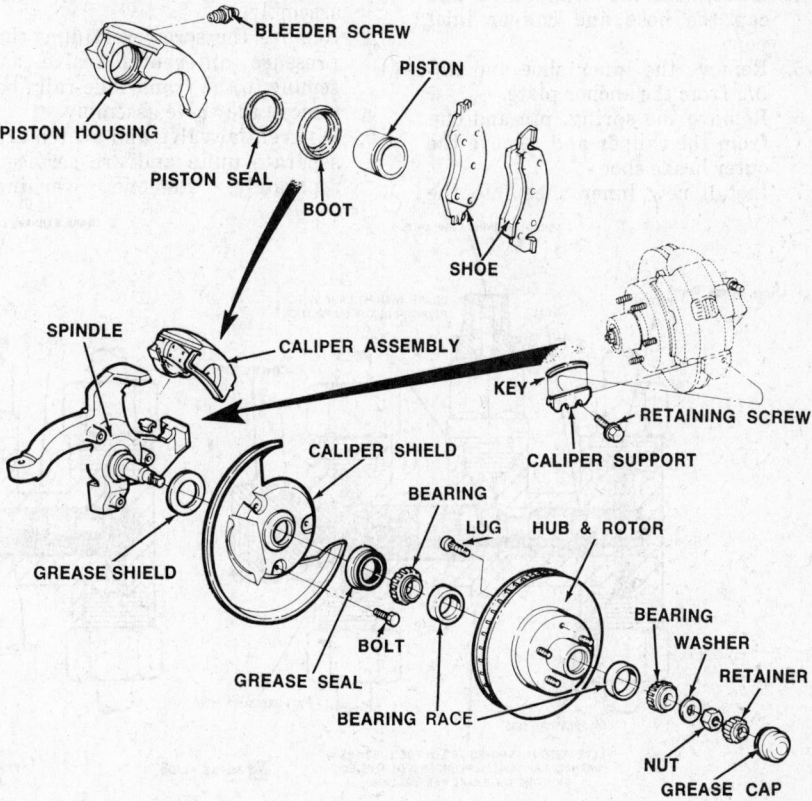

Exploded view of sliding caliper assembly (© Ford Motor Co.)

grit off the rotor breaking surface.

8. Place the caliper rail into the slide on the support and rotate the caliper into the rotor.
9. Position the key and spring between the caliper and support assembly and start in by hand. Note that the spring is between the key and caliper and that the spring tangs overlap the ends of the key. Use a break adjusting tool or screwdriver to hold up the caliper against the support assembly.
10. Using a hammer, drive the key and spring into position aligning the correct notch with the existing hole in the support.
11. Secure the key to the support with the key retaining screw. Tighten the screw to 20 ft-lbs.
12. Place new copper washer on the brake hose fitting and connect to the caliper inlet port.
13. Bleed the system, lower the vehicle, and refill the master cylinder if necessary.

Heavy Duty Two Piston Sliding Caliper

Removal and Installation
1. Raise the vehicle and remove the wheel and tire assembly.
2. Remove the four screws holding the caliper mounting plate and remove the plate.
3. Lift the caliper off the hub and rotor assembly.
4. Disconnect the brake hose and cap the hose and caliper inlet port.
5. Remove the inner shoe and lining from the anchor plate.
6. Remove the spring, pin and cup from the caliper and remove the outer brake shoe.
7. Install new inner shoe into the

anchor plate. Take care that the shoes do not fall out prior to installing the caliper.
8. Using a block of wood over the pistons and a large C-clamp, push the pistons to the bottom of the cylinder bore.
9. Place the outer shoe in the caliper assembly and install the retaining pin, spring and cup.
10. Install the caliper assembly over the rotor assembly and position in the anchor plate grooves.
11. Install the caliper hold down plate and tighten the attaching screws to 40 ft-lbs.
12. Install new copper washer on the brake hose fitting and connect the brake hose to the caliper inlet port.
13. Install the wheel and tire assembly and bleed the system.
14. Lower the vehicle and top off the master cylinder.

Differential Valve

Removal and Installation
1. Raise the vehicle on a hoist.
2. Disconnect the brake warning light from the pressure differential valve assembly switch. CAUTION: To prevent damage to the brake warning switch wire connector, expand the plastic lugs to allow removal of the shell-wire connector from the switch body.
3. Disconnect the brake hydraulic lines from the differential valve assembly.
4. Remove the screw retaining the pressure differential valve assembly to the frame side rail and remove the valve assembly.
5. Differential valve and switch are separate units and are serviced separately. Remove warning

light switch from valve to replace either unit.
6. To install, mount the pressure differential valve assembly on the frame side rail and tighten the attaching bolt.
7. Connect the brake hydraulic lines to the differential valve and tighten the tube nuts securely.
8. Connect the shell-wire connector to the brake warning light switch. Make sure plastic lugs on the connector hold the connector securely to the switch.
9. Bleed the brakes and centralize the pressure differential valve as described above.

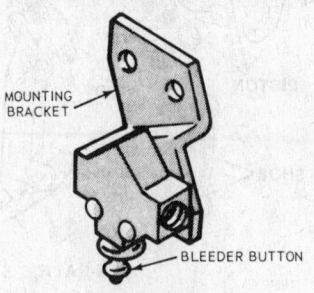

Brake metering valve
(© Ford Motor Co.)

Power Boosters

Reference
For power booster overhaul procedures see the General Repair Section.

Bendix Dash-Mounted Vacuum Booster

Removal and Installation
1. Remove retaining nuts and master cylinder from booster.
2. Loosen hose clamp and remove manifold vacuum hose from booster.
3. From inside the cab, remove the

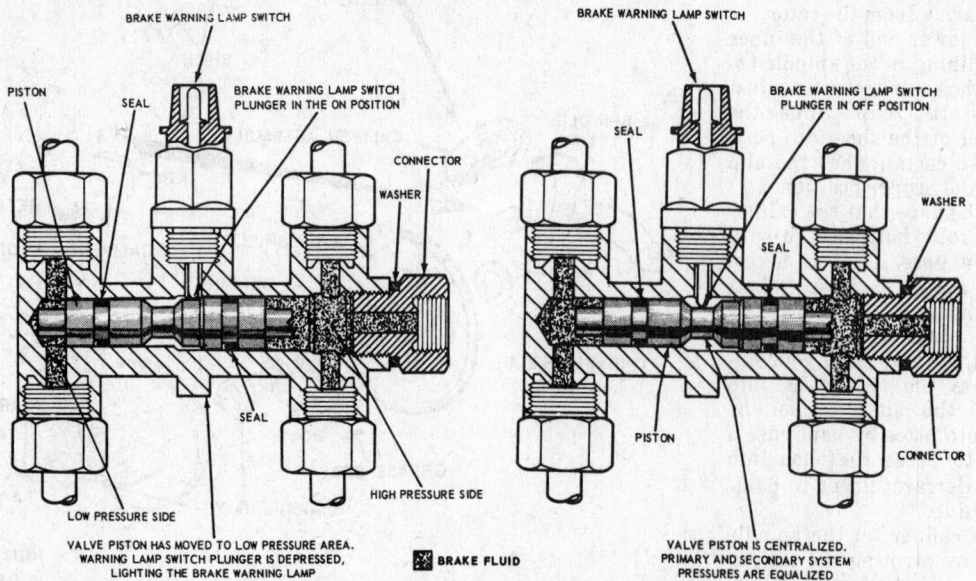

Brake differential valve—100-400 series (© Ford Motor Co.)

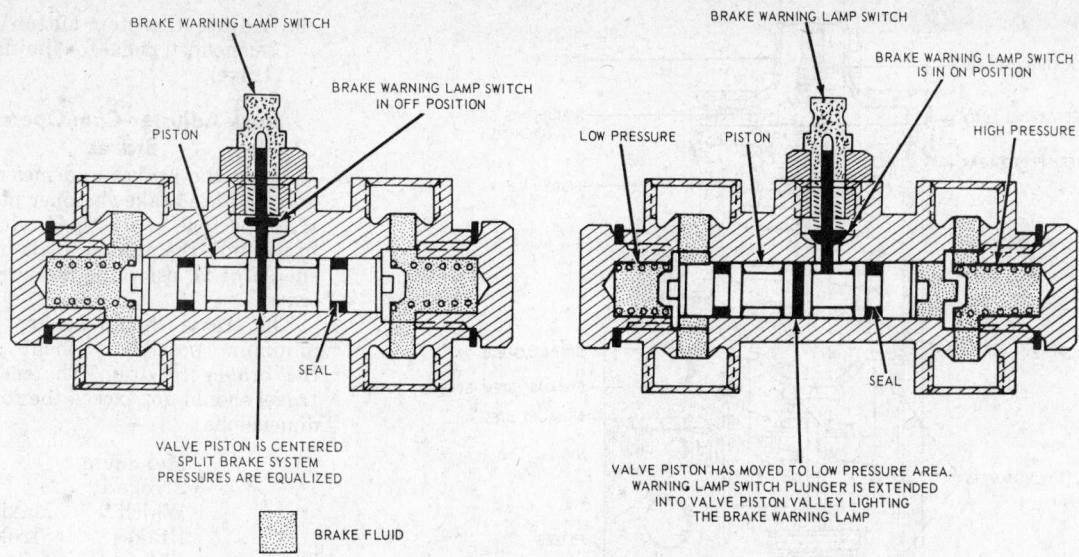

BRAKE WARNING LAMP SWITCH

BRAKE WARNING LAMP SWITCH

BRAKE WARNING LAMP SWITCH
IN OFF POSITION

PISTON

PISTON

BRAKE WARNING LAMP SWITCH
IS IN ON POSITION

LOW PRESSURE

HIGH PRESSURE

SEAL

SEAL

VALVE PISTON IS CENTERED
SPLIT BRAKE SYSTEM
PRESSURES ARE EQUALIZED

VALVE PISTON HAS MOVED TO LOW PRESSURE AREA.
WARNING LAMP SWITCH PLUNGER IS EXTENDED
INTO VALVE PISTON VALLEY LIGHTING
THE BRAKE WARNING LAMP

BRAKE FLUID

Brake differential valve—500-900 series w/split hydraulic brakes (© Ford Motor Co.)

attaching bolt, nut and plastic bushings and disconnect the booster pushrod from the brake pedal.

4. Remove nuts that retain the booster mounting bracket to the dash panel.

5. Remove the booster assembly from engine compartment.

6. To install, mount the booster and bracket assembly to the engine side of the dash panel by sliding the bracket mounting bolts and valve operating rod in through the holes in the dash panel.

7. From inside the cab, install the booster mounting bracket to dash panel retaining nuts.

8. Position the master cylinder to the booster assembly and install the retaining nuts.

9. Connect the manifold vacuum hose to the booster and secure with clamp.

10. From inside the cab connect the booster valve operating rod to the brake pedal with the attaching bolt, nut and plastic bushings.

11. Start engine and check operation of the brake system.

Midland-Ross Dash-Mounted Vacuum Booster

Removal and Installation

Removal and installation procedure for the Midland-Ross booster is the same as that described for the Bendix booster above.

Bendix Frame Mounted Vacuum Booster

Removal and Installation

1. Depress the brake pedal several times to remove all vacuum from the system.

2. Loosen the booster air inlet tube clamp and remove the tube.

3. Disconnect the hydraulic lines and vacuum lines from the booster.

4. Remove the booster mounting bolts and booster from the bracket.

5. Place the booster on the mounting bracket and secure with the mounting bolts. Use new lockwashers on the bolts.

6. Connect the hydraulic lines. Make sure that the connections are tight.

7. Connect the air inlet tube to the booster. Be sure that the hose clamp is tight.

8. Bleed the brake system.

9. Connect the vacuum line to the booster and tighten the clamp securely.

Air Brakes

Adjustments

Governor

Before adjusting the pressure settings of the governor, determine the accuracy of the dash gauge by checking the readings against an accurate test gauge. The cut-in cut-out setting is made at the adjusting screw.

1. With the engine running, build up pressure in the system and observe the pressure registered by the dash gauge.

2. If the pressure build-up continues beyond 125 psi before the governor cuts-out, remove the

CLAMP BAND

DIAPHRAGM

FRONT HOUSING

REAR HOUSING

FULCRUM RING

O-RING

SEAL

VALVE SPRING

FILTER

HUB

BRAKE PEDAL PUSH ROD

VALVE ASSEMBLY

O-RING

VALVE SEAT

PLUNGER CUSHION RING

VALVE SEAT RETAINING

PUSH ROD

DIAPHRAGM RETURN SPRING

COLLAR PLUNGER

PUSH ROD RETAINER

SNAP RING

REACTION LEVER ASSEMBLY

SPACER GUIDE

PLUNGER SPRING

Midland Ross dash mounted vacuum booster (© Ford Motor Co.)

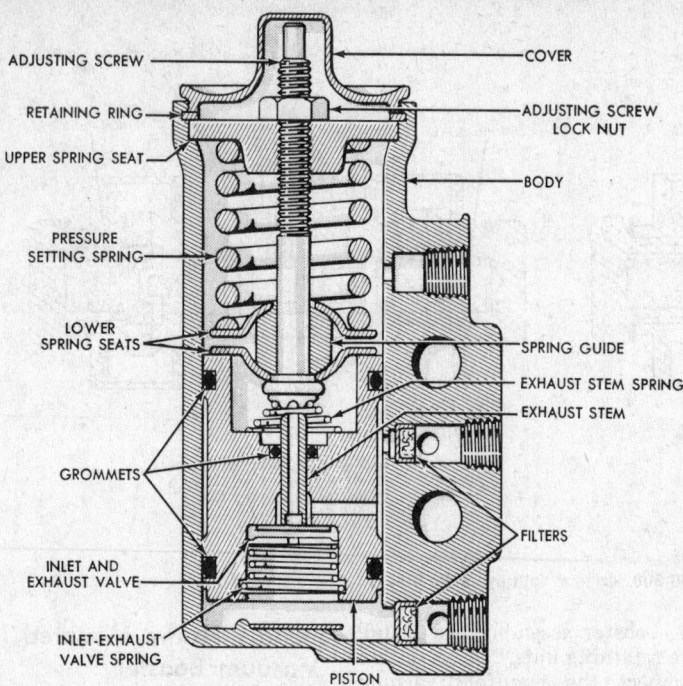

Compressor Governor—Typical (© Ford Motor Co.)

3. Adjust the stop button on the suspended pedal to eliminate free travel.

Slack Adjuster-Cam Operated Brakes

Apply the brakes and measure the travel of the brake chamber push rod. If the vehicle is equipped with Maxi brake unit, the minimum air pressure should be 90 psi while measuring the travel.

The travel should be kept to the minimum possible without causing the brakes to drag. The maximum travel should not exceed the following dimensions:

Type Number	Maximum Stroke at Which Brakes Should Be Adjusted	Maximum Stroke After Adjustment
#9	1⅜ in.	1.0 in.
#12	1⅜ in.	1.00 in.
#16	1⅜ in.	1.00 in.
#20	1¾ in.	1¼ in.
#24	1¾ in.	1¼ in.
#30	2.00 in.	1¼ in.

NOTE: Adjustment of the yoke on the brake chamber push rod should not be changed. When new, the yoke is adjusted so that the slack adjuster brake chamber push rod angle is slightly greater than 90° when the brakes are properly adjusted and the brakes are applied. Brake lining wear will not change this angle as long as the slack adjusters are kept adjusted to compensate for lining wear.

Front: This procedure applies only to trucks equipped with the S-cam operated brakes. Push rod travel which

cover from the top of the governor and loosen the locknut. Turn the adjusting screw clockwise to lower the cutout pressure, and counter clockwise to increase the cut-out pressure. After adjusting the cut-out pressure, tighten the adjusting screw locknut and install the cover.

NOTE: The air pressure range between the cut-out pressure (maximum) and the cut-in pressure (minimum) is fixed at about 20-25 psi and cannot be adjusted.

Foot Control Valve

To determine if the brakes are applying properly, proceed as follows:

1. Install a pressure gauge anywhere in the circuit between the control valve and brake chamber, or install the gauge in one of the extra service ports (upper row of ports).
2. Fully depress the brake pedal. The test gauge reading should approximate reservoir pressure as indicated by the dash gauge.

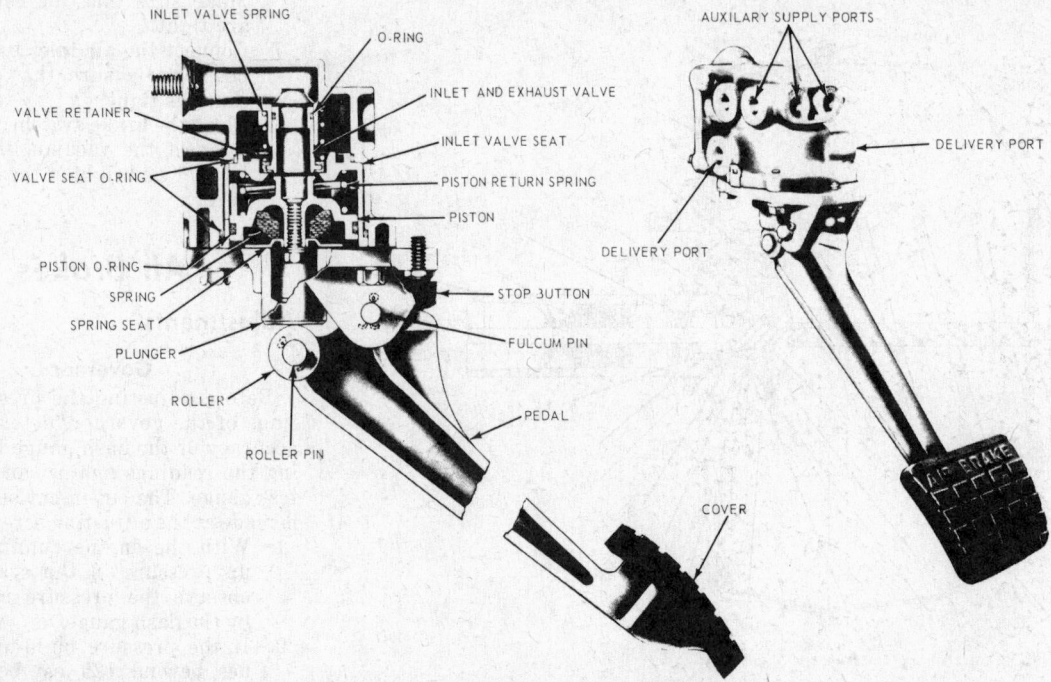

Foot control valve—typical (© Ford Motor Co.)

reaches or exceeds the maximum listed above indicates the need of adjustment. Turn the adjusting screw clockwise until the push for travels ¾ in. in, going from released to fully applied position. When making the adjustment, turn the screw in quarter turns.

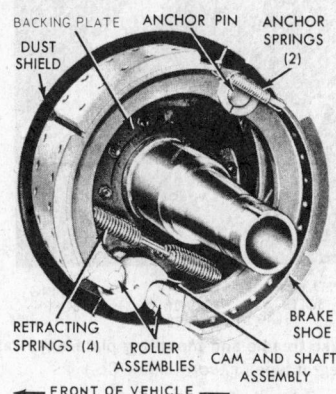

Cam type brake assembly (rear)
(© Ford Motor Co.)

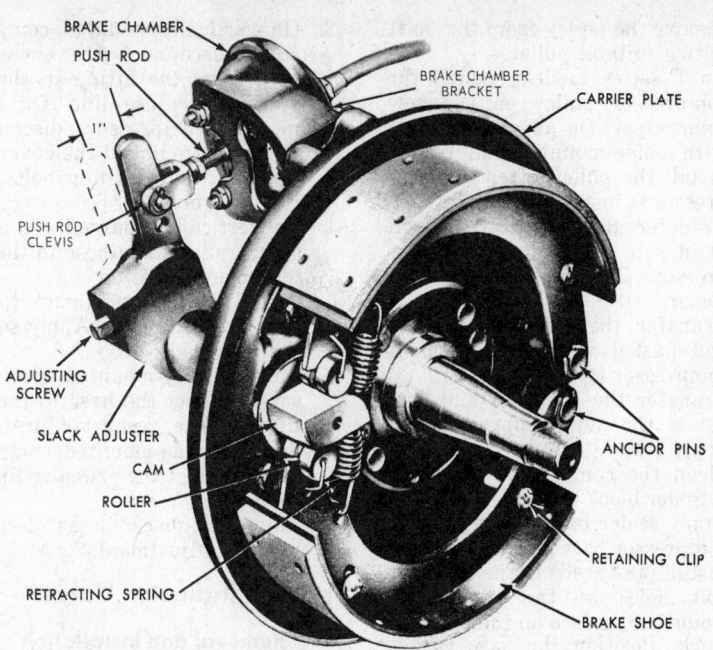

Cam Type brake assembly (front) showing slack adjuster adjustment
(© Ford Motor Co.)

Rear: This procedure applies to vehicles equipped with either standard or Maxi brake S-cam slack adjusters. Push rod travel which reaches or exceeds the maximum listed above indicates the need of adjustment. Depress the lock sleeve and turn the hex head of the wormshaft clockwise until the push rod travels one inch in, going from released to fully applied position. Be sure that the lock sleeve comes back out of the wormshaft so that the adjustment is locked.

NOTE: When adjusting either the front or rear slack adjuster, raise the wheels and make sure that there is no brake drag.

Air Compressor R & R

NOTE: The following operations are for Ford gasoline engine vehicles only. For R&R procedures on diesel engine models, refer to the "Diesel Engine" section in the General Repair portion of this manual.

Depending on the particular truck model and engine, compressors are mounted in various locations. The procedures for removal and installation differ between vertically mounted units and side mounted units.

The following procedures apply generally to liquid cooled compressors, but air cooled installations are similar.

Vertically Mounted Compressor—Removal and Installation

1. Open the reservoir drain cocks to exhaust air pressure from the system.
2. Drain the cooling system.
3. Disconnect the compressor air outlet line, the water inlet and outlet lines, the oil feed and return lines at the compressor.
4. Since the reservoir hose is difficult to remove at the governor, disconnect the hose at the fitting on the opposite end. If the hose is to be replaced, disconnect it from the compressor after compressor removal.
5. Remove the compressor-to-base plate bolts, then slide the compressor inward on the base plate. Disengage the drive belt and remove the compressor.
6. Transfer the pulley, Woodruff key and attaching nut to the new compressor.
7. Transfer the air outlet elbow and gasket, water outlet fittings, and oil inlet and outlet fittings to the new sealer. Apply sealer to all parts.
8. Transfer the governor and gasket to the new compressor and apply sealer. If the reservoir pressure hose is being replaced, connect it to the governor.
9. Transfer the air inlet filter and gasket to the new compressor and apply sealer.
10. Position the compressor on the base plate and engage the drive belt to the pulley.
11. Install the attaching bolts and slide the compressor away from the fan pulley until a ½ inch belt deflection is obtained. Tighten the bolts.
12. Connect the compressor air outlet line, water inlet line, and water outlet line to the compressor.
13. Connect the oil feed and return lines to the compressor.
14. Connect the hose from the governor to the fitting on the reservoir pressure line.
15. Fill the cooling system, close the reservoir drain cocks, start the engine and check for oil, air, or coolant leaks.

Side Mounted Compressor—Removal and Installation

1. Drain the cooling system, and open the reservoir drain cocks to exhaust the air pressure from the system.
2. Loosen the idler pivot bolt and adjusting bolt, remove the compressor drive belt.
3. Disconnect all oil, air, and coolant lines from the compressor and the reservoir pressure line from the governor.
4. Remove the bracket bolt at the cylinder head and the bracket bolt and washer at the manifold. Remove the bracket between the exhaust manifold and the cylinder head.
5. On vehicles with power steering, remove the attaching bolt, nut and lockwasher from the clip at the frame side rail. Remove the power steering hose retainer bolt, nut, and lockwasher at the frame crossmember and place the hoses out of the way.
6. Remove the front compresor-to-base attaching bolts. *There is very little clearance between the compressor base plate and the engine mount.*
7. Remove the remaining bolts.
8. Remove the air compressor from underneath the vehicle in a rearward direction. Discard the base gasket.
9. Remove the compressor pulley cotter pin and Woodruff key, and

remove the pulley from the shaft with a suitable puller.

10. On C-series models, install the compressor pulley on the new compressor. On all other models with a side-mounted compressor, install the pulley after the compressor is installed.

11. Transfer all air, water, and oil fittings to the replacement compressor. Coat the threads with sealer.

12. Transfer the air inlet strainer and gasket to the replacement compressor and apply sealer.

13. Transfer the governor and gasket to the new compressor and apply sealer.

14. Clean the compressor base and cylinder block base plate.

15. Apply sealer to both sides of the compressor base gasket.

16. Install pilot studs in the two rear bolt holes of the base plate mounting surface on the cylinder block. Position the base gasket on the base plate over the studs. From underneath the vehicle, install the compressor on the pilot studs and install the three attaching bolts. Remove the pilot studs and install the remaining bolts.

17. On vehicles with power steering, position the power steering hose on the frame rail and secure with the attaching bolt, washer, and nut.

18. Connect the water outlet line to the compressor.

19. Position the steering tube clips on the frame side rail and secure with the attaching nut, bolt, and lockwasher.

20. Connect the water inlet line to the compressor.

21. On all vehicles except C-series, loosen the power steering belt adjustment bolts and loosen the belt to provide clearance for installing the compressor pulley. Install the pulley on the shaft and secure with the nut and cotter pin.

22. Position the power steering belt on the pulleys and adjust the belt tension.

23. Position the air compressor drive belt on the pulleys and adjust the belt tension.

24. Connect the air outlet line to the compressor.

25. Connect the reservoir pressure line at the governor.

26. Fill the cooling system, close the air reservoir drain cocks, and build up pressure in the system. Check for air, oil, and coolant leaks.

Governor

Removal and Installation

1. Exhaust the air from the system.

2. On vertically-mounted compressors, disconnect the governor hose from the fitting in the reservoir pressure line. On side-mounted compressors, disconnect the pressure line at the governor.

3. Remove the attaching bolts and the governor.

4. On vertically-mounted compressors, transfer the hose to the replacement governor.

5. Install the governor, gasket, and the attaching bolts. Apply sealer to the threads.

6. On vertically-mounted compressors, connect the hose to the fitting on the reservoir pressure line. On side-mounted compressors, connect the pressure line to the governor.

7. Test the governor as detailed under "Adjustments".

Pressure Indicator Valve

Removal and Installation

1. Exhaust the air from the system.

2. Disconnect the wire at the buzzer switch.

3. Disconnect the air lines to the wiper control as required.

4. Unscrew the pressure indicator fitting and remove the assembly.

5. Install the pressure indicator valve assembly.

6. Connect the buzzer switch wire. Turn on the ignition switch to test that the buzzer and light are functioning properly before building up pressure in the system.

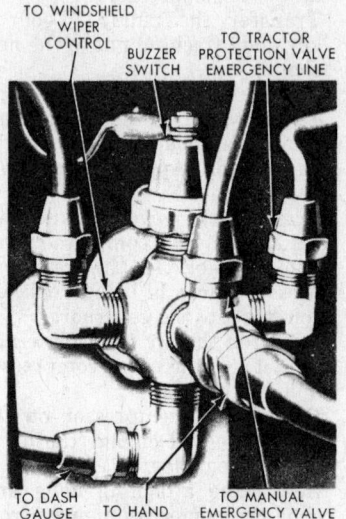

Pressure indicator valve (© Ford Motor Co.)

Foot Control Valve

Removal and Installation

1. Open the reservoir drain cocks to exhaust the air from the system.

2. Disconnect all but one line from the valve ports. Loosen, *but do not disconnect*, the remaining line. This will prevent the valve

Brake treadle and mounting plate—typical integral type (© Ford Motor Co.)

from falling when the attaching bolts are removed.

3. Remove the cotter pin and pivot pin that connect the brake tredle to the control valve mounting plate.

4. Remove the control valve attaching bolts.

5. Disconnect the remaining air line from the control valve, and remove the control valve.

6. If the valve is being replaced, transfer all brass fittings and the stop light switch to the new valve. Apply sealer to the threads before installation.

7. Remove the actuating button and rubber seal from the control valve mounting plate to allow installation of the new valve.

8. Position the new valve on the lower dash panel and mounting plate and install the attaching bolts.

9. Install the actuating button in the mounting plate bore, and install the rubber seal to the button and mounting seal.

10. Install the brake treadle on the control valve mounting plate with the pivot pin and cotter pin.

11. Connect the brake service lines to the upper ports of the valve.

12. Start the engine to build up pressure in the system. Check for leaks.

Quick Release Valve

Removal and Installation

1. Exhaust the air from the system.

2. Disconnect the air lines at the valve.

3. Remove the valve mounting bolts and the valve.

4. Install the valve and tighten the bolts and nuts.

5. Check the exhaust port to be sure that it is not plugged.

6. Connect the air lines.

Relay Valve

Removal and Installation

1. Block the wheels.
2. Exhaust the air from the system.
3. Disconnect the air lines from the relay valve.
4. Remove the valve mounting bolts and remove the valve.
5. Remove the insert (inlet and exhaust valve assembly), by removing the four exhaust cover cap screws and cover. Pull the insert out.
6. Clean and inspect the relay valve air lines. Replace any lines which are damaged or connecting hoses which are deteriorating or show signs of chafing.
7. Mount the relay valve and secure with the attaching bolts.
8. Connect the air lines to the valve.
9. Build up pressure in the system. Test the valve for correct operation and air leaks.

Brake Shoes

Reference

Removal and installation procedures for brake shoes can be found in the "Air Brake" section in the General Repair portion of this manual.

Anti-Skid System

Reference

Information on the trouble-shooting and repair of "Anti-Skid System" units will be found in the "Anti-Skid section of the General Repair portion of this manual.

Brake Service Air Chamber

Wedge Type Brakes—
Removal and Installation

1. Exhaust all air from the system.
2. Disconnect the air inlet line from the brake chamber.
3. Using a drift and a light weight hammer, loosen the spanner nut.
4. Unscrew the brake chamber from the wedge housing.
5. Check the position of the wedge in the plunger housing to be sure that the wedge assembly is properly seated.
6. Replace the automatic adjusting identification ring on the brake chamber tube. Thread the spanner nut into the power unit tube.
7. Screw the service chamber into the plunger housing until it bottoms (spanner nut loose).
8. Align the connection ports with the brake lines, if necessary, unscrewing the brake chamber not more than one full turn.
9. Connect the air lines.
10. Make and hold a full pressure brake application. Drive the spanner nut with a drift and

hammer until it is tight against the plunger housing. Release the brake pressure.
11. Check for leaks at all connections.

Cam Type Brakes—
Removal and Installation

1. Exhaust all air from the system.
2. Disconnect the air line at the brake chamber.
3. Disconnect the push rod clevis pin from the slack adjuster.
4. Remove the attaching nuts, and remove the brake chamber assembly.
5. Check and if necessary, cut the push rod of the new chamber to the same length as on the removed brake chamber.
6. If the chamber is being replaced, transfer the brake line fitting to the new brake chamber.
7. Position the brake chamber assembly on the mounting bracket and install the attaching nuts.
8. Install the clevis and cotter pin.
9. Connect the air line to the brake chamber.
10. Build up pressure in the system and inspect for leaks.
11. Adjust the brakes.

Slack Adjuster-Cam Type Brakes

Removal and Installation

1. Remove the clevis pin (on vehicles with a rear wheel spring brake, the spring brake must be released to remove the clevis pin) attaching the slack adjuster to the brake chamber push rod.
2. Remove the lock ring attaching the slack adjuster to the camshaft.
3. Mark the position of the slack adjuster on the camshaft, then slide the slack adjuster off the shaft.
4. Place the slack adjuster on the camshaft, aligning the locating marks. If there is excessive camshaft and play, install an additional spacer at the cam end of the camshaft. Install the lock ring.
5. Connect the brake chamber push rod to the slack adjuster by installing the clevis pin in the upper hole, and installing the cotter pin. The distance from the clevis to the mounting face of the chamber should be 2.60 in.
6. Lubricate the slack adjuster. Build up pressure in the system, check operation of brake assembly and for air leaks. Adjust the brakes.

Air-Hydraulic Intensifier

Master Cylinder Removal and Installation

1. Clean the dirt from around the

discharge fitting and the inlet fitting of the master cylinder.
2. Disconnect the inlet (hydraulic reservoir) line from the master cylinder inlet port and plug both the line and port.
3. Disconnect the discharge line and plug the descharge port of the master cylinder. The open brake line should be plugged to prevent contaminats from entering the line or fluid loss.
4. Unscrew the four $\frac{3}{8}$ in. screws holding the master cylinder onto the rotochamber. Remove the master cylinder.
5. Bolt the new master cylinder onto the rotochamber. Tighten the cap screws.
6. Remove the plug from the brake line and connect the line to the discharge port on the master cylinder.
7. Remove the plug from the reservoir line and connect it to the master cylinder inlet port.
8. Clean the dirt from around the hydraulic reservoir lid. Unfasten the bail wire from the reservoir lid and remove the lid. Add brake fluid until it is level with the tops of the four vertical ribs in the reservoir.
9. Connect a bleeder bottle to the bleed screw on the master cylinder and allow the unit to gravity bleed. Close the bleed screw when the fluid is flowing freely with no air bubbles.
10. If pressure bleeding is performed on the system, the warning switch on the rotochamber may be tripped and cause the brake light on the instrument panel to come on. After pressure bleeding is complete, manually reset the switch by pushing in the plunger located in the middle of the switch.
11. Fill the fluid reservoir, install the lid, and fasten the bail wire in place.
12. Check the installation for leaks as follows:
    a. Make sure that system pressure is at least 90 psi.
    b. Make a full brake application, hold for 5-10 seconds, then release.
    c. Check for fluid leaks around the discharge line, and bleeder screw on the master cylinder.

Intensifier Unit

Removal and Installation

1. Disconnect and cap hydraulic lines. Plug the port in the master cylinder.
2. Discharge the secondary system reservoir and exhaust chamber.
3. Remove the air line and cap to prevent the entry of dirt.

4. Remove the bracket mounting bolts and remove the assembly from the vehicle.

5. Mount the intensifier unit in the vehicle with the bracket attaching bolts and tighten securely.

6. Attach air lines and hydraulic lines, tighten securely.

7. Apply air pressure and hold. Leak test air line connections and intensifier with soap suds.

8. Pressure bleed front brakes at caliper bleed screws.

9. Fill the master cylinder reservoir to $\frac{1}{2}$-$\frac{1}{4}$ inch below the top of the unit with specified brake fluid.

Parking Brakes

Adjustment

Orscheln Lever

The Orscheln parking brake is the over center locking type. It is adjusted (in the fully released position) by turning the lever knob. When properly adjusted, it pulls over center with a distinct click. No other adjustment is normally required.

Cable Actuated Rear Wheel Type

Adjust service brakes before attempting to adjust the parking brake cables. Place parking brake lever in fully released position, then check for slack in the parking brake two rear cables. Cables are properly adjusted when the rear brake shoes are fully applied when parking brake is applied, and the brake shoes fully release when the parking brake lever is released.

To tighten cables, loosen locknut and tighten adjusting nut on equalizer. Tighten locknut when proper adjustment is obtained.

External Band Type

1. On cable controlled parking brakes, move the parking brake lever to the fully released position. On a vehicle with a rod type linkage, set the lever at the first notch.

2. Check the position of the cam to make sure that the flat portion is resting on the brake band bracket. If the cam is not flat with the bracket, remove the clevis pin from the upper part of the cam and adjust the clevis rod to allow the flat portion of the cam to rest on the brake band bracket. Install the clevis pin and cotter pin.

3. Remove the lock wire from the anchor adjusting screw, and turn the adjusting screw clockwise until a clearance of 0.100" is obtained between the brake lining and the brake drum at the anchor bracket. Install the lock wire in the anchor adjusting screw.

4. Adjust the clearance on the upper and lower halves of the band in a similar manner. See illustration for location of adjusting screws. Adjust for a 0.010" clearance between band and drum.

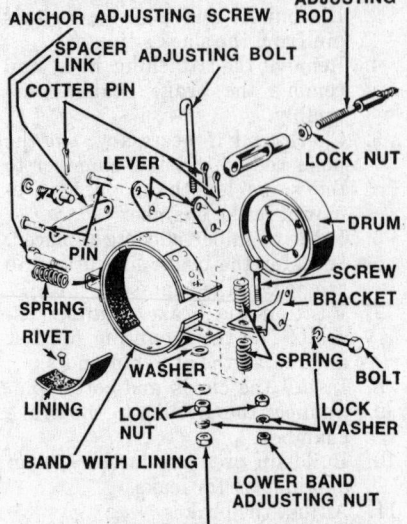

Exploded view of external band type parking brake (© Ford Motor Co.)

Internal Shoe Type 9" Drum

1. Release parking brake lever.

2. Remove cotter pin from the parking brake linkage adjusting clevis pin and remove the clevis pin.

3. Lengthen or shorten adjusting link by turning the clevis. There should be a 0.010" clearance between the drum and the band all the way around when the clevis pin is installed.

4. Install a new cotter pin in the clevis pin and check brake operation.

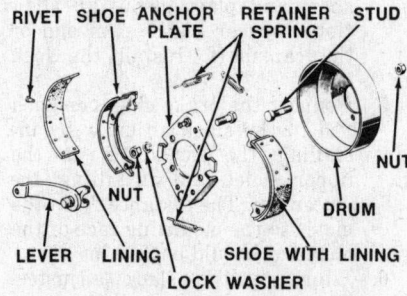

Exploded view of internal shoe type parking brake (© Ford Motor Co.)

12" Drum

There is no internal adjustment on this brake. Adjust the linkage as follows:

1. Remove clevis pin, loosen the nuts on the adjusting rod and turn clevis until a $\frac{1}{4}$"—$\frac{3}{8}$" free play is obtained at the brake lever with pin installed.

2. Tighten locknuts on adjusting rod and reinstall clevis pin.

Internal Shoe Type Parking Brake Drum and Shoe R & R

1. Remove the drive shaft. Disconnect the parking brake actuating lever from the linkage.

2. Remove the transmission spline flange and drum. Remove the bolts holding the carrier plate to the transmission housing. Slide the plate with the brake shoes and retaining springs off the transmission.

3. Remove the actuating lever, shoe retaining springs, and shoes.

4. Install the brake shoe lower retaining of the shoes.

5. Position the shoes and lower retaining spring on the back of the carrier plate and install the shoe upper retaining springs and the actuating lever. Place the assembly on the transmission with the lever properly positioned at the ball socket and shoe ends.

6. Install the brake mounting bolts and lockwashers.

7. Install the transmission drum, spline flange, nut, and cotter pin.

NOTE: If the drum mounting bolts are pressed into the companion flange the drum can be mounted during the following step.

8. Install the drive shaft and brake drum.

9. Connect the actuating lever to the parking brake linkage. Check the brake operation and adjust if necessary.

Transmission Mounted External Band Shoe R & R

1. Put the transmission in low gear and disconnect the driveshaft flange from the transmission.

2. Apply the parking brake and remove the nut attaching the transmission output shaft flange. Release the parking brake.

3. Disconnect the adjusting rod from the cam by removing the cotter and clevis pins.

4. Remove the cotter and clevis pins and remove the cam link from the cam.

5. Remove the lockwire and anchor adjusting screw.

6. Remove the brake band adjusting nuts and bolts.

7. Lift the brake band and lining from the drum.

8. The lining should be replaced if the lining is less than 1/32 in. off the top of the rivet.

9. Installation is the reverse of removal. Adjust the parking brake.

Fuel System

Note: For the 1978 and later Bronco, use F-150 4 x 4 procedures.

Reference

Information on application, and major adjustments can be found in the "Carburetor" sections of this manual.

Idle Mixture Adjustments

Idle Fuel Settings—Engine Off —1971-73

1. Set idle mixture screw(s) and limiter cap(s) to full counterclockwise position.
2. Back off idle speed adjusting screw until throttle plate(s) seat in throttle bore(s).
3. Be sure dashpot or solenoid (if so equipped) plunger is not interfering with the throttle lever. It may be necessary to loosen the dashpot to allow the throttle plate to seat in the throttle bore.
4. Turn the idle speed adjusting screw inward until it just makes contact with stop on throttle shaft and lever assembly, then turn screw inward 1½ turns to establish a preliminary idle speed adjustment.

Idle Fuel Settings—Engine Running —1971-73

1. Set parking brake, start engine and position idle screw on the intermediate step of the fast idle cam to obtain an engine speed of 1500 rpm. Let the engine run for at least 20 minutes so that engine and underhood temperatures are stabilized.
2. Check ignition timing and advance as described above in this section. These settings must be accurate.
3. On manual transmission model, idle setting is made with transmission in *neutral*. On automatic transmission models, set selector lever in DRIVE position and parking brake on, except as noted when using an exhaust gas analyzer.
4. Be sure choke plate is in full open position.
5. On carburetors equipped with a hot idle compensator or where the idle compensator is in the crankcase ventilation hose, be sure the compensator is seated to allow for proper idle adjustment.
6. Turn headlights on so that alternator is under load.
7. Turn air conditioner OFF for final idle speed adjustment.
8. Adjust curb idle rpm to specifications (see Specifications at the beginning of this section or in the Carburetor section of the General Repair Section). Use a tachometer and leave the air cleaner on. On Carter model YF IV with solenoid throttle modulator, turn solenoid plunger screw to obtain correct curb idle. Disconnect solenoid lead wire at bullet connector near the loom (*not* at solenoid), then adjust the throttle stop screw for 500 rpm. Reconnect solenoid lead wire and open throttle by hand: the throttle plunger should follow the throttle lever and remain in the fully extended position as long as the ignition is on and the solenoid energized. If it is impossible to adjust idle speed with the air cleaner on, remove it to adjust, then reinstall it and check speed, repeating the process until correct idle speed is obtained.
9. Turn idle mixture adjusting screw(s) inward to obtain the smoothest possible idle within range of limiter(s). On 2- and 4-barrel carburetors, turn idle mixture limiters and equal amount. Check for smoothness only with the air cleaner installed.

Additional Idle Speed and Mixture Procedures—1971-73

If a satisfactory idle is not obtained with the above normal procedures, make the following checks, making corrections if necessary:

1. Vacuum leaks, ignition wiring continuity, spark plugs, breaker point dwell angle, breaker point gap and initial ignition timing.
2. If a satisfactory idle condition is still not obtained with the checks and corrections of Step 1 above, check: fuel level, crankcase ventilation system, valve clearance and engine compression.
3. If the above procedures fail to produce a satisfactory idle condition, it may be due to a lean idle mixture. Check air-fuel mixture with an exhaust analyzer and adjust as described below.

Use of Exhaust Gas Analyzer —1971-73

1. Connect analyzer in accordance with manufacturer's instructions. All exhaust gas analyzers must be checked for calibration.
2. Observe reading with *air cleaner installed*. Refer to specifications below for correct air-fuel ratio.
3. Turn idle mixture adjusting screw(s) to obtain specified air-fuel ratio. On 2- and 4-barrel carburetors, turn the idle limiters equal amounts. Be sure to check idle speed frequently, correcting if necessary. Allow at least 10 seconds after each mixture adjustment for the analyzer to properly respond and stabilize.
4. If the air-fuel ratio is not to specifications, as shown by analyzer reading, it may be corrected by altering the controlled limits of the idle mixture system. See following paragraph for limiter cap replacement procedure.

Replacing Idle Limiter Caps —1971-73

1. Cut plastic cap with a knife or side-cutter pliers, then carefully pry limiter apart. On some carburetors it may be necessary to remove the carburetor to remove the limiters. On Holley 4-barrel carburetors, pry limiters out of metering block with a screwdriver.
2. After limiters are removed, set the carburetor to the correct fuel-ratio, using exhaust gas analyzer.
3. When air-fuel ratio is within specified value, install limiter caps. Install cap so that it is the maximum counterclockwise position with the tab of the limiter against the stop on the carburetor. Be careful not to turn the screw, installing cap with a straight, forward push.
4. Recheck air-fuel ratio with the analizer.

Idle Fuel Setting-Exhaust Gas Analyzer—1974-76

NOTE: The following operations are for vehicles without catalytic converters or with related partial catalyst equipped vehicles.

1. Operate the analyzer as recomended by the manufacturer's instructions.
2. Operate the engine for a minimum of 20 minutes at fast idle to normalize engine temperature.
3. Check and make sure that the engine timing and idle speed are as specified on the engine decal. All vacuum hoses must be connected.
4. If the vehicle is equipped with a dual spark delay valve, the DSDV must be disconnected and plugged.
5. Disconnect the evaporative emission purge line to the air cleaner. *NOTE: The air cleaner must be in position when the readings are taken for valid results.*
6. Disconnect the thermactor system air hose from the by-pass valve to the check valve at the check valve connection.
7. Place the transmission in neutral or park with the parking brake on. Start the engine and accelerate to 1500 rpm. Place your hand over the by-pass valve connec-

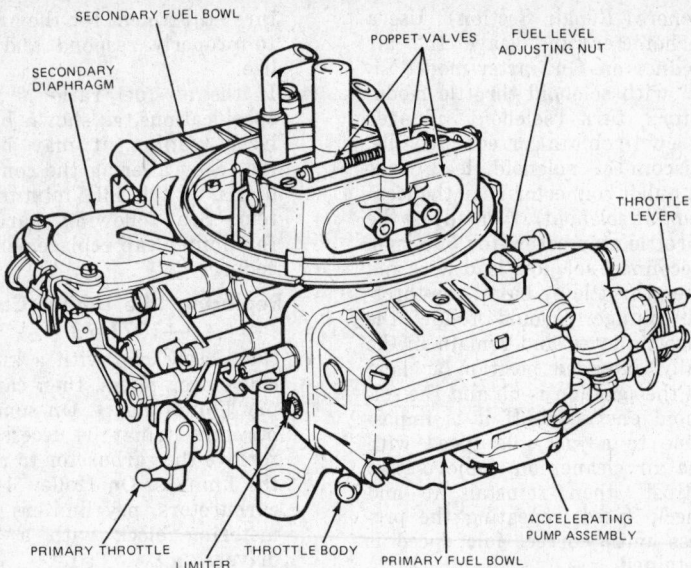

SECONDARY FUEL BOWL

SECONDARY DIAPHRAGM

POPPET VALVES

FUEL LEVEL ADJUSTING NUT

THROTTLE LEVER

PRIMARY THROTTLE LIMITER

THROTTLE BODY

PRIMARY FUEL BOWL

ACCELERATING PUMP ASSEMBLY

Holley 4 BBL—Adjustment locations (© Ford Motor Co.)

tion, air flow should be felt and heard.

8. With your hand held over the by-pass valve connection for 5 to 8 seconds, pinch off the vacuum hose at the by-pass valve to duplicate the air by-pass cycle. Release the pinched vacuum hose. Air flow should be felt and heard to diminish or stop for a short period of time then resume as before. The length of time required to resume normal flow cannot be specified since the time interval is dependant on the engine vacuum and length of time the vacuum line is pinched off. Air will be discharged through the exhaust ports in the side of the valve silencer cover. If this complete cycle does not occur, the valve must be replaced.

9. Insert the probe of the prepared "CO" analyzer into the tail pipe of a non-catalyst vehicle according to the manufacture's instructions.

10. With the brake pedal depressed, or the wheels blocked, increase the engine speed slightly and allow the throttle to return to the normal closed position. If the vehicle is equipped with an automatic transmission, place the selector in the drive position.

11. Observe the readings on the analyzer after allowing at least 10 seconds for the instrument to stabilize. *NOTE: All readings must be completed in 60 seconds to preclude incorrect readings caused by engine overheating.*

12. If the reading in step 11 is not within specifications, recheck the instrument according to the manufacture's instructions and adjust if required. Repeat steps 10 and 11, after stabi-

lization of the engine temperature.

13. If the readings are still not within specifications, proceed as follows:
    a. Remove the air cleaner.
    b. Adjust the idle mixture screws to provide the correct "CO" reading. If necessary, remove the limiter caps.
    c. If necessary, correct the idle speed immediately to specifications. The thermactor system must be connected each time the curb idle speed is adjusted.
    d. Install the air cleaner and tighten the nut.
    e. If necessary, re-stabilize the engine temperature by operating the vehicle at fast idle speed.

14. Repeat steps 10 thru 13 until the correct "CO" reading and idle speed has been obtained.

15. Install blue service limiter caps if caps were removed. The limiter caps should be installed so that they are against the stops and it is not possible to turn them counter clockwise.

16. Re-check the "CO" reading to be sure the reading has not changed during installation of the limiter caps.

17. Remove all test equipment and connect the evaporative emission hose and the thermactor air hose.

Idle Mixture Setting-Propane Enrichment Method—1975 and Later

NOTE: The following procedures are for vehicles without the Thermactor Air System. Remove the air cleaner when necessary to perform adjustments.

1. Bring the engine to normal operation temperature and connect a tachometer.

2. Disconnect the evaporative emission purge hose from the air cleaner. Disconnect the PCV closure hose from the air cleaner and plug the hose.

3. Adjust the curb idle speed to specifications on engine decal. *NOTE: With the transmission in neutral, run the engine at 2500 rpm for 15 seconds before each speed check. The idle speed must be adjusted with the air cleaner in place.*

4. Place the special gas tool into the air cleaner evaporative purge nipple. *With the engine idling, slowly open the propane valve until the engine speed reaches a maximum and then begins to drop. Note the maximum speed increase. If the speed will not drop, check the bottle gas supply. If necessary, repeat the operation with a new bottle gas supply.
    a. If the speed increase is within specifications, but not "O" rpm, proceed to step 5. If the speed increase is "O" and minimus specification is "O", proceed to step 4d.
    b. If the speed increase is higher than specification; †enrich the mixture without propane by turning the mixture limiter screws counterclockwise in equal amounts until the rpm increases as necessary. Example: If the increase was 80 rpm and the desired reset is 50 rpm, the mixture screws should be richened to attain a 30 rpm increase. Repeat steps 3 and 4.
    c. If the speed increase is lower than specifications proceed as follows; lean the mixture without propane by turning the mixture screws clockwise in equal amounts until the rpm decreases as necessary. Example: If the increase was "O" rpm and the desired reset increase is 20 rpm, the mixture screws should be leaned to attain a 20 rpm decrease. Repeat steps 3 and 4.
    d. If the speed increase is "O" rpm and the minimum speed gain specification is "O", perform the following speed drop test; Turn the mixture limiters counterclockwise to the maximum rich position. (If the limiters have been removed, do not enrich; assume the mixture screws are already set at the maximum rich position.) Lean the idle fuel mixture by turning the

screws clockwise equally as specified. Note the drop in engine rpm.

e. If the speed drop is equal to or greater than the specified minimum speed drop, return the mixture limiters to the maximum rich position or the mixture screws to the "assumed" maximum rich position. If the engine speed before mixture adjustment was 650 rpm and the speed drop specification is 100 rpm minimum, proceed to step 5 if the engine speed drops to at least 550 rpm or stalls.

f. If the speed drop is less than the specified minimum speed drop, leave the mixture limiters or screws in the adjusted position and repeat steps 3 and 4.

5. If the idle limiters were removed, install new blue service limiters at the maximum rich stop. Check the speed increase after installation of the limiters to be certain that the settings were not disturbed. If the setting is within specification, proceed to step 6, if not correct as required.

6. Remove the gas tool from the nipple and connect all system components that were removed.

7. Set the curb idle speed to specification if step 3 required an idle speed adjustment.

8. Turn off the engine and disconnect the tachometer.

* Every time propane is administered, place the transmission in the range specified on the engine decal.
† Remove the limiter caps with appropriate tool if required.

Idle Fuel Setting-Optimum Idle Method—1977 and Later

NOTE: This alternate method is to be used only when propane enrichment equipment is not available. The procedures are for vehicles without the Thermactor Air System. Remove the air cleaner when necessary to perform adjustments.

1. Bring the engine to normal operation temperature and connect a tachometer.

2. Disconnect the evaporative emission purge hose from the air cleaner.

3. Remove the idle mixture limiter

4. With the transmission in neutral, run the engine at 2500 rpm for 15 seconds.

5. Block the wheels or apply the brake. With the transmission in drive for automatic and neutral for manual, adjust the idle to curb idle RPM plus the optimum idle speed range RPM (if the specified optimum idle speed range is "O" rpm, simply adjust to the curb idle rpm).

6. With the transmission in drive for automatic and neutral for manual, adjust the idle mixture screws to the maximum idle rpm, leaving the screws in the leanest position that will maintain this "maximum idle rpm."

7. Repeat steps 4, 5, and 6 until further adjustment of the idle mixture screws does not increase the idle rpm.

8. If the specified optimum idle speed rpm is "O" proceed to step 10. Otherwise proceed to the next step.

9. With the transmission in drive for automatic and neutral for manual, turn the mixture screws equally in the lean direction until the curb idle rpm is obtained.

10. Install new blue service limiter caps at the maximum rich stops. Check the idle speed to insure that the limiter cap installation did not disturb the setting. Correct if necessary.

11. Turn off the engine, disconnect the tachometer, reinstall the system components, and make sure that the air cleaner attaching nut is tight.

Carburetor

Removal and Installation

1. Remove the air cleaner.

2. Remove the throttle cable or rod from the throttle lever. Disconnect the distributor vacuum line, EGR vacuum line, if so equipped, the inline fuel filter and the choke heat tube at the carburetor.

3. Disconnect the choke clean air tube from the air horn. Disconnect the choke actuating cable, if so equipped.

4. Remove the carburetor retaining nuts then remove the carburetor. Remove the carburetor mounting gasket, spacer (if so equipped), and the lower gasket from the intake manifold.

5. Before installing the carburetor, lean the gasket mounting surfaces of the spacer and carburetor. Place the spacer between two new gaskets and position the spacer and gaskets on the intake manifold. Position the carburetor on the spacer and gasket and secure it with the retaining nuts. To prevent leakage, distortion or damage to the carburetor body flange, snug the nuts, then alternately tighten each nut in a criss-cross pattern.

6. Connect the inline fuel filter, throttle cable, choke heat tube, distributor vacuum line, EGR vacuum line, and choke cable.

7. Connect the choke clean air line to the air horn.

8. Adjust the engine idle speed, the idle fuel mixture and anti-stall dashpot (if so equipped). Install the air cleaner.

Idle Speed Adjustment

1. Start the engine and run it until it reaches operating temperature.

2. If it hasn't already been done, check and adjust the ignition timing. After you have set the timing, turn off the engine.

3. Attach a tachometer to the engine.

4. Turn the headlights on to high beam.

5. On trucks with manual transmissions, engage the parking brake and place the transmission in Neutral; vehicles equipped with automatic transmission, engage the parking brake, and place the gear selector in Drive. Block the wheels.

6. Make sure that the choke plate is in the fully open position.

7. Adjust the engine curb idle rpm to the proper specifications. The tachometer reading must be taken with the carburetor air cleaner in place. If it is impossible to make the adjustment with

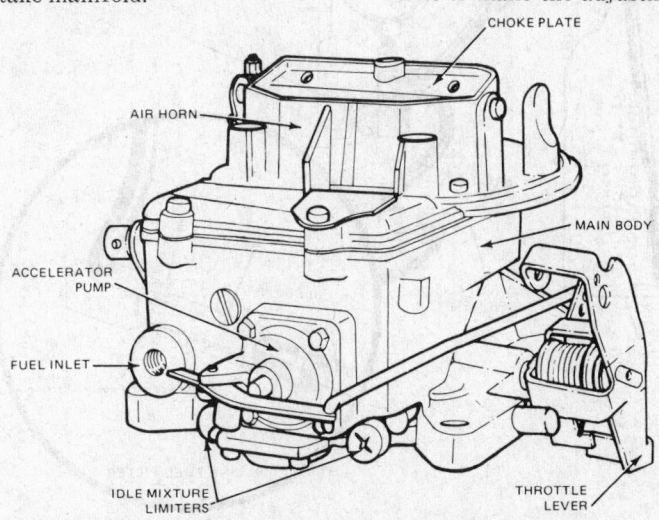

CHOKE PLATE

AIR HORN

MAIN BODY

ACCELERATOR PUMP

FUEL INLET

IDLE MIXTURE LIMITERS

THROTTLE LEVER

Motorcraft 2100 2BBL (Manual Choke)—Adjustment locations (© Ford Motor Co.)

163

the air cleaner in position, remove it and make the adjustment. Then replace the air cleaner and check the tachometer for the proper rpm reading.

On carburetors equipped with a solenoid throttle positioner, loosen the jam nut on the solenoid at the bracket and rotate the solenoid in or out to obtain the specified curb idle rpm. Disconnect the solenoid lead wire at the connector, set the automatic transmission in Neutral, then adjust the carburetor throttle stop screw to obtain 500 rpm. Connect the solenoid lead wire and open the throttle slightly by hand. The solenoid plunger will follow the throttle lever and remain in the fully extended position as long as the ignition is on and the solenoid energized.

Fuel Pump

All engines except the 330, 359, 361, 389, 391, 401, 475, 477 and 534 cu. in. are equipped with mechanical fuel pumps which are driven by an eccentric on the camshaft. On 6-cylinder engines it is located on the lower, left center of the block. On V-8 engines mechanical pumps are mounted on the left side of the cylinder front cover. Some fuel pumps have an integral filter. Other models have an in-line filter assembly. Filters are replaced (see owner's manual for specified replacement interval) by unscrewing the filter housing and replacing the old filter.

Testing Mechanical Fuel Pump

Install a new filter element before making the following tests. Tests are

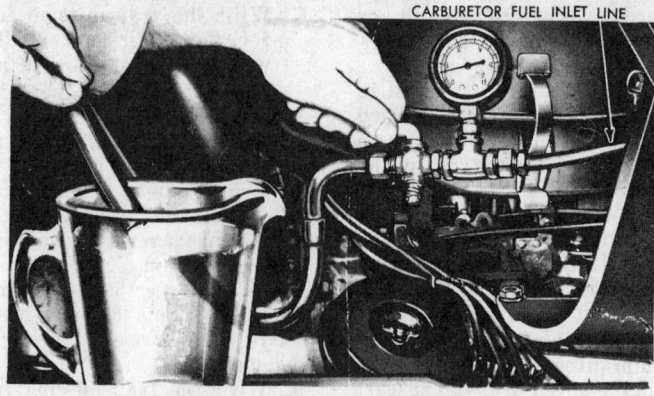

Testing electrical fuel pump (© Ford Motor Co.)

made with the fuel pump installed on the engine, with the engine warmed up to normal operating temperature, and with the engine idling at proper idle speed.

1. Remove air cleaner assembly and disconnect fuel inlet line at the carburetor. *Avoid fuel spillage due to risk of fire.*
2. Connect pressure gauge and fitting, flexible hose and restrictor clamp as illustrated.
3. Operate engine at idle speed. Momentarily open restrictor clamp to bleed system of air.
4. Close restrictor and note reading after pressure has stabilized. See Tune-Up Specifications at the beginning of this section for correct fuel pump pressure. If pressure is below specified value, fuel pump must be replaced or rebuilt.
5. If fuel pump is producing correct pressure, leave engine idling and proceed with volume test.
6. Open restrictor clamp and note time required to expell one pint

into the container. It should take no more than 30 seconds.
7. If it takes more than 30 seconds, check for restriction in the fuel line from the tank by connecting pump to an external fuel source. If the volume test still takes more than 30 seconds per pint, then the fuel pump must be replaced.

Testing Electric Fuel Pump

1. Remove air filter and disconnect the fuel inlet line at the carburetor.
2. Using 1/4" pipe fittings (smaller diameter will restrict the flow), connect the pressure gauge, gate valve and flexible hose as illustrated. Use a suitable container to collect expelled fuel.
3. Operate fuel pump with primer switch. *Be sure the battery is fully charged.* Adjust gate valve for a reading of 2 psi, then note time required to expell 1 quart of fuel.
4. If the time required to expell one quart exceeds that specified in the table below, repeat the same test at the outlet at the tank to establish whether or not there is restriction in the fuel line.

Engine	Time (Sec.)
330	40
359/361	37
389/391	34
401/475	36
477	33
534	30

5. If the fuel pump does not fill the quart container fast enough, the pump must be replaced as a unit.
6. Remove test equipment, connect fuel line and install air cleaner.

Mechanical Fuel Pump
Removal and Installation

1. Disconnect inlet and outlet lines from pump.
2. Remove pump mounting bolts and remove pump and gasket. Discard gasket.
3. To install, clean away all gasket material from mounting pad and

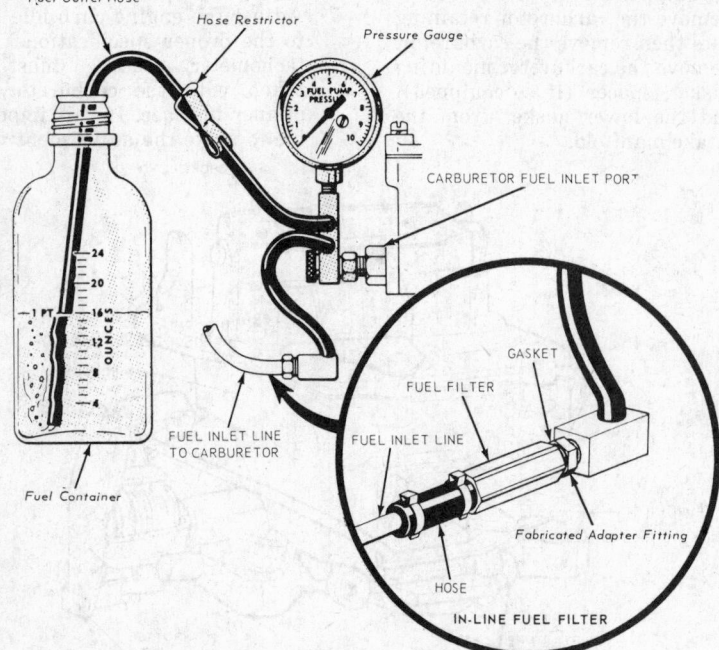

Testing mechanical fuel pump (© Ford Motor Co.)

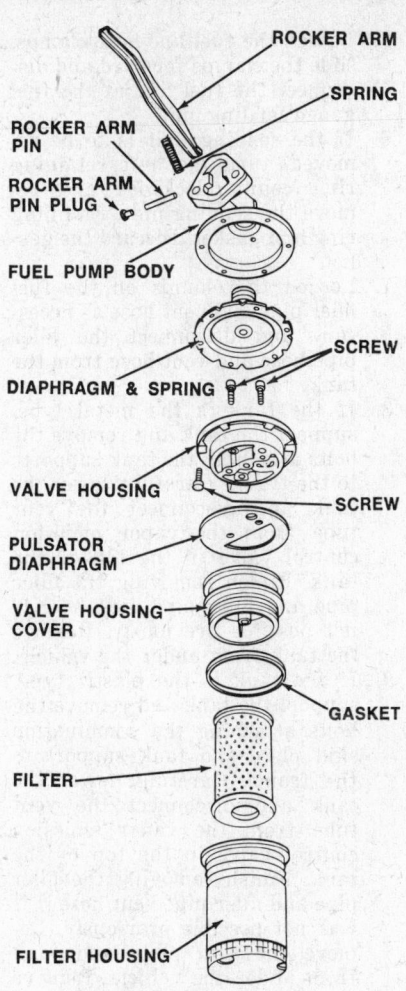

ROCKER ARM

SPRING

ROCKER ARM PIN

ROCKER ARM PIN PLUG

FUEL PUMP BODY

SCREW

DIAPHRAGM & SPRING

VALVE HOUSING

SCREW

PULSATOR DIAPHRAGM

VALVE HOUSING COVER

GASKET

FILTER

FILTER HOUSING

Typical mechanical fuel pump
. (© Ford Motor Co.)

pump flange. Apply sealant to new gasket and threads of bolts.

4. Position pump and gasket on the mounting pad, being sure the rocker arm is riding on the cam eccentric. Turn engine over until eccentric is on low side of stroke.
5. Install mounting bolts and tighten securely.
6. Connect fuel lines.
7. Operate engine and check for leaks.

Electric Fuel Pump
Removal and Installation
1. Disconnect the battery.
2. Disconnect the fuel outlet line at the pump cover.

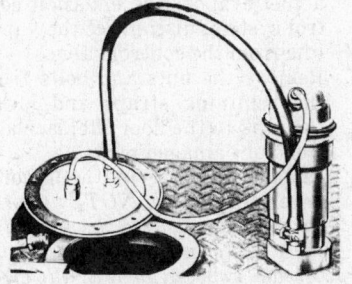

Electric fuel pump installation
(© Ford Motor Co.)

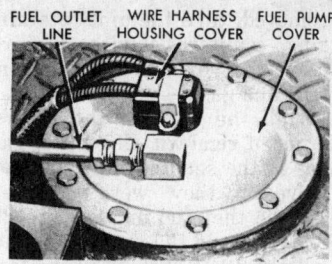

FUEL OUTLET LINE   WIRE HARNESS HOUSING COVER   FUEL PUMP COVER

Electric fuel pump wiring
(© Ford Motor Co.)

3. Remove wire harness housing cover from the fuel pump cover and disconnect the wires.
4. Remove fuel pump cover and reach down into the tank and disconnect the toggle clamp which secures the pump in the retaining bracket and remove the fuel pump.
5. Disconnect the fuel outlet tube and the wire assembly at the cover and at the pump.
6. To install, remove the old gasket from the cover and install a new gasket.
7. Connect the fuel outlet tube and the wire assembly to the pump cover and to the pump.
8. Install fuel pump in the bracket in the tank and fasten the toggle clamp.
9. Position the pump cover assembly on the tank and install the attaching bolts and wire clamps.
10. Connect the fuel line to the fitting on the cover.
11. Position the wiring harness in its housing on the cover and install the wire harness housing cover.
12. Connect the battery and operate the pump and truck engine to check for leaks.

Fuel Tank

Removal and Installation
Bronco
The following procedure can be used to remove either the main or auxiliary fuel tanks.
1. Insert a siphon through the filler neck and drain the fuel into a suitable container.
2. Raise the rear of the vehicle. If you are removing the auxiliary tank, raise the left side of the vehicle.
3. To avoid any chance of sparking at or near the tank(s), disconnect the ground cable from the vehicle battery. Disconnect the fuel gauge sending unit wire at the fuel tank.
4. Loosen the clamp on the fuel filler pipe hose at the filler pipe and disconnect the hose from the pipe.
5. Loosen the hose clamps, slide the clamps forward and disconnect

the fuel line at the fuel gauge sending unit.
6. If the fuel gauge sending unit is to be removed, turn the unit retaining ring, and gasket, and remove the unit from the tank.
7. Remove the strap attaching nut at each tank mounting strap, swing the strap down, and lower the tank enough to gain access to the tank vent hose.
8. Disconnect the fuel tank vent hose at the top of the tank. Disconnect the fuel tank-to-separator tank lines at the fuel tank.
9. Lower the fuel tank and remove it from under the vehicle.
To install the fuel tank:
10. Position the forward edge of the tank to the frame crossmember, and connect the vent hose to the top of the tank. Connect the fuel tank-to-separator tank lines at the fuel tank.
11. Position the tank and mounting straps, and install the attaching nuts and flat washers.
12. If the fuel gauge sending unit was removed, make sure that all of the old gasket material has been removed from the unit mounting surface on the fuel tank. Using a new gasket, position the fuel gauge sending unit to the fuel tank and secure it with the retaining ring.
13. Connect the fuel line at the fuel gauge sending unit and tighten the hose clamps securely. Install the drain plug. if so equipped.
14. Connect the fuel gauge sending unit wire to the sending unit.
15. Connect the filler pipe-to-tank hose at the filler pipe and install the hose clamp.
16. Connect the vehicle battery ground cable.
17. Fill the tank and check all connections for leaks.
18. Lower the vehicle.

Econoline-Frame and Body Mounted Tanks
1. Drain the tank with a siphon inserted through the fuel filler pipe.
2. Raise and secure the vehicle.
3. Disconnect the battery negative cable to avoid sparking at the tank. Disconnect the fuel gauge sending unit wire at the tank.
4. Loosen the clamps on the filler and vent hoses. Disconnect the hoses at the tank.
5. Disconnect the fuel line hose from the fuel gauge sending unit.
6. a. On body mounted tanks: Support the tank in position. Remove the nuts the mounting straps to the J-bolts. *NOTE: the J-bolts are attached to body brackets located at the rear of the tank.*

Disengage the straps from the J-bolts and the front body brackets. Lower the tank enough to gain access to the vapor valve.

b. On frame mounted tanks: Support the tank in position. Remove the nuts and bolts that attach the tank supports to the frame. Disengage the straps from the front frame support and the rear crossmember. Lower the tank enough to gain access to the vapor valve.

7. Disconnect the vapor hose from the vapor control valve.

8. Lower the fuel tank and remove it from the vehicle.

9. If the fuel gauge sending unit is to be removed from the tank, turn the unit retaining ring counterclockwise and remove the sending unit, retaining ring, and gasket.

10. If the vapor control valve is to be removed, pull it out of the grommet located in the top of the tank, and remove the grommet.

11. Install the tank in the reverse order of removal. Use a new sealing gasket at the sending unit if the unit was removed. Install a new grommet for the vapor control valve if the grommet was removed. Torque the body mounted J-bolt attaching nuts to 6-8½ ft lbs. and the frame mounted tank attaching bolts to 25-34 ft lbs. Fill the tank and check for leaks.

Econoline-Midship Tank

1. Insert a siphon through the fuel filler pipe and drain the fuel.

2. Raise and secure the vehicle.

3. Disconnect the battery negative cable to avoid sparking at the tank.

4. Disconnect the fuel gauge sending unit wire at the fuel tank.

5. Support the tank in position and remove the restrictor brace from the front of the tank. Disengage the mounting strap ends attached to the frame side rail. Remove the other end from the tank support by rotating the strap to disengage the L-shaped hook end.

6. Lower the tank enough to gain access to the vapor valve, fuel filler hose, fuel vent hose, and fuel line hose. Loosen the attaching clamps and disconnect the hoses.

7. Lower the tank and remove it from the vehicle.

8. If the fuel gauge sending unit is to be removed, turn the unit retaining ring counterclockwise and remove the sending unit, retaining ring and gasket.

9. If the vapor control valve is to be

removed, turn the unit retaining ring counterclockwise and remove the vapor valve, retaining ring and gasket.

10. Install the tank in the reverse order of removal. Use a new gasket at the sending unit or vapor valve if they were removed. Torque the stud end nuts to 26-31 ft lbs. and the restrictor brace bolts to 12-18 ft lbs. Fill the tank and check for leaks.

F-100, 150, 250, 350
In-Cab Fuel Tank

1. Siphon the fuel from the tank into a suitable container through the filler neck.

2. Move the seat to the full forward position and tilt the seat forward.

3. Disconnect the fuel gauge sending unit wire and fuel line from the tank. Disconnect the vapor vent chamber and vapor lines of the fuel evaporative emission control system.

4. Loosen the filler neck hose clamp at the tank end of the hose, and pull the filler neck away from the tank.

5. Remove the fuel tank retaining nuts and bolts and lift the tank out of the cab. If the tank is being replaced, remove the fuel gauge sending unit and install it in the new tank.

6. Install the fuel tank in the reverse order of removal.

F100, 150, 250, 350
In-Frame Fuel Tank

1. Drain the fuel from the tank into a suitable container by either removing the drain plug, if so equipped, or siphoning through the filler cap opening.

2. Disconnect the fuel gauge sending unit wire and fuel outlet line.

3. Disconnect the air relief tube from the filler neck and fuel tank.

4. Loosen the filler neck hose clamp at the fuel tank and pull the filler neck away from the tank.

5. Remove the retaining strap mounting nuts and bolts and lower the tank to the floor.

6. If a new tank is being installed, change over the fuel gauge sending unit to the new tank.

7. Install the fuel tank in the reverse order of removal.

F100, 150, 250, 350
Behind-the-Axle Fuel Tank

1. Raise the rear of the truck.

2. Disconnect the negative battery cable.

3. Disconnect the fuel gauge sending unit wire at the fuel tank.

4. Remove the fuel drain plug or siphon the fuel from the tank into a suitable container.

5. Loosen the fuel line hose clamps, slide the clamps forward and disconnect the fuel line at the fuel gauge sending unit.

6. If the sending unit is to be removed, turn the unit retaining ring counterclockwise and remove the sending unit, retaining ring and gasket. Discard the gasket.

7. Loosen the clamps on the fuel filler pipe and vent hose as necessary and disconnect the filler pipe hose and vent hose from the tank.

8. If the tank is the metal type, support the tank and remove the bolts attaching the tank supports to the frame. Carefully lower the tank and disconnect the vent tube from the vapor emission control valve in the top of the tank. Finish removing the filler pipe and filler pipe vent hose if not possible previously. Remove the tank from under the vehicle.

9. If the tank is the plastic type, support the tank and remove the bolts attaching the combination skid plate and tank support to the frame. Carefully lower the tank and disconnect the vent tube from the vapor emission control valve in the top of the tank. Finish removing the filler pipe and filler pipe vent hose if it was not possible previously. Remove the skid plate and tank from under the vehicle. Remove the skid plate from the tank.

10. Install the tank in the reverse order of removal.

F100, 350 and Bronco
Auxiliary Fuel Tank

NOTE: The Bronco auxiliary fuel tank is made of a high density plastic, and is not repairable. If this tank leaks, it must be replaced.

1. Insert a siphon through the filler neck and drain the fuel.

2. Disconnect the negative battery cable to avoid sparking when the tank is removed. Remove the skid plate if equipped.

3. Disconnect the fuel gauge sending unit wire from the unit.

4. Remove the clamps and disconnect the hoses connected to the tank. On vehicles equipped with a fuel evaporative emission control system, disconnect the vapor line from the control valve.

5. Remove the nuts and bolts from the retaining straps and lower the tank to the floor. Replace any worn or damaged parts.

6. Install the tank in the reverse order of removal. NOTE: On the Bronco, tightening the nuts for the J-bolts until there is 1¼ of thread exposed below the nut. This procedure applies to vehicles with and without a skid

plate. Fill the tank and check for leaks.

Medium & Heavy Duty Trucks Outside Frame Fuel Tanks

1. Drain the fuel by either removing the drain plug (if equipped), or by siphoning through the filler cap opening.
2. Disconnect the negative battery cable to avoid sparking during removal.
3. Disconnect the fuel gauge sending unit wire and fuel lines.
4. Disconnect the electric fuel pump wires if so equipped.
5. Disconnect the fuel return lines if so equipped.
6. In vehicles equipped with dual fuel tanks, disconnect the line connecting the tanks.
7. If the vehicle is equipped with a filler neck and connecting hose, loosen the hose clamp at the tank and disconnect the hose.
8. Support the tank in position and remove the tank retaining strap bolts and straps. Remove the fuel tank. Replace any worn or damaged parts.
9. If the fuel tank is to be replaced, remove the fuel gauge sending unit, shut off valve (if so equipped), and electric fuel pump (if equipped).
10. Install the tank in the reverse order of removal. Install new gaskets for fuel sending unit, shut off valve, and electric fuel pump. Make sure that the tank is properly aligned and that the insulators are properly centered under the straps and brackets. Fill the tank and check for leaks.

Medium & Heavy Duty Trucks Saddle Tanks

1. Drain both sides of the tank.
2. Disconnect the wires from the fuel gauge sending units, and the electric fuel pump (if equipped).

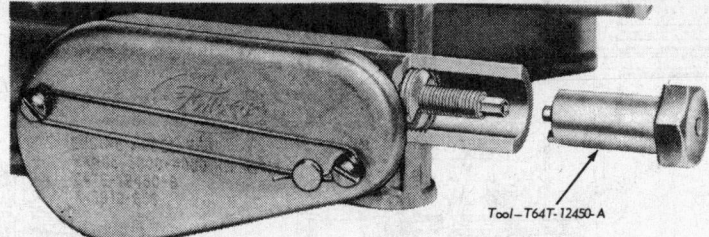

Altitude compensation adjustment-Velocity type governor (© Ford Motor Co.)

Disconnect the fuel lines.
3. Remove the nuts and bolts retaining the air line and hose bracket to the forward edge of the tank.
4. Remove the bolt, nut, flat washer and spring from each fuel tank mounting bracket.
5. Hook a lifting chain to the four corners of the fuel tank, near the mounting brackets and remove the tank with a hoist.
6. If a replacement is being installed, transfer all equipment to the new tank. Install new gaskets on transferred parts.
7. Install the tank in the reverse order of removal. Replace any damaged insulators. Fill the tank and check for leaks.

Governors

Adjusting Velocity Governors

1. Connect a tachometer to the engine, warm up the engine then read engine rpm at wide-open throttle. If governed speed is not within the range stamped on the governor plate, adjustment is required.
2. Remove the governor seal.
3. To increase rpm, turn the cap counterclockwise; to decrease the rpm turn it clockwise.
4. If the truck is to be operated at a consistent altitude, cut the seal wire and remove the adjusting

cap. *Do not rotate the cap during removal.* Use a mirror and light to observe the position of the slots in the adjusting bushing. *Do not disturb the center post or adjusting bushing*—if the tool does not engage the slots easily, remove the tool and realign it. For an increase in the average altitude of operation rotate the inserted tool the amount specified in the table below in the counterclockwise direction.

Aver. Operating Altitude—Feet	Amount of Tool Rotation
2000	1/3 turn (120°)
3000	1/2 turn (180°)
4000	2/3 turn (240°)
5000	5/6 turn (300°)
6000	1 turn (360°)

60° or 1/6 turn rotation is equivalent to one flat of the tool hex head.

5. Remove tool and install cap, but *do not turn the adjusting cap.*
6. Install a tachometer and check and adjust the no-load setting. It should be 3900 for no-load at altitude and 3600 for load at altitude. Load and no-load speed should be slightly above these speeds if the governor is being adjusted above anticipated operating altitude and slightly below if it is being adjusted below anticipated operating altitude.
7. If load rpm is below 3600 rpm at operating altitude, repeat Step 4 turning the tool counterclockwise. If load governed speed is above 3600 rpm, repeat Step 4 turning the tool clockwise.
8. Seal adjusting cap to the governor body using service governor seal wire.
9. If the engine is to be operated at varying altitudes, adjust the governor for 3800 rpm no-load for sea-level. Using the adjusting cap only, adjust the no-load speed for 4100 rpm at the anticipated altitude by turning the adjustment cap 1/4-turn (clockwise) for each 1000' difference between the adjusting and anticipated altitudes. If the maximum operating altitude of the truck is lower than the altitude at which the adjustment is being made, adjust the no-load speed to 4100 rpm with the adjusting cap.

MAINIFOLD VACUUM PASSAGE TO CARBURETOR POWER VALVE

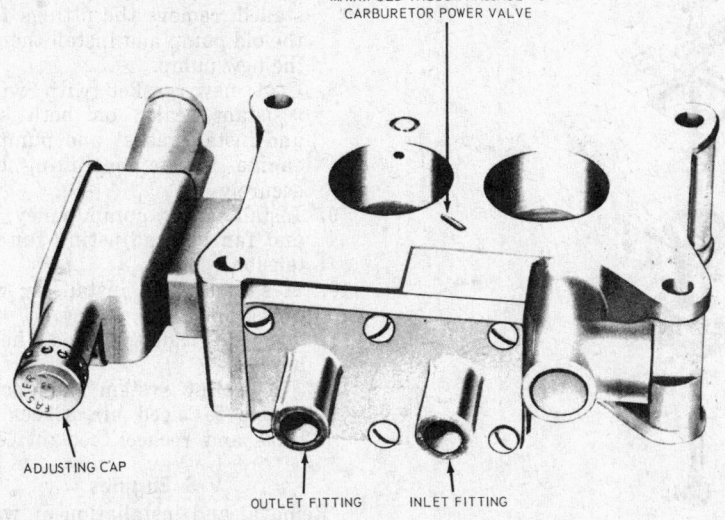

ADJUSTING CAP

OUTLET FITTING   INLET FITTING

Coolant heated velocity governor (© Ford Motor Co.)

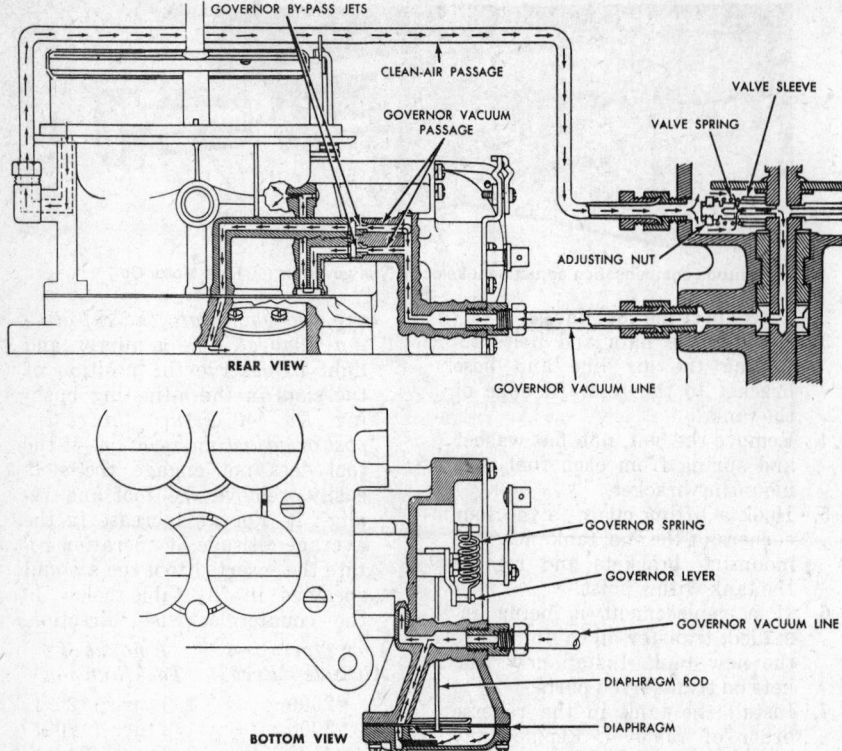

REAR VIEW

GOVERNOR BY-PASS JETS

CLEAN-AIR PASSAGE

GOVERNOR VACUUM PASSAGE

VALVE SLEEVE

VALVE SPRING

ADJUSTING NUT

GOVERNOR VACUUM LINE

GOVERNOR SPRING

GOVERNOR LEVER

GOVERNOR VACUUM LINE

DIAPHRAGM ROD

DIAPHRAGM

BOTTOM VIEW

Vacuum governor system (© Ford Motor Co.)

Adjusting Vacuum Governor

1. Warm up the engine until normal operating temperature is reached, then connect a tachometer.
2. Momentarily operate engine at wide-open throttle (governed speed) and note rpm reading.
3. If governed speed is not at the correct value (see Tune-Up Specifications at the beginning of this section), turn off the ignition switch and remove adjusting hole cover from the controlling unit (distributor).
4. Crank engine by hand until adjusting nut is aligned with access hole.
5. Turn adjusting nut clockwise to increase speed and counterclockwise to decrease speed. One full turn equals about 150 rpm.
6. Repeat above procedure until proper governed speed is obtained.
7. Install adjusting access hole cover and tighten securely.
8. Install new locking wire and lead seal.

Adjusting Mechanical Governor

1. Disconnect the throttle control rod at the carburetor.
2. Loosen the top nut on the primary spring adjusting eye bolt.
3. Tighten the bottom nut finger tight, then turn it in two additional turns to pre-load the spring. Tighten the top nut.

4. Move the throttle to the wide-open position and connect the governor throttle control rod to the carburetor control arm.
5. Adjust the governor throttle control rod so that the governor throttle control auxiliary lever is full forward, then back off

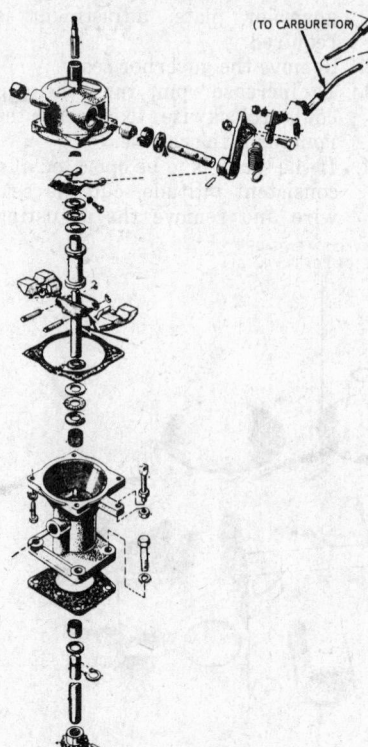

(TO CARBURETOR)

Mechanical governor (© Ford Motor Co.)

(shorten) the rod one full turn.

6. Check the throttle linkage to be sure that the throttle is wide-open when pedal is depressed to the floor. Be sure the rod or cable is attached to the proper hole in the throttle lever.
7. Check operation of choke plate for proper adjustment. On C- AND W-Series trucks the choke plate does not completely close when the dash knob is fully out.
8. To adjust speed, operate the engine (parking brake on) until normal operating temperature is reached. With throttle wide-open, adjust main spring (higher tension increases rpm and lower tension decreases rpm). Sensitivity of the governor can be sharpened by installing the governor spring in the hole closest to the lower arm pivot. Adjust governed speed after changing spring position.

Cooling System

Water Pump

Removal and Installation

6-Cylinder Engines

1. Drain cooling system.
2. Disconnect radiator lower hose and heater hose at the water pump.
3. Remove fan belt, fan and water pump pulley.
4. On trucks equipped with air compressors, remove the air compressor belt.
5. Remove water pump retaining bolts, then remove pump and gasket.
6. To install, clean gasket surfaces of pump body and engine block.
7. If a new water pump is being installed, remove the fittings from the old pump and install them on the new pump.
8. Coat new gasket with water-resistant sealer on both sides and install gasket and pump on engine. Tighten mounting bolts securely.
9. Install water pump pulley, fan and fan belt, adjusting fan belt tension.
10. If so equipped, install air compressor belt.
11. Connect radiator and heater hoses.
12. Fill cooling system and operate engine to bleed air. Check for leaks and recheck coolant level.

V-8 Engines

Removal and installation of water pumps on V-8 engines are essentially the same as described for 6-cylinder engines. On V-8 equipped Econolines, the radiator must be removed in order to replace the water pump. On late model F-100 and Bronco vehicles, there may be a radiator shroud which must be unbolted and hung on the fan to permit access to the lower radiator hose. Remove lower radiator hose, then shroud.

Heater Core

Heater Core Without Air Conditioning—Removal & Installation

F-100-350 1971-72

F-500-880 1971-78

1. Drain the cooling system and disconnect both heater hoses at the heater.
2. Remove the three nuts and washers that hold the heater assembly to the dash panel.
3. Remove the glove box.
4. Remove the screws that hold the air inlet duct to the cowl, pull the heater assembly from the dash panel, and disconnect both defroster nozzles from the heater.
5. Disconnect the wire lead to the blower motor and resistor.
6. Disconnect the wire and control cables from the heater, and remove the assembly to a bench.
7. Remove the core cover stop bracket. Remove the retaining clip and the heater core from the case.
8. Transfer the pad to the new core (no glue) and position the core in the heater case. Install the clip and heater core cover.
9. Position the heater assembly on the dash and cowl, and start one heater to dash retaining nut.
10. Connect the blower motor and resistor wire lead. Install the

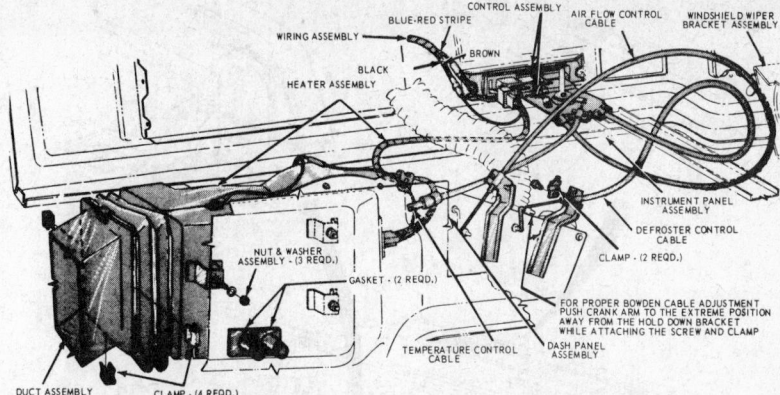

Bronco heater installation (© Ford Motor Co.)

duct to side cowl retaining screws.
11. Connect the control cables and the defroster nozzles to the heater assembly.
12. Install the remaining heater to dash nuts at the engine side of the dash panel and tighten all the nuts.
13. Position the pads on the core tubes, connect the heater hoses, and fill the cooling system.

F-100-350 1973-78

1. Disconnect the temperature and air door Bowden cables from the heater housing. *This must be done to prevent damage to the cables.*
2. Disconnect the wires from the blower resistor and the blower motor.
3. Remove the screws attaching the air inlet (vent) duct to the heater housing.
4. Disconnect the blower motor wires.
5. Drain the radiator and remove the heater hoses from the heater core.
6. Remove the heater stud retaining nuts and remove the heater assembly.

7. Remove the gasket between the heater hose ends and the dash panel at the core tubes.
8. Remove the heater core cover and gasket.
9. Pull the heater core and lower support out of the heater assembly.
10. Place the foam gaskets on the heater core and install the core in the heater assembly.
11. Install the core seal and cover plate.
12. Position the heater assembly in the vehicle and install the stud retaining nuts.
13. Connect the heater hoses to the heater core and fill the radiator.
14. Connect the blower motor wires.
15. Place the defroster nozzle on the heater assembly so that the defroster and heater openings are in the up position and there is no air leak around the seal.
16. Install the air inlet (vent) duct on the heater assembly. Push the duct firmly against the seal on the side cowl and tighten the attaching screws.
17. Connect the wires to the blower motor resistor and blower motor.
18. Connect the temperature and air door cables to the heater, and adjust the cables.
19. Install the gasket between the heater hose ends and the dash panel at the core ends.
20. Check the heater operation.

Bronco

1. Drain the cooling system.
2. Disconnect the heater hoses at the heater assembly.
3. Remove the nuts and star washers holding the heater assembly to the dash panel.
4. Disconnect the right and left defroster hoses at the plenum.
5. Disconnect the fresh air inlet at the cowl. Rest the heater assembly on the floor.
6. Disconnect the heat/defrost door cable at the door crank arm.
7. Disconnect the outside air door cable at the crank arm.
8. Disconnect the electrical wires at the connector.

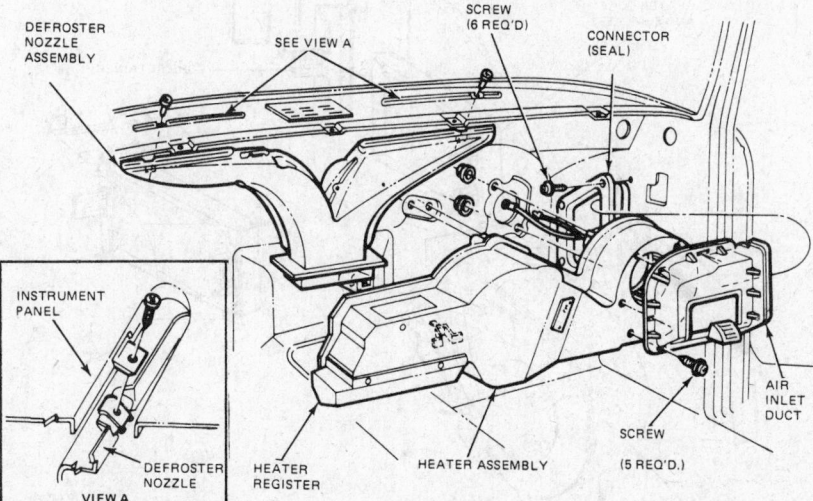

1973-78 F100-350 heater installation (© Ford Motor Co.)

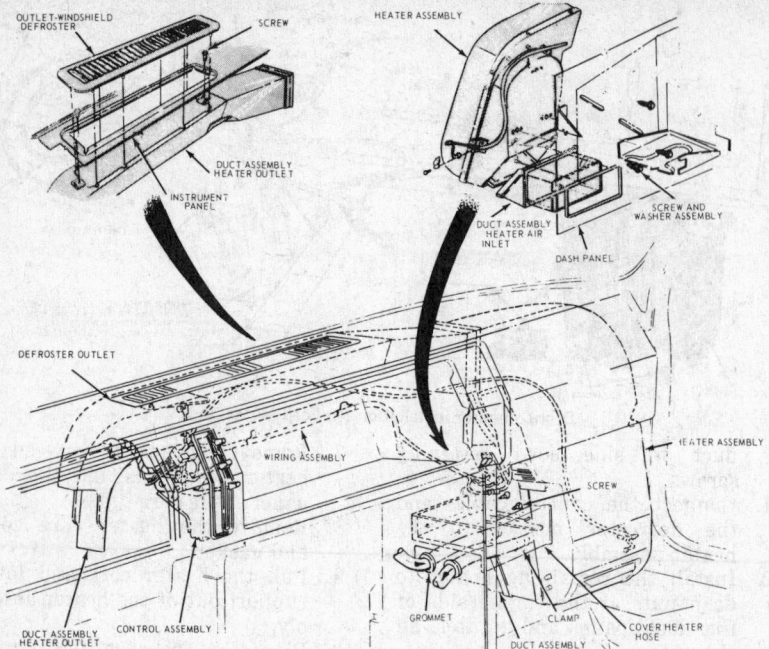

1971-74 Econoline heater installation (© Ford Motor Co.)

line up with the defroster and fresh air openings.

12. Connect the resistor and blower motor leads.
13. Connect the heater hoses, and fill the cooling system.
14. Install the battery. Check the operation of the heater.

Econoline 1975-78

1. Drain the cooling system and remove the battery.
2. Disconnect the resistor wire and the blower motor wire at the harness.
3. Remove the ground wire screw from the dash panel.
4. Disconnect the heater hoses from the core tubes. Remove the plastic strap retaining the heater hoses to the heater assembly.
5. Remove the five heater mounting screws inside of the vehicle.
6. Remove the heater assembly.
7. Cut the seal at the top and bottom edge of the heater core retainer, and remove the retainer.
8. Slide the core and seal assembly out of the heater case.
9. Insert the replacement heater core and seal assembly into the heater case, and install the attaching screws.
10. Install the heater assembly with the mounting screw from inside the vehicle.
11. Connect the heater hoses to the heater core and install the plastic strap.
12. Install the ground wire to the dash panel.
13. Connect the resistor and blower motor leads.
14. Install the battery.
15. Fill the cooling system and check the heater operation.

9. Remove the heater assembly from the vehicle.
10. Remove the screws holding the rear cover. Remove the clip retaining the cover. Remove the clip holding the heater core in the case, and remove the core.
11. Transfer the seals from the old heater core to the replacement core.
12. Position the heater core in the case. Install the retaining clip. Position the cover and install the screws.
13. Position the heater assembly on the floor of the vehicle. Connect the electrical wire connector.
14. Connect and adjust the heat/defrost door cable and the outside air door control cable.
15. Position the heater assembly on the dash panel and install the nuts and washers.
16. Place the heater core pads over the door and connect the hoses.
17. Connect the defroster hoses and the fresh air intake.
18. Fill the cooling system. Check the heater operation.

Econoline 1971-74

1. Remove the battery.
2. Drain the cooling system.
3. Disconnect the heater hoses at the heater.
4. Disconnect the heater resistor and motor leads.
5. From under the hood, remove the heater to dash mounting bolts. Move the heater out of position to gain access to the control cable and disconnect the cable.
6. Separate the two halves of the heater case.
7. Remove the heater core.

8. Transfer the core pads to the replacement heater core and position the core in the case.
9. Position both halves of the heater case together and install the screws and clip.
10. Place the heater controls in the OFF position. Place the heater on the wheel housing as near as possible to the installed position. Pull the air door closed (toward the rear of the vehicle) and connect the control cable.
11. Position the heater on the dash, and install the mounting bolts. Use an assistant and make certain that the housing openings

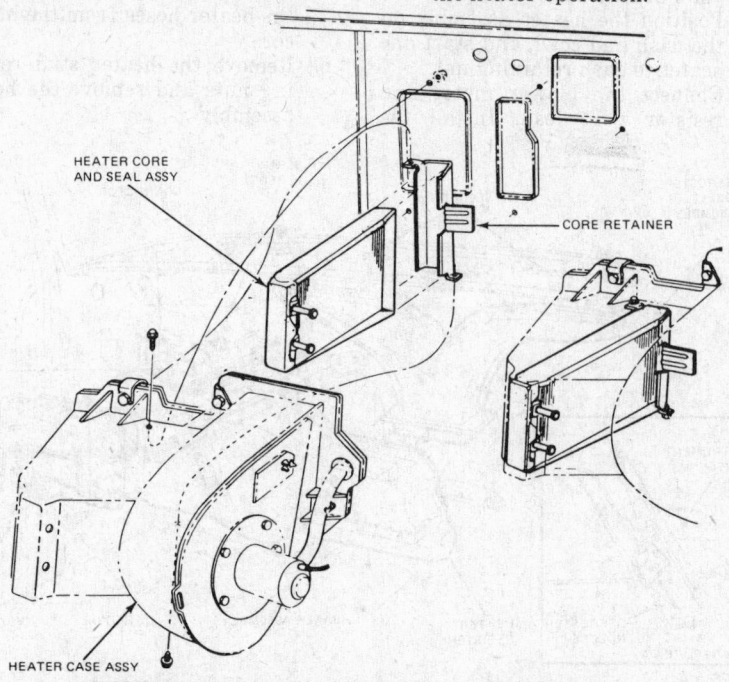

1975-78 Econoline heater core installation (© Ford Motor Co.)

C-Series
1. Drain the cooling system.
2. Disconnect the defroster hoses at the plenum chamber.
3. Disconnect the blower motor wires at the connectors. Remove the glove compartment liner to gain access to the heater wire connectors.
4. Disconnect the heater hoses at the heater.
5. Remove the heater lower mounting bolts, nuts and lockwashers.
6. Remove the upper mounting screws and move the heater assembly away from the dash. Disconnect the Bowden cable from the air inlet door and remove the heater assembly from the vehicle.
7. Remove the heater assembly top cover.
8. Remove the heater core mounting screws and remove the core from the housing.
9. Install the heater core assembly in the reverse order of removal.

L-Series
1. Drain the cooling system and disconnect the heater hoses at the heater core.
2. Disconnect the control cable at the top of the heater assembly.
3. Disconnect the motor wire leads, and remove the heater front cover.
4. Pull the heater core out of the heater assembly. Transfer the core pads to the replacement core and push the replacement core into the heater assembly.
5. Route the motor leads through the front of the cover assembly, position the cover on the heater assembly, and install the retaining screws.
6. Connect the blower motor leads.
7. Connect the control cable at the top of the heater assembly and adjust the cable.
8. Connect the heater hoses and fill the cooling system.
9. Check the heater operation.

W-Series
1. Drain the cooling system and unlock and tilt the cab.
2. Disconnect both heater hoses at the core and the control valve tubes.
3. Lower the cab.
4. Remove the top cover from the heater assembly, and the register from the side of the heater assembly.
5. Remove the cable retaining clip and disconnect the cable at the fresh air door.
6. Remove the cable from the fresh air door.
7. Remove the seal from the top of the blower housings.
8. Remove the clip holding the de-

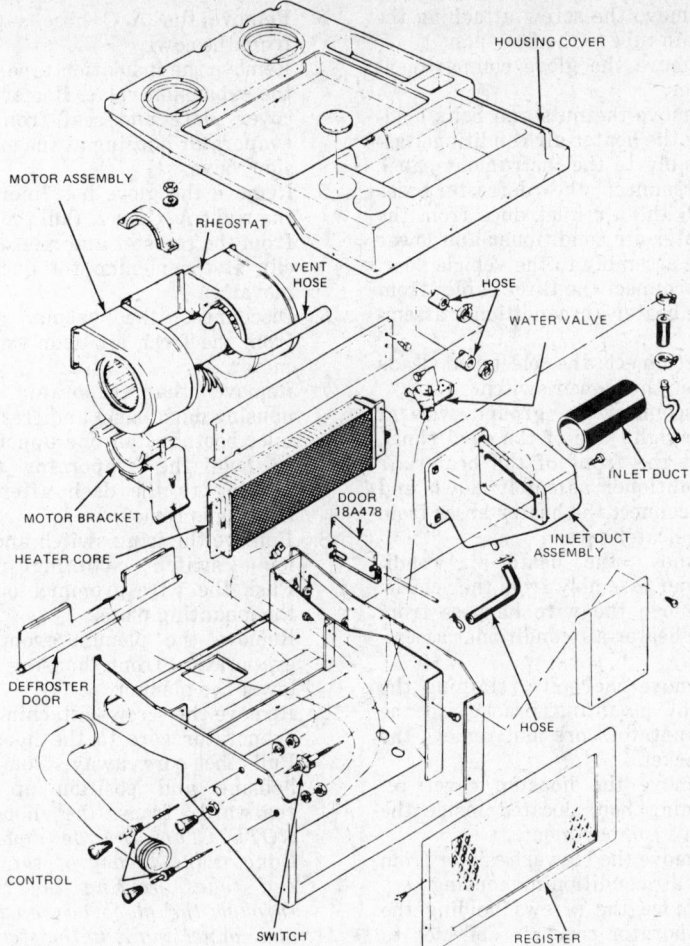

W-series heater assembly—exploded view (© Ford Motor Co.)

froster door cable to the heater housing, and disconnect the cable at the defroster door.
9. Remove the blower motor hold-down strap secured by tow nuts. One nut also holds the blower motor ground wire.
10. Disconnect the wires at the heater switch.
11. Remove the screws holding the blower housing to the heater housing and the center plate.
12. Lift out the blower housings, motor and cages as an assembly.
13. Remove the defroster door pivot plate.
14. Remove the heater core top retaining plate. Note the defroster cable routed between the plate and the heater core.
15. Remove the screws holding the heater core to the left side of the heater assembly.
16. Remove the clip from the switch and control valve cable and disconnect the cable at the valve.
17. Remove the screws securing the valve and bracket to the heater housing.
18. Lift the heater core, valve and bracket out of the heater housing as a unit.
19. Transfer the valve, bracket and hose to the replacement heater

core. (Disconnect the valve hose at the core tube.)
20. Install the heater core assembly in the reverse order of removal.

Heater Core With Air Conditioning—Removal and Installation

1971-72 F-100-350 and Bronco
1. Drain the cooling system
2. *Discharge the air conditioning system.*
3. Disconnect the battery ground cable.
4. Remove the carburetor air cleaner.
5. Remove the tape from the A/C hoses and clutch wire. CAUTION: Be sure that the system has been fully discharged before disconnecting any of the refrigerant hoses or lines.
6. Disconnect the refrigerant high pressure hose at the sight glass.
7. Disconnect the refrigerant low pressure line at the compressor service valve.
8. Remove the screws attaching the refrigerant hose seal and retainer to the dash panel.
9. Remove the bolts attaching the heater-air conditioner to the dash panel.

10. Remove the screw attaching the drain tube to the floor pan.
11. Remove the glove compartment liner.
12. Remove the nuts and bolts holding the heater-air conditioner assembly to the instrument panel. Disconnect the defroster hoses and the air inlet duct from the heater-air conditioner and lower the assembly to the vehicle floor.
13. Disconnect the three cables from the heater-air conditioner assembly.
14. Disconnect the electrical leads from the blower switch.
15. Disconnect the ground wire at the right side of the cowl panel.
16. Tilt the front of the heater-air conditioner assembly down and disconnect the heater hoses from the heater core.
17. Remove the heater-air conditioner assembly from the vehicle.
18. Remove the wire harness from the heater-air conditioner assembly.
19. Remove the bolts retaining the right mounting bracket to the evaporator core and remove the bracket.
20. Remove the housing cover retaining bolts located inside the glove compartment.
21. Remove the blower resistor from the air conditioning housing.
22. Remove the screws holding the evaporator core air deflector to the air inlet end of the housing.
23. Remove the housing cover retaining screws and remove the cover from the housing.
24. Remove the blower adapter plate attaching screws, and remove the blower and adapter plate.
25. Remove the screws holding the A/C door lever retainers to the door and remove the door from the bellcrank.
26. Remove the screws attaching the evaporator core air deflector to the housing and remove the assembly from the housing.
27. Remove the heater baffle from the housing.
28. Remove the heater core retainer screws and remove the heater core.
29. Remove the insulators from the heater core.
30. Install the heater core and heater-air conditioning assembly in the reverse order of removal. Be sure to leak test, evacuate, and charge the air conditioning system when installation procedures are completed. Test system operation.

1973-74 F-100-350 and Bronco
1. Disconnect the battery cable, remove the carburetor air cleaner, and drain the coolant system.
2. Remove the heater hoses from the heater core.

3. Remove the A/C hose support from the cowl.
4. Remove the insulation tape from the expansion valve. Remove the cover plate and seal from the evaporator housing at the expansion valve.
5. Remove the glove box liner and the right A/C duct. Pull the duct from the register and release the clip at the plenum for duct removal.
6. Disconnect the vacuum hose from the fresh air door vacuum motor.
7. Remove the evaporator rear housing dash panel and fresh air inlet boot. Install one upper nut to hold the evaporator front housing to the dash after the housing is removed.
8. Remove the icing switch and the icing switch mounting plate. Push the wire grommet out of the mounting plate.
9. Remove the plenum from the evaporator front housing and lower the plenum.
10. Remove the screws attaching the evaporator core to the housing. Pull the core away from the housing and position up and rearward from the housing. *NOTE: This can be done by tying twine to one of the core end tubes, routing the twine through the glove box opening, and connecting it to the steering wheel.*
11. Remove the heater core from the evaporator housing.
12. Install the heater core in the reverse order of removal. Be sure to test the operation of the heater-air conditioning system when assembly is completed.

1975-78 F-100-350 and Bronco
1. Disconnect the battery cable, remove the carburetor air cleaner, and partially drain the cooling system.
2. Remove the heater hoses from the heater core.
3. From under the hood, remove the A/C support bracket from the cowl.
4. Remove the insulation tape from the expansion valve. Then remove the cover plate and seal

from the evaporator housing at the expansion valve.
5. Remove the glove box liner and remove the A/C duct by pulling from the instrument panel register and releasing the clip at the plenum.
6. Disconnect the right hand cowl fresh air inlet vacuum hose from the fresh air inlet vacuum door vacuum motor.
7. Remove the evaporator rear housing from under the instrument panel. Remove the fresh air inlet tube from the evaporator rear housing and install one upper nut to retain the evaporator housing-to-dash after the rear housing is removed.
8. Disconnect the wires from the icing switch and pull the capillary tube out of the evaporator core. Remove the icing switch mounting plate and remove the plenum.
9. Remove the screws retaining the plenum-to-dash (above transmission tunnel) and the screws to evaporator case. Remove the plenum.
10. Install a piece of protective tape on the "A" pillar inner cowl panel, at the lower right corner of the instrument panel.
11. Remove the lower right instrument panel-to-"A" pillar bolt an lower the center instrument panel brace, bolt and nut.
12. Position the instrument panel rearward and install the "A" pillar bolt to hold the panel in the rearward position.
13. Remove the evaporator retaining screws.
14. Position the evaporator away from the case and secure it rearward and upward. Remove the evaporator sealing grommet.
15. Remove the heater core.
16. Install the heater core and evaporator housing components in the reverse order of removal. Be sure to test the operation of the heating and A/C systems after the installation is completed.

Econoline 1975-78
1. Disconnect the electrical leads from the resistor on the front

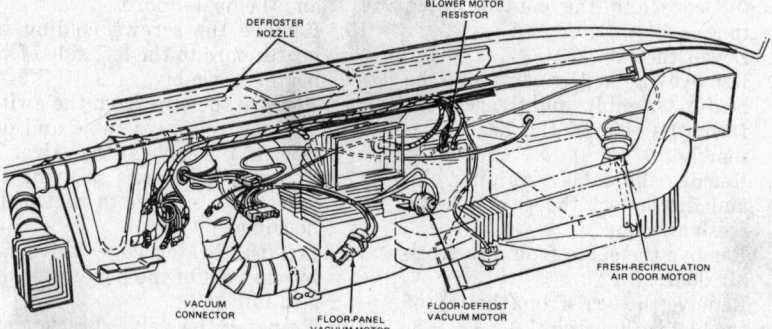

1975-78 F100-350 and Bronco heater-A/C installation (© Ford Motor Co.)

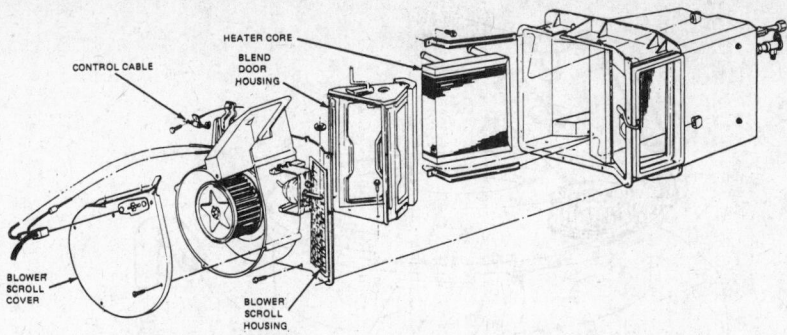

1975-78 Econoline with A/C-heater core installation (© Ford Motor Co.)

face of the A/C blower.

2. Disconnect the vacuum line from the fresh air/recirculated air door vacuum motor.
3. Remove the A/C blower cover.
4. Remove the push nut and washer from the fresh air/recirculated air door shaft.
5. Remove the control cable from the bracket and slide over the bracket. Remove the cable wire loop from the blend door shaft.
6. Remove the A/C blower motor housing.
7. Remove the blend door housing.
8. Partially drain the cooling system.
9. Remove the heater hoses from the heater core.
10. Remove the heater core retaining brackets.
11. Remove the heater core and seal assembly.
12. Install the heater core and related parts in the reverse order of removal. Check the operation of the heating and A/C system after installation is completed.

L-Series

1. Drain the cooling system.
2. Discharge the A/C system.
3. Disconnect the battery ground cable. **CAUTION: Before any refrigerant lines or hoses are disconnected, be sure that the air conditioning system is fully discharged.**

4. Disconnect the refrigerant low pressure line at the dash panel and remove the fitting retaining nut and lockwasher.
5. Remove the bolts holding the A/C-heater assembly to the dash panel.
6. Disconnect the drain tube at the bottom of the assembly housing.
7. Disconnect the heater hoses at the bottom of the assembly housing.
8. Disconnect the refrigerant high pressure hose at the side of the assembly housing.
9. Lower the assembly to the vehicle floor and disconnect the control cables from the A/C-heater assembly.
10. Disconnect the wires from the blower switch and remove the A/C-heater assembly from the vehicle.
11. Remove the wire harness from the A/C-heater assembly.
12. Remove the blower resistor from the A/C housing.
13. Remove the screws retaining the evaporator core air deflector to the air inlet end of the housing.
14. Remove the housing cover retaining screws and remove the cover.
15. Remove the evaporator core retaining screws and remove the evaporator core and hoses.
16. Remove the blower and adapter plate attaching screws, and re-

move the blower and adapter plate.

17. Remove the screws attaching the A/C door lever retainers to the door and remove the door from the bellcrank.
18. Remove the screws attaching the evaporator core air deflector to the housing and remove the assembly from the housing.
19. Remove the heater baffle from the housing.
20. Remove the heater core retainer attaching screws and remove the heater core.
21. Remove the insulators from the heater core.
22. Install the assembly in the reverse order of removal. Be sure to leak test, evacuate, and charge the A/C system when the installation is completed. Test the system operation.

Engine

Description

All Ford truck engines are of the conventional overhead valve design, either of six-cylinder in-line or V-8 configuration.

The 240 cu. in. (1971-74) and 300 cu. in. (1971-78) 6-cylinder engines are the large block units found in most medium and some heavy duty trucks. The 300 HD Six has a lower compression ratio to meet the higher torque requirements of heavier vehicles. These are the most common Ford power units.

The 330 cu. in. engine became the most popular V-8 option in medium and heavy duty trucks. The difference between the medium and heavy duty models is that the HD has a longer crankshaft in front.

The 302 cu. in. was the V-8 for both Econoline and Bronco models. The 351W was added in 1975, the 460 was added in 1973, and the 351M and 400 were added in 1977.

The large block Super Duty 401, 475, 477 and 534 cu. in. V-8 engines are used mainly in heavy trucks and will not be considered here.

Reference

Tune-up and general specifications for Ford truck engines may be found at the beginning of this section. Engine overhaul procedures can be found in the "ENGINE REBUILDING" section in the General Repair portion of this manual.

Emission Control

Crankcase Emission Controls

The crankcase emission control equipment consists of a positive crankcase ventilation (PCV) valve, a

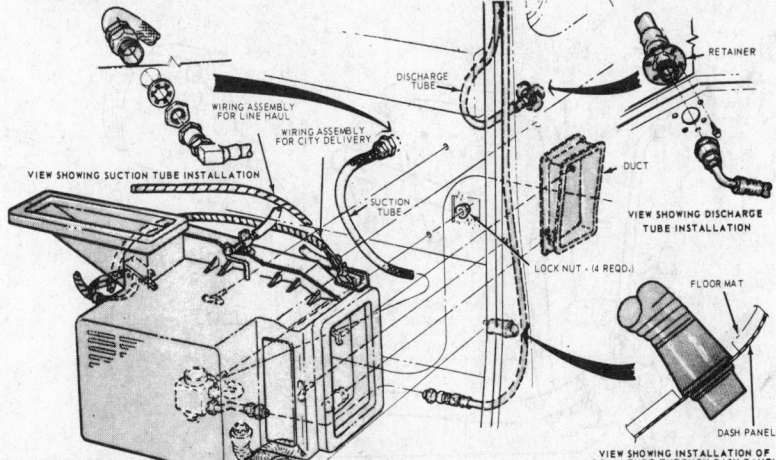

L-series heater-A/C installation (© Ford Motor Co.)

crankcase air filter that is vented to the air cleaner, and the hoses that connect the equipment.

When the engine is running, a small amount of the gases formed in the combustion chamber leak by the piston rings and enter the crankcase. The PCV system pulls these gases back into the intake manifold allowing fresh air to flow into the crankcase through the filter and filler cap. For service to the PCV system, refer to the "Emission Control" chapter in the General Repair section of this manual.

Evaporative Emission Control

The Evaporative Emission Control system consists of a sealed fuel tank, a vapor controlling orifice valve located in the top of the fuel tank, a pressure/vacuum relief fuel cap, and a carbon canister. This system is designed to limit the fuel vapors released into the atmosphere.

The open orifice valve is used to control the flow of fuel vapor and to minimize the amount of liquid gasoline entering the fuel vapor delivery line. The delivery line conducts the vapor forward to the carbon canister where the vapor is stored. During normal driving, the engine compartment mounted canister is purged of the fuel vapor by means of a hose connected to the air cleaner assembly. The vapors are drawn into the engine's induction system. The fuel cap is sealed and contains a vacuum and pressure relief valve. The vacuum valve relieves tank vacuum caused by consumption or cooling and the pressure relief valve prevents excessive fuel tank pressurization due to any system component failure or operation extremes.

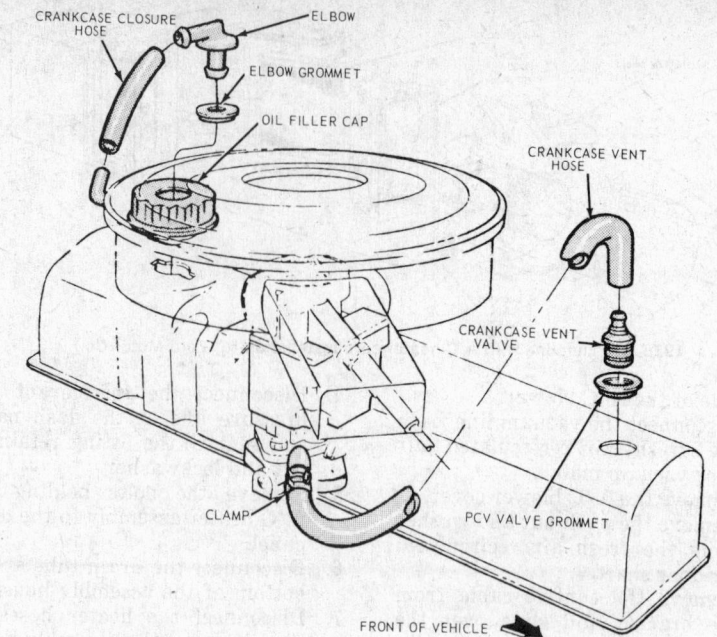

Closed crankcase ventilation system (© Ford Motor Co.)

Air Injector System— "Thermactor"

The air injection exhaust emission control system "Thermactor" consists of a air supply pump, external air manifold or cylinder head/exhaust manifold with internal air passages, air by-pass valve, check valve, and the hoses necessary to connect the components.

The air injection system reduces the carbon monoxide and hydrocarbon content of the exhaust gases by injecting fresh air into the hot exhaust gas stream as it leaves the combustion chamber. A pump supplies the air under pressure to the exhaust port near the exhaust valve by either an external air manifold or through an internal drilled passages in the cylinder head or exhaust manifold. The oxygen in the fresh air plus the heat of the exhaust gases cause further burning which converts the exhaust gases into carbon dioxide and water.

For service on the air injection system, refer to the "Emission Control" chapter in the General Repair section of this manual.

Catalytic Converter

The catalytic converter is a muffler type device installed in the vehicles exhaust system which contains a chemical catalyst. When the hot exhaust gas passes over and through the catalyst, it heats up to a high temperature and the chemical reaction which occurs breaks down the exhaust into harmless elements. For service on the catalytic converter,

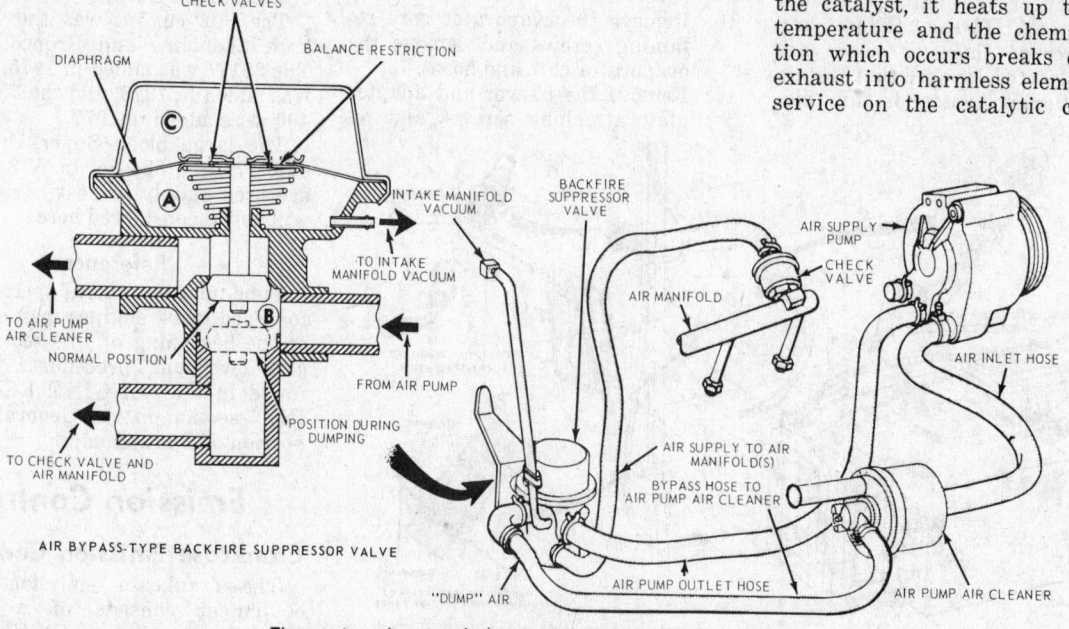

Thermactor exhaust emission control system (© Ford Motor Co.)

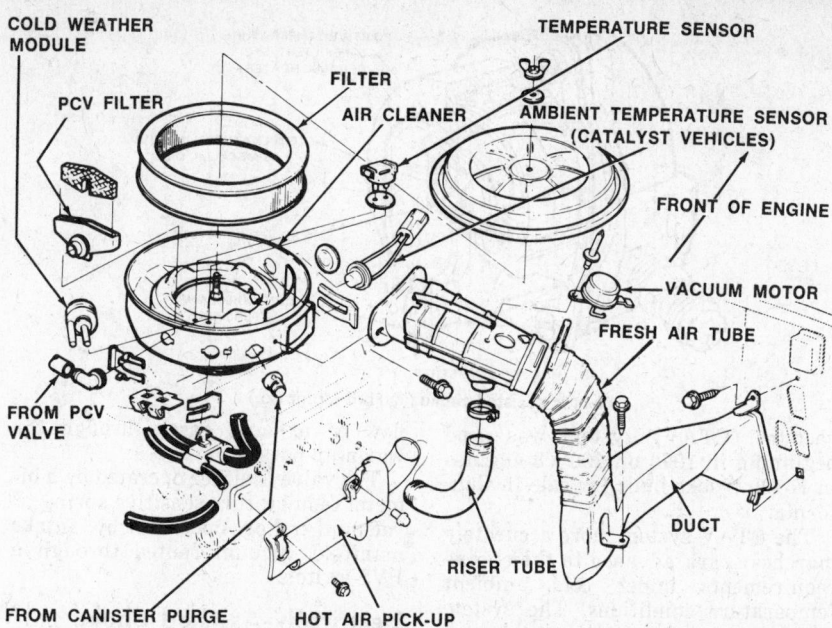

Typical thermostatically controlled air cleaner (TAC) (© Ford Motor Co.)

refer to the "Emission Control" chapter in the General Repair Section of this manual.

Thermostatically Controlled Air Cleaner System (TAC)

This system consists of a heat shroud which is integral with the right-side exhaust manifold, a hot air hose and a special air cleaner assembly equipped with a thermal sensor and vacuum motor and air valve assembly.

The purpose of TAC is to get hot air into the carburetor as soon as possible because the engine will emit less pollutants on a lean mixture.

Dual Diaphragm Distributor

The dual diaphragm distributor has two diaphragms which operate independently. The outer (primary) diaphragm makes use of carburetor vacuum to advance the ignition timing. The inner (secondary) diaphragm uses intake manifold vacuum to provide additional retardation of ignition timing during closed-throttle decleration and idle, resulting in the reduction of hydrocarbon emissions.

Ported Vacuum Switch Valve (PVS)

The PVS valve is a temperature sensing valve usually found on the distributor vacuum advance line, and is installed in the coolant outlet elbow. During prolonged periods of idle, or any other situation which causes engine operating temperatures to be higher than normal, the valve, which under normal conditions simply connects the vacuum advance diaphragm to its vacuum source within the carburetor, closes the normal source vacuum port and engages an

alternate source vacuum port. This alternate source is from the intake manifold which, under idle conditions, maintains a high vacuum. This increase in vacuum supply to the distributor diaphragm advances the timing, increasing the idle speed. The increase in idle speed causes a directly proportional increase in the operation of the cooling system. When the engine has cooled sufficiently, the vacuum supply is returned to its normal source, the carburetor.

These switches are used in several places in the emission systems. They can have anywhere from two to four ports and can be used to turn vacuum on and off, or to switch between two vacuum sources for a third delivery point.

Deceleration Valve

Some engines were equipped with a

distributor vacuum advance control valve (deceleration valve) which is used with dual diaphragm distributors to further aid in controlling ignition timing. The deceleration valve is in the vacuum line which runs from the outer (advance) diaphragm to the carburetor, the normal vacuum supply for the distributor. During deceleration, the intake manifold vacuum rises causing the deceleration valve to close off the carburetor vacuum source and connect the intake manifold vacuum source to the distributor advance diaphragm. The increase in vacuum provides maximum ignition timing advance, thus providing more complete fuel combustion and decreasing exhaust system backfire.

Exhaust Gas Recirculation System (EGR)

In this system, a vacuum-operated EGR flow valve is attached to the carburetor spacer (except on the 302 V8). A passage in the carburetor spacer mates with a hole in the mounting face of the EGR valve or the intake manifold. The EGR valve on the 302 V8 is located on the rear of the intake manifold. On all engines except the

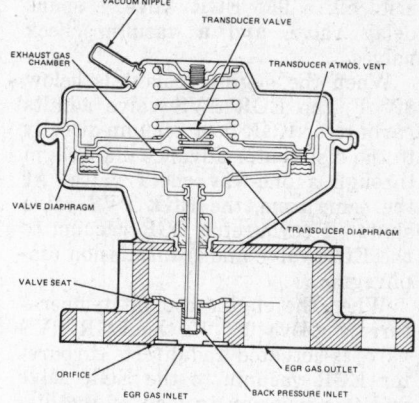

Integral transducer back pressure EGR (exhaust gas recirculation) valve (© Ford Motor Co.)

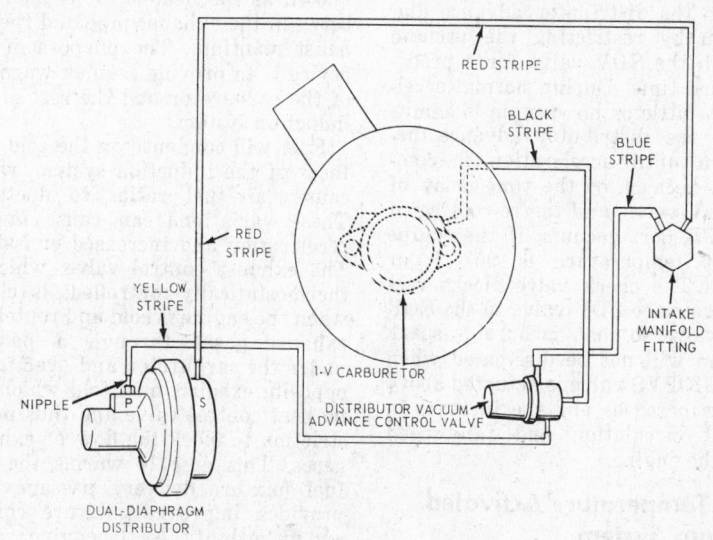

Ignition vacuum system with Imco emission (© Ford Motor Co.)

302 V8, the system allows exhaust gases to flow from the exhaust crossover, through the control valve and through the spacer into the intake manifold below the carburetor. For those engines where exhaust gases cannot be picked up from the exhaust crossover (6 cylinder) as described above, the gases are picked up from the choke stove located on the exhaust manifold or directly from the exhaust manifold. The exhaust gases are routed to the carburetor spacer through steel tubing.

The vacuum signal which operates the EGR valve originates at the EGR vacuum port in the carburetor. This signal is controlled by at least one, and sometimes, two series of valves.

A water temperature sensing valve (the EGR PVS) which is closed until the water temperature reaches either 60° F or 125° F, depending on application, is always used.

Another system working in conjunction with the EGR system is the EGR/CSC system. This system regulates both the distributor spark advance and operation of the EGR valve according to the temperature of the engine coolant. The system consists of: a 95° EGR valve, a spark delay valve, and a vacuum check valve.

When the engine coolant is below 82° F, the EGR PVS valve admits carburetor EGR port vacuum directly to the distributor advance diaphragm through a one-way check valve. At the same time, the EGR PVS valve shuts off carburetor EGR vacuum to the EGR valve and transmission diaphragm.

When the engine coolant temperature is above 95° F, the EGR PVS valve is actuated and directs carburetor EGR vacuum to the EGR valve and transmission instead of the distributor.

The spark delay valve (SDV) delays carburetor spark advance vacuum to the distributor advance diaphragm by restricting the vacuum through the SDV valve for a predetermined time. During normal acceleration, little or no vacuum is admitted to the distributor advance diaphragm until acceleration is completed, because of the time delay of the SDV valve, and the re-routing of the EGR port vacuum, if the engine coolant temperature is 95° F or higher. The check valve blocks vacuum from the SDV valve to the EGR PVS valve so that carburetor spark vacuum will not be dissipated when the EGR PVS valve is actuated above 95° F. increases and the increase in coolant circulation and fan speed cools the engine.

Cold Temperature Activated Vacuum System

The cold temperature activated

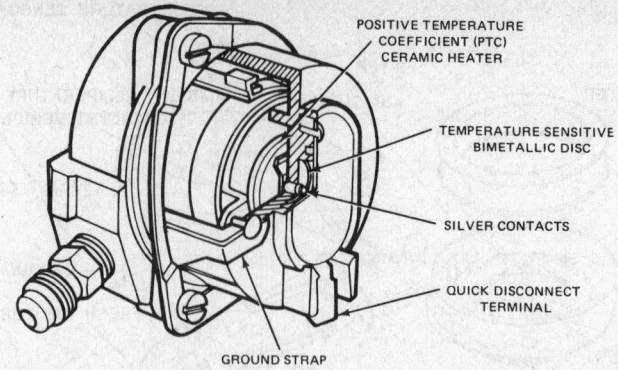

Electric assisted choke (© Ford Motor Co.)

POSITIVE TEMPERATURE COEFFICIENT (PTC) CERAMIC HEATER

TEMPERATURE SENSITIVE BIMETALLIC DISC

SILVER CONTACTS

QUICK DISCONNECT TERMINAL

GROUND STRAP

vacuum (CTAV) system was used beginning in 1974 on 460 V8 engines in F-100 trucks built for sale in California.

The CTAV system more accurately marches spark advance to the engine requirements under cold ambient temperature conditions. The system can select from two vacuum sources for spark advance depending on the ambient temperature: below 45° F—carburetor spark port vacuum, above 65° F—EGR vacuum. In between these two temperature ranges the system will select either source of vacuum, depending on the cycle it is in.

The CTAV system consists of an ambient temperature switch, a three-way vacuum switch, inline vacuum bleed, and a latching relay.

The vacuum from both the spark port of the carburetor and the EGR port is supplied to the three-way solenoid valve. The ambient temperature switch provides the signal that determines which of the sources will be selected. The latching relay provides for only one cycle each time the ignition switch is turned on.

Vacuum and Spring Controlled Heat Control Valve

The heat control valve, commonly known as the heat riser, is mounted between the exhaust pipe and the exhaust manifold. The purpose of the device is to provide a quick warm-up of the carburetor and the rest of the induction system.

Fuel will condense on the cold surfaces of the induction system, which causes air/fuel ratios to fluctuate. These variations can cause uneven acceleration and increased emissions. The exhaust control valve, which is thermostatically-controlled, is closed when the engine is cold and routes hot exhaust gases through a passage under the carburetor and over to the opposite exhaust manifold which has no heat control valve and thus no restriction to block the flow of exhaust gases. This quickly warms the air/fuel mixture delivery passages and provides improved mixture control and driveability. As the engine warms up, the valve opens and reduces the

flow of exhaust gases through the warm-up passage.

The valve is either operated by a bimetal temperature sensitive spring or vacuum motor operated by intake manifold vacuum routed through a PVS switch.

Electrically-Assisted Choke

Some pick-ups use an electrically heated choke thermostatic spring housing as an aid to fast choke release and better emission characteristics during engine warm-up. The heater operates from a lead off the alternator only when the engine is actually running.

The heater element only operates when ambient temperatures are above 60° F (when long periods of choke operation are not necessary for engine driveability). When temperatures are blow 60° F, the choke thermostatic spring is heated in the normal manner: via a tube running from an exhaust manifold heat stove.

Vacuum Delay Valves

Retard Delay Valves (RDV) and Spark Delay Valves (SDV)

Delay valves are found in many places on 1974 and later engines. The

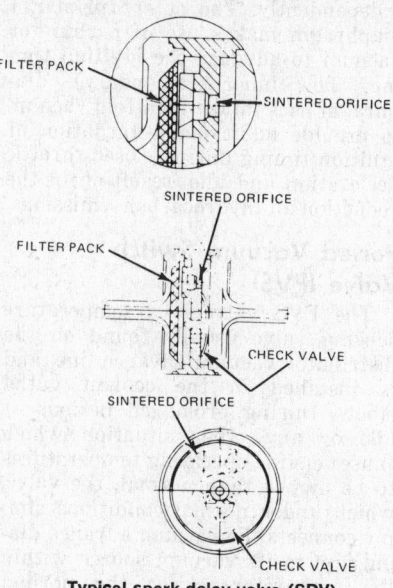

FILTER PACK

SINTERED ORIFICE

SINTERED ORIFICE

FILTER PACK

CHECK VALVE

SINTERED ORIFICE

CHECK VALVE

Typical spark delay valve (SDV)
(© Ford Motor Co.)

delay valves work to slow the air flow in the vacuum lines, thus providing closer control on vacuum operated equipment. The SDV is normally used to delay the opening function of a vacuum device and the RDV is used to delay the closing of a vacuum device. The delay valves have an interval sintered orifice; the check valve and filter pack must be installed in the correct direction in each system.

Engine Removal and Installation

Note: For the 1978 and later Bronco, use F-150 4 x 4 procedures.

Bronco 170 cu. in. 6-Cylinder

1. Drain cooling system and remove air cleaner.
2. Disconnect battery ground cable from the battery.
3. Disconnect upper and lower hoses at the engine, remove the four radiator mounting bolts, then remove radiator.
4. Disconnect the heater hose at the water pump and at the rear of the carburetor spacer.
5. Disconnect battery ground cable and alternator ground wire from the cylinder block.
6. Remove fan belt and alternator.
7. Disconnect the starter cable at starter, then remove the starter.
8. Remove U-clamp holding exhaust head pipe to the block, remove manifold stud retaining nuts, and remove pipe from the exhaust manifold.
9. Disconnect wiring at the coil and oil and temperature sending units.
10. Disconnect throttle and choke cables from the carburetor.
11. Remove windshield wiper vacuum hose from fuel pump.
12. Remove the crimped clamp, then the flex line from fuel pump.
13. Remove the retaining screws from the equalizer shaft bracket at the block and clutch housing, and remove the bracket.
14. Remove nut and washer from each of the engine mounts.
15. Loosen clutch housing to block bolts just enough so that they may be hand removed later.
16. Remove clutch housing cover retaining bolts.
17. Lower vehicle to the floor.
18. Position a jack under the transmission, and remove the clutch housing to block bolts.
19. Carefully remove engine.
20. To install, lower engine assembly carefully aligning the transmission shaft with the clutch disc splines.
21. Remove hoisting equipment and transmission jack.
22. Raise vehicle on a hoist.
23. Install lower clutch housing to block bolts securely.

24. Install clutch housing cover to clutch housing retaining bolts.
25. Install engine mount washers and retaining nuts, tightening securely.
26. Lubricate the clutch equalizer shaft and bracket and position the bracket and install the retaining bolts.
27. Lower vehicle.
28. Connect flex line (use a new clamp) and vacuum line to fuel pump.
29. Connect and adjust choke and throttle cables.
30. Connect the wiring at the coil and at the oil and temperature sending units.
31. If they were removed, install the water pump pulley, spacer and fan.
32. Place a new gasket on the exhaust inlet pipe, connect the pipe to the exhaust manifold, install the reaining nuts and tighten securely.
33. Install the U-bolt clamp to retain the inlet pipe to the bolck.
34. Install the starter and starter cable.
35. Install alternator and belt. Adjust belt.
36. Connect battery and alternator ground cables to block.
37. Connect heater hose at the rear of carburetor spacer and at the water pump.
38. Install radiator and, using a water-resistant sealer, connect upper and lower hoses to engine.
39. Connect battery ground cable and install air cleaner.
40. Fill cooling system and bleed all air.
41. Fill crankcase with lubricant.
42. Operate engine and check for fuel, lubricant and coolant leaks.

240, 300 Cu. In. 6-Cylinder

All Trucks Except 1973-78 Econoline

The engine and transmission are disattached in the following engine removal procedure.

1. Drain cooling system and crankcase.
2. Remove hood or tilt the cab.
3. Remove air cleaner (1973-76). On C-series, remove the oil filler tube.
4. Disconnect the battery positive cable.
5. Disconnect the heater hose from the water pump and coolant outlet housing.
6. Disconnect flexible fuel line from the fuel pump.
7. Remove the radiator following procedure described above.
8. Remove the cooling fan, water pump pulley and fan drive belt.
9. Disconnect the accelerator cable and the choke cable at the car-

buretor. Remove the cable retracting spring.
10. On a vehicle with power brakes, disconnect the vacuum line at the intake manifold.
11. On a vehicle with automatic transmission, disconnect the transmission kickdown rod at the bellcrank assembly.
12. Disconnect exhaust manifold from the muffler inlet pipe.
13. Disconnect the body ground strap and the battery ground cable at the engine.
14. Disconnect the engine wiring harness at the ignition coil, water temperature and oil pressure sending units.
15. Remove alternator mounting bolts and position the alternator out of the way, leaving wires attached.
16. On a vehicle with power steering, remove the power steering pump from the mounting brackets and position it right side up and to one side, leaving the line attached.
17. If equipped with an air compressor, bleed the air system and disconnect the two air lines at the compressor.
18. Raise vehicle and remove starter and, if so equipped, remove the automatic transmission fluid filler tube.
19. Remove engine rear plate upper right bolt.
20. On a vehicle with manual transmission, remove all the flywheel housing lower attaching bolts. Disconnect the clutch retracting spring.
21. On vehicles with an automatic transmission, remove the converter housing access cover assembly. Remove the flywheel to converter nut and secure the converter assembly in the housing. Remove the transmission oil cooler lines from the retaining clip at the engine. Remove the converter housing to engine lower attaching bolts.
22. On F-100, 250, remove the insulator to intermediate support bracket nut from each engine front support. On other vehicles, remove the engine from support insulator bolt.
23. Lower the vehicle and position a transmission jack under the transmission. Remove the remaining flywheel or converter housing to engine bolts.
24. Attach engine lifting hook and raise the engine slightly and carefully pull it from the transmission. Lift engine out of the chassis.
25. To install, place a new gasket on the muffler inlet pipe.
26. Lower engine carefully into

chassis. Make sure the studs on the exhaust manifold are aligned with the holes in the muffler inlet pipe and the dowels in the block engage the holes in the flywheel or converter housing.

27. On a vehicle with an automatic transmission, start the converter pilot into the crankshaft. Remove the retainer securing the converter in the housing.

28. On a vehicle with standard transmission, start the transmission input shaft into the clutch disc. It may be necessary to adjust the position of the transmission with relation to the engine if the transmission input shaft will not enter the clutch disc. If the engine hangs up after the shaft enters, turn the crankshaft slowly (with transmission in gear) until the shaft splines mesh with the clutch disc splines.

29. Install the converter or flywheel housing upper attaching bolts. Remove the transmission jack.

30. Lower the engine until it rests on the engine support(s) and remove the lifting hook.

31. On the F-100, 250, install engine left and right support insulator to intermediate support bracket retaining nuts, tightening securely. Install the front engine mount bolt and nut on other vehicles.

32. Install the transmission oil cooler lines bracket, if so equipped.

33. Install remaining converter or flywheel housing attaching bolts.

34. Connect clutch return spring.

35. Install starter and connect cable.

36. If so equipped, install transmission fluid filler tube bracket.

37. On a vehicle with automatic transmission, install transmission oil cooler lines in the bracket at the engine block.

38. Install exhaust pipe to exhaust manifold, tightening bolts securely.

39. Connect engine ground strap and battery ground cable.

40. On a vehicle with automatic transmission, connect the kickdown rod to the bellcrank assembly on the intake manifold.

41. Connect the accelerator linkage to the carburetor and install the retrackting spring. Connect the choke cable to the carburetor and hand throttle if so equipped.

42. On a vehicle with power brakes, connect the brake vacuum line to the intake manifold.

43. On a C-series vehicle, install the oil filler tube.

44. Connect coil primary wire, oil pressure and water temperature sending units, flexible fuel line, heater hoses and battery positive cable.

45. Install alternator on mounting bracket. On a vehicle with power steering, install the power steering pump on the mounting brackets.

46. Install water pump pulley, spacer, cooling fan and drive belt, tightening bolts securely.

47. Install radiator.

48. Connect air compressor lines.

49. If so equipped, connect oil cooler lines.

50. Install and adjust hood.

51. Fill and bleed cooling system, checking for leaks.

52. Adjust the carburetor idle speed and mixture.

53. Adjust clutch pedal free travel or automatic transmission control linkage and check transmission fluid level.

54. Install air cleaner.

Econoline—1973-78

1. Disconnect the battery and drain the cooling system. Remove the engine cover and the right hand seat. Remove the grille and bumper.

2. Remove the hood lock support bracket, right and left headlight doors, and grille.

3. Disconnect all hoses and lines to the radiator, and remove the battery deflector. Remove the radiator.

4. Disconnect
   A. The heater hoses, at the engine.
   B. Temperature, oil, and ignition wires.
   C. Starter solenoid, neutral safety switch, and back-up light wiring.

5. Remove the engine oil dipstick and oil filler tube. Remove hoses connecting the rocker cover and air filter. Remove the air cleaner and brackets.

6. Disconnect the choke and accelerator cables at the carburetor. Disconnect the auxiliary heater hose at the front heater. Disconnect the hoses at the right front of the engine, and position them out of the way.

7. Disconnect the fuel pump discharge line at the pump. Disconnect the alternator, and remove it from the brackets.

8. Disconnect the ground wires at the block, and the muffler inlet pipe at the manifold. Disconnect the modulator line at the intake manifold.

9. Put the vehicle on a hoist. Drain the crankcase and remove the oil filter.

10. Disconnect the starter wiring, and remove the starter.

11. Position an engine support bar to the chassis and engine, and adjust it.

12. On manual transmission vehicles:

A. Disconnect the driveshaft, and remove it. Install a plug in the transmission extension housing.

B. Disconnect the speedometer cable and housing, and secure the assembly out of the way.

C. Remove the nut and bolt holding the rear support to the crossmember. Raise the transmission, remove the mounting bolts, and remove the crossmember.

D. Remove the clutch equalizer arm bolts from the engine. Disconnect the retracting spring, and move the assembly away.

E. Remove the bolts connecting transmission and clutch, and remove the transmission.

13. On automatic transmission vehicles:

A. Remove the bolts connecting the adapter plate and inspection cover to the torque converter.

B. Unbolt and remove the transmission dipstick tube. Drain the transmission.

C. Remove the nuts attaching the converter to the flex plate. Disconnect the oil cooler and modulator lines at the transmission.

D. Disconnect the driveshaft at the companion flange.

E. Disconnect the speedometer cable and housing from the transmission. Disconnect the shift rod at the lever on the transmission. Jack the transmission up slightly.

F. Remove the nuts and bolts attaching the rear engine mount bracket to the crossmember. Remove the side support bolts, and remove the crossmember.

G. Secure the transmission to the jack with a safety chain, remove the remaining bolts attaching the transmission to the cylinder block, and remove the transmission from the vehicle.

14. Remove the nuts which attach the engine front support insulator, remove the bellcrank bolt from the block, and position it out of the way.

15. Lower the vehicle, and remove the fan spacer, and water pump pulley. Lift the engine from the vehicle with a lifting hook. Remove the clutch housing on vehicles with manual transmission.

16. Install the clutch housing on manual transmission vehicles. Hoist the engine into the vehicle, and allow it to rest on the front supports and support tool.

17. Raise the vehicle on a hoist.

18. On manual transmission equipped
    A. Raise the transmission and position it behind the clutch housing. Install the mounting bolts.
    B. Raise the transmission slightly further, position the crossmember to the chassis and rear support, and install the attaching bolts. Torque crossmember to body bolts to 20-30 ft-lbs.
    C. Remove the jack and engine support tool. Connect the shift linkage and the speedometer cable housing.
    D. Install and connect the drive shaft.
19. Install the engine front support insulator bolts, and torque to 45-55 ft-lbs.
20. Install the transmission bellcrank.
21. On automatic transmission equipped vehicles:
    A. Position the transmission against the block and install the mounting bolts. Torque to 23-33 ft-lbs.
    B. Position the crossmember to the rear mount bracket and frame side members. Install the attaching nuts and bolts, and torque to 20-30 ft-lbs.
    C. Remove the transmission safety chain, and remove the jack. Remove the engine support bar.
    D. Install the converter to the flex plate. Connect the vacuum and oil cooler lines to the transmission.
    E. Install the dipstick and tube into the transmission oil pan. Install the tube and vacuum line bracket attaching bolt to the block.
    F. Connect the driveshaft to the transmission companion flange. Connect the speedometer cable and housing to the transmission. Connect the shift rods to the transmission levers.
    G. Install the adapter plate and inspection cover.
22. Install the starter. Connect the muffler inlet pipe at the manifold.
23. Install the clutch equalizer arm bracket (on manual transmission models). Install the attaching bolts, and connect the retractor spring.
24. Lower the vehicle to the floor, install the remaining automatic transmission mounting bolts.
25. Install the starter ground wire and remaining starter attaching bolt. Connect the starter cable at the starter.
26. Install the water pump pulley, spacer, and fan assembly.
27. Install the alternator onto its mounting brackets, and install and tension the V-belt.
28. Connect the alternator wiring, and back-up light and neutral safety switch leads.
29. Connect the alternator and battery ground wires to the block.
30. Connect the transmission modulator line to the manifold.
31. Connect the fuel line from the tank to the fuel pump.
32. Connect the choke and accelerator cables to the carburetor. Install the dipstick tube, and bolt it to the cylinder head.
33. Connect remaining neutral safety switch and back-up light wires. Position the wiring harness, and connect the coil, oil, and temperature leads.
34. Connect the auxiliary heater hoses. Connect the oil filler tube to the rocker arm cover and install the retaining clamp. Install the filler tube bracket to the dash panel.
35. Install the radiator and battery deflector. Connect the radiator hoses and oil cooler lines.
36. Install the grille, head light and hood lock bracket.
37. Fill the crankcase and cooling system. Fill the automatic transmission.
38. Connect the positive battery cable, and operate the engine to check for leaks. Adjust idle speed and mixture, and automatic transmission linkage.
39. Install the air cleaner and brackets, engine front cover, and right front seat, grille and bumper.

302, 351M, 351W & 400 V8

F100-350, Bronco

Removal and installation procedure for the V-8 engine is the same as that for the 240, 300 cu. in. 6-cylinder described above except for the following.

1. Transmission shift rod must be disconnected and retracting spring removed.
2. A left and a right lifting bracket are used, the left bracket on the front of the left cylinder head and the right bracket on the rear of the right cylinder head. A sling is then attached to each bracket to lift out the engine.
3. When lowering the engine into the vehicle, make sure exhaust manifolds are properly aligned with the muffler inlet pipes.

Econoline

In this procedure the engine is removed without the transmission attached.

1. Remove engine cover.
2. Remove right front seat.
3. Drain cooling system.
4. Remove air cleaner and intake duct assembly (later models), including crankcase ventilation hose.
5. Disconnect battery and alternator ground cables at block.
6. Remove oil filler tube at dash panel and disconnect at the rocker arm cover.
7. Disconnect radiator upper and lower hoses at the radiator and automatic transmission cooler lines (if so equipped).
8. Remove radiator.
9. Disconnect heater hoses at the engine.
10. Remove fan, spacer, pulley and drive belt.
11. Disconnect accelerator linkage at accelerator shaft assembly on the left cylinder head.
12. Disconnect automatic transmission kickdown rod at the carburetor and vacuum line at the intake manifold, if so equipped.
13. Disconnect wiring harness from left rocker arm cover.
14. Remove upper nut attaching right exhaust manifolds to exhaust pipe.
15. Raise vehicle on hoist.
16. Drain crankcase and remove filter.
17. Disconnect fuel pump inlet line at the pump.
18. Disconnect oil dipstick tube bracket from exhaust manifold and oil pan.
19. On vehicles with standard transmission, remove bolts attaching the equalizer arm bracket to cylinder block and clutch housing (this includes clutch linkage disconnection and retracting spring).
20. Disconnect starter cable at starter and remove starter.
21. On a vehicle with standard transmission, disconnect driveshaft at the rear axle and remove driveshaft. Install plug in transmission end.
22. On a vehicle with automatic transmission, disconnect driveshaft at companion flange.
23. Disconnect speedometer cable and transmission linkage at transmission.
24. Position a transmission jack under transmission. Raise transmission and remove bolts attaching crossmember to chassis. Lower transmission slightly and remove bolt which attaches rear engine support to frame crossmember. Remove crossmember.
25. On a vehicle with standard transmission, remove bolts attaching transmission to clutch housing. Remove transmission.
26. On a vehicle with automatic transmission, remove the lower front cover from converter housing. Remove transmission dip-

Ford Trucks · Vans · Bronco

stick tube and drain transmission. Install plastic bag in transmission oil pan. Remove nuts attaching transmission converter to flywheel. Disconnect oil cooler lines and vacuum lines at transmission. Remove remaining bolts fastening transmission to engine. Remove transmission.

27. Position engine support bar (tool T65E-6000-J) to engine and chassis, or improvise a suitable support.
28. Disconnect exhaust pipes at exhaust manifold.
29. Remove front engine mount attaching nuts and washers.
30. Remove bellcrank bolt from side of engine block and position bellcrank aside.
31. Lower vehicle.
32. Remove bolts which fasten alternator and adjusting arm to the block and water pump. Set alternator aside.
33. Remove carburetor air horn stud. Disconnect fuel line at fuel pump.
34. Install engine lifting apparatus and remove engine through side door.
35. On vehicle with standard transmission, remove bolts which fasten adapter plate to clutch housing. Remove clutch housing from block.
36. To install, connect fuel line to fuel pump.
37. On models with manual transmission, position clutch housing to block and install mounting bolts. Install adapter plate to clutch housing.
38. Position alternator and adjusting arm to cylinder block and water pump. Install and tighten attaching bolts.
39. Lift engine and position to chassis and supporting tool.
40. Raise vehicle on hoist.
41. Install front engine support retaining nuts and washers, tightening securely.
42. On vehicle with automatic transmission, position bellcrank assembly to engine block and install attaching bolt, tightening securely. Position transmission to engine and install attaching bolts, tightening securely.
43. On vehicle with standard transmission, position transmission to clutch housing and install attaching bolt, tightening securely.
44. Remove engine support bar.
45. Position rear engine support crossmember to chassis and install retaining bolts. Position transmission with rear engine support attached to crossmember. Install bolt and nut. Tighten all bolts and nuts securely.
46. On vehicle with automatic transmission, install and tighten con-

verter-to-flywheel attaching nuts. Connect oil cooler and vacuum lines at transmission. Install transmission dipstick tube in pan. Install dipstick tube and vacuum line retaining bracket bolt to engine block.

47. Connect transmission shift linkages and speedometer cable.
48. On vehicle with automatic transmission, connect driveshaft to transmission companion flange.
49. On vehicle with standard transmission, remove plug from transmission and install driveshaft yoke into transmission. Connect rear end of driveshaft at rear axle.
50. Install starter and connect cable to starter.
51. Install the exhaust pipe to exhaust manifold, installing all bolts except upper nut on right exhaust manifold.
52. On a standard transmission, install the bolts connecting equalizer arm bracket to block and clutch housing (including clutch linkage connection and retracting spring).
53. Install oil filter, then oil dipstick tube bracket to oil pan and exhaust manifold.
54. Connect fuel line at fuel pump.
55. Lower the vehicle.
56. Install upper nut attaching right exhaust manifold to exhaust pipe.
57. Connect engine wire harness at left rocker arm cover.
58. Connect battery and alternator ground cables at block.
59. If applicable, connect automatic transmission vacuum line at intake manifold and transmission kickdown rod at carburetor.
60. Connect accelerator linkage at accelerator shaft assembly on left cylinder head.
61. Install drive belt, pulley, spacer and fan.
62. Connect heater hoses at engine, then install radiator and connect hoses.
63. Install oil filler tube by connecting to left rocker arm cover and dash panel.
64. Install air cleaner and intake duct assembly, including the crankcase ventilation hose.
65. Fill and bleed cooling system.
66. Fill crankcase and automatic transmission (if so equipped).
67. Install right front seat and engine cover.
68. Operate engine at fast idle and check for leaks.

460 V-8

1. Remove the hood.
2. Drain the cooling system, the radiator, and the cylinder block.
3. Disconnect the negative battery cable and remove the air cleaner assembly.

4. Disconnect the upper and lower radiator hoses and the transmission oil cooler lines from the radiator.
5. Remove the fan shroud from the radiator and remove the fan from the water pump. Remove the fan and shroud from the engine compartment.
6. Remove the upper support and remove the radiator.
7. If the truck is equipped with air conditioning, remove the compressor from the engine and position it out of the way. If the compressor must be removed completely, loosen the air conditioning service valves (disconnect) carefully to discharge the air conditioning system. Remove the compressor.
8. Remove the power steering pump from the engine, if so equipped, and position it to one side. Do not disconnect the fluid lines.
9. Disconnect the fuel pump inlet line from the pump and plug the line.
10. Remove the alternator drive belts and disconnect the alternator from the engine, positioning it aside.
11. Disconnect the ground cable from the right front corner of the engine.
12. Disconnect the heater hoses.
13. Remove the transmission fluid filler tube attaching bolt from the right side valve cover and position the tube out of the way.
14. Disconnect all vacuum lines at the rear of the intake manifold.
15. Disconnect the speed control cable at the carburetor, if so equipped. Disconnect the accelerator rod and the transmission kick-down rod and secure them out of the way.
16. Disconnect the engine wiring harness at the connector on the fire wall.
17. Raise the vehicle and disconnect the exhaust pipes at the exhaust manifolds.
18. Disconnect the starter cable and remove the starter. Bring the starter forward and rotate the solenoid outward to remove the assembly.
19. Remove the access cover from the converter housing and remove the fly-wheel-to-converter attaching nuts. Remove the lower converter housing-to-engine attaching bolts.
20. Remove the engine mount through bolts attaching the rubber insulators to the frame brackets.
21. Lower the vehicle and place a jack under the transmission to support it.
22. Remove the converter housing-

to-engine block attaching bolts (left-side).

23. Disconnect the coil wire and remove the coil and bracket assembly from the intake manifold.
24. Attach the engine lifting device and carefully lift the engine from the engine compartment.
25. Install the engine in the reverse order of removal.
   Tighten the alternator pivot bolt to 45-57 ft lbs and all the rest of the nuts and bolts as is outlined in Step 21 of the preceding "302 and 351 V8, etc. Removal and Installation" procedure.

330, 350, 360, 361, 389, 390, 391 Cu. In. V-8

All Models

1. On B, F and FT series trucks, remove engine hood assembly from vehicle; on C and CT vehicles, release cab lock and tilt the cab forward. On N and NT series vehicles, position a suitable support in front of the truck to accept the hood and fender assembly when it is fully forward, then raise the hood and fender assembly.
2. Disconnect the ground cable from the battery. On N and NT series trucks, disconnect and remove the battery.
3. Drain cooling system and crankcase.
4. On N and NT series trucks, disconnect the check cable assemblies and let the hood swing forward out of the way, resting on the support. Remove radiator to cowl support rod, disconnect water hoses and unbolt and remove radiator.
   On C and CT series trucks, remove the heater hoses from the radiator, transmission oil cooler lines (automatic transmission), disconnect upper and lower radiator hoses, disconnect and remove vent line between radiator and supply tank and hose between water outlet housing and supply tank, remove fan (leaving it lay in the shroud), then remove the radiator, shroud and fan as an assembly. Remove radiator supply tank from cab rear support.
   On B, F and FT series trucks, disconnect the upper and lower hoses from the engine and water pump, remove the fan, disconnect the transmission oil cooler hoses (if applicable), then unbolt and remove radiator.
5. On a vehicle with power steering, disconnect the power steering pressure line from the pump reservoir and return line from the pump housing. Drain oil,

then loosen and remove the power steering drive belt.
6. On N and NT series trucks, remove fan.
7. On N, NT, C and CT series truck, disconnect heater hoses at engine.
8. Remove air cleaner and, if applicable, vent hose from carburetor.
9. Disconnect choke and throttle cables and accelerator linkage.
10. Disconnect tachometer cable (and bracket, if so attached) and position out of the way.
11. Remove ignition coil.
12. On a vehicle with an air compressor, relieve pressure from the system and disconnect main line from the compressor and treadle valve. Remove drive belt if it is still in place.
13. Disconnect fuel line (from tank) at the fuel pump and cap line.
14. Disconnect wires from the alternator and remove wiring harness from engine (or disconnect from junction block).
15. Disconnect cable from starter and remove starter. Remove flywheel lower housing attaching bolts first on L and LT truck. Disconnect engine-to-body ground strap.
16. Unbolt exhaust pipes from right and left exhaust manifolds.
17. Disconnect vacuum lines from intake manifold.
18. Remove clutch return spring and hydraulic clutch slave cylinder attaching bolts on C and CT trucks.
19. Remove driveshaft center bearing retainer (except on C and CT series) and position a jack under transmission.
20. Remove flywheel housing cover.
21. On B, F and FT series trucks, remove lower clutch housing attaching bolts. Secure lifting apparatus. Remove front engine mount nuts. Remove remaining clutch housing bolts. Raise engine high enough to remove the bolts that attach the engine front mount bracket to the upper insulator. Carefully lift engine away from the transmission and chassis. On C and CT vehicles, remove flywheel housing to engine attaching bolts. Remove nuts and bolts from front engine mount and upper insulator. Attach lifting apparatus and remove engine from chassis. On N and NT trucks, remove flywheel upper housing to engine attaching bolts. Remove the bolt(s) attaching the front engine mounting plate to the front engine mount. Attach lifting apparatus. Raise engine sufficiently to remove the front support bracket-to-upper insulator bolt and nuts. Remove

engine.
22. To install, carefully lower engine into chassis, aligning clutch disc splines with transmission shaft and aligning front engine mount bracket with upper insulator.
23. Install flywheel housing to engine attaching bolts, tightening securely.
24. Install flywheel housing cover.
25. Lower engine and remove jack from under transmission, being careful to keep front engine mounting aligned.
26. Install and tighten front engine mount bolts and nuts. On N and NT trucks, install front engine mount to mounting plate.
27. Remove lifting apparatus.
28. Install driveshaft center bearing support to frame crossmember.
29. On vehicle with power steering, install power steering unit if it was removed.
30. Install starter motor and starter cable, attaching engine-to-frame ground cable if applicable.
31. Install left and right exhaust pipes to exhaust manifolds, using new gaskets.
32. Connect alternator wires and engine wiring harness wires. Secure engine wiring harness to bracket on engine.
33. On C and CT trucks, install hydraulic clutch slave cylinder to the flywheel housing and attach the clutch return spring.
34. Secure coil and bracket to cylinder head and connect all leads.
35. Connect vacuum line(s) to intake manifold.
36. Connect fuel line to fuel pump.
37. On C and CT trucks install radiator supply tank to cab rear support.
38. Connect accelerator linkage, choke and throttle cables and tachometer cable. Adjust linkage and cables if necessary.
39. Connect heater hoses.
40. On vehicles equipped with air system, install and connect all components.
41. Connect engine-to-body ground strap.
42. On N and NT series trucks, install fan. On B, F, FT, C and CT trucks, place fan in radiator, shroud, then install radiator, shroud and fan (loose) in vehicle. Secure all attaching bolts, insulators and radiator supports. Make sure belts are ready to be installed. Install fan and belts.
43. Connect radiator upper and lower hoses. On C and CT series trucks, connect hoses to radiator supply tank.
44. On vehicle with automatic transmission, connect transmission oil cooler lines to radiator lower tank.
45. On vehicle with power steering,

install and tighten the drive belt and connect power steering pressure line to the pump reservoir and return line to the pump housing.

46. On C and CT series trucks, attach the heater hoses, throttle cable, choke cable and tachometer cable to the radiator.
47. Install air cleaner.
48. Install (if removed) and connect battery cables.
49. Fill crankcase and cooling system.
50. On B, F and FT series trucks, install the hood.
51. On N and NT series trucks, raise hood and connect check cables.
52. Adjust clutch as required.
53. Operate engine and check for lubricant and coolant leaks.

Manifolds

Exhaust Manifold Removal and Installation

170, 200 Cu. In. 6-Cylinder

1. Remove air cleaner and, if so equipped (later models), hot air duct.
2. Disconnect exhaust pipe from manifold.
3. Remove retaining bolts and manifold.
4. To install, clean mating surfaces of head and manifold and scrape gasket material from manifold exhaust pipe flange and pipe.
5. Apply graphite grease to the mating surface of the exhaust manifold.
6. Install manifold, retaining bolts and tab washers, tightening bolts from the center out to 13-18 ft. lbs. Bend tabs to lock bolts.
7. Install exhaust pipe on manifold, using new gasket.
8. Install air cleaner and hot air duct.
9. Start engine and check for exhaust leaks.

302 and 351M Cu. In. V-8

1. Remove air cleaner, intake duct and crankcase ventilation hose as an assembly.
2. Remove air cleaner inlet duct attaching bolts (Bronco and F-100) and oil dipstick tube bracket on right exhaust manifold.
3. Disconnect exhaust pipes from manifolds.
4. Remove heat shield, if equipped.
5. Remove bolts, washers and manifolds.
6. To install, clean mating surfaces of cylinder head and manifolds. Clean out exhaust pipe flange of

manifolds and exhaust pipe.
7. Apply graphite grease to mating surfaces of manifolds.
8. Install manifolds, heat shield, tab washers, and bolts, tightening from the center out to 12-16 ft. lbs. torque. Bend tabs to lock bolts.
9. Install exhaust pipe to flange on manifold, using a new gasket.
10. Install air cleaner inlet duct (Bronco and F-100) and oil dipstick tube bracket on the right exhaust manifold.
11. Install air cleaner and intake duct, including crankcase ventilation hose.

351M and 400 Cu. In. V8

1. If the right manifold is being removed, remove the air cleaner, intake duct and heat stove. If the left manifold is being removed, remove the oil filter.
2. On vehicles equipped with column selector and automatic transmission, disconnect the selector lever cross shaft for clearance.
3. Disconnect the muffler inlet pipe or the catalytic converter at the exhaust manifold. Remove the spark plug heat shields.
4. Remove the exhaust manifold attaching bolts and remove the manifold.
5. Clean the mating surfaces of the exhaust manifold and cylinder head. Clean the mounting flange of the exhaust manifold and muffler inlet pipe or catalytic converter.
6. Apply a graphite grease to the mating surfaces of the exhaust manifold.
7. Position the exhaust manifold on the head and install the attaching bolts. Working from the center to the ends, tighten the bolts to specification.
8. Install the spark plug heat shields.
9. Install the spacer between the inlet pipe and the exhaust manifold.
10. Connect the muffler inlet pipe or the catalytic converter at the exhaust manifold. Tighten the nuts to specification.
11. If the left exhaust manifold is being installed, install the oil filter.
12. On vehicles with automatic transmissions and column selector, connect the selector cross shaft at the chassis and cylinder block.
13. If a right manifold is being installed, install the air cleaner heat stove.
14. Install the air cleaner and intake duct.
15. Start the engine and check for exhaust leaks.

330, 359, 360, 361, 389, 390, 391 and 460 Cu. In. V-8 —All Models

1. Remove air cleaner and disconnect exhaust pipes from manifolds.
2. On all engines except the 460, disconnect the power steering pump bracket from the cylinder block and move it out of the way. Position the pump so that the oil will not drain out. Remove dipstick and tube assembly.
3. Remove attaching bolts, washers of cylinder head, manifold, manifold pipe flange and exhaust pipe.
4. To install, clean mating surfaces.
5. Apply graphite grease to mating surface of manifold, then install manifold, tab washers and bolts. Tighten bolts from the center out to specifications. Bend tabs to lock bolts.
6. Install dipstick and tube assembly.
7. Install power steering pump bracket and adjust belt tension.
8. Connect exhaust pipes to manifolds, using new gaskets.
9. Install air cleaner.

Intake Manifold Removal and Installation

302 and 351 W Cu. In. V-8

1. Drain cooling system.
2. Remove air cleaner and intake duct assembly, including crankcase ventilation hose.
3. Disconnect accelerator rod, choke cable and automatic transmission kickdown rod (if applicable) at the carburetor. Remove the accelerator retracting spring, where so equipped.
4. Disconnect high tension lead and wires from the coil.
5. Remove spark plug wire from plugs and harness brackets, then remove distributor cap and spark plug wire assembly.
6. Disconnect fuel inlet line at carburetor.
7. Disconnect distributor vacuum hoses and remove distributor.
8. Remove heater hose, radiator hose and water temperature sending unit wire from manifold.
9. Remove water pump bypass hose from coolant outlet housing.
10. Disconnect crankcase ventilation hose from valve rocker cover.
11. Remove intake manifold and carburetor as an assembly, prying manifold from cylinder head if necessary. Throw away gaskets and bolt sealing washers.
12. When disassembling, identify all vacuum hoses before disconnecting them. Remove coolant outlet

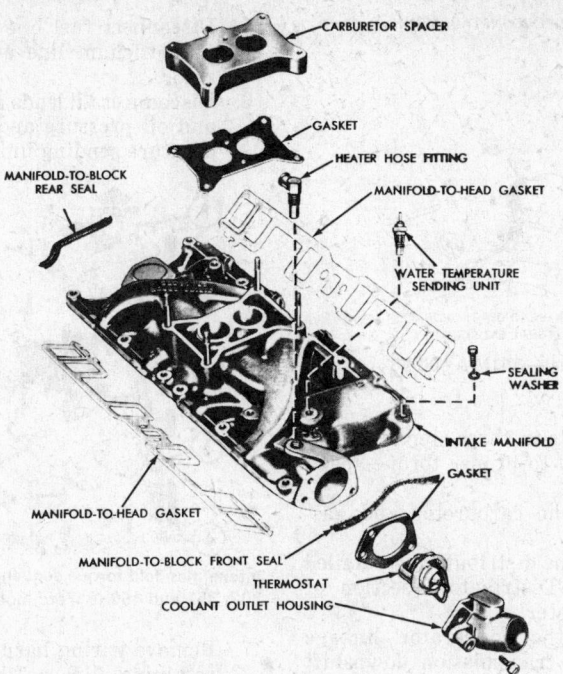

Intake manifold—302 and 351 V-8 (© Ford Motor Co.)

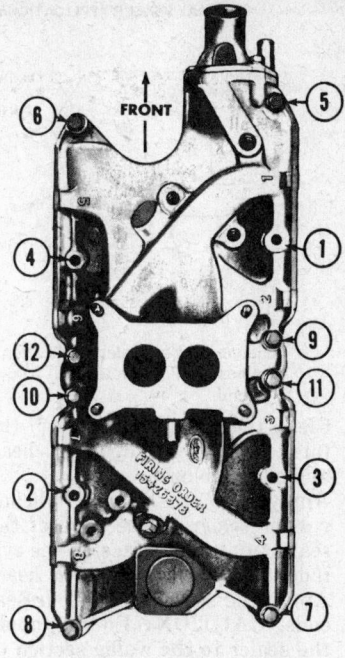

Intake manifold torque sequence— —302 V-8 (© Ford Motor Co.)

housing and gasket. Remove ignition coil and engine identification tag, temperature sending unit, carburetor, spacer, gasket, vacuum fitting, accelerator retracting spring bracket and choke cable bracket.

13. To install, first assemble manifold/carburetor unit by installing all components removed in Step 12 above, making sure vacuum lines are positioned correctly.

14. Clean all mating surfaces, using a suitable solvent to remove all oil. Apply block surfaces with adhesive sealer.

15. Position new gaskets and front and rear seals, using a nonhardening sealer at four gasket-seal junctions. Interlock gaskets with seal tabs and be sure all holes are aligned.

16. Carefully position manifold, making sure that gaskets and seals do not shift. Install bolts and new bolt seal washers, tightening in the sequence illustrated. *Retighten after engine has been operated until warmed up.* Torque bolts to 23-25 ft. lbs.

17. Install water bypass hose to coolant outlet housing, radiator upper hose and heater hose.

18. Install distributor as described in "Distributor Removal and Installation" above. Install distributor cap and spark plug wires, positioning wires in harness brackets on valve rocker covers.

19. Connect crankcase ventilation hose, high tension lead and coil wires, accelerator rod and retracting spring, choke cable and

automatic transmission kickdown rod (if applicable).

20. Fill and bleed cooling system.
21. Adjust ignition timing.
22. Connect vacuum hoses at distributor.
23. Operate engine until warmed up, checking for leaks.
24. Retorque manifold bolts.
25. Adjust transmission throttle linkage, if so equipped.
26. Install air cleaner and intake duct assembly including closed crankcase ventilation hose.

351M and 400 Cu. In. V8—All Models

1. Remove the air cleaner and intake duct. On vehicles with air conditioning, isolate and remove the compressor.
2. Disconnect the high tension lead and wires from the coil. Disconnect the engine harness and move it out of the way.
3. Disconnect the spark plug wires from the plugs. Remove the distributor cap and spark plug wire assembly.
4. Remove the air pump by pass valve and hose from the check valve.

5. Remove the carburetor fuel inlet line.
6. Remove the heater hoses from the retainers and position the hoses out of the way.
7. Remove the ignition coil, vacuum solenoid valve and bracket.
8. Disconnect the crankcase emission hose at the left rocker arm cover.
9. Disconnect the vacuum lines from the intake manifold.
10. Disconnect the distributor vacuum hose from the distributor. Remove the distributor as detailed in the "Distributor" section of this chapter. Block the distributor hole with a rag to prevent foreign material from entering the crankcase.
11. Disconnect the accelerator linkage and the transmission downshift linkage and position out of the way.
12. Remove the carburetor.
13. Remove the manifold attaching bolts, and remove the manifold. Remove and discard the manifold gasket and seals.
14. If the manifold assembly is to be disassembled, disconnect the vacuum hoses.

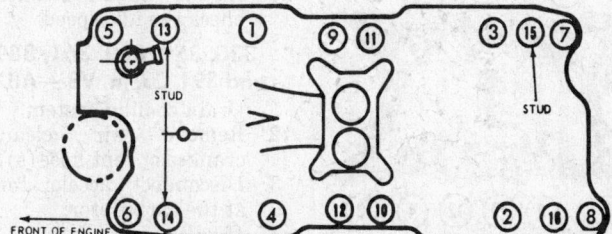

Intake manifold torque sequence—351 V-8 (© Ford Motor Co.)

TYPICAL SEALER APPLICATION AREAS FOR INTAKE MANIFOLD INSTALLATION

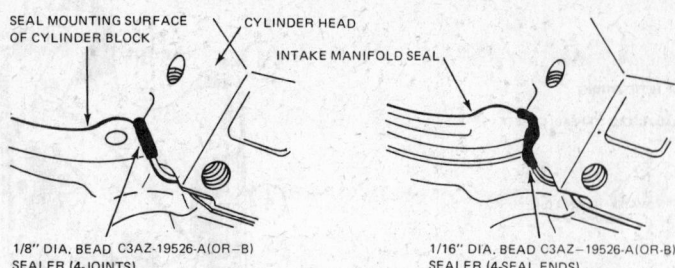

1/8" DIA. BEAD C3AZ-19526-A(OR-B) SEALER (4-JOINTS)

1/16" DIA. BEAD C3AZ-19526-A(OR-B) SEALER (4-SEAL ENDS)

Silicone rubber sealer application on intake manifold seal—351M & 400 V8 engines (© Ford Motor Co.)

15. Clean the mating surfaces of the intake manifold, cylinder head and engine block.
16. Apply a 1/8 in. bead of silicone rubber sealer at the edge of the seal mounting surface on the cylinder block and cylinder head. Apply the sealer at all four seal ends. CAUTION: Do not apply the sealer to the waffle section of the end seals as the sealer will rupture the seal material.
17. Position new seals on the cylinder block and press the seal locating extensions into the holes in the mating surface.
18. Apply a 1/6 in. bead of silicone rubber sealer to the outer end of each manifold seal, (above previous sealer application), along the full length of the seal end. CAUTION: Do not apply the sealer to the waffle section of the end seals as the sealer will rupture the seal material. The sealer sets up in 15 minutes so it is very important that the assembly be completed quickly.
19. Position the intake manifold gasket on the block and cylinder heads with the alignment notches under the dowels on the cylinder head. Be sure that the holes in the gasket are aligned with the holes in cylinder head.
20. Carefully lower the intake manifold into position on the cylinder block and heads.
21. Check and make sure that the holes in the manifold gaskets and the manifold are in alignment. Install the intake manifold attaching bolts. Tighten the in-

take bolts in three steps in sequence and to proper torque specifications.
22. Install the carburetor and gasket.
23. Install the distributor as detailed in the "Distributor" section of this chapter.
24. Install the accelerator linkage and the transmission downshift rod.
25. Install the vacuum solenoid valve and the ignition coil.
26. Connect the vacuum lines at the manifold. Install the air pump bypass air supply hose.
27. Position the wiring harness under the hold down clips on the left rocker arm cover and connect the wires to the ignition coil, water temperature sending unit and throttle solenoid.
28. Connect the PCV line at the left rocker arm cover.
29. Install the heater hoses in their retainers.
30. Connect the fuel inlet pipe to the carburetor.
31. Install the distributor cap and spark plug wires. Position the spark plug wires in the harness brackets on the rocker arm covers and connect the wires to the spark plugs.
32. Install the air conditioning compressor, if vehicle is so equipped.
33. Start the engine and check for leaks. Adjust the ignition timing and connect the distributor vacuum line.
34. When the engine temperature has stabilized, adjust the idle mixture and speed.
35. *Retorque the intake manifold in sequence and to specifications.*
36. Install the air cleaner and recheck the idle speed.

330, 359, 360, 361, 389, 390, and 391 Cu. In. V8—All Models

1. Drain cooling system.
2. Remove air cleaner and crankcase vent hose(s).
3. Disconnect accelerator linkage at the carburetor.
4. If so equipped, remove the accelerator cross shaft bracket from the intake manifold.

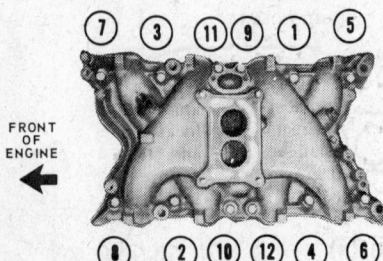

Intake manifold torque sequence—351M & 400 V8 engines (© Ford Motor Co.)

5. Disconnect fuel line and distributor vacuum line at the carburetor.
6. Disconnect all leads at the coil and oil pressure and water temperature sending units.

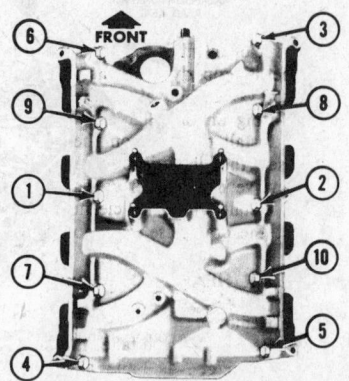

Intake manifold torque sequence—V-8 except 302, 351 and 460 (© Ford Motor Co.)

7. Remove wiring harness from the retaining clips on the left valve rocker cover.
8. Disconnect spark plug wires, remove wires from harness brackets, remove distributor cap and spark plug wire assembly.
9. Disconnect and remove distributor vacuum line.
10. Remove distributor.
11. Disconnect radiator upper hose(s) at the water outlet housing, heater hoses and, on C series vehicle, the coolant supply tank hose at the water outlet housing.
12. Disconnect water pump bypass hose(s) at the water pump.
13. Remove valve rocker covers and crankcase ventilation hoses.
14. Remove valve rocker shaft assembly as described below in the section, "Rocker Arm Shaft Assembly Removal and Installation."
15. Remove and identify pushrods so that they can be put back in their original positions.
16. Remove manifold attaching bolts.
17. Install eyebolts (5/16-18 thread) in the left front and right rear rocker arm cover screw holes and attach lifting sling.
18. Carefully lift out intake manifold, then remove seals and gaskets.
19. Remove water pump bypass hoses(s), water outlet housing, carburetor, gaskets spacer and water temperature sending unit.
20. Before installing, assemble components removed in Step 19. Use a new gasket and sealing compound when installing water outlet (thermostat) housing. Use electrical-conductive sealer when installing water temperature sending unit.

21. Thoroughly clean all manifold, cylinder head and block mating surfaces (use solvent to remove all traces of oil). Coat block seal surfaces with quick-setting seal adhesive and coat mating surfaces of cylinder heads and block with non-hardening Oil-resistant sealer. Position new seals on block and cylinder heads, *making sure they are properly aligned.* Position manifold gasket slots over the end tabs on the seals and coat these junctions with non-hardening sealer.

22. Install eyebolts in intake manifold and attach lifting sling, then carefully lower manifold onto engine. Position manifold by inserting distributor and check that seals and gaskets are still properly aligned and that all holes line up.

23. Install manifold attaching bolts, coating under side of bolt heads with non-hardening sealer. Tighten bolts to 32-35 ft. lbs. torque in the sequence illustrated. Retorque bolts after engine has been run and warmed up. Remove distributor, lifting sling and eyebolts.

24. Connect water pump bypass hose (s) to water pump, radiator upper hose, heater hoses, water temperature sending unit and, on C series trucks, coolant supply tank hose.

25. Apply lubriplate to both ends of pushrods and install them in their original positions. Install valve rocker shaft as described in "Rocker Arm Shaft Assembly Removal and Installation" below.

26. Install the distributor as described in "Distributor Removal and Installation" above.

27. Install rocker covers, using new gasket and sealer, tightening to 10-12 ft. lbs., waiting two minutes then torquing again.

28. Connect crankcase ventilation hoses.

29. Install carburetor fuel inlet line, spark plug wires, wiring harness, distributor vacuum line and distributor cap.

30. Connect oil pressure sending unit wire and coil wire and lead.

31. Install accelerator cross shaft bracket (if applicable) and accelerator rod.

32. Fill and bleed cooling system.

33. Install air cleaner and vent hose.

34. Start engine, then check ignition timing idle speed and idle fuel mixture, then let engine warm up.

35. Retorque intake manifold bolts to specifications.

460 V-8

1. Drain the cooling system and re-

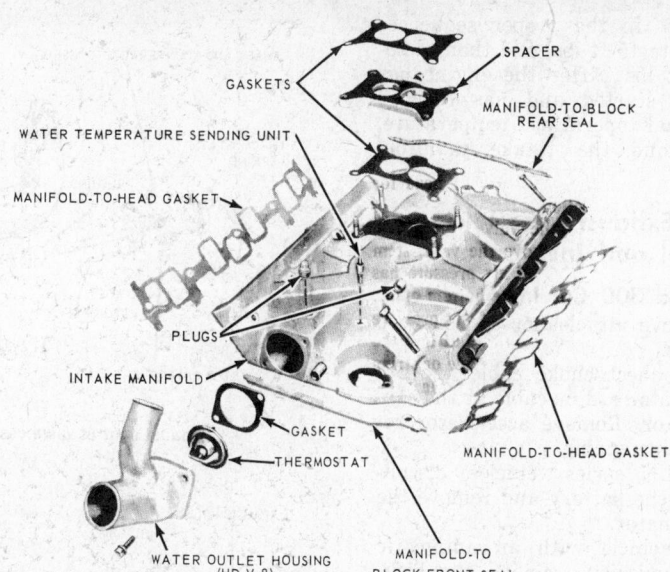

Typical V-8 intake manifold except 302, 351 and 460 V-8 (© Ford Motor Co.)

move the air cleaner assembly.

2. Disconnect the upper radiator hose at the engine.

3. Disconnect the heater hoses at the intake manifold and the water pump. Position them out of the way. Loosen the water pump by-pass hose clamp at the intake manifold.

4. Disconnect the PCV valve and hose at the right valve cover. Disconnect all of the vacuum lines at the rear of the intake manifold and tag them for proper reinstallation.

5. Disconnect the wires at the spark plugs, and remove the wires from the brackets on the valve covers. Disconnect the high-tension wire from the coil and remove the distributor cap and wires as an assembly.

6. Disconnect all of the distributor vacuum lines at the carburetor and vacuum control valve and tag them for proper installation. Remove the distributor and vacuum lines as an assembly.

7. Disconnect the accelerator linkage at the carburetor. Remove the speed control linkage bracket, if so equipped, from the manifold and carburetor.

8. Remove the bolts holding the accelerator linkage bellcrank and position the linkage and return springs out of the way.

9. Disconnect the fuel line at the carburetor.

10. Disconnect the wiring harness at the coil battery terminal, engine temperature sending unit, oil pressure sending unit, and other connections as necessary. Disconnect the wiring harness from the clips at the left valve cover and position the harness out of the way.

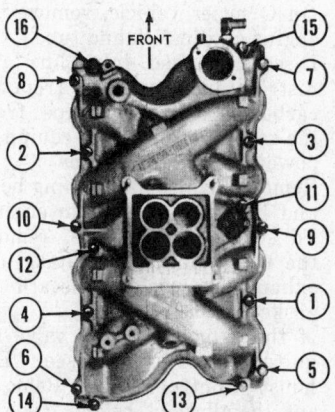

Intake manifold torque sequence— —460 V-8 (© Ford Motor Co.)

11. Remove the coil and bracket assembly.

12. Remove the intake manifold attaching bolts and lift the manifold and carburetor from the engine as an assembly. It may be necessary to pry the manifold away from the cylinder heads. Do not damage the gasket sealing surfaces.

13. Install the intake manifold in the reverse order of removal. Clean all gasket material from the mating surfaces of the manifold and cylinder heads and block. Glue the intake manifold end seals in place before installing the manifold. Use sealer at each end of the intake manifold-to-cylinder head gaskets for the full width of the gasket. When the manifold is placed on top of the engine run your fingers around the end seal areas to make sure that the end seals have not shifted. If they have, remove the manifold and reposition the seals. Tighten the intake manifold bolts in two

stages in the proper sequence; first to 15 ft lbs, and then to 25-30 ft lbs. After the engine has been started and has reached normal operating temperature, retorque the intake manifold bolts.

Intake/Exhaust Manifold Removal and Installation

240 and 300 Cu. In. 6-Cylinder

1. Remove air cleaner and hot air ducts.
2. Disconnect choke cable and accelerator rod or cable at the carburetor. Remove accelerator retracting spring.
3. On LN series vehicles, disconnect the battery and remove the alternator.
4. On vehicle with an automatic transmission, remove the kickdown rod retracting spring and remove the accelerator rod bellcrank assembly.
5. On C series vehicle, remove the engine oil dipstick and tube.
6. Disconnect fuel inlet line and distributor vacuum line from the carburetor, exhaust pipe from the manifold and, if so equipped, power brake vacuum line.
7. Remove manifold attaching bolts and lift manifolds from engine.
8. To separate manifolds, remove the nuts joining the intake and exhaust manifolds. Discard all gaskets.
9. If the exhaust control valve requires replacement, see "Exhaust Control Valve Removal and Installation" below.
10. To install, clean the mating surfaces of cylinder head and manifolds.
11. If a new manifold is to be used, remove the tube fittings on the old manifold and install them on the new one.
12. Before joining exhaust and intake manifolds, coat the mating surfaces lightly with graphite grease. Use a new gasket and tighten the nuts finger tight.
13. Coat the mating surfaces with graphite grease and install the manifold assembly on the cylinder head. Use a new intake man-

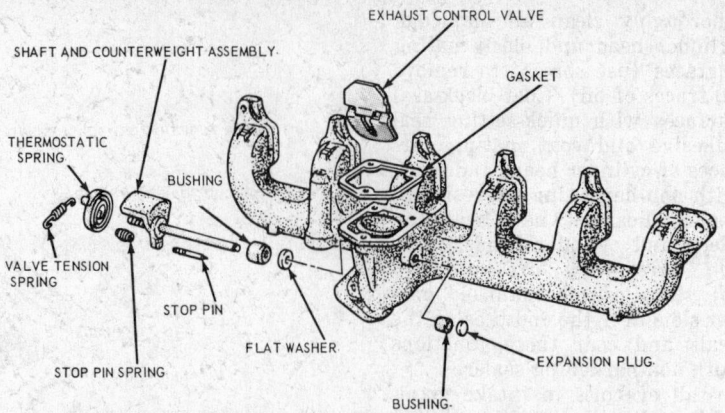

Exhaust control valve assy.—240 & 300 engines (© Ford Motor Co.)

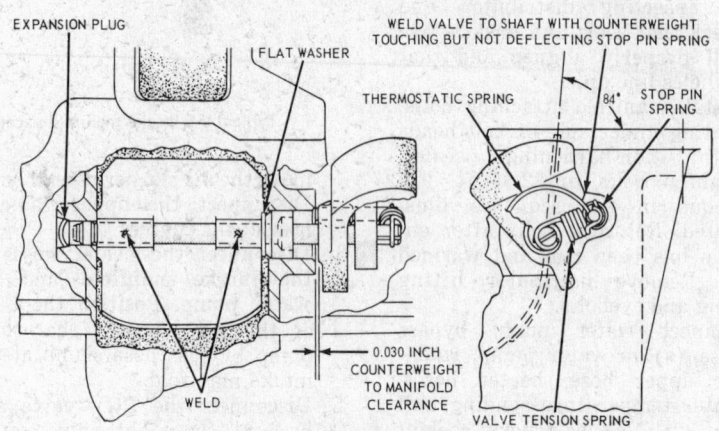

Exhaust valve plate position & counterweight clearance (© Ford Motor Co.)

ifold gasket. Tighten the bolts and nuts to 23-28 ft. lbs. torque in the sequence illustrated.
14. Tighten the intake to exhaust manifold stud nuts to 28-33 ft. lbs. torque.
15. Connect exhaust pipe to manifold, tightening nuts to 25-30 ft. lbs. torque.
16. Connect crankcase vent hose to intake manifold inlet tube, fuel inlet line and distributor vacuum line to carburetor, accelerator rod or cable and choke cable to carburetor. Install the accelerator retracting spring.
17. On LN series trucks, install the alternator and belts and connect the battery.
18. On C series trucks, install the dipstick and tube.
19. On a vehicle with an automatic transmission, install the bellcrank assembly and kickdown rod retracting spring. Adjust the transmission control linkage.
20. Install the air cleaner and hot air duct.
21. Adjust idle speed and idle fuel mixture.

Exhaust Control Valve Removal and Installation

240 and 300 Cu. In. 6-Cylinder

1. Separate intake and exhaust manifolds.
2. Remove valve tension spring, thermostatic spring and stop pin.
3. The valve shaft must be cut with a torch on each side of the valve plate. Remove valve plate and expansion plug.
4. Remove bushings and install new ones. There are two sizes of replacement bushings (OD) so make sure the right ones are used. Ream the ID of bushings to 0.51-0.253". The shorter bushing (front) is installed 0.010-0.015" below inside surface of the manifold and the longer bushing (rear) protrudes into

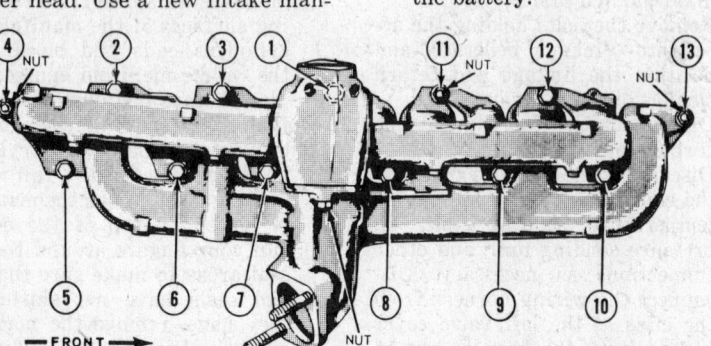

Manifold tightening sequence—240 & 300 engines (© Ford Motor Co.)

the manifold cavity 0.020″ (note that beveled end points inward).

5. Slide new shaft into the bushings, flat washer and valve plate. Note that the flat washer is between the valve plate and the long (rearward) bushing. Install a new stop pin spring on the stop pin.

6. Rotate the counterweight and shaft assembly clockwise until the counterweight contacts the stop pin spring, then place a 0.030″ feeler gauge between the counterweight and the manifold to maintain the specified clearance. Hold the valve plate at a 84 degree angle to the top surface of the manifold as illustrated and tack-weld the plate to the shaft, *using stainless steel welding rod.*

7. Check for free movement of the valve and install expansion plug in the bushing bore.

8. Install the thermostatic spring, positioned so that it will be necessary to wind the spring ½ turn clockwise to hook it over the stop pin.

9. Install a new valve tension spring on the exhaust control valve shaft and the stop pin.

Valve Rocker Arm Shaft Assembly

Removal and Installation

330, 359, 360, 361, 389, 390, and 391 Cu. In. Engine

1. Remove air cleaner, disconnect spark plug leads and remove leads from bracket on the valve rocker cover.

2. Remove crankcase ventilation hose from rocker cover, then remove rocker cover. On left rocker cover the wiring harness must be removed.

3. On right side, start at No. 4 cylinder (rearmost) and loosen the support bolts in sequence, two turns at a time. Remove the shaft assembly and baffle plate after all the bolts have been loosened. The same procedure is followed on the left bank, except that the bolt-loosening sequence starts with the No. 5 cylinder (foremost). **CAUTION: The above bolt-loosening procedure must be followed to avoid damage to the rocker arm shaft.**

4. To install, apply Lubriplate to the pad end of the rocker arms, to the tip of the valve stems and to both ends of the pushrods.

5. Rotate engine to 45 degrees *past* No. 1 cylinder TDC.

6. With the pushrods in place, position rocker arm shaft assembly

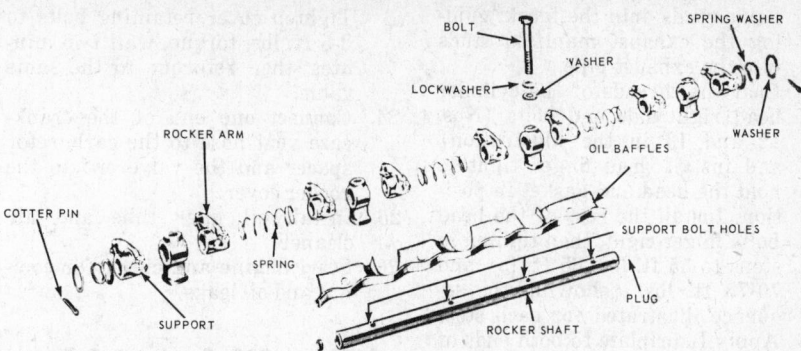

Valve rocker arm shaft assembly V8—typical (© Ford Motor Co.)

and baffle plate on the cylinder head such that *oil holes are on the bottom and identification notch is down and toward the front on the right bank and toward the rear on the left bank.* Tighten support bolts finger tight.

7. On the right bank, start at No. 4 cylinder and tighten the support bolts two turns at a time in sequence (4-3-2-1) until the supports are fully in contact with the cylinder head. Then tighten the support bolts to 40-45 ft. lbs. torque. The same procedure is followed on the left valve rocker arm shaft support bolts, starting with the No. 5 cylinder. This procedure allows time for the hydraulic lifter leakdown and thus prevents damage to pushrods, valves and rocker arms.

8. Check valve clearances and adjust if necessary.

9. Install rocker cover, using new gaskets and sealer.

10. Tighten cover retaining bolts to 10-12 ft. lbs, wait two minutes, then tighten to the same torque again.

11. Install crankcase ventilation regulator valve and hose(s), connect spark plug wires and crankcase vent hose, and install air cleaner.

Rocker arm shaft identification notch (© Ford Motor Co.)

Cylinder Head

Cylinder Head Removal and Installation

170, 200 Cu. In. 6-Cylinder

1. Drain cooling system, remove

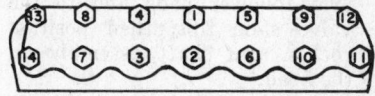

Cylinder head torque sequence—170, 200 cu. in. 6-cylinder (© Ford Motor Co.)

the air cleaner and oil filler tube, and disconnect the battery cable at cylinder head.

2. Disconnect exhaust pipe from manifold.

3. Disconnect accelerator rod retracting spring, choke control cable and accelerator rod at the carburetor, transmission kickdown rod (automatic transmission), accelerator linkage at bellcrank, fuel inlet line at fuel filter hose, distributor vacuum line at the carburetor and other vacuum lines as necessary for accessibility (identify them for proper reinstallation).

4. Remove upper radiator hose at the coolant outlet housing.

5. Disconnect the distributor vacuum line at the distributor and carburetor fuel inlet line at the fuel pump.

6. Disconnect spark plug wires at the plugs and temperature sending unit wire at the unit.

7. Remove PCV valve and hose from valve rocker cover and disconnect the other end of the hose from the intake manifold.

8. Remove the valve rocker arm cover, then remove the rocker arm shaft support bolts by loosening them two turns at a time in sequence.

9. Lift off rocker arm shaft.

10. Remove cylinder head bolts and cylinder head. *Do not pry.*

11. Before installing, clean gasket surfaces of cylinder head and engine block, install guide studs at each end of block, apply cylinder head gasket sealer evenly to both sides of the head gasket, and run the gasket down the guide studs into position on the engine block.

12. Put a new gasket on the flange of the exhaust pipe, then carefully lower the head down the

guide studs onto the block, guiding the exhaust manifold studs into the exhaust pipe.

13. Coat the threads of the cylinder head right side end bolts (Nos. 12 and 13 in the illustration) and install them finger tight to hold the head and gasket in position. Install the rest of the head bolts finger tight, then tighten in steps to 55 ft. lbs., 65 ft. lbs. and 70-75 ft. lbs., following the sequence illustrated for each step.

14. Apply Lubriplate to both ends of their original locations.

15. Apply Lubriplate to the rocker arm follower pads and to the valve stem tips, then position rocker arm shaft assembly on the head.

16. Install the rocker arm shaft support bolts and tighten them in sequence two turns at a time until they are torqued to 30-35 ft. lbs.

17. Check and adjust the preliminary (cold) lash as described in "Valve Clearance Adjustment" below.

18. Install lockwashers and nuts to exhaust manifold to exhaust pipe studs and tighten to 25-35 ft. lbs. torque.

19. Connect radiator upper hose at the coolant outlet housing.

20. Install vacuum lines(s) and fuel line.

21. Connect accelerator linkage at the bellcrank assembly, transmission kickdown rod, accelerator rod retracing spring, battery cable at the cylinder head, choke control cable and accelerator rod at carburetor, temperature sending unit wire and spark plug leads. Adjust choke cable.

22. Temporarily install rocker cover and PC valve, then operate engine until it is warmed up. Adjust final (hot) valve lash as described in "Valve Clearance Adjustment" below.

23. Clean rocker cover and head gasket surfaces and install cover using a new seal coated on both sides with oil-resistant sealer.

Tighten cover retaining bolts to 3-5 ft. lbs. torque, wait two minutes, then retorque to the same value.

24. Connect one end of the crankcase vent hose to the carburetor spacer and the valve end to the rocker cover.

25. Install oil filler tube and air cleaner.

26. Start engine and check for coolant and oil leaks.

240 and 300 Cu. In. 6-Cylinder

1. Drain cooling system and remove air cleaner and crankcase ventilation valve.

2. Disconnect the following: the vent hose at the intake manifold inlet tube, carburetor fuel inlet line, distributor vacuum line, choke cable, accelerator cable and heater hose at the coolant outlet elbow. Remove accelerator cable retracting spring.

3. On a vehicle with automatic transmission, disconnect the kickdown rod at the carburetor.

4. Disconnect upper radiator hose and exhaust pipe.

5. Remove the coil and the valve rocker cover and disconnect spark plug wires.

6. Loosen the rocker arm stud nuts so that the rocker arms can be twisted aside, then remove the pushrods, identifying each so that it may be installed in its original position.

7. Remove head bolts and install eyebolts for lifting (see illustration). Attach lifting apparatus and lift off cylinder head and manifold assembly. *Do not pry between head and block.*

8. Before installing, clean mating surfaces of block, cylinder head and exhaust pipe.

9. Position new gasket over the dowel pins on the cylinder block, then carefully lower head into place on the block. Remove lifting apparatus.

10. Oil the threads and install head bolts. Tighten the bolts in three

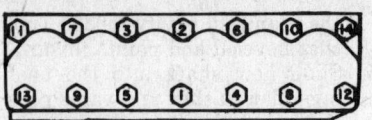

Cylinder head bolt tightening sequence—6 cyl.—240 & 300 engines
(© Ford Motor Co.)

steps, following the sequence illustrated in each step: first to 50-55 ft. lbs., then to 60-65 ft. lbs, and finally to 70-75 ft. lbs.

11. Connect exhaust pipe to manifold flange, using a new gasket and tighten the nuts to 25-30 ft. lbs.

12. Apply Lubriplate to both ends of the pushrods and install them in their original positions.

13. Apply Lubriplate to both the rocker arm fulcrum seat and the fulcrum seat socket of each rocker arm and install the rocker arms, tightening the stud nuts just enough to hold the pushrods. Adjust the valve lash as described in "Valve Clearance Adjustment" below.

14. Install the rocker cover, using oil-resistant sealer and a new gasket. Tighten the retaining bolts to 4-7 ft. lbs.

15. Connect the spark plug wires to the plugs, crankcase vent hose to the intake manifold, PCV valve in the valve rocker cover, fuel inlet line, distributor vacuum hose, accelerator cable and choke cable. Install accelerator cable retracting spring.

16. On a vehicle with automatic transmission, connect the kickdown rod to the carburetor.

17. Connect radiator upper hose to the coolant outlet housing and the heater hose to the coolant outlet housing, leaving the clamp loose.

18. Fill the cooling system and bleed. Then tighten the heater hose clamp.

19. Operate the engine until it is warmed up, checking for leaks.

20. Adjust engine idle speed and idle fuel mixture.

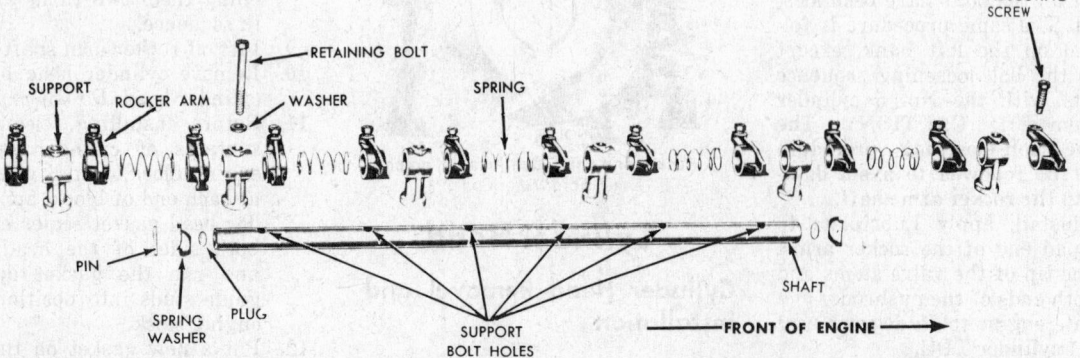

Rocker arm shaft assembly—6 cyl.—typical (© Ford Motor Co.)

302 and 351W Cu. In. V-8

1. Remove the intake manifold and carburetor as an assembly as described in "Intake Manifold Removal and Installation" above.
2. Remove rocker arm cover.
3. To remove right cylinder head, loosen alternator adjusting arm bolt and remove the alternator mounting bracket bolt and spacer. Swing alternator down out of the way. On Bronco and Econoline trucks, remove the coil and air cleaner inlet duct from the right head. To remove left cylinder head, remove accelerator shaft fastening bolts at the front of the head.
4. Disconnect exhaust pipe from the manifold.
5. Loosen rocker arm stud nuts and twist rocker arms so that the pushrods may be removed. Identify the pushrods when removing so that they may be reinstalled in their original locations.
6. Remove exhaust valve stem caps, on 302 engines.
7. Install cylinder head holding fixtures, remove head bolts and lift off head.
8. To install, clean all gasket surfaces of block, head and rocker cover. Position new head gasket over the dowels onto the block (do not use sealer on this composition gasket). Install head and remove holding fixture.
9. Install head bolts and tighten in three steps: first to 50 ft. lbs, then to 60 ft. lbs, and finally to 65-72 ft. lbs. Tighten in the sequence illustrated for each step.
10. Clean pushrods, blowing out oil passage, and check them for straightness. Lubricate pushrod ends, valve stem tips and rocker arm cups, fulcrum sets and followers. Install pushrods in their original positions, install exhaust stem caps and install rocker arms. Adjust the valve clearance as described in "Valve Clearance Adjustment" below.
11. Connect the exhaust pipe to the manifold, using new gasket and tightening nuts to 25-35 ft. lbs. torque.
12. On right cylinder head, position the alternator and install the attaching bolt and spacer, ignition coil (Bronco and Econoline) and air cleaner inlet duct. Adjust drive belt tension. On left cylinder head, install accelerator shaft assembly.
13. Install rocker cover using new gasket and tightening cover bolts to 3-5 ft. lbs.
14. Install intake manifold and carburetor assembly as described in "Intake Manifold Removal and Installation" above.

330, 359, 360, 361, 389, 390 and 391 Cu. In. V-8

Removal and installation of cylinder heads on the larger V-8's is essentially the same procedure as that described above for the 302 and 351W cu. in. engines. The intake manifold and carburetor assembly and rocker arm shaft assembly are removed first (see "Intake Manifold Removal and Installation" above). When installing new head gasket, note the word "front" on the gasket and install accordingly. The head bolts are tightened in three steps, first to 70 ft. lbs., then to 80 ft. lbs., and finally to 85-90 ft. lbs. Tighten in the sequence illustrated for each step.

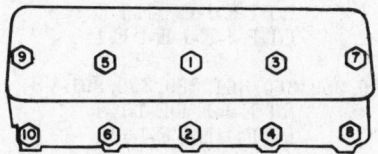

Cylinder head bolt tightening sequence— all V8 engines

351M and 400 Cu. In. V8

1. Remove the intake manifold and carburetor as an assembly. This procedure is described earlier in "Intake Manifold Removal and Installation."
2. Remove the rocker arm cover.
3. If the left cylinder head is being removed; isolate and remove the air conditioning compressor (if so equipped), disconnect the power steering pump bracket, and remove the drive belt from the pump pulley. Position the power steering pump out of the way and in a position that will prevent the fluid from draining out.
4. If the right cylinder head is being removed; remove the alternator mounting bracket through bolt and the air cleaner duct from the cylinder head assembly. Disconnect the ground strap from the rear of the cylinder head.
5. Disconnect the muffler inlet pipe or catalytic converter from the exhaust manifold.
6. Remove the rocker arm bolts, oil deflectors, fulcrum seats, rocker arms, and push rods in sequence so that they may be installed in there original positions.
7. Remove the cylinder head attaching bolts and lift the head off the cylinder block. Remove and discard the head gasket.
8. Clean the cylinder head, intake manifold, rocker arm cover and cylinder head mounting surface. If the cylinder head was removed to replace the head gasket, check the flatness of the head and cylinder block.
9. Position a new head gasket over the cylinder dowels on the block. Place the cylinder head on the block and install the attaching bolts.
10. The cylinder head bolts are tightened in three progressive steps. Tighten all the bolts in sequence to 75 ft-lbs., then to 95 ft-lbs. and finally to 105 ft-lbs. It will not be necessary to retorque the bolts after extended operation.
11. Clean the push rods and blow out the oil passages. Check the ends of the push rods for nicks, grooves, roughness or excessive wear. Visually check the push rods for straightness or check runout with a dial indicator. If the runout exceeds the maximum limit, replace the push rod. *Do not attempt to straighten a push rod.*
12. Lubricate and install the push rods in their original positions.
13. Lubricate and install the rocker arms, fulcrum seats, oil deflectors, and rocker arm bolts.
14. Connect the muffler inlet pipe or catalytic convertor to the exhaust manifold. Tighten the nuts to specification.
15. Install the alternator mounting bracket through bolt and air cleaner inlet duct on the right cylinder head. Connect the ground strap at the rear of the right cylinder head.
16. Install the power steering pump bracket and air conditioning compressor (if equipped) on the left cylinder head.
17. Adjust the drive belts to specification.
18. Apply sealer to the rocker arm cover gasket to hold the gasket in place. Position the gasket on the cover aligning the holes in cover and gasket. Install the rocker arm cover.
19. Install the intake manifold assembly as described earlier in this section, "Intake Manifold Removal and Installation."

460 V-8s

1. Remove the intake manifold and carburetor as an assembly.
2. Disconnect the exhaust pipe from the exhaust manifold.
3. Loosen the air conditioning compressor drive belt, if so equipped.
4. Loosen the alternator attaching bolts and remove the bolt attaching the alternator bracket to the right cylinder head.
5. Disconnect the air conditioning compressor from the engine and move it aside, out of the way. Do not discharge the air conditioning system, if possible.
6. Remove the bolts securing the power steering reservoir bracket

to the left cylinder head. Position the reservoir and bracket out of the way.

7. Remove the valve rocker arm covers. Remove the rocker arm bolts, rocker arms, oil deflectors, fulcrums and pushrods in sequence so that they can be reinstalled in their original positions.

8. Remove the cylinder head bolts and lift the head and exhaust manifold off the engine. If necessary, pry at the forward corners of the cylinder head against the casting bosses provided on the cylinder block. Do not damage the gasket mating surfaces of the cylinder head and block by prying against them.

9. Remove all gasket material from the cylinder head and block. Clean all gasket material from the mating surfaces of the intake manifold. If the exhaust manifold was removed, clean the mating surfaces of the cylinder head exhaust port areas and install the exhaust manifold.

10. Position the two long cylinder head bolts in the two rear lower bolt holes of the left cylinder head. Place a long cylinder head bolt in the rear lower bolt hole of the right cylinder head. Use rubber bands to keep the bolts in position until the cylinder heads are installed on the cylinder block.

11. Position new cylinder head gaskets on the cylinder block dowels. Do not apply sealer to the gaskets, heads, or block.

12. Place the cylinder heads on the block, guiding the exhaust pipe connections. Install the remaining cylinder head bolts. The longer bolts go in the lower row of holes.

13. Tighten all the cylinder head attaching bolts in the proper sequence in three stages: 75 ft lbs, 105 ft lbs, and finally, to 135 ft lbs. When this procedure is used, it is not necessary to retorque the heads after extended use.

14. Connect the exhaust pipes to the exhaust manifolds.

15. Install the intake manifold and carburetor assembly. Tighten the intake manifold attaching bolts in the proper sequence to 25-30 ft lbs.

16. Install the air conditioning compressor to the engine.

17. Install the power steering reservoir to the engine.

18. Apply oil-resistant sealer to one side of the new valve cover gaskets and lay the cemented side in place in the valve covers. Install the covers.

19. Install the alternator to the right cylinder head and adjust the alternator drive belt tension.

20. Adjust the air conditioning compressor drive belt tension.
21. Fill the radiator with coolant.
22. Start the engine and check for leaks.

Reference

For cylinder head overhaul procedures see General Repair Section.

Valve Train

Valve Arrangement

170, 200-Six: E-I-E-I-E-E-I-E-I-I-E

240, 300 Six: E-I-E-I-E-I-E-I-E-I-E-I

302, 351W 351M, 400, 460-V8:
RT I-E-I-E-I-E-I-E
LT E-I-E-I-E-I-E-I

330, 359, 360, 361, 389, 390, 391-V8:
RT E-I-E-I-E-I-E-I
LT E-I-E-I-E-I-E-I

401, 475, 477, 534-V8:
RT E-I-E-I-E-I-I-E-I-E
LT E-I-E-I-E-I-I-E-I-E

Valve Clearance (Lash) Adjustment

All engines used in full-size Ford products are equipped with hydraulic valve lifters except the 401, 475, 477, 534 cu. in. versions. Valve systems with hydraulic valve lifters operate with zero clearance in the valve train, and because of this the rocker arms are non-adjustable. The only means by which valve system clearances can be altered is by installing .060 in. over-or undersize pushrods; but, because of the hydraulic lifter's natural ability to compensate for slack in the valve train, all components of the valve system should be checked for wear if there is excessive play in the system.

Preliminary Valve Adjustment
6-Cylinder

1. Crank the engine until the TDC mark on the crankshaft damper is aligned with timing pointer on the cylinder front cover.
2. Scribe a mark on the damper at this point.
3. Scribe two more marks on the damper, each equally spaced from the first mark (see illustration).
4. With the engine on TDC of the compression stroke (mark A aligned with the pointer), back off the rocker arm adjusting nut until there is end-play in the pushrod. Tighten the adjusting nut until all clearance is removed, then tighten the adjusting nut one additional turn. To determine when all clearance is removed from the rocker arm, turn the pushrod with the

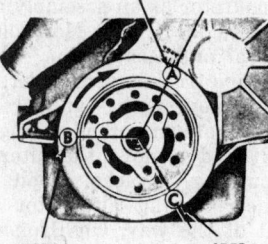

STEP 1—SET NO. 1 PISTON ON T.D.C. AT END OF COMPRESSION STROKE ADJUST NO. 1 INTAKE AND EXHAUST

STEP 4—ADJUST NO. 6 INTAKE AND EXHAUST

STEP 2—
ADJUST NO. 5
INTAKE AND
EXHAUST

STEP 3—
ADJUST NO. 3
INTAKE AND
EXHAUST

STEP 5—
ADJUST NO. 2
INTAKE AND
EXHAUST

STEP 6—
ADJUST NO. 4
INTAKE AND
EXHAUST

6-cylinder preliminary valve adjustment
(© Ford Motor Co.)

fingers. When the pushrod can no longer be turned, all clearance has been removed.

5. Repeat this procedure for each valve, turning the crankshaft ⅓ turn to the next mark each time and following the engine firing order of 1-5-3-6-2-4.

All V-8

NOTE: The early 302 V8 engine has rocker arm mounting studs which do not incorporate a positive stop shoulder on the mounting stud. These engines were originally equipped with this kind of stud. However, due to production differences, it is possible some early 302 engines may be encountered that are equipped with positive stop rocker arm mounting studs. Before adjusting the valves, verify that the rocker arm mounting studs do not incorporate a positive stop shoulder. On studs without a positive stop, the shank portion of the stud that is exposed just above the cylinder head is the same diameter as the threaded portion, at the top of the stud, to which the rocker arm retaining nut attaches. If the shank portion of the stud is of greater diameter than the threaded portion, this identifies it as a positive stop rocker arm stud and the adjustment specifications for the 351 engine with adjusting nuts should be used. Only the 302 and 351W engines require a preliminary valve adjustment. All other V8s use either a bolt and fulcrum (460) or rocker shafts (330, 360, 361, 390, 391).

302 and 351W
With Rocker Arm Adjusting Nuts

1. Crank the engine until #1 cylinder is at TDC of the compression stroke and the timing pointer is aligned with the mark on the crankshaft damper.
2. Scribe a mark on the damper at this point.
3. Scribe three more marks on the damper, dividing the damper into quarters.

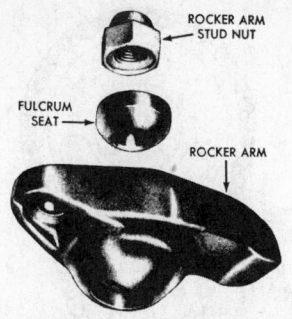

Stud/nut type rocker arm (© Ford Motor Co.)

4. With the first mark aligned with the timing pointer, adjust the valves on #1 cylinder by backing off the adjusting nut until the pushrod has free play in it. Then, tighten the nut until there is no free play in the rocker arm. This can be determined by turning the pushrod while tightening the nut; when the pushrod can no longer be turned, all clearance has been removed. After the clearance has been removed. tighten the nut an additional ¾ of a turn. (302 V8 w/o positive stop rocker arm studs).

5. Rocker arm adjusting nut tightening specifications are: 302 (with positive stop rocker arm studs and 351W—tighten the nut until it contacts the rocker shoulder, then torque it to 18-20 ft lbs.

6. Repeat this procedure for each valve, turning the crankshaft ¼ turn to the next mark each time and following the engine firing order.

330, 360, 361, 390, 391

These engines do not require a preliminary valve adjustment. In the event of cylinder head removal or some operation requiring that the valve train be disturbed, torque the rocker arm shaft supports to 40-45 ft lbs.

460

These engines use a bolt and fulcrum rocker arm and require no pre-

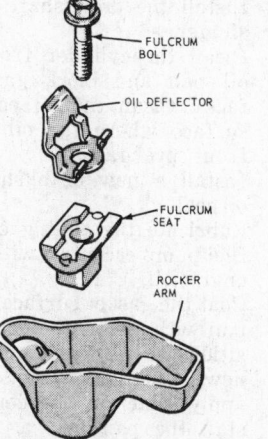

Bolt and fulcrum rocker arm (© Ford Motor Co.)

liminary valve adjustment. In the event that the valve train is disturbed, install the fulcrum, oil deflector, and tighten the bolt to 18-25 ft lbs.

Valve Overhaul

See the General Repair Section for complete valve overhaul procedures. All valves are removed by compressing the spring with a valve spring compressing tool, then removing the keepers from the end of the valve stem. Most models utilized an O-ring or cup type oil seal as illustrated. Valve spring, stem and seal specifications may be found in the Valve Specifications Table at the beginning of this section.

Valve rocker arm shaft assembly removal and installation procedures may be found immediately following "Intake Manifold Removal and Installation" above (330, 360, 361, 390 and 391 cu. in. V-8) and under "Cylinder Head Removal and Installation" (170 and 200 cu. in. 6-cylinder).

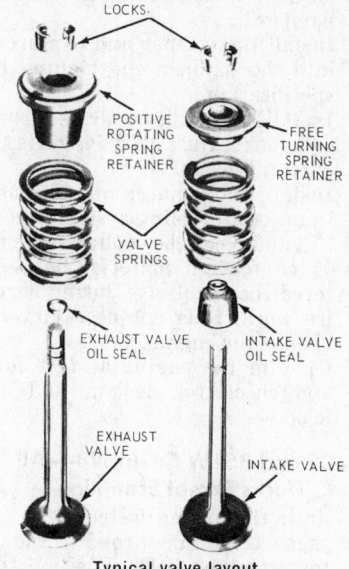

Typical valve layout
(© Ford Motor Co.)

Timing Gears and Chain

Timing (Front) Cover and Seal Removal and Installation

170 and 200 Cu. In. 6-Cylinder

1. Drain cooling system, disconnect the radiator upper hose at the coolant outlet elbow and remove the two radiator upper attaching bolts.
2. Raise vehicle and drain crankcase.
3. Remove splash shield (if applicable) and radiator as described above.
4. Remove drive belts, fan, pulley and crankshaft damper (use a suitable puller).

5. Remove front cover attaching bolts. Before removing the cover, cut the oil pan gasket. Remove cover.
6. Check timing chain deflection and camshaft endplay as described below. Endplay should be within 0.001-0.007" and chain deflection should not exceed 0.500".
7. Rotate crankshaft until sprocket timing marks are aligned, then remove camshaft sprocket attaching bolt and washer. Slide off sprockets and chain.

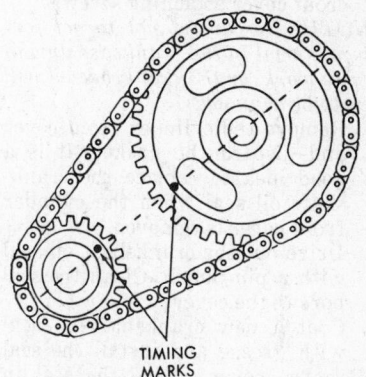

Timing mark alignment—170, 200 cu in.
6-cylinder (© Ford Motor Co.)

8. Drive out the old seal with a pin punch and clean out seal recess.
9. Coat new seal with grease and install using a suitable installing tool.
10. Clean and oil chain and sprockets, then install as an assembly with timing marks aligned as illustrated. Install camshaft sprocket retaining bolt and washer, tightening to 34-45 ft lbs.
11. Cut and install a piece of new gasket to go between front cover and oil pan, using sealer on all surfaces.
12. Install front cover and new seal, coating all surfaces with sealer. It may be necessary to force the cover downward to compress the new piece of oil pan gasket. Coat threads of cover attaching screws with oil-resistance sealer and install them finger-tight. While holding the front cover in alignment (using an aligning tool if available), tighten oil pan to cover attaching to 7-9 ft. lbs. Tighten the rest of the mounting bolts to the same torque.
13. Apply Lubriplate to the hub of the crankshaft and install damper, tightening attaching bolt to 85-100 ft. lbs.
14. Install fan, pulley, drive belt (adjust), radiator, radiator lower hose and splash shield.
15. Lower the vehicle and connect upper hose.
16. Fill and bleed cooling system.
17. Fill crankcase.

18. Operate engine and check for leaks.

240 and 300 Cu. In. 6-Cylinder

1. Drain the cooling system.
2. Remove the radiator and shroud.
3. Remove the alternator adjusting arm bolt, loosen the drive belt and swing the alternator arm aside. Remove the fan, drive belts and pulleys.
4. Remove the screw and washer from the end of the crankshaft, remove the crankshaft damper.
5. Remove the front oil pan and front cover attaching screws.

NOTE: Be careful not to get foreign material in the crankcase during service work, or the crankcase oil will have to be changed.

6. Remove the cylinder front cover and discard the gasket. It is a good idea to replace the crankshaft oil seal when the cylinder front cover is removed.
7. Drive out the crankshaft oil seal with a pin punch. Clean the seal bore in the cover.
8. Coat a new crankshaft oil seal with grease and install the seal in the cover. Drive the seal in until it is fully seated in the seal bore.
9. Cut the old front oil pan seal flush at the cylinder block/pan junction and remove the old seal material.
10. Clean all gasket surfaces.
11. Cut and fit a new pan seal flush to the cylinder block pan junction. Use the old seal as a pattern.
12. Coat the gasket surfaces of the block and cover with a oil resist-

ant sealer. Position a new front cover gasket on the cylinder block.
13. Align the pan seal locating tabs with the pan holes. Pull the seal tabs through until the seal is completely seated. Apply a silicone sealer to the block/pan junction.
14. Position the front cover assembly over the end of the crankshaft and against the cylinder block. Start the cover and pan attaching screws. Slide the cover alignment tool over the crank stub and into the seal bore of the cover. Install the alternator adjusting arm, tighten all attaching screws to specification.

NOTE: Tighten the oil pan screws first (compressing the pan seal) to obtain the proper alignment of the cover.

15. Lubricate the crank stub, damper hub I.D. and the seal rubbing surface with Lubriplate. Align the damper keyway with the key on the crankshaft and install the damper.
16. Install the washer and capscrew into the damper and tighten to specification.
17. Install the pulleys, drive belts, and fan. Adjust all drive belts to correct tension.
18. Install the radiator and shroud. Connect all cooling system hoses.
19. Fill and bleed the cooling system. If no foreign material has entered the crankcase during service work, it is not necessary to change the engine oil.
20. Operate the engine at fast idle and check for coolant and oil leaks.

302 and 351W Cu. In. V8—All Trucks Except Econoline

1. drain the cooling system.
2. Remove the fan shroud to radiator attaching bolts. Position the shroud over the fan.
3. Disconnect the radiator lower hose, heater hose and by-pass hose at the water pump. Remove the drive belts, fan, fan spacer, and pulley.
4. Remove the fan shroud.
5. Loosen the alternator pivot bolt and bolt attaching the alternator adjusting arm to the water pump.
6. Remove the crankshaft pulley from the crankshaft vibration damper. Remove the damper attaching bolt and washer. Install a puller on the vibration damper and remove the damper.
7. Disconnect the fuel pump outlet line from the fuel pump. Remove the fuel pump to one side with the flexible fuel line still attached.
8. Remove the oil dipstick and the

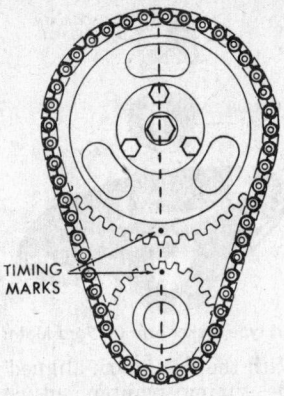

V8 engine timing mark alignment
(© Ford Motor Co.)

bolt attaching the dipstick to the exhaust manifold.

9. Remove the oil pan to cylinder front cover attaching bolts. Use a knife with a thin blade to cut the oil pan gasket flush with the cylinder block face prior to separating the cover from the cylinder block. Remove the cylinder front cover and water pump as an assembly.
10. Discard the cylinder front cover gasket. Remove the crankshaft front oil slinger.
11. Check the timing chain deflection. If the deflection exceeds specification, replace the chain and sprockets as follows:
   a. Crank the engine until the timing marks on the sprockets are correctly aligned.
   b. Remove the camshaft sprocket capscrew, washers, and fuel pump eccentric. Slide both sprockets and the timing chain forward and remove the chain and sprockets as an assembly.
   c. Position the sprockets and timing chain on the camshaft. Be sure that the timing marks are properly aligned.
   d. Install the fuel pump, eccentric, washers, and camshaft sprocket capscrew. Tighten the capscrew to specification.
12. Install the crankshaft front oil slinger.
13. Clean the cylinder front cover, oil pan and block gasket surfaces. Clean the oil pan gasket surface where the oil pan and front cover fasten.
14. Install a new crankshaft front oil seal.
15. Lubricate the timing chain and fuel pump eccentric with a heavy engine oil.
16. Coat the gasket surface of the oil pan with sealer, then cut and position the required sections of a new gasket on the oil pan and apply sealer at the corners. Install the pan seal as required. Coat the gasket surfaces of the

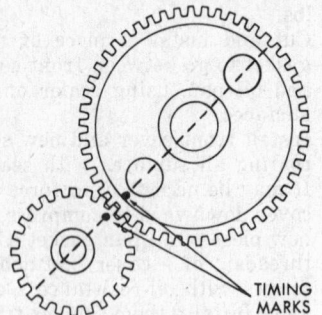

Timing mark alignment—6 cyl.— 240 & 300 engines

Tool—T64T-6306-A

Tool—T65L-6306-A

Installing camshaft gear
(© Ford Motor Co.)

block and cover with sealer, and position a new gasket on the block.

17. Position the cylinder front cover on the cylinder block. Use care when installing the cover to avoid seal damage or possible gasket dislocation.

18. Install the cylinder front cover to seal alignment tool.

19. It may be necessary to force the cover downward to slightly compress the pan gasket. This operation can be facilitated by using a suitable tool at the front cover attaching hole locations.

20. Coat the threads of the attaching bolts with a oil-resistant sealer and install the bolts. While pushing in on the alignment tool, tighten the oil pan to cover attaching bolts to specification. Tighten the cover to block attaching bolts to specification. Remove the alignment tool.

21. Apply Lubriplate or equivalent to the oil seal rubbing surface of the vibration damper inner hub to prevent damage to the seal. Apply a white lead and oil mixture to the front of the crankshaft for damper installation.

22. Line up the crankshaft vibration damper keyway with the key on the crankshaft. Install the vibration damper on the crankshaft. Install the capscrew and washer and tighten to specification. Install the crankshaft pulley.

23. Lubricate the fuel pump lever with heavy engine oil and install the pump using a new gasket. Connect the fuel pump outlet pipe.

24. Install the alternator pivot bolt and bolt attaching the alternator adjusting arm to the water pump.

25. Position the fan shroud over the water pump. Install the pulley, spacer and fan. Install and adjust the drive belts and adjust to specified tension. Connect the radiator, heater, and by-pass hoses. Position the fan shroud on the radiator and install the attaching bolts.

26. Fill and bleed the cooling system.

27. Run the engine at fast idle and check for coolant and oil leaks. Check the coolant level. Check and adjust the ignition timing.

28. Install the air cleaner and intake duct assembly including the crankcase ventilation hose.

302 and 351W Cu. In. V8-Econoline

1. Drain the radiator.
2. Remove the air conditioning idler pulley, bracket and drive belt if equipped.
3. Remove the upper radiator hose. Remove the fan and shroud as an assembly. Raise the vehicle on a hoist.
4. Loosen the thermactor and alternator drive belts.
5. Disconnect the lower radiator hose at the water pump. Disconnect the fuel line at the fuel pump and remove the pump. Lower the vehicle.
6. Remove the by-pass hose. Remove the power steering pump drive belt if equipped. Remove the water pump pulley and disconnect the heater hose at the water pump.
7. Remove the air condition compressor upper bracket and the power steering pump mount.
8. Remove the crankshaft pulley. Remove the oil pan to front cover bolts. Remove the front cover.
9. Clean the front cover, fuel pump, and damper. Lubricate the crankshaft front seal. Clean the gasket surface at the pan and trim the gasket. Clean the front cover gasket surface at the block.
10. Replace the oil seal in the front cover. Position the gasket on the front cylinder cover. Apply a silicone sealer to the oil pan and cylinder block junction. Cut the pan gasket and postion on pan and front cover.
11. Install the front cover, fuel pump, and crankshaft pulley.
12. Install the power steering pump and water pump by-pass hose. Connect the heater hose at the water pump.
13. Install the air conditioning compressor upper bracket, water pump pulley and power steering drive belt.
14. Install the alternator belt, thermactor belt, and fan/shroud assembly.
15. Adjust the power steering pump drive belt tension to specification.
16. Install the air conditioning drive belt idler pulley and bracket. Install the air conditioning drive belt and tighten to specification.
17. Install the upper radiator hose.
18. Raise the vehicle on a hoist. Install the fuel pump with a new gasket and connect the fuel line.
19. Install the lower radiator hose. Adjust the alternator and air injection pump drive belts to specified tension.
20. Drain the crankcase and replace the oil filter. Lower the vehicle.
21. Fill the crankcase and cooling system.
22. Start the engine and run at a fast idle, check for oil and coolant leaks.

351M and 400 Cu. In. V8

1. Drain the cooling system and disconnect the battery.
2. Remove the fan shroud attaching bolts and move the shroud to the rear.
3. Remove the fan and spacer from the water pump shaft.
4. Remove the air conditioner compressor drive belt lower idler pulley and the compressor mount to water pump bracket.
5. Loosen the alternator and power steering pump and remove the drive belts.
6. Remove the water pump pulley.
7. Remove the alternator and power steering pump brackets from the water pump and position them out of the way.
8. Disconnect the lower radiator and heater hose from the water pump.
9. Remove the crankshaft pulley from the crankshaft vibration damper. Remove the vibration damper attaching screw. Install a puller and remove the damper.
10. Remove the timing pointer.
11. Remove the bolts attaching the front cylinder cover to the cylinder block. Remove the front cover and water pump assembly.
12. Disconnect the fuel pump outlet line from the pump. Remove the fuel pump attaching bolts and lay the pump to one side with the flexible line still attached.
13. Discard the cylinder front cover gasket and oil pan seal.
14. Check the timing chain deflection.
15. If the timing chain deflection exceeds specification, proceed as follows:
    a. Crank the engine until the timing marks on the sprockets are aligned.

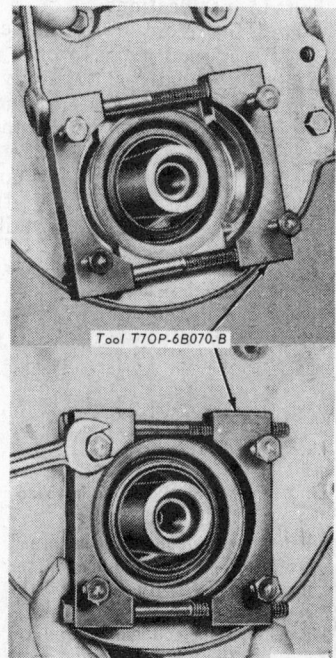

Tool T7OP-6B070-B

Removing the crankshaft front seal—351M & 400 engines (© Ford Motor Co.)

b. Remove the camshaft sprocket capscrew, washer, and two piece fuel pump eccentric. Slide both sprockets and the timing chain forward, and remove them as an assembly.

c. Position the sprockets and timing chain on the camshaft and crankshaft. Be certain that the timing marks on the sprockets are correctly aligned.

d. Install the two piece fuel pump eccentric, washers, and camshaft sprocket capscrew. Tighten the camshaft capscrew to specification. *Make sure that the outer fuel pump eccentric sleeve rotates freely.*

16. Coat a new fuel pump gasket with oil resistant sealer and position the fuel pump and gasket on the cylinder block with the fuel pump arm resting on the eccentric outer sleeve. Install the pump attaching bolt and nut and tighten to specification. Connect the fuel pump outlet line.

17. Remove the front crankshaft seal from the front cover. Clean the cylinder front cover and the engine block gasket surfaces.

18. Coat the gasket surfaces of the block and cover with sealer, and position a new gasket on the cylinder block alignment dowels.

19. Position the cylinder front cover and water pump assembly on the cylinder block alignment dowels.

20. Coat the threads of the attaching bolts with an oil resistant sealer and install the timing pointer and attaching bolts. Tighten the bolts to specifications.

21. Install the front cover oil seal into the cylinder front cover.

22. Apply lubriplate or its equivalent to the oil seal rubbing surface of the vibration damper inner hub to prevent damage to the seal. Apply a white lead and oil mixture to the front of the crankshaft for damper installation.

T70P–6B070–A

Installing the crankshaft front seal—351M & 400 V8 engines (© Ford Motor Co.)

23. Line up the crankshaft vibration damper keyway with the key on the crankshaft. Install the vibration damper on the crankshaft by pressing on with appropriate tool. Install the capscrew and washer, tighten to specification. Install the crankshaft pulley.

24. Connect the heater hose and the lower radiator hose to the water pump.

25. Install the air conditioner compressor to water pump bracket and lower idler pulley.

Tool—T52L-6306-AEE or 6306-AJ

Installing the crankshaft vibration damper —351M & 400 V8 engines (© Ford Motor Co.)

26. Position the alternator bracket and power steering pump bracket on the water pump and install the bolts.

27. Position the water pump pulley on the water pump shaft and install the drive belts.

28. Place the fan shroud over the pulley, and install the fan and spacer.

29. Position the fan shroud over the radiator and install the attaching bolts.

30. Adjust the drive belts to specification.

31. Raise the vehicle and remove the oil pan and install new gaskets and seals as described in "Oil Pan Removal and Installation" in this section.

32. Lower the vehicle. Fill the crankcase. Fill and bleed the cooling system. Connect the battery cable.

33. Operate the engine until normal operating temperature has been reached and check for oil or coolant leaks.

330, 359, 360, 361, 389, 390, and 391 Cu. In. V8

1. Drain the cooling system and the crankcase. Remove the air cleaner. Disconnect the battery ground cable and the distributor vacuum line.

2. Disconnect the upper radiator hose at the thermostat housing and the lower radiator hose at the water pump.

3. Disconnect the transmission oil

cooler lines if equipped.

4. Remove the radiator and support as an assembly. If the vehicle is equipped with an automatic radiator shutter, leave it attached to the radiator assembly.

5. Disconnect the heater hose at the water pump. Remove the water pump by-pass hose. On a high fan installation, remove the cooling fan and drive belt.

6. Remove the power steering pump and position it to one side, leaving the hoses attached.

7. If equipped with an air compressor, disconnect the air lines and remove the compressor.

8. Remove the alternator adjusting arm bolt at the alternator. Remove the drive belts. Disconnect the wiring and remove the alternator from the support bracket.

9. Remove the water pump and cooling fan drive belts. Remove the water pump and fan as an assembly. On C-series vehicles, remove the cooling fan from the crankshaft damper.

10. Remove the cap screw and washer from the end of the crankshaft. Install a puller on the crankshaft damper and remove the damper.

11. Disconnect the carburetor fuel inlet line at the fuel pump.

12. Remove the fuel pump attaching bolts and lay the fuel pump to one side with the flexible fuel line still attached.

13. Remove the bolts attaching the cylinder front cover to the cylinder block and oil pan. Using a knife with a thin blade, cut the oil pan gasket flush with the cylinder block to oil pan junction prior to separating the cover from the block.

14. Remove the cylinder front cover, alternator support bracket and adjusting arm, and the engine front support bracket.

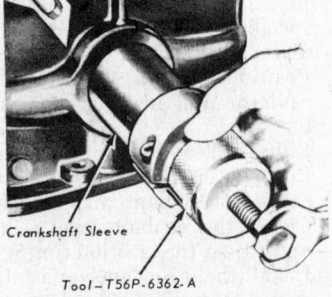

Crankshaft Sleeve

Tool—T56P-6362-A

Removing crankshaft sleeve— 330 engine (© Ford Motor Co.)

15. Discard the cylinder front cover gasket. Remove the oil slinger.

16. Check the timing chain deflection and camshaft end play.

17. If the timing chain deflection exceeds specification, proceed as follows:

a. Crank the engine until the

REFERENCE POINT   RIGHT SIDE OF CHAIN

TAKE UP SLACK ON LEFT SIDE, ESTABLISH
REFERENCE POINT. MEASURE DISTANCE **A.**
TAKE UP SLACK ON RIGHT SIDE. FORCE
LEFT SIDE OUT. MEASURE DISTANCE **B.**
DEFLECTION IS **A MINUS B.**

Checking timing chain deflection
(© Ford Motor Co.)

timing marks on the sprockets are correctly aligned.

b. Remove the camshaft sprocket capscrew and the fuel pump eccentric.

c. Slide both sprockets and the timing chain forward and remove the sprockets and timing chain as an assembly.

d. Clean the sprockets and chain replace any worn or damaged parts. Clean the crankshaft damper.

e. Position the sprockets and timing chain of the camshaft and crankshaft. Be sure that the timing marks are in alignment.

f. Install the fuel pump eccentric and the camshaft sprocket cap screw. Tighten the capscrew to specification. *If a new thrust plate was installed on the camshaft to bring end play within specifications, check the camshaft end play.*

18. Install the crankshaft front oil slinger.

19. Clean all oil pan and cylinder block to front cover gasket surfaces.

20. Coat the gasket surface on the oil pan with sealer. Cut and position the required section of a new gasket on the oil pan. Apply silicone sealer at the corners.

21. Coat the gasket surface of the cylinder block and front cover with sealer and position a new gasket on the cylinder block.

22. Position the cylinder front cover on the cylinder block. Be careful during installation of the cover to avoid dislocation or damage to the gasket.

23. Install the cylinder front cover-to-seal alignment tool in its proper position. It may be neces-

sary to force the cover downward to slightly compress the pan gasket. This operation can be faciliated by using a suitable tool at the cover attaching bolt hole locations in the cylinder block. Position the engine front support bracket and alternator mounting bracket and adjusting arm bracket on the cylinder front cover. Install the attaching bolts.

24. When pushing on the alignment tool, align the oil pan surfaces on the cylinder front cover and cylinder block. Tighten the attaching bolts to specification. Remove the alignment tool.

25. Clean the oil seal rubbing surface on the crankshaft sleeve with solvent and polish with crocus cloth. Examine for grooves, nicks, and burrs which could damage the seal. Lubricate the seal rubbing surface with grease and install the crankshaft sleeve.

26. Lubricate the damper hub and line the damper keyway with the key on the crankshaft. Install the damper on the crankshaft.

27. Install the damper capscrew and washer and tighten the capscrew to specification.

28. Install the power steering pump pulley on the damper and tighten the attaching bolts to specification.

29. Clean the water pump gasket surfaces. Coat the new gaskets with a water resistant sealer and position the gaskets on the cylinder front cover or water pump. Install the water pump and fan assembly. Tighten the attaching bolts to specification. On C-series vehicles, install the cooling fan on the crankshaft damper.

30. Install the alternator and the alternator drive belt. Adjust the alternator and water pump drive belts to the correct tension.

31. Install the power steering pump and drive belt, adjust the drive belt to the correct tension.

32. Install the air comrpessor and connect the air lines. Install the air compressor drive belt and adjust tension to specification.

33. On a high-fan installation, install the fan and drive belt and adjust the tension to specification.

34. Install the fuel pump using a new gasket.

35. Connect the carburetor fuel inlet line to the fuel pump.

36. Connect the heater hose to the water pump and install the by-pass hose.

37. Install the radiator, radiator support, and shutter assembly (if equipped) as an assembly. Connect the upper and lower radiator hoses.

38. Connect the transmission oil cooler lines and the battery ground cable.

39. If any coolant has entered the oil pan when separating the cylinder front cover from the block, it will be necessary to flush the crankcase.

40. Fill and bleed the cooling system.

41. Fill the crankcase with the correct grade and qualtity of oil.

42. Install the air cleaner and operate the engine at a fast idle to check for coolant or oil leak. Adjust the ignition timing and connect the vacuum line to the distributor.

460 V-8

1. Drain the cooling system and crankcase.

2. Remove the radiator shroud and fan.

3. Disconnect the upper and lower radiator hoses, and the automatic transmission oil cooler lines from the radiator.

4. Remove the radiator upper support and remove the radiator.

5. Loosen the alternator attaching bolts and air conditioning compressor idler pulley and remove the drive belts with the water pump pulley. Remove the bolts attaching the compressor support to the water pump and remove the bracket (support), if so equipped.

6. Remove the crankshaft pulley from the vibration damper. Remove the bolt and washer attaching the crankshaft damper and remove the damper with a puller. Remove the woodruff key from the crankshaft.

7. Loosen the by-pass hose at the water pump, and disconnect the heater return tube at the water pump.

8. Disconnect and plug the fuel inlet and outlet lines at the fuel pump, and remove the fuel pump.

9. Remove the bolts attaching the front cover to the cylinder block. Cut the oil pan seal flush with the cylinder block face with a thin knife blade prior to separating the cover from the cylinder block. Remove the cover and water pump as an assembly. Discard the front cover gasket and oil pan seal.

10. Transfer the water pump if a new cover is going to be installed. Clean all of the gasket sealing surfaces on both the front cover and the cylinder block.

11. Coat the gasket surface of the oil pan with sealer. Cut and position the required sections of a new seal on the oil pan. Apply sealer to the corners.

12. Coat the gasket surfaces of the cylinder block and cover with sealer and position the new gasket on the block.

13. Position the front cover on the cylinder block. Use care not to damage the seal and gasket or mislocate them.

14. Coat the front cover attaching screws with sealer and install them.

NOTE: it may be necessary to force the front cover downward to compress the oil pan seal in order to install the front cover attaching bolts. Use a screwdriver or drift to engage the cover screw holes through the cover and pry downward.

15. Assemble and install the remaining components in the reverse order of removal. Tighten the front cover bolts to 15-20 ft lbs, the water pump attaching screws to 12-15 ft lbs, the crankshaft damper to 70-90 ft lbs, the crankshaft pulley to 35-50 ft lbs, fuel pump to 19-27 ft lbs, the oil pan bolts to 9-11 ft lbs for the 5/16" screws and to 7-9 ft lbs for the 1/4" screws, and the alternator pivot bolt to 45-57 ft lbs.

Checking Timing Chain Deflection

To measure timing chain deflection, rotate crankshaft clockwise to take up slack on the left side of chain. Choose a reference point and measure distance from this point and the chain. Rotate crankshaft in the opposite direction to take up slack on the right side of the chain. Force the left (slack) side of the chain out and measure the distance to the reference point chosen earlier. The difference between the two measurements is the deflection.

Timing chain should be replaced if deflection measurement exceeds specified limit.

Camshaft Endplay Measurement

The fiber camshaft gears used on some engines is easily damaged if pried upon while the valve train load is on the camshaft. Loosen rocker arm nuts or rocker arm shaft support bolts before checking camshaft endplay.

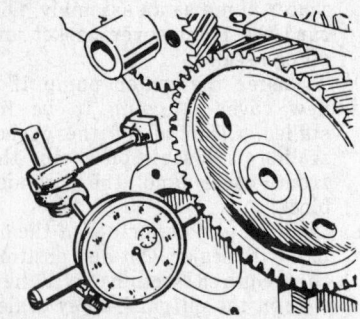

Checking camshaft gear runout

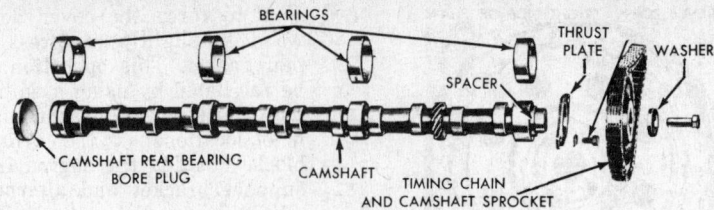

Camshaft—170 and 200 cu in. 6-cylinder (© Ford Motor Co.)

Push camshaft toward rear of engine, install and zero a dial indicator, then pry between camshaft gear and block to pull the camshaft forward. If endplay is excessive, check for correct installation of spacer. If spacer is installed correctly, then replace thrust plate.

Measuring Timing Gear Backlash

Use a dial indicator installed on block to measure timing gear backlash. Hold gear firmly against the block while making measurement. If excessive backlash exists, replace both gears.

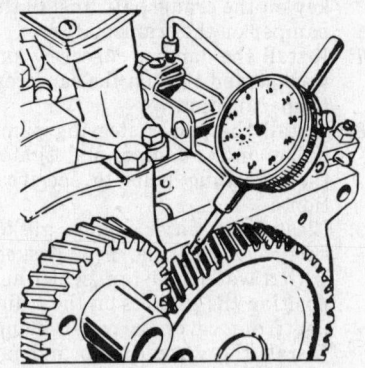

Checking timing gear backlash

Camshaft

Camshaft Removal and Installation

170 and 200 Cu. In. 6-Cylinder

1. Remove in order front cover, timing chain, cylinder head, distributor and fuel pump. Detailed removal and installation procedures for these components may be found in this section under each component heading.

2. Remove tappets with a magnet.

3. Remove dipstick, headlights, parking lights and the grill and hoodlock assembly.

4. Remove camshaft thrust plate and carefully slide camshaft from block.

5. Blow out rocker arm oil supply passages in the block with compressed air. Oil camshaft journals and apply Lubriplate to lobes. If a new camshaft is being used, transfer the spacer and dowel from old camshaft. Carefully slide camshaft into place.

6. Install thrust plate, tightening bolts to 12-15 ft. lbs.

7. Install timing sprockets, chain and all related front cover components as described in front cover removal and installation procedures above. Be sure to install a new front cover oil seal.

8. Install in reverse order all components removed, following instructions for each component group described topically in this section.

9. Start engine, adjust ignition timing, set idle speed and check for leaks.

240 and 300 Cu. In. 6-Cylinder

1. Remove the front cover following procedure described in "Front Cover Removal and Installation".

2. Remove the oil pan and oil pump as detailed in "Oil Pan Removal and Installation."

3. Remove air cleaner and crankcase vent tube at the rocker cover.

4. Disconnect accelerator cable, choke cable and hand throttle cable (if so equipped). Remove accelerator cable retracting spring.

5. If applicable, remove air compressor and power steering belts.

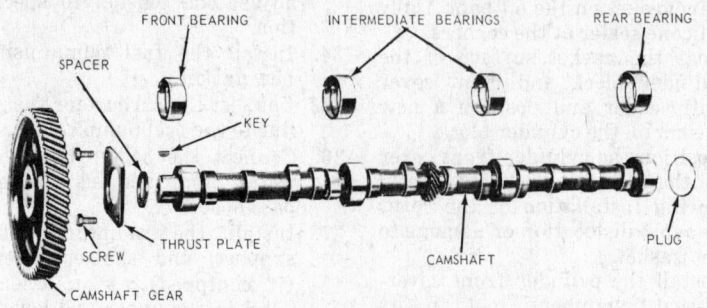

Camshaft—6 cyl.—240 & 300 engines (© Ford Motor Co.)

6. Disconnect oil filler hose from rocker cover.
7. Remove distributor cap and wiring as an assembly, then disconnect vacuum line and primary wire and remove distributor.
8. Remove fuel pump.
9. Remove valve rocker cover, loosen rocker arm stud nuts and move rocker arms to one side. Remove pushrods, identifying each so that they may be installed in their original locations.
10. Remove pushrod cover and valve lifters, identifying the position of each.
11. Turn crankshaft to align timing marks, remove camshaft thrust plate bolts and carefully pull camshaft and gear from block. Metal camshaft gear (300 HD) is bolted onto camshaft and fiber gear (240 and 300 LD) is pressed on and must be removed with an arbor press.

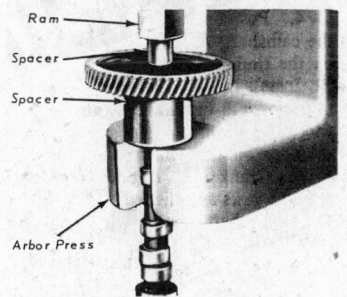

Removing fiber camshaft gear
(© Ford Motor Co.)

12. To install camshaft, oil journals and apply Lubriplate to lobes, then carefully install camshaft, spacer, thrustplate and gear as an assembly, making sure timing marks are aligned, then tightening thrustplate bolts to 19-20 ft. lbs. Do not rotate crankshaft until distributor is installed.
13. Install front cover, referring to "Front Cover Removal and Installation" for correct procedure.
14. Install valve lifters, then the pushrods in their original locations. Apply heavy engine oil to the lifters and Lubriplate to the pushrods.
15. Install in order the following components, referring to appropriate sections by topic for detailed instructions if necessary and using new gaskets with sealer: pushrod cover, valve rocker cover (adjust valve lash first), distributor (rotor in No. 1. cylinder firing position), fuel pump, distributor cap and wiring assembly, crankcase ventilation valve (in rocker cover), oil filler hose, accelerator cable and retracting spring, choke cable, hand throttle cable, front cylin-

der cover, oil pump, oil pan, water pump pulley, fan, belt, air compressor and power steering belts, radiator, hood latch, grill and air cleaner.
16. Fill crankcase.
17. Fill and bleed cooling system, checking for leaks.
18. Set the ignition timing, then connect distributor vacuum line.
19. Adjust carburetor idle speed and idle fuel mixture.

V-8 Engines

1. Remove the intake manifold and valley pan, if so equipped. On Econolines, remove the grill.
2. Remove the rocker covers, and either remove the rocker arm shafts or loosen the rockers on their pivots and remove the pushrods. The pushrods must be reinstalled in their original positions.
3. Remove the valve lifters in sequence with a magnet. They must be replaced in their original positions.
4. Remove the timing gear cover and timing chain and sprockets, as detailed in "Timing Cover and Seal Removal and Installation."
5. In addition to the radiator and air conditioning condenser, if so equipped, it may be necessary to remove the front grille assembly and the hood lock assembly to gain the necessary clearance to slide the camshaft out the front of the engine.
6. Remove the camshaft thrust plate attaching screws and carefully slide the camshaft out of its bearing bores. Use extra caution not to scratch the bearing journals with the camshaft lobes.
7. Install the camshaft in the reverse order of removal. Coat the camshaft with engine oil liberally before installing it. Slide the camshaft into the engine very carefully so as not to scratch the bearing bores with the camshaft lobes. Install the camshaft thrust plate and tighten the attaching screws to 9-12 ft lbs. Measure the camshaft end-play. If the end-play is more than 0.009 in.,

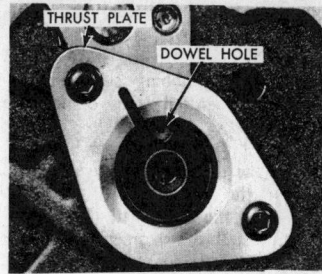

Camshaft thrustplate & spacer—330 engine
(© Ford Motor Co.)

replace the thrust plate. Assemble the remaining components in the reverse order of removal.

Camshaft Bearings and Valve Lifters

Reference

For detailed procedures for camshaft bearing replacement and hydraulic lifter service see the General Repair Section.

Pistons and Connecting Rods

Reference

Instructions for fitting of rings and rod bearings, ridge reaming and cylinder honing may be found in the General Repair Section.

Piston and Rod Removal and Installation

All Models

1. Drain cooling system and crankcase.
2. Remove the cylinder head.
3. Remove oil pan, oil pump pick-up tube and screen assembly and oil pump.
4. Turning crankshaft so that piston is at the bottom of its stroke, then ridge ream the top of the cylinder. *Never cut into the ring travel area in excess of 1/32" when removing ridges.*
5. Mark each rod bearing cap before removal so that it can be installed in its original location, then remove cap. Caps and rods are numbered on some models.

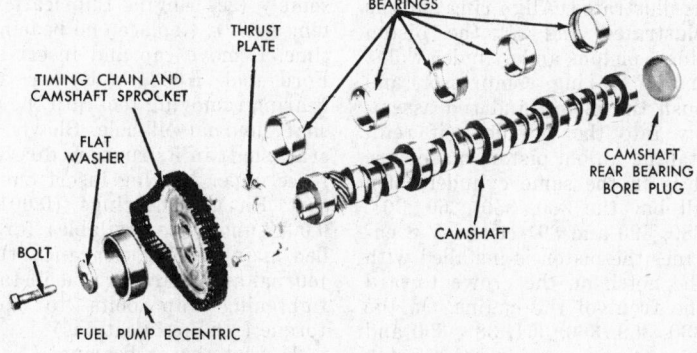

Typical camshaft—V8 engines (© Ford Motor Co.)

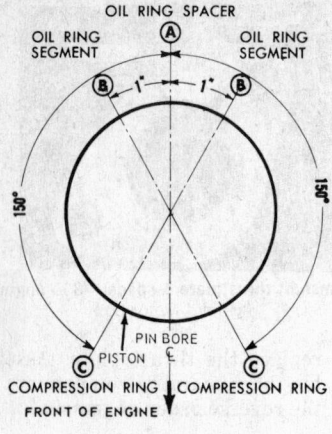

Piston ring gap spacing
(© Ford Motor Co.)

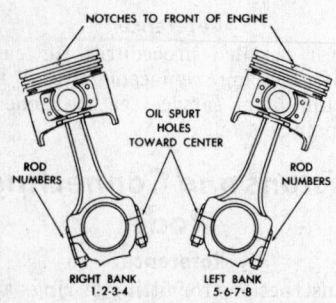

Piston-to-rod relationship—302 and 351 V-8
(© Ford Motor Co.)

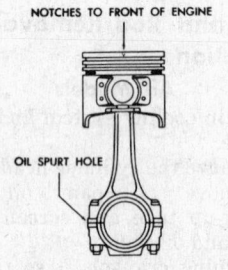

Piston-to-rod relationship—170 and 200 cu. in.
6-cylinder (© Ford Motor Co.)

6. Push connecting rod and piston assembly out the top of the cylinder.

7. Make sure piston is assembled in correct relation to the connecting rod, that is, that the notch on the top of the piston and the oil hole in the rod are positioned as illustrated. Align ring gaps as illustrated and oil the piston rings, pistons and cylinder walls.

8. Install a ring compressor and push the piston and rod assembly into the cylinder (if reinstalling an old piston, make sure it is in the same cylinder). On all but the 330, 359, 360, 361, 389, 390 and 391 cu. in. V-8 engines the piston is installed with the notch on the crown toward the front of the engine. On the 330, 359, 360, 361, 389, 390 and 391 cu. in. engines the notch faces in (toward "V").

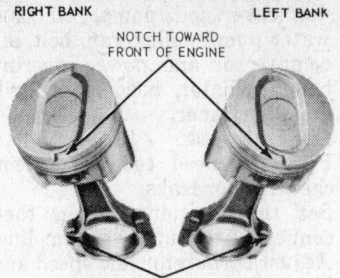

Piston and rod relationship—460 V8 engine
(© Ford Motor Co.)

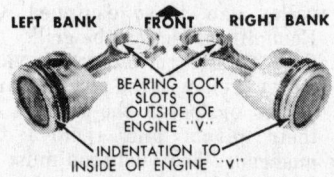

Piston and rod relationship—330, 359, 360, 361, 389, 390 and 391 V8 engines
(© Ford Motor Co.)

9. Fit rod bearings, apply oil to journals and bearings, then install bearings and cap, tightening cap bolts to specified torque (see Specifications at the beginning of this section).

10. Check rod bearing side clearance.

11. Thoroughly clean oil pump assembly, then prime it by filling and rotating shaft until pump is full. Install pump assembly.

12. Install oil pan, cylinder head and intake manifold (V-8 engines).

-13. Fill and bleed cooling system.

14. Fill crankcase.

15. Set ignition timing and operate engine to check for leaks.

16. Make final (hot) valve lash adjustment.

Main Bearings

Main Bearing Removal and Installation

All main bearings may be replaced with the engine in the vehicle with the exception of the 240 and 300 cu. in. rear main bearing. See the General Repair Section for main bearing fitting procedure.

Remove oil pan and oil pump assembly (see Engine Lubrication System below). Replace one bearing at a time. Remove cap and insert special Ford tool #6331 (-E) or similar bearing removing tool into the crankshaft journal oil hole. Slowly rotate crankshaft in its running direction to force upper bearing insert out of its seat. Fit new bearings (0.001″ and 0.002″ undersize available) for specified main bearing clearance, then oil journal and bearings and install cap, tightening cap bolts to specified torque (see Specifications).

To seat thrust bearing (No. 3 on 144, 170, 200 Six and V-8's; No. 5 on

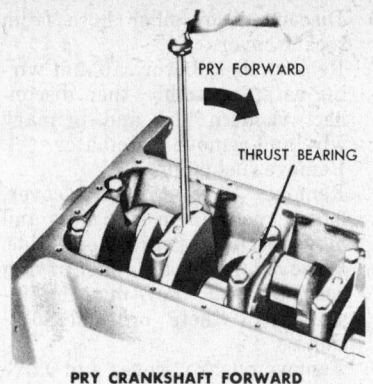

PRY CRANKSHAFT FORWARD

PRY CAP BACKWARD

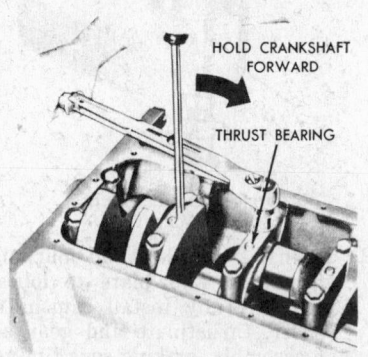

TIGHTEN CAP

Aligning thrust bearing (© Ford Motor Co.)

240, 300 Six), fit bearing inserts, then install bearings and cap, tightening cap bolts finger-tight. While prying crankshaft forward and bearing cap rearward (see illustration), tighten the cap bolts to specified torque.

If the rear main bearing is replaced, install a new rear oil seal as described below.

Rear Oil Seal Removal and Installation

All Engines Except 240 and 300 Cu. In. 6-Cylinder

On all models, remove the oil pan as described below. In some cases it may be necessary to remove the oil pump pick-up and screen or the whole pump assembly.

Use only the split-lip type crankshaft rear oil seal as a replacement.

1. Loosen all main bearing caps,

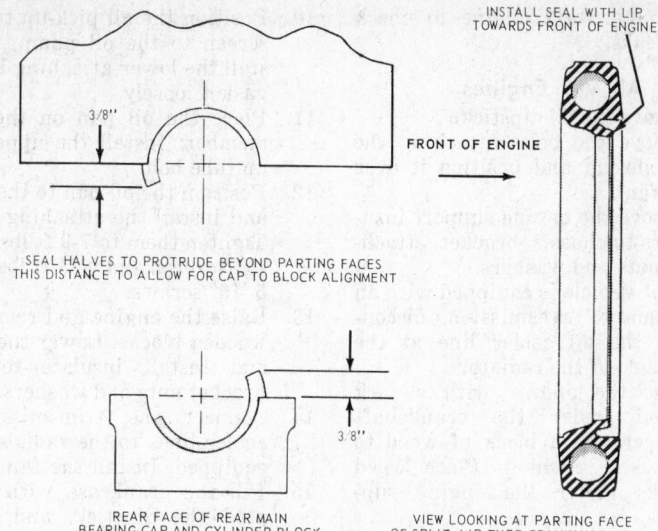

INSTALL SEAL WITH LIP
TOWARDS FRONT OF ENGINE

FRONT OF ENGINE

3/8"

SEAL HALVES TO PROTRUDE BEYOND PARTING FACES
THIS DISTANCE TO ALLOW FOR CAP TO BLOCK ALIGNMENT

3.8"

REAR FACE OF REAR MAIN
BEARING CAP AND CYLINDER BLOCK

VIEW LOOKING AT PARTING FACE
OF SPLIT, LIP-TYPE CRANKSHAFT SEAL

Crankshaft rear oil seal installation (© Ford Motor Co.)

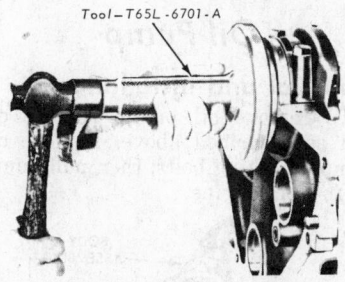

Tool—T65L-6701-A

Installing crankshaft rear oil seal
(© Ford Motor Co.)

lowering the crankshaft slightly, but not more than 1/32".
2. Remove the rear main bearing cap.
3. Remove the seal halves from cap and block. Use a seal removing tool on the block half or install a small metal screw in one end so that the seal may be pulled out. **CAUTION: Do not damage or scratch the crankshaft seal surfaces.**
4. If so equipped (not used on split-lip type), remove the oil seal retaining pin from the bearing cap.
5. Thoroughly clean seal grooves in block and cap with brush and solvent.
6. Dip seal halves in engine oil.
7. Carefully install upper half of seal with the lip facing toward the *front* of the engine until 3/8" is left protruding below parting surface. Be careful not to scrape seal.
8. Tighten all but the rear main bearing caps to specified torque.
9. Install lower seal half in the rear main bearing cap with the lip facing toward the *front* of the engine. Apply a light coat of oil-resistant sealer to the rear of the top mating surface of the cap. Do not apply sealer to the

area forward of the side seal groove.
10. Install rear main bearing cap and tighten bolts to specified torque.
11. Install side seals on the 330, 360, 361, 390 and 391 cu. in. V-8. Dip side seals in oil (do not use sealer as these seals expand when in contact with oil and sealer would retard or stop the expansion). Install seals in grooves, lightly tapping them if necessary. Do not cut off projecting ends. After allowing sufficient time for the seals to expand, squirt oil over the seal end and blow compressed air against the seals from inside the block to check for leakage.
12. Install oil pump and oil pan.
13. Fill crankcase and operate engine to check for leaks.

240 and 300 Cu. In. 6-Cylinder
1. Remove the starter.
2. Remove the transmission (see Transmission Removal and Installation). On standard transmission, remove pressure plate and cover assembly and the clutch disc.
3. Remove flywheel and engine rear cover plate.
4. Punch two holes with an awl on each side of the crankshaft just above the bearing cap to cylinder block splint line.
5. Install two sheet metal screws, then pry on both at once to remove seal. Be careful not to damage or scratch oil seal surface. Clean out seal recess in cap and block.
6. Lightly oil crankshaft and seal, then install seal with tool as illustrated. Carefully drive the seal straight in.
7. Install engine rear cover plate

and flywheel. Coat the flywheel attaching bolt threads with oil-resistant sealer and torque to 75-85 ft. lbs.
8. On standard transmission, install the clutch disc and pressure plate assembly (see Clutch Removal and Installation).
9. Install transmission.

Engine Lubrication

Oil Pan

Removal and Installation

170 and 200 Cu. In. 6-Cylinder
On some earlier models there is no frame crossmember under the oil pan. Therefore, some of the steps of the following procedure will not be applicable in certain cases.
1. Drain crankcase and cooling system and remove dipstick.
2. Remove fan and water pump pulley.
3. Disconnect the radiator upper and lower hoses, flex fuel line at the fuel pump and starter cable at the starter.
4. Remove the starter.
5. Remove nuts from both front engine support insulators and raise the front of the engine with a transmission jack and wood block. Remove crossmember from beneath the pan and install blocks between front support insulators and side rails. Lower and remove jack.
6. Remove the oil pan attaching bolts and pan.
7. Clean all gasket surfaces and remove seals from their grooves in the front cover and rear main bearing cap.
8. Position oil pan gasket, then pan front seal on front cover, making sure that the tabs on the seal are over the oil pan gasket. Position oil pan rear seal on the rear main bearing cap, making sure the tabs on the seal are over the oil pan gasket.
9. Install oil pan, tightening bolts to 7-9 ft. lbs.
10. Raise the engine with a transmission jack and remove the blocks. Install crossmember and lower the engine. Install washers and nuts on the insulator studs and tighten to 30-40 ft. lbs.
11. Install starter and connect starter cable.
12. Connect radiator hoses and fuel pump flex line.
13. Install water pump pulley, fan and drive belt. Adjust belt.
14. Install dipstick.

15. Fill and bleed cooling system.
16. Fill crankcase.
17. Operate the engine and check for leaks.

240 and 300 Cu. In. 6-Cylinder

1. Drain the crankcase.
2. On the F-100-250, also drain the cooling system.
3. Remove radiator from F-100-250 vehicles.
4. Raise vehicle on a hoist. On F-100-250 trucks disconnect and remove the starter.
5. On F-100-250, remove engine front support insulator to support bracket nuts and washers. Use a transmission jack to raise the front of the engine, then install blocks (1" thick) between the front support insulators and support brackets. Lower engine onto blocks and remove jack.
6. Remove the attaching bolts and oil pan. It may be necessary to remove the oil pump inlet tube and screen assembly in order to free the pan.
7. Remove the rear main bearing cap and front cover seals. Clean out the seal grooves and all gasket surfaces.
8. Apply oil-resistant sealer in the spaces between the rear main bearing cap and the block as illustrated. Install new rear cap seal, then apply a bead of sealer to the tapered ends of the seal.
9. Install new oil pan side gaskets with sealer and position the front cover seal.
10. Clean oil pump pick-up assembly and place it in the pan.
11. Position pan under the engine and install pick-up assembly.
12. Install pan and attaching bolts, tightening to 10-12 ft. lbs.
13. Raise engine enough with a jack and remove wood blocks. Lower engine and install washers and nuts on the support insulator studs, tightening to 40-60 ft. lbs. on LD trucks and 110-150 ft. lbs. on MD and HD trucks.
14. Install starter and starter cable on F-100-250 trucks.
15. Lower vehicle and install radiator if it was removed.
16. Fill crankcase and cooling sys-

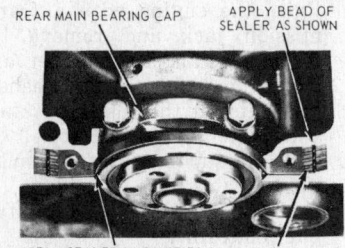

REAR MAIN BEARING CAP   APPLY BEAD OF SEALER AS SHOWN

APPLY SEALER IN CAVITIES PRIOR TO INSTALLING SEAL   OIL SEAL

Installing oil pan seal—240 & 300 engines
(© Ford Motor Co.)

tem and start engine to check for leaks.

All V-8 Engines

1. Remove the oil dipstick.
2. Remove the bolts attaching the fan shroud and position it over the fan.
3. Remove the engine support insulators-to-chassis bracket attaching nuts and washers.
4. If the vehicle is equipped with an automatic transmission, disconnect the oil cooler line at the left-side of the radiator.
5. Raise the engine with a jack placed under the crankshaft damper and a block of wood to act as a cushion. Place wood blocks under the engine supports.
6. Drain the crankcase. Remove the oil filter on the 460 V8.
7. Remove the oil pan attaching screws and lower the oil pan onto the crossmember. Remove the two bolts attaching the oil pump pick-up tube to the oil pump. Lower the pick-up tube and screen into the oil pan, then remove the pan.

NOTE: It may be necessary to turn the crankshaft to provide clearance between the crankshaft counterweights and the oil pan.

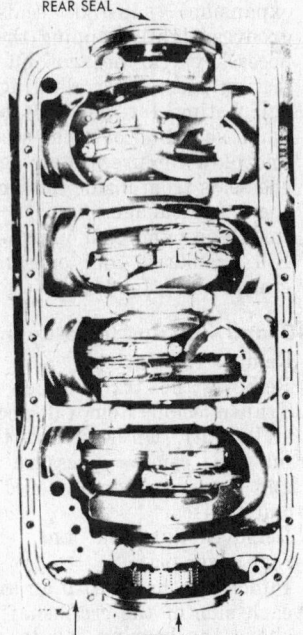

REAR SEAL

OIL PAN GASKET   FRONT SEAL

Oil pan gasket and seals—typical V-8
(© Ford Motor Co.)

8. Clean the oil pan and remove all gasket material from the mating surfaces. Repair and straighten any damage due to overtightening the retaining bolts.
9. Place the new oil pan gasket and end seals on the cylinder block. Use sealer to hold them in place.

10. Position the oil pick-up tube and screen to the oil pump, and install the lower attaching bolt and gasket loosely.
11. Place the oil pan on the crossmember. Install the upper pick-up tube bolt.
12. Position the oil pan to the engine and install the attaching screws. Tighten them to 7-9 ft lbs for the 1/4" screws, and 9-11 ft lbs for the 5/16" screws.
13. Raise the engine and remove the wooden blocks. Lower the engine and install insulator-to-chassis bracket nuts and washers.
14. Connect the transmission oil cooler lines to the radiator, if so equipped. Install the fan shroud.
15. Fill the crankcase with oil, install the dipstick, and operate the engine until it reaches normal operating temperature and check for leaks.

Oil Pump

Removal and Installation

To remove oil pump, remove oil pan as described above. Remove oil pump attaching bolts, then pull pump shaft from engine.

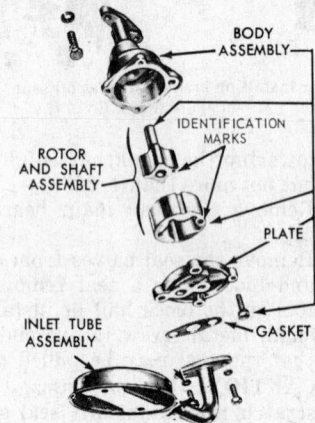

BODY ASSEMBLY

IDENTIFICATION MARKS

ROTOR AND SHAFT ASSEMBLY

PLATE

INLET TUBE ASSEMBLY

GASKET

Typical oil pump assembly
(© Ford Motor Co.)

Thoroughly clean all parts in solvent and dry with compressed air. Check the inside of the pump housing for obvious wear or scoring. Check mating surfaces of pump cover and rotors, replacing the cover if it is scored or grooved.

Feeler Gauge

Checking outer race to housing clearance
(© Ford Motor Co.)

Measure outer race to housing clearance and clearance (rotor end-play) between a straightedge and the rotor. The outer race, shaft and rotor are replaceable only as an assembly.

Measure the driveshaft to housing clearance by comparing shaft OD to housing bearing ID.

Inspect relief valve spring for collapsed or worn condition. Check the spring tension. Replace the spring if weak or worn.

Check relief valve piston and bore for scores and free operation.

Front Axle, Suspension

Solid I-Beam

Front Spring Removal and Installation

P-500 Series Trucks

1. Raise the vehicle frame until the weight is off the springs with the wheels still resting on the floor.
2. Remove spring U-bolts, nuts, an plate.
3. Remove the nut from the spring-to-frame bracket bolt and drive the bolt out of the spring bracket with a drift.
4. Remove the nut from the lower shackle bolt and drive the bolt out of the spring and shackle bars.
5. Remove the grease fittings from the shackle bolts.
6. Remove the spring, noting the position of any caster wedges.
7. Install new bushings in the spring and shackle bracket if required.

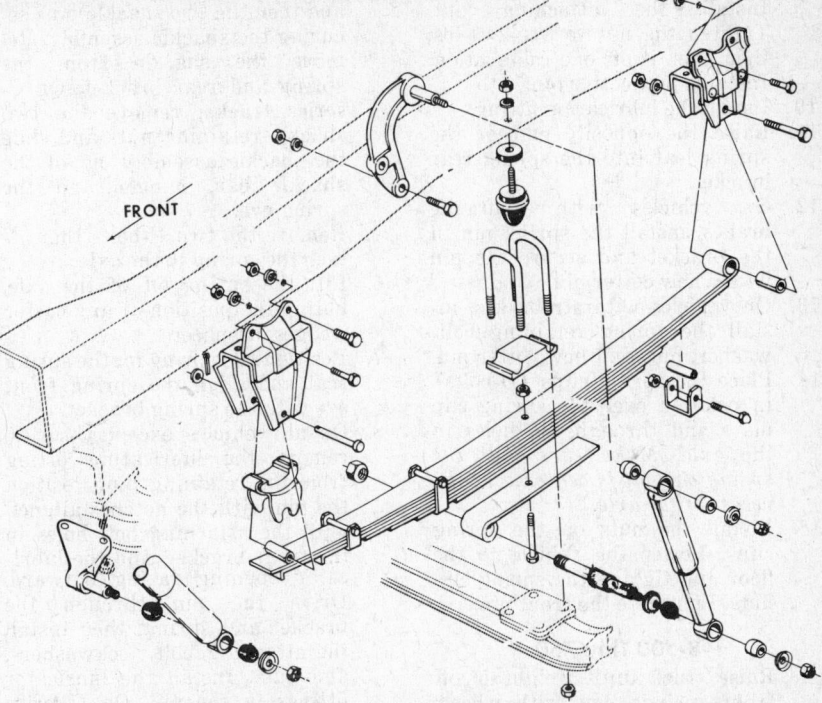

Front spring installation—F-B 600-750 (© Ford Motor Co.)

8. Install the caster wedges, if so equipped. Position the caster wedges with the thick edge in the same direction as they were before removal. Install the spring on the spring eye.
9. Position the spring eye between the shackle bars and install the lower shackle bolt and nut. *Do not tighten at this point.*
10. Position the spring eye in the bracket and install the bolt through the bracket and spring, with the bolt head toward the outside of the vehicle. *Install but do not tighten the nuts.*

Note: For the 1978 and later Bronco, use F-150 4 x 4 procedures.

11. Install the grease fittings in the shackle bolts.
12. Place the U-bolts and nuts in position over the spring plate and through the holes in the axle. *Make sure that the spring tie bolt is centered in the recess on the axle.*
13. Install the U-bolt nuts and lower the vehicle to the floor.
14. Tighten the lower shackle nut, spring bracket nut, and U-bolt nuts. *The nuts must be tightened in the order given.*

LN-500 to 750 Series Trucks

1. Raise the vehicle frame until the weight is off the front springs, with the wheels still touching the floor.
2. Remove the shock absorber.
3. Remove the cotter pin from the support bracket and remove the stud from the front bracket.
4. On trucks equipped with hydraulic brakes, remove the cotter pin and drive out the spring pin.
5. On trucks equipped with air brakes, remove the cotter pin and nut from the rear bracket shackle bolt and drive out the bolt.
6. Remove the nuts from the two spring clips (U-bolts) holding the spring on the axle.
7. Position the spring on the spring seat and align the spring eye with the spring bracket.
8. Prior to installation, coat the bushings with lubricant. Drive the stud through the bracket and the eye of the spring. *The lubrication opening in the stud should face inward.*

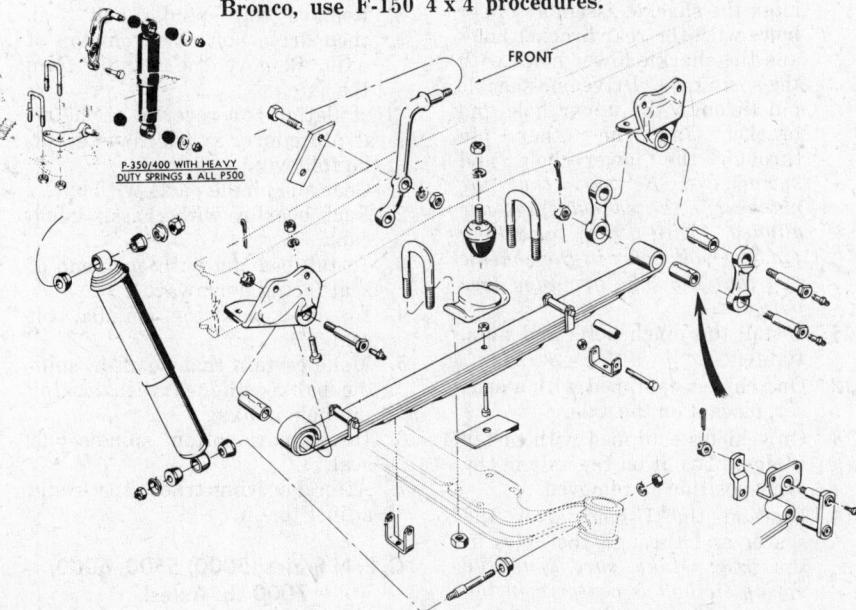

P-350/400 WITH HEAVY DUTY SPRINGS & ALL P500

FRONT

Front spring installation—P 350-500 (© Ford Motor Co.)

9. Install the attaching nut. Tighten the nut to 31-42 ft-lbs. then back it off one castellation. Install a new cotter pin.
10. Install the lubrication fitting.
11. Raise the opposite end of the spring leaf into the spring rear bracket.
12. On vehicles with hydraulic brakes, install the spring pin in the bracket and secure the pin with a new cotter pin.
13. On vehicles with air brakes, install the spring retaining bolt, washer, nut, and new cotter pin.
14. Place the spring clips (U-bolts) in position over the spring clip plate and through the holes in the axle. *Make sure that the spring tie bolt is centered in the recess of the axle.*
15. Install the nuts on the spring clips. Lower the vehicle to the floor and tighten the spring clip nuts. Lubricate the front pin.

F-8-500 Thru 750

1. Raise truck until weight is off front springs, but with wheels still touching ground.
2. Support front axle.
3. Remove attaching nuts and U-bolts.
4. Remove rear hanger cotter pin, nut, washers and bolt.
5. Remove securing cotter pin, then remove spring front hanger pin.
6. Remove spring.

Installation is a reversal of the removal procedure, in addition to noting the following:
1. Make certain that front U-bolt enters shock absorber lower bracket.
2. Raise or lower axle to align spring eye with bolt hole in rear hanger. Install bolt and one washer on each side of bracket; then, with the weight of the truck on the springs, tighten rear hanger bolt nut and install cotter pin.

NOTE: Do not back-off nut to align castellation with cotter pin hole. If necessary, tighten nut.

L-LT-LN-LNT-800-900, 8000-9000, C-CT And W Series

1. Raise the vehicle until the weight is off the front springs, with the wheels still touching the floor. Remove the shock absorber, if equipped.
2. On all vehicles except the C-series, remove the bolt securing the spring pin in the front bracket.
3. Remove the retaining pin from the bracket. On C-series trucks, remove the cotter pin and nut from the spring stud and drive the stud out of the front hanger bracket and spring eye.
4. On all vehicles except C-series, remove the four through bolts

and then the two shackle pins securing the shackle assembly. Remove the shackle from the spring and rear bracket. On C-series trucks, remove the two shackle retaining nuts and slide the shackle assembly out of the shackle bar, hanger and the spring eye.
5. Remove the two U-bolts that attach the spring to the axle.
6. Lift the spring off of the axle, noting the position of any caster wedges or spacers.
7. Position the spring on the spring seat and align the spring front eye with the spring bracket.
8. On all vehicles except C-series, remove the lubricating fitting from the retaining pin. Position the pin with the notches aligned with the attaching bolt holes in the front bracket. and the lubricator opening facing outward. Drive the pin through the bracket and spring, then install the attaching bolt, lockwashers, and nuts. Install the lubricator fittings in the pin. On C-series vehicles, remove the lube fittings from the retaining stud and drive the stud through the front hanger bracket and spring eye. Install the lube fitting on the outer end of the stud and the nut and cotter pin on the inner end.
9. On C-series vehicles, raise or lower the rear end of the spring as required to insert the shackle into the spring eye and rear bracket. Position the shackle bar to the inner side of the rear bracket, and drive the shackle assembly pins through the hanger bracket, spring eye and shackle bar. Install the two shackle retaining nuts and cotter pins.
10. On all vehicles except C-series, align the shackle assembly upper holes with the rear bracket holes and the shackle lower holes with the spring eye. Drive one shackle pin through the upper hole and bracket and the other pin through the lower hole and spring eye. *Be sure that the notches in the shackle pins are aligned with the attaching (pinch) bolt holes in the shackle and that the lube openings face outward.*
11. Install the pinch bolts and nuts, tighten.
12. On vehicles equipped with a spacer, place it on the axle.
13. On vehicles equipped with caster wedge, place it on the axle in the same position as removed.
14. Position the U-bolts over the spacer and through the holes in the axle. *Make sure that the spring tie bolt is centered in the recess of the axle, spacer or caster wedge.*

15. Install the shock absorber lower bracket to the underside of the axle, except C-series vehicles, with the spring clip (U-bolt) ends entering the holes in the bracket. Install the flat washers and lock nuts on the spring clip.
16. Lower the vehicle to the floor and tighten the clip nuts. Lubricate the shackle bolts.

Spindle R & R

P Series, C

1. Raise truck until wheels clear floor. Place a support under axle.
2. Remove wheel, drum, bearing and hub assembly. NOTE: It may be necessary to back-off brakes adjustment in order to remove drum.
3. Remove brake backing (carrier) plate, then remove spindle arm bolt and nut. Tie plate and arm to frame with wire.
4. Remove attaching nut, lockwasher and spindle bolt locking pin.

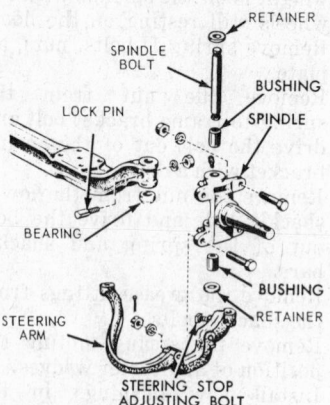

Spindle installation—P-Series
(© Ford Motor Co.)

5. Remove top spindle bolt seal, then drive bolt out from top of axle. Remove the spindle from the axle.

Installation is a reversal of the removal procedure, in addition to noting the following:
1. Coat all spindle parts with oil.
2. Pack bearing with chassis lubricant.
3. Install bearing with open end of seal facing downward.
4. Install a new top spindle bolt seal.
5. Make certain that notch in spindle bolt is aligned with locking pin hole in axle.
6. Install new bottom spindle bolt seal.
7. After lowering truck, check and adjust toe-in.

C, F, N Series (5000, 5500, 6000, 7000 Lb. Axles)

1. Raise truck and support front axle.

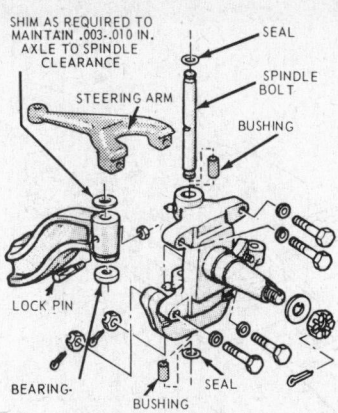

Spindle installation—5000-5500 lb. axles
(© Ford Motor Co.)

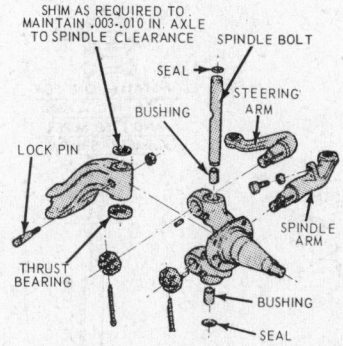

Spindle installation—6000-7000 lb. axles
(© Ford Motor Co.)

2. Remove wheel, hub, drum bearing.
3. Remove brake backing plate. Tie plate to frame with wire.
4. Disconnect spindle arm.
5. Remove top spindle bolt seal.
6. Remove attaching nut, then drive out spindle bolt locking pin.
7. Using a suitable drift, drive out spindle bolt from top of axle.
8. Remove spindle assembly.

Installation is a reversal of the removal procedure, in addition to noting the following:

1. Install thrust bearing with seal (retainer) lip downward.
2. Late model trucks, excluding F Series, are equipped with shims between top of axle and spindle. A clearance of .003-010″ between axle and spindle must be maintained on all models.
3. Excluding F Series, install spindle bolt with letter "T" upward.
4. Install new top and bottom spindle bolt seals.
5. On 5000 and 5500 lb. axles. excluding F Series, make certain that hardened flat washers, of the same type removed, are placed between backing plate and attaching bolt heads.
6. On 6000 lb. and larger axles, install spindle arm before backing plate.
7. Lubricate spindle bushings.

8. Adjust front wheel bearings and toe-in.

Spindle Bushing Replacement

1. Remove bronze bushings by driving them out with a drift slightly smaller than spindle bore. If a drift is not available, carefully drive a small center punch between the bushing and the spindle bore. Collapse the bushing, then remove.
2. Remove Delrin bushings, using a small center punch as described in Step 1.

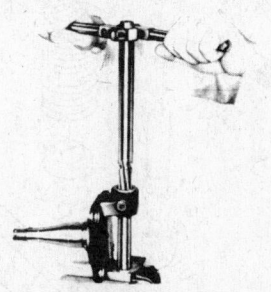

Reaming bronze bushing
(© Ford Motor Co.)

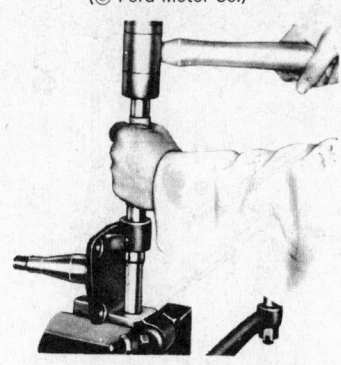

Removing bronze bushing
(© Ford Motor Co.)

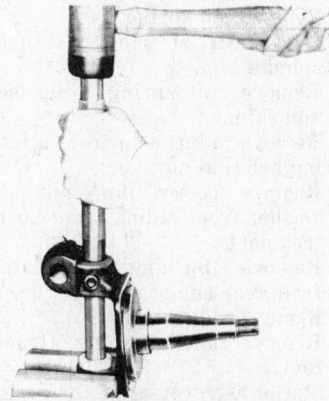

Installing bronze bushing
(© Ford Motor Co.)

3. Thoroughly clean spindle bores and make certain that lubrication holes are not obstructed in any way.
4. Place new bushing in spindle bore with lubricating holes properly aligned. Position open end of bushing oil groove toward axle.

5. Drive bronze bushing into spindle bore, using a drift as a pilot.
6. It is not necessary to drive Delrin bushings into spindle bores.
7. Install remaining bushing(s) in the same manner.
8. Ream bronze bushings .001 to .003″ larger than spindle bolt diameter.
9. DO NOT ream Delrin bushings.
10. After reaming bronze bushings, clean out spindle bore to remove metal shavings.
11. Apply a light coat of oil to all bushings before spindle assembly and installation.

Twin I-Beam

NOTE: F-100 and F-250 models with standard brakes utilize Delrin spindle bolt bushings.

NOTE: F-250, F-350 and E-300 models with BX or disc brakes are equipped with an integral arm and spindle.

Front Coil Spring R & R

Econoline 1971-75

1. Remove floor mat retainer from lower end of door opening.
2. Fold mat to one side, then remove attaching screws and shock absorber cover plate.
3. Remove spring upper retainer attaching screws, insulator and clamp.
4. Support frame side rails with a jack stand.
5. Position a floor jack under axle, then remove lower retainer attaching bolt, retainer and spring support.
6. Slowly lower axle to release tension, then remove spring.
7. Installation procedure is given in steps 8-14.
8. Position spring on axle with pigtails toward rear end of truck.
9. Position spring lower support and retainer. Install attaching bolt loosely.
10. Place upper insulator on spring.
11. Raise jack to apply light spring pressure.
12. Install upper retainer and clamp. *NOTE: Make certain that all upper retainer parts are correctly seated.*
13. Remove jack stand.
14. Install shock absorber cover plate and floor mat retainer.

Econoline 1976-78

1. Raise and support the front of the van. Support the front axle.
2. Disconnect the shock absorber from the lower bracket.
3. Remove the 2 upper spring attaching bolts from the upper spring seat. Remove the retainer.

4. Remove the nut from the lower spring retainer and remove the retainer.
5. Install a spring compressor.
6. Lower the axle carefully, and remove the spring.
7. Installation is the reverse of removal.

F-100 150, 250, 350 1971-74

1. Raise front end of truck and support frame with floor stands.
2. Position a jack under axle.
3. Disconnect shock absorber from lower bracket.
4. Remove attaching bolt, nut and rebound bracket.
5. On F-100 and F-250, perform the following:
   a. Remove attaching bolts and retainer.
   b. Remove attaching nut and lower spring retainer.
   c. Lower axle and remove spring.
6. On F-350, perform the following:
   a. Remove bolts attaching upper spring clip to seat. Remove clip.
   b. Remove attaching nut and lower spring retainer.
   c. Lower axle and remove spring.

Installation is a reversal of the removal procedure.

F100, 150, 250 & 350 1975-78

1. Raise the vehicle and support the axle with a jack.
2. Disconnect the shock absorber from the lower bracket.
3. Remove the two spring upper retainer attaching bolts from the top of the spring upper seat and remove the retainer.
4. Remove the nut attaching the spring lower retainer to the lower seat and axle, remove the retainer.
5. Lower the axle and remove the spring.
6. Place the spring in position and raise the front axle.
7. Position the spring lower retainer over the stud and lower seat. Install the attaching nut.
8. Position the upper retainer over the spring coil and against the spring upper seat. Install the two attaching bolts.
9. Tighten the upper retainer bolts and lower retainer attaching nut.
10. Connect the shock absorber to the lower bracket and install the rebound bracket.
11. Remove the jack and safety stands.

Radius Arm R & R
All Models

1. Raise front end of truck and support frame with floor stands.
2. Position a jack under axle or ap-

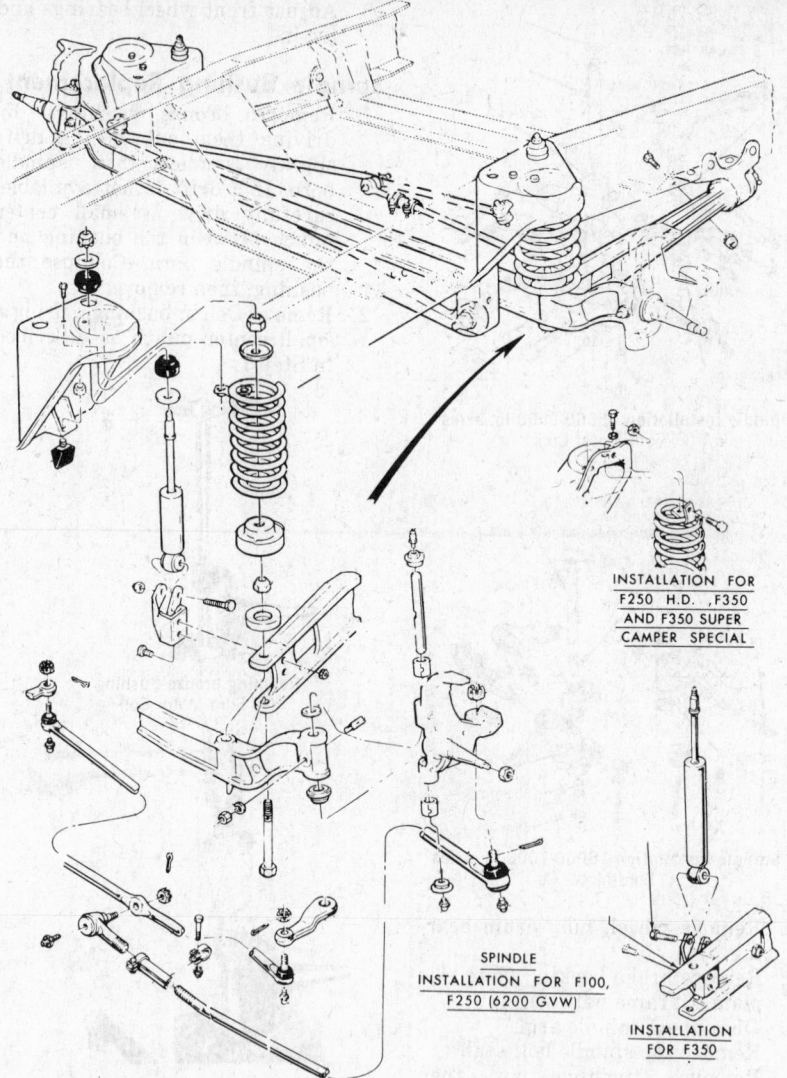

INSTALLATION FOR F250 H.D. ,F350 AND F350 SUPER CAMPER SPECIAL

SPINDLE INSTALLATION FOR F100, F250 (6200 GVW)

INSTALLATION FOR F350

Twin I-Beam front axle—exploded view—F100-350 shown (© Ford Motor Co.)

propriate wheel.
3. Remove coil spring.
4. Disconnect steering rod from spindle arm.
5. Remove coil spring lower seat and shim.
6. Remove radius arm front attaching bolt and nut.
7. Remove cotter pin, nut and washer from radius arm rear attachment.
8. Remove bushing (insulator) from rear end of arm.
9. Remove radius arm.
10. Remove inner bushing (insulator).

Installation is reversal of the removal procedure.

Refer to Wheel Alignment Specifications for toe-in, caster, camber and kingpin (spindle bolt) inclination values.

Front Drive Axle

Description

Bronco

The Dana Model 441F drive axle is used on Bronco trucks. The axle has an open yoke welded to the outer ends of the axle housing. Cardan-type universal joints transmit power to the front driving wheels.

F-100

The Dana Model 44-7F drive axle is used on F-100, F-150 four wheel drive trucks. The axle has an open spindle assembly, utilizing Cardan-type universal joints to transmit power to the front drive wheels.

F-250

Dana Models 44-6BF and 44-7BF-HD are used on early F-250 four wheel drive trucks. The axle has a spherical, closed spindle and enclosed universal joint.

Late model trucks use the Dana model 44-6CF or model 44-6CF-HD. These axles are identical to the other 44 series except for a few minor differences.

F-600

The Rockwell-Standard Single-Reduction Final Drive Axle is used on

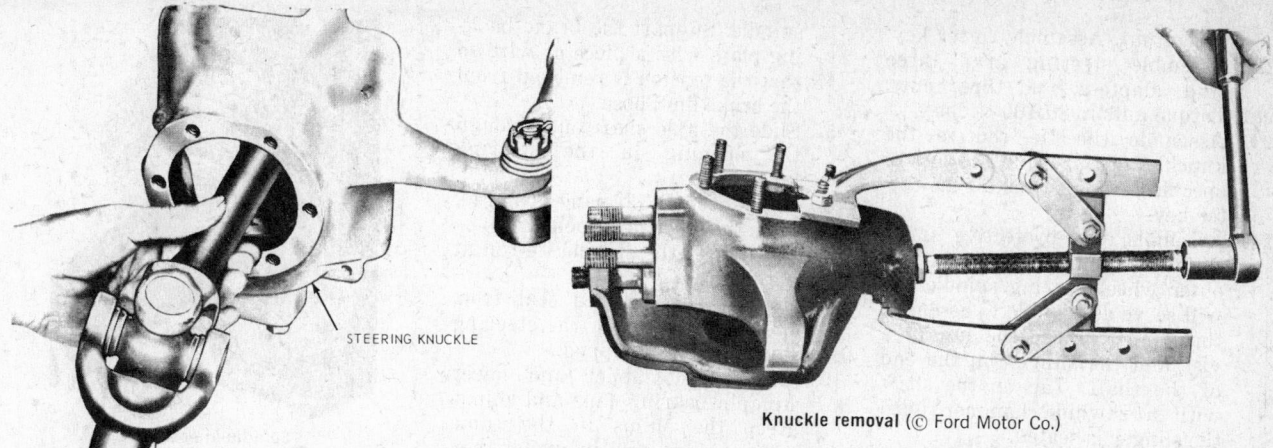

Knuckle removal (© Ford Motor Co.)

Removing axle shaft (© Ford Motor Co.)

F-600 trucks. The axle has a heavy duty spiral bevel or hypoid pinion and gear. The differential and gear assembly is carried in tapered roller bearings.

Refer to the General Repair Section for complete overhaul procedures on the front wheel drive axles.

With the exception of specialized components required for front drive application, the Dana and Rockwell-Standard models are identical to their rear axle counterparts (i.e. Dana 44-7F—Dana Model 44, Rockwell - Standard Single - Reduction—same, etc.). Spicer and Dana front drive axles are identical. The Dana Model 44IF, used on Bronco models, is similar to the other Dana axles except in size.

Axle Shaft and Steering Knuckle R & R

Bronco, F-100 and F-150

1. Raise vehicle on a hoist.
2. Remove front hub grease cap. Remove driving hub retaining snap-ring, then slide splined driving hub from between axle shaft and wheel hub.
3. Remove driving hub spring. *NOTE: If equipped with free-wheeling hubs, see hub removal.*
4. Remove lock nut, washer, and wheel bearing adjusting nut from spindle. Remove wheel, hub and drum as an assembly. The wheel outer bearing will be forced off the spindle at the same time. Remove wheel inner bearing cone.
5. Remove capscrews that attach brake backing plate and spindle

to steering knuckle. Remove brake backing plate and secure it to one side. Carefully remove the spindle.
6. Pull shaft assembly from axle housing, working universal joint through bore in steering knuckle.
7. To remove steering knuckle (housing), disconnect the steering connecting rod end from the steering knuckle and remove the bearing caps. Remove the steering knuckle.
8. Remove the three retaining nuts and remove the steering arm. Remove the cotter key from the upper ball socket.
9. Remove the nut from the upper ball socket. Remove the nut from the lower ball socket. *NOTE: Discard the nut from the bottom socket.*
10. Remove the knuckle from the yoke. If the top socket remains in the yoke it can be dislodged with a rawhide hammer.
11. Remove the bottom socket, the adjusting sleeve and the top socket from the knuckle. On F-100, F-150 remove the snap-ring before removing the socket.
12. Installation is given in steps 13 thru 30.
13. Place knuckle in vise and assemble bottom socket. Place new socket into knuckle making sure it is not cocked; place the driver over the socket; place forcing screw into driver as shown. Apply torque to screw and force socket into knuckle.
14. Make sure socket shoulder is seated against knuckle. Use a .0015″ feeler gauge between socket and knuckle. Feeler gauge is not to enter at minimum area of contact.
15. Assemble top socket into knuckle. Assemble holding plate onto backing plate screw. Tighten nuts snug. Place new socket into knuckle. Be sure socket is straight and not in a cocked position. Place driver over socket.
16. Make sure socket shoulder is

seated against knuckle. Use a .0015″ feeler gauge between socket and knuckle. Feeler gauge is not to enter at minimum area of contact.
17. Assemble new adjusting sleeve into top of yoke. Leave approximately two threads exposed. This will protect the threads in the yoke.
18. Assemble knuckle with sockets to yoke. Assemble new nut to bottom socket. Tighten nut finger loose. This will serve as a holding device.
19. Place spanner wrench and step plate over adjusting sleeve. Position puller, and turn forcing screw. This will pull the knuckle assembly into the yoke. With torque still applied, tighten the bottom nut on the socket. Torque nut to 70-90 ft. lbs. *NOTE: If the bottom stud should turn with the nut, add more torque to the puller forcing screw.*
20. Torque adjusting sleeve to 50 ft. lbs. Remove spanner wrench.
21. Assemble top socket nut. Torque to 100 ft. lbs. Line up cotter key hole of stud with the castellation or slot of the nut. Tighten nut when it is being lined up with the hole of the stud. Do not

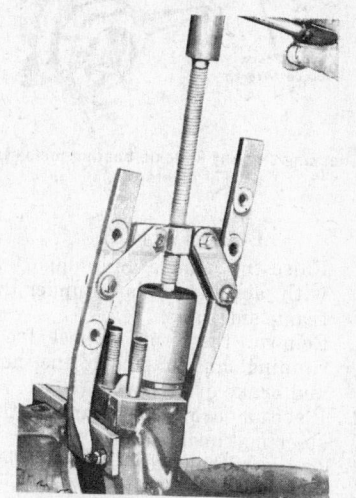

Top socket assembly
(© Ford Motor Co.)

loosen nut. Assemble cotter key.
22. Assemble steering arm, three stud adapters, and three nuts. Torque nuts to 80-100 ft. lbs.
23. Assemble the tie rod to the knuckle arm. Torque the nut to specifications and install the cotter key.
24. Assemble the protective inner slinger on to the axle shaft. The outer wheel bearing spindle nut will serve as a guide to assemble the slinger. Place the nut in a vise and the slinger on the end of the shaft. Tap on the shaft with a rawhide hammer until the slinger is seated.
25. Assemble the protective outer slinger on to the shaft. One of the wheel spindles will serve as a guide. Place the spindle in a vise. Do not clamp on the bearing diameters. Place the slinger on the shaft. Tap on the end of the shaft with a rawhide hammer until the slinger is seated. *NOTE: Take care not to damage the seal diameter of the slinger.*
26. Assemble the axle shaft joint assembly into the housing.
27. Place the spindle in a vise and install the needle roller bearings, using driver and rawhide hammer.
28. Assemble the grease seal into the spindle, flush with the spindle face.
29. After assembly of the needle bearing and oil seal, pack wheel bearing grease around the needle bearing and lip of the seal.
30. Assemble the axle shaft joint assembly, bronze spacer and spindle to the knuckle. *NOTE: The chamfer of the spacer should be inboard against the shaft.*

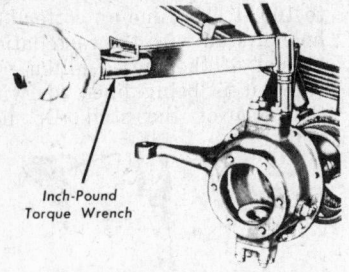

Inch-Pound Torque Wrench

Checking steering knuckle bearing preload
(© Ford Motor Co.)

F-250 1971-75

1. Raise the vehicle and support it with jackstands placed under the frame side rails.
2. Remove the front wheel free-running hub. Remove the hub and brake drum.
3. Place a drain pan under the steering knuckle assembly.
4. Remove the capscrews which hold the brake backing plate and spindle to the steering knuckle. Remove the backing plate and

spindle. Support the brake backing plate with a piece of wire so that the tension is removed from the brake fluid hose.
5. Slide the axle shaft out through the opening in the steering knuckle.
6. Disconnect the steering drag link from the steering knuckle.
7. Disconnect the spindle tie-rod at both ends.
8. Remove the unitized seal from the steering arm of the steering knuckle, if so equipped.
9. Remove the upper and lower kingpin bearing caps and shims. Keep the shims in the same order side-to-side in which they are removed for proper installation.
10. Remove the steering knuckle from the axle housing and remove the bearings.
11. If the bearing cups (races) are to be removed from the housing, tap them out with a hammer and drift.
12. Place a new steering knuckle seal over the axle housing tube, if the seal is being replaced. The seal is attached to the rear of the steering knuckle with the split facing up to prevent leakage. Do not attach the seal until the steering knuckle turning torque is checked.
13. Install and assemble the steering knuckle and axle shaft in the reverse order of removal. Install the shims under the kingpin bearing caps in the same order in which they were removed and tighten the cap attaching bolts to 30-40 ft lbs on the Dana Model 44-6CF and 80-90 ft lbs on the Dana Model 44-6CF-HD.
14. Place a torque wrench on the inside/forward bearing cap retaining bolt and check the torque required to turn the steering knuckle. The torque required should be 5-10 ft lbs on the 44-6CF and 10-15 ft lbs on the 44-6CF-HD. Adjust the number and thickness of the shims under the bearing cap to obtain the correct turning breakaway torque. *NOTE: Check the steering turning torque with the seal not yet attached to the back of the knuckle, the axle shaft removed, and the steering linkage unattached.*

F-250 1976-78

1. Raise and support the front of the truck.
2. Remove the wheel.
3. Remove the caliper from the rotor and suspend it from the frame.
4. Remove the dust cap, cotter pin, nut, washer and outer bearing and remove the rotor from the spindle.

SPINDLE POSITIONING SCREW

Spindle in position
(© Ford Motor Co.)

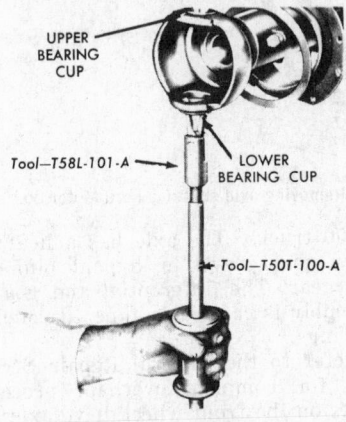

UPPER BEARING CUP

Tool—T58L-101-A

LOWER BEARING CUP

Tool—T50T-100-A

Removing king pin bearing cup
(© Ford Motor Co.)

5. Remove the inner bearing cone and seal.
6. Remove the axle shaft, working the U-joint through the bore of the steering knuckle. Be careful not to damage the seal.
7. At this point, the axle shaft, U-joints, spindle bore seals and bearings can be replaced without further disassembly.
8. Disconnect the steering connecting rod end from the steering knuckle.
9. Remove the cotter key from the upper ball socket. Loosen the nuts from the upper and lower ball sockets and discard the nuts.
10. Remove the knuckle from the yoke. If the top socket sticks in the yoke, it can be removed with a plastic mallet.
11. Remove and discard the bottom socket.
12. Remove and discard the adjusting sleeve.
13. Remove the top socket from the knuckle.
14. Place a new knuckle in a vise and install a new bottom socket. The socket shoulder should be seated against the knuckle so that a .0015 in. feeler blade cannot be inserted between the socket and the knuckle.

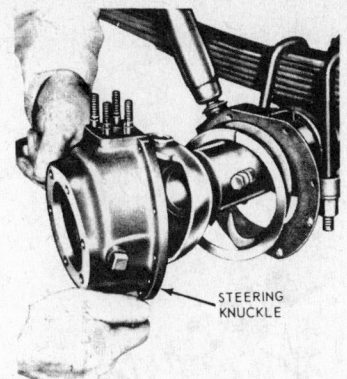

Removing or installing steering knuckle—
F-250
(ⓒ Ford Motor Co.)

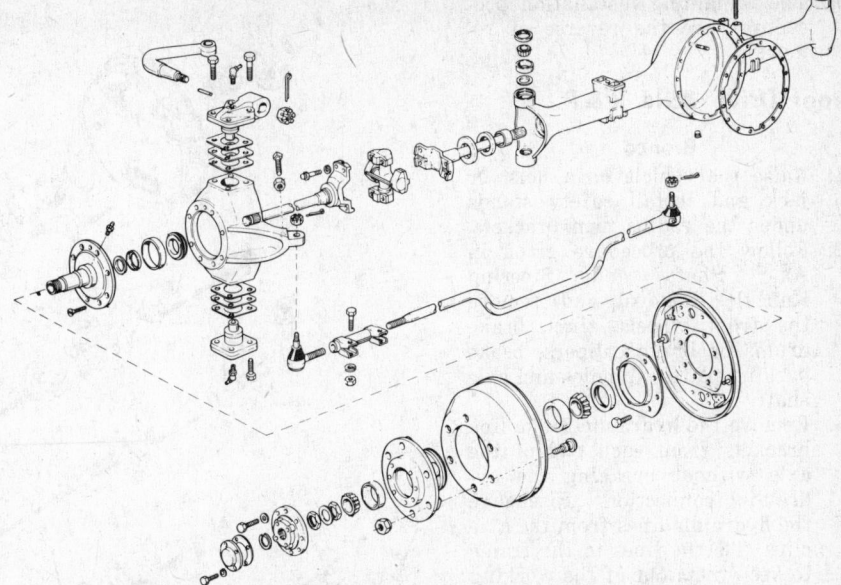

F-600 front driving axle housing—exploded view (ⓒ Ford Motor Co.)

15. Install the top socket as in Step 14.
16. Install a new adjusting sleeve, leaving about 2 threads exposed.
17. Assemble the knuckle to the yoke and tighten the nut finger-tight.
18. Pull the knuckle assembly into the yoke and tighten the bottom nut on the socket to 70-90 ft lbs.
19. Tighten the adjusting sleeve to 40 ft lbs.
20. Assemble the top socket nut. Tighten it to 100 ft lbs and align the keyway with the castellations on the nut. Tighten the nut to align the holes. Install a new cotter pin.
21. Test the steering knuckle turning effort with a spring scale hooked at the tie rod hole. If it is more than 26 lbs, the ball joints should be replaced.
22. Connect the steering rod to the steering knuckle and tighten the nut.
23. Further installation is the reverse of removal. Adjust the wheel bearings.

F-600

1. Raise and support the vehicle.
2. Remove the plug from the bottom of the axle housing and drain the lubricant. Replace the plug.
3. Disconnect the drive shaft at the pinion shaft.
4. Remove one front wheel from the hub and drum assembly.
5. Remove the outer hub cap retaining screws and remove the cap.
6. Remove the retaining lock ring.
7. Remove the retaining screws and remove the splined drive plate.
8. Bend the tab of the lock washer from the lock nut and remove the lock nut.
9. Remove the lockwasher and discard it. Remove the adjusting nut.
10. Remove the drum and hub assembly from the axle spindle.
11. Remove the brake backing plate retaining screws and carefully

lift out the backing plate and wire it up out of the way.
12. Remove the spindle positioning screw and remove the spindle.
13. Remove the axle shaft from the housing.
14. Remove the four upper bearing cap retaining screws.
15. Remove the bearing cap and steering arm and set them aside. Do not drop or lose the shims.
16. Remove the bearing from the socket and set it aside.
17. Remove the four lower bearing cap retaining screws.
18. Remove the cap and bearing from the socket and set them aside. Do not drop or lose the shims.
19. Remove the steering knuckle from the yoke.

20. Position the knuckle on the yoke.
21. Clean, lubricate and install the upper bearing.
22. Position the shims, bearing cap, and the steering arm on the knuckle. Install the retaining screws.
23. Clean, lubricate and position the lower bearing, shims and bearing cap. Install the retaining screws.
24. Tighten the capscrews to 185-235 ft-lbs. torque.
25. Check the steering knuckle bearing preload with a torque wrench. The preload should be between 11-15 ft-lbs.
26. Adjust the shim thickness at the bearing caps as required to bring the bearing preload to specification.

Axle shaft removal—F-600 (ⓒ Ford Motor Co.)

207

27. The remaining installation procedures are the reverse of removal.

Front Drive Axle R & R

Bronco

1. Raise the vehicle on a hoist or jack and install safety stands under the radius arm brackets.
2. Follow the procedure given in Axle Shaft and Steering Knuckle Removal and remove the front wheels, tires, brake drums or brake calipers, brake backing plates, spindles and axle shaft.
3. Remove the hydraulic brake line brackets from each end of the axle without breaking the hydraulic connection. Disengage the hydraulic lines from the axle clips. Tie the lines to the frame to keep them out of the working area.
4. Disconnect the steering tie rod at the knuckle connecting rod ends and tie it out of the working area. Disconnect the axle stablizer bar.
5. Disconnect the front drive shaft at the pinion companion flange and universal joint. Secure the drive shaft out of the working area.
6. Lower the vehicle onto the safety stands and place a jack under the axle to support it while disconnecting it from the radius arms.
7. Each radius arm and cap is numbered frcm 1 through 100 for proper assembly, since they are manufactured as matched pairs. Remove the bolts attaching the radius arms to the radius arm caps. Remove the rubber insulators and roll the axle form under the vehicle.
8. Installation is given in steps 9 thru 15.
9. Position the front drive axle under the vehicle, using a floor jack, and install the radius arms, insulators and caps to the axle. The numbers enscribed on cap and arm should be matched. Torque the attaching bolts to specifications, tightening them diagonally in pairs.
10. Raise the vehicle to working height and install the drive shaft to the pinion companion flange at the universal joint. Torque the universal joint U-bolt nuts to specifications.
11. Connect the axle stabilizer bar. Connect the steering tie rod to the steering knuckle by the steering connecting rod ends. Torque the attaching nuts to specifications, then install cotter pins.
12. Follow the procedure detailed in

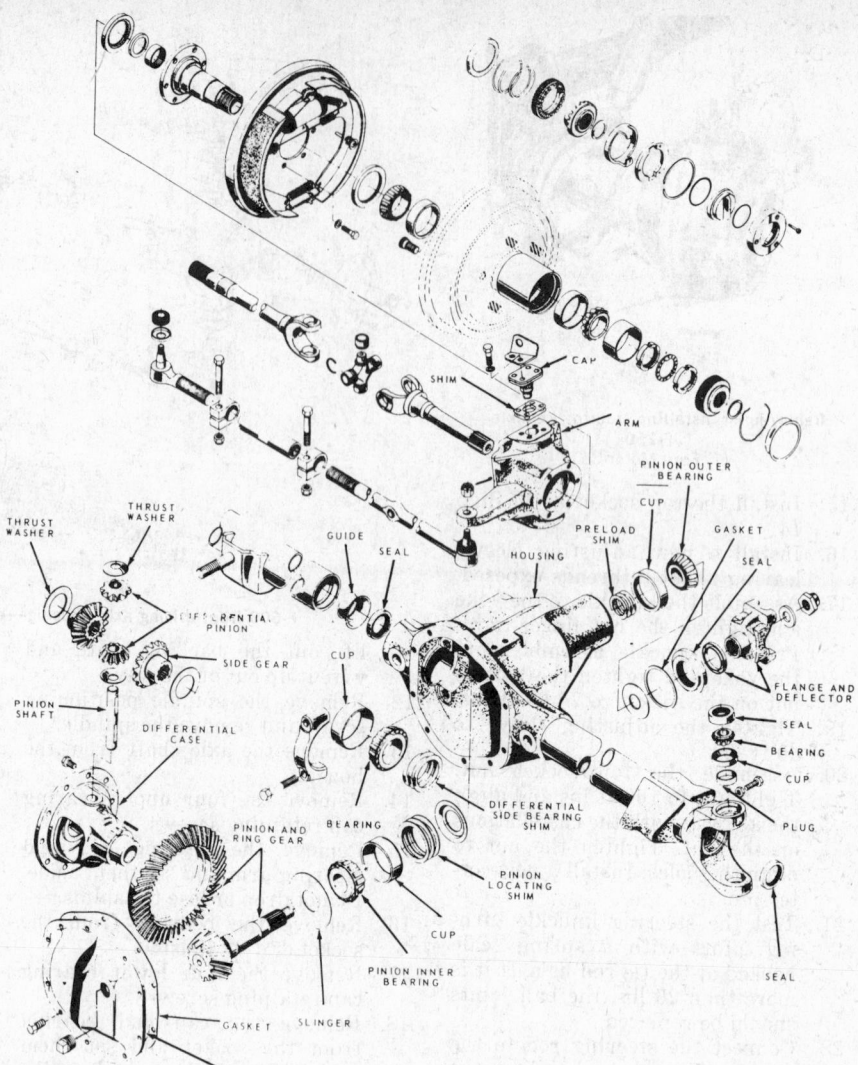

Front drive axle—F-100 and F-150 (© Ford Motor Co.)

Axle Shaft and Steering Knuckle Installation and install the axle shafts, spindles and brake backing plates.
13. Position the hydraulic brake lines and brackets, then install the retaining clips.
14. Install the front brake drums, wheels and tires. Adjust the front wheel bearings. Install the dust cap or locking hub cap and the wheel cover on each front wheel.
15. Lower the truck and fill the axle housing with the specified lubricant.

F-100 and F-150

1. Raise the vehicle on a hoist or jack and install safety stands under the radius arm brackets.
2. Follow the procedure detailed in Axle Shaft and Steering Knuckle Removal and remove the front wheels, tires, brake drums, brake carrier plates, spindles and axle shafts.
3. Remove the hydraulic brake line

brackets from each end of the axle without breaking the hydraulic connection. Disengage the hydraulic lines from the axle clips. Tie the lines to the frame.
4. Disconnect the steering tie rod at the spindle connecting rod ends. Disconnect the axle stabilizer bar.
5. Disconnect the front drive shaft at the pinion companion flange and universal joint. Secure the drive shaft out of the working area.
6. Lower the vehicle onto the safety stands and place a jack under the axle to support it while disconnecting it from the radius arms.
7. Each radius arm and cap is marked, since they are manufactured as matched pairs (parts are numbered 1 through 100). Remove the bolts attaching the radius to the radius arm caps. Remove the rubber insulators and roll the axle from under the truck.

8. Installation is given in steps 9 thru 15.
9. Position the front drive axle under the vehicle, using a floor jack, and install the radius arms, insulators and caps to the axle. Numbers on radius arm and cap should be matched. Torque the attaching bolts to specifications, tightening them diagonally in pairs.
10. Raise the vehicle to working height and install the drive shaft to the pinion companion flange at the universal joint. Torque the universal joint U-bolt nuts to specifications.
11. Connect the axle stablizer bar. Connect the steering tie rod to the spindle arms by means of the steering connecting rod ends. Torque the attaching nuts to specifications, then install the cotter pins.
12. Follow the procedure detailed in Axle Shaft and Spindle Arm Installation and install the axle shafts, spindles and brake backing plates.
13. Position the hydraulic brake lines and brackets, then install the retaining clips.
14. Install the front brake drums, wheels and tires. Adjust the front wheel bearings. Install the dust cap or locking hub cap and the wheel cover on each front wheel. NOTE: If a dust cap is used, install it with a coat of non-hardening sealer on the sealing surface.
15. Lower the truck and fill the axle housing with the specified lubricant.

F-250

1. Raise the vehicle on a hoist so that no weight is supported by the front axle.
2. Remove the hubs, rotors, brake carrier plates, axle shafts and steering knuckle as given in Axle Shaft and Steering Arm.
3. Disconnect both front axle shock absorbers at their lower ends.
4. Disconnect the front axle drive shaft at the pinion flange.
5. Support the front axle on a transmission jack, then remove the spring clip (U-bolt) nuts and the spring seats.
6. Lower the axle assembly and roll it from under the truck.
7. Installation is given in steps 8 thru 11.
8. Position axle under truck and raise it so that the spring clips and spring seats can be installed.
9. Connect the front axle shock absorbers.
10. Connect the front axle drive shaft.
11. Complete the assembly as given

in Axle Shaft and Steering Knuckle Installation.

F-600

1. Raise the vehicle and support with safety stands.
2. Remove the plug from the bottom of the axle housing and drain the lubricant.
3. Disconnect the driveshaft at the pinion shaft.
4. Remove one front wheel from the brake drum and hub assembly.
5. Remove the outer hub cap retaining screws and cap.
6. Remove the retaining lock ring.
7. Remove the retaining screws and the splined drive plate.
8. Bend the tab of the lockwasher from the locknut, then remove the locknut.
9. Remove lockwasher and discard. Remove the adjusting nut.
10. Remove the drum and hub assembly from the spindle.
11. Remove the brake backing plate retaining screws and carefully lift out the backing plate and wire it out of the way.
12. Remove the spindle positioning screw and spindle.
13. Remove the axle shaft from the housing.
14. Repeat steps 4 thru 13 to remove the opposite axle shaft.
15. Remove the carrier to housing stud nuts and washers. Loosen two top nuts and leave on studs to prevent carrier from falling.
16. Break carrier loose from axle housing with rawhide mallet. *NOTE: A roller jack should be positioned and fastened to the carrier at this point.*
17. Remove the top nuts and washers, then work the carrier free. A small pinch bar may be used to straighten the carrier in the housing bore. However, the end must be rounded to prevent identing the carrier flange. A roller jack may be used to facilitate removal of the carrier.
18. Installation is given in steps 21 thru 39.
19. Remove all traces of old gasket material from the carrier and housing surfaces, then position a new gasket over the housing mounting studs.
20. Using a roller jack, ease the carrier into position in the housing bore.
21. Position two washers and run two nuts part way on to the studs to hold the carrier; then align it properly with the housing.
22. Install the housing stud nuts and torque them to specifications.
23. Install one axle shaft in the housing.

24. Install the spindle and secure it with the positioning screw.
25. Position the brake backing plate, install the retaining screws and torque to specifications.
26. Carefully position the drum, and hub assembly on the axle spindle, and tighten the adjusting nut.
27. Adjust wheel bearing. Position a new lockwasher against the adjusting nut and apply a film of oil to the outer face of the lockwasher.
28. Run the lock nut against the lockwasher, then torque it to specifications.
29. Bend one tab of the lockwasher over the adjusting nut.
30. Bend other tab of the lockwasher (in the opposite direction) over the lock nut.
31. Apply Silastic Sealer to the front and rear mounting faces of the splined drive plate. Position the drive plate and install the plate retaining screws, then torque to specifications.
32. Install the retaining lock ring.
33. Position the outer hub cap and install the retaining screws, then torque to specifications.
34. Install the front wheel.
35. Connect the drive shaft at the pinion shaft.
36. Repeat steps 23 through 34 for the other axle shaft assembly.
37. Make sure that the housing drain plug has been installed.
38. Fill the axle housing with the correct grade and quantity of lubricant.
39. Lower the vehicle.

Wheel Bearing Adjustment

F-100, F-150, F-250, F-350, Econoline P-350-500—(2WD)

1. Remove the hub cap and hub grease cap.
2. Clean the end of the spindle and remove the cotter pin and nut-lock.
3. While rotating the wheel, torque the adjusting nut to 17-25 ft lbs. (40-55 ft lbs.—P-350-500), to seat the bearings.
4. Install the nut-lock so that the cotter pin hole in the spindle is aligned with the hole in the nut-lock.
5. Back the adjusting nut off, 2 slots and install the cotter pin. Be sure that the wheel rotates freely with no noticeable endplay.
6. Install the grease cap and hub cap.

4-Wheel Drive Except F600

1. Raise the vehicle and support with safety stands.
2. Back off the brake adjusting

screw, if necessary. Remove the wheel cover.

3. Remove the front hub grease cap. Remove the driving hub retaining snap-ring and slide the splined driving hub from between the axle shaft and the wheel hub. Remove the driving hub spacer. *NOTE: If equipped with free-wheeling hubs, see Free-Wheeling Hub Removal.*

4. Remove lock nut and lock ring from the spindle.

5. Tighten the bearing adjusting nut to 50 ft. lbs., while rotating the wheel back and forth to seat the bearings.

6. Continue rotating the wheel and then, loosen and re-torque the adjusting nut to 30-40 ft. lbs.

7. Back the adjusting nut off approximately 1/4 turn (90 degrees). Assemble the lock ring by turning the nut to the nearest notch where the dowel pin will enter.

8. Install the outer lock nut and torque to 50 ft. lbs. (80-100 ft. lbs—1973-78). Final endplay of the wheel on the spindle should be 0.001 to 0.010.

9. Install the driving hub, spacer, snap-ring and hub grease cap. Apply a thin coat of non-hardening sealer to the seating edge of the grease cap before installation. *NOTE: If equipped with free-wheeling hubs, installation refer to Free-Wheeling Hub.*

10. Adjust the brake.

11. Remove safety stands and lower the vehicle.

F-600

1. Remove the outer hub cap retaining screws, then remove the cap.

2. Remove the retaining lock ring.

3. Remove the retaining screws and the splined drive plate.

4. Bend the tab of the lockwasher away from the locknut, then remove the locknut.

5. Remove the lockwasher and discard.

6. While rotating the wheel back and forth to correctly seat the bearings, torque the adjusting nut to 50 ft. lbs.

7. Back-off the adjusting nut from 1/4 to 1/3 turn.

8. Position a new lockwasher against the adjusting nut and apply a film of oil to the outer face of the lockwasher.

9. Run the lock nut up against the lockwasher, then torque it to 100-150 ft. lbs.

10. Bend one tab of the lockwasher over the adjusting nut.

11. Bend other tap (in the opposite direction) over the lock nut.

NOTE: Use a blunt tool when bending the tabs.

12. Apply Silastic Sealer to the front and rear mounting faces of the spined drive plate.

13. Position the drive plate, install the retaining screws and torque to specifications.

14. Install the retaining lock ring.

15. Position the outer hub cap and install the retaining screws. Torque to specifications.

Free-Wheeling Lock-out Hubs

Free Wheeling Hub R&R

Bronco, F-100, F-150, F-250 (Except H.D.)

1. Remove the three screws (six screws on Bronco) attaching the lock-out actuating knob and retaining plate assembly to the wheel hub. Remove the actuating knob and retaining plate assembly and large O-ring.

2. Remove the large internal snap-ring, the outer clutch retaining ring and the actuating cam body.

3. While pressing inward against the axle shaft sleeve and ring assembly, remove the snap-ring that secures the axle sleeve and ring assembly to the axle shaft.

4. Remove the axle shaft sleeve and ring assembly and the inner clutch and bushing assembly.

5. Remove the pressure spring (or spacer) and spring retainer plate.

6. Installation is given in steps 7 thru 13.

7. Insert the spring retainer plate into the wheel hub with the flange side facing inward. Be sure spring retainer plate bottoms against the outer wheel hub bearing cap.

8. Position the pressure spring (or

spacer), with the large end seating against the spring retainer plate, inside the wheel hub.

9. Assemble the inner clutch ring and bushing assembly to the axle shaft sleeve and ring assembly, sliding both assemblies as a unit on to the axle shaft splines. Using a new axle shaft snap-ring, lock the axle shaft sleeve and ring assembly to the shaft.

10. Place the actuating cam body in position against the axle shaft sleeve and ring assembly, inside the wheel hub.

11. Install the outer clutch retaining ring and the internal snap-ring into position in the wheel hub. Make certain the snap-ring is well seated in the groove on the inside diameter of the hub.

12. Coat the large O-ring seal with O-ring lubricant. Install the large O-ring seal onto the actuating knob and retaining plate and place the knob and retaining plate assembly into position in the wheel hub.

13. Install the three (six on Bronco) knob and retaining plate assembly attaching screws and copper washers. Tighten screws securely.

F-250 Heavy-Duty

1. Remove the free-running hub screws and washers.

2. Loosen the gear hub housing and slide it away from the hub and drum assembly.

3. Remove and discard the inner metal gasket, remove the gear hub housing, and remove and discard the outer gasket. Wipe the exposed parts clean with a clean rag.

4. Remove the snap-ring while holding pressure on the clutch gear.

5. The actuator knob should be in the lock position while performing this operation. Ease the clutch gear and pressure spring out of the assembly.

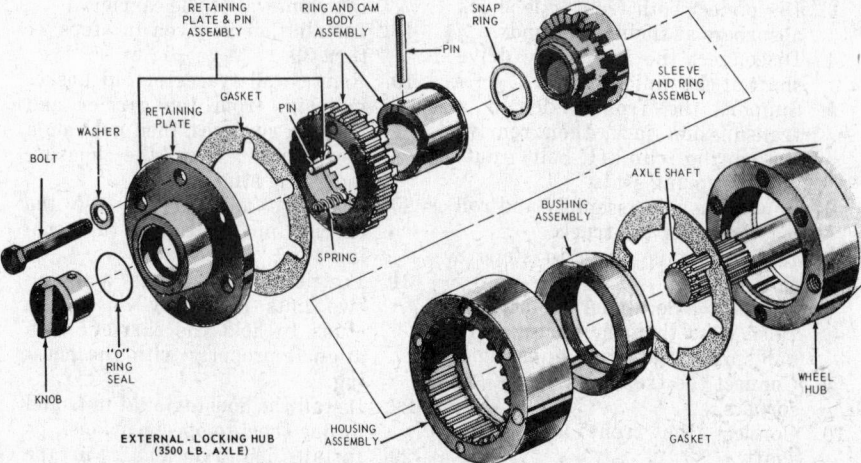

External lock-out hub—disassembled (© Ford Motor Co.)

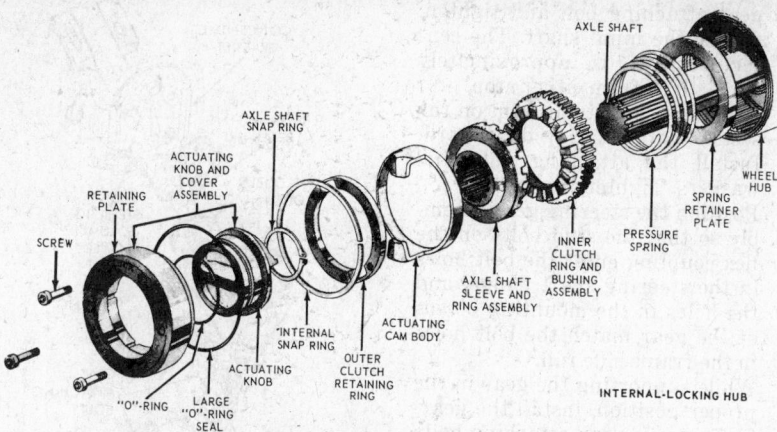

Internal lock-out hub—disassembled (© Ford Motor Co.)

6. Turn the actuator knob to the free position. Drive the cam lock pin out of the assembly with a drift.
7. Remove the actuating cam from the knob.
8. Remove the knob retainer snap-ring and remove the knob from the knob retainer.
9. Using a capscrew, pull out slightly on the axle shaft and remove the snap-ring which retains the bushing and inner clutch gear assembly.
10. Remove the inner clutch gear and the bushing behind it. Replace these two components as a set if excessive wear or damage is evident.
11. Inspect the splines of the axle shaft for nicks or burrs. Clean the threaded holes in the wheel hub.
12. Apply Moly XL Hi-Speed grease to the back face and thrust face of the bushing and the splines of the inner clutch gear.
13. Assemble the inner clutch gear into the bushing.
14. Install the bushing and inner clutch gear onto the axle shaft mating the splines of the axle and the gear.
15. Install a new snap-ring to retain the bushing and gear. It may be necessary to pull out the axle with a capscrew to gain clearance for the snap-ring to be installed. Make sure that the snap-ring is fully seated.
16. Apply a small amount of O-ring lubricant on the actuator knob and assemble the O-ring to the actuator knob.
17. Install the actuating knob into the knob retainer with the arrow pointing to the free position.
18. Install the knob retainer snap-ring.
19. Install the actuating cam onto the knob, aligning the ears of the cam with the slots of the retainer. Position the parts on a small block of wood.

20. Assemble the cam lock pin through the groove of the cam and the holes in the actuating knob. Be sure that the ends of the pin are flush with the outside diameter of the cam.
21. Turn the actuator knob to the lock position and apply a small amount of Moly XL Hi-Speed grease to both grooves of the cam.
22. Install the pressure spring and outer clutch gear. Compress the pressure spring by forcing the clutch gear down, then install the snap-ring. Be sure that the snap-ring is fully seated in the groove.
23. Turn the actuator knob to the free position. Assemble the six dished washers to the six retaining screws.
24. Install two screws with washers into the knob retainer to properly align the parts. Apply a small amount of Moly XL Hi-Speed grease to the outer spline and teeth of the outer clutch gear. Remove any excess lubricant from the gasket surface of the retainer.

25. Install a new outer retainer gasket. Assemble the gear hub housing by aligning the splines of the housing with those on the outer clutch gear. Then install a new inner metal gasket on the hub housing.
26. Position the free-running hub assembly to the axle and tighten the two installed retaining screws. Turn the actuator knob to the lock position.
27. Install the remaining four retainer screws with washers and tighten all screws in sequence to 30-36 ft lbs.

NOTE: The hubs may be hard to engage and disengage at first, but should loosen up after some use. Make sure that both the hubs are either engaged or disengaged before driving the truck.

Reference

Front wheel and steering alignment procedures may be found in the General Repair Section.

Steering Gear

Manual Steering Gear

Steering Gear R&R

Econoline 1971-74
1. Raise the vehicle on a hoist.
2. Remove the flex coupling lower attaching bolt.
3. Disconnect the pitman arm from the drag link.
4. Remove the three attaching bolts and remove the gear.
5. Remove the pitman arm attaching nut and remove the pitman arm.
6. To install, center the input shaft (approximately three turns from either stop).

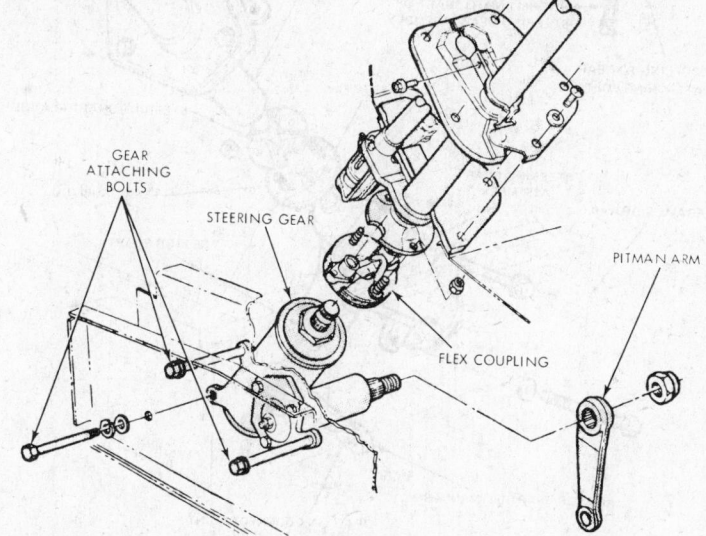

1971-74 Econoline manual steering gear installation (© Ford Motor Co.)

7. Install pitman arm pointing downward, tightening attaching nut securely.
8. Fill the steering gear to the proper level with lubricant.
9. Position the steering gear in the truck, aligning the input shaft splines to engage the flexible coupling. Install and tighten the three steering gear mounting bolts.
10. Connect drag link to pitman arm. Install attaching bolt and tighten securely. Install cotter pin.
11. Install flexible coupling attaching bolt and tighten securely.
12. Lower vehicle.

Econoline 1975-78

1. Raise the vehicle.
2. Disconnect the flex coupling from the steering shaft flange by removing the two attaching nuts.
3. Disconnect the drag link from the sector shaft arm.
4. Support the steering gear and remove the bolts and washers that attach the steering gear assembly to the frame side rail. Lower the steering gear from the vehicle.
5. Remove the coupling to gear attaching bolt from the lower half of the flex coupling and remove the coupling from the steering gear assembly.
6. Remove the pitman arm-to-sector shaft attaching nut and washer. Remove the pitman arm from the sector shaft.
7. Install the flex coupling on the worm shaft of the gear assembly. Install a new coupling-to-

gear attaching bolt and tighten.
8. Center the input shaft. The center position is approximately three turns from either stop.
9. Assemble the pitman arm on the sector shaft pointing downward. Install the attaching nuts and washers. Tighten the nuts.
10. Position the steering gear assembly so that the stud bolts on the flex coupling enter the bolt holes in the steering shaft flange, and the holes in the mounting bosses of the gear match the bolt holes in the frame side rail.
11. While supporting the gear in the proper position, install the gear-to-frame side rail attaching bolts and washers, tighten.
12. Connect the drag link to the pitman arm. Install the drag link ball stud nut and tighten the nut. Install the cotter pin.
13. Secure the flex coupling to the steering shaft flange with the two attaching nuts and tighten.

F-100, 150, 250, 350 (4x2)

1. Remove flex joint attaching bolt and remove the brake line bracket.
2. Raise the front of the vehicle.
3. Disconnect the pitman arm from the sector shaft.
4. Remove the steering gear attaching bolts and the gear.
5. Before installing gear, align the wheels and the sector shaft to the straight-forward position.
6. Install steering gear, tightening attaching bolts securely.
7. Install the brake line bracket to the gear cover studs.

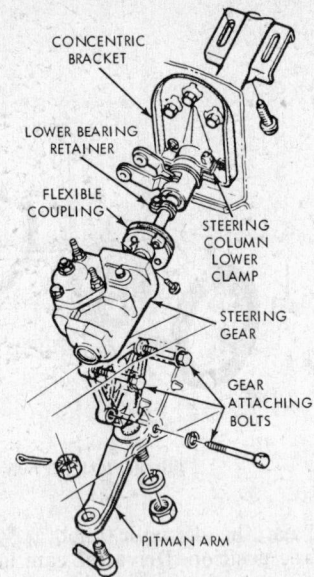

Steering gear installation— F100, 250 & 350, 4x2
(© Ford Motor Co.)

8. Connect the pitman arm to the sector shaft.
9. Install and tighten flex coupling bolt.
10. Remove the steering gear filler plug and housing lower cover bolt. Turn the steering wheel to the left to move the ball nut away from the filler hole. Fill the steering gear with lubricant (until lubricant comes out of the housing cover lower bolt hole). Install filler plug and cover bolt.

Bronco and F-100, 150, 250 (4x4)

1. Raise the vehicle on a hoist.
2. Remove the pitman arm.
3. Remove the three gear to frame attaching bolts, then lower the vehicle.
4. Remove the flex coupling clamp bolt at the steering gear input shaft and loosen the other clamp bolt (steering column). Remove the coupling from the steering gear input shaft. Discard clamp, bolt and nut.
5. Remove the steering gear from the vehicle.
6. When installing, first mount the steering gear to the frame, but do not tighten attaching bolts.
7. Install steering shaft flex coupling to the gear input shaft using a new clamp and bolt. Tighten bolt securely.
8. Install the flex coupling to the steering column shaft with a new clamp and bolt. Tighten bolt securely.
9. Raise the vehicle and tighten the steering gear attaching bolts securely.
10. Install the pitman arm on the steering gear sector shaft, tightening attaching nut securely.

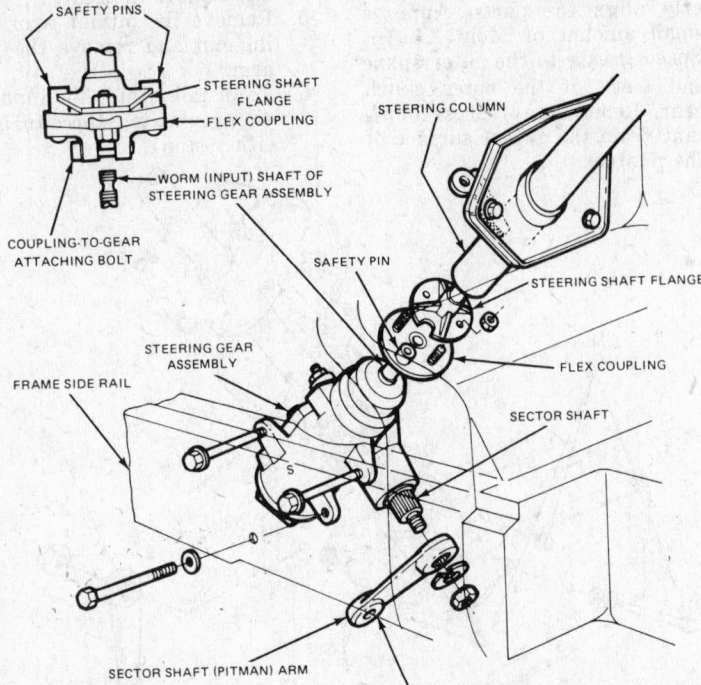

1975-78 Econoline manual steering gear installation (© Ford Motor Co.)

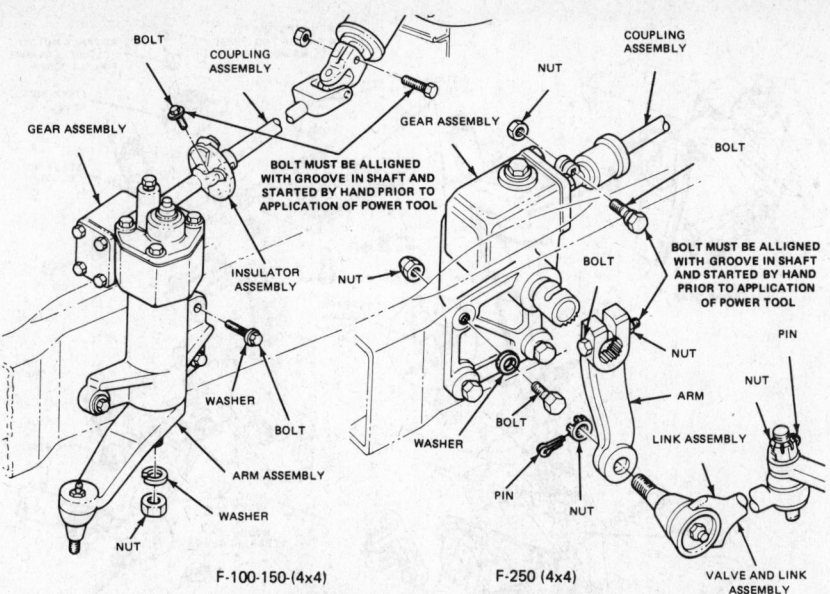

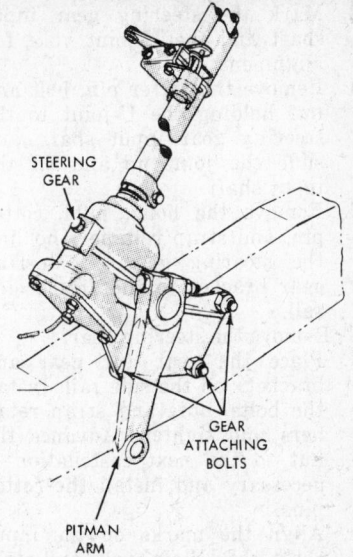

F-100, F-150, & F-250 manual steering gear installation (© Ford Motor Co.)

P-Series manual steering gear installation
(© Ford Motor Co.)

11. Lower the vehicle and fill the steering gear with lubricant.

P Series

1. Remove the steering wheel.
2. Remove the steering column bracket bolts from the instrument panel.
3. Raise the front of the vehicle 10 inches and support with safety stands.
4. Remove the sector shaft arm from the sector shaft and remove the steering gear attaching bolts. It may be necessary to spread the sector shaft opening in the arm.
5. Loosen the steering column

lower clamp and disconnect the horn wire.
6. Move the steering gear to the left and remove it from beneath the vehicle.
7. Place the steering gear in the steering column tube and partially tighten the lower clamp.
8. Install the steering gear attaching bolts finger tight and connect the horn wire.
9. Install the sector shaft arm and tighten the nut. Spread the split portion of the arm if necessary.
10. Loosely install the steering column tube bracket bolts.
11. Install the steering wheel.
12. Adjust the steering column tube

to provide a 1/16 inch clearance between the top of the tube and the steering wheel hub. Tighten the instrument panel bolts.
13. Tighten the lower clamp bolt and tighten the steering gear attaching bolts.
14. Lower the vehicle.

F-B Series—500 and Up

1. Loosen the steering column lower clamp and slide the clamp down the column.
2. Remove the pitman arm retaining bolt and nut from the sector shaft.
3. Loosen the steering column 9000 lb axles, remove the nuts, bolts, cotter pin, and strap retainer that hold the steering gear housing on the frame side member.
4. Remove the steering gear from under the vehicle.
5. Install the steering gear assembly from underneath the vehicle and tighten the mounting bolts, nuts, and strap retainer finger tight.
6. Connect the steering column lower clamp and tighten.
7. Tighten the nuts that attach the upper steering gear housing to the frame.
8. Install the bottom nut and cotter pin and tighten to 50-70 ft-lbs.
9. Install the pitman arm on the sector shaft and tighten.

L-N, L-LT-LNT- W-WT Series

1. Place the steering wheel and the front wheels in the straight ahead position.
2. Remove the bolt, nut and cotter pin holding the pitman arm to the sector shaft. Remove the pitman arm from the shaft.

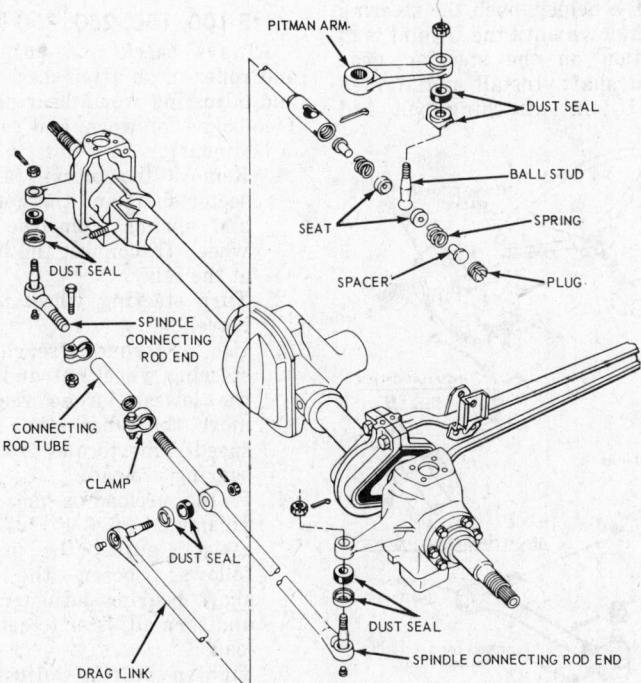

Steering linkage—F100 & 4x4 Bronco (© Ford Motor Co.)

3. Mark the steering gear input shaft and the U-joint yoke for alignment.

4. Remove the cotter pin, bolt and nut holding the U-joint to the steering gear input shaft, and slide the joint up and off the input shaft.

5. Remove the bolts, nuts, cotter pin, and strap retainers holding the steering gear or steering gear bracket to the frame side rail.

6. Remove the steering gear.

7. Place the gear (or gear and bracket) on the side rail. Install the bolts, nuts and strap retainers and tighten. Advance the nut to the next castellation if necessary and install the cotter pins.

8. Align the marks on the input shaft and U-joint yoke and place the steering shaft U-joint on the steering shaft U-joint on the steering gear input shaft and install a new nut and bolt. *NOTE: Be sure the bolt engages in the slot on the steering gear input shaft.* Tighten the nuts. Advance to the next castellation if necessary and install the cotter pin.

9. Place the pitman arm on the steering shaft and install the bolt and nut. On W and L series vehicles, align the slash marks on the pitman arm with the serrations on the sector shaft. Tighten the bolts. Advance the nut to the castellation if necessary and install the cotter pin.

C-Series

1. Tilt the cab up and remove the horn wire brush from the steering column just below the instrument panel bracket.

2. Remove the bolt that attaches the U-joint to the steering gear.

3. Turn the wheels all the way to the right. Remove the sector arm attaching nut and bolt. Pry the

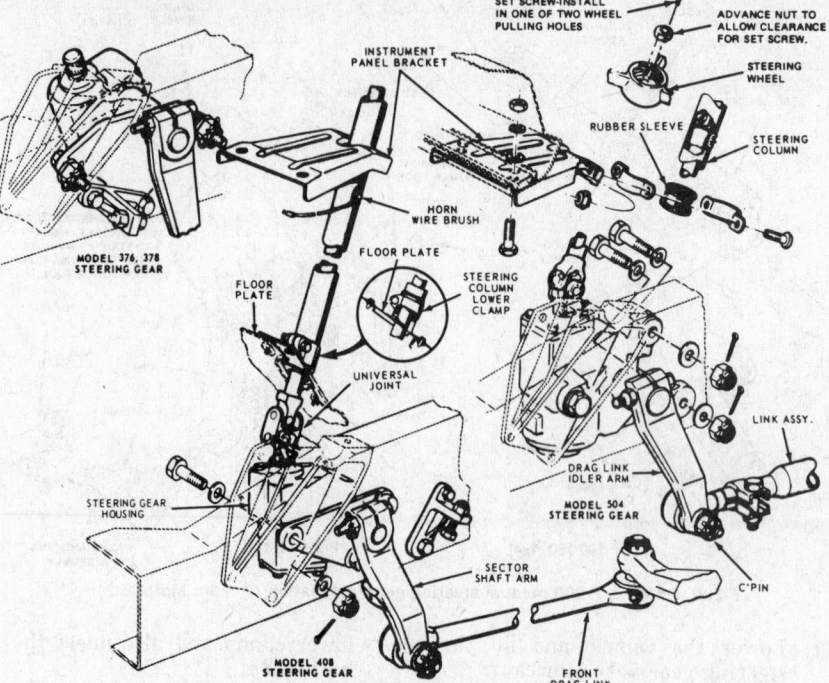

C-Series manual steering gear installation (© Ford Motor Co.)

sector arm off the sector shaft. Spread the sector shaft opening in the arm if necessary.

4. Remove the steering gear housing attaching bolts and remove the gear.

5. Place the steering gear housing on the frame side rail and install the attaching bolts. Tighten the nuts and install the cotter pins.

6. Center the steering gear input shaft. Turn the steering wheel to position the lower spoke in a vertical position.

7. With a helper, push the steering shaft down until the U-joint is in position on the steering gear input shaft. Install and tighten the U-joint attaching bolt and nut.

8. Install the sector shaft arm on the sector shaft. Make sure it is in line with the lock bolt slot in the sector shaft. Install and tighten the bolt and nut. Spread the sector shaft opening in the arm if necessary.

9. Connect the hornwire. Road test the vehicle in check for proper steering operation.

Worm Bearing Preload Adjustment

F-100, 150, 250, 350 (4x2)

Always check and adjust worm and roller mesh after check checking and adjusting worm bearing preload (see below for worm and roller mesh adjustment).

1. Remove the pitman arm from sector shaft and the horn button and spring from the steering wheel. Disconnect the horn wire at the relay.

2. Turn steering wheel to end of travel.

3. Use a torque wrench on the steering wheel nut and measure the lowest torque required to move the wheel at a constant speed. This torque is the worm bearing preload.

4. If the preload is not 4-5" lbs. (manual) or 3-4" lbs. (power assist), adjust the preload as follows: loosen the steering shaft bearing adjuster locknut and turn adjuster to set the preload.

5. Tighten bearing adjuster locknut, install the pitman arm and horn components.

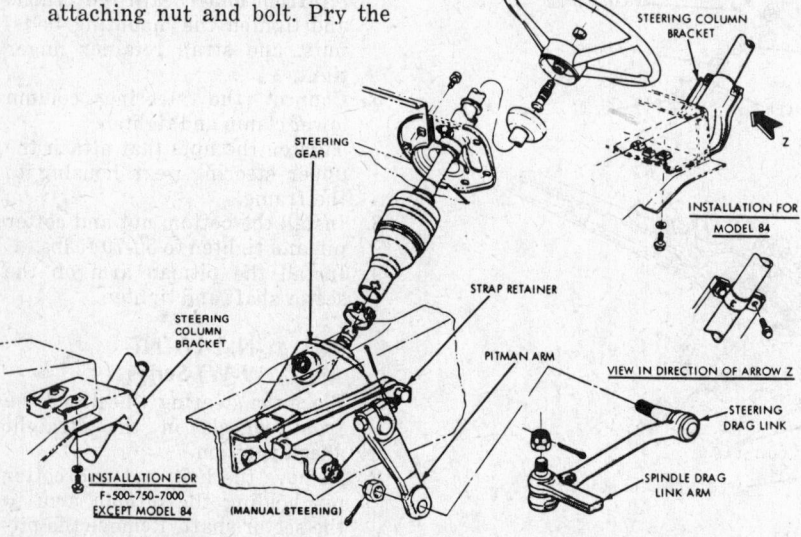

F-B 500-750 manual steering gear installation (© Ford Motor Co.)

F-100 (4x4)

Follow Steps 1 through 3 of the procedure for F-100, 250, 350 (4x2) above. If adjustment is necessary, remove or add shims between the worm shaft bearing retainer cover and the steering gear housing. If erratic readings are experienced, disconnect the steering shaft joint at the steering gear and check torque required to rotate the steering shaft. If that torque is measurable, add it to the preload specification.

Bronco

Remove the gear from the vehicle, loosen locknut and back off the mesh and roller adjusting nut. Check the torque required to rotate the input shaft 1½ turns either side of center. If bearing preload is not 5-10″ lbs., add or remove shims between the worm shaft bearing retainer cover and the steering gear housing until proper preload is obtained.

F and C-Series

Remove the steering gear and use a 12-point socket and torque wrench on the lower steering serrated shaft to measure preload at a constant speed rotation. If bearing preload is not within 9-11″ lbs. (C-series) or 5-9″ lbs. (F-series), add or remove gasket shims between the worm shaft bearing retainer and the gear housing. Adjust the worm and roller mesh preload. Install steering gear in vehicle.

Worm and Roller Mesh Adjustment

Always check and adjust worm bearing preload before making the mesh adjustment (see above).
1. Remove the steering gear from the vehicle.
2. Measure the torque required to move the gear through middle (straight-ahead) position with a torque wrench on the input shaft. The highest reading is used. If it is not 24-29″ lbs. (C-series), 14-22″ lbs. (F-500-750) or 12-21″ in. lbs. (F-100 4x4 and Bronco), loosen locknut and turn adjusting screw until correct mesh load is obtained. Tighten locknut.
3. Recheck mesh load, install steering gear in vehicle and fill gear with lubricant.

Steering Worm and Sector Adjustment

Econoline 1971-74

1. Remove the steering gear from the vehicle.
2. Loosen locknut on the sector shaft adjusting screw and turn adjusting screw out (counterclockwise) approximately three turns.

3. Using a torque wrench on the input shaft, rotate shaft about 1½ turns either side of center. If the preload is not 4-5″ lbs. (manual) or 3-4″ lbs. (power assist), loosen input shaft bearing adjuster locknut and tighten or loosen bearing adjuster to obtain the correct bearing preload. Tighten locknut and recheck preload.
4. Center the input shaft (turn against stops gently as it is easy to damage the ball return guides).
5. Turn sector shaft adjusting screw clockwise until an overcenter meshload of 9-10″ lbs. (manual) or 8-9″ lbs. (power assist) is obtained (measured at input shaft). Tighten the locknut.
6. Check the total gear lash by holding the sector shaft solid in the center position and pulling with the torque wrench 15 in. lbs. in each direction. If the travel of the wrench exceeds 1¼″, then the whole gear must be replaced.
7. Recheck total over-center preload, then install steering gear in vehicle.

Econoline 1975-78

1. Remove the steering gear.
2. Tighten the worm bearing adjuster plug until all end-play is removed. Loosen it ¼ turn.
3. Using a socket on an in. lb torque wrench, turn the worm shaft full right and back ½ turn.
4. Tighten the adjuster plug until 5-8 in. lbs is reached. Tighten the locknut to 85 ft. lbs.

5. Turn the worm shaft from stop to stop, counting the number of turns. Turn the shaft back exactly half the number of turns to the center position.
6. Turn the sector shaft adjuster screw clockwise to remove all lash between the ball nut and sector teeth. Tighten the locknut.
7. Using a socket and in. lb torque wrench, note the highest reading required to rotate the gear through the center position, which should be 16 in. lbs.
8. If necessary, adjust the sector shaft adjuster screw to obtain the proper torque and recheck it.

Power Steering Gear

Reference

For power steering gear overhaul procedures, see the General Repair Section.

Ford Integral Power Steering Gear

F-100, 250, 350 (4x2) F-150 (4x4)
Over Center Preload Adjustment

1. Disconnect pitman arm from the sector shaft.
2. Disconnect the fluid return line at the reservoir and cap the reservoir return line.
3. Place the end of the return line in a clean container and cycle the steering wheel in both directions to discharge the fluid from the gear.
4. Turn the steering wheel to 45-degrees from the left stop and measure the torque (at the

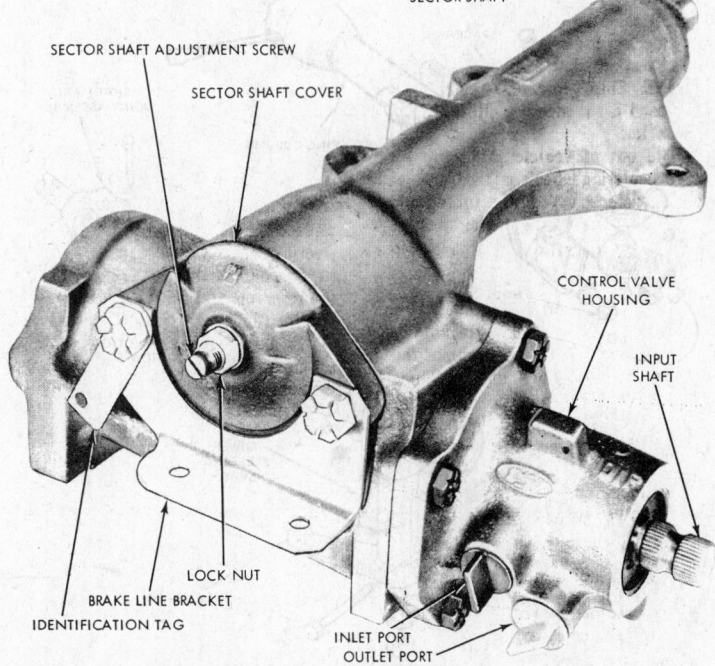

Ford integral power steering gear (© Ford Motor Co.)

steering wheel nut) required to turn ⅛-turn from there.

5. Determine the torque required to turn the steering gear through center position. Loosen adjusting screw locknut and turn adjusting screw to obtain a torque reading 11-12 in. lbs. greater than the torque 45-degrees from the stop.

6. Hold adjusting screw while tightening locknut.

7. Install the pitman arm and steering wheel hub cover.

8. Connect the fluid return line and fill reservoir with fluid.

Steering Gear Removal and Installation

1. Disconnect pressure and return lines from the steering gear, *being sure to tag them for identification.* Plug lines and ports.

2. Remove brake lines attached to bracket on the steering gear.

3. Remove the two bolts that secure the flex coupling to the steering gear and to the column steering shaft assembly.

4. Raise the vehicle.

5. Remove the pitman arm from the sector shaft, using a puller if necessary.

6. If vehicle has a standard transmission, remove the clutch release lever retracting spring.

7. Remove the steering gear attaching bolts and steering gear, working the steering gear free of the flex coupling.

8. To install, slide the flex coupling into place on the bottom of the steering shaft.

9. Set the steering wheel so that the spokes are horizontal and

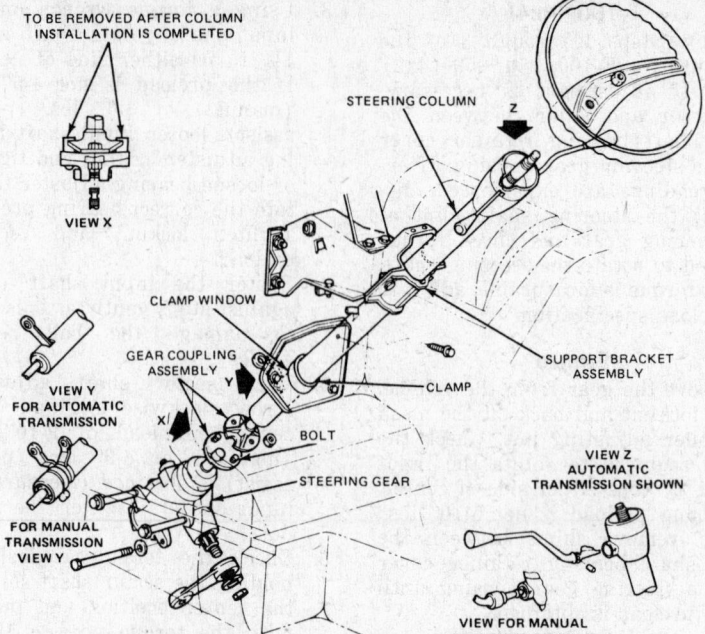

Econoline power steering gear installation (© Ford Motor Co.)

center the steering gear input shaft.

10. Slide the steering gear input shaft into the flex coupling. Install the three steering gear attaching bolts and tighten them securely.

11. With wheels in the straight ahead position, install the pitman arm on the sector shaft.

12. Install the flex coupling and tighten the bolts securely.

13. Connect fluid pressure and return lines to steering gear. Reinstall brake lines on bracket on steering gear.

14. Remove the coil wire, fill the power steering pump reservoir and, while engaging the starter, cycle the steering wheel to distribute the fluid. Add fluid if reservoir is not full.

15. Connect the coil wire, start the engine and check for leaks while cycling the steering wheel.

Integral Power Steering Gear (Saginaw)

Removal and Installation

Bronco

1. Disconnect the pressure and return lines and plug the lines and ports.

2. Raise the truck and disconnect the Pitman arm.

3. Remove the Pitman arm.

4. Unbolt the gear from the frame rails and lower the truck.

5. Remove the pinch-bolt from the flange and insulator.

6. Unbolt the horn and hold it outward.

7. Remove the attaching bolts and remove the gear from the truck, with the shaft and joint assemblies as a unit.

8. Remove the pinch-bolt from the shaft and joint.

9. Remove the shaft and joint from the gear.

10. Installation is the reverse of removal.

Econoline

1. Disconnect the pressure and return lines and plug the ports.

2. Raise the van and remove the drag link from the Pitman arm.

3. Unbolt the flex coupling from the steering shaft.

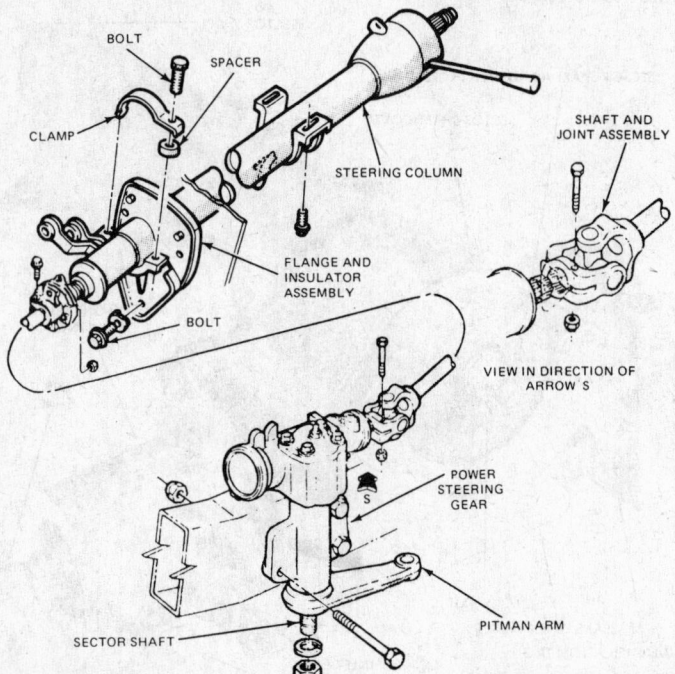

Bronco power steering gear installation (© Ford Motor Co.)

4. Support the gear and remove the attaching bolts.
5. Remove the pinch bolt from the flex coupling and remove the coupling from the gear.
6. Remove the Pitman arm from the sector shaft.
7. Installation is the reverse of removal.

Mesh Load Adjustment

1. On Econolines, remove the drag link from the Pitman arm and remove the horn pad. On Broncos, disconnect the Pitman arm from the sector shaft and remove the horn pad.
2. Disconnect the fluid return line and cap the reservoir return line. Put the end of the return line in a clean container and cycle the wheel several times to discharge fluid from the gear.
3. Using an in. lb torque wrench on the steering wheel nut, check the torque required to rotate the wheel through a 180° arc on each side of center. The new gear over center torque should be 4-8 in. lbs greater than the end readings but the total should not exceed 18 in. lbs. Used gears should be 4-5 in. lbs greater than the end reading, but should not exceed 14 in. lbs.
4. To adjust, make sure the Pitman shaft over center adjusting screw is backed all the way out. Turn it in ½ turn.
5. Rotate the shaft from one stop to the other. Count the number of turns and locate the center position. Check the combined preload on the ball and thrust bearing by rotating the shaft through the center of travel. Note the highest reading.
6. Tighten the adjusting screw until the torque wrench reads 3-6 in. lbs, higher than the reading in Step 5. The total should not exceed 14 in. lbs.
7. Hold the adjusting screw and tighten the locknut to 35 ft lbs.

Ross HF-54 and HF-64 Integral Power Steering Gears

Reference

Complete power steering gear overhaul procedures may be found in the General Repair Section.

F, B, L-500 through 9000 Series

Unloader Valve Adjustment

This adjustment is made for *right turn only* on the HF-54 gear and for *both turns* on the HF-64 gear.

Before making this adjustment, establish the straight-forward center of the steering system by driving the truck forward with hands off the wheel until the steering finds its own

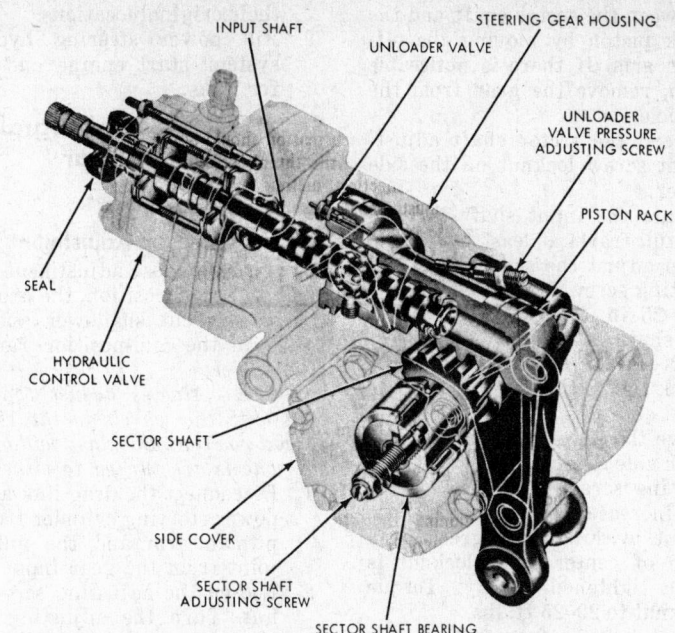

Ross model HF-64 integral steering gear (© Ford Motor Co.)

center. Stop the truck and mark the steering column to steering wheel with chalk or masking tape.

1. With wheels straight ahead engine warm and idling, turn the steering wheel 1¾ turns to the right (HF-54), 1½ turns (2 port HF-64) or 1¼ turns (4 port HF-64). Hold this position.
2. Loosen locknut and turn the unloader valve pressure adjusting screw until an audible hiss is heard. Tighten locknut.
3. While vehicle is moving, let the steering wheel center (straight-forward), then stop the truck and turn the wheel the prescribed number of turns (Step 2 above)

and listen for audible hiss. Repeat Step 2 if necessary. *NOTE: The pitman arm on the HF-64 must not contact the stop cast on the gear housing prior to contacting the unloader valve. The distance between the pitman arm and the stop should be 1/16" to ⅛" when the hiss is heard.*

4. To adjust the unloader valve for left turn (HF-64 ONLY), repeat Steps 1 through 3 above turning the steering wheel to the left.

Sector Shaft Adjustment

1. Disconnect the drag link from the pitman arm and center the steering wheel. Check for lash

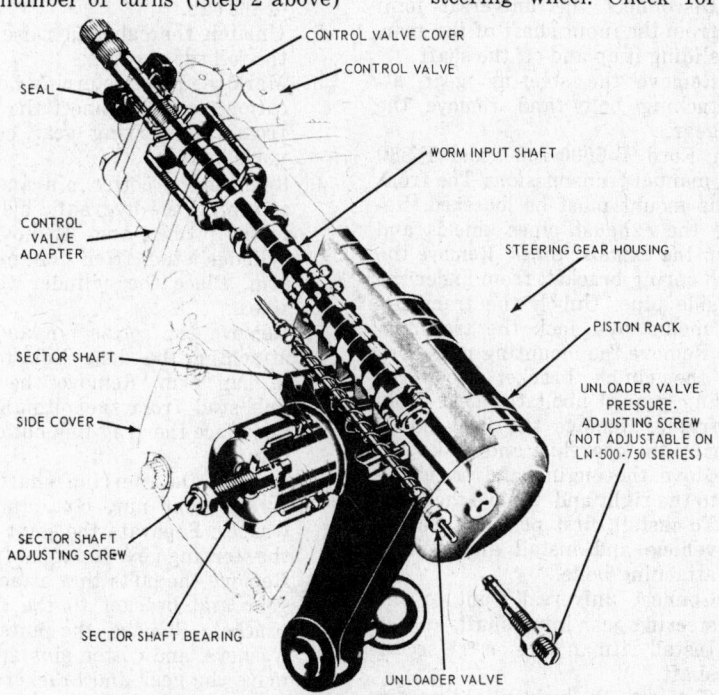

Ross model HF-54 integral steering gear (© Ford Motor Co.)

between the sector shaft and the rack piston by moving the pitman arm. If there is noticeable lash, remove the gear from the vehicle.

2. Loosen the sector shaft adjustment screw locknut on the side cover.

3. Rotate the input shaft through its full travel at least five times, then adjust the sector shaft adjusting screw for a 15–20 in. lb. (20-25 in. lbs. 1974-76) torque as shaft is rotated 90 degrees each side of center.

4. Back off adjusting screw one turn and note torque required to move the input shaft 90 degrees each side of center. Move the adjusting screw in to provide for an increase of 2-4 in. lbs. at a point within 45 degrees each side of center after locknut is first tightened snugly. Torque locknut to 20–25 ft. lbs.

5. Input torque of assembled gear (no fluid) should not exceed 15 in. lbs. over the full travel of output shaft.

6. Install steering gear in the vehicle.

7. Connect the drag link to the pitman arm.

8. Connect the pump lines and fill the system with fluid.

Steering Gear R & R

1. Position a drain pan under the steering gear and disconnect the pressure line from the gear, the return line from the gear and the turning hoses if so equipped. Mark all lines for identification.

2. Remove the pitman arm from the sector shaft.

3. Disconnect the universal joint from the input shaft of the gear, sliding it up and off the shaft.

4. Remove the steering gear attaching bolts and remove the gear.

On Ford B-6000-7000 and F-880 with manual transmission: The front engine mount must be loosened. Remove the exhaust pipes shields and loosen the exhaust pipes. Remove the clutch spring bracket, from under the left side pipe. Unbolt the transmission mount and jack the transmission. Remove the mounting pads. Unbolt the clutch bracket from the steering gear. Unbolt the gear from the frame. Remove the floor board covering the gearshift and brake levers. Move the engine and transmission to the right and remove the gear.

5. To install, first position gear in vehicle and install and tighten attaching bolts.

6. Connect universal joint to the steering gear input shaft.

7. Install pitman arm on the sector shaft.

8. Connect all hydraulic lines in their original locations.

9. Fill power steering hydraulic system, start engine and check for leaks.

Ross HPS-70 Semi-Integral Power Steering Gear

W Series

Mesh Load Adjustment

To provide close adjustment in the straight ahead position, the groove of the cam is cut shallower and narrower in the mid-position range of stud travel.

NOTE: Always adjust the mesh load with the wheels in the straight ahlead position. Backlash will occur if the wheels are turned to either side.

1. Disconnect the drag link and the power sterring cylinder from the pitman arm and the universal joint from the gear input shaft.

2. Loosen the adjusting screw lock nut. Turn the adjusting screw clockwise until a 15 in. lb. torque is required to turn the steering gear through mid-position.

3. Turn the adjusting screw counterclockwise ⅛ turn.

4. Holding the adjusting screw in place, tighten the locknut to 15-20 ft-lbs. to lock the adjustment.

5. Turn the gear from left to right with the engine off then check the adjustment. The torque at the input shaft should be 2-8 in-lb. for the full travel.

6. Connect the drag link and power sterring cylinder to the pitman arm, and the universal joint to the sterring gear input shaft.

Removal & Installation

1. Remove the grill from the front of the cab.

2. Unlatch the cab and raise it to the full tilt position.

3. Mark the power sterring hose location, and disconnect the hoses from the sterring gear control valve fittings.

4. Remove the cotter pin and nut attaching the hydraulic cylinder to the pitman arm. Remove the cylinder stud from the pitman arm. Place the cylinder to one side.

5. Remove the cotter pin and nut attaching the drag link to the pitman arm. Remove the drag link stud from the pitman arm and place the drag link out of the way.

6. Remove the sterring shaft universal joint nut, bolt, and flat washer. Separate the joint from the sterring gear input shaft.

7. Remove the bolts that attach the gear and bracket to the frame bracket. Remove the nuts, flat washers, and cotter pins and remove the gear and bracket from the vehicle.

8. Secure the gear and bracket in a vise and remove the pitman arm from the output shaft.

9. Remove the bolts and nuts attaching the gear to the bracket. Separate the gear from the bracket.

10. Place the gear on the bracket. Install and tighten the bolts and nuts.

11. Install the pitman arm on the output shaft with the notches on the output shaft and pitman arm aligned. Tighten the clamp bolt and install the cotter pin.

12. Place the steering gear and bracket on the frame bracket and install the bolts, washers, and nuts. Tighten the nuts and install the cotter pins.

13. Connect the drag link to the pitman arm. Tighten the nut on the pitman arm and install the cotter pin.

14. Connect the hydraulic cylinder to the pitman arm and tighten the nut. Install the cotter pin.

15. With the front wheels and steering gear in the straight ahead position, place the sterring shaft universal joint on the sterring gear input shaft. Install a new bolt, nut, and hardened washer, tighten the nut to 31-32 ft-lbs. *Be sure that the bolt passes through the slot on the sterring gear input shaft.*

16. Connect the hoses to the control valve.

17. Fill the fluid reservoir. Lower the cab. Start the engine and turn the sterring wheel from left to right three or four times to bleed the air from the system.

18. Install the grill on the front of the cab.

Clutch

Adjustment

Mechanical Clutch Linkage Adjustment

The clutch pedal free travel is the distance between the clutch pedal in the fully released position and the pedal position at which the clutch release fingers contact the clutch release bearing (this can be felt).

The clutch pedal total travel is the distance between the floor pan and the top of the pedal when the clutch is in the fully released position.

Only the clutch pedal free travel is adjusted. If the pedal free travel is not within 11/16″ to 1⅛″, adjust the clutch release rod until correct free travel is obtained.

Econoline

First the clutch pedal total travel is adjusted. It should be between

CLUTCH RELEASE ROD AND LEVER

Clutch adjustment—Econoline
(© Ford Motor Co.)

7½" and 7¾". Adjustment is made by loosening the locknut on the pedal stop eccentric bolt and turning the eccentric bolt until correct total travel is obtained. Tighten locknut.

Check the clutch pedal free travel. It should be 1⅛" to 1⅜". Adjustment is made at the clutch release rod turnbuckle.

Bronco

If the clutch pedal total travel is not within 6-11/16" and 15/16" (1971), or 6¾ to 7", move the clutch pedal bumper and bracket up or down to obtain required free travel.

Check and adjust pedal free travel. For 1971 Bronco's it is 1-7/32" to 1-27/64" and the 1972-78 is ¾" to 1½". Adjustment is made at the clutch release rod turnbuckle.

F- and N-Series

Total pedal travel is adjusted at the pedal bumper stop. Pedal free travel is made at the clutch release rod.

F-Series

1. Measure the clutch pedal free-play by depressing the pedal slowly until the free-play between the release bearing assembly and the pressure plate is removed. Note this measurement. The difference between this measurement and when the pedal is not depressed is the free-play measurement.
2. If the free-play measurement is less than ¾", the clutch linkage must be adjusted.
3. Loosen the two jam nuts on the release rod under the truck and back off both nuts several turns.
4. Loosen or tighten the first jam nut (nearest the release lever) against the bullet (rod extension) until a free-play measurement of ¾-1½" is obtained. A free-play measurement closer to 1½" is more desirable.

5. When the correct free-play measurement is obtained, hold the first jam nut in position and securely tighten the other nut against the first.
6. Recheck the free-play adjustment.

NOTE: Total pedal travel is fixed and is not adjustable.

Hydraulic Actuated Clutch Adjustment

Before adjusting hydraulic clutch, check the level of fluid in the master cylinder reservoir, filling to within ½" of the top if necessary. Bleed the hydraulic system as described below.

Bleeding Hydraulic Clutch

1. Attach a funnel to the bleeder screw of the slave cylinder by means of a transparent bleeder hose. The funnel must be higher than the master cylinder.
2. Pour fluid into the funnel as the system is filling, *being careful not to pour bubbles in the fluid.* Bleeder screw must be open.
3. Close the bleeder screw when master cylinder reservoir is full.
4. Check the slave cylinder pushrod total travel which should be 1⅛".
5. The clutch master cylinder relief port must be open in order to bleed the system. Clutch master cylinder pushrod lash of ¼" is required on the P-500 and 5000 series vehicles to permit the conventional type master cylinder piston to return to the piston stop ring and open the relief port. The clutch master cylinder cannot be adjusted on the P-350, 400, 3500 and 4000 series vehicles. In the case of tilt-cab (C-series) master cylinder, the O-ring seal on the end of the pushrod must not seat against the piston. Uncovering of the piston port is accomplished by adjustment of the pushrod to a piston lash of ¼". Forcing the clutch pedal to compress the pedal bumper more than normal by jamming a screwdriver between the pedal pad and the floor insures a complete uncovering of the relief port.

Hydraulic Clutch Adjustment

There are two adjustments for obtaining clutch pedal free travel. The initial pedal free travel should be ¼" for all models. On F, N, NT, and T Series vehicles the pedal height is first adjusted to 8-1/16" to 8-3/16" at the pedal stop. Turn eccentric bolt on pedal to obtain an initial pedal free travel of 3/16" to ⅜". Remove retracting spring, push slave cylinder pushrod as far forward as possible and release lever back until it con-

tacts the release fingers. Adjust the nut on the pushrod for a clearance of ¼" between pushrod adjusting nut and release lever. A total pedal travel of at least 2" should result.

On P and C Series trucks, the initial pedal free travel is that distance of pedal travel before the master cylinder pushrod contacts the piston. On C Series vehicles this adjustment is made with the eccentric bolt and on P Series it is made by rotating the master cylinder pushrod. Obtain an initial pedal free travel of ¼". With the retracting spring removed, the slave cylinder pushrod held completely forward and the release lever held in contact with the release fingers, turn adjusting nut on the slave cylinder pushrod to obtain a final pedal free travel of approximately 2".

Internal Clutch Adjustment

1. Remove inspection cover from the bottom of the clutch housing.
2. Disconnect the retracting spring and hold the release lever against the throwout bearing. Measure the distance between the throwout bearing and the clutch spring hub (synchromatic transmission) or clutch brake (non-synchromatic transmission). On vehicles with the 14" single plate clutch, this distance should be 11/32" to 13/32". On all Spicer clutches, also adjust the adjusting ring to obtain a ⅛" clearance between release yoke fingers and the throwout bearing.

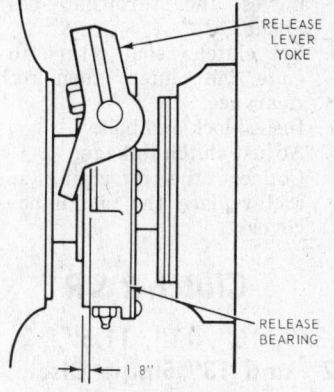

RELEASE LEVER YOKE

RELEASE BEARING

1.8"

Spicer clutch adjustment
(© Ford Motor Co.)

3. If adjustment is necessary, rotate the clutch assembly to get at the adjusting lockring and bolt. Remove the bolt and lockring.
4. With clutch pedal held or blocked in the released position, rotate the adjusting ring (use a pry bar) clockwise to move the throwout bearing toward the flywheel or counterclockwise to move the throwout bearing away from the flywheel. Rotating the adjusting ring one lug position

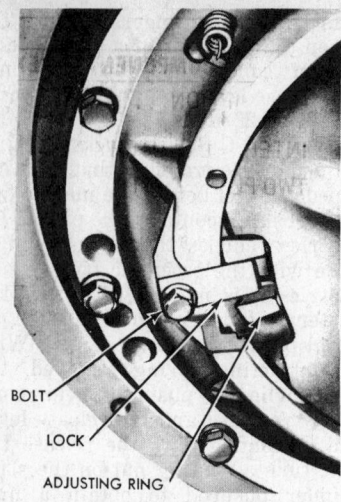

BOLT
LOCK
ADJUSTING RING

Internal clutch adjustment
(© Ford Motor Co.)

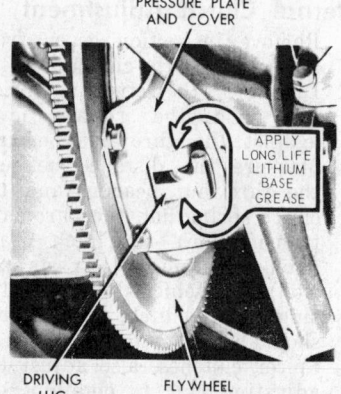

PRESSURE PLATE
AND COVER

APPLY
LONG LIFE
LITHIUM
BASE
GREASE

DRIVING
LUG   FLYWHEEL

Pressure plate lubrication point
(© Ford Motor Co.)

moves the throwout bearing about 1/32".

5. Let clutch pedal return to engage the clutch, then recheck clearance.
6. Install lock and bolt.
7. Adjust clutch linkage.
8. Connect the retracting spring and replace the clutch housing cover.

Clutch R & R

9.37", 10", 11", 11.5", 12" and 13" Single Disc

Removal & Installation

1. Disconnect the release lever retracting spring and pushrod at the lever.
2. If so equipped, remove the slave cylinder attaching bolts.
3. Remove the transmission.
4. If there is no dust cover on the flywheel housing remove the starter. Remove the release lever and bearing.
5. Mark the pressure plate and cover, remove the dust cover, release lever and bearing. If the flywheel housing has a dust

cover assembly and the flywheel so that they may be assembled in the same relative position.

6. Loosen the pressure plate attaching bolts evenly until the springs are loose then remove the bolts, pressure plate assembly and clutch disc. Do not remove the pilot bushing unless it is to be replaced.
7. To install, position the disc on the flywheel and install a pilot tool or spare transmission spline shaft.
8. Install the pressure plate assembly over the aligning tool and align the marks made during removal. Install the retaining bolts, tightening securely.
9. Remove pilot tool and apply a light coat of lithium-base grease to the hub splines of the clutch disc.
10. Apply lithium-base grease to the sides of the driving lugs.
11. Position throwout bearing and bearing hub on the release lever and install release lever on the trunnion in the flywheel housing.
12. Apply a light film of lithium-base grease to the release lever fingers and to the lever trunnion or fulcrum. Fill the angular groove of the release bearing hub with grease.
13. If removed, install the flywheel housing, tightening bolts securely.
14. Install the starter motor if it was removed.
15. Apply a light film of lithium-base grease to the transmission front bearing retainer and install the transmission assembly on the clutch housing, tightening attaching bolts securely.
16. Install the slave cylinder, if applicable.
17. Adjust the clutch linkage and install the clutch housing dust cover.

14" Single Plate Clutch

Removal & Installation

1. Disconnect the clutch pedal as-

sist spring, removing the left-hand exhaust pipe from the manifold if necessary.

2. Disconnect the release lever retracting spring and remove the slave cylinder attaching bolts, if applicable.
3. Remove the transmission.
4. Insert two ¾" wood blocks between the throwout bearing housing and the rear surface of the pressure plate assembly.
5. Remove the clutch to flywheel bolts and remove the flywheel ring and pressure plate assembly. Remove the clutch disc.
6. To install, position the clutch disc and pressure plate assembly on the flywheel and start the bolts. NOTE: the long hub of the clutch disc faces the rear.
7. Insert a spare transmission splined shaft or a disc aligning tool through the disc and into the pilot bushing. Tighten pressure plate bolts evenly and securely.
8. Remove the wood blocks and aligning tool or shaft.
9. Install the transmission.
10. Install the clutch slave cylinder.
11. Connect the release lever retracting spring and exhaust pipe.
12. Connect the clutch pedal assist spring.
13. Check internal clutch adjustment, linkage adjustment and correct if necessary.

13" Double Disc Clutch

Removal & Installation

1. Disconnect the clutch pedal reacting spring.
2. Remove the muffler inlet pipe if necessary.
3. Remove the bolts that attach the slave cylinder to the clutch housing.
4. Disconnect the slave cylinder push rod at the release lever. Disconnect the release lever retaining spring.
5. Remove the transmission as detailed in "Manual Transmission Removal and Installation."

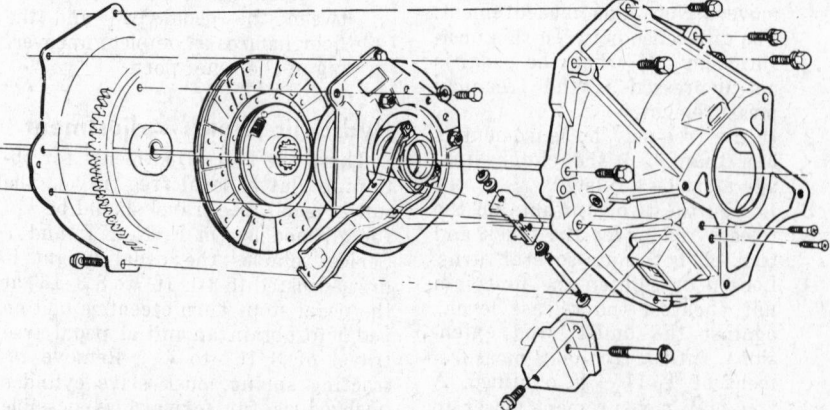

Typical single plate clutch installation—exploded view (© Ford Motor Co.)

6. Remove the dust cover from the clutch housing.

7. Remove the clutch release bearing and hub from the release lever.

8. Mark the center drive plate, pressure plate, cover assembly, and flywheel so that the parts may be installed in the same relative position.

9. Loosen the pressure plate and cover attaching bolts evenly until the pressure plate springs are expanded. Remove the bolts.

10. Remove the pressure plate and cover, the discs, and the intermediate pressure plate through the opening in the bottom of the clutch housing.

11. *Remove the pilot bearing only if replacement is necessary.*

12. Install the pilot bearing if it was removed.

13. Position the discs and second pressure plate on the flywheel so that the pilot tool can enter the clutch pilot bushing. *When installing the original pressure plate and cover assembly, align the assembly and flywheel according to the match marks made during the removal operations. The long end of the clutch disc hub must face the rear of the transmission.* The front (flywheel) side of the center drive plate is stamped "flywheel side" on one of the lugs.

14. Position the clutch discs, center drive plate, pressure plate and cover assembly on the flywheel and install the retaining bolts that fasten the assembly to the flywheel. Tighten the bolts to specification. Remove the clutch pilot tool.

15. Position the clutch release bearing and hub on the release lever.

16. Install the transmission assembly on the clutch housing and tighten the bolts to specification.

17. Install the slave cylinder and tighten the bolts.

18. Adjust the slave cylinder push rod. Connect the release lever reacting spring and (if removed) install the muffler left inlet pipe. Connect the clutch pedal reacting spring.

19. Lubricate the release bearing through the grease fitting with ½ oz of the specified lubricant.

20. Install the clutch housing dust cover.

14" and 15½" Double Disc Clutches

Removal & Installation

1. Disconnect the clutch pedal assist spring. If necessary, remove the left muffler inlet pipe.

2. Disconnect the release lever reacting spring. Remove the bolts

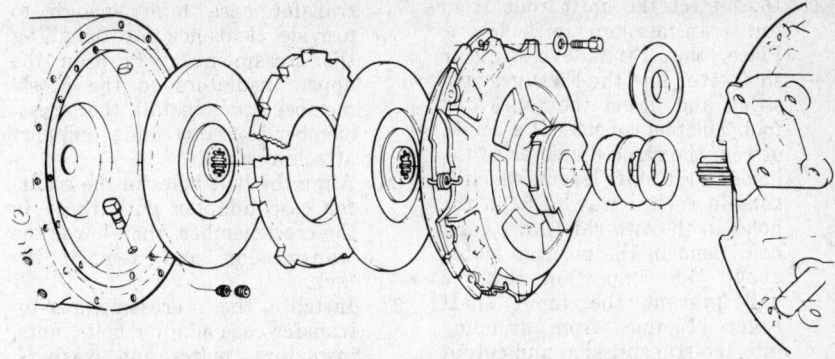

14 & 15½ Double disc clutch, typical installation—exploded view (© Ford Motor Co.)

that attach the slave cylinder. Remove the slave cylinder.

3. Remove the transmission as detailed in "Manual Transmission Removal and Installation." *NOTE: Pull the transmission straight out until it is clear of the clutch assembly. An unsupported, partially removed or installed transmission can spring or damage the clutch plates.*

4. Insert two ¾ in. wood blocks between the release bearing housing and the rear surface of the pressure plate assembly.

5. Remove the clutch-to-flywheel bolts and remove the flywheel ring and pressure plate assembly.

6. Remove the rear disc, intermediate plate, and the front disc from the engine. If necessary, remove the clutch drive pin retaining set screw and remove the four intermediate pressure plate drive pins.

7. Install the pressure plate drive pins in the flywheel if they were removed. Make sure they are a press fit. Oversize pins are available for service. Install the intermediate pressure plate and check the clearance between the pin heads and sots in tlhe pressure plate. A clearance of .006-.010 in. is required, measured on the same side of all pins. Replace the pins or plate if necessary. Remove the intermedidate plate, and using an adjustable square to maintain pin head alignment, install the pin set screws.

8. Place the front driven disc (shorter hub than rear disc) in the flywheel with the hub facing the rear. Set the intermediate pressure plate in the flywheel, aligning the slots with the pins. The side marked FRONT should face the engine. Position the read driven plate in the flywheel with the long end of the splined hub toward the rear. Insert the clutch disc aligning tool through the two discs and into the pilot bushing. Place the main (rear) pressure plate and flywheel ring assembly in position and start

the bolts. Tighten the pressure plate bolts evenly and tighten them to specification.

9. Remove the wood blocks from the release bearing housing and pressure plate. Remove the spline aligning tool.

10. Install the transmission.

11. If the vehicle is equipped with a slave cylinder, install the slave cylinder on the clutch housing. Connect the release lever reacting spring and the left-hand muffler inlet pipe if it was removed.

12. Connect the clutch pedal assist spring. Check the internal clutch adjustment, and make necessary correction.

Manual Transmission

Reference

For complete manual transmission overhaul procedures see the General Repair Section.

Transmission R & R

Ford 3.03 3-Speed

Bronco

1. Shift the transfer case into Neutral.

2. Remove the bolts attaching the fan shroud to the radiator support, if so equipped.

3. Raise the vehicle on a hoist.

4. Support the transfer case shield with a jack and remove the bolts that attach the shield to the frame side rails. Remove the shield.

5. Drain the transmission and transfer case lubricant. To drain the transmission lubricant, remove the lower extension housing-to-transmission bolt.

6. Disconnect the front and rear driveshafts at the transfer case.

7. Disconnect the speedometer cable at the transfer case.

8. Disconnect the T.R.S. switch, if so equipped.

9. Disconnect the shift rods from the transmission shift levers. Place the First-Reverse gear shift lever into the First gear position and insert the fabricated tool. The tool consists of a length of rod. the same diameter as the holes in the shift levers, which is bent in such a way to fit in the holes in the two shift levers and hold them in the position stated above. More important, this tool will prevent the input shaft roller bearings from dropping into the transmission and output shaft. THIS TOOL IS A MUST.
10. Support the engine with a jack.
11. Remove the two cotter pins, bolts, washers, plate and insulators that secure the crossmember to the transfer case adapter.
12. Remove the crossmember-to-frame side support attaching bolts.
13. Position a transmission jack under the transfer case and remove the upper insulators from the crossmember. Remove the crossmember.
14. Roll back the boot enclosing the transfer case shift linkage. Remove the threaded cap holding the shift lever assembly to the shift bracket. Remove the shift lever assembly.
15. Remove the two lower bolts attaching the transmission to the flywheel housing.
16. Reposition the transmission jack under the transmission and secure it with a chain.
17. Remove the two upper bolts setcuring the transmission to the flywheel housing. Move the transmission and transfer case rearward and downward out of the vehicle.
18. Move the assembly to a bench and remove the transfer case-to-transmission attaching bolts.
19. Slide the transmission assembly off the transfer case.
   To install the transmission:
20. Position the transfer case to the transmission. Apply an oil-resistant sealer to the bolt threads and install the attaching bolts. Tighten to 42-50 ft lbs.
21. Position the transmission and transfer case on a transmission jack and secure them with a chain.
22. Raise the transmission and transfer case assembly into position and install the transmission case to the flywheel housing.
23. Install the two upper and two lower transmission attaching bolts and torque them to 37-42 ft lbs.
24. Position the transfer case shift lever and install the threaded cap to the shift bracket. Reposition the rubber boot.
25. Raise the transmission and

transfer case high enough to provide clearance for installing the crossmember. Position the upper insulators to the crossmember and install the crossmember-to-frame side support attaching bolts.
26. Align the bolt holes in the transfer case adapter with those in the crossmember, then lower the transmission and remove the jack.
27. Install the crossmember-to-transfer case adapter bolts, nuts, insulators, plates and washers. Tighten the nuts and secure them with cotter pins.
28. Remove the engine jack.
29. Remove the fabricated tool and connect each shift rod to its respective lever on the transmission. Adjust the linkage.
30. Connect the speedometer cable.
31. Connect the T.R.S. switch. if so equipped.
32. Install the front and rear driveshaftshafts to the transfer case.
33. Fill the transmission and transfer case to the bottom of the filler hole with the recommended lubricant.
34. Position the transfer case shield to the frame side rails and install the attaching bolts.
35. Lower the vehicle.
36. Install the fan shroud, if so equipped.
37. Check the operation of the transfer case and the transmission shift linkage.

Econoline

1. Raise the vehicle on a hoist and drain the lubricant from the transmission by removing the drain plug if the vehicle is so equipped. For models without drain plugs, remove the lower extension housing-to-transmission bolt.
2. Disconnect the drive shaft from the flange at the transmission. Secure the front end of the drive shaft out of the way with lock wire.
3. Disconnect the speedometer cable from the extension housing and disconnect the gear shift rods from the transmission shift levers. Remove the wire to the TCS switch if so equipped.
4. Position a transmission jack under the transmission. Secure the transmission to the jack.
5. Raise the transmission slightly and remove the four bolts retaining the transmission support crossmember to the frame side rails. Remove the bolt retaining the transmission extension housing to the crossmember.
6. Remove the four transmission-to-flywheel housing bolts.
7. Position engine support bar

(Tool T65E-6000-J) to the frame.
8. Lower the transmission.
9. To install, make certain that the machined surfaces of the transmission case and the flywheel housing are free of dirt, paint and burrs.
10. Install a guide pin in each lower mounting bolt hole.
11. Start the input shaft through the release bearing. Align the splines on the input shaft with the splines in the clutch disc. Move the transmission forward on the guide pins until the input shaft pilot enters the bearing or bushing in the crankshaft. If the transmission front bearing retainer binds up on the clutch release bearing hub, work the release bearing lever until the hub slides onto the transmission front bearing retainer. Install the two transmission-to-flywheel housing upper mounting bolts and lock washers. Remove the two guide pins and install the lower mounting bolts and lock washers.
12. Raise the jack slightly and remove the engine support bar.
13. Position the support crossmember on the frame side rails and install the retaining bolts. Install the extension housing-to-crossmember retaining bolt.
14. Connect the gear shift rods and the speedometer cable.
15. Install the drive shaft and torque the attaching bolts to specification.
16. Fill the transmission to the bottom of the filler hole with the recommended lubricant.
17. Adjust the clutch pedal free travel and shift linkage as required.

F-100-250

1. Raise the vehicle and position safety stands.
2. Drain the transmission lubricant by removing the drain plug if the vehicle is so equipped. For models without drain plugs, remove the lower extension housing-to-transmission bolt.
3. Position a transmission jack under the transmission.
4. Disconnect the gear shift linkage at the transmission.
5. If the vehicle has a four-wheel drive, remove the transfer case shift lever bracket from the transmission.
6. Disconnect the speedometer cable.
7. Disconnect the drive shaft from the transmission.
8. Remove the transmission-to-clutch housing attaching bolts.
9. Move transmission to the rear until the input shaft clears the

clutch housing and lower the transmission. *Do not depress the clutch pedal while the transmission is removed.*

10. Before installing the transmission, apply a light film of lubricant to the clutch disc splines, release bearing inner hub surfaces, release lever fulcrum and fork and the transmission front bearing retainer. Exercise care to avoid contaminating the clutch disc with excessive grease.
11. Place the transmission on a transmission jack. Raise the transmission until the input shaft splines are in line with the clutch disc splines. The clutch release bearing and hub must be properly positioned in the release lever fork.
12. Install a guide stud in each lower clutch housing-to-transmission case mounting bolt and align the splines on the input shaft with the splines on the clutch disc.
13. Slide the transmission forward on the guide studs until it contacts the clutch housing.
14. Install the two transmission to flywheel housing upper mounting bolts and nuts. Remove the two guide studs and install the lower mounting bolts.
15. Connect the speedometer cable and the driven gear.
16. Install the drive shaft.
17. Connect each shift rod to its respective lever on the transmission.
18. If the vehicle is equipped with four-wheel drive, install the four-wheel drive shaft bracket.
19. Fill the transmission to the proper level with an approved lubricant.
20. Adjust the clutch pedal free travel and shift linkage as required.

Warner T-89F, T-87G

1. Raise and support the truck.
2. Drain the lubricant.
3. Support the transmission with a jack.
4. Disconnect the gearshift linkage at the transmission.
5. If the vehicle is a 4WD, remove the transfer case shift lever bracket from the transmission.
6. Disconnect the brake cable clevis at the cam and remove the cable clamp.
7. Disconnect the speedometer cable.
8. Remove the driveshaft and wire it aside.
9. Remove the transmission attaching bolts and remove the transmission.
10. Installation is the reverse of removal.

Warner T85N 3-Speed With Overdrive

F-100

The overdrive unit cannot be removed from the vehicle as a separate assembly. Drain both the transmission and the overdrive unit, and then remove the transmission and overdrive as a unit.

To remove the overdrive transmission, follow the same procedure as that for the standard three-speed transmission plus the following:
1. Disconnect the solenoid and governor wires at the connectors near the solenoid.
2. Remove the overdrive wiring harness from its clip on the transmission.
3. Disconnect the overdrive control cable.

To install the overdrive transmission, follow the same procedure as that for the standard transmission, plus the following:
1. Connect the overdrive control cable so that there is 1/4" clearance between the handle shank and dash bracket when the lever at the overdrive housing is against its rear stop.
2. Connect the solenoid and governor wires, and replace the overdrive wiring harness in its clip.
3. Fill only the transmission with lubricant.

Warner T-18, T-19 4-Speed

1. Disconnect the back-up light switch located at the rear of the gearshift housing cover.
2. Remove the rubber boot, floor mat, and the body floor pan cover, and remove the transmission shift lever. Remove the weather pad and pad retainer.
3. Raise the truck and position safety stands. Position a transmission jack under the transmission, and disconnect the speedometer cable.
4. If the truck is equipped with band-type parking brake, disconnect the brake cable clevis at the cam. Remove the brake cable conduit clamp.
5. Remove the front U-joint flange attaching bolts. Remove the bolts that attach the coupling shaft center support to the crossmember and wire the coupling shaft and drive shaft to one side. On F-100-350 Series trucks, remove the transmission rear support.
6. Remove the transmission attaching bolts.
7. Move the transmission to the rear until the input shaft clears the clutch housing, and lower the transmission.

Before installing the transmission,

apply a light film of lubricant to the clutch disc splines, release bearing inner hub surfaces, release lever fulcrum and fork, and the transmission front bearing retainer. Care must be exercised to avoid excessive grease from contaminating the clutch disc.
8. Place the transmission on a transmission jack, and raise the transmission until the input shaft splines are aligned with the clutch disc splines. The clutch release bearing and hub must be properly positioned in the release lever fork.
9. Install guide studs in the clutch housing and slide the transmission forward on the guide studs until it is in position on the clutch housing. Install the attaching bolts and nuts. Remove the guide studs and install the two lower attaching bolts.
10. Connect the speedometer cable and driven gear and parking brake clevis. Install the brake cable conduit clamp, and shift linkage.
11. Install the bolts attaching the coupling shaft center support to the crossmember.
12. Install the bolts attaching the front U-joint flange to the transmission output shaft flange. On F-100-350 Series trucks, install the transmission rear support.
13. Connect the back-up light switch.
14. Install the shift lever and lubricate the spherical ball seat with lubricant.
15. Install the weather pad and pad retainer. Install the floor pan cover, floor mat and boot.

New Process 435 4-Speed

1. On F-, LN- or B-Series truck, remove the rubber boot and floor mat.
2. On a F-, LN- or B-Series truck, remove the floor pan transmission cover plate. Remove the weather pad and pad retainer. It may be necessary first to remove the seat assembly.
3. Disconnect the back-up light switch located in the rear of the gearshift housing cover.
4. Raise the truck and position safety stands. Position a transmission jack under the transmission, and disconnect the speedometer cable.
5. Disconnect the parking brake lever from its linkage, and remove the gearshift housing. On a C-Series truck, disconnect parking brake cable and bracket at the transmission.
6. Disconnect the drive shaft. Remove the bolts that attach the coupling shaft center support to the cross-member and wire the coupling shaft and drive shaft to

one side. In F-100-350 Series Trucks, remove the transmission rear support.

7. Remove the two transmission upper mounting nuts at the clutch housing.

8. Remove the transmission attaching bolts at the clutch housing, and remove the transmission.

9. Before installing the transmission, apply a light film of lubricant to the clutch disc splines, release bearing inner hub surfaces, release lever fulcrum and fork, and the transmission front bearing retainer. Care must be exercised to avoid excessive grease from contaminating the clutch disc.

10. Place the transmission on a transmission jack, and raise the transmission until the input shaft splines are aligned with the clutch disc splines. The clutch release bearing and hub must be properly positioned in the release lever fork.

11. Install guide studs in the clutch housing and slide the transmission forward on the guide studs until it is in position on the clutch housing. Install the attaching bolts and nuts. Remove the guide studs and install the two lower attaching bolts.

12. Install the bolts attaching the coupling shaft center support to the crossmember.

13. Connect the drive shaft and the speedometer cable. On F-100-350 Series trucks, install the transmission rear support.

14. Connect the parking brake to the transmission.

15. Connect the back-up light switch.

16. On an F-, LN- or B-Series truck, install the weather pad, the pad retainer and the transmission cover plate. Install the seat assembly if it was removed.

17. On an F-, LN- or B-Series truck, install the weather pad, the pad retainer, the floor mat and rubber boot.

New Process 5 Speed

1. Remove the floor mat and the floor plate. Loosen the two nuts that secure the top of the transmission to the clutch housing studs. If the vehicle is not raised on a hoist, it may be necessary to remove the parking brake lever before the transmission can be removed.

2. Remove the shift lever and drain the transmission.

3. Disconnect the drive shaft or coupling shaft at the parking brake drum. If the vehicle is equipped with a coupling shaft support, disconnect the support

bracket and remove the coupling shaft.

4. Disconnect the parking brake adjusting rod and the speedometer cable.

5. Remove the dust cover from the bottom of the clutch housing. Position a transmission jack under the transmission. Remove the top nuts, and the bottom bolts and lockwashers that hold the transmission to the clutch housing.

6. Remove the transmission rearward until the input shaft splines clear the clutch housing. Be careful that the clutch release bearing and hub do not drop out of the release lever fork. Lower the transmission to the floor.

7. Place the transmission on the transmission jack and raise the transmission until the input shaft splines are aligned with the clutch disc splines. The clutch release bearing and hub must be properly positioned in the release lever fork. Slide the transmission forward until it is in position on the clutch housing.

8. Install the lock washers and the top nuts and the bottom bolts and tighten to specification.

9. Install the dust cover on the bottom of the clutch housing. Connect the speedometer cable.

10. Connect the parking brake adjusting rod.

11. Install the coupling shaft in the support bracket and connect the support bracket to the support plate. Connect the coupling shaft or drive shaft at the transmission. Install the shift lever.

12. Lower the vehicle to the floor and install the floor plate and the floor mat. Rill the transmission to the correct level with the correct lubricant.

13. Check the clutch pedal free travel and adjust if necessary.

Clark, Fuller, and Spicer 5 Speed Transmissions

1. Remove the floor mat and floor plate. On spicer transmissions, remove the 1st-reverse lockout plunger retainer, spring and plunger. Remove the gear shift lever housing and disconnect the parking brake lever (if so equipped). Cover the case opening to prevent foreign material from entering the case.

2. On C-series vehicles, shift the transmission into neutral, release the lock and tilt the cab forward. Remove the rear cross-shaft housing bolts. Tie the housing so that it does not fall. If necessary, raise the rear of the vehicle and install saefty stands to provide room for the removal of the transmission. Block the

wheels on W-series vehicles, remove the cross shaft housing bolts and tie the housing out of the way.

3. Drain the transmission. Disconnect the drive shaft or coupling shaft at the parking brake drum. If the vehicle is equipped with a coupling shaft support, disconnect the support bracket and remove the coupling shaft.

4. Disconnect the parking brake rod (if so equipped) and the speedometer cable. Remove the speedometer driven gear. Check the speedometer driven gear bushing in the mainshaft rear bearing cap on Spicer transmissions.

5. Disconnect the clutch linkage at the release arm. If the truck is equipped with a slave cylinder, disconnect the slave cylinder return spring, remove the slave mounting bolts, and set the slave cylinder out of the way.

6. Remove the dust cover from the bottom of the clutch housing.

7. a. *For Clark and Spicer transmissions;* position a transmission jack under the transmission. Remove the top nuts and the bottom bolts and lockwashers that attach the transmission to the clutch housing.

b. *For Fuller transmissions;* position a jack under the transmission and raise it slightly to relieve the pressure at the side rail mounting brackets. Remove the side rail bracket stud nuts, insulators and reinforcements. Remove the side rail brackets and the bolts that attach the clutch housing to the engine.

On W-series vehicles, remove the bolts and lockwashers that attach the transmission support to the frame side rail brackets. Mark the washers so that they can be re-installed in there original position. Remove the lock wire and the transmission support-to-transmission case bracket bolts, washers, and insulators. Remove the support and attach a chain hoist to the transmission. Remove the bolts that attach the clutch housing to the engine.

8. Move the transmission to the rear until it is clear and lower (or hoist on W-series vehicles) out of the chassis.

9. Install the transmission in the reverse order of removal.

NOTE: On Clark and Fuller transmissions when installing the gear shift lever housing, always lower it straight down onto the gear shift housing.

Fuller and Spicer Transmissions—6 Speed and Up

1. Bleed the air reservoir and disconnect the air line at the rear of the transmission. Drain the transmission lubricant.
2. Disconnect the drive shaft at the transmission companion flange.
3. Disconnect the speedometer cable at the transmission.
4. Clean the area around the cross shaft housing. Remove the housing and tie it out of the way. Cover the opening in the gear shft housing to prevent foreign material from entering.
5. Remove the shift control cable. Remove the retaining clamps for the speedometer and fuel line.
6. Disconnect the nylon cables at the air valve and shift cylinder. Tie the air cables next to the chassis frame to prevent damage to the cables when the transmission is removed.
7. Disconnect the clutch linkage at the clutch release arm. Remove all but two of the clutch attaching bolts at the engine. Remove the clutch housing dust cover.
8. Place a jack under the transmission and raise slightly to relieve the pressure at the side rail mounting brackets. Remove the bolts attaching the transmission support to the frame side rail brackets. After removing the lockwire, remove the transmission support-to-transmission case bolts, washers, insulators, and spacers. Remove the support.
9. Remove the remaining clutch housing attaching bolts. Pull the transmission rearward to clear the input shaft.
10. Raise the rear of the vehicle and support with safety stands. Remove the transmission from the vehicle.
11. Install the transmission in the reverse order of removal.

NOTE: If the pilot bearing was re-

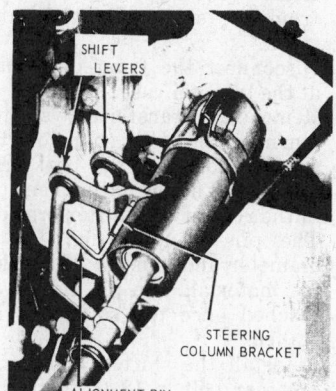

Gearshift linkage adjustment—Bronco
(ⓒ Ford Motor Co.)

moved, the new bearing must be pressed approximately 1/16 of inch beyond the flush position.

When installing the air lines to the cab control valve, relative to the piston air valve, keep the air lines with the protective cover at least four inches from the exhaust manifold. Heat will melt the protective cover and the air lines. Sharp bends in the air lines will restrict the required air pressure of 50-55 psi.

Gearshift Linkage Adjustment

Ford 3.03 3-Speed & T-85N 3-Speed

E and F-100, 150, 250 & Bronco

1. Place the shifter in the Neutral position and insert a gauge pin (3/16″ diameter) through the steering column shift levers and the locating hole in the spacer.
2. If the shift rods at the transmission are equipped with threaded sleeves, adjust the sleeves so that they enter the shift levers on the transmission easily with the shift levers in the Neutral position. Now lengthen the rods seven turns of the sleeves and insert them into the shift levers.

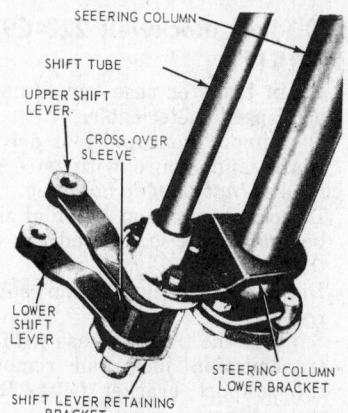

Gearshift linkage adjustment—Econoline
(ⓒ Ford Motor Co.)

3. If the shift rods, are slotted, loosen the attaching nut, make sure that the transmission shift levers are in the Neutral position, then retighten the attaching nuts.
4. Remove the gauge pin and check the operation of the shift linkage.

Warner T85N 3-Speed With Overdrive

1. Check the operation of the gear shift linkage for smooth operation.
2. To adjust the gear shift linkage, move the selector lever to the neutral position.
3. If the selector lever position

must be changed, disconnect the adjustable rods at the transmission, and install the gauge pin.
4. Loosen the lock nuts and adjust the swivels on the rods so that ends enter the levers freely, but do not put them in. Now adjust the swivels 7 turns to lengthen the rods by that amount. Connect the rods and lock the lock nuts.
5. Move the selector lever through all shift positions as a final check for smooth operation.
6. Adjust the position of the steering column tube if there is interference.

Overdrive Control Cable Adjustment

1. Loosen the overdrive cable lock nut at the overdrive control lever.
2. Position the overdrive cable until there is 1/4″ clearance between the cable control handle and bezel.
3. Move the overdrive control lever to the rear against the stop. When moving the lever, be careful not to move the cable out of position.
4. Tighten the control cable lock nut on the overdrive.

Overdrive Kickdown Switch Adjustment

1. With the accelerator shaft held in the full throttle position, slide the switch up or down on the elongated slots until the accelerator shaft is just touching the switch plunger. If at the closest position the plunger is still not touching the shaft, an adjustment not exceeding 4 turns counterclockwise may be made at the switch plunger.
2. The accelerator shaft can now be returned to the normal position and an additional adjustment of 5 turns counterclockwise performed at the switch plunger.

Transfer Case

Reference

For transfer case overhaul procedures see the General Repair Section.

Transfer Case R & R

Bronco (Dana 20)

1. Shift the transfer case into neutral and remove the fan shroud. Raise vehicle, support transfer case shield, then remove shield.
2. Drain the transmission and transfer case lubricant then dis-

connect driveshafts from transfer case.

3. Disconnect speedometer cable and shift rods.

4. A tool must be placed in the transmission shift levers to keep the input shaft roller bearings from falling out.

5. Remove the frame crossmember from side frame and transfer case, first raising transmission and removing adapter insulators.

6. Disconnect the shift rod from the transfer case shift lever bracket.

7. Remove the shift lever bracket to transfer case adapter bolt and let the assembly hang.

8. Position a transmission jack under the transfer case, remove the transfer case to transmission attaching bolts, then pull the transfer case back until it clears the transmission output shaft.

9. To install, place the transfer case in position and install the attaching bolts, tightening securely.

10. Install the shift lever to transfer case adapter. Connect the shift rod to the shift lever bracket.

11. Raise the transmission and transfer case high enough to provide clearance for installing the crossmember. Position the upper insulators on the crossmember, install the crossmember, then lower the transmission and install the bolts, tightening securely.

12. Remove fabricated tool from shift levers and install the shift rods.

13. Connect the speedometer cable.

14. Install the forward and rear driveshafts.

15. Fill the transmission and transfer case with lubricant.

16. Install transfer case shield.

F-100, 150 (4x4) (Dana 21)

1. Raise the vehicle on a hoist and disconnect the front and rear driveshafts.

2. Disconnect the shift rod from transfer case shift lever.

3. Remove transmission extension housing to transfer case attaching bolts and remove the transfer case.

4. To install, position transfer case and new gasket to extension housing and install attaching bolts.

5. Connect shift rod to shift lever.

6. Connect the front and rear driveshafts.

7. Lower vehicle and check for proper operation.

F-250 (4x4) (Dana 24)

1. Raise the vehicle on a hoist and drain transfer case lubricant.

2. Disconnect front and rear driveshafts, speedometer cable and shift rods.

3. Position a transmission jack under the transfer case and remove mounting bolts. Remove transfer case.

4. To install, raise transfer case on the jack and install nuts on mounting studs.

5. Connect shift rod, speedometer cable and driveshafts.

6. Fill transfer case to filler plug with lubricant.

F-100 and F-250 (New Process 205)

1. Drain the Transfer case and remove the rear drive shaft.

2. On F-250 models remove the front drive shaft from the transfer case. On F-100 models remove the front drive shaft and remove the transmission adapter to transfer case bolts.

3. Disconnect the shift rod and speedometer cable. Disconnect the reaction bar.

4. Support the transfer case with a jack and remove the transfer case mounting bolts.

5. Remove the transfer case from the vehicle.

6. To install reverse the removal procedures.

F-600 (4x4) (Rockwell 223-C9 & 223 C11)

1. Drain transfer case and disconnect speedometer cable.

2. Disconnect front and rear driveshafts and secure with wire or rope *so that they do not hang.*

3. Disconnect parking brake at the bellcrank and secure rod out of the way.

4. Disconnect de-clutch and shift rods.

5. Support the transfer case with a transmission jack and remove the support bracket bolts. Remove transfer case.

6. Install with the use of a transmission jack. Install and tighten mounting bolts.

7. Connect and adjust parking brake rod at the bellcrank.

8. Connect de-clutch and shift rods.

9. Connect front and rear driveshafts.

10. Connect speedometer.

11. Fill transfer case with lubricant.

New Process Model 203

1. Drain the transfer case by removing the power take-off lower bolts and the front output rear cover lower bolts.

2. Disconnect the front axle driveshaft from the flange at the transfer case.

3. Disconnect the shift rods from the transfer case.

4. Disconnect the speedometer

cable and lockout light switch wire from the transfer case rear output shaft housing.

5. On F-250 pick-ups, disconnect the front input and output shafts from the transfer case flanges. On F-100 trucks, remove the transfer case-to-transmission adapter attaching bolts. Disconnect the rear axle driveshaft at the transfer case flange.

6. Position a transmission jack under the transfer case and secure it with a chain.

7. Remove the transfer case mounting bolts and remove the unit from the vehicle. On F-250 models, remove the stabilizer bar-to-transfer case attaching bolt before lowering the case.

8. Install the transfer case in the reverse order of removal.

Transfer Case Shift Linkage Adjustment

New Process Model 205— F-250

Manual Transmission

Adjust the length of the shift rod between the transfer case and the shift lever with the lever in 4WD-Low so that the distance between the rear face of the transmission and the shift lever-to-rod clevis pin is 3.94 to 3.82".

Automatic Transmission

Adjust the length of the shift rod between the transfer case and the shift lever with the lever in 4WD-Low so that the distance between the upper surface of the automatic transmission extension housing and the upper horizontal edge of the shift lever is 0.640 to 0.600".

New Process 203 Full-Time 4WD

F-100, F-150

1. Place the shift lever in the Low position.

2. Remove the adjusting stud nuts from the shift rods under the truck.

3. Disconnect the front driveshaft at the transfer case.

4. Remove the transfer case shifter.

5. Unsnap the top snap of the splash boot, peel the boot back and install a new adjustment pin in the rear of the shifter bracket. The pin is 1.5" long, 1/4" in diameter and must be of breakable material. The pin cannot be reached once the shifter is installed.

6. Reinstall the shifter. tightening the rear bolt to 12-17 ft lbs and the front bolt to 70-90 ft lbs.

7. Move the lower lever on the transfer case completely forward

and the upper lever completely rearward.

8. Install new adjusting stud nuts and tighten them to 20-25 ft lbs.
9. Connect the front driveshaft.

F-250

1. Move the transfer case shifter to the low position.
2. Loosen the adjusting stud nuts.
3. Unsnap the top splash boot snap and peel the boot back. Install a new adjustment pin in the rear of the shifter bracket. The pin is the same size as previously mentioned for the F-100. If the pin is made of steel, it must be removed after tightening the adjusting studs.
4. Move the lower lever on the transfer case completely forward and the upper lever completely rearward.
5. Install new adjusting nuts and tighten them to 15-20 ft lbs.

Automatic Transmission

Reference

For complete automatic transmission overhaul procedures see the General Repair Section.

Shift Linkage Adjustment

Shift Linkage Adjustment

F-100, 250, 350 1971

1. Remove and discard shift rod trunnion retainer clip and grommet.
2. Install a new grommet and position selector lever against stop in D position.
3. Shift lever on transmission in D position (third from rear).
4. Adjust shift rod length until trunnion fits easily into the column shift lever, then lengthen rod seven additional turns.
5. Install trunnion in column shift lever and secure with new retainer clip.
6. Tighten trunnion locknut.
7. Check operation of selector lever in all positions.

P-350, 400, 500—1971

1. Disconnect shift rod from column shift lever.
2. Position selector lever against the stop in N position.
3. Shift lever on transmission to N position (fourth from the rear).
4. Rotate trunnion until it is easily inserted in column lever, then turn trunnion two full turns to lengthen rod.

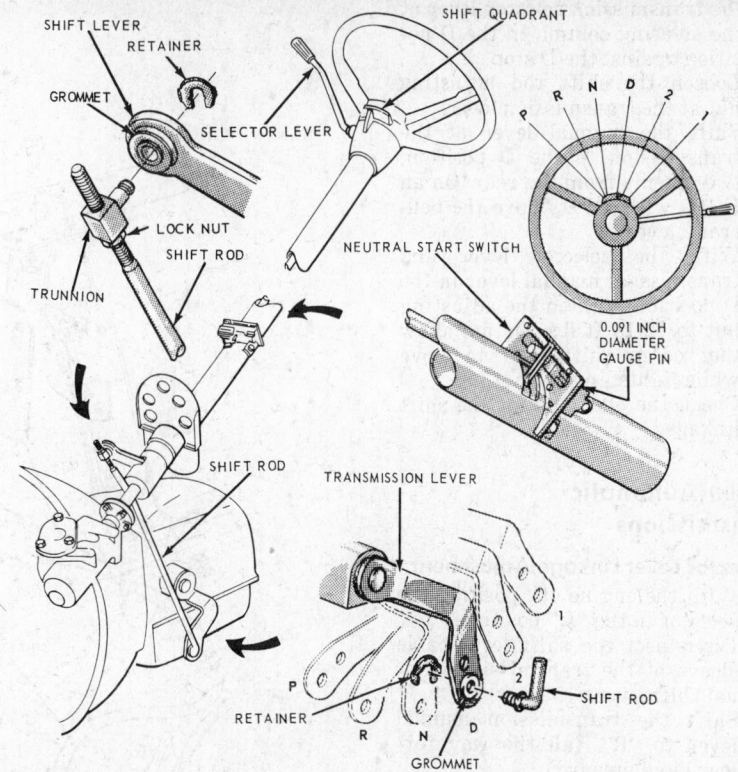

Manual linkage adjustment F-100-250-350 (© Ford Motor Co.)

5. Connect trunnion to column lever.
6. Check operation in all shift positions, adjusting trunnion one turn at a time if necessary. Relocate quadrant to selector indicator if necessary.

Econoline 1971

1. Position selector lever in drive position.
2. Loosen the locknut on the shift rod at the transmission and disconnect the manual shift rod.
3. Place transmission lever in drive position.
4. Adjust trunnion so that it fits into the lower column lever, then rotate the trunnion four additional turns to lengthen rod.
5. Connect trunnion and tighten the locknut.

6. Check operation of linkage in all positions, adjusting the rod length one turn at a time if necessary.
7. Adjust selector indicator to quadrant if necessary.

1972

1. With the engine stopped, place the transmission selector lever in the D position against the D stop.
2. Shift the manual lever at the transmission into the D position, second detent position from the rear.
3. With the transmission manual lever in the D position, connect the trunnion to the manual lever.
4. Turn the nuts on each side of the trunnion against the face of the trunnion finger-tight.
5. While holding the upper nut stationary tighten the lower nut to 12-18 ft lbs.

1973-78

1. With the engine stopped, place

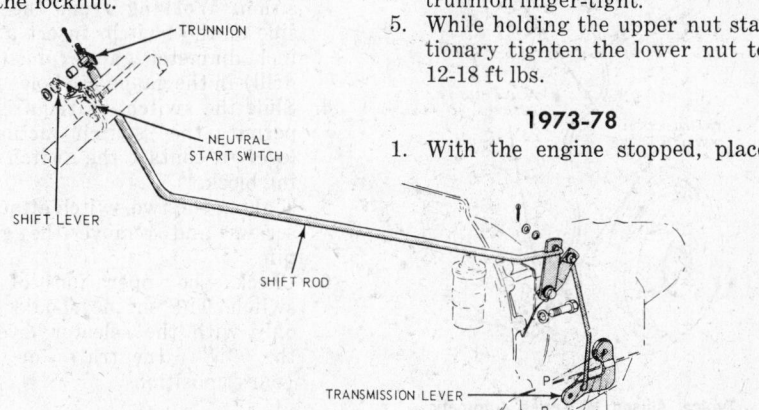

Manual linkage adjustment P-350-400-500 (© Ford Motor Co.)

the transmission selector lever at the steering column in the D position against the D stop.

2. Loosen the shift rod adjusting nut at the transmission lever.
3. Shift the manual lever at the transmission to the D position, two detents from the rear. On an F-100 with 4WD, move the bell-crank lever.
4. With the selector lever and transmission manual lever in the D position, tighten the adjusting nut to 12-18 ft lbs. Do not allow the rod or shift lever to move while tightening the nut.
5. Check the operation of the shift linkage.

Allison Automatic Transmissions

Selector Lever Linkage Adjustment

1. With the engine off, position the selector in the "R" position.
2. Disconnect the shift lever cable sleeve at the transmission manual shift lever.
3. Shift the transmission manual lever to "R" (all the way forward and upward).
4. With the manual lever in the "R" position, adjust the sleeve until it freely enters into the hole in the manual lever.
5. Connect the sleeve to the manual lever with the flat washers, spring washers, and cotter pin.
6. Operate the shift lever in all positions to make certain that the manual lever at the transmission is in full detent at all gear ranges. It may be necessary to re-adjust the sleeve slightly to obtain the detent position in all drive ranges.

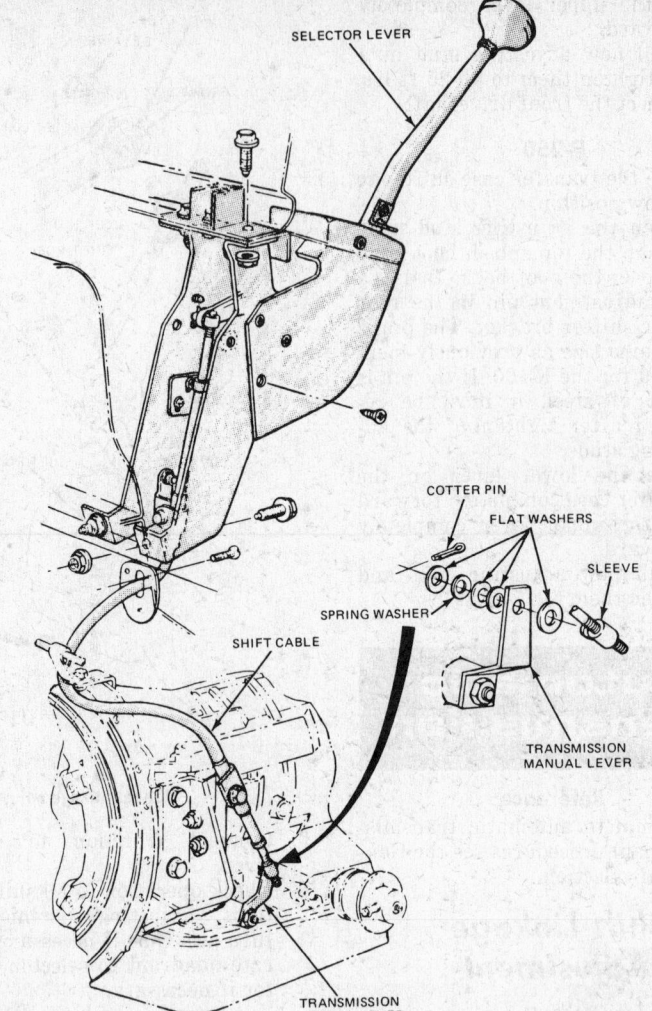

Typical Allison AT-540 automatic transmission selector lever linkage (© Ford Motor Co.)

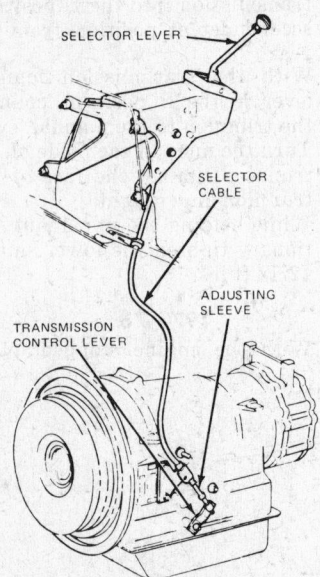

Typical Allison MT-series automatic transmission selector lever linkage (© Ford Motor Co.)

Neutral Safety Switch Adjustment

F-100-350

1. Loosen the two neutral start switch attaching screws.
2. Hold the selector lever against the neutral stop.
3. Move the sliding block on the neutral switch to the neutral position. Working from the rear side of the switch, insert a .091 inch diameter gauge pin (#43 drill) in the gauge pin hole.
4. Slide the switch, as required, to permit the switch actuating lever to contact the switch sliding block.
5. Tighten the two switch attaching screws and remove the gauge pin.
6. Check the operation of the switch. The engine should start only with the selector lever in the "N" (neutral) or "P" (park) positions.

E-100-350 and Bronco

1. With the manual linkage prop-

erly adjusted, loosen the two switch attaching bolts.
2. Place the transmission selector lever in neutral. Rotate the switch and insert .091 in. (#43 drill) gauge pin into the gauge pin holes. The gauge pin has to be inserted a full 31/64 in. into the three holes in the switch.
3. Tighten the two neutral start switch attaching bolts and remove the gauge pin from the switch.
4. Check the operation of the switch. The engine should start only with the selector lever in the "N" (neutral) or "P" (park) position.

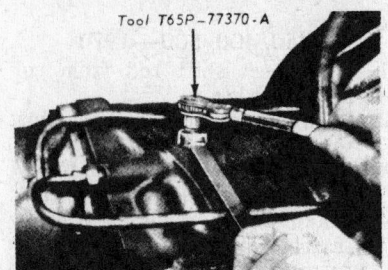

FMX rear band adjustment (© Ford Motor Co.)

P-500

1. Place the selector lever in neutral.
2. Loosen the two neutral start switch attaching bolts.
3. Move the neutral start switch until the switch lever is in the neutral detent position. Tighten the attaching screws.
4. Check the operation of the switch in each selector lever position. The engine should start only with the selector lever in the "N" (neutral) or "P" (park) positions.

B-500-700 and F-500-600

1. Check the starter at all selector lever positions. The circuit must be open at all positions except "N" (neutral).
2. To adjust, loosen the neutral switch to bracket attaching screws.
3. Position the switch so that the starter circuit is closed when the selector lever is in the "N" (neutral) position.)

Band Adjustments

C-4 Automatic Transmission

Intermediate Band Adjustment

1. Clean adjusting screw, apply penetrating lubricant and remove and discard locknut.

Adjusting intermediate band
(© Ford Motor Co.)

Adjusting low-reverse band
(© Ford Motor Co.)

2. Install a new locknut (loosely) and torque adjusting screw to 10 ft. lbs., then *back off 1 3/4 turns.*
3. Hold adjusting screw and tighten locknut.

Low-Reverse Band Adjustment

The adjusting procedure is exactly the same as that described for the intermediate band above except the *adjusting screw is backed off 3 full turns.*

C-6 Automatic Transmission

Intermediate Band Adjustment

F-100, 150, 250, 350

The adjusting procedure is exactly the same as that described for the C4 intermediate band adjustment above except that the adjusting screw is *backed off 1 1/2 turns.*

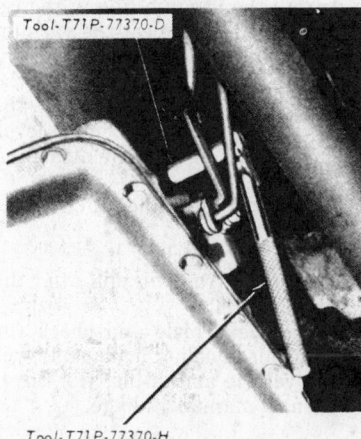

C6 intermediate band adjustment
(© Ford Motor Co.)

MX-HD Automatic Transmission

P-350, 400, 500 and F-350

Front Band Adjustment

1. Disconnect filler tube and drain fluid.
2. Remove and clean transmission pan and screen. Discard pan gasket.
3. Loosen locknut and adjusting screw, then insert a 1/4" spacer between the servo piston stem and the adjusting screw.
4. Tighten adjusting screw to 10 in. lbs., then remove spacer and *tighten adjusting screw an additional 3/4 turn.*
5. Hold adjusting screw and tighten locknut.
6. Install screen and pan using a new gasket.
7. Install the filler tube and fill transmission with fluid.

Rear Band Adjustment

1. Clean adjusting screw threads and apply oil.
2. Loosen locknut and tighten adjusting screw to 10 ft. lbs., then *back off adjusting screw 1 1/2 turns.*
3. Hold adjusting screw and tighten locknut.

FMX Automatic Transmission

Front Band Adjustment

1. Remove the transmission oil pan.

2. Loosen the front servo adjusting screw locknut.
3. Pull back on the actuating rod, and insert a 1/4" spacer between the adjusting screw and the servo piston stem.
4. Tighten the adjusting screw to 10" lbs.

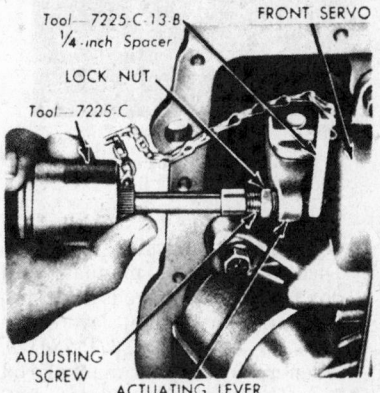

FMX front band adjustment (© Ford Motor Co.)

5. Remove the spacer and tighten the adjusting screw an additional 3/4 turn.
6. Hold the adjusting screw stationary and tighten the locknut. Tighten the locknut to 20-25 ft lbs.
7. Install the oil pan and a new gasket in the reverse order of removal.

Rear Band Adjustment

1. Remove all dirt away from the adjusting screw threads then oil the threads.
2. Loosen the rear band adjusting screw locknut.
3. Tighten the adjusting screw to 10 ft lbs.
4. Back off the adjusting screw *exactly 1 1/2 turns.*
5. Hold the adjusting screw stationary and tighten the adjusting screw locknut to 35-40 ft lbs.
6. Reinstall the oil pan and a new gasket.

Automatic Transmission R & R

C4 Automatic Transmission

F-100, 150, 250

1. Raise the vehicle and disconnect the transmission fluid filler tube from the pan. Drain the transmission fluid.
2. At the front lower edge of the converter housing, remove the cover attaching bolts and remove the dust cover. Remove the splash shield at the control levers.
3. Remove the drive shaft or coupling shaft. Remove the converter drain plug. Allow the converter to drain and install the drain plug.

Converter drain plug location
(© Ford Motor Co.)

FLYWHEEL   DRAIN PLUG HOLE

4. Disconnect the oil cooler lines from the transmission.
5. Disconnect the manual and downshift linkage rods from the transmission control levers. Be sure to use tool T67P-7341-A when removing the manual rod from the transmission lever.
6. Remove the speedometer gear from the extension housing.
7. Remove the four converter to flywheel attaching nuts. Disconnect the starter cable. Remove the three starter to converter housing attaching bolts. Remove the starter.
8. Disconnect the vacuum line from the diaphragm unit and the vacuum line retaining clip.
9. Position the transmission jack to support the transmission. Install the safety chain to hold the transmission on the jack.
10. Remove the two engine rear support crossmember-to-frame attaching bolts.
11. Remove the two engine rear support-to-extension housing attaching bolts.
12. Raise the transmission and remove the rear support. Remove the six converter housing-to-engine attaching bolts.
13. Move the transmission away from the engine. Lower the transmission and remove it from under the vehicle.
14. To install secure the transmission in a transmission jack. Align the transmission with the engine and move it into place. Install the converter housing-to-engine attaching bolts. Make sure that the converter studs are entered in the flywheel.
15. Install the rear support. Install the rear support-to-extension housing attaching bolts.
16. Position the starter into the converter housing and install the three attaching bolts. Install the starter cable.
17. Remove the transmission jack. Install the four converter to flywheel attaching nuts.
18. Connect the transmission filler tube to the transmission pan.

Connect the oil coolers lines to the transmission.
19. Install the speedometer driven gear in the extension housing.
20. Connect the transmission linkage rods to the transmission control levers. Be sure to use tool T67P-7341-A to install a new grommet into the manual lever, and to install the manual linkage rod into the ground.
21. Install the drive shaft or coupling shaft.
22. Install the vacuum line in the retaining clip. Connect the vacuum line to the diaphragm unit.
23. At the front lower area of the converter housing, install the lower cover and the control lever dust shield. Install the attaching bolts.
24. Secure the fluid filler tube to the pan.
25. Lower the vehicle.
26. Fill the transmission to the proper level.
27. Raise the vehicle and check for transmission fluid leakage. Lower the vehicle and adjust the throttle and manual linkage.

Econoline

1. Working from inside the vehicle, remove the engine compartment cover.
2. Disconnect the neutral start switch wires at the plug connector.
3. If the vehicle is equipped with a V8 engine, remove the flex hose from the air cleaner heat tube.
4. Remove the upper converter housing-to-engine attaching bolts (three bolts on 6-cylinder engines; four bolts on 8-cylinder engines).
5. On V8 engines, remove the upper muffler inlet pipe-to-exhaust manifold flange nut (right side of engine).
6. Raise the vehicle on a hoist.
7. On V8 engines, remove the three remaining muffler inlet pipe-to-exhaust manifold flange nuts and allow the exhaust pipe to hang.
8. Disconnect the transmission filler tube at the pan and drain the transmission fluid.
9. At the front lower edge of the converter housing, remove the dust cover attaching bolts and remove the cover.
10. Remove the converter-to-flywheel attaching nuts. As the flywheel is being rotated, remove the converter drain plug and drain the fluid from the converter.
11. Disconnect the drive shaft from the transmission companion flange and position it out of the way.
12. Remove the bolt retaining the

fluid filler tube to the engine and remove the tube.
13. Disconnect the starter cable at the starter. Remove the starter-to-converter housing attaching bolts and remove the starter.
14. Position the engine support bar (Tool T65E-6000-J) to the side rail and engine oil pan flanges.
15. Disconnect the cooler lines from the transmission. Disconnect the vacuum line from the vacuum diaphragm unit.
16. Remove the speedometer driven gear from the extension housing.
17. Disconnect the manual and downshift linkage rods from the transmission control levers.
18. Install the converter drain plug. If the converter is not going to be cleaned, torque the drain plug to specification.
19. Position a transmission jack to support the transmission. Install the safety chain to hold the transmission on the jack.
20. Remove the bolt and nut securing the rear mount to the crossmember. Remove the four bolts retaining the crossmember to the side rails. Then, with the transmission jack, raise the transmission and remove the crossmember.
21. Remove the remaining converter housing-to-engine attaching bolts. Lower the transmission and remove it from under the vehicle.
22. Position the transmission on the jack and secure the transmission and converter to the jack with the safety chain.
23. Raise the transmission and guide the transmission and converter into position. The converter-to-flywheel retaining studs must line up with the holes in the flywheel. The converter hub must enter the end of the crankshaft.
24. Install the converter engine attaching bolts. Install the converter-to-flywheel attaching nuts.
25. Install the crossmember. Install the rear mount-to-crossmember attaching bolt and nut.
26. Remove the safety chain and remove the jack from under the vehicle. Remove the engine support bar.
27. Connect the cooler lines to the transmission. Connect the vacuum line to the vacuum diaphragm unit.
28. Install the speedometer driven gear into the extension housing.
29. Connect the transmission linkage rods to the transmission control levers.
30. Connect the transmission filler tube to the transmission pan. Secure the tube to the engine with the attaching bolt.

31. Install the converter dust cover.
32. Position the starter into the converter housing and install the attaching bolts. Install the starter cable.
33. Install the drive shaft.
34. If equipped with a V8 engine, install the muffler inlet pipe on the exhaust manifolds and install and torque the three retaining nuts.
35. Lower the vehicle.
36. On V8 engines, install and torque the upper muffler inlet pipe-to-exhaust manifold flange nut.
37. Install the upper converter housing-to-engine attaching bolts.
38. On V8 engines, install the flex hose to the air cleaner heat tube.
39. Connect the neutral start switch wires at the plug connector.
40. Fill the transmission to the proper level with the specified fluid.
41. Raise the vehicle and check for transmission fluid leakage. Lower the vehicle and adjust the throttle and manual linkage.
42. Install the engine compartment cover.

C-6 Automatic Transmission

1. Drive the vehicle on a hoist, but do not raise at this time.
2. Remove the two upper converter housing-to-engine bolts.
3. Remove the bolt securing the fluid filler tube to the engine cylinder head.
4. Raise the vehicle and drain the fluid from the transmission and converter.
5. Disconnect the coupling shaft or driveshaft from the transmission companion flange and position it out of the way.
6. Disconnect the speedometer cable from the bearing retainer.
7. Disconnect the throttle and manual linkage rods from the levers at the transmission.
8. Disconnect the oil cooler lines from the transmission.
9. Remove the vacuum hose from the vacuum unit. Remove the vacuum line retaining clip.
10. Disconnect the cable from the terminal on the starter motor. Remove the three attaching bolts and remove the starter motor.
11. Remove the four flywheel attaching nuts. Place a wrench on the crankshaft pulley attaching bolt to turn the converter to gain access to the nuts.
12. Remove the two engine rear support crossmember-to-frame attaching bolts.
13. Remove the two engine rear support-to-extension housing attaching bolts.
14. Remove the eight bolts securing

the No. 2 crossmember to the frame side rails.
15. Raise the transmission with a transmission jack and remove both crossmembers.
16. Secure the transmission to the jack with the safety chain.
17. Remove the remaining converter housing-to-engine attaching bolts.
18. Move the transmission away from the engine. Lower the transmission and remove it from under the vehicle.
19. Install the converter on the stator support.
20. Secure the transmission to the jack with the safety chain.
21. Rotate the flywheel to place the two converter mounting stud holes in a vertical position.
22. Rotate the converter so that studs and drain plug are in alignment with those in the flywheel.
23. Move the transmission toward the cylinder block until they are in contact. Install attaching bolts.
24. Remove the transmission jack safety chain from around the transmission.
25. Position the No. 2 crossmember to the frame side rails. Install the attaching bolts.
26. Position the engine rear support crossmember to the frame side rails. Install the rear support to extension housing mounting bolts.
27. Lower the transmission and remove the jack.
28. Secure the engine rear support crossmember to the frame side rails with the attaching bolts.
29. Connect the vacuum line to the vacuum diaphragm making sure that the metal tube is secured in the retaining clip.
30. Connect the oil cooler lines to the transmission.
31. Connect the throttle and manual linkage rods to their respective levers on the transmission.
32. Connect the speedometer cable to the bearing retainer.
33. Secure the starter motor in place with the attaching bolts. Connect the cable to the terminal on the starter.
34. Install a new O-ring on the lower end of the transmission filler tube and insert the tube in the case.
35. Secure the converter-to-flywheel attaching nuts. Use a wrench on the crankshaft pulley attaching nut to rotate the flywheel. Do not use a wrench on the converter attaching nuts to rotate it.
36. Install the converter housing dust shield and secure it with the attaching bolts.

37. Connect the coupling shaft drive shaft.
38. Adjust the shift linkage.
39. Lower the vehicle. Then install the two upper converter housing-to-engine bolts.
40. Position the transmission fluid filler tube to the cylinder head and secure with the attaching bolt.
41. Fill the transmission to the correct level with the specified lubricant. Start the engine and shift the transmission thru all ranges, then re-check the fluid level.

MX-HD and FMX Automatic Transmission

1. Drive the vehicle onto a hoist, but do not raise it at this time.
2. After removing the converter access hole covers, remove the two upper bolts and lockwashers which attach the converter housing to the engine.
3. Raise the vehicle, and remove the cover from the lower front side of the converter housing.
4. Remove the converter drain plug. Drain the converter and reinstall the plug. If desired, the converter may be drained after the unit has been removed' from the vehicle.
5. Disconnect the fluid filler tube from the transmission pan.
6. When the fluid has stopped draining from the transmission, remove the flywheel to converter bolts. Wedge the converter to hold it in place when the transmission is removed.
7. Disconnect the starter cable from the starter, and disconnect the transmission to body ground cable from the transmission. Remove the starter.
8. On a transmission that has an oil cooler in the radiator, disconnect the fluid lines. On models equipped with a sidemounted cooler, disconnect the radiator-to-cooler coolant lines at the cooler. Plug the lines to prevent coolant loss from the radiator.
9. Disconnect the manual and throttle linkages from the transmission.
10. Remove the vacuum hose from the vacuum unit. Remove the vacuum line retaining clip.
11. Disconnect the speedometer cable from the extension housing, and remove the drive shaft and/or coupling shaft.
12. Remove both engine rear support bolts.
13. With the transmission jack, raise the engine and transmission high enough to remove the engine rear supports.
14. Lower the engine against a floor

stand or engine support bar so that the converter housing is clear of the cross member when all weight is off the transmission jack.

15. Remove the remaining converter housing to engine attaching bolts.

16. Remove the flywheel to converter attaching bolts. Then move the assembly toward the rear and lower it, leaving the flywheel attached to the crankshaft. If additional clearance is needed, tilt the rear of the assembly upright slightly and to the rear (enough to allow removal of the six flywheel to crankshaft bolts). Move the assembly to the rear, and remove it.

17. If the converter has been removed from its housing, position the converter in the housing, and install a wedge to prevent the converter from slipping out of the housing.

18. Rotate the converter until the studs are in the vertical position. Position the flywheel on the crankshaft flange and install the six attaching bolts.

19. Raise the transmission and converter with the jack to align it with the engine. Remove the wedge from the converter housing. Carefully move the transmission toward the engine and at the same time engage the converter studs with the holes in the flywheel.

20. Install the converter engine lower bolts.

21. Raise the engine and transmission with the transmission jack, and remove the engine support stand or support bar.

22. Place the engine rear supports in position on the cross member.

23. Lower the engine and transmission against the supports, and at the same time install the support bolts.

24. Rotate the flywheel, and tighten the attaching bolts.

25. Install the four converter-to-flywheel attaching nuts. Install the converter drain plug and the access plate.

26. Connect the oil cooler inlet and outlet lines to the transmission or side-mounted cooler.

27. Install the vacuum line and retaining clip. Install the vacuum hose on the vacuum unit.

28. Coat the universal joint knuckle with transmission fluid, and install the drive shaft and/or coupling shaft.

29. Connect the speedometer cable to the transmission.

30. Connect the manual linkage rod to the transmission manual lever.

31. Connect the throttle linkage to the transmission throttle lever.

32. Install the linkage splash shield.

33. Install the starter motor. Connect the transmission to frame ground cable to the transmission.

34. Connect the fluid filler tube to the pan.

35. Lower the hoist. Then install the upper two converter housing-to-engine bolts.

36. Install the access hole covers, and position the floor mat.

37. Fill the transmission with fluid.

38. Check the transmission, converter assembly, and oil cooler lines for fluid leaks. Then adjust the transmission control linkage.

Allison AT-540 Automatic Transmission

1. Remove the right and left door sill scuff plates. Move the floor mat out of the way. Remove the bolts securing the transmission access cover to the floor pan and remove the cover.

2. On a one-piece drive shaft, disconnect the shaft from the yoke at the parking brake drum. On vehicles with a two-piece drive shaft, disconnect the coupling shaft from the yoke at the parking brake drum. Remove the center support bearing bracket. Move the forward end of the drive shaft out of the way.

3. Disconnect the parking brake linkage and remove the parking brake drum from the output shaft flange.

4. Remove the parking brake handle and brackets from the transmission case.

5. Place a drain pan under the transmission and drain the fluid.

6. Disconnect the speedometer cable from the rear of the transmission

7. Disconnect the oil cooler lines at the transmission.

8. Disconnect the vacuum line from the vacuum modulator.

9. Remove the vacuum modulator retainer bolt and remove the retainer. Remove the vacuum modulator valve actuating rod from the case.

10. Disconnect the shift cable from the manual selector lever at the transmission.

11. Remove the two bolts holding the shift cable bracket to the transmission and position the cable and bracket out of the way.

12. Remove the inspection cover from the bottom front side of the flywheel housing.

13. Remove the six converter to flywheel attaching bolts. It may be necessary to turn the engine over manually.

14. Remove the four upper and two lower converter housing to fly-

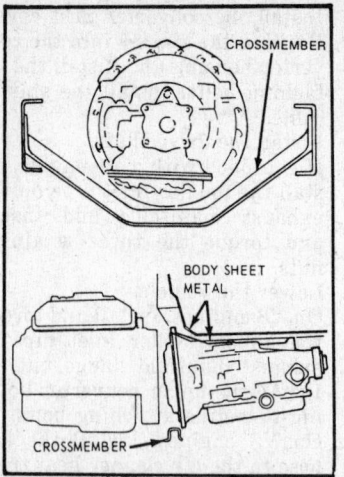

Removal or installation of Allison AT-540 transmission (© Ford Motor Co.)

wheel attaching bolts.

15. Remove the bolts and nuts securing the two rear engine supports to the cross member. Remove the lower insulators and flat washers.

16. Loosen the front engine support bolts.

17. Position a transmission jack under the transmission and secure the transmission to the jack with a chain.

18. Place a jack under the engine. Raise both jacks to relieve the pressure on the cross member.

19. Remove the two upper engine rear support insulators from the crossmember.

20. Remove the bolts securing the engine rear support brackets to the transmission and remove the brackets.

21. With both jacks supporting the engine and transmission, remove the remaining converter housing to flywheel attaching bolts.

22. Pull the transmission rearward until the converter housing touches the body sheet metal. Then, tilt the rear of the transmission upward until the bottom of the converter housing clears the crossmember. Lower the transmission assembly and remove it from the underside of the vehicle. If necessary, raise the rear of the vehicle to permit the transmission to clear the chassis.

NOTE: Prior to installing the transmission, it is mandatory that the flexplate be checked for runout. The maximum T.I.R. should be .020 in. In addition, the flywheel housing transmission mouting face alignment inspection should be made on the engine flywheel housing transmission surface. The face and bore runout should not exceed .008 in T.I.R.

23. Position the torque converter on the transmission, engaging the turbine shaft with the stator and

the hub with the oil pump drive gear.

24. Place the transmission on a jack and secure with a chain.

25. Raise the converter and transmission of the jack and move the transmission assembly into position over the cross member. Align the holes in the converter with the holes in the flywheel.

26. Install six converter housing-to-flywheel attaching bolts, three on each side. Tighten the bolts to specification.

27. Lower the engine jack and remove it from the vehicle.

28. Position the engine rear support brackets on the side of the transmission and secure with the attaching bolts.

29. Position the two upper engine rear support insulators on the crossmember. Position the lower insulators and flat washers on the cross member and install the rear support-to-cross member bolts and nuts. Tighten the bolts to specification.

30. Lower the transmission jack and remove it from the vehicle.

31. Install the four upper and two lower conver housing-to-flywheel attaching bolts.

32. Tighten the front engine support bolts to specification.

33. Install the six converter to flywheel attaching bolts. Tighten the bolts to specification.

34. Install the inspection cover on the bottom front side of the flywheel housing.

35. Position the shift cable bracket on the transmission and install the attaching bolts.

36. Connect the shift cable to the manual lever on the transmission.

37. Install the vacuum modulator actuating rod and vacuum modulator into the case. Install the vacuum modulator retainer and secure with the attaching bolt.

38. Connect the vacuum line to the vacuum modulator.

39. Connect the oil cooler lines to the transmission.

40. Connect the speedometer cable.

41. Install the fluid filler tube on the oil pan.

42. Position the parking brake handle and bracket on the transmission and secure with the attaching bolts.

43. Install the parking brake drum and connect the parking brake linkage.

44. Install the drive shaft, center support bearing bracket and the coupling shaft. Connect the coupling shaft to the parking brake drum and tighten all bolts.

45. Position the transmission access cover on the floor pan and secure with the attaching screws.

46. Add enough ATF to the transmission to bring the fluid level to the full mark on the transmission.

47. Check the transmission, converter assembly and oil cooler for leaks.

Allison MT-Series Automatic Transmissions

1. Remove the filler tube from the oil pan and drain the oil from the transmission.

2. Install a protective plug in the drain hole.

3. On vehicles with a one piece drive line, disconnect the shaft from the pinion shaft flange and from the yoke at the parking brake drum. *Do not remove the bolt that attaches the flange to the transmission output shaft.* On vehicles with two piece drive shafts, disconnect the coupling shaft from the yoke at the parking brake drum, and then remove the center support bearing bracket. Move the forward end of the drive shaft out of the way.

4. Disconnect the speedometer cable from the rear of the transmission.

5. Disconnect the throttle control rod and the selector lever cable from the levers on the left side of the transmission housing. Remove the cable clamp bracket from the transmission.

6. Disconnect the parking brake linkage.

7. Disconnect the oil cooler lines from the fittings on the retarder valve body, then remove the forward fitting. Plug the lines and valve body openings. If desired, the lines and valve body can be drained, but they should be plugged before removing the transmission assembly from the vehicle.

8. Remove the dust shield from the bottom front side of the flywheel housing.

9. Remove the nuts and flat washers that hold the converter pump cover to the engine flywheel. The nuts and washers can be reached through an opening in the lower right side of the flywheel housing. The flywheel must be turned to remove all the nuts and washers. *The ignition system should be disconnected during this operation to prevent accidental starting of the engine.*

10. Cut the lock wires and remove the two bolts and nuts that hold the converter housing to the frame crossmember.

11. Cut the lock wire and remove the two bolts, washers, and insulators from the top of the transmission rear support and cross member.

12. Place an engine support under the rear of the engine, then raise the engine to take the weight off the cross member.

13. Support the transmission with a 1000-lb. transmission jack. The jack should be placed so that the oil pan does not support the entire weight of the transmission. Fasten a safety chain over the top of the transmission and to both sides of the jack.

14. Remove the two bolts from each end of the transmission rear support crossmember, and remove the crossmember.

15. Remove the bolts and lockwashers that attach the converter housing the flywheel housing.

16. Move the transmission assembly away from the engine until the converter clears the crossmember. If necessary raise the floor pan slightly to permit the converter housing to clear the crossmember. *NOTE: If the engine, flywheel housing has been replaced, check the housing alignment and flywheel shim adjustment before installing the torque converter and transmission.*

The flywheel housing transmission mounting face alignment inspection should be made on the engine flywheel housing transmission surface. The bace bore runout should not exceed .008 in T.I.R.

17. Raise the transmission assembly on the jack, then move the unit into position over the crossmember that supports the converter. Align the studs in the converter with the holes in the flywheel. Install the two guide studs in the flywheel housing, then push the unit forward so that the converter studs enter the holes in the flywheel.

18. Install the bolts, except the fluid filler tube support bracket bolt, and lockwashers that attach the converter housing to the flywheel housing. Torque the bolts to 23-28 ft-lbs.

19. Install six new self locking nuts and flat washers to attach the converter to the engine flywheel. Torque the nuts to 34-40 ft-lbs.

20. Install the dust shield on the bottom front side of the flywheel housing.

21. Install the transmission rear support crossmember on top of the frame brackets, then install the bolts on the cross member and tighten to 40-45 ft-lbs. torque.

22. Lower the jack, and remove it from under the vehicle. Remove the engine support.

23. Install the nuts and bolts that hold the converter housing to the frame crossmember and tighten

the bolts to 70-91 ft-lbs. torque. Install the lock wire.

24. Install the bolts, washers, and insulators at the transmission rear 60 ft.-lbs. torque and install the support. Tighten the bolts to 40- lock wire.

25. Install the oil cooler lines.

26. Install the selector lever cable bracket on the left side of the transmission and tighten the attaching screws to 8-10 ft.-lbs. torque.

27. Connect the selector lever cable to the lever on the left side of the housing.

28. Connect the selector lever cable to the lever on the left side of the housing.

28. Connect the fluid filler tube to the oil pan, then install the tube support bracket bolt.

29. Push the breather hose across the top of the transmission and connect the hose to the fitting above the left PTO plate. Tighten the hose clamp.

30. Connect the parking brake linkage.

31. If the exhaust system was removed or disconnected during transmission removal, install and connect the parts.

32. Install the drive shaft, center bearing support bracket, and the coupling shaft. Connect the coupling shaft to the parking brake drum. Tighten all nuts and bolts.

33. Connect the speedometer cable to the rear of the transmission.

34. Adjust the selector lever linkage.

35. Check all fluid line connections for tightness, and lower the vehicle to the floor.

36. Add enough ATF to the converter and transmission to bring the fluid level up to the full mark on the dipstick.

Drive Shaft

Driveshaft R & R

Single Snap Ring U-Joint Type E-100 to 350, F100 to 350 and P Series (2WD)

1. Disconnect the driveshaft from the rear axle flange.

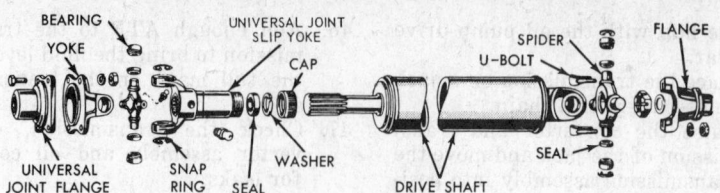

Drive shaft components—115" wheelbase (© Ford Motor Co.)

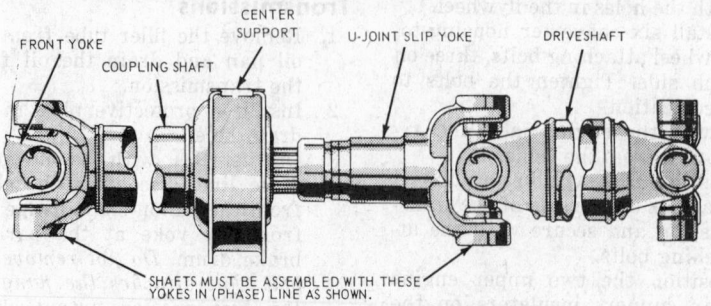

SHAFTS MUST BE ASSEMBLED WITH THESE YOKES IN (PHASE) LINE AS SHOWN.

Drive shaft components—132" wheelbase (© Ford Motor Co.)

2. On a 117" wheelbase vehicle, disconnect the driveshaft rear flange from the flange on the rear axle.

3. If vehicle has a coupling shaft, slide the driveshaft off the coupling splines.

4. Working from the center support nearest to the rear of the vehicle, remove the two attaching bolts and support the bearing.

5. On a vehicle with more than one coupling shaft, disconnect the rear shaft from the front one.

6. Remove the remaining center support attaching bolts and support the bearing.

7. Remove the transmission coupling shaft flange attaching nuts and remove the shaft and center bearing(s) as an assembly.

8. Thoroughly clean all driveshaft components before installing.

9. To install, connect the front flange or joint to the transmission flange.

10. Secure the center bearing to the frame bracket, tightening the bracket attaching bolts securely.

11. If vehicle has more than one coupling shaft, connect the rear shaft to the forward one, then install the remaining center support.

12. Connect the rear universal to the

rear axle flange, tightening nuts or bolts securely.

13. Be sure all driveshaft and coupling shaft yokes are in phase.

Double Cardan U-Joint Type Bronco, F150 and F250 (4WD)

1. Disconnect the double cardan joint from the transfer case flange and at the rear axle. Remove axle.

2. Remove the front axle by disconnecting the double cardan joint from the flange at the transfer case and the universal joint flange at the front axle. Remove the driveshaft.

3. Install axles connecting the universal joint ends before the cardan ends.

Single Bearing Cap, Bolt/ Bolted End Cap U-Joint Type Medium and Heavy Duty Trucks

1. Disconnect the drive shaft from the flange at the rear axle.

2. If working on a vehicle equipped with a coupling shaft, slide the drive shaft off the coupling shaft splines. On some Mechanics, Rockwell, and heavy duty Spicer assemblies, remove the attaching bolts.

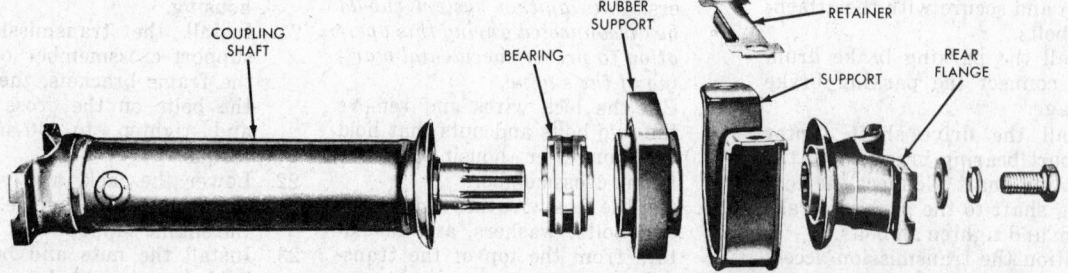

Coupling shaft and center support bearing (© Ford Motor Co.)

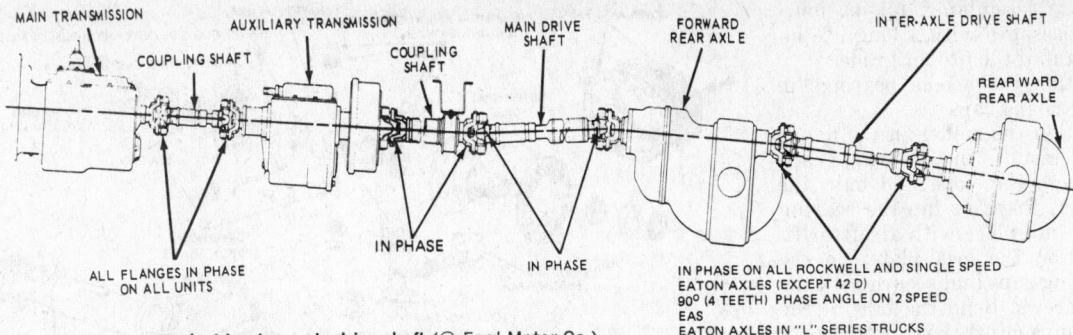

Typical tandem axle drive shaft (© Ford Motor Co.)

3. Working from the center support nearest to the rear of the vehicle, remove the two attaching bolts and support bearing. Different types of support brackets are used. Some have elongated mounting holes. This permits close adjustment.

4. If working on a vehicle equipped with more than one coupling shaft, disconnect the rear shaft from the front one.

5. Remove the remaining center support attaching bolts and support the bearing.

6. Remove the nuts that attach the coupling shaft flange to the transmission, and remove the shaft and center bearing assembly.

7. Thoroughly clean all drive shaft components before installation.

8. Connect the front flange or joint of the drive shaft or coupling shaft to the flange on the transmission.

9. Secure the center bearing to the frame bracket with the center support and attaching bolts. Note that L-series vehicles are adjustable with two-piece center support brackets. Tighten all bolts.

10. If working on a vehicle with more than one coupling shaft, connect the rear shaft to the forward one, then install the remaining center support. Make sure all splines are properly lubricated.

11. Connect the U-joint to the rear axle flange and tighten the nuts or bolts.

12. Make sure that all drive shaft and coupling shaft yokes are properly in phase.

U-Joints

E-100, 150, 250, F-100, 150, 250, 350, P-350-500, Bronco

Disassembly and Assembly

1. Mark the position of the spiders, the center yoke, and the centering socket as related to the stud yoke which is welded to the front of the driveshaft tube. The spiders must be assembled with the

bosses in their original positions to provide proper clearance.

2. Remove the snap-rings that secure the bearings in the front of the center yoke.

3. Position the driveshaft in a vise so that the bearing caps that are pressed into the center yoke can be pressed or driven out with a drift and hammer. Do this for all of the spiders.

4. Clean all the serviceable parts in cleaning solvent. If you are using a repair kit, install all of the parts supplied with the kit.

NOTE: If the driveshaft is damaged in any way, replace the complete driveshaft to insure a balanced assembly.

5. Assemble the U-joints in the reverse order of disassembly.

Bearing Cap and Bolt Type U-Joint

1. Remove the cap screws attaching the bearing caps to the U-joint flange and yoke. Remove the bearing caps and bearings from the spider.

2. Remove the grease seals and retainers from the spider.

3. Clean the assembly thoroughly and check for damage or wear.

4. Pack the recess in the spider with the proper grade of grease.

5. Install the grease seals on the spider.

6. Position the needle bearings in the bearing cap, then position the caps on the spider. Place the

spider in the yokes, and then install the bearing caps.

7. Lubricate the U-joints with the proper grade of grease.

Bolted End Cap Type U-Joint

1. Bend the tangs on the lock plates away from the capscrews.

2. Remove the capscrews and lock plates holding the bearing caps to the U-joint flange. Remove the bearing caps and bearings from the flange and spider.

If the bearing caps are integral, tap the bearing cap lightly clockwise and, using a screw driver, pry under first one end of the bearing cap and then the other until the bearing comes out of the yoke. Turn the joint over, and remove the opposite bearing in the same way.

If the bearing cap and bearing are separate, the bearing cap will come off with a light tap. Remove the bearings by first tapping with a round soft drift on the exposed face of one bearing until the opposite bearing comes out. Turn the joint over and tap the exposed end of the journal cross pin until the opposite bearing is free.

3. Remove the spider. All burrs on the bearing cap or yoke must be removed before replacing the spider in the yoke.

4. Clean all parts thoroughly and inspect for damage or wear.

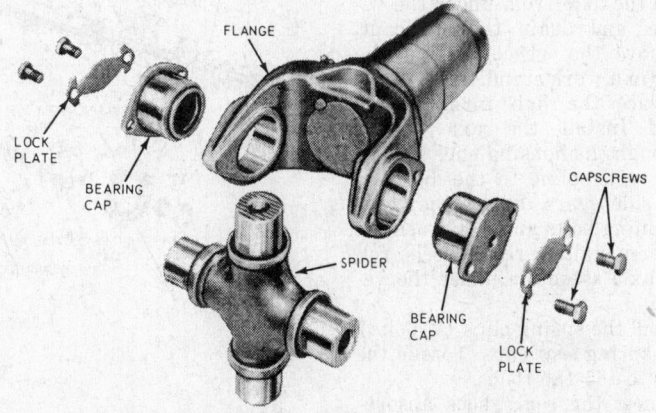

Bolted end cap type U-joint—exploded view (© Ford Motor Co.)

5. Before assembling, fill the journal passages with a long life lubricant of the proper grade.
6. Position the needle bearings in the bearing caps.
7. Position the spider in the flange and install the bearing caps through the yoke and onto the spider. Press or tap the bearing caps into place with a soft drift.
8. Position the lock plates on the bearing caps and secure with the capscrews. Bend the tabs of the lock up against the capscrews.
9. Lubricate the U-joint with a long life grease of the proper grade.

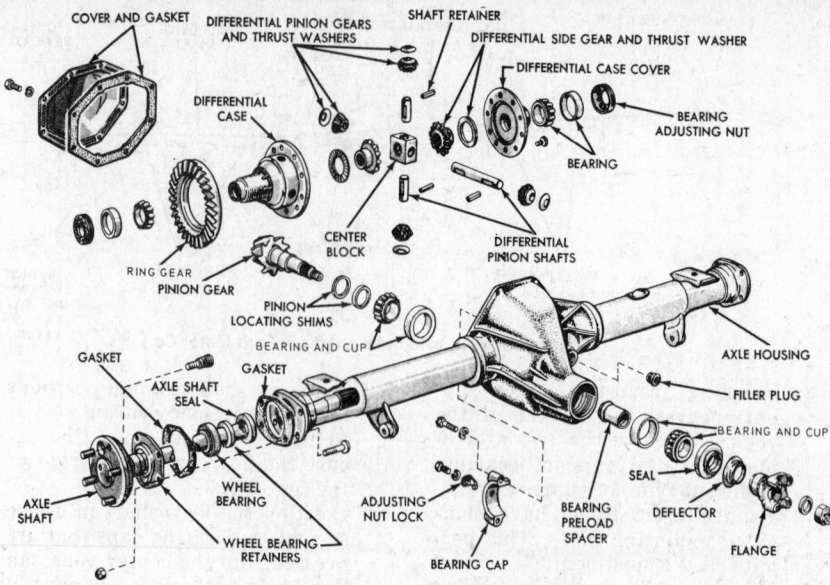

Ford integral carrier differential (© Ford Motor Co.)

Rear Axles,

Reference

See the Unit Repair Section for overhaul procedures for rear axles.

Integral Carrier Axle (Dana) R & R

E250, E350, F100, F250, F350, P Series

1. Loosen the wheel stud nuts and the axle shaft retaining bolts.
2. Disconnect the rear shock absorbers from the spring seat caps. Then raise the rear end of the vehicle frame until the weight is off the rear springs. Place safety stands under the frame in this position.
3. Disconnect the flexible hydraulic line at the frame and disconnect the axle vent hose at the axle connection.
4. Disconnect the parking brake cable (if so equipped) at the equalizer, and remove the cables from the cable support brackets.
5. Disconnect the drive shaft from the rear U-joint flange.
6. Remove the nuts from the spring clips (U-bolts), and remove the spring seat caps.
7. Roll the axle from under the vehicle, and drain the lubricant. Remove the wheels. Mount the axle in a work stand.
8. Replace the hub inner grease seal. Install the axle shafts through the housing ends so that they will spline to the differential side gears. Install the shaft retaining bolts and lock washers.
9. After installing rear wheels, roll the axle assembly under the vehicle.
10. Install the spring clips (U-bolts) and spring seat caps. Torque the nuts to 165-185 ft-lbs.
11. Connect the rear shock absorbers.

12. Lower the vehicle to the floor. Connect the drive shaft to the rear universal joint flange.
13. Connect and adjust the parking brake cables (if so equipped).
14. Connect the hydraulic brake hose and bleed the brakes. Also connect the axle vent hose to the axle fitting.
15. Fill the axle with the proper grade and amount of lubricant.

Ford Removable Carrier Axle

Bronco, E100, F200, F100, and F150

Carrier Assembly R & R

1. Raise the vehicle on a hoist and remove the two rear wheel and tire assemblies.
2. Remove the two brake drums (3 Tinnerman nuts at each drum) from the axle shaft flange studs.

If difficulty is experienced in removing the drums, back off the brake shoes.

3. Working through the hole provided in each axle shaft flange, remove the nuts that secure the rear wheel bearing retainer plate. Pull each axle shaft assembly out of the axle housing using axle shaft remover, Tool 4235-C. Care must be exercised to prevent damage to oil seal, if so equipped. Any roughing or cutting of the seal element during removal or installation can result in early seal failure. Install a nut on one of the brake carrier plate attaching bolts to hold the plate to the axle housing after the shaft has been removed. Whenever a rear axle shaft is replaced, the wheel bearing oil seals must be replaced. Remove the seals with tool 1175-AB.
4. Make scribe marks on the drive

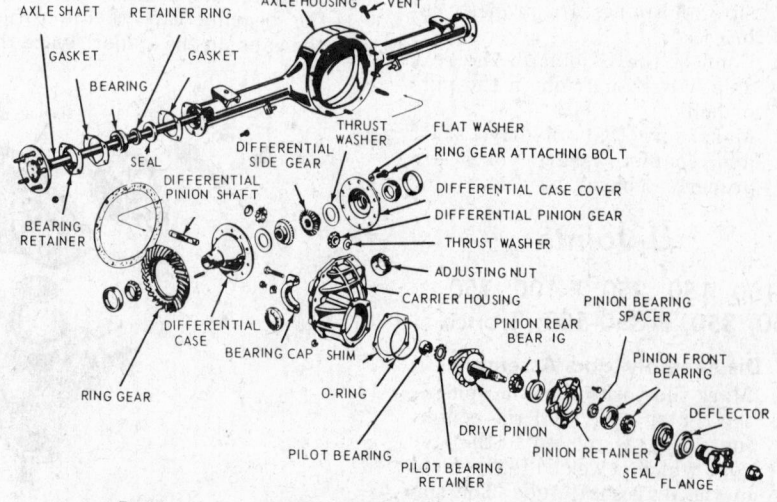

Ford banjo type differential—exploded view (© Ford Motor Co.)

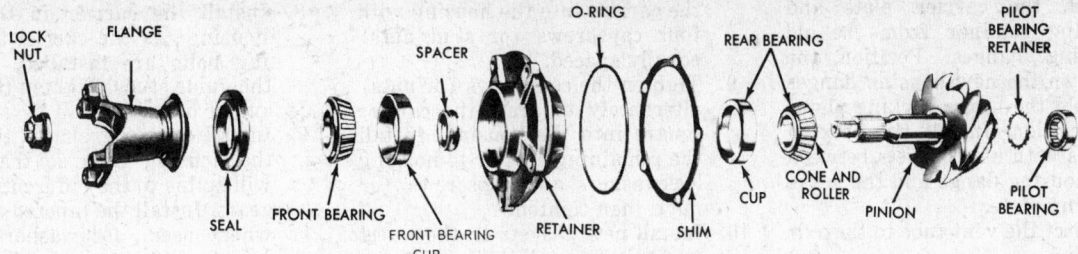

Pinion and bearing retainer components (© Ford Motor Co.)

shaft end yoke and the axle U-joint flange to insure proper position at assembly. Disconnect the drive shaft at the rear axle U-joint, remove the drive shaft from the transmission extension housing. Install oil seal replacer tool T57P-7657-A in the housing to prevent transmission leakage.

5. Place a drain pan under the carrier and housing, remove the carrier attaching nuts, and drain the axle. Remove the carrier assembly from the axle housing.

6. Synthetic wheel bearing seals must not be cleaned, soaked or washed in cleaning solvent. Clean the axle housing and shafts using kerosene and swabs. To avoid contamination of the grease in the sealed ball bearings, do not allow any quantity of solvent directly on the wheel bearings. Clean the mating surfaces of the axle housing and carrier.

7. Position the differential carrier on the studs in the axle housing using a new gasket between carrier and housing. Install the carrier-to-housing attaching nuts.

8. Remove the oil seal replacer tool from the transmission extension housing. Position the drive shaft so that the front U-joint slip

yoke splines to the transmission output shaft.

9. Connect the drive shaft to the axle U-joint flange, aligning the scribe marks made on the drive shaft end yoke and the axle U-joint flange during the removal procedure. Install the U-bolts and nuts and torque to specifications.

10. Wipe a small amount of an oil-resistant sealer on the outer edge of each seal before it is installed. Do not put any of the sealer on the sealing lip. Install the oil seals in the ends of the rear axle housing with tool shown in illustration.

11. Install the two axle shaft assemblies in the axle housing. Care must be exercised to prevent damage to the oil seals. The shorter shaft goes into the left side of the housing. When installing an axle shaft, place a new gasket between the housing flange and the brake backing plate, and carefully slide the axle shaft into the housing so that the rough forging of the shaft will not damage the oil seal. Start the axle splines into the differential side gear, and push the shaft in until the bearing bottoms in the housing.

12. Install the bearing retainer plates on the attaching bolts on the axle housing flanges. Install and tighten the nuts on the bolts.

13. Install the two rear brake drums and the drum attaching nuts.

14. Install the rear wheel and tire assemblies.

15. If the rear brake shoes were backed off, adjust the brakes.

16. Fill the rear axle with lubricant.

Axle Housing R & R

1. Remove the carrier assembly from the axle housing as outlined in the foregoing procedure.

2. Position safety stands under the rear frame members, and support the axle housing with either a floor jack or hoist.

3. Disengage the brake line from the clips that retain the line to the axle housing.

4. Disconnect the vent tube from the rear axle housing.

5. Remove the brake backing plate assemblies from the axle housing, and support them with wire. Do not disconnect the brake line.

6. Disconnect each rear shock absorber from the mounting bracket stud on the axle housing.

7. Lower the rear axle slightly to reduce some of the spring tension. At each rear spring, remove the spring clip (U-bolt) nuts, spring clips, and spring seat caps.

8. Remove the rear axle housing from under the vehicle. If the axle housing is new, install a new vent. The hose attaching portion must face toward the front of the vehicle.

9. Install new rear wheel bearing oil seals in the ends of the rear axle housing. If leather-type wheel bearing service seals are to be installed, soak the new rear wheel bearing oil seals in SAE 10 oil for 1/2 hour before installation.

10. Position the rear axle housing under the rear springs. Install the spring clips (U-bolts), spring seat caps, and nuts. *Torque the spring clip nuts evenly.*

11. If a new axle housing is being installed, remove the bolts that

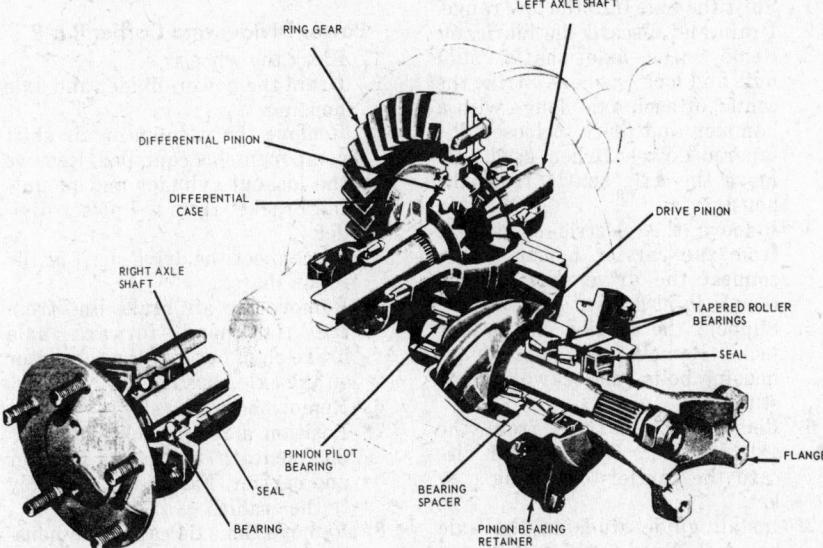

Ford banjo (removable carrier) type differential (© Ford Motor Co.)

attach the carrier plate and bearing retainer from the old housing flanges. Position the bolts in the new housing flanges to hold the brake backing plates in position. Install the backing plates with new gaskets between the housing flange and the brake backing plate.

12. Connect the vent tube to the axle housing.
13. Position the brake line to the axle housing, and secure with the retaining clips.
14. Raise the rear axle housing and springs enough to allow connecting the rear shock absorbers to the mounting bracket studs on the axle housing.
15. Install the carrier assembly and the two axle shaft assemblies in the axle housing as outlined in this section.

Single-Speed, Single-Reduction Axles

Differential Carrier R & R

1. Drain the axle lubricant. Remove the axle shaft and the stud nuts and lockwasher.
   If tapered dowels are installed in the axle shaft flange, hold a short drift firmly in the center of icant.
2. Remove the axle shaft and the stud nuts and lockwasher.
   If tapered dowels are installed in the axle shaft flange, hold a short drift firmly in the center of the flange, and then strike it sharply to loosen the dowels. Remove the dowels and then remove the axle shaft.
   Tapered dowels are not used at the axle shaft flange on most of the larger vehicles, on these flanges, two threads are provided. Use one or both puller threads to remove the axle shafts.
3. Disconnect the drive shaft at the rear U-joint pinion flange.
4. Support the carrier assembly with a roller jack, then remove the carrier-to-housing stud nuts or bolts, and lock washers.
5. Tighten the puller screws, when provided in the carrier, until the carrier is losened from the housing, and then back off the puller screws. Remove and discard the gasket.
6. Place the carrier on a transmission stand.
7. Position a new gasket on the axle housing. *A dry gasket will prevent creeping between the carrier and housing under heavy loads.*
8. Place the differential carrier assembly on a roller jack, and roll the carrier into position. Start

the carrier into the housing with four capscrews (or stud nuts) equally spaced.

9. Tighten the capscrews (or nuts) alternately to draw the carrier square into the housing. Install the remaining carrier-to-housing lockwashers and capscrews (or nut), then tighten.
10. Install new gaskets on the wheel hugs, then install the axle shafts. Install the dowels, lockwashers, and stud nuts.
11. Connect the drive shaft at the rear U-joint flange.
12. Fill the axle with the proper grade of lubricant.

Rear Axle Housing R & R

1. Disconnect the drive shaft and remove the axle shafts and differential carrier as detailed above.
2. Disconnect the rear shock absorbers from the spring seal caps or shock absorber brackets. Raise the rear of the vehicle frame until the weight is off the rear springs. Place safety stands under the frame.
3. Disconnect the flexible hydraulic brake hose or air lines at the frame or frame cross member.
4. Remove the nuts from the spring clips (U-bolts), and remove the spring seat caps and shock absorber brackets.
5. Roll the axle housing and rear wheels from underneath the vehicle.
6. Remove the wheels from the axle housing.
7. Install the axle housing in the reverse order of removal.

Two Speed, Double Reduction Axles

Differential Carrier R & R

1. Shift the axle into the low range.
2. Drain and discard the lubricant.
3. Remove the axle shafts, stud nuts and lock washers. Strike the center of each axle flange with a hammer and drift to loosen the tapered dowels, when used. Remove the axle shafts from the housing.
4. Remove the electric shift unit from the carrier housing. Disconnect the drive shaft at the rear U-joint flange.
5. Support the carrier on a roller jack. Remove the carrier-to-housing bolts and lockwashers or stud nuts.
6. Remove the carrier from the axle housing. Remove and discard the carrier-to-housing gasket.
7. Install guide studs in the axle housing. Place a new gasket over the guide studs.

8. Install the carrier in the axle housing. As the carrier-to-housing bolts are installed, remove the guide studs. Tighten the nuts and or bolts.
9. Install the axle shafts through the housing ends so that they will spline to the differential side gears. Install the tapered dowels, when used, lockwashers, and axle shaft flange stud nuts.
10. Install the electric shift unit. Connect the drive shaft at the rear U-joint flange.
11. Fill the axle with the proper grade lubricant up to the bottom of the filler hole in the rear cover, then add one pint of axle lubricant at the lubricant channel filler plug.
12. Road test the vehicle.

Rear Axle Housing R & R

1. Disconnect the drive shaft and remove the axle shafts and differential carrier as detailed above.
2. Disconnect the rear shock absorbers from the spring seal caps of shock absorber brackets. Raise the rear of the vehicle frame until the weight is off the rear springs. Place safety stands under the frame.
3. Disconnect the flexible hydraulic brake hose or air lines at the frame or frame crossmember.
4. Remove the nuts from the spring clips (U-bolts), and remove the spring seat caps and shock absorber vrackets.
5. Roll the axle housing and rear wheels from underneath the vehicle.
6. Remove the wheels from the axle housing.
7. Install the axle housing in the reverse order of removal.

Eaton Tandem Axles

Power Divider and Carrier R & R

1. Block the wheels.
2. Drain the power divider and axle housing.
3. Remove the vacuum or air shift components as equipped. Remove the lockout cylinder and mounting bracket from the power divider.
4. Disconnect the drive shaft at the input shaft.
5. Remove the air brake line from the right-hand forward axle brake chamber and the connector on the axle housing.
6. Remove both axle shafts.
7. Position a roller jack, with cradle, beneath the power divider and carrier. Fasten the assembly to the cradle.
8. Remove the axle carrier to housing stud nuts.
9. Disconnect the inter-axle drive

shaft at the power divider output shaft flange.

10. Hold the output shaft flange, and remove the flange nut. Pull the flange out.
11. Carefully remove the power divider and carrier from the housing. As the power divider and carrier are moved out of the housing, the output shaft and roller bearing inner race must slide out of the straight rollers in the axle housing cover.
12. Remove the power divider and carrier from under the vehicle.
13. Clean the inside of the axle housing.
14. Remove the power divider output shaft flange seal from the axle housing rear cover. Remove the flange spacer and the roller bearing from the cover.
15. Mount the power divider and carrier in a roller jack.
16. Place a new gasket on the axle housing. Install the gasket dry to prevent leakage due to creeping caused by sealing compound.
17. Start the carrier chosing on the axle housing studs and pull it into position with the stud nuts. Tighten the nuts.
18. Install the roller bearing outer race and cage in the housing cover. Place the spacer next to the bearing.
19. Install a new seal in the housing cover for the output shaft flange.
20. Start the flange on the output shaft. Hold the flange and pull it into position with the flange nut and flat washer. Tighten the nut and install a new cotter pin.
21. Connect the inner-axle drive shaft. Connect the drive shaft at the input shaft flange.
22. Install and connect the power divider lockout unit. Install the air line to the brake chamber.
23. Place new gaskets on the wheel hub studs and install the axle shafts. Tighten the stud nuts.
24. Fill the power divider and axle housing with the proper grade lubricant.

Rear Axle Carrier R & R
For removal and installation of the rear axle carrier, follow the procedure outlined in the Single-Speed, Single-Reduction Axle or Two-Speed, Double Reduction Axle.

Rockwell Tandem Axles

Inter-Axle Differential and Forward Carrier R & R
1. Disconnect the forward drive shaft at the inter-axle differential input shaft flange.
2. Disconnect the rear drive shaft at the through shaft flange.
3. Remove the plug from the bottom of the axle housing and drain the lubricant. Drain the lubricant from the inter-axle differential cover by removing the bottom plug.
4. Remove the axle shaft, stud nuts and lockwashers.
   If tapered dowels are installed in the axle shaft flange, hold a short drift firmly in the center of the axle shaft flange, and then strike it sharply to loosen the dowels and remove the axle shaft.
   Tapered dowels are not used at the axle shaft flange on most of the larger trucks. On these flanges, two puller threads are provided. Use one or both puller threads to remove the axle shafts.
5. Remove the attaching screws and lockwashers, and remove the shift housing assembly.
6. Remove the shift lever attaching nut, and lift out the button, shift lever, cap, and spring.
7. Remove the through-shaft, cage and flange assembly. To free the through-shaft cage from the housing bore, it may be necessary to tap the yoke with a soft mallet. While the through-shaft assembly is being drawn out from the rear of the housing, work the sliding clutch splines by hand at the shift lever oepning. When the through-shaft clears the opening, lift out te clutch.
8. Remove all the carrier-to-housing nuts except the two nuts, which should be loosened but left on the studs to prevent the carrier from falling. Break the carrier loose with a rawhide mallet.
9. Place a roller jack under the carrier, and remove the two top nuts. Work the carrier free. A small pinch bar may be used to straighten the carrier in the housing bore, but the end of the bar should be rounded to prevent indenting the carrier flange.
10. Clean the inside of the axle housing and install a new gasket over the housing studs.
11. Mount the inter-axle differential and carrier assembly on a roller jack and move the assembly into position. Start the carrier into the housing with nuts and flat washers, equally spaced on the housing. Tighten the nuts enough to draw the carrier into the housing.
12. Remove the nuts and flat washers, install the lockwashers and stud nuts. Tighten the nuts.
13. Insert the through-shaft and cage, with a new gasket, into the cage bore in the rear of the axle. Move the through-shaft in until the forward end of the shaft is even with the shift lever opening.
14. Slide the splined clutch collar over the forward end of the shaft, through the shift opening, and ease the shaft through and into the forward side gear of the inter-axle differential. At the same time, pass the splined clutch collar through and onto the through-shaft clutch splines.
15. Install the through-shaft cage retaining screws and lockwashers, tighten the screws.
16. Install the shift lever spring, cup, and lever over the shift lever bolt. Properly locate the lever inner yoke in the clutch groove at this time.
17. Install the shift lever button and nut. Tighten the nut and install a new cotter pin.
18. Position the shift shaft housing with a new gasket, on the carrier. Be sure that the shift lever outer yoke is properly located in the shift shaft collar groove.
19. Install the shift housing capscrews and tighten.
20. Install the axle shafts and the retaining nuts and lockwashers.
21. Remove the fill plugs on the axle housing and inter-axle differential cover and fill the assembly with the proper grade of lubricant. Install the plugs.
22. Connect the rear drive shaft at the shrough shaft flange.
23. Connect the forward drive shaft at the inter-axle differential input shaft flange.

Rear Axle Carrier R & R
For removal and installation of the rear axle carrier, follow the procedures outlined in Single-Speed, Single-Reduction Axle R&R above.

Axle Shaft

Ford Removable Carrier Type

Removal and Installation
1. Raise and support the vehicle and remove the wheel/tire assembly from the brake drum.
2. Remove the nuts which secure the brake drum to the axle flange, then remove the drum from the flange.
3. Working through the hole provided in each axle shaft flange, remove the nuts which secure the wheel bearing retainer plate.
4. Using an axle puller, pull the axle shaft assembly out of the axle housing.
NOTE: The brake backing plate must not be dislodged. Install one nut to hold the plate in place after the axle shaft is removed.
   To replace the axle shaft:
5. Remove the one nut which holds the brake backing plate and

carefully slide the axle shaft into the housing, so that the rough forgings on the shaft will not damage the oil seal.

6. Start the axle splines into the side gear and push the shaft in until the bearing bottoms in the housing.
7. Install the bearing retainer plate and the nuts which secure it.
8. Install the brake drum and the drum attaching nuts.
9. Install the wheel/tire assembly and lower the vehicle.

Dana Integral Carrier Type

1. Remove the lockbolts and lockwashers which hold the axle flange to the hub and drum assembly.
    NOTE: *It is not necessary to raise the vehicle to remove the axle shafts.*
2. Carefully slide the axle shaft out of the axle housing.
3. Clean the mating surfaces of the axle flange and the hub and drum assembly.
4. Position a new gasket on the axle flange and carefully slide the axle shaft into the axle housing. When the splined end of the axle shaft reaches the side gear, gently rotate the shaft until it is inserted into the side gear.
5. Position the gasket between the axle flange and the hub and drum and install the lockbolts and lockwashers.

Rear Spring

Rear Spring (Single Axle) R & R

E-100, 200—1971-72

1. Raise the vehicle frame until the

weight is off the rear spring, with the wheels touching the floor.
2. Remove the nuts from the spring clips (U-Bolts) and drive the clips out of the spring seat cap. If so equipped, remove the auxiliary spring and spacer.
3. Remove the spring to bracket nut and bolt at the front of the spring.
4. Remove shackle lower nut and bolt at the rear of the spring, and remove the spring and shackle assembly from the rear bracket. Remove the spring and compensating shackle from the rear bracket.
5. Remove the shackle from the spring. If the bushing in the spring or the bushing in the shackle are worn or damaged, remove them.
6. If required, install new bushings in the spring and the shackle.
7. Position the spring in the shackle, and install the upper shackle to spring bolt and nut with the bolt head facing outboard.
8. Position the front end of the spring in the bracket and install the bolt and nut.
9. Position the shackle in the rear bracket and install the bolt and nut.
10. Position the spring on top of the axle with the spring tie bolt centered in the hole provided in the seat. If so equipped, install the auxiliary spring and spacer.
11. Install the spring clips, spring seat cap, and nuts.
12. Lower the vehicle to the floor, tighten the spring bolt nuts and

spring clip nuts (U-bolts), tighten the front spring bolt and nut, the rear shackle bolt and nut and the compensating shackle bolt and nut.

E-100, 150, 200—1973-78

1. Raise the rear end of the vehicle and support the chassis with safety stands. Support the rear axle with a floor jack or hoist.
2. Disconnect the lower end of the shock absorber from the bracket on the axle housing.
3. Remove the two U-bolts and plate.
4. Lower the axle and remove the upper and lower rear shackle bolts.
5. Pull the rear shackle assembly and rubber bushings from the bracket and spring.
6. Remove the nut and mounting bolt that secures the front end of the spring. Remove the spring assembly from the front shackle bracket.
7. Install new rubber bushings in the rear shackle bracket and in the rear eye of the replacement spring.
8. Position the spring assembly and connect the front eye of the spring to the front shackle bracket by installing the front mounting bolt and nut. Do not tighten the nut.
9. Mount the rear end of the spring by inserting the upper bolt of the rear shackle assembly through the eye of the spring and lower bolt through the rear spring hanger.
10. Position the spring center bolt to the pilot hole in the axle and in-

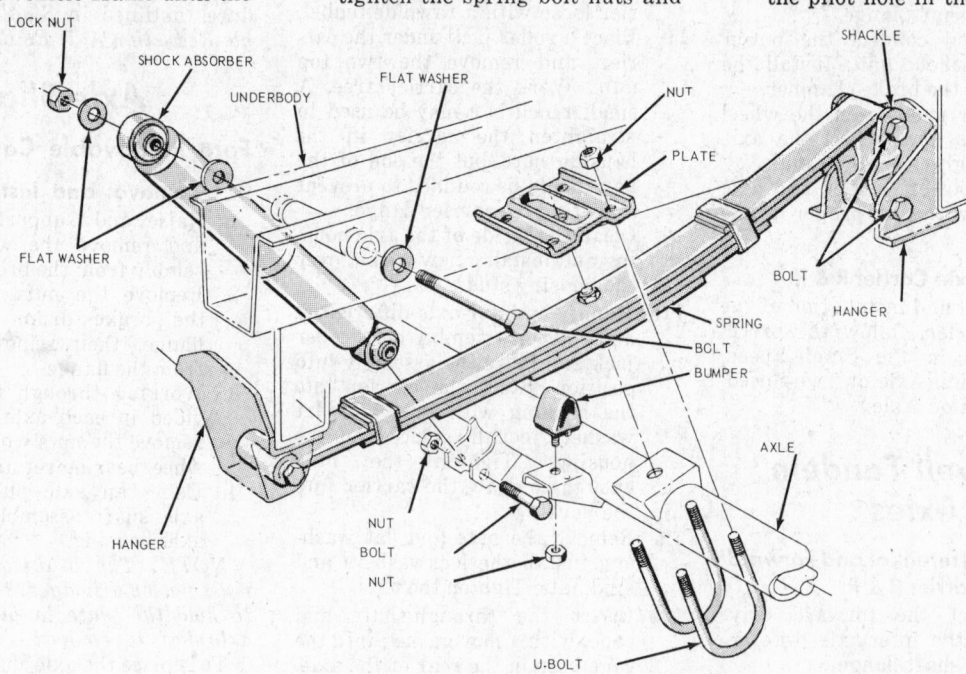

Rear spring E-100—E-200 (© Ford Motor Co.)

stall the plate. Install the U-bolts through the plate. Do not tighten the attaching nuts at this time.

11. Raise the axle with a floor jack or hoist until the vehicle is free of the stands and connect the lower end of the shock absorber to the bracket on the axle housing.

12. Tighten the spring front mounting bolt and nut, the rear shackle nuts and the U-bolt nuts.

13. Remove the safety stands and lower the vehicle.

E-250, 300, 350

1. Raise the rear end of the vehicle and support the chassis with safety stands. Support the rear axle with a floor jack or hoist.

2. Disconnect the lower end of the shock absorber from the bracket on the axle housing.

3. Remove the two spring clips (U-bolts) and the spring clip cap.

4. Lower the axle and remove the spring front bolt from the hanger.

5. Remove the two attaching bolts from the rear of the spring. Remove the spring and the shackle.

6. Assemble the upper end of the shackle to the spring with the attaching bolt.

7. Connect the front of the spring to the front bracket with the attaching bolt.

8. Assemble the spring and shackle to the rear bracket with the attaching bolt.

9. Place the spring clip plate over the head of the center bolt.

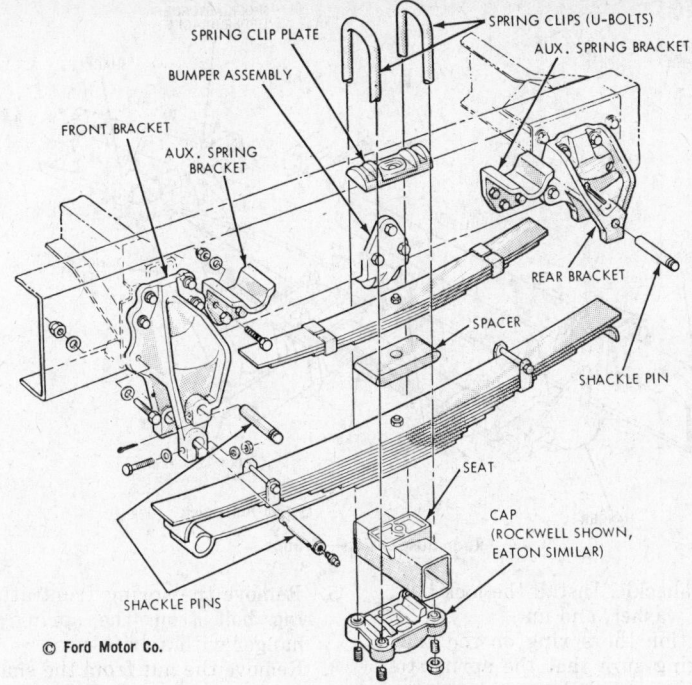

© Ford Motor Co.

Rear Spring Installation

10. Raise the axle with a jack and guiding it so that the center bolt enters the pilot hole in the pad on the axle housing.

11. Install the spring clips, cap and attaching nuts. Tighten the nuts snugly.

12. Connect the lower end of the shock absorber to the lower bracket.

13. Tighten the spring front mounting bolt and nut, the rear shackle nuts and spring clip nuts.

14. Remove the safety stands and lower the vehicle.

F-100, 150, F-250 4-Wheel Drive and P-Series

1. Raise the vehicle frame until the weight is off the rear springs with the wheels still touching the floor.

2. Remove the nuts from the spring clips (U-bolts) and drive the clips out of the spring seat cap and remove the spring clip plate. If the truck is so equipped, remove the auxiliary spring and spacer.

3. Remove the shackle pin locking bolts from each end of the spring (bolt and nut on F-250 4 x 4 front eye).

4. Working from the inner side of the frame, insert a drift in the hole provided in the frame for removing the shackle pin. Drive the shackle pin out of each spring bracket (bolt on F-250 4 x 4 front eye).

5. Remove the spring and shackle from the vehicle. On F-100-250 4 x 4 models, remove the spring to axle spacer.

6. Drive out the remaining shackle pin from the rear spring eye and remove the shackle from the spring.

7. Position the shackle to the rear spring eye.

8. Install the shackle pin through the shackle and spring eye with the lubricator fitting on the shackle pin facing outward.

9. Line up the shackle pin lock bolt groove with the lock bolt hole in

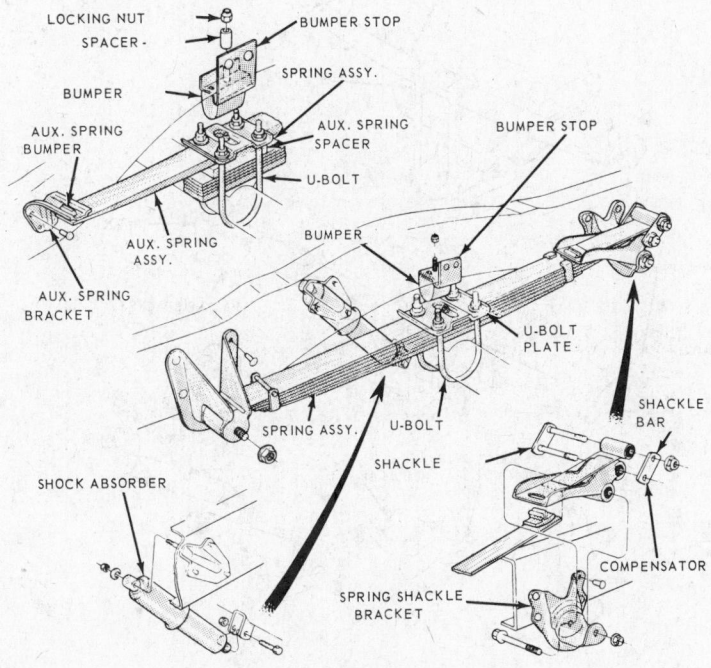

Rear spring F-100-250 (© Ford Motor Co.)

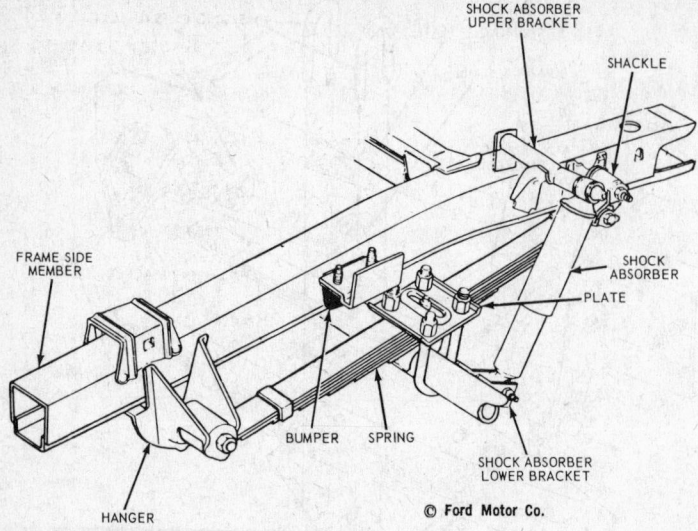

SHOCK ABSORBER UPPER BRACKET

SHACKLE

SHOCK ABSORBER

FRAME SIDE MEMBER

PLATE

BUMPER   SPRING

SHOCK ABSORBER LOWER BRACKET

HANGER

© Ford Motor Co.

Rear Suspension—Bronco

the shackle. Install the lock bolt, lock washer, and nut.

10. Position the spring on the axle, making sure that the spring tie bolt is in the hole provided in the axle spring seat or spacer. On F-100-250 4 x 4 models, first install the spacer between the spring seat and the spring, making sure the spacer dowel is positioned in the pilot hole of the axle spring seat.

11. Install a shackle pin through the shackle and rear bracket, with the lubricating fitting on the shackle pin facing outward. Line up the pin groove with the lock bolt hole in the bracket, and install the lock bolt, lock washer, and nut as before.

12. Repeat this operation to install a shackle pin at the front bracket and spring eye (bolt on F-250 4 x 4).

13. If so equipped, install the auxiliary spring and spacer. Place the spring clip plate on top of the spring at the tie bolt, and put the spring clips over the spring assembly and axle.

14. Position the spring seat cap, and install the nuts on the spring clips.

15. Lower the vehicle to the floor and tighten the spring clip nuts. Lubricate the fittings on the shackle pins.

Bronco

1. Raise the vehicle by the axles and install safety stands under the frame.

2. Disconnect the shock absorber from the axle.

3. Remove the U-bolt attaching nuts and remove the two U-bolts and the spring clip plate.

4. Lower the axle to relieve spring tension and remove the nut from the spring front attaching bolt.

5. Remove the spring front attaching bolt from the spring and hanger with a drift.

6. Remove the nut from the shackle to hanger attaching bolt and drive the bolt from the shackle and hanger with a drift and remove the spring from the vehicle.

7. Remove the nut from the spring rear attaching bolt. Drive the bolt out of the spring and shackle with a drift.

8. Position the shackle (closed section facing toward front of vehicle) to the spring rear eye and install the bolt and nut.

9. Position the spring front eye and bushing to the spring front

hanger, and install the attaching bolt and nut.

10. Position the spring rear eye and bushing to the shackle, and install the attaching bolt and nut.

11. Raise the axle to the spring and install the U-boats (when an axle cap is not used, the U-bolt shank should contact the leaf edges) and spring clip plate. Align the spring leaves.

12. Tighten the U-bolt nuts and the spring front and rear attaching bolt nuts. The U-bolts should contact the spring assembly edges or axle seat.

13. Connect the shock absorber to the axle and tighten the nut.

14. Remove the safety stands and lower the vehicle.

F-350

1. Raise the vehicle frame until the weight is off the rear springs with the wheels still touching the floor.

2. Remove the nuts from the spring U-bolts.

3. Drive the U-bolts out of the shock absorber lower bracket and the spring cap and remove the U-bolts.

4. Remove the spacer from the top of the spring.

5. If equipped with auxiliary springs, remove the auxiliary spring and spacer.

6. Remove the shackle to bracket bolt and nut from the rear of the spring.

7. Remove the spring to hanger bolt and nut from the front of

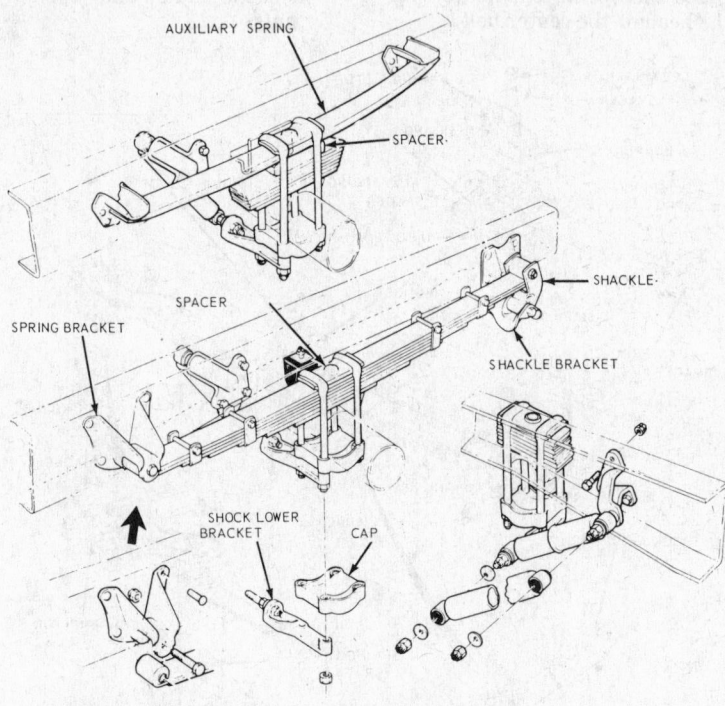

AUXILIARY SPRING

SPACER

SHACKLE

SPACER

SHACKLE BRACKET

SPRING BRACKET

SHOCK LOWER BRACKET

CAP

Rear spring F-350 (© Ford Motor Co.)

the spring and remove the spring.

8. Remove the shackle to spring bolt and nut and remove the shackle from the spring.

9. Position the shackle to the spring and install the attaching bolt and nut. The bolt must be installed so the nut is away from the frame.

10. Position the spring to the spring front hanger and install the attaching bolt and nut.

11. Position the shackle to the bracket and install the attaching bolt and nut.

12. Align the spring toe bolt with the pilot hole in the axle spring seat and, if so equipped, install the auxiliary spring and spacer.

13. Position the spacer on top of the spring and install the U-bolts over the spacer, spring and axle.

14. Position the spring cap and shock lower bracket to the axle and U-bolts. Install the U-bolt attaching nuts.

15. Lower the vehicle and tighten the front spring bracket bolt and nut and the rear shackle bolts and nuts.

C-Series

1. Raise the vehicle frame until the weight is off the rear springs with the wheels still touching the floor.

2. Remove the nuts from the spring clips (U-bolts) and drive the clips out of the spring seat cap. If so equipped, remove the auxiliary spring and spacer.

3. Remove the shackle pin locking bolts from each spring bracket.

4. A hole is provided in the frame opposite each spring bracket for removing the shackle pin. Insert a drift from the inside of the frame through these holes and drive the shackle pin out of each bracket.

5. Remove the spring and shackle assembly from the truck. Separate the spring from the shackle by removing the locking bolt and driving out the shackle lower pin from the shackle and spring eye.

6. Remove the lubricating fittings from the shackle pins.

7. Align the upper bore of the shackle with the holes in the rear bracket. Drive the shackle upper pin through the shackle and bracket with the pin lubricator hole facing outward.

8. Line up the shackle pin groove with the locking bolt hole in the bracket and install the locking bolt, washer and nut.

9. Install the spring seat and the wedge (if so equipped) between the axle and the spring. Position the spring on the axle, being

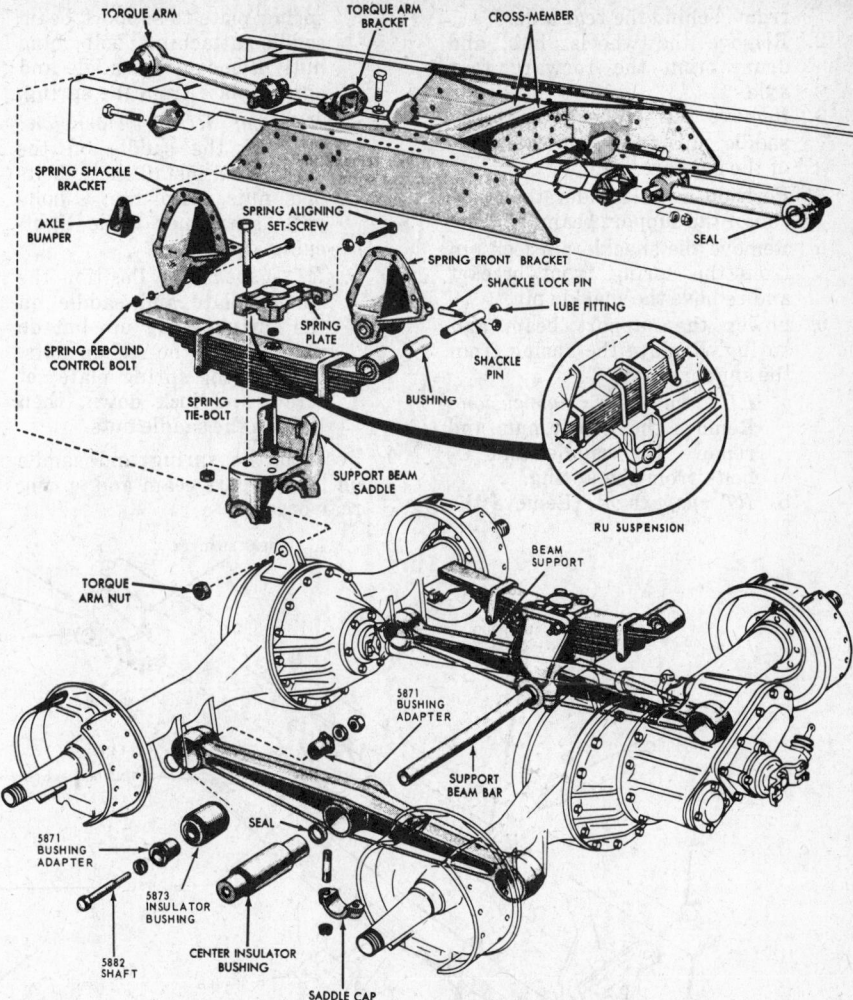

Hendrickson tandem suspension—Typical (© Ford Motor Co.)

sure that the spring tie bolt is in the hole provided in the axle or spring seats. If so equipped, install the auxiliary spring and spacer.

10. Drive the shackle lower pin through the shackle and spring rear eye. Install the locking bolt, washer and nut as before. Repeat the operation to install shackle pin through the spring front bracket and eye.

11. Place the spring clip plate on top of the spring at the tie bolt, and put the spring clips over the spring assembly and the axle.

12. Install the spring seat cap on the spring clips and install the spring clip nuts on the clips.

13. Lower the vehicle to the floor and tighten the spring clip nuts.

L-LN-B-W Series

1. Lift the vehicle until the weight is off of the rear spring but the wheels still touch the ground.

2. Remove the nuts from the U-bolts and drive the bolts out of the spring seat caps.

3. Remove the auxiliary spring and spacer.

4. Support the spring and remove

the front shackle pins. The lower pin is held in place by a lock bolt and the upper is held by a cotter pin.

5. Remove the cotter pin in the rear shackle pin and remove the pin. Remove the spring from the vehicle.

6. To install place the spring in the rear shackle bracket and install the shackle pin and cotter pin.

7. With a jack or C-clamp press the front eye of the spring up into the bracket until the eye is lined up with the hole in the bracket. Install the shackle pins in the spring and secure them in place with the lock bolt or the cotter pin.

8. Install the spring seat and the wedge (if so equipped) between the axle and spring and install the U-bolts over the axle and install the nuts on the bolts after installing the spring seat cap.

9. Lower the vehicle and tighten the nuts on the U-bolts securely.

Rear Spring (Hendrickson Tandems) R & R

1. Raise the rear of the vehicle and position the blocks under the

frame behind the rear axle.

2. Remove the wheels, hub, and drum from the forward rear axle.

3. Remove the support beam bar saddle caps from the lower side of the support beam.

4. Position a jack under the front end of the support beam.

5. Remove the shackle pin lock pin from the spring front bracket and remove the shackle pin.

6. Lower the support beam and spring. Remove the spring from the support beam.

7. a. *RU and RUE suspension*: Remove the U-bolt nuts and remove the saddle and U-bolts from the spring.

   b. *RT suspension*: Remove the spring plate-to-support beam saddle attaching bolts and nuts. Remove the saddle and spring plate from the spring.

8. a. *RU and RUE suspension*: Position the saddle on the spring and install the U-bolts and nuts. Tap the U-bolts with a hammer while tightening the nuts.

   b. *RT suspension*: Position the spring plate and saddle on the spring. Snug up, but do not tighten, the saddle nuts. Tighten the spring plate set screw and lock down, then tighten the saddle nuts.

9. Position the spring and saddle on the support beam and spring rear bracket.

10. Raise the support beam and spring, and position the spring on the front spring bracket.

11. Align the spring with the front bracket and install the shackle pin.

12. Install the shackle pin lock pin and tighten the lock nut.

13. Install the hub, drum, and wheel on the forward rear axle.

14. Remove the jack from the support beam and the blocks from the rear of the frame. Lower the saddle on the support beam bar center insulator bushing and install the saddle caps.

15. Install the saddle caps and tighten. *NOTE: The weight of the vehicle must be on the suspension when the saddle cap attaching nuts are tightened.*

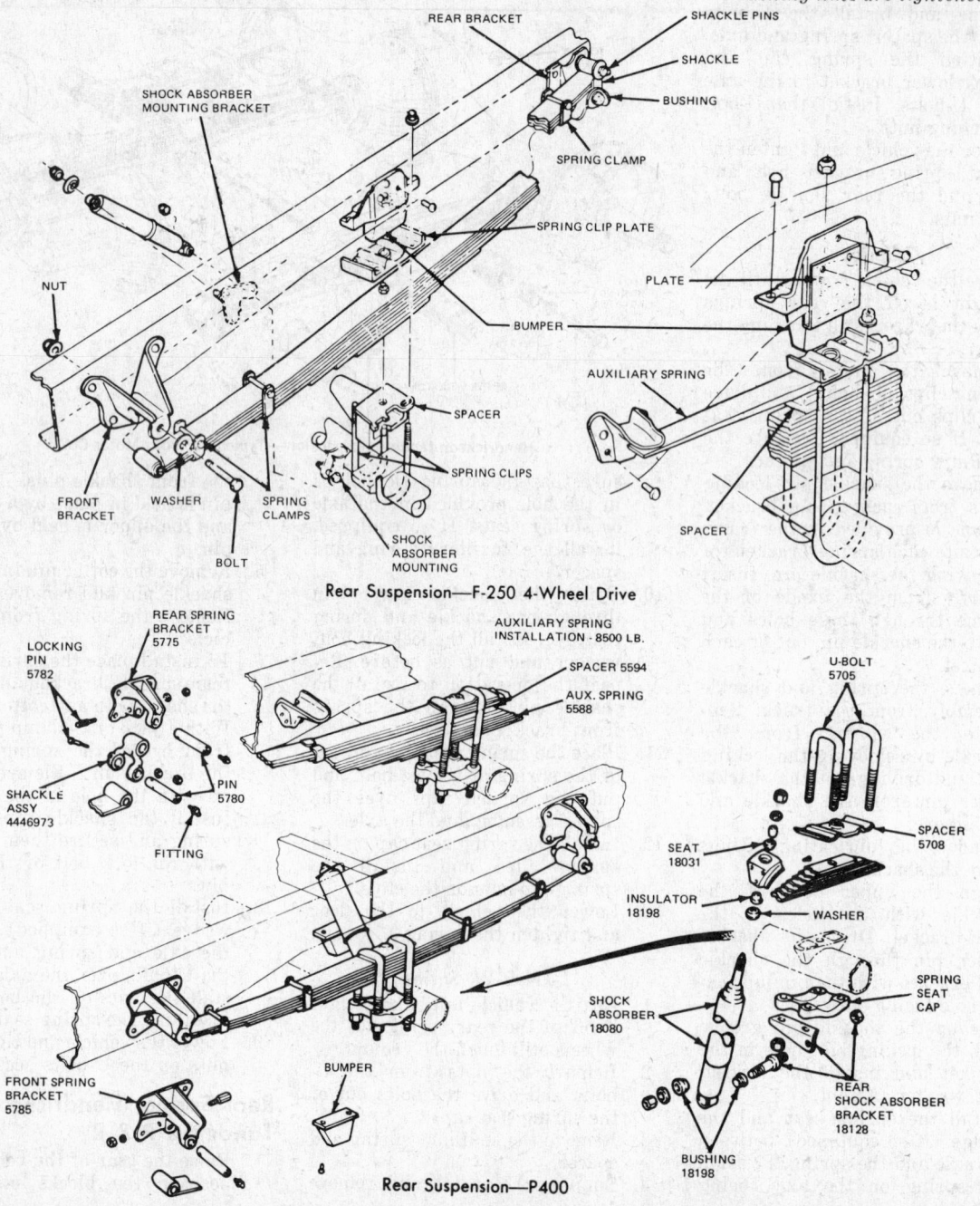

Rear Suspension—F-250 4-Wheel Drive

Rear Suspension—P400

Index

TUNE-UP SPECIFICATIONS

NOTE: Due to the many small variations made to the basic engines to compensate for light duty use, heavy duty use, light and heavy duty emission control equipment, federal engines, California engines, and altitude engines, (over 4000 feet), it is most important for the service man to determine the engine used and consult the underhood emission control specification sticker and the following specifications.

CU. IN. DISPLACEMENT (cu. in.)	YEAR	SPARK PLUG GAP (in.)	DISTRIBUTOR POINT DWELL (deg.)	DISTRIBUTOR POINT GAP (in.)	IGNITION TIMING (DEGREES)	CRANKING COMP. PRESS	VALVE CLEARANCE Int. Exh.	GOV. RPM NO LOAD (rpm)	PUMP FUEL PRESS (psi)	IDLE SPEED* (rpm) STD.	IDLE SPEED* (rpm) AUTO.
SIX CYLINDER											
250	1971	0.035	31-34	0.019	4B	130	Hyd.		3.5-4.5	550	500
	1972	0.035	31-34	0.019	4B②	130	Hyd.		3.5-4.5	700	600
	1973	0.035	31-34	0.019	6B③	130	Hyd.		3.5-4.5	700	600④
	1974	0.035	31-34	0.019	8B⑩	130	Hyd.		3.5-4.5	850⑤	600
	1975	0.060	Elec.	Elec.	10B	130	Hyd.		3.5-4.5	900N	550DR
	1976	0.060	Elec.	Elec.	10B	130	Hyd.		3.5-4.5	900N	550DR
	1977-78	0.035	Elec.	Elec.	12B	130	Hyd.		3.5-4.5	—	550⑥
	1977-78	0.035	Elec.	Elec.	8B	130	Hyd.		3.5-4.5	750	—
HD①	1977-78	0.035	Elec.	Elec.	6B	130	Hyd.		3.5-4.5	600	600N
292	1971	0.035	31-34	0.019	4B	130	Hyd.		3.5-4.5	550	500
	1972	0.035	31-34	0.019	4B	130	Hyd.		3.5-4.5	700	700
	1973	0.035	31-34	0.019	4B⑧	130	Hyd.		3.5-4.5	700⑨	700⑨
	1974	0.035	31-34	0.019	8B	130	Hyd.		3.5-4.5	700	700
	1975	0.060	Elec.	Elec.	8B	130	Hyd.		3.5-4.5	600	600
	1976	0.060	Elec.	Elec.	8B	130	Hyd.		3.5-4.5	600	600
	1977-78	0.035	Elec.	Elec.	8B	130	Hyd.	3800-4000	3.5-4.5	600	600N
305C (V6)	1971-73	0.040	31-34	0.019	7-½B	125	0.012-0.018	3600	5.0-7.0	550	550
	1974	0.035	31-34	0.019	7-½B	125	0.012-0.018	3600	5.0-7.0	600	600
351	1971-72	0.035	31-34	0.019	5B	125	0.012-0.018	3900	5.0-6.5	550	550
379 (V6)	1973	0.035	31-34	0.019	8B	125	0.012-0.018	4000	5.0-7.0	550	550
	1974	0.035	31-34	0.019	6B	125	0.012-0.018	4000	5.0-7.0	575	575
432 (V6)	1973	0.035	31-34	0.019	8B	125		3200	5.0-7.0	525	525
	1974	0.035	31-34	0.019	6B	125	0.012-0.018		5.0-7.0	525	525
200 Fed.	1978				Refer to Underhood Specification Sticker						
231 Calif.	1978				Refer to Underhood Specification Sticker						
EIGHT CYLINDER											
305	1976-78	0.045	Elec.	Elec.	8B	150	Hyd.		5.0-6.5	600	500DR
HD①	1976-78	0.045	Elec.	Elec.	6B	150	Hyd.		5.0-6.5	700	700N
307	1971 (200hp)	0.035	28-32	0.019	4B⑫	150	Hyd.		5.0-6.5	600	550
	1971 (215hp)	0.035	28-32	0.019	4B	150	Hyd.		5.0-6.5	550	500
	1972	0.035	28-32	0.019	4B⑬	150	Hyd.		5.0-6.5	900⑭	600
	1973-75	0.035	29-31	0.019	4B⑮	150	Hyd.		5.0-6.5	900⑯	600
350	1971	0.035	28-32	0.019	4B⑳	150	Hyd.		5.0-6.5	600	550
	1972	0.035	29-31	0.019	4B⑳	150	Hyd.		5.0-6.5	750	600
	1973	0.035	29-31	0.019	㉑	150	Hyd.		5.0-6.5	900㉒	600㉒
	1974	0.035	29-31	0.019	㉓	150	Hyd.		5.0-6.5	900㉔	600㉔
	1975 (2bbl)	0.060	Elec.	Elec.	6B	150	Hyd.		5.0-6.5	—	600
	1975	0.060	Elec.	Elec.	6B㉖	150	Hyd.	3600	5.0-6.5	800㉗	600㉗
	1976	0.060	Elec.	Elec.	6B	150	Hyd.	3600	5.0-6.5	800	600
	1977-78	0.045	Elec.	Elec.	8B⑦	150	Hyd.	3600	5.0-6.5	700	500DR
HD①	1977-78	0.045	Elec.	Elec.	8B⑩	150	Hyd.	3600	5.0-6.5	700	700N
366	1971	0.035	28-32	0.019	8B	150	Hyd.		5.0-6.5	500	500
	1972	0.035	28-32	0.019	8B	150	Hyd.		5.0-6.5	550	550
	1973	0.035	28-32	0.019	8B	150	Hyd.		5.0-6.5	550㉕	550㉕

TUNE-UP SPECIFICATIONS

CU. IN. DISPLACE-MENT (cu. in.)	YEAR	SPARK PLUG GAP (in.)	DISTRIBUTOR POINT DWELL (deg.)	POINT GAP (in.)	IGNITION TIMING (DEGREES)	CRANKING COMP. PRESS	VALVE CLEARANCE Int. Exh.	GOV. RPM NO LOAD (rpm)	PUMP FUEL PRESS (psi)	IDLE SPEED* (rpm) STD.	IDLE SPEED* (rpm) AUTO.
	1974	0.035	28-32	0.019	8B	150	Hyd.		5.0-6.5	600	600
	1975 (Fed.)	0.035	28-32	0.019	8B	150	Hyd.		5.0-6.5	700	700
	1975 (w/AT-475)	0.035	28-32	0.019	8B	150	Hyd.	3600	5.0-6.5	700	700 600㉗
	1975-76 (Calif.)	0.060	Elec.	Elec.	8B	150	Hyd.		5.0-6.5	700	700
366	1977-78	0.045	Elec.	Elec.	8B	150	Hyd.	3750	5.0-6.5	700	——
400	1973-74	0.035	29-31	0.019	8B	160	Hyd.		7.0-8.5	——	600
	1975-76 (Fed.)	0.060	Elec.	Elec.	4B	150	Hyd.		5.0-6.5	700	700
	1975-76 (Calif.)	0.060	Elec.	Elec.	2B	150	Hyd.		5.0-6.5	700	700
	1977-78	0.045⑪	Elec.	Elec.	4B⑰	150	Hyd.		5.0-6.5	700	700N
396	1971	0.035	28-32	0.019	4B	150	Hyd.		7.0-8.5	700	600
402	1972	0.035	28-32	0.019	8B	150	Hyd.		7.0-8.5	700	600
427	1971	0.035	28-32	0.019	8B	150	Hyd.		7.0-8.5	750	600
	1972	0.035	28-32	0.019	8B	150	Hyd.		7.0-8.5	500	500
	1973	0.035	28-32	0.019	8B	150	Hyd.		7.0-8.5	550	550
	1974	0.035	28-32	0.019	8B	150	Hyd.		7.0-8.5	550㉕	550㉕
	1975-76 (Calif.)	0.060	Elec.	Elec.	8B	150	Hyd.		7.0-8.5	600	600
	1975-76 (Fed.)	0.035	28-32	0.019	8B	150	Hyd.		7.0-8.5	700	700
	1977-78	0.045	Elec.	Elec.	8B	150	Hyd.	4200	5.0-6.5	700	——
454	1971-72	0.035	29-31	0.019	10B	160	Hyd.		7.0-8.5	900	600
	1973 (Fed.)	0.035	28-32	0.019	10B	150	Hyd.		7.0-8.5	900④	600④
	1973 (Calif.)	0.035	28-32	0.019	㉘	150	Hyd.		7.0-8.5	900④	600④
	1974	0.035	29-31	0.019	10B㉘	150	Hyd.		7.0-8.5	800④	600④
	1975-76 (Fed.)	0.060	Elec.	Elec.	16B㉘	150	Hyd.		7.0-8.5	——	650④
	1975-76 (Calif.)	0.060	Elec.	Elec.	8B	150	Hyd.		7.0-8.5	600	600
	1977-78	0.045	Elec.	Elec.	4B	150	Hyd.		5.0-6.5	——	600DR
HD①	1977-78	0.045	Elec.	Elec.	8B	150	Hyd.		5.0-6.5	700	700N

—— Not Applicable
Elec.—Electronic Ignition
Hyd.—Hydraulic valve lifters
N—Neutral
DR—Drive
B—BTDC
①—HD—6201 lbs GVW or over
②—TDC on K-20 Suburban California only
③—4B All C-K 20 except Suburban; all C-P 30 series and all G-30 except Sportvan
④—700 rpm All C-K 20 except Suburban; all C-P 30 series and all G-30 except Sportvan
⑤—600 rpm All C-K 20 except Suburban; all C-P 30 series and all G-30 except Sportvan
⑥—W/AC and High Altitude—600DR
⑦—W/Distributor number 1103254—6B
⑧—8B California
⑨—600 rpm California
⑩—W/Distributor number 1103250—2B
⑪—8B w/automatic transmission
⑫—8B w/automatic transmission
⑬—8B 10 series w/automatic transmission only
⑭—W/Distributor number 1103250—2B
    W/RBL-12-6 spark plugs—0.060
⑮—8B All 10 series, C-K 20 Suburban, G-20 & 30 Sportvans w/auto transmission TDC All others
⑯—600 rpm All C-K 20 except Suburban; all C-P 30 series and all G-30 except Sportvan
⑰—W/Distributor number 1103250—2B
⑱—Not used

⑲—Not used
⑳—8B w/automatic transmission
㉑—C-20 Suburban 2B
    All 10 series, K-20 Suburban, G-20, and G-30 Sportvan:
      w/manual transmission 8B
      w/automatic transmission 12B
    All others 4B
㉒—All C-K 20 except Suburban; all C-P 30 series and all G-20 & 30 except Sportvan 700 rpm
㉓—Federal except C-K 10 & 20 Suburban and G-20 & 30 Sportvan 8B
    Federal C-K 10 & 20 Suburban and G-20 & 30 Sportvan:
      w/automatic transmission 12B
      w/automatic transmission (except Suburban) 8B
      Suburban (w/manual transmission) 6B
    California:
      w/automatic transmission 8B
      Suburban (w/manual transmission) 68
      w/manual transmission 4B
㉔—All C-K 20 except Suburban; all C-P 30 series and all G-20 & 30 except Sportvan 600 rpm
㉕—750 rpm California
㉖—2B California
㉗—700 rpm California
㉘—All 10 series, C-K 20 Suburban, G-20 and G-30 Sportvan
㉘—All 10 series, C-K 20 Suburban, G-20 and G-30 Sportvan 10B
    All C-K 20 except Suburban, all C-P 30 Series and all G-30 except Sportvan:
      w/manual transmission 5B
      w/automatic transmission 8B
㉙—All C-K 20 except Suburban, all C-P 30 series and all G-30 except Sportvan 8B

FIRING ORDER AND ROTATION

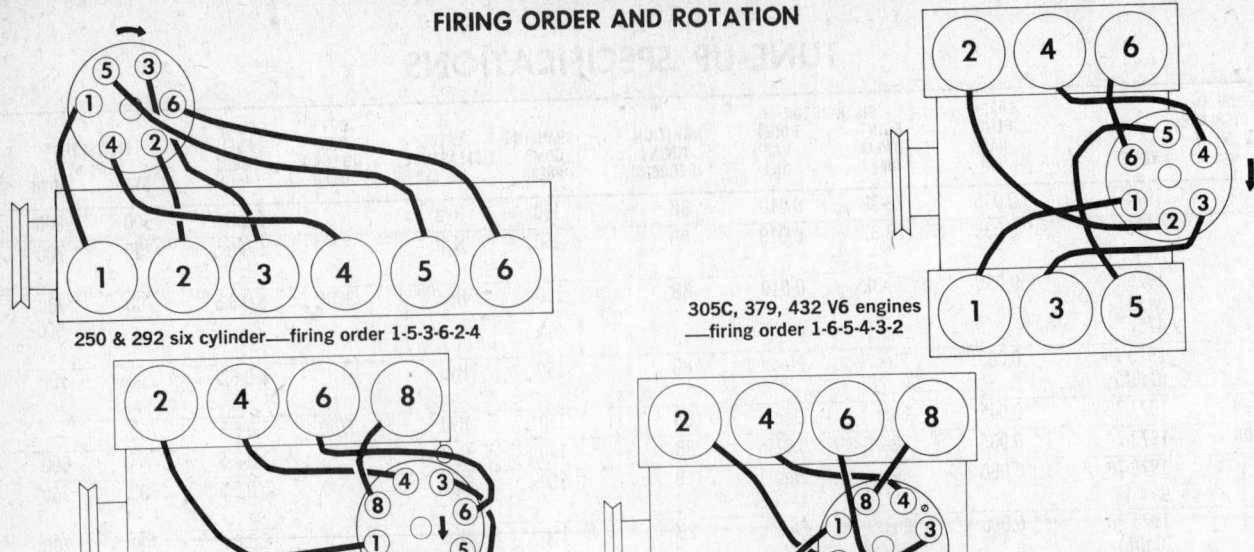

250 & 292 six cylinder—firing order 1-5-3-6-2-4

305C, 379, 432 V6 engines
—firing order 1-6-5-4-3-2

307, 350, 366, 402, 427, 454 engines—firing order 1-8-4-3-6-5-7-2

305, 350, 366, 400, 454 engines
—firing order 1-8-4-3-6-5-7-2

GENERAL ENGINE SPECIFICATIONS

CU. IN. DISPLACE-MENT	YEAR	BORE X STROKE	FIRING ORDER	DEVELOPED HORSEPOWER @ R.P.M.	DEVELOPED TORQUE @ R.P.M.	COM-PRESSION RATIO	CARBU-RETOR	VALVE LIFTER TYPE	NORMAL OIL PRESSURE
SIX CYLINDER									
250	1971	3.88 x 3.53	153624	145 @ 4200	230 @ 1600	8.5-1	1V	Hyd.	40-60
	1972	3.88 x 3.53	153624	110 @ 3800	185 @ 1600	8.5-1	1V	Hyd.	40-60
	1973-74	3.88 x 3.53	153624	100 @ 3600	175 @ 2000	8.25-1	1V	Hyd.	40-60
	1973-74	3.88 x 3.53	153624	105 @ 3800	185 @ 1200	8.25-1	1V	Hyd.	40-60
	1975-76	3.88 x 3.53	153624	110 @ 3800	195 @ 1600	8.3-1	1V	Hyd.	40-60
	1977-78	3.88 x 3.53	153624	165 @ 4000	270 @ 1600	8.0-1	1V	Hyd.	40-60
	1971	3.88 x 4.13	153624	125 @ 3600	225 @ 2400	8.0-1	1V	Hyd.	40-60
	1972	3.88 x 4.13	153624	120 @ 3600	215 @ 2000	8.0-1	1V	Hyd.	40-60
	1973-74	3.88 x 4.13	153624	120 @ 3600	215 @ 2000	8.0-1	1V	Hyd.	40-60
	1975-78	3.88 x 4.13	153624	130 @ 3600	225 @ 2000	8.0-1	1V	Hyd.	40-60
V-6 ENGINES									
305	1971	4.25 x 3.58	165432	165 @ 4000	270 @ 1600	7.75-1	2V	Mech.	60
	1972	4.25 x 3.58	165432	125 @ 3600	225 @ 1600	7.75-1	2V	Mech.	60
	1973-74	4.25 x 3.58	165432	120 @ 3600	220 @ 1600	7.75-1	2V	Mech.	60
351	1971-72	4.56 x 3.58	165432	195 @ 3600	314 @ 1600	7.5-1	2V	Mech.	60
379	1973-74	4.56 x 3.86	165432	170 @ 3600	266 @ 1600	7.5-1	2V	Mech.	60
432	1973-74	4.88 x 3.86	165432	190 @ 3200	331 @ 1600	7.5-1	2V	Mech.	60
EIGHT CYLINDER									
305	1976	3.74 x 3.48	18436572	140 @ 3800	245 @ 2000	8.5-1	2V	Hyd.	40-60
	1977-78	3.74 x 3.48	18436572	145 @ 3800	245 @ 2400	8.5-1	2V	Hyd.	40-60
307	1971	3.88 x 3.25	18436572	200 @ 4600	300 @ 2400	8.5-1	2V	Mech.	60
	1972	3.88 x 3.25	18436572	130 @ 4000	230 @ 2400	8.5-1	2V	Hyd.	30
	1973-75	3.88 x 3.25	18436572	115 @ 3600	205 @ 2000	8.5-1	2V	Hyd.	30
350	1971	4.0 x 3.48	18436572	250 @ 4600	350 @ 3000	8.5-1	2V	Hyd.	30
	1972	4.0 x 3.48	18436572	165 @ 4000	280 @ 2400	8.5-1	2V	Hyd.	30
	1973	4.0 x 3.48	18436572	145 @ 4000	255 @ 2400	8.5-1	2V	Hyd.	30
	1974	4.0 x 3.48	18436572	145 @ 4000	250 @ 2400	8.5-1	2V	Hyd	40
	1975	4.0 x 3.48	18436572	145 @ 3500	250 @ 2200	8.5-1	2V	Hyd.	30
		4.0 x 3.48	18436572	160 @ 4000	265 @ 2400	8.5-1	2V	Hyd.	40
		4.0 x 3.48	18436572	155 @ 3800	250 @ 2800	8.5-1	4V	Hyd.	40
		4.0 x 3.48	18436572	160 @ 3800	250 @ 2400	8.5-1	4V	Hyd.	40

GENERAL ENGINE SPECIFICATIONS

CU. IN. DISPLACE-MENT	YEAR	BORE X STROKE	FIRING ORDER	DEVELOPED HORSEPOWER @ R.P.M.	DEVELOPED TORQUE @ R.P.M.	COM-PRESSION RATIO	CARBU-RETOR	VALVE LIFTER TYPE	NORMAL OIL PRESSURE
	1976	4.0 x 3.48	18436572	145 @ 3800	250 @ 2200	8.5-1	2V	Hyd.	40
		4.0 x 3.48	18436572	165 @ 3800	260 @ 2400	8.5-1	4V	Hyd.	40
		4.0 x 3.48	18436572	185 @ 4000	275 @ 2400	8.5-1	4V	Hyd.	40
		4.0 x 3.48	18436572	210 @ 5200	255 @ 3600	9.0-1	4V	Hyd.	40
	1977-78	4.0 x 3.48	18436572	170 @ 3800	270 @ 2400	8.5-1	4V	Hyd.	40-60
		4.0 x 3.48	18436572	160 @ 4000	265 @ 2400	8.5-1	4V	Hyd.	40-60
366	1971	3.94 x 3.76	18436572	230 @ 4000	340 @ 2400	8.0-1	4V	Hyd.	50-75
	1972	3.94 x 3.76	18436572	211 @ 4000	312 @ 2800	8.0-1	4V	Hyd.	40-55
	1973	3.94 x 3.76	18436572	200 @ 4000	310 @ 2800	8.0-1	4V	Hyd.	40-55
	1974	3.94 x 3.76	18436572	200 @ 4000	310 @ 2800	8.0-1	4V	Hyd.	40-55
	1975	3.94 x 3.76	18446572	200 @ 4000	305 @ 2800	8.0-1	4V	Hyd.	40-55
	1976-78	3.94 x 3.76	18436572	195 @ 4000	290 @ 2800	8.0-1	4V	Hyd.	40
		3.94 x 3.76	18436572	200 @ 4000	305 @ 2800	8.0-1	4V	Hyd.	40
396	1971	4.94 x 3.76	18436572	310 @ 4800	400 @ 3200	9.0-1	4V	Hyd.	50-75
400	1973-74	4.13 x 3.75	18436572	150 @ 3200	295 @ 2000	8.5-1	2V	Hyd.	40
		4.13 x 3.75	18436572	180 @ 3800	290 @ 2400	8.5-1	4V	Hyd.	40
	1975-76	4.13 x 3.75	18436572	175 @ 3600	305 @ 2000	8.5-1	4V	Hyd.	40
	1977-78	4.13 x 3.75	18436572	175 @ 3600	290 @ 2800	8.5-1	4V	Hyd.	40
402	1972	4.94 x 3.76	18436572	210 @ 4000	320 @ 2800	8.5-1	4V	Hyd.	50-75
427	1971	4.25 x 3.76	18436572	260 @ 4000	405 @ 2600	8.0-1	4V	Hyd.	45-55
	1972	4.25 x 3.76	18436572	230 @ 4000	360 @ 2400	8.0-1	4V	Hyd.	45-55
	1973	4.25 x 3.76	18436572	230 @ 4000	360 @ 2800	8.0-1	4V	Hyd.	45-55
	1974	4.25 x 3.76	18436572	230 @ 2800	360 @ 2800	8.0-1	4V	Hyd.	40-55
	1975	4.25 x 3.76	18436572	220 @ 4000	360 @ 2400	8.0-1	4V	Hyd.	40-55
	1976-78	4.25 x 3.76	18436572	200 @ 4000	360 @ 2400	8.0-1	4V	Hyd.	40
454	1971-72	4.25 x 4.0	18436572	270 @ 4000	390 @ 3200	8.25-1	4V	Hyd.	40
	1973	4.25 x 4.0	18436572	240 @ 4000	355 @ 2800	8.25-1	4V	Hyd.	40
	1974	4.25 x 4.0	18436572	230 @ 4000	350 @ 2800	8.25-1	4V	Hyd.	40
	1975	4.25 x 4.0	18436572	245 @ 4000	350 @ 2400	8.25-1	4V	Hyd.	40
		4.25 x 4.0	18436572	245 @ 4000	355 @ 3000	8.25-1	4V	Hyd.	40
		4.25 x 4.0	18436572	245 @ 4000	375 @ 2800	8.25-1	4V	Hyd.	40
	1977-78	4.25 x 4.0	18436572	245 @ 3800	365 @ 2800	8.15-1	4V	Hyd.	40

CRANKSHAFT BEARING JOURNAL SPECIFICATIONS

CU. IN. DISPLACE-MENT	YEAR	MAIN BEARING JOURNALS		SHAFT END PLAY	THRUST ON NO.	CONNECTING ROD BEARING JOURNALS		END PLAY
		JOURNAL DIAMETER	OIL CLEARANCE			JOURNAL DIAMETER	OIL CLEARANCE	
SIX CYLINDER								
250	1971-74	2.2983-2.2993	.0003-.0029	.002-.006	Rear	1.999-2.000	.0007-.0027	.0085-.0135[10]
	1975-78	2.2893-2.2993	.0003-.0029	.002-.006	Rear	1.999-2.000	.0007-.0027	.0007-.0016
292	1971-72	2.2983-2.2993	.0008-.0034	.002-.006	5	2.099-2.100	.0007-.0027	.0009-.0014
	1973-78	2.2983-2.2993	.0008-.0034	.002-.006	5	2.099-2.100	.0007-.0027	.0006-.0017
V-6 ENGINES								
305	1971-74	3.1237-3.1247[2]	.0013-.0039	.003-.008	3	2.8112-2.8122	.0013-.0039	.006-.011
351	1971-72	3.1237-3.1247[2]	.0013-.0039	.003-.008	3	2.8112-2.8122	.0013-.0039	.006-.011
379	1973-74	3.1247-3.1237[1]	.0023-.0039[2]	.003-.008	—	2.8112-2.8122	.0015-.0035	.006-.011
432	1973-74	3.1247-3.1237[1]	.0023-.0039[2]	.003-.008	—	2.8112-2.8122	.0015-.0035	.006-.011
EIGHT CYLINDER								
305	1976-78	[17]	[18]	.002-.006	5	1.8682-1.8692	.0013-.0035	.008-.0014
307	1971-73	2.4684-2.4693[11]	.0011-.0023[12]	.002-.006	5	2.199-2.200	.0013-.0035	.008-.014
350	1971-74	2.4484-2.4493[11]	.0011-.0023[12]	.002-.006	5	2.199-2.200	.0013-.0035	.008-.014
	1975	[19]	[20]	.002-.006	5	2.199-2.200	.0013-.0035	.008-.0014
	1978	[17]	[21]	.002-.006	5	2.099-2.100	.0013-.0035	.008-.0014
366	1971-78	2.7481-2.7490[6]	.0013-.0025[7]	.006-.010	5	2.1988-2.1998	.0007-.0030	.019-.025
396	1971	2.7481-2.7490[6]	.0013-.0025[7]	.006-.010	5	2.1988-2.1998	.0007-.0028	.019-.025
400	1973-78	[22]	[20]	[23]	5	2.099-2.100	.0013-.0035	.008-.0014

CRANKSHAFT BEARING JOURNAL SPECIFICATIONS

CU. IN. DISPLACE-MENT	YEAR	MAIN BEARING JOURNALS				CONNECTING ROD BEARING JOURNALS		
		JOURNAL DIAMETER	OIL CLEARANCE	SHAFT END PLAY	THRUST ON NO.	JOURNAL DIAMETER	OIL CLEARANCE	END PLAY
402	1972	2.7487-2.7496[13]	.0013-.0025[14]	.006-.010	5	2.1985-2.1995	.0009-.0025	.013-.023
427	1971-78		.0013-.0025[9]	.006-.010	5	2.1988-2.1998	.0007-.0028	.019-.025
454	1971-72		.0013-.0025[16]	.006-.010	5	2.199-2.200	.0009-.0025	.015-.021
	1973-77	2.7481-2.7490[15]	.0013-.0025[16]	.006-.010	5	2.1985-2.1995	.0009-.0025	.013-.023

1—Not used
2—Rear No. 4—3.1229-3.1239.
3—Not used
4—Rear—3.1229-3.1239 in.
5—No. 4—.0031-.0047 in.
6—Rear 2.7478-2.7488.
7—Rear .0015-.0031.
8—Rear 2.7473-2.7483.
10—1971-72 .0009-.0014, 1973-74 .0006-.0017.
11—No. 5—2.4479-2.4488.
12—No. 1—.0008-.0020, No. 5—.0017-.0033.
13—Applies to No.'s 1-2. For No.'s 3-4, figures are 2.7481-2.7490. For No. 5, figures are 2.7473-2.7483.

14—Applies to No.'s 2-3-4. For No. 1, figures are .0007-.0019. For No. 5, figures are .0019-.0033.
15—Applies to No.'s 2-3-4. For No. 1 figures are 2.7485-2.7494. For No. 5, figures are 2.7478-2.7488.
16—For No. 5, figures are .0024-.0040.
17—#1—2.4484-2.4493 #2, 3, 4—2.4479-2.4490 #5—2.4479-2.4488
18—Auto.—#1,—.0019-.0031 #2, 3, 4—.0013-.0025 #5—.0023-.0033
Man.—#1, 2, 3, 4—.0013-.0025 #5,—.0023-.0033

19—#1, 2, 3, 4—2.4484-2.4493 #5—2.4479-2.4488
20—#1—.0008-.0020 #2, 3, 4—.0011-.0023 #5—.0017-.0033
21—#1, 2, 3, 4—.0013-.0025 #5—.0023-.0035
22—#1, 2, 3, 4—2.6484-2.6493 #5—2.6479-2.6488
23—2 bbl.—.002-.006 4 bbl.—.006-.0010

VALVE SPECIFICATIONS

CU. IN. DISPLACE-MENT	YEAR	LASH (HOT) INCHES INT.	EXH.	ANGLE DEGREES FACE	SEAT	STEM DIA. INCHES INT.	EXH.	STEM CLEARANCE INTAKE	EXHAUST	VALVE LIFT INCHES	VALVE SPRING LBS. @ INCHES OPEN	CLOSED	FREE LENGTH INCH
						SIX CYLINDER							
250	1971-72	0 + 1 Turn		45	46	.341	.341	.0010-.0027	.0015-.0032	.390	186 @ 1.27	59 @ 1.66	1.90
	1973	0 + 1 Turn		45	46	.341	.341	.0010-.0027	.0010-.0027	.390	186 @ 1.27	59 @ 1.66	1.90
	1974	0 + 1 Turn		45	46	.341	.341	.0010-.0027	.0015-.0032	.390	186 @ 1.27	59 @ 1.66	1.90
	1975-78	0 + ¾ Turn		45	46	.341	.341	.0010-.0027	.0015-.0032	.390	175 @ 1.26	82 @ 1.06	2.08
292	1971-72	0 + 1 Turn		45	46	.341	.341	.0010-.0027	.0015-.0032	.400	179 @ 1.30	89 @ 1.69	1.90
	1973	0 + 1 Turn		45	46	.341	.341	.0010-.0027	.0010-.0027	.400	179 @ 1.30	89 @ 1.69	1.90
	1974-78	0 + 1 Turn		45	46	.341	.341	.0010-.0027	.0015-.0032	.400	179 @ 1.30	89 @ 1.69	1.90
						V-6 ENGINES							
305	1971-74	.012H	.018H	[7]	[7]	.341	.340	.001-.003	.002-.004	.454	204 @ 1.50	80 @ 1.92	2.27
351	1971-72	.012H	.018H	[7]	[7]	.341	.341	.001-.003	.002-.004	.454	204 @ 1.50	80 @ 1.92	2.27
379	1973-74	.012	.018	—	30	.373	.438	.0015-.003	.0019-.0036	.454[2]	203 @ 1.50	80 @ 1.92	2.27
432	1973-74	.012	.018	—	30	.373	.438	.0015-.003	.0019-.0036	.454[2]	203 @ 1.50	80 @ 1.92	2.27
						EIGHT CYLINDER							
305	1976-78	0 + ¾ Turn		45	46	—	—	.0010-.0027	.0010-.0027	—	200 @ 1.25	80 @ 1.70	2.03
350	1971-74	0 + 1 Turn		45	46	.341	.341	.0010-.0027	.0010-.0027	[4]	200 @ 1.20	80 @ 1.61	2.03
	1975-78	0 + ¾ Turn		45	46	—	—	.0010-.0027	.0010-.0027	—	12	12	13
366	1971-78	0 + 1 Turn		[1]	46	.371	.371	.001-.003	.001-.003	.400	220 @ 1.40	90 @ 1.80	2.05
400	1973-75	0 + 1 Turn		45	46	—	—	.0010-.0027	.0012-.0027	—	200 @ 1.25	80 @ 1.70	2.03
	1976-78	0 + ¾ Turn		45	46	—	—	.0010-.0027	.0010-.0027	—	200 @ 1.25	80 @ 1.70	2.03
402	1972	0 + 1 Turn		45	46	.372	.372	.0010-.0027	.0012-.0027	[9]	215 @ 1.48[10]	90 @ 1.80	2.05
427	1972-78	0 + 1 Turn		[5]	46	.371	.371	.001-.003	.001-.003	[6]	220 @ 1.40	90 @ 1.88	2.05
454	1971-78	0 + 1 Turn		45	46	—	—	.0010-.0027	.0012-.0029	—	305 @ 3.05	80 @ 1.88	2.12

1—Intake 45, Exhaust 46.
2—Not used
3—Not used
4—Intake 390, Exhaust 410.
5—Not used

6—Intake .398, Exhaust .430.
7—Inlet 30. Exhaust 45.
8—Not used
9—Intake .398, Exhaust .430
10—Inner spring 90 @ 1.28

11—Not used
12—Exhaust—Closed-80 @ 161, Open-189 @ 120 Intake—Closed-80 @ 170, Open-200 @ 125
13—Exhaust—1.91 Intake—2.03

RING GAP SPECIFICATIONS (in)

ENGINE CU. IN.	YEAR	TOP COMPRESSION	BOTTOM COMPRESSION	OIL CONTROL
250	1971-78	.010-.020	.010-.020	.015-.055
292	1971-78	.010-.020	.010-.020	.015-.055
305 V6	1971-74	.017-.027	①	no gap
351 V6	1971-72	.017-.027	①	no gap
379 V6	1973-74	.022-.032	②	no gap
432 V6	1973-74	.024-.034	③	no gap
200 V6	1978	—	—	—
231 V6 Calif.	1978	—	—	—
305	1976-78	.010-.020	.010-.020	.015-.055
307	1971-75	.010-.020	.010-.020	.015-.055
350	1971-78	.010-.020	.013-.025	.015-.055
366	1971-78	.010-.020	.010-.020	.010-.023
396	1971	.010-.020	.010-.020	.010-.030
400	1973-78	.010-.020	.010-.020	.010-.035
402	1972	.010-.020	.010-.020	.015-.055
427	1971-78	.010-.020	.010-.020	.010-.023
454	1971-78	.010-.020	.010-.020	.010-.030
350 Diesel	1978 (Olds)	—	—	—

① —2nd compression—.015-.025
   3rd compression—.015-.025
② —2nd compression—.022-.032
   3rd compression—.015-.025
③ —2nd compression—.025-.034
   3rd compression—.015-.025

RING SIDE CLEARANCE (in)

ENGINE CU. IN.	YEAR	TOP COMPRESSION	BOTTOM COMPRESSION	OIL CONTROL
250	1971-78	.0012-.0027	.0012-.0032	.0000-.0050
292	1971-78	.0020-.0040	.0020-.0040	.0005-.0055
305 V6	1971-74	.0030-.0050	①	.0010-.0040
351 V6	1971-72	.0030-.0050	①	.0010-.0040
379 V6	1973-74	.0030-.0045	②	.0025-.0040
432 V6	1973-74	.0030-.0045	②	.0025-.0040
200 V6	1978	—	—	—
231 V6 Calif.	1978	—	—	—
305	1976-78	.012-.027	.012-.027	.0000-.0050
307	1971-75	.0012-.0027	.0012-.0032	.0000-.0050
350	1971-78	.0012-.0032	.0012-.0032	.0000-.0050
366	1971-78	.0018-.0032	.0018-.0032	.0020-.0035
396	1971	.0012-.0032	.0012-.0032	.0012-.0060
400	1973-78	.0012-.0032	.0012-.0032	.0000-.0050
402	1972	.0017-.0032	.0017-.0032	.0005-.0065
427	1971-78	.0018-.0032	.0018-.0032	.0020-.0035
454	1971-78	.0017-.0032	.0017-.0032	.0005-.0065
350 Diesel	1978 (Olds)	—	—	—

① —2nd compression—.0030-.0045
   3rd compression—.0025-.0040
② —2nd compression—.0025-.0040
   3rd compression—.0025-.0040

TORQUE SPECIFICATIONS

CU. IN. DISPLACEMENT	YEAR	CYLINDER HEAD BOLTS FT. LBS.	ROD BEARING BOLTS FT. LBS.	MAIN BEARING BOLTS FT. LBS.	CRANKSHAFT BALANCER BOLT FT. LBS.	FLYWHEEL CRANKSHAFT BOLTS FT. LBS.	MANIFOLDS INTAKE FT. LBS.	MANIFOLDS EXHAUST
SIX CYLINDER								
250	1971-78	95	30-35	60-70	Pressed on	55-75	35	35
292	1971-78	95	35-45	60-70	Pressed on①	110	35	35
V-6 ENGINES								
305-351	1971-74	65-72	50-55	130-140③	200-210	100-105	20-25	15-20
401	1971-74	65-72	50-55	130-140③	200-210	100-105	20-25	15-20
432	1973-74	65-72	50-55	130-140③	200-210	100-105	20-25	15-20
487	1971-72	65-72	50-55	130-140③	200-210	100-105	20-25	15-20
EIGHT CYLINDER								
305	1976-78	65	45	70④	Pressed on①	60	30	20
307	1971-73	65	45	70	Pressed on①	60	30	20
350	1971-78	65	45	70	70-80	55-65	30	20
366	1971-78	75-85	67-73②	100-110	80-90	60-70	25-35	25-35
400	1974-78	80	50	110	85	65	30	20
402	1971-72	80	50	110	85	65	30	20
427	1971-78	75-85	50	100-110	80-90	60-70	30	25-30
454	1972-78	80	50	110	85	65	30	20

① With bolt-60 ft. lbs.
② ⅜ nut—45-55 ft. lbs.
③ Rear—55-65 ft. lbs.
④ Outer bolt on engine with four bolt caps—65 ft. lbs.

WHEEL ALIGNMENT SPECIFICATIONS

YEAR	MODEL	CASTER (Deg.)	CAMBER (Deg.)	TOE-IN (In.)	KING PIN (Deg.)
		1500 thru 3500 Series			
1971-72	K1500	+4	+1-3	⅛	—
	CP1500, C1550	+¼-¼	0+½	⅛-¼	—
	G1500-2500	+3¼	+¼-1¾	3/32-3/16	7¼
	K1500-2500	+3¼	+½-1½	3/32-3/16	—
	P2500-3500	−¼+¼	0+½	⅛-¼	—
1973	GA10-30	±1	+¼	±1/16	—
	CA, PA10	±1	+¼	±1/16	—
	CA, PA20-30	±1	+¼	±1/16	—
	K10-20	±1	+1½	±1/16	—
	GE, GS1500, 2500, 3500	±½	±½	±1/16	—

Note: For 1974 to present Alignment Specifications for series 150 thru 3500 see the chart & illustration following these tables.

YEAR	MODEL	CASTER (Deg.)	CAMBER (Deg.)	TOE-IN (In.)	KING PIN (Deg.)
		4000 thru 6500 Series			
1971-72	CP4500	+2-3	+½-1½	⅛-3/16	7¼
	CS5500	+2-3	+½-1½	3/32-3/16	7¼
	CM6500①	+2-3	+½-1½	3/32-3/16	7¼
	M6500	+2-3	0±¼	⅛-7/32	6
	CM6500③	+2-3	0±¼	⅛-7/32	6
	T5500, T6500①	+¾-1¾	+1-2	⅛-¼	7
	T5500, T6500③	+¾-1¾	0±¼	⅛-¼	6
1973	C4500	2½±½	1±½	⅛-3/16	7¼
	S/C5500	2½±½	1±½	3/32-3/16	7¼
	C/M6500①	2½±½	1±½	3/32-3/16	7¼
	C/M6500③	2½±½	+¼	⅛-7/32	5¾
	T5500①	1¼±½	1½±½	⅛-¼	7
	T5500③	1¼±½	¼±½	⅛-¼	5¾
	T6500①	1¼±½	1½±½	⅛-¼	7
	T6500③	1¼±½	¼±½	⅛-¼	5¾
	T6500②	1¼±½	¼±½	⅛-¼	5¾
1974	P40-45	2½±½	1±½	1/16-⅛	7¼
	C-50	2½±½	1±½	1/16-⅛	7¼
	S/C-60	2½±½	1±½	1/16-⅛	7¼
	C/M-65①	2½±½	1±½	1/16-⅛	7¼
	C/M-65②③	2½±½	−0¼	⅛-¼	L-5¾, R-6¼
	T-60①	1¼±½	1½±½	1/16-⅛	7
	T-60④	1¼±½	−¼±½	⅛-¼	L-5¾, R-6¼
	T-65①	1¼±½	1½±½	1/16-⅛	7
	T-65④	1¼±½	−¼±½	⅛-¼	L-5¾, R-6¼
	T-65②	1¼±½	−¼	⅛-¼	L-5¾, R-6¼
1975-78	CE/CS-50, 60	+1½-2½	+1±2	1/16-⅛	7
	SE-60	+1½-2½	+1±2	1/16-⅛	7
	CE/ME-65①	+1½-2½	+1±2	1/16-⅛	7
	CE/ME-65④	+1½-2½	−¼±¾	1/16-⅛	L-5¾, R-6¼
	CE/ME-65②	+1½-2½	−¼±¾	1/16-⅛	5¾
	TE-60, 65①	+3½-4½	+1±2	1/16-⅛	7
	TE-60, 65④	+3½-4½	−¼±¾	1/16-⅛	L-5¾, R-6¼
	TE-60, 65②	+3½-4½	−¼±¾	1/16-⅛	5¾

①—W/F070 Axle
②—W/F120 Axle
③—W/F160 Axle
④—W/F090 Axle

⑤—Manual Steering
⑥—Power Steering
⑦—Left +½° to +1½°
   right 0° to +1°

Caster—1974 to Present 1500 thru 3500 Series

All caster specifications are given assuming a frame angle of zero. Therefore, it will be necessary to know the angle of the frame (whether "up" in rear or "down" in rear) before a corrected caster reading can be determined. Camber and toe can be read "as is" from the alignment equipment.

How to Determine Caster

1. With the vehicle on a level surface, determine the frame angle "B" in the illustration, using a bubble protractor or clinometer.
2. Draw yourself a graphic as shown in the illustration that is representative of the frame angle (either "up" in rear or "down" in rear).
3. Determine the caster angle from the alignment equipment and draw a line that is representative of the caster reading.
4. To determine an "actual (corrected) caster reading" with various frame angles and caster readings one of the following rules applies:
   a. A "down" in rear" frame angle must be subtracted from a positive caster reading.
   b. An "up in rear" frame angle must be added to a positive caster reading.
   c. A "down in rear" frame angle must be added to a negative caster reading.
   d. An "up in rear" frame angle must be subtracted from a negative caster reading
5. Add or subtract as necessary to arrive at the corrected caster angle.
6. Measure dimension "A" (bump stop bracket to frame) and check the specifications for that dimension.
7. Correct the actual caster angle, as arrived at in Step 4, as necessary to keep within the specifications by adding or subtracting shims from the front or rear bolt on the upper control arm shaft.

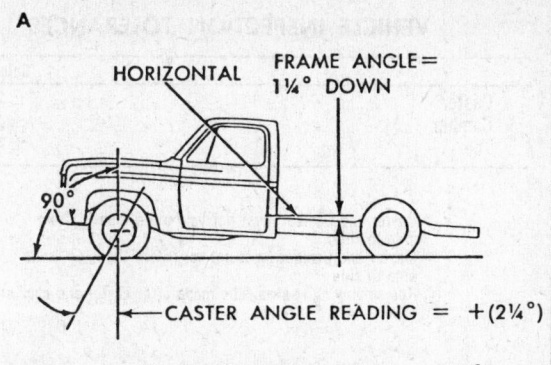

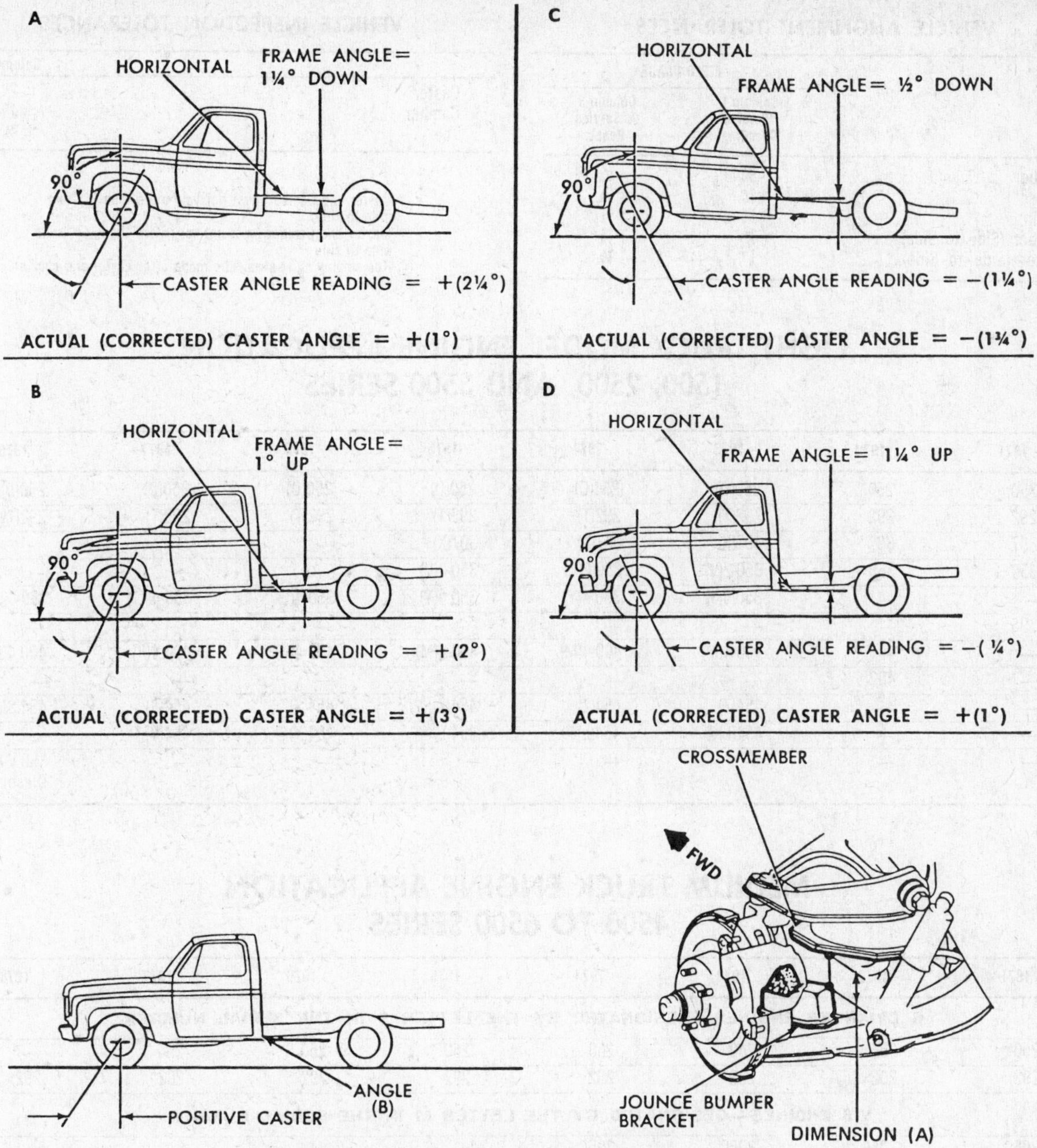

A
HORIZONTAL
FRAME ANGLE= 1¼° DOWN
90°
CASTER ANGLE READING = +(2¼°)
ACTUAL (CORRECTED) CASTER ANGLE = +(1°)
C
HORIZONTAL
FRAME ANGLE= ½° DOWN
90°
CASTER ANGLE READING = −(1¼°)
ACTUAL (CORRECTED) CASTER ANGLE = −(1¾°)
B
HORIZONTAL
FRAME ANGLE= 1° UP
90°
CASTER ANGLE READING = +(2°)
ACTUAL (CORRECTED) CASTER ANGLE = +(3°)
D
HORIZONTAL
FRAME ANGLE= 1¼° UP
90°
CASTER ANGLE READING = −(¼°)
ACTUAL (CORRECTED) CASTER ANGLE = +(1°)
ANGLE (B)
POSITIVE CASTER
CROSSMEMBER
FWD
JOUNCE BUMPER BRACKET
DIMENSION (A)

Determining Caster—Typical—1974-76—Series 1500 thru 3500

WHEEL ALIGNMENT SPECIFICATIONS

1974 To Present—1500 thru 3500 Series

CASTER

Dimension "A" in inches	2½″	2¾″	3″	3¼″	3½″	3¾″	4″	4¼″	4½″	4¾″	5″
GA10-30 PA 30	+2¼°	+2°	+1½°	+1¼°	+1°	+¾°	+½°	+¼°	0°	−¼°	−½°
CA PA 10			+2°	+1½°	+1¼°	+1°	+¾°	+½°	+¼°	0°	−½°
CA PA 20-30	+1½	+1¼°	+1°	+¾°	+½°	+¼°	0°	−¼°	−½°	−¾°	−1°
K10-20	+4° —no provision for resetting										
PA 10-20	—	—	—	+1½°	+1¼°	+1°	+¾°	+½°	+¼°	0°	−¼°
CAMBER											
CA GA PA 10-20-30										+¼°	
KA 10-20 .... No provision for resetting										+1½°	
TOE-IN (TOTAL)											
CA GA PA 10-20-30 K 10-20										3/16″†	
K 10-20										0°	

* See Column 1, 2 or 3 under Vehicle Alignment Tolerances for applicable tolerances.
† K-10 with full time four wheel drive toe-in = 0″.

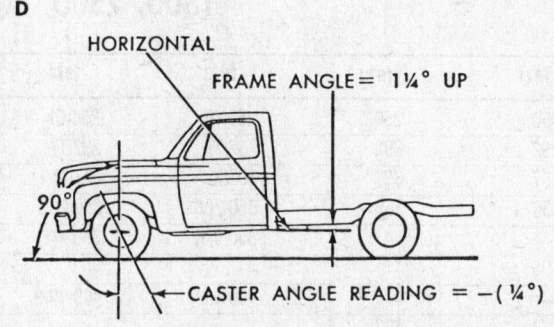

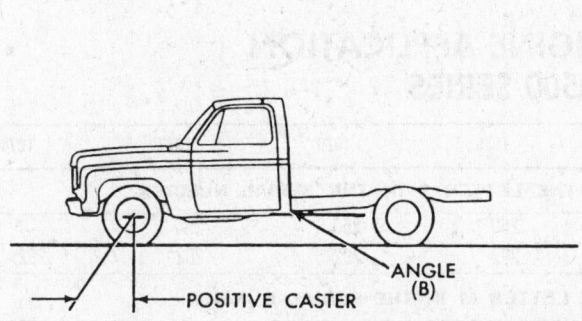

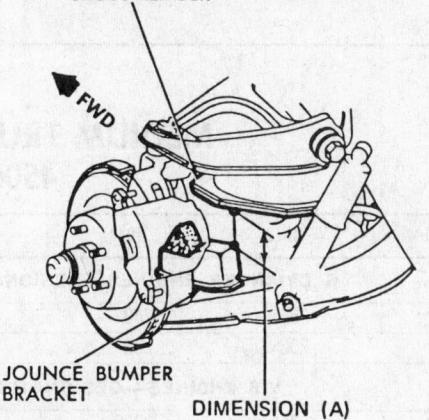

VEHICLE ALIGNMENT TOLERANCES

	FIELD USAGE	
	Column 1 *Service Checking	Column 3 @ Service Reset
Camber	±¾°	±½°
Caster	±1°	±½°
① Toe-in	±⅛"	±¹⁄₁₆"
Camber (Side to Side)	1°	½°
Caster (Side to Side)	1°	½°

VEHICLE INSPECTION TOLERANCES

	Column 2
Caster	± 2°
Camber	± 1½°
Toe	± ⅜"

* Caster and Camber must not vary more than 1° from side to side.
@ Caster and Camber must not vary more than ½° from side to side.
① Toe setting must always be made after caster and camber.

LIGHT TRUCK MODEL ENGINE APPLICATION
1500, 2500, AND 3500 SERIES

1971	1972	1973	1974	1975	1976	1977	1978
250	250	250(Q)	250(Q)	250(Q)	250(D)	250(D)	250(D)
292	292	292(T)	292(T)	292(T)	292(T)	292(T)	292(T)
307	307	307(X)	307(X)	307(X)	—	—	—
350	350	350-2(V)	350(2(V)	350-2(V)	—	—	—
—	—	350-4(Y)	350-4(Y)	350-4(Y)	350-4(L)	350-4(L)	350-4(L)
396	—	—	—	—	—	—	—
—	—	400-4(M)	400-4(M)	400-4(M)	400-4(U)	400-4(R)	400-4(R)
—	402	—	—	—	—	—	—
—	—	454(Z)	454(Z)	454(Z)	454(S)	454(S)	454(S)
—	—	454(L)②	454(L)②	454(L)②	454(Y)②	454(Y)②	—
—	—	—	—	—	—	—	350 V-8 (—) Diesel (Olds)

MEDIUM TRUCK ENGINE APPLICATION
4500 TO 6500 SERIES

1971	1972	1973	1974	1975	1976	1977	1978
6 CYLINDER ENGINES—DESIGNATED BY THE LETTER S IN THE SERIAL NUMBER.							
250	250	250	250	250	250	250	—
292	292	292	292	292	292	292	292
V-6 ENGINES—DESIGNATED BY THE LETTER M IN THE SERIAL NUMBER.							
305C	305C	305C	305C	—	—	—	—
351C	351C	—	—	—	—	—	—
—	—	379	379	—	—	—	—
—	—	432	432	—	—	—	—
V-8 ENGINES—DESIGNATED BY THE LETTER E IN THE SERIAL NUMBER.							
307	307	—	—	—	—	—	—
350	350	350	350	350	350	350	350
366	366	366	366	366	366	366	366
396	—	—	—	—	—	—	—
—	402	—	—	—	—	—	—
—	—	427	427	427	427	427	427
DIESEL ENGINES—DESIGNATED BY THE LETTERS D, G, AND Y IN THE SERIAL NUMBER.							
DH478(G)	DH478(G)	DH478(G)	DH478(G)	—	—	—	—
—	—	—	—	3208(Y)	—	—	—
—	—	—	—	4-53(D)	4-53(D)	4-53(D)	4-53(D)
—	—	—	—	—	—	—	4-53T(D)

①—231 V-6 Standard in California only
②—Heavy Duty Model
③—Heavy Duty Emission Controls

Light Duty Trucks

For the model years 1971-72, the engine types and applications are identified by a three letter code. The code is stamped on a pad, located on the engine at the rear of the distributor on the six cylinder engines, and on a pad forward of the right side cylinder head on the V-8 engines. Utilize this code when ordering parts for these engines.

Beginning with the 1973 model year, the engine type is included as the third designation in the serial number and is represented as a letter. The vehicle identification plate containing the serial number, is mounted on the left door lock pillar on the convention cabs, on the left door hinge pillar post for steel tilt cabs, and on the inside face of the dash and toe panel on the step vans. The three letter engine code is retained on the engines for precise parts ordering, and their locations on the engines remain the same.

Medium Trucks— 4500 to 6500 Series

A general code is used in the serial number for the engine identification.

D — In-line 4-53 Detroit Diesel
E — V-8 Gasoline engines
G — GMC V-6 Diesel
M — V-6 Gasoline engines
S — In-line Gasoline engines
Y — Caterpillar Diesel V-8 3208

For positive engine identification, refer to the three letter code on the engine block and to a plate located on the right valve cover. The engine Emission Control Certificate can also be used in the identification procedure. The following is an example of a vehicle identification number and the important break-down of the number.

Distributor

Standard Ignition

The distributor used on inline engines is driven from the engine camshaft by spiral cut gears and is located on the right side of the engine. A gasket is used between the distributor flange and cylinder block. The distributor is held in place by a hold-down clamp and cap screw. The lower end of the distributor shaft is tongued and fits a slot in the upper end of the oil pump shaft to drive the oil pump.

The distributor used on V6 and V8 engines is mounted on top center of cylinder block at the rear end, and is driven from the camshaft by spiral cut gears. A gasket is used between distributor flange and engine block. The distributor is held in place by a hold-down clamp and cap screw. The lower end of the distributor shaft has a hexagonal opening that fits the end of the oil pump shaft to drive the oil pump. Model number is stamped on distributor housing.

Distributor R & R

Removal

1. Locate number one cylinder spark plug wire and mark the position of the wire tower on the distributor cap and body.
2. Remove the distributor cap, primary wire from the coil terminal, and the vacuum advance line from the advance unit.
3. With the use of the starter, rotate the engine until the crankshaft pulley timing marks are aligned with the pointer or timing mark tab, located on the timing cover housing.
4. The rotor segment should point toward the mark previously made on the distributor housing. If the rotor segment points 180° degrees away from the mark, rotate the engine one complete revolution and re-align the timing marks.
5. Note the position of the vacuum advance unit in relation to the engine. Remove the hold-down bracket and bolt, and lift the distributor upward until the spiral gear disengages from the camshaft gear. Remove the distributor from the engine.
6. The rotor will move a few degrees as the gears disengage. Mark the second position of the rotor segment on the distributor housing to aid in reassembly.

NOTE: Keep the distributor in the upright position so that oil from the distributor shaft will not run out onto the breaker plate and points, or the electronic units within the distributor.

Installation

1. Lubricate the distributor drive gear with engine oil and install a new distributor flange gasket.
2. Turn the rotor segment to point toward the mark made on the housing after the gears were disengaged. Insert the distributor into the engine while observing the previous position of the vacuum advance unit, relative to the engine.
3. As the gears engage, the rotor will rotate a few degrees and the rotor segment will align with the mark on the housing, that repre-

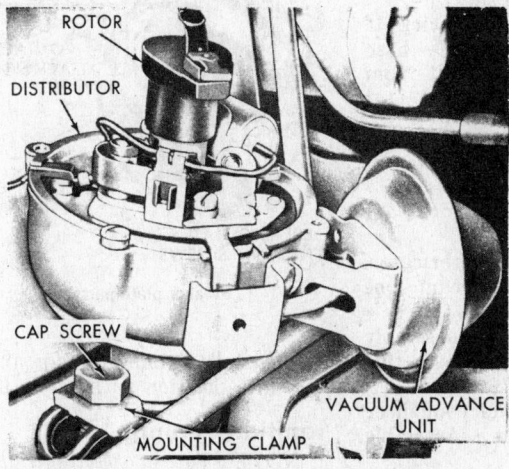

Distributor with cover removed—inline engine (typical)
(© G.M.C.)

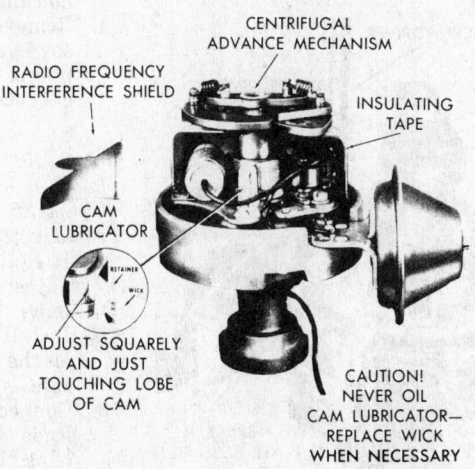

Distributor view—V8 engine (typical)
(© G.M.C.)

sents the number one cylinder spark plug wire.

4. Press down on the distributor to seat it fully against the block. If necessary, engage the starter several times to make certain the oil pump shaft is engaged.

5. Rotate the distributor body until the points begin to open with the rotor segment pointing to the number one cylinder position. Install the hold-down bracket and bolt and snug them in place.

6. Replace the primary wire to the coil terminal, the vacuum line to the advance unit, and the distributor cap to the distributor body. Start the engine, adjust the dwell and the ignition timing.

Locating Number One Firing Position

If the engine has been cranked with the distributor out, locating number one firing position can be accomplished by one of the two methods outlined below:

a. Remove number one spark plug and place a finger over the hole and crank the engine until compression can be felt. Continue cranking until the timing mark on the crankshaft pulley aligns with the pointer or the timing tab.

b. Remove the rocker arm cover over number one cylinder and crank the engine until the number one intake valve begins to close. Continue to crank the engine until the timing marks align between the crankshaft pulley and the pointer or tab.

Relocate the oil pump shaft to a position to accept the distributor gear, and install the distributor as outlined in the removal and installation section.

Contact Point Replacement
Cleaning

Dirty breaker points should be cleaned with a few strokes of a fine-cut contact file. File should be kept free of grease, dirt and should not be used on other metals. Never use emery cloth to clean breaker points. Do not attempt to file point surface smooth, just remove scale or dirt.

NOTE: Highly pitted or burnt points are often caused by improper condenser capacity.

Inline, V6 Engines—Removal

1. Remove distributor cap and place it away from work area.
2. Lift off rotor.
3. Pull primary and condenser lead wires from quick-disconnect terminal.
4. Remove attaching screws and lift breaker point set from plate.
5. Remove oil, dirt and smudge from breaker plate.

Installation

1. Carefully remove protective covering from points and place set on breaker plate. Install attaching screws.
2. Connect primary and condenser leads to terminals. Assemble clips "back to back". Do not push on spring.
3. Apply a slight amount of petroleum jelly to breaker cam and a few drops of S.A.E. #20 oil to top of shaft.
4. Check points for alignment and breaker arm spring for proper tension.
5. Set point gap to specifications.
6. Install rotor and distributor cap.

V8 Engines—Removal

1. Remove distributor cap and place it away from work area.
2. Remove two screws attaching rotor to weight assembly, then pull primary and condenser lead wires from quick-disconnect terminal.
3. Remove two screws breaker plate, point set to distributor housing, and lift out set.
4. Remove cam lubricating wick, if so equipped, with long nose pliers. Clean old lubricant from cam surface.

Installation

NOTE: Breaker point set is replaced as a complete assembly; point alignment and spring tension are pre-adjusted.

1. If equipped with cam lubricator, adjust wick to touch cam lobe only.
2. Install new contact set assembly on the plate and attach with two screws.
3. Connect primary and condenser leads to terminals. Assemble clips "back to back".
4. Install rotor on weight assembly with two screws and washers.

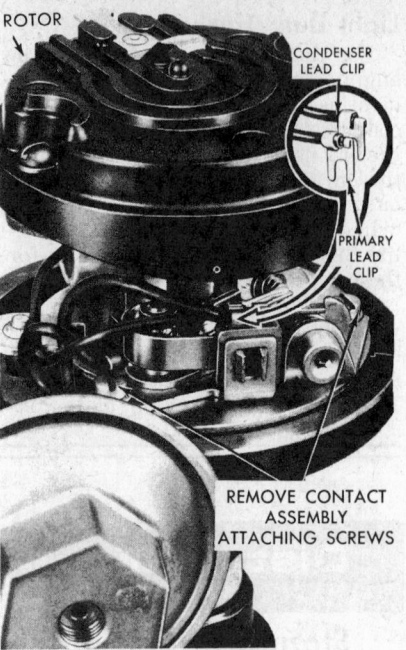

Distributor point set replacement
(© G.M.C.)

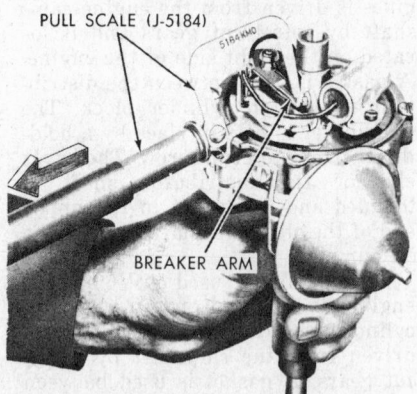

Checking breaker arm spring tension
(© G.M.C.)

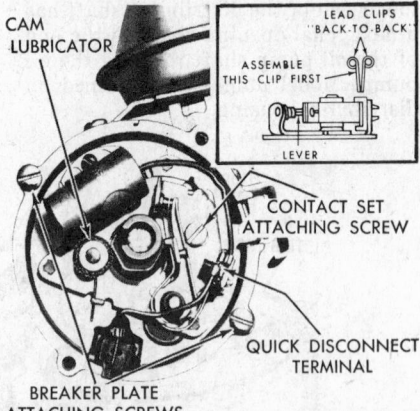

Breaker plate parts
(© G.M.C.)

5. Install distributor cap and lock into position with screw latches.

Ignition Timing

NOTE: To use a timing light, disconnect vacuum advance line to carburetor and tape open end. Carbure-

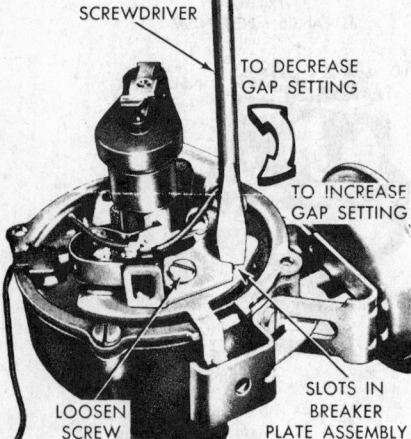

Adjusting point gap, typical—In-line and V6 engines (© G.M.C.)

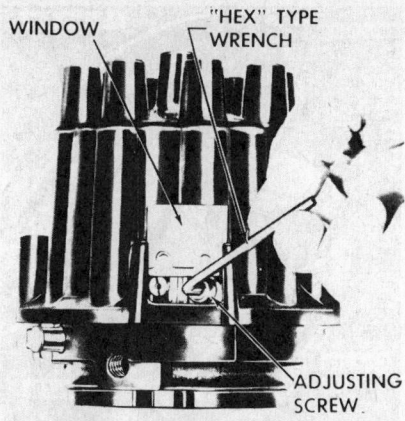

WINDOW   "HEX" TYPE WRENCH

ADJUSTING SCREW

Adjusting dwell angle—V8 engine (© G.M.C.)

tor trouble can affect ignition timing adjustments. Without a power timing unit, an accurate method of setting timing with engine stopped is by using a jumper light.

1. Remove distributor cap and rotor, set breaker gap to specifications.
2. Rotate engine until No. 1 cylinder is at firing position (timing mark on crankshaft pulley aligned with timing tab).
3. Connect jumper light between distributor ignition terminal and ground.
4. Turn on ignition.
5. Loosen distributor and move in normal rotation until light goes out (points closed), then slowly turn distributor back until light just comes on. Tighten distributor.

Reference

Refer to Troubleshooting section for ignition problem analysis.

High Energy Ignition

Beginning with 1975 and 1976 models, most GMC truck engines use a High Energy Ignition system. Two types are used. V8 distributors and 1978 and later sixes combine all ignition components in one unit. The coil is in the distributor cap and connects directly to the rotor. The 6 cylinder distributor through 1977 has an externally mounted coil. Both units operate in basically the same manner, except that the module and pick-up coil replace the conventional breaker points. The module automatically controls the dwell, stretching it with increased engine speed. The system also features a longer spark duration due to the greater amount of energy stored in the primary coil.

The centrifugal and vacuum advance mechanisms are basically the same type of unit as on a conventional ignition distributor.

The electronic module is serviced by complete replacement.

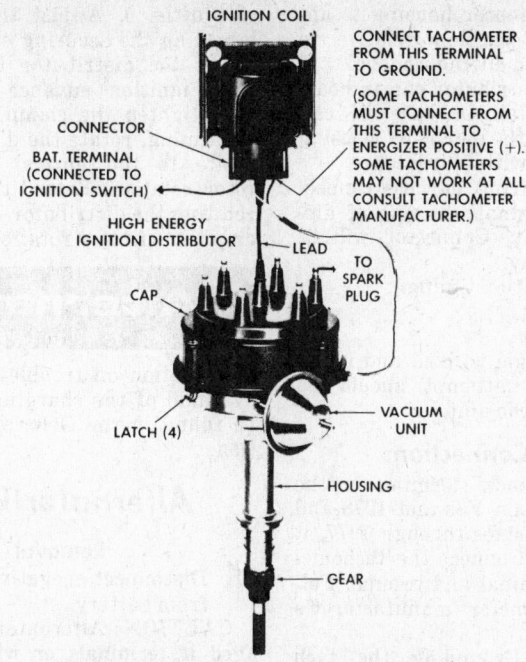

IGNITION COIL

CONNECT TACHOMETER FROM THIS TERMINAL TO GROUND.

(SOME TACHOMETERS MUST CONNECT FROM THIS TERMINAL TO ENERGIZER POSITIVE (+). SOME TACHOMETERS MAY NOT WORK AT ALL. CONSULT TACHOMETER MANUFACTURER.)

CONNECTOR

BAT. TERMINAL (CONNECTED TO IGNITION SWITCH)

HIGH ENERGY IGNITION DISTRIBUTOR

LEAD

TO SPARK PLUG

CAP

VACUUM UNIT

LATCH (4)

HOUSING

GEAR

HEI Distributor—In-line and V6 engines (© G.M.C.)

WARNING: *Do not remove the spark plug wires with the engine running. Severe shock could result.*

Distributor R & R—All Models

Removal

1. Disconnect wiring harness connectors at side of distributor cap.
2. Remove distributor cap and position out of way.

3. Disconnect vacuum advance hose from vacuum advance mechanism.
4. Scribe a mark on the engine in line with rotor. Note approximate position of distributor housing in relation to engine.
5. Remove distributor hold-down nut and clamp.
6. Lift distributor from engine.

Installation

1. Install distributor using same procedure as for standard distributor.
2. Install distributor hold-down clamp and snugly install nut.

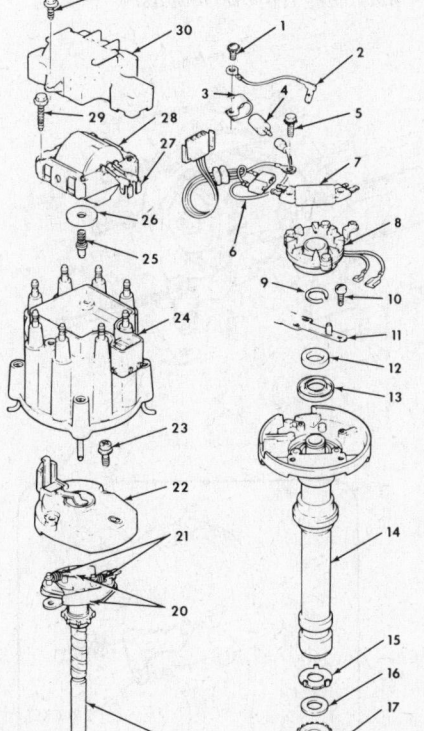

1 Screw
2 Wiring lead
3 Capacitor clamp
4 Capacitor
5 Screw
6 Wiring harness and module leads
7 Module
8 Pick-up coil magnet assembly
9 Thin "C" washer
10 Screw
11 Plastic retainer
12 Felt washer
13 Felt retainer
14 Housing
15 Thrust washer
16 Shim
17 Driven pin
18 Roll pin
19 Shaft
20 Weights
21 Springs
22 Rotor
23 Screw
24 Cap
25 Spring and button assembly
26 Seal
27 Coil tower terminals
28 Coil
29 Screw
30 Cover
31 Screw

HEI Distributor—V8 engine (© G.M.C.)

3. Move distributor housing to approximate position relative to engine noted during removal.

4. Position distributor cap to housing with tab in base of cap aligned with notch in housing and secure with four latches.

5. Connect wiring harness connector to terminals on side of distributor cap. Connector will fit only one way.

6. Adjust ignition timing.

Dwell Angle

The dwell angle is fixed and is not adjustable. No attempt should be made to adjust the unit.

Tachometer Connections

There is a "tach" terminal on the distributor cap on V8s and 1978 and later sixes. On sixes through 1977, it is on the coil. Connect the tachometer to this terminal and ground. Follow the tachometer manufacturer's instructions.

CAUTION: Grounding the tach terminal could damage the H.E.I. ignition module.

Ignition Timing

On H.E.I. systems, connect the timing light in parallel at the No. 1 tower on the distributor cap. Disconnect the distributor spark advance hose and plug the vacuum opening. Start the engine and run it at idle speed. Aim the timing light at the degree scale just over the harmonic balancer. The markings on the scale are in 2° increments with the greatest number of markings on the *before*

side of the 0. Adjust the timing by loosening the securing clamp and rotating the distributor until the desired ignition advance is achieved, then tighten the clamp. To advance the timing, rotate the distributor opposite to the normal direction of rotor rotation. Retard the timing by rotating the distributor in the normal direction of rotor rotation.

Alternator

Information on trouble-shooting and overhaul of the charging system can be found in the General Repair section.

Alternator R & R

Removal

1. Disconnect negative (-) cable from battery.

CAUTION: Alternator will be damaged if terminals or wiring is accidentally shorted or grounded with negative (−) cable connected to battery.

2. Depress lock and pull connector out of socket on generator. Remove rubber boot from "BAT" terminal and remove terminal nut. Disconnect wire from "GRD" terminal and remove clip.

NOTE: On 130 amp alternator, remove nuts and washers from harness leads at alternator terminals. Remove harness clip from alternator, then pull leads from terminals.

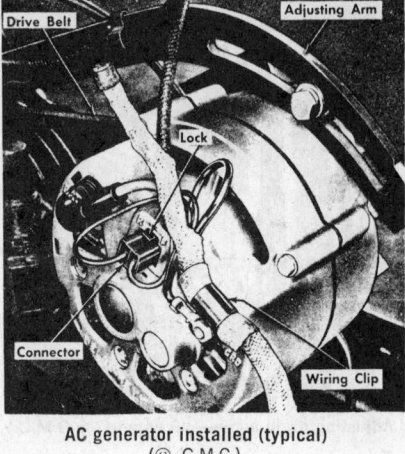

AC generator installed (typical)
(© G.M.C.)

Integral AC generator (typical)
(© G.M.C.)

3. Loosen alternator mounting bolts and adjusting arm pivot bolt, then remove drive belt(s).

4. Remove alternator mounting bolts and adjusting arm pivot bolt. Lift alternator assembly from engine.

Installation

CAUTION: Make certain negative (−) cable is disconnected from battery.

1. Attach alternator to mounting bracket and install adjusting arm. Tighten lock nuts.

2. Install drive belt(s) and adjust to specifications. Torque lock nuts and mounting bolts to specifications.

3. Push connector into socket, making certain that it locks; place clip on "GRD" terminal and connect ground wire. *NOTE: On 130 amp alternator, connect leads to respective terminals, install attaching nuts and washers.*

4. Install harness clip.

5. Attach red wire to "BAT" terminal on generator and fit rubber boot.

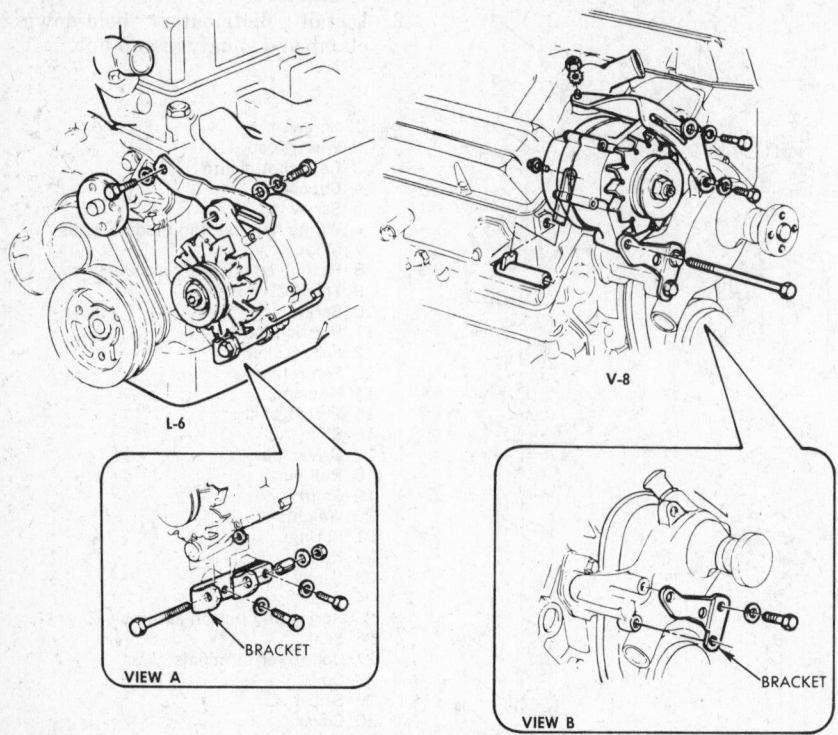

Delcotron installation
(© G.M.C.)

Voltage Regulator

Non-Integral

Removal

1. Disconnect negative (-) cable from battery.
2. Carefully remove wiring harness connector from regulator.
3. Remove regulator mounting screws; lift off regulator assembly.

Installation

CAUTION: Make certain negative (—) cable is disconnected from battery.

1. Place regulator ground wire on mount and install attaching screws.
2. Lift regulator terminal latch and insert wiring harness connector. Make certain connector is locked.
3. Connect negative (-) cable to battery.

Adjustment

1. Connect (POS) voltmeter lead to battery (POS) terminal on regulator and (NEG) voltmeter lead to ground on regulator.
2. Adjust engine speed to approximately 1500 rpm, turn heater to medium speed and turn all other electrical load "OFF." Disconnect negative cable from battery.

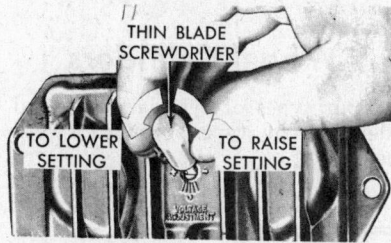

Adjusting voltage setting (typical)
(© G.M.C.)

3. Place a thermometer approximately ¼″ from regulator cover and operate engine 15 minutes.
4. Compare voltmeter reading with those given in specifications.
5. If voltmeter is not within limits listed in specifications, remove plug from regulator cover and insert thin-bladed screwdriver into adjustment screw. *NOTE: After two notches in either direction there is a positive stop. Forcing screw beyond normal stop will destroy regulator.*
6. For an undercharged battery, turn adjusting screw one notch clockwise.
7. For an overcharged battery, turn adjusting screw one notch counterclockwise.
8. Check battery condition after a service period of reasonable length.

9. If regulator cannot be adjusted to within limits listed in specifications, repair or replace the regulator.

Starter

Information on trouble-shooting and overhaul of the starter motor can be found in the General Repair Section.

The starter motor used on all models is the over-running clutch type, having an enclosed shift lever and solenoid plunger mechanism, within the extended drive end housing. A compression type lever return spring is used to operate the over-running clutch. Oil impregnated bronze bushings and oil saturated wicks provide the lubrication to the armature shaft at the commutator end and the nose housing.

Removal

1. Remove ground strap from negative (-) post on battery.
2. Disconnect wires from terminals on starter solenoid. Reinstall nuts as each wire is disconnected.
3. Loosen front bracket mounts where applicable. Remove bolts, nuts and washers attaching starter to flywheel housing.
4. Remove starter and spacer (when used).

Installation

1. Install spacer (when used) and position starter against flywheel housing.
2. Install bolts, nuts and washers and torque to specifications.
3. Connect wires to proper terminals on starter solenoid and tighten attaching nuts.

4. Connect ground strap to negative (-) post on battery.

Refer to General Section for starter motor overhaul.

Brakes

Master Cylinder

Specific information will be found in the General Brake Section on adjustment, bleeding, master cylinder and wheel cylinder overhaul procedures and troubleshooting.

Refer to the Power Brake Section for details concerning power brakes.

Three types of master cylinders are used on GMC trucks covered by this manual.

1. Single barrel firewall and frame mounted type.
2. Double barrel firewall mounted type.
3. Dual single barrel double reservoir type.

The removal and installation of the master cylinders are covered in this chapter, while the disassembly and assembly is covered in the hydraulic brake section.

Single Barrel Firewall Mounted Type

Removal

1. Clean the area at the fitting and place a dry cloth under the line to absorb any fluid leakage.
2. Disconnect the hydraulic line at the cylinder and cover the end to prevent any foreign material from entering the system.
3. Disconnect the pushrod from the pedal.
4. Remove the two nuts and washers holding the cylinder to the firewall and remove the cylinder.

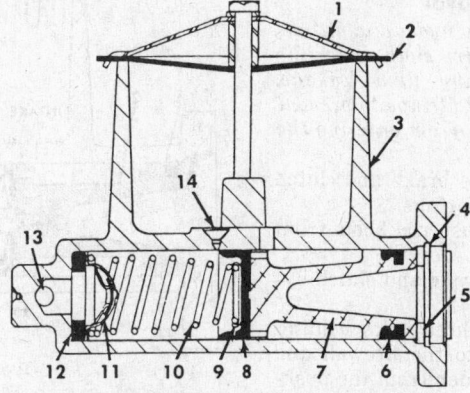

Typical single barrel master cylinder (© G.M.C.)

1 Cover Assembly	8 Primary Cup
2 Cover Gasket	9 Spring Retainer
3 Reservoir Body	10 Return Spring
4 Snap Ring	11 Check Valve
5 Stop Plate	12 Check Valve Seat
6 Secondary Cup	13 Outlet Port
7 Piston Assembly	14 By-Pass Port

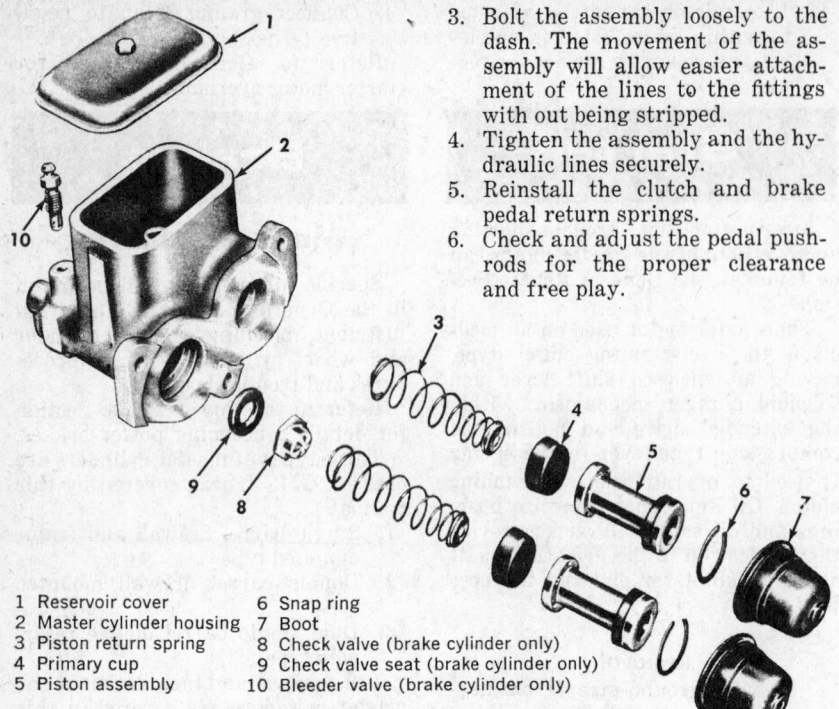

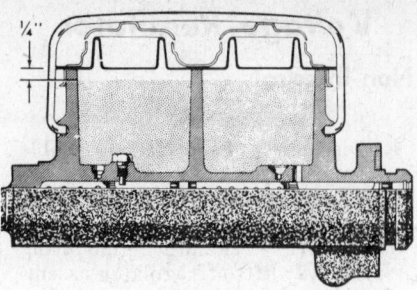

Correct Fluid Level (© G.M.C.)

3. Bolt the assembly loosely to the dash. The movement of the assembly will allow easier attachment of the lines to the fittings with out being stripped.
4. Tighten the assembly and the hydraulic lines securely.
5. Reinstall the clutch and brake pedal return springs.
6. Check and adjust the pedal pushrods for the proper clearance and free play.

1 Reservoir cover
2 Master cylinder housing
3 Piston return spring
4 Primary cup
5 Piston assembly
6 Snap ring
7 Boot
8 Check valve (brake cylinder only)
9 Check valve seat (brake cylinder only)
10 Bleeder valve (brake cylinder only)

Typical double barrel master cylinder (© G.M.C.)

Installation

1. Position the master cylinder at the dash and align the pushrod to the pedal.
2. Attach the nuts and washers, but do not tighten.
3. Install the fluid lines, and tighten the attaching nuts and the lines.
4. Check and adjust the pedal pushrod for 1/8 in. clearance between the rod and the piston of the master cylinder.
5. Fill the cylinder with brake fluid and bleed the system.

Twin or Double Barrel Type

Removal

NOTE: Wipe the hydraulic fittings clean and place a dry cloth under the lines to absorb any fluid leakage. Cover the lines and fittings to prevent any foreign matter from entering the system.

1. Disconnect the brake and clutch pedal return springs.
2. Remove the push rod boots from the cylinders.
3. Remove the brake and clutch hydraulic lines.
4. Remove the three bolts holding the cylinder to the firewall and slide the cylinder from the pushrods.

Installation

1. Place new pushrod rubber boots over the cylinder tubes.
2. Hold the cylinder next to the firewall and insert the pushrods, making sure the rods are centered in the pistons.

7. Fill the reservoirs and bleed both systems.

Dual Reservoir Type

Removal

1. Clean the area at the fittings and place a dry cloth under the lines to absorb any fluid leakage.
2. Disconnect both lines at the cylinder and cover them to prevent foreign matter from entering the system.
3. Unbolt the cylinder and remove,

while allowing the pushrod to hang loose.

Installation

1. Install a new boot on the push rod.
1. Install a new boot on the pushrod and install the boot on the cylinder.
3. Fasten the cylinder loosely to the attaching point. This freedom of the assembly will allow both hydraulic lines to be started easily.
4. Tighten the mounting bolts and the lines securely. Check for the proper pedal rod free play.
5. Fill the reservoir and bleed the system.

NOTE: To identify the reservoir for the conventional front and rear brakes, it is advisable to check the brake line routing from the master cylinder. On master cylinders with a large and small reservoir, the large reservoir is used for the disc brake system.

Wheel Cylinders

Four types of cylinders are used, identified by type of brake system used.

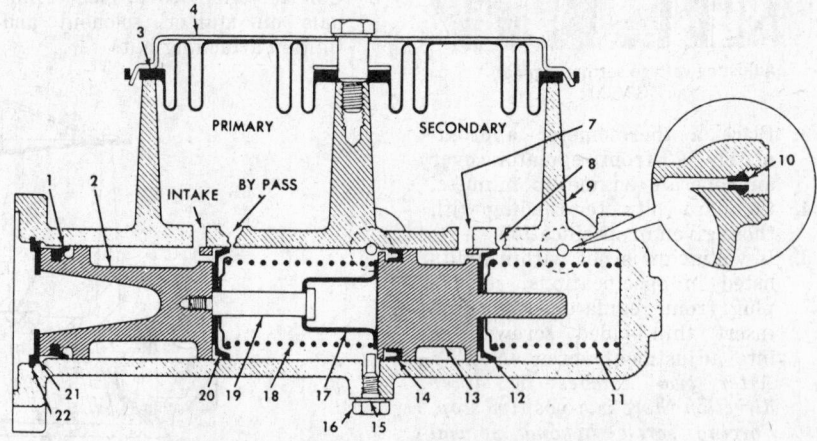

Typical split system master cylinder (© G.M.C.)

1 Primary piston seal cup
2 Primary piston
3 Cover seal
4 Reservoir cover
5 Gasket
6 Cover bolt
7 Intake port
8 By-pass port
9 Reservoir housing
10 Tube seat
11 Secondary piston return spring
12 Secondary piston pressure cup
13 Floating secondary piston
14 Secondary piston seal cup
15 Gasket
16 Stop bolt
17 Primary return spring retainer
18 Primary return spring
19 Primary piston stop pin
20 Primary piston pressure cup
21 Stop plate
22 Retainer ring

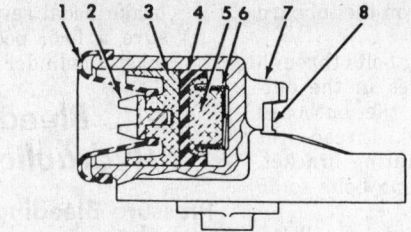

Type F and FA—front wheel cylinder

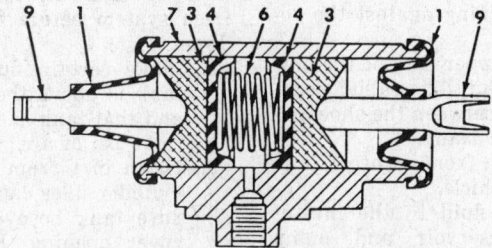

Type FR-3 and FR-3A—rear wheel cylinder (© G.M.C.)

1 Boot	4 Piston cup	7 Cylinder
2 Brake shoe guide	5 Cup filler	8 Brake shoe anchor slot
3 Piston	6 Piston spring	9 Push rod

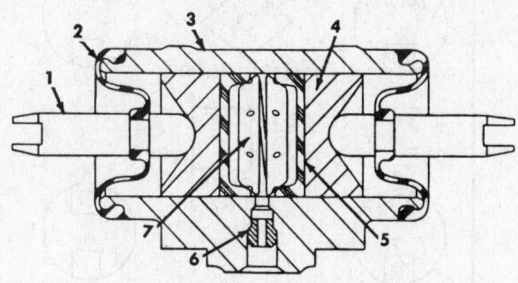

Twin action wheel cylinder (© G.M.C.)

1 Connecting rod	5 Cup
2 Seal	6 Connector insert
3 Wheel cylinder body	7 Spring assembly
4 Piston	

Duo servo—One double-end cylinder mounted at toe ends of shoes.

Twinplex—Two double-end cylinders mounted between shoes at toe and heel.

Wagner F (FA)—Two single end cylinders (single piston, single direction) mounted so as to be an anchor for one and powering other.

Wagner FR3 (FR3A)—Two double-end cylinders mounted between shoes.

Wheel Cylinder

Removal

1. Jack up axle and support. Remove wheel and drum. *NOTE: To remove drum it may be necessary to back off brake adjustment, also if rear drum, release hand brake cable if so equipped.* CAUTION: **To gain access to adjusting starwheel, a knockout lanced area is located in web of drum. After knocking out metal be sure to clean all metal particles from brake compartment. A new cover plug must be installed.**

2. Release shoe return springs and spread shoes to clear wheel cylin-

der links. Make sure any lubricant or brake fluid does not get on facings by covering same.

3. At front—Disconnect metal line from flexible hose and remove hose if accessible or remove hose later after cylinder is removed. At rear—Disconnect metal line from cylinder.

4. Remove shield over cylinder and connecting line between cylinders, if so equipped.

5. Remove cap screws and washers holding cylinder to backing plate. Remove cylinder being careful of any fluid spillage.

Installation

1. Clean mounting surface and reverse above procedures.

2. Bleed and readjust brakes. Check pedal before moving vehicle.

NOTE: Twinplex — Upper and lower cylinders are not interchangeable due to position of connector tube openings. Upper cylinder has threaded bleeder valve opening drilled at outer edge of bore.

Wagner F & FA—*Two wheel cylinders (upper and lower) are identical,*

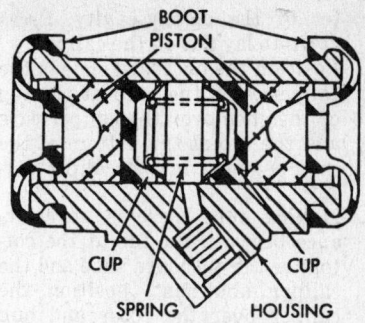

Duo-servo wheel cylinder (© G.M.C.)

however cylinders on right and left brakes have opposite castings.

Wagner FR3 & FR3A—*Upper and lower cylinders on both right and left brakes are interchangeable.*

Disc Brakes

Information on repair and overhaul of the disc brake components can be found in the General Repair Section.

Removal

1. Check the master cylinder fluid level and if full, siphon approximately ⅔ of the fluid from the reservoir and discard.

2. Raise the front of the vehicle and remove the front wheels.

3. Using a "C" clamp type tool, push the pistons back into the caliper bores.

4. Remove the two mounting bolts which attach the caliper to the support.

5. Lift the caliper assembly off the rotor and hub assembly.

6. Remove the inboard shoe, dislodge and remove the outboard shoe.

7. Position the caliper on the front suspension arm so that the brake hose does not support the weight of the caliper.

8. Remove the shoe support spring from the piston.

9. Remove the two sleeves from the inboard ears of the caliper.

10. Remove the four rubber bushings from the grooves of the caliper ears.

Installation

1. Lubricate the new sleeves, rubber bushings, the bushing grooves and the ends of the mounting bolts.

2. Install the new rubber bushings in the caliper ears.

3. Install the new sleeves in the inboard ears of the caliper. Position the sleeve so that the end towards the shoe and lining assembly is flush with the machine surface of the ear.

4. Install the shoe support spring and the inboard shoe in the cen-

ter of the piston cavity. Press down to lay flat on the caliper.

5. Position the outboard shoe in the caliper with the ears at the top of the shoe over the caliper ears and the tab at the bottom of the shoe engaged in the caliper cutout.

6. Making sure there is no clearance between the tab at the bottom of the outboard shoe and the caliper abutment, position the caliper over the rotor and hub assembly, lining up the hole in the caliper ears with the holes in the mounting brackets.

7. Insert the mounting bracket bolts through the sleeves in the inboard caliper ears and through the mounting bracket, making sure that the bolts pass under

the retaining ears on the inboard shoe.

8. Push the mounting bolts through to engage the holes in the outboard shoes and the outboard caliper ears, and thread the bolts into the mounting bracket. Torque the mounting bolts to 35 ft. lbs.

9. Pump the brake pedal to fill the piston cavity and to force the shoes and lining against the rotors.

10. Bend both upper ears of the outboard shoes until no radial clearance exists between the shoe and the caliper housing.

11. Install the front wheels and lower the vehicle.

12. Add brake fluid to the master cylinder reservoir and pump

brake pedal several times to assure a firm pedal. Recheck the master cylinder fluid level.

Bleeding Hydraulic Brakes

Pressure Bleeding

CAUTION: Stop engine and relieve vacuum or exhaust pressure from system before following procedures.

1. Make certain fluid in pressure tank is above the petcock outlet and that tank is charged with 40 to 50 psi of air.

2. Clean dirt from around master cylinder filler cap. Connect pressure tank hose to filler cap or cover opening. Bleed air from

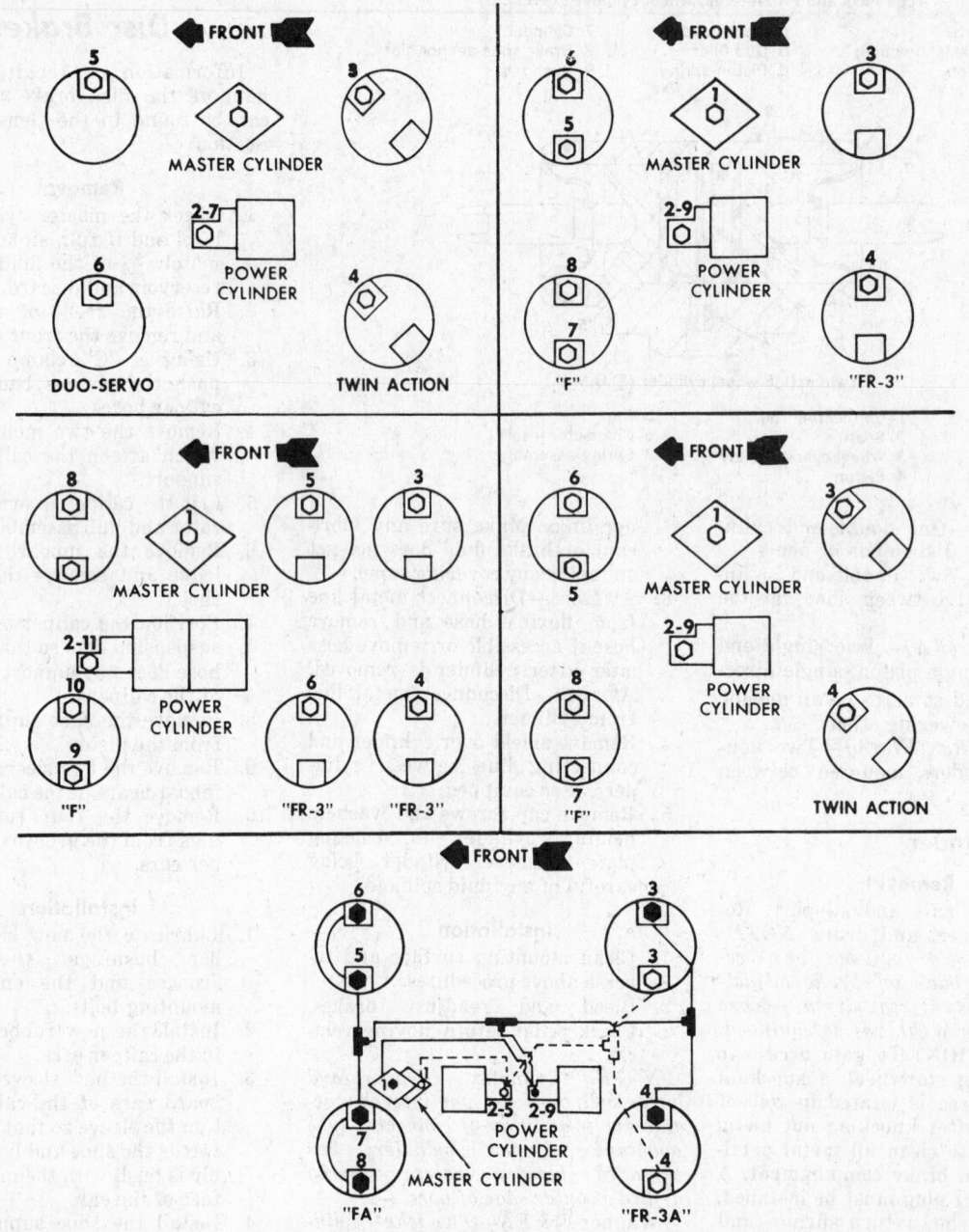

Brake bleeding sequence (© G.M.C.)

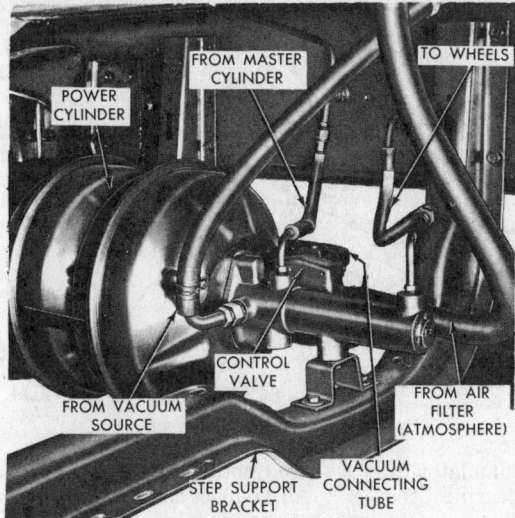

Power cylinder (conventional)
(© G.M.C.)

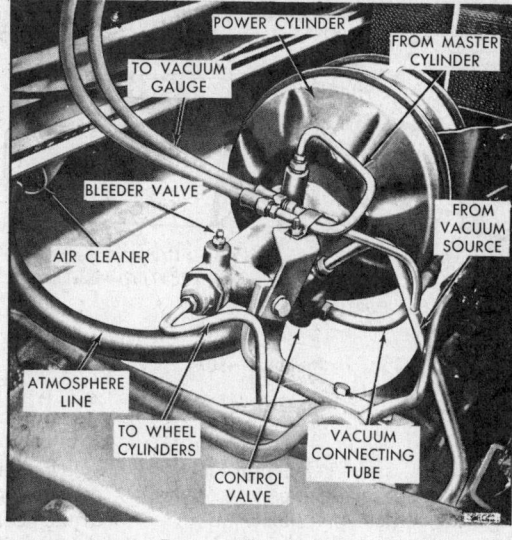

Power cylinder (tilt cab)
(© G.M.C.)

hose before tightening connection. Open valves at both ends.

3. Bleed slave cylinder and control valve first. (when used). Slip end of bleeder hose over bleeder valve No. 1 and place other end in glass jar containing enough hydraulic fluid to cover end of hose. Open bleeder valve with wrench and observe flow of fluid. On Models 4000 and 5000, start engine and make at least two power brake applications with bleeder valve open to force air out of slave cylinder. Close bleeder valve as soon as bubbles stop and fluid flows in solid stream. Stop engine and relieve vacuum from system.

4. Bleed valve No. 2 (on power cylinder control valve), then bleed wheel cylinders in sequence. Repeat bleeding operations at power cylinder. On Models 4000 and 5000, repeat power brake applications with engine running as in Step (3).

5. If, after bleeding, the pedal "feel" is not satisfactory, inspect residual check valve in the master cylinder and the check valve in the power cylinder piston. Improper operation of either or both of these valves will result in same pedal "feel" as air in the system, since malfunction permits recirculation of fluid through compensating line and back to master cylinder reservoir. Refer to applicable procedures for repair.

Manual Bleeding

Manual bleeding follows the same procedures as pressure bleeding, except that brake fluid is forced through lines by pumping the brake pedal instead of by air pressure. Fluid in master cylinder must be replenished after bleeding at each

valve. Brake pedal should be pumped up and down slowly, and should be on downstroke as valve is closed.

Split System ("S" Models)

The system consists of a dash mounted master cylinder and two power cylinders mounted on the frame. The main system consists of the front wheel brakes and one cylinder on each rear wheel brake. The secondary system consists of one cylinder on each secondary wheel brake. Each system must be bled separately.

Vacuum
Power Brakes

Power Cylinder

Removal

1. For easier accessibility, it is recommended that cab step be removed on conventional models and that cab be tilted forward on tilt cab models.

2. Clean away road dirt and grease to prevent contamination of vacuum or hydraulic systems.

3. Have a container available to catch hydraulic brake fluid which will flow from system. *Do not re-use this fluid.*

4. Disconnect all hydraulic, vacuum and atmospheric lines and hoses from power cylinder. Plug vacuum line.

5. On light duty models, disconnect the push rod from the brake pedal. Remove bolts and nuts fastening cylinder to frame and support brackets.

6. Remove power cylinder.

Installation

1. Place power cylinder in position and fasten with nuts and bolts to frame and support brackets. Connect the push rod to the brake pedal.

2. Connect all hydraulic, vacuum, and atmospheric lines and hoses to power cylinder.

3. Bleed master cylinder and vacuum power cylinder as directed under "Bleeding Brakes." If ONLY the power cylinder has been removed, it should not be necessary to bleed the wheel cylinders IF the master cylinder and power cylinders are bled first AND lines to wheel cylinder have not been disturbed.

Hydro-Boost
Power Brakes

Specific information will be found in the Power Brake portion of the General Repair Section on adjustments, overhauling and troubleshooting.

The Hydro-boost system was designed to eliminate the need for the remote frame mounted boosters. It utilizes power steering fluid in place of engine vacuum to provide a power assist that operates a dual master cylinder brake system. A spring accumulator is used in conjunction with the hydraulic brake booster. The accumulator is a sealed hydraulic cylinder with a port at each end. On 1974 models the accumulator and booster are mounted separately. In 1975 all "C" model vehicles equipped with the hydro-boost incorporate an accumulator which is integral with the booster.

CAUTION: The accumulater contains a spring compressed under high pressure. Any attempt to disassemble or cut could cause personal injury.

Hydraulic Brake Booster
Removal and Installation

Motor Home Chassis

1. Make sure all pressure is discharged from the accumulator by

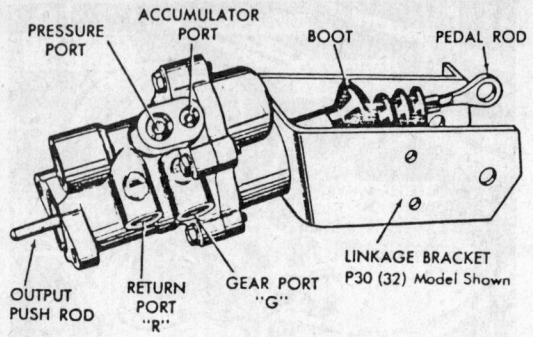

Hydro-boost—1974.

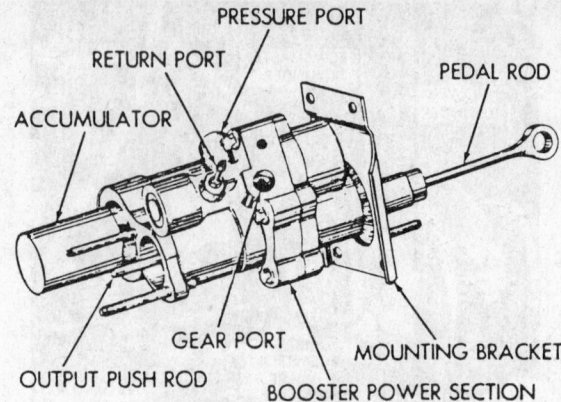

Hydro-Boost—1975—Cabs, Suburban.

depressing and releasing the brake pedal several times.

2. Raise the vehicle on a hoist.
3. Clean all the dirt from the booster at the hydraulic line connections and master cylinder.
4. Remove the nuts and lockwashers that secure the master cylinder to the booster and support bracket. Support the master cylinder leaving the hydraulic lines attached to the master cylinder.
5. Disconnect and plug the hydraulic lines from the booster ports.
6. Remove the cotter pin, nut, bolt and washers that secure the operating lever to the vertical brake rod.
7. Remove the six nuts, lockwashers and bolts that secure the booster linkage bracket to the front and rear support brackets, then slide the booster off the rear support studs and remove the booster from the vehicle.
8. Remove the cotter pin, nut, washer and bolt that secures the operating lever to the pedal rod.
9. Remove the brake pedal rod lever nut and bolt and then remove the lever, sleeve and bushings.
10. To install reverse the removal procedures. Bleed the booster-power steering hydraulic system and check the brake pedal and stoplamp switch adjustment.

Conventional Cab & Suburban

1. Make sure all pressure is dis-

charged from the accumulator by depressing and releasing the brake pedal several times.

2. Remove the nuts and lockwashers that secure the master cylinder to the booster and support bracket. Support the master cylinder leaving the hydraulic lines attached to the master cylinder.
3. Remove the booster pedal push

rod cotter pin and washer, and disconnect the push rod from the brake pedal.

4. Remove the booster support bracket.
5. Remove the booster bracket to dash panel or support bracket nuts and remove the booster assembly.
6. To install reverse the removal

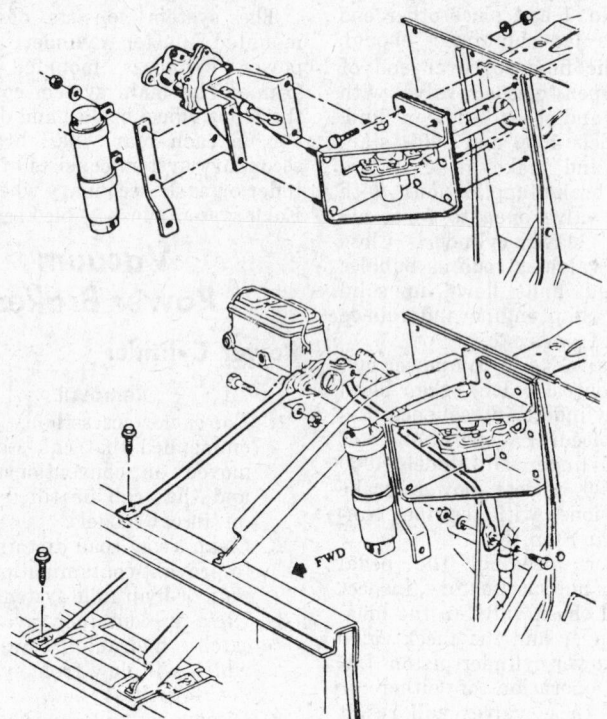

Booster Installation—forward control chassis exc. Light Duty Vans.

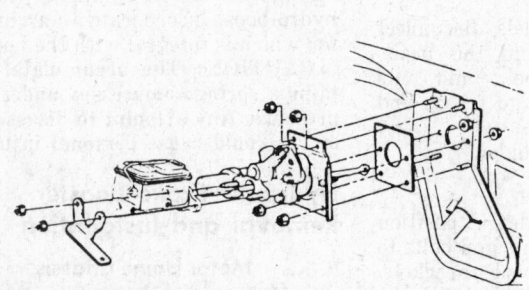

Booster Installation—Cab and Suburban.

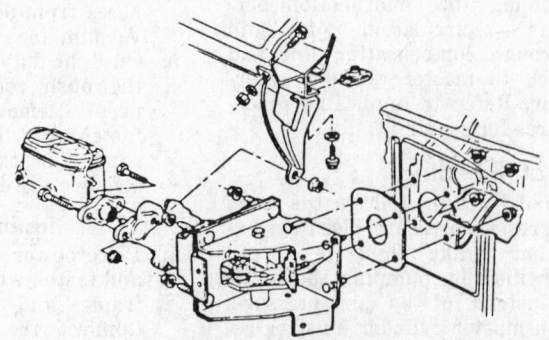

Booster Installation—Vans

procedure. Bleed the booster-power steering hydraulic system and check brake pedal and stop-lamp switch adjustment.

Forward Control Chassis (Vans)

1. Make sure all pressure is discharged from the accumulator by depressing and releasing the brake pedal several times.
2. Clean all the dirt from the booster at the hydraulic line connections and master cylinder.
3. Remove the nuts and lockwashers that secure the master cylinder to the booster and the support bracket. Support the master cylinder leaving the hydraulic lines attached to the master cylinder.
4. Remove the booster pedal push rod cotter pin and washer and disconnect the push rod from the booster bracket pivot lever.
5. Remove the booster upper braces.
6. Remove the booster bracket nuts and remove the booster.
7. To install reverse the removal procedure. Bleed the booster-power steering hydraulic system and check the brake pedal and stop lamp switch adjustment.

Bleeding Hydro-Boost System

1. Fill the power steering pump to the proper level.
2. Start the engine for approximately two seconds. Check the fluid and add if necessary.
3. Repeat step 2 until the fluid level remains constant.
4. Raise the front end of the vehicle so that the tires are clear of the ground.
5. Start the engine and run approximately 1500 rpm. Depress and release the brake pedal several times then turn the steering wheel right and left, lightly contacting the wheel stops.
6. Turn off the engine and check the fluid level in the reservoir and add fluid if necessary.
7. Lower the vehicle, start the engine and run at approximately 1500 rpm. Depress and release the brake pedal several times and turn the steering wheel to full right and left.

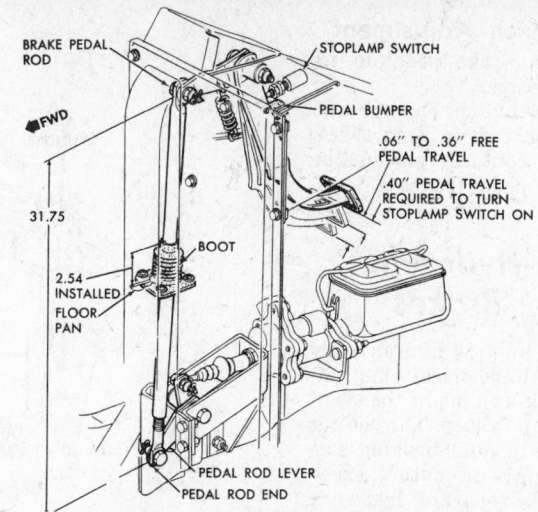

Brake pedal and stop lamp adjustments—Motor Home Chassis.

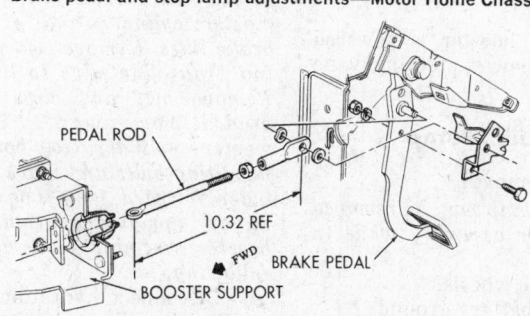

Brake pedal adjustment—Vans

8. Turn the engine off and check the fluid level in the reservoir. Add fluid if necessary.

NOTE: If the fluid is extremely foamy, or there is an erratic pedal feel allow vehicle to stand a few moments with the engine off and repeat the above procedure.

Brake Pedal Adjustment

Motor Home Chassis

1. With brake pedal pull back spring installed, brake pedal hard into bumper, brake master cylinder assembly and brake pedal rod lever at full return, install the preassembled brake pedal rod assembly (rod end at boot) and adjust to 31.75 in.
2. Turn the brake pedal rod end and adjust the free pedal travel to .06 in. to .36 in.

3. Fasten the boot to the floor pan assembly and compress the boot to 2.54 in. installed height.

Forward Control Chassis—(Vans)

1. Adjust the pedal rod to 9.90 in. and install in the vehicle.

NOTE: The brake pedal push rod is not adjustable on cab and Suburban models.

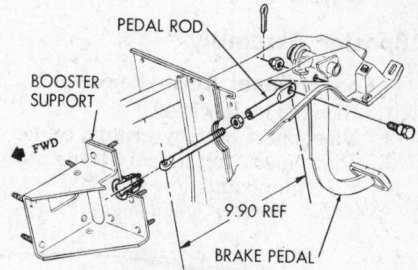

Brake pedal adjustment
Forward Control Chassis.

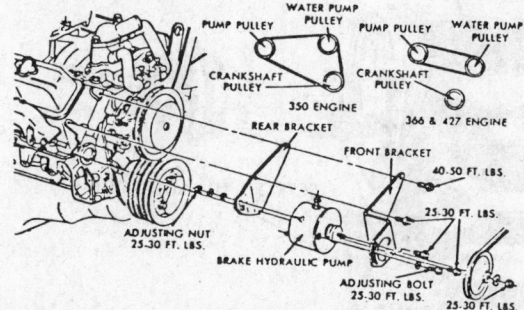

Brake Hydraulic pump installation—(V-8)

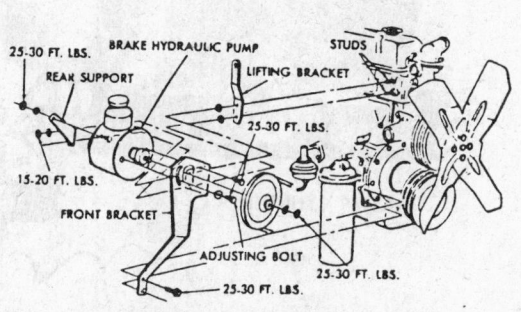

Brake Hydraulic pump installation—(L-6)

Stoplamp Switch Adjustment

1. Release the brake pedal to its normal position.
2. Loosen the switch locknut and rotate the switch in its bracket. Electrical contact should be made when pedal travel is 3/8 in-5/8 in.

Electro-Hydraulic Power Brakes

The booster unit is hydraulically operated with an electrical pump attached as a back-up unit in the event of primary pump failure. The booster system features a dual braking system, increased pressure output, excellent pedal feel, separate hydraulic pump, and an electrical powered back-up system.

The hydraulic booster is powered by a standard power steering vane type pump.

Electro-Hydraulic Pump

Removal

NOTE: The pump may be removed from the booster assembly while in the vehicle.

1. Block vehicle wheels.
2. Disconnect battery ground.
3. Disconnect E.H. pump lead.
4. Position a container to catch fluid and remove bracket-to-booster 9/16 in. hex-head bolt.
5. Remove two 9/16 in. hex-head mounting bolts. Remove pump and two O-rings.

Installation

1. To install reverse the removal procedure. Use new O-rings and torque the two 3/8-16 x 1 1/8 in mounting bolts to 16-30 ft. lbs.

Booster Assembly

Conventional Cab—Removal

1. Block vehicle wheels.
2. Disconnect battery ground cable.
3. Disconnect electrical leads to E.H. pump and flow switch.

NOTE: Booster head may be re-moved from vehicle without removing master cylinder and disconnecting brake lines. Remove two nuts retaining brake line clips to line support. Remove bolt and clips in booster head. Remove four 9/16 in. hex head master cylinder to booster head mounting bolts and move master cylinder forward from booster. Secure in an upright position. Complete booster assembly removal procedure following:

4. Disconnect hydraulic lines from booster. Use a suitable container to catch fluid. DO NOT reuse fluid.
5. Remove cotter key from push rod pin. Remove nut and bolt from pedal push rod eye. Thread push rod and nut from booster push rod.
6. Remove line and hose supports from booster head.

NOTE: Use care when proceeding to next two steps. This is a heavy unit, approximately 50 pounds including master cylinder, and should be handled as such. After removing from vehicle, use care in handling so that flow switch or E.H. pump is not damaged and bail or cover on master cylinder is not damaged.

7. Remove two 3/8-24 thread hex-head nuts and washers from inside cab at dash panel.

8. Remove upper two mounting bolts and washers from booster head at dash. Remove booster assembly.
9. Remove four 9/16 in. hex-head bolts which attach brake master cylinder to booster head (Fig. 8). Remove master cylinder, gasket, and brake line support.

Installation

1. To install reverse the removal procedure. Torque the four 3/8-16 x 1 3/8 in. hex-head bolts attaching the master cylinder to the booster head to 16-30 ft lbs. The two hex-nuts on the 3/8-24 in. studs are torqued to 25-30 ft lbs. The two mounting bolts are torqued to 25-30 ft lbs.

Tilt Cab—Removal and Installation

1. Block vehicle wheels.
2. Disconnect battery ground cable.
3. Disconnect electrical lead to E.H. pump.
4. Disconnect hydraulic lines from booster and master cylinder. Use a suitable container to catch fluid. Do not reuse fluid.
5. Loosen nuts on each end of push rod extension. Turn extension until free of booster push rod. Bellcrank ball joint may have to

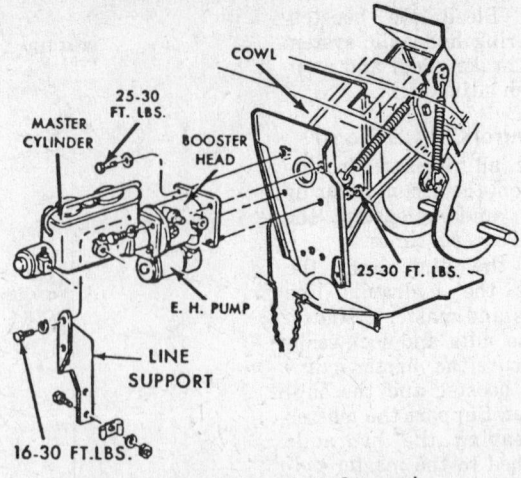

Booster assembly mounting—Conv. cab.

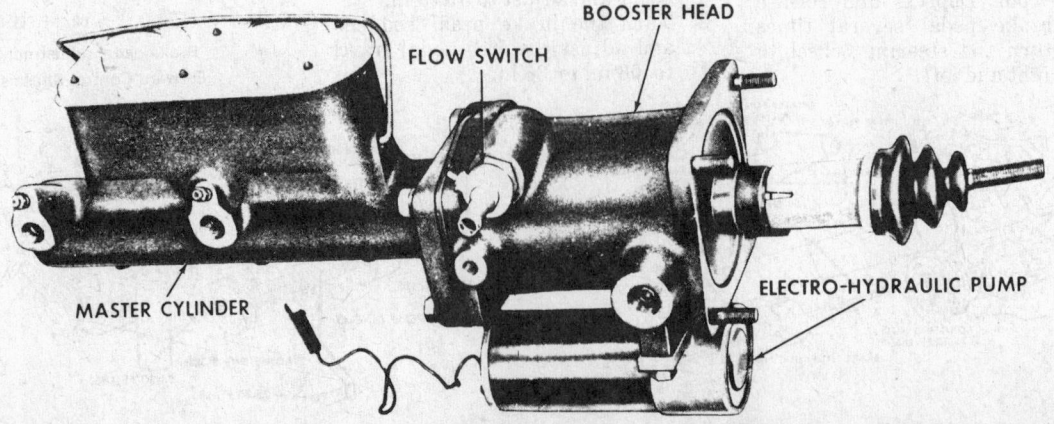

Electro-hydraulic brake booster and master cylinder (© G.M.C.)

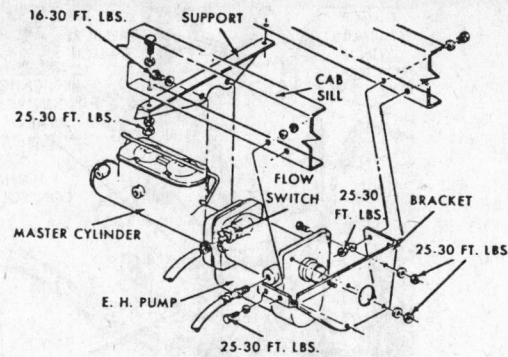

16·30 FT. LBS. SUPPORT
CAB SILL
25·30 FT. LBS.
FLOW SWITCH 25·30 FT. LBS. BRACKET
MASTER CYLINDER
25·30 FT. LBS.
E. H. PUMP
25·30 FT. LBS.

Booster assembly mounting—Tilt cab.

be removed to facilitate turning extension.

6. Remove two 9/16 in. hex-head bolts and nuts retaining support to cab sill.
7. Remove two 9/16 in. hex-head bolts and nuts retaining bracket to sill.
8. Remove booster assembly. Move push rod end forward and down while revolving unit to clear E.H. pump, support, and bracket.
9. Remove four 9/16 in. hex-head bolts which attach brake master cylinder to booster head and remove master cylinder and gasket.

Flow Switch

Removal

1. Position container to catch fluid from disconnected hose at flow switch.
2. Loosen outer hose clamp at switch and remove hydraulic line from hose. Drain fluid into previously placed container. DO NOT reuse fluid.
3. Loosen inner hose clamp at switch and remove hose.
4. Disconnect electrical lead to switch.
5. Remove switch using a one-inch thin blade wrench.

Installation

1. Use a new O-ring seal and torque to 20-30 ft lbs.

Brake Hydraulic Pump

V8 Engine—Removal

1. Disconnect E.H. pump electrical lead.
2. Disconnect hydraulic lines at pump. Cap or otherwise restrict flow of fluid from lines.
3. Remove hose to reservoir (if used).
4. Remove pulley.
5. Remove adjusting nut and washers at rear of pump at bracket slot.
6. Remove adjusting bolt and washers from front of pump at bracket slot.

7. Remove front mounting bolt and washers and remove pump.

Installation

1. To install reverse the removal procedure. Use a belt tension gauge and adjust the belt to 90-100 lbs (new belt).

Six Cylinder Engine—Removal

1. Disconnect E.H. pump electrical lead.
2. Disconnect hydraulic lines at pump. Cap or otherwise restrict flow of fluid from lines.
3. Remove pulley.
4. Remove adjusting bolt and washer from slot at front bracket.
5. Remove nut and washer from support at rear of pump.
6. Remove mounting bolt at front bracket and remove pump.

Installation

1. To install reverse the removal procedure. Use a belt tension gauge and adjust the belt to 90-100 lbs (new belt).

NOTE: See the "Power Brake" section of the General Repair section.

Air Brakes

Information on repair and overhaul of air brake components can be found in the General Repair Section.

Application Valve

Conventional—Removal

1. Relieve air pressure from system.
2. Disconnect all air lines from valve. On "L" models, disconnect hose from exhaust tube.
3. Remove valve as follows:
   a. All except SPA5000: Remove three attaching bolts and valve assembly.
   b. SPA5000: Remove bolts attaching valve and treadle mounting plate to toeboard and then remove the complete assembly from the truck. Remove three bolts attaching valve to mounting plate and remove valve.

Installation

a. Except SPA5000 models: Position valve assembly on mounting bracket with pushrod inserted into piston cup. On dash mounted units, the exhaust opening (with filter screen) must be down; on "L" models, exhaust tube must be toward right side of cab. Attach valve to mounting bracket with three bolts. Check pushrod or stop screw adjustment and correct if necessary.
b. SPA5000: Models: Connect valve assembly to mounting plate, with pushrod inserted into plate. Attach with three bolts and lock washers. Check for free travel between end of pushrod and piston cup. If necessary, remove treadle pin, then remove treadle stop bumper and add shims under bumper as necessary to remove clearance. Install treadle pin and secure with cotter pin. Install the complete assembly on toeboard and attach with three bolts, lock washers, and nuts.

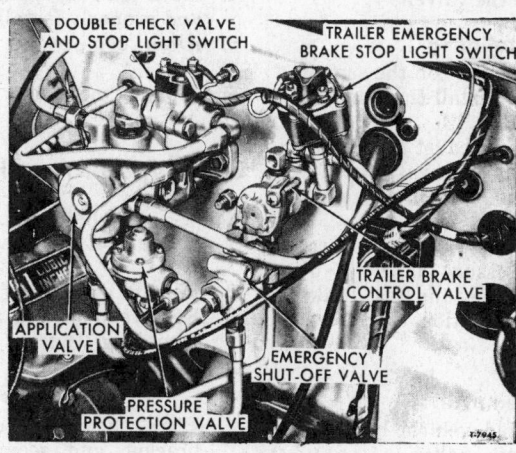

DOUBLE CHECK VALVE AND STOP LIGHT SWITCH
TRAILER EMERGENCY BRAKE STOP LIGHT SWITCH
TRAILER BRAKE CONTROL VALVE
APPLICATION VALVE
EMERGENCY SHUT-OFF VALVE
PRESSURE PROTECTION VALVE

Application valve—model 6500 (typical)
© G.M.C.

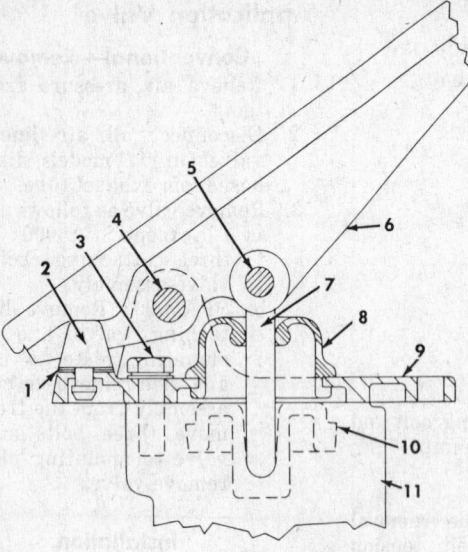

Brake treadle installation (SPA5000)
(© G.M.C.)

1 Shims	6 Treadle
2 Treadle stop bumper	7 Push rod
3 Mounting plate	8 Push rod boot
to application	9 Mountain plate
valve bolt	10 Application valve
4 Treadle pin	piston cup
5 Push rod pin	11 Application vlave

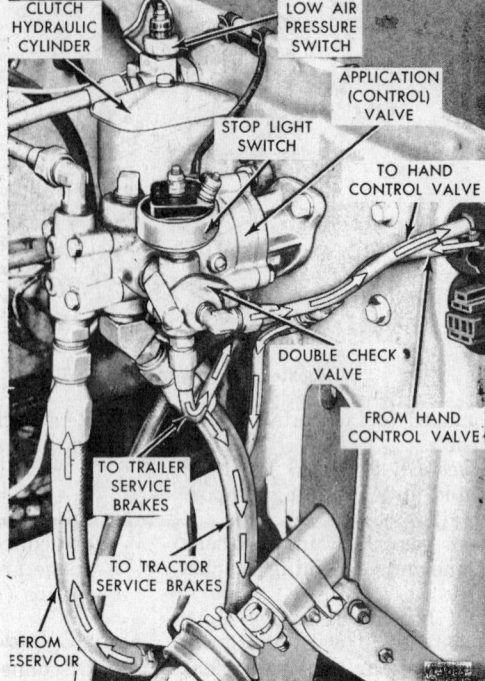

Application valve—models 1000-4500 (typical)
(© G.M.C.)

Connect air lines to valve. When installing connector fittings in valve, use sealing compound on threads. *Keep sealing compound off first two threads of fittings.* Sealing compound inside the valve could foul valve seats and block compensating port. On "L" models, connect hose to exhaust tube. Build up air pressure in system and test application valve operation.

Tilt Cab 5000-6500 Models—
Removal

1. Block the vehicle wheels, drain air pressure from the brake system, and disconnect the air lines from the application valve.
2. Disconnect the exhaust hose from the exhaust port of the valve.
3. Remove the bolts which attach the valve to the support bracket, and remove the valve.

Installation

1. Position the valve on the support bracket and install the attaching bolts.
2. Connect the exhaust hose to the exhaust port of the valve.
3. Connect the air lines to the valve.
4. Operate the engine until operating air pressure is built up, check for leaks, and test brakes for proper operation.

Compressor
Removal

1. Block or hold vehicle by means other than air brakes.
2. Drain air from system, usually at reservoirs.

3. If water cooled, drain cooling system.
4. Disconnect ALL lines (air, water and oil).
5. Loosen belt adjusting stud and remove drive belt.
6. Remove mounting bolts, remove compressor.

Installation

1. Run engine briefly to clear and check oil supply lines. Clean oil return lines and passages. Check coolant supply and lines (if used).
2. Clean mounting surface and replace gasket, be sure oil holes in gasket are aligned.
3. Install compressor with mounting cap screws loose, compressor will be movable to allow fittings on lines to be started.
4. Make sure air cleaner is cleaned and properly installed.
5. Align compressor, check drive belt.
6. Tighten mounting bolts and adjust.

Cam Type Brake Chamber
Removal

1. Disconnect air hose from chamber.
2. Remove clevis pin from yoke.
3. Remove nuts and washers from mounting studs.
4. Remove chamber.

Installation

1. Place chamber on mounting bracket and secure with stud nuts and lock washers.
2. Connect air hose.

3. Install slack adjuster yoke clevis pin, after adjusting for minimum travel. (Angle made by push rod and slack adjuster lever should not be less than 90°, brakes applied). Lock yoke with locking nut. Push rod travel should be as short as possible without brakes dragging.
4. Check for leaks, and possible brake shoe adjustment.

Wedge Type Brake Chamber
Removal

1. Block or hold vehicle by means other than air brakes.
2. Disconnect air hose from chamber.
3. Remove lock washer tangs from notches in spanner nut and spider housing.
4. Loosen spanner nut and unscrew air chamber from housing.

Installation

1. Screw air chamber in spider housing until it "bottoms," then back off (no more than one turn) until chamber air port aligns with air hose. The plastic guide will assure proper position of wedge. Lock brake chamber in position with spanner nut and lock washer.
2. Start engine, charge air system, check for leaks. Pump brake pedal to allow automatic adjusters to adjust brakes.

NOTE: When brakes are equipped with "fail-safe" brake chambers, cage power spring before starting any disassembly or removal of wheels or drums to avoid possible injury. When a vehicle is disabled, due to low or lost air pressure, cage power spring before attempting to move the vehicle. Cage the power spring by rotating the release bolt approximately 18 to 21 turns clockwise. Caging and uncaging can be made easier by applying air pressure, 65 lbs needed (if possible). This takes spring load off release bolt.

CAUTION: Before removing or caging brake chamber, block wheels since parking brake will not be applied.

Slack Adjusters

Removal

1. Remove clevis pin at lever.
2. Remove lock ring and washer on splined camshaft (some front slack adjusters are held on by a retaining screw).
3. Slide adjuster off splined shaft.

Installation

1. Make sure spacer washers are in place (if used).
2. Slide slack adjuster on splined camshaft, lock in place with snap ring and washer. If held on by retaining screw, allow .010 in. end play. Stake edge of screw to lock.
3. Connect yoke clevis pin. Angle made by push rod and slack adjuster lever should NOT be less than 90° brakes applied. Push rod travel should be as short as possible without brakes dragging.
4. Lubricate adjuster.

Parking Brakes

Adjustment

NOTE: Parking brake adjustment can be accomplished only when service brakes are in adjustment.

Foot Pedal Type

1. Jack up rear wheels.
2. Apply parking brake 1 notch from fully released position.
3. Loosen equalizer check nut and tighten the adjusting nut until moderate drag is felt when rear wheels are rotated.
4. Tighten check nut securely.

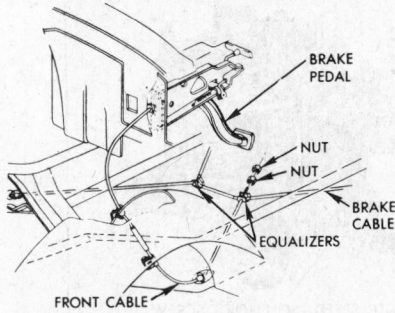

Parking brake system ("C"-"K" models)
(© G.M.C.)

5. Fully release parking brake and rotate rear wheels. No drag should be present.

Orscheln Lever Type

1. Turn adjusting knob on parking brake lever counter-clockwise to stop.
2. Apply parking brake.
3. Jack up rear wheels.
4. Loosen lock nut at intermediate cable equalizer and adjust front nut to give light drag at rear wheels.
5. Re-adjust parking brake lever knob to give definite snap-over-center feel.
6. Fully release parking brake and rotate rear wheels. No drag should be present.

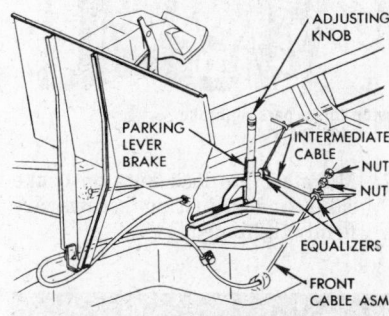

Parking brake system (P2500-3500 models)
(© G.M.C.)

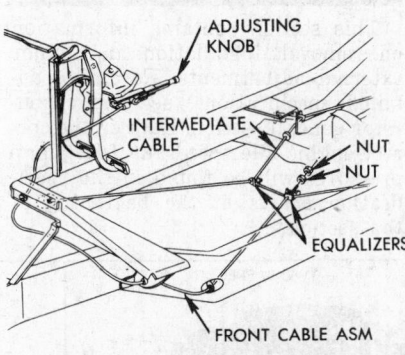

Parking brake system (P1500 models)
(© G.M.C.)

Driveshaft Type (Drum On)

1. Jack up at least one rear wheel. Block wheels and release hand brake.
2. Remove cotter pin and clevis pin connecting pull rod to relay lever. *NOTE: It may be necessary to knock out lanced area in brake drum with punch and hammer to gain entry to adjusting screw through brake drum.*
3. Rotate brake drum to bring an access hole into line with adjusting screw at bottom of shoes.
4. Expand shoes by rotating adjusting screw with screw driver. Move outer end of screw driver away from drive shaft. Continue adjustment until shoes are tight against drum and drum cannot be rotated by hand. Back off ad-

justment ten notches and check drum for free rotation.

5. Place parking brake lever in fully released position. Take up slack in brake linkage by pulling back on cable just enough to overcome spring tension. Adjust clevis of pull rod or front cable to line up with hole in relay levers.
6. Insert clevis pin and cotter pin, tighten clevis locknut.
7. Install new hole cover in drum to prevent dirt contamination.
8. Lower rear wheels. Remove jack and wheel blocks.

Driveshaft Type (Drum Off)

1. With parking brake drum off, check diameter of drum clearance surface.
2. Turn tool to the opposite side and fit over brake shoes by turning star wheel until gauge just slides over linings.
3. Rotate gauge around brake shoe lining surface to insure proper clearance.
4. Install driveshift flange at mainshaft.
5. Lower rear wheels. Remove jack and wheel blocks.

Stopmaster "Fail-Safe" Type

Stopmaster "Fail-Safe" parking brakes are used as standard equipment on some models and as optional equipment on others. When Stopmaster service brakes with "Fail-Safe" feature are used, no other parking brake system is required.

Anchorlok Type

The "Anchorlok" brake chamber is used as optional equipment on some models. This chamber incorporates a spring applied, air pressure released parking/emergency brake.

Tru-Stop (Disc) Type

This type brake is used only on "W" models when equipped with 3-speed or 4-speed auxiliary transmission. The ventilated brake disc is mounted between propeller shaft flange and auxiliary transmission shaft companion flange. Brake shoes are mounted in opposed positions with brake disc between, as shown in illustration. When brake is applied, the shoes are forced against disc. Brake should be adjusted before a full application requires parking brake lever to be pulled to travel limit.

1. Disconnect brake cable clevis (5) from brake lever (6).
2. Tighten adjusting nut (16) until spring (14) exerts enough pressure to bring lever (6) against front lever arm (11).
3. Insert a 1/32" shim between rear shoe lining and brake disc.
4. Tighten adjusting nut (16) until

1 Brake support
 bracket
2 Parallel adjusting
 screws
3 Front lever arm pin
4 Pin retaining screw
5 Brake cable clevis
6 Brake lever
7 Brake shoe pin
 retainer
8 Brake shoe pin
9 Front brake shoe
10 Rear brake shoe
11 Front lever arm
12 Brake disc
13 Tension spring
14 Spring
15 Rear lever arm
16 Adjusting nut
17 Tie Rod

Tru-stop (disc) parking brake
(© G.M.C.)

Carburetors

Various carburetors are designed to meet requirements of engine, transmission and vehicle; therefore carburetors may look alike but are not always interchangeable. All carburetors have conventional float, idle, low speed, power or high speed and accelerating circuits with either manual or automatic chokes.

NOTE: Some symptoms indicate carburetor trouble but in reality are ignition related. Before any extensive repairs on carburetor, check first—heat riser, intake manifold and ignition.

Removal

1. Remove air cleaner.
2. Disconnect fuel, vacuum, spark control and governor lines.
3. Disconnect choke and hand throttle controls.
4. Disconnect throttle and automatic transmission linkage at carburetor. Remove pull back spring.
5. Remove mounting nuts and washers.
6. Lift carburetor off manifold and drain. Discard gasket.

Installation

1. Clean carburetor mounting surface.
2. Install new gasket on manifold. Be sure vacuum port and gasket slots are aligned.
3. Place carburetor on manifold, reconnect finger-tight all lines before (evenly) tightening carburetor mounting nuts.

front shoe lining is firmly against disc, yet allowing removal of shim.

5. Make certain tension spring (13) is in place. Turn adjusting screws (2) so that both linings are parallel with disc. This provides 16¼" clearance between front and rear shoe linings and brake disc at all points.
6. Make certain parking brake lever is in fully released position. Adjust clevis (5) on brake cable to permit installation of clevis pin through clevis and brake lever (6) without changing position of lever. Install clevis pin and cotter pin.

7. Make certain lock nuts on brake cable and adjusting screws are firmly tightened.

Fuel System

This section contains information on removal, installation and minor external adjustments. For more detailed specifications, see the carburetor general section. Data on the correct engine idle speed and fuel pump pressures will be found in the specification charts at the beginning of this section.

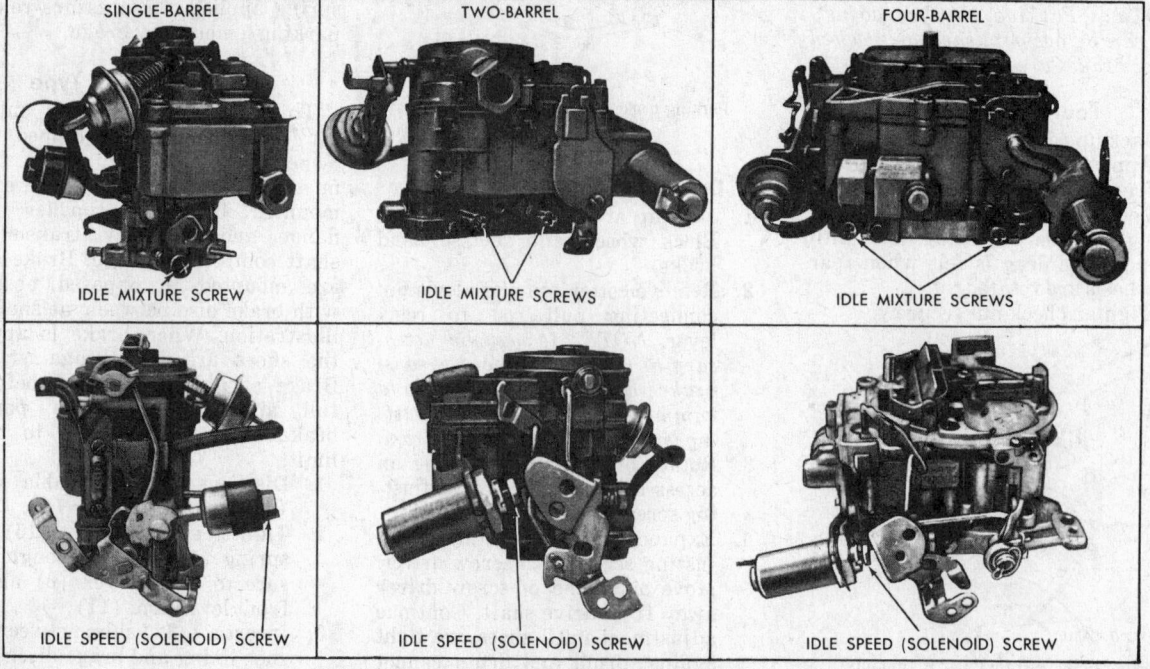

Typical idle speed and mixture screw locations (© G.M.C.)

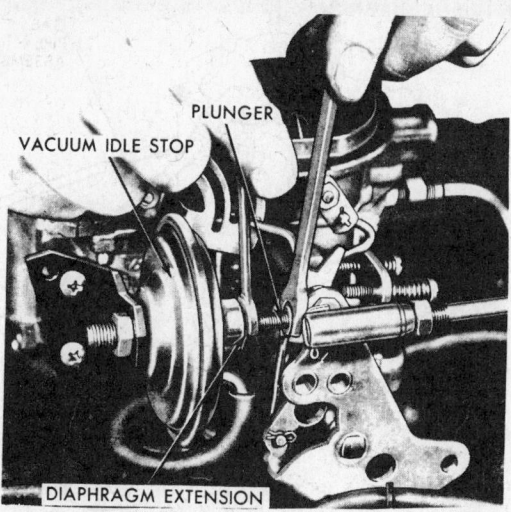

Adjusting Vacuum Idle Stop (© G.M.C.)

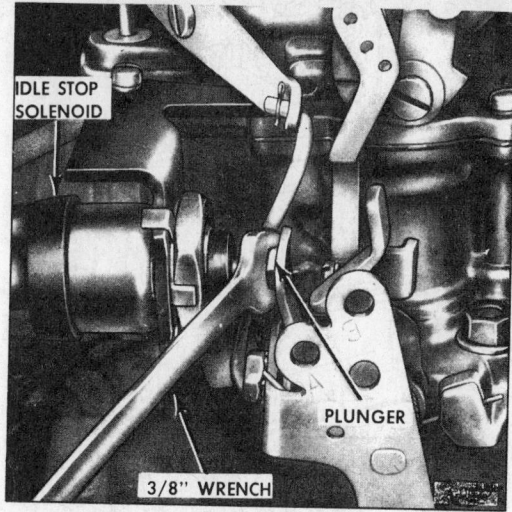

Adjusting idle stop solenoid—Bendix Stromberg Carb.

4. Tighten all lines.
5. Reconnect choke and hand throttle controls, replace throttle and automatic transmission linkage reconnect all vacuum and electrical lines. Connect pull back spring. Check choke and throttle operation.
6. Install air cleaner, check element.
7. Start engine and warm to operating temperature. During warm-up time, torque intake manifold.
8. Adjust idle mixture and idle speed screws, be sure choke is fully open.

CAUTION: Do not force idle mixture screw against seat, this will damage needle. On transmission controlled spark engines (TCS), fast idle adjustment must be set with electrical leads disconnected at solenoid and transmission in neutral.

NOTE: If possible (safely), fill carburetor bowl before installing. This will save time and battery drain as well as reducing possible backfiring.

CAUTION: Dirt is greatest troublemaker for carburetors, check all filters. Use starter briefly to clear fuel lines (before reconnecting) of any metal flakes that are always present

when fuel lines (metal) are disturbed.

NOTE: For Idle Speed and Mixture adjustments see the underhood tune-up specifications sticker. For all other adjustments, refer to the appropriate carburetor in the General Repair Section.

Fuel Pump

Mechanical Fuel Pump

Removal

1. Disconnect all inlet and outlet pipes from fuel pump.
2. Remove fuel pump mounting bolts.
3. Remove fuel pump and gasket. *NOTE: On V8 engines, remove fuel pump adapter and gasket if pushrod is to be removed.*
4. Transfer fittings if new pump is to be installed.

Installation

1. On V8 engines, install fuel pump pushrod and adapter.
2. Install fuel pump, using new gasket and sealer on mounting bolt threads.
3. Connect fuel lines to pump.
4. Start engine and check for leaks.

NOTE: On V8 engines use mechanical fingers or heavy grease to hold push rod up while installing pump.

Electric Fuel Pump

Removal

1. Disconnect battery ground cable.
2. Disconnect pump wiring harness from connector.
3. Disconnect fuel outlet fitting from hose.
4. Remove cap screws and washers.
5. Rotate pump 90° counterclockwise and lift out.

Installation

1. Insert pump outlet line into tank and connect to fitting.
2. Carefully install fuel pump and cap screws with washers. Tighten to specifications.
3. Connect pump harness to connector.
4. Connect battery ground cable.

Governor

All models covered by this manual utilize single-throat, dual-throat or hydraulic governors. Governors are adjusted for correct maximum speed and sealed at the factory.

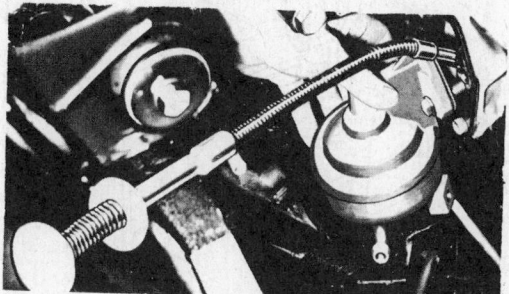

Installing V8 engine fuel pump
(© G.M.C.)

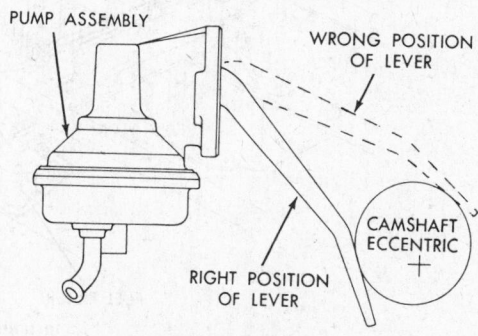

Installing V6 engine fuel pump

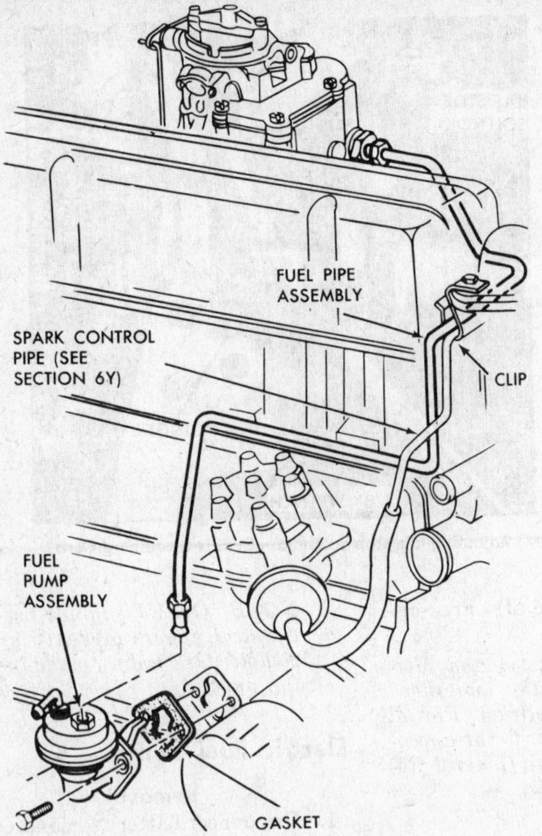

SPARK CONTROL
PIPE (SEE
SECTION 6Y)

FUEL PIPE
ASSEMBLY

CLIP

FUEL
PUMP
ASSEMBLY

GASKET

Fuel pump—inline engine (typical)
(© G.M.C.)

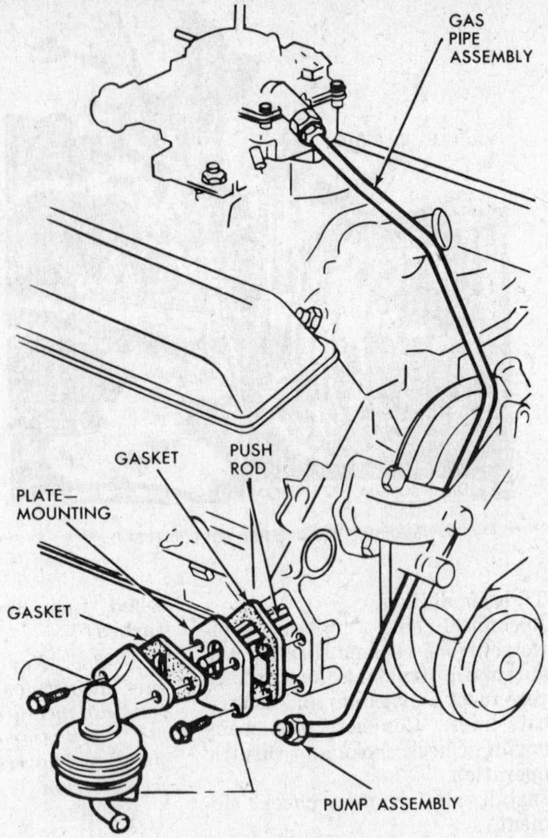

GAS
PIPE
ASSEMBLY

GASKET

PUSH
ROD

PLATE-
MOUNTING

GASKET

PUMP ASSEMBLY

Fuel pump—V6 engine (typical)
(© G.M.C.)

Adjustments

Governor adjustments are pre-set at the factory. However, minor adjustments to satisfy local conditions may be accomplished by turning adjusting cap counterclockwise for higher speed and clockwise for lower speed. One turn on the adjusting cap will vary speed 300-400 rpm or 4-5 mph.

Fuel Tank

NOTE: The following procedure is intended only as a guide. It will vary according to truck model and tank type.

Removal and Installation

1. On trucks with dual tanks check shut-off valve position.

2. Be sure ignition switch is off or battery ground is disconnected.
3. Remove seat back rest if cab mounted.
4. Drain tank. If tank does not have a drain plug, disconnect gas line and use opening to drain tank. If not accessible, siphon fuel. Do not siphon by mouth, use equipment for that purpose

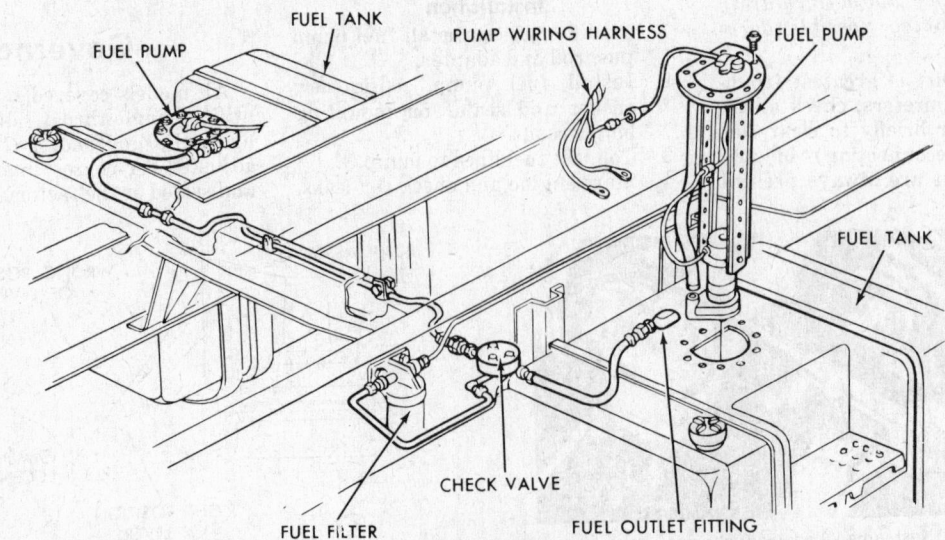

FUEL PUMP

FUEL TANK

PUMP WIRING HARNESS

FUEL PUMP

FUEL TANK

CHECK VALVE

FUEL FILTER

FUEL OUTLET FITTING

Installing electric fuel pump
(© G.M.C.)

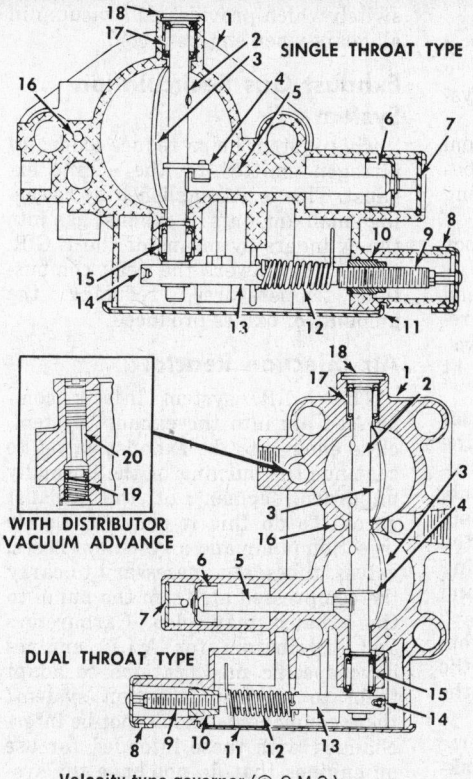

SINGLE THROAT TYPE

WITH DISTRIBUTOR VACUUM ADVANCE

DUAL THROAT TYPE

Velocity type governor (© G.M.C.)

1 Governor body
2 Valve shaft
3 Throttle valve
4 Throttle valve arm
5 Stabilizer piston rod
6 Stabilizer piston
7 Stabilizer piston plug
8 Adjusting screw cap
9 Adjusting screw
10 Adjusting screw bushing
11 Governor cover
12 Operating spring
13 Cam ribbon
14 Cam ribbon clip
15 Roller bearing
16 Vacuum by-pass passage
17 Roller bearing
18 Shaft plug
19 Transfer valve spring
20 Transfer valve

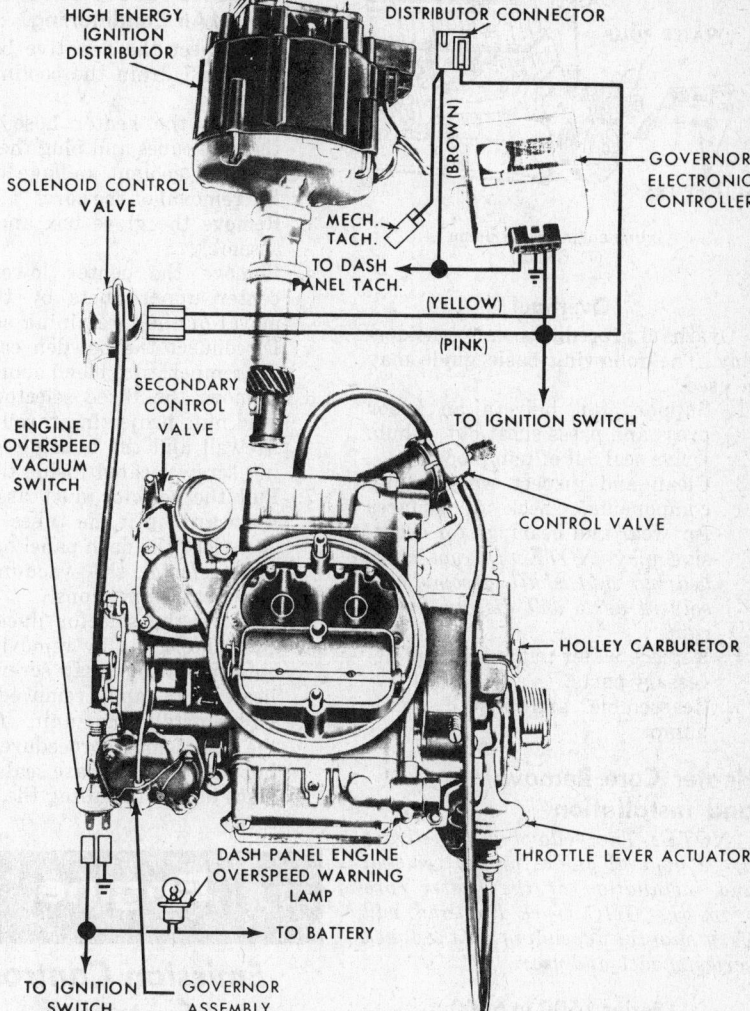

Holley governor control valve and electronic controller with HEI system (© G.M.C.)

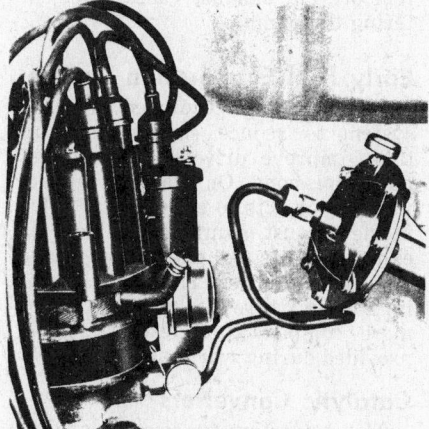

Governor mounting (© G.M.C.)

or make a siphon hose as shown below and use air pressure if available.

5. Remove filler neck, cap and vent hose.
6. Disconnect tank gage wire and any ground lead.
7. Remove tank support straps or mounting bolts. Remove tank.
8. Clean all lines, check filters. (Blow clean only after disconnecting other end of line).

9. To install, reverse removal procedures.

CAUTION: Do not use drop cord in area. A bulb breakage could have disastrous effects. Use only safety cans for fuel storage. Do not over-tighten lines, as this could distort or twist and lead to leaks.

Cooling System

Water Pump

Removal

1. Drain cooling system.
2. Remove fan spacers (when used) and pulley(s) from water pump drive hub.
3. Remove all hoses connected to water pump.
4. Remove mounting bolts and washers.
5. Remove pump and gasket.

Installation

Reverse removal procedure, installing new water pump gasket.

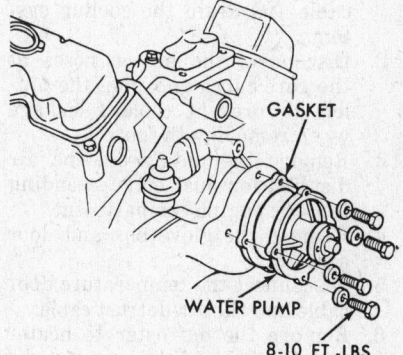

V6 engine water pump (© G.M.C.)

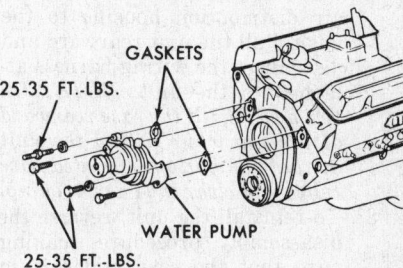

V8 engine water pump (© G.M.C.)

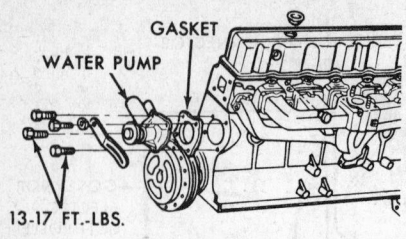

GASKET

WATER PUMP

13-17 FT.-LBS.

Inline engine water pump
(© G.M.C.)

Overhaul

Overhaul procedures vary considerably. The following basic guide may be used.

1. Support fan hub in an arbor press and press shaft out of hub.
2. Drive seal out of pump body.
3. Clean and inspect water pump components. Check all surfaces for wear and bearings for excessive play. *NOTE: Do not clean bearing and shaft assembly in solvent as it will dissolve lubricant.*
4. Replace water pump seal and necessary parts.
5. Reassemble and install water pump.

Heater Core Removal and Installation

NOTE: The following procedures are a general guide for the removal and installation of the heater core from the GMC truck line and will vary slightly depending on the truck series, model, and year.

Series 1500 to 6500 w/o Air Conditioning

1. Disconnect the battery ground cable and drain the cooling system.
2. Disconnect the heater hoses at the core tubes and plug the outlets to prevent coolant leakage when removing the core.
3. Remove the nuts from the air distribution duct bolts extending into the engine compartment.
4. Remove the glove box and door assembly.
5. Disconnect the temperature door cable and the air-defrost cable.
6. Remove the defroster to heater distribution air duct screw and the floor outlet duct.
7. Remove the screws holding the air distribution housing to the dash. Pull the unit rearward and disconnect the wiring harness attached to the unit. *NOTE: On van models, tilt the case rearward at the top while lifting the unit so that the bottom mounted core tubes will clear the dash opening.*
8. To reinstall the unit, reverse the disassembly procedure, making sure that the case sealer is in place before installing the heater core.

Series 1500 to 6500 w/Air Conditioning

1. Disconnect the negative battery cable and drain the cooling system.
2. Remove the heater hoses from the core tubes and plug the tubes to avoid coolant spillage during the removal of the core.
3. Remove the glove box and door assembly.
4. Remove the center lower and center upper ducts by the removal of their retaining screws.
5. Disconnect the bowden cable at the temperature blend door.
6. Remove the three selector duct stud nuts projecting through the firewall and the housing retaining screw located inside the cab.
7. Pull the selector duct assembly rearward until the tubes of the core clear the dash panel opening and remove the vacuum and electrical connections.
8. Remove the selector duct from the vehicle and by removing the core mounting strap screws, the heater core can be removed.
9. To reinstall the unit, reverse the disassembly procedure, making sure that the case sealer is in place before installing the heater core.

Engine

Emission Control Systems

Emission control systems are designed to control the emissions of Hydrocarbons (HC), Carbon Monoxide (CO), and Oxides of Nitrogen (NOx), at the levels specified by the Federal and State Governments. Emission control systems vary in their usage, in relationship to the engine, transmission and series application. The units are covered in the Emission Control Section of the General Repair Section.

Evaporative Emission System

This system was designed to reduce fuel vapor emissions that are normally vented into the atmosphere from the gas tank and the carburetor fuel bowl, through the use of a carbon canister and liquid-vapor separator.

Transmission Controlled Spark System

This system controls emissions by eliminating the vacuum advance in the low gears. The control is accomplished by the use of a solenoid vacuum switch jhich regulates the vacuum to the distributor. This system also incorporates a thermal override switch which provides full vacuum in all gears when engine is cold.

Exhaust Gas Recirculation System

This system helps reduce oxides of nitrogen emitted by the engine exhaust. This is accomplished by releasing small amounts of exhaust gas into the cylinders by means of the E.G.R. valve. This lowers the peak combustion temperatures, reducing the amounts of oxides produced.

Air Injection Reactor

The A.I.R. system injects compressed air into the exhaust system, close enough to the exhaust valves to continue the burning of the normally unburned segment of the exhaust gases. To do this it employs an air injection pump and a system of hoses, valves, tubes, etc., necessary to carry the compressed air from the pump to the exhaust manifolds. Carburetors and distributors for A.I.R. engines have specific modifications to adapt them to the air injection system; these components should not be interchanged with those intended for use on engines that do not have the system.

A diverter valve is used to prevent backfiring. The valve senses sudden increases in manifold vacuum and ceases the injection of air during fuel-rich periods. During coasting, this valve diverts the entire air flow through the muffler and during high engines speeds, expels it through a relief valve. Check valves in the system prevent exhaust gases from entering the pump.

Early Fuel Evaporation System

1975 models are equipped with this system to reduce engine warm-up time, improve driveability, and reduce emissions. On start-up, a vacuum motor acts to close a heat valve in the exhaust manifold which causes exhaust gases to enter the intake manifold heat riser passages. Incoming fuel mixture is then heated and more complete fuel evaporation is provided during warm-up.

Catalytic Converters

The converters are used to oxidize hydrocarbons (HC) and carbon monoxide (CO). The catalysts are made of noble metals (platinum and palladium) which are bonded to either a monolithic (one-piece) element or to individual pellets. The catalyst causes the HC and CO to break down without taking part in the reaction; hence, a catalyst life of 50,000 miles is expected.

Some engines equipped with the converters require an air injection pump to supply air for the reaction; others will not.

Engine Assembly

Inline Engines In Conventional Models

Removal
1. Drain radiator.
2. Disconnect battery.
3. Remove hood and attaching parts.
4. Remove grille and radiator support braces.
5. Remove radiator and heater hoses.
6. Remove radiator and grille assembly.
7. Disconnect fuel line at fuel pump.
8. Remove air cleaner and cover carburetor to protect it from dirt.
9. Disconnect choke control and accelerator linkage.
10. Disconnect exhaust pipe from manifold.
11. Disconnect wiring harness and battery cable.
12. Remove hand brake lever and gearshift lever from transmission.
13. Disconnect driveshaft from transmission flange.
14. Attach lifting equipment, remove mounting bolts and rear crossmember.
15. Lift out engine and transmission assembly.

Installation
1. Attach lifting equipment and lower assembly into chassis. Install support crossmember and engine mountings.
2. Connect driveshaft.
3. Install handbrake and gearshift levers.
4. Connect exhaust pipe to manifold.
5. Connect wiring harness and carburetor control linkage.
6. Install hood and attaching parts.
7. Install air cleaner, connect fuel line.
8. Fill cooling system.
9. Fill crankcase with oil to the proper level.
10. Install battery.
11. Start engine and check for leaks.

Inline Engines in Tilt Cab Models

Removal
1. Drain cooling system.
2. Disconnect battery cables.
3. Tilt cab forward; remove radiator and shroud.
4. Disconnect shift linkage at control island.
5. Disconnect throttle and choke controls at carburetor.
6. Disconnect parking brake cable and housing.
7. Remove control island mounting bolts; swing control island forward.
8. Disconnect hoses and remove surge tank.
9. Remove right and left island supports.
10. Disconnect cab safety lock, remove cab rear support.
11. Disconnect electrical wiring from engine units.
12. Disconnect all fuel, heater, oil lines from engine.
13. Disconnect exhaust pipe from manifold.
14. Remove engine fan and pulley.
15. Remove rocker arm cover and attach lifting brackets at cylinder head bolts.
16. Attach hoist and take up slack.
17. Remove engine mounting bolts and bolts attaching transmission to engine. Support transmission.
18. Move engine forward until it is disengaged from transmission.
19. Lift engine from the chassis.

Installation
Install engine by reversing the removal procedure. Be certain to maintain cleanliness and to avoid damaging the engine parts. After engine is installed, check operation of all control linkages, fill cooling system and check for leaks. Fill crankcase with oil to the proper level.

V6 Engines in Conventional Models

Removal
1. Drain radiator.
2. Disconnect battery.
3. Remove hood and attaching parts.
4. Remove grille and radiator braces.
5. Remove radiator and heater hoses.
6. Disconnect oil cooler lines (when used).
7. Remove grille, radiator, and front bumper.
8. Disconnect fuel line.
9. Disconnect air lines (when used).
10. Disconnect engine ground strap.
11. Disconnect exhaust pipes from manifolds.
12. Disconnect accelerator and choke controls from the carburetor.
13. Disconnect tachometer drive and oil gauge pressure line (when used).
14. Attach lifting equipment and take up slack.
15. Remove clutch housing to flywheel housing bolts.
16. Move engine forward to disengage transmission.
17. Lift engine from the chassis.

Installation
Install engine by reversing the removal procedure. Be certain to maintain cleanliness and to avoid damaging the engine parts. After engine is installed, check operation of all control linkages, fill cooling system and check for leaks. Fill crankcase with oil to the proper level.

V6 Engines in Tilt Cab Models

Removal
1. Drain radiator.
2. Disconnect battery.
3. Disconnect oil cooler lines (when used).
4. Disconnect electrical wiring and cables from the engine units.
5. Disconnect engine ground strap.
6. Disconnect accelerator and choke controls at the carburetor.
7. Disconnect transmission control rods, surge tank hoses, air cleaners and hoses.
8. Remove control island and rear cab support.
9. Disconnect exhaust pipes from manifolds.
10. Disconnect clutch control cylinder, parking brake control and speedometer drive from the rear of the transmission.
11. Disconnect cooling system and heater hoses.
12. Disconnect air lines from compressor (when used).
13. Disconnect driveshaft from transmission.
14. Attach lifting equipment and take up slack.
15. Remove bolts from front and rear mountings.
16. Lift assembly from the chassis.

Installation
Install engine by reversing the removal procedure. Be certain to maintain cleanliness and to avoid damaging the engine parts. After engine is installed, check operation of all control linkages, fill cooling system and check for leaks. Fill crankcase with oil to the proper lever.

V8 Engines in Conventional Models

Removal
1. Disconnect battery cables and remove battery.
2. Drain cooling system.
3. Disconnect air intake hose (when used).
4. Disconnect all wiring to engine units as necessary.
5. Disconnect radiator and heater hoses.
6. Disconnect fuel line.
7. Remove front end sheet metal, including hood.
8. Remove engine fan and drive belts.
9. Remove power steering pump.
10. Remove air compressor and disconnect air lines (when used).
11. Remove air cleaner and discon-

nect accelerator and choke cables at the carburetor.
12. Disconnect exhaust pipes from manifolds.
13. Disconnect clutch and transmission controls as necessary.
14. Remove valve covers and attach lifting brackets at cylinder head bolts.
15. Attach lifting equipment and take up slack.
16. Remove bolts from front and rear mountings.
17. Remove bolts attaching transmission to engine.
18. Move engine forward to disengage transmission.
19. Lift engine from the chassis.

NOTE: Engines coupled to an automatic transmission must be removed as a unit. Disconnect all transmission controls accordingly. Transmission may be removed from engine assembly after power plant removal.

Installation

Install the engine by reversing the removal procedure. Be certain to maintain cleanliness and to avoid damaging the engine parts. After engine is installed, check operation of all control linkages, fill cooling system and check for leaks. Fill crankcase with oil to the proper level.

NOTE: Engines coupled to an automatic transmission must be installed as a unit.

V8 Engines in Tilt Cab Models
Removal

1. Remove radiator, support and shroud.
2. Disconnect shift linkage at control island, accelerator and choke cables at carburetor.
3. Disconnect parking brake control.
4. Disconnect surge tank hoses.
5. Remove surge tank and control island rear support.
6. Remove cab safety lock control and cab rear support.
7. Disconnect all wiring to engine units.
8. Disconnect fuel line.
9. Disconnect heater hoses and vacuum/air lines (when used).
10. Disconnect engine ground straps.
11. Disconnect exhaust pipes from manifolds.
12. Disconnect clutch control.
13. Remove engine fan and pulley.
14. Remove valve covers and install lifting brackets at cylinder bolts.
15. Attach lifting equipment and take up slack.
16. Remove bolts attaching transmission to engine.
17. Remove bolts from front and rear mountings.
18. Move engine forward to disengage transmission.
19. Lift engine from the chassis.

Installation

Install the engine by reversing the removal procedure. Be certain to maintain cleanliness and to avoid damaging engine parts. After engine is installed, check operation of all control linkages, fill cooling system and check for leaks. Fill crankcase with oil to the proper level.

Manifolds

Inline Engine
Removal

1. Remove air cleaner.
2. Disconnect both throttle rods at bellcrank, remove throttle return spring.
3. Disconnect fuel and vacuum lines, choke cable at carburetor.
4. Disconnect crankcase ventilation hose.
5. Disconnect exhaust pipe at manifold flange. Discard packing.
6. Remove heat stove (when used).
7. Remove attaching bolts and clamps and manifold.

Installation

1. Clean all surfaces.

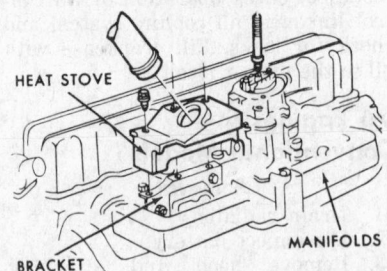

Manifold heat stove (inline engine)
(© G.M.C.)

2. Place new gasket over manifold end studs on head.
3. Position manifold and install bolts and clamps while holding manifold in place.
4. Torque bolts to specifications.
5. Connect exhaust pipe to manifold using new packing.
6. Connect crankcase ventilation hose.
7. Connect fuel and vacuum lines.
8. Connect choke cable, throttle rods, and install throttle return spring.
9. Install air cleaner.

V6 and V8 Engines
Intake Manifold Removal

1. Drain radiator and remove air cleaner.
2. Disconnect battery cables, radiator and heater hoses, water pump by-pass, accelerator linkage, choke control, and fuel line at carburetor. If necessary, remove crankcase ventilation lines and spark advance connections.
3. Where necessary, remove the distributor cap, mark the rotor position with chalk, and remove the distributor.

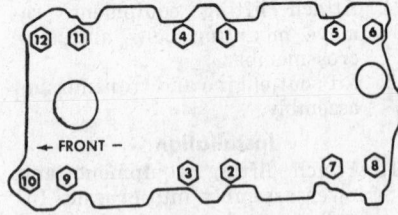

Intake manifold bolt tightening sequence (350 V8 engine)
(© G.M.C.)

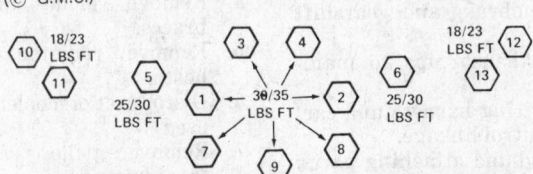

In-line engine manifold torque (typical) (© Chevrolet Div., G.M. Corp.)

Intake manifold gaskets and seals (V8 engine)
(© G.M.C.)

4. Remove, as necessary, the oil filler bracket, air cleaner bracket, air compressor and bracket, coil, accelerator return spring, and bracket, and accelerator bellcrank.
5. Remove the attaching bolts, and remove the manifold.

Intake Manifold Installation

1. Clean all surfaces.
2. Install manifold seals on block and gaskets on cylinder heads.
3. Install manifold and torque bolts to specifications.

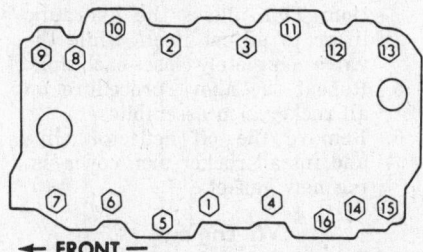

Intake manifold bolt tightening sequence
(366, 427 V8 engines)
(© G.M.C.)

4. To finish installation, reverse removal procedure. Fill cooling system and check for leaks.

V6 and V8 Engines

Exhaust Manifold Removal

1. Remove carburetor bracket as required. Remove generator and bracket from manifold, as required. *NOTE: On large V8 engines, remove sparkplugs.*
2. Disconnect exhaust pipe from manifold.
3. On 1975-78 models remove the carburetor heat choke tube assembly.
4. Bend back bolt lock tabs and remove manifold bolts.
5. Remove exhaust manifolds.

Exhaust Manifold Installation

1. Clean all surfaces.
2. Install bolts with locks and torque in proper sequence to specifications.
3. Connect exhaust pipe to manifold with new packing.
4. On 1975-78 models install the

carburetor heat choke tube with a new gasket.
5. Install generator and bracket and carburetor bracket as re- *NOTE: Install spark plugs on large engines.*

Cylinder Head

Inline Engines

Removal

1. Remove manifold assembly.
2. Remove valve mechanism.
3. Drain cooling system (block).
4. Remove fuel and vacuum lines from retaining clip and disconnect wires at temperature sending units.
5. Disconnect radiator hose at water outlet and ground strap at cylinder head.
6. Remove coil.
7. Remove cylinder head bolts, cylinder head and gasket.

Installation

1. Clean all surfaces and make certain there are no nicks or deep

Cylinder head gasket installed (inline engine)
(© G.M.C.)

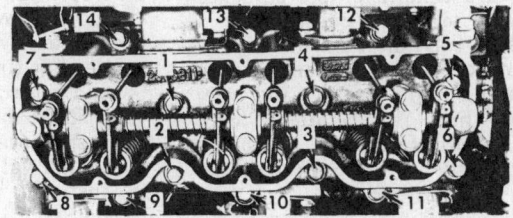

Cylinder head bolt tightening sequence
V6 engines
(© G.M.C.)

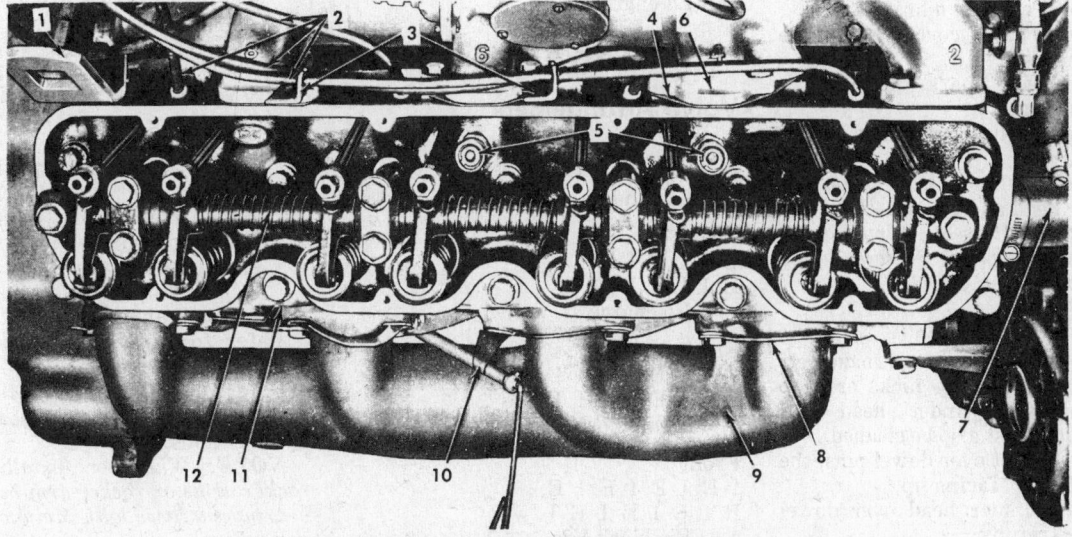

Exhaust manifold installed (V8 engine) (© G.M.C.)

1 Engine lifting bracket	8 Exhaust manifold bolt locks
2 Spark plug wires	9 Exhaust manifold
3 Plug wire supports	10 Dip stick tube clip
4 Intake manifold gasket	11 Exhaust manifold gasket
5 Crankcase ventilation valves	12 Rocker arms, shaft, and
6 Intake manifold	brackets
7 Water outlet hose	

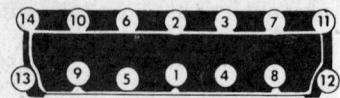

6 cylinder

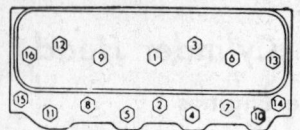

Big block V8s

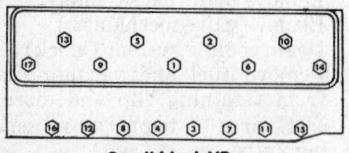

Small block V8s

Cylinder head torque sequence (© G.M.C.)

scratches. Cylinder head bolt threads must also be cleaned.

2. Place gasket over dowel pins with the bead up.
3. Place cylinder head over dowel pins carefully.
4. Apply sealer to head bolts and tighten down finger tight.
5. Tighten cylinder head bolts a little at a time in proper sequence and to torque specifications.
6. Install valve mechanism.
7. Connect wires to temperature sending units, connect fuel vacuum lines to retaining clip.
NOTE: Make certain to follow proper cylinder head tightening sequence.

V6 Engines

Removal

1. Drain cooling system.
2. Remove exhaust manifold.
3. Remove valve covers and valve mechanism.
4. Disconnect spark plug wires from plugs.
5. Remove water outlet and intake manifolds.
6. Remove cylinder head bolts, cylinder head and gasket.
NOTE: If lifters are removed, always install them in the same bores they were removed from.

Installation

1. Clean all surfaces and make certain there are no nicks or deep scratches. Cylinder head bolt threads must also be cleaned.
2. Place gasket over dowel pins, the word "top" facing up.
3. Place cylinder head over dowel pins carefully.
4. Apply sealer to head bolts and tighten down finger tight.
5. Tighten cylinder head bolts a little at a time in proper sequence and to torque specifications.
6. Install exhaust manifolds, using new gaskets.

7. Install intake manifold and water outlet manifold, using new gaskets.
8. Lubricate and install valve mechanism. Install valve cover with new valve cover gasket after valve adjustment.

V8 Engines

Removal

1. Drain cooling system (block).
2. Remove intake manifold.
3. Remove exhaust manifolds.
4. Remove valve mechansim.
5. Remove cylinder head bolts, cylinder head and gasket.

Installation

1. Clean all surfaces and make certain there are no nicks or deep scratches. Cylinder head bolt threads must also be cleaned.
2. Coat both sides of gasket with a thin coat of sealer. *NOTE: Use no sealer on a composition steel-asbestos gasket.*
3. Place gasket over dowel pins with bead facing up.
4. Place cylinder head over dowel pins carefully.
5. Apply sealer to head bolts and tighten down finger tight.
6. Tighten cylinder head bolts a little at a time in proper sequence and to torque specifications.
7. Install exhaust manifolds.
8. Install intake manifold.
9. Install valve mechanism and adjust.

Reference

Refer to the specifications area for proper head bolt tightening sequences and torques for all Inline, V6 and V8 engines.

Valve Train

Valve Arrangement

Front
E I I E E I I E E I I E
6 cylinder

Front
E I I E I E
E I E I I E
V6 engine

Front
E I I E E I I E
E I I E E I I E
Small block V8s

Front
I E I E I E I E
E I E I E I E I
Big block V8s

Adjustments

Inline and V8 Engines

1. Remove rocker arm cover.
2. With engine running at idle speed, install oil deflector clips to prevent oil splatter.
3. Back off rocker arm stud nut at one rocker arm until it begins to clatter, then tighten it slowly until the clatter just stops. This is zero lash position.
4. Turn nut down 1/4 turn and pause for 10 seconds. Repeat this procedure until nut has been turned down for one complete revolution. This allows the hydraulic lifter to adjust itself while the valve completely closes each time.
5. Repeat the above procedure on all rocker arm assemblies.
6. Remove the oil deflector clips and install rocker arm cover using new gasket.

V6 Engines

1. Run engine until it reaches normal operating temperature.
2. Remove rocker arm covers.
3. Using a feeler gauge and a box end wrench, adjust exhaust and intake valve clearance to specifications.
4. Repeat the above procedure for all rocker arms, and check to make certain that all valve rocker arms are receiving sufficient lubrication.

Valve Rocker Arm

Inline and V8 Engines—Removal

1. Remove rocker arm cover.
2. Remove rocker arm nuts, rocker arm balls, rocker arms and push rods.
NOTE: Place assemblies in a specific order so that they may be installed in their previous positions.

Installation

1. Install push rods, making certain they seat in the lifter socket.
2. Install rocker arms, rocker arm balls and rocker arm nuts. Tighten all rocker arm nuts until all lash is eliminated.
3. Adjust valves.
Refer to Engine Rebuilding Section for more details.
Refer to valve specifications for valve spring, stem and seat angle specifications.
NOTE: Whenever installing new rocker arms or rocker arm balls, coat bearing surfaces with Molykote or its equivalent.

Engine (Cu. In.)	Camshaft Part No. & Valve Lift	Gears Properly Indexed	One Tooth Adv.	One Tooth Ret.
250	3864896-.388	.014"±.004"	.0351"	.0055"
292	3848000-.405	.016"±.004"	.0379"	.0068"

Valve Rocker Arms and Shaft

V6 Engines—Removal

1. Remove rocker arm covers.
2. Loosen rocker arm shaft bracket bolts until spring pressure is fully relieved from rocker arms.
3. Lift off rocker arm shaft and brackets as an assembly.
4. Remove push rods and place them in a specific order so that they may be installed in their original positions.

Installation

1. Install push rods in their proper positions.
2. Install rocker arm shaft and bracket assembly.
3. Set clearance between rocker arm shaft end brackets and adjacent rocker arms to specifications. Set initial valve clearance to 0.014"-intake, 0.022"-exhaust.
4. Lubricate rocker arms with engine oil, temporarily install rocker arm covers.
5. Run engine until operating temperature is reached, then stop engine.
6. Adjust valves.

Checking Engine Valve Timing

Inline Engine

1. Remove the valve rocker arm cover and the push rod front cover.
2. Loosen the nut and the No. 2 intake valve rocker arm, swing the rocker arm away from the push rod and remove the push rod and valve lifter.
3. Temporarily install a flat face mechanical lifter in place of the hydraulic lifter.
4. Turn the crankshaft until the no. 2 exhaust valve opens and the notch on the pulley dampener is aligned with the "O" mark on the timing pointer.
5. Position a deal indicator to measure lifter movement and set the indicator at zero. Turn the crankshaft to 360 degrees and read the indicator. On engines that are correctly timed the indicator will read as follows:
   250 engines . . . . 0.014 in.
   292 engines . . . . 0.016 in.
   If the reading is not as above, reset the indicator at zero and turn the crankshaft 360 degrees, then read the indicator again. If the reading is now correct, the engine is timed properly. The following chart shows indicator readings with gears properly indexed for each engine and indicator readings resulting from improperly indexed gears.
6. If after checking an out-of-time condition exists, remove the engine front cover and check for proper indexing of timing marks on gears.

V6 Engine

1. Remove the left-hand rocker cover.
2. Facing the front of the engine turn the engine clockwise to the top dead center No. 1 mark at the crankshaft pulley on the compression stroke. Both the intake and the exhaust valve on the No. 1 cylinder will then be closed.
3. On all V6 engines adjust the clearance to exactly 0.099 in. at the No. 1 exhaust valve (front valve).
4. Turn the engine clockwise until the No. 1 exhaust valve opens and begins to close. then with fingers, try turning the push rod of the No. 1 exhaust valve as the engine is cranked slowly. When the push rod rotates with finger pressure, the 5-degree (before top dead center mark) on the pulley should be at the pointer. This will be about one revolution from the starting point. If the push rod can be rotated at any point between the 10-degree mark and the top dead center No. 1 mark, the valve timing is correct. After making the timing check be sure to adjust the exhaust valve clearance to 0.018 in.

NOTE: If the timing chain has been installed improperly, there will be a 15 degree out-of-time condition for each mismatched tooth on the sprocket.

V8 Engine

1. Remove the rocker arm covers.
2. Turn the crankshaft so that the timing mark is at the "0" mark on the pointer and the No. 1 cylinder is ready to fire.
3. Move the crankshaft, and the exhaust valve in the No. 6 cylinder will just close and the intake valve will just begin to open.
4. If the exhaust valve or the intake valve is open when the pointer is at the "0" timing mark the camshaft is out of time.
5. When the No. 6 cylinder is ready to fire and the pointer is at the "0" timing mark the No. 1 cylinder exhaust valve will have just closed and the intake valve will begin to open.

Timing Gears

Timing Case Cover

Removal and Installation

1. Remove the damper and/or pulley. If necessary, remove the radiator.
2. Remove the bolts from the front cover to block and remove the two bolts from the oil pan to the front cover, if needed.
3. Carefully remove the front cover, discard the old gasket and oil seal.
4. To reassemble, clean the gasket surface and install a new gasket. Replace the oil seal in the cover.
5. Install the cover to the block and install the bolts.
6. Reinstall the damper and/or pulley, and install the radiator, if removed.

Timing Chain

V6 and V8 Engines

Removal

1. Drain and remove radiator.
2. Remove torsional damper.
3. Remove oil pan (see oil pan R & R).
4. Remove water pump or pulley if necessary.
5. Remove timing case cover.
6. Align timing marks.
7. Remove (3) camshaft sprocket bolts.
8. Remove sprocket and chain.

Timing chain and alignment marks (V8 engine)
(© G.M.C.)

NOTE: sprocket is a light press fit on-camshaft, especially at dowel pin. If sprocket is tight, tap lightly with plastic hammer on lower edge of sprocket.

9. Clean all gasket and seal areas and inspect.
10. Check teeth on both sprockets for wear.

Installation

1. Suspend chain on camshaft sprocket with timing mark in approximate position.
2. Place chain over crankshaft sprocket and position camshaft sprocket on dowel. Recheck timing marks.
3. Draw camshaft sprocket in place using the three mounting bolts. Torque to specifications. **CAUTION: do not drive sprocket on camshaft as plug at rear of camshaft can be dislodged.**
4. Lubricate chain with engine oil, make sure oil slinger is in place.
5. Reverse removal procedures.
6. Fill radiator, add engine oil and start engine. Check for leaks.

Check for Worn Chain

1. Check fan belt tension, adjust if too loose.
2. Remove distributor cap, loosen spark plugs.
3. Move fan blades until rotor moves. Mark distributor housing and balancer at pointer.
4. Move fan in opposite direction until rotor moves, remark both units.
5. Marks in excess of 4° apart, 2 graduations (usually) on balancer, indicate excess wear.

Inline Engines

Crankshaft Gear Removal

1. Attach gear puller to crankshaft gear (puller screw holes are provided on gear).
2. Turn puller screw and remove gear.
3. Check condition of timing gear key; replace if necessary.

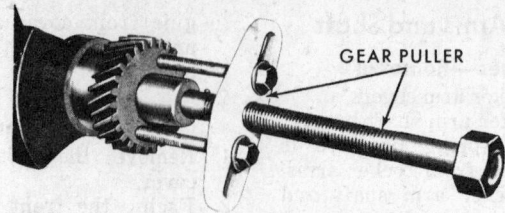

Crankshaft gear removal (inline engine)
(© G.M.C.)

Installation

1. Install key in crankshaft and coat gear seat with oil.
2. Align key way in gear to key on crankshaft with timing mark toward front of crankshaft.
3. Drive gear into place.

Camshaft Gear Removal

1. Gear is press fit on camshaft. Place in arbor press and apply pressure to front end of camshaft.
2. Press camshaft out of gear and remove key and spacer.

Installation

1. Install spacer and key, apply transmission oil on gear seat.
2. Support camshaft at journal, then position gear with timing mark forward and keyway aligned with key in camshaft.
3. Press gear on to camshaft until gear hub stops at spacer.
4. Measure clearance at thrust plate with a feeler gauge. A clearance of 0.001-0.005" is necessary for proper lubrication.

Crankshaft Pulley or Damper

Removal

1. Drain radiator; remove hoses.
2. Remove radiator.
3. Loosen fan belt and remove any accessory drive.
4. Use puller to remove pulley or damper.
   NOTE: On some early models, pulley is bolted to hub. Remove pulley first then, using pulley holes, remove hub. More recent models have the hub

bonded to inertia weight. Care must be exercised when using a puller.

Timing Cover Oil Seal

Replacement

1. Remove torsional damper—See Damper R & R.
2. Pry old seal out, be careful not to mar crankshaft or bend cover.
3. Install new seal (lip of seal toward block), tap lightly into position. Coat seal with oil.
4. Reverse removal procedures.

Camshaft

Removal and Installation

Inline, V6, and V8 engines

1. Drain and remove the radiator assembly.
2. Remove the front end sheetmetal and/or grille.
3. Remove the damper and/or pulley.
4. Remove the timing case cover. Remove the two bolts at the front of the oil pan.
5. Remove the rocker arm covers, the rocker arms, and push rods.
6. On 6 cylinder inline models, remove the push rod covers and remove the lifters. On the V6 and V8 engines, remove the intake manifold, and remove the lifters. *NOTE: The lifters may be kept in their respective bores with the use of spring clothes pins, snapped around the lifter housing.*
7. Align the timing gear marks on

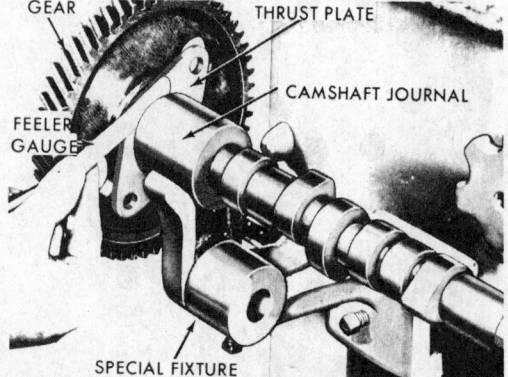

Measuring camshaft thrust plate clearance (inline engine)
(© G.M.C.)

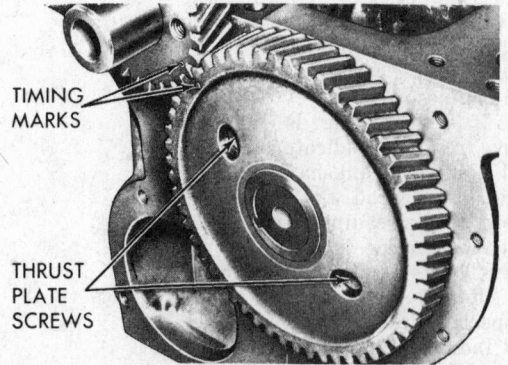

Timing gear marks and thrust plate screws (inline engine)
(© G.M.C.)

the camshaft and crankshaft gears for easier assembly when replacing the camshaft.

8. Remove two thrust plate screws by reaching through the two access holes in the camshaft gear.
9. Remove the camshaft and gear assembly by pulling and turning the shaft so the lobes will not strike and mar the camshaft bearings.
10. Use a press to remove the cam gear from the camshaft, using caution not to damage the thrust plate with the Woodruff key.
11. Clean all gasket surfaces and install the gear on the camshaft and the camshaft assembly into the engine in the reverse order of the disassembly.

Piston and Connecting Rod

Removal

1. Ream out top of bore if a piston ring travel ridge is present.
2. Position crankshaft so that a pair of connecting rods can be moved without interference.
3. Remove nuts from connecting rod cap bolts, remove rod cap and lower bearing half.
4. Mark all parts for reassembly.
5. With the aid of a length of wood or a hammer handle, push upward on the connecting rod to remove the piston from the cylinder bore.

Installation

1. Clean and coat piston pin, rings, and cylinder bore with engine oil.
2. Stagger piston ring gaps.

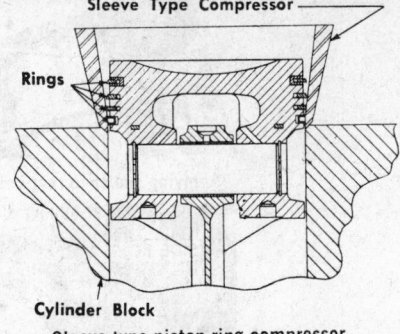

Sleeve type piston ring compressor
(© G.M.C.)

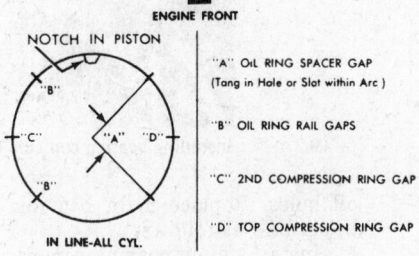

Piston ring gap locations (inline engine)
(© G.M.C.)

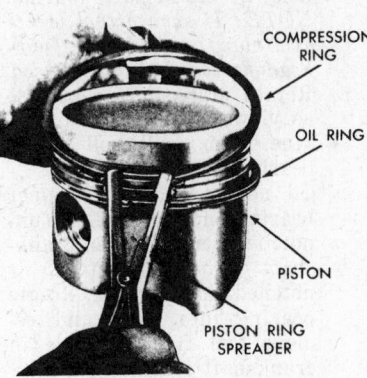

Installing piston rings with spreader tool
(© G.M.C.)

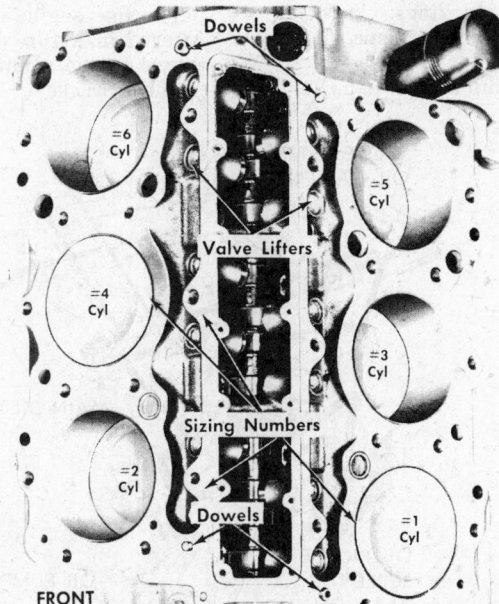

Cylinder numbering and size selection (V6 engine)
(© G.M.C.)

3. Place connecting rod bearing halves in rod and cap.
4. Use a ring compressor to install piston and connecting rod assembly in cylinder bore.
5. Install cap and lower bearing half. Lubricate nut threads and torque to specifications.
6. Check bearing clearances.

Refer to General Engine Rebuilding Section for further details on piston and connecting rod assembly and disassembly.

Piston Rings

Piston rings are removed and installed with the use of a spreader tool. Use care not to break the rings.

Crankshaft Bearings Main and Rod Bearings

Main bearings and connecting rod bearings are replaceable inserts, precision fit and held in place by locking tangs. Excessive bearing clearances reduce oil pressure. Never replace the lower half of any bearing without replacing the upper half. Do not file any bearing cap. Make sure, on main bearings, the upper half oil hole is aligned. Be certain oil passages in the crankshaft are open. Mark rod caps and upper forgings, also main caps and block, in numerical order to aid in reassembly. Rod bearings are available in standard size as well as .001″, .002″, .010″ and .020″ undersizes. Main bearings are furnished in standard size and .001″, .002″, .009″, .010″, .020″ and .030″ undersizes.

Rod Bearing Replacement

1. Drain crankcase oil and remove oil pan. See Oil Pan Removal.
2. Remove connecting rod bearing cap.
3. Wipe bearing shell and crankpin clean of oil.
4. Inspect bearings for evidence of wear or damage. (Bearings showing the above should not be installed).
5. Measure crankpin for out-of-round or taper with a micrometer. If within specifications measure bearing clearance with plastigage or its equivalent. See Engine Rebuilding—General Section.
6. Install bearing in connecting rod and cap.
7. Coat bearing surface with oil, install rod cap and torque nuts to specifications.
8. Rotate crankshaft after bearing adjustment to be sure bearings are not too tight. Check side clearance.
9. Install oil pan gaskets, seals and oil pan. See Oil Pan Installation.

GMC Trucks · Vans · Jimmy

Main Bearing Replacement

NOTE: Main bearings may be replaced with or without removing crankshaft.

(Engine in Vehicle)

1. Drain crankcase oil and remove oil pan. See Oil Pan Removal.
2. Remove oil pump.
3. Loosen or remove spark plugs for easier crankshaft rotation.
4. Starting with rear main bearing, remove bearing cap and wipe oil from journal and cap.
5. Inspect bearings for evidence of wear or damage.
6. Measure bearing clearance with plastigage or its equivalent. The crankshaft should be supported at damper and flywheel to remove clearance from upper bearing. Total clearance can then be measured between lower bearing and journal. See Engine Rebuilding—General Section.
7. Remove bearing shell from cap.
8. On inline engine crankshaft, rear main bearing has no oil hole. Replace rear main bearing upper half as follows:
   a. Use a small drift punch and hammer to start upper bearing half rotating out of block.
   b. Use a pair of pliers (with taped jaws) to hold bearing thrust surface to oil slinger and rotate crankshaft to remove bearing.
   c. Oil new selected size upper bearing and insert plate (unnotched) between crankshaft and indented or notched side of block.
   d. Use pliers as in removing to rotate bearing into place. The last ¼ movement may be done by holding just the slinger with pliers or tapping into place with a drift punch.
9. All other crankshaft journals (inline and "V" models) have

oil holes. Replace main bearing upper half as follows:
   a. Install a main bearing removing and installing tool in oil hole in crankshaft journal. *NOTE: If such a tool is not available, a cotter pin (with head flattened) may be used.*
   b. Rotate crankshaft clockwise as viewed from front of engine. This will roll upper bearing out of block.
   c. Oil new selected size upper bearing and insert plain (unnotched) end between crankshaft and indented or notched side of block. Rotate bearing into place and remove tool from oil hole in crankshaft journal.
10. Oil new lower bearing and install in bearing cap.
11. Install main bearing cap according to markings. Torque bearing cap bolts to specifications.
12. Install oil pump and oil pan. Refill crankcase. Install and tighten spark plugs.

Applying sealer to rear bearing cap area (© G.M.C.)

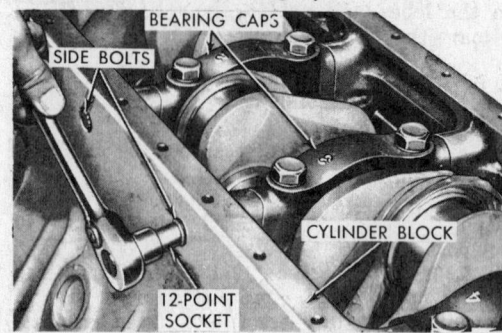

Installing bearing cap side bolts (V8 engine) (© G.M.C.)

Engine Lubrication

Oil Pan R & R

Series 1500-3500

6 Cyl.—Removal

1. Disconnect battery ground cable.
2. Raise vehicle on a hoist and disconnect starter at engine block—leave electrical connections intact and position starter out of way.
3. Remove bolts securing engine mounts to crossmember brackets—then, using a suitable jack with a flat piece of wood to protect oil pan, raise engine sufficiently to insert 2 in. x 4 in. wood block between engine mounts and crossmember brackets.

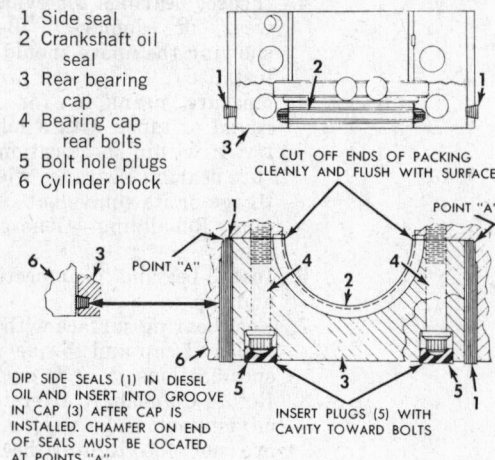

1 Side seal
2 Crankshaft oil seal
3 Rear bearing cap
4 Bearing cap rear bolts
5 Bolt hole plugs
6 Cylinder block

CUT OFF ENDS OF PACKING CLEANLY AND FLUSH WITH SURFACE

POINT "A"

DIP SIDE SEALS (1) IN DIESEL OIL AND INSERT INTO GROOVE IN CAP (3) AFTER CAP IS INSTALLED. CHAMFER ON END OF SEALS MUST BE LOCATED AT POINTS "A"

INSERT PLUGS (5) WITH CAVITY TOWARD BOLTS

Cross section of rear seals (© G.M.C.)

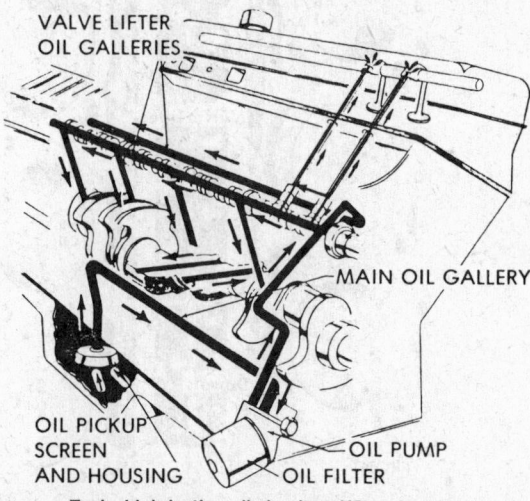

Typical lubrication oil circuits—V6 engines (© Oldsmobile Div., G.M. Corp.)

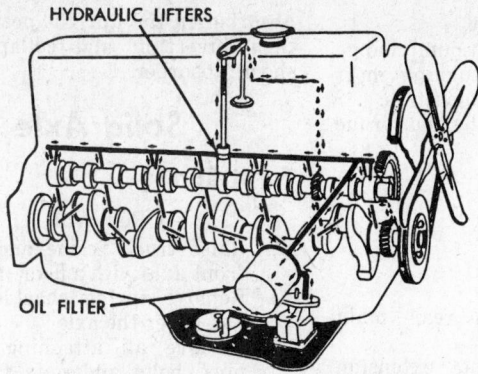

Lubrication oil circuits—In-line engines (© G.M.C.)

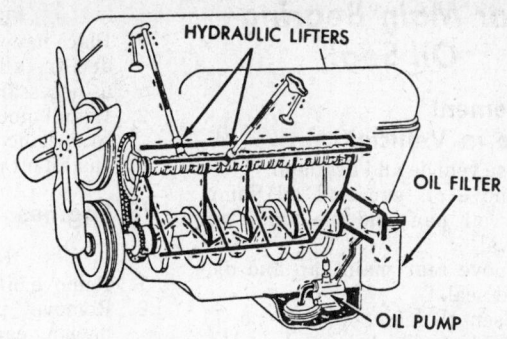

Typical lubrication oil circuits—Mark IV-V8 engines (© G.M.C.)

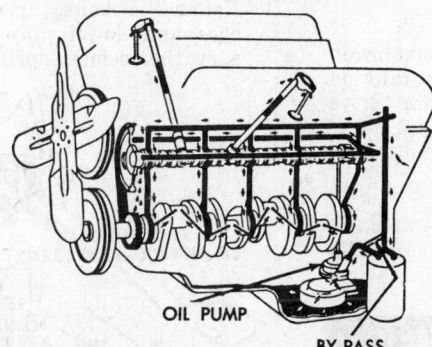

Typical lubrication oil circuits—small V8 engines (© G.M.C.)

4. Drain engine oil and remove flywheel (converter) cover.
5. Remove engine mount spacer on C Series vehicles.
6. Remove oil pan bolts and withdraw oil pan from engine.

Installation
1. Discard old gaskets and seals, thoroughly clean all gasket sealing surfaces.
2. Install new rear seal in rear main bearing cap.
3. Install new front seal on crankcase front cover.
4. Install new side gaskets on cylinder block.
   NOTE: DO NOT USE SEALER.
5. Position oil pan to block, making sure that seals and gaskets remain in place, install and torque pan screws to specifications.
6. Raise engine as outlined above and remove blocks used to support engine.
7. Lower engine, install and torque mount-to-crossmember bracket bolts.
8. Install starter and flywheel (converter) cover.
9. Install engine mount spacer on C Series vehicles.
10. Fill engine with specified quantity of oil, then start engine and check for leaks.

V8—Removal
1. Drain engine oil.
2. Remove oil dip stick and tube.
3. On vehicles so equipped remove exhaust crossover pipe.

4. On vehicles equipped with automatic transmission remove converter housing under pan.
5. Remove starter brace and inboard bolt, swing starter aside.
6. Remove oil pan and discard gaskets and seals.

Installation
1. Thoroughly clean all gasket and seal surfaces on oil pan, cylinder block, crankcase front cover and rear main bearing cap.
2. Install new oil pan side gaskets

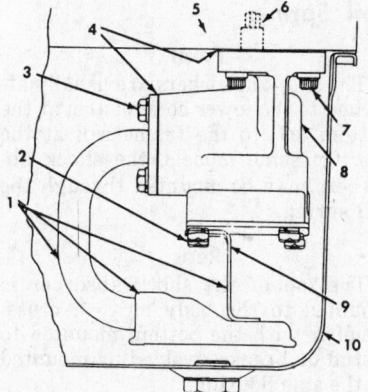

Engine oil pump installation (© G.M.C.)

1 Suction tube and screen assy.
2 Bracket bolt (20-25 ft.-lbs.)
3 Flange bolts and lock washers
4 Gaskets
5 Cylinder block
6 Pump drive shaft
7 Pump to block bolts (30-35 ft.-lbs.)
8 Oil Pump
9 Bracket
10 Oil pan

on cylinder block using gasket sealer as a retainer. Install new oil pan rear seal in rear main bearing cap groove, with ends butting side gaskets. Install new oil pan front seal in groove in crankcase front cover with ends butting side gaskets.
3. Install oil pan and torque bolts to specifications.
4. Install starter brace and attaching bolts. Torque bolts to specifications.
5. Install converter housing under pan (if removed).
6. Install exhaust crossover pipe (if removed).
7. Install oil dip stick tube and dip stick.
8. Fill with oil, start engine and check for leaks.

Series 4500-6500
Removal
1. Clean all dirt and accumulated material from oil pan attaching bolts and drain plug.
2. Drain oil out of crankcase.
3. Remove oil pan bolts, then remove oil pan. Scrape off any portions of gaskets which adhere to oil pan flange or bolting flange on engine block and front cover. Gasket at front cover is neoprene type. Remove seal at crankshaft rear bearing cap.

Installation
1. Install seal at rear bearing cap.
2. Install front seal on timing gear cover, pressing tips into holes in cover.
3. Use grease or cement to hold side gaskets in place on cylinder block. Side gasket tabs must index with front seal on timing gear cover.
4. Install oil pan.

Oil Cooler Service

Should foreign matter be suspected of contaminating oil system, back flush oil cooler and lines, using cleaning solvent and compressed air. Do not exceed 100 psi.

Rear Main Bearing Oil Seal

Replacement (Engine in Vehicle)

1. Raise vehicle and drain oil.
2. Remove oil pan and oil pump. See oil pan and oil pump removal.
3. Remove rear main cap and discard seal.
4. Loosen all mains (except no. 1) and block crankshaft down for maximum clearance at rear main.
5. Use wooden dowel, so as not to mar journal, tap lightly until seal can be gripped and removed. Rotating shaft may help.
6. Wick type seal: Use "Chinese Finger" type of tool, or insert a piece of soft wire through seal approximately ¼ from end, then wrap end with wire. Insert wire through seal opening in crankcase and around crankshaft. Start upper half of seal in place, lubricated with light sealer. Pull seal into place (rotating shaft may help) until centered. Cut each end of seal so that ¼" protrudes. Install new lower half of oil seal in cap and roll or pack into position. Cut ends flush with cap parting face. Replace cap. Molded type seal: Insert new upper half, lubricated with light sealer, into channel and apply firm pressure with hammer handle until seal is centered. (Lip of seal facing toward front of engine.) Install lower half of oil seal in cap (lip toward front), lubricate lip with oil and install cap.
7. Torque all mains, check for drag.
8. Replace oil pump, oil pan and add oil.
9. Start engine and check for leaks.

Oil Pump R & R

Inline Engines

Removal

1. Remove oil pan.
2. Remove oil suction pipe bolt and two bolts attaching the pump flange to engine.
3. Remove oil pump and screen.

Installation

1. Install oil pump, aligning drive shaft with distributor tang. Install suction pipe support bolt.
2. Install oil pan.

V6 Engines

Removal

1. Remove oil pan.
2. Remove two mounting bolts and oil pump.

Installation

1. Place new gasket on pump flange and install pump, turning shaft as necessary to engage gears.
2. Install mounting bolts and torque to specifications.
3. Install oil pan.

V8 Engines

Removal

1. Remove oil pan.
2. Remove pump to rear main bearing cap bolt.
3. Remove pump and extension shaft.

Installation

1. Install pump and extension shaft, aligning slot with tang on lower end of distributor drive shaft.
2. Install attaching bolt and torque to specifications.
3. Install oil pan.

NOTE: To prime the oil pump, fill the gear cavity with petroleum jelly or with engine oil.

Front Axle, Suspension

Shock Absorbers

Shock absorbers are used to dampen the rebound of the two types of springs used on the vehicles, the coil and leaf springs.

Leaf Type

The top of the shock absorbers are mounted to the frame and the bottom is usually mounted to the U bolt bracket at the axle area or to a bracket welded to the axle housing.

Coil Spring

Front

The shock absorbers are usually attached to the lower control arm at the bottom and to the frame rail at the top. On some models, the shock absorbers may be mounted through the coil spring.

Rear

The top of the shock absorber is mounted to the body or to a crossmember with the bottom mounted to a stud or bracket welded or mounted on the axle housing.

Removal and Installation

The removal and replacement is accomplished by the removal of the attaching retainers located at the top and bottom of the shock absorber, and withdrawing the shock. Replacement is the reverse of removal. Air should be purged from the shock absorber by extending with the shock absorber in the upright position, and then inverting and collapsing the shock absorber.

Solid Axle

Front Spring

Removal

1. Raise truck frame and support front axle with a floor jack.
2. Remove spring shackle U-bolts and lower the axle.
3. Remove all attaching retainer pins, bolts and nuts from rear hanger.
4. Remove U-bolts, spacers, tow eyes and dowel pins as necessary, then remove spring.

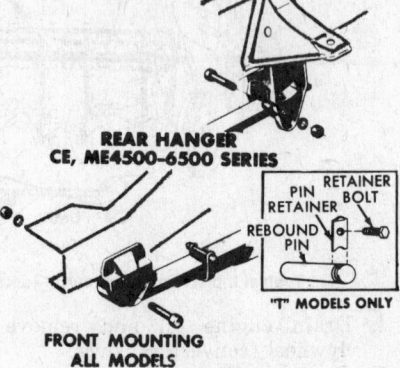

Front spring mounting (typical)
(© G.M.C.)

Installation

1. Install rear end of spring with all attaching parts.
2. Raise front end of spring and install attaching parts.
3. Raise front axle to spring and install attaching U-bolts, nuts and washers.
4. Torque all bolts and nuts to specifications.
5. Lower vehicle.

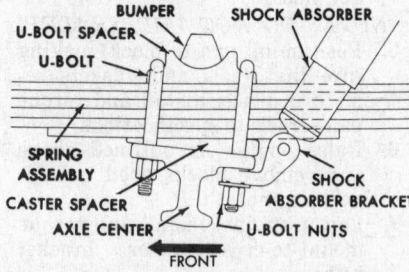

Front spring center mounting (tilt cab models)
(© G.M.C.)

King Pins and Bushings

Removal

1. Jack up axle and remove hubs and bearings.
2. Remove steering arm from knuckle.
3. Remove king pin draw key by using a brass drift and hammer.
4. Remove king pin cap screws, caps and gaskets.

5. Drive out king pin using a brass drift and hammer.
6. Remove bearings, shims and O-ring.

Installation

1. Clean all parts and coat king pin with S.A.E. #10W oil.
2. Install thrust bearing and knuckle, then raise to take up all clearance.
3. Check thrust bearing minimum clearance with specifications and correct with shims, if necessary.
4. Align king pin holes and partially install king pin.
5. Drive king pin into place, aligning milled slot with draw key (when used).
6. Install required king pin bushings and nuts and torque to specifications. Install dust caps and gaskets.
7. Install remaining parts, lubricate according to specifications, then install bearings and hubs.

Coil Spring Suspension

Coil Spring

Removal

1. Raise vehicle and place jack stands under frame, allowing control arms to hang free.
2. Disconnect lower end of shock absorber.
3. Install chain through spring and lower control arm as a safety precaution.
4. Raise cross-shaft to relieve load; remove U-bolts securing shaft to crossmember.
5. Slowly lower control arm until the coil spring can be removed.
6. Remove coil spring.

Installation

1. Install spring and then lift control arm. Position cross-shaft with crossmember and install attaching U-bolts.
2. Connect lower end of shock absorber.
3. Torque all nuts to specifications.
4. Lower vehicle.

Lower Control Arm

Removal

1. Remove coil spring.
2. Remove cotter pin from lower ball stud and loosen stud nut.
3. Loosen lower ball stud in steering knuckle, then remove nut.
4. Remove control arm.

Installation

1. Install lower ball stud through steering knuckle and tighten nut.
2. Install spring and control arm.
3. Torque stud to specifications and install cotter pin.

Upper Control Arm

Removal

1. Raise vehicle and support lower control arm.
2. Remove wheel assembly.
3. Remove cotter pin from upper ball stud and loosen stud nut.
4. Loosen upper ball stud in steering knuckle, then remove nut.
5. Remove nuts attaching control arm to crossmember bracket. Remove control arm.

Installation

1. Install control arm in position on bracket. Make certain camber and caster shims are in proper order.
2. Install ball joint stud, nut and cotter pin.
3. Remove lower control arm support and lower vehicle.

Adjustments

Refer to the wheel alignment specifications for necessary information.

Ball Joint Inspection

Lower Ball Joint

1. Support the control arms at the wheel hub and at the drum.
2. Measure the distance between the tip of the ball stud and the tip of the grease fitting below the ball joint.
3. Remove the support under the control arm, and measure the distance between the tip of the ball stud and the tip of the grease fitting, as in 2.
4. Subtract the smaller measurement from the larger. If the difference exceeds .094″ (or 3/32″), the ball joint must be replaced.

Upper Ball Joint

1. Check the stud for perceptible lateral shake.
2. Attempt to twist the stud in its socket with the fingers. If either test produces positive results, replace the ball joint.

Upper Ball Joint

Removal

1. Put the vehicle on a hoist. Support the lower control arm, using a floor jack, if necessary.
2. Remove the cotter pin from the upper ball stud. Loosen the stud nut just two turns.
3. Remove the brake caliper assembly and wire it to the frame for clearance.
4. Install a special tool between the ball studs, as shown.
5. Extend the bolt from the special tool to loosen the ball stud in the steering knuckle. When the stud is loose, remove the tool and stud nut.
6. Center punch the rivet heads. Drill out the rivets.
7. Remove the assembly.

Installation

1. Install a new service ball joint with fasteners supplied, torquing nuts to 45 ft lbs.
2. Install the ball stud in the steering knuckle, and install the stud nut, torquing the nuts, as specified below:
   10 Series—40-60 ft lbs
   20, 30 Series—80-100 ft lbs, plus additional torque required to cotter pin—not to exceed 130 ft lbs
3. Install a new cotter pin.
4. Install the lube fitting, and lubricate the joint.
5. Install the brake caliper assembly and tire and wheel.

Lower Ball Joint

Removal

1. Put the vehicle on a hoist. Support the lower control arm using a floor stand, if necessary.
2. Remove the tire and wheel.
3. Remove the lower stud cotter pin and loosen the stud nut just two turns.
4. Remove the brake caliper assembly and wire it to the frame for clearance.

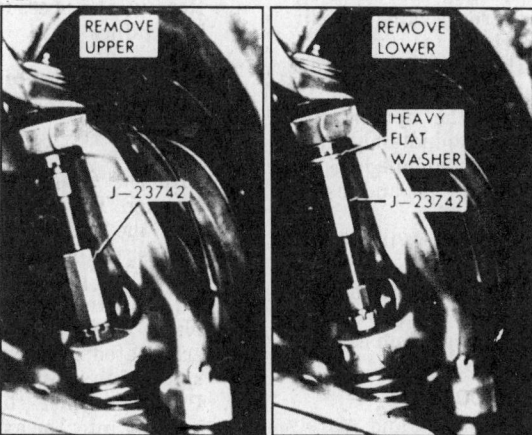

Use of a special tool to remove upper and lower ball joint studs (© G.M.C.)

5. Install a special tool between the ball studs, as shown.

6. Extend the bolt from the special tool to loosen the ball stud in the steering knuckle. When the stud is loosened, remove the tool and ball stud nut.

7. Pull the brake disc and knuckle assembly off the ball stud and support the upper arm with a block of wood.

8. If working on a 20 or 30 Series vehicle, cut a piece of 3″ water pipe to a length of $2\frac{5}{8}$″. Install the special tools and, on 20 and 30 Series vehicles the pipe, as shown in the illustration.

9. Press the ball joint out of the control arm with the hex head screw. Remove the ball joint.

Installation

1. Start the new joint into the control. Then, install the special installation tools shown. Make sure bleed vent in rubber boot is inward.

2. Turn the hex head screw just far enough to seat the new joint in the control arm.

3. Lower the upper arm and start the lower ball stud into the steering knuckle.

4. Install the brake caliper assembly.

5. Install the ball stud nut and torque to 80-100 ft lbs. Then continue to torque the nut until the cotter pin hole is in alignment. The maximum torque must not exceed 130 ft lbs.

6. Install a lube fitting and lubricate the new joint. Install the tire and wheel.

Wheel Bearing Adjustment

1500-3500 Series

1. Remove the hub cap or wheel disc from the wheel.

2. Remove the dust cap from the hub.

3. Remove the cotter pin from the spindle and spindle nut.

4. Tighten the spindle nut to 12 ft lbs. while turning the wheel assembly forward by hand to fully seat the bearings.

5. Back off the nut to the "just loose" position.

6. Hand tighten the spindle nut then loosen the spindle nut until either hole in the spindle lines up with a slot in the nut and install a new cotter pin.

7. Measure the looseness in the hub assembly. If properly adjusted the end play will be from .001-.005 in.

8. Install the dust cap, hub cap and lower the vehicle.

4500-6500 Series

1. Raise the vehicle and safely support the axle.

2. Remove the hub cap and gasket.

3. Remove the cotter pin from the adjusting nut.

4. Tighten the adjusting nut to 50 ft. lbs. while rotating the wheel in both directions.

5. Back off the nut 1/3 turn and install a new cotter pin. This adjustment should result in bearing end play of .001-.007 in.

6. Install the dust cap and wheel and lower the vehicle.

Four Wheel Drive Axle

The front axle is a hypoid type gear unit equipped with ball joint steering knuckles, and is powered through a transfer case which may be one of two types. A full-time four wheel drive unit (New Process 203 transfer case) is used mainly with V8 engines and automatic transmission. The other type is a conventional part-time four wheel drive system.

A yoke and a trunnion universal joint as part of the drive axle, allows a continous power flow to each wheel, regardless of the turning angle.

Free-wheeling hubs are available on the front wheels except those vehicles equipped with the full time 4 wheel drive transfer case.

For repairs to the hypoid gear unit, refer to the General Repair Section.

Free Wheeling Hub

1500 Series—Removal

1. Turn the actuator lever to the "Lock" position. Raise the vehicle on a hoist.

2. Remove the six retaining plate bolts. Remove the retaining plate actuating knob and O-ring.

3. Remove the internal snap ring, outer clutch retaining ring, and actuating cam body.

4. Relieve the pressure on the axle shaft snap ring and remove it.

5. Remove the axle shaft sleeve and clutch ring assembly and the inner clutch ring and bushing assembly.

6. Remove the pressure spring and its retainer plate.

Installation

1. Lubricate all parts with a high speed grease.

2. Install the spring retainer plate with the flange side facing the bearing. The retainer plate goes over the spindle nuts and seats against the outer bearing cup.

3. Install the pressure spring, large o.d. against spring retainer plate. When the spring is properly seated, it extends beyond the spindle nuts approx. $\frac{7}{8}$″.

4. Place the inner clutch ring and bushing into the axle shaft sleeve and clutch ring. Install the entire

assembly onto the axle shaft, press in on it, and install the axle shaft snap ring. Note: Installing a 7/16 x 20 bolt in the axle shaft end and pulling it outward during installation may be of help.

5. Install the actuating cam body with the cams facing outward, the outer clutch retaining ring, and internal snap ring.

6. Install the O-ring on the retaining plate. Install the actuating knob and retaining plate with the knob in "Lock" position. The grooves in the knob must fit into the actuator cam body.

7. Install the cover bolts and seals. Turn the knob to the "Free" position and check for proper operation.

2500 Series—Removal

1. Raise the vehicle on a hoist. Turn the hub key knob to "Free" position.

2. Remove the allen head bolts, and remove the hub cap assembly and gasket from the wheel hub. Remove the exterior sleeve extension housing and gasket.

3. Turn the hub key knob to the locked position. Drive out the key knob retainer roll pin.

4. Remove the outer clutch gear assembly.

5. Remove the lock ring. Remove the slotted adjustment sleeve.

6. Remove the spring, and then remove the lock ring securing the plastic key knob to the hub retainer cap.

7. Remove the O-ring from the plastic hub key knob.

8. Remove the snap ring from the end of the axle shaft. Pull the internal clutch gear and its collar.

Installation

1. Apply a high speed grease to both faces of the bushing, the splines, the teeth of the inner and outer clutch gears, and to the actuating cam.

2. Install the internal clutch collar and gear. Install the lock ring at the end of the axle shaft.

3. Lubricate the O-ring and install it in the groove of the plastic hub key knob. Insert the knob into the retainer cap.

4. Install the lock ring into the hub retainer cap. Push outward on the plastic knob to ensure that the lock ring is fully engaged, and correct its position as necessary.

5. Install the slotted adjustment sleeve with the two tabs outward.

6. Install the key knob retaining roll pin with the knob in the locked position. Install the spring.

7. Place the outer clutch gear as-

sembly on top of the spring, compress the spring, and install the lock ring at the sleeve end. Turn the key knob to the "Free" position.

8. Install a ⅜" bolt 5 inches long into one of the hub housing bolt holes.
9. Install the new exterior sleeve extension housing gasket and the housing and retainer cap assembly and gasket. Install the allen head bolts, and tighten securely.
10. Turn the hub key knob to the locked position and check for proper engagement. Install the wheel and tire.

Axle Assembly

Removal

1. Disconnect the drive shaft from the front axle. Raise the vehicle far enough to take the weight off the front springs and place jack stands under the truck.
2. Disconnect the connecting rod at the steering arms.
3. Disconnect the brake hoses at the frame fittings, and cover all open ends.
4. Disconnect the shock absorbers at the axle brackets.
5. Disconnect the axle vent tube clip at the differential housing.

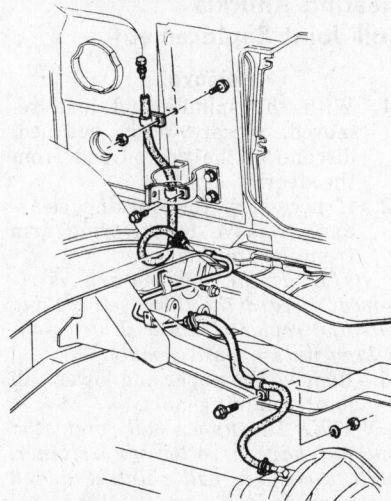

Axle Vent Hose Routing—Typical (© G.M.C.)

6. Unfasten the U-bolts, raise the truck further, as necessary, and roll the axle out from underneath.

Installation

1. With truck on axle stands, roll the axle under the truck. Lower the truck until axle and truck are in proper relative positions. Again support the vehicle with axle stands.
2. Attach the shock absorbers to the axle brackets. Connect the brake hoses to the frame fittings and fill and bleed the brake system.

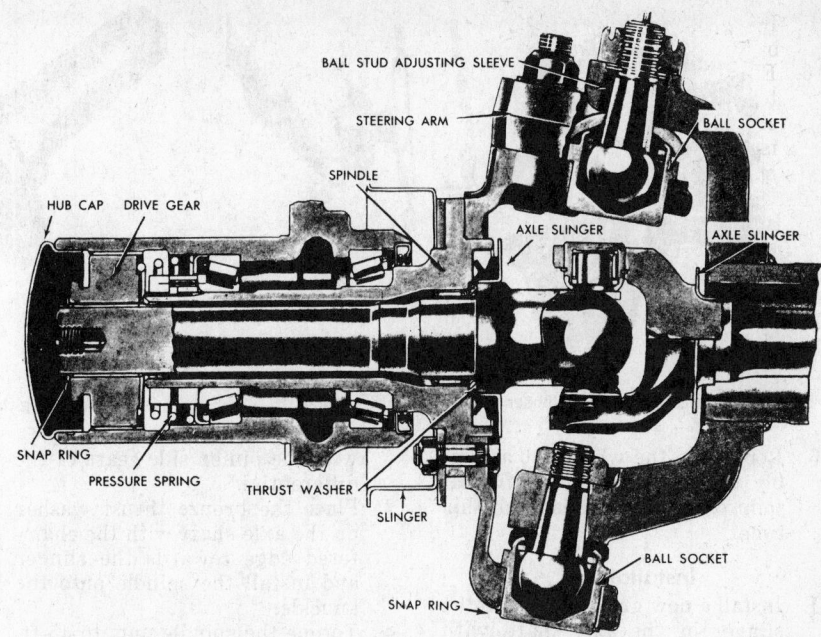

Steering knuckle and hub assembly—4 wheel drive—without free-wheeling hubs (© G.M.C.)

3. Attach the steering connecting rod at the steering arms.
4. Connect the drive axle to the front differential.

Axle Shaft Assembly

Removal

1. Remove the free-wheeling hubs as outlined, if so equipped.
2. Remove the wheel bearing outer lock nut, lock ring, and inner adjusting nut.
3. Remove the hub assembly from the spindle.

4. Remove the spindle retaining bolts and tap the end of the spindle with a soft faced hammer, to separate the spindle from the knuckle.
5. Remove the axle shaft and joint assembly by pulling outward on the shaft.

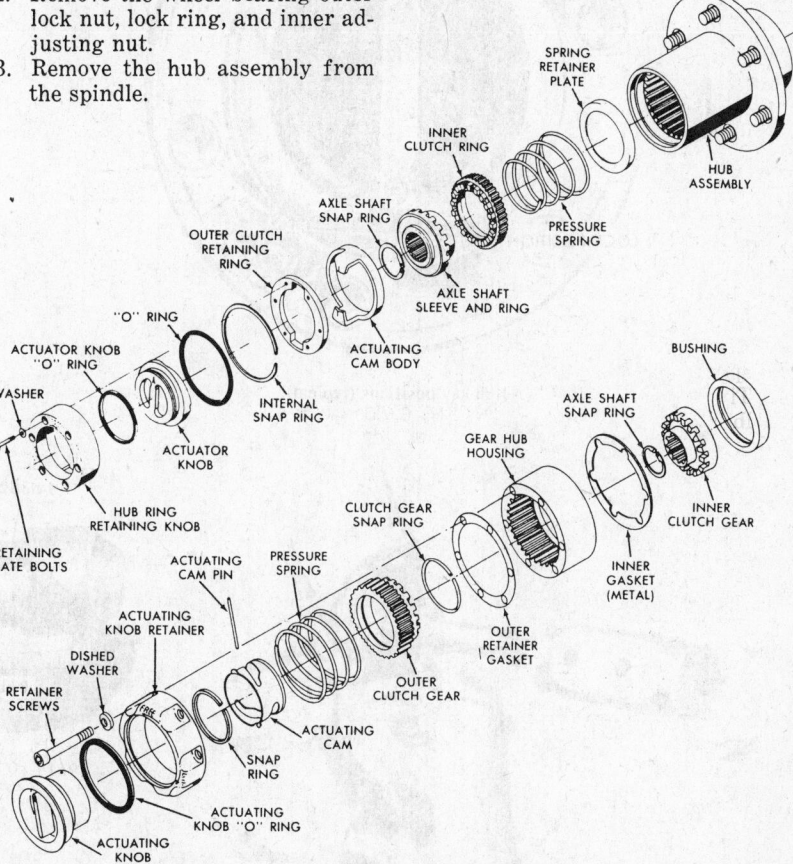

Exploded views of free-wheeling hub assemblies (© G.M.C.)

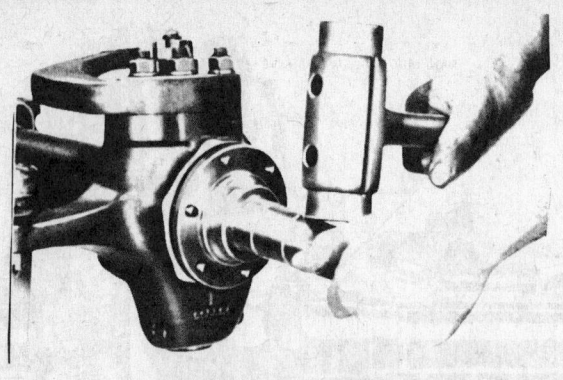

Tapping spindle to loosen from knuckle (© G.M.C.)

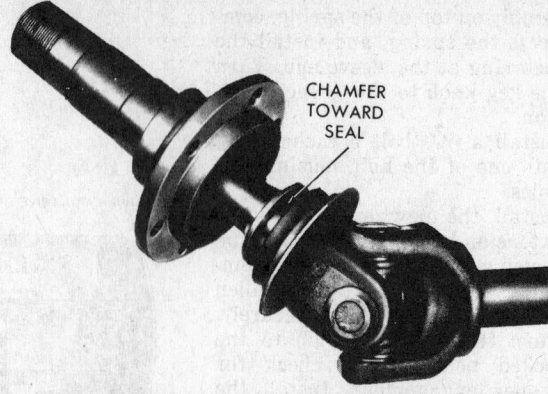

CHAMFER
TOWARD
SEAL

Assemblying thrust washer, seal, and spindle (© G.M.C.)

6. Repairs to the wheel hub assembly and to the axle universal joint can be accomplished at this time.

Installation

1. Install a new grease seal onto the slinger of the axle shaft, with the lip of the seal facing toward the spindle.
2. Install the axle shaft into the housing and engage the splines with the pinion side gears of the differential.
3. Place the bronze thrust washer on the axle shaft with the chamfered edge towards the slinger and install the spindle onto the knuckle.
4. Torque the spindle nuts to 45 ft. lbs. and assemble the hub to the spindle. Torque the inner adjusting nut to 50 ft. lbs. while rotating the hub. Back off the inner nut and retorque to 35 ft. lbs. while still rotating the hub. To complete, back off the inner nut an additional 3/8 of a turn maximum.
5. Assemble the lock washer and the outer lock nut to the spindle. Torque the outer lock nut to 50 ft. lbs. minimum. The hub should have .001 to .010 inch end play.
6. If the vehicle is equipped with free-wheeling hubs, refer to the installation procedure for correct installation and if not equipped, install the hub cap assembly.

Steering Knuckle Ball Joint Replacement

Removal

1. With the spindle and axle removed, as previously outlined, disconnect the tie rod end from the steering arm.
2. If necessary for working clearance, remove the steering arm from the knuckle.

NOTE: If the steering arm is removed, discard the three self-locking nuts and replace them with new self-locking nuts upon assembly.

3. Remove the upper and lower ball joint retaining nuts.

NOTE: The upper ball joint stud and nut have a cotter pin retainer, while the lower ball point stud and nut have none.

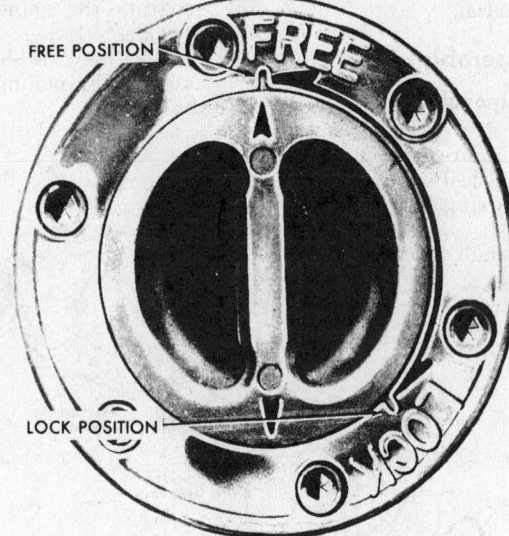

FREE POSITION

FREE

LOCK POSITION

Hub key positions (typical)
(© G.M.C.)

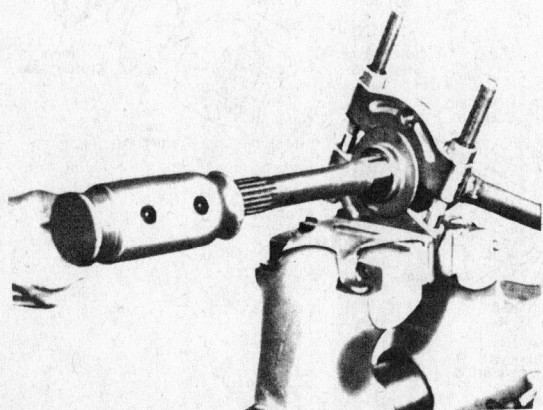

Removing axle shaft slingers (© G.M.C.)

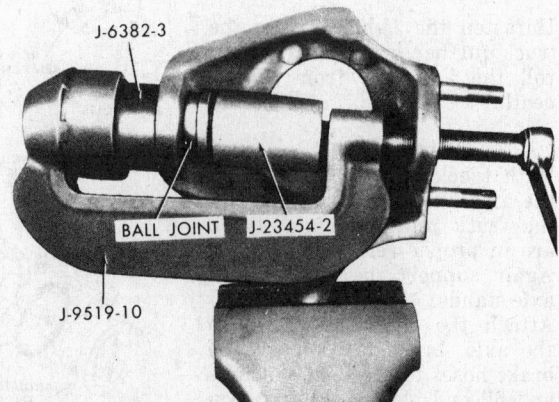

J-6382-3

BALL JOINT   J-23454-2

J-9519-10

Use of a "C" clamp type tool to install lower ball joint in knuckle (© G.M.C.)

4. With a wedge type tool, separate the lower ball joint stud from the knuckle. Repeat this operation for the upper ball joint stud.

5. Remove the snap ring retainer from the lower ball joint. With the aid of a "C" clamp tool, press the lower ball joint from the knuckle.

NOTE: The lower ball joint must be removed before any service can be performed on the upper ball joint.

6. With the aid of the "C" clamp tool, press the upper ball joint from the knuckle. Replacement of the knuckle can be accomplished at this point in the disassembly.

Installation

1. Press the lower ball joint into the knuckle with the aid of the "C" type tool and install the snap ring retainer on the lower ball joint.

2. Install the upper ball joint into the knuckle with the aid of the "C" clamp tool.

3. Position the ball joint studs in their respective openings on the yoke and install the new nuts finger tight.

NOTE: The castellated nut is placed on the upper ball joint stud.

4. Torque the lower ball joint stud nut to 70 ft. lbs. while exerting upward pressure on the knuckle.

5. Torque the upper ball joint stud adjusting sleeve to 50 ft. lbs. using a spanner type socket.

6. Torque the upper ball joint stud nut to 100 ft. lbs. Apply additional torque if necessary to align the cotter pin hole in the nut and stud.

7. Reassemble the steering arm, if removed, tie rod ends, spindle, axle and hub as outlined previously.

8. Torque the steering arm nuts to 90 ft. lbs. and the tie rod nut to 45 ft. lbs. and install the cotter pin.

Wheel Bearing Adjustment

Refer to the axle shaft removal and installation section.

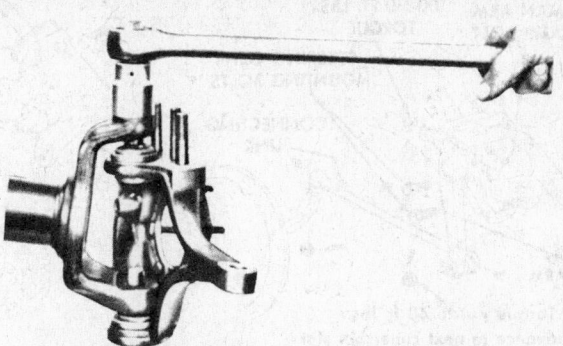

Removing upper ball socket stud retaining nut (© G.M.C.)

Ball Joint Adjustment

Adjustment of the ball joint is necessary only when there is excessive play in the steering, persistent loosening of the tie rod, or wear on the tires.

1. Raise vehicle on hoist and place jack stands just inside of front springs.

2. Disconnect tie rod and connecting rod to allow each steering knuckle to move independently.

3. Apply torque wrench, at top of knuckle, to one of the steering arms attaching and stud nuts and check torque necessary to turn knuckle.
   Maximum torque is:

15 ft.-lbs.	Axle 44-5B
20 ft.-lbs.	Axle 44- B

NOTE: Knuckle should turn smoothly through turning arc and have no vertical end play.

IMPORTANT: Temperature will affect turning torque considerably so allowances should be made.

4. If torque is too low perform the following procedures:
   a. Remove upper ball stud cotter pin and lock nut.
   b. Torque stud adjusting sleeve to 50 ft.-lbs.
   c. Install lock nut and new cotter pin.
   d. Recheck torque necessary to turn steering knuckle. If torque is still too low retorque sleeve to 60 ft.-lbs. and recheck.

NOTE: If torque specifications for turning of knuckle can not be reached, replace required parts.

Front Spring Removal and Installation

Refer to the beginning of the Front Suspension section under "Front Spring Removal and Installation." and follow the outline to remove and install the leaf springs of the front wheel drive units.

Steering Gear

Standard Steering Gear

Medium and Heavy Trucks

Removal

1. Disconnect steering linkage from pitman arm.
2. Scribe alignment marks on worm shaft and clamp yoke for reassembly.
3. Remove bolts attaching clamp yoke or coupling to steering gear worm shaft.
4. Remove pitman arm nut and washer, then use a puller to remove arm.
5. Remove attaching bolts, nuts and washers, then remove steering gear.

Torquing upper ball socket stud nut (© G.M.C.)

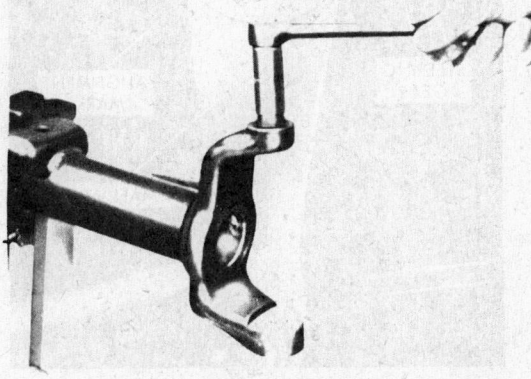

Loosening upper ball stud adjusting sleeve (© G.M.C.)

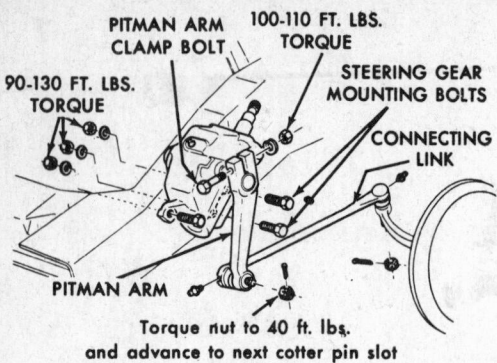

PITMAN ARM CLAMP BOLT

100-110 FT. LBS. TORQUE

90-130 FT. LBS. TORQUE

STEERING GEAR MOUNTING BOLTS

CONNECTING LINK

PITMAN ARM

Torque nut to 40 ft. lbs. and advance to next cotter pin slot

Steering gear and shaft (with flexible coupling)
(© G.M.C.)

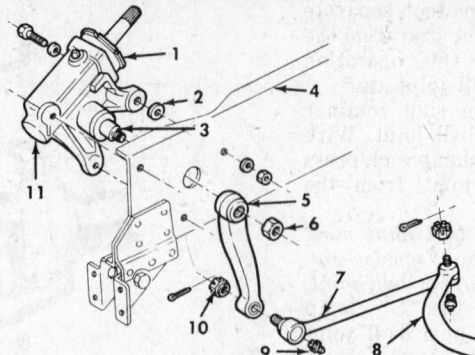

1 Worm shaft adjuster lock nut
2 Spacer washer
3 Pitman shaft
4 Frame
5 Pitman arm
6 Pitman shaft nut
7 Drag link
8 Steering arm
9 Lubrication fitting
10 Drag link stud nut
11 Steering gear

Steering gear and linkage—"T" models (typical)
(© G.M.C.)

Installation

1. Turn steering wheel to straight-ahead position and center steering gear.
2. Position the steering gear, matching the alignment marks.
3. Install attaching bolts, washers and nuts.
4. Connect clamp yoke or coupling to steering gear worm shaft with attaching parts and torque to specifications.
5. Align pitman arm and shaft; press arm onto shaft and torque to specifications.

Adjustments

Before proceeding to adjust steering gear, check lubricant, front end alignment, tire pressures, shock absorbers, king pins, pitman arm and all mounting bolts for looseness, wear or incorrect adjustment. These are common causes for steering wheel shimmy, clunking or chucking movements.

Worm Bearing Adjustment

1. Loosen adjuster lock nut and turn screw clockwise until there is no perceptible play.
2. Tighten adjuster lock nut.

Pitman Shaft Lash Adjustment

1. Center steering wheel and mark its position.
2. Loosen adjuster nut and turn screw clockwise to remove all backlash between gear teeth.
3. Check steering rim pull.

Light Duty Trucks

Removal

1. Drive the vehicle a short distance and move the wheels to the straight ahead position.
2. On P models, remove the lower universal joint pinch bolt. On C and K models, remove the bolts from the flexible coupling to steering shaft flange. Mark the relationship between the univer-

sal yoke and the wormshaft.
3. Mark the relationship between the pitman arm and the pitman shaft, and remove the pitman shaft nut or pitman arm pinch bolt.

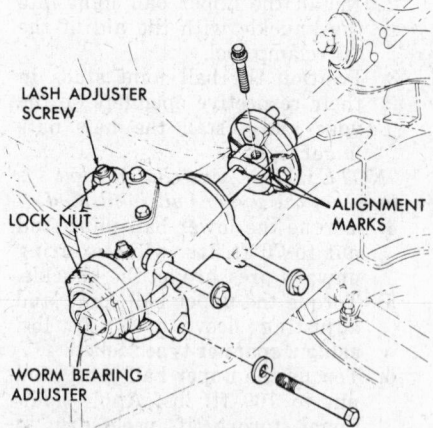

LASH ADJUSTER SCREW

LOCK NUT

ALIGNMENT MARKS

WORM BEARING ADJUSTER

Steering gear adjustment points (typical)
(© G.M.C.)

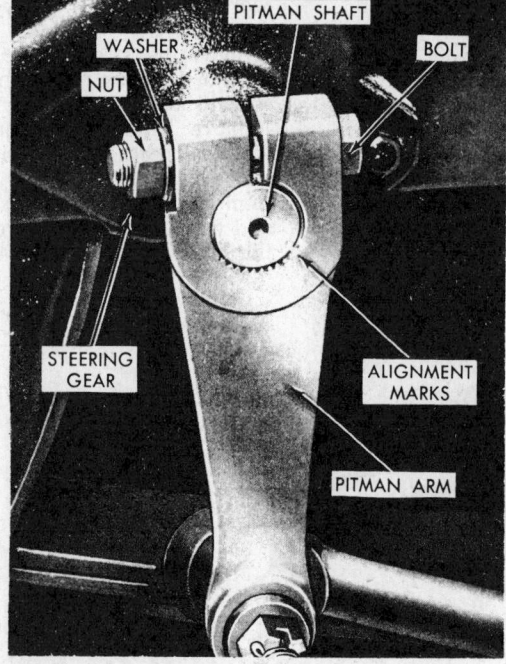

WASHER
NUT
PITMAN SHAFT
BOLT
STEERING GEAR
ALIGNMENT MARKS
PITMAN ARM

Pitman arm installed (typical)
(© G.M.C.)

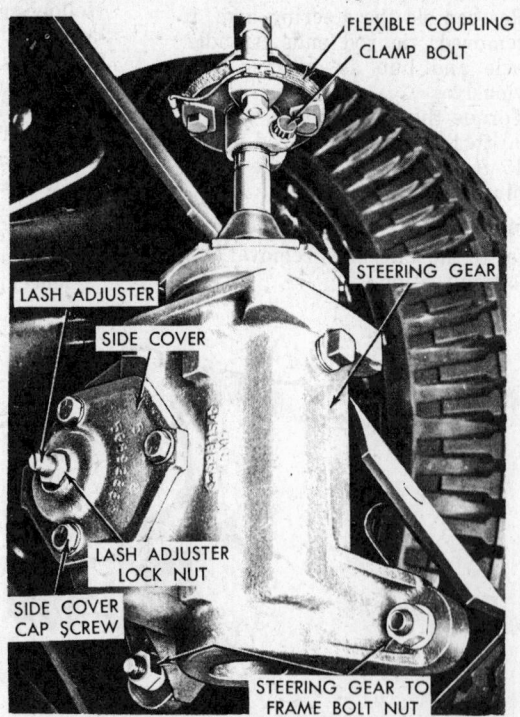

FLEXIBLE COUPLING CLAMP BOLT
STEERING GEAR
LASH ADJUSTER
SIDE COVER
LASH ADJUSTER LOCK NUT
SIDE COVER CAP SCREW
STEERING GEAR TO FRAME BOLT NUT

Steering gear and linkage—convential models (typical) (© G.M.C.)

4. Remove the pitman arm from the shaft with a special puller.
5. Remove the bolts between steering gear and frame. Remove the steering gear.
6. On C and K models, remove the flexible coupling pinch bolt and remove the coupling from the wormshaft.

Installation

1. On C and K models:
   a. Install the flexible coupling onto the steering gear wormshaft. The flats in the coupling and on the shaft must line up. Then, push the coupling onto the shaft until the wormshaft bottoms on the coupling reinforcement. Install the pinch bolt. *Note that the coupling bolt must pass through the shaft undercut.*
   b. Position the steering gear, guiding the coupling bolt into the steering shaft flange.
   c. Install the steering gear to frame bolts. Torque to 65 ft-lbs.
   d. In cases where plastic spacers are used on the flexible coupling alignment pins, make sure they are bottomed on the pins, torque the flange bolt nuts, and then remove the spacers.
   e. If plastic spacers are not used, center the pins in the steering shaft flange slots, and install and torque the flange bolt nuts.
2. On P models:
   a. Position the steering gear, guiding wormshaft into the U-joint assembly. Line up the marks made at removal. In cases where a new steering gear is being installed, line up the mark on the wormshaft with the slit in the yoke of the U-joint.
   b. Install the steering gear to frame bolts, and torque to 65 ft-lbs.
   c. Install and torque the universal joint pinch bolt. Make sure the bolt passes through the shaft undercut.
3. Install the pitman arm onto the shaft, lining up marks made at removal. Install the shaft nut or pinch bolt and torque to 140 ft-lbs.

Power Steering Gear

Medium and Heavy Trucks

Integral Type 553-DU—Removal

1. Mark steering gear worm shaft and clamp yoke or coupling for reassembly.
2. Remove connecting rod from pitman arm.

3. Remove attaching pinch bolt, nuts and washers from pitman arm; press pitman arm from shaft.
4. Drain as much fluid as possible from steering gear.
5. Disconnect all tubes from the control valve ports. Plug all tubes and cover all ports to prevent any dirt from entering the system.
6. Remove attaching bolts, nuts and washers from the steering gear and control valve assembly. Remove steering gear.

Installation

1. Center steering wheel and steering gear.
2. Install steering gear by reversing the removal procedure. Make certain to match all alignment marks and to torque all nuts to specifications.
3. Bleed the system and fill reservoir to the proper level.

Adjustments—Type-553-DU

The only adjustment that can be made on the vehicle is the over-center adjustment.
1. Disconnect pitman arm from shaft, marking the alignment positions.
2. Loosen pitman shaft adjusting screw nut and turn screw out to its limit of travel.
3. Disconnect battery ground cable.
4. Remove horn button.
5. Center steering wheel.
6. Check combined ball and thrust bearing preload by using an in. lbs. torque wrench. Note the highest reading.
7. Tighten pitman shaft adjusting screw and torque steering shaft nut to specifications.
8. Install horn button and connect battery ground cable.
9. Connect pitman arm to shaft, making certain to match alignment marks.
NOTE: There are no on-vehicle adjustments of the type 710-D steering gears.
Refer to the General Repair Section for overhaul procedures.

Integral Type 710-D Removal

1. Center steering gear by positioning tires straight ahead and remove Pitman arm clamp bolt.
2. Install Pitman arm puller and remove Pitman arm.
NOTE: It may be necessary to spread the Pitman arm clamp bosses slightly to remove arm. Install a dial indicator as shown. Insert a wedge shaped tool and spread clamp bosses .004 in.
3. Remove pot joint to stub shaft clamp bolt (conventional cab) or carden joint to stub shaft clamp bolt (school bus model). Remove

steering column plastic cap and metal cover at dash. Remove column clamp cap screws. Pull steering shaft up until shaft coupling clears stub shaft.
4. Remove steering gear mounting bolts. It will be necessary to use two technicians to remove ear bolt as shown.
5. Disconnect lines from steering gear and plug lines. Turn gear in a vertical position (stub shaft up) and work gear down between frame and inner fender panel and remove gear.
6. Remove adapter to gear bolts.

Installation

1. Install adapter to gear with washers under bolt heads and torque bolts to specifications.
NOTE: One adapter plate to frame bolt and washer (lower forward bolt as shown) must be installed before adapter is bolted to the gear.
2. Holding gear in the vertical position (stub shaft up) push gear up between inner fender panel and frame and position ear on adapter plate over frame and push the bolt previously installed into frame and install washer and nut only finger tight. With gear loose, reach between gear and frame and remove line plugs and install line fittings to gear.
3. Install steering gear to frame mounting bolts with washers and torque to specifications.
4. With one technician inside cab centering steering wheel and one at steering gear with steering gear centered, install pot joint or carden joint over stub shaft. Push steering shaft down until coupling lines up with the cross groove in stub shaft. Install clamp bolt and adjust pot joint to 3.08 in. Tighten steering column clamp cap screws, replace metal cover panel and column plastic cap.
5. Install Pitman arm to Pitman shaft, install clamp bolt and torque to specifications.
6. Bleed the system and fill reservoir to the proper level.

Light Duty Trucks

Removal

1. Disconnect the hoses at the gear and secure the open ends in a raised position.
2. Cap the open ends of the hoses and plug the openings of the steering unit.
3. Remove the flexible coupling to flange bolts on G, C, and K models, or the U-joint pinch bolt on P models. Mark the relationship between the universal yoke and stub shaft.
4. Mark the relationship between

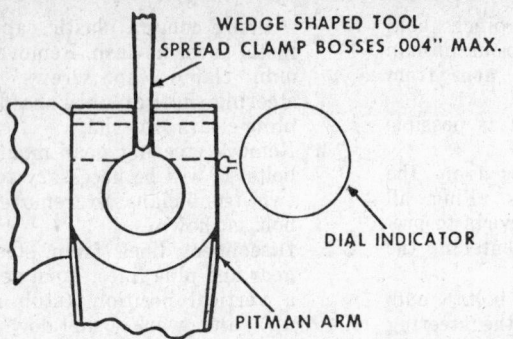

WEDGE SHAPED TOOL
SPREAD CLAMP BOSSES .004" MAX.

DIAL INDICATOR

PITMAN ARM

Pitman arm removal—Dial indicator and wedge shaped tool installed.

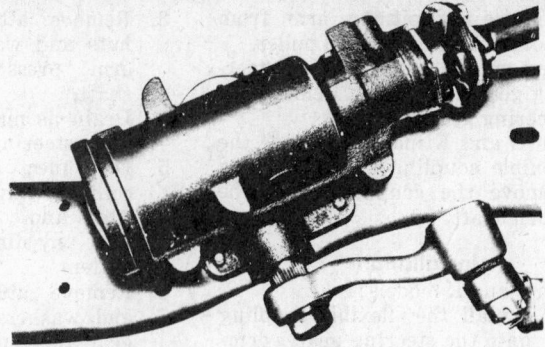

Power steering gear—models 1500-3500 (typical) (© G.M.C.)

the pitman arm and shaft. Remove the shaft nut or pinch bolt from the pitman arm.

5. Remove the arm from the shaft with a special puller.

6. Remove the steering gear mounting bolts, and remove the gear. On G, C, and K models, also remove the pinch bolt from the flexible coupling, and remove the coupling.

Installation

1. Where applicable, install the flexible coupling, aligning the flats in coupling and on shaft. Make sure the stub shaft bottoms on the coupling reinforcement. Install the pinch bolt and torque to 18 in lbs. Make sure the bolt passes through the shaft undercut.

2. Position the steering gear, guiding the coupling bolt into the shaft flange.

3. Install the mounting bolts and torque to 65 ft-lbs. (110 ft-lbs on G series).

4. If plastic spacers are used in the are bottomed on the pins, tighten flexible coupling, make sure they the flange bolt nuts to 18 in-lbs (20 ft-lbs on P series), and then remove the spacers. Where plastic spacers are not used, center

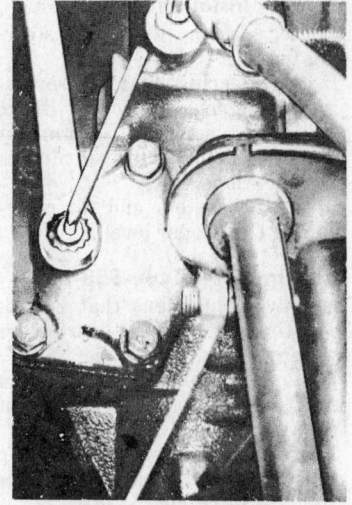

Over center adjustment
(© G.M.C.)

the pins in the steering shaft flange slots and install the bolt nuts and torque as above.

5. On P Models:

a. Position the steering gear. guiding the stub shaft into the U-joint assembly and lining up marks made at removal or, with a new unit, lining up the mark on the

stub shaft with the mark on the universal yoke.

b. Install the gear to frame bolts and torque to 65 ft-lbs.

c. Install the U-joint pinch bolt and torque to 20 ft-lbs. Make sure the bolt passes through the shaft undercut.

6. Install the pitman arm, lining up the marks made at removal. Install the shaft nut or pinch bolt and torque to 180 ft-lbs on C series, 90 ft-lbs on K series, and 180 ft-lbs on G and P series.

7. Remove the plugs and caps from the fluid fittings and reinstall both hoses.

Power Cylinders

Power cylinders are used to assist in lowering the steering effort for the vehicle operator. Two types are used, side mounted and axle mounted.

The side mounted unit is attached to the frame side rail at one end and to the pitman arm at the other end. The axle mounted cylinder is attached to the front axle at one end and to the steering tie rod at the other.

A control valve mounted on the steering gear housing, directs oil pressure from the belt driven oil

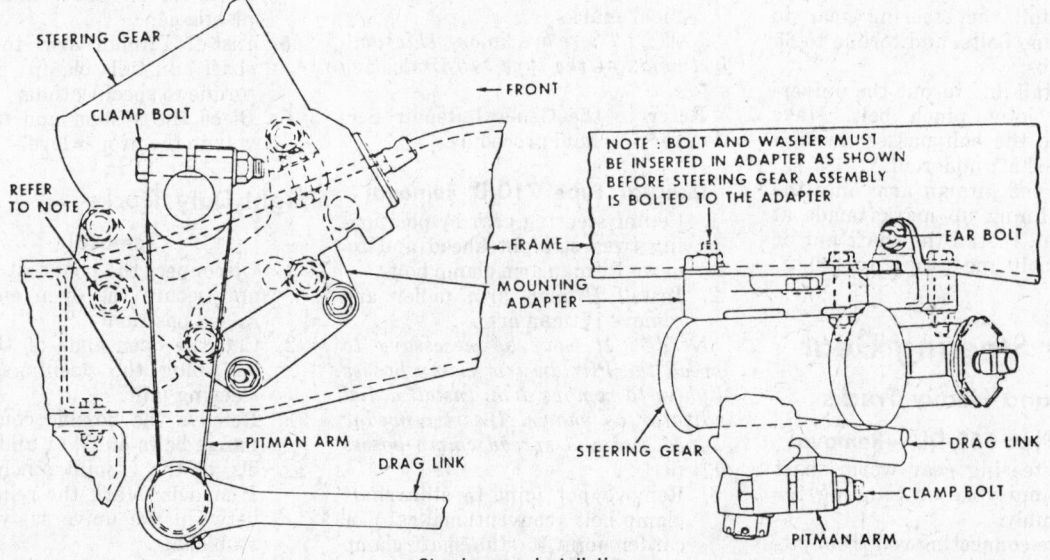

STEERING GEAR

CLAMP BOLT

REFER TO NOTE

← FRONT

PITMAN ARM

DRAG LINK

NOTE : BOLT AND WASHER MUST BE INSERTED IN ADAPTER AS SHOWN BEFORE STEERING GEAR ASSEMBLY IS BOLTED TO THE ADAPTER

FRAME

MOUNTING ADAPTER

EAR BOLT

STEERING GEAR

DRAG LINK

CLAMP BOLT

PITMAN ARM

Steering gear installation.

pump to either the right or left sides of the piston within the power cylinder, depending upon the turn being made. If a hydraulic failure occurs with this type unit, the steering reverts to manual with no hydraulic assist.

Overhaul of the power cylinder is detailed in the Power Steering Section of the General Repair Section.

Removal and Installation

Removal and installation of the power cylinder is accomplished by the removal of the power cylinder hydraulic lines, and the attaching bolts at each end of the cylinder. A container should be placed under the lines to catch the fluid that will drain from the lines.

To install the cylinder, reverse the procedure, fill the power steering reservoir, and bleed the system of air.

Transfer Case

Removal

1. Raise and support vehicle on hoist. Drain transfer case.
2. Disconnect speedometer cable, back-up lamp and TCS switch.
3. Remove skid plate and crossmember supports as necessary.
4. Disconnect rear prop shaft from transfer case and tie up away from work area.
5. Disconnect front prop shaft from transfer case and tie up shaft away from work area .
6. Disconnect shift lever rod from shift rail link. On full time 4 wheel drive models, disconnect shift levers at transfer case.
7. Remove transfer case to frame mounting bracket bolts.
8. Support transfer case and remove bolts attaching transfer case to transmission adapter.
9. Move transfer case to rear until input shaft clears adapter and lower assembly from vehicle.

Installation

1. Support transfer case in suitable stand and position case to transmission adapter. Install bolts attaching case to adapter and torque to 45 ft. lbs.
2. Remove stand as required and install bolts attaching transfer case to frame rail. Bend lock tabs after assembly.
3. Install connecting rod to shift rail link or connect shift levers to transfer case, as applicable.
4. Connect front prop shaft to transfer case front output shaft.
5. Connect rear prop shaft to transfer case rear output shaft.
6. Install crossmember support and skid plate, if removed.

7. Connect speedometer cable, backup lamp and TCS switch.
8. Fill transfer case to proper level with lubricant.
9. Lower and remove vehicle from hoist.

CAUTION: Check and tighten all bolts to specified torques.

NOTE: Before connecting prop shafts to companion flanges be sure locknuts are torqued to specifications.

For overhaul—see General Section.

Manual Transmission

Reference

Information on the overhaul of standard transmissions will be found in the General Repair Section.

NOTE: The following procedures for removing and installing the transmission are intended as a guide only. Procedures will vary according to optional equipment and individual truck requirements.

Transmission R & R

Except "K" Models

Removal

1. Remove floor mat and transmission floor pan cover, then place transmission in neutral and remove gearshift lever and control tower. *NOTE: On trucks equipped with SM465 or New Process transmission, remove gearshift lever only.*
2. Place a clean cloth over the transmission opening to prevent the entrance of dirt.
3. Disconnect back-up light switch at transmission and, where applicable, TCS connection.
4. Drain transmission lubricant.
5. Disconnect speedometer cable at transmission and remove parking brake controls.
6. Disconnect driveshaft from transmission.
7. Remove power take-off unit (when used) and cover opening.
8. Position a jack or dolly under the transmission and make certain that all attaching lines and brackets are disconnected.
9. Support rear of engine when mounts are located on the clutch housing.
10. Remove transmission to flywheel housing bolts.
11. Move transmission straight back from the engine, keeping mainshaft in alignment. **CAUTION: Do not let the weight of the transmission hang on the clutch disc hub.**
12. When transmission is free of the

engine, lower jack or dolly and remove transmission.

Installation

1. Apply a light coating of high temperature grease to main drive gear bearing retainer and splined portion of driveshaft. **CAUTION: Do not apply an excessive amount of grease as it will be thrown onto the clutch facings.**
2. To complete transmission installation, place transmission in fourth gear, reverse the removal procedures and torque mounting bolts to specifications. Fill transmission with lubricant to the proper level.

"K" Models (3 Speed)

Removal

1. Drain transfer case and transmission.
2. Disconnect speedometer cable and TCS connections at transmission.
3. Disconnect driveshaft at U-joint.
4. Remove attaching shift assembly bolt, then push assembly to one side.
5. Supporting transfer case in cradle, remove attaching bolts and case.
6. Disconnect shift control rods from transmission.
7. Support engine rear and remove adapter mount bolts.
8. Remove top 2 transmission to flywheel housing mount bolts and install guide pins.
9. Remove remaining mount bolts and slide transmission straight back on guide pins until it is disengaged from engine.
10. Remove transmission and adapter as an assembly.

Installation

To install transmission, reverse the removal procedures and fill with lubricant to the proper level.

"K" Models (4 Speed)

Removal

1. Remove transfer case shift lever retainer.
2. Remove floor mat, shift lever, console and heater duct (when used).
3. Remove transmission floor cover and disconnect shift lever link assembly.
4. Remove all attaching wiring and clamps.
5. Support engine. Drain transmission and transfer case.
6. Disconnect front and rear driveshafts at transfer case.
7. Remove transmission and transfer case to frame bolts and place

jack or dolly under transmission assembly.

8. Remove frame crossmember.

9. Remove flywheel housing cover and exhaust crossover pipe (when used).

10. Remove transmission to flywheel housing bolts. Using guide pins, slide assembly back until it disengages the engine.

11. Remove transmission assembly.

Installation

To install the transmission, reverse the removal procedure and torque mounting bolts to specifications. Fill transmission with lubricant to the proper level.

5 Speed and Larger Transmissions

Removal

1. Remove the floor mat and transmission floor pan cover. Remove the gear shift lever and control tower assembly, with the transmission in the neutral position. Cover the opening in the transmission to keep foreign material from entering.

NOTE: On vehicles equipped with the New Process or SM465 Transmission, remove the gear shift lever only.

2. Disconnect the electrical wiring from the back-up lamp switch and remove the speedometer cable from the transmission adapter.

3. Drain the transmission assembly.

4. Disconnect the clutch linkage and the parking brake lever and controls, if used, from the rear of the transmission.

5. Match-mark the drive shaft, yoke, and/or the universal joint and remove the drive shaft from the rear of the transmission.

6. Remove the power take-off unit and controls, if equipped.

7. Position a jack or dolly under the transmission and support the weight.

NOTE: On vehicles having the rear engine mountings attached to the clutch housing, support the weight of the engine with a suitable dolly or jack and remove the mountings.

8. Remove the bolts holding the transmission to the engine as follows:

   a. Apron type flywheel housing (has sheet metal pan covering the entire lower portion of the clutch housing). Remove the bolts retaining the transmission to the flywheel housing.

   b. S.A.E. #2 type flywheel housing (has a separate clutch housing used in addi-

tion to the flywheel housing). Remove the clutch housing to flywheel housing mounting bolts.

9. Pull the transmission assembly straight away from the engine to clear the clutch hub with the input shaft and lower the transmission and remove from under the vehicle.

Installation

1. Shift the transmission into high gear.

2. Place the transmission on a dolly or a jack and move into position under the vehicle.

3. Raise the transmission and align the assembly with the engine, and move the unit forward. Align the transmission main drive gear shaft with the clutch hub by rotating the transmission output shaft or yoke.

4. Move the transmission forward until the flywheel housing seats on the engine or the clutch housing.

5. On the Apron type flywheel housing, install and torque the bolts to 55 to 65 ft. lbs.

6. On the S.A.E. type flywheel housing, install and torque the bolts to 25 to 30 ft. lbs.

7. Reinstall the rear mountings, if removed, and remove the engine support. Tighten the rear engine mounting bolts.

8. Reinstall the power take-off and controls, if equipped.

9. Connect the drive shaft and align the match-marks.

10. Connect the brake control cable or rod to the parking brake unit, if equipped.

11. Connect the speedometer cable, clutch control linkage, and any other equipment that may have been removed.

12. Shift the transmission into neutral and install the gear shift lever and control tower assembly.

13. Connect the back-up light switch wires at the transmission.

14. Fill the transmission with lubricant and adjust the clutch linkage as needed.

15. Install the transmission floor pan and the floor mat.

Shift Linkage Adjustment

Column Shift

1. Disconnect central rods and align both second/third and first/reverse shifter tube levers in neutral position.

2. Install gauge in holes provided to maintain alignment.

3. Position relay levers so that gearshift is in neutral position.

4. Connect control rods to tube levers and then remove gauge.

5. Move gearshift through pattern to check adjustment.

Tilt Cab Models

1. Place transmission selector and shift levers in "neutral" position.

2. Adjust selector and shift rods to provide 90° angularity at the lower end of the gearshift lever to the control island panel by rotating adjustable clevis on each rod to the desired position and then connecting rods and tightening lock nuts.

3. Check adjustment by moving the gearshift through the pattern.

4. Install new cotter pins and lubricate linkage.

   Refer to the General Repair Sec-

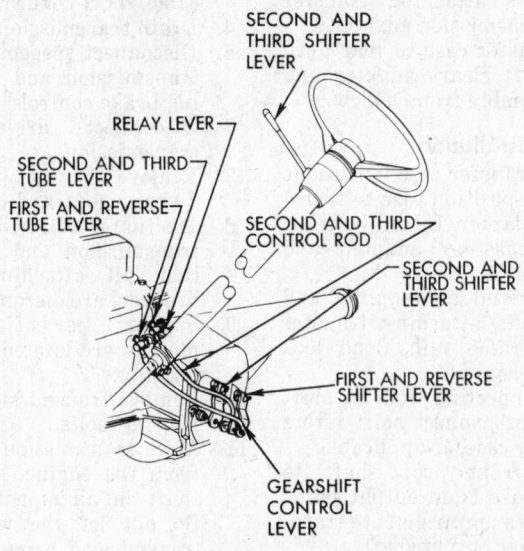

SECOND AND THIRD SHIFTER LEVER
RELAY LEVER
SECOND AND THIRD TUBE LEVER
FIRST AND REVERSE TUBE LEVER
SECOND AND THIRD CONTROL ROD
SECOND AND THIRD SHIFTER LEVER
FIRST AND REVERSE SHIFTER LEVER
GEARSHIFT CONTROL LEVER

3-speed column shift controls
(© G.M.C.)

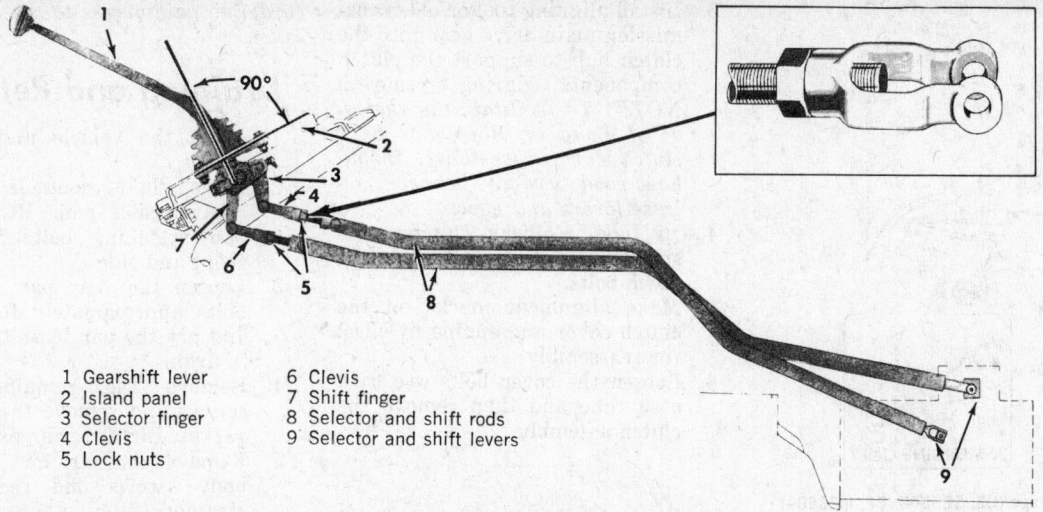

1 Gearshift lever   6 Clevis
2 Island panel   7 Shift finger
3 Selector finger   8 Selector and shift rods
4 Clevis   9 Selector and shift levers
5 Lock nuts

Manual transmission control linkage—tilt cab models (typical) (© G.M.C.)

tion for complete overhaul procedures, illustrations and out of truck adjustments.

Clutch

Clutches used on inline engines are single (driven) disc with either of two types of pressure (drive) plates, diaphragm or coil spring. "V" engines use a single disc with coil spring pressure plate. Heavy duty clutches use two discs. The two plate clutch consists of three basic assemblies, the cover with rear pressure plate assembly, front pressure plate and two discs. The front pressure plate, located between the two driven discs, has two friction surfaces and is coupled to the rear pressure plate through steel drive straps bolted at each of its four driving bosses. Diaphragm spring covers operate with light pedal pressure while coil spring levers combine operating ease and high torque capacity. The operating controls are either mechanical or hydraulic. Discs have torsion spring centers.

Free Pedal Travel

This adjustment is for the amount of pedal travel (measured at pedal) before clutch release bearing contacts the levers, or fingers of a coil spring cover or the diaphragm spring of a diaphragm clutch cover. This is called free-play. With normal clutch wear the amount of free-play is reduced and in time this will cause the release bearing to be in constant contact with the cover. This in turn will cause clutch disc slippage resulting in premature failure of disc and release bearing. It is necessary to maintain sufficient free pedal travel for clutch efficiency and long life.

Cover Assemblies

While no wear adjustment is needed (except Spicer), original settings at time of manufacture must be retained for good clutch operation. If a diaphragm clutch cover fails to release properly, after first checking pedal travel and pedal lash and linkage for looseness, replace diaphragm retracting springs. This can be done in the vehicle. If trouble still persists the diaphragm is probably overstressed.

With coil spring covers, the finger or lever adjusting screw nut is locked (staked) at time of manufacture and should not be disturbed unless rebuilding. Some Spicer heavy duty covers are internally adjusted for wear.

Adjustment

Mechanical Pedal Free Play

1. Disconnect return spring and loosen nuts.
2. Apply approximately 5 lbs. force to pushrod in direction of letter "G" in the illustration.

3. Move lever until pedal makes contact with stop.
4. Turn nut as necessary to obtain a clearance of 3/16-1/4" between nut and swivel.
5. Tighten lock nut and connect return spring.
6. Check pedal free play. The clutch pedal free play should be ¾"-1" on all models.

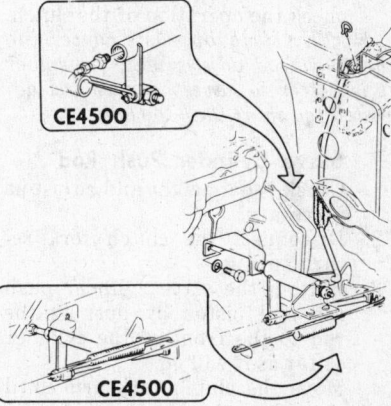

Clutch linkage (CE, CS4500; CS, SS5500)
(© G.M.C.)

Linkage adjustment
(all models, except V6, 427 engines)
(© G.M.C.)

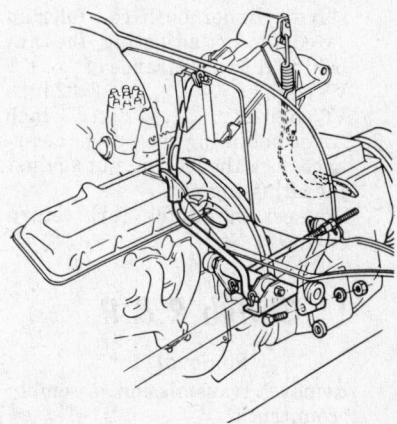

Clutch linkage (CE, ME6500—427 V8 engine)
(© G.M.C.)

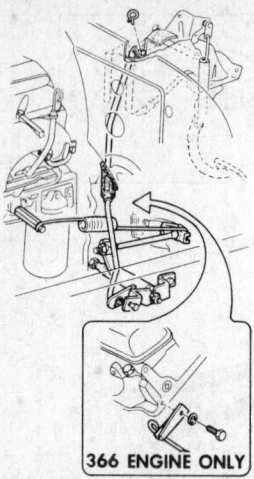

Clutch linkage (CE, SE5500; CE, ME6500)
(© G.M.C.)

Hydraulic Controls

Master Cylinder Push Rod

1. Loosen the adjusting nut on the push rod. Turn the rod in and out with the aid of pliers, so that a ⅛ inch movement of the clutch pedal at the pad end is necessary before the end of the push rod contacts the piston in the master cylinder.
2. Tighten the adjusting nut and check the operation of the clutch.
NOTE: If doubt exists concerning the push rod adjustment, remember it is better to have the push rod adjusted too short than too long.

Slave Cylinder Push Rod

1. Check pedal height and adjust as necessary.
2. Disconnect the clutch fork return spring.
3. Bottom the slave cylinder push rod and piston by pushing the rod to the front of the cylinder as far as it will go.
4. Move the clutch fork arm until the release bearing contacts the pressure plate levers or the diaphragm springs.
5. Adjust the clearance between the slave cylinder push rod fulcrum (wedge) by adjusting the nut behind it to a clearance of:
   V8 engines ....13/32 inch
   V6 engines ....½ inch
6. After obtaining the proper clearance, lock the locking nut against the adjusting nut.
7. Connect the clutch fork return spring.

Clutch R & R

Removal

1. Remove transmission assembly from truck.
2. Remove clutch release fork from ball stud.

3. Install aligning tool or old transmission main drive gear into the clutch hub to support the clutch components during removal.
NOTE: To facilitate the removal of Long or Borg and Beck clutch cover assemblies, install hardwood wedges between release levers and cover.
4. On Lipe-Rollway clutches, install three flat washers and hold down bolts.
5. Make alignment marks on the clutch cover and engine flywheel for reassembly.
6. Loosen the cover bolts one turn at a time and then remove the clutch assembly.

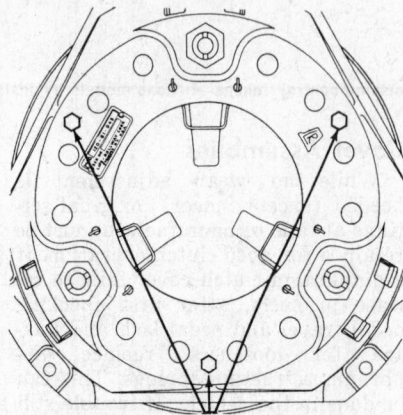

HOLD-DOWN BOLTS ¼"-20 x 2¼"
Use of hold-down bolts (Lipe-Rollway)
(© G.M.C.)

Installation

1. Install clutch assembly.
2. Install clutch cover assembly using the alignment tool or old transmission main drive gear. Make certain alignment marks match.
3. Install cover to flywheel bolts and torque slowly one turn at a time to specifications.
4. Remove aligning tool and wood wedges or hold-down bolts.
5. Connect clutch release mechanism. Lubricate clutch release bearing sparingly and apply a small amount of high temperature grease to recess in release fork. Tighten ball stud to specifications.
6. Install transmission assembly.
Refer to General Repair Section for clutch overhaul procedures.

Automatic Transmission

NOTE: The automatic transmissions covered in this section are Allison and Turbo Hydra-Matic. Removal and installation procedures are covered separately for each. It may be necessary to remove air tanks, fuel tanks, and optional equipment in

order to gain access to the transmission.

Draining and Refilling

1. Raise the vehicle and support safely.
2. Place a fluid receptacle under the transmission pan. Remove the pan attaching bolts from the front and side.
3. Loosen the rear pan attaching bolts approximately four turns and pry the pan loose to allow it to drain.
4. Remove the remaining pan screws and remove the pan and gasket. Discard the gasket.
5. Remove the strainer to valve body screws and remove the strainer (filter) and gasket and discard.
6. Install the new strainer and gasket and install the strainer to valve body screws and tighten.
NOTE: The 400/475 transmissions use a filter retainer bolt, oil filter assembly, O-ring seal and an intake pipe. The filter and O-ring seal must be replaced when the fluid is changed.
7. Install a new gasket on the pan and install the pan. Tighten the pan bolts to 12 ft. lbs. torque. Connect and tighten the filler tube.
8. Lower the vehicle and install 2.5 quarts of transmission fluid into the engine and start the engine.
9. Move the selector lever through the detents for each range. Add fluid to bring the level to ¼ inch below the ADD mark on the dipstick.

Powerglide

Shift Linkage Adjustment

Manual linkage adjustment and neutral safety switch are important from a safety standpoint. The neutral safety switch should be adjusted so that the engine will start in Park and Neutral positions only. With selector lever in Park position, the parking pawl should freely engage and prevent the vehicle from moving. The pointer on the indicator quadrant should line up with the range indicators in all ranges.

Check for Adjustment

With engine off, lift gearshift lever toward steering wheel, allow gearshift lever to be positioned in Drive (D) by the transmission detent. DO NOT use the indicator pointer to position the gearshift lever. Pointer is adjusted (or set) after linkage adjustment is made. Gearshift lever now won't engage reverse range unless lifted. A properly adjusted link-

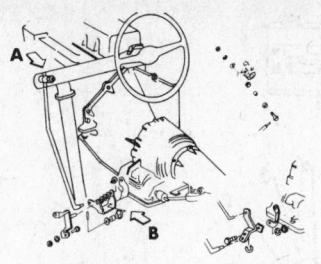

CA 100, 200

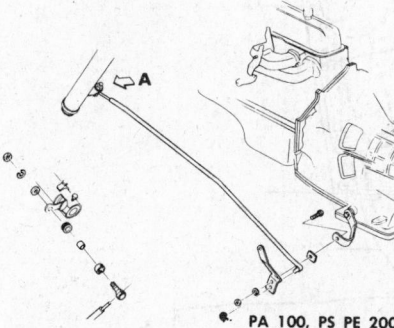

Powerglide shift linkage adjustments—typical (© GMC)

age will prevent the gearshift lever from moving beyond the Neutral detent and drive detent unless lever is raised to pass over mechanical stop in steering column.

Linkage Adjustment

1. Place the gearshift lever in Drive (D) position, as determined by detent.
2. Loosen adjustment swivel at mast jacket lever and rotate transmission lever so that it contacts the Drive stop in the steering column.
3. Tighten the swivel and recheck the adjustment.
4. Readjust indicator pointer, if necessary, to agree with transmission detent positions.
5. Readjust neutral safety switch, if necessary.

Low Band Adjustment

1. Raise vehicle and support on stands.
2. Place gearshift lever in Neutral.
3. Remove protective cap from adjustment screw on transmission.
NOTE: Powerglide throttle link-be necessary to remove rear mounting bolts from crossmember and move transmission slightly to passenger side for additional clearance.
4. Loosen adjusting screw locknut 1/4 turn and hold in this position during adjusting procedure.
5. Using an inch-pound torque wrench, adjust band to 70 inch pounds and back off four complete turns for a band which has been in service for 6000 miles or more, or three full turns for one with less than 6000 miles.

6. Tighten adjusting screw locknut to specified torque.
CAUTION: The back-off amount is not an approximate figure, it must be exact.

Throttle Linkage Adjustments

NOTE: Powerglide throttle linkage adjustments are made when carburetor throttle valve(s) is in closed position (hot idle) and transmission throttle lever is back against its internal stop. If shift occurs too early, shorten rod a turn or two and road test. If shift occurs too late, lengthen rod.

6 Cylinder Models

1. Remove air cleaner and disconnect carburetor return spring.
2. Rotate lever (A) to the wide open throttle position and T.V. lever through detent. (Depress accelerator fully).
3. Hold slot on rod (B) against pin on lever (A) and adjust swivel on rod so that it freely enters hole on lever (C).
4. Hold rod (B) perpendicular to pin on lever (A) and then tighten swivel nut. Rod must be 1/64"-1/16" off lever stop with transmission lever against internal stop. Connect carburetor return spring.
5. Check for throttle linkage freedom and proper operation.

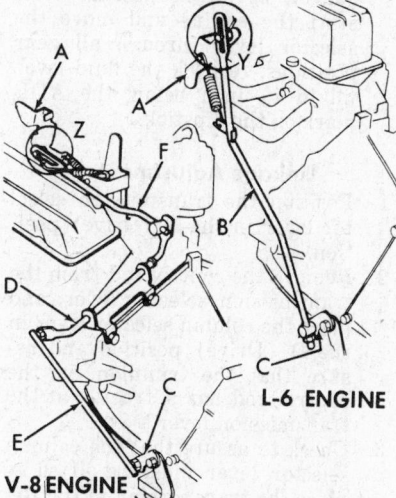

L-6 ENGINE

V-8 ENGINE

Powerglide T.V. valve linkage adjustments typical (© GMC)

V8 Models

1. Remove air cleaner.
2. Disconnect accelerator linkage at carburetor.
3. Disconnect accelerator / return and throttle valve rod return springs.
4. With right hand, pull T.V. upper rod (F) forward until transmission lever (C) is through detent. With left hand, rotate carburetor lever (A) to wide open throttle position. Carburetor must reach

wide open throttle position at the same time pin or ball stud on lever (A) contacts end of slot in upper T.V. rod.
5. Adjust swivel on upper end of T.V. rod (F) to obtain setting described in step 4 (approximately 1/32").
6. Connect and adjust accelerator linkage. Connect carburetor return springs.
7. Check for throttle linkage freedom and proper operation.

Powerglide Transmission R & R

Removal

1. Raise vehicle on hoist or support on stands.
2. Drain transmission if equipped with drain plug (or remove pan to drain, then replace pan using several bolts). Fluid can be drained after transmission removal, if desired.
3. On models with long fill pipe, remove dipstick and pipe.
4. Disconnect cooler lines (if equipped), vacuum modulator hose and speedometer cable at transmission. Support lines out of work area.
5. Disconnect ground cable, T.C.S. switch, neutral switch and backup lamp switch (if used).
6. Remove shift and throttle valve controls from transmission.
7. Disconnect drive shaft from rear of transmission.
8. Position transmission jack and secure transmission to jack with safety chain.
9. Raise jack slightly to relieve engine rear mount on transmission extension and remove transmission support crossmember (Vans—remove thru-bolt).
NOTE: Use care to remove any shims (if used) at crossmember. It is vital that exactly the same number be reinstalled.
10. Remove flywheel cover or converter under pan, scribe flywheel-converter marks for reassembly, remove flywheel to converter attaching bolts.
11. Support engine at oil pan rail with a jack to support engine weight when transmission is removed.
12. Vans—Remove linkage cross shaft frame bolts and accelerator linkage at bellcrank before lowering engine. Lower engine to allow rear mount housing to clear support bracket.
13. Lower rear of transmission slightly so that the upper transmission to engine bolts can be reached using a universal socket and a long extension. Remove upper bolts. **CAUTION: On V8 engines do not lower too far be-**

cause of distributor—firewall interference.

14. Remove remaining transmission mounting bolts.
15. Remove transmission by moving it slightly to the rear and downward, remove from under vehicle.

NOTE: Watch converter when moving transmission back. If it does not move with transmission pry it free of flywheel before proceeding.

CAUTION: Keep front of transmission up to prevent the converter from falling out. Install a converter hold tool or improvise.

For Overhaul Procedures—See General Repair Section.

Installation

1. Reverse removal procedures.
   NOTE: If reusing converter align scribed marks, align as closely as possible the white stripe painted on engine side of flywheel outer rim (heavy side) with the blue stripe painted on end of converter (light side) to maintain balance.
2. Refill transmission and check for proper operation and leaks.

Turbo Hydra-Matic 250

Removal and Installation

1. Disconnect the negative battery cable and raise the vehicle and support safely.
2. Disconnect the detent downshift cable at the carburetor and release the parking brake.
3. Remove the drive shaft, speedometer cable, modulator vacuum line and the cooler lines at the transmission.
4. Disconnect the shift linkage at the transmission and support the transmission with a suitable jack.
5. Disconnect the rear mount from the cross-member and remove the crossmember from the frame.
6. Remove the converter housing under pan.
7. Remove the converter to flywheel bolts.
8. Lower the engine-transmission assembly and remove the transmission to engine bolts.
9. Remove the filler tube from the transmission.
10. Support the rear of the engine and slide the transmission rearward, down and out of the vehicle.
    NOTE: Loss of the converter can occur if the transmission front is lower than the rear.
11. The installation is in the reverse of the removal procedure.
12. Fill the transmission with 2.5

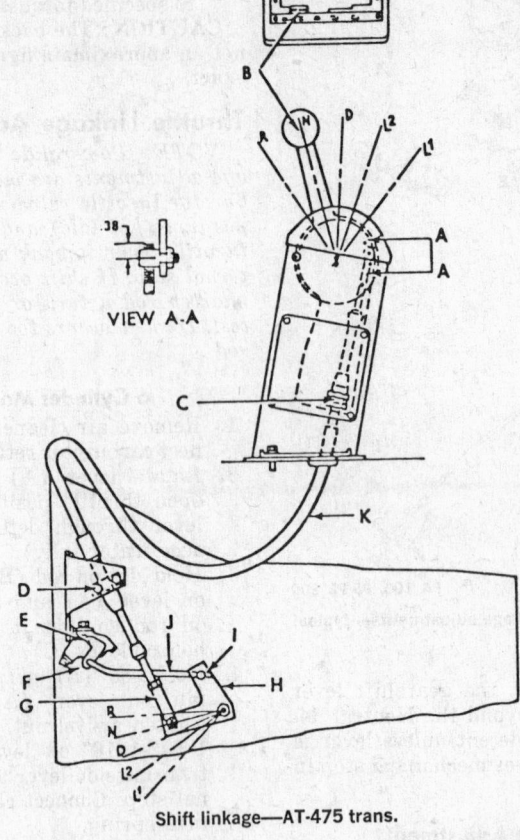

VIEW A-A

Shift linkage—AT-475 trans.

quarts of transmission fluid, start the engine and move the selector lever through all gear positions, recheck the fluid level, fill to 1/4 inch below the ADD mark on the dipstick.

Linkage Adjustment

1. Position the transmission selector lever in the D (Drive) position.
2. Remove the control rod from the transmission selector lever, and place the column selector lever in the D (Drive) position and assure that the trunnion on the control rod has a free fit in the transmission lever.
3. Check to assure that the column selector lever must be lifted to place the transmission in the Reverse and Low ranges.
   NOTE: Do not use the indicator pointer to position the shift selector lever.
4. Install the control rod retainer.
5. Adjust the indicator needle by loosening and repositioning the cable end mounted to the steering housing shift tube.
6. Adjust the neutral start switch to allow the engine to start in Park and Drive only, by loosening the attaching screws and moving the switch either right or left. Tighten the attaching screws.

Intermediate Band Adjustment

1. Place the selector lever in Neutral and raise the vehicle and support safely.
2. Loosen the adjusting screw lock nut and tighten the adjusting screw to 30 inch pounds torque and back off exactly three turns.
3. Holding the adjusting screw in position, tighten the lock nut to 15 ft. lbs. torque.
4. Lower the vehicle.

Downshift Cable Detent

Adjustment

With the snap lock disengaged, position the carburetor to the wide open throttle position and push the snap lock downward until the top is flush with the cable, thereby adjusting the cable length.

Turbo Hydra-Matic 350, 400, 475

Linkage Adjustment— Light Duty Trucks

1971-78 Cabs, Suburbans, 4-wheel drive
1974-78 Forward Control exc. Vans

1. Place gearshift lever in Drive (D), as determined by transmission detent. Obtain Drive

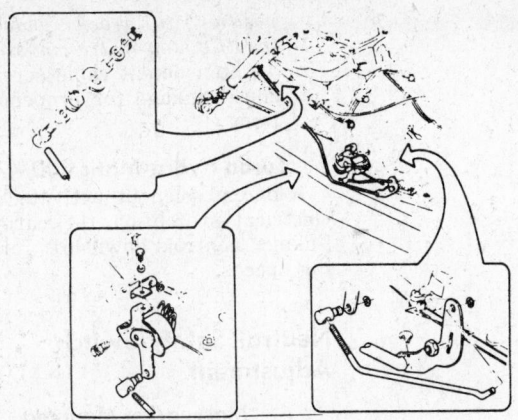

Manual selector linkage—"P" models (Turbo Hydra-Matic)
(© G.M.C.)

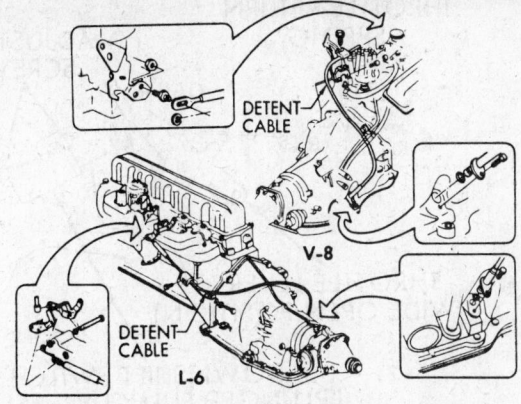

Detent cable adjustment (Turbo Hydra-Matic 350)
(© G.M.C.)

position by rotating transmission lever counterclockwise to detent, then clockwise two detent positions to Drive.

2. Loosen adjustment swivel at mast jacket lever and rotate transmission lever so that it contacts the Drive stop in the steering column.
3. Tighten swivel and recheck adjustment.
4. Readjust indicator pointer, if necessary, to agree with transmission detent positions.
5. Readjust Neutral safety switch if necessary.

1971-73—Forward Control
1971-74—Vans

1. Set transmission lever in Drive position. Obtain Drive position by rotating transmission lever counter-clockwise to Low detent, then clockwise two detent positions to Drive.
2. Attach control rod to lever and shaft assembly with retainers.
3. Assemble ring, washers, grommet, swivel, retainer and nut loosely on shaft.
4. Insert control rod in swivel and retainer, attach opposite end to tube and lever assembly.
5. Set tube lever assembly in Drive position and tighten nut.

NOTE: When above procedure is adhered to, the following conditions

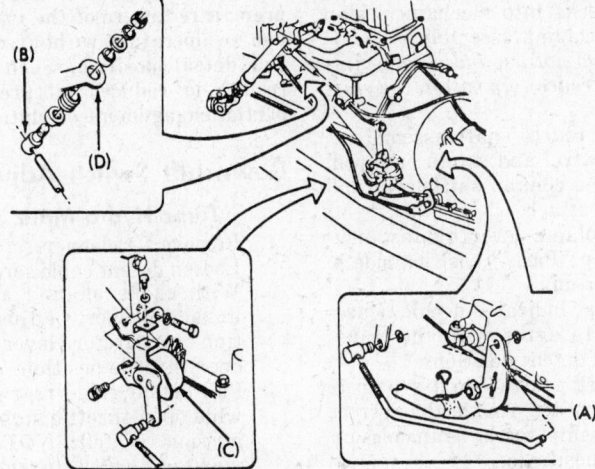

Linkage adjustment—Turbo Hydra-matic—1970-76 Cabs.
1974-76 Forward control exc. Vans

must be met by manual operation of the gearshift lever. From Reverse to Drive position travel, the transmission detent feel must be noted and related to indicated position on dial. When in Drive and Reverse position, pull lever toward steering wheel and then release. It must drop back into position with no restriction.

1975-78 Vans

1. The shift tube and lever assembly must be free in the mast jacket.

2. Set transmission lever in "neutral" position by one of the following optional methods. NOTE: Obtain "neutral" position by moving transmission lever counter-clockwise to "LI" detent, then clockwise three detent position to "neutral" or obtain "neutral" position by moving transmission lever clockwise to the "park" detent then counter-clockwise two detents to "neutral."'

3. Set the column shift lever in

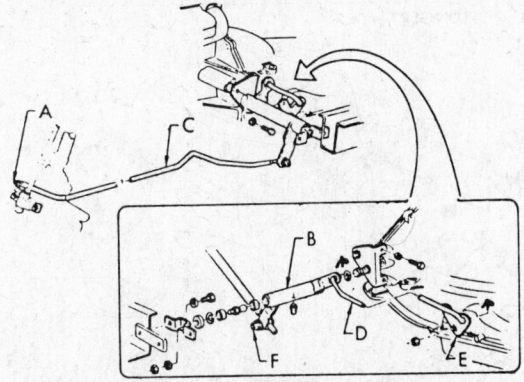

Linkage adjustment—Turbo-Hydra-Matic 1970-73
Forward control, 1970-74 Vans.

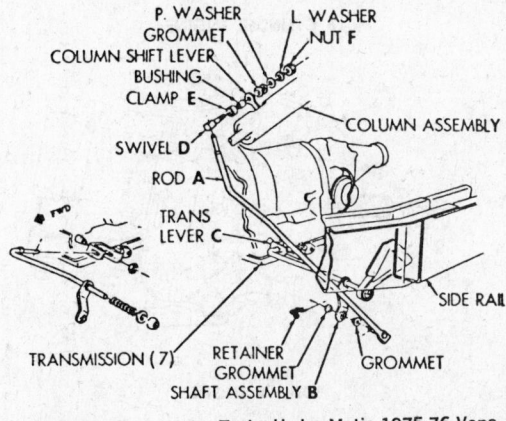

Linkage adjustment—Turbo-Hydra-Matic 1975-76 Vans.

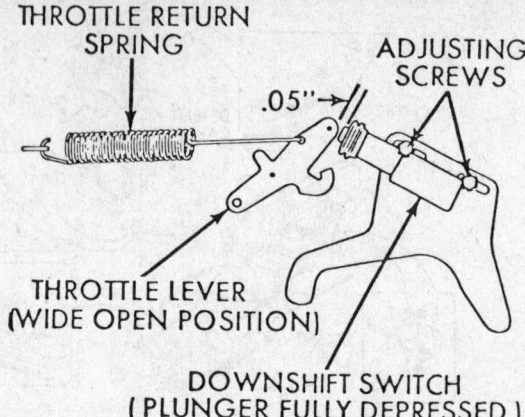

Downshift linkage—AT-475 trans.

"neutral" position. This is obtained by rotating shift lever until it locks into mechanical stop in the column assembly.

NOTE: Do not use indicator pointer on a reference to position the shift lever.

4. Attach rod to shaft assembly.
5. Slide swivel and clamp onto rod, align the column shift lever and loosely attach.
6. Hold column lever against neutral stop "Park"' position side.
7. Tighten nut.
8. Readjust indicator needle if necessary to agree with the transmission detent positions.
9. Readjust neutral start switch if necessary to provide the correct relationship to the transmission detent positions.

CAUTION: Any inaccuracies in the above adjustments may result in premature failure of the transmission due to operation without controls in full detent position. Such operation results in reduced oil pressure and partial engagement of clutches.

Downshift Switch Adjustment

Turbo Hydra-Matic 350

1. Remove air cleaner.
2. Loosen detent cable screw.
3. With choke off and accelerator linkage properly adjusted, position carburetor lever in wide open throttle position.
4. Pull detent cable rearward until wide open throttle stop in transmission is felt. *NOTE: Cable must be pulled through detent position to reach wide open throttle stop in transmission.*
5. Tighten detent cable screw and check linkage for proper operation.

Turbo Hydra-Matic 400/475

A detent solenoid, activated by an electrical switch on the carburetor linkage, controls downshift for passing speeds.

Neutral Safety Switch Adjustment

Mast Jacket Mounted

1. Place gearshift lever in Neutral (N).
2. Loosen retainer screws holding switch, install 3/32" drill (or pin) through hole in lower switch arm and bracket. Adjust position of switch until engine turns over (with ignition switch in start).

Transmission Mounted

1. Place gearshift lever in Neutral (N), loosen transmission lever extension bolt.
2. Pin switch lever in Neutral position with 3/32" drill or pin.
3. Install rod into switch lever, adjust swivel on rod to allow free entry of rod into lever.
4. Secure rod with retainer, tighten transmission lever extension bolt.
5. Check adjustment by testing for

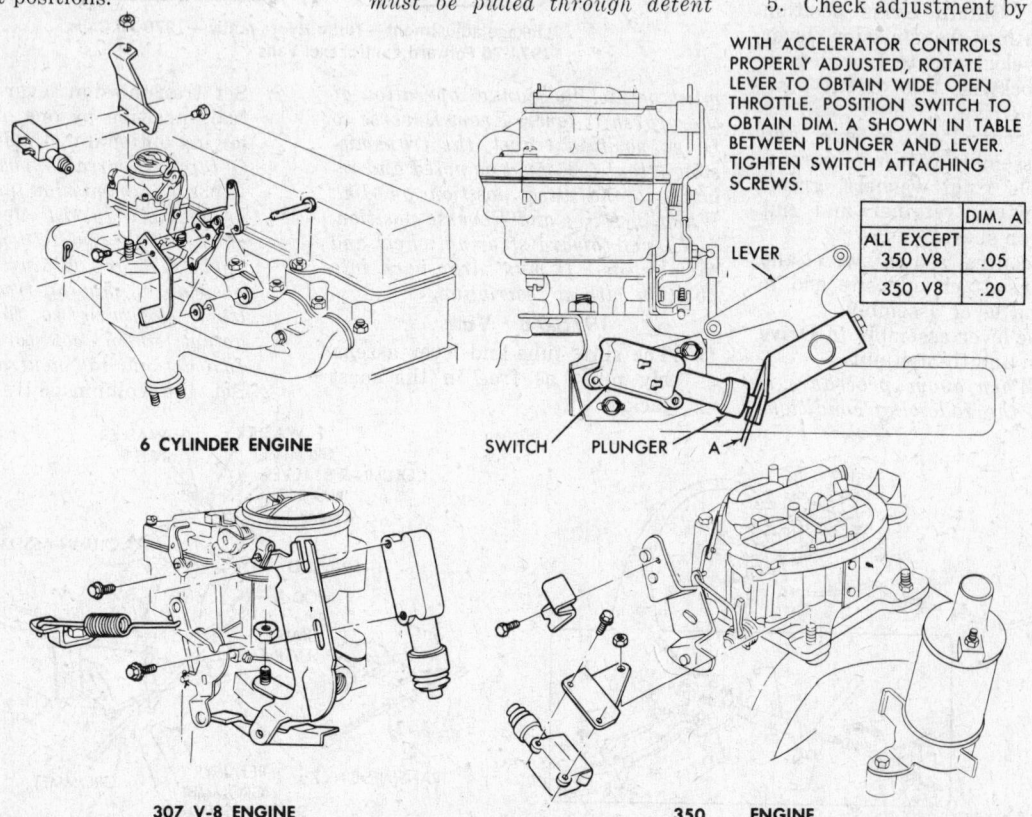

WITH ACCELERATOR CONTROLS PROPERLY ADJUSTED, ROTATE LEVER TO OBTAIN WIDE OPEN THROTTLE. POSITION SWITCH TO OBTAIN DIM. A SHOWN IN TABLE BETWEEN PLUNGER AND LEVER. TIGHTEN SWITCH ATTACHING SCREWS.

	DIM. A
ALL EXCEPT 350 V8	.05
350 V8	.20

Detent switch adjustment (Turbo Hydra-Matic 400) © G.M.C.

cranking in both Neutral and Park.

Transmission R & R

Turbo Hydra-Matic 350 (Except 4 Wheel Drive)

Removal

1. With vehicle on hoist drain by removing pan or drain plug (if so equipped). *NOTE: Fluid can be drained after transmission is removed if so desired.*
2. Remove the vacuum modulator line and speedometer cable from the transmission and secure out of way.
3. Remove detent cable and manual control lever from the transmission.
4. Disconnect the drive shaft and remove.
5. Place suitable jack or other support under transmission and secure transmission to it.
6. At transmission extension disconnect rear engine mount and then remove support crossmember.
7. Remove converter underpan. Place marks on the flywheel and converter to insure proper installation and then remove the flywheel-to-converter bolts.
8. Support engine at oil pan rail with jack capable of supporting the weight of the engine when transmission is removed.
9. Lower rear of the transmission slightly so that upper, housing-to-engine, transmission bolts can be reached with a long extension and universal socket. Remove upper bolts. *NOTE: Have an assistant watch upper engine parts to make sure everything clears when rear of transmission is being lowered.*
10. Remove remaining transmission-to-housing bolts.
11. Remove transmission by moving it slightly to the rear and downward. Remove from under vehicle.

CAUTION: Watch converter when removing transmission to be sure that it moves with transmission. If it does not move, pry it free from flywheel before proceeding any further.

NOTE: On those vehicles so equipped, disconnection of the catalytic converter may be necessary to provide adequate clearance for transmission removal.

NOTE: Keep transmission front upward when removing transmission to prevent converter from falling out. Install converter holding tool after removal from engine.

For overhaul procedures—see General Repair Section.

Installation

1. Mount transmission on transmis-

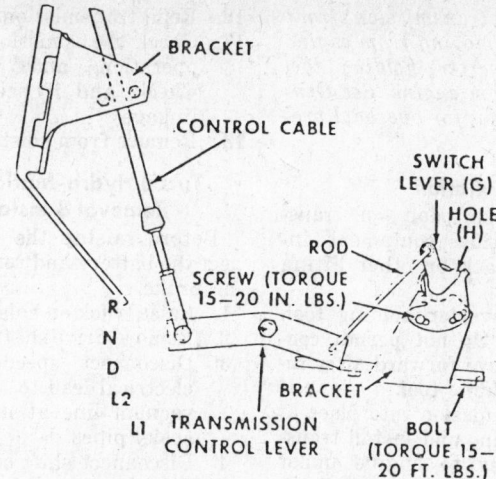

Neutral safety and back-up lamp switch—1975-76 AT-475 trans.

sion lifting equipment installed on jack or other lifting device.
2. Remove converter holding tool. **CAUTION: Do not permit converter to move forward after removal of holding tool.**
3. Raise transmission into place at rear of engine and install transmission case to engine upper mounting bolts, then install remainder of the mounting bolts.
4. Remove support from beneath engine, then raise rear of transmission to final position.
5. If scribed during removal, align scribe marks on flywheel and converter cover. Install converter to flywheel attaching nuts and bolts.
6. Install converter underpan.
7. Reinstall transmission support crossmember to transmission and frame.
8. Remove transmission lift equipment.
9. Connect drive shaft to transmission.
10. Connect shift control lever rod and detent cable to transmission.
11. Connect vacuum modulator line, and speedometer drive cable to transmission.
12. Lower vehicle.
13. Refill transmission.
14. Check transmission for proper operation and for leakage. Check and, if necessary, adjust linkage.
15. Remove vehicle from hoist.

Turbo Hydra-Matic 350— (4 Wheel Drive) Removal

1. With vehicle on hoist drain by removing pan or drain plug (if so equipped). *NOTE: Fluid can be drained after transmission removal if so desired.*
2. Remove shift lever and rod from transfer case.
3. Remove speedometer cable and vacuum modulator line from

transmission and secure out of way.
4. Disconnect detent cable and manual control lever rod from transmission.
5. Remove front and rear drive shafts from transfer case.
6. Place jack or other support under transfer case and remove transmission-to-adapter case bolts.
7. Place jack or other support under transmission and secure transmission to it.
8. Remove transfer case-to-frame bracket bolts and remove transfer case.
9. On V8 engines remove exhaust crossover pipe.
10. Disconnect and remove rear transmission crossmember.
11. Remove converter underpan. Place marks on the flywheel and converter to insure proper installation and then remove the flywheel-to-converter bolts.
12. Support engine at oil pan rail with jack capable of supporting engine weight when transmission is removed.
13. Lower rear of transmission slightly so that upper housing-to-engine transmission bolts can be removed with a long extension and universal socket. Remove upper bolts. *NOTE: Have an assistant watch upper engine parts to make sure everything clears when transmission is lowered.*
14. Remove the remaining transmission-to-housing bolts.
15. Remove the transmission by moving it slightly to the rear and downward. Remove from under vehicle.

CAUTION: Watch converter when removing transmission to make sure converter moves with transmission. If it does not, pry it loose from flywheel before proceeding any further.

NOTE: Keep transmission front upwards when removing from engine to prevent converter holding tool after removal from engine. See General Repair Section for overhaul procedures.

Installation

1. Mount transmission on transmission lifting equipment installed on jack or other lifting device.
2. Remove converter holding tool. **CAUTION: Do not permit converter to move forward after removal of holding tool.**
3. Raise transmission into place at rear of engine and install transmission case to engine upper mounting bolts, then install remainder of the mounting bolts.
4. Remove support from beneath engine, then raise rear of transmission to final position.
5. If scribed during removal, align scribe marks on flywheel and converter cover. Install converter to flywheel attaching bolts.
6. Install flywheel cover.
7. Place transfer case and adapter assembly at rear of transmission on suitable lift equipment and install transfer case to frame bracket attaching bolts.
8. Reinstall transmission to transfer case adapter attaching bolts and remove lift equipment.
10. Connect front and rear drive shafts to transfer case.
11. Install exhaust system cross pipe.
12. Connect manual control lever rod and detent cable to transmission.
13. Connect vacuum modulator line and speedometer drive cable to transmission.
14. Assemble rod on transfer case shift lever before installing rod to transfer case shift linkage.
15. Lower vehicle.

16. Refill transmission.
17. Check transmission for proper operation and for leakage. Check, and if ncessary, adjust linkage.
18. Remove from hoist.

Turbo Hydra-Matic 400/475 Removal & Installation

Before raising the truck, disconnect the battery and release the parking brake.
1. Raise truck on hoist.
2. Remove drive shaft.
3. Disconnect speedometer cable, electrical lead to case connector, vacuum line at modulator, and cooler pipes.
4. Disconnect shift control linkage.
5. Support transmission with transmission jack.
6. Disconnect rear mount from frame crossmember.
7. Remove two bolts at each end of frame crossmember and remove crossmember.
8. Remove converter under pan.
9. Remove converter to flywheel bolts.
10. Loosen exhaust pipe to manifold bolts approximately ¼ inch, and lower transmission until jack is barely supporting it.
11. Remove transmission to engine mounting bolts and remove filler tube at transmission.
12. Raise transmission to its normal position, support engine with jack and slide transmission rearward from engine and lower it away from vehicle.

CAUTION: Use converter holding tool when lowering transmission or keep rear of transmission lower than front so as not to lose converter.

The installation of the transmission is the reverse of the removal with the following additional steps.

Before installing the flex plate to converter bolts, make certain that the weld nuts on the converter are flush

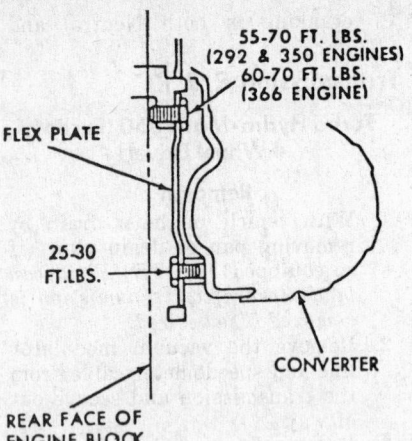

55-70 FT. LBS. (292 & 350 ENGINES)
60-70 FT. LBS. (366 ENGINE)
FLEX PLATE
25-30 FT.LBS.
CONVERTER
REAR FACE OF ENGINE BLOCK

Flex plate installed (in line and V-8 engines) AT-475

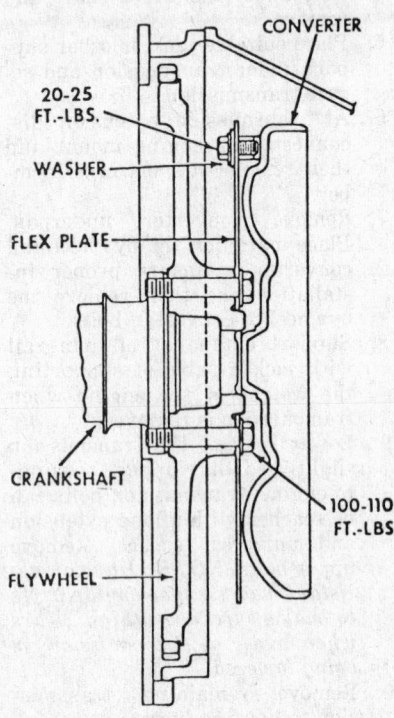

CONVERTER
20-25 FT.-LBS.
WASHER
FLEX PLATE
CRANKSHAFT
100-110 FT.-LBS
FLYWHEEL

Flex plate installation (V-6 engine)—AT-475

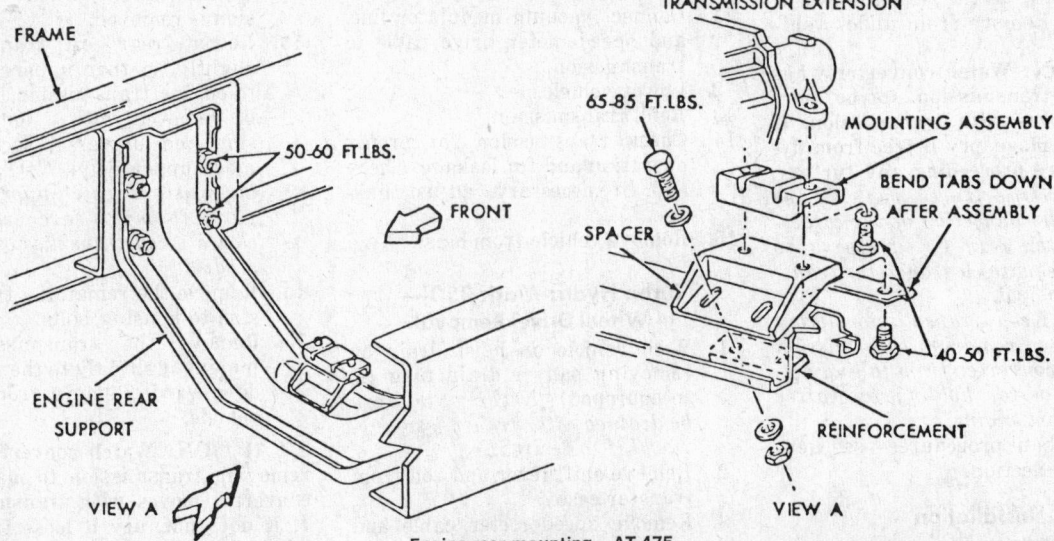

FRAME
50-60 FT.LBS.
FRONT
ENGINE REAR SUPPORT
VIEW A

TRANSMISSION EXTENSION
65-85 FT.LBS.
MOUNTING ASSEMBLY
BEND TABS DOWN AFTER ASSEMBLY
SPACER
40-50 FT.LBS.
REINFORCEMENT
VIEW A

Engine rear mounting—AT-475

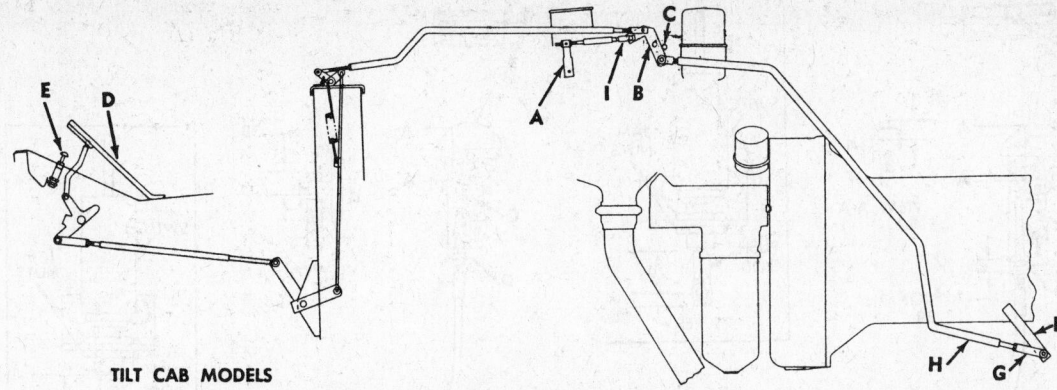

TILT CAB MODELS

TV linkage adjustment—Allison—MT-40, AT-41.

with the flex plate and the converter rotates freely by hand in this position. Then, hand start all three bolts and tighten finger tight before torquing to specification. This will insure proper converter alignment.

NOTE: After installation of transmission check linkage for proper adjustment and check for leaks.

Allison—AT540, MT640, MT650 Automatic Transmission

Shift Linkage Adjustment

With engine off and vehicle wheels blocked, shift the selector lever through each drive range while feeling for full engagement in the transmission. The transmission shift control linkage should fully engage all transmission detent positions in the transmission just before the range selector lever hits the stops incorporated in the range selector cover. Note the position of the selector lever after each shift. The transmission detent "R" (Reverse) should not engage until the selector lever is completely out of the neutral notch. If the selector lever is not properly located or operating, adjust the linkage.

Gas and Diesel Engines

1. Disconnect clevis from transmission shift lever by removing cotter pin and clevis pin.
2. Disconnect control cable from anchor point at bracket on transmission and remove cable retainer clips from underside of cab.
3. Remove four cross recess screws retaining range selector cover to tower or bracket, depending on truck model. Lift range selector assembly out to inspect cable attachment at range selector lever and hanger.
4. Check cable at trunnion for dimension "A". This dimension is necessary to allow for proper cable length at clevis. If adjustment is necessary, remove cable from range selector lever and hanger assembly as follows:
   a. Remove 1/2 inch locknut from trunnion.
   b. Disconnect cable from anchor point on hangers by removing two locknuts and U-bolt.

NOTE: Locknuts can be removed and reinstalled up to six times before their replacement becomes necessary.

   c. Mark location of trunnion. Remove trunnion with attached cable core from lever.

CAUTION: The trunnion has been placed in the proper hole at the factory. Any change in the location could cause vehicle operation to be dangerous.

5. With control cable removed, loosen jam nut at trunnion.

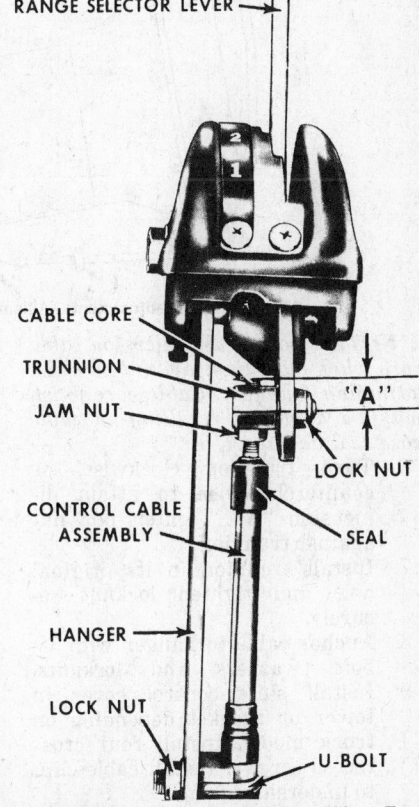

Trunion and cable core assembly—Allison Trans.

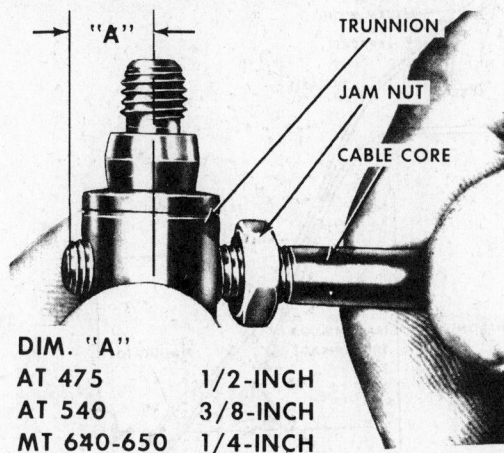

DIM. "A"
AT 475   1/2-INCH
AT 540   3/8-INCH
MT 640-650   1/4-INCH

Trunion adjustment—Allison Trans.

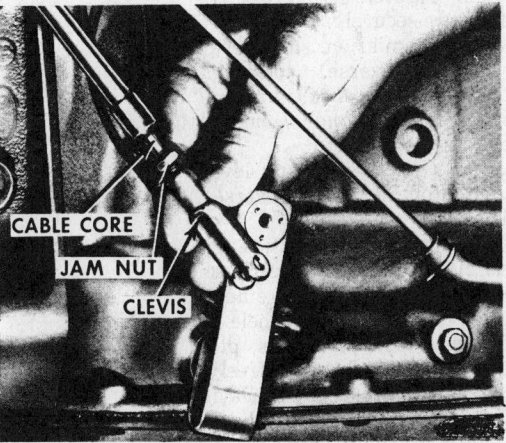

Clevis adjustment—Allison Trans.

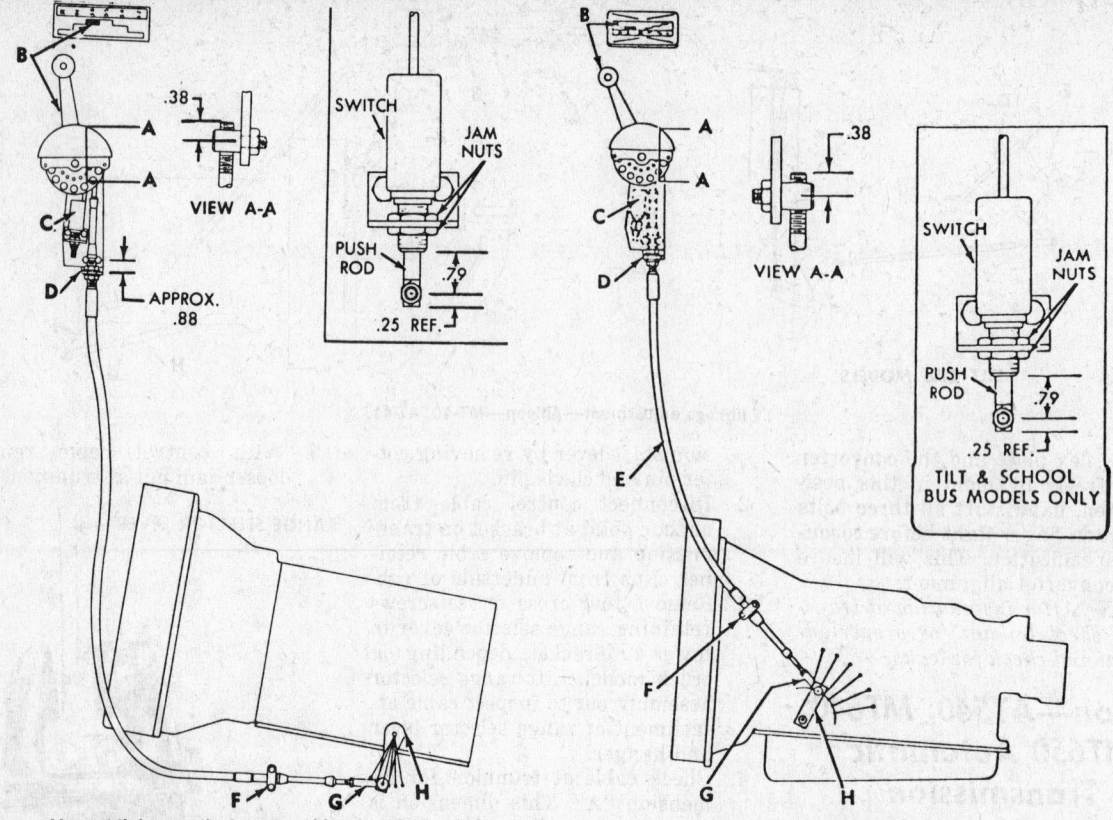

Manual linkage adjustment—Allison—MT-40, AT-41

Shift linkage—AT-540 trans.

NOTE: Do not use pliers on cable core when loosening jam nut or when adjusting trunnion. Cable core finish may be damaged resulting in cable core seal damage.

6. Turn trunnion clockwise, or counterclockwise to attain dimension "A". Tighten jam nut against trunnion.
7. Install trunnion in its original hole and tighten locknut securely.
8. Anchor cable to hanger with U-bolt, washers and locknuts.
9. Install shift control cover in tower or bracket depending on truck model. Install four cross recess screws. Install cable clips to underside of cab.
10. Anchor cable to bracket at transmission. Tighten two cross recess screws securely.
11. Locate transmission shift lever in "R" (Reverse) position. *NOTE: "R" position is full counterclockwise movement of lever.*
12. Locate range selector lever against stop in "R" (Reverse) position.
13. Loosen jam nut at clevis. Turn clevis clockwise or counterclockwise on threaded cable core until holes in clevis align with hole in shift lever. Install clevis pin, note that pin should enter freely, if it does not, adjust clevis slightly by ½ turn in each direction until pin does enter freely.
14. Turn clevis one full turn clock-

wise to allow for cable backlash.
15. Connect clevis to shift lever with clevis pin and a new cotter pin, but do not spread cotter pin at this time.
16. Move the range selector lever through all drive ranges. The transmission detents should fully engage just before the range selector lever hits the stops incorporated in the shift control cover.

Throttle Valve Linkage Adjustment Diesel Engines

1. Apply parking brake and block vehicles's driving wheels.

2. Disconnect upper TV cable end from accelerator cross shaft.
3. Push upper cable end toward transmission modulator until seated against the stop.
4. Assemble slotted clevis end to cable with locknut.
5. With accelerator linkage in "IDLE" position, upper end of the slotted clevis must be against the clevis pin on the cross shaft lever.
6. Shorten cable at slotted clevis by turning clevis three complete turns clockwise to locate lower end of the cable in actuator ⅛-inch from idle stop.
7. Tighten locknut securely.

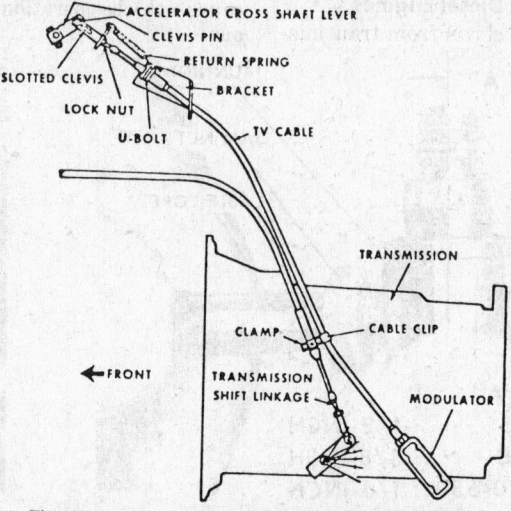

Throttle valve linkage (diesel engine)—Allison Trans.

8. Assemble slotted clevis to cross shaft lever with clevis pin and a new cotter pin.
9. Attach TV return spring to bracket.

Neutral Safety and Backup Lamp Switch Adjustment

Allison (AT 540) Transmission Tilt & Schoolbus Models

NOTE: "Shift Linkage Adjustment" should be performed as described previously, prior to adjustment of the neutral safety and back-up lamp switch.

1. Block driving wheels, apply parking brake, and perform the following to prevent the vehicle from accidentally starting.
NOTE: Pull the secondary wire out of center socket in the distributor cap and ground wire to prevent possible damage to coil.
2. Move selector lever to "N" (Neutral) position. Then, loosen jam nuts and adjust length of push rod.
3. With switch push rod properly adjusted, tighten jam nuts securely.
4. Check each range position of shift linkage to make sure the starter does not operate with the selector lever in any position other than "N." Have assistant check for proper operation of back-up lights with selector lever in "R". If necessary, readjust switch.
5. Reconnect secondary wire to distributor cap.
NOTE: The neutral safety and back-up lamp switches are non-adjustable on the Allison-MT-640, MT-650 transmission and all diesel engines with automatic transmissions.

Fluid and Filter Changes

AT and MT Models

1. Have the transmission at normal operating temperature (160 to 220 degrees F), and the transmission in neutral.
2. Remove the fill tube from the pan or the drain plug from the right side of the transmission pan. Allow the fluid to drain into a large container. *NOTE: Do not allow fluid to spill and splash. Burns can result.*
3. Remove the pan bolts, loosen and remove the pan and gasket. Discard the gasket.
4. Remove the one screw that retains the filter, remove the filter and discard. *NOTE: Later models will have a suction tube that separates from the filter. Retain the tube for use with the new filter. The tubes will have a sealring. Replace with a new sealring upon installation.*

5. To clean or replace the governor feed screen, the valve body must be removed on the AT models, and the screen taken from the governor feed bore. Refer to the valve body removal and installation procedures outlined in the General Repair Section. The MT Models have discontinued the use of the primary governor screen, and if one is found in the governor feed tube area of the valve body, discard it.
6. The MT Models have the main governor screen located in the rear cover. Early models will have a screen to be cleaned and reinstalled while the later models will have a replaceable cartridge type filter. *NOTE: The screen or filter is inserted into the rear cover open end first.*
7. Retain the screen or filter with the plug or cap.
8. Install the new filter, suction tube and sealring in the sump area and secure with the retaining bolt. Torque to 10 to 15 ft. lbs. for the MT models and 10 to 13 ft. lbs for the AT models.
9. Install the pan with a new gasket. Torque the pan bolts to 10 to 15 ft. lbs. for the MT models and 10 to 13 ft. lbs. for the AT models.
10. Install the filler tube or drain plug into the transmission or pan.
11. Install 10 quarts of transmission fluid into the AT models, and 15 quarts into the MT models. Start the engine, check for leaks, move the selector through the gear positions, and recheck the fluid level of the transmission and refill to the full mark on the dipstick.

Transmission R & R

AT540   Removal

NOTE: It may be necessary to remove the air tanks, fuel tanks, special equipment, etc., on some vehicles to provide clearance before the transmission is removed.

1. Block vehicle so that it cannot move. Disconnect ground strap from battery negative (—) post. Remove the spark plugs so the engine can be turned over manually.
2. Remove the level gauge (dipstick). Drain transmission by disconnecting filler tube at right side of transmission pan. Remove bracket holding filler tube to transmission and remove filler tube from vehicle. Replace dipstick in tube and cover the pan opening to prevent entry of foreign material.
3. Disconnect cooler lines from fittings on right side of trans-

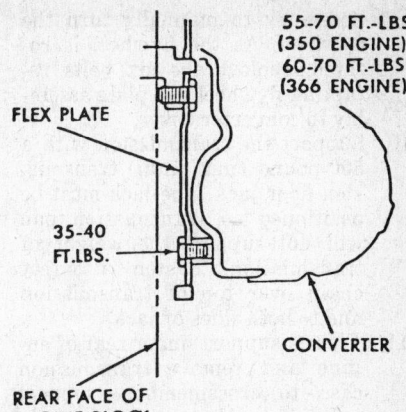

55-70 FT.-LBS (350 ENGINE)
60-70 FT.-LBS (366 ENGINE)
FLEX PLATE
35-40 FT.LBS.
CONVERTER
REAR FACE OF ENGINE BLOCK

Flex plate installation (V-8 engine)—AT-540.

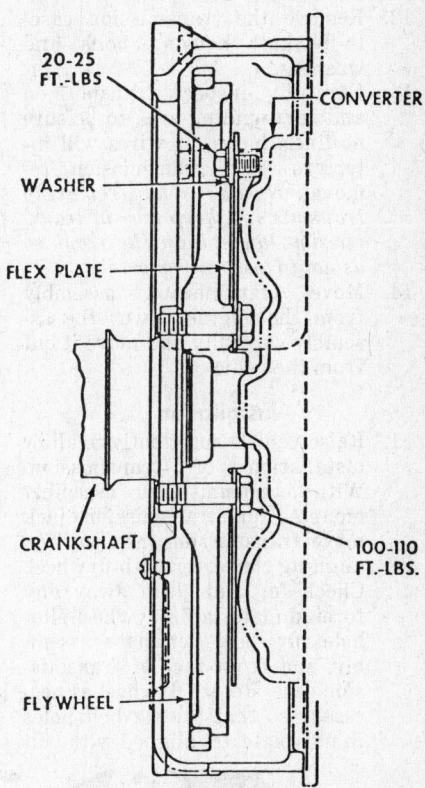

20-25 FT.-LBS
CONVERTER
WASHER
FLEX PLATE
CRANKSHAFT
100-110 FT.-LBS.
FLYWHEEL

Flex plate installation (V-6 engine)—AT-540.

mission case. Plug line ends and case openings with lint-free material.
4. Disconnect the range selector cable from shift lever at left-side of transmission.
5. Disconnect vacuum modulator line from modulator. Also, on conventional cab models, disconnect wiring from neutral safety and back-up lamp switches.
6. Disconnect the speedometer shaft fitting from adapter at rear of transmission.
7. Disconnect the drive shaft from transmission.
8. Disconnect the mechanical parking brake linkage at the right side of the transmission (if used).
9. Through the opening in the flywheel housing, use a pry-bar, as

necessary to manually turn the flywheel. As the flywheel is rotated, remove the six bolts retaining flywheel flex plate assembly to converter cover.

10. Support the transmission with a 500-pound (minimum) transmission floor jack. The jack must be positioned so transmission pan will not support the weight of transmission. Fasten a safety chain over top of transmission and to both sides of jack.

11. Place a support under rear of engine and remove transmission case - to - crossmember support bolts. Raise the engine to remove weight from the engine rear mounts.

12. Remove the transmission case-to-flywheel housing bolts and washers.

13. Carefully inspect transmission and surrounding area to be sure no lines, hoses, or wires will interfere with transmission removal. *NOTE: When removing transmission, keep rear of transmission lower than the front so as not to lose converter.*

14. Move transmission assembly from the engine, lower the assembly carefully and move it out from the vehicle.

Installation

1. Raise vehicle sufficiently to allow installation of transmission. With transmission assembly mounted on transmission jack move transmission into position aligning converter with flywheel. Check for and clean away any foreign material in flywheel pilot hole, flywheel flex-plate assembly, and front face of transmission case. Rotate flywheel as necessary so that the six bolt holes in flex-plate are aligned with bolt holes in converter cover. Carefully move transmission assembly toward engine so flex-plate-to-converter cover bolts can be loosely installed and so that pilot on transmission converter enters pilot hole in center of flywheel.

2. Install bolts and washers that attach transmission case to flywheel housing. Tighten bolts to 25-30 foot-pounds torque.

3. Tighten the six flex-plate-to-converter cover bolts to 35-40 foot-pounds torque

4. Carefully lower engine and transmission assembly onto engine rear mounts. Tighten engine rear mounting bolts to 60-70 foot-pounds torque. Then bend lock tabs down over head of each bolt. Remove lifting equipment from beneath vehicle.

5. Remove plugs from cooler lines and transmission case fittings. Be sure fittings are clean and lint-free, then connect cooler lines to transmission.

6. Install filler tube and bracket on right side of transmission. Install level gauge (dipstick).

7. Connect the speedometer shaft fitting to adapter at rear of transmission.

8. Connect drive shaft to transmission.

9. Connect parking brake linkage (if used) at side of transmission.

10. Connect the range selector cable to shift lever at left side of transmission.

11. Connect the vacuum modulator line to modulator. Also, on conventional cab models, connect wiring to neutral safety and back-up lamp switches. *NOTE: Make sure the ignition switch is in the off position before proceeding to the next step.*

12. Install spark plugs and connect battery ground strap, previously disconnected.

13. Connect any other lines, hoses, or wires which were disconnected to aid in transmission removal.

14. Adjust the shift linkage. (see "Shift Linkage Adjustment")

15. Refill the transmission with the proper lubricant.

MT640, MT650—Removal

NOTE: It may be necessary to remove the air tanks, fuel tanks, special equipment, etc., on some vehicles to provide clearance before the transmission is removed.

1. Block vehicle so that it cannot move. Disconnect ground strap from battery negative (−) post. Remove the spark plugs so the engine can be turned over manually.

2. Remove the level gauge (dipstick). Drain transmission by disconnecting filler tube at right side of transmission pan. Remove bracket holding filler tube to transmission and remove filler tube from vehicle. Replace dipstick in tube and cover the pan opening to prevent entry of foreign materials.

3. Disconnect cooler lines from fittings on right side of transmission case. Plug line ends and case openings with lint-free material.

4. Disconnect shift cable from shift lever at left side of transmission.

5. Disconnect vacuum modulator line from modulator. Also, disconnect wiring from back-up lamp switch (right side of transmission) and neutral safety switch (left side of transmission).

6. Disconnect the speedometer shaft fitting from adapter at

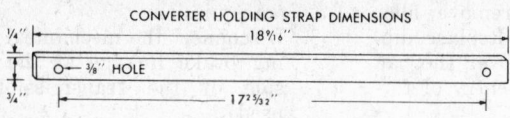

CONVERTER HOLDING STRAP DIMENSIONS

Converter holding strap
(© G.M.C.)

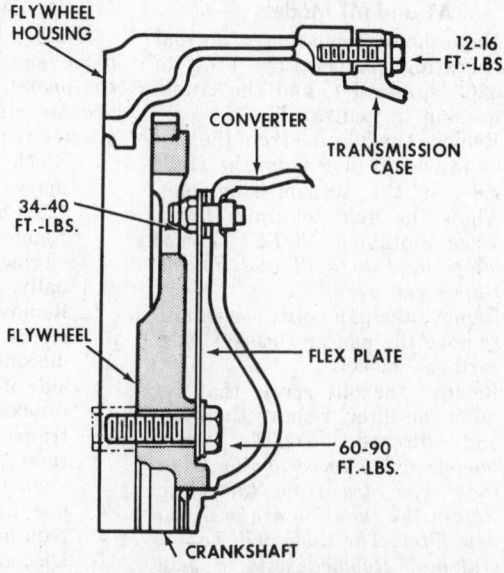

Flex plate installation—MT—640, MT-650.

7. Disconnect the drive shaft from transmission.
8. Disconnect the mechanical parking brake linkage at the right side of the transmission (if used).
9. Through access openings in the flywheel housing, use a pry bar, as necessary to manually turn the flywheel. As the flywheel is rotated, remove the six nuts retaining flex-plate assembly to converter cover.
10. Support the transmission with a 750-pound (minimum rating) transmission floor jack. The jack must be positioned so transmission pan will not support the weight of transmission. Fasten a safety chain over top of transmission and to both sides of jack.
11. Place a support under rear of engine and remove transmission case - to - crossmember support bolts. Raise the engine to remove weight from the engine rear mounts.
12. Remove the transmission case-to-flywheel housing bolts and washers.
13. Carefully inspect transmission and surrounding area to be sure no lines, hoses, or wires will interfere with transmission removal.
    NOTE: *When removing transmission keep rear of transmission lower than the front so as not to lose converter.*
14. Move transmission assembly from the engine, lower the assembly carefully and move it out from the vehicle.

Installation

1. Raise vehicle sufficiently to allow installation of transmission. With transmission assembly mounted on transmission jack move transmission into position aligning converter with flywheel. Check for and clean away any foreign material in flywheel pilot hole, flex-plate assembly, and front face of transmission case. Rotate flywheel as necessary so that the six studs in converter cover are aligned with holes in flex plate. Carefully move transmission assembly toward engine so flex-plate-to-converter cover nuts can be loosely installed and so that pilot on transmission converter enters pilot hole in center of flywheel.
2. Install bolts and washers that attach transmission case-to-flywheel housing. Tighten bolts to 12-16 foot-pounds torque.
3. Tighten the six flex-plate-to-converter cover nuts to 34-40 foot-pounds torque.
4. Carefully lower engine and transmission assembly onto engine rear mounts. Tighten engine rear mounting nuts to 190-210 foot-pounds torque. Remove lifting equipment from beneath vehicle.
5. Remove plugs from cooler lines and transmission case fittings. Be sure fittings are clean and lint-free, then connect cooler lines-to-transmission.
6. Install filler tube and bracket on right side of transmission. Install level gauge (dipstick).
7. Connect the speedometer shaft fitting to adapter at rear of transmission.
8. Connect drive shaft to transmission.
9. Connect parking brake linkage (if used) at side of transmission.
10. Connect shift cable to shift lever at left side of transmission.
11. Connect the vacuum modulator line to modulator. Also, connect wiring to neutral safety switch (left side of transmission) and back-up lamp switch (right side of transmission). NOTE: *Make sure the ignition switch is in the off position before proceeding to the next step.*
12. Install spark plugs and connect battery ground strap, previously disconnected.
13. Connect any other lines, hoses, or wires which were disconnected to aid in transmission removal.
14. Adjust the shift linkage. (see "Shift Linkage Adjustment")
15. Refill the transmission.

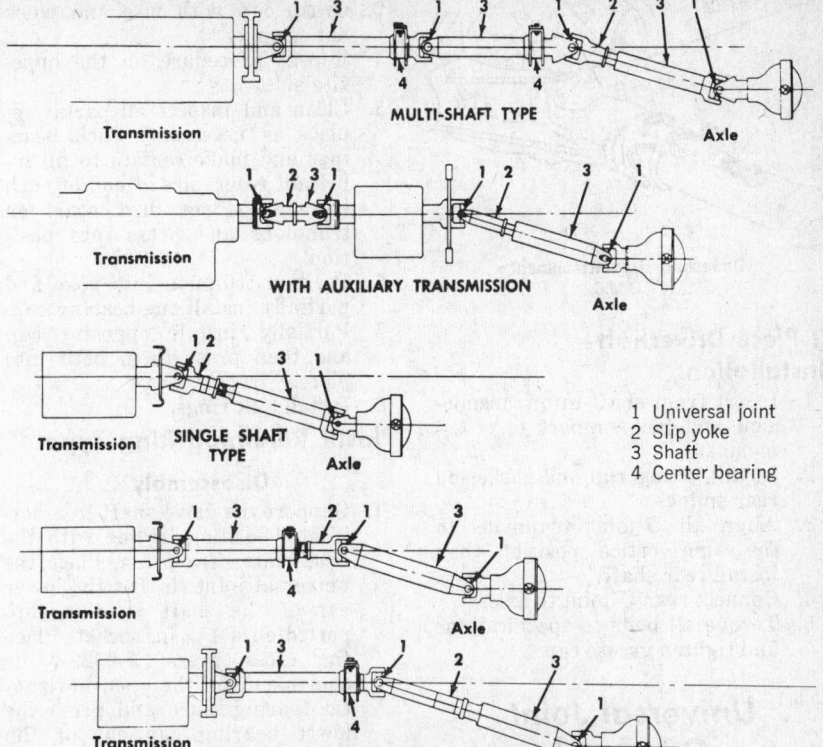

1 Universal joint
2 Slip yoke
3 Shaft
4 Center bearing

Driveshaft combinations (typical) (© G.M.C.)

Drive Shaft, U Joints

Drive Shaft R & R

Removal

1. Raise vehicle.
2. Mark shaft and companion flange alignment.
3. Disconnect rear universal joint by removing attaching U-bolt or strap.
4. Remove bearing support bolts as necessary.
5. Slide driveshaft forward to disengage trunnion, then back to disengage transmission.

1 Piece Driveshaft—Installation

1. Slide shaft into transmission and connect rear U-joint.
2. Torque bolts to specifications.

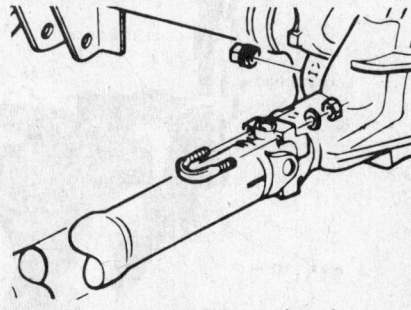

Driveshaft U-bolt attachment
(© G.M.C.)

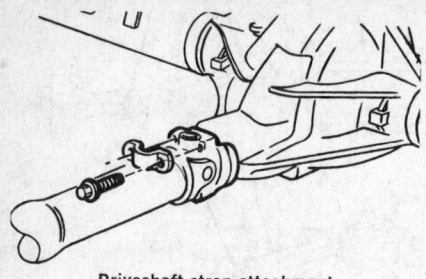

Driveshaft strap attachment
(© G.M.C.)

2 Piece Driveshaft— Installation

1. Insert front shaft into transmission and bolt support to crossmember.
2. Install grease cap and gasket on rear splines.
3. Align all U-joint trunnions in the same vertical position, then install rear shaft.
4. Connect rear U-joint to axle.
5. Torque all bolts to specifications and tighten grease cap.

Universal Joint Overhaul

Snap Ring Type

1500-3500 Models

1. Remove bearing lock rings.
2. Support yoke in an arbor press and apply pressure on trunnion until bearing cup is almost out. *NOTE: Bearing cup cannot be fully pressed out.*

3. Grasp cup with vise and work out of yoke.
4. Repeat procedure on the opposite side.
5. Clean and inspect all parts; replace as necessary. Pack bearings and make certain to fill lubricant reservoirs at end of each trunnion. Place dust seals on trunnions and press into position.
6. Position trunnion into yoke and partially install one bearing cup.
7. Partially install opposite cup and then press them both into place.
8. Install lock rings.

Plastic Retaining Ring Type

Disassembly

1. Support the drive shaft in a horizontal position in line with the base plate of a press. Place the universal joint so that the lower ear of the shaft yoke is supported on a 1⅛ in. socket. Place the cross press, J-9522-3 or equivalent, on the open horizontal bearing cups, and press the lower bearing cup out of the yoke ear as shown in the illustration. This will shear the plastic retaining the lower bearing cup.
2. If the bearing cup is not completely removed, lift the cross and insert Spacer J-9522-5 or equivalent, between the seal and bearing cup being removed.
   Complete the removal of the bearing cup, by pressing it out of the yoke.

Pressing out bearing cup (Plastic retaining ring type)

3. Rotate the drive shaft, shear the opposite plastic retainer, and press the opposite bearing cup out of the yoke as before, using Spacer J-9522.
4. Disengage the cross from the yoke and remove.

NOTE: Production universal joints cannot be reassembled. There are no bearing retainer grooves in production bearing cups. Discard all universal joint parts removed.

5. Remove the remains of the sheared plastic bearing retainer from the ears of the yoke. This will aid in reassembly of the service joint bearing cups. It usually is easier to remove plastic if a small pin or punch is first driven through the injection holes.

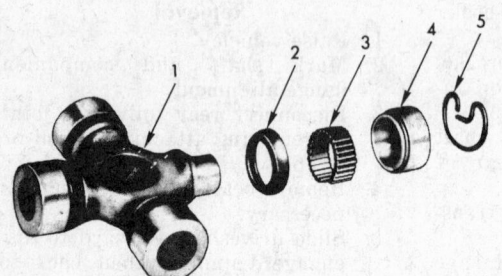

1 Trunnion
2 Seal
3 Bearings
4 Cap
5 Snap ring

Universal joint repair kit—(snap ring type)

Bearing cap removal—(Snap ring type)

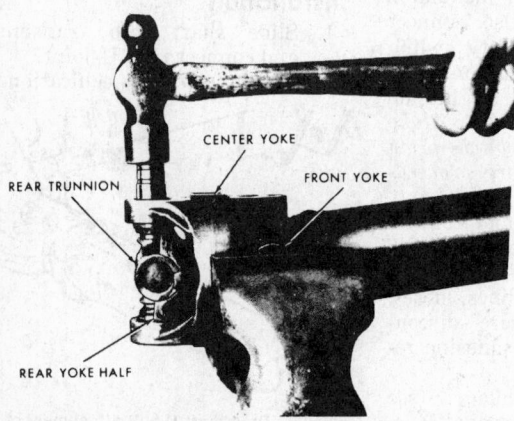

Driving out bearing caps (Snap ring type)

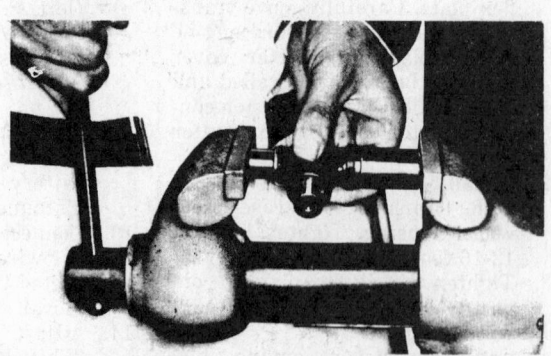

Installing U-joint trunnion (Snap ring type)

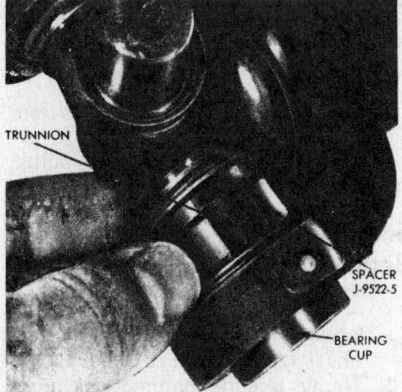

Using spacer to remove bearing cup (Plastic retaining ring type)

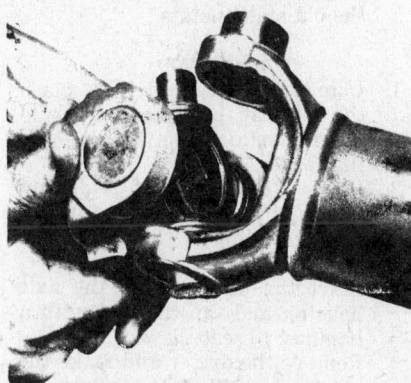

Installing trunnion into yoke (Plastic retaining ring type)

Installing plastic snap rings.

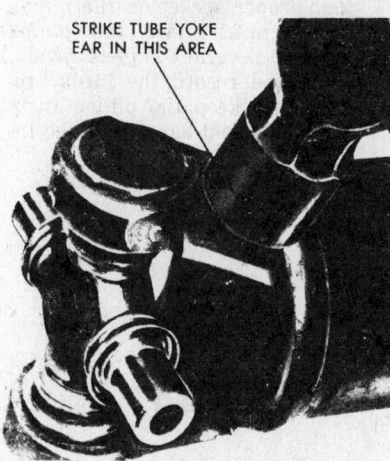

Seating plastic snap rings.

6. If the front universal joint is being serviced, remove the pair of bearing cups from the slip yoke in the same manner.

Reassembly

1. A universal joint service kit is used when reassembling this joint. This kit includes one pre-greased cross assembly, four service bearing cup assemblies with seals, needle rollers, washers, grease and four bearing retainers.
2. Make sure that the seals are in place on the service bearing cups to hold the needle rollers in place for handling.
3. Remove all of the remains of the sheared plastic bearing retainers from the grooves in the yokes. The sheared plastic may prevent the bearing cups from being pressed into place, and this prevents the bearing retainers from being properly sealed.
4. Install one bearing cup part way into one side of the yoke, and turn this yoke ear to the bottom.
5. Insert cross into yoke so that the trunnion seats freely into bearing cup as shown in illustration.
6. Install opposite bearing cup part way. Make sure that both trunnions are started straight and true into both bearing cups.
7. Press against opposite bearing cups, working the cross all of the time to check for free movement of the trunnions in the bearings. If there isn't, stop pressing and recheck needle rollers to determine if one or more of them has been tipped under the end of the trunnion.
8. As soon as one bearing retainer groove clears the inside of the yoke, stop pressing and snap the bearing retainer into place as shown in the illustration.
9. Continue to press until the opposite bearing retainer can be snapped into place. If difficulty is encountered, strike the yoke

firmly with a hammer to aid in seating bearing retainers. This springs the yoke ears slightly.
10. Assemble the other half of the universal joint in the same manner.
11. Check the freedom of rotation of both sets of trunnions of the cross. If too tight, again rap the yoke ears as described above. This will loosen the bearings and help seat the bearing retainers.

4500-7500 Models

1. Remove bearing retaining snap rings or U-bolts.
2. Strike one side of yoke with hammer to remove bearing. Repeat on opposite side.
3. Tilt journal and remove yoke.
4. Remove remaining bearings in the same manner as step (2).
5. Clean all parts with cleaning fluid. Make certain lubricating passages in journal cross are clear, and all old lubricant has been removed. Check for wear and any missing bearing rollers. If any excessive wear is noted, discard bearings and journal; replace with new parts. Lubricate as recommended in specifications.
6. Install lubrication fitting in journal.
7. Install journal in yoke, then install bearings using a mallet to tap them in place.
8. Install bearing retaining snap rings or U-bolts and torque nuts (when used) to specifications.

Drive Axles

Axle Shaft

Semi-Floating—8⅞ Axle

Removal

1. Remove the brake drum.
2. Drain lubricant from the differ-

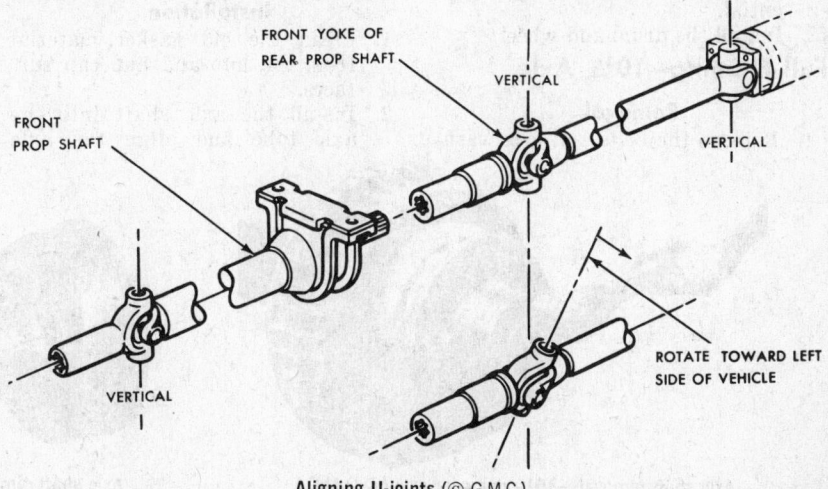

Aligning U-joints (© G.M.C.)

ential and remove the housing cover.

3. Remove the differential pinion shaft lock screw, pinion shaft and axle shaft spacer.

4. Push the axle shaft in and remove the "C" washer from the inner end of the axle shaft.

5. Remove the axle shaft from the housing.

Installation

NOTE: If a new axle shaft is to be installed.

1. Position the axle shaft gasket to the axle shaft flange.

2. Apply heavy shellac or paint to both sides of the gasket and axle shaft oil deflector.

3. Install the axle shaft oil deflector over the gasket aligning the oil pocket with the notch in the flange.

4. Insert six special axle shaft bolts and force the heads down to the deflector.

5. Peen the end of the shoulder on the bolts into the countersink around the bolt holes in the flange.

6. Slide the axle shaft into place. **CAUTION: Exercise care that the splines on the end of the shaft do not cut the axle shaft oil seal and that they engage with the splines of the differential side gears.**

7. Install the "C" washer on the inner end of the shaft.

8. Pry the shafts apart so that the "C" washers are seated in the counterbore in the differential side gears and install the pinion gears.

9. Select the proper axle shaft spacer to give free fit to .014″ maximum clearance between the end of the axle shaft and the spacer.

10. Install the spacer and pinion shaft, locking in place with the special screw.

11. Install the axle housing cover and gasket and refill the differential.

12. Install the drum and wheel.

Full Floating—10½ Axle

Removal

1. Remove the bolts and lock wash-

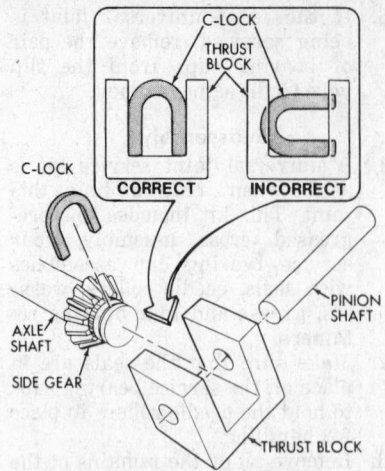

Correct C lock position (© G.M.C.)

ers from the axle shaft flange to hub.

2. Tap on the axle shaft flange with a soft-faced hammer to loosen the flange from the hub.

3. Grip the axle flange and pull the axle from the axle tube.

Installation

1. Clean the hub surface and the axle flange.

2. Install a new gasket over the axle shaft and install the axle shaft into the axle tube.

3. Engage the axle splines with the splines of the differential side gear, align the bolt holes in the axle flange with the bolt holes in the wheel hub and push the axle flange to mate with the hub surface.

4. Install the bolts and lock washers. Torque the bolts to 90 ft. lbs.

Full Floating—12¼ Axle

Removal

1. Remove the bolts from the hub cap and install a side hammer assembly into the tapped hole in the axle shaft flange.

2. Remove the axle shaft with the aid of the slide hammer.

Installation

1. Clean the old gasket material from the hub and hub cap surfaces.

2. Install the axle shaft into the axle tube and align the axle

splines of the differential side gear, and align the splines of the axle flange with the splines of the wheel hub.

3. Tap the axle shaft into position on the hub.

4. Install the hub cap and bolts. Torque the bolts to 15 ft. lbs.

Semi-Floating Axle Shaft Bearing or Oil Seal

Removal

1. Remove the wheel, drum and axle shaft (see axle shaft removal).

2. Using a slide hammer, remove the bearing, bearing retainer and oil seal.

3. Inspect the bore and dress out the old stake points.

Installation

1. Using the proper driver, place the oil seal, bearing and inside bearing retainer on the driver in that order.

2. Place a light coat of sealer on the outside of the seal to insure proper sealing of the seal in the housing bore.

3. Start the bearing into the axle housing and tap the tool with a hammer to seat the parts.

4. Remove the driver and stake the oil seal in place with a punch.

5. Assemble the axle shafts (see Axle Shaft Installation).

Pinion Flange, Oil Deflector and/or Oil Seal

Replacement

1. Raise the vehicle and support the frame on stand jacks, allow the axle to drop for clearance and expand the brake shoes on one wheel to lock the wheel.

2. Check the free wheel for freedom of rotation.

3. Separate the rear universal, tape the trunnion bearings to the joint and lower the rear of the propeller shaft

4. Using a one inch torque wrench, and proper socket on the pinion flange nut, rotate the pinion through several complete revolutions and record the torque required to keep the pinion turning. If the old flange is to be in-

Axle shaft removal—10½ ring gear axle (© G.M.C.)

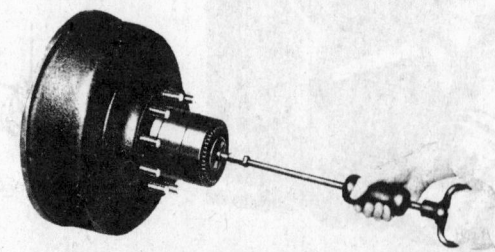

Axle shaft removal—12¼ ring gear axle (© G.M.C.)

stalled, mark the pinion and flange for reassembly in the same relative position.

5. Hold the pinion flange, remove the pinion flange nut and special washer. Discard the nut and use a new one upon reassembly.
6. Remove the pinion flange Pry the old oil seal out of the case.
7. Inspect the pinion flange for smooth oil seal surface or worn drive splines. Replace if necessary.
8. Install a new flange oil deflector if the deflector is damaged.
9. Soak the new seal in light engine oil before installation. Wipe the outside of the seal and coat the outside with sealer.
10. Install the new seal using the proper driver.
11. Install the pinion flange, aligning the marks on the pinion and flange if the old flange is being used. If the flange does not go on the shaft easily, pull the flange on the shaft using a special tool. Remove the special tool and install the special washer and new nut.
12. Tighten the nut to remove end play and continue alternately tightening and checking preload with an inch pound torque wrench until it is the same as recorded in step 4.
13. Readjust the brake on the locked wheel.
14. Connect the propeller shaft, lower the vehicle and road test for leaks and noise.

Rear Axle Assembly

3500 and 5500 lbs Capacity Axle (Models 1500-3500)

Removal

1. Raise truck and support rear axle to relieve load from springs, tie-rod, and shock absorbers.
2. Disconnect tie-rod at axle (when used).
3. Disconnect driveshaft. *NOTE: Secure bearing caps to trunnion with tape.*
4. Disconnect shock absorbers.
5. Disconnect vent hose.
6. Disconnect brake hose on axle housing. Remove brake drum and disconnect parking brake cable.
7. Make certain coil springs (when used) are compressed, then remove U-bolts, spacers and clamp plates.
8. Withdraw axle assembly.

Installation

To install axle assembly, reverse the removal procedure, bleed brake system, adjust parking brake and torque all bolts to specifications.

11,000 lb. Capacity Axle (Models 1500-3500)

Removal

1. Raise truck, support frame side rails and remove rear wheels.
2. Disconnect driveshaft. *NOTE: Secure bearing caps to trunnion with tape.*
3. Disconnect brake hose and shock absorbers.
4. Remove drum and disconnect parking brake.
5. Support axle, remove spring U-bolts and then withdraw assembly.

Installation

To install axle assembly, reverse the removal procedure, bleed brake system, adjust parking brake and torque all bolts to specifications.

11,000 lb. Capacity Axle (Models 4500-7500)

Removal

1. Raise rear of truck and support frame rails.
2. Disconnect brake lines and electrical wiring.
3. Disconnect driveshaft and torque or radius rods (when used).
4. Remove spring U-bolts and withdraw axle assembly.

Installation

To install axle assembly, mount wheels and tires, roll axle under truck and reverse the removal procedure. Bleed brake system, fill axle with lubricant to proper level and torque all bolts to specifications.

Refer to General Repair Section for complete overhaul procedures and out of truck adjustments.

Tandem Axle Suspension

Description

The Hendrickson type tandem axle suspension uses equalizing beams to tie the front rear axle to the rear axle and to permit independent vertical movement of each axle as required by the road surface. The torque rods are used to maintain proper drive line alignment and to stablize the driving and braking forces. Bolts are used on some models to hold the spring to the top saddle pad, while U-bolts are used on other models.

NOTE: When major overhaul is required, the complete tandem axle should be removed as a unit. The torque rods, springs, equalizing beams and other parts may be removed separately as required.

CAUTION: Block the vehicle securely before removal of the assembly to avoid rolling or pivoting at the equalizer beams when the torque rods

are disconnected. The use of a helper is suggested, along with proper lifting tools so that personal injury does not occur.

Tandem Axle Removal and Installation

1. Block all wheels and disconnect all applicable brake lines or hoses, differential lock lines, or electrical wiring from the rear axles.
2. Remove the rebound bolts from the rear spring brackets.
3. Remove all nuts, washers from the front spring brackets.
4. If equipped with ball stud torque rods, remove the stud nuts and tap each ball stud loose with a soft hammer. Remove the ball studs from the axle brackets.
5. If equipped with straddle mount torque rods, remove the mounting bolts from the rear axle bracket.
6. Support the rear axle differential with a floor jack, and disconnect the propeller shaft from the forward rear axle.
7. Using a suitable hoist, raise the rear of the frame high enough to clear the tandem axle assembly. Roll the assembly out from under the frame.
8. The installation is in the reverse of the removal procedure.

Locking Differentials

Refer to General Repair Section for complete overhaul procedures.

Rear Suspension

Leaf Spring

1500-3500 Models

Removal

1. Raise truck.
2. Loosen, but do not remove, spring to shackle retaining nut.
3. Remove shackle to spring hanger attaching bolt.
4. Remove attaching spring to front hanger bolt.
5. Remove U-bolts and then remove spring.

Installation

To install leaf spring assembly, reverse the removal procedure and torque nuts and bolts to specifications.

NOTE: Torque nuts and bolts after lowering the truck.

4500-7500 Models—Single or Tandem Axle Springs

Removal and Installation

1. Raise the rear of the vehicle,

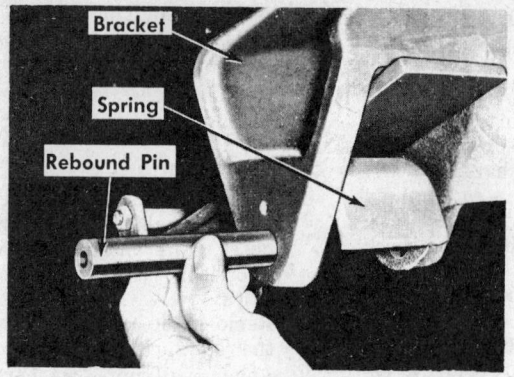

Installing rebound pin (typical) (© G.M.C.)

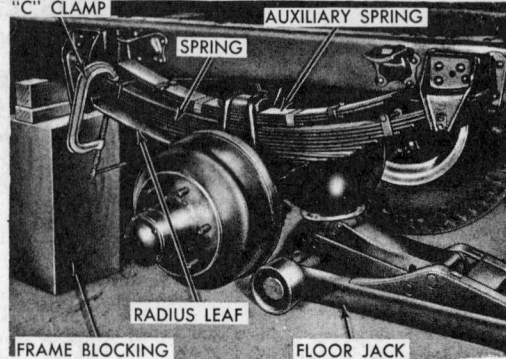

Using C-clamp at radius leaf (© G.M.C.)

place floor jacks under the axle(s) and remove the dual wheels from the hubs to facilitate the removal of the spring eye pin and to expose the other nuts and bolts.

2. Remove the saddle cap stud nuts and/or spring U-bolts.
3. Remove the rebound pin locks or retainers, and then remove the rebound pins.
4. Remove the eye bolts or radius lead pin clamp bolts, then remove the lubrication fitting from the inner end of the pin, if equipped.
5. Remove the pins from the springs and lower the axle(s) or raise the frame until the spring will clear the brackets.
6. Remove the spring from the vehicle.
7. Installation is the reverse of removal.
8. Torque the U-bolts or saddle cap stud nuts to specifications after the vehicle is lowered to the floor.

Coil Spring

1500-3500 Models

Removal

1. Raise truck on hoist and support rear axle.
2. Disconnect shock absorber at control arm bracket.
3. Remove upper and lower spring clamps.
4. Lower control arm enough to remove spring.

Installation

1. Position clamp in spring.
2. Install spring and clamp loosely; torque upper clamp nut to specifications.
3. Connect shock absorber and torque to specifications.
4. Torque lower clamp nut to specifications.
5. Remove axle support and lower truck.

Auxiliary Springs

Some models are equipped with auxiliary springs. They are mounted

Rear leaf spring suspension (© G.M.C.)

to brackets on the frame and are usually held to the main spring by long U-bolts. The purpose of these springs is to provide additional stability under unusual operating conditions.

Shock Absorbers

Shock absorbers used are non-adjustable and non-repairable. Maintenance operations are limited to periodic tightening and replacement of rubber mounting grommets. If a shock absorber is worn, the complete unit must be replaced.

Tie Rod

CP1500-C2500 Models

Removal

1. Raise truck.
2. Remove attaching nuts and bolts, then remove rod.

Installation

1. Position rod ends in mounting brackets, insert pivot bolts and install nuts finger tight.
2. Lower truck and torque pivot nuts to specifications.

Control Arm

CP1500-C2500 Models

Removal

1. Raise truck and relieve load on springs by supporting axle.

2. Place jack under control arm.
3. Remove spring clamp bolt from underside of arm.
4. Remove U-bolt nuts; disconnect shock absorber bracket, and lower end of control arm.
5. Disconnect parking brake lever, then remove pivot bolt and control arm.

Installation

To install control arm, reverse the removal procedure. Torque spring clamp and shock absorber BEFORE lowering truck, torque pivot bolts AFTER.

Stabilizer Shaft

P3500 Model

Removal

1. Raise truck and support axle to relieve load on U-bolts.
2. Remove the forward U-bolt retaining nuts, and remove shaft anchors.
3. Remove shaft retaining brackets and then remove shaft.

Installation

To install stabilizer shaft, reverse the removal procedure. Alternately torque all nuts and bolts to specifications.

NOTE: Make certain that both ends of shaft protrude equally from the anchors.

Index

TUNE-UP SPECIFICATIONS

| CYL — CID | YEAR | SPARK PLUG GAP | DISTRIBUTOR | | IGNITION TIMING DEGREES | CRANKING COMPRESSION PRESSURE | VALVE CLEARANCE | | GOVERNOR NO LOAD RPM | FUEL PRESS. | IDLE SPEED | |
			POINT DWELL	POINT GAP			IN	EXH			STD	AUTO
4-196	1971	.030	68-72	.019⑤	TDC	143	0	0	4000⑦	5	475	—
4-196	1972	.030	26-32	.017	TDC	143	0	0	4000⑦	5	475	—
4-196	1975-78	.035	24-34	①	0⑥	143	0	0	4000⑦	5	575	600
6-232	1971-74	.035	32	.016⑧	TDC	130	0	0	4000⑦	5	725	650
6-258	1972-74	.035	32	.016⑧	TDC	130	0	0	4000⑦	5	675	675
8-304	1971-72	.030	30	.019	TDC	145	0	0	3900⑦	5	600	550N
8-304	1973	.030	30	.019	TDC	145	0	0	3900⑦	5	700	700N
8-304	1974-78	.030	30	.017①	TDC	145	0	0	3900⑦	5	700	650N
8-345	1971-75	.030	30	.019①⑤	TDC②	143	0	0	3800	5	600⑪	600N
8-345	1976-78	.035	30	①	5B⑩	143	0	0	3800	5	650⑩	650⑩
8-392	1971	.027	30	.019	TDC	143	0	0	3600	5	700	600D
8-392	1972-75	.030⑫	30	.019①⑤	TDC⑬	140	0	0	3600	5	700	600D
8-400	1973-74	.035	30	.016	5B	140	0	0	3400	5	700	
8-401	1971-74	.027	30	.017	7B	140	0	0	3400	5	700	—
8-404(399)	1975-78	.030	24-34	①	9B	140	0	0	3600	5	525	575
6-406	1971-74	.030	32	.019	5B	120	.025H	.025H	2450	5	475	—
8-446	1975-78	.030	24-34	①	5B	145	0	0	3600	5	525	—
6-450	1971-74	.030	32	.019	5B	120	.025H	.025H	2800	5	525	—
8-478	1971-74	.027	32	.017	10B	135	0	0	3400	5	550	550
6-501	1971-74	.030	32	.019	5B	120	.025H	.025H	2800	5	525	—
8-537	1975-78	.030	24-34	①	7B	140	0	0	3400	5	550	—
8-549	1971-78	.027	32	.017	7B④	140	0	0	3200	5	475	—
8-605	1975-78	.030	24-34	①	7B	140	0	0	3400	5	500	550

① .008 on breakerless ignition
② LPG—10B
　　1975 Low compression—5B
③ with thermoquad—5B
④ FTV models—9B
⑤ 1972 -.016 on light duty vehicles
⑥ With automatic transmission—5BTDC
⑦ Maximum recommended RPM
⑧ 1971-73—.019 inch
⑨ Idle speed shutdown—525-575N

⑩ 1976-78—Equipped with distributor numbers—
　　461270-C91
　　461271-C91
　　461272—C91
　　Idle speed (AT and ST) 650-700
　　Timing—TDC
⑪ 1973—700 RPM
　　1974—800 RPM⑨
　　1975—650 RPM
⑫ 1975—.035 inch
⑬ Equipped with carter carburetor and distributor number 519861—5 BTDC

FIRING ORDER AND ROTATION

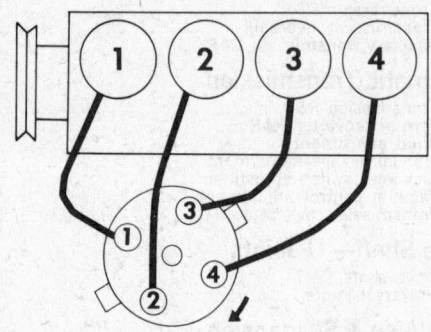

4-cyl Firing Order
1-3-4-2
4-196

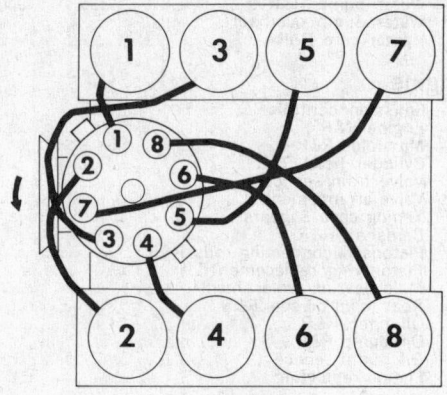

Firing Order
1-2-7-3-4-5-6-8
MV 404, 446

International Trucks · Scout · Travelall · Traveler

FIRING ORDER AND ROTATION

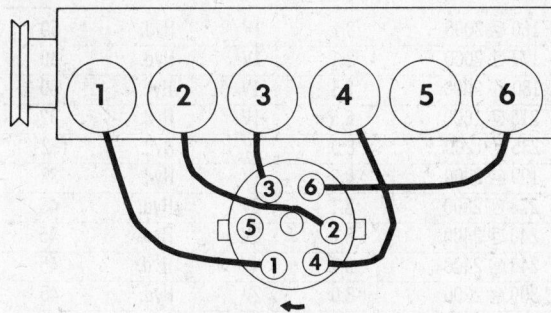

6-cyl Firing Order
1-5-3-6-2-4
RD 406, 450, 501

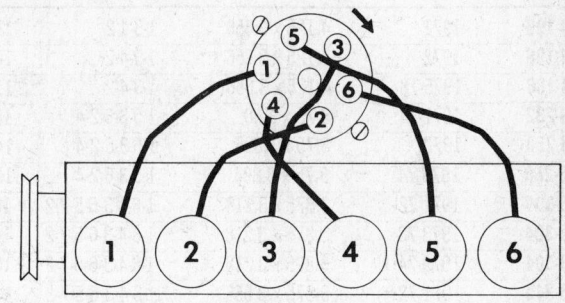

6-cyl Firing Order
1-5-3-6-2-4
P-6 232, 258

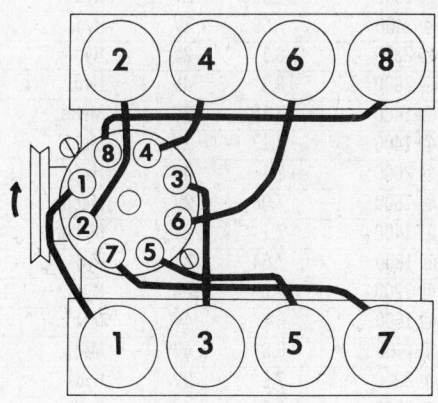

8-cyl Firing Order
1-8-4-3-6-5-7-2
VS 401, 478, 549

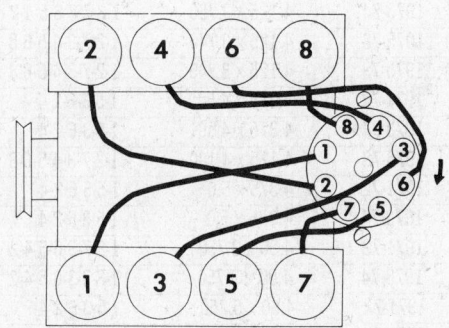

8-cyl Firing Order
1-8-4-3-6-5-7-2
V 537, 605

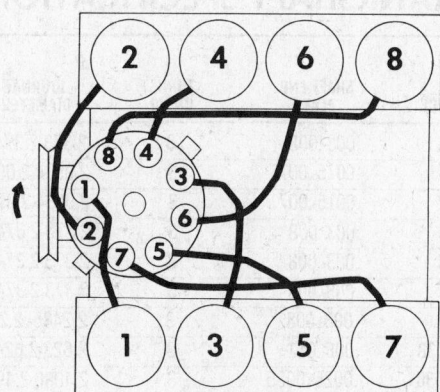

8-cyl Firing Order
1-8-4-3-6-5-7-2
V 304, 345, 392

GENERAL ENGINE SPECIFICATIONS

CYL — CID	YEAR	BORE x STROKE	FIRING ORDER	SAE HORSEPOWER @ RPM	SAE TORQUE @ RPM	COMP. RATIO	CARBU- RETOR	VALVE LIFTER	OIL PRESSURE
4-196	1971	4.125 x 3.656	1-3-4-2	111 @ 4000	180 @ 2000	8.1	1V	Hyd.	50
4-196	1972	4.125 x 3.656	1-3-4-2	103 @ 4000	177 @ 2000	8.1	1V	Hyd.	50
4-196	1975-78	4.125 x 3.656	1-3-4-2	111 @ 4400	180 @ 2000	8.1	1V	Hyd.	50
6-232	1971-74	3.75 x 3.50	1-5-3-6-2-4	145 @ 4300	215 @ 1600	8.5	1V	Hyd.	32
6-258	1972	3.75 x 3.895	1-5-3-6-2-4	140 @ 3800	235 @ 1200	8.0	1V	Hyd.	32
6-258	1973-74	3.75 x 3.895	1-5-3-6-2-4	115 @ 3800	199 @ 2000	8.0	1V	Hyd.	32
8-304	1971-72	3.875 x 3.218	1-8-4-3-6-5-7-2	193 @ 4400	273 @ 2800	8.1	2V	Hyd.	45
8-304	1973-78	3.875 x 3.218	1-8-4-3-6-5-7-2	147 @ 3900	240 @ 2400	8.2	2V	Hyd.	45
8-304	1973-74	3.875 x 3.218	1-8-4-3-6-5-7-2	153 @ 3900	246 @ 2400	8.2	4V	Hyd.	45
8-345	1971-72	3.875 x 3.656	1-8-4-3-6-5-7-2	197 @ 4000	309 @ 2200	8.0	2V	Hyd.	45
8-345	1973-78	3.875 x 3.656	1-8-4-3-6-5-7-2	157 @ 3800	266 @ 2400	8.1	2V	Hyd.	45
8-345	1973-75	3.875 x 3.656	1-8-4-3-6-5-7-2	163 @ 3800	273 @ 2400	8.1	4V	Hyd.	45
8-392	1971-72	4.125 x 3.656	1-8-4-3-6-5-7-2	236 @ 4000	357 @ 2800	8.0	4V	Hyd.	45
8-392	1973-74	4.125 x 3.656	1-8-4-3-6-5-7-2	191 @ 3600	299 @ 2800	8.0	2V	Hyd.	45
8-392	1973-75	4.125 x 3.656	1-8-4-3-6-5-7-2	194 @ 3600	308 @ 2800	8.0	4V	Hyd.	45
8-400	1973-74	4.166 x 3.680	1-8-4-3-6-5-7-2	211 @ 4000	326 @ 2800	8.25	2V	Hyd.	50
8-401	1971-72	4.125 x 3.750	1-8-7-3-6-5-4-2	206 @ 3600	355 @ 1800	7.69	2V	Hyd.	55
8-401	1973-74	4.125 x 3.750	1-8-7-3-6-5-4-2	186 @ 3400	322 @ 2400	7.69	2V	Hyd.	55
8-404 (399)	1975-78	4.125 x 3.740	1-2-7-3-4-5-6-8	188 @ 3600	311 @ 2300	8.1	2V	Hyd.	55
8-404 (399)	1975-78	4.125 x 3.740	1-2-7-3-4-5-6-8	210 @ 3600	336 @ 2800	8.1	4V	Hyd.	55
6-406	1971-72	4.375 x 4.50	1-5-3-6-2-4	193 @ 3200	373 @ 1600	7.13	2V	Mech.	50
6-406	1973-74	4.375 x 4.50	1-5-3-6-2-4	160 @ 2750	336 @ 1400	7.13	2V	Mech.	50
8-446	1975-78	4.125 x 4.180	1-2-7-3-4 5-6-8	235 @ 3600	385 @ 2600	8.1	4V	Hyd.	55
6-450	1971-72	4.375 x 5.0	1-5-3-6-2-4	202 @ 3000	422 @ 1600	7.06	2V	Mech.	50
6-450	1973-74	4.375 x 5.0	1-5-3-6-2-4	168 @ 2600	372 @ 1400	7.1	2V	Mech.	50
8-478	1971-72	4.50 x 3.750	1-8-7-3-6-5-4-2	234 @ 3600	431 @ 1800	7.64	2V	Hyd.	50
8-478	1973-74	4.50 x 3.750	1-8-7-3-6-5-4-2	209 @ 3400	384 @ 2200	7.6	2V	Hyd.	55
6-501	1971-72	4.50 x 5.250	1-5-3-6-2-4	215 @ 3000	451 @ 1600	6.8	4V	Mech.	40
6-501	1973-74	4.50 x 5.250	1-5-3-6-2-4	181 @ 2600	406 @ 1400	6.8	4V	Mech.	40
8-537	1975-78	4.625 x 3.750	1-8-7-3-6-5-4-2	208 @ 3200	415 @ 2000	7.5	2V	Hyd.	55
8-537	1975-78	4.625 x 3.750	1-8-7-3-6-5-4-2	236 @ 3200	429 @ 2200	7.5	4V	Hyd.	55
8-549	1971-72	4.50 x 4.312	1-8-7-3-6-5-4-2	257 @ 3400	505 @ 2000	7.6	4V	Hyd.	55
8-549	1973-78	4.50 x 4.312	1-8-7-3-6-5-4-2	227 @ 3200	446 @ 2000	7.6	4V	Hyd.	55
8-605	1975-78	4.625 x 4.50	1-8-7-3-6-5-4-2	227 @ 3200	446 @ 2000	7.5	4V	Hyd.	55

CRANKSHAFT SPECIFICATIONS

CYL — CID	MAIN BEARING JOURNALS JOURNAL DIAMETER	OIL CLEARANCE	SHAFT END PLAY	THRUST ON NO.	CONNECTING ROD JOURNALS JOURNAL DIAMETER	OIL CLEARANCE	ROD SIDE CLEARANCE
4-196	2.7484-2.7494	.001-.004	.003-.008	3	2.373-2.374	.0011-.0036	.004-.011
6-232	2.4981-2.500	.001-.002	.0015-.007	3	2.0934-2.0955	.001-.002	.008-.010
6-258	2.498-2.500	.001-.002	.0015-.007	3	2.0934-2.0955	.001-.002	.008-.010
8-304	2.7484-2.7494	.001-.004	.003-.008	3	2.373-2.374	.0011-.0036	.008-.016
8-345	2.7484-2.7494	.001-.004	.003-.008	3	2.373-2.374	.0011-.0036	.008-.016
8-392	2.7484-2.7494	.001-.004	.003-.008	3	2.373-2.374	.0011-.0033	.010-.018
8-400	2.7474-2.7489	.001-.002	.003-.008	3	2.2485-.2.2464	.001-.002	.006-.018
8-401	3.124-3.125	.0008-.0038	.006-.011	3	2.623-2.624	.0017-.0042	.010-.018
8-404(399)	3.1228-3.1236	.0010-.0036	.0025-.0085	3	2.4980-2.4990	.0011-.0036	.008-.020
6-406	3.2495-3.2505	.0013-.0043	.006-.015	7	2.751-2.752	.0012-.0037	.007-.013
8-446	3.1228-3.1236	.0010-.0036	.0025-.0085	3	2.4980-2.4990	.0011-.0036	.008-.020
6-450	3.2495-3.2505	.0013-.0043	.006-.015	7	2.751-2.752	.0012-.0037	.007-.013
8-478	3.124-3.125	.0008-.0038	.006-.011	3	2.623-2.624	.0017-.0042	.010-.018
6-501	3.2495-3.2505	.0013-.0043	.006-.015	7	2.751-2.752	.0012-.0037	.007-.013
8-537	3.1235-3.1245	.0015-.0035	.006-.012	3	2.623-2.624	.011-.036	.008-.018
8-549	3.124-3.125	.0008-.0038	.006-.011	3	2.623-2.624	.0017-.0042	.010-.018
8-605	3.1235-3.1245	.0015-.0035	.006-.012	3	2.623-2.624	.011-.036	.008-.018

VALVE SPECIFICATIONS

CYL — CID	ANGLE FACE	ANGLE SEAT	STEM DIA. IN	STEM DIA. EX	STEM-TO-GUIDE CLEARANCE IN	STEM-TO-GUIDE CLEARANCE EX	VALVE LIFT IN.	VALVE SPRING TENSION LBS @ INCHES	SPRING FREE LENGTH
4-196	⑥	⑥	.372	.415	.001-.0035	.0015-.004	①	188 @ 1.428	2.065
6-232	⑩	⑪	.371	.371	.001-.003	.001-.003	.254	190 @ 1.437	2.265
6-258	⑩	⑪	.371	.371	.001-.003	.001-.003	.254	190 @ 1.437	2.265
8-304	45	45	.372	.371	.001-.0035	.0015-.004	①	180-195 @ 1.429	2.065
8-345	45	45	.372	.371	.001-.0035	.0015-.004	①	180-195 @ 1.429	2.065
8-392	⑥	⑥	.372	.414	.001-.0035	.0015-.004	①	180-195 @ 1.429	2.065
8-400	⑥	⑥	.372	.372	.001-.003	.001-.003	.286	210-226 @ 1.365⑧	2.200⑨
8-401	⑤	⑤	.435	.434	.0015-.0035	.0025-.0045	.426	113-121 @ 1.663⑩	2.562⑦
8-404(399)	45	45	.372	.372	.0012-.0028	.0017-.0024	.435	188 @ 1.429	2.065
6-406	②	②	.435	.434	.0015-.004	.002-.0045	.449	133-141 @ 1.703③	2.562④
8-446	45	45	.372	.372	.0012-.0028	.0017-.0024	.435	188 @ 1.429	2.065
6-450	②	②	.435	.434	.0015-.004	.002-.0045	.449	133-141 @ 1.703③	2.562④
8-478	⑤	⑤	.435	.434	.0015-.0035	.0025-.0045	.426	113-121 @ 1.663⑩	2.562⑦
6-501	②	②	.435	.434	.0015-.004	.002-.0045	.449	133-141 @ 1.703③	2.562④
8-537	②	②	.435	.434	.0016-.0034	.002-.0037	.465	200 @ 1.397	2.075
8-549	⑤	⑤	.435	.434	.0015-.0035	.0015-.0045	.426	113-121 @ 1.663⑩	2.562⑦
8-605	②	②	.435	.434	.0016-.0034	.002-.0037	.465	200 @ 1.397	2.075

① Intake .440; Exh. 395
② Intake 15; Exh. 45
③ Inner 83-88 @ 1.50
④ Inner 2.343
⑤ Intake 15; Exh. 45
⑥ Intake 30; Exh. 45
⑦ Inner 2.281
⑧ Exhaust 210-226 @ 1.183
⑨ Exhaust 2.00
⑩ Inner 79-87 @ 1.538
⑪ Face—Intake 30°
   Exhaust 44°
⑫ Seat—Intake 29°
   Exhaust 45°

PISTON RING SPECIFICATIONS

Cyl — CID	RING GAP Compression	RING GAP Oil Control	RING CLEARANCE Compression	RING CLEARANCE Oil Control
4-196	.013-.023	.013-.028	.0015-.003	.002-.0035
6-232	.010-.020	.015-.055①	.0015-.0035	.000-.005
6-258	.010-.020	.015-.055①	.0015-.0035	.000-.005
8-304	.010-.020	.015-.055①	.0015-.003	.000-.0084
8-345	.010-.020	.015-.055①	.0015-.003	.000-.0084
8-392	.013-.023	.013-.028	.0015-.003	.002-.0035
8-400	.010-.020	.0015-.030①	.0015-.003	.002-.0035
8-401	.013-.025	.013-.028	.0035-.005	.001-.003
8-404	.013-.023	.013-.023	.002-.004	.002-.004
6-406	.025-.035	.013-.028	.0035-.005	.002-.0035
8-446	.013-.023	.013-.023	.002-.004	.002-.004
6-450	.025-.035	.013-.028	.0035-.005	.002-.0035
8-478	.013-.025	.013-.028	.0035-.005	.001-.003
6-501	.013-.023	.013-.028	.0035-.005	.002-.0035
8-537	.012-.022②	.012-.022	.002-.004	.002-.004
8-549	.013-.025	.013-.028	.0035-.005	.001-.003
8-605	.012-.022②	.012-.022	.002-.004	.002-.004

① Spring spacer or steel rails—no gap at joint
② 2nd compression—.014-.024

Bolt Torque Specifications (Unlisted)

Many bolts are used that have no torque specification listed. Refer to the following charts for the classification of the bolts and the allowable torque for each bolt class.

Note that torque specifications given in the chart are based on the use of clean and dry threads. Reduce the torque by 10 percent when threads are lubricated with oil and by 20 percent if new plated bolts are used.

S.A.E. CLASSIFICATION

SAE GRADE NUMBER	1 or 2	5	6 or 7	8
Capscrew Head Markings				
Manufacturer's marks may vary. Three-line markings on heads, for example, indicate SAE Grade 5. 	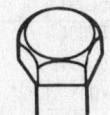			

Usage	Used Frequently	Used Frequently	Used at Times	Used at Times
Quality of Material	Indeterminate	Minimum Commercial	Medium Commercial	Best Commercial
Capscrew Body Size (Inches)—(Thread)	Torque	Torque	Torque	Torque
	Ft-Lb	Ft-Lb	Ft-Lb	Ft-Lb
1/4-20	5	8	10	12
-28	6	10		14
5/16-18	11	17	19	24
-24	13	19		27
3/8-16	18	31	34	44
-24	20	35		49
7/16-14	28	49	55	70
-20	30	55		78
1/2-13	39	75	85	105
-20	41	85		120
9/16-12	51	110	120	155
-18	55	120		170
5/8-11	83	150	167	210
-18	95	170		240
3/4-10	105	270	280	375
-16	115	295		420
7/8-9	160	395	440	605
-14	175	435		675
1 -8	235	590	660	910
-14	250	660		990

TORQUE SPECIFICATIONS

CYL —CID	HEAD BOLTS	ROD CAP BOLTS	MAIN CAP BOLTS	CRANKSHAFT BALANCER BOLT	FLYWHEEL BOLTS	MANIFOLDS IN.	EX.
4-196	90-100	40-45	75-80	100-110	45-55	40-45	40-45
6-232,258	80-85	25-30	75-85	70-80	100-110	20-25	20-25
8-304	90-100	45-55	75-85	100-110	45-55	40-55	40-45
8-345,392	90-100	45-55	75-85	100-110	45-55	40-55	40-45
8-400	110	35-40	95-105	50-60	100-110	40-45	20-30
8-401	90-100	60-70	100-110	Press-Fit	90-100	25-30	25-30
8-404(399)	100-110	38-44	90-100	80-100	—	—	15-20
6-406	100-110	65-75	100-110	Press-Fit	150-160	25-30	25-30
8-446	100-110	38-44	90-100	80-100	—	—	15-20
6-450	100-110	65-75	100-110	Press-Fit	150-160	25-30	25-30
8-478	90-100	60-70	100-110	Press-Fit	90-100	25-30	25-30
6-501	100-110	65-75	100-110	Press-Fit	150-160	25-30	25-30
8-549	90-100	60-70	100-110	Press-Fit	90-100	25-30	25-30
8-537, 8-605	80-90	65-70	125-135	260-290	110-120	30-38	30-38

FRONT END ALIGNMENT SPECIFICATIONS

AXLE	CASTER LEVEL FRAME	CAMBER	TOE-IN	KING PIN INCLINATION
FA-1	2° ± 1° (Not Metro) 2° ± ½° (Metro)	1½° ± ½°	$^1/_{16}$ ± $^1/_{16}$	4°
FA-3	0°	1°	$^3/_{16}$	8½°
FA-10	3° ± ½°	1½° ± ½°	$^1/_{16}$ ± $^1/_{16}$	4°
FA-13	0°	1°	$^3/_{16}$	8½°
FA-12	2° ± 1° (Not Motor Home) 4° ± ½° (Motor Home)	½° ± ½°	$^1/_{16}$ ± $^1/_{16}$	4°
FA-28	1° ± 1° (Not Motor Home or Metro) 4° ± ½° (Motor Home) 3° ± ½° (Metro)	½° ± ½°	$^1/_{16}$ ± $^1/_{16}$	4°
FA-44	0°	1°	$^3/_{16}$	8½°
FA-48	2° ± 1° (Not Motor Home) 4° ± ½° (Motor Home)	½° ± ½°	$^1/_{16}$ ± $^1/_{16}$	4°
FA-54	1	0°	$^1/_{16}$	8°
FA-57	0 ± 2	2°	0-$^3/_8$	0°
FA-59	0 ± 2	2°	0-$^3/_8$	0°
FA-68	2° ± 1°	½° ± ½°	$^1/_{16}$ ± $^1/_{16}$	4°
FA-69	3° ± ½° (Cargostar Manual Strg.) 4½° ± ½° (Cargostar Power Strg.)	½° ± ½° ½° ± ½°	$^1/_8$ ± $^1/_{16}$ $^1/_8$ ± $^1/_{16}$	4° 4°
FA-71	2° ± ½°	½° ± ½°	$^1/_{16}$ ± $^1/_{16}$	LT 4¼° RT 4½°
FA-72	2° ± ½°	½° ± ½°		4°
FA-73	2° ± ½°	½° ± ½°	$^1/_{16}$ ± $^1/_{16}$	LT 4¼° RT 4½°
FA-74	3° ± ½° (Manual Strg.) 4½° ± ½° (Power Strg.)	½° ± ½°	$^1/_8$ ± $^1/_{16}$	LT 4¼° RT 4½°
FA-78	0 ± 2	2°	0-$^3/_8$	0°
FA-91	2° ± 1° (Not CO-190) 2° ± ½° (CO-190)	½° ± ½°	$^1/_{16}$ ± $^1/_{16}$	4°
FA-98	2° ± (Not Fleetstar A) 2° ± ½° (Fleetstar A 4x2) 4° ± ½° (Fleetstar A 4x4)	½° ± ½°	$^1/_{16}$ ± $^1/_{16}$	4°
FA-99	3° ± ½° (Cargostar Manual Strg.) 4½° ± ½° (Cargostar Power Strg.)	½° ± ½°	$^1/_8$ ± $^1/_{16}$	4°
FA-101	2° ± ½° (Not Fleetstar 6x4)	½° ± ½°	$^1/_{16}$ ± $^1/_{16}$	LT 4¼° RT 4½°
	4° ± ½° (Fleetstar 6x4)	½° ± ½°	$^1/_{16}$ ± $^1/_{16}$	LT 4¼° RT 4½°
FA-103	3° ± ½° (Manual Strg.) 4½° ± ½° (Power Strg.)	½° ± ½°	$^1/_8$ ± $^1/_{16}$	LT 4¼° RT 4½°
FA-109	Lt. ¼ + ½ Rt. 0 + ½	1°	$^1/_{16}$	4½ R.H. 4¼ L.H.
FA-112	3½	0°	$^1/_{16}$	1°
FA-136	4	0°	$^1/_{16}$	5½°
FA-139	Lt. ¼ ± ½ Rt. 0 ± ½	1°	$^1/_{16}$	4½ R.H. 4¼ L.H.
FA-182	4	0°	$^1/_{16}$	5½°
FA-228	4	0°	$^1/_{16}$	5½°
FA-309	Lt. ¼ ± ½ Rt. 0 ± ½	1°	$^1/_{16}$	4½ R.H. 4¼ L.H.
FA-329	3	Lt. ¼ + ½ Rt. 0 + ½	$^1/_{16}$	4½ R.H. 4¼ L.H.
FA-339	3	Lt. ¼ + ½ Rt. 0 + ½	$^1/_{16}$	4½ R.H. 4¼ L.H.

ENGINE APPLICATION BY YEAR—GASOLINE

CID	YEARS	CID	YEARS	CID	YEARS
4-196	1971-72, 1975-78			8-404	1975-78
6-232	1971-74	8-304	1971-78	8-446	1975-78
6-258	1972-74	8-345	1971-78	8-478	1971-74
6-406	1971-74	8-392	1971-75	8-537	1975-78
6-450	1971-74	8-400	1973-74	8-549	1971-78
6-501	1971-74	8-401	1971-74	8-605	1975-78

ENGINE APPLICATIONS

SERIES	AXLES	ENGINES— HP RATING
SCOUT SERIES (with or without top)		
Scout II	4x2	G 86-158
Scout II	4x4	G 86-158
MULTI-STOP SERIES		
MS-1210	4x2	G 158-163
MS-1510	4x2	G 158-163
PICKUP MODELS		
150 Bonus Load, Regular	4x2	G 140-163
200 Bonus Load, Regular	4x2	G 140-163
150 Bonus Load, Regular	4x4	G 140-163
200 Bonus Load, Regular	4x4	G 140-163
TRAVELALL		
150	4x2	G 187-196
200	4x2	G 187-196
150	4x4	G 187-196
200	4x4	G 187-196
LIGHT-DUTY SERIES		
150	4x2	G 140-196
200	4x2	G 140-196
500	4x2	G 140-196
150	4x4	G 140-196
200	4x4	G 140-196
LOADSTAR SERIES		
1600	4x2	G 157
1700	4x2	G 190
1750	4x2	D 150-175
1800	4x2	G 205-235
1850	4x2	D 170-210
1600	4x4	G 147
1700	4x4	G 157
F-1800	6x4	G 205-235
F-1850	6x4	D 170-210

SERIES	AXLES	ENGINES— HP RATING
FLEETSTAR A SERIES		
2010A	4x2	G 209-227
2050A	4x2	D 170-216
2070A	4x2	D 228-290
F-2010A	6x4	G 209-227
F-2050A	6x4	D 170-216
F-2070A	6x4	D 228-290
TRANSTAR SERIES		
4270	4x2	D 260-350
4370	4x2	D 290-350
F-4270	6x4	D 260-430
F-4370	6x4	D 290-450
PAYSTAR SERIES		
5050	4x4	D 190-210
5070	4x4	D 228-304
F-5050	6x4	D 190-210
F-5070	6x4	D 228-400
F-5050	6x6	D 190-210
F-5070	6x6	D 228-400
CARGOSTAR SERIES		
CO-1610B	4x2	G 157
CO-1710B	4x2	G 205
CO-1810B	4x2	G 205-235
CO-1850B	4x2	D 170-190
CO-1910B	4x2	G 209-227
CO-1950B	4x2	D 170-210
COF-1810B	6x4	G 205-235
COF-1910B	6x4	G 209-227
COF-1950B	6x4	D 170-210
CO TRANSTAR II SERIES		
CO-4070B	4x2	D 290-350
COF-4070B	6x4	D 290-450

G = Gasoline
D = Diesel

Distributor

Distributor Removal

1. Remove distributor cap.
2. Mark position of rotor by scribing a mark on the distributor body.
3. Mark position of distributor body on the engine block or mounting bracket.
4. If so equipped, remove vacuum advance line, governor lines, and tachometer cable.
5. Loosen clamp screw or hold-down bolt and remove distributor from engine.
6. To install, reverse the above procedure, making sure to align marks which position rotor and distributor body.
7. If the engine has been disturbed (crankshaft position unknown), rotate engine until engine is in No. 1 firing position (No. 8 on V-304, 345, and 392 engines). Compression stroke may be felt by placing finger tightly over spark plug hole as engine is rotating. Holding distributor shaft so that rotor is in No. 1 (or No. 8 on V-304, 345, 392) firing position, insert distributor, rotating slightly until it drops into keyed position.
8. Adjust ignition timing before tightening clamp screw or hold-down bolt.

Breaker Points Replacement

1. Unsnap and remove distributor cap.
2. Remove rotor and dust cover if so equipped.
3. Loosen nut retaining primary and condenser leads and remove leads.

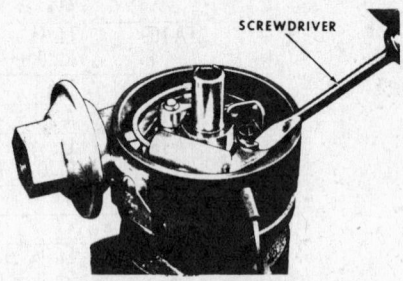

Adjusting point gap—slots
(© International Harvester Co.)

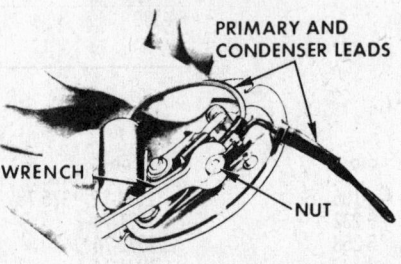

Removing primary condenser
(© International Harvester Co.)

4. Remove breaker assembly mounting screws.

5. Remove breaker assembly, carefully freeing conductor spring. Do not lose eccentric screw which may fall out.

6. To install, follow the above procedure in reverse order, being sure to include eccentric screw.

7. Lightly lubricate cam with distributor cam grease.

8. Set breaker gap and ignition timing.

Breaker Point Setting

Internal Adjustment Type

1. Remove distributor cap.

2. Loosen distributor housing clamp screw or hold-down bolt.

3. Rotate distributor until cam positions breaker points at maximum gap.

4. Inspect alignment of breaker points and, if necessary, gently bend the stationary contact support (never bend arm) with needle-nose pliers.

5. Loosen lock screws which mount breaker assembly and adjust gap using a small screwdriver in slots of contact bracket and upper plate. On models without slots and with eccentric adjusting screw instead, rotate adjusting screw until proper gap is obtained.

6. Replace cap and set ignition timing.

External Adjustment Type

1. With engine idling, raise window in distributor cap and in-

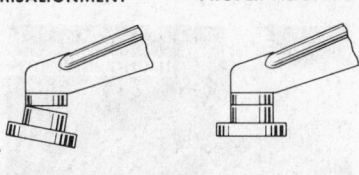

MISALIGNMENT   PROPER ALIGNMENT

CORRECT MISALIGNMENT BY BENDING FIXED CONTACT SUPPORT NEVER BEND BREAKER LEVER

LOCK SCREW

ECCENTRIC SCREW

Adjusting point gap—eccentric screw
(© International Harvester Co.)

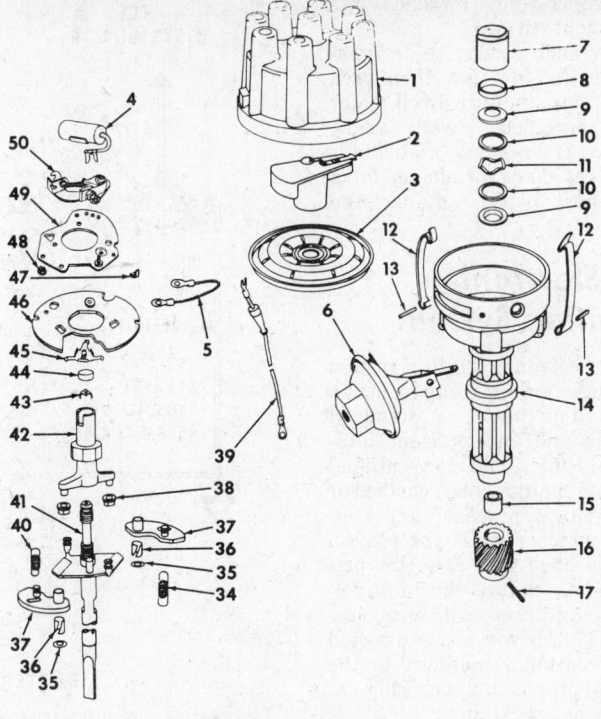

STANDARD DISTRIBUTOR

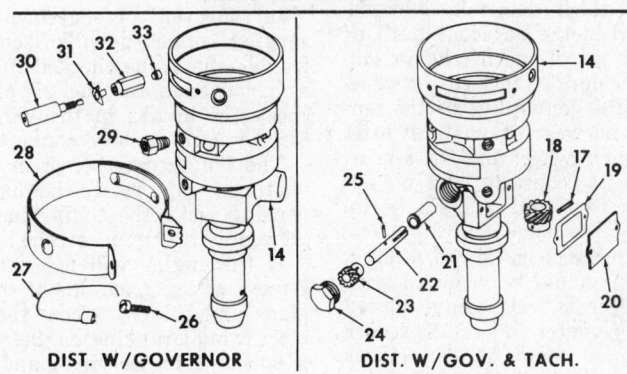

DIST. W/GOVERNOR   DIST. W/GOV. & TACH.

1 Cap, assembly	26 Screw, governor clamp
2 Rotor, assembly	27 Wire, seal
3 Cover, dust, assembly	28 Clamp, governor
4 Condenser, assembly	29 Fitting
5 Cable, ground, assembly	30 Housing, with valve, governor
6 Chamber, diaphragm, assembly	31 Plate, lock weight
7 Bushing, drive shaft, upper	32 Weight, governor valve counter
8 Retainer, seal	33 Plug, governor valve
9 Seal, Drive shaft	34 Spring, weight (primary)
10 Washer, drive shaft	35 Washer
11 Spacer, drive shaft	36 Bushing
12 Clamp, cap	37 Plate, weight, assembly
13 Pin, roll	38 Block, slider
14 Housing	39 Cable, primary
15 Bushing, drive shaft, lower	40 Spring, weight (secondary)
16 Gear, drive shaft	41 Shaft, drive, assembly
17 Pin, roll	42 Cam, breaker
18 Gear, tachometer drive	43 Retainer, cam
19 Gasket, cover plate	44 Wick, shaft oil
20 Plate, cover	45 Spring, retainer
21 Bushing, tachometer drive shaft	46 Plate, lower breaker
22 Shaft, driven, tachometer	47 Retainer, diaphragm rod
23 Gear, driven, tachometer	48 Button
24 Plug, tachometer driven shaft	49 Plate, upper breaker
25 Pin, roll	50 Point set

Exploded view—Model 1510 distributors (© International Harvester Co.)

sert proper "hex" wrench into adjustment screw.

2. Turn wrench clockwise until engine begins to miss, then back off ½ turn, or until dwell meter reads specified dwell angle. *NOTE: To remove distributor cap, press down in slot of latch head with a screwdriver and twist.*

Electronic Ignition System

The I.H. electronic ignition system contains three major components, a breakerless distributor, a standard ignition coil, and an electronic ignition control unit. The conventional cam, ignition points, and condensor are replaced by the sensor and trigger wheel, which signals the control box when to open and close the primary circuit to induce the high voltage in the ignition coil secondary circuit. This high voltage is directed in the conventional manner, to the rotor, distributor cap, spark plug cables, and to the spark plugs.

Dwell angle is determined by the angle between the adjacent teeth of the trigger wheel and by the air gap between the ends of the trigger wheel teeth and the center line of the sensor. Since no wearing surfaces exist on the trigger wheel and the sensor, dwell remains constant and no adjustment is required, after the initial sensor air gap is made.

To obtain the proper dwell reading, the air gap should be adjusted with the use of a brass feeler gauge, placed between the center line of the sensor and a aligned tooth of the trigger wheel. Attach a *modified* dwell meter to the distributor circuitry in a conventional manner, and operate the engine at curb idle. If the dwell is within specifications, the trigger wheel to sensor gap is satisfactory. If the dwell is out of specifications, stop the engine and move the sensor toward the trigger wheel to decrease dwell, and move the sensor away from the trigger wheel to increase dwell.

NOTE: Dwell is affected approximately ½ degree per each .001 inch of sensor movement.

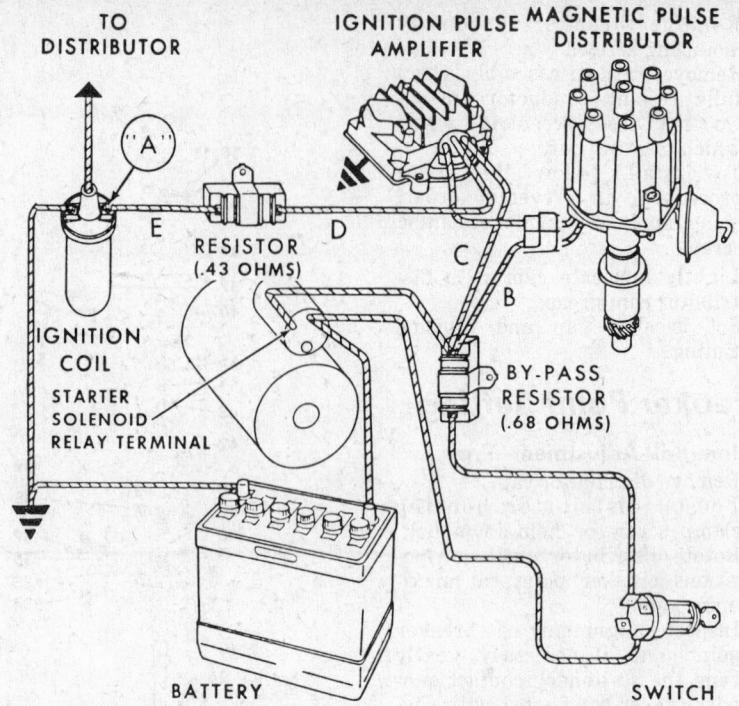

Typical transistor ignition wiring (© International Harvester Co.)

In the event of engine misfire or surging, other possible sources of trouble should be checked first, such as carburetion and fuel supply. Then check for breaks in the wiring and for corroded or loose connections.

The timing may be set in the conventional way: rotate the distributor housing while the timing marks are viewed with a timing strobe light.

If the engine will not run at all, remove a lead from one of the spark plugs and hold ½" from the engine block while cranking engine. If there is no spark, check wiring and connections.

CAUTION: Never disconnect the high voltage lead between the coil and distributor and never disconnect more than three spark plugs at a time unless the ignition switch is off.

To make compression checks, disconnect the harness plug at the control box or disconnect the lead at the negative terminal of the coil.

Ignition Timing

See tune-up specification table at the beginning of this section for timing settings. Timing light is con-

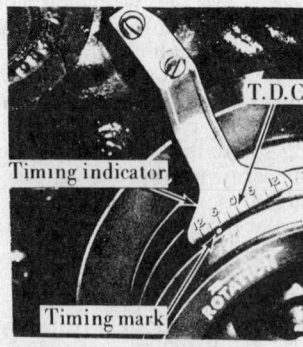

Ignition timing marks—OHV4 and V8 engines
(© International Harvester Co.)

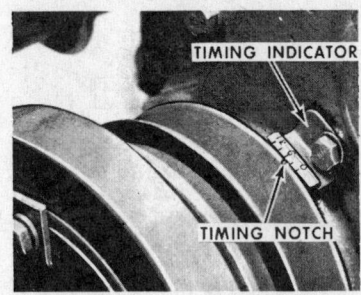

Ignition timing marks RD engine
(© International Harvester Co.)

nected to the No. 1 spark plug lead (No. 8 on V-304, 345, 392).

Alternator

Precautions—

Rectifiers and regulators in alternator systems are easily damaged by incorrect polarity. Observe the following precautions when wiring and testing circuits:

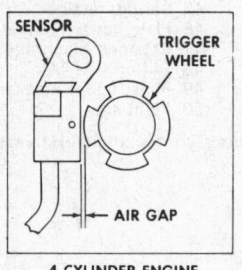

Air gap location—trigger wheel to sensor (© International Harvester Co.)

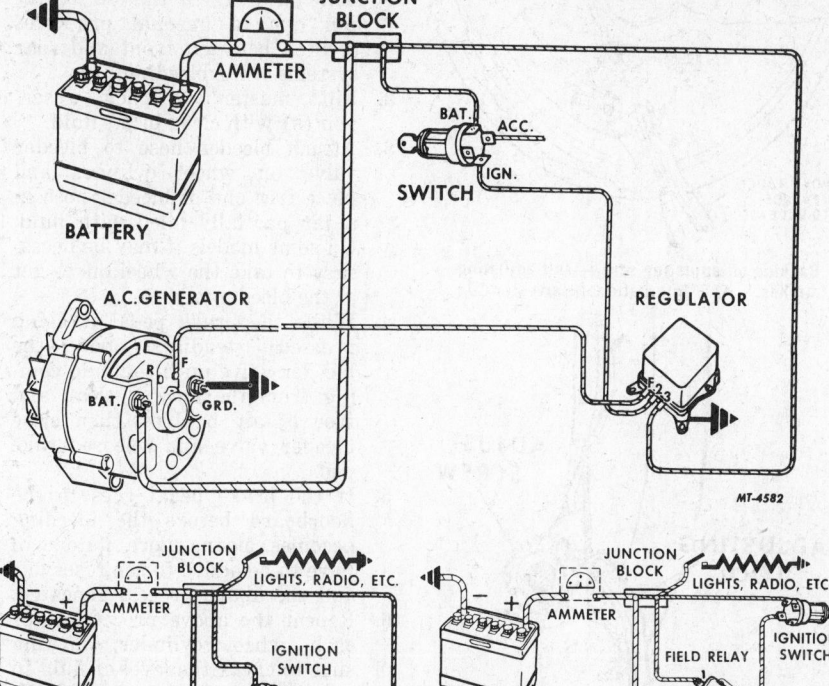

MT-4582

Circuit with Vibrating Contact Type Regulator

Circuit with Transistor Type Regulator

Typical circuits used for alternator charging system (© International Harvester Co.)

1. Always be certain of battery polarity.
2. Always connect booster battery negative to negative and positive to positive.
3. Never ground alternator output terminal.
4. When adjusting voltage regulator, be careful not to short adjusting tool.
5. Before making any tests, turn off ignition switch and disconnect battery ground.
6. Never use a fast charge with the battery connected unless charging unit is equipped with a special alternator protector.
7. Never try to polarize the alternator regulator, this will cause severe damage to the regulator and alternator.

Alternator Removal

1. Disconnect the negative battery cable.
2. Remove the wire terminals from the rear of the alternator.
3. Loosen the adjusting strap and pivot bolts. Push inward on the alternator to loosen the belt and slip it off the pulley.

4. Remove the adjusting strap and pivot bolts, and remove the alternator from the engine.
5. Installation is in the reverse of the removal. Adjust the belt to have no more than 1/2 inch deflection on the longest span of the belt.

Voltage Regulators

Two types of voltage regulators are used to control the output of the alternators. One type is the internal unit, mounted with-in the alternator, and the second is an external type, normally mounted on the inner fender panel or the firewall.

Voltage Regulator Removal

External Type

1. Disconnect clamp lead at the negative terminal of battery.
2. Disconnect the wiring harness connector at regulator terminals.
3. Remove mounting screws and regulator unit from vehicle.
4. To install, reverse the above procedure.
5. Reconnect cable clamp to battery terminal, checking polarity first.

Internal Type

1. Remove the alternator as outlined.
2. Mark and separate the front housing from the rear housing.
3. Remove the diode trio screws and nuts, and remove the trio assembly.
4. Remove the two remaining screws in the regulator, and remove the brush holder and the regulator from the rear housing.
5. Installation is the reverse order of the removal, assuring that the insulated sleeves are installed on the proper screws, during installation.

Starter

For servicing and overhauling starter motors, see Electrical Diagnosis Section of General Section.

Starter Motor—Removal

1. Disconnect cable clamp from negative terminal of battery.
2. Disconnect cable and wire leads from terminals of solenoid assembly, identifying leads with tags. If the solenoid is not mounted directly on the starter motor, disconnect the cable from the solenoid to the motor at the motor terminal.
3. Remove starter motor mounting bolts or stud nuts.
4. Pull starter assembly forward to clear housing and remove starter.
5. To install, reverse the above procedure, installing new tang lockwashers where removed.

Brakes

Hydraulic Brakes

Late model trucks are equipped with a dual hydraulic brake system in which there are separate hydraulic systems for the front and rear brakes. In this dual system a warning light switch operates a warning light on the dashboard when there is a pressure failure in either the front or rear system. A power system may be employed to reduce the effort applied to the brake pedal. See General Repair Section for hydraulic brake service and overhaul.

Master Cylinder R & R

1. Disconnect hydraulic lines from master cylinder.
2. Disconnect master cylinder pushrod at brake pedal and

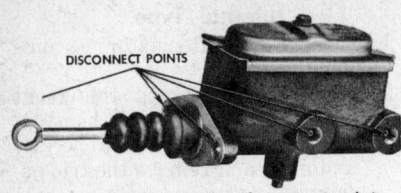

Tandem master cylinder disconnect points
(© International Harvester Co.)

remove nuts securing cylinder to dash panel.

3. If master cylinder is mounted on power unit, remove nuts securing master cylinder to power unit and remove cylinder from vehicle.

4. Installation is the reverse of the above procedure.

5. Bleed system.

Resetting the Warning Light Switch

Once a difference of 85-150 psi pressure between the front and rear systems has activated the warning light switch, it will not go off by itself and must be manually reset.

1. Clean switch and disconnect wire from terminal.

2. Unscrew and completely remove switch from body. This will allow the pistons to center and hold the switch in "off" position.

3. Screw switch back into body and reconnect wire to terminal.

4. *NOTE: If fluid is in the switch cavity, press brake pedal to see if pistol O-ring seals are leaking. If there is leakage, the O-rings must be replaced.*

5. Warning light switch should be checked periodically for proper function and the presence of foreign matter and dirt.

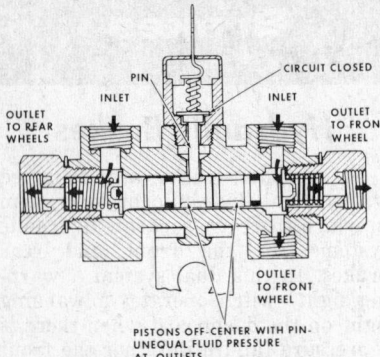

Warning light switch—circuit closed
(© International Harvester Co.)

Adjusting Brake Shoes

1. Remove rubber dust cover from access hole.

2. Using an adjusting tool or screwdriver, turn star screw until shoes drag on the drum.

3. Rotate star screw back from drag position until drag is com-

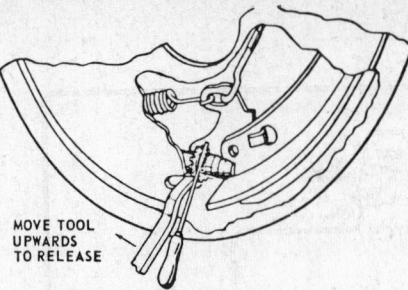

Backing off adjusting screw—self adjusting brakes (© International Harvester Co.)

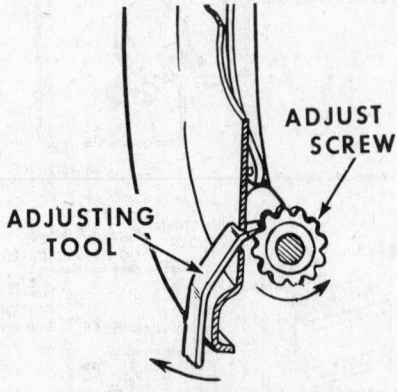

Adjusting brakes
(© International Harvester Co.)

pletely eliminated.

4. On brakes equipped with automatic adjusters it will be necessary to hold the adjusting lever away from the star wheel with a screwdriver while the adjustment is made.

Bleeding Hydraulic Brakes

1. Before bleeding the brake system, disconnect electrical wire from warning light switch and remove any foreign material or dirt accumulation around warning light switch. Then remove the switch from body. The switch must be removed to prevent shearing of the end of the pin due to unequal pressures created between front and rear systems while bleeding.

2. Fill master cylinder reservoir(s) with clean brake fluid.

3. Attach bleeder hose to bleeder valve on wheel cylinder and place free end of bleeder hose in a jar partially filled with fluid. On some models it may be necessary to take the wheel off to get at the bleeder valve.

4. While the brake pedal is being pressed steadily, open the bleeder valve until the fluid coming from the hose is clean and free of air bubbles, then close bleeder valve and release brake pedal.

5. If the brake pedal goes to the floorboard before the bleeding becomes clean, more fluid will have to be added to the reservoir and the above process repeated.

6. Repeat the above procedure for each wheel cylinder, making sure to check the level of fluid in the reservoir frequently. *NOTE: On models equipped with power boosters, the booster must be bled first.*

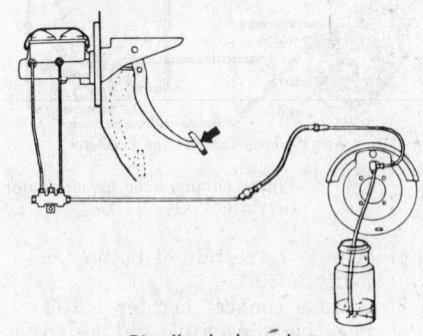

Bleeding brake system
(© International Harvester Co.)

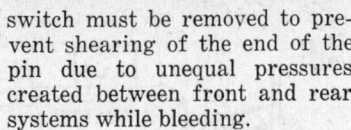

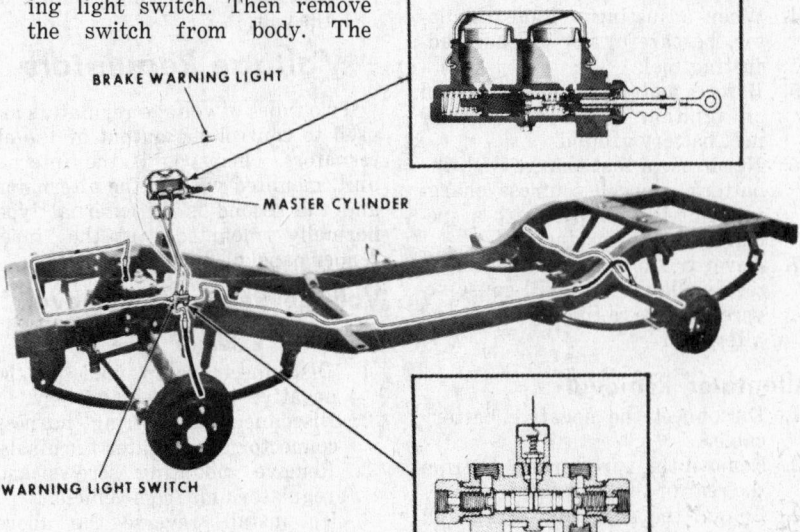

Typical standard dual brake system (© International Harvester Co.)

Wheel Cylinder (Front Wheel)

Removal

1. Raise the front of the vehicle and support it safely.
2. Remove the front wheel assembly, including the drum, to expose the brake shoes.
3. Remove the brake shoes from the brake support plate.
4. Loosen or remove the hydraulic brake line. Remove the wheel cylinder from the brake support plate.

NOTE: The brake hose can be loosened at the cylinder and removed when the cylinder is loose from the support plate, or can be removed from the pipe and fitting at the frame rail, and withdrawn with the wheel cylinder from the brake support plate, to be separated later.

Installation

1. Connect the brake hose to the wheel cylinder and install the wheel cylinder to the brake support plate.
2. Install the brake lining and secure.
3. Install the drum and wheel assembly.
4. Adjust the wheel bearings and brakes.
5. Bleed the hydraulic system. Refill the master cylinder.

NOTE: The warning switch should be disconnected and removed from its seat in the switch until after the bleeding operation is completed.

Wheel Cylinder (Rear Wheel)

Removal

1. Raise the rear of the vehicle and support it safely.
2. Remove the axle shaft (if full floating), and the rear wheel assembly.
3. If necessary, remove the drum separately.
4. Remove the brake lining from the brake support plate.
5. Remove the hydraulic line from the wheel cylinder.
6. Remove the wheel cylinder attaching bolts and remove the cylinder.

Installation

1. Install the wheel cylinder on the brake support plate.
2. Install the hydraulic line to the wheel cylinder.
3. Install the brake shoes on the brake support plate.
4. Install the drum, if removed separately.
5. Install the wheel assembly and the axle shaft (if removed).

6. Bleed the system and refill the master cylinder. Refer to the *NOTE* under the front cylinder installation concerning the warning switch removal during bleeding.

Disc Brakes

The disc brakes are the sliding caliper, single piston type, and are used on the front wheels in combination with drum type brakes on the rear.

Removal

1. Raise the front of the vehicle and support it safely.
2. Remove the front wheels from the hub.
3. Remove approximately a third of the fluid from the large reservoir of the master cylinder, to avoid leakage of fluid when the pistons are forced back into the calipers.
4. Position a large "C" clamp over the caliper and engage the rear of the caliper with the shoe of the clamp, and place the screw on the outboard disc pad. Tightening the screw will cause the piston to be forced deeper in the bore, by the movement of the caliper.
5. Remove the key retaining screw and drive the support key and support spring from the caliper and support, using a brass drift and a light hammer.
6. Remove the caliper from the support bracket and support the assembly on a wire.
CAUTION: Do not support the assembly by the brake hose.
NOTE: It is not necessary to remove the brake hose from the caliper when only replacing the disc pads, and therefore it would not be necessary to bleed the caliper when reinstalled.
7. Remove the disc pads from the calipers.

Installation

1. Position the new disc pads into the calipers, using a new anti-rattle spring clip, and position it on the inboard pad.
2. Place the caliper assembly over the rotor and engage the anchor bracket.
3. Position the caliper support spring and support key between the bottom edge of the caliper and the anchor bracket.
4. With the use of a brass drift and hammer, drive the key and spring assembly into position and install the key retaining screw.
5. Refill the master cylinder as needed, apply the brakes several times to seat the pads, and recheck the master cylinder fluid level.

6. Install the wheels and lower the vehicle.

Brake Pedal Adjustment

There are no provisions available for the adjustment of the brake pedal height. However, it should be checked to determine if sufficient height exists. Corrections can only be made by replacement of parts, alignment, or straightening of the affected parts. To determine if sufficient pedal height exists, open a wheel cylinder bleed valve to simulate a failed system, and depress the brake pedal. The pedal should not contact the floor board during this test.

NOTE: Close the bleeder valve before releasing the brake pedal. The brake warning light switch will have to be reset after the test is completed.

Stoplight Switch Adjustment

No stoplight switch adjustments are provided. If the stop lamps are inoperative, a defective switch, defective bulbs, loose or broken connections, or an improper positioned switch would be indicated. A mechanical type switch is located on the brake pedal, at the pushrod bolt location, while the hydraulic type switch is located on or near the master cylinder, and operated by hydraulic pressure.

The air brake system on the straight truck models, utilizes two air operated switches. One switch is used on the primary brake system, and the second switch is used on the secondary brake system, to provide stop lamps in case of a failure in one of the systems. The tractor type models use a single switch, but it is mounted on a double check valve, and is operated by air being passed through from both the primary and secondary brake systems.

Vacuum Power Brakes

The vacuum power cylinder used with the single and dual hydraulic brake systems assists braking differently than a power booster in that its activation results directly from the foot pedal and not from the master cylinder. There are single and tandem dual types. Both types are mounted on the engine side of the firewall.

Vacuum Power Brake Cylinder R & R

1. Disconnect vacuum hose from check valve.
2. Disconnect hydraulic lines from master cylinder.

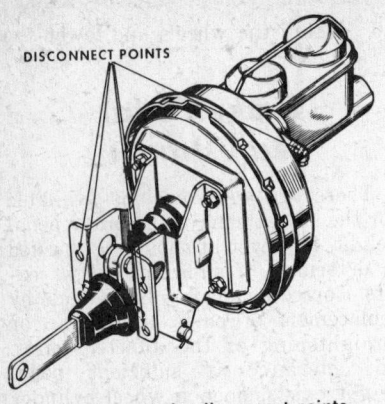

DISCONNECT POINTS

Power cylinder disconnect points
(© International Harvester Co.)

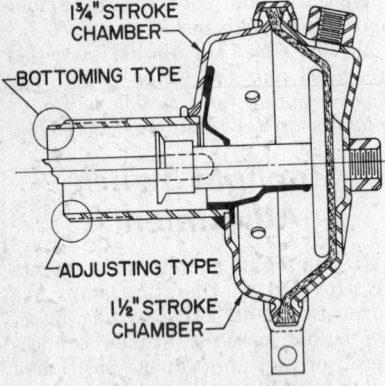

1¾" STROKE CHAMBER

BOTTOMING TYPE

ADJUSTING TYPE

1½" STROKE CHAMBER

Power brake air chamber
(© International Harvester Co.)

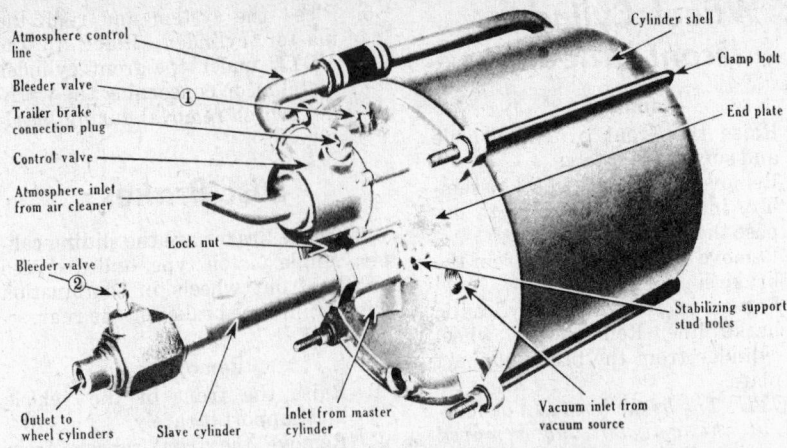

Atmosphere control line
Bleeder valve ①
Trailer brake connection plug
Control valve
Atmosphere inlet from air cleaner
Bleeder valve ②
Lock nut
Outlet to wheel cylinders
Slave cylinder
Inlet from master cylinder
Cylinder shell
Clamp bolt
End plate
Stabilizing support stud holes
Vacuum inlet from vacuum source

Typical single cylinder power booster (© International Harvester Co.)

Vacuum Brake Boosters

Vacuum boosters add pressure to the hydraulic brake system. The activation and amount of pressure is controlled by a hydraulic line from the master cylinder. Disassembly and service procedures for vacuum boosters may be found in the General Repair Section.

Booster R & R

1. On units lacking integral air filter, remove air inlet hose (from engine air cleaner).
2. Disconnect vacuum inlet tube (coming from engine manifold).
3. Disconnect hydraulic line from master cylinder and hydraulic line going to wheel cylinders.
4. Remove mounting bolts and lift out vacuum unit.
5. To install, reverse the above procedure.
6. Bleed complete hydraulic system, starting with cylinder on vacuum unit as described below.

Bleeding Vacuum Booster Systems

1. The booster must be bled before proceeding to wheel cylinders.
2. All vacuum boosters have a bleed valve on the control valve as indicated in the figures by the number "1". This valve must always be bled first.
3. On boosters having an additional bleeder valve on the hydraulic cylinder indicated by the number "2", bleed this cylinder after control valve.
4. Bleed wheel cylinders as described in preceding section.

Trailer Brake Hand Control Adjustment

The advance control valve is used in conjunction with the Hydrovac and trailer brake system to vary the initial braking of the trailer.

1. Place advance plate in full released position (rotate counterclockwise).
2. While coasting on smooth road at 20 mph, apply valve by rotating clockwise until a slight drag is felt from the trailer brakes.
3. Rotate advance plate to where it just touches the valve operating handle. This releases the brakes on the trailer but sets the advance effect.
4. Leaving advance valve controls as set above, gently apply tractor brakes to check the "advance" of the trailer brakes.

Parking Brake Adjustment

Rear Brake Type

1. Loosen locknut on the equalizer rod and turn front nut forward several turns.
2. Turn the locknut (rear) forward just enough to remove any slack but not so much that the brake shoes lift of their anchors.

3. Disconnect pedal link from pedal from inside the cab.
4. Remove bolts which mount the bracket to the firewall.
5. To install reverse the above procedure.
6. Bleed master cylinder output ports while connecting tube nuts which secure lines to ports.
7. Bleed hydraulic brake system.

Atmosphere control line
Clamp bolt
Bleeder screw (1)
Integral air cleaner
Bleeder screw (2)
Cylinder shell
End plate
Vacuum inlet from vacuum source
Outlet to wheel cylinders
Inlet from master cylinder

Power cylinder with integral air cleaner (© International Harvester Co.)

NOTE: Hydraulic Switch shown on Hydrovac. Some installations provide for Hydraulic Switch located at Master Cylinder.

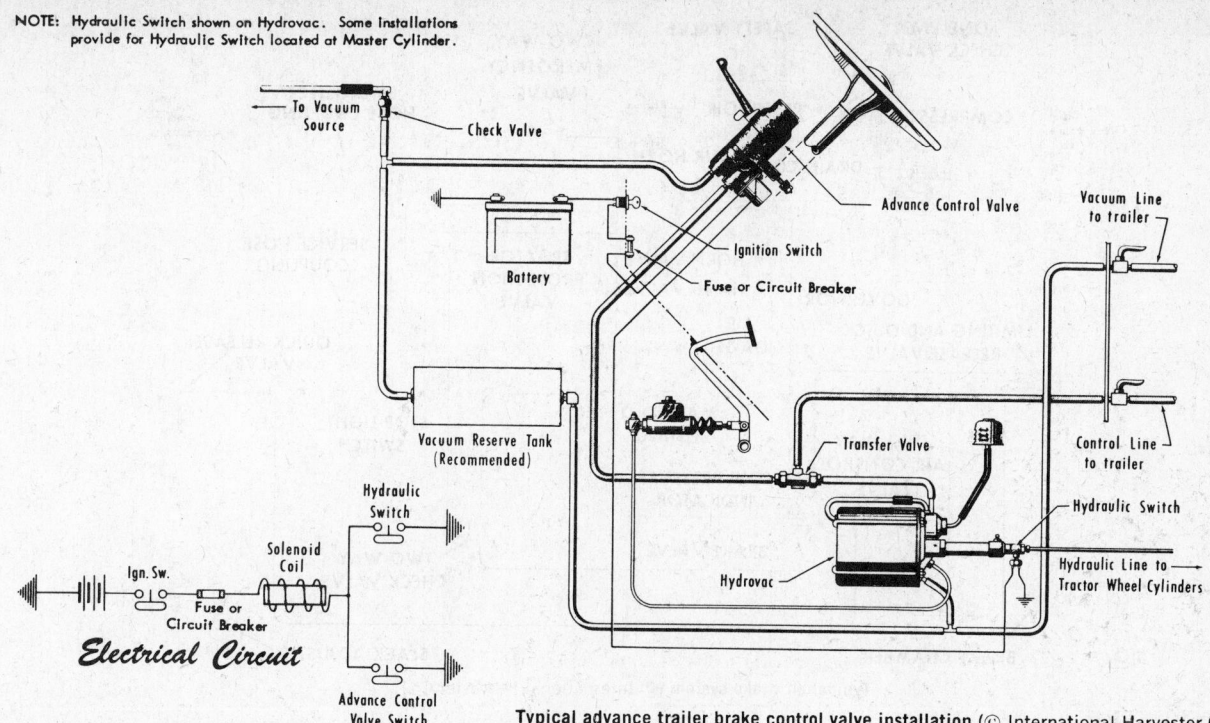

Electrical Circuit

Typical advance trailer brake control valve installation (© International Harvester Co.)

3. Tighten both nuts against the equalizer.

Drive Shaft Band Type

1. Leaving parking brake lever in the extreme release position, check that the cam lever is resting squarely on upper brake band bracket. This is adjusted by removing the clevis pin and readjusting yoke.
2. Adjust screw nut (1) until a clearance of .020-.030" is reached.
3. Adjusting nuts (4) on bolt (5), obtain .020-.030" clearance on lower half of lining and drum.
4. Adjust nuts (2) on bolt (3) for a clearance of .020-.030" clearance on top half of lining.
5. Lock all adjustment with lock nuts.

Enclosed Drum Type— Driveshaft

1. Block the vehicle wheels.
2. Disconnect the clevis pin from the bellcrank and clevis at the brake assembly.
3. Move the hand brake control

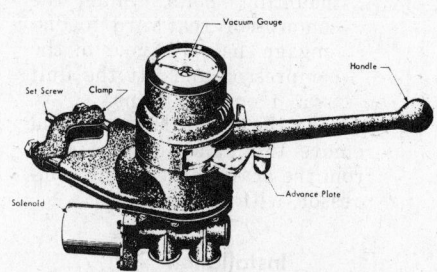

Advance control valve assembly
(© International Harvester Co.)

lever approximately $\frac{1}{4}$ to $\frac{3}{8}$ inch in the apply direction to compensate for the allowable freeplay.
4. Move the bellcrank lever in the apply direction until contact is made with the brake cam, without any brake shoe movement.
5. Adjust the clevis until the hole in the clevis aligns with the mating hole in the positioned bellcrank lever.
6. Assemble the clevis and bellcrank lever. Tighten the clevis locknut.
7. Recheck the freeplay and lever adjustment.

NOTE: If the vehicle is equipped with the Orschelm type parking brake lever, (over-center type), rotate the adjusting knob on the end of the lever, while in the released position, to attain a force of 90 lbs. to apply the parking brake. When properly adjusted, a distinct click will be heard when pulled over center.

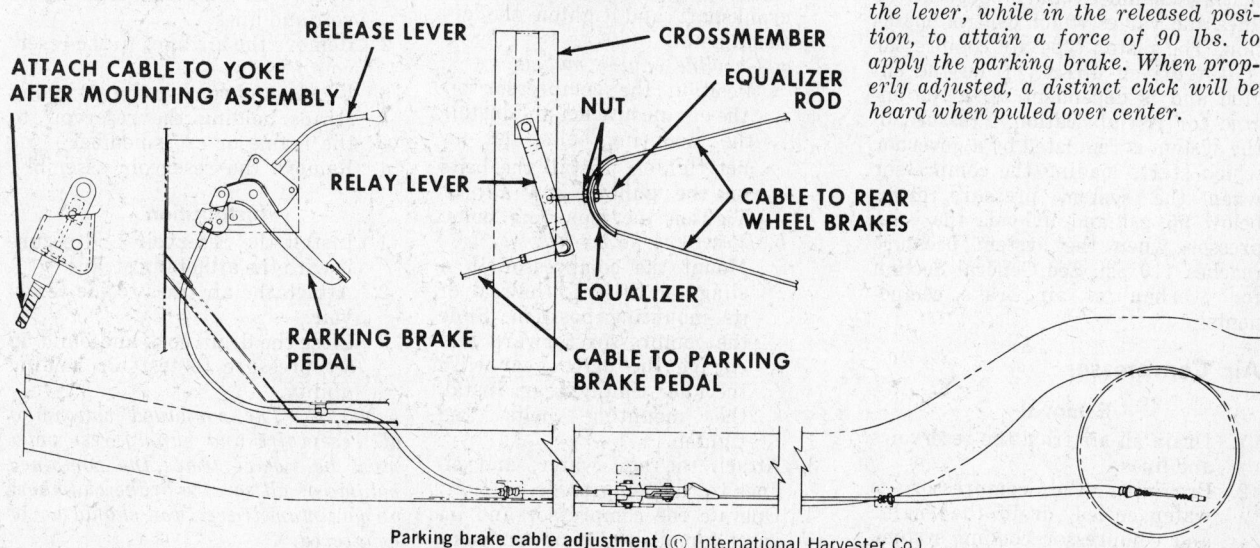

Parking brake cable adjustment (© International Harvester Co.)

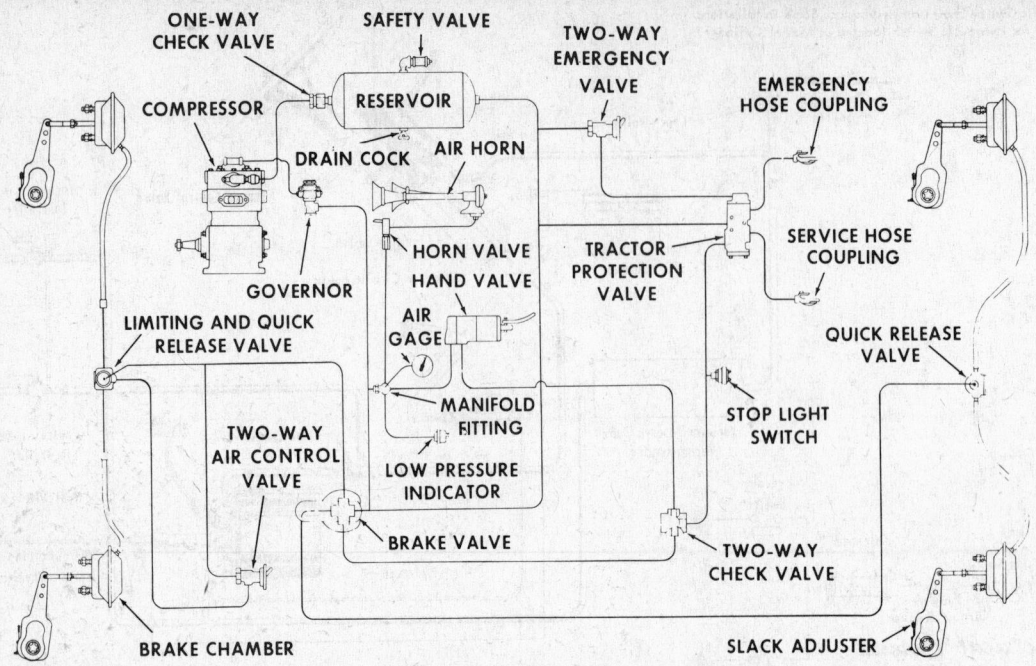

Typical air brake system (© International Harvester Co.)

Spring Actuated—
Tandem Type—Air

This unit is a spring actuated type parking brake consisting of a tandem-type cylinder, connected to the brake shoes, through the air brake slack adjuster and brake camshaft. The cylinder assembly is divided into two sections. One section is the regular air brake chamber, and the second is the spring actuated chamber, containing a powerful spring, compressed by air pressure and applied by the operator with the use of a control valve. The brake adjustment is controlled by the adjustment of the slack adjuster for the regular brakes.

Air Brake System

Air brake systems are composed of a compressor, a reservoir, brake actuating chambers and a network of lines and valves which control operation. The piston-type air compressor is belt driven directly from the engine and is dependent upon the engine for its lubrication. Pressure in the system is regulated by a governor which starts loading the compressor when the system pressure drops below 95 psi and unloads the compressor when the system pressure reaches 110 psi. See General Section for overhaul of air brake components.

Air Compressor

Removal

1. Drain all air from the reservoirs and lines.
2. Providing the compressor is water cooled, drain the engine and compressor cooling system.
3. Disconnect all air, water, and oil lines from the compressor.
4. a. *Gasoline engine models*
   Remove the compressor mounting bolts and remove the belt or belts from the pulleys. Remove the compressor.
   b. *Diesel engine models*
   Attach a lifting sling to the compressor and remove the mounting bolts. Slide the compressor rearward to disengage the drive gear of the compressor and lift the unit away from the engine.
5. Remove the crankshaft nut and remove the pulley or the gear from the crankshaft of the compressor, with puller.

Installation

1. Install the compressor crankshaft pulley or gear on the crankshaft and tighten the attaching nut.
2. a. *Gasoline engine models*
   Position the compressor on the engine bracket and install the mounting bolts, but do not tighten. Install the belts on the pulleys and adjust. Tighten the mounting bolts.
   b. *Diesel engine models*
   Mount the compressor in a sling and lift it to the rear of its mounting position. Slide the compressor forward and engage the drive gear with the idler timing gear. Install the mounting bolts and tighten.
3. Attach the air, water, and oil lines to the compressor.
4. Operate the compressor and inspect for oil, water or air leaks.

Compressor Governor Valve

Removal and Installation

The compressor governor valve can be either the remote-mounted or compressor-mounted type. The air system must be drained before any attempt is made to remove either type valve. If the governor valve is the compressor-mounted type, the reservoir line must be disconnected, and then the attaching bolts removed. If the governor valve is the remote-mount type, the unloader line and the reservoir line must be disconnected from the valve, and then the attaching bolts removed. Installation is the reverse of removal. Test the air system before the vehicle is placed in service.

Reservoir

Removal

1. Drain the air from the reservoir and lines.
2. Remove the air lines to the reservoir.
3. Loosen and remove the attaching straps holding the reservoir to the frame or crossmember.
4. Remove the reservoir assembly.

Installation

1. Install the reservoir and secure it with the attaching straps.
2. Attach the air lines to the reservoir.
3. Close the drain cock and build up air pressure to test air holding ability.

NOTE: The combined volume of all reservoirs and supply reservoirs must be twelve times the combined volume of all service brake chambers at maximum travel, and should never be altered.

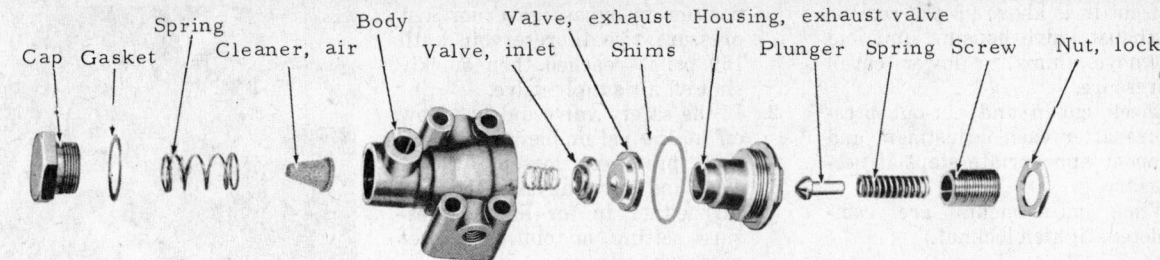

Air compressor governor—early models (© International Harvester Co.)

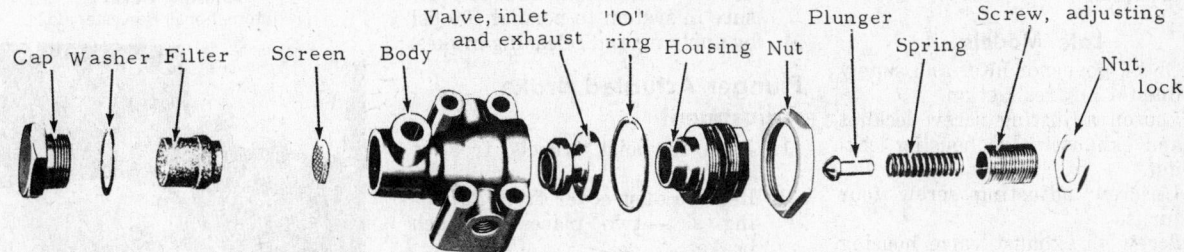

Air compressor governor—late models (© International Harvester Co.)

Control and Check Valves

Removal and Installation

NOTE: Before any valve, line, or fitting is loosened or removed, all air must be drained from the system. Personal injury can result if these precautions are not adhered to.

The safety valve, pressure gauge or gauges, low pressure indicator, stop light switch, automatic reservoir drain valve, check valves, inversion valve, and quick release valves are located in the lines and may be bolted to the frame or crossmember. To remove or install the above mentioned valves, is simply a matter of disconnecting and connecting the lines from the valve, and removing or installing any attaching bolts, and repairing or replacing the affected valve. The lines must be maintained in their proper order so as not to interchange the primary and secondary air systems.

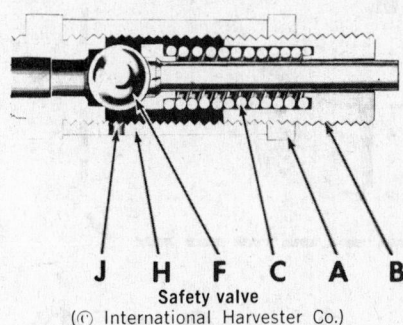

J H F C A B
Safety valve
(© International Harvester Co.)

Brake Valve

Removal

1. Drain the air from the primary and secondary air systems.
2. Dsisconnect all supply and delivery lines at the brake valve. Mark each line to assist in reassembly.
3. Remove the fittings from the

valve and mark for reassembly.
4. Remove the valve from the vehicle.
   a. *Suspended pedal valves.*
      Remove the attaching nuts on the engine fire wall side and remove valve.
   b. *Treadle type valve*
      Remove three capscrews on the outer bolt circle of the mounting plate and lift the valve upward. The valve can be removed from the mounting plate by the removal of capscrews from the inner bolt circle.

Installation

1. The installation of the brake valve is the reverse of removal. Connect all fittings and lines in their proper order, and test the brake operation before vehicle movement.

Spring Brake Two-way Control Valve

Push and Pull—Dash Mounted

Removal

1. Drain the air from the system.
2. Loosen locknut and remove the knob and nut.
3. Disconnect the air lines at the valve.
4. Loosen and remove the valve mounting nut, and name plate.
5. Remove the valve from the rear of the instrument panel.

Installation

1. The installation is in the reverse order of the removal. Test the system before vchicle operation.

Flip Switch—Dash Mounted

Removal

1. Drain air from the system.

2. Remove the air lines from the valve.
3. Remove the machine screw securing the valve to the instrument panel and remove the valve.

Installation

1. Installation is the reverse of removal. Test the system before vehicle operation.

Air Compressor Governor Adjustment

Early Models

1. Check governor filter and supply line for any restriction.
2. Loosen adjusting screw locknut and exhaust valve housing.
3. Remove exhaust valve housing with adjusting screw locknut, and shims as a complete unit.
4. Replace exhaust valve housing, adjusting screw, locknut and three shims into governor body and tighten.
5. Turn adjusting screw until it sticks out 3/8" from exhaust valve housing.
6. Start truck engine and build up pressure in air brake system until it reaches 115 psi then shut off engine. If governor cuts out before 115 psi then turn adjusting screw out one complete turn and repeat this step.
7. With pressure holding at 115 psi slowly turn adjusting screw out until governor cuts out (dull pop).
8. Start truck engine and slowly bleed pressure until governor cuts in. Cut-in should be between 93-98 psi.
9. If cut-in is below 93 psi, remove exhaust valve housing unit and add shims (one shim equals 4 psi) to adjust cut-in and install as in step 4.

329

10. If cut-in is above 98 psi, remove exhaust valve housing unit and remove shims to lower cut-in pressure.
11. Check cut-in and cut-out pressure after each adjustment and repeat appropriate steps if necessary.
12. When adjustments are completed, tighten locknut.

Air Compressor Governor Adjustment

Late Models

1. Check governor filter and supply line for any restriction.
2. Loosen adjusting screw locknut and exhaust valve housing locknut.
3. Unscrew adjusting screw four turns.
4. Screw in exhaust valve housing until it bottoms (do not tighten—seats are easily ruined).
5. Back off exhaust valve housing ¾ turn.
6. While holding exhaust valve housing, turn in adjusting screw three turns after it has made contact with the spring. A slight resistance should be felt when contact is made. If this contact cannot be felt, turn adjusting screw until it sticks out ⅜" from exhaust valve housing.
7. Start engine and build up air pressure to 115 psi and shut off engine. If governor cuts out before 115 psi, turn adjusting screw out one turn and repeat this step.
8. With pressure holding at 115 psi, turn in adjusting screw until governor cuts out (dull pop).
9. Start truck engine and bleed pressure down until governor cuts in. Cut-in pressure should be 93-98 psi.
10. If cut-in pressure is below 93 psi, hold adjusting screw and turn out exhaust valve housing (1/6 turn equals 5 psi).
11. If cut-in pressure is above 98 psi, hold adjusting screw and turn in exhaust valve housing.
12. Repeat steps until proper cut-in pressure is reached.
13. Check cut-out pressure and adjust if necessary.
14. Tighten adjusting screw locknut and exhaust valve housing locknut.

Air Reservoir Safety Valve Adjustment

1. Connect an accurate air pressure gauge to the emergency line at the rear of the truck and open emergency line valve. With truck engine running, turn air supply valve to the air supply position to bypass governor. Let pressure rise in reservoir until 150 psi is reached then quickly shut off air supply valve.
2. If the safety valve did not blow off at 150 psi or blew off before that pressure, loosen locknut (A) and turn adjusting screw (B) either in for higher pressure setting or out for lower pressure setting.
3. When adjustment is complete, tighten locknut and reduce pressure in system to normal 100 psi by applying and releasing brakes.

Plunger Actuated Brake Adjustment

1. Jack or hoist wheels free of ground.
2. Remove dust cover from adjusting slot—two places on each brake.
3. Turn the star wheel until heavy drag on drum is developed.
4. Back off bolt barely past light drag.

Adjusting brakes
(© International Harvester Co.)

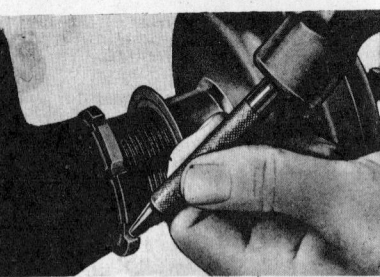

Loosening collect nut—air brake power unit
(© International Harvester Co.)

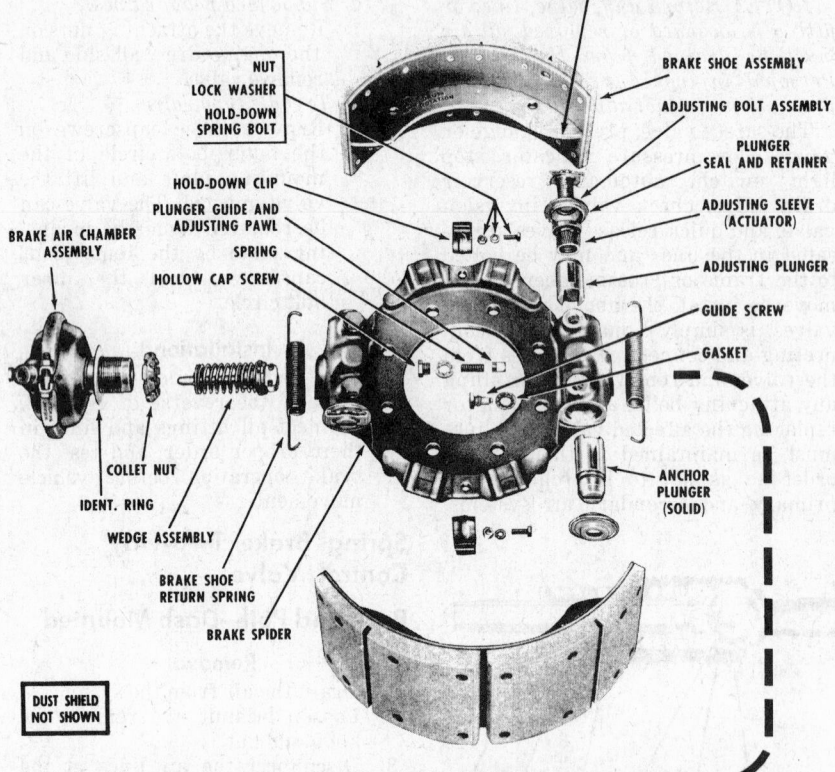

LONG RADIUS
NUT
LOCK WASHER
HOLD-DOWN SPRING BOLT
BRAKE SHOE ASSEMBLY
ADJUSTING BOLT ASSEMBLY
PLUNGER SEAL AND RETAINER
HOLD-DOWN CLIP
PLUNGER GUIDE AND ADJUSTING PAWL
ADJUSTING SLEEVE (ACTUATOR)
BRAKE AIR CHAMBER ASSEMBLY
SPRING
HOLLOW CAP SCREW
ADJUSTING PLUNGER
GUIDE SCREW
GASKET
COLLET NUT
IDENT. RING
WEDGE ASSEMBLY
ANCHOR PLUNGER (SOLID)
BRAKE SHOE RETURN SPRING
BRAKE SPIDER
DUST SHIELD NOT SHOWN

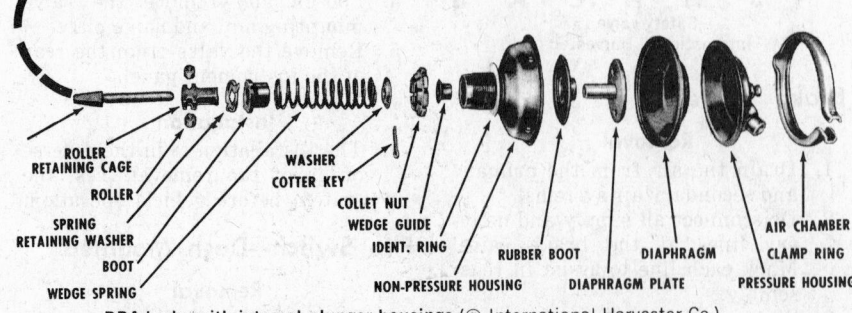

ROLLER RETAINING CAGE
ROLLER
SPRING RETAINING WASHER
BOOT
WEDGE SPRING
WASHER
COTTER KEY
COLLET NUT
WEDGE GUIDE
IDENT. RING
NON-PRESSURE HOUSING
RUBBER BOOT
DIAPHRAGM
DIAPHRAGM PLATE
AIR CHAMBER CLAMP RING
PRESSURE HOUSING

RDA brake with integral plunger housings (© International Harvester Co.)

5. Replace dust covers in adjusting slots.
6. Repeat for other brakes.
7. If brakes are equipped with automatic adjusters, check drum to lining clearance. If it is more than .060″, adjust brakes manually until they can be serviced. See Unit Repair section for servicing air brakes.

Air Power Unit Adjustment

1. Determine whether the power unit is the adjustable or bottoming type. Bottoming units have an identification tag fastened to the clamp ring bolt of the air chamber. Adjustable units have no identification markings. Loosen collet nut.
2. Bottoming units automatically provide optimum useful chamber stroke and need only be screwed in until they bottom.
3. Adjustable units are adjusted manually by screwing the unit in until the wedge is just starting to lift the plungers off the abutment seats at the first movement of the diaphragm.
4. After screwing in air chamber unit to proper depth, tighten collet nut to lock position.

Air Brake Slack Adjusters Adjustment

Cam actuated air brakes should be adjusted for lining wear every 2000 miles. Adjustment is made by turning a worm screw on a gear which positions stack adjuster angle.

1. With wheel free to rotate, disconnect pushrod from the slack adjuster to determine whether or not it is in fully released position.
2. Reinsert clevis pin through bottomed pushrod and slack adjuster arm, adjusting worm gear if necessary.
3. Holding the locking sleeve in, adjust worm screw until shoes drag against drum, then back off enough to eliminate drag.

4. Angle that slack adjuster makes when brake is fully applied should not "go over" 90° point.
5. If the slack adjuster goes over the 90° point the maximum force will not be exerted and the pushrod must be adjusted as follows:
6. Carefully disconnect slack adjuster from pushrod—it may snap into the air chamber with considerable force.
7. Loosen locknut on pushrod clevis and thread clevis onto pushrod towards air chamber several turns.
8. Connect pushrod and clevis with pin.
9. Check pushrod-to-slack adjuster angle again as the brake is applied to make sure that it is not still going over the 90° point.
10. Readjust if necessary.
11. When adjustment is correct tighten locknut on pushrod clevis and install cotter pin which secures clevis pin.

Fuel System

The carburetors used on the International Harvester truck engine models, vary from one barrel to four barrels. Due to emission control regulations, different internal components and adjustments are necessary for each carburetor model, regardless of the similarity of the carburetor exteriors. When replacing or overhauling a carburetor, it is most important the model number is referred to. This number can be found either on a metal tag, fastened to a bowl cover screw, or embossed on the carburetor casting.

The Emission Control Information Label should be referred to, for the correct specifications necessary for the idle mixture and speed adjustments.

For carburetor overhaul specifications, refer to the Unit Repair Section.

Fuel Tanks

The location and size of the fuel tanks vary as the requirements of the vehicle vary. The removal and installation procedure will depend upon the location and size of the tanks.

When left and right tanks are used a fuel selector switch is used and is located on the floor panel or the instrument panel.

Carburetor Removal

Single-Barrel Holley Model 1920

1. Remove the air cleaner, fuel lines, vacuum lines and any other hoses and linkage attached to the carburetor.
2. Remove the attaching bolts from the base of the carburetor and remove the carburetor from the manifold. Remove and discard the old gasket under the base of the carburetor.
3. To install reverse the removal procedure making sure to install a new gasket under the carburetor base.

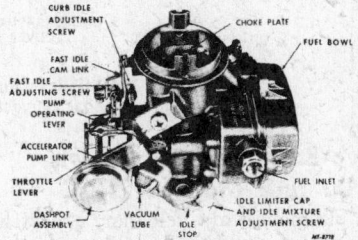

Model 1920 carburetor
(© International Harvester Co)

Single-Barrel Holley Model 1940

1. Remove the air cleaner, fuel lines, vacuum lines and any other lines or linkage attached to the carburetor.
2. Remove the attaching bolts from the base of the carburetor and remove the carburetor from the manifold. Remove and discard the old gasket from under the carburetor.
3. To install reverse the removal procedure making sure to install a new gasket under the carburetor base.

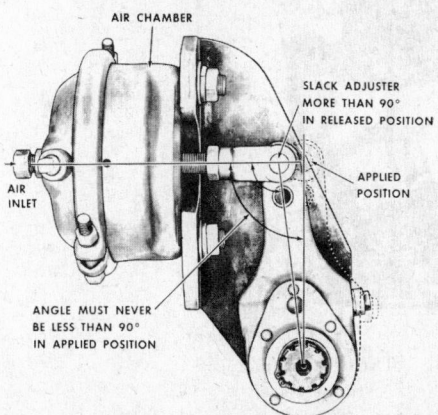

Foundation air brake slack adjuster
(© International Harvester Co.)

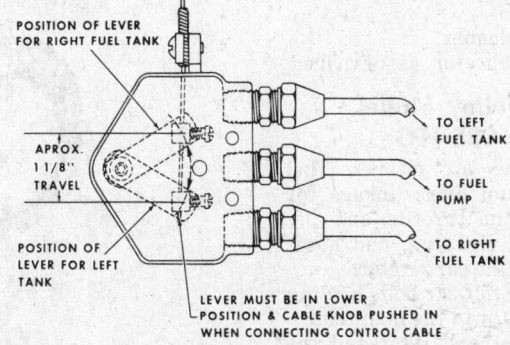

Fuel tank selector valve cable positions (© International Harvester Co.)

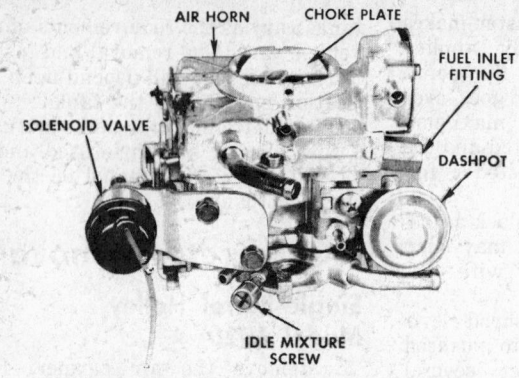

Carburetor model 1940 (© International Harvester Co.)

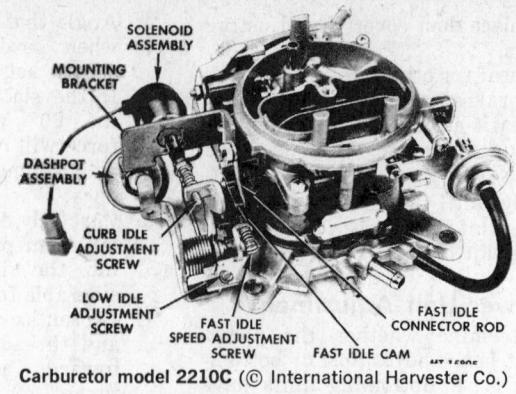

Carburetor model 2210C (© International Harvester Co.)

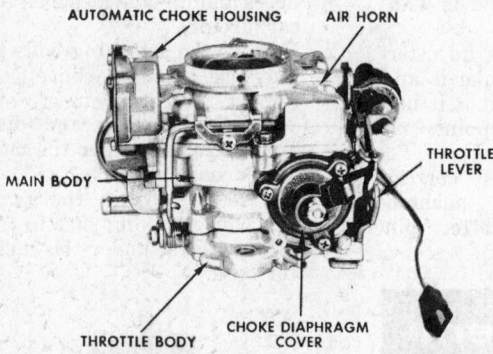

Model 1940 Holley carburetor
(© International Harvester Co.)

Model 1940 Holley carburetor with dashpot
(© International Harvester Co.)

Two-Barrel Holley Models 2100 and Model 2210-C

1. Remove air cleaner, throttle linkage and choke cable.
2. Disconnect fuel line and distributor and governor vacuum lines.
3. Remove bolts from mounting studs and lift off carburetor.
4. To install, clean manifold mating surface and install a new flange gasket.
5. Install carburetor but do not tighten down stud nuts.
6. Connect fuel line and vacuum lines.
7. Tighten nuts on mounting studs in an alternating fashion so that flange gasket compresses evenly for a good seal.
8. Connect throttle linkage and choke cable, making sure that choke plates are fully open when the choke knob is pushed in.
9. Check throttle for complete travel.
10. Install air cleaner.
11. Adjust carburetor as described.

Two-Barrel Holley Model 852-FFG and 885-FFG

1. Remove the air cleaner, fuel lines, vacuum lines, linkage for the choke and throttle, and any other lines, linkages, and hoses attached to the carburetor.
2. Remove the nuts or bolts holding the carburetor to the manifold.
3. Remove the carburetor and the base gasket. Clean any gasket

particles from the base and manifold surface.
4. To install the carburetor, reverse the removal procedure.
5. Adjust the idle speed and mixture as outlined.

Two-Barrel Holley Models 2300 and 2300G

1. Remove air filter and disconnect fuel line, distributor and governor vacuum lines and throttle and choke linkages.
2. Remove mounting stud nuts and lift off carburetor.

3. Remove flange gasket and discard.
4. To install, clean manifold mating surface and install a new flange gasket.
5. Operate choke and throttle levers to be sure they are functioning properly.
6. Install carburetor and mounting stud nuts, but do not tighten nuts.
7. Connect fuel line, vacuum lines and throttle and choke linkage.
8. Tighten down mounting stud nuts in an alternating criss-cross

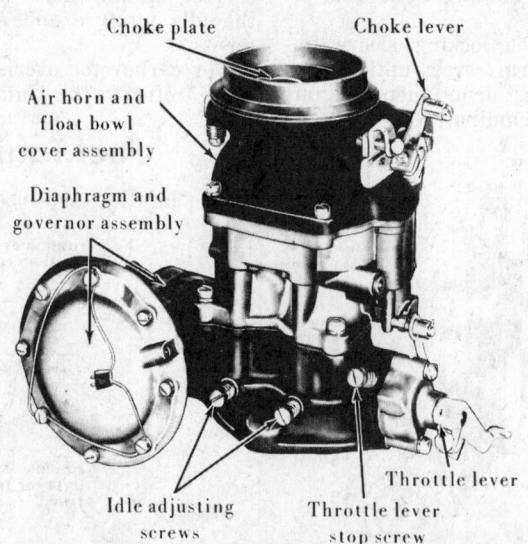

Model 2110G Holley 2 BBL carburetor
(© International Harvester Co.)

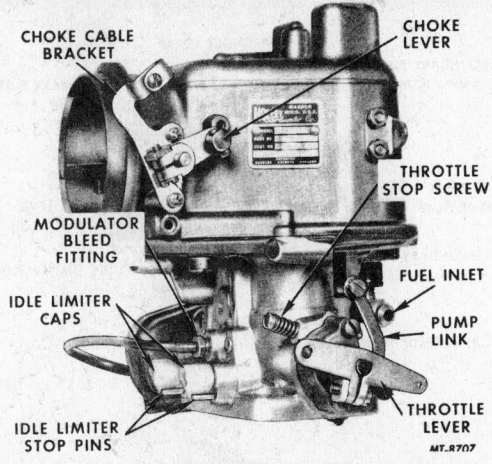

Carburetor model 885-FFG (© International Harvester Co.)

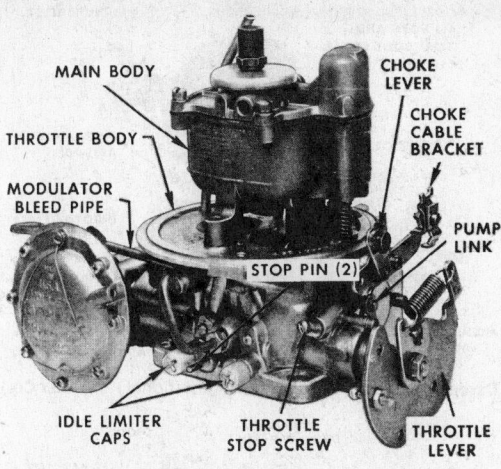

Carburetor model 2140G (© International Harvester Co.)

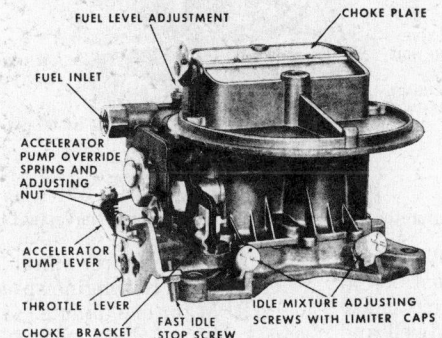

Carburetor model 2300 (© International Harvester Co.)

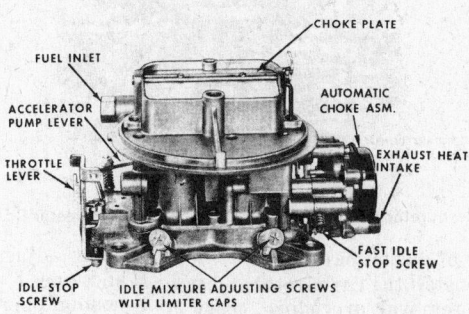

Carburetor model 2300G (© International Harvester Co.)

pattern to make sure that flange gasket is compressed evenly.

9. Check to see that choke plate is fully open and dashboard knob is in when connecting choke.

10. Make all adjustments described in text.

11. Install air cleaner.

Four-Barrel Holley Model 2140G and 2140SG

1. Remove the air cleaner, throttle linkage, vacuum lines, fuel lines, choke control cable, and other hoses and linkages attached to the carburetor.

2. Remove the attaching bolts or nuts from the base of the carburetor, and remove the carburetor from the manifold.

3. Discard the base gasket and clean the base and manifold surfaces of gasket particles.

4. To reinstall the carburetor, reverse the removal procedure, using a new base gasket.

5. Adjust the carburetor as outlined.

Four-Barrel Holley Model 4150, 4150C, 4150G

1. Remove the air cleaner, throttle linkage and choke linkage, fuel lines, vacuum lines, and other hoses and linkages attached to the carburetor.

2. Remove the nuts or bolts holding the carburetor to the intake manifold.

3. Lift the carburetor from the manifold and remove the gasket from the base of the carburetor and from the manifold surface.

4. To install the carburetor, reverse the removal procedure.

5. Adjust the idle mixture and speed as outlined.

Carter Thermo-Quad Carburetor

1. Remove the air cleaner, throttle linkage, vacuum hoses, fuel lines, and any other hoses and linkages attached to the carburetor.

2. Remove the bolts or nuts holding the carburetor to the manifold, and remove the carburetor from the intake manifold.

3. Discard the base gasket and clean the base and manifold sur-

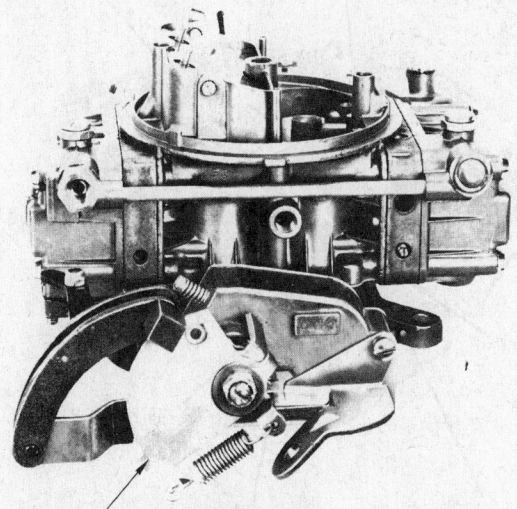

Model 4150G carburetor with automatic transmission operating cam
(© International Harvester Co.)

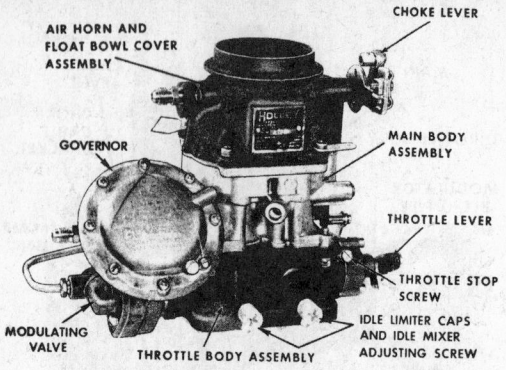

Carburetor model 852-FFG (© International Harvester Co.)

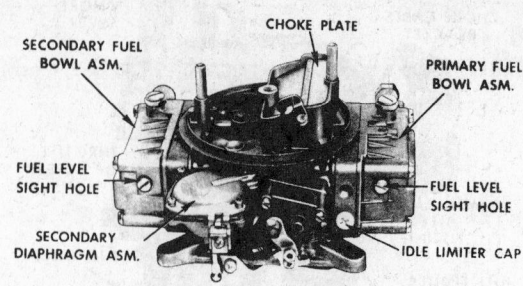

Carburetor model 4150 (© International Harvester Co.)

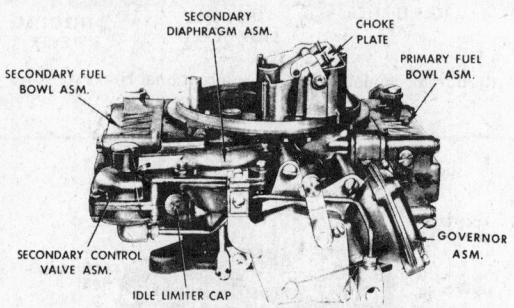

Carburetor model 4150G (© International Harvester Co.)

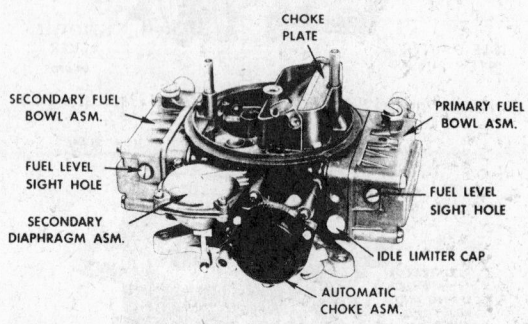

Carburetor model 4150C (© International Harvester Co.)

face of gasket particles.
4. To install the carburetor, reverse the removal procedure, using a new base gasket.
5. Adjust the idle speed and air mixture as outlined.

Idle Mixture and Speed Adjustment

To comply with the mandated emission control requirements, certain procedures must be followed when adjusting the air/fuel mixture and speed. The engine must be at normal operating temperature, choke open, air cleaner installed, dwell and ignition timing correct, and the parking brake applied. The following procedures apply to all carburetors, with minor deviations possible, depending upon the carburetor used.

Observe the following precautions when adjusting the idle mixture and speed.
1. Do not idle the engine for longer than three minutes at a time.

2. After each three minute interval, increase the engine speed to 2000 RPM for one minute.
3. Continue with the idle adjustment and repeat step 2 as necessary.

Preliminary Idle Setting— (After carburetor overhaul)
1. Connect a calibrated tachometer to the engine.
2. Connect a test vacuum gauge to the engine intake manifold.

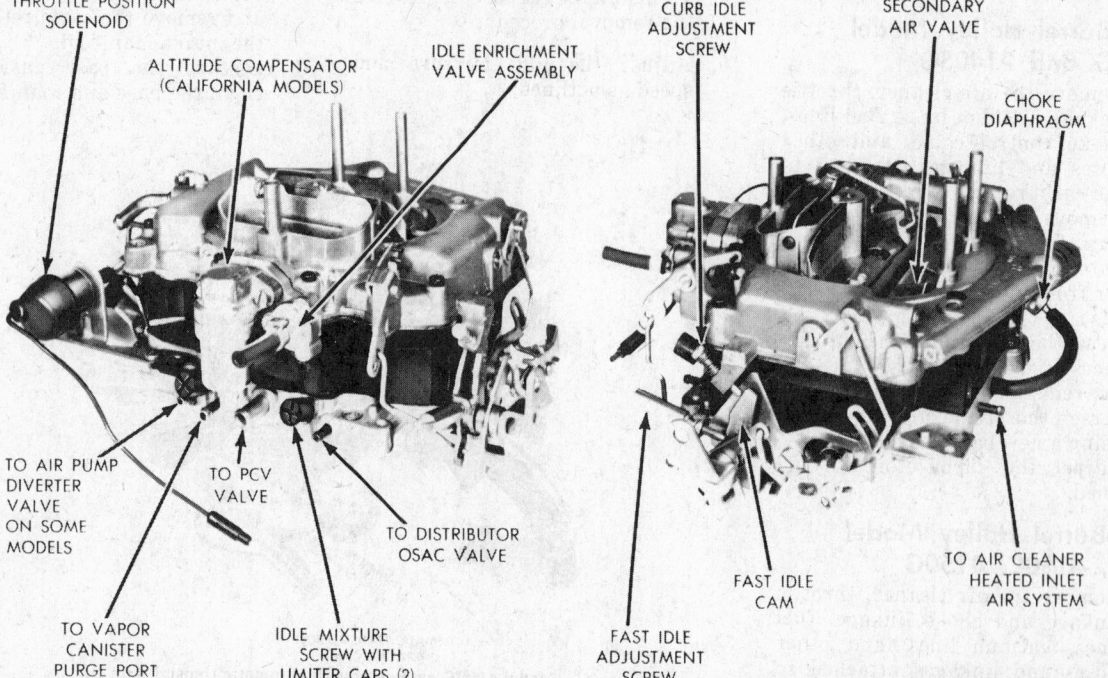

Carburetor model—Carter Thermo-Quad (© International Harvester Co.)

3. Operate the engine at a fast idle speed to bring the operating temperature to normal.

4. Adjust the carburetor to the specified idle speed. Refer to the chart at the beginning of this section.

5. Adjust the idle mixture screw(s) and idle speed screw to obtain "lean best idle" at the specified speed. *NOTE: "Lean best idle" is the point at which intake manifold vacuum starts to drop due to leanness.*

6. Install the colored (service) plastic cap(s) with the tab fully turned counterclockwise against the stop.

7. Adjust the idle speed to specifications.

8. Make final idle adjustments to obtain the recommended idle setting.

Idle Adjustment— Lean Drop Method

1. Connect a calibrated tachometer to the engine.

2. Rotate the idle adjusting screw(s) counterclockwise against the stops.

3. Adjust the idle speed to give an engine speed 25 RPM higher than the specified idle speed.

4. Rotate the idle mixture screw(s) clockwise slowly and equally (if two) until the specified speed is obtained.

5. If the engine is rough or the specified idle speed cannot be attained, remove the limiter cap(s) and continue the adjustment as outlined in step 4, until the specified RPM is attained and the engine is smooth.

6. Install new plastic limiter cap(s) with the tab fully counterclockwise against the stop.

7. Readjust as necessary to maintain the specified RPM.

Idle Adjustment— Exhaust Analyzer Method

When exhaust analyzer equipment is used, the following procedure is recommended to be used to adjust the idle mixture and speed. The test equipment must give accurate readings in the 0-5% Carbon Monoxide (CO) range.

1. Connect a calibrated tachometer to the engine and insert the exhaust analyzer into the exhaust pipe. *NOTE: Refer to the manufacturers instructions for complete connection procedures.*

2. Operate the engine for fifteen minutes at fast idle speed (approximately 1000-1200 RPM), to bring engine to normal operating temperature and to stabilize the temperature of the exhaust analyzer.

4. Calibrate the test equipment as per the manufacturers instructions. *NOTE: If the combustion analyzer does not respond to changes in the mixture quality, check for leaks or restrictions in the sample lines. The thermal conductivity instruments used in the analyzer are both temperature and pressure sensitive, and require a definite sample flow. Refer to the manufacturer's instructions as necessary.*

5. Adjust the idle mixture screw(s) counterclockwise against the tab stop.

6. Adjust the idle speed screw to obtain the specified idle speed.

7. Observe the analyzer dial and adjust the idle mixture screw(s) clockwise by 1/16 turn increments to obtain the specified idle mixture setting and readjust the idle speed as necessary.

8. If the idle speed and mixture cannot be obtained, remove the idle limiter cap(s). *NOTE: To prevent damage to the mixture screw(s) or seat, file or grind the side of the plastic cap. Do Not Pry Cap Off.*

9. With the engine operating, adjust the mixture screw(s) to obtain the "lean best idle" at the specified idle speed. *NOTE: "Lean best idle" is the point at which maximum manifold vacuum begins to drop due to leanness.*

10. Install new plastic limiter cap(s) with the tab fully counterclockwise against the stop.

11. Readjust the idle mixture screw(s) to obtain the recommended CO setting.

NOTE: After completing the idle adjustment procedure, if unsatisfactory idle operation still exists, a recheck of the ignition system, crankcase ventilation system, timing advance system, air induction system, exhaust gas recirculation system, or hot idle compensation system should be made.

Fuel Pumps

The fuel pumps used on the gasoline engines are of two types.

a. Mechanical type—This type is mounted on the engine block and is operated by a special eccentric on the camshaft.

b. Electric type—This type is mounted in the fuel tank and

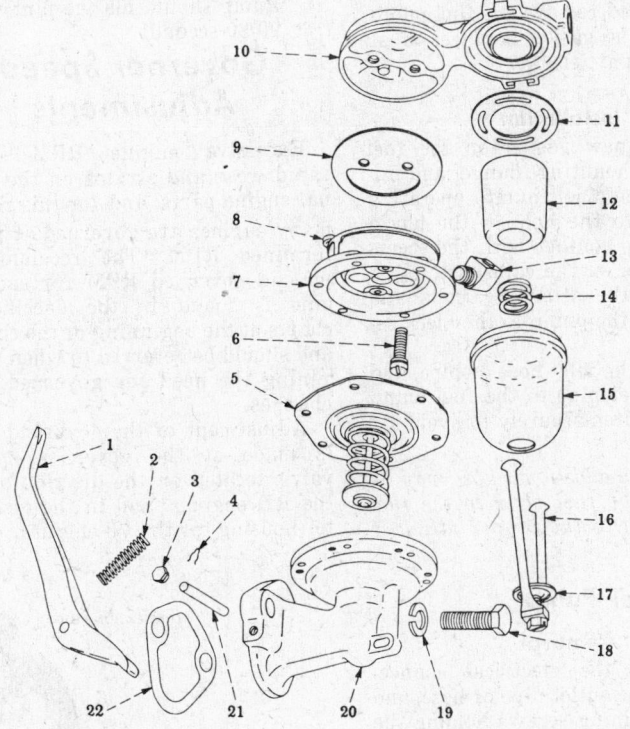

1 Lever, cam
2 Spring, cam lever return
3 Plug, cam lever shaft seal
4 Pin, spring, cam lever shaft retaining
5 Diaphragm, assembly
6 Screw, with lockwasher, assembly
7 Housing, valve assembly
8 Screw, with lockwasher, assembly
9 Diaphragm, air dome
10 Air dome and filter cover
11 Gasket, filter bowl
12 Filter, glazed ceramic or paper
13 Elbow
14 Spring, filter
15 Bowl, filter
16 Retainer, with screw assembly
17 Washer, Filter bowl retaining
18 Bolt, hex head
19 Lockwasher
20 Pump body
21 Pin, cam lever
22 Gasket, pump-to-crankcase

Exploded view—mechanical fuel pump—typical (© International Harvester Co.)

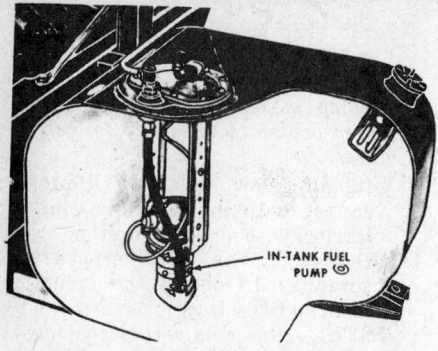

Cutaway view—in-tank fuel pump mounted on adjustable supports
(© International Harvester Co.)

is supported by an adjustable hanger assembly, therefore, making it adaptable to all I. H. fuel tank depths. A spring loaded latch is normally used to permit easy motor replacement.

Mechanical Fuel Pump

Removal

1. Remove the fuel inlet pipe or hose and the outlet fuel pipe to the carburetor from the fuel pump fittings.
2. Remove the attaching bolts from the fuel pump housing to engine block and remove the fuel pump.
3. Clean the gasket surfaces of all gasket particles.

Installation

1. Install new gasket on the fuel pump mounting flange and install the fuel pump operating arm into the hole in the block, and into contact with the eccentric lobe on the camshaft.
2. Install the attaching bolts and tighten the pump to the block securely.
3. Install the inlet hose or pipe, and the outlet pipe to the fuel pump and tighten securely to avoid air leaks.

NOTE: An additional hose may be used from the fuel filter to the fuel pump. Install to the proper fitting if so equipped.

Electric Fuel Pump

Removal

1. Remove the electrical connections, the outlet pipe or hose, and the retaining screws holding the assembly to the tank.
2. Withdraw the pump/support assembly from the tank, being careful not to allow dirt to enter the tank hole.
3. With the unit out of the tank, remove the pump from the support assembly.
4. Remove the gasket and any particles from the gasket surfaces.

Installation

1. Install the pump assembly into the support arms and retain it securely.
2. Using a new gasket, insert the pump/support assembly into the tank and secure it with the attaching screws. Tighten securely to avoid air or gasoline leaks.
3. Install the outlet pipe or hose, and the electrical connections to the assembly.

Fuel Pump Pressure Test

1. Disconnect fuel line at carburetor inlet and attach pressure gauge between the inlet and disconnected line.
2. Start engine and take reading. Consult Tune-up Specifications at the beginning of this section for correct pump pressure.
3. When engine is stopped, the pressure should remain constant or very slowly return to zero.

Fuel Pump Capacity Test

1. Disconnect fuel line from the fuel pump.
2. Connect a piece of hose to the line so that fuel can be directed into a measuring container.
3. Start engine and note time it takes to fill a pint container. Pump should fill one pint within 20-30 seconds.

Governor Speed Adjustments

Excessive engine RPM causes rapid wear and strains on the internal engine parts, and for this reason, many engines are governed at predetermined RPM. The recommended no-load governed RPM for each engine is found in the specification charts at the beginning of the chapter and should be referred to when determining the need for governed RPM changes.

Adjustment of the governed speed is made at the governor spinner valve, located on the distributor, for the RD engines, and in the distributor housing for the V8 engines.

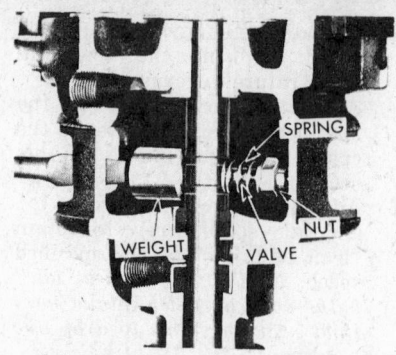

Adjusting spinner governor—distributor
(© International Harvester Co.)

RD Engine Governor Adjusting Procedure

1. Attach a calibrated tachometer to the engine.
2. Accelerate the engine and note the governed top speed and compare with the specification charts.
3. If necessary to adjust the governor, remove the seal wire and the adjusting hole plug from the governor housing.
4. With the ignition off, rotate the engine until the adjusting screw in the end of the spinner valve appears at the plug hole.
5. Insert a screwdriver and engage the adjusting screw.
6. Turn the adjusting screw clockwise to increase the governed speed, or counterclockwise to decrease governed speed. *NOTE: One turn of the screw will affect governed speed approximately 150 RPM.*
7. Again accelerate the engine, and observe the governed speed. Readjust the governor as needed.
8. Install the adjusting screw hole plug and install a new seal wire.

V8 Engine Governor Adjustment Procedure

1. Connect a calibrated tachometer to the engine.
2. Accelerate the engine and observe the governed speed at no-load. Refer to the specification

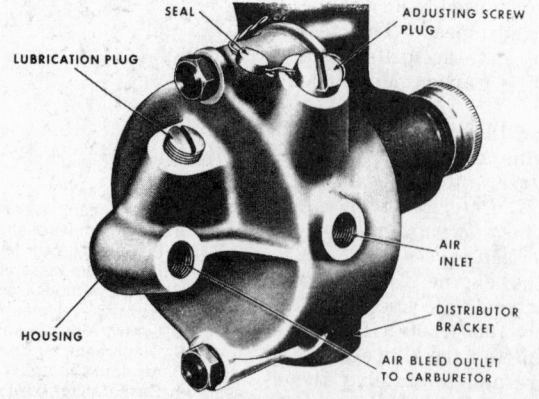

RD Engine model governor control housing (© International Harvester Co.)

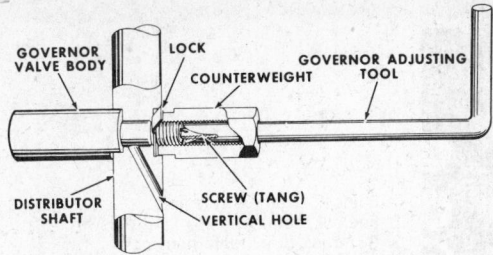

GOVERNOR VALVE BODY — LOCK — GOVERNOR ADJUSTING TOOL — COUNTERWEIGHT — DISTRIBUTOR SHAFT — SCREW (TANG) — VERTICAL HOLE

Governor valve adjustment—model 1530 distributor (© International Harvester Co.)

charts at the beginning of this chapter.

3. If necessary, adjust the governor by removing the seal wire, governor clamp and gasket from the distributor housing.

4. With the ignition off, rotate the engine until the adjusting screw hole appears in the opening of the distributor housing.

5. With a 1/8 Allen wrench, remove the adjusting hole plug from the governor.

6. With the use of a special tool (IH*SE-2072-2) inserted through the governor adjusting hole, engage the adjusting screw tang. *NOTE: The adjusting screw is of a special design and cannot be adjusted with a screwdriver or any device, other than the special adjusting tool.*

7. Turning the adjusting screw clockwise to decrease governed speed, or counterclockwise to increase governed speed. *NOTE: 1/4 turn of the adjusting screw will affect the governed speed approximately 100 RPM.*

8. When specified governed speed has been attained, reinstall the plug in the adjusting hole and tighten securely.

9. Reinstall governor clamp and gasket on the distributor housing and install a new seal wire.

Carburetor Adjustment

LPG (propane-butane) RD-501, V-304, 345, 401, & 549 Engines

Setting the idle fuel adjustment automatically gives the correct mixture for part and full throttle operation.

CAUTION: Liquefied petroleum gas is extremely flammable. Observe all safety precautions regardless of the nature of work being performed. No work is to be done on fuel tanks except by qualified concerns who normally service such containers.

1. Set throttle for fast idle by turning throttle stop screw in from closed position 3 or 4 turns.

2. Start engine and adjust idle adjusting screw in drag link (either end) until engine runs smoothly. Turning screw in (clockwise) enriches mixture.

3. Adjust throttle stop screw for idle of 600 rpm.

4. Readjust idle adjusting screw in drag link for maximum engine speed. If engine speed starts to go above 900 rpm, set back speed with throttle stop screw to 600 rpm and continue adjusting drag link screw for maximum engine speed.

5. Adjust idle to 400 rpm and replace cotter pin in drag link adjusting screw to lock adjustment.

Carburetor Adjustment

LPG (Propane-Butane) Carburetor Adjustment

All OHV 6 Engines Except PT 6 & RD 501

1. Loosen locknut on starting adjusting screw and turn screw clockwise until it bottoms, then back it out the required number of turns (see table). Tighten locknut.

2. Turn idle adjusting screw on the regulating unit clockwise until it bottoms gently, then back it out the required number of turns (see table).

3. On BD engines, loosen the large lock nut on the economizer unit and turn load adjusting screw clockwise to extreme position, then turn counterclockwise the required number of turns (see table). Tighten locknut. On BD engines (engine not running), disconnect vacuum line from economizer before setting load adjustment. Then loosen locknut and turn entire economizer assembly clockwise until it bottoms in carburetor, then counterclockwise the specified number of turns (see table). Temporarily plug the vacuum line and start engine. After the engine is warmed up, set hand throttle at about 2/3 governed speed. Screw economizer body in until engine speed starts to drop and then out until the engine speed starts to drop again, then set in between the two extremes and tighten locknut. For part throttle adjustment, loosen small locknut on economizer cover and turn economizer adjusting screw clockwise the required number of turns from bottom position (see table) and tighten locknut.

LPG Carburetor Adjustment Specifications

Engine	RD-406	RD-450
Starting Adjusting Screw (number turns off seat)	1	2
Economizer (load adjustment) (number turns off seat)	5 4-11/16	4-3/4 3-3/4

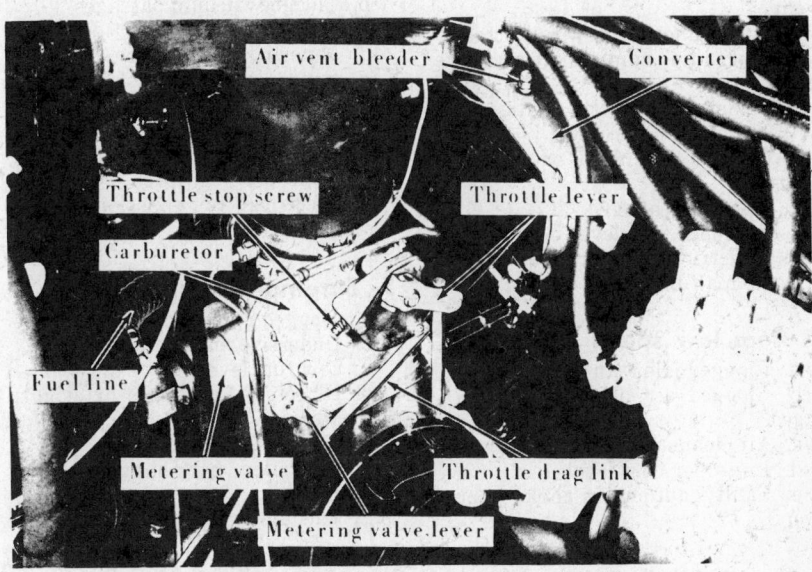

Air vent bleeder — Converter — Throttle stop screw — Throttle lever — Carburetor — Fuel line — Metering valve — Throttle drag link — Metering valve lever

LPG carburetion—V401, 461 & 549 engines (© International Harvester Co.)

1 Fuelock-strainer, assembly
2 Fuelock, connection to ignition switch
3 Governor, diaphragm housing
4 Governor, spinner box air line
5 Drag, link idle adjusting screw
6 Spacer (insulator)
7 Throttle, control rod
8 Metering, valve lever
9 Throttle, lever

LPG carburetor
(© International Harvester Co.)

Carburetor
Diagnosis Service

The following diagnosis and troubleshooting chart can be used as a general guide to determine the causes of carburetor related problems. When the problem has been isolated to a particular component and more information is needed, refer to the related chapter within this section, or to the Unit Repair Section.

1. Rough Idle or Stalling—Engine Hot or Cold

a. Binding linkage, choke valve, or choke piston
b. Disconnected or broken choke control cable
c. Incorrect choke thermostat adjustment
d. Fast idle linkage and cam not properly adjusted
e. Idle mixture screw(s) out of adjustment
f. Idle speed screw out of adjustment
g. Air cleaner air flow restricted
h. Hot idle compensator valve stuck.
i. Secondary throttle plates open (4V carburetors)
j. Clogged air bleed or idle passages
k. Vacuum leakage
l. Improper float level
m. Electrical or emission control systems malfunction

2. Poor low speed operation

a. Clogged idle transfer slots
b. Clogged air bleed or idle passages
c. Air cleaner air flow restricted
d. Improper float level
e. Faulty automatic choke operation

f. Improper use of hand controlled choke
g. Vacuum leakage
h. Electrical or emission control system malfunction

3. Poor Engine Acceleration

a. Improper acceleration pump stroke
b. Inoperative or missing pump discharge check valve, ball, or needle
c. Damaged or worn pump diaphragm or piston
d. Leaking gaskets
e. Defective fuel pump
f. Clogged discharge jets
g. Electrical or emission control systems malfunction

4. Poor High Speed Operation

a. Defective fuel pump or clogged fuel filter
b. Clogged vacuum passages
c. Power valve stuck
d. Metering rods stuck
e. Improper size or obstructions in the main jets
f. Restricted air supply to air cleaner
g. Improper float level
h. Electrical or emission control system malfunction

5. Surging—Cruising Speeds

a. Clogged main jets
b. Undersize main jets
c. Low fuel level
d. Defective fuel pump or clogged fuel filter
e. Blocked air bleeds
f. Restricted air supply to air cleaner
g. Vacuum leakage

h. Metering rods out of adjustment
i. Power valve sticking
j. Electrical or emission control system malfunction

6. Stalling When The Accelerator is Closed Quickly

a. Improperly adjusted or defective throttle modulator or dash pot
b. Clogged air bleed or idle passages
c. Vacuum leakage
d. Throttle plates not closing

7. Governor Not Operating—No Engine Speed Control

a. Seal broken—Governor maladjusted
b. Vacuum leakage or lines broken
c. Modulator diaphragm leaking or broken
d. Sticking governor spinner valve

8. Governor Cuts Off At Low Speeds—Erratic Operation

a. Clogged distributor governor filter
b. Restricted vacuum lines
c. Sticking governor spinner valve
d. Restrictions in the spinner shaft or housing

9. Engine Surges At And Below Governed Speed

a. Vacuum leakage or restriction
b. Carburetor jets or air bleeds clogged
c. Spinner valve sticking
d. Electrical or emission control systems malfunction

Cooling System

The cooling system is a closed type, utilizing a two valve pressure cap. One valve is used to relieve excessive pressure from the system, and the second valve is used to allow atmospheric air to enter the system during the cooling down period. The engine temperature is controlled in two ways. One method is by a thermostat, located on the front of the engine block or cylinder head. The second method is by a shutter assembly, mounted on the grille of the vehicle, and controlled by heat sensitive switches mounted on the engine and radiator. The coolant is forced through the engine and radiator by the water pump, located on the front of the engine, which is belt driven by the crankshaft pulley.

CAUTION: To avoid personnel injury, remove the pressure cap from the radiator in two steps. Loosen the cap to its first notch and allow the pressure to escape through the overflow pipe. After the pressure has been released, press on the cap and continue to turn until the prongs on the cap disengage from the radiator neck.

Heater Core Removal and Installation

Without Air Conditioning Except Scout

Due to the many body styles, a general description of the heater core removal and installation will be outlined. Utilize the operations as necessary for the vehicle being serviced. Use caution when removing the heater box assembly, to avoid breakage to the box due to hidden screws or bolts.

1. Drain the cooling system and remove the negative battery cable.
2. Disconnect the heater hoses at the heater core tubes.
3. Disconnect the heater electrical wiring and the control cables. *NOTE: Cables may have to be disconnected at the control panel and removed with the heater box.*
4. Remove the heater distribution manifold at the heater box.
5. Remove the heater box attaching bolts or nuts from the cab side and from the engine side of the firewall.
6. Remove the heater box assembly from the vehicle.
7. Remove the heater box back cover or the outer door and shaft. Remove the heater core.
8. The installation of the heater assembly is the reverse of removal.

Fill the cooling system and check the operation of the heater system.

With Air Conditioning (Blend Air System) Except Scout

NOTE: This air conditioning/ heater system unit is located under the passenger seat as a compact unit, for ease in servicing.

1. Close the heater hose shutoff valves at engine, if equipped, or drain the cooling system. Disconnect the negative battery cable.
2. Remove the heater hoses from the heater core tubes.
3. Remove the floor mounted blower motor well cover, and disconnect the wiring at the connectors.
4. Loosen the unit mounting stud nuts. *NOTE: Lower the cab on the CO models before removing the mounting stud nuts.*
5. Remove the floor mat from the passenger side.
6. Remove the floor panel on the passenger side to expose the air duct.
7. Remove the rubber seals from the fresh air duct and treated air duct.
8. Remove the hose connecting the bunk duct to the blower housing (if equipped).
9. Remove the thumb screws securing the passenger seat and heater cover plate assembly. Remove the seat and cover assembly.
10. Disconnect the control cables at the heater assembly.
11. Disconnect the electrical wiring from the heater assembly.
12. Remove the unit mounting stud nuts and washers. Remove the heater assembly from the cab.
13. Remove the sponge rubber seals on the heater tubes and set aside for later installation.
14. The installation is in the reverse of the removal. Replace the coolant lost during removal, and operate the system and check for leakage.

Scout

1. Drain the cooling system and remove the negative battery cable.
2. Remove the heater hoses from the heater core outlets.
3. Remove the windshield washer bottle from the firewall.
4. Remove the cover plate from the heater box and remove the heater core from the housing.
5. Remove the core end cover. *NOTE: Do not damage the core fins during the removal and installation procedure.*
6. Installation is the reverse of removal. Fill the cooling system and check the operation of the heater system.

Water Pump Removal

1. Drain cooling system.
2. If radiator shrouds hinder access they must be removed before proceeding (large V-8's and cab-forward models).
3. Loosen alternator pivot bolts and adjusting bolt on bracket to relieve tension on the fan belt and remove belt from water pump pulley. On V(S)-401, 478, 549, 537, 605 engines all accessories are driven by belts from the water pump pulley and are removed by loosening the alternator and power steering pump adjusting brackets and pivots. The idler pulley on the belts between the crankshaft and water pump pulleys is then loosened so that these belts may be removed.
4. On V(S)-401, 478, 549, 537, 605 engines the fan blades and pulley are removed.
5. Remove all pipes and hoses connected to the water pump.
6. Remove mounting bolts or stud nuts and water pump. On all models except 4-196 and V-304, 345, 392, only the front half of the pump housing is removed for water pump servicing.
7. Installation is the reverse of the above procedure. Be sure to install new gaskets and, if applicable, new O-rings on pipe end fittings.

Water Pump Service

Engine Models 232, 258 MV 404, 446, V537, V605

This water pump is nonadjustable, has a packless seal and must be serviced as a complete unit: in the event of malfunction the whole assembly is replaced.

1. Remove pump as described in the preceding section.
2. Clean impeller cavity before installing new pump.
3. Spin shaft on new pump to be sure it rotates freely.
4. Install as described in preceding section, using a new gasket.

Water Pump Disassembly

Engine Models 4-196 and 4-196E

1. Separate the housing from the pump body. *NOTE: The housing holds the impeller and shaft assembly.*
2. Unless special tools, number SE 1950 and SE 1950-1 installers or their equivalents are used, the following measurements should be taken and recorded for the reassembly of the pump.
   a. Measure the distance from

Water pump location—MV 404, 446 engines
(© International Harvester Co.)

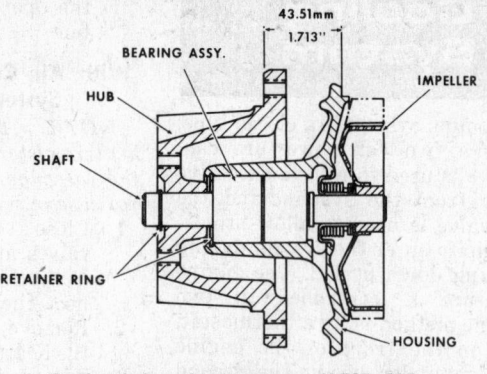

Sectional view of water pump—V 537, 605
engines (© International Harvester Co.)

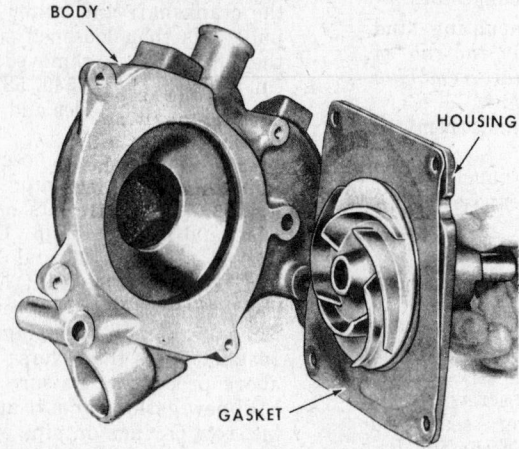

Separation of water pump body and housing—4-196 engine
(© International Harvester Co.)

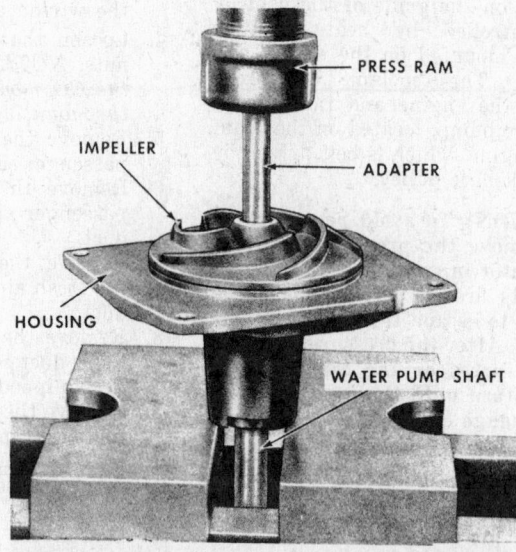

Removal of water pump shaft and bearing assembly from
the pump housing (© International Harvester Co.)

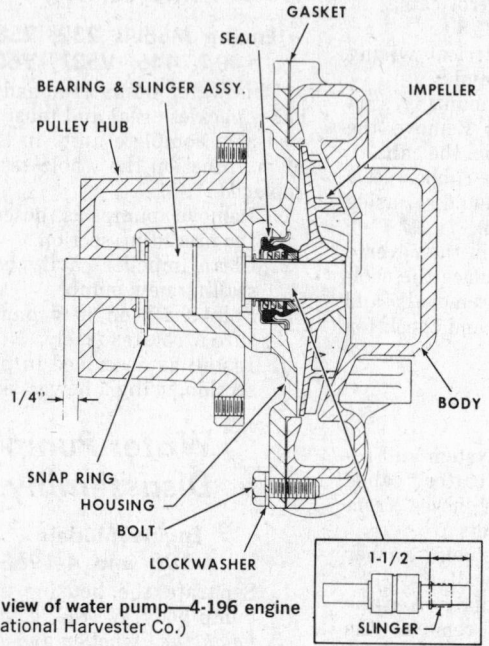

Sectional view of water pump—4-196 engine
(© International Harvester Co.)

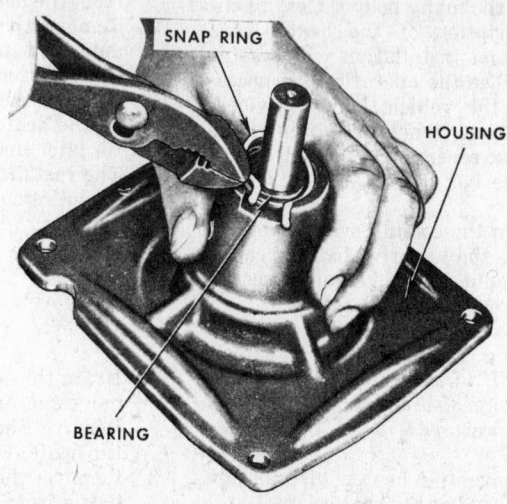

Removing shaft and bearing assembly snap-ring from water
pump (© International Harvester Co.)

the pump hub to the end of the shaft and record the measurement.

b. Measure the distance from the impeller hub to the end of the shaft and record the measurement.

3. Remove the snapring from the groove in front of the shaft bearing assembly.

4. Press the shaft and bearing assembly from the housing and from the impeller. *NOTE: The bearing and shaft assembly are replaced as an assembly only.*

5. Remove the seal from the rear of the housing with the aid of a drift and hammer.

Installation

1. Press a new seal into the housing bore.

2. Install new slinger on the bearing shaft, if not equipped, 1½

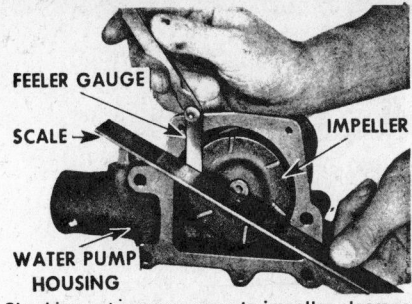

FEELER GAUGE

SCALE →

IMPELLER

WATER PUMP HOUSING

Checking water pump cover to impeller clearance
(© International Harvester Co.)

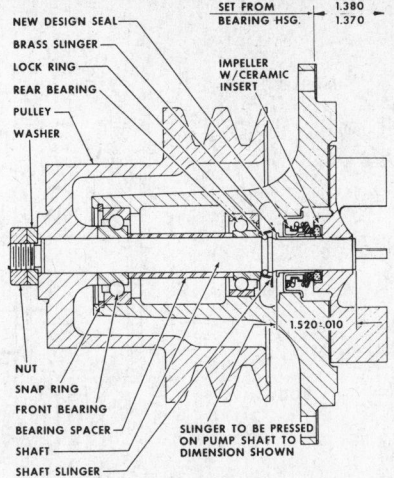

NEW DESIGN SEAL
BRASS SLINGER
LOCK RING
REAR BEARING
PULLEY
WASHER

SET FROM BEARING HSG.   1.380 / 1.370

IMPELLER W/CERAMIC INSERT

NUT
SNAP RING
FRONT BEARING
BEARING SPACER
SHAFT
SHAFT SLINGER

1.520-.010

SLINGER TO BE PRESSED ON PUMP SHAFT TO DIMENSION SHOWN

Sectional view of water pump—RD 406, 450, 501 engines (© International Harvester Co.)

inch from the rear end of the shaft to the forward edge of the slinger.

3. Press the bearing and shaft assembly into the housing and install the snapring.

4. Using special tool SE 1950 or equivalent, or by using the measurement taken before removal, press the impeller in place on the shaft.

5. Using special tool SE 1950-1 or equivalent, or by using the measurement taken during the removal, press the pulley hub in position on the shaft.

6. Using a new gasket, install the pump housing to the pump body and secure with the attaching bolts.

Engine Models RD 406, 450, 501

1. Remove the nut and washer holding the pulley to the shaft and remove the pulley from the shaft with the aid of a puller.

2. Remove the snapring from the front of the water pump shaft front bearing.

3. Press the shaft and bearing from the impeller and the pump housing.

4. Press the shaft out of the bearing, spacer and slinger. Do not lose the two half-moon lock rings from under the slinger.

5. Remove the seal assembly from the housing through the back of the pump.

6. Use a drift and hammer, and carefully drive the seal from the pump housing and discard.

Installation

1. Install a new pump seal into the housing bore and press into place.

2. Place the slinger on the shaft and press into position 1 33/64 inch on the shaft, measuring from the rear of the shaft. Pack the bearings with short fiber wheel bearing grease and place the rear bearing, spacer, and front bearing onto the shaft.

3. Place the two half-moon lock rings in the groove on the shaft and press the shaft into the bearings and spacer until the rear

bearing rests firmly against the slinger and lock rings.

4. Install the shaft and bearing assembly into the housing of the pump and hold it into place by inserting the snapring in the groove in front of the front bearing. *NOTE: Fill the pump housing with an ounce and one-half of the short fiber grease before installing the shaft and bearing assembly.*

5. Press the pulley onto the front end of the shaft and secure with the washer and nut.

6. Support the shaft at the front end and press the impeller onto the rear of the shaft.

7. Locate the impeller at 1.520 ± .010 inch, measuring from the shaft end to the edge of the impeller vane.

8. Check for freedom of operation and install the pump, using a new gasket.

Engine Models VS 401, 478, 549 series, V 304, 345, 392

1. Remove the pump housing from the pump body on engine models V 304, 345, 392.

2. Remove the pulley from the shaft, and the bearing and shaft snapring.

3. Remove the impeller from the bearing shaft and the shaft from the housing by the use of a press.

4. Using a drift and hammer, remove the seal and discard.

5. Remove the impeller seat and bushing from the housing and discard.

Installation

1. Install new bushing and seat into the housing bore.

2. Install a new seal into the housing.

3. Press the bearing and shaft assembly into the housing until the bearing bottoms into the counterbore. *NOTE: Due to variations in the pump housing, bearings and shaft assembly, and the impeller, the impeller cannot be pressed to the correct location on the shaft during the first press application. The following procedure must be used to achieve the correct location of the impeller.*

4. With the use of SE 2086 special tool kit or its equivalent, select a .060 shim stop and place it on the impeller end of the shaft and press the impeller on the shaft until the press ram bottoms on the shim. *NOTE: The purpose of the shim stop is to limit the travel of the pump shaft in the impeller, so that an interference will exist between the impeller and the pump housing.*

5. Place the pump assembly on the front cover of the engine and in-

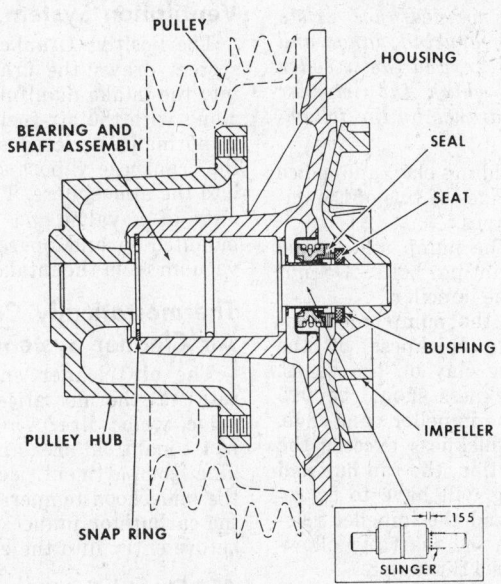

PULLEY

HOUSING

BEARING AND SHAFT ASSEMBLY

SEAL

SEAT

BUSHING

IMPELLER

PULLEY HUB

SNAP RING

.155

SLINGER

Sectional view of water pump—VS 401, 478, 549 engines
(© International Harvester Co.)

341

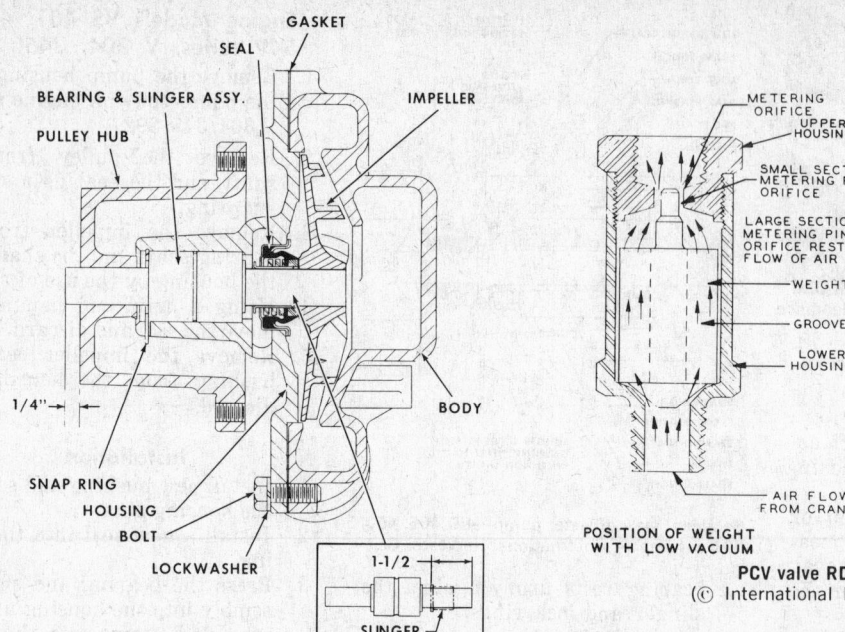

Water pump—V-266, 304, 345 & 392 engines
(© International Harvester Co.)

PCV valve RD engines
(© International Harvester Co.)

stall two bolts finger tight.

6. With the use of a feeler gauge, determine the clearance between the housing flange and the front cover.

7. Record the clearance and add to the specification of .015 inch running clearance. Subtract the total of the two from the original .060 shim thickness and record.

8. From the special tool kit SE 2086 or its equivalent, select the shims to provide the thickness of the above recorded figure, and place the shims on the shaft end. Press the impeller further onto the shaft until the ram bottoms on the shim stop. This operation should provide the proper operating clearance for the impeller.

NOTE: *If no clearance exists between the housing flange and the front cover, and the impeller turns free, check the impeller running clearance by the following method.*

a. Use molding clay and place on the edges of two of the impeller vanes.

b. Install the pump and torque the mounting bolts. Do not rotate the impeller.

c. Remove the pump and measure the thickness of the moulding clay on the vanes. The thickness should be .015 above the impeller vane edge.

d. If the thickness exceeds the specification, the fan hub and snapring will have to be removed and the impeller relocated on the shaft by following steps 4 through 8.

9. Position the gasket on the pump body and install the pump assem-

bly. Torque the mounting bolts to specified torque.

Engine

Emission Control Systems

Emission control systems are designed to control the emissions of Hydrocarbons (HC), Carbon Monoxide (CO), and Oxides of Nitrogen (NOx) at the levels specified by the Federal and State governments. Emission control Systems vary with engine, transmission, and series applications.

Positive Crankcase Ventilation System

The Positive Crankcase ventilation system draws the crankcase vapors into the intake manifold to be burned along with the air-fuel mixture. This is normally a closed system so that the crankcase vapors are not emitted into the atmosphere. The system consists of a valve and hose routings, mounted to and operated by engine vacuum from the intake manifold.

Thermostatically Controlled Air Cleaner System

The air cleaner snorkel incorporates a thermostatically controlled valve, which directs air from the exhaust manifold area and from the engine compartment, depending upon the underhood temperature, to insure the carburetor induction air is warm before entry into the engine.

Air Guard System

This system is used to inject air

into the exhaust ports to mix with the hot unburned gases, and to further burn the combustion mixture and reduce the emissions of hydrocarbons and carbon monoxide into the atmosphere. The system includes an air pump, a diverter valve, hose routings, and air injector manifolds and tubes.

Exhaust Gas Recirculation System

This system is used to meter exhaust gases into the combustion chambers to dilute the intake charge, thereby reducing the peak temperature of the gases and limit the formation of the oxides of nitrogen that form as the result of the high temperature during the combustion process. The system consists of a exhaust gas recirculating valve which connects the intake manifold to the exhaust manifold, and is operated by vacuum and temperature.

Fuel Tank Vapor Emission Control System

A closed fuel tank vent system is used to prevent fuel vapors from entering the atmosphere. The system consists of a two-way relief valve filler cap, which is closed to the atmosphere under normal operating conditions and opens when pressure exceeds 0.75 to 1.50 PSI, or vacuum exceeds 15 to 25 inches. A liquid check valve is used to route the vapors and collect any liquid before they are drawn into the fuel vapor storage cannister. The vapors are then drawn into the intake manifold through the air cleaner assembly. The amount of vapor drawn from the cannister is relative to the air volicity through the air cleaner snorkel.

Vacuum Throttle Modulating System

This system is used to reduce the

emissions of hydrocarbons during rapid throttle closure at high speeds. It consists of a deceleration valve and a throttle modulating diaphragm located on the carburetor base to allow the throttle to remain slightly open and admit more air into the combustion chambers to lean out the overrich mixture. The decel valve and the modulator diaphragm are operated by engine vacuum signals.

Electric Choke

This system is used to assist in maintaining an open choke butterfly during cruising conditions, when vacuum may not be sufficient to draw enough heated air from the manifold to the choke assembly. When the engine cylinder head temperature is below 130 degrees, the electric choke is inoperative and the choke operates in the normal manner. Above the stated temperature, the electric choke is in operation.

Engine Removal

The following is an outline of general engine removal. Removal procedure will vary from truck to truck due to the variety of body models and accessory equipment. Before lifting out engine be certain that everything has been disconnected. Remove anything that might be in the way of the actual lifting.

1. Drain water from radiator and engine block.
2. Drain crankcase oil.
3. Disconnect battery ground cable and remove cable clamp from hot terminal of battery.
4. Remove all water hoses to radiator and heater.
5. Remove fan blades and fan shroud.
6. Remove any radiator cross-brace rods or brackets.
7. Remove radiator mounting bolts and lift out radiator. WARNING: On vehicles with LPG fuel systems observe all safety precautions. Be sure shop procedures are in compliance with

Engine lifting sling—V8 engines
(© International Harvester Co.)

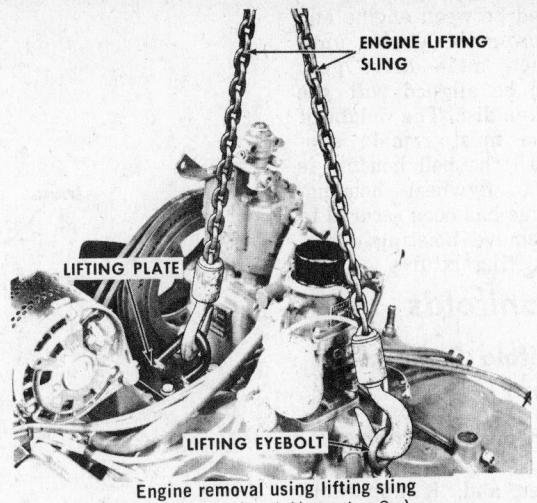

Engine removal using lifting sling
(© International Harvester Co.)

local fire regulations. Close all tank valves and exhaust fuel from lines before working on fuel system.
8. On conventional chassis, remove hood hinge bracket mounting bolts and remove hood assembly. On CO models, tilt cab forward and prop securely. On cab-forward models, remove all front end sheet metal: bumper, fenders, radiator shell and disconnect any wiring that goes to these parts.
9. Disconnect and remove air filter from engine. Remove breather hose from air cleaner, if applicable.
10. Disconnect fuel pump inlet line.
11. Remove vacuum lines from manifold and all other components, and lines from air compressor, and air pump, if applicable.
12. Disconnect throttle linkage, choke control wire and hand throttle control wire, if applicable. On V-304, 345, 392 engines the carburetor must be removed for the fitting of the lifting fixture.
13. If so equipped, disconnect wire from heater control valve.
14. Disconnect all wiring from engine:
    a. Water temperature gauge sender.
    b. Oil pressure gauge sender.
    c. Generator wires.
    d. Primary ignition wire to resistor.
    e. Starter solenoid wires and battery cable.
15. On V(S)-401, 478, 549 engines, loosen alternator belts and disconnect alternator strap bracket to swing alternator away from thermostat housing. Then remove thermostats and housing.
16. If so equipped, remove tachometer drive at the distributor on the small V-8—or at the rear of the block on the big V-8.

17. Disconnect exhaust pipes at manifolds.
18. If so equipped, remove automatic transmission filler tube, freon compressor lines and disconnect power steering pump line and hose.
19. Install lifting fixtures and suitable sling. On the V(S)-401, 478, 549 engines an eye bolt is installed at rear of intake manifold and a plate where the thermostat was mounted. On V-304, 345, 392 models the lifting fixture is mounted on the intake manifold where the carburetor was removed. On 400 engines, lifting eyes are installed in predrilled holes at front right and left rear of intake manifold. On six cylinder engines two brackets are secured by the two extreme head bolts and sling attached.
20. Connect hoisting equipment to lifting fixture and hoist enough to support engine.
21. Remove bell housing mounting bolts. On the PT-6 and V-8 engines the flywheel housing front cover is removed before the flywheel housing is removed from crankcase. On vehicles with RD engines the floor panel is removed to provide access to the bolts securing bell housing.
22. Disconnect clutch linkage.
23. Remove front engine mounting bolts. On some models it is easier to unbolt the mount from the frame crossmember.
24. Remove side engine mount bolts.
25. In hoisting out engine, first pull engine forward to clear clutch assembly from transmission, then tilt front up and carefully out of the chassis. **CAUTION: Avoid damaging clutch driven disc.**
26. Installation of the engine is in general the reverse of the above described procedure. Be careful when installing that wires are

not pinched between engine and frame. Lower the engine until transmission main drive gear spline can be aligned with the clutch driven disc. The weight of the engine must remain supported until the bell housing is secured to flywheel housing. After engine has been secured to chassis, remove hoisting equipment and lifting fixtures.

Manifolds

Intake Manifold Removal & Installation

4 Cylinder and V8 Engines

1. If engine is in vehicle, remove air cleaner and, if applicable, governor vacuum line.
2. Disconnect throttle linkage, choke cable and fuel line.
3. Remove carburetor.
4. On V-304, 345, 392, 400 engines, disconnect hose from thermostat housing and bracket for spark plug wires.
5. On 4 cylinder models, remove coil, coil mounting bracket and ignition resistor from intake manifold.
6. Remove positive crankcase ventilation pipe and vacuum line.
7. Remove mounting bolts, manifold and gasket.
8. Installation is the reverse of the above procedure. Install new gaskets and tighten the mounting bolts from the center out, torquing to 40-45 ft. lbs.

Exhaust Manifold Removal & Installation

4 Cylinder and V8 Engines

1. Disconnect exhaust pipe from manifold.
2. Unbolt exhaust manifold from head.
3. Remove manifold.
4. Installation is the reverse of the above procedure. Install new manifold-to-head gasket and new manifold-to-pipe gasket.

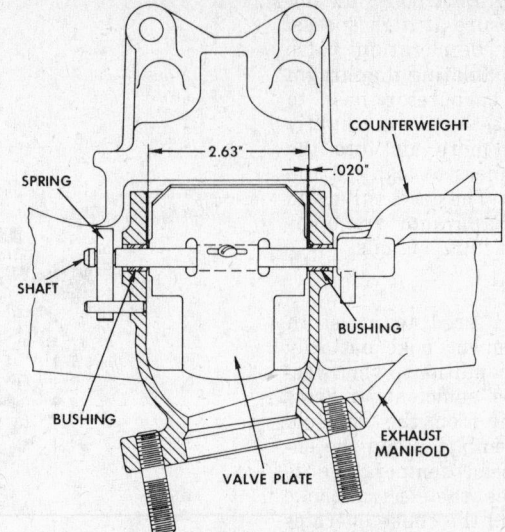

Manifold heat control valve—PT-6 engines
(© International Harvester Co.)

5. Torque manifold-to-head bolts to 25-30 ft.lb.

6 Cylinder Engines

1. Remove exhaust pipe from manifold.
2. Disconnect PCV line and vacuum lines from intake manifold.
3. Disconnect throttle linkage, fuel line and choke cable from carburetor.

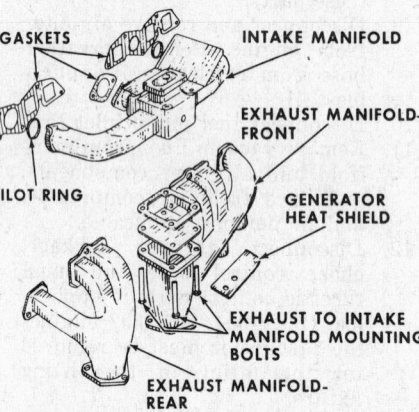

Exploded view—intake and exhaust manifolds—RD 406, 405, 501 engines
(© International Harvester Co.)

4. Remove air filter and carburetor.
5. Remove exhaust/intake manifold unit and gaskets.
6. Separate intake from exhaust manifold.

NOTE: The RD engines use a two piece exhaust manifold and one piece intake manifold, bolted together as a unit, with only the rear half of the exhaust manifold being separate and removable by itself.

7. The PT-6, engines are equipped with a manifold heat control valve in the exhaust manifold. To remove this valve, first note position of counter-weight in relation to valve plate. Remove thermostatic spring from end of shaft. With a hacksaw blade or cutting torch, cut the shaft on both sides of the valve plate and remove plate and shaft pieces, being careful not to ruin bushings. If bushings need replacement, remove and install, spacing them .3120-.3130". With valve plate in the "heat on" position, insert shaft, hold counterweight in the correct position and secure

Intake manifold removal—V8 engine (© International Harvester Co.)

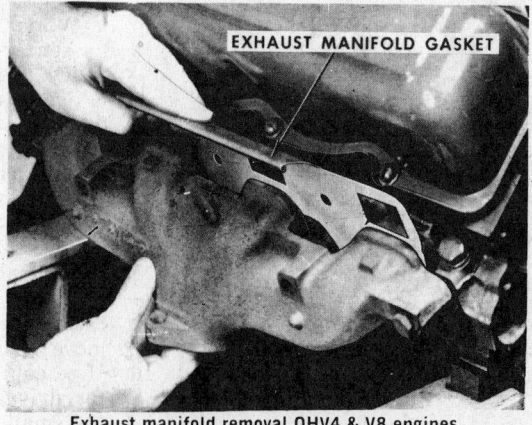

Exhaust manifold removal OHV4 & V8 engines
(© International Harvester Co.)

plate with screw or tack weld. Install thermostatic spring and hook spring over stop pin. Lubricate with a mixture of penetrating oil and graphite.

8. Intake and exhaust manifolds must be assembled properly to insure alignment. Position intake manifold to exhaust manifold using new gasket.

9. Install nuts and bolts holding the two manifolds together but do not tighten.

10. Position manifolds to cylinder head with intake manifold pilot rings in place and install mounting bolts, but do not tighten.

11. Tighten nuts and bolts which hold manifolds together, torquing to 25-28 ft. lbs.

12. Tighten manifold mounting bolts, torquing to 25-28 ft. lbs.

Cylinder Head

Cylinder Head Removal & Installation

6 Cylinder Engines

1. If the engine is to remain in the vehicle, first do the following: Drain cooling system, disconnect water hose(s) and disconnect the fuel line from the carburetor. On the PT-6 engine, positive crankcase ventilation and vacuum advance lines must also be removed. In some cases, depending on the model, various accessory or other equipment may have to be removed.

2. Disconnect spark plug wires and, if applicable, remove wires from bracket on cylinder head. Identify cylinder number of each wire with tags or marks.

3. Disconnect ignition coil wires and remove coil from cylinder head.

4. Remove manifolds as described in preceding section.

5. On BD engines, disconnect water by-pass hose at cylinder head and remove water pump as described in preceding section.

6. Remove rocker arm cover bolts and remove rocker arm cover and gasket. If there are any dowel sleeves in brackets, note their location.

7. Remove push rods and keep them in order so that they may be replaced in the same locations.

8. Remove cylinder head bolts, marking any odd length or oil feed bolts for proper reinstallation.

9. Remove cylinder head and gasket.

10. Installation is the reverse order of the above instructions. Be sure to note the following steps where applicable.

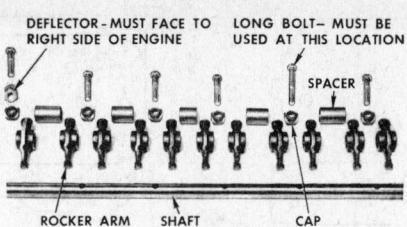

Rocker arm assembly—1st type—P-6 engines
(© International Harvester Co.)

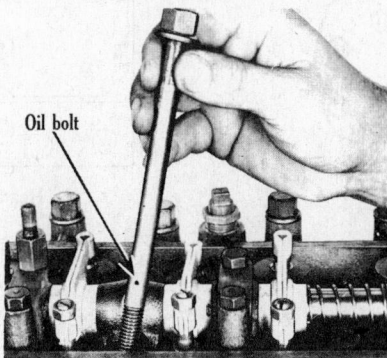

Rocker arm shaft oil feed bolt—RD engines
(© International Harvester Co.)

11. Align new head gasket with bolt holes and carefully place cylinder head into position without damaging or shifting gasket.

12. Install all short head bolts and flat washers, but do not tighten.

13. On the PT-6 engine, apply No. 2 Permatex or equivalent sealant to the threads of the long (4¼") bolt at position "11". On this engine, proceed to tighten the head bolts to 80-85 ft. lbs. torque in the sequence illustrated. Tighten in approximately 20 ft. lb. steps until proper torque readings are obtained.

14. Install push rods into original positions.

15. Install rocker arm assembly, installing any dowel sleeves which were removed. On BD engines, rocker arm bolts serve as head bolts and all should be tightened in the sequence illustrated at this point. Tighten to 85-95 ft. lbs. torque.

16. On RD engines, leave the long oil bolt out and tighten only the short head bolts to 100-110 ft. lbs. torque.

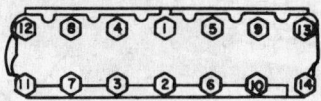

Cylinder head bolt tightening sequence— PT-6-232 engines

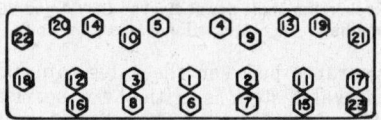

Cylinder head bolt tightening sequence— RD engines

17. On the PT-6 engine, the long rocker arm assembly mounting bolt is placed in the fifth (next to rear) position. Tighten rocker arm assembly mounting bolts to 20-23 ft. lbs. torque, inner first, then outer.

NOTE: On later PT-6 engines, individual rocker arms are used for each value, with a double bridge used as bearing surface and retainer.

18. On RD engines, insert drilled oil bolt in special oil connector and tighten rocker arm shaft mounting bolts to 25-30 ft. lbs. torque. Be sure to install those bolts with the cover mounting studs in the correct position. Note lock washers which key into end brackets.

19. Adjust rocker to valve clearance. See the following section, "Valve Clearance Adjustment," for correct procedure. The PT-6 engine has hydraulic lifters and cannot be adjusted. See Unit Repair Section for hydraulic lifter service operations.

20. Reinstall any components which were removed and have not yet been put back. Replace rocker cover gasket if necessary.

V8 and 4 Cylinder Engines

1. Remove intake and exhaust manifolds as described in preceding section. On V-8's, this may entail removal of the air compressor and air compressor mounting bracket.

2. Head removal is facilitated by the use of a lifting sling which is attached with bolts in the intake manifold mounting bolt holes.

3. Remove cylinder head covers and gaskets.

4. Loosen rocker arm shaft bracket bolts and remove the rocker arm assembly.

NOTE: Be sure to remove and keep track of the two dowel sleeves on the end brackets of the rocker arm assembly.

5. Remove pushrods, marking them so that they may be installed in their same locations.

6. Remove spark plug wires.

7. Remove cylinder head bolts.

8. When lifting off cylinder, do not lose the two locating dowel sleeves.

9. Installation is basically the reverse of the above procedure, with the exception of the following additional steps.

10. Be sure to use a new head gasket and to reinstall dowel sleeves when positioning the head and mounting the rocker assembly. Reinstall pushrods in their original locations.

11. On 4-196 and V-304, 345, 392, turn engine crankshaft until leading edge of balance weight

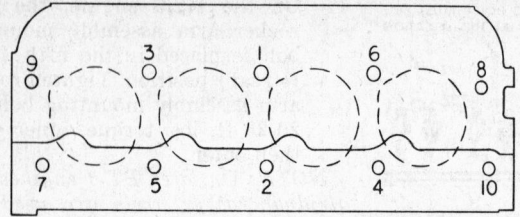

Cylinder head bolt tightening sequence—4-196, V 304, 345, 392, MS 406, 446 engines (© International Harvester Co.)

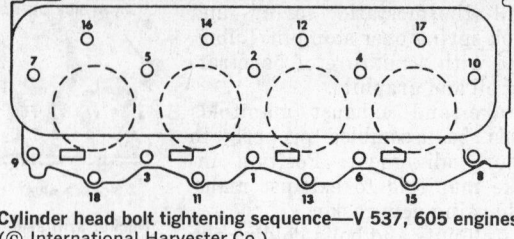

Cylinder head bolt tightening sequence—V 537, 605 engines (© International Harvester Co.)

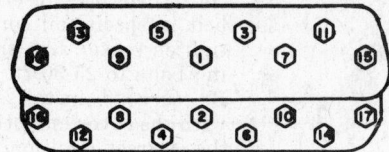

Cylinder head bolt tightening sequence— 401, 561 & 549 engines

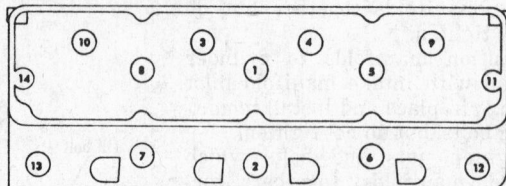

Head bolt torque sequence—V8-400

on crankshaft pulley is aligned with the zero degree mark on the timing indicator before installing rocker arm assembly.

12. On 4-196 engines, be sure to install rocker assembly so that the oil feed shaft bracket is third from the rear.

13. On V-304, 345, 392 engines, install rocker arm assembly so that the notches at the end of the shaft are facing upward. Oil feed brackets are third from the rear on the right (even numbers) bank and third from the front on the left (odd numbers) bank.

14. On the V(S)-401, 478, 549 engines, rocker arm assembly mounting bolts serve as head bolts and are tightened in the head bolt torque sequence. Torque head bolts in the sequence pattern illustrated to 80-90 ft. lbs. torque.

CAUTION: Do not use a power wrench on heads of engines with hydraulic lifters. Torque head bolts slowly so that the leakdown of the lifters may relieve strain from the valve train.

15. On 4-196 and V-304, 345, 392, 400 engines, tighten the head bolts in the sequence illustrated to 110 ft. lb. on 400 engines and 90-100 ft. lb. on all others.

16. Retorque head bolts to the specified torque after 1000 miles of operation.

17. Install rocker covers and any other equipment removed for head work. Replace rocker cover gasket if necessary.

Valves

Valve Rotators

RD Engines

Exhaust valves on RD engines have a special stem and valve cap which permit the valve to rotate for an instant during each cycle. The

Correct pulley location for installing pushrods—OHV-4 & small V8 engines (© International Harvester Co.)

Cylinder head dowel sleeve location— OHV-4 and V8 engines (© International Harvester Co.)

Rocker arm shaft dowel sleeve—OHV-4 and V8 engines (© International Harvester Co.)

clearance between the valve cap and the valve stem is critical for service life of the valve. The clearance is measured with a special gauge as illustrated. Grind bottom of valve cap

to decrease clearance. To increase clearance, grind top of valve stem. If valve keys are removed they must be reinstalled with the wear on the top side.

4 Cylinder and V8 Engines

Rotators are used between the valve springs and the cylinder heads on both the intake and exhaust valves on the VS 401, VS 478, VS 549, V 537, and V 605 engines. On the 4-196, MV-404, MV-446, V-304, V-345, V-392 engines, rotators are used between the valve spring and the cylinder head on the exhaust valve only.

NOTE: Keep the valves and their related parts together so they may be reinstalled in their respective positions.

Valve Train Service

The PT-6, 4-196 and V-8 engines utilized hydraulic lifters for which there is no lash adjustment. Excess noise in the valve train of these engines indicates that service is required. Instructions for servicing hydraulic lifters may be found in the General Repair Section.

Valve removal, service, and installation procedures may be found in the Engine Rebuilding Section. See Specifications table at the beginning of this section for valve spring and valve seat angle specifications.

E E E E
I I I I
4-196 Engine

E I I E E I I E E I I E
RD 406, 450, 501, P-6 232, 258 Engines

I E I E I E I E
E I E I E I E I
MV 404, 446, V 537, 605, V 304, 345, 392 Engines

E I I E E I I E
E I I E E I I E
VS 401, 478, 549 Engines

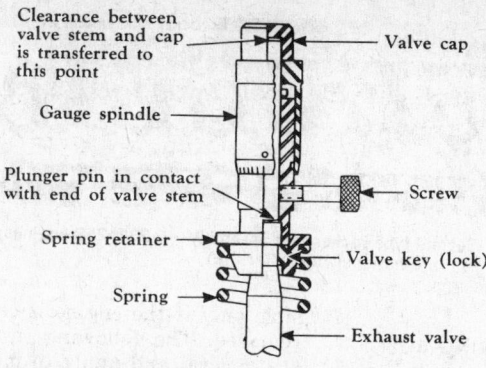

Checking rotating valve cap clearance—RD engines (© International Harvester Co.)

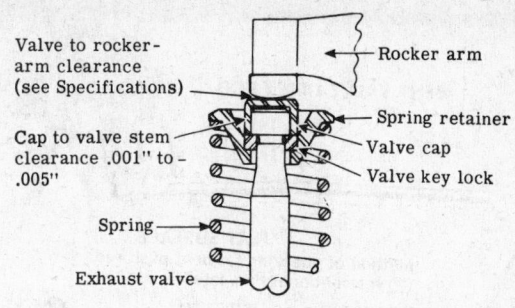

"Slo-Roto" valve rotators—RD engines (© International Harvester Co.)

Valve Adjustment

RD Engines

1. Remove valve rocker cover and gasket.
2. Rotate crankshaft until No. 1 cylinder is at top dead center of the compression stroke (both valves closed). Timing marks on the crankshaft pulley of RD engines should be aligned for zero degrees (TDC).
3. Clearance is adjusted on both valves. Loosen lock nut.
4. Holding the proper feeler gauge between adjusting screw on the end of the rocker arm and the valve stem, turn adjusting screw until specified clearance is obtained (feeler gauge just held snugly).
5. While holding adjusting screw, tighten lock nut.
6. Recheck clearance.
7. Rotate crankshaft 1/3 turn so that No. 5 cylinder is at top dead center of compression stroke and repeat Steps 3 to 6.
8. Repeat for each cylinder following the sequence of the firing order.
9. Replace rocker cover, using a new gasket if necessary.
10. Valve clearance should be rechecked with engine at normal running temperature and after 500 miles when new or reground valves have been installed.

Rocker Arm Removal & Installation

1. Remove rocker cover and gasket.
2. Remove rocker arm assembly mounting bolts and flat washers.
3. Remove rocker assembly.
4. If applicable, remove clip-ring and retainer to disassemble

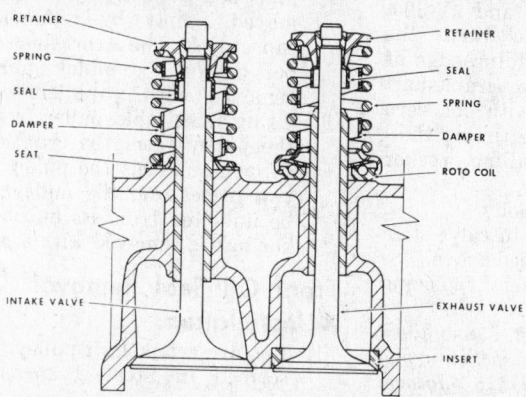

Rotor used under exhaust valve spring only—MV 404, 446, 4-196, V 304, 345, 392 engines (© International Harvester Co.)

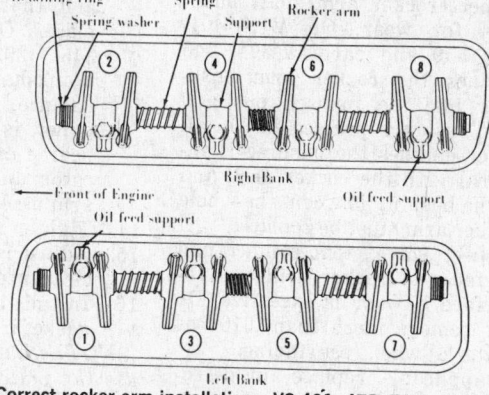

Correct rocker arm installation—VS 401, 478, 549 engines (© International Harvester Co.)

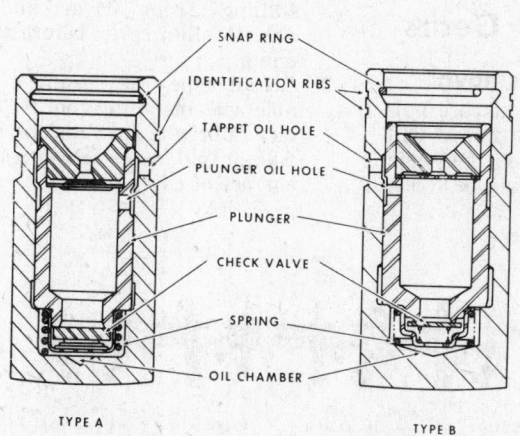

Sectional view of two types of hydraulic valve lifters (© International Harvester Co.)

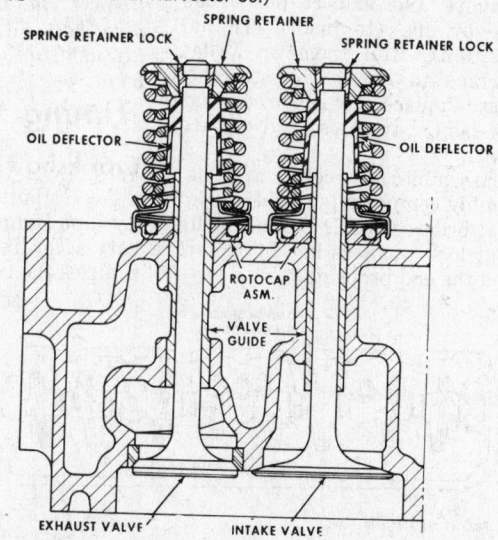

Rotors used under valve springs—intake and exhaust valves— –VS 401, 478, 549 engines (© International Harvester Co.)

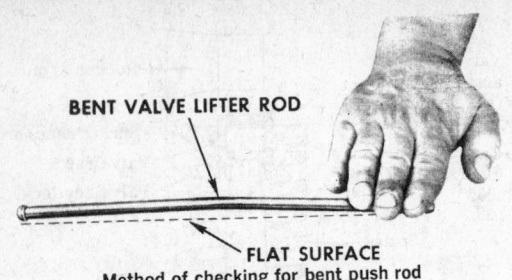

Method of checking for bent push rod
(© International Harvester Co.)

BENT VALVE LIFTER ROD
FLAT SURFACE

Second type rocker arm assembly—6-232, 258 engines
(© International Harvester Co.)

ROCKER ARM BRACKETS
ROCKER ARMS

rocker components. Be sure to keep all parts in order so that they may be replaced in their original positions.

5. Clean all parts thoroughly, making sure that oil passages are clear. If necessary to remove plugs from ends of shaft, drill a hole in one plug, knock out the other with a steel rod, then knock out the drilled plug.

6. Inspect shaft for wear and warpage. Replace bent or worn shaft.

7. On engines without hydraulic lifters (all 6 cylinder except the PT-6), inspect rocker arm adjusting screws for wear at the contact surface and for damaged threads. Replace any that are defective.

8. Inspect rocker arm shaft bushings for wear. On V(S)-401, 478, 549 and early V-304, 345 engines the rocker arm bushings may be pressed out and new bushings pressed in. On all other engines the bushing is integral with the rocker arm, and if the bushing is worn, the whole rocker arm must be replaced.

9. Inspect valve stem contact pad surfaces of rocker arm and resurface if wear is excessive. Do not remove more than .010″ of material when resurfacing.

10. If applicable, replace any defective tension springs.

11. Remove and inspect push rods one by one (to insure original position). Roll them on a flat surface to check for straightness. Replace any pushrods that are bent, have loose ends or are worn.

12. Reassemble all rocker arm assembly components in their original order. On RD engines, note that lock washers key into shaft and the end brackets.

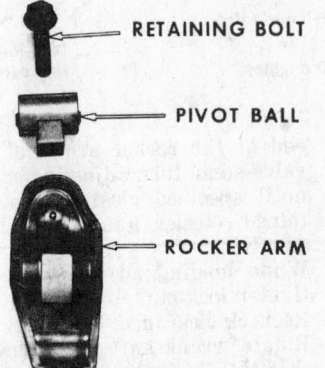

Rocker arm assembly—MV 404, 446, engines
(© International Harvester Co.)

RETAINING BOLT
PIVOT BALL
ROCKER ARM

13. Install rocker arm assembly, making sure that the oil feed bracket is in the proper position and that dowel sleeves are in place. On 4-196 and V-304, 345, 392 engines turn the crankshaft until leading edge of balance weight on crankshaft pulley is aligned with the zero degree mark on the timing indicator before installing rocker arm assembly.

14. Tighten mounting bolts.

15. Adjust rocker arm to valve stem clearance as described above.

16. Install rocker cover, replacing gasket if necessary.

NOTE: Later P-6, 232, 258 engines use the individual type rocker arms, as do the MV 404, and 446 engines. Inspect the bearing surface of the pivot ball and bridge to determine the extent of wear and need for replacement.

Timing Case & Gears

Crankshaft Pulley Removal

Accessibility of the crankcase pulley and front (timing) cover will vary according to the model. On some vehicles the timing case will be acces-

sible only if the engine is completely removed. The following instructions are general and apply to most front cover repairs and service.

1. Drain cooling system.
2. Disconnect radiator hoses and remove radiator. In some cases the radiator shroud and truck hood must be removed.
3. Loosen front engine mounts and jack up engine enough to provide access to the crankshaft pulley with a puller.
4. Loosen and remove fan belts and remove fan blades.
5. Remove crankshaft pulley retaining bolt. On V(S)-401, 478, 549 engines, scribe a line alongside the timing indicator and remove the indicator from front cover. The line is for reassembly purposes. The 400 engine has a pulley which is removed simply by removing 4 capscrews. The vibration damper behind the pulley must be removed with a puller.
6. Using a suitable puller, remove the pulley from the crankshaft. On some models the pulley is in two pieces and the pulley must be unbolted from its hub before the hub is removed with a puller.

Front Oil Seal Removal & Installation

1. Remove crankshaft pulley as described in Steps 1 through 6 above.
2. Remove seal. It is preferable to use an appropriate seal puller. Use a new gasket when installing front cover and be sure to align cover before tightening.
3. Install a new seal using a suitable seal installing tool if possible. Lubricate first and be careful not to damage seal or seating surface of cover.

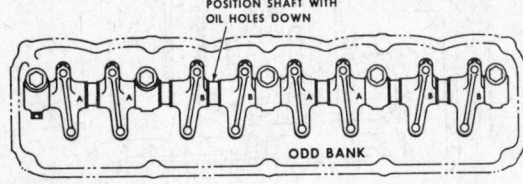

POSITION SHAFT WITH OIL HOLES DOWN
ODD BANK
FRONT OF ENGINE

Correct rocker arm positioning—V 537, 605 engines
(© International Harvester Co.)

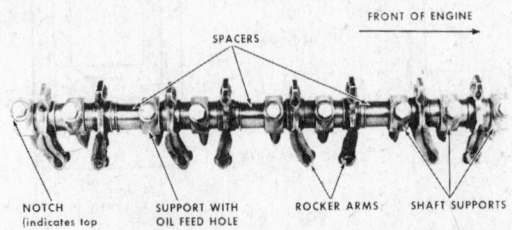

FRONT OF ENGINE
SPACERS
NOTCH (indicates top of shaft)
SUPPORT WITH OIL FEED HOLE
ROCKER ARMS
SHAFT SUPPORTS

Rocker arm assembly—4-196 (© International Harvester Co.)

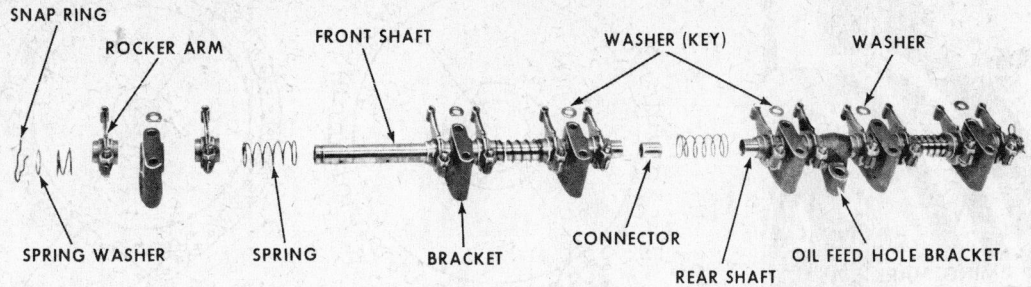

Rocker arm assembly—RD engines (© International Harvester Co.)

4. Install crankshaft pulley, fan belt and fan blades.
5. Lower engine and tighten mounting bolts.
6. Install radiator, shroud, hoses and whatever else was removed.
7. Fill cooling system.

Timing Gear Removal

Timing gears can be removed without disassembling the engine. In some cases, however, the engine must be removed.

1. Remove crankcase pulley as described in Steps 1 through 6 above.
2. Remove engine front cover.
3. Rotate engine to align timing marks on crankshaft gear and camshaft gear.
4. To remove either gear, remove bolt and washer. Use a suitable puller.

NOTE: On the P-6 engines, simultaneously loosen the camshaft gear while pulling the gear from the crankshaft, and remove both gears and the timing chain as an assembly. Normal replacement would include the three parts, due to running wear patterns. On all other engines, replace both the cam gear and crankshaft gear, due to being serviced in matched sets.

5. Use a suitable installing tool to install gears. Lubricate with engine oil and insert key in shaft to align gear. Align timing marks as illustrated. Be careful not to damage threads on shaft. Install and tighten retaining bolt.

6. Rotate engine to check that gears are not binding.
7. Check gear backlash with a dial indicator. It should be within .004-.007" on OHV6 engines and .0005-.0045" on OHV4 and V8 engines.
8. Use a new gasket when installing front cover and be sure to align cover before tightening. On some models there is an oil slinger on the crankshaft.
9. Install crankshaft pulley, belt and fan blades. Tighten retaining bolt.
10. Lower engine and tighten engine mounting bolts.
11. Install radiator and hoses.
12. Fill cooling system.

Camshaft Removal & Installation

On most International truck models, it is possible to remove the camshaft with the engine remaining in the vehicle. However, the body grille work, radiator, A/C condensor (if equipped), hood, bumper, and braces must be removed to allow clearance for the camshaft to be withdrawn from the engine block. In some cases, it would be more advantageous to remove the engine from the vehicle to replace the camshaft. The decision would depend upon the individual mechanic and his shop facilities.

1. Remove the cylinder head on the P-6 engines, and the intake manifolds on the V8 engines, and the rocker covers on all engines.
2. Remove the rocker arms or assemblies, pushrods and tappets.

NOTE: On the RD engines, the tappets are of the mushroom type and must be removed from the bottom. In order for the tappets to be out of the way during camshaft removal, the engine, if out of the vehicle, must be turned upside down. If the engine is in the vehicle the tappets must be raised and held to allow the camshaft to clear when being withdrawn from the block.

3. Remove the distributor and mechanical fuel pump.
4. Remove the oil pan and oil pump, if necessary.

NOTE: RD Series, VS 401, 478, 549 engines have camshaft driven oil pumps which must be removed before the camshaft is removed.

5. Remove the crankshaft pulley as previously described.
6. Remove the front timing cover, gasket and seal.
7. Remove the two screws securing the camshaft thrust flange to the block.

NOTE: On the P-6 engines, remove the camshaft bolt and washer retaining the gear to the shaft. Remove the crankshaft and camshaft gears and chain as an assembly.

8. Remove the camshaft and gear. To prevent nicking and damaging the camshaft or bearings, use a camshaft removal and installation tool, which is an extension on the front of the camshaft to act as a handle.
9. When installing the camshaft and gear, coat the bearing surfaces and lobes with lubricant and use the installing tool if pos-

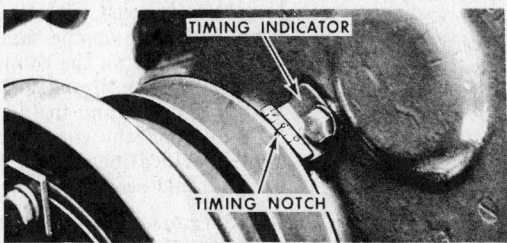

Timing mark alignment for valve adjustment
(© International Harvester Co.)

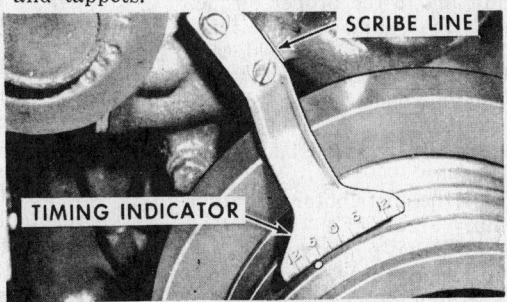

Marking exact position of timing indicator—V8 engines
(© International Harvester Co.)

349

Typical timing marks—4-196, V 304, 345, 392 engines (© International Harvester Co.)

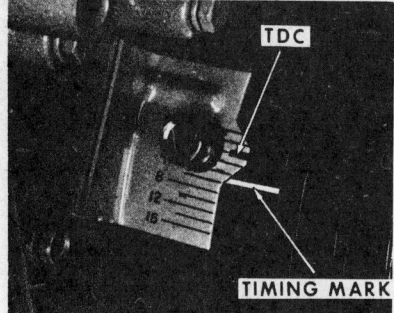

Typical timing marks—MV 404, 446, engines (© International Harvester Co.)

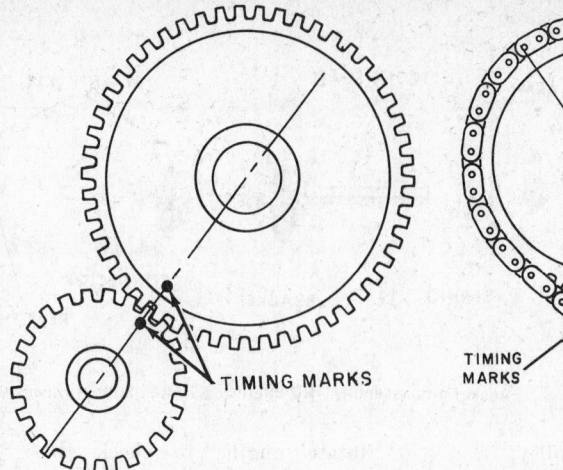

Timing gear alignment marks except PT6 engines

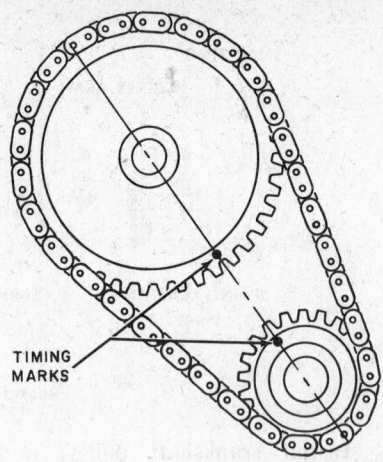

Timing gear alignment marks PT6 engines

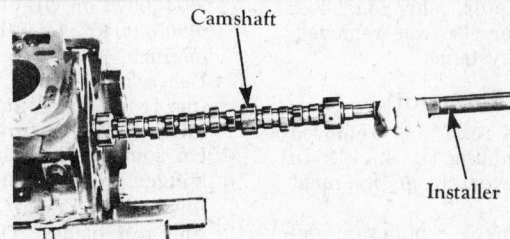

Use of a camshaft remover and installer
(© International Harvester Co.)

sible, to aid in the installation of the camshaft. Make sure the gear timing marks align properly.

NOTE: For the P-6 engine chain installation, refer to the timing gear removal and installation procedure.

NOTE: FTVS 549 engines should have the camshaft gear retarded one tooth from normal position.

10. Working through the two large holes in the camshaft gear, install the two thrust flange screws and tighten to proper torque specifications.
11. Check timing gear backlash. If the end play exceeds the allowable limits, replace the thrust flange.
12. Place the oil slinger over the end of the crankshaft.
13. Install the front cover, using a new seal and gasket. Align the cover before tightening the bolts to the specified torque.
14. Install the crankshaft pulley, tightening to the proper torque.
15. Install the cylinder head, if removed, the intake manifold, tappets, pushrods, and rocker arms. Torque all bolts to the specified torque.
16. Install the fan pulley, blades and belts.
17. Install the distributor and fuel pump.
18. Install the oil pump and oil pan.
19. If the engine was raised, lower and tighten the engine mounts.
20. Complete the assembly as necessary for the removed body parts.

21. Start the engine, time it to specifications, and check for proper operation.

Pistons and Connecting Rods

For piston and connecting rod overhaul procedures see Engine Rebuilding General Section.

Piston Removal & Installation

1. Remove oil pan and oil pump. On some models this may require loosening the engine mounts and jacking up the engine until spacer blocks can be installed. When removing oil pump on big V8's turn crank until counter weight is out of the way.
2. Remove cylinder head. See "Cylinder Head R&R" above.
3. Using a ridge reamer, remove the ridge from the top of the cylinders.
4. Rotate the crankshaft until journal is in lowermost position for removal of connecting rod assemblies. Remove the cap and push the connecting rod and piston up through the cylinder bore. Replace the cap and bearing inserts on the rods so the numbered sides match. The numbers indicate position.
5. To install pistons and connecting rods, rotate crankshaft until No.

1 crankpin is at the bottom of its stroke. Correctly seat rod bearing insert in rod then dip piston assembly in clean oil to lubricate rings. Using a ring compressor, install piston and rod in cylinder. Push piston in, do not strike. On the PT-6 engines, the notch on the top of piston perimeter faces toward the front of the engine (number on connecting rod toward the camshaft). On 4-196 and all V8 engines, except the 400, the piston assembly is installed with the word "UP" toward the top (camshaft) side of the engine block. On the 400, the piston is installed with the notch facing front. On all OVH 6 except the PT-6, install piston assembly with the arrow stamped on piston crown toward camshaft.

6. Place lower half of bearing insert in rod cap and lubricate with oil. Assemble bearing cap to connecting rod with the number side of cap on the same side as the number on the connecting rod. Lubricate threads of bolts with engine oil and install bolts, tightening to the correct torque (see Specifications at the beginning of this section).
7. Rotate crankshaft and repeat installation procedure with the rest of the pistons and connecting rods.
8. Install oil pump and oil pan,

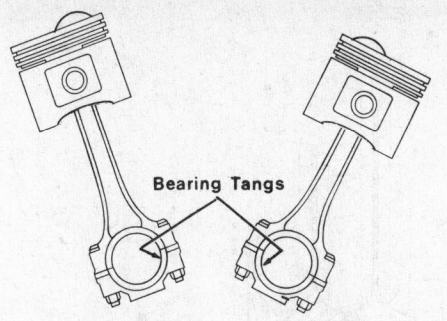

Right Bank
2, 4, 6, 8

Left Bank
1, 3, 5, 7

Correct assembly of piston & rods—
V8 engines
(© International Harvester Co.)

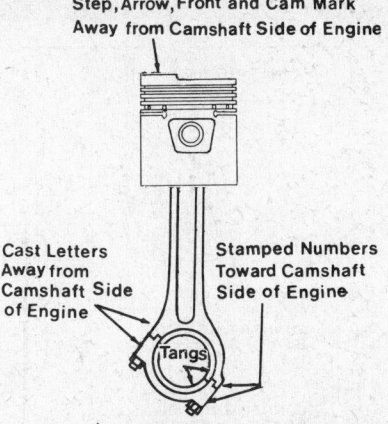

Step, Arrow, Front and Cam Mark
Away from Camshaft Side of Engine

Cast Letters Away from Camshaft Side of Engine

Stamped Numbers Toward Camshaft Side of Engine

Tangs

Piston & rod assembly—RD engines
(© International Harvester Co.)

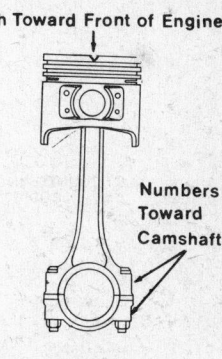

Notch Toward Front of Engine

Numbers Toward Camshaft

Piston & rod assembly PT-6—232 engines

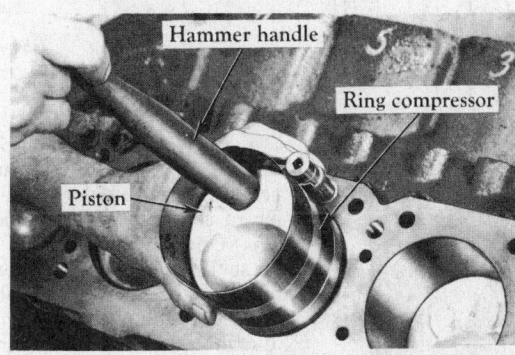

Piston installation
(© International Harvester Co.)

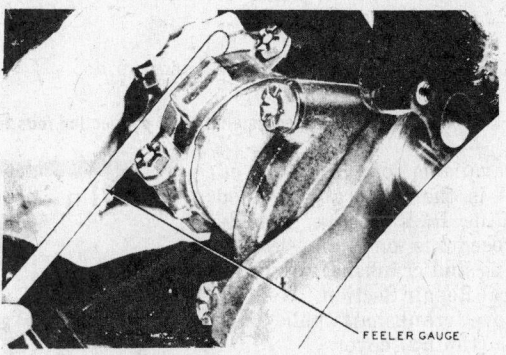

Checking connecting rod end clearance—
typical (© International Harvester Co.)

using a new pan gasket if necessary.
9. Install cylinder head as described in "Cylinder Head R&R" above.
10. If engine was raised, remove spacers and lower engine. Tighten engine mount bolts.

Piston Ring Replacement

1. Remove pistons as described above.
2. Remove both compression rings and three-piece oil ring.
3. Using rings which correspond to the piston size (standard or oversize), check rings for gap clearance and ring-to-groove side clearance.
4. Install rings on piston with a suitable ring expander tool.
5. Install piston assembly as described above.

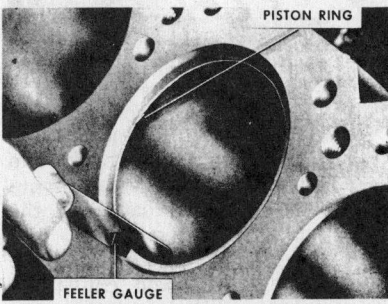

Checking piston ring gap
(© International Harvester Co.)

Engine Lubrication

Oil Pan Removal and Installation

The engine and mounts may have to

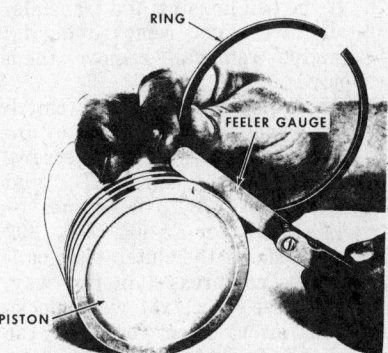

Checking ring groove clearance
(© International Harvester Co.)

be loosened from the crossmember and lifted, and spacer blocks installed between the mounts and crossmember, to gain clearance to remove the oil pan from the engine. Other engine applications may only require the removal of steering linkage to gain sufficient clearance. Be sure to clean all old gasket material from the oil pan and the block before installing the oil pan and new gasket.

Main Bearing Replacement

On most models it is possible to replace main bearings without removing the engine from the vehicle. However, it is easier to do a better job with the engine removed and, if facil-

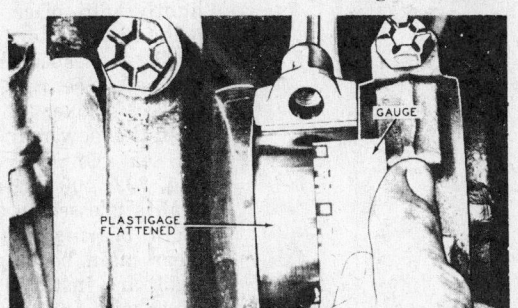

Checking connecting rod bearing clearance with plastigage—typical (© International Harvester Co.)

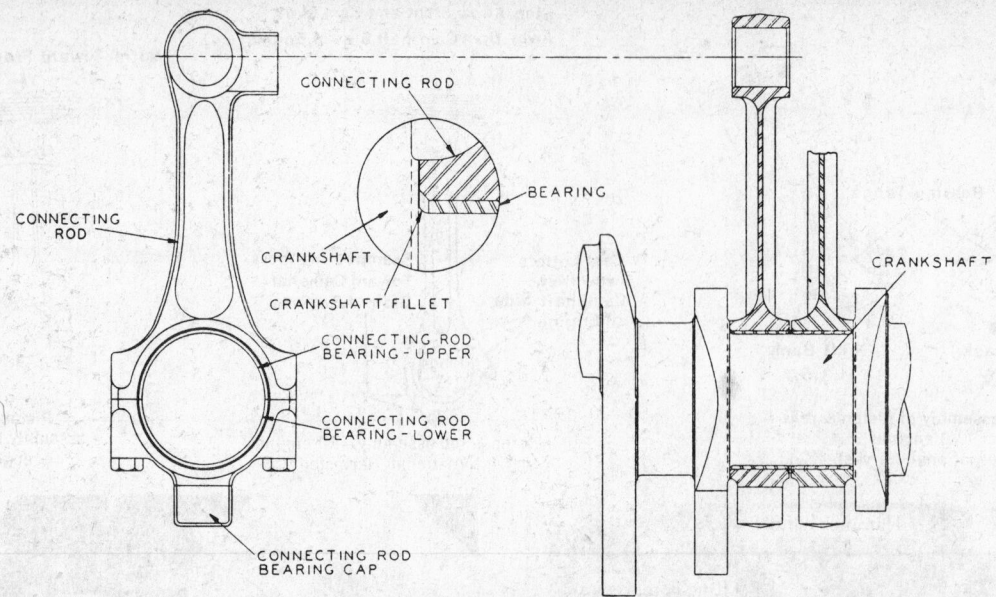

Proper installation of connecting rods to crankshaft (© International Harvester Co.)

ities are available for pulling the engine, this is the preferable method. See "Engine R&R" above. For detailed procedures on main bearing, rod bearing and crankshaft servicing see General Repair Section.

1. Remove crankshaft pulley and front (timing) cover.
2. Remove cylinder head(s) and piston assemblies.
3. If the bell housing and transmission were not removed during engine removal, remove them now.
4. On many engines the clutch plate may be compressed by installing three cap screws (3/8"-16x2½" for RD and V(S)-401, 478, 549 engines or 3/8"-16x2" for V-304, 345, 392 engines). If the clutch plate cannot be compressed in this way, cut three ½"x1"x3" wood blocks and insert them between the clutch fingers and back plate. Loosen backing plate mounting bolts slightly if it is difficult to insert wood blocks. A third alternative procedure for compressing

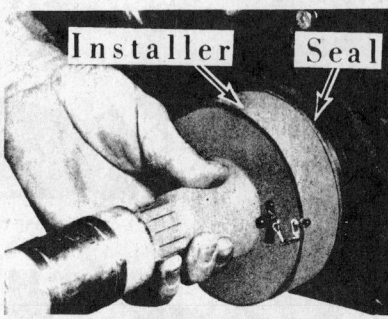

Installing rear main oil seal
(© International Harvester Co.)

the clutch plate is to insert three retaining clips as illustrated. The clutch plate is compressed during removal and installation of the clutch to prevent warpage.

5. Remove the backing plate mounting bolts and clutch assembly.
6. Remove flywheel bolts and pull off flywheel.
7. On RD engines remove the upper and lower rear main bearing oil seal retainers.
8. Remove main bearing caps. On

OHV4, V-304, 345, 392, engines use a rear main bearing cap puller to remove rear main bearing cap. On RD engines the lower oil seal retainer must be removed before the cap can be unbolted. Note that the caps are numbered and should be reinstalled in their original positions.

9. When installing new main bearings make sure that the oil holes are properly aligned and that bearing tangs are fitted into tang recesses. Thoroughly clean all surfaces and coat lightly with oil. Be sure to align timing marks when positioning crankshaft. On the PT-6 engine main bearing caps have an arrow which indicates front of engine. On the OHV4 and V-304, 345, 392, 400 engines the numbered sides of the main bearing caps face the left side of the engine. On the RD engines the numbered side of the bearing caps face the camshaft side of the engine. On V(S)-401, 478, 549, 537, 605 engines the bearing caps are installed with the numbered side facing to the right side of the engine. When tightening main bearing caps, first tap them lightly into place, then tighten bolts in an alternating manner until the specified torque is reached. See Specifications for correct torque.

10. Install a new rear main bearing oil seal. On OHV4 and V-304, 345, 392, 400 engines the round seal is pressed in after the rear main bearing cap is installed. Rear main bearing cap side oil seals are installed on these engines with an installer tool made from a piece of ⅛" welding rod. Puddle a ball on the end of the

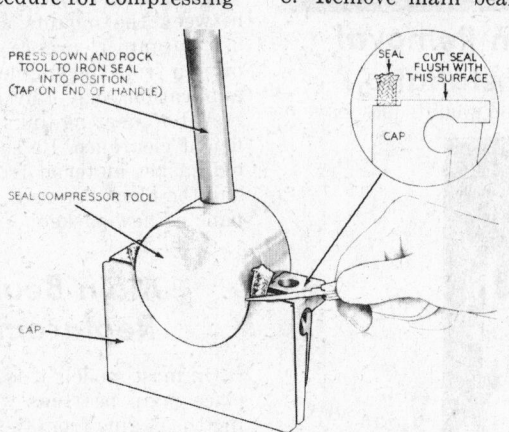

Installation of oil seal in rear main bearing cap
(© International Harvester Co.)

rod and file the ball to approximately 5.32″ diameter. On RD engines the rear upper seal and retainer are installed before the crankshaft and the lower main bearing lower oil seal is installed after the rear main bearing cap, using new gaskets on each side of the lower seal retainer. On V(S)-401, 478, 549 engines the rear main bearing cap side seals are marked for right and left side. On PT-6, and V(S)-401, 478, 549 engines, the top and bottom rear main bearing oil seals are made of wick type material which fits into slots in the block and bearing cap. These must be cut to the correct length while the seal is being held in place with a seal compressing tool as illustrated.

11. Check main bearing clearance and crankshaft endplay and compare to clearance limits listed in Specifications at the beginning of this section. See Engine Rebuilding General Section for clearance measurement and service procedures.
12. Reassemble engine following Steps 1 through 7 in reverse order. Be sure to align clutch driven disc with transmission shaft or clutch aligning tool before tightening clutch plate mounting bolts.

Rear Main Bearing Seal Replacement

OHV 4, V 304, 345, 392, 400, 536, 605 Engines

The rear main bearing cap seal can be replaced with the engine in the chassis, but the transmission, clutch assembly, and flywheel must be removed to gain access to the seal.

1. Remove the transmission, clutch assembly, and the flywheel.
2. Remove the engine oil pan.
3. With a slide hammer with a screw end adapter pierce the seal and remove it from the recess in the cap and block.
4. Lubricate the new seal, seat it squarely with a seal installer tool

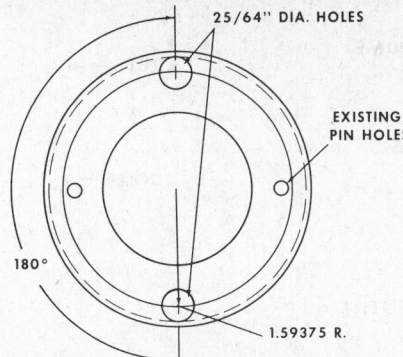

Rework of pilot tool—SE 1942-2 for MV 404, 446, engine rear main bearing oil seal installation (© International Harvester Co.)

.085 inch from the rear face of the block.
NOTE: Production installed seals are seated flush with the rear of the block.

5. Install the bearing cap side seals with the use of a ⅛ inch welding rod, 8 inches long, with a 5/32 inch puddled ball on the end. Cut off any excess side seal, flush with the oil pan block surface.
6. Install the oil pan, flywheel, clutch assembly, and transmission.

PT 6, RD, VS 401, 478, 549 Engines

1. Drain crankcase and remove oil pan.
2. Remove rear main bearing cap.
3. Remove oil seal from bearing cap and clean cap thoroughly.
4. Loosen all remaining main bearing cap mounting bolts.
5. Using a brass drift and hammer, tap the upper seal until sufficient seal is protruding on the other side to permit pulling it out with pliers.
6. Wipe crankshaft seal surface clean and coat lightly with oil.
7. Coat crankcase surface of upper seal with soap and the lip of seal with No. 40 engine oil.
8. Install upper seal with lip toward front of engine.
9. Coat mating surfaces of crankcase and cap with No. 2 Permatex or equivalent, back surface of seal with soap and lip of seal with No. 40 engine oil.

10. Seat seal firmly into seal recess of cap and apply No. 2 Permatex or equivalent to both chamfered edges of the rear main bearing cap.
11. Install main bearing halves and cap.
12. Tighten all main bearing cap mounting bolts to specified torque.
13. Install oil pan.

MV 404, 446 Engines

The rear main bearing oil seal is a pressed fit into a retainer plate, which is bolted to the rear of the engine block.

1. Remove the transmission, clutch assembly, and flywheel.
2. Remove the capscrews holding the retainer to the engine block, and remove the retainer.
3. Press out the old seal from the retainer, clean the retainer seal pocket, and press the new seal in place on the retainer.
NOTE: The seal must be installed from the crankcase side of the retainer, flush with the seal bore inner surface.
4. With the use of tool SE-1942-2 pilot, or its equivalent, install the rear oil seal and retainer with gaskets on the engine block.
NOTE: When using the SE-1942-2 pilot tool to install the seal with the engine in the vehicle, drill two 25/64 inch diameter holes in the pilot, 180 degrees apart, 90 degrees from each existing hole. This will allow the use of two ⅜ x 4 inch pilot studs to serve as a safety measure to retain the pilot on the crankshaft.
5. Replace the flywheel, clutch assembly, and transmission.

Oil Filter Replacement

On all models except the RD-6 and the 400 engines, the oil filter unit is on the left side of the engine block. On the PT-6 and 400 engines, the filter is located on the right side of the block. All engines except the RD series and the VS 401, 478, and 549 series, use a spin-on type oil filter, which is replaced as a complete unit, using a strap wrench to remove it from the engine. Follow the instruc-

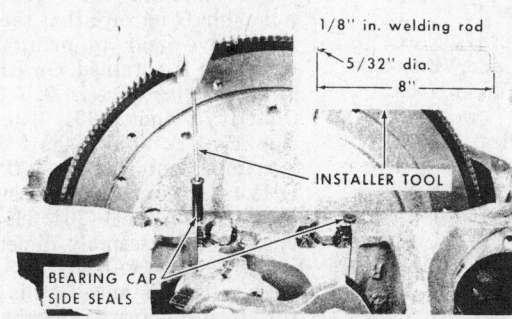

Installation of rear main bearing cap side seals—V 304, 345, 392 engines (© International Harvester Co.)

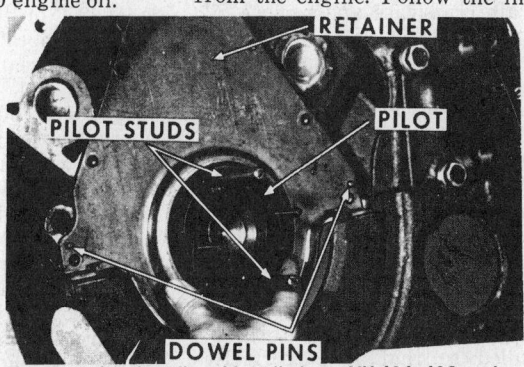

Rear main bearing oil seal installation—MV 404, 406 engines (© International Harvester Co.)

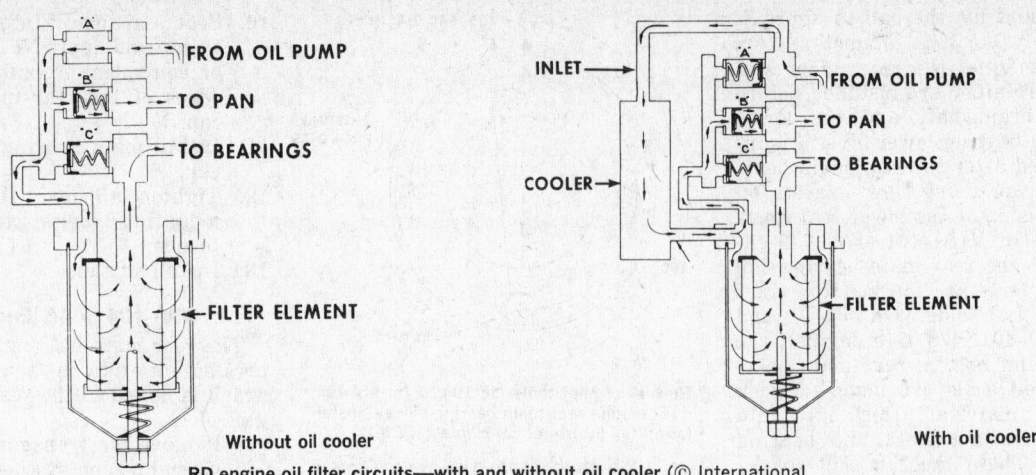

FROM OIL PUMP
TO PAN
TO BEARINGS
FILTER ELEMENT

INLET
COOLER
FROM OIL PUMP
TO PAN
TO BEARINGS
FILTER ELEMENT

Without oil cooler   With oil cooler

RD engine oil filter circuits—with and without oil cooler (© International Harvester Co.)

tions printed on the filter assembly to install.

The RD and VS series use a paper type filter, inserted in the filter shell, with the shell bolted to the filter base. To replace the filter, follow this procedure.

1. Remove drain plug from bottom of filter body, drain oil and replace drain plug.
2. Loosen filter body retaining bolt and remove filter body and element. Check condition of body to base gasket and replace if necessary.
3. Wash filter body with cleaning solvent, being sure to remove all sediment.
4. Install new filter element onto filter base with seal end away from base. Be sure element is fully seated onto base.
5. Install oil filter body and bolt with spring, making sure body seats evenly on gasket. Tighten filter body retaining bolt to 30-35 ft. lbs. torque on all engines.
6. Start engine and run for at least 5 minutes until oil is warm and check for leaks.
7. Check crankcase oil level. Lubricant capacity of oil filter is about one quart.

Oil Pump Removal and Installation

1. Drain crankcase and remove oil pan.
2. Remove oil pump mounting bolts and pull straight down on pump to remove.
3. When installing oil pump, guide pump shaft into position and rotate shaft until tang of drive gear is engaged. On PT-6 and V(S)-400, 401, 478, 549 engines install a new oil pump gasket when installing. On RD engines the oil pump shaft drives the distributor shaft and must be installed so that it is correctly timed to the crankshaft. Rotate

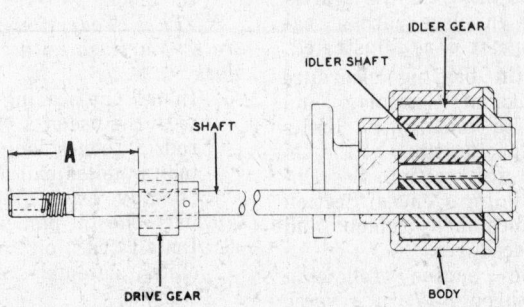

IDLER GEAR
IDLER SHAFT
SHAFT
A
DRIVE GEAR
BODY

Oil pump drive gear installation—VS 401, 478, 549 engines

| | 2-7/8" | Front mounted distributor |
| "A" | 1-15/16" | Rear mounted distributor |

(© International Harvester Co.)

the crankshaft until No. 1 cylinder is in firing position. On RD engines the oil pump is installed so that the slot in the top of the shaft is at a 60-degree angle to the side of the engine.
4. Tighten oil pump mounting bolts to:
    25-30 ft. lb. for all exc. 400.
    400-55 in. lb.
5. Install oil pan and fill crankcase.

Oil Pump Service

1. Thoroughly clean oil pump. Do not disturb or remove pickup tube unless absolutely necessary.
2. Remove pump cover bolts and pump cover.
3. Check gear to body clearance. If it is not within .0025-.0055" on the OHV6 engines or .0007-.0027" on the OHV4 and V8 engines, obtain new parts.
4. Check gear backlash. If it exceeds .011" on OHV4 and V8 engines or .006" on OHV6 engines, replace gears.
5. Check pump shaft clearance in bore. If it exceeds .004" on OHV6 engines or .003" on V8 and OHV4 engines, replace the whole pump assembly.
6. Remove relief valve and spring. Remove any burrs and clean. Be sure to install with bevelled or pointed end in seat. Check that valve moves freely in bore.

7. Check body and gear clearance. This is the distance between the pump gears and the pump cover, except on the PT-6 where it is the distance the gears protrude beyond the pump body. Adjustment of this clearance is made by the addition or removal of cover gaskets. On V(S)-401, 478, 549 engine oil pumps the clearance must be .0015-.009". On the OHV4 and V-304, 345, 392 engines the clearance is .0015-.006". On all OHV6 except the PT-6 (which is .000-.004" *above* pump body) the clearance is .0025-.0055".

On the V 537, 605 engines, the body to gear clearance is .0007 to .0027 inch, while the clearance on the MV 404, 446 engine oil pumps are .0014 to .0054 inch.

8. When installing drive gears on pump shaft be sure that the correct drive gear to pump body clearance is obtained. On RD engines the clearance is .025-.035". On OHV4 and V-304, 345, 392 engines the oil pump shaft sleeve is crimped onto the shaft. On the OHV4 the assembly dimension is .200" and on the V-304, 345, 392 engines the assembly dimension is .375". On V(S)-401, 478, 549 engines the distance "A" is $2\frac{7}{8}$" for front-mounted distributors, 1-15/16" for rear-mounted distributors.

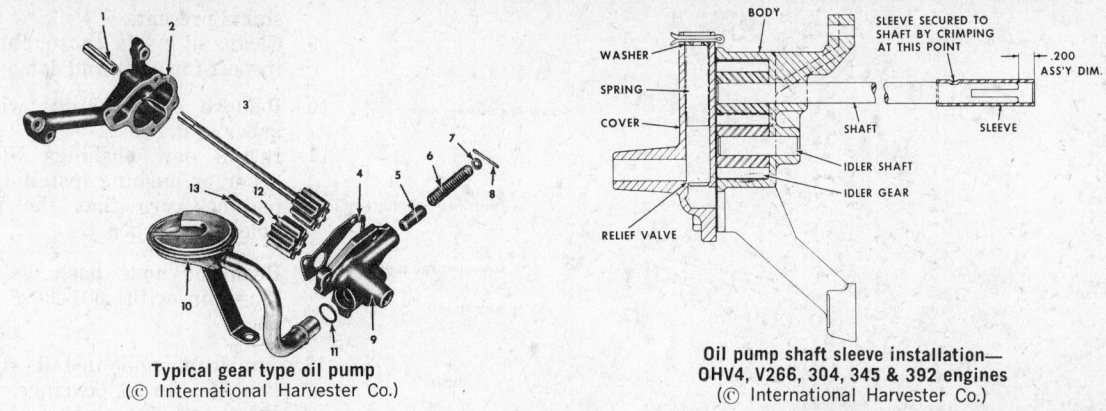

Typical gear type oil pump
(© International Harvester Co.)

Oil pump shaft sleeve installation—
OHV4, V266, 304, 345 & 392 engines
(© International Harvester Co.)

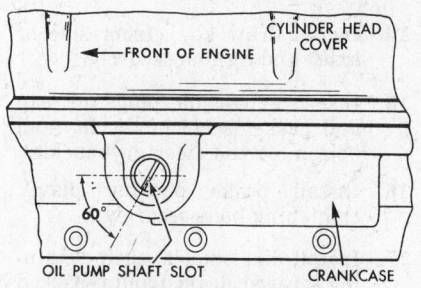

Oil pump installation—RD engines
(© International Harvester Co.)

Oil filter bypass valve & spring—
PT-6 engines
(© International Harvester Co.)

Lubrication System Priming

The recommended procedure to prime the internal parts and the oil pump is to attach a bearing leak detector or similar tool to a suitable fitting on the oil gallery, located on the left side of the engine block. (right side on the PT-6 and 400 engines). Inject enough oil into the engine to fill the oil filter and the various passage ways for the lubrication system. Disconnect the primary coil wire and turn the engine over, while the priming operation is in process. Do not overfill the crankcase when this method is used. This type of priming will minimize the possibility of scuffing or heat build-up in the areas of friction, which could cause premature engine failure.

Front Suspension

Front I-Beam Suspension

Shock Absorber Removal And Installation

1. Raise the vehicle and support safely.
2. Remove the retaining nuts and washers from the upper and lower attaching bolts.
3. Remove the shock absorber from the bolts.
4. Install the new shock absorber on the bolts and position the rubber grommets at the shock absorber eyes.
5. Install the retaining nuts and washers on the upper and lower attaching bolts and tighten securely.
6. Lower the vehicle.

Spring Removal

1. Disconnect shock absorber at lower mount.
2. Raise front of vehicle just enough to take weight off spring.
3. Unbolt U-bolts, tapered caster wedge and U-bolt.
4. Remove lube fittings from spring mountings.
5. Remove spring pins and spring.
6. To install, reverse the above procedure, mounting fixed end of spring first.

King Pin and Bushings Replacement

1. Remove spindle nuts and spindle bearing retaining nuts.
2. Remove wheels, inner bearings and grease retainers from spindles.
3. Remove dirt shields.
4. Remove bolts holding backing plates and place backing plate assemblies over ends of axle I-beam.

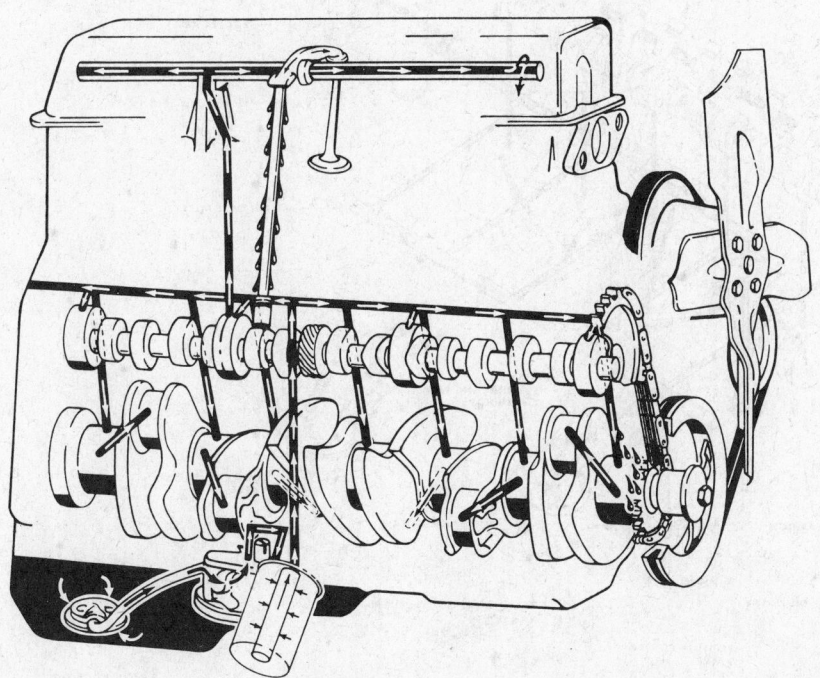

Engine lubrication system PT-6 engines (© International Harvester Co.)

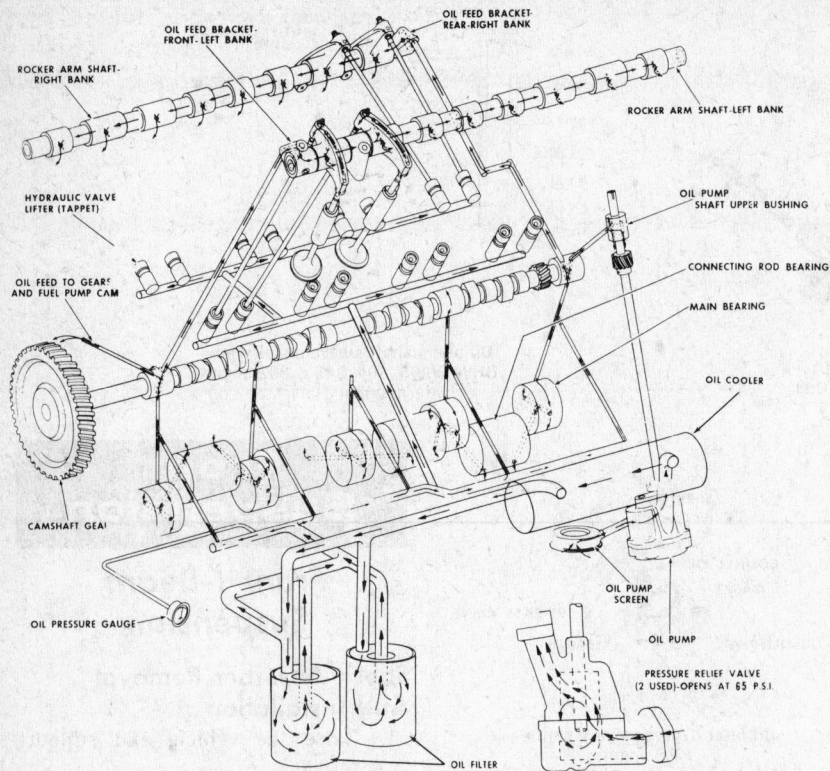

Lubricating system circuits—V 537, 605 engines (© International Harvester Co.)

shims present.

9. Clean all parts thoroughly and inspect for wear and damage.

10. Remove old bushings with an arbor or drift.

11. Install new bushings with an arbor or bushing installing tool, making sure that the grease holes are aligned.

12. Ream or hone bushings to fit king pin with .001-.002" clearance.

13. Lubricate and install steering knuckle, thrust bearings, spacer shims and king pins.

14. Install draw key (front side of axle) and tighten securely.

15. Insert expansion plugs or cap and gasket seals in the top and bottom of the steering knuckles.

16. Install brake backing plates, tightening bolts securely.

17. Install dirt shields, their retaining screws, cleaned and repacked wheel bearings and new grease seals.

18. Install wheel and spindle nuts, rotating wheel while tightening nut until slight drag is felt. Back off to the first castellation and install new cotter pin.

19. Lubricate and check and align front wheels if necessary.

5. Remove tapered draw keys holding the knuckle pins.

6. Remove expansion plugs or cap and gasket from the top and bottom of steering knuckles. (Remove expansion plugs by drilling a hole in one of the plugs and driving king pin with a punch to remove the other).

7. Drive out king pin.

8. Remove steering knuckles, thrust bearings and any spacer

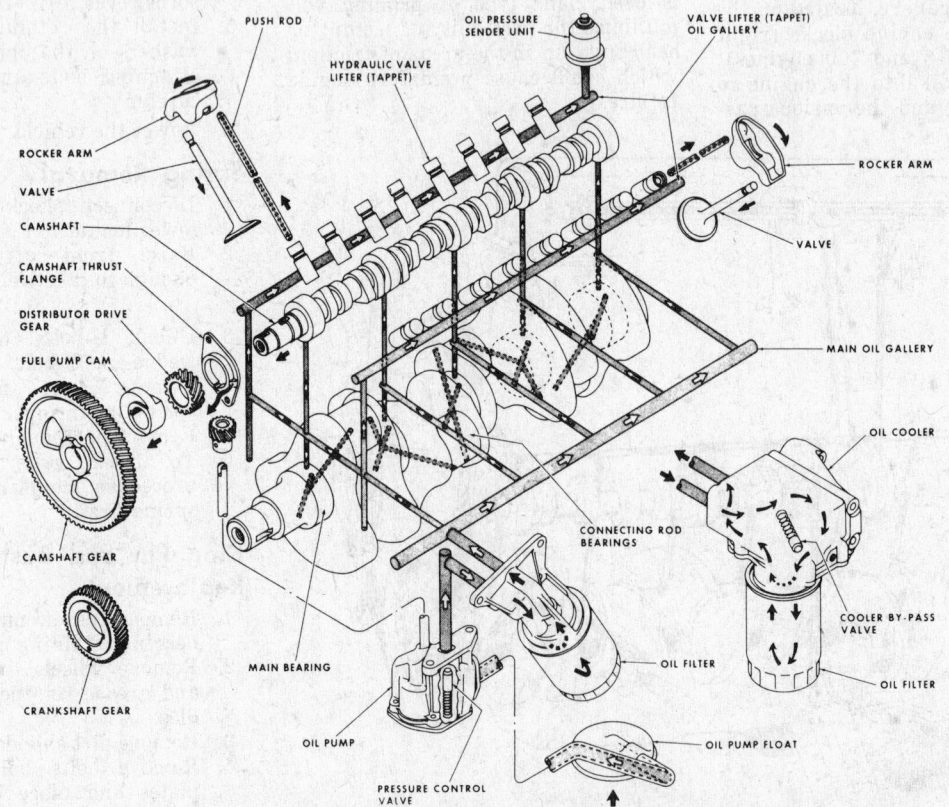

Lubricating system circuits—MV 404, 446 engines (© International Harvester Co.)

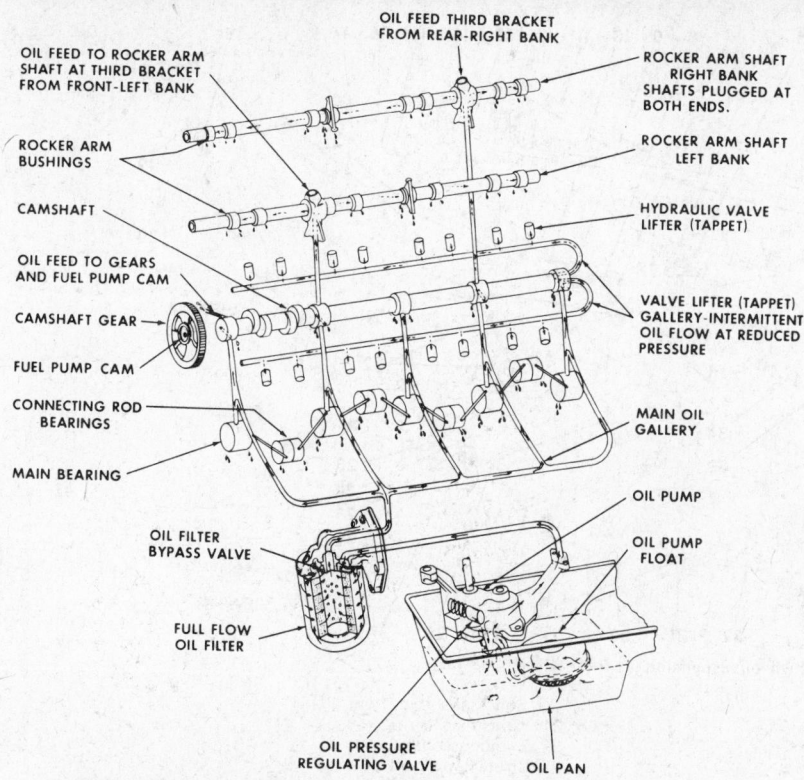

OIL FEED THIRD BRACKET FROM REAR-RIGHT BANK

OIL FEED TO ROCKER ARM SHAFT AT THIRD BRACKET FROM FRONT-LEFT BANK

ROCKER ARM SHAFT RIGHT BANK SHAFTS PLUGGED AT BOTH ENDS.

ROCKER ARM BUSHINGS

ROCKER ARM SHAFT LEFT BANK

CAMSHAFT

HYDRAULIC VALVE LIFTER (TAPPET)

OIL FEED TO GEARS AND FUEL PUMP CAM

CAMSHAFT GEAR

VALVE LIFTER (TAPPET) GALLERY-INTERMITTENT OIL FLOW AT REDUCED PRESSURE

FUEL PUMP CAM

CONNECTING ROD BEARINGS

MAIN BEARING

MAIN OIL GALLERY

OIL PUMP

OIL FILTER BYPASS VALVE

OIL PUMP FLOAT

FULL FLOW OIL FILTER

OIL PRESSURE REGULATING VALVE

OIL PAN

Lubricating system circuits—V 304, 345, 392 engines. Typical of MS 401, 478, 549 engines (© International Harvester Co.)

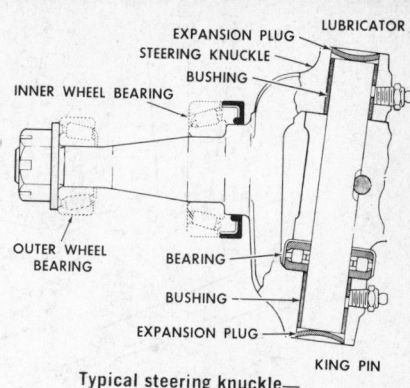

EXPANSION PLUG   LUBRICATOR
STEERING KNUCKLE
BUSHING
INNER WHEEL BEARING
OUTER WHEEL BEARING
BEARING
BUSHING
EXPANSION PLUG
KING PIN

Typical steering knuckle— expansion type seal plugs (© International Harvester Co.)

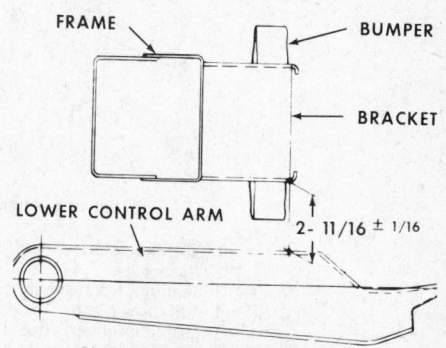

FRAME   BUMPER
BRACKET
LOWER CONTROL ARM   2- 11/16 ± 1/16

Checking front suspension height. (© International Harvester Co.)

Torsion Bar Front Suspension

Shock Absorber Removal and Installation

1. Remove the shock absorber upper retaining nut, washer, and rubber grommet from the top of the upper control arm.
2. Raise the vehicle and support it in such a manner so as not to cover the bottom of the lower arm with a floor jack or jack stands.
3. Remove the two shock absorber retaining bolts from the bottom of the lower control arm and withdraw the shock absorber.
4. To install new shock absorber, position the washer and rubber grommet on the extended shock absorber rod, and position the shock absorber up through the lower control arm, and engage the hole in the upper control arm with the extended rod.
5. Install the two bottom retaining bolts and tighten the shock absorber securely to the lower control arm.
6. Carefully lower the vehicle, so as not to lose the shock absorber rod from the upper control arm hole.
7. Install the upper rubber grommet, washer, and nut on the rod and tighten until the rubber grommet squeezes out slightly.

NOTE: Follow the manufacturers recommendation concerning the installation of lock nuts or self-locking nuts.

Torsion Bar Removal

Right and left torsion bars are not interchangeable. The bars are marked with an "L" or "R" on one end and the bars should always be installed with the marked end towards the rear of the vehicle. There is an arrow indicating the direction of wind-up on the end of the bar.

1. Jack up the vehicle by the frame crossmember and release the load from the torsion bar by loosening the retainer lever adjusting bolt.
2. Remove retainer lever adjusting bolt and slide retainer lever from end of torsion bar.
3. Remove torsion bar by sliding it rearward. CAUTION: Do not nick or scratch torsion bars— this may create a fracture.

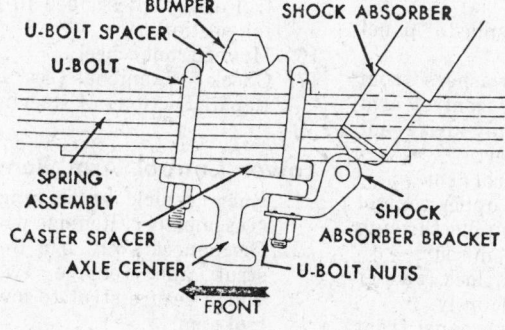

BUMPER
U-BOLT SPACER
SHOCK ABSORBER
U-BOLT
SPRING ASSEMBLY
SHOCK ABSORBER BRACKET
CASTER SPACER
AXLE CENTER
U-BOLT NUTS
FRONT

Typical front axle mounting (© International Harvester Co.)

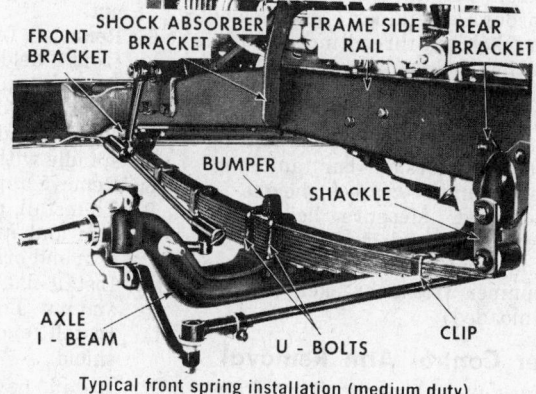

SHOCK ABSORBER BRACKET   FRAME SIDE RAIL   REAR BRACKET
FRONT BRACKET
BUMPER
SHACKLE
AXLE I-BEAM
U-BOLTS
CLIP

Typical front spring installation (medium duty) (© International Harvester Co.)

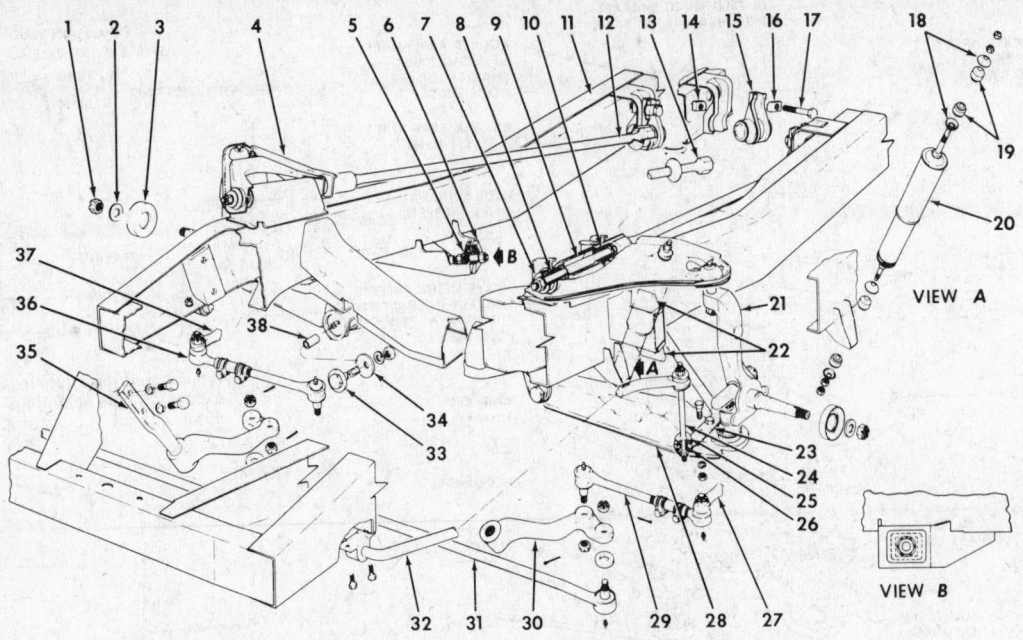

Torsion bar front wheel suspension (© International Harvester Co.)

1 Nut, hex., slotted	20 Shock absorber, front
2 Washer	21 Knuckle, steering
3 Seal, oil, front wheel	22 Bumper, control arm
4 Arm, upper, asm.	23 Strut, lower control arm
5 Cushion, rubber, strut	24 Link, sway bar
6 Washer, lower arm strut	25 Retainer, sway bar link
7 Washer	cushion
8 Bushing, upper control arm, front	26 Cushion, rubber, sway bar link
9 Spindle, upper control arm	27 Arm, steering, left
10 Bushing, upper control arm, rear	28 Arm, lower, asm.
11 Frame	29 Link, vertical, left (or tie rod)
12 Bar, torsion	30 Arm, pitman
13 Seal, torsion bar	31 Rod, tie, asm.
14 Nut, adjusting, torsion bar	32 Bar, sway
15 Lever, retainer, torsion bar	33 Bolt, hex.-hd.
16 Washer, adjusting, torsion bar	34 Cam, lower control arm
17 Bolt, hex.-hd.	35 Arm, idler, asm.
18 Washer, retaining	36 Link, vertical, right (or tie rod)
19 Cushion, rubber	37 Arm, steering, right
	38 Spacer, lower control arm

4. To install torsion bar, position torsion bar in upper control arm, observing right and left side and rearward direction as indicated above.

5. Install retainer lever on end of torsion bar and position bar nut in bracket on frame so that torsion bar adjusting bolt may be installed.

6. Insert bolt in bar washer, then through retainer lever and bracket and thread into bar nut.

7. Adjust height by lowering vehicle to ground (check for correct tire pressure), bouncing front end up and down, then turning bolt on torsion bar adjusting lever until correct height is achieved. Measure height between top of lower control arm and lower edge of rubber bumper frame bracket (vehicle unloaded).

Upper Control Arm Removal

1. Jack up vehicle by front frame crossmember until front wheels are off the ground.

2. Remove wheel and torsion bar (see preceding section).

3. Disconnect top mount of shock absorber.

4. Remove cotter pin, nut and dust seal (cut away seal) from lower ball joint.

5. Drive out lower ball joint stud (do not damage threads) or use special ball stud remover and nut.

6. Remove fender splash panel front shield.

7. Remove nut and washers from front end of upper control arm spindle and carefully drive out spindle with hammer.

8. Remove upper control arm.

9. To install, position upper control arm and install spindle through arm and bracket from rear.

10. Install flat washer, lock washer and nut. Tighten securely.

11. Install fender splash panel front shield.

12. Install new dust seal on ball stud, then line up shock absorber with hole in control arm and position ball stud into steering knuckle. Use jack to raise lower control arm until ball stud is well into steering knuckle.

13. Install nut on ball stud. Tighten securely.

14. Install top mounting of shock absorber, tightening just enough to squash rubber cushion slightly.

15. Install torsion bar on upper control arm as described in preceding section.

16. Mount front wheel.

17. Check alignment (see General Repair Section) of steering.

Lower Control Arm Removal

1. Raise vehicle by jacking frame crossmember. Remove wheel.

2. Disconnect sway bar link from strut and remove two bolts which secure strut to lower control arm.

3. Cut away dust seal from lower ball stud and remove cotter pin and nut from lower ball stud.

SPECIFICATIONS

IH MODEL	FA1 FA10	FA12 FA28 FA48	FA68 FA72	FA91 FA98 FA99	FA69
IH CODE	02001 02010	02012 02028 02048	02068 02072	02091 02098 02099	02069
Knuckle Pin Bushing Diameter (Inch)	.8625 .8615	1.111 1.110	1.236 1.235	1.236 1.235	1.236 1.235
Knuckle Pin Diameter (Inch)	.861 .860	1.110 1.109	1.234 1.233	1.234 1.233	1.234 1.233
Knuckle Pin Length (Inch)	5-7/16	6-1/4	6-3/4	7-21/32	6-3/4
Steering Knuckle Spindle Diameter At Inner Bearing (Inch)	Early Prod. 1.312 1.311 Late Prod. 1.374 1.371	Early Prod. 1.562 1.561 Late Prod. 1.687 1.686	1.749 1.748	2.1248 2.1240	1.874 1.873
At Outer Bearing (Inch)	Early Prod. .812 .811 Late Prod. .843 .842	.937* .936	1.124 1.123	1.312 1.311	1.124 1.123

*FA12 .9998
        .9988

SPECIFICATIONS

IH Model	FA-71	FA-73	FA-74	FA-101	FA-103
IH Code	02071	02073	02074	02101	02103
Knuckle Pin Bushing Diameter (inch)	1.360	1.360	1.360	1.360	1.360
Knuckle Pin Diameter (inch)	1.358	1.358	1.358	1.358	1.358
Knuckle Pin Length (inch)	7.97	7.97	7.97	7.97	7.97
Steering Knuckle Spindle Diameter: At Inner Bearing (inch)	1.967	1.967	1.967	2.164	2.164
At Outer Bearing (inch)	1.374	1.374	1.374	1.374	1.374

4. Either drive out ball stud with a soft hammer while supporting control arm or use special ball stud remover and nut.
5. Disconnect tie rod end from either side of vehicle.
6. Remove nut, lockwasher, cam and bolt from lower control arm and frame bracket.
7. Remove control arm and spacer in bushing.
8. To install, place new dust seal on ball stud and position ball stud into steering knuckle.
9. Tighten nut on ball stud and install cotter pin.
10. Position spacer in bushing and, while holding control arm in position, install bolt from front. Install cam, lockwasher, and nut, tightening to 81-135 ft. lbs. torque.
11. Mount strut to lower control arm tightening bolts securely.
12. Connect tie rod end.
13. Position cushion and retainer on sway bar link and place sway

bar link into strut with cushion, retainer and nut. Tighten nut until cushion is slightly squished. Insert cotter pin.
14. Mount front wheel.
15. Check alignment (see General Repair Section) and tighten nut on strut to 120-150 ft. lbs. torque and camber adjusting bolt nut to 81-135 ft. lbs. torque.

Ball Joint Inspection

Upper Ball Joint

The upper ball joint is a loose fit when not connected to the steering knuckle.

1. Use a floor jack or position the vehicle on a frame contact lift and raise the vehicle until the wheels fall to the full down position.
2. Grasp the tire at the top and bottom and move the tire in and out. The radial end play should not exceed .180 inch. If so, replace the ball joint.

Lower Ball Joint

The lower ball joint is spring loaded in its socket and this minimizes looseness and compensates for normal wear.

1. Locate a floor jack or position the vehicle on a frame contact lift and raise the vehicle until the wheels fall to the full down position.
2. Grasp the tire at the top and bottom and move the tire in and out. Any movement at the ball joint socket and stud indicates wear and the loss of preload, and the ball joint should be replaced.

Ball Joint Removal and Installation

Refer to "Independent Front Suspension-Coil Spring," for the procedures necessary to remove and install the ball joints.

Independent Front Suspension— Coil Spring

Shock Absorber Removal and Installation

1. Remove the shock absorber upper retaining nut, washer, and rubber grommet from the top of the upper control arm.
2. Raise the vehicle and support it in such a manner so as not to cover the bottom of the lower arm with a floor jack or jack stands.
3. Remove the two shock absorber retaining bolts from the bottom of the lower control arm and withdraw the shock absorber.
4. To install new shock absorber, position the washer and rubber grommet on the extended shock absorber rod, and position the shock absorber up through the lower control arm, and engage the hole in the upper control arm with the extended rod.
5. Install the two bottom retaining bolts and tighten the shock absorber securely to the lower control arm.
6. Carefully lower the vehicle, so as not to lose the shock absorber rod from the upper control arm hole.
7. Install the upper rubber grommet, washer, and nut on the rod and tighten until the rubber grommet squeezes out slightly.

NOTE: Follow the manufacturers recommendation concerning the installation of lock nuts or self-locking nuts.

Ball Joint Inspection— Coil Spring Suspension

Upper Ball Joint

The upper ball joint stud is spring loaded in its socket and this minimizes looseness and compensates for normal wear.

1. Locate a floor jack under the lower control arm on the outboard side and raise the vehicle so that the wheels clear the floor.
2. Grasp the tire at the top and bottom and move the tire in and out. If any perceptible lateral or vertical movement is noted, the ball joint should be replaced.

Lower Ball Joint

The lower ball joints are a loose fit when not connected to the steering knuckle.
1. Locate a floor jack under the lower control arm on the outboard side and raise the vehicle until the wheels clear the floor.
2. Grasp the tire at the top and bottom and move the tire in and out. The radial play should not exceed .250 inch. If so, the ball joint should be replaced.

Coil Spring Removal and Installation

1. Raise the front of the vehicle and support safely on floor stands.
2. Remove the wheels and tires.
3. Remove the caliper from the rotor assembly and support to avoid damage to the brake hose.
4. Remove the hub assembly from the steering knuckle.
5. Remove the brake shield from the knuckle.
6. Disconnect the sway bar link, if equipped.
7. Remove the axle bumper and wheel stop bracket from the lower control arm.
   NOTE: This also disconnects the lower control arm rod assembly from the control arm.
8. Remove the shock absorber.
9. Position the spring compressor screw into the shock absorber upper mounting hole in the crossmember. Position the puller hooks under the lower second spring coil and turn the spring compressor screw until coil spring unseats from the lower control arm.
10. Remove the cotter pins from the upper and lower ball joint studs, and loosen the lower nut approximately two turns.
    NOTE: The cotter pin is removed from the upper ball joint stud to allow a special tool to be placed over the upper stud and nut.
11. With the aid of special tool SE-2493 (ball joint stud remover) or its equivalent, and by placing it over the upper stud at its hex end, extend the screw to contact the lower ball joint stud.
12. Apply pressure by turning the screw out from the tool. Tap the steering knuckle lightly to loosen the stud from the knuckle.

13. Remove the tool and remove the nut from the lower ball joint stud. Separate the ball joint stud from the knuckle.
14. Loosen the spring compressor and relieve the spring of all tension. Remove the spring from the vehicle.
15. Position new spring into the crossmember and the lower control arm.
    Note: Turn the coil spring to line up the bottom coil with the seal groove in the lower control arm.
16. Install the hooks of the spring compressor under the second coil of the spring. Tighten the compressor until the lower ball joint stud can be installed into the steering knuckle.
17. Position a hydraulic jack under the lower control arm and raise the arm until the bottom ball joint stud will enter the steering knuckle. Install the nut and torque to specifications.
18. Install the cotter pins in both the upper and lower ball joint studs. Confirm the position of the spring in the lower arm.
19. Remove the spring compressor tool and install the shock absorber.
20. Install the lower control arm rod assembly and axle bumper and wheel stop bracket to the lower control arm.
21. Install the brake shield on the steering knuckle.
22. Install the rotor-hub assembly.
23. Install the caliper on the rotor assembly.
24. Install the wheel and tire assembly.
25. Remove the floor stands and lower the vehicle to the floor.
26. Depress the brake pedal to force the disc pads against the rotor.

Upper Control Arm and Ball Joint Removal and Installation

Follow the procedure outlined under "Coil Spring Removal And Installation," except that the special tool SE-2493 or its equivalent, is used to apply pressure to the upper ball joint stud after loosening the nut approximately two turns. Follow the procedure as outlined.
1. After removing the ball joint stud from the steering knuckle, remove the upper control arm from the frame rail.
2. Place the upper control arm in a vise and with a ball joint remover socket, remove the ball joint from the upper arm.
3. Lubricate the threads in the arm and place the new ball joint into the arm.
4. With the use of the ball joint socket, tighten the ball joint to

specifications in the upper arm.
5. Inner upper arm bushings may be replaced by pressing the old bushings out and pressing the new bushings into the upper arm.
6. Install the upper arm onto the frame rail and tighten securely.
7. Follow the procedure outlined under the "Coil Spring Removal and Installation" to complete the operation.

Lower Control Arm and Ball Joint Removal and Installation

Follow the procedure outlined under "Coil Spring Removal and Installation" to remove the coil spring from the suspension. After the coil spring is removed, follow this procedure to remove the lower control arm and ball joint.
1. After removing the lower ball joint stud from the steering knuckle, remove the lower control arm from the frame bracket.
2. Place the lower control arm in a vise.
3. Remove the ball joint from the lower control arm with the aid of tool SE-2494-2 ball joint remover and installer, or an equivalent tool.
4. Lubricate the threads in the arm and place a new ball joint into the control arm.
5. With the use of tool SE-2494-2 or equivalent, tighten the ball joint into the arm.
6. Inner lower arm bushings may be replaced by pressing the old bushings out and pressing the new bushings into place.
7. Install the lower arm onto the frame rail bracket and tighten securely.
8. Follow the procedure outlined under "Coil Spring Removal and Installation" to complete the operation.

Front Wheel Bearing Adjustment

(All Non-Driving Axles)

1. While rotating the wheel and hub assembly, adjust the spindle nut to 30 ft. lbs. (50 ft. lbs. on Loadstar and Cargostar), then back off nut 1/4 turn.
2. **Series 100, 200**—Finger tighten and insert lock so that cotter pin can be inserted with out backing off nut.
   Loadstar and Cargostar—If the lock or cotter pin can be installed at this position, do so, if not, tighten to the nearest locking position and insert the cotter pin or lock.
   NOTE: Bent type lockwashers must have one tab bent over the adjusting nut. With double locknuts,

tighten jam nut to 100-200 ft. lbs. and bend one tab of the lockwasher over the jam nut.

All other series—Finger tighten and if possible, insert cotter pin. If not able to install cotter pin, back off the nut to the nearest hole and insert the cotter pin.

NOTE: When using the cotter pin as a lock, the long tang should be bent over the spindle end. Clip the remaining tang, leaving enough stock to bend down against the side of the nut.

Front Drive Axle

For Service on Transfer Case and Differential—see General Repair Section.

Description

The front drive axles incorporate hypoid gears and use both spherical and ball joint wheel-end steering knuckles. The axle shaft assemblies are full floating and may be removed without disassembling the steering knuckles. Two types of axle shafts are used, one, a drive flange arrangement, bolted to the hub, and a second, mated to an internally splined gear, which in turn is splined and mated to the wheel hub, to transmit the driving torque to the front wheels.

Leaf Spring—Removal and Installation

1. Raise the vehicle and support on the frame rails behind the front springs with floor stands.
2. Remove the shock absorber from the spring.
3. Remove the U-bolts, spring bumpers and retainer, or the U-bolt seat.
4. Remove the lubricators, if used.
5. Remove the nuts from the shackles and bracket pins.
6. Slide the spring off the bracket and shackle pins.
7. Remove the spring from the vehicle.
8. Installation is the reverse of removal. Tighten all nuts and bolts securely.

Front Drive Axle Removal

1. Jack up truck until load is re-

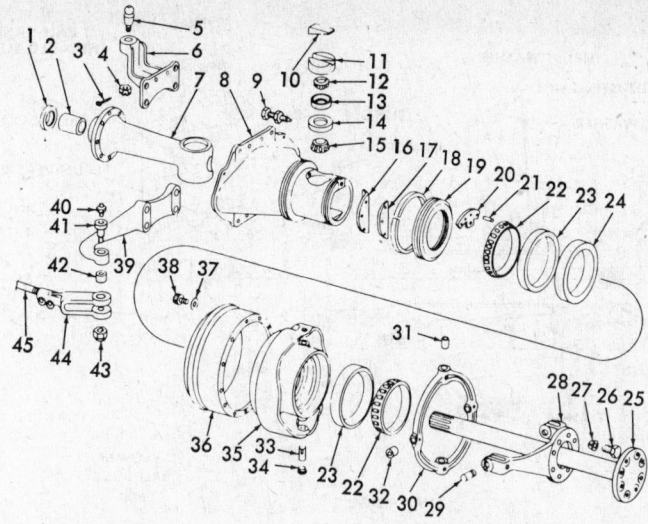

1 Seal axle shaft
2 Bushing, axle shaft
3 Pin, cotter
4 Nut
5 Ball, steering arm
6 Bracket, ball stud
7 End, stub
8 Spindle
9 Screw, adjusting
10 Wedge, adjusting
11 Cap, upper brg.
12 Bearing cone
13 Bearing cup
14 Bearing cup
15 Bearing cone
16 Seal
17 Plate, retaining
18 Seal, wheel
19 Nut, bearing adjusting
20 Plate lock
21 Pin, lock plate
22 Bearing
23 Bearing
24 Ring, clamp
25 Shaft, left axle
26 Bolt, hex-hd.
27 Dowel, shaft flange
28 Yoke, power
29 Pin, ring to yoke
30 Ring, compensating
31 Bushing, ring
32 Bushing, yoke
33 Pin, hub to ring
34 Plug, pipe
35 Hub
36 Drum
37 Washer, lock
38 Bolt, hex-hd.
39 Arm, steering
40 Lubricator
41 Bolt, tie-rod end
42 Bushing, steering
43 Nut, tie-rod end bolt
44 Yoke, tie-rod
45 Rod, tie

Front drive axle (drive gear type) (© International Harvester Co.)

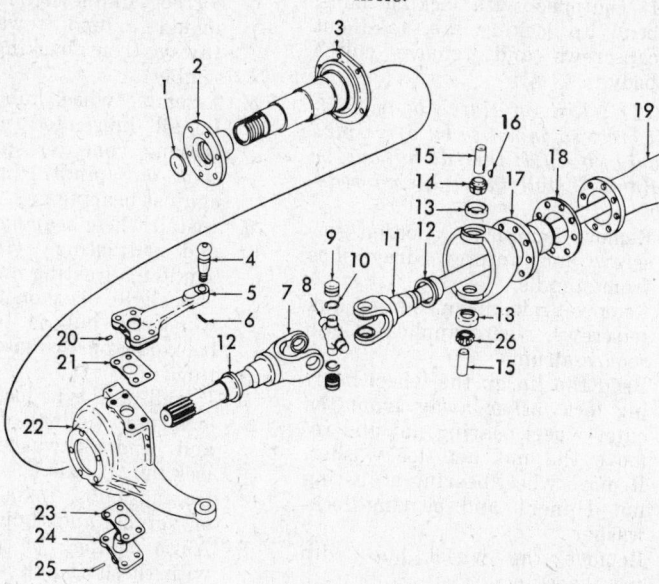

1 Plug, expansion
2 Flange, wheel drive
3 Knuckle, wheel
4 Ball, steering arm
5 Arm, steering
6 Pin, cotter
7 Shaft, axle outer
8 Spider
9 Bearing, trunnion
10 Ring, snap
11 Shaft, axle inner
12 Bushing, knuckle
13 Bearing trunnion
14 Bushing, steering knuckle
15 Pin, king cone
16 Key, woodruff
17 Yoke, trunnion
18 Gasket, yoke mounting
19 Housing, axle
20 Pin
21 Gasket
22 Knuckle, steering
23 Shim
24 Cap, king pin bearing
25 Pin
26 Cone, bearing

Front drive axle (© International Harvester Co.)

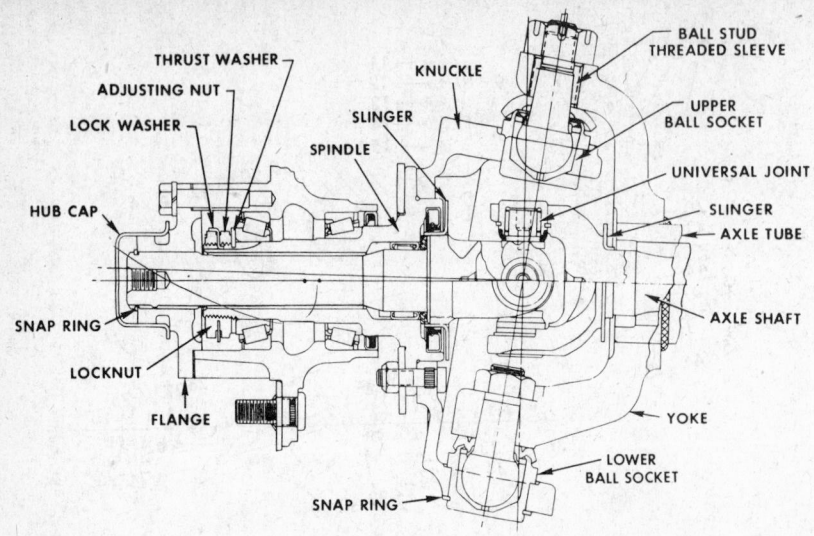

THRUST WASHER
ADJUSTING NUT
LOCK WASHER
SPINDLE
HUB CAP
SNAP RING
LOCKNUT
FLANGE
SNAP RING
SLINGER
KNUCKLE
BALL STUD THREADED SLEEVE
UPPER BALL SOCKET
UNIVERSAL JOINT
SLINGER
AXLE TUBE
AXLE SHAFT
YOKE
LOWER BALL SOCKET

Front drive axle with 40° steer (© International Harvester Co.)

moved from springs and block up frame to safely hold weight.
2. Drain lubricant from main housing and, if applicable, from wheel end housings.
3. Disconnect brakes.
4. Disconnect drag link from ball stud bracket.
5. Disconnect drive shaft from pinion shaft yoke.
6. Supporting axle with a portable floor jack, remove spring U-bolts.
7. Roll axle assembly out from under truck.
8. To install, reverse the above procedure.

Front Drive Axle Adjustments

Preload on the knuckle bearings of these front axles must be maintained at all times. Check for looseness each time knuckle is lubricated.
1. Jack up front end of truck until off-center weight of the wheel is relieved (wheel just barely touching ground).
2. Remove wheel and wheel adapter from hub.
3. Disconnect tie rod and drag link.
4. Remove axle shaft.
5. To remove play (check for play by pushing and pulling on top and bottom of knuckle) and increase preload drag, turn adjusting bolt into back of knuckle. Preload should read (spring scale hooked into end of steering arm) 12 lbs.

Front Wheel Bearing Adjustment

1. Remove wheel and adapter from hub.
2. Remove axle shaft or internal gear, and adjusting nut lock plate.
3. Tighten nut until just against bearing.
4. Rotate the wheel forward and

backward until a slight drag can be felt. Turn nut back to the first lock hole to obtain about a ½ hole relief.
5. Bearing adjustment is correct when no play can be felt when pushing and pull at top and bottom of wheel.

Reference for overhaul procedures see International Single Reduction Rear Axle in General Repair Section.

Axle Shaft and Universal Joint

(Axles having drive flange)

Removal

1. Raise vehicle, support with floor stands and remove wheel from vehicle.
2. Remove grease cap and snap ring from end of axle shaft.
3. Remove drive flange cap screws, lock-washer, flange and gasket. If equipped with locking hubs, bend up locking tab, take out capscrews and remove clutch body.

NOTE: Lift off clutch body holding it erect so as not to let drive pins fall out of body. If they do fall out, be certain to install them during reassembly.

Remove hub body. Loosen setscrew and unscrew drag shoe from spindle.
4. Remove brake drum countersunk setscrews, where applicable and remove drum.
5. Bend the lip on the wheel bearing lockwasher away from the outer wheel bearing nut and remove the nut and lockwasher. Remove wheel bearing adjusting nut (inner) and bearing lockwasher.
6. Remove the wheel hub with wheel bearing.
7. Remove backing plate and wheel spindle retaining bolts and lock-

washers. Support backing plate to prevent damage to brake hose if hose is not disconnected.
8. Remove wheel bearing spindle with bushing. If spindle bushing requires replacing, press out bushing using an adapter of correct size. An alternate method of bushing removal is the use of a cape chisel or punch to collapse the bushing.
9. Pull axle shaft and universal joint assembly out of axle housing.

Installation

1. Insert axle shaft and universal joint assembly into axle housing. Position splined end of axle shaft into differential pinion gear and push into place.
2. If wheel bearing spindle bushing was removed, press new bushing into spindle using an installer tool or adapter of proper size. Lubricate ID of bushing with chassis lube when installed to provide initial lubrication. Bushing should be pressed in until bushing flange is seated against shoulder in spindle. Assemble wheel spindle and backing plate to steering knuckle. Secure with six (6) bolts and lockwashers and tighten to specifications. Connect hydraulic brake fluid line if disconnected.
3. Pack wheel bearings using a pressure lubricator or by carefully working lubricant into bearing cones by hand. Slide lubricated inner wheel bearing on spindle until it stops against spindle shoulder.
4. Apply thin coating of lubricant specified for wheel bearings to seal lip and install seal into wheel hub using an adapter of correct diameter. Lip of seal should extend towards wheel (away from backing plate assembly).
5. Assemble wheel hub on spindle. Install lubricated outer wheel bearing cone on spindle. Push cone on spindle until it rests against bearing cup.
6. Install wheel bearing lockwasher and adjusting (inner) nut. Tighten adjusting nut until there is a slight drag on the bearings when the hub is turned; then back-off approximately one-sixth turn.
7. Install tang-type lockwasher and lock nut (outer). Tighten nut and bend lockwasher tang over lock nut. If axle is equipped with locking hubs, install drag shoe on spindle and tighten setscrew.
8. Align splines of drive flange with those of axle shaft and secure drive flange and new gasket to wheel hub with capscrews and

lockwashers. Tighten capscrews securely. If equipped with locking hubs, lightly lubricate hub body and clutch using a light grade chassis lubricant and install new gasket, hub body, snap ring and hub clutch. Be certain that all drive pins are positioned in locking hub clutch when clutch is installed. Secure hub clutch to wheel hub with capscrews and lock. Tighten to specifications and bend tang over head of capscrew.

9. Install snap ring and grease cup if not equipped with locking hubs.
10. Assemble brake drum and wheels to wheel hub. Bleed and adjust brakes.

CAUTION: **Be certain that master cylinder is full of brake fluid after completing bleeding operation.**

(Axles having drive gear)

Removal

1. Raise and support vehicle with floor stands placed under frame rails. Remove wheel from vehicle.
2. Lightly tap alternately around edge of hub cap with hammer and screwdriver or similar tool until hub cap is removed.
3. If axle is equipped with locking hubs, remove the eight (8) socket-head setscrews securing hub clutch assembly to wheel hub assembly.

NOTE: Drive pins may fall out of hub clutch when separated from wheel hub assembly. Be certain to replace them during installation.

4. Remove retaining ring from wheel hub if equipped with locking hubs.
5. Remove snap ring from axle shaft.
6. Pull drive gear out of wheel hub. If difficulty is encountered in removing drive gear, obtain a screwdriver or similar tool having the end bent approximately 90° with the handle. Insert end of tool into groove in drive gear and withdraw gear. If necessary, move wheel alternately backward and forward to aid removal of gear.
7. Remove retaining ring and locking hub body, if so equipped.
8. Using Wheel Bearing Adjusting Nut Wrench, remove wheel bearing outer nut and slide lock ring off of axle shaft. Again using wrench, remove wheel bearing inner nut.
9. Pull drive gear spacer out of wheel hub.
10. Remove brake drum from wheel hub and slide wheel hub assembly off of spindle.

NOTE: Do not allow tapered roller bearings to drop on floor as bearings may be damaged.

11. Remove screws retaining grease guard to backing plate. Take off grease guard and gasket.
12. Remove the six (6) bolts securing wheel spindle and backing plate to steering knuckle. Pull spindle with bushing off of axle shaft. If spindle bushing requires replacing, press or drive out bushing using an adapter of correct size. An alternate method of bushing removal is the use of a cape chisel or punch to collapse the bushing.
13. Pull axle shaft and universal joint assembly out of axle housing.

Installation

1. Proceed with steps 1 through 5 *of Axle Shaft and Universal Joint Installation* (Axles having drive flange).
2. Insert drive gear spacer over spindle and against outer wheel bearing cup.
3. Position wheel bearing inner adjusting nut Wheel Bearing Adjusting Nut wrench with pin in nut extending toward handle end of wrench. Install nut on spindle and tighten until it is snug against outer wheel bearing; then loosen adjusting nut 1/4 turn. Align tang on adjusting nut lock ring with groove in wheel spindle. Slide ring on spindle and index pin on adjusting nut with hole in lock ring. If pin will not index with hole in lock ring, turn adjusting nut to the left (Loosen) until it will index.

NOTE: When attempting to index pin with hole in lock ring, turn nut very slightly since adjusting nut should be locked with first hole in lock ring past 1/4 turn lose. Position wheel bearing outer nut in adjusting nut wrench and install on spindle. Tighten nut securely.

4. Align splines on axle shaft and splines in wheel hub with those of drive gear. Insert drive gear on axle shaft. Push gear into hub until it rests against drive gear spacer.

NOTE: Groove on side of gear must be toward hub cap.

5. If axle is equipped with locking hubs, lightly lubricate locking hub body using a light grade chassis lubricant. Align splines and insert hub body into wheel hub.
6. Install snap ring on end of axle shaft.
7. Place retaining ring in groove in wheel hub, if equipped with locking hub.
8. If applicable, lightly grease hub clutch assembly using a light grade chassis lubricant. Be sure that all eight (8) drive pins are positioned in the locking hub clutch. Assemble hub clutch to hub body and secure with eight (8) socket head setscrews.
9. Position hub cap on wheel hub and lightly tap alternately around cap until flange is against edge of hub.
10. Assemble brake drum and wheel to wheel hub. Bleed and adjust brakes.

CAUTION: **Be certain that the master cylinder is full of brake fluid after bleeding operation.**

Steering Knuckle (Spherical)

Removal

1. Remove drag link at steering arm and tie-rod at steering knuckle.
2. Remove oil seal retaining bolts from inner flange of steering knuckle and remove oil seals.
3. Remove bolts and lockwashers securing king pin lower bearing cap. Remove bearing cap and shim pack. Retain shim pack for use during reassembly.
4. Remove capscrews or self-locking nuts, which ever is applicable, securing steering arm or upper bearing cap to steering knuckle.
5. Lift steering arm assembly and knuckle until bronze bearing cone will clear ball yoke. Separate steering knuckle from ball yoke.

NOTE: Do not allow lower tapered roller bearing cone to drop on floor during removal of steering knuckle.

6. Support steering knuckle and with a long brass drift, drive or press king pin out of bronze bearing cone.

NOTE: Be careful not to damage end of king pin during removal of cone.

Installation

1. Assemble steering arm to knuckle using original shim pack. Install self-locking nuts or capscrews and tighten securely.
2. Coat king pin and bronze bearing cone ID and OD with chassis lubricant to prevent galling. Align serrations of new bronze bearing cone with serrations of king pin and press cone on king pin.

NOTE: Make sure the cone is pressed all the way on or against the shoulder.

3. With bronze cone and tapered roller bearing pre-lubricated, place tapered roller bearing cone into cup at lower end of ball yoke. While retaining lower bearing cone in position, assemble steering knuckle to ball yoke.

Seat bronze cone into cup at upper end of ball yoke.

4. Lubricate lower king pin with chassis lubricant and using original shim pack, install lower bearing cap to knuckle securing with bolts and lockwasher. Tighten bearing cap bolts securely.

5. Assemble opposite knuckle proceeding with instructions similar to those outlined above.

6. Individually check knuckle bearing preload by placing a torque wrench on any one (1) of the steering arm or bearing cap bolts or nuts. Read the starting torque (not rotating torque). Remove or add shims at the lower bearing cap until the specified preload is obtained. Knuckle bearing preload should be checked without ball joint oil seal, drag link or tie-rod installed.

7. Assemble the knuckle oil seals with the split on top. Knuckle retainer plate must be adjacent to ball yoke; followed by rubber seal, felt seal and metal retainer. Install seal retainer bolts and tighten securely.

8. Connect tie-rod to steering knuckle and tighten nut. Connect drag link to steering arm ball. Install cotter keys.

Cleaning, Inspection

All parts of the wheel end assembly should be thoroughly cleaned and dried with compressed air or a lint-free clean cloth.

Inspect all parts for wear, cracks or other damage. Replace all oil seals, felts and gaskets to prevent lubricant leakage.

Steering Knuckle (Ball Joint)

Removal

1. With the vehicle safely supported, remove the wheel and brake drum.

2. Remove the backing plate and the spindle from the knuckle.

NOTE: If necessary, tap the spindle lightly with a soft hammer to loosen it from the knuckle bolts. The spindle oil seal, needle bearings, and bronze spacer can be removed and replaced at this time.

3. Remove the axle from the housing.

NOTE: The slingers can be removed from the axle by using pullers or tapping the axle through the slingers.

4. Disconnect and remove the tie rod from the steering arm.

5. Remove the cotter pin from the upper ball socket stud and remove the nut.

6. Remove the nut from the lower-ball socket stud and discard.

NOTE: This nut is of a special torque design and should only be used one time.

7. Remove the lower ball socket snap ring (used on 4 x 4 applications only), and unseat the upper and lower ball socket studs with a lead hammer or with a puller tool arrangement, to separate the knuckle from the yoke.

NOTE: If the upper ball socket stud remains in the yoke flange, remove it by striking it on the stud with a soft hammer.

8. With the aid of puller tools or a press and ram, remove the bottom ball socket.

9. Reverse the knuckle and remove the upper ball socket.

10. With the aid of a special socket, remove the threaded sleeve in the top flange of the yoke.

Installation

1. Assemble the lower ball socket into the knuckle with a press and ram or a puller tool arrangement, making sure that the ball socket is firmly seated against the knuckle. Install the snap ring on the 4 x 4 application.

2. Assemble the upper ball socket into the knuckle with a press and ram or a puller type tool arrangement, making sure that the ball socket is firmly seated against the knuckle. *NOTE: Use a .0015 inch feeler gauge blade between the socket and knuckle. The blade should not enter at the minimum area of contact.*

3. Install new threaded sleeve into the top flange of the yoke, leaving approximately two threads exposed.

4. Install the knuckle assembly to the yoke, using a new nut on the lower ball socket stud. Torque the lower nut to 80 ft. lbs.

5. With the use of a special socket, torque the threaded sleeve to 50 ft. lbs. in the upper yoke flange.

6. Install the top ball socket stud nut and torque to 100 ft. lbs. Align the cotterpin holes between the stud and the castellated nut. Do not loosen nut to align the holes. Install the cotter pin.

7. Assemble the tie rod to the steering arm.

8. Assure that slingers are properly installed on the axle shaft and install the shaft into the housing.

9. Position the spindle over the axle end with the bronze bushing in place.

10. Install the backing plate and torque the nuts to 30 ft. lbs.

11. Install the hub and wheel assembly, and lower the vehicle.

Checking Ball Sockets For Looseness

To check the ball sockets for excessive looseness, raise the vehicle and attach a dial indicator to the lower yoke or axle tube and set the indicator against the knuckle or lower ball socket, with a loaded pressure so as to read in both directions. Grasp the wheel at the top and bottom and move the wheel inward and outward. If the total indicator reading exceeds .020 inch, both the upper and lower ball sockets should be replaced.

Front Drive Locking Hubs

Two types of locking hubs are

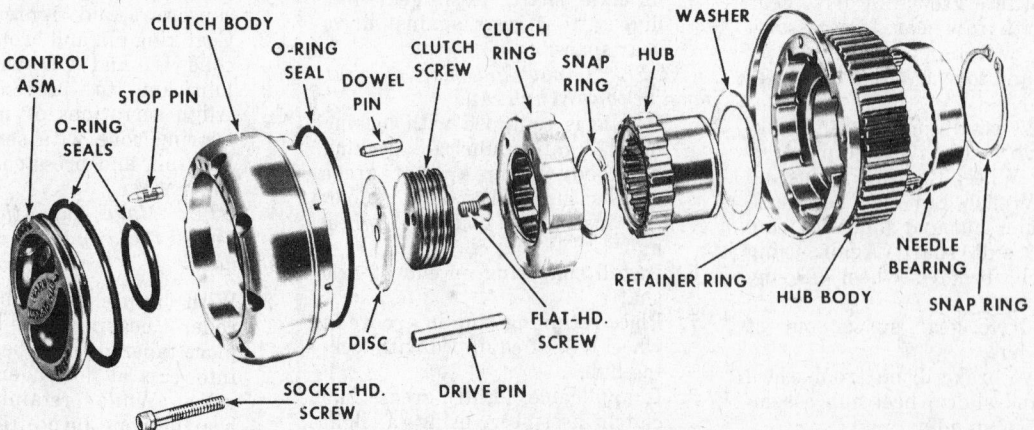

Exploded view—manual type locking hub (© International Harvester Co.)

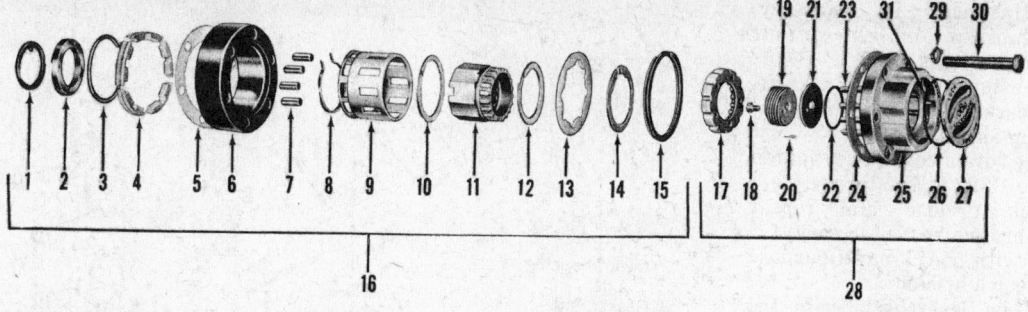

1 Washer, spindle lock	9 Cage, roller	17 Ring, clutch	25 Body, clutch
2 Shoe, drag	10 Ring, lock	18 Screw, flat head	26 "U" ring, oil seal
3 Spring, friction shoe	11 Hub, axle shaft	19 Screw, clutch	27 Control, assembly
4 Shoe, friction	12 Ring, lock	20 Pin, dowel	28 Body, clutch assembly
5 Gasket	13 Washer, thrust	21 Disc	29 Washer, lock
6 Body, hub	14 Ring, lock	22 "U" ring, oil seal	30 Bolt
7 Roller	15 Ring, lock	23 Pin, drive	31 Pin, stop
8 Spring, centering	16 Body, hub assembly	24 Gasket, clutch	

Exploded view—Lock-O-Matic hub (© International Harvester Co.)

used: manual and Lock-O-Matic. Manual locking hubs are either engaged or disengaged, depending on how they are set. Lock-O-Matic hubs, when in "free" position, automatically engage axle and wheel when forward torque is applied by the axle shaft. Thus, whenever front wheel drive is disengaged at the transmission, the wheels free wheel. "Lock" position is required only when engine braking control on the front wheel is desired.

Front Locking Hubs Removal

1. Bend up tabs on mounting bolt lock washers.
2. Remove six (eight) mounting bolts using a thin-walled socket or appropriate hex wrench (externally splined type).
3. When clutch body is lifted off, immediately tilt it up so that the drive pins do not fall out.
4. Remove lock ring holding hub body onto axle shaft and pull off hub body.
5. Remove drag shoe (Lock-O-Matic only) from axle spindle by loosening hex-head set screw and unscrew drag shoe.
6. To install, reverse the above procedure.

Steering Linkage

See General Repair Section for steering alignment procedures. See specifications at the beginning of this section for steering alignment specifications.

Tie Rods

Tie rods are of three-piece construction: rod and two end assemblies. The end assemblies are threaded into the end of the tie rod and adjustment is made by turning them either in or out to shorten or lengthen the tie rod. When tightening the clamp it is important to make sure that the end assembly is threaded in far enough so that the clamping action of the clamp is right over the end pieces. Ball studs are integral in the end assemblies.

When disconnecting ball studs, loosen ball stud nut, then strike the nut with one hammer while another larger hammer is backing up the nut.

Steering Gear

For manual steering gear overhaul, see the Unit Repair Section.

Steering Gear Removal

1. Loosen collar clamp at bottom of steering wheel column.
2. Remove nut or loosen clamp bolt which secures steering arm to lever shaft, removing steering arm from lever shaft using a suitable puller if necessary.
3. Remove mounting bolts and steering gear assembly.
4. To install, reverse the above procedure, taking special care not to bind steering column if there is no universal joint.

Steering Gear Adjustment

Twin Stud Levershaft Type

1. Free steering gear of all load by disconnecting drag link from steering arm and loosening bracket clamp on steering gear jacket tube.
2. To adjust end play on cam (ball thrust bearings), loosen lock nut and adjusting screw, then unscrew four upper cover (steering column) bolts.
3. Remove (cut) or add shims, replacing and tightening down upper cover to test drag. Drag should be just slight enough so that steering wheel can be moved from lock to lock with one finger.
4. To adjust lever shaft cams for backlash, place steering wheel in middle (straight-forward) position and turn adjusting screw until very slight drag is felt through mid-position range.
5. Tighten lock nut and give final test for drag.
6. Install drag link onto levershaft and tighten clamp on steering gear jacket tube.

Triple Roller Steering Gear

1. To check cam preload, disconnect linkage from steering arm and turn steering wheel through entire range, noting lash area.
2. Check the preload in lash area. It should be 7-13 inch-pounds.

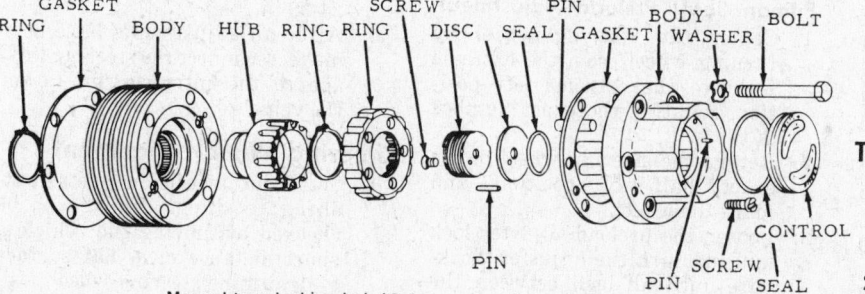

Manual type locking hub (© International Harvester Co.)

3. If adjustment is necessary, drain lubricant and remove four lower cover bolts.

4. Remove top shim with a knife, being careful not to mutilate remaining shims.

5. Replace lower cover and tighten bolts to 13-22 ft. lbs. torque.

6. Recheck preload and repeat above procedure if necessary.

7. Refill with SAE multi-purpose type gear lubricant.

8. To adjust levershaft, raise the vehicle and centralize the steering.

9. Rotate the steering 180° through center and check preload:
   S-161: 2-3⅜ lbs.
   S-108, S-165: 1 11/16-3¼ lbs.
   S-109: 2⅝-3⅞ lbs.

10. To adjust:
   a. Loosen lock nut.
   b. Turn slotted adjusting screw clockwise to increase preload; counterclockwise to decrease.
   c. when preload is within specifications, hold adjusting screw and tighten lock nut to 16-20 ft. lb.

Cam (Worm) and Roller Type

1. Disconnect linkage from pitman arm and drain lubricant.

2. Loosen locknut (1) at housing side cover, then turn adjusting screw (6) counterclockwise one turn to assure release of levershaft preload.

3. Turn steering wheel tube to about center of travel and check bearing preload (should register 9/16 to 1⅛ ft. lbs. @ 9″ radius).

4. If preload needs adjustment, drain lubricant and remove four housing cover bolts and housing.

5. Remove one shim.

6. Replace housing cover and bolts, torquing bolts to 18-22 ft. lbs.

7. Check preload and repeat the above procedure if necessary.

8. Adjust levershaft preload by turning adjusting screw (6). Correct levershaft preload is ¾ to 1¼ ft. lbs. @ 9″ radius (over cam preload).

9. Refill with SAE-90 SP type lubricant.

Recirculating Ball Type

Worm Bearing Adjustment

1. Raise the vehicle and disconnect the linkage from the pitman arm.

2. Loosen the adjuster lock nut and turn the worm bearing adjuster plug clockwise.

3. Using a spring scale attached to the steering wheel, pull at a right angle to the wheel spoke, and measure the pull required to keep the wheel moving.

4. Turn the adjuster plug until a pull of 14-18 in. lbs. is obtained

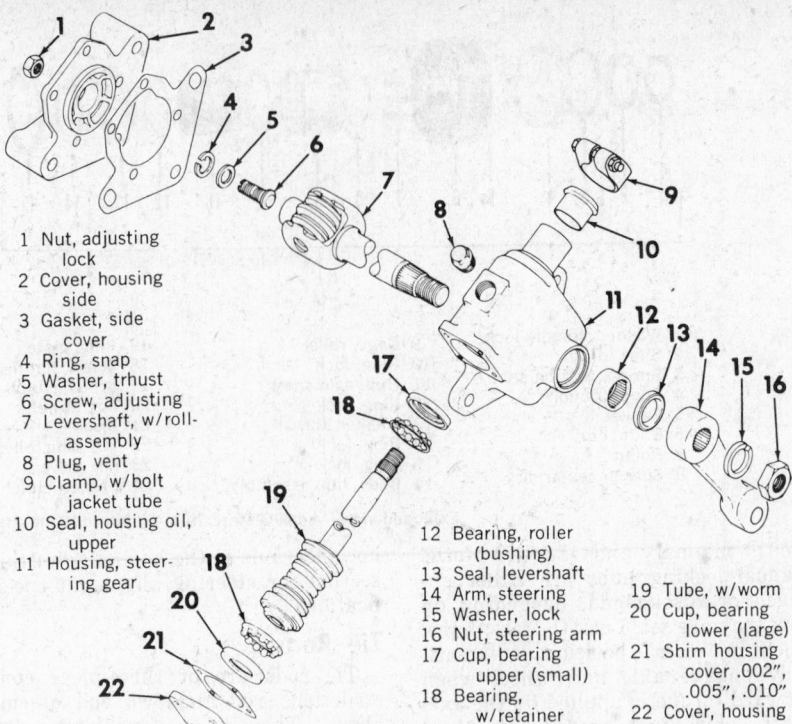

1 Nut, adjusting lock
2 Cover, housing side
3 Gasket, side cover
4 Ring, snap
5 Washer, thrust
6 Screw, adjusting
7 Levershaft, w/roll assembly
8 Plug, vent
9 Clamp, w/bolt jacket tube
10 Seal, housing oil, upper
11 Housing, steering gear
12 Bearing, roller (bushing)
13 Seal, levershaft
14 Arm, steering
15 Washer, lock
16 Nut, steering arm
17 Cup, bearing upper (small)
18 Bearing, w/retainer
19 Tube, w/worm
20 Cup, bearing lower (large)
21 Shim housing cover .002″, .005″, .010″
22 Cover, housing

Cam (worm) and roller type steering gear
© International Harvester Co.

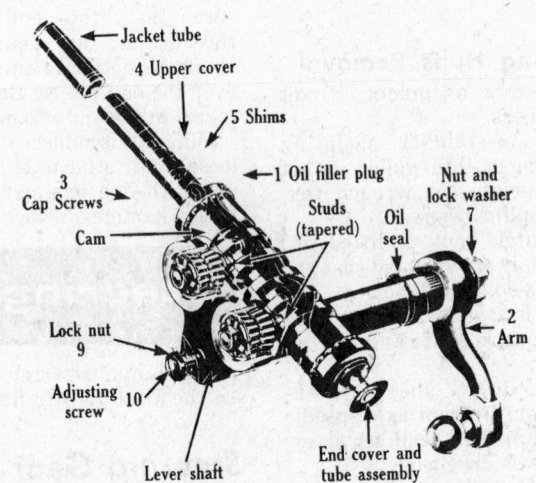

Twin stud levershaft steering gear © International Harvester Co.

on the spring scale.

5. Tighten the adjuster plug lock nut and recheck worm bearing preload. Readjust if necessary.

Pitman Shaft Preload Adjustment

1. Center the steering wheel by turning wheel from the extreme right to the extreme left position, counting the exact number of turns.

2. Return the steering wheel to the exact half-way position and mark the wheel.

3. Loosen the preload adjuster lock nut and turn the adjuster clockwise until all lash between the gears is removed.

4. Tighten the locknut and as outlined previously, with the aid of a spring scale, pull the steering wheel through the center position. The pull pressure should be 24-30 in. lbs.

5. After all adjustments have been made, reconnect the steering linkage to the pitman arm. Lower the vehicle.

Steering Wheel Alignment

1. Set front wheels in a straight-ahead position. This can be checked by driving the vehicle a short distance on a flat surface to determine steering wheel position when vehicle is following a

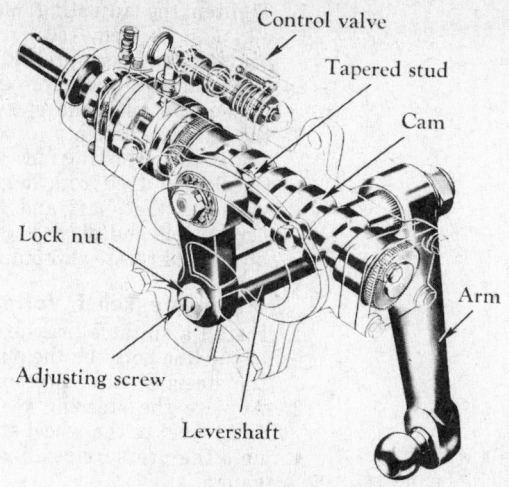

Power steering gear with toggle type
integral valve
(© International Harvester Co.)

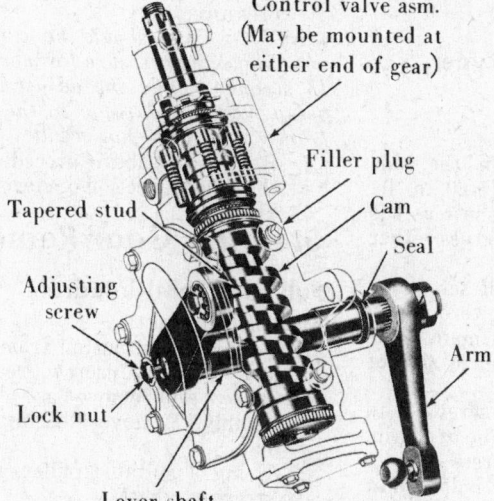

Model S-36 power steering gear
with integral concentric valve
(© International Harvester Co.)

1 Seal, housing upper oil	12 Plug, housing filler
2 Housing	13 Spacer, steering gear mounting
3 Bushing, housing	14 Levershaft
4 Seal, housing side oil	15 Screw, levershaft adjusting
5 Arm, steering	16 Washer, thrust
6 Bearing, upper cup	17 Ring, snap
7 Bearing, cone	18 Gasket, housing side cover
8 Tube, with cam	19 Cover, housing side
9 Bearing, lower cup	20 Nut, hex. jam
10 Shim, housing lower cover	21 Bracket, steering gear mounting
11 Cover, housing lower	

Triple roller steering gear
(© International Harvester Co.)

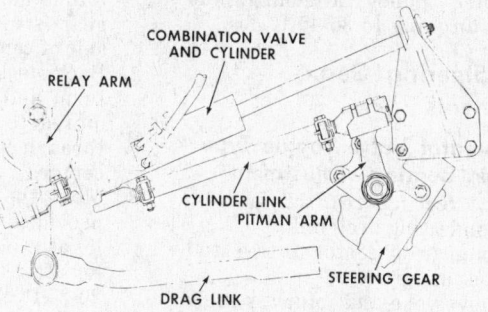

In-line type power steering booster
(© International Harvester Co.)

straight path.

2. Raise vehicle and check number of turns required from center point to extreme right and left. The number of turns should be the same in each direction.

3. If step 2 moves wheels off of straight-ahead, loosen adjusting sleeve clamps on both left and right-hand tie rods, then turn both sleeves an equal number of turns in the same direction to bring gear back on high point.

Power Steering Gear

For overhaul of power steering systems, see General Section.

In-Line Booster Removal

1. Disconnect and plug hydraulic lines from valve and cylinder unit.

2. Loosen clamp bolts and disconnect cylinder link from cylinder.

3. Remove nut and lockwasher from piston rod and remove rod from frame bracket.

4. Loosen clamp bolt and unscrew cylinder assembly from pivot on relay.

5. Installation is the reverse of the above procedure. Be sure to center steering wheel and wheels before tightening clamp bolts of cylinder link. Bleed hydraulic system.

Hydraulic Cylinder Removal

1. Disconnect hydraulic lines from cylinder and plug lines.

2. Unbolt piston rod from frame bracket, noting approximate position of clamp.

3. Disconnect cylinder assembly from steering linkage.

4. Installation is the reverse of the above procedure. Approximate original position of clamp on piston rod, center adjusting steering wheel if necessary.

Separate Control Valve Removal

1. Disconnect and plug hydraulic lines from control valve.

2. Loosen clamp bolts at each end of control valve and remove valve.

3. Installation is the reverse of removal. Since the control valve in this type of power steering system serves, in a sense, as a relay arm, it must be adjusted to center the steering wheel for straight-forward running. Tighten clamp bolts after adjustment is made.

Hydraulic Pump Removal

1. Disconnect hydraulic lines at pump. When hoses are disconnected, secure them in a raised position to prevent leakage. Plug fittings of pump.

2. Remove drive pulley attaching nut.

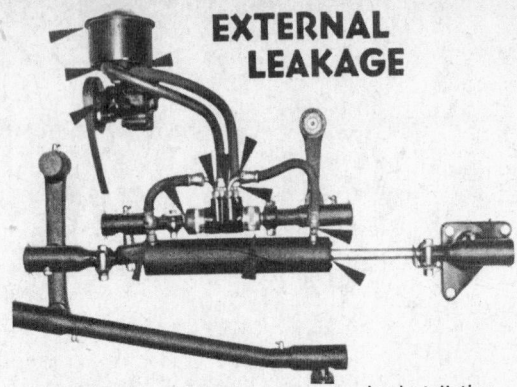

EXTERNAL LEAKAGE

Separate control valve type power steering installation
(© International Harvester Co.)

3. Loosen bracket-to-pump mounting bolts, and remove pump belt.
4. Slide pulley from shaft. **CAUTION: Do not hammer pulley off shaft as this will damage the pump.**
5. Remove bracket-to-pump bolts and take off pump assembly.
6. Installation is the reverse of the above removal procedure. Do not tighten mounting bolts or pulley nut until installation is complete. Move pump until belt is tight, then tighten mounting bolts. Tighten pulley attaching nut last, torquing to 35-45 ft. lbs.

Power Steering Gear Adjustments

Semi-Integral Valve Toggle Type Thrust Bearing Adjustment

1. Free steering gear of load by disconnecting drag link.
2. Turn gear off center to free stud in the cam groove.
3. Remove yoke and joint assembly from stub shaft.
4. Remove key from stub shaft.
5. Remove upper cover.
6. Reassemble actuator housing screws with a ⅜" thick spacer under each head to hold actuator and cam assembly in the gear.
7. Remove the adjusting nut, lock washer, thrust washer and thrust housing.
8. Clean threads on nut and camshaft so that nut can be run freely by hand.
9. Reassemble nut, washers, and bearing.
10. Tighten nut to 10 ft. lb., then back off 10-20°. Bend locknut.
11. Attach cover and other parts.

Stud in Cam Groove Adjustment

1. Tighten side cover adjusting screw until a very slight drag can be felt when turning through mid-position. If no drag is felt, remove shims until it can be felt. Back off adjusting nut 1/16 turn and lock.

2. If a drag can be felt without removing shims, it will be necessary to add shims, then remove enough shims to reestablish a very slight drag.
3. After adjustment, back off nut 1/16 turn and lock.

Fully Integral Valve Type Adjustment

Sector Shaft

1. Adjust the screw in the side cover to provide a 15-20 in. lb. torque at the input shaft as the gear is moved 90 degrees either side of center.
2. Back out the adjusting screw one turn and note the torque required to move the input shaft through 90 degrees each side of center.
3. Move the adjusting screw in to provide a rise in torque of 2-4 in. lb. at a point 45 degrees off center after the jam nut is locked at 20-25 ft. lbs.
NOTE: Input torque of the completely assembled gear unit less oil, should not exceed 15 in. lbs. over full travel of 95 degrees at the output shaft.

Valve Thrust Bearings

NOTE: The upper housing must be removed for this adjustment procedure.

1. Tighten the adjusting nut until all play is removed from the bearings and zero preload exists.
2. Back off the nut approximately 20 degrees and bend one tang of the lock washer into a matching slot in the adjusting nut.
3. Check for free rotation of the valve on the shaft and for any perceptible end play. The assembly should rotate at 3-5 in. lbs.

Pressure Relief Valve

1. Install a suitable pressure gauge in the line between the pump and the steering valve pressure port.
2. Actuate the steering to provide full travel to the wheel stops and note the pressure reading on the gauge.
3. Adjust the pressure relief screw in a clockwise direction to provide a pressure of at 400 psi below the maximum operating pressure.
NOTE: Care should be exercised not to hold the pressure for more than 15 seconds, while the adjustment is being made or damage to the pump from excess heat can result.

4. Repeat the above procedure for the other direction of steering.

Steering Gear Removal

Full and Semi-Integral Steering Gear

1. Remove horn button from steering wheel. Unscrew retaining screws and remove base plate assembly. Remove steering wheel nut.
2. Using a suitable puller, remove steering wheel.
NOTE: Where the steering column is the jointed type, the steering column and wheel need not be removed. Loosen nut on steering column shaft collar and remove steering column shaft from gear. Retain woodruff key.
3. Using suitable puller, remove pitman arm from levershaft.
4. Identify hydraulic connecting

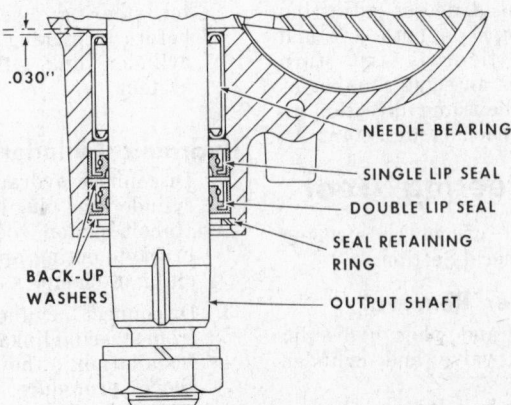

.030"

NEEDLE BEARING
SINGLE LIP SEAL
DOUBLE LIP SEAL
SEAL RETAINING RING
BACK-UP WASHERS
OUTPUT SHAFT

Correct installation of output shaft seals and bearing
S-281, 282
(© International Harvester Co.)

lines by tagging and marking the valve ports to which they are connected. Disconnect hydraulic lines from valve. Plug all openings.

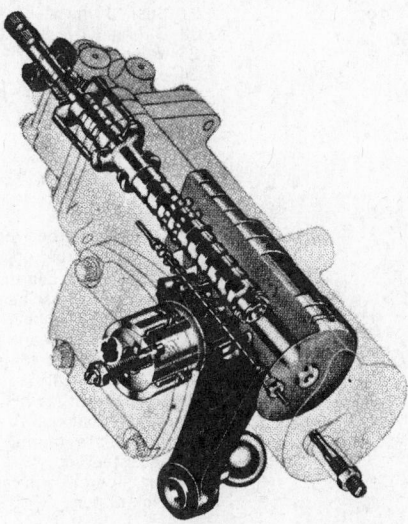

Types S-276, 299 Integral power steering gear

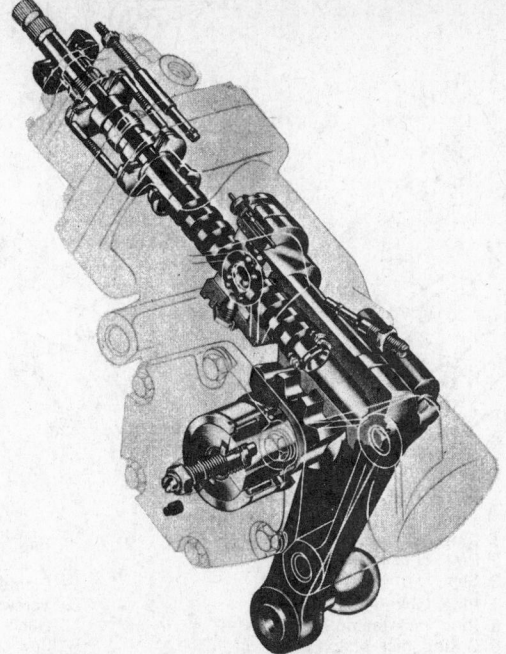

Types S-261, 277, 301, 302, 320, 321, 322, 323, 325 Integral power steering gear

5. Remove mounting flange bolts and remove gear from chassis.
6. To install, mount steering gear in chassis and fasten securely.
7. Place woodruff key in stud end of steering gear and install collar of steering column shaft. Secure with bolt and nut.
8. Center steering gear. Center steering wheel. Set front wheels straight ahead.
9. Connect drag link to ball on steering arm.
10. Install steering arm on levershaft of gear. If arm does not line up with splines of shaft, turn steering wheel to the right or left 1/4 turn until it does.
11. Secure arm to levershaft with lockwasher and nut. Tighten nut to 250 ft. lbs.

12. Install hydraulic lines to control valve.
13. Fill the steering gear housing with the specified lubricant on the semi-integral power steering units.
14. Fill the Power steering reservoir with the recommended fluid and bleed the system.

Output Shaft Preload Adjustment

Integral Rotary Valve Type

1. Position steering in the straight-ahead position. Check for lash by moving steering wheel. If there is steering wheel movement without moving the steering arm, over-center adjustment must be made.

2. Disconnect steering arm and remove horn ring.
3. Position steering wheel at center of travel, then turn 1/2 turn off center.
4. Using an inch pound torque

Rotary valve power steering gear with horizontal output shaft
(© International Harvester Co.)

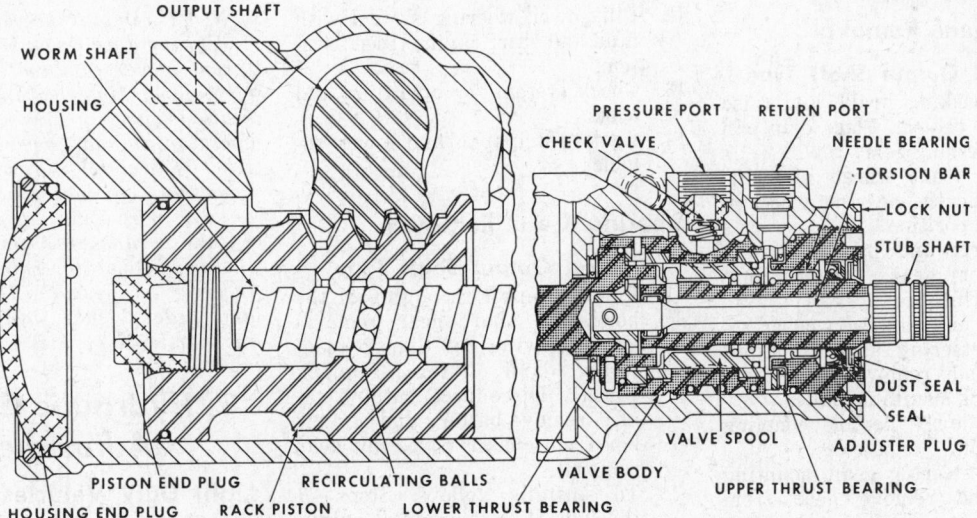

Rotary valve type power steering gear (© International Harvester Co.)

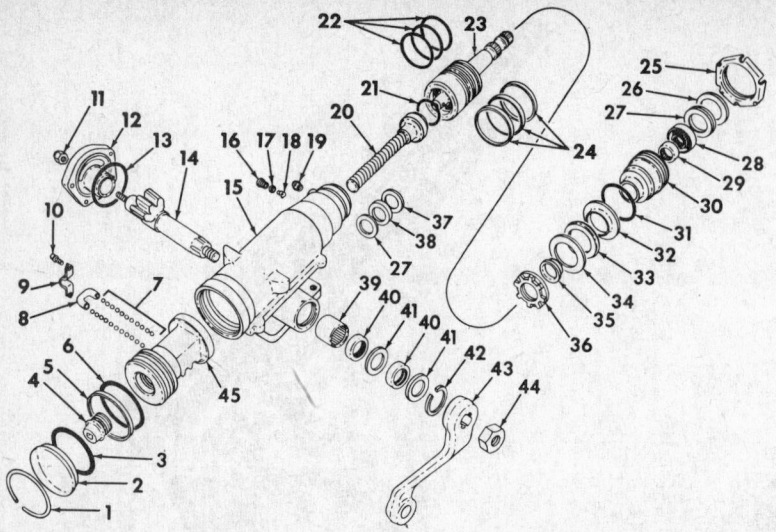

13 Seal, O-ring side cover
14 Shaft, output
15 Housing, gear
16 Seat, pressure connector
17 Valve, poppet check
18 Spring, poppet check valve
19 Seat, return connector
20 Shaft, worm
21 Seal, O-ring
22 O-Ring, valve body back-up
23 Valve, control
24 Ring, valve body
25 Nut, lock adjuster plug
26 Ring, retainer
27 Seal, dust
28 Seal, oil
29 Bearing, Needle
30 Plug, adjuster
31 Seal, O-ring
32 Race, thrust bearing, upper
33 Bearing, thrust upper
34 Race, thrust bearing, upper
35 Retainer, thrust bearing
36 Spacer, thrust bearing
37 Race, thrust bearing, lower
38 Bearing, thrust lower
39 Bearing, needle
40 Seal, output shaft
41 Washer, back-up
42 Ring, seal retaining
43 Arm, steering
44 Nut, steering arm retaining
45 Rack piston

1 Ring, retaining
2 Plug, housing end
3 Seal, O-ring
4 Plug, rack piston end
5 Ring, rack piston
6 O-Ring, rack piston backu-up

7 Ball, recirculating
8 Guide, ball return
9 Clamp, ball return guide
10 Screw, clamp
11 Nut, lock output shaft adjuster
12 Cover, housing side

Integral rotary valve type power steering gear (© International Harvester Co.)

wrench and socket on the steering wheel retaining nut, determine the torque required to rotate the shaft slowly through a 20-degree arc. Turn gear to center and take a second reading. If second torque reading is 4-8 inch pounds in excess of first reading, no preload adjustment is necessary.

5. If adjustment is required, loosen adjuster screw locknut and turn screw until second reading exceeds first reading by 4-8 inch pounds.

6. Tighten locknut while holding adjusting screw in place to 27-37 ft. lbs. torque.

7. Recheck torque reading after adjustment is made.

8. Install steering arm, tightening nut to 120-125 ft. lbs. torque.

9. Install horn ring.

Steering Gear Removal

Horizontal Output Shaft Type

1. Apply parking brake and raise front of vehicle. Place drip pan under steering gear.

2. Position front wheels straight-ahead and tie steering wheel in centered position.

3. Disconnect and plug hydraulic lines from gear. Tag the lines for identification.

4. Remove nut and lockwasher securing steering arm to output shaft and remove arm from shaft with a suitable puller.

5. Remove lower flexible coupling clamp bolt.

6. Remove steering gear mounting bolts and remove gear from chassis.

7. To install, set gear assembly on center and position gear in chassis.

8. Insert gear stub shaft into lower flexible coupling and install and tighten gear mounting bolts to 55-60 ft. lbs. torque.

9. Install and tighten lower flexible coupling bolt clamp to 30-35 ft. lbs. torque.

10. With steering wheel centered and front wheels straight ahead, place steering arm on output shaft by matching master serrations of arm with shaft. Secure arm to shaft with lockwasher and nut, tightening nut to 120-125 ft. lbs. torque. Untie steering wheel.

11. Remove plugs and connect hydraulic lines to proper ports, tightening connections to 20-30 ft. lbs. torque.

12. Fill power steering system with fluid and start engine. Bleed system.

13. Check system for operation and leaks.

14. Remove drip pan and lower vehicle.

Steering Gear Removal

Vertical Output Shaft Type

1. Follow Steps 1 through 6 of the horizontal shaft gear removal above, leaving out Step 2 (centering and tying steering wheel). Disconnect battery cables, remove battery and battery box from chassis to permit gear removal.

2. To install, follow Steps 7 through 11 of horizontal output

shaft gear installation described above.

3. Install battery box, battery and cables.

4. Follow Steps 12 through 14 of horizontal output shaft installation described above.

Power System Bleeding

1. Fill pump reservoir to correct level with fluid.

2. Start engine and turn steering wheel through entire travel two or three times. This will permit air to escape and be replaced with fluid.

3. Check fluid level and refill if necessary.

Clutch

NOTE: Due to the various medium and heavy duty truck models and different power combinations used, the clutch operating clearances and specifications are given in the owner-operator manual, accompaning each truck. The clutch release bearing may be actuated by mechanical, hydraulic, or a combination of air over hydraulic means, and for this reason the owner-operator manual should be referred to regarding the free travel and pedal height measurement for the particular model being serviced.

Hydraulic Clutch Adjustment

Light Duty Vehicles

1. Check clutch pedal height. If it

is not approximately 7¾" from the floorboard (measured at right angles from floorboard), loosen two bolts on the clutch pedal stop bracket and move bracket either way until proper pedal height is achieved. Tighten bracket mounting bolts.

2. Clutch pedal push rod to master cylinder piston clearance is adjusted by loosening the locknut on the pushrod and turning the rod either in or out until 3/16" pedal stroke is obtained before clutch pedal push rod contacts master cylinder. Tighten locknut on pushrod and recheck pedal stroke.

3. Release bearing to clutch lever (finger) clearance is adjusted at the slave cylinder pushrod. Measure stroke of clutch pedal required to produce contact of clutch release bearing to clutch release levers. If it is not 1⅞" ± ⅛", loosen locknut on slave cylinder pushrod and rotate pushrod either in or out until proper pedal travel is obtained. Tighten locknut on pushrod and recheck pedal free travel.

Medium and Heavy Duty Vehicles

1. Clutch pedal pushrod stroke adjustment is made by disconnecting clutch pedal from pushrod (remove yoke pin). Holding master cylinder pushrod out snugly against the stop in the

Hydraulic clutch master cylinder—1600 thru 1890 models (© International Harvester Co.)

Clutch and brake master cylinder installation (© International Harvester Co.)

end of the cylinder and making sure that the spring is holding the clutch pedal against its stop, adjust yoke on pushrod until yoke can be connected to pedal. Install yoke pin and cotter pin. Tighten locknut on yoke. A clutch pedal free travel of 5/16" should result.

2. Clutch finger to bearing clearance is adjusted at the slave cylinder pushrod. Clutch pedal should travel 1½" before clutch bearing makes contact with re-

lease fingers. Loosen locknut on slave cylinder pushrod and turn pushrod to obtain correct pedal travel. Tighten locknut and recheck pedal travel.

Mechanically Controlled Linkage or Cable Adjustment

Light Duty Vehicles

1. Measure and correct the clutch pedal height to approximately 9 inches. *NOTE: On some models it may be necessary to increase the clutch pedal height setting slightly over the amount specified, in order to obtain complete clutch release.*

2. Disconnect the return spring on release fork.

3. Loosen the nut on the cable or linkage rod.

4. Hold the pedal assembly against the pedal stop and lengthen or shorten the rod or cable to obtain zero clearance at the release bearing face and the pressure plate fingers.

5. After obtaining zero clerance, lengthen or shorten cable or linkage to obtain 3/32 inch between the bearing face and the fingers of the pressure plate.

6. Tighten nut on the cable or linkage rod.

7. Reconnect the return spring.

Medium and Heavy Duty Vehicles

External Adjustment

NOTE: Refer to the owner-operator manual for the correct free travel and pedal height for the particular model being serviced.

1. In general, the mechanically controlled clutch should have 1½ inch of free travel before the clutch begins to disengage.

2. Pedal clearance or free travel in the linkage must be sufficient to prevent the clutch from being partially disengaged. A clearance of ⅛ inch should be maintained between the yoke fingers and the release bearing wear pads.

3. Slave cylinder air assist units must have an average of ½ inch clutch release bearing travel for poper release.

Internal Adjustment

1. Remove the clutch housing inspection cover.

2. Inspect the running clearance between the release fork and the bearing housing for ⅛ inch clearance.

3. Inspect the clearance between the release bearing and the clutch brake (if equipped with a

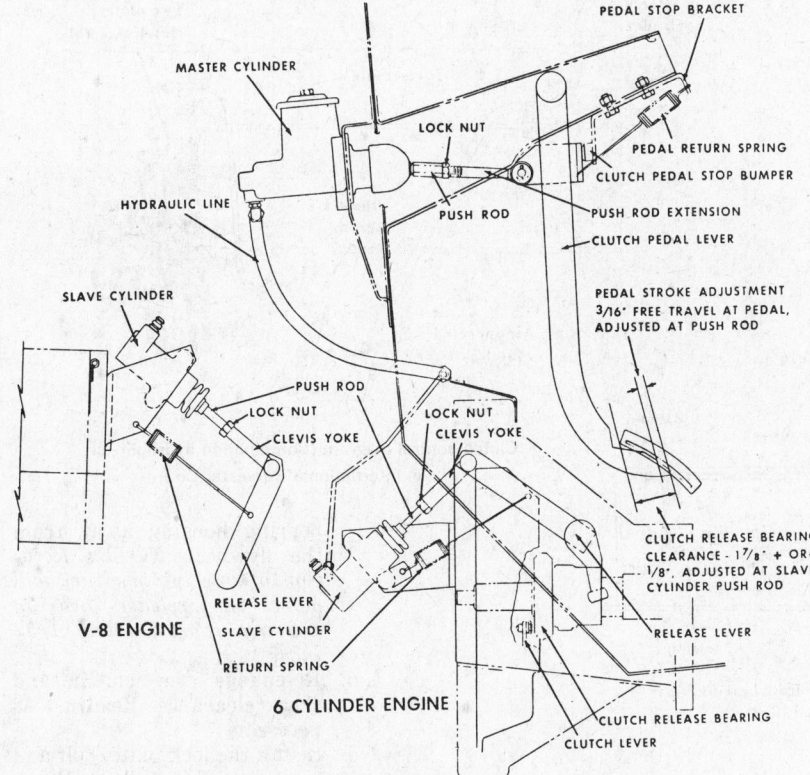

Hydraulic clutch control system (© International Harvester Co.)

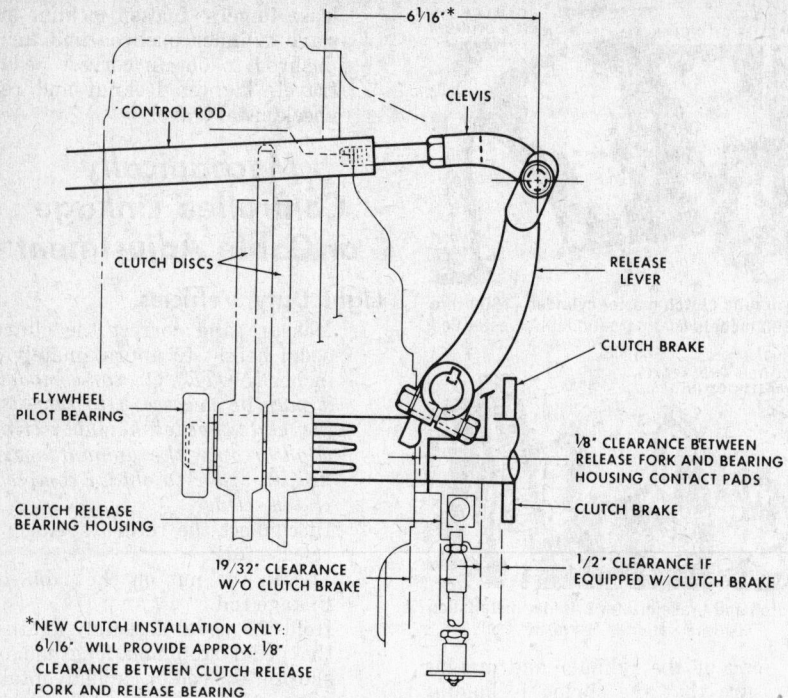

Clutch clearances—heavy duty vehicles with two clutch discs
(© International Harvester Co.)

rod to check the release bearing to release lever clearance.

b. Vehicles with out clutch brake — Weld 19/32 inch stock material to a handle to check the gap between the release bearing and the clutch cover hub. The same 1/8 inch rod as used above, can be used to check the clearance between the release bearing and the release lever.

4. If the clearances are more or less than specified, readjust as follows:

a. Rotate the engine flywheel until the adjusting ring lock is exposed. Remove the lock bolt and pry the lock free of the adjuster ring. **CAUTION: Lock is spring loaded.**

b. Release clutch by blocking the pedal in the depressed position.

c. Turn the clutch adjusting ring counterclockwise to move the release bearing housing towards the flywheel and clockwise to move the

clutch brake), or the clearance between the clutch cover hub and the release bearing housing (if not equipped with a clutch brake).

NOTE: Gauges of proper thickness can be fabricated locally.

a. Vehicles with clutch brake— Weld 1/2 inch stock material to a handle to check the gap between the release bearing and the clutch brake assembly and use a 1/8 inch wire

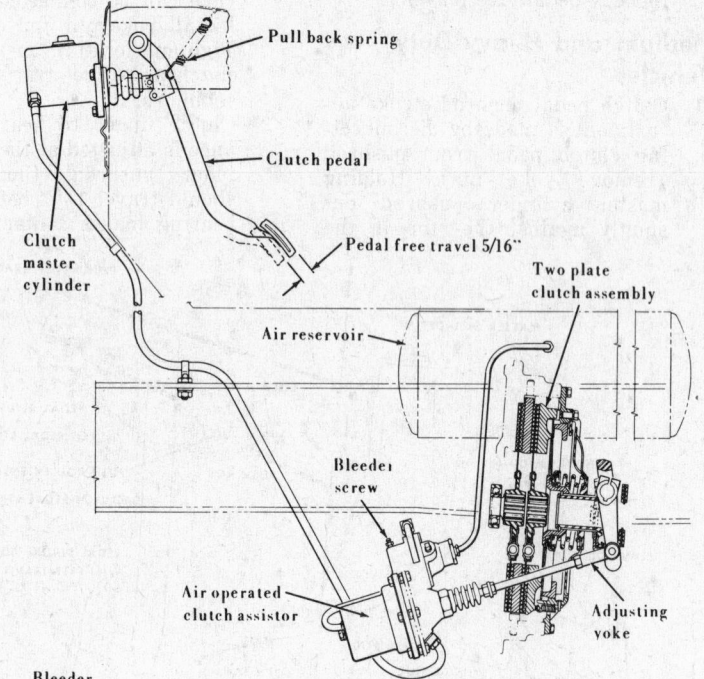

Clutch control diagram (Combination hydraulic-air)
(© International Harvester Co.)

bearing housing away from the flywheel. *NOTE: Rotation of one lug position will move the release bearing housing approximately 1/32 inch.*

d. Re-engage the clutch and check clearance. Readjust as necessary.

e. Install the lock plate, bolt and washer, and install the clutch housing cover plate.

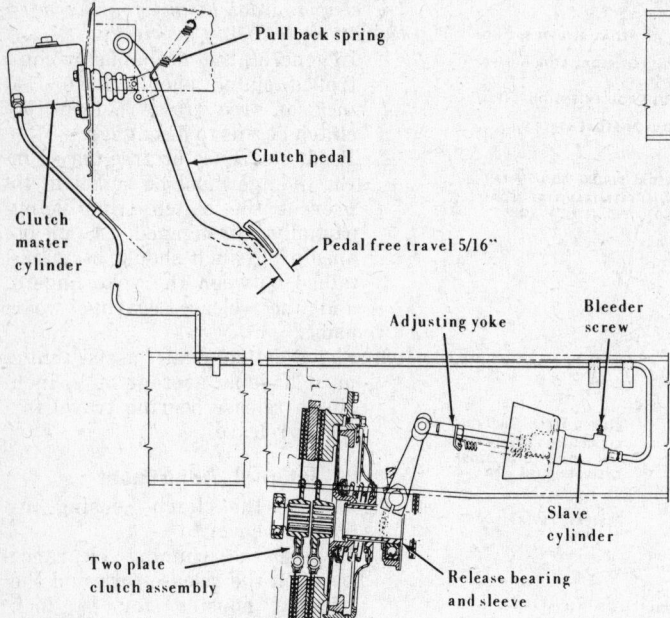

Clutch control diagram (Hydraulic) (© International Harvester Co.)

Clutch Master Cylinder Removal

1. Remove hydraulic line from master cylinder and remove yoke pins connecting pedal to master cylinder pushrod. On models with dual cylinders (brake-/clutch integral unit), disconnect brake hydraulic line, stoplight switch wire and brake pedal. If the fluid reservoir is separate from the cylinder unit, disconnect reservoir fluid line.
2. Remove cylinder assembly mounting bolts and remove cylinder.
3. To install, first mount the cylinder unit, then connect hydraulic lines.
4. Adjust clutch pedal as described above.
5. Bleed hydraulic clutch system.

Slave Cylinder Removal

1. Disconnect slave cylinder pushrod from clutch release lever.
2. Disconnect hydraulic line.
3. Unbolt and remove slave cylinder unit.
4. To install, reverse Steps 1 through 3.
5. Adjust clutch pedal as described above.
6. Bleed hydraulic clutch system.

Bleeding Hydraulic Clutch System

1. Fill fluid reservoir with hydraulic brake fluid.
2. Remove dust cover from bleeder screw on slave cylinder and open bleeder screw approximately ¾ turn.
3. Attach a short bleeder tube to bleeder screw and place the other end in clear container filled with brake fluid.
4. Pump clutch pedal slowly through full stroke repeatedly until only clear (no air bubbles) fluid flows from bleeder hose.
5. Tighten bleeder screw on down stroke of clutch pedal and remove bleeder tube. Replace rubber bleeder screw dust cover.
6. Refill fluid reservoir if necessary.

Clutch Removal

1. Remove transmission. Extreme care should be taken to support the transmission until it is completely removed so that the main shaft splines will clear the driven member. For transmission removal procedures see "Transmission—R&R" immediately following this section.
2. Remove flywheel housing cover.
3. Disconnect clevis, yoke from clutch release lever.
4. Compress clutch assembly. On the 13", 14", 15" and 10" (9 spring) clutches, the pressure plate is drilled and tapped so that three retaining cap screws and flat washers may be installed. Tighten the cap screws until flat washers and cap screw heads are seated on the back plate. On the 11", 12" and 10" (6 spring, open back plate type) clutches, three retaining spacers are used to hold the clutch assembly compressed during removal. Slightly loosen the back plate to flywheel mounting screws to wedge the retaining spacers into place. On the 10" six spring (full back plate type) clutch, three ⅝"x3"x¼" hardwood blocks are used to compress the clutch during removal. Loosen back plate to flywheel retaining screws enough to wedge the blocks between the back plate inner flange and release fingers.
5. Remove back plate to flywheel screws and remove back plate assembly and driven disc.
6. When removing the clutch assembly, observe that the balance mark (spot of white paint) on the back plate flange is located as near as possible to the balance mark ("L") stamped on the flywheel face. These balance marks should be located in the same relative position at clutch installation. If there are no marks, scribe a line to indicate correct position.
7. To install clutch, position the clutch driven member so that the long portion of the hub is toward the rear (all except the 10" 6 spring open back plate type, which may be fitted either way). Clutch must be compressed for correct installation.
8. Place clutch assembly over the driven member on the flywheel so that the balance mark (spot of white paint) is as near as possible to the flywheel balance mark ("L"). Loosely install two or three back plate to flywheel mounting screws.
9. Using a clutch aligning arbor or transmission main drive gear shaft to hold the driven member in place, complete installation of the remaining back plate to flywheel mounting screws and lockwashers. Tighten capscrews alternately and evenly.
10. Remove retaining capscrews, wood blocks or retaining spacers which were used to hold the clutch compressed.
11. Install transmission as described in "Transmission—R&R".
12. Connect linkage to clutch release lever.
13. Install flywheel housing cover.
14. If the vehicle is equipped with a hydraulic system, bleed the sys-

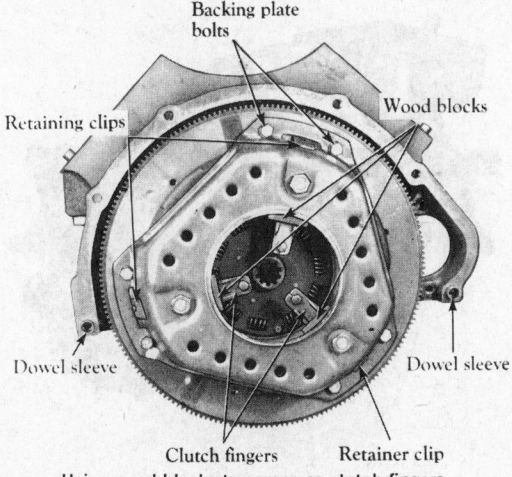

Using wood blocks to compress clutch fingers-
(© International Harvester Co.)

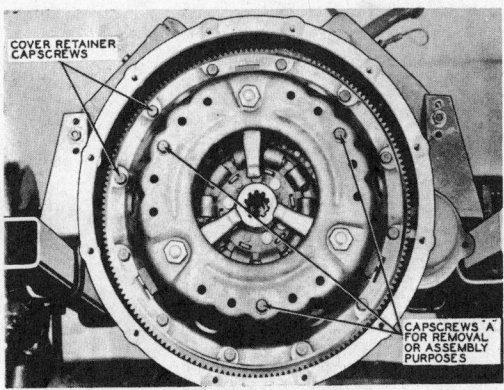

Clutch compressing cap screws
(© International Harvester Co.)

tem and refill the reservoir with hydraulic fluid.

15. If the vehicle is equipped with the manual type linkage or cable, adjust as outlined.

Transfer Case

Transfer cases may be mounted to the rear of the transmission and connected directly to the output shaft of the transmission by a coupler, or may be mounted to a crossmember and connected to the transmission by a driveshaft.

For transfer case overhaul procedures, see the Unit Repair Section.

Linkage and Cable Adjustment

Shifter rods connect the shift arms of the transfer case to the shift lever arms. Non-adjustable and adjustable links are used on the various models of vehicles. To insure the proper alignment of the rods to the arms, use the following procedure.

Linkage Adjustment

1. Place the shift lever in the neutral position.
2. Remove the shift control rod at the transfer case.
3. Assure that the shift arm of the transfer case is in the center or neutral position.
4. If the control rod is adjustable, position the trunnion or clevis to align with the hole in the shift arm of the transfer case.
5. If the control rod is non-adjustable and the rod does not line up with the hole in the shift arm of the transfer case, replacement or bending will be necessary for the control rod.
6. Reconnect the control rod to the transfer case shift arm and check for proper operation.

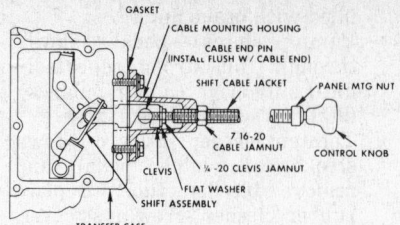

Exploded view—shift cable control
(© International Harvester Co.)

Cable Adjustment

A pull on the cable will engage the gears. To disengage, merely push the control cable in. To adjust, follow this procedure.

1. Pull the control cable knob out approximately two inches and block in this position.
2. Loosen the cable mounting housing jam nut.
3. Remove the two cable mounting housing bolts.
4. Unscrew the cable mounting housing away from the transfer case housing.
5. Confirm the inner clevis is positioned in the engaged position.
6. Turn the cable housing down the cable jacket to a snug fit against the gasket on the transfer case mounting boss. Install the two retaining screws.
7. Turn the jam nut down the cable jacket and secure against the cable mounting housing.
8. Remove the control cable knob block and operate the cable assembly to check the shifter operation.

Cable Removal and Installation

Removal

1. Leave the control knob pushed in.
2. Loosen the cable jam nut at the cable mounting housing on the transfer case and turn it back to

the end of the threads.
3. Remove the two bolts holding the cable mounting housing to the transfer case.
4. Unscrew the housing all the way to the jam nut.
5. Pull the cable mounting housing forward until the inner cable jam nut is clear.
6. Loosen the inner cable jam nut at the shift clevis and unhook the cable end pin from the clevis.
7. Position the shift cable to obtain working clearance.

Installation

1. Turn the jam nut and the cable mounting housing to the bottom of the thread end.
2. With the transfer case shifter assembly in the fully engaged position, install the cable mounting housing gasket to the case. *NOTE: The clevis is pulled out to engage.*
3. Connect the cable end pin to the shifter assembly and secure with the jam nut. *NOTE: Confirm that the pin is installed flush with the cable end.*
4. Block the control knob out approximately two inches.
5. Turn the cable mounting housing down the cable jacket to a snug fit against the gasket on the transfer case mounting boss.
6. Secure the cable mounting housing with the two mounting bolts to the transfer case.
7. Turn the jam nut down the cable jacket and lock against the cable mounting housing.
8. Remove the block from the shift control cable knob and operate the cable to check the shifter operation.

Transfer Case Removal and Installation

Frame Mounted

1. Drain the transfer case and dis-

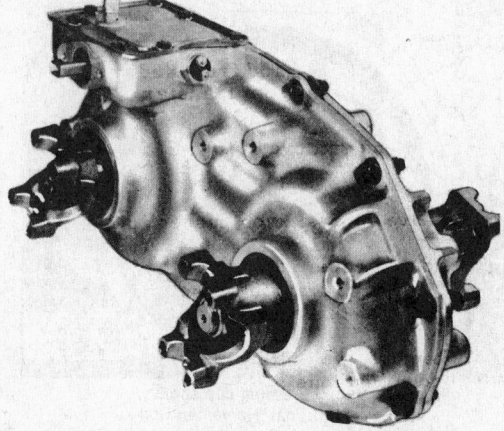

Frame mounted transfer case—typical (© International Harvester Co.)

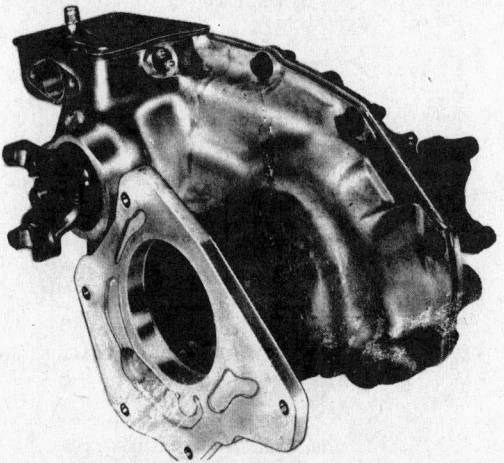

Transmission mounted transfer case—typical (© International Harvester Co.)

connect the rear axle drive shaft at the transfer case.

2. Disconnect the front drive shaft at the transfer case.

3. Disconnect the speedometer cable, and indicator light switch wire, if equipped.

4. Disconnect the shift linkage. If erquipped with a shift cable, refer to the cable removal and installation outlined previously.

5. Place a transmission jack under the transfer case and remove the mounting bolts from the frame to case.

6. Remove the transfer case from the vehicle.

7. The installation of the transfer case is the reverse of removal.

Transmission Mounted

1. Disconnect the rear driveshaft at the transfer case and drain the case assembly.

2. Disconnect the front driveshaft at the transfer case.

3. Disconnect the speedometer cable and the indicator light switch wire, if equipped.

4. Disconnect the shift linkage or cable. If equipped with a shift cable, refer to the removal and installation of the cable outlined previously.

5. Place a transmission jack under the transfer case and remove the flange bolts holding the transfer case to the transmission.

6. Pull the transfer case rearward to disengage the transmission output shaft from the coupler.

7. Lower the transfer case and remove from the vehicle.

8. The installation of the transfer case is the reverse of removal.

Manual Transmission

For manual transmission overhaul procedures see the Manual Transmission General Section.

Shift Linkage

Different types of transmissions are used which may require the shift linkage to be either mounted in the transmission and controlled by a shift lever, or to have a shift lever mounted remotely with linkage rods connecting the lever to the transmission. No adjustment is provided when the linkage is mounted in the transmission. When the shift lever is remotely mounted, the connecting rods have adjustment provisions. The adjustments are made with the shift control and the transmission arms in the neutral position, and the control rods adjusted to enter either the transmission arms or the shift lever arms with a free fit. Normally the control rods

are threaded and trunnions and jam nuts are used to position the rods.

Manual Transmission Removal

Removal and installation of manual transmissions will vary in detail, depending on which vehicle is being serviced. The following general procedure includes the basic steps common to all models.

1. Access to the transmission may be improved by removing cab floor panels.

2. Raise vehicle on a hoist or jack up and support with jack stands.

3. Drain the transmission lubricant.

4. Disconnect drive shaft at the transmission. If the vehicle is equipped with a transfer case which is not mounted directly to the transmission, disconnect the shaft between the transfer case and transmission at the yoke. If the vehicle is equipped with a transfer case which is mounted directly to the transmission, it must be removed with the transmission as a unit and the forward and rear drive shafts must be disconnected. Secure shaft out of the way with wire.

5. Disconnect shift linkage from transmission shift levers. If the vehicle is equipped with a transfer case which is mounted directly to the transmission, disconnect the shift linkage from the transfer case shift levers.

6. If the vehicle is equipped with a transmission mounted handbrake, disconnect the handbrake cable at the relay lever.

7. Disconnect speedometer cable from the transmission.

8. Remove the clutch slave cylinder from its mount. Do not disconnect the hydraulic line from the slave cylinder. Secure the push rod to the slave cylinder to avoid ejection of the internal components. Keeping the hydraulic system sealed eliminates the necessity of bleeding.

9. On some models it may be necessary to remove the starter motor.

10. Support the rear of engine by means of a hydraulic jack.

11. Remove the transmission mounting bolts and insulators at the engine rear crossmember. If possible, remove the rear engine crossmember. Remove gear shift lever and housing from top of transmission if applicable.

12. Attach suitable hoisting equipment or jack to transmission and raise enough to support the transmission assembly.

13. Remove top transmission to

clutch housing bolts and install transmission guide pins.

14. Remove remaining transmission to clutch housing bolts.

15. Carefully pull transmission rearward, keeping it in line until the main drive gear shaft is clear of ther clutch. **CAUTION: Extreme care must be exercised to insure that the weight of the transmission does not rest on the hub of the clutch driven disc.**

16. Depending on vehicle model, either lift the transmission up through the floorboard and out the right door or lower it with a jack.

17. Installation is the reverse of the above procedure.

18. Fill transmission with fluid.

Auxiliary Transmission Linkage Adjustment

1. Disconnect the shift rods at the lever assembly.

2. Position the transmission shafts in neutral.

3. Position the shift control lever so that the slots in the rails are exactly opposite each other with the lever inclined to the rear.

4. Adjust each rod clevis until the pins enter the shift control rail and clevis easily.

5. Install the clevis pins and retainers.

6. Check the shift operation.

Auxiliary Transmission Removal and Installation

1. Drain the lubricant from the auxiliary transmission.

2. Remove the pins and retainers from the shift rods and remove the rods from the rails.

3. Disconnect the front drive shaft universal joint and tie out of the way.

4. Disconnect the rear drive shaft universal joint and tie the shaft out of the way.

5. Remove the speedometer cable, if equipped.

6. Remove the parking brake linkage from the output shaft parking brake drum.

7. Place a transmission jack under the auxiliary transmission, remove the mounting bolts, and remove the unit from the vehicle.

8. Installation is the reverse of removal.

Automatic Transmission

Identification

Two types of automatic transmissions are used in the I. H. vehicles. The models T-39, T-49 and T-409 use a cast iron case with a removable aluminum converter housing. The transmission model T-407 uses a complete aluminum case and converter housing, cast as a unit. The rear extension housings may differ from one transmission to another, due to the vehicle application of engine and drive train. Some models may use a parking brake mounted on the rear extension housing. All transmissions use a parking pawl type lock, controlled by the shift linkage. The selector lever may be mounted on the floor or on the steering column, along with an indicator quadrant for the gear position.

For automatic transmission overhaul procedures, see the Automatic Transmission General Section in the Unit Repair Section.

Transmission Models T-39, T-49, T-409

Removal

1. Raise the vehicle with a hoist.
2. Disconnect the fluid filler tube at the pan, and drain the fluid. Loosening the tube clip capscrew at the extension on the starting motor will permit rotating filler tube for transmission removal.
3. Disconnect the vacuum line at vacuum unit located at rear of transmission.
4. Disconnect the speedometer cable at speedometer adapter on transmission.
5. Disconnect the hand brake cable at drive shaft brake if chassis is so equipped.
6. Disconnect the shift linkage at manual shift lever on transmission.
7. Disconnect the two oil cooler lines on right side of transmission if chassis is so equipped.
8. Disconnect the drive shaft at transmission companion flange. *NOTE: Wire the end of the drive shaft to the frame to permit transmission removal.*
9. Place the hydraulic hoist with a suitable transmission lift cradle in position under the transmission oil pan. Adjust the hoist to align the cradle to the transmission oil pan flange so that the weight of the transmission case is supported by the hoist.
10. Remove the transmission case to

Pilot studs

Pilot stud location—automatic transmission
(© International Harvester Co.)

converter housing upper capscrews and install two pilot studs into the capscrew holes.
11. Remove the transmission case to converter lower capscrews.
12. With the hydraulic hoist and cradle adjusted so the transmission case is in alignment with the converter housing, pull the transmission rearward with the hydraulic hoist to disengage the transmission from the converter housing and converter assembly. Lower transmission and remove from the vehicle.
13. To install, place two transmission pilot studs in the upper transmission to converter housing mounting screw holes.
14. Mount the transmission on a jack and position it under vehicle.
15. Rotate engine until the front pump drive lugs on the converter are in a vertical position.
16. Rotate the front pump until the slots in the pump drive gear are in a vertical position.
17. Apply lubricant similar to lubriplate to seal the surface of converter impeller cover hub.
18. Being extremely careful to align the turbine shaft splines with the turbine hub splines and the converter impeller lugs with the slots in the front pump drive gear, raise transmission and move it forward into the converter housing and converter.
19. Install the transmission to converter housing lower mounting screws. Remove two pilot studs and install upper mounting screws. Tighten all mounting screws securely.
20. Install oil cooler lines on the right side of the transmission if so equipped.
21. Connect the shift linkage at the manual shift lever on the transmission.
22. Connect the hand brake cable to the drive shaft brake if so equipped.
23. Connect the speedometer cable to the transmission.

24. Connect the vacuum line to the vacuum unit located at the rear of transmission.
25. Connect the fluid filler tube to the oil pan, tightening securely. Also tighten tube clip capscrews at the extension on the starter motor.
26. Connect the drive shaft to the transmission companion flange. Tighten mounting screws securely and lock with lock plates.
27. Lower vehicle to floor and fill transmission with type "A" automatic transmission fluid.
28. Road test vehicle to check performance and shift points.

Torque Converter Removal

1. Remove transmission as described above.
2. Disconnect and remove starter motor.
3. If vehicle has conventional chassis, remove floormat and transmission floor opening cover.
4. Install a rear engine support or support the rear weight of the engine with a jack.
5. Remove converter housing to crossmember mounting bolts, lower insulators and retainers.
6. Unbolt and remove rear engine crossmember.
7. Remove eight capscrews and lockwashers which attach converter housing to crankcase adapter and remove converter housing. On some models it may be necessary to lower the engine to provide clearance for converter removal.
8. Remove six nuts which attach the converter assembly to the flywheel assembly and remove converter.
9. To install, carefully place converter into position (do not damage bolt mounting threads) and install six nuts which attach converter to converter drive plate, but do not tighten nuts at this time.
10. After thoroughly cleaning crankcase converter housing adapter, install converter housing and engage the dowels being careful not to damage the dowels or the converter housing.
11. Install eight converter housing mounting capscrews and lockwashers.
12. Install rear engine crossmember.
13. Install upper and lower insulators and retainers, the converter to crossmember mounting bolts and lockwashers. Lower converter housing and engine. With insulators and retainers firmly seated, hand tighten bolts, then give bolts one-half additional

turn and lock.

14. Remove rear engine support or jack.
15. If vehicle has conventional chassis, install floormat and transmission floor opening cover.
16. Install starter motor and wiring.
17. Install transmission as described above.
18. Rotate engine and converter assembly through two complete revolutions to center converter (to rotate, remove spark plugs and pry on drive plate ring gear).
19. Tighten converter to the converter drive plate attaching nuts.
20. Install converter housing adapter cover.

Transmission Fluid

1. Transmission fluid should be changed and band adjusted every 15,000 miles.
2. Remove converter housing front plate.
3. Remove one of the converter drain plugs, then rotate converter 180 degrees and remove other converter drain plug.
4. Disconnect fluid filler tube at the transmission pan.
5. Drain fluid and remove pan. Clean pan.
6. Connect filler tube to pan and tighten securely.
7. Install drain plugs in converter cover and tighten them to 7-10 ft. lbs. torque.
8. Install converter housing front plate.
9. Add five quarts of type "A" automatic transmission fluid through filler tube.
10. Run engine at idle (do not race) for about two minutes, then add five more quarts of fluid. Let engine idle until it reaches normal operating temperature.
11. Move selector lever through all positions, then place it in "P" (park). Check fluid level and add enough fluid to bring level up to the "F" (full) mark on indicator.

Front Band Adjustment

1. Drain fluid from transmission and remove pan (disconnect filler tube from pan).
2. Loosen front servo adjusting screw locknut two full turns and check the adjusting screw for free rotation.
3. Pull back on the actuating rod and insert gauge block of the front band adjusting tool (SE-1910) between the servo piston stem and adjusting screw. Tighten adjusting screw until adjusting tool handle overruns. Holding adjusting screw stationary, tighten locknut to 20-25 ft. lbs. torque. Remove gauge block.
4. Install fluid screen and pan, using a new gasket. Connect filler tube to pan.
5. Refill transmission as described above, adding new fluid if necessary.

Rear Band Adjustment

1. If vehicle is conventional model, remove floor mat and transmission cover plate from floor board.
2. Clean adjusting screw threads thoroughly and oil threads.
3. Loosen rear band adjusting locknut.
4. Using tool SE-1909, tighten adjusting screw until wrench overruns, at 10 ft. lbs. torque. *NOTE: If adjusting screw is tighter than overrun of wrench, loosen screw and retighten.*
5. Back off adjusting screw one and one-half turns, then hold adjusting screw stationary and tighten locknut to 25-40 ft. lbs. torque. **CAUTION: Severe damage may result if the adjusting screw is not backed off exactly one and one-half turns.**
6. Install transmission cover plate and floor mat to floor board.

Shift Linkage Adjustment

1. With engine off, disconnect the manual shift rod from the selector lever on the steering column and the transmission lever on the conventional chassis, or bellcrank on the Metro chassis.
2. Position selector lever in "D" and place transmission manual lever in the "D" detent (second from the top of the transmission).
3. Position the manual shift rod into the ball joint on the steering column. The opposite end of the rod should be installed in the transmission shift lever on the conventional chassis and secured with a washer and cotter pin. On the Metro chassis, the rod yoke should be positioned on the bellcrank and secured with the clevis pin, washer and cotter pin.
4. Tighten ball joint nut at the steering column lever.
5. Move the selector lever through all positions, checking the alignment of the pointer in all positions.

Kickdown Switch Adjustment

1. On the conventional chassis the kickdown switch is located in the toeboard under the throttle pedal. On the Metro chassis it is mounted on a bracket under the toeboard and operated by a pad welded to one of the throttle linkage rods.
2. Loosen two mounting nuts and turn them either direction until a clearance of 1/4" is obtained between switch and throttle pedal on the conventional chassis and between switch and throttle linkage pad on the Metro chassis.

Vacuum Control Adjustment

1. Connect a tachometer to the engine.

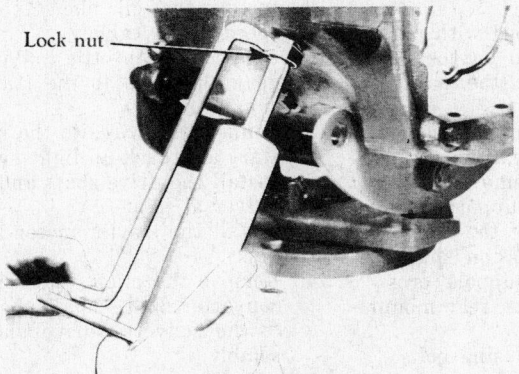

Adjusting rear band
(© International Harvester Co.)

Adjusting front band
(© International Harvester Co.)

377

2. Remove the 1/8 inch pipe plug located on the left front of the transmission case. Install a pressure gauge line connection at this point, then connect a pressure gauge to the line and place gauge in cab.

3. Start engine and move selector lever to "D" (drive) position. Apply hand brake and accelerate engine until 1000 rpm is reached. Pressure reading on gauge should be 82-88 psi.

4. If correct pressure is not obtained, loosen locknut on vacuum control unit (located at rear of transmission) and turn vacuum unit clockwise to increase pressure or counterclockwise to decrease pressure. Adjust for proper pressure, then tighten locknut. CAUTION: Do not operate engine over 10 seconds at any one time while performing the above.

Neutral Safety Switch Adjustment

1. Place the selector lever in the "N" position and loosen the two capscrews securing the switch to the steering jacket tube.

2. Using a 3/32 pin punch as an aligning tool, insert the pin into the hole on the face of the switch.

   NOTE: If necessary, rotate the switch until the pin enters freely into the hole in the switch.

3. Secure the two capscrews holding the switch to the steering jacket tube and remove the pin punch.

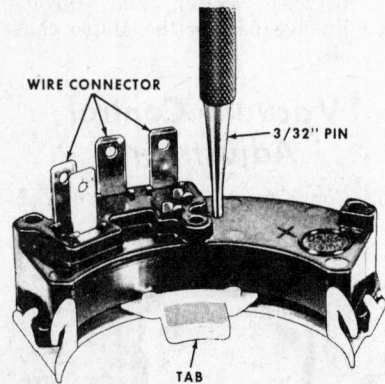

WIRE CONNECTOR

3/32" PIN

TAB

Using a 3/32" pin punch to adjust the safety starter switch (© International Harvester Co.)

4. Attempt to start the vehicle in all positions. If the engine will start in the "P" and the "N" positions and not in any of the others, the switch is correctly adjusted.

   Caution: Block the wheels before testing the switch adjustment.

Transmission Model T-407

Removal

NOTE: The transmission and converter must be removed as a unit assembly. Damage can result to the converter drive plate, pump bushing, or to the pump seal, if the converter is allowed to remain on the converter drive plate.

1. Connect a remote switch to the starter solenoid so that the engine can be rotated from under the vehicle.

2. Disconnect the coil high tension cable.

3. Raise the vehicle and support safely.

4. Remove the engine rear crossmember on 4 x 4 vehicles, if necessary.

5. Remove the cover plate from the front of the converter housing to provide access to the converter drain plug and mounting bolts.

6. Rotate the engine to bring the drain plug to the six o'clock position. Drain the converter and loosen the pan bolts to drain the transmission.

7. Mark the converter and drive plate to aid in the assembly. Rotate the engine to locate the converter-to-drive plate bolts and remove the bolts.

8. Disconnect the negative battery cable and remove the starter motor assembly.

9. Disconnect the wires from the back-up light and neutral start switch.

10. Disconnect the gearshift cable or rod and bellcrank from the transmission.

11. Disconnect the throttle rod from the left side of the transmission.

12. Disconnect the cooler lines at the transmission and remove the filler tube.

13. Disconnect the speedometer cable, and move cable away from the transmission.

14. Disconnect the front universal joint and secure the shaft out of the way.

15. On vehicles equipped with parking brake mounted on the rear extension, remove the parking brake cable.

16. On vehicles equipped with dual exhaust, the left exhaust system may have to be removed.

17. Install an engine support fixture to hold the rear of the engine.

18. Raise the transmission slightly, and remove the support crossmember holding the rear mount assembly.

19. Remove all bell housing bolts.

20. Carefully move the transmission assembly rearward off the block

dowels and disengage the converter hub from the end of the crankshaft. Place a converter holding tool on the bell housing to hold the converter in place.

21. Lower the transmission assembly and remove the transmission from the vehicle.

22. To remove the converter assembly from the transmission, remove the holding tool and carefully slide the converter out of the transmission.

Installation

1. Rotate the pump rotors with tool SE-2402 or its equivalent, so that the lugs on the pump inner rotor are vertical.

2. Position the converter so that the impeller shaft slots are vertical and carefully slide the converter assembly over the input shaft and reaction shaft. Make sure that the converter slots fully engage the pump inner rotor lugs.

   NOTE: The surface of the converter front cover lug should be at least 1/2 inch to the rear of a straightedge, placed on the face of the bell housing, when the converter is pushed all the way into the transmission.

3. Install the converter holding tool to hold the converter in place.

4. Position the transmission on a jack assembly and move the unit under the vehicle.

5. Rotate the converter to align the previously made marks on the drive plate and converter.

6. Raise the transmission and align with the engine. Install a pilot stud to aid in the alignment of the converter to the drive plate. Carefully work the transmission assembly forward over the engine block dowels with the converter hub entering the crankshaft opening.

7. Install the converter housing bolts and tighten to specified torque.

8. Install the crossmember and mount at the rear of the transmission. Remove the engine support fixture.

9. Install the oil filler tube and speedometer cable.

10. Connect the throttle rod and the gear shift rod to the transmission levers.

11. Connect the wires to the neutral start and back-up light switch.

12. Install the drive shaft and front universal joint.

13. Install the starter motor assembly.

14. Remove the pilot stud from the converter and install the bolts to the converter-drive plate assembly.

15. Install the cooler lines to the transmission.

16. Install the converter access plate on the front of the converter housing.
17. If the left exhaust system was removed, replace the pipes and brackets.
18. Install the parking brake cable and adjust, if equipped with the extension housing parking brake assembly.
19. Adjust the shift and throttle linkage.
20. Fill the transmission and connect the negative battery cable, if not done, and start the engine. Recheck the fluid level and refill as necessary.

Transmission Fluid Drain and Refill

1. Raise the vehicle on a jack or hoist. Support safely.
2. Place a large drain container under the transmission oil pan.
3. Loosen the pan bolts and tap one corner of the pan to break it loose, allowing the fluid to drain.
4. Remove the access plate from the front of the converter housing. Remove the converter drain plug and allow the fluid to drain.
5. Remove and clean the pan, remove the fluid filter and discard.
6. Install a new filter assembly on the valve body and tighten the screws securely.
7. Using a new pan gasket, install the pan and tighten the bolts securely.
8. Install and tighten the converter drain plug.
9. Install the converter housing access plate.
10. Install six quarts of transmission fluid into the transmission. Start the engine and allow to run for two minutes. Check the fluid level and add enough oil to bring the level to the "ADD ONE PINT" mark.
11. Recheck the level after moving the selector lever through all the gear positions and after the transmission has reached normal operating temperature. The level should be between the "FULL" mark and the "ADD ONE PINT" mark.

Kickdown Band Adjustment

NOTE: The kickdown band is located on the left side of the transmission case near the throttle lever shaft.
1. Loosen the locknut and back off approximately five turns.
2. Tighten the adjusting screw to 10 ft. lbs.
3. Back off the adjusting screw 2¼ turns with the 6 and 8 cylinder engines. Hold the adjusting

screw in position and tighten the lock nut to 29 ft. lbs.

Low and Reverse Band Adjustment

1. Raise the vehicle, support safely, drain the transmission fluid, and remove the pan.
2. Loosen the lock nut on the adjusting screw.
3. Tighten the adjusting screw to 10 ft. lbs.
4. Tighten the lock nut to 30 ft. lbs.
5. Install the pan using a new pan gasket.
6. Fill the transmission with fluid, start the engine and recheck the level. Add as necessary.

Back-up Light and Neutral Start Switch

No provisions are made for any adjustments of the back-up light and neutral start switch. The neutral start circuit is controlled by the inner terminal and the back-up light circuits are controlled by the two outside terminals.

The replacement of the switch is accomplished by unscrewing the switch from the transmission case, and screwing a new switch into the case. Since fluid leakage will occur when removing the switch, fluid must be added after the new switch is installed.

Shift Linkage

Adjustable Cable Control

1. Install cable conduit anchor clamps at both ends.
2. Install swivel on the control lever so that a distance of .55 inch exists from the end of the cable to the opposite side of the trunnion. Tighten the jam nut securely.
3. With the control in PARK position and transmission lever in the full rearward position (PARK detent), adjust the yoke so that the rod end pin installs freely and secure the yoke nut and install the cotter pin.

Column Shift

1. Assemble all linkage parts, but leave the upper control rod bolt loose.
2. Place the selector lever in DRIVE position.
3. Move the shift control lever on the transmission to the DRIVE position.
4. Tighten the upper bolt on the control rod to 14-16 ft. lbs.
5. Check the adjustment as follows.
   a. Shift effort must be free and detents feel crisp. All gate stops must be positive.

b. Key start must only occur in the PARK or NEUTRAL positions.
   c. Detent positions must be in proper relationship to the transmission lever positions.

Throttle Valve Linkage Adjustment

1. With the engine off and an assistant holding the accelerator pedal to the floor, check for full carburetor throttle plate opening.
2. If necessary, adjust the throttle cable and pedal floor stop to obtain wide open throttle.
3. If necessary, adjust the idle speed of the engine with the use of a tachometer and with the engine at normal operating temperature and the carburetor off the fast idle cam. Adjust the curb idle speed, (throttle stop solenoid activated) with the transmission in neutral and the air conditioning in the OFF position.

NOTE: Be sure that carburetor is not being held open by a deceleration valve dashpot, solenoid valve, or a vacuum throttle modulator valve.

CAUTION: All components in the throttle control and transmission linkage system must operate freely with absolutely no sticking, excessive friction, or interference from other chassis components.

6-258 Engine with 1940 Carburetor

1. Hold the throttle valve control rod on the transmission in the forward position while adjusting the length of the throttle push rod.
2. Adjust the link at the upper end of the control rod until the rear end of the slot in the adjusting link contacts the pin in the bellcrank.
3. Tighten the lock nut against the throttle valve control rod link. Position the link on the bellcrank pin and install the washer and cotter pin. Connect the control rod return spring.
4. Check the throttle valve control rod linkage for freedom of movement and for full return to the forward position upon movement of the bellcrank.
5. Road test the vehicle. If no kickdown is present, adjust the throttle valve control rod one turn to move the lever at the transmission farther to the rear.
6. Repeat the road test. If no kickdown is obtainable as yet, repeat the procedure as outlined in paragraph 5 above.

V-304, V-345, V-392 and V-400 Engines With 2210, 2300, and 4150 Holley Carburetors

Two types of throttle linkages are used on these carburetors to transmissions. They are designated as FIRST RELEASE and MODIFIED type linkages. The differences are;

    a. FIRST RELEASE—has both the throttle push rod and the throttle valve control rod piloted by guide holes in the throttle valve rod bracket. The throttle push rod is capped by a lock nut and acorn nut.

    b. MODIFIED TYPE—has only the throttle valve control rod piloted in the throttle valve rod bracket. The throttle push rod is piloted in the spring clip plate which is welded to the throttle valve control rod.

FIRST RELEASE—Adjustment

1. On the first release type throttle valve linkage, loosen the jam nut of the acorn adjusting nut and turn the acorn nut clockwise until clearance between the acorn nut and the spring clip is obtained.
2. Move the throttle valve linkage control rod to its forward most position. Adjust the acorn nut so it just rides against the spring clip on the throttle valve control rod and tighten the jam nut.
3. Check the throttle linkage for freedom of movement by pushing the linkage to its full rearward position and making sure its complete return to the full forward position.

MODIFIED LINKAGE—Adjustment

1. Loosen the jam nut and turn the adjusting nut clockwise until clearance between adjusting nut and spring clip is obtained.
2. Move the control rod to its forward most position. Adjust the nut so it just rides against the spring clip on the throttle valve control rod and tighten the jam nut.
3. Check the throttle linkage for freedom of movement by pushing the linkage to its full rearward position and making sure of its complete return to the full forward position.

V-345 and V-392 Engines With Carter Thermo-Quad Carburetor

NOTE: Two types of throttle linkages are used as with the V-304, V-345, V-392, and V-400 Carburetors.

FIRST RELEASE—Adjustment

This procedure is the same as for the V-304, V-345, V-392, and V-400.

MODIFIED LINKAGE—Adjustment

1. Loosen the jam nut and turn the adjusting nut clockwise until the adjusting nut just contacts the spring clip.
2. Move the control rod to its forward most position. Adjust the nut so it just rides against the spring clip on the throttle valve control rod and tighten the jam nut.
3. Check the throttle linkage for freedom of movement and return to its full forward position when moved.

Drive Shaft, U Joints

Driveshaft Assembly

It is imperative that all components of the drive train be tight to insure balance. Check companion flanges at the axles and transmission, center bearing mounts and engine mounts.

When assembling drive train, lubricate slip joint (splined) assemblies and universal joint bearings. Make sure that universal joints are kept on parallel planes by observing the arrows stamped on the shaft end and slip yoke.

Removal and Installation

1. Beginning at the front or the rear of the shaft, bend the lock strip tabs, if equipped, away from the bolts in the trunnion flanges of the universal joints. Mark the shaft, universal joint, and yoke for proper reassembly.
2. Remove the bolts from the trunnion flange to yoke and support the ends of the shaft.
3. If equipped with a center bearing assembly, use a floor jack or other means to support the shaft assembly and center bearing during removal. Remove the bolts from the center bearing to cross-

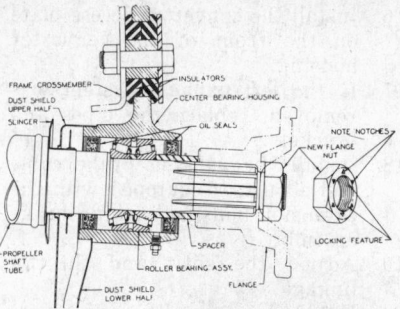

Silent spin type center driveshaft bearing
(© International Harvester Co.)

member and lower the shaft assembly.

NOTE: Some drive shafts will have a slip joint or universal joint near the center bearing assembly. Separating the shaft at these points will lighten the weight and control awkwardness of handling the complete shaft assembly during the disassembly and assembly.

4. The installation of the shaft is in the reverse of the removal procedure. Make sure all indexing marks are properly aligned.

Universal Joints

To remove universal joints, bend down tabs on bearing bolt lock plate and remove four bolts at each universal joint. Joint and shaft assembly must be removed as a unit to service "R" type trunnion bearings. Do not disassemble drive shaft from slip yoke unless these parts are to be replaced. On vehicles equipped with "CL" type trunnion bearings, the universal joint may be unbolted from both the drive shaft and the companion flange and the whole shaft assembly does not have to be removed to service one joint.

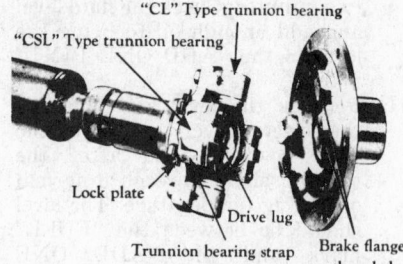

CL type universal joint
(© International Harvester Co.)

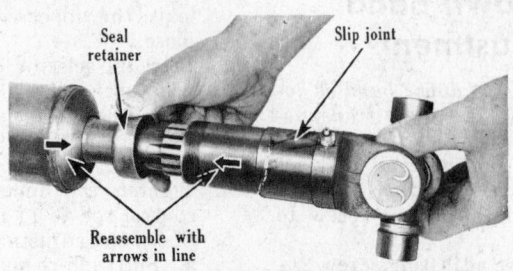

Correct slip joint assembly
(© International Harvester Co.)

When replacing trunnion bearings, remove retaining clips then carefully drive out one, then the other, bearing. Use new packing washers (seals) when reassembling.

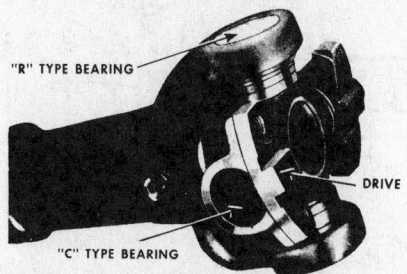

"R" TYPE BEARING

DRIVE

"C" TYPE BEARING

CR type universal joint
(Ⓒ International Harvester Co.)

Rear Axles, Suspension

For overhaul procedures and out of truck adjustments see Rear Axle General Section.

Rear Axle Assembly Removal

1. Jack and block up truck until load is removed from springs and rear wheels are clear of the ground.
2. Drain differential housing.
3. Disconnect brake lines and parking brake cables (where used).
4. On two speed differentials, disconnect control wires or air hoses from the shift mechanism.
5. Disconnect driveshaft at rear axle companion flange.
6. Support differential on portable floor jack and take off U-bolts at springs.
7. Roll out axle from under truck.
8. Installation is the reverse of the above procedure. Be sure to bleed hydraulic brake systems. Remove axle housing breather valve and clean thoroughly with solvent.

Rear Axle Shaft Removal

Semi-Floating Type

1. Remove wheel and nut from axle shaft end.
2. Remove hub and drum assembly with a suitable puller.
3. Unbolt and remove brake backing plate and bearing retainer.
4. Pull axle shaft and bearing using suitable puller. Bearings are pressed on.
5. When installing axle shaft assembly, use new oil seals and be careful not to damage seals.
6. Lightly tap bearing cap into axle housing.

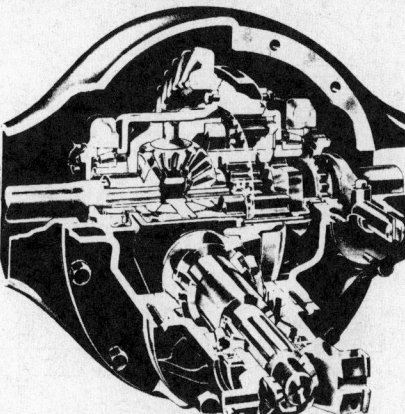

Two speed planetary gear type axle final drive
(Ⓒ International Harvester Co.)

7. Install shims on end of axle housing flange and insert backing plate bolts to retain shims.
8. Install backing plate, bearing retainer and seal retainer. Tighten nuts to specified torque.
9. Install wheel hub. Grasp wheel hub and pull outward to be sure that axle shaft is withdrawn as far as possible.
10. Check axle shaft end play as follows: Mount dial indicator on stationary location at right-hand side of axle assembly. Position indicator against end of axle shaft and check shaft end play. If end play is not within .006″, shims must be added or removed between backing plate and axle housing flange.
11. Place key in axle shaft and install hub and drum assembly on shaft, securing with washer and nut. Tighten nut and install cotter pin.

Full-Floating Type

1. Axle shaft is removed without taking off wheels. Remove axle shaft nuts from studs in the wheel hub.
2. Install puller screws in the two tapped holes provided in the axle shaft flange.
3. Turn in puller screws until axle shaft is loose, then pull axle.
4. Installation is the reverse of the above procedure. Be sure puller screws are removed.

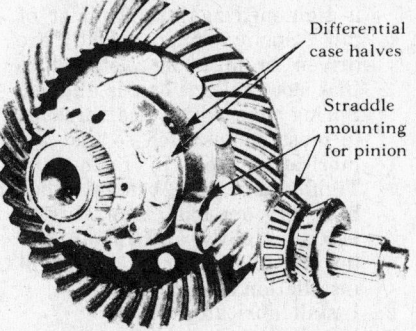

Differential case halves

Straddle mounting for pinion

Single reduction type axle final drive
(Ⓒ International Harvester Co.)

Full-Floating Type With Tapered Dowel Mounting

1. Remove flange nuts from studs of wheel hub.
2. Using a heavy hammer, strike sharply on the center of the flange of the axle shaft. This will unseat and loosen tapered dowels.
3. Remove tapered dowels.
4. Push axle flange back into position against wheel hub and strike again with a hammer to spring axle shaft away from the wheel hub. Do not pry on flange.
5. When installing axle shaft, make sure there is between the axle shaft driving flange and lockwasher. Dowel must not be "sunken in."

Axle Shaft Bearing and Oil Seal

Removal and Installation

The only axle type that has a bearing mounted on the shaft is the semi-floating axle. Follow the procedure outlined under the Axle Shaft Removal to expose the seals and bearings for replacement.

1. The bearing must be pressed from the axle shaft and a new one pressed back on the shaft to seat against the bearing shoulder on the axle.
2. Lubricate the bearing with wheel bearing grease and fill the roller cavities completely.
3. Remove the axle tube inner seal by the use of a seal puller or by engaging the lip of the seal with the end of the axle shaft and prying outward on the seal.
4. Using a seal installer or its equivalent, seat the seal against the shoulder within the axle tube.
5. During assembly of the brake support plate, install the outer seal.

NOTE: Lubricate the lips of the seals during installation.

Oil Seal Installation and Wheel Bearing Adjustment

Full Floating Axle

1. Raise the rear of the vehicle and support safely. Keep both rear wheels parallel with the floor surface.
2. Remove the axle as outlined under Rear Axle Shaft Removal.
3. Remove the wheel bearing lock nut by bending the lock tab away from the lock nut shoulders. Remove the lock nut and lock.
4. Place a dolly under the rear wheels and maneuver the truck

height so that the wheels are neither hanging nor supporting the truck.

5. Remove the adjusting nut and outer wheel bearing. Pull the wheel assembly outward and off the housing tube.

6. Remove the oil seal and inner wheel bearing from the truck side of the hub assembly.

7. Clean the hub of any old grease and replace the bearings and races as necessary. Repack the bearings if they are to be used again. Pack wheel bearing grease in the hub cavity between the inner and outer bearings.

8. Install the wheel hub oil seal and clean the hub and drum of any grease droppings.

9. Install the wheel assembly on the housing tube, keeping the hub assembly parallel with the housing tube.

10. Install the outer bearing and adjuster nut. Raise the vehicle and wheel assembly upward to clear the dolly.

11. Tighten the adjusting nut to 50 ft. lbs. while rotating the wheels to seat the bearings.

12. Back off the adjusting nut ¼ turn and install the lock washer and lock nut. Tighten the lock nut to 150 ft. lbs. and bend the lock tabs to secure the nut.

    NOTE: Assemblies using doweled adjusting nuts and pierced wheel bearing lock nut, require 200-300 ft. lbs. torque on the outer nut.

13. Install the axle shaft and bolt to the wheel assembly.

14. Lower the vehicle and check the level of lubricant in the differential.

Locking Differentials

For overhaul procedures of differentials with "NoSPIN" and "Powr-Lok" locking units see Rear Axle portion of General Section.

Rear Suspension

Description

The rear springs used on the IH vehicles are classified as leaf, air, and rubber block types. The heavier the vehicle load requirement, the heavier the spring assemblies would be to carry the load. Care should be exercised in the removal and installation of the spring assemblies so that personal injury can be avoided. The suspensions using shock absorbers will normally be found on the light and medium duty vehicles. The removal and installation of the shock absorb-

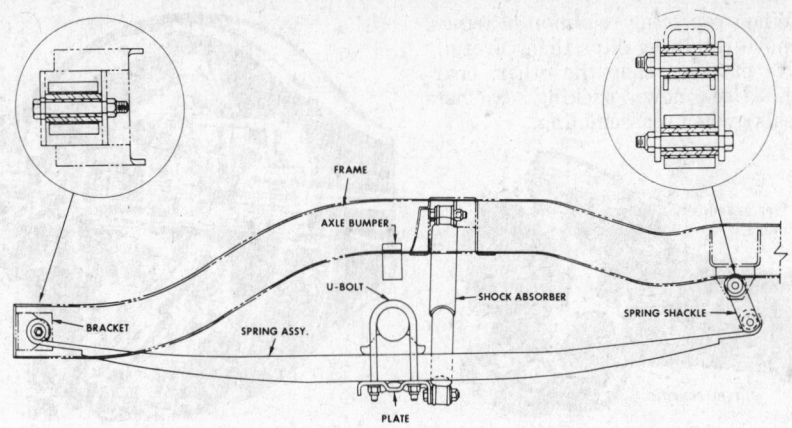

Rear spring installation diagram—light and medium duty vehicles
(© International Harvester Co.)

ers require only the removal and installation of the retaining nuts or bolts, with the possibility of raising the vehicle for working clearance.

The following procedure can be used as a general outline for the removal and installation of leaf springs. The air and rubber block suspension, along with the tandem leaf springs, are outlined in this section.

Rear Spring Removal

1. Place floor jack under truck frame and raise truck sufficiently to relieve weight from spring to be removed.

2. Remove shock absorbers where used.

3. Remove U-bolts, spring bumper and retainer or U-bolt seat.

4. Remove lubricators (not used where springs are equipped with rubber bushings).

5. Remove nuts from spring shackle pins or bracket pins.

6. Slide spring off bracket pin and shackle pin.

7. If spring is rubber bushed, bushing halves may be removed from each side of spring and shackle eye.

Installation

1. Install pivot end of spring first. Align shackle end to other frame bracket. When installing nuts on spring pins which are welded or pressed in, be sure that washer is tightened against shoulder of pin. Spring pins which are driven in must be installed so that slot for lock bolt is aligned. Spring pins which are threaded in must be installed so that the lubrication hole is facing up. Tighten pin into bracket, then back off one-half turn. Install locknut tightly and install cotter pin. Turn pin out to permit installation of cotter pin.

2. Install lubricators.

3. Install U-bolt seat or retainer and U-bolts. Install U-bolt nuts,

but do not tighten.

4. Install shock absorber where used.

5. Lower vehicle.

6. Tighten U-bolt nuts securely.

Air Suspension

The air suspension system was developed to improve the ride characteristics of highway transport vehicles. The major components, whether used on single or tandem axles, are as follows: Trailing arms, frame hanger brackets, shock absorbers, track bars, air springs, axle connections, air leveling valve.

Spring Height Adjustment

1. Start engine, and wait until the air pressure indicates near maximum pressure of 85-90 lbs.

2. Measure the distance from the top of the trailing arm to the bottom of the frame at the rear of the air spring on one side only.

3. If the distance is greater than 12¼ inches, (± ⅜ inch), shorten the linkage from the height control valve arm to the axle, wait 15 seconds and remeasure. Repeat the procedure if necessary.

4. If the distance is less than 12¼ inches, (± ⅜), lengthen the linkage from the height control valve arm to the axle, wait 15 seconds and remeasure. Repeat the procedure if necessary.

5. Repeat the adjustment procedure on the opposite side of vehicle.

Air Spring Pressure Balance

Air spring pressure imbalance on an unloaded IH air suspension system is a normal condition and causes no harm to the suspension components so long as chassis remains unloaded. If, however, the imbalance continues after the chassis is loaded, do not operate the vehicle until the cause is located and corrected. These causes can

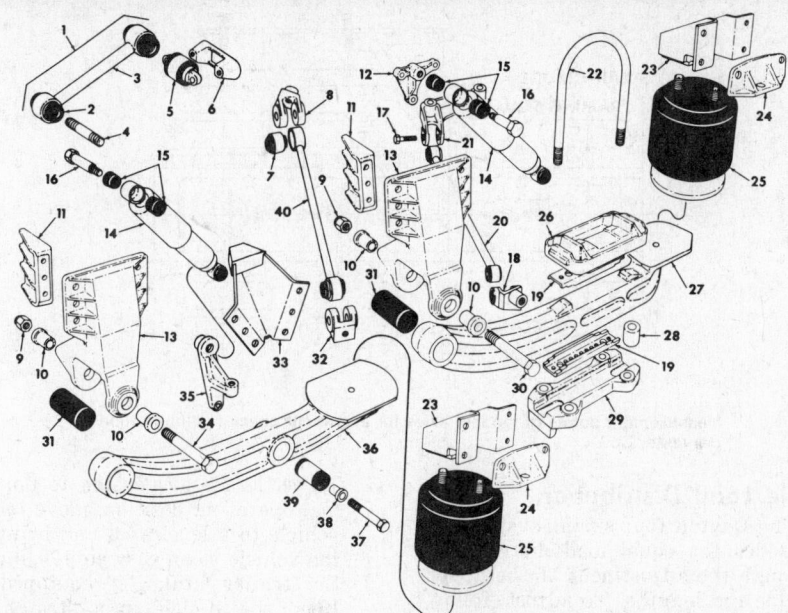

4. Lower the arm and the air spring away from the axle.

5. Installation is the reverse of removal.

Equalizing Beam Suspension (Hendrickson)

Tandem drive axles require a special suspension which permits flexibility between the axles, the equalizing beam suspension. Semi-elliptic springs are used mounted on saddle assemblies above the equalizer beams and pivoted at the front end on spring pins and brackets. The rear end of the springs have no rigid attachment to the spring brackets, but are free to move forward and backward to compensate for spring deflection.

NOTE: As options, airspring assemblies and four point rubber mounted suspensions are available in place of the leaf spring type.

There are two approaches to servicing the suspension system. One is the removal and installation of individual parts. Removal and overhaul of the entire unit can also be done.

CAUTION: When complete removal is performed, be careful when disconnecting the torque rods, springs, or rubber cushions from the frame since the axle assemblies will be free to roll or pivot at the equalizer beam ends. Use jacks and other equipment and block the vehicle securely to prevent injury to personnel and damage to the unit.

1 Rod, torque, assembly	21 Yoke, right rear trac bar mounting
2 Bushing, torque rod	22 U-bolts with nuts
3 Tube, torque rod	23 Bracket, air spring mounting
4 Stud, torque rod mounting	24 Bracket, air spring mounting
5 Shaft, torque rod	25 Spring, air ride, assembly
6 Bracket, torque rod mounting	26 Saddle, axle mounting
7 Bushing, trac bar	27 Arm, rear trailing, assembly
8 Yoke, right front trac bar mounting	28 Spacer, saddle
9 Nut, hex lock, 1 NC	29 Bracket, saddle
10 Adapter, bushing	30 Bolt, hex head, 1 NC x 8
11 Bracket, crossmember mounting	31 Bushing, trailing arm
12 Bracket, rear shock absorber mounting	32 Yoke, left front trac bar mounting
13 Bracket, trailing arm mounting	33 Bracket, shock absorber mounting
14 Absorber, shock, assembly	34 Bolt, hex head, 1 NC x 10
15 Bushing, shock absorber mounting	35 Bracket, Front, lower, shock absorber mounting
16 Bolt, hex head, 3/4 NC x 3	36 Arm, front trailing
17 Bolt, hex head, 3/4 NC x 4	37 Bolt, hex head, 3/4 NF x 7 3/4
18 Yoke, left rear trac bar mounting	38 Adapter, bushing
19 Pad, rear axle mounting	39 Bushing, front trailing arm center
20 Bar, rear trac, assembly	40 Bar, trac, assembly front

Exploded view—air suspension—typical (© International Harvester Co.)

be incorrect spring height adjustment, air leaks, plugged or pinched air lines, defective leveling control valve, or loose or broken parts.

Air Spring Removal and Installation

1. Block the vehicle wheels and release the air pressure build-up from the air brake system.

2. Support the vehicle in the raised position to relieve any pressure on the air spring.

3. Remove the front or rear trailing arm, depending upon the air spring affected, by removing the attaching bolts at the front and center of the arm.

Four Spring Suspension (Dayton)

The four spring suspension system is used to distribute the load over a greater area of the frame rail. The six torque rods of the Dayton four spring suspension serve a dual purpose. The rods provide a means of

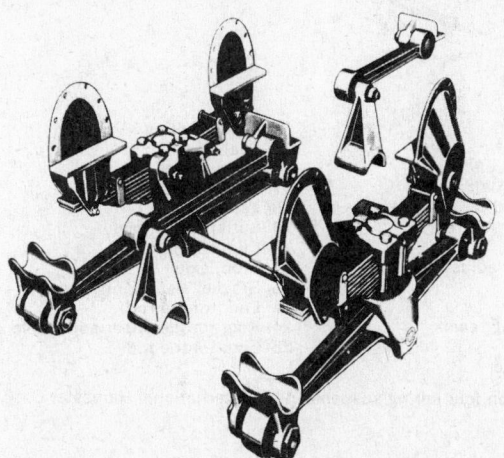

Equalizing beam suspension (leaf spring type)
(© International Harvester Co.)

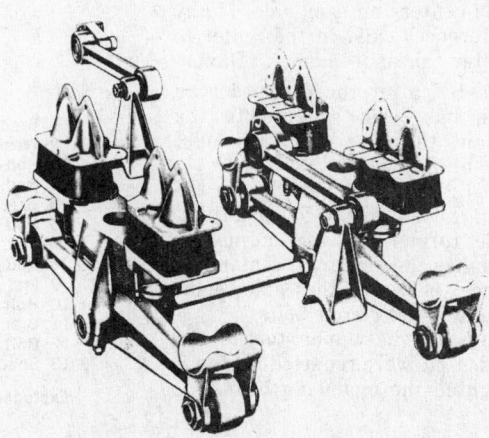

Equalizing beam suspension (four point rubber mounted)
(© International Harvester Co.)

suspension alignment as well as permiting the axles to accept complete drive line torque. The torque rods consist of two non-adjustable and four adjustable units.

The removal and installation of parts can be accomplished by the removal of the individual parts or the removal of the complete unit, as with the Equalizing Beam Suspension.

Axle Alignment

1. Clamp a straight edge to the top of the frame rail ahead of the forward rear axle. Use a framing square against the straight-edge and the outside surface of the frame siderail to insure the straightedge is perpendicular to the frame.
2. Suspend a plumb bob from the straightedge in front of the tire and on the outboard side of the forward rear axle.
3. Position a bar with pointers that can be engaged in the center holes of the rear axles.
4. Measure the distance between cord of the plumb bob and the pointer on the forward axle and record (Dimension A).
5. Position the plumb bob and bar on the opposite side of the vehicle and measure as outlined in paragraph 4. Record the result.
6. Any difference in dimensions from side to side must be equalized if the difference exceeds .0625 inch.
7. Equalize the dimensions by loosening the clamp bolts on the lower adjustable torque rod on the forward rear axle and adjusting the length of the torque rod. Tighten the clamp bolts.

   NOTE: Remove one end of the left and right upper torque rods on the forward rear axle to relieve any stresses which may be present due to an improperly adjusted torque rod, before adjusting the lower torque rods.

8. Reposition the bar pointers to the axle centers on each side. If any differences exist in the center to center measurement, (Dimension B), after the forward rear axle has been squared to the frame, the rear rear axle must also be aligned.
9. To align the rear axle, loosen the clamp bolts on the lower adjustable torque rod and adjust to equalize the center to center distance between the axle ends. Tighten the clamp bolts.
10. Reinstall the upper torque rod ends that were removed in step 7. Tighten the mounting bolts.

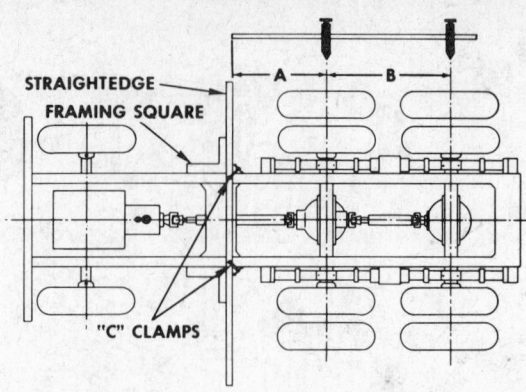

Measurement points of tandem axles for alignment check (© International Harvester Co.)

Axle Load Distribution

The Dayton four spring suspension provides for equal load distribution through the adjustment of the upper torque rod lengths. To adjust, follow this procedure.

1. Disconnect the forward and rear upper torque rods at the frame crossmembers.
2. If the vehicle is equipped with an adjustable fifth wheel, position it in the normal operating location.
3. Apply the maximum rated load on the suspension assembly to obtain the full deflection of the leaf springs when adjusting the torque rods.
4. To settle the suspension to normal operating position, move the vehicle to a level area and bring the vehicle to an easy stop, using the trailer brake, if equipped. Keep the vehicle in a straight ahead position.
5. Loosen the torque clamp bolts and lengthen or shorten the torque rods as required to obtain bolt hole alignment for easy installation of the bolts in the torque rod ends.
6. Tighten the mounting bolts and the torque rod clamp bolts.
7. No further adjustment should be required.

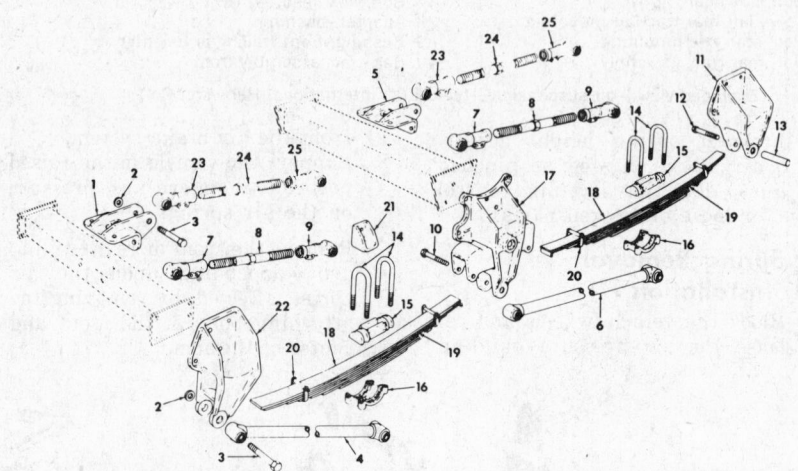

1 Bracket, torque rod
2 Washer, torque rod
3 Bolt and nut, torque rod
4 Rod, torque, lower left front
5 Bracket, torque rod
6 Rod, torque, lower left rear
7 End, torque rod
8 Rod, torque, lower adjustable
9 End, torque rod
10 Bolt, shoulder
11 Bracket, rear spring rear
12 Bolt, shoulder
13 Spacer, spring roller
14 U-bolt, spring
15 Seat, U-bolt
16 Plate, U-bolt
17 Bracket, Equalizer
18 Leaf, spring
19 Spring assembly
20 Clip, spring
21 Stop, axle
22 Bracket, rear spring front
23 End, torque rod
24 Rod, torque upper adjustable
25 End, torque rod

Exploded view—Dayton four spring suspension (© International Harvester Co.)

Index

TUNE-UP SPECIFICATIONS

When analyzing compression test results, look for uniformity among cylinders rather than specific pressures.

Year	Engine No. Cyl Displacement (cu. in.)	hp	Spark Plugs Type	Gap (in.)	Distributor Point Dwell (deg)	Point Gap (in.)	Ignition Timing (deg) ▲	Valves Intake Opens (deg) ■	Fuel Pump Pressure (psi)	Idle Speed (rpm) • Man Trans	Auto Trans
'71	4—134	75	J8	.030	40	.020	5B	9	3	600	N.A.
	6—232	140	N-14Y	.035	32	.016	[1]	12½	5	650[5]	700[5]
	6—258	150	N-14Y	.035	32	.016	5B	12½	4-5½	600	700
	6—225	160	44S	.035	30	.016	5B	24	5	550	N.A.
	8—350	230	H-14Y	.035	30	.016	[4]	24	5	650[5]	700[5]
	8—304	210	N-12Y	.035	30	.016	2½B	18½	4-5½	650	700
	8—360	245	N-12Y	.035	30	.016	2½B	18½	4-5½	650	750
'72	6—232	100	N-12Y	.035	32	.016	5B[6]	12½	4-5	700	600
	6—258	110	N-12Y	.035	32	.016	3B[6]	12½	4-5	700	600
	8—304	150	N-12Y	.035	30	.016	5B[6]	14¾	5-6½	750	650
	8—360	175	N-12Y	.035	30	.016	5B[6]	14¾	5-6½	750	650
'73-'74	6—232	100	N-12Y	.035	32	.016	5B[6]	12½	4-5	700[7]	—
	6—258	258	N-12Y	.035	32	.016	3B[6]	12½	4-5	700[7]	550
	8—304	150	N-12Y	.035	30	.016	5B[6]	14¾	5-6½	750	700
	8—360	175	N-12Y	.035	30	.016	5B[6]	14¾	5-6½	750	700
	8—360	195	N-12Y	.035	30	.016	5B[6]	14¾	5-6½	750	700
	8—401	225	N-12Y	.035	30	.016	5B[6][8]	25½	5-6½	750[9]	700
'75	6—232	100	N-12Y	.035	Electronic		5B	12	4-5	700(600)	—
	6—258	110	N-12Y	.035	Electronic		3B	12	4-5	700[10](600)	550
	8—304	150	N-12Y	.035	Electronic		5B	14¾	5-6½	750	
	8—360	175	N-12Y	.035	Electronic		2-5B	14¾	5-6½	750	700
	8—360	195	N-12Y	.035	Electronic		2-5B	14¾	5-6½	750	700
	8—401	215	N-12Y	.035	Electronic		2-5B	25½	5-6½	750	700
'76-'78	6—232	90	N-12Y	.035	Electronic		8B	12	4-5	600	—
	6—258	95	N-12Y	.035	Electronic		6B[11]	12	4-5	600	550(700)
	8—304	120	N-12Y	.035	Electronic		5B[12]	14¾	5-6½	750	700
	8—360	175	N-12Y	.035	Electronic		5B[13]	14¾	5-6½	750	700
	8—401	215	N-12Y	.035	Electronic		5B[13]	25½	5-6½	750	700

NOTE: Figures in parentheses are for California engines
▲ With vacuum advance disconnected
■ All figures before TDC (BTDC)
• With manual transmission in Neutral and automatic transmission in Drive
B Before top dead center (BTDC)
[1] W/o emission control, 5B on dist. model 1110340, 0 on dist. model 110444, 0 w/ emission control
[2] 5B w/o emission control, 0 w/ emission control
[3] 650/700 rpm w/o emission control
[4] 0 on dist. model 1111330, 1111474, and 1111938, 5B on model 11116964

[5] 100 rpm less w/o emission control
[6] At 550 rpm in 1973; 700 rpm in 1974-75
[7] 700 rpm for CJ, 600 rpm Commando and Wagoneer
[8] 2.5° B on heavy-duty engine (painted red)
[9] 650 rpm on heavy-duty engine (painted red)
[10] 650 rpm w/EGR
[11] 8B w/Automatic transmission
[12] 10B w/Automatic transmission; 5B in California
[13] 8B w/Automatic transmission; 5B in California

FIRING ORDER AND ROTATION

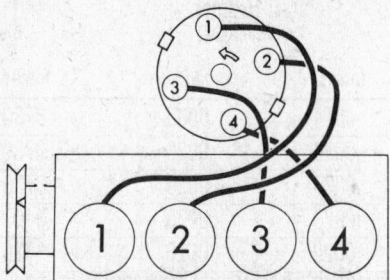
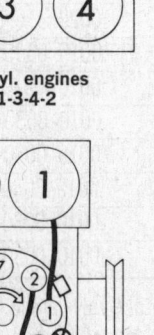

Jeep F head 134 4 cyl. engines
Engine firing order: 1-3-4-2

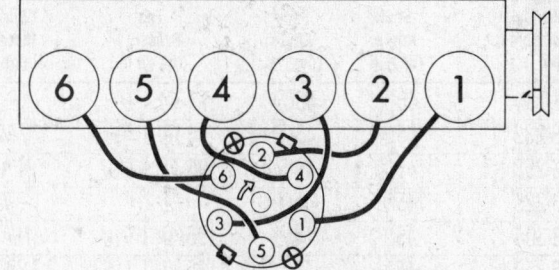
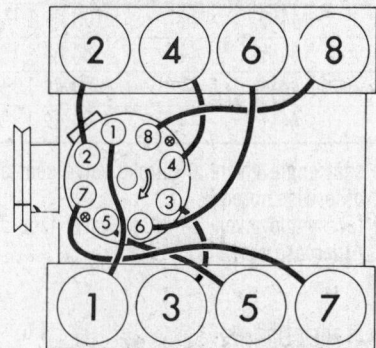

Jeep 232, 258 6 cyl. engines
Engine firing order: 1-5-3-6-2-4

Jeep 304, 360, 401 V8 engines
Engine firing order: 1-8-4-3-6-5-7-2

Jeep (Buick) 350 V8 engine
Engine firing order: 1-8-4-3-6-5-7-2

GENERAL ENGINE SPECIFICATIONS

Year	Engine Cu In. Displacement	Carburetor Type	Advertised Horsepower @ rpm ■	Advertised Torque @ rpm (ft lbs) ■	Bore and Stroke (in.)	Advertised Compression Ratio	Oil Pressure @ 30 mph (psi)
'71	4-134	1 bbl	75 @ 4000	114 @ 2000	3.125 x 3.375	7.4:1/6.9:1/7.8:1	35
	6-232	1 bbl	145 @ 4300	215 @ 1600	3.750 x 3.500	8.5:1	50
	6-258	1 bbl	150 @ 3800	240 @ 1800	3.750 x 3.895	8.0:1/7.6:1	37
	6-225	2 bbl	160 @ 4200	235 @ 2400	3.750 x 3.400	9.0:1	33
	8-350	2 bbl	230 @ 4400	350 @ 2400	3.800 x 3.850	9.0:1	37
	8-304	2 bbl	210 @ 4400	300 @ 2600	3.750 x 3.440	8.4:1	37
	8-360	2 bbl	245 @ 4400	365 @ 2600	4.080 x 3.440	8.5:1	37
'72-'73	6-232	1 bbl	100 @ 3600	185 @ 1800	3.895 x 3.500	8.0:1/7.6:1	37
	6-258	1 bbl	110 @ 3500	195 @ 2000	3.750 x 3.500	8.0:1/7.6:1	37
	8-304	2 bbl	150 @ 4200	245 @ 2500	3.750 x 3.440	8.4:1	37
	8-360	2 bbl	175 @ 4000	285 @ 2400	4.080 x 3.440	8.5:1	37
	8-360	4 bbl	195 @ 4400	295 @ 2900	4.080 x 3.440	8.5:1	37
'74-'78	6-232	1 bbl	100 @ 3600①	185 @ 1800②	3.750 x 3.500	8.0:1	37
	6-258	1 bbl	110 @ 3500③	195 @ 2000④	3.750 x 3.900	8.0:1	37
	8-304	2 bbl	150 @ 4200⑤	245 @ 2500⑥	3.750 x 3.440	8.4:1	37
	8-360	2 bbl	175 @ 4000⑦	285 @ 2400⑧	4.080 x 3.440	8.3:1	37
	8-360	4 bbl	195 @ 4400⑨	295 @ 2900⑩	4.080 x 3.440	8.3:1	37
	8-401	4 bbl	215 @ 4200	320 @ 2800	4.165 x 3.680	8.4:1	37

■ Beginning in 1972, horsepower and torque are SAE net figures. They are measured at the rear of the transmission with all accessories installed and operating. Since the figures vary when a given engine is installed in different models, some are representative rather than exact.

① 90 @ 3050 1976-78
② 170 @ 2000 1976-78
③ 95 @ 3050 1976-78
④ 180 @ 2100 1976-78
⑤ 120 @ 3200 1976-78
⑥ 220 @ 2200 1976-78
⑦ 140 @ 3300 1976-78
⑧ 251 @ 1600 1976-78
⑨ 180 @ 3600 1976-78
⑩ 280 @ 2800 1976-78

VALVE SPECIFICATIONS

Year	Engine No. Cyl. Displacement (cu in.)	Seat Angle (deg) •	Face Angle (deg) ■	Spring Test Pressure (lbs @ in.)	Spring Installed Height (in.)	STEM TO GUIDE Clearance (in.) Intake	Exhaust	STEM Diameter (in.) Intake	Exhaust
'71	4-134	45①	46②	73 @ 1 27/32	⑥	.0014	.0035	.3730	.3710
	6-232	45③	45③	90 @ 1 13/16	2 13/64⑦	.0020	.0020	.3730	.3730
	6-258	45③	44	195 @ 1 7/16	N.A.	.0020	.0020	.3720	.3720
	6-225	45④	45④	⑤	1 23/32	.0020	.0025	.3407	.3407
	8-304	45	44½	195 @ 1 7/16	N.A.	.0020	.0020	.3720	.3720
	8-350	45	45	185 @ 1 11/32	1 23/32	.0015-.0035	.0015-.0035⑪	.372	.378⑩
	8-360	45	44½	195 @ 1 7/16	N.A.	.0020	.0020	.3720	.3720
'72-'78	6-232	44½	44	100 @ 1 13/16⑨	2 15/64⑦⑧	.0010-.0030	.0010-.0030	.3720	.3720
	6-258	44½	44	100 @ 1 13/16⑨	2 15/64⑦⑧	.0010-.0030	.0010-.0030	.3720	.3720
	8-304	44½	44	84 @ 1 13/16⑨	2 7/32⑦⑧	.0010-.0030	.0010-.0030	.3720	.3720
	8-360	44½	44	84 @ 1 13/16⑨	2 7/32⑦⑧	.0010-.0030	.0010-.0030	.3720	.3720
'74-'78	8-401	44½	44	84 @ 1 13/16⑨	2 7/32⑦⑧	.0010-.0030	.0010-.0030	.3720	.3720

• Exhaust valve seat angle given; all intake valve seat angles are 30° unless otherwise noted
■ Exhaust valve face angle given; all intake valve face angles are 29° unless otherwise noted
— Not applicable
N.A. Not available
① Intake valve seat angle 45°
② Intake valve face angle 46°
③ Intake valve angle 30°
④ Intake valve angle 45°
⑤ Intake: 64 @ 1 11/16; exhaust 64 @ 1 5/8
⑥ Free length of intake spring 1 31/32 in.; exhaust 2½ in.
⑦ Free length
⑧ 1974 and later only; other years N.A.
⑨ Without rotators
⑩ .375-371 (head end)
⑪ .00250-.0045 (head end)

TORQUE SPECIFICATIONS
All readings in ft lbs

Engine No. Cyl. Displacement (cu in.)	Cylinder Head Bolts	Rod Bearing Bolts	Main Bearing Bolts	Crankshaft Balancer Bolt	Flywheel to Crankshaft Bolts	MANIFOLD Intake	Exhaust
4-134	60-70	35-45	65-75	60-70	35-41	29-35	29-35
6-232, 258	95-115③	26-30	75-85	50-64	95-120④	37-47②	20-30②
6-225	65-80	30-40	95-120	140-160	50-65	25-35	15-20
8-350	①	35	110	140-180	60	50	18
8-304, 360	100-120⑤	26-30⑥	90-105⑦	48-64⑧	95-120④	37-47⑨	20-30⑥
8-401	100-120	35-40⑥	90-105⑦	48-64⑧	95-120④	37-47⑨	20-30⑥

① Metal gasket 75; composition 80
② 20-25 for 1970-72, 18-28 for 1974-'77
③ 80-85 for 1970-72
④ 100-110 for 1970-72
⑤ 105-115 for 1970-72
⑥ 30-35 for 1976-77
⑦ 95-100 for 1970-72
⑧ 50-60 for 1970-72
⑨ 40-45 for 1970-72

PISTON RING SPECIFICATIONS

	Engine	Ring Gap Top Compression	Bottom Compression	Oil Control	Ring Side Clearance Top Compression	Bottom Compression	Oil Control	Piston to Bore Clearance
'71	4-134		.007-.017		.002-.004	.0015-.0035	.001-.0025	.003 (selective)
	V6-225	.010-.020	.010-.020	.015-.035	.002-.0035	.003-.005	.0015-.0085	.0005-.0011①
	8-350	.010-.020	.010-.020	.015-.055	.002-.004	.002-.004	.000-.005	.0009-.0015①
'71-'78	6-232 6-258	.010-.020	.010-.020	.010-.025	.0015-.003	.0015-.003	.001-.008	.0009-.0017
	8-304	.010-.020	.010-.020	.010-.025	.0015-.0035	.0015-.003	.0011-.008	.0010-.0018
	8-360	.010-.020	.010-.020	.015-.045	.0015-.0035	.0015-.0035	.000-.007	.0012-.0020
	8-401	.010-.020	.010-.020	.0015-.0055	.0015-.003	.0015-.0035	.000-.007	.0010-.0018

① measured at skirt bottom

CRANKSHAFT AND CONNECTING ROD SPECIFICATIONS

All measurements are given in inches

Engine No. Cyl. Displacement (cu in.)	CRANKSHAFT				CONNECTING ROD		
	Main Brg. Journal Dia	Main Brg. Oil Clearance	Shaft End-Play	Thrust on No.	Journal Diameter	Oil Clearance	Side Clearance
4-134	2.333	.0019	.005	1	1.9375	.0014	.007
6-225	2.4995	.0009	.006	2	2.0000	.0021	.010
6-232	2.4986-2.5001	.001-.003	.0015-.0065	3	2.0934-2.0955	.001-.002	.005-.014①
6-258	2.4986-2.5001	.001-.002	.0015-.0065	3	2.0934-2.0955	.001-.002	.005-.014①
8-350	2.9995	.0010	.006	3	2.0000	.0012	.010
8-304	2.7489-2.7474②	.001-.003	.003-.008	3	2.0934-2.0955	.001-.002	.006-.018
8-360	2.7489-2.7474②	.001-.003	.003-.008	3	2.0934-2.0955	.001-.002	.006-.018
8-401	2.7489-2.7474②	.001-.003	.003-.008	3	2.2464-2.2485	.001-.002	.006-.018

① .008-.010 1973-only
② Rear main, 2.7479-2.7464

WHEEL ALIGNMENT

Model	CASTER Pref. Setting (deg)	CAMBER Pref. Setting (deg)	Toe-In (in.)	King-Pin Inclination (deg)	WHEEL PIVOT RATIO Inner Wheel	Outer Wheel
CJ-5, CJ6, CJ-7, DJ-5, DJ-6, CJ-5A, CJ-6A	3	1°30′	3/64-3/32	7½①	20	20
Jeepster, Commando	3	1°30′	3/64-3/32	7½	N.A.	N.A.
Wagoneer, Cherokee	3②	1°30′	3/64-3/32	7½①	N.A.	N.A.

N.A. Not available
① 8½° in 1974-'75
② 4° in 1974-78

TRUCK MODEL AND ENGINE APPLICATION

(All Jeep engines are gas engines)

Truck Model	Cu. In.	Carb.	H.P.	Torque (ft lbs)
CJ-5, CJ-6 and CJ-7	134-4 Cyl.	1 bbl	75	114
	225-V6	2 bbl	160	235
	232	1 bbl	100①	185②
	258	1 bbl	110③	195④
	304-V8	2 bbl	150⑤	245⑥
Wagoneer	232⑦	1 bbl	145	215
	232	1 bbl	100①	185②
	258	1 bbl	110③	195④
	304-V8	2 bbl	150⑤	245⑥
	350-V8	2 bbl	230	350
	360-V8	2 bbl	175	285
	360-V8	4 bbl	195	295
	401-V8	4 bbl	215	320

Truck Model	Cu. In.	Carb.	H.P.	Torque (ft lbs)
Cherokee	258	1 bbl	110③	195④
	360-V8	2 bbl	175	285
	360-V8	4 bbl	195	295
	401-V8	4 bbl	215	320
Jeepster and Commando	134-4 Cyl.	1 bbl	75	114
	225-V6	2 bbl	160	235
	232	1 bbl	100	185
	258	1 bbl	110	195
	304-V8	2 bbl	150	245
J-100 Truck	232	1 bbl	100①	185②
	258	1 bbl	110③	195④
	304-V8	2 bbl	150⑤	245⑥
	350-V8	2 bbl	230	350
	360-V8	2 bbl	175	285
	360-V8	4 bbl	195	295
	401-V8	4 bbl	215	320

① 90 HP in 1976-78
② 170 ft lbs in 1976-78
③ 95 HP in 1976-78
④ 180 ft lbs in 1976-78
⑤ 120 HP in 1976-78
⑥ 220 ft lbs in 1976-78

Distributor

Refer to the Electrical General Repair section for detail procedures on the different distributors.

Starting in the 1975 model year all engines use the American Motors Breakerless Inductive (BID) Ignition System. This system consists of five major components: an electronic ignition control unit, an ignition coil, a distributor, high tension spark plug wires, and spark plugs. This system uses a conventional coil, spark plug wires, and spark plugs. The main difference in components from a conventional system is the points and condensor are eliminated from the distributor and are replaced by a trigger wheel and an electromagnetic sensor, and an electronic control unit is added.

Distributor R & R

1. Remove the high-tension wires from the distributor cap terminal towers, noting their positions to assure correct reassembly. For diagrams of firing orders and distributor wiring, refer to the front of this section.
2. Remove the primary lead from the terminal post at the side of the distributor.
3. Disconnect the vacuum tube if there is one.
4. Unlatch the two distributor cap retaining hooks and remove the distributor cap.
5. Note the position of the rotor in relation to the base. Scribe a mark on the base of the distributor and on the engine block to facilitate reinstallation. Align the marks with the direction the metal tip of the rotor is pointing.
6. Remove the screw that holds the distributor to the engine.
7. Lift the distributor assembly from the engine.

If the engine has not been disturbed, install the distributor as follows:

1. Insert the distributor shaft and assembly into the engine. Line up the mark on the distributor and the one on the engine with the metal tip of the rotor. Make sure that the vacuum advance diaphragm is pointed in the same direction as it was pointed originally. This will be done automatically if the marks on the engine and the distributor are lined up with the rotor.

NOTE: On the F4 the distributor shaft fits into a slot in the end of the oil pump shaft.

2. Install the distributor hold-down bolt and clamp. Leave the screw loose enough so that you can

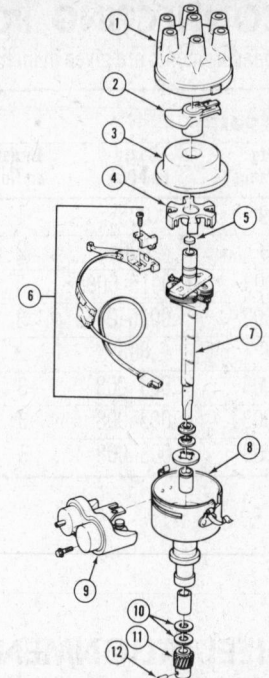

Electronic ignition distributor (BID)—exploded view (© Jeep Corp.)

1. DISTRIBUTOR CAP
2. ROTOR
3. DUST SHIELD
4. TRIGGER WHEEL
5. FELT WICK
6. SENSOR ASSEMBLY
7. SHAFT ASSEMBLY
8. HOUSING
9. VACUUM CONTROL
10. SHIM
11. DRIVE GEAR
12. PIN

move the distributor with heavy hand pressure.

3. Connect the primary wire to the distributor side of the coil. Install the distributor cap on the distributor housing. Secure the distributor cap with the spring clips or the screw type retainers, whichever is used.
4. Install the spark plug wires. Make sure that the wires are pressed all of the way into the top of the distributor cap and firmly onto the spark plugs.

NOTE: Design of the V6 engine requires a special form of distributor cam. The distributor may be serviced in the regular way and should cause no more problems than any other distributor, if the firing pin is thoroughly understood. The distributor cam is not ground to standard six cylinder indexing intervals. This particular form requires that the original pattern of spark plug wiring be used. The engine will not run in balance if number one spark plug wire is inserted into number six distributor cap tower, even though each wire in the firing sequence is advanced to the next distributor tower. There is a difference between the firing intervals of each succeeding cylinder through the 720° engine cycle.

5. Adjust the point cam dwell and set the ignition timing.

If the engine has been turned while the distributor has been removed, or if the marks were not drawn, it will

be necessary to initially time the engine. Follow the procedure below:

1. It is necessary to place the No. 1 cylinder in the firing position to correctly install the distributor. To locate this position, some engines have marks placed on the flywheel while other engines have marks placed on the timing gear covers and crankshaft pulleys. The flywheel marks may be viewed through a covered opening directly in back of the starting motor by loosening the hole cover and sliding it to one side.
2. Remove the No. 1 cylinder spark plug. Turn the engine until the piston in No. 1 cylinder is moving up on the compression stroke. This can be determined by placing your thumb over the spark plug hole and feeling the air being forced out of the cylinder. Stop turning F-head engines when either the 5° mark on the flywheel is in the middle of the flywheel inspection opening, or the marks on the crankshaft pulley and the timing gear cover are in alignment.
3. Oil the distributor housing lightly where the distributor bears on the cylinder block.
4. Install the distributor so that the rotor, which is mounted on the shaft, points toward the No. 1 spark plug terminal tower position when the cap is installed. Of course you won't be able to see the direction in which the rotor is pointing if the cap is on the distributor. Lay the cap on the

Timing mark locations on six cylinder and V8 AMC engines (© Jeep Corp.)

top of the distributor and make a mark on the side of the distributor housing just below the No. 1 spark plug terminal. Make sure that the rotor points toward that mark when you install the distributor.

5. When the distributor shaft has reached the bottom of the hole, move the rotor back and forth slightly until the driving lug on the end of the shaft enters the slot, which is cut in the end of the oil pump gear on the F4, or when the drive gears of the distributor and cam mesh on the other engines, and until the distributor assembly slides down into place.

On models that have a gear on the end of the distributor shaft and a gear on the end of the oil pump drive, these gears have to mesh with the same teeth as originally installed when the distributor is inserted into the engine. Once again, the marks that were placed on the engine and the base of the distributor housing come into play. If the distributor shaft gear and the oil pump drive gear are but one tooth off from what they are supposed to be, the engine will not run correctly.

6. When the distributor is correctly installed, the breaker points should be in such a position that they are just ready to break contact with each other. This is accomplished by rotating the distributor body after it has been installed in the engine. Once again, line up the marks that you made before the distributor was removed from the engine.

7. Install the distributor hold-down screw and the hold-down bracket. Be sure that the models that have vacuum advance units are free to turn in the mounting socket. Note that the vacuum advance control of some distributors is connected directly to the plate on which the points are mounted. When this is the case, the plate must be free to turn rather than the distributor body.

8. Install the spark plug into the No. 1 spark plug hole and continue from Step 3 of the distributor installation procedure.

Contact Points R & R

1. Remove the distributor cap by releasing the hold-down screws or clamps. Remove the rotor.

2. Release the primary and condensor wire from the point set. Remove the holding screws from the point set and remove the points from the distributor.

3. Clean the breaker plate with a clean cloth, to remove any dirt or oil.

4. Align the pilot pin on the new point set with the pilot hole in the breaker plate and install the point set. Lubricate the cam lobe or the cam wick.

5. Connect the primary and condensor leads to the point set.

6. Rotate the engine until the contact set rubbing block is resting on the high point of a cam lobe.

7. Adjust the point gap to specifications and tighten the locking screw.

8. Replace the rotor and cap, start the engine to check the point dwell and the ignition timing.

Ignition Timing

All Engines

1. Locate the timing marks on the crankshaft pulley and the front of the timing case cover.

2. Clean off the timing marks, so that you can see them.

3. Use chalk or white paint to color the mark on the scale that will indicate the correct timing, when aligned with the mark on the pulley or the pointer. It is also helpful to mark the notch in the pulley or the tip of the pointer with a small dab of color.

4. Attach a tachometer to the engine.

5. Attach a timing light to the engine.

6. Disconnect the vacuum lines to the distributor at the distributor and plug the vacuum lines. Disconnect the TCS switch if so equipped. Loosen the distributor lock-bolt just enough so that the distributor can be turned with a little resistance.

7. Check to make sure that all of the wires clear the fan and then start the engine.

8. Adjust the idle to the correct specification.

9. With the timing light aimed at the pulley and the marks on the engine, turn the distributor in the direction of rotor rotation to retard the spark, and in the opposite direction of rotor rotation to advance the spark. Align the marks on the pulley and the engine with the flashes of the timing light.

Alternator

Refer to the Electrical General Repair Section for detailed alternator test and overhaul procedures.

CAUTION: Since the AC generator and regulator are designed for use on only one polarity system, the following precautions must be observed:

a. The polarity of the battery, generator and regulator must be matched and considered before making any electrical connections in the system.

b. When connecting a booster battery, be sure to connect the negative battery terminals together and the positive battery terminals together.

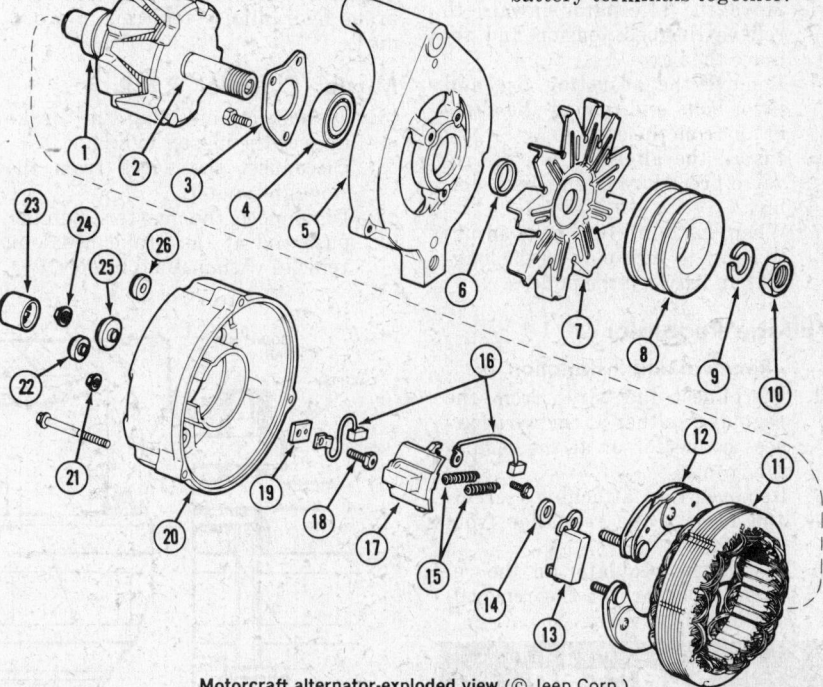

Motorcraft alternator-exploded view (© Jeep Corp.)

1. ROTOR	10. NUT	19. BRUSH TERMINAL INSULATOR
2. STOP RING	11. STATOR	20. REAR HOUSING
3. FRONT BEARING RETAINER	12. RECTIFIER ASSEMBLY	21. GRD TERMINAL NUT
4. FRONT BEARING	13. RADIO NOISE SUPPRESSION CAPACITOR	22. FIELD INSULATOR (ORANGE)
5. FRONT HOUSING	14. INSULATOR CAPACITOR	23. REAR BEARING
6. FRONT BEARING SPACER	15. BRUSH SPRING	24. BAT TERMINAL NUT
7. FAN	16. BRUSH SET	25. BATTERY INSULATOR (RED)
8. PULLEY	17. BRUSH HOLDER	26. STATOR INSULATOR (BLACK)
9. LOCKWASHER	18. BRUSH TERMINAL SCREW	

c. When connecting a charger to the battery, connect the charger positive lead to the battery positive terminal. Connect the charger negative lead to the battery negative terminal.

d. Never operate the AC generator on open circuit. Be sure that all connections in the circuit are clean and tight.

e. Do not short across or ground any of the terminals on the AC generator.

f. Do not attempt to polarize the AC generator.

g. Do not use test lamps of more than 12 V for checking diode continuity.

h. Avoid long soldering times when replacing diodes or transistors. Prolonged heat is damaging to these units.

i. Disconnect the battery ground terminal when servicing any AC system. This will prevent the possibility of accidentally reversing polarity.

Alternator

Removal and Installation

1. Remove the negative battery cable from the battery.
2. Remove the wire terminals attached to the rear of the alternator.
3. Loosen the bolt holding the adjusting bar and the pivot bolt at the opposite side of the alternator.
4. Move the alternator inward to relieve the belt tension and remove the belt.
5. Remove the adjusting bar and pivot bolts and remove the alternator from the engine.
6. Install the alternator in the reverse procedure of the disassembly.
7. When installing the belt, adjust to allow 1/2 inch play on the longest run between the pulleys.

Voltage Regulator

Removal and Installation

1. Disconnect the wires from the regulator, either at the wire harness connector or at the regulator frame.
2. Remove the attaching screws and remove the regulator from the vehicle.
3. Install the regulator in the reverse procedure of the removal, and attach all wires.

Starter

Reference

Refer to the Electrical General Repair Section for detailed starter test and overhaul procedures.

Starter R & R

232 and 258 Sixes

Remove the oil filler pipe and disconnect the battery and solenoid leads from the starter. From underneath the vehicle, remove the bolts that attach the starter to the bellhousing and lift out the starter.

Install in the reverse of removal.

304 V8, 360 V8, 401 V8

Disconnect the battery lead and solenoid lead from the starter. From underneath the vehicle, remove the attaching bolts and lift out the starter. Install in the reverse of removal.

225 V6

Disconnect the negative battery cable from the battery. Note the locations of the wiring connections and disconnect the electrical leads from the starter. Remove the capscrew which secures the starter motor to the angle bracket on the side of the engine. Remove the two capscrews which secure the drive end of the starter motor to the cylinder block and remove the starter. Install the starter in the reverse order of removal.

Brakes

Reference

Refer to the Brakes General Repair Section for detail troubleshooting and brake hydraulic system repair procedures.

Master Cylinder R & R

1. Disconnect and plug the brake lines at the master cylinder.
2. Disconnect the wires from the stoplight switch.
3. Disconnect the master cylinder push rod at the brake pedal on vehicles with manual brakes.

4. Remove all attaching bolts and nuts and lift the master cylinder from the vehicle.
5. Install the master cylinder in the reverse order of removal and bleed the hydraulic system.

Power Unit R & R

1. Clean the master cylinder and booster unit.
2. Remove the cotter and clevis pins securing the booster pushrod to the pedal linkage.
3. Disconnect the vacuum hose from the booster check valve.
4. Disconnect the fluid lines from the master cylinder. Plug the ends and catch any escaping fluid. *Do not reuse brake fluid.*
5. Disconnect the stoplight wires from the switch.
6. Remove the attaching nuts, booster unit assembly, and block spacers.
7. Remove the attaching nuts and separate the master cylinder from the booster.

To install the booster unit, reverse the removal procedure and bleed the brakes.

CAUTION: Do not pressure-bleed power-assisted brake systems.

Wheel Cylinder R & R

1. Raise and support the vehicle and remove the brake drums and brake shoes.
2. Disconnect the brake line. Do not bend the line away from the wheel cylinder. When the cylinder is removed from the support plate, the line will separate from the wheel cylinder easily.
3. Remove the wheel cylinder mounting bolts and remove the wheel cylinder from the brake backing plate.
4. Clean the wheel cylinder mounting surface on the brake support plate. Clean the brake line fitting and threads.

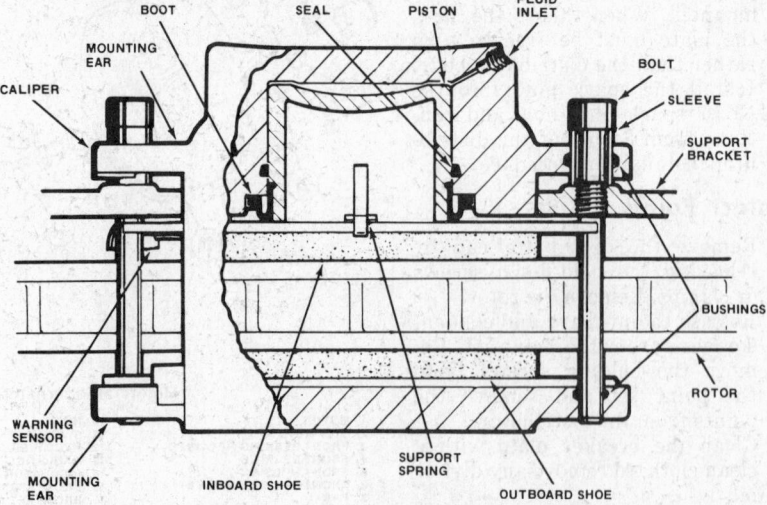

Single piston caliper—exploded view (© Jeep Corp.)

5. Start the brake line fitting into the wheel cylinder and attach the wheel cylinder to the support plate and tighten the brake line fitting. Tighten the wheel cylinder mounting bolts to 18 ft lbs.

Disc Brakes

Floating caliper, single piston type disc brakes are used on the front wheels of the Jeep vehicles, and consists of three assemblies, the caliper assembly, the hub and rotor assembly, and the support and shield assembly.

The caliper is attached to the support and shield assembly and upon hydraulic pressure application, the piston within the caliper is forced outward and pushes the inboard shoe against the rotor face. The reaction force moves the caliper body and the outboard shoe against the opposite rotor face, causing a pinching action of the two brake shoes against the rotor and bringing the vehicle to a stop. Brake adjustment is not needed because wear is automatically compensated for by the sliding movement of the caliper and the increased piston extension.

Brake Shoes

Replace the brake shoes when the linings are worn within 1/32 inch of the shoe or rivets.

Removal

1. Raise the front of the vehicle and support. Remove the wheel and tire assembly.
2. Remove approximately 2/3 of the brake fluid from the front section of the master cylinder.
3. Using a C-clamp, bottom the piston in its bore by placing the solid end of the clamp on the back of the caliper and the screw end contacting the metal part of the out board shoe, and tighten the clamp screw. *NOTE: This procedure backs the brake shoes off the rotor surface, easing the lining replacement.*
4. Remove both allen head mounting screws from the caliper to

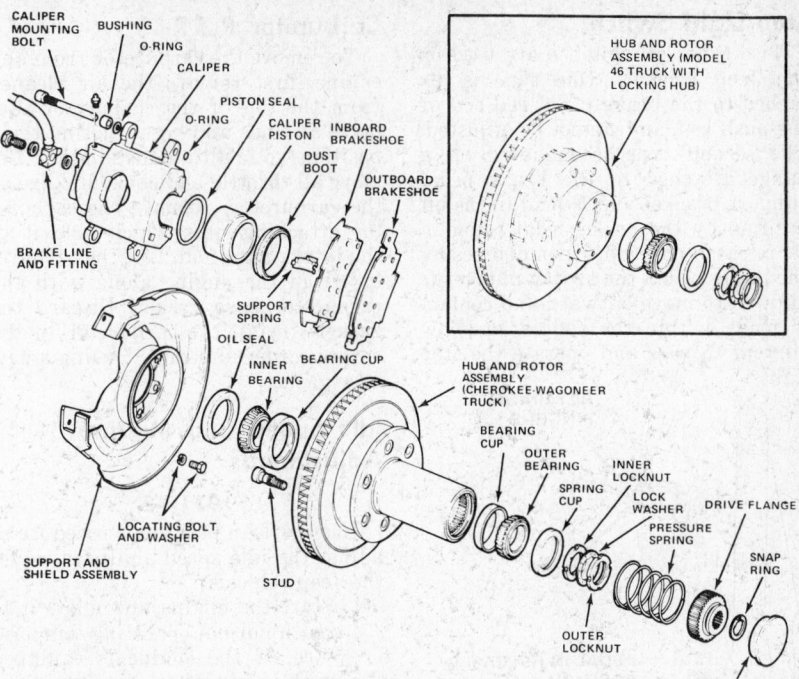

Disc brake assembly—1977-78 Cherokee, Wagoneer, and truck (© Jeep Corp.)

support, and lift the caliper off the rotor. *NOTE: Hang the caliper by a wire hook or tie it to the frame, to avoid allowing the brake hose to support the weight of the caliper assembly.*

5. Remove both disc brake shoes from the caliper, and note the position of the support spring on the inboard shoe for later installation, and remove the spring from the shoe.
6. Remove the sleeves and rubber bushings from the ears of the calipers.

Cleaning and Inspection

Clean the sliding surfaces of the caliper and clean any dirt from the mounting bolts, clips or keys. Inspect the boot on the piston for signs of cracks, cuts of any other damage. Check to see if there is any signs of fluid leakage around the seal on the piston. This will show up in the boot.

If there is any indication of a fluid leak, the entire caliper will have to be overhauled. *NOTE: Refer to the Brake General Repair Section for caliper overhaul procedures.*

Installation

1. Clean all mounting holes, bolts, and bushing grooves. Lubricate and install new bushings and sleeves. *NOTE: Sleeves should be installed in the inboard mounting ears of the caliper, and positioned so the sleeve end facing the shoe and lining, is flush with the machined surface of the mounting ear.*
2. Install the support spring on the inboard shoe. Place the single tang end of the spring over the notch in the shoe.
3. Install the two brake shoes into the caliper and assure that both shoes are fully seated by having the ears of the shoes resting on the ears of the caliper.
4. Position the caliper over the rotor and install the mounting bolts. Assure that the bolts pass through the holes of the outboard shoes and caliper ears, and that the retaining ears of the inboard shoes are over the bolts. Torque the bolts to 35 ft. lbs.
5. Fill the master cylinder and pump the brake pedal to seat the shoes to the rotor.
6. Bend the upper ears of the outboard shoe until the radial clearance between the shoe and the caliper is eliminated.
7. Install the wheel and tire. Lower the vehicle and test the brakes before moving the vehicle.

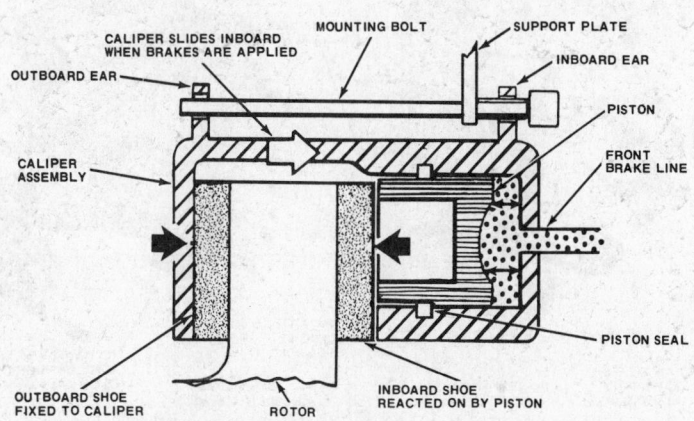

Operation of disc brake—typical (© Jeep Corp.)

Stop Light Switch

Two types of switches are used on the Jeep vehicles. One type is attached to the brake pedal rod end of the push rod, and cannot be adjusted. The second type is mounted on a flange attached to the brake pedal support bracket and is held in the off position by the brake pedal being in its released position. Upon depressing the brake pedal, the switch plunger is allowed to move outward and contact is made within the switch to allow current to pass and operate the stop lights.

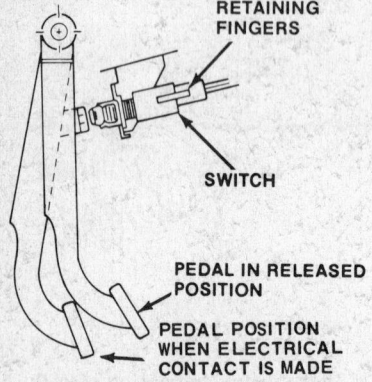

Adjustable stop light switch operation (© Jeep Corp.)

Switch Adjustment

1. Release the brake pedal, unhook the retaining fingers of the wire connector from the switch and remove the wire harness.
2. Adjust the switch by turning it in or out of the mounting bracket. The switch should operate after ⅜ to ⅝ inch of brake pedal travel.
3. Connect the wire harness and recheck the switch operation.

Parking Brake

Adjustment

1. Make sure that the hydraulic brakes are in satisfactory adjustment.
2. Raise the rear wheels off the ground and disengage the parking brake pedal.
3. Loosen the locknut on the brake cable adjusting rod, located directly behind the frame center crossmember.
4. Spin the wheels and tighten the adjustment until the rear wheels drag slightly. Loosen the adjustment until there is no drag and the wheels spin freely.
5. Tighten the locknut to lock the adjusting nut.

Fuel System

Reference

Refer to the Carburetor General Repair Section for exploded views of carburetors and specifications.

Carburetor R & R

To remove the carburetor from any engine, first remove the air cleaner from the top of the carburetor. Remove all lines and hoses, noting their positions to facilitate installation. Remove all throttle and choke linkage at the carburetor. Remove the carburetor attaching nuts which hold it to the intake manifold. Lift the carburetor from the engine along with the carburetor base gasket. Discard the gasket. Install the carburetor in the reverse order of removal, using a new base gasket.

Idle Speed and Mixture Adjustments

1971-72

Use the 'lean best idle' procedure to adjust the idle speed and mixture on the Jeep vehicles.

1. Start the engine and allow it to reach normal operating temperature. If the engine is equipped with an air pump, disconnect the bypass valve air inlet hose.
2. Adjust the idle speed to the specified rpm by turning either the idle adjustment screw or the throttle stop solenoid.

NOTE: If the idle adjustment procedure is not completed within 3 minutes, run the engine at 2000 rpm for 1 minute to stabilize the engine temperature.

3. Turn the mixture screw(s) counterclockwise (richer until loss of rpm is noticed, then turn mixture screw(s) clockwise (leaner), past original starting point, counting the number of the screws(s).
4. Continue turning mixture screw(s) clockwise until engine looses rpm due to an overly lean mixture.
5. Return the screw(s) to the mid-

point between the two extremes and the highest rpm reading.
6. Turn the mixture adjusting screw(s) clockwise (leaner) to the point where the engine rpm just begin to drop, then counterclockwise the minimum amount to obtain the previously established highest rpm. This is the 'lean best idle'.

NOTE: This procedure should first be attempted with the limiter caps installed on the mixture adjustment screw(s). If a satisfactory idle cannot be obtained with the caps in place, then carefully remove them by threading a sheet metal screw into the center of the cap and make the adjustment as outlined above. Once the 'lean best idle' has been established, install service limiter caps with the tabs positioned against the full rich stops.

1973-78

The procedure for adjusting the idle speed and mixture on 1973-78 Jeep vehicles is called the lean drop procedure and is made with the engine operating at normal operating temperature and the air cleaner in place as follows:

1. Turn the mixture screws to the full rich position with the tabs on the limiters against the stops. Note the position of screw head slot inside limiter cap slots.
2. Remove idler limiter caps.
3. Remove limiter caps by threading a sheet metal screw in center of cap and turning clockwise. Discard limiter caps.
4. Reset adjustment screws to same position noted before limiter caps were removed.
5. Start engine and allow it to reach normal operating temperature.
6. Adjust idle speed to 30 rpm

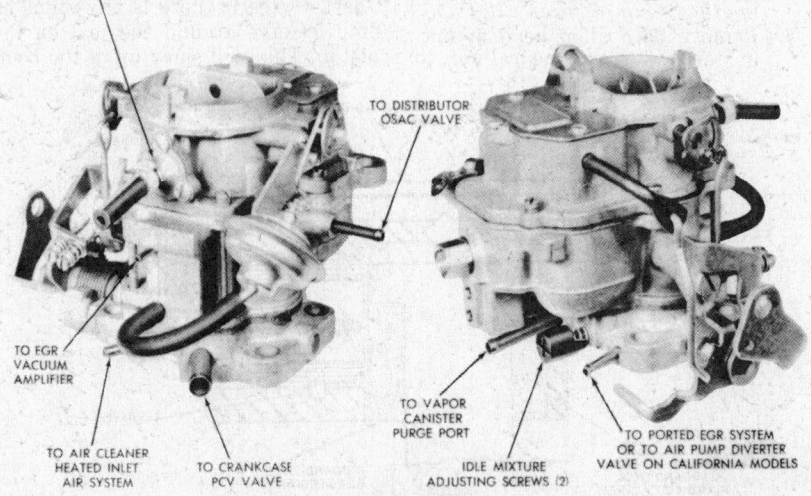

Carter BBD carburetor (© Chrysler Corp.)

above the specified rpm in the Tune-Up Specifications.

  a. On 6 cylinder engines with throttle stop solenoid, turn nut on solenoid plunger in or out to obtain specified rpm. This is done with solenoid wire connected.

  b. On V8 engines with throttle stop solenoid, turn hex screw on throttle stop solenoid carriage to obtain specified rpm. This is done with solenoid wire connected.

  c. Tighten solenoid locknut, if so equipped.

  d. Disconnect solenoid wire and adjust curb idle speed screw to obtain idle speed of 500 rpm.

  e. Re-connect solenoid wire.

7. Starting from the full rich position, as was determined before limiter caps were removed, turn mixture adjusting screws clockwise (leaner) until a loss of engine speed is noticed.

8. Turn screws counterclockwise (richer) until the highest rpm reading is obtained at lean best idle setting. The lean best idle setting is on the lean side of the highest rpm setting without changing rpm.

9. If the idle speed changed more than 30 rpm during the mixture adjustment procedure, reset the idle speed to 30 rpm above the specified rpm with idle speed adjusting screw or the throttle stop solenoid and repeat the mixture adjustment.

10. The final adjustment is to turn the mixture adjusting screws clockwise until engine rpm drop as follows:

1973-74

6 Cylinder Automatic	20 rpm
6 Cylinder Manual	35 rpm
V8 All	40 rpm

1975

6 Cylinder Automatic	25 rpm
6 Cylinder Manual with EGR and Catalytic Converter	35 rpm
6 Cylinder Manual with EGR only	50 rpm
V8 Automatic	20 rpm
V8 Manual	40 rpm

1976-78

6 Cylinder Automatic	25 rpm
6 Cylinder Manual	50 rpm
V8 Automatic	20 rpm
V8 Manual	100 rpm

11. Install new limiter caps over mixture adjusting screws with tabs positioned against full rich stops. Be careful not to disturb idle mixture setting while installing caps.

Fuel Pump R & R

All Engines

1. Disconnect the fuel lines leading to the carburetor and from the fuel tank.
2. Remove the two attaching bolts that hold the fuel pump to the engine and lift the fuel pump off of the engine.
3. Before installing the fuel pump, make sure that all of the mating surfaces are clean.
4. Cement a new gasket to the mating surface of the fuel pump.
5. Position the fuel pump on the cylinder block so that the cam lever of the pump rests on the camshaft.
6. Secure the pump to the engine with the two bolts and lock washers.
7. Connect the fuel lines to the fuel pump.

Fuel Tank R & R

1971 CJ-5 and CJ-6

The fuel tank on these models is installed under the driver's seat.

To remove the fuel tank, first make sure that the tank is either completely drained or that the level is at least below any of the vent lines or filler openings so that when these lines are disconnected fuel will not run out.

Remove the driver's seat from the vehicle.

Disconnect all of the ventline hoses, the fuel gauge electrical lead, the fill hose and the fuel outlet line at the tank.

Remove the tank hold down screws from the mounting brackets, or the hold down strap, and lift the tank from the vehicle.

If there is still gas in the tank, be careful not to spill any fuel when lifting it out of the vehicle. Also, empty the tank of all fuel and flush it with water before soldering or welding the tank.

Install the tank in the reverse order of removal.

1972-78 All Models

The fuel tank on 1972-78 models is attached to the frame by brackets and bolts. The brackets are attached to the tank at the seam flange.

Before removing the fuel tank, make sure the level of the fuel inside the tank is at least below any of the various hoses connected. It is best to either drain or siphon the majority of fuel out of the tank to make it easier to handle while removing it.

To remove the tank, loosen all of the clamps retaining hoses to the tank and disconnect the hoses from the tank. It may be necessary to remove the fuel tank-to-mounting bracket screws and lower the tank slightly to gain access to some of the connecting hoses. Disconnect the tank from the mounting brackets, if not already done, and lower the tank from under the vehicle. Be careful not to spill any fuel in the tank while removing it.

Empty the tank of all fuel and flush it with water before soldering or welding the tank.

Install the fuel tank in the reverse order of removal.

Cooling System

Water Pump R & R

F4

1. Drain the coolant from the system by opening the petcock.
2. Remove the fan belt, fan, and fan pulley.
3. Remove the bolts which attach

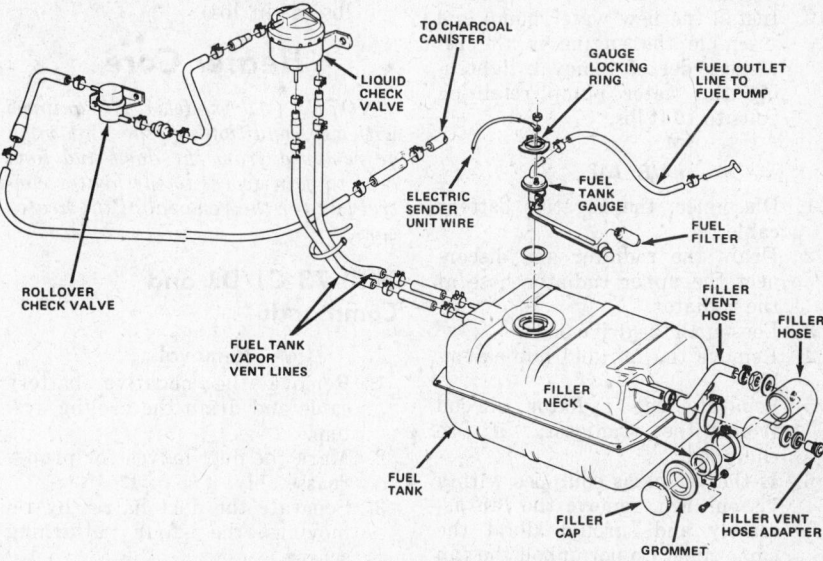

Fuel tank and vent lines—1977 CJ-5 and CJ-7 (© Jeep Corp.)

the water pump to the block and remove the pump.
4. Install the pump in the reverse order of removal, using a new gasket.

225 V6

1. Drain the cooling system.
2. Remove the fan belt and the fan and pulley from the hub on the water pump.
3. Disconnect the hoses from the water pump.
4. Remove the capscrews which secure the water pump to the engine and remove the pump from the vehicle.
5. Clean all of the mating surfaces, install new gaskets with sealer applied and install the water pump in the reverse order of removal.

232 and 258 Sixes

1. Drain the cooling system at the radiator.
2. Disconnect the radiator and heater hoses from the water pump.
3. Loosen the alternator adjustment strap screw, upper pivot bolt, and remove the drive belt.
4. If the vehicle is equipped with a radiator shroud, separate the shroud from the radiator to facilitate removal and installation of the cooling fan and hub.
5. Remove the cooling fan and hub assembly.
6. Remove air conditioning intermediate idler pulley and mounting bracket, if so equipped.
7. Remove the power steering pump front mounting bracket, if so equipped.
8. Remove the water pump and gasket from the engine.
9. Clean all the old gasket material from the gasket surface of the engine.
10. Install the new water pump and assemble the engine in the reverse order of removal, tightening the water pump retaining bolts to 13 ft lbs.

V8 All

1. Disconnect the negative battery cable.
2. Drain the radiator and disconnect the upper radiator hose at the radiator.
3. Loosen all the drive belts.
4. Remove the fan and hub assembly.
5. Separate the radiator shroud from the radiator, if so equipped.
6. If the vehicle is equipped with a viscous fan, remove the fan assembly and shroud all at the same time. Do not unbolt the fan blades.

NOTE: The studs in the water pump may back out of the water pump while removing the nuts, preventing the fan assembly from clearing the water pump. If this happens, install a double nut on the stud(s) and remove the studs.

7. If the vehicle is equipped with air conditioning, install a double nut on the air conditioning compressor bracket to water pump stud and remove the stud. Removal of this stud eliminates removing the compressor mounting bracket.
8. Remove the alternator and mounting bracket assembly and place it aside. Do not disconnect the alternator wires.
9. Remove the two nuts attaching the power steering pump to the rear half of the pump mounting bracket, if so equipped.
10. Remove the two bolts attaching the front half of the bracket to the rear half.
11. Remove the remaining upper bolt from the inner air pump support brace, loosen the lower bolt and drop the brace away from the power steering front bracket.
12. Remove the front half of the power steering bracket from the water pump mounting stud.
13. Disconnect the heater hose, bypass hose, and lower radiator hose at the water pump.
14. Remove the water pump and gasket from the timing chain cover and clean all old gasket material from the gasket surface of the timing chain cover.
15. Install the new water pump and assemble the remaining components in the reverse order of removal, tightening the water pump-to-engine block screws to 25 ft lbs and the water pump-to-timing case cover screws to 4 ft lbs (48 in. lbs).

Heater Core

NOTE: (All models) If equipped with air conditioning, the unit must be removed from the dash and lowered to gain access to the heater control box for the removal of the heater core.

1971-73 CJ/DJ and Commando

Removal

1. Remove the negative battery cable and drain the cooling system.
2. Mark the duct halves for proper reassembly.
3. Separate the duct halves by removing the four attaching screws.
4. Remove the heater core.

Installation

1. The assembly is in the reverse of the disassembly.
2. When the assembly is complete, test the operation of the heater and controls.

1971-73 Wagoneer—Truck

Removal

1. Drain the cooling system and disconnect the negative battery cable.
2. Disconnect the temperature control cable at the heater housing.
3. Disconnect the heater hoses and remove the wire plug connector from the heater resistor.
4. Remove the four nuts that retain the heater core and duct to the firewall. *NOTE: Two of the nuts are located inside the vehicle to the right of the transition duct.*
5. Remove the duct assembly and scribe a mark on the two halves for proper assembly.
6. Separate the duct halves by removing the four attaching screws.
7. Remove the heater core.

Installation

1. The assembly is in the reverse of the disassembly.
2. Upon completion of the assembly, test the heater assembly for proper operation.

1974-78 CJ Models

Removal

1. Remove the battery, drain the cooling system, and disconnect the heater hoses.
2. Disconnect the damper door control cable.
3. Disconnect the blower motor wire harness and ground wire at the switch and instrument panel.
4. Iemove the glove box; water drain hose and defroster duct hose.
5. Disconnect the heater to air deflector duct at the heater housing.
6. Remove the nuts from the heater housing studs, protruding into the engine compartment.
7. Remove the heater housing assembly from the vehicle and remove the core from the housing.

Installation

1. The assembly is in the reverse of the disassembly.
2. When the assembly is completed, test the heater for proper operation.

1974-78 Cherokee—Wagoneer —Truck

Removal

1. Remove the negative battery cable and drain the cooling system.

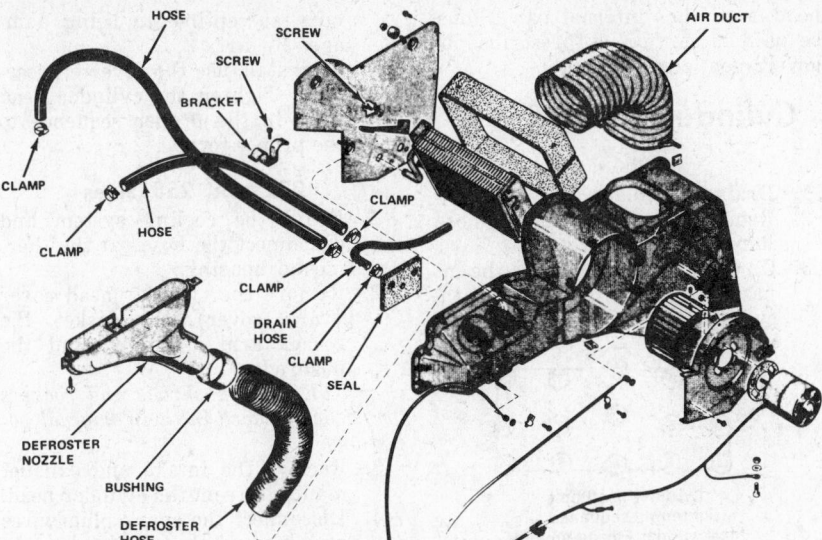

Typical heater—defroster assembly—CJ models (© Jeep Corp.)

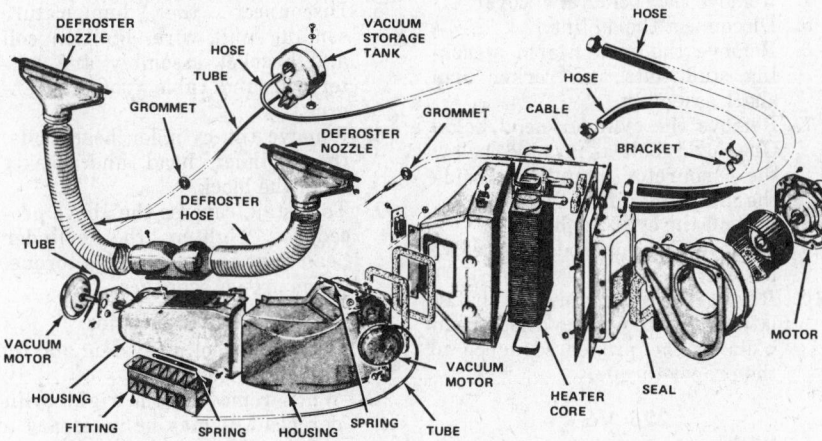

Typical Heater—defroster assembly—Cherokee, Wagoneer, Truck (© Jeep Corp.)

2. Disconnect the temperature control cable from the blend air door.
3. Remove the heater hoses and blower motor resistor wires.
4. Remove the heater core housing to dash panel attaching screws or nuts, projecting into the engine compartment.
5. Remove the heater housing assembly from the vehicle.
6. Separate the halves of the housing, after scribing a mark on the two halves, remove the core retaining screws and remove the heater core.

Assembly

1. The assembly is in the reverse of the disassembly.
2. When the assembly is completed, test the heater for proper operation.

Engine

Reference

Refer to the Engine General Repair Section for troubleshooting and rebuilding procedures.

Emission Control Systems

Emission control systems are designed to control the emissions of Hydrocarbons (HC), Carbon monoxide (CO), and Oxides of Nitrogen (NOx), at the levels specified by Federal and State governments. Emission control systems vary in their usage, in relationship to engine, transmission, and series applications.

Overhaul of the units are covered in the General Repair Section.

Air Guard System

This system is used to inject air into the exhaust ports, to mix with the hot unburned gases, and to further burn the combustion mixture and reduce the hydrocarbons and carbon monoxide emissions into the atmosphere. The system consists of a belt driven air pump, a diverter valve, air injector manifolds, air injector tubes, and connecting hoses.

Exhaust Gas Recirculation System (EGR)

This system is used to meter exhaust gases into the combustion chambers to dilute the intake charge, thereby reducing the peak temperature of the gases and limit the formation of the Oxides of Nitrogen that forms as a result of the high temperature during the combustion process.

The exhaust gases introduced is inert and much cooler than the combustion temperature. Since the exhaust gas will not burn, peak combustion temperatures are lowered.

The system consists of a exhaust gas recirculating valve which connects the exhaust manifold to the intake manifold. An exhaust back pressure sensor is used on some models. A coolant override temperature switch and various lengths of vacuum hoses complete the system.

Catalytic Converter

A catalytic converter is used as part of the exhaust system to have the exhaust gases pass through and undergo a chemical reaction which changes the hydrocarbons and carbon monoxide into harmless carbon dioxide and water before it is emitted into the atmosphere. Beads of Alumina, covered with Platinum and Palladium are used as the catalyst. Unleaded gasoline must be used as leaded fuels poisons or spoils the catalyst used in the converter.

Fuel Tank Vapor Emission Control System

A closed fuel tank vent system is used to prevent fuel vapors from entering the atmosphere. The raw vapors are routed into the intake system and is burned along with the fuel-air mixture.

The system consists of a two-way relief valve filler cap, which is closed to the atmosphere pressures under normal operating conditions and opens when pressure exceeds 0.75 to 1.50 PSI, or a vacuum of 15 to 25 inches. A liquid check valve is used to route the vapors and collect any liquid before the vapors are drawn into and stored in the fuel vapor storage cannister, until they are drawn into the intake manifold through the carburetor air cleaner assembly. The amount of vapors drawn from the cannister is relative to the air velocity passing through the air cleaner snorkel.

Positive Crankcase Ventilation System (PCV)

The positive crankcase ventilation system draws the crankcase vapors into the intake manifold, to be burned along with the air-fuel mixture. This is normally a closed system so that the crankcase vapors are not emitted into the atmosphere. The system consists of hose routings, a ventilation valve, connected to and operated by engine vacuum from the intake manifold.

Thermostatically Controlled Air Cleaner System (TAC)

The thermostatically controlled air cleaner system operates to avoid the induction of cold air into the carburetor and intake manifold before the engine reaches normal operating temperature. When the engine is first started, a thermostatically controlled valve in the air cleaner snorkel closes to outside air and exhaust manifold heater air is directed into the air cleaner assembly, carburetor, and intake manifold. As the engine warms, and the surrounding air temperature increases, the valve in the air cleaner snorkel opens and admits air from the engine compartment, while closing off the heated manifold air.

Vacuum Throttle Modulating System (VTM)

The VTM system is used to reduce the emission off hydrocarbons during rapid throttle closure at high speeds and consists of a deceleration valve and a throttle modulating diaphragm located on the carburetor base to allow the throttle to remain slightly open and admit more air into the combustion chambers to lean out the overrich mixture, during the rapid throttle release. The decel valve and modulator diaphragm are operated by engine vacuum signals.

Transmission Controlled Spark System (TCS)

The TCS system is used to reduce the emissions of oxide of nitrogen by lowering the peak combustion temperatures during the power stroke by not allowing vacuum to be routed to the distributor vacuum advance unit during low speed operation, thereby not allowing the advance unit to operate. This system is controlled by switches located and operated by the transmission gear selector, or oil pressure, directed to a switch, at a predetermined speed.

Spark Coolant Temperature Override Switch (Spark CTO)

This system is used to override the TCS system to improve driveability during the warmup period by providing full distributor vacuum advance operation until the temperature reaches 160° F within the cooling system. The system then reverts to the transmission controlled spark system.

Engine Modification

The design of certain engine components are directly related to the approved emission standards. The correct combination of engine components, such as the camshaft, carburetor, ignition distributor, cylinder

head, and other internal parts, must be used in service, as prescribed by government certification.

Cylinder Head R & R

F4

1. Drain the coolant.
2. Remove the upper radiator hose. Remove the carburetor.
3. On early engines, remove the bypass hose from the front of the cylinder head.

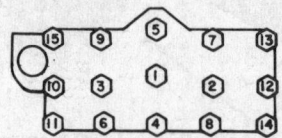

Cylinder head bolt tightening sequence— four-cylinder F-head engines

4. Remove the rocker arm cover.
5. Disconnect the oil line.
6. Remove the rocker arm attaching stud nuts and rocker arm shaft assembly.
7. Remove the cylinder head bolts. One head bolt is located below the carburetor mounting, inside the intake manifold.
8. Lift off the cylinder head.
9. Remove the pushrods and valve lifters.
10. Reverse the procedure for installation. Tighten the cylinder head bolts in the proper sequence to the proper torque.

225 V6

1. Remove the intake manifold.
2. Remove the rocker cover.
3. Remove the exhaust pipes at the flanges.
4. Remove the alternator in order to remove the right head.
5. Remove the dipstick and power steering pump, if so equipped, in order to remove the left head.
6. Remove the valve cover and the rocker assemblies. Mark these parts so that they can be reinstalled in exactly the same positions.
7. Unbolt the head bolts and lift off the cylinder head(s). It is very important that the inside of the engine be protected from dirt. The hydraulic lifters are particu-

larly susceptible to being damaged by dirt.
8. To install, use the reverse procedure. Tighten the cylinder head bolts in the proper sequence to the proper torque.

232 and 258 Sixes

1. Drain the cooling system and disconnect the hoses at the thermostat housing.
2. Remove the cylinder head cover (valve cover), the gasket, the rocker arm assembly, and the pushrods.
NOTE: The pushrods and rockers must be replaced in their original positions.
3. Remove the intake and exhaust manifold from the cylinder head.
4. Disconnect the spark plug wires and remove the spark plugs to avoid damaging them.
5. Disconnect the temperature sending unit wire, ignition coil and bracket assembly and battery ground cable from the engine.
6. Remove the cylinder head bolts, the cylinder head and gasket from the block.
7. To install, reverse the above procedure. Tighten the cylinder head bolts to the specified torque, in the proper sequence.

All AMC V8 Engines

1. Drain the cooling system and the cylinder block.
2. When removing the right cylinder head, it may be necessary to remove the heater core housing from the firewall.
3. Remove the valve cover(s) and gasket(s).
4. Remove the rocker arm assemblies and push rods.
NOTE: The valve train components must be replaced in their original positions.
5. Remove the spark plugs to avoid damaging them.
6. Remove the intake manifold with the carburetor still attached.
7. Remove the exhaust pipes at the flange of the exhaust manifold. When replacing the exhaust pipes, it is advisable to install new gaskets at the flange.
8. Loosen all of the drive belts.

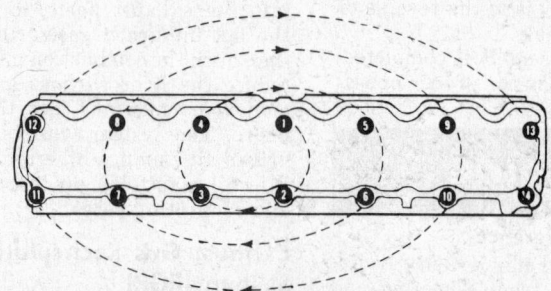

Cylinder head bolt tightening sequence—232 and 258 Sixes.

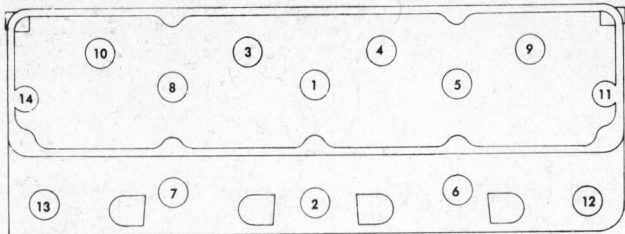

Cylinder head bolt tightening sequence—V8—304, 360 & 401 engines

9. Disconnect the battery ground cable and the alternator bracket.
10. Disconnect the air pump and power steering pump brackets from the left cylinder head.
11. Remove the cylinder head bolts and lift the head(s) from the cylinder block.
12. Remove the cylinder head gasket(s) from the head(s) or the block.
13. To install, reverse the above procedure.

NOTE: Apply an even coat of sealing compound to both sides of the new head gasket only. Wire brush the cylinder head bolts, then lightly oil them prior to installation. First, tighten all bolts to 80 ft lbs, then tighten them to the specified torque. Follow the correct tightening sequence.

350 V8

1. Drain the cooling system at the radiator and at the engine block.
2. Disconnect the battery cables from the battery and remove the alternator and alternator bracket assembly from the engine.
3. Remove the air cleaner assembly.
4. Remove the rocker covers and gaskets and remove the rocker assemblies together with the pushrods.

NOTE: The pushrods must be reinstalled in the same positions from which they were removed. Identify each pushrod so that this can be accomplished.

5. Disconnect the exhaust pipes from the exhaust manifolds by removing the nuts from the attaching bolts of the clamp. Remove the donut gasket.
6. Remove the distributor.
7. Remove the cap screws, lockwashers, nuts, and flat washers that secure the intake manifold to the cylinder heads.
8. Remove the intake manifold with the carburetor attached from the

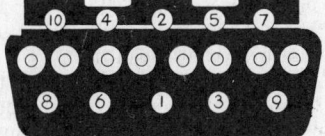

Cylinder head bolt tightening sequence 350 V8 engine

engine. Remove the gaskets and discard them.

9. Remove the bolts holding the cylinder head to the block. Removing one head bolt on the left bank frees the oil dipstick and tube which may now be lifted off. There is a rubber bushing on the lower end of the tube; take care not to lose the rubber bushing.
10. Lift the heads off the block and remove and discard the cylinder head gaskets. Remove all traces of gasket material from all mating surfaces on both the heads and the intake manifold.
11. Install the cylinder heads and assemble the engine in the reverse order of removal and disassembly. Torque the head bolts and intake manifold bolts in the correct sequence and to the proper specifications.

Engine Mounts R & R

F4

The front of the engine is supported by two rubber insulator mountings attached to the frame side rail brackets. The rear of the engine-transmission assembly is supported by a rubber insulator mounting under the rear of the transmission on the frame center crossmember. This crossmember is bolted to the frame side rails so that it can be dropped when removing the transmission or engine-transmission assembly. The rubber insula-

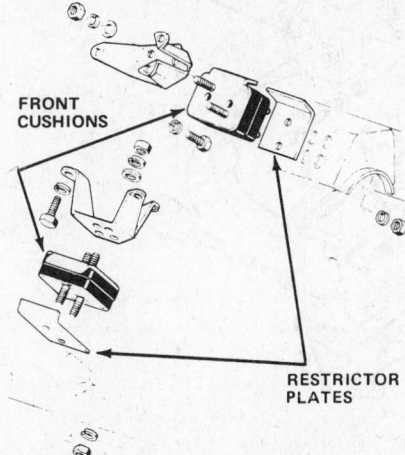

FRONT CUSHIONS

RESTRICTOR PLATES

Typical V8 engine mounts-AMC engines (© Jeep Corp.)

tors allow free side and vertical oscillation to effectively neutralize engine vibration at the source.

The rubber insulator mountings should be inspected for separation and deterioration by jacking the power plant away from the frame, near the supports. Vibration cannot be effectively absorbed by separated or worn insulators. They should be replaced if faulty.

225 V6

The engine-transmission unit is mounted to the chassis at three points by rubber pads. The two front mounts are bolted to the engine cylinder block and the frame members. These mounts support most of the engine weight, and absorb vibration which would otherwise be caused by changes in engine output torque. The single rear mount is placed between the transmission and the transmission support. It supports part of the engine and transmission weight, and locates the rear of the engine with respect to the centerline of the vehicle.

232 and 258 Sixes and All V8

Resilient rubber mounting cushions support the engine and transmission at three points. A cushion is located at each side on the center line of the engine, with the rear supported by a cushion between the transmission extension housing and the rear engine support crossmember.

Replacement of the cushion may be accomplished by supporting the weight of the engine or transmission at the area of the cushion.

Manifolds R & R

F4

The intake manifold is cast as an integral part of the cylinder head.

The exhaust manifold is removed by removing the five nuts from the manifold studs, removing the two bolts that hold the manifold to the exhaust pipe, and lifting the manifold off of the mounting studs. Remove the center and two end gaskets from the cylinder block. Install in the reverse order of removal, using new gaskets.

NOTE: On engines with the AIR exhaust emission control devices, the rubber hose leading from the air pump to the air injection manifold must be disconnected in order to remove the exhaust manifold.

Intake Manifold

225 V6

1. Disconnect the crankcase vent hose, exhaust manifold line, choke heat tube, distributor vacuum hose, and the fuel line at the carburetor.
2. Disconnect the two distributor leads from the ignition coil.

3. Disconnect the wiring harness from the coolant temperature sending unit.
4. Remove the 10 capbolts which attach the intake manifold to the cylinder heads.
5. Remove the intake manifold and carburetor as an assembly.
6. Clean all the mating surfaces of the intake manifold and cylinder heads, removing all traces of gasket material.
7. Install a new rubber intake manifold seal at the front and rear rails of the cylinder block. Be sure that the pointed ends of the seals fit snugly against the block and cylinder heads.
8. Set the intake manifold in place on the cylinder block between the cylinder heads. Thread two capbolts through the manifold into each cylinder head as guide bolts. Lift the manifold slightly and insert each of the two gaskets into position between the manifold and corresponding cylinder head. Be certain that the gasket is installed with its 3 apertures aligned with the ports of the head and manifold. One gasket should be installed in position on the right-side and the other gasket reversed and installed on the left-side.
9. Install the manifold attaching bolt in the open bolt hole at the right center side of the intake manifold. During manufacture, this open bolt hole is held to close tolerances so that the bolt in this location serves to locate the manifold front and rear.
10. Install the remaining intake manifold attaching bolts. The

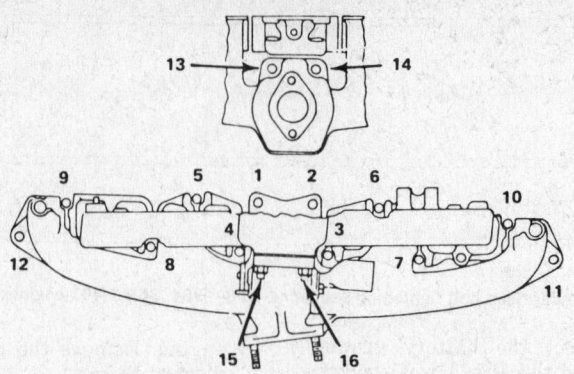

232 and 258 Six intake/exhaust manifold torque sequence.

longer bolts are installed in the two forward holes on either side.
11. Tighten the intake manifold bolts to the proper torque in the proper sequence.
12. Assemble the remaining components to the engine in the reverse order of removal.

232 and 258 Sixes

The intake and exhaust manifold are mounted externally on the left side of the engine and are attached to the cylinder head. They are removed as a unit. On later engines, an exhaust gas recirculation valve is mounted on the side of the intake manifold.

1. Remove the air cleaner and carburetor.
2. Disconnect the accelerator cable from the accelerator bellcrank.
3. Disconnect the PCV vacuum hose from the intake manifold.
4. Disconnect the distributor vacuum hose and electrical wires at the TCS solenoid vacuum valves.

5. Remove the TCS solenoid vacuum valve and bracket from the intake manifold. In some cases it might not be necessary to remove the TCS unit.
6. If so equipped, disconnect the EGR valve and back pressure sensor hoses.
7. Remove the power steering mounting bracket and pump and set it aside without disconnecting the hoses.
8. Remove the EGR valve and backpressure sensor, if so equipped.
9. Disconnect the exhaust pipe from the manifold flange.
10. Remove the manifold attaching bolts, nuts and clamps.
11. Separate the intake manifold and exhaust manifold from the engine as an assembly, and discard the gasket.
12. If either manifold is to be replaced, they should be separated at the heat riser area.
13. Clean the mating surface of the manifolds and the cylinder head before replacing the manifolds. Replace them in reverse order of the above procedure with new gasket. Tighten the bolts and nuts to the specified torque in the proper sequence.

350 V8

1. Drain the cooling system.
2. Remove the carburetor air cleaner. Disconnect all tubes and hoses from the carburetor. Disconnect and remove the ignition coil.
3. Disconnect the temperature indicator wire from the sending unit.
4. Disconnect the accelerator and transmission linkage at the carburetor.
5. Slide the front thermostat bypass hose clamp back on the hose. Disconnect the upper radiator hose at the outlet.
6. Disconnect the heater hose at the temperature control valve inlet. Force the end of the hose down to allow the coolant to drain from the intake manifold.

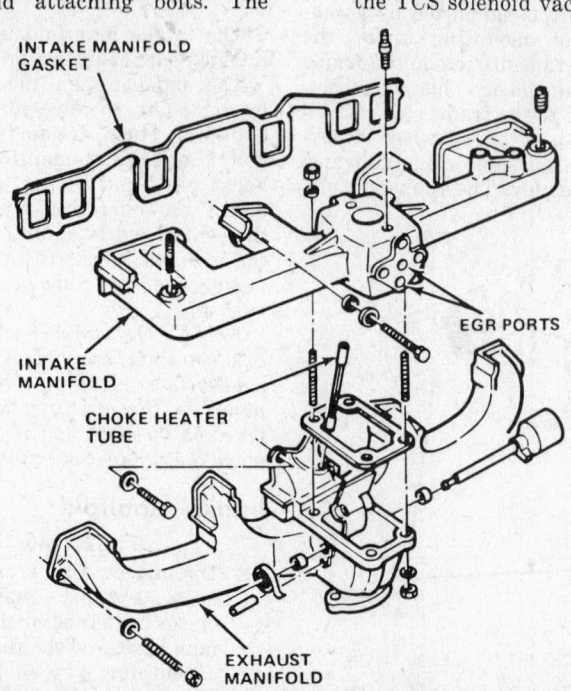

INTAKE MANIFOLD GASKET

EGR PORTS

INTAKE MANIFOLD

CHOKE HEATER TUBE

EXHAUST MANIFOLD

Exploded view of the intake and exhaust manifold on 232 and 258 Sixes.

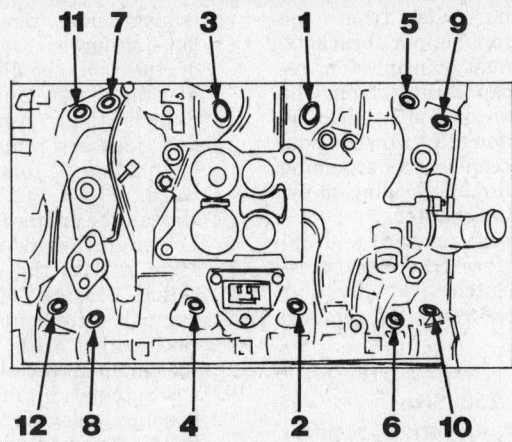

Intake manifold torque sequence 350 V8 engine

7. Remove the intake manifold-to-head attaching bolts.

8. Remove the intake manifold and carburetor as an assembly, by sliding the assembly rearward to disengage the thermostat bypass hose from the water pump. Remove all traces of gasket material from the mating surfaces of the intake manifold and the cylinder heads and block.

9. Install the intake manifold in the reverse order of removal, tightening the attaching bolts in the correct sequence and to the proper torque.

304, 360 and 401 V8s

1. Drain the coolant from the radiator.

2. Remove the air cleaner assembly.

3. Disconnect the spark plug wires. Remove the spark plug wire brackets from the valve covers, and the bypass valve bracket.

4. Disconnect the upper radiator hose and the by-pass hose from the intake manifold. Disconnect the heater hose from the rear of the manifold.

5. Disconnect the ignition coil bracket and lay the coil aside.

6. Disconnect the TCS solenoid vacuum valve from the right side valve cover.

7. Disconnect all lines, hoses, linkages and wires from the carburetor and intake manifold and TCS components as required.

8. Disconnect the air delivery hoses at the air distribution manifolds.

9. Disconnect the air pump diverter valve and lay the valve and the bracket assembly, including the hoses, forward of the engine.

10. Remove the intake manifold after removing the cap bolts that hold it in place. Remove and discard the side gaskets and the end seals.

11. Clean the mating surfaces of the intake manifold and the cylinder head before replacing the intake manifold. Use new gaskets and tighten the cap bolts to the correct torque. Install in reverse order of the above procedure.

NOTE: There is no specified tightening sequence for the V8 304, 360 and 401 intake manifold. Begin at the center and work outward.

Exhaust Manifold

225 V6

1. Remove the five attaching screws, one nut, and exhaust manifold(s) from the side of the cylinder head(s).

2. Use a new gasket when replacing the exhaust manifolds. Make sure that the mating surfaces of the manifold and the cylinder head are clean. Tighten the manifold nuts and bolts to the correct torque.

232, 258 Sixes

The intake and exhaust manifolds of the 232 and 258 cu in. Sixes must be removed together. See the procedure for removing and installing the intake manifold.

304, 360 and 400 V8

1. Disconnect the spark plug wires.

2. Disconnect the air delivery hose at the distribution manifold.

3. Remove the air distribution manifold and the injection tubes.

4. Disconnect the exhaust pipe at the manifold.

5. Remove the exhaust manifold attaching bolts and washers along with the spark plug shields.

6. Separate the exhaust manifold from the cylinder head.

7. Install in reverse order of the above procedure. Clean the mating surfaces and tighten the attaching bolts to the correct torque.

350 V8

1. Disconnect the exhaust pipe at the exhaust manifold/exhaust pipe joining flange by removing the retaining nuts.

2. Remove the donut gasket and save it, if it is in good condition. If not, discard and replace it.

3. Remove the bolts attaching the exhaust manifold to the cylinder head, leaving one upper bolt toward the center until last.

4. Remove all traces of gasket material from the exhaust manifold and cylinder head mounting surfaces.

NOTE: The sheet metal heat stove is removed with the left exhaust manifold.

Engine Assembly R & R

F4

1. Drain the cooling system.

2. Disconnect the battery ground cable.

3. Remove the air cleaner and disconnect the breather hose at the oil filter.

4. Disconnect the choke and throttle controls.

5. Disconnect the fuel line and windshield wiper hose at the fuel pump.

6. Remove the radiator stay bar, if so equipped.

7. Remove the radiator and heater hoses.

8. Remove the fan blades, fan hub, radiator, and shroud.

9. Remove the starter motor.

10. Disconnect:
    a. The alternator or generator.
    b. The ignition primary wire at coil.
    c. The oil pressure and temperature sending units.
    d. The exhaust pipe from manifold.
    e. The engine ground strap.

11. Attach a lifting device to the engine. Unbolt and remove the front engine supports.

12. Remove the flywheel housing bolts.

13. Pull the engine forward until the clutch clears the flywheel housing. Lift the engine from the vehicle.

14. Install the engine by reversing the removal procedure.

225 V6

1. Remove the hood if necessary.

2. Disconnect the battery ground cable.

3. Remove the air cleaner.

4. Drain the coolant.

5. Disconnect the radiator hoses.

6. Remove the radiator support bars.

7. On the Universal series, remove the radiator. On the Jeepster series, disconnect the headlamp wiring from the block on the left fender, the horn wiring from the horn, the oil cooler lines if

equipped with automatic transmission, and remove the front fenders, radiator, and grille as a unit.

8. Disconnect the engine wiring from the connectors on the firewall.
9. Remove the starter motor.
10. Disconnect the fuel hoses at the right frame rail. Plug the hoses.
11. Disconnect the throttle and choke.
12. Disconnect the exhaust pipes.
13. Place a jack under the transmission and support the weight.
14. Remove the front motor mount bolts.
15. Support the engine with a lifting device.
16. Remove the flywheel housing bolts.
17. Raise the engine slightly and slide the engine forward until the engine is free of the transmission shaft. Remove the engine.
18. Install in reverse order of the above procedure.

350 V8

1. Remove the hood.
2. Remove the air cleaner.
3. Drain the radiator and cylinder block.
4. Disconnect the radiator and heater hoses.
5. If equipped with an automatic transmission, disconnect the oil cooler lines at the radiator. Remove the radiator.
6. Remove the fan, belt, and hub.
7. Drain the engine oil and remove the filter.
8. Disconnect the temperature-sender lead, pressure-sender lead, coil, starter solenoid, and alternator and distributor leads.
9. Disconnect the accelerator cable at the carburetor throttle shaft lever and at the cable support bracket.
10. Disconnect the heater system vacuum valve hose at the intake manifold.
11. Disconnect the flexible fuel line from the frame-to-crankcase at the frame end. Plug the end of the hose.
12. Disconnect the exhaust pipes at both manifolds.
13. Remove the air conditioning compressor (if so equipped). Do not disconnect the hoses.
14. If equipped with a manual transmission, disconnect the linkage at the clutch. Disconnect the clutch cross-shaft support brackets at the flywheel housing and frame.
15. Install a suitable lifting fixture on the engine.
16. Support the transmission on a jack.

17. Remove the nuts from the engine-to-front support brackets.
18. With a manual transmission, remove the cap screws from the transmission-to-clutch housing. With an automatic transmission, remove the cap screws attaching the transmission housing-to-flywheel housing adapter.
19. Pull the engine forward and upward until free from the transmission or clutch.
20. Install by reversing the removal procedure.

232, 258 Sixes

NOTE: This operation requires discharging the air conditioning system. This requires special tools and skills for safety reasons. It should not be attempted by untrained persons.

1. Remove the hood after marking the hinge locations. The hood need not be removed on the CJ series.
2. Remove the air cleaner.
3. Drain the coolant. Disconnect the radiator hoses. Disconnect automatic transmission cooler lines from the radiator. If there is a radiator shroud, remove it, then remove the radiator.
4. Remove the fan.
5. Remove and set aside the power steering pump and belt. Do not disconnect the hydraulic lines.
6. Bleed the compressor refrigerant charge. See the note at the start of this procedure. Remove the condenser and receiver assembly.
7. Disconnect all wires, lines, linkage, and hoses from the engine.
8. Drain the oil and remove the filter.
9. Remove both engine front support cushion-to-frame retaining nuts.
10. Disconnect the exhaust pipe at the support bracket and the manifold.
11. Support the engine with the lifting equipment.
12. Remove the front support cushion and bracket assemblies from the engine.
13. Remove the transfer case lever boot, the floor mat, and the transmission access cover.
14. On automatic transmissions, remove the upper bolts holding the bellhousing to the engine adapter plate. On manual transmissions, remove the upper bolts holding the clutch housing to the engine.
15. Remove the starter.
16. On automatics, remove the two adapter plate inspection covers. Mark the relationship of the converter to the flex plate and remove the converter-to-flex plate bolts. Remove the rest of the bolts holding the bellhousing to the adapter plate. On manual

transmissions, remove the clutch housing lower cover and the rest of the bolts holding the clutch housing to the engine.
17. Support the transmission with a floor jack and remove the engine by pulling it forward and upward.

To install the engine:
18. Lower the engine into place and align it with the bellhousing or clutch housing. Make sure the manual transmission clutch shaft aligns with the splines of the clutch driven plate.
19. On automatics, install the bellhousing-to-engine adapter plate bolts. On manuals, install the clutch housing-to-engine bolts. Torque the bolts to 25-28 ft lbs at the top and 40-45 ft lbs at the bottom.
20. Remove the floor jack.
21. Align the marks made in step 16 and install the converter-to-flex plate bolts, torquing them to 21-23 ft lbs.
22. Install the two engine adapter plate inspection covers or the clutch housing lower cover.
23. Replace the starter.
24. Install the front support cushion and bracket assemblies to the engine, torquing the bolts to 25-30 ft lbs. Lower the engine onto the frame supports. Install the front support cushion retaining nuts, torquing them to 25-30 ft lbs.
25. Connect the exhaust pipe at the support bracket and manifold. A new manifold seal is advisable.
26. Install the oil filter.
27. Replace all the items removed in step seven.
28. Replace the air conditioning condenser and receiver assembly and recharge the system.
29. Replace the power steering pump and belt. Install the fan and tighten the bolts to 15-25 ft lbs.
30. Replace and reconnect the radiator. Replace the oil cooler lines. Fill the cooling system.
31. Fill the crankcase and replace the air cleaner. Install the transmission access cover, floor mat, and transfer case lever boot. Replace the hood.

304, 360 and 401 V8

NOTE: This operation requires discharging the air conditioning system. This requires special tools and skills. For safety reasons, it should not be attempted by untrained persons.

The engine is removed without the transmission and bellhousing.

1. On the Commando, Cherokee, and Wagoneer, the hood must be removed. Mark the hinge locations at the hood panel for alignment during installation. Remove the hood from the hinges.

2. Remove the air cleaner assembly.
3. Drain the cooling system and disconnect the upper and lower radiator hoses. If equipped with automatic transmission, disconnect the cooler lines from the radiator.

NOTE: If the vehicle is equipped with a radiator shroud, it is necessary to separate the shroud from the radiator to facilitate removal and installation of the radiator and engine fan.

4. Remove the radiator.
5. Remove the engine fan.
6. If equipped with power steering, remove the pump from the engine and lay it aside. Do not disconnect the hoses.
7. If equipped with air conditioning, turn both service valves clockwise to the front seated position. Bleed the compressor refrigerant charge by slowly loosening the service valve fittings. Disconnect the condenser and evaporator lines from the compressor. Disconnect the receiver outlet at the disconnect coupling. Remove the condenser and receiver assembly.
8. Remove the battery and tray on CJ models only (1972-75).
9. On Wagoneers and Cherokees, remove the heater core housing and charcoal canister from the firewall.
10. Disconnect all wires, lines linkage, and hoses which are connected to the engine.
11. If equipped with automatic transmission, disconnect the transmission filler tube bracket from the right cylinder head. Do not remove the filler tube from the transmission.
12. Remove both engine front support cushion-to-frame retaining nuts.
13. Support the weight of the engine with a lifting device.
14. On CJ and Commando models, remove the transfer case shift lever boot, floor (if so equipped), and transmission access cover.
15. Remove the upper bolts which secure the transmission bellhousing to the engine adapter plate on vehicles equipped with automatic transmission. If equipped with manual transmission, remove the upper bolts which secure the clutch housing to the engine.
16. Disconnect the exhaust pipes at the exhaust manifolds and support bracket.
17. Remove the starter motor.
18. Support the transmission with a floor jack.
19. If equipped with automatic transmission, remove the two engine adapter plate inspection covers. Mark the assembled posi-

tion of the converter and flex plate and remove the converter-to-flex plate cap screws. Remove the remaining bolts which secure the transmission bellhousing to the engine adapter plate.
20. If equipped with manual transmission, remove the clutch housing lower cover and the remaining bolts which secure the clutch housing to the engine.
21. Remove the engine by pulling upward and forward.

NOTE: If equipped with power brakes, care must be taken to avoid damaging the power unit while removing the engine.

To install the engine:
22. Lower the engine slowly into the engine compartment and align with the transmission bellhousing (automatic transmission) or clutch housing (manual transmission). On manual transmissions, make certain the clutch shaft is aligned properly with the splines of the clutch driven plate.
23. Install the transmission bellhousing-to-engine adapter plate bolts (automatic transmission) or the clutch housing to engine bolts (manual transmission). Tighten the bolts to the specified torque. Remove the floor jack which was used to support the transmission.
24. If equipped with automatic transmission, align the marks previously made on the converter and flex plate, install the converter-to-flex plate cap screws and tighten to the specified torque.
25. Install the two engine adapter plate inspection covers (automatic transmission) or the clutch housing lower cover (manual transmission).
26. Install the starter motor.
27. Lower the engine onto the frame supports, remove the lifting device and install the front support cushion retaining nuts. Tighten the nuts to the specified torque.
28. Connect the exhaust pipes at the exhaust manifolds and support bracket.
29. If equipped with automatic transmission, connect the transmission filler tube bracket to the right cylinder head.
30. On Wagoneers and Cherokees, install the heater core housing and charcoal canister to the firewall.
31. If removed, install the battery and tray.
32. Connect all wires, lines, linkage and hoses which were previously disconnected from the engine.
33. If removed, install the air conditioning condenser and receiver assembly. Connect the receiver outlet to the disconnect coupling.

Connect the condenser and evaporator lines to the compressor. Purge the compressor of air.

NOTE: Both service valves must be open before the air conditioning system is operated.

34. If equipped with power steering, connect the power steering, pump to the engine.
35. Install the engine fan and tighten the retaining bolts to the specified torque.
36. Install the radiator and connect the upper and lower hoses. If equipped with automatic transmission, connect the cooler lines.
37. Fill the cooling system to the specified level.
38. Install the air cleaner assembly.
39. Start the engine. Check all connections for leaks. Stop the engine.
40. If removed, install the transmission access cover, floor mat and transfer case shift lever boot.
41. If removed, install the transmission access cover, floor mat and transfer case shift lever boot.

Valve Arrangement

F4 Engine
Intake valves located in cylinder head.
Exhaust valves located in cylinder block.

AMC 6 Cylinder Engines
E I I E E I I E E I I E → Front

V6 Engine
E I I E I E
E I E I I E → Front

V8 Engines
E I I E E I I E
E I I E E I I E → Front

Valve Adjustment

F4
NOTE: While all valve adjustments must be made as accurately as possible, it is better to have the valve adjustment slightly loose than slightly tight, as a burned valve may result from overly tight adjustments.

The only engine in which the valves can be adjusted is the F-Head 4 cylinder. The V6-225, 232 and 258 sixes, 304, 350, 360 and 401 V8s all have hydraulic lifters and are not adjustable in service.

NOTE: The engine must be cold when the valves are adjusted on the F-Head engine.

1. Start the engine and let it run until it has reached operating temperature. Remove the valve cover and check all the cylinder head bolts to make sure they are tightened to the correct torque specifications.

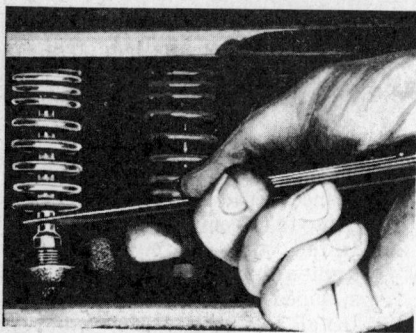

Measuring the clearance between the exhaust valve and tappet adjusting screw on an F4 engine

2. While the engine is cooling, remove the side valve spring cover. Be careful not to burn yourself on the exhaust pipe; it may still be hot.

3. After the engine has cooled to ambient temperature, turn the engine until the lifter for the front intake valve is down as far as it will go. The lifter should be resting on the center of the heel (back) of the cam lobe for that valve. You can observe the position of the lifter by looking through the side valve spring cover opening. Put the correct size feeler gauge between the rocker arm and the valve stem for that particular intake valve. There should be a very slight drag on the feeler gauge when it is pulled through the gap. If there is a slight drag, you can assume that the valves are at the correct setting. If the feeler gauge gannot pass between the rocker arm and the valve stem, the gap between them is too small and must be increased. If the gauge can be passed through the gap without any drag, the gap is too large and must be decreased. Loosen the locknut on the top of the rocker arm (pushrod side) by turning it counterclockwise. Turn the adjusting screw clockwise to lessen the gap and counterclockwise to increase the gap. When the gap is correct, turn the locknut clockwise to lock the adjusting screw. Follow this procedure for all of the intake valves, making sure that the lifter is all the way down for each adjustment.

4. Turn the engine so that the first exhaust valve is completely closed and the lifter that operates that particular valve is all the way down and on the heel of the cam lobe that operates it.

5. Insert the feeler gauge between the stem of the valve and the adjusting screw. If there is a slight drag on the feeler gauge, the adjustment is correct. If the ad-

justment is incorrect, turn the adjusting screw clockwise to increase the gap and counterclockwise to decrease the gap.

6. When all the valves have been adjusted, scrape the block and valve covers clean of all old gasket material, install new gaskets, and bolt the covers in place.

Rocker Arm Assemblies

F4 Engine

Removal

1. Remove the cylinder head cover and gasket.
2. Remove the retaining nuts from the rocker arm shaft support studs.
3. Lift and remove the rocker arm assembly from the cylinder head.
4. Slide the rocker arm shaft brackets, the four rocker arm assemblies, and the two springs from the shaft.
5. Remove the two rocker arm shaft lock screws from the two

remaining shaft brackets, and remove the brackets.
6. If needed, the valve adjustment screws and lock nuts can be removed from the rocker arms.

Installation

1. Replace the valve adjustment screws and lock nuts, if removed from the rocker arms.
2. Install the two center shaft brackets and align the tapped holes in the brackets with the drilled holes in the top of the shaft, and install the lock screws. Tighten securely. *NOTE: The lock screw points should engage and enter the drilled holes in the shaft.*
3. Slide the rocker arm onto the shaft, so that the adjusting screw end of the arm is angled away from the center bracket, and the adjusting screw is on the same side of the shaft as the mounting hole in the bracket.
4. Repeat step 3 on the opposite end of the shaft and install the

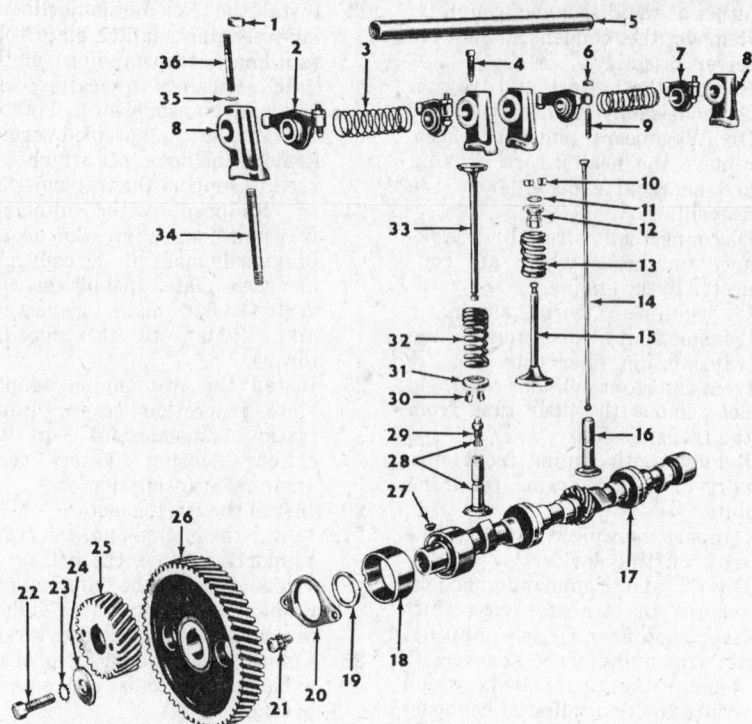

Exploded view—valves, camshaft, and timing gears—F4 engine (© Jeep Corp.)

1 Nut	19 Camshaft thrust plate spacer
2 Left rocker arm	20 Camshaft thrust plate
3 Rocker arm shaft spring	21 Bolt and lock washer
4 Rocker shaft lock screw	22 Bolt
5 Rocker shaft	23 Lock washer
6 Nut	24 Camshaft gear washer
7 Right rocker arm	25 Crankshaft gear
8 Rocker arm shaft bracket	26 Camshaft gear
9 Intake valve tappet adjusting screw	27 Woodruff key no. 9
10 Intake valve upper retainer lock	28 Exhaust valve tappet
11 Oil seal	29 Tappet adjusting screw
12 Intake valve spring upper retainer	30 Spring retainer lock
13 Intake valve spring	31 Roto cap assembly
14 Intake valve push rod	32 Exhaust valve spring
15 Intake valve	33 Exhaust valve
16 Intake valve tappet	34 Rocker shaft support stud
17 Camshaft	35 Washer
18 Camshaft front bearing	36 Rocker arm cover stud

rocker shaft on to the cylinder head.

5. Align the push rods to the rocker arms, and install the nuts onto the end bracket studs, and torque to 30-36 ft. lbs.
6. Adjust the engine valves as previously outlined.
7. Install the cylinder head cover and new gasket. Torque the retaining nuts to 3-5 ft. lbs.

AMC 6 Cylinder Engines

Removal (1st type)

1. Remove the cylinder cover and gasket.
2. Loosen the shaft retaining bolts evenly until the shaft is loose, and then remove all the retaining bolts from the supports and lift the shaft assembly from the cylinder head.
3. Remove the roll pin and spring washer from one end of the shaft, and remove the rocker arms, spacers, retainers and bolts, and the oil deflector.
4. Maintain the parts in the order of the disassembly.

Installation

1. Assemble the rocker arms, spacers, retainers and bolts, and the oil deflector in the same order as removed. *NOTE: The oil holes in the rocker arm shaft must face the cylinder head.*
2. Install the rocker arm and shaft assembly onto the cylinder head and install the retaining bolts. Align the push rods with the rocker arms and tighten the retaining bolts evenly, by starting at the center and working outward.
3. Torque the bolts to 20-23 ft. lbs.
4. Install the cylinder head cover with a new gasket and torque the bolts to 45-55 in. lbs.

AMC 6 Cylinder Engines

Removal (2nd Type)

1. Remove the cylinder head cover and gasket.
2. Remove the two capscrews at each bridged pivot, a turn at a time, to avoid cocking and breaking the bridge.

V6 rocker arm arrangement (© Jeep Corp.)

Rocker arm assembly—6 cylinder—2nd type (© Jeep Corp.)

3. Remove the bridge and rocker arm from the cylinder head.

Installation

1. Install the bridge and rocker arms in the same order as the removal, making sure the push rods are aligned to the rocker arms.
2. Install the capscrews and tighten alternately, a turn at a time, to avoid breaking the bridges. Torque the capscrews to 21 ft. lbs.
3. Install the cylinder head cover and a new gasket. Torque the bolts to 45-55 in. lbs.

V6 Engine

Removal

NOTE: The removal and installa-

tion is described for one side only. Follow the same procedure to remove and install the opposite side.

1. Remove the cylinder head cover and gasket.
2. Remove the rocker arm shaft bracket retaining bolts and lift the rocker arm assembly from the cylinder head.
3. Remove the cotter pins, flat washers, spring retaining rings and one rocker arm from each end of the shaft.
4. Remove the two bolts from the outer shaft support and remove the two supports, two rocker arms, two spacer springs, and the two remaining rocker arms from the shaft.
5. Remove the center support bolt and remove the support from the shaft. *NOTE: The three shaft supports are identical and are interchangeable.*

CAUTION: Two different types of rocker arms are used, three of each type on each rocker arm assembly, and are not interchangeable. One face of each rocker arm has a notch and the notch must touch a shaft support when assembled correctly.

CAUTION: Oil ports of the rocker arm shaft must coincide with the oil return passages of the rocker arms, to allow an oil return path from the cylinder head to the crankcase. A notch is located at one end of the rocker arm shaft and when the arms are properly installed, the notch will be at the front of the right rocker arm shaft, and at the rear of the left rocker arm shaft.

1. Position the center support on the shaft and install the retaining bolt.
2. Install the center pair of rocker arms, so that the notched faces touch the support.
3. Install the front and rear spacer springs and one rocker arm of

Exploded view—rocker arm and shaft—6 cylinder—1st type (© Jeep Corp.)

the pair, so that the notched face of the arm is facing outward.

4. Install the outer shaft supports on the shaft, compress the springs to position the supports, and install the retainer bolts to hold the support in place.

5. Install the two remaining rocker arms, one on each end of the shaft with the notched face touching the shaft supports.

6. Install the spring retainer rings, flat washers, and secure each end with a new cotter pin.

7. Install the rocker arm assembly on the cylinder head, align the push rods to the rocker arms, and tighten the support bolts.

8. Torque the support bolts to 25-35 ft. lbs.

9. Install the cylinder head cover and new gasket. Torque the cover screws to 3-5 ft. lbs.

AMC V8 Engines

Removal (1st Type)

NOTE: The removal and installation is described for one side only. Follow the same procedure to remove and install the opposite side.

1. Remove the cylinder head cover and gasket.

2. Remove the retaining nuts from the individually mounted rocker arms, and remove the rocker arm and pivot ball from the rocker arm retaining stud.

Installation

1. Install the rocker arm and pivot ball on the retaining stud.

2. Align the push rod to the rocker

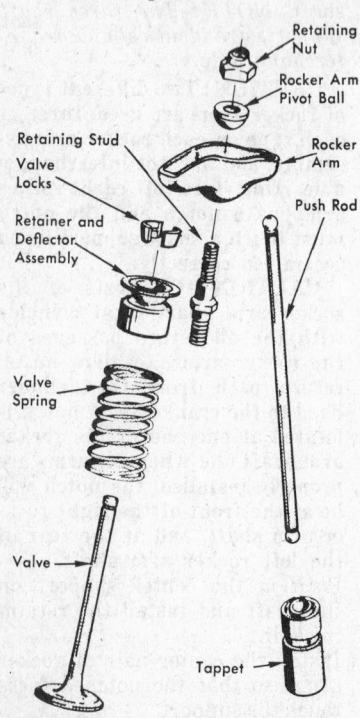

Exploded view—valve train assembly—V8—1st type (© Jeep Corp.)

arm and torque the retaining nut to 20-25 ft. lbs.

3. Install the cylinder head cover and new gasket. Torque the bolts to 20-30 in. lbs.

AMC V8 Engine

Removal (2nd Type)

1. Remove the cylinder head cover and gasket.

2. Loosen the bridged pivot capscrews a turn at a time, so as not to break the bridge.

3. Remove the rocker arm and bridge assembly from the cylinder head.

Installation

1. Install the rocker arms and bridge assembly on the cylinder head, and align the push rods.

2. Install the capscrews and tighten each one a turn at a time to avoid breaking the bridge. Tighten the capscrews to 19 ft. lbs. torque.

V8 rocker arm-bridged pivot assembly—2nd type (© Jeep Corp.)

3. Install the cylinder head cover with a new gasket and torque the cover bolts to 50 in. lbs.

350 V8 Engine

NOTE: The removal and installation is described for one side only. Follow the same procedure to remove and install the opposite side.

Removal

1. Remove the cylinder head cover and gasket.

2. Remove the rocker arm shaft retaining bolts, and lifting the shaft assembly from the cylinder head.

3. With the use of a hammer and chisel, split the end cap of the shaft, and remove the rocker arms and springs.

Installation

NOTE: The drill mark on the rocker arm shaft must be facing up and toward the rear on the left cylinder head, and toward the front on the right cylinder head.

1. Lubricate all parts and install the rocker arms and springs on the shaft. *NOTE: Each set of rocker arms must be offset to each other.*

2. Install new end caps on the shaft.

3. Install the rocker shaft on the cylinder head and install the retaining bolts.

4. Align the push rods to the rocker arms, and tighten the retaining bolts evenly. Torque the bolts to 30 ft. lbs.

5. Install the cylinder head cover and a new gasket. Torque the bolts to 4 ft. lbs.

Timing Gear Cover and Oil Seal R & R

F-Head

1. Remove the drive belts and crankshaft pulley.

2. Remove the attaching bolts, nuts and lock washers that hold the timing gear cover to the engine.

3. Remove the timing gear cover.

4. Remove the timing pointer.

5. Remove the timing gear cover gasket.

6. Remove and discard the crankshaft oil seal from the timing gear cover.

7. Replace in reverse order of the above procedure. Replace the crankshaft oil seal. Use a new timing gear cover gasket.

225 V6

1. Remove the water pump and crankshaft pulley.

2. Remove the two bolts that attach the oil pan to the timing chain cover.

3. Remove the five bolts that attach the timing chain cover to the engine block.

4. Remove the cover and gasket.

1 Rocker arm
2 Drill mark—left bank
3 Offset position
4 Drill mark—right bank

Rocker arm positioned on shaft—350 V8 engine (© Jeep Corp.)

5. Remove the crankshaft front oil seal.
6. From the rear of the timing chain cover, coil new packing around the crankshaft hole in the cover so that the ends of the packing are at the top. Drive in the new packing with a punch. It will be necessary to ream out the hole to obtain clearance for the crankshaft vibration damper hub.

232, 258 Sixes

1. Remove the drive belts, engine fan and hub assembly, the accessory pulley and vibration damper.
2. Remove the oil pan to timing chain cover screws and the screws that attach the cover to the block.
3. Raise the timing chain cover just high enough to detach the retaining nibs of the oil pan neoprene seal from the bottom side of the cover. This must be done to prevent pulling the seal end tabs away from the tongues of the oil pan gaskets which would cause a leak.
4. Remove the timing chain cover and gasket from the engine.
5. Use a razor blade to cut off the oil pan seal end tabs flush with the front face of the cylinder block and remove the seal. Clean the timing chain cover, oil pan, and cylinder block surfaces.
6. Remove the crankshaft oil seal from the timing chain cover.
7. Install in reverse order of the above procedure. It will be necessary to cut the same amount from the end tabs of a new oil pan seal as was cut from the original seal, before installing the new gasket.

304, 360 and 401 V8

1. Remove the negative battery cable.
2. Drain the cooling system and disconnect the radiator hoses and by-pass hose.
3. Remove all of the drive belts and the fan and spacer assembly.
4. Remove the alternator and the front portion of the alternator bracket as an assembly.
5. Disconnect the heater hose.
6. Remove the power steering pump, and/or the air pump, and the mounting bracket as an assembly. Do not disconnect the power steering hoses.
7. Remove the distributor cap and note the position of the rotor. Remove the distributor. (See the Engine Electrical Section.)
8. Remove the fuel pump.
9. Remove the vibration damper and pulley.
10. Remove the two front oil pan bolts and the bolts which secure the timing chain cover to the engine block.
    NOTE: The timing gear cover retaining bolts vary in length and must be installed in the same locations from which they were removed.
11. Remove the cover by pulling forward until it is free of the locating dowel pins.
12. Clean the gasket surface of the cover and the engine block.
13. Pry out the original seal from inside the timing chain cover and clean the seal bore.
14. Drive the new seal into place from the inside with a block of wood until it contacts the outer flange of the cover.
15. Apply a light film of motor oil to the lips of the new seal.

Installation of the timing gear cover-to-oil pan seal on AMC V8 engines (© Jeep Corp.)

16. Before reinstalling the timing gear cover, remove the lower locating dowel pin from the engine block. The pin is required for correct alignment of the cover and must either be reused or a replacement dowel pin installed after the cover is in position.
17. Cut both sides of the oil pan gasket flush with the engine block with a razor blade.
18. Trim a new gasket to correspond to the amount cut off at the oil pan.
19. Apply seal to both sides of the new gasket and install the gasket on the timing case cover.
20. Install the new front oil pan seal.

21. Align the tongues of the new oil pan gasket pieces with the oil pan seal and cement them into place on the cover.
22. Apply a bead of sealer to the cutoff edges of the original oil pan gaskets.
23. Place the timing case cover into position and install the front oil pan bolts. Tighten the bolts slowly and evenly until the cover aligns with the upper locating dowel.
24. Install the lower dowel through the cover and drive it into the corresponding hole in the engine block.
25. Install the cover retaining bolts in the same locations from which they were removed, tightened to 25 ft lbs.
26. Assemble the remaining components in the reverse order of removal.

350 V8

1. Drain the cooling system and remove the radiator, shroud, fan, pulleys and drive belts.
2. Remove the crankshaft pulley, fuel pump and distributor.
3. Remove the alternator and the power steering pump, if necessary.
4. Loosen and slide rearward the front clamp on the thermostat by-pass hose.
5. Remove the vibration damper.
6. Remove the bolts attaching the timing chain cover to the cylinder block and oil pan. Remove the timing chain cover assembly and gasket.
7. Use a punch to drive the old seal out of the cover. Drive the seal from the front toward the rear of the timing chain cover.
8. Tap the new seal into place from the rear of the timing case cover. If a rope seal is used, position the joint of the two ends at the top, then drive in a new shedder with a suitable punch. Stake the shedder in at least 3 places to secure it in position. Size the packing by rotating a hammer handle or similar smooth tool around the inside of the seal to obtain clearance for the crankshaft hub.
9. If the oil pump has not been removed from the timing chain cover, remove the attaching screws, oil pump cover and gasket from the timing chain cover.
10. Completely pack the space around the oil pump gears with petroleum jelly. There must be no air space left inside the pump. Secure the oil pump cover and a new gasket to the timing chain cover with the 5 slotted attaching screws. Torque the screws alternately and evenly to 8-12 ft lbs.

CUT HERE

CUT HERE

Trim the timing gear cover-to-oil pan seal as shown on 232 and 258 Sixes.

NOTE: Unless the oil pump gears are packed with petroleum jelly, the pump may not prime itself when the engine is started, thus oil will not be pumped throughout the engine for lubrication.

11. The gasket surfaces of the cylinder block and the timing chain cover must be smooth and clean before positioning the gasket in place on the engine. Use sealer on the engine and timing chain cover mating surfaces.

12. Position the timing chain cover to the engine block. Be sure that the dowel pins on the engine engage the holes in the cover before installing the bolts.

13. Install the attaching bolts and tighten them evenly to 30 ft lbs.

Timing Chain or Gears and Tensioner R & R

F4

1. Remove the timing gear cover.
2. Use a puller to remove both the crankshaft and the camshaft gear from the engine after removing all attaching nuts and bolts.
3. Remove the Woodruff keys.

Installation is as follows:

1. Install the Woodruff key in the longer of the two keyways on the front end of the crankshaft.
2. Install the crankshaft timing gear on the front end of the crankshaft with the timing mark facing away from the cylinder block.
3. Align the keyway in the gear with the Woodruff key and then drive or press the gear onto the crankshaft firmly against the thrust washer.
4. Turn the camshaft or the crankshaft as necessary so that the timing marks on the two gears will be together after the camshaft gear is installed.
5. Install the Woodruff key in the keyway on the front of the camshaft.
6. Start the large timing gear on the camshaft with the timing mark facing out.

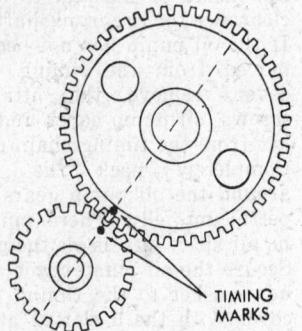

TIMING MARKS

Aligning timing marks—4 cylinder
(© Willys Corp.)

NOTE: Do not drive the gear onto the camshaft as the camshaft may drive the plug out of the rear of the engine and cause an oil leak.

7. Install the camshaft retaining screw and torque it to 30-40 ft lbs. This will draw the gear onto the camshaft as the screw is tightened. Standard running tolerance between the timing gears is 0.000 to 0.002 in.

225 V6

1. Remove the timing chain cover.
2. Make sure that the timing marks on the crankshaft and the camshaft sprockets are aligned. This will make installing the parts easier.

NOTE: It is not necessary to remove the timing chain dampers (tensioners) unless they are worn or damaged and require replacement.

3. Remove the front crankshaft oil slinger.
4. Remove the bolt and the special washer that hold the camshaft distributor drive gear and fuel pump eccentric at the forward end of the camshaft. Remove the eccentric and the gear from the camshaft.
5. Alternately pry forward the camshaft sprocket and then the crankshaft sprocket until the camshaft sprocket is pried from the camshaft.
6. Remove the camshaft sprocket, sprocket key, and timing chain from the engine.
7. Pry the crankshaft sprocket from the crankshaft.

Install as follows:

1. If the engine has not been disturbed proceed to step Number 4 for installation procedures.
2. If the engine has not been disturbed turn the crankshaft so that number one piston is at top dead center of the compression stroke.
3. Temporarily install the sprocket key and the camshaft sprocket on the camshaft. Turn the camshaft so that the index mark of the sprocket is downward. Remove the key and sprocket from the camshaft.
4. Assemble the timing chain and sprockets. Install the keys, sprockets, and chain assembly on the camshaft and crankshaft so that the index marks of both the sprockets are aligned.

NOTE: It will be necessary to hold the spring loaded timing chain damper out of the way while installing the timing chain and sprocket assembly.

5. Install the front oil slinger on the crankshaft with the inside diameter against the sprocket (concave side toward the front of the engine).

6. Install the fuel pump eccentric on the camshaft and the key, with the oil groove of the eccentric forward.
7. Install the distributor drive gear on the camshaft. Secure the gear and eccentric to the camshaft with the retaining washer and bolt.
8. Torque the bolt to 40-55 ft lbs.

232, 258 Sixes

1. Remove the drive belts, engine fan and hub assembly, accessory pulley, vibration damper and timing chain cover.
2. Remove the oil seal from the timing chain cover.
3. Remove the camshaft sprocket retaining bolt and washer.
4. Rotate the crankshaft until the timing mark on the crankshaft sprocket is closest to and in a center line with the timing pointer of the camshaft sprocket.
5. Remove the crankshaft sprocket, camshaft sprocket and timing chain as an assembly. Disassemble the chain and sprockets.

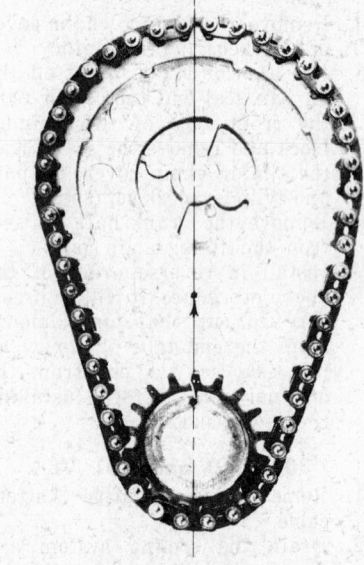

Alignment of the timing chain sprockets timing marks on 232 and 258 Sixes.

Installation is as follows:
1. Assemble the timing chain, crankshaft sprocket and camshaft sprocket with the timing marks aligned.
2. Install the assembly to the crankshaft and the camshaft.
3. Install the camshaft sprocket retaining bolt and washer and tighten to 45-55 ft lbs.
4. Install the timing chain cover and in a new oil seal.
5. Install the vibration damper, accessory pulley, engine fan and hub assembly and drive belts. Tighten the belts to the proper tension.

304, 360 and 401 V8s

1. Remove the timing chain cover and gasket.
2. Remove the crankshaft oil slinger.
3. Remove the camshaft sprocket retaining bolt and washer, distributor drive gear and fuel pump eccentric.
4. Rotate the crankshaft until the timing mark on the crankshaft sprocket is adjacent to, and on a center line with, the timing mark on the camshaft sprocket.
5. Remove the crankshaft sprocket, camshaft sprocket and timing chain as an assembly. Disassemble the chain and sprockets.

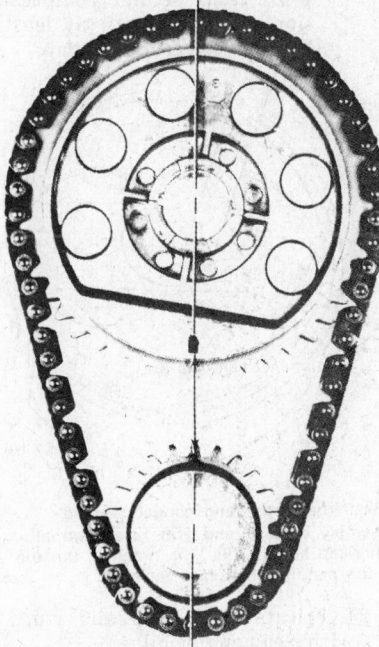

Alignment of the timing chain sprockets timing marks on the American Motors V8s.

Installation is as follows:
1. Assemble the timing chain, crankshaft sprocket and camshaft sprocket with the timing marks on both sprockets aligned.
2. Install the assembly to the crankshaft and the camshaft.
3. Install the fuel pump eccentric, distributor drive gear, washer and retaining bolt. Tighten the bolt to 25-35 ft lbs.
4. Install the crankshaft oil slinger.
5. Install the timing chain cover using a new gasket and oil seal.

350 V8

1. Remove the timing chain cover, and rotate the crankshaft until the timing marks on both the crankshaft and camshaft sprockets are aligned.
2. Remove the crankshaft oil slinger from the crankshaft.
3. Remove the fuel pump eccentric from the face of the camshaft sprocket by removing the bolt,

lockwasher and flat washer. Remove the distributor drive gear.
4. Pry the camshaft sprocket and the crankshaft sprocket forward a little at a time until the camshaft sprocket is free from the camshaft. Remove the timing chain from both sprockets.
5. Continue prying the crankshaft sprocket forward until it is free of the crankshaft.

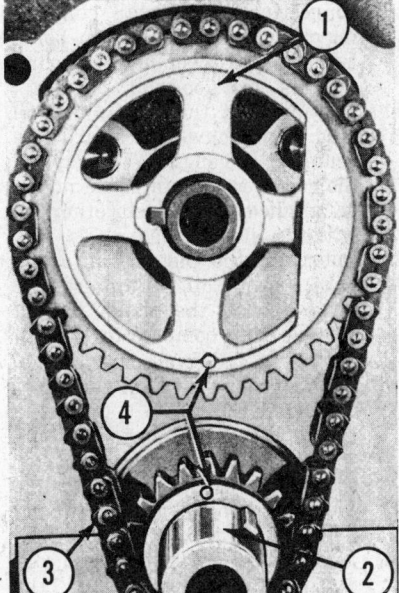

1 Camshaft sprocket 3 Timing chain
2 Crankshaft sprocket 4 Timing marks

Alignment of the timing chain sprockets timing marks on the 350 V8.

To install the timing chain and sprockets:
6. If the engine has been disturbed while the timing chain was removed from the engine or the timing marks on both the sprockets were not aligned prior to being removed, then it is necessary to establish the correct valve timing.
7. Index the "0" marks on the camshaft and crankshaft sprockets on an imaginary line drawn vertically through the centerline of each sprocket. See Step 1.
8. Adjust the positioning of the crankshaft and camshaft so that the sprockets can be installed in the position described in Step 7.
9. Without disturbing the positioning of the indexed sprockets and chain assembly, lift up the indexed assembly and slide the sprockets onto their shafts. Make sure the longer hubs of both sprockets face the block.
10. To check the assembly, rotate the crankshaft until the timing mark on the camshaft sprocket is on a horizontal line at either the 3 or 9 o'clock position. Count the number of links or pins on the

timing chain between the two timing marks. There should be 10 links or 20 pins between the timing marks.
11. On the 350 V8, install the fuel pump eccentric and the distributor drive gear. Install the bolt and washer and torque to 50-55 ft lbs.
12. Install the oil slinger on the crankshaft with the flange facing outward. Install any timing chain cover studs that may have been removed and install the timing chain cover.

Pistons and Connecting Rods R & R

F4

1. Remove the cylinder head and the oil pan.
2. Remove the ridge from the tops of the cylinder bores, using a ridge reamer.
3. Remove the oil pan and, one at a time, remove the connecting rod caps and push the piston assemblies out the top of the block. Number all pistons, connecting rods, and caps on removal.

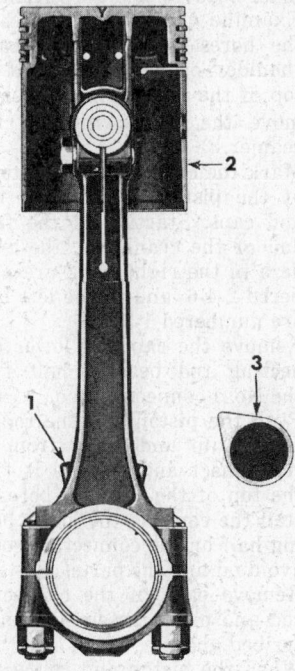

1 Oil spray hole
2 Piston skirt T-slot
3 Relative position of the camshaft

Piston and connecting rod assembly for the F4 engine.

4. Remove the old rings with a ring expander.
5. Release the piston pin lockscrews and force out the pins.
6. Check the cylinder bores for distortion, taper, and other evidence of excessive wear. Bore and hone as necessary.
7. Assemble the piston to the rod

by pushing in and locking the pin.

8. Using a ring expander, install the piston rings. Install the bottom (oil) ring first, center ring second, and top ring last.

9. Coat all the bearing surfaces and rings, and the piston skirt, with engine oil.

10. Turn down the crankpin of the cylinder being worked on.

11. Make sure that the gaps in the rings are not in line.

12. Using a ring compressor, install the piston and rod assembly into the cylinder with the connecting rod identifying number toward the camshaft and oil spray hole away from the camshaft. Carefully tap down until the rod bearing is solidly seated on the crankpin.

13. Install the cap and lower bearing shell. Torque to the specified figure.

14. Install all piston assemblies in the same manner.

15. Install the cylinder head and oil pan.

225 V6

1. Remove the intake manifold, cylinder head(s) and the oil pan.

2. Examine the cylinder bores. If the bores are worn so that a shoulder or ridge exists at the top of the piston ring travel, remove the ridges with a ridge reamer.

3. Mark the cylinder number on all of the pistons, connecting rods and caps. Starting at the front end of the crankcase, the cylinders in the right bank are numbered 2-4-6, and in the left bank are numbered 1-3-5.

4. Remove the cap and lower connecting rod bearing half from the No. 1 connecting rod.

5. Push the piston and the rod assembly up and away from the crankshaft and remove it from the top of the cylinder bore. Install the cap and the lower bearing half on the connecting rod to avoid mixing the parts.

6. Remove each of the connecting rod and piston assemblies as described above.

7. When the piston and connecting rod assembly is properly installed, the oil spurt hole in the connecting rod will face the camshaft. The rib on the edge of the bearing cap will be on the same side as the conical boss on the connecting rod web. These marks on the rib and boss will be toward the other connecting rod on the same crankpin. The notch on the piston will face the front of the engine.

8. Be certain that the cylinder bores, pistons, connecting rod bearings and crankshaft journals are clean. Coat all the bearing surfaces with engine oil.

9. Before installing a piston and connecting rod assembly into its bore, rotate the crankshaft so that the corresponding crankpin is moved downward, away from the cylinder bore.

10. Remove the bearing cap from the connecting rod. With the upper bearing half seated in the connecting rod, install connecting rod guides on the cap bolts. Lengths of rubber hose covering the bolts do a good job.

11. Be sure that the gap in the oil ring faces upward toward the center of the engine. The gaps of the compression rings are not to be aligned with each other or with the oil ring gap.

12. Lubricate the piston and rings with engine oil. Compress the rings around the piston with a suitable ring compressor. Install the piston and connecting rod assembly from the top of the cylinder bore.

13. Install the connecting rod bearing cap with the lower half of the bearing shell in place on the connecting rod. Tighten the nuts to the proper torque.

14. Install the other piston and connecting rod assemblies in the same manner.

15. Check the end clearance between the connecting rods on each crankpin with a feeler gauge.

16. Assemble the oil pan, cylinder head(s) and intake manifold in the reverse order of removal.

232 and 258 Sixes

1. Remove the cylinder head cover and gasket.

2. Remove the rocker arm assemblies. If the engine is equipped with the individual bridged pivot type of rockers, back off each capscrew a turn at a time to prevent breaking the bridge.

3. Remove the push rods.

4. Remove the cylinder head and gasket.

5. Position the pistons one at a time near the bottom of their stroke and use a ridge reamer to remove any ridge from the top end of the cylinder walls.

6. Drain the engine oil.

7. Remove the oil pan and gaskets.

8. Remove the connecting rod bearing caps and inserts and retain them in the same order as removed. The connecting rods and caps are stamped with the number of the cylinder when they are installed.

9. Remove the connecting rod and piston assemblies through the top of the cylinder bores.

NOTE: Be careful the connecting

rod bolts do not scratch the connecting rod journals or cylinder walls. Short pieces of rubber hose can be slipped over the rod bolts to prevent damage.

10. After thoroughly cleaning the cylinder bores, apply a light film of clean engine oil to the cylinder bores with a clean cloth.

11. Position the piston rings on the pistons with:

   a. Oil ring spacer gap on centerline of piston skirt.
   b. Oil ring rail gaps 180° apart on center line of piston pin.
   c. Second compression ring gap 180° from top oil rail gap.
   d. First compression ring gap 180° from second compression ring gap with at least 30° between each ring gap.

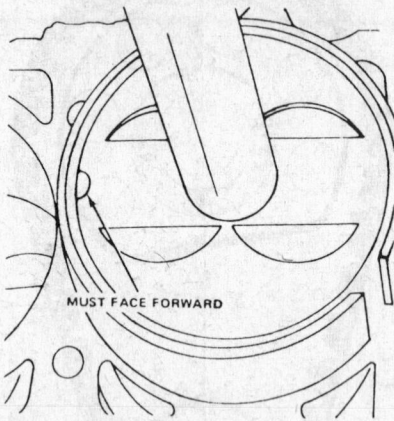

MUST FACE FORWARD

Install the piston and connecting rod assembly on 232 and 258 Sixes and all American Motors V8s. The notch in the top of the pistons must face forward.

12. Lubricate the piston and rings with clean engine oil.

13. With the notch in the top of the piston facing forward and the oil squirt hole in the connecting rod facing the camshaft, use a piston ring compressor to install the connecting rod and piston assemblies through the top of the cylinder bores. Make sure you have the little pieces of hose over the connecting rod bolts so as not to scratch the cylinder bores or the crankshaft journals.

14. Install the connecting rod bearing caps and inserts in the same order as removed. Tighten the retaining nuts to 28 ft lbs.

15. Install the oil pan with new gaskets. Tighten the drain plug.

16. Install a new gasket and the cylinder head.

17. Install push rods.

18. Install the rocker assemblies.

19. Install the cylinder head cover and gasket.

20. Fill the crankcase with new oil.

V8 All

1. Remove the cylinder head covers.

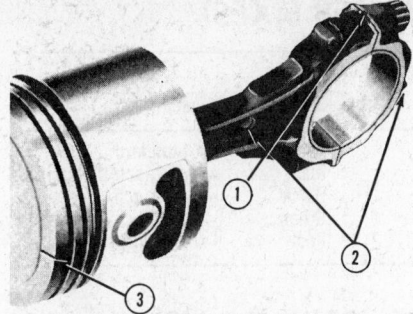

Piston and connecting rod assembly for the 350 V8 (right side).

1. Oil squirt hole facing camshaft
2. Boss on rod and cap facing forward (rearward on left side assembly)
3. Notch in piston facing forward

2. Remove the rocker arms and bridged pivot assemblies, loosening each capscrew one turn at a time to avoid breaking the bridge.
3. Remove the push rods.
4. Remove the intake manifold assembly.
5. Remove the cylinder head and gasket.
6. Position the pistons, one at a time, near the bottom of their stroke and use a ridge reamer to remove any ridge from the top end of the cylinder walls.
7. Drain the engine oil.
8. Remove the oil pan.

Location of the connecting rod identifying numbers and the oil squirt hole on the AMC V8 engines. The numbers and the squirt hole are on the same side of the connecting rod on the 232 and 258 six cylinder AMC engines. In both cases, the oil squirt hole must face the camshaft when installed in the engine (© Jeep Corp.)

9. Remove the connecting rod bearing caps and inserts. Keep these parts in the same order as they are removed. The connecting rods and caps are stamped with the number of the cylinder from which they are removed.
10. Remove the connecting rod and piston assemblies through the top of the cylinder bores. Be careful not to scratch the cylinder bores or the crankshaft connecting rod bearing journals with the connecting rod bolts. Install short pieces of rubber hose over the bolts for protection.

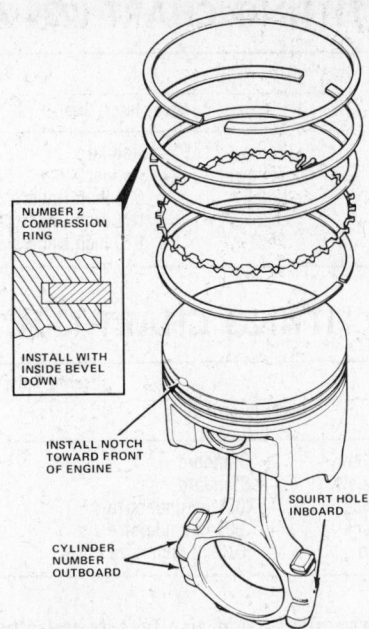

NUMBER 2 COMPRESSION RING

INSTALL WITH INSIDE BEVEL DOWN

INSTALL NOTCH TOWARD FRONT OF ENGINE

SQUIRT HOLE INBOARD

CYLINDER NUMBER OUTBOARD

AMC V8 engine piston and connecting rod assembly (© Jeep Corp.)

11. After thoroughly cleaning the cylinder bores, apply a light film of clean engine oil to the bores with a clean cloth.
12. Arrange the piston ring gaps around the piston as follows:
    a. Oil spacer gap on centerline of either skirt face.
    b. Oil rail gaps 180° apart and in line with piston pin centerline.
    c. Second compression ring gap 180° from top oil rail gap.
    d. First compression ring gap 180° from second compression ring gap.
13. Lubricate piston and ring surfaces with clean engine oil.
14. With the notch on the top of the pistons facing forward and the oil squirt hole in the connecting rod facing the camshaft, use a ring compressor to install the piston assemblies through the top of the cylinder bores. Place the lengths of rubber hose over the connecting rod bolts so to protect the cylinder bores and crankshaft bearing journals.
15. Install the connecting rod bearing caps and inserts in the same order as removed. Tighten the nuts to the proper torque (see Torque Specifications).
16. Install the oil pan with new gaskets and tighten the drain plug.
17. Install the cylinder heads and gaskets.
18. Install the push rods and rocker assemblies.
19. Install the intake manifold assembly.
20. Install the cylinder head covers with new gaskets.
21. Fill the crankshaft with new oil.

Crankshaft Main Bearings

F4

The bearings are positioned and prevented from rotating in their supports in the cylinder block by dowel pins. Dowel pins are used in both the center and the rear bearing caps. No dowel pins are used in the front bearing cap because the bearing has a flange. The front main bearing takes the end thrust of the crankshaft. The main bearings are of premium type which provides long bearing life. They are replaceable and when correctly installed, provide proper clearance without filing, boring, scraping, or shimming. Crankshaft bearings can be removed from this engine only with the engine out of the vehicle. Crankshaft bearings must be replaced as a complete set of three bearings, each bearing consisting of two halves. Main bearings are available in the standard size and the following undersizes:

.001 in., .002 in., .010 in., .012 in., .020 in., .030 in.

The .001 in. and .002 in. undersize main bearings are for use with standard size crankshafts having slightly worn journals. The .010 in., .020 in., and .030 in. undersize bearings are for use with undersize crankshafts in those sizes. The .012 in. undersize bearings are for use with .010 in. undersize crankshafts having slightly worn journals. Bearing sizes are rubber stamped on the reverse side of each bearing half.

350 V8

The crankshaft bearings consists of two halves which are neither alike nor interchangeable. One half is carried in the corresponding main bearing cap; the other half is located between the crankshaft and cylinder block. The upper (cylinder block) half of the bearing is grooved to supply oil to the connecting rod bearings, while the lower (bearing cap) half of the bearing is not grooved. The two bearing halves must not be interchanged. All crankshaft bearings except the thrust bearing and the rear main bearing are identical. The thrust bearing (No. 3) is longer and it is flanged to take crankshaft end thrust, and the rear main upper bearing groove does not extend the full length of the bearing. When the bearing halves are placed in cylinder block and bearing cap, the ends extend slightly beyond the parting surfaces. When cap bolts are tightened, the halves are clamped tightly in place to ensure positive seating and to prevent turning. The ends of bearing halves must never be filed flush with parting surface of crankshaft or bearing cap.

MAIN BEARING FITTING CHART (232 AND 258 SIXES)

Crankshaft Main Bearing Journal Color Code and Diameter in Inches (Journal Size)	Bearing Color Code			
	Upper Insert Size		Lower Insert Size	
Yellow — 2.5001 to 2.4996 (Standard)	Yellow	— Standard	Yellow	— Standard
Orange — 2.4996 to 2.4991 (0.0005 Undersize)	Yellow	— Standard	Black	— .001-inch Undersize
Black — 2.4991 to 2.4986 (0.001 Undersize)	Black	— .001-inch Undersize	Black	— .001-inch Undersize
Green — 2.4986 to 2.4981 (0.0015 Undersize)	Black	— .001-inch Undersize	Green	— .002-inch Undersize
Red — 2.4901 to 2.4896 (0.010 Undersize)	Red	— .010-inch Undersize	Red	— .010-inch Undersize

MAIN BEARING FITTING CHART (304, 360, AND 401 V8)

Crankshaft Main Bearing Journal Color Code and Diameter (Journal Size)	Bearing Color Code			
	Upper Insert Size		Lower Insert Size	
Yellow — 2.7489 to 2.7484 in.	Yellow	— Standard	Yellow	— Standard
Orange — 2.7484 to 2.7479 in.	Yellow	— Standard	Black	— .001-in. undersize
Black — 2.7479 to 2.7474 in.	Black	— .001-in. undersize	Black	— .001-in. undersize
Green — 2.7474 to 2.7469 in.	Black	— .001-in. undersize	Green	— .002-in. undersize
Red — 2.7389 to 2.7384 in.	Red	— .010-in. undersize	Red	— .010-in. undersize

If the thrust bearing shell is disturbed or replaced it is necessary to line up the thrust surfaces of the bearing shell before the cap bolts are tightened. To do this, move the crankshaft fore and aft the limit of its travel several times (the last movement fore) with the bearing cap bolts finger tight.

Crankshaft bearings are the precision type which do not require reaming to size or other fitting. Shims are not provided for adjustment since worn bearings are readily replaced with new bearings of proper size. Bearings for service replacement are furnished in standard size and undersizes. Under no circumstances should crankshaft bearing caps be filed to adjust for wear in the old bearings.

232 and 258 Sixes

The crankshaft main bearings are steel-backed, micro-babbit, precision type. Each bearing is selectively fitted to its respective journal to obtain the desired operating clearance. In production, the select fit is obtained by using various sized color coded vearing inserts. The color code appears on the edge of the insert. Bearing size is not stamped on the inserts used in production.

The main bearing journal size is identified in production by a color coded paint mark on the adjacent cheek toward the rear end of the crankshaft, except for the rear main journal which is on the crankshaft rear flange.

When required, different sized upper and lower bearing inserts may be used as a pair. A standard size insert is sometimes used in combination with a 0.001 in. undersize insert to reduce clearance by 0.0005 in.

NOTE: Never use bearing inserts in pairs with greater than 0.001 in. difference in size. When replacing inserts, all the odd size inserts must be either on the top (in the block) or the bottom (in the main bearing cap).

Service replacement bearing inserts are available as pairs in the following sizes: standard, 0.001, 0.002, 0.010, and 0.012 in. undersize. The size on these service replacement bearings is stamped on the back of the inserts.

NOTE: The 0.012 in. undersize insert is not used in production.

304, 360 and 401 V8

The main bearing caps are numbered from front to rear 1 through 5, with an arrow indicating forward. The upper inserts are grooved while the lower inserts have a smooth surface on the 304 and 360 V8. The 401 V8 has a groove in both the upper and lower insert.

Each bearing is a select fit to its respective journal to obtain the desired operating clearance. In production the select fit is obtained by using various sized color coded bearing inserts. The bearing code color appears on the edge of the insert. The bearing size is not stamped on production inserts.

The main bearing journal size is identified in production by a color coded paint mark on the adjacent cheek toward the rear end of the crankshaft except for the rear main journal. The paint mark for the rear main journal is on the crankshaft rear flange.

When required, different sized upper and lower bearing inserts may be used as a pair to reduce clearance by 0.0005 in.

NOTE: When using upper and lower inserts of different sizes, install all the same size inserts together either on the top (block) or bottom (bearing cap). Never use bearing inserts with greater than 0.001 in. difference in pairs.

Service replacement bearings are available as pairs in the following sizes: standard, 0.001, 0.002, 0.010, and 0.012 in. undersize. The size on the service replacement inserts is stamped on the back of the inserts.

NOTE: The 0.012 in. undersize inserts are not used in production.

Connecting Rod Bearings

F4

The connecting rod bearings, like the crankshaft main bearings, are of the replaceable type. When correctly installed, the bearings provide proper clearance without filing, boring, scraping, or shimming.

Main bearings with maximum wearing surfaces are obtained through the use of offset connecting rods. When the rods are installed, the offset is placed away from the nearest main bearing.

The oil spray hole should be on the "follow" side or away from the camshaft, toward the right side of the vehicle. Because of the offset and oil spray hole, No. 1 and 2 or No. 3 and 4 connecting rods cannot be interchanged for if they are reversed, the oil spray hole will be on the wrong side. No. 1 and 3 or No. 2 and 4 can be interchanged.

Connecting rod bearings should be replaced as a complete set. Each bearing consists of two halves. Connecting rod bearing sets are available in standard size and the following undersize:

.001 in., .002 in., .010 in., .012 in., .020 in., .030 in.

The .001 and .002 in. undersize bearings are for use with standard size crankshafts having slightly worn crankpins that do not require grinding. The .012 in. undersize bearings

CONNECTING ROD BEARING FITTING CHART (232 AND 258 SIXES) (IN.)

Crankshaft Connecting Rod Journal Color and Diameter in Inches (Journal Size)	Bearing Color Code			
	Upper Insert Size		Lower Insert Size	
Yellow — 2.0955 to 2.0948 (Standard)	Yellow	— Standard	Yellow	— Standard
Orange — 2.0948 to 2.0941 (0.0007 Undersize)	Yellow	— Standard	Black	— .001 Undersize
Black — 2.0941 to 2.0934 (0.0014 Undersize)	Black	— .001-Inch Undersize	Black	— .001 Undersize
Red — 2.0855 to 2.0848 (0.010 Undersize)	Red	— .010-Inch Undersize	Red	— .010 Undersize

CONNECTING ROD BEARING FITTING CHART (304, 360 AND 401 V8) (IN.)

Crankshaft Connecting Rod Journal Color and Diameter in Inches (Journal Size)	Bearing Color Code			
	Upper Insert Size		Lower Insert Size	
304-360 CID Engines				
Yellow — 2.0955 to 2.0948 (Standard)	Yellow	— Standard	Yellow	— Standard
Orange — 2.0948 to 2.0941 (0.0007 Undersize)	Yellow	— Standard	Black	— .001 Undersize
Black — 2.0941 to 2.0934 (0.0014 Undersize)	Black	— .001-inch Undersize	Black	— .001 Undersize
Red — 2.0855 to 2.0848 (0.010 Undersize)	Red	— .010-inch Undersize	Red	— .010 Undersize
401 CID Engine				
Yellow — 2.2485 to 2.2478 (Standard)	Yellow	— Standard	Yellow	— Standard
Orange — 2.2478 to 2.2471 (0.0007 Undersize)	Yellow	— Standard	Black	— .001 Undersize
Black — 2.2471 to 2.2464 (0.0014 Undersize)	Black	— .001-inch Undersize	Black	— .001 Undersize
Red — 2.2385 to 2.2378 (0.010 Undersize)	Red	— .010-inch Undersize	Red	— .010 Undersize

are for use with slightly worn crankshafts that have been previously ground for .010 in. undersize bearings.

Should it be necessary to replace the bearings due to wear, replacement of piston rings and piston pins is also recommended.

232 and 258 Sixes, 304, 360 and 401 V8

the connecting rod bearings are steel-backed aluminum-alloy, precision type.

Each bearing is selectively fitted to its respective journal to obtain the desired operating clearance. In production, the select fit is obtained by using various sized, color coded bearing inserts. The color code appears on the edge of the insert. Bearing size is not stamped on production inserts.

The rod journal size is identified in production by a color coded paint mark on the adjacent cheek or counterweight toward the rear end of the crankshaft.

When required, different sized upper and lower bearing inserts may be used as a pair, thus reducing clearance by 0.0005 in.

NOTE: Never use a pair of bearing inserts with more than 0.001 in. difference in size.

Service replacement bearing inserts are available as pairs in the following sizes: standard, 0.001, 0.002, 0.010, and 0.012 in. undersize. The bearing size is stamped on the back

of the service replacement bearing inserts.

NOTE: The 0.002 and 0.012 in. undersize inserts are not used in production.

Engine Lubrication

Rear Main Bearing Oil Seal R & R

F4

1. Drain the oil from the engine and remove the oil pan.
2. Remove the screws and lockwashers that attach the rear main bearing cap to the cylinder block.
3. Remove the bearing cap with a lifting bar to pry the cap off of the mounting dowels. Do not apply too much pressure to the cap. Lift the cap off a little at a time from one side to the other until the cap is removed.
4. Remove the two pieces of rear main bearing cap packing out of position between the side of the bearing cap and the cylinder block.
5. Remove the lower half of the oil seal from the bearing cap and install a new half.

NOTE: Coat the seal with oil to facilitate installation.

6. Remove the upper half of the seal by pushing it out with a suitable tool.
7. Coat the new seal with oil and, using a wire to pull it around the crankshaft, install the new upper half of the seal.

NOTE: For better sealing, make the joints of the seal off-set from the joints of the rear main bearing cap and the cylinder block.

8. Reinstall the bearing cap, being careful not to bend the dowel pins and making sure that the seal is inserted correctly. Tighten to the correct torque.
9. Apply a small amount of plastic type gasket cement to the rubber packings to be inserted into the holes between the bearing cap and the crankcase. Insert the packings in these holes. Do not trim the packings. They are trimmed to the correct length at the time of manufacture. They will protrude about ¼ in. from the crankcase. When the oil pan is installed, it will force them tightly into the holes and seal any spaces between the bearing cap and the crankcase.
10. Reinstall the oil pan and fill the crankcase with the correct amount of oil.

225 V6

1. Remove the oil pan and the crankshaft. When the rear main bearing cap is removed, remove

the fabric packing type seal from the radius portion of the cap and the neoprene seals from the grooves on the side of the cap.

2. With an ice pick or similar tool pry the upper half of the seal from the block.

3. Place the new seal in the groove in the rear main bearing cap with both ends protruding above the parting surface of the cap. Force the seal into the groove by rubbing down with a hammer handle or similar tool until the seal projects above the groove no more than 1/16 in. Cut the ends off flush with the mating surface of the cap with a razor blade. Lubricate the seal with engine oil just before installing it in the engine.

4. Lubricate all of the main bearings and install the crankshaft. Install all the main bearing caps and tighten them to the correct torque.

5. Dip the neoprene main bearing cap side seals in kerosene (any light oil will do) for 1½ minutes. This will cause the seals to swell; so you will have to work quickly. Install the seals into the bearing cap grooves. The protruding ends of the seals should then be squirted with kerosene (or oil), wiped off, and peened over with a hammer to be sure of a seal at the upper mating surface of the cap and cylinder block.

NOTE: Do not cut the seals to length before installing them.

6. Attach the connecting rods to the crankshaft and install the oil pan in the reverse order of removal.

232 and 258 Sixes, 304, 350, 360, and 401 V8

1. Remove the engine from the vehicle.

2. Remove the timing chain cover and the crankshaft timing gear.

3. Remove the oil pan, oil float support and the oil float.

4. Slide the crankshaft thrust washer and all of the end-play adjusting shims off the front end of the crankshaft.

5. Move the two pieces of the rear main bearing cap packing away from the side of the bearing cap and the cylinder block.

6. You will now be able to see the marks on the bearing caps and the cylinder block for bearing number and position.

7. Remove the screws and lockwashers that attach the main bearing caps to the cylinder block. Use a lifting bar beneath the ends of each bearing cap. Be careful not to exert too much pressure which could damage the cap or the dowels they fit onto. Lift each cap evenly on both sides until free of the dowels. If there is any reason to believe that any of the dowels have become bent during bearing cap removal, remove them and install new dowels. Remove the connecting rod caps and bearings.

8. Remove the crankshaft.

9. Remove the upper half of the rear main bearing oil seal from the cylinder block and the lower half from the oil seal groove in the rear main bearing cap.

10. Install the main bearing caps and bearings on the cylinder block in their original positions.

11. Reassemble the engine in reverse order of the above procedure.

NOTE: It is possible to replace the rear main rubber seal without removing the crankshaft. The procedure is as follows:

1. Loosen the crankshaft cap bolts and lower the crankshaft not more than 1/32 in. Remove the rear main bearing cap. Do not turn the crankshaft while it is loosened.

2. Push the seal around the groove until it can be gripped and removed with a pair of pliers.

3. Coat the new seal with clean motor oil and slide it into the groove until ⅜ in. protrudes from the groove. On American Motors engines push the seal around until its ends are flush with the block.

4. Install the other half of the seal in the bearing cap with ⅜ in. protruding from the opposite side. This is so that the juncture of the two halves of the seal is not at the same point as the juncture of the bearing cap and the cylinder block. On American Motors engines install the seal so that the ends are flush.

5. Install the rear main bearing cap and tighten all the caps to the proper torque. Be sure that the caps have not fallen out of place and that they are tightened in a straight manner.

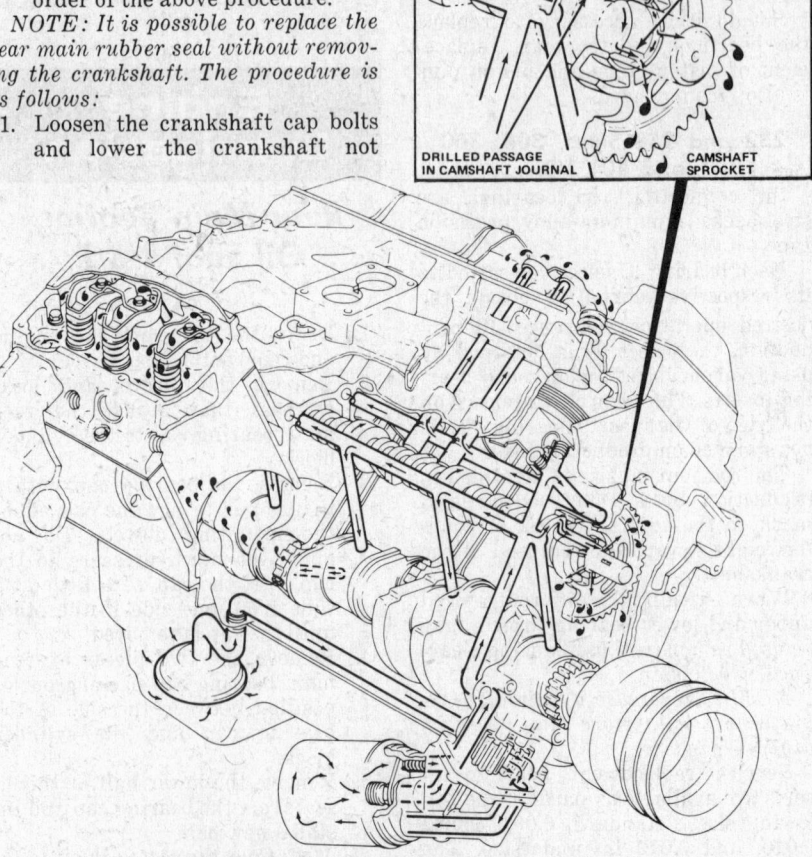

Typical V8 engine lubrication system (© Jeep Corp.)

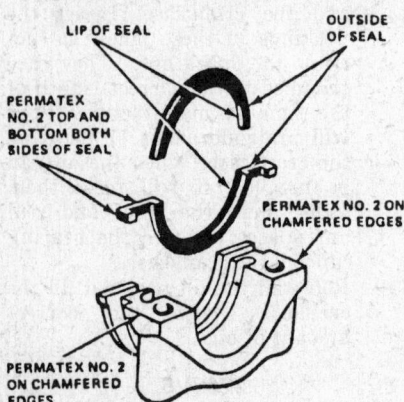

Rear main oil seal on 232 and 258 Sixes and American Motors V8s.

Oil Pan R & R

Due to the variety of engines used in Jeep vehicles, removal and installation procedures for the oil pan could range from simply removing the attaching screws to removing front end components for access.

On CJ and Commando models equipped with 232 and 258 Sixes, remove the starter, drain the oil pan and disconnect the right engine support cushion bracket from the engine block and raise the engine with a jack placed under the transmission bell housing. The oil pan can then be removed with sufficient clearance.

On Jeep vehicles equipped with V8 engines, simply drain the oil pan, remove the starter motor, and remove the oil pan.

In all cases, install the oil pan in the reverse order of removal.

Oil Pump

The oil pump is located on the timing gear cover on the V6 and V8 engines, and does not require the removal of the oil pan to service the oil pump. On the 6 cylinder inline engine, the oil pump is located within the oil sump cavity, and the oil pan must be removed to gain access to the oil pump. The F4 engine has the oil pump located on the lower left side of the engine block, and can be serviced by the removal from its location. Each pump has a pressure relief valve assembly to control the oil pressure produced, and is located within the oil pump body.

F4 Engines

Removal

1. Locate No. 1 cylinder on its firing stroke at TDC by the alignment of the crankshaft pulley and the distributor rotor. *NOTE: Mark the rotor position on the distributor housing for reference upon the installation of the oil pump.*
2. Remove the capscrews from the pump body to cylinder block.
3. Slide the oil pump assembly and shaft out of the engine block, noting the position of the oil pump shaft notch for its alignment upon installation.
4. Remove the three capscrews from the cover, and remove the cover from the pump body.
5. Remove the oil pump drive gear by removal of the retaining pin. *NOTE: It is necessary to file off one end of the retaining pin before driving it through the gear and shaft, or damage may result to the gear and shaft.*

6. Remove the outer rotor and inner rotor, and the shaft from the oil pump body.
7. Remove the pressure relief valve and spring from the pump body.
8. Inspect the pump body, rotor, shaft, relief valve, and spring for abnormal wear, distortion, galling, or scoring.

Installation

1. Assemble the inner rotor and shaft, and the outer rotor into the pump body. Match the rotors together with one lobe of the inner rotor pushed as far as possible into the notch of the outer rotor. Measure the clearance between the lobes of the rotors. The clearance should be .010 inch or less.
2. Measure the clearance between the outer rotor and the pump

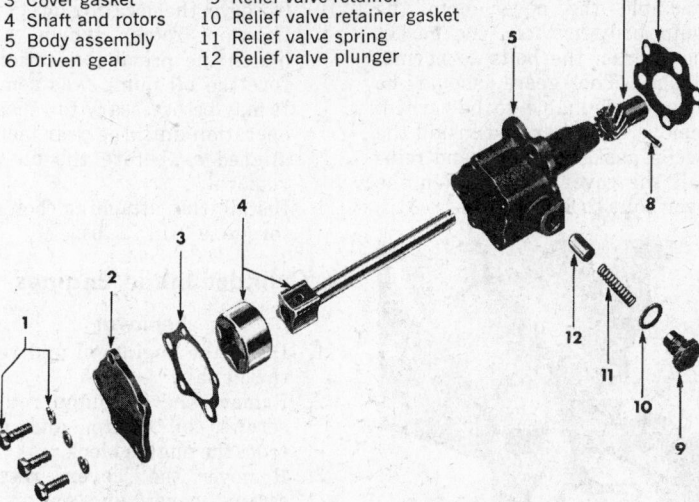

1 Cover screw
2 Cover
3 Cover gasket
4 Shaft and rotors
5 Body assembly
6 Driven gear
7 Pump gasket
8 Gear retaining pin
9 Relief valve retainer
10 Relief valve retainer gasket
11 Relief valve spring
12 Relief valve plunger

Exploded view—F4 engine oil pump assembly (© Jeep Corp.)

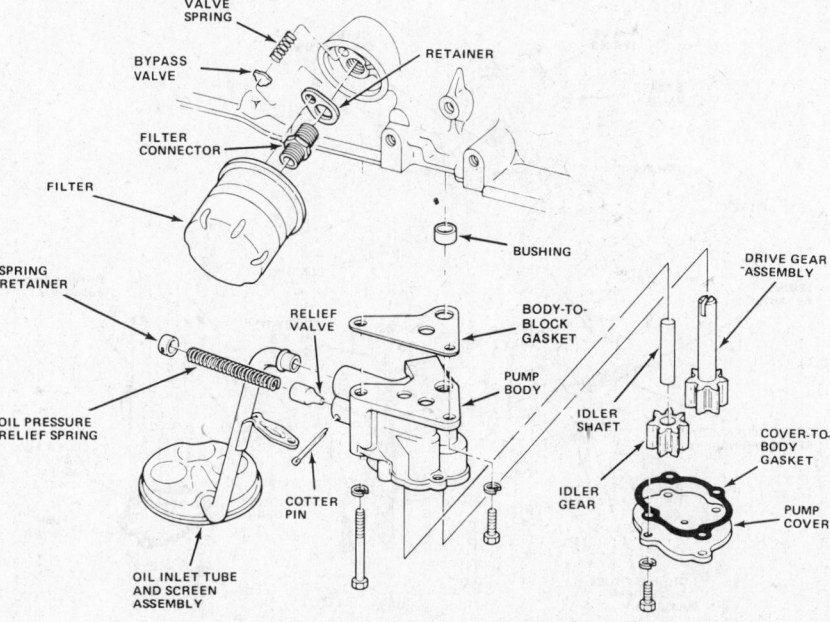

Oil filter and pump assembly on six cylinder AMC engines (© Jeep Corp.)

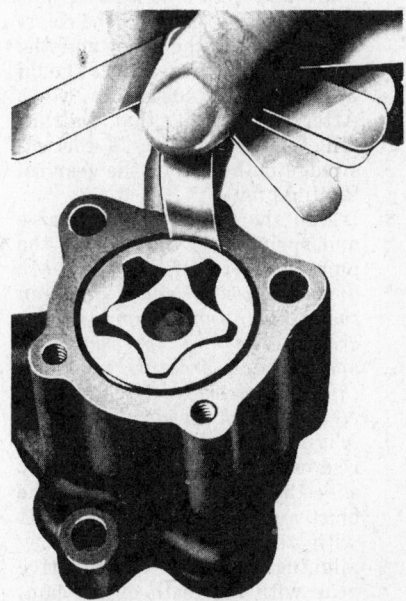

Measuring outer rotor to oil pump body—F4 engine (© Jeep Corp.)

body. The clearance should not exceed .012 inch.

3. Assemble the cover onto the pump body *without* the gasket and tighten the bolts to normal tension. The gears should be locked and unable to be turned.

4. Remove the cover and install the special gasket in place and reinstall the cover, and tighten the cover bolts to 8-12 ft. lbs.

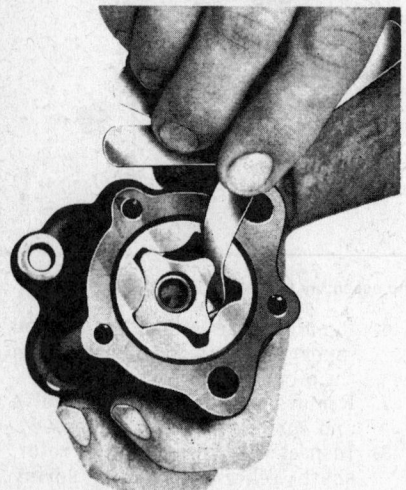

Measuring oil pump rotors—F4 engine
(© Jeep Corp.)

5. The rotors should be free to rotate, assuring the end float of the rotors are less than .004 inch, or the thickness of the special gasket. *NOTE: If the assembly measurements are correct, remove the cover and fill the cavity with petroleum jelly to aid in the priming of the pump. Retorque the cover bolts to the specified torque.*

6. Install the drive gear onto the shaft and retain it with a new retaining pin. Measure the clearance between the gear and the pump body. The clearance should be between .022-.051 inch.

7. After determining that the running gear clearance is correct, swedge both ends of the gear retaining pin.

8. Install the pressure relief valve and spring into the bore of the pump body. *NOTE: Shims are available to be added between the retainer and the spring to increase the oil pressure. If springs are present, removing them will decrease the pressure. Never stretch the spring.*

9. Align the pump shaft notch to its removed location, and carefully slide the pump assembly into the block and engage the pump shaft with the distributor shaft, by aligning the teeth of the drive gear with the teeth of the camshaft gear and allowing a partial turning of the pump shaft as the

pump body is bottomed to the block.

10. Observe the location of the distributor rotor, which should match the previous location, before the oil pump was removed. It may be necessary to repeat the operation until the gear teeth are aligned as before the oil pump removal.

11. Install the attaching bolts and torque to 9-14 ft. lbs.

6 Cylinder Inline Engines

Removal

1. Drain the engine oil and remove the oil pan.
2. Remove the oil pump retaining screws, oil pump, and gasket from the engine block.
3. Remove the cover retaining screws, cover, and gasket from the pump body.
4. Measure the gear end clearance between the gears and the face of the oil pump body. The clearance should be .002-.006 inch.
5. Measure the gear lobe clearance to the pump body sides. The clearance should be .0025-.005 inch.
6. Remove the gears and shaft from the body.
7. Remove the cotter pin, spring retainer, spring, and oil pressure relief valve from the pump body. *NOTE: The oil inlet tube must be removed to gain clearance for the removal of the relief valve. A new pick-up tube assembly must be replaced upon installation, to assure an air tight seal.*

Installation

1. Install the oil pressure relief valve, spring retainer, and cotter pin.

Method of obtaining gear to body clearance measurement (© Jeep Corp.)

Method of obtaining gear end clearance measurement—typical (© Jeep Corp.)

2. Install new inlet tube into the pump body, making sure that the support bracket is properly lined up.

3. Install the idler shaft, Idler gear and drive gear assembly. *NOTE: Fill the pump gear cavity with petroleum jelly prior to the installation of the pump cover, to insure self priming.*

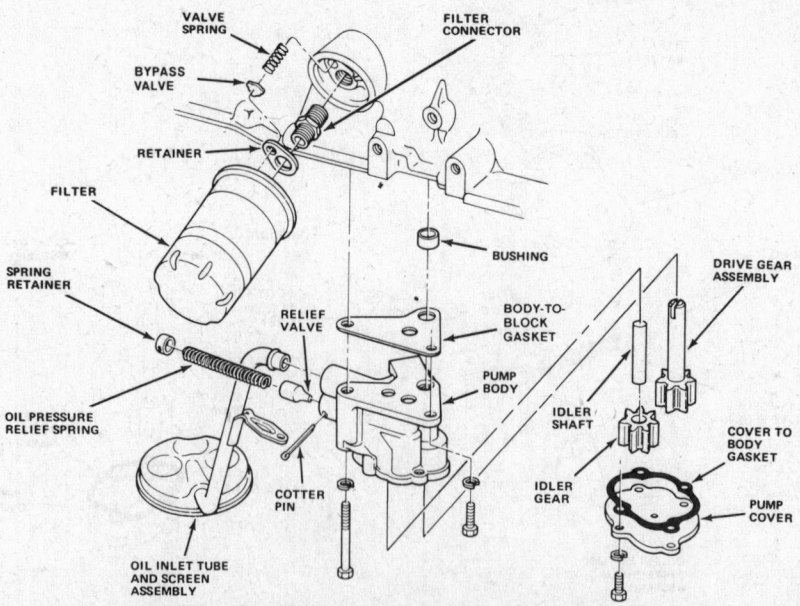

Exploded view—AMC 6 cylinder oil pump assembly (© Jeep Corp.)

4. Install the pump cover and new gasket. Tighten the cover bolts to 70 in. lbs.

5. Install the oil pump assembly and a new gasket on the engine block. Torque the screws to 10 ft. lbs. for the short screws, and 17 ft. lbs. for the long screws.

6. Install the oil pan, using a new gasket and seals.

V6 and V8 Engines

Removal

1. Remove the oil pump cover from the timing chain cover and remove the oil pump gears and shaft.

2. Remove the oil pressure relief valve from the body.

3. Inspect the gears for abnormal wear, chips, looseness on the shafts, galling, and scoring.

4. Inspect the cover and cavity for breaks, cracks, distortion, and abnormal wear.

5. Install the gears into the pump cavity, and with the use of a straight edge and feeler gauge, check the gear to housing clearance.

6. The allowable clearance is .002 to .006 inch and the clearance between the gears and cavity side is .0025 to .0005 inch.

7. If the clearances measure out of the allowable span, the timing chain cover and gears should be replaced.

Installation

1. Install the pressure relief valve, if previously removed.

2. Install the gears into the gear cavity, and pack the cavity with petroleum jelly to insure the self priming of the pump. *NOTE: Never use chassis or wheel bearing grease to pack the gear cavity.*

3. Install the gear cover, using a new gasket. Tighten the retaining screws to 55 in. lbs. on the AMC V8 engines, and 8 to 12 ft. lbs. on the 350 V8 and V6 engines.

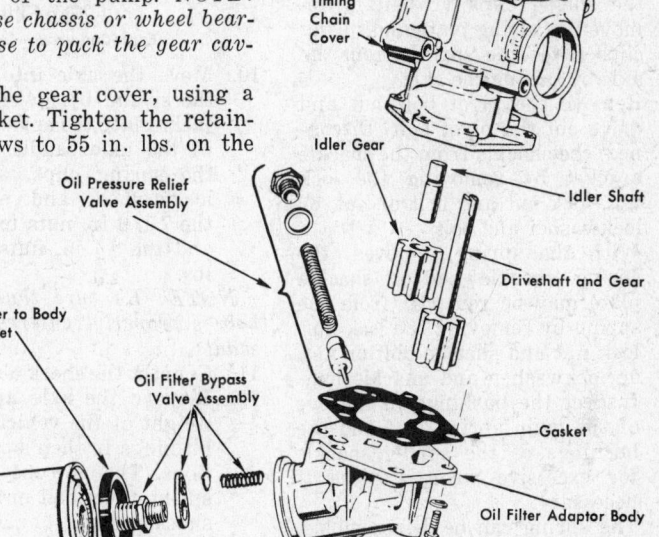

Exploded view—oil pump assembly—AMC V8 engine (© Jeep Corp.)

1 Oil dipstick
2 Oil pan baffle
3 Oil pan gasket
4 Oil pan
5 Drain plug gasket
6 Drain plug
7 Oil pump screen
8 Oil suction housing, pipe and flange
9 Oil suction pipe gasket
10 Oil pump idler gear
11 Valve by-pass and cover assy.
12 Oil pressure valve

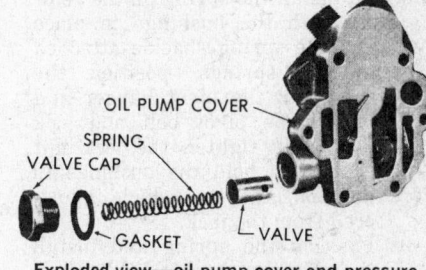

Exploded view—oil pump cover and pressure relief valve—350 V8 engine
(© Buick Motor Div., GM Corp.)

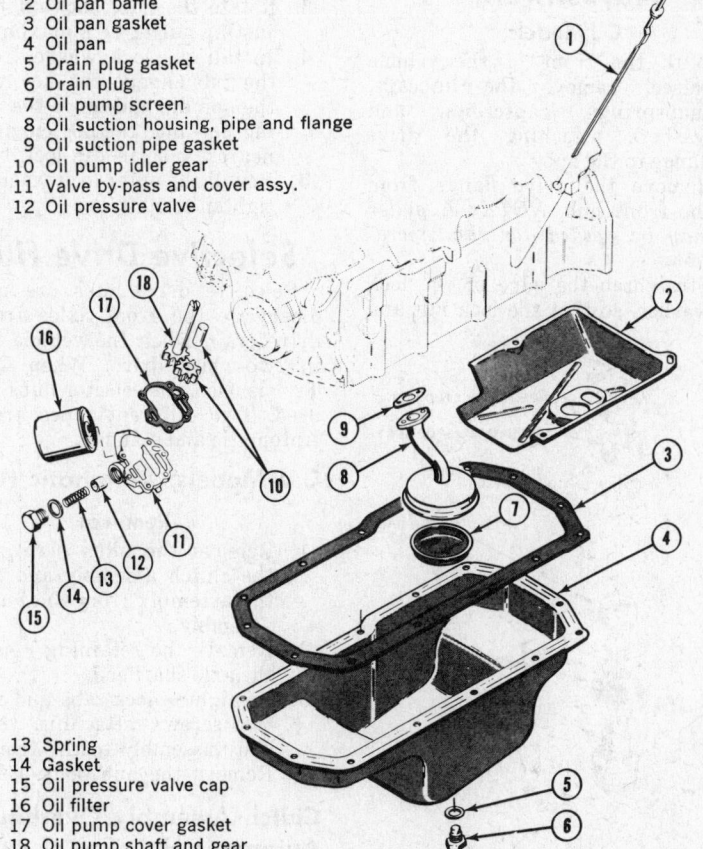

13 Spring
14 Gasket
15 Oil pressure valve cap
16 Oil filter
17 Oil pump cover gasket
18 Oil pump shaft and gear

Oil pan and pump assembly—V6 engine—Typical of 350 V8 engine (© Jeep Corp.)

Front Axle, Suspension

Reference

Refer to the Drive Axles General Repair Section for application, troubleshooting overhaul, and specifications.

Shock Absorbers

The upper ends of the shock absorbers are attached to the frame side rails with mounting brackets and pins. The lower ends are attached to the axle or to the spring by mounting brackets. The shock absorbers are not refillable or adjustable and must be replaced if defective.

Jeep Trucks · Wagoneer · Cherokee ·

Front Spring R & R

1. Raise the vehicle with a jack under the axle. Place a jackstand under the frame side rail. Then lower the axle jack so the load is relieved from the spring and the wheels rest slightly on the floor.
2. Disconnect the shock absorber from the spring clip plate.
3. Remove the nuts which secure the spring clips (U-bolts). Remove the spring plate and spring clips. Free the spring from the axle by raising the axle.
4. Remove the pivot bolt nut and drive out the pivot bolt. Disconnect the shackle from the shackle bracket by removing the lock nut, lock nut and bolt or nut, or lockwasher and bolt.
5. With the spring removed, the spring shackle and/or shackle plate may be removed from the spring by removing the lock nut, lock nut and shackle bolt or nut, or lockwasher and shackle bolt.
6. Inspect the bushings in the eye of the main spring leaf and the bushings of the spring shackle for excessive wear. Replace if necessary.
7. The spring can be disassembled, for replacing an individual spring leaf, by removing the clips and the center bolt.
8. To install the spring on the vehicle with the bushings in place and the spring shackle attached to the springs, position the spring in the pivot hanger and install the pivot bolt and lock nut. Only tighten the lock nut enough to hold the bushings in position until the vehicle is lowered from the jack.
9. Position the spring and install the shackle, shackle bolts, shackle plate if applicable, lockwasher, and nut. Only finger tighten the nuts at this time.

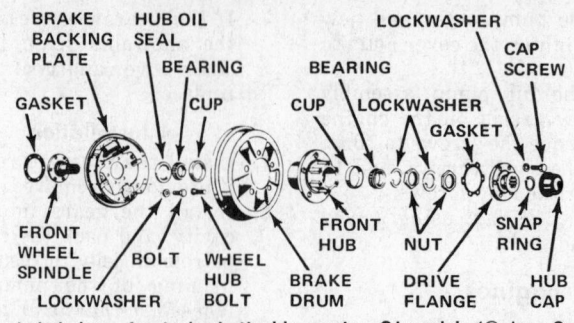

Exploded view—front wheel attaching parts—CJ models (© Jeep Corp.)

10. Move the axle into position on the spring by lowering the axle jack. Place the spring center bolt in the axle saddle hole. Install the spring clips, spring plate, lockwashers and nuts. Torque the 7 1/6 in. nuts to 36-42 ft lbs and the 1/2 in. nuts to 45-65 ft lbs.

NOTE: Be sure that the center bolt is properly centered in the axle saddle.

11. Connect the shock absorber.
12. Remove the axle and allow the weight of the vehicle to seat the bushings in their operating positions. Then torque the 7/16 in. spring pivot bolt nuts and spring shackle nuts to 25-40 ft. lbs. Torque the 5/8 in. shackle nuts to 55-75 ft lbs.

Wheel Bearing Adjustment
CJ Models

1. With the front of the vehicle raised, remove the hubcaps, snapsprings, capscrews, and washers attaching the drive flange to the hub.
2. Remove the drive flange from the front hub. *NOTE: A puller may be needed for this operation.*
3. Straighten the edge of the lock washer, so that the lock nut and

lock washer can be removed.
4. With a special wrench or equivalent, tighten the adjusting nut until the wheel binds, and back off approximately 1/6 turn, so that the wheel turns freely without any lateral shake.
5. Install the lock washer and lock nut on the housing end, tighten the lock nut and crimp the lock washer edge over the lock nut.
6. Assemble the drive flange, bolts, and the hub cap. Assure that the gasket is properly installed between the hub and the flange.

Cherokee, Wagoneer, and Trucks

1. Remove the hub caps, snapring, drive gear pressure spring, outer lock nut, and lock washer.
2. Loosen the inner wheel bearing nut, and then retighten to 50 ft. lbs. torque. (inner nut has peg on outer side).
3. Rotate the hub, back off the adjusting nut 1/4 turn maximum.
4. Install the lock washer to have the tab engage the key way in the spindle and move the adjusting nut until the peg engages the nearest hole in the lock washer.
5. Install the outer lock washer and tighten to 50 ft. lbs.

Selective Drive Hubs

Selective drive hubs are used to disengage the front axles from the drive train when the vehicle is used in two-wheel drive. When Quadra-Trac is used, the selector hubs are not used. Two different types are used, Automatic and manual.

CJ Models, Automatic Hubs
Removal

1. Remove the allen screws from the clutch assembly and remove the assembly from the hub body assembly.
2. Remove the retaining ring from the axle shaft end.
3. Straighten lock tabs and remove the screws attaching the hub body assembly to the front hub. Remove the hub body assembly.

Clutch Assembly Overhaul— Automatic Hubs

1. Push out the control dial, turn

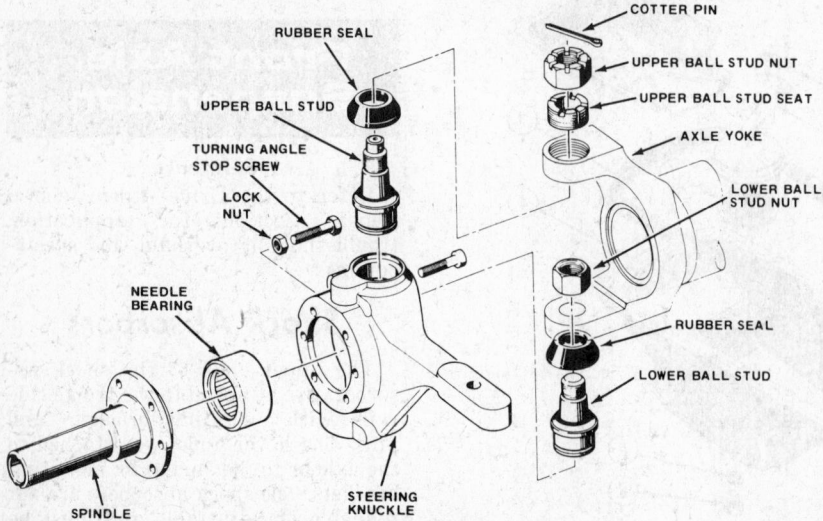

Model 44 steering knuckle-exploded view (© Jeep Corp.)

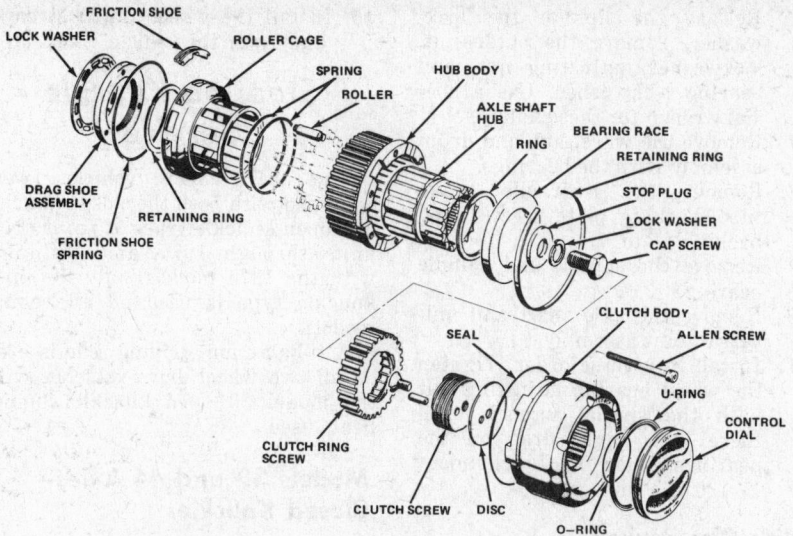

Automatic type—front hub—exploded view (© Jeep Corp.)

the unit over and remove the cluster ring and disc.

2. Clean and inspect all parts for damage. Replace "U" ring and "O" ring seals on the control dial.

3. Install the control dial assembly and install the disc. *NOTE: Lubricate "O" ring and inside of cap.*

4. Rotate the control dial to "FREE" position. Install the clutch ring and thread to the bottom.

5. Turn back the clutch ring until the holes align and install in the body assembly.

6. Turn the control from "FREE" to "LOCK" and check the operation.

Body Assembly

Overhaul

1. Remove the friction shoe spring, retaining ring, and separate the hub body from the roller clutch.

2. Remove the centering spring, and spirolock ring. Separate the cage and axle shaft hub.

3. Clean and inspect all parts, coat lightly with grease.

4. Install the friction shoe on the cage, (avoid stretching the shoe) and lubricate the friction shoes liberally, and install.

Installation

1. Position the gasket and body assembly on the wheel hub.

2. Install the tab lock washer and screws. Torque the screws to 40-45 ft. lbs. and secure with the lock washer tabs.

3. Install the retaining ring on the axle shaft end.

4. Place the gasket and cap assembly on the body assembly and torque the allen headed screws to 6-8 ft. lbs.

Cherokee and Truck

Removal

1. Remove the six allen screws and remove the clutch assembly.

2. Remove the capscrews, lock washers, and stop-ring on the axle shaft end.

3. Remove the retaining ring and slide the hub body off the axle end.

Clutch Assembly—Automatic Hubs

Overhaul

1. Remove the taper headed screw from the clutch screw unit.

2. Push out the control dial, turn the assembly over and push out the clutch and clutch screw.

3. Clean and inspect all parts. Replace the "U" ring and "O" ring seals on the control dial.

4. Lubricate the clutch ring and thread the clutch screw into the clutch ring until the ring raises slightly.

5. Install the taper headed screw and stake it in place. *NOTE: If*

new parts are used, drill a 3/16 inch hole through the clutch screw into the thick webbing on the control dial 5/8 inch deep. Install a pin and stake into place.

6. Turn the dial from "FREE" to "LOCK" and check the operation.

Hub Body—Automatic Hubs

Overhaul

1. Remove the friction shoe spring and the retaining ring.

2. Clean and inspect all parts.

3. Lubricate the bearing race lightly.

4. Place the cage into the body and pack rollers with chassis lub.

5. Place the body over the axle shaft hub, carefully, and install the retaining ring.

Installation

1. Carefully install the friction shoe spring and lubricate the shoes with chassis lube, and slide the body assembly into the hub. *NOTE: The body assembly will stop about 1/4 inch from full position. Allow the body to slide to full position by pushing the assembly to expand the friction shoes over the drag shoe nut.*

2. Install the retaining ring to hold the body assembly to the hub.

3. Install the screw, lock washer, and stop ring in the axle end and torque to 35-40 ft. lbs.

4. Install the clutch assembly to the body assembly with the allen headed screws, and torque to 4-6 ft. lbs.

5. Rotate the wheel and check for freedom of movement.

Manual Hubs

Removal

1. Remove the allen screws from the cap and remove the cap from the clutch hub.

2. Remove the hub body bolts from the front hub.

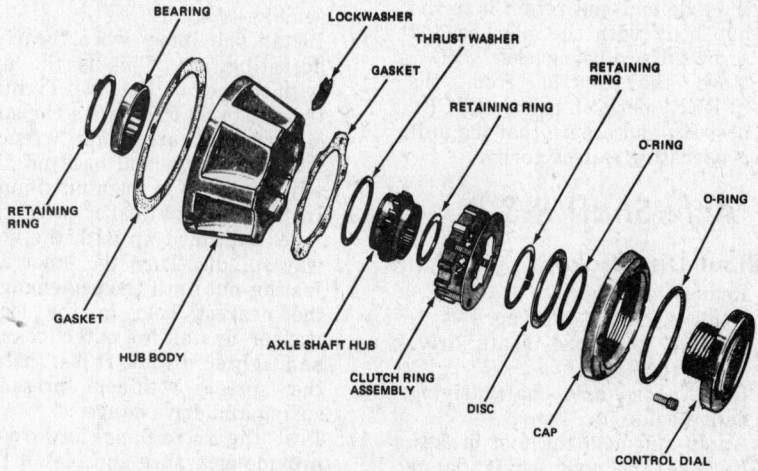

Manual type—front hub—exploded view (© Jeep Corp.)

3. Remove the retaining ring and pull the clutch ring assembly and axle shaft hub body from the axle. *NOTE: The clutch ring assembly cannot be disassembled. The control assembly and the clutch screw cannot be separated.*

Disassembly

1. Remove the snap ring from the hub bore.
2. Remove the needle bearing and thrust washer from the clutch hub body and the axle shaft hub, noting the side of the hub body from which the axle shaft hub is removed.

Assembly

1. Install the axle shaft hub into the hub body from the same side as it was removed.
2. Install the needle bearings in the hub and install the snap ring.

Installation

1. With a new gasket, install the hub body assembly onto the wheel hub and over the axle shaft, and install the attaching bolts.
2. Install the retaining ring in the groove at the end of the axle shaft hub.
3. Lubricate the bearing side and the grooves of the control assembly and install the new "O" rings.
4. Insert the clutch ring assembly into the hub body and retain it with the snap ring. *NOTE: Try the clutch ring for a free sliding fit on the drive pins. If a binding occurs, lift the unit out and reposition it. If the binding still exists, remove and examine for damage.*
5. Position the control assembly with the dowel pin into the face of the clutch body, so that the arrow stops on the dot marked "FREE".
6. Install the control assembly over the axle end and retain it to the hub body with the allen headed screws. Use a new gasket.
7. Move the control from the "FREE" position to the "LOCK" position and assure that the unit is operating satisfactorily.

Axle Shaft R & R

Without Disc Brakes

1. Remove the wheel.
2. Remove the hub dust cap.
3. Remove the axle shaft drive flange snap ring.
4. Remove the axle shaft driving flange bolts.
5. Apply and hold the foot brakes. Remove the axle shaft flange with a puller.

6. Release the lip on the lockwasher, remove the outer nut, lockwasher, adjusting nut, and bearing lockwasher. Use a special wrench for these nuts.
7. Remove the wheel hub and drum assembly with the bearings.
8. Remove the hydraulic brake tube, backing plate screws, and backing plate.
9. Remove the spindle and spindle bearing.
10. Remove the axle shaft and universal joint assembly.
11. Install in reverse order. Tighten the wheel bearing adjusting nut with the special wrench until there is a slight drag on the bearings, then back off about 1/4-1/6 of a turn.

With Disc Brakes

1. Raise and support the vehicle.
2. Remove the wheel and dust cover.
3. Remove the axle shaft snap ring, drive flange, pressure spring and spring retainer. If the drive flange is stuck to the shaft, use a screwdriver to pry it out.
4. Use the special nut wrench to remove the wheel bearing locknut, lockring, and wheel bearing adjusting nut.
5. Remove the two bolts securing the brake caliper assembly to the disc brake shield and move the caliper assembly aside.
6. Remove the rotor and hub assembly. The spring retainer and outer wheel bearing will slide out as the hub assembly is removed.
7. Remove the nuts and bolts attaching the spindle and disc brake shield.
8. Remove the spindle and disc brake shield. It may be necessary to tap the spindle lightly to free it.
9. Remove the axle shaft.
10. Install the new axle shaft, spindle, and bearing assembly.
11. Install the hub, brake shield, rotor, and the brake caliper assembly.
12. Install the inner wheel bearing adjusting nut. This is the nut with the peg on the side. Tighten the nut to 50 ft lbs with the special wheel bearing nut wrench. Rotate the hub and back off the adjusting nut 1/4 turn maximum.
13. Install the lockwasher with the inner tab lined up with the keyway spindle. Turn the inner adjusting nut until the peg engages the nearest hole in the lockwasher. Install the outer locknut and tighten it to 50 ft lbs. Install the spring retainer, pressure spring and drive flange.
14. Push the drive flange inward to provide clearance and install the axle shaft snap ring.

15. Install the wheel and dust cover and lower the vehicle.

Steering Knuckle Service

The AMC Jeep vehicles were equipped with both the closed knuckle and open knuckle types of front drive units through 1973, and beginning with the 1974 models, only the open knuckle type is used on all vehicle models.

A tubular non-driving axle is used on all two wheel drive vehicles, with the model 30 open knuckle components used.

Models 30 and 44 Axles— Closed Knuckles

Steering Knuckle and Pivot Pins

Removal

NOTE: Replacement of the bearings require the removal of the hub and drum assembly, axle shaft, spindle, steering tie rod, and the steering knuckle. To complete the following procedure , it is assumed that the above have been removed, with the exception of the steering knuckle.

1. Remove the eight screws holding the seal retainer in place.
2. Remove the four screws from the upper bearing cap and the four screws from the lower bearing cap. Remove the caps.
3. Remove the steering knuckle from the axle.
4. The bearings and cups can be replaced with suitable tools. Do not loose the shims from under the top bearing cap.

Installation

1. Position the upper and lower bearings in the cups and place the knuckle on the axle end.
2. Place the old shim pack in position under the upper bearing cap to determine the preload of the knuckle pivot bearings. Tighten the bearing cap screws to 25-40 ft. lbs. torque for the model 30 axle, and 70-90 ft. lbs. torque for the model 44 axle.
3. Without the seal in place, use a spring scale, graduated to 25 lbs., and attach it to the steering arm hole for the tie rod stud, and pull the knuckle through its arc and record the reading on the spring scale.
4. The preload reading should be 12 to 16 lbs. Remove or add shims under the upper bearing cap to obtain the proper preload. *NOTE: Shims are available in .003," .005", .010", and .030" thicknesses.*
5. Install the oil seal and oil seal

felt. Install the backing ring assembly wth the two retainer split halves, and tighten the eight screws to 10 to 25 ft. lbs.

6. Assemble the Axle shaft and components as previously outlined.

Models 30 and 44 Axles—Open Knuckles

Steering Knuckle

Removal

1. Follow the Axle Removal procedure as outlined previously.
2. Remove the steering rods from the steering arm.
3. Remove the lower ball stud nut. *NOTE: This nut is a self-locking nut and should be discarded and replaced with a new nut upon assembly.*
4. Remove the cotter pin from the upper ball stud nut and loosen it to the top of the stud in a flush manner. With the aid of a lead hammer, unseat the upper and lower ball studs from the yoke.
5. Remove the knuckle assembly from the axle.
6. With the use of a suitable too. remove the upper ball stud seat from the axle yoke.

Ball Joints

Removal and Replacement

1. With the aid of a puller or a press, position the tool to force the *lower* ball joint from the knuckle.
2. Position the puller or the press to force the upper ball joint from the knuckle.
3. To install the ball joints, press the *lower* joint into place and follow with the upper ball joint.

Installation

1. Install the upper ball stud seat into the axle yoke, until the top of the seat is flush with the top of the yoke.
2. Install the knuckle assembly onto the axle yoke by inserting the ball studs into their respective holes in the yoke. Install the new lower stud nut and tighten to 70-90 ft. lbs. torque.
3. Tighten the upper ball stud seat to 60 ft. lbs. torque, and install the upper ball stud nut and torque to 100 lbs. torque.
4. If the cotter pin holes do not align, tighten the nut until the pin can be installed. Never loosen the nut to install the cotter pin.
5. Continue with the assembly as previously outlined.

Upper Ball Joint Adjustment

Adjustment of the upper ball joint is necessary only when there is excessive play in the steering, persistent loosening of the steering linkage, or abnormal wear of the tires.

Adjustment Procedure

1. Raise the vehicle on a hoist or jack and support with stands if necessary.
2. Disconnect the steering linkage from the left and right knuckle assemblies.
3. Attach a torque wrench and socket to a steering arm stud and check the torque needed to move the knuckle through its arc.
4. Maximum Torque is:
   12-16 lbs.   Model 30 Axle
   15-20 lbs.   Model 44 Axle
   NOTE: The knuckle should turn smoothly through the turning arc and have no vertical end play.
5. If the torque is too low, perform the following procedures.
   a. Remove the upper ball stud cotter pin.
   b. Torque the stud adjusting sleeve to 50 ft lbs.
   c. Install the lock nut and a new cotter pin.
   d. Recheck the torque necessary to turn the knuckle through its arc, and if the torque is still too low, retorque the sleeve to 60 ft. lbs. and recheck.
   e. If the torque specifications can not be obtained, it will be necessary to replace parts in the knuckle assembly.
   Important: Temperature will affect the turning torque, therefore, allowances in the toruqe reading should be made.

Front Axle Assembly

Removal

1. Raise and support the front of the vehicle, supporting the weight at the rear of the front spring.
2. Remove the wheel covers and wheels.
3. Index the propeller shaft to the differential yoke for the proper alignment upon installation.
4. Disconnect the steering linkage from the steering knuckles.
5. Disconnect the shock absorbers and breather tube from the axle housing.
6. Remove the brake drums and backing plates, or the brake calipers, hub and rotor, and the brake shield.
7. Remove the spring clips and the spring clip plates.
8. Support the assembly on a jack and loosen the nuts securing the rear shackles, but do not remove the bolts.
9. Remove the front spring shackle bolts and rest the front of the spring on the floor.
10. Pull the jack and axle housing from underneath the vehicle.

Installation

1. Support the axle on a jack and slide the assembly under the vehicle, and position it over the springs.
2. Raise the front of the springs and install the front shackle bolts, but do not tighten.
3. Position the axle on the springs and install the spring clips and spring clip plates.
4. Tighten the front and rear shackle bolts.
5. On disc brake models, install the brake shield, hub and rotor, and brake calipers. On the drum brake models, install the backing plates, hubs and drums.
6. Connect the breather tube and shock absorbers.
7. Connect the steering linkage at the steering knuckles.
8. Align the indexing marks and install the propeller shaft.
9. Install the wheels and tighten. Install the wheel covers.
10. Lower the vehicle and check the wheel alignment and turning angle.

Steering Gear

Manual Steering Gear

Reference

Refer to the Manual Steering Gear General Repair Section for troubleshooting and overhaul procedures.

Manual Steering Gear R & R

1. Remove the bolt and nut attaching the coupling to the wormshaft, disconnecting the steering gear from the lower steering shaft.
2. Disconnect the steering arm from the connecting rod using a puller.
3. Remove the upper steering gear-to-frame bracket bolt.
4. Remove the two lower steering gear-to-frame bracket bolts and remove the steering gear.
5. Install in the reverse order.

Power Steering Gear

References

Refer to the Power Steering Gear General Repair Section for troubleshooting and overhaul procedures. Power steering pump services are also covered in the general repair section.

Power Steering Gear R & R

1. Disconnect the hoses from the return port and pressure port. Raise the hoses above the level of the pump to prevent any further lose of power steering fluid.
2. On CJ models, disconnect the intermediate shaft coupling at the steering gear stub shaft. On the other models, disconnect the flexible coupling at the intermediate shaft.
3. Remove the pitman arm nut, lockwasher, and remove the pitman arm using a puller.
4. Remove the mounting bolts attaching the steering gear assembly to the frame, and remove the steering gear assembly.
5. To install the steering gear, mount the steering gear on the frame and install the attaching bolts, tightening them to 65 ft lbs.
6. Install the pitman arm on the pitman shaft and install the lockwasher and pitman arm nut. Tighten the nut to 190 ft lbs.
7. On CJ models, connect the stub shaft to the intermediate shaft. Tighten the clamp bolt to 40 ft lbs.
8. On the other models, install the flexible coupling on the stub shaft, if it was removed, and tighten the clamp to 30 ft lbs. Connect the intermediate shaft to the flexible coupling and tighten the attaching bolts and nuts to 20 ft lbs.
9. Connect the hoses to the gear and tighten the hose fittings to 30 ft lbs.
10. Check the level of the fluid in the reservoir and add as necessary.

Power Steering Pump R & R

Before working on the power steering pump, clean the exterior of the pump and the reservoir assembly.

NOTE: On all engines except the 1971-1972 232 Six, the power steering pump need not be removed to service the pressure relief/flow control valve. The valve is located behind a pressure union on all the other engines.

1. Loosen the drive belt tension adjustment bolt and remove the belt (power steering and air pump, if so equipped).
2. Place a receptacle under the pump/reservoir assembly and disconnect the pressure and return hoses from the pump. Fluid will drain out of the pump and hoses. Lay the ends of the hoses up higher than the steering gear to prevent all of the fluid from draining out. Cover the ends of the hoses to prevent dirt from entering.
3. Remove either the bolts that hold the pump to the mounting

bracket or the bolts that hold the bracket and pump to the engine, whichever is easiest. Remove the pump from the engine.
4. Install the pump in the reverse order of removal and install the correct type and amount of fluid.

Clutch

Description

1971 CJ Models

Vehicles with the F-head engine have a 8½ or 9¼ in. clutch plate. The clutch is either a three pressure spring, three fingered Auburn clutch or a six pressure spring, three fingered Rockford clutch.

Jeeps with the V6 engine have a General Motors diaphragm type clutch.

1971 Commando, Wagoneer, and Truck

The clutch installed in Commandos is either a 3-pressure-springed 3-fingered, 9¼ in. diameter Auburn clutch installed behind the F-Head engine with 1150 lbs of plate pressure or, a diaphragm type, 10.4 in. diameter G.M. clutch installed behind the V6 engine with 1600 lbs of plate pressure.

The clutch installed in the Wagoneers are as follows:

Behind the 232 Six, there is a Borg & Beck, 10½ in. diameter, 9-pressure-springed, 3-fingered clutch with 1350 lbs of plate pressure.

The clutch installed behind the 350 V8 is a G.M. diaphragm type, 11 in. in diameter with 2450-2750 lbs of plate pressure.

All Models 1972 and Later

A single plate, dry disc-type clutch

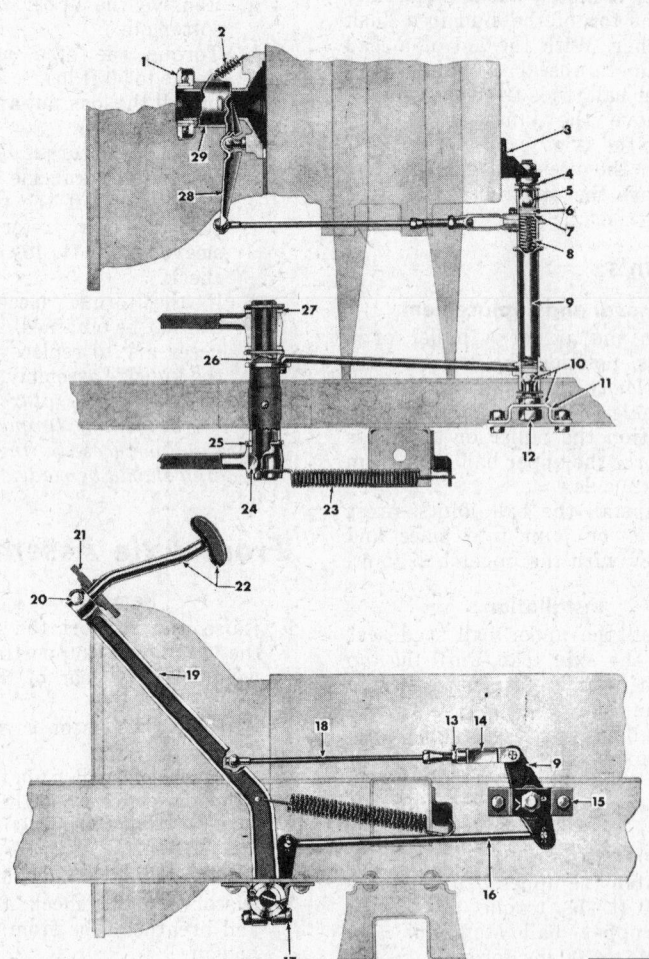

1 Throwout bearing	11 Frame bracket	21 Draft pad
2 Throwout bearing carrier spring	12 Ball stud nut	22 Pedal pad and shank
3 Bracket	13 Yoke locknut	23 Retracting spring
4 Dust seal	14 Adjusting yoke	24 Pedal to shaft key
5 Ball stud	15 Bolt	25 Washer
6 Pad	16 Pedal release rod	26 Pedal shaft
7 Retainer	17 Pedal clamp bolt	27 Master cylinder tie bar
8 Control tube spring	18 Control cable	28 Control lever
9 Control lever and tube	19 Clutch pedal	29 Bearing carrier
10 Ball stud and bracket	20 Screw and lockwasher	

Clutch linkage on CJ models prior to 1972.

is used with two types of steel covers; a 10½ in. diameter direct spring pressure type and an 11 in. diameter semi-centrifugal type. Both types are similar in function in that they both apply direct spring pressure to the pressure plate. But, the semicentrifugal cover also utilizes six rollers that are forced outward by centrifugal force to apply extra force to the pressure plate and provide extra positive clutch engagement.

Clutch Pedal Free-Play Adjustment

CJ 1971

When sufficient wear occurs, the pedal clearance must be adjusted to 1-1½ in. The free pedal clearance is adjusted by lengthening or shortening the clutch fork cable. To make this adjustment, loosen the jam nut on the cable clevis and lengthen or shorten the cable to obtain the proper clearance at the pedal pad, then tighten the jam nut.

1972 CJ and Commando

The clutch pedal has an adjustable stop located on the pedal support bracket directly behind the instrument cluster.

Adjust the stop to provide the specified clearance between the top of the pedal pad and the closest point on the bare floor pan. The distance must be 8 in.

1. Lift up the clutch pedal against the pedal support bracket stop.
2. Unhook the clutch fork return spring.
3. Loosen the ball adjusting nut until some cable slack exists.
4. Adjust the ball adjusting nut until the slack is removed from the cable and the clutch throwout bearing contacts the pressure plate fingers.

1971 Wagoneer

1. Disconnect the adjustable rod from the clutch pedal.
2. Adjust the clutch pedal stop bolt

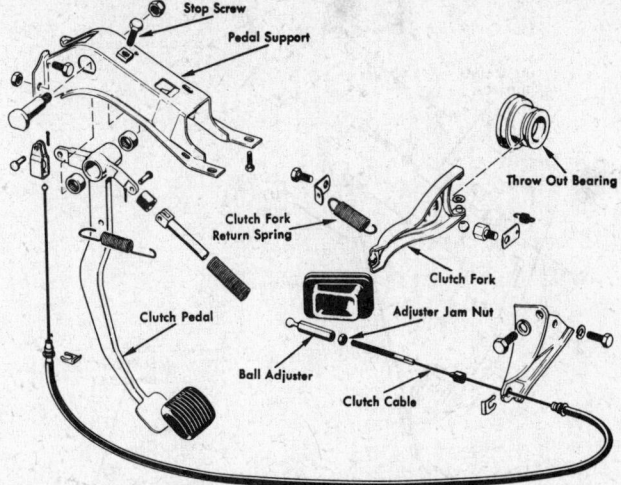

Clutch linkage on a 1972 Wagoneer (© Jeep Corp.)

for a positive overcenter position. This is done by turning the bolt in or out of the crossmember to obtain ¼-½ in. of clutch pedal travel before the overcenter spring assists the pedal to the floorboard. The pedal may be ⅜-⅝ in. higher than the pad on the brake pedal. This is a normal condition.

3. With the clutch pedal resting against the stop bolt, adjust the length of the adjustable rod by turning the ball joint in or out on the rod, so that when the adjustable rod is connected, the arm of the clutch pedal cross-shaft is at a 30° angle with a 350 V8 and 49° with a 232 Six and 327 V8, in relation to the top of the left frame side rail.
4. To adjust clutch pedal free-play, adjust the cross-shaft-to-throwout lever link, so that the clutch pedal can be depressed 1 in. before the clutch starts to disengage.

NOTE: There is a special tool available for setting the angle of the arm of the clutch pedal cross-shaft as outlined in Step 3 of the above procedure. Tool W-341 is for the vehicle equipped with a 350 V8 and gives the 30° angle and tool W-317 is for the

vehicle equipped with a 232 Six and gives the 49° angle.

1972 Wagoneer

1. The clutch pedal has an adjustable stop located on the pedal support bracket, directly behind the instrument cluster. Adjust the stop to provide 8⅜ in. between the top of the pedal pad and the closest point, 90° on the bare floor pan.
2. Lift the clutch pedal up against the pedal support bracket stop.
3. Unhook the clutch fork return spring.
4. Loosen the ball adjusting nut until some cable slack exists.
5. Adjust the ball adjusting nut until the slack is removed from the cable and the clutch throwout bearing contacts the pressure fingers.
6. Back off the ball adjusting nut ¾ of a turn to provide free-play.
7. Tighten the jam nut.
8. Hook the clutch fork return spring.

All Models 1973-78

1. Adjust the bellcrank outer support bracket to provide about ⅛ in. of bellcrank end play.
2. Lift up the clutch pedal against the pedal stop.
3. On the clutch push rod (pedal-to-bellcrank), adjust the lower ball pivot assembly onto or off the rod, as required, to position the bellcrank inner lever parallel to the front face of the clutch housing (slightly forward from vertical).

NOTE: There is no lower ball pivot assembly on 1976 CJ models, so step 3 does not apply.

4. Adjust the clutch fork release rod (bellcrank-to-release fork) to obtain the maximum specified clutch pedal free play of ¾ in. on 1973-74 models and 1 in. on 1975-76 models.

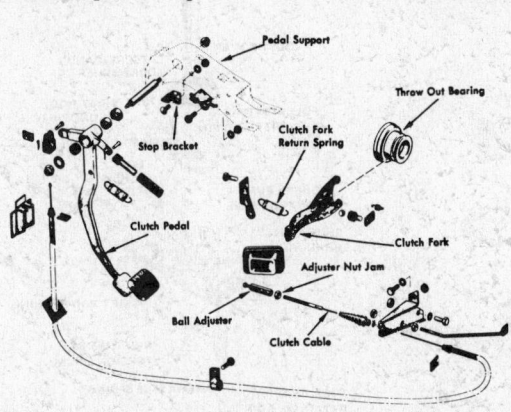

Clutch linkage for 1972 Commandos and CJs (© Jeep Corp.)

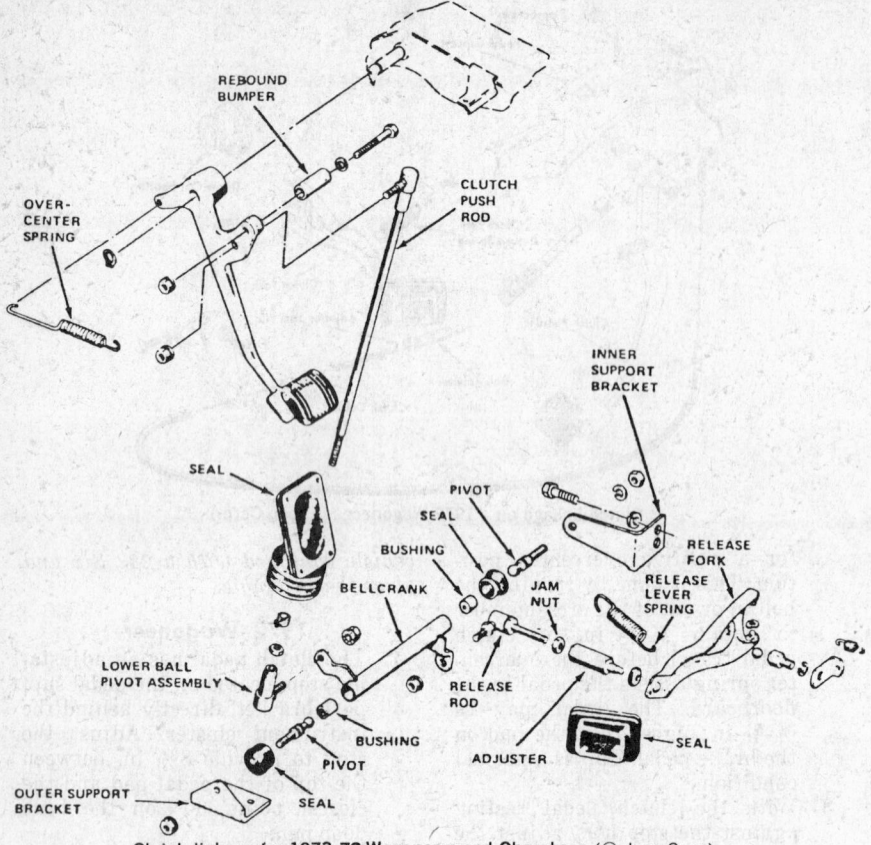

Clutch linkage for **1973-78 Wagoneers and Cherokees** (© Jeep Corp).

Clutch R & R

All Models

1. Remove the transmission.
2. Remove the starter, throwout bearing and sleeve assembly and clutch housing.
3. Mark the clutch cover, pressure plate and flywheel with a center punch to insure correct alignment during assembly.
4. Loosen the attaching screws in sequence, one or two turns at a time until the spring tension on the cover is released. This is to prevent the clutch cover from becoming warped, which could result in clutch chatter when reinstalled.
5. Install the clutch in the reverse order of removal, tighten the attaching bolts of the cover in sequence, one or two turns at a time to 40 ft lbs.

NOTE: The clutch pedal is not to be depressed until the transmission has been installed.

Transfer Case

Reference

Refer to the Transfer Case General Repair Section for troubleshooting and overhaul procedures.

Transfer Case—Spicer Model 20

Removal

1. Remove the shift lever knob, boot, and shift lever.
2. Raise and support the vehicle safely.
3. Mark the propeller shafts for reference at assembly and disconnect them from the yokes.
4. Disconnect the parking brake cable at the equalizer bar, and disconnect the speedometer cable from the transfer case.
5. Remove the transfer case to transmission bolts and install a

guide bolt on each side to aid in the removal and installation.
6. Remove the transfer case and gasket.

Installation

1. Install a new gasket on the transmission.
2. Shift the transfer case into 4WD low and install the case assembly on the guide bolts.
3. Rotate the transfer case output shaft until the transmission main shaft gear engages the rear output shaft gear of the transfer case.
4. Slide the transfer case forward until the two units mate flush.
5. Install one upper bolt, remove the dowel guide bolts and install the remaining bolts. Torque to 30 ft. lbs.
6. Connect the speedometer cable and parking brake cable.
7. Install the propeller shafts after aligning the indexing marks.
8. Fill the unit with gear lube, and lower the vehicle.
9. Install the transfer case shift lever, boot, and knob.

Transfer Case— Quadra-Trac

Removal

1. Lift and support the vehicle safely.
2. Index the marks on the front and rear yokes and propeller shafts for proper alignment during assembly.
3. Disconnect both the front and rear propeller shafts.

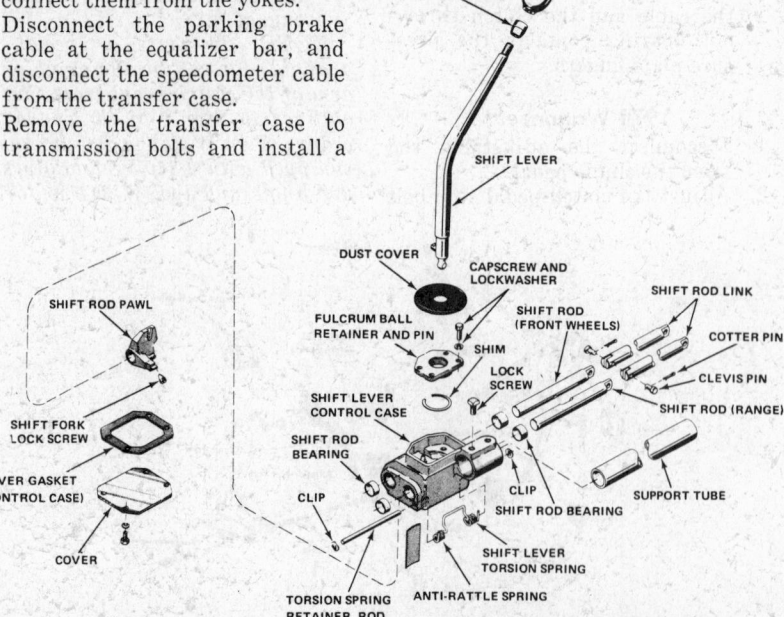

Exploded view—transfer case shift controls—**Cherokee and truck** (© Jeep Corp.)

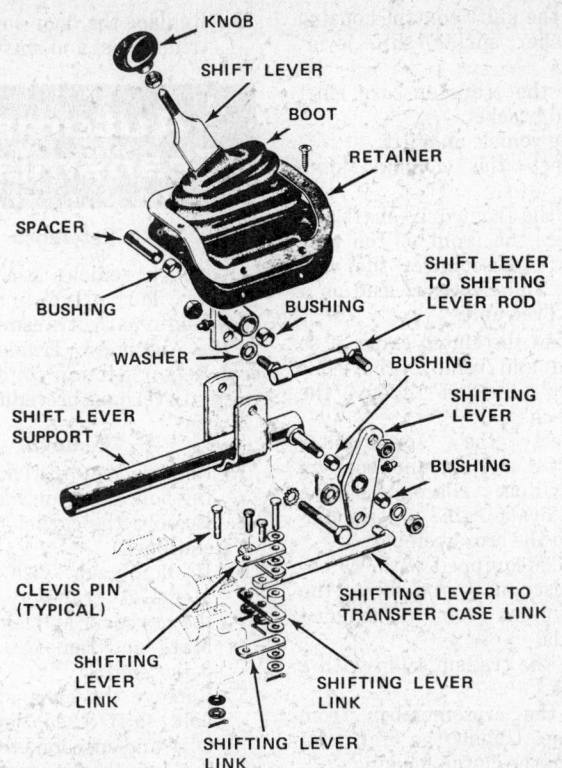

KNOB
SHIFT LEVER
BOOT
RETAINER
SPACER
SHIFT LEVER
TO SHIFTING
LEVER ROD
BUSHING
BUSHING
WASHER
BUSHING
SHIFT LEVER
SUPPORT
SHIFTING
LEVER
BUSHING
CLEVIS PIN
(TYPICAL)
SHIFTING LEVER TO
TRANSFER CASE LINK
SHIFTING
LEVER
LINK
SHIFTING LEVER
LINK
SHIFTING LEVER
LINK

Exploded view—transfer case shift controls—C J models (© Jeep Corp.)

The reduction unit is actuated by a shift cable and can be adjusted in the following manner:

1. Loosen the nut which clamps the cable to the shift lever pivot. Be sure that the cable can move freely in the pivot.
2. Move the reduction shift lever to the most rearward detent position (Hi-Range position).
3. Push the Low Range lever inward until it stops. Pull the Low Range lever out slightly, no more than 1/16 in.
4. Tighten the cable clamp nut at the reduction unit shift lever.

NOTE: This procedure only applies to the Quadra-Trac transfer case equipped with a reduction unit.

Manual Transmission

Reference

Refer to the Manual Transmission General Repair Section for application, troubleshooting and overhaul.

Linkage Adjustments

Jeepster—Console Shift

1. Remove the plug from the hole in the left side of the console. If there is no hole, cut a 1⅛ in. dia. hole or remove the console.
2. Lift the shift tower rubber cover and remove the plug in the shift tower. Move the selector lever to the neutral position.
3. Loosen the adjusting nuts at the transmission.
4. Insert a 3/16 in. dia. rod through the holes in the console and shift tower, and through the aligning holes in the two shift levers. Check to see that the transmission shift levers are in neutral positions.
5. Torque the adjusting nuts to 15-20 ft lbs.
6. Remove the adjusting rod, replace the plugs, and check the shifting action. If the selector lever interferes with the console, relocate the console.

Wagoneer—Column Shift

1. Put the selector lever in Neutral position.
2. Loosen the shift rod adjusting nuts at the transmission. Place the shift levers in Neutral position.
3. Insert a 3/16 in. dia. rod through the remote control shift levers and housing at the bottom of the steering column.
4. Torque the adjusting nuts to 15-20 ft lbs. Remove the adjusting rod.

---

4. Mark the vacuum diaphragm control for identification during the assembly, and then disconnect the vacuum hoses, wiring, and speedometer cable.
5. Disconnect the parking brake cable guide from the pivot on the right frame side.
6. Remove the two front side transfer case to transmission bolts, and install a guide bolt into the upper hole.
7. Remove the two rear side bolts, holding the transfer case to the transmission, and install a guide bolt into the upper hole.
8. Move the transfer case rearward until the unit is free of the transmission output shaft and guide pins. Lower the assembly to the floor.
9. Remove all gasket material from the rear of the transmission.

Installation

1. Install a new gasket on the rear of the transmission.
2. Install the guide bolts in the upper transmission adapter and transfer case, if they were removed.
3. Raise the transfer case, engage the guide bolts, and move the case assembly forward to the transmission. Make sure a flush fit is achieved.
4. If necessary, rotate the transfer case rear output shaft yoke until the drive hub splines align with the transmission output shaft.
5. Install front and rear attaching

bolts, and remove the guide bolts during this operation.
6. Attach the exhaust pipe bracket support, if removed.
7. Align the propeller shaft and indexing marks on the yokes and attach the propeller shafts.
8. Connect the speedometer cable, wiring, and vacuum hoses.
9. Connect the parking brake cable guide to the pivot bracket on the right frame side.
10. Install the specified lubricant, and lower the vehicle.

Linkage Adjustments

Spicer Model 20

The shifter rails of the transfer case lever assembly connect to the shifter rails of the transfer case either directly or through non-adjustable links. The linkage should be lubricated periodically.

Warner Quadra-Trac

Since the Quadra-Trac system is a "full time 4WD" system, and is constantly engaged in 4WD, there is no "shift linkage" as such. There are two features which can be operated manually concerning the transfer case: the "Lock-Out" feature and the engagement of the optional "Low Range Reduction Unit."

Since the "Lock-Out" feature is a vacuum actuated unit, there are no external adjustments that can be made other than making sure that all vacuum lines are in place, connected and not damaged in any way.

Transmission R & R

The transmission and transfer case can be removed as a unit. These instructions apply to both three and four-speed transmissions.

1971 CJ

1. Drain the transmission and transfer case.
2. Remove the floor pan inspection plate.
3. Remove the shift lever and housing, or disconnect the remote control rods, depending on the model.
4. If the vehicle has a power takeoff, remove the shift lever.
5. Disconnect the front and rear driveshafts from the transfer case. Disconnect the power takeoff driveshaft.
6. Disconnect the speedometer cable.
7. Disconnect the handbrake cable.
8. Disconnect the clutch release cable or rod.
9. Place the jacks under the engine and transmission, protecting the oil pan with a wooden block.
10. Remove the rear crossmember.
11. Unbolt the transmission from the flywheel housing.
12. Force the transmission to the right to disengage the clutch control lever tube ball joint.
13. Lower the jacks. Slide the transmission and transfer case rearward until the clutch shaft clears the flywheel housing.
14. Lower the transmission jack. Remove the assembly from beneath the vehicle.
15. To install, reverse the procedure.

1971 Jeepster

The procedure for these models is the same as that for the Universal series, with the substitution of the following steps:

1. Remove the right front seat, floor mat, and floorboard center section. Disconnect the back-up switch wires.
2. Remove the rear crossmember. Remove the transmission and transfer case stabilizer brackets.

1971 Wagoneer

The procedure for these models is the same as that for the Universal series, with the substitution of the following step:

1. Remove the transmission access cover.

All Models—1972 and Later

1. Remove the shift lever knobs, trim rings, and boots.
2. On three-speed floorshift models, remove the floor covering. Remove the floor pan section from above the transmission. Remove the shift control and lever assembly. On four-speed models,

remove the shift control housing cap, washer, spring, shift lever, and pin.
3. Remove the transfer case shift lever and bracket.
4. Raise the vehicle on a lift.
5. Disconnect the column shift rods.
6. Remove the front driveshaft and disconnect the front of the rear driveshaft. Disconnect the vacuum line and electrical lead on a Quadra-Trac unit.
7. Disconnect the clutch cable, if so equipped and remove the cable mounting bracket from the transfer case.
8. Disconnect the speedometer cable, TCS switch, and back-up light switch. Disconnect the parking brake cable if it is connected to the crossmember.
9. On models equipped with V8 engines, disconnect and lower the exhaust pipes from the exhaust manifolds.
10. Support the transmission with a floor jack.
11. Unbolt the crossmember from the frame. Unbolt the transmission from the clutch housing.
12. Lower the transmission slightly and move it to the rear to disengage the clutch shaft. Remove the unit.

To replace the transmission:
13. Place the wave washer and throw-out bearing and sleeve assembly in the fork. Center the bearing over the release levers.
14. Slide the transmission into place, being careful to align the transmission splines with those on the clutch plate. Bolt the transmission to the clutch housing and torque the bolts to 55 ft lbs.
15. Bolt the crossmember to the frame. Torque the bolts to 30-35 ft lbs. Remove the jack.
16. On models equipped with V8 engines, connect the exhaust pipes to the exhaust manifolds.
17. Connect the speedometer cable, back-up light switch, TCS switch, parking brake cable, clutch cable and cable mounting bracket.
18. Install the driveshafts. Flange bolts should be torqued to 25-45 ft lbs.
19. Replace the column shift linkage.
20. Lower the vehicle.
21. Replace the transfer case shift lever and bracket.
22. On four-speeds, install the lever pivot pin, shift lever, spring. washer, and control housing cap. On three-speeds, install the shift control and lever assembly. Set the gears and the cover in the neutral position. Install the cover, placing the shift forks into the sleeves. Torque the cover bolts to 8-15 ft lbs.

23. Replace the floor covering, boots, trim rings, and shift lever knobs.

Automatic Transmission

Reference

All Jeep vehicles use the General Motors Turbo-Hydramatic 400 3-speed automatic transmission. Refer to the Automatic Transmission General Repair Section for troubleshooting and overhaul procedures.

Removal

1. Remove the dipstick tube attaching bolt at the engine block.
2. Remove the carpet trim ring (if equipped).
3. If equipped with the Spicer Model 20 transfer case, remove the top cover and lever.
4. Mark and remove the rear propeller shaft.
5. Remove the exhaust pipe clamp bolt, shift lever, detent solenoid wire and speedometer cable.
6. Support the transmission and remove the rear cross member.
7. Remove the exhaust pipes.
8. Mark and remove the front propeller shaft from the transfer case end.
9. Remove the vacuum lines and the oil cooler lines from the transmission.
10. When equipped with the Quadra-Trac transfer case, remove the lockout signal wire and the diaphragm control hoses.
11. Disconnect the low range cable (if equipped), and the converter housing splash pan.
12. Mark the converter and flywheel for alignment at assembly and remove the converter to flywheel bolts.
13. Remove the converter housing to engine bolts and remove the transmission.

Installation

1. Install the transmission to the engine and install the retaining bolts. Torque the bolts to 28 ft. lbs.
2. Align the indexing marks made during the removal and install the converter to flywheel bolts. Torque to 33 ft. lbs.
3. Install the converter splash pan, the oil cooler lines, and the vacuum line.
4. If equipped with the Quadra-Trac transfer case, install the diaphgram control hoses, and the lockout signal switch wire.
5. Connect the low range cable (if equipped).
6. Index the front propeller shaft to yoke marks, and install the

propeller shaft at the transfer case end.

7. Install the exhaust pipes, but do not tighten the attaching bolts.
8. Install the rear crossmember and remove the jack support.
9. Install the exhaust pipe clamp bolt, shift lever, speedometer cable, and detent solenoid wire.
10. Install the rear propeller shaft, after indexing the marks made during the removal. Tighten all exhaust system bolts.
11. If equipped with the Spicer Model 20 transfer case, install the top cover and shift lever.
12. Install the carpet trim ring (if equipped).
13. Install the dipstick tube to engine bolt.
14. Install specified oil into the unit and inspect for any leakages.

Linkage Adjustment

1. Place the steering column gearshift lever in the Neutral position.
2. Raise the vehicle on a hoist.
3. Loosen the locknut on the gearshift rod trunnion just enough to permit movement of the gearshift rod in the trunnion.
4. Place the outer range selector lever at the transmission, fully into the Neutral detent position and tighten the locknut at the trunnion to 9 ft lbs.
5. Lower the car and operate the steering column gearshift in all ranges. The car should start in Park and Neutral only and the column gearshift lever engage properly in all detent positions.

Neutral Switch Adjustment

1. Apply the parking brake.
2. Check and adjust the manual linkage, if necessary.
3. Remove the Neutral switch from the steering column.
4. Place the selector lever in Park and lock the steering column.
5. Move the switch actuating lever until it is aligned with the letter "P" stamped on the back of the switch.
6. Insert a 3/32 in. drill in the hole located below the letter "N" stamped on the back of the switch.
7. Move the switch actuating lever until it stops against the drill.
8. Position the switch on the steering column, install the attaching screws and remove the drill.
9. Check the operation of the switch. The engine should start in Park and Neutral only. The backup light should glow only in the Reverse position.

Transmission Oil Pan

Removal

1. Raise the vehicle and support safely.
2. Position a drain pan under the transmission and remove the oil pan bolts, except the four corner ones.
3. Loosen the corner bolts and pry the oil pan loose from the transmission case.
4. Allow the oil to drain from the corners of the oil pan, while tilting the pan to remove as much oil as possible.
5. Carefully remove the corner bolts and the oil pan from the transmission case.
6. Remove the oil filter, oil pan gasket, and intake tube "O" ring seal.

Installation

1. Place a new "O" ring on the intake tube, and install the tube in place.
2. Install a new filter and retain it to the control valve assembly with the attaching bolt.
3. Install the oil pan and gasket. Torque the pan bolts to 12 ft. lbs.
4. Lower the vehicle and fill the transmission with the specified fluid. (Dexron or equivalent).

Band Adjustment

No provisions are made for the external band adjustments of this transmission. Only during the assembly, can a different sized pin be installed in the rear band apply system, to compensate for lining wear.

Detent Adjustment

An electrical detent solenoid is located on the control valve assembly of the transmission and is activated by an electrical switch, located either on the carburetor or under the accelerator pedal, and is energized by depressing the pedal to the bottom of its travel, which causes the transmission to downshift at speeds below 70 MPH. The only adjustment needed is

to insure the switch contacts engage when the pedal is depressed. This can be accomplished by movement of the switch or of the throttle linkage.

U Joints, Drive Lines

Drive Shaft R & R

In order to remove the front and rear driveshafts, unscrew the attaching nuts from the universal joint's U-bolts, remove the U-bolts and slide the shaft forward or backward toward the slip-joint. The shaft can then be removed from the end yokes and removed from under the vehicle. Install the driveshaft in the reverse order.

NOTE: Some driveshafts are marked at the slip-joints with arrows on the spline and sleeve yoke. When installing the driveshaft, align the arrows to have the yokes at the front and rear of the shaft in the same parallel plane.

U-Joint Overhaul

Snap-Ring Type Disassembly and Repair

1. Remove the snap-rings.
2. Press on the end of one bearing until the opposite bearing is pushed from the yoke arm.
3. Turn the joint over. Press the first bearing back out of the arm by pressing on the exposed end of the journal shaft. Repeat this operation for the other two bearings, then lift out the journal assembly by sliding it to one side.
4. Wash all parts in solvent and inspect for wear. Replace all worn parts.
5. Install new gaskets on the journal assembly. Make certain that the grease channel in each journal trunnion is open.
6. Pack the bearing cones one-third full of grease and install the rollers.
7. Assemble in the reverse order of

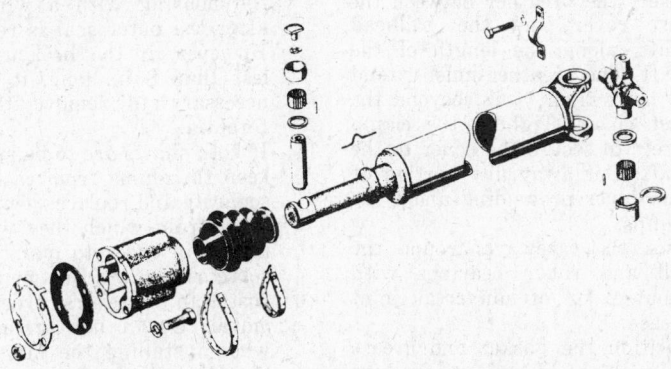

Exploded view of the Ball and Trunnion type universal joint.

disassembly. If the joint binds when assembled, tap the arms lightly to relieve any pressure on the bearings at the end of the journal.

U-Bolt Type Disassembly and Repair

Remove the attaching U-bolts to release one set of bearing races. Slide the driveshaft into the yoke flange to remove the races. The rest of the disassembly and repair procedure is the same as that given above for the snap-ring type of cross and roller joint. The correct U-bolt torque is 15-20 ft lbs.

Ball and Trunnion Disassembly and Repair

1. Clamp the shaft firmly in a vise.
2. Bend the grease cover lugs away from the universal joint body. Remove the cover and gasket.
3. Remove the two clamps from the dust cover. Push the joint body toward the driveshaft tube. Remove two each; centering buttons, spring washers, ball and roller bearings, and thrust washers, from the trunnion pin.
4. Press the trunnion pin from ballhead.
5. If the ballhead is bent out of alignment or if the trunnion pin bore is worn or damaged, replace the driveshaft.

To reassemble:

6. Secure the larger end of the dust cover to the joint body with the larger of two clamps. Install the smaller clamp. Fit the cover over the ballhead shaft.
7. Push the universal joint cover toward the driveshaft tube. Press the trunnion pin into the centered position. If the trunnion pin is not centered, imbalance will result.
8. Install the thrust washers, ball and roller bearings, spring washer, and centering buttons on the trunnion pin. Compress the centering buttons. Move the joint body to hold the buttons in place.
9. Insert the breather between the dust cover and the ballhead shaft, along the length of the shaft. The breather must extend no more than ½ in. beyond the dust cover. Tighten the clamp screw to secure the cover to the shaft. Cut away any portion of dust cover protruding under the clamps.
10. Pack the raceways around the ball and roller bearings with about 2 oz of universal joint grease.
11. Position the trasket and grease cover on the body. Bend the lugs of the cover into the notches of

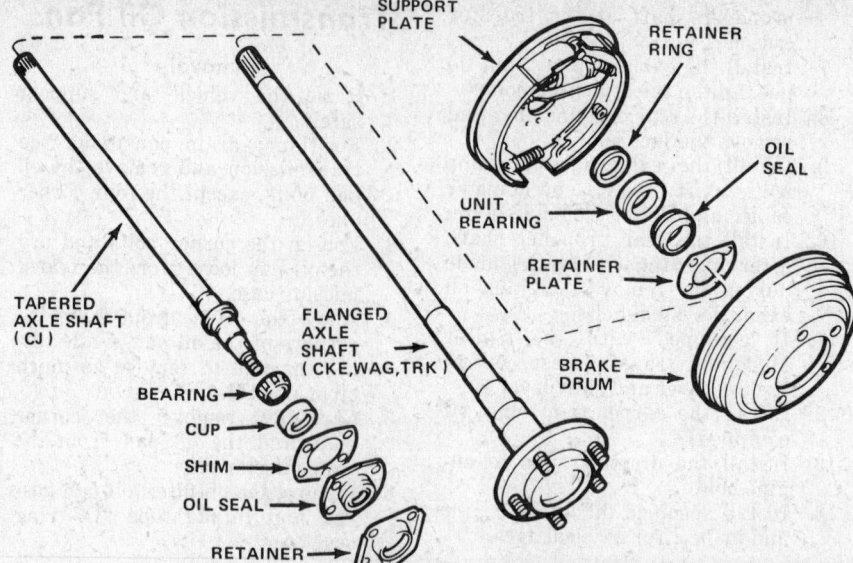

Typical flanged and tapered axle assembly (© Jeep Corp.)

the body. Move the body back and forth to distribute grease in the raceways.

Drive Axles

Refer to the Drive Axle General Repair Section for application, troubleshooting, and overhaul procedures.

Axle Shaft R & R

Tapered Shaft—1971 CJ

1. Jack up the wheel and remove the hub cap.
2. Remove the axle shaft nut.
3. Use a puller to remove the wheel hub and key.
4. Remove the screws which attach the brake dust shield, grease and bearing retainers, and brake assembly. Remove the shield and retainer.
5. Pull out the axle shaft with a puller, being careful not to lose the adjusting shims. Should the axle shaft be broken, the inner end can usually be drawn out of the housing with a wire loop, after the outer seal is removed. However, if the broken end is less than 8 in. long, it will be necessary to remove the differential.
   If both shafts are to be removed, keep the shims from each shaft separate and replace them on the shaft from which they were removed in order to maintain the correct bearing adjustment.
6. Install in the reverse order of removal. Use a new grease seal when installing the hub assembly. Install the hub assembly. then the key.

Tapered Shaft—Wagoneer, Commando, and Cherokee

1. Jack up the vehicle and remove the hub cap.
2. Remove the wheel.
3. Remove the axle nut dust cap.
4. Remove the axle shaft cotter pin, castle nut and flat washer.
5. Back-off the brake adjustment.
6. Use a puller to remove the wheel hub.
7. Remove the screws attaching the brake dust protector, grease and bearing retainers, brake assembly and shim to the housing.
8. Remove the hydaulic line from the brake assembly.
9. Remove the dust shield and oil seal.
10. Use a puller to remove the axle shaft.
11. Install the axle shaft in the reverse order of removal, using a new grease seal and installing the hub assembly before the woodruff key.

NOTE: Should the axle shaft be broken, the inner end can usually be drawn out of the housing with a wire loop after the outer oil seal is removed. However, if the broken end is less than 8 in. long, it usually is necessary to remove the differential assembly.

Axle Shaft Bearing
Removal and Installation (Axle Out)

1. With the aid of a combination puller, remove the bearing from the axle shaft *NOTE: If a puller is not available, place the threaded end of the axle on a heavy block of wood and with the aid of an assistant, drive the bearing from the axle shaft with a punch and hammer. Contact the inner race only with the punch.*

2. The new bearing can be installed with the use of a combination puller, or with the use of a length of pipe, fitted to the diameter of the inner bearing race, and slipped over the axle end to contact and drive the bearing to its seat on the axle shaft.

3. Lubricate the bearing with wheel bearing grease, making sure the grease fills the cavities between the bearing rollers.

Inner Oil Seal

Removal and Installation (Axle Out)

1. Insert the splined end of the axle shaft into the inner seal, hooking the axle end to the seal, and prying downward. The seal will move outward from the housing tube.

2. Install a new seal into the housing tube with the aid of a seal installer or its equivalent.

4. Seat the seal and lubricate the lip. *NOTE: The lip of the seal should point towards the center of the axle housing.*

5. The outer seal is installed during the assembly of the brake support plate.

1976—CJ Models Only and later

1. With the wheel on the ground, remove the axle shaft cotter pin and nut. Loosen the wheel nuts.

2. Raise and support the rear of the car, preferably with jackstands under the axle housing.

3. Remove the wheel.

4. Remove the drum retaining screws, 3 per drum.

5. Remove the drum from the hub. If the brake shoes hold the drum, the brake adjustment will have to be backed off slightly.

6. Attach a puller to the wheel bolts and pull off the hub. **CAUTION: Don't use a knockout type puller. It could damage the rear wheel bearings or the differential.**

7. Disconnect the parking brake cable at the equalizer. The equalizer is where the single cable from the parking brake pedal joins the double cable from the rear wheels.

8. Disconnect the brake tube at the wheel cylinder and remove the brake support plate assembly (backing plate), oil seal, and shims (left-side only).

9. Use a puller to remove the axle shaft and bearing.

10. Remove and discard the axle shaft inner oil seal.

11. The bearing cone is pressed onto the shaft. A hydraulic press must be used to remove it.

12. Before installation, pack the axle shaft bearings with high quality grease. Place a healthy glob of grease in the palm of one hand and force the edge of the bearing into it so that grease fills the bearing. Do this until the whole bearing is packed. Grease packing tools are available which make this task much easier.

13. Press the axle shaft bearings onto the axle shafts with the small diameter of the cone toward the outer end of the shaft. **CAUTION: Always press on the inner bearing race.**

14. Coat the inner axle shaft seal with light oil.

15. Coat the outer surface of the metal seal retainer with sealant.

16. Use a seal driver to install the inner oil seal in the axle housing.

17. Install the axle shaft(s), turning them as necessary to fit the splines into the differential.

18. Install the outer bearing cup.

19. Apply sealant to the axle housing flange and brake support plate mounting areas. Install the original shims in their original locations, oil seal assembly, and brake support plate. Tighten the retaining bolts to 35 ft lbs.

NOTE: The oil seal and retainer go on the outside of the brake support plate.

20. Axle shaft end-play can be measured by installing the hub retaining nut on the shaft so it can be pushed and pulled with relative ease. Strike the end of each axle shaft with a lead hammer to seat the bearing cups against the support plate. Mount a dial indicator on the left side support plate with the stylus resting on the end of the axle shaft. Check the end-play while pushing and pulling on the axle shaft. End play should be within 0.004-0.008 in., with 0.006 in. ideal. Add shims to increase end play and subtract whims to decrease end-play. Remove the hub retaining nut when finished checking end-play.

NOTE: When a new axle shaft is installed, a new hub must also be installed. However, a new hub can be installed on an original axle shaft if the serrations on the shaft are not worn or damaged. The procedures for installing an original hub and a new hub are different.

21. Install an original hub in the following manner:
   a. Align the keyway in the hub with the axle shaft key.
   b. Slide the hub onto the axle shaft as far as possible.
   c. Install the axle shaft nut and washer.
   d. Install the drum, drum retaining screws, and wheel.
   e. Lower the vehicle onto its wheels and tighten the axle shaft nut to 250 ft lbs. If the cotter pin hole is not aligned, tighten the nut to the next castellation and install the pin. Do not loosen the nut to align the cotter pin hole.

22. Install a new hub in the following manner:
   a. Align the keyway in the hub with the axle shaft key.
   b. Slide the hub onto the axle shaft as far as possible.
   c. Install two well-lubricated thrust washers and the axle shaft nut.
   d. Install the brake drum, drum retaining screws, and wheel.
   e. Lower the vehicle onto its wheels.
   f. Tighten the axle shaft nut until the distance from the outer face of the hub to the outer end of the axle shaft is 1-19/64 in. Pressing the hub onto the axle to the specified distance is necessary to form the hub serrations properly.
   g. Remove the axle shaft nut and one thrust washer.
   h. Install the axle shaft nut and tighten it to 250 ft lbs. If the cotter pin hole is not aligned, tighten the nut to the next castellation and install the pin. Do not loosen the nut to install the cotter pin.

23. Connect the brake line to the wheel cylinder and bleed the brake hydraulic system and adjust the brake shoes.

Flanged Shaft—Wagoneer, Commando, Cherokee, Truck, and 1972-75 CJ Models

Removal and Installation

1. Raise the vehicle and support safely. Remove the wheel.

2. Remove the brake drum spring lock nuts and remove the drum.

3. Remove the axle shaft flange cup plug by piercing the center with a sharp tool and prying it out.

4. Using the access hole in the axle shaft flange, remove the nuts which attach the brake support plate and retainer to the axle tube flange.

5. Remove the axle shaft from the housing with an axle puller.

6. Remove the inner oil seal from the axle housing tube. Install a new seal in the tube. *NOTE: Lip of seal must be facing towards the center of the differential.*

7. Mount the axle in a vise, and with a chisel, cut the bearing retaining ring, and drive the ring off the axle shaft.

8. Using a hacksaw, cut through the oil seal and remove from the axle shaft. Do not damage the

seal contact surface while cutting.

9. With the aid of a puller or its equivalent, remove the bearing from the shaft.

10. Install the retainer plate on the axle shaft.

11. Apply wheeel bearing grease to the oil seal cavity and between the seal lips and install seal on the axle shaft seal seat. *NOTE: Outer face of seal must face the axle flange.*

12. Pack wheel bearing with grease and install on the axle shaft. *NOTE: Cup rib ring should be facing the axle flange.*

13. Install the retainer ring on the axle shaft, and press both the retainer and the bearing on the shaft at the same time, until both are seated against the shaft shoulder.

14. Install the axle shaft into the housing bore, being careful not to damage the inner seal.

15. Lubricate the outer surface of the bearing cup before installing into the bearing bore.

16. Tap the flanged end of the axle to position it into the bearing bore.

17. Attach the axle shaft retainer and brake support plate to the axle tube flange, and secure with the nuts and lockwashers.

18. Install the brake drum, spring type locknuts, and rear wheels.

19. Remove the safety stands and lower the vehicle.

Full-Floating Axle Shaft— Truck Only

It is not necessary to raise the rear wheels in order to remove the rear axle shaft on full-floating rear axles.

1. Remove the axle flange nuts, lock washers, and split washers retaining the axle shaft flange.

2. Remove the axle shaft from the axle housing.

3. Clean the axle flange mating area on the hub and axle, removing all old gasket material.

4. Install a new flange gasket onto the hub studs.

5. Insert the axle shaft into the housing. It may be necessary to rotate the axle shaft to align the shaft splines with the differential gear splines and the flange attaching holes with the hub studs.

6. Install the split washers, lockwashers, and flange nuts. Tighten the nuts securely.

Wheel Bearing Adjustment

Full Floating Axle

1. Raise the vehicle so that the

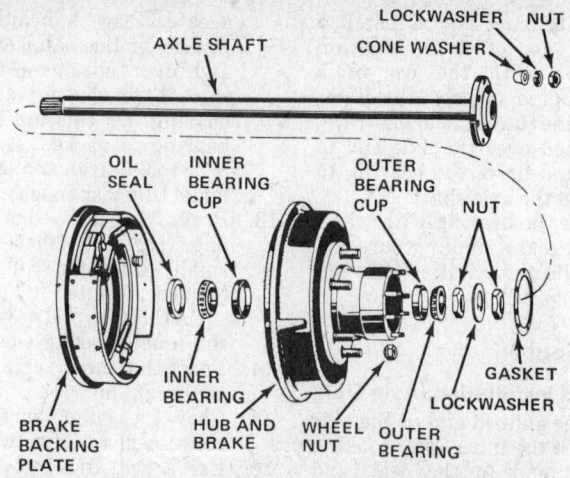

Axle shaft and wheel attaching parts-full floating axle (© Jeep Corp.)

wheel can be rotated. Support the vehicle safely.

2. Remove the axle shaft.

3. Straighten the lip of the lock washer and remove the lock washer and lock nut.

4. Tighten the adjusting nut and rotate the wheel until binding exists. Back off the adjusting nut 1/6 turn until the wheel rotates freely without any lateral shake.

5. Replace the lock washer and tighten the lock nut, bending the lip of the lock washer over the lock nut.

6. Install the axle with a new gasket and tighten the axle nuts securely.

Pinion Oil Seal R & R

Semi-Floating Axle with Tapered Shaft

1. Raise and support the vehicle and remove the rear wheels and brake drums.

2. Mark the driveshaft and yoke for reassembly and disconnect the driveshaft from the rear yoke.

3. With a socket on the pinion nut and an in. lb torque wrench, rotate the drive pinion several revolutions. Check and record the torque required to turn the drive pinion.

4. Remove the pinion nut. Use a flange holding tool to hold the flange while removing the pinion nut. Discard the pinion nut.

5. Mark the yoke and thee drive pinion shaft for reassembly reference.

6. Remove the rear yoke with a puller.

7. Inspect the seal surface of the yoke and replace it with a new one if the seal surface is pitted, grooved, or otherwise damaged.

8. Remove the pinion oil seal.

9. Before installing the new seal, coat the lip of the seal with rear axle lubricant.

10. Install the seal, driving it into place with the proper driving tool.

11. Install the yoke on the pinion shaft. Align the marks made on the pinion shaft and yoke during disassembly.

12. Install a new pinion nut. Tighten nut until end play is removed from the pinion bearing. Do not overtighten.

13. Check the torque required to turn the drive pinion. The pinion must be turned several revolutions to obtain an accurate reading.

14. Tighten the pinion nut to obtain the torque reading observed during disassembly (step 3) plus 5 in. lbs. Tighten the nut minutely each time, to avoid overtightening. Do not loosen and then re-tighten the nut.

NOTE: If the desired torque is exceeded a new collapsible pinion spacer sleeve must be installed and the pinion gear preload reset. Refer to the General Repair Section and Overhaul procedures for this operation.

15. Install the driveshaft, aligning the index marks made during disassembly. Install the rear brake drums and wheels.

Semi-Floating and Full-Floating Axles with Flange Shaft

1. Raise and support the vehicle.

2. Mark the driveshaft and yoke for reference during assembly and disconnect the driveshaft at the yoke.

3. Remove the pinion shaft nut and washer.

4. Remove the yoke from the pinion shaft, using a puller.

5. Remove the pinion shaft oil seal.

6. Install the new seal with a suitable driver.

7. Install the pinion shaft washer and nut. Tighten the nut to 210 ft lbs on the semi-floating axles

and 260 ft lbs on the full-floating axles.

8. Align the index marks on the driveshaft and yoke and install the driveshaft. Tighten the attaching bolts or nuts to 16 ft lbs.
9. Remove the supports and lower the vehicle.

Rear Axle Assembly

Removal

1. Raise the vehicle and place jack stands forward of the rear springs.
2. Remove the rear wheels.
3. Place an indexing mark on the rear yoke and propeller shaft, and disconnect the shaft.
4. Disconnect the shock absorbers from the axle tubes.
5. Disconnect the brake hose from the tee fitting on the axle housing.
6. Disconnect the parking brake cable at the frame mounting.
7. Support the axle on a jack, remove the spring clips, and remove the axle assembly from under the vehicle.

Installation

1. Position the axle assembly under the vehicle, supported by a floor jack.
2. Align the springs to the axle spring pads, and install the spring clips, and tighten securely.
3. Attach the brake line to the tee at the top of the axle housing.
4. Connect the parking brake cable at the frame mounting.
5. Install the shock absorbers to the axle housing.
6. Align the indexing marks and attach the propeller shaft to the rear yoke.
7. Bleed the brakes and install the rear wheels.
8. Fill the rear with lubricant, remove the supports, and lower the vehicle.

Rear Suspension

Spring R & R

Mounted Below the Axle

1. Raise the vehicle and support the axle.
2. Disconnect the shock absorber and stabilizer bar, if so equipped.
3. Remove the U-bolts and tie plates.
4. Disconnect the front and rear ends of the spring and remove the spring.
5. The spring can be disassembled by removing the spring rebound clips and the center bolt.

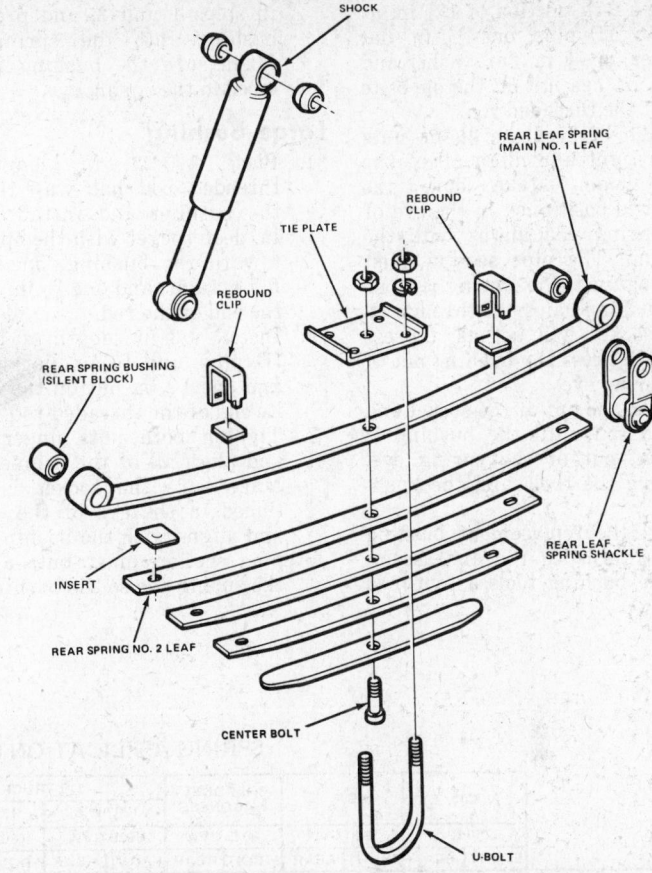

Typical rear spring and attaching parts-exploded view (© Jeep Corp.)

6. Mount the spring in the vehicle, but do not tighten the pivot bolts.
7. Align the spring center bolt and install the tie plate and U-bolts.
8. Connect the shock absorber and stabilizer bar, if so equipped.
9. Remove the axle support and lower the vehicle.
10. Tighten the pivot bolts with the weight of the vehicle on the springs to 45 ft lbs on CJ models and 75 ft lbs on all other models. Tighten 9/16 in. U-bolt nuts to 100 ft lbs, 1/2 in. nuts to 55 ft lbs and 7/16 in. U-bolt nuts to 40 ft lbs.

Mounted Above the Axle

1. Raise the vehicle and support the frame ahead of the axle.
2. Remove the U-bolts.
3. Unclip the axle vent hose from the frame.
4. Disconnect the shock absorber.
5. Remove the spring pivot bolts.
6. Lower the axle enough so the spring can be turned over and remove the spring. The spring can be disassembled by removing the rebound clips and center bolt.
7. Mount the spring in the vehicle and install the pivot bolts and nuts.
8. Raise the axle, align the spring center bolt, and install the U-bolts.

9. Connect the shock absorber, vent hose, and remove the supports and lower the vehicle.
10. Tighten the spring pivot bolts with the weight of the vehicle on the springs. Tighten the pivot bolts to 75 ft lbs. Tighten 9/16 in. U-bolt nuts to 100 ft lbs, 1/2 in. nuts to 55 ft lbs, and 7/16 in. U-bolt nuts to 40 ft lbs.

Spring Bushing R & R

Small Bushing

1. Install an 8 in. length of threaded rod halfway through the bushing and place a 1⅛ in. socket with the open end toward the bushing, one ½ in. flat washer, and one ⅜ in. hex nut on one end of the rod.

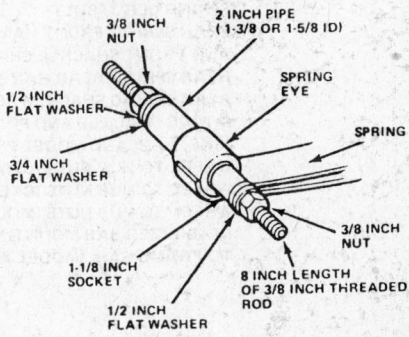

Set-up for removing small spring eye bushings.

2. Place a 2 in. section of $1\frac{5}{8}$ in. or $1\frac{3}{8}$ in. ID pipe, one $\frac{3}{4}$ in. flat washer, or $\frac{1}{2}$ in. flat washer and on $\frac{3}{8}$ in. hex nut on the opposite end of the threaded rod.
3. Tighten both of the $\frac{3}{8}$ in. nuts finger-tight and align all of the components. Make sure the socket is positioned in the eye of the spring and aligns with the bushing. The pipe section must butt against the spring eye so the bushing can pass through it. The socket will act as a press and will press the bushing out of the spring eye.
4. Tighten the nut at the socket end of the rod until the bushing is pressed out of the spring eye. Remove the tools and the bushing.
5. Install the replacement bushing on the threaded rod and assemble the bushing tools as outlined

in steps 1 and 2, and press the bushing into the spring eye. Make sure the bushing is centered in the spring eye.

Large Bushing

1. Place $\frac{1}{2} \times 11$ in. length of threaded rod half-way through the bushing and install 1-1/16 in. deep socket with the open end toward the bushing, one $\frac{1}{2}$ in. flat washer, and one $\frac{1}{2}$ in. nut on the end of the rod.
2. Install a 3 in. length of $1\frac{1}{2}$ in. ID pipe, one $\frac{1}{2}$ in. flat washer and one $\frac{1}{2}$ in. nut on the opposite end of the threaded rod.
3. Tighten both nuts finger tight and align all of the components. Make sure the socket is positioned in the eye of the spring and aligns with the bushing. The pipe section must butt against the spring eye so the bushing can

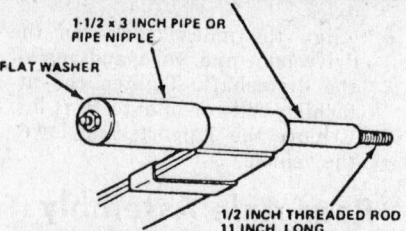

Set-up for removing large spring eye bushings.

pass through it. The socket will act as a press ram and press the bushing out of the spring eye.
4. Tighten the nut at the socket and press the bushing out of the spring eye.
5. Install the new bushing on the threaded rod and assemble the bushing tools as outlined in steps 1 and 2. Press the bushing into the spring eye until it is centered in the eye.

SPRING APPLICATION CHART

	CJ5		CJ6		CHEROKEE-WAGONEER		J10 TRUCK 120" W.B.		132" W.B.		J20 TRUCK 132" W.B.					
	3750 GVW		3900 GVW		6025 GVW		6025 GVW		6025 GVW		6500 GVW		7200 GVW		8000 GVW	
	FRONT	REAR	FRONT	REAR	FRONT	REAR	FRONT	REAR	FRONT	REAR	FRONT	REAR	FRONT	REAR	FRONT	REAR
STANDARD SPRING; NUMBER OF LEAVES	7	5	9	5	4	6	4	6	4	2	2	2	2	2	2	3
SPRING RATE (#/IN.²)	190	155/230	210	155/230	200	160/260	200	160/260	200	340	260	340	260	340	260	510
LENGTH	46	46	46	46	47	52	47	52	47	57	47	57	47	57	47	57
WIDTH	1-3/4	1-3/4	1-3/4	1-3/4	2-1/2	2-1/2	2-1/2	2-1/2	2-1/2	2-1/2	2-1/2	2-1/2	2-1/2	2-1/2	2-1/2	2-1/2
OPTIONAL SPRING; NUMBER OF LEAVES	10	10	10	10	2	6	2	6	2	—	3	3	3	3	3	—
SPRING RATE (#/IN.²)	270	276	270	276	260	230	260	230	260	—	330	510	330	510	330	—
LENGTH	46	46	46	46	47	52	47	52	47	—	47	57	47	57	47	—
WIDTH	1-3/4	1-3/4	1-3/4	1-3/4	2-1/2	2-1/2	2-1/2	2-1/2	2-1/2	—	2-1/2	2-1/2	2-1/2	2-1/2	2-1/2	—

Torque Specifications

ALL TORQUE VALUES ARE GIVEN IN FOOT-POUNDS WITH DRY FITS UNLESS OTHERWISE SPECIFIED

	SIZE	TORQUE
SHOCK - UPPER ATTACHMENT	7/16 - 20	25 - 40
SHOCK - LOWER ATTACHMENT	1/2 - 20	40 - 55
SPRING - PIVOT BOLTS (CJ MODELS)	7/16 - 14	35 - 50
SPRING - SHACKLE BOLTS (CJ MODELS)	7/16 - 14	35 - 50
SPRING CLIP U-BOLT	9/16 - 18	85 - 105
SPRING CLIP U-BOLT	1/2 - 20	45 - 65
SPRING CLIP U-BOLT	7/16 - 20	36 - 42
REAR SPRING FRONT HANGER SHAFT NUT AND FRONT SHACKLE SHAFT	5/8 - 18	45 - 65
REAR SPRING REAR HANGER SHAFT NUT	5/8 - 18	45 - 65
REAR SPRING SHACKLE BOLT (MODEL 46 TRK)	5/8 - 18	45 - 65
SPRING SHACKLE AND PIVOT BOLTS (CKE, WAG, AND MODEL 25 - 45 TRK)	9/16 - 12	50 - 100
WHEEL TO HUB NUTS (CJ MODELS)	1/2 - 20	65 - 90
WHEEL TO HUB NUTS (CKE, WAG, AND MODEL 25 - 45 TRK)	7/16 - 20	65 - 85
WHEEL TO HUB NUTS (MODEL 46 TRK)	9/16 - 18	110 - 150
STABILIZER BAR MOUNTING BRACKET TO FRAME RAIL (MODEL 46 TRK)	7/16 - 20	25 - 40

Index

Electrical Diagnosis

To satisfy the growing trend toward organized engine diagnosis and tune-up, the following gauge and meter hook-ups, as well as diagnosis procedures are covered. The most sophisticated tune-up and diagnostic facilities are no more than a complex of the basic gauges and meters in common, everyday use. Therefore, to understand gauge and meter hook-ups, their applications and procedures, is to be equipped with the know-how to perform the most exacting diagnosis.

Know Your Instruments

Ohmmeter

An ohmmeter is used to measure electrical resistance in a unit or circuit. The ohmmeter has a self-contained power supply. In use, it is connected across (or in parallel with) the terminals of the unit being tested.

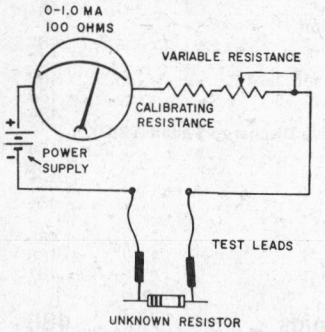

Ammeter

An ammeter is used to measure current (amount of electricity) flowing through a unit, or circuit. Ammeters are always connected in the line (in series) with the unit or circuit being tested.

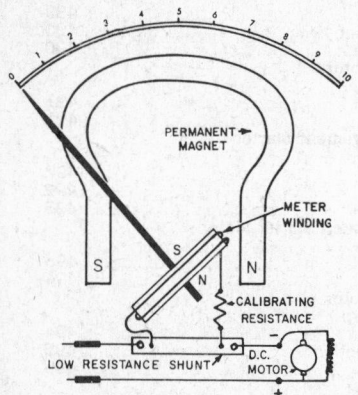

Voltmeter

A voltmeter is used to measure voltage (electrical pressure) pushing the current through a unit, or circuit. The meter is connected across the terminals of the unit being tested.

The meter reading will be the difference in pressure (voltage drop) between the two sides of the unit.

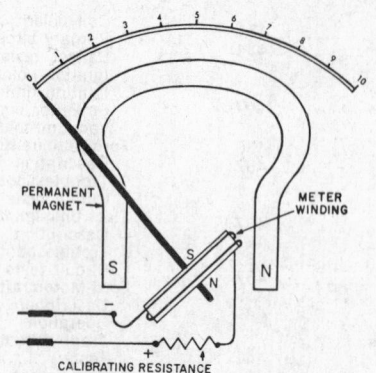

Testing the Starting Motor

Testing the Starter Circuit

The starter circuit should be divided and tested in four separate phases:

1. Cranking voltage check.
2. Amperage draw.
3. Voltage drop—grounded side.
4. Voltage drop—battery side.

NOTE: The battery must be in good condition for this test to have significance. To accurately check battery condition, use equipment designed to measure its capacity under a load. Instructions accompanying the equipment should be followed.

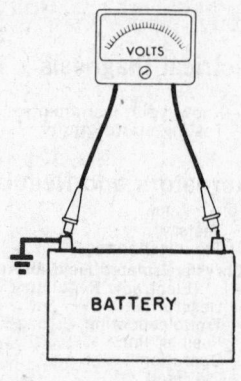

Cranking Voltage

Connect voltmeter leads to prods tapped into the battery posts (observe polarity and reverse meter leads if necessary). Remove the high tension wire from the distributor cap and ground it to prevent starting. With electronic ignition, disconnect the control box harness from the distributor. Now, turn the key. Observe both voltmeter reading and cranking speed. The cranking speed should be even, and at a satisfactory rate of speed, with a voltmeter reading of at least 9.6 volts for 12-volt systems.

Amperage Draw

The amount of current the starter motor draws is usually (but not always) associated with the mechanical problems involved in cranking the engine. (Mechanical trouble in the

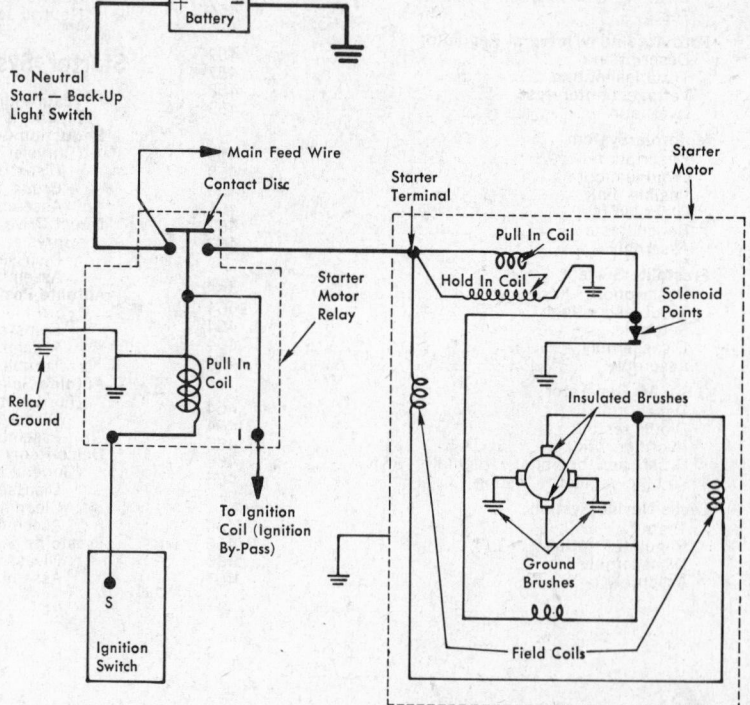

Positive engagement starter circuits (© Jeep Corp.)

engine, frozen or worn starter parts, misaligned starter or starter components, etc.) Because starter motor amperage draw is directly influenced by anything restricting the free turning of the engine, or starter, it is important that the engine and all components be at operating temperatures.

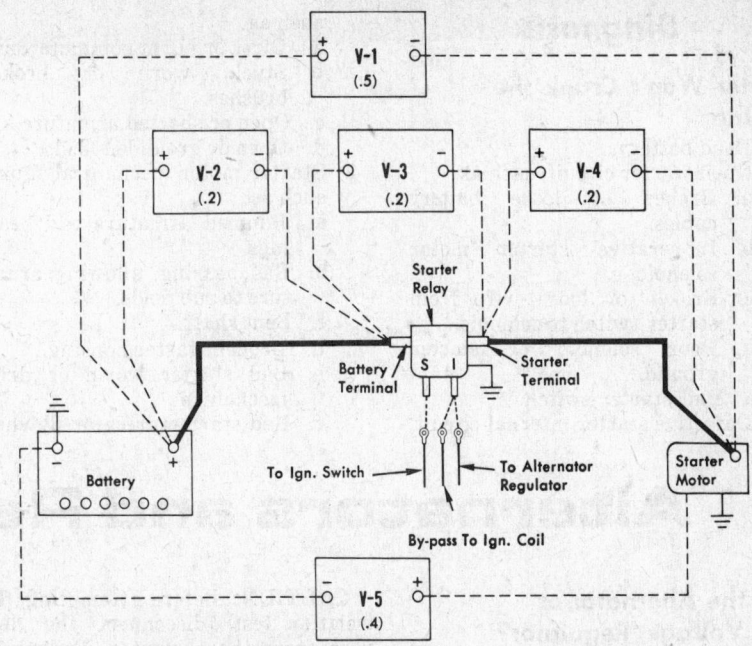

Starter cable resistance tests—typical
(Maximum voltage resistance noted per test) © Jeep Corp.)

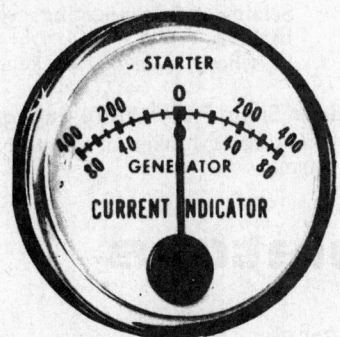

Starter current Indicator

To measure starter current draw, remove the high tension wire from the center of the distributor cap and ground it. With electronic ignition, disconnect the control box harness from the distributor. A very simple and inexpensive starter current indicator is available at auto stores. This indicator is an induction type gauge and shows, without disconnecting any wires, starter current draw.

Place the yoke of the meter directly over the insulated starter supply cable (cable must be straight for a minimum of 2 in.). Close the starter switch for about 20 seconds, watch the meter dial and record the average reading. If the indicator swings in the wrong direction, reverse the position of the meter.

The cranking amperage draw can vary from 150 to 400 amperes, depending on the engine size, engine compression, and starter type.

NOTE: When starter specifications are not available, average starter draw amperage can be derived from testing a like starter unit, known to be operating satisfactorily.

More accurate but complex equipment is available from many manufacturers. This equipment consists of a combination voltmeter, ammeter, and carbon pile rheostat. When using this equipment, follow the equipment manufacturer's procedures and recommendations.

High amperage and lazy performance would suggest an excessively tight engine, friction in the starter or starter drive, grounded starter field or armature.

Normal amperage and lazy performance suggest high resistance, or possibly poor connections somewhere in the starter circuit.

Low amperage and lazy or no performance suggest battery condition poor, bad cables or connections along the line.

Voltage Drop—Grounded Side

With a voltmeter on the 3-volt scale, without disconnecting any wires, connect negative test lead of the voltmeter to a prod secured in the grounded battery post. The positive test lead is connected to a cleaned, bare metal portion of the starter motor housing. Close the starter switch and note the voltmeter reading. If the reading is the same as battery reading, the ground circuit is open somewhere between the battery and the starter. In many cases the reading will be very small. The reading shown will indicate voltage drop (loss) between battery ground post and starter housing. The drop should not exceed 0.2 volt. If the voltage drop is above the specified amount, the next step is to isolate and correct the cause. It can be a bad cable or connection anywhere in the battery-to-starter ground circuit. A check of this type should progress along the various points of possible trouble, between the battery ground post and the starter motor housing, until the trouble spot has been located.

NOTE: due to the design of the Chrysler reduction gear starter, testing is limited to measuring voltage drop to starter cable connection.

Voltage Drop—Battery Side

Bad starter cranking may result from poor connections or faulty com-

ponents of the battery or hot phase of the starter motor circuit. To check this phase of the circuit, without disconnecting any wires, connect one lead of a voltmeter to a prod secured in the hot post of the battery and the other voltmeter lead to the field terminal of the starting motor. The meter should be set to the 16-20 volt scale. Before closing the starter switch, the voltmeter reading will be that of the battery. After closing the starter switch, change the selector on the voltmeter to the 3-volt scale. With a jumper wire between the relay battery terminal and the relay starter switch terminal, crank the engine. If the starting motor cranks the engine, the relay (solenoid) is operating.

While the engine is being cranked, watch the voltmeter. It should not register more than 0.5 volt. If more than this, check each part of the circuit for voltage drop to isolate the trouble, (high resistance).

Without disturbing the voltmeter-to-battery hook-up, move the free voltmeter lead to the battery terminal of the relay (solenoid), and crank the engine. The voltmeter should show no more than 0.1 volt.

If this reading is correct, move the same voltmeter lead to the starting motor terminal of the relay (solenoid). While the engine is being cranked, the voltmeter should show no more than 0.3 volt. If it does, the trouble lies in the relay.

If the reading is correct, the trouble is in the cable or connections between the relay and the starting motor.

Diagnosis

Starter Won't Crank the Engine

1. Dead battery.
2. Open starter circuit, such as:
   a. Broken or loose battery cables.
   b. Inoperative starter motor solenoid.
   c. Broken or loose wire from starter switch to solenoid.
   d. Poor solenoid or starter ground.
   e. Bad starter switch.
3. Defective starter internal circuit, such as:
   a. Dirty or burnt commutator.
   b. Stuck, worn or broken brushes.
   c. Open or shorted armature.
   d. Open or grounded fields.
4. Starter motor mechanical faults, such as:
   a. Jammed armature end bearings.
   b. Bad bearing, allowing armature to rub fields.
   c. Bent shaft.
   d. Broken starter housing.
   e. Bad starter worm or drive mechanism.
   f. Bad starter drive or flywheel driven gear.
5. Engine hard or impossible to crank such as:
   a. Hydrostatic lock, water in combustion chamber.
   b. Crankshaft seizing in bearings.
   c. Piston or ring seizing.
   d. Bent or broken connecting rod.
   e. Seizing of connecting rod bearing.
   f. Flywheel jammed or broken.

Starter Spins Free, Won't Engage

1. Sticking or broken drive mechanism.

Alternators and Regulators

Is it the Alternator or the Voltage Regulator?

The first step in diagnosing troubles of the charging system, is to identify the source of failure. Does the fault lie in the alternator or the regulator? The next move depends upon preference or necessity; either repair or replace the offending unit.

It is just as easy to separate an alternator, electrically, from the AC regulator as it is to separate its counterpart, the DC generator from its regulator.

AC generator output is controlled by the amount of current supplied to the field circuit of the system.

Unlike the DC generator, an AC generator is capable of producing substantial current at idle speed. Higher maximum output is also a possibility. This presents a potential danger when testing. As a precaution, a field rheostat should be used in the field circuit when making the following isolation test. The field rheostat permits positive control of the amount of current allowed to pass through the field circuit during the isolation test. Unregulated alternator capacity could ruin the unit.

NOTE: most manufacturers of precision gauges offer special test connectors, in sets, that will adapt to the leads and connections of any AC charging system.

CAUTION: Before attempting the isolation test, disconnect the field wire from the regulator. Failure to take this precaution can cause instant burning and permanent damage to the regulator.

Isolation Test

(By-passing the Regulator)

1. Connect voltmeter leads to two prods driven into the battery posts.
2. Disconnect field wire from the FLD terminal of the voltage regulator.
3. Connect one lead of a field rheostat to the undisturbed IGN terminal of the regulator, and the other field rheostat lead to the wire that was removed from the FLD terminal of the regulator.
4. With field rheostat turned to the low side of the scale, (high resistance) start the engine and adjust throttle to about 2,000 rpm.
5. Slowly move field rheostat control knob to decrease resistance, (allowing more current to flow through the field circuit) until voltmeter reading slightly exceeds manufacturers' specifications.

NOTE: under load conditions, observe the alternator for arcing or any other evidence of malfunction.

6. If alternator performs satisfactorily, repair or replace the regulator. Conversely, if the voltmeter reading is zero, or below specifications, repair or replace the alternator.

Alternator Test Plans

The following is a procedure pattern for testing the various alternators and their control systems.

There are certain precautionary measures that apply to alternator tests in general. These items are listed in detail to avoid repetition when testing each make of alternator, and to encourage a habit of good test procedure.

1. Check alternator drive belt for condition and tension.
2. Disconnect battery cables, check physical, chemical, and electrical condition of battery.
3. Be absolutely sure of polarity before connecting any battery in the circuit. Reversed polarity will ruin the diodes.
4. Never use a battery charger to start the engine.
5. Disconnect both battery cables when making a battery recharge hook-up.

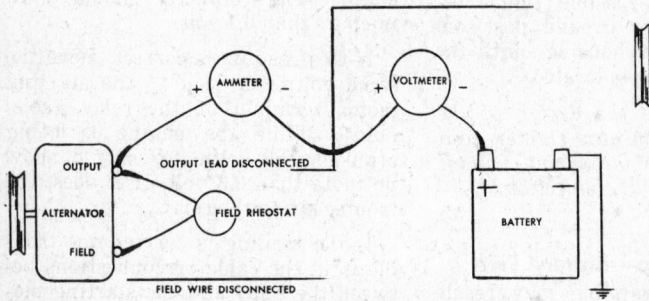

Plan to check charging circuit resistance

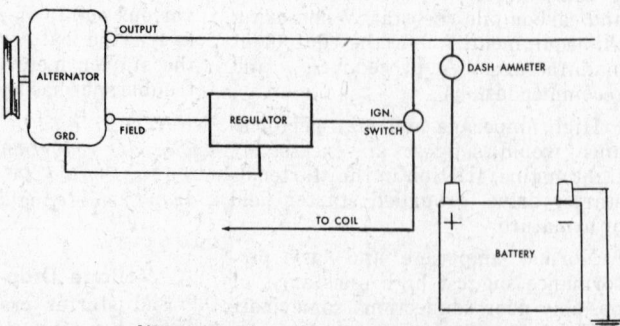

Alternator system with ammeter in the circuit

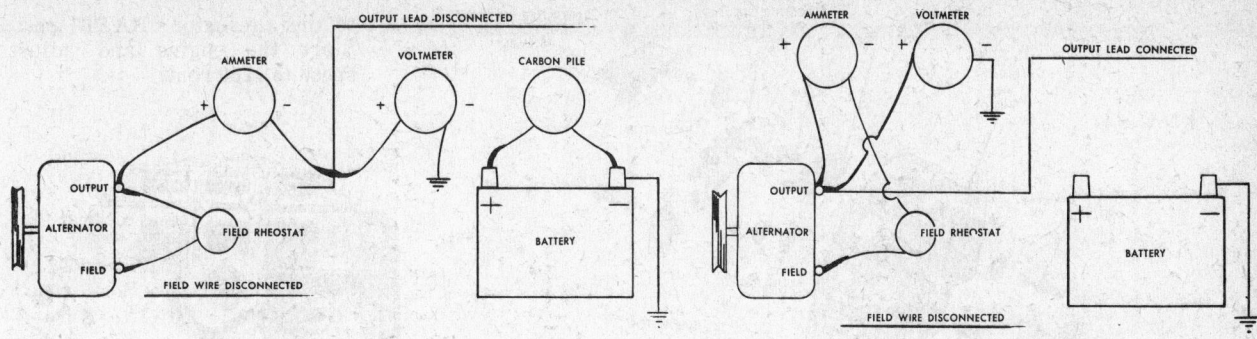

Plan to check current output

Plan to check field current draw

6. Be sure of polarity hook-up when using a booster battery for starting.
7. Never ground the alternator output or battery terminal.
8. Never ground the field circuit between alternator and regulator.
9. Never run any alternator on an open circuit with the field energized.
10. Never try to polarize an alternator.
11. Do not attempt to motor an alternator.
12. The regulator cover must be in place when taking voltage limiter readings.
13. The ignition switch must be in off position when removing or installing the regulator cover.

14. Use insulated tools only to make adjustments to the regulator.
15. When making engine idle speed adjustments, always consider potential load factors that influence engine rpm. To compensate for electrical load, switch on the lights, radio, heater, air conditioner, etc.

Diagnosis

Low or No Charging

1. Blown fuse.
2. Broken or loose fan belt.
3. Voltage regulator not working.
4. Brushes sticking.
5. Slip ring dirty.
6. Open circuit.
7. Bad wiring connections.
8. Bad diode rectifier.

9. High resistance in charging circuit.
10. Voltage regulator needs adjusting.
11. Grounded stator.
12. May be open rectifiers (check all three phases).
13. If rectifiers are found blown or open, check capacitor.

Noisy Unit

1. Damaged rotor bearings.
2. Poor alignment of unit.
3. Broken or loose belt.
4. Open diode rectifiers.

Regulator Points Burnt or Stuck

1. Regulator set too high.
2. Poor ground connections.
3. Shorted generator field.
4. Regulator air gap incorrect.

Chrysler Alternators

Chrysler Isolated Field Alternator (Electronic Regulator)

The Chrysler isolated field alternator derives its name from its construction. Both of the brushes are insulated from ground and there is no heat sink connection, thereby isolating the internal field.

Troubleshooting

NOTE: see the "Alternator Test Plans" section before proceeding further. Make sure that the continuous running blower, if equipped, is disconnected. This blower will run with the key turned on even if the blower controls are off unless disconnected.

Fusible Links

Chrysler Corporation trucks have a single fusible link which is connected between the starter relay and the junction block. Failure of this link will cause all electrical systems to stop functioning.

Charging System Operation

NOTE: if the current indicator is to give an accurate reading, the bat-

tery cables must be of the same gauge and length as the original equipment.

1. With the engine running and all electrical systems off, place a current indicator over the positive battery cable.
2. If a charge of about 5 amps is

recorded, the charging system is working. If a draw of about 5 amps is recorded the system is not working. The needle moves toward the battery when a charge condition is indicated and away from the battery when a

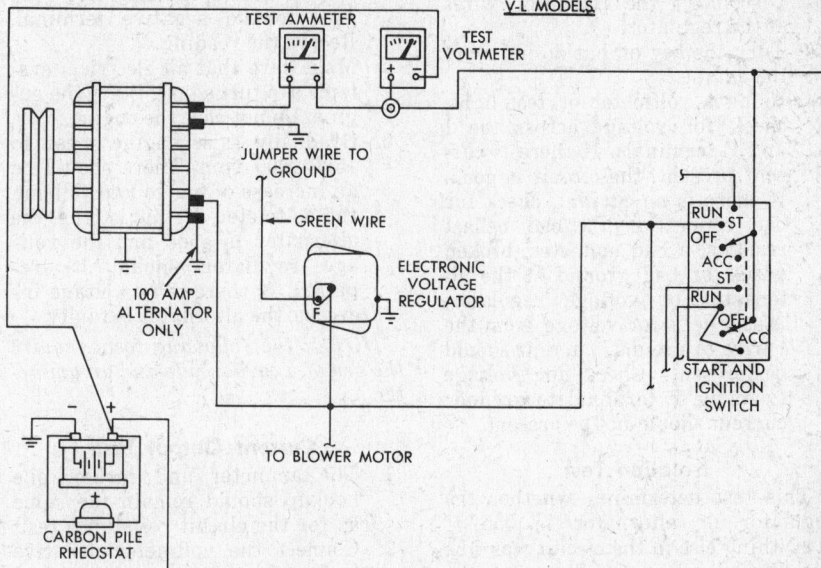

Typical resistance test connections—Chrysler charging system (© Chrysler Corp.)

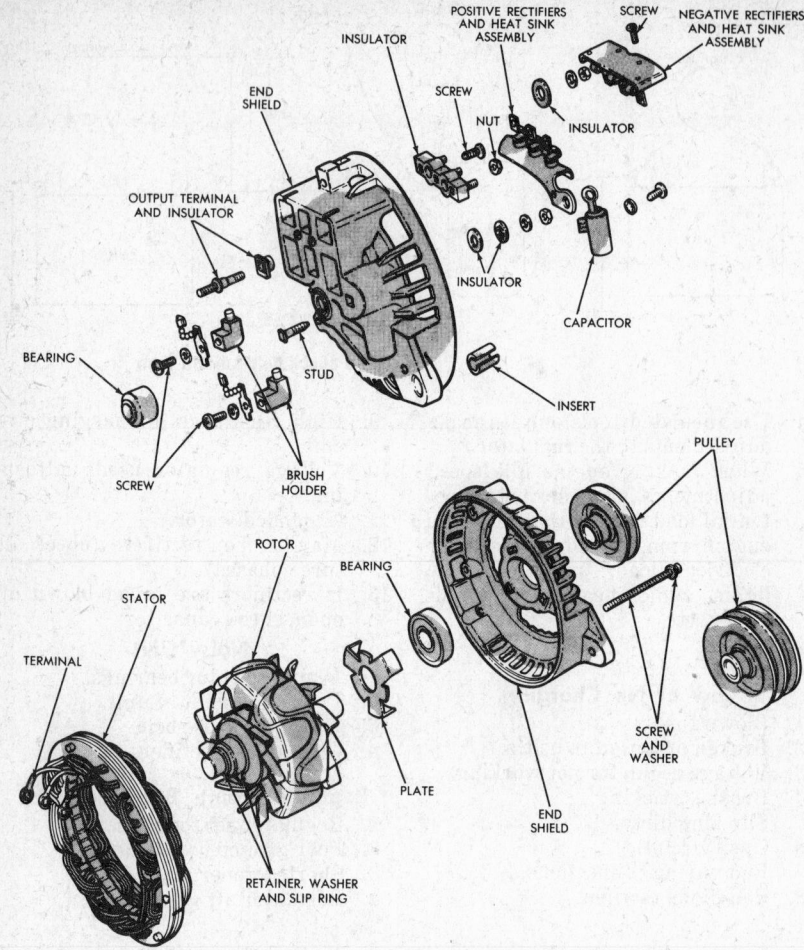

Typical Chrysler alternator—exploded view (© Chrysler Corp.)

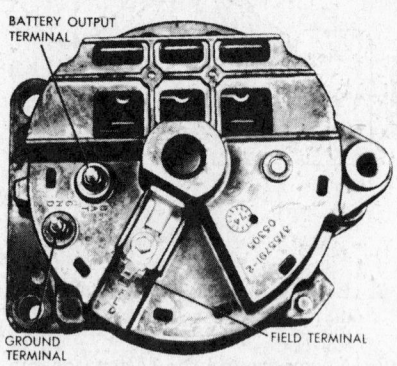

View of rear housing—100 ampere Chrysler alternator (© Chrysler Corp.)

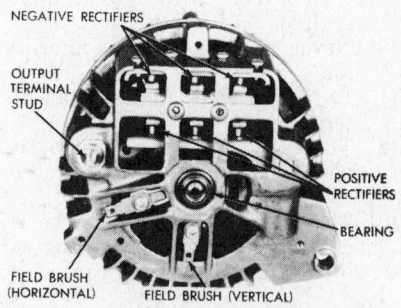

View of rear housing (except 100 ampere) —Chrysler alternator (© Chrysler Corp.)

4. Start the engine and adjust speed to 1250 rpm.

draw condition is indicated. If a draw is indicated, proceed to the next testing procedure. If an overcharge of 10-15 amps is indicated, check for a faulty regulator.

Ignition Switch-to-Regulator Circuit Check

1. Disconnect the regulator wires at the regulator.
2. Turn the key on but do not start the engine.
3. Using a voltmeter or test light, check for voltage across the I and F terminals. If there is current present, the circuit is good. If there is no current, check for bad connections, a bad ballast resistor, a bad ammeter, broken wires, or bad ground at the alternator or voltage regulator. Also, check for voltage from the I wire to ground; current should be present. Check for voltage from the F terminal to ground; current should not be present.

Isolation Test

This test determines whether the regulator or alternator is bad if everything else in the circuit was OK.
1. Disconnect, at the alternator, the wire that runs between one of

the alternator field connections and the voltage regulator.
2. Run a jumper wire from the disconnected alternator terminal to ground.
3. Connect a voltmeter to the battery. The positive voltmeter lead connects to the positive battery terminal, and the negative lead goes to the negative terminal. Record the reading.
4. Make sure that all electrical systems are turned off. Start the engine. Do not race the engine.
5. Gradually raise engine speed to 1500-2000 rpm. There should be an increase of one to two volts on the voltmeter. If this is true, the alternator is good and the voltage regulator should be repaired. If there is no voltage increase, the alternator is faulty.

NOTE: the following tests require the use of a carbon pile and an ammeter.

Current Output Test

1. The ammeter and carbon pile hookup should remain the same as for the circuit resistance test.
2. Connect the voltmeter negative lead to the battery negative post.
3. Move the positive voltmeter lead

to the alternator "BATT" post.

5. Note voltmeter and ammeter readings. Maintain a 15 volt reading by adjusting the carbon pile control.
6. Compare ammeter reading with manufacturer's specifications The reading should be no less than specified, 3 amps.
7. If below specifications, internal trouble is indicated. Remove the alternator for further testing.

Electronic Voltage Regulator Test

1. Make sure battery terminals are clean and battery is charged.
2. Connect the positive lead of a test voltmeter to ignition Terminal No. 1 of the ballast resistor.
3. Connect the negative voltmeter lead to a good *body* ground.
4. Start engine and allow it to idle at 1250 rpm, all lights and accessories turned off. Voltage should be as follows:

Ambient Temp. 1/4 in. from Regulator	Voltage
-20°F.	14.3-15.3
80°F.	13.8-14.4
140°F.	13.3-14.0

5. If the voltage is *below* specifications, check the following:

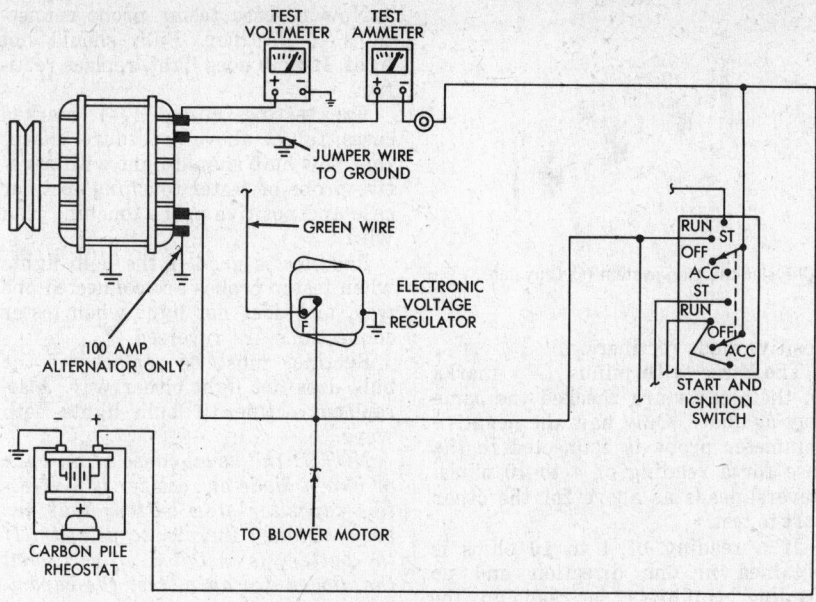

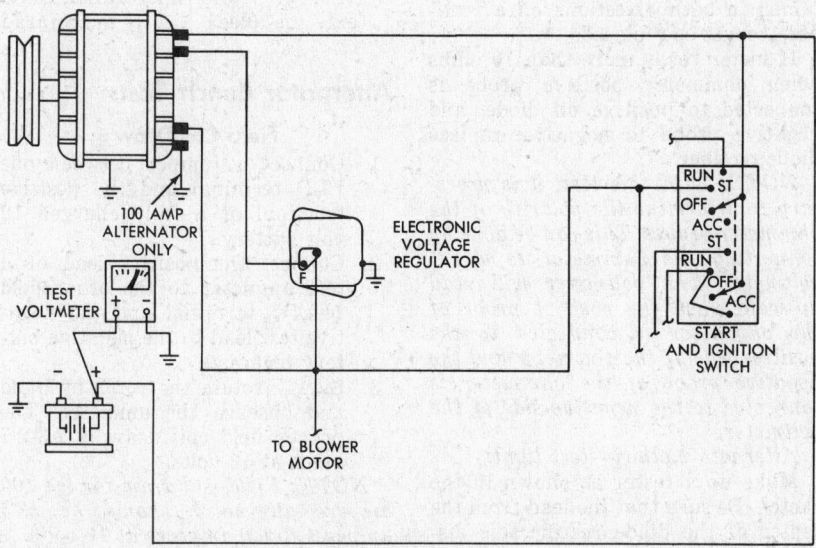

a. Voltage regulator ground—check voltage drop between regulator cover and ground.

b. Harness wiring—disconnect regulator plug (ign. switch off), then turn on ign. switch and check for battery voltage at the terminal having the blue and green leads. *Wiring harness must be disconnected from the regulator when checking individual leads.* If no voltage is present in either lead, the problem is in the truck wiring or alternator field.

6. If Step 5 tests showed no malfunctions, install a new regulator and repeat Step 4.

7. If voltage is *above* specifications (Step 4), or fluctuates, check the following:

a. Ground between regulator and body, and between body and engine.

b. Ignition switch circuit between switch and regulator.

8. If voltage is still more than 1/2 volt above specifications, install a new regulator and repeat Step 4.

Chrysler Overhaul and Internal Testing

Alternator disassembly, repair and assembly procedures are basically the same for all Chrysler alternators. Certain variations in design, or in-production modifications, could require slightly different procedures that should be obvious upon inspection of the unit being serviced.

Disassembly

To prevent damage to the brush assemblies they should be removed before proceeding with the disassembly of the alternator. The insulated brush is mounted in a plastic holder that positions the brush vertically against one of the slip rings.

1. Remove the retaining screw, flat washer, nylon washer and field terminal and carefully lift the plastic holder containing the spring and brush assembly from the end housing.

2. The ground brush is positioned horizontally against the remaining slip ring and is retained in the holder that is integral with the end housing. Remove the retaining screw and lift the clip, spring and brush assembly from the end housing.

CAUTION: The stator is laminated, don't burr the stator or end housings.

3. Remove the through bolts and pry between the stator and drive end housing with a thin blade screwdriver. Carefully separate the drive end housing, pulley and rotor assembly from the stator and rectifier housing assembly.

4. The pulley is an interference fit on the rotor shaft. Remove with a puller and special adapters.

5. Remove the three nuts and washers and, while supporting the end frame, tap the rotor shaft with a plastic hammer and separate the rotor and end housing.

6. The drive end ball bearing is an interference fit with the rotor shaft. Remove the bearing with puller and adapters.

NOTE: further dismantling of the rotor is not advisable, as the remainder of the rotor assembly is not serviced separately.

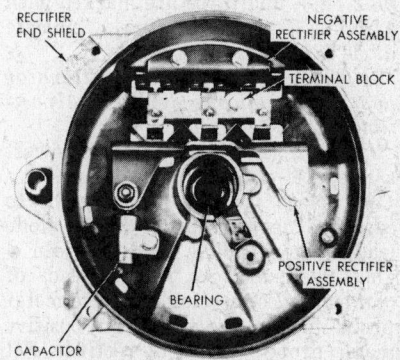

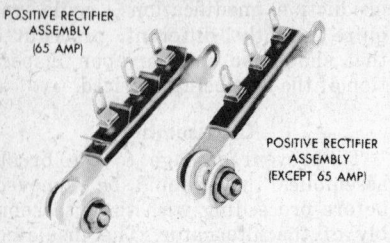
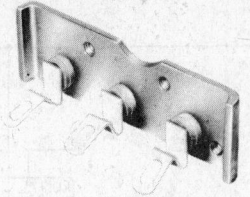

Positive and negative rectifier identification—Chrysler charging system (© Chrysler Corp.)

7. Remove the DC output terminal nuts and washers and remove terminal screw and inside capacitor (on units so equipped).

8. Remove the insulator.

NOTE: three positive rectifiers are pressed into the heat sink and three negative rectifiers in the end housing. When removing the rectifiers, it is necessary to support the end housing and/or heat sink to prevent damage to these castings. Another caution is in order relative to the diode rectifiers. Don't subject them to unnecessary jolting. Heavy vibration or shock may ruin them.

  a. Cut rectifier wire at point of crimp.

  b. Support rectifier housing.

NOTE: the factory tool is cut away and slotted to fit over the wires and around the bosses in the housing. Be sure that the bore of the tool completely surrounds the rectifier, then press the rectifier out of the housing.

NOTE: the roller bearing in the rectifier end frame is a press fit. To protect the end housing it is necessary to support the housing with a tool when pressing out the bearing.

Bench Tests

Testing Silicon Diode Rectifiers With Ohmmeter

Preferred method—rectifiers open in all three phases.

Disassemble the alternator and separate the wires at the Y-connection of the stator.

There are six diode rectifiers mounted in the back of the alternator. Three of them are marked with a plus (+), and three are marked with a minus (−). These marks indicate diode case polarity.

NOTE: The 100 ampere alternator has twelve silicone diodes. Six positive and six negative.

To test, set ohmmeter to its lowest range. If case is marked positive (+), place positive meter probe to case and negative probe to the diode lead. Meter should read between 4 and 10 ohms. Now, reverse leads of ohmmeter, connecting negative meter probe to positive case and positive meter probe to wire of rectifier. Set meter on a high range. Meter needle should move very little, if any (infinite reading). Do this to all three positive diode rectifiers.

The three with minus (−) marks on their cases are checked the same way as above. Only now the negative ohmmeter probe is connected to the case for a reading of 4 to 10 ohms. Reverse leads as above for the other part to test.

If a reading of 4 to 10 ohms is obtained in one direction and no reading (infinity) is read on the ohmmeter in the other direction, diode rectifiers are good. If either infinity or a low resistance is obtained in both directions on a rectifier, it must be replaced.

If meter reads more than 10 ohms when ohmmeter positive probe is connected to positive on diode, and negative probe to negative, replace diode rectifier.

NOTE: with this test, it is necessary to determine the polarity of the ohmmeter probes. This can be done by connecting the ohmmeter to a DC voltmeter. The voltmeter will read up-scale when the positive probe of the ohmmeter is connected to the positive side of the voltmeter and the negative probe of the ohmmeter is connected to the negative side of the voltmeter.

Alternate method—test light.

Make up a tester as shown in the sketch. Be sure that the lead from the center of the diode rectifiers is disconnected.

To test rectifiers with plus (+) case, touch positive probe of tester to case and minus (−) probe to lead wire of rectifier. Bulb should light if rectifier is good. If bulb does not light, replace rectifier.

Rectifier test with test light and 12 volt battery (© Chrysler Corp.)

Now reverse tester probe connections to rectifier. Bulb should not light. If bulb does light, replace rectifier.

For testing minus (−) marked cases, follow above procedure, except that now bulb should light with negative probe of tester touching rectifier case and positive probe touching lead wire.

Rectifier is good if the bulb lights when tester probes are connected one way, and does not light when tester connections are reversed.

Rectifier must be replaced if the bulb does not light either way. Also, replace rectifier if bulb lights both ways.

NOTE: the usual cause of an open or blown diode or rectifier is a defective capacitor or a battery that has been installed in reverse polarity. If the battery is installed properly and the diodes are open, test the capacitor.

Capacitor capacity:
  (int. installed)
  ....158 microfarad, min.
  (ext. installed) ....5 microfarad

Alternator Bench Tests

Field Coil Draw

1. Connect a jumper between one FLD terminal and the positive terminal of a fully charged 12 volt battery.

2. Connect the positive lead of a test ammeter to the other field (FLD) terminal and the negative test lead to the negative battery terminal.

3. Slowly rotate the rotor by hand and observe the ammeter: The proper field coil draw is 2.3-2.7 amps at 12 volts.

NOTE: Field coil draw for the 100 ampere alternator should be 4.75 amperes to 6.0 amperes at 12 volts.

Field Circuit Ground Test

1. Touch one test lead of a 110 volt AC test bulb to one of the alternator brush (field) terminals and the other test lead to the end shield.

2. If the lamp lights, remove the field brush assemblies and separate the end housing by removing the three thru-bolts.

3. Place one test lead on a slip ring and the other on the end shield.

4. If the lamp lights, the rotor assembly is grounded internally and must be replaced.

5. If the lamp does not light, the cause of the problem was a grounded brush.

Grounded Stator

1. Disconnect the diode rectifiers from the stator leads.

2. Test from stator leads to stator core, using a 110-volt test lamp.

Test lamp should not light. If it does, stator is grounded and must be replaced.

Low Output

(About 50% output accompanied with a growl-hum caused by a shorted phase or a shorted rectifier.)

Perform Steps 1, 2 and 3 (rectifier open in all three phases). If the rectifiers are found to be within specifications, replace the stator assembly.

Current Output Too High (No Control) Caused by Open Rectifier or Open Phase

Perform Steps 1, 2 and 3 (rectifier open in all three phases). If the rectifier tests satisfactorily, inspect the stator connections before replacing the stator.

Assembly

1. Support the heat sink or rectifier end housing on circular plate.
2. Check rectifier identification to be sure the correct rectifier is being used. The part numbers are stamped on the case of the rectifier. They are also marked, red for positive and black for negative.
3. Start the new rectifier into the casting and press it in squarely.

CAUTION: Do not start rectifier with a hammer or it will be ruined.

4. Crimp the new rectifier wire to the wires disconnected at removal, or solder (using a heat sink with rosin core solder).
5. Support the end housing on tool so that the notch in the support tool will clear the raised section of the heat sink, then press the bearing into position with tool SP-3381, or equivalent.

NOTE: new bearings are pre-lubricated, additional lubrication is not required.

6. Insert the drive end bearing in the drive end housing and install the bearing plate, washers and nuts to hold the bearing in place.
7. Position the bearing and drive end housing on the rotor shaft

Rotor test for short or open circuits—Chrysler charging system (© Chrysler Corp.)

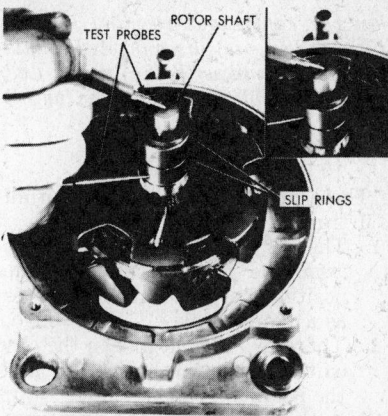

Rotor test for ground—Chrysler charging system (© Chrysler Corp.)

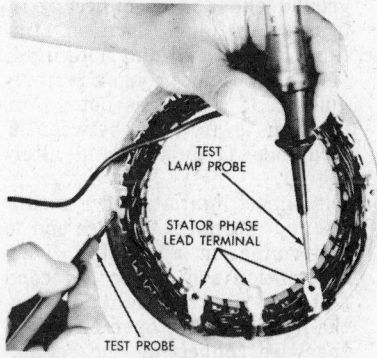

Stator test for ground—Chrysler charging system (© Chrysler Corp.)

and, while supporting the base of the rotor shaft, press the bearing and housing in position on the rotor shaft with an arbor press and arbor tool.

CAUTION: Be careful that there is no cocking of the bearing at installation; or damage will result. Press the bearing on the rotor shaft until the bearing contacts the shoulder on the rotor shaft.

8. Install pulley on rotor shaft. Shaft of rotor must be supported so that all pressing force is on the pulley hub and rotor shaft.

NOTE: Do not exceed 6,800 lbs. pressure. Pulley hub should just contact bearing inner race.

9. Some alternators will be found

to have the capacitor mounted internally. Be sure the heat sink insulator is in place.

10. Install the output terminal screw with the capacitor attached through the heat sink and end housing.
11. Install insulating washers, lockwashers and locknuts.
12. Make sure the heat sink and insulator are in place and tighten the locknut.
13. Position the stator on the rectifier end housing. Be sure that all of the rectifier connectors and phase leads are free of interference with the rotor fan blades and that the capacitor (internally mounted) lead has clearance.
14. Position the rotor assembly in the rectifier end housing. Align the through bolt holes in the stator with both end housings.
15. Enter stator shaft in the rectifier end housing bearing, compress stator and both end housings manually and install through bolts, washers and nuts.
16. Install the insulated brush and terminal attaching screw.
17. Install the ground screw and attaching screw.
18. Rotate pulley slowly to be sure the rotor fan blades do not hit the rectifier and stator connectors.

General Motors Alternators

Delcotron 5.5 Series 1D and 6.2 Series 2D

Description

The Delcotron continuous output AC generator consists of two major parts—the stator and the rotor. The stator is composed of many turns of wire on the inside of a laminated core that is attached to the generator frame. The rotor is mounted on bearings at each end. Two brushes carry current through slip rings to the field

coils, which are wound on the rotor shaft.

The 5.5 Series 1D Delcotron is similar in operation to the 6.2 Series 2D perforated Stator Delcotron. Where differences exist, the two units are mentioned separately.

Six diodes, mounted on internal heat sinks, change the AC current output into DC current. This current is controlled by the alternator regulator. The regulator is a double-contact unit combined with a field relay or a

triple-contact unit containing an indicator lamp relay as well as the field relay and voltage relay. Transistor regulators were also used in production intermittently.

On high-output Delcotron units, the regulator incorporates a field discharge diode.

Troubleshooting

NOTE: see the "Preliminary Charging System Inspection" section before proceeding further. Make sure that the continuous running blower, if equipped, is disconnected. This blower will run with the key on and even if the blower control is off, it is not disconnected.

Fusible Links

There are four fusible links on all GM trucks.

1. The 14 gauge wire that runs from the junction block to the positive battery terminal serves as a fusible link. .
2. There is a second link in the circuit between the horn relay and the ignition switch.
3. A third link is in the wire running to the No. 3 voltage regulator terminal. Its purpose is to protect the regulator contacts and the alternator field circuit.
4. The fourth link is connected between the main junction block and the horn relay.

These links must be inspected before proceeding with troubleshooting

Charging System Operation

NOTE: if the current indicator is to give an accurate reading, the battery cables must be the same gauge and length as the original equipment.

1. With the engine running and all electrical systems turned off, place a current indicator over the positive battery cable.
2. If a charge of about 5 amps is recorded, the charging system is working. If a draw of about 5 amps is recorded, the system is not working. The needle moves toward the battery when a charge condition is indicated, and away from the battery when a draw condition is indicated. If a draw is indicated, proceed with further testing. If an excessive charge (10-15 amps) is indicated, check for an overcharge, caused by a faulty regulator.

Indicator Light Circuit Testing

The indicator lights is important in AC charging systems, for it provides initial field excitation current to the alternator. The light goes out when the field relay closes, which applies

battery current to both sides of the bulb. If the light does not go on when key is turned, the bulb could be faulty, there could be an open circuit in the wiring or a positive diode in the alternator could be shorted to ground.

1. Disconnect plug from regulator and connect a test light between terminal No. 4 (in plug) and ground. Turn on ignition switch and observe the light. If light does not go on, check bulb, socket or wiring between switch and regulator plug. If light goes on, check regulator, wiring between regulator F terminal and alternator, or Delcotron itself.
2. Disconnect jumper wire at ground end and reconnect to F terminal in plug. Turn on ignition for a second and note light. If light goes on, problem is in regulator. If light does not go on, problem is in wire between F terminals (regulator and alternator).
3. Disconnect light at plug F terminal and reconnect the free end to F terminal at alternator. Turn on ignition switch for a second and note light. If light goes on, the problem is an open circuit in the wire connecting the regulator and alternator F terminals. If light does not go on, the alternator field windings are defective.

If the indicator light does not extinguish when engine is started, check for a loose drive belt, faulty field relay, faulty alternator, open parallel resistance wire (usually shows up at idle). If the light stays on with the key turned off, an alternator positive diode is shorted to ground.

Isolation Test

1. Disconnect the wiring harness from the voltage regulator. With a jumper wire connect the F wire to the no. 3 wire in the wire harness plug.
2. Connect a voltmeter across the battery terminals, the positive voltmeter lead to the positive

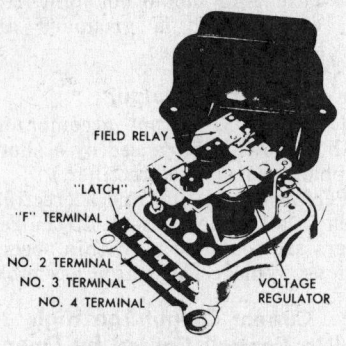

Mechanical voltage regulator
(© Chevrolet Div., G.M. Corp)

battery terminal, and the negative lead to the negative terminal. Record the reading.

3. Start the engine. Do not race the engine.
4. Gradually raise engine speed to 1500-2000 rpm. The reading on the voltmeter should increase one to two volts over the initial reading. If there is no increase in the reading, repair the alternator. If there is an increase in the voltmeter reading, replace the regulator.

Field Relay Test

1. Connect a voltmeter between the No. 2 terminal and the ground on the regulator.
2. Turn ignition switch on; do not start the engine. Voltmeter should read battery voltage.
3. If voltmeter reads zero, check circuit connecting regulator terminal No. 2 and Delcotron R terminal.
4. Start engine and run at 1,500-2,000 rpm. If voltage exceeds closing voltage (field relay), and light remains on, field relay is faulty and must be checked.

Field Relay Adjustment

1. Connect a voltmeter between No. 2 regulator terminal and ground.
2. To adjust, connect a 50 ohm rheostat between wiring harness terminal No. 3 and regulator terminal No. 2, after disconnect-

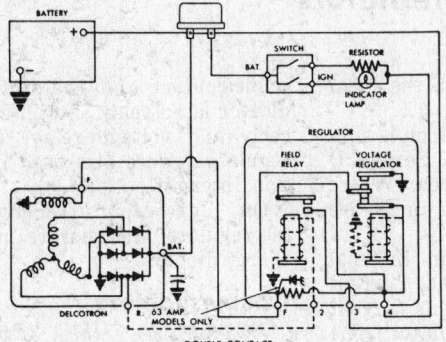

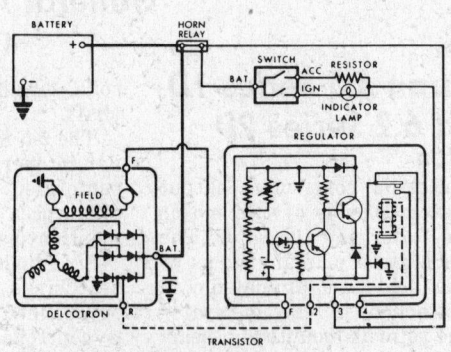

Voltage regulator circuit diagrams (© Chevrolet Div., G.M. Corp)

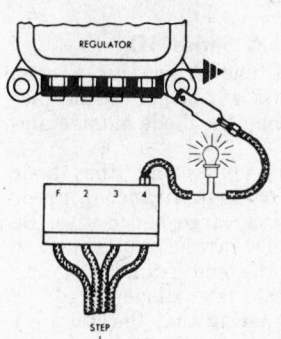
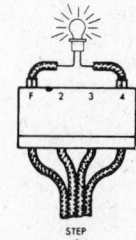
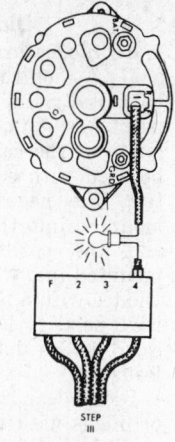

Initial field excitation circuit test hook up (© Chevrolet Div., G.M. Corp)

Adjusting field relay closing voltage
(© Chevrolet Div., G.M. Corp)

ing the spade lug on the end of the No. 2 regulator terminal wire. Connect a voltmeter between regulator terminal No. 2 and ground, then turn the resistor to "open" position, turn off ignition switch and slowly decrease resistance until relay closes (noting voltage at this point). Voltage can be adjusted by bending heel iron.

Field Circuit Resistance Testing

The resistance wire is an integral part of the ignition wiring harness. The wire cannot be soldered; any connections must be made using crimp-type connectors. Resistance is 10 ohms, 6¼ watts.
1. Connect a voltmeter between the wiring harness terminal No. 4 and ground.
2. Turn on ignition switch, needle must indicate or resistor is open.

Delcotron Current Output Test

NOTE: disconnect battery ground cable while making test connections, then reconnect cable after completing Step 5. Disconnect battery ground cable again before removing test set-up. This test yields the same information as the isolation test but requires the use of an ammeter and a carbon pile.

1. Disconnect lead from BAT. terminal of Delcotron.
2. Hook an ammeter to the lead just disconnected, and to the BAT. terminal of the Delcotron.
3. Hook up the voltmeter leads to the BAT. terminal and a good ground on the alternator.
4. Disconnect the lead from the FR. terminal of the Delcotron.
5. Hook up a jumper wire between BAT. and F terminals of the Delcotron.
6. With a carbon pile load control hooked up to the battery posts, start the engine and set engine to 1,500 rpm, while adjusting carbon pile to obtain 14 volts. With a 6.2 in. alternator, only

600-800 rpm is required.
CAUTION: Be careful not to exceed the recommended regulator voltage setting. This is controlled by the carbon pile load.
7. Ammeter should read within 10% of rated output, as stamped on frame of each unit.

Alternator Overhaul (5.5 and 6.2 Delcotron)

Disassembly—5.5 Series 1D
1. Remove four bolts.
2. Separate drive end frame and rotor from stator assembly by prying with screwdriver. Note that separation is between stator frame and drive end frame.
3. Place tape over slip ring end frame bearing to seal dirt.
4. Lightly clamp rotor in vise to remove shaft nut.
CAUTION: Do not distort rotor by overtightening vise.
5. After nut removal, take off washer, pulley, fan and collar.
6. Separate drive end frame from rotor shaft.
7. Remove three stator lead attaching nuts and separate stator from end frame.
8. Remove screws, brushes and

holder assembly.
9. Remove BAT., GND., and attaching screw terminals, then remove heat sink.

Disassembly—6.2 Series 2D
1. Clamp drive end mounting flange in a vise, remove the two screws that secure the cover to the brush holder and remove the cover.
2. Remove the nut that holds the indicator light wire to the blade connector post; disconnect wire lead from post.
3. Remove the two screws that hold the condenser and brush holder to rear end frame, then remove brush holder. Allow condenser to remain with alternator.
4. Remove three slip ring end frame bolts and tab nuts, then carefully pry end frame and case apart, working evenly around the circumference.
5. Remove the three drive end frame bolts and tab nuts, then remove end frame, rotor and pulley as an assembly.
6. Remove shaft nut, washer, pulley and Woodruff key from rotor shaft, then slide rotor from end frame.

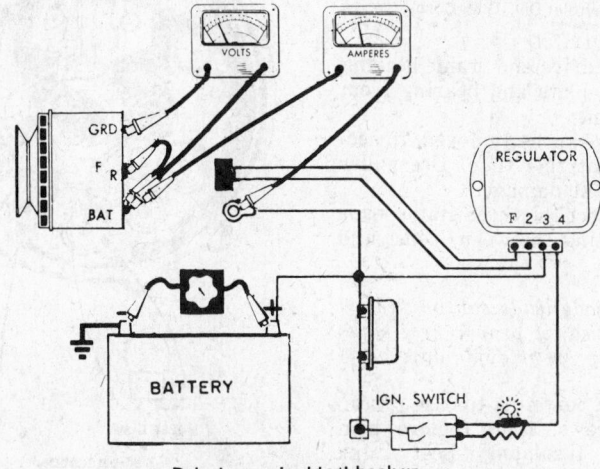

Delcotron output test hook up
(© Chevrolet Div., G.M. Corp)

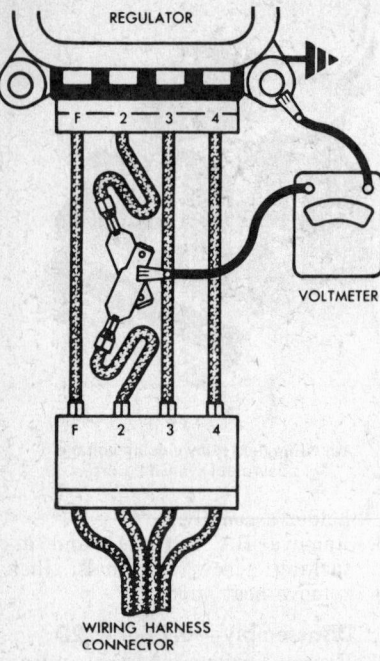

Testing field relay
(© Chevrolet Div., G.M. Corp)

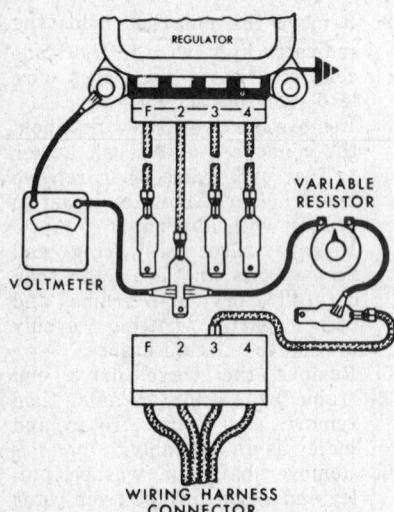

Testing field relay closing voltage
(© Chevrolet Div., G.M. Corp)

7. Remove drive end frame bearing retainer plate and bearing from end frame.
8. Bearing can be removed, if necessary, at this time. Use puller to prevent damage.
9. Disconnect the three stator leads by cutting between coils and diodes.

NOTE: diode leads can be cleaned and unsoldered, if proper heat sinks are used to prevent diode damage.

10. Remove heat sink-to-case retaining screws, then remove heat sinks. Insulated heat sink (BATT. terminal) holds positive diode.

Diode Tests

All diodes are marked with either a + or − on the head or are marked with *red* paint for + diodes, *black* paint for − diodes to identify the polarity of the case. On a generator to be used with a negative ground system, the negative case diodes are mounted into the slip ring end frame and the positive case diodes are mounted into the insulated heat sink. Diodes with a negative case have positive polarity leads, whereas positive case diodes have negative polarity leads.

Diodes can be checked for shorts or opens with an ohmmeter.

With the stator leads disconnected, connect one ohmmeter test prod to the diode lead and the other test prod to the heat sink. Reverse the test prods and note the ohmmeter readings. The meter should read high ohms in one direction, low ohms in the other. If both readings are the same, either both high or both low, the diode is faulty and must be replaced. A 1½ volt test light also will indicate a faulty diode. It will light in one direction and not in the other if the diode is good. If it lights in both directions, or in neither direction, the diode is bad.

Diode Replacement

Early Delcotrons had screwed-in diodes, as does the extremely heavy-duty 6.6 Series 4D Delcotron. Models covered here use pressed-in diodes exclusively. If there is any doubt about the year and model of the Delcotron being serviced, not the diode construction—screwed-in diodes have hexagonal heads and the later, pressed-in, units have straight sides with no hex head. Old-style, screwed-

in diodes have both right-and left-hand threads. Plus (+) diodes have left-hand and minus (−) have right-hand threads.

5.5 Series 1D

1. Support end frame on a deep socket with a larger inside diameter than the diode outside diameter.
2. Carefully *press* out the diode with a brass drift and an arbor press, or a large bench vise. Be extremely careful so as not to distort the end frame.
3. Select a new diode (red or black), noting that the red (+) diodes go into the heat sink and the black (−) diodes into the end frame.
4. Support end frame on a flat, smooth surface around diode hole and carefully *press* the new diode into position. Diode must be square when starting or both diode and frame will be ruined.

6.2 Series 2D

1. Cut leads connected to diode stem as close as possible to stem, then support end frame as for 5.5 Delcotron and press out diode.
2. Select diode with proper color marking (same as for 5.5 Delcotron), then press new diode into position.
3. Scrape enough insulation from diode stem and leads to ensure good contact, then install a sleeve over diode and place the T-clip from diode package over diode stem.
4. Place the flexible lead and stator lead (if applicable) into the T-clip, crimp clip and solder with rosin core solder only,

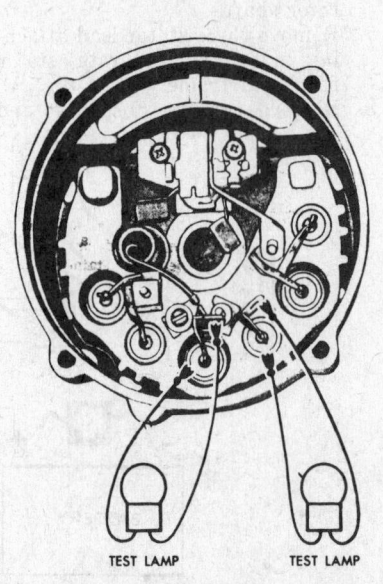

5.5" DELCOTRON

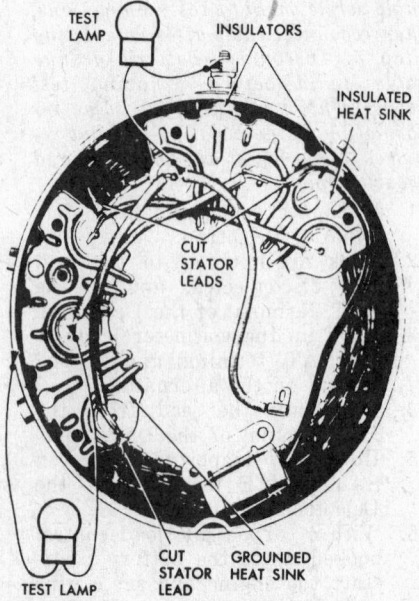

6" DELCOTRON

Testing diodes (© Chevrolet Div., G.M. Corp)

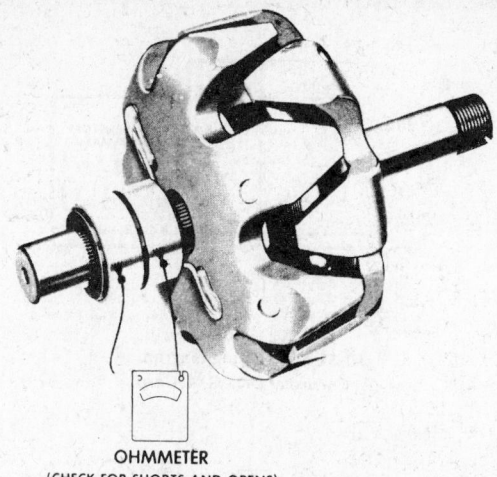

OHMMETER
(CHECK FOR SHORTS AND OPENS)

Checking rotor for grounds or opens
all Delcotron models
(© Chevrolet Div., G.M. Corp)

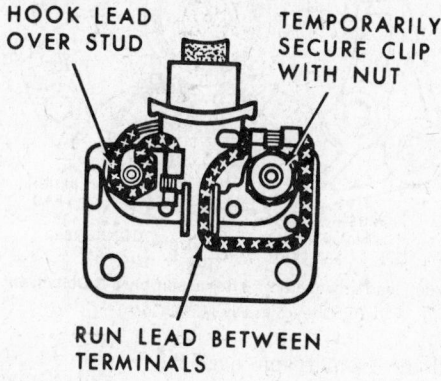

HOOK LEAD OVER STUD

TEMPORARILY SECURE CLIP WITH NUT

RUN LEAD BETWEEN TERMINALS

Brush lead arrangement during assembly 6-2
Delcotron (© Chevrolet Div., G.M. Corp)

using heat sinks (pliers) to avoid destroying diode.

5. Tape the leads together to prevent vibration damage.

Rotor Checks—All Models

The rotor may be checked electrically for grounded, open, or shorted field coils.

To check for grounds, connect a 110-volt test light from either slip ring to the rotor shaft or to the laminations. If the lamp lights, the field windings are grounded.

To check for opens, connect the leads of a 110-volt test light to each slip ring. If the lamp fails to light, the windings are open.

The windings are checked for short-circuits by connecting a battery and ammeter in series with the two slip rings. Note the ammeter reading.

An ammeter reading greater than that specified indicates shorted windings.

Since the field windings are not serviced separately, the rotor assembly must be replaced if the windings are defective.

Stator Checks—All Models

Stator windings may be checked for grounded, open, or shorted windings. If a 110-volt test lamp lights when connected from any stator lead to the stator frame, the windings are

grounded. If the lamp fails to light when successively connected between each pair of stator leads, the windings are open.

A short circuit in the stator windings is difficult to locate without laboratory equipment, due to the low resistance of the windings. However, if all other electrical checks are normal and the generator fails to supply the rated output, shorted stator windings are indicated.

Slip Ring Servicing and Replacement—All Models

Slip rings which are rough or out of round should be trued in a lathe to .001 in. maximum indicator reading. Remove only enough material to make the rings smooth and round. Finish with 400 grit or finer polishing cloth and blow away all dust.

Slip rings which must be replaced can be removed from the shaft with a gear puller, after the leads have been unsoldered. The new assembly should be pressed on with a sleeve which just fits over the shaft; this will apply all the pressure to the inner slip ring collar and prevent damage to the outer slip ring. Only pure tin solder should be used when reconnecting field leads.

Brush Replacement—All Models

The extent of brush wear can be determined by comparison with a new brush. If brushes are one-half

worn, they should be replaced.

1. Remove brush holder assembly from end frame by removing two holder assembly screws.
2. Place springs and brushes in the holder and insert straight wire or pin into holes at bottom of holder to retain brushes.
3. Attach holder assembly onto end frame.

Assembly—5.5 Series 1D

1. Install stator assembly into slip ring end frame and locate diode connectors over the relay, diode and stator leads. Tighten terminal nuts.
2. Install rotor into drive end frame.
3. Install fan, spacer, pulley washer and nut.
4. Install Allen wrench (5/16 in.) into end of shaft to hold drive shaft, then tighten pulley nut to 40-50 ft. lbs. using a crowsfoot wrench (15/16 in.) and torque wrench.
5. Assemble slip ring end frame and stator assembly to drive end frame and rotor.
6. Install four through bolts and tighten securely.

Assembly—6.2 Series 2D

1. Install stator assembly into slip ring end frame and locate diode connectors over the relay, diode and stator leads. Tighten terminal nuts.

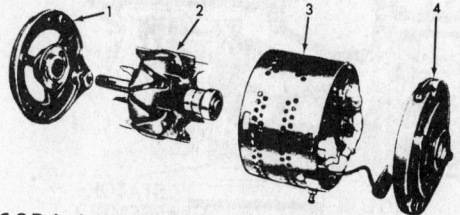

6-2 Delcotron assembly sequence: 1) drive end frame, 2) rotor 3) stator 4) end frame
(© Chevrolet Div., G.M. Corp)

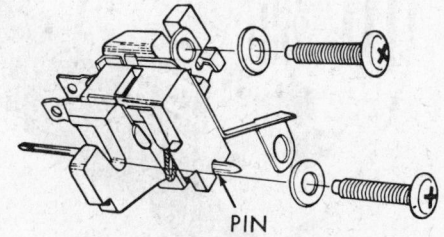

PIN

Brush Assembly—5-5 Delcotron
(© Chevrolet Div., G.M. Corp)

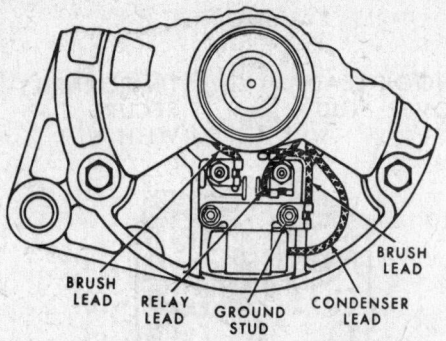

Brush lead arrangement after assembly 6-2 Delcotron
(© Chevrolet Div., G.M. Corp)

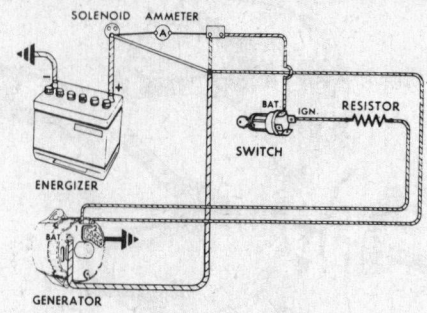

10-S1 basic wiring diagram
(© Chevrolet Div., G.M. Corp)

2. Install the front frame over the rotor.

3. Install the front frame over the key, pulley, washer and nut.

4. Clamp pulley in a padded-jaw vise and tighten shaft nut to 50-60 ft. lbs.

CAUTION: Do not clamp rotor, or segments will be distorted.

5. Position rotor and drive end frame assembly into slip ring end frame and stator. Install through bolts and tighten securely.

6. Push the brushes into the holder and secure the leads.

7. Attach brush assembly and condenser to the end frame with left-hand hex stud only.

8. Arrange the leads with the right-hand brush lead connected under the right-hand hex stud.

9. Attach terminal cover with two screws, making sure not to pinch the leads.

Delcotron 10-SI Series 100 (General Motors Corp.)

This system is an integrated AC generating system containing a built-in voltage regulator. Removal and replacement is essentially the same as for the standard AC generator.

The regulator is mounted inside the slip ring end frame. All regulator components are enclosed in an epoxy molding, and the regulator cannot be adjusted. Rotor and stator tests are the same as for the 5.5 Delcotron, covered previously.

Troubleshooting

NOTE: see the "Alternator Test Plans" section before proceeding further. Make sure that the continuous running blower, if equipped, is disconnected. This blower will run with the key on even if the blower control is off, unless disconnected.

Charging System Test—Low Charging Rate

1. After battery condition, drive belt tension, and wiring terminals and connections have been checked, charge the battery fully and perform the following test:

2. Connect a test voltmeter between the alternator BAT. terminal and ground, ignition switch on. Connect the voltmeter in turn to alternator terminals No. 1 and No. 2, the other volt-

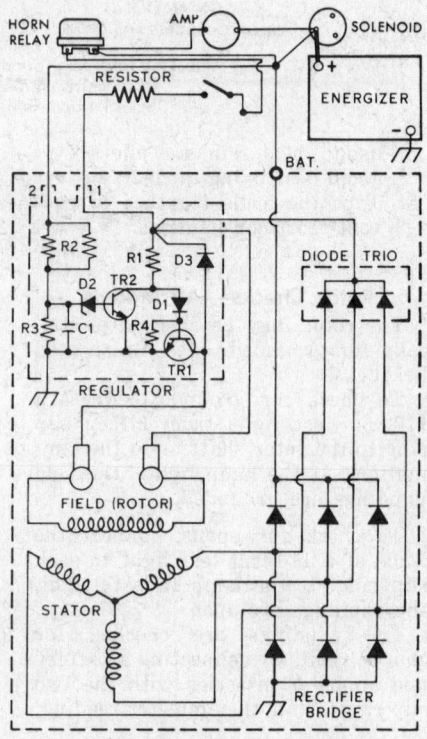

10-S1 charging circuit schematic
(© Chevrolet Div., G.M. Corp)

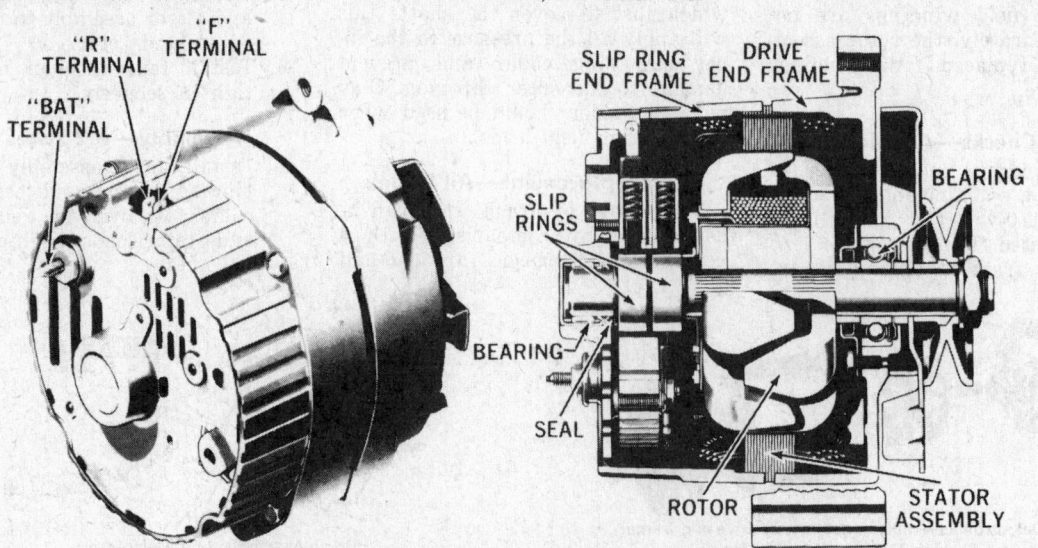

10-S1 Delcotron (© Chevrolet Div., G.M. Corp)

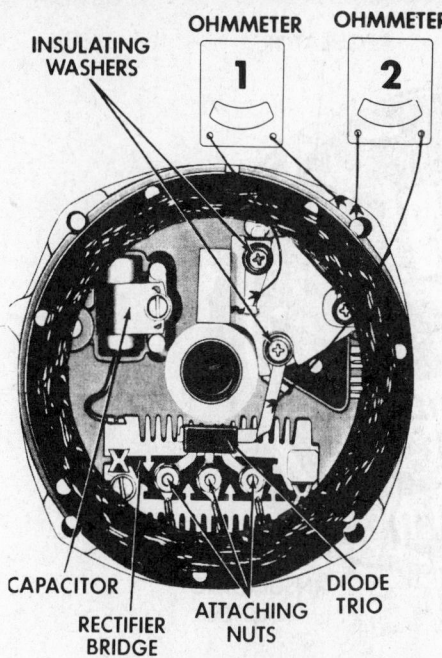

Brush lead clip ground test—
10-S1 Delcotron
(© Chevrolet Div., G.M. Corp)

meter lead being grounded as before. A zero reading indicates an open circuit between the battery and each connection at the alternator. If this test discloses no faults in the wiring, proceed to Step 3.

3. Connect the test voltmeter to the alternator BAT. terminal (the other test lead to ground), start the engine and run at 1,500-2,000 rpm with all lights and electrical accessories turned on. If the voltmeter reads 12.8 volts or greater, the alternator is good and no further checks need be made. If the voltmeter reads less than 12.8 volts, ground the field winding by inserting a screwdriver into the test hole in the end frame.

CAUTION: Do not force tab more than ¾ in. into end frame.

   a. If voltage increases to 13 volts or more, the regulator unit is defective.

   b. If voltage does not increase significantly, alternator is defective.

Alternator Output Test

1. Connect a test voltmeter, ammeter and 10 ohm 6 watt resistor into the charging circuit. Do not connect the carbon pile to the battery posts at this time.

2. Increase alternator speed and observe voltmeter—if voltage is uncontrolled with speed and increases to 16 volts or more, check for a grounded brush lead clip as covered previously. If brush lead clip is not grounded,

the voltage regulator is faulty and must be replaced.

3. Connect the carbon pile load to the battery terminals.

4. Operate the alternator at moderate speed and adjust the carbon pile to obtain maximum alternator output as indicated on the ammeter. If output is within 10% of rated output as stamped on the alternator frame, alternator is O.K. If ouput is not within specifications, ground the alternator field by inserting a screwdriver into the test hole in the end frame. If output now is within 10% of rating, replace the voltage regulator; if still not within specifications, check field winding, diode trio, rectifier bridge and stator, as described later. Disassembly of alternator up to and including Step 6 is necessary.

Disassembly and Assembly

1. Place alternator in a vise, clamped by the mounting flange only.

2. Remove the four through bolts and separate the slip ring end frame and stator assembly from the drive end and rotor assembly, using a screwdriver to pry the two sections apart. Use the slots provided for the purpose.

3. Place a piece of tape over the slip ring end frame bearing to prevent entry of dirt; also tape shaft at slip ring end to prevent scratches.

4. Clean brushes, if they are to be reused, with trichloroethylene or carbon tetrachloride solvent. Use these solvents only in an adequately ventilated area.

5. Remove the stator lead nuts and separate the stator from the end frame.

6. Remove the screw that secures the diode trio and remove diode trio.

NOTE: at this point, test the rotor, rectifier bridge, stator and diode trio if these tests are necessary.

7. Remove the rectifier bridge hold-down screw and the BAT. terminal screw, then disconnect condenser lead. Remove rectifier bridge from end frame.

8. Remove the two securing screws and brush holder and regulator assemblies. Note the insulating sleeves over the screws.

9. Remove the retaining screw and condenser from the end frame.

10. Remove the slip ring end frame bearing, if it is to be replaced, using the procedure given later in this section.

11. Remove the pulley nut, washer, pulley, fan and spacer from the rotor shaft, using a 5/16 in. Allen key to hold the shaft while loosening the nut.

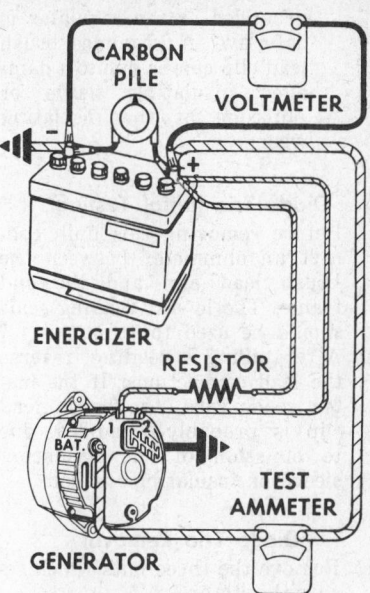

10-S1 Delcotron output test hook up
(© Chevrolet Div., G.M. Corp)

12. Remove rotor and spacers from drive end frame assembly.

13. Remove drive end frame bearing retainer plate, screws, plate, bearing, and slinger from end frame, if necessary.

14. To assemble, reverse order of disassembly. Pulley nut must be tightened to 40-50 ft. lbs.

Cleaning and Inspection

1. Clean all metal parts, except stator and rotor assemblies, in solvent.

2. Wipe off bearings and inspect them for pitting or roughness.

3. Inspect rotor slip rings for scoring. They may be cleaned with 400 grit sandpaper (not emery), rotating the rotor to make the rings concentric. Maximum out-of-true is 0.001 in. If slip rings are deeply scored, the entire rotor must be replaced as a unit.

4. Inspect brushes for wear; minimum length is ¼ in.

Charging System Test—High Charging Rate

1. With the battery fully charged, connect a voltmeter between alternator terminal no. 2 and ground. If the reading is zero, no. 2 circuit from the battery is open.

2. If no. 2 circuit is OK, but an obvious overcharging condition still exists, proceed as follows:

   a. Remove the alternator and separate the end frames.

   b. Connect a low-range ohmmeter between the brush lead clip and the end frame, as illustrated (test no. 1), then reverse the lead connections. If both readings are zero, either the brush lead clip is grounded or the regulator is defective. A grounded brush lead clip can be due to a damaged insulating sleeve or omission of the insulating washer.

Diode Trio Initial Testing

1. Before removing this unit, connect an ohmmeter between the brush lead clip and the end frame. The lowest reading scale should be used for this test.
2. After taking a reading, reverse the lead connections. If the meter reads zero, the brush lead clip is probably grounded, due to omission of the insulating sleeve or insulating washer.

Diode Trio Removal

1. Remove the three nuts which secure the stator.
2. Remove stator.
3. Remove the screw which secures the diode trio lead clip, then remove diode trio.

NOTE: The position of the insulating washer on the screw is critical; make sure it is returned to the same position on reassembly.

Diode Trio Testing

1. Connect an ohmmeter, on lowest range, between the single brush connector and one stator lead connector.
2. Observe the reading, then reverse the meter leads. Repeat this test with each of the other two stator lead connectors. The readings on each of these tests should NOT be identical, there should be one low and one high reading for each test. If this is not the case, replace the diode trio.

CAUTION: Do not use high voltage on the diode trio.

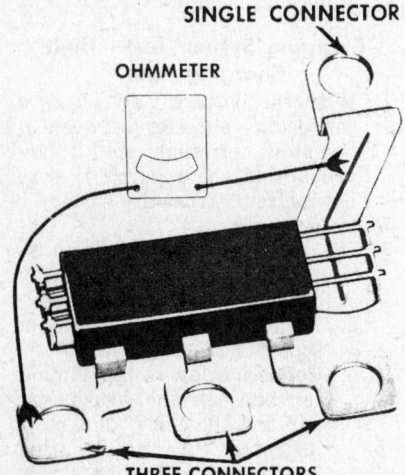

SINGLE CONNECTOR

OHMMETER

THREE CONNECTORS

Testing diode trio—10-S1 AC Generator

BRUSH HOLDER   **REGULATOR**

INSULATED HEAT SINK

GROUNDED HEAT SINK

INSULATING WASHER

OHMMETER

Rectifier bridge testing—10-S1 Delcotron

(© Chevrolet Div., G.M. Corp)

Rectifier Bridge Testing

1. Connect an ohmmeter between the heat sink (ground) and the base of one of the three terminals. Then, reverse the meter leads and take a reading. If both readings are identical, the bridge is defective and must be replaced.
2. Repeat this test with the remaining two terminals, then between the INSULATED heat sink (as opposed to the GROUNDED heat sink in previous test) and each of the three terminals. As before, if any two readings are identical, on reversing the meter leads, the rectifier bridge must be replaced.

Rectifier Bridge Removal

1. Remove the attaching screw and the BAT. terminal screw.
2. Disconnect the condenser lead.
3. Remove the rectifier bridge.

NOTE: The insulator between the insulated heat sink and the end frame is extremely important to the operation of the unit. It must be replaced in exactly the same position on reassembly.

Brush and/or Voltage Regulator R & R

1. Remove two brush holder screws and stator lead to strap nut and washer, brush holder screws and one of the diode trio lead strap attaching screws.

NOTE: The insulating washers must be replaced in the same position on reassembly.

2. Remove brush holder and

Brush holder—10-S1 Generator

brushes. The voltage regulator may also be removed at this time, if desired.

3. Brushes and brush spring must be free of corrosion and must be undamaged and completely free of oil or grease.
4. Insert spring and brushes into holder, noting whether they slide freely without binding. Insert wooden or plastic toothpick into bottom hole in holder to retain brushes.

NOTE: The brush holder is serviced as a unit; individual parts are not available.

5. Reassemble in reverse order of disassembly.

Slip Ring End Frame Bearing and Seal R & R

1. With stator removed, press out bearing and seal, using a socket or similar tool that fits inside the end frame housing. Press from outside to inside, supporting the frame inside with a hollow cylinder (large, deep socket) to allow the seal and bearing to pass.

2. The bearings are sealed for life and permanently lubricated. If a bearing is dry, do not attempt to repack it, as it will throw off the grease and contaminate the inside of the generator.

3. Using a flat plate, press the new bearing from the outside toward the inside. A large vise is a handy press, but care must be exercised so that end frame is not distorted or cracked. Again, use a deep socket to support the inside of the end frame.

4. From inside the end frame, insert seal and press flush with housing.

5. Install stator and reconnect leads.

Alternators
40 SI (100 ampere)
2600 JB (105 ampere)
2700 JB (130 ampere)
(General Motors Corp.)

These alternators feature a fully adjustable built-in, solid state voltage regulator and six silicone diodes mounted in heat sinks, to convert the alternating current, produced in the Delta wound stator, to direct current. These units use a capacitor to assist in suppressing transient voltage which could effect diode operation. A diode trio is used to supply field current to the rotor.

Troubleshooting

NOTE: refer to the "Alternator test plans" section before proceeding with any alternator tests. Because of the high output of amperage that can be produced, extreme care must be exercised in the testing and handling of these units.

40 SL Alternator

Charging System Tests—High and Low Rated Output

1. After checking the battery condition drive belt tension, and

2600 JB and 2700 JB alternators—typical
(© Chev Div., G.M.C. Corp.)

wiring terminals, charge the battery fully and perform the following tests.

2. Disconnect the negative battery cable and connect an ammeter between the positive battery post and the Battery terminal of the alternator. Connect a voltmeter between the battery terminal of the alternator and to ground. Reconnect the negative battery

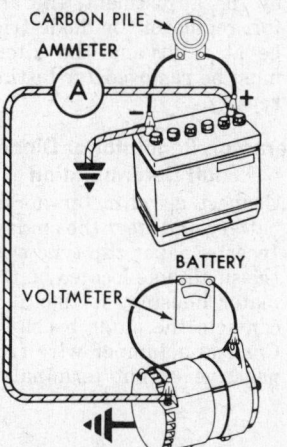

Test connections for 40 SI Delco-Remy
(©G.M.C. Corp.)

cable. The voltmeter should register battery voltage.

3. Connect a carbon pile control across the battery posts. Start the engine and operate at approximately 2000 RPMs. Turn on the vehicles' accessories and adjust the carbon pile control to obtain maximum current output.

4. If the amperage reading is within 10% of the rated output, the alternator is not defective. If the amperage is not within the 10% range, the alternator should be removed for testing and repairs.

NOTE: With the carbon pile control in the off position, and the engine operating at 2000 RPMs, the voltage reading should increase to 15 volts. If needed, the voltage can be adjusted up or down scale by raising the voltage regulator cap and relocating it in one of the new positions, which are indicated by; LO (low), 2 (med. low), 3 (med. high), and HI (high). Recheck the voltage reading after movement of the cap.

Disassembly

1. Remove the cover plate from the rear housing.

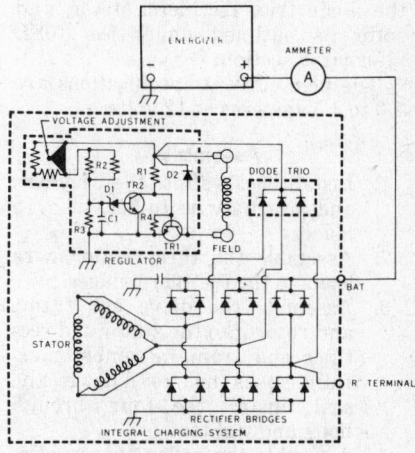

Charging circuit—model 40 SI Delco-Remy
(© G.M.C.)

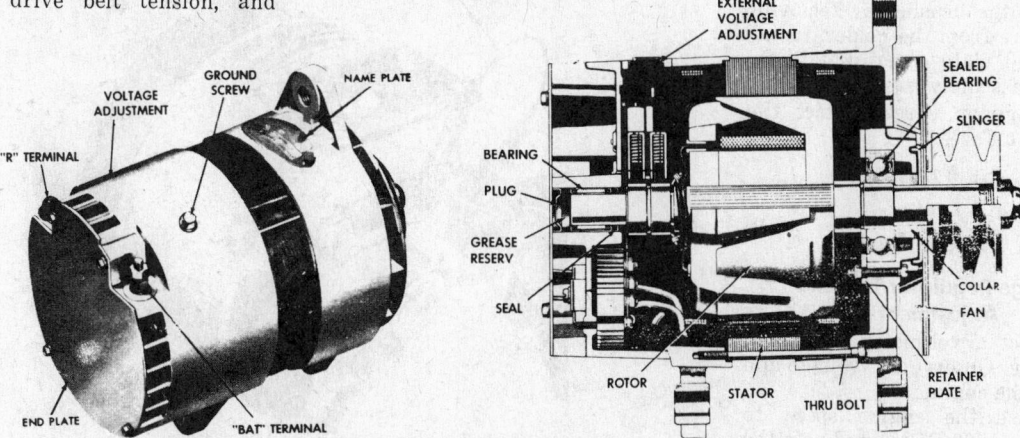

Typical 100 ampere alternator—model 40 SI Delco-Remy (© G.M.C.)

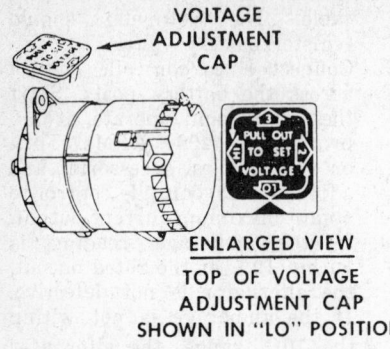

VOLTAGE
ADJUSTMENT
CAP

ENLARGED VIEW
TOP OF VOLTAGE
ADJUSTMENT CAP
SHOWN IN "LO" POSITION

Voltage adjustment cap—model 40 SI
Delco-Remy (© G.M.C.)

2. Remove the pulley nut, pulley, fan, slinger, and spacer collar.
3. Remove the four through bolts. Separate the rear housing from the stator and drive end frame by inserting a screw driver in the stator slots.
   NOTE: Expect the brushes to fall from their holders. Do not break them.
4. Disconnect the stator leads from the rectifier bridge and remove the stator from the frame.

Testing

Follow the procedure for testing the diode trio, rectifiers, stator, and rotor as outlined under the 10SL alternator section.

The field current specifications are 4.0 to 4.5 amperes at 12 volts.

Assembly

1. Install brushes into their holders and insert a wire to hold them in place.
2. Assemble the three stator wire leads to the rectifier bridges.
3. Assemble the drive end frame and rotor into the stator and rectifier end frame assembly carefully so as not to damage the seal. Install the four through bolts and tighten.
4. Assemble the collar spacer, slinger, fan, and pulley. While holding the rotor shaft, torque the pully nut to 72 ft. lbs.
5. Relase the brushes by removing the wire from the holder. Install the end plate cover. Rotate the rotor assembly to insure that no windings or wires contact the rotor surface.

2600 JB and 2700 JB Alternators
Charging System Tests

Voltage Regulator Test and Adjustment

1. Connect a voltmeter across the battery. Observe the voltage and start the engine.
2. Increase the engine speed to 1000 to 1500 RPMs and note the voltage reading. An increase to

Voltage regulator adjustment (© Chev. Div., G.M.C. Corp.)

13.6 to 14.2 volts should occur.
3. If the voltage reading is out of specifications, attempt to bring the voltage within the OK range by rotating the regulator adjusting screw back and forth.
4. If the voltage cannot be lowered by adjustment, the voltage regulator may be defective and should be replaced.
5. If the voltage cannot be raised by the adjustment, the alternator, regulator, or diode trio may be at fault and the alternator must be removed for testing and repairs.

Alternator, Regulator or Diode Trio Fault Determination

1. Connect a voltmeter across the battery and start the engine.
2. Insert a paper clip type wire into the small hole located on the regulator housing, so that it firmly contacts the outer brush holder.
3. Connect a jumper wire from the negative output terminal to the

stiff wire inserted in the regulator housing.
4. With the engine operating between 1500 to 2000 RPMs the voltmeter reading should rise above battery voltage. This indicates the alternator is good and the fault lies with the voltage regulator and/or the diode trio.
5. If the voltage reading does not increase above battery voltage, the alternator must be removed for testing and repairs.

Checking Output—Alternate Method

An alternate method used to check current output from each phase is to construct a test lamp from a two filament sealed beam bulb, connected in such a manner that the two filaments are in parallel.

The lamp should light with equal brilliancy when each phase is tested. If a dimmer light is observed between two phases, a defective diode, stator, or power diode is indicated and fur-

JUMPER
LEAD

Testing alternator with jumper wire (© Chev. Div., GMC Corp.)

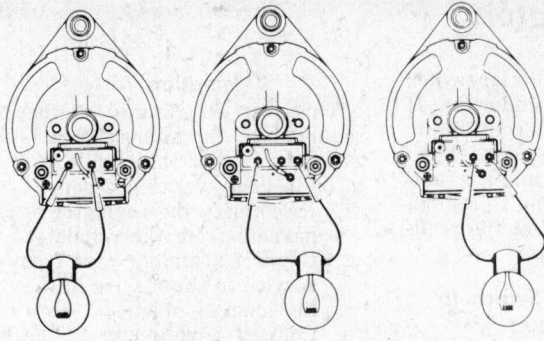

Alternate method of testing alternator output with the use of a test lamp tool (© Chev. Div., G.M.C. Corp.)

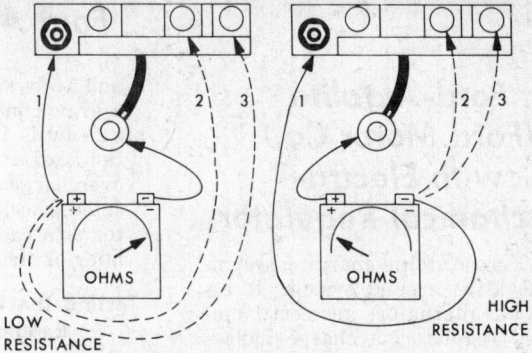

Method of testing diode trio (© Chev. Div., G.M.C. Corp.)

ther testing would have to be done.

NOTE: Point to point resistance checks with an ohmmeter on the circuit board of the voltage regulator can be misleading and inconclusive, due to the circuitry in parallel of the capacitors, resistors, diodes and transistors. To check a regulator, it is advisable to install it on an alternator known to be good, if possible. If in doubt, replace the regulator with a new unit.

Disassembly

1. Remove the voltage regulator panel from the regulator housing carefully, so as not to loose the brush springs.
2. Remove the red and black wire leads from the regulator panel. Note their positions.
3. Remove the blue lead from the diode trio assembly within the regulator housing and remove the regulator panel.
4. Remove the pulley nut, pulley, fan and spacer. Remove the through bolts and nuts and separate the rear stator housing from the front drive housing assembly.
5. Upon separation, the rotor can be removed from the front hous-

ing, and the stator leads removed from the rectifier bridges for further testing of the diodes and the stator windings. The diode trio can be removed from the outside of the regulator housing by the removal of the three retaining nuts.

Testing

Follow the procedure for testing the diode trio, rectifiers, stator, and rotor as outlined under the 10 SI alternator section.

The field current specifications are 2.8 amperes at 12 volts for the 2600 JB alternator, and 5.0 to 6.0 amperes at 12 volts for the 2700 JB alternator.

Assembly

1. Assemble the rotor into the front drive housing.
2. Connect the stator leads to the rectifier bridges and assemble the front drive housing and rotor into the stator and rear housing assembly.
3. Install the three through bolts and nuts and tighten. Rotate the rotor to insure free movement.
4. Install the diode trio to the regulator housing and secure with attaching nuts.

5. Install the brushes and springs. Compress the springs and retain them with a stiff 1/16 inch wire, placed through the hole in the regulator housing.
6. Attach the red and black leads to the regulator panel and position it on the housing and secure it loosely with the four attaching screws.
7. Remove the wire retainer from the brush springs, and tighten the regulator retaining screws.
8. Attach the blue regulator leads to the diode terminal.
9. Assemble the pulley, fan, and spacer to the rotor shaft. Torque the pulley nut to 70-80 ft. lbs.

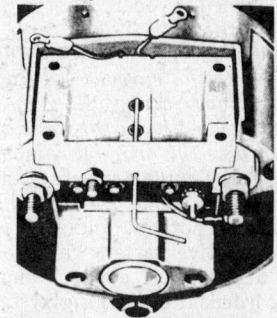

Holding brushes and springs in place with a 1/16 wire tool (© Chev. Div., G.M.C. Corp.)

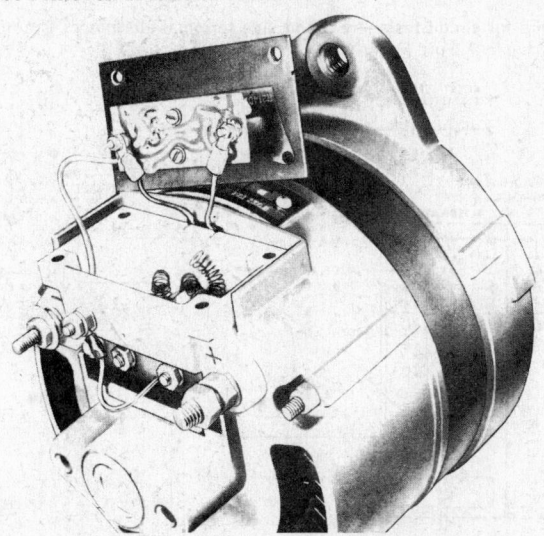

Voltage regulator panel—underside shown (© Chev. Div., G.M.C. Corp.)

Removal of diode trio (© Chev. Div., G.M.C. Corp.)

Ford Alternators

Ford-Autolite (Ford Motor Co.) with Electro-Mechanical Regulator

The Ford-Autolite charging system is a negative ground system. It includes an alternator, an electro-mechanical regulator, a charge indicator, and a storge battery.

NOTE: Late model Ford systems have replaced the electro-mechanical regulator with either a non-adjustable transistorized regulator or an adjustable transistorized regulator. The adjustable transistorized unit used with high output systems has a single, voltage limit adjusting screw under the cover. Do not use a metal screwdriver for adjustment.

Troubleshooting

NOTE: see the "Alternator Test Plans" section before proceeding further.

Fusible Links

1. Check the fusible link located between the starter relay and the alternator. Replace the link if it is burned or open.

Charging System Operation

NOTE: if the current indicator is to give an accurate reading, the battery cables must be of the same gauge and length as the original equipment.

1. With the engine running, and all electrical systems turned off, place a current indicator over the positive battery cable.
2. If a charge of about 5 amps is recorded, the charging system is working. If a draw of about 5 amps is recorded, the system is not working. The needle moves toward the battery when a charge condition is indicated,

and away from the battery when a draw condition is indicated. If a draw is indicated, continue to the next testing procedure. If an overcharge of 10-15 amps is indicated, check for a faulty regulator or a bad ground at the regulator or the alternator.

Testing the Ignition Switch to Regulator Circuit

1. Disconnect the regulator wiring harness from the regulator.
2. Turn on the key. Using a test light or voltmeter, check for voltage between the I wire and ground. Check for voltage between the A wire and ground. If voltage is present at this part of the system, the circuit is OK. If there is no voltage at the I wire, check for a burned-out charge indicator bulb, a burned-out resistor, or a break or short in the wiring. If there is no voltage present at the A wire, check for a bad connection at the starter relay or a break or short in the wire.

Isolation Test

This test determines whether the regulator or the alternator is faulty, after the rest of the circuit is found to be in good working order.

1. Disconnect the regulator wiring harness from the regulator.
2. Connect a jumper wire from the A wire to the F wire in the wiring harness plug.
3. Connect a voltmeter to the battery. The positive voltmeter lead goes to the positive terminal and the negative lead to the negative terminal. Record the reading on the voltmeter.
4. Turn off all of the electrical systems and start the engine. Do not race the engine.
5. Gradually increase engine speed to 1500-2000 rpm. The voltmeter reading should increase above the previously recorded battery voltage reading by at least one to two volts. If there is no increase, the alternator is not working correctly. If there is an increase,

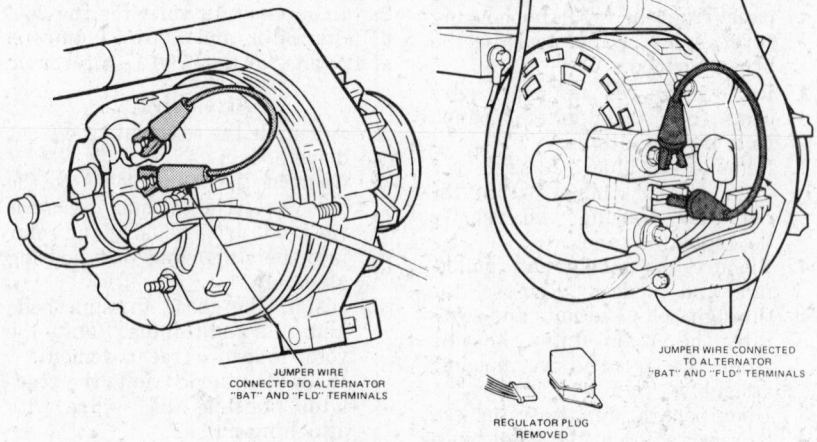

Location of jumper wire for circuit tests—rear and side terminal alternators shown
(© Ford Motor Co.)

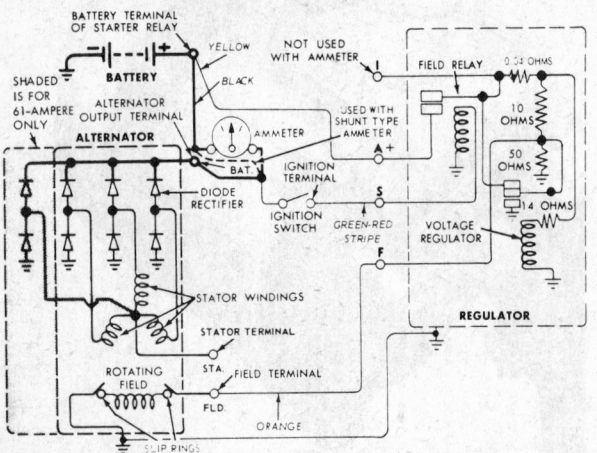
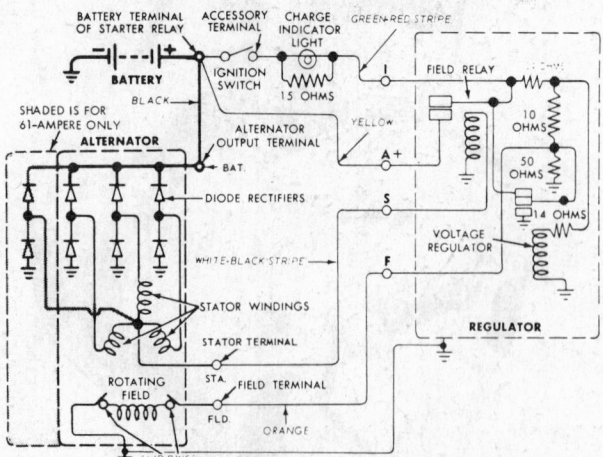

Charging system schematic with electro-mechanical regulator and charging light
(© Ford Motor Co)

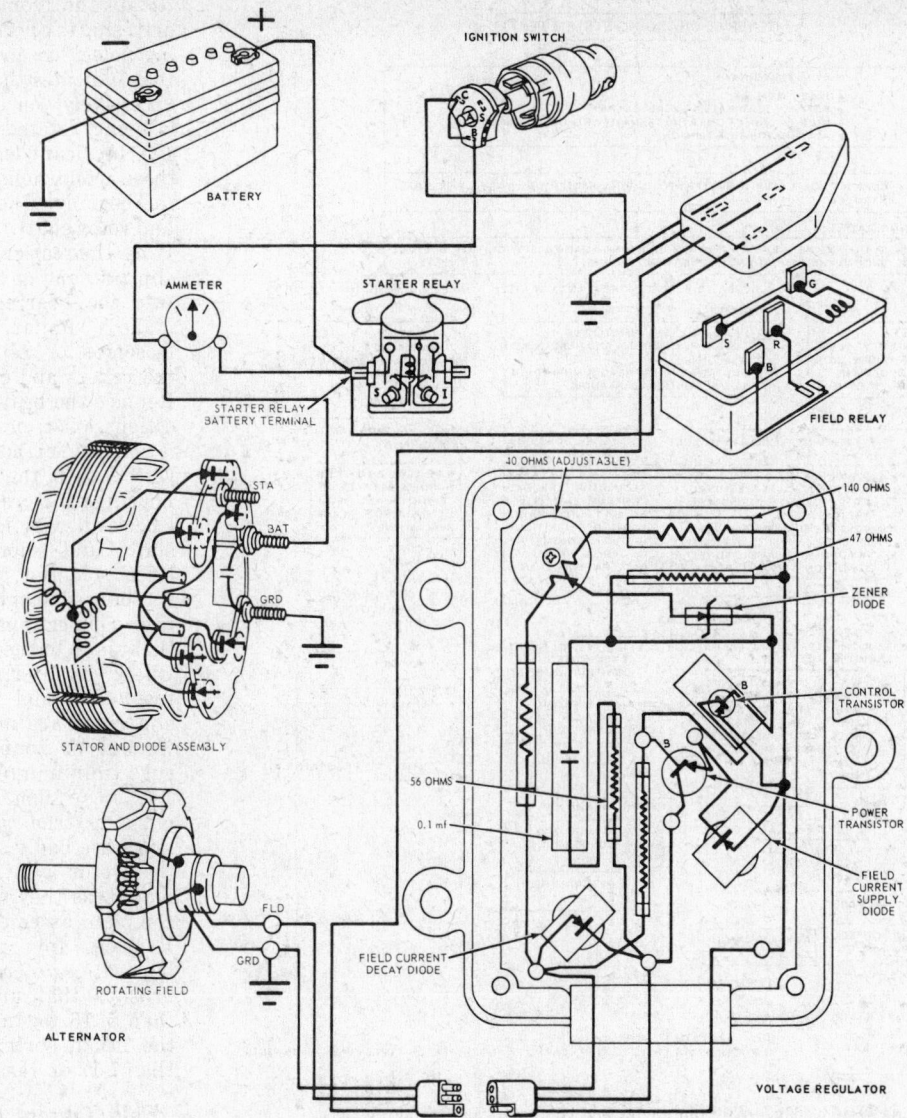

Charging system schematic with transistor regulator and ammeter (© Ford Motor Co)

the voltage regulator needs to be replaced.

Overhaul

Disassembly—Except 65, 70, 90 Amp Alternators

1. Mark both end housings with a scribe mark for assembly.
2. Remove the three housing through bolts.

3. Separate the front housing and rotor from the stator and rear housing.
4. Remove the nuts from the rectifier to rear housing mounting studs, and remove the rear housing.
5. Remove the brush holder mounting screws and the holder, brushes, springs, insulator, and terminal.

6. If replacement is necessary, press the bearing from the rear end housing, support housing on inner boss.
7. If rectifiers are to be replaced, carefully unsolder the leads from the terminals.

CAUTION: Use only a 100-watt soldering iron. Leave the soldering iron in contact with the diode terminals only long enough to remove the

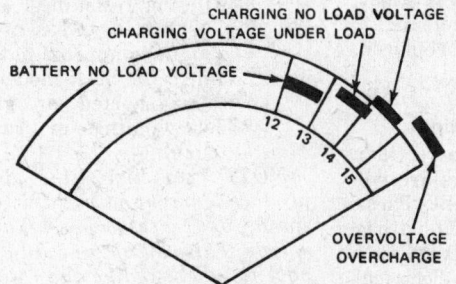

VOLTMETER TEST
TYPICAL VOLTAGE BANDS SHOWN

Voltmeter reading isolation test (© Ford Motor Co)

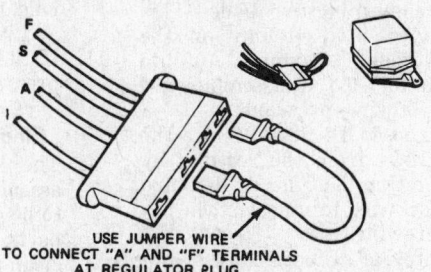

USE JUMPER WIRE
TO CONNECT "A" AND "F" TERMINALS
AT REGULATOR PLUG

USE OF JUMPER WIRE AT REGULATOR PLUG
TO TEST ALTERNATOR FOR NORMAL OUTPUT AMPS
AND FOR FIELD CIRCUIT WIRING CONTINUITY

Isolation test jumper wire (© Ford Motor Co)

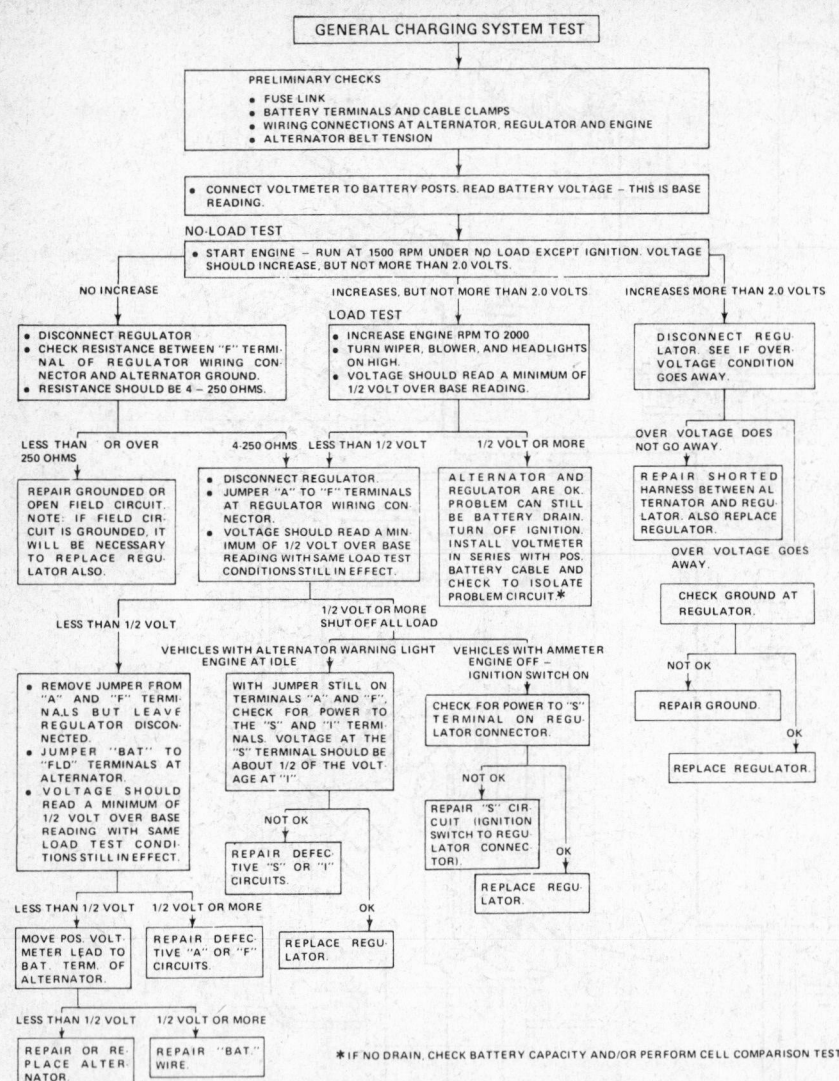

General charging system tests—With ohmmeter and voltmeter (© Ford Motor Co.)

2. Rotate the front bearing on the driveshaft. Check for any scraping noise, looseness or roughness that indicates that the bearing is excessively worn. As the bearing is being rotated, look for excessive lubricant leakage. If any of these conditions exist, replace the bearing. Check rear bearing and rotor shaft.

3. Place the rear end housing on the slip ring end of the shaft and rotate the bearing on the shaft. Make a similar check for noise, looseness or roughness. Inspect the rollers and cage for damage. Replace the bearing if these conditions exist, or if the lubricant is missing or contaminated.

4. Check both the front and rear housings for cracks.

5. Check all wire leads on both the stator and rotor assemblies for loose soldered connections, and for burned insulation. Solder all poor connections. Replace parts that show burned insulation.

6. Check the slip rings for damaged insulation and runout. If the slip rings are more than 0.0005 in. out of round, take a light cut (minimum diameter limit 1.22 in.) from the face of the rings to true them. If the slip rings are badly damaged, the entire rotor will have to be replaced, as they are serviced as a complete assembly.

7. Replace any parts that are burned or cracked. Replace brushes that are worn to less than 5/16 in. in length. Replace the brush spring if it had less than 7-12 oz. tension.

wires. Use pliers as temporary heat sinks in order to protect the diodes.

8. There are various types of rectifier assembly circuit boards installed in production. One type has the circuit board spaced away from the diode plates and the diodes are exposed. Another type consists of a single circuit board with integral diodes; and still another has integral diodes with an additional booster diode plate containing two diodes. This last type is used only on the eight-diode, 61-amp. Autolite alternator. To disassemble, use the following procedures:
   a. Exposed Diodes—remove the screws from the rectifier by rotating bolt heads 1/4 turn clockwise to unlock, then unscrewing.
   b. Integral Diodes—press out the stator terminal screw, making sure not to twist it while doing this. Do not remove grounded screw.
   c. Booster Diodes—press out the

stator terminal screw about 1/4 in., then remove the nut from the end of the screw and lift screw from circuit board, making sure not to twist it as it comes out.

9. Remove the drive pulley and fan. On alternator pulleys with threaded holes in the outer end of the pulley, use a standard puller for removal.

10. Remove the three screws that hold the front bearing retainer, and remove the front housing.

11. If the bearing is to be replaced, press from housing.

Cleaning and Inspection

1. The rotor, stator, diode rectifier assemblies, and bearings are not to be cleaned with solvent. These parts are to be wiped off with a clean cloth. Cleaning solvent may cause damage to the electrical parts or contaminate the bearing internal lubricant. Wash all other parts in solvent and dry them.

Field Current Draw Test

NOTE: alternator must be removed from the truck.

1. Connect a test ammeter between the alternator frame and the positive post of a 12-volt test battery.

2. Connect a jumper wire between the negative test battery post and the alternator field terminal.

3. Observe the ammeter:
   a. Little or no current flow indicates high brush resistance, open field windings, or high winding resistance.
   b. Current in excess of specifications (approximately 2.9 amps. for most models) indicates shorted or grounded field windings, or brush leads touching.

NOTE: sometimes the alternator produces current output at low engine speeds, but ceases to put out at higher speeds. This can be caused by centrifugal force expanding the rotor windings to the point where they short to ground. Place in a test stand and check field current draw while spinning alternator.

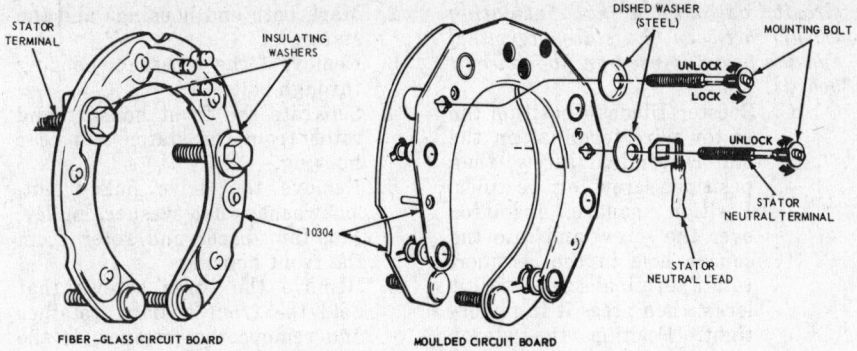

Rectifier assembly (© Ford Motor Co)

FIBER –GLASS CIRCUIT BOARD

MOULDED CIRCUIT BOARD

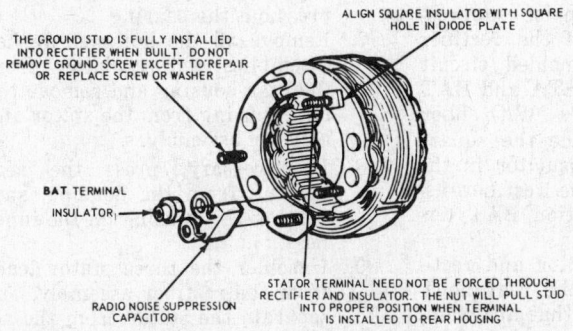

Terminal insulators—fiber circuit board (© Ford Motor Co)

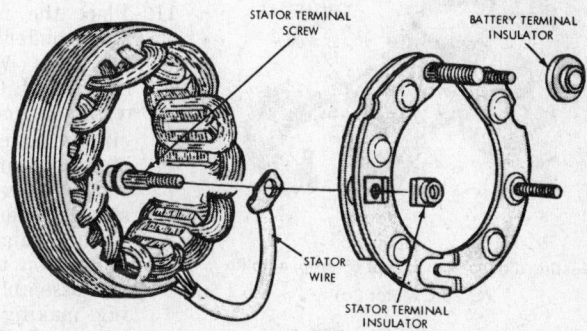

Stator and rectifier assembly—61 amp. booster diode model
(© Ford Motor Co)

press a new one into rear housing.

7. Assemble brushes, springs, terminal and insulator in the brush holder, retract the brushes and insert a short length of 1/8 in. rod or stiff wire through the hole in the holder to hold the brushes in the retracted position.

8. Position the brush holder assembly in the rear housing and install mounting screws. Position brush leads to prevent shorting.

9. Wrap the three stator winding leads around the circuit board terminals and solder them using only rosin core solder and a 100-watt iron. Position the stator neutral lead eyelet on the stator terminal screw and install the screw in the rectifier assembly.

10. A. Exposed Diodes—insert the special screws through the wire lug, dished washers and circuit board. Turn 1/4 turn counterclockwise to lock in place.
   B. Integral Diodes—insert the screws straight through the holes.

Diode Tests

Disassemble the alternator. Disconnect diode assembly from stator and make tests. To test one set of diodes, contact one ohmmeter probe to the diode plate and contact each of the three stator lead terminals with the other probe. Reverse the probes and repeat the test. All six tests (eight for 61 amp. Autolite eight-diode models) should show a reading of about 60 ohms in one direction and infinite ohms in the other. If two high readings, or two low readings, are obtained after reversing probes the diode is faulty and must be replaced.

Stator Tests

Disassemble the stator from the alternator assembly and rectifiers. Connect test ohmmeter probes between each pair of stator leads. If the ohmmeter does not indicate equally between each pair of leads, the stator coil is open and must be replaced.

Connect test ohmmeter probes between one of the stator leads and the stator core. The ohmmeter should not show any reading. If it does show continuity, the stator winding is grounded and must be replaced.

Assembly—Except 65, 70, 90 Amp Alternators

1. Press the front bearing into the front housing boss, putting pressure on outer race only. Install bearing retainer.

2. If the stop ring on the driveshaft was damaged, install a new stop ring. Push the new ring onto the shaft and into the groove.

3. Position the front bearing spacer on the driveshaft against the stop ring.

4. Place the front housing over the shaft, with the bearing positioned in the front housing cavity.

5. Install fan spacer, fan, pulley, lockwasher and retaining nut and tighten nut to 60-100 ft. lbs. holding the drive shaft with an Allen key.

6. If rear bearing was removed,

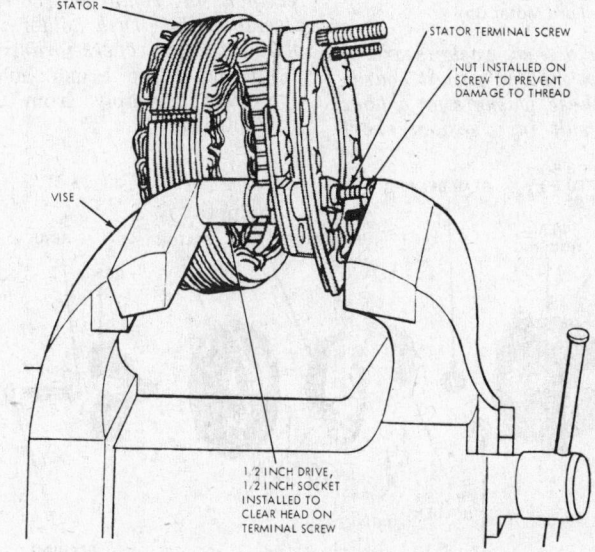

Stator Terminal screw removal—61 amp. booster diode model
(© Ford Motor Co)

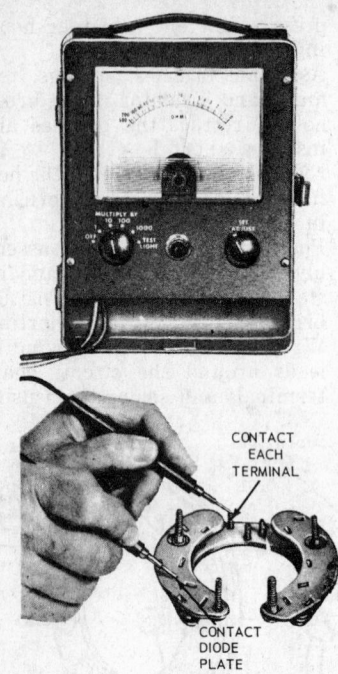

Testing diodes—all except 65 amp. autolite
(© Ford Motor Co)

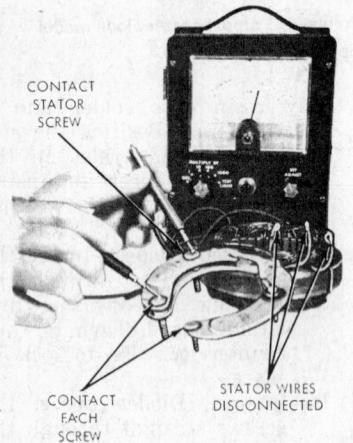

Testing diodes—65 amp. Autolite
(© Ford Motor Co)

NOTE: the dished washers are to be used on the molded circuit boards only. Using these washers on a fiber board will result in a serious short

circuit, as only a flat insulating washer between the stator terminal and the board is used on fiber circuit boards.

    C. Booster Diodes—position the stator wire terminal on the stator terminal screw, then position screw on rectifier. Position square insulator over the screw and into the square hole in the rectifier, rotate terminal screw until it locks, then press it in finger-tight. Position the stator wire, then press the terminal screw into the rectifier and insulator with a vise.

11. Place the radio noise suppression condenser on the rectifier terminals. With molded circuit board, install the STA and BAT terminal insulators. With fiber circuit board, place the square stator terminal insulator in the square hole in the rectifier assembly, then position BAT terminal insulator.

    Position the stator and rectifier assembly in the rear housing, making sure that all terminal insulators are seated properly in the recesses. Position STA, BAT and FLD insulators on terminal bolts; install nuts.

12. Clean the rear bearing surface of the rotor shaft with a rag, then position rear housing and stator assembly over rotor. Align matchmarks made during disassembly and install through bolts. *Remove brush retracting wire and place a dab of silicone sealer over the hole.*

Disassembly (typical) 65, 70, 90 Ampere Alternators

NOTE: When disassembling the side terminal alternator, the brush holder would be removed after the rectifier is removed. During the assembly, the brush holder would be installed in the reverse order.

1. Remove the brush holder and cover assembly from the rear housing.

2. Mark both end housings and the stator.
3. Remove the three housing through bolts.
4. Separate the front housing and rotor from the stator and rear housing.
5. Remove the drive pulley nut, lockwasher, flat washer, pulley, fan, fan spacer and rotor from the front housing.
6. Remove the three screws that hold the front bearing retainer and remove the retainer. If the bearing is damaged or has lost its lubricant, support the housing close to the bearing boss and press out the bearing.
7. Remove all the nut and washer assemblies and insulators from the rear housing and remove the rear housing from the stator and rectifier assembly.
8. If necessary, press the rear bearing from the housing, supporting the housing on the inner boss.
9. Unsolder the three stator leads from the rectifier assembly, and separate the stator from the assembly. Use a 200-watt soldering iron.
10. Perform a diode test and an open and grounded stator coil test.

Cleaning and Inspection

Nicks and scratches may be removed from the rotor slip rings by turning down the slip rings. Do not go beyond the minimum diameter limit of 1.22 in. If the slip rings are badly damaged, the entire rotor must be replaced. The rectifier also is serviced as an assembly. See Lower Ampere Alternator Section for test procedures.

Assembly—65, 70, 90 AMP

1. If the front bearing is being replaced, press the new bearing into the bearing boss, putting pressure on the outer race only. Install the bearing retainer and tighten the retainer screws until the tips of the retainer touch the housing.

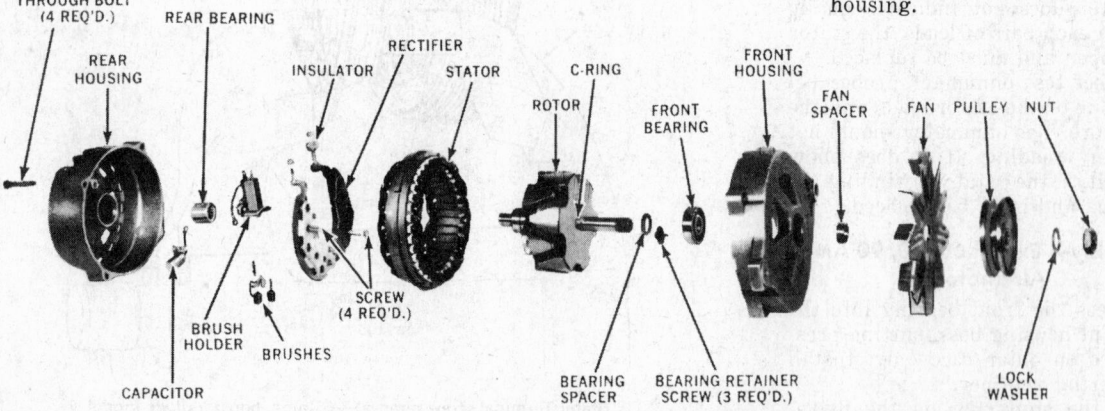

Exploded view—side terminal alternator (© Ford Motor Co)

Exploded view—70 ampere rear terminal alternator—Typical 90 ampere alternator
(© Ford Motor Co.)

2. Position the rectifier assembly to the stator, wrap the three stator leads around the diode plate terminals and solder them using a 200-watt soldering iron.
3. If the rear housing bearing was removed, press in a new bearing from the inside of the housing, putting pressure on the outer race only.
4. Install the BAT-GRD insulator, and position the stator and rectifier assembly in the rear housing.
5. Install the STA (purple) and BAT (red) terminal insulators on the terminal bolts and install the nut and washer assemblies. *Make certain that the shoulders on all insulators, both inside and outside of the housing, are seated properly before tightening the nuts.*
6. Position the front housing over the rotor and install the fan spacer, fan, pulley, flat and lockwasher and nut on the rotor shaft.
7. Wipe the rear bearing surface of the rotor shaft with a clean rag.
8. Position the rotor with the front housing into the stator and near housing assembly, and align the matchmarks made during disassembly. Seat the machined portion of the stator core into the step in both housings and install the through bolts.
9. If the field brushes have worn to less than ⅜ in., replace both brushes. Hold the brushes in position by inserting a stiff wire into the brush holder.
10. Position the brush holder assembly into the rear housing and install the three mounting screws. Remove the brush retracting wire and put a dab of silicone cement over the hole.

Brush Replacement—65, 70, 90 AMP

1. Remove the brush holder and cover assembly from the rear housing.
2. Remove the terminal bolts from the brush holder and cover assembly, then remove the brush assemblies.
3. Position the new brush terminals on the terminal bolts and assemble the terminals, bolts, brush holder washers and nuts. The insulating washer mounts under the FLD terminal nut. The entire brush and cover assembly also is available for service.
4. Depress the brush springs in the brush holder cavities and insert the brushes on top of the springs. Hold the brushes in position by inserting a stiff wire in the brush holder as shown. Position the brush leads as shown.
5. Install the brush holder and cover assembly into the rear housing. Remove the brush retracting wire and put a dab of silicone cement over the hole.

Autolite Alternator with Integral Regulator

Description

Some vehicles are equipped with an Autolite alternator having an integral regulator mounted to the rear end housing. The regulator is a hybrid unit featuring use of solid state integrated circuits. These circuits may consist of transistors, diodes and resistors. The unusual feature of this type of micro-electronic circuit is that the entire circuit is within a silicone crystal approximately ⅛ in. square. Because of the small size of the circuit, it is not repairable or adjustable and must be replaced as a unit if found to be defective. It should be noted that the size of the regulator housing is dictated only by the fact that some means of connecting the regulator to the alternator is necessary. Overhaul is the same as for other Autolite alternators.

Troubleshooting

NOTE: see the "Alternator Test Plans" section before proceeding further.

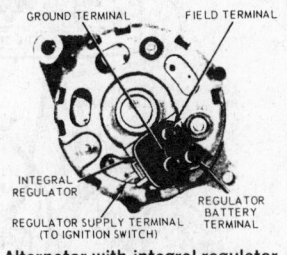

Alternator with integral regulator
(© Ford Motor Co)

Fusible Links

1. Check the fusible link located between the starter relay and the alternator. Replace the link if it is burned or open.

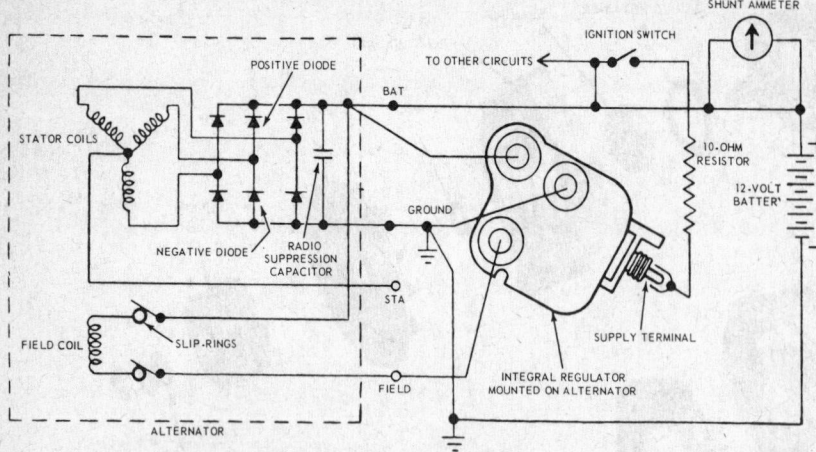

Charging system schematic with integral regulator (© Ford Motor Co)

Output Test

1. Place transmission in Neutral or Park.
2. Remove the positive battery cable and install a battery adapter switch in the line.
3. Attach one lead of a test voltmeter to the negative battery post and the other test lead to the circuit side of the adapter switch.
4. Connect a test ammeter to each side of the adapter switch, so that charging current will go through the ammeter when the switch is opened.
5. Connect a jumper wire between the alternator frame and the integral regulator field terminal (cover plug removed).
6. Close adapter switch, start engine and open adapter switch.
7. Running engine at 2,000 rpm, observe voltmeter and ammeter. At 15 volts indicated, the ammeter should read 50-57 amps. If so, and there is still a no-charge condition, the regulator is proba-

bly faulty and must be replaced. An output 2-8 amps. below 50 amps. usually indicates an open diode rectifier, while an output 10-15 amps. below minimum specifications usually indicates a shorted diode. An alternator with a shorted diode usually will whine at idle speed.

Field Test (Voltmeter)

1. Turn ignition switch to OFF position.
2. Remove wire from regulator supply terminal.
3. Remove cover plug from regulator field terminal and connect one test voltmeter lead to this terminal. A $\frac{1}{4}$ ohm resistor should be in the circuit.
4. Connect the other test voltmeter lead to a good engine ground.
5. The voltmeter should read 12 volts. If *no* voltage is present, the field circuit is open or grounded.
6. If voltmeter reads more than I

volt, but still less than battery voltage, there is probably a partial ground in the alternator field circuit and the circuit should be checked with an ohmmeter.

Field Test (Ohmmeter)

1. Disconnect battery ground cable; remove alternator from truck.
2. Remove the regulator from the alternator (covered later).
3. Make the ohmmeter tests as illustrated. If any of the tests indicates a field circuit problem, disassemble the alternator to further isolate the trouble.
   a. Contact each ohmmeter probe to a slip ring. Resistance should be 4-5 ohms. A higher reading indicates a damaged slip ring soldered connection or a broken wire. A lower reading indicates a shorted wire or slip ring assembly.
   b. Contact one ohmmeter probe to a slip ring and the other probe to the rotor shaft. Any reading other than infinite ohms indicates a short to ground. If neither of these tests (A and B) isolates the trouble, the brushes or brush assembly are the probable cause.

Voltage Limiter Test

1. Check the battery specific gravity. If it is not at least 1.230, charge the battery or install a charged battery for the test.
2. Make sure all lights and accessories are turned off, including such items as dome lights.
3. Make the test connections as illustrated.

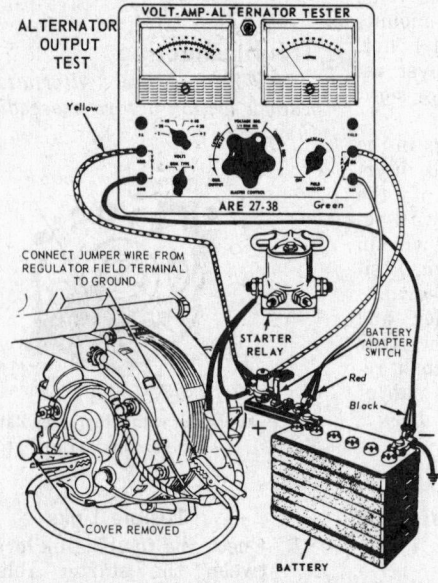

Output test hook-up integral regulator alternator
(© Ford Motor Co)

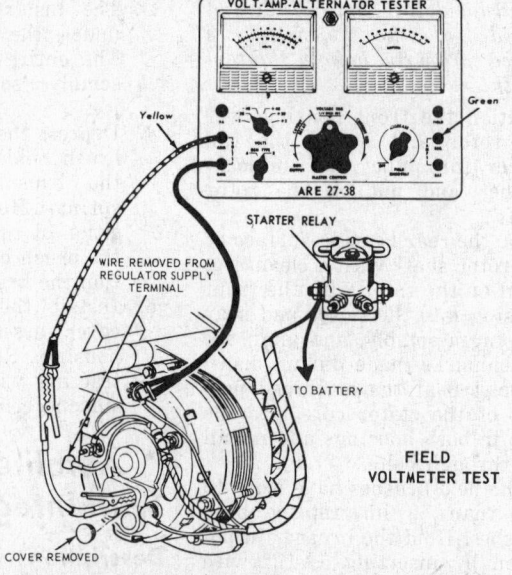

Field voltmeter test hook-up—integral regulator alternator
(© Ford Motor Co)

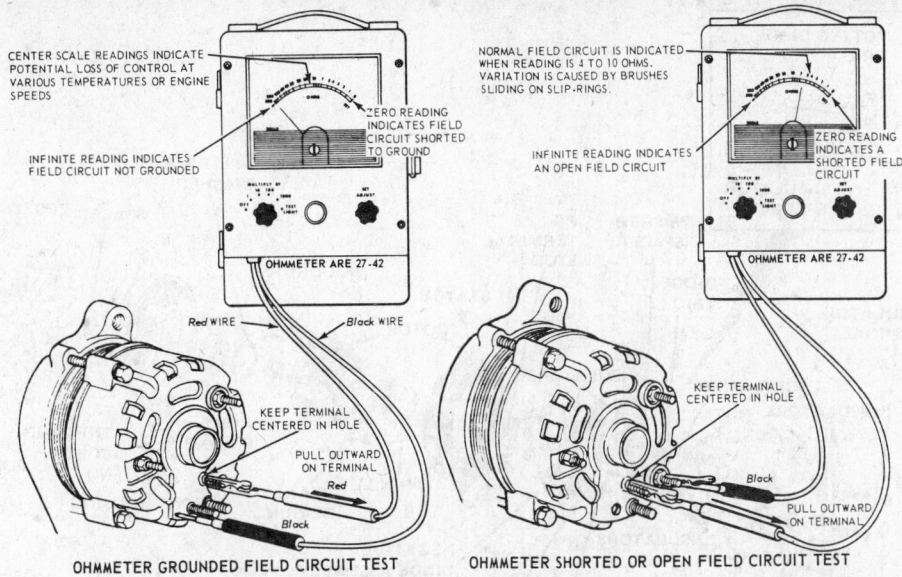

Field circuit test hook-up with ohmmeter—integral regulator alternator (© Ford Motor Co)

4. Place transmission in Neutral or Park, close battery adapter switch and start the engine.

5. Open the battery adapter switch and operate engine at 2,000 rpm for 5 minutes. The voltmeter should read 13.3-15.3 volts.

6. If voltage does not rise above 12 volts, perform a regulator supply voltage test to determine whether or not the regulator is getting voltage from the battery. Before replacing a regulator, check the wiring of the entire charging system for shorts, opens, or high resistance connections.

Regulator Supply Voltage Test

The regulator is "turned on" by the application of battery voltage through a 10 ohm resistor wire. If the supply circuit is defective, the regulator will not function and the alternator will not put out current.

1. Connect a 12-volt test light or voltmeter between the regulator supply lead and ground.

2. Turn on the ignition switch. The test light should glow or the voltmeter indicate. If not, the supply circuit should be checked back to the battery, especially the resistance wire.

Overhaul

The overhaul procedures for the alternator are the same as for the Ford Autolite electro-mechanical alternator.

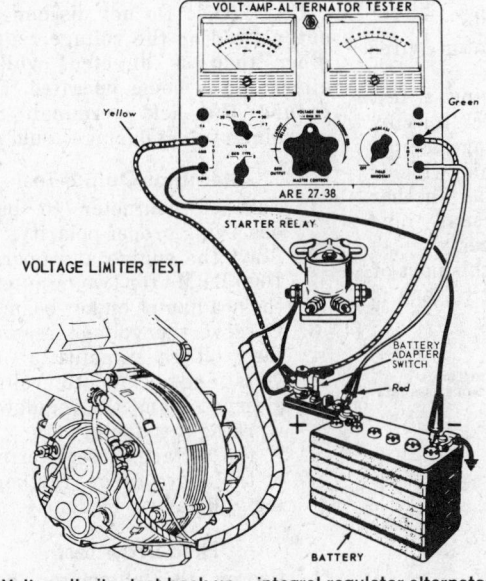

Voltage limiter test-hook-up—integral regulator alternator
(© Ford Motor Co)

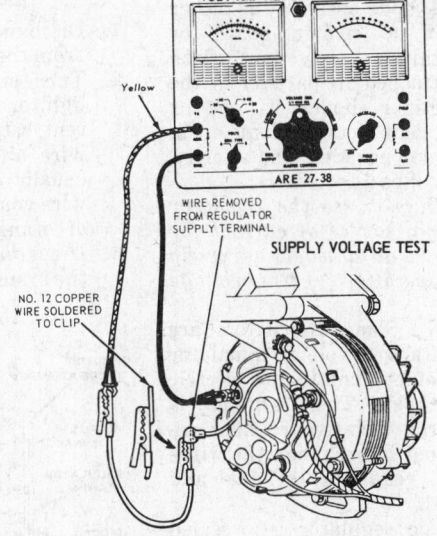

Supply voltage test hook-up—integral regulator alternator
(© Ford Motor Co)

Motorola System

The Motorola alternator is an electro-mechanical device producing alternating current, which is changed to direct current by the rectifier diodes, accomplished by the characteristics of the diodes to allow current to flow in one direction only.

A three phase stator winding is used, a "Wye" type for the 37 ampere

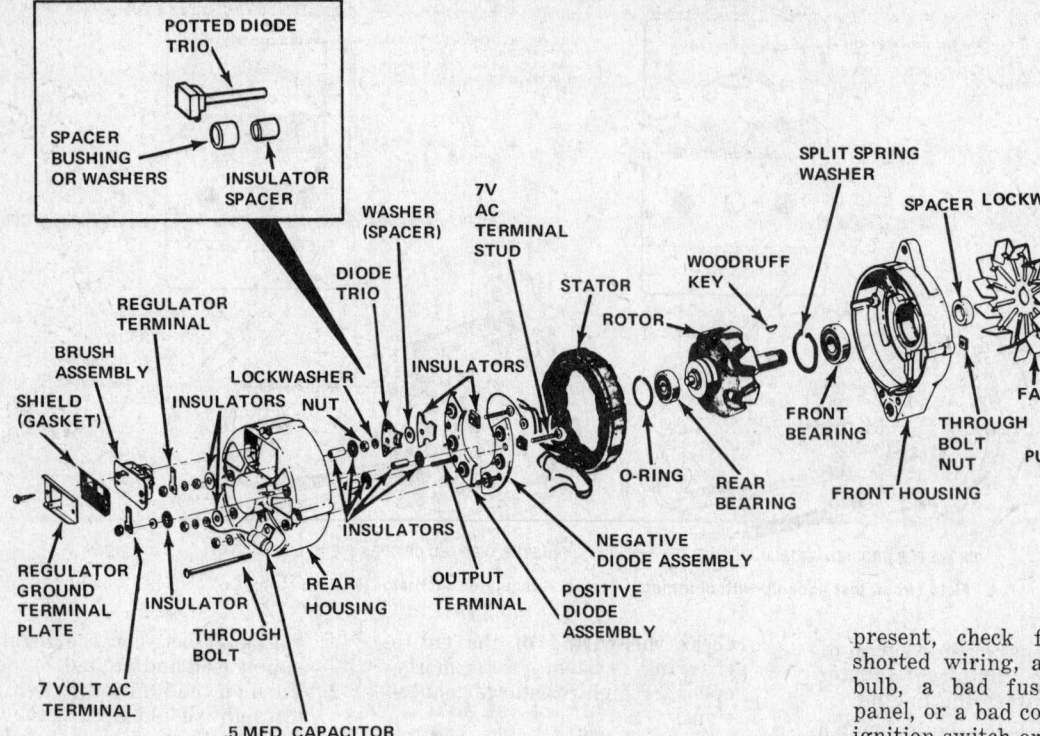

POTTED DIODE TRIO

SPACER BUSHING OR WASHERS — INSULATOR SPACER

REGULATOR TERMINAL

BRUSH ASSEMBLY

SHIELD (GASKET)

INSULATORS

LOCKWASHER NUT

WASHER (SPACER)

DIODE TRIO

7V AC TERMINAL STUD

STATOR

ROTOR

WOODRUFF KEY

SPLIT SPRING WASHER

SPACER LOCKWASHER

NUT

FAN

INSULATORS

O-RING

REAR BEARING

FRONT BEARING

THROUGH BOLT NUT

PULLEY

FRONT HOUSING

REGULATOR GROUND TERMINAL PLATE

INSULATOR

THROUGH BOLT

REAR HOUSING

INSULATORS

OUTPUT TERMINAL

NEGATIVE DIODE ASSEMBLY

POSITIVE DIODE ASSEMBLY

7 VOLT AC TERMINAL

.5 MFD CAPACITOR

Motorola alternator—exploded view (© Jeep Corp.)

rated alternator, and a "Delta" type for the 51 and 55 ampere rated alternators.

A field diode assembly is used to provide the excitation current to the rotor (field) windings when the alternator is operating and is sensed and regulated by the voltage regulator to control the output of the alternator.

The field diode assembly is either mounted on a circuit board or encased within a epoxy "pot" with the leads attached in parallel to the positive rectifier diodes. If one or more of the field diodes become open, shorted, or downgraded, the alternator output will be affected.

NOTE: Do not use the regulator terminal for a source of current for any reason. To do so would adversely affect the operation of the voltage regulator.

CAUTION: Some alternators are equipped with a 7 volt terminal for the supply of current to the electric automatic choke. This terminal is located on the negative rectifier assembly. Do not interchange the wires between the regulator terminal and this terminal.

The voltage regulator is a sealed unit and requires no adjustment. Replacement of the unit is required if the regulator becomes defective.

Troubleshooting

NOTE: see the "Alternator Test Plans" section before proceeding further.

Fusible Link Test

There are many fuse links in the truck however the fuse link located in the wiring between the battery terminal of the horn relay to the main wire harness is the only one that concerns the charging system. This link protects the entire wiring harness. If it fails, all the electrical systems will fail to function.

Testing the Ignition Switch to Regulator Circuit

1. Disconnect the regulator wires from the regulator.
2. Turn on the key. Using a test light or voltmeter, check for current between the voltage supply wire and ground. This wire is usually orange and has another wire connected to it, usually blue or orange with a tracer.
3. If current is present, this part of the system is OK. If no voltage is

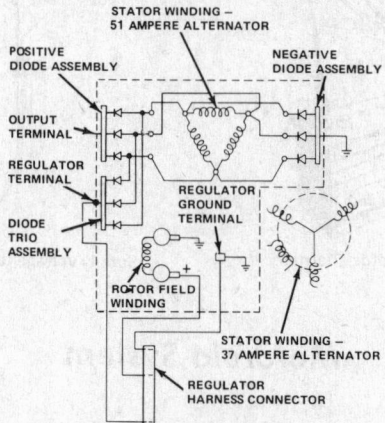

STATOR WINDING – 51 AMPERE ALTERNATOR

POSITIVE DIODE ASSEMBLY

NEGATIVE DIODE ASSEMBLY

OUTPUT TERMINAL

REGULATOR TERMINAL

REGULATOR GROUND TERMINAL

DIODE TRIO ASSEMBLY

ROTOR FIELD WINDING

STATOR WINDING – 37 AMPERE ALTERNATOR

REGULATOR HARNESS CONNECTOR

Motorola alternator internal circuits (© Jeep Corp.)—**51 and 55**

present, check for broken or shorted wiring, a bad indicator bulb, a bad fuse in the fuse panel, or a bad connection at the ignition switch or on the battery side of the starter relay.

Alternator Tests (In vehicle)

NOTE: Various types of charging system testers are available to perform the tests necessary to determine if the system or components are defective. Follow the manufactures instructions for the tester being used, as the following charging system tests are generalized.

CAUTION: Do not disconnect the output lead or the voltage regulator, other than as directed, while the alternator is being operated. Do not ground the field terminal. Severe charging system damage could result.

Alternator Output Test

1. Connect voltmeter to battery, observing proper polarity.
2. Start the engine and operate at 1000 RPMs for two minutes with the headlamps on low beam.
3. Observe the voltage reading. If the voltage remains above 13 volts and below 15 volts, the alternator and the regulator are working satisfactorily.
4. If the voltage is registering out of the above range, further testing will have to be done.

Field Draw Test

1. Loosen the alternator belt so that the rotor can be turned by hand.
2. Connect ammeter leads between the positive battery post and the positive brush post on the alternator.
3. The ammeter should register a reading within a range of 1½ to 3 amperes and if by turning the

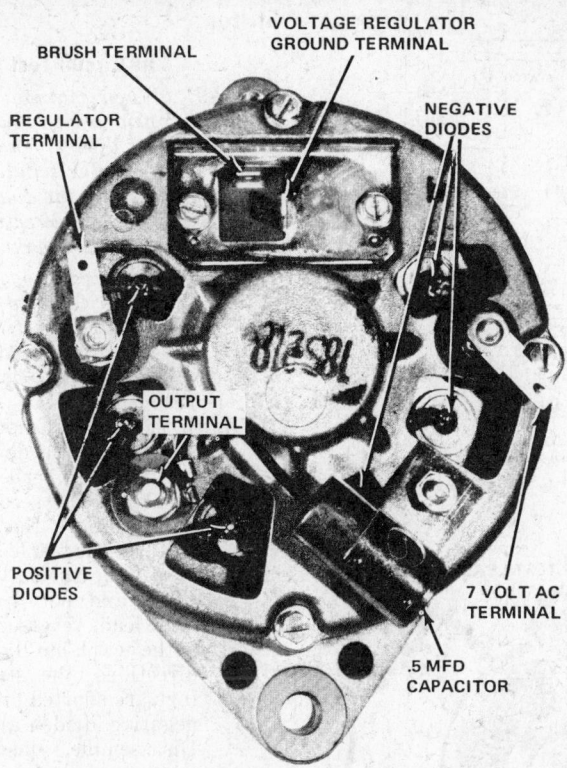

Terminal locations—Motorola alternator (© Jeep Corp.)

rotor by hand, the reading varies within the scale, the brushes and the slip rings require cleaning or repairs.

4. If the readings are too high or too low, the alternator should be removed and disassembled for further tests and repairs.

Regulator Bypass Test

1. Connect a volt meter to the battery, observing the proper polarity. Disconnect the voltage regulator.
2. Start the engine and allow to idle.
3. Connect an ammeter lead between the positive battery post to the alternator positive brush terminal.
4. Increase the engine speed while observing the voltage reading. A reading of 16 volts should be obtained, if the alternator is not defective.

NOTE: Do not allow the voltage to increase over 16 volts, as damage to the charging system can result.

Field Diode Assembly Test

NOTE: A shorted or open field diode assembly will cause reduced alternator output and require unit disassembly and removal of the diode assembly for testing. A downgrading of one or more of the diodes will cause the dash indicator bulb to glow dimly, but will normally not effect the alternator output.

1. Start the engine and operate at idle speed.

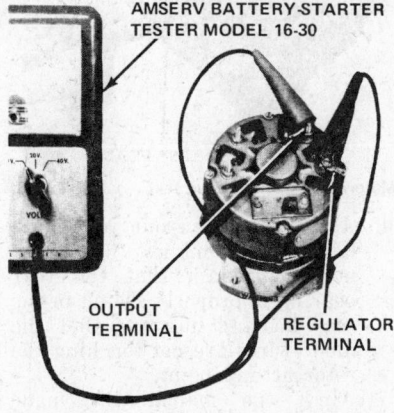

Motorola field diode test (© Jeep Corp.)

2. With the voltmeter adjusted to the low scale, connect the leads to the alternator output terminal and the negative lead to the regulator terminal.
3. Turn the blower motor to the high position and turn the headlamps to the high beam position for approximately two minutes of operation. This causes the diode assembly to heat up due to the electrical load.
4. Observe the reading on the voltmeter. A range of 0 to 0.2 volts indicates the diode assembly is good. A reading above 0.2 volts indicates the downgrading of the diode assembly although it is not necessary to replace the assembly unless the reading is over 0.6 volts.

5. A pulsating reading on the meter indicates a positive diode of the rectifier or a soldered connection is breaking down under heat, and the alternator will have to be disassembled for testing repairs.
6. If the reading is over 0.6 volts and the alternator output was deemed satisfactory in the earlier tests, a bench test of the diode assembly will have to be made.
7. If the dash indicator bulb remains on dimly after a satisfactory diode assembly test has been made, inspect the following locations for loose or corroded connections.
   a. alternator output terminal
   b. starter relay battery terminal
   c. ignition switch
   d. fuse panel
   e. instrument harness connections
   f. instrument cluster printed circuits
   g. indicator bulb socket
   h. main wiring harness connectors

Alternator

Disassembly

1. Remove the two self-tapping screws and the cover. Pull the brush assembly straight up to clear the locating pins, then lift out the brush assembly.
2. Scribe a matchmark across the front housing, stator, and rear housing. Remove the four thrubolts and nuts, then carefully separate the rear housing and stator from the front housing using two screwdrivers in the slots provided.

CAUTION: Do not insert screwdrivers deeper than 1/16 in., to avoid damaging stator winding.

3. Remove the four locknuts and insulating washers that hold the stator and diode assembly, then separate the assembly from the rear housing. Avoid bending the stator wires—do not unsolder the wires without using pliers as a heat sink.
4. There is no reason to remove the rotor from the front housing unless there is a defect in the field coil or front bearing. Front and rear bearings are lubricated for life and sealed and, as a rule, do not go bad unless the drive belt has been adjusted with too much tension. If the rotor must be removed, use a puller to remove the front drive pulley, then unseat the split-ring washer using long-nose pliers through the front housing to compress the washer while pulling on the rotor. Tap the rotor shaft lightly to remove the rotor and front

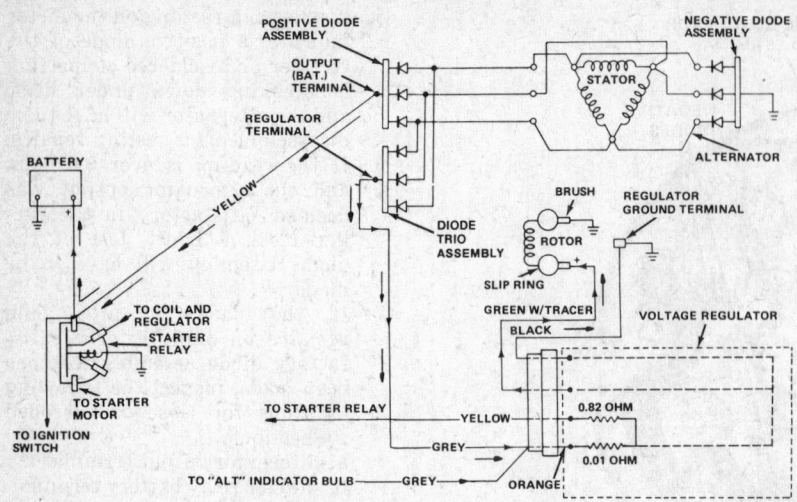

Motorola charging system (© Jeep Corp.)

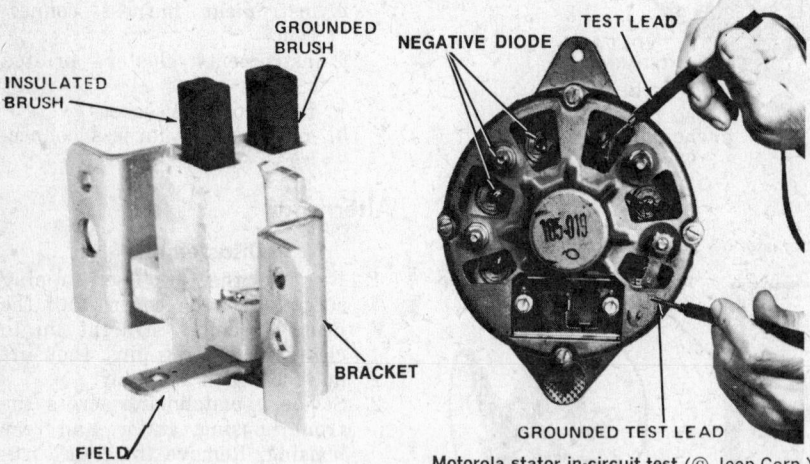

INSULATED BRUSH

GROUNDED BRUSH

NEGATIVE DIODE

TEST LEAD

BRACKET

FIELD TERMINAL

GROUNDED TEST LEAD

Motorola brush assembly—typical
(© Jeep Corp.)

Motorola stator in-circuit test (© Jeep Corp.)

bearing, then reach in and remove the split-ring washer. Bearings must be removed using a puller and new bearings must be pressed into place.

Assembly

1. Clean the bearing and the inside of the bearing hub in the front housing, then gently seat the bearing using a socket of appropriate size and a small hammer.
2. Insert the split-ring washer into the hub of the front housing and seat the washer in its groove. Be extremely careful doing this, because the bearing seal is easily damaged.
3. The front bearing now must be seated against the shoulder on the rotor shaft. Install the fan and pulley spacer, then the Woodruff key, fan and pulley. Using a 7/16 in. socket or equivalent tool to fit inside the rear bearing race, apply pressure to drive the bearing against the shoulder of the rotor shaft.

4. Assemble the front and rear housing assemblies by hand, making certain that the rear bearing is properly seated in the rear housing hub and that the diode wires are not touching the rotor at any point.
5. Align the matchmarks made during disassembly, then spin the rotor to make sure sufficient clearance exists between it and the diode wires. Install the through bolts and tighten them evenly, using only a hand wrench. Continue assembly in reverse of disassembly.

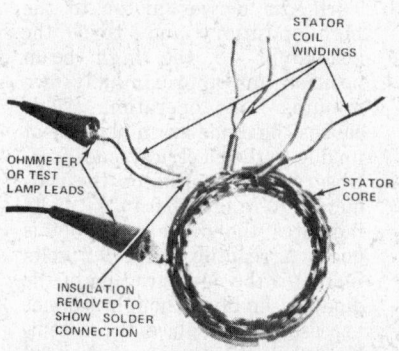

STATOR COIL WINDINGS

OHMMETER OR TEST LAMP LEADS

STATOR CORE

INSULATION REMOVED TO SHOW SOLDER CONNECTION

Motorola stator winding short test (© Jeep Corp.)

Stator

In-Circuit Test

NOTE: When making the in-circuit test, consideration must be given to the rectifier diodes, which are connected to the stator windings. When properly polarized, the diode will conduct current in one direction only. A shorted diode would make the stator appear to be shorted also, so if during this test, a defect is noted, the stator windings and the rectifier diodes must be tested individually. Do not use a 120 volt test lamp as the diodes will be damaged.

1. With the use of a diode continuity light tool or a dc test lamp, connect one test lead to a diode terminal and the second lead to ground. Observe the test lamp and reverse the test leads.
2. The test lamp should light in one direction and not in the other with the leads reversed.
   a. If the test lamp lights in both directions, the stator windings are shorted or one of the negative diodes are shorted. Disassemble, unsolder, and test.
   b. If the test lamp does not light in either direction, all three rectifiers in the negative assembly are indicated to be open. Disassemble, unsolder, and test.

Out of Circuit Tests

To prepare for out of the circuit tests, the stator and diode assemble must be removed from the rear housing. Unsolder the stator leads from the diode stems. Upon reassemble, be certain that the same leads are soldered to the diodes in the same location as removed.

Stator Short Test

1. With the use of a test lamp or ohmmeter, test the windings of the stator by attaching one lead to the stator core and probing the stator leads with the other test lead.

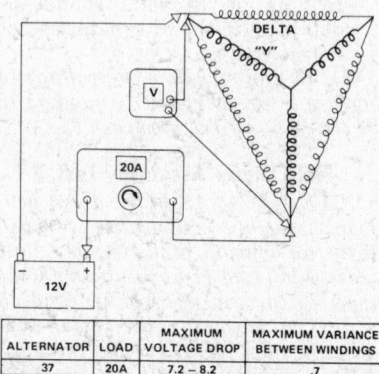

ALTERNATOR	LOAD	MAXIMUM VOLTAGE DROP	MAXIMUM VARIANCE BETWEEN WINDINGS
37	20A	7.2 – 8.2	.7
51	20A	5.5 – 6.5	.6

Motorola stator load test
(© Jeep Corp.)—51 and 55

2. The test lamp will light and the ohmmeter will register if a short circuit exists between the windings and the core. The short circuit must be found or the stator unit be replaced.

Stator Load Test

To test the stator coil windings for short circuits or high resistance, the following tools are needed. A fully charged 12 volt battery, a voltmeter, an ammeter, and a variable load control.

1. Connect the negative battery lead to any one of the three stator leads.
2. Connect the positive battery lead to one lead of the variable load control.

NOTE: If the load control has a built-in ammeter, the other load control lead would be connected to either of the two remaining stator leads. If the ammeter is a separate unit, the remaining load control lead would be connected to the positive ammeter lead and the negative ammeter lead would be connected to one of the two remaining stator leads. (series connection)

3. Connect the voltmeter leads between the two stator leads being tested, (parallel connection) and adjust the variable load control to draw 20 amperes. Allow the windings to warm up for 15 seconds and note the reading on the voltmeter scale. The reading should not exceed 8.2 volts for a 37 ampere rated alternator, or exceed 6.5 volts for the 51 and 55 ampere rated alternators.
4. Stop the current flow to the coil and disconnect the test leads from the stator leads and reconnect them to the remaining stator leads and test the circuits as outlined in paragraph 3. Continue with the test for the third set of windings.
5. Note the variance between the windings. It should not exceed 0.7 volt for the 37 ampere alternator or 0.6 volt for the 51 and 55 ampere alternator.

Rectifier Diode Test

1. With diodes unsoldered, use a commercial type tester and follow the manufacturer's test procedure or make up a heavy load tester as illustrated.

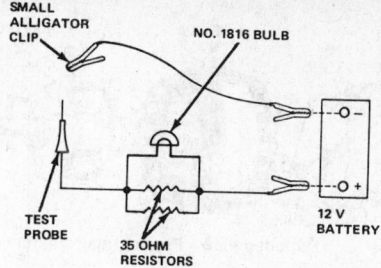

Rectifier diode tester (produces 15 AMP load)

NOTE: A 15 ampere load is necessary to properly test the rectifier diodes for heat related defects.

2. With the use of the heavy load tester probes, connect them to the diode so the test bulb is lighted.
3. Maintain the test load on the diode for 1 to 3 minutes. If the light flickers or goes out, the diode is defective.
4. If the light remains on after three minutes, immediately reverse the test leads. If the test bulb lights, the diode is defective.
5. Test the remaining diodes in the same manner.

NOTE: The diodes are normally not replaced separately, but are replaced as a positive or negative rectifier bridge assembly.

CAUTION: When soldering the stator wires to the diodes, it is advisable to use a set of needle nose pliers attached to the diode stem, to act as a heat sink to avoid heat damage to the diodes.

Field Diode Assembly (Diode Trio)

Two types of diode assemblies are used. The board and the potted type and both are tested with the same procedure.

1. With the diode assembly removed, use a commercial type tester and follow the manufacturer's test procedures or make a load tester as illustrated.
2. Connect the test leads to one of the diodes so that the test bulb is lighted.

NOTE: A one ampere load is needed to properly test the field diode assembly for heat related defects.

3. Maintain a load on the diode for approximately one minute to detect any heat failure.

Field diode tester (produces 1 AMP load)

4. Reverse the test leads and if the test bulb would light, the diode is defective. Test the remaining diodes as outlined.

Rotor Winding Tests

With the rotor removed, use a test probe connected in series with a 110 volt test lamp. Place one probe on a slip ring and the other probe on the rotor core. The rotor is shorted if the test bulb lights.

To test for shorted windings, use a fully charged 12 volt battery, an ammeter, a voltmeter, a variable rheostat, and test probes.

With the use of the test probe leads, place the rheostat and ammeter in series with the battery. Connect one test probe to one slip ring and the other test probe to the other slip ring. Place the voltmeter in parallel with the slip rings.

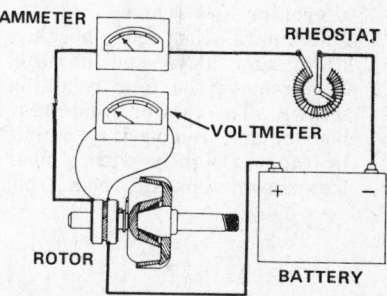

Motorola rotor winding test (© Jeep Corp.)

Slowly reduce the resistance of the rheostat to zero and with full battery voltage, (12.6 ± 0.2 volt), applied to the rotor coil, the field current should register between 1.8 to 2.5 amperes. Excessive ampere draw would indicate shorted windings and low ampere draw would indicate open windings of the rotor.

The Prestolite System

Prestolite alternators incorporate an *isolation diode,* mounted as a component part of the internal positive heat sink assembly. Such alternators are almost identical to late model

Motorola units in operation. Test procedures for the Motorola alternator also apply to the diode-equipped Prestolite.

Troubleshooting

NOTE: see the "Alternator Test Plans" section before proceeding further.

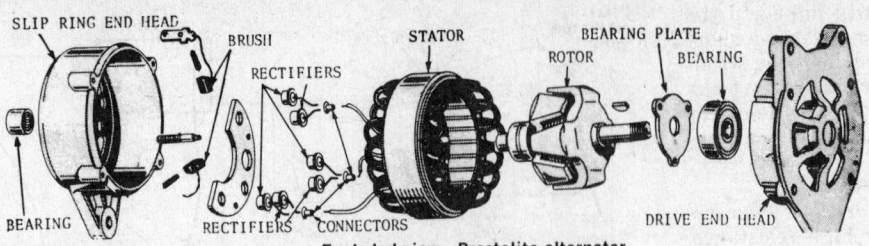

Exploded view—Prestolite alternator

Fusible Link Test

See the Motorola system section for the fuse link test.

Charging System Operation

See the Motorola system section for the "Charging System Operation" tests.

Testing the Ignition Switch-to-Regulator Circuit

1. Disconnect the regulator wires from the regulator.
2. Turn on the key. Using a test light or voltmeter, check for current between the I terminal and ground and the L terminal ground. If voltage is present, this part of the system is OK. If no voltage is present, check for broken or shorted wires, a bad indicator bulb, a bad ammeter (if so equipped), or bad connections.

Alternator Disassembly

1. Remove the two brush mounting screws and cover, then tip the brush assembly away from the alternator and remove.
2. Matchmark the rear housing, stator and drive end housing, then remove the four retaining screws. The stator and rear housing are removed as a unit by tapping lightly with a fiber hammer to separate them from the front housing.
3. The rotor should not be removed unless it or the front bearing is defective. To remove the rotor under these conditions, first remove the pulley nut and pulley (using a two-jaw puller), then remove the fan, Woodruff key and spacer. The rotor is removed from the front housing using a three-paw puller.
4. The front bearing is easily removed, after taking out the retaining ring, by pressing it out in a large vise using sockets to support the housing from the rear.

Stator Coil Test—Diode Type

1. Using a No. 57 bulb, connected in series with a 12-volt battery, as a test light, touch one test lead to the connection of the three stator windings and the other test lead to each stator lead that is connected to the diodes. If the bulb does not light, the winding is open.
2. To test for a grounded stator, use a 110-volt test lamp. First disconnect the diodes from the stator leads, then touch one test lead to the stator core and the other test lead to each of the three stator leads. If the test lamp lights, the winding is grounded.

NOTE: if all other components are O.K. and alternator still does not work, it can be assumed that the stator windings are internally shorted. This type of short is impossible to detect by using the previous test. Diode tests are the same as for the Motorola alternator.

Alternator Assembly

1. Press the front bearing into the front housing, making sure the dust seal faces the rotor. Install the bearing retaining snap-ring, then press the shoulder of the shaft against the inner bearing race using a tool that fits over the shaft and against the race. Install the spacer, Woodruff key, fan and pulley, then install lockwasher and pulley nut.
2. Install the diode heat sink, negative diodes and stator. Solder any stator to diode connections that were unsoldered, using pliers as a heat sink to prevent overheating.
3. Install the rotor and front drive housing to stator and rear housing, aligning matchmarks made during disassembly. Install the four retaining screws, then the brush holder assembly and retaining screws.
4. Make sure the stator leads and brush holder assembly clear the rotor and that the rotor can be spun by hand without binding.

C.S.I. AC Generator

The C.S.I. system is an integrated AC generating system containing a built-in voltage regulator. Removal and replacement is essentially the same as for the standard AC generator. Specialized service procedures are as follows:

Diode Trio Initial Testing

1. Before removing this unit, easily identified in the illustration, connect an ohmmeter between the brush lead clip and the end frame. The lowest reading scale should be used for this test.
2. After taking a reading, reverse the lead connections. If the meter reads zero, the brush lead clip is probably grounded, due to omission of the insulating sleeve or insulating washer.

Diode Trio Removal

1. Remove the three nuts which secure the stator.
2. Remove stator.
3. Remove the screw which secures the diode trio lead clip, then remove diode trio.

NOTE: The position of the insulating washer on the screw is critical; make sure it is returned to the same position on reassembly.

Diode Trio Testing

1. Connect an ohmmeter, on lowest range, between the single brush connector and one stator lead connector.
2. Observe the reading, then reverse the meter leads. Repeat this test with each of the other two stator lead connectors. The

readings on each of these tests should NOT be identical, there should be one low and one high reading for each test. If this is not the case, replace the diode trio.

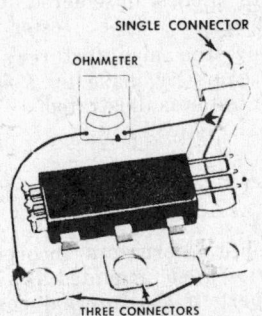

Testing diode trio—C.S.I. AC generator

CAUTION: Do not use high voltage on the diode trio.

Rectifier Bridge Testing

1. Connect an ohmmeter between the heat sink (ground) and the base of one of the three terminals. Then, reverse the meter leads and take a reading. If both readings are identical, the bridge is defective and must be replaced.
2. Repeat this test with the remaining two terminals, then between the INSULATED heat sink (as opposed to the GROUNDED heat sink in previous test) and each of the three terminals. As before, if any two readings are identical, on reversing the meter leads, the rectifier bridge must be replaced.

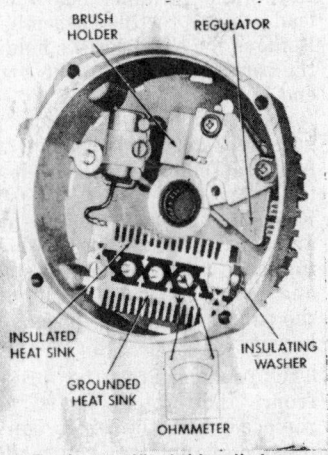

Testing rectifier bridge diodes— C.S.I. AC generator

Rectified Bridge Removal

1. Remove the attaching screw and the BAT. terminal screw.
2. Disconnect the condenser lead.
3. Remove the rectifier bridge.

NOTE: The insulator between the insulated heat sink and the end frame is extremely important to the operation of the unit. It must be replaced in exactly the same position on reassembly.

Brush and/or Voltage Regulator R & R

1. Remove two brush holder screws and stator lead to strap nut and washer, brush holder screws and one of the diode trio lead strap attaching screws.

NOTE: The insulating washers must be replaced in the same position on reassembly.

2. Remove brush holder and brushes. The voltage regulator may also be removed at this time, if desired.
3. Brushes and brush springs must be free of corrosion and must be undamaged and completely free of oil or grease.
4. Insert spring and brushes into holder, noting whether they slide

Brush holder—C.S.I. generator

freely without binding. Insert wooden or plastic toothpick into bottom hole in holder to retain brushes.

NOTE: The brush holder is serviced as a unit; individual parts are not available.

5. Reassemble in reverse order of disassembly.

Voltage Regulator Testing

NOTE: The voltage regulator must be tested with the C.S.I. unit still in place.

Grounding tab for voltage regulator test—C.S.I. AC generator

1. Disconnect battery ground strap.
2. Connect an ammeter in series with the BAT. terminal of the generator and the lead removed from that terminal.

3. Reconnect battery ground strap, then turn on all accessoris to place a load on the system.
4. Connect a carbon pile across the battery terminals.
5. Ground the field winding by inserting an insulated screwdriver into the test hole in the alternator frame and depressing the tab. Do not push the tab into the generator more than 1 in.
6. Run the engine at moderate rpm, equivalent to 30-40 mph in high gear, and adjust the carbon pile to obtain maximum current output.
7. If the output is within 10% of the rated output of the alternator and the system does not charge properly, the voltage regulator is defective and must be replaced.

Generator—C.S.I.
Amps.—cold rating 55
Output @ rpm ....30 @ 2000
Output @ rpm ....55 @ 5000
Field current draw 4.0-4.5
Regulator—C.S.I.
Model 1116368
Normal range .....13.5-16.0 volts

Slip Ring End Frame Bearing and Seal R & R

1. With stator removed, press out bearing and seal, using a socket or similar tool that fits inside the end frame housing. Press from outside to inside, supporting the frame inside with a hollow cylinder (large, deep socket) to allow the seal and bearing to pass.
2. The bearings are sealed for life and permanently lubricated. If a bearing is dry, do not attempt to repack it, as it will throw off the grease and contaminate the inside of the generator.
3. Using a flat plate, press the new bearing from the outside toward the inside. A large vise is a handy press, but care must be exercised so that end frame is not distorted or cracked. Again, use a deep socket to support the inside of the end frame.
4. From inside the end frame, insert seal and press flush with housing.
5. Install stator and reconnect leads.

The Leece-Neville System

The Leece-Neville charging systems use varied types of alternator housings and controls. The types most commonly encountered on light and medium trucks and vans will be illustrated and described.

105 Ampere Alternator (early type)

With the use of an adjustable carbon pile load control, field rheostat control, and voltamp tester, the rated ampere output and field current draw

can be tested with the alternator mounted on the engine. If any indication of malfunction is determined, the alternator should be removed and the internal circuits tested.

Upon disassembly, the testing procedure for the internal circuits fol-

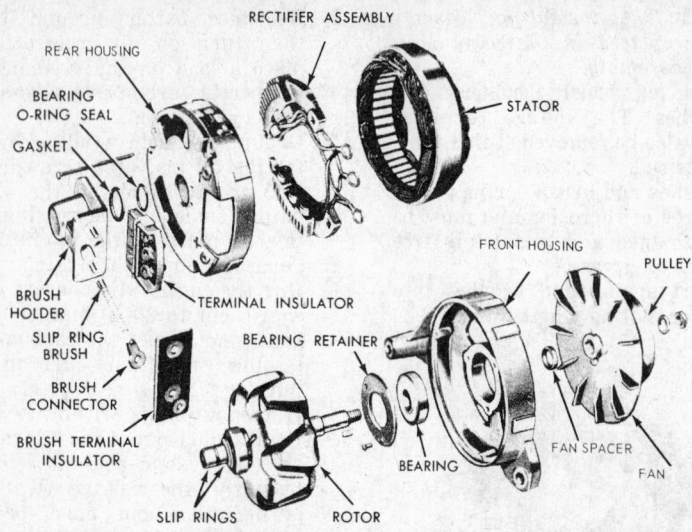

Exploded view 65, 70, 105 Leece-Neville Alternator (© Ford Motor Co.)

lows the outline described in the Autolite (Ford) type alternator section.

Rated Output Test

1. With battery disconnected, ignition switch in off position, and the wire disconnected at the alternator B+ terminal, hook an ammeter between the alternator B+ terminal and the wire that was just disconnected from this terminal.
2. Disconnect wire from the alternator F terminal.
3. With the field rheostat adjusted to open position, connect its leads to the alternator F terminal and the alternator B+ terminal.
4. Hook up a voltmeter between the alternator B+ terminal and ground.
5. Reconnect battery cables.
6. Connect a carbon pile load control between the battery posts.
7. With a tachometer connected to the engine, start the engine and set its speed at the recommended rpm.
8. While watching both ammeter and voltmeter, adjust the carbon pile load control to maintain 15 volts. When field rheostat control is fully closed, note voltmeter and ammeter readings.
9. Adjust field rheostat to open position and carbon pile control to off.
10. Compare readings with specifications. Readings should be at least equal to rated output specified by the manufacturer. If output complies with specifications, proceed to the voltage regulator test. If readings are below rated output, it indicates possible internal troubles.

Field Current Draw Test

1. Disconnect battery ground cable.
2. Disconnect carbon pile, volt-

meter, ammeter and field rheostat from the previous check.
3. Reconnect the regular circuit wire to the alternator B+ terminal.
4. With the field rheostat adjusted to the open position, connect one of its leads to the alternator B+ terminal and the other lead to a test ammeter lead.
5. Connect the remaining ammeter lead to the alternator F terminal.
6. Connect one voltmeter lead to the alternator F terminal and the remaining voltmeter lead to ground.
7. Reconnect battery ground cable, start the engine, and run it at about 1,000 rpm.
8. With field rheostat in closed position, read the ammeter and volt-

meter for a very brief period.
9. Compare these readings with specifications.
10. If readings are low, it indicates trouble in slip rings or brushes.
11. If readings are too high, it indicates trouble in the rotor field windings.
12. If readings are as specified on an alternator which did not deliver its rated output, look for trouble in the stator or diodes.

Disassembly

1. Remove the mounting brackets. Remove the pulley nut, pulley, fan, key, and spacer from rotor shaft.
2. Remove the dust shield. Remove three screws holding the six stator and terminal board wire leads to the rectifier assemblies.
3. Remove the three screws holding the terminal board to the brush end housing.
4. Remove the field lead from the brush holder and remove the brush assembly from the housing.
5. Remove the four through bolts and separate the drive end housing and rotor from the stator and brush end housing. Remove the stator from the brush end housing.
6. The bearings can be pressed from the rotor shaft and replaced. The slip rings can be pressed from the rotor after removal of the field wire.
7. Remove the two circular copper collector plates from the brush end housing.

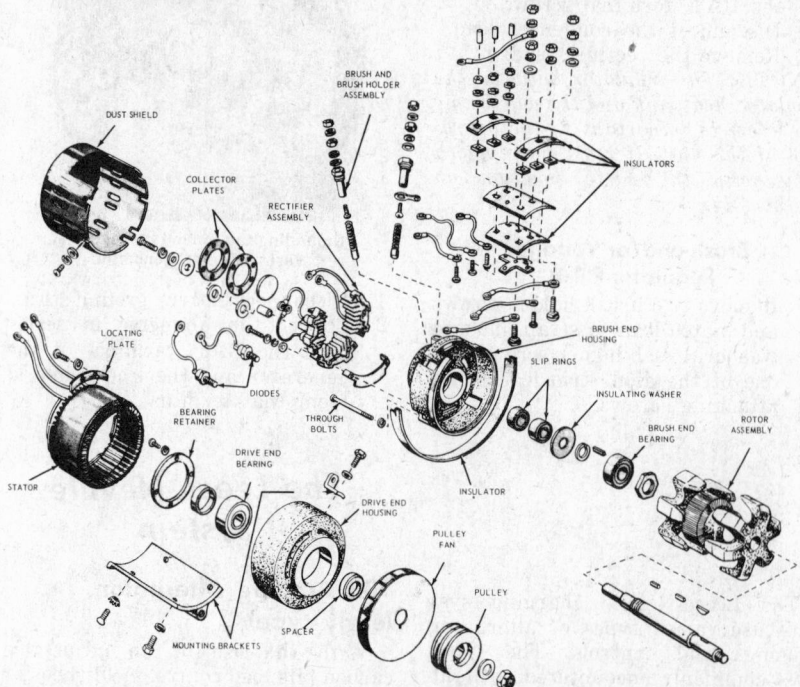

Leece-Neville 105 Ampere alternator—Exploded view (early Model) (© Ford Motor Co.)

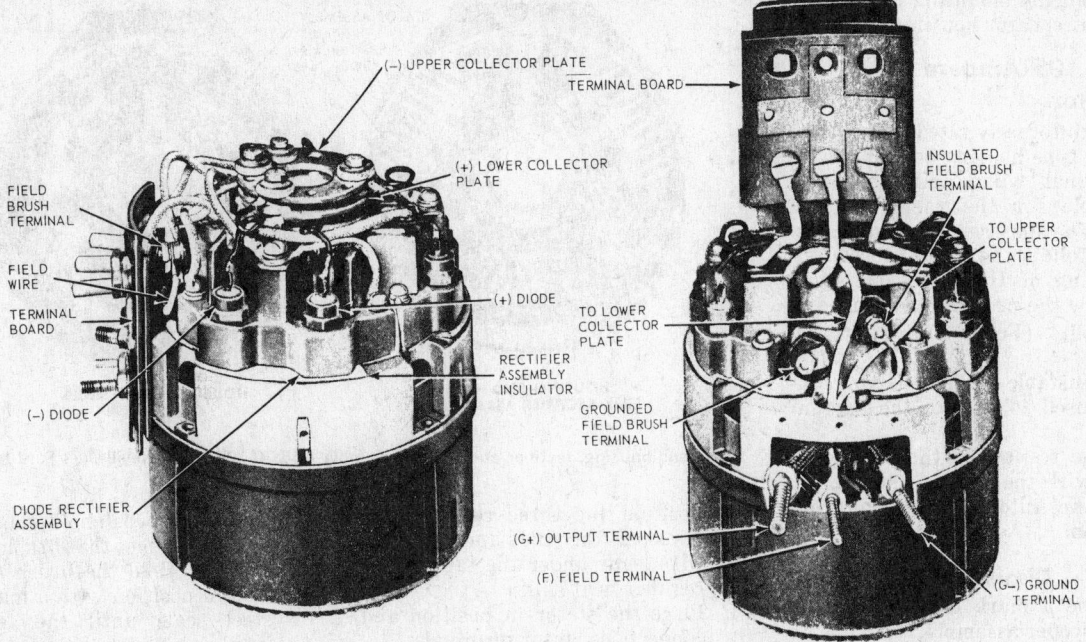

Collector plates and wire lead assembly—Leece-Neville 105 ampere (© Ford Motor Co.)

NOTE: The retaining screws hold the diode lead wires and terminal board output leads. Tag the wires for identification during assembly.

8. Remove the terminal board and the rectifier assemblies from the brush end housing.
9. Testing of the internal circuits can now be accomplished.

Assembly

1. Replace any necessary parts and assemble the terminal board and wires.
2. Position the two stator winding insulators in the brush end housing.

3. Install the rectifier assemblies and insulators into the brush end housing.
4. Position the terminal board and wire assembly, the two circular collector plates, insulators, and washers into the brush end housing and install the retaining screws with the diode leads and output leads in their respective locations as previously tagged.
5. Position the stator to the brush end housing and connect the assembly to the rotor and drive

end housing. Install the four through bolts and tighten.
6. Install the brushes, springs, and brush holders into the housing and connect the field wire to the brush holder.
7. Position the terminal board leads and stator leads to their respective rectifiers as previously tagged.
8. Install the dust shield. Install the rotor spacer, fan, key, and pulley. Torque the pulley nut to 45-50 ft. lbs.

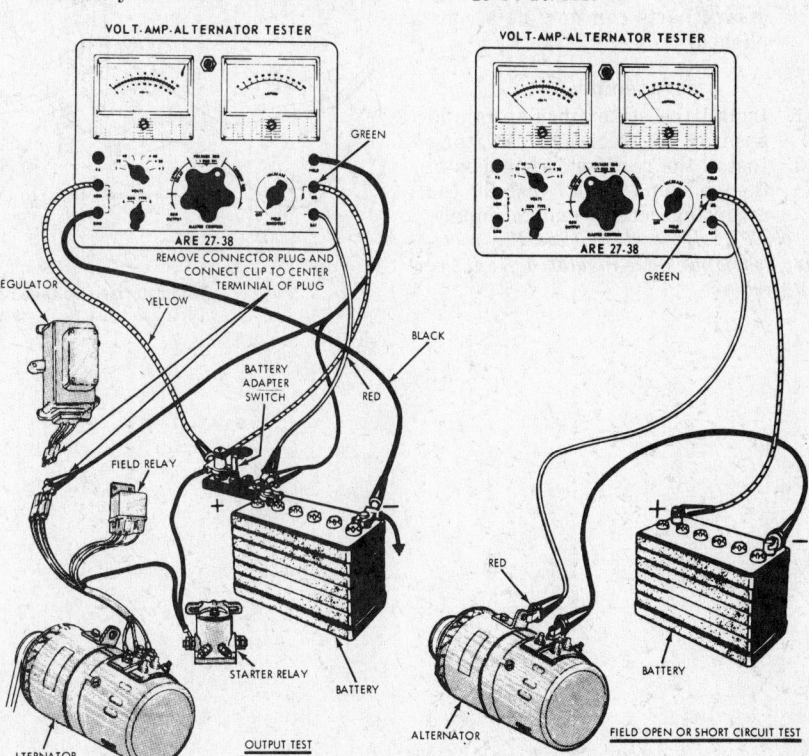

Testing field relay—Leece-Neville system
(early model) (© Ford Motor Co.)

Alternator tests-105 ampere alternator (early model) (© Ford Motor Co.)

9. Install the mounting brackets to the alternator housing.

65, 70, 105 Ampere Alternators

These differently rated units utilize the same type housing assembly with the internal wiring circuits being responsible for the varied current output. The disassembly and assembly will follow the same procedures. The testing of the internal circuits will follow the procedures outlined in the Autolite (Ford) alternator section.

An adjustable transistorized regulator is used to control the current output.

For the testing of the alternator circuits with the unit on the vehicle, refer to the Autolite (Ford) alternator section.

Disassembly

1. Scribe a mark on the housings for proper assembly.
2. Remove the pulley nut, pulley, fan, key, and spacer from the rotor shaft.
3. Remove the brush terminal insulator, springs, and brushes. Remove the brush holder assembly.
4. Remove the through bolts and separate the brush end housing from the drive end housing.
5. Remove the stator lead retaining nuts and remove the stator from the brush end housing.
6. Remove the rectifier assemblies and stator terminal insulators.
7. Testing and replacement of necessary parts can now be accomplished.

Assembly

1. Install the stator insulators and position the rectifier insulators.
2. Install the rectifier assemblies in the housing and install the mounting screws and terminals.

NOTE: Be certain that the rectifier assemblies are insulated from the end frame.

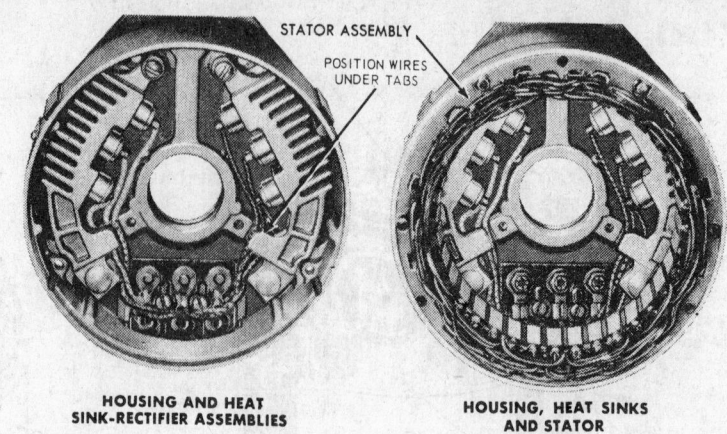

Brush end housing, rectifier and stator assembly—65,70,105 Leece-Neville (© Ford Motor Co.)

3. Position the three rectifier terminals to the studs and route the wire leads under the tabs of the rectifier heat sinks.
4. Place the stator in position and connect the stator terminals.
5. Assemble the stator and brush end housing to the rotor and drive end housing, while aligning the previously scribed marks.
6. Install the brush holder assembly and assemble the brushes and springs. Hold the brush connectors in position with a machinist steel scale until the terminal insulator is installed.
7. Install the spacer, key, fan, and pulley on the rotor shaft and torque the pulley nut to 40 50 ft. lbs.

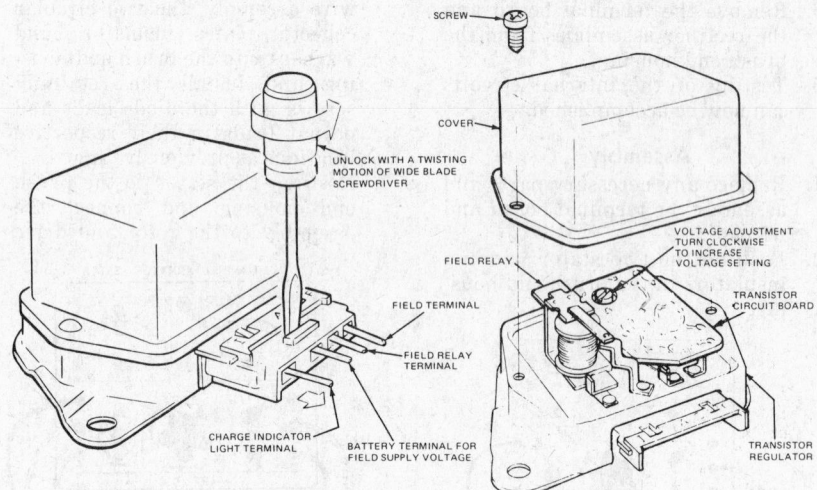

Transistorized voltage regulator (© Ford Motor Co.)

Conventional Ignition Systems

The Ignition System is divided into two circuits; a low voltage or primary circuit, and a high voltage or secondary circuit.

The primary circuit carries current, (usually modified for ignition by a resistor, on 12-volt systems) at battery voltage. It includes the battery, ignition-starter switch, starter relay, ignition ballast resistor, primary winding of the coil, condenser, contact points, and ground.

The secondary circuit begins with the ignition coil. Secondary voltage is a product of the coil and emerges from the secondary terminal and flows through a cable to the distributor cap. It is distributed by the rotor, through the distributor cap and cables, to the spark plugs, and to ground.
CAUTION: Secondary circuit pressure could reach as high as 30,000 volts.

Coil Polarity

Coil polarity is predetermined and must match the circuit polarity of the system being tested. It is an established fact that the electron flow through the spark plug is better from the hotter center plug electrode-to-ground than by the opposite route, from ground-to-center electrode. Therefore, negative ground polarity has been established as standard. There is about a 14% difference in required voltage of the two polarity designs at idling speed. This differential increases with engine speed.

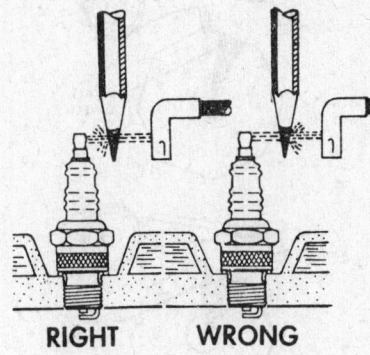

RIGHT   WRONG

Correct coil polarity can be checked on the truck by connecting a voltmeter negative lead to the ignition coil secondary wire, and the positive voltmeter lead to engine ground. If the voltmeter reading is up-scale, polarity is correct: if voltmeter reading is down-scale, polarity is reversed.

Lately, automotive batteries are designed with the battery posts on the same side of the battery, opposed to the earlier diagonal post design. Therefore, terminal size and cable length will discourage improper battery installation. This results in the

battery and distributor terminals of the coil being the most likely points of possible reversal of polarity.

Another tentative, but less precise, method is to hold a regular carbon-cored wooden lead pencil in the gap between a disconnected spark plug wire and ground. It is possible to observe the direction of spark flow, from wire-to-pencil-to-ground when polarity is correct.
CAUTION: Hold the pencil with a heavy glove or you may observe the spark flow along your fingers.

Primary Circuit Test

A quick, tentative check of the 12 volt ignition primary circuit (including ballast resistor) can be made with a simple voltmeter, as follows:

1. With engine at operating temperature, but stopped, and the distributor side of the ignition coil grounded with a jumper wire, hook up a voltmeter between the ignition coil (switch side) and a good ground.
2. Jiggle the ignition switch (switch on) and watch the meter. An unstable needle will indicate a defective ignition switch.
3. With ignition switch on (engine stopped) the voltmeter should read 5.5 to 7 volts for 12-volt systems.

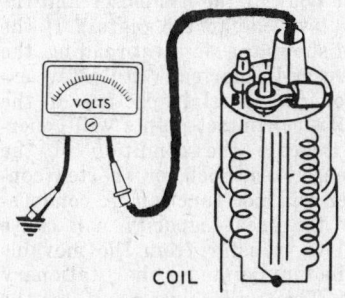

COIL

4. Crank the engine. Voltmeter should read at least 9 volts during cranking period.
5. Now remove the jumper wire from the coil. Start the engine. Voltmeter should read from 9.0 volts to 11.5 volts (depending upon generator output) while running.

Coil Resistance, with Ohmmeter —Primary Circuit

To check ignition coil resistance, primary side, switch ohmmeter to low scale. Connect the ohmmeter leads across the primary terminals of the coil and read the low ohms scale.

Coils requiring ballast resistors should read about 1.0 ohm resistance. 12-volt coils, not requiring external ballast resistors, should read about 4.0 ohms resistance.

Coil Resistance, with Ohmmeter —Secondary Circuit

To check ignition coil resistance, secondary side, switch ohmmeter to high scale. Connect one test lead to the distributor cap end of the coil secondary cable. Connect the other test lead to the distributor terminal of the coil. A coil in satisfactory condition should show between 4 K and 8 K on the scale. Some special coils (Mallory, etc.) may show a resistance as high as 13 K. If the reading is much lower than 4 K, the coil probably has shorted secondary turns. If the reading is extremely high (40 K or more) the secondary winding is either open, there is a bad connection at the coil terminal, or resistance is high in the cable.

If both primary and secondary windings of the coil test good, but the ignition system is still unsatisfactory, check the system further.

Ballast Resistor

Some sort of ballast resistor is used with most trucks equipped with 12-volt ignition systems. This resistance may be built into the ignition coil, or it may be a special wire of specific resistance, comprising the primary ignition circuit.

To provide a greater safety margin of sufficient voltage for high speed operation, a special ignition coil is used with whatever type of ballast resistance is used. Other reasons for

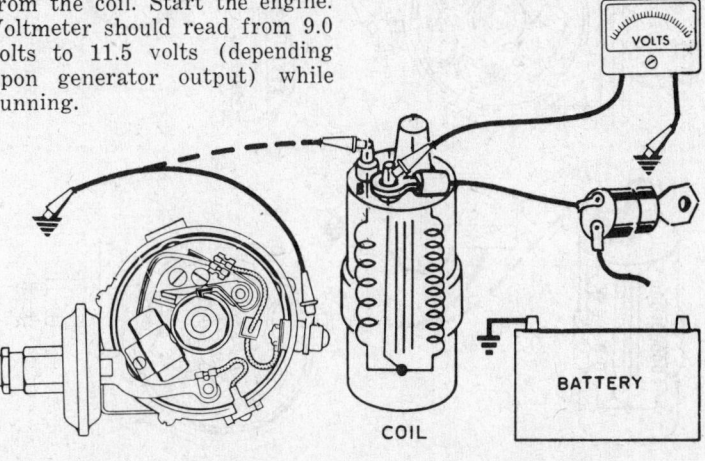

BATTERY

COIL

Ballast resistor

ballast resistance are to limit to a safe maximum the primary current flow through the coil and through the distributor contact points. This helps protect the contact points at slow engine speed when they are closed for a longer period of time. The resistor also protects against excessive build-up of primary current when the ignition switch is on with the engine stopped and ignition points closed.

On some systems, the resistor is removed from the ignition circuit during engine cranking, then with the ignition connected directly to battery voltage. This keeps ignition voltage as high as possible while cranking. The by-pass type system can have the by-pass factor built into the ignition switch, or it may be part of the starter solenoid.

Primary Circuit—Distributor Side

With the voltmeter on the 16-20 volt scale, connect one voltmeter lead to ground. Connect the other voltmeter lead to the distributor side of the coil. Remove the high tension wire from the coil and ground it. Close ignition switch and slowly bump the starter switch to open and close the points. When the distributor contacts make and break during cranking, the voltmeter reading should be from one-third to one-half battery voltage. Normally, with engine stopped and with points open,

the reading will be the same as battery voltage. Furthermore, with the engine stopped and the points closed and in good condition, the reading will be close to zero.

If while cranking, the voltmeter reading remains zero or close to it, the trouble may be one or more of the following:

A. No current at distributor. Disconnect the distributor primary wire from the top of the coil. Now, take a voltmeter reading from the distributor terminal of the coil to make sure that the current is going through the circuit.

B. Points are not opening because of mechanical (points or cam) failure or maladjustment. Dual points in parallel, one set not opening.

C. The movable point, the stud at the primary distributor wire terminal, or the pigtail wire is grounded.

D. The condenser has a dead short. An ohmmeter check of the condenser will show this condition. Connect one test lead of an ohmmeter to the body of the condenser and the other test lead to the pigtail. If the meter shows the slightest reading, the condenser is shorted. With a few exceptions, a visual inspection of the distributor contact points will generally indicate the condition of the condenser. An open, or shorted condenser will not function. A condenser of too great capacity will cause metal to transfer from the movable distributor point to the stationary point. This will cause a pit on the

movable point. An under-capacity condenser, causes metal to leave the stationary point and build up on the movable point.

Any excessive resistance in either the primary or secondary circuits will upset the sensitive balance of the ignition system and cause the ignition points to pit.

Ignition Point Dwell

It is very important that point dwell be adjusted to exact specifications before any attempt is made to time the engine.

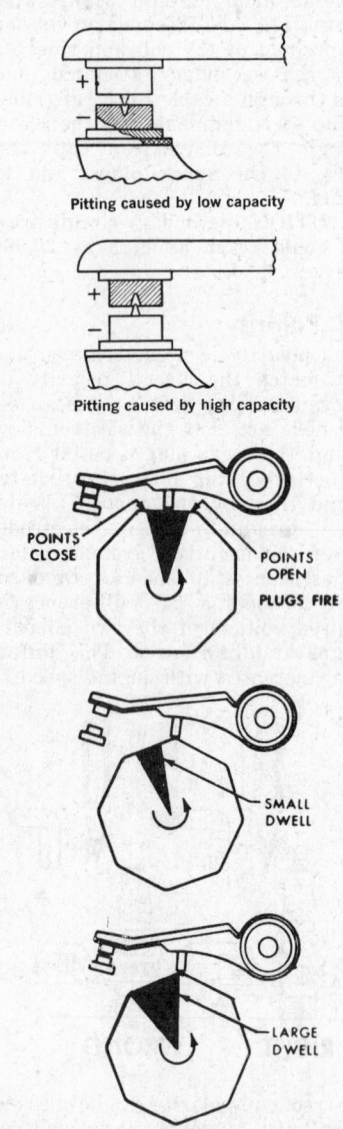

Pitting caused by low capacity

Pitting caused by high capacity

Point dwell (cam angle) is the degree value for the closed attitude of ignition points for each make-and-break period of a distributor cycle. It is that phase of ignition system functioning during which the coil becomes saturated (builds up to voltage capacity) for its next discharge at the moment of point opening.

Some current production truck engines demand in excess of 23,000 spark plug firings per minute to fire

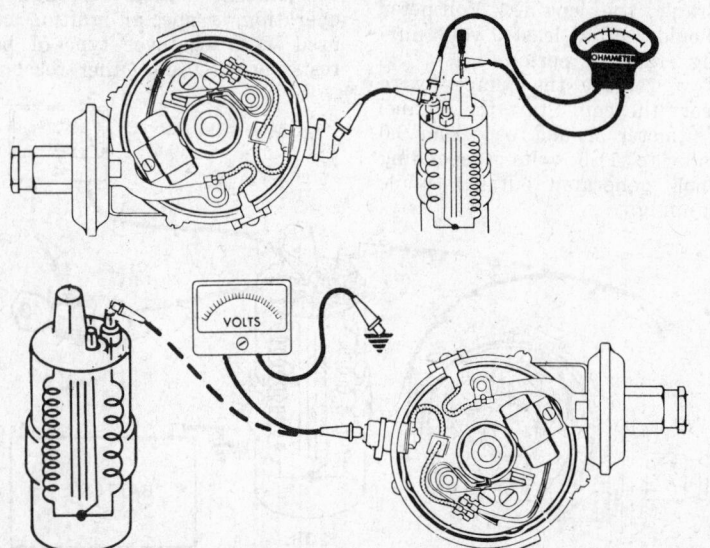

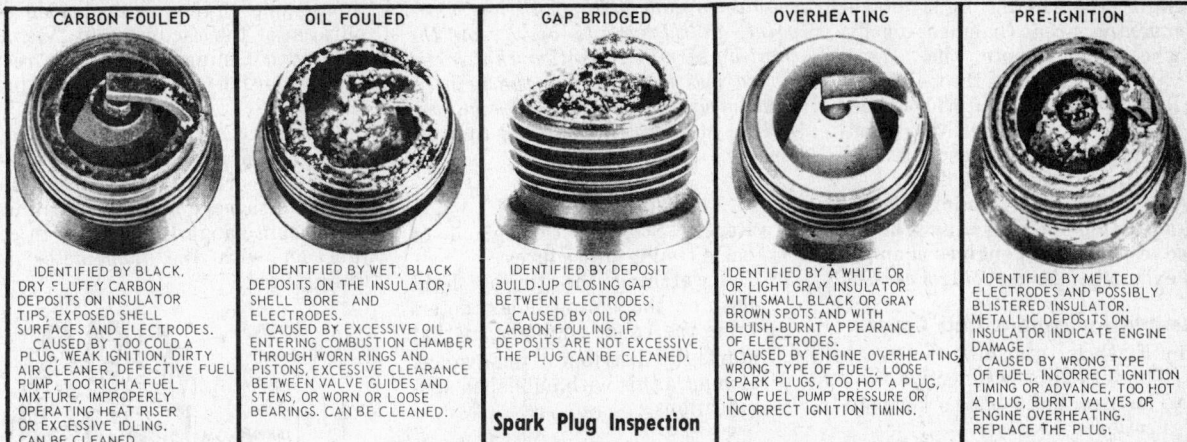

CARBON FOULED	OIL FOULED	GAP BRIDGED	OVERHEATING	PRE-IGNITION
IDENTIFIED BY BLACK, DRY FLUFFY CARBON DEPOSITS ON INSULATOR TIPS, EXPOSED SHELL SURFACES AND ELECTRODES. CAUSED BY TOO COLD A PLUG, WEAK IGNITION, DIRTY AIR CLEANER, DEFECTIVE FUEL PUMP, TOO RICH A FUEL MIXTURE, IMPROPERLY OPERATING HEAT RISER OR EXCESSIVE IDLING. CAN BE CLEANED.	IDENTIFIED BY WET, BLACK DEPOSITS ON THE INSULATOR, SHELL BORE AND ELECTRODES. CAUSED BY EXCESSIVE OIL ENTERING COMBUSTION CHAMBER THROUGH WORN RINGS AND PISTONS, EXCESSIVE CLEARANCE BETWEEN VALVE GUIDES AND STEMS, OR WORN OR LOOSE BEARINGS. CAN BE CLEANED.	IDENTIFIED BY DEPOSIT BUILD-UP CLOSING GAP BETWEEN ELECTRODES. CAUSED BY OIL OR CARBON FOULING. IF DEPOSITS ARE NOT EXCESSIVE, THE PLUG CAN BE CLEANED.	IDENTIFIED BY A WHITE OR OR LIGHT GRAY INSULATOR WITH SMALL BLACK OR GRAY BROWN SPOTS AND WITH BLUISH-BURNT APPEARANCE OF ELECTRODES. CAUSED BY ENGINE OVERHEATING, WRONG TYPE OF FUEL, LOOSE SPARK PLUGS, TOO HOT A PLUG, LOW FUEL PUMP PRESSURE OR INCORRECT IGNITION TIMING.	IDENTIFIED BY MELTED ELECTRODES AND POSSIBLY BLISTERED INSULATOR. METALLIC DEPOSITS ON INSULATOR INDICATE ENGINE DAMAGE. CAUSED BY WRONG TYPE OF FUEL, INCORRECT IGNITION TIMING OR ADVANCE, TOO HOT A PLUG, BURNT VALVES OR ENGINE OVERHEATING. REPLACE THE PLUG.

Spark Plug Inspection

their cylinders. This places a tremendous demand upon the ignition system, particularly the aspect of coil build up (saturation) and discharge time.

While it is true that ignition points can be adjusted by using a thickness gauge, the results, even when using new points, are sometimes inconclusive. Point gap is incidental to particular distributor cam shape and could be misleading. This is one reason for the use of a dwell meter.

Another point in favor of the dwell meter is its ability to detect high resistance, (oxidized points, poor connections, etc.). The dwell meter, a modified voltmeter, often includes a band on the extreme high end of the scale to indicate excessive point resistance. Follow instrument manufacturer's instructions to get the most out of your particular equipment.

The most informative procedure is to use both methods (dwell meter and point gap) then compare the two. Many times the comparison is surprising, and leads to the location of previously unnoticed distributor troubles.

Using the Dwell Meter

1. With distributor vacuum control line disconnected and plugged, turn the meter selector switch to the eight lobe position (eight cylinder engines) or the six lobe position (six cylinder engines). On four cylinder engines, follow instrument makers' instructions, or select the eight lobe position

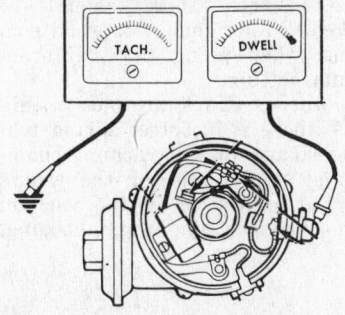

and double the reading for eight cylinder engines.
2. Connect one tach-dwell meter lead to the coil terminal of the distributor and the other meter lead to ground.
3. Start the engine and operate it at idle speed. Note reading on dwell meter. On eight cylinder engines, (single contacts) dwell should read 26°-32°. Double contacts should read 26°-32° (each set), or 34°-40° combined. Six cylinder engines should show a dwell of 36°-45° and four cylinder engines, a dwell in the area of 40°. These are tentative figures and cover a wide latitude. It is therefore, urgent that manufacturers' specifications be followed without exception.

An excessive variation in dwell, (over 3°) as engine speed is increased usually indicates a worn distributor shaft, bushing, breaker plate or secondary circuits.

NOTE: on some Auto-Lite or Ford distributors, a pivoted, movable type breaker plate is used. This pivoted plate, operated by the vacuum control unit, carries the contacts and rotates on its own center, independent of the distributor cam center. This design affects a running dwell variation of as much as 12°. To check this type distributor, hook up the distributor vacuum control line. Increase engine rpm and observe dwell changes at various engine rpm and throttle attitudes.

NOTE: experience dictates that all distributor adjustments are best performed with the use of a good off-car distributor tester.

Dwell information at idle speed is given for each engine in the Tune-up Specification Table in the truck Section.

Ignition Timing

Ignition timing is a term applied to the relationship of piston travel and moment of spark in a gas engine.

Due to the many variables involved, such as compression ratio, temperature, humidity, elevation, fuel octane value, engine condition, work load, etc. published timing data must be considered approximate; some tolerance permitted.

Ignition timing consists of basic (prime) timing and dynamic (variable) timing.

It is very important that point dwell be correct before setting timing.

Basic Timing

Basic timing can be checked quite accurately by using one of the many timing lights, (strobe-flashers) available. A timing light, when properly connected to No. 1 spark wire, (or the exact opposite cylinder in firing sequence of any multiple cylinder, four stroke cycle automotive engine) will indicate the moment of ignition for that cylinder.

NOTE: Some International Harvester engines are timed on No. 8 cylinder.

Index markings may be on either the rotating member of the crankshaft, (vibration damper or flywheel) with the pointer stationary, or the index may be on an engine stationary member, with the pointer or scribe mark rotating.

NOTE: because ignition timing is directly affected by distributor contact spacing, points should be adjusted to specifications before timing is attempted.

1. Unless otherwise stated by the manufacturer, the distributor control vacuum line should be disconnected, and plugged, to prevent fuel induction disturbance.
2. Hook up the timing light, (power or otherwise) according to the equipment instructions.
3. With engine at operating temperature and adjusted to function smoothly, run engine at low idle. Use a tachometer and be sure the rpm is below the speed of governor advance influence.

4. Shine the timing light on the indexing area (balancer or flywheel) and note the degree value indicated by the pointer.

5. Rotate the distributor body one way or the other until the pointer appears to correspond with the index value published.

CAUTION: Power timing (on the road adjusting for ping cannot be tolerated, especially on engines equipped with exhaust emission control devices.

Possible Indications and Causes

If The Flash Is Intermittent:
A. The test light is defective.
B. The test light has a bad connection.
C. Distributor points are bad or badly out of adjustment.
D. Distributor grounding is poor.
E. Distributor cap is cracked or tracking.
F. Spark plug gaps too small.
G. Broken or badly worn rotor.

If Pointer Appears To Move On The Index Scale (Unfixed At Constant Engine Speed):
A. Distributor governor weights loose or with broken springs.
B. Distributor shaft or bushing worn.
C. Rotor loose or broken.
D. Distributor base plate loose.
E. Cam lobes worn.

Dynamic Timing

To accurately check and calibrate dynamic timing through all attitudes of engine operation, more sophisticated equipment than the common strobe-light is needed. A distributor tester, an oscillograph, or one of the more complex timing lights equipped with an advance value index is needed.

It is possible, however, to determine to some degree, the action of both governor advance and vacuum control mechanisms with a tachometer and a timing light.

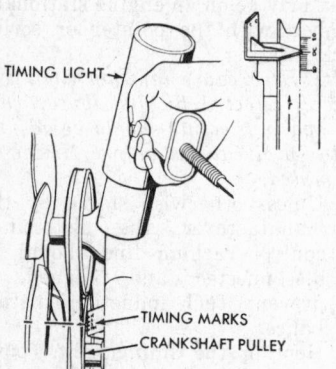

NOTE: before checking dynamic timing with a timing light, extend the index graduations on the timing member involved, by about 30°. This should be done with chalk or white paint, in increments of 5°, on the rotating member, whether that member carries the index or the pointer. Some measurements and extreme care will be necessary in making this extension.

Governor Control

1. Repeat Steps 1 through 4 of basic timing procedures.
2. By watching the timing light flash on the timing index, determine the exact engine rpm that starts the distributor to advance. Compare this with published specifications.

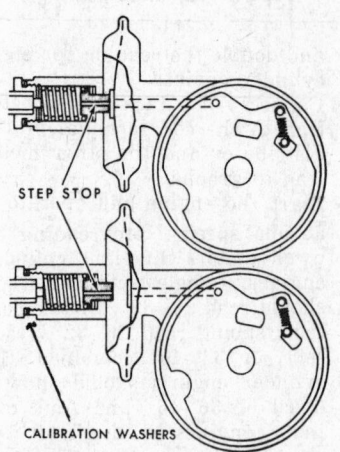

3. It is equally important that distributor advance progresses steadily with engine speed. It is just as important that a decrease in engine speed will smoothly and gradually return the index pointer to its original position.
4. After checking the indications against specifications, turn the engine off and make corrections, if necessary.

NOTE: dynamic timing cannot be accurately checked using the above method; therefore it is recommended that no attempt be made to modify advance curves (especially on exhaust emission equipped engines) unless the proper distributor test facilities are available.

Vacuum Control

Vacuum control action can be observed and evaluated by using a tachometer and a timing light.

This type of spark control, whether used as the only means of control or used in conjunction with a governor type mechanism, operates through a spring loaded vacuum chamber. This chamber is attached to the side of the distributor, then, through linkage, to the breaker plate (or pick-up assembly of transistorized distributors).

Carefully metered vacuum is piped to one side of the spring loaded diaphragm of the vacuum unit. Vacuum controlled timing is then the result of differential (vacuum-spring) pressures.

In the case of vacuum-only controls, (Ford Loadmatic, etc.) metering is more critical; therefore, a manometer or a very accurately calibrated vacuum gauge is required, in conjunction with the tachometer and timing light.

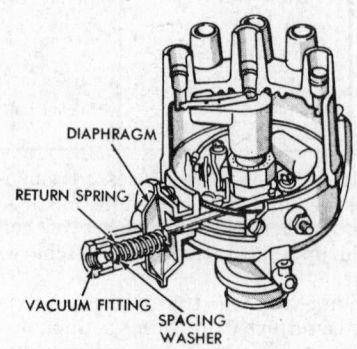

To Check Vacuum Control:

1. Hook up a tachometer and the timing light in the conventional manner.
2. Connect a good vacuum gauge or a manometer into the vacuum line between the carburetor and distributor.
3. With engine at operating temperature and adjusted to function smoothly, run engine at low idle.
4. Shine timing light on the indexing area and observe the vacuum reading and timing light index relationship.
5. Compare these readings with published vacuum advance data.

Indications and Causes

If timing is not within degree range as specified vacuum reading, a faulty vacuum or mechanical control mechanism, or Loadmatic control mechanism is defective.

If all parts are good, adjustments in control valve can sometimes be made by changing the calibration washers between the vacuum chamber spring and the spring retaining nut. Adding washers will decrease the amount of advance. Removing washers will increase the advance. After one vacuum setting has been adjusted, the others should be checked. Do not change original rpm setting when going to a different vacuum setting.

If other settings are not within limits, there is incorrect spring tension, leakage in the vacuum chamber and/or the line, or the wrong stop has been used in the vacuum chamber of the diaphragm housing.

Electronic Ignition Systems

Ford-Autolite Breaker Point System

This transistorized system uses conventional breaker points, but does not use a condenser. The only external components that serve to distinguish this system from a conventional ignition are an external ballast resistor, a tachometer connecting block, a cold-start relay and an amplifier (transistor switching device).

The major hurdle to increasing primary ignition circuit voltage in point-type is that the points burn very easily. This system uses a transistor to by-pass that weakness.

The design of the main transistor in this system allows it to conduct current from the wire running into it to the wire running out of it—if it is connected to a complete electrical circuit. However, as the current passes through the transistor, it breaks that current down into two paths, one of high voltage and one of low voltage. This transistor is called a PNP transistor because of its three component parts. The top part of the transistor is called a collector (C), the middle part the base (B), and lower part the emitter (E). The two currents that form inside the transistor are the high current one, or power current, which runs from E to C and the low current, or the switching current, which runs from B to E.

The power current is connected to the primary side of the ignition coil and the switching current is connected to the ignition points. This allows high current to energize the pri-

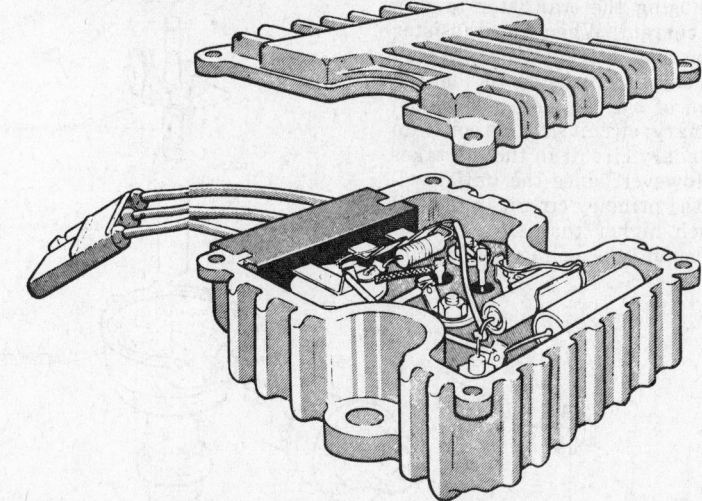

Autolite amplifier assembly (© Ford Motor Co.)

mary circuit in the coil while permitting a much smaller current to pass through the points.

In order for the transistor to pass current, both the power circuit and the switching circuit must make a complete circuit. This has made adaptation of this transistor system to an automotive ignition system relatively simple.

When the ignition key is turned on and the breaker points are closed, current passes from the battery, through the ignition switch, to the amplifier which contains the PNP transistor. As the current from the battery passes through the transistor,

the two transistor circuits are connected as follows: the power current is connected to the coil, and the switching current is connected to the points. Since the points are closed, both transistor circuits are complete and the primary side of the ignition

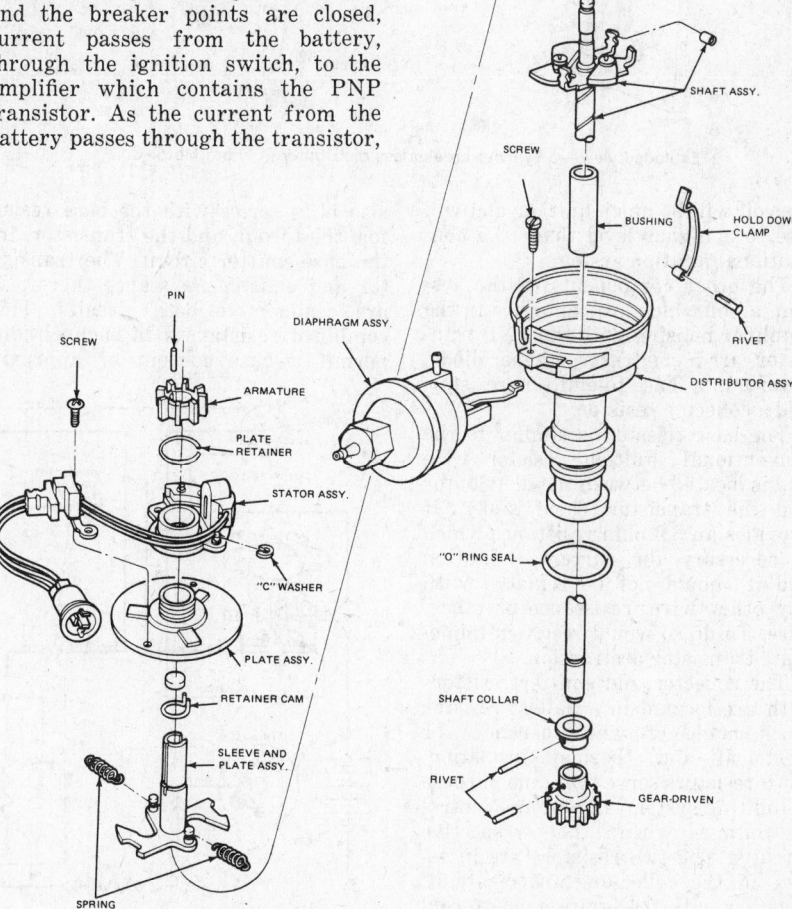

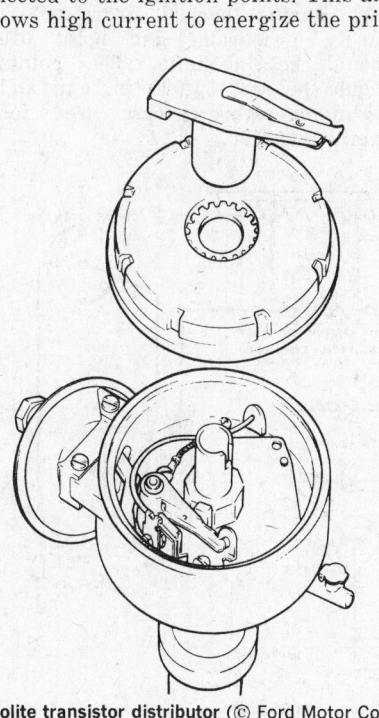

Autolite transistor distributor (© Ford Motor Co.)

Exploded view—8 cylinder breakerless distributor (© Ford Motor Co.)

coil builds up a magnetic field. When the ignition points open, the switching circuit in the transistor (B to E) opens causing the transistor to stop passing current. When the transistor stops passing current, the primary ignition circuit breaks down, and the induction of the magnetic field from the primary circuit in the coil into the secondary circuit in the coil takes place. However, since the initial voltage in the primary circuit in the coil was much higher than in a conventional system, the voltage buildup in

mately 1.0 amp. and a collector current of approximately 12 amps.

A tach block is included in the circuit for attaching tachometer and dwellmeter leads. In the conventional system, these leads are connected to the distributor primary lead and ground, but in the transistorized circuit, the connection of the leads in this manner would jump the contact gap, contributing to a current buildup in the base circuit and in the collector circuit which would overheat and burn out the transistor. The area surrounding the collector terminal is colored *red* for the meter red lead while the area surrounding the emitter terminal is colored *black* for the meter black lead.

A cold start relay is incorporated into the circuit at the starter relay, interrupting the conventional battery-to-coil lead. The purpose of this is to furnish additional current to the coil primary windings during situations when the starter draw is excessive. The cold start relay contacts normally are closed; only opening during the cranking cycle. However, when the available battery voltage drops below a predetermined value during the cranking cycle, they again close, bypassing the ignition resistor and furnishing full battery current to the system.

The distributor differs from the conventional distributor only in the absence of the condenser and in the highly polished breaker cam. Because one of the big advantages of the transistor ignition is long breaker point life, wear on the rubbing block must be reduced to a minimum. Because the current at the breaker points is so small, the amount of pitting that occurs during normal operation is hardly measurable and point life should be indefinite. The points should be set to .020 in. gap and high-temperature grease used for cam lubrication.

Exploded view—6 cylinder breakerless distributor (© Ford Motor Co.)

the coil will be much quicker and will rise to a higher level than in a conventional ignition system.

The other components of the system, all of which are contained in the amplifier housing with the PNP transistor, are a condenser, a zener diode, a toroid, a base-to-emitter resistor, and a collector resistor.

The base reisstor is similar to the conventional ignition resistor wire and is located between the distributor and the transistor (heat sink). It provides an 8.0 ohm resistance which is necessary for current limitation and it should not be replaced with any other wire, resistance or otherwise. To do so would result in immediate transistor destruction.

The collector and emitter resistors both are located in a ballast resistor block made of white ceramic for electrical and thermal insulation. Both resistors serve the same purpose —limiting system current and control of voltages within their respective circuits. The two resistors are in series in the collector-emitter secion, together with the ignition coil, toroid and transistor. The emitter resistor

also is in series with the base resistor, the toroid, and the transistor, in the base-emitter circuit. The transistor and emitter resistance therefore are common to both circuits. The combined resistances in each circuit permit a base current of approxi-

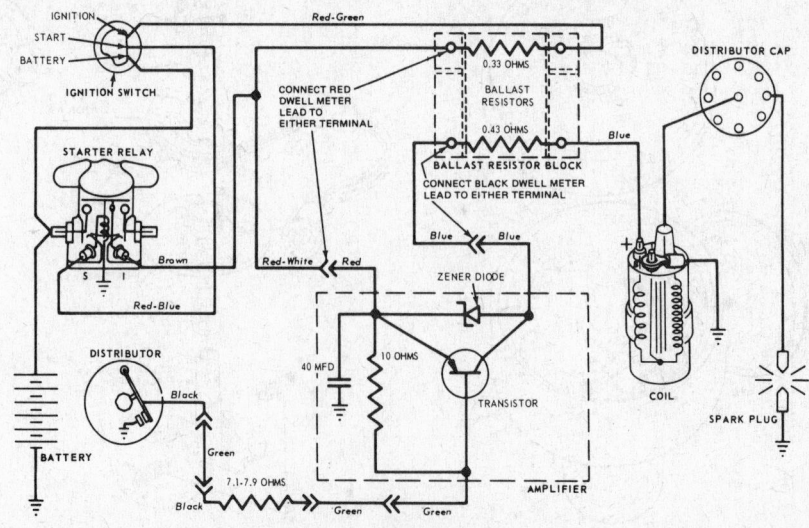

Transistor ignition circuits (© Ford Motor Co.)

When testing the transistor ignition distributor in a test machine, incorporate a condenser into the primary-to-ground circuit using a jumper wire. This will prevent point pitting or oxidation during testing.

CAUTION: When connecting an in-car tachometer to the Ford System, always shunt the tachometer leads that go to the coil IGN terminal and ignition switch with a 10 in. length of Ford ignition resistor wire, part No. COLF-12250-A, to prevent tachometer damage. The higher current draw of the transistor system can ruin a tach if this precaution is not taken.

Troubleshooting

Ignition problems are caused by a failure in the primary or secondary circuit, or incorrect ignition timing. Isolate the trouble as follows:

1. Remove the coil high tension lead from the distributor cap.
2. Disconnect the brown wire from the starter relay "I" terminal and the red and blue wire from the starter relay "S" terminal.
3. Turn the ignition switch on.
4. While holding the high tension lead approximately 1/4 in. from the engine block, crank the engine by using a remote starter switch between the starter relay "S" and battery terminals.

If the spark is good, the trouble lies in the secondary (high voltage) circuit. If there is no spark or a weak spark, the trouble is in the primary (low voltage) circuit.

A breakdown or energy loss in the primary circuit can be caused by:
1. Defective primary wiring.
2. Improperly adjusted, contaminated or defective distributor points.
3. Defective amplifier assembly.

The trouble can be isolated by performing a primary circuit test.

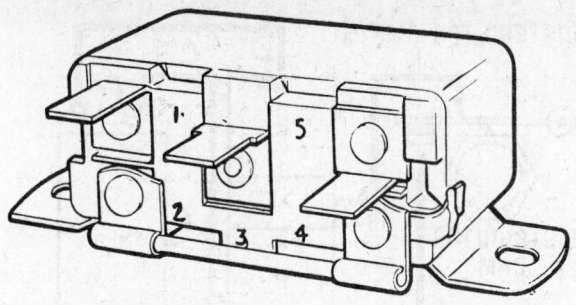

Autolite cold start relay (© Ford Motor Co.)

A breakdown or enery loss in the secondary circuit can be caused by:
1. Fouled or improperly adjusted spark plugs.
2. Defective high voltage wiring.
3. High voltage leakage across the coil, distributor cap, or rotor.

To isolate a problem in the secondary circuit, turn the ignition switch off, remove the remote starter switch from the starter relay, install the coil high tension lead in the distributor cap, the red and blue wire to the starter relay (this goes on the "S" terminal) and the brown wire to the starter relay (this goes on the "I" terminal) and perform a secondary circuit test.

Primary Circuit Tests

CAUTION: Do not use any other procedure, conventional short-cut, or connect test equipment in any other manner than described, or extensive damage can be caused to the transistor ignition system.

Connect a dwell meter to the tachometer block. Connect the black lead to the black (large) terminal and the red lead to the red (small) terminal.

With the remote starter switch installed and the ignition switch on, ground the coil high tension wire and crank the engine and observe the dwell reading.

0° Dwell

1. The distributor points are contaminated or are not closing.
2. An open circuit in the distributor lead to the amplifier.

To determine which item listed is causing the trouble, proceed as follows:

Disconnect the distributor lead at the bullet connector and connect a voltmeter red lead to the red (small) tach block terminal and the voltmeter black lead to the distributor lead from the distributor. *Do not connect the voltmeter to the lead from the amplifier.* Crank the engine and note the voltmeter reading.

If a steady indication of voltage is obtained, the trouble is in the distributor lead to the amplifier. Absence of any voltage indication on the voltmeter shows that there is an open circuit between the distributor lead and the breaker point ground.

0-45° Dwell

1. The transistor and the primary circuit are functioning properly.
2. The trouble could be in the secondary circuit.

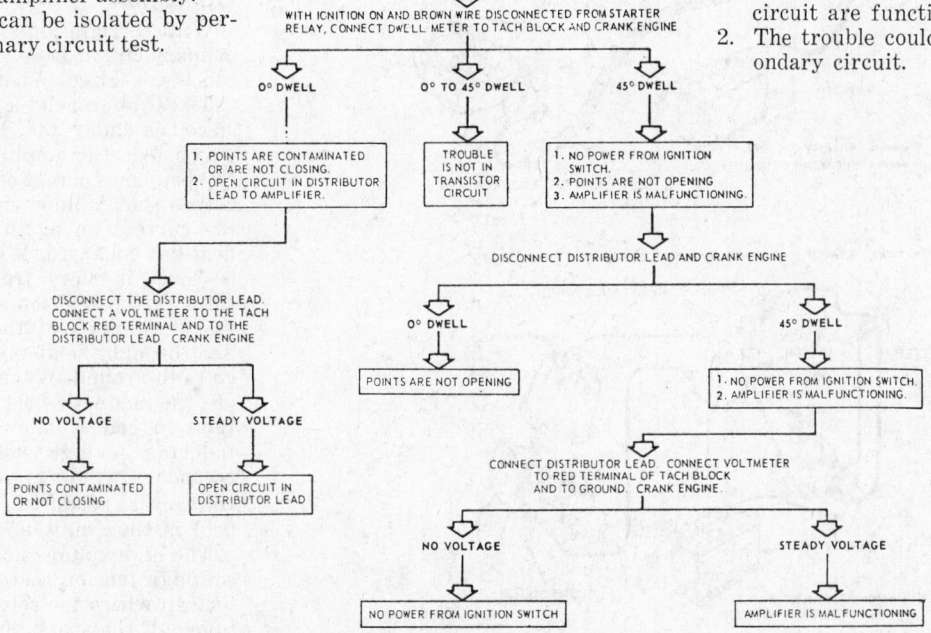

Autolite transistor ignition troubleshooting chart (© Ford Motor Co.)

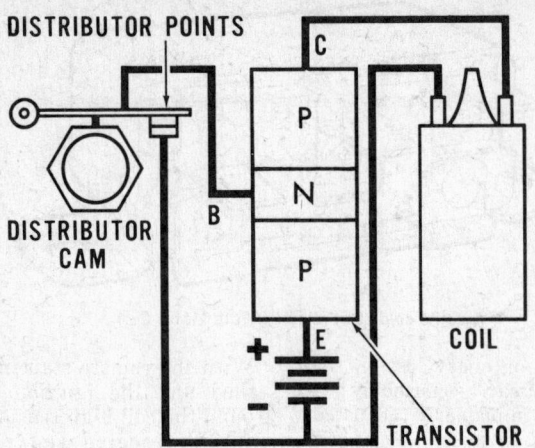

Primary circuit connected to PNP transistor (© Ford Motor Co.)

45° Dwell

1. No power from the ignition switch.
2. The distributor points are closed and not opening.
3. Defective amplifier assembly.

To determine which of the three items listed is causing the trouble, proceed as follows:

Disconnect the distributor lead at the bullet connector, and crank the engine. If the dwellmeter indicates 0° dwell, the distributor points are not opening. If 45° dwell is indicated, the amplifier is malfunctioning or there is no power from the ignition switch.

Use a voltmeter or test light to determine if the transistor (amplifier assembly) is at fault. Connect the voltmeter to the red-green lead terminal of the ballast resistor and to ground. Crank the engine.

Absence of any voltage indication on the voltmeter shows there is an open circuit, or no power between the ignition switch and the amplifier. The ballast resistor could be defective. Replace it with a good ballast resistor, and repeat the test.

A steady indication of voltage on the voltmeter indicates either a defective amplifier or the coil to amplifier lead is defective or improperly connected to the ballast resistor. Proceed as follows:

1. Disconnect the amplifier at the quick disconnect.
2. Connect an ohmmeter across the outside terminals of the amplifier side of the quick disconnect.
3. Reverse the ohmmeter leads.

If a very high resistance is obtained one way and a very low or *zero* resistance is obtained the other way, the amplifier is not defective. Check the coil to amplifier wiring for a loose connection or defective wiring.

After a repair has been made, run through the test again to check for any other malfunctions.

Secondary Circuit Tests

Use conventional system test procedures.

Ford-Motorcraft Solid-State Ignition System

The Ford-Motorcraft Solid-State Ignition System is a pulse triggered, breakerless, transistor controlled ignition system. The system utilizes most of the standard ignition components, but substitutes an amplifier module and magnetic pickup assembly for the conventional ignition contact points.

Operation

With the ignition switch "on", the primary circuit is on and the ignition coil is energized. When the armature "spokes" approach the magnetic pickup coil assembly, they induce a voltage which tells the amplifier to turn the coil primary current off. A timing circuit in the amplifier module will turn the current on again after the coil field has collapsed. When the current is "on", it flows from the battery through the ignition switch, the primary windings of the ignition coil, and through the amplifier module circuits to ground. When the current is off, the magnetic field built up in the ignition coil is allowed to collapse, inducing a high voltage into the secondary windings of the coil. High voltage is produced each time the field is thus built up and collapsed.

The high voltage flows through the coil high tension lead to the distributor cap where the rotor distributes it to one of the spark plug terminals in the distributor cap. This process is re-

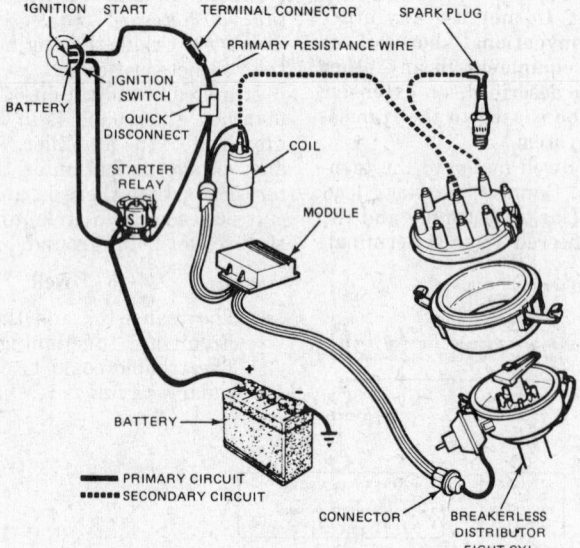

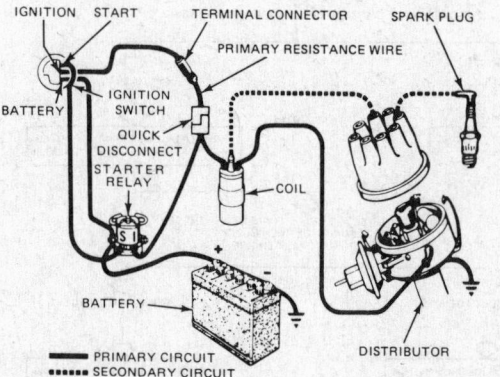

Circuit routings for conventional and breakerless ignition systems (© Ford Motor Co.)

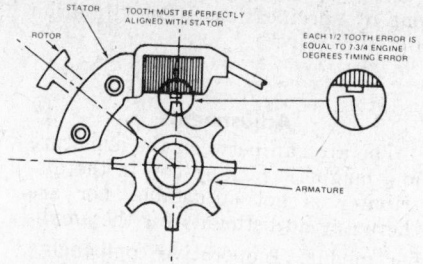

Breakerless armature alignment in relationship with ignition timing (© Ford Motor Co.)

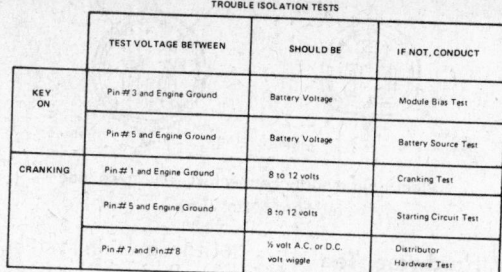

TROUBLE ISOLATION TESTS

	TEST VOLTAGE BETWEEN	SHOULD BE	IF NOT, CONDUCT
KEY ON	Pin # 3 and Engine Ground	Battery Voltage	Module Bias Test
	Pin # 5 and Engine Ground	Battery Voltage	Battery Source Test
CRANKING	Pin # 1 and Engine Ground	8 to 12 volts	Cranking Test
	Pin # 5 and Engine Ground	8 to 12 volts	Starting Circuit Test
	Pin # 7 and Pin # 8	½ volt A.C. or D.C. volt wiggle	Distributor Hardware Test

	TEST RESISTANCE BETWEEN	SHOULD BE	IF NOT, CONDUCT
KEY OFF	Pin # 7 and Pin # 8	400 to 800 ohms	Magnetic Pick-up (Stator) Test
	Pin # 6 and Engine Ground	0 ohms	
	Pin # 7 and Engine Ground	more than 70,000 ohms	
	Pin # 8 and Engine Ground	more than 70,000 ohms	
	Pin # 3 and Coil Tower	7000 to 13000 ohms	Coil Test
	Pin # 5 and Pin # 4	1.0 to 2.0 ohms	
	Pin # 5 and Engine Ground	more than 10.0 ohms	Short Test
	Pin # 3 and Pin # 4	1.0 to 2.0 ohms	Resistance Wire

Solid state ignition system diagnosis (© Ford Motor Co.)

peated for every power stroke of the engine.

Ignition system troubles are caused by a failure in the primary and/or the secondary circuit; incorrect ignition timing; or incorrect distributor advance. Circuit failures may be caused by shorts, corroded or dirty terminals, loose connections, defective wire insulation, cracked distributor cap or rotor, defective pick-up coil assembly or amplifier module, defective distributor points, fouled spark plugs, or by improper dwell angle.

If an engine starting or operating trouble is attributed to the ignition system, start the engine and verify the complaint. On engines that will not start, be sure that there is gasoline in the fuel tank and that fuel is reaching the carburetor. Then locate the ignition system problem by an oscilloscope test or by a spark intensity test.

Primary Circuit Testing

A breakdown or energy loss in the primary circuit can be caused by: defective primary wiring, loose or corroded connections, inoperative or defective magnetic pick-up coil assembly, or defective amplifier module.

A complete test of the primary circuit consists of checking the circuits in the ignition coil, the magnetic pick-up coil assembly and the amplifier module. Wiring harness checks will be included as a part of basic component circuit tests.

Always inspect connectors for dirt, corrosion or poor fit before assuming you have spotted a possible problem.

Troubleshooting

Make sure that the battery is fully charged before beginning tests. Perform a Spark Intensity Test. If no spark is observed, make sure that the high tension coil wire is good. Discon-

nect the three-way and four-way connectors at the electronic module.

The first trouble isolation test will be conducted on the harness terminals, with the electronic module disconnected from the circuit. The pin numbers shown in the schematic correspond to those shown in the trouble isolation test table.

Make the following tests using a sensitive volt-ohmmeter. These tests will direct you to the proper follow-up test to determine the actual problem.

If the circuit checks good at all these test points, connect a known good electronic module in place of the vehicle module and again perform the spark intensity test. If the substitution corrects the malfunction again reconnect the vehicle module and perform the spark intensity test. If the malfunction still exists, the problem is in the module and it must be replaced. If the problem is gone, it may be in the wiring connectors.

If the substitute module does not correct the problem, reconnect the original module and make repairs elsewhere in the system.

Module Bias Test

Measure the voltage at Pin 3 to engine ground with the ignition key "on". If the voltage observed is less than battery voltage, repair the voltage feed wiring to the module for running conditions (re-wire).

Battery Source Test

1. Connect the voltmeter leads from the battery terminal at the coil to engine ground, without disconnecting the coil from the circuit.
2. Install a jumper wire from the DEC terminal of the coil to a good engine ground.
3. Turn the lights and all accessories off.
4. Turn the ignition switch "on".
5. If the voltmeter reading is between 4.9 and 7.9 volts, the primary circuit from the battery is satisfactory.
6. If the voltmeter reading is less than 4.9 volts, check the following:
   a. The primary wiring for worn insulation, broken strands, and loose or corroded terminals.
   b. The resistance wiring for defects.
7. If the voltmeter reading is greater than 7.9 volts, the resistance wire should be replaced after verifying a defect.

Cranking Test

Measure the voltage at Pin 1 to engine ground with the engine cranking. If the voltage observed is not 8 to 12 volts, repair the voltage feed to the module for starting conditions (white wire).

Starting Circuit Test

If the reading is not between 8 and 12 volts, the ignition by-pass circuit is open or grounded from either the starter solenoid or the ignition switch to Pin 5. Check the primary connections at the coil.

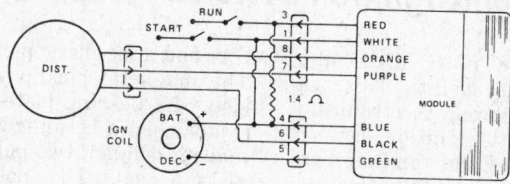

Electronic module schematic—solid-state ignition (© Ford Motor Co.)

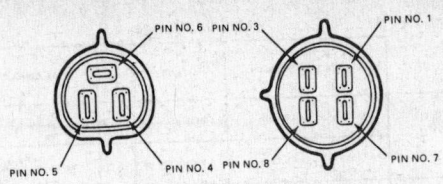

PIN NO. 6 PIN NO. 3   PIN NO. 1

PIN NO. 5   PIN NO. 4 PIN NO. 8   PIN NO. 7

Electronic module connectors—harness side
(© Ford Motor Co.)

Distributor Hardware Test

1. Disconnect the three-wire weatherproof connector at the distributor pigtail.
2. Connect a D.C. voltmeter on a 2.5 volt scale to the two parallel blades. With the engine cranking, the meter needle should oscillate.
3. Remove the distributor cap and check for visual damage or misassembly.
   a. Sintered iron armature (6 or 8-toothed wheel) must be tight on the sleeve, and the roll pin aligning the armature must be in position.
   b. Sintered iron stator must not be broken.
   c. Armature must rotate when the engine is cranked.
4. If the hardware is alright, but the meter doesn't oscillate, replace the magnetic pick-up assembly.

Magnetic Pick-up Tests

1. Resistance of pick-up coil measured between two parallel pins in the distributor connector must be 400-800 ohms.
2. Resistance between the third blade (ground) and the distributor bowl must be zero ohms.
3. Resistance between either parallel blade and engine ground must be greater than 70,000 ohms.
4. If any test fails, the distributor stator assembly is defective and must be replaced.
5. If the above readings are not the same as measured in the original test, check for a defective harness. If the readings are the same, proceed.
6. If these tests check alright, the signal generator portion of the distributor is working properly.

Ignition Coil Test

The breakerless ignition coil must be diagnosed separately from the rest of the ignition system.

1. Primary resistance must be 1.0-2.0 ohms, measured from the BAT to the DEC terminals.
2. Secondary resistance must be 7,000-13,000 ohms, measured from the BAT or DEC terminal to the center tower of the coil.
3. If resistance tests are alright, but the coil is still suspected, test the coil on a coil tester by following the test equipment manufacturer's instructions for a standard coil. If the reading differs from the original test, check for a defective harness.

Module Test 1975 and later

1. Unplug the electronic module connector, but don't remove the existing module from the car.
2. Connect a module which is known to be good to the connector. There is no need to attach the module to the car in order to have it work.
3. Start the engine; if it starts and operates correctly go on to the next step. If it won't start and run the trouble is somewhere else. Check and repair the wiring and other systems, as required.
4. If the engine started in step 3, reconnect the original module and try to start the engine again. If the engine won't start, replace the module.
5. If the engine starts in step 4, the original module is not defective. Check all the wiring and connections in the ignition system.

Short Test

If the resistance from Pin 5 to ground is less than 10 ohms, check for a short to ground at the DEC terminal of the ignition coil or in the connection wiring to that terminal.

Resistance Wire Test

Replace the resistance wire if it is out of specifications (See "Ignition Resistor Wire").

Adjustments

The air gap between the armature and magnetic pick-up coil in the distributor is not adjustable, nor are there any adjustments for the amplifier module. Inoperative components are simply replaced. Any attempt to connect components outside the vehicle may result in component failure.

Component Replacement

Magnetic Pick-up Assembly Removal and Installation

1. Remove the distributor cap and rotor and disconnect the distributor harness plug.
2. Using a small gear puller or two screwdrivers, lift or pry the armature from the advance plate sleeve. Remove the roll pin.
3. Remove the large wire retaining clip from the base plate annular groove.
4. Remove the snap-ring which secures the vacuum advance link to the pick-up assembly.
5. Remove the magnetic pick-up assembly ground screw and lift the assembly from the distributor.
6. Lift the vacuum advance arm off the post on the pick-up assembly and move it out against the distributor housing.
7. Place the new pick-up assembly in position over the fixed base plate and slide the wiring in position through the slot in the side of the distributor housing.
8. Install the fine wire snap-ring securing the pick-up assembly to the fixed base plate.
9. Position the vacuum advance arm over the post on the pick-up assembly and install the snap-ring.
10. Install the grounding screw through the tab on the wiring harness and into the fixed base plate.
11. Install the armature on the advance plate sleeve making sure that the roll pin is engaged in the matching slots.
12. Install the distributor rotor cap.
13. Connect the distributor wiring plug to the vehicle harness.

Delco Remy Electronic Ignition Systems

Delco-Remy Magnetic Pulse System

Components

The Delco-Remy magnetic pulse, fully transistorized ignition system uses a magnetic pulse distributor having no breaker points. This system switches power electronically rather than with ignition contact points. Instead of the familiar cam and breaker plate assembly, this distributor uses a rotating iron timer core and a magnetic pickup assembly. The magnetic pickup assembly consists of a bearing plate on which are sandwiched a ceramic ring-type permanent magnet, two pole pieces and a pick-up coil. The pole pieces are doughnut shaped steel plates with

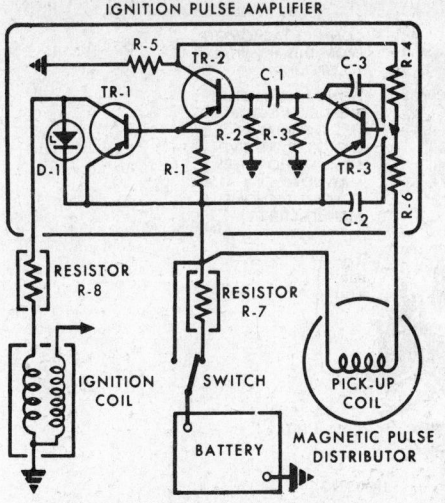

IGNITION PULSE AMPLIFIER

Magnetic pulse amplifier schematic
(© Chevrolet Div., G.M. Corp.)

accurately spaced internal teeth, one tooth for each cylinder of the engine.

A critically important part is the iron timer core. It has a number of equally spaced projections or vanes and is attached to, and rotates with, the distributor shaft.

The transistor control unit, the switchbox of the system, is mounted in an aluminum case and contains three transistors, a zener diode, a condenser and five small resistors. The zener diode is a circuit protection device. Remaining components control and switch ignition-coil current electronically; there are no moving parts in the control unit.

The ignition coil is of standard design except for a special winding. The external primary resistor is a ceramic type, similar to those used on various conventional systems.

Operation

The ignition primary circuit is connected from the battery, through the ignition switch, through the ignition pulse amplifier assembly, through the primary side of the ignition coil, and back to the amplifier housing where it is grounded externally. The secondary circuit is the same as in conventional ignition systems: the secondary side of the coil, the coil wire to the distributor, the rotor, the spark plug wires and the spark-plugs.

The magnetic pulse distributor is also connected to the ignition pulse amplifier. As the distributor shaft rotates, the distributor rotating pole piece turns inside the stationary pole piece. As the rotating pole piece turns inside the stationary pole piece, the eight teeth on the rotating pole piece align with the eight teeth on the stationary pole piece eight times during each distributor revolution (two crankshaft revolutions since the distributor runs at one-half crankshaft speed). As the rotating pole piece

teeth move close to, and align with, the teeth on the stationary pole piece, the magnetic rotating pole piece induces voltage into the magnetic pole piece through the stationary pole piece. This voltage pulse is sent to the ignition pulse amplifier from the magnetic pole piece. When the pulse enters the amplifier, it signals the ignition pulse amplifier to interrupt the ignition primary circuit. This causes the primary circuit to collapse and begins the induction of the magnetic lines of force from the primary side of the coil into the secondary side of the coil. This induction provides the required voltage to fire the spark plugs.

The advantages of this system are that the transistors in the ignition pulse amplifier can make and break the primary ignition circuit much faster than conventional ignition points, and higher primary voltage can be utilized since this system can be made to handle higher voltage without adverse effects, whereas ignition breaker points cannot. The shorter switching time of this system allows longer coil primary circuit saturation time and longer induction time when the primary circuit collapses. This increased time allows the primary circuit to build up more current and the secondary circuit to discharge more current.

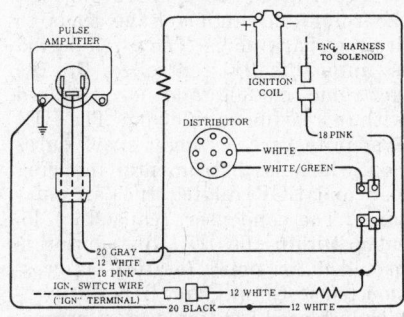

Magnetic pulse system circuit diagram
(© Chevrolet Div., G.M. Corp.)

Troubleshooting

Cautions

1. Don't use 18 volts or 24 volts for emergency starting.
2. Never crank engine with coil high-tension lead or more than three spark plug leads disconnected.
3. Don't short circuit between coil positive terminal and ground.
4. On any repair that necessitates replacement of control unit or ignition resistor, perform complete charging system check before releasing the unit. Basic cause of trouble may be high or uncontrolled charging rate.

Engine Surge or Intermittent Miss

Since there are so many possible causes for this problem, all other pos-

sible defects must be ruled out before the specialized components of the electronic ignition system are judged defective.

As a general rule, a miss or surge that is caused by an ignition problem will be much more pronounced than a similiar problem that is caused by carburetion. Also, carburetion is usually affected by temperature more than the ignition system is. A carburetor or intake manifold vacuum leak is often compensated for by the choke when the engine is cold. When the engine warms up and the choke is released, the engine surge will show up.

If the ignition system is found to be the source of the problem, first check all connections in the system to make sure that they are *clean and tight*. Check the coil and spark plug high-tension wires with an ohmmeter to be sure they have the correct resistance. Check the inside and outside of the distributor cap and the tower on the ignition coil for cracks which would allow the high voltage intended for the spark plugs to short to ground.

If none of the above checks uncovers a defective component, the distributor pick-up coil leads may be reversed in the connector, or the pickup coil itself may have an intermittent open.

Engine Will Not Start or Is Hard to Start

1. Disconnect a spark plug wire from one spark plug and hold the wire 1/4 in. from a good ground with a pair of insulated pliers.
2. Crank the engine over and observe whether a spark jumps from the plug wire to ground.
3. *If spark occurs*, the problem is not in the ignition system.
4. *If spark does not occur*, reconnect the spark plug wire that was disconnected and connect a tachometer between the positive (+) coil primary terminal and the pink wire in the three-wire connector to the ignition pulse amplifier.
5. Crank the engine over and observe the tachometer.
6. *If the tachometer needle deflects* while cranking the engine, perform "Ignition Distributor Test" to locate the problem.
7. *If the tachometer needle does not deflect* while cranking the engine, perform "Circuit Resistance Test" to pinpoint the problem.

Circuit Resistance Test Ignition Distributor Check

1. Disconnect the distributor leads from the engine wiring harness.
2. Connect the two leads of an ohmmeter to the distributor

leads at the connector.

3. Rotate the magnetic pick-up assembly in the distributor through full vacuum advance travel and read the ohmmeter. If the reading is not within a range of 500-700 ohms, replace the magnetic pick-up assembly.

4. If the reading is within the 500-700 ohms range, disconnect one ohmmeter lead from the distributor connector and connect it to a good ground. If the reading is less than infinity (needle moves to end of scale), replace the magnetic pick-up assembly.

5. If the reading is infinite, and there was no spark when the spark plug wire was disconnected from the plug, the amplifier is defective.

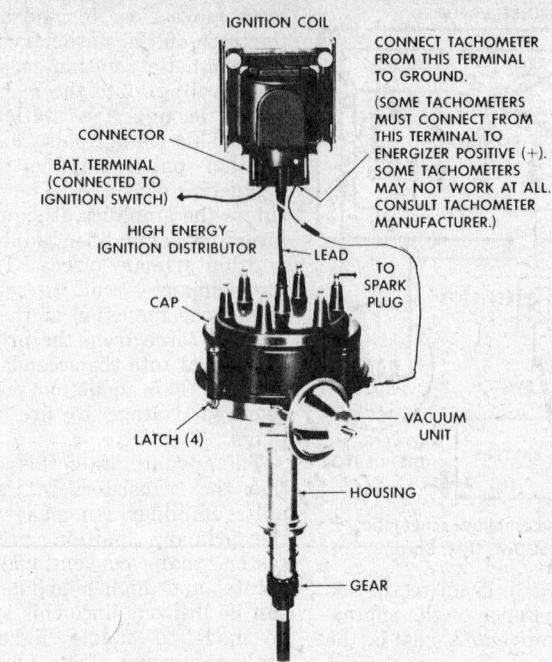

Connector details for inline six-cylinder HEI system (© Chevrolet Div., G.M. Corp.)

Delco-Remy High Energy Ignition (HEI) System

Components

The Delco-Remy High Energy Ignition (HEI) System is a breakerless, pulse triggered, transistor controlled, inductive discharge ignition system.

It is similar in operation to the Magnetic Pulse System. There are only nine external electrical connections; the ignition switch feed wire, and the eight spark plug leads. On V8 engines, the ignition coil is located within the distributor cap, connecting directly to the rotor.

Operation

The magnetic pick-up assembly located inside the distributor contains a permanent magnet, a pole piece with internal teeth, and a pick-up coil. When the teeth of the rotating timer core and pole piece align, an induced

voltage in the pick-up coil signals the electronic module to open the coil primary circuit. As the primary current decreases, a high voltage is induced in the secondary windings of the ignition coil, directing a spark through the rotor and high voltage leads to fire the spark plugs. The dwell period is automatically controlled by the electronic module and is increased with increasing engine rpm. The HEI System features a longer spark duration which is instrumental in firing lean and EGR diluted fuel/air mixtures. The condenser (capacitor) located within the HEI distributor is provided for noise (static) suppression purposes only and is not a regularly replaced ignition system component.

Major Repair Operations (Distributor in Engine)

Ignition Coil Replacement V8 Engines

1. Disconnect the feed and module wire terminal connectors from the distributor cap.
2. Remove the ignition set retainer.
3. Remove the 4 coil cover-to-distributor cap screws and the coil cover.
4. Remove the 4 coil-to-distributor cap screws.
5. Using a blunt drift, press the coil wire spade terminals up out of distributor cap.
6. Lift the coil up out of the distributor cap.
7. Remove and clean the coil spring, rubber seal washer and coil cavity of the distributor cap.
8. Coat the rubber seal with a dielectric lubricant furnished in the replacement ignition coil package.
9. Reverse the above procedures to install.

Six Cylinder Engines

On 6 cylinder engines, a separate ignition coil is used. To remove and install it, proceed as follows:

1. Remove the ignition switch-to-coil lead from the coil.
2. Unfasten the distributor leads from the coil.
3. Remove the screws which secure the coil to the engine and lift it off.

Installation is the reverse of removal.

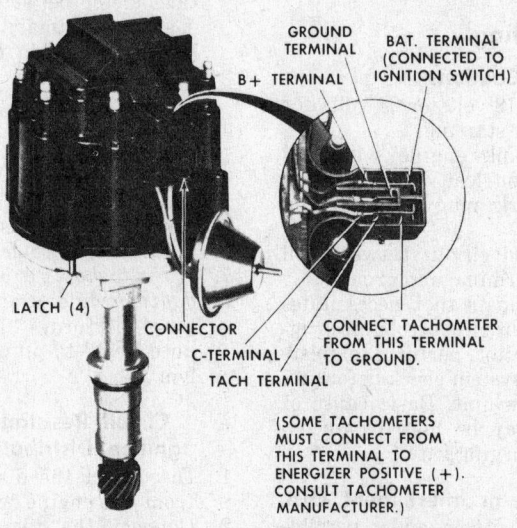

Connector details for V8 HEI system (© Chevrolet Div., G.M. Corp.)

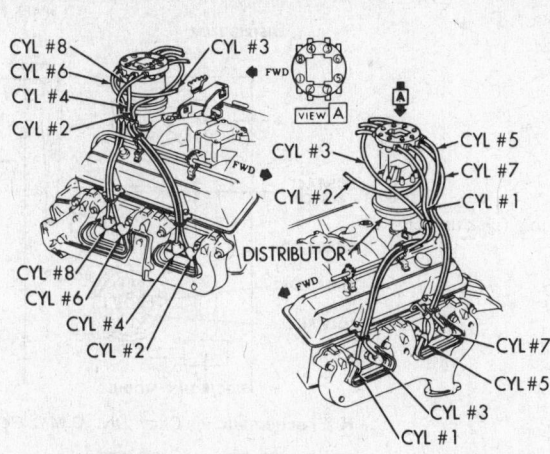

V8 HEI secondary wiring (© Chevrolet Div., G.M. Corp.)

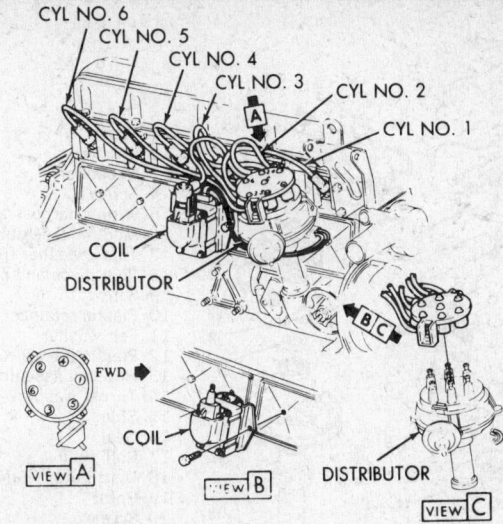

Inline six-cylinder HEI secondary wiring (© Chevrolet Div., G.M. Corp.)

Distributor Cap Replacement

1. Remove the feed and module wire terminal connectors from the distributor cap.
2. Remove the retainer and spark plug wires from the cap.
3. Depress and release the 4 distributor cap-to-housing retainers and lift off the cap assembly.
4. Remove the 4 coil cover screws and cover (V8 only).
5. Using a finger or a blunt drift, push the spade terminals up out of the distributor cap (V8 only).
6. Remove all 4 coil screws and lift the coil, coil spring and rubber seal washer out of the cap coil cavity (V8 only).
7. Using a new distributor cap, reverse the above procedures to assemble being sure to clean and lubricate the rubber seal washer with dielectric lubricant.

Rotor Replacement

1. Disconnect the feed and module wire connectors from the distributor.
2. Depress and release the 4 distributor cap to housing retainers and lift off the cap assembly.
3. Remove the two rotor attaching screws and rotor.
4. Reverse the above procedure to install.

Vacuum Advance Unit Replacement

1. Remove the distributor cap and rotor as previously described.
2. Disconnect the vacuum hose from the vacuum advance unit. Remove the module.
3. Remove the two vacuum advance retaining screws, pull the advance unit outward, rotate and disengage the operating rod from its tang.
4. Reverse the above procedure to install.

Module Replacement

1. Remove the distributor cap and rotor as previously described.
2. Disconnect the harness connector and pick-up coil spade connectors from the module (note their positions).
3. Remove the two screws and module from the distributor housing.
4. Coat the bottom of the new module with dielectric lubricant. Reverse the above procedure to install. Be sure that the leads are installed correctly.

Distributor Removal

1. Disconnect the ground cable

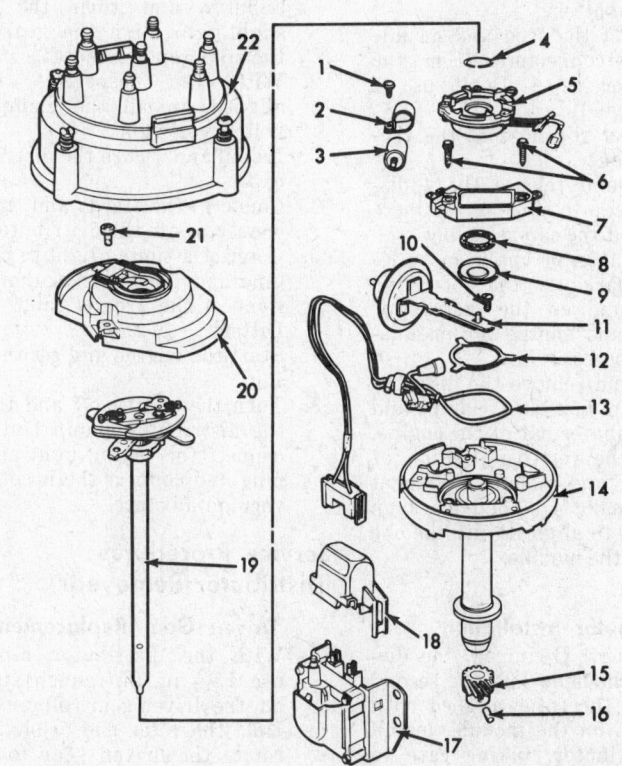

1 Screw	12 Retainer (wire harness)
2 Bracket	13 Wire harness assembly
3 Capacitor	14 Housing assembly
4 Thin c-washer (retainer)	15 Gear
5 Pole piece and plate assembly (pick up coil)	16 Roll pin
6 Screw	17 Ignition coil
7 Module assembly	18 Cover
8 Felt washer	19 Distributor shaft assembly
9 Plastic grease retainer seal	20 Rotor
10 Screw	21 Screw
11 Vacuum control assembly	22 Distributor cap

H E I distributor—6 cylinder (© Chev. Div., G.M.C. Corp.)

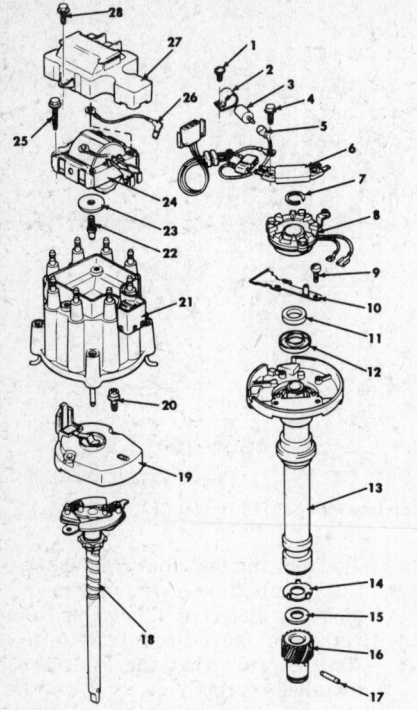

1 Screw
2 Bracket
3 Capacitor
4 Screw
5 Wiring harness assembly
6 Module assembly
7 Thin c-washer (retainer)
8 Pole piece and plate assembly (pick up coil)
9 Screw
10 Plastic retainer
11 Felt washer
12 Plastic grease retainer seal
13 Housing assembly
14 Thrust washer
15 Shim
16 Gear
17 Roll pin
18 Distributor shaft assembly
19 Rotor
20 Screw
21 Distributor cap
22 Resistor brush and spring
23 Seal
24 Ignition coil
25 Screw
26 Ground lead
27 Cover
28 Screw

H E I distributor—V-8 (© Chev. Div., G.M.C. Corp.)

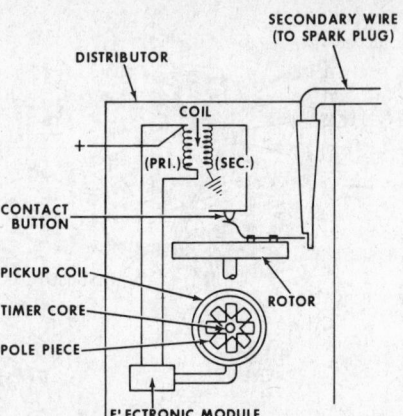

H E I schematic (© Chev. Div., G.M.C. Corp.)

Terminals on distributor cap assembly
(© Chev Div., G.M.C. Corp.)

from the battery.

2. Disconnect the feed and module terminal connectors from the distributor cap. (Don't use a screwdriver).
3. Disconnect the hose at the vacuum advance.
4. Depress and release the 4 distributor cap-to-housing retainers and lift off the cap assembly.
5. Using crayon or chalk, make locating marks on the rotor and module and on the distributor housing and engine for installation purposes.
6. Loosen and remove the distributor clamp bolt and clamp, and lift distributor out of the engine. Noting the relative position of the rotor and module alignment marks, make a second mark on the rotor to align it with the one mark on the module.

Distributor Installation

1. With a new O-ring on the distributor housing and the second mark on the rotor aligned with the mark on the module, install the distributor, taking care to align the mark on the housing with the one on the engine. It may be necessary to lift the dis-

tributor and turn the rotor slightly to align the gears and the oil pump driveshaft.

2. With the respective marks aligned, install the clamp and bolt finger-tight.
3. Install and secure the distributor cap.
4. Connect the feed and module connectors to the distributor cap.
5. Connect a timing light to the engine and plug the vacuum hose.
6. Connect the ground cable to the battery.
7. Start the engine and set the timing.
8. Turn the engine off and tighten the distributor clamp bolt. Disconnect the timing light and unplug and connect the hose to the vacuum advance.

Service Procedures (Distributor Removed)

Driven Gear Replacement

1. With the distributor removed, use a 1/8 in. pin punch and tap out the driven gear roll pin.
2. Hold the rotor end of shaft and rotate the driven gear to shear any burrs in the roll pin hole.
3. Remove the driven gear from the shaft.

4. Reverse the above procedure to install.

Mainshaft Replacement

1. With the driven gear and rotor removed, gently pull the mainshaft out of the housing.
2. Remove the advance springs, weights and slide the weight base plate off the mainshaft.
3. Reverse the above procedure to install.

Pole Piece, Magnet or Pick-up Coil Replacement

1. With the mainshaft out of its housing, remove the 3 retaining screws, pole piece and magnet and/or pick-up coil.
2. Reverse the removal procedure to install making sure that the pole piece teeth do not contact the timer core teeth by installing and rotating the mainshaft. Loosen the 3 screws and realign the pole piece as necessary.

Chrysler Electronic Ignition

Components

This system consists of a special pulse-sending distributor, an electronic control unit, a two-element bal-

last resistor, and a special ignition coil.

The distributor does not contain breaker points or a condenser, these

parts being replaced by a distributor reluctor and a pick-up unit.

Operation

The ignition primary circuit is connected from the battery, through the ignition switch, through the primary side of the ignition coil, to the control unit where it is grounded. The secondary circuit is the same as in conventional ignition systems: the secondary side of the coil, the coil wire to the distributor, the rotor, the spark plug wires, and the spark plugs.

The magnetic pulse distributor is also connected to the control unit. As the distributor shaft rotates, the distributor reluctor turns past the pick-up unit. As the reluctor turns past the pick-up unit, each of the eight teeth on the reluctor pass near the pick-up unit once during each distributor revolution (two crankshaft revolutions since the distributor runs at one-half crankshaft speed). As the reluctor teeth move close to the pick-up unit, the magentic rotating reluctor induces voltage into the magnetic pick-up unit. This voltage pulse is sent to the ignition control unit from the magnetic pick-up unit. When the pulse enters the control unit, it signals the control unit to interrupt the ignition primary circuit. This causes the primary circuit to collapse and begins the induction of the magnetic lines of force from the primary side of the coil into the secondary side of the coil. This induction provides the required voltage to fire the spark plugs.

The advantages of this system are that the transistors in the control unit can make and break the primary ignition circuit much faster than con-

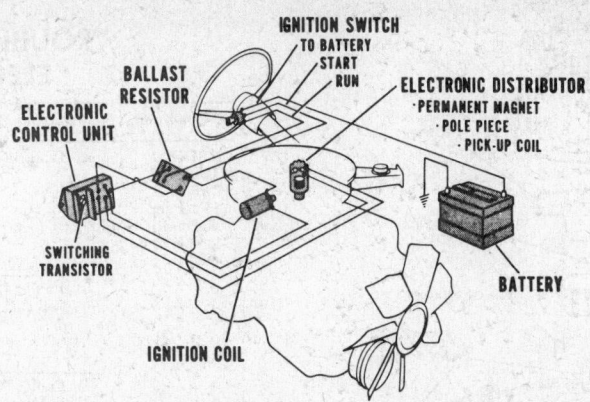

Chrysler electronic ignition system (© Chrysler Corp.)

ventional ignition points can, and higher primary voltage can be utilized, since this system can be made to handle higher voltage without adverse effects, whereas ignition breaker points cannot. The quicker switching time of this system allows longer coil primary circuit saturation time and longer induction time when the primary circuit collapses. This increased time allows the primary circuit to build up more current and the secondary circuit to discharge more current.

Pick-up Coil Replacement

1972-74

1. Remove the distributor.
2. Remove the pick-up coil mounting screw.
3. Remove the wires from the retainers on the upper plate and dis-

tributor housing.
4. Remove pick-up coil from the upper plate.
5. Position the pick-up coil on the pivot of the upper plate and install the mounting screw. Do not tighten.
6. Insert the wires into the appropriate retainers in the distributor.
7. Install the distributor.
8. Set the air gap.

1975 and later

1. Remove the distributor from the engine.
2. Using two small pry-bars or screwdrivers (maximum 7/16 in. wide), pry the reluctor off the shaft from the bottom.

CAUTION: Do not damage the teeth on the reluctor.

3. Unfasten the vacuum advance-

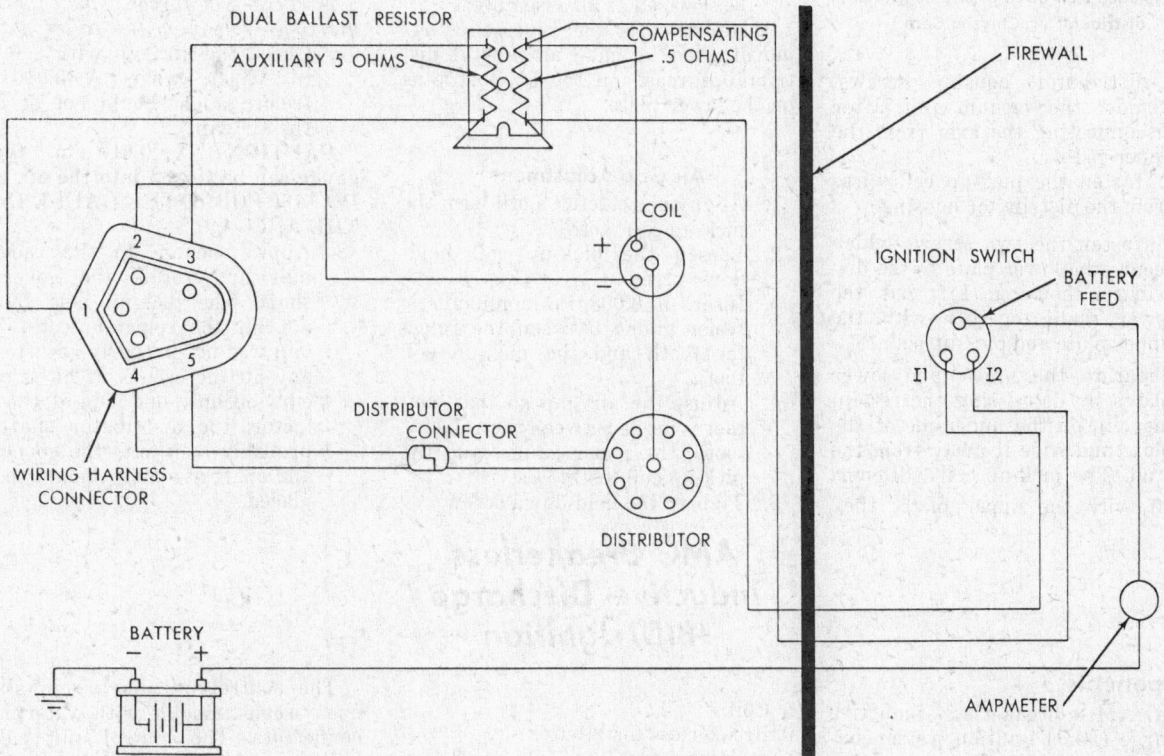

Chrysler electronic ignition system schematic (© Chrysler Corp.)

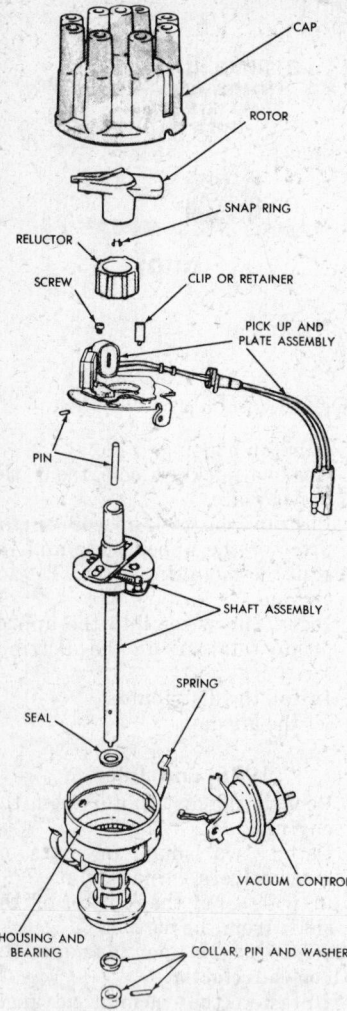

Exploded view of V8 electronic ignition distributor (© Chrysler Corp.)

TROUBLESHOOTING CHRYSLER ELECTRONIC IGNITION

Condition	Possible Cause	Correction
ENGINE WILL NOT START (Fuel and carburetion known to be OK)	a) Dual Ballast	Check resistance of each section: Compensating resistance: .50-.60 ohms @ 70°-80°F Auxiliary Ballast: 4.75-5.75 ohms Replace if faulty. Check wire positions.
	b) Faulty Ignition Coil	Check for carbonized tower. Check primary and secondary resistances: Primary: 1.41-1.79 ohms @ 70°-80°F Secondary: 9,200-11,700 ohms @ 70°-80°F Check in coil tester.
	c) Faulty Pickup or Improper Pickup Air Gap	Check pickup coil resistance: 400-600 ohms Check pickup gap: .010 in. feeler gauge should not slip between pickup coil core and an aligned reluctor blade. No evidence of pickup core striking reluctor blades should be visible. To reset gap, tighten pickup adjustment screw with a .008 in. feeler gauge held between pickup core and an aligned reluctor blade. After resetting gap, run distributor on test stand and apply vacuum advance, making sure that the pickup core does not strike the reluctor blades.
	d) Faulty Wiring	Visually inspect wiring for brittle insulation. Inspect connectors. Molded connectors should be inspected for rubber inside female terminals.
	e) Faulty Control Unit	Replace if all of the above checks are negative. Whenever the control unit or dual ballast is replaced, make sure the dual ballast wires are correctly inserted in the keyed molded connector.
ENGINE SURGES SEVERELY (Not Lean Carburetor)	a) Wiring	Inspect for loose connection and/or broken conductors in harness.
	b) Faulty Pickup Leads	Disconnect vacuum advance. If surging stops, replace pickup.
	c) Ignition Coil	Check for intermittent primary.
ENGINE MISSES (Carburetion OK)	a) Spark Plugs	Check plugs. Clean and regap if necessary.
	b) Secondary Cable	Check cables with an ohmmeter, or observe secondary circuit performance with an oscilloscope.
	c) Ignition Coil	Check for carbonized tower. Check in coil tester.
	d) Wiring	Check for loose or dirty connections.
	e) Faulty Pickup Lead	Disconnect vacuum advance. If miss stops, replace pickup.
	f) Control Unit	Replace if the above checks are negative.

to-distributor housing screws. Remove the vacuum unit, after disconnecting the arm from the upper plate.

4. Unfasten the pick-up coil wires from the distributor housing.

5. Unfasten the two screws which secure the lower plate to the distributor housing. Lift out the lower plate together with the upper plate and pick-up coil.

6. Separate the upper and lower plates by depressing the retaining clip on the underside of the plate and slide it away from the stud. The pick-up coil will come off with the upper plate; they cannot be separated; they must be serviced as an assembly.

Installation is the reverse of removal. Place a small amount of distributor grease on the support pins on the lower plate.

Air Gap Adjustment

1. Align one reluctor tooth with the pick-up coil tooth.
2. Loosen the pick-up coil hold-down screw.
3. Insert a 0.008 in. nonmagnetic feeler gauge between the reluctor tooth and the pick-up coil tooth.
4. Adjust the air gap so that contact is made between the reluctor tooth, the feeler gauge, and the pick-up coil tooth.
5. Tighten the hold-down screw.

AMC Breakerless Inductive Discharge (BID) Ignition

Components

The AMC breakerless inductive discharge (BID) ignition system consists of five components:
Control unit

Coil
Breakerless distributor
Ignition cables
Spark plugs

6. Remove the feeler gauge.
NOTE: No force should be required in removing the feeler gauge.
7. Check the air gap with a 0.010 in. feeler gauge. A 0.010 in. feeler gauge should not fit into the air gap.
CAUTION: A 0.010 in. feeler gauge can be forced into the air gap. DO NOT FORCE THE GAUGE INTO THE AIR GAP.
8. Apply vacuum to the vacuum unit and rotate the governor shaft. The pick-up pole should not hit the reluctor teeth. The gap was not properly adjusted if any hitting occurs. If hitting occurs on only one side of the reluctor, the distributor shaft is probably bent, and the governor and shaft assembly should be replaced.

The control unit is a solid-state, epoxy-sealed module with waterproof connectors. The control unit has a built-in current regulator, so no separate ballast resistor or resistance wire is needed in the primary circuit. Battery voltage is supplied to the ignition coil positive (+) terminal when the ignition key is turned to the "ON" or "START" position; low voltage is also supplied by the control unit.

The coil used with the BID system requires no special service. It works just like the coil in a conventional ignition system.

The distributor is conventional, except for the lack of points, condenser and cam. Advance is supplied by both a vacuum unit and a centrifugal advance mechanism. A standard cap, rotor, and dust shield are used.

In place of the points, cam, and condensor, the distributor has a sensor and trigger wheel. The sensor is a small coil which generates an electromagnetic field when excited by the oscillator in the control unit.

Standard spark plugs and ignition cables are used.

Operation

When the ignition switch is turned on, the control unit is activated. The control unit then sends an oscillating signal to the sensor which causes the sensor to generate a magnetic field. When one of the trigger wheel teeth enters this field, the strength of the oscillation in the sensor is reduced. Once the strength drops to a predetermined level, a demodulator circuit operates the control unit's switching transistor. The switching transistor is wired in series with the coil primary circuit; it switches the circuit off when it gets the demodulator signal.

From this point on, the BID ignition system works in the same manner as a conventional ignition system.

Troubleshooting

1. Check all of the BID ignition system electrical connections.
2. Disconnect the coil-to-distributor high tension lead.
3. Hold the end of the lead ½ in. away from a ground. Crank the engine. If there is a spark, the trouble is not in the ignition system.
4. If there was no spark in step 3, connect a test light with a No. 57 bulb between the positive coil terminal (+) and a good ground. Have an assistant turn the ignition switch to "ON" and "START" (Do not start the engine). The bulb should light in both positions; if it doesn't, the fault lies in the battery-to-coil circuit. Check the ignition switch and related wiring.
5. If the test light lit in step 4, disconnect the coil-to-distributor leads at the connector and connect the test light between the

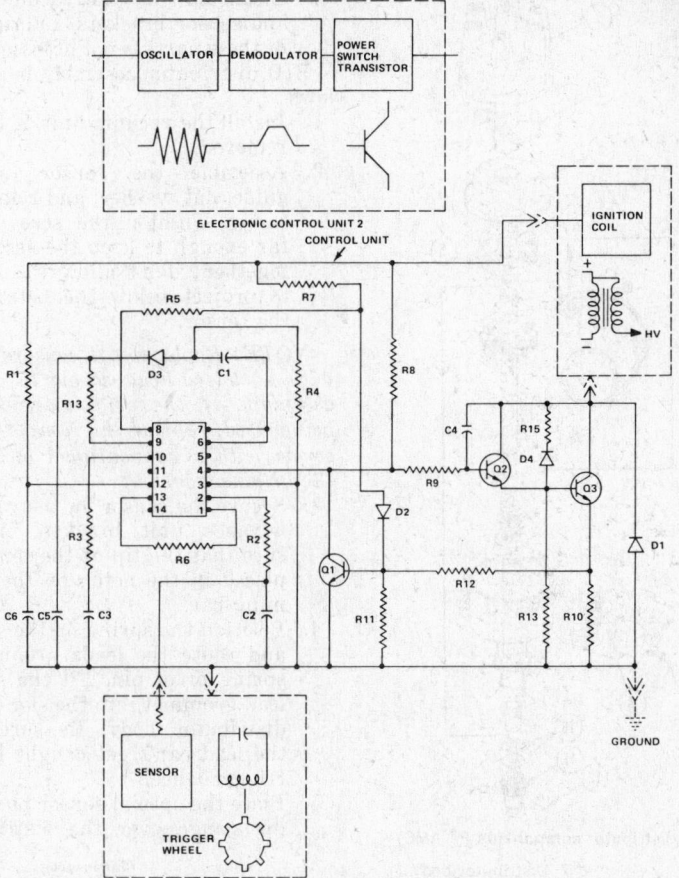

BID system schematic (© AMC)

positive (+) and negative (—) coil terminals.
6. Turn the ignition switch on. If the test light doesn't come on, check the control unit's ground lead. If the ground lead is in good condition, replace the control unit.
7. If the bulb lights in step 6, leave the test light in place and short the terminals on the coil-to-distributor connector together with a jumper lead, (connector separated) at the coil side of the connector. If the light stays on, replace the control unit.
8. If the test light goes out, remove it. Check for a spark, as in step 2, each time that the coil-to-distributor connector terminals are shorted together with the jumper lead. If there is a spark, replace the control unit; if there is no spark, replace the coil.

Coil Testing

Test the coil with a conventional coil checker or an ohmmeter. Primary resistance should be 1-2 ohms and secondary resistance should be 8-12 kilohms. The open output circuit should be more than 20 kilovolts. Replace the coil if it doesn't meet specifications.

Sensor Testing

Check the sensor resistance by connecting an ohmmeter to its leads. Resistance should be 1.8 ohms (±10%) at 77° F. Replace the sensor if it doesn't meet these specifications.

Distributor Overhaul

NOTE: If you must remove the sensor from the distributor for any reason, it will be necessary to have the special sensor positioning gauge in order to align it properly during installation.

1. Scribe matchmarks on the distributor housing, rotor, and engine block. Disconnect the leads and vacuum lines from the distributor. Remove the distributor. Unless the cap is to be replaced, leave it connected to the spark plug cables and position it out of the way.
2. Remove the rotor and dust cap.
3. Place a small gear puller over the trigger wheel, so that its jaws grip the inner shoulders of the wheel and not its arms. Place a thick washer between the gear puller and the distributor shaft to act as a spacer; do not press against the smaller inner shaft.
4. Loosen the sensor hold-down screw with a small pair of needle-nosed pliers; it has a tamper-

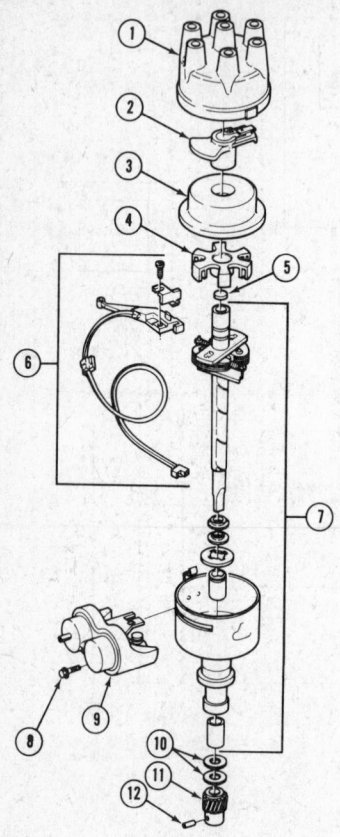

BID distributor components (© AMC)

1 Cap	7 Distributor body
2 Rotor	8 Vacuum unit screw
3 Dust shield	9 Vacuum advance unit
4 Trigger wheel	10 Shim
5 Felt lubricator	11 Drive gear
6 Sensor assembly	12 Pin

proof head. Pull the sensor lead grommet out of the distributor body and pull out the leads from around the spring pivot pin.

5. Release the sensor securing spring by lifting it. Make sure that it clears the leads. Slide the sensor off the bracket. *Remember, a special gauge is required for sensor installation.*

6. Remove the vacuum advance unit securing screw. Slide the vacuum unit out of the distributor. Remove it only if it is to be replaced.

7. Clean and dry the vacuum unit and sensor brackets. Lubrication of these parts is not necessary.

BID distributor assembly is as follows:

1. Install the vacuum unit, if it was removed.

2. Assemble the sensor, sensor guide, flat washer, and retaining screw. Tighten the screw only far enough to keep the assembly together; don't allow the screw to project below the bottom of the sensor.

NOTE: Replacement sensors come with a slotted-head screw to aid in assembly. If the original sensor is being used, replace the tamper-proof screw with a conventional one. Use the original washer.

3. Secure the sensor on the vacuum advance unit bracket, making sure that the tip of the sensor is placed in the notch on the summing bar.

4. Position the spring on the sensor and route the leads around the spring pivot pin. Fit the sensor lead grommet into the slot on the distributor body. Be sure that the lead can't get caught in the trigger wheel.

5. Place the special sensor positioning gauge over the distributor

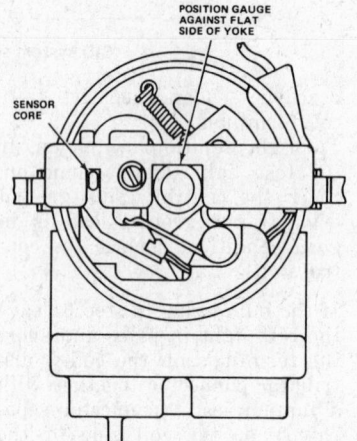

Using the special gauge to align the BID sensor coil (© AMC)

shaft, so that the flat on the shaft is against the large notch on the gauge. Move the sensor until the sensor core fits into the small notch on the gauge. Tighten the sensor securing screw with the gauge in place (through the round hole in the gauge).

6. It should be possible to remove and install the gauge without any side movement of the sensor. Check this and remove the gauge.

7. Position the trigger wheel on the shaft. Check to see that the sensor core is centered between the trigger wheel legs and that the legs don't touch the core.

8. Bend a piece of 0.050 in. gauge wire, so that it has a 90° angle and one leg ½ in. long. Use the gauge to measure the clearance between the trigger wheel legs and the sensor boss. Press the trigger wheel on the shaft until it just touches the gauge. Support the shaft during this operation.

9. Place 3 to 5 drops of SAE 20 oil on the felt lubricator wick.

10. Install the dust shield and rotor on the shaft.

11. Install the distributor on the engine using the matchmarks made during removal and adjust the

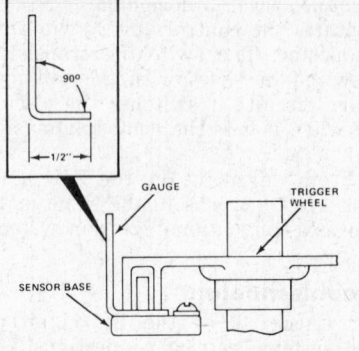

Fabricated BID trigger wheel clearance gauge (© AMC)

timing. Use a new distributor mounting gasket.

IH Electronic Ignition

There are two versions of the system. The first, with a black control box, was used on 1974 V8 models, while the second, with a gold control box, was introduced in 1975 on four-cylinder and V8 models. The two versions of the system are very similar in appearance; the main external difference is that the gold control box combines all wiring into a single plug.

The system uses a standard ignition coil with no ballast resistor or resistance wire, since the control box regulates the primary, low-voltage,

current. The distributor part of the system consists of a metal detecting sensor and a toothed trigger wheel in place of the distributor cam. The only adjustment is for sensor to trigger wheel air gap. The control box components are permanently sealed in a waterproof and vibration resistant compound. Most of these systems use vacuum spark advance in addition to mechanical advance.

Disassembly and overhaul of the distributor is very similar to that for International V8 point-type distributors.

Sensor to Trigger Wheel Air Gap Adjustment

1. Align the trigger wheel so that one tooth is aligned with the centerline of the sensor. The tooth should be at right angles to the flat side of the sensor.

2. The gap should be .014 in. for 1974 units and .008 in. for 1975 and later models.

3. A distributor machine or a dwell/tachometer calibrated for electronic ignition systems can be used to measure dwell.

NOTE: Most ordinary dwell/tach-

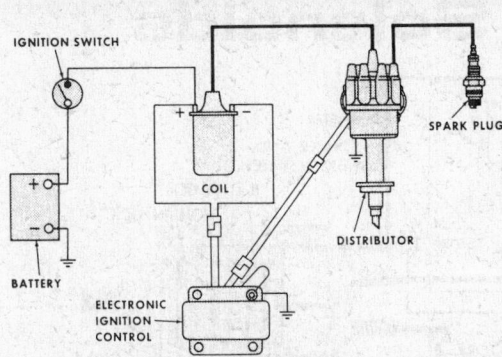

IH electronic ignition system schematic (© IH)

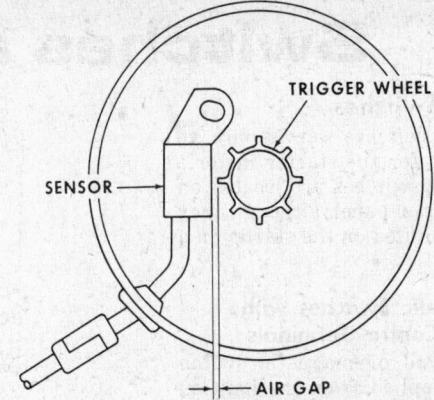

Sensor to trigger wheel air gap (© IH)

ometers will give a reading, but this will not be accurate.

On 1974 models, dwell should be 27-30 degrees at curb idle; on a distributor machine the reading will be 22-28 degrees at 2000 distributor rpm with 12-13 volts primary input. On 1975 and later models, dwell should be 26-32 degrees at curb idle and also at 300 distributor rpm with 12-13 volts primary input.

4. Adjust the gap and dwell by moving the sensor. Move the sensor toward the trigger wheel to decrease dwell or away to increase dwell. .001 in. of sensor movement equals about ½ degree of dwell change.

Troubleshooting

1. Make sure that the battery is fully charged, delivering 12-13 volts. Make sure that all wiring, connections, and mounting bolts are in good condition.

2. Disconnect one spark plug wire and insert an extension of some sort into the boot. Hold the wire with insulated pliers and a heavy glove so that there is a gap of about ¼-½ in. between the extension and a ground. If the spark jumps the gap when the engine is cranked with the starter, the system is in good condition. If not, replace the wire and go on.

3. Detach the coil wire from the center of the distributor cap. Attach one end of a jumper wire to a ground and the other around the coil wire (don't pierce the insulation) ¼-½ in. from the metal tip. If there is a spark when the engine is cranked, the distributor cap, rotor, or spark plug and coil wires may be faulty. If not, keep the jumper in place and go on.

4. Disconnect the primary wiring plug near the distributor and plug a tester switch, part no. SE-2503, into the wiring harness. The switch replaces the distributor sensor in the circuit. Turn the ignition switch on and press the tester switch button. If there is a spark at the jumper, the sensor is defective and must be replaced. If not, go on.

5. Disconnect the primary wiring plug near the control box and install the tester switch. Turn the ignition switch on and press the tester switch button. If there is a spark at the jumper now, the primary wiring harness is defective. If not, go on.

6. Connect a voltmeter between the coil negative terminal and a ground. Voltage should be 12-13 volts. A low reading indicates high resistance between the battery and the coil, probably due to defective wires or ignition switch.

7. Connect the voltmeter between the coil negative terminal and a ground. With the ignition switch on, voltage should be 5-8 volts. A lower or higher reading indicates a bad coil. Press the tester switch button; voltage should go up to 12-13 volts and go back down when the button is released. If the voltage doesn't go up and down, the control box is faulty and must be replaced. If the voltage goes up and down but there is no spark at the jumper, the coil is defective.

8. Reconnect the system and make a final check for spark at the plug wire, as in Step 2.

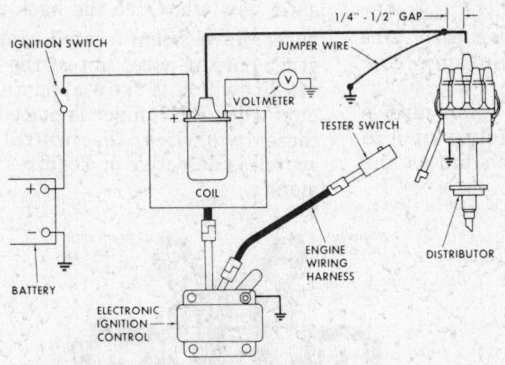

Voltmeter connected to coil negative terminal (© IH)

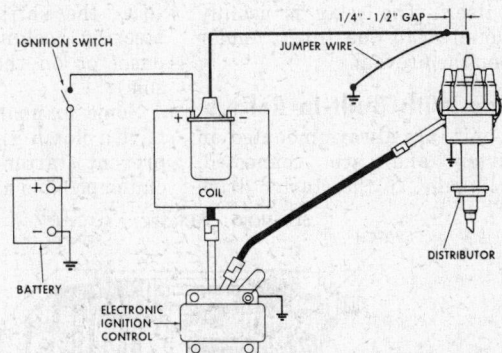

Spark gap for testing electronic ignition system (© IH)

Switches and Solenoids

Magnetic Switches

Magnetic switches serve only to make contact for the starter motor. Usually, such switches are located on the inner fender panel, although they are found mounted on the starter in a few cases.

Magnetic Switches with Two Control Terminals

On this type of magnetic switch current is supplied from the ignition switch or transmission neutral button to one of the magnetic switch control terminals. The other control terminal is connected to the transmission neutral safety switch (on the transmission) where it is grounded.

Magnetic Switches with Ignition Resistor By-Pass Terminals

All normally use a magnetic switch with a single control terminal. The second terminal is an ignition resistor by-pass terminal.

Solenoids Without Relays

This type of starter solenoid is always mounted on the starter. Makes electrical contact for the starter and pulls the starter and drive clutch into mesh with the flywheel. The Chrysler reduction gear starter has this solenoid embodied in the starter housing.

There is only one control terminal on the solenoid.

The ignition by-pass terminal is usually marked R or IGN, if it is used.

Solenoids With Separate Relays

The solenoid itself is always mounted on the starter. In addition to making contact for the starter, it also pulls the starter drive clutch gear into mesh with the flywheel. A single control terminal is used on the solenoid itself. The relay is usually found mounted to the inner fender panel or on the firewall.

Solenoids With Built-In Relays

These units are always mounted on the starter and are connected, through linkage, to the starter drive

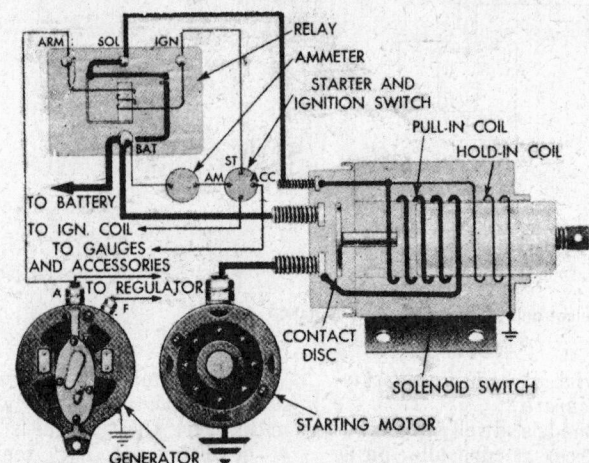

Pictorial drawing of solenoid with a separate relay

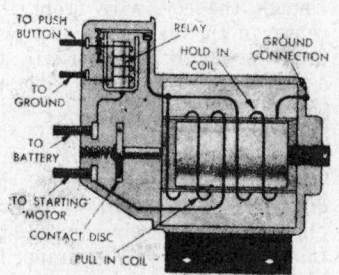

Pictorial drawing of solenoid with a built-in relay

clutch. The relay portion is a square box built into and integral with the front end of the solenoid assembly.

Neutral Safety Switches

The purpose of the neutral safety switch is to prevent the starter from cranking the engine except when the transmission is in neutral or park.

On some trucks the neutral safety switch is located on the transmission. It serves to ground the solenoid or magnetic switch, whichever is used.

On other trucks the neutral safety switch is located either at the bottom of the steering column, where it contacts the shift mechanism, on the steering column, underneath the dash, or on the shift linkage (console).

Some manual transmission models have a clutch linkage safety switch to prevent starter operation unless the clutch pedal is depressed.

On most trucks the neutral safety switch and the backup light switch are combined into a single switch mechanism.

Troubleshooting Neutral Safety Switches—Quick Test

If the starter fails to function and the neutral safety switch is to be checked, a jumper can be placed across its terminals. If the starter then functions the safety switch is defective.

In the case of neutral safety switches with one wire, this wire must be grounded for testing purposes. If the starter works with the wire grounded, the switch is defective.

Neutral Safety Switch—Back-Up Light Switch

When the neutral safety switch is built in combination with the back-up light switch, the easiest way to tell which terminals are for the back-up lights is to take a jumper and cross every pair of wires. The pair of wires which light the back-up lamps should be ignored when testing the neutral safety switch. Once the back-up light wires have been located, jump the other pair of wires to test the neutral safety switch. If the starter functions only when the jumper is placed across these two wires, the neutral safety switch is defective or requires adjustment.

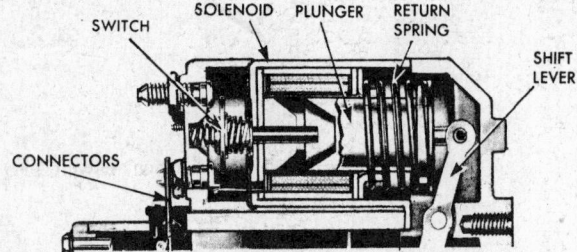

Starter solenoid mounted on starter motor

Schematic diagram of a magnetic switch with two control terminals

Starting Systems

Reduction-Gear Starter Motor
(Chrysler Corporation)

The housing is die-cast aluminum. A 3.5 to 1 reduction, combined with the starter to ring gear ratio, results in a total gear reduction of about 45 to 1.

NOTE: the high-pitched sound is caused by the higher starter speed.

The positive shift solenoid is enclosed in the starter housing and is energized through the ignition switch. When ignition switch is turned to start, the solenoid plunger engages drive gear through a shifting fork. At the completion of travel, the plunger closes a switch to revolve the starter.

The tension of the spring-type shifting prevents a butt-tooth lock up and motor will not start before total shift.

An overrunning clutch prevents motor damage if key is held on after engine starts.

No lubrication is required due to Oilite bearings.

Disassembly

1. Support assembly in a vise equipped with soft jaws. Do not clamp. Care must be used not to distort or damage the die cast aluminum.
2. Remove the thru-bolts and the end housing.
3. Carefully pull the armature up and out of the gear housing, and the starter frame and field assembly. Remove the steel and fiber thrust washer.

NOTE: on eight cylinder engines the starting motors have the wire of the shunt field coil soldered to the brush terminal. Six cylinder engines have the four coils in series and do not have a wire soldered to the brush terminal. One pair of brushes is connected to this terminal. The other pair of brushes is attached to the series field coils by means of a terminal

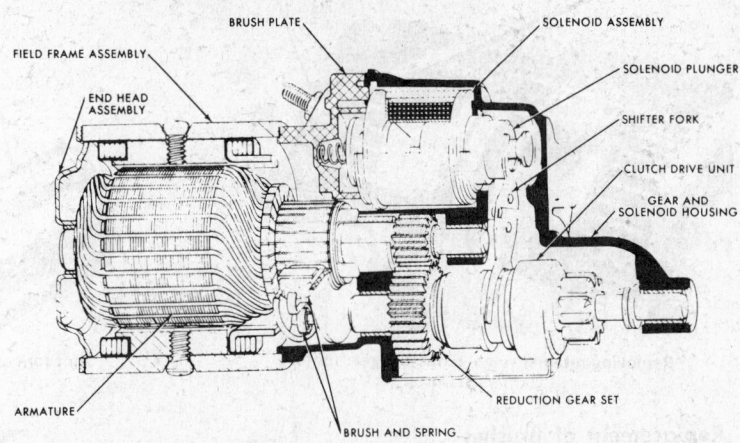

Reduction gear starter motor (© Chrysler Corp)

screw. Carefully pull the frame and field assembly up just enough to expose the terminal screw and the solder connection of the shunt field at the brush terminal. Place two wood blocks between the starter frame and starter gear housing to facilitate removal of the terminal screw and unsoldering of the shunt field wire at the brush terminal.

4. Support the brush terminal with a finger behind terminal and remove screw.
5. On eight cylinder engine starters unsolder the shunt field coil lead from the brush terminal and housing.
6. The brush holder plate with terminal, contact and brushes is serviced as an assembly.
7. Clean all old sealer from around plate and housing.
8. Remove the brush holder attaching screw.
9. On the shunt type, unsolder the solenoid winding from the brush terminal.
10. Remove 11/32 in. nut, washer and insulator from solenoid terminal.
11. Remove brush holder plate with brushes as an assembly.

12. Remove gear housing ground screw.
13. The solenoid assembly can be removed from the well.
14. Remove nut, washer and seal from starter battery terminal and remove terminal from plate.
15. Remove solenoid contact and plunger from solenoid and remove the coil sleeve.
16. Remove the solenoid return spring, coil retaining washer, retainer and the dust cover from the gear housing.
17. Release the snap-ring that locates the driven gear on pinion shaft.
18. Release front retaining ring.
19. Push pinion shaft toward the rear and remove snap-ring, thrust washers, clutch and pinion, and two shift fork nylon actuators.
20. Remove driven gear and friction washer.
21. Pull shifting fork forward and remove moving core.
22. Remove fork retainer pin and shifting fork assembly. The gear housing with bushings is serviced as an assembly.

Removing terminal screw—reduction gear motor
(© Chrysler Corp)

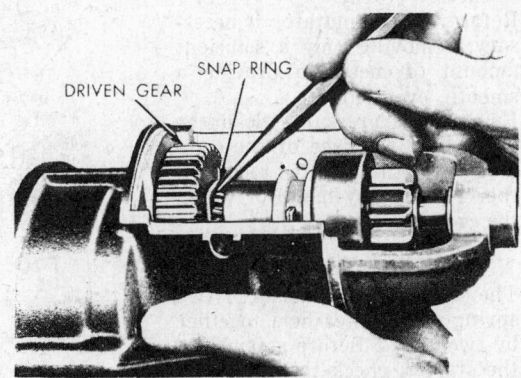

Removing drive gear snap-ring—reduction gear motor
(© Chrysler Corp)

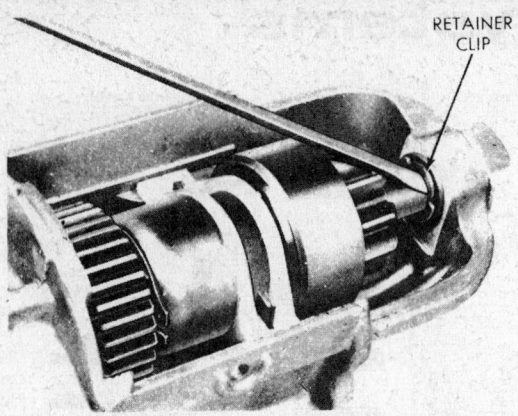

Removing retainer ring—reduction gear motor
(© Chrysler Corp)

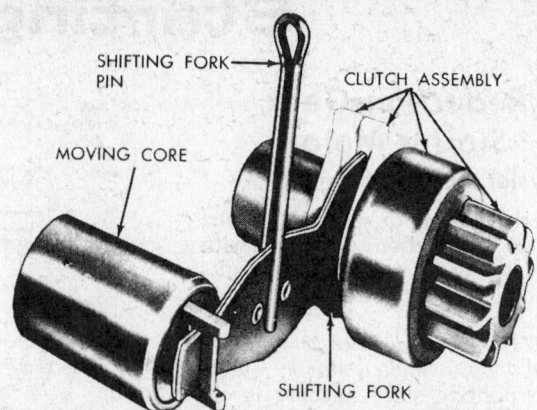

Shift fork and clutch arrangement—reduction gear motor
(© Chrysler Corp)

Replacement of Brushes

1. Brushes that are worn more than one-half the length of new brushes, or are oil-soaked, should be replaced.
2. When resoldering the shunt field and solenoid lead, make a strong, low-resistance connection using a high-temperature solder and resin flux. Do not use acid or acid-core solder. Do not break the shunt field wire units when removing and installing the brushes.

Starter Clutch and Pinion Gear Inspection

1. Do not immerse the starter clutch unit in a cleaning solvent. The outside of the clutch and pinion must be cleaned with a cloth so as not to wash the lubricant from the inside of the clutch.
2. Rotate the pinion. The pinion gear should rotate smoothly and in one direction only. If the starter clutch unit does not function properly, or if the pinion is worn, chipped, or burred, replace the starter clutch unit.

Commutator Inspection

1. Inspect the commutator and the surface contacted by the brushes when the starter is assembled, for flat spots, out-of-roundness, or excessive wear.
2. Reface the commutator if necessary, removing only a sufficient amount of metal to provide a smooth, even surface.
3. Using light pressure, clean the grooves of the face of the commutator with a pointed tool. Neither remove any metal or widen the grooves.

Assembly

1. The shifter fork consists of two spring steel plates held together by two rivets. Before assembling the starter, check the plates for side movement. After lubricating between the plates with a small

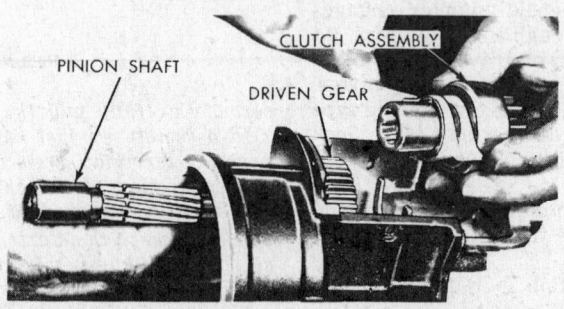

Removing clutch assembly—reduction gear motor
(© Chrysler Corp)

amount of SAE 10 engine oil, they should have about 1/16 in. side movement to insure proper pinion gear engagement.
2. Position the shift fork in the drive housing and install the shifting fork retainer pin. One tip of the pin should be straight and the other bent at a 15 degree angle away from the housing. The fork and retainer pin should operate freely after bending the tip of the pin.
3. Install the solenoid moving core and engage the shifting fork.
4. Place the pinion shaft into the drive housing and install the friction washer and drive gear.

5. Install the clutch and pinion assembly, thrust washer, and retaining washer.
6. Engage the shifting fork with the clutch actuators.
CAUTION: The friction washer must be positioned on the shoulder of the splines of the pinion shaft before the driven gear is positioned.
7. Install the driven gear snap ring.
8. Install the pinion shaft retaining ring.
9. The starter solenoid return spring can now be inserted in the moveable core.
10. Install the solenoid contact plunger assembly into the solenoid and reform the double wires so they

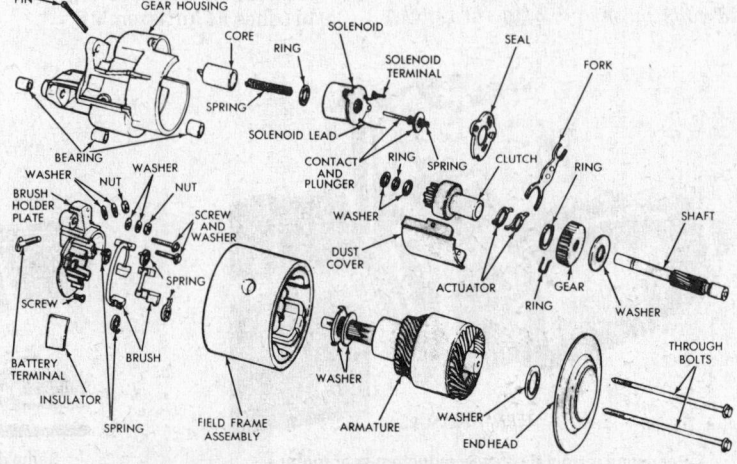

Reduction gear motor—exploded view (© Chrysler Corp)

490

can be curved around the contactor. This will allow the terminal stud to enter the brush holder properly.

CAUTION: The contactor must not touch these double wires after assembly is complete.

11. Assemble the battery terminal stud in the brush holder.
12. Position the seal on the brush holder plate.
13. Run the solenoid lead wire through the hole in the brush holder and attach the solenoid stud, insulating washers, flat washer, and nut.
14. Wrap the solenoid lead wire tightly around the brush terminal post and solder it.
15. Fix the brush holder to the solenoid attaching screws.
16. Gently lower the solenoid coil and brush plate into the gear housing.
17. Position the brush plate assembly into the starter gear housing, install the nuts, and tighten.
18. Solder the shunt coil lead wire to the starter brush terminal.
19. Install the brush terminal screw.
20. Position the field frame on the gear housing and start the armature into the housing, carefully engaging the splines on the shaft with the reduction gear by rotating the armature.
21. Install the fiber thrush washer and the steel washer on the armature shaft.
22. Replace the starter end housing and starter through bolts; tighten securely.

Direct Drive Starter Motor

(Chrysler Corporation)

This starter can be identified by the externally mounted solenoid bolted to the case.

Disassembly

1. Remove through bolts and tap commutator end head from frame.
2. Remove thrust washers from armature shaft.
3. Lift brush holder springs and remove brushes from holders.
4. Remove brush holder plate.
5. Disconnect the field coil wires at the solenoid connector, and remove the solenoid screws.
6. Remove solenoid and boot.
7. Drive out shift fork pivot pin.
8. Remove drive end pinion housing and spacer washer.
9. Remove shift fork from starter drive.
10. Slide overrunning clutch pinion gear toward commutator, drive stop retainer toward clutch pinion gear and remove the now-exposed snap-ring.
11. Remove overrunning clutch drive from armature shaft.
12. If field coils are good, stop disassembly at this point. If field coils must be replaced, remove ground brushes terminal screw and remove brushes, terminal and shunt

wire. Remove pole shoe screws, using a ratchet-type impact driver and special wide screwdriver blade, then remove field coils.
13. Replacement of the brushes, inspection of the starter clutch and pinion, and inspection of the commutator procedures are the same as the reduction-gear starter procedures.

Assembly

1. Install field coils into frame, if removed.
2. Lubricate armature shaft and splines with engine oil.
3. Install starter drive, stop retainer, lock ring and spacer washer.
4. Install shift fork, with *narrow* leg of fork toward commutator.
5. Install pinion housing onto armature shaft, indexing shift fork with slot in housing.
6. Install shift fork pivot pin.
7. With clutch drive, shift fork, and pinion housing assembled onto the armature, slide armature into frame until pinion housing indexes with slot.
8. Install solenoid and boot, tight-

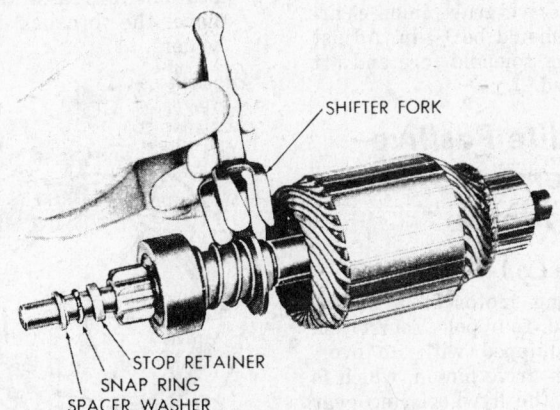

Removing shift fork—direct drive motor
(© Chrysler Corp)

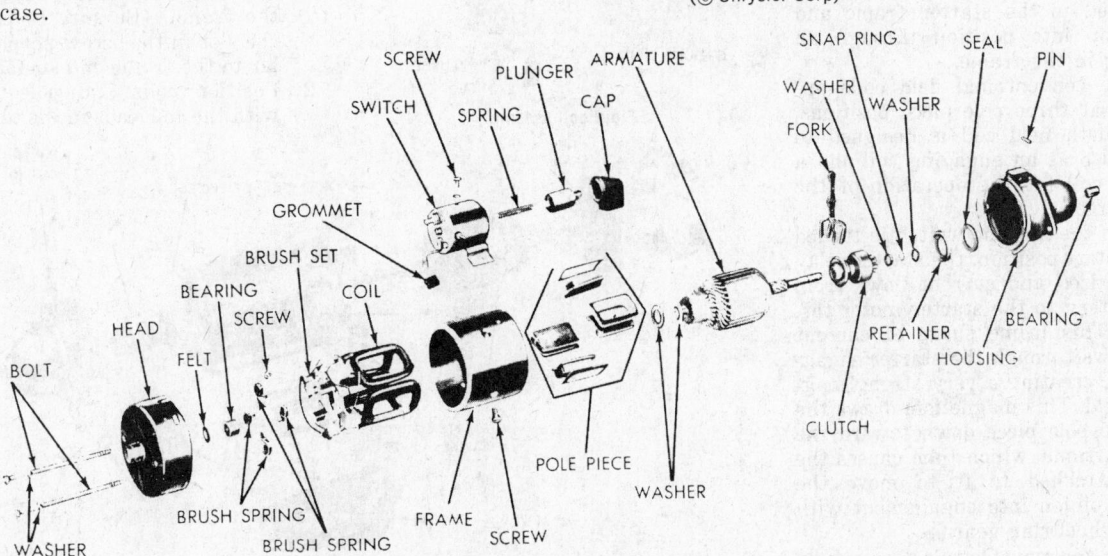

Chrysler direct drive motor—exploded view (© Chrysler Corp)

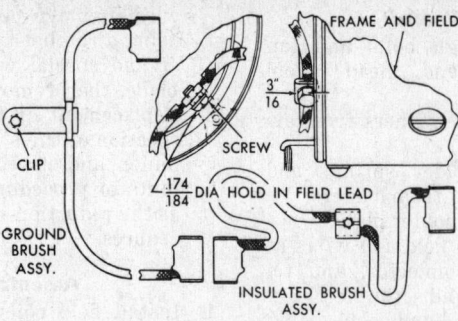

Brush lead arrangement—Chrysler direct drive motor
(© Chrysler Corp)

ening bolts to 60-70 in. lbs.

9. Conect field coil wires to solenoid connector, making sure they do not touch frame.
10. Install brush holder plate, indexing tang in frame hole.
11. Place brushes in holders, making sure field coil wires do not interfere.
12. Install thrust washers on commutator end of armature shaft to obtain a maximum of 0.010 in. end-play.
13. Install commutator end head and through bolts. Tighten bolts to 40-50 in. lbs.
14. Measure drive gear pinion clearance; it should be ⅛ in. Adjust by moving solenoid fore and aft as required.

Autolite Positive Engagement Starter Motor

(Ford Motor Co.)

This starting motor is a series-parallel wound, four pole, four brush unit. It is equipped with an over-running clutch drive pinion, which is engaged with the flywheel ring gear by an actuating lever, operated by a movable pole piece. This pole piece is hinged to the starter frame and can drop into position through an opening in the frame.

Three conventional field coils are located at three pole piece positions. The fourth field coil is designed to serve also as an engaging coil and a hold-in coil for the operation of the drive pinion.

When the ignition switch is turned to the start position, the starter relay is energized and current flows from the battery to the starter motor terminal. This prime surge of current first flows through the starter engaging coil, creating a very strong magnetic field. This magnetism draws the movable pole piece down toward the starter frame, which then causes the lever attached to it to move the starter pinion into engagement with the flywheel ring gear.

When the movable pole shoe is fully seated, it opens the field coil, ground-

ing contacts, and the starter is then in normal operation. A holding coil is used to hold the movable pole shoe in the fully seated position during the engine cranking operation.

Trucks, equipped with automatic transmissions have a starter neutral switch circuit control. This is to prevent operation of the starter if the selector lever is not in Neutral or Park.

Disassembly

1. Remove brush cover band and starter drive gear actuating lever cover. Observe the brush lead locations for reassembly, then remove the brushes from their holders.

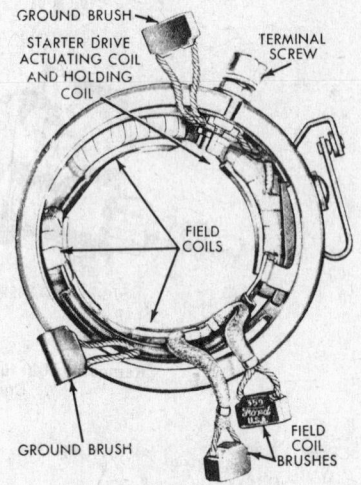

Field coil assembly

NOTE: factory brush length is ½ in.; wear limit is ¼ in.

2. Remove the through bolts, starter drive gear housing and the drive gear actuating lever return spring.
3. Remove the pivot pin retaining the starter gear actuating lever and remove the lever and the armature.
4. Remove the stop ring retainer. Remove and discard the stop ring holding the drive gear to the armature shaft; then remove the drive gear assembly.
5. Remove the brush end plate.
6. Remove the two screws holding the ground brushes to the frame.
7. On the field coil that operates the starter drive gear actuating lever, bend the tab up on the field retainer and remove the field coil retainer.
8. Remove the three coil retaining screws. Unsolder the field coil leads from the terminal screw, then remove the pole shoes and coils from the frame (use a 300 watt iron).
9. Remove the starter terminal nut, washer, insulator and terminal from the starter frame.
10. Check the commutator for runout. If the commutator is rough, has flat spots, or is more than 0.005 in. out of round, reface the commutator. Clean the grooves in the commutator face.
11. Inspect the armature shaft and the two bearings for scoring and excessive wear. Replace if necessary.
12. Inspect the starter drive. If the gear teeth are pitted, broken, or excessively worn, replace the starter drive.

Assembly

1. Install starter terminal, insulator, washers and retaining nut in the frame. (Be sure to position the slot in the screw perpendicular to the frame end surface.)
2. Position coils and pole pieces, with the coil leads in the terminal

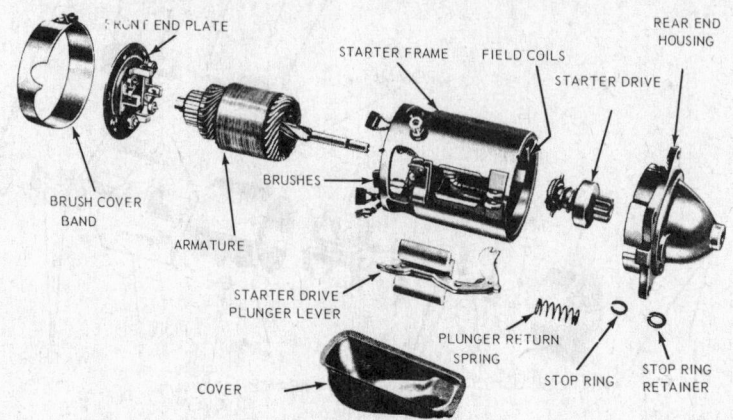

Starter motor

Starter Cranks Engine Slowly

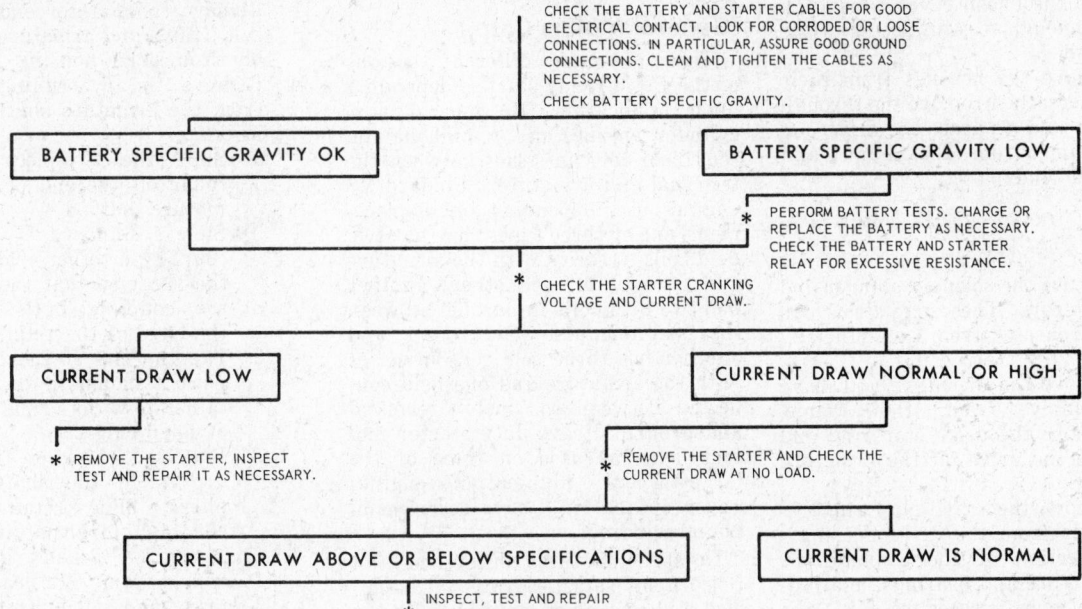

CHECK THE BATTERY AND STARTER CABLES FOR GOOD ELECTRICAL CONTACT. LOOK FOR CORRODED OR LOOSE CONNECTIONS. IN PARTICULAR, ASSURE GOOD GROUND CONNECTIONS. CLEAN AND TIGHTEN THE CABLES AS NECESSARY.
CHECK BATTERY SPECIFIC GRAVITY.

BATTERY SPECIFIC GRAVITY OK | **BATTERY SPECIFIC GRAVITY LOW**

* PERFORM BATTERY TESTS. CHARGE OR REPLACE THE BATTERY AS NECESSARY. CHECK THE BATTERY AND STARTER RELAY FOR EXCESSIVE RESISTANCE.

* CHECK THE STARTER CRANKING VOLTAGE AND CURRENT DRAW.

CURRENT DRAW LOW | **CURRENT DRAW NORMAL OR HIGH**

* REMOVE THE STARTER, INSPECT TEST AND REPAIR IT AS NECESSARY.

* REMOVE THE STARTER AND CHECK THE CURRENT DRAW AT NO LOAD.

CURRENT DRAW ABOVE OR BELOW SPECIFICATIONS | **CURRENT DRAW IS NORMAL**

* INSPECT, TEST AND REPAIR THE STARTER AS NECESSARY.

screw slot, then install the retaining screws. As the pole screws are tightened, strike the frame several sharp hammer blows to align the pole shoes. Tighten, then stake the screws.

3. Install solenoid coil and retainer and bend the tabs to hold the coils to the frame.
4. Solder the field coils and solenoid wire to the starter terminal, using rosin-core solder and a 300 watt iron.
5. Check for continuity and ground connections in the assembled coils.
6. Position the solenoid coil ground terminal over the nearest ground screw hole.
7. Position the ground brushes to the starter frame and install retaining screws.
8. Position the brush end plate to the frame, with the end plate boss in the frame slot.
9. Lightly Lubriplate the armature shaft splines and install the starter drive gear assembly in the shaft. Install a new retaining stop ring and stop ring retainer.
10. Position the fiber thrust washer on the commutator end of the armature shaft, then position the armature in the starter frame.
11. Position the starter drive gear actuating lever to the frame and starter drive assembly, and install the pivot pin.
NOTE: fill drive gear housing bore ¼ full of grease.
12. Position the drive actuating lever return spring and the drive gear housing to the frame, then install

and tighten the through bolts. Do not pinch brush leads between brush plate and frame. Be sure that the stop ring retainer is properly seated in the drive housing.
13. Install the brushes in the brush holders and center the brush springs on the brushes.
14. Position the drive gear actuating lever cover on the starter and install the brush cover band with a new gasket.
15. Check starter no-load amperage draw.

Autolite Solenoid Actuated Starter Motor

(Ford Motor Co.)

This starter motor, usually used with late-model 460 engines, is a four-brush, four-field, four-pole wound unit. The frame encloses a wound armature, which is supported at the drive end by caged needle bearings and at the commutator end by a sintered copper bushing. The four pole shoes are retained to the frame by one pole screw apiece, and on each pole shoe is wound a ribbon-type field coil connected in series-parallel.

The solenoid is mounted to a flange on the starter drive housing, which encloses the entire shift mechanism and solenoid plunger. The solenoid utilizes two windings—a pull-in winding and a hold-in winding.

Disassembly

1. Disconnect the copper strap from the solenoid starter termi-

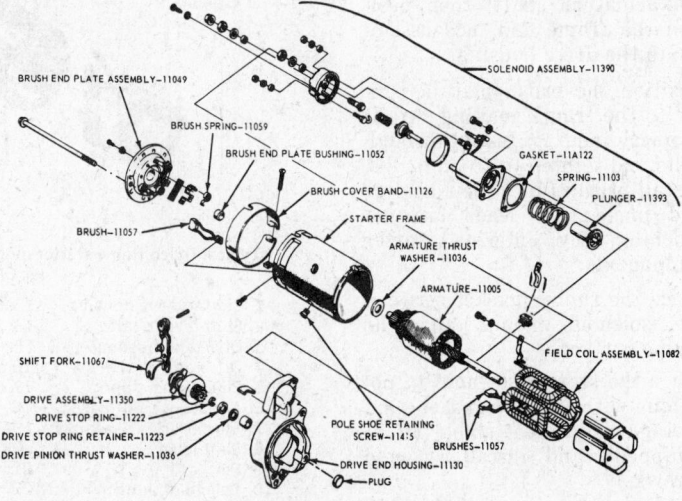

Ford solenoid starter motor (© Ford Co)

nal, remove the remaining screws and remove the solenoid.

2. Loosen the retaining screw and slide the brush cover band back far enough to gain access to the brushes.

3. Remove the brushes from their holders, then remove the through bolts and separate the drive end housing from the frame and brush end plate.

NOTE: factory brush length is ½ in., wear limit ¼ in.

4. Remove the solenoid plunger and shift fork. These two items can be separated from each other by removing the roll pin.

5. Remove the armature and drive assembly from the frame. Remove the drive stop ring and slide the drive off the armature shaft.

6. Remove the drive stop ring retainer from the drive housing.

7. Inspection of the commutator, armature and bearings, and pinion gear procedures is the same as the positive engagement starter procedures.

Assembly

1. Lubricate the armature shaft splines with Lubriplate, then install drive assembly and a new stop ring.

2. Lubricate shift lever pivot pin with Lubriplate, then position solenoid plunger and shift lever assembly in the drive housing.

3. Place a new retainer in the drive housing. Apply a small amount of Lubriplate to the drive end of the armature shaft, then place armature and drive assembly into the drive housing, indexing the shift lever tangs with the drive assembly.

4. Apply a small amount of Lubriplate to the commutator end of the armature shaft, then position the frame and field assembly to the drive housing.

5. Position the brush plate assembly to the frame, making sure it properly indexes. Install through bolts and tighten to 45-85 in. lbs.

6. Install brushes into their holders and make sure leads are not touching any interior starter components.

7. Place the rubber gasket between the solenoid mount and the frame surface.

8. Place the starter solenoid in position with metal gasket and spring, install heat shield (if so equipped) and install solenoid screws.

9. Connect copper strap and install cover band.

Delco-Remy Starter Motor

(General Motors Corp.)

There are many different versions of the Delco-Remy starter, depending upon application. In general, six-cylinder engines use a unit having four field coils in series between the terminal and armature. Standard V8 engines use, depending on displacement, one of three types: one has two field coils in series with the armature and parallel to each other; another has two field coils in parallel between the field terminal and ground, and another has three field coils in series with the armature and one field connected between the motor terminal and ground. Heavy-duty starter motors, such as used on some of the largest G.M. high-output engines (over 400 cu. in.) have series compound windings.

In spite of these differences, all Delco-Remy starters are disassembled and assembled in essentially the same manner.

Disassembly

1. Disconnect the field coil connectors from the motor solenoid terminal.

NOTE: on models so equipped, remove solenoid mounting screws.

2. Remove the through bolts.

3. Remove commutator end frame, field frame and armature assembly from drive housing.

4. Remove the overruning clutch from the armature shaft as follows:

   a. Slide the two-piece thrust collar off the end of the armature shaft.

   b. Slide a standard ½ in. pipe coupling or other spacer onto the shaft so that the end of the coupling butts against the edge of the retainer.

   c. Tap the end of the coupling with a hammer, driving retainer towards armature end of snap-ring.

   d. Remove snap-ring from its groove in the shaft using pliers. Slide retainer and clutch from armature shaft.

5. Disassemble brush assembly from field frame by releasing the V-spring and removing the support pin. The brush holders, brushes and springs now can be pulled out as a unit and the leads disconnected.

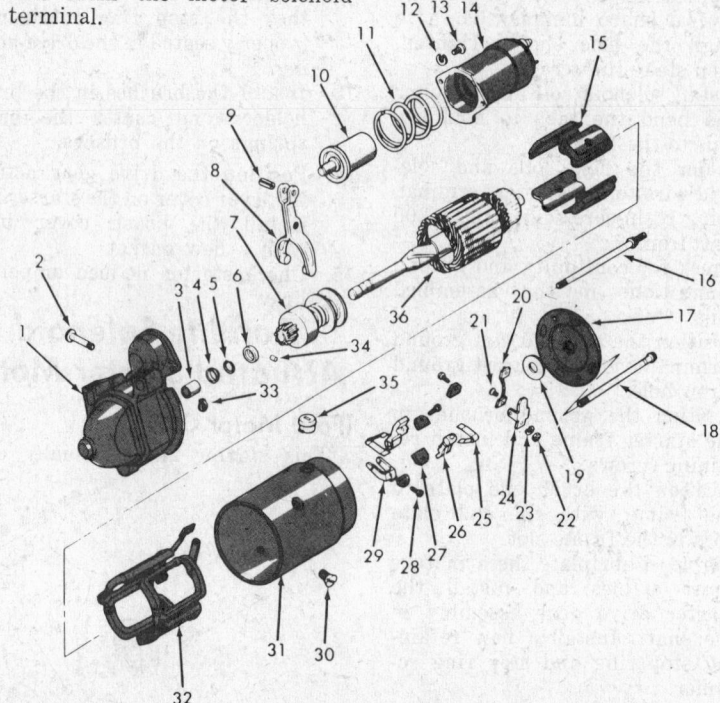

Typical Delco-Remy starter motor exploded view—light duty Chevrolet illustrated
(© Chevrolet Div., G.M. Corp)

1	Starter drive housing	13	Screw	25	Brush spring
2	Shift lever shaft	14	Solenoid switch assembly	26	Brush holder
3	Drive end bushing	15	Pole shoes	27	Brush
4	Drive end washer	16	Through bolt	28	Screw
5	Pinion ring stop	17	End frame	29	Ground brush holder
6	Armature shaft collar	18	Through bolt	30	Screw
7	Starter drive assembly	19	Washer	31	Field frame
8	Shift lever	20	Brush lead	32	Field coil assembly
9	Pin	21	Screw	33	Ring
10	Solenoid plunger	22	Nut	34	Pin
11	Solenoid return spring	23	Washer	35	Field frame grommet
12	Washer	24	Brush support pin	36	Armature

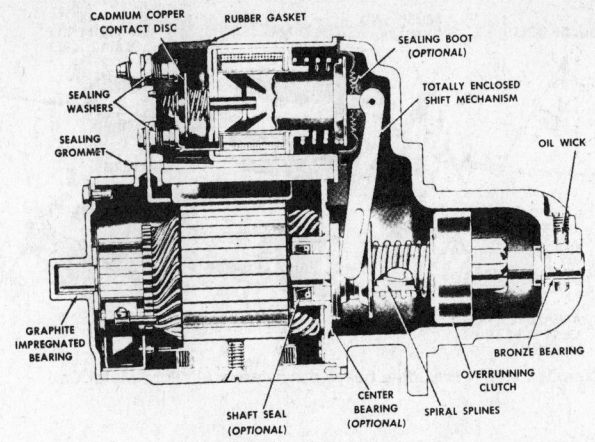

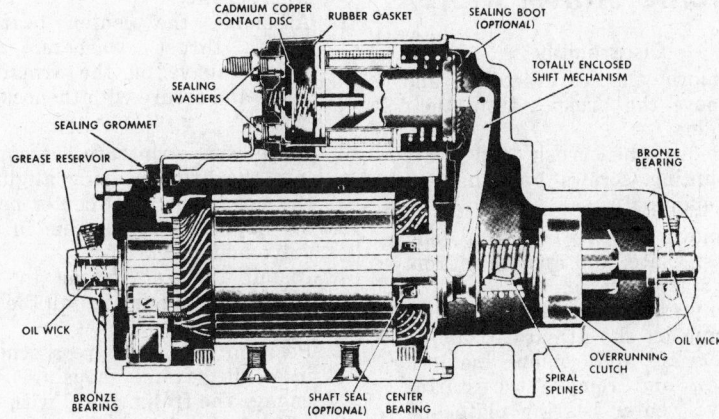

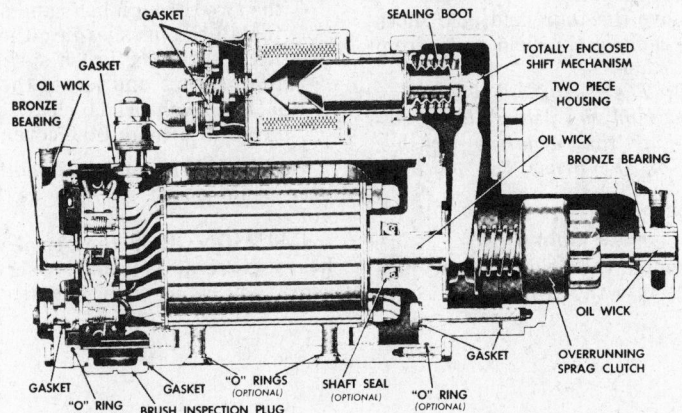

Three types of Delco-Remy starter motors (© G.M.C. Corp.)

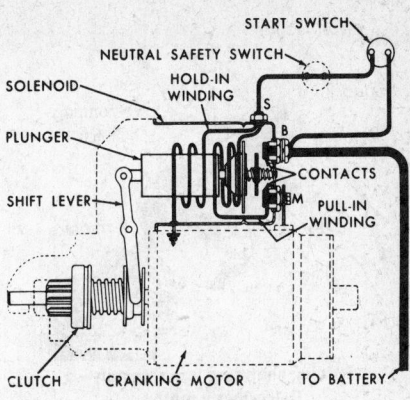

Delco-Remy solenoid windings (© G.M.C. Corp.)

Assembly

1. Install brushes into holders. Install solenoid, if so equipped.
2. Assemble insulated and grounded brush holder together using the V-spring and position the assembled unit on the support pin. Push holders and spring to bottom of support and rotate spring to engage the slot in support. Attach ground wire to grounded brush and field lead wire to insulated brush, then repeat for other brush sets.
3. Assemble overrunning clutch to armature shaft as follows:
   a. Lubricate drive end of shaft with silicone lubricant.
   b. Slide clutch assembly onto shaft with pinion outward.
   c. Slide retainer onto shaft with cupped surface facing away from pinion.

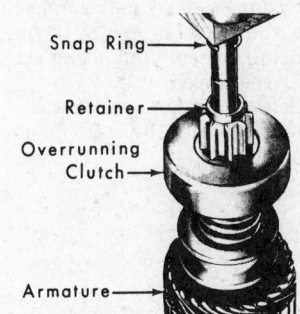

Forcing snap ring over armature shaft—
Delco-Remy motor
(© Chevrolet Div., G.M. Corp)

   d. Stand armature up on a wood surface, commutator downwards. Position snap-ring on upper end of shaft and drive it onto shaft with a small block of wood and a hammer. Slide snap-ring into groove.
   e. Install thrust collar onto shaft with shoulder next to snap-ring.
   f. With retainer on one side of snap-ring and thrust collar on the other side, squeeze together with two sets of pliers until ring seats in retainer.

6. On models so equipped, separate solenoid from lever housing.

Cleaning and Inspection

1. Clean parts with a rag, but do not immerse the parts in a solvent. Immersion in a solvent will dissolve the grease that is packed in the clutch mechanism and damage the armature and field coil insulation.
2. Test overrunning clutch action. The pinion should turn freely in the overrunning direction and must not slip in the cranking direction. Check pinion teeth to see that they have not been chipped, cracked, or excessively worn. Replace the unit if necessary.
3. Inspect the armature commutator. If the commutator is rough or out of round, it should be turned down and undercut.

CAUTION: Undercut the insulation between the commutator bars by 1/32 in.

This undercut must be the full width of the insulation and flat at the bottom; a triangular groove will not be satisfactory. Some starter motor models use a molded armature commutator design and no attempt to undercut the insulation should be made or serious damage may result to the commutator.

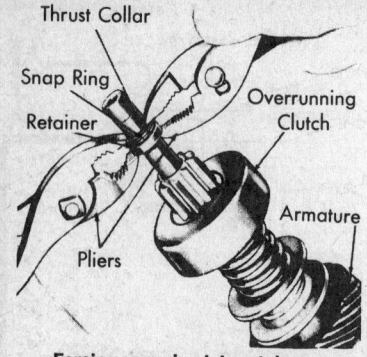

Forcing snap ring into retainer—
Delco-Remy motor
(© Chevrolet Div., G.M. Corp)

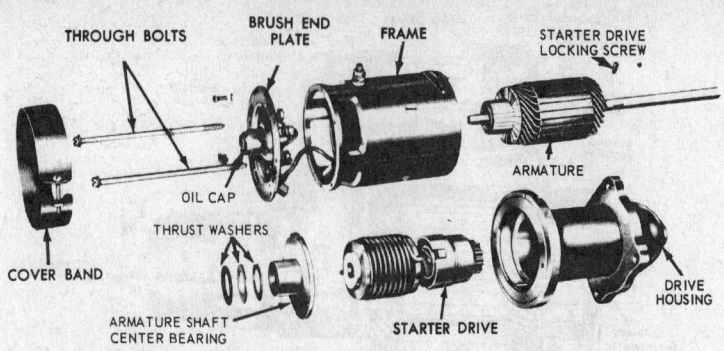

Exploded view—Prestolite heavy duty starter (© Ford Motor Co.)

On models without thrust collar, use a washer. Remember to remove washer before continuing.

4. Lubricate drive end bushing with silicone lubricant, then slide armature and clutch assembly into place, at the same time engaging shift lever with clutch.

5. Position field frame over armature and apply sealer (silicone) between frame and solenoid case. Position frame against drive housing, making sure brushes are not damaged in the process.

6. Lubricate commutator end bushing with silicone lubricant, place a leather brake washer on the armature shaft and slide commutator end frame onto shaft. Install through bolts and tighten to 65 in. lbs.

7. Reconnect field coil connector/s to the solenoid motor terminal. Install solenoid mounting screws, if so equipped.

8. Check pinion clearance; it should be 0.010-0.140 in. on all models.

Prestolite Starter Motor

Disassembly

1. Remove the cover band and remove the brushes from their holders.

2. Remove the brush end plate mounting screws and the two through bolts.

3. Remove the drive housing, end brush plate, and armature from the starter frame.

4. Compress the starter drive spring on the armature side of the shaft and remove the lock screw and remove the starter drive, center bearing plate and thrust washers.

5. Remove the four field pole shoes and remove the field coils from the frame.

NOTE: The positive brushes can be replaced on the field coils by soldering, and the negative brushes replaced on the brush end plate by riveting.

Assembly

1. Assemble the field coils and pole shoes into the frame and secure with screws.

2. Assemble the center bearing plate, thrust washers, and starter drive on the armature shaft and secure with the locking screw.

3. Place the armature assembly into the drive housing aligning the slot in the shaft center bearing support with the pin in the drive housing.

4. Install the end frame to the frame housing and install the six mounting screws.

5. Position the armature assembly into the frame housing and engage the frame dowel with the bolt of the drive frame. Install the two through bolts and secure.

6. Install the brushes into the holders. Center the brush springs on the brushes and locate the insulated brush leads clear of the armature. Install the cover band.

CAUTION: Secondary circuit pressure could reach as high as 30,000 volts.

CAUTION: Hold the pencil with a heavy glove or you may observe the spark flow along your fingers.

Index

HYDRAULIC BRAKE SYSTEM TROUBLE DIAGNOSIS

Condition	Possible Cause	Correction
Insufficient brakes	1. Improper brake adjustment.	1. Adjust brakes.
	2. Worn lining.	2. Replace brake lining and adjust brakes.
	3. Sticking brakes.	3. Lubricate brake pivots and support platforms.
	4. Brake valve pressure low.	4. Inspect for leaks and obstructed brake lines.
	5. Slack adjuster to diaphragm rod not adjusted properly.	5. Adjust slack adjuster.
	6. Master cylinder low on brake fluid.	6. Fill master cylinder and inspect for leaks.
Brakes apply slowly	1. Improper brake adjustment or lack of lubrication.	1. Adjust brakes and lubricate linkage.
	2. Low air pressure.	2. Check belt tension and compressor for output. Adjust as necessary.
	3. Brake valve delivery pressure low.	3. Check valve pressure and clean or replace as necessary.
	4. Excessive leakage with brakes applied.	4. Inspect all fittings and lines for leaks and repair as necessary.
	5. Restriction in brake line or hose.	5. Clean or replace brake line or hose.
Spongy pedal	1. Air in hydraulic system.	1. Fill and bleed hydraulic system.
	2. Swollen rubber parts due to contaminated brake fluid.	2. Clean hydraulic system and recondition wheel cylinders and master cylinder.
	3. Improper brake shoe adjustment.	3. Adjust brakes.
	4. Brake fluid with low boiling point.	4. Flush hydraulic system and refill with proper brake fluid.
	5. Brake drums ground excessively.	5. Replace brake drums.
Erratic brakes	1. Linings soaked with grease or brake fluid.	1. Correct the leak and replace brake lining.
	2. Primary and secondary shoes mounted in wrong position.	2. Match the primary and secondary shoes and mount in proper position.
Chattering brakes	1. Improper adjustment of brake shoes.	1. Adjust brakes.
	2. Loose front wheel bearings.	2. Clean, pack and adjust wheel bearings.
	3. Hard spots in brake drum.	3. Grind or replace brake drums.
	4. Out-of-round brake drums.	4. Grind or replace brake drums.
	5. Grease or brake fluid on lining.	5. Correct leak and replace brake lining.
Squealing brakes	1. Incorrect lining.	1. Install correct lining.
	2. Distorted brakedrum.	2. Grind or replace brake drum.
	3. Bent brake support plate.	3. Replace brake support plate.
	4. Bent brake shoes.	4. Replace brake shoes.
	5. Foreign material embedded in brake lining.	5. Replace brake shoes.
	6. Dust or dirt in brake drum.	6. Use compressed air and blow out drums and support plate and shoes.
	7. Shoes dragging on support plate.	7. Sand support plate platforms and lubricate.
	8. Loose support plate.	8. Tighten support plate attaching nuts.
	9. Loose anchor bolts.	9. Tighten anchor bolts.
	10. Loose lining on brake shoes or improperly ground lining.	10. Replace brake shoes and cam-grind lining.
Brakes fading	1. Improper brake adjustment.	1. Adjust brakes correctly.
	2. Improper brake lining.	2. Replace brake lining.
	3. Improper type of brake fluid.	3. Drain, flush and refill hydraulic system.
	4. Brake drums ground excessively.	4. Replace brake drums.
Dragging brakes	1. Improper brake adjustment.	1. Correct adjust brakes.
	2. Distorted cylinder cups.	2. Recondition or replace cylinder.
	3. Brake shoe seized on anchor bolt.	3. Clean and lubricate anchor bolt.
	4. Broken brake shoe return spring.	4. Replace brake shoe return spring.
	5. Loose anchor bolt.	5. Adjust and tighten anchor bolt.
	6. Distorted brake shoe.	6. Replace defective brake shoes.
	7. Loose wheel bearings.	7. Lubricate and adjust wheel bearings.
	8. Obstruction in brake line.	8. Clean or replace brake line.
	9. Swollen cups in wheel cylinder or master cylinder.	9. Recondition wheel or master cylinder.
	10. Master cylinder linkage improperly adjusted.	10. Correctly adjust master cylinder linkage.
Hard pedal	1. Incorrect brake lining.	1. Install matched brake lining.
	2. Incorrect brake adjustment.	2. Adjust brakes and check fluid.
	3. Frozen brake pedal linkage.	3. Free up and lubricate brake linkage.
	4. Restricted brake line or hose.	4. Clean out or replace brake line hose.
Wheel locks	1. Loose or torn brake lining.	1. Replace brake lining.
	2. Incorrect wheel bearing adjustment.	2. Clean, pack and adjust wheel bearings.
	3. Wheel cylinder cups sticking.	3. Recondition or replace the wheel cylinder.
	4. Saturated brake lining.	4. Reline front, rear or all four brakes.

HYDRAULIC BRAKE SYSTEM TROUBLE DIAGNOSIS

Condition	Possible Cause	Correction
Brakes fade (high speed)	1. Improper brake adjustment. 2. Distorted or out of round brake drums. 3. Overheated brake drums. 4. Incorrect brake fluid (low boiling temperature). 5. Saturated brake lining.	1. Adjust brakes and check fluid. 2. Grind or replace the drums. 3. Inspect for dragging brakes. 4. Drain flush and refill and bleed the hydraulic brake system. 5. Reline brakes as necessary.

Hydraulic Brakes

General Information

Servicing the hydraulic system is chiefly a matter of adjustments, replacement of worn or damaged parts and correcting the damage caused by grit, dirt or contaminated brake fluid. It is highly important to make sure the brake system is clean and tightly sealed when a brake job is completed and that only approved heavy duty brake fluid is used.

The approved heavy duty type brake fluid retains the correct consistency throughout the widest range of temperature variation, will not affect rubber cups, helps protect the metal parts of the brake system against failure and assures long trouble-free brake operation.

Never use brake fluid from a container that has been used for any other liquid. Mineral oil, alcohol, anti-freeze, or cleaning solvents, even in very small quantities, will contaminate brake fluid. Contaminated brake fluid will cause piston cups and the valve in the master cylinder to swell or deteriorate.

Brake adjustment is required after installation of new or relined brake shoes. Adjustment is also necessary whenever excessive travel of pedal is needed to start braking action.

Low Pedal

Normal brake lining wear reduces pedal reserve. Low pedal reserve may also be caused by the lack of brake fluid in the master cylinder. The wear condition may be compensated for by a minor brake adjustment. Check fluid level in master cylinder and add as required.

Fluid Loss

If the master cylinder requires constant addition of hydraulic fluid, fluid may be leaking past the piston cups in the master cylinder or brake cylinders, the hydraulic lines; hoses or connections may be loose or broken. Loose connections should be tightened, or other necessary repairs or parts replacement made and the hydraulic brake system bled.

Fluid Contamination

To determine if contamination exists in the brake fluid, as indicated by swollen, deteriorated rubber cups, the following tests can be made.

Place a small amount of the drained brake fluid into a small clear glass bottle. Separation of the fluid into distinct layers will indicate mineral oil content. Be safe and discard old brake fluid that has been bled from the system. Fluid drained from the bleeding operation may contain dirt particles or other contamination and should not be reused.

Brake Adjustment

Normally self adjusting brakes will not require manual adjustment but in the event of a brake reline it may be advisable to make the initial adjustment manually to speed up adjusting time.

Automatic Adjuster Check

Place vehicle on a hoist, with a helper in the driver's seat to apply brakes. Remove plug from rear adjustment slot in each brake support plate to observe adjuster star wheel. Then, to exclude possibility of maximum adjustment; that is, the adjuster refuses to operate because the closest possible adjustment has been reached; the star wheel should be backed off approximately 30 notches. It will be necessary to hold adjuster lever away from star wheel to allow backing off of the adjustment.

Spin the wheel and brake drum in reverse direction and apply brakes vigorously. This will provide the necessary inertia to cause the secondary brake shoe to leave the anchor. The wrap up effect will move the secondary shoe, and cable will pull the

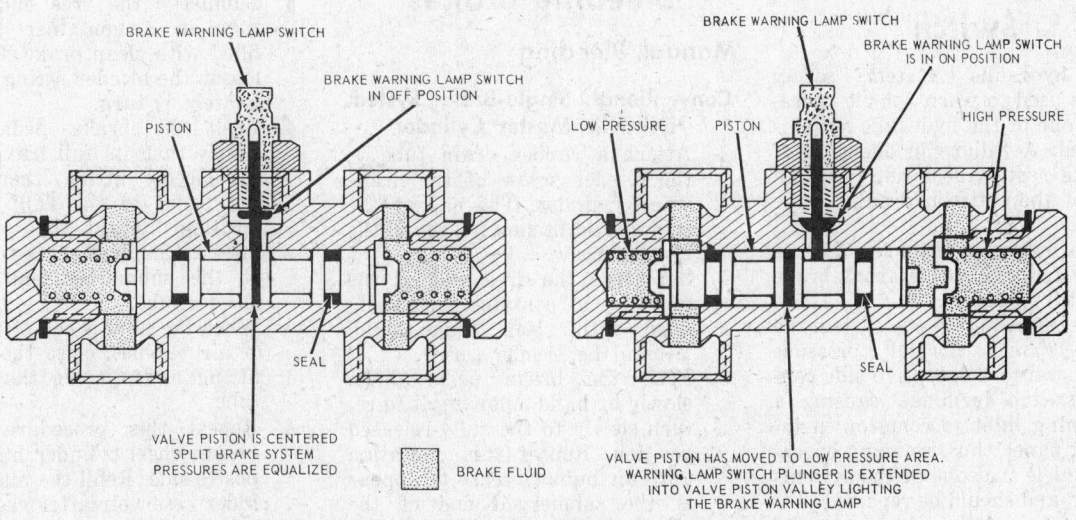

Differential valve system—with split hydraulic brakes (© Ford Motor Co.)

Hydraulic Brakes

adjuster lever up. Upon release of brake pedal, the lever should snap downward, turning star wheel. Thus, a definite rotation of adjuster star wheel can be observed if automatic adjuster is working properly. If by the described procedure one or more automatic adjusters do not function properly, the respective drum must be removed for adjuster servicing.

Hydraulic Line Repair

Steel tubing is used in the hydraulic lines between the master cylinder and the front brake tube connector, and between the rear brake tube connector and the rear brake cylinders. Flexible hoses connect the brake tube to the front brake cylinders and to the rear brake tube connector.

When replacing hydraulic brake tubing, hoses, or connectors, tighten all connections securely. After replacement, bleed the brake system at the wheel cylinders and at the booster, if so equipped.

Brake Tube

If a section of the brake tube becomes damaged, the entire section should be replaced with tubing of the same type, size, shape, and length. *Copper tubing should not be used in the hydraulic system.* When bending brake tubing to fit the frame or rear-axle contours, be careful not to kink or crack the tube.

All brake tubing should be double flared to provide good leak-proof connections. Always clean the inside of a new brake tube with clean isopropyl alcohol.

Brake Hose

A flexible brake hose should be replaced if it shows signs of softening, cracking, or other damage.

When installing a new brake hose, position the hose to avoid contact with other truck parts.

Pressure Differential Switch

The hydraulic system safety switch is used to warn vehicle operator that one of the hydraulic systems has failed. A failure in one part of the brake system does not result in failure of the entire hydraulic brake system.

As an example, failure of rear brake system will leave front brake system still operative.

As pressure falls in one system the other system's normal pressure forces piston to inoperative side contacting switch terminal, causing a red warning light to come on in instrument panel, thus, warning operator of vehicle that one of the systems has failed and should be repaired.

The safety switch body is mounted in a vertical position, with the brake tubes connected to opposite sides.

The component parts of the switch body are not serviced. However, terminal unit can be removed if a malfunction occurs and a new terminal unit installed.

Centralizing the Pressure Differential Valve

1. Turn the ignition switch to the ACC or ON position. Loosen the pressure differential valve inlet tube nut of the system that remained operative, or the side opposite the system that was bled last. Operate the brake pedal carefully and gradually until the pressure differential valve is returned to a centralized position and the brake warning light goes out. Tighten the tube nut.
2. Check the fluid level in the master cylinder reservoirs and fill them to within 1/4" of the top with the specified brake fluid.
3. Turn the ignition switch to the OFF position.

With Split Hydraulic Brakes

The pressure differential valve used with the split hydraulic brake system has a self-centering spring. Use the following procedure to reset the valve:
1. Remove the switch connector wire.
2. Remove the threaded hex-shaped electrical switch body from the center of the valve. This allows the valve centering springs to re-position the valve.
3. Install the electrical switch and connect the wire.
4. Apply the brakes a few times and check the operation of the warning light. The light should go on with the ignition switch in the START position only.

Bleeding Brakes

Manual Bleeding

Conventional, Single-Brake System Hydraulic Master Cylinder

1. Attach a rubber drain tube to the bleeder screw of the brake wheel cylinder. The end of the tube should fit snugly around the bleeder screw.
2. Submerge the free end of the tube in a container partially filled with clean brake fluid. Loosen the bleeder screw.
3. Push the brake pedal down slowly by hand, allowing it to return slowly to the fully-released position. Repeat this operation until air bubbles cease to appear at the submerged end of the tube.

4. When the fluid is completely free of air bubbles, close the bleeder screw and remove the drain tube.
5. Repeat this procedure at each brake cylinder. Refill the master cylinder reservoir after each brake cylinder is bled and when the bleeding operation is completed.

Dual-Brake System Hydraulic Master Cylinder

The primary and secondary hydraulic brake systems are individual systems and are bled separately. Bleed the longest line first on the individual system being serviced. *During the complete bleeding operation, do not allow the reservoir to run dry.* Keep the master cylinder reservoirs filled with the specified brake fluid. *Never use brake fluid that has been drained from the hydraulic system.*

1. Bleed the master cylinder at the outlet port side of the system being serviced.
   NOTE: On a master cylinder without bleed screws, loosen the master cylinder to hydraulic line nut. Operate the brake pedal slowly until the brake fluid at the outlet connection is free of bubbles, then tighten the tube nut to the specified torque. Do not use the secondary piston stop screw located on the bottom of the master cylinder to bleed the brake system. Loosening or removing this screw could result in damage to the secondary piston or stop screw. Operate the brake pedal slowly until the brake fluid at the outlet connection is free of air bubbles, then tighten the bleed screw.
2. Position a suitable 3/8" box wrench on the bleeder fitting on the brake wheel cylinder. Attach a rubber drain tube to the bleeder fitting. The end of the tube should fit snugly around the bleeder fitting.
3. Submerge the free end of the tube in a container partially filled with clean brake fluid, and loosen the bleeder fitting approximately 3/4 turn.
4. Push the brake pedal down slowly thru its full travel. Close the bleeder fitting, then return the pedal to the fully-released position. Repeat this operation until air bubbles cease to appear at the submerged end of the bleeder tube.
5. When the fluid is completely free of air bubbles, close the bleeder fitting and remove the bleeder tube.
6. Repeat this procedure at the brake wheel cylinder on the opposite side. Refill the master cylinder reservoir after each wheel cylinder is bled.

When the bleeding operation is complete, the master cylinder fluid level should be filled to within ¼" from the top of the reservoirs.

7. Centralize the pressure differential valve.

Master Cylinder Service

Bendix Tandem

Disassembly

1. Clean the outside of the master cylinder assembly. Remove the residual pressure valves.
2. Remove the tube seats by installing "easy outs" firmly into the seats. Tap lightly with a hammer to loosen, remove seats.
3. Slide clamp off master cylinder cover and remove the cover and its gasket. Drain the brake fluid from the master cylinder.
4. Remove the snap ring from the open end of the cylinder with snap ring pliers. Remove the washer from cylinder bore.
5. Remove the front piston retaining screw. Carefully remove the rear piston assembly.
6. Remove the front piston assembly.

Cleaning and Inspection

1. Clean all parts with a suitable solvent and dry with filtered compressed air. Wash cylinder bore with clean brake fluid and check for damage or wear.
2. If cylinder bore is lightly scratched or shows slight corrosion it can be cleaned with crocus cloth. Heavier scratches or corrosion can be removed by honing, providing that diameter of cylinder bore is not increased by more than .002 inch. If master cylinder bore does not clean up at .002 inch when honed, the master cylinder should be replaced.
3. If master cylinder pistons are badly scored or corroded, replace them with new ones. All caps and seals should be replaced when rebuilding a master cylinder.

Assembly

NOTE: Before assembly of master cylinder, dip all parts in clean brake fluid and place on clean paper. Assembling master cylinder dry can damage rubber seals.

1. Coat master cylinder bore with brake fluid and carefully slide the front piston into cylinder body.
2. Slide the rear piston assembly into the cylinder bore. Compress pistons and install the front piston retaining screw.
3. Position washer in cylinder bore and secure with snap ring.
4. Install the residual pressure valve and spring in the outlet port and install tube seats firmly.

Chrysler Tandem

Disassembly

1. Clean the outside of the master cylinder assembly. Remove the master cylinder cover and drain the brake fluid.
2. Remove the front piston retainer screw from inside reservoir and the snap ring from the outer end of cylinder bore. Slide the rear piston assembly out of cylinder bore.
3. Tamp the master cylinder assembly lightly on bench, open end down, to remove the front piston and spring. If the front piston sticks in cylinder bore, use air pressure to force it out.
4. Remove the front piston compression spring from the cylinder bore.
5. Remove the tube seats by installing "easy outs" firmly into the seats. Tap lightly with a hammer to loosen, remove the seats.
6. Take note of the position of cup lips and remove them from pistons. DO NOT remove the center cup of the rear piston. If this cup is damaged or worn, install a new rear piston assembly.

Cleaning and Inspection

1. Clean all parts with a suitable solvent and dry with filtered compressed air. Wash cylinder bore with clean brake fluid and check for damage or wear.
2. If cylinder bore is slightly scrached or shows slight corrosion it can be cleaned with crocus cloth. Heavier scratches or corrosion can be removed by honing, providing that the diameter of the cylinder bore is not increased by more than .002 inch. If master cylinder bore does not clean up at .002 inch when honed, the master cylinder should be replaced.
3. If master cylinder pistons are bady scored or corroded, replace them with new ones. All caps and seals should be replaced when rebuilding a master cylinder.

Assembly

NOTE: Before assembly of master cylinder, dip all parts in clean brake fluid and place on clean paper. Assembling a master cylinder dry can damage the new seals.

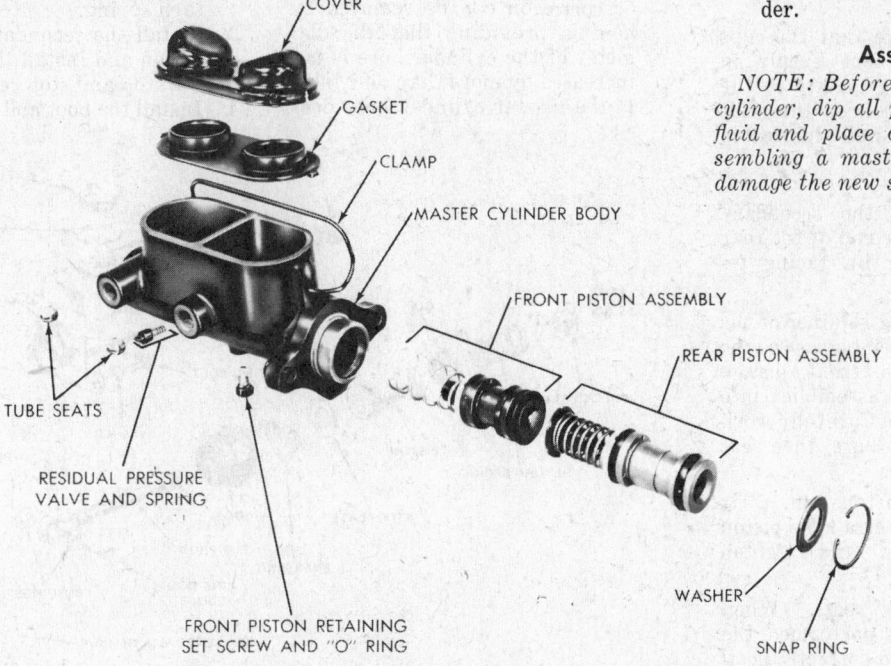

COVER
GASKET
CLAMP
MASTER CYLINDER BODY
FRONT PISTON ASSEMBLY
REAR PISTON ASSEMBLY
TUBE SEATS
RESIDUAL PRESSURE VALVE AND SPRING
FRONT PISTON RETAINING SET SCREW AND "O" RING
WASHER
SNAP RING

Bendix tandem master cylinder—exploded view (© Chrysler Corp.)

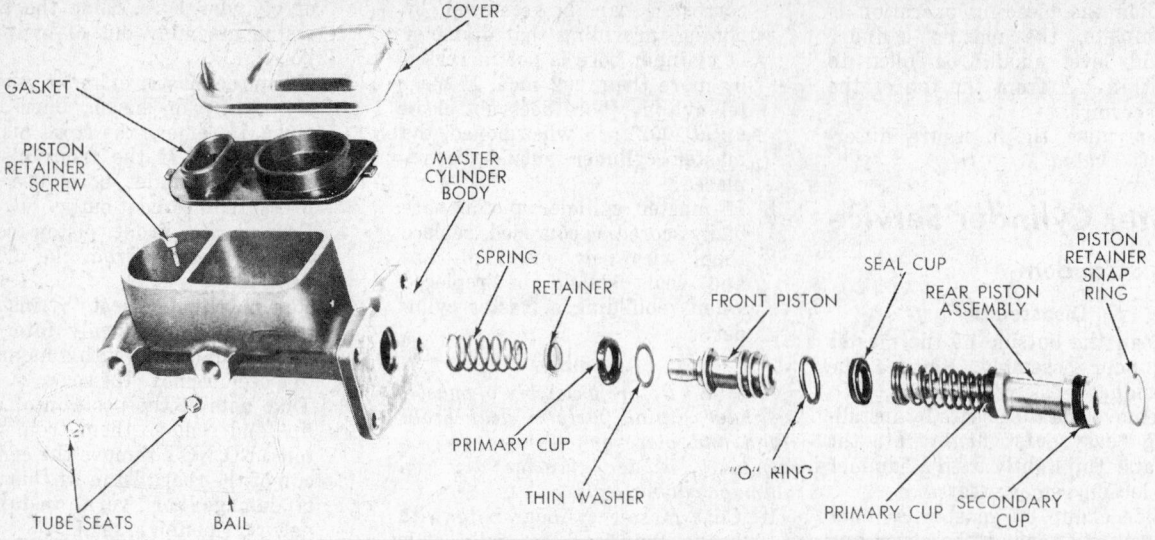

Chrysler tandem master cylinder—exploded view (©Chrysler Corp.)

1. Carefully work the primary cup on the end of the front piston with the lip facing away from the piston.

2. Carefully, work the second seal cup over the rear end of the piston and into the second land. Be sure that the lip of the cup is facing the front of piston.

3. Carefully work the rear secondary cup over the piston and into the rear land. The lip must face the rear of the piston.

4. Slide the cup retainer over the stem of the front piston with the beveled side away from the piston cup.

5. Position the small end of the pressure spring into the retainer, then slide the assembly into the master cylinder bore.

CAUTION: Be sure that the cups enter the cylinder bore evenly in order that the sealing quality of the cups is not damaged, keep the seals and cylinder bore well lubricated with brake fluid.

6. Carefully work the secondary cup over the rear end of the rear piston with the lip facing towards the front.

7. Center the spring retainer of the rear piston assembly over the shoulder of the front piston. Push the piston assemblies into the cylinder bore. Carefully work the cup lips into bore, then seat piston assemblies.

8. Holding the pistons in their seated position, install the piston retaining screw and tighten securely.

9. Install new tube seats. (When bench bleeding is performed, the tube seats will be correctly positioned.)

Wagner Single

Disassembly

1. Clean the outside of the master cylinder assembly. Remove the master cylinder cover and drain the fluid.

2. Remove the boot, stop retainer, piston stop and the end plug.

3. Remove the piston, cups, return spring and the valve assembly.

Cleaning and Inspection

1. Clean all parts with a suitable solvent and dry with filtered compressed air. Wash cylinder bore with clean brake fluid and check for damage and wear.

2. If cylinder bore is slightly scratched or shows slight corrosion it can be cleaned with crocus cloth. Heavier scratches or corrosion can be removed by honing, providing that the diameter of the cylinder bore is not increased by more than .002 inch. If the master cylinder bore does

not clean up at .002 inch when honed, the master cylinder should be replaced.

3. If the master cylinder piston is badly scored or corroded, replace it with a new one. All caps and seals should be replaced when rebuilding a master cylinder.

Assembly

NOTE: Before assembly of master cylinder, dip all parts in clean brake fluid and place on clean paper. Assembling a master cylinder dry can damage the new seals.

1. Install the check valve and spring in the cylinder, with the valve facing toward the outlet of the master cylinder.

2. Install the cylinder cup in the cylinder with the open end of the cup over the closed end of the return spring.

3. Install the secondary cup on the piston and install the piston, piston stop and stop retainer.

4. Install the boot and push rod.

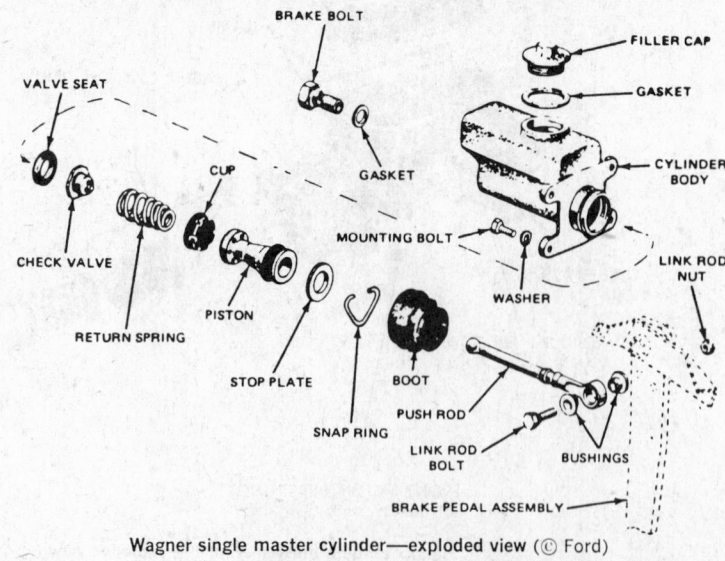

Wagner single master cylinder—exploded view (© Ford)

Wagner Tandem

Disassembly

1. Clean the outside of the master cylinder. Remove the cylinder cover screw or spring retaining clip. Lift off the cover and the diaphragm gasket and pour off excess brake fluid. Use the push rod to stroke the cylinder forcing fluid from the cylinder through the outlet ports.
2. Loosen and remove the piston stop screw and gasket from the right hand side of the cylinder.
3. Pull back the push rod boot and remove the snap ring from the groove in the end of the cylinder bore.
4. Remove the push rod and stop plate from the master cylinder.
5. Remove the internal parts from the master cylinder. If the parts will not slide out apply air pressure at the secondary outlet port. If after applying air, parts still do not move easily, check bore carefully for extensive damage which may eliminate possibility of rebuilding master cylinder.

Inspection and Repair

1. Clean all parts in clean brake fluid. Inspect the parts for chipping, excessive wear or damage. Replace them as required. When using a master cylinder repair kit, install all the parts supplied.
2. Check all recesses, openings and internal passages to be sure they are open and free of foreign matter. Passages may be probed with soft copper wire, 0.020″ OD, or smaller.
3. Minor scratches or blemishes in the cylinder bore can be removed with crocus cloth or a clean up hone. Do not oversize the bore more than 0.007″.

Assembly

1. Dip all parts except the master cylinder in clean hydraulic brake fluid of the specified type.
2. Install the rear rubber cup on the secondary piston with the cup lip facing the rear. All other cups face the front or closed end of the cylinder.
3. Assemble and install the secondary piston spring, front cup, and the secondary piston.
4. Install the piston stop screw and gasket, making sure the screw enters the cylinder behind the rear of the secondary piston.
5. Assemble and install the primary piston and push rod parts.
6. Locate the stop plate in the seat in the bore and engage the snap ring into the groove at the rear of the cylinder.
7. Install the push rod boot onto the push rod and the groove of the cylinder housing.
8. Bleed the master cylinder.

Bench Bleeding the Master Cylinder

Before the master cylinder is installed on the vehicle, the unit should be bled.

1. Support the master cylinder body in a vise, and fill both fluid reservoirs with the specified brake fluid.
   CAUTION: Do not tighten the vise too tightly on the master cylinder as this can cause damage to the cylinder which can not be repaired.
2. Loosely intall plugs in the front and rear brake outlet bores. Depress the primary piston several times until air bubbles cease to appear in the brake fluid.
3. Tighten the plugs and attempt to depress the piston. The piston travel should be restricted after all air is expelled.
4. Remove the plugs. Install the cover and diaphragm gasket assembly, and make sure the cover screw is tightened securely.

Midland-Ross Tandem (Removable Reservoirs Type)

Disassembly

1. Clean the outside of the cylinder and remove the filler cap and gasket (diaphragm). Pour out any brake fluid that may remain in the reservoir. Stroke the push rod several times to remove fluid from the cylinder bore.
2. Remove the reservoir retainers, washers, and reservoir from the master cylinder body.
3. Remove the two rubber washers from the reservoir and the two O-rings from the reservoir retainers.
4. Remove the snap ring, spring retainer and push rod spring.
5. Unscrew the retainer bushing counterclockwise and remove the push rod, retainer bushing, seal retainer and primary piston from the master cylinder.
6. Remove the primary piston from the push rod and discard it.
7. Remove the seal retainer, and retainer bushing from the push rod. Remove the two lip seals and the two Orings from the retainer bushing.
8. Unscrew the end cap counterclockwise and remove the end cap and secondary piston assembly from the master cylinder.

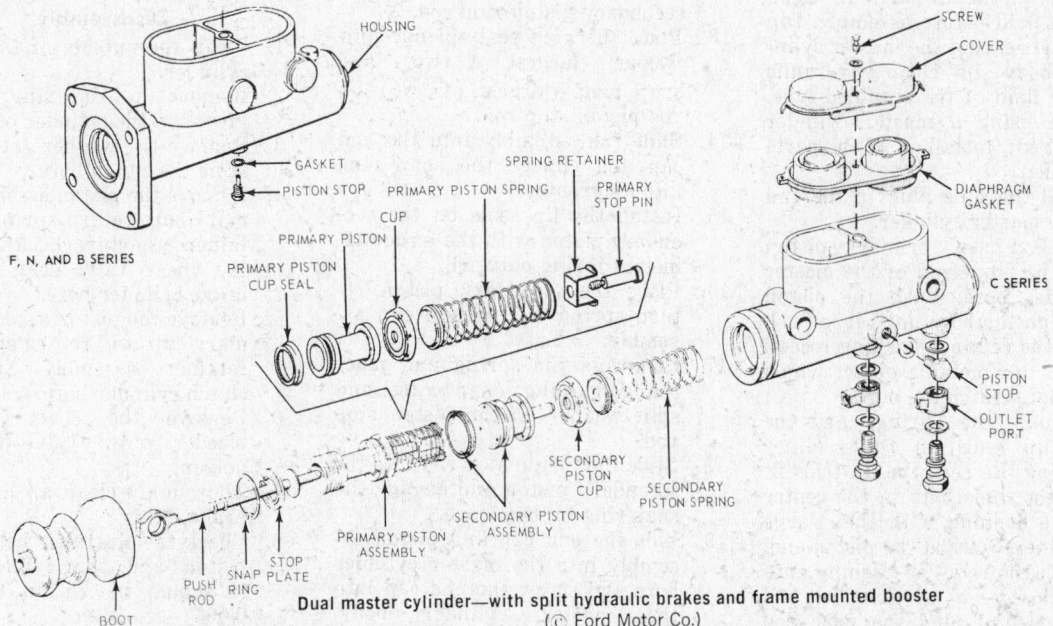

Dual master cylinder—with split hydraulic brakes and frame mounted booster
(ⓒ Ford Motor Co.)

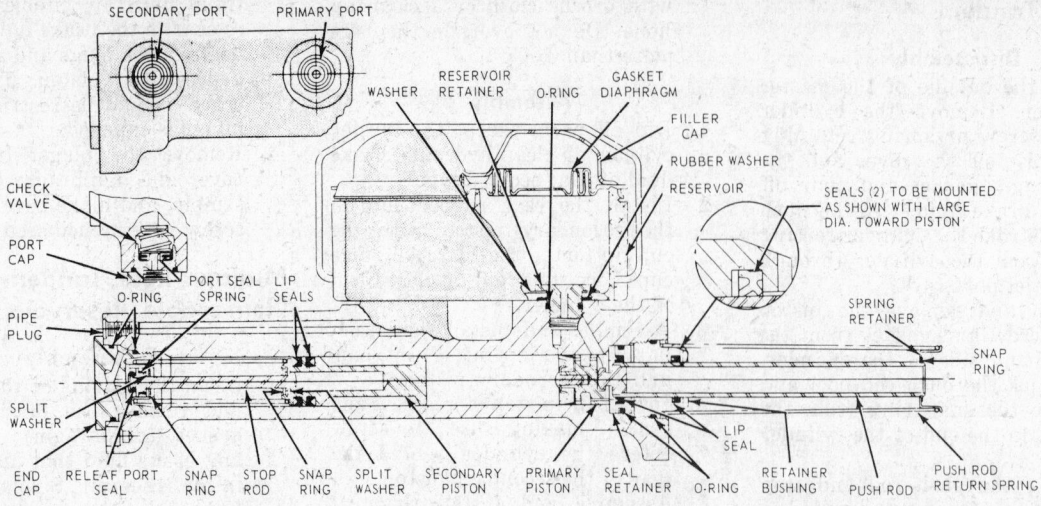

SECONDARY PORT   PRIMARY PORT

WASHER   RESERVOIR RETAINER   O-RING   GASKET DIAPHRAGM

FILLER CAP

RUBBER WASHER

RESERVOIR

SEALS (2) TO BE MOUNTED AS SHOWN WITH LARGE DIA. TOWARD PISTON

CHECK VALVE

PORT CAP

PORT SEAL   LIP SEALS

O-RING   SPRING

PIPE PLUG

SPLIT WASHER

SPRING RETAINER

SNAP RING

END CAP   RELEAF PORT SEAL   SNAP RING   STOP ROD   SNAP RING   SPLIT WASHER   SECONDARY PISTON   PRIMARY PISTON   SEAL RETAINER   O-RING   LIP SEAL   RETAINER BUSHING   PUSH ROD   PUSH ROD RETURN SPRING

Dual master cylinder—with midland ross dash mounted booster (© Ford Motor Co.)

9. Remove the snap ring from the secondary piston and remove the piston and return spring from the end cap and stop rod assembly.
10. Remove the two lip seals from the piston.
11. Remove the snap ring from the end cap and remove the secondary piston stop rod, relief port seal spring, the two snap rings and the two split washers from the end cap.
12. Remove the relief port seal from the secondary piston stop rod.
13. Remove the O-rings from the end cap.
14. Remove the primary and secondary port caps and discard.
15. Remove the check valves and springs from the ports.
16. Remove the pipe plug from the end of the master cylinder.

Assembly

1. Wash all metal parts in clean brake fluid before assembly. Dip all parts except the master cylinder body in clean hydraulic brake fluid of the specified type. When using a master cylinder repair kit, install all of the parts supplied.
2. Install the pipe plug in the end of the master cylinder.
3. Install a new primary piston into the front end of the master cylinder bore. Push the piston through the bore until it is flush with the retainer bushing recess. Use a non-metallic object which will not scratch the bore.
4. Assemble the O-rings and the two lip seals on the retainer bushing. Be sure the lip seals fit into the undercuts in the center of the bushing with their large diameters toward the piston end.
5. Install the retainer bushing onto the closed end of the push rod and push it onto the push rod

approximately half way. Be sure the lip seal at the piston end of the retainer bushing remains in the undercut portion of the retainer bushing.
6. Install the seal retainer onto the closed end of the push rod with the raised lip toward the retainer bushing.
7. Insert the push rod into the master cylinder bore and hook the push rod onto the primary piston.
8. Slide the seal retainer into the recess in the master cylinder bore.
9. Screw the retainer bushing into the master cylinder body and tighten to 15-20 ft-lbs torque.
10. Install the push spring with the large end toward the master cylinder and install the spring retainer and snap ring.
11. Install the O-rings on the end cap.
12. Install the relief port seal on the secondary piston stop rod.
13. Place the port seal spring, split washer (largest of two), and snap ring (largest of two) on the piston stop rod.
14. Slide the assembly into the end cap and engage the snap ring into its groove.
15. Install the lip seals on the secondary piston with the large diameters facing outward.
16. Place the secondary piston return spring on the end cap assembly.
17. Compress the spring and place the remaining snap ring and split washer on the piston stop rod.
18. Slide the piston stop rod into the secondary piston and engage the snap ring in its groove.
19. Slide the end cap and piston assembly into the master cylinder bore and screw the end cap into the master cylinder body.

Tighten the cap to 15-20 ft-lbs torque.
20. Install washers on the reservoir retainer and place the retainers in the mounting holes of the reservoir.
21. Place the rubber washers and O-rings on the retainers.
22. Place the reservoir and retainer assembly on the master cylinder body and tighten the retainers to 15-20 ft-lbs torque.
23. Replace the springs and check valves in the output ports of the cylinder.
24 Replace the primary and secondary port caps. Tighten to 15-20 ft-lbs torque.
25. Install the mounting seal on the flange of the master cylinder. Install the filler cap and gasket (diaphragm).

Single and Double Barrel- G.M.C.

Disassembly

1. Clean the outside of the master cylinder.
2. Remove the snap ring from the groove in the cylinder bore.
3. Remove the washer (stop plate) from the clutch bore.
4. Remove the piston assembly, primary cup, return spring and retainer assembly, check valve, and the check valve seat from the brake cylinder bore.
5. Remove the piston assembly, primary cup, and return spring and retainer assembly from the clutch cylinder bore.
6. Remove the cover and the bleeder screw valve from the housing.
7. Thoroughly clean all parts with brake fluid.
8. Check the clearance between the piston and the cylinder wall. It should be within 0.001" to 0.005"

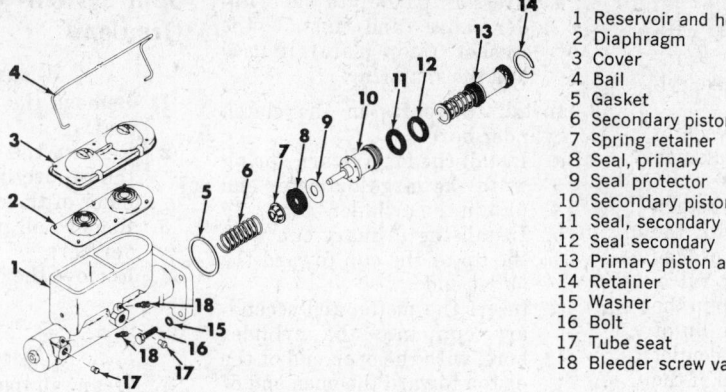

1 Reservoir and housing
2 Diaphragm
3 Cover
4 Bail
5 Gasket
6 Secondary piston spring
7 Spring retainer
8 Seal, primary
9 Seal protector
10 Secondary piston
11 Seal, secondary
12 Seal secondary
13 Primary piston assembly
14 Retainer
15 Washer
16 Bolt
17 Tube seat
18 Bleeder screw valve

Split system master cylinder (© G.M.C.)

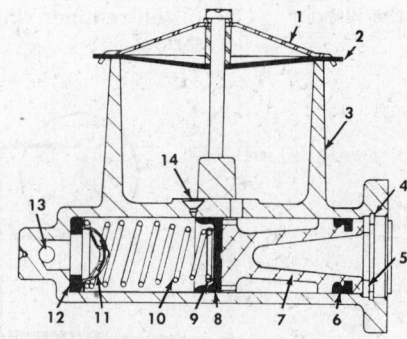

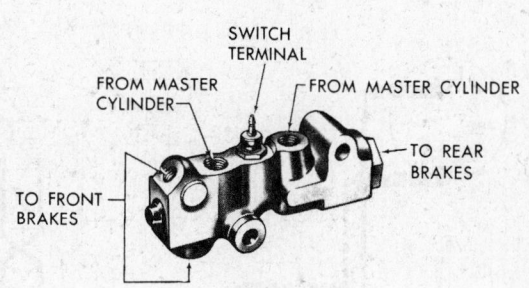

SWITCH
TERMINAL

FROM MASTER
CYLINDER

FROM MASTER CYLINDER

TO FRONT
BRAKES

TO REAR
BRAKES

Combination Valve

1 Cover assembly
2 Cover gasket
3 Reservoir body
4 Snap ring
5 Stop plate
6 Secondary cup
7 Piston assembly
8 Primary cup
9 Spring retainer
10 Return spring
11 Check valve
12 Check valve seat
13 Outlet port
14 By-pass port

Single barrel master cylinder (© G.M.C.)

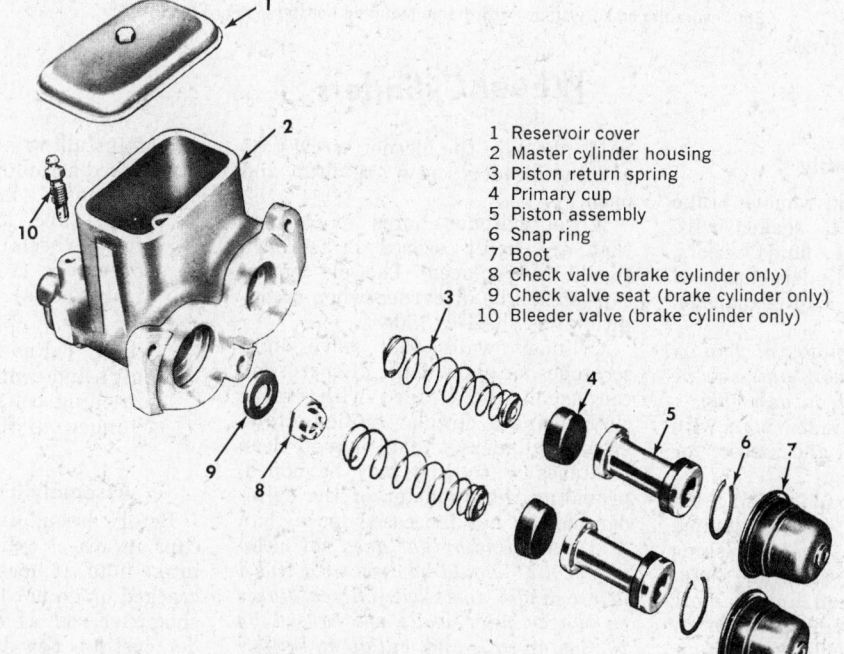

1 Reservoir cover
2 Master cylinder housing
3 Piston return spring
4 Primary cup
5 Piston assembly
6 Snap ring
7 Boot
8 Check valve (brake cylinder only)
9 Check valve seat (brake cylinder only)
10 Bleeder valve (brake cylinder only)

Double barrel master cylinder (© G.M.C.)

Hydraulic Brakes

Assembly

1. Coat all internal parts with brake fluid.
2. Install the parts in the brake cylinder bore.
   a. Install the check valve seat and then the check valve in the cylinder bore.
   b. Install the short return spring in the bore with the large diameter end of the spring over the check valve.
   c. Install the primary cup in the cylinder bore with the lip of the cup toward the outlet end. Make sure the end of the return spring seats inside the cup.
   d. Insert the piston and secondary cup assembly into the cylinder bore with the open end of the piston toward the open end of the cylinder bore.

e. Press all parts into the cylinder bore and install the washer (stop plate) if used and the snap ring.
3. Install the parts in the clutch cylinder bore.
   a. Install the long return spring with the large diameter end first in the cylinder bore.
   b. Install the primary cup with the lip of the cup toward the outlet end.
   c. Insert the piston and secondary cup into the cylinder bore, with the open end of the piston toward the open end of the cylinder.
   d. Press all parts into the cylinder bore and install the washer (stop plate) if used and the snap ring.
4. Install the cover and the bleeder screw.

Split System—G.M.C. (Tandem)

Disassembly

1. Remove the cover and reservoir seal.
2. Remove the retaining ring from the groove in the end of the cylinder of the cylinder bore.
3. Remove all parts from the cylinder bore.
4. Remove the bleeder screw valves.

Assembly

1. Clean all parts in clean brake fluid.
2. Leave a coating of brake fluid on all internal parts and install parts in the cylinder bore using new rubber seals.
3. Install retainer ring and bleeder screws.

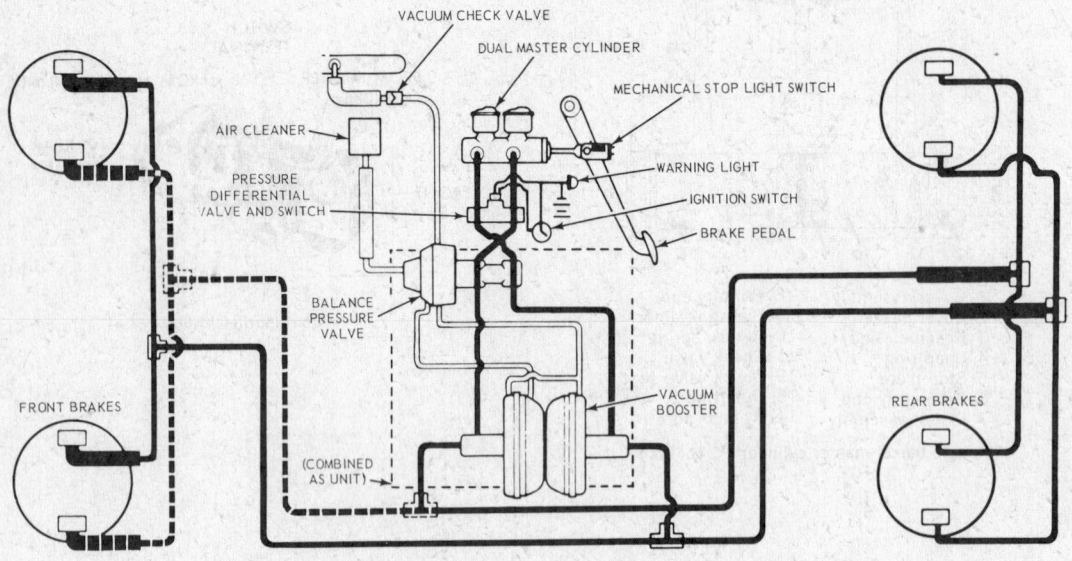

Split hydraulic brake system—with frame mounted booster (© Ford Motor Co.)

Wheel Cylinders

Disassembly

1. In case of a leak, remove brake shoes (replace if soaked with grease or brake fluid), boots, piston wheel cylinder cups and wheel cylinder cup expansion spring.
   NOTE: A slight amount of fluid on boot may not be a leak, but may be preservative oil used on assembly.
2. Wash wheel cylinder bore with clean brake fluid and inspect for scoring or pitting.

Use extreme care in cleaning the wheel cylinder after reconditioning. Remove all dust or grit by flushing the cylinder with clean brake fluid; wipe dry with a clean lintless cloth and clean a second time with brake fluid. Dry the wheel cylinder with air pressure, then flush with clean brake

fluid. (Be sure the bleeder screw port and the bleeder screw are clean and open.)

Wheel cylinder bores or pistons that are badly scored or corroded should be replaced. The old piston cups should be discarded when reconditioning wheel cylinders.

Cylinder walls that have light scratches, or show signs of corrosion, can usually be cleaned with crocus cloth, using a circular motion. However, cylinders that have deep scratches or scoring may be honed, providing the diameter of the cylinder bore is not increased more than .002". *A cylinder that does not clean up at .002" should be discarded and a new cylinder installed. (Black stains on the cylinder walls are caused by the piston cups and will do no harm.)*

Should inspection reveal the neces-

sity of installing a new wheel cylinder proceed as follows:
1. Disconnect brake hose from brake tube at frame bracket (front wheels) or disconnect the brake tube from wheel cylinder (rear wheels).
2. Disconnect brake hose from wheel cylinder (front wheels only) and remove wheel cylinder attaching bolts, then slide wheel cylinder out of backing plate.

Assembly (Front or Rear)

Before assembling pistons and new cups in wheel cylinder, dip them in brake fluid. If boots are deteriorated, cracked or do not fit tightly on brake shoe push rod, as well as wheel cylinder casting, new boots should be installed.

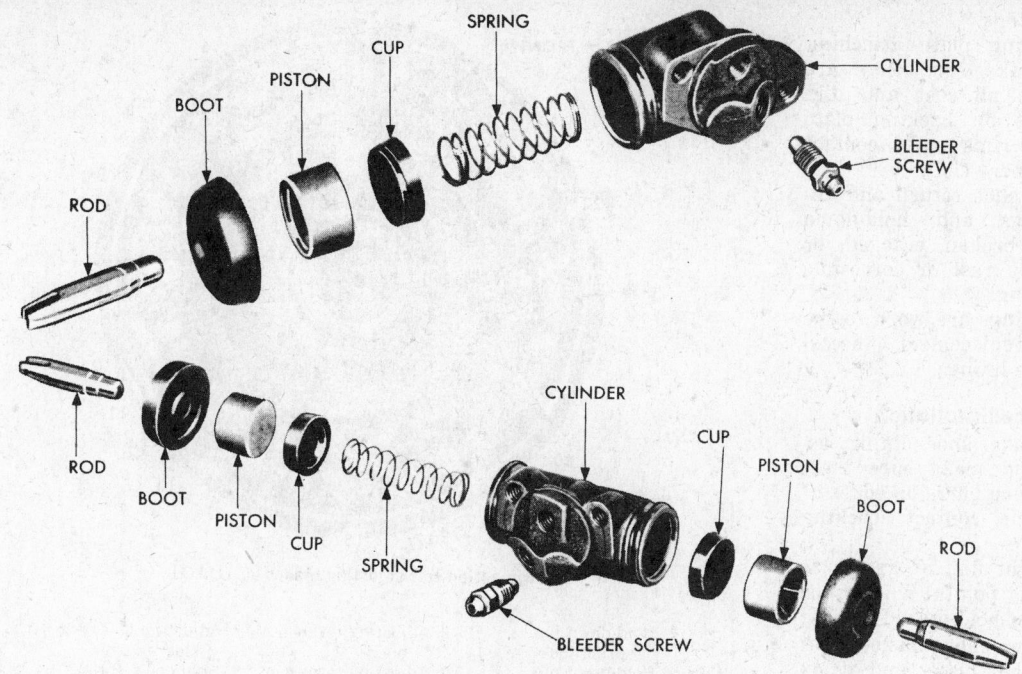

Wheel cylinder (typical) (©️ Chrysler Corp.)

1. Wash wheel cylinder with clean brake fluid and wipe dry.
2. Install expansion spring in cylinder. Install wheel cylinder cups in each end of cylinder with open end of cups facing each other.
3. Install wheel cylinder pistons in each end of cylinder with recessed end of pistons facing open ends of cylinder.
4. Install boots over ends of cylinder. Keep assembly compressed with aid of a brake cylinder clamp until brake shoes are assembled.

Non-Servo Type

This brake is a non-servo, floating shoe type brake. Upper ends of shoes extend through wheel cylinder boots and contact inserts in wheel cylinder pistons. Shoe ends are held firmly against pistons by the brake shoe return spring. Lower ends of shoes are held against a fixed anchor plate by the anchor spring. Hold-down spring at center of each shoe holds shoes in alignment. Lining-to-drum clearance adjustment is made through eccentric cam type adjusting studs.

Brake Shoe Removal

1. Back off brake adjustment, then remove brake drum.
2. Remove brake shoe return spring. Spread upper end of shoes until they are clear of wheel cylinders and hold-down springs, then disengage shoes from anchor plate at bottom. Remove anchor spring from shoes.
3. Do not depress brake pedal while shoes are removed.

Cleaning and Inspection

1. Clean all dirt out of brake drum. Inspect drum for roughness, scoring, or out-of-round. Replace or recondition drum as necessary.

Brake Service

2. Carefully pull lower edge of each wheel cylinder boot away from cylinder and note whether interior is excessively wet with brake fluid. Excessive fluid indicates leakage past piston cups, requiring overhaul of wheel cylinder.

NOTE: A slight amount of fluid is nearly always present and acts as a

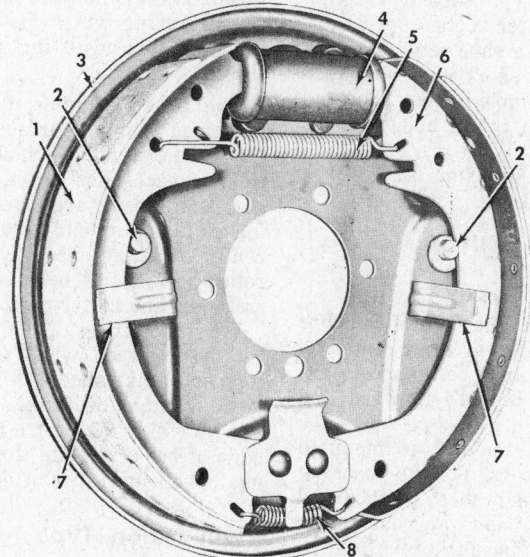

Non servo type brake installed (©️ G.M.C.)

1 Secondary shoe	5 Shoe return spring
2 Adjusting cam	6 Primary shoe
3 Backing plate	7 Shoe hold-down spring
4 Wheel cylinder	8 Shoe anchor spring

lubricant for pistons.

3. Check backing plate attaching bolts to make sure they are tight. Clean all rust and dirt from ledges on backing plate where shoe rims make contact using fine emery cloth.
4. Inspect the shoe return and anchor springs and hold-down springs. If broken, cracked, or weakened by rust or corrosion, replace springs.
5. If brake linings are worn to the extent that replacement is necessary, replace linings.

Brake Shoe Installation

1. Inspect brake shoe lining assemblies and make sure there are no nicks or burrs on edges of shoes which contact backing plate.
2. Apply a light film of grease at the following points: where shoe webs contact hold-down springs; where anchor ends of shoe webs contact anchor plate; and at six places where shoe rims contact ledges on backing plate.
3. Install hold-down springs on backing plate. Hook anchor spring into slot at bottom of each shoe. Swing upper ends of shoes apart and position at backing plate, with lower ends of shoe webs engaging anchor plate and with anchor spring behind extension on anchor plate. *NOTE: The shoe with the shorter lining must be to the rear of the vehicle.*
4. Swing shoes up into position with center of shoe webs engaging hold-down springs, and with upper ends inserted through wheel cylinder boots.
5. Install brake shoe return spring, being sure short end is hooked into slotted hole in rear shoe and long end in round hole in front shoe.
6. Install brake drum and wheel. Adjust brakes.

Wagner

Twin Action Type

Twin-action brake is a four-anchor type. Brake shoes are self-centering in operation, and both shoes are self-energizing in both forward and reverse.

Two wheel cylinders are mounted on opposite sides of the backing plate. One brake shoe is mounted above wheel cylinders and one below. Sliding pivot type anchor is used at front end of upper shoe and at rear end of lower shoe. Adjustable anchor is used at front end of lower shoe and at rear end of upper shoe. Four shoe return springs hold shoe ends firmly against anchors when brakes are released.

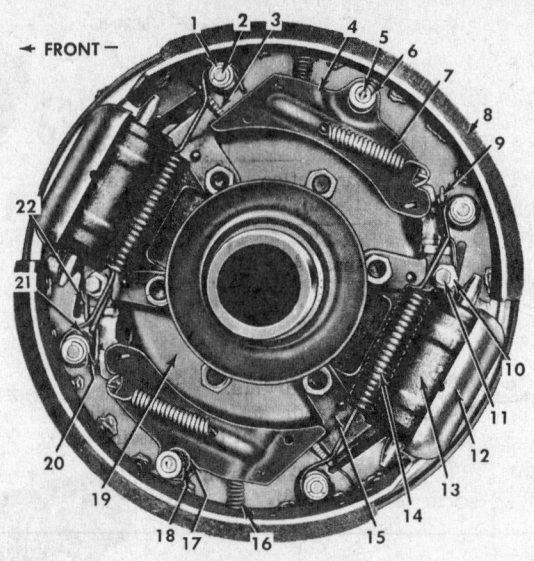

Twin action self adjusting brakes (© G.M.C.)

1 Hold-down pin spring lock	9 Adjusting lever pivot	17 Adjusting lever pin sleeve
2 Hold-down pin	10 Adjusting lever cam	18 Hold-down spring
3 Adjusting screw	11 Adjusting lever bolt	19 Brake backing plate
4 Adjusting lever	12 Wheel cylinder sheild	20 Hold-down pin retainer
5 Adjusting lever pin spring	13 Wheel cylinder	21 Hold-down pin spring
6 Hold-down spring cup	14 Brake shoe return spring	22 Adjusting lever link
7 Lever override spring	15 Brake shoe anchor	
8 Brake shoe and lining	16 Lever return spring	

Anchor brackets are steel forgings, attached to flange on axle housing in conjunction with the backing plate. At adjustable anchor end of each shoe, shoe web bears against flat head of adjusting screw which threads into anchor bracket. The adjusting screw heads are notched and are rotated for brake adjustment through access holes in backing plate. A lock spring which fits over anchor bracket holds adjusting screw in position.

The brake backing plate has six machined bearing surfaces, three for each shoe, against which the inner edge of each shoe bears. Two brake shoe guide bolts are riveted to backing plate and extend through holes in center of brake shoe web. Shoes are retained on guide bolts by flat washers, nuts, and cotter pins.

Wheel cylinder push rods make contact between wheel cylinder pistons and brake shoes.

Inner edge of brake drum has a groove which fits over a flange on the edge of backing plate, forming a seal against the entrance of dirt and mud.

Twin-Action Type Rear Brake

Brake Shoe Removal

1. Remove brake drums. *NOTE: If brake drums are worn severely, it may be necessary to retract the adjusting screws.*

2. Remove the brake shoe pull back springs. *NOTE: Since wheel cylinder piston stops are incorporated in the anchor brackets, it is not necessary to install wheel cylinder clamps when the brake shoes are removed.*
3. Loosen the adjusting lever cam cap screw, and while holding the star wheel end of the adjusting lever past the star wheel, remove the cap screw and cam.
4. Remove the brake shoe hold down springs and pins by compressing the spring and, at the same time, pushing the pin back through the flange plate toward the tool. Then, keeping the spring compressed, remove the lock ("C"-washer) from the pin with a magnet.
5. Lift off the brake shoe and self-adjuster lever as an assembly.
6. The self-adjuster lever can now be removed from the brake shoe by removing the hold-down spring and pin. Remove lever return spring also. *NOTE: The adjusting lever, override spring and pivot are an assembly. It is not recommended that they be disassembled for service purposes unless they are broken. It is much easier to assemble and disassemble the brake leaving them intact.*
7. Thread the adjusting screw out of the brake shoe anchor and

remove and discard the friction spring.

8. Clean all dirt out of brake drum. Inspect drums for roughness, scoring or out-of-round. Replace or recondition drums as necessary.

9. Carefully pull lower edges of wheel cylinder boots away from cylinders. If brake fluid flows out, overhaul of the wheel cylinders is necessary. *NOTE: A slight amount of fluid is nearly always present and acts as a lubricant for the piston.*

10. Inspect flange plate for oil leakage past axle shaft oil seals. Install seals if necessary.

Brake Shoe Installation

1. Put a light film of lubricant on shoe bearing surfaces of brake flange plate and on threads of adjusting screw.

2. Thread adjusting screw completely into anchor without friction spring to be sure threads are clean and screw turns easily. Then remove screws, position a new friction spring on screw and reinstall in anchor.

3. Assemble self-adjuster assembly and lever return spring to brake shoe and position adjusting lever link on adjusting lever pivot.

4. Position hold-down pins in flange plate.

5. Install brake shoe and self-adjuster assemblies onto hold down pins. Insert ends of shoes in wheel cylinder push rods and legs of friction springs. *NOTE: Make sure the toe of the shoe is against the adjusting screw.*

6. Install cup, spring and retainer on end of hold-down pin. With spring compressed, push the hold-down pin back through the flange plate toward the tool and install the lock on the pin.

7. Install brake shoe return springs.

8. Holding the star wheel end of the adjusting lever as far as possible past the star wheel, position the adjusting lever cam into the adjusting lever link and assemble with cap screw.

9. Check the brake shoes for being centered by measuring the distance from the lining surface to the edge of the flange plate. To center the shoes, tap the upper or lower end of the shoes with a plastic mallet until the distances at each end become equal.

10. Locate the adjusting lever .020″ to .039″ above the outside diameter of the adjusting screw thread by loosening the cap screw and turning the adjusting cam. *NOTE: To determine .020″ to .039″, turn the adjusting screw 2 full turns out from the*

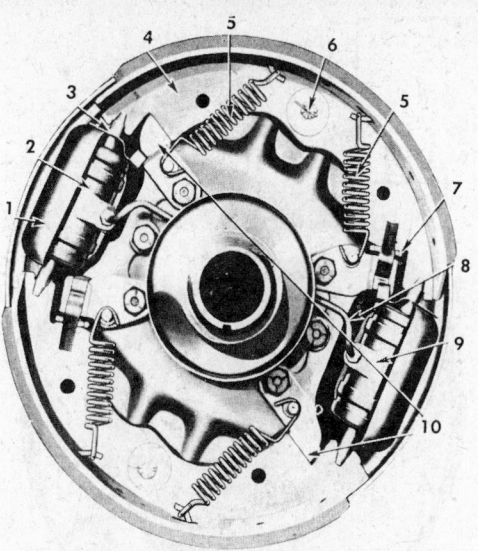

Twin action type brake installed (© G.M.C.)

1 Heat shield	6 Brake shoe guide bolt
2 Front wheel cylinder	7 Adjusting screw
3 Dust shield	8 Hydraulic line
4 Brake shoe	9 Rear wheel cylinder
5 Brake shoe return spring	10 Brake shoe anchor

fully retracted position. Hold a .060″ wire gauge at a 90° angle with the star wheel edge of the adjusting lever. Turn the adjusting cam until the adjusting lever and threaded area on the adjusting screw just touch the wire.

11. Secure the adjusting cam cap screw and retract the adjusting screw.

12. Install brake drums and wheels.

13. Adjust the brakes by making several forward and reverse stops until a satisfactory brake pedal height results.

Wagner

Type "F"

Two identical brake shoes are arranged on backing plate so that their toes are diagonally opposite. Two single-end wheel cylinders are arranged so that each cylinder is mounted between the toe of one shoe and the heel of the other. The two wheel cylinder pistons apply an equal amount of force to the toe of each shoe. Each cylinder casting is shaped to provide an anchor block for the brake shoe heel.

Each shoe is adjusted by means of an eccentric cam which contacts a pin pressed into brake shoe web. Each cam is attached to the backing plate by a cam and shoe guide stud which protrudes through a slot in the shoe web and, in conjunction with flat washers and C-washers, also serves as a shoe hold-down. Two return springs are connected between the shoes, one at each toe and heel.

With vehicle moving forward, both shoes are forward acting (primary shoes), self-energizing in forward direction of drum rotation. With vehicle in reverse, both shoes are reverse acting since neither is self-energized in the reverse direction of drum rotation.

Brake Shoe Removal

1. Remove both brake shoe return springs, using brake spring pliers.

2. Remove C-washer and flat washer from each adjusting cam and hold-down stud. Lift shoes off backing plate.

Cleaning and Inspection

1. Clean all dirt out of brake drum. Inspect drum for roughness, scoring, or out-of-round. Replace or recondition brake drum as necessary.

2. Inspect wheel bearings and oil seals.

3. Check backing plate attaching bolts to make sure they are tight. Clean all dirt off backing plate.

4. Inspect brake shoe return springs. If broken, cracked, or weakened, replace with new springs.

5. Check cam and shoe guide stud and friction spring on backing plate for corrosion or binding. Cam stud should turn easily with a wrench but should not be loose. If frozen, lubricate with kerosene or penetrating oil and work free.

6. Examine brake shoe linings for wear. Lining should be replaced

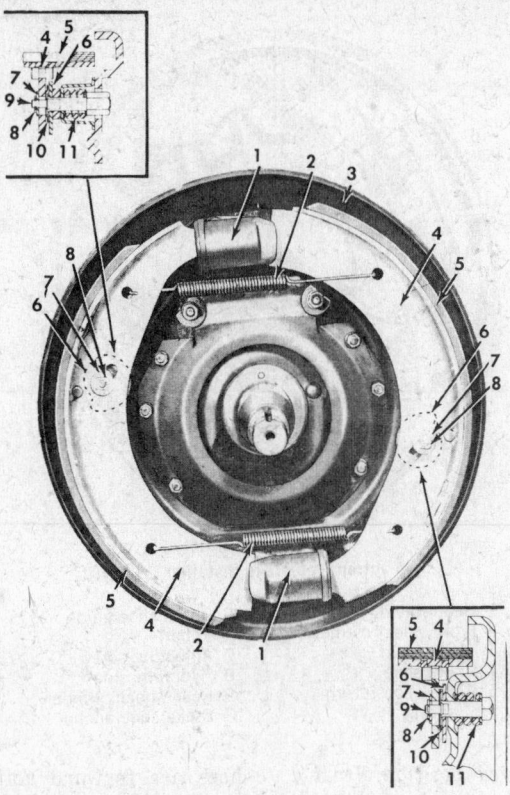

Type "F" brake assembly (Ⓒ G.M.C.)

1 Wheel cylinder
2 Brake shoe return
  spring
3 Backing plate
4 Brake shoe
5 Brake lining
6 Brake shoe adjusting
  cam

7 Brake shoe guide
  washer
8 Brake shoe guide
  C-washer
9 Adjusting cam and
  shoe guide stud
10 Shoe guide anti-
  rattle washer
11 Adjusting cam spring

if worn down close to rivet heads.

Brake Shoe Installation

1. Install anti-rattle spring washer on each cam and shoe guide stud, pronged side facing adjusting cam.
2. Place shoe assembly on backing plate with cam and shoe guide stud inserted through hole in shoe web; locate shoe toe in wheel cylinder piston shoe guide and position shoe heel in slot in anchor block.
3. Install flat washer and C-washer on cam and shoe guide stud. Crimp ends of C-washer together.
4. After installing both shoes, install brake shoe return springs. To install each spring, place spring end with short hook in toe of shoe, then using brake spring pliers, stretch spring and secure long hook end in heel of opposite shoe.
5. install hub and brake drum assembly.
6. Adjust brake.
7. After checking pedal operation, road test vehicle.

Wagner

Type "FA"

Brake Shoe Removal

1. Block brake pedal in up position. Raise vehicle off ground and support.
2. Remove the brake drums. Disconnect the shoe retaining springs and hold down clips and lift off shoes.
3. Unhook the wedge actuating coil spring from the wedge.
4. Unhook the lever actuating spring from the shoe web, work the spring coil off the lever pivot pin and slide the spring "U" hook off the contact plug-lever pin.
5. Pull the adjuster lever from the opposite side of the shoe web, the contact plug through the shoe table and lift off the wedge washer, wedge and the wedge guide.
6. Clean all parts with the exception of the brake shoe linings in a suitable solvent. Inspect all components for wear or damage. Replace all parts that are in questionable condition.

Brake Shoe Installation

1. Install the automatic adjuster, contact plug flush with the lining surface.
2. Position the wedge guide on the back side of the shoe with serrations facing away from the shoe table.
3. Position the wedge on the shoe with the serrations against matching serrations on the wedge guide with the slot aligned on the lever pivot pin hole.
4. Working from the drum side of the shoe, insert the contact plug, with the guide shank through the hole in the shoe table and over wedge guide and wedge.
5. Insert the adjuster lever pins through the shoe web from the opposite side, guiding actuating (center) pin into the mating hole of the contact plug shank.
6. Install the wedge washer over the shoulder of the pivot pin. Slide the "U" hook of adjuster spring on the pin over the contact plug shank.
7. Attach the end of the wedge actuating spring to the "U" hook of the adjusting spring. Position the coil of the adjuster torsion spring over the pivot pin and pull spring hook over the edge of shoe web.
8. Connect the wedge actuating spring on the raised hook of wedge fork.
9. Fully retract the wedge against the lever pivot pin, pressing upon contact plug to permit this movement. If the plug protrudes more than 0.005 in. above lining, clamp shoe in vise so that jaws press against adjuster lever. With a file, press down on the plug until it is even with the brake lining. Exercise caution when filing so as not to create a flat spot on the brake lining. If the fully extended plug is more than 0.005 below the surface of the lining, replace with a new contact plug.
10. Locate the shoe on hold-downs. Install the retracting springs, long ends of springs are at the ends of shoes.

Initial Adjustment

1. Fully release the manual cams.
2. Center each shoe by sliding up or down on its anchor slot until the leading and trailing edges of the lining are equal distant from the inner curl of the support plate.
3. Install the wheel and drum.
4. Rotate the manual adjuster cam in the direction of forward drum rotation, while rotating the drum in the same direction, until the shoe slightly drags on the brake

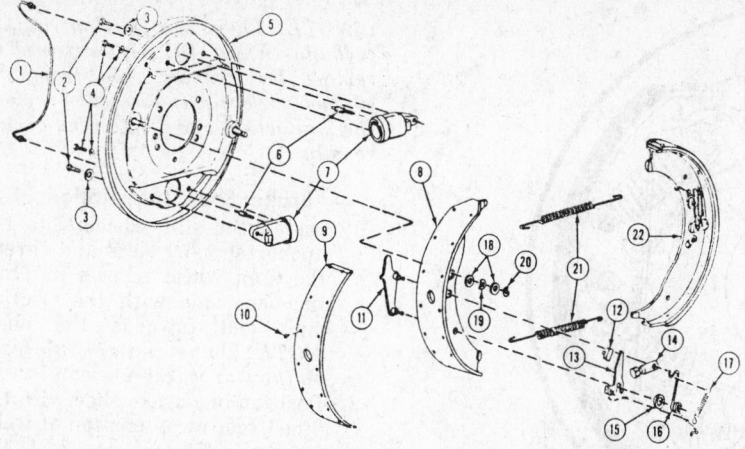

Wagner Type "FA" brake assembly—exploded view (© Chrysler Corp.)

1 Connector tube	8 Brake shoe	16 Adjuster wedge spring
2 Cylinder anchor bolt	9 Brake lining	17 Adjuster torsion spring
3 Anchor bolt washer	10 Rivet	18 Shoe guide washer
4 Screw & lockwasher	11 Automatic adjuster lever	19 Shoe guide wave washer
5 Support plate	12 Adjuster wedge guide	20 Shoe guide 'C' washer
6 Bleeder screw	13 Automatic adjuster wedge	21 Shoe retracting spring
7 Wheel cylinder	14 Drum contact plug	22 Complete shoe assembly
	15 Wedge retainer washer	

drum. Back off adjuster until drag is just relieved. Use only sufficient adjustment torque to obtain drag that will just allow turning the wheel by hand (approximately 120 to 130 inch pounds) as excessive torque may damage the adjuster mechanism.

5. Adjust other manual adjuster in the same manner, forward to tighten, and reverse to relieve drag.

6. Lower vehicle and road test. Automatic adjusters should operate from this point and additional manual adjustment should not be necessary.

Wagner

Type "FR-3"

Each brake is equipped with two double-end wheel cylinders which apply hydraulic pressure to both the toe and the heel of two identical, self-centering shoes. The shoes anchor at either toe or heel, depending upon the direction of rotation. Each adjusting screw is threaded into or out of its support by means of an adjusting wheel. Adjusting wheels are accessible through adjusting slots in the backing plate.

Brake Shoe Removal

1. Remove hub and brake drum assembly.

2. Install wheel cylinder clamps to hold pistons in cylinders.

3. Remove brake shoe return springs.

4. Remove lock wires, nuts, and washers from brake shoe guide bolts, then remove brake shoe assemblies.

5. Remove screws attaching adjust-

ing wheel lock springs to anchor supports. Thread each adjusting screw from the shoe side of its anchor support by turning adjusting wheels, then lift adjusting wheels out of slots in anchor supports.

Brake Shoe Installation

1. *Install adjusting screws and wheels in anchor supports dry; use no lubricant.* Insert each adjusting wheel in slot in anchor support, insert threaded end of adjusting screw in anchor support, then turn adjusting wheel to thread adjusting screw into anchor support. Insert anchor pins into holes in anchor supports, with slots in pins facing slots in supports.

1 Wheel cylinder heat shield
2 Upper wheel cylinder
3 Anchor pin
4 Return spring (short)
5 Guide washer
6 Guide bolt
7 Guide bolt nut lock wire
8 Brake shoe and lining assembly
9 Return spring (long)
10 Adjusting wheel lock spring
11 Adjusting wheel
12 Lower wheel cylinder
13 Backing plate

2. Install brake shoes with cutaway end of shoe web next to adjusting screw and with ends of shoes engaging slots in wheel cylinder push rods and anchor pins. Install flat washer and nut on each brake shoe guide bolt. Tighten nuts finger-tight, then back off nuts only far enough to allow movement of shoes without binding.

3. Install brake shoe return springs, hooking one end of each spring in brake shoe web, then hook other end over anchor pins.

4. Remove wheel cylinder clamps.

5. Install hub and brake drum assembly.

6. Adjust brakes.

7. After checking pedal operation, road test vehicle.

Type "FR-3A" & "FR-5A"

The FR-3A brake system is basically the same as the FR-3 except in that the FR-3A employs a self-adjuster assembly and the FR-3 does not. Each shoe is individually adjusted to compensate for wear by a link-crank system. This serves to maintain a high, firm brake pedal at all times.

Brake Shoe Removal

1. Remove the hub and brake drum assemblies.

2. Remove the two springs for the automatic adjusters.

3. Remove the two long crank links from the adjuster assemblies by turning back the star wheel cranks until their slots align with the crank link "U" hooks. Lift out the links and then slide the "S" hooks out of the adjuster cranks.

4. Remove the short crank links by rotating the adjuster cranks until the link "U" hooks clear the

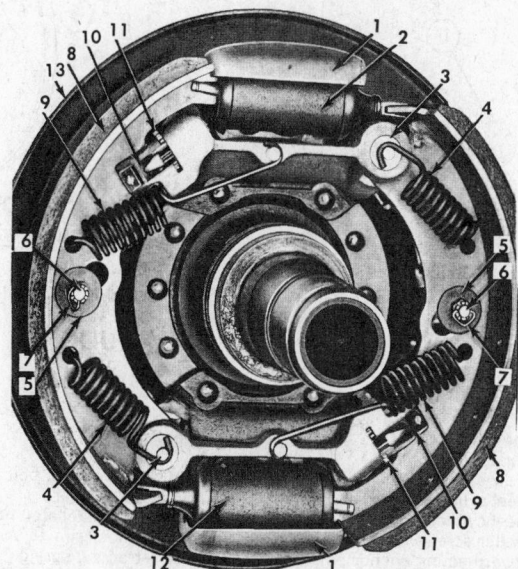

Type "FR3" brake assembly (© G.M.C.)

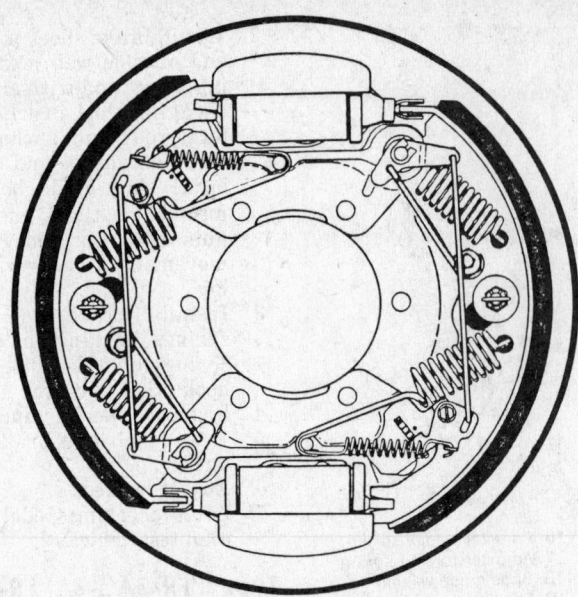

Wagner Type "FR-3A" and "FR-5A" brake assembly (© Chrysler Corp.)

port slots.

NOTE: There is a friction ring on each star wheel. DO NOT attempt to remove the friction ring from the star wheel screw. If necessary replace the star wheel and friction as an assembly.

Brake Shoe Installation

1. Install the star wheels into the anchor support slots and thread the star wheel screws in from the shoe side with the friction ring end towards the shoe. *NOTE: Do not put any lubricant on the star wheel screws.*
2. Position one brake shoe with the "toe" (cut away portion of web) in the adjuster slot and the "heel" in the anchor pin slot of the anchor supports. Install the brake shoe hold down bolt, washer and hold down nut. Tighten the nut finger tight and back off one turn and insert the lock wire in the nut.
3. Install the other brake shoe in the same manner as described in step 2.
4. Install the long return springs in the shoe and hook them over the anchor pins. Then install the short springs in the same manner.
5. Install the adjuster eccentrics on the brake shoes and fasten them with a self-tapping screw, only

eccentrics on the brake shoe webs, then remove the small "U" hooks from the adjuster cranks.

5. Spread the C-washers for the adjuster cranks and remove the cranks.
6. Remove the hold down bolt which holds the star wheel crank to the anchor support and remove the crank.
7. Remove the adjuster eccentric screw and eccentric from the brake shoe.

8. Remove the two long and the two short brake shoe return springs.
9. Remove the lock wires, hold-down nuts and washers from the hold down bolt and remove the brake shoes from the backing plate.
10. Thread each star wheel screw out of the anchor support from the shoe side of the support and lift the star wheels from the sup-

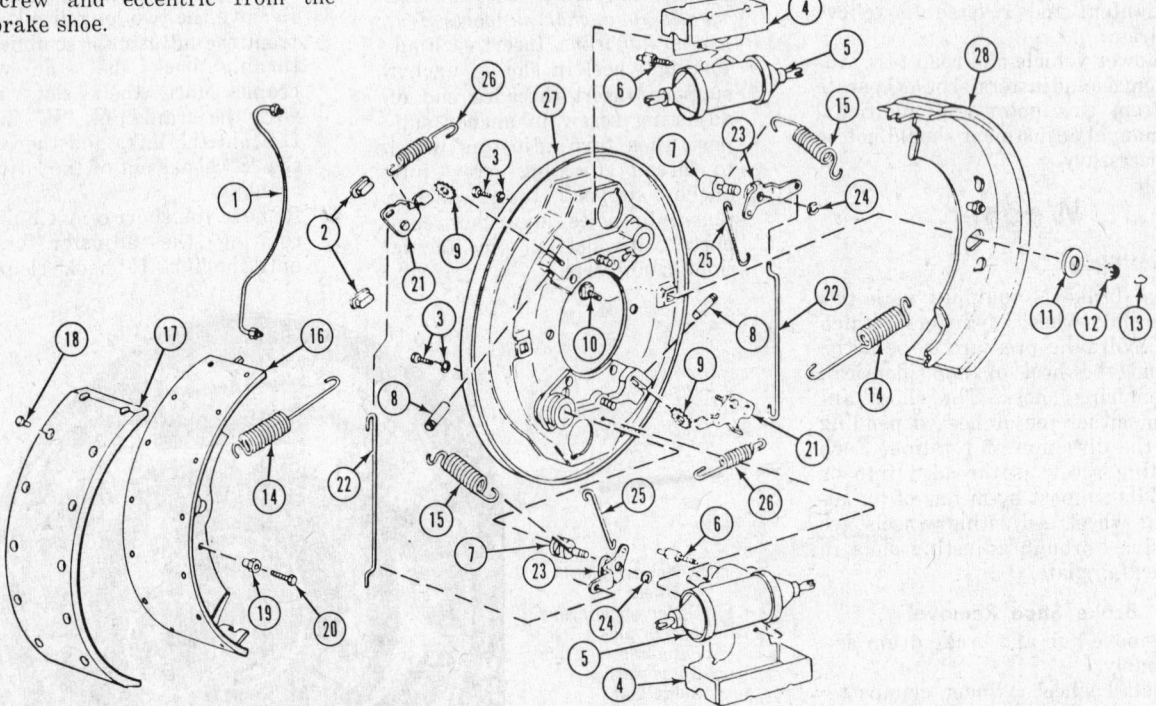

1 Connector tube	15 Short shoe retracting spring
2 Adjuster slot cover	16 Brake shoe
3 Screw and lockwasher	17 Brake lining
4 Wheel cylinder cover	18 Rivet
5 Wheel cylinders	19 Automatic adjuster hex eccentric
6 Bleeder screws	20 Eccentric self-tapping screw
7 Forward-acting anchor pin	21 Starwheel crank assembly
8 Starwheel adjuster screw	22 Long starwheel crank link
9 Shoe adjuster starwheel	23 Automatic adjuster anchor crank
10 Shoe hold-down carriage bolt	24 Anchor crank 'C' washer
11 Shoe hold-down washer	25 Short anchor crank link
12 Shoe hold-down castellated nut	26 Automatic adjuster spring
13 Hold-down nut lockwire	27 Support plate
14 Long shoe retracting spring	28 Complete shoe assembly

Wagner Type "FR-3A" and "FR-5A"—exploded view (© Chrysler Corp.)

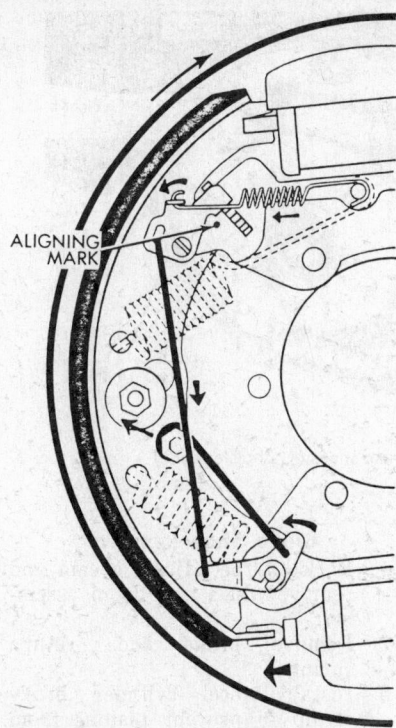

Wagner Type "FR-3A" and "FR-5A"—alignment marks (© Chrysler Corp.)

make the screw finger tight to allow for final adjustment.

6. Install the adjuster cranks on the anchor pins with the long arm towards the shoe and the bushing towards the backing plate so that they rotate easily while resting against the return springs hooks.

7. Install the C-washer for the adjuster crank and crimp in place.

8. On the anchor support place the star wheel and fasten with the crank bolt.

9. On each adjuster crank assembly install the short links small hook into the short arm of the crank from the lower side and hook the other end of the link around the eccentric on the shoe web.

10. Install the long link "S" hook into the long arm of the adjuster crank from the upper side. Then rotate the star wheel crank so the slot lines up with the "U" hook on the long link. Insert the hook and rotate the crank back to its adjusting position.

11. Install the adjuster springs with the short hooks on the star wheel crank fingers and the long hook on the outer groove of the anchor pin on the wheel cylinder side.

12. Adjust the brakes in the following manner:
   a. Center the shoes on the backing plate.
   b. On the shoe web, rotate the eccentrics until the linkage aligns the star wheel crank pawl with the center line of the star wheel screw.
   c. When they are aligned, tighten the self-tapping screw to 19 ft lbs torque.
   d. Install hub and drum assembly and remove the plugs from the slots in the backing plate.
   e. Adjust the brakes in the normal manner to achieve the required amount of drag on each wheel.

13. Road test the vehicle to check for proper braking action.

Bendix

Duo Servo Type

Removing Front Brake Shoes

With the vehicle elevated on a hoist, jack or suitable stands remove front wheel and drums.

1. Remove brake shoe return springs. (*Note how secondary spring overlaps primary spring.*)

2. Remove brake shoe retainer, spring and nails.

3. Slide eye of automatic adjuster cable off anchor and unhook from lever. Remove cable, cable guide and anchor plate.

4. Disconnect lever spring from lever and disengage from shoe web. Remove spring and lever.

5. Remove the primary and secondary brake shoe assemblies and adjusting star wheel from support. Install wheel cylinder clamps to hold pistons in cylinders.

Removing Rear Brake Shoes

1. With the vehicle elevated on a hoist, jack or suitable stand, loosen parking brake equalizer nut, remove rear wheel, and drum retaining clips. Remove drum.

2. Remove brake shoe return springs. (Note how secondary spring overlaps primary spring.)

3. Remove brake shoe retainers, springs and nails.

4. Slide eye of automatic adjuster cable off anchor and then unhook from lever. Remove cable, cable guide and anchor plate.

5. Disconnect lever spring from lever and disengage from shoe web. Remove spring and lever.

6. Spread anchor ends of the primary and secondary shoes and remove parking brake strut and spring.

7. Disengage parking brake cable from parking brake lever and remove brake assembly.

8. Remove the primary and secondary brake shoe assemblies and adjusting star wheel from support. Install wheel cylinder to hold pistons in cylinders.

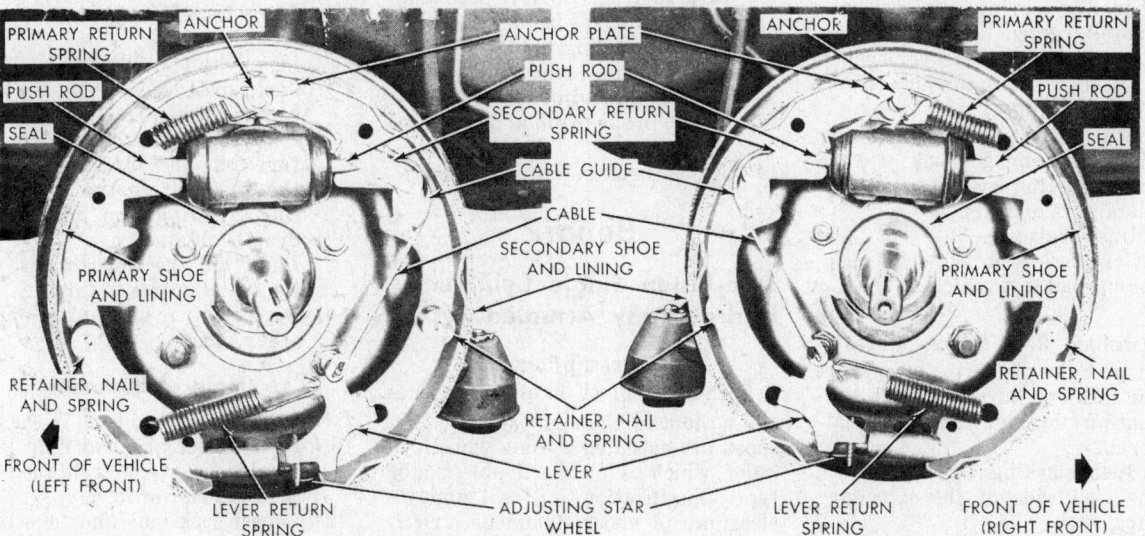

Brake assemblies front (© Chrysler Corp.)

Removing or installing parking brake strut and spring—rear
(© Chrysler Corp.)

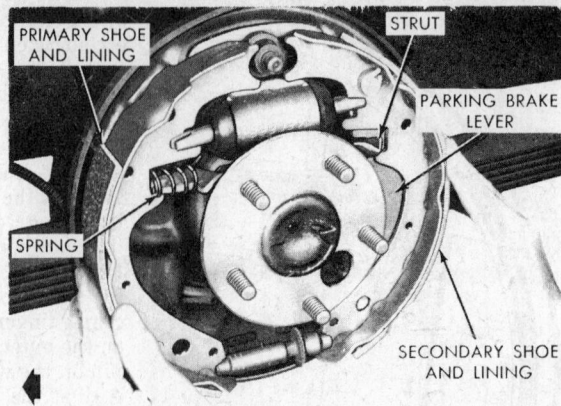

Installing brake shoes (© Chrysler Corp.)

Installing Front Brake Shoes

1. Match a primary with a secondary brake shoe and place them in their relative position on a work bench.
2. Lubricate threads of adjusting screw and install it between the primary and secondary shoes with star wheel next to secondary shoe. The star adjusting wheels are stamped "R" (right side) and "L" (left side), and indicate their location on vehicle.
3. Overlap anchor ends of primary and secondary brake shoes and install adjusting spring lever.
4. Spread anchor ends of brake shoes to maintain adjusting lever and spring in position.
5. Holding brake shoes in their relative position, place brake shoe assembly on support and over the anchor pin.
6. Install nails, cups, springs and retainers.
7. Install anchor pin plate.
8. Install cable guide in the secondary shoe and place the "eye" of adjusting cable over anchor pin.
9. Install return spring in primary shoe. Slide spring over anchor.
10. Install return spring in secondary shoe and slide over anchor. (Be sure the secondary spring overlaps primary).
11. Place adjusting cable cover guide and engage hook of cable into adjusting lever.
12. Install brake drum.
13. Adjust brakes.
14. After checking brake pedal operation, road test vehicle.

Installing Rear Brake Shoes

1. Inspect the platforms of support for nicks or burrs. Apply a thin coat of lubricant to support platforms.
2. Attach parking brake lever to the back side of the secondary shoe.
3. Place the secondary and primary shoe in their relative position on

a work bench.
4. Lubricate threads of adjusting screw and install it between the primary and secondary shoes with star wheel next to secondary shoe. The star adjusting wheels are stamped "R" (right side) and "L" (left side), and indicate their location on vehicle.
5. Overlap anchor ends of the primary and secondary brake shoes and install adjusting spring and lever.
6. Hold the brake shoes in their relative position and engage parking brake cable into parking brake lever.
7. Install parking brake strut and spring between the parking brake lever and primary shoe.
8. Place brake shoes on the support and install retainer nails, springs and retainers.
9. Install anchor pin plate.
10. Install "eye" of adjusting cable over anchor pin and install return spring between primary shoe and anchor pin.
11. Install cable guide in secondary shoe then install secondary return spring. (Be sure secondary spring overlaps primary.)
12. Place adjusting cable in groove of cable guide and engage hook of cable into adjusting lever.
13. Install brake drum and retaining clips.
14. Adjust brakes.

Bendix

Two-Piston Single Cylinder Hydraulically Actuated Type

Description

Both shoes pivot on anchor pins at the bottom of the support plate. The shoes are actuated by one wheel cylinder which is of the double piston type. Specification for heel and toe clearance of shoes should be strictly followed to obtain efficient brake operation.

Brake Shoe Removal

1. Back off the adjusting cam and remove wheel and drum assembly.
2. Remove brake shoe return spring.
3. Install wheel cylinder brake clamp to prevent pistons from being forced out of cylinder.
4. Remove "C" washer, oil washer and retainer, guide spring retainer and guide spring from anchor bolts to remove brake shoes.

Brake Shoe Installation

1. Install brake shoes, oil washers and retainers on anchor bolts and secure with "C" washers.
2. Install brake return spring.
3. Install wheel and drum assembly.

Adjustments

Since tapered brake lining is thicker at the center than at the ends, the adjustment procedures outlined in the paragraphs that follow must be performed in order to assure maximum braking efficiency.

Minor Adjustment

1. Jack up truck so that one wheel can be rotated freely.
2. Then, while rotating that wheel forward and backward, bring the shoe out to the drum with the adjusting cam until a light drag is obtained.
3. Back off the adjustment until the wheel is free to turn.
4. Repeat this procedure on the other shoe.

Major Adjustment

1. Inspect the fluid level in the master cylinder and add fluid if the level is $\frac{3}{8}$" to $\frac{1}{2}$" from the top of the reservoir or lower.
2. Loosen lock nuts and turn brake shoe anchor bolts to the fully released position.

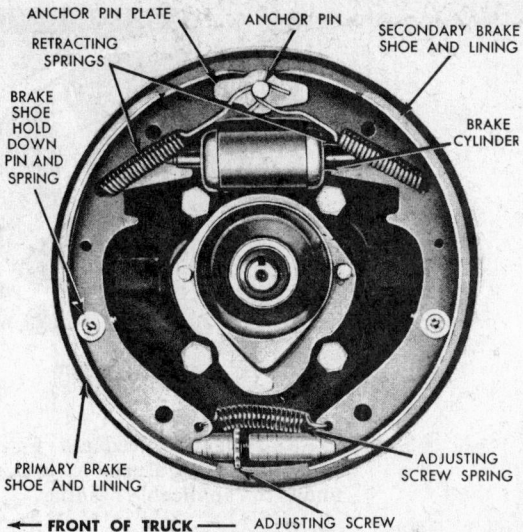

Duo-servo single anchor brake assembly
(© Ford Motor Co.)

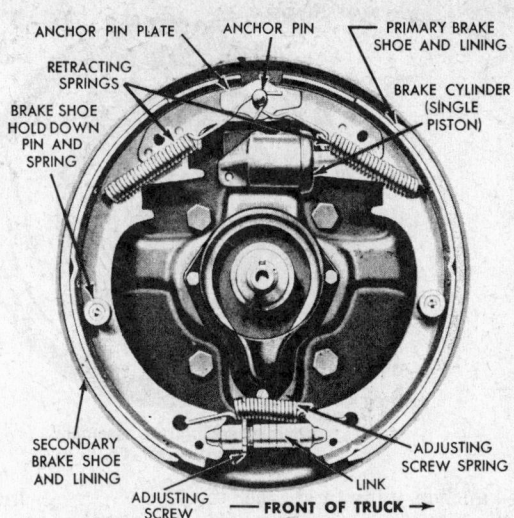

Uni-servo single anchor brake assembly
(© Ford Motor Co.)

3. Adjust the anchor bolt and cam and the minor adjustment cam at the top of the shoe to give equal clearance at the toe and heel. Make sure that sufficient center contact is maintained to produce a slight drag.

4. Lock anchor adjusting nut. After adjusting the clearance on one shoe, repeat the procedure on the other shoe. Then apply the brakes a couple of times to make sure adjustment is up to specifications.

NOTE: Whenever cams are adjusted, check brakes by applying pressure on the brake pedal a couple of times so as to make sure wheel drag has not increased, since the spring loaded cams may cause shoe adjustment to change by shifting position. Wheel should only have a slight drag at room temperature.

Bendix

Single Anchor Brake Shoe Replacement

Brake Shoe Removal

1. Remove the wheel and drum. *Do not push down the brake pedal after the brake drum has been removed.* On a truck equipped with a vacuum or air booster, be sure the engine is stopped and there is no vacuum or air pressure in the system before disconnecting the hydraulic lines.

2. Clamp the brake cylinder boots against the ends of the cylinder, and remove the brake shoe retracting springs from both shoes.

3. Remove the anchor pin plate.

4. Remove the hold-down spring cups and springs from the shoes, and remove the shoes and the

adjusting screw parts from the carrier plate. *Do not let oil or grease touch the brake linings.* If the shoes on a rear brake assembly are being removed, remove the parking brake lever, link, and spring with the shoes. Unhook the parking brake cable from the lever as the shoes are being removed.

5. If the shoes are from a rear brake assembly, remove the parking brake lever from the secondary shoe.

Brake Shoe Installation

1. Coat all points of contact between the brake shoes and the other brake assembly parts with Lubri-plate or a similar lubricant. Lubricate the adjusting screw threads.

2. Place the adjusting screw, socket, and nut on the brake shoes so that the star wheel on the screw is opposite the adjusting hole in the carrier plate. Then install the adjusting screw spring.

3. Position the brake shoes and the adjusting screw parts on the carrier plate, and install the hold-down spring pins, springs, and cups. When assembling a rear brake, connect the parking brake lever to the secondary shoe, and install the link and spring with the shoes. Be sure to hook the parking brake cable to the lever.

4. Install the anchor pin plate on the pin.

5. Install the brake shoe retracting springs on both shoes. *The primary shoe spring must be installed first.*

6. Remove the clamp from the brake cylinder boots.

7. Install the wheel and drum.

8. Bleed the system and adjust the brakes. Check the brake pedal operation after bleeding the system.

Bendix

Double Anchor Brake Shoe Replacement

1. Remove the wheel and drum. *Do not push down the brake pedal after the brake drum has been removed.* On trucks equipped with vacuum boosters, be sure the engine is stopped and there is no vacuum in the system before disconnecting the hydraulic lines.

2. Clamp the brake cylinder boots against the ends of the cylinder, and remove the brake shoe retracting springs from both shoes.

3. At each shoe, remove the 2 brake shoe retainers and washers from the hold-down pins and remove the spring and pin from the carrier plate. Remove the anchor pin retainers and remove the shoes from the anchor pins. *Do not allow grease or oil to touch the linings.*

4. Clean all brake assembly parts. If the adjusting cams do not operate freely apply a small quantity of lubricating oil to points where the shaft of the cam enters the carrier plate. Wipe dirt and corrosion off the plate.

5. Clean the ledges on the carrier plate with sandpaper. Coat all points of contact between the brake shoes and the other brake assembly parts with high temperature grease.

6. Position the brake shoes on the carrier plate with the heel (lower) end of the shoes over

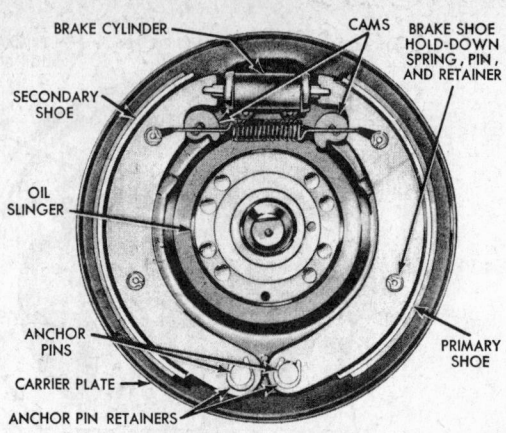

Double anchor brake assembly
(© Ford Motor Co.)

Anchor adjustment
(© Ford Motor Co.)

the anchor pins and the toe (upper) end of the shoes engaged in the brake cylinder link. Install the hold-down spring pins, spring, washers and retainers.

7. Install the anchor pin retainers and then install the brake shoe return spring.
8. Turn the brake shoe adjusting cams to obtain maximum clearance for brake drum installation.
9. Install the wheel and drum assembly.
10. Bleed the brake system and adjust the brakes.
11. Check brake pedal operation and road test.

Bendix

Two-Cylinder Brake Shoe Replacement

Removal

1. Remove the wheel, and then remove the drum or the hub and drum assembly. *Mark the hub and drum to aid assembly in the same position.* On trucks equipped with vacuum or air boosters, be sure the engine is stopped and there is no vacuum or air pressure in the system before disconnecting the hydraulic lines.
2. Clamp the brake cylinder boots against the ends of the cylinder and remove the four brake shoe retracting springs.
3. Remove the brake shoe guide bolt cotter pin, nut, washer, and bolt from both shoes, and remove the shoes from the carrier plate.
4. Remove the clamp-type adjusting wheel lock from the anchor pin support, and unthread the adjusting screw and wheel assembly from the anchor pin support.

Installation

1. Clean the carrier plate ledges with sandpaper. Coat all points of contact between the brake shoes and other brake assembly parts with high temperature grease.
2. Thread the adjusting screw and wheel assembly into the anchor pin support and install the clamp-type adjusting wheel lock. Thread the adjusting wheel into the support so that the brake shoe will rest against the adjusting wheel end.
3. Place the brake shoe over the two brake shoe anchor pins, insert the ends in the brake cylinder links, and install the shoe guide bolt, washer, and nut. Finger tighten the nut, then back off one full turn, and install the cotter pin.
4. Install the four retracting springs.
5. Remove the cylinder clamps, install the drum or the hub and drum assembly, then install the wheel assembly. Align the marks on the hub and drum during installation.
6. Bleed and adjust the brakes.
7. Check pedal operation and road test.

Bendix

Brake Shoe Adjustment

The brake drums should be at normal room temperature, when the brake shoes are adjusted. If the shoes are adjusted when the shoes are hot and expanded, the shoes may drag as the drums cool and contract.

A minor brake adjustment reestablishes the brake lining-to-drum clearance and compensates for normal lining wear.

A major brake adjustment includes the adjustment of the brake shoe anchor pins as well as the brake shoes. Adjustment of the anchor pin permits the centering of the brake shoes in the drum.

Adjustment procedures for each type of brake assembly are given under the applicable heading.

Minor Adjustment

The brake shoe adjustment procedures for the uniservo single anchor brake assmebly are the same as those for the duo-servo single anchor type.

A major brake adjustment should be performed when dragging brakes are not corrected by a minor adjustment, when brake shoes are relined or replaced, or when brake drums are machined.

Duo-Servo Single Anchor Brake

The duo-servo single-anchor brake is adjusted by turning an adjusting screw located between the lower ends of the shoes.

1. Raise the truck until the wheels clear the floor.
2. Remove the cover from the adjusting hole at the bottom of the brake carrier plate, and turn the adjusting screw inside the hole to expand the brake shoes until they drag against the brake drum.
3. When the shoes are against the drum, back off the adjusting screw 10 or 12 notches so that the drum rotates freely without drag.
4. Install the adjusting hole cover on the brake carrier plate.

Duo-servo brake shoe adjustment
(© Ford Motor Co.)

5. Check and adjust the other three brake assemblies. When adjusting the rear brake shoes, check the parking brake cables for proper adjustment. Make sure that there is clearance between the ends of the parking brake link and the shoes.

6. Apply the brakes. If the pedal travels more than halfway down between the released position and the floor, too much clearance exists between the brake shoes and the drums. Repeat steps 2 and 3 above. Internal inspection and/or bleeding may be necessary.

7. When all brake shoes have been properly adjusted, road test the truck and check the operation of the brakes. *Perform the road test only when the brakes will apply and the truck can be safely stopped.*

Single Anchor Pin
Major Adustment

A major brake adjustment should be made when dragging brakes are not corrected by a minor adjustment, when brake shoes are relined or replaced, or when brake drums are machined.

1. Raise the truck until the wheel clears the floor.

2. Rotate the drum until the feeler slot is opposite the lower end of the secondary (rear) brake shoe.

3. Insert a 0.010-inch feeler gauge through the slot in the drum. Move the feeler up along the secondary shoe unit it is wedged between the sedondary shoe and the drum.

4. Turn the adjusting screw (star wheel) to expand the brake shoes until a heavy drag is felt against the drum. Back off the adjusting screw just enough to establish a clearance of 0.010", between the shoe and the drum at a point 1½" from each end of the secondary shoe. This adjustment will provide correct operating clearance for both the primary and secondary shoes. If the 0.010" clearance cannot be ob-

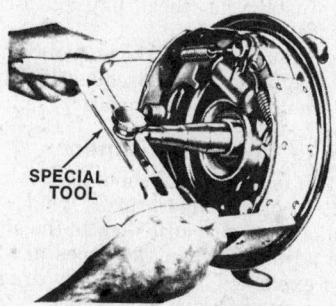

SPECIAL
TOOL

Measuring brake shoes
(Ⓒ Ford Motor Co.)

SPECIAL TOOL

Measuring brake drum
(Ⓒ Ford Motor Co.)

tained at both ends of the secondary shoe, the anchor pin must be adjusted.

5. To adjust the anchor pin setting, loosen the anchor pin nut just enough to permit moving the pin up or down by tapping the nut with a soft hammer. *Do not back the nut off too far or the shoes will move out of position when the nut is tightened.* Tap the anchor pin in a direction that will allow the shoes to center in the drum and provide an operating clearance of 0.010". Torque the anchor pin nut to 80-100 ft-lbs. Recheck the secondary shoe clearance at both the heel and toe ends of the shoe.

6. When all brake shoes and anchor pins have been properly adjusted, road test the truck and check the operation of the brakes. *Perform the road test only when the brakes will apply and the truck can be safely stopped.*

Double Anchor Pin
Major Adjustment

1. Raise the truck until the wheels clear the floor.

2. Rotate the drum until the feeler slot is opposite the lower (heel) end of the secondary (rear) brake shoe.

3. Insert a 0.007" feeler gauge through the slot in the drum. Move the feeler up along the secondary shoe until it is wedged between the shoe and the drum.

4. Loosen the secondary shoe anchor pin nut. Turn the secondary shoe anchor pin until the brake shoe-to-drum clearance at a point 1½" from the heel end of the shoe is 0.007". Remove the feeler gauge.

5. Rotate the drum until the feeler slot is opposite the upper (toe) end of the secondary brake shoe.

6. Insert a 0.010" feeler gauge through the slot in the drum. Move the feeler gauge down along the secondary shoe until it is wedged between the shoe and the drum. Turn the adjusting cam, to expand the brake shoe, until a heavy drag is felt against the drum.

7. Turn the anchor pin until the brake shoe-to-drum clearance at a point 1½" from the toe end of the shoe is 0.010". Remove the feeler gauge.

8. Torque the anchor pin nut to 80-100 ft-lbs. Recheck the heel and toe clearances.

9. Using the preceding secondary brake shoe adjustment procedure as a guide, adjust the primary brake shoe-to-drum clearance.

10. Road test the truck and check the operation of the brakes. *Perform the road test only when the brakes will apply and the truck can be safely stopped.*

Kelsey Hayes

Front Brake Shoes

Removal

1. Raise the vehicle until the wheel clears the floor. Remove the wheel, drum and hub assembly.

2. Clamp the wheel cylinder boots against the ends of the cylinder.

3. Remove the brake shoe retracting springs from both shoes.

4. Remove the adjusting lever link, anchor plate and the adjusting lever spring.

5. Remove the hold down spring cups, springs and the adjusting lever.

6. Remove the brake shoes and adjuster screw assembly from the backing plate.

Installation

1. Clean all brake dust from the brake assembly parts with a *clean dry* rag.

2. Coat all points of contact between the shoes and other brake parts with high temperature grease.

3. Coat the adjuster screw with high temperature grease before assembly. Thread the adjuster screw into the adjuster screw sleeve.

4. Position the brake shoes on the backing plate and install the adjusting lever, hold down pins, springs and cups.

5. Position the adjuster screw assembly on the brake shoes so that the star wheel is opposite the adjusting slot in the backing plate. Install the adjusting lever spring.

6. Install the anchor plate and adjusting lever link.

7. Install the secondary brake shoe retracting spring.

8. Install the primary brake shoe retracting spring.

9. Remove the clamp from the wheel cylinder boots.

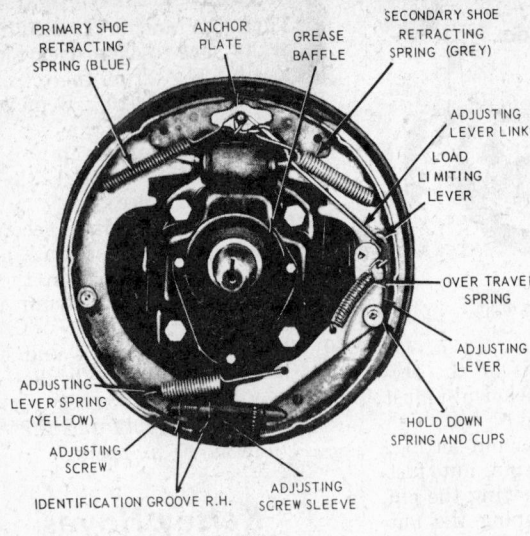

Kelsey hayes hydraulic brake assembly
(© Ford Motor Co.)

Two cylinder brake—unequal length springs
(© Ford Motor Co.)

10. Install the wheel, drum and hub assembly.
11. Adjust the brakes. Subsequent adjustment will be automatic.

Rear Brake Shoes

Removal

1. Raise the truck until the wheel clears the floor.
2. Remove the wheel, hub and drum assembly.
3. Clamp the brake cylinder boots against the ends of the cylinder with brake piston clamps.
4. *Note the two different types of brake shoe retracting springs and remove the springs.*
5. Remove the brake shoe hold down post cotter key, nut, and shoe hold down washer.
6. Loosen and remove the eccentric adjuster bolt, lock washer, eccentric and adjusting link.

7. Remove the shoe and lining assembly from the backing plate.
8. Remove the anchor block spring and slide the adjuster assembly from the shoe web.
9. Remove the adjuster star wheel and screw from the adjuster block. Unthread the star wheel from the adjuster screw.

Installation

1. Wipe all brake dust from the brake assembly parts with a *clean dry* rag. Coat all points of contact between brake shoes and other parts with high temperature grease.
2. Coat the adjuster screw and the inside of the adjuster block with high temperature grease.
3. Thread the adjuster screw onto the star wheel and insert the adjuster screw assembly into the adjuster block. Maintain a

2.12-2.18 inch dimension from the end of the adjuster block to the adjuster screw web slot.
4. Install the adjuster assembly onto the shoe web and attach the anchor block spring.
5. Place the brake shoe over the retracting spring toggle pin and insert the ends of the shoe in the wheel cylinder links.
6. Install the shoe hold down washer and nut. Do not install the cotter pin.
7. Install the four brake shoe retracting springs. Make sure the retracting springs are installed. On 15 x 5″ brakes the inner hook ends face the wheel cylinders. On 15 x 4″ brakes the inner hook ends face the center of the axle.
8. Install the adjusting link, eccentric, lockwasher and adjuster bolt. Do not tighten.
9. Remove the brake piston clamps.
10. Tighten the shoe hold down nut until there is 0.015-0.025″ clearance between the shoe and hold down washer with the shoe held against the backing plate. Install the cotter pin.
11. Center the shoes on the backing plate. Using a ½″ wrench, rotate the adjuster eccentric until the adjusting lever is at the index mark. Tighten the eccentric adjuster bolt to specification.
12. Install the wheel, hub and drum assembly.
13. Adjust the brake to obtain a slight drag. Subsequent adjustments will be automatic.

Brake Shoe Adjustment

The brake drums should be at normal room temperature, when the brake shoes are adjusted. If the shoes are adjusted when the shoes are hot and expanded, the shoes may drag as the drums cool and contract.

The brake shoes are automatically adjusted when the vehicle is driven

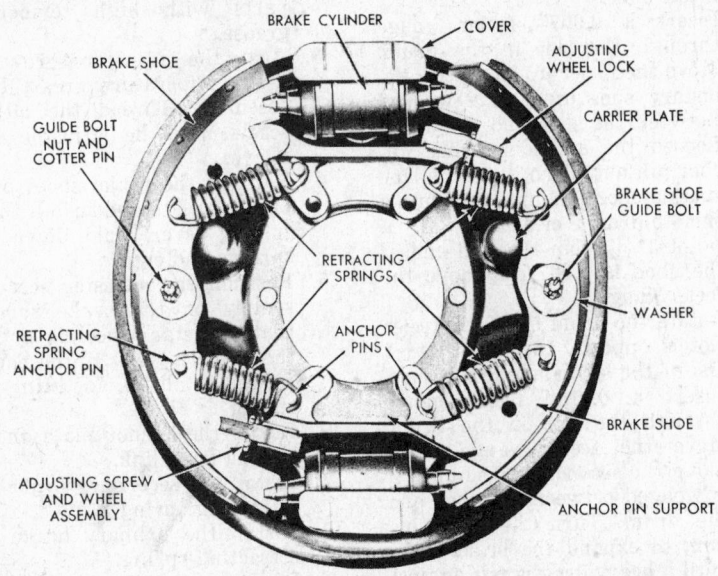

Two cylinder brake—equal length springs (© Ford Motor Co.)

Hydraulic Brakes

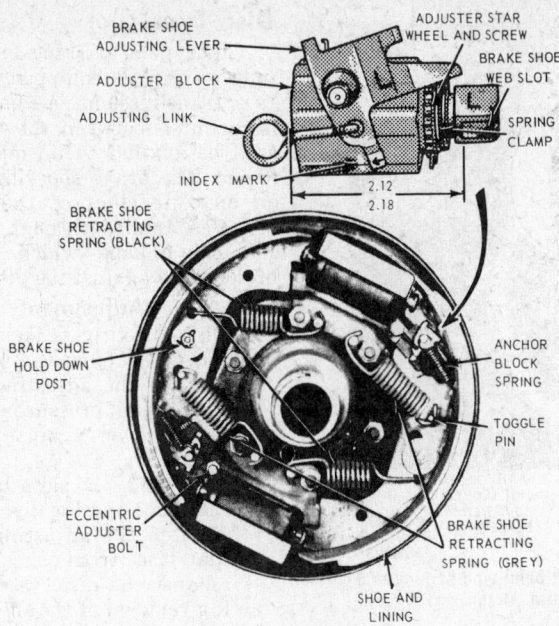

LEFT REAR
Kelsey hayes rear brake assembly
(© Ford Motor Co.)

in reverse and the brakes applied. A manual adjustment is required only after the brake shoes have been relined or replaced. *The manual adjustment is performed while the drums are removed, using the tool and the procedure detailed below.*

When adjusting the rear brake shoes, check the parking brake cables for proper adjustment. Make sure that the equalizer operates freely.

To adjust the brake shoes:

1. Use special tool (see illustration) and adjust to the inside diameter of the drum braking surface.
2. Reverse the tool as shown in illustration and adjust the brake shoes to touch the gauge. The gauge contact points on the shoes must be parallel to the vehicle with the center line through the center of the axle. Hold the automatic adjusting lever out of engagement while rotating the adjusting screw, to prevent burring the screw slots. Make sure the adjusting screw rotates freely.
3. Apply a small quantity of high temperature grease to the points where the shoes contact the carrier plate, being careful not to get the lubricant on the linings.
4. Install the drums. Install the retaining nuts and tighten securely.
5. Install the wheels on the drums and tighten the mounting nuts to specification.
6. Complete the adjustment by applying the brakes several times while backing the vehicle.
7. After the brake shoes have been

properly adjusted, check the operation of the brakes by making several stops while operating in a forward direction.

Self Adjusting Brake— Two-Cylinder Front

Two-cylinder front brakes are adjusted by means of exposed, hexhead, self-locking cam adjusters. The brakes are to be manually adjusted initially. Subsequent adjustment is automatic. To adjust this brake:

1. Raise the vehicle and check the front brakes for drag by rotating the wheels.
2. Adjust one shoe by rotating the wheel backward and forward while turning the cam hex-head with a wrench. Bring the shoe out to the drum until a light drag is felt. *Do not apply excessive force on the hex head cam, as automatic adjuster parts can be damaged.* Back off the adjustment until the wheel turns freely. Adjust the other cam on the same wheel in the same manner.
3. Adjust the other front wheel brake using the procedure above.
4. Apply the brakes and recheck the adjustment.

Self Adjusting Brakes—Rear

The brake shoes are automatically adjusted when the vehicle is driven in reverse and the brakes applied. A manual adjustment is required only after the brake shoes have been relined or replaced.

The two-cylinder brake assembly brake shoes are adjusted by turning adjusting wheels reached through

slots in the backing plate.

Two types of two-cylinder brake assemblies are used on truck rear wheels. The assemblies differ primarily in the retracting spring hookup, and in the design of the adjusting screws and locks. However, the service procedures are the same for both assemblies.

The brake adjustment is made with the vehicle raised. Check the brake drag by rotating the drum in the direction of forward rotation as the adjustment is made.

1. Remove the adjusting slot covers from the backing plate.
2. Turn the rear (secondary shoe) adjusting screw inside the hole to expand the brake shoe until a slight drag is felt against the brake drum.
3. Repeat the above procedure on the front (primary) brake shoe.
4. Replace the adjusting hole covers.
5. Complete the adjustment by applying the brakes several times while backing the vehicle.
6. After the brake shoes have been properly adjusted, check the operation of the brakes by making several stops while operating in a forward direction.

Parking Brakes

Internal Shoe Type

Adjustment

Nine-Inch Diameter Drum

1. Release the parking brake lever in the cab.
2. From under the truck, remove the cotter pin from the parking brake linkage adjusting clevis pin. Remove the clevis pin.
3. Lengthen the parking brake adjusting link by turning the clevis. Continue to lengthen the adjusting link until the shoes seat against the drum when the clevis pin is installed.
4. Remove the clevis pin and shorten the linkage adjustment until there is 0.010″ clearance between the shoes and the drum. The measurement should be taken at all points around the

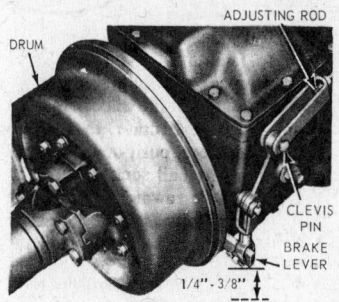

Shoe type parking brake
(© Ford Motor Co.)

519

drum with the clevis pin installed.

5. Install a new cotter pin in the clevis retaining pin and check the brake operation.

Twelve-Inch Diameter Drum

There is no internal adjustment on this brake. Adjustment is made on the linkage. Remove the clevis pin, loosen the nuts on the adjusting rod, and turn the clevis on the rod until a ¼-⅜" free play is obtained at the brake lever. Tighten the nuts, and connect the clevis to the bellcrank with the clevis pin.

External Band Type

Adjustment

1. On cable-controlled parking brakes, move the parking brake lever to the fully released position. On a vehicle with a rod-type linkage, set the lever at the first notch.

2. Check the position of the cam to make sure the flat portion is resting on the brake band bracket. If the cam is not flat with the bracket, remove the clevis pin from the upper part of the cam, and adjust the clevis rod to allow the flat portion of the cam to rest on the brake band bracket. Install the clevis pin and cotter pin.

3. Remove the lock wire from the anchor adjusting screw, and turn the adjusting screw clockwise until a clearance of 0.010" is established between the brake lining and the brake drum at the

Typical external band type parking brake
(Ⓒ Ford Motor Co.)

anchor bracket. Install the lock wire in the anchor adjusting screw.

4. Loosen the lock nut on the adjusting screw for the lower half of the brake band, and adjust the screw to establish a 0.010" clearance between the lining and the brake drum at the lower half of the brake band. Tighten the lock nut.

5. Turn the upper band adjusting rod nut until a 0.010 clearance is established between the upper half of the band and the drum.

6. Apply and release brake several times to insure full release.

Disc Type

This type of parking brake is used only on models equipped with auxiliary transmissions. A ventilated disc is mounted between the drive shaft and the auxiliary transmission shaft flange. The brake shoes are mounted on opposite sides of the disc. The brake is applied through use of mechanical linkage which forces the brake shoes against the disc.

Adjustment

1. Disconnect the brake cable or the rod clevis from the brake lever.

2. Tighten the adjusting nut until the spring pressure pushes the brake lever against the front lever arm.

3. Insert a 1/32" shim between the rear shoe and the disc.

4. Tighten the adjusting screw so that the front shoe is firmly against the disc, yet still allowing removal of the shim.

5. Remove the shim and adjust the parallel adjusting screws so that both linings are parallel to the disc. This should provide 1/64" clearance between the shoes and disc.

6. Making sure the brake lever in the cab is fully released, adjust the clevis on the brake cable as much as necessary to allow installation of clevis pin through the clevis and brake lever without changing position of the lever. Install the pin and cotter pin.

7. Make sure all lock nuts and adjusting screws are tightened securely.

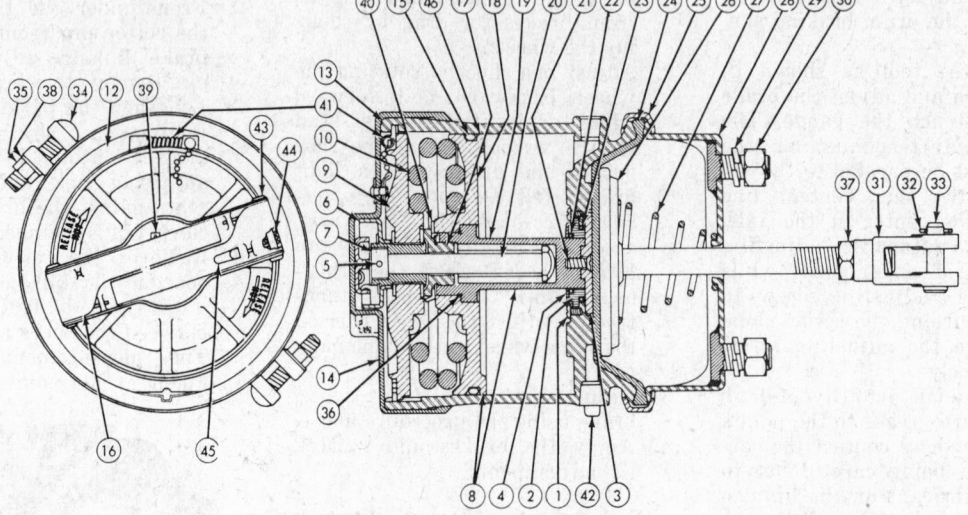

Spring type parking brake (Ⓒ Chrysler Corp.)

1 Screw	12 Retaining ring	24 Diaphragm	36 Seal "O" ring
2 Bushing	13 Cylinder housing	25 Clamp band	37 Jam nut
3 Plunger	14 Protective boot	26 Push rod	38 Washer
4 Cap	15 Seal "O" ring	27 Spring	39 Snap ring
5 Pin	16 Handle W/O stop	28 Non-pressure housing	40 Tooth washer
6 Cross bar	17 Piston nut	29 Lock washer	41 Lock spring
7 Nut	18 Piston	30 Nut	42 Pipe plug
8 Piston-cover-spring (sub-assy).	19 Piston shaft	31 Yoke	43 Rivet
9 Cover (sub assy.)	20 Relase bolt	32 Cotterpin	44 Spring
10 Spring	21 Screw	33 Clevis pin	45 Handle with stop
11 Retaining ring (sub assy.)	22 Seal "O" ring	34 Bolt	46 Piston-cover-spring
	23 Seal "O" ring	35 Nut	

DISC BRAKES—TROUBLE DIAGNOSIS

CAUSE	CORRECTION
1. Master cylinder fluid level low.	1. Fill to proper level with approved fluid. Note, fluid level drops as disc brake linings wear.
2. Poor quality brake fluid (low boiling point) in system.	2. Drain hydraulic system and fill with approved
3. Air in hydraulic system.	3. Bleed hydraulic system and refill with approved fluid.
4. Hoses soft or weak (expanding under pressure).	4. Replace defective hoses.
	combination valve and all cups and seals in complete brake s
1. Power brake malfunctioning.	1. Check and repair power unit.
2. Linings soiled with brake fluid, oil or grease.	2. Replace shoes and linings.
3. Lines, hoses or connections dented, kinked, collapsed, clogged or disconnected.	3. Repair or replace defective parts.
4. Master cylinder cups swollen.	4. Drain hydraulic system, flush system with brake fluid and replace combination valve and all cups and seals in complete brake system.
5. Master cylinder bore corroded or rough.	5. Repair or replace master cylinder.
6. Caliper pistons frozen or seized.	6. Disassemble caliper and free pistons (replace if necessary).
7. Caliper cylinder bores corroded or rough.	7. Disassemble caliper and remove corrosion or roughness, or replace caliper.
8. Pedal push rod and linkage binding.	8. Free and lubricate.
9. Metering valve not working.	9. Replace combination valve.

GRABBING OR PULLING (Severe Reaction To Pedal Pressure and Out of Line Stops)

CAUSE	CORRECTION
1. Linings soiled with brake fluid, oil or grease.	1. Replace shoes and linings.
2. Caliper loose.	2. Tighten caliper mounting bolts to specified torque.
3. Lines, hoses or connection dented, kinked, collapsed or clogged.	3. Repair or replace defective parts.
4. Master cylinder bore corroded or rough.	4. Repair or replace master cylinder.
5. Caliper pistons frozen or seized.	5. Disassemble caliper and free pistons (replace if necessary).
6. Caliper cylinder seals soft or swollen.	6. Drain hydraulic system, flush system with brake fluid and replace all cups and seals in complete brake system.
7. Caliper cylinder bores corroded or rough.	7. Disassemble caliper and remove corrosion or roughness, or replace caliper.
8. Pedal linkage binding (and suddenly releasing).	8. Free and lubricate linkage.
9. Metering valve not functioning properly.	9. Replace combination valve.

DISC BRAKES—TROUBLE DIAGNOSIS—Cont'd

FADING PEDAL (Pedal Falling Away Under Steady Pressure)

CAUSE	CORRECTION
1. Poor quality brake fluid (low boiling point) in system.	1. Drain hydraulic system and fill with approved fluid.
2. Hydraulic connections loose; lines or hoses ruptured (causing leakage).	2. Tighten or replace defective parts.
3. Master cylinder cup worn or damaged. (primary, secondary or both).	3. Repair master cylinder.
4. Master cylinder bore corroded, worn or scored.	4. Repair or replace master cylinder.
5. Caliper cylinder seals worn or damaged.	5. Replace seals.
6. Caliper cylinder bores corroded, worn or scored.	6. Disassemble caliper and remove corrosion or scoring, or replace caliper.
7. Bleed screw open.	7. Close bleed screw and bleed hydraulic system.

NOISE AND CHATTER (May Be Accompanied By Brake Roughness and Pedal Pumping)

CAUSE	CORRECTION
1. Disc has excessive lateral runout.	1. Replace or machine disc.
2. Disc has excessive thickness variations (out of parallel).	2. Replace or machine disc.
3. Disc has casting imperfections.	3. Replace disc.
4. Car creeping or moving slowly with brakes applied (may produce groan or crunching noise).	4. Increase or decrease pedal effort slightly.
5. Squeal, during application.	5. A small amount of high-pitched squeal is inherent in disc brake design and must be considered normal. Some relief may be obtained with service package backing.

DRAGGING BRAKES (Slow or Incomplete Release of Brakes)

CAUSE	CORRECTION
1. Lines, hoses or connections dented, kinked, collapsed or clogged.	1. Repair or replace defective parts.
2. Master cylinder compensating port restricted by swollen primary cup.	2. Drain hydraulic system, flush system with brake fluid and replace combination valve and all cups and seals in complete brake system.
3. Residual pressure check valve in lines to front wheels.	3. Remove check valve.
4. Caliper pistons frozen or seized.	4. Disassemble caliper and free pistons (replace if necessary).
5. Caliper cylinder seals swollen.	5. Drain hydraulic system, flush system with clean brake fluid and replace combination valve and all cups and seals in complete brake system.
6. Caliper cylinder bores corroded or rough.	6. Disassemble caliper and remove corrosion or roughness, or replace caliper.
7. Hydraulic push rod on power brake out of adjustment or binding (causing primary cup to restrict master cylinder compensating port).	7. Adjust or free and lubricate.

Floating Caliper Disc Brakes—Dual Piston

This disc brake is a floating caliper design with two pistons on one side of the rotor. It is a two piece unit consisting of the caliper and cylinder housing. The caliper is mounted to the anchor plate on two mounting pins which travel in bushings in the anchor plate. The bushings and pins are protected by boot type seals.

Two brake shoe and lining assemblies are used in each caliper, one on each side of the rotor. The shoes are identical and are attached to the caliper with two mounting pins.

The cylinder housing contains the two pistons. The pistons are fitted with an insulator on the front and a seal on the back lip. A friction ring is attached to the back of the piston with a shouldered cap screw. The pistons and cylinder bores are protected by boot seals which are fitted to a groove in the piston and attached to the cylinder housing with retainers. The cylinder assembly is attached to the caliper with two cap screws and washers.

The anchor plate is bolted directly to the spindle. It positions the caliper assembly over the rotor forward of the spindle.

Disc Brake Shoe Adjustment

The front disc brake assembly is designed so that it is inherently self-adjusting and requires no manual adjustment.

Automatic adjustment for lining wear is achieved by the piston and friction ring sliding outward in the cylinder bore. The piston assumes a new position in the cylinder and maintains the correct adjustment.

Front Disc Brake Shoe and Lining

Replace shoe and lining assemblies when lining is worn to a minimum of 1/16" in thickness (combined thickness of shoe and lining 1/4" minimum).

Removal

1. Remove the shoe and lining mounting pins, anti-rattle springs and old shoe and lining assemblies.

Installation

1. Remove the master cylinder cover.
2. Loosen the piston housing-to-caliper mounting bolts sufficiently to permit the installation of new shoe and lining assemblies. *Do not move pistons.*

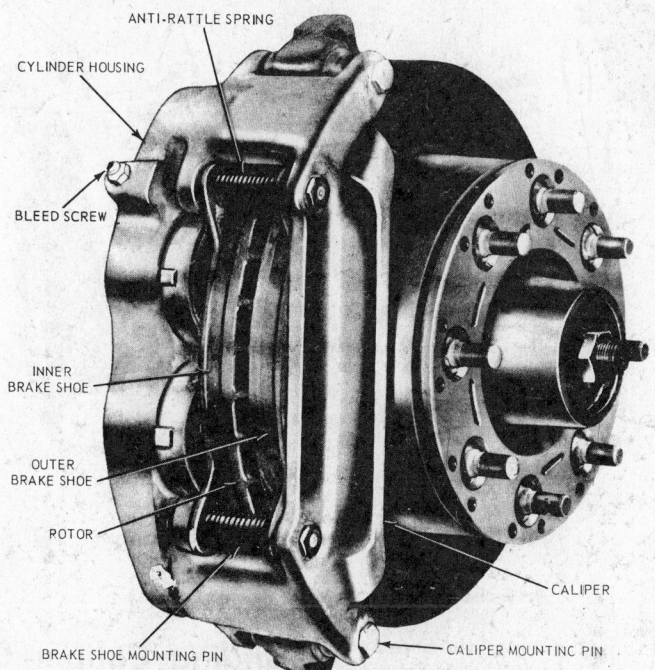

Floating caliper disc brake (© Ford Motor Co.)

3. Install new shoe and lining assemblies. Install the brake shoe mounting pins and anti-rattle springs. *Be sure that the spring tangs are located in the holes provided in the shoe plates.*
4. Torque the brake shoe mounting pins to 17-23 ft lbs.
5. Reset the pistons to the correct location in the cylinders by placing shims or feeler gauges of .023 to .035" thickness between the shoe plate of the outboard shoe and lining assembly and the caliper; then, retighten the piston housing-to-caliper mounting bolts. *Keep the cylinder housing square with the caliper.*
6. Loosen the piston housing-to-caliper mounting bolts and remove the shims.
7. Torque the piston housing-to-caliper mounting bolts to 155 to 185 ft lbs.
8. Check the master cylinder reservoirs.
9. Install the master cylinder cover.

Disc Brake Caliper

Removal

1. Remove the wheel and tire assembly.
2. Remove the pins and nuts retaining the caliper assembly to the anchor plate.
3. Disconnect the brake hose from the caliper and remove the caliper.

Installation

1. Connect the brake hose to the caliper.
2. Position the caliper assembly to the anchor plate and install the retaining pins and nuts. Torque the nuts to specifications.
3. Install drum and wheel and bleed brake system.

If the caliper assembly is leaking, the piston assemblies must be removed from the piston housing and replaced. If the cylinder bores are scored, corroded or excessive wear is evident, the piston housing must be replaced. *Do not hone the cylinder bores.* Piston assemblies are not available for oversize bores. The piston housing must be removed from the caliper for replacement.

Disassembly

1. Remove the two pins and nuts retaining the caliper to the support. Disconnect the flexible brake hose and plug the end to prevent brake fluid leakage.
2. Remove the boot retainers and remove the dust boots from the pistons and cylinder housing.
3. Position the caliper assembly in a vise.
4. Place a block of wood between the caliper and the cylinders, and apply low pressure air to the brake hose inlet. One piston will be forced out.
5. Reverse the piston and install it by hand pressure back into the cylinder bore far enough to form a seal. Block the reversed piston

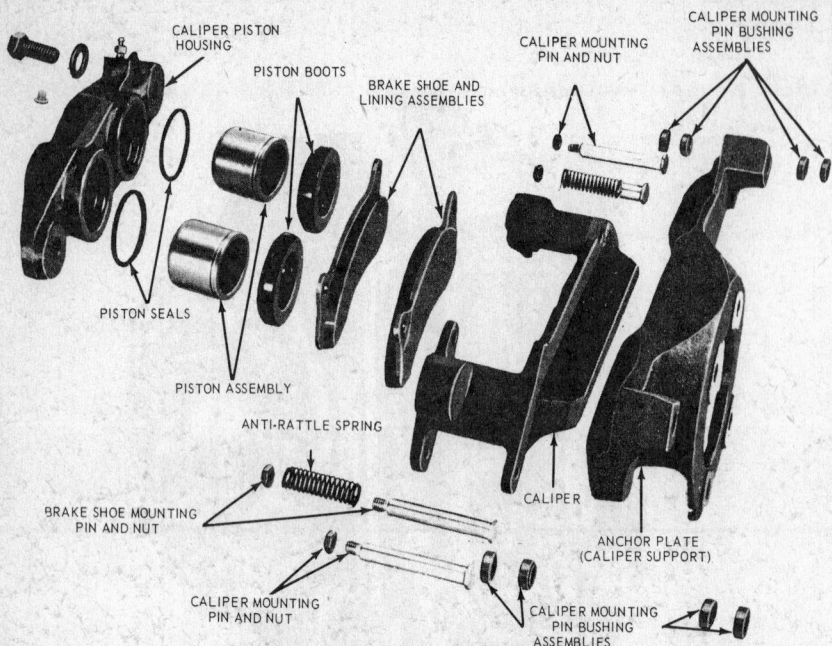

Front disc brake disassembled (© Ford Motor Co.)

from moving out of the bore and place the wooden block between the remaining piston and the caliper.

6. Force out the second piston with low pressure air. *Care should be taken as the piston is forced out of the bore.*

7. Remove the two bolts and separate the caliper from the cylinder housing.

Assembly

The piston assembly and dust boots are not to be reused. A new set is to be used each time the caliper is assembled.

1. Apply a film of clean brake fluid in the cylinder bores and on the piston assemblies. *Do not apply brake fluid on the insulators.*

2. Start the piston assemblies into the cylinder bores using firm hand pressure. *Exercise care to avoid cocking the piston in the bore.*

3. Lightly tapping with a rawhide mallet, seat each piston assembly until the friction ring bottoms out in the cylinder bore.

4. Install the piston dust boots and retainers.

5. Position the piston housing on the caliper and install the piston housing-to-caliper mounting bolts and washers. Torque the piston housing-to-caliper mounting bolts to 155 to 185 ft lbs.

6. Install the flexible brake hose.

7. Bleed the brake system and centralize the pressure differential valve.

Do not move the vehicle after working on the disc brakes until a firm brake pedal is obtained.

Sliding Caliper Disc Brakes—Single Piston

This caliper is a one piece type with a single piston on the inboard side. The piston is made of steel and is plated to resist wear and corrosion.

The piston has a square cut seal which provides for a seal between the piston and the caliper cylinder wall. A rubber dust boot located in a groove in the cylinder helps keep contamination from the piston and cylinder wall.

The caliper is mounted on an adapter which is mounted on the steering knuckle.

Disc Brake Adjustment

No adjustment is required on this unit other than applying the pedal several times after the unit has been worked on. This is to seat the shoes and after this is done the hydraulic pressure maintains the proper clearance between the brake shoes and the rotor.

Brake Shoe Removal

Replace the brake shoes when the linings are worn within 1/32" of the shoe or the rivets.

Removal

1. Remove the master cylinder cover and if the cylinder is more than 1/3 full remove the fluid necessary to make the cylinder only 1/3 full. This is done to prevent any overflow from the cylinder when the piston is pushed into the bore of the caliper.

2. Raise vehicle on hoise and remove the front wheels.

3. Compress the piston back into the bore by using a large C-clamp and compressing the unit until the piston bottoms in the bore.

4. Remove the two retaining bolts that hold the caliper into the support. If the caliper has retaining clips remove the retaining clips and anti-rattle springs. If the caliper has key type retainers, remove the key retaining screws, and using a hammer and drift, punch drive the key out of the caliper.

5. Slide the caliper off the rotor disc. Be careful not to damage the dust boot on the piston when removing the caliper.

NOTE: Do not let the caliper hang with the brake hose supporting the

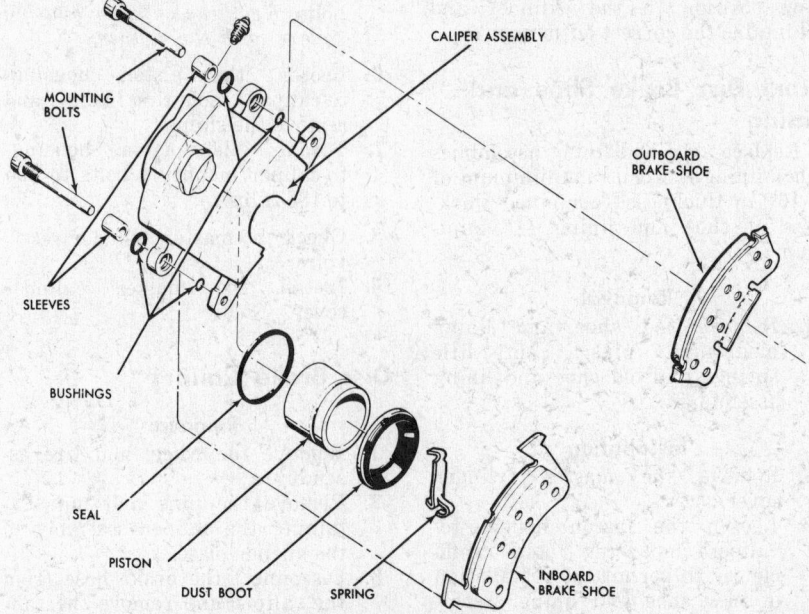

Bolt Mounted Caliper (© Chevrolet Div. GM Corp)

weight. This can cause damage to the hose which could result in a loss of brakes. Set the caliper on the front suspension arm or tie rod.

6. Remove the outer shoe from the caliper. It may be necessary to tap the shoe to loosen it from the caliper. Remove the inner shoe from the caliper or spindle assembly depending on where the shoe stays.
7. Remove the shoe support spring from the piston.

Cleaning and Inspection

Clean the sliding surfaces of the caliper and clean any dirt from the mounting bolts, clips or keys.

Inspect the boot on the piston for signs of cracks, cuts or other damage. Check to see if there is signs of fluid leaking around the seal on the piston. This will show up in the boot. If there is an indication of a fluid leak, the entire caliper has to be disassembled and the seal replaced.

Installation

1. Make sure that the piston is fully bottomed in the cylinder bore and install the outboard shoe in the recess of the caliper. NOTE: On shoes with anti-rattle springs be sure to install the spring before installing the shoe in the caliper.
2. Place the outer shoe on the caliper and press it into place with finger pressure.
3. Position the caliper on the rotor and carefully slide it down into position over the rotor.
4. Install the caliper mounting bolts and torque them to 35 ft lbs. On models with retaining clips install the anti-rattle springs and the retaining clips and torque the retaining screws to 200 in lbs. On models with key type retainers press down the caliper and install the key in its slot and drive it in place with a hammer and drift. Install the retaining screw and torque to 12-18 ft lbs.
5. Install the wheels and lower the vehicle. Check the master cylinder fluid level and add any fluid necessary to bring it up to the proper level.
6. Pump the brake pedal several times until a firm brake pedal is established. Road test the vehicle to check for proper operation.

Disc Caliper

Removal

1. Remove the cover on the master cylinder and check if the fluid level is 1/3 full. If it is more than 1/3 full remove the necessary amount to bring the level down.

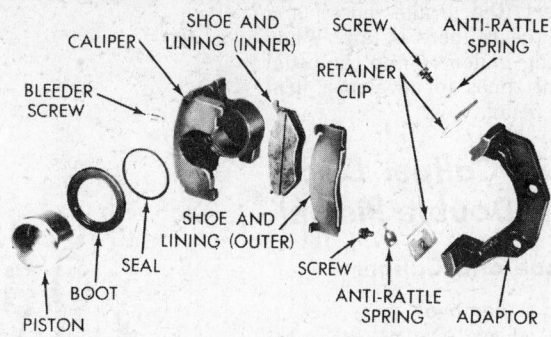

Caliper Assembly Retaining Clip Type (© Chrysler Corp)

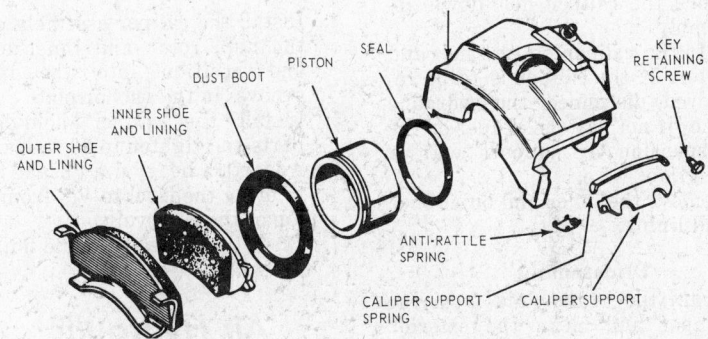

Disc Brake Caliper Retainer Key Type (© Ford Motor Co)

This step is necessary to avoid overflow from the master cylinder when the piston is compressed into the cylinder bore.

2. Raise the vehicle and remove the wheel.
3. Compress the piston into the caliper bore and remove the brake hose from the caliper. Tape the end of the hose to prevent dirt from entering the line.
4. Remove the caliper retaining bolts, clips or wedges and remove the caliper from the vehicle.

Disassembly

1. Clean the outside of the caliper with clean brake fluid and drain any fluid from the caliper.
2. Remove the piston from the caliper by connecting the hydraulic line to the caliper and gently stroking the brake pedal. This will push the piston from the caliper bore.
3. With care remove the boot from the caliper piston bore.
4. Remove the piston seal from the caliper bore using a piece of wood or plastic.
NOTE: DO NOT use a metal tool to remove the seal. This can damage the bore or burr the edges of the seal groove.
5. Remove the bleeder valve.

Cleaning and Inspection

1. Clean all the parts with clean brake fluid and blow out all the passages in the caliper.
NOTE: When ever the caliper is disassembled discard the boot and piston seal. These parts must not be reused.

2. Inspect the outside of the piston for signs of wear, corrosion, scores or any other defects. If any defects are detected replace the piston.
3. Check the caliper bore for the same defects as the piston. However, the bore can be cleaned up to a point with crocus cloth. If there are any marks that will not clean up with the cloth the caliper must be replaced.

Assembly and Installation

1. Lube the caliper bore and the piston with clean brake fluid and position the seal for the piston in the cylinder bore groove.
2. Install the dust boot into the groove in the piston with the fold faces toward the open end of the piston.
3. Install the piston in the bore being careful not to unseat the piston seal in the bore.
4. With the piston bottomed in the cylinder position the boot in the groove in the caliper. Make sure that the retaining ring in the seal is pressed down evenly around the cylinder.
5. Install the bleeder screw in the caliper and install the caliper back on the vehicle.

6. Connect the brake hoses and bleed the calipers of air. When bleeding is done pump the pedal several times to develop a firm brake pedal.

Sliding Caliper Disc Brakes (Double Piston)

Brake Shoe and Caliper

Removal

1. Drain about 2/3 of the total brake fluid from the reservoir.
2. Jack up the vehicle and remove the front wheels.
3. Remove the four screws and remove the caliper hold-down assembly.
4. Lift the caliper off the hub and rotor. If the caliper is to be removed, disconnect the hydraulic line; if not, lay the caliper on the suspension or support with a length of wire.
5. Remove the inner and outer shoe and lining.

Disassembly

1. Drain the brake fluid from the caliper and clean the exterior with clean brake fluid.
2. Place a small block of wood under the caliper pistons and place a protective pad over the exterior. Remove the pistons by directing compressed air into the caliper fluid outlet.
3. Remove and discard piston boots.
4. Remove the piston seals from the groove in the caliper bore.

Assembly

1. Clean all parts in clean brake fluid and blow dry.
2. Dip the new piston seal in clean brake fluid and install it into the cylinder groove.

NOTE: Be sure that the seal is not rolled or twisted in the groove.

3. Install the dust boot in the cylinder groove.
4. Coat the outside diameter of the piston with clean brake fluid. Use something plastic or wood and gradually work the dust boot around the piston.
5. Press the piston straight into the caliper bore until it bottoms. Position the boot in the piston groove.

Installation

1. Install a new shoe and lining assembly into the anchor plate.
2. Push the pistons to the bottom of the piston bore. Place a small block of wood over both pistons and boots. Push the pistons to the bottom of the bores with a C-clamp.
3. Install the outer shoe and lining onto the caliper and install the shoe hold-down spring and pin.

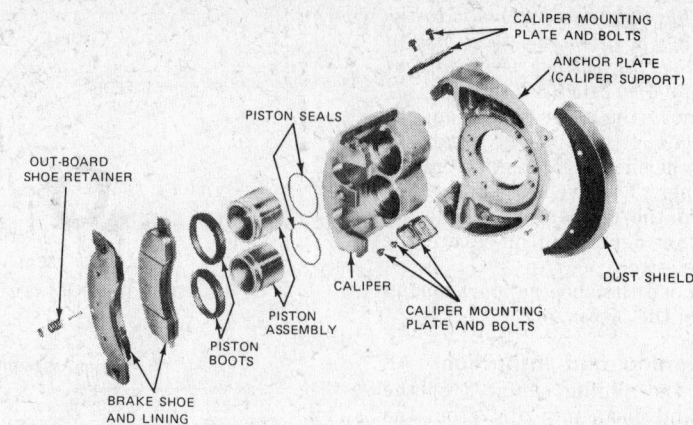

Sliding caliper disc brakes (Double piston) (© Ford Motor Co.)

4. Install the caliper assembly over the hub, rotor and inner shoe, and position into the inner grooves in the anchor plate.
5. Install the caliper hold-down parts and tighten to 40 ft. lbs.
6. Add extra heavy duty brake fluid to bring the level to ¼" from the top of the reservoir.
7. Bleed the system and add fluid as necessary.

Air Hydraulic Intensifier

Description

The intensifier is used to convert air pressure into mechanical force on the hydraulic system which operates the disc brakes.

Air pressure is applied to a diaphragm in the roto-chamber. The diaphragm, through a push rod, activates a master cylinder. The master cylinder transmits hydraulic pressure to the brake caliper pistons. The air hydraulic intensifier is used in conjunction with the sliding caliper (double piston) disc brake system used on some heavy duty models.

Disassembly

NOTE: Soak chamber in solvent for 24 hours if there are signs of rust accumulation.

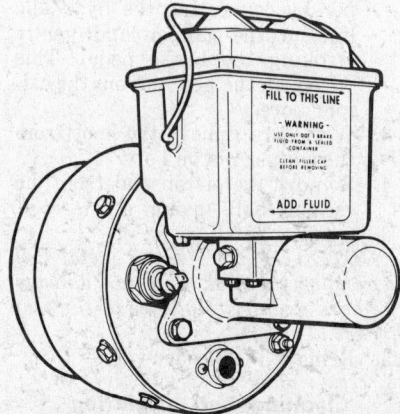

Air-hydraulic intensifier © Ford Motor Co.

1. Remove the master cylinder cap and separate the master cylinder from the chamber cover.

NOTE: Do not get cleaning solvent on master cylinder seals.

2. Remove the chamber cover cap screws and cover. Remove the spring and spring guides if so equipped.
3. Remove the nuts at the air inlet end that secure the diaphragm outer clamp to the body.
4. Grasp the push rod and pull the diaphragm and clamp assembly out of the body.
5. Unroll the diaphragm.
6. Remove the outer clamp.
7. Remove the nuts from the inside of the guide.
8. Separate the inner clamp, diaphragm, push plate and rod assembly and guide.

Assembly

1. Clean and inspect all parts.
2. Drain the master cylinder and inspect for wear. Do not repair. Install a new master cylinder if necessary.
3. Place the small diameter end of the diaphragm in the inner diaphragm clamp.
4. Install the diaphragm guide through the diaphragm and over the clamp studs and against the diaphragm bead.
5. Install the push rod plate and rod assembly against the guide and over the studs. Then install the nuts and washers.
6. Place the assembly inside of the outer clamp and roll the outer bead of the diaphragm back and over the end of the outer clamp.
7. Lubricate the inside wall of the body and the diaphragm surface with a silicone lubricant or equivalent.
8. Slide the assembled parts into the body so that the diaphragm outer bead fits snugly against the shoulder in the body and the outer clamp studs are through the holes in the body.
9. Install the nuts and lockwashers.

POWER BOOSTER TROUBLE DIAGNOSIS

Condition	Possible Cause	Correction
Vacuum leak (booster in released position)	1. End plate, center plate or control valve body gaskets leak. 2. Distortion of end plate. 3. Misalignment of control valve poppet. 4. Loose vacuum cylinder bolts. 5. Loose control valve body screws. 6. Large control valve poppet spring not centered in spring retainer.	1. Recondition booster unit. 2. Replace end plate. 3. Disassemble, clean and correctly reassemble. 4. Coat vacuum cylinder bolts lightly with a suitable sealing compound and tighten to specified torque. 5. Tighten control valve body screws to specified torque. 6. Disassemble unit and correctly reassemble.
Vacuum leak (booster in applied position)	1. Leak at control valve poppet and seat. 2. Dry or faulty piston leather packing. 3. Faulty control valve diaphragm assembly.	1. Clean and inspect poppet and seat for damage and repair as necessary. 2. Clean and lubricate piston leather or replace. 3. Replace faulty parts.
External hydraulic leaks	1. Gasket ("O" ring) leaking at hydraulic end plate joint. 2. Fluid leaking at copper gasket under hydraulic cylinder end cap.	1. Disassemble clean and replace ("O" ring) gasket and reassemble. 2. Remove end cap and inspect copper gasket and seat install new copper gasket.
Internal hydraulic leak at low pressures	1. Control valve hydraulic piston cup failure. 2. Faulty push rod seal.	1. Recondition control valve unit. 2. Replace push rod seal.
Internal leaks at high pressure	1. Fluid passing copper gasket under hydraulic fitting in control valve. 2. Inspect cups and seals of master cylinder for cuts and scores. 3. Inspect cups of the control valve hydraulic piston.	1. Clean and inspect gasket and fitting, replace if faulty. 2. Hone master cylinder and replace cups and seals. 3. Replace faulty cups.
Hydraulic pressure buildup (without added input)	1. Check hydraulic piston check valve and slot for foreign material under valve.	1. Clean or replace valve and seats as condition indicates.
Failure to release	1. Weak vacuum cylinder piston return spring. 2. Dry vacuum piston leather packing. 3. Swollen rubber cups due to inferior or contaminated brake fluid. 4. Damaged or dented vacuum cylinder shell. 5. Dirty or sticky control valve piston.	1. Replace vacuum cylinder piston return spring. 2. Lubricate vacuum piston leather packing. 3. Flush hydraulic system and recondition or replace all cylinders. 4. Replace vacuum cylinder shell. 5. Recondition control valve assembly.
Failure of booster to operate within specified pressures	1. Rusty, dirty or distorted vacuum cylinder shell. 2. Dry or worn vacuum cylinder leather packing. 3. Swollen rubber cups due to inferior brake fluid. 4. Worn or scored hydraulic cups. 5. Dirt, rust or foreign matter in any component of the system.	1. Clean or replace vacuum cylinder shell. 2. Recondition and lubricate the vacuum booster. 3. Recondition the master cylinder. Replace brake fluid. 4. Recondition the master cylinder. 5. Recondition and lubricate the brake booster assembly.
Loss of fluid	1. Fluid leaking past cup in master cylinder. 2. Brake wheel cylinders leaking. 3. Loose hydraulic hose connectors. 4. Leaking stop light switch.	1. Recondition master cylinder or replace. 2. Recondition or replace wheel cylinders. 3. Inspect and tighten all hydraulic connections. 4. Replace stop light switch.
Presence of brake fluid on hy-power vacuum cylinder	1. Piston cup or push rod seal leaking.	1. Recondition master cylinder.
Pedal kicks back against foot when brakes are applied	1. Vacuum leakage. 2. Dirt under control valve or damaged seat. 3. Weak or broken spring.	1. Inspect and correct vacuum leak. 2. Clean and recondition booster assembly. 3. Replace spring.
Engine runs unevenly at idle with brakes released	1. Vacuum leakage. 2. Dirt under control valve disc or damaged seat. 3. Defective spring.	1. Inspect and tighten all vacuum fittings. 2. Clean control valve or replace. 3. Replace defective spring.
Engine runs evenly and pedal is hard with brakes applied	1. Control valve piston assembly not seating on vacuum disc. 2. Defective control valve plate and diaphragm. 3. Defective pressure plate and diaphragm.	1. Clean or replace control valve piston assembly. 2. Replace control valve plate and diaphragm. 3. Replace pressure plate and diaphragm.

Power Brakes

POWER BOOSTER TROUBLE DIAGNOSIS

Condition	Possible Cause	Correction
Brakes are slow to release Note: First, jack up truck and determine whether or not the wheels are dragging.	1. Incorrect pedal linkage adjustment. 2. Compensating port of master cylinder plugged. 3. Brake shoes sticking. 4. Weak brake shoe return spring. 5. Booster control valve piston sticking. 6. Booster air filter clogged. 7. Control valve diaphragm return spring missing. 8. Defective check valve in slave cylinder piston. 9. Dirt under atmospheric valve disc.	1. Adjust and lubricate pedal linkage. 2. Clean master cylinder with compressed air. 3. Free up and lubricate brake shoes. 4. Replace brake shoe return spring. 5. Clean booster control valve piston and lubricate. 6. Clean air filter in mineral spirits. 7. Install new control valve return spring. 8. Recondition slave cylinder pistons. 9. Clean atmospheric valve.
Brake pedal is hard at different intervals	1. Defective manifold check valve. 2. Slave cylinder piston sticking due to dirt or inferior brake fluid. 3. Brake booster air cleaner clogged.	1. Clean or replace manifold check valve. 2. Clean and recondition slave cylinder. 3. Clean air cleaner in mineral spirits and blow dry with compressed air.

Power Brake Boosters

Brake System Preliminary Checks

Always check the fluid level in the brake master cylinder reservoir(s) before performing the test procedures. If the fluid level is not within 1/4" of the top of the master cylinder reservoirs, add the specified brake fluid.

Push the brake pedal down as far as it will go. If the pedal travels more than halfway between the released position and the floor, adjust the brakes. If the vehicle is equipped with automatic brake adjusters, several sharp brake applications while backing up may be necessary to adjust the brakes.

Road test the vehicle and apply the brakes at a speed of about 20 mph to see if the vehicle stops evenly. If not, the brakes should be adjusted. *Perform the road test only when the brakes will apply and the vehicle can be safely stopped.*

Dual Brake Warning Light System Tests

1. Turn the ignition switch to the ACC or ON position. If the light on the brake warning lamp remains on, the condition may be caused by a shorted or broken switch, grounded switch wires or the differential pressure valve is not centered. Centralize the differential pressure valve. If the warning light remains on, check the switch connector and wire for a grounded condition and repair or replace the wire assembly. If the condition of the wire is good, replace the brake warning lamp switch.
2. Turn the ignition switch to the start position. If the brake warning lamp does not light, check the light and wiring and

replace or repair wiring as necessary. When both brake systems are functioning normally, the equal pressure at the pressure differential valve during brake pedal application keeps the valve centered. The brake warning light will be on only when the ignition key is in the start position.
3. If the brake warning lamp does not light when a pressure differential condition exists in the brake system, the warning lamp may be burned out, the warning lamp switch is inoperative or the switch to lamp wiring has an open circuit. Check the bulb and replace it, if required. Check the switch to lamp wires for an open circuit and repair or replace them, if required. If the warning lamp still does not light, replace the switch.

Power Brake Function Test

With the engine stopped, eliminate all vacuum from the system by pumping the brake pedal several times. Then push the pedal down as far as it will go, and note the effort required to hold it in this position. If the pedal gradually moves downward under this pressure, the hydraulic system is leaking and should be checked by a hydraulic pressure test.

With the brake pedal still pushed down, start the engine. If the vacuum system is operating properly, the pedal will move downward. If the pedal position does not change, the vacuum system is not operating properly and should be checked by a vacuum test.

Vacuum Booster Check Valve Test

Disconnect the line from the bottom of the vacuum check valve, and connect a vacuum gauge to the valve.

Start the engine, run it at idle speed, and check the reading on the vacuum gauge.

The gauge should register 17-19" with standard transmission and 14-15" in Drive range if equipped with an automatic transmission. Stop the engine and note the rate of vacuum drop. If the vacuum drops more than one inch in 15 seconds, the check valve is leaking. If the vacuum reading does not reach 18" or is unsteady, an engine tuneup is needed.

Remove the gauge and reconnect the vacuum line to the check valve.

Vacuum Booster Test—Bendix Piston Type

Disconnect the vacuum line from the booster end plate. Install a tee fitting in the end plate, and connect a vacuum gauge (No. 1) and vacuum line to the fitting. Install a second vacuum gauge (No. 2) in place of the pipe plug in the booster control valve body.

Start the engine, and note the vacuum reading on both gauges. If both gauges do not register manifold vacuum, air is leaking into the vacuum system. If both gauges register manifold vacuum, stop the engine and note the rate of vacuum drop on both gauges. If the drop exceeds one inch in 15 seconds on either gauge, air is leaking into the vacuum system. Tighten all vacuum connections and repeat the test. If leakage still exists, the leak may be localized as follows:
1. Disconnect the vacuum line and gauge No. 1 from the booster.
2. Connect vacuum gauge No. 1 directly to the vacuum line. Start the engine and note the gauge reading. Stop the engine and check the rate of vacuum drop. If gauge No. 1 does not register manifold vacuum, or if the vacuum drop exceeds 1" in 15 seconds, the leak is in the vacuum

line or check valve connections.

3. Reconnect vacuum gauge No. 1 and the vacuum line to the tee fitting. Start the engine, and run it at idle speed for one minute. Depress the brake pedal sufficiently to cause vacuum gauge No. 2 to read from zero to 1 inch of vacuum. Gauge No. 1 should register manifold vacuum of 17-19" with standard transmission and 14-16" in Drive range if equipped with an automatic transmission. If the drop of vacuum on gauge No. 2 is slow, the air cleaner, or air cleaner line, may be plugged. Inspect and if necessary, clean the air cleaner.

4. Release the brake pedal and observe the action of gauge No. 2. Upon releasing the pedal, the vacuum gauge must register increasing vacuum until manifold vacuum is reached. The rate of increase must be smooth, with no lag or slowness in the return to manifold vacuum. If the gauge readings are not as outlined, the booster is not operating properly and should be removed and overhauled.

Vacuum Booster Test— Diaphragm Type

This procedure can be used to test all diaphragm boosters which are equipped with a pipe thread outlet on the atmosphere portion of the diaphragm chamber.

Remove the pipe plug from the rear half of the booster chamber, and install a vacuum gauge. Start the engine and run it at idle speed. The gauge should register 18-21" of vacuum.

1. With the engine running, depress the brake pedal with enough pressure to show a zero reading on the vacuum gauge. Hold the pedal in the applied position for one minute. Any downward movement of the pedal during this time indicates a brake fluid leak. Any kickback (upward movement) of the pedal indicates brake fluid is leaking past the hydraulic piston check valve.

2. With the engine running, push down on the brake pedal with sufficient pressure to show a zero reading on the vacuum gauge. Hold the pedal down, and shut the engine off. Maintain pedal position for one minute. A kickback of the pedal indicates a vacuum leak in the vacuum check valve, in the vacuum line connections, or in the booster.

Bleeding Vacuum-Hydraulic Booster Systems

1. Eliminate vacuum in the booster

by depressing the brake pedal several times while the engine is not running.

2. On trucks not equipped with reservoir tanks, disconnect the manifold tube at the booster side of the manifold check valve (engine not running).

3. Alternately loosen the brake tube at each unit until all air is expelled. Booster slave-cylinder is bled first.

CAUTION: Where air pressure brake bleeding equipment is used to bleed brakes, do not use more than 25-30 psi.

NOTE: A piston stop is provided in the slave cylinder to eliminate the possibility of damaging the return spring while bleeding the system. This damage occurs only when bleeding the brakes with a vacuum present in the booster system.

Hydraulic Tandem Brake Unit

Disassembly

1. Disconnect hydraulic and vacuum by-pass tube from valve body.

2. Remove control valve air inlet fitting from control valve body.

3. Remove control valve body and valve parts from end plate.

4. Make a special tool. NOTE: If this to be a regular service this tool is recommended. For one time or emergency, a vise, "C" clamps, and a guide tube 10" long may be used.

5. Insert tool through end plate opening, and force vacuum cylinder piston forward.

6. Attach flange of tool to end plate with three valve body cover screws.

7. Loosen slave cylinder check nut, and remove slave cylinder.

8. Compress push rod pin retaining spring, remove retainer pin, then remove hydraulic piston from push rod.

9. Hold end cap in a vise, and remove hydraulic cylinder from cap.

10. Loosen vacuum hose clamps, then slide both hoses on the vacuum tube toward center plate.

11. Remove hydraulic by-pass tube from rear end plate, then remove return spring compression tool from end plate.

12. Remove the nuts and studs from power cylinder, then disassemble end plates, cylinder shells and center plate assembly.

13. Force center plate and vacuum piston together, and insert a rod through hole in piston rod to hold piston return spring in the compressed position.

14. Place assembly ring over piston, then remove piston assembly, but keep piston parts assembled in assembly ring. After vacuum tubes and tee fittings have been removed from center plate, position plate on a flat surface.

15. Remove fast application valve cover.

16. To disassemble the diaphragm assembly, hold valve shaft with a screw driver, and remove nut.

17. Lift retainer and diaphragm off valve shaft.

18. Turn center plate upside down, then remove valve seat plate screws and plate, gasket, valve, and spring from center plate.

19. Position front end plate assembly on a flat surface with flat side down.

20. Remove the "O" ring seal, snap ring and retainer washer, push rod seal spring and flange washer, push rod rubber cup seal, and guide washer from end plate.

21. Drive push rod leather seal out of end plate.

22. Position end plate in a holding fixture, then remove hydraulic valve fitting with a 1-7/8" socket wrench.

23. Push hydraulic piston out of valve fitting, and remove gasket from fittings.

Clean all metal parts in a suitable cleaning fluid. After cleaning, wash all the hydraulic system parts in alcohol.

Examine the bore of the cylinder shells for rust and corrosion, and polish with fine steel wool or crocus cloth if necessary.

If the cylinders are badly pitted or scored, install new cylinders.

If felt type wicks are worn, replace them with cotton type wicks.

Use overhaul kit and install ALL parts contained. Do not gamble on ANY old parts that the kit replaces.

Assembly

1. Install nut on piston rod with flat side of nut upward.

2. Position larger diameter piston plate on piston rod with chamfered side of hole at top. Guide rubber seal ring over threads of piston rod.

3. Place assembly ring on a flat surface, then install leather packing, with lip side upward; and smaller diameter piston plate with chamfered side of hole downward in the ring.

4. Cut a new piece of wick to the required length, then place it against inner face of leather packing lip.

5. Assemble expander ring against wick with gripper points upward, and hook notched end of

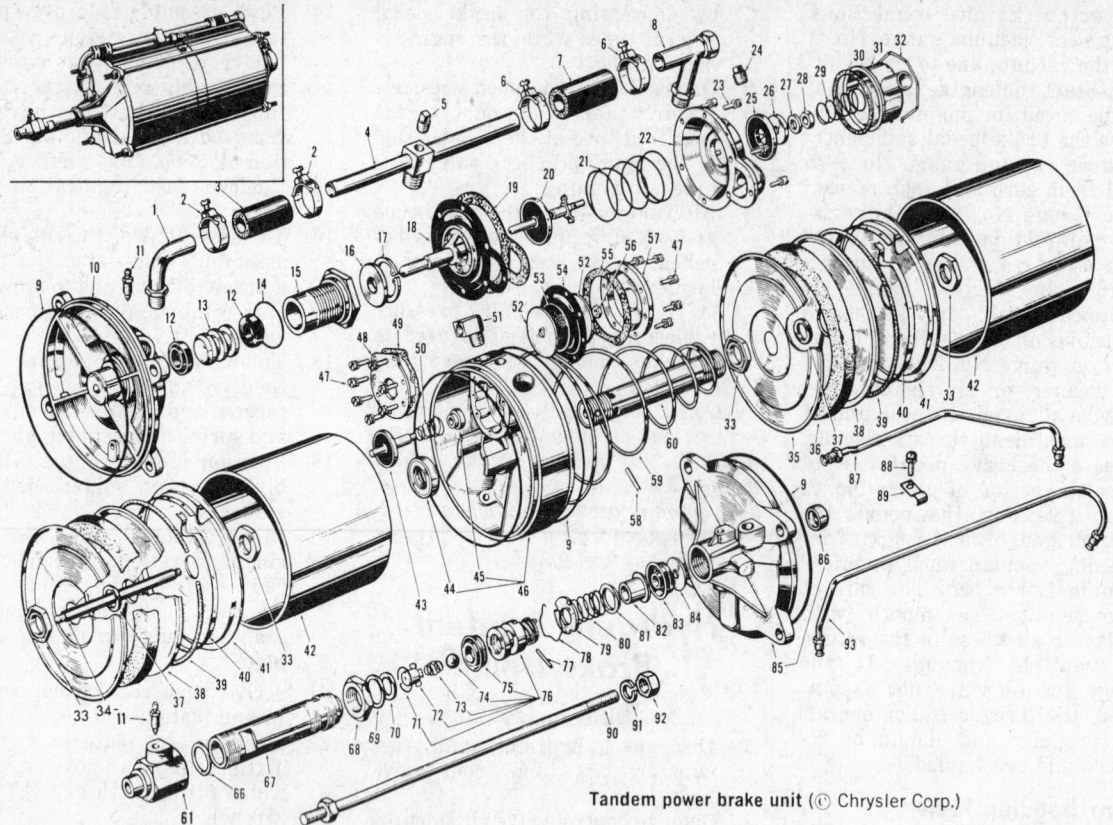

Tandem power brake unit (© Chrysler Corp.)

1 Tube & bushing
2 Clamp—hose
3 Hose—vacuum ¾x2½"
4 Tube # fitting—vacuum
5 Plug—pipe (⅛")
6 Clamp—hose (1")
7 Hose—vacuum 1"x3"
8 Tee—1" hose ¾" pipe, male & female
9 Gasket—center plate & end plate
10 Plate—cylinder end
11 Valve—bleeder
12 Cup—hydraulic valve piston
13 Piston—hydraulic valve
14 Seal—hyd. valve fitting
15 Fitting—hyd. valve piston
16 Washer—stop
17 Ring—retainer
18 Diaphragm & plates—control valve
19 Gasket—poppet (valve body)
20 Shaft & vac. poppet—control valve
21 Spring—poppet valve
22 Body—control valve

23 Screw & lockwasher
24 Plug—(⅜")
25 Seal—poppet valve
26 Valve—poppet
27 Washer—poppet valve
28 Nut—Hex (#6-32)
29 Spring—poppet valve (small)
30 Gasket—A. C. tube & cover
31 Tube & cover—air inlet
32 Snap ring—air inlet tube & cover
33 Nut—piston rod
34 Push rod—piston
35 Plate—piston (outer)
36 Packing—piston
37 Seal—vacuum piston
38 Plate—piston (inner)
39 Wick—piston
40 Ring—piston expander
41 Plate—retainer (piston felt & exp. ring)
42 Shell—vac. cyl.
43 Shaft & seal (fast application valve)
44 Seal—center plate piston rod
45 Seal—Center plate poppet

valve
46 Center plate & seals
47 Screw & lockwasher fast application valve
48 Seat—poppet (fast application valve)
49 Gasket—poppet seat
50 Spring—return—valve poppet
51 Elbow—inverted flared tube
52 Plate—diaphragm
53 Gasket—diaphragm plate
54 Diaphragm—F. A. valve
55 Gasket—cover (F. A. valve)
56 Nut—thin hex. check (¼"-28)
57 Cover—fast application valve
58 Pin—retainer (push rod)
59 Spring—return (vac. piston)
60 Piston rod & thrust cup
61 Cap—end—hydraulic cylinder
66 Gasket—end cap
67 Tube—hydraulic cylinder
68 Nut—hyd. cyl. tube
69 Seal—hyd. cyl. tube nut
70 Seal—hyd. cyl.
71 Snap ring—ball retainer
72 Retainer—ball

73 Spring—ball return
74 Ball (hyd. piston check valve)
75 Cup—hyd. piston
76 Piston—hydraulic
77 Pin—retainer, hydraulic piston
78 Snap ring
79 Washer—stop
80 Spring—retainer
81 Sleeve—retainer
82 Retainer—seal
83 Hyd. seal
84 Washer—guide
85 End plate & seal
86 Seal—oil (end plate)
87 Tube—vacuum by-pass
88 Screw & lockwasher—clip attached
89 Clip—tube (¼")
90 Stud—cylinder attaching
91 Lockwasher—cylinder attaching
92 Nut—hex. (½"-20) cylinder attaching
93 Tube—hyd. by-pass

spring under the clip near opposite end of spring. Position cut of retainer plate over loop of the spring.

6. Hold piston parts in the assembly ring, assemble them on end of piston rod, then install nut on tip of piston assembly. Tighten nut until it is flush with end of rod. Stake nuts securely at two places.

7. Clamp staked nut firmly in a vise, and tighten nut on opposite side of piston plate solidly against piston plate.

8. Press the fast application valve stem and push rod seals into cen-

ter plate. The application valve seal must be flush with bottom of hole. The push rod seal should rest against the shoulder of center plate. Position center plate. Then place valve spring on top of seal with the small end at top.

9. Install the bullet-nosed tool at threaded end of valve shaft, and insert valve shaft through seal. Position gasket on center plate.

10. Place valve seat plate, with seat side downward, on gasket, and install screws and lockwashers.

11. Turn center plate over. Place lower diaphragm plate on valve shaft with rounded edge at top,

then place diaphragm gasket at top of plate. Position diaphragm on top of gasket so screw holes and the bypass hole index with the identical holes in center plate.

12. Install the other diaphragm plate with rounded edge facing diaphragm.

13. Install valve shaft nut on valve shaft. Use a screw driver to prevent shaft from turning, and tighten nut. Stake nut securely at opposite points.

14. Position cover gasket and cover plate, then install screw and lockwashers.

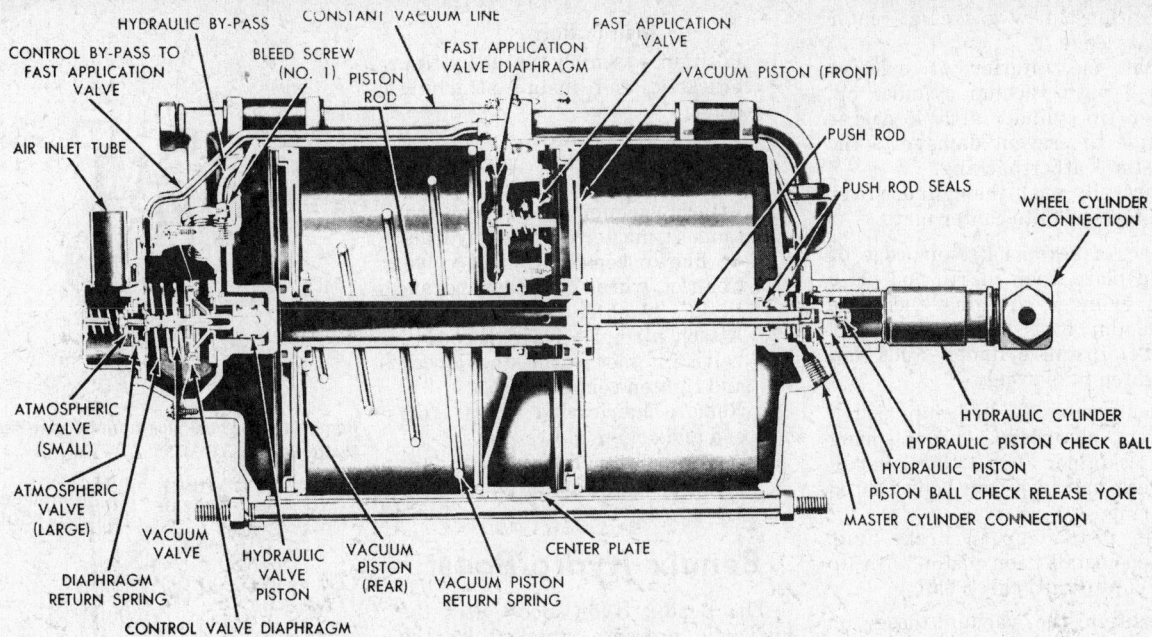

CONTROL BY-PASS TO FAST APPLICATION VALVE — HYDRAULIC BY-PASS — CONSTANT VACUUM LINE — FAST APPLICATION VALVE — FAST APPLICATION VALVE DIAPHRAGM — BLEED SCREW (NO. 1) — PISTON ROD — VACUUM PISTON (FRONT) — PUSH ROD — PUSH ROD SEALS — WHEEL CYLINDER CONNECTION — AIR INLET TUBE — ATMOSPHERIC VALVE (SMALL) — ATMOSPHERIC VALVE (LARGE) — DIAPHRAGM RETURN SPRING — VACUUM VALVE — HYDRAULIC VALVE PISTON — CONTROL VALVE DIAPHRAGM — VACUUM PISTON (REAR) — VACUUM PISTON RETURN SPRING — CENTER PLATE — HYDRAULIC CYLINDER — HYDRAULIC PISTON CHECK BALL — HYDRAULIC PISTON — PISTON BALL CHECK RELEASE YOKE — MASTER CYLINDER CONNECTION

Tandem power brake (sectional view) (© Chrysler Corp.)

15. Place piston return spring over piston rod with small end of spring at bottom.
16. Carefully guide piston rod through leather seal in center plate, with piston stop flanges of center plate facing upward. Press center plate down against spring, and insert a rod in piston rod. Thread piston rod nut on piston rod, with flat side of nut upward to limit of threads.
17. If forward piston was disassembled to replace leather piston packing, cotton wicking, or other parts, assemble the piston parts in the ring and turn assembly ring over.
18. Remove larger piston plate and "O" ring seal.
19. With assembly ring still in place, guide the remaining piston parts over end of push rod and against piston nut. Carefully install "O" ring seal over threads of piston rod.
20. Place the larger diameter piston plate on piston rod with chamfered side of hole toward "O" ring seal.
21. Assemble large end of push rod in end of piston rod and install retainer pin. Install piston rod nut on end of piston rod with flat side downward. Tighten nut until it is flush with face of piston rod, then stake nut securely at opposite points.
22. Hold piston rod nut in a vise or with a wrench, and tighten inner nut securely against piston. Care must be exercised when tightening inner nut to prevent expander spring retainer plate from shifting.
23. Remove assembly ring, then remove rod holding return spring compressed. Install a new copper gasket in end cap.
24. The hydraulic cylinder must be assembled with milled flats next to end cap. Tighten hydraulic cylinder solidly in end cap, then thread check nut on hydraulic cylinder up to the limit of the threads.
25. Install check nut seal (if used) in groove of cylinder tube. Install bleeder screw in cap.
26. Press push rod leather seal into hydraulic cylinder bore of front end plate from inner side of plate with lip of seal toward outer end of the plate. Install push rod seal parts.
27. The chamfered side of stop washer is down, lip of cup is up, flat side of washer is next to cup, and small end of spring is down. Place washer against spring. Install snap ring in inner groove of end plate.

Compressor tool for vacuum piston return spring
(© Chrysler Corp.)

2¼"
⅝"
1¹⁄₁₆"
1⁷⁄₃₂"
1½"
1⁷⁄₃₂"
1"
3"
½" DIAMETER
½"
3 HOLES ¾₁₆" DIAMETER
6"
⅛"
3"
WELD HERE

28. Install stop washer with flat side in control valve hydraulic fitting. Install stop washer retaining ring.
29. Dip hydraulic piston cups in brake fluid, and assemble them on the hydraulic piston with lips of cups positioned away from each other. Insert piston into the fitting with open end of piston toward stop washer.
30. Install a new gasket on the hydraulic fitting (copper gasket on fitting without the groove, and a rubber seal gasket on fitting with the groove). Install the hydraulic fitting in end plate with a 1-⅞" socket wrench. Tighten fitting equipped with a rubber gasket firmly, and fitting equipped with a copper gasket to 324-330 foot-pounds.
31. Assemble vacuum control parts in control body. Install a new lead washer.
32. Hold slave cylinder end cap in a vise, and thread cylinder into end plate. Install T-fitting and tubes on center plate.
33. Position an end plate gasket on the plate, place cylinder shell on end plate, and coat interior of cylinder with vacuum cylinder oil.
34. Dip cylinder piston on packing in vacuum cylinder oil and allow the excess oil to drain off the wickings.
35. Position a gasket on ledge of center plate, then carefully guide push rod through seal in front end plate. At the same time, align the vacuum tube in end plate with vacuum tube on center plate. Slide hose in place to contact the two vacuum tubes.

36. Position a new gasket at center plate ledge.
37. Coat the interior of cylinder shell with vacuum cylinder oil, then tip cylinder at a 45 degree angle to prevent damage to the piston leather packing.
38. Carefully push the cylinder over piston and onto center plate.
39. Place a new gasket on ledge of end plate, then install end plate on cylinder, aligning end plate vacuum tube and center plate tube. Install cylinder studs and tighten nuts evenly.
40. To assemble the hydraulic piston parts, place large end of spring in retainer cup, then install check ball in piston body behind spring.
41. Dip piston cup in brake fluid, then install it on piston with lip of cup toward check ball.
42. Position the vacuum hoses on tubes, and tighten hose clamps firmly.
43. Connect hydraulic by-pass tube to front and rear end plate.
44. Remove slave cylinder from end plates, then insert and attach return spring compressing tool.
45. Assemble hydraulic piston on push rod.
46. Make certain lock ring is positioned over the retainer pin. Install hydraulic gasket in the plate. Carefully guide the hydraulic cylinder over piston cup, and thread cylinder into end plate.
47. Adjust cylinder 7¾", measuring between points shown in illustration.
48. Align bleeder screw in end cap with bleeder screw in control valve.
49. Remove spring compressing tool. After cylinder length adjustment is completed, tighten cylinder check nut solidly.
50. Install guide pins, made from 8-32 x 2½" machine screws with the heads cut off, in end plate.
51. Install diaphragm with diaphragm stem inserted into hydraulic control piston hole. Place diaphragm return spring and control valve body on top of diaphragm.
52. Remove guide pins, one at a time, and replace each guide pin with an attaching screw and a new lock washer. Tighten screws progressively and firmly.
53. Install air inlet fitting in control body, then install retainer.
54. Install vacuum by-pass tube.
55. Inspect assembly to see that all bolts, nuts, screws, washers, and plugs are in place, and that all tubes, clamps, and fittings are firmly tightened.

Installation

1. Position assembly on mounting brackets, and install attaching bolts.
2. Tighten bolts firmly.
3. Connect stop light wires and hydraulic lines to stop light switch.
4. Attach vacuum hose to booster.
5. Connect master cylinder hydraulic line to booster control valve.
6. Connect wheel cylinder hydraulic line to booster end cap.
7. Attach air inlet hose to control valve air inlet fitting, then check and tighten connections.
8. Remove lubricating plugs from end and center plates.
9. Add vacuum cylinder oil to level of filler holes, install plugs, then bleed hydraulic system.

Bendix Hydro-Boost

The Bendix Hydro-Boost uses the hydraulic pressure supplied by the power steering pump to provide a power assist to brake application.

Disassembly

1. Place the booster in a vise with the bracket end up. Using a hammer and chisel, cut the bracket nut that holds the linkage bracket to the booster assembly. The nut should be cut at the open slot in the booster cover threads. Care must be exercised to avoid damage to the threads. Spread the nut and remove the bracket.
2. Remove the pedal boot by pulling it off over the pedal rod eyelet.
3. Position pedal rod removing tool around the pedal rod. The tool should be resting on the booster cover. Insert a punch through the pedal rod from the lower side of the special tool. Push the punch through until it rests on the higher side of the tool. Push up on the punch to shear the pedal rod retainer; remove the pedal rod
4. Remove the grommet from the groove near the end of the pedal rod and from the groove in the input rod.
5. Disengage the tabs of the spring retainer from the ledge inside

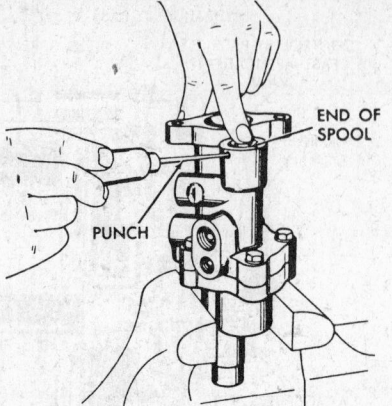

Removing the spool plug from booster-Bendix Hydro-Boost (© G.M.C.)

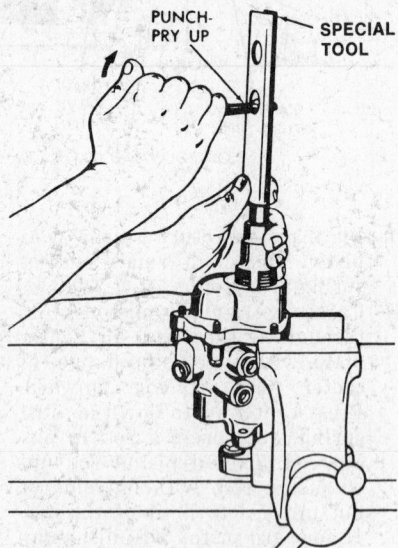

Removing the booster pedal rod-Bendix Hydro-Boost (© G.M.C.)

the opening near the master cylinder mounting flange of the booster. Remove the retainer and piston return spring from the opening.
6. Pull straight out on the output push rod to remove the push rod and push rod retainer from inside the booster piston.
7. Press in on the spool plug, and insert a small punch into the hole on top of the housing. This unseats one side of the spool plug snap ring from the groove in the bore. Remove the snap ring.

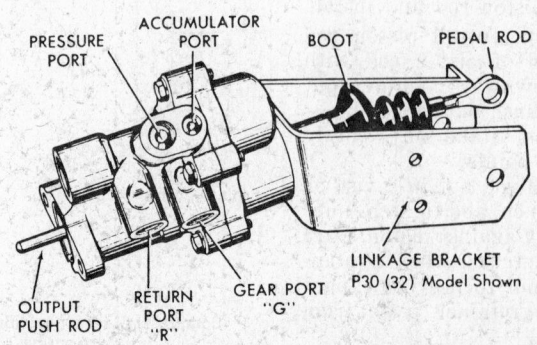

Bendix Hydro Boost (Typical) (© G.M.C.)

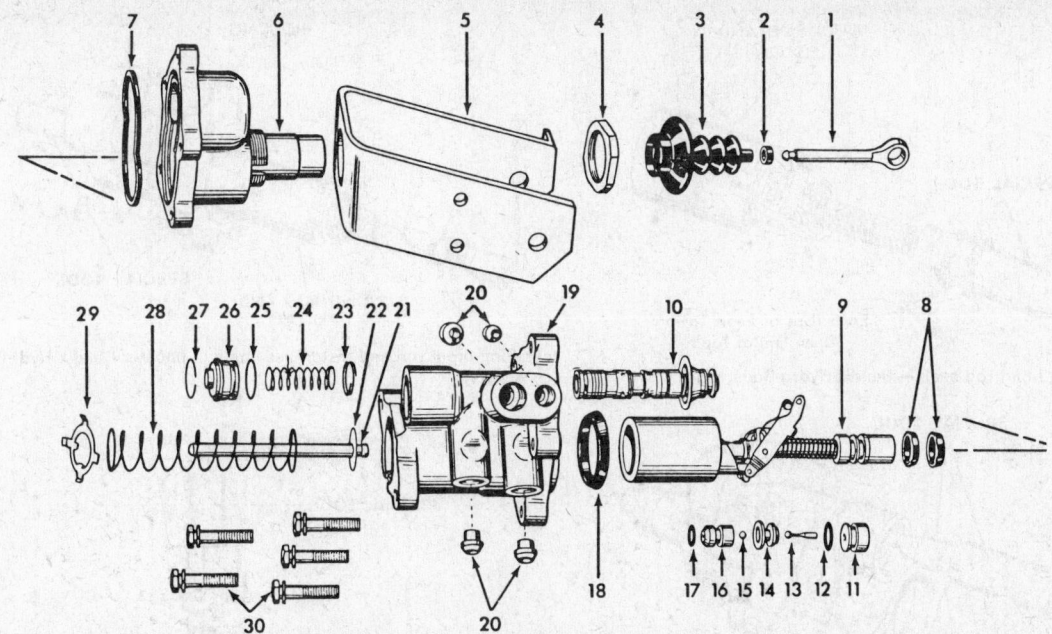

Bendix Hydro-Boost—exploded view (© G.M.C.)

1	Pedal push rod	7	Cover to housing seal	13	Spacer
2	Pedal push rod grommet	8	Input rod seals	14	Spacer
3	Pedal push rod boot	9	Input rod and piston assy.	15	Check valve ball
4	Bracket nut	10	Spool assembly	16	Accumulator check valve
5	Linkage bracket	11	Plunger seat	17	'O' ring
6	Booster cover	12	'O' ring	18	Piston seal

19	Booster housing	25	Plug 'O' ring
20	Tube seat inserts	26	Spool plug
21	Output push rod	27	Snap ring
22	Push rod retainer	28	Piston return spring
23	Spiral snap ring	29	Spring retainer
24	Spool spring	30	Housing to cover bolts

8. Remove the spool plug from the bore with a pair of pliers. Remove the 'O' ring from the plug and discard. Remove the spool spring from the bore.

9. Place the booster cover in a soft-faced vise and remove the cover retaining bolts. Remove the booster assembly from the vise and separate the booster cover from the housing. Remove the large seal ring and discard.

10. Press in on the end of the spool assembly, and use a spiral snap ring removing tool to remove the snap ring from the forward groove in the spool. Discard the snap ring.

11. Remove the input rod and piston assembly, and the spool assembly from the booster housing.

12. Remove the input rod seals from the input rod end, and the piston seal from the piston bore in the housing. Discard the seals.

13. Remove the plunger, seat, spacer and ball from the accumulator valve bore in the flange of the booster housing. Remove the 'O' ring from the seat and discard.

14. Thread a screw extractor into the opening in the check valve in the bottom of the accumulator valve bore, and remove the check valve from the bottom of the bore. Discard the check valve and 'O' ring.

NOTE: Using a screw extractor damages the seat in the check valve. A new check valve, 'O' ring and valve

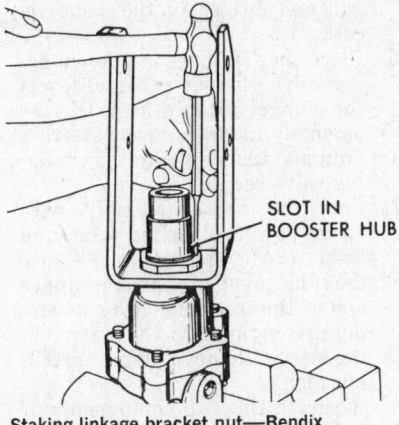

SLOT IN BOOSTER HUB

Staking linkage bracket nut—Bendix Hydro-Boost (© G.M.C.)

Engage notched end of tool under first coil, then rotate to remove the remaining coils from the snap ring groove.

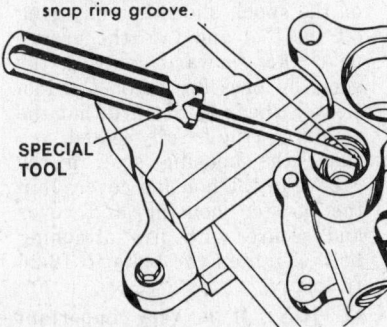

SPECIAL TOOL

Removing spiral snap ring—Bendix Hydro-Boost (© G.M.C.)

must be installed whenever the check valve is removed from the accumulator valve bore.

15. Using a ¼" or a 5/16" spiral flute type screw extractor, remove the tube seats from the booster ports.

Cleaning & Inspection

1. Clean all parts in a suitable solvent.

2. Inspect the valve spool and the valve spool bore for any damage or ware. Discoloration of the spool or bore is normal, particulary in the grooves. If any damage is noted, replace the valve spool and housing.

NOTE: The clearance between the valve spool and the bore is very important. Because of this, the valve spool and housing are to be replaced only as an assembly.

3. Inspect the input rod and piston assembly for any damage or ware. Replace any defective components.

4. Inspect the piston bore in the housing for any damage or ware. If defective, replace the booster housing and spool valve assembly.

Assembly

CAUTION: Parts must be kept VERY clean. If there is any reason to doubt the cleanliness of the components, re-wash before assembly.

Lubricate all seals and metal friction points with power steering fluid before assembly. Whenever the booster is disassembled, be sure that seals, tube inserts, spiral snap ring, check valve and ball are replaced.

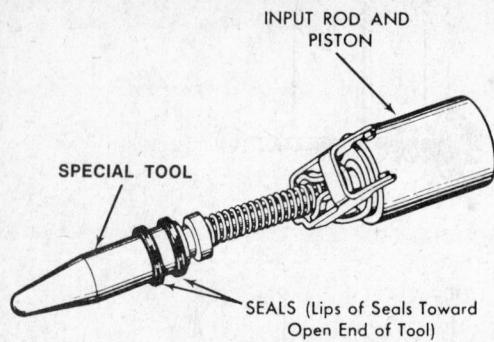

Installing input rod seals—Bendix Hydro-Boost (© G.M.C.)

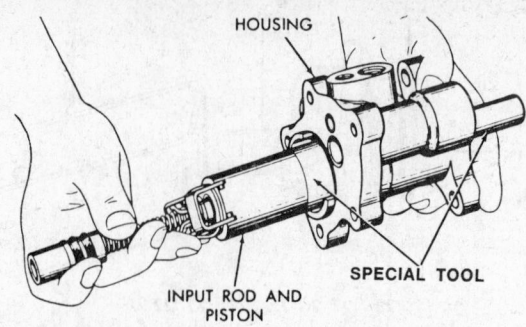

Installing input rod and piston assembly in booster-Bendix Hydro-Boost (© G.M.C.)

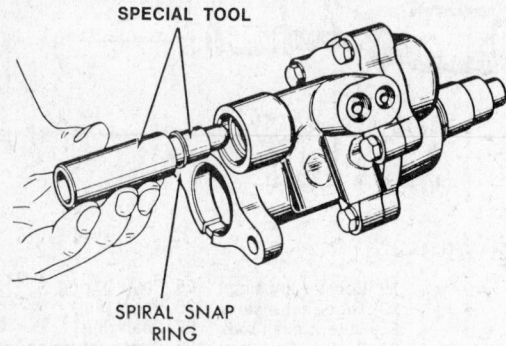

Installing spiral snap ring—Bendix Hydro-Boost (© G.M.C.)

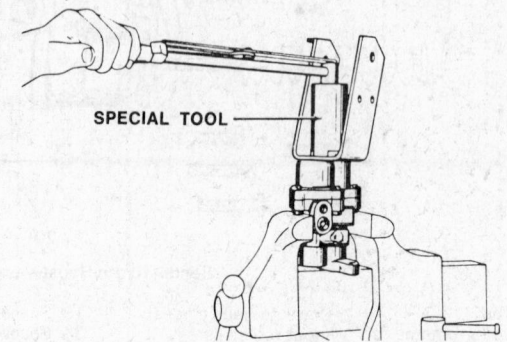

Installing linkage bracket nut (typical)—Bendix Hydro-Boost (© G.M.C.)

1. Position a tube seat in each booster port and screw a spare tube nut in each port to press the seat down into the port. Do not tighten the tube nuts in the port as this may deface the seats. Remove the spare tube nuts and check for aluminium chips in the ports. Be sure that there is no foreign matter in the ports.

2. Coat the piston bore and piston seal with clean power steering fluid. Assemble the seal in the piston bore. The lip of the seal must be towards the rear (away from the master cylinder mounting flange). Be sure that the seal is fully seated in the housing.

3. Lubricate the input rod end, input rod seals and the seal installer tool with clean power steering fluid. Slide the seals on the tool with the lip of the cups towards the open end of the tool. Slide the tool over the input rod end end down to the second groove; then slide the forward seal off the tool and into the groove. Assemble the other seal in the first groove. Be sure that both seals are fully seated.

4. Lubricate the piston and piston installing tool with clean power steering fluid. Insert the large end of the tool into the piston and the tool and piston into the piston bore, through the seal.

5. Position the 'O' ring on the accumulator check valve and coat the assembly with clean power steering fluid. Insert the check valve in the accumulator valve recess

in the housing flange. Place the ball and spacer in the same recess.

6. Place the 'O' ring on the changing valve plunger seat and insert the plunger into the seat. Dip the assembly in clean power steering fluid and insert it into the changing valve recess.

7. Coat the spool assembly with clean power steering fluid and insert in the spool bore. Be sure that the pivot pins on the upper end of the input rod lever assembly are engaged in the groove in the sleeve. Remove piston installing tool.

8. Separate the two components of the snap ring installation tool and place the spiral snap ring on the tool. Insert the rounded end of the installer into the spool bore. While pressing on the rear of the spool, slide the snap ring off the tool and into the groove near the forward end of the spool by pressing in on the tool sleeve. Check to be sure that the retaining ring is fully seated.

9. Place the housing seal in the groove in the housing cover. Join the booster housing and cover and secure with five attaching bolts. Tighten the bolts to 18-26 ft. lbs.

CAUTION: It is very important that the same cover attaching bolts are used as they are designed for the booster only. If they are damaged, replace with the same part numbers.

10. Place an 'O' ring on the spool plug. Insert the spool spring and

the spool plug in the forward end of the spool bore. Press in on the plug and position the snap ring in its groove in the spool valve bore.

11. Place the linkage bracket on the booster assembly. The tab on the inside of the large hole in the bracket should fit into the slot in the threaded portion of the booster cover.

12. Install the bracket nut with a staking groove outward on the threaded portion to the booster cover. Use special tool and tighten to 95-120 ft. lbs.

13. Insert a small punch into the staking groove of the nut, at the slot in the booster cover, and with a hammer stake the nut in place. Be sure that the threads on the nut are deformed so the nut will not loosen.

14. Position a new boot and grommet on the pedal rod. Moisten the grommet and insert the grommet end of the pedal rod into the input rod of the booster. When the grommet is fully seated, the pedal rod will rotate freely.

15. Install the boot on the booster cover.

Bendix Master Vac

Removal

1. Disconnect clevis at brake pedal to push rod.
2. Remove vacuum hoses from power cylinder.

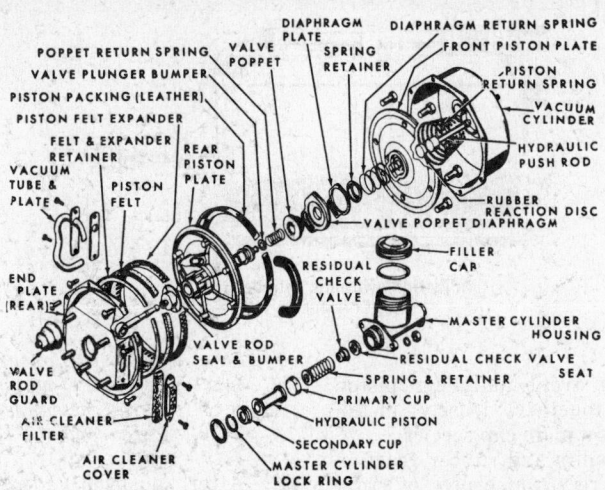

Bendix master vac unit (© Ford Motor Co.)

3. Disconnect hydraulic line from master cylinder.
4. Remove the four attaching nuts and lock washers that hold the unit to the firewall. Remove the power brake unit.

Disassembly

2. Remove four master cylinder to vacuum cylinder attaching nuts and washers.
2. Separate master cylinder from vacuum cylinder, then remove the rubber seal from the outer groove at end of master cylinder.
3. Remove the push rod from the power section. (Do not disturb adjusting screw.)
4. Remove push rod boot and valve operating rod.
5. Scribe alignment marks across the rear shell and vacuum cylinder. Remove all but two of the end plate attaching screws (opposite each other). Hold down

on the rear shell while removing the two remaining screws to prevent the piston return spring from expanding.
6. Scribe a mark across the face of the piston, to index the mark on the rear shell, and remove rear shell with vacuum piston and piston return spring.
7. Remove vacuum hose from vacuum piston and from vacuum tube on inside of rear shell. Separate rear shell from vacuum piston.
8. Remove air cleaner and vacuum tube assembly, and air filter from the rear shell.

9. Spring the felt retaining ring enough to disengage ring from grooves in bosses on rear piston plate.
10. Remove piston felt and expander ring from piston assembly.
11. Remove six piston plate attaching screws and separate front piston plate and piston packing from piston plate.
12. Remove valve return spring, floating control valve and diaphragm assembly, valve spring and diaphragm plate. Separate floating control valve spring-retainer and control valve diaphragm from control valve.
13. Remove rubber reaction disc and shim (if present) from front piston plate. *NOTE: Do not remove the valve operating rod and valve plunger from the rear piston plate unless it is necessary to replace defective parts. Normally, the next two steps can be omitted.*
14. When it is necessary to replace the valve operating rod or valve plunger, remove valve rod seal from groove in piston plate and pull seal over end of rod.
15. Hold piston with valve plunger side down and inject alcohol into valve plunger through opening around valve rod. This will wet the rubber lock in the plunger.

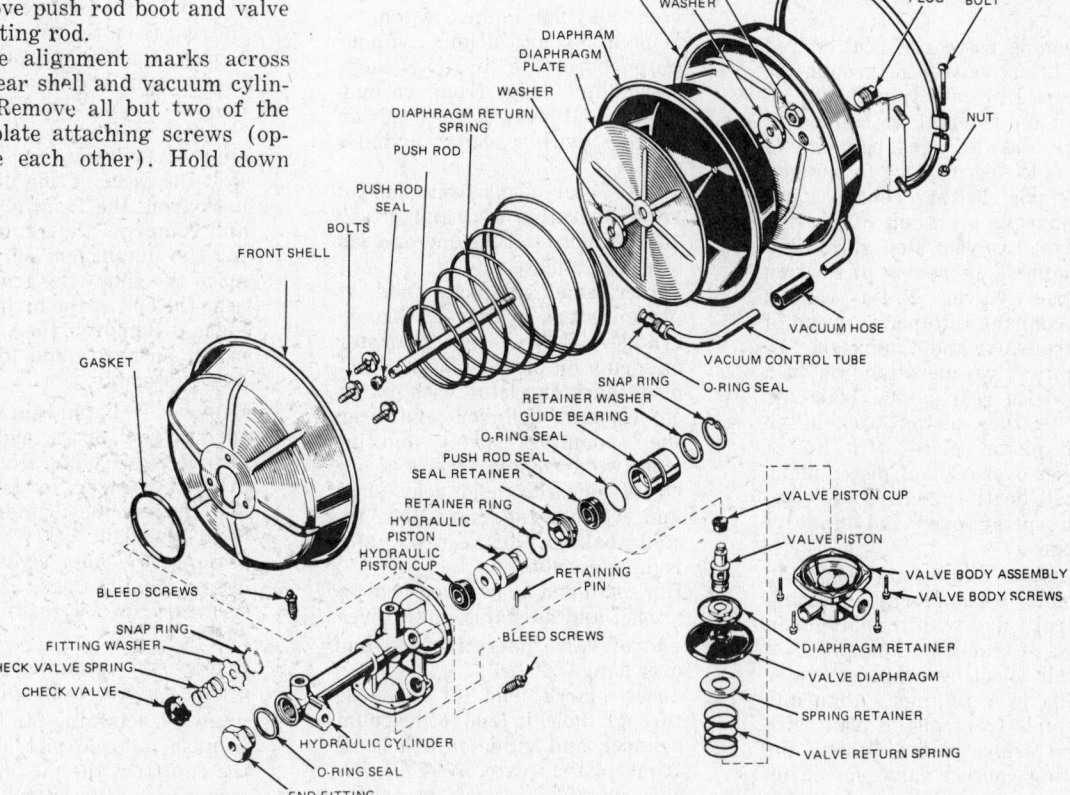

Bendix Single Diaphragm, Frame-Mounted Booster (© Ford Motor Co)

Then drive or pry valve plunger off the valve rod. *NOTE: If master cylinder is not to be rebuilt, omit Steps 16-19*

16. Remove snap-ring from groove in base at end of master cylinder.
17. Remove piston assembly, primary cup, retainer spring, and check-valve from master cylinder.
18. Remove filler cap and gasket from master cylinder body.
19. Remove secondary cup from master cylinder piston.

Cleaning Note

After disassembly, cleaning of all metal parts in satisfactory commercial cleaner solvent is recommended. Use only alcohol or Declene on rubber parts or parts containing rubber. After cleaning and drying, metal parts should be rewashed in clean alcohol or Declene before assembly.

Assembly

Steps 1-5 apply to a completely disassembled master cylinder. Otherwise, omit these steps (1-5).

1. Coat bore of master cylinder with brake fluid.
2. Dip secondary cup in brake fluid and install on master cylinder piston.
3. Dip other piston parts in brake fluid and assemble the piston. Install piston.
4. Install snap-ring into groove of cylinder.
5. Use new gasket and install filler cap.
6. Assemble valve rod seal on rod and insert valve rod through the piston. Dip valve plunger in alcohol and assemble to ball end of valve rod. Be sure ball end of rod is locked in place in plunger.
7. Assemble floating control valve diaphragm over end of floating control valve. Be sure diaphragm is in recess of floating control valve. Press control valve spring retainer over end of control valve and diaphragm.
8. Clamp valve operating rod in a vise with rear piston plate up. Lay leather piston packing on rear piston plate with lip of leather over edge of piston plate.
9. Install floating control valve return spring over end of valve plunger.
10. Assemble diaphragm plate to diaphragm and assemble floating control valve with diaphragm in recess of rear piston plate.
11. Install floating control valve spring over retainer. Align and assemble front piston plate with rear piston plate. Center the floating control valve spring on front piston plate and center valve plunger stem in hole of

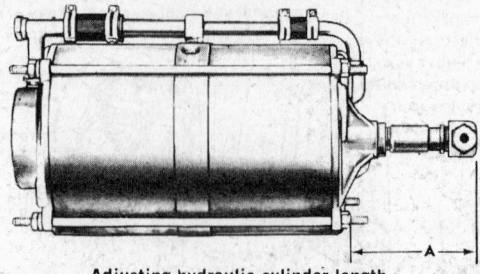

Adjusting hydraulic cylinder length
(© Chrysler Corp.)

piston.
12. Holding front and rear piston plates together, loosely install six piston plate cap screws.
13. Install shim and rubber reaction disc in recess at center of front piston plate. *NOTE: A piston assembling ring is handy in assembling the piston.*
14. Place the assembling tool over piston packing, turn piston assembly upside down and assemble the expander ring against inside lip of leather packing. Saturate felt with Vacuum Cylinder Oil or shock absorber fluid—type A, then assemble in expander ring. Assemble retainer ring over bosses on rear piston plate. Be sure retainer is anchored in grooves of piston plate.
15. Assemble air cleaner filter over vacuum tube of air cleaner and attach air cleaner shell in position with screws.
16. Slide vacuum hose onto vacuum inlet tube of piston and align hose to lay flat against piston.
17. Wipe a coat of vacuum cylinder oil on bore of cylinder. Remove assembling ring from vacuum piston and coat leather piston packing with vacuum cylinder oil.
18. Install rear shell over end of valve operating rod and attach vacuum hose to tube end on each side of end plate.
19. Center small diameter end of piston return spring in vacuum cylinder. Center large diameter of spring on piston. Check alignment mark on piston with marks on vacuum cylinder and rear shell, compress spring and install two attaching screws at opposite sides to hold rear shell and cylinder together. Now, install balance of screws and tighten evenly.
20. Dip small end of pushrod boot in alcohol and assemble guard over end of valve operating rod and over flange of shell.
21. Insert large end of pushrod through hole in end of vacuum cylinder and guide into hole of front piston plate. *NOTE: Before going on with assembly, check the distance from the*

outer end of the pushrod to the master cylinder mounting surface on the vacuum cylinder. This measurement should be 1.195-1.200".
22. After pushrod adjustment is correct, replace rubber seal in groove on master cylinder body.
23. Assemble master cylinder to the vacuum cylinder at four studs. Replace lock washers and nut and securely tighten.

Bendix Single Diaphragm Type Frame Mounted

Disassembly

1. Remove the booster unit and hydraulic cylinder from frame mounting bracket.
2. Scribe marks across front and rear shells and across flange of hydraulic cylinder. Disconnect the control tube nut from the control valve seat and remove the seal from the tube.
3. Remove the clamp band from the booster unit and disassemble the rear shell. *NOTE: The plug in the rear shell should be removed only if it is damaged.*
4. Roll the bead of the diaphragm back from the front shell flange and compress the return spring for the diaphragm slightly. Remove the snap ring from groove near the end of the hydraulic cylinder. Remove the hydraulic parts, push rod and diaphragm as an assembly.
5. Remove diaphragm return spring from piston end of push rod. Remove bolts securing hydraulic cylinder to front shell and remove the cylinder gasket from the shell. *NOTE: The diaphragm assembly should be removed from the push rod only if necessary to remove damaged parts.*
6. Remove the retaining ring from groove in hydraulic piston and press the retaining pin from hole in push rod and piston. Remove the cup from the piston and if a new seal is to be installed in end of push rod, remove old push rod

seal. *NOTE: Be careful to avoid damaging push rod.*

Carefully slide the seal retainer, seal, O-ring, guide bearing, retainer washer and snap ring from push rod.

7. Scribe marks across the flanges of valve body and housing and remove the four attaching bolts. Remove the valve body and remove cups from the control valve piston.

Assembly

1. Install check valve, spring, washer and snap ring in hydraulic cylinder end fitting. Next assemble O-ring seal and end fitting on the hydraulic cylinder.

2. Install cups, back to back, on control valve piston. Then assemble piston, diaphragm retainer and valve diaphragm. *NOTE: Make sure inner bead of diaphragm is seated in the piston groove.*

3. Install the spring retainer, with the flange down, on the spring in the valve body. Install the piston and diaphragm assembly on the retainer and press the outer bead of the diaphragm into the groove in the valve body.

4. Coat the valve piston with clean brake fluid and assemble the piston in the control valve cylinder bore. Align the scribe marks on the valve cylinder body and housing and attach with bolts. Torque bolts to 40-60 in. lbs.

5. If installing new seal on push rod, place new seal on clean block of wood, with the rubber side down. Place the push rod vertically on the seal stem and strike the threaded end with a soft mallet to seat the seal stem

in the push rod. *NOTE: Make sure the shoulders of the push rod and seal are in contact.*

6. Slide snap ring, retaining washer, guide bearing with O-ring seal in outer groove of guide, seal cup and seal retainer on push rod.

7. Attach the piston to the push rod with retaining ring and pin. Dip the piston cup in clean brake fluid and install on piston.

8. Install new gasket in groove at flange end of hydraulic cylinder and install cylinder on front shell with the hold down bolts.

9. Install the diaphragm return spring, with the large coil first, against the diaphragm plate. Lubricate the cylinder bore with clean brake fluid and carefully insert the piston, cups and seals into cylinder bore. Roll back the edge of the diaphragm and press against the diaphragm to compress the return slightly. When push rod and parts are installed all the way into the cylinder bore, install the retaining snap ring. *NOTE: Be sure the snap ring is seated properly before releasing pressure on the spring.*

10. Coat both sides of the diaphragm lightly with talcum powder or silicone lubricant. Align scribe marks made on front and rear shells and press the rear shell flange and diaphragm bead into position against the front shell flange.

11. Install the clamp band on the shells and secure with bolt. Install a new seal on the control vacuum tube and assemble the

tube and hose onto the rear shell tube and tighten nut.

12. Reinstall on vehicle and check for vacuum leaks and road test to check for proper operation.

NOTE: Be sure to bleed all air from hydraulic cylinder and brake lines before attempting to road test vehicle.

Bendix Piston Type Frame Mounted

Disassembly

1. Remove all vacuum and hydraulic lines to booster and remove unit from vehicle.

2. Scribe marks across end plate and vacuum cylinder, also across the control valve body and flange on the end plate.

3. Clamp the end nut of the hydraulic cylinder in a soft jawed vise and unscrew the lock nut on the control tube and remove the tube and O-ring seal.

4. Remove the four bolts securing the valve body to the end plate and remove the control valve body, valve return spring and diaphragm from the end plate.

5. From the valve body remove the snap ring, tube and cover, gasket, the two poppet return springs and the valve seal. From the valve housing remove the valve poppet seal.

6. Remove the four hook bolts that hold the shell to the end plate and separate the shell from the end plate.

7. Compress the return spring and hold in place with two hook type clamps.

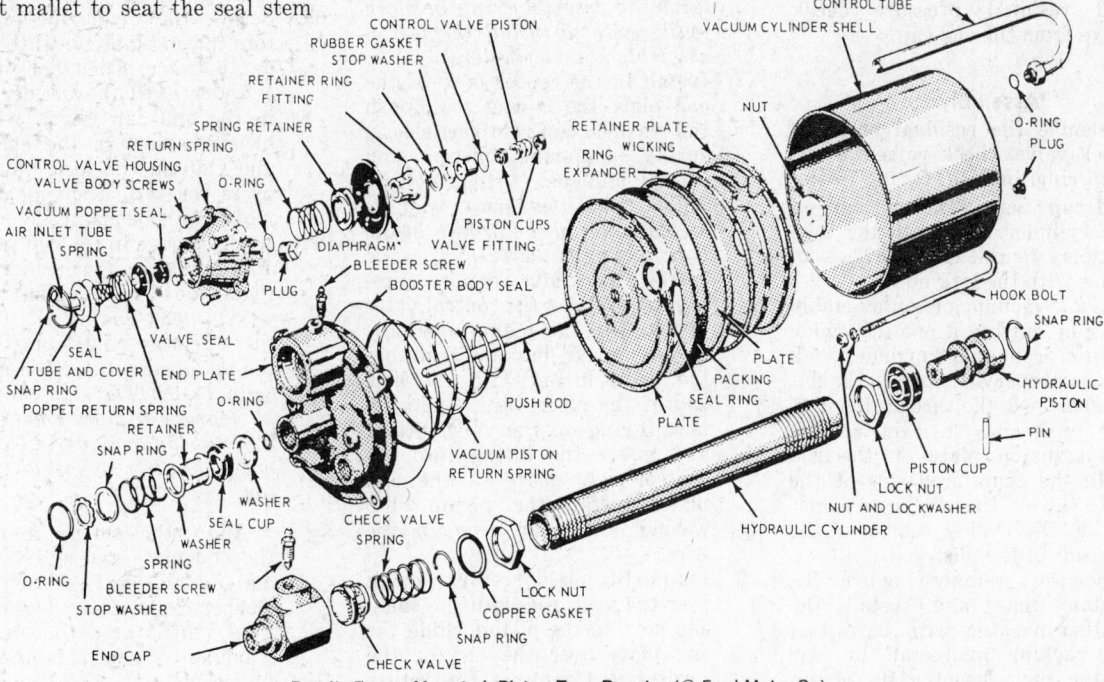

Bendix Frame-Mounted, Piston-Type Booster (© Ford Motor Co)

8. Loosen the lock nut on the hydraulic cylinder and separate the cylinder from the end plate. Remove the retaining ring from the hydraulic piston and press out the retaining pin in the piston. Remove the piston from the push rod and remove the piston cup from the piston.

9. Compress the piston return spring and remove the hook clamps. Separate the piston and push rod assembly from the end plate, and remove the return spring from the push rod.

10. From the end plate remove the seal O-ring, stop washer, snap ring, spring, sleeve, retainer, seal cup, push rod washer and small seal O-ring. On the opposite of the end plate remove the large seal O-ring.

11. From the control valve opening in the end plate, remove the retaining snap ring, piston stop washer and valve fitting. Remove the O-ring seal from the valve fitting.

12. Remove the hydraulic piston from the valve fitting and remove the piston cups from the piston.

13. To prevent damaging the push rod when disassembling the vacuum power piston, clamp the push rod in a soft jawed vise. Next remove the nut, piston felt retainer, packing ring, packing wick, rear plate, packing, front plate and washer from the push rod.

14. To remove the end cap on hydraulic cylinder, loosen the lock nut on the cylinder and remove the cap. Remove the seal O-ring, snap ring, check valve spring and residual pressure check valve from the end cap.

Assembly

1. Assemble the residual pressure check valve, check valve spring, snap ring and seal O-ring in the end cap. Screw the end cap on the cylinder tube until the tube bottoms in the cap and lock in place with the lock nut.

2. Make a vacuum piston assembly ring by cutting a one inch wide section from an old cylinder shell of the proper size. Install the flat washer over the threaded end of the push rod, then install the front piston plate on the rod with the chamfered side of the hole away from the washer. Guide the O-ring seal over the threads of the push rod.

3. Place the assembly ring over the piston plate and install the leather packing with the lip of the packing up. Install the rear piston plate keeping the cham-

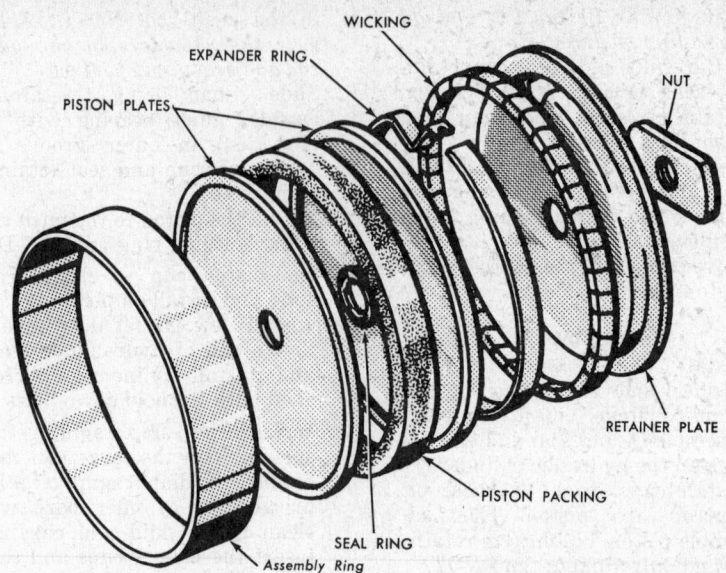

Bendix Vacuum Piston (© Ford Motor Co)

fered side of the hole next to the seal.

4. Cut the packing wick to the required length and saturate with a good quality oil. Install the wick against the inner lip of the leather packing.

5. Install the packing expander ring inside the wick with the gripping point up. Put the notch, at the loop end of the ring, under the clip on the opposite end of the ring. Install the piston retainer with the cutout portion over the loop of the expander ring.

6. Install the nut on the threaded end of the push rod and tighten securely. *NOTE: Be careful the piston retainer does not shift when tightening the nut. Also leave the assembly ring in place until ready to install the piston assembly in the cylinder.*

7. Install in the center bore of the end plate the O-ring seal, push rod washer, cup retainer, sleeve, spring, snap-ring, piston stop washer and seal O-ring. *NOTE: Make sure the snap ring is seated fully, in the groove in the bore of the end plate.*

8. Install the valve piston cups, back to back, on the control valve piston. Dip the piston assembly in clean brake fluid and install in the valve fitting with the hole end of the piston first. Install a new O-ring on the valve fitting and insert the fitting into the control valve bore in the end plate. Install the piston stop washer and snap ring in the bore.

9. Install the piston return spring over the push rod with the small end next to the piston. Slide the end plate over the end of the push rod. Compress the return

spring slightly to project the push rod approximately 2 inches through the end plate, and hold in place with two hook type clamps. Install the hydraulic piston on the end of the push rod and attach with the retainer pin. Slide the snap ring into its groove to hold the pin in place.

10. Dip the piston cup in clean brake fluid and install the cup on the piston with the lip away from the end plate. Next dip the piston assembly in clean brake fluid and coat the inside of the hydraulic cylinder with fluid. Guide the hydraulic cylinder over the piston being careful not to turn the lip of the seal cup backwards when installing.

11. Screw the hydraulic cylinder into the end plate until the cylinder bottoms. Then back off the cylinder until the bleeder screw in the end cap aligns with the bleeder screw in the end plate and tighten the lock nut. Compress the return spring and remove the hook clamps.

12. In the groove in the end plate install the large O-ring seal. Coat the inside of the vacuum cylinder with a good quality oil. Remove the assembly ring from the piston and insert the piston into the cylinder. *NOTE: Tip the piston 45 degrees or more when sliding it into the cylinder. This will make it easier to install the piston.*

Align the scribe marks on the end plate and cylinder shell and install the hook bolts, lock washers and nuts. Tighten the nuts a little at a time to avoid warping the end plate or the cylinder shell.

13. Install the valve diaphragm and spring retainer on the diaphragm plate. Install the valve poppet seal over the end of the valve seal and place the assembly in the valve body.

14. Install the poppet return springs with the small end of the small spring over the button on the valve seal. In the recess in the valve body, install the O-ring seal and attach the tube and cover with the snap ring.

15. In the groove in the control vacuum port of the valve body install the small O-ring seal.

16. Install the valve diaphragm and spring retainer assembly on the end plate flange and position the control valve return spring on the retainer. Position the control valve housing on the spring and secure with the four hold down bolts. *NOTE: Be sure to align the scribe marks made on the housing and end plate before assembling the housing to the end plate.*

17. Install the O-ring seal on the end of the vacuum tube and install the tube with one end in the control valve body and the other end secured to the rear of the shell with the tube nut.

18. Install the unit on the vehicle and check for vacuum leaks and road test for proper operation of unit.

NOTE: Be sure to bleed all air from the hydraulic cylinder and brake lines before road testing the vehicle.

Bendix Dual Diaphragm Type

This unit features a direct pedal connection to a vacuum unit mounted on the firewall, with the master cylinder directly mounted to booster.

The booster chamber contains two diaphragms and is under constant engine vacuum. When brakes are applied, the control valve is opened to allow atmospheric pressure behind both diaphragms. This provides the power boost to the master cylinder.

This vacuum-suspended system provides reserve against fade. Pedal linkages are eliminated, no additional vacuum storage tanks are needed.

NOTE: Do not attempt to disassemble the booster. It is serviced only by the dealer.

Bendix Tandem Diaphragm Type Frame Mounted

Disassembly

1. Remove all hydraulic and vacuum lines attached to booster and hydraulic cylinder. Then remove the unit from frame brackets and remove from vehicle.

2. Scribe marks across both clamp rings onto the shell surfaces, also across top of hydraulic cylinder flange onto the front shell. Scribe marks across control valve body and housing below hydraulic cylinder.

3. Disconnect the control tube and nut from control valve port and remove the three hose clamps and tee. Remove the seal ring from the control tube.

4. Remove the rear clamping ring and remove the rear shell.

5. Remove the front clamping ring and separate the front, center and rear shells and remove the diaphragm assemblies.

6. Clamp the hydraulic cylinder in a soft jawed vice being careful to avoid damaging cylinder.

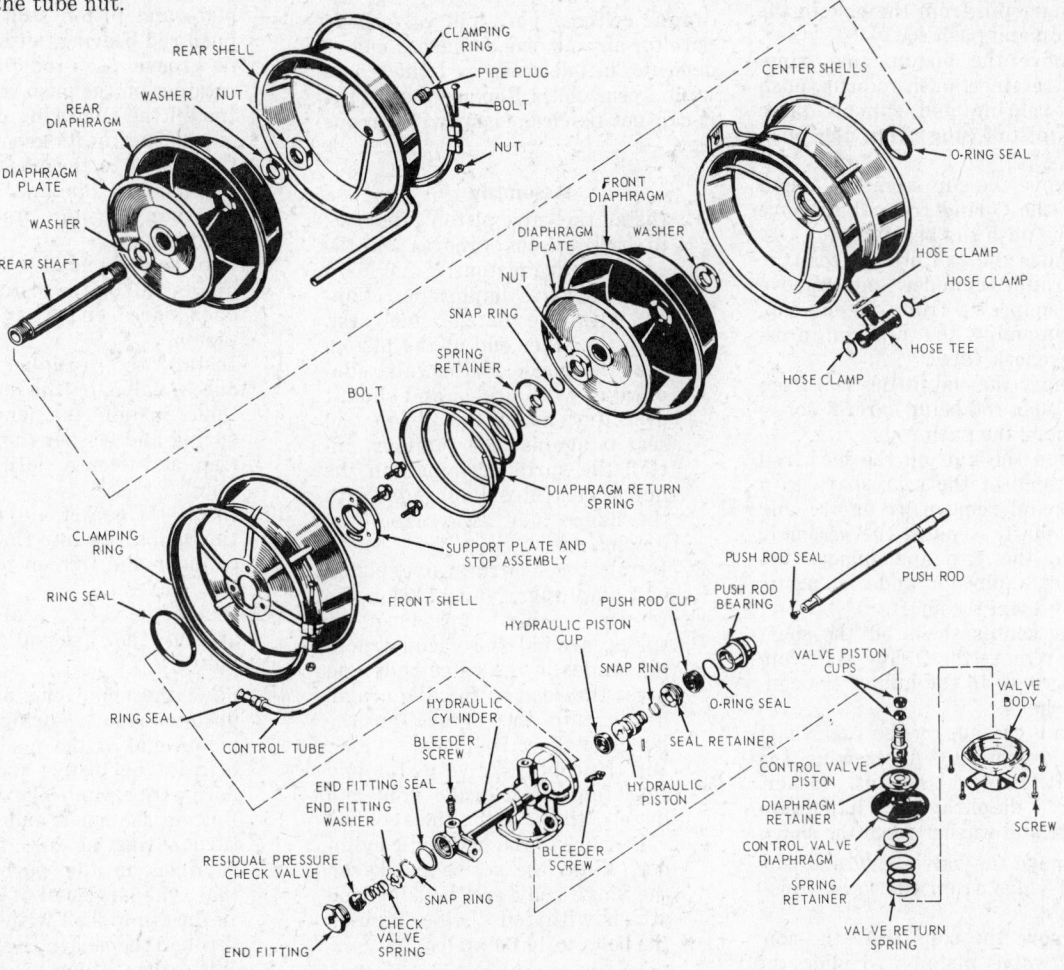

Bendix Tandem Diaphragm—Frame-Mounted Booster (© Ford Motor Co)

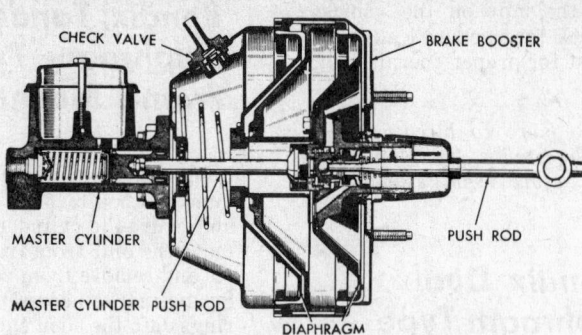

CHECK VALVE

BRAKE BOOSTER

MASTER CYLINDER

PUSH ROD

MASTER CYLINDER PUSH ROD

DIAPHRAGM

Cutaway view of brake booster and master cylinder
(© Chrysler Corp.)

7. Press on the spring retainer to compress the diaphragm return spring and remove the three bolts that hold the hydraulic cylinder to the front shell and support plate. Then carefully release the pressure on the return spring and pull the push rod and hydraulic piston from the cylinder.

8. Remove the return spring from the piston end of the push rod. *NOTE: Do not remove the spring retainer except to replace damaged parts.*

9. Remove the snap ring from the groove in the piston and press plunger pin from the hole in the piston and push rod.

10. Remove the piston, snap ring, seal retainer, push rod cup, push rod bearing and support plate and piston stop from the push rod.

11. Remove the cup from the piston and the O-ring from the groove in the push rod bearing.

12. Remove the end fitting from the hydraulic cylinder and remove the snap ring from the end cap. Disassemble the residual pressure check valve.

13. Remove the end fitting seal from the push rod being careful not to damage the push rod.

14. Clamp the nut, on the push rod seat end of the rear shaft, in a vise and remove by unscrewing the shaft. Remove the assembly from the vise and remove the front diaphragm and diaphragm plate from the shaft.

15. Slide center shells off the shaft and remove the O-ring seal from its groove in the hub of the center shells.

16. Clamp the nut on the rear shaft and remove by unscrewing the shaft. Remove the nut, washer, rear diaphragm, diaphragm plate and washer from the shaft.

17. Remove the valve body and control valve from the valve housing.

18. Remove the cups from the control valve piston and slide the control valve diaphragm and re-

tainer off the opposite end of the piston.

Cleaning and Inspection

Clean all metal parts in clean metal parts cleaner. Discard any old parts that are to be replaced with new ones. Clean all hydraulic parts in clean brake fluid. Check the diaphragms for cracks, tears and kinks and replace any diaphragms that are questionable. Inspect all metal and plastic parts for nicks, cracks, scores or burrs and replace any damaged parts. Check the shells for cracked or broken welds, dents or cracks. *DO NOT attempt to disassemble the center shell assembly.* Inspect the hydraulic cylinder bore and valve body bore for any surface damage. Remove deposits, pitted areas or light scores with crocus cloth. Replace the part if it can not be cleaned up with crocus cloth.

Assembly

1. Install the valve piston cups back to back in the grooves on the control valve piston.

2. Slide the valve diaphragm retainer, with the flange side first, onto the other end of the piston. Wet the inside of the valve diaphragm with alcohol and slide it over the end of the piston and seat it against the retainer. Install the spring retainer on the hub of the valve diaphragm with the flange side away from the diaphragm.

3. Install the control valve piston and diaphragm assembly on the return spring. Position the spring around the vacuum poppet guides in the valve body and press the bead of the diaphragm firmly into the groove on the flange of the valve body.

4. Dip the control valve piston and cups into clean brake fluid and install them into the control valve bore in the hydraulic cylinder. Align the scribe marks on the valve body and housing and attach with four bolts. Tighten the bolts to 40-60 in. lbs.

5. Assemble the washer, rear diaphragm plate, diaphragm and

washer onto the rear shaft with the holes in the shaft towards the diaphragm.

6. Install the nut on the end of the shaft and tighten to 10-15 ft. lbs. Stake the nut in two places to prevent any movement.

7. Install the O-ring seal in its groove inside the hub of the center shells. With a silicone lubricant, coat the seal and bearing and the outer surface of the rear shaft.

8. Insert the front end of the rear shaft through the middle of the center shells.

9. Install the washer, front diaphragm and diaphragm plate on the end of the shaft. Screw on retaining nut and tighten to 10-15 ft. lbs. Stake the nut in two places to prevent any movement.

10. To install a new push rod seal, in end of push rod, place new seal face down on a clean block of wood. Place the push rod upright on the seal stem and strike end of rod with a soft mallet to seat the seal. *NOTE: Be sure that the shoulders of the seal and push rod are in contact.*

11. Dip all hydraulic parts, push rod and push rod bearing in clean brake fluid. Install the support plate and piston stop assembly, push rod bearing, with O-ring in its groove, push rod cup and seal retainer on the push rod.

12. Install the snap ring on the piston but not in its groove. Attach the piston to the push rod with the plunger pin and then slide the snap ring into its groove in the piston.

13. Dip the piston cup in brake fluid and install on the piston with the open flared end away from the piston.

14. Install the residual pressure check valve in the end fitting, then install the check valve spring and washer in the end fitting and secure with the snap ring.

15. Install the gasket onto the end fitting and screw into the hydraulic cylinder and tighten to 50-85 ft. lbs.

16. Install the O-ring seal in groove around the hydraulic cylinder flange.

17. Slide the small end of the diaphragm return spring over the piston end of the push rod. Lubricate the piston and cylinder bore with clean brake fluid.

18. Bottom the small end of the return spring against the spring retainer on the push rod and place the large coil of the spring in the front shell with the piston through the hole in the shell.

19. With the return slightly compressed, guide the piston, seal re-

tainer, push rod cup, and push rod bearing into the cylinder bore. Seat the cylinder flange against the front shell, make sure the O-ring is in place. Place the support plate and stop plate on the opposite side of the shell and secure the stop assembly and front shell to the cylinder with the three securing bolts. Release the pressure on the spring.

20. On the front and rear diaphragm beads put a light coat of talcum powder or silicon lubricant.
21. Guide the rear shaft onto the push rod and align the scribe marks made on the front shell flanges. Press the shells together and seat the bead of the diaphragm all the way around in the shell flanges.
22. Install the clamp ring on the shell flanges and align the scribe marks. Tighten the clamp screw to 30-40 in. lbs.
23. Align the scribe marks on the rear shells and press them together making sure the diaphragm bead is in the shell flange all the way around. Install the clamp ring, aligning the marks, and tighten the clamp bolt to 30-40 in. lbs.
24. Install the hose tee to the control tube on the rear shell and to hose nipple on the center shell.
25. Install the seal ring on the end of the control tube and nut assembly and attach the tube to the hose tee with a hose clamp. Screw the nut onto the control valve port and tighten to 80-120 in. lbs.
26. Install unit on vehicle and test for vacuum leaks and road test for proper operation.

NOTE: Be sure to bleed hydraulic cylinder of all air before attempting to road test vehicle.

Bendix Single Diaphragm Booster

Disassembly

1. Scribe a line across the front and rear housings for reassembly.
2. Pull the piston rod from the front housing and remove the seal.
3. Attach a holding fixture to the front housing and clamp the base in a vise with the power section up.
4. Loosen the locknut and remove the pushrod device and locknut.
5. Remove the mounting bracket from the rear housing.
6. Remove the dust boot retainer, dust boot and silencer from the diaphragm plate extension.
7. The edge of the rear housing contains twelve lances. Four of

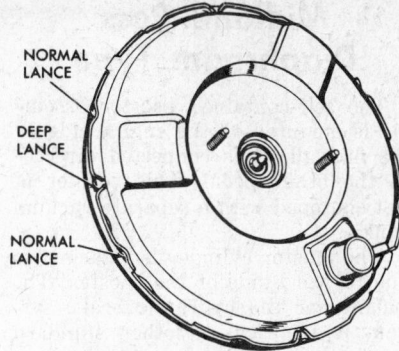

NORMAL LANCE

DEEP LANCE

NORMAL LANCE

Lances in the rear housing (© G.M.C.)

these lances (one in each quadrant) are deeper than the other lances. The metal that forms the four deep lances must be partially straightened so that the lances will clear the cutouts in the front housing.

NOTE: If the metal tabs that form the deep lances crack or break during straightening, the housing must be replaced.

8. Place a spanner wrench over the studs on the rear housing and attach with nuts and washers.
9. Press down on the spanner wrench and rotate the rear housing clockwise to separate the two housings.

NOTE: It may be necessary to tap the rear housing lightly with a plastic hammer to loosen.

10. Lift the rear housing assembly from the unit.
11. Use a small screwdriver and carefully remove the air filter element from the diaphragm plate extension.
12. Separate the diaphragm plate assembly from the rear housing and disassemble the plate assembly.
13. Remove the rolling diaphragm from the groove in the diaphragm plate hub.

NOTE: Protect the diaphragm from oil, and nicks.

14. Hold the diaphragm plate in a horizontal position and depress the push rod approximately 1/16" and rotate the piston so the air valve lock will fall from its location. Remove the air valve pushrod assembly and the reaction disc.
15. If a new seal is needed, support the outer surface of the rear housing and drive out the seal with a suitable tool.

NOTE: Do not reuse old seal once it has been removed.

16. Remove the check valve and grommet from the front housing and discard.
18. Remove the front housing from the holding fixture.

Assembly

1. Use clean brake fluid and thoroughly clean all reusable brake parts.
2. Inspect all rubber parts and replace if nicked, cut or damaged.
3. When rebuilding, make sure that no grease or mineral oil comes in contact with any of the rubber parts.
4. Install a new check valve grommet in the front housing.
5. Position and secure the holding fixture to the front housing and place in a vise.
6. Place the rear housing on a block of wood, stud side down, and position the housing seal in the center hole. Using the special installing tool seat the seal in the recess of the rear housing.
7. Assemble the diaphragm plate assembly:
   a. Apply a silicone lubricant to the outside diameter of the diaphragm plate and extension, to the bearing surfaces of the air valve and to the

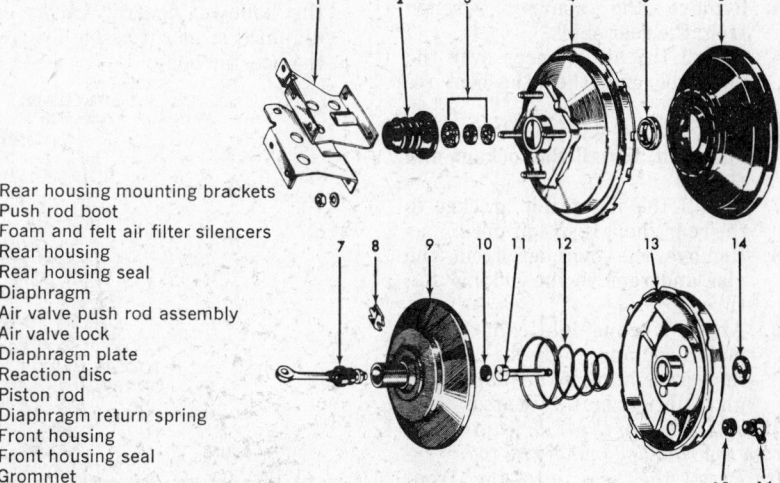

1 Rear housing mounting brackets
2 Push rod boot
3 Foam and felt air filter silencers
4 Rear housing
5 Rear housing seal
6 Diaphragm
7 Air valve push rod assembly
8 Air valve lock
9 Diaphragm plate
10 Reaction disc
11 Piston rod
12 Diaphragm return spring
13 Front housing
14 Front housing seal
15 Grommet
16 Check valve

Bendix single diaphragm booster (© G.M.C.)

outer edge of the valve poppet. Insert the air valve and push-rod assembly in the extension of the diaphragm plate.

b. Depress the pushrod slightly and install the air valve. Make sure the lock indexes and retains the air valve.

c. Install the rolling diaphragm in the groove of the diaphragm plate.

d. Apply silicone lubricant to the surface of the reaction disc and position the disc in the center bore of the diaphragm plate. Use the piston rod to seat the disc in the bore.

NOTE: It is important that the disc be fully seated before removing the piston rod.

8. Apply silicone lubricant to the inside diameter of the rear housing seal and the diaphragm bead contact surface of the rear housing. Install the diaphragm plate assembly in the rear housing.

9. Position the air filter element over the pushrod and into the diaphragm plate extension. Install the air filter retainer.

10. Attach the base of the holding fixture to the front housing and clamp the base in a vise with the power section up.

11. Place a spanner wrench over the studs on the rear housing.

12. Place a diaphragm plate return spring in the front housing and position the rear housing assembly on the front housing with the small end of the spring downward. Align the scribe marks and lock in place.

13. Press down on the spanner wrench and rotate the rear housing counterclockwise to assemble the two housings.

NOTE: Bend the lances in on the rear housing. If the tangs crack or break, it will be necessary to replace that half of the housing.

14. Remove the spanner wrench from the rear shell.

15. Install the air silencer over the push rod end, then the boot retainer.

16. On vehicles with a clevis type push rod, install the locknut and clevis.

17. Install the mounting bracket to the rear shell, if so equipped.

18. Remove the cylinder from the vise and remove the holding fixture.

19. Apply silicone lubricant to the piston rod and guide the rod into the center bore of the diaphragm plate until it is fully seated.

NOTE: Keep the lubricant away from the rounded end of the rod.

20. Press the seal into the front housing until it is bottomed in the recess of the housing.

Midland Ross Diaphragm Type

The self-contained booster assembly is mounted on the engine side of the firewall. It is connected directly to the brake pedal. This booster is not equipped with a separate vacuum tank.

The master cylinder is attached to the forward side of the booster. The balance of the hydraulic brake system is identical to other standard service brakes.

Booster Repairs

1. Separate master cylinder from booster body.

2. Remove air filter cover and hub and the filter from the booster body.

3. Remove the vacuum manifold mounting bolt, manifold, gaskets and vacuum check valve from the booster body.

4. Disconnect the valve operating rod from the lever by removing its retaining clip, washers, and pivot pin.

5. Disconnect the lever from the booster end plate brackets by removing its retaining clip, washers, and pivot pin.

6. Remove two brackets from the end plate.

7. Remove the rubber boot from the valve operating rod.

8. To remove the bellows, control valve, and diaphragm assemblies, remove large C-ring that holds the rear seal adapter assembly to the booster end plate.

9. Scribe matching lines on the booster body and the end plate. Then remove the ten retaining screws. Tap the outside of the plate with a soft hammer and separate the plate from the booster body.

10. Push the bellows assembly into the vacuum chamber and remove the bellows, control valve, and diaphragm as an assembly from the booster body.

11. Remove the outer O-ring from the control valve hub.

12. To disassemble the bellows, pushrod, and control valve assemblies, remove the large bellows retaining ring, bellows, bellows retainer, and support ring from the diaphragm and valve assembly.

13. Remove the retainer and support ring from the bellows.

14. Remove pushrod assembly, the reaction lever and ring assembly, and the rubber reaction ring from the control valve hub.

15. Remove the reaction cone and cushion ring from the pushrod assembly. Then disassemble the reaction levers from the ring.

16. Remove the two plastic plunger guides from the control valve plunger. Then remove the retainer that holds the reaction load ring and atmospheric valve on the control valve hub.

17. Slide the reaction load ring and atmospheric valve from the control valve hub.

18. Separate the control valve hub and the plunger assembly from the diaphragm by sliding the plunger and rear seal adapter from the rear of the hub. Then remove the hub outer O-ring from the front side of the diaphragm.

19. To disassemble the control valve plunger, remove the hub rear seal adapter from the valve plunger assembly, and remove the seal from the adapter.

20. Remove the O-rings, the seal, and the fiber gaskets from the plunger.

21. If the plunger assembly needs to be replaced, hold the plunger and pull out the valve operating rod with pliers. Do not separate the operating rod and plunger unless the plunger is to be replaced.

Assembly

1. If valve operating rod was removed for replacement of

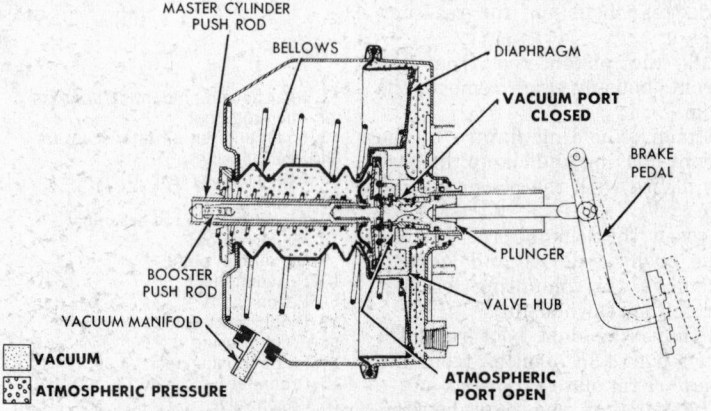

MASTER CYLINDER PUSH ROD
BELLOWS
DIAPHRAGM
VACUUM PORT CLOSED
BRAKE PEDAL
BOOSTER PUSH ROD
VACUUM MANIFOLD
PLUNGER
VALVE HUB
ATMOSPHERIC PORT OPEN
VACUUM
ATMOSPHERIC PRESSURE

Midland Diaphragm Type Booster Applied

plunger, install a new rubber bumper and spring retainer on the rod before installing it on the replacement plunger. Then push the rod firmly until it bottoms in the plunger.

2. Install fiber gaskets, plunger seal, and the two O-rings on the plunger assembly.

3. Install the valve hub rear seal in the adapter assembly with the sealing lip toward the rear. Then slide the adapter assembly onto the plunger with the small diameter end of the hub toward the rear.

4. To assemble the control valve, pushrod, and bellows assemblies, install the hub outer O-ring. Then install the plunger with the seal adapter and the hub on the diaphragm. To do this, hol the hub on the front side of the diaphragm and insert the plunger assembly in the hub from the rear side of the diaphragm.

5. Install atmospheric valve and then the reaction load ring onto the plunger and hub. Compress the valve spring, and install the load ring retainer into the groove of the plunger.

6. Install two plastic plunger guides into their grooves on the plunger.

7. Install rubber reaction ring into the valve hub so that the ring locating knob indexes in the notch in the hub, with the ring tips toward the front.

8. Assemble the reaction lever and ring assembly, and install the assembly into the valve hub.

9. Install the reaction cone and cushion ring on the pushrod. Then install the pushrod assembly on the valve hub so that the plunger indexes in the rod.

10. Assemble the bellows, retainer, and support ring. The ring should be positioned on the middle fold of the bellows.

11. Position the bellows assembly on the diaphragm, and secure it with the retaining ring. Make sure the retaining ring is fully seated.

12. Install the bellows, control valve, and diaphragm assemblies with a screwdriver, moving the booster body retaining screw tapping channel just enough to provide a new surface for the self-tapping attaching screws.

13. Install the diaphragm, the control valve components, and the bellows as an assembly into the booster body. (Be sure the lip of the diaphragm is evenly positioned on the retaining radius of the booster body.) Pull the front lip of the bellows through the booster body, and position it around the outer groove of the body.

14. Install O-ring in the front side of the end plate, and locate the plate on the booster body. Align the scribed lines, compress the two assemblies together with a clamp. Then install all ten self-tapping attaching screws.

15. Install the large C-ring onto the rear seal adapter at the rear side of the end plate.

Pushrod Adjustment

The pushrod has an adjusting screw to maintain the correct relationship between the control valve plunger and the master cylinder piston after the booster is completely assembled. If this screw is not properly adjusted, the brakes may drag.

To check adjustment of the screw, make a gauge to the dimensions shown. Place this gauge against the master cylinder mounting surface of the booster body. The pushrod screw should be adjusted so that the end of the screw just touches the inner edge of the slot in the gauge.

Booster Installation

1. Install rubber boot on the valve operating rod.

2. Position the two mounting brackets on the end plate, and install on retaining nuts.

3. Connect the lever assembly to the lower end of the mounting brackets with its pivot pin. Then install the spring washer and retaining clip.

4. Connect the valve operating rod to the upper end of the lever with its pivot pin, washer, and retaining clip.

5. Install the vacuum check valve, the vacuum manifold, the two gaskets, and the mounting bolt. Torque the mounting bolt to 8-10 ft lbs.

Midland Diaphragm Type Frame Mounted

The Midland frame mounted booster is a remote type, without mechanical operation, utilizing vacuum to boost the hydraulic pressure between master and wheel cylinder.

Removal

1. Remove all hydraulic lines from the booster unit hydraulic cylinder.

2. Remove all vacuum lines from the booster unit and remove the support bracket bolts.

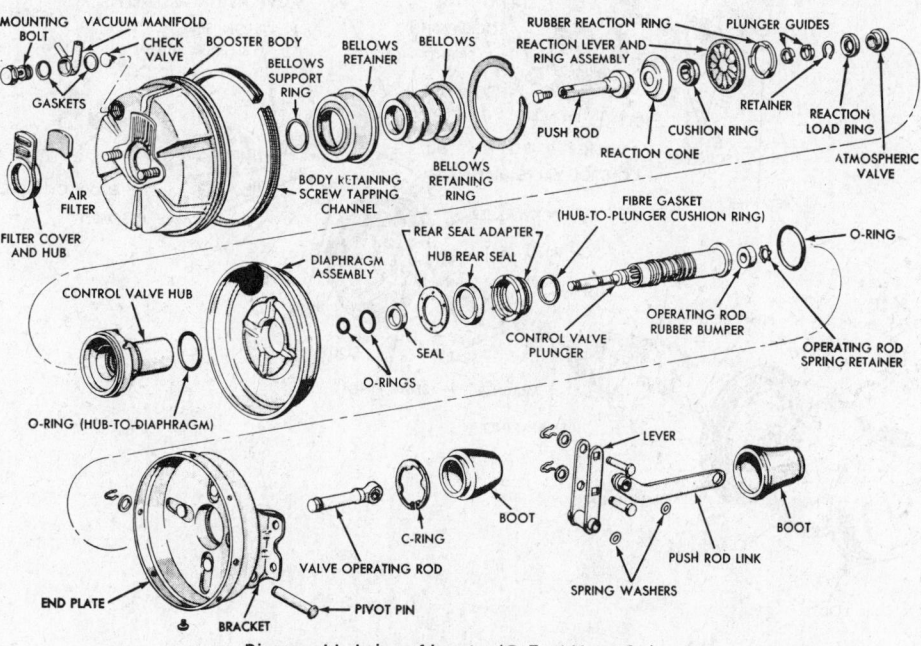

Disassembled view of booster (© Ford Motor Co.)

3. Remove the unit from the vehicle and place on a clean work bench.

Disassembly

1. Remove the control tube from the control valve body and the rear body.
2. Scribe marks across the diaphragm body and across the flanges of the slave cylinder body and the control valve body.
3. Remove the body clamp carefully, and remove the rear body and diaphragm with the return spring.
4. Remove the push rod, spring retainer and collar from the return spring.
5. Scribe a line across the valve body cover and the valve body, and remove the valve body cover and gasket.
6. Remove the valve body, spring, and the piston and diaphragm assembly from the slave cylinder.
7. From the end of the slave cylinder remove the end plug, copper gasket, spring, spring seat and spring retainer.
8. Remove the piston cup and piston assembly from the cylinder. *NOTE: If the assembly does not fall free from the cylinder it may be pushed out by inserting the push rod through the bushing.*
9. From the hydraulic piston remove the check valve, check valve retainer and the return spring.
10. Hold the cylinder in a soft jawed vise and remove the push rod bushing, lockwasher and front body.
11. Remove the gasket, rubber seal

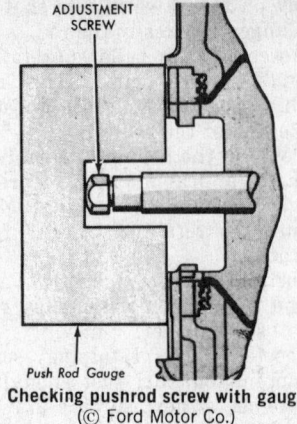

Checking pushrod screw with gauge
(© Ford Motor Co.)

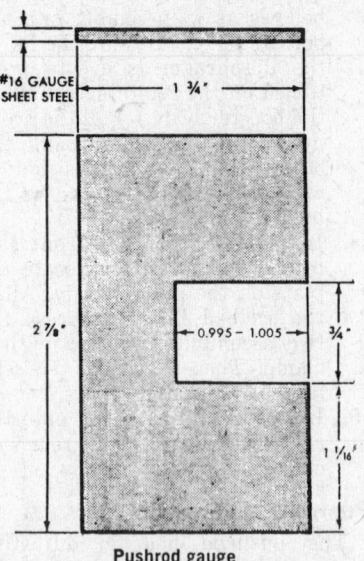

Pushrod gauge
(© Ford Motor Co.)

and transfer bushing from the slave cylinder body. From the bushing remove the two push rod bushing snap rings, and remove the washer and two seals. From the outside of the push rod bushing remove the O-ring seal.

12. From the lower end of the control valve piston remove the seal, also remove the seal from the piston boss.
13. Remove the retaining nut from the piston boss and remove the diaphragm plate and control valve diaphragm.
14. Remove the screw, lockwasher, spacer, spring, disc, and the seal from the control valve body.

Assembly

1. Install new spring in the control valve body and assemble the spring and spacer in the valve body. Secure with the screw and locknut.
2. Secure the control valve diaphragm and plate in place with the attaching nut.
3. On the control valve piston install the piston seal.
4. In the hydraulic piston install the check valve spring, check valve and retainer, making sure that the valve floats free in the bore and does not bind.
5. On the front end of the slave cylinder body, install the transfer bushing, seal and gasket.
6. In the push rod bushing install the push rod seals, washer and snap rings. *NOTE: Install the push rod seals with the open end of the seal towards the slave cylinder body.*

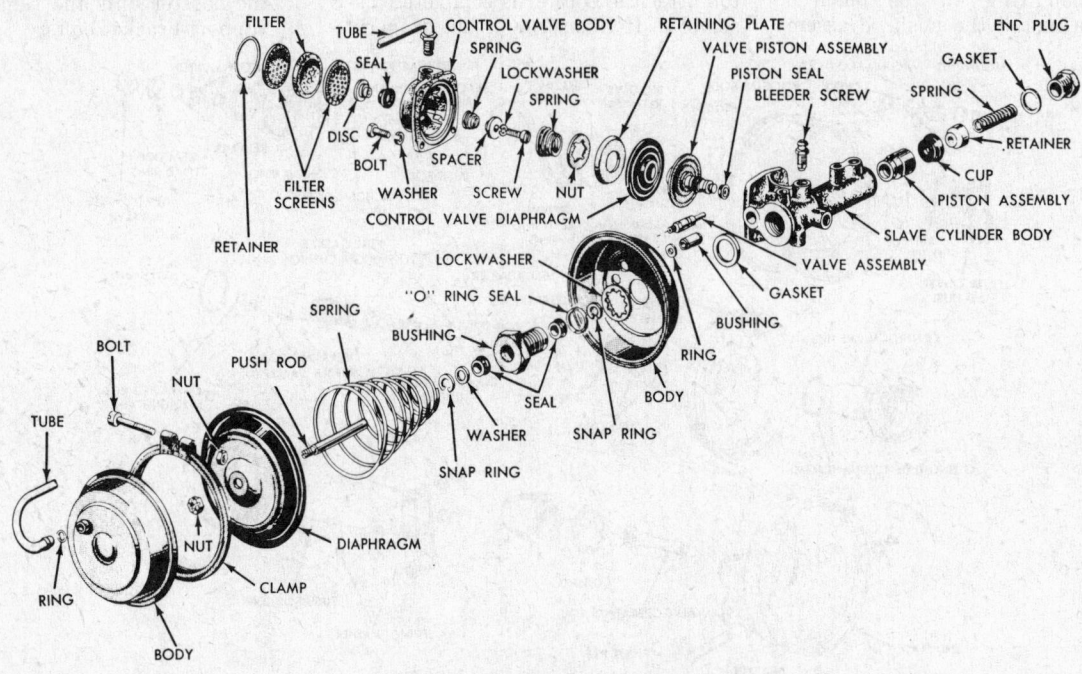

Midland Vacuum Booster (© Ford Motor Co)

Install the lockwasher over the end of the rod bushing and install the bushing seal.

7. With the slave cylinder mounted in a vise, position the front body over the end of the cylinder, inserting the transfer bushing in the front body.

8. Thread push rod bushing in place and tighten securely, making sure the front body seats squarely on the slave cylinder body.

9. Coat the piston bore of the slave cylinder with brake fluid, also the hydraulic piston, seals, spring retainer and spring.

10. With the recessed end towards the push rod bushing, install the hydraulic piston in the slave cylinder bore. On top of the piston, install the piston cup, large spring retainer and spring. Install the spring seat in the spring coils.

11. Install a new copper gasket on the end plug and screw the plug into the cylinder tightening securely.

12. Dip the control valve piston and diaphragm in brake fluid and position the control valve spring on the diaphragm, making sure that the small end of the spring is over the piston boss.

13 With the control valve body positioned over the spring, and with the scribe marks aligned, secure the valve body to the slave cylinder with the four attaching bolts.

14. Install a new gasket on the valve body cover and secure the cover to the valve body with the four attaching bolts.

15. Install the collar over the threaded end of the push rod and position the retainer on the spring. Insert the rod and collar through the coils of the spring and the retainer. Install the diaphragm over the threaded end of the push rod and secure in place with the push rod nut. Coat the push rod with brake fluid.

16. Install the return spring assembly over the push rod bushing.

17. Install the rear body on the diaphragm aligning the scribe marks on the front and rear shells. Making sure the bead of the diaphragm is properly placed between the body halves, compress the return spring and install and tighten the ring clamp band.

18. Install the by-pass tube and install the unit on the vehicle.

19. Check the unit for vacuum leaks and road test the vehicle for proper operation of the unit.

NOTE: Bleed all air from the hydraulic cylinder and lines before road testing the vehicle.

Kelsey-Hayes Diaphragm Type

Identification

The Kelsey-Hayes power brake unit can be identified by the twistlock method of locking the housing and cover together, plus the white-colored vacuum check valve assembly.

Removal

1. With engine off, apply brakes several times to equalize internal brake pressure.
2. Disconnect hydraulic line from master cylinder.
3. Disconnect vacuum hose from power brake check valve.
4. Disconnect power brake from brake pedal (under instrument panel).
5. Disconnect power brake unit from dash panel.
6. Remove power brake and master cylinder assembly from the vehicle.

Disassembly

1. Separate master cylinder from power brake unit.
2. Remove master cylinder pushrod and air cleaner plate.
3. Mount the power unit in a vise with the master cylinder attaching-studs up.
4. Scribe an index line across the housing and cover for reassembly reference.
5. Pry out the housing lock. Do not damage the lock, as it must be used at assembly.
6. Remove check valve from cover by prying out of rubber grommet.
7. Place parking brake flange holding tool over the master cylinder mounting studs.
8. Rotate the tool and cover in a counterclockwise direction. Then, separate the cover from the housing. This will expose the power piston return spring and diaphragm.
9. Lift out the power piston return spring. Remove the brake unit from the vise.
10. Remove power piston by slowly lifting the piston straight up.

11. Remove air cleaner, guide seal and seal retainer from the cover.
12. Remove the block seal from the center hole of the housing, using a blunt drift. (Don't scratch the bore of the housing, it could cause a vacuum leak.)

Power Piston Disassembly

1. Remove power piston diaphragm from the power piston. Keep it clean.
2. Remove screws that attach the plastic guide to the power piston. Remove guide and place to one side.
3. Remove the power piston square seal ring, reaction ring insert, reaction ring and reaction plate.
4. Depress operating rod slightly, then remove the Truarc snapring.
5. Remove control piston by pulling the operating rod.
6. Remove the O-ring seal from the end of the control piston.
7. Remove the filter elements and dust felt from the control piston rod.

Cleaning and Inspection

Thoroughly wash all metal parts in a suitable solvent and dry with compressed air. The power diaphragm, plastic power piston and guide should be washed in a mild soap and water solution. Blow dust and all cleaning material out of internal passages. All rubber parts should be replaced, regardless of condition. Install new air filters at assembly. Inspect all parts for scoring, pits, dents or nicks. Small imperfections can be smoothed out with crocus cloth. Replace all badly damaged parts.

Assembly

When assembling, be sure that all rubber parts, except the diaphragm and the reaction ring are lubricated with silicone grease.

1. Install control piston O-ring onto the piston.
2. Lubricate and install the control piston into the power piston. Install the Truarc snap-ring into its groove. Wipe all lubricant off the end of the control piston.

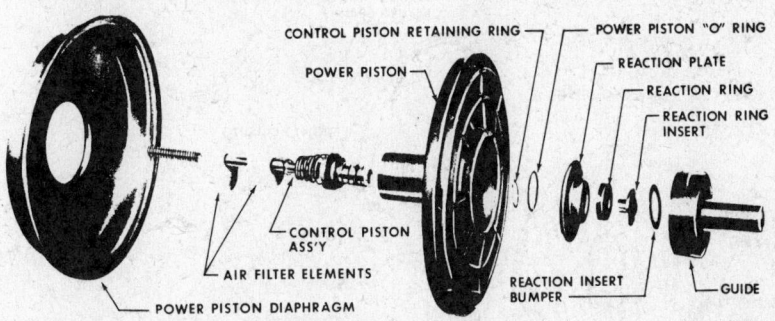

Power piston assembly (© G.M. Corp.)

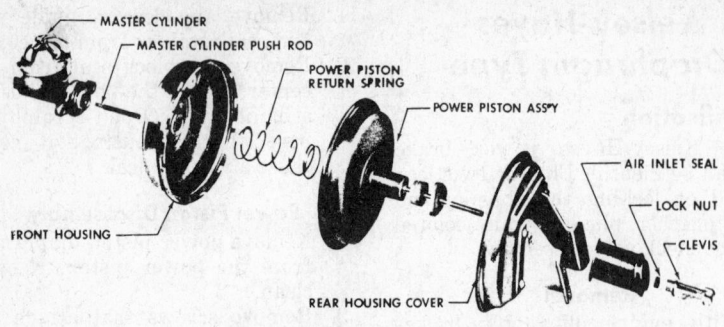

Power brake unit (© G.M. Corp.)

3. Install air filter elements and felt seal over the pushrod and down past the retaining shoulder on the rod. Install the power piston square seal ring into its groove.

4. Install the reaction plate in the power piston. Align the three holes with those in the power piston.

5. Install the rubber reaction ring in the reaction plate. Do not lubricate this ring.

6. Lubricate outer diameter of the reacton insert and install in the reaction ring.

7. Install reaction insert bumper into the guide.

8. Place guide on the power piston, align the holes with the aligning points on the power piston. Install retaining screws and torque to 80-100 in. lbs.

9. Install diaphragm on power piston; be sure that the diaphragm is correctly seated in the power piston groove.

10. With the housing blocked to prevent damage, install the block seal in the housing.

11. Install a new cover seal on the retainer and lubricate thoroughly, inside and out, with silicone grease, then install in the cover bore. Install new air filter.

12. Lubricate check valve grommet and install the vacuum check valve.

13. Mount the power unit in a vise, with master cylinder attaching studs up.

14. Apply a light coating of silicone grease to the bead, *outer edge only*, of the power piston diaphragm.

15. Install the power piston assembly in the housing with the operating rod down.

16. Install the power piston return spring into the flange of the guide.

17. Place the cover over the return spring and press down on the cover. At the same time, pilot the guide through the seal.

18. Rotate the cover to lock it to the housing. Be sure the scribe lines are in correct index and that the diaphragm is not pinched during assembly.

19. Install the housing lock on one of the long tangs of the housing.

20. Remove the power unit from the vise.

21. Install the master cylinder pushrod and air cleaner plate, then install the master cylinder on the studs. Install attaching nuts and washers. Torque to 200 in. lbs.

Installation

1. Install the power brake seal to the firewall.

2. Install power brake unit onto firewall and torque the attaching nuts to 200 in. lbs.

3. Install pushrod to brake pedal attaching bolt. Torque to 30 ft. lbs.

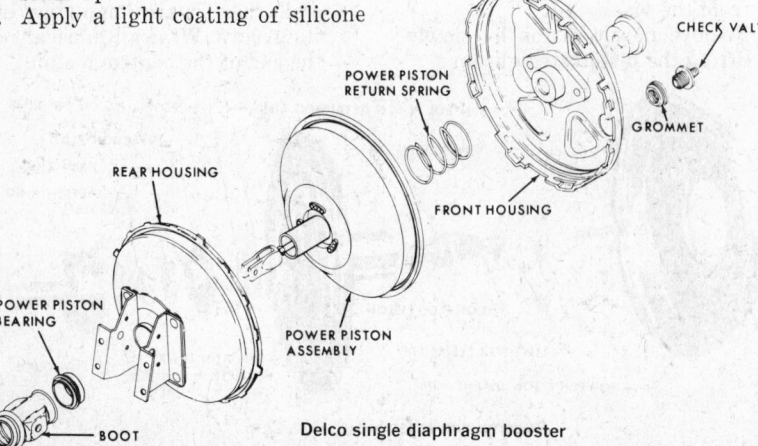

Delco single diaphragm booster

4. Install vacuum hose onto the power brake unit.

5. Attach the hydraulic tube and fill the master cylinder. Bleed hydraulic system.

6. Adjust stop-light switch if necessary.

Delco Single Diaphragm Booster

Disassembly

1. Scribe a mark on the bottom center of front and rear housings for reassembly.

2. Attach a base tool to the front housing and clamp the base in a vise with the power section up.

3. Separate the front and rear housings by securing a spanner wrench to the bracket. Press down on the wrench and rotate rear housing counterclockwise to the unlocked position. Loosen the housing carefully as it is spring loaded.

4. Remove the spanner wrench, then lift the rear housing and power piston assembly from the unit. Remove the return spring.

5. Remove the silencer by removing the retaining ring on the push rod.

6. Remove the seal, vacuum check valve and grommet from the front housing.

7. Remove the power piston assembly from the rear housing.

8. Remove the silencer from the neck of the power piston tube.

9. Remove the lock ring from the power piston.

10. Remove the reaction retainer, piston, plate, levers, bumper and spring.

11. Place a power piston wrench in a vise and position the assembly so that the three lugs on the tool fit into the three notches in the piston.

12. Press down on the support plate and rotate it counterclockwise until it separates from the power piston.

13. Remove the diaphragm from the support plate.

14. Position the power piston, tube down, in a tool fabricated from a piece of wood 2″ x 4″ x 8″ long with a $1\frac{3}{8}$″ hole in the center clamped in a vise.

15. Remove the snap ring on the air valve.

16. Using the power pump and press plate insert the power piston, tube down, in a press plate and remove the air valve assembly using a $\frac{3}{8}$″ drive extension as a remover.

17. Remove the floating control valve assembly from the push rod. Use a new one when rebuilding.

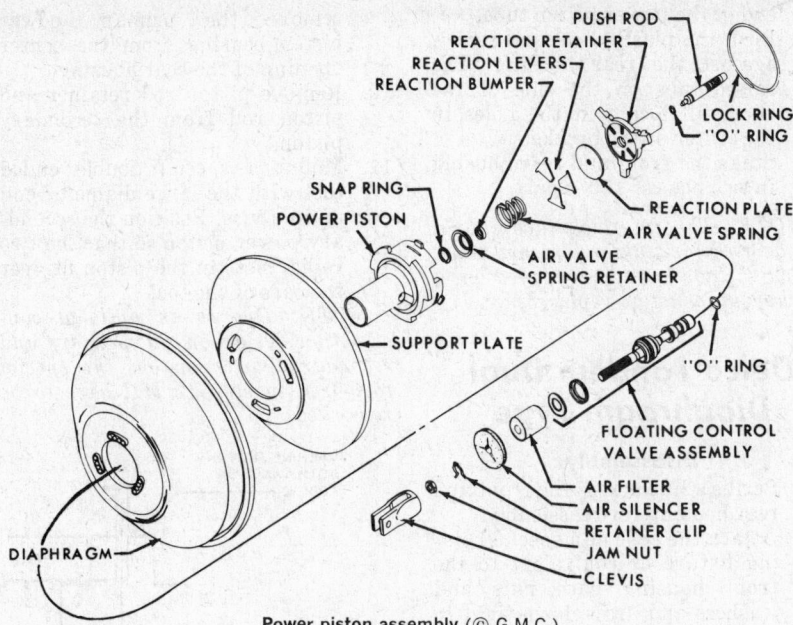

Power piston assembly (© G.M.C.)

18. Push the master cylinder push rod from the center of the reaction retainer.
19. Remove the O-ring from the groove in the master cylinder piston rod.

Assembly

1. Use clean brake fluid and thoroughly clean all reusable brake parts.
2. Inspect all rubber parts and replace if nicked, cut or damaged.
3. When rebuilding make sure that no grease or mineral oil comes in contact with any of the rubber parts.
4. Install a new vacuum check valve using a new grommet.
5. Position a new front housing seal so that the flat surface of the cup lies against the bottom depression in the housing.
6. Place a new O-ring in the groove on the master cylinder piston rod, wipe a thin film of silicone lubricant on the "O" ring.
7. Insert the master cylinder piston rod through the reaction retainer so the round end protrudes from the end of the tube on the reaction retainer.
8. Place the power piston wrench in a vise and position the power piston on the wrench so that the three lugs fit into the notches.
9. Position a new "O" ring on the air valve on the second groove from the push rod end.
10. Place a new floating control valve on the push rod-air valve assembly so that the flat face of the valve will seat against the valve seat on the air valve. *NOTE: The old floating control valve assembly must be replaced with a new one since the force required to remove it distorts component parts.*
11. Wipe a thin film of silicone lubricant on the control valve and the "O" ring on the air valve.
12. Press the air valve push rod assembly, air valve first, onto its seat in the tube of the power piston.
13. Place the control valve retainer over the push rod so that the flat side seats on the floating control valve.
14. Press the floating control valve and its retainer onto the power piston tube by use of an installer tool and pushing down by hand.
15. After the floating control valve is seated, position the push rod limiter washer over the push rod and down onto the valve.
16. Stretch the air filter element over the end of the push rod and press it into the the power piston tube.
17. Assemble the power piston diaphragm to the support plate. The raised flange of the diaphragm is pressed through the hole in the center of the support plate. *NOTE: Be sure that the edge of the center hole fits into the groove in the flange of the diaphragm.*
18. Pull the diaphragm away from the outside diameter of the support plate so that the support plate can be gripped with both hands.
19. With the power piston still positioned on the holding tool in a vise, coat the bead of the diaphragm that contacts the power piston with silicone lubricant.
20. Place the support plate and diaphragm assembly over the tube of the power piston with the locking tangs facing downward.

NOTE: The flange of the power piston will fit into the groove on the power piston.

21. Press down and rotate the support plate clockwise, until the lugs on the power piston come against the stops on the support plate.
22. Turn the assembly over and place tube down in a tool, fabricated from a piece of wood 2" x 4" x 8" long with a 1⅜" hole in the center, clamped into a vise.
23. Replace the snap ring into the groove of the air valve.
24. Place the air valve spring retainer on the snap ring and assemble the reaction bumper into the groove in the end of the air valve.
25. Position the air valve return spring, large end down, on the spring retainer.
26. Place the three reaction levers into position with the wide ends in the slots of the power piston and the narrow ends resting on top of the air valve return springs.
27. Position the reaction plate (with the numbered side up) on top of the reaction levers. Press down on the plate until the large ends of the reaction levers pop up so the plate rests flat on the levers and is centered.
28. With the round end of the master cylinder piston rod up, and with the reaction retainer held toward the top of the piston rod, place the small end of the piston rod in the hole in the center of the reaction plate. Line up the ears on the reaction retainer with the notches in the power piston and push the reaction retainer down until the ears seat in the notches.
29. With pressure on the reaction retainer, position the large lock ring down over the master cylinder push rod.
30. There is a lug on the power piston which has a raised divider in the center. One end of the lock ring goes under the lug and on one side of the divider.
31. As you work your way around the power piston, the lockring goes over the ear of the reaction retainer and under a lug on the power piston until the other end of the lock ring is seated under the lug with the raised divider. *NOTE: Make certain both ends of the lock ring are securely under the large lug.*
32. Place a new power piston bearing in the center of the rear housing so the flange on the center hole of the housing fits into the groove of the power piston bearing. The large flange on the power piston bearing will be on

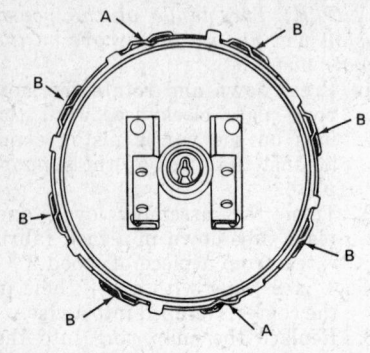

A- STAKED TABS
2 PLACES 180°
APART

B – OPTIONAL
STAKING
LOCATIONS

Housing locking tabs (© G.M.C.)

UNSTAKED TAB SOCKET

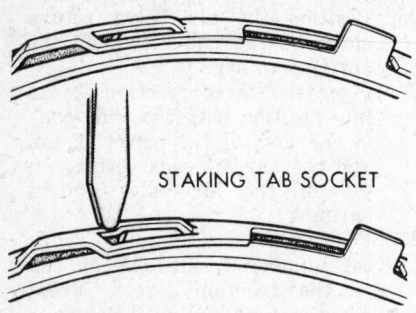

STAKING TAB SOCKET

Staking housing tabs (© G.M.C.)

the stud side of the housing. Coat the inside of the bearing with silicone lubricant.

33. Place the air silencer over the holes on the tube of the power piston. Wipe the tube with silicone lubricant.

34. Attach the holding fixture to the front housing and clamp the base in a vise.

35. Place the power piston return spring over the insert in the front housing.

36. Lubricate the inside diameter of the support plate seal, the reaction retainer tube, and the beaded edge of the diaphragm with silicone lubricant.

37. Place the rear housing assembly over the front housing assembly and align the scribe marks of the two housings so they will match when in the locked position.

38. Place a spanner wrench on the rear housing and tighten the nuts and washers to the bolts.

39. Press down on the spanner wrench and twist the rear housing clockwise until fully locked.

NOTE: Do not break the studs loose in the rear housing or put pressure on the power piston tube when locking the housings.

40. Remove the spanner wrench and the holding fixture from the front housing.

41. Push the felt silencer over the pushrod and seat it against the

end of the power piston tube.

42. Push the plastic boot and seat it against the rear housing. The raised tabs on the side of the boot will locate in the holes in the center of the brackets.

43. Stake the front and rear housing in two places: 180° apart.

NOTE: The interlock tabs should not be used for staking a second time. When all tabs have been staked once, the housing must be replaced.

Delco Tandem Dual Diaphragm Type

Disassembly

1. Scribe a line across the front and rear housing for reassembly.

2. Attach the base of a special holding fixture or equivalent to the front housing with nuts and washers and draw down tight to eliminate damage to the studs. Clamp the base in a vise with the power section up.

3. On vehicles with a straight mounting bracket place a spanner wrench over the studs on the rear housing and attach with nuts and washers.

4. On vehicles with a tilted mounting bracket there is a special tool placed inside the mounting bracket with the spanner wrench placed on top.

5. Press down on the spanner wrench and rotate the rear housing counterclockwise to separate the two housings. Remove the special tools.

6. Remove the power piston return spring, and remove and discard the vacuum check valve and grommet from the front housing.

7. Remove the front housing seal.

8. Remove the boot retainer and boot from the rear housing and remove the felt silencer from inside the boot.

9. Remove the power piston group from the rear housing and

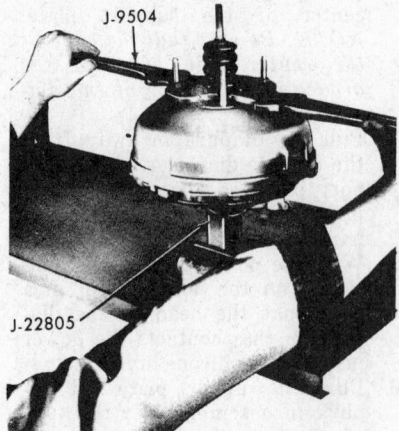

J-9504

J-22805

Unlocking front and rear housings (© G.M.C.)

remove the primary power piston bearing from the center opening of the rear housing.

10. Remove piston rod retainer and piston rod from the secondary piston.

11. Mount a special double ended tool with the large diameter end up in a vise. Position the secondary power piston so that the two radial slots in the piston fit over the ears of the tool.

NOTE: Due to an optional construction design on the primary and secondary power pistons the special tool used in step 11 will have to be reworked.

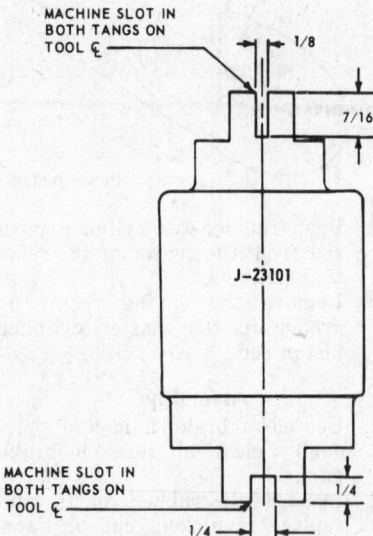

MACHINE SLOT IN
BOTH TANGS ON
TOOL ₵

1/8

7/16

J–23101

MACHINE SLOT IN
BOTH TANGS ON
TOOL ₵

1/4

1/4

Reworking of tool for optional power piston design (© G.M.C.)

12. Fold back the primary diaphragm from the outside diameter of the primary support plate. Grip the edge of the support plate and rotate it counterclockwise to unscrew the primary power piston from the secondary power piston.

13. Remove the housing divider from the secondary power piston bearing from the housing divider.

14. The secondary power piston should still be positioned on the special double ended tool. Fold back the secondary diaphragm from the outside diameter of the secondary support plate. Rotate the support plate clockwise to unlock the secondary power piston.

15. Remove the secondary diaphragm from the secondary support plate.

16. Remove the reaction piston and disc from the center of the secondary power piston by pushing down on the end of the piston.

17. Remove the air valve spring from the end of the valve, if not removed earlier.

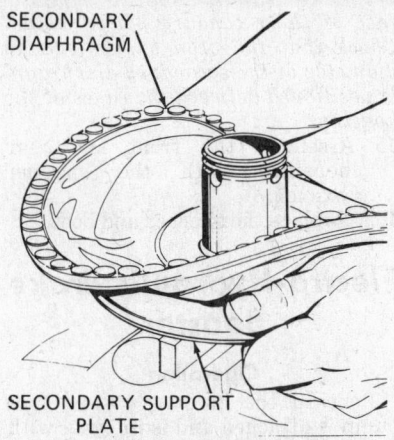

SECONDARY
DIAPHRAGM

SECONDARY SUPPORT
PLATE

Locking or unlocking the secondary support
plate and secondary power piston
(© G.M.C.)

18. Remove the primary diaphragm and piston using the same procedure as the secondary with the exception of turning the support plate counterclockwise to unlock it.
19. Remove the air filter from the tubular section of the primary power piston.
20. Remove the power head silencer from the neck of the power piston tube.
21. Remove the rubber reaction bumper from the end of the air valve.
22. Using snap ring pliers, remove the retaining ring from the air valve.
23. Remove the air valve push rod assembly.
    a. The recommended method would be to place the primary power piston in an arbor press and press the air valve push rod assembly out the bottom of the power piston tube using a rod not larger than ½" in diameter.
    b. An alternate method would

be to insert a heavy, round shanked screwdriver on both sides of the pushrod and pull the air valve-push rod assembly straight out.
24. Remove the "O" ring seal from the air valve.

Assembly
1. Use clean brake fluid and thoroughly clean all reusable brake parts.
2. Inspect all rubber parts and replace if nicked, cut or damaged.
3. When rebuilding, make sure that no grease or mineral oil comes in contact with any of the rubber parts.
4. Install a new vacuum check valve and a new grommet in the front housing.
5. Place a new seal in the front housing so that the flat surface lies against the bottom of the depression in the housing.
6. Reassemble the power piston group.
7. Lubricate the inside and outside diameter of the "O" ring seal with silicone lubricant and place on the air valve.
8. Wipe a thin film of silicone lubricant on the large and small outside diameter of the floating control valve. If the floating control valve needs replacement, it will be necessary to replace the complete air valve-push rod assembly.
9. Place the air valve end of the air valve push rod assembly into the tube of the primary power piston. Manually press the air valve push rod assembly so that the floating control valve bottoms on the tube section of the primary power piston.
10. Place the inside diameter of the floating control valve retainer on

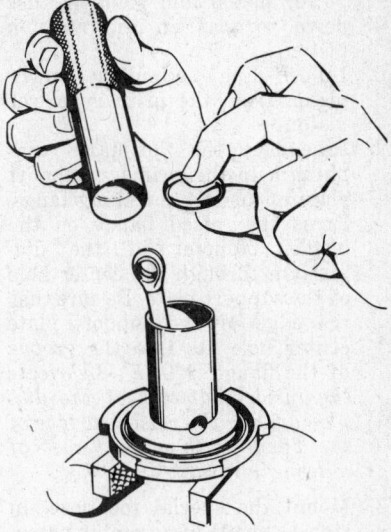

Installing the floating control valve retainer
(© G.M.C.)

the outside diameter of the special installer. Place it over the pushrol so that the closed side of the retainer seats on the floating control valve. Using the installer manually press the retainer and floating control valve to seat in the tube.
11. Stretch the filter element over the pushrod and press it into the piston tube.
12. Place the retaining ring into the groove in the air valve using snap ring pliers.
13. Position the rubber reaction bumper on the end of the air valve.
14. Determine the correct reaction piston and apply a light coat of silicone lubricant to the outside diameter of the rubber reaction disc.
15. Place the rubber reaction disc in the large cavity of the secondary

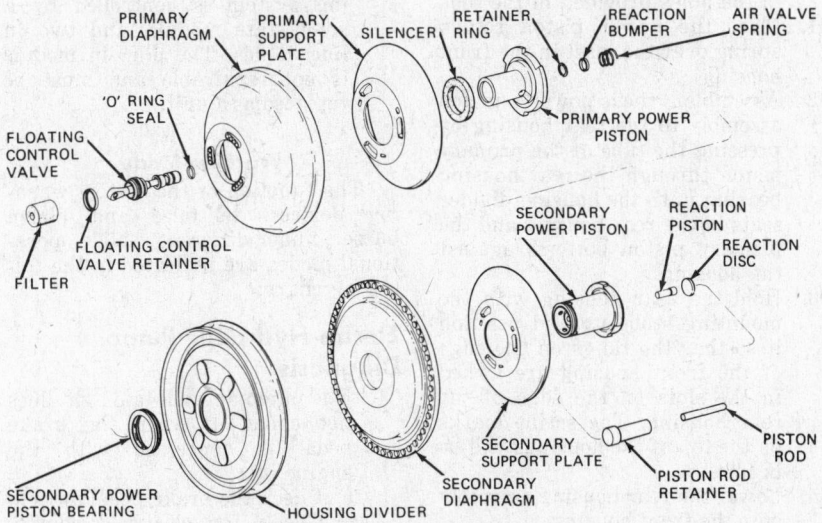

Power piston group (© G.M.C.)

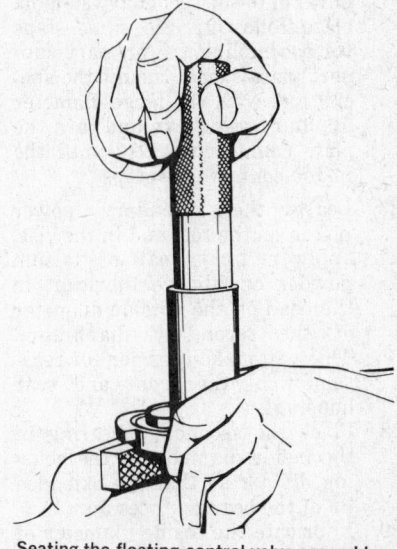

Seating the floating control valve assembly
(© G.M.C.)

power piston and push the disc down to seat on the reaction piston.

16. Unlock the secondary power piston from the primary power piston.

17. Assemble the primary diaphragm to the primary support plate opposite the locking tangs. Press the raised flange on the inside diameter of the diaphragm through the center hole of the support plate. Be sure that the edge of the support plate center hole fits into the groove of the flange. *NOTE: Lubricate the inside diameter of the diaphragm and the raised surface of the flange with a light coat of silicone lubricant.*

18. Mount the special tool used in step 11 of the disassembly procedures in a vise with the small end up. Position the primary power piston so that the two radial slots in the piston fit over the ears (tangs) of the tool.

19. Fold the primary diaphragm away from the outside diameter of the primary support plate.

20. Place the primary support plate and diaphragm assembly over the tube of the primary piston. Make sure the locking tangs are facing down.

21. Press down and rotate the support plate clockwise until the tabs on the piston contact the stops.

22. Place the power head silencer on the tube of the piston so that the holes at the base of the tube are covered.

23. Coat the outside of the tube with silicone lubricant.

24. Remove the primary piston assembly from the special tool and lay it aside.

25. Assemble the secondary diaphragm to the secondary support plate following the same steps for assembling the primary support plate except mount the special tool with the large diameter up, and press down and turn the plate counterclockwise until the piston contacts the stops.

26. Leave the secondary power piston on the tool and in the vise.

27. Apply a light coat of talcum powder or silicone lubricant to the bead on the outside diameter of the secondary diaphragm. This will make it easier for reassembly of the front and rear housing.

28. Place the secondary bearing in the inside diameter of the housing divider so that the extended lip of the bearing faces up.

29. Lubricate the inside diameter of the bearing with silicone lubricant.

30. Using a special protector tool or equivalent, position the secondary bearing on the threaded end of the secondary power piston.

31. Hold the housing divider so that the six oblong protrusions on the middle of the divider are facing up. Press the divider down over the tool and onto the piston tube so it rests against the support ring. Remove the bearing protector tool.

32. Pick up the primary power piston assembly and fold the primary diaphragm away from the outside diameter of the support plate.

33. Place the small end of the air valve return spring on the air valve so that it contacts the air valve retaining ring.

34. Position the primary power piston. Make sure that the air valve return spring seats down over the raised center section of the secondary piston.

35. Rotate the secondary power piston clockwise into the threaded portion of the primary piston. Tighten to 5-15 ft. lbs.

36. Fold the primary diaphragm back into position.

37. Cover the outside diameter of the piston rod retainer with a light coat of silicone lubricant.

38. Insert the master cylinder piston rod retainer into the secondary power piston so that the flat end bottoms against the rubber reaction disc.

39. Place the new primary piston bearing in the rear housing center hole. The thin lip of the bearing will protrude to the outside of the housing. Coat the inside diameter of the bearing with silicone lubricant.

40. Mount the holding fixture in a vise and position the front housing so that the housing studs fit in the holes provided in the tool.

41. Place the power piston return spring over the inset in the front housing.

42. Assemble the power piston assembly to the rear housing by pressing the tube of the primary piston through the rear housing bearing until the housing divider seats in the rear housing and the primary piston bottoms against the housing.

43. Hold the rear housing with the mounting studs up and position it so that the tangs on the edge of the front housing are locked in the slots on the edge of the rear housing. The scribe marks on the top of the housings will be in line.

44. Lower the rear housing assembly onto the front housing. *NOTE: The power piston spring must seat in the depression in the* face of the secondary power piston. Check that the bead on the outside diameter of the secondary diaphragm is positioned between the edges of the housing.

45. Assembly the front and rear housings with the spanner wrench.

46. Replace the silencer and boot.

Electro-Hydraulic Brake Booster

Operation

Beneath the booster a vane type pump is attached and is integral with a 12 volt DC electric Motor. If the vehicle engine was not operating or a hose or belt was broken the pump and motor would serve as a reserve power source to provide boost pressure. The electric pump draws fluid from the low pressure side of the booster piston and delivers it to the high pressure side. The electric pump provides boost pressure of approximately one-half of the primary system pressure.

Electrical Control Circuits

The electric pump operation is controlled by a relay which is operated by a flow switch located in the booster outlet to sense the fluid flow. A pedal switch also controls the electric pump operation whenever the brake pedal is depressed and the engine is not operating. The system is monitored by two dash mounted telltale lamps and a buzzer. The two lamps will be marked to:

1. Warn of failure of the primary system.
2. Warn of failure of the reserve system.
3. To make the driver aware that the reserve is in operation, the dash lamp will light and the buzzer will sound. The monitoring system is controlled by a solid state module, and two in line diodes. The plug in module is not repairable and must be replaced as a unit.

Warning Mode

The function of the system warning devices, tell-tales and alarm buzzer, under different vehicle operational modes are indicated in the following chart:

Electro-Hydraulic Pump Diagnosis

1. The pump and tell-tale light does not come on when the brake pedal is depressed with the engine off:
   a. Check the brake pedal switch.
   b. Check for electrical continuity through the pump flow switch.

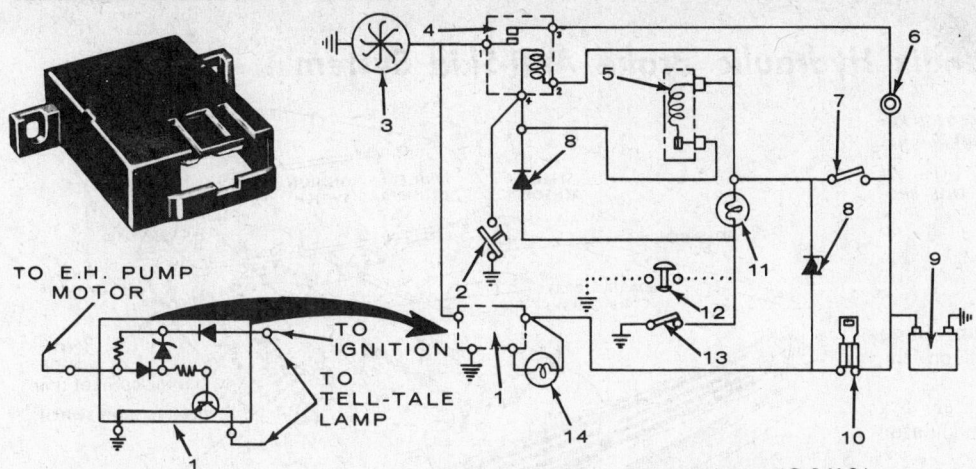

1 Sensor
2 Flow switch
3 Pump
4 Relay
5 Buzzer
6 Starter solenoid, Ignition bus bar (Tilt Cab)
7 Pump switch
8 Diode
9 12 volt battery
10 Ignition switch
11 Tell-Tale no. 1
12 Pressure differential switch
13 Pedal travel (safety) switch
14 Tell-Tale no. 2

TO E.H. PUMP MOTOR

TO IGNITION

TO TELL-TALE LAMP

Hydraulic brake booster electrical diagram (© G.M.C.)

c. Check for voltage to the ignition side of the relay coil.
d. Check for voltage at the battery connection to the relay.
e. Check for an open at the ignition diode.
f. Check for voltage at the pump terminal at the relay.
2. The engine is off and the pump is operating but the light is not on when the brake pedal is depressed.
   a. Check the voltage at the warning light bulb.
   b. Replace the bulb.
3. The engine and the pump are off, but when the brake pedal is depressed, the light is on.
   a. Check the voltage at the pump motor.
   b. Replace pump.
4. The accessories, radio, heater, wipers etc. operate when the brake pedal is depressed and the engine off.
   a. Check the ignition diode for a short.
5. The pump and warning light stay on after the engine is started.
   a. Check for air in the boost systems.
   b. Check to see if the flow switch is shorted or in the stuck position.
   c. Check to see if the relay is in the closed position.

*BRAKE
+BRAKE ELECT. HYD. BOOST

DIAGNOSIS OF EH PUMP

PROBLEM	POSSIBLE CAUSE	CORRECTION
EXCESSIVE PUMP NOISE (gurgle, chatter, etc.)	1. Trapped air in pump.	1. Depress brake pedal lightly with the engine off for thirty seconds and release. Recheck and should the problem persist, repeat above procedure after a three minute waiting period. NOTE: This noise will diminish upon continued use of the brakes under normal driving conditions.
INOPERATIVE PUMP	(A) Non-functioning motor.	1. Check electrical connection between motor lead wire and wiring harness. If loose, corroded, or disconnected, clean and secure connection. 2. Check grounding of pump housing to booster. The pump housing must be securely bolted to the booster to properly ground the motor. 3. Replace EH Pump.
	(B) Low or no voltage at motor connection of wiring harness.	1. Check condition of battery and battery terminals. Correct an abnormally low battery condition and/or clean battery terminals if necessary. 2. Check electrical leads at battery terminal of starter or ignition bus bar—not corroded or loose.
OIL LEAK AT BOOSTER AND EH PUMP MATING SURFACE.	1. Damaged or missing O-rings at pressure and/or return port.	1. Replace two O-rings.
OIL LEAK FROM PUMP END PLATE	1. Damaged or missing end plate seal.	1. Replace EH Pump assembly.
OIL LEAK FROM EH PUMP MOTOR	1. Damaged shaft seal.	1. Replace EH pump assembly.

MODE	TELL-TALE #1*	TELL-TALE #2+	BUZZER
1. Engine off—ignition off			
A. No brake apply	off	off	off
B. Brake apply	on	off	on
2. Engine off—ignition on with or without brake apply—(bulb check)	on	on	on
3. Engine off—ignition on start with or without brake apply	on	on	on
4. Engine on with or without brake apply	off	off	off
5. Engine on—primary boost interrupted with or without brake apply	on	on	on
6. Engine on—open circuit in EH pump motor with or without brake apply	off	on	off

Bendix Hydraulic Brake Anti-Skid System

NOTE: Before attempting any repair operations on the anti-skid system, determine the problem area by tests outlined at the end of this section.

Pressure Modulator

Removal

1. Disconnect the vacuum hose from the vacuum inlet on the modulator bypass valve.
2. Disconnect the modulator hydraulic lines from the modulator hydraulic cylinder.
3. Disconnect the modulator electrical leads.
4. Loosen the nut that secures the rear of the modulator to the mounting bracket. Loosen the outer nuts on the two lower 'J' bolts, these nuts secure the front of the modulator to the bracket.
5. Lift the modulator upward to disengage the modulator bolts from the slots in the mounting bracket and remove the modulator from the vehicle.

CAUTION: Do not attempt to disassemble the modulator. The heavy-duty diaphragm return spring is compressed inside the modulator, and any attempt to remove the retaining bolts from the end plate could result in injury.

Installation

1. Position the modulator, with the bypass tube up, on the mounting bracket so that the two lower bolts and the stud on the rear of the modulator fit into the slots in the mounting bracket.
2. Tighten the nut on the stud at the rear of the modulator. Tighten the outer nuts on the two lower 'J' bolts.
3. Connect the brake lines to the hydraulic cylinder.
4. Connect the vacuum hose to the vacuum inlet on the bypass valve.

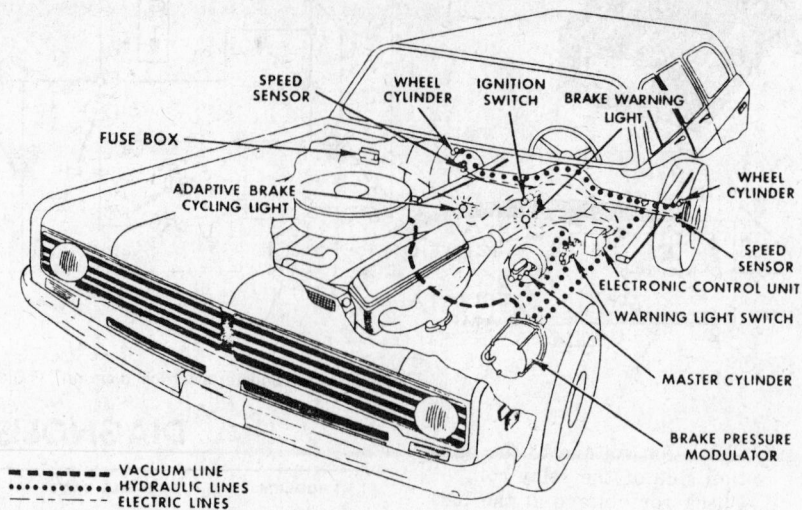

Typical Bendix system installation (© IHC)

5. Connect the electrical leads to the modulator.
6. Proceed to the "Testing & Trouble Shooting" section to make sure the modulator is operating properly.

Electronic Control Unit (Computer)

Removal

1. Turn off the vehicle power to the control unit.
2. Disconnect and lable the control unit electrical connectors.

3. Remove the control unit mounting screws and remove the control unit.

Installation

1. Position the control unit over the mounting holes and secure with the retaining screws.
2. Connect the electrical leads.
3. Proceed to the "Testing & Trouble Shooting" section to make sure the control unit is working properly.

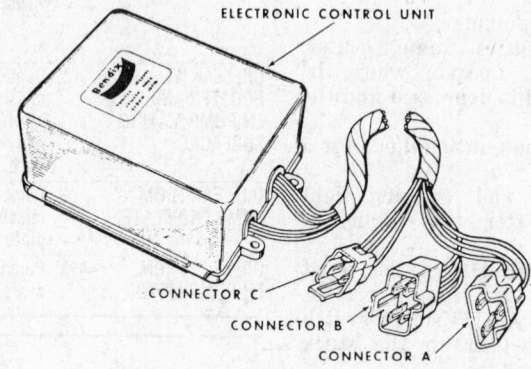

Electronic Control Unit—Bendix System (© IHC)

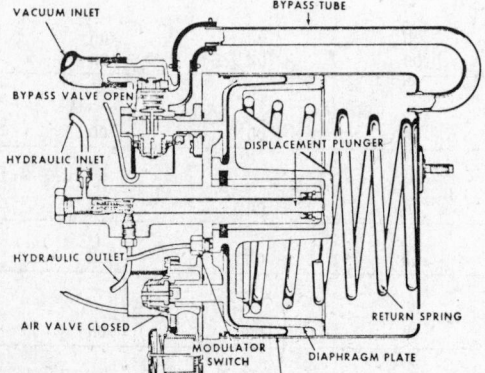

Pressure modulator-cross section view—Bendix System (© IHC)

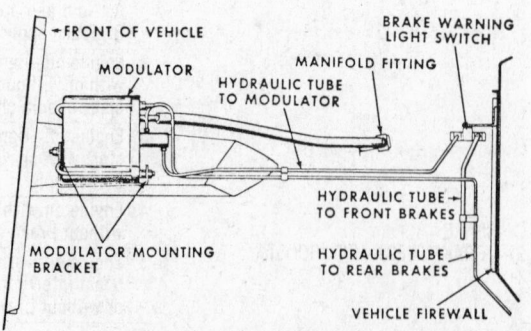

Modulator vacuum and hydraulic lines—Bendix System (© IHC)

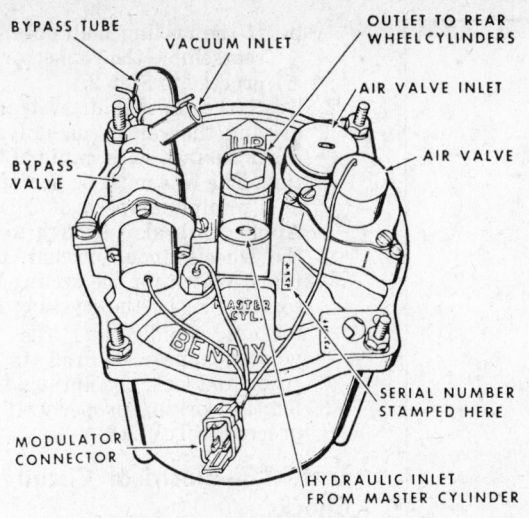

Pressure modulator—Bendix (© IHC)

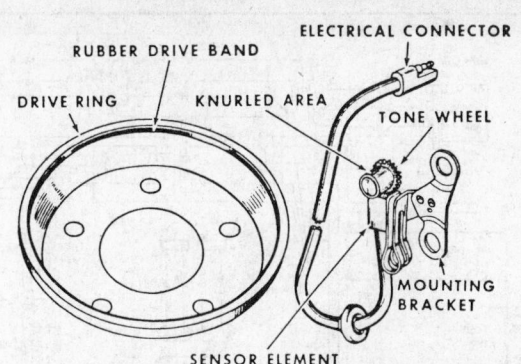

Drive ring and speed sensor components—Bendix System (© IHC)

Wheel Speed Sensor

Removal

1. Raise the vehicle and secure. Remove the wheel and brake drum assembly.
2. Remove the straps that hold the sensor leads in place. Disconnect the wheel speed sensor leads.
3. Press the wheel speed sensor inward (toward the vehicle), to reduce the pressure on the knarled shaft, and remove the drive ring assembly.
4. Remove the nut which secures the sensor to the brake backing plate and remove the sensor.
5. Loosen the sensor lead grommet in the backing plate, and pull the sensor lead through the hole.

Installation

1. Insert the sensor lead through the hole in the backing plate and seat the grommet.
2. Position the sensor mounting bracket with the lower front brake mounting hole aligned with the lower hole in the sensor bracket. Align the upper sensor bracket hole with the upper forward mounting hole. Hold the bracket in place with the lower nut.
3. Press the speed sensor inward

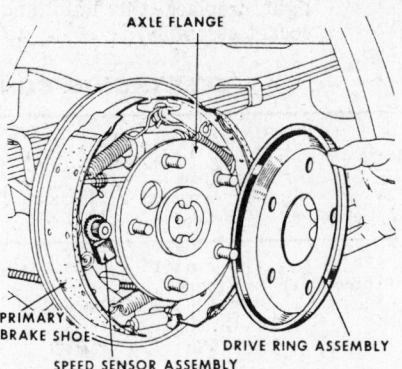

Removing the drive ring—Bendix System (© IHC)

and position the drive ring on the hub.
4. Connect the sensor electrical lead and secure with new strap.
5. Install the wheel and drum assembly. Lower the vehicle.
6. Proceed to the "Testing & Trouble Shooting" section to make sure the sensor is working properly.

Testing & Trouble Shooting Bendix Hydraulic Anti-Skid System

NOTE: The following equipment is required to perform tests on this system:

a. V.O.M. capable of measuring: AC voltage in the .10-5.0 range; DC voltage in the 6-18 range; Resistance accurately in the 0-10 ohm, 100-1,000 ohm, 5,000-100,000 ohm ranges.
b. Vacuum gauge that measures vacuum from 0-25 inches Hg.
c. Pressure gauge capable of measuring hydraulic pressure from 0-1,000 psi.

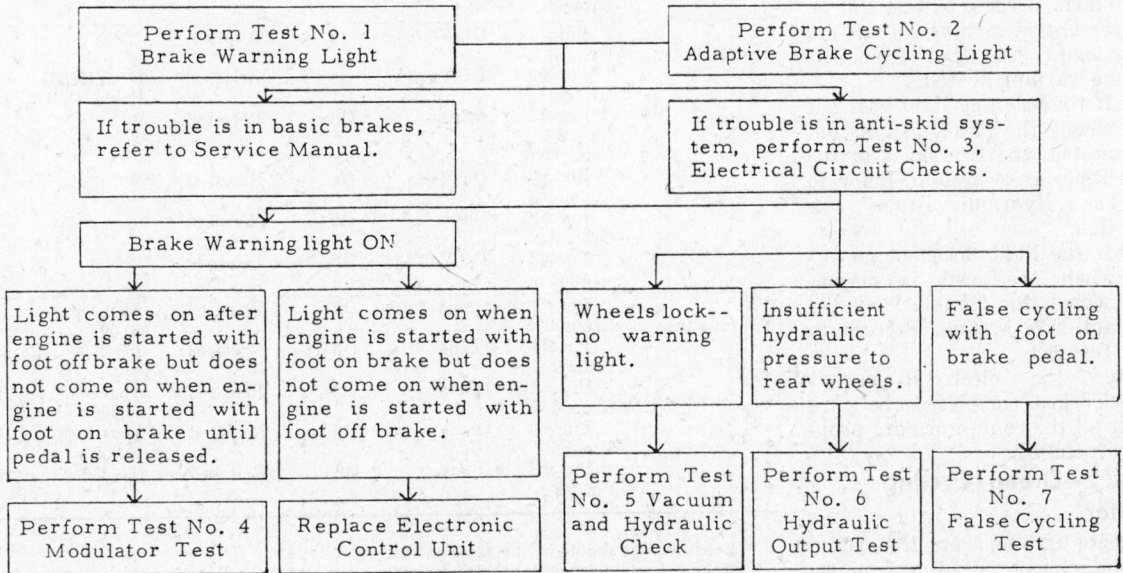

Trouble shooting chart—Bendix System (© IHC)

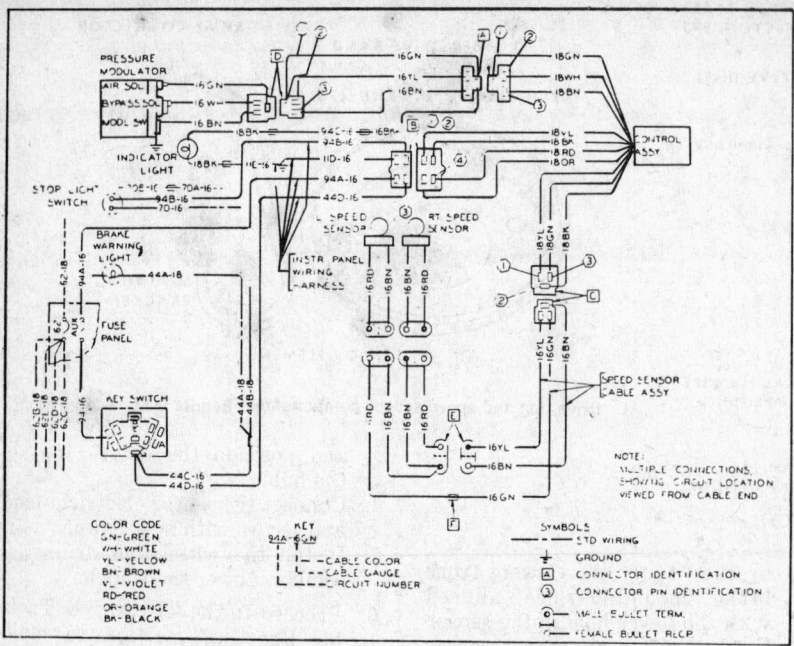

Electrical circuit diagram—Bendix System (© IHC)

COLOR CODE
GN—GREEN
WH—WHITE
YL—YELLOW
BN—BROWN
V—VIOLET
RD—RED
OR—ORANGE
BA—BLACK

KEY
94A—GN
│ └── CABLE COLOR
│ └── CABLE GAUGE
└── CIRCUIT NUMBER

SYMBOLS
━ STD WIRING
⏚ GROUND
Ⓐ CONNECTOR IDENTIFICATION
① CONNECTOR PIN IDENTIFICATION
MALE BULLET TERM.
FEMALE BULLET RECP.

NOTE:
MULTIPLE CONNECTIONS
SHOWING CIRCUIT LOCATION
VIEWED FROM CABLE END

Test #1—Red Brake Warning Light

1. *If the brake warning light is not on,* turn the ignition switch to the "start" position. The light should come on while the engine is cranking.

   a. If the warning light does not come on when the engine is cranking, replace the bulb and check the warning lamp circuit operation.

   b. If the light stays on after the bulb has been replaced, proceed to Step 2 to determine if the problem is in the anti-skid system or if the trouble is in the basic brake system.

   c. If the light comes on when the engine is cranking and goes off when the engine starts, proceed to Test #3.

2. *If the brake warning light is on,* disconnect the lead from the brake warning switch.

   a. If the warning light goes out when the lead is disconnected, the trouble is in the basic brake system. Refer to the "Hydraulic Brake" section.

   b. If the light does not go out when the lead is disconnected, the trouble is in the anti-skid system. Proceed to Test #3.

After the electrical circuit checks are completed, be sure that all the connectors are properly installed.

Test #2—Green Cycling Indicator

1. Locate and separate the cycling light connector leading from the controller (connector "A").

Apply 12 volts directly from the battery positive terminal to the bulb side of the connector.

   a. If the cycling bulb does not light, replace the bulb and socket assembly.

   b. If the cycling bulb does light reassemble the connector and proceed to Step 2.

2. Test the anti-skid system by driving the vehicle on gravel or other surface with minimal traction. The test must be performed at a minimum speed of 20 mph. Apply the brakes enough to lock the wheels under normal conditions. Watch for the cycling light to flicker. If the cycling light flickers rapidly and the rear wheels appear to roll to low speed and lock, the anti-skid system is working properly. If not, proceed to Test #3.

Test #3—Electrical Circuit Checks

1. *Check at the electronic control unit.*

   NOTE: Make the measurements listed in the chart that follows. These measurements are made at the control unit connectors. The connectors should be disconnected except those measurements indicated by a double asterisk (**). All readings must be made on the terminals on the side of the connector attached to the control unit assembly and the speed sensor assemblies. If any reading is not as

ELECTRONIC CONTROL UNIT CHECKS

Connector	Take Measurement Between	Measure For	Ignition Switch Position	Brake Pedal Position	Reading	Improper Reading Refer to Test #3; Step #
"A" (Three Pins)	Pin #1 -Ground	Ohms	Off	Released	3.0-50	3
	Pin #2 -Ground	Ohms	Off	Released	3.0-5.0	3
	Pin #3 -Ground	Ohms	Off	Released	0	2
	**Pin #2 -Ground	DC Volts	On	Applied	0	8
	**Pin #1 -Ground	DC Volts	On	Applied	0	8
"B" (Four Pins)	Pin #1 -Ground	DC Volts	Off	Applied	12*	6
	***Pin #2 -Ground	DC Volts	Off	Released	0 w/Head-lights ON	7
	**Pin #2 Pin #2 -Con. "C"	Ohms	Off	Released	0	8
	Pin #3 -Ground	DC Volts	On	Released	12*	9
	Pin #2 -Ground	Ohms	Off	Released	0	4
	Pin #4 -Ground	DC Volts	On	Released	12*	5
"C" (Three Pins)	Pin #1 -Ground	Ohms	Off	Released	Above-50,000	10
	Pin #1 -Pin #2	Ohms	Off	Released	250-350	10
	Pin #2 -Ground	Ohms	Off	Released	Above-50,000	10
	Pin #3 -Ground	Ohms	Off	Released	Above-50,000	10
	Pin #3 -Pin #2	Ohms	Off	Released	250-350	10

* Measurement should be the same as battery voltage.
** Make the measurement with all electrical connectors installed.
*** Make the measurement with the headlights on.

specified, refer to the step indicated for additional readings or corrections.

2. *Modulator switch circuit.* Make the measurements in the sequence listed below to identify the defective component. When it appears that the faulty component has been found, reconnect all connectors except at the control unit and repeat the measurements in step 1 that produced the incorrect reading. *This remeasurement should be made prior to replacing any components to insure that the trouble still exists.*
   a. Disconnect connector "D" (located at the modulator).
   b. On the modulator lead side of the connector, measure the resistance Pin #3 and ground. If zero ohms is measured, the modulator cable is defective and must be replaced. If the measurement is above zero ohms, the modulator or the modulator switch is defective and must be replaced.

3. *Air valve and bypass valve circuits.* Make the measurements in the sequence listed below to identify the defective component. When it appears that the faulty component has been found, reconnect all connectors except at the control unit and repeat the measurements in step 1 that produced the incorrect reading. *This remeasurement should be made prior to replacing any components to insure that the trouble still exists.*
   a. Disconnect connector "D" (located at the modulator).
   b. On the modulator lead side of the connector, measure the resistance between the pin where the incorrect reading was recorded and ground.
   c. If the reading is 3.0-5.0 ohms, the modulator cable assembly is defective and must be replaced.
   d. If the reading is still incorrect, the modulator must be replaced.

4. *Electronic Control Unit ground circuit.*
   a. Disconnect the body ground wire under the instrument panel.
   b. Clean the eye on the end of the wire and ground point on wire where the wire is attached.
   c. Reinstall the ground wire. Remeasure the resistance between Pin #2, connector 'B', on the brake harness side and ground.
   d. If the reading is zero ohms, the ground connection was faulty and the problem has

been corrected.
   e. If the reading is above zero ohms, the brake harness is faulty and must be replaced.

5. *Power supply circuit.* Make the measurements in the sequence listed below to identify the defective component. When it appears that the faulty component has been found, reconnect all connectors, except at the control unit, and repeat the measurements in step 1 that produced the incorrect reading. *This measurement should be made prior to replacing any components to insure that the trouble still exists.*
   a. Check the 10 amp fuse on the fuse panel in line between the ignition switch and the power connector, replace the fuse if defective.
   b. If the fuse is good, be sure that the circuit is properly connected at the ignition switch and that battery voltage is present at the ignition switch terminal.
   c. If the fuse is good, connections and battery voltage are correct at the switch terminals, but there is no power at Pin #4, Connector "B", with the ignition switch "on", the brake harness is faulty and must be replaced.

6. *Stop light to Electronic Control Unit circuit.* Make the measurements in the sequence listed below to identify the defective component. When it appears that the defective component has been found, reconnect all connectors, except at the control unit, and repeat the measurements in step 1 that produced the incorrect reading. *This measurement should be made prior to replacing any components to insure that the trouble still exists.*
   a. With the brakes applied, measure the DC voltage between the output (stop light) side of the stop light switch and ground.
   b. If the reading is correct (battery voltage), the brake harness is defective and must be replaced.
   c. If the reading is still incorrect (below battery voltage), measure the DC voltage between the battery side of the stop light switch and ground.
   d. If the reading is correct (battery voltage), adjust the switch position or replace the switch as necessary.
   e. If the reading is incorrect (below battery voltage), check the wiring to the stop light switch and repair or replace the wiring as necessary.

7. *Voltage feedback.* Make the measurements in the sequence listed below to identify the defective component. When it appears that the faulty component has been found, reconnect all connectors, except at the control unit, and repeat the measurements in step 1 that produced the incorrect reading. *This measurement should be made prior to replacing any components to insure that the trouble still exists.*
   a. Depress and release the brake pedal and check the operation of the brake lights.
   b. If the stop lights are not functioning normally, refer to the "Hydraulic Brake" section. After the defect has been found and corrected, repeat the measurement of DC voltage between Pin #1 Connector "B" and ground (with the headlights on).
   c. If the stop lights function properly and the reading is still above zero volts, check the ground connections in both tail light-stop light assemblies, correct as necessary.

8. *Electronic Control Unit circuits.* If the reading is above zero volts, the electronic control unit is defective and must be replaced.

9. *Warning light circuit.* Make the measurements in the sequence listed below to identify the defective component. When it appears that the defective component has been found, reconnect all the connectors, except at the control unit, and repeat the measurement in step 1 that produced the incorrect reading. *This measurement should be made prior to replacing any components to insure that the trouble still exists.*
   a. Check the circuit connection at the ignition switch.
   b. Measure the DC voltage between the circuit terminal on the ignition switch and ground.
   c. If the reading is correct (battery voltage) but the battery voltage is not present at Pin #3, Connector "B", the brake harness is defective and must be replaced.

10. *Speed sensor circuit.* Make the measurements in the sequence listed below to identify the defective component. When it appears that the defective component has been found, reconnect all the connectors, except at the control unit, and repeat the measurements in step 1 that produced the incorrect reading. *This measurement should be made prior*

to replacing any components to insure that the trouble still exists.

a. Separate the connector at the wheel speed sensor.
b. On the wheel speed sensor side of the connector, repeat the measurement from step 1 that produced the incorrect reading.
c. If the reading is now within limits, the speed sensor cable is defective and must be replaced.
d. If the reading is still incorrect, separate the speed sensor cable connector in the line with the incorrect reading.
e. On the speed sensor side of the connector, measure the resistance of the two terminals and between each terminal and ground.
f. If the readings are within limits (see chart) the speed sensor harness is defective and must be replaced.
g. If any of the readings are incorrect, the corresponding speed sensor is defective and must be replaced.

Test #4 Modulator Operation

1. With the engine running, disconnect the "D" connector. (modulator lead).
2. On the modulator side of the connector, measure the resistance between Pin #3 and ground.
3. If the reading is above zero ohms, replace the modulator.

Test #5 Vacuum and Hydraulic Test

1. Check the vacuum supply as follows
   a. Visually inspect all vacuum hoses in the engine compartment for damage. Make sure all hose connections are tight. Replace or tighten all hoses as necessary.
   b. Start the engine and listen for a hissing sound indicating a vacuum leak. Repair as necessary.
   c. Disconnect all vacuum lines at the modulator, and connect a vacuum gauge between the modulator and the vacuum line.
   d. With the engine running, the gauge should read normal engine vacuum. If reading is not within limits, recheck for vacuum leak or tune the engine as necessary.
2. Check the operation of the air valve and the bypass valve as follows:
   a. Disconnect connector "D" (modulator electrical lead).
   b. Start the engine and note the reading on the vacuum gauge.

CAUTION: The air and bypass valves should not be energized for more than two minutes, with a five minute cooling off period to follow. If the valves are energized longer, the solenoids will overheat and be damaged.

   c. Connect a jumper lead from the positive pole of the battery to Pin #1 on the modulator side of the connector.
   d. The reading on the vacuum gauge should drop. If the reading remains constant, the modulator is defective and should be replaced.
   e. Connect a second jumper wire from the positive side of the battery to Pin #2 on the connector (the jumper wire to Pin #1 must still be in place).
   f. The vacuum gauge reading should jump to the reading noted in step 2b above.
   g. If the vacuum gauge reading does not increase, the modulator is defective and must be replaced.
   h. If the vacuum gauge reading does increase, proceed to step 3.
3. Check the operation of the hydraulic shut-off valve as follows:
   a. Install a hydraulic pressure gauge in the modulator output port.
   b. Bleed the air from the hydraulic system and the gauge.
   c. Start the engine and depress the brake pedal to obtain 500-600 psi on the gauge.
   d. With the brake pedal depressed, connect jumper wires from the battery positive terminal to Pins #1 and #2 of the electrical connector, modulator side.
   e. The hydraulic pressure should drop to zero. If not, the modulator is faulty and should be replaced.
   f. Hold the pedal depressed for two minutes, with the jumper wires connected. The hydraulic pressure must not increase.
   g. Release the brake pedal, disconnect the jumper wires from the battery and turn the engine off.
   h. If the hydraulic pressure did rise during the two minutes the pedal was depressed, the modulator is defective and must be replaced.
4. If the results of all the steps in this test were normal, the electronic control unit is faulty and must be replaced.

Test #6 Hydraulic Pressure Test

1. Check the output of the hydraulic system as follows:
   a. Connect a hydraulic pressure gauge in the output side of the modulator.
   b. Bleed the air from the hydraulic gauge and the hydraulic system.
   c. Disconnect the modulator electrical lead.

CAUTION: The air and bypass valves should not be energized for more than two minutes, with a five minute cooling off period to follow. If the valves are energized longer, the solenoids will overheat and be damaged.

   d. Connect a jumper wire from the positive battery terminal to the #2 pin on the modulator side of the electrical connector.
   e. Start the engine and depress the brake pedal to obtain 600 psi on the hydraulic pressure gauge.
   f. If the pressure does not rise when the brake pedal is depressed, the modulator is defective and must be replaced.

Index
Gasoline Engines Rebuilding

Diesel Engines

Index

Diesel Engines

This section describes, in detail, the procedures involved in rebuilding a typical gasoline engine. The procedures specifically refer to an inline engine. However, they are basically identical to those used in rebuilding engines of nearly all designs and configurations.

If an engine runs quietly and efficiently after it has been rebuilt, it's no accident. An engine can be expected to give many miles of dependable service only if the proper reconditioning procedures and checks are performed carefully and clearances are kept within the recommended specifications. The following outline can serve as a checklist of the many procedures and parts that can effect the performance of a rebuilt engine.

Oil Pressure

Engine oil pressure is a function of all of the pressure fed bearing surfaces in the engine. All of the bearing clearances must be correct for good oil pressure and maximum service life. Listed here are some of the check points.

- Main bearings and journals—clearance, taper, and roundness
- Connecting rod bearings and journals—clearance, taper, and roundness
- Camshaft and bearings—clearance
- Rocker arms—arm and shaft wear, or ball and seat wear
- Oil pump clearances—(a new pump is a good investment)
- Fuel pump—A leaking diaphragm can allow fuel to enter the oil pan and dilute the oil.
- Carburetor—A leaking float valve or high float level can cause a rich mixture or fuel drip when the engine is not running. Either of these conditions can cause oil dilution and decrease the oil pressure.

Compression

Compression in an engine is determined by the correct fit and sealing efficiency of the piston and rings against the cylinder walls, the quality of the seal between the valve and its seat, and the seal between the cylinder head, head gasket and block. Here are some important check points.

- Valve seat and face—Machine face and seat
- Valve guides—The reconditioned seat and face won't hold up long if the valve stem clearance isn't within specifications.
- Valve seals—Oil reaching the valve seat will become a solid when it combines with the heat in the combustion chamber. This solid (carbon) will build up and eventually keep the valve from seating.
- Cylinder walls—Check for taper, out-of-round and hone to proper cross hatch pattern.

- Pistons—Check all dimensions. A poorly fitted piston will shorten the life of the new rings.

Engine Noises

Engine noises are not only annoying, but indicate conditions inside the engine that can limit the service life of the engine or shut it down completely. Generally noises are caused by too much clearance between parts or loss of oil supply. Engine noises can be caused by any of the following parts.

- Push rods—Check the condition of the ends and straightness.
- Rocker arms—Push rod seat, pivot points, and valve stem contact point.
- Valve lifters—leakdown rate, flatness of the face.
- Camshaft—lobe or journal wear.
- Valve springs—check height and pressure.
- Timing chain—stretch and wear.
- Timing gears—check wear.

Cooling System

Maintaining engine temperatures within the specified range is critical to the life of a rebuilt engine. Until new parts mate properly with each other, excessive heat can cause permanent damage or substantially reduce the service life of the reconditioned engine. If the engine is operated at temperatures below normal, the oil may not properly lubricate all of the parts. Some parts that should be checked during the rebuilding process are:

- Coolant passages—Should be free of rust and corrosion deposits.
- Core plugs—all plugs should be replaced during the rebuilding process.
- Hoses—Should be free of cracks, hard spots, and oil softened spots.
- Thermostat—Check for opening and closing at the specified temperature.
- Radiator—Check for leaks, and rust or corrosion deposits.
- Pressure cap—Should hold specified pressure, also check the gasket and vent valve operation.

The section is divided into two parts. The first, Cylinder Head Reconditioning, assumes that the cylinder head is removed from the engine, all manifolds are removed, and the cylinder head is on a workbench. The camshaft should be removed from overhead cam cylinder heads. The second section, Cylinder Block Reconditioning, covers the block, pistons, connecting rods and crankshaft. It is assumed that the engine is mounted on a work stand, and the cylinder head and all accessories are removed.

Procedures are identified as follows:

Unmarked—Basic procedures that must be performed in order to suc-

cessfully complete the rebuilding process.

Starred (*)—Procedures that should be performed to ensure maximum performance and engine life.

Double starred (**)—Procedures that may be performed to increase engine reliability. These procedures are usually reserved for extremely heavy-duty usage.

In many cases, a choice of methods is also provided. Methods are identified in the same manner as procedures. The choice of method for a procedure is at the discretion of the user.

The tools required for the basic rebuilding procedure should, with minor exceptions, be those included in a mechanic's tool kit. An accurate torque wrench, and a dial indicator (reading in thousandths) mounted on a universal base should be available. Bolts and nuts with no torque specification should be tightened according to size (see chart). Special tools, where required, all are readily available from the major tool suppliers. The services of a competent automotive machine shop must also be readily available.

When assembling the engine, any parts that will be in frictional contact must be pre-lubricated, to provide protection on initial start-up. Any product specifically formulated for this purpose may be used. NOTE: *Do not use engine oil.* Where semi-permanent (locked but removable) installation of bolts or nuts is desired, threads should be cleaned and coated with a liquid locking compound. Studs may be permanently installed using a stud mounting compound.

Aluminum has become increasingly popular for use in engines, due to its low weight and excellent heat transfer characteristics. The following precautions must be observed when handling aluminum engine parts:

—Never hot-tank aluminum parts.

—Remove all aluminum parts (identification tags, etc.) from engine parts before hot-tanking (otherwise they will be removed during the process).

—Always coat threads lightly with engine oil or anti-seize compounds before installation, to prevent seizure.

—Never over-torque bolts or spark plugs in aluminum threads. Should stripping occur, threads can be restored according to the following procedure, using Heli-Coil thread inserts:

Tap drill the hole with the stripped threads to the specified size (see chart). Using the specified tap (NOTE: *Heli-Coil tap sizes refer to the size thread being replaced, rather than the actual tap size*), tap the hole for the Heli-Coil. Place the insert on the proper installation tool (see chart). Apply pressure on the insert while winding it clockwise into the hole, until the top of the insert is one turn below the surface. Remove the installation tool, and break the installation tang from the bottom of the insert by moving it up and down.

If the Heli-Coil must be removed, tap the removal tool firmly into the hole, so that it engages the top thread, and turn the tool counter-clockwise to extract the insert.

Snapped bolts or studs may be removed, using a stud extractor (unthreaded) or locking pliers (threaded). Penetrating oil will often aid in breaking frozen threads. In cases where the stud or bolt is flush with, or below the surface, proceed as follows:

Drill a hole in the broken stud or bolt, approximately ½ its diameter. Select a screw extractor of the proper size, and tap it into the stud or bolt. Turn the extractor counter-clockwise to remove the stud or bolt.

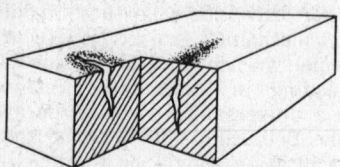

Heli-Coil installation (© Chrysler Corp.)

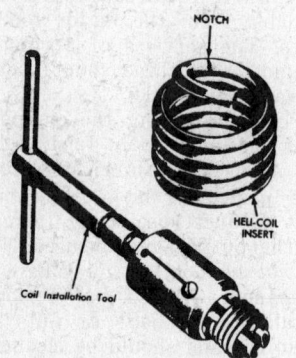

Magnaflux® indication of cracks

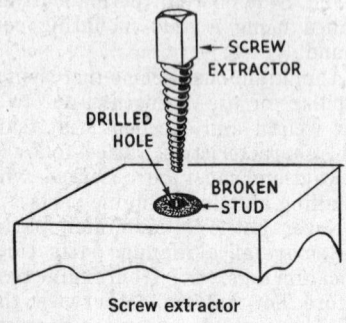

Heli-Coil and installation tool

Screw extractor

Magnaflux® and Zyglo® are inspection techniques used to locate material flaws, such as stress cracks. Magnafluxing® coats the part with fine magnetic particles, and subjects the part to a magnetic field. Cracks cause breaks in the magnetic field, which are outlined by the particles. Since Magnaflux® is a magnetic process, it is applicable only to ferrous materials. The Zyglo® process coats the material with a fluorescent dye penetrant, and then subjects it to blacklight inspection, under which cracks glow brightly. Parts made of any material may be tested using Zyglo®. While Magnaflux® and Zyglo® are excellent for general

HELI-COIL SPECIFICATIONS

HELI-COIL INSERT			DRILL	TAP	INSERT. TOOL	EXTRACT-ING TOOL
THREAD SIZE	PART NO.	INSERT LENGTH (IN.)	SIZE	PART NO.	PART NO.	PART NO.
1/2 -20	1185-4	3/8	17/64(.266)	4 CPB	528-4N	1227-6
5/16-18	1185-5	15/32	Q(.332)	5 CPB	528-5N	1227-6
3/8 -16	1185-6	9/16	X(.397)	6 CPB	528-6N	1227-6
7/16-14	1185-7	21/32	29/64(.453)	7 CPB	528-7N	1227-16
1/2 -13	1185-8	3/4	33/64(.516)	8 CPB	528-8N	1227-16

TORQUE (FT. LBS.)*

U.S.

Bolt Diameter (inches)	Bolt Grade (SAE)				Wrench Size (inches)	
	1 and 2	5	6	8	Bolt	Nut
1/4	5	7	10	10.5	3/8	7/16
5/16	9	14	19	22	1/2	9/16
3/8	15	25	34	37	9/16	5/8
7/16	24	40	55	60	5/8	3/4
1/2	37	60	85	92	3/4	13/16
9/16	53	88	120	132	7/8	7/8
5/8	74	120	167	180	15/16	1
3/4	120	200	280	296	1-1/8	1-1/8
7/8	190	302	440	473	1-5/16	1-5/16
1	282	466	660	714	1-1/2	1-1/2

METRIC

Bolt Diameter (mm)	Bolt Grade				Wrench Size (mm)
	5D	8G	10K	12K	Bolt and Nut
6	5	6	8	10	10
8	10	16	22	27	14
10	19	31	40	49	17
12	34	54	70	86	19
14	55	89	117	137	22
16	83	132	175	208	24
18	111	182	236	283	27
22	182	284	394	464	32
24	261	419	570	689	36

*—Torque values are for lightly oiled bolts. CAUTION: Bolts threaded into aluminum require much less torque.

inspection, and locating hidden defects, specific checks of suspected cracks may be made at lower cost and more readily using spot check dye. The dye is sprayed onto the suspected area, wiped off, and the area is then sprayed with a developer. Cracks then will show up brightly. Spot check dyes will only indicate surface cracks; therefore, structural cracks below the surface may escape detection. When questionable, the part should be tested using Magnaflux® or Zyglo®.

Cylinder Head Reconditioning

PROCEDURE	METHOD
Remove the rocker arms:	Remove the rocker arms with shaft(s) or balls and nuts. Wire the sets of rockers, balls and nuts together, and identify according to the corresponding valve.
Remove the valves and springs:	Compress the valve springs with a spring compressor, remove the retaining locks, valve retainers and springs, and the valve seals. Check the condition of the valve stem tips and retainer grooves. Remove any burrs so that the valve guides won't be damaged as the valves are removed. Remove the valves and place them in a numbered holder so that they can be installed in the same position.
Check the valve stem-to-guide clearance:  **Checking the valve stem-to-guide clearance**	Clean the valve stem with lacquer thinner or a similar solvent to remove all gum and varnish. Clean the valve guides using solvent and an expanding wire-type valve guide cleaner. Mount a dial indicator so that the stem is at 90° to the valve stem, as close to the valve guide as possible. Move the valve off its seat, and measure the valve guide-to-stem clearance by moving the stem back and forth to actuate the dial indicator. Measure the valve stems using a micrometer, and compare to specifications, to determine whether stem or guide wear is responsible for excessive clearance.
De-carbon the cylinder head and valves: 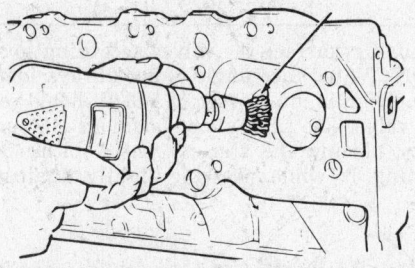 **Removing carbon from the cylinder head**	Chip carbon away from the valve heads, combustion chambers, and ports, using a chisel made of hardwood. Remove the remaining deposits with a stiff wire brush. *NOTE: Ensure that the deposits are actually removed, rather than burnished.*
Hot-tank the cylinder head:	Have the cylinder head hot-tanked to remove grease, corrosion, and scale from the water passages. *NOTE: In the case of overhead cam cylinder heads, consult the operator to determine whether the camshaft bearings will be damaged by the caustic solution.*
Degrease the remaining cylinder head parts:	Using solvent (i.e., Gunk), clean the rockers, rocker shaft(s) (where applicable), rocker balls and nuts, springs, spring retainers, and keepers. Do not remove the protective coating from the springs.
Check the cylinder head for warpage: 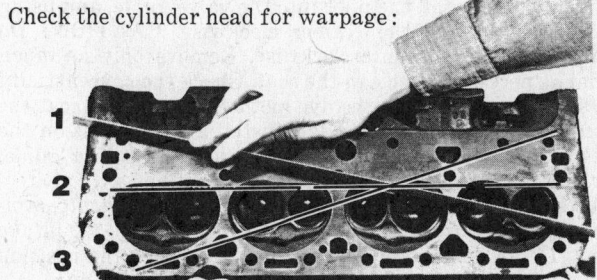 1 & 3 CHECK DIAGONALLY 2 CHECK ACROSS CENTER **Checking the cylinder head for warpage**	Place a straight-edge across the gasket surface of the cylinder head. Using feeler gauges, determine the clearance at the center of the straight-edge. Measure across both diagonals, along the longitudinal centerline, and across the cylinder head at several points. If warpage exceeds .003″ in a 6″ span, or .006″ over the total length, the cylinder head must be resurfaced. *NOTE: If warpage exceeds the manufacturers maximum tolerance for material removal, the cylinder head must be replaced.* When milling the cylinder heads of V-type engines, the intake manifold mounting position is altered, and must be corrected by milling the manifold flange a proportionate amount.

Cylinder Head Reconditioning

PROCEDURE	METHOD

*Knurling the valve guides:

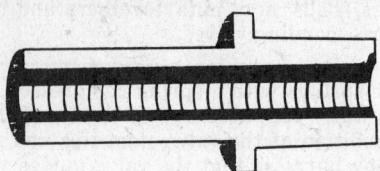

Cut-a-way view of a knurled valve guide

*Valve guides which are not excessively worn or distorted may, in some cases, be knurled rather than replaced. Knurling is a process in which metal is displaced and raised, thereby reducing clearance. Knurling also provides excellent oil control. The possibility of knurling rather than replacing valve guides should be discussed with a machinist.

Replacing the valve guides: *NOTE: Some worn valve guides can be resized by knurling, or the guide can be reamed to accept a valve with an oversized stem.*

Valve guide removal tool

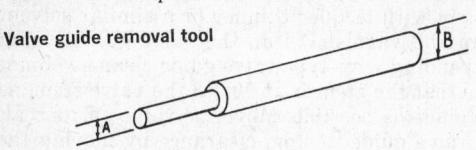

A-VALVE GUIDE I.D.
B-SLIGHTLY SMALLER THAN VALVE GUIDE O.D.

WASHERS

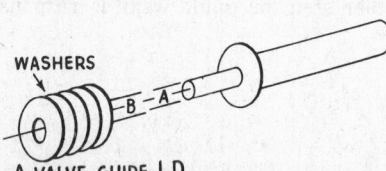

A-VALVE GUIDE I.D.
B-LARGER THAN THE VALVE GUIDE O.D.

Valve guide installation tool with washers used during installation

Depending on the type of cylinder head, valve guides may be pressed hammered, or shrunk in. In cases where the guides are shrunk into the head, replacement should be left to an equipped machine shop. In other cases, the guides are replaced as follows: Press or tap the valve guides out of the head using a stepped drift (see illustration). Determine the height above the boss that the guide must extend, and obtain a stack of washers, their I.D. similar to the guide's O.D., of that height. Place the stack of washers on the guide, and insert the guide into the boss. *NOTE: Valve guides are often tapered or beveled for installation.* Using the stepped installation tool (see illustration), press or tap the guides into position. Ream the guides according to the size of the valve stem.

Replacing valve seat inserts:

Replacement of valve seat inserts which are worn beyond resurfacing or broken, if feasible, must be done by a machine shop.

Resurfacing (grinding) the valve face:

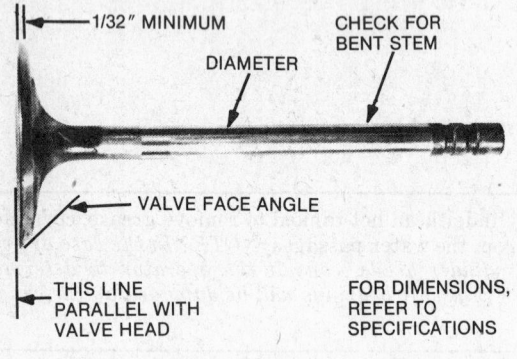

1/32" MINIMUM
DIAMETER
CHECK FOR BENT STEM
VALVE FACE ANGLE
THIS LINE PARALLEL WITH VALVE HEAD
FOR DIMENSIONS, REFER TO SPECIFICATIONS

Critical valve dimensions

Using a valve grinder, resurface the valves according to specifications. **CAUTION: Valve face angle is not always identical to valve seat angle.** A minimum margin of 1/32" should remain after grinding the valve. The valve stem should also be squared and resurfaced, by placing the stem in the V-block of the grinder, and turning it while pressing lightly against the grinding wheel.

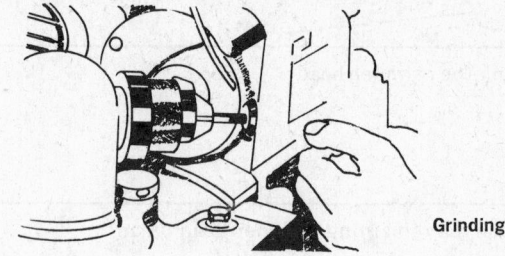

Grinding a valve

Resurfacing the valve seats using reamers:

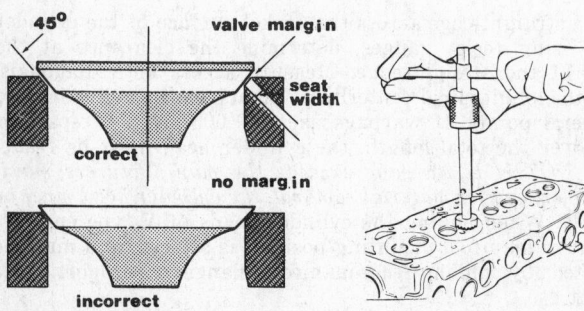

45°
valve margin
seat width
correct
no margin
incorrect

Valve seat width and centering

Reaming the valve seat (© Outboard Marine Corp.)

Select a reamer of the correct seat angle, slightly larger than the diameter of the valve seat, and assemble it with a pilot of the correct size. Install the pilot into the valve guide, and using steady pressure, turn the reamer clockwise. **CAUTION: Do not turn the reamer counterclockwise.** Remove only as much material as necessary to clean the seat. Check the concentricity of the seat (see below). If the dye method is not used, coat the valve face with Prussian blue dye, install and rotate it on the valve seat. Using the dye marked area as a centering guide, center and narrow the valve seat to specifications with correction cutters. *NOTE: When no specifications are available, minimum seat width for exhaust valves should be 5/64", intake valves 1/16".* After making correction cuts, check the position of the valve seat on the valve face using Prussian blue dye.

Cylinder Head Reconditioning

PROCEDURE	METHOD

*Resurfacing the valve seats using a grinder :

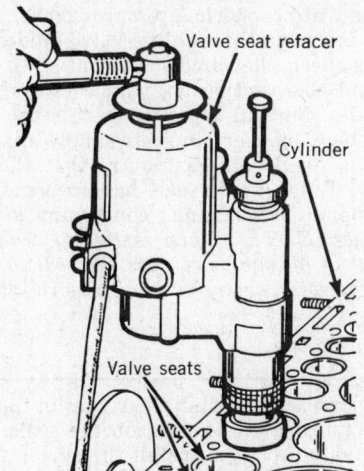

Valve seat refacer

Cylinder

Valve seats

Grinding a valve seat (© Subaru)

Select a pilot of the correct size, and a coarse stone of the correct seat angle. Lubricate the pilot if necessary, and install the tool in the valve guide. Move the stone on and off the seat at approximately two cycles per second, until all flaws are removed from the seat. Install a fine stone, and finish the seat. Center and narrow the seat using correction stones, as described above.

Checking the valve seat concentricity :

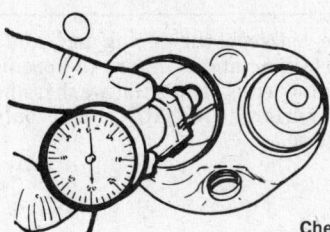

Check the valve seat for concentricity using a dial guage

Coat the valve face with Prussian blue dye, install the valve, and rotate it on the valve seat. If the entire seat becomes coated, and the valve is known to be concentric, the seat is concentric.

*Install the dial gauge pilot into the guide, and rest of the arm on the valve seat. Zero the gauge, and rotate the arm around the seat. Run-out should not exceed .002″.

---

*Lapping the valves: *NOTE: Valve lapping is done to ensure efficient sealing of resurfaced valves and seats. Valve lapping alone is not recommended for use as a resurfacing procedure.*

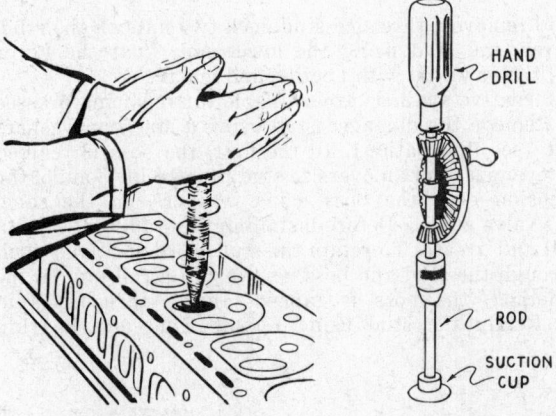

HAND DRILL

ROD

SUCTION CUP

Hand lapping the valves   **Home made mechanical valve lapping tool**

*Invert the cylinder head, lightly lubricate the valve stems, and install the valves in the head as numbered. Coat valve seats with fine grinding compound, and attach the lapping tool suction cup to a valve head (*NOTE: Moisten the suction cup*). Rotate the tool between the palms, changing position and lifting the tool often to prevent grooving. Lap the valve until a smooth, polished seat is evident. Remove the valve and tool, and rinse away all traces of grinding compound.

Fasten a suction cup to a piece of drill rod, and mount the rod in a hand drill. Proceed as above, using the hand drill as a lapping tool. **CAUTION: Due to the higher speeds involved when using the hand drill, care must be exercised to avoid grooving the seat. Lift the tool and change direction of rotation often.

Check the valve springs :

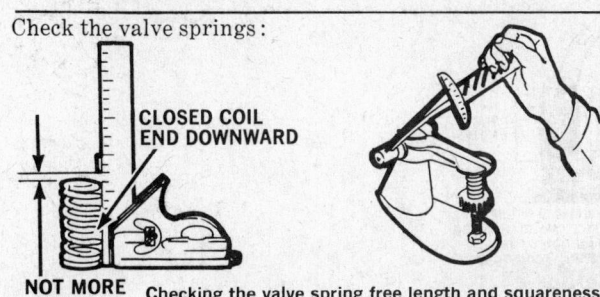

CLOSED COIL END DOWNWARD

NOT MORE THAN 1/16″   **Checking the valve spring free length and squareness**

Place the spring on a flat surface next to a square. Measure the height of the spring, and rotate it against the edge of the square to measure distortion. If spring height varies (by comparison) by more than 1/16″ or if distortion exceeds 1/16″, replace the spring.

**In addition to evaluating the spring as above, test the spring pressure at the installed and compressed (installed height minus valve lift) height using a valve spring tester. Springs used on small displacement engines (up to 3 liters) should be ± 1 lb. of all other springs in either position. A tolerance of ± 5 lbs. is permissible on larger engines.

Cylinder Head Reconditioning

PROCEDURE	METHOD

*Install valve stem seals:

RETAINER

SPRING

INTAKE VALVE

SEAL

Valve stem seal installation (© Chrysler Corp.)

*Due to the pressure differential that exists at the ends of the intake valve guides (atmospheric pressure above, manifold vacuum below), oil is drawn through the valve guides into the intake port. This has been alleviated somewhat since the addition of positive crankcase ventilation, which lowers the pressure above the guides. Several types of valve stem seals are available to reduce blow-by. Certain seals simply slip over the stem and guide boss, while others require that the boss be machined. Recently, Teflon guide seals have become popular. Consult a parts supplier or machinist concerning availability and suggested usages. *NOTE: When installing seals, ensure that a small amount of oil is able to pass the seal to lubricate the valve guides; otherwise, excessive wear may result.*

Install the valves:

Lubricate the valve stems, and install the valves in the cylinder head as numbered. Lubricate and position the seals (if used, see above) and the valve springs. Install the spring retainers, compress the springs, and insert the keys using needlenose pliers or a tool designed for this purpose. *NOTE: Retain the keys with wheel bearing grease during installation.*

Checking valve spring installed height:

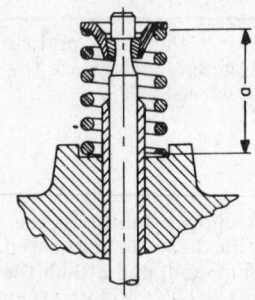

Valve spring installed height dimension (© Porsche)

Measure the distance between the spring pad and the lower edge of the spring retainer, and compare to specifications. If the installed height is incorrect, add shim washers between the spring pad and the spring. **CAUTION: Use only washers designed for this purpose.**

Replacing rocker studs:

Reaming the stud bore for oversize rocker studs (© Buick Div. G.M. Corp.)

In order to remove a threaded stud, lock two nuts on the stud, and unscrew the stud using the lower nut. Coat the lower threads of the new stud with Loctite, and install.

Two alternative methods are available for replacing pressed in studs. Remove the damaged stud using a stack of washers and a nut (see illustration). In the first, the boss is reamed .005-006″ oversize, and an oversize stud pressed in. Control the stud extension over the boss using washers, in the same manner as valve guides. Before installing the stud, coat it with white lead and grease. To retain the stud more positively drill a hole through the stud and boss, and install a roll pin. In the second method, the boss is tapped, and a threaded stud installed. Retain the stud using Loctite Stud and Bearing Mount.

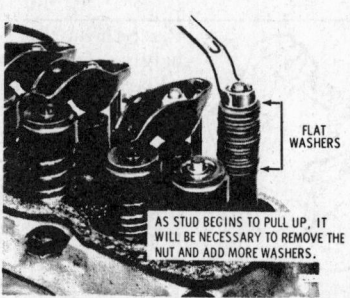

FLAT WASHERS

AS STUD BEGINS TO PULL UP, IT WILL BE NECESSARY TO REMOVE THE NUT AND ADD MORE WASHERS.

Extracting a pressed in rocker stud (© Buick Div. G.M. Corp.)

Cylinder Head Reconditioning

PROCEDURE	METHOD
Inspect the rocker arms, balls, studs, and nuts (where applicable): 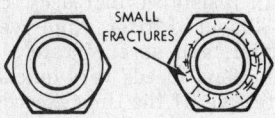 **Stress cracks in rocker nuts** (© Ford Motor Co.)	Visually inspect the rocker arms, balls, studs, and nuts for cracks, galling, burning, scoring, or wear. If all parts are intact, liberally lubricate the rocker arms and balls, and install them on the cylinder head. If wear is noted on a rocker arm at the point of valve contact, grind it smooth and square, removing as little material as possible. Replace the rocker arm if excessively worn. If a rocker stud shows signs of wear, it must be replaced (see below). If a rocker nut shows stress cracks, replace it. If an exhaust ball is galled or burned, substitute the intake ball from the same cylinder (if it is intact), and install a new intake ball. *NOTE: Avoid using new rocker balls on exhaust valves.*
Inspect the rocker shaft(s) and rocker arms (where applicable): 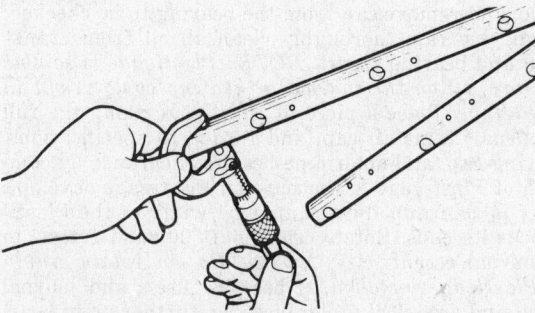 **Checking the rocker arm shaft O.D.** (© Chrysler Corp.) 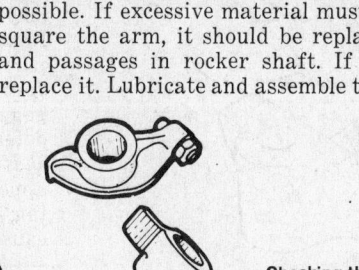 **Checking the rocker arms** (© Chrysler Corp.)	Remove rocker arms, springs and washers from rocker shaft. *NOTE: Lay out parts in the order they are removed.* Inspect rocker arms for pitting or wear on the valve contact point, or excessive bushing wear. Bushings need only be replaced if wear is excessive, because the rocker arm normally contacts the shaft at one point only. Grind the valve contact point of rocker arm smooth if necessary, removing as little material as possible. If excessive material must be removed to smooth and square the arm, it should be replaced. Clean out all oil holes and passages in rocker shaft. If shaft is grooved or worn, replace it. Lubricate and assemble the rocker shaft.
Inspect the pushrods:	Remove the pushrods, and, if hollow, clean out the oil passages using fine wire. Roll each pushrod over a piece of clean glass. If a distinct clicking sound is heard as the pushrod rolls, the rod is bent, and must be replaced. Check the ends of each push rod for wear or galling. *NOTE: If the end of a push rod is worn the rocker arm that it was operating is also worn and must be replaced or it will ruin the new push rod.*
Inspect the valve lifters: 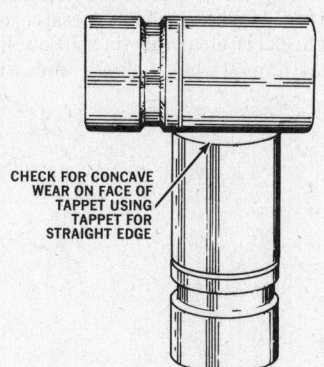 CHECK FOR CONCAVE WEAR ON FACE OF TAPPET USING TAPPET FOR STRAIGHT EDGE **Checking the lifter face**	Remove lifters from their bores, and remove gum and varnish, using solvent. Clean walls of lifter bores. Check lifters for concave wear as illustrated. If face is worn concave, replace lifter, and carefully inspect the camshaft. Lightly lubricate lifter and insert it into its bore. If play is excessive, an oversize lifter must be installed (where possible). Consult a machinist concerning feasibility. If play is satisfactory, remove, lubricate, and reinstall the lifter. *NOTE: A valve lifter that is worn concave (dished) indicates a worn camshaft. It is poor practice to install new lifters against a worn camshaft.*

Engine Rebuilding

Cylinder Block Reconditioning

PROCEDURE	METHOD

*Testing hydraulic lifter leak down:

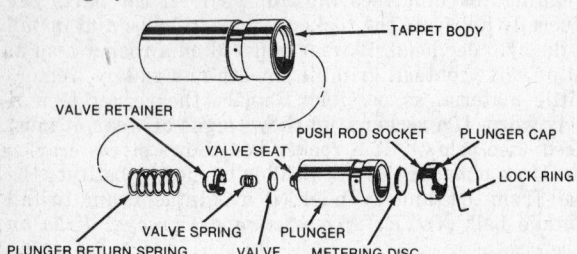

Exploded view of a typical hydraulic lifter (© American Motors Corp.)

Submerge lifter in a container of kerosene. Chuck a used pushrod or its equivalent into a drill press. Position container of kerosene so pushrod acts on the lifter plunger. Pump lifter with the drill press, until resistance increases. Pump several more times to bleed any air out of lifter. Apply very firm, constant pressure to the lifter, and observe rate at which fluid bleeds out of lifter. If the fluid bleeds very quickly (less than 15 seconds), lifter is defective. If the time exceeds 60 seconds, lifter is sticking. In either case, recondition or replace lifter. If lifter is operating properly (leak down time 15-60 seconds), lubricate and install it.

Checking the main bearing clearance:

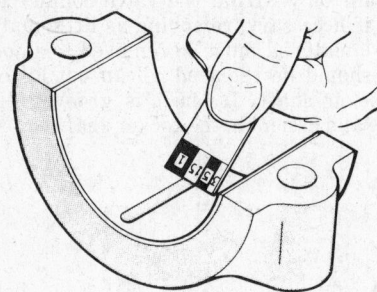

Measuring plastigauge to determine bearing clearance
(© Chrysler Corp.)

Invert engine, and remove cap from the bearing to be checked. Using a clean, dry rag, thoroughly clean all oil from crankshaft journal and bearing insert. *NOTE: Plastigage is soluble in oil; therefore, oil on the journal or bearing could result in erroneous readings.* Place a piece of Plastigage along the full length of journal, reinstall cap, and torque to specifications. Remove bearing cap, and determine bearing clearance by comparing width of Plastigage to the scale on Plastigage envelope. Journal taper is determined by comparing width of the Plastigage strip near its ends. Rotate crankshaft 90° and retest, to determine journal eccentricity. *NOTE: Do not rotate crankshaft with Plastigage installed.* If bearing insert and journal appear intact, and are within tolerances, no further main bearing service is required. If bearing or journal appear defective, cause of failure should be determined before replacement.

*Remove crankshaft from block (see below). Measure the main bearing journals at each end twice (90° apart) using a micrometer, to determine diameter, journal taper and eccentricity. If journals are within tolerances, reinstall bearing caps at their specified torque. Using a telescope gauge and micrometer, measure bearing I.D. parallel to piston axis and at 30° on each side of piston axis. Subtract journal O.D. from bearing I.D. to determine oil clearance. If crankshaft journals appear defective, or do not meet tolerances, there is no need to measure bearings; for the crankshaft will require grinding and/or undersize bearings will be required. If bearing appears defective, cause for failure should be determined prior to replacement.

Checking the connecting rod bearing clearance:

Connecting rod bearing clearance is checked in the same manner as main bearing clearance, using Plastigage. Before removing the crankshaft, connecting rod side clearance also should be measured and recorded.

Check connecting rod side clearance:

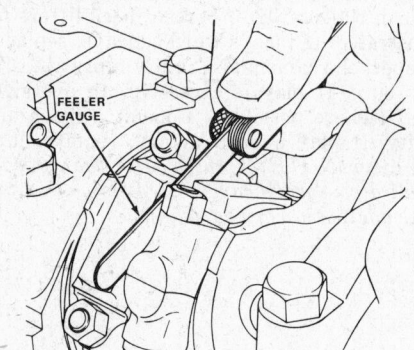

Connecting rod side clearance measurement (© Jeep Corp.)

Determine the clearance between the sides of the connecting rods and the crankshaft, using feeler gauges. If clearance is below the minimum tolerance, the rod may be machined to provide adequate clearance. If clearance is excessive, substitute an unworn rod, and recheck. If clearance is still outside specifications, the crankshaft must be welded and reground, or replaced.

Cylinder Block Reconditioning

PROCEDURE	METHOD

Remove the ridge from the top of the cylinder:

RIDGE CAUSED BY CYLINDER WEAR

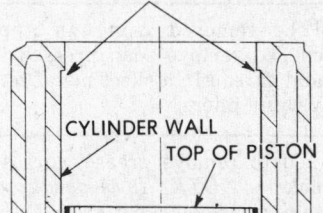

CYLINDER WALL
TOP OF PISTON

Cylinder bore ridge (© Pontiac Div. G.M. Corp.)

In order to facilitate removal of the piston and connecting rod, the ridge at the top of the cylinder (unworn area; see illustration) must be removed. Place the piston at the bottom of the bore, and cover it with a rag. Cut the ridge away using a ridge reamer, exercising extreme care to avoid cutting too deeply. Remove the rag, and remove cuttings that remain on the piston. **CAUTION: If the ridge is not removed, and new rings are installed, damage to rings will result.**

Removing the crankshaft:

Rod number and squirt hole location (© Jeep Corp.)

Using a punch, mark the corresponding main bearing caps and saddles according to position (i.e., one punch on the front main cap and saddle, two on the second, three on the third, etc.). Using number stamps, identify the corresponding connecting rods and caps, according to cylinder (if no numbers are present). Remove the main and connecting rod caps, and place sleeves of plastic tubing over the connecting rod bolts, to protect the journals as the crankshaft is removed. Lift the crankshaft out of the block.

Removing the piston and connecting rod:

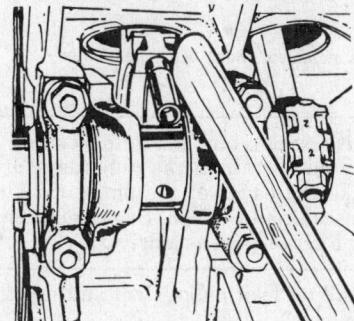

Removing the piston

Invert the engine, and push the pistons and connecting rods out of the cylinders. If necessary, tap the connecting rod boss with a wooden hammer handle, to force the piston out. **CAUTION: Do not attempt to force the piston past the cylinder ridge** (see above).

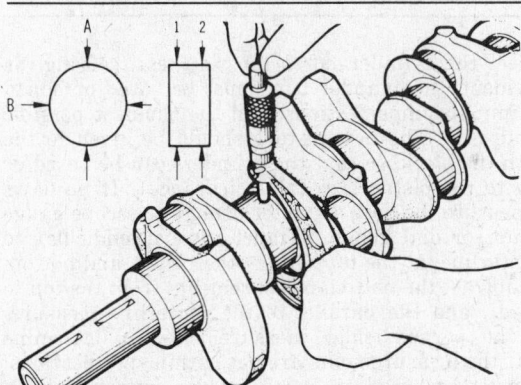

Measuring the crank pin O.D. (© Chrysler Corp.)

Service the crankshaft:

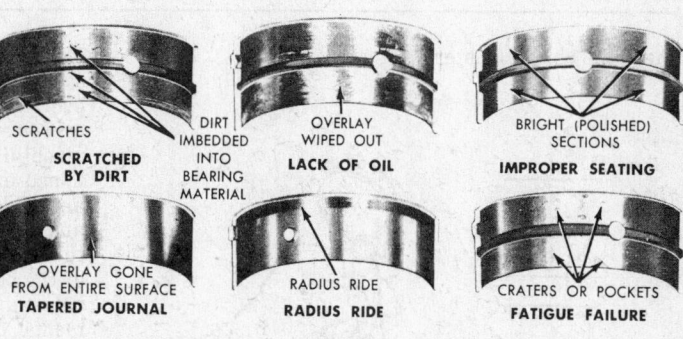

Causes of bearing failure (© Ford Motor Co.)

Ensure that all oil holes and passages in the crankshaft are open and free of sludge. If necessary, have the crankshaft ground to the largest possible undersize.

****Have the crankshaft Magnafluxed, to locate stress cracks. Consult a machinist concerning additional service procedures, such as surface hardening (e.g., nitriding, Tuftriding) to improve wear characteristics, cross drilling and chamfering the oil holes to improve lubrication, and balancing.

Cylinder Block Reconditioning

PROCEDURE	METHOD
Removing freeze plugs :	Drill a hole in the center of the freeze plugs, and pry them out using a screwdriver or drift.
Remove the oil gallery plugs :	Threaded plugs should be removed using an appropriate (usually square) wrench. To remove soft, pressed in plugs, drill a hole in the plug, and thread in a sheet metal screw. Pull the plug out by the screw using pliers.
Hot-tank the block :	Have the block hot-tanked to remove grease, corrosion, and scale from the water jackets. *NOTE: It is good practice to replace the camshaft bearings whenever a cylinder block has been hot tanked.*
Check the block for cracks : 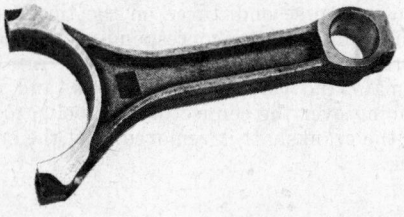 Connecting rod as seen in natural light before magnetic inspection 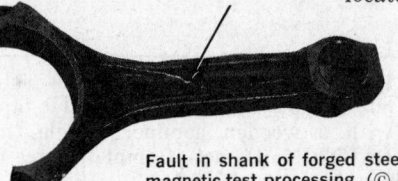 Fault in shank of forged steel connecting rod is revealed under black light after magnetic test processing. (© Magnaflux Corp.)	Visually inspect the block for cracks or chips. The most common locations are as follows : Adjacent to freeze plugs. Between the cylinders and water jackets. Adjacent to the main bearing saddles. At the extreme bottom of the cylinders. Check only suspected cracks using spot check dye (see introduction). If a crack is located, consult a machinist concerning possible repairs. **Magnaflux the block to locate hidden cracks. If cracks are located, consult a machinist about feasibility of repair.
Install the oil gallery plugs and freeze plugs :	Coat freeze plugs with sealer and tap into position using a piece of pipe, slightly smaller than the plug, as a driver. To ensure retention, stake the edges of the plugs. Coat threaded oil gallery plugs with sealer and install. Drive replacement soft plugs into block using a large drift as a driver. **Rather than reinstalling lead plugs, drill and tap the holes, and install threaded plugs.
Check the bore diameter and surface : 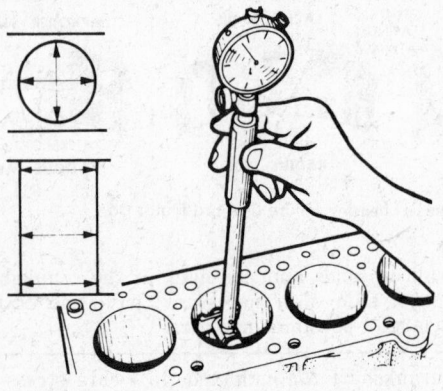 Checking the cylinder bore size (© Chrysler Corp.)	Visually inspect the cylinder bores for roughess, scoring, or scuffing. If evident, the cylinder bore must be bored or honed oversize to eliminate imperfections, and the smallest possible oversize piston used. The new pistons should be given to the machinist with the block, so that the cylinders can be bored or honed exactly to the piston size (plus clearance). If no flaws are evident, measure the bore diameter using a telescope gauge and micrometer, or dial guage, parallel and perpendicular to the engine centerline, at the top (below the ridge) and bottom of the bore. Subtract the bottom measurements from the top to determine taper, and the parallel to the centerline measurements from the perpendicular measurements to determine eccentricity. If the measurements are not within specifications, the cylinder must be bored or honed, and an oversize piston installed. If the measurements are within specifications the cylinder may be used as is, with only finish honing (see below). *NOTE: Prior to submitting the block for boring, perform the following operation(s).*

Cylinder Block Reconditioning

PROCEDURE	METHOD

Check the bore diameter and surface:

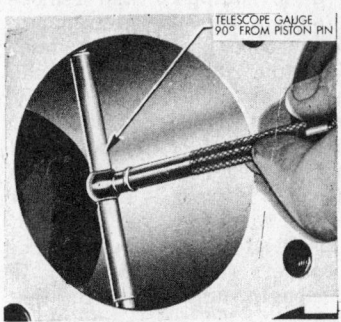

Measuring the cylinder bore with a telescopic gauge
(© Buick Div. G.M. Corp.)

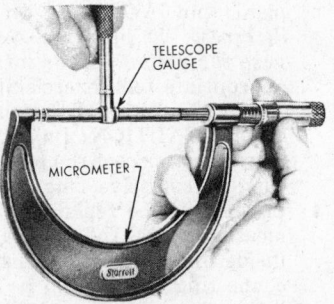

Determining the cylinder bore size by measuring the telescopic gauge with a micrometer (© Buick Div. G.M. Corp.)

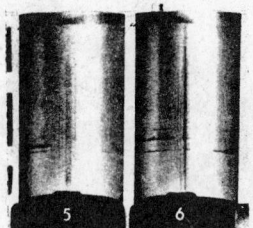

5, 6 Score marks caused by a split piston skirt. Damage is not serious enough to warrant reboring

7. Ring seized longitudinally, causing a score mark 1 3/16" wide, on the land side of the piston groove. The honing pattern is destroyed and the cylinder must be rebored

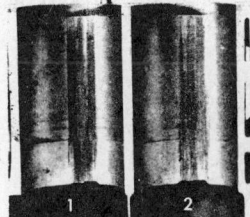
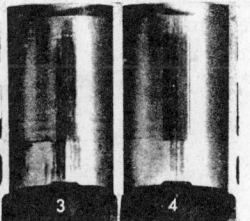

1, 2, 3 Piston skirt seizure resulted in this pattern. Engine must be rebored

4. Piston skirt and oil ring seizure caused this damage. Engine must be rebored

Cylinder wall damage (© Daimler-Benz A.G.)

8. Result of oil ring seizure. Engine must be rebored

9. Oil ring seizure here was not serious enough to warrant reboring. The honing marks are still visible

---

Check the block deck for warpage:

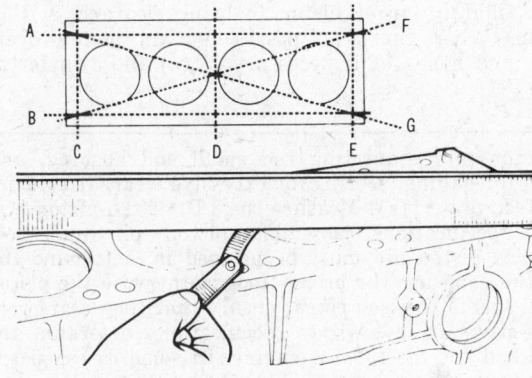

Using a straightedge and feeler gauges, check the block deck for warpage in the same manner that the cylinder head is checked. If warpage exceeds specifications, have the deck resurfaced. *NOTE: In certain cases a specification for total material removal (Cylinder head and block deck) is provided. This specification must not be exceeded.*

Checking the cylinder block for distortion (© Chrysler Corp.)

Check the cylinder block bearing alignment:

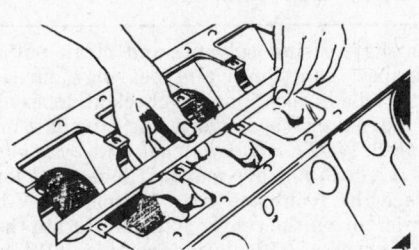

Remove the upper bearing inserts. Place a straightedge in the bearing saddles along the centerline of the crankshaft. If clearance exists between the straightedge and the center saddle, the block must be alignbored.

Checking main bearing saddle alignment

Cylinder Block Reconditioning

PROCEDURE	METHOD

Clean and inspect the pistons and connecting rods:

Removing the piston rings

Using a ring expander, remove the rings from the piston. Remove the retaining rings (if so equipped) and remove piston pin. *NOTE: If the piston pin must be pressed out, determine the proper method and use the proper tools; otherwise the piston will distort.* Clean the ring grooves using an appropriate tool, exercising care to avoid cutting too deeply. Thoroughly clean all carbon and varnish from the piston with solvent. **CAUTION: Do not use a wire brush or caustic solvent on pistons.** Inspect the pistons for scuffing, scoring, cracks, pitting, or excessive ring groove wear. If wear is evident, the piston must be replaced. Check the connecting rod length by measuring the rod from the inside of the large end to the inside of the small end using calipers (see illustration). All connecting rods should be equal length. Replace any rod that differs from the others in the engine.

*Have the connecting rod alignment checked in an alignment fixture by a machinist. Replace any twisted or bent rods.

*Magnaflux the connecting rods to locate stress cracks. If cracks are found, replace the connecting rod.

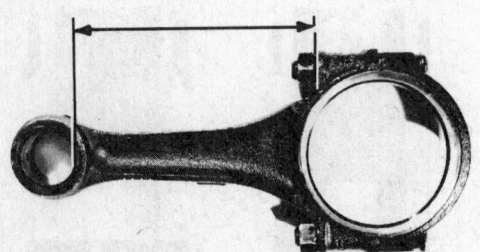

Connecting rod length checking dimension

Ring Groove Cleaner

Cleaning the piston ring grooves (© Ford Motor Co.)

Fit the pistons to the cylinders:

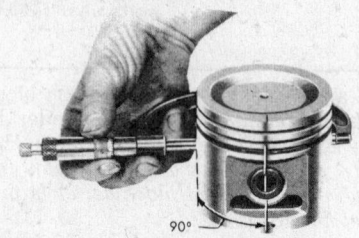

90°

Measuring the piston for fitting (© Buick Div. G.M. Corp.)

Using a telescope gauge and micrometer, or a dial gauge, measure the cylinder bore diameter perpendicular to the piston pin, 2½″ below the deck. Measure the piston perpendicular to its pin on the skirt. The difference between the two measurements is the piston clearance. If the clearance is within specifications or slightly below (after boring or honing), finish honing is all that is required. If the clearance is excessive, try to obtain a slightly larger piston to bring clearance within specifications. Where this is not possible, obtain the first oversize piston, and hone (of if necessary, bore) the cylinder to size.

Assemble the pistons and connecting rods:

Installing the piston pin lock rings

Inspect piston pin, connecting rod small end bushing, and piston bore for galling, scoring, or excessive wear. If evident, replace defective part(s). Measure the I.D. of the piston pin. If within specifications, assemble piston pin and rod. **CAUTION: If piston pin must be pressed in, determine the proper method and use the proper tools; otherwise the piston will distort.** Install the lock rings; ensure that they seat properly. If the parts are not within specifications, determine the service method for the type of engine. In some cases, piston and pin are serviced as an assembly when either is defective. Others specify reaming the piston and connecting rods for an oversize pin. If the connecting rod bushing is worn, it may in many cases be replaced. Reaming the piston and replacing the rod bushing are machine shop operations.

Clean and inspect the camshaft:

BEARING JOURNALS

FUEL PUMP DRIVE ECCENTRIC

DISTRIBUTOR DRIVE GEAR

Checking the camshaft for straightness (© Chevrolet Motor Div. G.M. Corp.)

Degrease the camshaft, using solvent, and clean out all oil holes. Visually inspect cam lobes and bearing journals for excessive wear. If a lobe is questionable, check all lobes as indicated below. If a journal or lobe is worn, the camshaft must be reground or replaced. *NOTE: If a journal is worn, there is a good chance that the bushings are worn.* If lobes and journals appear intact, place the front and rear journals in V-blocks, and rest a dial indicator on the center journal. Rotate the camshaft to check straightness. If deviation exceeds .001″, replace the camshaft.

Cylinder Block Reconditioning

PROCEDURE	METHOD

Clean and inspect the camshaft:

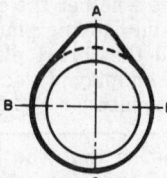

Camshaft lobe measurement (© Ford Motor Co.)

*Check the camshaft lobes with a micrometer, by measuring the lobes from the nose to base and again at 90° (see illustration). The lift is determined by subtracting the second measurement from the first. If all exhaust lobes and all intake lobes are not identical, the camshaft must be reground or replaced.

Replace the camshaft bearings:

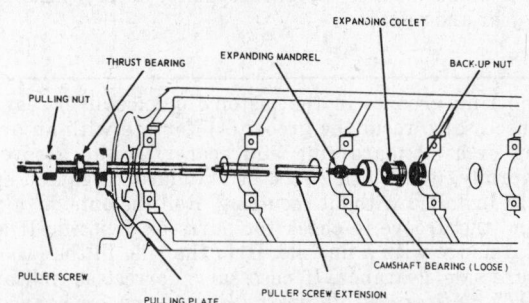

Camshaft bearing removal and installation tool (typical) (© Ford Motor Co.)

If excessive wear is indicated, or if the engine is being completely rebuilt, camshaft bearings should be replaced as follows: Drive the camshaft rear plug from the block. Assemble the removal puller with its shoulder on the bearing to be removed. Gradually tighten the puller nut until bearing is removed. Remove remaining bearings, leaving the front and rear for last. To remove front and rear bearings, reverse position of the tool, so as to pull the bearings in toward the center of the block. Leave the tool in this position, pilot the new front and rear bearings on the installer, and pull them into position: Return the tool to its original position and pull remaining bearings into position. *NOTE: Ensure that oil holes align when installing bearings.* Replace camshaft rear plug, and stake it into position to aid retention.

Finish hone the cylinders:

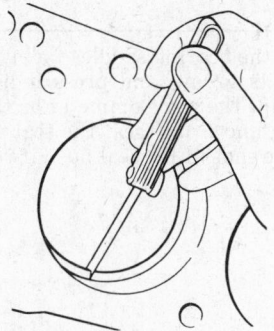

Finish honed cylinder (© Chrysler Corp.)

Chuck a flexible drive hone into a power drill, and insert it into the cylinder. Start the hone, and move it up and down in the cylinder at a rate which will produce approximately a 60° cross-hatch pattern (see illustration). *NOTE: Do not extend the hone below the cylinder bore.* After developing the pattern, remove the hone and recheck piston fit. Wash the cylinders with a detergent and water solution to remove abrasive dust, dry, and wipe several times with a rag soaked in engine oil.

Check piston ring end-gap:

Checking ring end gap (© Outboard Marine Co.)

Compress the piston rings to be used in a cylinder, one at a time, into that cylinder, and press them approximately 1″ below the deck with an inverted piston. Using feeler gauges, measure the ring end-gap, and compare to specifications. Pull the ring out of the cylinder and file the ends with a fine file to obtain proper clearance. **CAUTION: If inadequate ring end-gap is utilized, ring breakage will result.**

Install the camshaft:

Liberally lubricate the camshaft lobes and journals, and slide the camshaft into the block. **CAUTION: Exercise extreme care to avoid damaging the bearings when inserting the camshaft.** Install and tighten the camshaft thrust plate retaining bolts.

Cylinder Block Reconditioning

PROCEDURE	METHOD

Check camshaft end-play :

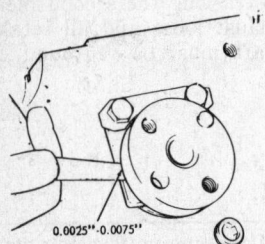 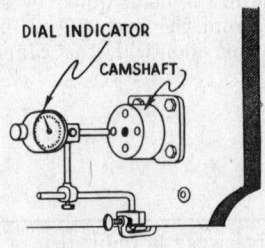

Checking camshaft end-play with a feeler gauge

0.0025"-0.0075"

Checking camshaft end-play with a dial guage

Using feeler gauges, determine whether the clearance between the camshaft boss (or gear) and backing plate is within specifications. Install shims behind the thrust plate, or reposition the camshaft gear and retest endplay. In some cases, adjustment is by replacing the thrust plate.

*Mount a dial indicator stand so that the stem of the dial indicator rests on the nose of the camshaft, parallel to the camshaft axis. Push the camshaft as far in as possible and zero the gauge. Move the camshaft outward to determine the amount of camshaft endplay. If the endplay is not within tolerance, install shims behind the thrust plate, or reposition the camshaft gear and retest.

Install the piston rings :

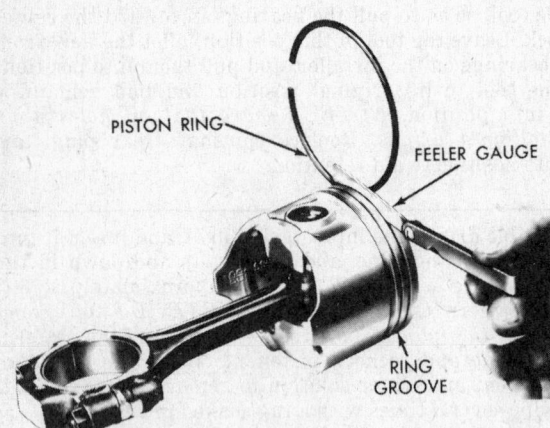

PISTON RING

FEELER GAUGE

RING GROOVE

Checking ring side clearance (© Chrysler Corp.)

Inspect the ring grooves in the piston for excessive wear or taper. If necessary, recut the groove(s) for use with an over-width ring or a standard ring and spacer. If the groove is worn uniformly, overwidth rings, or standard rings and spacers may be installed without recutting. Roll the outside of the ring around the groove to check for burrs or deposits. If any are found, remove with a fine file. Hold the ring in the groove, and measure side clearance. If necessary, correct as indicated above. *NOTE: Always install any additional spacers above the piston ring.* The ring groove must be deep enough to allow the ring to seat below the lands (see illustration). In many cases, a "go-no-go" depth gauge will be provided with the piston rings. Shallow grooves may be corrected by recutting, while deep grooves require some type of filler or expander behind the piston. Consult the piston ring supplier concerning the suggested method. Install the rings on the piston, lowest ring first, using a ring expander. *NOTE: Position the ring markings as specified by the manufacturer (see car section).*

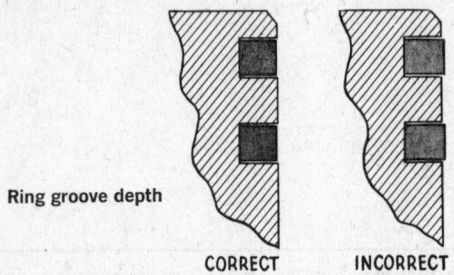

Ring groove depth

CORRECT   INCORRECT

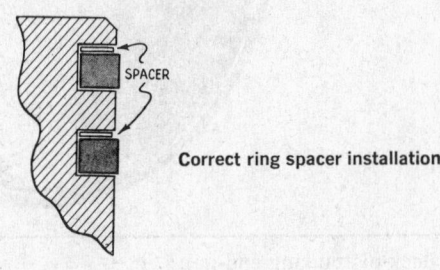

SPACER

Correct ring spacer installation

Install the rear main seal (where applicable) :

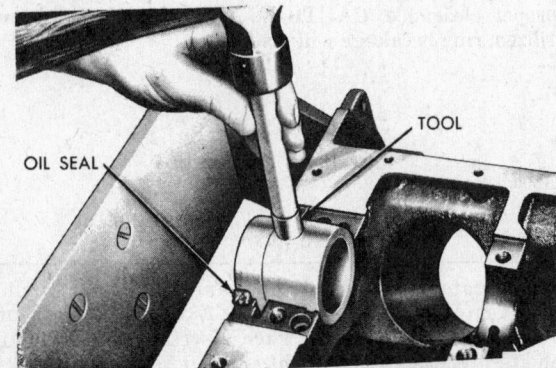

OIL SEAL

TOOL

Seating the rear main seal

Position the block with the bearing saddles facing upward. Lay the rear main seal in its groove and press it lightly into its seat. Place a piece of pipe the same diameter as the crankshaft journal into the saddle, and firmly seat the seal. Hold the pipe in position, and trim the ends of the seal flush if required.

Cylinder Block Reconditioning

PROCEDURE	METHOD

Install the crankshaft:

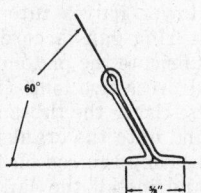

Home made bearing roll-out pin (© Pontiac Div. G.M. Corp.)

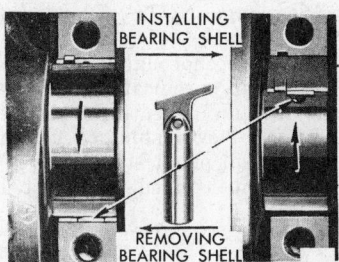

Removal and installation of upper bearing insert using a roll-out pin (© Buick Div. G.M. Corp.)

Thoroughly clean the main bearing saddles and caps. Place the upper halves of the bearing inserts on the saddles and press into position. *NOTE: Ensure that the oil holes align.* Press the corresponding bearing inserts into the main bearing caps. Lubricate the upper main bearings, and lay the crankshaft in position. Place a strip of Plastigage on each of the crankshaft journals, install the main caps, and torque to specifications. Remove the main caps, and compare the Plastigage to the scale on the Plastigage envelope. If clearances are within tolerances, remove the Plastigage, turn the crankshaft 90°, wipe off all oil and retest. If all clearances are correct, remove all Plastigage, thoroughly lubricate the main caps and bearing journals, and install the main caps. If clearances are not within tolerance, the upper bearing inserts may be removed, without removing the crankshaft, using a bearing roll out pin (see illustration). Roll in a bearing that will provide proper clearance, and retest. Torque all main caps, excluding the thrust bearing cap, to specifications. Tighten the thrust bearing cap finger tight. To properly align the thrust bearing, pry the crankshaft the extent of its axial travel several times, the last movement held toward the front of the engine, and torque the thrust bearing cap to specifications. Determine the crankshaft end-play (see below), and bring within tolerance with thrust washers or new thrust bearings.

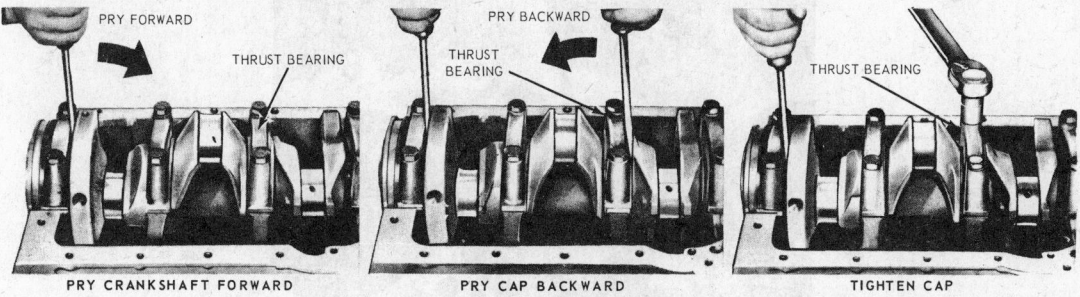

Aligning the thrust bearing (© Ford Motor Co.)

Measure crankshaft end-play:

Mount a dial indicator stand on the front of the block, with the dial indicator stem resting on the nose of the crankshaft, parallel to the crankshaft axis. Pry the crankshaft the extent of its travel rearward, and zero the indicator. Pry the crankshaft forward and record crankshaft end-play. *NOTE: Crankshaft end-play also may be measured at the thrust bearing, using feeler gauges (see illustration).*

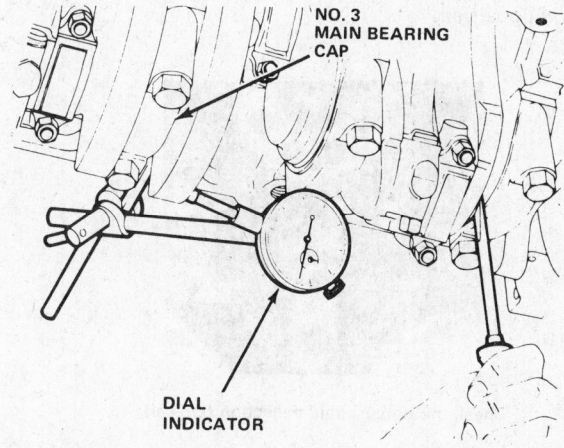

Checking crankshaft end-play with a dial indicator

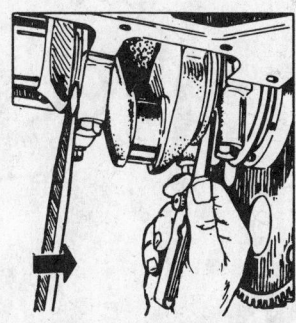

Checking crankshaft end-play with a feeler gauge

Cylinder Block Reconditioning

PROCEDURE	METHOD

Install the pistons:

Press the upper connecting rod bearing halves into the connecting rods, and the lower halves into the connecting rod caps. Position the piston ring gaps according to specifications (see car section), and lubricate the pistons. Install a ring compresser on a piston, and press two long (8") pieces of plastic tubing over the rod bolts. Using the tubes as a guide, press the pistons into the bores and onto the crankshaft with a wooden hammer handle. After seating the rod on the crankshaft journal, remove the tubes and install the cap finger tight. Install the remaining pistons in the same manner. Invert the engine and check the bearing clearance at two points (90° apart) on each journal with Plastigage. *NOTE: Do not turn the crankshaft with Plastigage installed.* If clearance is within tolerances, remove *all* Plastigage, thoroughly lubricate the journals, and torque the rod caps to specifications. If clearance is not within specifications, install different thickness bearing inserts and recheck. **CAUTION: Never shim or file the connecting rods or caps.** Always install plastic tube sleeves over the rod bolts when the caps are not installed, to protect the crankshaft journals.

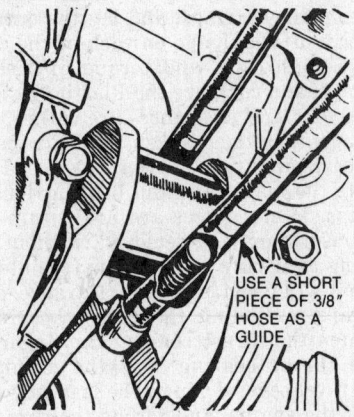

USE A SHORT PIECE OF 3/8" HOSE AS A GUIDE

Tubing used as a guide when installing a piston

Installing a piston

Inspect the timing chain:

Visually inspect the timing chain for broken or loose links, and replace the chain if any are found. If the chain will flex sideways, it must be replaced. Install the timing chain as specified. *NOTE: If the original timing chain is to be reused, install it in its original position.*

TORQUE WRENCH

3/16 INCH

Measuring timing chain strech

REFERENCE POINT

Check the timing chain deflection (typical)

Cylinder Block Reconditioning

PROCEDURE	METHOD
Check timing gear backlash and runout:	Mount a dial indicator with its stem resting on a tooth of the camshaft gear (as illustrated). Rotate the gear until all slack is removed, and zero the indicator. Rotate the gear in the opposite direction until slack is removed, and record gear backlash. Mount the indicator with its stem resting on the edge of the camshaft gear, parallel to the axis of the camshaft. Zero the indicator, and turn the camshaft gear one full turn, recording the runout. If either backlash or runout exceed specifications, replace the worn gear(s).

Checking camshaft gear backlash

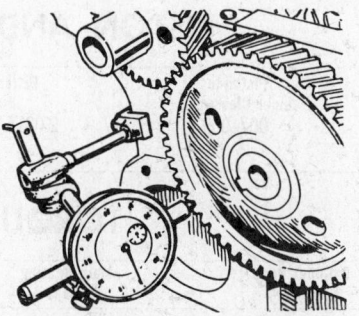

Checking camshaft gear runout

Completing the Rebuilding Process

Following the above procedures, complete the rebuilding process as follows:

Fill the oil pump with oil, to prevent cavitating (sucking air) on initial engine start up. Install the oil pump and the pickup tube on the engine. Coat the oil pan gasket as necessary, and install the gasket and the oil pan. Mount the flywheel and the crankshaft vibration damper or pulley on the crankshaft. *NOTE: Always use new bolts when installing the flywheel.* Inspect the clutch shaft pilot bushing in the crankshaft. If the bushing is excessively worn, remove it with an expanding puller and a slide hammer, and tap a new bushing into place.

Position the engine, cylinder head side up. Lubricate the lifters, and install them into their bores. Install the cylinder head, and torque it as specified in the car section. Insert the pushrods (where applicable), and install the rocker shaft(s) (if so equipped) or position the rocker arms on the pushrods. If solid lifters are utilized, adjust the valves to the "cold" specifications.

Mount the intake and exhaust manifolds, the carburetor(s), the distributor and spark plugs. Adjust the point gap and the static ignition timing. Mount all accessories and install the engine in the car. Fill the radiator with coolant, and the crankcase with high quality engine

Break-in Procedure

Start the engine, and allow it to run at low speed for a few minutes, while checking for leaks. Stop the engine, check the oil level, and fill as necessary. Restart the engine, and fill the cooling system to capacity. Chcek the point dwell angle and adjust the ignition timing and the valves. Run the engine at low to medium speed (800-2500 rpm) for approximately ½ hour, and retorque the cylinder head bolts. Road test the car, and check again for leaks.

Follow the manufacturer's recommended engine break-in procedure and maintenance schedule for new engines.

GENERAL ENGINE SPECIFICATIONS

Engine Model	Number of Cylinders	Bore x Stroke	Piston Cu. In. Displ.	Maximum HP @ RPM	Maximum Torque @ RPM	Firing Order	Gov. No. Load RPM	Low Idle
25000 MK II	6	5¼ x 6½	844	378 @ 2000	1075 @ 1400	1-5-3-6-2-4	2300	550

CRANKSHAFT AND BEARING SPECIFICATIONS

Engine Model	Main Bearing Journal Dia.	Main Bearing Clearance	Rod Bearing Journal Dia.	Rod Bearing Clearance	Crankshaft Endplay
25000 MK II	3.9945-3.9960	.0021-.0048	3.3715-3.3730	.0026-.0056	.007-.015

PISTON AND RING SPECIFICATIONS

Engine Model	Piston-to Liner Clearance	Piston Pin Dia.	Ring End Gap 1st	2nd	3rd	4th	Liner Inside Dia.
25000 MK II	.007-.009	2.0015-2.0017	.019-.036	.016-.038	.017-.038	.016-.035	5.871-5.873

TORQUE SPECIFICATIONS

Engine Model	Cylinder Head Capscrews				Main Bearing Cap		Crankshaft Pulley	Flywheel Capscrews	Exhaust Manifold	Connecting Rod
25000 MK II	⅝-11 185	¾-10 390	⅞-9x6⁷⁄₁₆ 400	⅝-11x6 185			325	200	85	socket head 245 195

VALVE SPECIFICATIONS

Valve Lash Intake (in.)	Exhaust (in.)	Face Angle (deg.)	Seat Angle (deg.)	Stem-to-Guide Clearance Intake (in.)	Exhaust (in.)	Valve Spring Free Length (in.)	(Intake and Exhaust) Installed Height (in.)	Pressure (lb. @ in.)
.018	.023	30	30	.0010-.0015	.0020-.0025	2.625	2.200	133 @ 1.626

Cooling System

Thermostat Removal

1. Remove capscrews from upper end of water by-pass tube and remove it from thermostat cover.

2. Unbolt and remove outlet elbow and gasket.
3. Remove the thermostat cover, and thermostats and gaskets from the manifold.
4. Clean and inspect cover, gaskets and seals.

Thermostat Installation

1. Position gasket on manifold and install thermostats.
2. Position thermostat cover on manifold and secure with capscrews and lockwashers.
3. Place the radiator inlet elbow and gasket in position at the front of the thermostat cover and secure with capscrews and lockwashers.
4. Using a new gasket, position the top of the water by-pass tube on the thermostat cover and secure with capscrews and lockwashers.

Fuel System
Robert Bosch

Injection Pump Removal

1. Make sure the FPI mark on the pump coupling hub is aligned with the timing pointer on the pump.
2. Disconnect the engine stop and speed control linkage from the injection pump.
3. Shut off fuel supply.
4. Disconnect fuel inlet and outlet tubes and fuel return and injection lines. Disconnect lube oil feed and drain tubes.

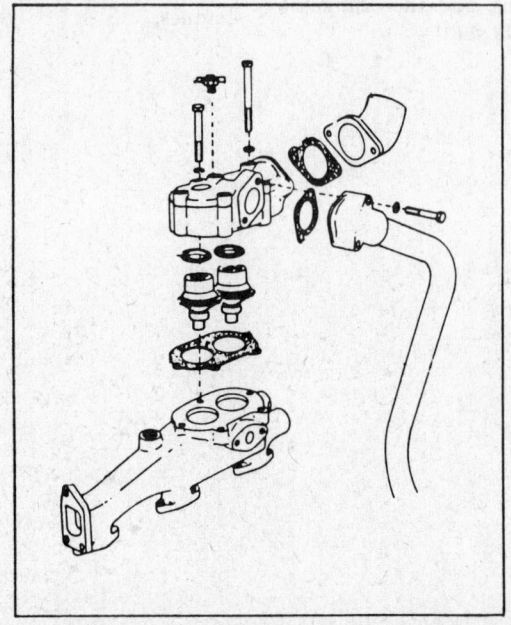

Thermostat Assemblies

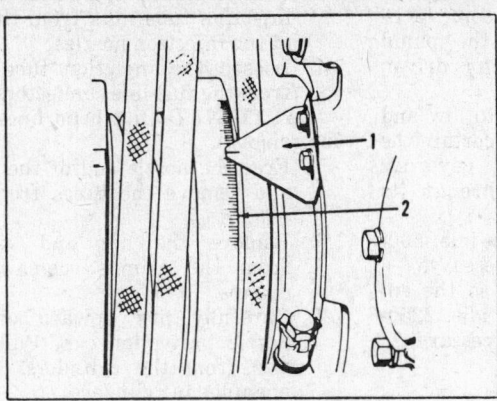

Front end timing pointer

1 Pointer
2 Timing marks

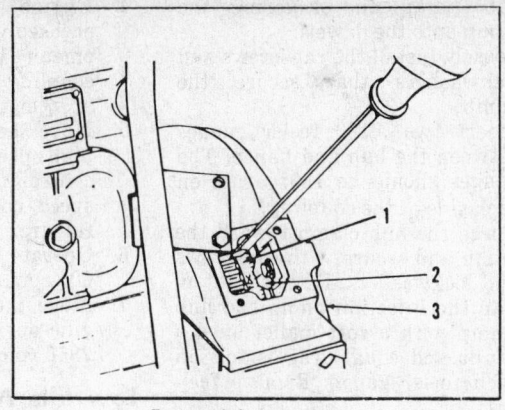

Rear end timing pointer

1 Flywheel timing marks
2 Pointer
3 Flywheel housing

5. Remove capscrews from injection pump drive coupling flange.
6. Remove capscrew which secures hub clamp to the shaft of the accessory drive assembly. Force the hub clamp as far as possible toward the front of the engine. This will facilitate removing the pump from the dowels in the pump mounting bracket.
7. Remove the capscrews and lockwashers that secure the injection pump to the mounting bracket.
8. Lift the injection pump to clear the dowels and remove it and the coupling as an assembly from the mounting bracket.

Injection Pump Installation and Timing

1. Depending on the production date, the timing pointer is located either on the front gear housing cover or in the flywheel housing.

2. If the engine has been rotated since the pump was removed, position number one piston on TDC compression.
3. Rotate the engine in normal direction of rotation until the timing pointer lines up with the mark denoting 34° BTDC at 2100 rpm.
4. Make certain that all the slack is out of the timing gears and fuel pump drive. If the correct mark is passed, back up several notches past the mark and approach it in the normal direction of rotation. Never back up to the mark for a setting.
5. If coupling was removed from pump:
   a. Position key in injection pump camshaft keyway.
   b. Position hub on camshaft. Install lockwasher and nut; torque nut to 85-90 ft lbs;
   c. Current production couplings

have a dust shield which installs in a groove on the flange with the peak of the shield toward the front of the engine.
   d. Position the two inserts in the flange.
   e. Assemble the flange to the hub with rubber inserts separating the flange and hub fingers.
   f. Work the cover over the rubber inserts and secure to flange with capscrews and lockwashers.
6. Position the FPI mark on the coupling hub, opposite the pointer on the pump.
7. Install the pump and coupling as an assembly on the pump mounting bracket by holding the pump assembly at a slight angle and positioning the bore in the coupling flange on the pilot of the

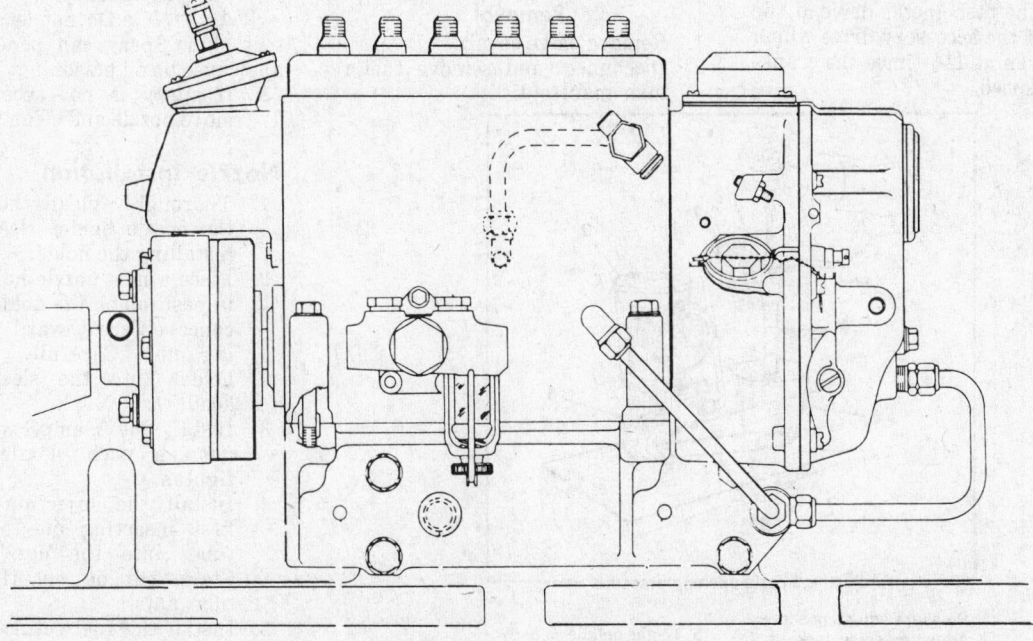

Robert Bosch injection pump

hub clamp and lowering the pump onto the dowels.

8. Loosely install the capscrews and lockwashers that secure the pump.

9. Insert two .010″ feeler gauges between the hub and flange. The gauges should be 180° apart on each side of the coupling.

10. Force the hub clamp toward the pump and secure with capscrews and lockwashers. Do not tighten.

11. Tap the injection pump and hub clamp with a soft mallet until a slight and equal drag is felt on each feeler gauge. Remove feelers.

12. Torque the pump-to-bracket capscrews to 18-23 ft lbs.

13. Torque the clamp-to-accessory drive capscrews to 39 ft. lbs.

14. Align the FPI mark on the flange with the pointer on the pump.

15. Torque the clamp-to-flange capscrews to 29 ft lbs.

16. Install lube oil supply and drain tube assemblies.

17. Install fuel supply and return tube assemblies.

18. Install high pressure lines.

19. Connect the engine stop and control linkages.

20. Install the timing cover, if so equipped.

Governor

Engine Speed Adjustment

1. Run engine to normal operating temperature.

2. If engine is not equipped with a tachometer, one may be installed on the front end of the engine crankshaft.

3. The engine speed may be checked from the tachometer drive at the rear of the accessory drive which is driven at 1½ times the crankshaft speed.

4. Engine speed may also be checked with a stroboscopic tachometer by marking the pump coupling flywheel which is driven at ½ engine speed.

5. Move the accelerator to low and high speeds and make certain the control moves the governor speed control lever through its full arc.

6. Operate engine at low idle, 500-600 rpm. Adjust if necessary.

7. Move the accelerator so the engine operates at high idle, 2250-2350 rpm. Adjust if necessary.

Low Idle Adjustment

1. With engine running, loosen the jam nut on the low idle adjusting screw.

2. Hold the governor speed control lever so the stop contacts the low idle adjusting screw. Turn the idle adjusting screw as necessary to increase or decrease idle speed.

3. When correct idle speed, 500-600 rpm, is achieved, hold the low idle adjusting screw and tighten the jam nut.

High Idle Adjustment

1. Remove the flat head screw seal from the securing link and remove the link to gain access to the high idle adjusting screw.

2. With engine running, loosen jam nut on the high idle screw. Hold governor speed control lever so control lever shaft stop contacts the high idle screw. Turn adjusting screw to increase or decrease high speed.

Injector Nozzles

Removal

1. Remove valve cover.

2. Disconnect and remove fuel return manifold.

3. Using a tubing socket, loosen the injection tube nuts from the top of the injection nozzles.

4. Loosen the injection tube nuts from the fuel line connectors.
   CAUTION: Do not bend lines during removal.

5. Free the nozzle end of the tubes and remove the tubes from the engine.

6. Remove the nut and washer from the clamps; remove the clamps.

7. Carefully pry upward on the nozzle protection cap. Pull nozzles from the cylinders. Cover openings in cylinders.

Testing and Adjusting

1. Open the nozzle tester valve and loosen the filler cap.

2. Connect the nozzle to the tester.

3. Operate the tester handle and observe the tester gauges. The nozzle should pop at 4100-4150 psi.

4. To adjust pop pressure, remove the protection cap from the upper end of the nozzle and loosen the adjusting screw locknut.

5. Operate the tester handle and turn the screw until the proper pop point is reached. Hold the screw and tighten the locknut to 75 ft lbs.

6. Operate the pressure handle until the gauge shows about 3900 psi. Check nozzle tip for leakage. If leakage is observed, relieve the pressure by loosening one of the line nuts and remove the valve and valve body for cleaning and inspection.

7. Operate the handle at about 100 strokes per second and observe the spray pattern.
   WARNING: Do not let spray contact skin. Spray can penetrate skin and cause blood poisoning.

8. If spray is not acceptable, remove nozzle and clean tip.

Nozzle Installation

1. Thoroughly clean the inside of the nozzle holder sleeves before installing the holder.

2. Place a new nozzle holder gasket in position on the holder with the concave face toward the retaining nut. Carefully insert the holder into the sleeve in the head.

3. Install the clamp, washer and nut on each nozzle. Do not tighten.

4. Install the injection tubes by first inserting one end of the tubes into the nozzle holders. Start, but do not tighten, the tube nuts.

5. Install the fuel return manifold.

6. Tighten the nozzle holder clamp nuts to 43-48 ft lbs.

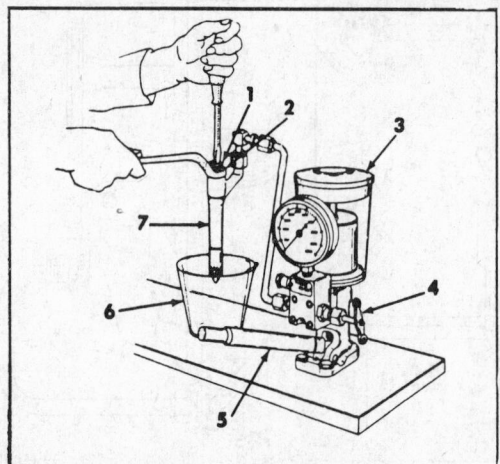

Adjusting nozzle opening pressure

1 Pressure adjusting screw
2 Adjusting screw locknut
3 Nozzle tester
4 Valve handle
5 Tester handle
6 Spray collector
7 Nozzle holder assembly

7. Tighten the injection tube nuts and the fuel return manifold nuts.

Flow Timing the Injection Pump

NOTE: A hand primer pump capable of supplying 2-3 psi. to the pump gallery is necessary.

1. Position the engine crankshaft so that the timing marks indicate #1 piston TDC compression.
2. Remove the injection pump top cover and position it out of the way. Clean all external surfaces of the pump and lines.
3. Temporarily position the cover on the pump and remove the #1 injection line.
4. Remove the delivery valve holder from the pump.
5. Remove the fill piece, spring, delivery valve gasket, and delivery valve from the top of the barrel and plunger assembly.
6. Remove the fluted delivery valve from the valve body and reinstall the valve body and old gasket with the holder.

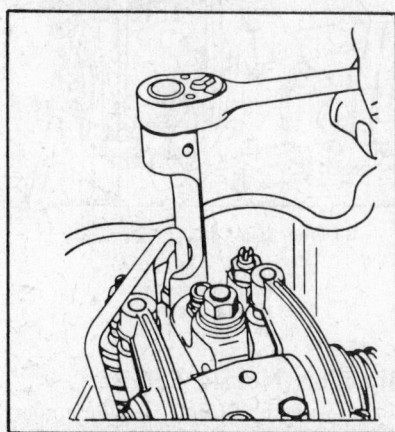

Removing fuel injection tube

7. Loosen the capscrews securing the injection pump coupling hub clamp to the coupling flange. Place the stop lever in the run position and the speed control lever in the high speed position. Turn the pump coupling clockwise several degrees. Rotation is determined from pump drive end.
8. Rotate the injection pump camshaft counterclockwise while operating the hand primer. Camshaft should be rotated very slowly. As the #1 plunger starts to close off the intake port of the barrel the fuel flow will be reduced. Continue turning the camshaft until the fuel stops. A bulge of fuel will appear on top of the valve holder. Wipe this away and continue to operate the primer. If the bulge appears

again, it means that the fuel inlet port in the barrel is not completely closed. A very slight rotation of the pump camshaft will now cause the plunger to close the inlet port. When the port is completely closed, the FPI mark on the pump coupling will align with the pointer on the injection pump, indicating proper timing of the pump to the engine. Repeat this step several times.

9. Securely tighten the hub clamp-to-injection pump coupling flange retaining capscrews.
10. Remove the delivery valve holder, copper gasket, and valve body from the pump. Wash all components in clean fuel oil.
11. Place the delivery valve into the valve body and position this assembly on top of the plunger and barrel.
12. Place a new copper gasket on the top of the delivery valve assembly and insert the fill piece into the spring. Position this assembly on top of the delivery valve, fill piece up.
13. Lubricate a new O-ring and position it on the valve holder. Screw the holder into the pump and torque to 50-55 ft. lbs. Install the top cover, connect the high pressure line and tighten the nut.

Aneroid Adjustment

NOTE: Prior to adjustments on early models, make certain the stop screw is turned until it bottoms. Current production pumps have the stop screw bottomed. The stop screw is locked in position by a locknut within the aneroid housing. A setscrew is screwed into the stop screw threads. A soft metal seal plug seals the setscrew. This plug can be removed with a sharp instrument.

1. Remove the aneroid housing side cover.
2. Back off the pump shutoff lever stop screw until clearance is obtained between the lever and the stop screw, when the lever is completely in the stopped position.
3. With the injection pump shutoff in the stopped position, check the clearance between the end of the rack and the aneroid lever gap adjustment screw. The current specified clearance is 10-10½ mm.
4. To adjust clearance, loosen the lever adjusting screw locknut and turn the screw to gain specified clearance.
5. Connect shop air to aneroid.
6. Place pump shutoff lever in the run position.
7. Slowly apply air pressure from zero until the linkage begins to move. This movement must occur

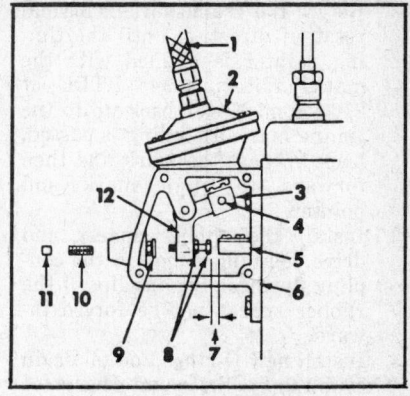

Aneroid assembly

1 Hose	7 Gap-10mm
2 Diaphragm cover	8 Gap adjusting screw
3 Guide bushing	9 Stop screw
4 Pin & snap ring	10 Setscrew
5 Rack	11 Seal plug
6 Aneroid housing	12 Lever

between 6.3-6.6 psi for all units.
8. Turn the guide bushing with a screwdriver to adjust.

NOTE: It is important that the aneroid lever and gap adjustment screw does not interfere with the movement of the rack to the specified full load position.

9. With the swivel position on the fuel stop, observe the rack end. It must not be contacting the aneroid lever gap adjusting screw.
10. With the shutoff lever held in the stopped position, turn the stop screw until it contacts the shutoff lever. Then continue to turn the screw an additional ½ turn.

Fuel System American Bosch

Injection Pump Removal

1. Align the FPI mark on the pump coupling flywheel and the timing pointer on the pump.
2. Disconnect engine stop and speed control linkage from the pump.
3. Shut off the fuel supply.
4. Disconnect the inlet and outlet hoses. return line and injection lines.
5. Remove capscrews from the pump drive coupling flange, and the capscrews which secure the injection pump to the mounting bracket. Remove the pump and coupling assembly.

Installation and Timing

1. If the engine was rotated, make sure the #1 piston is on TDC compression. Timing marks may be located on the damper or the flywheel depending on engine production date. The timing hole cover must be removed to gain access to flywheel timing marks.

2. Rotate the crankshaft in normal rotation direction until the timing pointer is aligned with the mark indicating 34° BTDC at 2100 rpm. Never back up to the timing mark. If timing is passed, back up past the mark and then forward to align mark and pointer.

3. Install the rubber spacer and drive coupling flange to the coupling flywheel. Sealing lips of the rubber spacer not be forced inward.

4. Install new O-rings and oil drain coupling in the counterbores of the oil passages in the pump mounting bracket.

5. Align the FPI mark on the coupling flywheel with the timing pointer.

6. Position the pump and drive coupling assembly on the mounting bracket. Slide the pump forward and install, but do not tighten the capscrews and washers.

7. Insert two .010″ feeler gauges on the sides between the coupling hub and flange, 180° apart.

8. Tap the pump forward with a soft mallet until a slight and equal drag is felt at each gauge. Torque the mounting bracket capscrews to 44-49 ft. lbs. for 7/16″-14 × 1⅛″ capscrews and 40-45 ft. lbs. for the 7/16″-14 × 1⅝″ capscrews. Remove feeler gauges.

9. Make sure that the FPI timing mark on the coupling flywheel is aligned with the timing pointer. If not, turn the coupling flywheel to obtain proper alignment.

10. Torque the two drive coupling capscrews to 29 ft. lbs.

11. Connect the inlet and outlet hoses, fuel return and injection lines.

12. Install timing cover, if equipped.

Governor

Engine Speed

See engine speed adjustment under Robert Bosch procedures.

Governor Adjustments

1. Disconnect the accelerator linkage from the governor speed control lever so the lever can be moved by hand.

2. Remove the speed adjusting screw access cover from the governor.

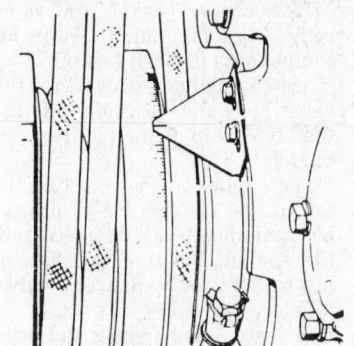

Front end timing pointer

3. With engine running, loosen low idle adjusting screw jam nut. Hold governor speed control lever so lever shaft stop plate contacts the low idle adjusting screw. Turn the idle adjusting screw to attain 500-600 rpm. Tighten jam nut.

4. Loosen high idle screw jam nut. Hold governor speed control lever so lever shaft stop contacts the high idle screw. Turn screw to attain 2250-2350 rpm. Tighten jam nut.

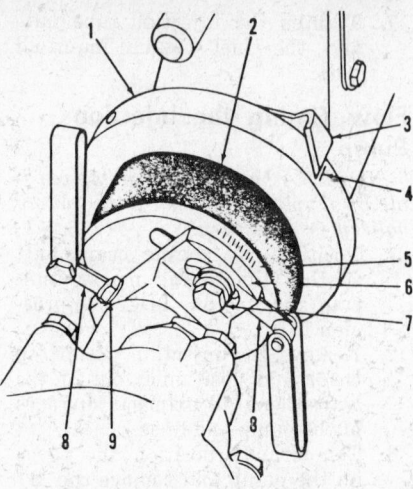

Injection pump alignment

1	Coupling flywheel	6	Drive coupling hub
2	Rubber spider	7	Capscrew
3	Pump timing pointer	8	Feeler gauge
4	FPI timing mark	9	Hub capscrew
5	Drive coupling flange		

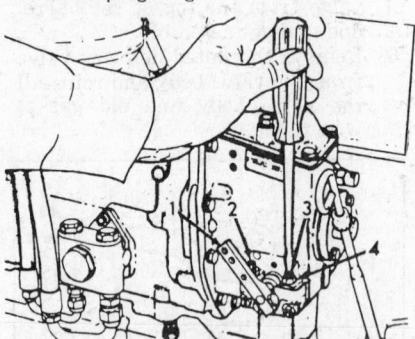

Governor speed adjustment points

1 Speed control lever
2 High idle screw
3 Low idle screw
4 Jam nut

Injection Nozzle Service

See Injector Nozzle section under Robert Bosch procedures.

Flow Timing the Injection Pump

NOTE: A hand primer pump capable of supplying 2-3 psi to the fuel oil gallery is necessary.

1. Align the FPI timing marks with #1 piston at compression.

2. Clean all external pump and line surfaces.

3. Remove #1 nozzle injection line from the pump.

4. Remove holder, delivery valve spring and delivery valve from the valve body at the top of the plunger and barrel assembly.

5. Reinstall the holder and nut. Tighten nut securely.

6. Loosen the pump coupling hub capscrews, place the stop lever in the run position, place the speed control lever in the high speed position and turn the coupling several degrees clockwise as viewed from the pump drive end.

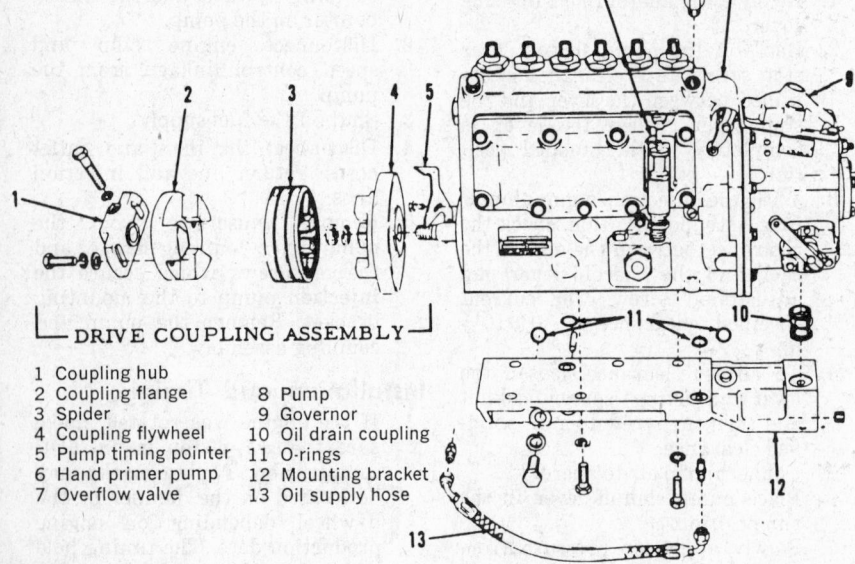

┗ DRIVE COUPLING ASSEMBLY ┛

1	Coupling hub	8	Pump
2	Coupling flange	9	Governor
3	Spider	10	Oil drain coupling
4	Coupling flywheel	11	O-rings
5	Pump timing pointer	12	Mounting bracket
6	Hand primer pump	13	Oil supply hose
7	Overflow valve		

American Bosch pump

7. Operate the hand primer pump and rotate the pump camshaft counterclockwise very slowly. As #1 plunger starts to close off the intake port of the barrel, the fuel volume from the delivery valve holder will be reduced.

8. When the fuel stops flowing a bulge of fuel will appear on top of the delivery valve holder. Wipe off the bulge and continue to operate the primer pump. If the bulge appears again, it indicates that the inlet port in the barrel is not completely closed. A slight rotation of the pump camshaft will now cause the plunger to close the inlet port. At this point the FPI mark on the pump coupling will align with the pointer on the pump.

9. Repeat this procedure several times.

10. Securely tighten the coupling hub-to-coupling flange capscrews.

11. Remove the holder nut and holder, wash all components and reassemble. Torque the holder nut to 60-65 ft. lbs. Connect the injection line.

Lubrication System

Full-Flow Filters

Removal & Installation

1. Clean filter head and bodies.
2. Drain filters.
3. Remove centerbolts, filter bodies and elements.
4. Install new elements and gaskets, position filter bodies and torque center bolts to 45-50 ft lbs.

By-Pass Filter

Removal & Installation

1. Clean the filter cover.
2. Drain filter housing.
3. Remove the cover clamp ring and lift the cover from the filter housing.
4. Unscrew and remove the T-handle hold-down from the center tube. Remove the filter element.
5. Clean interior of housing and install drain.
6. Insert new element and press down firmly. Be sure that the T-handle hold-down orifice plug hole is open. Install hold-down and tighten securely.
7. Install cover gasket and cover.

Lubricating Oil Pump Specifications

Disassemble oil pump and check the following fits and clearances:
Pump driving gear-to-
cover, running   .010-.029"

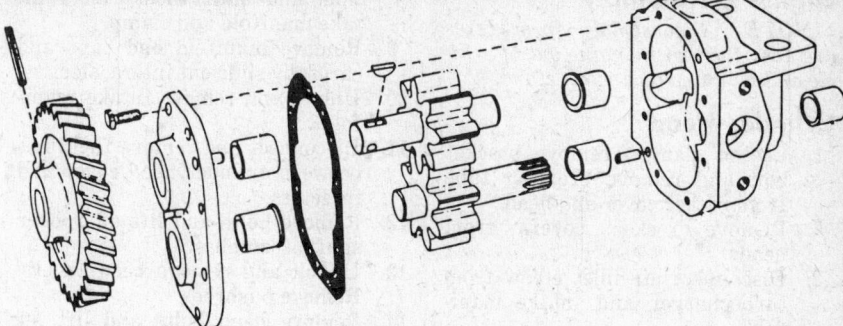

Oil pump-exploded view

Gear-to-pumpbody radial clearance	.00225-.00325"
Gear side clearance	.005-.007"
Shaft bushings— I.D.	.7495-.7505"
Gear shafts—O.D.	.7475-.7480"
Shaft-to-bushing clearance	.0015-.0030"

Valve Adjustment

1. Operate the engine until normal operating temperature is reached.
2. Remove the rocker cover.
3. Crank the engine until the #6 cyl. intake valves begin to open.
4. Loosen the locknut on #1 cyl. bridge adjusting screw. Back off screw one turn.
5. Press down firmly on the center of the bridge and turn the ad-

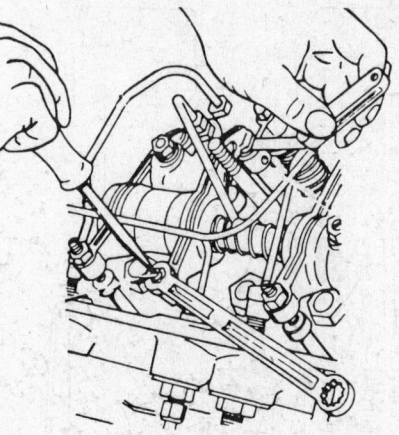

Valve lash adjustment

justing screw until it makes contact with the valve stem. Tighten the locknut.

6. Loosen the rocker arm adjusting screw locknut and adjust rocker arm to bridge clearance. Clearance is: intake—.015", exhaust—.020". Tighten lock nut.

Air Compressor

Cylinder Head Removal and Installation

1. Drain coolant from compressor crankcase.
2. Remove discharge valve capnuts, springs and valves from top of air compressor cylinder head.
3. Remove hose assembly from compressor cylinder head.
4. Remove discharge flange and gasket from head.
5. Unbolt and remove cylinder head. It may be necessary to tap cylinder head with a soft mallet to loosen.
6. Remove inlet valve springs from head and mark for replacement.
7. Clean loose carbon and foreign matter from head with a safe solvent. Head may be scraped to remove hard deposits.
8. Rinse head in solvent and blow dry.
9. Clean spring and discharge valve parts in solvent and blow dry.
10. Reassemble parts and head in reverse order of removal. Use new gaskets.
11. Refill cooling system.

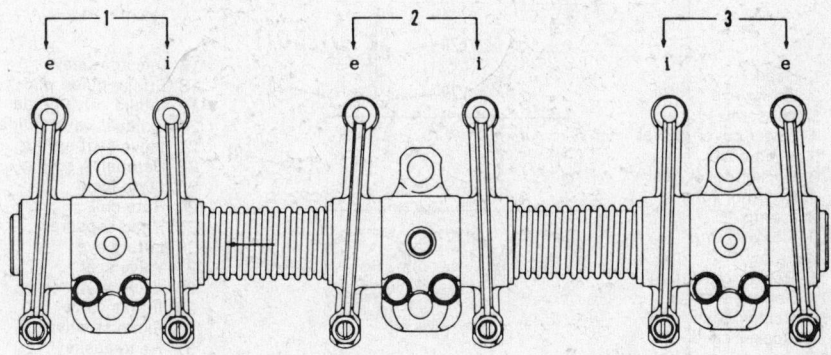

Valve arrangement—front head; rear head identical

Engine Disassembly

NOTE: Disassembly procedures are described with engine out of vehicle and mounted on a work stand.

Cylinder Head

1. Loosen clamps, remove bracket bolt and remove breather tube from rocker cover and head.
2. Remove rocker covers from heads.
3. Disconnect air inlet elbow from turbocharger and intake manifold.
4. Remove bracket bolts and disconnect and remove oil inlet and outlet lines from turbocharger.
6. Remove high pressure fuel lines and fuel drain line from head covers and pump.
7. Remove exhaust manifold end sections by unbolting and swiveling out of connecting sleeves. Unbolt and remove center section.
8. Disconnect and remove water inlet and outlet hoses from intake manifold and clamp.
9. Remove manifold end caps and carefully slide out intercooler.
10. Unbolt and remove intake manifold.
11. Disconnect and remove fuel lines between cylinder head cover and injectors.
12. Remove bolts and lift off rocker shaft assemblies.
13. Unbolt and remove head covers. Remove pushrods.
14. Remove head bolts and lift off heads.

Cylinder Block

15. Remove fire rings from block.
16. Remove lifter covers from block, unbolt and remove lifter brackets and remove lifters.
17. Remove fan, hub and housing assemblies.
18. Disconnect lines from oil cooler and unbolt and remove cooler and bracket assemblies.
19. Disconnect all lines and hoses, remove water pump pulley, remove pump attaching bolts and remove pump assembly. Remove clamp and idler assembly.
20. Unbolt and remove coolant inlet and outlet manifolds.
21. For injection pump and governor removal, see Fuel Section.
22. Disconnect coolant inlet and outlet lines, and oil inlet line from oil compressor. Remove compressor-to-bracket bolts and lift off compressor.

NOTE: Compressor may be either a right or left hand mount.

23. Remove crankshaft pulley bolt and 6 pulley-to-damper bolts. Using a suitable puller, remove the pulley and damper.
24. Unbolt and remove the timing case cover. Remove the oil pan.
25. Unbolt and remove the accessory drive unit.
26. Remove camshaft drive gear. Remove camshaft retainer and key; carefully slide camshaft from block.
27. Unbolt and remove timing case and gasket.
28. Remove the eight flywheel capscrews and carefully lift off flywheel.
29. Remove the eight flywheel housing bolts and carefully remove housing.
30. Remove crankshaft rear seal and O-ring. Remove camshaft rear O-ring. Remove the oil pump.
31. Remove connecting rod nuts and lower bearing halves. Remove carbon ridge from liner and push piston and rod assembly out of block.
32. Remove lower main bearing cap bolts and lift out main caps.
33. Lift crankshaft from block with hoist.
34. Using a wood block, drive out the liners.
35. Remove upper main bearing halves.
36. Unbolt and remove piston cooling nozzle jets.

Engine Assembly

1. Push cylinder liners, with new seals into place in the block. Protrusion should be .002-.005".
2. Install piston cooling nozzles making sure that the nozzles spray toward the center of the piston. Torque to 35 ft lbs.
3. Install upper main bearing halves.
4. Carefully lower crankshaft into position.
5. Install lower main bearing halves with a piece of Plastigage to check clearance. Remove caps and check Plastigage width. If satisfactory (.0021-.0048) coat bearings and journals with SAE 30 oil and install caps. Torque

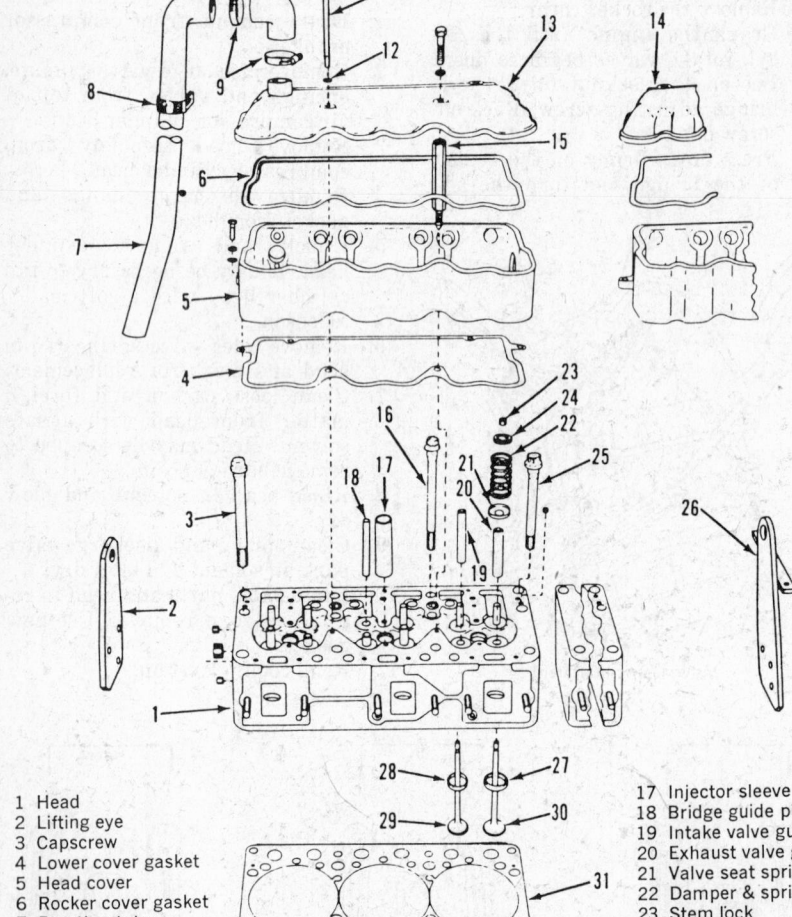

1 Head
2 Lifting eye
3 Capscrew
4 Lower cover gasket
5 Head cover
6 Rocker cover gasket
7 Breather tube
8 Clamp
9 Elbow clamp
10 Elbow
11 Rocker cover bolt
12 Sealing washer
13 Rocker cover-rear
14 Rocker cover—front
15 Rocker cover bolt stud
16 Head capscrew

17 Injector sleeve
18 Bridge guide pin
19 Intake valve guide
20 Exhaust valve guide
21 Valve seat spring
22 Damper & spring
23 Stem lock
24 Roto-coil
25 Head capscrew
26 Lifting eye
27 Valve seat
28 Valve seat
29 Intake valve
30 Exhaust valve
31 Head gasket
32 Liner gasket

Rear cylinder head, exploded view

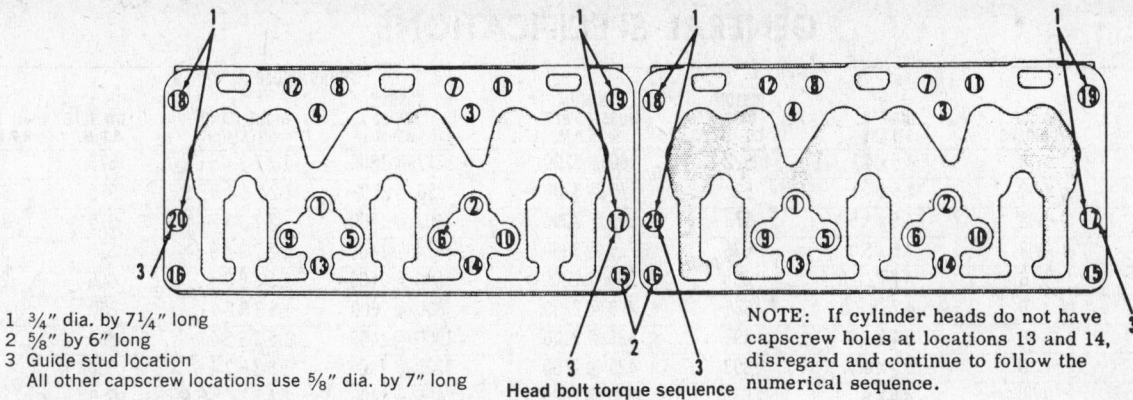

1 ¾" dia. by 7¼" long
2 ⅝" by 6" long
3 Guide stud location
   All other capscrew locations use ⅝" dia. by 7" long

Head bolt torque sequence

NOTE: If cylinder heads do not have capscrew holes at locations 13 and 14, disregard and continue to follow the numerical sequence.

⅞-9 × 6- 7/16 bolts to 400 ft lbs and ⅝-11 × 6 bolts to 185 ft lbs.

6. Install piston and rod assemblies taking care to avoid damage to the cooling nozzles. Position rods on crankshaft journals.

7. Using Plastigage, check rod cap-to-crankshaft clearance. If satisfactory (.0026-.0056") coat journal and bearing with SAE 30 oil and install caps. Torque to 245 ft lb for ⅝-18 × 3¾ socket head and 195 ft lbs for ⅝-18 nut.

8. Install rear oil seal and O-ring, and camshaft O-ring.

9. Install flywheel housing and tighten bolts.

10. Install flywheel and torque bolts to 200 ft lbs.

11. Install timing case and new gasket.

12. Coat camshaft with SAE-30 oil and carefully slide into place. Install retainer, key and gear.

13. Install and secure accessory drive unit.

14. Install timing case cover and gasket.

15. Install oil pump discharge tube and suction tube assemblies.

16. Install oil pan.

17. Using a suitable press, install crankshaft pulley and damper. Torque center bolt to 325 ft lbs.

18. Mount air compressor on bracket. Connect coolant and oil lines.

19. For injection pump and governor installation, see Fuel Section.

20. Install coolant inlet and outlet manifolds.

21. Position water pump assembly in place, install retaining bolts and clamp. Install and secure idler assembly. Connect all hoses and lines.

22. Position oil cooler and secure bracket. Connect lines and hoses.

23. Assemble and secure fan, hub and housing.

24. Install lifters and brackets. Position and secure lifter covers.

25. Install fire rings on block face.

Cylinder Head

26. For injector installation see Fuel Section.

27. Position heads on block with new head gaskets.

28. Torque head bolts in sequence to 185 ft lbs for ⅝-11 capscrews and 390 ft lbs for ¾-10 capscrews.

29. Install head covers and torque to 38-42 ft lbs.

30. Install rocker shaft assemblies and tighten capscrews in order from front cover.

31. Install fuel lines between injectors and head covers.

32. Install air intake manifold. Carefully insert intercooler and install end caps. Connect coolant lines.

33. Install exhaust manifold center section. Insert end section into connector sleeves and bolt to head.

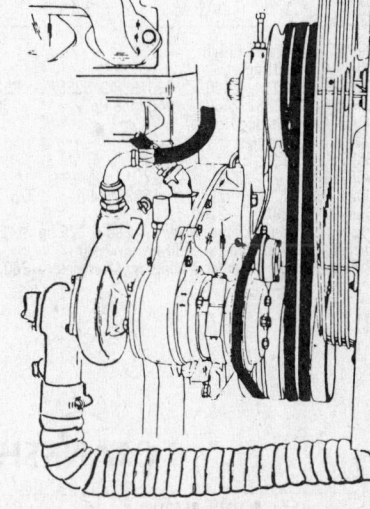

Water pump, idler and fan and hub mounting

34. Install high pressure fuel lines and drain line between head covers and pump.

35. Position turbocharger on manifold and torque nuts to 18-21 ft lbs.

36. Connect oil inlet and outlet lines to turbocharger.

37. Connect air inlet elbow from turbocharger to manifold.

38. Check valve adjustment as previously described. Install rocker covers.

39. Install breather tube.

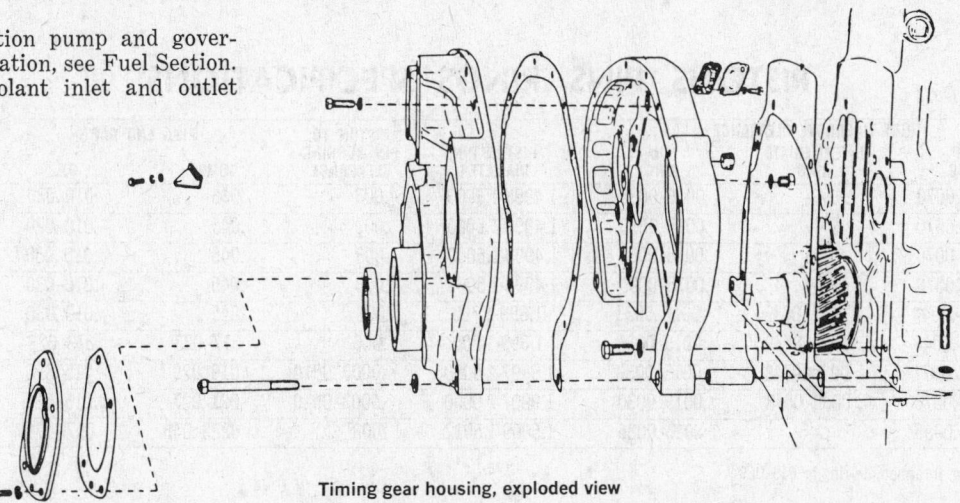

Timing gear housing, exploded view

GENERAL SPECIFICATIONS

ENGINE MODEL	NO. OF CYLINDERS	BORE & STROKE	PISTON DISP. CU. IN.	MAXIMUM HORSEPOWER @ R.P.M.	TORQUE SAE @ R.P.M.	FIRING ORDER RIGHT HAND ROTATION	LOW IDLE R.P.M.	HIGH IDLE R.P.M.
1140	(V)8	4.5 x 4.1	522	150 @ 3200	277 @ 1800	1-2-7-3-4-5-6-8	575	3380
1145	(V)8	4.5 x 4.1	522	175 @ 3200	326 @ 1700	1-2-7-3-4-5-6-8	575	3410
1150	(V)8	4.5 x 4.5	573	200 @ 3000	403 @ 1600	1-2-7-3-4-5-6-8	575	3255
1160	(V)8	4.5 x 5.0	636	225 @ 2800	474 @ 1400	1-2-7-3-4-5-6-8	575	3060
1673C	6	4.75 x 6.0	638	250 @ 2200	690 @ 1600	1-5-3-6-3-4	600	2530
1674	6	4.75 x 6.0	638	270 @ 2200	805 @ 1600	1-5-3-6-2-4	650	1
1693T	6	5.4 x 6.5	893	325 @ 2100	1000 @ 1450	1-5-3-6-2-4	550	2
1693TA	6	5.4 x 6.5	893	425 @ 2100	1275 @ 1400	1-5-3-6-2-4	550	3
3208	(V8)	4.5 x 5	636	210 @ 2800	485 @ 1400	1-2-7-3-4-5-6-8	650	3030
3406	6	5.4 x 6.5	893	4	6	1-5-3-6-2-4	600	5
3408	(V)8	5.4 x 6	1099	7	N.A.	1-2-7-3-4-5-6-8	650	3030

1. Serial Numbers and High
   Idle R.P.M. Settings:
   94B1-94B1060   2500
   94B1061-94B1681 2480
   94B1682-94B2557 2520
   94B2558-Up 2505
2. Serial Numbers and High
   Idle R.P.M. Settings:
   65B1-65B781   2190
   65B782-Up 2305
3. 65B1-65B781   2230   65B2918-Up 2375
   65B782-65B2917 2260
4. Precombustion Chamber Models—280 & 325 hp @ 2100
   Direct Injection Models—280 hp @ 2100
   Precombustion Chamber Models w/Aftercooler—360 hp @ 2100

5. 280 PC—2260
   280 DI—2300
   325 PC—2285
   360 PC—2300
6. 280 PC—840 @ 1400
   280 DI—865 @ 1385
   325 PC—970 @ 1400
   360 PC—1080 @ 1400
7. Direct injection—360 & 400 @ 2100
   Precombustion chambers w/aftercooling—450 @ 2100

CRANKSHAFT AND BEARING SPECIFICATIONS

ENGINE SERIES	MAIN BEARING JOURNAL DIAMETER	MAIN BEARING CLEARANCE	SHAFT END THRUST ON BEARING NO.	SHAFT END PLAY	ROD BEARING JOURNAL DIAMETER	ROD BEARING CLEARANCE
1140	3.4995-3.5005	.0015-.0045	4	.003-.009	2.7496-2.7504	.0015-.0045
1145	3.4995-3.5005	.0015-.0045	4	.003-.009	2.7496-2.7504	.0015-.0045
1150	3.4995-3.5005	.0015-.0045	4	.003-.009	2.7496-2.7504	.0015-.0045
1160	3.4995-3.5005	.0015-.0045	4	.003-.009	2.7496-2.7504	.0015-.0045
1673C	3.499-3.500	.0030-.0059	7	.006-.019	2.999-3.000	.0004-.0007
1674	3.499-3.500	.0030-.0059	7	.011-.018	2.999-3.000	.0004-.0007
1693T	4.4995-4.5005	.0035-.0066	7	.006-.018	3.5395-3.5405	.0029-.0060
1693TA	4.4995-4.5005	.0035-.0066	7	.006-.018	3.5395-3.5405	.0029-.0060
3208	3.4990-3.5000	.002-.005	4	.003-.009	2.7496-2.7504	.0015-.0045

PISTONS, PINS, RINGS SPECIFICATIONS

ENGINE SERIES	RING-TO-GROOVE CLEARANCE TOP RING	INTERMEDIATE RING	OIL RING	PISTON PIN DIAMETER	PISTON TO PIN BUSHING CLEARANCE	RING END GAP COMP.	OIL	CYLINDER LINER INSIDE DIAMETER
1140	.0045-.0070	—	.0015-.0035	1.4998-1.5000	.003	.045	.010-.030	4.5000-4.5010
1145	.0045-.0070	—	.0015-.0035	1.4998-1.5000	.003	.045	.010-.030	4.5000-4.5010
1150	.0045-.0070	—	.0015-.0035	1.4998-1.5000	.003	.045	.010-.030	4.5000-4.5010
1160	.0045-.0070	—	.0015-.0035	1.4998-1.5000	.003	.045	.010-.030	4.5000-4.5010
1673C	.0022-.0044	.0025-.0039	.0015-.0033	1.699-1.700	.006	.045	.010-.030	4.750-4.752
1674	.0030-.0044	.0025-.0039	.0015-.0033	1.699-1.700	.006	.017-.023	.013-.023	4.750-4.752
1693T	.0057-.0071	.0030-.0048	.0015-.0030	1.9997-2.0000	.0003-.0010	.019-.029	.015-.025	5.400-5.402
1693TA	.0055-.0073	.0030-.0048	.0015-.0030	1.9997-2.0000	.0003-.0010	.021-.027 1	.015-.025	5.400-5.402
3208	.0030-.0055	—	.0015-.0035	1.5009-1.5012	.003	.0225-.045	.0175-.045	4.500-4.5010

1. Applies to top ring. Intermediate ring is .019-.029

ENGINE TORQUE SPECIFICATIONS

ENGINE MODEL	CYLINDER HEAD FT. LBS.	INTAKE MANIFOLD FT. LBS.	EXHAUST MANIFOLD FT. LBS.	CONN. ROD FT. LBS.	MAIN BEARING CAPS FT. LBS.	DAMPER TO CRANKSHAFT BOLT FT. LBS.	FLYWHEEL TO CRANKSHAFT FT. LBS.	OIL PUMP MOUNTING BOLTS FT. LBS.	OIL PAN BOLTS FT. LBS.
1140	①	32	32	30②	③	265	55	18	15
1145	①	32	32	30②	③	265	55	18	15
1150	①	32	32	30②	③	265	55	18	15
1160	①	32	32	30②	③	265	55	18	15
1673	④	—	32	30⑤	175	230	100	—	—
1674	④	—	32	30⑤	175	230	100	—	—
1693	⑥	—	70	⑦	75⑧	148	—	—	—
3208	①	—	32	30②	30⑧	460	55	18	15

① Large bolts: 60 then 95
   Small bolts: 32
② Plus 60°
③ See text under engine assembly
④ Numbered bolts: 115 then 175
   Lettered bolts: 22 then 32
⑤ Plus 90°
⑥ 200 then 330
⑦ 50 plus 180°
⑧ Plus 120°

Caterpillar Diesel Engines

Trouble Diagnosis

Hard Starting or Failure to Start
1. No fuel getting to engine—clogged filters, empty tank, plugged or kinked lines.
2. Fuel shutoff solenoid not energized on 1100 engines, energized on 1673C and 1674 engines.
3. Fuel shutoff solenoid stuck.
4. Fuel transfer pump delivering less than 10 PSI.
5. Injection timing incorrect.
6. Fuel injection pump drive slipping (1100 Series only).

Misfiring
1. Fuel injection lines installed in improper firing order.
2. Defective injection nozzle or pump.
3. Incorrect valve lash.
4. Sticking valves.
5. Incorrect injection timing.
6. Fuel transfer pump delivering less than 10 PSI.
7. Faulty high pressure fuel line.
8. Poor compression.
9. Air in fuel system.

Excessive Black or Gray Smoke
1. Insufficient combustion air—clogged air cleaner or manifold, malfunctioning turbocharger.
2. Faulty fuel nozzle.
3. Incorrect injection timing.
4. Incorrect fuel ratio control setting or injector rack setting.

Excessive White or Blue Smoke
1. Oil consumption due to:
   a. Worn valve guides.
   b. Scored rings or liners.
   c. High crankcase oil level.
   d. Misfiring.

Lack of Power
1. Improperly adjusted accelerator linkage.
2. Failed fuel nozzle.
3. Improper grade of fuel.
4. Turbocharger clogged or dragging.
5. Air induction leaks.
6. Improper injection timing.
7. Excessive valve lash.
8. Fuel supply pressure less than 10 PSI.
9. Faulty timing advance unit.

Low Oil Pressure
1. Lubricating oil diluted with fuel.
2. Excessive clearances in bearings in crankshaft or timing gears.
3. Defective oil pump or relief valve.
4. Crankcase oil level excessive.
5. Oil temperature too high—faulty cooler or cooler relief valve.

Coolant Temperature Too High
1. Combustion gases leaking into coolant.
2. Low coolant level.
3. Faulty water pump.
4. Faulty thermostat.
5. Injection timing incorrect.
6. Pinched shunt line (1100 Series).

Troubleshooting-General Notes
When troubleshooting for misfiring, operate the engine at the speed where missing is most noticeable. Loosen each injector pump fuel line nut, one at a time. That cylinder (or those cylinders) which effect performance the least should be tested for a defective fuel nozzle or pump first.

Fuel transfer pump output should be 10-20 PSI at cranking speed. At full load, the rating is 25 PSI, and at high idle, 30 PSI.

In testing for poor performance or smoke, remember that they can result from either fuel system defects or engine mechanical defects. An air intake restriction across the air cleaner of 30″ of water, or an exhaust system restriction of more than 15″ of water

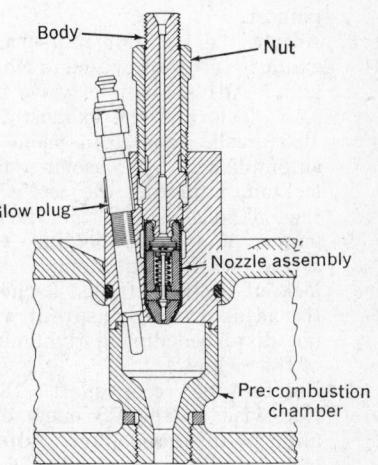

Injection valve and pre-combustion chamber
(© Caterpillar Tractor Co.)

can cause smoke and poor performance because of lack of oxygen and compression. If fuel system inspection reveals no problem on cylinders that misfire, a compression test and measurement of intake and exhaust manifold pressures might be the next step.

Compression Check
1. Remove the fuel injection nozzle of the cylinder to be checked, leaving the pre-combustion chamber in place.
2. Rotate the crankshaft until the piston of the cylinder to be tested is at TDC on the compression stroke.
3. Adapt an air pressure hose to the pre-combustion chamber, using a threaded fitting or rubber adapter.
4. Apply approximately 100 PSI pressure and listen for air leakage at the air cleaner inlet, exhaust outlet, and crankcase breather. On turbocharged engines, it may be necessary to remove the air inlet and exhaust connections to detect leakage.

Air through the air intake indicates a leaky intake valve, air through the exhaust indicates a leaky exhaust valve, and air through the crankcase breather indicates problems with piston, rings, or liner. If there is valve leakage, check the valve clearance.

Engine Tune-Up

Valve Adjustments

1100 and 3208 Series Engines

1. Remove valve covers.
2. Rotate the crankshaft in a clockwise direction until the piston of No. 1 cylinder is at TDC on the compression stroke. The TDC-1 mark on the damper or pulley will align with the timing pointer.
3. Adjust the lash for intake and exhaust valves of cylinders No. 1 and 2. Adjust the inlet valve for .015" lash, and the exhaust for .025" lash. Lash adjustment is accomplished by loosening the locknut, adjusting the screw to the dimension of the feeler gauge, and then holding the screw while retightening the locknut to 19-25 ft lbs. Recheck the adjustment to ensure it was not disturbed during tightening of the locknut.
4. Rotate the crankshaft 180° clockwise, so the VS mark will align with the pointer. Adjust lash for all valves of cylinders 7 and 3.
5. Rotate the crankshaft 180° clockwise again, until the TDC mark on the damper or pulley aligns with the pointer. Adjust the lash of all valves on cylinders 4 and 5.
6. Again rotate the crankshaft 180° clockwise (the VS mark will again align on engines so

equipped). Adjust the lash for cylinders 6 and 8.
7. Replace the valve covers.

1673 Series Engines

1. Remove the valve cover. Rotate the flywheel so the timing pointer is at TC 1-6 cyl. on the flywheel. Observe the positions of the valves for cylinders 1 and 6.
2. If No. 1 cylinder is on compression stroke, adjust the clearances of the inlet valves on cylinders 1, 2, and 4, and the clearances of the exhaust valves on cylinders 1, 3, and 5. If No. 6 cylinder is on compression stroke, adjust the clearances of the inlet valves of cylinders 3, 5, and 6, and the exhaust valve clearances for cylinders 2, 4, and 6. Clearances are: exhaust—.025", intake—.015".
3. Following this, rotate the crankshaft 360° so the cylinder at the opposite end of the engine is on compression stroke, and adjust valves as described above.

1674 Series Engines

1. Remove the valve cover. Rotate the flywheel so the timing pointer is at TC 1-6 cyl. on the flywheel. Observe the positions of the valves for cylinders 1 and 6.
2. If No. 1 cylinder is on compression stroke, adjust the clearances of the inlet valves on cylinders 1, 2, and 4, and the clearances of the exhaust valves on cylinders 1, 3, and 5. If No. 6 cylinder is on compression stroke, adjust the clearances of the inlet valves of cylinders 3, 5, and 6, and the exhaust valve clearances for cylinders 2, 4, and 6. Adjustments are made by turning the adjusting screw clockwise until the

button just contacts the tip of the valve stem (can no longer be wiggled), and then turning backward 10 clicks (.020") for exhaust valves, and four clicks (.008") for intake valves.
3. Following this, rotate the crankshaft 360" so the cylinder at the opposite end of the engine is on compression stroke, and adjust as described above.

1693 and 3406 Series Engines

1. Follow generally the adjusting procedures given for 1674 Series engines, adjusting the inlet valve cam followers to .018" and the exhaust cam followers to .030".

Injection Pump Timing

1100 Series

1. Remove the timing slot plug. Insert a timing pin.
2. Very slowly rotate the crankshaft clockwise until the timing pin drops into the camshaft timing slot.
3. Remove the timing hole plug. Insert a timing bolt and attempt to thread it into the timing gear. The cover retaining bolt to the right of the timing hole, which is shown in the illustration, may be used.
4. If the bolt threaded into the timing gear in the last step, the engine is in time. Otherwise, proceed as described below.
5. Remove the tach drive adaptor on the front cover.
6. On hydraulically governed engines, remove the tachometer driveshaft with a special deep well socket. Then, with a Puller Group, thread a special bolt assembly into the camshaft, avoiding the use of excessive force. Install another special bolt into the gear carrier or adapter, and tighten it until the gear carrier comes loose.
7. On engines with "max-min" governors, loosen the timing advance retaining bolt with a 9/16" deep socket. Tighten the bolt until it starts to turn with effort as the taper drive of the carrier starts separating from the camshaft.
8. Turn the crankshaft clockwise until the bolt will thread into the timing gear and is centered in the hole through the front cover.
9. On hydraulically governed engines, torque the gear retaining bolt to 30-34 ft lbs, and on other engines, torque it to 33-37 ft lbs.
10. Recheck the timing by removing the timing pin and bolt, rotating the engine two revolutions, and replacing them, making further adjustments as necessary.
11. Remove the cover retaining bolt

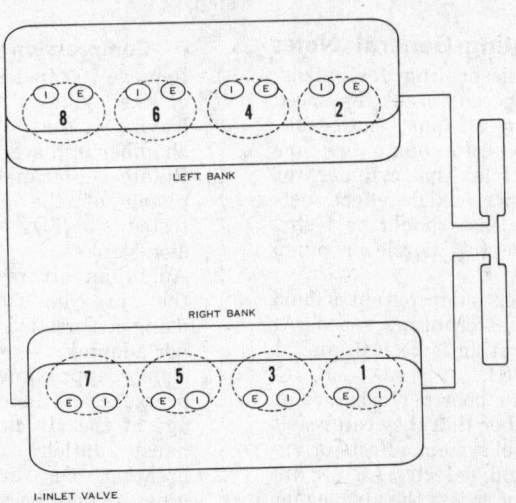

I-INLET VALVE
E-EXHAUST VALVE

Locations of cylinders and valves (© Caterpillar Tractor Co.)

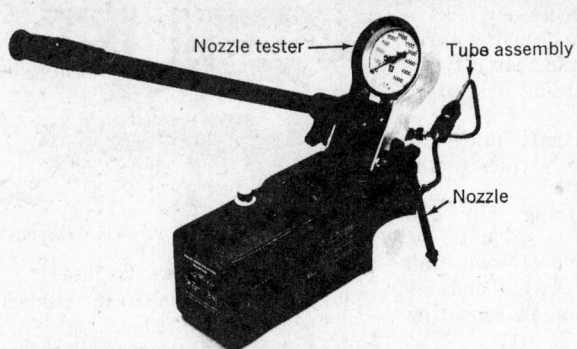

Nozzle tester → Tube assembly

Nozzle

Nozzle testing setup (© Caterpillar Tractor Co.)

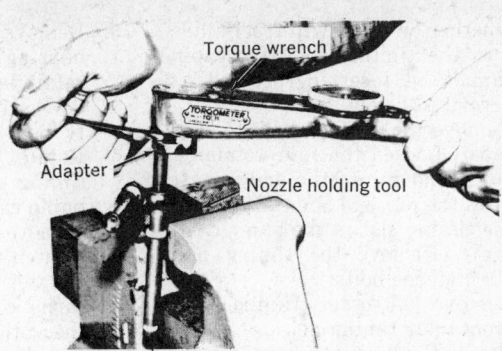

Torque wrench

Adapter

Nozzle holding tool

Tightening nozzle locknut (© Caterpillar Tractor Co.)

from the timing hole, and install it back into the cover. Remove the timing pin from the timing slot. Install the plugs into the timing hole and fuel pump housing opening. Install the tach drive adapter.

Injection Pump Timing— 3208 Series

1. Remove the bolt from the timing pin hole.
2. Turn the crankshaft clockwise until the pin drops in the notch in the pump camshaft.
3. Remove the timing hold plug and put a bolt through the front cover and into the threaded hole in the timing gear. The timing should be correct if the bolt threads easily.
4. If not, remove the tachometer drive cover and disconnect the tachometer drive shaft and washer from the camshaft for the injection pumps.
5. Pull the drive gear from the injection pump camshaft.
6. Turn the crankshaft clockwise until the bolt threads into the hole in the timing gear. Timing should now be correct.
7. Install the washer and tachometer drive shaft. Tighten tachometer drive shaft to 80 ft lbs. for early models and 110 ft lbs. for later models.

1673C Series

1. Remove the injection pump housing. Remove the valve and timing pointer covers.
2. Rotate the crankshaft counterclockwise 60 degrees, and then continue rotating until the TC1-6 cyl. mark on the flywheel aligns with the pointer and both valves of No. 1 cylinder are closed.
3. Attempt to install a special timing plate onto the rear face of the accessory drive housing with the dowels aligned and bolts installed as shown.
4. If the plate can be installed as shown, the engine is in time. Otherwise, proceed as described below.

5. Remove the small front cover from the timing gear housing. Loosen, but do not remove, the gear retaining nut.
6. Separate the gear from the accessory drive shaft. Rotate the shaft in either direction as necessary to permit installation of the timing plate, and install it.
7. Torque the gear retaining nut to 90-110 ft lbs.
8. Remove the timing plate and reinstall the valve cover and pump housing.

1674 Series

1. Remove the valve and timing covers. Rotate the crankshaft counterclockwise 30 degrees, and then continue rotating until the TC1 and 6-6 cyl. mark is aligned with the pointer in the flywheel

housing. Do not turn the crankshaft backwards anywhere in the procedure.

NOTE: There are two timing marks on the flywheel of engines 94B3472 and up. See the accompanying illustration for the correct mark.

2. If all four valves on No. 1 cylinder are not closed, turn the crank another revolution. Install the timing bolt into the flywheel to make sure flywheel and housing are aligned. Also make sure the timing pin will drop into the timing hole between the inlet valves of clynders No. 5 and No. 6.
3. Remove the timing pin and cover shown in the illustration. Insert the timing pin back into the hole it came out of, and determine

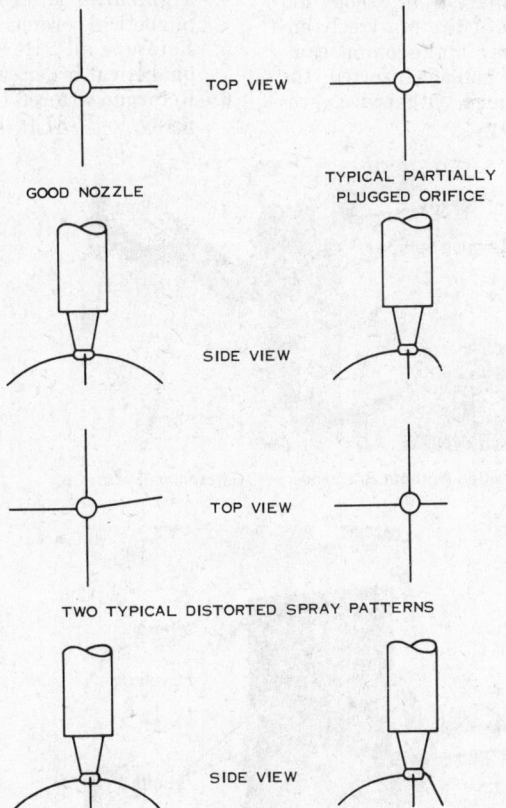

TOP VIEW

GOOD NOZZLE

TYPICAL PARTIALLY PLUGGED ORIFICE

SIDE VIEW

TOP VIEW

TWO TYPICAL DISTORTED SPRAY PATTERNS

SIDE VIEW

Nozzle spray pattern appearance (© Caterpillar Tractor Co.)

Caterpillar • Diesel Engines

whether or not it will freely slide into the timing slot. If the pin cannot be inserted, proceed as described below.

4. Remove the accessory drive gear cover. Loosen the four retaining bolts, and rotate the drive shaft until the pin will slide freely into the timing slot in the pump camshaft. Remove the timing pin. Tighten the bolts.
5. Remove all other timing pins from their timing holes.
6. Rotate the flywheel counterclockwise two complete turns, without going backwards. Insert the timing pin into the pump camshaft again, and make sure alignment is correct, repeating the timing procedure, if necessary.
7. Replace valve and timing hole covers and timing pin.

1693 Series

1. Remove the valve cover and rotate the flywheel clockwise 30 degrees.
2. Rotate the flywheel counterclockwise until the timing bolt can be installed into the flywheel. Do not back up during the procedure.
3. Ensure that all valves for No. 1 cylinder are closed, and turn the flywheel exactly another 360 degrees if they are not. Install the timing bolt and plug back into their normal locations.
4. Insert the timing pin into the pump camshaft timing hole and check to see if the pin freely enters the slot in the pump camshaft. If it can be inserted, the pump is timed. Otherwise, proceed as below.

5. Remove the accessory drive housing cover. Loosen the three retaining bolts, and then retighten the front coupling bolt to 10 ft lbs.
6. Rotate the camshaft until the timing pin squarely rides in the pump camshaft slot.
7. Remove the timing pin and tighten the bolts to 66-74 ft lbs. Rotate the flywheel two full turns counterclockwise and recheck the adjustment, repeating it as necessary.
8. Reinstall the timing pin and cover.

Torquing Head Bolts

1100 and 3208 Series Engines

1. Tighten bolts 1-18, in numerical order, to 50-70 ft lbs.
2. Tighten bolts 1-18, in numerical order, 90-100 ft lbs.
3. Retorque 1-18, in numerical order, to 90-100 ft lbs.
4. Torque 19-22, in numerical order, to 27-37 ft lbs.

1673C Series Engines

1. Tighten numbered bolts, in numerical order to 115 lb ft.
2. Retorque numbered bolts, in numerical order, to 170-180 ft lbs.
3. Retorque all numbered bolts, in numerical order, to 170-180 ft lbs.
4. Tighten all lettered bolts, in alphabetical order, to 22 ft lbs.
5. Retorque all lettered bolts, in alphabetical order, to 27-37 ft lbs.
6. Retorque lettered bolts, alphabetically, to 27-37 ft lbs.

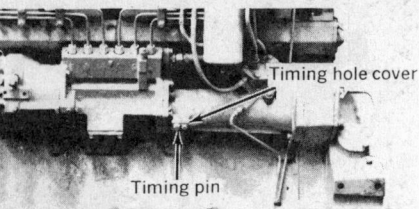

Timing pin location (© Caterpillar Tractor Co.)

1674 Series Engines

1. Torque all bolts, in numerical order, to 135 ft lbs.
2. Retorque, in numerical order, to 180-190 lb.
3. Retorque, in numerical order, to 180-190 ft lbs.

1693 Series Engines

1. Tighten all bolts, in numerical order, to 200 ft lbs.
2. Retorque, in sequence, to 320-340 ft lbs.
3. Retorque, in sequence, to 320-340 ft lbs.

Fuel System

Fuel Injection Pump Housing and Governor

Removal and Installation

1100 Series with Hydraulic Governor

Removal

1. Remove the plug from the timing pin hole in the fuel injection pump housing. Install the timing pin.
2. Rotate the crankshaft clockwise until the timing pin drops into the slot in the pump camshaft.
3. Disconnect the electrical shutoff solenoid wire.

Timing bolt location (without Brakesaver) (© Caterpillar Tractor Co.)

Timing pin location (© Caterpillar Tractor Co.)

Coupling bolt location (© Caterpillar Tractor Co.)

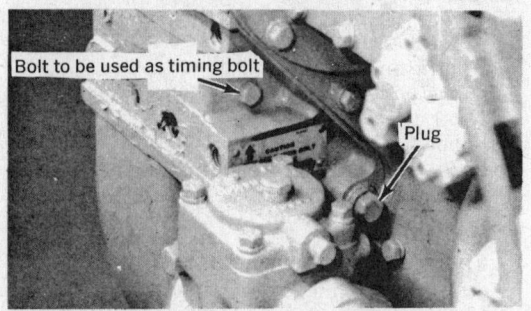

Timing bolt location (with Brakesaver) (© Caterpillar Tractor Co.)

588

4. Remove the tach drive adapter housing. Remove the adapter shaft with a socket.

5. Thread a special bolt assembly into the camshaft, avoiding the use of force. Then, install another special bolt into the gear carrier, and tighten this bolt with a wrench until the gear carrier pops loose. Remove the puller bolts.

6. Remove the timing hole cover from the front cover and insert a 5/16-18 bolt. 2½" long.

7. Rotate the crankshaft in a clockwise direction until the bolt mentioned in the above step can be threaded into the timing gear and is centered in the hole.

8. Remove the injection pump housing retaining bolts, and remove the pump and governor with a hoist.

Installation

1. Position the pump with a hoist and install the mounting bolts.

2. Install the tach drive adapter shaft. Tighten the retaining nut to 30-34 ft lbs.

3. Remove the timing pin and bolt. Rotate the crankshaft, in a clockwise direction, two full revolutions. Install the timing pin and bolt. If both cannot be installed at once, retime the pump as described in "Injection Pump Timing".

4. Remove the timing bolt, and reinstall the timing hole cover.

5. Remove the timing pin from the slot in the injection pump camshaft, and install the plug. Install the tach drive housing, and reconnect the fuel shutoff solenoid wire.

1100 Series with "Max-Min" Governor

Removal

1. Remove the timing hole plug from the injection pump housing and install the timing pin. Rotate the crankshaft clockwise until the pin drops into the timing slot in the pump camshaft.

2. Disconnect the fuel shutoff solenoid wire. Remove the tach drive adapter housing.

3. Back off the retaining bolt for the automatic timing advance unit.

4. Loosen the automatic timing advance drive at the pump camshaft. Back off the retaining bolt well through the area where it turns easily, until the tapered portion starts to separate from the camshaft.

5. Remove the plug from the timing hole in the front cover. Insert a 5/16"-18 bolt about 2½" long.

6. Rotate the crankshaft clockwise

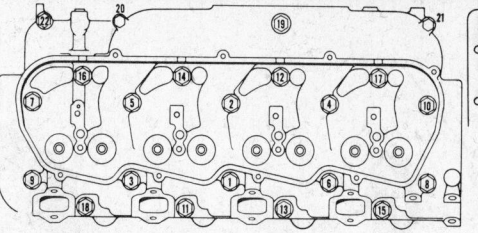

1100 Series head bolt torquing pattern
(© Caterpillar Tractor Co.)

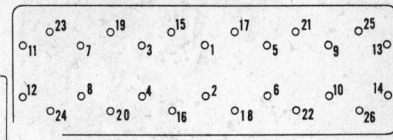

1693 Series head bolt torquing sequence
(© Caterpillar Tractor Co.)

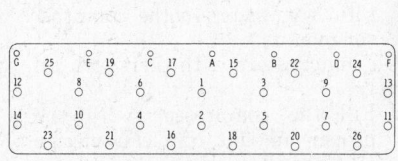

1673 Series head bolt torquing pattern
(© Caterpillar Tractor Co.)

1674 Series head bolt torquing pattern
(© Caterpillar Tractor Co.)

until the bolt can be threaded into the timing gear and is centered in the hole.

7. Remove the pump housing retaining bolts, and remove the pump and governor.

Installation

1. Position the governor and pump and install the retaining bolts.

2. Install the retaining bolt for the automatic advance unit, and torque to 33-37 ft lbs.

3. Remove the timing pin and bolt, rotate the crankshaft two revolutions in a clockwise direction, and attempt to reinstall them. If either cannot be reinstalled, retime the pump as described in "Injection Pump Timing".

4. Remove the timing bolt and install the hole plug. Remove the timing pin from the slot in the pump camshaft and install the plug in the hole.

5. Install the tach drive adapter housing, and connect the wire to the fuel shutoff solenoid.

3208 Series

1. Remove the air cleaner, injection lines and intake manifold.

2. Remove the plug from the cover in the pump housing. Turn the engine clockwise until the timing pin drops into the notch in the pump camshaft.

3. Remove the tachometer drive housing and disconnect the tachometer drive shaft.

4. Pull the drive gear free of the camshaft.

5. Remove the plug from the timing hole in the front cover and install a 5/16"-18NC bolt 2½" long. Turn the crankshaft until the bolt can be installed in the timing gear.

6. Remove fuel line and bolts from pump and remove pump and governor assembly.

1600 Series Engines

Removal

1. Rotate the crankshaft clockwise to bring it to the position where No. 1 piston is at TDC on the compression stroke (see Injection Pump Timing).

2. Remove the turbocharger lubrication and oil drain lines.

3. Disconnect the fuel ratio control sensing line, the governor linkage, and the shutoff solenoid wire. If applicable, disconnect the wiring from the Brakesaver switch.

4. Disconnect the fuel injection lines and fuel supply line at the housing.

5. Remove the injection pump mounting bolts, and remove the assembly with a hoist.

Installation

1. Position the assembly on the engine and install the mounting bolts.

2. Install fuel lines and torque retaining nuts to 25-35 ft lbs.

3. Connect the fuel supply line at the housing. Connect the governor control linkage, and, where applicable, the wiring to the Brakesaver switch.

4. Install the turbocharger oil supply and drain lines.

5. Check and, as necessary, correct the pump camshaft timing as described in the section on "Injection Pump Timing".

Air Compressor Troubleshooting

Insufficient Operating Pressure

1. Slipping drive belt.
2. Excessive leakage in system.
3. Dirty intake strainer.
4. Restriction in system.
5. Leaking valves.
6. Excessive wear.

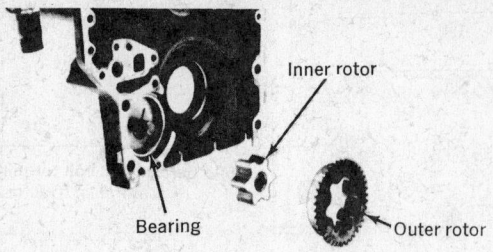

1100 Series oil pump—rotors removed (© Caterpillar Tractor Co.)

Checking rotor tip clearance (© Caterpillar Tractor Co.)

Excessive Noise

1. Improper lubrication.
2. Loose drive pulley.
3. Damaged oil pump
4. Loose mounting brackets.
5. Restriction in line.

Compressor Passes Oil

1. Dirty air strainer.
2. Flooded oil return line.
3. Back pressure from engine crankcase.
4. Excessive oil pressure.
5. Worn oil seals.
6. Piston rings improperly installed.

Air Leakage Test

1. Build up pressure in the system until the governor cuts out. Stop the engine.
2. Listen for air escaping at the intake. If air is escaping, squirt some oil around the unloader pistons. If there is no leakage, the air may be escaping at the discharge valves.
3. If the location of the leak has not been determined, disconnect the discharge line and supply air back through the discharge port. If air movement can be heard, the discharge valves are leaking.

Removal

NOTE: Perform these procedures with caution, as only general procedures are available.

1. Carefully block the vehicle's wheels.
2. Drain the entire brake system.
3. If the compressor is water cooled, drain the engine cooling system, and the compressor head and block.
4. Disconnect air, water, and oil lines.
5. Remove the mounting bolts. Remove the compressor.
6. Remove the pulley with a gear puller.

Installation

On compressors lubricated by the engine:

1. Clean the oil supply line. Operate the engine long enough to determine that an adequate oil flow is present.
2. Clean the oil return line or the return passages in the compressor mounts.

On compressors with their own oil supply:

1. Fill the compressor with the proper amount of clean engine oil.

Then, on all units:

1. Install a new gasket, ensuring that oil holes are properly aligned, where applicable.
2. Inspect the pulley. Replace or reinstall.
3. Clean air cleaner. Install properly. Where engine air cleaner is used, make sure connections are tight.
4. Inspect air and water lines. Clean or replace, as necessary.
5. Reconnect air lines, using a new gasket for the discharge line.
6. Align the pulley, install and tension the belt.
7. Tighten mounting bolts securely and evenly.
8. Operate to check for leakage of any kind or noisy operation.

Oil Pan
1100 and 3208 Series Engines

Removal

1. Drain crankcase.
2. Remove oil pan mounting bolts. Remove the pan.

Installation

1. Clean oil pan and cylinder block gasket surfaces thoroughly.
2. Install the new gasket.
3. Put the pan in position and install the mounting bolts, torquing to 14-20 ft lbs.
4. Refill crankcase (12 qts.).

1600 Series Engines

Removal

1. Drain the crankcase.
2. Remove the oil gauge level and its guide. Remove the Brakesaver oil line, where applicable.
3. Remove the pan mounting bolts, and remove the pan.

Installation

1. Position the new gasket on the engine. Install pan and mounting bolts.

2. Install the oil level gauge and its guide. Where applicable, install the Brakesaver oil line.
3. Refill crankcase.
   clearance on the drive gear side of .002"-.004".
3. Install shafts and integral rotors into rear housing half.
4. Install separator plate. Install main pump rotors.
5. Assemble the housings, and install the housing bolts.
6. Install the drive gear, and install the nut, torquing to 100 ft lbs. Continue torquing the nut just until cotter pin grooves are in proper alignment.
7. Install a new cotter pin in the drive gear mounting nut.
8. Install the suction bell.

Intake Manifold
1100 and 3208 Series

Removal

1. Remove the injection lines, and plug or cap all open ends.
2. Disconnect crankcase ventilator hose.
3. Remove intake manifold bolts, and remove the manifold. Tape engine inlet ports.

Installation

1. Remove sealing tape from intake ports. Clean gasket surfaces thoroughly.
2. Install the gaskets. Position the manifold on the engine block.
3. Tighten the bolts numbered 3 and 4 to 10-20 ft lbs.
4. Tighten *all* bolts to 27-37 ft lbs.
5. Reconnect the crankcase ventilation hose.
6. Uncap and reconnect all injector lines, torquing to 25-35 ft lbs.

Aftercooler Core
1693 Series

Removal

1. Drain the cooling system.
2. Remove the turbocharger outlet pipe and elbow.
3. Disconnect the aftercooler vent line.
4. Remove the cover from the front of the aftercooler top cover.

5. Remove the aftercooler top cover.
6. Remove the mounting bolts from the water inlet collar, and pull the collar out of the housing.
7. Remove the core from the housing.

Oil Pump

1100 and 3208 Series Engines

Disassembly

1. Remove the timing gear cover and oil pickup.
2. Remove the cover retaining bolts. Remove the cover.
3. Remove the outer and inner rotors.
4. Check size of bearing to determine whether or not replacement is required. Bearing must be 2.802"-2.806". Remove the bearing if it is not to specifications.

Assembly

1. Clean all parts thoroughly in a safe solvent. Thoroughly lubricate all parts with 30 weight oil.
2. Install the bearing in the timing cover, using a special driver.
3. Install outer and inner rotors.
4. Check the rotor tip clearance to make sure it falls within the permissible range of .002"-.006". Up to .009" is permitted with used parts.
5. Install the cover and mounting bolts and locks, torquing the bolts to 13-23 ft lbs.
6. Check end clearance to ensure that it is within the permissible range of .002"-.006".

1673 C

Disassembly

1. Remove the cover retaining bolts, and remove the cover.
2. Remove the pump drive gear with a puller.
3. Remove rotors and shafts and idler gear and shaft.

Assembly

1. Clean all parts thoroughly in an approved solvent.
2. Check to make sure the edges of the drive shaft bearings are .020" beneath finished surfaces of the pump body.
3. Check to make sure idler gear shaft diameter is 1.2479"-1.2485", with a bearing clearance of .0020"-.0036".
4. Check pump rotor shafts to ensure their diameters are .7404"-.7410", and that bearing clearance is .0010"-.0026". Check that rotor bearings are recessed

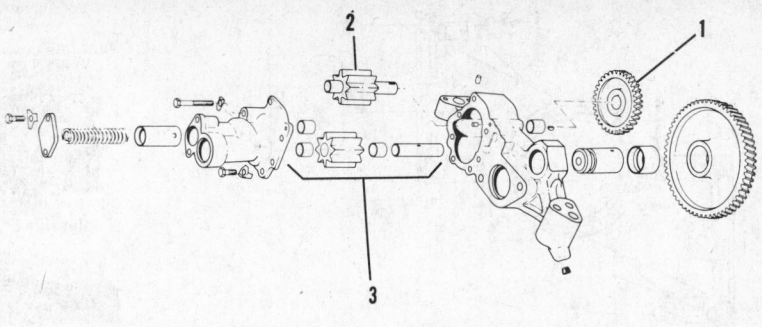

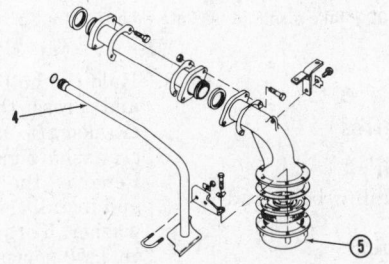

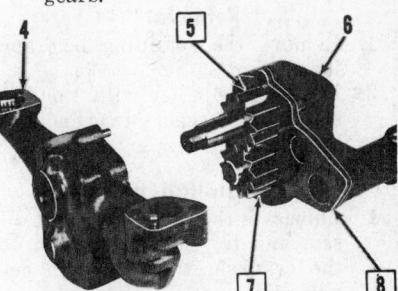

Oil pump, 1673C (© Caterpillar Tractor Co.)

.030" from finished gear surfaces.
5. Install rotors and shafts. Install cover, and torque retaining bolts. Install idler gear and shaft.
6. Heat the drive gear to nearly 750°F. and install.
7. When assembling the unit to the engine, make sure the plate on the bottom of the tube assembly is level with the bottom of the cover.

1674 Series

Disassembly

1. Remove the suction bell.
2. Remove cotter pin, nut and washer from the end of driving gear, and pull gear off with a special puller.
3. Remove housing bolts, and remove outer housing and bushings.
4. Separate the two housing halves.
5. Remove shafts and remaining bushings.

Assembly

1. Clean all parts thoroughly in an approved solvent.
2. Check bearings and shafts for a clearance of .002"-.003".
3. Install shafts and bushings.
4. Assemble the front and rear housing halves.
5. Install housing front cover so the dowel matches the dowel hole in the body. Install housing bolts.
6. Install drive gear and install and torque the mounting nut to 60 ft lbs. Torque nut further to align slots in it for installation of the cotter pin.
7. Install new cotter pin.

1693 Series

Disassembly

1. Remove the suction bell.
2. Remove the cotter pin, nut, and washer from the drive gear shaft. Pull the gear with a puller.
3. Remove housing bolts and separate housing halves.
4. Remove outer gears and separator plate.
5. Remove remaining shafts and gears.

Oil pump, 1693 (© Caterpillar Tractor Co.)

Assembly

1. Clean all parts in an approved solvent.
2. Check all bearings and shafts for a clearance of .005" or less. Check rotors for a rotor to cover

Installation

1. Put the core in position in the housing.
2. Push the water inlet line collar into the housing, and install the mounting bolts.
3. Install the top cover.
4. Install the front cover and mounting bolts.
5. Connect the vent line and install the turbocharger outlet and elbow.
6. Refill the cooling system.

Caterpillar • Diesel Engines

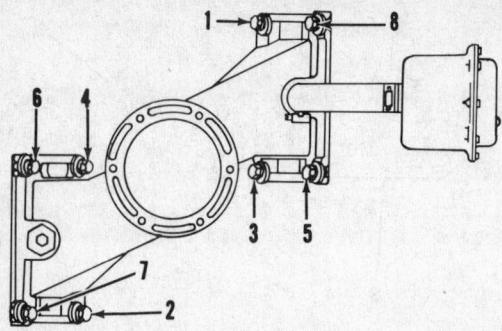

Tightening sequence, 1100 intake manifold (© Caterpillar Tractor Co.)

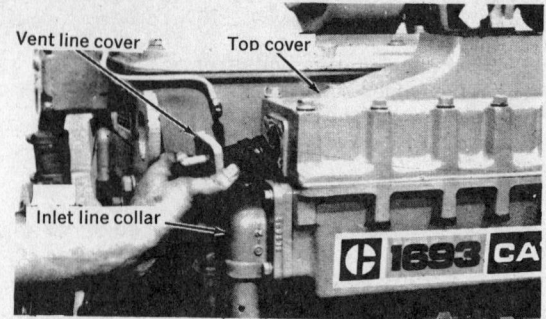

Aftercooler—1693 Series (© Caterpillar Tractor Co.)

Exhaust Manifold

1100 and 3208 Series

Removal

1. Remove the retaining bolts and locks.
2. Remove the manifold.

Installation

1. Clean the gasket mounting surfaces, and position the gaskets on the block.
2. Install the retaining bolts and torque to 27-37 ft lbs.
3. If necessary, turn manifold bolts up to 30° farther to make one edge parallel to each locking tab.
4. Bend lock tabs over flats.

Crankshaft Pulley and Vibration Damper

1100 and 3208 Series

Removal

1. Remove the retaining bolt and its washer.
2. Remove the pulley with a puller, step plate, washers (2), and two bolts.

Installation

1. Lubricate the lip of the front oil seal and the sealing surface of the crankshaft pulley with engine oil.
2. Start the pulley onto the crankshaft. Install a hardened nut, thrust bearing assembly, and one small and four large washers, as shown.

3. Hold the bolt to prevent turning, and press the pulley onto the crankshaft until it contacts the crankshaft gear.
4. Remove the installation tools, and install the retaining bolt and washer. Torque to 230-300 ft lbs on 1100 series and 400-520 ft lbs on 3208 series.

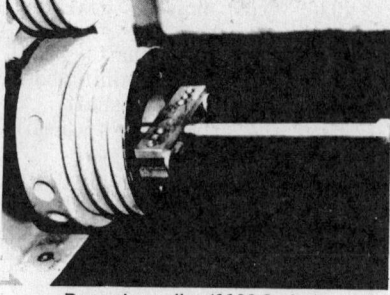

Removing pulley (1100 Series)
(© Caterpillar Tractor Co.)

1673 C

Removal

1. Remove the pulley mounting bolts. Remove the pulley hub with two ⅜″ NC forcing screws.
2. Remove damper-to-hub fastening bolts, and separate hub, pulley, and damper.

Installation

1. Replace faulty parts and assemble hub, damper, and pulley, aligning bolt holes.
2. Install damper-to-hub fastening bolts.

3. Reposition damper assembly on the crankshaft, and employ mounting bolt to press unit back in place. Lubricating the inside surface of the hub, and tapping the assembly periodically with a hard rubber mallet during tightening of the bolt with facilitate installation.
4. Torque the mounting bolt to 210-250 ft lbs.

1674 Series

Removal

1. Remove the hub mounting bolts, and remove the hub assembly, retaining tab lock and spacer.
2. Remove pulley-to-hub bolts and washers, and damper-to-hub bolts and washers, separate parts, and replace faulty components.

Installation

1. Assemble hub, pulley, and damper, and install pulley-to-hub bolts and washers and damper-to-hub bolts and washers.
2. Install assembly, spacer, tab lock, and mounting bolts, torquing bolts to 130-170 ft lbs.
3. Tap bolt head with a hammer, and retorque. Bend tab locks to retain bolts.

1693 Series

Removal

1. Remove alternator and water pump drive belts.

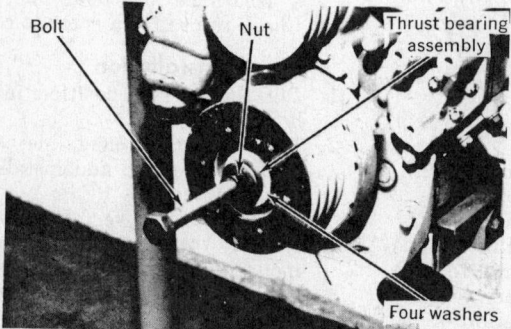

Installing pulley (1100 Series) (© Caterpillar Tractor Co.)

Crankshaft pulley and vibration damper removal (1693 Series)
(© Caterpillar Tractor Co.)

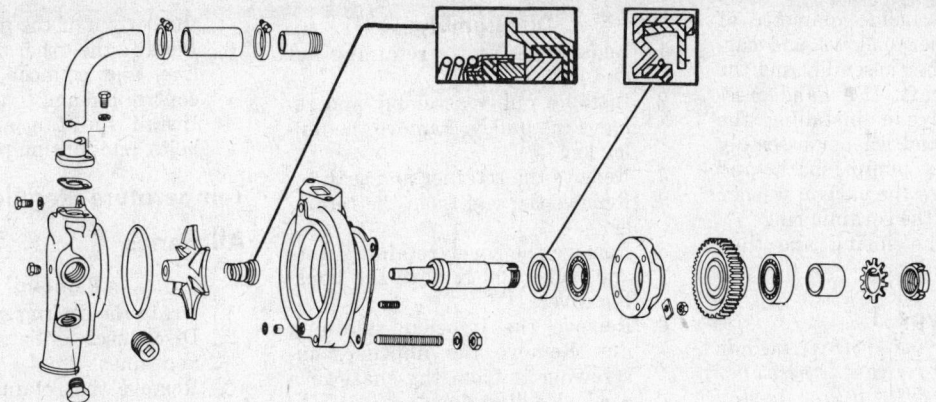

Exploded view of water pump (1673C Series) (© Caterpillar Tractor Co.)

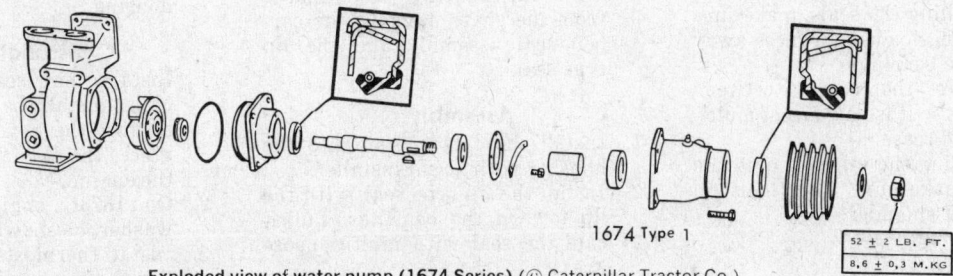

1674 Type 1

52 ± 2 LB. FT.
8,6 ± 0,3 M. KG

Exploded view of water pump (1674 Series) (© Caterpillar Tractor Co.)

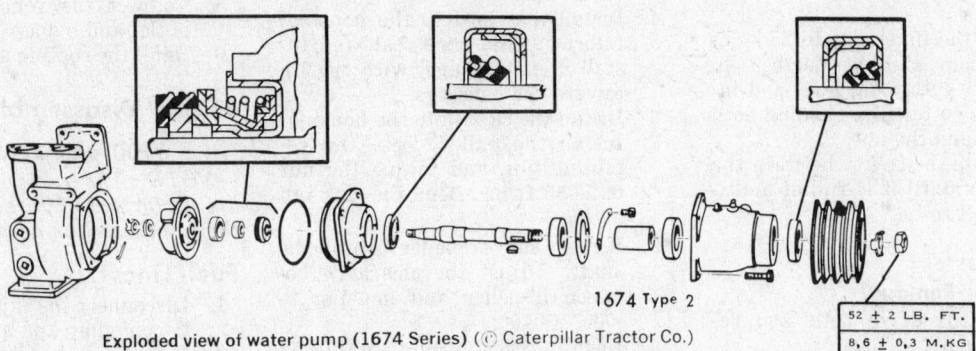

1674 Type 2

52 ± 2 LB. FT.
8,6 ± 0,3 M. KG

Exploded view of water pump (1674 Series) (© Caterpillar Tractor Co.)

2. Remove pulley mounting bolts, and remove the pulley.
3. Remove the vibration damper retaining bolts, lock, and plate. Install a suitable spacer using the retaining plate and bolts.
4. Install a hydraulic press, as shown, and remove the assembly.

Installation

1. Position the damper on the crankshaft. Install the retaining plate, bolts, and lock.
2. Torque the retaining bolts to 137-159 ft lbs, tap the bolts with a hammer, and retorque. Bend the retaining lock.
3. Install the pulley, install and tension the V-belts.

Turbocharger
Removal and Installation
1600 Series

Removal

1. Disconnect the oil supply and drain lines.

2. Remove the air discharge pipe and elbow, and the inlet pipe.
3. Attach a hoist to support the unit, remove the mounting bolts and nuts, and remove the unit with the hoist.

Installation

1. Hoist the unit into place on the engine. Install the mounting bolts and nuts, torquing them to 36-44 ft lbs.
2. Connect air inlet piping, air outlet piping and the outlet elbow.
3. Install oil supply and drain lines.

Water Pump

1673 Series

NOTE: These instructions include an exploded view for ease in assembly and disassembly without step-by-step procedures, and notes on special procedures.

1. In removing bearings:
   a. Remove the bearings and gear with a special push-

puller, bearing pulling attachment, step plate, and ratchet box wrench.
2. In removing impeller:
   a. Use a ½" NC forcing screw to remove the impeller.
3. In installing bearings:
   a. Heat the bearings to 300°F. before installation. Then, do not install shaft assembly in the pump housing until the gear and bearings have cooled to room temperature.
4. In installing the impeller:
   a. Press the unit onto the shaft until it bottoms on the pump shaft shoulder.
5. In installing the seal:
   a. Wet the outside diameter of the rubber cup and ceramic ring assembly, and the seal bore of the pump housing. Use hand pressure only in installation of the rubber and ceramic seats in the pump housing bore. Make sure both seats are square.

b. Wet the inside diameter of the rubber bellows and carbon washer assembly and the pump shaft. Use hand pressure only in installing the bellows and washer assembly onto the pump shaft, and make sure the carbon washer contacts the ceramic ring.

c. Install the spring and then the impeller.

1674 Series Type 1

NOTE: These instructions include an exploded view for ease in assembly and disassembly without step-by-step procedures, and notes on special procedures.

1. In installing the seals, make sure the spring loaded lip faces away from the bearings.
2. Lubricate the cavity between bearings so it is half full of multipurpose grease.
3. In pressing the impeller onto the shaft, make sure it bottoms on the shaft shoulder.

Type 2

1. In installing the seals, make sure the spring loaded lip faces the bearings.
2. Torque the inner nut to 25-35 ft lbs. Then, tighten further to align the cotter pin hole, and install the cotter pin, bending both legs around the nut.
3. Lubricate the cavity between the bearings until it is full of multipurpose grease.

1693 Series

Removal

1. Loosen the drive belts and remove them.
2. Remove the thermostat bypass line, and disconnect the governor linkage.
3. Disconnect the water inlet line from the bottom of the water pump and the aftercooler water supply line. On Brakesaver engines, disconnect the cooler hose at the elbow.
4. Install a 3/8"-16NC forged eyebolt in the pump and attach a hoist.
5. Remove the attaching bolts and remove the pump.

Disassembly

1. Remove the pulley retaining nut and lock.
2. Install a puller assembly and remove the pulley. Remove the pulley key.
3. Remove the retainer and seal.
4. Remove the seal from the retainer.
5. Remove the cover retaining nuts and bolts, and remove the housing cover.
6. Remove the impeller retaining nut. Remove the impeller, unscrewing it from the shaft in a clockwise direction.
7. Remove the shaft assembly. Remove the bearings and spacer from the shaft. Remove the carbon seal assembly and the lip type seal.

Assembly

1. Install the carbon seal into the housing with a seal installer.
2. Install the lip type seal, with the lip toward the bearings. Lubricate the seal with multipurpose grease.
3. Install the bearings and spacer on the shaft.
4. Install the shaft in the housing.
5. Lubricate the cage seal, and install it in the cage, with the lip toward the bearings.
6. Install the cage onto the housing.
7. Install the pulley, lock, and retaining nut, and torque the nut to 50-55 ft lbs. Bend the lock tab over.
8. Screw the impeller onto the shaft. Adjust the clearance between impeller and housing to .005"-.0015".
9. Install the impeller retaining nut, and torque to 50-55 ft lbs. Strike the pulley end of the shaft, and check impeller clearance while rotating the shaft.
10. Install the cover, retaining bolts, and nuts.

Installation

1. Attach a hoist to the pump and hoist the pump into position on the front of the engine. Install the pump retaining bolts.
2. Connect the aftercooler water supply line and the inlet hose at the bottom of the pump.
3. Install the thermostat bypass line, and connect the governor control linkage.
4. Install the fan and alternator belts, and tension properly.

Temperature Regulator

All Series

Removal

1. Drain the cooling system.
2. Disconnect outlet, bypass, and vent lines.
3. Remove the retaining bolts, and remove the housing.
4. Remove thermostat(s) from the housing.

Installation

1. Install a new seal. On 1673C engines, install the seal with the lip toward the cover. On 1647 engines, the lip faces the front of the engine.
2. On 1673C engines, install the washer, as shown.
3. Install thermostat(s).
4. Put the housing into position and install the retaining bolts.
5. Connect the vent line, and the outlet and bypass hoses.
6. Refill the cooling system.

Engine Disassembly

1100 and 3208 Series

NOTE: The following procedures are performed with the engine out of the vehicle and on a work stand.

Fuel Lines

1. Disconnect the injection lines between pump and heads.
2. Disconnect left fuel return line.
3. Disconnect the inlet and outlet lines from the fuel filter base.
4. Disconnect the inlet and outlet lines from the fuel transfer pump.
5. Disconnect the fuel shutoff solenoid lines.
6. Remove all disconnected lines from engine.
7. Unbolt and remove fuel transfer pump.

Injection Nozzles

8. Remove valve covers, fuel return

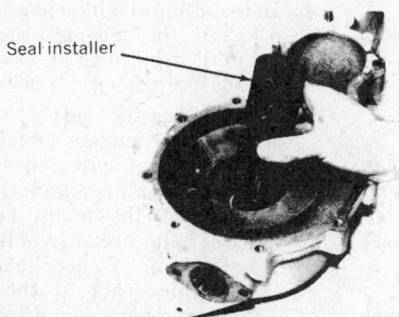

Installing water pump seal (1693 Series) (© Caterpillar Tractor Co.)

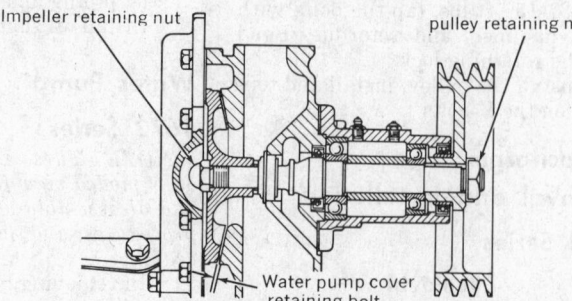

Water Pump Cross Section (1693 Series) (© Caterpillar Tractor Co.)

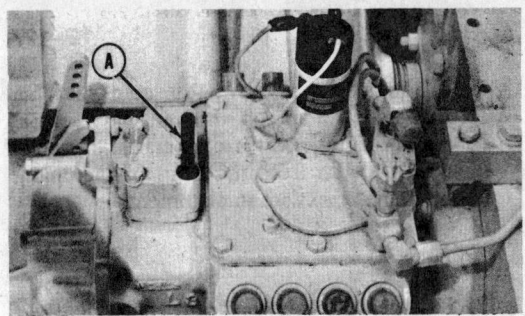

Timing pin in housing 3208 series

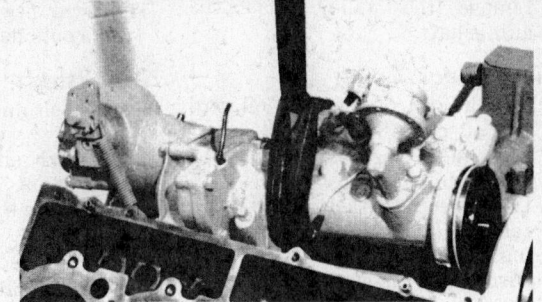

Removing housing—1100 and 3208, typical

manifold and rocker arm assembly.

9. Remove fuel injection nozzle clamp.
10. Disconnect injection nozzle and line from adapter.
11. Remove adapter and lift nozzle from head. Do not pry nozzle!
12. Remove carbon seal nozzle dam.

Automatic Timing Advance Unit

13. To remove automatic timing advance unit on engine equipped with hydraulic governor and engine camshaft mounted timing advance nut:
    a. Remove timing slot plug and insert timing pin.
    b. Turn crankshaft clockwise (front view) until timing pin drops into slot in pump camshaft.
    c. Remove retaining screw and washer from engine camshaft and remove timing unit.
14. To remove automatic timing advance unit from engines equipped with hydraulic governor and fuel pump camshaft mounted timing advance unit:
    a. Remove timing slot plug and insert timing pin.
    b. Turn the crankshaft clockwise (front view) until the timing pin drops into slot in pump camshaft.
    c. Remove tachometer drive adapter shaft—a $5/8$" deep socket helps.
    d. Using a suitable puller, remove the advance unit from the camshaft.
15. To remove automatic timing advance unit from engines equipped with a "Max-Min" governor:
    a. Remove timing slot plug and insert timing pin.
    b. Rotate engine crankshaft clockwise (front view) until the timing pin drops into slot in fuel pump camshaft.
    c. Loosen retaining bolt, then tighten when the taper drive on the carrier starts to separate from the injection pump camshaft.
    d. Remove the automatic timing advance unit.

Injection Pump and Governor

16. To remove injection pump housing and governor on engines with hydraulic governor:
    a. Remove the plug from the timing pin hole in the fuel injection pump housing. Install the timing pin.
    b. Rotate the crankshaft clockwise until the timing pin drops into the slot in the pump camshaft.
    c. Disconnect the electrical shutoff solenoid wire.
    d. Remove the tach drive adapter shaft with a socket.
    e. Thread a special bolt assembly into the camshaft, avoiding the use of force. Then, install another special bolt into the gear carrier, and tighten this bolt with a wrench until the gear carrier pops loose. Remove the puller bolts.
    f. Remove the timing hole cover from the front cover and insert a 5/16-18 bolt, $2\frac{1}{2}$" long.
    g. Rotate the crankshaft in a clockwise direction until the bolt mentioned in the above step can be threaded into the timing gear and is centered in the hole.
    h. Remove the injection pump housing retaining bolts, and remove the pump and governor with a hoist.
17. To remove injection pump housing and governor on engines with "Max-Min" governor:
    a. Remove the timing hole plug from the injection pump housing and install the timing pin. Rotate the crankshaft clockwise until the pin drops into the timing slot in the pump camshaft.
    b. Disconnect the fuel shutoff solenoid wire. Remove the tach drive adapter housing.
    c. Back off the retaining bolt for the automatic timing advance unit.
    d. Loosen the automatic timing advance drive at the pump camshaft. Back off the re-

taining bolt well through the area where it turns easily, until the tapered position starts to separate from the camshaft.
    e. Remove the plug from the timing hole in the front cover. Insert a 5/16-18 bolt about $2\frac{1}{2}$" long.
    f. Rotate the crankshaft clockwise until the bolt can be threaded into the timing gear and is centered in the hole.
    g. Remove the pump housing retaining bolts, and remove the pump and governor.

Intake Manifold

18. Remove the injection lines and plug or cap all open ends.
19. Disconnect the crankcase ventilator hose.
20. Remove the intake manifold bolts, and remove the manifold. Tape engine inlet ports.

Exhaust Manifold

21. Remove the retaining bolts and locks.
22. Remove the manifold.

Rocker Shafts

23. On engines with through head adapters:
    a. Remove shaft retaining bolts.
    b. Remove shaft assembly and pushrods.

Cylinder Heads

24. Remove the water sleeve clamps from between the cylinder head and timing gear cover.
25. Using tool 8S6692 or equivalent, move water sleeve into front cover.
26. Attach lifting device to head, unbolt and remove head.

Pistons & Connecting Rods

27. Remove ridge from cylinders.
28. Turn crankshaft so that journal for piston being removed is at top of travel.
29. Remove the rod cap.
30. Remove piston from cylinder.

Crankshaft Pulley

31. Remove pulley bolt and washer.
32. Using a suitable pulley puller,

remove the pulley from the crankshaft.

Timing Gear Cover

33. Unbolt and remove the oil pan and oil pump suction bell.
34. Disconnect the water temperature sending unit.
35. Disconnect the tachometer drive cable and remove the drive housing.
36. Disconnect the heater hoses from the timing gear cover.
37. Remove oil cooler water outlet elbow from rear of timing gear cover.
38. Remove cover bolts and mark each bolt as to location.
39. Attach a hoist to the front cover and pry cover off engine.

Camshaft Gears (on Engines With Fuel Injection Pump Mounted Timing Advance)

40. Using puller 1P2320, or equivalent, remove the small outer gear.
41. Remove the spacer immediately behind the gear.
42. Install puller 1P2321, or equivalent, and remove the large inner gear.

Camshaft (Gear Installed)

43. Remove cam followers with a magnet. Mark location.
44. Turn crankshaft clockwise (front view) and align crankshaft gear timing mark with timing mark on large camshaft gear.
45. Remove camshaft thrust pin from rear of block.
46. Carefully withdraw camshaft from block.
47. Remove bearings with suitable puller.

Flywheel, Rear Oil Seal and Wear Sleeve

48. Unbolt and remove flywheel.
49. Unbolt and remove flywheel housing.
50. Using a puller, remove rear oil seal.

51. Wear sleeve can be removed with an outside puller.

Crankshaft

52. Unbolt and remove main bearing caps.
53. Attach sling to crankshaft and hoist out of block.
54. Remove upper bearing halves.

Oil Pump

55. Remove pump cover.
56. Remove outer rotor and inner rotor. Check bearing size. Bearing should be 2.80″ ± .002″.

Fuel Nozzle Service

1100 and 3208 Series

1. Remove loose carbon from the tip, carbon seal dam groove, and the body below this groove with a special wire brush. Avoid brushing more than the minimum required to remove deposits, and this can remove the protective coating.

   NOTE: Be careful to avoid contact of the fuel spray with clothing or skin. Direct nozzle discharge into the collector, as fuel spray can penetrate the skin and cause infection.

2. Connect the nozzle to a tester, using one hand squeezing two wrenches, in order to avoid damage to the nozzle and tester. Only perfectly clean fuel may be used.
3. Close the gauge protector valve and the on-off valve. Open the isolator valve on the pump. Operate the tester at 60 strokes/minute until the nozzle is adequately flushed.
4. Open the gauge protector valve and raise the pressure slowly until the nozzle valve opens, and note the pressure. The pressure should be 2,400-2,850 PSI with a used valve, and 2,750-2,850 with a new valve.
5. Remove the nozzle from the tester, install it in a special nozzle holding tool, and loosen the adjusting screw locknut. Then reinstall the nozzle into the tester, tip downward.

6. Turn the lift adjusting screw in until it bottoms, and note the fraction of a turn required. If it can be turned more than ¾ of a turn, it is damaged. Back out the screw far enough to prevent bottoming when the pressure adjusting screw is turned.
7. Adjust the opening pressure to 2,550-2,650 PSI with a used nozzle, 2,750-2,950 with a new nozzle, but do not tighten the locknut.
8. Pump fuel through the nozzle and hold the pressure adjusting screw in place. Slowly turn the lift adjusting screw clockwise until the valve ceases to open. Raise the pressure an additional 200 PSI to ensure valve is bottomed.
9. Back out the lift screw ⅝-⅞ of a turn.
10. Remove the nozzle from the tester and place it in a holding tool. Secure the tool in a vise, avoiding clamping any part of the nozzle. Hold the adjusting screw, and tighten the locknut to 70-75 in. lbs.
11. Mount the nozzle in the tester and recheck the opening pressure.
12. Wrap the nozzle body to absorb leakage from the leak-off boot. Point the tip into the fuel collector and operate the pump rapidly. Dry the tip of the nozzle.
13. Raise the pressure to within 200-300 PSI of the opening pressure. The nozzle should leak three drops or less in 15 seconds.
14. Close the gauge protector valve and the on-off valve, and open the pump isolator valve. Point the nozzle tip into the fuel collector.
15. Operate the tester at 60 strokes per minute, and observe the fuel spray. Fuel should be finely atomized, and in four straight, equally spaced streams.
16. Loosen the collector nuts and reposition the nozzle tip slightly above the horizontal. Operate the tester, bringing the pressure to

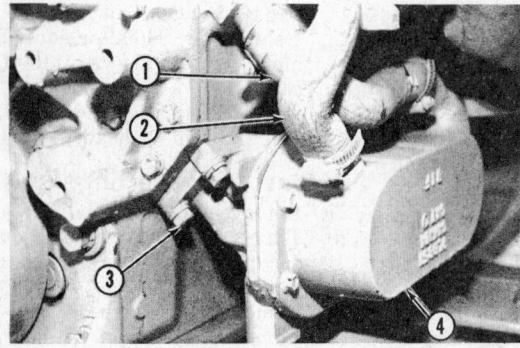

Oil cooler

1 Inlet water line   3 Retaining bolts
2 Outlet water line   4 Oil cooler

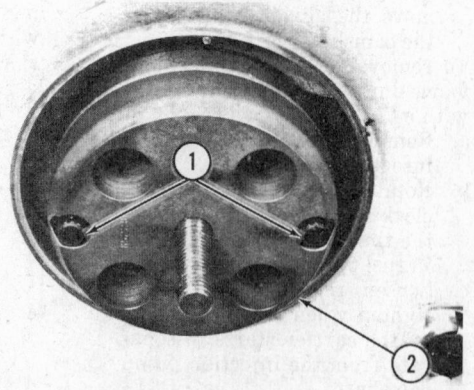

Installing wear sleeve

1 Bolts   2 Pilot

1400-1600 PSI. Observe the leakage from the return at the top of the nozzle. Leakage should be 1-10 drops in 15 seconds after the first drop fall with No. 2 diesel oil.

Engine Assembly

Oil Cooler

1. Clean block surfaces and install oil cooler unit. Torque retaining bolts to 32 ft. lb.
2. Connect inlet and outlet hoses.

Oil Pump

3. Coat parts with SAE 30.
4. Install inner and outer rotors.
5. Check rotor tip clearance .004-.009".
6. Install cover and torque mounting bolts to 18 ft. lbs.
7. Check end clearance .002-.006 in.

Crankshaft and Main Bearings

8. Install upper bearing halves and coat with SAE 30.
9. Lower crankshaft into place.
10. Lubricate lower halves with SAE 30 and install.
11. Oil bolt threads and torque as follows:
    1140 (36B1-36B1923) 1145 (97B1-97B6154) 1150 (96B1-96B6580) 1160 (95B1-95B13691) torque from front to rear to 60 ft. lb., then to 175 ft. lb. 1140 (36B1924 and up) 1145 (97B6155 and up) 1150 (96B6581 and up) 1160 (95B13692 and up) and 3208 Series torque from front to rear to 30 ft. lb., then an additional 120°.
12. Check crankshaft end play .006-.012".

Flywheel, Rear Oil Seal and Wear Sleeve

13. Press wear sleeve into place with chamfer outward.
14. With seal installer, first apply gasket cement to the O.D. of the metal shell, lubricate the seal lip with SAE 30, position seal with spring lip inward.
15. Using a new gasket, install flywheel housing and tighten nuts.
16. Position flywheel on crankshaft so that marks are aligned.
17. Coat bolt threads with Permatex Form—A-Gasket.
18. Install retaining bolts and torque to 55 ft. lb.
    NOTE: The "V" cut on the lock should align with the mark on the flywheel.

Camshaft

19. Install bearings with pilot type installer.
20. Coat the bearing surfaces with SAE 30.
21. Install camshaft, aligning gear timing mark with mark on crankshaft gear.

22. Install thrust pin and torque to 35 ft. lb.
23. If gears were removed, heat both gears to 600°F and press on shaft; aligning timing marks.

Timing Gear Cover

24. Lubricate water sleeve O-rings.
25. Position new cover gasket on block.
26. Install water sleeves in cover (1140, 145 and 1150 use silver sleeves; 1160 uses gold).
27. Position cover on engine and install retaining bolts.
28. Connect oil cooler water elbow to back of cover.
29. Using tool 8S6692, or equivalent, slide water sleeves into heads and install clamps.
30. Install oil pump suction bell, oil pan gasket and oil pan. Torque pan bolts to 17 ft. lb.
31. Connect water temperature sending unit and install tachometer drive housing and cable.

Crankshaft Pulley

32. Coat pulley sealing surface with SAE 30.
33. With suitable installer press pulley onto crankshaft.
34. Remove installer and tighten retaining bolt to 265 ft. lb. on 1100 Series and 460 ft. lb. on 3208 Series.

Pistons & Connecting Rods

35. Turn crankshaft so that journal worked on is at the top of its travel.
36. Lubricate all parts with SAE 30.
37. Install piston and rod assembly with the piston notch toward engine center.
38. Install rod cap, and torque bolt to 30 ft. lb. +60°.

Cylinder Head

39. If removed, install valves.
40. Install new gasket on block.
41. Lower head into place with a hoist.
42. Use anti-seize on bolt threads and tighten in sequence.
    1 thru 18 to 60 ft. lb. then to 95 ft. lb. Tighten bolts 19 thru 22 to 32 ft. lb.
43. Slide water sleeve into head and install clamp.

Rocker Arm Assembly

44. Lubricate all parts with SAE 30.
45. Install pushrods and position shaft assembly on head.
46. Install retaining bolts and nuts and torque to 18 ft. lb.
47. Install fuel return line.
48. Adjust valves to: inlet .015 cold; exhaust .025 cold.
49. Install valve cover. Torque bolts to 120 in. lb.

Exhaust & Intake Manifolds

50. Position new gasket and exhaust manifold on engine and torque bolts to 32 ft. lb. Bend lock tabs over bolts. Position intake manifold and gasket on engine, connect one fuel injection line and torque bolts nearest center to 15 ft. lb., then torque all bolts to 32 ft. lb.

Fuel Injection Pump & Governor (Engines with "Max-Min" Governor)

51. Position pump and governor in place as a unit.
52. Install retaining bolts and tighten.

Fuel Injection Pump & Governor (Engines with Hydraulic Governor)

53. Install injection pump and governor assembly and tighten retaining bolts.
54. Install the tachometer drive adapter shaft and tighten retaining nut to 32 ft. lb.

Automatic Timing Advance Unit (Engines Equipped with "Max-Min" Governor)

55. Lubricate unit through opening in end plates before installing on engine.
56. Position automatic timing advance unit on the fuel injection pump camshaft.
57. Install retaining bolt and torque to 35 ft. lb.

Automatic Timing Advance Unit (Engines Equipped with Hydraulic Governor & Camshaft Mounted Timing Advance Unit.)

58. Align holes in weights with dowels in gear and install the timing unit.
59. Align pin in washer with hole in camshaft and install washer.
60. Install retaining screw and tighten to 108 in. lb. Stake screw in two places.
61. Check gear and weight assembly end clearance .003-.027".

Automatic Timing Advance Unit (Enginess Equipped with Hydraulic Governor and Fuel Pump Camshaft Mounted Timing Advance Unit.)

62. Apply SAE 30 through openings in end plates of the timing unit prior to installation.
63. Position the timing unit on the injection pump camshaft.
64. Install the tachometer drive adapter shaft; tighten nut to 32 ft. lb.

1 Water pipe
2 Fan and pulley mounting bracket
3 Water by-pass line
4 Compressor water return line
5 Turbocharger-to-inlet manifold air pipe
6 Turbocharger lubricating oil supply line
7 Turbocharger oil drain line
8 Exhaust elbow
9 Turbocharger
10 Fuel lines (six)
11 Valve cover
12 Cylinder head

Fuel Return Manifold (Engines Without Through-Head adapters)

65. Position Unit on engine and connect boots to top of injection nozzles.
66. Connect boot to bleed off line.

Fuel Return Manifold (Engines with Through-Head Adapters)

67. Position unit on engine and connect boots to top of injection nozzles.
68. Connect boot to bleed off line. Install clamps and tighten to 18 ft. lb.

Fuel Injection Nozzles (Without Through-head Adapters)

69. Clean nozzle bore, fittings and gasket surfaces.
70. Install a new compression seal and carbon seal.
71. Install a new grommet.
72. Lubricate the exterior surface of the cap, pull grommet over cap. Rotate grommet so the flat surface will face the valve cover when nozzle is installed.
73. Coat grommet groove with 7M7260 Liquid Gasket, or equivalent, remove caps and install nozzle into head with a twisting motion.
74. Assemble and tighten clamp; connect fuel lines to nozzles and hand-tighten nuts.
75. Install rocker assembly. Torque bolts to 18 ft. lb. Tighten fuel line fitting after lines have been purged.

Fuel Injection Nozzles (Engines with Through-head Adapters)

76. Clean nozzle core and fittings.
77. Install a new compression seal and carbon seal dam.
78. Install a new O-ring on the nozzle and adapter.
79. Install the nozzle in the head with a twisting motion.
80. Install adapter and tighten injection nozzle-to-adapter nut to 30 ft. lb. Hand tighten line fitting.
81. Install the nozzle clamp.
82. Install the fuel return manifold

and rocker assembly. Install bolts and tighten to 18 ft. lb.

Fuel Transfer Pump

83. Clean pump mounting surface.
84. Position pump, install bolts and tighten to 32 ft. lb.

Fuel Injection Lines

85. Remove caps, lay on lines as a unit and hand tighten fittings. When lines are purged, tighten fittings. On engines with through-head adapter, tighten retaining nuts to 30 ft. lb.

Engine Disassembly 1673

NOTE: Disassembly is performed with engine out of vehicle and on work stand.

1. Disconnect the turbocharger oil supply and drain lines.
2. Remove the air discharge pipe and elbow and the inlet pipe.
3. Attach a hoist to support the unit, remove the mounting bolts and lift off the unit.
4. Remove alternator.
5. Disconnect and remove fuel lines.
6. Remove rocker housing rear cover.
7. Remove rocker housing retaining bolts, and lift off housing.
8. Remove water pipe, and exhaust manifold.
9. Remove fan and pulley mounting bracket assembly.
10. Remove water by-pass line, and air compressor coolant line.
11. Remove air compressor water return line.
12. Remove head bolts and attach lifting fixture to head. Lift head off engine.
13. Unbolt and remove rocker assembly.
14. Compress springs and remove keepers; lift out valves.
15. Remove pushrod and keep in order of removal.
16. Lift tappets out of block with magnet.
17. Remove injectors from head; do not pry.
18. Remove pulley hub bolt and with

2-bolt puller, remove pulley from crankshaft.
19. Remove bolts, attach hoist and pry off front and rear covers.
20. Using suitable pullers, remove auxiliary drive shaft gear and camshaft gear.
21. Remove fuel transfer pump.
22. Remove idler gear.
23. Remove auxiliary driveshaft retainer and seal and carefully slide shaft out of block.
24. Remove camshaft front and rear retainers and carefully slide camshaft out of block.
25. Remove bolts, attach hoist to flywheel and lift off.
26. Unbolt and remove flywheel housing.
27. Using outside puller, remove oil seal and wear sleeve.
28. Remove oil pan.
29. Remove cylinder ridge.
30. Unbolt and remove rod bearing caps.
31. Push piston and rod assemblies out of cylinders.
32. Unbolt and remove main bearing caps, attach a hoist and lift out crankshaft. Remove liners with a puller.
33. Remove governor cover, plate and housing.
34. Remove injection pump bolts and using extractor 8S2244 or equivalent, remove pump.
35. Pull off oil pump drive gear and remove pump.
36. Unbolt and remove oil cooler.
37. Remove attaching bolts and lift off water pump.

Engine Assembly

1. Install water pump and tighten bolts.
2. Install oil cooler taking care to avoid damaging element.
3. Install oil pump and tighten mounting bolts. Heat gear to 750°F before installing.
4. Check end clearance—.0015-.0035". The edges of the shaft bearings should be .020" below surface of pump body and cover.
5. When installing fuel pump, sight down pump and align notches in bonnet and barrel with the mark

1 Fuel injection nozzles
2 Fuel lines
3 Fuel priming pump
4 Governor
5 Injection pump housing
6 Fuel filter
7 Transfer pump
8 Tachometer drive

1 Fuel injection lines (six)
2 Exhaust pipe connection
3 Fuel priming pump
4 Governor
5 Inject on pump housing

180″ from the pump gear segment center tooth. Position the pump so that the notches align with the guide pins in the housing bore. Align the pump gear segment center tooth with the fuel rack center notch. Keep a downward force on the pump and install the bushing until it's flush with the housing.

6. Install fuel transfer pump and tachometer drive.
7. Assemble and install governor on engine front.
8. Check orifice in piston cooling nozzles with a .090 drill rod.
9. Install upper main bearing halves, coat with SAE 30 and install thrust plates with words "Block Side" next to block.
10. Install upper lower caps and torque bolts to 175 ft. lb.
11. Install cylinder liners coating seals and bore with liquid soap prior to insertion. Drive liner into place with a hammer and block of wood. Attach push-puller and torque to 50 ft. lb. Liner protrusion must be .001-.005″.
12. Install piston and rod assembly, with crankshaft journal up.
13. Install rod bearing cap and torque to 30 ft. lb. +90°.
14. Install oil seal and wear sleeve with a seal installer.
15. Assemble flywheel housing and flywheel torque all bolts to 100 ft. lb. Trim housing gasket square with bottom of block.

16. Install front plate and gasket. Trim gasket.
17. Install camshaft and tighten fasteners.
18. Install camshaft gear aligning "C" on gear with "C" on crankshaft gear.
19. Install idler gear; bearing must be centered in hub.
20. Install accessory drive shaft and gear.
21. Rotate the shaft in the direction necessary to install the 8S5417 Timing Plate. Tighten the gear retaining nut to 100 ft. lb. and remove the timing plate. The fuel injection pump camshaft will be in time with the engine crankshaft when the pump housing is installed.
22. Install the fuel transfer pump.
23. Install front and rear covers. Tighten bolts, trim gaskets.
24. Install shim between trunnion hub and support halves to obtain a .002″ fit. New bushing must be heated to 250° in oil for 2 hrs. before installing.
25. Install pulley and tighten bolt to 230 ft. lb. Tap bolt head with hammer and re-torque.
26. Install damper.
27. Assemble valves and rocker assemblies in head.
28. Install tappets in block.
29. Install injectors in head, torque to 150 ft. lb. Coat mating surfaces with liquid soap.
30. Position gasket on head. Lower head into place. Install head

bolts and torque large numbered bolts as follows:
    Step 1—115 ft. lb.
    Step 2—175 ft. lb.
Tighten small lettered bolts as follows:
    Step 1—22 ft. lb.
    Step 2—32 ft. lb.
31. Install pushrods in order as removed.
32. Install following parts: valve cover—torque to 6 ft. lb., fuel lines, turbocharger oil drain line, exhaust elbow, turbocharger oil supply line, turbocharger-to-inlet manifold air pipe, air compressor water return line, water bypass line, fan and pulley mounting bracket, water pipe.
33. Install exhaust manifold and torque to 32 ft. lb.

Engine Disassembly—1674

NOTE: Disassembly is performed with engine out of vehicle and on workstand.

1. Remove alternator.
2. Remove turbocharger oil drain line.
3. Remove camshaft housing rear cover.
4. Remove camshaft housing retaining bolts, attach hoist and lift off housing.
5. Unbolt retainers and remove camshaft gears.
6. Remove stake mark and drive out retaining pin. Slide camshafts out of heads.

Securing top plate to block around one liner

1 Air sensing line
2 Retaining bolts
3 Cover

Fuel ratio control

Fuel shut-off solenoid

Accessory drive housing

7. Disconnect and remove fuel injection lines.
8. Disconnect turbocharger exhaust pipe.
9. Remove air compressor coolant line.
10. Remove cylinder head bolts, attach hoist and lift off head.
11. Compress springs, remove keepers and lift out valves.
12. Remove precombustion chambers.
13. Remove pulley and damper bolts; lift off as an assembly. Remove trunnion hub retainer and pull off hub.
14. Unbolt support halves, unbolt and pry off front cover.
15. Unbolt flywheel and remove with hoist.
16. Unbolt and remove flywheel housing.
17. Remove rear cover.
18. Remove ridge in cylinder.
19. Remove rod caps and push piston and rod assemblies out of block.
20. Remove lower main bearing halves and hoist crankshaft out of block. Remove upper bearing halves.
21. Remove cylinder liners with a puller.
22. Remove cylinder block top plate.
23. Remove fuel lines from governor, remove attaching bolts and lift off.

24. Attach hoist to injection pump, remove bolts and lift off unit.
25. Remove bolts, locks and plate and remove variable timing unit.
26. Remove attaching bolts and lift out oil pump.
27. Remove water temperature regulator cover from head.
28. Remove aftercooler from head.

Engine Assembly—1674

1. Install aftercooler on head; tighten bolts.
2. Install water temperature regulator with seal lip toward engine.
3. Install oil pump by aligning dowel in cover with dowel hole in body, tighten nut to 60 ft lbs., back off and retighten to align cotter pin.
4. When installing injection pump, sight down pump to align notches in the bonnet and barrel with the pump gear segment center tooth. Position the pump to align notches with the guide pins in the housing bore. Align the pump gear segment center tooth with the fuel rack center notch. Install pump keeping a downward force on the pump and install bushing until flush with the housing. Torque pump retaining bolts to 150 ft lbs.
5. Remove injection pump cam-

shaft timing pin, rotate camshaft to TDC #1 and install pin.
6. Install governor and automatic advance unit.
7. Install upper main bearing halves in block and coat with SAE 30. Lower crankshaft into position.
   NOTE: If crankshaft gear was removed, it must be heated to 400° for installation.
8. Install lower bearing caps and torque to 175 ft lbs.
9. Install top plate and gasket. Install liners by coating seals, liners and bores with liquid soap and driving liners into place with a hammer and wood block.
10. Install piston and connecting rod assemblies, coat bearing surfaces with SAE 30 and install caps. Torque to 30 ft lbs. = 90°.
11. Install rear cover and new gasket; cut gasket flush with bottom of block.
12. Install flywheel housing and flywheel; torque bolts to 100 ft lbs.
13. Install front cover and new gasket; cut gasket flush with bottom of block.
14. Install shims between trunnion hub and upport halves to obtain a .002-.003" fit. Install hub bolts and tighten to 150 ft lbs., tap with hammer and retighten.

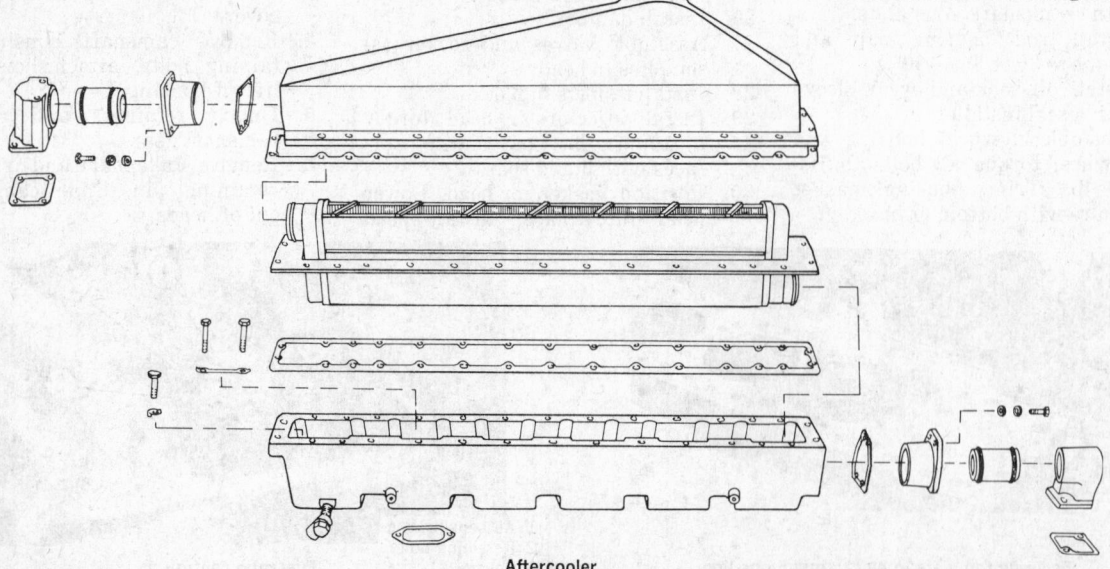

Aftercooler

Camshaft alignment

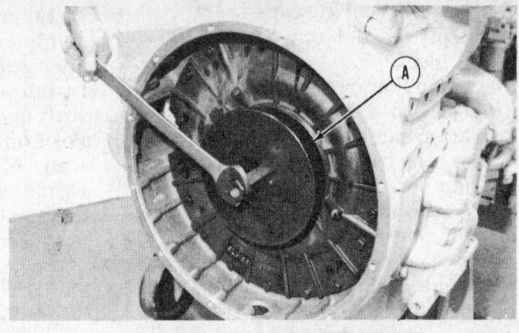

Installing flywheel rotor seal with tool (A)

15. Install pulley and damper.

NOTE: Support bushing must be heated in oil at 250° for 2 hrs.

16. Install precombustion chambers in head and torque to 150 ft lbs. Zinc coated washer is installed first, then glow plug. Tighten glow plug to 120 in lbs.

17. Install valves and injectors; assemble valve springs. Injector nozzle nut torque is 105 ft lbs. Nozzle should be finger tight in body.

18. Position new head gasket on block and lower head into place. Install bolts and torque in sequence, to 135 ft lbs.; then to 185 ft lbs.

NOTE: If cylinder block top plate was removed, install a new gasket between the plate and block.

19. Install air compressor coolant line.

20. Install exhaust manifold and exhaust pipe-to-turbocharger pipe.

21. Install turbocharger and make all connections. Torque mounting bolts to 40 ft lbs. and clamp bolts to 120 in lbs.

22. Install fuel injection lines and torque connections to 30 ft lbs.

23. Install camshafts in housing. Shafts are not interchangeable and are markeed "IN" and "EX."

24. Install camshaft gears and retainers.

25. Install camshaft housing and rear cover.

26. Install turbocharger oil drain line.

27. Install alternator.

Engine Disassembly—1693

NOTE: Disassembly is performed with engine out of vehicle and mounted in work stand.

1. Remove oil cooler water outlet elbow.

2. Remove oil outlet line.

3. Disconnect oil inlet line.

4. Unbolt and remove unit.

5. Remove alternator and water pump drive belts.

6. Remove water temperature regulator bypass line.

7. Disconnect governor control linkage.

8. Disconnect water pump inlet hose and after cooler water supply line. Remove mounting bolts and lift off water pump.

NOTE: On engines with Brakesaver do not disconnect the cooler elbow. Disconnect the cooler hose from the elbow.

9. Disconnect air sensing line from the fuel ratio control.

10. Remove the bolts and use the cover to turn the adjusting bolt 5 turns.

11. Remove the two mounting bolts and pull out and down on the unit to remove.

12. Disconnect the wiring from the fuel shut-off solenoid.

13. Unbolt and remove the unit.

14. Rotate the crankshaft clockwise (front view) to #1 TDC.

15. Remove the turbocharger oil supply line and drain line.

16. Disconnect the Brakesaver wiring where applicable.

17. Disconnect the fuel injection lines from the pump housing.

18. Disconnect the fuel inlet line from the pump housing.

19. Unbolt and remove the pump and governor.

20. Disconnect the fuel supply line and fuel transfer pump drain line from the accessory drive housing.

21. Remove the oil level gauge guide mounting bolt from the fuel priming pump.

22. Remove the accessory drive housing cover and loosen coupling bolts. Remove the tachometer drive assembly.

23. Attach a hoist and remove the accessory drive housing mounting bolts. Remove the unit.

24. Unbolt and remove the oil pan. On Brakesaver engines, remove the oil line.

25. Remove oil pump.

26. Remove lines from water temperature regulator.

27. Unbolt and remove unit.

28. Remove turbocharger air outlet pipe and elbow, and air inlet pipe.

29. Remove mounting bolts and lift off unit.

30. Disconnect aftercooler vent line.

31. Remove front and top covers from aftercooler.

32. Remove water inlet collar from aftercooler.

33. Remove the aftercooler core from housing.

34. Remove the valve cover.

35. Bring #1 to TDC compression.

36. Remove the camshaft driveshaft cover and pull out drive shaft.

NOTE: On Brakesaver engines it is necessary only to remove the two rear plugs to remove driveshaft.

37. Remove the flow plug wiring harness and all inner fuel injection lines.

38. Remove all 15 bolts and locks, pry camshafts off of dowels and lift out.

39. Remove outer fuel injection lines.

40. Remove valve cover base.

41. Remove the air compressor water line.

42. Remove the oil filter base mounting plate.

43. Remove all cylinder head mounting bolts.

44. Attach a hoist and lift off cylinder head.

45. Pry off spacer plate using 6 forcing screws 3/8-16 NC.

Liner installation

46. Remove rod bearing caps.
47. Push piston and rod assemblies out of block.
48. Using a puller, remove liners.
49. Unbolt and remove pulley and vibration damper spacer.
50. Using a hydraulic puller, remove the damper.
51. Pry out front seal.
52. Unbolt and remove front cover.
53. Remove crankcase breather tube.
54. Disconnect all lines from air compressor.
55. Remove mounting bolts and lift off unit.
56. Attach hoist to flywheel and lift off.

NOTE: On engine with Brakesaver, make two holes in metal part of seal and remove seal. Remove rotor housing bolts, housing and rotor. Remove ring seal from rotor. Remove seal carrier and wear ring from rotor.

57. Remove baffle retaining bolts and locks and remove the baffle and rear seal. On engines with Brakesaver, remove seal ring and carrier from crankshaft.
58. Remove fuel filter housing assembly.
59. Remove the tachometer drive and starter.
60. Unbolt and remove flywheel housing. On engines with Brakesaver, remove Brakesaver control valve, oil filter mounting base, starter and flywheel housing.
61. Remove camshaft idler gear, auxiliary drive idler gear, cluster idler gear, pump idler gear, camshaft drive gear and injection pump drive gear.
62. Unbolt and remove timing plate.
63. Remove main bearing caps and hoist out the crankshaft. Remove the upper main halves.

Engine Assembly

1. Install upper mains, coat with SAE 30 and lower crankshaft into place.
2. Install lower main halves, coated with SAE 30 and torque to 75 ft lbs. plus 120°.
3. Install new timing plate gasket and plate. Trim gasket.
4. Install camshaft drive gear and align plug with master spline on camshaft drive collar.
5. Install camshaft gear drive shaft.
6. Install camshaft idler gear with V aligned with V on camshaft idler gear.
7. Install cluster idler gear and align V with camshaft.

8. Install accessory drive idler gear and line up V marks on camshaft idler gear and small cluster gear.
9. Install accessory drive gear and scavenge pump drive gear.
10. Position flywheel housing and install bolts. Torque to 100 ft lbs.
11. Install starter and tachometer drive.
12. Install fuel filter assembly.

NOTE: On engines with Brakesaver: position and install flywheel housing, starter, oil filters, Brakesaver control valve and tachometer drive. Install rear seal and wear ring, rotor and rotor housing.

13. Position flywheel and torque bolts to 100 ft lbs.
14. Position air compressor and drive gear assembly and install mounting bolts. Connect water and air lines.
15. Install the crankcase breather tube.
16. Install front cover, lubricate front seal lip with SAE 30 and coat O.D. of seal with Locktite. Install seal, lip first.
17. Position the damper on the crankshaft. Install retaining plate, bolts and lock. Torque bolts to 148 ft lbs., tap with hammer and retorque.
18. Install pulley.
19. Coat underside of liner flange with Form-A-Gasket. Coat seals with liquid soap.
20. Install liners and drive into position with hammer and wood block.
21. Check liner protrusion—.002-.0076 in four places.
22. Lubricate rings, pistons, liner and rod bearings with SAE 30.
23. Install piston and rod assembly with piston mark front.
24. Install bearing cap and torque nuts to 60 ft lbs. plus 120°.
25. Install spacer plate and gasket.
26. Install head gasket and water seals.
27. Lower head into position, lubricate and install bolts.
28. Torque head bolts in sequence to 200 ft lbs., then to 330 ft lbs., back off and retighten to 330 ft lbs.
29. Install aftercooler inlet and outlet lines.
30. Install oil filter base plate.
31. Connect the air compressor water line. Connect the fuel ratio control sensing line.
32. Connect the turbocharger supply and drain lines.

33. Connect the outer fuel injection lines to the valve cover base and install base on head. Connect glow plug wiring harness. Torque fuel lines to 30 ft lbs.
34. Adjust all camshaft followers to maximum clearance.
35. Align camshaft phasing gear timing marks and install camshafts on engine.

NOTE: #1 piston must be at TDC when installing camshafts.

36. Install inner fuel injection lines and torque to 30 ft lbs.
37. Install camshaft driveshaft and cover. Camshaft driveshaft should have blind spot on top.
38. Install timing bolt in cover.
39. Adjust valves: inlet .018" exhaust .030".
40. Install valve cover.
41. Install aftercooler core, push water inlet collar into housing and install mounting bolts. Install top cover and front cover. Connect vent line to front cover.
42. Install air outlet pipe to front cover.
43. Position turbocharger on engine and torque bolts to 40 ft lbs.
44. Connect inlet pipe and outlet pipe and elbow. Connect supply and drain lines.
45. Install temperature regulators in housing. Attach housing to engine. Connect vent and water lines.
46. Install oil pump.
47. Install oil pan and oil level gauge and guide. On engines with Brakesaver, connect the oil line.
48. Position accessory drive and install mounting bolts. Connect fuel transfer pump drain and supply lines.
49. Position fuel injection pump and governor and install mounting bolts. Connect injection lines and torque to 30 ft lbs. Connect fuel inlet line.
50. Connect fuel ratio control sensing line. Connect Brakesaver wiring.
51. Position fuel shutoff solenoid and install nuts. Connect wire.
52. Install fuel ratio control. Connect air sensing line.
53. Attach hoist and position water pump. Install bolts. Connect aftercooler supply line and inlet hose. Install temperature regulator bypass line and governor control linkage.
54. Install oil cooler and make oil connections.

GENERAL SPECIFICATIONS

ENGINE MODEL	BORE & STROKE	PISTON DISP. CU. IN.	MAXIMUM HORSEPOWER @ R.P.M.	TORQUE @ R.P.M.	FIRING ORDER		LUBE OIL PRESSURE @ GOVERNED SPEED
					RIGHT HAND ROTATION	LEFT HAND ROTATION	
V6-140	4⅝ x 3½	352	140 @ 3300	247 @ 1900	1-4-2-5-3-6	—	55 @ 3300
V-352-C	4⅝ x 3½	352	140 @ 3300	247 @ 1800	1-4-2-5-3-6	—	55 @ 3300
V-352-HT	4⅝ x 3½	352	134 @ 3000	264 @ 1800	1-4-2-5-3-6	—	55 @ 3300
V6-140-HT	4⅝ x 3½	352	140 @ 3300	268 @ 1800	1-4-2-5-3-6	—	55 @ 3300
V6-155	4⅝ x 3¾	378	155 @ 3300	289 @ 1900	1-4-2-5-3-6	—	55 @ 3300
V8-185	4⅝ x 3½	470	185 @ 3300	329 @ 1900	1-5-4-8-6-3-7-2	—	55 @ 3300
V-470-C	4⅝ x 3½	470	185 @ 3300	325 @ 1800	1-5-4-8-6-3-7-2	—	55 @ 3300
V-470-HT	4⅝ x 3½	470	178 @ 3000	354 @ 1800	1-5-4-8-6-3-7-2	—	55 @ 3300
V8-185-HT	4⅝ x 3½	470	185 @ 3300	354 @ 1800	1-5-4-8-6-3-7-2	—	55 @ 3300
V8-210	4⅝ x 3¾	504	210 @ 3300	387 @ 1900	1-5-4-8-6-3-7-2	—	55 @ 3300
V-555-225	4⅝ x 4⅛	555	225 @ 3300	425 @ 1800	1-5-4-8-6-3-7-2	—	40 @ 3300
V-555-240	4⅝ x 4⅛	555	240 @ 3300	445 @ 1900	1-5-4-8-6-3-7-2	—	40 @ 3300
VT-555-225	4⅝ x 4⅛	555	225 @ 3300	425 @ 1800	1-5-4-8-6-3-7-2	—	40 @ 3300
VT-555-240	4⅝ x 4⅛	555	240 @ 3300	445 @ 1900	1-5-4-8-6-3-7-2	—	40 @ 3300
NHE-225	5½ x 6	855	225 @ 1950	668 @ 1463	1-5-3-6-2-4	1-4-2-6-3-5	50 @ 1950
NH-250	5½ x 6	855	250 @ 2100	690 @ 1575	1-5-3-6-2-4	1-4-2-6-3-5	50 @ 2100
Super-250	5½ x 6½	927	250 @ 2100	710 @ 1575	1-5-3-6-2-4	1-4-2-6-3-5	50 @ 2100
NTC-230	5½ x 6	855	230 @ 2100	805 @ 1600	1-5-3-6-2-4	1-4-2-6-3-5	50 @ 2100
NH-230	5½ x 6	855	230 @ 2100	625 @ 1600	1-5-3-6-2-4	1-4-2-6-3-5	50 @ 2100
Formula 230	5½ x 6	855	230 @ 1900	805 @ 1300	1-5-3-6-2-4	1-4-2-6-3-5	40 @ 1900
NTC-250	5½ x 6	855	250 @ 2100	855 @ 1575	1-5-3-6-2-4	1-4-2-6-3-5	40 @ 2100
NHC-250	5½ x 6	855	250 @ 2100	685 @ 1575	1-5-3-6-2-4	1-4-2-6-3-5	40 @ 2100
Formula 250	5½ x 6	855	250 @ 1900	850 @ 1300	1-5-3-6-2-4	1-4-2-6-3-5	40 @ 2000
NTE-235	5½ x 6	855	235 @ 2100	650 @ 1575	1-5-3-6-2-4	1-4-2-6-3-5	50 @ 2100
NT-6	5½ x 6	855	250 @ 2100	695 @ 1575	1-5-3-6-2-4	1-4-2-6-3-5	50 @ 2100
NHHTC-250	5½ x 6	855	250 @ 2100	830 @ 1575	1-5-3-6-2-4	1-4-2-6-3-5	50 @ 2100
NHH-250	5½ x 6	855	250 @ 2100	658 @ 1575	1-5-3-6-2-4	1-4-2-6-3-5	40 @ 2100
NHCT-270	5½ x 6	855	270 @ 2100	740 @ 1575	1-5-3-6-2-4	1-4-2-6-3-5	40 @ 2100
Power Torque-270	5½ x 6	855	270 @ 1950	825 @ 1575	1-5-3-6-2-4	1-4-2-6-3-5	40 @ 2100
NT-280	5½ x 6	855	280 @ 2100	780 @ 1575	1-5-3-6-2-4	1-4-2-6-3-5	50 @ 2100
NTC-290	5½ x 6	855	290 @ 2100	825 @ 1575	1-5-3-6-2-4	1-4-2-6-3-5	40 @ 2100
NTCC-290	5½ x 6	855	290 @ 2100	835 @ 1575	1-5-3-6-2-4	1-4-2-6-3-5	40 @ 2100
NHTCC-290	5½ x 6	855	290 @ 2100	845 @ 1575	1-5-3-6-2-4	1-4-2-6-3-5	40 @ 2100
NHHTC-290	5½ x 6	855	290 @ 2100	855 @ 1575	1-5-3-6-2-4	1-4-2-6-3-5	40 @ 2100
Formula 290	5½ x 6	855	290 @ 1900	930 @ 1300	1-5-3-6-2-4	1-4-2-6-3-5	40 @ 1900
Power Torque-300	5½ x 6	855	300 @ 2300	875 @ 1600	1-5-3-6-2-4	1-4-2-6-3-5	40 @ 2100
NT-310	5½ x 6	855	310 @ 2100	855 @ 1575	1-5-3-6-2-4	1-4-2-6-3-5	40 @ 2100
NT-320	5½ x 6	855	320 @ 2100	900 @ 1575	1-5-3-6-2-4	1-4-2-6-3-5	40 @ 2100
Power Torque-330	5½ x 6	855	330 @ 2300	930 @ 1500	1-5-3-6-2-4	1-4-2-6-3-5	40 @ 2100
NT-335	5½ x 6	855	335 @ 2100	925 @ 1575	1-5-3-6-2-4	1-4-2-6-3-5	40 @ 2100
NTC-335	5½ x 6	855	335 @ 2100	930 @ 1575	1-5-3-6-2-4	1-4-2-6-3-5	50 @ 2100
NTCC-335	5½ x 6	855	335 @ 2100	930 @ 1500	1-5-3-6-2-4	1-4-2-6-3-5	50 @ 2100
NHHTC-335	5½ x 6	855	335 @ 2100	838 @ 1575	1-5-3-6-2-4	1-4-2-6-3-5	40 @ 2100
NHHTCC-335	5½ x 6	855	335 @ 2100	945 @ 1500	1-5-3-6-2-4	1-4-2-6-3-5	40 @ 2100
NTC-350	5½ x 6	855	350 @ 2100	975 @ 1575	1-5-3-6-2-4	1-4-2-6-3-5	40 @ 2100
NTCC-350	5½ x 6	855	350 @ 2100	990 @ 1600	1-5-3-6-2-4	1-4-2-6-3-5	40 @ 2100
Formula 350	5½ x 6	855	350 @ 1900	1060 @ 1300	1-5-3-6-2-4	1-4-2-6-3-5	40 @ 1900
NTA-370	5½ x 6	855	370 @ 2100	1015 @ 1575	1-5-3-6-2-4	1-4-2-6-3-5	50 @ 2100
NT-380	5½ x 6	855	380 @ 2300	855 @ 1600	1-5-3-6-2-4	1-4-2-6-3-5	40 @ 2300
NTA-380	5½ x 6	855	380 @ 2300	855 @ 1500	1-5-3-6-2-4	1-4-2-6-3-5	40 @ 2300
NTA-400	5½ x 6	855	400 @ 2100	1000 @ 1575	1-5-3-6-2-4	1-4-2-6-3-5	40 @ 2100
NTC-400	5½ x 6	855	400 @ 2100	1150 @ 1500	1-5-3-6-2-4	1-4-2-6-3-5	40 @ 2100
NTA-420	5½ x 6	855	420 @ 2300	1045 @ 1725	1-5-3-6-2-4	1-4-2-6-3-5	40 @ 2300
V-903	5½ x 4¾	903	280 @ 2600	700 @ 1575	1-5-4-8-6-3-7-2	—	50 @ 2600
VT-903	5½ x 4¾	903	300 @ 2400	795 @ 1600	1-5-4-8-6-3-7-2	—	50 @ 2400
V-903-C	5½ x 4¾	903	307 @ 2600	706 @ 1800	1-5-4-8-6-3-7-2	—	50 @ 2600
VT-903-C	5½ x 4¾	903	350 @ 2600	848 @ 1800	1-5-4-8-6-3-7-2	—	50 @ 2600

GENERAL SPECIFICATIONS - Cont'd.

ENGINE MODEL	BORE & STROKE	PISTON DISP. CU. IN.	MAXIMUM HORSEPOWER @ R.P.M.	TORQUE @ R.P.M.	FIRING ORDER RIGHT HAND ROTATION	FIRING ORDER LEFT HAND ROTATION	LUBE OIL PRESSURE @ GOVERNED SPEED
Formula 903	5½ x 4¾	903	290 @ 2200	795 @ 1600	1-5-4-8-6-3-7-2	—	50 @ 2200
N-927	5½ x 6½	927	280 @ 2100	775 @ 1500	1-5-3-6-2-4	1-4-2-6-3-5	40 @ 2100
KT-450	6¼ x 6¼	1150	450 @ 2100	1350 @ 1500	1-5-3-6-2-4	1-4-2-6-3-5	50 @ 2100
KTA-525	6¼ x 6¼	1150	525 @ 2100	1650 @ 1300	1-5-3-6-2-4	1-4-2-6-3-5	50 @ 2100
KTA-600	6¼ x 6¼	1150	600 @ 2100	1650 @ 1600	1-5-3-6-2-4	1-4-2-6-3-5	50 @ 2100

CRANKSHAFT AND BEARING SPECIFICATIONS

ENGINE MODEL	PART NUMBER	MAIN BEARING JOURNAL DIAMETER	MAIN BEARING CLEARANCE	ROD BEARING JOURNAL DIAMETER	ROD BEARING CLEARANCE	CRANKSHAFT END PLAY
V-6	159970	3.249-3.250	.0015-.0045	2.499-2.50	.0015-.0045	.004-.014
	176760	3.499-3.500	.0015-.0045	2.499-2.50	.0015-.0045	.004-.014
	181130	3.249-3.250	.0015-.0045	2.499-2.50	.0015-.0045	.004-.014
	183570	3.499-3.500	.0015-.0045	2.499-2.50	.0015-.0045	.004-.014
	195240	3.499-3.500	.0015-.0045	2.499-2.50	.0015-.0045	.004-.014
V-8 378, 504	156340	3.249-3.250	.0015-.0045	2.499-2.50	.0015-.0045	.004-.014
	175050	3.499-3.500	.0015-.0045	2.499-2.50	.0015-.0045	.004-.014
	181140	3.249-3.250	.0015-.0045	2.499-2.50	.0015-.0045	.004-.014
	182360	3.499-3.500	.0015-.0045	2.499-2.50	.0015-.0045	.004-.014
	194840	3.499-3.500	.0015-.0045	2.499-2.50	.0015-.0045	.004-.014
V-8 555	All	3.499-3.500	.0015-.0050	2.749-2.750	.0015-.0050	.004-.014
6-855, 927	All	4.4985-4.50	.0015-.0050	3.1235-3.125	.0015-.0045	.007-.017
V-8 903	All	3.749-3.750	.0015-.0050	3.124-3.125	.0015-.0050	.005-.015
KT-1150	All	5.4985-5.5000	.0026-.0065	3.9985-4.0000	.0020-.0050	.004-.016

VALVE SPECIFICATIONS

ENGINE MODEL	FACE ANGLE (DEG.)	SEAT ANGLE (DEG.)	STEM DIAMETER (IN.)	STEM-GUIDE CLEARANCE (IN.)	VALVE SPRING FREE LENGTH (IN.)	PRESSURE (LB @ IN.)	VALVE LASH INTAKE	VALVE LASH EXHAUST
V-378, 504, 555	30	30	.3785-.3795	.0015-.0022	1.953	203-221 @ 1.329	.012	.022
855, 927	30	30	.4500-.4510	.0022-.0025	①	②	.014H	.027H
903	30	30	.4500-.4510	.0020-.0022	③	④	.012	.025
1150	30	30	.4945-.4955	.0016	3.349	266-294 @ 1.908	.014	.027

① outer: 2.890 inner: 2.685
② outer: 147-163 @ 1.724 inner: 108-130 @ 1.765
③ outer: 2.350 inner: 2.090
④ outer: 124-136 @ 1.287 inner: 80-90 @ 1.237

PISTON SPECIFICATIONS

ENGINE SERIES	PART NO. PISTON	PART NO. LINER	PISTON TO LINER CLEARANCE	PISTON PIN DIAMETER
V6 352, 378 V8 470, 504, 555	173800, 185310, 200800	179710	.0085-.0110	1.3738-1.3740
6-855, 927	All	All	.0125-.013	1.9988-1.9990
V8-903	All	All	.0095-.0120	1.7488-1.7490
KT-1150	All	All	.0115-.0120	2.3988-2.3990

PISTON RING GAP

ENGINE SERIES	RING PART NO.	GAP
V6-352, 378 V8-470, 504, 555	154700	.013-.023
	171690	.013-.023
	171700	.010-.020
	201645	.010-.030
	201650	.013-.023
	208990	.025-.035
6-855, 927	147670	.023-.033
	132880	.019-.029
	168680	.028-.038
	194610	.010-.020
V-8 903	216981	.017-.027
	194600	.013-.023
	194610	.010-.020
KT-1150	Top Ring	.025-.040
	Center Ring	.025-.040
	Oil Ring	.015-.030

TORQUE SPECIFICATIONS

V6—140 V6—155 V8—555
V8—185 V8—210

For each step, tighten in specified pattern to the figure indicated. All figures are in ft./lbs. unless indicated inch/lbs. or degrees of rotation. Degrees of rotation refers to rotation of nut or capscrew.

MAIN CAPSCREWS	SIDE CAPSCREWS	CON ROD NUTS	HEAD BOLTS	FLYWHEEL HOUSING	OIL PAN
1. 55-65	35-40	25-30	25	70-75	snug
2. 115-125	F, G 110-120,	50-55	50-55		$5/16$″—15-17
3. 175-185	Others 100-110	loosen fully	75-80		$3/8$″—25-30
4. loosen fully		20-25	110-115		
5. repeat 1, 2, 3.		30°			
6.		30°			

FLYWHEEL	INTAKE MANIFOLDS	CRANKSHAFT PULLEY	VIBRATION DAMPER	FAN HUB (LOW MOUNTING)	FAN HUB (HIGH MOUNT)
1. 30-35	10	20	90-100	with spacer 20-22	50
2. 60-70	20	40		w/o spacer 30-32	60°
3. 100-105	30-32	60			
4. #195198 caps only—140-145①		80			
5.		90-100			

FAN HUB (INTERMEDIATE MOUNT)	FAN HUB (ECCENTRIC)	INJECTOR PLUNGER ADJUSTMENT	VALVE CROSSHEAD ADJUSTMENT LOCKNUT	VALVE CLEARANCE LOCKNUTS
78-85	300	60 inch/lbs.	25-30	40-45

① Serial #F040000 and later.

6—855, 927

For each step, tighten in specified pattern to the figure indicated. All figures are in ft./lbs. unless indicated inch/lbs. or degrees of rotation. Degrees of rotation refers to rotation of nut or capscrew.

MAIN CAPSCREWS	CONNECTING ROD NUTS	CYLINDER HEAD BOLTS 4⅞″ & 5⅛″ BORES	CYLINDER HEAD BOLTS 5½″ BORES
1. 150	60	25	25
2. 300-310	120	125	125
3. loosen completely	140	325	225
4. 140	loosen completely	425	280-300
5. 30°	25	460-480	
6.	50-55		
7.	60°		

CAM FOLLOWER HOUSINGS	INJECTOR HOLD DOWN CAPSCREWS	INJECTOR INLET & DRAIN CONNECTIONS 4⅞″—5⅛″ BORES	INJECTOR INLET & DRAIN CONNECTIONS 5½″ BORES
1. 15-20	4-5	20-25	4-5
2. 30-35	7-8		7-8
3.	10-12		NEWER TYPE (PT TYPES D, A, OR B)
4.	NYLOCK CAPSCREWS 12-14		10-12

VALVE CROSSHEAD ADJUSTING NUTS	ROCKER LEVER HOUSING 4⅞″—5⅛″	5½″ BORE	OIL FILTER	FUEL FILTER
25-30	65-75	55-75	20-35	20-25

KT, KTA—1150

MAIN CAPSCREWS	CONNECTING ROD NUTS	CYLINDER HEAD BOLTS (CADMIUM PLATED)	CYLINDER HEAD BOLTS (LUBRITED)	FLYWHEEL
1. 190	70	40	40	40
2. 440	140	110	140	100
3. Loosen Completely	210	180	240	140
4. 190	Loosen Completely	250	350	200
5. 440	70			
6.	140			
7.	210			

EXHAUST MANIFOLDS	CRANKSHAFT ADAPTER	VIBRATION DAMPER	INJECTOR HOLDDOWN BOLTS	VALVE CROSSHEAD ADJUSTING NUTS
40-45	320-340	65-75	4-5	25-30
			7-8	
			10-12	

6—855, 927 (CONT)

SECOND ORDER COUNTERBALANCER (4 CYL.) HOUSING	DRIVE ASSEMBLY	FLYWHEEL HOUSING	OIL PAN	FLYWHEEL		CRANKSHAFT FLANGE IF CAST IRON PART #139145
1. 50-55	30-35	150	50-60	Snug	180-200	90-100
2.				190-200		
3.						

VIBRATION DAMPER	ACCESSORY DRIVE PULLEY	FAN HUB SHAFT TO BRACKET NUT	EXHAUST MANIFOLD WITH LOCKPLATE	WITHOUT LOCKPLATE	OIL COOLER-FILTER ASSEMBLY
60	90-100	400-500	25	40	40-45

INJECTOR PLUNGER ADJUSTING SCREW	INJECTOR PLUNGER ADJUSTING NUT	VALVE ADJUSTMENT LOCKNUTS	ROCKER COVER
1. 15° beyond contact	30-40	30-40	75-95
2. Loosen 1 Turn			
3. 48 inch/lbs.			

TORQUE SPECIFICATIONS- Cont'd.

V—903

	PIPE PLUGS			MAIN CAPSCREWS	SIDE CAPSCREWS	CON ROD NUTS	CAMSHAFT THRUST PLATE
	1/8"	3/8"	1/2"				30-32
1.	5-15	30-45	45-55	50	50-55	50-55	
2.				100	100-110	105-115	
3.				150-160	140-150	loosen completely	
4.				250		30-32	
5.				300-350		60-65	
6.				loosen completely		30-32	
7.				45-50		60-65	
8.				85-90		30°	
9.				135-140		30°	
10.				170-180			
11.				30°			
12.				30°			

V—903

	INJECTOR HOLD DOWN CAPSCREWS	VALVE CROSSHEAD ADJUSTING SCREWS	CYLINDER HEAD	GEAR HOUSING	CRANKSHAFT ADAPTER	FLYWHEEL HOUSING	OIL PAN	
							5/16"	3/8"
1.	30-35	25-30	Snug	30-32	75	Snug	16-18	25-30
2.			50		150	50-55		
3.			100		225			
4.			150		300			
5.			200		330-350			
6.			250					
7.			280-300					

	FLYWHEEL	INTAKE MANIFOLDS	WATER PUMP	VIBRATION DAMPER	FAN HUB	ALTERNATOR PULLEY	WATER HEADER COVER PLATE
1.	50-60	10	30-35	200-205	50	60	30-35
2.	100-110	20					
3.	150-165	30-32					
4.	200-210						

	LUBE OIL COOLER/FILTER	EXHAUST MANIFOLDS	VALVE ADJUSTING SCREWS	INJECTOR ROCKER LOCKNUT
1.	30-35	15	40-45	25-30
2.		30		
3.		45-50		

Trouble Diagnosis

The satisfactory performance of a Diesel engine depends on two items of foremost importance, (1) Sufficiently high compression pressure, and (2) The injection of the proper amount of fuel at the right time.

The first one of these items depends mainly on pistons, piston rings, valves and valve operating mechanism; the second item depends on injectors and their operating mechanism, and fuel system.

Lack of engine power, uneven running, excessive vibration, and tendency to stall when idling may be caused by either a compression loss or faulty injector operation.

The causes of trouble symptoms may be varied; therefore a hit-and-miss search should be avoided. A proper diagnosis of symptoms is an orderly process. An "orderly process" means to check the most probable common cause first; then proceed with the nextmost probable cause.

Hard Starting or Failure To Start

1. Fuel shut off valve closed or fuel tank empty.
2. Inferior quality fuel.
3. Restricted fuel lines.
4. Fuel pump pressure regulation faulty.
5. Plugged injector spray holes.
6. Broken fuel pump drive shaft.
7. Gear pump gears scored or worn.
8. Injector inlet or drain connections loose.
9. Water in fuel.
10. Air leaks in fuel suction line.
11. Incorrect injector timing.
12. Valve leakage.
13. Restricted air intake.
14. Engine in need of overhaul.
15. Incorrect valve timing.

Engine Runs but Misses

1. Restricted fuel lines.
2. Water in fuel or poor quality fuel.
3. Air leaks in fuel suction line.
4. Injectors improperly adjusted or plugged.
5. Low compression, intake or exhaust valves leaking.
6. Leaking supercharger air connection.
7. Restricted drain line.
8. Stuck injector plunger.
9. Improper valve and injector adjustments.

Excessive Smoke

1. Restricted fuel system drain lines.
2. Plugged injector spray holes.
3. Inferior quality fuel.
4. Engine fuel rate too high.
5. Injectors improperly adjusted.
6. Intake manifold or cylinder head gasket leak.
7. Restricted air intake.
8. High exhaust back pressure.
9. Broken or worn piston rings.
10. Engine in need of overhaul.
11. Incorrect valve timing.
12. Worn or scored cylinder liners or pistons.
13. Engine overloaded.

Low Power or Loss of Power

1. Inferior quality fuel.
2. Water in fuel.
3. Fuel suction line leaking.
4. Restricted fuel lines.
5. Low fuel pressure.
6. Plugged injector spray holes.
7. Dirty fuel filters or screens.
8. Improper valve and injector adjustments.
9. Improperly adjusted throttle linkage.
10. High speed governor set too low.
11. Air in system.
12. Sticking stop control (fuel shut off).
13. Restricted air intake.
14. High exhaust back pressure.
15. Intake manifold or cylinder head gasket leakage.
16. Low compression—intake or exhaust valve leakage.
17. Broken or worn piston rings.
18. Incorrect bearing clearances.
19. Worn or scored cylinder liners or pistons.
20. Engine in need of overhaul.
21. Incorrect valve timing.
22. Dirty air cleaner.
23. Overheating engine.

Excessive Fuel Consumption

1. Inferior quality fuel.

2. Restricted fuel system drain lines.
3. Fuel rate set too high.
4. Fuel leaks—external or internal.
5. Plugged injector spray holes.
6. Injectors not adjusted properly.
7. Cracked injector body or cup.
8. Restricted air intake.
9. High exhaust back pressure.
10. Engine overloaded.
11. Incorrect bearing clearances.
12. Engine in need of overhaul.

Excessive Oil Consumption

1. External or internal oil leaks.
2. Cylinder oil control not working.
3. Wrong grade oil for climatic conditions.
4. Broken or worn piston rings.
5. Engine in need of overhaul.
6. Worn or scored cylinder liners or pistons.

Low Lubrication Oil Pressure

1. Oil suction line restricted.
2. Oil pump or pressure regulator valve not working properly.
3. Crankcase oil level too low.
4. Wrong grade of oil for conditions.
5. Insufficient coolant.
6. Worn water pump.
7. Coolant thermostat not working.
8. Loose fan belts.
9. Clogged coolant passages.
10. Clogged oil cooler.
11. Radiator core openings restricted.
12. Air in cooling system.
13. Insufficient radiator capacity.
14. Leaking coolant hoses, connections or gaskets.
15. Incorrect bearing clearances.
16. Engine in need of overhaul.
17. Engine overloaded.

Coolant Temperature Too High

1. Low coolant level.
2. Air leaks in suction line.
3. Low oil level.
4. Engine overloaded.

5. Injectors not properly adjusted.
6. Injector pipe partially clogged.
7. Faulty injectors.
8. Injector timing too early.
9. Worn or scored cylinder liners or pistons.
10. Broken valve springs.
11. Crankshaft vibration damper faulty.
12. Excessive crankshaft end clearance.
13. Flywheel loose or unbalanced.
14. Broken or worn piston rings.
15. Incorrect bearing clearances.
16. Engine in need of overhaul.
17. Broken tooth in engine gear train.
18. Loose mounting bolts.

Fuel System

Description and Operation

The Cummins PT fuel system delivers fuel to each cylinder of the engine in equal, pre-determined amounts; thus PT is an abbreviation for pressure-time. The fuel system consists of the supply tank(s), fuel filter, fuel pump, governor, injectors, fuel shut-off valve, and the fuel supply and drain lines.

The fuel pump is coupled to the air compressor which is driven from the engine gear train. The fuel pump main shaft turns at engine crankshaft speed, and drives the gear pump, governor and tachometer shaft.

The fuel pump consists of a gear pump to draw fuel from the supply tank and deliver it to the injectors under pressure, and a governor and throttle to control fuel pressure to the injectors. A pulsation damper prolongs gear pump life, as it

C-160 naturally aspirated diesel engine
(© Cummins Engine Co.)

Cummins V6 diesel engine
(© Cummins Engine Co.)

C-180 supercharged diesel
(© Cummins Engine Co.)

Cummins Diesel Engines

smoothes the flow of fuel through the pump.

The gear pump receives fuel from the supply tank and pumps the fuel under pressure, varying with engine speed, through the filter screen and governor to the throttle and to the injectors. The filter screen, mounted in the top of the fuel pump, prevents damage to the fuel pump and injectors by filtering from the fuel any foreign particles that may be present. To provide fuel pump cooling a small amount of fuel is bled from the fuel pump and returned to the fuel tank. Thus, a circulation of cooling fuel is provided.

At idle speed, the main throttle shaft passage is closed, allowing fuel to flow around the throttle shaft to the idle port in the governor barrel. At engine speeds above idle, the fuel flows through the main throttle shaft hole and enters the main fuel port in the governor barrel.

When the throttle lever is moved to the idle position and the vehicle is moving in excess of idle speed, a small amount of throttle leakage is provided to purge the injectors of air. The PT fuel pump governor controls both idle and high speed. On the 855 CID engines except the NH-250, the fuel manifold is a tube mounted externally on the engine, with equal length branches (tubing) to the injectors. On the V-6, V-8, and NH-250 engines the manifolds are drilled passages, one in each cylinder head, that are aligned with a passageway in the injector body. These fuel manifolds are connected to the pressure side of the fuel pump by a common fuel line.

Fuel circulates through the injector at all times, except during a short period following injection into the cylinder. From the inlet fuel flows down the inlet passage of the injector, around the injector plunger, between the body end and cup, up the drain passage to the drain connections and manifold and back to the supply tank.

As the plunger comes up the injector feed passage is opened and fuel flows through the metering orifice into the cup. At the same time, fuel flows past the cup and out the drain orifice. The amount of fuel which enters the cup is controlled by the fuel pressure against the metering orifice. Fuel pressure is controlled by the fuel pump.

The plunger, during injection, comes down until the metering orifice is closed and the fuel in the cup is injected into the cylinder. While the plunger is seated in the cup, all fuel flow in the injector is stopped.

On the earlier 855 CID engines, except the NH-250, a flanged injector is used. The fuel manifold is connected to the inlet port of the injector flange.

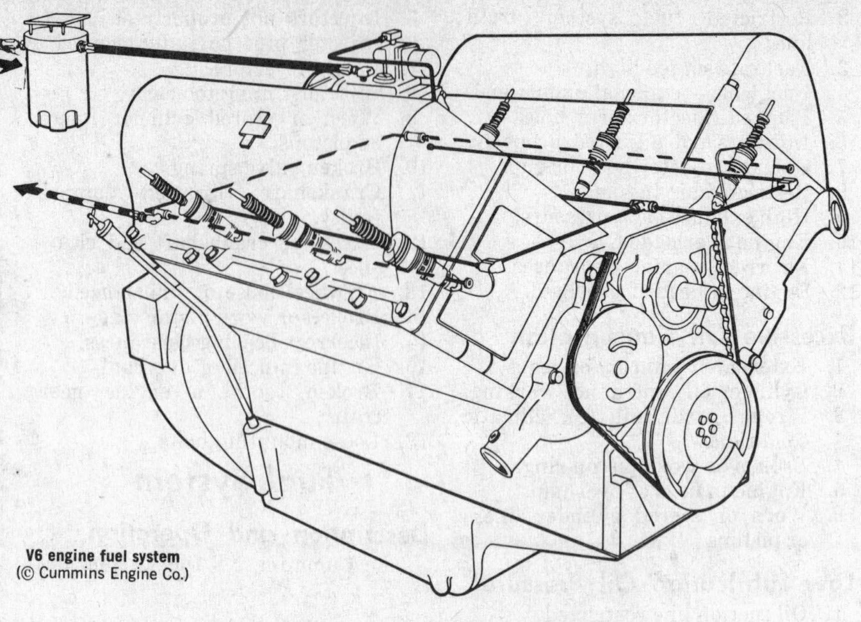

V6 engine fuel system
(© Cummins Engine Co.)

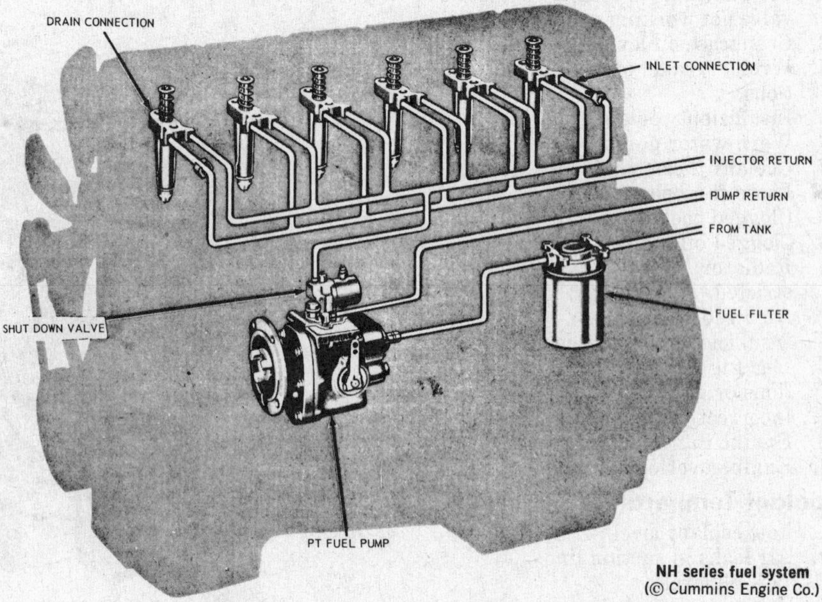

NH series fuel system
(© Cummins Engine Co.)

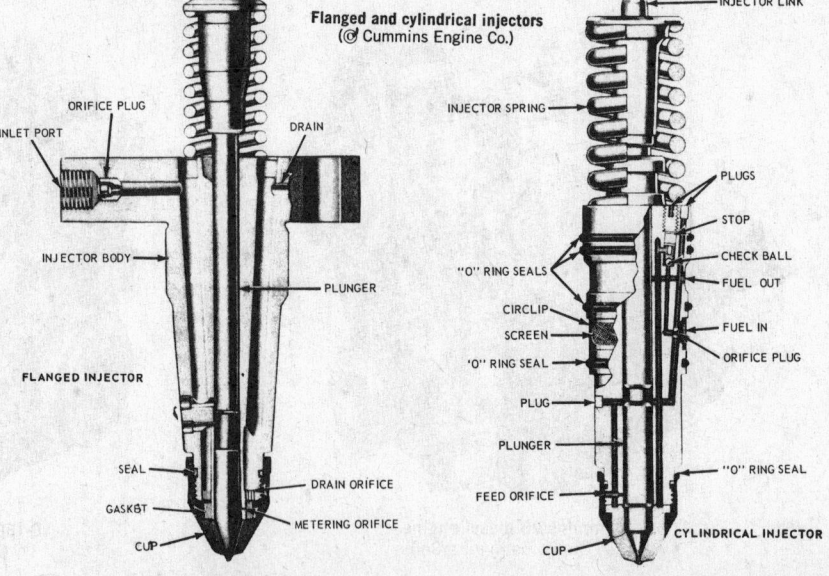

Flanged and cylindrical injectors
(© Cummins Engine Co.)

608

A removable inlet screen is installed in each inlet port for final fuel filtration. The drain port is located adjacent to the inlet port. The drain ports are connected by a drain mainfold to return surplus fuel to the supply tank.

The V-6, V-8, NH-250 and later 855 CID and 1150 cid engines utilize a cylindrical injector. Two radial grooves (annuli) on the injector body index with two drilled passages in each cylinder head, which serve as fuel inlet and drain mainfolds. The inlets annuli contains a fine mesh screen for final fuel filtration. A fuel crossover drain tube, at the flywheel end of the engine connects the two cylinder head drain passages and a common line returns the unused fuel to the fuel tank.

The fuel grooves around the cylindrical injectors are separated by O-rings which seal against the cylinder head injector bore. This forms a leak-proof passage between the injectors and the cylinder head injector bore surface. The injectors contain an orifice plug which can be varied in inside diameter to achieve identical fuel delivery through all injectors.

Adjustments and Repairs

Before Making Adjustments

NOTE: The following applies only to in-vehicle adjustment.

1. Run engine to operating temperature; fuel temperature should be above 110° F. (43° C).
2. Engine parts should be in good condition; timing, valves and injectors properly adjusted.
3. Guages and tachometers should be of known accuracy.
4. Throttle control linkage should be adjusted so that full throttle is obtained and when released,

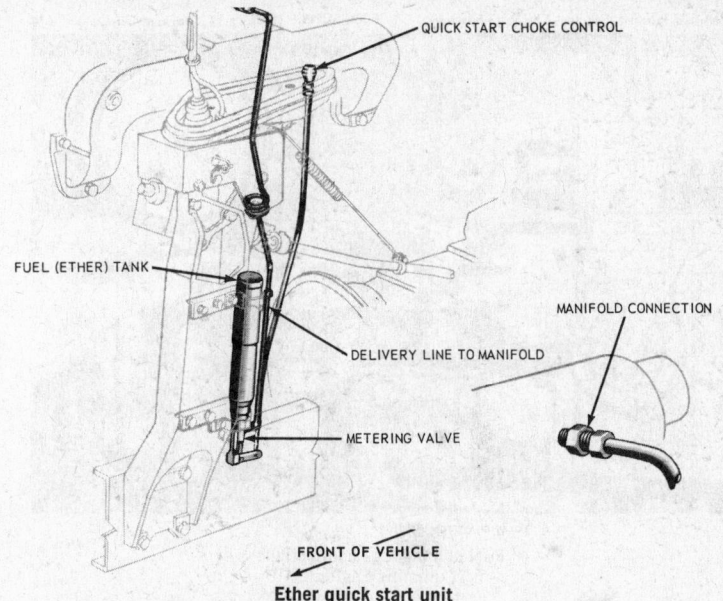

Ether quick start unit
(© Cummins Engine Co.)

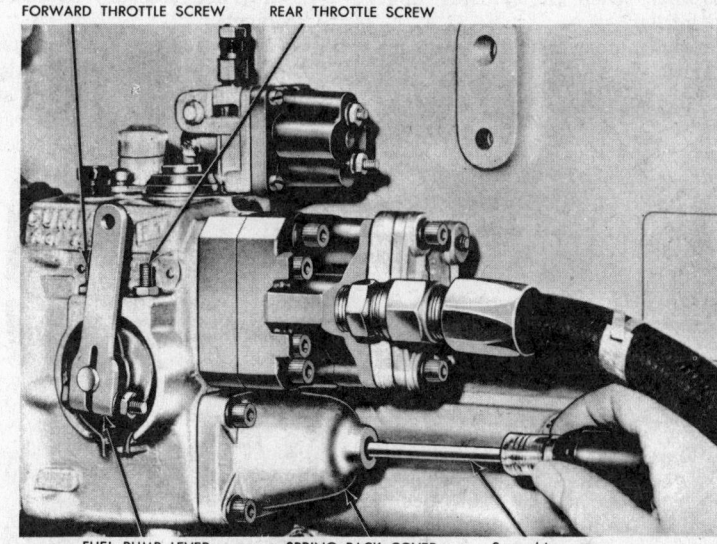

Fuel pump adjustments
(© Cummins Engine Co.)

the throttle is stopped by the front throttle adjusting screw.

NOTE: Pump should be properly adjusted and calibrated prior to installation on vehicle engine. Very little adjustment should be required after installation. The most often made adjustment is the idle setting since this depends on vehicle load.

Adjustments

Idle Speed

1. Remove pipe plug from spring pack cover.
2. The adjusting screw is held in place by a spring clip.
3. Turn screw in to increase or out to decrease speed.
4. If spring pack sealing tool is not used, some air will collect in the spring pack causing a temporary

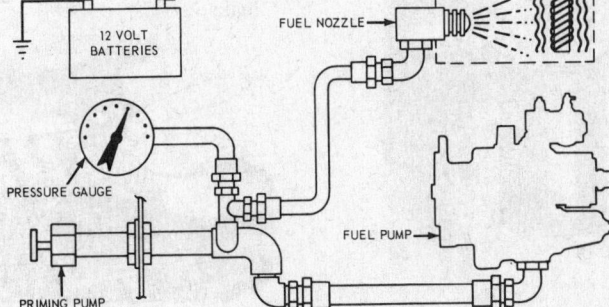

Preheater operation
(© Cummins Engine Co.)

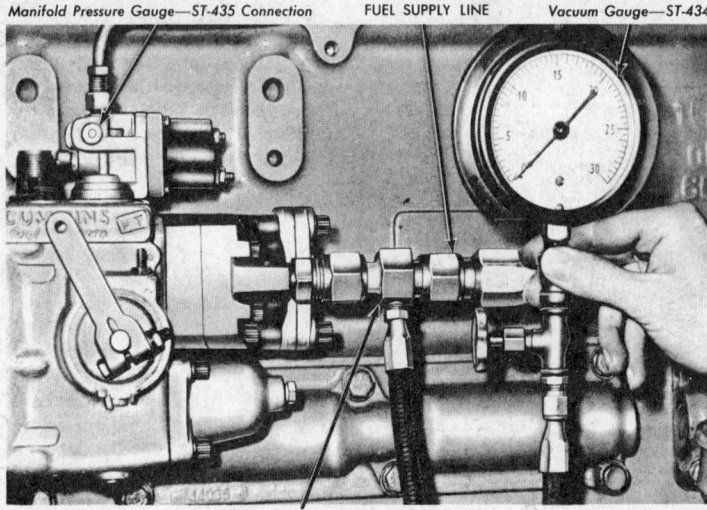

Fuel pump gauge connections
(© Cummins Engine Co.)

rough idle condition. This will stabilize when the housing fills with fuel.

5. Replace pipe when idle speed is correct.

NOTE: The following are factory recommended idle speeds intended as reference points. Some deviations can be made as long as extreme care is taken to avoid causing related problems.

V6, V8: 625 ± 25
V, VT-903: 550 ± 25
855, 927: 600 ± 20
K, KT: 600 ± 25

High Idle

1. Operate engine to purge all air from fuel system.
2. With transmission in neutral, open throttle and hold fully open.
3. Add or remove shims under the governor high speed spring until high idle is 10-12% greater than governor cutoff speed.

Throttle Leakage

1. Operate engine to purge all air from system.
2. Adjust throttle control linkage so that pump throttle just contacts the front throttle stop screw when throttle is closed.
3. Place transmission in neutral, open throttle fully and let engine run at high idle—no load.
4. Using a stop watch, check the time required for throttle release to 1000 rpm movement. Repeat several times.
5. If engine begins to stall upon deceleration, increase throttle leakage.
6. Note position of leakage adjusting screw. Turn screw in while checking engine operation until deceleration time is increased 1 to 2 seconds.

7. If engine decelerates too slowly, it may be necessary to decrease leakage.
8. Note position of adjusting screw. Back out screw as engine decelerates until engine tends to stall. Turn screw in until deceleration time is increased 1 to 2 seconds.

Typical chassis mooted fuel filter V6 engine
(© Cummins Engine Co.)

Engine mounted fuel filter NH series
(© Cummins Engine Co.)

Fuel Pump Screen

Remove the retainer cap from the fuel pump. Remove the spring and lift out the filter screen assembly.

Remove the top screen retainer from the filter screen assembly. Wash the screen and magnet in cleaning solvent and dry with compressed air. Install the top screen retainer.

Install the filter screen, *with the hole down*, in the fuel pump.

Install the spring and retainer cap. Torque the retainer cap to specifications.

Fuel Pump Removal

In-Line 6 Cyl. Series Engines

1. Disconnect the accelerator return spring. Disconnect the hand throttle control, tachometer cable and accelerator rod at the fuel pump. Disconnect the lead to the fuel shut-off valve.
2. Disconnect the primer fuel pump line and fuel supply line at the full pump. Remove the fuel outlet line from the shut-off valve and fuel manifold. Disconnect the fuel bleed line.
3. Remove the fuel pump mounting bolts. Remove the fuel pump, gasket and pump drive rubber buffer.

Installation

1. Clean the gasket surfaces of the fuel pump and air compressor.
2. Install the pump, drive rubber buffer. Position a new fuel pump gasket on the air compressor and install the fuel pump. Torque the mounting bolts to specifications.
3. Connect the primer fuel pump line and fuel supply line to the fuel pump. Connect the fuel outlet line to the fuel manifold and fuel shut-off valve. Connect the fuel bleed line.
4. Connect the accelerator rod, hand throttle control and tachometer cable to the fuel pump. Connect the fuel shut-off valve lead. Connect the accelerator return spring.
5. Check the accelerator linkage and adjust if necessary.
6. Start the engine and check for leaks.

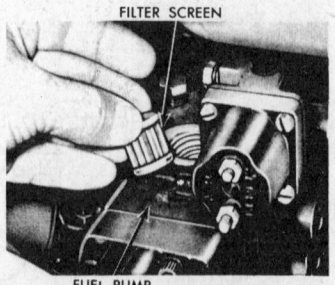

FILTER SCREEN

FUEL PUMP

Fuel filter screen removal
(© Cummins Engine Co.)

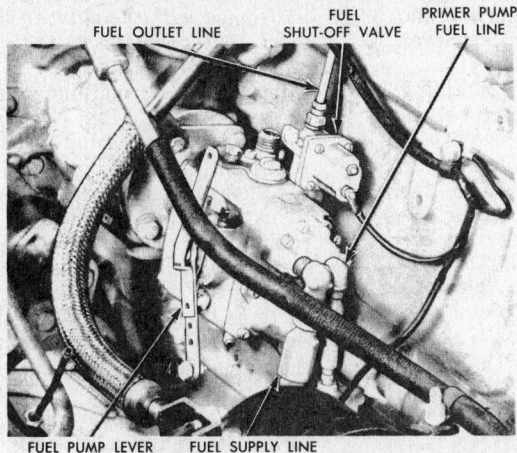

Typical fuel pump installation NH series
(© Cummins Engine Co.)

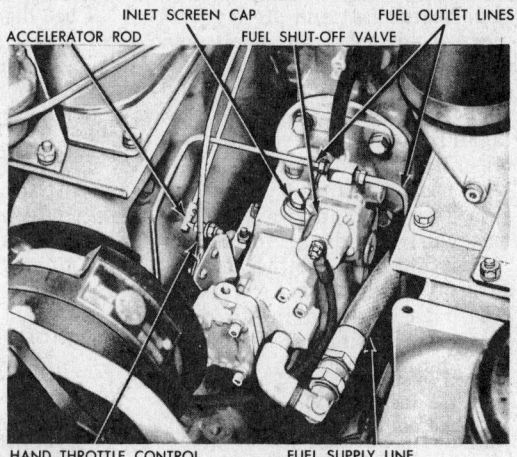

Typical fuel pump installation V6 engines
(© Cummins Engine Co.)

Fuel Pump Removal

V-6 and V-8 Engines

1. Disconnect the air cleaner inlet air tube from the air intake connection on the right intake manifold. Remove the air intake connection and air cross-over connection from the intake manifolds.
2. Disconnect the fuel shut-off valve lead. Disconnect the hand throttle control, accelerator rod and tachometer cable from the fuel pump.
3. Disconnect the fuel supply line and the fuel outlet line(s) from the fuel pump and shut-off valve. Disconnect the fuel bleed line.
4. Remove the fuel pump to air compressor mounting bolts, and remove the fuel pump. Remove the gasket and the pump drive rubber buffer.

Installation

1. Clean the gasket surfaces of the fuel pump and air compressor.
2. Install the pump drive rubber buffer. Use a new gasket and install the fuel pump on the air compressor. Torque the mounting bolts to specifications.
3. Connect the fuel outlet line(s) to the fuel shut-off valve. Connect the fuel supply line to the fuel pump. Connect the fuel bleed line.
4. Connect the accelerator rod, hand throttle control and tachometer cable to the fuel pump. Connect the fuel shut-off valve lead. Check the accelerator linkage and adjust if necessary.
5. Install the air cross-over connection and air intake connection on the intake manifolds. Torque the retaining bolts to specifications. Connect the air cleaner inlet air tube to the air intake connection.
6. Start the engine and check for air and fuel leaks.

Fuel Pump Hydraulic Governor Drive

The fuel pump/hydraulic governor drive is used whenever a hydraulic governor is needed for a particular engine application. The governor mounts atop the housing and the fuel pump is coupled behind the housing.

Disassembly

1. Remove governor drive assembly from governor reservoir. Remove and discard gasket.
2. On governors with 2:1 gear ratio, press out governor drive shaft from gear end. Remove snap ring, ball key and collar.
3. On governors with 3:1 gear ratio, press on shaft from gear end to separate all units from housing.
4. Remove snap ring and ball key; then press out shaft.
5. Remove capscrews and lockwashers and separate drive gear and support assembly from reservoir.

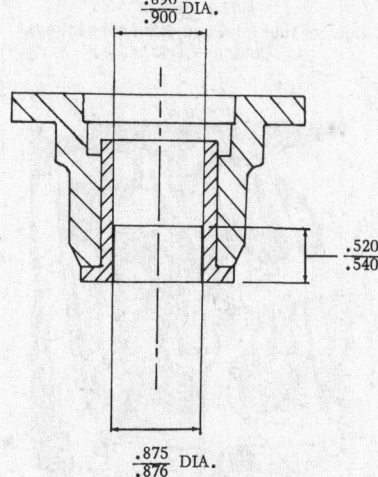

Bearing and retainer assembly
(© Cummins Engine Co.)

6. Remove drain plug, dipstick, vent plug and elbow.
7. Remove shaft locknut and washer from drive shaft.
8. Use puller to remove coupling, spacer and governor drive gear.
9. Press on small end of shaft to remove shaft from support.
10. Press on large end of shaft to remove drive gear. Remove keys from shaft key way.
11. Remove snap ring from support. Invert housing and press out rear bearing and oil seal. Discard oil seal.

Cleaning and Inspection

1. Clean all parts in approved cleaning solvent and dry thoroughly with clean cloth or air jet.
2. Check bearing for worn race or rough action.
3. Check gears for chipped or broken teeth or uneven wear.
4. Check governor shaft housing oil holes. Make certain they are open.
5. Inspect support and reservoir for cracks, breaks, or rough mating surfaces.

Assembly

1. Lubricate outside of oil seal and press into housing from large end. Open end of seal must be down. Check to see that oil holes in housing are open.
2. Invert housing. Coat outside of rear bearing with lubricant and press bearing into housing to shoulder. Insert snap ring, flat side down.
3. Lubricate large end of shaft. Place key in shaft and press shaft into gear against shoulder. Place flat side, not beveled side, of gear hub against shoulder.
4. Press small end of shaft assembly into large end of housing. Coat shaft first with lubricant so that oil seal will not be damaged.

5. Lubricate shaft and press governor drive gear on small end against bearing in housing.
6. Insert key; then press on coupling. Shoulder of coupling goes against gear.
7. Install flat washer and shaft lock nut.
8. Place reservoir in vise with governor drive studs up. Install dipstick, vent plug, weatherhead fitting, and drain plug.
9. Place gaskets and install drive gear and housing assembly to serial number side of reservoir with four capscrews and lockwashers. Large oil hole in housing must be at top.
10. On governors with 2:1 gear ratio:
    a. Drop thrust washer into housing.
    b. Lubricate ball key and insert in drive shaft; install snap ring.
    c. Line up ball key with thrust washer, invert assembly, and press on gear. Allow .003/.006 inch end play. Check with feeler gauge.
11. On governor with 3:1 gear ratio:
    a. Press governor drive shaft into washer flush with bottom side. Note relief in washer to start shaft.
    b. Press shaft assembly into cylinder until shoulder on shaft is flush with end of cylinder.
    c. Slide this assembly into governor drive housing so that flatwasher rests on bronze bushing.
    d. Invert assembly and install ball key, collared washer and snap ring.
    e. Press on end of cylinder until flat washer is against bronze bushing.
    f. Press gear into position allowing .003/.006 inch end play. Check with feeler gauge.

Fuel Injectors

Removal

Early 855, 927 CID Series Engines (except NH-250) Flange Type

1. Remove the rocker housing cover(s).
2. Disconnect the fuel manifold and fuel return line from the injector inlet and drain connections. Remove the fuel inlet screen (Fig. 13). Remove the fuel inlet and drain connections from the injector.
3. Loosen the injector rocker lever lock nut and unscrew the adjusting screw until the push rod can

be disengaged. Disengage the push rod and tip the rocker lever away from the injector.
4. Remove the injector retaining screws and lift the injector from the cylinder head. *Do not bruise the injector tip. Do not turn the injector bottom side up or the plunger will fall out. Leave the plunger in the injector as the plungers are not interchangeable.*

Installation

1. Insert the injector in the cylinder head. *Be sure the plunger is*

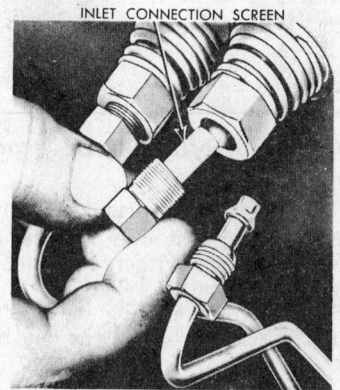

Removing injector inlet screen
(© Cummins Engine Co.)

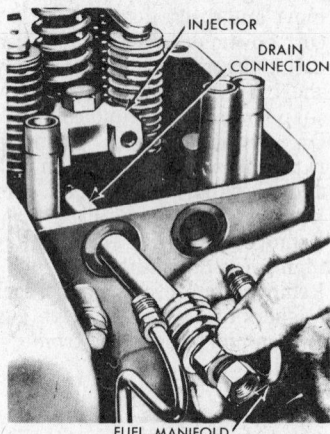

Injector fuel inlet and drain connections
(© Cummins Engine Co.)

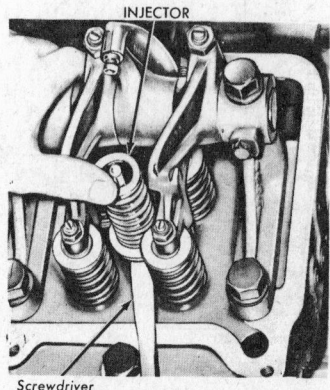

Typical cylindrical injector removal
(© Cummins Engine Co.)

positioned with the plunger class mark centered between the inlet and drain connections.

2. Lubricate the retaining screw threads. Start, *but do not tighten,* the retaining screws. Align the injector by installing the inlet and drain connections 2 turns in the injector. Torque the retaining screws alternately to specifications.
3. Torque the inlet and drain connections to specifications. Install the fuel inlet screen and connect the fuel manifold and fuel return line.
4. Position the push rod under the rocker lever and adjust the injector rocker lever lash to specifications.
5. Install the rocker housing cover(s). Partially tighten the cover screws.
6. Operate the engine until the oil temperature is 140°F, and adjust the injector lash to specifications.

Removal

V-6, V-8, and Later in-line 6 Cyl. Cylindrical Type

1. On some model trucks, it will be necessary to remove the right side air cleaner for right cylinder bank operations.
2. On some model trucks, it will be necessary to disconnect the brace rod from the left side of the transmission remote control mounting support and position it aside. Disconnect the accelerator pedal rod at the cross-shaft on the engine.
3. Remove the valve covers.
4. Remove the power steering pump and drive bolts from the engine. Leave the hoses attached and position the power steering pump out of the way.
5. Loosen the lock nut and back-off the injector rocker lever adjusting screw until the push rod can be disengaged. Hold the push rod aside and tip the rocker lever away from the injector to clear the injector link.
6. Remove the injector clamp retaining screws and the clamps. Pry the injector from the cylinder head. Remove the O-rings.
Avoid bruising the injector tip. Do not turn the injector upside down as the plunger might fall out. Injector plungers are not inter-changeable.

Installation

1. Clean the injector sleeve with a cloth-wrapped stick.
2. Install new O-rings on the injector, and coat the O-rings with lubricant.
3. Install the injector in the cylin-

der head. *Be sure the fuel inlet hole is toward the camshaft.* Seat the injector in the sleeve by giving it a quick, hard push with a wooden hammer handle. A snap should be heard as the injector cup seats in the sleeve.

4. Install the hold-down clamp and torque the retaining screws to specifications.

5. Depress the injector springs with a screwdriver (use as a lever with blade under rocker lever shaft), and engage the injector plunger link with the rocker lever. At the same time engage the push rod with the rocker lever adjusting screw.

6. Adjust the injector lash to specifications. Tighten the lock nut. Install the cylinder head cover valve cover if removed and partially tighten the cover screws.

7. Install the right air cleaner, if removed.

8. Operate the engine until the oil temperature is 140°F. Remove the valve cover or cylinder head cover and adjust the injector lash to specifications.

9. Install the power steering pump and drive belts. Adjust the belt tension to specifications.

10. Connect the accelerator pedal rod to the cross shaft. Connect the brace rod to the transmission remote control mounting support.

Injector Adjustment (Exc. 903 Cid V-8)

1. Turn engine, in direction of rotation, to No. 1 top-center firing position. In this position, intake and exhaust valves for No. 1 cylinder are closed.

2. Continue rotating crankshaft until valve set mark on accessory drive pulley aligns with timing marks on gear case cover. (1-6 VS on 6 cylinder, 1-4 VS on 4-cylinder engines.) When these

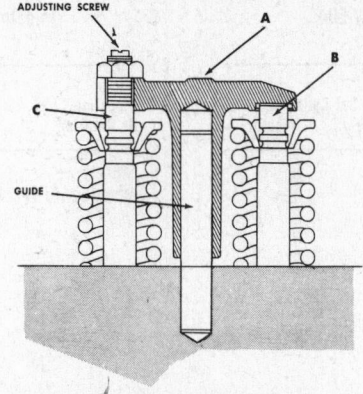

Alignment of crosshead-stem and guide
(© Cummins Engine Co.)

marks align, the engine is in position to adjust valves and injector plunger for No. 1 cylinder.

3. Check intake and exhaust valve position of cylinder to be adjusted. See that valves are closed and rocker levers free.

4. Check threads on adjusting screw and nut. See that they are clean, well-oiled and free-turning.

5. Turn injector adjusting screw down until plunger contacts cup. Turn adjusting screw an additional 15° to squeeze oil from cup.

6. Slack off adjusting screw one turn.

7. Tighten adjusting screw to 70°F. setting. Use a torque wrench equipped wtih a screwdriver adaptor, graduated in inch-pound divisions, and having a maximum capacity of 150 inch-pounds.
   Setting at 70°F. 48 inchpounds.
   Setting at 140°F. 60 inchpounds.

8. Lock screw in place with jam nut.

9. Make final adjustment when en-

gine is at operating (140°F.) temperature.

Injector Torque Settings (in.-lbs.)

	60°-70°F.	140°F.
855, 927 CID	48	72
V6, V8 exc. 903	60	60
KT-1150 CID	48	72

Injector Adjustment V-903 and VT-903

NOTE: Engine oil must be 140-160°F. for this adjustment.

1. Turn the engine until the proper "VS" mark on the vibration damper lines up with the pointer. Make sure both valve rockers for each of the cylinders to be adjusted are free. If rockers are not loose, the engine must be turned over 360 degrees.

2. Support a dial indicator on the engine so that the indicator extension just sits on the top of the injector plunger flange for the cylinder to be adjusted. Make sure the indicator extension is not against the rocker lever.

3. Turn the rocker lever forward until the injector plunger is bottomed in the cup, and all oil is squeezed out.

4. Allow the injector to rise, and then bottom it again. Set the indicator to zero. Again, allow the injector to rise, and bottom it. Make sure the indicator still reads zero.

5. Release the injector completely, and note the reading; .179-.182 is desirable.

6. After adjusting the cylinders listed by each "VS" mark on the vibration damper, turn the engine over to the next mark, adjust those cylinders as described above, and continue in this manner until all eight have been adjusted.

Valve set marks
(© Cummins Engine Co.)

Adjusting injector plunger
(© Cummins Engine Co.)

Cummins · Diesel Engines

Injector and Valve Set Positions for V-504, V-555, VT-555 Series Engines				
Bar In Direction Of Rotation	Pulley Position Old	New	Set Cylinder Valve	Injector
Start	8 & 2 VS	B	—	5
Advance To	1 & 6 VS	A	—	4
Advance To	5 & 3 VS	C	5	8
Advance To	4 & 7 VS	D	4	6
Advance To	8 & 2 VS	B	8	3
Advance To	1 & 6 VS	A	6	7
Advance To	5 & 3 VS	C	3	2
Advance To	4 & 7 VS	D	7	1
Advance To	8 & 2 VS	B	2*	—
Advance To	1 & 6 VS	A	1*	—

* denotes valves are to be set at end of sequence.

Injector and Valve Set Positions For the Left Hand V-8 Engine				
Bar in Direction Of Rotation	Pulley Position Old	New	Set Cylinder Valve	Injector
Start	2 & 8 V.S.	B	—	3
Advance To	7 & 4 V.S.	D	—	6
" "	3 & 5 V.S.	C	3	8
" "	1 & 6 V.S.	A	6	4
" "	2 & 8 V.S.	B	8	5
" "	7 & 4 V.S.	D	4	1
" "	3 & 5 V.S.	C	5	2
" "	1 & 6 V.S.	A	1	7
" "	2 & 8 V.S.	B	2	—
" "	7 & 4 V.S.	D	7	—

Crosshead Adjustment

It is necessary to adjust the crossheads before making valve adjustments. Some crossheads pilot over a solid guide while others pilot within a tubular guide.

1. Loosen valve crosshead adjusting screw locknut and back off screw one turn.
2. Use light finger pressure at the rocker lever contact surface to hold crosshead in contact with valve stem nearest the push rod.
3. Turn adjusting screw down until it contacts its mating valve stem.
4. For new crossheads and guides, advance adjusting screw one-third of one hex (20 degrees) to straighten the stem in its guide and to compensate for slack in threads. With worn crossheads and guides, it may be necessary to advance the screw as much as 30 degrees in order to straighten the stem in its guide.
5. Hold the adjusting screw in this position and tighten locknut to 25/30 foot-pounds torque.
6. Check clearance between crosshead and valve spring retainer with wire gauge. There must be a minimum of .025 inch clearance at this point.

Valve Adjustment

1. Warm engine until oil temperature is 140-160 degrees F. For each two cylinders, turn the engine over and position it at the proper "VS" mark. If the rockers are not loose at this position, turn the engine another 360 degrees.
2. Adjust crossheads as described above.
3. Adjust valves in the standard manner, to the following clearances: Valve Clearance Setting in Inches

Engine Series	Intake	Exhaust
855/927 cid	.014	.027
V6, V8 exc		
903 cid	.012	.022
V-903 cid	.012	.025
K, KT 1150 cid	.014	.027

NOTE: Adjust valves in firing order, except on small V6 and V8. On those engines, follow chart below.

Supercharger

Preliminary Inspection

Before disassembling the supercharger the following inspection procedures must be observed.
1. Check for excessive oil at the air outlet port indicating broken piston ring oil seals.
2. Check radial clearance in bearings:
   a. Remove pump-end cover and force fuel oil through the oil inlet to flush *all* lubricating oil from bearings.
   b. Install an indicator gauge on the rotor timing gear outer diameter.
   c. Check total radial movement by moving gear from side to side. If movement exceeds .003 inch, disassemble unit for a complete bearing inspection.
3. Check rotor shaft end play.
   a. Install a dial indicator at end of rotor shaft.
   b. Push shaft back and forth and note indicator reading.
   c. If total end play exceeds .005 inch, disassemble supercharger and inspect thrust faces.
   d. Perform this operation on both rotor shafts.
4. Check timing gear backlash.
   a. Install an indicator gauge to supercharger housing and check rotor timing gear backlash.
   b. If backlash exceeds .004 inch, a new set of gears must be installed.
5. Insert a feeler gauge into the inlet port between the housing and rotor lobe. Minimum clearance is .005 inch.
6. Check clearance between rotor lobes through the air inlet port. Minimum clearance is .006 inch.

If the preliminary inspection does not indicate any of the above wear

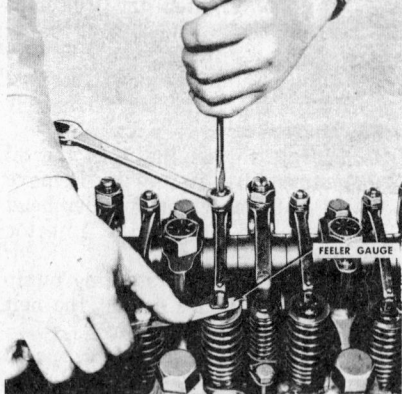

Adjusting valves
(© Cummins Engine Co.)

conditions, the supercharger may be reinstalled on the engine.

NOTE: Use a new gasket when replacing the pump-end cover.

If one or more of the above wear conditions are noted, disassemble the supercharger for a complete inspection or return to factory for a complete rebuild.

Disassembly

1. Raise the lockwasher flange clear of slot in the shaft nut and remove shaft nut with a shaft nut wrench.
2. Wedge a piece of soft metal between timing gear teeth to lock gears while loosening shaft nuts.
3. Remove shaft nut lockwashers from the rotor shafts and slide the outboard bearing journal from the drive rotor shaft. Remove expansion plugs if damaged.
4. Pull the water pump coupling half from the driven rotor shaft with a gear puller.
5. Remove capscrews and lock washers securing end cover and gasket. Discard gasket. Tap cover with a rubber mallet to loosen from housing dowels.
6. Remove capscrews and lockwires securing bearing cages to end plate. Remove bearing cages by rotating and prying.

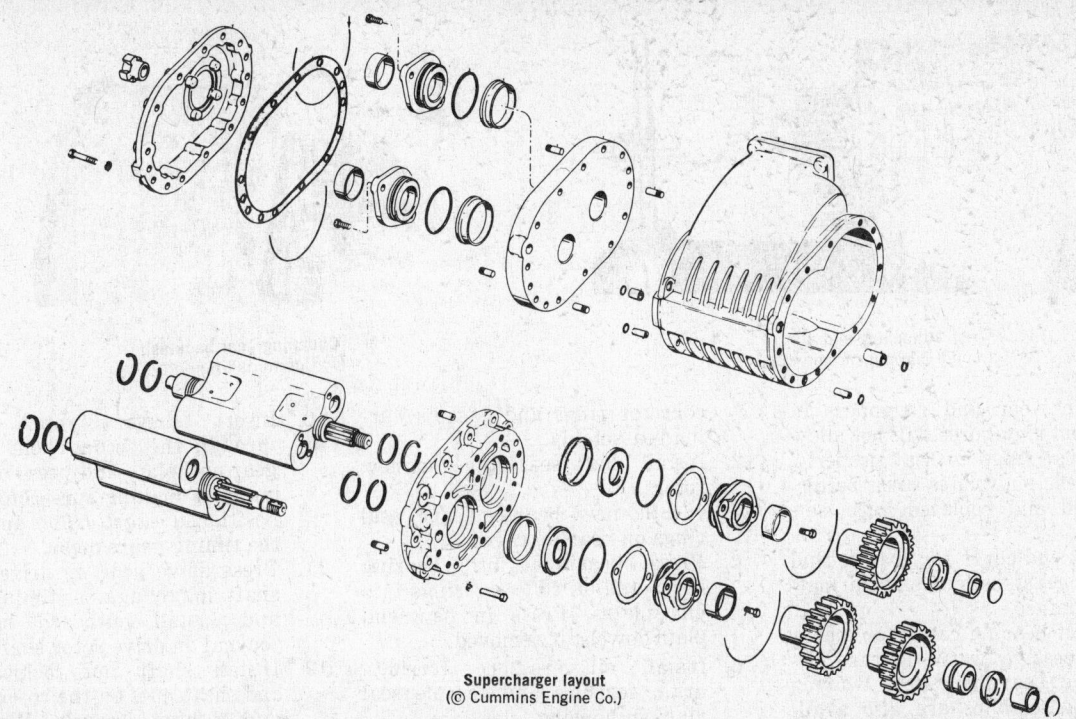

Supercharger layout
(© Cummins Engine Co.)

7. Remove bearing cage seal rings and press bearings from bearing cages. *Mark bearing cages to assure replacement in identical position from which removed.*

8. If damaged, pry oil collector rings from end plate and remove dowels from end plate.

9. Remove capscrews from front end plate and remove end plate and rotor assembly from housing.

10. Remove oil pressure ferrules and "O" rings.

11. Press drive gear and timing gears from the rotors and end plate assembly and extract the rotors from the end plate assembly. *NOTE: The rotors must be rotated into position so the lobe of the rotor being pressed out does not catch on the shaft of the other rotor.*

12. Remove lockwires and capscrews securing bearing cages to the gear end plate.

13. Turn the bearing cages sufficiently to clear the end plate and pry them free. Remove shims and bearing cage seal rings from bearing cages and press out bearings. Mark bearing cages to assure identical replacement.

14. Remove thrust washers from the gear end plate.

15. Remove oil collector rings and dowels, if damaged.

16. Remove and discard all shaft seal rings.

Inspection

1. Clean all parts in approved cleaning solvent.

2. Check housing for cracks, nicks, obstructed oil passages and stripped threads. Remove nicks or scratches with a handstone. Be sure all mating surfaces are smooth.

3. Inspect rotors for undue wear, burrs, pits, scratches, or other damage. Check threaded ends, shoulders, spline bearing surfaces, and seal ring grooves for wear. Small burrs and imperfections on the rotor lobes can be dressed down with a handstone. If the lobe surfaces or rotor shafts are badly scored or damaged, discard rotor set.

4. Inspect timing gears, drive gear and all gear hubs for cracks and broken or worn teeth, damaged splines and excessive wear. **CAUTION: Rotors, timing gears and the driven gear are not interchangeable. If one of these components is worn or damaged, it must be replaced as part of a complete set of rotors and gears. Do not attempt to replace a single component that is not a part of the original or a completely new set.**

5. Examine end plates for cracks or damaged seal ring bores. Check for snug dowel fits. Ex-

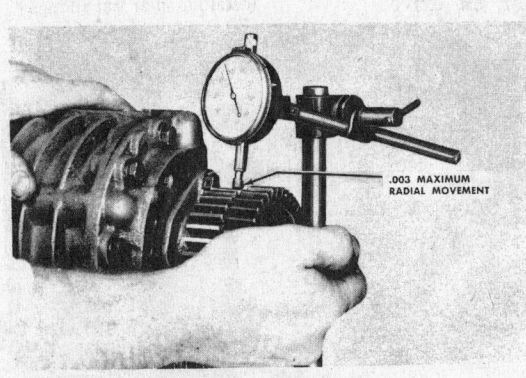

.003 MAXIMUM RADIAL MOVEMENT

Checking radial bearing clearance
(© Cummins Engine Co.)

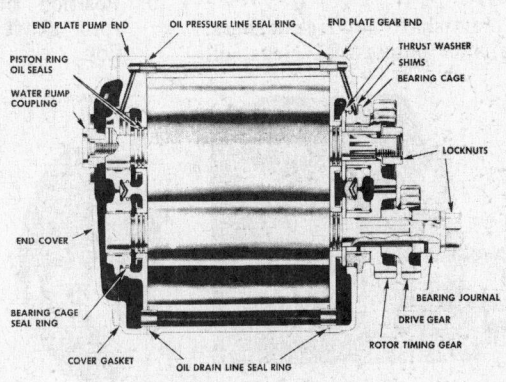

END PLATE PUMP END — OIL PRESSURE LINE SEAL RING — END PLATE GEAR END
PISTON RING OIL SEALS
WATER PUMP COUPLING
THRUST WASHER
SHIMS
BEARING CAGE
LOCKNUTS
END COVER
BEARING JOURNAL
DRIVE GEAR
ROTOR TIMING GEAR
BEARING CAGE SEAL RING
COVER GASKET — OIL DRAIN LINE SEAL RING

Supercharger—cross section
(© Cummins Engine Co.)

Cummins • Diesel Engines

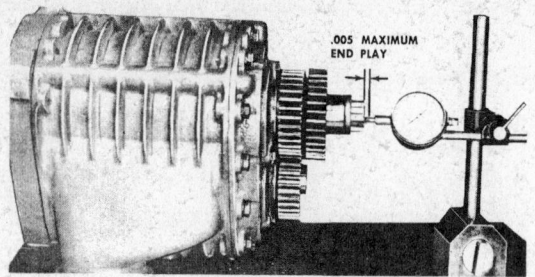

Checking end play
(© Cummins Engine Co.)

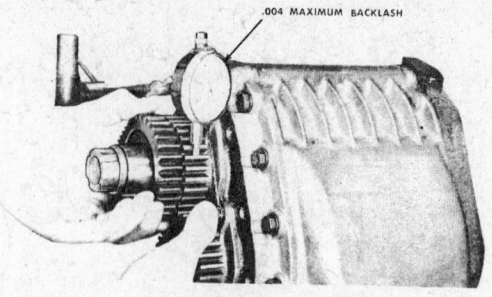

Checking gear backlash
(© Cummins Engine Co.)

cessive wear and roughness in the seal ring bores will not allow the seal rings to seat properly, thus the end plates must be discarded and replaced with new parts.

6. Check end cover for cracks. End cover *must* have a smooth mating surface.

7. Inspect bearing cages for cracks or excessive wear. Discard defective bearing cages. Bearing cage assemblies are also available in .010 and .020 inch *undersize.*

8. Inspect bearings for scratches or scoring. Check bearing bores with inside micrometers. If bore exceeds 1.3765 inch, discard bearings.

9. Examine thrust washers for burrs, cracks and wear. Discard damaged or worn thrust washers.

10. Check outboard bearing journal for nicks, scoring or excessive wear.

11. Inspect water pump drive coupling half for excessive wear.

12. Check end thrust by measuring bearing cage width and distance from thrust face of gear to end of gear hub. Bearing cage width must be .002/.005 inch less than length of gear hub. If difference between these dimensions is greater than .005 inch, install new gears, thrust washers, and bearing cages.

Assembly

1. Use new shaft nut lockwashers, expansion plugs, seal rings, oil collector rings and gaskets during reassembly.

2. Install shaft seal rings on new rotors.

3. Install new bearing cage seal rings on bearing cages.

4. Press bearings into bearing cages and oil collector rings into end plates. Press in new end plate dowels, if removed.

5. Install oil pressure ferrules, drain ferrules and ferrule seal rings in housing.

6. Position thrust washers, shims, and bearing cages into gear end plate; secure with capscrews and lockwires. *NOTE: If old bearing cages are reused, replace in same hole from which removed. Use same shims as originally removed from respective cages. Use .005 inch total shim thickness under each bearing cage. Shim thickness may have to be adjusted when rotor end clearance is checked.*

7. Press water pump coupling on the driven rotor shaft extension so that the shaft end is flush with the water pump coupling counterbore.

8. Lubricate the gear end plate seal ring seats and bearings with clean, light engine oil and insert splined shaft end of driven rotor in the bottom hole in the gear end plate. NOTE: Do not damage seal rings while pressing into gear end plate.

9. Position timing gear on drive rotor shaft and press into position.

10. Insert driven rotor shaft through the second hole in the gear end plate and press on timing gear making sure rotors are positioned *exactly* 90° apart as the timing gears mesh.

11. Press drive gear on drive rotor shaft snugly against timing gear and install outboard bearing journal on drive rotor shaft.

12. Install shaft nut lockwashers and shaft nuts on the rotor shaft and tighten securely. Wedge a piece of soft metal or rag between the gear teeth to lock gears while tightening locknuts.

13. Position the expansion plugs on locknuts and tap into place.

14. Push rotors towards gear end plate and check end clearance. End clearance should be .003/.004 inch. Add or remove

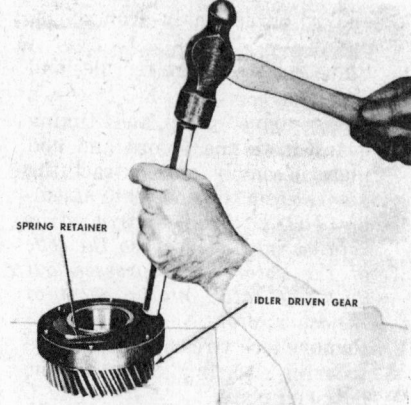

Installing idler leaf springs
(© Cummins Engine Co.)

Removing rotor shaft locknuts
(© Cummins Engine Co.)

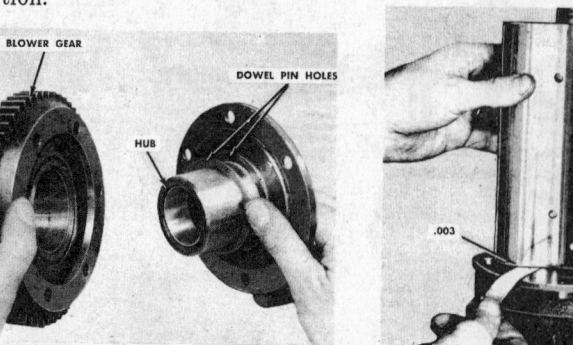

Assembling supercharger idler gears
(© Cummins Engine Co.)

Checking rotor end clearance
(© Cummins Engine Co.)

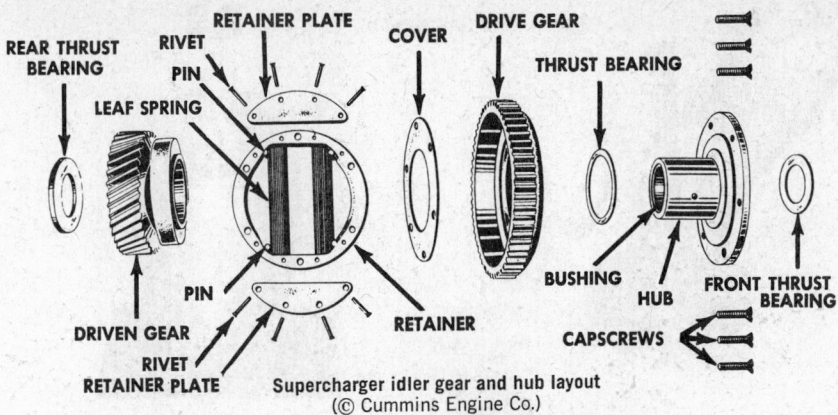

Supercharger idler gear and hub layout
(© Cummins Engine Co.)

shims under bearing cage mounting flange to obtain correct end clearance. If shims must be changed, remove locknuts and gears as described in disassembly instructions.

15. Position housing over rotors against end plate and secure with lockwashers and capscrews.
16. Lubricate bearings and seal rings in end plate and position end plate to housing. *Do not damage shaft seal rings.* Tap into place.
17. Position end cover and new gasket to end plate; secure with lockwashers and capscrews.
18. Turn drive gear by hand to check for smooth, free operation.

Supercharger Idler Gear Assembly

It should not be necessary to rebuild this unit unless it has become noisy or loose, thus indicating broken leaf springs or worn bushings.

If disassembly is necessary, proceed as follows:

1. Remove capscrews and idler drive hub, blower gear, thrust washer, and spring retaining cover from assembly.
2. Carefully pry and remove idler leaf springs, 4 spring retaining rings, and separate idler driven gear from spring retainer.

Repair

1. Replace all broken or damaged idler leaf springs.
2. Check two supercharger idler gear bushings and three idler hub bushings.
   a. New idler gear bushings are 1.501/1.500 inside diameter, bushings pressed in place.
   b. New idler hub bushings are 1.501/1.500 inside diameter, idler bushings pressed in place.
   c. Idler gear shaft is 1.4985/1.4975 outside diameter.
3. Check thrust washer for wear. Replace as needed.

Assembly

1. Place spring retainer over idler

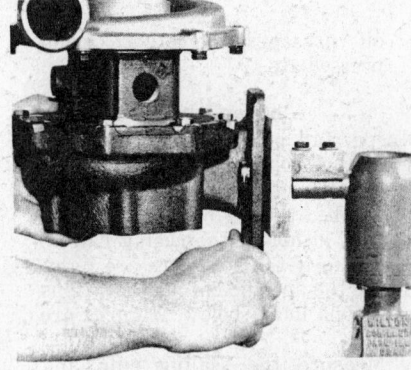

Supercharger mounting fixture
(© Cummins Engine Co.)

driven gear cam and drive idler leaf springs and four pins in place. **CAUTION: Idler leaf springs are made from spring steel and will cause a serious injury if allowed to fly loose as they are forced between gear cam and spring retainer.**

2. Assemble thrust washer to idler gear hub.
3. Mount retainer plate, blower gear and idler gear hub and fasten with six capscrews.
4. Peen capscrews in place.

Gear Train

1. Check all drive and driven gears, crankshaft, camshaft, and idler gears for cracks, chipping or signs of wear.
2. Replace all gears showing visible deep wear pattern or other tooth damage.

NOTE: The extent of gear wear can be determined by checking gear backlash after all gears are assembled.

Gear Case Cover

Disassembly

1. Remove crankshaft oil seal and fuel pump and compressor drive oil seal.
2. Remove bearing race at fuel pump drive, if used.

Inspection and Repair

1. Check bearings, if used; replace

if rough or worn. This is a two piece bearing, one-half remains on drive shaft with outer race only in gear cover.

2. Check fuel pump and compressor drive shaft bushing in cover. If replacement is necessary knock out old bushing and press in new one with suitable driver.
3. A replaceable steel trunnion bushing, may be used to repair worn trunnions. To use bushing:
   a. Machine gear case trunnion to 4.747/4.750 inches outer diameter.
   b. Press bushing over machined trunnion with chamfered side of bushing toward gear case cover.

Assembly

Do not install oil seals until gear case cover is ready to be installed on engine.

Turbocharger

The information listed below covers procedures for cleaning turbocharger compressor wheels and checking bearing clearances.

T-35 and T-50

Disassembly of Turbocharger

1. Clean the turbocharger exterior, and mark compressor casing, turbine casing, bearing housing and clamps to facilitate assembly.
2. Remove fittings from the lube oil inlet and outlet ports.
3. Unbolt turbocharger V-clamps, discarding nuts, washers, and bolts. Remove V-clamps.
4. Remove compressor casing and O-ring from bearing housing, and discard O-ring.
5. Remove the turbine casing from the bearing housing. If necessary, the turbine casing may be pressed off by supporting the casing at the center, and putting pressure on the turbine end of the shaft, while heating the tur-

Removing front cover plate
(© Cummins Engine Co.)

Removing compressor wheel

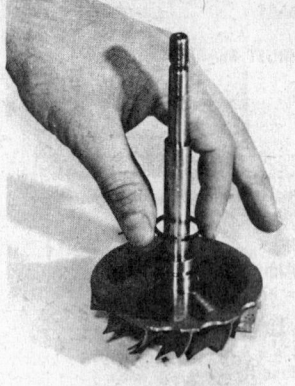

Installing sealing ring

Installing compressor wheel

bine casing evenly with a torch. Be careful not to drop the bearing housing assembly.

6. Place the bearing housing and rotor assembly on a press, supporting the housing in a specially designed bearing support. Put cardboard or soft wood under the support. Press the rotor shaft through the compressor wheel, using a rod small enough to enter the bore of the wheel. Be careful not to drop any parts.

7. Remove the compressor wheel from the housing. On T-50, remove the heat shield and insulation pad from the rotor shaft. Discard the insulation pad if it is damaged.

8. Seat a soft mandrel against the floating bearing, and drive the oil seal plate from the bearing housing.

9. Remove the oil seal plate from the bearing housing, and remove the seal plate O-ring, and discard it.

10. On T-35, remove the thrust washer and floating bearing from the bearing housing. On T-50, remove the floating bearing and Teflon insert, and discard the insert. Discard the thrust washer if it is scored.

11. Remove the sealing rings from the grooves in the rotor shaft, and discard them.

Cleaning

1. Immerse parts, so they do not touch each other, in a carburetor solvent that *will not damage aluminum, or Ni-Resist alloys*. A soft brush may be used to help remove carbon.

2. Flush out oil passages in the bearing housing from the drain end, to remove dirt loosened by the solvent.

3. Continue to soak parts for at least 12 hours, and as long as 24 hours if possible. Then, repeat the flushing of oil passages.

4. Steam clean all parts, especially oil passages. Dry with compressed air.

Inspection

Inspect all parts for cracks, burning, distortion, scoring, burrs, chips, etc. If compressor or turbine wheels have rubbed the casing, they must be replaced, although the casing can be cleaned up if the scratches are light.

Make the specific checks of various dimensions listed below. Figures are in inches.

1. Measure the sealing ring and bearing bores in the bearing housing. Limits are 3.527-3.533 for the T-35, and 5.049-5.055 for the T-50.

3. On T-35, check distance from oil seal plate stop to turbine stop on casing surface. Distance should be 2.239-2.241″.

4. On T-50 make the measurement from the compressor stop to the bearing stop.
T-50   2.989-2.992

5. Check bearing flange for wear:

	Thickness	Width
T-35	.199-.201	1.97-2.00
T-50	.138-.142	1.920-1.950

6. Check bearing dimensions:

	I.D.	O.D.
T-35	.533-.5335	.873-.874
T-50	.752-.7525	1.272-1.273

Assembly

1. Hold thrust washer on shoulder of shaft with a piece of tubing. Check floating bearing end clearance on shaft.
T-35: .001-.007″
T-50: .006-.014″

2. Position sealing ring in groove on turbine end of rotor shaft.

3. Lubricate bearing housing bore on T-50 and install insulation pad, heat shield and insert rotor shaft in housing.

4. Lubricate rotor shaft and floating bearing. Insert floating bearing over shaft and into bearing housing bore. On T-50, position Teflon bearing insert over flange of floating bearing.

5. On T-35 position new thrust washer over turbine shaft with lapped side toward compressor end of shaft and against floating bearing flange. Align mark on thrust washer with balance mark on end of rotor shaft. Balance mark on washer must face outward.

6. Lubricate oil seal bores and insert sealing sleeve assembly.

7. Lubricate and install new O-ring housing with vaseline and place in seal plate. Coat bearing bore oil seal plate over shaft. Align so that retainer of oil seal plate straddles sides of floating bearing flange. Push oil seal plate in bearing housing until plate seats on shoulder. Stops on oil seal plate should be turned so they seal on bearing flange 90° from disassembled position.

CAUTION: On T-35, check alignment mark of thrust washer with mark on end of rotor shaft before installing sleeve.

8. Support rotor and position assembly in a press.

9. Lubricate rotor shaft and position compressor wheel on shaft. On T-35, align the balance mark of compressor wheel with balance mark on shaft end. Press wheel onto shaft down to shoulder. A clearance of .0007″ is acceptable on T-50.

10. On T-35, install self-locking nut on shaft. Tighten to 120-130 in. lb.

11. On T-50, install self-locking nut on shaft. Tighten to 20-24 ft. lb.

12. Coat turbine casing bore with anti-seige compound. Insert bearing housing and rotor in bore. Align scribed marks. Assembly seats on shoulders.

13. Install new O-ring in bearing housing and coat lightly with engine oil.

14. Position compressor casing over bearing and shaft. Align marks.

15. Position V-bands around casings. Center V-band openings over oil supply and drain ports. Secure bands; torque nuts to 30 in. lb. It is helpful to tap V-

Checking bearing clearance
(© Cummins Engine Co.)

bands while torquing nuts. Clearance between V-band clamp and casing should be .035″ MIN.
CAUTION: Do not over-torque or retighten clamp.
16. Check radial clearance at turbine end and compressor end by:
   a. Push shaft toward side of bore.
   b. Check minimum distance between tip of vanes & bore.

	Compressor end	Turbine end
T-35	.0011-.0409	.0023-.0397
T-50	.005-.033	.011-.043

17. Check total end clearance with a dial indicator.
   T-35   -.004-.015
   T-50   -.006-.017
18. Turn rotor by hand to check for interference.

T-506

Currently, T-506 turbochargers utilize a universal main casing which can be rotated 360° to simplify mounting arrangement and facilitate positioning oil drain.

Formerly, a stationary front cover was used.
1. Clean exterior of turbocharger.
2. Attach turbocharger to Assembly Stand.

Cleaning and Inspection

1. Clean compressor wheel.
   a. Use Bendix cleaner or equivalent; NEVER use a caustic solution which will attack aluminum.
   b. Use nylon or hog bristle brush; NEVER use wire brush.
   c. Immerse compressor wheel end in cleaning solution.
   d. Dry thoroughly with compressed air.
2. Check bearing clearances. Use a dial indicator to indicate side and end-play of rotor shaft. Check as follows:
   a. Fasten a dial indicator to turbine casing.
   b. Place indicator point against hub of turbine wheel.

   c. Force turbine wheel up and down, or sideways; note total indicator reading.
   d. Remove one capscrew from front plate (compressor wheel end) and replace with a long stud bolt.
3. Remove capscrews and lockplates from exhaust casing on turbine end; remove universal main casing or front cover plate.
   e. Attach an indicator to long stud bolt and register indicator point against flat on end of rotor shaft.
   f. Push shaft sideways, noting total indicator reading.
   g. Move indicator point to end of shaft.
   h. Check end-play of rotor assembly.
   i. Compare readings with limits shown in following table.

Type	Min. Radial Clearance	Rotor End Play
T-506	.008	.003/.011
T-350	.0135	.003/.008

If radial or end clearance exceeds the maximum limits, send turbocharger to a factory approved rebuild station for bearing replacement.

T-350

Disassembly

Before disassembling, clean exterior thoroughly. Mark housing position with a scribe mark.
1. Attach turbocharger to Assembly Stand.
2. Remove capscrews and lockplates from the exhaust outlet connection. Remove connection and gasket.
3. Remove capscrews, lockwashers and flatwashers from bearing housing to compressor casing. Lift off compressor casing.
4. Remove capscrews and lockplates from the bearing or main housing to turbine casing. Lift bearing housing assembly from turbine casing.

Cleaning

1. Immerse parts in Turco Super-Carb or similar solvent. CAUTION: Never use a caustic solution or solvent that may attack aluminum, stellite or ni-resist alloys.
2. Allow parts to soak to remove carbon. A soft bristle brush may be used to remove heavy deposits. *Never use wire or stiff bristle brush.*
3. Steam clean thoroughly to remove all carbon and grease. Apply steam liberally to oil passages in the bearing housing.
4. Blow off excess water and dry with compressed air.

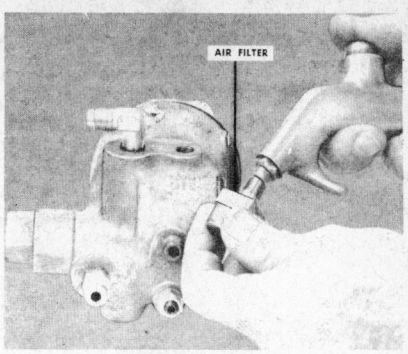

Cleaning aneroid control air filter
(© Cummins Engine Co.)

Inspection and Assembly

1. Place a new "O" ring on compressor end of bearing housing and coat with oil.
2. Install compressor casing aligning scribe marks made at disassembly.
3. Check impeller radial clearance between impeller vanes and compressor casing. Impeller wheel must be pushed to side and minimum clearance checked. See Table for clearance.
4. Check turbine wheel radial clearance between turbine wheel vanes and turbine casing. Turbine wheel must be pushed to side and minimum clearance checked. It should be .0135.
5. Use a dial indicator and check rotor end play. See Table for end play.
6. Turn rotor by hand to be sure no internal interference is present.
7. Install pipe fittings in bearing housing as required.
8. Install exhaust connection using Never-Seez anti-gall compound on the capscrew threads.
9. Cover all inlets and outlets to prevent dirt entry.

Aneroid Control

Cleaning and Lubrication

1. Clean air filter assembly by reverse flushing with compressed air.
2. Fill aneroid control cavity with clean lubricating oil.

Fuel Pressure Setting

1. Attach a pressure gauge into the injector supply line.
2. Operate engine until oil is 140° F.
3. Disconnect the air line between aneroid bellows and intake manifold.
4. Operate engine at full throttle at BHP setting speed and adjust bottom screw in aneroid until fuel pressure is reduced as needed for the particular altitude.

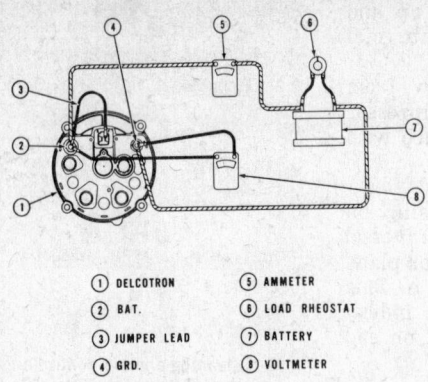

(1) DELCOTRON (5) AMMETER
(2) BAT. (6) LOAD RHEOSTAT
(3) JUMPER LEAD (7) BATTERY
(4) GRD. (8) VOLTMETER

Alternator Output—Bench Test
(© Cummins Engine Co.)

(1) STEEL FRAME MEMBER
(2) BODY GROUNDED TO ENGINE
(3) GENERATOR
(4) CUTOUT RELAY
(5) AMMETER
(6) BATTERY
(7) CLAMP
(8) TINNED

Basic Generator Wiring Cut-Out Relay
(© Cummins Engine Co.)

Alternator

Self-rectifying A.C. generators are designed and constructed to give long periods of trouble-free service with a minimum amount of maintenance. The rotor is mounted on ball bearings, and each bearing has a grease reservoir which eliminates the need of periodic lubrication. Only two brushes are required to carry current through the two slip rings to the field coils which are wound on the four pole rotor. The brushes are extra long and under normal operating conditions will provide long periods of service.

The stator windings are assembled on the inside of a laminated core that forms the generator frame. Six rectifier diodes are mounted in the slip ring end frame, and are connected to the stator windings through connectors mounted internally in two nylon holders, or a separately mounted rectifier, and they act to change the generator A.C. voltages to a D.C. voltage which appears at the "BAT" terminal on the generator.

Even though the generator is constructed to give long periods of trouble-free service, a regular inspection procedure should be followed to obtain the maximum life from the generator.

Inspect the generator as follows:

1. Inspect the terminals for corrosion and looseness.
2. Inspect the wiring for frayed insulation.
3. Check mounting bolts for proper tension.
4. Inspect the belt for excessive wear and replace if necessary.
5. Check belt tension and alignment.
6. Inspect slip rings and brushes through the end frame assembly. If the slip rings are dirty, clean them up with a 400 grain or finer polishing cloth. To clean, hold the cloth on the rings with engine turning alternator rotor. Blow dust away thoroughly. Rings that are worn or out of round must be trued on a lathe. Brushes worn close to the holder must be replaced.
7. Check the output of the alternator on a test bench, making electrical connections as shown in the illustration. Operate it at the specified speed and check for specified output, adjusting the field rheostat as necessary.

Alternator Safety Precautions

1. Make sure that type, polarity, and part numbers of alternator and regulator are matched ac-

cording to manufacturer's specifications before installation.
2. Make sure battery and alternator polarity are correct (+ to + and − to −). Make sure polarity is correct when installing a booster battery, also.
3. Never operate an alternator when the circuit is open (without a load).
4. Never attempt to polarize an alternator.

Generator

The shunt generator converts mechanical energy to electrical energy, supplying current for electrical equipment by replacing electricity consumed from the battery.

The shunt generator requires external current regulation in the form of a current regulator, voltage regulator, and cut-off relay, which control generator output under all operating conditions.

If the generator produces no output, check for sticking brushes, a gummed or burned commutator, or weakened brush springs. If output is low, check for sticking brushes, low spring tension, or a rough, out of round, dirty, or burned commutator. If necessary, turn down the commu-

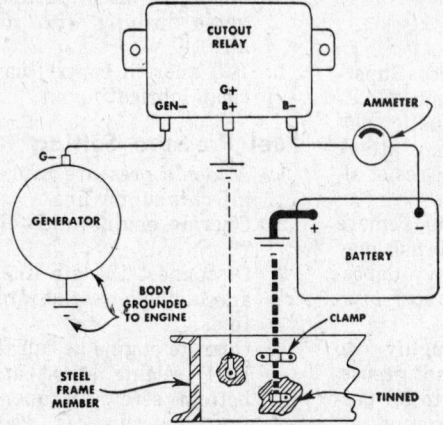

Basic generator wiring cut-out relay
(© Cummins Engine Co.)

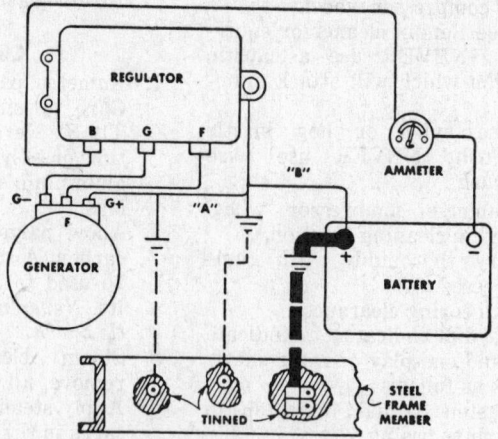

Basic generator wiring-leece-neville
(© Cummins Engine Co.)

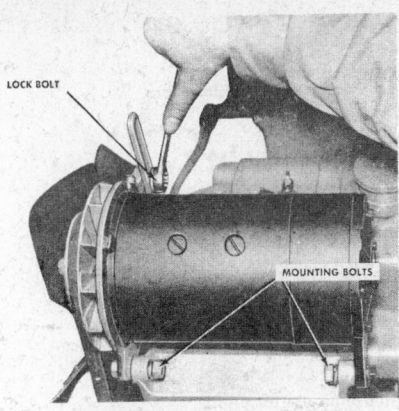

Removing belt driven generator
(© Cummins Engine Co.)

tator in a lathe, and undercut mica if the unit has carbon brushes.

If the generator's output is excessive, the problem is that the field circuit is grounded or the field is shorted. If terminal insulation is not defective, the problem is internal.

Removal

1. Disconnect wiring leads from generator.
2. On belt driven models, remove drive belt. On some C and J models it may be necessary to disconnect No. 1 cylinder injector inlet and drain manifolds to remove generator.
3. Remove belts from generator bracket and lift off generator and bracket.
4. Remove bracket from generator.

Installation

1. Mount bracket to generator, checking that the bracket is mounted to generator so that oil cup and passages are in proper position.
2. On gear driven generators, install rubber buffer between generator coupling and drive coupling and connect the two units.
3. Mount generator and bracket to block.
4. On belt driven models, install drive belt.
5. Connect generator wiring leads to proper terminals.

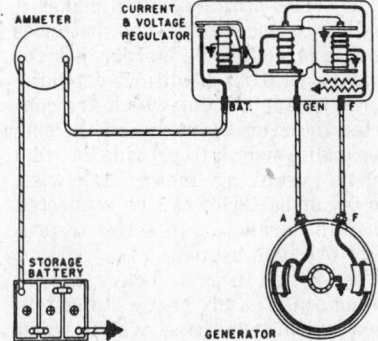

Leece neville current and voltage regulator wiring
(© Cummins Engine Co.)

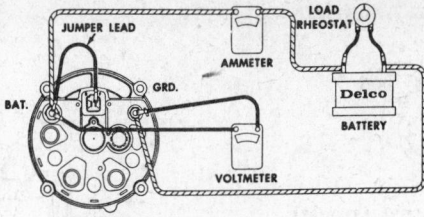

Alternator output—bench test
(© Cummins Engine Co.)

Starter Motors

Air Starter

An improved air starting system with increased starting motor torque is now used on current production engines without increasing required air pressure above 120 psi. The air piping system reduces the drive torque engagement of starter drives and flywheel ring gears, therefore, improving service life.

A specially hardened flywheel ring gear is now used and is available for service. On all new installations make sure the letter "W" appears on the ring gear near the part number. This indicates the ring gear is especially hardened.

Air Starter Removal

1. Relieve air system before disconnecting air supply to starter.
2. Depress starter handle to bleed off air in starter.
3. Disconnect main union between starter handle and motor.
4. Unbolt starter from flywheel housing.
5. Support starter and slide it back to disengage starter drive from flywheel ring gear.

Installation

1. Start one capscrew in mounting flange to hold starter during installation.
2. Some applications use a spacer between mounting flange and flywheel housing. Be certain spacer is in place.
3. Slide starter into flywheel housing opening, engaging the drive with flywheel ring gear.
4. Install remaining mounting bolts and tighten securely.
5. Connect air supply line to starter.

Electric Starter

The starter motor used on Cummins engines is a special overload motor capable of delivering high horsepower. In order to obtain this power, it is necessary to build the cranking motor with a minimum of resistance so a large current will be taken through it. The starter motor should be used for short periods only—30 seconds maximum to avoid the possibility of burning out.

Troubleshooting Cranking Motor Problems

1. If cranking is poor, turn on headlights and close starter switch. If lights go out, the problem is most likely a bad connection, while if they dim but still burn, the battery is run down or the starter or engine is mechanically defective. If the lights do not dim at all, the cranking motor or switch may be open.
2. To make a more thorough analysis, check the specific gravity of each battery cell, and, if gravities are satisfactory, check the battery connections and cables and cranking motor switch. If supply of current to the motor is satisfactory, inspect the brushes and commutator. The brushes should contact the commutator snugly, and the commutator should be clean and smooth. Remove the unit for turning down of the commutator or further testing and examination if required.

Removal

1. Disconnect battery cable and all electrical leads from starter.
2. Unbolt and remove starter motor and spacer (if used).

Installation

1. Before installation starter should be checked for type used. There are several different type drives used that must be matched with flywheel ring gears.
2. Place the spacer (if used) on starter.
3. Install starter to flywheel mounting bolts and tighten securely.

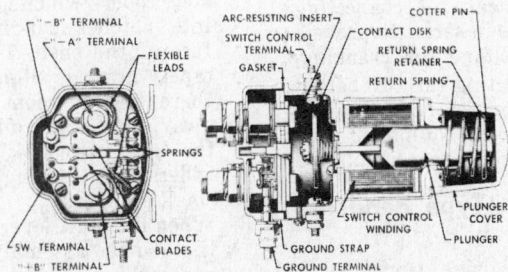

Series parallel switch cross section
(© Cummins Engine Co.)

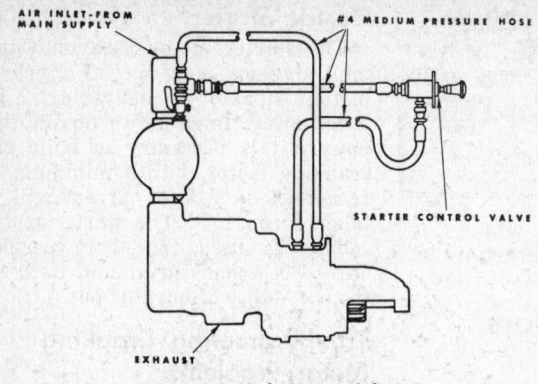

Ingersoll rand air starter piping
(© Cummins Engine Co.)

Bendix westinghouse air starter piping
(© Cummins Engine Co.)

4. Connect wires and cables to starter and battery.

Starter Motor Controls

1. Because of extremely high current flow during cranking, a positive means of connecting and disconnecting the battery and cranking motor must be used. The switch used must have contacts heavy enough to carry the current without burning. A manually operated switch mounted on the floor board or cranking motor frame is the simplest type.

2. Some applications with Bendix drive use a magnetic switch which when energized draws in a plunger and causes a contact disc to make contact between two terminals to complete the circuit from the battery to cranking motor. The magnetic switch winding is usually energized by a push button.

3. Some applications with the overrunning clutch, or Dyer type drive, use a somewhat larger magnetic switch called a solenoid switch. Here, the plunger not only thrusts against a contact disc to close the battery-to-cranking motor circuit, but is also linked to the shift lever so that the drive pinion is shifted into mesh with the flywheel teeth by the solenoid action.

4. Cummins engines require a comparatively high voltage to assure adequate cranking. The series-parallel system is designed to provide a means of connecting two batteries in series to provide increased voltage for cranking, and reconnecting the two batteries in parallel for normal operation of electrical equipment after starting has been accomplished.

Friction Clutch Type Bendix Drive

This type of drive performs in much the same manner as other Bendix drives except it uses a series of spring-loaded clutch plates which slip momentarily during the shock of engagement to relieve the shock and prevent it from being carried back through the cranking motor. The slipping stops as engagement is completed so that cranking torque is transmitted from the cranking motor armature through the drive pinion to the flywheel ring gear.

Overrunning Clutch Drive

1. The overrunning clutch is designed to provide positive meshing and disengagement of the drive pinion and flywheel ring gear. It uses a shift lever which slides the clutch and drive pinion assembly along the armature shaft so it can be meshed and disengaged as required. The clutch transmits cranking torque from cranking motor to the engine flywheel but permits the drive pinion to overrun, or run faster than, the armature after the engine is started. This protects the armature from excessive speed during the brief interval that the drive pinion remains in mesh.

2. The overrunning clutch consists of a shell and sleeve assembly which is splined internally to match splines on the armature shaft. Thus, both the shell and sleeve assembly and armature shaft must turn together. A pinion and collar assembly fit loosely into the shell, and the collar is in contact with four matched steel rollers which are assembled into notches cut in the inner face of the shell. These notches taper inward slightly so that there is less room in the end away from the rollers than in the end with the rollers. The rollers are spring loaded by small plungers.

3. When the shift lever is operated, the clutch assembly is moved endways along the armature shaft so the pinion meshes with the flywheel ring gear. If the teeth should butt instead of mesh, the clutch spring compresses so the pinion is spring loaded against the ring-gear teeth. When the armature begins to rotate, meshing takes place at once. Completion of the shift lever movement closes the cranking motor switch so the armature begins to rotate.

4. When the engine begins to operate, it attempts to drive the cranking motor armature through the pinion, faster than the armature is rotating. This causes the pinion to rotate with respect to the shell so that it overruns the shell and armature. The rollers are turned back toward the larger section of the shell notches where they are free, and thus, permit the pinion to overrun. This protects the armature until the automatic controls take over so that the shift lever is released, causing the shift lever spring to pull the overrunning clutch drive pinion out of mesh from the engine flywheel ring gear. This shift lever movement also opens the cranking motor switch so that the armature stops rotating.

5. The overrunning clutch pinion requires the same ring gear as the Bendix pinion.

Series-Parallel Switch

The series-parallel switch makes it possible to use two 12-volt batteries which are connected in parallel for normal operating conditions after the engine is started, but which are connected in series by means of the series-parallel switch to provide 24 volts for the cranking motor. Likewise, two 6-volt batteries can be connected either in parallel or in series to provide a 6-12 volt system.

The switch uses a heavy copper contact disk and heavy tungsten-faced main terminals, which resist the effects of the arcs that occur when the circuits are broken. The main cranking current is carried through these contacts and terminals. In addition, there are contacts in the terminal plate assembly which complete the parallel connections between the batteries for normal operation and also complete the connections that energize the cranking motor solenoid in the cranking position.

After cranking has been accomplished and the series-parallel switch is released, the two batteries again become connected in parallel to provide 12-volt operation of the equipment. In the mechanically operated switch a quick-break mechanism, consisting of a pair of triggers and a cam, causes the contact disk to be snapped away from the stationary contacts very quickly so that there is a very small amount of arcing.

Due to the high voltage of the starting circuit and the great amount of power available from the two batteries, it is essential that every precaution be taken to avoid short circuits or grounds. All wires should be of sufficient size to carry the electrical load to which they are subjected without overheating. Stranded wire and cable should be used throughout to reduce the possibility of breakage because of vibration. All connections should be clean and tight and all terminal clips should be soldered to the wires or cables. Only rosin flux should be used to solder electrical connections. All wires should be adequately insulated and supported at enough points to prevent movement and consequent chafing through of the insulation. Rubber boots, rubber tape or friction tape and shellac should be applied to cover all exposed terminals and clips. This will prevent accidental grounding of an exposed terminal which could cause serious damage to the system.

Cooling System
Water Pump
855, 927 Cid Engines
Removal
1. Drain the cooling system.
2. Remove the fan. Remove the alternator drive belt.
3. Remove the water pump retaining ring capscrews, and disengage the water pump drive belts. Remove the retaining ring and water pump body. Remove and

discard the gasket. *On some engines, the fan hub bracket is an integral part of the water pump retaining ring.*

Installation
1. Clean the gasket surfaces on the engine and the water pump.
2. Position the retaining ring and a new copper gasket on the pump. Position the pump in the engine block, and install, but do not tighten, the pump retaining ring bolts.
3. Install the alternator and water pump drive belts on the water pump pulley. Tighten the water pump drive belt and tighten the pump retaining ring bolts alternately starting on side away from belts.
4. Place the alternator drive belt on the alternator pulley and adjust. Tighten the alternator adjusting arm and mounting bolts.
5. Install the fan.
6. Fill and bleed the cooling system. Operate the engine and check for coolant leaks.

V-6 & V-8 Engines
Removal
1. Drain the cooling system.
2. Remove the alternator drive belt. Loosen the fan drive belt adjusting screw, and remove the drive belts from the fan and water pump pulleys.
3. Loosen the radiator lower hose at the water pump. Loosen the hose clamps on the by-pass hose connecting the water pump to the thermostat housing. Disconnect the heater and corrosion resistor hoses at the water pump.
4. Disconnect the temperature and warning light sending unit leads at the sending units on the thermostat housing. Disconnect the radiator upper hose and the radiator supply tank vent line at the thermostat housing. *Leave the heater hose and shutterstat air hoses connected.*
5. Remove the thermostat housing from the cylinder head and position it out of the way.
6. Remove the water pump from the engine and radiator lower hose. Discard the pump gasket.

Installation
1. Clean the gasket surfaces of the thermostat housing, water pump, cylinder head and cylinder block.
2. Apply water-resistant sealer to a new water pump gasket and position it on the water pump housing. Insert the water pump lower elbow in the radiator lower hose, and install the water pump on the engine block. Torque the pump attaching screws to specifications.
3. Use water-resistant sealer and position a new thermostat housing gasket on the cylinder head. Install the thermostat housing; be sure the thermostat housing to water pump by-pass hose and clamps are in place. Tighten the by-pass hose clamps and the mounting bolts.
4. Connect the radiator upper hose and the radiator supply tank vent line to the thermostat housing. Connect the temperature and warning light sending unit leads.
5. Connect the heater and corrosion resistor hoses to the water pump. Tighten the radiator lower hose clamp.
6. Install the fan and water pump drive belts and the alternator drive belt.
7. Fill and bleed the cooling system. Operate the engine and check for coolant leaks.

1150 cid Engines
Removal & Installation
1. Close valve above water filter, remove and discard filter.
NOTE: Valve is closed when plunger is down and locked.
2. Remove tube assembly connecting water pump to cooler housing.
3. Remove by-pass tube between pump and thermostat.
4. Unbolt pump and pull straight back, parallel with engine to remove.
5. Install in reverse of above.

Water Pump—Overhaul
Supercharger Driven
Disassembly
1. Remove mounting capscrews, lockwashers, gasket and cover from water pump body.
2. Pull drive coupling from shaft.
3. Remove outer snap-ring from coupling end of pump body.
4. Support pump body on its mounting flange in an arbor press and press out bearing and shaft assembly from impeller and pump housing.
5. Press shaft seal out of pump body.

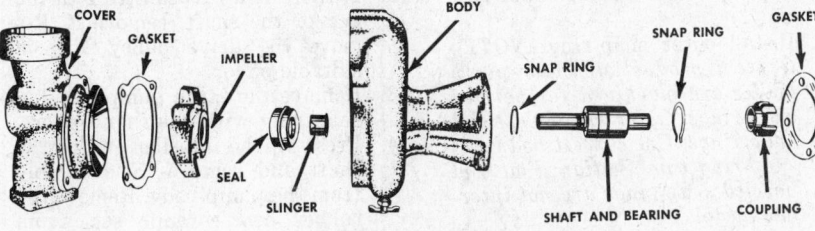

Supercharger driven water pump
(© Cummins Engine Co.)

623

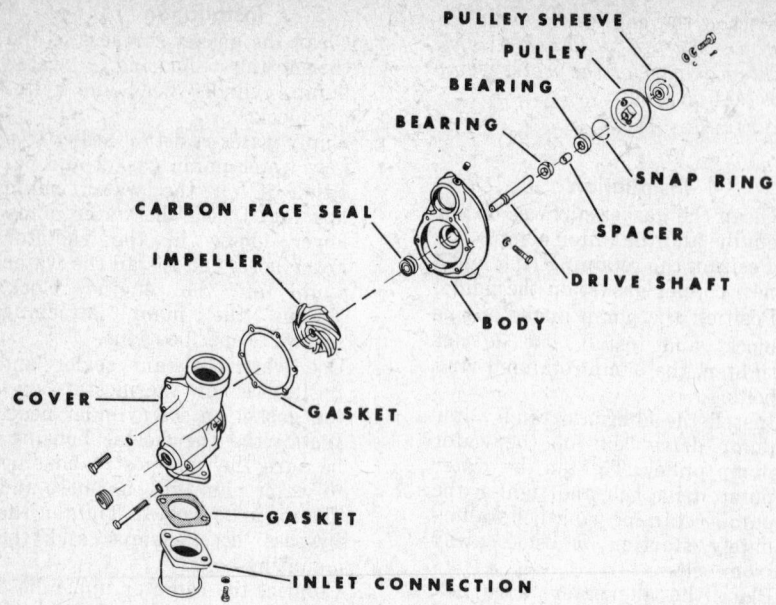

PULLEY SHEEVE
PULLEY
BEARING
BEARING
SNAP RING
CARBON FACE SEAL
SPACER
IMPELLER
DRIVE SHAFT
BODY
COVER
GASKET
GASKET
INLET CONNECTION

Belt driven water pump
(© Cummins Engine Co.)

6. Remove inner snap-ring from housing.

Inspection

1. Inspect water pump bearing(s). Replace bearing(s) with rough or worn races.
2. Inspect water pump impeller. Replace if cracked or corroded to extent that it will interfere with circulation.
3. Measure impeller bore and shaft outer diameter. There must be a minimum of .0015 inch press-fit between shaft and impeller; replace if necessary.
4. Inspect water pump body seal face.
5. Inspect water pump mounting parts for cracks. Replace as necessary.
6. Examine carbon seal carefully to make sure it is not cracked or chipped.
7. Inspect ceramic seat.

Replacing Ceramic Seat

1. Strike damaged seat with a sharp tool perpendicular to axis of impeller.
2. Scrape adhesive from counterbore, be careful not to damage counterbore surface.
3. Check counterbore of impeller. It must be 1.042/1.046 in diameter and square to the axis of .624/.625 inch bore and within .002 inch total indicator runout.
4. Wash and de-grease impeller in an approved cleaning solvent and air dry. *The bore must be chemically clean to insure a good strong water tight bond.*
5. Apply Bonding Film in counterbore. Care must be used to prevent contamination of film by dust, dirt, oil, moisture or finger-

prints or use of film more than six months old.
6. Remove bonding film liner, it may be necessary to pick liner to start separation.
7. Place ceramic seat in counterbore with identification mark (dimple) against adhesive. Rotate seat to insure a continuous bond.
8. Apply a 15 pound weight on ceramic seat and place assembly in temperature controlled oven at 345°/355°F. for 55 to 65 minutes. NOTE: The temperature and time must be closely controlled to obtain proper bond.
9. Check ceramic seat after curing for squareness to the axis of the .625/.625 inch bore. It must be within .004 inch total indicator reading. Check for scratches, holes, chips or cracks. All unusable parts must be discarded and usable assemblies protected against further damage.

Assembly

1. Install inner snap-ring.
2. Support pump body on its cover mounting face and press bearing and shaft assembly in place against inner snap-ring. When pressing bearings onto shaft, apply load on bearing inner race only.
3. Install outer snap-ring. *NOTE: Current production water pump bodies are machined for beveled snap-rings. Older models use flat snap-rings. Use correct body and snap-ring combination. Flat and beveled snap-rings are not interchangeable.*
4. Press drive coupling onto shaft and flush with end of shaft.

5. Support pump on shaft and install slinger over shaft until flange is 1.805/1.825 from end of shaft.
6. Apply sealer to the bottom side or seat of seal driving flange.
7. Press seal in place, apply force on driving lip only. Check carbon for damage before going further.
8. Clean seal with lint free cloth to remove dirt, oil or grease.
9. Press impeller on shaft. Face of impeller hub should be .872/.878 inch below cover mounting face on body. **CAUTION: Use extreme care when pressing impeller to prevent damaging ceramic seal.**
10. Assemble lockwashers, capscrews and new gasket to water pump cover.
11. Mount cover to pump body and turn shaft to be sure that it is free.

Thermostats

1. Using new gasket, install new or tested thermostat in thermostat housing. *NOTE: If thermostat with "V" notch is used, plug vent hole in thermostat housing.*
2. If water-cooled turbocharger is used, install new "O" ring on water by-pass coupling.
3. Install thermostat with vent hole *at top.* Failure to do so may result in air lock and incomplete coolant circulation.
4. Install thermostat housing with washers and capscrews.
5. Connect air compressor and turbocharger water outlet lines, if used.
6. Connect water outlet hose. *NOTE: The by-pass to water pump hose connection on supercharged engines must be assembled with a shield to protect it from exhaust manifold heat. Four cylinder engines use a small thermostat. Always replace same type thermostat as removed.*

Water Pump Overhaul V-6 and V-8 (470 and 504)

Disassembly

1. Pull the impeller from the shaft with a suitable puller.
2. If there is a grease fitting in the end of the shaft, remove it. Remove the drive pulley with a suitable puller.
3. Remove the large pump housing snap ring with snap ring pliers.
4. Press on the impeller end of the shaft and remove the assembly from the pump body. Remove the rubber boot/ceramic seal from the shaft and discard it, if the pump is so equipped.

Removing Impeller
(© Cummins Engine Co.)

5. If the bearing is still in the pump body, press it out.
6. Remove the carbon face seal from the pump body. Remove the flinger from the shaft. Then, remove the shaft from the large bearing, pressing on the inner race of the bearing. If the assembly uses a double row bearing and spacer, use a suitable puller to pull the spacer and bearing from the shaft.
7. Remove the small snap ring from the shaft.
8. Press the single row bearing, if the assembly is so equipped, from the shaft over the pulley end, pressing on the inner race as much as possible.

Inspection
1. Inspect the water pump bearings and replace if they do not appear to be capable of lasting until the next overhaul.
2. Inspect the impeller, and replace if cracking or corrosion will interfere with circulation.
3. Measure impeller bore and shaft outside diameter for proper fit. There must be a minimun .001″ press fit between them.
4. Inspect the pump mounts for cracks, and replace as necessary.
5. Examine the carbon face seal for cracks or chipping. Replace if necessary. If the seat requires replacement, it is recommended that the impeller be machined to permit installation.

Assembly
1. Lightly lubricate the pump shaft with clean engine oil.
2. Press the bearing onto the shaft, shielded side toward impeller, until the inner race seats against the shoulder.
3. Apply sealant to the inside diameter of the pump housing at the small bearing seat.
4. Fill the cavity of the bore about 60% with an approved grease.
5. Press the shaft and small bearing into the housing until the bearing seats.

6. Then, press the large bearing over the shaft and into the housing until it seats against the shoulder on the shaft.
7. Install a snap ring in the bore of the body to secure the bearings.
8. Slide the slinger downward until it seats on the shaft bearing shoulder.
9. Press the carbon face seal into the housing with a mandrel until it seats.
10. Support the shaft on the impeller end. Press the pulley onto the shaft until it rests snugly against the bearing inner race.
11. Lubricate the inside diameter of the rubber seat with a water soluble lubricant. No lube must touch the seal face of the ceramic ring. Install the seat assembly over the pump shaft until the polished side of the ceramic ring bottoms against the pump seal.
12. Supporting the shaft on the pulley end, press the impeller on until there is .010″-.020″ clearance between the vanes and the pump body. Turn the shaft to check for free rotation.

Right Hand NH Series

Disassembly
1. Remove the hugnut and washer (if used) which hold the pulley in place on the shaft.
2. Using a special puller, remove the pulley from the shaft.
3. Compress and remove the snap ring.
4. Place the pump body and shaft in a press and press the shaft and bearing assembly from the body by exerting force on the impeller end of the shaft.
5. Drive the carbon face seal out of the body with a suitable mandrel.
6. Install the halves of a bearing removal tool around the bearing spacer in the grease channel, press the shaft through the bearing and spacer.
7. Remove the inner bearing snap ring.
8. Remove the inner bearing by supporting it and pressing the shaft out of it.

Inspection
1. Inspect the bearings for worn races or damaged shields, and replace as necessary.
2. Inspect the bearing spacer and replace if worn or damaged.
3. Inspect the impeller and replace if cracks or corrosion are likely to retard the flow of coolant.
4. Inspect the ceramic seat, and replace if damaged.
5. Measure the impeller bore and shaft diameter to check for the presence of a minimum .001″

press fit between them. Replace parts as necessary.
6. Inspect the straightness of the shaft, check for galling on the press fit surfaces, and worn threads, and replace or repair as necessary.
7. Inspect the pulley grooves for wear by installing a new belt in the grooves. The belt should protrude 1/16-″1/8″, and should not bottom in the groove. Replace the pulley if the belt does not fit properly or if the groove is chipped or cracked.
8. Measure the pulley bore and shaft diameter to check for the minimum required .001″ clearance. Replace the pulley as necessary.
9. Inspect the pump body, checking for cracks, damage from a spun bearing, or stopped up weep holes, and clean, repair, or replace as necessary.

Assembly
1. Lubricate the shaft and bearings with an approved grease.
2. Press the inner bearing over the shaft until it bottoms on the shoulder, using a suitable mandrel pressing on the inner race.
3. Install the snap ring and the bearing spacer.
4. Using a mandrel as described in step 2, press the outer bearing onto the shaft until it seats against the bearing spacer.
5. Check to see that both bearings turn freely, and then press the shaft assembly into the pump body, supporting the body on the seal surface of the impeller end. Press on the outer races of the bearings until they seat within the body.
6. Press the carbon face seal into the pump body until it bottoms, using a special mandrel. Make sure to keep grease off the face of the seal.
7. With the pump supported on the pulley end of the shaft, press the impeller onto the end of the shaft.
8. Place the pump on the impeller end of the shaft and measure the clearance between impeller vanes and the water pump bodly. Clearance should be .0015″-.035″.
9. Install the snap ring in the pump body in front of the outer bearing.
10. Support the assembly on the impeller end of the shaft (not on the impeller) and press the pulley onto the shaft until the hub contacts the outer bearing inner race.
11. Install the washer (if used), and install and tighten the nut.

12. Grease the pump bearings before installation.

855 and 927 NTA Series Engines

Disassembly

1. Remove inlet housing to water pump housing mounting bolts.
2. Remove the inlet housing and discard the O-ring.
3. Using a suitable puller, remove the pulley and wear sleeve assembly. Remove the idler pulley assembly.
4. Remove the large snap ring which secures the large bearing in the water inlet housing.
5. Using a special puller, pull the pump impeller off the shaft. Then remove the ceramic seat and rubber seal and discard them.
6. Support the housing, and press the shaft bearing assembly on the impeller end from the inlet housing. Remove the bearings from the shaft.
7. Remove the carbon face seal and the small oil seal from the housing with a suitable mandrel.
8. Remove the large pipe plug from the idler pulley. Remove the small snap ring, and pulley assembly from the shaft.
9. Remove the oil seal, snap ring, and the bearing from the idler pulley.

Inspection

1. Check the impeller for cracks and corrosion which would interfere with coolant flow, and replace if necessary.
2. Inspect bearings for wear or roughness and replace as necessary.
3. Measure the impeller bore, pulley bore, and shaft diameter. There should be a minimum .001″ press fit, and the shaft should be straight and smooth. Replace the shaft if necessary.
4. Inspect the pulley grooves for wear and the wear sleeve for wear or grooves.
5. Inspect the housings for cracks, bearing spin damage, and cleanliness of the weep hole. Clean or replace as necessary.
6. Inspect the bypass valve retainer disc and spring.
7. Replace any unserviceable parts.

Assembly

1. Press the bearing into the pulley with a special mandrel. Install the snap ring (flat side toward bearing) and press the seal in until it is flush with the surface of the pulley, using a special seal driver.
2. Now press the pulley assembly onto the shaft and install the snap ring.

3. Install the pipe plug into the pulley. Remove the small pipe plug and install a grease fitting. Fill the unit ½ to 2/3 full with an approved grease.
4. With a mandrel and press, press the small bearing onto the impeller end of the shaft and the large bearing onto the pulley end of the shaft. Bearings should be pressed on until they seat against the shoulders.
5. Using a special mandrel, press the small seal into the water inlet housing until it is flush with the bore to .015″ indented.
6. Insert the shaft assembly into the water inlet housing. With a suitable mandrel, apply pressure to the outer race of the large bearing to press the assembly in until it seats. Install the retaining ring.
7. Support the impeller end of the shaft and, using a seal driver, press the large oil seal into the inlet housing.
8. Mount the idler pulley bracket onto the pump assembly and secure it.
9. Support the impeller end of the shaft, and press the drive pulley all the way onto the end of the shaft.
10. Use a special mandrel to press the carbon face seal into the inlet housing until it seats. Place the ceramic seat and rubber seal onto the shaft.
11. Support the drive end of the shaft and install the impeller. Press the impeller on until there is .020″-.040″ clearance between the impeller and cavity.
12. Install new grease and relief fittings where required, and fill the cavity ½ to 2/3 full with an approved type of grease.
13. Lubricate a new O-ring with clean oil, and then install it in the groove of the water housing inlet.
14. Install the water housing inlet onto the main pump housing, carefully aligning the capscrew holes to avoid damaging the O-ring. Install the lockwashers and capscrews.

V-903 and VT-903

Disassembly

1. If there is a grease fitting on the end of the shaft, remove it.
2. With an appropriate puller, remove the impeller and drive pulley from the shaft.
3. Using a special puller with the puller jaws reversed, remove the oil seal from the body of the pump, and discard it. Remove the snap ring from the pump housing with snap ring pliers.

4. Support the body of the pump on the side opposite from the impeller, and press on the impeller end of the shaft to remove the bearing and shaft assembly from the pump body. If the small bearing remains in the body, press it out.
5. Remove the carbon face seal from the pump body. Press the oil seal out of the body and discard it.
6. Press the shaft out of the large bearing. Support it on the inner race of the bearing during the operation.

Inspection

1. Inspect the bearings for roughness or worn races.
2. Inspect the impeller for cracks or corrosion that might interfere with circulation.
3. Measure the impeller bore and the outside diameter of the shaft. There must be .0015″ clearance.
4. Inspect the pump body parts for cracking.
5. Examine the carbon face seal carefully for cracks or chips.
6. Inspect the ceramic ring and rubber seat for damage.
7. Replace all unsatisfactory or questionable parts.
8. If there is a wear sleeve on the drive pulley, remove it as follows:
   a. Procure a 3/16″ straight shank round punch and grind a 15 degree angle on the end.
   b. Secure the pulley in a vise (exercise care in tightening, to prevent cracking the pulley).
   c. Use the punch to drive the sleeve from the pulley hub by placing the punch through the puller holes in the pulley. Discard the sleeve.

Assembly

1. Lubricate the shaft outside diameter and the inside diameter of the bearings with an approved grease. Then, pack the bearings with grease.
2. If a small bearing is used, press it onto the shaft, with the shielded side toward the impeller end of the shaft, until the inner race seats against the shoulder. Press on the inner race of the bearing.
3. Press the large bearing onto the shaft via the inner race, with the shielded side toward the impeller end of the shaft, until the inner race seats against the shoulder.
4. Press a new oil seal into the body (steel face toward impeller end of body) with a special seal driving mandrel, making sure the seal is not pressed in below the face next to the small bearing.

5. Apply a liquid lead sealer to the press fit area of the seal outside diameter. Then, press the new carbon face seal into the body using a special mandrel, or applying force on the driving lip of the seal.

6. Press the shaft assembly into the pump body bore via the outer bearing race until the race seats in the body. Install the snap ring.

7. Using a special mandrel, press a new oil seal into the pump body until it seats.

8. Press a new wear sleeve into the drive pulley with a sleeve driving mandrel until there is .000"-.015" indentation.

9. Support the impeller end of the shaft. Press the pulley on until it seats against the inner race of the bearing.

10. Lightly coat the inside diameter of the rubber seat with a water soluble lubricant. Then, slide the seat assembly over the pump shaft, with the polished side of the ring toward the carbon face, until the ring contacts the face.

11. Support the pulley end of the shaft, and press the impeller on until there is .010"-.020" clearance between the impeller vanes and pump body.

12. Install the grease fitting and fill the cavity about 60% full of an approved lubricant.

NOTE: When installing the pipe plug, use sealant and reduce torque on the 3/4" plug to 25-30 ft-lbs.

1150 cid Engines

Disassembly

1. Inspect driveshaft for wear.
2. Remove inlet assembly and discard O-ring.

3. If impeller shroud insert is eroded or worn, replace by:
   a. Pull insert from roll pin in inlet housing; discard insert.
   b. Insert new roll pin in inlet housing.
   c. Lubricate O-ring and position in groove in inlet mating surface.
   d. Install insert over roll pin, press to mating surface of inlet housing.

4. Pull impeller from shaft, remove and discard seal seat.

5. Remove retaining ring, support drive end of pump body, and press out bearing and shaft.

6. Pull carbon face seal and oil seal from body.

7. Inspect bearings and shafts. Remove retaining ring to replace outer bearing.

8. Check impeller bore and shaft diameter. A minimum .00." press fit should exist between shaft and bore.

Assembly

1. Lubricate shaft and press bearings onto shaft to shoulders. Secure with retaining ring.

2. Support body at inlet end and press seal into bore until flush within .030" below shoulder. Seal lip is toward bearing surface area.

3. Lubricate shaft seal mating surface, position bearing and shaft in bore and press outer bearing into bore until seated.

4. Support body at drive end, press carbon face seal into position and install seal seat over shaft.

5. Press impeller onto shaft; impeller-to-body clearance.

6. Lubricate O-ring with clean engine oil, install on inlet housing and secure housing to body.

NOTE: Until filter is installed, inlet housing shut-off valve should be "OFF."

Coolant Manifold

Removal

In-line Engines

Remove engine compartment covers or tilt cab whichever be the case.

1. Drain the engine cooling system.

2. Disconnect the high temperature warning light and the coolant temperature sending unit leads at the coolant manifold.

3. Close the shut-off valve at the shutterstat, and disconnect the air hoses at the shutterstat.

4. Disconnect the heater hose from the coolant manifold. Disconnect the radiator upper hose and the air compressor coolant line from the thermostat housing. Loosen the hose clamps on the cooler coolant elbow.

5. Remove the retaining bolts; then, remove the coolant manifold as an assembly, or in sections if necessary. Discard the gaskets.

6. If the coolant manifold was removed as an assembly, separate the sections. Remove the coupling O-ring seals.

Installation

1. Clean the gasket surfaces of the coolant manifold and cylinder heads. Clean the O-ring seal surfaces on the coolant manifold and couplings.

2. Apply lubricant to new coupling O-ring seals, and assemble the coolant manifold sections, couplings and seals.

3. Position new gaskets coated with water-resistant sealer on the coolant manifold. Position the manifold on the cylinder heads, and install and tighten the retaining bolts.

4. Position the hose on the oil cooler coolant elbow and the thermostat housing, and tighten the hose clamps. Connect the radiator upper hose and the air compressor coolant line to the thermostat housing. Connect the heater hose to the coolant manifold.

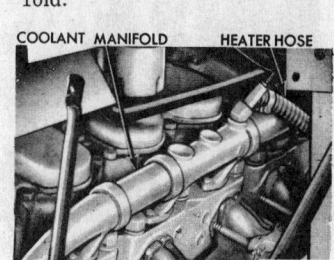

Typical coolant manifold
(© Cummins Engine Co.)

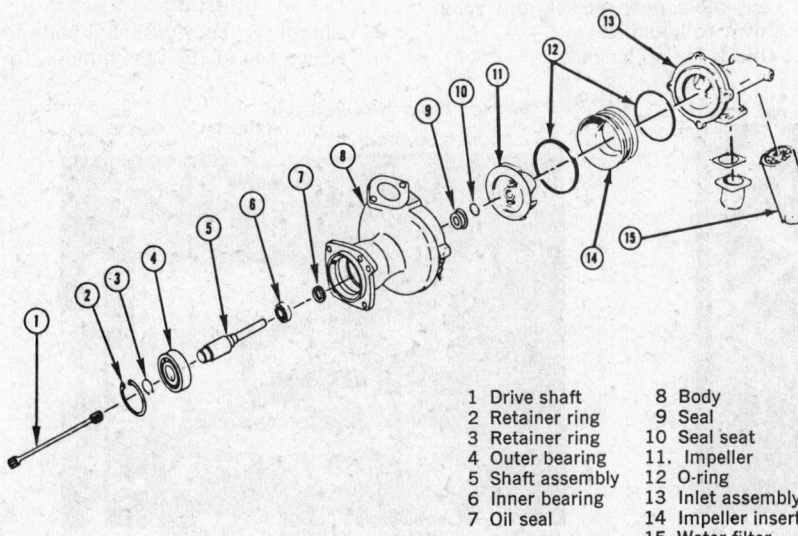

1	Drive shaft	8	Body
2	Retainer ring	9	Seal
3	Retainer ring	10	Seal seat
4	Outer bearing	11.	Impeller
5	Shaft assembly	12	O-ring
6	Inner bearing	13	Inlet assembly
7	Oil seal	14	Impeller insert
		15	Water filter

Water pump—1150 cid engines

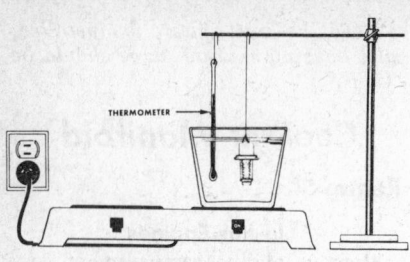

Testing thermostat
(© Cummins Engine Co.)

5. Connect the air hoses to the shutterstat.
6. Connect the high temperature warning light and the coolant temperature sending unit leads at the coolant manifold.
7. Fill and bleed the cooling system. Operate the engine and check for coolant and air leaks.

Lubricating Oil Cooler

855, 927 Cid

Disassembly

1. Remove top exposed "O" ring under the brass retainer being careful not to scratch or mar the sealing surface on the element.
2. Clean out lube oil and contaminants trapped in housing by forcing cleaner through the oil ports.
3. Remove the element from housing, by inserting two 7/32 inch rods 8 inches long into the outside row of tubes opposite each other. *NOTE: The rods should not drag bottom of housing.*
4. Place a flat bar on top of housing and bundle faces between rods and rotate element in housing to unseal lower "O" ring. *NOTE: While turning element lift up gradually on rods to free "O" rings. When up about 5/32 inch, element should be free to lift out of housing.*

5. Discard "O" rings and retainer.

Cleaning and Inspection

1. Clean cooler core immediately after removal with approved cleaning solvent that will not harm non-ferrous metal.
2. Inspect element for corrosion or cracks where tubes are welded to end plates.
3. Clean housing in approved cleaning solvent and inspect for cracks or damage.
4. Clamp the cooler assembly in a fixture and connect an air hose.
5. Immerse in a water tank and apply 1-4 psi to the water side. Check for air bubbles.
6. Apply 35-40 psi to the oil side, and repeat the inspection for air leaks.

Repair

1. To repair damaged tubes insert a smaller outside diameter tube inside damaged tube. Cut and flare ends, then solder securely. **CAUTION: Do not damage adjacent tubes with heat while soldering.**
2. Do not restrict more than 5% of total number of tubes in this manner. If more than 5% of tubes are defective, discard element.

Assembly

1. Lubricate rubber "O" ring and place in groove at bottom of housing.
2. Push element into housing, aligning index marks on housing and element.
3. Press second "O" ring around top of element with a wooden block to assure equal pressure around ring circumference.
4. Place new retainer ring over rubber "O" ring.
5. Assemble new gasket and front cover to housing.
6. Check for oil leaks.

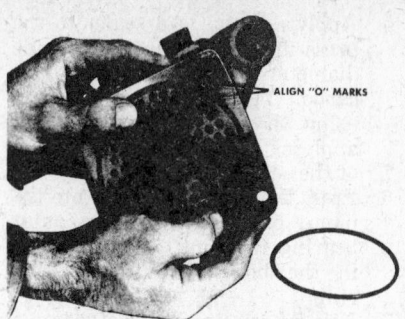

ALIGN "O" MARKS

Installing oil cooler element
(© Cummins Engine Co.)

V6 and V8

Disassembly

1. Remove the cooler head and gasket from the housing.
2. Remove the retainer ring from the housing.
3 Remove the top O-ring carefully, avoiding damage to the sealing surface.
4. Use mineral spirits to clean out the lube oil and contaminants. Force the cleaner through the oil ports to accomplish this.
5. Insert two 7/32" x 8" rods into outside tubes opposite each other. Do not insert the rods far enough for them to rub the inside of the housing.
6. Place a flat bar between the rods and rotate the element, while lifting on the rods, to free the lower O-ring.
7. After the tube bundle has risen up about 5/32", lift the element out of the housing. If resistance is encountered, a special puller may be made for pulling the unit out using the two threaded holes in the end of the tube bundle. Discard the O-rings and retainer.

Testing

1. Clamp a rubber gasketed plate to either end of the tube bundle. In-

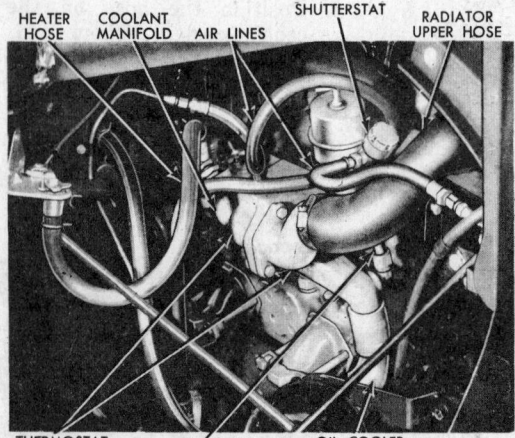

HEATER HOSE · COOLANT MANIFOLD · AIR LINES · SHUTTERSTAT · RADIATOR UPPER HOSE

THERMOSTAT HOUSING · AIR COMPRESSOR COOLANT LINE · OIL COOLER COOLANT ELBOW

Shutterstat and thermostat housing connections N series
(© Cummins Engine Co.)

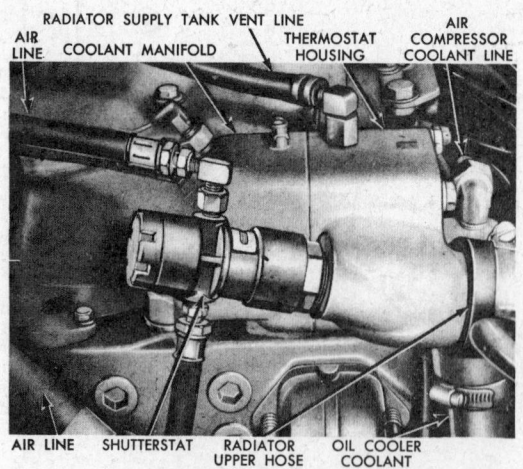

RADIATOR SUPPLY TANK · VENT LINE · AIR LINE · COOLANT MANIFOLD · THERMOSTAT HOUSING · AIR COMPRESSOR COOLANT LINE

AIR LINE · SHUTTERSTAT · RADIATOR UPPER HOSE · OIL COOLER COOLANT HOSE

Shutterstat and thermostat housing connections H series
(© Cummins Engine Co.)

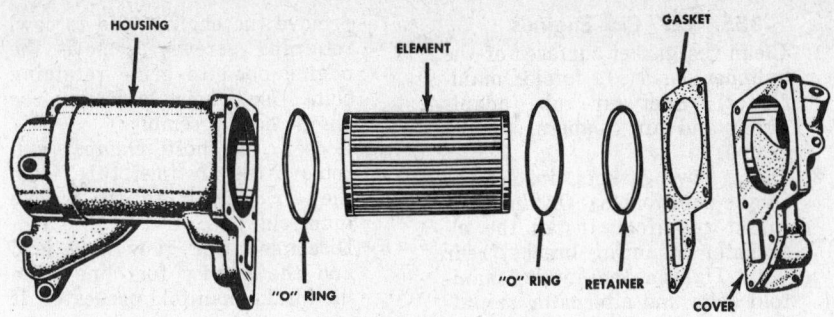

Lubricating oil cooler layout
(© Cummins Engine Co.)

stall an air connection in one of the plates.

2. Attach an air hose to the connection. Immerse the tube bundle and apply about 40 psi air pressure.
3. Check for the presence of air bubbles. Repair as described below if they are present.
4. Inspect the support for broken or cracked welds, and repair or replace as necessary.

Repair

1. Damaged tubes may be repaired by inserting a smaller diameter tube into the damaged tube, cutting and flaring the ends, and then soldering securely. No more than 5% of the tubes may be restricted in this manner. If more than 5% of the tubes require such a repair, discard the unit.

Assembly

1. Lubricate a new O-ring with engine oil and install it in the groove at the bottom of the housing. Make sure the ring is fully seated, and that it seats in the groove without being damaged or twisted. Make sure the sealing area on the element is not scratched.
2. Push the element carefully into the housing, aligning the index marks on the housing and element. If unusual resistance is felt, check the O-ring to make sure it has not come out of the groove.
3. Carefully install the other O-ring around the top of the element. Install the retainer ring.
4. Place a new gasket on the cooler/filter head, and secure the head snugly with the capscrews and lockwashers.
5. With the assembly on a flat work surface, and the inlet and outlet flat on top of the plate, check to make sure that inlet and outlet are on exactly the same plane. Torque the capscrews to 28-31 ft-lbs.

Suction Tube

1. Remove the mounting capscrews

from flange mounted suction tube to lube pump body.

2. Remove retaining clip and screen from suction bell.
3. Clean screen and tube; soak in solvent, dry with compressed air.
4. Inspect parts for damage and replace as needed.
5. Reassemble screen.
6. When "O" rings are used on the suction tubes, install new ones.

Corrosion Resistor

Disassembly

1. Remove capscrews and washers from resistor cover. Remove cover, gaskets, and upper plate.
2. Lift element from resistor body.
3. Lift out lower plate (sacrificial metal plate), and spring.
4. Remove drain plug and sump plate from body.

Inspection and Repair

1. Buff or polish sacrificial plate to remove scale and expose metal. Discard plate if less than 50% of metal surface can be exposed by polishing.
2. Inspect resistor body and cover for cracks or leaks.
3. Inspect spring, drain plug and capscrews for wear. Replace as necessary.

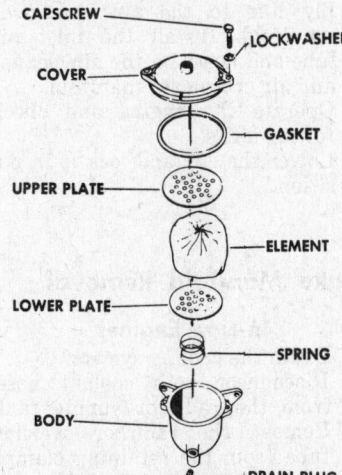

Corrosion resistor—layout
(© Cummins Engine Co.)

Assembly

1. Install sump plate and install spring in resistor housing.
2. Remove transparent bag from new element. Install lower plate, new element and upper plate.
3. Install new gasket to resistor cover. Secure cover to housing with capscrews and washers.
4. Install pipe plug.

1150 cid Engines

Disassembly and Inspection

1. Remove capscrews from elements and remove elements from housing.
2. Clean elements immediately with a safe solvent.
3. Inspect all parts for wear or cracks.
4. Plug openings in elements and place elements in water tank. Apply 60 psi air to check for leaks.

Assembly

1. Lubricate O-rings and install elements in housing.
2. If hold-down clamps are used, snug-tighten the cap screws.
3. Install cooler to block and tighten capscrews to 30-35 ft. lb.

Manifolds

Intake Manifold Removal

V-6 Engines

1. Remove the power steering pump mounting bolts and position the pump out of the way where necessary.
2. Loosen the clamp and remove the inlet air flex tube from the air cross-over manifold. Disconnect the air compressor supply line at the air-cross-over manifold. Loosen the hose clamps on the cross-over manifold connection hose and remove air manifold.
3. On the right intake manifold, remove the alternator bracket rear bolt and loosen the 2 front bolts.

   On the left intake manifold, remove the alternator adjusting arm to left intake manifold bolt.
4. Remove the intake manifold(s).

855, 927 cid—Engines

1. Open the air reservoir drain cock and exhaust the system. Disconnect the inlet and outlet lines from the shutterstat.
2. On a right intake manifold, remove the corrosion resistor from the air cleaners support frame, and attach it to the cowl with wire. Disconnect the air compressor supply line from the

right air cleaner. For one intake manifold, remove the support frame braces from the air cleaner support. For both intake manifolds, remove the air cleaners and support frame as an assembly. Remove the air cleaner(s), support frame and inlet air tube(s) assembly from the intake manifold.

3. Remove the power steering pump mounting bolts and tie the pump out of the way where necessary.

4. On the right intake manifold, remove the alternator bracket rear bolt and loosen the 2 front bolts. On the left intake manifold, remove the alternator adjusting arm to left intake manifold bolt.

5. Remove the intake manifold(s).

V-8 Engines

1. Release the cab locks, tilt the cab forward and lock it in place.
2. Remove the inlet air tube and hoses from the air cleaner and air cross-over manifold. Disconnect the air compressor supply line from the air cross-over manifold. Loosen the hose clamps on the cross-over manifold connection hose.
3. Remove the air cross-over manifold(s) and gasket(s). Remove the intake manifold(s).

Intake Manifold Installation

V-6 Engines

1. Clean the gasket surfaces of the cylinder head(s), intake manifold(s) and air cross-over manifold.
2. Using new gaskets, install the intake manifold(s). On the right intake manifold, install the alternator mounting bracket rear bolt. Tighten the intake manifold bolts and alternator mounting bracket bolts in alternate steps. Position a new gasket on the inlet port flange of the intake manifold(s), and install the air cross-over manifold. On the left intake manifold, attach the alternator adjusting arm.
3. Connect the air compressor supply line to the air cross-over manifold. Connect the inlet air flex tube to the air cross-over manifold and tighten the hose clamp.
4. Install the power steering pump. Adjust the alternator and power steering pump drive belt.
5. Operate the engine and check for air leaks.

855, 927 Cid Engines

1. Clean the gasket surfaces of the cylinder head(s), intake manifold(s), inlet air tube adapter(s) and air cleaners support frame.
2. Using new gaskets, install the intake manifold(s). On the right intake manifold, install the alternator mounting bracket rear bolt. Tighten the intake manifold bolts and alternator mounting bolts in alternate steps. On the left intake manifold, attach the alternator adjusting arm.
3. Position new gasket(s) on the inlet port flange of the intake manifold(s). Place new gasket(s) between the inlet air tube adapter(s) and the base of the air cleaner support(s). Install the air cleaner(s) and support frame assembly (ies) on the intake manifold(s).
4. Connect the air compressor supply line to the right air cleaner. Install the corrosion resistor on the air cleaners support frame. Connect the inlet and outlet lines to the shutterstat.
5. Install the power steering pump on the left intake manifold. Adjust the alternator and power steering pump drive belt.
6. Operate the engine and check for air leaks.

V-8 Engines

1. Clean the gasket surfaces of the cylinder head(s), intake manifold(s) and air cross-over manifold(s).
2. Using new gasket(s), install the intake manifold(s) and tighten the retaining bolts in alternate steps. Use new gasket(s) and install the air cross-over manifold(s). Tighten the clamps on the manifold connection hose.
3. Connect the air compressor supply line to the air cross-over manifold. Install the inlet air tube and hoses on the air cleaner and air cross-over manifold.
4. Operate the engine and check for air leaks.
5. Lower the cab and lock it in position.

Intake Manifold Removal

In-Line Engines

1. Drain the cooling system.
2. Disconnect both coolant hoses from the radiator supply tank. Remove the radiator overflow tube from the retaining clamps. Disconnect the windshield washer hose at the "Y" connection.
3. In most cases it is necessary to

remove the engine hood to cowl retaining screws. Remove the engine hood to grille retaining bolts. Use a hoist to remove the engine hood assembly.

4. Loosen the hose clamps, and remove the air inlet tube from the air cleaner to the intake manifold.
5. Disconnect the glow plug lead and the primer fuel line from the intake manifold preheater, if so equipped.
6. Disconnect the compressor air inlet line from the compressor, and remove it from the intake manifold. Disconnect the air compressor coolant line from the compressor and the engine coolant manifold. Remove the retaining clip from the rocker cover; then remove the compressor coolant line.
7. Remove the engine compartment covers from the cab floor.
8. Remove the retaining bolts; then remove the intake manifold.

Intake Manifold Removal

Tilt Cab Models

1. Release the cab locks and tilt the cab forward.
2. Loosen the hose clamps, and remove the air inlet tube from the intake manifold and air cleaner. Remove the air cleaner assembly.
3. Disconnect the glow plug lead and the primer fuel line from the intake manifold preheater, if so equipped.
4. Remove the air compressor air inlet line from the intake manifold and air compressor. Remove the parking brake retaining clip from the intake manifold.
5. Remove the retaining bolts; then remove the intake manifold.

Intake Manifold Installation

In-line Engines

1. Clean the gasket surfaces on the intake manifold and cylinder heads.
2. Using new gaskets, install the intake manifold and tighten the manifold bolts in sequence.
3. Install the compressor coolant line on the air compressor and the engine coolant manifold. Attach the retaining clip to the front rocker cover. Install the compressor air inlet line on the compressor and intake manifold.
4. Connect the primer fuel line and glow plug lead to the intake manifold preheater, if so equipped.
5. Install the air inlet tube on the intake manifold and air cleaner, and tighten the hose clamps.
6. If removed, use a hoist to posi-

tion the engine hood assembly on the cowl and grille. Install and tighten the hood to cowl screws and hood to grille bolts.

7. Position the radiator overflow tube in the retaining clamps, and tighten the clamps. Connect the coolant hoses to the radiator supply tank. Connect the windshield washer hose.
8. Install the engine compartment covers on the cab floor.
9. Fill and bleed the cooling system.
10. Start the engine and check for coolant and air leaks.

Intake Manifold Installation

Tilt Cab Models

1. Clean the gasket surfaces on the intake manifold and cylinder heads.
2. Using new gaskets, install the intake manifold. Tighten the manifold bolts in sequence.
3. Install the parking brake retaining clip on the intake manifold. Install the air compressor air inlet line on the compressor and the intake manifold.
4. Connect the primer fuel line and glow plug lead to the intake manifold preheater, if so equipped.
5. Install the air cleaner assembly. Install the air inlet tube on the intake manifold and air cleaner, and tighten the hose clamps.
6. Lower the cab and lock it in position.

Cylinder Head

Removal

855, 927 Cid Engines

1. Remove the fuel manifold, coolant manifold, exhaust manifold, intake manifold, and rocker housing(s).
2. Remove the injector and valve push rods in order, and place them in a rack for installation in

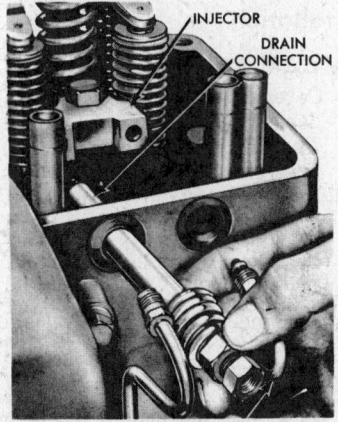

Removing injector drain connection
(© Cummins Engine Co.)

the same position from which they were removed.
3. Remove the compression release lever return spring and retaining bolt.
4. Remove the injector inlet and drain connections.
5. Remove the injector hold-down bolts and lift or pry the injectors from the cylinder heads. *Do not turn the injectors upside down, thus avoiding dropping the plungers.*
6. Loosen and remove the cylinder head bolts. Using a lifting eye and a hoist, remove the cylinder head(s), and place them on a bench, with the valves horizontal. Remove the grommet retainers. Remove and discard the grommets, O-rings and gaskets.

Installation

855, 927 Cid Engines

1. Carefully remove excess carbon deposits from the cylinder head(s). *Do not use a scraper or wire brush in the injector sleeve area, to prevent distortion or mutilation of the copper injector sleeves.* Clean the O-ring, grommet and gasket surfaces of the cylinder head(s) and cylinder block. Clean the gasket surfaces of the intake and exhaust manifolds.
2. Check the cylinder liner protrusion of all liners.
3. Press new grommet retainers, with the small end up, in the coolant passages of the cylinder block. Position a new cylinder head gasket on the engine block and cylinder head dowels so that the gasket side marked "TOP" is facing up. Use a gasket with white water hole grommets if block water holes are free of erosion. Use a gasket with black water hole grommets where there is evidence of water hole erosion, if sleeves have not been installed. Install the grommets on the grommet retainers, and press the grommets into the gasket holes. *Avoid all contact be-*

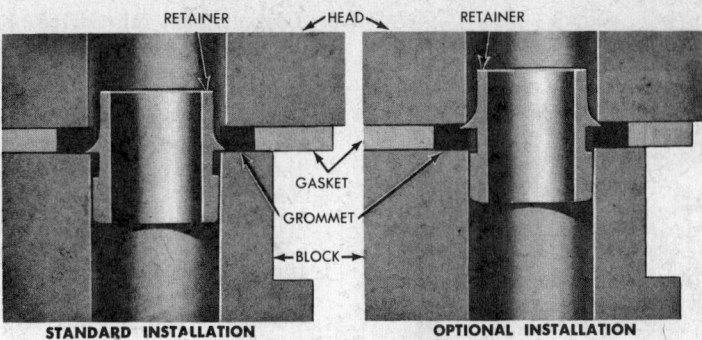

Installing grommets and retainers
(© Cummins Engine Co.)

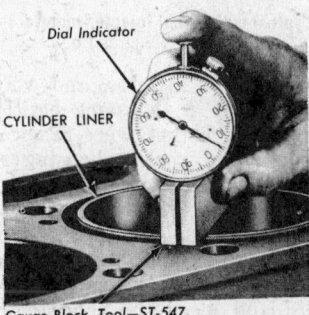

Checking cylinder liner protrusion
(© Cummins Engine Co.)

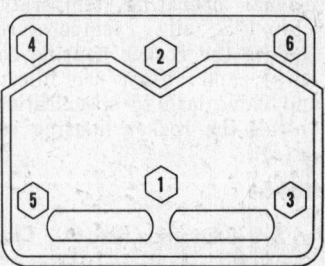

Cylinder head bolt torque sequence
(© Cummins Engine Co.)

tween grommets and fuel or lubricating oil to prevent damage due to swelling. The optional installation of the grommet retainer above the grommet may become necessary when the cylinder head as 15/32 inch coolant holes or when the holes are eroded.
4. Using lifting eye and a hoist, install the cylider head(s).
5. Lubricate the cylinder head bolt threads and washers with rust preventive. Install the head bolts or caps and torque, in sequence, as specified in the specifications charts.
6. Install the injectors, following the procedures described in this section. *Screw the inlet and drain connections into the injectors about 3 turns to align the injectors before tightening the injector hold-down bolts.*
7. Install the compression release lever return spring and retaining bolt.

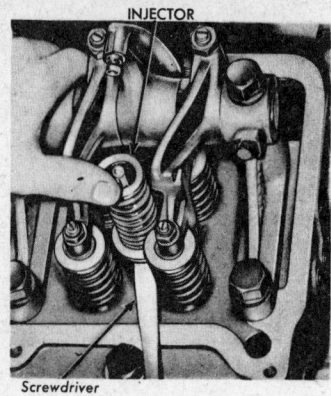

Typical cylindrical injector removal
(© Cummins Engine Co.)

8. Install the injector and valve push rods in the same position from which they were removed.

9. Install the rocker housing(s), intake manifold, exhaust manifold, coolant manifold and fuel manifold, and adjust the valve cross-heads and injector and valve lash.

10. Fill and bleed the cooling system. Operate the engine and check for fuel, oil and coolant leaks. When the engine reaches normal operating temperature (140°F oil temperature), remove the rocker housing cover(s), and adjust the injector and valve lash to specifications. Install the rocker housing cover(s).

Removal

V-6 & V-8 Engines (Exc 903 Cid)

1. Drain the cooling system and exhaust the air reservoir. Disconnect the battery cables.

2. Remove the engine hood assembly where necessary.

3. Disconnect the headlamp wiring harness and remove the grille and headlamp assembly(ies). Remove the front fender(s) where necessary.

4. Loosen the clamps and remove the steering gear shaft to steering wheel shaft universal joint. Position the steering gear shaft out of the way where necessary.

5. Remove the radiator brace rod from the left side of the radiator support and frame rail where necessary.

6. Remove the alternator mounting and adjusting bolts. Leave the leads attached and position the alternator out of the way.

7. On the left cylinder head, remove the power steering pump and drive belts from the engine. Leave the hoses attached and position the power steering pump out of the way. On the right cylinder head, remove only the steering pump drive belts.

8. Loosen the fan drive belts. Remove the cooling fan and hub assembly, spacers and drive belts from the engine.

9. On a right cylinder head, disconnect the shutterstat air hoses at the shutterstat. Disconnect the heater hoses at the thermostat housing and the water pump. Disconnect the coolant temperature and warning light sending unit leads. Loosen the water pump to thermostat housing by-pass hose clamp. Remove the thermostat housing and shutterstat assembly.

10. On the right cylinder head, remove the dipstick tube assembly. Remove the air inlet duct from the air cleaner and air cross-over manifold.

11. Disconnect the accelerator rod and the tachometer cable at the fuel pump. Disconnect the service brake air hose at the brake treadle valve. Remove the air hose from the brake treadle valve to the air reservoir line.

12. On some models it may be necessary to disconnect the grille to cab brace rods at the cab. Remove the cab front support to chassis bolts and nuts. Attach a sling and hoist to the cab supports and raise the cab approximately 4 inches.

13. On the left cylinder head, disconnect the air compressor, supply line at the left air cross-over manifold. Loosen the hose clamps and remove the air cross-over manifold(s) and gasket(s).

14. Remove the valve cover(s) and gasket(s) and cylinder head cover(s) and gasket(s.

15. On the right cylinder head, remove the alternator mounting bracket. Remove the intake manifold(s) and gasket(s).

16. Disconnect the fuel supply line at the front of the cylinder head(s). Disconnect the fuel return line at the rear of the cylinder head(s).

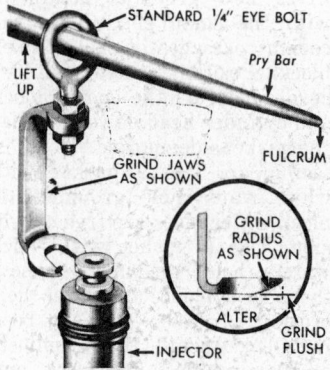

Injector removal
(© Cummins Engine Co.)

17. On the right cylinder head, disconnect the air compressor coolant line at the coolant cross-over tube. Remove the coolant cross-over tube(s) from the cylinder head(s).

18. Remove the push rod cover(s) and gasket(s).

19. Remove the exhaust manifold bolts and separate the exhaust manifold(s) from the cylinder head(s). Leave the manifold(s) attached to the resonator inlet pipe(s).

20. Loosen the lock nuts and back off all of the injector and valve rocker lever adjusting screws. Remove the rocker lever support screws. Remove the injector and valve push rods and place them in a rack for installation in the same position from which they were removed. Remove the rocker lever assembly(ies) with the cylinder head(s). Loosen the crosshead adjusting screw nuts and remove the crossheads, on four valve engines.

21. Remove the injector hold-down clamp screws and clamps. Pry the injectors loose from the cylinder head. *It may be necessary to use a tool such as for removal of the rear injectors.* Remove the injectors and O'rings. *Do not remove the injector plunger, since they are not interchangeable between injectors.*

22. Remove the remaining cylinder head bolts. If the rear bolts cannot be removed due to interference with the cab supports, split a 1½-inch length of rubber hose and install it around the bolt. This will hold the bolt above the bottom of the cylinder head.

23. Install a lifting bracket on the cylinder head, and use a hoist to remove the cylinder head(s). Mark heads "left" or "right".

24. Remove the cylinder head gaskets, grommets, grommet retainers and O-rings. Remove the exhaust manifold gasket(s).

Installation

V-6 & V-8 Engines (Exc 903 Cid)

1. Carefully remove excess carbon deposits from the cylinder head. *Do not use a scraper or wire brush in the injector sleeve area. to prevent distortion or mutilation of the copper injector sleeves.* Clean all gasket surfaces. Clean the injector bores of the cylinder head with a clean rag and stick. Remove all liquid and dirt from capscrew holes. Make sure cylinder walls are clean and well lubricated. Make sure breather vent tube assemblies (if used) are in place. Coat short capscrews with rust pre-

ventive and install until just snug in the sequence shown.

2. Position new head gasket(s) on the cylinder block and dowels with the side marked "TOP" up. Press new grommet retainers, with the small end up, in the coolant passages of the cylinder block. Install the grommets on the grommet retainers, and press the grommets into the gasket holes. *Avoid all contact between grommets and fuel or lubricating oil to prevent damage due to swelling.* The optional installation of the grommet retainer above the grommet may be necessary when the cylinder head has 15/32-inch coolant holes or when the holes are eroded. With a lifting bracket secured to the head with capscrews, hoist the head in position over the block and lower it in place over the ring dowels. Coat the short capscrews with rust preventive and install. Tighten in the proper sequence until snug.

3. Make sure the injectors are clean and that the button screen around the inlet is in place. Use a clean cloth and clean the injector sleeves. Install new O-rings on injector bodies and lubricate with clean engine oil. Start each C type injector into the bore so the ball check retainer plug is at one o'clock position and guide it by hand until alignment and lack of binding are ensured. Place a clean, blunt object on each injector and snap it onto its seat. In the case of type D injectors, align the injector so the button screen is toward the center of the Vee. Seat D injectors with a T-handle made from 5/8" bar stock with a 5/16" hole 1/2" deep drilled in the bottom and a type D injector link braised into the hole. Apply force with the T handle apparatus positioned on top of the injector plunger. Install type D injector links, if removed. Both types of injectors then require installation of hold-down clamps, washers, and capscrews torqued to 30-35 ft-lbs. Make sure the injector plunger moves freely after installation of clamps.

4. Install crosshead retainers where used, and insert crossheads over crosshead guides. Run adjusting screws down until they touch valve stems, and torque locknuts to 25-30 ft-lbs.

5. Install the push rods in the same position from which they were removed. Lubricate the bolts with rust preventive. Install the bolts and tighten them until snug. Then torque all head bolts

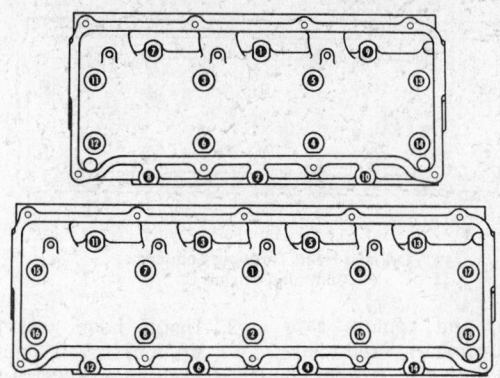

Cylinder Head Tightening Sequence
(© Cummins Engine Co.)

according to the specifications charts in the proper sequence.

6. Adjust the crossheads and the injector and valve lash. See Injector section for injector lash adjustment instructions. Valve clearance at room temperature is .012 intake, .022 exhaust.

7. Using new gaskets, install the exhaust manifold(s). Torque the bolts in 15 ft-lb increments to specifications.

8. Use new gasket(s) and install the push rod cover(s).

9. Install the coolant cross-over tube(s) on the cylinder head(s). On the right cylinder head, connect the air compressor coolant line to the coolant cross-over tube.

10. Connect the fuel return line at the rear of the cylinder head(s). Connect the fuel supply line at the front of the cylinder head(s).

11. Using new gasket(s), install the intake manifold(s). On the right cylinder head, install the alternator mounting bracket.

12. Coat the upper surface of new gaskets and position the gaskets on the cylinder head cover(s) and valve cover(s). Install the cylinder head cover(s) and valve cover(s). *Partially tighten the valve cover screws.*

13. Install the air cross-over manifold(s) with new gasket(s). Tighten the screws and hose clamps. On the left cylinder head, connect the air compressor supply line to the left air cross-over manifold.

14. Install the inlet air duct on the air cross-over manifold, and tighten the hose clamp. On the right cylinder head, install the dipstick tube assembly.

15. Install the thermostat housing and shutterstat assembly. Tighten the hose clamps. Connect the coolant temperature and warning light sending unit leads. Connect the radiator

upper hose and heater hoses to the thermostat housing and water pump. Connect the shutterstat air hoses to the shutterstat.

16. Install the cooling fan and hub assembly, spacers and drive belts.

17. Install the alternator, but do not install the drive belt at this time.

18. Connect the battery cables. Fill and bleed the cooling system. Operate the engine until the oil temperature is 140°F. Remove the valve cover(s) and adjust the injector and valve lash to specifications. Install the valve cover(s) and tighten the screws in alternate steps.

19. Install the alternator drive belt.

20. Lower the cab. Install and tighten the cab front support to chassis bolts, insulators and nuts. Connect the grille to cab brace rods to the cab.

21. Install the air hose to the brake treadle valve and the air reservoir line. Connect the service brake air hose to the brake treadle valve. Connect the tachometer cable and accelerator rod to the fuel pump. Adjust the accelerator linkage.

22. On the left cylinder head, install the power steering pump. Install the drive belts and adjust the belt deflection to specifications.

23. Install the radiator left brace rod.

24. Connect the steering gear shaft to the steering wheel shaft.

25. Install the front fender(s). Install the headlamp and grille assembly(ies). Connect the headlamp wiring harness.

26. Install the engine hood assembly.

27. Operate the engine and check for air, oil and coolant leaks.

Removal

V-903 and VT-903 Engines

1. Label cylinder head covers as to which side of the engine they are

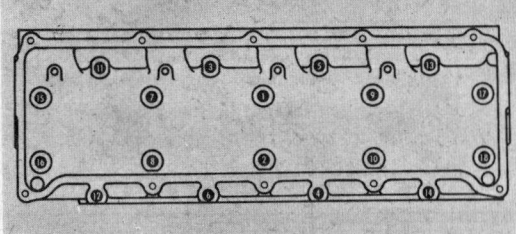

Cylinder Head Torquing Sequence
(© Cummins Engine Co.)

mounted on, and remove capscrews, lockwashers and flatwashers holding the covers on the heads. Remove the covers, discarding the gaskets.

2. Remove capscrews and associated parts from the push rod cavity covers, remove the covers and discard the gaskets.

3. Loosen all rocker lever adjusting nuts and back out all the adjusting screws.

4. Secure all the rocker levers together with rubber bands or a similar device.

5. Remove the rocker assembly mounting capscrews and washers.

6. Lift off the rocker assemblies.

7. Remove the pushrods.

8. Loosen all crosshead adjusting screw nuts and lift the crossheads off the guides.

9. Remove the injector hold down yoke capscrews, and remove the yokes.

10. Remove the injectors with a special tool. Place injectors in a clean rack carefully, wtih the injectors labeled as to which cylinder they belong in.

11. Attach a special lifting fixture to the head. Remove all remaining capscrews and flatwashers.

12. Insert guide studs in the block for easiest removal. Lift the heads off the block with an adequate hoist via the lifting fixture.

13. Tag heads as to which side of the engine they belong on, remove guide studs, and discard gaskets and grommets.

Installation

V-903 and VT-903 Engines

Valve guides, valves, and springs should be installed on heads and all liquid and dirt should be removed from the capscrew holes before installation.

1. Plug breather vent tube holes if heads are used on a turbocharged engine. Clean the mating surfaces and make sure cylinders are clean and coated with lubricating oil. Make sure there are no lime deposits in the water passages that could interfere with installation of grommet retainers.

2. Install head gasket, "Top" side up, and install a new O-ring over each ring dowel. Install an oversize gasket if head has been resurfaced.

3. Install new grommets in water passages.

4. Install a new ring dowel and O-ring in the oil passage.

5. Lower the head with the lifting bracket and a hoist until it moves down into place over the ring dowels.

6. Lubricate the capscrews and washers with clean oil, position the washers over the capscrews with the chamfered edge down, and position the O-rings, coated with engine oil, over the threads.

7. Install the capscrews and tighten just until snug in the sequence shown.

8. Make sure the injector sleeves are free of chips or carbon particles. If cleaning is required, perform it with a rag wrapped around a 3/4" wooden dowel soaked in solvent.

9. Lubricate the O-rings with engine oil.

10. Lower the injector into the head, aligned so the button screen faces the center of the Vee.

11. With a clean, blunt object, push each injector down into place by exerting pressure on the body. The injector will snap into place.

12. Install links, hold down rings, and clamps, flatwashers, and capscrews. Torque the capscrews to 30-35 ft-lbs. Make sure the hold down clamps do not contact the crosshead stems. Make sure the injector plunger moves freely.

13. Install the crossheads on their guides.

14. Adjust the crossheads by holding each firmly down on the mating valve stem and turning the adjusting screw down until it just touches the other stem. Torque all adjusting screw locknuts to 25-30 ft-lbs.

15. Position both rocker assemblies to the cylinder heads. Coat the capscrews with rust preventive and install in the heads, tightening just until snug.

16. Torque all the cylinder head capscrews in the sequence shown according to the specifications charts.

17. Install push rods and engage adjusting screws with the push rod sockets.

18. Position the pushrod cavity covers onto the heads with new gaskets. Install flatwashers, lockwashers, and capscrews.

Cylinder Head Removal 1150 cid

1. Remove intake manifold.

2. Remove rocker lever covers.

3. Loosen valve and injector adjusting screw locknuts and back out adjusting screws one or two turns.

4. Remove rocker lever capscrews and pull straight up on lever to clear ring dowels.

5. Remove valve crossheads.

NOTE: Mark push rods, injector links, rocker levers, and all such parts which establish a wear pattern.

6. Remove injector plunger links and push rods.

7. Remove water transfer tube retaining ring between #5 and #6 cylinders.

8. Drive transfer tube into #5 housing far enough to clear #6.

9. Remove capscrews from rocker housing and lift off housing.

10. Remove transfer tube from #5; discard O-rings.

11. Repeat this procedure for the remaining housings.

NOTE: The thermostat must be removed before #1 housing is removed.

12. Remove by-pass tube, turbocharger supply line and air compressor drain tube from housing.

13. Remove housing support bracket.

14. Remove thermostat-to-gear housing capscrews.

15. Cock thermostat housing clear of fan hub support.

16. Pull thermostat housing and transfer tube from rocker housing.

17. Remove remaining housings.

Removing injector plunger links

Removing thermostat housing

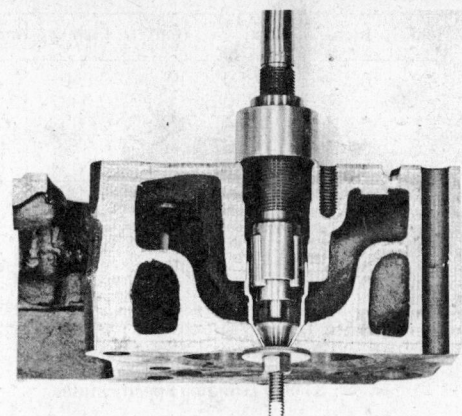

Rolling Injector Sleeve—Upper Portion
(© Cummins Engine Co.)

18. Remove injector hold-down plate.
19. Remove injectors with suitable tool.
20. Remove cylinder head capscrews; lift heads straight up.

Installation

1. Place new head gasket over dowels in block. Gaskets are marked "TOP".
2. Place water, oil, and push tube grommets into position in each cylinder head gasket, on the camshaft side of the engine, requires a gasket.
3. Fix head in hoist and position over block.
4. Lower head onto block, taking care not to disturb grommets.
5. Lubricate head bolts with rust-proof oil. Before snugging capscrews, coat underside of capscrew head with SAE 140 w.
   NOTE: Washers must be placed under capscrews with rounded side up.
6. Tighten capscrews to recommended torque.

Rebuilding Cylinder Head

Valve Guides V6 and V8

1. Where replacement is required, drive out valve guides from underside of cylinder head. Clean new guides sonically, or with a 45 caliber piston cleaning brush. All sealer must be removed. Then, flush with solvent to remove residue.
2. Install new valve guides with special mandrel. Oil mist hole (if used) must face exhaust manifold. If proper mandrels are unavailable, press guides in until protrusion is .775″-.800″ above surface (.695″-.710″ on 903).
3. If reaming of valve guide is necessary to install valves ream valve guide from bottom side of cylinder head, using a drill press and floating tool holder. Use lu-

bricating oil or soluble oil and water solution for a good finish.
CAUTION: Special care must be used to avoid breaking carbide tips. Sharpen carbide tipped tools on a diamond impregnated wheel. New guides are to be reamed to .3800″-.3817″ (.4520″-.4532″ on 903).

Valve Guides 855 and 927

1. Drive out faulty guides from the bottom of the head. If reaming is required to clean up guide bore, use a special reamer to ream the bore to .760″-.761″. Make sure all burrs are removed, and that the corner break does not exceed .015″.
2. If a damaged guide bore does not clean up, use a special oversize reamer, and ream to .765″-.766″. Install an oversize guide, and make sure that all burrs are removed, and that the corner break does not exceed .015″. If necessary, ream valve spring guide hole to .768″-.773″.
3. If a special mandrel is unavailable, press the guide in until it protrudes 1.315″-1.325″.
4. If guide must be reamed for insertion of valve stem, ream the guide from the bottom with a special reamer using lube oil or soluble oil and water solution. Do not ream beyond .445″.

Valve Guides 1150 cid

1. Drive out guides from bottom.
2. Drive in new guide with guide driver.
3. Check guide dimensions:
Installed diameter—inside .4961-.4971
Assembled height   1.325-1.340

Valve Crosshead Guides

1. Remove guides to be replaced, using Puller which contains different size collets to fit the various guides. NOTE: After Engine Serial No. 298655, engines are equipped with "solid" pin-type crosshead guides and crossheads with hollow stems.

Old and new-style crossheads and guides may be used in the same cylinder head.
2. Press in new guides.
NOTE: A special solid guide has been released to permit the new hollow crossheads in cylinder heads machined for the larger diameter hollow guide. If no spacers are available, press in the new guide until it protrudes 2.040″-2.060″ on V6 and small V8, 2.090″-2.110″ on 903. On 855 and 927 engines, if the special installation mandrel is not available, press the new guides in until they protrude 1.860″-1.880″.

Injector Sleeves

Removal

1. Remove worn sleeves by cutting them from cylinder head with a ⅜ inch gouge chisel (sometimes called a muffler sleeve cutting tool) and driving out from lower end. Remove O-ring, if used.
NOTE: Remove all foreign material from injector sleeve sealing area.
A special sleeve remover tool may be used as follows:
   a. Install tool into sleeve bore with teeth of forming collar resting on sleeve and inside bore. Do not allow hammer to strike underside of knob.
   b. Hold down knob and hammer on nut until forming collar bottoms out.
   c. Hold up on the rod, and turn down nut until it bottoms on forming collar.
   d. Slide hammer a light blow against the knob to raise and seat the extractor tip within the injector sleeve.
   e. Retighten the nut against the top of the forming collar.
   f. Repeatedly strike the knob with a hammer and pull the injector sleeve from the head.
   g. Back the nut out all the way.
   h. Tap the end of the injector sleeve until the forming collar is free or touches the nut.

Item	Inch.	[mm]	Item	Inch	[mm]
1.	0.375	[9.5250]	10.	*0.010	[0.2540]
2.	1.250	[31.7500]	11.	0.0937	[2.3812]
3.	0.3136	[7.9654]	12.	0.969	[24.6126]
4.	0.199	[5.0546]		0.971	[24.6634]
	0.209	[5.3086]	13.	0.0034	[0.0863]
5.	0.304	[7.7216]		0.0054	[0.1371]
	0.306	[7.7724]	14.	0.0015	[0.0381]
6.	0.3250	[8.2550]		0.0025	[0.0635]
	0.3300	[8.3820]	15.	0.040 R	[1.0160] R
7.	0.020	[0.5080]		0.060	[1.5240]
			16.	15 deg. Angle Relief	
8.	0.250	[6.3500]	17.	30 deg. Angle Relief	
9.	0.0937	[2.3812]	*Land		

ST-920 Grinding Specifications

i. Rotate the sleeve ⅓ turn until the indentations match the flats on the extractor tip. Remove the sleeve.

Installation

1. Lightly lube a new O-ring with engine oil.
2. Install the O-ring in the groove. Installation may be facilitated by sitting the head on one edge so the injector bore is horizontal and using two small rods to feed the seal ring through the bore to the groove.
3. Support the head adequately around the injector opening to avoid cracking it during the following operations. Place the sleeve in the injector bore and drive it in until it bottoms, using a special driver.
4. Remove the driver and install a special hold-down tool. Torque the nut to 35-40 ft-lbs.
5. Install a special sleeve driver into the sleeve over the hold down tool. Apply two light blows to the driver to embed the beads into the sleeve. Retighten the hold-down tool nut to 35-40 ft-lbs.
6. Seal the upper portion of the injector sleeve area with a special expanding roller until 75 in/lbs of torque are required.
7. Remove all the tools and install a special driver. Reseat the sleeve with two light blows of a 2 lb. lead hammer.
8. Seal the sleeve in the lower, tapered seating area with a special angle roller tool.
9. Cut the seat with a special sleeve cutter. Cutter must conform to dimensions shown in illustration. Provide a solid stream of cutting oil during the operation. A special pilot is available for adapting this tool to a drill press. Alternately insert injector to measure tip protrusion (from flat head surface) and cut the head. Final protrusion should be .095"-.115" on V6 and small V8,

.075"-.090" on 903, and .060"-.070" on 855 and 927 engines.
10. Install the injector and determine seating surface with Prussian Blue. A band .060" wide must appear all the way around the injector beginning about .275" from the bottom of the head surface.
NOTE: The manufacturer suggests a water pressure test of the cylinder head to check the seal of injector sleeves.

Replace Valve Seat Inserts

1. Remove loose or excessively worn valve seat inserts.
2. Enlarge counterbore to next oversize. Inserts are available in standard .010, .020, .030, and .040 inch oversizes.
3. Use Valve Seat Insert Tool to hold and drive cutters which come in sets. This tool must be driven by an electric motor. On all V6 and V8 engines, restore counter bore and pocket depth at the valve counterbore area (see chart). Restore chamfer by grinding with a 35° stone. Chamfer should be 2.0000"-2.0200" on V6 and small V8, and 2.3000"-2.3200" on 903 V8.
4. Machine counterbore .006/.010 inch deeper than insert height to permit peening of head to hold insert.
5. Install valve seat insert.
6. Peen insert in head with peening tool. A ¼-inch diameter round-end punch may be used.

Grind Valve Seats

1. Check condition of grinding equipment.
   a. Mandrels must be straight and of proper size to fit in reamed valve guides.
   b. Bushings in the grinder must be clean and must fit properly on guide mandrel.
   c. Bearings of drive unit must be in good condition.
2. Dress stone to 30° from horizontal.
3. Grind valve seats, holding seat-

ing motor as nearly vertical as possible. A severe angle will cause the seat to be out-of-true depending upon the amount of wear in the grinder bearings, mandrel bushines, etc., even though the grinder has a universal joint.
4. Check valve seat width which should be .063"-.125".
   a. If ground seat is wider than the maximum ⅛ inch, stock can be removed from points "A" and "B" with specially dressed valve seat grinder stones.
   b. Narrowing should not extend beyond chamfer on seat insert. Chamfer provides for peen metal.
5. Dress wheel for final finish.
6. Finish grind with light touches of stone against face.
7. Check guide alignment with eccentrimeter.
   a. Run-out should not exceed .002 inch total indicator reading.
   b. The gauge must be a perfect fit on pilot mandrel.

Grind Valves

1. Check valve grinder setting by using a new valve and an indicator gauge.
   a. Chuck valve on guide area of stem. Relieved portions on both ends of guide area are not necessarily concentric to guide area of stem.
   b. Indicate on ground face of valve.
   c. Turn valve and mark high spot on head of valve.
   d. Rechuck the valve 180° from first position.
   e. Repeat "b" and "c". If the high spots are same for both "a" and "d" positions, the valve is warped. If high spots occur in different positions, chuck is out of alignment. Runout should not exceed .001 inch.
2. Check bearings of machine.
3. The grinding wheel must be the proper grade and properly dressed to avoid chatter and grind marks.
4. Wet grind valves to an exact 30° angle from horizontal.
5. Valves and seats properly ground with precision equipment should not require lapping to effect an airtight seal; however, a small amount of lapping is permissible if necessary in order to pass vacuum test.
6. Check rim thickness. If rim is thinner than specified, valve is not suitable for use because of the danger of burning and cupping.

	Intake	Exhaust
V6, V8	.130″	.130″
In-line	.105″	.105″

7. Check valve in a finish reamed guide and against a newly ground valve seat face. Pencil mark valve as shown in illustration, drop into position and rotate 10 degrees.

8. A true seat will be indicated if all pencil marks are broken. If pencil marks are not broken, the valve seat tools need dressing or the machine has not been properly adjusted. A good seat must:
1. Have no grinding marks.
2. Be at a true 30° angle.
3. Have a grind width within limits.
4. Have proper guide-to-stem clearance.

Assemble Valves and Springs
1. Lubricate valve stems. Insert valves.
2. Place cylinder head face down on a wooden bench to prevent marring milled surface.
3. Place plastic sleeve over valve stem, and then install seal. Press seal into place with a special driver until it is properly seated.

Valve Seat Insert Data — Inch [mm]

Insert Part No.	ST Cutter	Oversize Diameter	Thickness	Insert Outside Diameter	Cylinder Head Inside Diameter	Insert Thickness	Counterbore Depth In Head
154390	ST-258	Std.	Std.	1.690/1.691 [42.9260/42.9514]	1.687/1.688 [42.8498/42.8752]	0.196/0.201 [4.9784/5.1054]	0.351/0.356 [8.9154/9.0424]
154395	ST-258	0.005 [0.1270]	Std.	1.695/1.696 [43.0530/43.0784]	1.692/1.693 [42.9768/43.0022]	0.196/0.201 [4.9784/5.1054]	0.351/0.356 [8.9154/9.0424]
154391	ST-258-1	0.010 [0.2540]	Std.	1.700/1.701 [43.1800/43.2054]	1.697/1.698 [43.1038/43.1292]	0.196/0.201 [4.9784/5.1054]	0.351/0.356 [8.9154/9.0424]
154392	ST-258-2	0.020 [0.5080]	0.005 [0.1270]	1.710/1.711 [43.4340/43.4594]	1.707/1.708 [43.3578/43.3832]	0.201/0.206 [5.1054/5.2324]	0.356/0.361 [9.0424/9.1694]
154393	ST-258-3	0.030 [0.7620]	0.010 [0.2540]	1.720/1.721 [43.6880/43.7134]	1.717/1.718 [43.6118/43.6372]	0.206/0.211 [5.2324/6.3594]	0.361/0.366 [9.1694/9.2964]
154394	ST-258-4	0.040 [1.0160]	0.015 [0.3810]	1.730/1.731 [43.9420/43.9674]	1.727/1.728 [43.8658/43.8912]	0.211/0.216 [5.3594/5.4864]	0.366/0.371 [9.2964/9.4234]

Caution: Be sure to measure insert before machining head or installing insert in head.

Intake Valve Seat Insert Data (If Swirl Plates Are Used) — Inch [mm]

Insert Part No.	ST Cutter	Oversize Diameter	Thickness	Insert Outside Diameter	Cylinder Head Inside Diameter	Insert Thickness	Counterbore Depth In Head
195000	ST-258	Std.	Std.	1.690/1.691 [42.9260/42.9514]	1.687/1.688 [42.8498/42.8752]	0.176/0.181 [4.4704/4.5974]	0.351/0.356 [8.9154/9.0424]
195001	ST-258-1	0.010 [0.2540]	Std.	1.700/1.701 [43.1800/43.2054]	1.697/1.698 [43.1038/43.1292]	0.176/0.181 [4.4704/4.5974]	0.351/0.356 [8.9154/9.0424]
195002	ST-258-2	0.020 [0.5080]	0.005 [0.1270]	1.710/1.711 [43.4340/43.4594]	1.707/1.708 [43.3578/43.3832]	0.181/0.186 [4.5974/4.7244]	0.356/0.361 [9.0424/9.1694]
195003	ST-258-3	0.030 [0.7620]	0.010 [0.2540]	1.720/1.721 [43.6880/43.7134]	1.717/1.718 [43.6118/43.6372]	0.186/0.191 [4.7244/4.8514]	0.361/0.366 [9.1694/9.2964]
195004	ST-258-4	0.040 [1.0160]	0.015 [0.3810]	1.730/1.731 [43.9420/43.9674]	1.727/1.728 [43.8658/43.8912]	0.191/0.196 [4.8514/4.9784]	0.366/0.371 [9.2964/9.4234]

Caution: Be sure to measure insert before machining head or installing insert in head.

Valve Seat Insert Data — Inch [mm]

Insert Part No.	ST Cutter	Oversize Diameter	Thickness	Insert Outside Diameter	Cylinder Head Inside Diameter	Insert Thickness	Counterbore Depth in Head
127930	ST-662	Std.	Std.	2.0025/2.0035 [50.8635/50.8889]	1.9995/2.0005 [50.7873/50.8127]	0.277/0.282 [7.0358/7.1628]	0.437/0.442 [11.0998/11.2268]
127935	ST-662	0.005 [0.1270]	Std.	2.0075/2.0085 [50.9905/51.0159]	2.0045/2.0055 [50.9143/50.9397]	0.277/0.282 [7.0358/7.1628]	0.437/0.442 [11.0998/11.2268]
127931	ST-662-1	0.010 [0.2540]	Std.	2.0125/2.0135 [51.1175/51.1425]	2.0095/2.0105 [51.0413/51.0667]	0.277/0.282 [7.0358/7.1628]	0.437/0.442 [11.0998/11.2268]
127932	ST-662-2	0.020 [0.5080]	0.005 [0.1270]	2.0225/2.0235 [51.3715/51.3969]	2.0195/2.0205 [51.2953/51.3207]	0.282/0.287 [7.1628/7.2898]	0.442/0.447 [11.2268/11.3538]
127933	ST-662-3	0.030 [0.7620]	0.010 [0.2540]	2.0325/2.0335 [51.6255/51.6509]	2.0295/2.0305 [51.5493/51.5747]	0.287/0.292 [7.2898/7.4168]	0.447/0.452 [11.3538/11.4808]
127934	ST-662-4	0.040 [1.0160]	0.015 [0.3810]	2.0425/2.0435 [51.8795/51.9049]	2.0395/2.0405 [51.8033/51.8287]	0.292/0.297 [7.4168/7.5438]	0.452/0.457 [11.4808/11.6078]

Caution: Be sure to measure insert before machining head or installing insert in head.

Cummins Diesel Engines

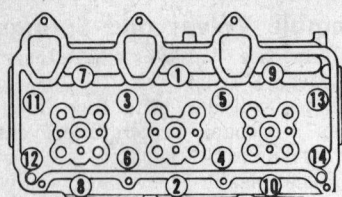

Cylinder head bolt torque sequence
(© Cummins Engine Co.)

4. Assemble lower valve spring guides on valve guides.

5. Assemble springs. *NOTE: Reground valve heads seat deeper in cylinder head causing valve stem to protrude further above the guide. This allows valve spring to extend beyond length limits and causes weak spring action. Therefore, up to two spacers may be used to reduce valve spring length.* **CAUTION: Too many spacers will cause the compressed spring to become a solid sleeve.**

6. Assemble upper valve spring guide. Insert half-collets.

7. Use Valve Spring Compressor to compress valve spring. Use new half-collets.

Test Valves and Seats for Leakage

A vacuum tester to check valves and seats for leakage is available. It consists of a vacuum pump, vacuum gauge and suction cup. Use with any

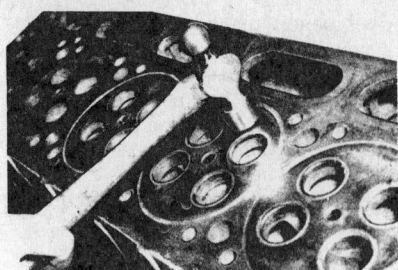

Checking for loose valve seat inserts
(© Cummins Engine Co.)

6-volt battery source or 110-volt electrical outlet as required. Follow manufacturers procedures for operation.

CAUTION: Never vacuum test cylinder head with injectors installed. Installation of injectors while head is removed from block could cause misalignment of valves in valve seat and result in leakage during the vacuum test which would not necessarily occur during actual engine operation.

Valves and seats must be dry and clean.

NOTE: It is possible to mistake leakage around the valve seat insert for valve seat leakage. If this type of leakage is suspected apply grease around the outside edge of the insert to make a grease seal. Perform the vacuum test and inspect the grease seal for a break indicating air leakage between the wall of counterbore and valve seat insert. If a leak around valve seat insert is found, correction

is required before continuing with the test.

Install Crossheads

1. Insert valve crossheads in crosshead guides, or assemble valve crossheads over crosshead guides, depending upon design.

2. Install crosshead retainer to keep crosshead in position and insure correct valve-to-crosshead and rocker lever contact.

Check Vent Holes

Check the vent hole at the top of the cylinder head and at the front of every other rocker shaft bearing. *On all supercharged and turbocharged engines the vent holes must be plugged with a 1/8 inch pipe plug. On naturally aspirated engines, a vented plug must be installed.*

Compression Release

Disassembly

1. Remove mounting stud nuts and lockwashers.

2. Lift compression release housing, shaft, lever and spring from cylinder head.

3. Inspect and install a new packing gland, if necessary.

Assembly

1. Reassemble compression release to cylinder head, being careful to let one end of spring ride

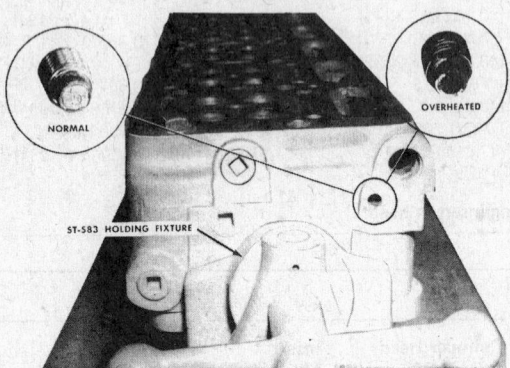

Check fuse plug for overheating
(© Cummins Engine Co.)

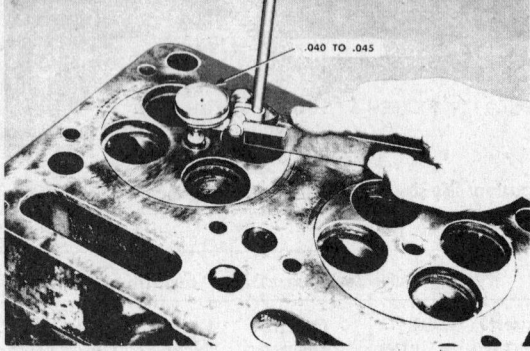

Measuring injector tip protrusion
(© Cummins Engine Co.)

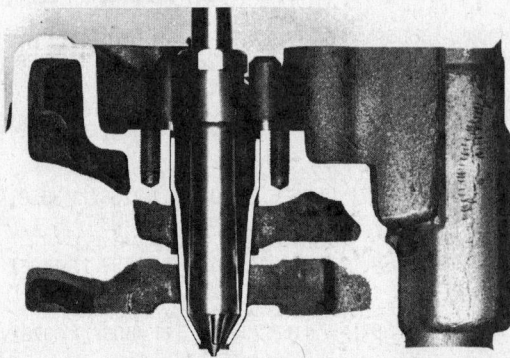

Sealing lower end of injector sleeve
(© Cummins Engine Co.)

Sealing upper end of injector sleeve
(© Cummins Engine Co.)

against housing with other end held against upper stud bolt.

2. Clamp lever in place so that open position will lift valves 1/16 inch maximum.

Regrooving cylinder head with special tool
(© Cummins Engine Co.)

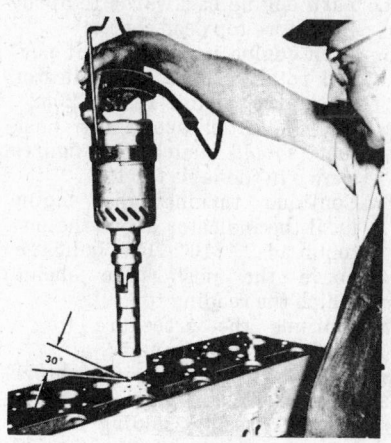

Refacing valve seats
(© Cummins Engine Co.)

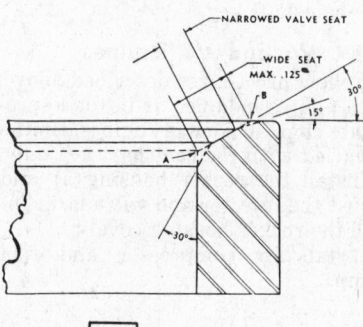

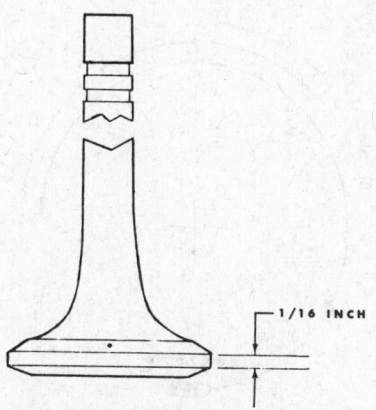

Minimum valve head rim thickness
(© Cummins Engine Co.)

Rocker Assembly and Housing

Rocker Removal

In-line Engines

1. Remove the rocker housing covers. Remove the lubricating oil pipe caps and gaskets.
2. Loosen the locknuts and back off the adjusting screws two or three turns on all rocker levers.
3. Remove the rocker housing retaining bolts and lift the rocker housing assemblies from the engine by grasping the ends of the injector rocker. Leave the compression lever attached to the linkage on No. 1 cylinder.

Installation

In-Line Engines

1. Clean the gasket surfaces of the cylinder heads, rocker housings and covers.
2. Using a new housing gasket, position the rocker housing(s) on the cylinder heads with the ball-ends of the rocker levers fitting into their respective push rod sockets. Install and torque the housing bolts in sequence to specifications. Install the compression release arm with the front rocker housing assembly.
3. Using new gaskets, install the lubricating oil pipe caps and torque to specifications.
4. Adjust the valve crossheads and the injector and valve lash.
5. Coat a new cover gasket with oil-resistant sealer and position it on the rocker housing covers. *Care must be taken in positioning the gasket to the rocker cover to prevent covering the crankcase vent or leaving a gap which would permit oil leakage.* Position the cover and gasket on the rocker housing, and install the retaining bolts and washers.
6. Operate the engine and check for leaks. When engine oil temperature is 140°F. remove the rocker housing covers and adjust the injector and valve lash.

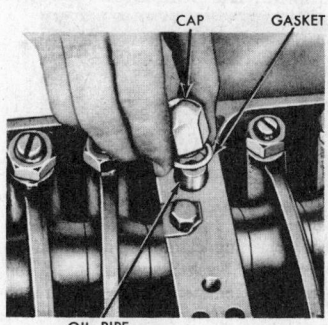

Removing oil pipe cap
(© Cummins Engine Co.)

Rocker Lever Assembly Removal

V-6 & V-8 Engines

Remove the valve cover.

Loosen the lock nuts and back off all of the valve and injector rocker lever adjusting screws. Hold the rocker levers together with rubber bands. Remove the rocker lever support screws and remove the rocker lever assembly(ies).

Installation

V-6 & V-8 Engines

Clean all parts before reassembly.

1. Back off the rocker lever adjusting screws. Lubricate the valve and injector rocker lever pads and adjusting screws with engine oil. Install the rocker lever assembly(ies) and torque the mounting screws to specifications (cylinder head bolt torque).
2. Adjust the crossheads and injector and valve rocker lash.
3. Install the valve cover(s), *and partially tighten the cover screws.*
4. Connect the accelerator rod to the fuel pump. Operate the engine until the oil temperature is 140°F. *Be sure the air reservoir drain cock is open.* Remove the valve cover(s) and adjust the injector and valve rocker lash. Install the valve cover(s) and cover screws. Tighten the cover screws in alternate steps.

Cam Follower Housing Assembly Removal

In-line Engines

Remove the rocker housing assembly(ies).

1. Remove the valve and injector push rods, and place them in a rack for installation in the proper sequence.
2. Remove the cam follower housing assembly for cylinders Nos. 5 and 6. Discard the gasket.
3. Remove the fuel pump. Remove the cam follower housing assembly for cylinders Nos. 3 and 4. Discard the gasket.
4. Remove the air compressor.
5. Remove the cam follower housing assembly for Cylinders Nos. 1 and 2. Discard the gasket.

Inspect and clean all parts before assembly.

NOTE: Do not discard metal spacer used with gasket on some models.

Installation

In-line Engines

1. Clean the gasket surfaces of the cam follower housings and cylinder block. Using new gaskets, in-

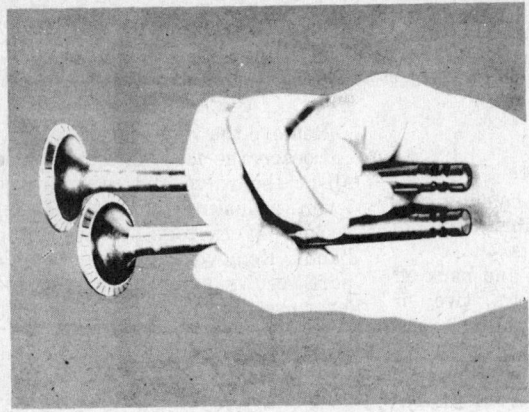

Pencil marks on valves
(© Cummins Engine Co.)

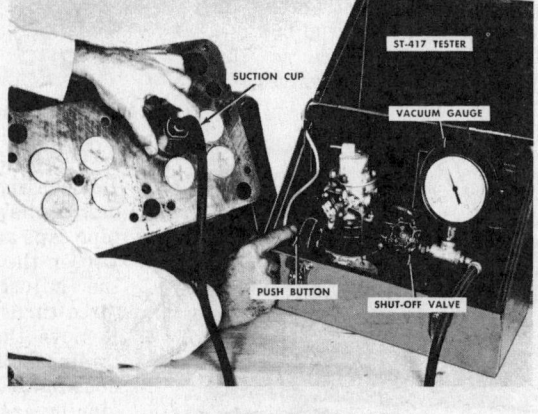

Vacuum testing valves for leaks
(© Cummins Engine Co.)

stall the cam follower housing assembly(ies), and tighten the retaining screws.

2. Install the valve and injector push rods in the same sequence that they were removed.

Valve Adjustment

The same crankshaft position used in setting the injectors is used for setting the intake and exhaust valves. Set valves after injectors are adjusted.

1. Loosen the locknut and adjusting screw. Insert a feeler gauge between the rocker lever and top of the valve stem or crosshead and turn screw down until the lever just touches the feeler gauge.

Valve clearances in-line except 1150 cid:

Valve	70°F	140°F
Intake	.016	.014
Exhaust	.029	.027

Valve clearances—1150 cid:

Valve	70°F	140°F
Intake	.014	.014
Exhaust	.027	.027

V6, V8 except	INT.	EXH.
903:	.012	.022
V8 903:	.012	.025

2. Lock adjusting screw in position with jam nut. Recheck valve clearance after tightening jam nut.

3. Continue turning crankshaft in direction of rotation, performing adjustments, until all valves are adjusted correctly.

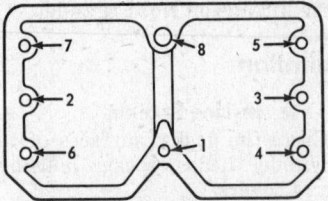

4-7/8" AND 5-1/8" BORE ENGINES
Rocker housing torque sequence
(© Cummins Engine Co.)

4. Make final adjustment when engine is at operating (140°F.) temperature.

5. Continue adjustments, each cylinder in firing order, until all cylinders are adjusted. This will require two complete revolutions of the crankshaft.

Time the Engine

NOTE: For complete timing specifications, see the chart at the beginning of this section.

855, 927, Engines

1. Install the special timing tool in injector sleeve. The short rod is engaged with the injector tube push rod, and the long rod rests on top of the piston.

2. Turn engine in normal direction of rotation until piston stops rising. Install indicator above piston so it is compressed to within .010" of full compression and zero the dial. Continue turning engine until the rod reaches the 90° mark on the left side of the tool.

3. Install the indicator above the

push tube using the procedure described in 2.

4. Turn engine backwards to about 45° before top dead center.

5. Turn engine in direction of normal rotation until the indicator above the piston reads .2032". The indicator above the push tube should match the figures shown in the table for 19°.

6. Continue turning the engine until the indicator above the piston reads .0816". The indicator above the push tube should match the reading for 12°.

7. Continue the procedure for a reading of .0143" and 5°.

8. If downward travel of pushtube is greater than limits, timing is slow. Correct by adding gaskets between follower housing and block on engines with right hand rotation, or by changing the camshaft key.

V6 and V8 Engines

Follow procedures described above, with figures modified as in the appropriate chart. On these engines, timing is varied with the camshaft key only.

Install the rocker housing(s) and adjust the injector and valve lash. Install the rocker housing covers.

Install air compressor and fuel pump.

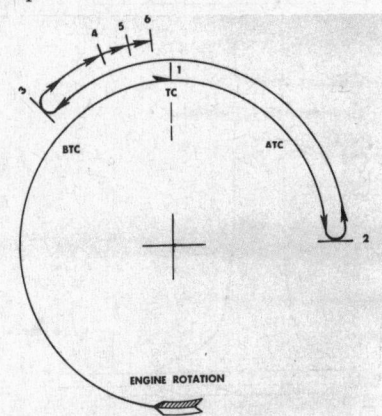

Injection Timing Procedure
(© Cummins Engine Co.)

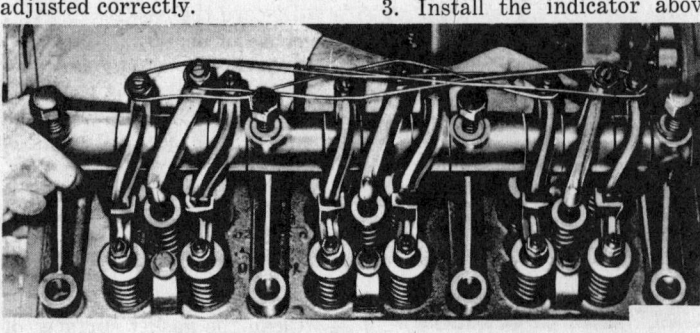

Rocker lever removal
(© Cummins Engine Co.)

Operate the engine and check for leaks. When the engine reaches normal operating temperature (140°F oil temperature), adjust the valve and injector lash to specifications.

1150 cid Engines

1. Adjustments to timing are made by changing camshaft keys.

Timing fixture ST-593 or equivalent and support block ST-593-40 or equivalent are used to obtain precise timing of push tube and piston travel.

2. Position timing fixture in injector well. Engage push rod indicator in injector push tube socket. Hand tighten hold-downs evenly.
3. If adaptor block is used, attach block to housing then secure rod indicator to block. Mount tools straight in cylinder and over push rod tube. Loosen indicators in their supports to prevent damage when barring engine.
4. Turn crankshaft in direction of rotation to bring piston to be checked to TDC firing position.
5. Position piston indicator to compress stem within .010″ of inner travel stop. Secure indicator.

Camshaft Key Data — Inch [mm]

Current Key Part Number	Replaces Key Part Number	Amount Of Offset	Push Rod Indicator Travel	Assembly Instructions	Timing Change	Relative Timing Affect
S-313	—	0.000 [0.0000]	0.0000 [0.0000]	—	—	—
200715	162055 & 172363	0.007 [0.1778]	0.0025 [0.0635]	Arrow toward rear of engine / Arrow toward front of engine	Adv. / Ret.	1/2° / 1/2°
200716	162056 & 172364	0.013 [0.3302]	0.0046 [0.1168]	Arrow toward rear of engine / Arrow toward front of engine	Adv. / Ret.	1° / 1°
200717	161154 & 172365	0.019 [0.4826]	0.0068 [0.1727]	Arrow toward rear of engine / Arrow toward front of engine	Adv. / Ret.	1-1/2° / 1-1/2°
200718	162057 172366	0.025 [0.6350]	0.0089 [0.2260]	Arrow toward rear of engine / Arrow toward front of engine	Adv. / Ret.	2° / 2°
200719	162058 172367	0.031 [0.7874]	0.0110 [0.2794]	Arrow toward rear of engine / Arrow toward front of engine	Adv. / Ret.	2-1/2° / 2-1/2°
200721	162059 172368	0.037 [0.9398]	0.0132 [0.3352]	Arrow toward rear of engine / Arrow toward front of engine	Adv. / Ret.	3° / 3°

Injection Timing Specifications

Model	Crank Angle	Piston Travel Inch [mm]	Push Rod Travel — Inch [mm] Nominal	Fast	Slow
V-903	21.0 deg BTC	-0.2032	-0.049	-0.046	-0.052
	13.5 deg BTC	-0.0816	-0.025	-0.022	-0.028
	5.5 deg BTC	-0.0143	-0.007	-0.004	-0.010
VT-903	21.0 deg BTC	-0.3032	-0.054	-0.051	-0.057
	13.5 deg BTC	-0.0816	-0.0295	-0.027	-0.032
	5.5 deg BTC	-0.0143	-0.0115	-0.0095	-0.0135

Injection Timing

Engine Model	Crank Angle (Degrees)	Piston Travel (Inches)	Push Tube Travel (Inches) Nominal	Fast	Slow
HRC-4, NHC-4, H-6, HR, HRF, NH-180, NH-195, NH-220	19° BTC	— 0.2032	— 0.0295	— 0.0275	— 0.0315
	12° BTC	— 0.0816	— 0.0150	— 0.0130	— 0.0165
	5° BTC	— 0.0143	— 0.0037	— 0.0020	— 0.0050
NHE-180, NHE-195	19° BTC	— 0.2032	— 0.0285	— 0.0275	— 0.0295
	12° BTC	— 0.0816	— 0.0140	— 0.0130	— 0.0150
	5° BTC	— 0.0143	— 0.0028	— 0.0020	— 0.0037
NH-250, NHE-225, NHD-230 NHC-250, NHCT-270-1,2,3 (before Serial No. 569212)	19° BTC	— 0.2032	— 0.0305	— 0.0295	— 0.0335
	12° BTC	— 0.0816	— 0.0160	— 0.0150	— 0.0185
	5° BTC	— 0.0143	— 0.0044	— 0.0037	— 0.0062
NT, NT-265, NT-280, NT-280 NT-200, NFT, HS, HRS, NHS, NHRS, NT-310 (Automotive and Marine)	19° BTC	— 0.2032	— 0.0335	— 0.0315	— 0.0355
	12° BTC	— 0.0816	— 0.0185	— 0.0165	— 0.0200
	5° BTC	— 0.0143	— 0.0062	— 0.0045	— 0.0075
NTO, NRT, NRTO, NTC-260, NTC-280, NTC-300, NTC-320, NTC-335, NT-380, NHCT-270-4* (Automotive and Marine)	19° BTC	— 0.2032	— 0.0360	— 0.0340	— 0.0380
	12° BTC	— 0.0816	— 0.0205	— 0.0190	— 0.0220
	5° BTC	— 0.0143	— 0.0080	— 0.0065	— 0.0095
NRT, NRTO, NT-335, NT-380 (Off-Highway)	19° BTC	— 0.2032	— 0.0415	— 0.0395	— 0.0435
	12° BTC	— 0.0816	— 0.0250	— 0.0230	— 0.0265
	5° BTC	— 0.0143	— 0.0115	— 0.0100	— 0.0130

Note: Procedures and values for timing horizontal engines are identical with corresponding standard models.

*NHCT-270 and NHCT-270HT after Serial No. 569212.

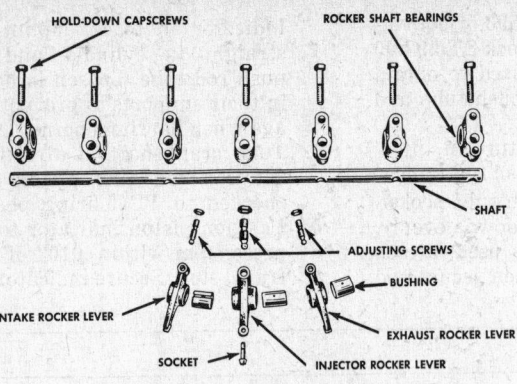

Rocker levers, shafts and bearings
(© Cummins Engine Co.)

Tappets and push rods
(© Cummins Engine Co.)

6. Check for exact O TDC with dial indicator.
7. Turn crankshaft in direction of rotation to 90 ATDC. Position pushrod indicator on follower to .020″ from inner travel stop. Secure indicator.
8. Turn crankshaft back to a position 40 B TDC.
9. Slowly turn engine in direction of rotation until piston indicator is positioned at .0032″ before 0. This is equivalent to .2032″ BTC.
10. Pushrod travel indicator should read:

Timing Code	Piston Travel	Push rod travel
AE	-.2032	.107-.109
AM	-.2032	.124-.128

Gear Case Cover and Seal

In-line Engines
Removal

1. Drain the cooling system and the crankcase.
2. When necessary release the cab locks and tilt the cab forward.
3. Remove the radiator and shutter assembly. Remove the front crossmember (radiator support) from the frame where necessary.
4. Remove the oil pan.
5. Remove the cooling fan. Remove all drive belts. Remove the fan hub and bracket assembly. Remove the fan drive pulley from the compressor crankshaft.
6. Remove the power steering pump bracket to engine bolts and position the pump and bracket assembly out of the way.
7. Remove the crankshaft damper capscrews, and pull the damper off the crankshaft. It may be necessary to tap the damper with a soft hammer in order to loosen it from the crankshaft. Remove the crankshaft flange capscrew and retainer, or nut and lock-

plate on some engines, and use puller to remove the flange.
8. Use a jack and wood block under the front of the engine to support it. Remove the engine front support retaining bolts, spacers and insulators. Remove the support bracket cap bolts and bracket cap. Raise the engine and remove the support bracket and insulators as an assembly.
9. Remove the compressor crankshaft pulley nut. Remove the compressor crankshaft pulley. Remove the crankshaft key.
10. If the gear case cover has an outboard camshaft bearing, remove it from the cover. Install guide studs in two of the capscrew holes. Remove the gear case retaining bolts. Using a soft hammer, loosen the gear case cover from the cylinder block, and remove the cover. Remove the seals from the cover. Remove the camshaft thrust-plate, O-ring seal and spacers, if so equipped.

Installation

1. Clean the oil pan and gear case cover with solvent. Be sure all gasket surfaces and oil seal recesses are clean. Remove and clean the oil pump inlet screen.
2. Apply oil-resistant sealer to the OD of the new cover seals (crankshaft and compressor shaft) and install the seals in the gear case cover. Use oil-

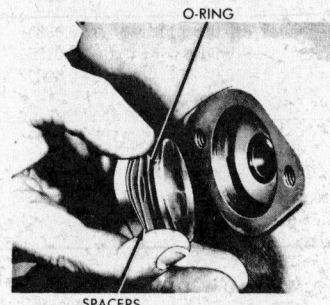

Installing camshaft thrust plate
(© Cummins Engine Co.)

resistant sealer to cement a new gasket to the gear case cover.
3. Apply lubricant to the oil seal lips and the seal surface of the compressor shaft. Install the gear case cover and tighten the retaining bolts.
4. Install the crankshaft and fuel pump-compressor drive shaft oil seals with special mandrels. Make sure crankshaft oil seal bore is concentric with shaft within .010″. Relocate cover, if necessary, to ensure concentricity, clearance between seal and shoulder on gear case must be at least .030″. Front face of seal must be square with crankshaft axis within .010″ to 45-50 ft-lbs. Make sure bottom surface is within .004″ of flush with bottom of block.
5. Position the engine front support and insulator assembly on the frame crossmember, and lower the engine onto the support bracket. Install the support bracket cap, lower insulators, spacers, and retaining bolts. Torque the bracket to frame bolts, and then the bracket cap bolts to specifications.
6. If equipped with a camshaft thrust plate, install it as follows: Hold the thrust plate, without spacers, in position against the camshaft, and use feeler gauges to check the clearance between the thrust plate and the gear case cover. Using micrometers, select enough spacers to bring the camshaft end clearance within specifications (.001″-.005″ on engines with thrust plates, .008″-.013″ on engines with outboard bearing supports). Place the spacers on the thrust plate. Apply Lubriplate to a new O-ring and install it on the thrust plate. Install the thrust plate on the gear case cover and tighten the screws.
7. Coat the seal surfaces of the crankshaft flange with lubricant and install the flange on the crankshaft. Install the flange

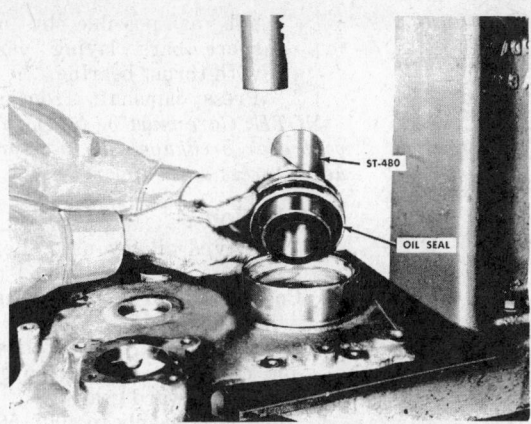

Installing crankshaft seal in gear case cover
(© Cummins Engine Co.)

Installing gear case cover
(© Cummins Engine Co.)

lockplate and lock nut and torque to specifications. Bend the lockplate to secure the nut. Install the crankshaft damper on the crankshaft flange and tighten the retaining screws.

8. Install the fan drive pulley on the compressor crankshaft.

9. Install the power steering pump and bracket assembly on the engine.

10. Install the fan hub and bracket assembly. Torque the screws to specifications. Position the cooling fan, alternator and power steering pump drive belts in the proper pulleys. Install the cooling fan.

11. Install the oil pan.

12. Install the front crossmember, and install the radiator and shutter assembly if removed.

13. Fill and bleed the cooling system. Fill the crankcase. Operate the engine and check for oil and coolant leaks.

Cylinder Front Cover Oil Seal Removal

V-6 & V-8 Engines

1. Drain the cooling system. Disconnect the battery cables.

2. Remove the shutter and radiator.

3. Remove the alternator, cooling fan and power steering pump drive belts.

4. Remove the cooling fan. Remove the crankshaft damper and the crankcase pulley.

5. Punch two holes in the cylinder front cover oil seal. Install a self-tapping screw in each

punched hole, and pry against the screws to remove the oil seal.

Installation

Apply oil-resistant sealer to the OD of a new oil seal. Lubricate the oil seal lips and the crankshaft with engine oil. Install the seal.

Reverse the removal procedure.

Camshaft

For best operation we do not recommend camshaft lobe regrinding.

Journal Wear

New camshaft journal diameter is 1.996"-1.997" on V-6 and small V-8, 2.496"-2.497" on V8 903, and 1.997"-2.496"-2.497" on V8 903, 1.997"-1.998" on 855 and 927 engines, and 2.996-2.997 on 1150 engines.

1. Replace camshaft if journals are smaller than specified.

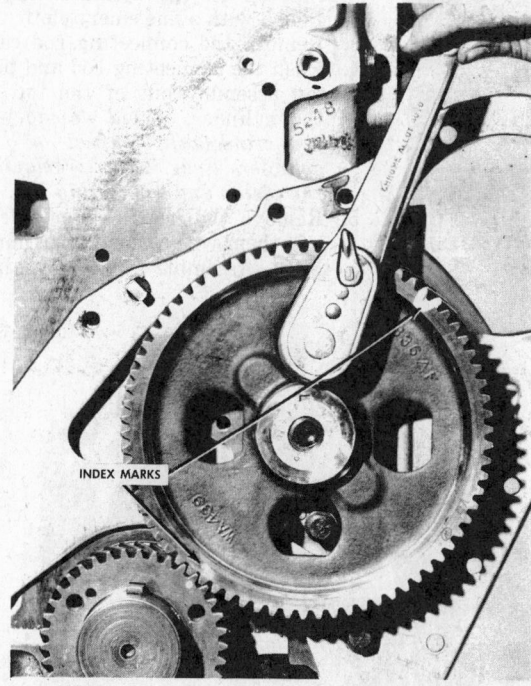

Tightening camshaft retaining bolts
(© Cummins Engine Co.)

INDEX MARKS

.007 TO .019

OIL SLINGER

Checking clearance between thrust washer
(© Cummins Engine Co.)

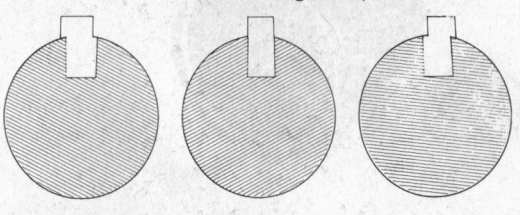

RETARD FROM STRAIGHT   STRAIGHT   ADVANCE FROM STRAIGHT

FROM GEAR CASE END OF CAMSHAFT
Camshaft gear keys from gear case end
(© Cummins Engine Co.)

ROTATE DURING REMOVAL

Removing camshaft
(© Cummins Engine Co.)

2. Replace any camshaft with scuffed, scored or cracked injector or valve lobes. Check by magnetic inspection for possible cracks.

Camshaft Gear and Thrust Bearing

The interference fit camshaft gear is keyed to the camshaft. The type of keys used are shown in illustration. The key controls engine timing limits, it may be straight or offset as shown.

The current camshaft thrust bearing is mounted with two instead of three capscrews; interchangeable with older bearing mounted with three capscrews.

Do not remove camshaft gear unless absolutely necessary. If inspection shows it is necessary to remove gear, do so by one of two following methods:

Camshafts with three capscrews mounted thrust bearing.

Removal

1. Heat gear to 300°-400°F.
2. Insert gear puller jaws through access holes in gear.
3. Tighten puller bolt until gear is free of camshaft.
4. Camshafts with two capscrew mounted thrust bearing.
   a. Place camshaft and gear in press. Support gear as near

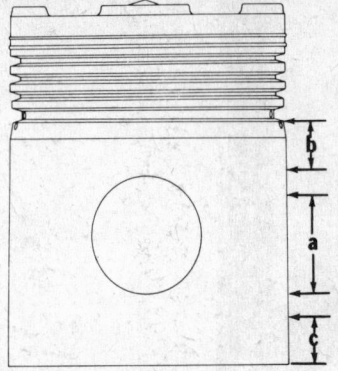

a. Pin Bore Area
b. Area Below Ring Groove
c. Piston Skirt

Piston Check Points (N20171)
(© Cummins Engine Co.)

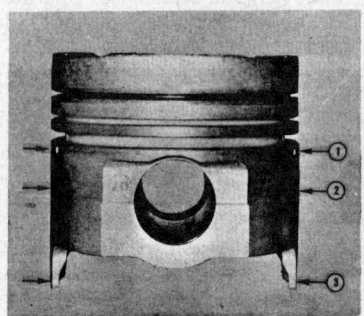

Piston Dimensions Check Points (V10167)
(© Cummins Engine Co.)

Piston Dimensions Check Points (V50125)
(© Cummins Engine Co.)

hub as possible by using short bars laying parallel with thrust bearing.
   b. Press camshaft from gear.
NOTE: Care must be taken to prevent gear breakage due to improper use of press and support bars.

Installation

1. Note type of key used. Replace key.
2. Coat gear hub area of camshaft with Lubriplate.
3. Install camshaft thrust bushing on camshaft with grooves in bushing toward gear.
4. Heat gear evenly to 400°/500°F. with heating torch (not a cutting torch).
5. Press on new camshaft gear, aligning gear keyway with camshaft key.
6. Check between camshaft flange and camshaft gear for .0015" maximum clearance for new thrust bearing. Check timing.

Pistons and Connecting Rods

Pistons and Connecting Rods Removal

Typical of All Engines

1. Drain the cooling system and the crankcase and remove the exhaust manifold(s), intake manifold(s), cylinder head or valve cover(s), cylinder head(s), oil pan, and oil pump.
2. Scrape all of the carbon from the top of the cylinder liners. Polish with a fine emery cloth.
3. Remove the connecting rod cap.
4. Push the connecting rod and piston assembly out of the top of the cylinder. *Avoid damage to the crankshaft journal or the cylinder liner when removing the piston and connecting rod.*
5. Remove the bearings and identify them with the cylinder numbers. Assemble the connecting

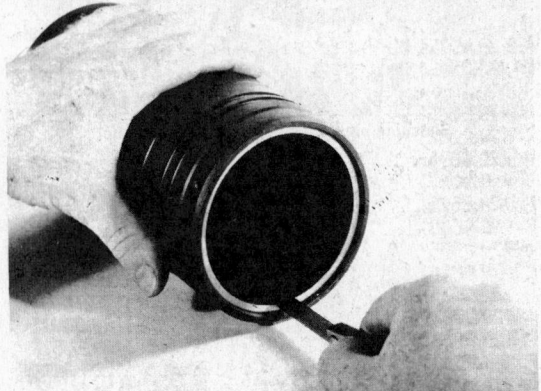

Checking piston ring gap
(© Cummins Engine Co.)

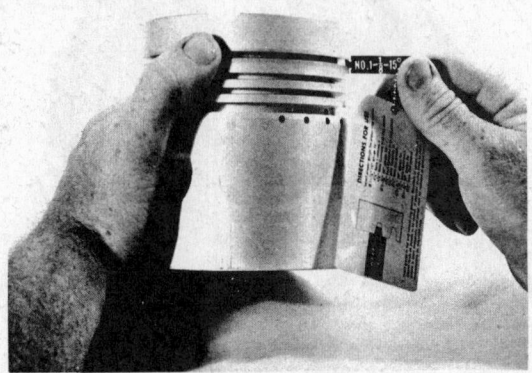

Checking ring groove wear
(© Cummins Engine Co.)

rod, cap, bolts and nuts as each rod and piston assembly is removed.

Disassembly

1. Using piston ring expander, remove the piston rings.
2. Remove the piston pin retaining rings. *Heat the pistons in hot water; then remove the piston pins. Do not use any force but pressure from the fingers to remove piston pins.*
3. Remove carbon deposits from the piston surfaces. Clean gum or varnish from the piston pins, piston skirt and piston rings with solvent. *Do not use a caustic cleaning solution or a wire brush to clean pistons. Make sure the oil ring holes are clean. Do not use a ring groove cleaning tool or a broken piston ring to clean the ring grooves. Use a soft bristle brush and cleaning solvent.*
4. Consult parts catalog for use of proper pistons and rings. Chrome plated compression rings are used only in the top groove, and should never be filed or used with chrome plated liners. Inspect the pistons, piston pins and piston rings as follows:
   a. Insert each ring into its cylinder liner, positioning it with the head of the piston. Measure the ring gap with a feeler gauge, and ensure that gap compares with figures in specifications section.
   b. Check the top and second ring grooves with the special gauge for each type piston. The shoulders of the gauge must not touch ring groove lands, or piston is not reusable. If the proper gauge is not available, use a segment of a new ring and a .006" feeler gauge. If gauge does not require force, do not use the piston.
   c. With pistons at 70-90°F., measure the skirt diameter at right angles to the pin bore. Measure at points shown in the appropriate illustration, and compare with specifications. Pistons must not be worn to less than 6.235" on 1150 cid engines, 5.483" on 855, 927 engines, 5.4850" on V8-903 engines. Use the chart below for V6 and small V8 engines.

Piston Part #	Check Points:	Wear Limit
173800	1	4.6130
185310	2, 3	4.6110
200800	2, 3	4.6110

   d. Check the piston pin bore at 70°F. It should be 2.3990 on 1150, 1.999" on 855, 927 engines, 1.750" on V8-903 engines, and 1.3479" on other engines.
5. Check piston pin diameter. It should be 2.398 on 1150 1.9978 on 855, 927 engines, 1.7478" on V8-903 engines and 1.3733" on V6 and small V8 engines. Permissible out of round limit is .001" on in-line engines, .0005" and V6 and V8 engines.

Assembly

1. Install one piston pin retaining ring in the piston. Heat the piston in hot water at 210°F for 15 min. and install the piston pin. Install the remaining piston pin retainer.
2. Install the piston rings on the piston, using a piston ring, expander. The side of the piston ring marked "TOP" must be toward the top of the piston. If new piston rings are installed, fit them to the piston and the cylinder bore. *Never use chrome-plated piston rings in a chrome-plated cylinder liner.*

Installation

1. Oil the piston rings, pistons and cylinder walls with clean engine oil. *Be sure to install the pistons in the same cylinders from which they were fitted. The numbers on the connecting rod and bearing cap must be on the same side when installed in the cylinder bore. If a connecting rod is ever transposed from one block or cylinder to another, new bearings should be fitted and the connecting rod should be numbered to correspond with the new cylinder number. On V6 and small V8 engines, install piston with cylinder or piston number, or word "out," toward exhaust side of engine.*
2. Make sure the ring gaps are properly spaced around the circumference of the piston but not in line with the piston pin. Install the piston ring compressor on the piston and push the piston in with a hammer handle until it is slightly below the top of the cylinder liner. *If a band-type ring compressor is used, make certain the inner band does not slip down and bind the piston, or ring breakage will occur.* Be sure to guide the connecting rods to avoid damaging the crankshaft journals. *Install the piston with the number on the rod toward the camshaft on in-line engines, or outside of block on V type engines.*
3. Turn the crankshaft throw to the bottom of its stroke. Install the connecting rod bearings, and check the clearance of each bear-

ing following the procedure under "Connecting Rod Bearing Replacement."
4. After the bearings have been checked and found satisfactory, apply a light coat of engine oil to the journals and bearings. Install the connecting rod cap. Torque the bolts to specifications. Lock the tab washers.
5. After the piston and connecting rod assemblies have been installed, check the connecting rod bearing side clearance.
6. Install the oil pump, oil pan, cylinder head(s), cylinder head cover(s), intake manifold(s), and exhaust manifold(s).
7. Adjust the valve crossheads; then adjust the injector and valve lash. Torque the rocker lever lock nuts to specifications.
8. Fill the cooling system and crankcase. Operate the engine until the oil temperature is 140°F. Remove the cylinder head covers and adjust the injector and valve lash. Check for oil, air, fuel and coolant leaks. Install the cylinder head covers.

Cylinder Liners

Cylinder Liner Removal

Typical—All Engines

1. Drain the cooling system and the crankcase.
2. Remove the following: coolant manifold, exhaust manifold, fuel manifold, intake manifold, rocker housing assembly(ies), cylinder head(s), oil pan and piston(s) and connecting rod(s).
3. Remove the cylinder liner, using liner puller (or equivalent). If more than one liner is removed, identify them for installation in the same position as removed.

Inspection and Repair

Inspect the cylinder liners and cylinder liner counterbore (crankcase) and repair as necessary. *Before dis-*

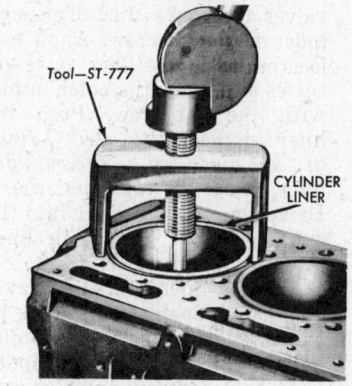

Tool—ST-777

CYLINDER LINER

Removing cylinder liner
(© Cummins Engine Co.)

Measuring cylinder liner counterbore depth
(© Cummins Engine Co.)

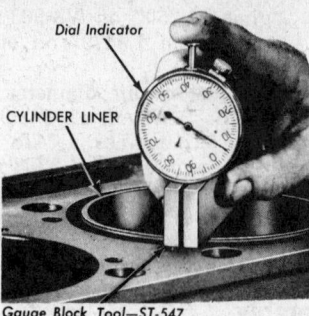

Checking cylinder liner protrusion
(© Cummins Engine Co.)

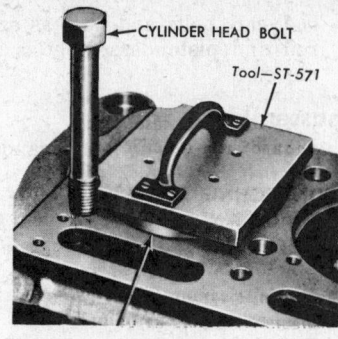

Aligning cylinder liner valve recess
(© Cummins Engine Co.)

carding a cylinder liner worn beyond the maximum wear limit, determine if it can be ground oversize for use with the next oversize piston. If it is necessary to grind the ID of one liner, inspect all liners for conformance to wear limits.

Check the diametrical clearance of the cylinder liner to cylinder bore in the liner O-ring seal area. *It must be within specifications.*

Installation

1. Check the cylinder liner protrusion by measuring the liner flange and liner counterbore. If necessary, install shims around the liner to provide the specified protrusion.

ENGINE	Protrusion
V6, V8	To eng. "F" .004-.007; After eng. "F" .006-.009
855, 927	.003-.006"
V8-903	.003-.0065"
K-1150	.003-.006"

2. Lubricate new cylinder liner O-rings with clean engine oil, and install them in the liner seal grooves. Be sure the O-rings are not twisted in the grooves.
3. Use engine oil to lubricate the machined portions of the liner bore in the cylinder block.
4. Carefully position the cylinder liner in the same cylinder bore from which it was removed. *Do not use beaded liners with steel-asbestos type head gasket.* Align valves as follows: Install one cylinder head capscrew. Align tool locators with the liner valve recesses and index the notch in tool with the capscrew. Push the liner in place. *Be sure O-rings are not displaced as the cylinder liner passes the top counterbore.*
5. Drive the press-fit liner into the cylinder block bore with liner driver and a hammer.
6. Check the liner protrusion at 4 equidistant points; it must be within specifications. Use a dial bore gauge to check the cylinder liner bore for roundness at several points within range of the piston travel. If the liner bore is

more than 0.002 inch out-of-round, remove the liner and check for possible binding condition which would cause distortion of the liner.

7. Using news gaskets, install the piston(s) and connecting rod(s), cylinder head(s), oil pan and rocker housing assembly(ies), and perform the necessary adjustments for the valve crossheads and injector and valve lash. Install the intake manifold, exhaust manifold, coolant manifold and fuel manifold.
8. Fill the cooling system and crankcase. Operate the engine and check for fuel, oil, coolant and air leaks. When the engine reaches normal operating temperature, check and adjust the injector and valve lash. Use the following specifics for liner counterbore.

Check Cylinder Liner Counterbore

1. Remove all corrosion and carbon deposits. These would affect readings obtained from block gauge.
2. Set gauge on block and depress indicator stem until indicator point touches block. Set dial to "0".
3. Position indicator in gauge block so stem moves about .10 inch to reach "0" on gauge. This leaves

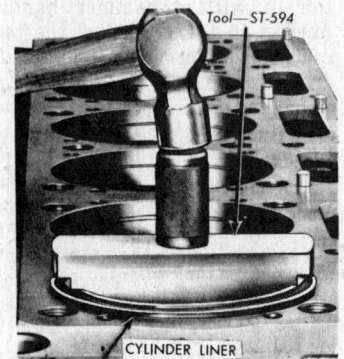

Installing cylinder liner
(© Cummins Engine Co.)

maximum amount of indicator travel available.

4. With indicator at "0", position gauge on deck so that indicator stem is over liner ledge.
5. Depress indicator stem till its point contacts liner ledge.
6. Read depth of counterbore shown on indicator.
7. Check each liner edge at four or more equi-distant locations.
NOTE: The liner ledge must not be "cupped" more than .00075 inch. ("Cupped" refers to the condition where that part of the cylinder liner ledge next to the liner is the highest point of the ledge surface). The depth must not vary more than .001 inch.
8. The counterbore at top of a new block is .30925/.31050 inch deep. If counterbore exceeds these limits oversize liners are available. *Only top ⅛ inch of counterbore provides press-fit.*
9. An installed cylinder liner must protrude .003-.006 inch above milled surface of block on in-line engines, .003-.006 inch on V8-903 engines, and .004-.007 inch on V6 and small V8 engines. To check protrusion *without* pressing liner into block:
   a. Measure liner flange outside bead with micrometer. *Do not include bead on top of liner flange in measurement.*
   b. Measure counterbore depth with dial indicator depth gauge.
   c. If measurement difference between two is not in proper range, add shims or counterbore block as necessary to achieve correct cylinder liner protrusion. **CAUTION: The following operation must be preceded by careful counter-bore depth measurement and calculation of shim thickness to be used after counterboring.** The .007 inch thick shim is the thinnest available. *Always check shim thickness before installing shim. Use as few shims as possible. Use one thick*

Checking counterbore depth
(© Cummins Engine Co.)

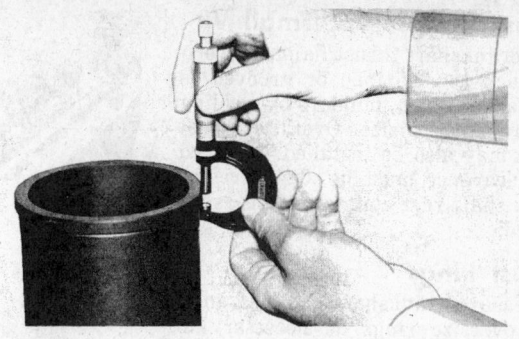

Measuring cylinder liner flange
(© Cummins Engine Co.)

shim in place of two thinner shims. Never use shims thinner than .007 inch.

Counterbore Cylinder Liner Ledge

(A special tool for this is available from Cummins)

1. Loosen adjustable tool head lock collar and remove tool holder.
2. Place upper adapter in position and fasten with capscrews.
3. Reposition tool holder, tightening adjustable tool head lock collar finger tight.
4. Install lower adaptor plate on tool with capscrews.

Installing Tool in Block

1. Raise tool bit until tip is flush with tool holder. *Tool bit set screws are not to be loosened to raise or lower tool bits.*
2. Make sure top and bottom adjustable locating pins are retracted and flush with housing before inserting tool in bore. The socket head screws on upper adaptor plate, above pins control this movement.
3. Insert tool into block bore so locating surfaces are on water jacket side of block. Position top of upper adaptor plate 1/32 inch below counterbore.
4. Tighten adjustable locating pins against bore with socket head wrench and extension.
5. Insert wooden block between capscrews inserted in block to prevent tool head from rotating counter-clockwise.
6. Insert handle in socket and rotate counterclockwise to loosen adjustable tool head lock collar.
7. Move tool head out until tool bit contacts counterbore edge.
8. Rotate collar clockwise, by hand, to lock.
9. Lower tool bit until it engages counterbore and remove wooden block. Rotate head clockwise to see if it binds.
10. Replace wooden block between capscrews. Check tool head posi-

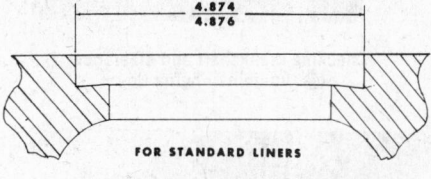

FOR STANDARD LINERS

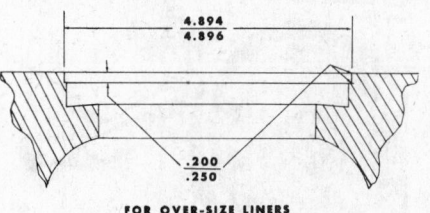

FOR OVER-SIZE LINERS

Oversize press-fit liner counterbore
(© Cummins Engine Co.)

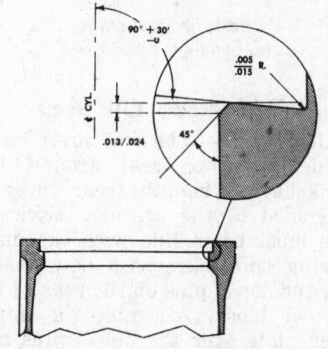

Cross section of new block counterbore
(© Cummins Engine Co.)

tion then lock in place by turning handle clockwise.

Counterbore block

1. Screw tool bit down until it touches counterbore seat.
2. Insert handle and rotate slowly, clockwise, until tool bit stops cutting.
3. If necessary, continue cutting by feeding tool bit down. Feed dial graduations are approximately .002 inch. Cut only ½ graduation at a time. Remove only enough metal to clean up counterbore.
4. Check counterbore depth with dial indicator depth gauge. Measure cylinder liner flange.

Do not include head on liner.
5. Subtract flange thickness from counterbore depth and calculate thickness of shim needed to provide proper liner protrusion above block.

Lapping Counterbore in Block

(Non-Press-fit liners only)

If the cylinder liner flanges and counterbore are not perfectly smooth, the counterbore may be lapped.

1. Place Grade A grit lapping compound on counterbore ledge.
2. Insert cylinder liner.
3. Apply light and even pressure while lapping. Rotate liner one complete revolution, then lap in 30° strokes. Add a few drops of lube oil each 2 or 3 revolutions to keep lapping compound moist.
4. *Remove all lapping compound from counterbore and liner flange.*
5. Apply light coat of prussian blue and check seat. *A full seat must be indicated on both counterbore and liner flange.*
6. Repeat lapping operation as necessary.

Crankshaft

Journals

Crankshafts which are scored or scratched should be reground to standard undersize to fit available standard main bearings and connecting rod bearings. Main bearings are available in .010, .020, .030, and .040 inch undersize. Undersize connecting rod bearings are available in corresponding undersizes.

Oil Passages

1. Use a rod and rag, just as you would to clean a rifle barrel, to check and completely clean oil passages in crankshaft.
2. Replace pipe plugs.
   a. Coat threads with sealer.
   b. Tighten to 5 ft. lbs.
   c. Stake to prevent loosening.

Thrust Flange—V6, Small V8

If crankshaft thrust flange is worn as much as .005 inch or grooved as deep as .005 inch, build up by welding and regrind to size. Oversize thrust rings may also be installed. Installed end clearance must be .014″ or less. Mark shaft rear counterweight as required.

Thrust Rings

Measure crankshaft end clearance. Use oversize rings as necessary to create .022″-.035″ clearance on 855 engines, .004″-.016″ on 1150 engines, .005″-.015″ clearance on 903 engines, and .022″-.030″ clearance on 927 engines. Mark ring size and location on rear counterweight.

Crankshaft Gear In-Line Engines

If it is necessary to remove crankshaft gear:
1. Attach puller to gear.
2. Apply 75/100 ft. lbs. on puller screw.
3. Heat gear with a heating torch—not a cutting torch—to 300°/400°F. The gear will expand, making it easier to pull.

To assemble gear:
1. Heat gear evenly to 400°/500°F with a heating torch (not a cutting torch).
2. Drive onto crankshaft with tubing.

Crankshaft Gear V-Type Engines

If the gear is damaged, remove it with a hammer and chisel. Place end of chisel between teeth above the key-way and strike several times to loosen the gear.

General Inspection

1. Inspect journals and thrust flange for scoring or scratching and have the shaft reground if they are present.
2. Gauge crankshaft journals for conformity to specifications.
3. Check all journals for out-of-round exceeding .002″. Regrind to correct excessive out-of-round.

Checking crankshaft end clearance
(© Cummins Engine Co.)

Removing rear cover
(© Cummins Engine Co.)

Crankshaft Rear Oil Seal

All Engines—The rear cover has a braided asbestos seal around the crankshaft. When the rear cover is assembled to the cylinder block its bore must be in line with the main bearing bore. Cement a new gasket over the dowel pins on the rear of the cylinder block. Assemble the upper cover plate over the dowel pins and tighten the four cup screws. With crankshaft pushed toward the cover plate, check the clearance between the crankshaft flange and the opposing face of the cover place.

Clearance should be a minimum of .004″. The clearance of .004″ to .006″ can be obtained by using gaskets between the cover plate and the cylinder block.

The clearance between the crankshaft and bore of the cover plate should be .009 to .011″ (.006″ on in-line) around the entire circumference. This clearance is provided when the original rear cover is installed at the factory and maintained by locating dowels. This can be checked after the old asbestos seal has been removed by using a .009″ feeler gauge between the shaft and upper cover.

Remove the cover plate and install new asbestos seals into the recess of both upper and lower cover plates. New seals must be beveled on three sides of each end 13/20″ by 45°, leaving the inside edge square. The ends of the seals should extend 1/32″ above the joining surfaces of the cover plates. Seals should be soaked in lubricating oil an hour or more before installing. Make sure they are well seated in their grooves.

Cement new connecting gaskets to the lower cover plate. Assemble the lower plate to the upper plate and bolt together tightly. Install capscrews to cover plates and pull up evenly to the cylinder block over the dowel pins.

Main and Connecting Rod Bearings

All Cummins main and connecting rod bearings are precision parts with shell thickness, bearing material and bearing crush accurately calculated and held to close tolerances. Both standard and undersized bearings are provided with recommended clearance for the oil film. *Under no circumstances should any attempt be made to scrape or ream these bearings nor should they be lapped or filed to increase oil clearances.*

A properly fitted bearing, after a reasonable period of service, will appear dull gray; indicating it is running on an oil film. Bright spots indi-

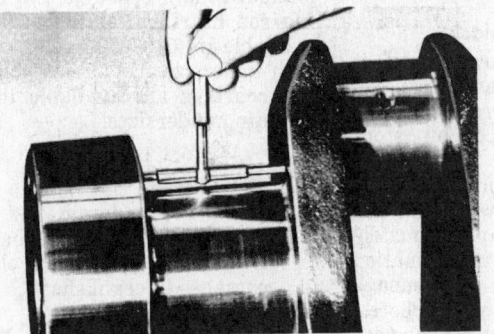

Checking thrust flange wear
(© Cummins Engine Co.)

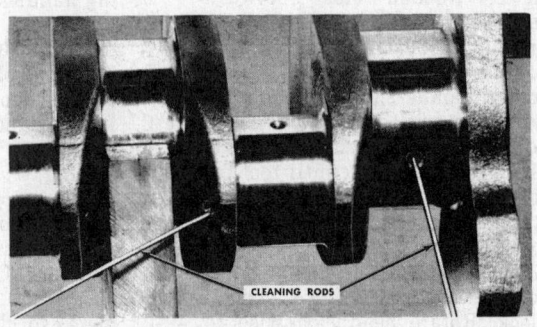

CLEANING RODS

Cleaning oil passages
(© Cummins Engine Co.)

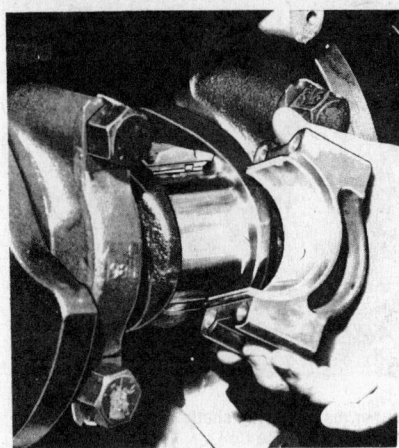

Installing connecting rod gap
(© Cummins Engine Co.)

cate metal-to-metal contact; black spots excessive clearance.

CAUTION: Never use shims with main or connecting rod bearing shells.

Main Bearings

All current engines have grooveless lower main bearing shells and grooved upper shells. Grooveless lower main bearing shells can be installed in engines with serial numbers previous to Serial No. 113353 provided the crankshaft journals do not have prominent ridges formed from wear with the grooved type lower shells. If the ridge is higher than .001 inch, it will be necessary to install grooved shells until the crankshaft is reground.

Main bearing shells are held in place by a locking tang or lip on each shell which mates with milled recesses in the block and cap opposite the camshaft side of the engine.

NOTE: If grooved shells are to be reused, they must be installed in the same position as removed.

Inspection

Inspect connecting rod and main bearing shells for the following defects:
1. Fatigue, corrosive damage, or pitting in the loaded area.
2. Evidence of overheating, i.e., badly discolored or burnished surface.
3. Excessive lead wiping.
4. High spot due to dirt behind bearing.
5. Cracks on the bearing surface or on back of shell.
6. Excessive edge loading or fretting.

Check scratches on the bearing surface. Scratches deep enough to result in ridges on either side of scratch will cause contact during operation and lead to engine failure.

It is not necessary to discard shells due to *minor* discoloration, eroded

lead plate, slight scratches or imperfection outside the load area.

Gauge bearing shells with a ball point micrometer or equivalent. Replace if worn beyond limit shown below.

855, 927	.1215″
903	.15300″
1150	.1690″
V6, V8	
Serial Prefix D	.12275″
Serial Prefix F, G	.12250

Connecting Rod Bearing Replacement

Typical—All Engines

Bearing clearance can be checked by using micrometers or by using Plastigage as outlined in the following procedure. If the Plastigage method is employed, use new standard-size bearings.

Worn limits on bearing shell thickness are as shown below:

855	.071″
1150	.1230″
927	.093″
903	.09295″
V6, V8	.007″

1. Drain the crankcase. Remove the oil pan; if necessary, remove the oil pump.
2. Remove the cap from the connecting rod to which new bearings are to be fitted. Push the piston part way up the cylinder.
3. Remove the bearings from the connecting rod and cap. Clean the crankshaft journal, connecting rod and bearing cap.
4. Install new bearings in the connecting rod and cap. Pull the connecting rod assembly down firmly on the crankshaft journal.
5. Place a piece of Plastigage on the lower bearing surface, the full width of the cap and about ¼ inch off-center.
6. Install the cap and torque the connecting rod nuts to 140 ft-lbs. *Do not turn the crankshaft while the Plastigage is in place.*
7. Remove the cap. then using the Plastigage scale, check the width of the Plastigage. When checking the width of the Plastigage, check at the widest point in order to get the minimum clearance. The difference between the two readings is the taper.
8. Check the bearing side clearance; it must be to specifications. After the bearing clearance has been checked and found to be within the specified limits, clean the Plastigage from the crankshaft journal. *If standard-size bearings or the next undersize bearing does not provide the specified clearance, then crankshaft grinding or replacement is indicated.*

9. Apply Lubriplate to the journal and bearings, then install the connecting rod cap. Install new lock plates and torque the rod nuts to specifications. Bend the tabs on the lock plates to secure the nuts.
10. Repeat the procedure for the remaining connecting rods that require new bearings.
11. Install the oil pump. Install the oil pan.
12. Fill the crankcase. Operate the engine at fast idle and check for oil leaks.

Main Bearing Replacement

In-line Engines

Bearing clearance can be checked by using micrometers or by using Plastigage as outlined in the following procedure. If the Plastigage method is employed, use new standard-size bearings.

1. Drain the crankcase. Remove the oil pan.
2. Remove the main bearing cap to which new bearings are to be installed. *Replace one bearing at a time, leaving the other bearings securely fastened.*
3. Remove the thrust washers. Rotate the crankshaft with barring tool, and force the upper bearing out of the block.
4. Remove the bearing from the cap. Clean the crankshaft journal and bearing inserts.
5. All upper main bearings are grooved and drilled for lubrication. On 855,927 engines, numbers 1, 3 and 5 are alike and Nos. 2, 4 and 6 are alike. Number 7 has the oil groove off center, and it must not be interchanged with the other bearings. On 1150 engines, nos. 2, 3, 4, 5, 6 are interchangeable; nos. 1 & 7 are interchangeable. The wide portion of the bearing is installed toward the flywheel end.
6. Apply lubricant to the upper main bearing, and insert it between the crankshaft journal and crankcase web. Rotate the crankshaft until the bearing seats itself. *The groove on each bearing for the dowel ring must match with the counterbore at the capscrew hole on the exhaust manifold side of the block.*
7. Install the upper thrust rings with the babbit sides *next to the crankshaft flanges.* The upper thrust rings are not doweled to the block. The doweled lower halves prevent them from turning.
8. Replace the cap bearing. The lower main bearings are plain with no grooves or drillings.
9. Support the crankshaft so that its weight will not compress the

Cummins Diesel Engines

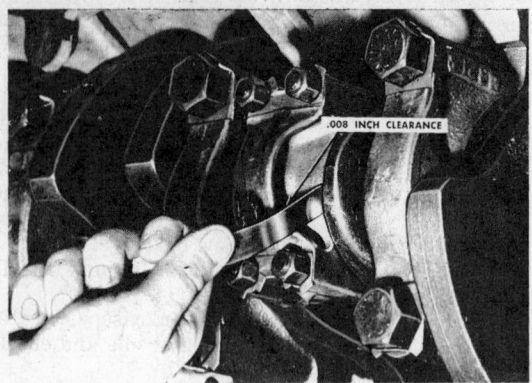

Checking rod to crankshaft clearance
(© Cummins Engine Co.)

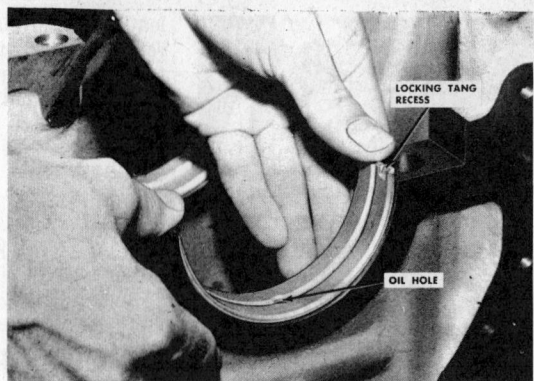

Laying upper main bearing shells
(© Cummins Engine Co.)

Plastigage and provide a false reading. Position the support so that it will bear against the counterweight adjoining the bearing which is being checked.

10. Place a piece of Plastigage on the bearing surface the full width of the bearing cap and ¼ inch off-center.

11. On the No. 7 (rear) main bearing, install the lower thrust rings. *The lower thrust rings must be held by the dowels in the main bearing cap.*

12. Install the cap and torque the bolts to 150 ft-lbs. *Do not turn the crankshaft while the Plastigage is in place.*

13. Remove the cap, then using the Plastigage scale, check the width of the Plastigage. When checking the width of the Plastigage, check at the widest point in order to get the minimum clearance. Check at the narrowest point in order to get the maximum clearance. The difference between the two readings is the taper.

14. After the bearing clearance has been checked and found to be within the specified limits, clean off the Plastigage. *If standard-size bearings or the next under-size bearing does not provide the specified clearance, then crank-*

shaft grinding or replacement is indicated.

15. Apply lubricant to the bearing and journal. Install the bearing cap, using new lock plates.

16. Torque the bolts to specifications. Bend the tabs on the lock plates to secure the bolts.

17. Repeat the procedure for the remaining bearings that require replacement.

18. Install the oil pan.

19. Fill the crankcase. Operate the engine at fast idle and check for oil leaks.

V-6 & V-8 Engines

Bearing clearance can be checked by using micrometers or by using Plastigage as outlined in the following procedure. If the Plastigage method is employed, use new standard-size bearings.

1. Drain the crankcase. Remove the oil pan and oil pump.

2. Remove the main bearing cap side bolts from both sides of the engine block.

3. Remove the cap bolts from the main bearing to which new bearings are to be installed. Use a small pry-bar to loosen the bearing cap. Remove the bearing cap. *Replace one bearing at a time,*

leaving the other bearings securely fastened.

4. Rotate the crankshaft with the barring tool and force the upper bearing out of the block. On the rear main bearing, remove the upper thrust half-rings from the block. On the rear main bearing, remove the thrust half-rings cap.

5. Remove the bearing from the from the bearing cap dowels. Clean the crankshaft journal and bearing inserts.

6. Only the upper main bearings are grooved and drilled for lubrication. Numbers 1 and 4 are alike and Nos. 2 and 3 are alike.

7. Apply Lubriplate to the upper main bearing, and insert it between the crankshaft journal and crankcase web. Rotate the crankshaft until the bearing seats itself. *The groove on each bearing for the dowel ring must match with the counterbore at the capscrew hole on the exhaust manifold side of the block.*

8. Install the upper thrust rings with the *grooved side next to the crankshaft flange.* The upper thrust rings are not doweled to the block; the doweled lower halves prevent them from turning.

9. The lower main bearings are plain with no grooves or drill-

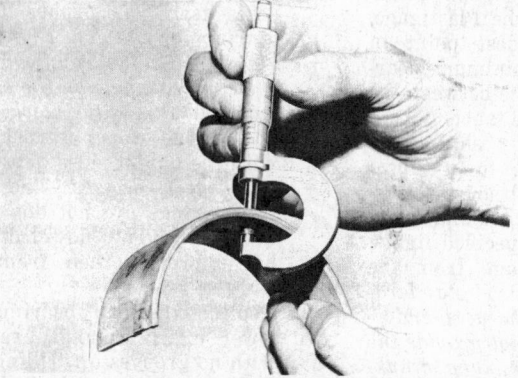

Measuring main bearing shells
(© Cummins Engine Co.)

Installing thrust rings
(© Cummins Engine Co.)

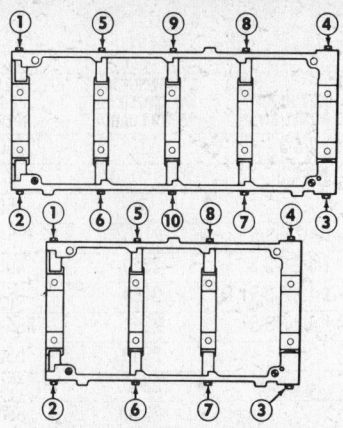

Main bearing side cap screw torque sequence
(© Cummins Engine Co.)

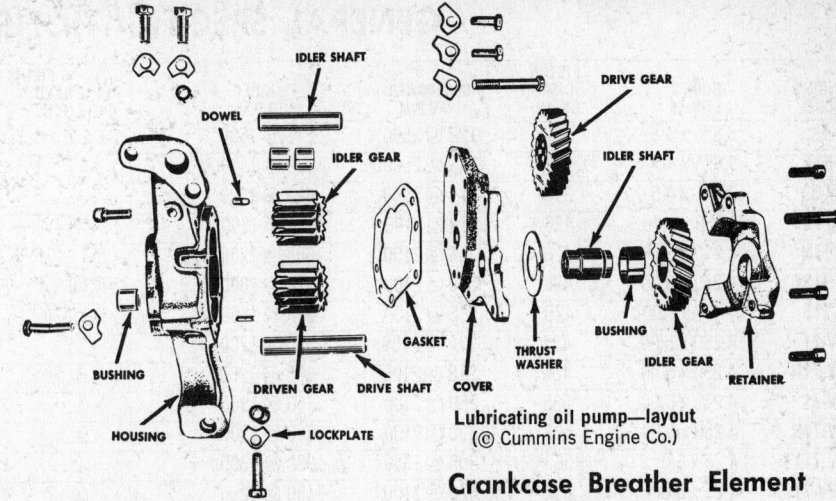

Lubricating oil pump—layout
(© Cummins Engine Co.)

ings. Numbers 1 and 4 are alike and Nos. 2 and 3 are alike.

10. Support the crankshaft so that its weight will not compress the Plastigage and provide an erroneous reading. Position the support so that it will bear against the counterweight adjoining the bearing which is being checked.

11. Place a piece of Plastigage on the bearing surface the full width of the bearing cap and ¼ inch off-center.

12. On the rear main bearing, install the lower thrust rings on the dowels in the main bearing cap.

13. Install the cap and torque the bolts to 180 ft-lbs. *Do not turn the crankshaft while the Plastigage is in place.*

14. Remove the cap, then using the Plastigage scale, check the width of the Plastigage. When checking the width of the Plastigage, check at the widest point in order to get the minimum clearance. Check at the narrowest point in order to get the maximum clearance. The difference between the two readings is the taper.

15. After the bearing clearance has been checked and found to be within the specified limits, clean off the Plastigage. *If standard-size bearings or the next under-size bearing does not provide the specified clearance, then crankshaft grinding or replacement is indicated.*

16. Apply Lubriplate to the bearing and journal. Install the bearing cap, using new lock plates.

17. Torque the bolts to specifications. Bend the tabs on the lock plates to secure the bolts.

18. Repeat the procedure for the remaining bearings that require replacement.

19. Torque the main bearing side capscrews to specifications, following the sequence.

20. Install the oil pump and oil pan.

21. Fill the crankcase. Operate the engine at fast idle and check for oil leaks.

Engine Oil Pump

In-Line Engines

The standard oil pump in these engines is of the two-gear type. In servicing these pumps, note the following:

1. The ball bearing outer race must not turn in the ball bearing cage.

2. The gears must show no excessive wear, or be scored or otherwise damaged.

3. The drive shaft must be replaced if it has been turning in the inner race of the ball bearing; if it is worn smaller than .8735″ where it bears against the bracket bushing, or if it is worn smaller than .8705″ on the body end.

4. The bracket bushing should not be used if its inside diameter is worn larger than .887″.

5. Idler pins should be replaced if worn smaller than .873″.

6. Replace bracket or cover if finished surfaces are scored or visibly worn larger than 2.301″ inside diameter. Maximum gear to pocket clearance should never exceed .008″ or .004″ on a side.

7. The by-pass valve in the body should not open below 75 lbs.

8. When assembling the pump, the pump gear should be pressed on the shaft so there is a clearance of .0015″ between the gear and bracket.

Typical oil filter and luber-finer installation
(© Cummins Engine Co.)

Crankcase Breather Element Replacement

The crankcase breather contains a chemically-treated paper element that must be changed at the specified interval. *Do not attempt to clean the element.*

Remove the wing nut, washer and gasket. Lift out the cover and element. Discard the element. Clean the breather housing and cover with a clean cloth. Install a new element and the cover on the housing. Install the gasket, washer and wing nut.

Oil Filter Replacement

The standard full-flow oil filter is remote-mounted. The optional by-pass oil filter (Luber-Finer) is also remote-mounted.

Full-Flow Oil Filter

To change the standard oil filter element, remove the element from the case and wipe it with clean cloths. *Do not wash the filter bag.*

Reverse the bag and inspect it. If bearing metal or grit is found, immediately inspect the connecting rod and main bearings. Thoroughly wash all parts, except the bag. Install new gaskets and filter bag. If the filter bag is incorrectly installed, the studs may pierce the filter bag.

By-Pass Filter

To change the by-pass filter element (Luber-Finer): Drain the oil, remove the cover, unscrew the T-handle pack hold-down assembly and lift out the pack(s). After cleaning the housing, replace the drain plug, install new pack(s), and replace the hold-down assembly.

Check the O-ring gasket for cleanliness and correct position. Replace the O-ring if it is damaged. Replace the cover and clamping ring. Tighten the capscrew until the clamping lugs come together.

Add an extra 3¼ gallons of oil to the crankcase to fill the filter case and element. Loosen the cover vent plug and start the engine. When the oil reaches the vent, close the plug.

GENERAL SPECIFICATIONS

ENGINE MODEL	BORE & STROKE	PISTON DISP. CU. IN.	HORSEPOWER* @ R.P.M.	TORQUE* @ R.P.M.	FIRING ORDER* RIGHT HAND ROTATION	FIRING ORDER* LEFT HAND ROTATION	GOVERNOR SPEED NO LOAD	IDLE R.P.M. SPEED
4-53	3.875 x 4.5	212	122 @ 2500	271 @ 1500	1-3-4-2	1-2-4-3	①	450
4-53T	3.875 x 4.5	212	170 @ 2500⑤	402 @ 1800⑥	1-3-4-2	1-2-4-3	①	①
6V-53	3.875 x 4.5	318	171 @ 2500	423 @ 1500	③	③	①	450
6-71E	4.250 x 5.00	426	219 @ 2100	574 @ 1600	1-5-3-6-2-4	1-4-2-6-3-5	2150	385-400
6-71N	4.250 x 5.00	426	218 @ 2100	605 @ 1200	1-5-3-6-2-4	1-4-2-6-3-5	2340	400
6-71SE	4.250 x 5.00	426	210 @ 2100	575 @ 1600	1-5-3-6-2-4	1-4-2-6-3-5	2150	385-400
6-71T	4.250 x 5.00	426	235 @ 2100	630 @ 1600	1-5-3-6-2-4	1-4-2-6-3-5	2150	385-400
6V-71	4.250 x 5.00	426	210 @ 2100	565 @ 1200	③	③	2340	550
6V-71N	4.250 x 5.00	426	218 @ 2100	605 @ 1200	③	③	2340	550
8V-71	4.250 x 5.00	568	280 @ 2100	750 @ 1200	③	③	2340	550
8V-71N	4.250 x 5.00	568	290 @ 2100	800 @ 1200	③	③	2340	550
8V-71TT	4.25 x 5.0	568	305 @ 1950	1038 @ 1300	③	③	2300	550
12V-71	4.25 x 5.0	852	420 @ 2100	1130 @ 1200	③	③	2300②	550
6V-92TT	4.84 x 5.0	552	270 @ 1950	958 @ 1200	③	③	①	500④
8V-92TT	4.84 x 5.0	736	365 @ 1950	1223 @ 1400	③	③	①	500④

① See service plate
② 1973 engines—2100
③ See Firing order diagrams
④ City buses—400
⑤ 155 with 5A55 injectors
⑥ 379 with 5A55 injectors
* Ratings are average; engines may be rated according to specification

CRANKSHAFT AND BEARING SPECIFICATIONS

ENGINE SERIES	MAIN BEARING JOURNAL DIAMETER	MAIN BEARING CLEARANCE	SHAFT END THRUST ON BEARING NO.	SHAFT END PLAY	ROD BEARING JOURNAL DIAMETER	ROD BEARING CLEARANCE
53(T)	2.999-3.000	.0013-.0042	Rear	.004-.011	2.449-2.500	.0015-.0046
V-53	3.499-3.500	.0025-.0057	Rear	.004-.011	2.749-2.750	.0016-.0046
71(T)	3.499-3.500	.0014-.0044	Rear	.004-.014	2.749-2.750	.0014-.0044
V-71(T)	4.499-4.500	.0012-.0046	Rear	.004-.011	2.999-3.000	.0014-.0044
6V-92TT	4.499-4.500	.0016-.0050	Rear	.004-.011	2.999-3.000	.0010-.0040
9V-92TT	4.499-4.500	.0016-.0050	Rear	.004-.011	2.999-3.000	.0010-.0040

PISTONS, PINS, RINGS SPECIFICATIONS

ENGINE SERIES	PISTON CLEARANCE TOP OF SKIRT	PISTON CLEARANCE BOTTOM OF SKIRT	PISTON PIN DIAMETER	PISTON TO PIN BUSHING CLEARANCE	RING END GAP COMP.	RING END GAP OIL	CYLINDER LINER INSIDE DIAMETER
53, V53	.0052-.0083	.0052-.0083	1.3746-1.3750	.0025-.0034	.020-.046	.010-.025	3.8752-3.8767
71, V71	.004-.0078	.004-.0078	1.4996-1.5000	.0025-.0034	.018-.043	.008-.025	4.2495-4.2511
71T, V71T	.006-.0098	.006-.0098	1.4996-1.5000	.0025-.0034	.018-.043	.008-.025	4.2495-4.2511
V-92TT	.0055-.0093	.0055-.0093	1.4996-1.5000	.0025-.0034	.025-.040	.025-.040	4.8395-4.8411

ENGINE TORQUE SPECIFICATIONS

ENGINE SERIES	CYLINDER HEAD FT. LBS. CAST IRON BLOCK NUTS	CYLINDER HEAD FT. LBS. CAST IRON BLOCK BOLTS	MAIN BEARING BOLTS FT. LBS.	CONN. ROD NUTS FT. LBS.	ROCKER SHAFT FT. LBS.	VALVE BRIDGE GUIDE FT. LBS.	CRANKSHAFT END BOLT FT. LBS.
53	——	170-180	120-130	50-55	50-55	——	200-220
V-53	——	170-180	120-130	24-28	50-55	——	200-220
71(T)	175-185①	175-185	180-190②	60-70③	90-100	46-50	290-310
V-71(T)	175-185	175-185	180-190	65-75	90-100	46-50	290-310
V-92TT		230-240	230-240	60-70	90-100④	46-50	290-310

①—Aluminum block nuts 140-160.
②—Aluminum block 120-140.
③—Castellated nut 65-75.
④—Load limit or power control bolts—75 to 85

FIRING ORDER AND ROTATION

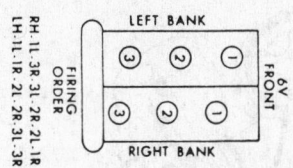
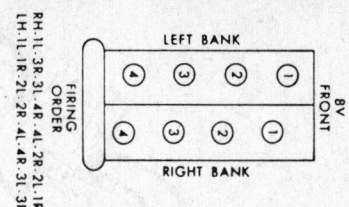
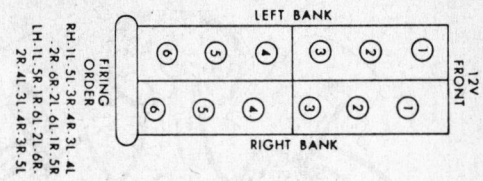

Trouble Diagnosis

The satisfactory performance of a Diesel engine depends on two items of foremost importance, (1) Sufficiently high compression pressure, and (2) The injection of the proper amount of fuel at the right time.

The first one of these items depends entirely on pistons, piston rings, valves and valve operating mechanism; the second item depends on injectors and their operating mechanism, and fuel system.

Lack of engine power, uneven running, excessive vibration, and tendency to stall when idling may be caused by either a compression loss or faulty injector operation.

The causes of trouble symptoms may be varied; therefore a hit-and-miss search should be avoided. A proper diagnosis of symptoms is an orderly process of diagnosing the symptoms. An "orderly process" means to check the most probable common cause first; then proceed with the next probable cause.

Low Starting R.P.M.

1. Poor or shorted electrical connections.
2. Undercharged or defective battery.
3. Faulty starter or loose starter connections.
4. Infrequent oil changes—change oil.

Low Compression

1. Exhaust valves sticking or burned.
2. Compression rings worn or broken.
3. Cylinder head gasket leaking.
4. Improper valve clearance adjustment.
5. Blower not functioning.
6. Emergency shutoff valve partially or completely closed.

Fuel Problems

1. Injector racks not in full position.
2. Fuel tank or fuel filter empty.
3. Improper grade and type fuel.
4. Diluted fuel.
5. Improperly filtered or restricted fuel flow.
6. Air in fuel system.
7. Faulty fuel pump.

Uneven Running or Frequent Stalling

1. Cylinders cutting out due to incorrect valve adjustment.
2. Insufficient fuel supply.
3. Faulty injectors.
4. Improper injector timing.
5. Incorrect rack setting.
6. Injector tip spray leaks.
7. Cylinder pressures low.
8. Governor instability.
9. Binding injector rack.
10. Faulty adjustments—perform tune up.

11. Improper grade and type of fuel.
12. Diluted fuel.
13. Improperly filtered or restricted fuel flow.
14. Below normal cooling system temperature.

Detonation

1. Oil picked up by air stream.
2. Oil level too high in air cleaners.
3. Oil accumulation in air box.
4. Defective blower to cylinder block gasket.
5. Leaking blower oil seals.
6. Low coolant temperature.
7. Faulty injectors.
8. Improper injector timing.
9. Injector check valve leaking.
10. Injector spray tip holes enlarged.
11. Broken injector spray tip.

Lack of Power

1. Governor gap setting incorrect.
2. Rack setting incorrect.
3. Injector timing incorrect.
4. Valve clearance incorrect.
5. Insufficient fuel.
6. Air cleaners damaged or clogged.
7. Cylinder liner air inlet ports clogged.
8. Low engine compression.
9. Blower air intake obstructed.
10. Excessive exhaust rack pressure.

General Information

The Diesel engine is an internal combustion power unit, in which the heat of fuel is converted into work in the cylinder of the engine.

Diesel engines differ from gasoline engines principally in the method used to introduce and ignite the fuel. Gasoline engines draw a mixture of fuel and air through the carburetor into the combustion chamber, where it is compressed, then ignited by an electric spark. In the Diesel engines, air alone is compressed in the cylinder; then, a charge of fuel is sprayed into the cylinder, after the air has been compressed, and ignition is accomplished by the heat of compression.

The 2-Cycle Engine

In the two-cycle engine, intake and exhaust take place during part of the compression and power strokes as shown in illustrations. A two-cycle engine, therefore, does not function as an air pump, so an external means of supplying the air is provided. A specially designed blower, on the top of the engine, forces cleaned air into the cylinders to expel the exhaust gases and fill the cylinders with fresh air for combustion.

A series of ports cut into the circumference of the cylinder wall (liner), above the piston, in its lowest position, admits air from the blower into the cylinder when top face of piston uncovers the ports.

The swirling flow of air toward the exhaust valve produces a scavenging effect, leaving the cylinders full of clean air when the piston rises and covers the inlet ports.

As piston continues on upward stroke, exhaust valves close and the charge of fresh air is subjected to the final compression. Air in cylinder is heated to approximately 1000°F. while being compressed.

Shortly before the piston reaches its highest position, the required amount of fuel is sprayed into the combustion space by the unit fuel injector. The intense heat generated during the high compression of the air ignites the fine fuel spray immediately, and the combustion con-

653

(Detroit 2-cycle) Diesel Engines

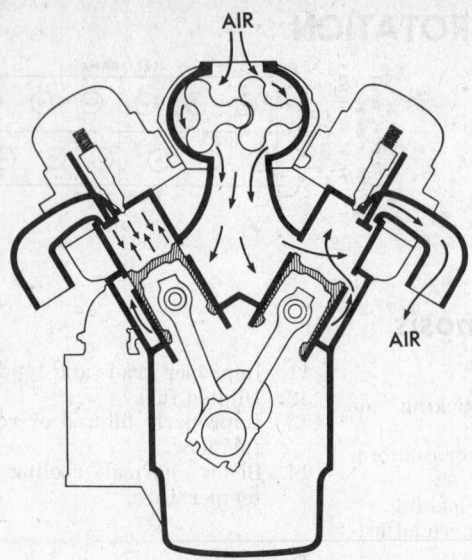

**AIR ENTERING COMBUSTION CHAMBER
THROUGH CYLINDER LINER PORTS**

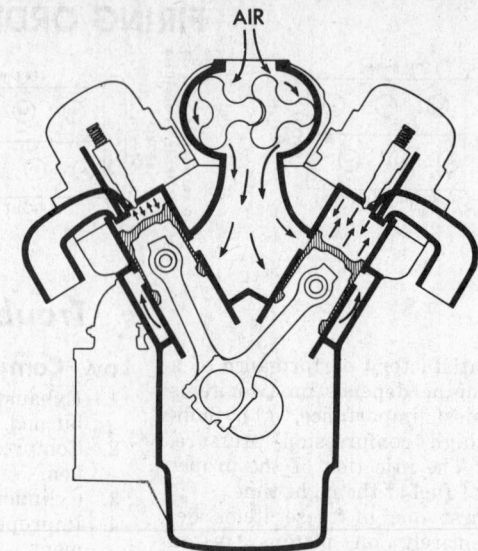

**AIR BEING COMPRESSED WITH
THE EXHAUST VALVE CLOSED**

Operation of GM Detroit two cycle diesel engine (© G.M.C.)

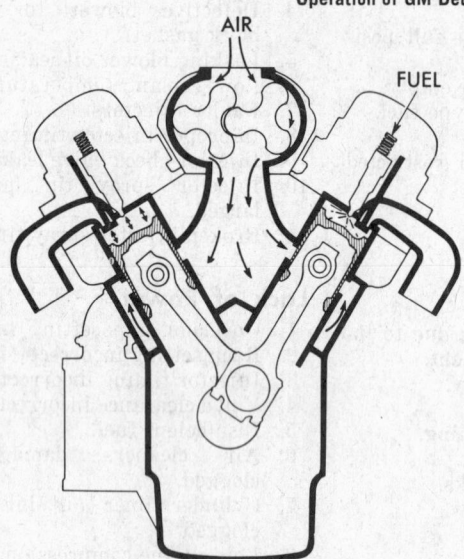

**CHARGE OF FUEL BEING INJECTED
INTO COMBUSTION CHAMBER**

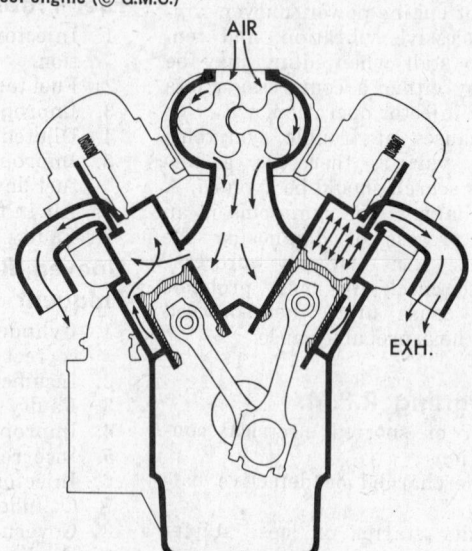

**EXHAUST TAKING PLACE AND CYLINDER ABOUT
TO BE SWEPT WITH CLEAN SCAVENGING AIR**

Operation of GM Detroit two cycle diesel engine (© G.M.C.)

tinues as long as the fuel spray lasts. The resulting pressure forces the piston downward to provide a power stroke.

As piston nears the bottom of the downward stroke, exhaust valves are opened and burnt gases are released. Still further downward movements of piston uncovers liner ports and cycle is repeated.

This entire combustion cycle is completed in each cylinder for each revolution of the crankshaft, or, in other words, two strokes; hence, the "two stroke cycle."

Diesel Engine Data

On in-line 53 Series engines, the model and serial numbers are stamped on the right side of the block at the upper rear corner. On V-type 53 Series engines, the model and serial numbers are located on the top right-hand front corner of the block, as viewed from the rear.

On in-line 71 Series engines, serial and model numbers are stamped on a pad located on the block directly in front of the blower and below the exhaust manifold. On V-71 Series engines, engine model and serial numbers are stamped on the block at the right side near the rear end plate.

On 92 Series engines, the engine serial number and engine model number are stamped on the rear of the cylinder block.

When ordering engine parts, order by part number and description, and refer to engine model and serial numbers.

Right and left rotation of engine is indicated by the letters "R" or "L" in the model designation. Truck engines are "R" and coach engines are left "L" rotation.

Certain arrangement of standard accessories are indicated by the letters "A," "B," "C" or "D." Both truck and coach engines are identified by the letter "C," indicating that the starter is on the left and oil cooler on the right as indicated in figure.

Right and left side of engine is determined by standing at rear (transmission) end.

Right and left rotation of engine is determined by standing at front (opposite transmission) end. If crankshaft rotates clockwise engine is right-hand rotation and is indicated by the letter "R" in model, or if rotation is anti-clockwise engine is left-hand rotation and is indicated by the letter "L" in model.

Minimum and Maximum Engine RPM

The following information is the approved minimum and maximum full-load governor setting for current engines used in truck models. Do not permit engine operation with governor settings higher than indicated for full load setting.

92 Series

Maximum (Full Range) ......See Emission Sticker
Maximum (Dual Range Reduced Speed ....See Emission Sticker
Minimum ....500 (EPA)

71 Series

Maximum (Full Range) 2100
Maximum (Dual Range Reduced Speed) 1900
Minimum 400

53 Series

Maximum 2600-2800
(see nameplate)
Minimum 500-600
NOTE: Above RPM are taken from blower using tachometer head —1549207 which is corrected to compensate for 2.05 to 1 ratio of blower. to crankshaft. When corrected tachometer is not available, above RPM can be increased by 5%, or 50 RPM per thousand.

Tune Up and Adjustments

Engine Tune-Up

NOTE: Before tune-up procedure is started it is important that air cleaners and fuel filters are serviced as described in applicable maintenance manuals. Crankcase breather tube and air box drains must be clean and unobstructed. Air box drains may be cleaned with compressed air.
CAUTION: Remove or at least loosen air box hand hole cover, otherwise blower or end plate gaskets may be damaged by excessive air pressure.

Tune-Up Procedures

To completely tune-up an engine, all adjustments except bridge balancing adjustment and exhaust valve cold setting must be performed only after engine has reached its normal operating temperature.

Results obtained from an engine tune-up are usually unsatisfactory, unless a step-by-step, systematic, and

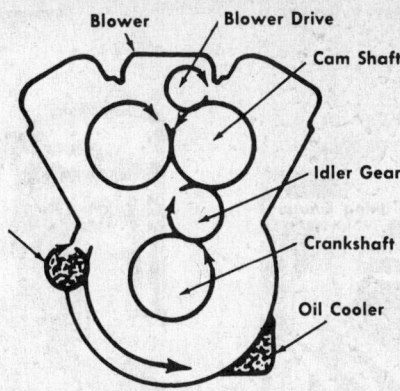

MODEL RC (TRUCK)
Engine rotation and accessory location
(© G.M.C.)

orderly approach is used. Proceed in the following sequence:

53 Series

1. Adjust exhaust valve clearance.
2. Time fuel injectors.
3. Adjust the governor gap.
4. Position the injector rack control levers.
5. Adjust maximum no-load speed.
6. Adjust idle speed.
7. Adjust the buffer screw.

71 Series

1. Bridge Balancing Adjustment
2. Exhaust Valve Lash Adjustment
3. Time Fuel Injectors
4. Back-out Buffer Screw
5. Loosen Throttle Delay Cylinder
6. Adjust Idle Speed-Preliminary
7. Adjust Low Speed Gap
8. Adjust Injector Control Racks
9. Adjust Maximum No-Load Speeds
10. Adjust Idle Speed—Final
11. Adjust Buffer Screw
12. Adjust Throttle Delay Cylinder (Coach)

92 Series

1. Adjust exhaust valves.
2. Time fuel injectors.
3. Adjust governor gap.
4. Position rack control levers.
5. Adjust maximum no-load speed.
6. Adjust idle speed.
7. Adjust bumper screw.
8. Adjust throttle booster spring—variable speed governor only.

NOTE: Whenever a push rod has been disconnected from the clevis, the push rod must be screwed back into place until end of push rod appears through the clevis. If this is not done, the piston may hit the head of the valve when the engine is being turned, due to the small clearance, between the valves and piston head at the piston's upper position, or an injector plunger may "Bottom" before being properly adjusted.

Clearance between exhaust valve stem and bridge with engine at operating temperature is important and should be maintained. Too little

clearance causes a loss of compression, misfiring of cylinders and eventual burning of valves and valve seats. Too much clearance results in noisy operation of the engine, especially in the low speed range.

Bridge Balancing Adjustment

The exhaust valve bridge assembly is adjusted and the adjustment screw locked securely at the time the cylinder head is installed on the engine. Until wear occurs with the operation of the engine, no further adjustment is required on the exhaust valve bridge.

1. Remove injector fuel jumper lines, then remove rocker arm shaft brackets. Lift rocker arms and swing back to provide access to valve bridge.
2. Remove bridge and spring (when used) from guide. NOTE: Use of bridge springs has been discontinued on latest engines.
3. Loosen adjusting screw lock nut.
4. Install bridge on bridge guide without spring.
5. Press straight down on the pallet surface of bridge. Turn adjusting screw until it just touches the valve stem, then turn screw an additional $\frac{1}{8}$ to $\frac{1}{4}$ turn and tighten lock nut finger tight.
6. Remove bridge and place in a vise. With screwdriver, hold screw from turning and tighten lock nut on the adjusting screw. Complete the operation by tightening the lock nut to 25 ft. lbs. torque, being sure that screw does not turn. IMPORTANT: Do not tighten while on engine, as binding may damage bridge, guide, and valve.
7. Apply engine oil to bridge and bridge guide.
8. Reinstall the bridge in its original position without the bridge spring.
9. Place a .0015" feeler under each end of the bridge. NOTE: Feeler used at inner end of bridge must be narrow enough to fit in bridge locating groove. Pressing down on the pallet surface of the bridge, both feelers must be tight. If both feelers are not tight, readjust the screw as previously instructed.
10. Remove the bridge and reinstall in its original position with the bridge spring (when used) in place.
11. Adjust remaining bridges as instructed in previous paragraphs.
12. If cylinder head has been removed, reinstall on the engine. Tighten cylinder head stud nuts to their specified torque before assembling the rocker shaft

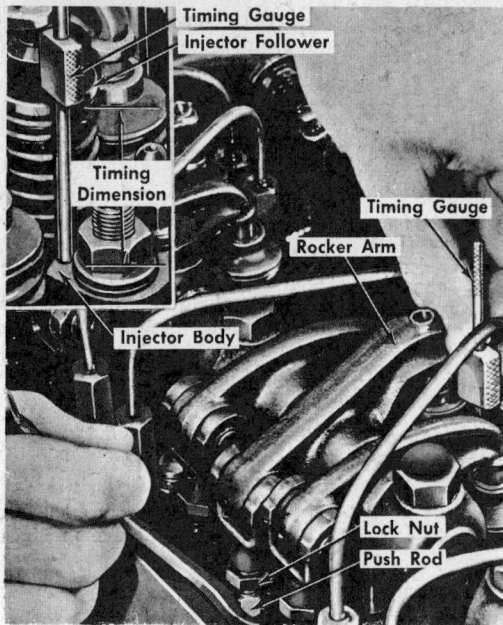

Timing fuel injector
(© G.M.C.)

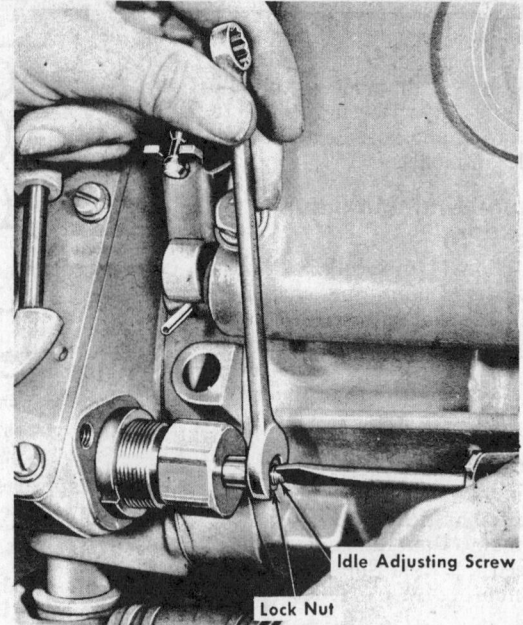

Engine idle speed adjustment
(© G.M.C.)

brackets to the head. Otherwise valves may be damaged.

13. Install the rocker arm assemblies, *being sure valve bridges are properly positioned on the inner valve stems.*
14. Tighten rocker shaft bracket bolts to their recommended torque.
15. Reconnect fuel jumper lines to injectors and connectors.

Valve Lash Adjustment

Valve lash adjustment must always be made with the engine COLD.

All valves may be lashed in firing order sequence during one full revolution of the crankshaft.

Valve lash must always be adjusted at the push rod. Do not disturb bridge adjusting screw.

1. Place governor throttle control lever in the NO FUEL position.
2. Rotate crankshaft until the injector follower is fully depressed on the cylinder being adjusted.
3. Loosen valve push rod lock nut.
4. On 71 and 92 Series engines, place a .017" feeler gauge be-

tween valve stem and valve bridge adjusting screw on engines with bridge spring, or between bridge and rocker arm on engines not using bridge spring. Adjust push rod to obtain a smooth "pull" on the feeler gauge.

On 53 Series engines, place a .027" feeler gauge between the rocker arm bridge. Adjust push rod to obtain a smooth pull on the feeler gauge.

5. Remove feeler gauge. Hold push rod with wrench and tighten push rod lock nut.
6. Check and adjust remaining valves in manner described in previous paragraphs.
7. With engine at operating temperature (160°F. to 180°F.), recheck clearances. At this time, on 71 Series engines, a .013" feeler gauge should pass between the end of the valve stem and the adjusting screw on the bridge (or between rocker arm and the bridge), and a .015" feeler gauge should not pass through. On 53

Series engines, the figures are .023" and .025". Readjust push rod if necessary.

Injector Timing

To properly time the injector, the injector follower-guide must be adjusted to a definite height.

1. Set governor control lever in the NO FUEL (off) position.
2. Turn engine crankshaft manually or by means of the starter, until the exhaust valve rocker arms are fully depressed, for that particular cylinder.
3. The injector is identified by a colored tab stamped onto the injector body. Injector is timed with 1.460" gauge. With 9290 injectors and turbocharger and 9295 injectors use a 1.484" gauge.
4. Place the injector timing gauge in the hole provided on top of the injector body with one of the "flats" toward the injector.
5. Adjust the injector rocker arm by loosening lock nut and turning the push rod with an end

Adjusting valve bridge
(© G.M.C.)

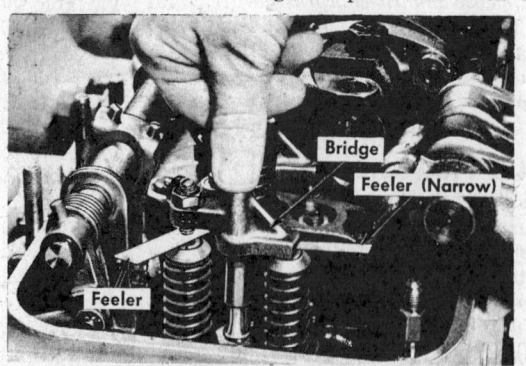

Valve bridge adjustment check
(© G.M.C.)

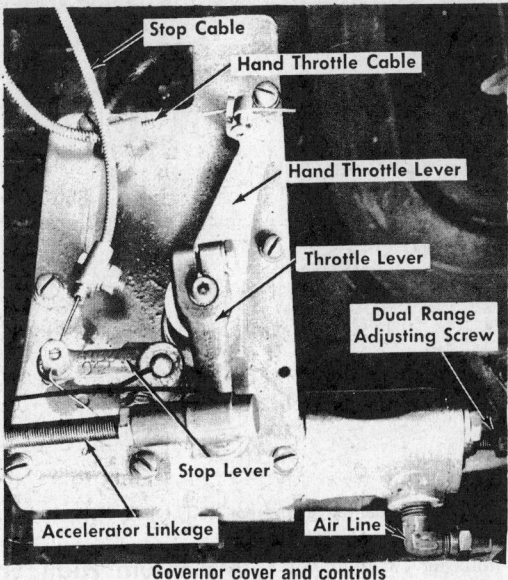

Stop Cable
Hand Throttle Cable
Hand Throttle Lever
Throttle Lever
Dual Range Adjusting Screw
Stop Lever
Accelerator Linkage
Air Line

Governor cover and controls
(© G.M.C.)

wrench, until the bottom of the timing gauge head will just pass over (drag lightly) the top of the injector follower guide.

6. Hold push rod from turning and tighten lock nut. Recheck adjustment with injector timing gauge and readjust if necessary.

7. Repeat the adjustment for all other injectors.

Governor Adjustments—

The governor used on these engines is a limiting speed double weight type, which controls the inejctor control racks.

The following adjustments must be performed in the sequence given, but not until the valves and injectors are in proper adjustment as previously described.

IMPORTANT: Tachometer drive ratio at blower is 2.05 times engine speed and rotation is counterclockwise. Make allowances accordingly or use corrected tachometer head when performing engine tune-up procedures.

Buffer Screw Adjustment

Adjust buffer screw so it extends at least ⅜" beyond the locknut.

Idle Speed Adjustment

1. Remove governor high speed spring cover to governor housing.

2. Start engine and operate at idle speed while observing RPM at tachometer.

3. To adjust, loosen lock nut and turn idle adjusting screw IN to increase or OUT to decrease speed. Lock adjusting screw with lock nut when idling speed is approximately 15 RPM below desired setting.

Low Speed Spring Gap Adjustment

When necessary to adjust, proceed as follows:

1. Remove governor control housing cover.

2. Start engine and run engine at 800-1000 RPM by manual operation of the differential lever or injector control tube lever. WARNING: Do not overspeed engine.

3. Measure gap between low speed spring cap and plunger. If gap is not between .001" to .002", loosen lock nut and turn adjusting screw as necessary to obtain desired gap. Tighten lock nut, then recheck gap.

4. Stop engine, then reinstall governor control housing cover and lever assembly to governor housing.

Positioning Injector Racks

The position of the injector control racks must be properly adjusted in relation to the governor.

The letters R or L indicate injector location in right or left cylinder bank when viewed from rear of engine. Cylinders are numbered starting at front of the engine.

NOTE: These instructions cover V-type engines, primarily. The same basic procedure applies to in-line engines, however.

1. Disconnect any linkage attached to the governor throttle control lever.

2. Loosen idle speed adjusting screw until ½" of threads project from lock nut.

3. Remove rocker covers from cylinder heads.

4. Loosen all inner and outer injector control rack adjusting screws in right and left cylinder bank. Be sure all injector rack

control levers are free on the injector control tubes. NOTE: On coaches equipped with Hydraulic transmission, loosen two nuts attaching throttle delay clamp to injector control tube.

5. Move governor throttle control lever to FULL FUEL position and hold in that position with light finger pressure. Turn *inner* adjusting screw of #1L injector rack control lever until clevis pin binds, then tighten *outer* adjusting screw. NOTE: When the setting is correct, the injector rack will be snug on the pin of the rack control lever and still maintain a snug (not loose or tight) fit at clevis pin, while throttle is in FULL FUEL position and inner and outer adjusting screws are tight.

6. At right bank, repeat procedure previously described in paragraph 5 for left bank.

7. Set #2 and #3 injector racks to #1 rack for their respective heads.

8. Turn idle speed adjusting screw in until it projects 3/16" from the lock nut to permit starting the engine.

High Speed No-Load Adjustment

Some governors are equipped with a device for reducing the engine RPM when in transmission or rear axle high speeds. These governors are referred to as "Dual Range Type" and are used on some truck engines.

Except Dual Range—71 and 92 Series Engines

1. Loosen the high speed spring retainer lock nut using a spanner wrench.

2. Start engine and operate at maximum speed, while observing engine RPM at tachometer.

3. Turn high speed adjuster IN to increase or OUT to decrease speed until desired no-load setting is obtained.

4. Hold adjuster, then tighten lock nut, using spanner wrench.

5. Recheck and if necessary readjust after tightening lock nut.

53 Series Engines

1. With engine at normal operating temperature, and no load, place the speed control lever in the maximum speed position. Note the engine speed.

2. Stop the engine. If it is necessary to adjust the speed, remove the high speed spring retainer with a special tool and withdraw the high speed spring and plunger assembly. Perform the operation very carefully to avoid

causing the low speed spring and cap to fall into the governor.

3. Separate the spring and plunger and add or remove shims as required to correct the no-load speed. Adding .010″ in shims will increase engine speed about 10 rpm.
4. Operate the engine and recheck the no-load speed.

Dual Range

Two types of pistons have been used on Dual Range governors. Early type incorporated a piston with a threaded sleeve, while latest type sleeve is positioned in piston by a shim pack.

Maximum High Speed Adjustment

(Early Type)

1. Start engine and operate at full speed. Be sure air is supplied to governor. No load speed is 125-150 RPM above full load speed.
2. Remove two bolts attaching piston housing to governor control housing. Then remove housing and piston assembly.
3. Remove piston and sleeve assembly with seal ring from piston housing.
4. Measure distance from end of piston to end of sleeve. This dimension should be approximately 25/32″.
5. Should it be necessary to increase engine RPM the dimension should be increased. Also, when necessary to decrease engine RPM the dimension should be decreased. NOTE: Each full turn of piston and sleeve will change engine approximately 100 RPM.
6. Install piston and sleeve assembly in piston housing, then in-

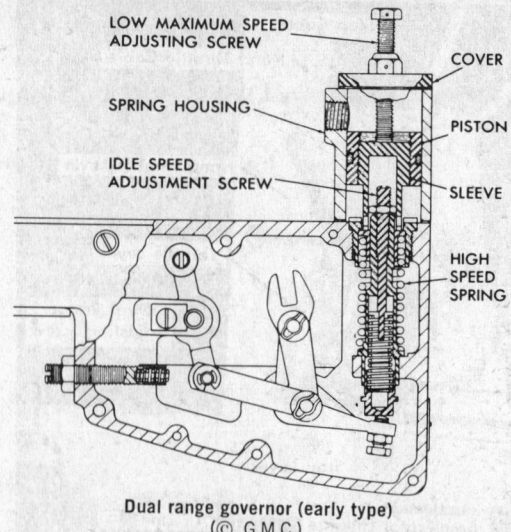

Dual range governor (early type)
(© G.M.C.)

stall piston housing to governor control housing.
7. Check for proper setting and readjust as necessary.

Reduced High Speed Adjustment

(Early Type)

1. Disconnect wire at governor air supply solenoid so as to prevent air reaching governor.
2. Start engine and operate at full speed while observing RPM at tachometer. Proper adjustment is approximately 200 RPM below maximum no-load speed.
3. Loosen lock nut and turn adjusting screw as necessary to obtain proper reduced high speed RPM. Lock screw with nut when proper adjustment is obtained.
4. Should it be necessary to increase engine RPM turn adjusting bolt in to increase, or out to decrease engine RPM.
5. Reconnect wire at solenoid when adjustment is completed.

Maximum High Speed Adjustment

(Late Type)

1. Start with a pack of four .100″ and then .010″ shims. Turn low maximum speed screw out until it extends 1¼″.
2. Disconnect air line so as to prevent air reaching governor piston.
3. Start engine and operate at full speed. No load speed is 125-150 RPM above full load speed.
4. Turn the low maximum speed adjusting screw inward until the desired high maximum no-load speed is obtained.
5. Stop engine, then remove spring housing and piston assembly.
6. Measure distance from inner edge of piston to inner edge of housing, when piston is against adjusting screw.
7. Remove sleeve from piston, then remove shims equal to the distance the piston is from the edge of housing.

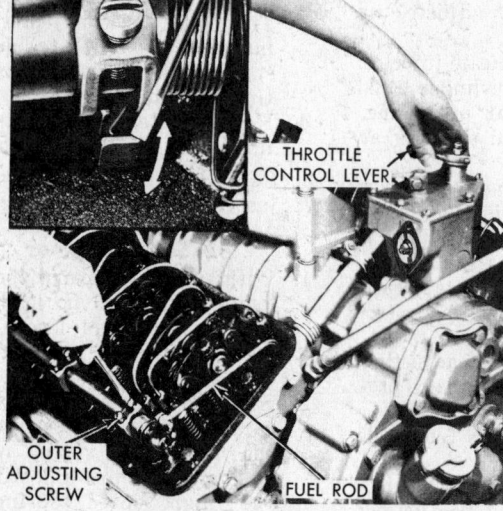

Adjusting No. 3L injector rack control lever
(© G.M.C.)

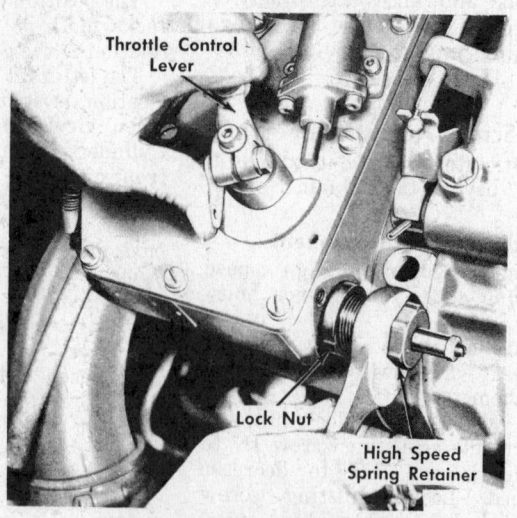

Engine high speed no-load adjustment (except dual range)
(© G.M.C.)

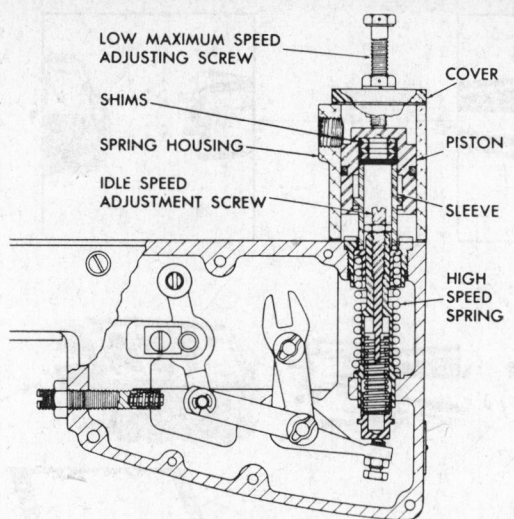

Governor springs—dual range
(© G.M.C.)

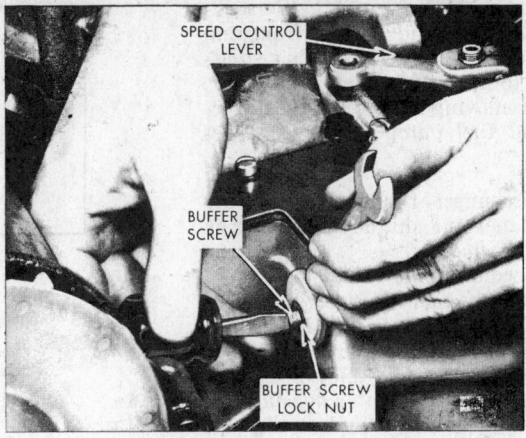

Buffer spring screw adjustment
(© G.M.C.)

8. Install sleeve in the piston, then install piston assembly in spring housing. Install spring housing assembly onto the governor.
9. Start engine and operate at full throttle with air applied at governor. Observe tachometer.
10. Remove air pressure from governor and stop engine. Remove or install shims as required to obtain desired high maximum no-load speed. Remove shims to decrease or add shims to increase engine speed. Each .010" of shims removed or added will change engine speed approximately 10 RPM.

Reduced High Speed Adjustment

(Late Type)

1. Disconnect air line to prevent air reaching governor piston.
2. Start engine and operate at full throttle while observing RPM at tachometer. Proper adjustment is approximately 200 RPM below maximum no-load speed.
3. Turn low maximum speed adjusting screw as necessary to obtain proper reduced high speed. Turn screw in to increase, or out to decrease engine RPM.
4. Reconnect air line when adjustment is completed.

Idle Speed Adjustment

Final

The engine must be at operating temperature when making the final idle speed adjustment.

1. Start engine and operate at idle speed while observing RPM at tachometer.
2. If necessary to adjust, loosen lock nut and turn idle adjusting screw IN to increase or OUT to decrease. Lock adjusting screw with lock nut when idle speed is

approximately 15 RPM below desired setting.
3. Install governor spring housing assembly to governor housing.

Buffer Spring Adjustment

With the idle speed adjusted, and engine at operating temperature adjust buffer spring screw as follows:

1. Turn buffer screw IN until "surge" or "roll" is eliminated and engine idles evenly. NOTE: Do not increase engine idle speed more than 15 RPM with the buffer screw. On 53 Series engine, buffer spring may increase idle speed up to 25 rpm.
2. Hold buffer screw and tighten lock nut.

Throttle Delay Adjustment

The following instructions outline a procedure to follow when adjusting the throttle delay cylinder, used on coaches having hydraulic transmission.

1. Adjustment is made with the engine stopped.
2. Loosen two bolts attaching clamp to injector control tube.
3. Temporarily install a gauge between injector body and shoulder on injector rack. Gauge is .404" (approx. 13/32") thick and can be made locally.
4. Position gauge as shown and exert a light pressure on the control tube in the direction of FULL FUEL.
5. Align the piston so it is flush with the cylinder edge.
6. Tighten the two clamps bolts. then remove gauge.
7. Rotate the injector control rack from "OFF" to "FULL" position to be sure it does not bind.
NOTE: Each time the injector racks are set the throttle delay must be first taken out of adjustment; injector racks are then set and the throttle delay readjusted.

Blower

Description And Operation

In the scavenging process employed in GM two cycle engines, air is forced into cylinders by blower which thoroughly sweeps out all of the burnt gases through the exhaust valve ports, and also helps to cool internal engine parts, particularly exhaust valves. At the beginning of compression stroke, therefore, each cylinder is filled with fresh, clean air, which permits highly efficient combustion.

The blower designed especially for efficient diesel operation, supplies the fresh air needed for combustion and scavenging. Its operation is similar to that of a gear-type pump. On V type engines, two hollow three-lobe rotors revolve with very close clearances in a housing bolted to the top deck of the cylinder block, between the two banks of cylinders. On in-line engines, the blower is mounted on the side of the block. To provide continuous and uniform displacement of air, the rotor lobes on 71 and 92 Series engines are made with a helical (spiral) form.

Two timing gears, located on the drive end of the rotor shafts, space the rotor lobes with a close tolerance; therefore, as the lobes of the two rotors do not touch at any time, no lubrication is required. However, build-up of oil and carbon deposits on the rotor lobes improves the sealing clearance and tends to increase the air flow.

Oil seals located in the blower end plates prevent air leakage and also keep the oil, used for lubricating the timing gears and rotor shaft bearings from entering the rotor compartment.

Blower Removal—92 Series Engines

The engine governor components

are assembled in a combination governor housing and blower front end plate cover. The fuel pump is also attached to the front end of the blower. When removing the blower the governor and fuel pump will also be removed.

1. Disconnect tubing between air cleaner and shutdown housing or turbocharger.
2. Disconnect the exhaust manifold from the turbocharger.
3. Disconnect the air inlet hose from the compressor.
4. Remove the oil inlet line from the top of the center housing and the oil outlet line from the bottom.
5. Unbolt and lift off the turbocharger.
6. Disconnect the shutdown wire from the air shutoff cam pin handle.
7. Remove the air shutdown housing.
8. Remove the adapter and blower screen.
9. Disconnect the oil pressure line from the blower.
10. Loosen the hose clamp on the blower drive support-to-blower seal.
11. Disconnect the tachometer drive cable from the adapter at the rear of the blower.
12. Remove the flywheel housing cover at the blower drive support.
13. Remove the blower drive shaft.
14. Drain engine.
15. Remove thermostat by-pass tube.
16. Disconnect fuel pump lines and fuel return crossover.
17. Remove breather pipe.
18. Disconnect throttle control rods from governor.
19. Remove rocker covers.
20. Remove governor cover.
21. Remove fuel rods.
22. Slide fuel rod cover tube hoses out of way.
23. Remove blower retaining bolts.
24. Attach a hoist to the blower, lift it slightly and move it forward to detach it from the seal at the drive end.

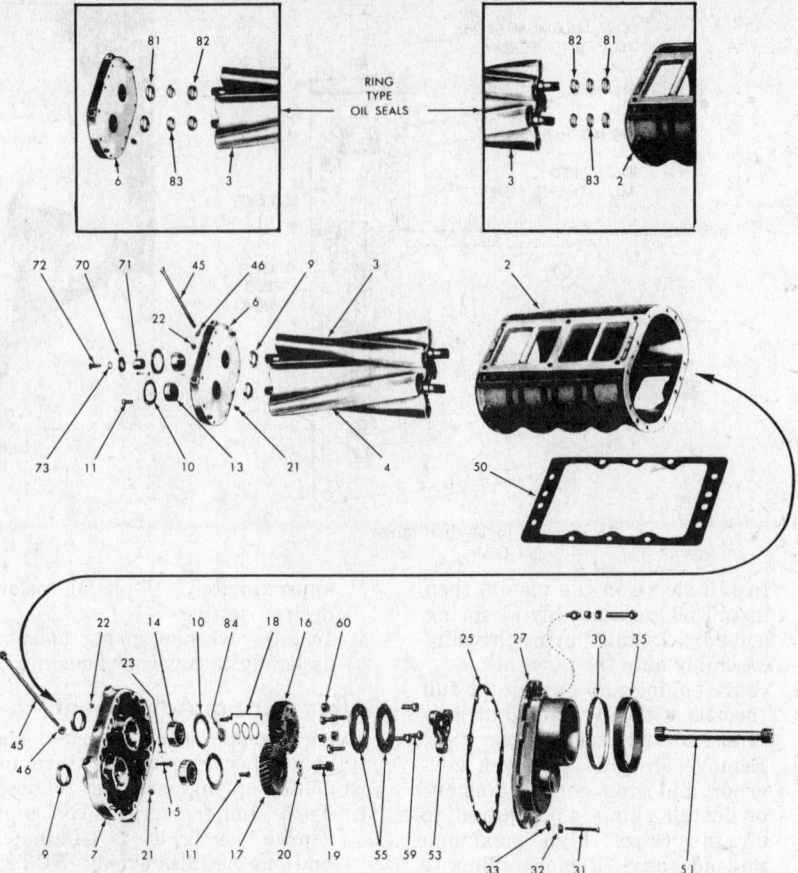

Exploded view of governor

2. Housing--Blower	19. Bolt--Blower Gear	50. Gasket
3. Rotor Assy.--R. H. Helix	20. Washer	51. Shaft--Blower Drive
4. Rotor Assy.--L. H. Helix	21. Strainer--End Plate Oil Passage	53. Drive Hub
6. Plate--Blower Front End	22. Plug--Oil Passage	55. Plate--Flexible
7. Plate--Blower Rear End	23. Orifice--Oil Passage	59. Bolt
9. Seal--Oil	25. Gasket--Cover	60. Spacer
10. Retainer--Blower Bearing	27. Cover--Rear End Plate	70. Disc--Fuel Pump Drive
11. Lockscrew-Retainer	30. Clamp--Hose	71. Spacer--Fuel Pump Drive Disc
13. Bearing--Blower Rotor Front (Roller)	31. Bolt--Cover	72. Bolt--Drive Disc
14. Bearing--Blower Rotor Rear (Ball)	32. Lock Washer	73. Lock Washer
15. Screw--Fillister Head	33. Washer--Special Flat	81. Collar--Blower End Plate
16. Gear--R. H. Helix Rotor	35. Hose--Blower Drive Cover	82. Carrier--Seal Ring
17. Gear--L. H. Helix Rotor	45. Bolt--Blower Mounting	83. Ring--Seal (Piston Type)
18. Shim--Blower Gear	46. Lock Washer	84. Spacer

Blower Removal—V-71 Engines

The engine governor components are assembled in a combination governor housing and blower front end plate cover. The fuel pump is also attached to the front end of the blower. Therefore, when removing the blower assembly from the engine, the governor and fuel pump will also be removed at the same time.

1. Remove the manifold and blower screen.
2. Loosen oil pressure line fitting from rear of blower to blower drive support and slide fitting back on tube.
3. Loosen hose clamp on blower drive support-to-blower seal.

4. Remove flywheel housing cover at blower drive support.
5. Remove snap ring and withdraw blower drive shaft from blower.
6. Remove by-pass tube between thermostat housings.
7. Remove fuel inlet and outlet lines to fuel pump. Also, remove the fuel return cross-over tube between cylinder heads.
8. Remove rocker cover(s).
9. Remove governor cover.
10. Disconnect and remove fuel rods.
11. Loosen hose clamps and slide governor fuel rod cover hose back on each side of governer.
12. Remove bolt and washer through top of each end plate which secures blower to cylinder block.

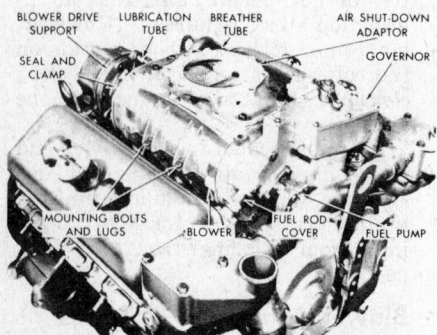

Blower mounting 92 series engine

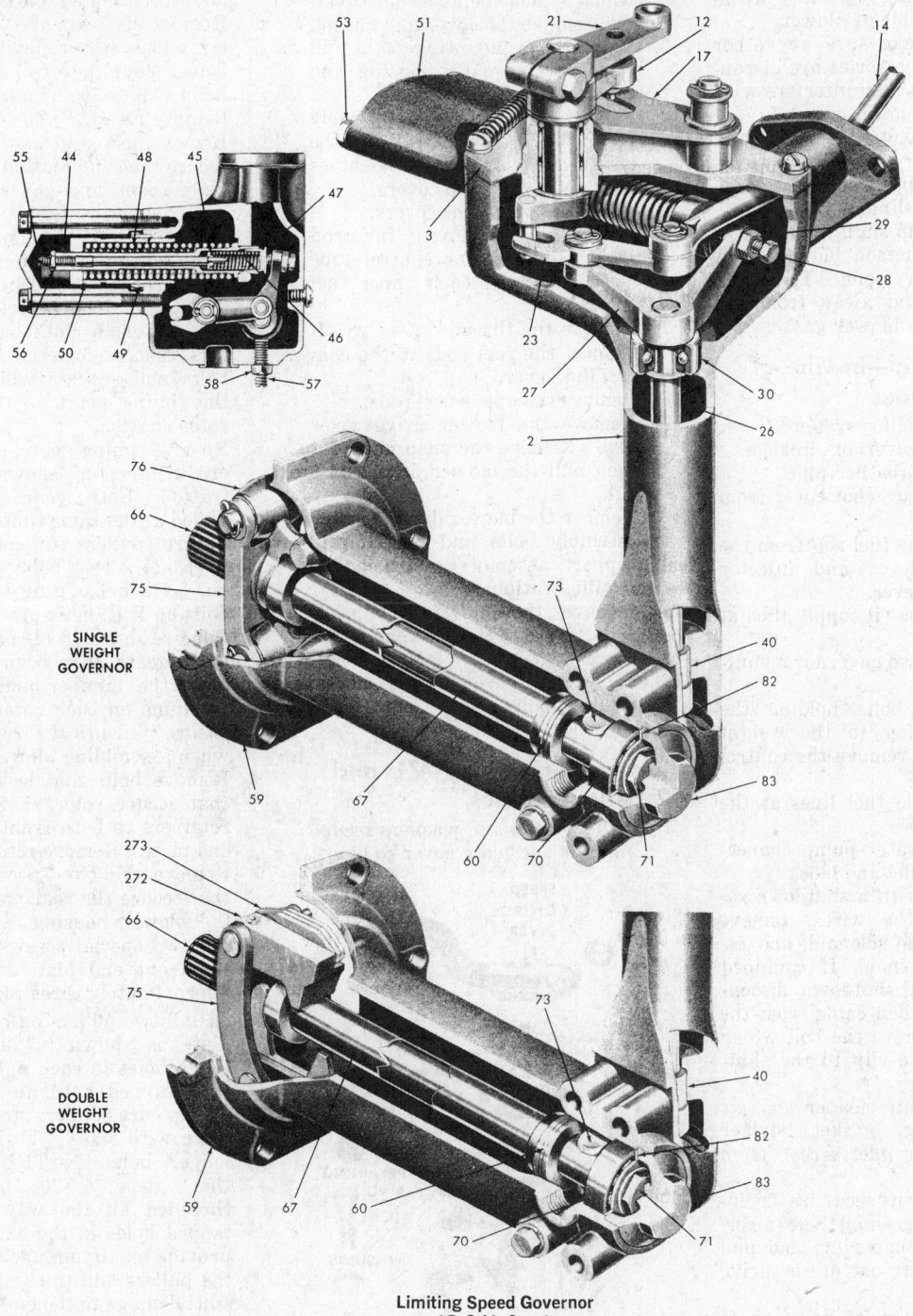

SINGLE
WEIGHT
GOVERNOR

DOUBLE
WEIGHT
GOVERNOR

Limiting Speed Governor
(© G.M. Corp)

2. Housing--Governor
   Control
3. Cover--Governor
5. Screw
6. Lock Washer
12. Lever--Throttle Shaft
14. Rod--Fuel
17. Cam
21. Lever--Speed Control
23. Lever--Differential
26. Shaft--Operating
27. Lever--Operating
   Shaft

28. Screw--Gap Adjusting
29. Lock Nut
30. Bearing--Operating
   Shaft
40. Bushing--Operating
   Shaft
44. Plunger--High Speed
   Spring
45. Seat--Low Speed Spring
46. Spring--Low Speed
47. Cap--Low Speed Spring
48. Spring--High Speed
49. Lock Nut--Retainer

50. Retainer--High Speed
   Spring
51. Cover--High Speed
   Spring Retainer
53. Bolt
55. Screw--Idle Speed
   Adjusting
56. Lock Nut
57. Screw--Buffer
58. Lock Nut
59. Housing--Weight
60. Bearing--Riser Thrust

66. Shaft Assy.--Weight
67. Riser--Governor
70. Bearing--Shaft End
71. Bolt--Retaining
73. Fork--Operating
   Shaft
75. Carrier--Weight
76. Weight Assy.--Single
   Weight Governor
82. Gasket
83. Plug--Weight Housing
272. Weights--Low Speed
273. Weights--High Speed

13. Remove two blower-to-block bolts, retaining lugs and washers on each side of blower.
14. Check to make sure any other tubing or accessories are disconnected which may interfere with removal of blower.
15. Thread eyebolts in diagonally opposite air inlet manifold-to-blower tapped holes and attach lifting sling.
16. Lift blower up slightly and move forward to detach blower from seal at drive end. Then lift blower up and away from engine. Remove blower gasket.

Blower Removal—In-Line 71 Series Engines

1. Drain the cooling system.
2. Disconnect governor linkage.
3. Remove the breather pipe.
4. Remove the governor cover from the housing.
5. Disconnect the fuel rod from the differential lever and injector control tube lever.
6. Disconnect the oil supply tube at the housing.
7. Remove the two governor mounting bolts.
8. Remove the bolts holding the control housing to the weight housing, and remove the control housing.
9. Disconnect the fuel lines at the fuel pump.
10. Loosen the water pump connections at the inlet and block.
11. If equipped with a shutdown solenoid, tag the wires, remove them from the solenoid, and remove the solenoid. If equipped with a manual shutdown, disconnect the Bowden cable from the lever and remove the bolt which holds the wire clip to the shutdown housing.
12. Remove the air cleaner and air inlet housing, gasket, striker plate, and air inlet screen from the blower.
13. Remove the bolts securing the flywheel housing small hole cover.
14. Remove the snap ring, and pull the drive shaft out of the drive assembly.
15. Loosen the blower drive shaft cover seal at the blower drive gear hub support.
16. Remove the bolts and washers that secure the blower to the block. Slide the blower slightly forward, pull the drive shaft cover off the seal, and lift the blower away from the block.

Blower Removal—53 Series Engines

1. Disconnect the governor control linkage. Remove the governor cover screws and washers and remove the cover.

2. Remove the screws and washers which attach the governor cover, and remove the cover and gasket.
3. Remove the mounting bolts and remove the spring housing and gasket from the governor.
4. Remove the spring assembly from the governor. Loosen the hose clamps and slide the hoses back on the fuel rod covers.
5. Remove both rocker covers.
6. Disconnect the lower fuel rod from each injector control tube lever and from each upper fuel rod.
7. Remove the threaded pins which connect the fuel rods to the control link lever.
8. Remove the upper fuel rods.
9. Remove the blower drive cover plate. Remove the snap ring and then pull the blower drive shaft out.
10. Remove the blower drive support assembly bolts and remove the support assembly as shown in the illustration.
11. Remove the blower mounting bolts, and lift the blower off the block.
12. Remove the governor housing attaching bolts, and remove the governor housing.

Blower Disassembly—71 and 92 Series Engines

1. Remove bolts and washers securing rear end plate cover to blow-

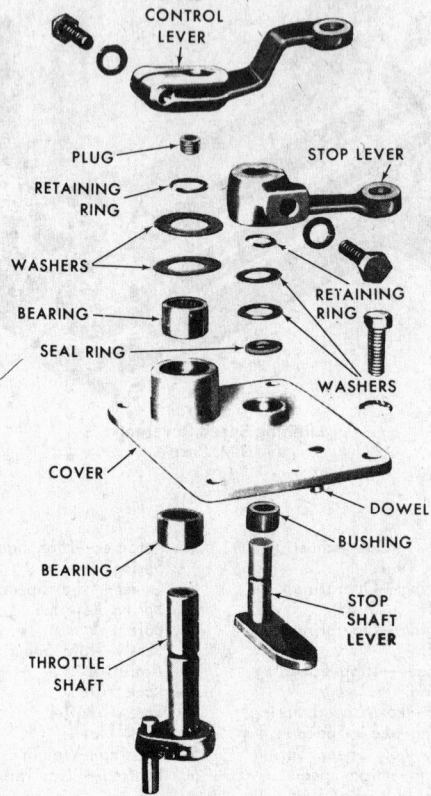

Governor Details
(© G.M. Corp)

er, including tachometer drive adaptor. Remove gasket.
2. Remove six bolts and lock washers which secure flexible blower drive coupling to right-hand helix blower timing gear. Remove retainer and coupling.
3. Remove bolts and washers which secure the combination front end plate cover and governor housing to blower. Remove housing and gasket. Fuel pump may be removed with governor housing. Remove fuel pump driving disc, spacer and driving fork.
4. Place a clean cloth between rotors, then remove the Allen head bolts and spacers which secure the timing gears to the blower rotor shafts.
5. Remove timing gears, using pullers installed as shown in illustration. Both gears must be pulled at the same time.
6. Secure pullers to gears with 5/16″-24 x 1½″ bolts (two bolts on L.H. helix gear and three bolts on R.H. helix gear).
7. Remove shims from rotor shafts, after gears have been removed. Note the number and thicknss of shims on each rotor shaft to ensure identical replacement when assembling blower.
8. Remove bolts and lock washers that secure rotor shaft bearing retainers to both front and rear end plates. Remove retainers.
9. Remove the two special screws that secure the rear end plate to the blower housing, and loosen the two special screws securing the front end plate to housing approximately three turns.
10. Install two pullers on blower end plate as shown in illustration. Align holes in each puller flange with tapped holes in end plate and secure pullers to the end plate with six ¼″-20 x 1¼″ or longer bolts. NOTE: Be sure that the ¼″-20 bolts are threaded all the way into the tapped holes in the end plate to provide maximum anchorage for the pullers and to eliminate possible damage to the end plate.
11. Turn the two puller screws uniformly clockwise to withdraw end plate and bearings from blower housing and rotors as shown in illustration.
12. Remove blower front end plate in the same manner as described above for rear plate.
13. Remove rotors from blower housing.
14. New seals should be installed. The seals may be removed from the end plates at the same time the individual bearings are removed.
15. Insert long end of remover

Blower Mounting 6-53
(© G.M. Corp)

Removing Blower Drive Support
(© G.M. Corp)

through seal and bearing and bring ram of press down on opposite end, forcing bearing and seal out of end plate.

Blower Disassembly 53 Series Engines

1. Remove the bolts, washers, and reinforcement plates which retain the front end plate and cover. Remove the end plate cover and gasket from the end plate.
2. Place a clean rag between the rotors. Remove the bolts which hold the blower drive cam retainer and blower drive spring support to the rotor gear. Separate the retainer and support from the gear.
3. Remove the governor drive plate from the opposite rotor gear.
4. Remove the bolts and blower drive cam pilots from both gears.
5. Install pullers, and remove both drive gears simultaneously. Mark the left hand rotor for identification during reassembly. Remove the shims and rotor gear spacers, and place each part with the proper gear for proper positioning at reassembly.
6. Remove the three bolts from the front end plate and remove the thrust plate and spacers. Remove the bolts that retain the thrust washers. Tap the end plate lightly with a soft mallet to remove it from the dowel pins on the blower housing.
7. Remove the rotors from the housing and then remove the rear end plate as the front end plate was removed.
8. Remove the oil seals from the end plates.
9. Drive the cam from the blower drive spring support with a brass drift. This permits the drive springs and spring seats to fall free.

Blower Inspection

After the blower has been disassembled, wash all parts thoroughly in clean fuel oil, blow dry with compressed air, and inspect as follows:

1. Races and balls or rollers of bearings for indications of corrosion or pitting.
2. Lip type oil seals for scoring, charring, or hardening. Replacement is recommended in all cases.
3. Blower rotor lobes and sealing rib for burrs or scoring.
4. Rotor shaft serrations and bearing surfaces for wear or burrs. On 53 Series engines, rotor shaft bearing surface outside diameter should be .8595"-.8600".
5. Inside of blower housing for burrs and scoring.
6. The finished inside face of end plates must be smooth and flat. On 53 Series engines, end plate shaft bore inside diameter should be .8610"-.8615".
7. Blower timing gears must be replaced as a set when worn or damaged to the point where the backlash exceeds .004".
8. Replace blower drive shaft if bent or serrations are worn badly.
9. Make sure blower drive coupling springs (pack) and cam are intact and not worn.
10. Clean out all oil holes and cavities.
11. If blower timing gear serrations are worn, replace gears.

Blower Assembly—71 and 92 Series Engines

1. Place the end plate, finished side up, on wood blocks on the bed of the arbor press.
2. Start the oil seal straight into the bore with sealing edge downward. Insert short end of seal installer into seal and use ram of press to press seal down until the flange of the tool contacts the end plate. Repeat for remaining seals.
3. The front end plate may be identified by the fact that the dowel pins extend on either side. The bottom of the plate has six holes. Start the plate dowels into the dowel holes in the front end of the blower. Then, tap the end plate lightly with a rawhide hammer to fit the plate to the

housing. Install the special screws.
4. In installing rotors, remember the following points:
   a. The lobes and gear teeth for the driving rotor form a right-hand helix.
   b. The lobes and gear teeth for the driven rotor form a left-hand helix.
   c. Serrations omitted on the gear end of the rotor shaft and in the gear should be in alignment.
   d. Rotors must be installed in the housing with both serrations pointing toward the left when viewing blower from gear end (see illustration).
5. Install the rotors into the housing and front end plate using the oil seal pilot tools as shown in the illustration. Be sure to place rotors as in d. above. The right hand helix rotor is marked "gear end" on one end, while the left hand helix rotor should be positioned so the serrated shaft is at the gear end.
6. Remove the oil seal pilot tools.
7. Install the oil seal pilots over the serrated ends of the rotor shafts.
8. Within the six holes at the bottom, start the end plate dowels into the dowel holes of the housing. Tap the end plate lightly to bring it into position.
9. Install the special screws snugly and remove the oil seal pilot tools.
10. Lube the inside diameter of each front roller bearing with light engine oil. Start the bearings straight onto the rotor shaft with the numbered face facing out.
11. Tap each bearing into the end plate with the installer as shown in the illustration.
12. Install bearing retainers with bolts and lockwashers, and torque the bolts to 7 to 9 ft-lb.
13. Install the rear bearings in the same manner, using the special tool designed for their installation. Install the retainers in exactly the same manner.
14. Follow all applicable precautions listed in step 4 in performing

Removing oil seal and bearing
(© G.M.C.)

Installing oil seal in end plate
(© G.M.C.)

this step. The punch mark on the end of each rotor shaft will assist in aligning the gears on the shafts.

15. Reposition in their original positions any shims removed from the back side of the gears. Rotate the two rotors until the two omitted serrations are in alignment and facing to the left.

16. Mesh the teeth of the rotor gears so that the omitted serrations are aligned and face in the same direction.

17. With the gears in mesh, start them onto the rotor shafts with the omitted serrations in line with those on the rotors. Make sure the right hand helix gear is on the right hand helix rotor, etc. Use ½"-20x1¼" bolts and plain washers, install and turn the bolts uniformly to pull the gears onto the shaft.

18. Remove the bolts and washers used in drawing the gears onto the shafts. Install the special spacers and Allen head gear retaining bolts and tighten the retaining bolts to 55–65 ft lbs. Lubricate the bolts before installation.

19. At this stage of the blower assembly, the blower rotors must be timed.
   NOTE: Before timing blower, install four 5/16"-18 x 2½" bolts and ⅜" spacers temporarily at each end to secure end plates to housing. After timing, remove spacers and bolts.
   a. The blower rotors, when properly positioned in the housing, run with a slight clearance between the lobes. This clearance may be varied by moving one of the helical gears in or out on the shaft relative to the other gear.
   b. If the right-hand helix gear is moved out, the right-hand helix rotor will turn counterclockwise when viewed from the gear end. If the left-hand helix gear is moved out, the left-hand helix rotor will turn clockwise when viewed from

Installing roller bearings in front end plate
(© G.M.C.)

the gear end. This positioning of the gear, to obtain the proper clearance between the rotor lobes, is known as blower timing.
   c. Moving the gears OUT or IN on the rotors is accomplished

by adding or removing shims between the gear hub and the bearing.
   d. The clearance between rotor lobes may be checked with various thickness feeler ribbons ½" wide. When measuring clearances of more than .005", laminated feelers that are made up of .002", .003" or .005" feeler stock are more practical and suitable than one single thick feeler gauge. Clearances should be measured from both the inlet and outlet sides of the blower.
   e. Time rotors to have from .002" to .006" clearance between the TRAILING edge of the right-hand helix rotor and the LEADING edge of the left-hand helix rotor ("CC" clearance) measured from both the inlet and outlet sides as shown in illustrations. If possible, keep this clearance to the minimum (.002"). Then check clearance between LEADING edge of right-hand helix rotor and TRAILING edge of left-hand helix rotor ("C" clearance)

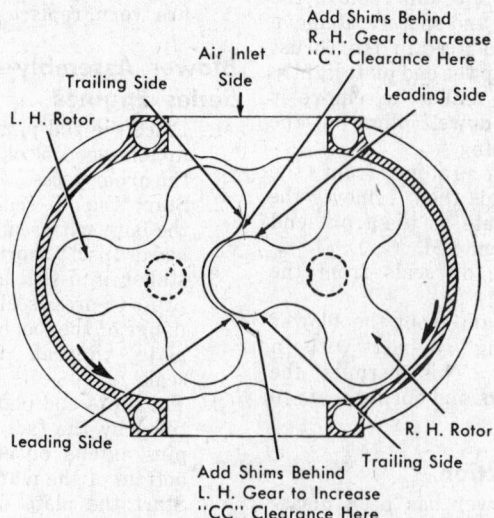

VIEW FROM GEAR END OF BLOWER

Proper location of shims for correct rotor clearances
(© G.M.C.)

Removing blower gears
(© G.M.C.)

for the minimum clearance of .012″. Rotor-to-rotor measurements should be taken 1″ from each end and at center of blower.

f. Having determined the amount one rotor must be revolved to obtain the proper clearance, add shims back of proper gear to produce the desired result. When more or fewer shims are required, both gears must be removed from the rotors.

g. Placing a .003″ shim in back of a rotor gear will revolve the rotor .001″.

h. Install the required thickness of shims back of the proper gear and next to the bearing inner race and reinstall both gears. Recheck clearances between rotor lobes.

i. Determine minimum clearances at points "A" and "B" shown in illustration. Insert feelers, between end plates and ends of rotors. This operation must be performed at both ends of each lobe, making 12 measurements in all.

20. Set drive coupling support on two wood blocks as shown in illustration.

21. Install spring seats in grooves of coupling support.

22. Apply grease to springs to hold the leaves together and slide the two spring packs, consisting of 21 leaves per pack, into support.

23. Place blower drive cam over end of tool, insert tool between spring packs and press cam into place.

24. Attach flexible blower drive coupling and retainer to right-hand helix gear and secure with six bolts and lock washers.

25. Place gasket on blower rear end plate.

26. Attach rear end plate cover and secure with bolts and lock washers.

27. Attach tachometer drive adaptor, if used.

28. Install fuel pump driving disc and spacer.

29. Place gasket on blower front end plate.

30. Install governor assembly (combination governor and front end plate cover) on front end plate and secure with bolts and washers.

NOTE: Make certain that fuel pump driving fork is in registration with fuel pump driving disc. Governor weight shaft is splined to front left-hand helix rotor shaft.

Blower Assembly—53 Series Engines

1. Using an installing tool, install new oil seals in both blower end plates. Install the seals so the lip of the seal faces the inside of the counterbore.

2. Place the front end plate (the front plate has tapped holes for the thrust washer plate bolts and thrust washer lube oil holes) on two wood blocks. Install the rotors with the lobes meshed and the gear ends upward.

3. Install the blower housing (see illustration).

4. Place the rear end plate over the rotor end shafts (see illustration). Then, bolt the end plates in place.

5. Attach the two thrust washers to the front end of the blower with the two retaining bolts. Tighten 5/16″–24 bolts to 24–30 ft lbs, and 3/8″–24 bolts to 54–59 ft lbs.

6. Attach the three spacers and the thrust plate to the front of the blower with the thrust plate bolts, torquing the bolts to 7 to 9 ft lbs. Check the clearance between the thrust plate and the washers. It should be .001″–.003″.

7. Position the rotors so the missing serrations where the gears fit are 90° apart. See the illustration.

8. Install shims and spacers in the counterbores of the rear faces of the rotor gears. Place the gears on the ends of the shafts with the missing serrations aligned with those in the shafts. Tap the gears lightly with a soft hammer to seat them. Rotate the gears until the punch marks line up. Reposition the gears as necessary if the marks do not line up.

9. Wedge a cloth between the rotors. Install the gear retaining bolts and plain washers and turn the bolts in uniformly until the gears are fully installed. Then, remove the bolts and washers.

10. Place the pilot in the counterbore of the gears, install the 12 point bolt in the right hand rotor shaft, and install the hex head

Removing or installing bearing retainers
(© G.M.C.)

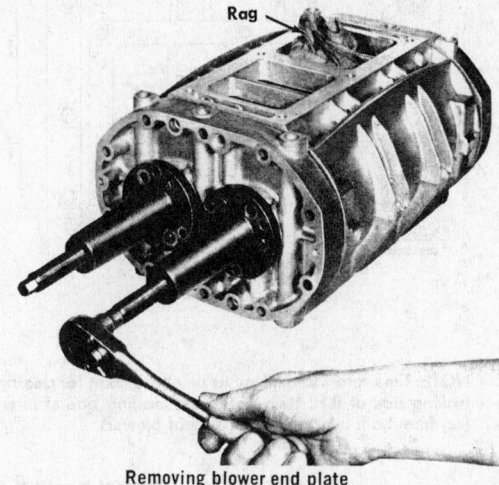

Removing blower end plate
(© G.M.C.)

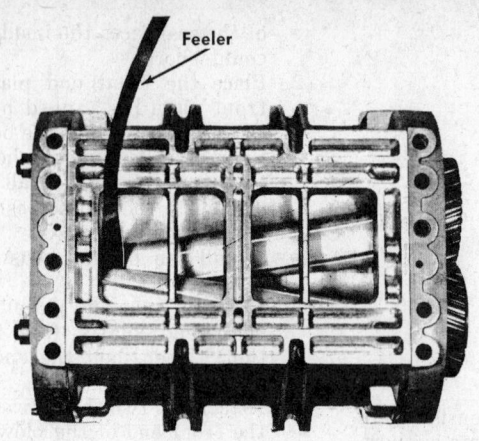
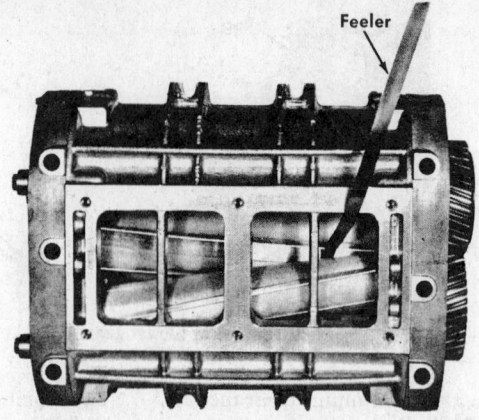

Measuring "CC" and "C" clearances (© G.M.C.)

Installing blower timing gears
(© G.M.C.)

Measuring end clearance between rotor and end plate
(© G.M.C.)

Using pilot tools to install rotors
(© G.M.C.)

bolt in the left hand rotor shaft. Tighten the bolts to 20-25 ft lbs.

11. Check the backlash between the blower gears. It should be .0005"-.0025" with new gears. Gears must be replaced if backlash exceeds .0035".

12. Check the blower rotor timing and other clearances with a feeler gauge. If necessary, change the rotor-to-rotor clearance by moving one of the gears in or out on its shaft through addition or removal of shims. Measurements should be made using two or three gauge blades combined so as to prodce the proper size gauge. Take measurements from both the inlet and outlet sides of the blower, and across the entire length of each rotor lobe. Measure rotor-to-rotor clearance with the rotors at a 90° angle to each other.

The clearance between rotors should be at least .010". Rotor-to-housing clearance should be .004" on the outlet side and .010 on the inlet side.

NOTE: During measurement of

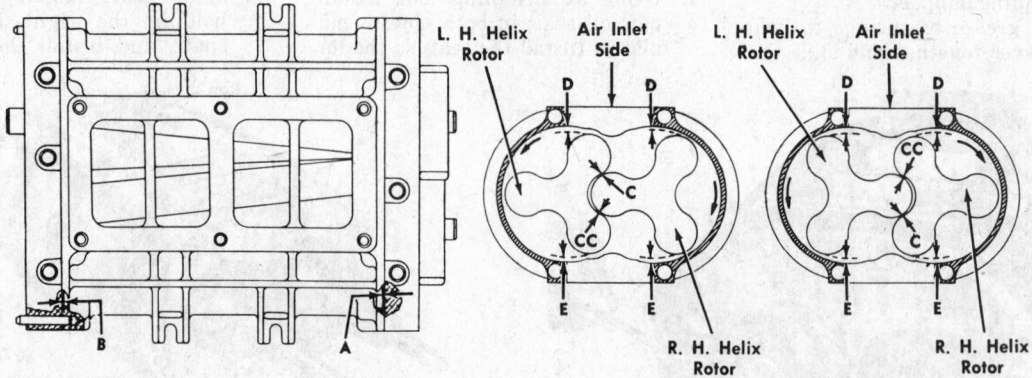

NOTE: Time rotors to dimensions on above chart for clearance between trailing side of R.H. Helix Rotor and leading side of L.H. Helix Rotor (cc) from both inlet and outlet side of blower.

VIEWS FROM GEAR END OF BLOWER

	A	B	C	CC	D	E
MIN.	.007"	.014"	.012"	.002"	.016"	.004"
MAX.				.006"		

Chart of minimum blower clearances (© G.M.C.)

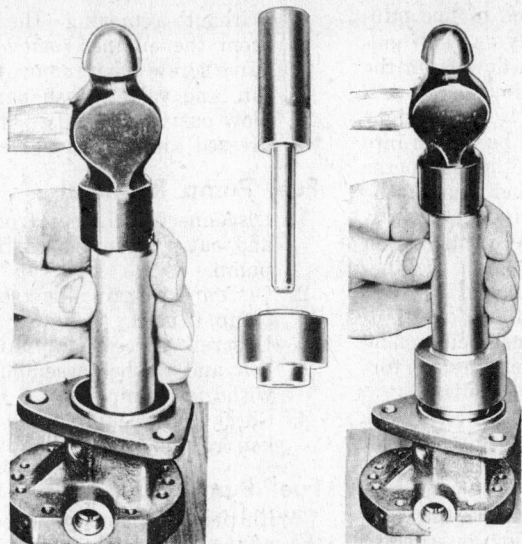

Installing inner and outer seals in pump body
(© G.M.C.)

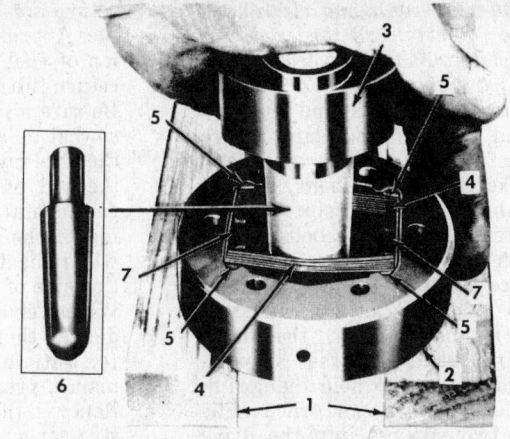

Inserting blower drive cam between springs— using spreader tool
(© G.M.C.)

1 Wood blocks
2 Drive spring support
3 Blower drive cam
4 Blower drive spring
5 Spring seat
6 Spring spreader tool
7 Spring seat

Inserting Blower Drive Cam
(© G.M. Corp)

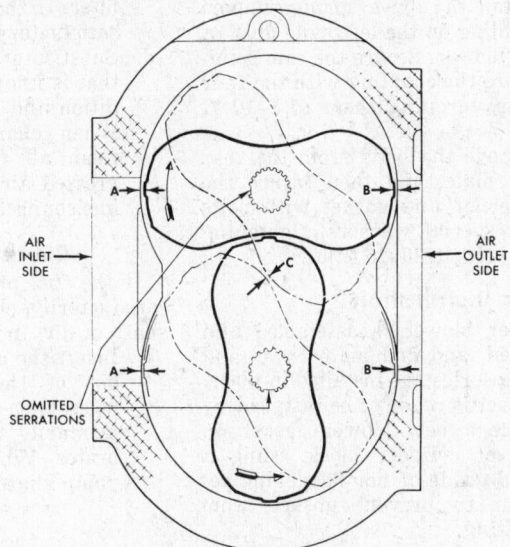

Minimum Blower Rotor Clearance
(© G.M. Corp)

	A	B	C
3-53	.0075"	.004"	.010"
4-53	.0075"	.004"	.010"
6V-53	.010"	.004"	.010"

BLOWER ROTOR END CLEARANCES (Minimum)

Engine	Front End Plate	Rear End Plate
3-53	0.006"	0.008"
4-53	0.006"	0.009"
6V-53	0.008"	0.012"

Measuring Lobe to Housing Clearance
(© G.M. Corp)

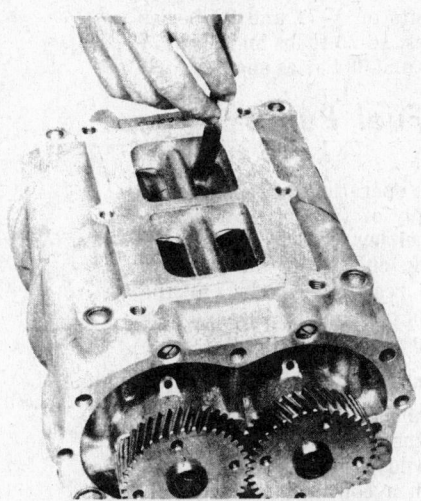

Measuring Lobe to End Plate Clearance
(© G.M. Corp)

front and rear rotor end clearances, push the rotor toward the appropriate end plate to make the clearance as tight as possible.

13. Remove the bolts and washers from the front end plate. Secure the front end plate cover and gasket with the bolts, special washers, and reinforcement plates. Tighten the bolts to 20-25 ft lbs.
14. Place the drive spring support on two wood blocks (see illustration) and position the drive spring seats in the support. Apply grease to hold the spring packs together, and then slide the two springs into the drive spring seats.
15. Place the blower drive cam over the end of a special tool, and insert the tool between the springs. Then, press the cam into place between the springs.
16. Install the drive spring support coupling on the left-hand gear of the blower. Secure the cam retainer to the coupling with the four bolts, torquing them to 8–10 ft lbs.
17. Remove the bolts from the rear end plate, and then secure the governor and gasket with bolts and special washers, tightening the bolts to 20–25 ft lbs.

Blower Installation

1. After blower is assembled and timed, and end plate cover and accessories are installed in place, blower is ready to be installed.
2. Place a new blower gasket on top of cylinder block, using a good grade of non-hardening cement to prevent gasket from shifting.
3. Lower the blower into position, reversing the removal procedure.
4. Secure the blower to the block with bolts, retaining lugs, and washers. Tighten the bolts uniformly to 25 ft-lbs. in 5 ft-lb increments on V-71 and 92 Series engines, 16-20 ft lbs on in-line 71 Series and 53 Series engines.

Fuel Pump

Tests

If engine operation indicates insufficient supply of fuel to the injectors and the fuel level is not low in the supply tank, check the fuel flow between the restricted fitting in the fuel return passage in the cylinder head and the fuel supply tank.

Fuel Flow Check

1. Disconnect the fuel return line from the fitting at the fuel tank and hold the open end of the pipe in a convenient receptacle.
2. Start and run the engine at 1200 r.p.m. and measure the fuel flow

return for a period of one minute. Approximately one half gallon of fuel should flow from the return tube per minute.
3. Be sure all connections are tight so that no air will be drawn into the fuel system; then immerse the end of the fuel line in the fuel container. Air bubbles rising to the surface of the liquid will indicate a leak on the suction side of the pump.
4. Whenever the fuel flow check indicates there is insufficient flow for satisfactory engine performance, proceed as follows:
a. Renew the element in the strainer as outlined in respective applicable vehicle manual.
b. Start the engine and run it at 1200 r.p.m. to check the fuel flow. If the fuel flow is still unsatisfactory continue as follows.
c. Renew the element in the fuel filter. If the fuel flow is still unsatisfactory continue as follows.
d. Substitute another fuel pump that is known to be in good condition and again check the flow. When changing a fuel pump, clean all fuel lines with compressed air and be sure all fuel line connections are tight.

Check Fuel Pump

If the fuel pump fails to function satisfactorily, check for broken pump shaft, or dirt in relief valve.
1. Insert the end of a wire through one of the pump body drain holes, then crank the engine momentarily and see if wire vibrates. Vibration will be felt if pump shaft rotates.

2. Without removing the pump from the engine, remove relief valve screw, then remove spring, pin, and valve. Wash parts and blow out valve cavity with compressed air. Install valve parts.

Fuel Pump Removal

1. Disconnect fuel lines from inlet and outlet openings of the fuel pump.
2. Disconnect drain line from fuel pump, if used.
3. Unscrew three pump attaching bolt and washer assemblies and withdraw pump.
4. Check drive coupling and if broken, replace.

Fuel Pump Disassembly

With the fuel pump removed from the engine, disassemble the pump as follows:
1. Remove eight cover bolts, then withdraw the pump cover away from the pump body and off the two cover dowels.
2. Withdraw drive shaft, drive gear and gear retaining ball as an assembly from the pump body.
3. Remove drive gear, if necessary, from shaft being careful not to misplace the locking ball.
4. Remove driven gear and shaft as an assembly from pump body.
5. Remove valve screw, holding hand on screw to relieve valve spring tension.
6. Remove spring, pin, and valve from valve cavity in pump body.
7. If inspection indicates oil seals require replacing, remove in manner illustrated, by clamping

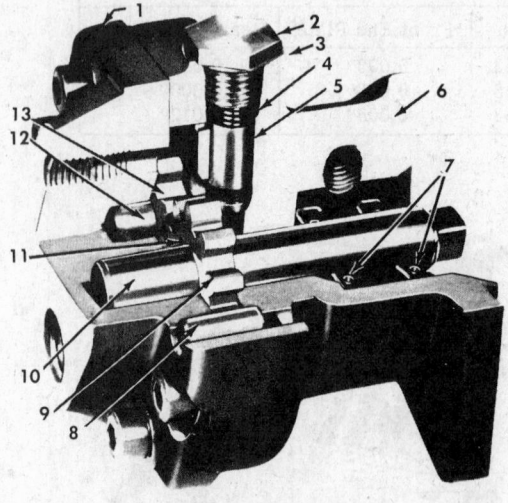

Fuel pump assembly
(© G.M.C.)

1 Pump cover	8 Dowel pin
2 Valve plug	9 Drive gear
3 Gasket	10 Drive shaft
4 Valve spring	11 Retaining ball
5 Relief valve	12 Driven shaft
6 Pump body	13 Driven gear
7 Oil seals	

pump body in a bench vise and screwing threaded end of tool shaft into outer oil seal (seal nearest to bolting flange). Then tap pilot end of shaft with hammer thus removing seal. Repeat this operation to remove the inner seal.

Inspection

1. When the fuel pump has been disassembled, all parts should be washed in clean fuel oil, blown dry with compressed air and inspected.
2. Oil seals once removed from the pump body should be discarded and replaced with new seals. Lips of oil seals must fit snug around the pump shaft.
3. Pump gear teeth should be checked for scoring or chipping. If gear teeth are scored or chipped, they should be replaced.
4. Mating faces of the pump body and cover must be flat and smooth and fit tightly together.
5. The relief valve must be free from score marks and must fit its seat in the pump body. If the relief valve is scored and cannot be cleaned up with crocus cloth, the valve must be replaced.

Fuel Pump Assembly

1. Place inner oil seal on pilot of installer with lip of seal facing shoulder on handle.
2. With pump supported on wood blocks, insert pilot of installer into pump body so seal starts straight into pump flange, then drive seal into place in counterbore of flange until it bottoms.
3. Place adaptor on pilot end of installer with shorter end of adaptor against shoulder on installer. Position outer oil seal on adaptor with lip of seal facing adaptor, then insert pilot of installer into pump so seal starts straight in pump flange and drive seal into pump body until the shoulder of adaptor contracts body.
4. Clamp pump body in soft jaws of bench vise with relief valve cavity up. Lubricate the outside diameter of relief valve and place valve in cavity with hollow end up. Insert spring inside the valve and pin inside of spring. With gasket in place next to head of valve screw place screw over spring and thread into pump body.
5. Install fuel pump drive gear, if removed, over plain end of drive shaft with slot in gear facing plain end of shaft. This operation is very important. Press gear beyond locking ball retaining hole. Then place ball in hole and press gear back until end of slot contacts ball.
6. Lubricate pump shaft and insert square end of shaft into opening at gear side of pump body and through the two oil seals.
7. Place driven gear shaft and gear assembly in pump body with chamfered end of gear teeth facing pump body.
8. Lubricate gears and shafts with clean engine oil.
9. Apply a thin coat of reputable sealer on face of pump cover outside of gear pocket area, then place cover against pump body with two dowel pins in cover entering holes in pump body. The cover can be installed in only one position over the two shafts.
10. Secure cover in place.
11. After assembly, rotate pump shaft by hand to make certain that parts rotate freely. If binding exists, it may be necessary to tap corner of pump cover with a hammer to relieve binding.

Fuel Pump Installation

1. Affix a new gasket to pump body and locate pump drive coupling over square end of fuel pump drive shaft.
2. Install fuel pump on engine and secure.
3. Connect inlet and outlet fuel lines to the fuel pump.
4. Connect drain tube, if used, to pump body.

Crown Valve Fuel Injectors

Description and Operation

The cross section of the injector illustrated, shows the various fuel injector parts. Fuel oil supplied to injector enters the dropforged steel body at the top through the filter cap. After passing through the filter element inlet passage, the fuel oil fills the supply chamber between bushing and spill deflector. The plunger operates up and down in this bushing, the bore of which is connected to fuel supply by two funnel-shaped ports.

Motion of injector rocker arm is transmitted to plunger by follower which bears against plunger spring. Follower is positioned in body by a follower stop pin. In addition to up and down motion, the plunger can be rotated, in operation, around its axis by gear, which is in mesh with the control rack. The fuel is metered by an upper helix and lower helix which are machined into lower end of plunger. The relation of these helixes to the two ports changes with the rotation of the plunger. As plunger moves downward, fuel oil in high-pressure cylinder or bushing is first displaced through the ports back into supply chamber until lower edge of plunger closes the lower port. Remaining oil is then forced upward through center passage in plunger into recess between upper helix and the lower cut-off from which it can still flow back into supply chamber until helix closes upper port. Rotation of plunger, by changing position of helix, retards or advances the closing of the ports and beginning and ending of injection period, at the same time increasing or decreasing desired amount of fuel which remains under plunger for injection into the cylinder.

Illustration shows the various plunger positions from NO INJECTION to FULL INJECTION.

Illustration shows four positions for downward travel of plunger, rack fixed. On downward travel of plunger, the metered amount of fuel is forced through center passage of the valve assembly, through check valve, and against spray tip valve. When sufficient fuel pressure is built up, valve is forced off its seat and fuel is forced through small orifices in spray tip and atomized into the combustion chamber.

Constant circulation of fresh, cool fuel through the injectors, which renews the fuel supply in the chamber, helps to maintain even operating temperatures of injectors, and also effectively removes all traces of air which might otherwise accumulate in system and interfere with accurate metering of fuel.

The fuel injector outlet opening, which returns the excess fuel oil supplied by fuel pump, is directly adjacent to inlet opening.

Refer to illustration for injector identifying marks. The correct injector must be used and type can be determined by checking injector in engine for number which appears on color tag pressed into body.

Injector Mounting

The injectors are mounted in the cylinder heads, with their spray tips projecting slightly below top of inside surface of combustion chambers. A clamp, bolted to cylinder head and fitting into a machined recess in each side of injector body, holds injector in place in a water-cooled copper tube which passes through cylinder head. A dowel pin in injector body registers with a hole in cylinder head for accurately locating injector assembly.

A copper tube is installed in cylinder head with a seal ring at flanged upper end. The lower end is peened into a recess of the cylinder head. The tapered lower end of injector seats in copper tube, forming a tight seal to withstand the high pressure inside combustion chamber.

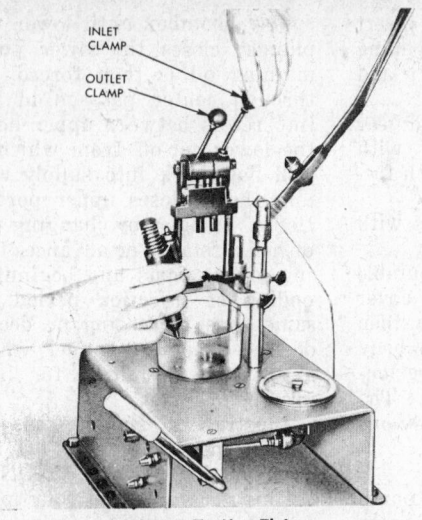

Injector Testing Fixture
(© G.M. Corp)

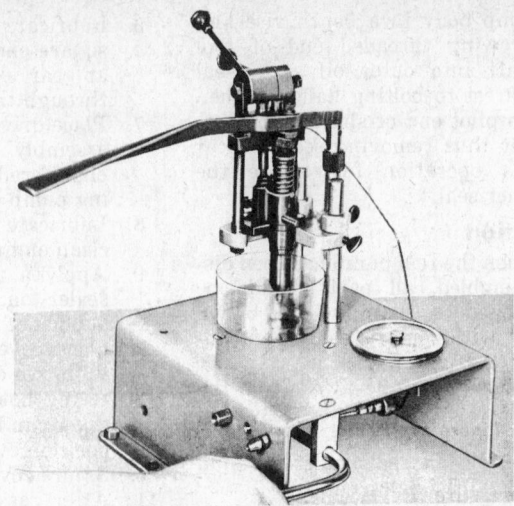

Pumping Up Test Fixture
(© G.M. Corp)

Injector Removal

If it becomes necessary to remove one of the fuel injectors for inspection or replacement, follow the procedure given below:

1. Remove valve rocker cover.
2. Remove fuel lines from both the injector and the fuel connectors. *Immediately after removal of fuel lines from an injector, the two fuel feed fittings should be protected to prevent dirt entering injector.*
3. If necessary, crank engine with the starter or a bar at flywheel ring gear until the three rocker arm clevis pins, at outer end of arms, are in line.
4. Loosen two rocker arm bracket bolts holding brackets to cylinder head and swing rocker arm assembly over away from valves and injector.
5. Remove injector hold-down bolt, special washer, and injector clamp.
6. Using tool as illustrated, pry injector from its seat.
7. Lift injector from seat, at the same time disengage control rack linkage.

Injector Tests

Tests using a popping fixture and a comparator may be made if the injector shows no external damage. If these tests and a visual inspection of the plunger show satisfactory condition and performance, the injector may be put back into service without further disassembly and work. Perform all the tests without disassembling the unit. Then, evaluate the test results, and repair the unit as necessary. Rebuilt injectors must pass all tests before being returned to service.

NOTE: Fuel injectors operate at extreme pressure for efficient atomization. Fuel leaving an injector nozzle is moving so fast that it will easily

penetrate the skin and cause severe infection. Therefore, extreme care must be taken to ensure that fuel spray is discharged well away from any portion of the skin.

1. Place the upper end of the injector against a bench, as shown in the illustration, and move the rack back and forth while fully depressing the follower. Any binding indicates that internal parts are dirty or damaged.
2. Mount the injector in a testing and popping fixture, as shown in the illustration, with the injector dowel in the proper slot. An

adaptor plate must be used for offset body injectors. Position the support plate and popping handle so they are at the proper height.

3. Close the clamp with the inlet tube on the injector. Operate the pump handle until all air has been purged, and close the outlet clamp.
4. Put the rack in the full fuel position and pump the handle of the test fixture with smooth, even strokes until the injector pops (sprays fuel). Take note of and record the pressure at which this

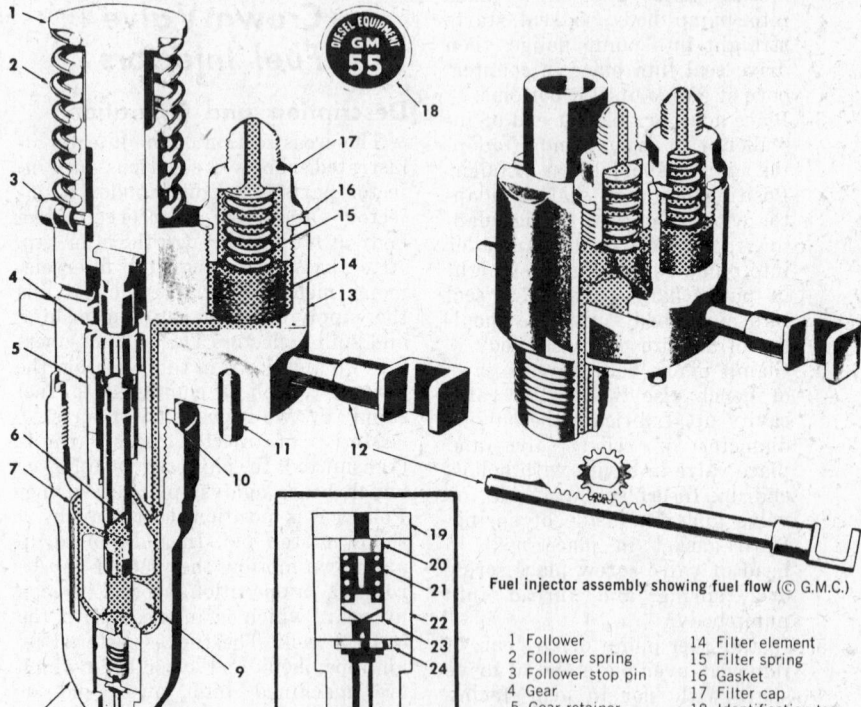

Fuel injector assembly showing fuel flow (© G.M.C.)

1 Follower	14 Filter element
2 Follower spring	15 Filter spring
3 Follower stop pin	16 Gasket
4 Gear	17 Filter cap
5 Gear retainer	18 Identification tag
6 Plunger	19 Valve seat
7 Spill deflector	20 Valve
8 Injector nut	21 Valve spring
9 Plunger bushing	22 Valve cage
10 Seal ring	23 Valve stop
11 Dowel	24 Check valve
12 Rack	25 Spray tip
13 Injector body	

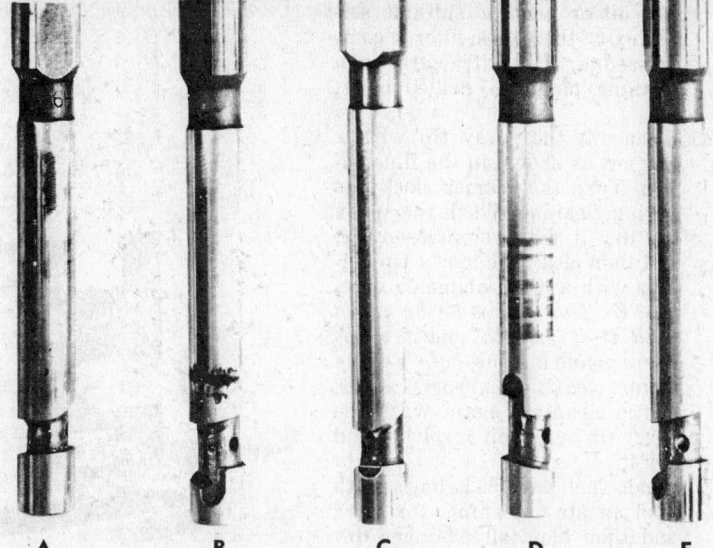

A. Tight rack causing binding in up and down movement.

B. Dirt in fuel. This shows advanced stages of abrasive matter in fuel.

C. Chipped at lower helix.

D. High pressure scoring caused by a plugged tip or wrong size tip being installed.

E. The condition shown can be caused by either lack of fuel at high speeds or water in fuel.

Damaged and Usable Plungers
(© G.M. Corp)

occurs. The pressure must be 450–850 psi, or the unit is faulty.

5. With the pressure just below the pop off point, close the fuel shut off valve and note and time the pressure drop. A drop from 450 to 250 psi must not take less than 40 seconds.

6. If the time for the pressure drop is less than 40 seconds, thoroughly dry the outside of the injector with compressed air, and then open the fuel valve and operate the pump to maintain test pressure. A leak at the injector rack opening indicates a poor bushing-to-body fit. A leak around the spray tip or seal ring indicates a loose injector nut, damaged seal ring, or a brinneled surface on the nut or spray tip. A leak at the filter cap indicates that the cap is loose or a damaged filter cap gasket. A dribble of fuel (if more than a drop or two) indicates a damaged surface or dirt.

7. Thoroughly dry the injector with compressed air and check and tighten the fuel connections.

8. Put the rack in the FULL FUEL position and lock the popping handle with the lock provided. Operate the pump to maintain full test pressure.

9. Adjust the screw in the popping handle so it will depress the plunger just far enough to close both ports in the injector bushing (at this point, the injector spray decreases appreciably and the pressure will rise).

10. Pump up and maintain 1400–2000 psi. Check for leaks and refer to the "Insufficient Injector Holding Time Chart" for evaluation of the causes.

11. Loosen the adjusting screw in the popping handle, pump up pressure to near the pop pressure, and operate the popping handle several times while watching the spray pattern very carefully. Fuel should be discharged from every orifice and should form an even spray pattern. Otherwise, the spray tip requires cleaning.

12. Remove the plunger from the injector as follows:
A. Mount the injector, right side up, in a holding fixture. Compress the follower spring and, with a screwdriver, raise the spring above the stop pin and withdraw the pin. Allow the plunger spring to rise gradually.
B. Remove the injector, turn it upside down, and catch the spring and plunger.

13. Inspect plunger, comparing it with plungers shown in the illustration. Check for a chip in the area of the bottom helix and lower portion of the upper helix. Reinstall the plunger and go on with the following test. If the plunger is chipped, replace the plunger and bushing after performing the last test.

14. Install the injector in a GM Diesel injector comparator and seal it firmly. Pull the rack out to the NO FUEL position.

15. Turn on the comparator operating switch. Then, move the rack to the FULL FUEL position.

16. Wait 30 seconds, and then press the fuel flow starting button.

17. After fuel flow stops (after 1000 strokes), pull the rack to the NO FUEL position, turn the comparator off, and reset the counter.

18. Observe the reading on the vial, and refer to the chart on the comparator to see if the volume

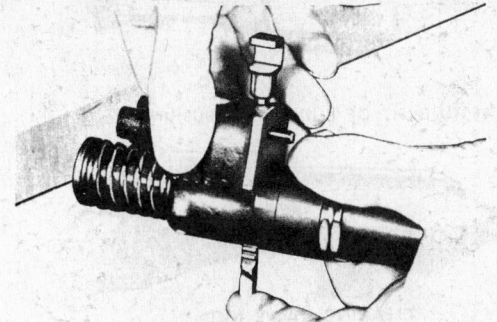

Checking the Freeness of the Rack and Plunger
(© G.M. Corp)

SCREW DRIVER

Removing Follower Stop Pin
(© G.M. Corp)

671

of fuel falls within specified limits.

Injector Disassembly

If required, the injector may be disassembled in the following manner:

1. Support injector upright in injector assembly fixture, and remove filter caps, springs, filter elements and gaskets. NOTE: Whenever injector is dsassembled, filter elements and gaskets should be discarded and replaced with new filters and gaskets.
2. Compress follower spring. Then, using a screwdriver, raise spring above stop pin and withdraw pin. Allow follower spring to rise gradually.
3. Refer to illustration for details and remove plunger follower, spring and plunger as an assembly.
4. Reverse the injector in the fixture and loosen nut from injector body.
5. Remove the spray tip and valve parts off bushing and place in a clean receptacle until ready for assembly.
6. When an injector has been in use for some time, the spray tip, even though clean on the outside, may not be pushed readily from the nut with the fingers. In this event, support the nut on a wood block and drive the tip down through the nut, as shown in illustration.
7. Remove spill deflector, and seal ring from injector nut.
8. Remove plunger bushing, gear retainer, and gear from injector body.
9. Withdraw injector control rack from injector body.

Injector Service

1. Wash all injector parts in clean fuel oil or another suitable solvent. Dry them with filtered compressed air. Use extra care when cleaning passages, drilled holes, etc.
2. Clean out the spray tip with a reamer, as shown in the illustration. Turn the reamer clockwise during cleaning. Wash the spray tip, dry it with compressed air, and then clean the spray tip orifices with a .005" diameter wire. *NOTE: The wire must be honed until it is smooth and free of burrs, using a stone, before using it for cleaning purposes.* After orifice cleaning, again wash the spray tip in fuel oil or solvent and dry it.
3. Clean fuel and rack holes with appropriate cleaning brushes, and then blow all passages dry with compressed air.
4.
5. When handling the injector plunger, never touch the finished surfaces with the fingers. Wash the plunger and its bushing in clean fuel oil and dry it with compressed air. Wrap tissue paper around the bushing cleaner tool and out the bushing bore, as shown in the illustration. Then, submerge the parts in clean fuel oil to protect them from corrosion during the remaining steps of injector service.
6. Inspect the control rack and rack gear teeth for excessive wear or damage, and check for excessive wear in the bore of the gear. Replace damaged or worn parts.
7. Inspect the ends of the spill deflector for sharp edges. Remove sharp edges with a medium stone.
8. Inspect the plunger spring for defects and check in a spring tester for a free length of 1.659" (1.504"-92 Series), and compres-

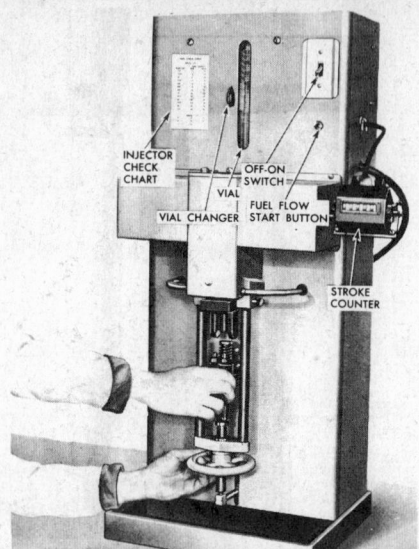

Injector Comparator
(© G.M. Corp)

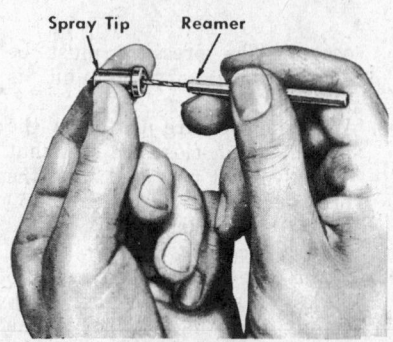

Reaming Spray Tip
(© G.M. Corp)

sion to 1.028" with a load of 53-59 lbs (70 lbs-92 Series). Replace the spring if less than 49 lbs. will bring it to the 1.028" length.

9. Check the seal ring area of the injector body for burrs and scratches. Check the surface

CLEANING BUSHING

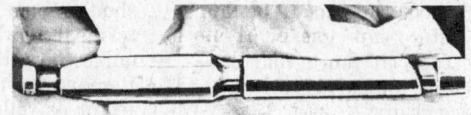

POLISHING PLUNGER WITH TISSUE

DRYING BUSHING WITH FILTERED AIR

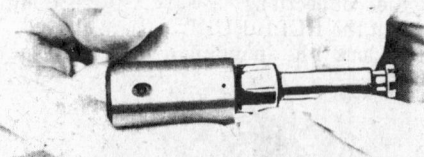

TESTING FIT OF PLUNGER IN BUSHING

CLEANING BUSHING WITH TISSUE AND TOOL

CLEANING VALVE NUT

Cleaning Plunger, Bushing, and Nut
(© G.M. Corp)

which contacts the injector bushing for scratches or scuff marks. Lap this surface as described below if necessary.

10. Inspect the plunger or any damage like that shown in the illustration. Replace it and the bushing as a set if such damage exists. Inspect the plunger for damage at the portion which rides inside the gear. If sharp edges are found, remove them with a 500 grit stone. Wash the plunger after stoning.

11. Inspect the plunger bushing for cracks or chipping. Slip the plunger into the bushing and check for free movement. If wear or scoring prevents free movement, replace the plunger and bushing with a new set.

12. Examine the spray tip seating surface of the injector nut for nicks, burrs, or brinelling. Reseat the surface if damage is mild, or replace the nut if it is severe.

13. Inspect the injector valve spring for wear or breakage, and replace if either is evident. When the spring is compressed to length of 240″, the load should be $4\frac{3}{4}$.-$5\frac{3}{4}$ lbs. If compression to .240″ requres less than $4\frac{1}{4}$ lbs., replace the spring.

14. Inspect the sealing surfaces of the spray tip and valve parts shown in the illustration. Inspect the parts for any imperfection whatever with a magnifying glass, and replace as necessary.

15. Inspect the injector body, bushing, spray tip, and valve assembly seating surfaces, and, if necessary, lap them as described below. Whenever reinstalling used valve parts, all the sealing surfaces must be lapped as described below except those on the crown valve. This is also good practice on new parts.

16. Clean the lapping blocks, using compressed air only. Spread 600 grit dry lapping powder on one of the blocks.

17. Place the part to be lapped flat on the block. Move the part back and forth in a figure eight motion, applying just sufficient pressure to keep the part flat on the block.

Draw the part across a clean piece of tissue paper placed on a flat surface every four or five passes to clean it. Inspect it after each cleaning to ensure that it is lapped just enough to ensure production of a flat surface.

18. Wash the part in cleaning solution and dry it with compressed air.

19. Apply lapping powder to a second lapping block and move the part lightly across the block just a few times to give it a smooth surface. Wash the part in cleaning solution and dry it with compressed air.

20. Place the part on a third lapping block and, without lapping powder, repeat the process very lightly just a few times.

21. Inspect the edge of the hole in the crown valve seat. Use a magnifying glass and check to make sure the seat is a true circle and has a perfectly smooth surface. If the edge is imperfect, lap it as described below:

Mount a deburring tool in a drill motor and place a small amount of lapping powder and oil in a mixture on the tool. Place the valve seat over the pilot of the tool and start the motor. Gently grab the valve seat with the fingers and touch it lightly against the rotating tool. When the seat is uniform, flat lap the seat slightly. Clean the part and examine the width of the edge. The width of the chamfer at the edge of the hole should be .002″-.005″. A greater width will lower the pop pressure of the injector.

22. Wash all parts that have been lapped in a suitable solvent and dry with compressed air. Clean the inside of the injector bushing by wrapping clean tissue around a bushing cleaner tool and rotating it inside the bushing as shown in the illustration.

Injector Assembly

Before starting to assemble an injector, it is necessary to have an extremely clean bench on which to work and place the parts. Refer to illustrations for proper relative posi-

tions of the injector parts and the methods for their assembly prior to proceeding with assembly. Since the plunger and bushing are matched parts, they must be considered as one piece and, if one is replaced, both must be replaced.

Assemble Injector Filters

New filters and gaskets should always be used when reassembling injectors.

1. Holding the injector body right side up, place a filter in each of the fuel cavities in the top of the injector body. Note that the fuel filters have a dimple in one end. When assembling the filters, always have the dimple at the bottom.

2. Place a spring above each filter, a new gasket up against the shoulder of each filter cap, lubricate the threads, and tighten the filter cap in place in the injector body to a torque of 65 to 75 ft.-lbs., using a 9/16″ deep socket wrench. It is important that the filter caps be tightened securely so as to compress the gaskets and effect a good seal with the injector body. Also, when the caps are tightened, they compress the filter springs which hold the filters securely in place so all fuel entering the injector is properly filtered.

3. Install covers on injector filter caps to prevent any dirt particles from entering injector. Be sure covers are clean.

Assemble Rack and Gear

When rack and gear are assembled, the marked tooth of the gear must be engaged between the two marked teeth on the rack.

1. Hold the injector body bottom end up and slide the rack through the proper hole in the body. The two marked teeth can then be observed when looking into the bore for the gear from the bottom of the injector body. The injector rack can be placed in the injector body in only one position and have the tooth marks show in the opening for the gear.

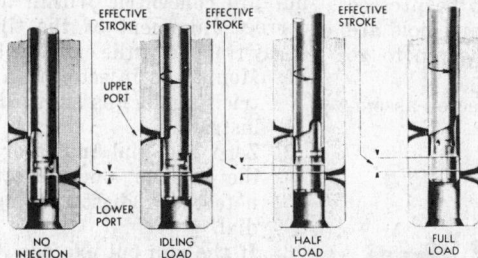

Fuel Metering From No Injection To Full Injection, Produced By Rotating Plunger with Control Rack

Fuel metering produced by rotating plunger
(© G.M.C.)

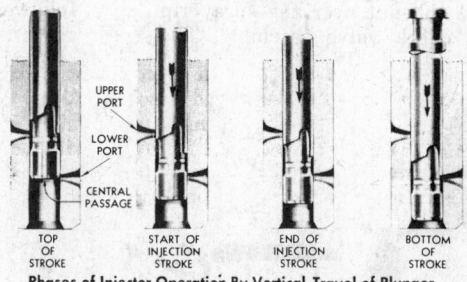

Phases of Injector Operation By Vertical Travel of Plunger

Injector operation by vertical travel of plunger
(© G.M.C.)

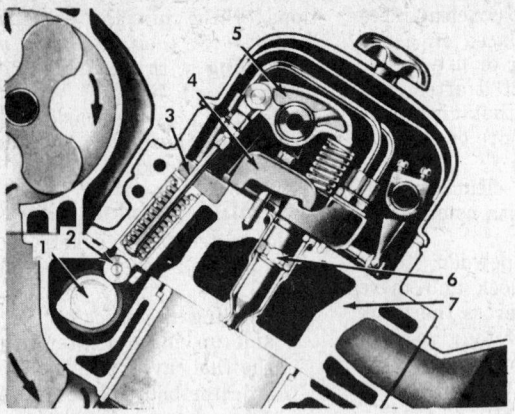

Injector installed
(© G.M.C.)

1 Camshaft
2 Cam follower assy.
3 Push rod
4 Hold-down clamp

5 Rocker arm
6 Injector assy.
7 Cylinder head

Injector removal
(© G.M.C.)

2. Holding the rack in position so the tooth marks show, slide the gear into proper engagement with the rack.
3. Slide gear retainer down on top of gear; then place plunger bushing down onto retainer with locating pin in bushing guided into slot of injector body.

Assemble Injector Valve Parts

Refer to the illustration and assemble as described below.
1. Support the body of the injector in an injector assembly fixture with the bottom end up.
2. Locate the seal ring on the shoulder of the body. Slide the spill deflector over the barrel of the bushing.
3. Place the valve seat on the end of the bushing. Insert the stem of the valve into one end of the valve spring and the valve stop into the opposite end. Lower the valve cage over the assembly so that the stop seats in the cage. Position the valve cage on the valve seat.
4. Locate the check valve on the center of the cage. Place the spray tip over the check valve and against the cage. Then, lubricate the threads and carefully pilot the nut over the spray tip and check valve assembly. The

tip will slide through the hole in the small end of the nut.
5. Screw the nut into place, making sure the valve assembly doesn't shift. If they do shift, turn the end of the spray tip with the fingers while screwing the nut onto the body by hand. A wrench must not be used to tighten the nut at this time!

Assemble the Plunger and Follower

1. Slide the head of the plunger onto the follower. Insert the assembly through the plunger spring and onto the injector.
2. Invert the injector in the fixture, so the connector is upward, and push the rack all the way in. Insert the free end of the plunger into the top of the injector body.
3. Start the stop pin into position in the injector body so the bottom coil of the follower spring rests on the flange on the stop pin (see illustration). With the follower slot and the injector body hole in alignment, and the flat side of the plunger positioned to engage the flat side of the gear, press down on the top of the follower and press the follower stop pin into position. The follower will have to be pressed down until the slot and hole are aligned and allow the pin to go in.
4. Again invert the injector assem-

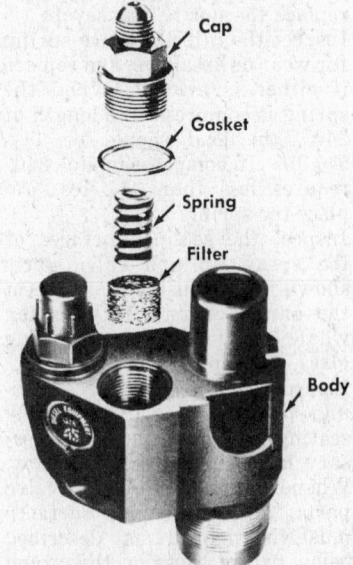

Injector body, filters, springs and cap
(© G.M.C.)

bly, and, using a socket and torque wrench drive, tighten the nut to 55-56 ft lbs. (75-85 on 92 Series).

Check Spray Tip Concentricity

This step must be performed to ensure that the spray tip and injector nut are concentric within .008" for correct alignment of the tip in the spray tip hole of the cylinder head.
1. Mount the injector in a concentricity gauge as shown in the illustration.
2. Zero the dial indicator. Rotate the injector one full turn and note the highest reading on the dial.
3. If the run out exceeds .008", remove the injector from the gauge, loosen the injector nut, re-center the spray tip, and retorque the nut.

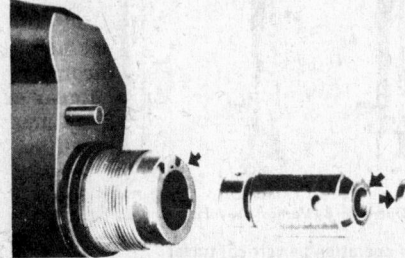

Sealing Surfaces Which Require Lapping
(© G.M. Corp)

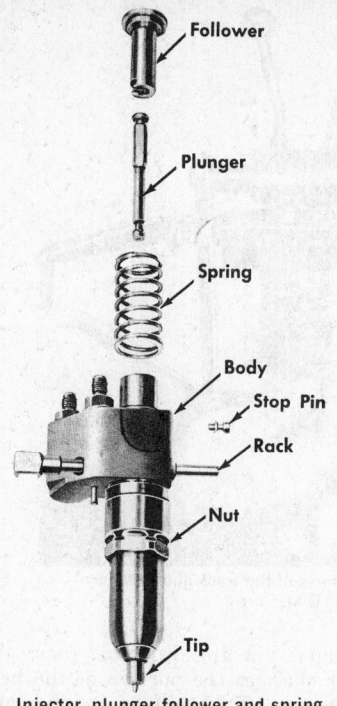

Injector, plunger follower and spring
(© G.M.C.)

tube in the cylinder head. If not, the injector may be cocked resulting in a fractured spray tip.

Injector tube bevel seat reamer should be used to clean the carbon out of the tube before installing the injector. Refer to "Reaming" in "CYLINDER HEAD" section. Care must be exercised to remove ONLY the carbon so that the proper clearance between the injector body and cylinder head is maintained. Pack the flutes of the reamer with heavy grease.

1. Insert the injector into the injector tube with the locating dowel registering with the locating hole in the cylinder head and with the pin on the injector control tube lever registering with the injector control rack lever.

2. Place injector clamp in position, then install washer and bolt. Tighten bolt to 20-25 ft.-lbs. *NOTE: Check injector rack for free movement. Excess torque will cause the injector control rack to stick or bind. Also, make sure clamp does not interfere with operation of valve or injector springs.*

3. Move rocker arm assembly into position and secure rocker arm bracket bolts to torque of 90 to 100 ft.-lbs.

4. Connect fuel pipes to injector fuel connectors, torquing the connections to 12-15 ft.-lbs.

5. As a precautionary measure against any possibility of scoring injector parts upon initial installation due to lack of lubrication, any entrapped air should be bled from the injector before tightening the connections on the fuel outlet side of the injector. This may be accomplished by cranking the engine briefly with the injector rack in the NO FUEL position and then tightening the fuel pipe connection.

6. When installing injector fuel lines, connections should be tightened only enough to prevent leakage of fuel. Excess tightening may result in the flared end

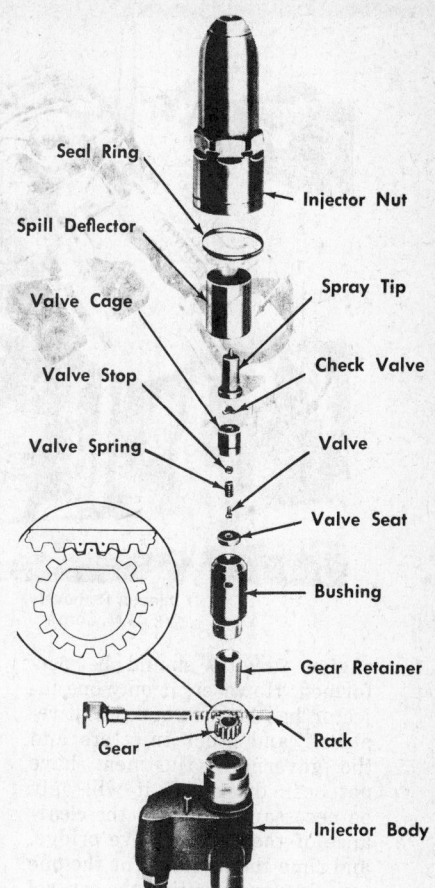

Injector body, nut, rack, bushing and spray tip details
(© G.M.C.)

Injector Testing

Before placing a reconditioned injector in service, all of the tests (except the visual inspection of the plunger) previously outlined in this section under "Injector Tests" must be performed again.

If an injector is not to be used immediately, caps should be installed on the injector caps. The injector test oil remaining in the injector after the fuel output test will serve as a rust preventive while the injector is in storage.

Injector Installation

Before installing a new or reconditioned injector in an engine, the carbon deposits must be removed from the beveled seat of the injector hole

4. If repeated attempts fail to center the tip within the specified runout, the assembly of the injector is unsatisfactory.

of the fuel line being twisted or fractured. A torque wrench adapter is essential to properly tighten fuel line connector nut to 12-15 ft. lbs. torque. Use of this adapter and torque wrench is highly recommended.

7. When all injector lines are installed, the engine should be run long enough (during tune-up procedure) to check for leaks. Should leaks occur, the connections should be tightened only enough to stop the leak.

8. Following installation of the injectors in the engine, a complete

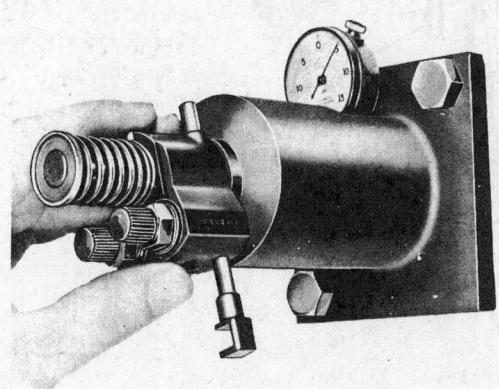

Checking Spray Tip Concentricity
(© G.M. Corp)

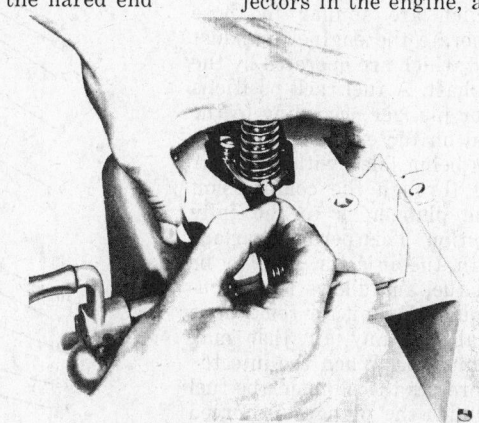

Installing Follower Stop Pin
(© G.M. Corp)

Injector Removal
(© G.M. Corp)

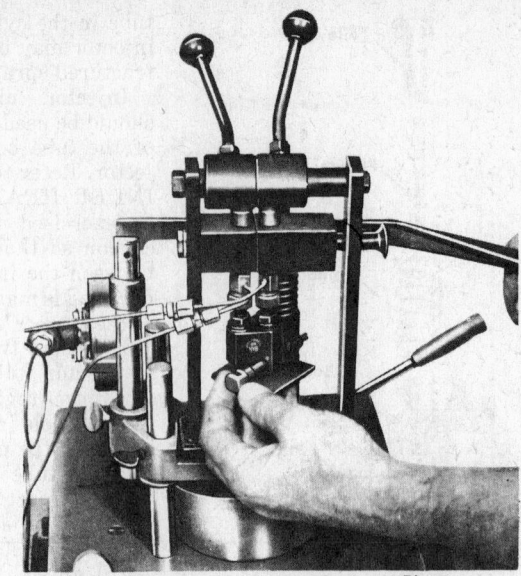

Checking the Freeness of the Rack and Plunger
(© G.M. Corp)

engine tune-up should be performed. However, if only one injector has been removed and replaced, and other injectors and the governor adjustment have not been disturbed, it will only be necessary to adjust the clearance of the valves, valve bridge, and time the injector for the one cylinder, and position the control rack on the cylinder involved to correspond with that of the other racks.

Needle Valve Fuel Injectors

The injectors used in Detroit Diesel Engines are of the unit type, which means that the injector requires only that fuel be supplied to it under pressure, and performs all metering, injection, and atomization functions without the help of a complex distributor type injector pump and the accompanying high pressure fuel lines. This type of design insures a minimum of problems with leakage and uneven fuel distribution.

The injector plunger is operated by a follower, rocker lever, pushrod, and tappet which are similar to those used to operate the engine's exhaust valves, and which are operated by the same camshaft. A fuel rack positions the injector plunger according to the load placed on the engine. While the cylinder is being filled with fresh air and going through the compression stroke, the plunger is in the fully raised position. This permits various passages in the injector unit to be filled with fuel and allows fuel circulation through the unit for removal of excess heat and any air that may have accumulated. When the injector rocker depresses the plunger, the fuel trapped under the plunger is forced down through the check valve. Pressure then builds up in the passages to

the needle valve, lifts that valve off its seat, and allows fuel to be injected through the spray tip in atomized form.

The injector plunger contains passages which allow fuel to pass between the area below the plunger and the helical cutout on the upper part of the plunger. Since both the upper and lower ports in the plunger bushing connect with the low pressure fuel supply system, injection will only occur during that portion of the plunger stroke during which both ports are covered. Rotation of the

plunger via the injector rack and gear changes the portion of the helical cutout which effects opening and closing of the ports (see illustration), thus controlling the amount of fuel injected. The high pressure required to force fuel through the needle valve insures that injection will occur only when both ports are fully blocked off. Thus, metering is precise, and plenty of pressure is available for efficient atomization.

The injectors are mounted in the heads so their spray tips will protrude through the lower surface of the head

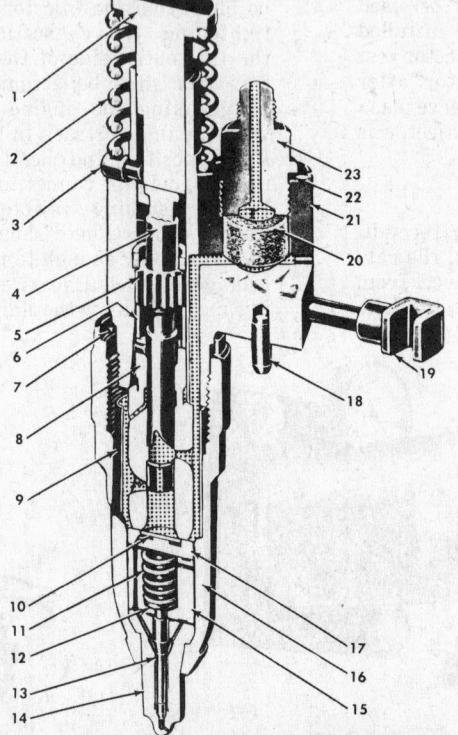

1	Follower
2	Follower Spring
3	Follower Stop Pin
4	Plunger
5	Gear
6	Gear Retainer
7	Seal Ring
8	Plunger Bushing
9	Spill Deflector
10	Check Valve
11	Valve Spring
12	Spring Seat
13	Needle Valve
14	Spray Tip
15	Spring Cage
16	Body Nut
17	Check Valve Cage
18	Dowel
19	Control Rack
20	Fuel Filter
21	Injector Body
22	Gasket
23	Filter Cap

Needle Valve Type Fuel Injector
(© G.M. Corp)

and into the combustion chamber. A water cooled copper tube surrounds the injector and permits it to be removed without draining the cooling system. A mounting clamp is bolted to the head and fits into a machined recess located on either side of the injector body. A dowel pin in the body of the injector registers with a hole in the head for accurately locating the assembly.

Injector Removal

1. Remove the valve cover. Remove the fuel lines from the injector(s) to be removed, immediately covering the two fuel feed fittings with shipping caps.
2. Crank the engine over until the three clevis pins at the outer ends of the rocker arms are in line.
3. Loosen the bracket bolts holding the brackets to the head and swing the rocker assembly over, away from the valves and injector.
4. Loosen the injector clamp bolt, and remove the bolt, washer, and clamp.
5. Loosen both injector rack control lever adjusting screws and slide the lever away from the injector.
6. Free the injector from its seat and lift it from the head.
7. Cover the injector holes to keep dirt out of the head. Clean the outside of the injector and dry it with compressed air.

Injector Tests

Inspect the exterior of the injector. If there is no damage, the tests following should be performed to determine whether or not the injector requires overhauling. If the injector passes all these tests and examination of the plunger shows no damage, it may be returned to service. If the plunger is visibly damaged or the unit fails to pass any of the tests, the unit requires overhaul. All tests should be performed before an overhauled injector is put back into service.

1. Mount the injector in the tester, locating the dowel in the proper slot in the adaptor plate. Position the injector support and the handle support at the proper height.
2. Place the handle on top of the injector follower and then close the inlet and outlet clamps to hold the injector in the tester. With the control rack in the NO FUEL position, push the handle down to depress the follower as far as it will go. Then, release the pressure on the handle. Begin moving the control rack back and forth and continue doing so while gradually allowing the follower to rise to its fully raised position. Any binding in the rack indicates internal damage or dirt.
3. Dry the injector thoroughly with compressed air. Pump up the tester, maintaining a pressure of 1600–2000 psi. Inspect for leaks at the filter cap gaskets, body plugs, injector nut seal ring, and rack hole and spray tip orifices.

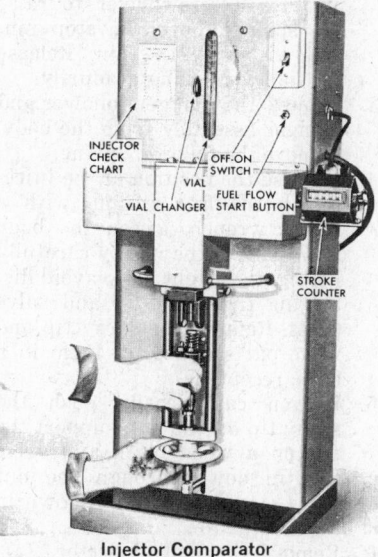

Injector Comparator
(© G.M. Corp)

4. Mount the injector valve parts on a tool body mounted in the adapter in the tester. Operate the pump handle until the spray tip valve has opened several times. Then, operate the handle smoothly at about 40 strokes per minute and note the opening pressure of the needle valve. The pressure should be 2000-3200 psi, the action should be sharp, and a finely atomized spray should be produced. If the opening pressure is low or the atomization is poor, the needle valve and tip assembly must be replaced.
5. Actuate the pump handle several times and then maintain 1500 psi for 15 seconds. Inspect the spray tip for leakage. There should be no fuel droplets, although a very slight wetting is permissible.
6. Cease pumping and time the pressure drop from 1500 to 1000 psi. The drop should require at least five seconds, timed with a watch. If the pressure drop takes less than five seconds, replace the valve and tip assembly.
7. Put the injector rack in FULL FUEL position, and pump the handle to obtain 200-250 psi. Close the fuel shut-off valve. Operate the tester handle at a rate of about 40 times per minute and observe the spray pattern. All spray orifices should be open and injecting evenly. The beginning and ending of injection should be sharp and all the fuel should be well atomized. Sluggish action, dribbling, or plugging of orifices indicate that disassembly and inspection are required.
8. Remove the injector from the tester and support it in a holding fixture. Compress the follower spring and, with a screw driver, raise the spring above the stop pin and withdraw the pin. Gradually release the spring.
9. Lift the follower and plunger from the injector body as an assembly. Inspect the plunger for chipping like that shown in figure 41 C. If plunger is chipped, the injector should be reconditioned before going on to the next test.
10. Invert the injector in the assembly fixture, and push the control rack all the way in. Place the follower spring and stop pin on the injector body so the bottom coil of the spring rests on the narrow flange of the pin. Slide the head of the plunger into the slot in the

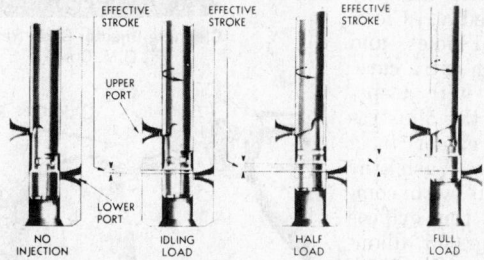

Fuel Metering From No Injection To Full Injection, Produced By Rotating Plunger with Control Rack

How Metering is Accomplished by Rotating the Plunger
(© G.M. Corp)

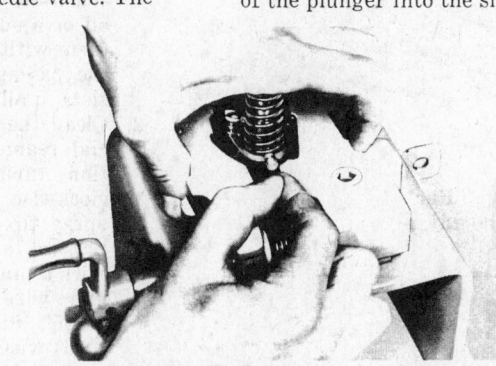

Installing the Follower Stop Pin
(© G.M. Corp)

(Detroit 2-cycle) Diesel Engines

follower as shown in the illustration. Position the follower and plunger over the injector body and align the slot in the follower with the stop pin hole in the body. Align the flat side of the plunger with the flat in the gear and lower the plunger and follower into the injector body until the follower rests on the spring. Press down on the follower and press the stop pin into the slot in the follower until the spring drops into the slot in the stop pin.

11. Mount the injector in the comparator and turn the wheel to clamp it and the adaptor in position. Make sure the counter is preset to 1,000 strokes.
12. Put the rack in the NO FUEL position. Turn the power switch on.
13. Push the injector rack to the FULL FUEL position. Allow the unit to run for about 30 seconds for purging of air. Then, push the fuel flow start button.
14. After the fuel flow stops (1,000 strokes), pull the rack out to the NO FUEL position. Then turn the unit off and reset the counter.
15. Note the reading on the fuel collection vial and compare it with the chart on the comparator. If the quantity is not within specified limits, the unit requires servicing.

Injector Disassembly

1. Support the injector in the assembly fixture and remove the filter caps, gaskets, and filter elements. Discard used filters and gaskets.

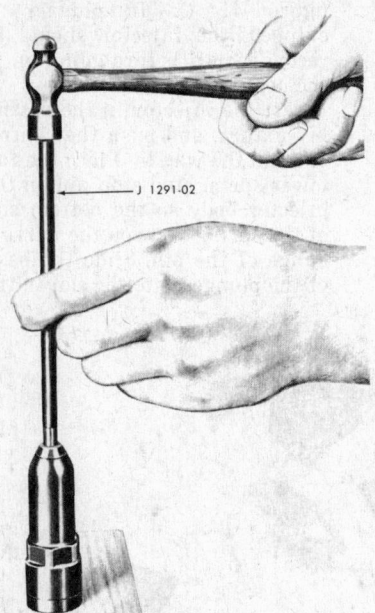

Removing the Spray Tip from the Injector Nut
(© G.M. Corp)

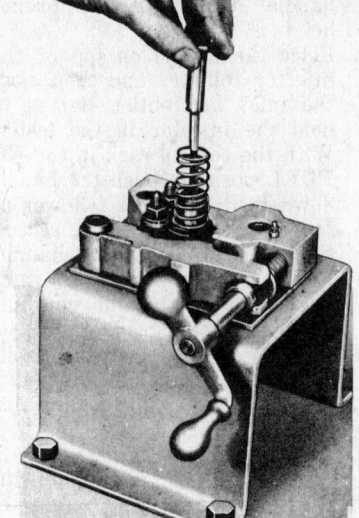

Removing Follower Spring and Plunger
(© G.M. Corp)

2. Compress the follower spring and use a screwdriver to raise the spring above the stop pin, and withdraw the pin. Release the follower spring gradually.
3. Remove the plunger follower and plunger assembly from the body. Remove the follower spring.
4. Reverse the position of the injector in the fixture and, with a socket wrench, loosen the body nut. Remove the nut by carefully lifting it straight up to avoid dislodging the spray tip and valve parts. Remove the spray tip and valve parts and place them in a clean receptacle.
5. If you can't easily push the spray tip off the nut, support the nut on a wood block and drive the tip down through the nut using a special tool, as shown in the illustration.
6. Remove the spill deflector. Lift the bushing straight out of the body.
7. Remove the body from the holding fixture. Invert the body and catch the gear and its retainer. Withdraw the rack and the seal ring.

Injector Service

1. Wash all the parts in clean fuel oil or a suitable solvent and dry them with compressed air. Clean the passages, drilled holes, and slots in all parts with extra care.
2. Clean the spray tip with a special reamer, as in the illustration, turning the reamer in a clockwise direction. Wash the spray tip and dry it with compressed air. Clean the orifices with a pin vise using .005" diameter wire for tips with .005"-.0055" diameter holes, and a .006" wire for tips with .006" diameter holes. Hone the end of the wire to remove burrs and

taper the end 1/16" back from the tip with a stone. Allow the wire to protrude 1/8" from the holder. After cleaning, wash the spray tip in clean fuel oil and dry it with compressed air. Note: During reaming of the spray tip, avoid contacting the needle valve seat.
3. Clean and brush all injector body passages with an appropriate brush. Dry all passages with compressed air.
4. Insert a reamer into the top of the injector body, turning it clockwise a few turns. Then, remove the reamer and check the face of the ring to make sure the reamer has been contacting it. Repeat the procedure until the reamer makes contact over the entire surface of the ring. Clean up the opposite side of the ring similarly.
5. Insert a .375" straight fluted reamer inside the ring bore in the injector body. Turn the reamer clockwise to remove any burrs inside the ring bore. Then, wash the injector body in clean fuel and dry with compressed air.
6. Use the two types of carbon removing tools to remove carbon deposits from the spray tip seat and the lower end of the injector nut. The tool to be used should be inserted carefully and turned clockwise. Be careful not to remove metal or set up burrs.
7. Wash the injector nut in clean fuel oil and dry with compressed air. Wash the plunger and bushing with clean fuel oil and dry

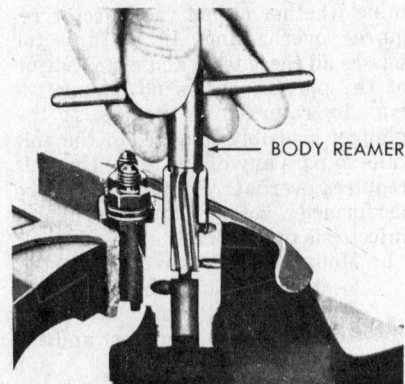

← BODY REAMER

Cleaning Injector Body Ring
(© G.M. Corp)

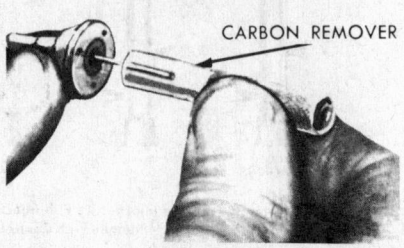

CARBON REMOVER

Cleaning the Spray Tip
(© G.M. Corp)

CLEANING BUSHING

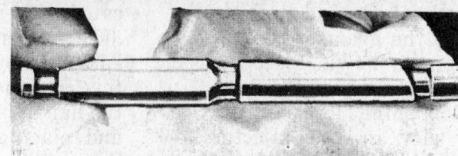

POLISHING PLUNGER WITH TISSUE

DRYING BUSHING WITH FILTERED AIR

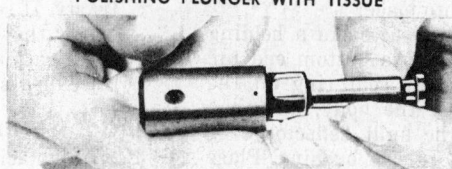

TESTING FIT OF PLUNGER IN BUSHING

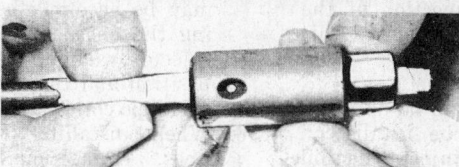

CLEANING BUSHING WITH TISSUE AND TOOL

CLEANING VALVE NUT

Cleaning plunger, bushing and nut
(© G.M.C.)

with compressed air, avoiding contact with the finished plunger surfaces.

8. Submerge all parts in a receptacle containing clean fuel oil. Keep all parts of each injector assembly together.

9. Inspect the teeth on the control rack and control rack gear for excessive wear or damage. Check for excessive wear in the bore of the gear. Replace any damaged or worn parts.

10. Inspect the ends of the spill deflector for sharp edges or burrs which could create burrs on the injector body or nut. Remove burrs with a medium stone.

11. Inspect the follower spring and check it with a spring tester. Free length should be approximately 1.659″. Load should be at least 48 lbs. when compressed to 1.028″. Replace the spring if it

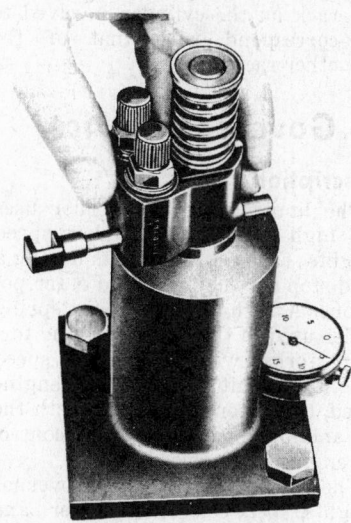

Checking Spray Tip Concentricity
(© G.M. Corp)

does not pass both of these tests.

12. Check the seal ring area for burrs or scratches. Check the surface which contacts the injector bushing for scratches, scuff marks, etc. Lap this surface if it is faulty.

13. Inspect the plunger for scoring, chipping, or excessive wear. Check for sharp edges where the plunger rides in the gear. If sharp edges exist, remove them with a 500 grit stone. Wash the plunger after stoning. Inspect the plunger bushing for cracks or chipping. Check the locating pin in the bushing and replace it if it is damaged or sheared off. Check the plunger and bushing for free movement. Replace badly worn, chipped, or scored plungers and bushings with a new assembly.

14. Examine the spray tip seating surface of the injector nut for nicks, burrs, or brinelling. Reseat the seating surface, if necessary, or replace the nut if it is severely damaged.

15. Inspect the injector valve spring and replace it if it is worn or broken.

16. Inspect the arrowed sealing surfaces shown in figure 53. Examine the surfaces under a magnifying glass for burrs, nicks, erosion, cracks, chipping, or excessive wear, and replace as necessary. Check the spray tip for enlarged orifices and replace it if necessary.

17. Examine the needle valve for wear, scoring, or damage to the quill where it contacts the spring seat. If the needle valve is scored or damaged, replace the spray tip assembly. Lap all surfaces indicated by arrows as described

below. Lap also the sealing surfaces of new parts to remove burrs or nicks picked up during handling.

18. Clean three lapping blocks with compressed air only. Spread 600 grit dry lapping powder on two of the blocks.

19. Place each part flat on the block. Using a figure eight motion, move the part back and forth across the block, maintaining just sufficient pressure to keep the part completely in contact with the block.

20. After each five passes, clean the part with tissue paper by placing the paper on a flat surface and drawing the part across it. Lap the part just until it is flat—avoid excessive lapping. When the part is flat, wash it in solvent and dry with compressed air.

21. Repeat the lapping process very briefly on the second block. Then, wash and dry the part as before.

22. Lap the part very, very lightly on the third block without the use of powder. Wash again and dry with compressed air.

Injector Assembly

1. Insert the new filters, slotted end up, in each of the two fuel cavities. Place new gaskets on the filter caps, coat the threads with oil, and install the caps, torquing them to 65-75 ft-lbs. Install clean shipping caps on the filter caps.

2. With the injector body bottom end up, slide the rack through the hole in the body. Look into the bore and slide the rack back and forth until you can see the two drill marks shown in the illustration. Hold the rack in this position while placing the gear

in the injector body and meshing the teeth of the gear so the drill marks align as shown. Place the gear retainer on top of the gear, align the bushing locating pin with the slot in the injector body, and slide the end of the bushing into place.

3. Mount the injector in a holding fixture with the bottom end up. Install a new seal ring on the shoulder of the body, and then position the spill deflector over the barrel of the bushing. Place the check valve on top of the center of the bushing, and then place the check valve cage over the valve and against the bushing.

4. Insert the spring seat inside the valve spring, and insert the assembly into the spring cage with the spring seat going first. Place the spring cage, spring seat, and valve spring assembly, with the valve spring down, on top of the check valve cage. Insert the needle valve (tapered end down) into the spray tip.

5. Place the spray tip and needle valve on top of the spring cage with the small end of the needle valve in the hole in the spring cage. Lube the threads of the injector nut, and then carefully position the nut and screw it in place by hand, watching carefully to make sure the valve assembly does not shift. If the valve parts are not centrally located, assembly may be eased by turning the end of the spray tip while screwing the nut onto the body.

6. Invert the injector in the fixture, and push the control rack all the way in. Place the follower spring and stop pin on the body so the bottom coil of the spring rests on the narrow flange of the pin. Slide the head of the plunger into the slot in the follower, and then position the plunger and follower over the injector body and align the slot in the follower with the stop pin hole in the body. Align the flat side of the plunger with the flat in the gear, and lower the plunger and follower straight into the gear and injector body until the follower rests on the spring.

7. Press down on the follower and press the stop pin into the slot in the follower until the spring drops into the stop pin slot. Invert the injector and use a special socket wrench and torque drive to tighten the nut to 75-85 ft.-lbs.

8. Place the injector in a concentricity gauge. Zero the dial indicator. Rotate the injector 360° and note the total run-out indi-

cated. If the run-out exceeds .008", attempt to recenter the spray tip by removing the injector from the gauge and loosening the nut. After retorquing the nut and placing the unit back into the gauge, recheck the concentricity. If several attempts fail to bring the concentricity within specifications, check the entire injector assembly.

Injector Testing

Before placing a reconditioned injector in service, all of the tests (except the visual inspection of the plunger) previously outlined in this section under "Injector Tests" must be performed again.

If an injector is not to be used immediately, caps should be installed on the injector caps. The injector test oil remaining in the injector after the fuel output test will serve as a rust preventive while the injector is in storage.

Injector Installation

Before installing a new or reconditioned injector in an engine, the carbon deposits must be removed from the beveled seat of the injector hole tube in the cylinder head. If not, the injector may be cocked resulting in a fractured spray tip.

Injector tube bevel seat reamer should be used to clean the carbon out of the tube before installing the injector. Refer to "Reaming" in "CYLINDER HEAD" section. Care must be exercised to remove ONLY the carbon so that the proper clearance between the injector body and cylinder head is maintained. Pack the flutes of the reamer with heavy grease.

1. Insert the injector into the injector tube with the locating dowel registering with the locating hole in the cylinder head and with the pin on the injector control tube lever registering with the injector control rack lever.

2. Place injector clamp in position, then install washer and bolt. Tighten bolt to 20-25 ft.-lbs. *NOTE: Check injector rack for free movement. Excess torque will cause the injector control rack to stick or bind. Also, make sure clamp does not interfere with operation of valve or injector springs.*

3. Move rocker arm assembly into position and secure rocker arm bracket bolts to torque of 90 to 100 ft.-lbs.

4. Connect fuel pipes to injector and fuel connectors, torquing the connectors to 12-15 ft.-lbs.

5. As a precautionary measure against any possibility of scoring injector parts upon initial installation due to lack of lubrication, any entrapped air should be bled from the injector before tightening the connections on the fuel outlet side of the injector. This may be accomplished by cranking the engine briefly with the injector rack in the NO FUEL position and then tightening the fuel pipe connection.

6. When installing injector fuel lines, connections should be tightened only enough to prevent leakage of fuel. Excess tightening may result in the flared end of the fuel line being twisted or fractured. A torque wrench adapter is essential to properly tighten fuel line connector nut to 12-15 ft.-lbs. torque. Use of this adapter and torque wrench is highly recommended.

7. When all injector lines are installed, the engine should be run long enough (during tune-up procedure) to check for leaks. Should leaks occur, the connections should be tightened only enough to stop the leak.

8. Following installation of the injectors in the engine, a complete engine tune-up should be performed. However, if only one injector has been removed and replaced, and other injectors and the governor adjustment have not been disturbed, it will only be necessary to adjust the clearance of the valves, valve bridge, and time the injector for the one cylinder, and position the control rack on the cylinder involved to correspond with that of the other racks.

Governor Service

Description

The limiting speed governor uses two high speed and two low speed weights; each set of weights having a fixed stop for the inner and outer positions. A high and low speed spring is also used in this governor; the former works with the high speed weights to limit the maximum engine speed, the latter works with both the low and high speed weights to control the engine idling speed.

The travel of the governor weights, between their inner and outer positions, is transferred to the injector racks by a system of cams

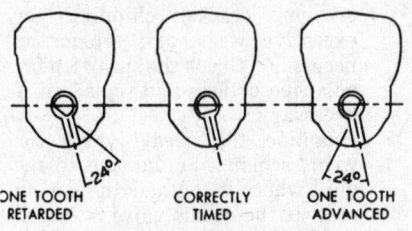

ONE TOOTH RETARDED   CORRECTLY TIMED   ONE TOOTH ADVANCED

Injector rack to gear giming
(© G.M.C.)

(Detroit 2-cycle) Diesel Engines

and levers over which a manual control can be used.

The governor, which is mounted at front of blower is divided into two main assemblies in separate housings. These assemblies are: Governor control housing and governor weight and lower housing, and the cover.

The two sets of weights are carried on a horizontal weight shaft inside the governor weight housing. The weight carrier shaft is mounted on an annular ball bearing at one end and opposite end is supported inside and driven by hollow blower rotor shaft. The blower end of governor shaft is serrated and engages with corresponding serrations inside blower shaft, which drives the governor shaft and weights.

Control mechanism transfers motion of governor weights to the injector racks. This mechanism consists of a vertical shaft mounted inside a housing, with a fork or yoke fixed at lower end, an operating lever fixed at upper end, and a high and low speed governing spring with suitable adjustments. The vertical shaft is mounted on annular ball bearings at upper and lower ends.

The motion of the governor weights is transferred to the vertical shaft through a movable riser on the weight carrier shaft and the fork on the lower end of the vertical shaft. This motion is, in turn, transferred to the injector control tube by means of the operating and differential levers on the upper end of the vertical shaft.

The cover assembly serves as a carrier for the throttle control lever, stop lever, and covers the top of the control housing.

The lower portion of governor is lubricated by means of a slinger attached to front end of blower lower rotor shaft. This slinger dips into a well of oil dammed up in blower housing cover and throws the oil onto all parts within governor weight housing. The upper portion of the governor, including the vertical shaft bearings and the control mechanism, is lubricated partly by splash from oil slinger on weight carrier shaft and partly by return oil through the vertical housing from cylinder head.

Limiting Speed Governor Removal—92 Series

1. Remove all accessories attached to the head, governor or engine front, which may interfere with removal.
2. Disconnect governor linkage.

3. Remove the governor housing cover.
4. Remove the fuel rods from the control link operating lever assembly and the injector control tube levers as follows:
   a. Remove the rocker covers.
   b. Remove the right bank fuel rod by removing the control link operating lever pin and the control tube lever clevis pin.
   c. Remove the left bank fuel rod by removing the clevis pin in the control tube lever and lift the connecting pin up out of the control link operating lever about ¾".
5. Loosen water by-pass tube clamps and remove tube.
6. Remove fuel lines from injection pump and crossover line from heads.
7. Remove fuel rod cover tube hoses.
8. Note washer locations and remove the bolts securing the governor and fuel pump to the blower.
9. Tap to loosen and remove governor and fuel pump.

In-Line 71 Engines

1. Disconnect governor control lever linkage.

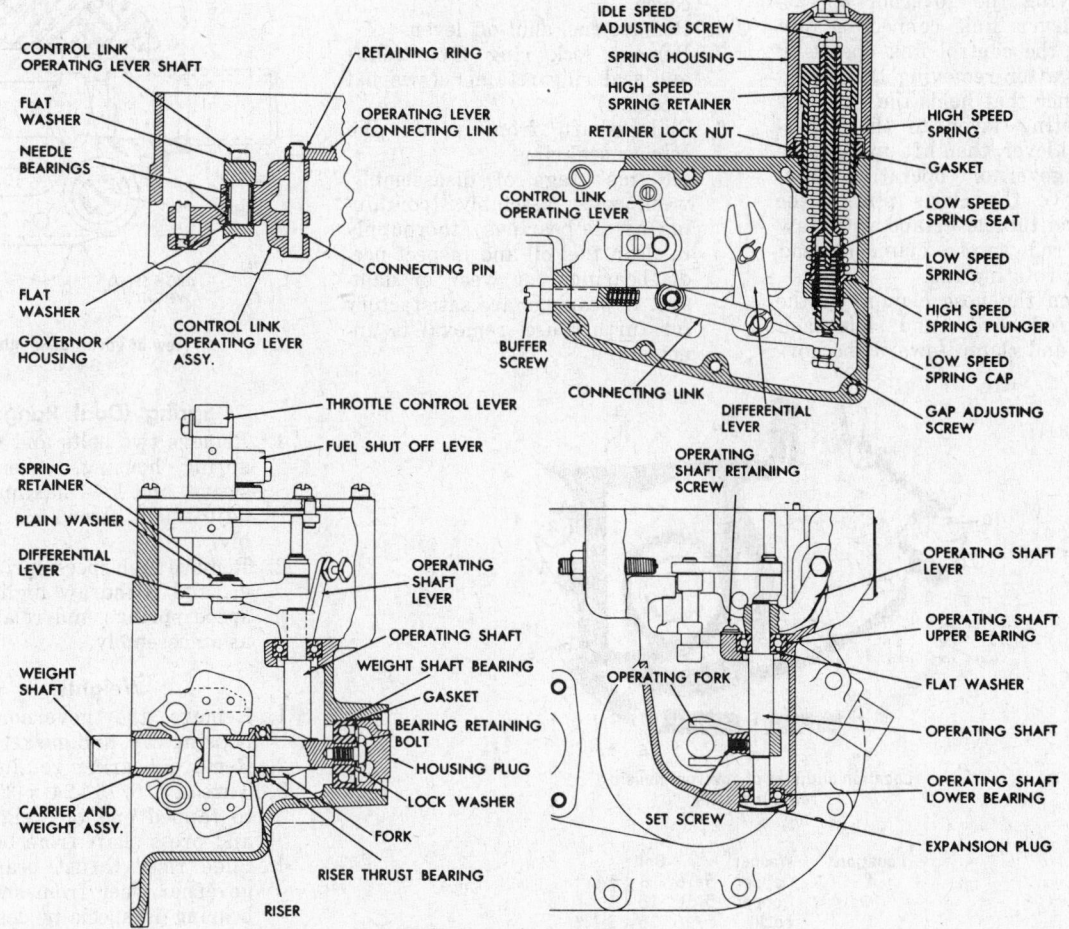

Limiting speed governor (Ⓒ G.M.C.)

681

2. Remove the breather pipe.
3. Remove the screws and lock-washers and pull the governor cover and gasket off the housing.
4. Disconnect the fuel rod from the differential lever (see illustration) and the injector control tube lever.
5. Disconnect the oil tube where it enters the weight housing.
6. Remove the governor-to-cylinder head bolts.
7. Remove the bolts and lock washers attaching the control housing to the weight housing and remove the control housing from the cylinder head and weight housing.
8. Remove the governor weight housing bolts, and pull the housing off the blower.

V-71 Models

1. Remove injector and rocker arm covers.
2. Remove cover and gasket from governor housing.
3. Remove the right bank fuel rod by removing the screw type pin, in the control link operating lever, and the clevis pin at the control tube lever and withdraw the fuel rod across cylinder head while rotating rod.
4. Remove the left bank fuel rod by removing the governor operating lever link connecting pin from the control link operating lever after removing the spring retainer that holds the governor operating lever to the differential lever, then lift upwards on the governor operating link. Remove the clevis pin at the control tube lever and withdraw fuel rod across cylinder head while rotating rod.
5. Loosen the hose clamps on the fuel rod covers and slide each hose and clamp toward the governor.

6. Loosen hose clamps on water by-pass tube. Slide hoses and clamps onto by-pass tube and remove tube from engine.
7. Disconnect and remove the fuel lines from the fuel pump.
8. Remove retaining bolts and slide the governor and fuel pump assembly straight forward, from the blower dowels. Remove governor-to-blower gasket. NOTE: The fuel pump drive coupling may stay with either the fuel pump or the blower rotor shaft. Remove the drive coupling.
9. Remove three bolts and remove the fuel pump and gasket from the governor housing.

Governor Disassembly 71 & 92 Series Engines

Cover

With cover removed from the control housing, disassemble the governor as follows:

1. Loosen governor throttle control lever retaining bolt, and lift control lever from throttle shaft.
2. Remove tapered pin from spacer. Lift lever spacer and seal ring retainer from throttle shaft. Withdraw shaft from cover.
3. Remove seal ring from governor cover.
4. Remove fuel shut-off lever.
5. Remove lock ring from shaft and seal ring retainer (two flat washers).
6. Slide shaft from cover and remove seal ring.
7. At this stage of disassembly, wash cover assembly (containing needle bearings) thoroughly in clean fuel oil and inspect needle bearings for wear or damage. If bearings are satisfactory for further use, removal is unnecessary.

8. Inspect fuel shut-off shaft bushing. If bushing is damaged, replace governor cover.
9. If needle bearing removal is necessary, then press both bearings out of cover.

Spring (Except Dual Range)

1. Place control housing in soft jaw vise.
2. Remove two bolts attaching high speed spring retainer cover to housing, then remove cover.
3. Loosen lock nut. Remove high speed spring retainer, idle adjusting screw, high speed spring, spring plunger, low speed spring, spring seat, and spring cap as an assembly.

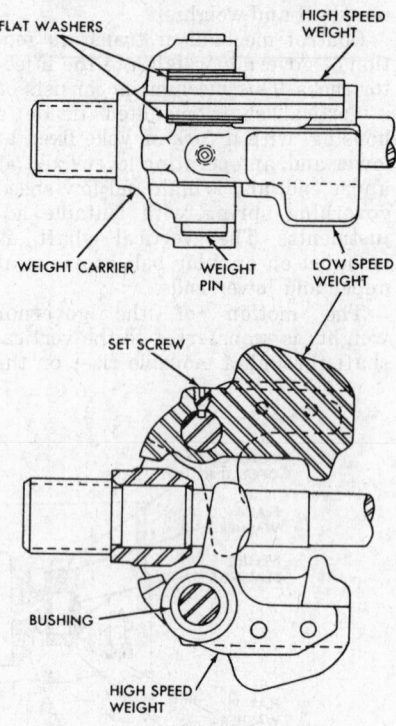

View of governor weights
(© G.M.C.)

Spring (Dual Range)

1. Remove two bolts and withdraw spring housing, cover, piston, sleeve, and low maximum speed adjustment screws as an assembly.
2. Remove high speed spring retainer, and withdraw high and low speed springs and related parts as an assembly.

Weights

1. Remove the governor weight housing cap, and gasket.
2. Remove bearing retaining bolt.
3. Thread a 5/16"-24 x 3" bolt into tapped end of weight shaft, and press shaft from bearing.
4. Slide riser thrust bearing and governor riser from shaft. This bearing is specially designed to absorb thrust load; therefore

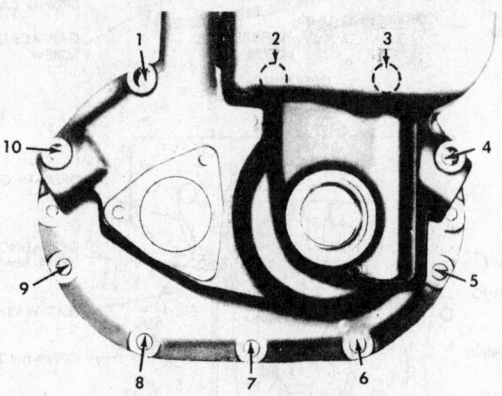

Location and size of governor housing retaining bolts
(© G.M.C.)

Location	Washer	Bolt
4	Copper	5/16"-18 x 5-1/4"
10	Copper	5/16"-18 x 4"
1	Lock	5/16"-18 x 3-3/4"
2-3-5-6-7-8-9	Lock	5/16"-18 x 2-1/4"

looseness between the mating parts does not indicate excessive wear.

5. Remove weight shaft bearing from governor housing.
6. Mark low and high speed weights and carrier with a center punch for identification, also note position of flat washers so that the parts can be replaced in original position.
7. Using a 3/32" Allen wrench remove Allen set screw from low speed weights. Withdraw pins and governor weights. NOTE: If necessary the bushings in the high speed weights may be removed at this time.
8. If required, the weight carrier may be pressed from governor weight shaft and a new carrier installed.

Operating Shaft

1. Lift differential lever off of pin of operating shaft lever after removing retainer clip and washer.
2. Remove bearing retaining screw, flat washer, and lock washer.
3. Remove expansion plug from bottom of control housing.
4. Loosen operating fork set screw.
5. Support control housing bottom side up on bed of press and insert a 9/16" open end wrench between operating fork and governor housing. Using a brass rod, press operating shaft from operating fork. Withdraw operating shaft, operating lever, and upper bearing as an assembly from control housing.
6. Support operating shaft lever and upper bearing on bed of arbor press. Using a brass rod, press operating shaft from operating lever and upper bearing.

Link Operating Lever

1. Remove retaining screw from governor housing and withdraw governor control lever shaft lock clip.
2. Slide governor control link operating lever shaft from governor housing. Remove governor control link operating lever assembly and two flat washers.
3. Remove the needle bearings, if necessary.

Governor Inspection

Inspect all bearings to be sure that they are satisfactory for further use.

Inspect spring seats, plungers, adjusting screws, lock nuts, and other parts of control housing for defects that might affect governor operation.

Inspect operating shaft and shaft bearing for excessive wear. If excessive wear is noted, a new bearing or shaft must be installed.

Examine riser thrust bearing for excessive wear, flat spots or corrosion. If any of these conditions exist, a new thrust bearing assembly must be installed.

Inspect roller bearings and throttle shaft for excessive wear or flat spots. If one or both conditions exist, new bearings and throttle shaft must be installed.

If new bearings are installed in the governor cover, the lower bearing should be flush with the lower end of the bearing boss. The upper bearing must be pressed in approximately 1/8" below the top surface of the upper bearing boss.

Examine weight carrier pins and bushings for excessive wear and flat spots. If either of these conditions exist, new parts must be installed.

Weights should be assembled and checked for free movement.

Governor Assembly

Cover

1. Place the cover on the bed of an arbor press with the inner face of the cover facing down. Start upper needle bearing straight into bearing bore of cover with number on bearing up. Then press bearing into cover.
2. Reverse cover on bed of press (inner face of cover up). Start second bearing straight into bearing bore of cover with number on bearing up. Place a flat washer over pilot of remover tool, insert pilot of tool into bearing and under ram of press, then press bearing down into cover until washer contacts cover.
3. Lubricate throttle shaft needle bearings with multi-purpose grease. Then insert throttle shaft through bearings.
4. Insert seal ring over shaft and
6. Press spacer down on shaft until holes in spacer are in line with shaft and against seal ring.
5. Start spacer over throttle shaft bearing. Place retainer over with holes in lever and shaft for hole in shaft.
into counterbore against upper tapered pin in alignment.
7. Insert tapered pin in hole of spacer, then drive pin into place.
8. Position throttle control lever on throttle shaft and tighten retaining bolt.
9. Slide fuel shut-off lever shaft through bushing and place lever, under governor cover, on right side of pin (viewed from blower end of governor.)
10. Place seal ring on fuel shut off lever shaft and seal ring retainer (two flat washers) and retain with lock ring.
11. Install fuel shut-off lever on shaft and secure with clamping bolt.

Installation

1. Place gasket on fuel pump and mount fuel pump on governor housing.
2. Place fuel pump drive fork on fuel pump shaft. Rotate pump shaft until fork aligns with drive on blower rotor.
3. Slide governor and fuel pump assembly straight on dowels of blower housing. Align splines of weight shaft with splines in blower rotor and fuel pump coupling with drive while assembling governor to blower.
4. Insert bolts and washers at locations as indicated and tighten alternately and evenly.
5. Reconnect fuel lines at fuel pump.
6. Install water by-pass tube between thermostat housing with the use of hoses and hose clamps.
7. Slide each hose on the fuel rod covers and tighten hose clamps.
8. Assemble the governor operating lever link connecting pin to the governor operating lever connecting link as follows:
a. Insert one control link operating lever retaining ring on inner groove of pin.
b. Insert connecting link over end of pin and against retaining ring.
c. Insert one control link operating lever retaining ring on outer groove of pin.
9. Insert fuel rods between cylinder heads and governor as follows:
10. Install right fuel rod:
a. Insert fuel rod through fuel rod cover and into control link operating lever.
b. Align rod end with opening in the control link operating lever and insert pin. Screw pin into postion and tighten securely.
11. Install left fuel rod:
a. Place on shim on differential lever.
b. Slide the operating lever link connecting pin into the control link operating lever and at the same time slide the connecting link over the pin of the differential lever.
c. Secure the assembly by placing shim over pin of differential lever and retain with the spring retainer.
12. Place governor housing cover-gasket on housing and install governor cover. Be sure governor control lever assembly enters slot of differential lever and shut down lever, on under side of cover, is between the stop pin and the differential pin. Tighten screws securely.
13. Reconnect control linkage to throttle and stop control levers at top of cover.
14. Perform tune-up.

Governor Service—53 Series

1. Disconnect the governor control linkage.
2. Remove the attaching screws and washers, pull the governor cover and lever assembly off the governor, and remove the cover gasket.
3. Remove the two bolts and washers, and remove the spring housing and gasket from the governor.
4. Referring to the illustration, remove the spring retainer with a special wrench and pull the spring assembly out of the governor.
5. Loosen the clamps on the fuel rod cover hoses, and then slide the hoses back.
6. Remove the rocker covers.
7. Disconnect the lower fuel rods at the injector control tube levers and at the threaded portions where they connect with the upper fuel rods.
8. Remove the threaded pins that connect the fuel rods to the control link lever. Then, remove the upper fuel rods.

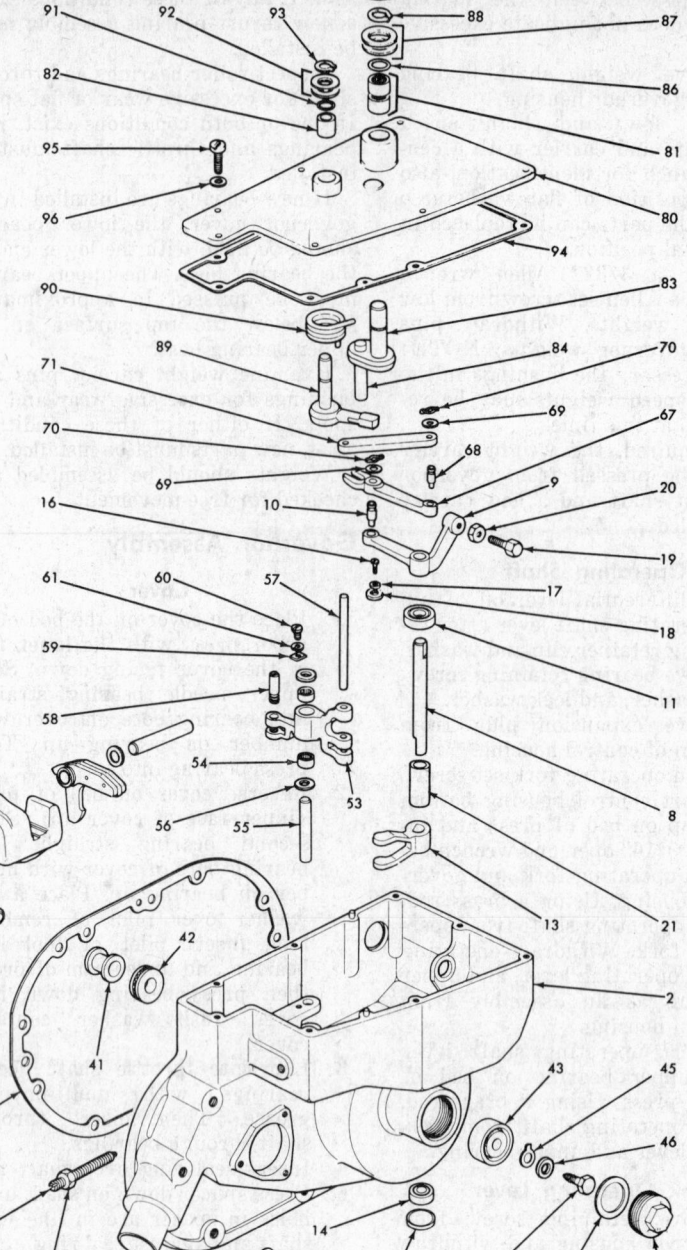

Limiting speed governor

2. Housing--Governor	26. Weight--Low Speed	57. Pin--Fuel Rod Connecting (Short)	81. Bearing--Speed Control Shaft
3. Gasket--Housing to Blower	27. Weight--High Speed	58. Pin--Fuel Rod Connecting (Long)	82. Bushing--Stop Lever Shaft
4. Bolt--Housing to Blower	29. Pin--Weight	59. Clip--Operating Lever Shaft Lock	83. Shaft--Speed Control Lever
5. Lock Washer	30. Flat Washer	60. Screw--Lock Clip	84. Pin--Fulcrum Lever
8. Shaft--Governor Operating	31. Screw--Weight Pin Set	61. Lock Washer	85. Spacer--Speed Control Shaft
9. Lever--Operating Shaft	40. Shaft--Weight Carrier	65. Screw--Buffer	86. Ring--Control Shaft Ring
10. Pin--Shaft Lever	41. Riser--Governor	66. Lock Nut--Buffer Screw	87. Washer--Seal Ring Retainer
11. Bearing--Operating Shaft (Upper)	42. Bearing--Riser Thrust	67. Lever--Governor Differential	88. Snap Ring--Speed Control Shaft
13. Fork--Operating Shaft	43. Bearing--Weight Carrier Shaft End	68. Pin--Differential Lever	89. Shaft--Stop Lever
14. Bearing--Operating Shaft (Lower)	44. Bolt--Bearing Retainer	69. Washer--Differential Lever and Connecting Link Flat	90. Spring--Stop Lever Shaft Return
15. Plug--Expansion	45. Lock Washer--Special	70. Retainer--Spring	91. Ring--Stop Shaft Seal
16. Screw--Bearing Retaining	46. Flat Washer	71. Link--Operating Lever Connecting	92. Washer--Seal Ring Retainer
17. Lock Washer	47. Plug--Governor Housing	80. Cover--Governor Housing	93. Snap Ring--Stop Shaft
18. Flat Washer	48. Gasket--Housing Plug		94. Gasket--Governor Housing Cover
19. Screw--Gap Adjusting	53. Lever--Control Link Operating		95. Screw--Housing Cover
20. Lock Nut	54. Bearing--Operating Lever		96. Lock Washer
21. Spacer--Operating Shaft	55. Shaft--Operating Lever		
25. Carrier--Governor Weight	56. Washer--Operating Lever Shim		

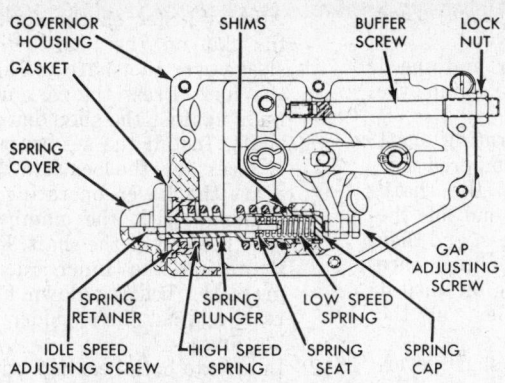

Governor Spring Assembly
(© G.M. Corp)

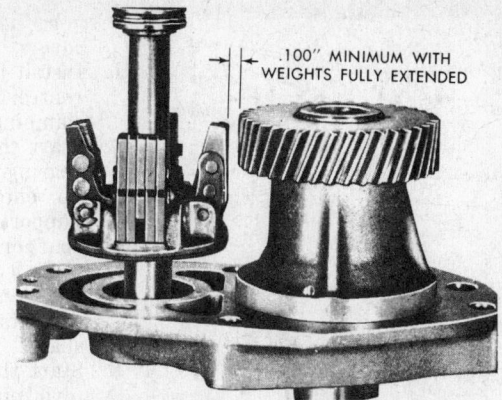

.100" MINIMUM WITH WEIGHTS FULLY EXTENDED

Minimum Clearance Between Drive Gear and Weights
(© G.M. Corp)

9. Remove the blower drive support as described in the blower removal instructions.
10. Check the clerance between the drive gear and each weight, with the weight fully extended, as shown in the illustration. Clearance must be .10".
11. Remove the governor weight shaft and carrier from the drive support, prying if necessary.
12. Remove the blower as described in the appropriate section. Remove the attaching bolts, and pull off the governor housing and gasket.

Disassembly

1. Remove the pipe plug from the throttle shaft.
2. Loosen the bolt, and remove the speed control lever.
3. Remove the throttle shaft retaining ring and retaining washers. and pull the throttle shaft out of the cover. Remove the cover seal ring.
4. Loosen the clamp, and remove the speed control lever.
5. Remove the retaining ring and two retaining washers and withdraw the stop lever shaft.
6. Remove the seal ring from the cover.
7. Use the fuel oil to clean the governor cover. Inspect the needle bearings for wear. If the bearings are badly worn, use an arbor press to remove them with a special bearing remover.
8. Remove the low speed spring cap, spring, and spring seat.
9. Depress the high speed spring and remove the idle speed adjusting screw locknut. Then, pull the high speed spring and shims from the assembly. Remove the idle speed adjusting screw from the spring plunger.
10. Remove the buffer screw and spring.
11. Remove the spring pin and washer from the control link lever, and pull off the control link lever and washer.
12. Inspect the bearings. If replace-

ment is required, place the control link lever on an arbor press supported by a sleeve. Press the bearings out of the lever with a special tool, as shown in the illustration.
13. Remove the spring pin and washer from the operating shaft lever, and remove the differential lever.
14. Remove the plug from the bottom of the governor housing. If applicable, remove the set screws from the governor operating fork.
15. Remove the retaining screw and washer from the operating shaft upper bearing. Invert the governor housing and place it on wooden blocks on the bed of an arbor press. Using a rod small enough to pass through the bearing, press down on the shaft until the bearing is free, and remove the bearing.
16. Place an open end wrench between the operating fork and the housing, and a rod on the end of the operating shaft, as shown in the illustration. Press the shaft out of the fork. Then, withdraw the shaft, operating shaft lever, and bearings.
17. Press the shaft off the operating shaft and upper bearing.
18. Remove the governor weight pin retaining rings. Drive the pins out with a punch held against

J 21967

SPACER

Removing the Governor Cover Bearing
(© G.M. Corp)

the grooved end of each pin. Remove the governor weights.
19. Press the shaft from the weight carrier as shown in the illustration.
20. Slide the riser and bearing assembly from the shaft.
21. Remove the snap ring and thrust washer from the blower drive gear shaft. Slide the shaft and gear off the blower drive support.
22. Press the drive gear off the shaft and remove the key.
23. Tap the governor weight shaft bearing to remove it from the blower drive support. If the bearing will not move, drive the plug from the support and, with a spacer against the outer race of the bearing, tap the bearing off the support.

Inspection

1. Clean all parts in fuel oil and dry them with compressed air.
2. Inspect all bearings and replace if they are pitted or corroded. Check ball bearings by revolving them slowly by hand, and replace any that have rough or tight spots.
3. Examine the riser thrust bearing for wear, flat spots, or corrosion. Install a new assembly if any are indicated.
4. Inspect the control link lever, needle bearings, and control link lever pin. The pin must project 1.055"-1.060" above the boss. Replace worn parts.
5. Examine the weight carrier, weights, and pins. Replace any worn parts.
6. Inspect the governor springs, spring seat, spring cap, plunger, spring retainer, adjusting screws, and housing for wear,, and replace parts as necessary.

Assembly

1. Place the cover on the bed of an arbor press, inner face down. Start a needle bearing, number side up, into the bearing bore in the cover. Then, insert a special

(Detroit 2-cycle) Diesel Engines

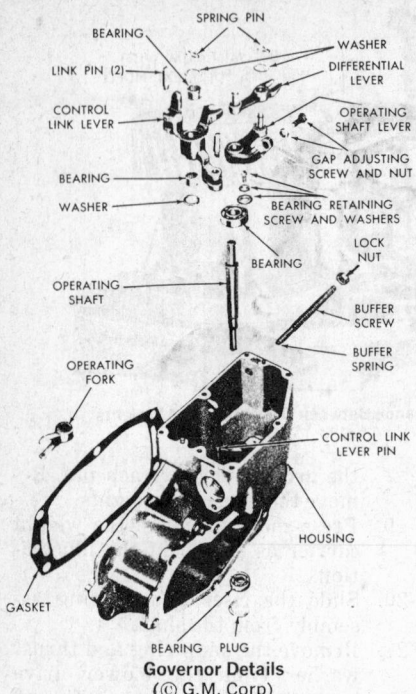

Governor Details
(© G.M. Corp)

bearing installer and press the bearing in until the shoulder of the tool touches the cover. *Do not use an impact tool for installation.*

2. Invert the cover and install the second bearing in precisely the same way.
3. Install the pipe plug in the throttle shaft.
4. Pack the needle bearings with grease. Slide the throttle shaft assembly through the bearings. The fulcrum lever pin must be seated in the slot on the underside of the cover.
5. On the top of the upper bearing, install a new seal ring, two seal retaining washers, and the retaining ring.
NOTE: A seal ring back up washer is used in place of the

lower washer on some governor covers.

6. Install the stop lever and speed control lever, and tighten the clamping bolts.
7. Start the upper operating shaft bearing, with the numbered side up, onto the end of the shaft. Support the lower end of the shaft on an arbor press, and then place a sleeve on the inner race and press the bearing on until it contacts the shoulder on the shaft.
8. Start the operating shaft lever, pivot pin up, onto the end of the shaft with the flat on the shaft registering with the flat in the lever bore. Then, use a sleeve to press the lever until it rests tightly against the bearing.
9. Insert the lever and shaft through the top of the governor housing. Position the operating fork so it is over the lower end of the shaft, with the finished cam surfaces facing the rear of the governor.
10. Support the operating shaft and governor housing on the bed of an arbor press with the upper end of the shaft resting on a steel block (see illustration).

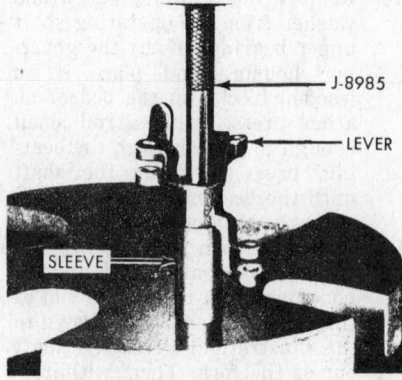

Removing Bearings from Control Link Lever
(© G.M. Corp)

Align the flat on the fork with the flat on the shaft. Place a sleeve over the shaft and against the fork. Press the fork until it rests against the shoulder on the shaft. Install the set screw and, if necessary, the lock screw.
11. Start the lower operating shaft bearing, with the number up, onto the end of the shaft. Place a sleeve on the inner race and press the bearing down until it rests against the shoulder in the housing.
12. Lubricate both bearings with engine oil.
13. Apply sealant around a new expansion plug and install it in the housing.
14. Install the upper operating shaft bearing retaining screw and washer.
15. Place the differential lever over the pivot pin in the operating shaft lever. Install the washer and spring pin.
16. Install the gap adjusting screw and locknut in the operating shaft lever.
17. Place a steel spacer on the bed of an arbor press, and position the control link lever on top of the spacer.
18. Install bearings on either side of the lever, number side up, with a special tool, as shown in the illustration.
19. Place the washer on the control link lever pin in the housing. Pack the needle bearings with grease and install the lever, tapped end of the link pin holes down, over the pin in the governor housing. Install the washer and spring pin.
20. Install the buffer screw so it extends 9/16"-5/8" out of the housing and lock it in place with the lock nut.
21. Lube the governor weight shaft with engine oil and slide the

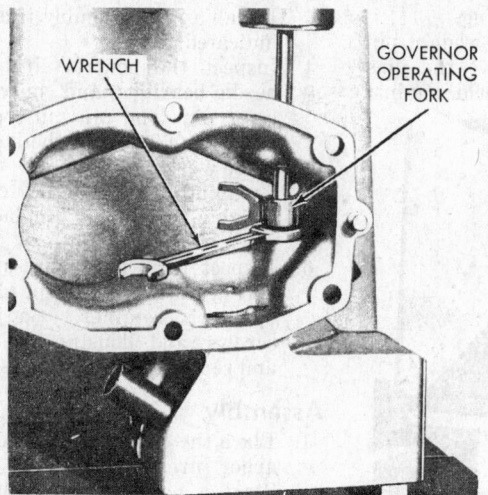

Removing Governor Operating Fork
(© G.M. Corp)

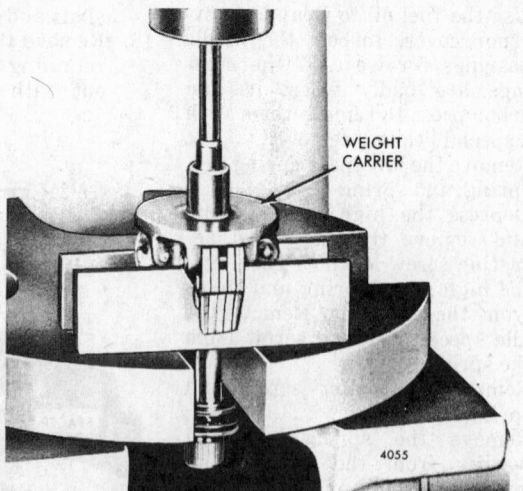

Removing Shaft from Weight Carrier
(© G.M. Corp)

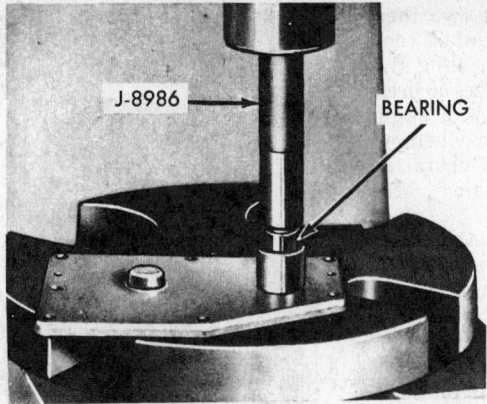

Installing Governor Cover Bearings
(© G.M. Corp)

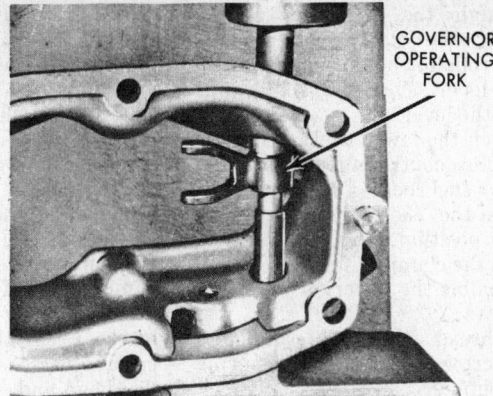

Installing Governor Operating Fork
(© G.M. Corp)

riser assembly over the shaft, bearing end toward the serrated end of the shaft. Pack the bearing with grease.

22. Use an installer as illustrated and press the shaft into the weight carrier.

23. Position the low speed weights (longer cam arm) on opposite sides of the weight carrier. Install the weight pins and retaining rings. The grooved end of the weight pins should be pushed through the smaller hole in the carrier and through the weight, and the knurled end should then be driven in just until the retaining ring can be installed. Install the high speed weights similarly.

24. Place the blower drive support, inner face up, on the bed of an arbor press. Start the governor weight shaft bearing, numbered side up, into the bore of the support. Place a suitable sleeve against the outer race and press the bearing against the shoulder of the blower drive support.

25. Install the steel thrust washer and snap ring on the end of the blower drive gear shaft.

26. Lubricate the blower drive gear

Installing Drive Gear
(© G.M. Corp)

shaft with engine oil and install it in the blower drive support.

27. Install the key in the shaft. Install the blower drive support in an arbor press.

28. Lubricate the inner surface of the blower drive gear. Start it onto the shaft, with the keyway in the gear properly aligned with the shaft key. Place a spacer on the gear and then press it onto the shaft until the clearance between the lower surface of the

gear and the blower drive support is .005".

29. Support the inner race of the bearing in the blower drive support. Start the weight end of the governor weight shaft onto the bearing. Press the shaft straight in until the shoulder contacts the inner race of the bearing.

30. Apply sealant to the cup plug, and press the plug in until it is flush with the blower drive support.

31. Check the clearance between the fully extended governor weights and the blower drive gear. It must be at least .100".

Installation

1. Position a new gasket on the governor housing, and then put the housing in position against the blower rear end plate. Secure the housing with the bolts and lock washers.

2. Install the blower and governor assembly as described in the appropriate section.

3. Install the blower drive support assembly as described in the appropriate section.

4. Insert the upper fuel rods

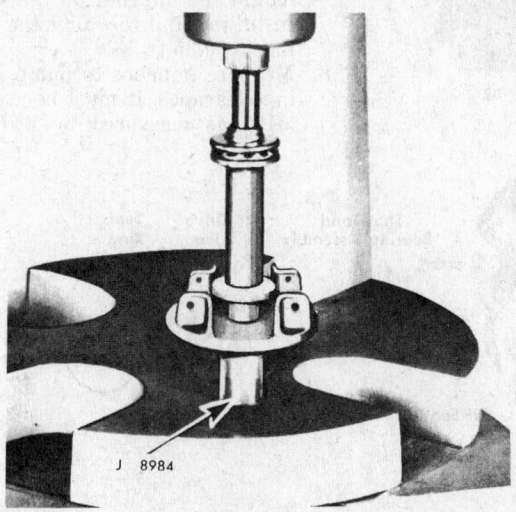

Installing Bearings in Control Link Lever
(© G.M. Corp)

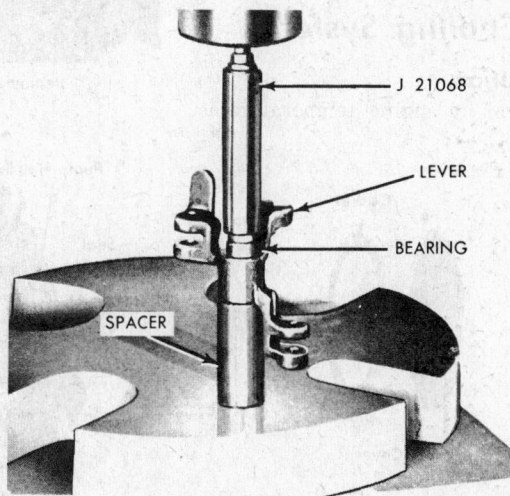

Installing Weight Carrier
(© G.M. Corp)

through the fuel rod covers, hoses, and clamps. Attach the fuel rods to the governor control link lever. Thread the link pins into the lever.

5. Attach the lower fuel rods to the injector control tube levers and upper fuel rods.

6. Slide the fuel rod cover hoses into position and secure them with the clamps.

7. Assemble the governor spring as follows:
   a. thread the idle adjusting screw into the spring plunger.
   b. Install the original shims over the spring plunger.
   c. Place the high speed spring over the spring plunger.
   d. Lube the spring and plunger with engine oil. Place the spring retainer over the plunger and secure it with a locknut when about 1/4" of the idle adjusting screw extends beyond the locknut.
   e. Lube and then insert the spring seat, low speed spring, and spring cap into the open end of the spring plunger.
   f. Start the retainer and spring assembly into the governor housing.

8. Position a new gasket on the governor housing and install the cover and lever assembly. The control link lever must engage the pin on the differential lever, the pin in the stop lever shaft must enter the slot in the differential lever, and the pin in the stop lever shaft must engage between the stop on the underside of the cover and the vertical extension of the control link lever.

9. Secure the cover with the screws and lockwashers.

10. Perform a complete engine tune up. Then, connect the linkage to the governor control linkage.

Cooling System

Description

Control of engine temperature is accomplished by means of two thermostats controlling the flow of cooling liquid within a sealed cooling system. A sealed system utilizes a pressure valve which maintains a slight pressure within the system when engine is running at normal operating temperature, thus raising the efficiency of the system.

Coolant is drawn from the lower portion of the radiator by the water pump and is forced through the oil cooler housing and into the cylinder block. From the cylinder block the coolant passes up through the cylinder head and, when the engine is at normal operating temperature, through the thermostat housing and into the upper portion of the radiator. Then the coolant passes down a series of tubes, where the coolant temperature is lowered by the air stream created by the revolving fan.

During the engine warm-up period, action of thermostats directs flow of cooling liquid back to the water pump. This arrangement permits circulation of liquid within the

Removing retaining ring
(© G.M.C.)

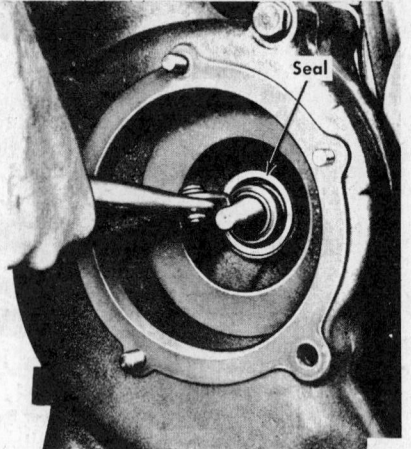

Removing water pump seal
(© G.M.C.)

engine, warming the engine and shortening the warm-up period. As engine reaches normal operating temperature, thermostats open gradually, allowing cooling liquid to flow through radiator, thus maintaining an efficient operating temperature.

Water Pump Service—71 Series Seal Replacement

The water pump seal can be replaced by removing the impeller without removing the pump from the engine.

1. Remove the water pump cover and gasket.
2. Remove nut and washer securing impeller to pump shaft.
3. Remove the impeller from the shaft with a special puller.
4. Pry the seal from the housing with a pair of pliers.
5. Using a hollow tool placed over the outer casing, tap the new seal into the pump housing.
6. Inspect the ceramic seat for cracks and for a good bond to the impeller. Replace impeller if necessary.
7. Apply oil to the ceramic seat and mount the impeller on the shaft, installing and torquing the locknut to 30-35 ft-lbs.
8. Measure impeller to pump housing clearance. It must be at least .015" as measured by a feeler

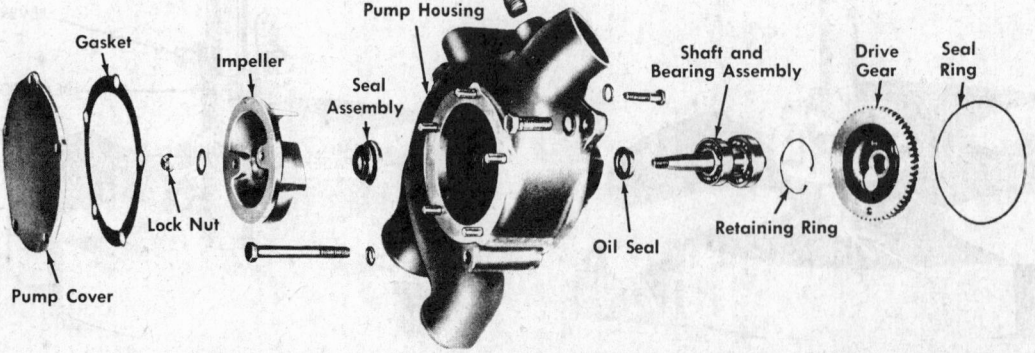

Gasket   Impeller   Pump Housing   Shaft and Bearing Assembly   Drive Gear   Seal Ring

Seal Assembly   Lock Nut   Oil Seal   Retaining Ring

Pump Cover

Water pump components (© G.M.C.)

Pressing drive gear from shaft
(© G.M.C.)

gauge placed in the water inlet opening.

9. Position a new gasket and locate the pump cover on the studs. Install the mounting nuts.

Removal

1. Drain cooling system by opening drain cocks at corners of cylinder block.
2. Remove hose clamps and slide hoses off the pump body.
3. Remove the three bolts that secure the pump to the balance weight cover, then lift the pump away from the engine being careful not to damage teeth of water pump drive gear.

Disassembly

1. Remove the cover and impeller.
2. Remove bearing retaining snap ring by placing pliers through one of the holes in the water pump drive gear.

3. Push the pump shaft, bearings and gear from housing.
4. Press gear from shaft.
5. Inspect the pump seals and remove only if cracked or worn. New seals should be used for replacement.

Inspection

1. Wash all the pump parts thoroughly in clean fuel oil and blow dry with compressed air, then inspect.
2. Ball bearings and races should be examined for indications of corrosion or pitting.

Assembly

1. Apply a film of lube oil to the outside diameter of the seal and place it so the lip is facing the bearings. Tap the seal into the pump housing.
2. Mount pump body in an arbor press with impeller side upward. Using a suitable hollow tool (to slide over seal face and spring) place on outer casing and press water seal into pump housing.
3. Apply film of lube oil to shaft, then slide shaft and bearings in pump housing. Install bearing retaining ring. *NOTE: Early engines have pump gear and camshaft gear with 59 and 42 teeth respectively. Latest pump gear and camshaft gear have 92 and 66 teeth respectively. Early and latest type gears are not interchangeable.*
4. Place water pump in arbor press. Support shaft on impeller end and press the drive gear onto shaft, against the inner race of bearing. **CAUTION: Support should be placed directly between shaft and base of press. Pump housing or studs should not be allowed to touch press while gear is being installed.**
5. Apply a thin film of oil to the ce-

ramic seat of impeller and place on tapered shaft. Place the washer and lock nut on shaft and tighten to 30-35 ft-lbs. torque.
6. Measure the clearance between the impeller and pump housing. There should be a minimum clearance of .015″. This can be checked by placing a feeler gauge in the water outlet opening.
7. Using a new gasket between cover and body tighten nuts securely.

Installation

1. Mount the pump on the engine and tighten the three bolts.
2. Slide the hose into position on the pump and tighten the clamps.
3. Fill the cooling system with clean fresh water.

Water Pump Service—92 Series

Seal Replacement— On Vehicle

1. Remove pump cover and gasket.
2. Remove the lock nut and washer and pull the impeller.

NOTE: Prevent damage to the ceramic insert by positioning the impeller with the insert up.

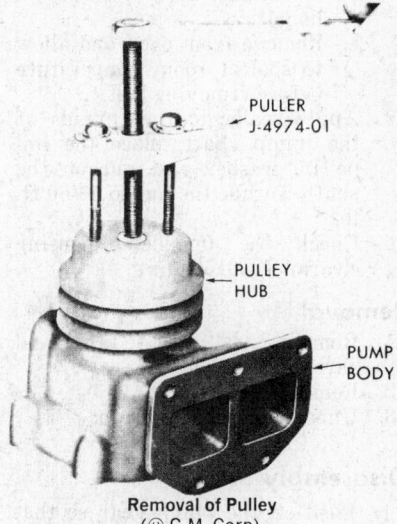

Removal of Pulley
(© G.M. Corp)

3. Place seal puller over the seal and into the two slots.
4. Position the sleeve over the puller. Insert the slotted end of the arm through the seal puller and position the cover stud in the slot of the arm.
5. Finger-tighten the set screw to hold the arm in position.
6. Pull the seal by turning the screw clockwise.
7. Press a new seal into place in the seal cavity.
8. Inspect the ceramic impeller in-

Removing water pump seal
(© G.M.C.)

ARM J 22150-5
SLEEVE J 22150-4
PULLER J 22150-2
4394

PULLER J-4974-01
PULLEY HUB
PUMP BODY

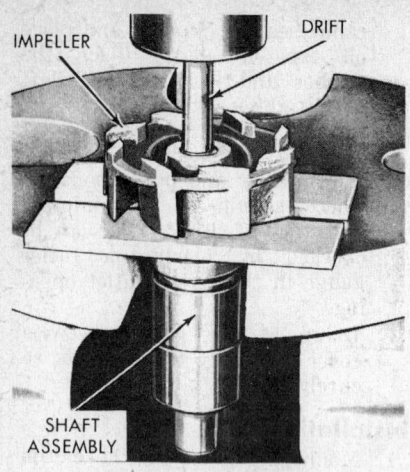

Removing Shaft from Impeller
(© G.M. Corp)

sert. If damaged, replace as follows:

a. Bake the used insert and impeller at 500° F for one hour to remove insert. Clean the bond area with a wire brush and alcohol. Dry thoroughly.

b. Place the adhesive washer in the bond area with the insert on top.

c. Clamp the insert and impeller together with a ⅜" bolt and nut and two smooth ⅛" bolt and nut and two smooth ⅛" thick washers.

d. Place the assembly in an oven and bake at 350°F for 1 hour.

e. Remove from oven and allow to cool at room temperature before removing bolt.

9. Apply sealer to the threads of the pump shaft, place the impeller, washer and nut on the shaft. Torque the nut to 45-50 ft. lb.

10. Check the impeller-to-housing clearance: .015" min.

Removal

1. Remove radiator, fan and shroud.
2. Remove hoses.
3. Unbolt and remove pump.

Disasembly

1. Position the pump gear so that the slot is over the ends of the bearing retaining ring. Remove the ring.
2. Remove the pump cover and gasket.
3. Remove the impeller lock nut.
4. Pull the impeller.
5. Press out the shaft, bearings and pump gear assembly.
6. Press the shaft out of the gear.
7. Press the shaft out of the large bearing.
8. Press the shaft out of the small bearing.
9. Remove the water seal as de-

scribed in the on-vehicle operation.

10. Remove the oil seal.

Assembly

1. Lubricate bearing bores and shaft. Install bearings.
2. Press the shaft into the housing.
3. Install the retaining ring.
4. Press the gear flush on the shaft end.
5. Coat the oil seal lip with SAE 30 and insert into the pump housing. Tap into place.
6. Install the water seal as described in the on-vehicle operation.
7. Install impeller and check impeller-to-housing clearance: .015" minimum.

Installation

1. Fit the seal ring to the housing and mount the pump on the engine. Install and tighten mounting bolts.
2. Check gear lash at impeller puller holes: .0015-.0045". Lash is adjusted by repositioning the pump.
3. Install water pump cover and new gasket. Check nylon bolt inserts to prevent leaks. Never substitute an ordinary bolt for the cadmium plated bolts used.

Water Pump Service—53 Series
Removal

1. Drain the cooling system by opening all drain cocks. Loosen the adjustment and then remove the water pump belts.
2. Loosen the hose clamps and then slide the water by-pass tube.
3. Remove the water pump mounting bolts and remove the pump.

Disassembly

1. Note the position of the pulley and then remove it with a puller as shown in the illustration.
2. Remove the water pump cover.
3. Press the shaft and bearing, seal, and impeller out of the pump housing by applying pressure *on the outer race of the bearing* with a special remover.
4. Press the end of the shaft out of the impeller, as shown in the illustration, using special plates and a holder.
5. Remove the seal assembly from the pump shaft.

Inspection

1. Clean all parts in fuel oil, with the exception of the bearing and shaft assembly, and dry with compressed air.
2. Examine the impeller for damage and for excessive wear on the face contacting the seal, and replace it if it is faulty.

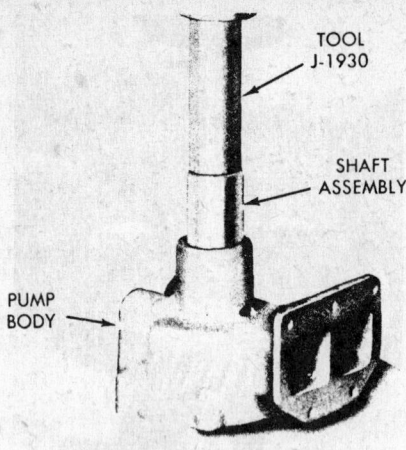

Pressing Shaft into Pump
(© G.M. Corp)

3. Check the bearing for a feeling of roughness when rotated, and for pitting, corrosion or other evidence of damage. Replace if faulty.

Assembly

1. Using a special installer as shown in the illustration, applying pressure to the *outer bearing race*, press the shaft and bearing into the pump body until the race is flush with the outer face of the body.
2. Coat the outside diameter of a new seal with sealing compound. Support the face of the body and the outer race of the bearing, apply pressure to the outer seal flange only to force the flange on until it contacts the body. Wipe the face of the seal with a chamois.
3. Place the pulley end of the shaft on the bed of an arbor press. Press the impeller onto the shaft until it is flush with the large end of the body.
4. Place the pulley on the bed of an arbor press. Using a heavy rod between the press ram and the impeller end of the shaft, press the shaft into the pulley until the pulley is in its original position.
5. Bolt the cover and a new gasket onto the pump body, tightening the bolts to 6-7 ft lbs.
6. Run the pump dry at 1,200 rpm for at least 30 seconds to seat the seal.

Installation

1. Position a new gasket onto the body flange, and bolt the water pump to the oil cooler housing.
2. Install the water-pump-to-by-pass-hose and clamps.
3. Install and adjust the drive belts.
4. Close all drains, refill the system. and operate the engine to check for leaks.

Thermostat

Description

The temperature of the coolant is controlled by two thermostats, one for each cylinder head. On truck engines, thermostats are installed in separate housing attached directly to cylinder head. On coach engines both thermostats are installed in a single housing attached to right cylinder head.

Removal

1. Drain cooling system to necessary level.
2. On 71 Series engines, remove the hose connections from the thermostat housing cover. On 53 and 92 Series engines, remove the cross-over by-pass tube and disconnect the by-pass tube that goes to the water pump.
3. On 71 and 92 Series engines, remove the bolts which secure the cover to the thermostat housing. On 53 Series engines, remove the bolts and the thermostat housings.
4. Remove the thermostat and clean seat for thermostat in housing.
5. Inspect seal pressed in thermostat housing. Remove if necessary.

Installation

1. Place new gasket on the thermostat housing. If the seal was removed, replace the seal with the closed end toward the thermostat.
2. Set the thermostat in housing and replace the cover or housing and hoses.
3. Fill cooling system and check for leaks.

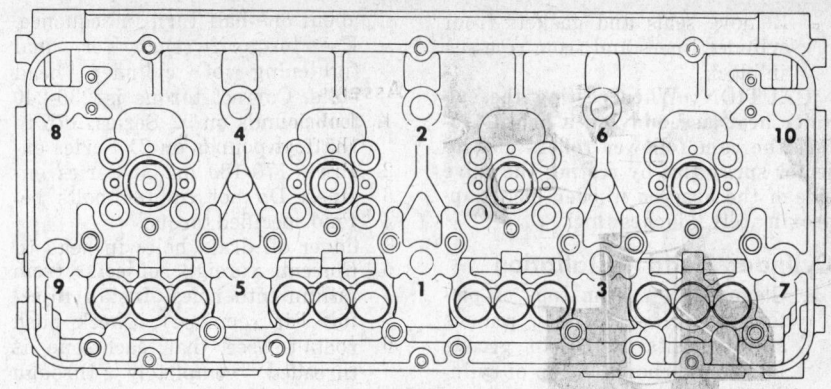

4-53 CYLINDER HEAD

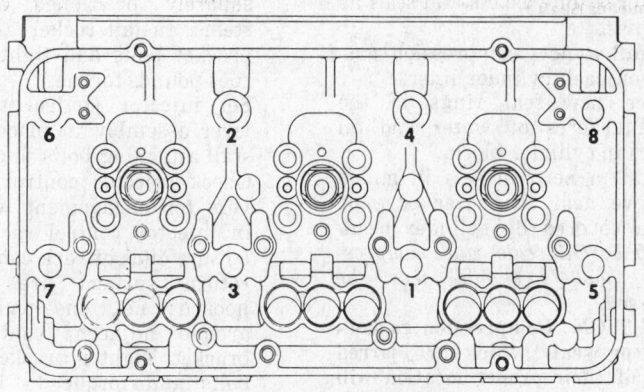

3-53 AND 6V-53 CYLINDER HEAD

Cylinder Head Bolt Tightening Sequence
(© G.M. Corp)

Engine Repair Section

Cylinder Head

The cylinder head and/or gasket at either right-hand or left-hand cylinder bank may be replaced in manner described below:

Cylinder Head Removal

1. Drain cooling system. Remove engine access cover from inside cab.
2. Disconnect battery. Remove air cleaners.
3. Disconnect exhaust pipe from exhaust manifold.
4. Remove exhaust manifold.
5. At front of cylinder head remove thermostat housing and accessory brackets (when used).
6. Disconnect fuel lines from fittings at front of cylinder head. When removing left-hand head, disconnect fuel return line from fitting at rear of head.
7. Remove rocker arm cover from head.
8. Remove cover from governor housing. Remove nut from fuel rod, then remove pin attaching link to tube lever and remove link.
9. Inside governor housing, remove fuel rod.
10. Loosen hose clamps, then move hoses as necessary to permit removal of rod cover (tube).
11. Remove injector rack control tube and bracket assembly.
12. If the cylinder head is to be overhauled, remove the injectors at this point. Remove cylinder head bolts in gradual stages to relieve spring pressure. When bolts have been removed, lift cylinder head assembly off cylinder block.

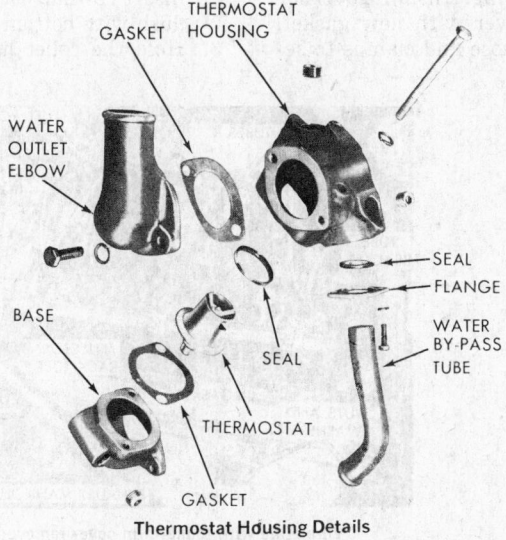

GASKET
THERMOSTAT HOUSING
WATER OUTLET ELBOW
SEAL
FLANGE
BASE
SEAL
WATER BY-PASS TUBE
THERMOSTAT
GASKET

Thermostat Housing Details
(© G.M. Corp)

[Detroit 2-cycle] Diesel Engines

Remove seals and gaskets from cylinder liners and from recesses in block.

CAUTION: When resting the cylinder head assembly on a bench protect the cam follower rollers and injector spray tips by resting the valve side of the head on wooden blocks approximately 2 inches thick.

Cylinder Head Installation

1. Clean carbon from tops of pistons and, if necessary, remove any deposits found in groove and counterbores in top of cylinder block.
2. Install new cylinder head compression gaskets and seals as follows:
   a. Install a new compression gasket on each cylinder liner.
   b. Place new seal rings in the counterbores of water and oil holes in cylinder block.
   c. Install a new oil seal in milled groove near the outer edge of area covered by cylinder head. *NOTE: Oil seals, and compression gaskets should never be re-used.*
   CAUTION: Compression gaskets and/or seals which are jarred out of their proper location will allow leaks and "blow-by" which will result in poor performance and damage to engine.
3. Make sure each pushrod is screwed far enough into its clevis that the end projects all the way through.
4. Wipe bottom of cylinder head clean, then lower cylinder head onto guide studs and down into contact with block. Lubricate threads and pressure area of bolt heads with S.A.E. #30 engine oil, then start all cylinder head bolts. Beginning at camshaft side of cylinder head tighten head bolts lightly to overcome tension of cam follower springs.
5. Tighten cylinder head bolts about one-half turn in sequence. Use torque wrench for final tightening of cylinder head bolts. Correct torque is 230-240 foot-pounds on 92 Series, 175 to 185 foot-pounds on 71 Series engines, 170-180 on 53 Series engines. Do not tighten bolts beyond specified torque.
6. Cover oil drain holes in head to prevent foreign objects from falling into holes. Install injectors (if removed). Check push rods to see that each one is threaded completely through clevis, then tip rocker arms into position with valve bridges squarely positioned at valve stems. Install rocker arm shaft bracket bolts and tighten to 50 foot-pounds torque.
7. Set injector control tube and lever assembly in place and install attaching bolts finger tight. Check injector control tube levers for engagement with slots in injector control racks. Also, be sure ends of rack control tube return springs are properly hooked; i.e., one end hooked around adjacent control tube bracket. Tighten bracket to head bolts and torque to 10 to 12 foot-pounds.
8. Try operating the injector control tube to determine if return spring rotates tube back to "No-Fuel" position after tube is manually moved to "Full-Fuel" position. If there is binding present, strike the control tube brackets lightly with soft hammer to correct any misalignment of tube bearings.
9. Install fuel rod through opening in top of governor housing; pass rod through hoses, clamps, and tubular cover (used at left-hand head). Attach fuel rod to governor lever with screw inside governor housing. Install governor housing cover with new gasket. Assemble hose and clamps to secure fuel rod cover.
10. Install fuel link between fuel rod and lever on control tube.
11. Connect fuel line at fitting at rear of left-hand cylinder head.
12. Mount thermostat housing on front of cylinder head.
13. Install exhaust manifold. Tighten manifold stud nuts to 25 to 40 foot-pounds.
14. Connect exhaust pipe. Fill cooling system and check for leaks.
15. Adjust exhaust valve clearance. Set injector timing.
16. Start engine and operate until normal operating temperature is reached, then recheck cylinder head bolt torque and make final check of exhaust valve clearance.
17. Install rocker arm cover on cylinder head.

Valve Seats

Inspect valve seats to determine if they are fit for further use. Inspection should include cleanliness, concentricity, flatness, and presence of cracks. If exhaust valve seat inserts are loose in cylinder head, or if for any reason due to wear pitting, etc., replacement of valve seat is necessary, oversize (.010") exhaust valve seat inserts are available.

Valve Seat Insert Counterbore Dimensions

	Diameter	Depth
53 Series	1.159-1.160	.300 -.312
71 Series	1.260-1.261	.338 -.352
92 Series	1.440-1.4410	.3395-.3505

The valve seat inserts are shrunk into the cylinder head and, therefore, must be replaced as outlined in the following procedure to avoid damage to the cylinder head:

Valve Seat Removal

1. Place cylinder head on its side on workbench.
2. Place collet of tool inside valve insert so that bottom of collet is flush with bottom of insert.
3. Hold the collet handle and turn

Governor cover housing removed showing fuel rod connecting screws
(© G.M.C.)

1 Fuel rod to R.H. cylinder head
2 Governor springs
3 Fuel rod-to-lever screws
4 Governor housing
5 Fuel rod to L.H. cylinder head
6 Fuel rod operating lever

V6-engine with rocker arm cover removed
(© G.M.C.)

692

Fuel rod and link installation
(ⓒ G.M.C.)

1 Governor housing	7 Link pin
2 Hose and clamp	8 Injector control tube
3 Control rod cover	lever
4 Fuel control rod	9 Control tube spring
5 Rod Nut	10 Control tube
6 Lower link	11 Cylinder head (L.H.)

Installing cylinder head on engine
(ⓒ G.M.C.)

the T handle to expand collet cone and insert is held securely by the tool.

4. Insert drive bar of tool through valve guide.

5. Tap the drive bar once or twice to move the insert about 1/16" away from its seat in the cylinder head.

6. Loosen the collet cone and move the tool into the insert slightly so that the narrow flange at the bottom of the collet is below the valve seat insert.

7. Tighten the collet cone and continue to drive the insert out of the cylinder head.

Valve Seat Installation

1. Wash the counterbores in the cylinder head for the valve seat inserts with trichloroethylene or other good solvent. Also, wash the valve inserts with the same solvent. Dry the counterbores and inserts with compressed air.

2. Inspect the counterbores for cleanliness, concentricity, flatness and cracks.

3. Immerse the cylinder head for at least 30 minutes in water heated to 180°F. to 200°F. At the same time, cool the valve seat insert as

much as possible (dry ice may be used for this purpose).

4. Rest the cylinder head, bottom side up, on bench and lay an insert in counterbore—valve seat up. If the temperature of the two parts are allowed to become nearly the same, installation may become difficult and damage to the parts may result.

5. Using tool, drive insert down tight into counterbore.

6. Grind the valve seat insert and check it for concentricity in relation to the valve guide.

Recondition Valve Seat

Exhaust valves which are to be reused may be refaced, if necessary, using standard refacing procedure. Before installing either a new or used valve, examine the valve insert in the cylinder head for proper valve seating. The proper angle for the seating face of both the valve and valve insert is 30°. The angle of the valve seat insert must be exactly the same as the angle of the valve face so as to provide proper seating of the valve.

Grind the inserts as follows:

1. First apply a 30° grinding wheel on valve seat insert.

2. Use a 60° grinding wheel to open

the throat of the insert.

3. Then grind top surface with a 15° wheel to narrow width of the seat from 1/16" to 3/32" (71 and 92 Series), or 3/64"-5/64" (53 Series). The 30° face of the insert may be adjusted relative to the center of the valve face with the 15° and 60° grinding wheels.

CAUTION: Do not permit grinding wheels to contact the cylinder head when grinding the inserts.

4. After valve inserts have been ground, the cylinder head should be thoroughly cleaned. Then check the concentricity of the valve seats relative to the valve guides. Total runout for a good valve seat should not exceed .002". If total runout exceeds .002" check for bent valve guide before regrinding insert.

5. After valve seats have been ground, the position of the contact area between the valve and valve insert should be determined in the following manner:

a. Apply a light coat of Prussian Blue or similar paste to the valve seat insert.

b. Lower stem of valve in the valve guide and "bounce" valve

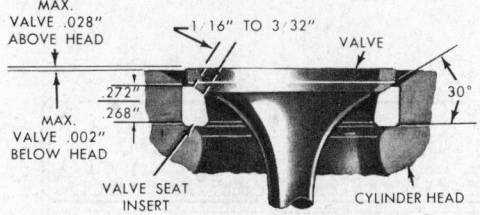

CURRENT 2 VALVE—FORMER 4 VALVE

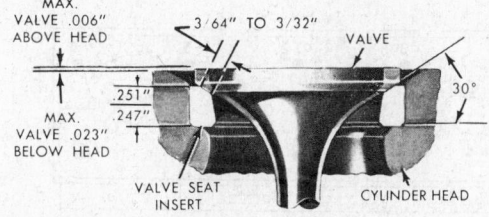

CURRENT 4 VALVE

Exhaust Valve, Insert, and Seat Dimensions
(ⓒ G.M. Corp)

on seat. *Do not rotate valve.* This procedure will show (on the valve face) the area of contact. The most desirable area of contact is at the center of the valve face.

Injector Tube

Whenever inspection indicates that injector copper tube in cylinder head requires replacement, the operation must be performed with the greatest of care and with special tools in the manner following:

Injector Tube Removal

1. Attach cylinder head holding plates to the cylinder head or support the cylinder head on its side with exhaust ports down. Place the injector tube installer into injector tube.
2. Insert pilot through the small opening in injector tube and thread pilot into hole in end of installer.
3. Tap end of pilot with hammer to loosen injector tube. Lift tube, pilot, and installer from cylinder head.

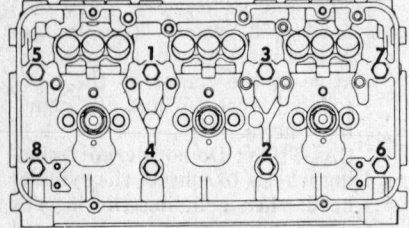

Cylinder head tightening sequence V6-engine
(© G.M.C.)

Inspection

Installation of the injector tube requires careful preparation and inspection to make sure that the area of the cylinder head is clean so that when tube contacts the cast iron cylinder head, foreign material will not prevent tube seating at bottom or sealing at top.

Injector Tube Installation

1. Place injector tube sealing ring in cylinder head injector tube counterbore.
2. Place drive in injector tube, insert pilot through the small opening of injector tube and

thread pilot tapped end of driver.
3. Slip the injector tube into injector bore and drive into place. The flange at upper end of tube will seat on the seal ring and into the counterbore of cylinder head when the injector tube is properly positioned.
4. With the injector tube properly positioned in cylinder head, flare lower end of injector tube, as follows:
   a. Turn cylinder head bottom side up, remove pilot and thread flaring tool into tapped end of driver.
   b. Then, using a socket and torque wrench, apply approximately 30 ft.-lb. torque on flaring tool.
   c. Remove injector tube installing tools and proceed to ream the injector tube.

Injector Tube Reaming

After an injector tube has been installed in a cylinder head, it must be hand reamed, spot faced, and hand reamed again, each step using its own special tool, as described below:

1. Place cylinder head right side up and clean out the injector tube.
2. Place a few drops of cutting oil on the reamer flutes and carefully position the reamer in the injector tubes.
3. Turn the reamer clockwise, withdrawing it frequently for removal of chips until its lower shoulder contacts the injector tube. Clean out all the chips.
4. Turn the head upside down, insert the pilot of the cutting tool into the small hole of the injector tube, and attach a socket drive.
5. Rotate the tool clockwise until enough stock has been removed that the lower end of the tube is from flush to .005″ below the finished surface of the head.
6. Wash the interior of the injector tube with clean fuel oil and dry with compressed air.
7. Lubricate the bevel seat of the tube with light cutting oil. Seat the reamer into the tube so it contacts the bevel seat.
8. Turn the reamer steadily without any downward force, and then remove it. Blow out the chips and observe the cutting pattern.

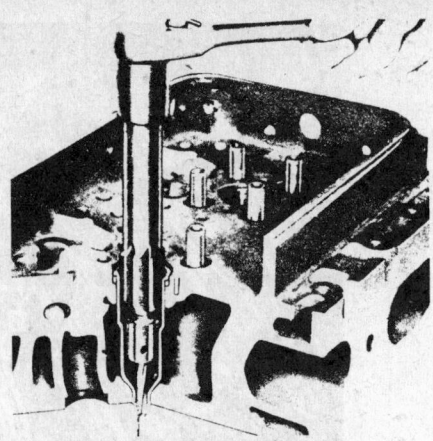

Reaming Injector Tube Bevel Seat
(© G.M. Corp)

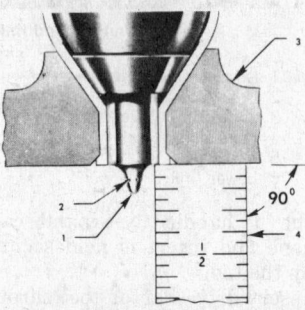

Checking Spray Tip Location in Relation to Head Surface (© G.M. Corp)
1 Spray Tip
2 Orifice - Injector Spray Tip
3 Cylinder Head
4 Straight Edge - Steel

9. Continue reaming carefully, removing the reamer, blowing out the chips, and inspecting the cut frequently. Test the reaming job by installing the injector after each inspection of the cutting. The job is complete when the shoulder of the spray tip is flush to .015″ on 53 and 71 Series and .005″ on 92 Series below the fire deck of the cylinder head.

Valve Guides

Cleaning And Inspection

After cleaning, inspect the valve guides for fractures, scoring or excessive wear and replace if necessary. The inside diameter of a new valve guide is .3125″ to .3135″ which will produce a valve stem-to-guide clearance, with a new valve, of .002″

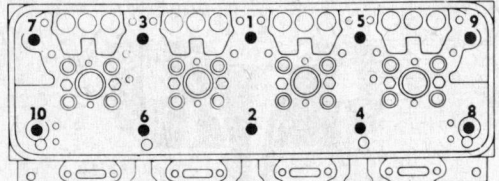

Cylinder head tightening sequence in-line 4 cyl. and V8 engines
(© G.M.C.)

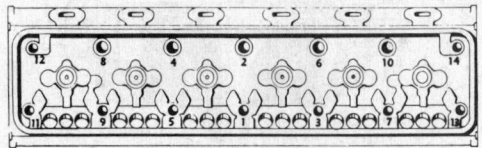

Cylinder head tightening sequence in-line 6 cyl. engine
(© G.M.C.)

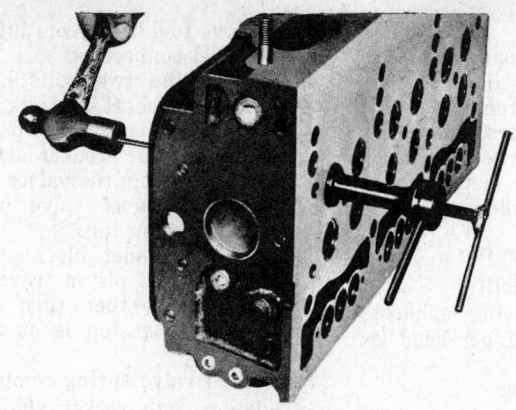

Removing valve seat insert
(© G.M.C.)

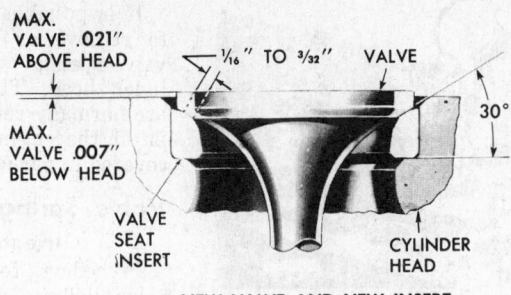

MAX. VALVE .021" ABOVE HEAD

MAX. VALVE .007" BELOW HEAD

VALVE SEAT INSERT

1/16" TO 3/32"   VALVE

30°

CYLINDER HEAD

NEW VALVE AND NEW INSERT

Installation of valve and valve seat insert
(© G.M.C.)

to .004". If the clearance exceeds .005", install a new guide and/or valve.

Valve Guide Removal

1. Support the cylinder head (bottom side up) on wood blocks 2" above the workbench.
2. Drive valve guide out of cylinder head.

Valve Guide Installation

Rest the bottom of the cylinder head on a workbench, and install guide as follows:

1. Remove exhaust valve bridge guides.
2. Insert the threaded end of the valve guide into installer. Valve guides must be installed with internal threads toward top to provide for lubrication of valve stems.
3. Locate the valve guide squarely in the top of cylinder head and tap gently to start it into place.
4. With the guide properly started, drive in until its upper end projects .88" on 53 and 71 Series and .67-.71" on 92 Series above the top of the cylinder head. This dimension is automatically provided by the valve guide installer tool.
   NOTE: Service valve guides are "pre-finished" and, therefore, do not require reaming.
5. Install threaded valve bridge guides. Tighten to 45-50 ft. lbs. Latest type guides are pressed into cylinder head.

Valve Bridge and Guide

Early engines used an exhaust valve bridge guide which was threaded into cylinder head. Threaded bridge guides were superseded by a straight guide which is pressed into cylinder head. Valve bridge springs have now been eliminated on latest engines.

A combination of soft steel guide and a soft steel bridge must be avoided, otherwise premature wear will occur.

Valve bridge springs can be removed and discharged on any engine whenever desirable; however, it is suggested that they be removed at time of overhaul. On engines used in trucks and parlor coaches heavier exhaust valve springs must be used when bridge springs are removed.

Bridge Guide Removal

1. Threaded guide can be identified by the hexagon section at the base of the guide and can be removed by using a thin wall socket.
2. On pressed guides, file or grind two notches 1/16" deep on opposite sides of the guide and about 1¼-1½" from upper end on 92 Series engines, ¾" from upper end on 71 Series engines, and about 1½" from upper end on 53 Series engines.
3. Place tool spacer over end of guide, then place remover tool over guide with two screws in alignment with two notches. Tighten set screws securely on 53 Series engines, place the other spacer required for the operation over the guide remover.
4. Thread nut onto remover tool and turn to pull guide from cylinder head.
5. Should guide be broken, drill a hole about ½" deep in end of guide with #3 (.2130") drill. Tap the guide with ¼"-28 bottoming tap. Thread adapter into guide, then attach slide hammer to adapter and pull guide.

Bridge Guide Installation

1. Threaded bridge guide may be installed, using thin wall socket.
2. Start pressed guide into cylinder head (undercut end first).
3. Use installer tool to drive guide into place.
4. Use of installer tool will properly position guide to its proper height of 2.04" above cylinder head.

Cylinder Head Assembly

New service replacement cylinder heads are equipped with valve guides, bridge guides, valve seat inserts, exhaust manifold studs, water nozzles, injector hole tubes and necessary plugs.

CAUTION: When installing plugs in fuel manifolds, apply a small amount of sealant to threads of plugs only. Work sealant into threads and wipe off excess with clean, lint-free cloth so that sealant will not be washed into the fuel system and result in serious damage to the injectors.

When a new cylinder head is to be used, the parts listed above should be removed from the old head and installed in the new head. If the old cylinder head is to be reused, the parts listed below should be installed in the old head prior to installation of the head to the cylinder block.

1. Exhaust valves and springs, cam followers and springs, valve and injector rocker arms and shafts should be assembled to cylinder head.
2. Using new steel washers, install fuel connectors and tighten to 35-40 ft.-lbs. torque.
3. Install fuel injectors as described in the appropriate section.

Valve and Injector Mechanism

Description

Rocker Arms

Valve and injector rocker arms for each cylinder operate on a separate shaft supported by two cast iron brackets. A single bolt fastens each bracket securely to top of cylinder head.

Cam Followers

Contact between cam followers and cam is made by a hardened steel roller. The roller is equipped with a steel-backed bronze bushing running directly on a pin in lower end of cam

695

Checking valve seat concentricity
(© G.M.C.)

follower. A separate coil spring, located inside of the hollow cam follower, is held in place in cylinder head by a spring seat retainer. A dropforged steel guide is provided for each set of three cam followers. This guide, located on bottom of cylinder head, keeps the rollers in line with cams and also serves as a retainer during assembly and disassembly.

Maintenance

Some service operations may be accomplished on cylinder head, valves, and injectors assembly without removing the cylinder head assembly from the block. These operations are:

1. Valve Lash.
2. Valve Spring Replacement.
3. Valve Bridge Adjustment.
4. Rocker Arm, Shaft or Shaft Bracket Replacement.
5. Injector Replacement.
6. Cam Follower Spring Replacement.
7. Cam Follower Replacement.
8. Push Rod Replacemet.

Valve Lash and Bridge Adjustment

CAUTION: **Whenever a push rod has been disconnected from the push rod clevis, the rod must be screwed back into the clevis, flush with the top of the threaded portion of the clevis before valve lash is checked.** *If this is not done, before engine is turned, the piston may hit head of valve due to the small clearance between valves and piston head at the piston upper position.*

Correct valve lash is important due to high compression pressures. Too little clearance causes a loss of compression, missing cylinder, and eventual burning of valves and valve seats. Too much clearance between valve stem and valve rocker arm results in noisy operation of engine, especially in idling range.

NOTE: It is recommended when lashing valves to set them first cold; then start engine and warm up to normal operating temperature. Reset valve lash after engine has warmed up to operating temperature.

Valve Spring

It is possible, if occasion requires, to remove or replace the exhaust valve springs without removing cylinder head. The springs, however, are normally removed when the head is off the engine. Both methods are covered in paragraph following.

Valve Spring Removal
(Head installed)

Procedure for removing exhaust valve springs while cylinder head is installed is as follows:

1. Remove rocker cover.
2. Bring valve and injector rocker arms in line horizontally.
3. Remove fuel lines from injectors and fuel connectors.
4. Remove the two bolts holding rocker arm shaft brackets to cylinder head then remove brackets and shaft. Lift rocker arms up and away from the valves.
5. Remove exhaust valve bridges and bridge springs.
6. Remove cylinder block air box cover so that piston travel may be observed, then turn crankshaft until piston is at top of stroke.
7. Thread valve spring compressor adaptor into rocker shaft bolt hole in cylinder head. Apply

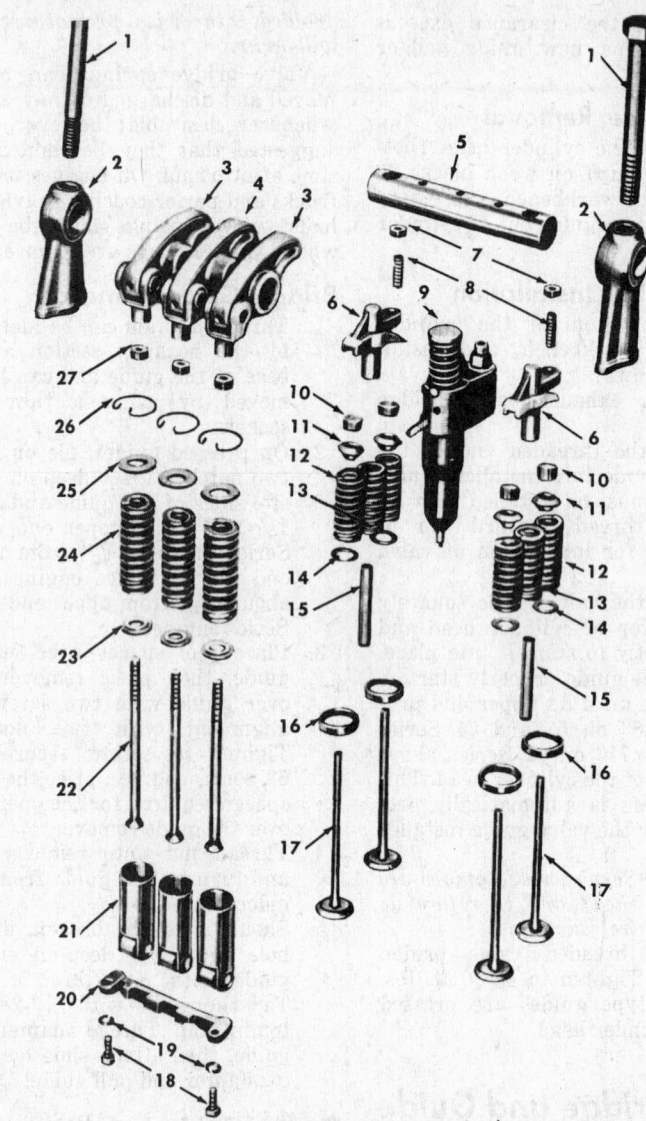

Valves, injector and operating mechanism
(© G.M.C.)

1 Bracket bolt	11 Valve spring cap	20 Cam follower guide
2 Rocker shaft bracket	12 Valve spring	21 Cam follower
3 Valve rocker arm	13 Bridge spring (Early models)	22 Push rod
4 Injector rocker arm	14 Valve spring seat	23 Spring seat—lower
5 Rocker shaft	15 Bridge guide (Pressed Type)	24 Push rod spring
6 Valve bridge	16 Valve seat insert	25 Spring seat—Upper
7 Screw nut	17 Exhaust valve	26 Spring seat retainer
8 Adjusting screw	18 Guide bolt	27 Lock nut
9 Injector assembly	19 Lock washer	
10 Valve spring lock		

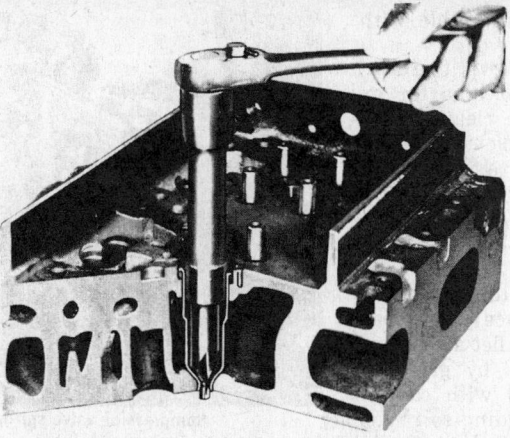

Reaming injector tube bevel seat for injector body nut
(© G.M.C.)

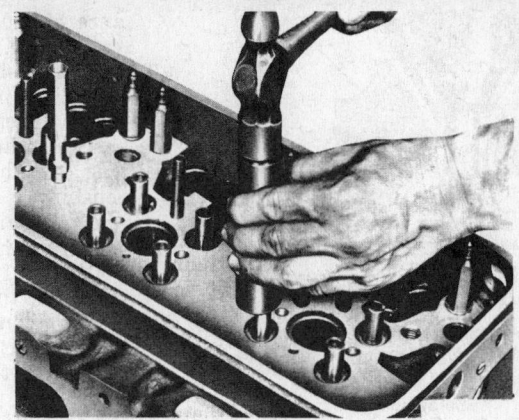

Installing exhaust valve guide
(© G.M.C.)

pressure to end of valve spring compressor handle to compress valve spring. Remove the two-piece tapered spring lock.

8. Release tool and remove valve spring cap, valve spring and spring seat.

Valve Spring Removal

(Head removed)

To remove the exhaust valve spring with the cylinder head removed from the engine, use the following procedure:

1. Support the cylinder head on wood blocks 2″ above the workbench so as to keep the cam followers clear of workbench.
2. Remove fuel lines from injector and fuel connectors.
3. Remove the two bolts holding rocker arm shaft brackets to cylinder head. Lift rocker arms and brackets up and away from injector.
4. Remove fuel injector.
5. Remove exhaust valve bridges and bridge springs.
6. Remove exhaust valve spring from cylinder head as previously outlined. In addition, use a block

of wood under cylinder head to support exhaust valve.

Inspection

After removing valve spring, clean with fuel oil and dry with compressed air. Check spring for pitted or fractured coils. Then, using spring tester, test spring load.

Valve Spring Testing Specifications

Valve Spring Testing Specifications	
53 Series	25 lbs. @ 1.93″
71 Series-2 valve	135 lbs. @ 1 49/64″
71 Series-4 valve	79 lbs. @ 1.416″
92 Series	25 lbs. @ 1.80″

Valve Removal

With the cylinder head removed from the engine and valve springs removed, number each valve and adjacent area prior to removal to facilitate later assembly to the same seat.

Then withdraw valves from cylinder head.

Inspection

1. Scrape the carbon from the valve stem, wash with clean fuel

oil, and check for scratches or scuff marks.
2. Valve faces should be smooth, and free from ridges and pits. Carbon on the face of the valve indicates a faulty seat and a resultant leak or "blow-by." If any of these conditions exist, a new valve should be installed or the valve refaced to an angle of 30°. *NOTE: The valve face angle must be identical to the valve seat angle.*
3. Valve heads should be square with the valve stem and should not be warped.

Valve Installation

With the cylinder head thoroughly cleaned, the valve guides checked or replaced and the valves and valve inserts ground, install the valves as follows:

1. Apply a light coat of engine oil to the valve stems and install them in the cylinder head.
2. Secure the valves in place temporarily with masking tape. Then turn head over, resting valve heads on wood block or board and install spring seats, valve springs, valve spring caps

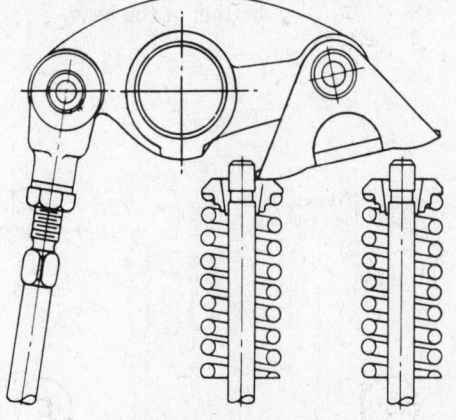

BRIDGE IMPROPERLY POSITIONED

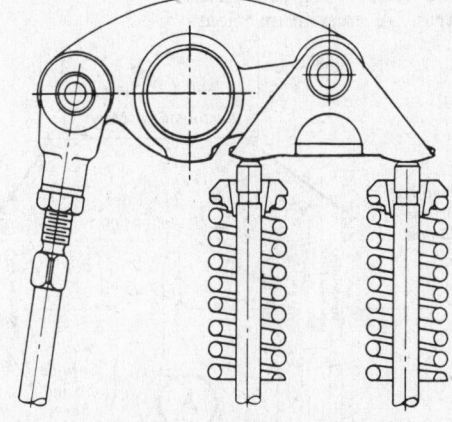

BRIDGE PROPERLY POSITIONED

Relationship of Exhaust Valve Bridge and Stems on 53 Series Engines
(© G.M. Corp)

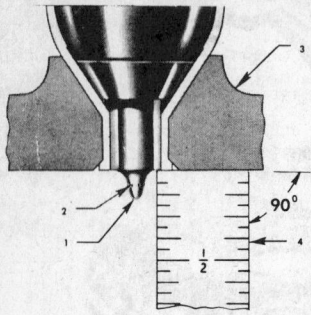

Checking location of injector spray tip
in relation to surface of cylinder head
(© G.M.C.)

1 Spray tip
2 Orifice—injector
   spray tip
3 Cylinder head
4 Straight edge-steel

and locks by reversing the procedure for removal.

NOTE: After valves have been installed, check to make sure the heads of the valves do not project more than .021″ beyond surface of cylinder head.

Rocker Arms and Shaft

Removal

1. Remove valve rocker cover.
2. Remove fuel lines from injector and fuel connectors.
3. Bring push rod ends—outer ends—or injector and valve rocker arms in line horizontally.
4. Remove the two bolts which hold the rocker arm shaft brackets to the cylinder head. Remove brackets and shaft.
5. Loosen the lock nut at upper end of push rod, next to clevis, and unscrew the rocker arm from the push rod.

Inspection

1. Make certain that oil passages in rocker arm shaft and bracket bolts are open and clean.
2. Inspect rocker arm shaft and bushings inside the rocker arms for excessive wear. Clearance should be .001″ to .0025″ with new parts. A maximum clear-

ance of .004″ is allowable with worn parts.

3. Check push rod clevis to rocker arm bushings and pins for excessive wear. The clearance be-between the steel clevis pin bushing and the bronze rocker arm bushing is .0015″ to .003″. The side clearance between the clevis and rocker arm should be from .008″ to .017″.
4. Examine the injector rocker arm pallet (contact face) for wear. The rocker arm pallet may be refaced up to .010″ by grinding. However, proceed with caution when surface grinding so rocker arms are not overheated. All radii and finish should be as close to original grind as possible.

Installation

1. Install cylinder head, if removed.
2. Apply clean engine oil to outer surface of rocker arm shaft and install rocker arm shaft and/or rocker arms by reversing the sequence of operations for removal.
3. Tighten rocker arm bracket bolts to 90-100 ft lb. torque on 71 and 92 Series engines, 50-55 ft lb. on 53 Series engines. Then check position of exhaust valve bridges. Valve stem on camshaft

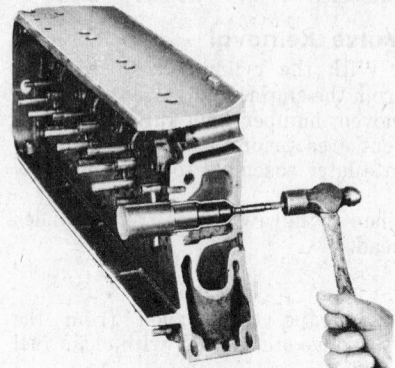

Removing injector tube
(© G.M.C.)

Compressing valve spring
(© G.M.C.)

side of cylinder head should register with recess in bridge guide.

4. Connect fuel lines from injectors to fuel connectors, torquing them to 12-15 ft lbs.
5. Adjust exhaust valve clearances and time fuel injectors.
6. Start engine. While engine is running in preparation for final tune-up, inspect the fuel lines for leaks. If leaks at the connections are discovered, tighten connecting nuts carefully. Should fuel oil leak into the crankcase and dilute the lubricating oil, the engine bearings will be seriously damaged.
7. Perform final engine tune-up after engine reaches normal operating temperature.

Cam Follower

Cam followers may be removed from either the top or bottom of the cylinder head. When followers are removed from the bottom, removal of the cylinder head will be necessary; when they are removed from the top, cylinder head removal is unnecessary. If the cylinder head is off the engine for any reason, the cam followers can best be removed from the bottom of the head.

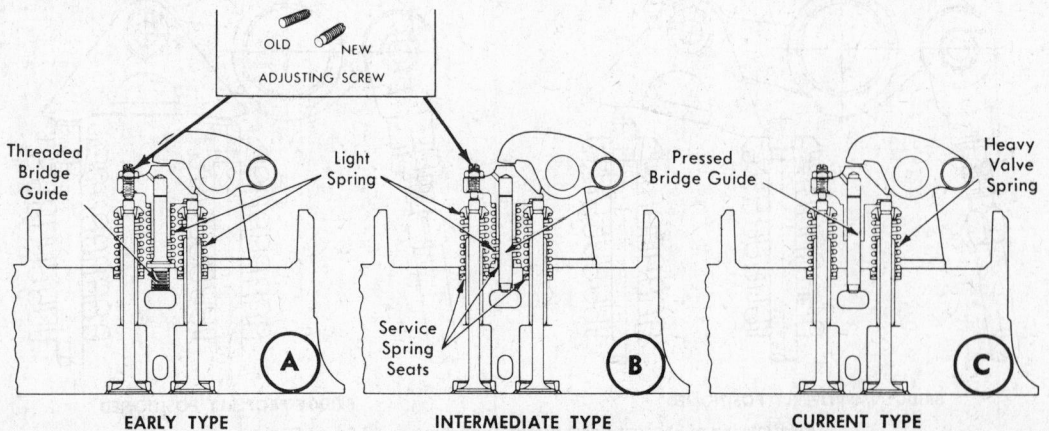

OLD   NEW
ADJUSTING SCREW

Threaded
Bridge
Guide   Light
               Spring   Pressed
                                          Bridge Guide   Heavy
                                                                 Valve
                                                                 Spring

Service
Spring
Seats

A   B   C

EARLY TYPE   INTERMEDIATE TYPE   CURRENT TYPE

Exhaust valve bridge and guide (© G.M.C.)

Cam Follower Removal

(Head removed)

1. Remove cylinder head.
2. Rest cylinder head on its side and remove the two bolts that secure cam follower guide to cylinder head. Remove guide.
3. Pull the cam followers from the bottom of the cylinder head.
4. Remove fuel lines from injector and fuel connectors.
5. Loosen the lock nuts at upper end of push rods, and unscrew push rods from the rocker arm clevises.
6. Pull push rod and spring assemblies from bottom of cylinder head.
7. Remove push rod lock nut, upper spring seat, spring and lower spring seat from each push rod for cleaning and inspection.
8. The upper push rod spring seat retainers remain in the cylinder head. If the head is to be changed, these retainers must be removed, if not, they may be left in place.

Cam Follower Removal

(Head installed)

A push rod, push rod spring, spring seats and cam follower may be removed from the top of the cylinder head.

1. Remove rocker cover.
2. Remove fuel lines from injector and fuel connectors.
3. Remove rocker arm brackets and rocker arm shaft.
4. Loosen lock nut at upper end of push rod, next to clevis, and unscrew rocker arm from push rod to be removed.
5. Run nut out on push rod so that remover may be inserted between the nut and the upper spring seat, with lower end of tool resting on upper spring seat. On 53 Series engines, remove the locknut and install a flat washer under it.
6. Screw nut down on upper end of push rod, thus compressing push

rod spring, relieving spring seat retainer.
7. Remove retainer from cylinder head with a screwdriver or similar tool.
8. Unscrew nut at outer end of push rod, thus releasing spring.
9. Pull push rod, spring, spring seats and cam followers out through top of cylinder head.

Replace Roller And Pin

1. Lock holding tool securely in vise and place cam follower in the groove in top of tool with follower pin resting on top of corresponding size plunger.
2. With suitable drift, drive pin from follower. Exercise caution in removing cam follower body and roller from holder, as follower pin is seated on top of spring loaded plunger.
3. Position follower body in groove of holding tool with the proper size tool plunger extending through roller pin hole in one of the legs of the follower body.
4. Coat new roller bushing and pin with engine oil.
5. With roller assembly placed properly on tool, align pin with hole in follower and carefully drive pin into the assembly until end of pin is centered in legs of follower.
6. Check side clearance between roller and follower body. This clearance should be .015″ to .023″.

Cam Follower Inspection

1. Clean the cam follower carefully and inspect it and the cam for excessive wear or scoring. Replace parts as necessary.
2. Check the cam follower spring for a free length of 2.62″ on 53 Series engines, and 2⅝″ on 71 and 92 Series engines. Replace if free length is not approximately correct.
3. Using a spring tester and torque wrench, check 92 series springs for a load of 250 lbs @ 2.1406″, 71 Series springs for a load of

172 lbs. @ 2.116″, 53 Series springs for a load of 250 lbs. @ 2-9/64″. Replace springs which do not meet these minimum specifications.
4. Check the cam follower-to-head clearance, which must be .006″ maximum, and replace followers as necessary.
5. Measure the clearance between roller bushing and pin (.010″ maximum) and the side clearance between the roller and follower (.015″-.023″) and replace parts which do not meet specifications.

Cam Follower Installation

(Head removed)

To assemble the cam follower and push rod assembly through the bottom of the cylinder head proceed as follows:

1. Assemble over the push rod, in order listed, lower spring seat, push rod spring, upper spring seat, and push rod lock nut. On 53 Series engines, replace lower spring seats with the newer, serrated design.
2. With the upper spring seat retainer in place in the cylinder head, slide the push rod assemblies into position from bottom of cylinder head.
3. Screw push rod lock nut down on upper end of push rod as far as possible, and screw push rod into clevis until end of rod is flush with or above inner side of clevis.
4. Note the oil hole in the bottom of the follower. With this oil hole pointing away from the valves, so that the hole is not covered by the follower guide, slide the followers into position from bottom of head. NOTE: To insure initial lubrication of the follower roller pin, immerse cam follower assemblies in clean lubricating oil for at least five minutes before installing them in cylinder head.
5. Secure follower guide to bottom of cylinder head to hold group of cam followers in place. Tighten guide bolts to 12-15 ft lbs. Then check to be sure there is clearance between the cam follower legs and the cam follower guide. Clearance is .005″ on 53 Series engines.

Cam Follower Installation

(Head installed)

1. With the oil hole in the bottom of the follower over the trailing side of roller (cam follower oil hole toward outside of cylinder head), insert follower into bore in head. *NOTE: To insure ini-*

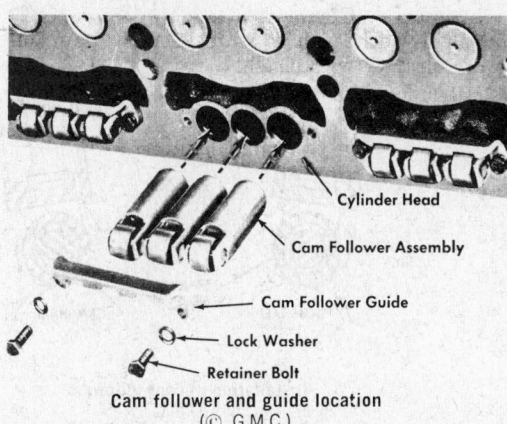

Cylinder Head
Cam Follower Assembly
Cam Follower Guide
Lock Washer
Retainer Bolt

Cam follower and guide location
(© G.M.C.)

tial lubrication of the follower roller pin, immerse cam follower assemblies in clean lubricating oil for at least five minutes before installing them in cylinder head.

2. Install lower spring seat, push rod spring and upper spring seat on push rod in the order listed and set the assembly down into the cam follower. Use serrated lower spring seats on 53 Series engines.

3. Start nut on outer end of push rod, and install tool beneath nut on push rod and against upper spring seat. Then screw nut down on push rod until spring is compressed sufficiently to permit retainer to be inserted in cylinder head. Partially collapse retainer and install in groove in the cylinder head.

4. Unscrew nut at outer end of push rod and remove tool.

5. Screw lock nut as far as possible down on push rod, then screw rocker arm clevis down on push rod until the end of the push rod is flush with or above inner side of clevis.

6. Observe that the injector rocker arm (the center arm of the group) is slightly different from the exhaust valve rocker arms; the boss for the shaft on the valve rocker arms is longer on one side of the arm than on the other. The extended boss of valve rocker arms must face the injector rocker arm.

Camshaft

Removal

Whenever an engine is being completely reconditioned or the camshafts, camshaft gears, bearings or thrust washers need replacing, the camshafts must be removed from the engine in the following manner:

92 Series

1. Drain engine and mount in overhaul stand.

2. Remove heads.
3. Remove flywheel and housing.
4. Remove water pump.
5. Remove from balance weight cover.
6. Remove the retaining plate bolts and plates.
7. Wedge on a clean rag between the gears and remove the gear retaining nut from each shaft. On left hand rotation engines, remove the lock bolt and washer from the right camshaft.
8. Pull the camshaft vibration damper and hub.
9. Remove the water pump drive gear from the right camshaft.
10. Remove the key and spacer from each shaft.
11. Remove intermediate bearing lock screws from the block.
12. Remove end bearing retaining bolts.
13. Pull each shaft assembly from the block.

71 Series

1. Remove all accessories and assemblies with their attaching parts necessary to mount engine on overhaul stand.
2. Remove cylinder head.
3. Remove flywheel and flywheel housing.
4. Remove water pump.
5. Remove balance weight cover.
6. Remove bolts securing retainer plates to camshaft timing gears. Then remove retainer plates.
7. Wedge a clean rag between gears. Using a socket wrench, remove nuts from both ends of each camshaft.
8. Attach puller to camshaft pulley. Use adaptor between end of camshaft and puller screw to protect end of camshaft.
9. Remove water pump drive gear from front end of right bank camshaft.
10. Remove Woodruff key and spacer from forward end of both camshafts.
11. Remove lock screw from the top of cylinder block at each cam-

Removing the Camshaft Gear
(© G.M. Corp)

shaft intermediate bearing.

12. Remove three bolts which secure each camshaft end bearing assembly to rear of engine. Rotate camshaft gear as required to reveal bearing bolts through hole in web of gear.
13. Withdraw camshaft, bearings, and gear as an assembly from rear end of cylinder block.
14. Remove three bolts from each camshaft front end bearing. Then, withdraw bearings from cylinder block and front end plate. If the bearings cannot be withdrawn readily by hand, remove by prying under bearing flange.

53 Series

1. Drain the cooling system.
2. Remove accessories and assemblies and their attaching parts as required to permit mounting the engine in an overhaul stand. Mount the engine securely in the stand, and release the lifting sling.
3. Remove the cylinder head(s) as described in the appropriate section.
4. Remove the flywheel and housing as described in the appropriate section.
5. Remove the gear nut retainer plate bolts (if used) and remove the retainer plates.

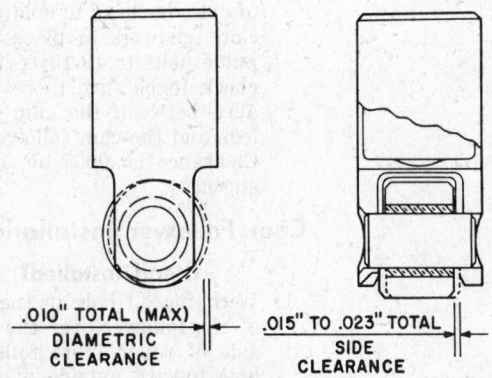

.010" TOTAL (MAX)
DIAMETRIC
CLEARANCE

.015" TO .023" TOTAL
SIDE
CLEARANCE

Cam roller wear clearance
(© G.M.C.)

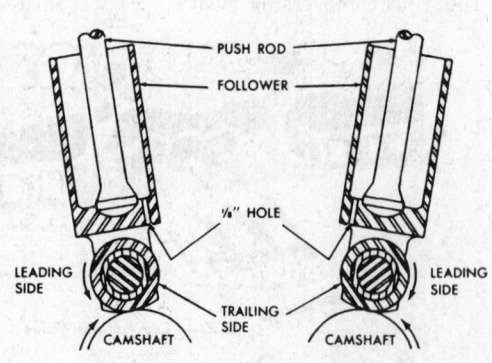

PUSH ROD
FOLLOWER
⅛" HOLE
LEADING SIDE
TRAILING SIDE
LEADING SIDE
CAMSHAFT
CAMSHAFT

Installation of cam followers
(© G.M.C.)

Removing camshaft gear retaining nut
(Ⓒ G.M.C.)

Removing camshaft bearing bolts
(Ⓒ G.M.C.)

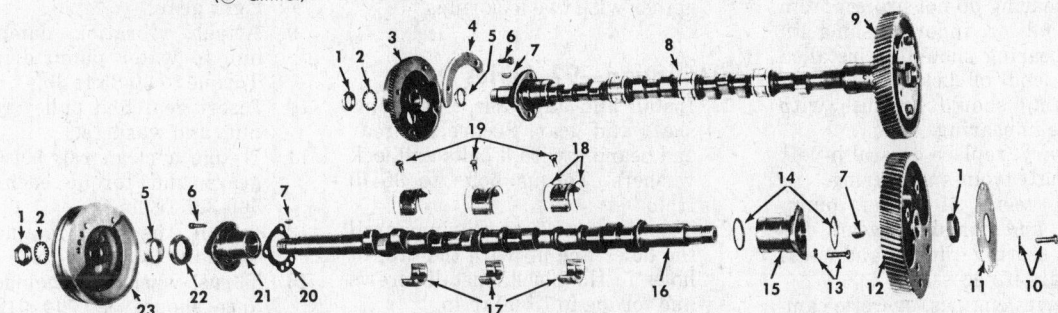

Camshaft components (Ⓒ G.M.C.)

1 Gear retaining nut	9 Timing gear-right	16 Camshaft—left
2 Lock washer	10 Retainer bolt and lock washer	17 Intermediate bearing
3 Water pump drive gear	11 Nut retainer	18 Lock ring
4 Balance weight	12 Timing gear-left	19 Set screw
5 Spacer	13 Retainer bolt and lock washer	20 Gasket
6 Weight bolt	14 Thrust washer	21 Front bearing
7 Woodruff key	15 Rear bearing	22 Oil seal
8 Camshaft—right		23 Camshaft pulley

6. Wedge a clean rag between the camshaft gears, and then remove the gear mounting nuts.
7. Force the balance weight off the end of each shaft, using heavy screwdrivers between the weight and upper front cover.
8. Remove the upper front cover-to-block attaching bolts. Tap the cover and dowel assembly away from the block, remove the wood-ruff keys and oil seal spacers, and clean all traces of old gasket material off the block.
9. Remove the oil slingers from the front ends of both shafts.
10. Remove the thrust washer retaining bolts securing the shaft thrust washers to the block by inserting a socket through the hole in the web of the gear.
11. Withdraw the shaft, thrust washer, and gear from the rear of the block.

Disassembly—92 Series
1. Pull camshaft gear.
2. Remove rear end bearing and thrust washers.
3. Remove intermediate bearing lock rings.
4. Remove end plugs. This is accomplished by drilling and tapping the plugs.

Disassembly—71 Series
1. Remove gear from camshaft.
2. Slide camshaft rear end bearing and thrust washers off camshaft
3. Remove lock rings from camshaft intermediate bearings then separate and remove two halves of each bearing.

Disassembly—53 Series
1. Place the assembly in an arbor press, with the gear suitably supported. Place a wood block under the lower end of the shaft.
2. Place a short piece of ¾″ O.D. brass rod between the end of the camshaft and the ram of the press. Force the camshaft out of the gear.
3 Remove the thrust washer, key and spacer.

Inspection
Be sure that oil holes in camshaft end bearings are clean.

Inspect cams and journals and replace the camshaft if scored or worn. Inspect cam followers.

Examine both faces of camshaft rear end bearing and thrust washers. If either face is scored or if thrust washers are worn excessively, replace washers. New standard size thrust washers are .120″ to .122″

thick. Examine surfaces of gear and camshaft which the thrust washers contact; if these surfaces are scratched but not severely scored, they may be smoothed down with an oil stone. However, if score marks are too deep to be removed or parts are badly worn, new parts must be used.

Clearance between the thrust washer and thrust shoulder of the camshaft, using new parts, is from .004″ to .012″, or a maximum of .018″ with used parts. Thrust washers are available in .005″ and .010″ oversize and may be used to reduce excessive clearance within the specified limits of .004″ to .012″.

On 71 and 6V-92 Series engines, the clearance between new camshaft end journals and their respective new bearings is from .0025″ to .004″, or a maximum of .006″ with worn parts. On 8V-92 Series engines the clearance is .0035 to .005″ or a maximum of .006″ On 53 Series engines, clearance is .0045″-.006″ with new parts, or .008″ with worn parts. End bearings are available in .010″ or .020″ undersize for use with shafts which have worn or been reground and the clearances exceed the specified limits.

On 71 and 92 Series engines, the clearance between the camshaft journals and the intermediate bearings is from .0025" to .005" with new parts, or a maximum of .009" with worn parts. Camshaft intermediate bearings are available in .010" and .020" undersize for use with worn or reground shafts in which the clearances exceed the .0025" to .005" limits. Bushings must be finished bored to a 20 R.M.S. finish after installation. Also inside diameter of bushing must be square with rear face of bearing within .0015" total indicator reading. Camshaft rear end bearing bushings must project .045" to .055" from each end of bearing. Bushings in camshaft front end bearing do not project from bearing. End of inner bushing in front end bearing should be installed flush with end of bearing; end of outer bushing should be flush with counterbore in bearing.

If necessary, replace oil seal in left bank camshaft front end bearing.

Examine teeth of water pump drive gear and camshaft gears for evidence of scoring, pitting and wear. Replace gears if necessary.

On 53 Series engines, oversize camshaft and balance shaft bearings are available in sets .010" oversize on the outside diameter to permit re-use of a block having scored bearing bores. To use the oversize bearings, camshaft and balance shaft bores must be line bored to the dimensions shown below:

End Bearings   2.385"-2.386"
Intermediate Bearings 2.375"-2.376"

In removal and installation of bearings, end bearings must be removed prior to removal of intermediate bearings. Reverse this procedure during installation.

Assembly—92 Series

1. Install new end plugs.

2. Apply heavy grease to the steel face of each thrust washer and place a thrust washer against each end of the two camshaft rear end bearings. The steel face of each washer must be against the bearing.

3. Lubricate the rear camshaft bearing journal and slide a rear end bearing on each shaft.

4. Install the camshaft gear with a press.

5. Lubricate the intermediate bearing journals and place the two halves of each bearing on a camshaft journal and lock halves together with two lock rings.

Installation—92 Series

1. Install and align the right camshaft and gear. Secure the rear end bearing with 3 bolts and lock washers. Torque bolts to 35-40 ft. lb.

2. Turn intermediate bearings until the holes line up with the tapped holes in the block. Install screws and torque to 15-20 ft. lb.

3. Install the left camshaft the same way.

4. Fix a new gasket to the front end bearing and seal. Lubricate the journal and slide the bearing on the left bank with the bolting flange toward the outer end. Install the bolts and torque to 35-40 ft. lb.

5. Install the right bank camshaft front end bearing, without the oil seal. Secure the bearing and torque to 35-40 ft. lb.

6. Install the spacers—polised one on the left.

7. Install both keys.

8. Install the pulley on the left shaft and the water pump drive

Removing or installing camshaft assembly
(© G.M.C.)

gear and external weight on the right gear.

9. Attach vibration damper and hub to water pump drive gear. Torque to 30-35 ft. lb.

10. Insert gear and pulley retaining nuts and washers.

11. Wedge a clean rag between the gears and torque each nut to 300-325 ft. lb.

12. Install the nut retainers and torque to 35-39 ft. lb.

13. Thrust washer-to-shoulder clearance should be .004-.012" when checked. Gear backlash is .003-.008".

Assembly—71 Series

Assemble parts on camshaft as follows:

1. Apply heavy cup grease to the steel face of camshaft thrust washers and install thrust washers against each end of camshaft rear bearing. Be sure steel face of washer is next to bearing.

2. Lubricate the camshaft rear bearing journal and slide rear bearing onto camshaft journal with bolting flange of bearing toward outer (gear) end of shaft.

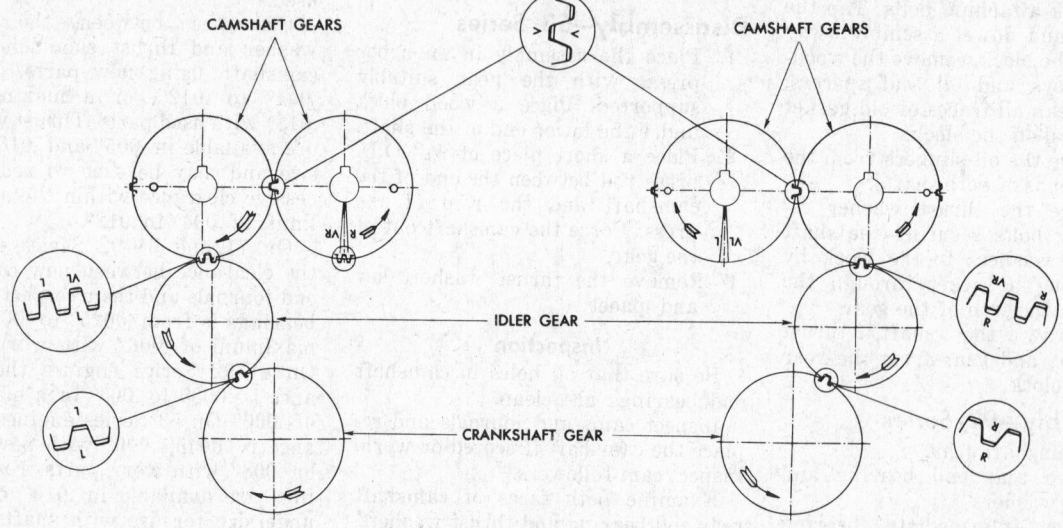

CAMSHAFT GEARS · CAMSHAFT GEARS · IDLER GEAR · CRANKSHAFT GEAR

LEFT-HAND ROTATION ENGINES   RIGHT-HAND ROTATION ENGINES

Gear train and timing marks (© G.M.C.)

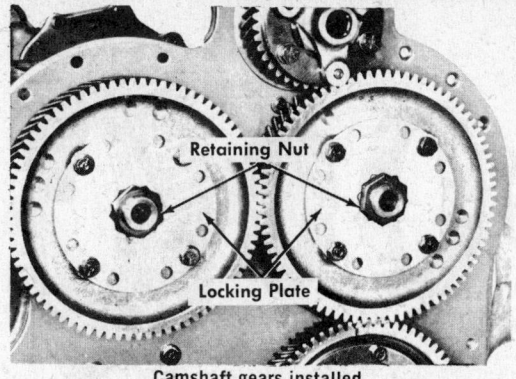

Camshaft gears installed
(© G.M.C.)

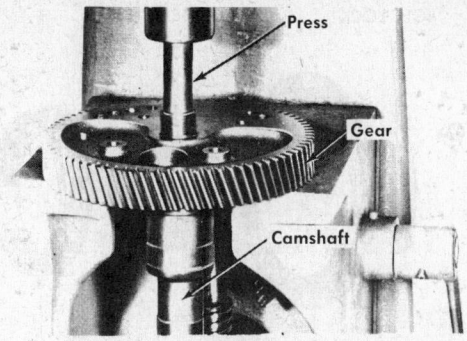

Pressing gear off camshaft
(© G.M.C.)

3. Install timing gear on camshaft.

4. Lubricate the camshaft intermediate bearing journals, then place the two halves of each intermediate bearing on camshaft journal and lock halves together with two lock rings. Assemble lock rings with gap in upper bearing and ends equal distance (approximately ½ inch) above split line of bearing.

Installation—71 Series

1. Insert forward end of assembled camshaft through opening in rear end plate until the first intermediate bearing enters bore. Continue to work the camshaft and bearings into the cylinder block until gear teeth are about to engage the teeth of mating gear. Use care not to damage cam lobes when installing shaft.

2. Then, with the timing marks on the mating gears in alignment, slide camshaft gear into mesh.

3. Secure camshaft rear bearing to cylinder block. Rotate camshaft gear as required to install bearing bolts through hole in web of gear. Tighten bolts to 35-40 ft.-lb. torque.

4. Turn camshaft intermediate bearings so that locking holes in bearings align with holes in top of cylinder block and secure in place with lock screws.

5. Install opposite camshaft as outlined in preceding paragraphs.

6. Using a new bearing to end plate gasket, install left bank camshaft front end bearing—the one with the oil seal—in the cylinder block. Secure bearing in place. Tighten bolts to 35-40 ft.-lbs. torque.

7. Install right bank camshaft front end bearing—the one without the oil seal—in the cylinder block. Tighten bearing retaining bolts to 35-40 ft.-lb. torque.

8. Lubricate outside diameter of camshaft front spacer and slide spacer onto front end of left bank camshaft.

9. Install camshaft front spacer on right bank camshaft.

10. Install Woodruff keys in keyways in front end of each camshaft.

11. Install pulley on front end of left bank camshaft and water pump drive gear on right bank camshaft.

12. Slip an internal tooth lock washer over front end of each camshaft. Install star nuts on forward end of both camshafts. *NOTE: Early engines have water pump gear, and camshaft gear with 59 and 42 teeth respectively. Latest pump gear and camshaft gear have 92 and 66 teeth respectively. Early and latest type gears are not interchangeable.*

13. Start nuts on rear end of each camshaft.

14. Wedge a clean rag between the camshaft gears to prevent their rotation; then, using a 1½" socket wrench, tighten nut on each end of both camshafts to 300-325 ft.-lb. torque.

15. Secure camshaft timing gear nuts with gear nut retainers, bolts and lock washers. Tighten gear nut retainer bolts to 35-39 ft.-lb. torque.

16. Check clearance between thrust washer and thrust shoulder of camshaft. Clearance should be .004" to .012", or a maximum of .018" with used parts.

17. Check backlash between mating

Camshaft gear installed
(© G.M.C.)

gears. Backlash should be .003" to .008", and should not exceed .010" between worn gears.

18. Install flywheel housing and other parts or assemblies, that were removed from the engine.

Installation—53 Series

1. Identify the timing gear end of the camshaft, stamped "R.H.-R. BANK" and "V7" or "R.H.-L. BANK" and "V7". The gear must be installed on this end of the camshaft.

2. Place the rear spacer over the timing gear end of the shaft and install the key.

3. Lube the thrust washer with engine oil and install it over the gear end of the shaft and spacer.

4. Start the camshaft gear over the end of the shaft with the key and keyway in line.

5. Support the shaft in an arbor press and place a sleeve between the gear and the ram of the press. Press the gear on until it is tight against the spacer.

6. Measure the clearance between the camshaft thrust washer and the camshaft. It should be .008"-.015" with new parts, with a maximum used parts figure of .021".

7. Install the gear retainer nut finger tight.

8. Lubricate the bearings and shafts with engine oil and slide the shaft assemblies into the block. Align the timing marks on the gears.

9. Install the oil slinger.

10. Put a new gasket onto the engine front cover, install the cover onto the engine, and secure it with bolts and lock washers. Torque the bolts to 35 ft lbs.

11. Apply cup grease to the outside diameters of the oil seal spacers and slide them onto the shafts.

12. Secure the thrust washers in place and tighten the bolts to 30-35 ft lbs.

13. Install the keys and then install the pulleys, tightening the retaining nuts to 300-325 ft lbs. If a balance pulley does not slip

(Detroit 2-cycle) Diesel Engines

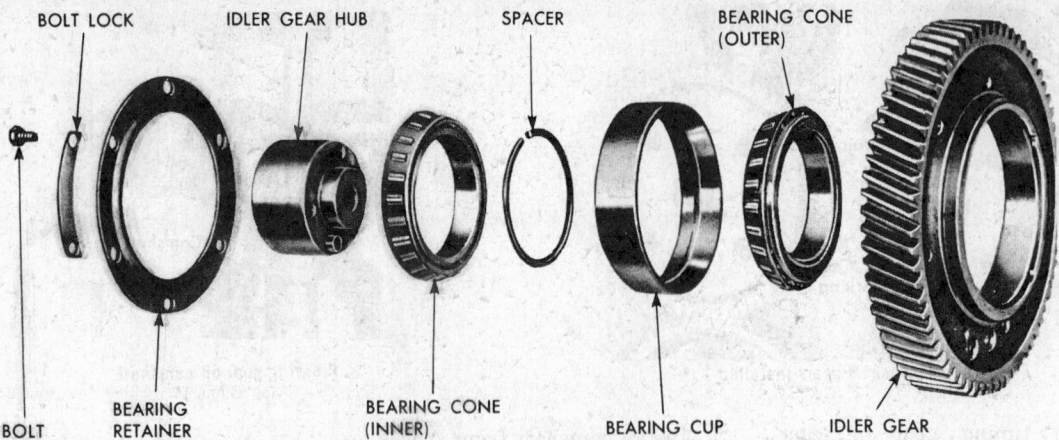

BOLT LOCK   IDLER GEAR HUB   SPACER   BEARING CONE (OUTER)

BOLT   BEARING RETAINER   BEARING CONE (INNER)   BEARING CUP   IDLER GEAR

Idler gear and hub details (© G.M.C.)

easily onto the end of the shaft, loosen the thrust washer retaining bolts, support the rear end of the shaft, and place a hollow sleeve over the end of the shaft to permit the pulley to be tapped into position.

14. Install the gear nut retainer plates (if used) and torque the bolts to 35-39 ft lbs.
15. Check the clearance between the thrust washer and gear. The clearance should be .008"-.015" with new parts, or up to .019" with used parts.
16. Check the backlash between mating gears. It must be .003"-.005", or up to .007" with used parts.
17. Replace cylinder heads and flywheel housing as described in the appropriate section.

Gear Train

Description

The gear train consists of a crankshaft gear, an idler gear, two camshaft gears and a blower drive gear.

The crankshaft gear is bolted to a flange at the rear end of the crankshaft. The idler gear is mounted on a stationary hub on either the right or left side of the engine depending upon the engine rotation. The camshaft gears are pressed on and keyed to their respective shafts and each is secured by a nut and gear nut retainer.

The two camshaft gears mesh with each other and run at the same speed as the crankshaft gear. Since the former two gears must be in time with each other, and the two as a unit, in time with the crankshaft gear, timing marks have been stamped in the face of the gears to facilitate correct gear train timing. When assembling the engine, it is important to remember engine rotation. Then line up the appropriate timing marks on the gears as each gear is installed on the engine.

Crankshaft Gear

The crankshaft timing gear is bolted to a flange at the rear end of the crankshaft and drives the camshaft gears, as well as the blower drive gear, through an idler gear. One gear attaching bolt hole is offset so that the gear can be attached in only one position.

Crankshaft Gear Removal—71 and 92 Series

1. Remove six bolts retaining gear to crankshaft.

2. Remove gear using a suitable gear puller.

Installation

1. Position gear on rear end of crankshaft with all six bolt holes in gear aligned with tapped holes in the crankshaft flange. Since one bolt hole is offset, the gear can be attached in only one position on the crankshaft.
2. Align timing mark on crankshaft gear with corresponding mark on the idler gear.
3. Install gear to crankshaft bolts, then tighten bolts to specified torque (35-39 ft.-lb.)
4. Check backlash between mating gears. Backlash should be .003" to .008", and should not exceed .010" between worn gears.

Crankshaft Gear Removal—53 Series

1. Attach a bar type puller with plates to the crankshaft gear with long bolts. Adjust the lengths of the bolts so they are even.
2. Turn the center screw of the puller and pull the gear off the crankshaft.

Press
Hub
Bearing
Gear

Pressing hub from idler gear bearing
(© G.M.C.)

Idler gear installed
(© G.M.C.)

Checking idler gear bearing pre-load
(© G.M.C.)

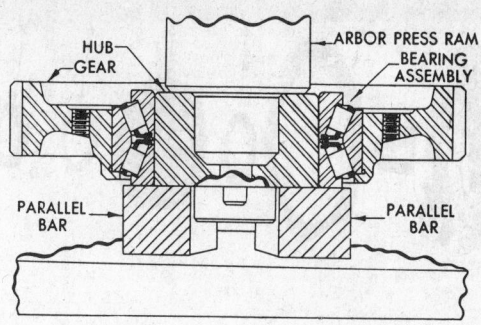

Pressing hub into idler gear bearing
(© G.M.C.)

Installation

1. Start the gear over the end of the crankshaft with the timing marks aligned and the gear keyway aligned with the key in the shaft.
2. Hold a heavy hammer against the head of the bolt in the front end of the crankshaft. Place an installer against the rear face of the timing gear. Use a heavy hammer to drive the gear up against the shoulder on the crankshaft.

Camshaft Gears

Since the two camshaft gears must be in time with each other, timing marks are stamped on the rim of both gears. Also, since these two gears as a unit must be in time with the crankshaft, timing marks are located on the idler and crankshaft gears.

Camshaft Gear Removal

1. Remove camshafts from engine.
2. Place one of the camshaft and gear assemblies in arbor press with gear suitably supported. Lay a wood block under the lower end of the camshaft.
3. Then, using a short piece of one inch (¾″ on 53 Series) O.D. brass rod beween end of camshaft and ram of press, force camshaft out of camshaft gear.
4. If necessary, remove Woodruff

key from camshaft. Remove the thrust washer and spacer on 53 Series engines.

Installation

NOTE: On 53 Series engines, install gear on end marked "R.H.-R Bank" or "R.H.-L Bank."

1. Install camshaft rear end bearing and thrust washers on camshaft. Apply heavy cup grease to the steel face of camshaft thrust washers and place a washer on each end of camshaft rear bearing. Steel face of washer is next to bearing. On 53 Series engines, install the rear camshaft spacer and key.
2. Lubricate the camshaft rear bearing journal and slide rear

Installing idler gear assembly
(© G.M.C.)

bearing onto camshaft journal with bolting flange of bearing toward outer (gear) end of shaft.
3. Start camshaft gear over end of camshaft with key in shaft registering with keyway in gear.
4. Press gear tight against shoulder on shaft in manner illustrated in Fig. 5. If available, camshaft gear installer may be used for this purpose. On 53 Series engines, measure the thrust washer-to-camshaft clearance. It should be .008″-.0015″ with new parts, or up to .021″ with used parts.
5. Start camshaft gear retaining nut on camshaft. Tighten nut after shaft is installed in cylinder block.
6. Install camshaft in cylinder block.

Idler Gear

The idler gear is mounted on a double row, tapered roller bearing, which in turn is supported on a stationary hub. A bolt passes through the hub and rear end plate. A dowel in the hub correctly positions the hub and prevents the hub from rotating.

The idler gear bearing cup has a light press fit in the gear and is held in place by a retainer. The idler gear bearing cones are pressed onto the gear hub and do not rotate. A spacer separates the two bearing cones.

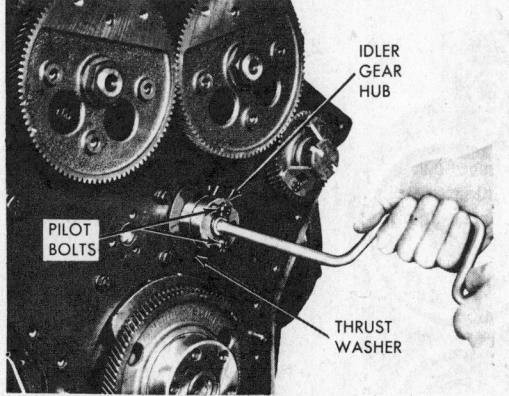

Installing the Idler Gear Hub
(© G.M. Corp)

Installing the Idler Gear
(© G.M. Corp)

1 Support to plate bolt
2 Copper washer
3 Blower drive gear support
4 Gasket
5 Blower drive gear
6 Thrust bearing
7 Thrust washer
8 Lock plate
9 Drive gear support nut
10 Plate to hub bolt and lockwasher
11 Spring plate
12 Plate to gear bolt and lockwasher
13 Gear hub
14 Snap ring

Blower drive gear and support components (© G.M.C.)

A left-hand helix gear is provided for right-hand rotation engines, and a right-hand helix gear is provided for left-hand rotation engines.

An idler gear hole spacer (dummy hub) is used on the side opposite the idler gear.

Removal—71 and 92 Series

1. Remove bolt and special washer. Then remove idler gear, hub and bearing assembly from the engine.
2. Remove idler gear hold spacer (dummy hub) from the opposite side in the same manner.
3. Remove bearing retainer. *NOTE: Component parts of idler gear bearing are mated; therefore, match-mark the parts during disassembly to assure they will be reassembled in their original positions.*
4. Place idler gear and bearing assembly in arbor press. While rotating the idler gear to prevent brinelling of bearing, press the hub out of the bearing.
5. Tap bearing cup (outer race) from idler gear by using a brass drift alternately at four notches provided in shoulder of gear.

Assembly

Pay attention to matching marks previously made on parts to assure their reassembly in same positions from which they were removed and proceed as follows:

1. Support idler gear, with shoulder down, on bed of arbor press, locate bearing cup (outer race) at bearing bore of gear, and press cup into place tight against shoulder in gear.
2. Support one bearing cone, numbered side down, on bed of arbor press and lower idler gear and bearing cup assembly down over the bearing cone.
3. Lay spacer ring on face of bearing cone.
4. Place second bearing cone, numbered side up, in idler gear and bearing cup assembly and against spacer ring.
5. Position the idler gear hub over the bearing cones so that the oil

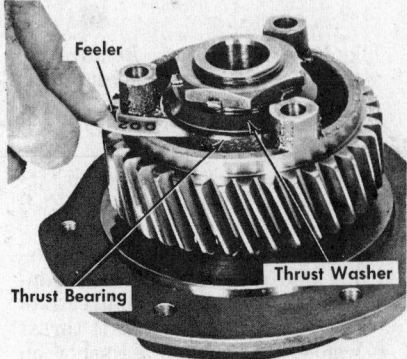

Feeler

Thrust Bearing

Thrust Washer

Checking clearance between thrust washer and thrust bearing (© G.M.C.)

hole in hub is 180° from gap in spacer ring.

6. Press hub into idler gear bearing cones, while rotating the gear to seat rollers properly between cones, until face or hub which will be adjacent to cylinder block end plate is flush with the corresponding face of bearing cone. Bearing cones should be supported so as not to load the bearing rollers during this operation.
7. Prior to installing and securing the bearing retainer, check the pre-load of the bearing assembly as outlined below.

Then, when the torque required to rotate the idler gear is within specified limits, secure bearing retainer to the idler gear with three bolt locks.

Blower drive gear assembly installed (© G.M.C.)

Pre-Load Check

If the mating crankshaft and camshaft gears are not already mounted on the engine, the torque required to rotate the idler gear may be checked by mounting the idler gear in position on the engine, using a steel plate 4″ square and ⅜″ thick against the hub and cone as follows under heading "Engine Mounted Method."

However, if the crankshaft and camshaft gears are on the engine, a suitable fixture, which may be held in a vise, can be made locally to accomplish a "Bench Test Method" of determining torque. Three plates, a ½″-13 x 1-¾″ bolt and a plain washer are used with a ½″-13 nut and plain washer for mounting. One of the plates is used to take the place of a flywheel housing, and the other two plates, the cylinder block. "Engine-mounted" conditions are simulated by tightening the nut to 80-90 ft.-lb. torque and tightening the three plate-to-hub attaching bolts to 25-40 ft.-lb. torque. The components of the fixture may be made from steel stock.

The idler gear bearing should be cleaned and lubricated with clean light engine oil prior to the pre-load test. Idler gear assemblies which include new bearings should be "worked in" by grasping the gear firmly by hand and rotating the gear back and forth several times.

Installation—71 and 92 Series

1. Position the crankshaft gear and camshaft gear so that the match marks will align with those on the idler gear.
2. With these marks in alignment, start the idler gear into mesh with the crankshaft gear and the camshaft gear, and simultaneously rotate the gear hub so that the hollow pin at the inner face of the hub nearly registers with the hole in the end plate.
3. Then simultaneously roll the idler gear into position, and align the hollow dowel with the hole in the end plate. Gently tap the hub until the hub seats against the end plate.
4. After making sure that the hub is tight against the end plate, se-

cure the idler gear assembly in place with a ½"-13 bolt and special washer. Tighten bolt to 80-90 ft.-lb. torque.
5. Lubricate the idler gear and bearing liberally with clean engine oil.
6. Check backlash between mating gears. Backlash should be .003" to .008", and should not exceed .010" between worn gears.

Removal—53 Series

1. Remove the idler gear thrust washer from the hub.
2. Slide the idler gear straight back off the hub.
3. Remove the bolt securing the idler gear hub to the block. Remove the hub and idler gear thrust washer assembled.

Installation—53 Series

1. Place the inner thrust washer on the forward end of the idler gear hub with the flat in its inner diameter over the flat on the end of the hub. Make sure the oil grooves face the idler gear.
2. Place the small end of the idler gear hub through the end plate and into the counterbore of the block.
3. Insert two ⅜"-16 bolts through the idler gear hub and thread them into the block to ensure alignment of the holes when installing the flywheel housing.
4. Insert the special bolt through the center of the idler gear hub and thread it into the block. Torque it to 40-45 ft lbs., and then remove the two bolts installed in the step above.
5. Lubricate the idler gear hub and idler gear bearings with clean oil.
6. Position the crankshaft gear and camshaft or balance shaft gear so they will align with the marks on the idler gear. Then, install the idler gear.
7. Apply a thin layer of cup grease to the inner face (with oil grooves) of the outer idler gear thrust washer. Place the washer over the hub with the oil grooves facing the gear and the flat over the flat on the end of the hub.
8. Check the backlash between the mating gears. It should be .003"-.005" between new gears, with a maximum clearance of .007" between used gears.

Blower Drive Gear and Support Assembly —71 and 92 Series

The blower drive gear, is mounted on the blower drive gear support, which is attached to the cylinder block rear end plate. This gear is driven by the right cylinder bank camshaft gear.

Removal

1. Remove blower.
2. Remove two blower drive support-to-cyclinder block rear end plate attaching bolts with copper washers and tap the assembly away from the end plate, using care not to damage the gear teeth.

Disassembly

1. Clamp blower drive gear support in soft jaws of bench vise.
2. Remove three spring plate to gear bolts and lock washers.
3. Remove spring plates and blower drive gear hub as an assembly from gear. If necessary, spring plates may be removed from hub.
4. Straighten lugs on blower drive gear support nut lock plate, then remove nut.
5. Remove lock plate, blower drive gear thrust washer, thrust bearing and gear from support.

Inspection

Examine faces of thrust bearings and thrust washer and, if scored or worn excessively replace.

Check inside diameter of blower drive gear bearings for wear; also check outside diameter of support hub on which gear bearings ride. The inside diameter of new bearings in 1.6260" to 1.6265". The clearance between bearings and support should be .001" to .0025" with new parts and a maximum of .005" with used parts. Install new parts, whenever clearance exceeds the maximum.

If new bearings are installed in blower drive gear, outer end of bearings must be pressed in flush with sides of gear and bore of bearings should be square with bolt boss faces and pitch diameter of teeth of gear within .003" total indicator reading, after reaming bushings to twenty micro-inches.

Assembly

1. With blower drive support clamped in soft jaws of bench vise, position one of the blower drive gear thrust bearings on support so that tangs on bearing register with holes in support.
2. Lubricate hub of support, bearings in gear, both thrust bearings and the blower drive gear, support thrust washer with clean engine oil.
3. Slide gear on hub with flat side of gear down.
4. Install second thrust bearing on support with tangs on bearing facing up.
5. Position blower drive support

thrust washer on support so that slots in thrust washer register with tangs on bearing.
6. Install gear on support with lock washer and nut. Tighten nut to 50-60 ft.-lb torque and bend lugs on lock washer against flats on nut to secure nut.
7. Check clearance between blower drive gear support thrust washer and blower drive gear thrust bearing. Clearance should be .005" to .010", and should not exceed .012" between worn parts.
8. If spring plates were removed from blower drive gear hub, attach spring plates to hub with three bolts and lock washers. Tighten bolts to 35-39 ft.-lb. torque.
9. Assemble spring plates and hub to gear with three bolts and lock washers. Tighten bolts to 35-39 ft.-lb. torque.

Blower Drive Gear and Support Assembly Installation

1. Using a new gasket, attach gear and support assembly to cylinder block rear end plate with two bolts and copper washers. Tighten bolts to 25-30 ft.-lb. torque.
2. Check backlash between blower drive gear and camshaft gear. Backlash should be .003" to .008" with worn parts.
3. Remove four fly-wheel housing-to-blower drive support bolts. Then install blower.

NOTE: Refer to appropriate sections of Governor Service—53 Series for Blower Drive Gear and Support Assembly service for that series.

Piston and Piston Rings

Inspection

Excessively worn or scored pistons, rings and cylinder liners indicate improper maintenance or operating conditions which should be corrected as quickly as possible.

Piston

1. A careful examination of the piston should be made for scoring, cracks, worn or damaged ring grooves, plugged oil drain holes, or indications of excessive heat such as may melt the tin coating on the pistons. Pistons which are heavily scored, overheated, cracked or have excessively widened ring grooves should be discarded. Pistons which are only slightly scored may be cleaned up and reused.
2. Examine the inside of the piston closely for cracks across the struts and discard piston if such cracks are present. The pistons may be checked for cracks for

use of the various methods of magnetic particle inspection.

3. Inspect the top of the piston crown for burned spots or other indications of overheating such as carbon formation in the underside of the piston crown. Such spots or carbon coating on underside of piston may indicate a lack of sufficient oil spray on the underside of piston via the drilled passage in the connecting rod.

Piston Pin Bushing

1. Inspect and measure the inside diameter of the piston pin bushings in each piston and, also inspect and measure the piston pin. With new parts, the piston pin-to-bushing clearance should be from .0025" to .0034". However, with worn parts, a maximum clearance of .010" is allowable.

A .010" oversize piston pin is available to aid in obtaining the proper clearance with worn bushings.

Piston Pin Bushing Removal

1. Place piston in holding fixture.
2. Drive bushing out of piston.
3. Revolve piston 180° in fixture and remove the second piston pin bushing in the same manner.

Installation

1. When installing bushings in a piston, locate the split or joint at the bottom, that is away from the piston head.
2. Insert spacer in hole in fixture and place piston on fixture so that spacer protrudes into bushing bore.
3. Insert installer in a bushing; then slide the bushing and installer inside piston and position over inside edge of lower bushing bore.

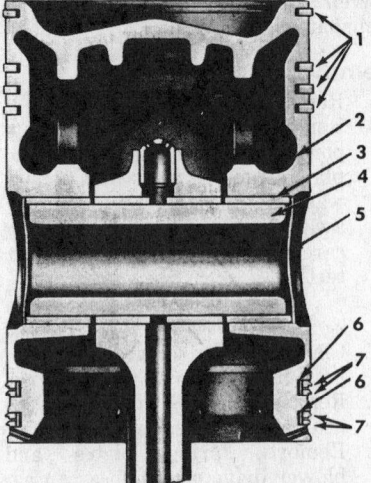

Piston and ring assembly
(© G.M.C.)

1	Compression rings	5	Retainer
2	Piston	6	Expander
3	Bushing	7	Oil ring
4	Piston pin		

4. Insert handle through upper bushing bore and into installer.
5. Using hammer on upper end of handle, drive bushing into piston bushing bore until bushing bottoms on spacer.
6. Turn piston over in fixture and install second bushing in a similar manner.

Reaming

The bushing must be reamed after pressing into place.

1. Clamp piston bushing reaming fixture in vise, then insert guide bushing in fixture and secure with set screw.
2. Place piston assembly in fixture and insert pilot end of reamer, through clamping bar, piston pin bushings, and into guide bushing.
3. With piston, fixture and reamer in alignment, tighten wing nuts securely.

4. Turning reamer in a clockwise direction only, with uniform motion, both when reaming and withdrawing reamer, ream the piston pin bushings.
5. Remove piston from fixture. Blow out chips resulting from reaming and check bushings for correct reamed inside diameters, as below:

71 and
92 Series—1.5025"-1.5030"
53 Series—1.3773"-1.3780"

Piston Ring Installation

With the connecting rod assembly inspected and assembled to the piston, be certain to check proper location of piston rings assembled on the piston.

1. Assemble compression rings on piston, and stagger ring gaps around piston. When installing compression or oil control rings, do not spread the rings more than is necessary to slip them on piston to avoid overstressing rings. **CAUTION: The top compression ring is a tapered face ring. Therefore, this ring must be installed with the side marked "T" or "TOP" facing up.**

2. Install oil control rings by hand with the scraping edge down. First install expander being careful not to overlap ends of expander. Second, install top oil ring with gap 180° from gap in expander. Check to be sure expander ends are not overlapped. Third install bottom oil ring with gap 45° from gap of top oil ring. Recheck to be sure ends of expander are not overlapped. *NOTE: Upper oil control ring, used on "N" series truck and all coach engines, has chrome plated wiping edge, and must be installed on top of other rings.*

3. Install the second set of oil con-

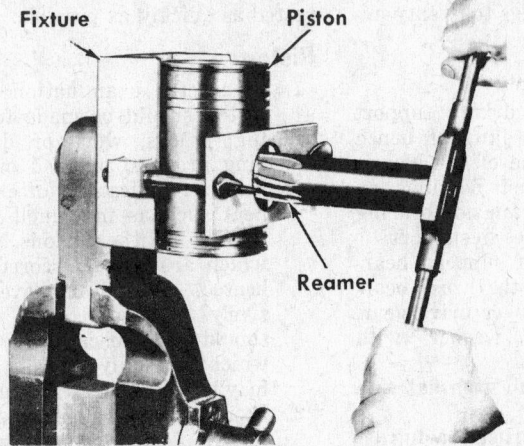

Reaming piston pin bushing

(© G.M.C.)

Removing or installing piston rings
(© G.M.C.)

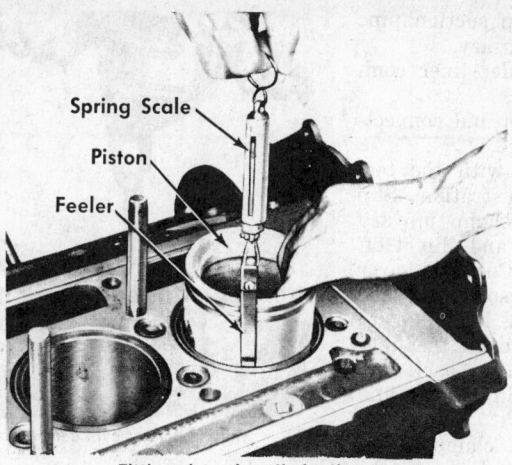

Fitting piston in cylinder liner
(© G.M.C.)

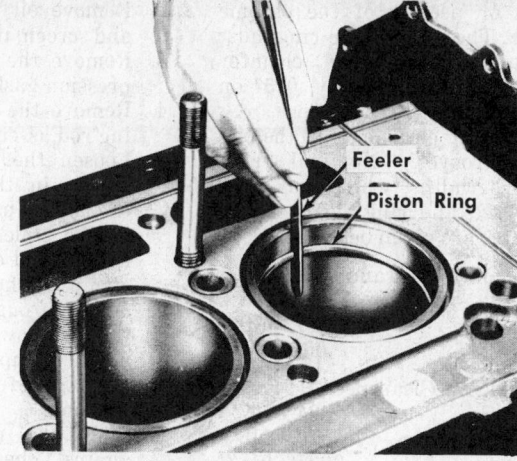

Measuring piston ring gap
(© G.M.C.)

trol rings and their expander as above.

Fitting Pistons

1. Measurements of pistons and bore for pistons in cylinder liners should be taken at room temperature (70°F.).
2. Measurements should be taken on the piston skirt lengthwise and crosswise of the piston pin. On 92 Series engines, piston diameter is 4.4650"-4.4750" above and below the seal ring groove. On 71 Series engines, piston diameter is 4.2428"-4.245", and the piston taper should not exceed .0005" from bottom compression ring groove to bottom of piston. On 53 Series engines, the diameter is 3.8693"-3.8715" measured 1.750" or more from the piston top.
3. On 92 Series engines, piston-to-liner clearance should be .0055-.0093" with a .012" clearance allowable. On 71 Series engines piston-to-liner clearance should be from .004" to .0078". With used parts, a maximum clearance of .012" is allowable. On 53 Series engines, piston-to-liner clearance is .0037"-.0074", not to exceed .010". This clearance should be checked at four places, 90° apart, around the circumference of the piston.
4. Cylinder liner bores should be round within .002" and straight within .001" when fitted into the cylinder block. Measure cylinder liner bores with a suitable gauge, both lengthwise and crosswise of the piston pin throughout the entire length of the cylinder liner.
5. Check the piston-to-liner clearance while the piston is held upside down in the liner. The liner should be in place in the cylinder block, otherwise inaccurate readings will be obtained. This clearance should always be checked

when installing either new or used pistons and liners.
6. A feeler gauge set may be used for this purpose. The set consists of a feeler gauge pack, spring scale, and swivel connection. The spring scale, attached to the feeler gauge pack by means of the swivel connection, is used to measure the force in pounds required to withdraw the feeler from between the piston and liner.
7. Select the feeler ribbon with a

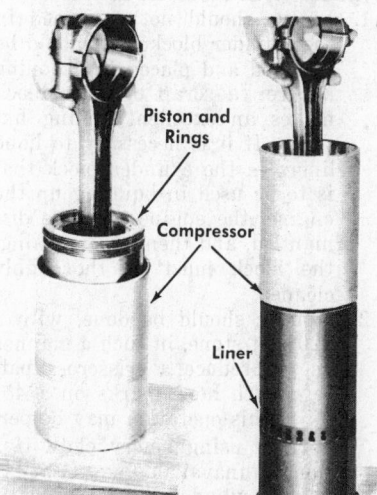

Installing piston assembly in ring compressor and cylinder assembly
(© G.M.C.)

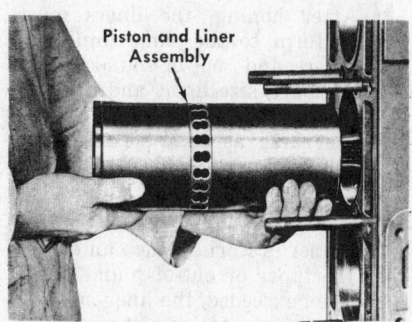

Installing piston assembly and liner in cylinder block
(© G.M.C.)

thickness that will require approximately six pounds pull to remove. The clearance will be .001" greater than the thickness of the feeler used, i.e., a feeler .004" thick indicates a clearance of .005" when withdrawn at a pull of six pounds. Feeler ribbon used for checking this clearance must be perfectly flat and free of nicks and scratches.
8. If any bind between the piston and liner is detected, remove the piston and liner and inspect them for burrs.
9. Service pistons are available in standard size, and in .010", .020", and .030" oversize. These within very close limits. Therefore, one or more oversize pistons may be installed in an engine without causing damage to the engine. Since cylinder liners are furnished in standard inside diameter only, the installation of oversize pistons becomes necessary only when cylinder liners have been rehoned.
10. All new piston rings must be installed whenever a piston is removed, whether or not a new liner and/or piston is installed.

Fitting Piston Rings

1. Insert one ring at a time inside the liner and far enough down in the bore to be on the wiping area of the rings when piston is installed. To assure that the ring is parallel with top of liner, use a piston to crowd the ring into liner bore; then measure gap with a feeler gauge.
2. If the piston ring gap is below the specified limits, it may be increased by stoning the ends of the ring. Stone both ends of the piston ring in such a direction that the stone will cut from the outside (chrome plated) surface of ring toward the inside surface. This will prevent any chip-

709

(Detroit 2-cycle) Diesel Engines

ping or peeling of the chrome plate. The ends of the ring must remain square and chamfer must be approximately .015″ on outer edge.

3. Check ring clearances in the piston grooves. Nominal ring groove widths, which may vary ± .001″, and ring clearances in grooves are given below

Ring-to-Groove-Clearance—92 Series
#1 .001″-.005″
#2 .010″-.013″
#3 .004″-.007″

Ring-to-Groove-Clearance—71 Series
#1 .004″-.007″
#2—N Series .010″-.013″
#2—Others .0095″-.013″
#3, #4—N Series .004″-.007″
#3—Others .0075″-.011″
#4—Others .055″-.009″
Oil Rings .0015″-.0055″

Ring-to-Groove Clearances—53 Series
#1 .0026″-.0049″
#2 .0070″-.0100″
#3, #4 .0050″-.0080″
Oil Rings .0015″-.0055″

When an oversize piston is installed, piston rings of the same oversize must also be installed.

Cylinder Liners

Description

The cylinder liners used in Detroit Diesel engines are of the replaceable dry type, made of hardened alloy cast iron and are a slip fit in the cylinder block.

A flange at the top of each liner fits into a counterbore in the cylinder block and rests on a cast iron insert to provide accurate alignment and positioning of the liners. Compression is sealed with an individual laminated compression gasket for each cylinder.

Removal

1. Remove the cylinder head assembly. Remove the oil pan.

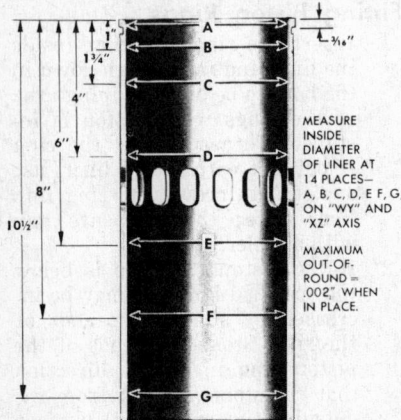

Cylinder liner measurement diagram
(© G.M.C.)

2. Remove oil pickup suction pipe and screen, if necessary.
3. Remove the cylinder liner compression gasket.
4. Remove the piston and connecting rod assembly.
5. Loosen the liner with the tool shown in the illustration. Slip the lower puller clamp upward on the puller rod and slip it off the tapered cone. Cock the lower clamp on the rod so it will slide down through the liner. Lower the rod down through the liner so the clamp will drop back onto the tapered cone in a horizontal position.
6. Slide the upper clamp down against the top edges of the liner.
7. Strike the upset head on the upper end of the puller rod sharply with the puller weight. thus releasing the liner from the cylinder block.

Honing Liner

Cylinder liners are available in standard size inside diameter only, therefore installation of oversize pistons becomes necessary when liners are honed oversize. Also, liners are available in oversize outside dimensions which must be used whenever cylinder block bores are honed.

1. Liners should not be honed in the cylinder block, but should be removed and placed in a honing fixture (a scrap cylinder block makes an excellent honing fixture). If it is necessary to hone liners in the cylinder block that is to be used in building up the engine, the engine must be dismantled, and then, after honing, the block must be thoroughly cleaned.
2. Honing should be done, with a 120 grit stone, in such a manner as to produce a crisscross pattern with hone marks on a 45° axis. This operation may be performed using emery cloth if a hone is unavailable.
3. When cylinder liner has been honed, inspect to be sure liner is free of burrs over its entire area.
4. After honing, the liners must conform to the same limits on taper and out-of-round as a standard size liner, and the piston-to-liner clearance must be the same as that specified for standard parts.

Taper and Out-of-Round

If a liner is worn to such an extent that the taper or out-of-round limits have been exceeded, the liner must be honed out for oversize pistons or the liner must be replaced. Permissible taper is .001″, out of round .002″.

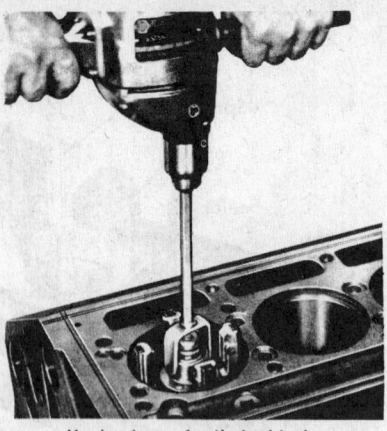

Honing bore of cylinder block
(© G.M.C.)

Remove Ridge

The inside diameter of a used liner measured just below the top of the ring travel must not exceed the unworn diameter at the top by more than .004″. A larger step than this may cause interference with new compression rings. In order to prevent ring breakage caused by this difference in diameter, hone the liner just enough to remove this ridge.

Break Glaze

Never install new piston rings in a used liner unless the glaze in the liner is broken.

Even though the liner taper and out-of-round are within specifications, the glaze must be removed by working a hone up and down the full length of the liner a few times.

Honing Cylinder Block

The fit of a liner depends to a great extent upon the condition of the bore before honing. Distortion may cause low spots that will clean up. A low spot is not objectionable above the ports if it does not exceed 1¼″ or the size of a half dollar. Below the ports, larger spots are permissible.

Roughing and finishing stones should be fairly coarse. A No. 80 grit stone may be used for roughing, and a No. 120 grit stone is satisfactory for finishing.

The following inspection should be made before honing cylinder block:

1. Clean bore and inlet port opening.
2. Measure the bore for high spots and the most narrow section.
3. Liners are fitted from .0005″ to .0025″ loose. A clearance of .0005″ produces a slip fit and .0024″ clearance allows the liner to slide freely into place. Refer to chart for standard and oversize cylinder liner bore in cylinder block.

Rough Honing

Insert hone in bore and adjust stones snugly to the most narrow section.

(Detroit 2-cycle) Diesel Engines

Cylinder liner removal
(© G.M.C.)

Checking bore at cylinder block
(© G.M.C.)

Start hone and "feel out" bore for high spots. These will cause an increased drag on the stone. Move hone up and down bore with short overlapping strokes about 1" long: Concentrate on the high spots in the first cut. Do not hone as long at the ports as in the rest of the bore—this area, as a rule, cuts away more rapidly.

When bore is fairly clean, remove hone to inspect the stones and measure bore. Decide carefully which spots must be honed most. To move the hone from top to bottom of bore will not correct an out-of-round condition. To remain in one spot too long may cause bore to become tapered. Where and how much to hone can be judged by feel. A heavy cut in distorted bore produces a more steady drag on the hone than a light cut and so makes it difficult to feel the high spots. Therefore, use a light cut with frequent stone adjustment.

Finish Honing and Fitting Liner

Rough hone cylinder bore until the liner can be pushed from 3" to 4" into the bore, or until bore is within approximately .005" of the diameter of the liner. Do not expect finishing stones to remove more than .001" of stock, or to true up the bore to any extent.

Work the finish hone with short, rapid strokes up and down the bore. Use light tension on the stones and hone only enough to allow the liner to enter the bore either with a light push fit or a free slip fit.

Cylinder Liner Installation

Install the cylinder liner in the block and measure the distance from top of liner flange to top of block to assure proper sealing of the cylinder head compression gaskets as follows:
1. Clean liner and be sure bore and counterbore in block is clean.
2. Check cylinder liner counterbore depth.

92 Series
Depth .... .4755"-.4770"
Variation ....none
71 Series
Depth .... .4770"-.4795"
Variation ....0015"*
53 Series
Depth .... .300"-.302"
Variation ....0015"
Adjacent liners to vary no more than .001" gauged on longitudinal cylinder centerline.

3. Place a cylinder liner insert in counterbore in cylinder block. *NOTE: Cylinder liner inserts are available in .004" and .008" thinner sizes as well as standard size for both standard to .005" oversize and .010", 020" and .030" oversize outside diameter cylinder liners. Matching oversize inserts should be installed whenever oversize (O.D.) liners are used.*

4. Push liner into cylinder block by hand, until liner flange rests on insert. If the liner does not slide freely into place, withdraw, turn 90° and insert liner again. Each liner should slide freely in-

to place to assure there will be no distortion.
5. Clamp liner in place with cylinder liner hold-down clamps.
6. Measure the distance from the top of liner flange to top of block, using a dial indicator. When in place, top of liner flange should be from .045" (.0465" on 53 Series) to .050" below top of cylinder block, and there must not be over .001" (.0015" on 53 Series) difference in height below top of block between any two adjacent liners when gauged on longitudinal cylinder centerline.
6. Install new cylinder liner compression gaskets.

Connecting Rod and Bearings

Bearing Removal

(Connecting Rod Installed)

When removal of connecting rod bearing shells becomes necessary, they may be removed as follows:
1. Remove the oil pan.

Installing cylinder liner in block
(© G.M.C.)

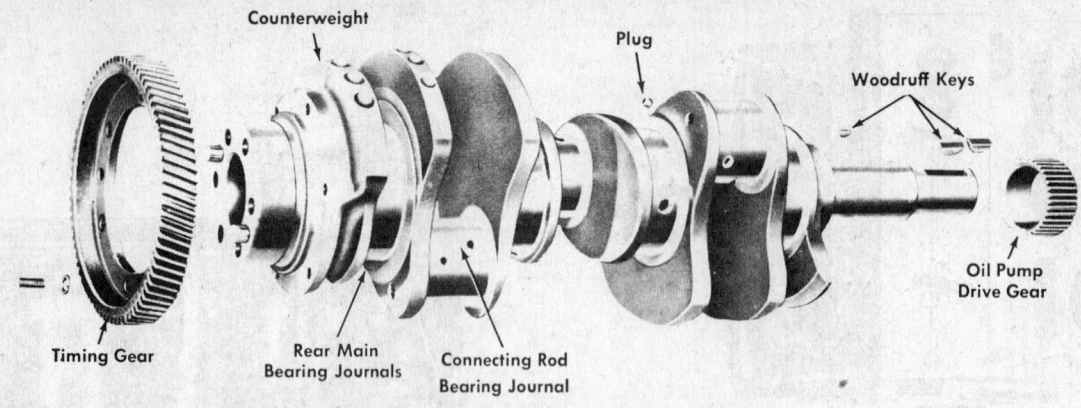

Crankshaft and gears (© G.M.C.)

2. Disconnect and remove the lubricating oil pump suction pipe and screen assembly from main bearing cap.
3. Remove one connecting rod bearing cap. Push connecting rod and piston assembly up into cylinder liner far enough to permit removal of upper bearing shell.
4. Install bearing shells before another cap is removed.

Bearing Installation

1. Rotate crankshaft until connecting rod journal is at the bottom of its travel, wipe the journal clean and lubricate with clean engine oil.
2. Install upper bearing shell—the one without an oil groove (on 53 Series engines, this shell has a short groove and a hole at each parting line.)
3. With tang of upper bearing shell in groove of connecting rod, pull piston and rod assembly down until upper rod bearing seats firmly on crankshaft journal.
4. Place lower bearing shell—the one with the oil groove—in bearing cap, with tang of shell in groove of cap, and lubricate with clean engine oil.
5. Note identifying marks on cap and rod and assemble cap to rod. Tighten connecting rod bolt nuts to 60-70 ft lbs. on 92 Series engines, 65-75 ft lbs. torque on 71 Series engines, 50-55 ft lbs. on 53 Series engines.
6. Install lubricating oil pump suction pipe and screen assembly.
7. Install oil pan and fill crankcase with recommended lubricating oil to "FULL" mark on dipstick.
8. If new bearings were installed, operate engine on run-in schedule.

Assembly of Rod and Piston

(Conecting Rod to Piston)

Apply clean engine oil to the piston pin and bushings and proceed with assembly of the rod to piston in the following manner:

1. Rest piston in holding fixture. Place a new piston pin retainer in either side of piston. Position installer tool over retainer and strike tool just hard enough to cause a deflection in the piston pin retainer and even seating of retainer. Driving too hard against the tool may force the piston bushing inward, thus reducing the piston pin end clearance.
2. Place upper end of connecting rod between piston pin bosses in line with piston pin holes and slide pin into position. The piston pin should slip readily into position without forcing if clearances are correct.
3. Install second piston pin retainer as outlined above. After piston pin retainers have been installed, it is advisable to visually check for piston pin end clearance by cocking rod on pin and shifting pin in its bushings.

Piston and Rod Installation

1. With piston assembled to connecting rod and piston ring in place, apply clean engine oil to piston, rings, and inside of piston ring compressor.
2. Place compressor on wood bench or table with tapered side of compressor up.
3. Stagger the piston ring gaps properly on the piston, being careful that oil ring expanders are not overlapped.
4. Start the top of the piston straight into compressor, then push piston down until it contacts wood block.
5. Place the ring compressor, piston, and connecting rod assembly on the liner. On 53 Series engines, align numbers on rod and cap with the match-mark on the liner. Replacement connecting rods must be stamped with number stamped on original rod. Carefully push the piston and rod assembly down into the liner until piston is free of the ring compressor, using care to avoid ring breakage. Remove the compressor. Install hold-down clamps to keep other liners in place on 53 Series engines.
6. Rotate the crankshaft until connecting rod journal is at bottom of its travel, wipe the journal clean and lubricate journal with clean engine oil.
7. Lubricate and install upper bearing shell—the one without an oil groove—in the connecting rod. On 53 Series engines, this shell has a short groove at each parting line.
8. Note that each connecting rod and its cap are stamped on one side—1L, 1R, 2L, 2R, etc. These numbers and letters are for identification purposes and indicate the particular cylinder in which each is used. These positions should always be maintained when rebuilding an engine. On 53 Series engines, apply permanent anti-freeze to the inner surface of the seal ring and install it.
9. Position the piston, rod and liner assembly in its proper location according to marks on liner and with connecting rod numbers and letters toward outside of the cylinder block. Slide the liner, pistons, and rod assembly straight into the block bore until the flange of liner seats on the insert in cylinder block counterbore. CAUTION: The centerline of the connecting rod is slightly offset. Therefore, when installing to crankshaft, be sure that the narrow side of the two connecting rods on the crankshaft journal are together to avoid cocking the rod.
10. Lubricate and place lower bearing shell—the one with the oil groove—in connecting rod cap with tang of shell in groove of cap.
11. Assemble cap to rod with identifying marks on cap and rod adjacent to each other. Tighten

connecting rod bolt nuts to 60-70 ft lbs. on 92 Series engines, 65-75 ft.-lbs. torque on V71 Series engines, or 50-55 ft.-lbs. on 53 Series engines. Phosphate coated, 53 Series nuts should be torqued to 45-50 ft.-lbs.

12. Check connecting rod side clearance. The clearance between each pair of connecting rods should be .008″ to .016″ with new parts with V-type engines, .008″-.012″ on In-line engines.
13. Remove cylinder liner hold-down clamps.

Crankshaft and Main Bearings

Crankshaft Removal

When necessary to remove crankshaft the operation may be performed as follows:

1. Drain oil, then remove the oil pan.
2. Remove the lubricating oil pump screen and suction pipe assembly.
3. Remove hydraulic fan assembly.
4. Remove flywheel and flywheel housing.

Removing main bearing cap
(© G.M.C.)

Removing rear main bearing upper shell
(© G.M.C.)

5. Remove crankshaft front cover. On 53 Series engines, remove the crankshaft pulley.
6. Remove connecting rod bearing caps. Note that each cap and rod is stamped with a number corresponding with the cylinder number.
7. Remove main bearing caps. Note that each cap is stamped with a number which corresponds with number stamped on cylinder block.
8. Lift crankshaft, timing gear, oil pump drive spline as an assembly from crankshaft.
9. Remove main bearing inserts from crankcase and bearing caps, also thrust washers at rear main bearing. Carefully identify each bearing so that it can be replaced in its original location.

Main Bearings Removal

(Crankshaft Installed)

When removal of main bearing shells becomes necessary, and the removal of the crankshaft is not required or undesirable, this operation may be performed by using the following procedure.

NOTE: Remove bearing shells one at a time, inspect, and reinstall or replace, replacing all shells if only one or more need replacing.

All crankshaft main bearing journals except the rear are drilled for an oil passage.

1. Remove the oil pan to expose the main bearing caps.
2. Remove the lubricating oil pump intake pipe assembly. *NOTE: In text following, main bearing stabilizers are mentioned in several instances. Early model engines used these parts which have since been discontinued and any reference made to stabilizers should be disregarded on latest engines.*
3. Remove only one main bearing cap stabilizer at a time, then place washers on main bearing bolts just removed, equal to thickness of stabilizer and reinstall bolts. Remove remaining stabilizer(s) in a like manner.
4. Remove main bearing caps using temporarily installed cap screws and pinch bars.
5. To remove all except rear main bearing shell, insert a 5/16″ x 1″ bolt with a 1/2″ diameter and a 1/16″ thick head (made from a standard bolt) into the crankshaft journal oil hole, then revolve the shaft to the right (clockwise) and roll the bearing shell out of position. The head of the bolt should not extend beyond the outside diameter of the shell.
6. Two-piece thrust washers are used each side of the rear main bearing. The lower half of these washers will be removed when removing the rear main bearing cap; upper half can be removed upon removal of bearing cap by pushing on end of washer with a small rod, thus forcing washer around and out on opposite side of bearing.
7. Remove rear main bearing upper shell by driving on the edge of the bearing shell with a small curved rod, at the same time revolving the crankshaft, thus rolling the shell from its position.

Main Bearings Installation

(Crankshaft Installed)

Make sure parts are clean. Apply clean engine oil to each crankshaft journal and install main bearing shells by reversing the sequence of operations given for removal.

Upper and lower main bearing shells are not alike; the upper shell is grooved and drilled for lubrication—the lower shell is not. Be sure to install the grooved and drilled shells in the cylinder block and plain shells in the bearing caps, otherwise the oil feed to the upper end of the connecting rods will be blocked off.

1. When installing the upper main bearing shells with crankshaft in place, start the end of the shell having no tang around the crankshaft journal, so that when shell is in place the tang will fit into groove in the shell support.

2. Assemble crankshaft thrust washer before installing rear main bearing cap. Clean both halves of thrust washer carefully, removing any burrs from the seats—*the slightest particle of dirt between washers and crankshaft may decrease clearance beyond limits.*

Slide upper halves of thrust washers into place in their grooves; then assemble lower halves over dowel pins in bearing cap. NOTE: Main bearing caps are bored in position and marked 1, 2, 3, etc. They must be replaced in their original positions.

3. With lower bearing shell installed in bearing cap, install cap and draw bolts up snug. Then rap cap sharply with soft hammer to seat cap properly and draw cap bolts uniformly tight, 230-240 ft lbs. on 92 Series engines, to 180-190 ft lbs. torque on 71 Series engines, 120-130 ft lbs. on 53 Series engines.

NOTE: If the bearings have been installed properly, the crankshaft will turn freely with all main bearing caps bolted tightly.

4. Install main bearing cap stabilizers by reversing their removal procedure. Tighten main bearing bolts to specified torque and 7/16"-14 x 1⅛" stabilizer cylinder block bolts, if used, to 70-75 ft-lbs. torque.

5. Check crankshaft end play at thrust washers. This clearance should be from .004" to .011" with new parts or a maximum of .018" with used parts. Insufficient clearance will usually de-

1 Main bearing cap bolts
2 Main bearing caps
3 Main bearing—lower
4 Rear main bearing thrust washer
5 Main bearing—upper
6 Crankcase

Upper and lower main bearings, thrust washers and seals
(© G.M.C.)

note misalignment of the rear main bearing. In such case, loosen and retighten the rear main bearing cap. If lack of clearance is still present, dirt or a burr on inner face of one or more of the thrust washers may be the cause.

6. Install lubricating oil pump intake pipe assembly.

7. Install oil pan with new gasket. Fill crankcase to proper level.

Crankshaft Installation

If shells previously used are to be used again, install them in the same

Checking crankshaft end play
(© G.M.C.)

locations from which they were removed.

When a new crankshaft is installed, all new main and connecting rod (upper and lower) bearing shells should also be installed.

With oil pump drive gear, and crankshaft timing gear, assembled on the front and rear ends respectively of the crankshaft, the crankshaft may be installed as follows:

1. Install upper grooved bearing shells in the cylinder block.

2. Apply clean engine oil to all crankshaft journals and set crankshaft in place so that timing marks on the crankshaft timing gear and the gear train idler gear match.

3. Install the upper halves of the rear main bearing thrust washers on each side of the bearing, and the doweled lower halves on each side of the rear main bearing cap.

4. With lower main bearing shells installed in bearing caps, install caps and stabilizers in their original position and draw bolts up snug. Then rap caps sharply with a soft hammer to seat them properly, and draw bearing cap bolts uniformly tight, starting with center cap and working alternately towards both ends of block, to 230-240 ft lbs. on 92

Crankshaft — Oil Seal

Crankshaft rear oil seal installed
(© G.M.C.)

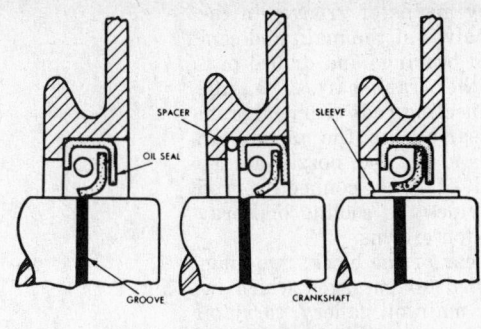

Use of oil seal spacer and sleeve
on grooved crankshaft
(© G.M.C.)

Series engines, 180-190 ft-lbs. torque on 71 Series engines, 120-130 ft-lbs. on 53 Series engines. Tighten 7/16"-14" x 1⅛" stabilizer to cylinder block bolts, if used, to 70-75 ft-lbs. torque.

NOTE: If bearings have been installed properly, the crankshaft will turn freely with all main bearing cap bolts drawn to specified torque.

5. Check crankshaft end play at thrust washers. This clearance should be .004" to .011" with new parts or a maximum of .018" with used parts. Insufficient clearance will usually denote misalignment of the rear main bearing. In such case, loosen the bearing cap, then retighten as described above. If lack of clearance is still present, either dirt or a burr on the inner face of one or more of the thrust washers may be the cause.
6. Lubricate and install connecting rod bearings.
7. Install flywheel housing and flywheel.
8. Install crankshaft front cover and oil pump assembly.
9. Affix new gasket to bolting flange and secure oil pan in place with bolts and lock washers.
10. Fill crankcase to the proper level with recommended oil.

Oil Seal Removal

If inspection reveals a worn or damaged oil seal, it must be replaced. Remove the oil seal as follows:
1. Remove flywheel housing as directed in respective section of this manual.
2. Support crankshaft flywheel housing on wood blocks. Then drive out old oil seal.

Oil Seal Sleeve Installation

When the oil seal spacer can no longer be used, an oil seal sleeve may be pressed on the crankshaft to provide a replaceable wear surface at the point of contact with the rear oil seal. The oil seal sleeve may be used with either the single lip or double lip type oil seal, and can also be used in conjunction with the seal spacer.

However, oversize oil seals must be used with the oil seal sleeve. Installation of the oil seal sleeve is performed as follows:
1. Stone all high spots off the circumference of the crankshaft.
2. Coat the area of the shaft where the seal will be positioned with shellac or equivalent sealer.
3. Press the sleeve onto the shaft. The sleeve must be *pressed squarely on the shaft.*
4. Wipe off all excess sealer.

Rear Oil Seal Installation

1. Support inner face of flywheel housing on a flat surface.
2. Install rear oil seal spacer, if used, against shoulder in oil seal counterbore in housing.
3. Apply a non-hardening sealant to the outer edge of the oil seal and position the seal with the lip pointing toward the inner face (toward shoulder in counterbore) of the housing.
4. Drive oil seal against seal spacer or shoulder in oil seal counterbore of housing.
5. If flywheel housing has not been removed from engine, lubricate lip of seal with Lubriplate or equivalent. Slip seal over crankshaft and start squarely into counterbore of housing. Using a hardwood block and hammer, tap seal alternately around edge of casing until seal is in place.
6. Remove all excess sealant from housing and oil seal.
7. Coat lip of oil seal lightly with Lubriplate or equivalent.
8. When installing flywheel housing on engine, use oil seal expander to avoid damage to lip of seal.

Engine Lubrication System

Illustrations show the flow of oil through engine lubricating system including the various components such as the oil pump, full flow oil filter, oil cooler, pressure regulator, and by-pass valve.

The oil pump is placed in the crankshaft front cover and consists of a pair of spur gears, one large and one small, which mesh together and ride in a cavity inside the crankshaft front cover. The large gear is concentric with and splined to a pump drive hub on the front end of the crankshaft. The pump idler gear is much smaller and runs on a bushing and hardened steel shaft pressed into the crankshaft cover.

Oil is drawn by suction from the oil pan through the intake screen and pipe to the oil pump where it is pressurized. The oil then passes from the pump into a short gallery in the cylinder block to the oil cooler adaptor plate. At the same time, oil from the pump is directed to a spring-loaded pressure relief valve mounted on the cylinder block. This valve discharges excess oil directly to the oil sump when the pump pressure exceeds 100 p.s.i.

From the oil cooler adaptor plate, the oil passed into the full flow filter through the oil cooler (truck engines only) and then back into the cylinder block where a short vertical oil gallery and a short diagonal oil gallery carry the oil to the main longitudinal oil gallery through the middle of the block. On engines used in coaches, oil filter is mounted at crankcase front cover and oil from pump goes directly to filter and returns to cylinder block. Valves are also provided to by-pass the oil filter and oil cooler should either one become plugged.

Stabilized lubricating oil pressure is maintained within the engine at all speeds, regardless of the oil temperature, by means of a pressure regulator valve located at the end of a vertical oil gallery connected to the main oil gallery. This vertical gallery is located at the front of the cylinder block on the side opposite the cooler. When the oil pressure at the valve exceeds 50 p.s.i., the regulator valve opens, discharging excess oil back into the sump.

From the main oil gallery, the pressurized oil flows through drilled passages to each main bearing then passes to an adjacent pair of connecting rods by means of grooves in the unloaded halves of the main and connecting rod bearings and drilled passages in the crankshaft. The rifle drilled connecting rods carry oil from the rod bearings to the piston pin bushings and to the nozzle at the upper end of each connecting rod which provides the cooling oil spray for the piston crowns.

At the rear of the block, two diagonally drilled oil passages which intersect the main oil gallery, carry oil to the two rear camshaft end bearings. Oil is then conducted through the rifle drilled camshaft to the intermediate and front end bearings. Oil from the camshaft intermediate bearings is directed against the camshaft lobes and cam rollers which run in an oil bath. This oil from the intermediate bearings provides lubrication of the cam lobes immediately after starting the engine when the oil is cold and before camshaft bearing oil flow and oil drainage from the cylinder head have had time to build up.

The diagonally drilled oil passage on the right side at the rear of the block intersects with a vertical passage to carry oil to the right bank cylinder head. A short gallery also intersects with this diagonal passage to lubricate the idler gear bearing. Another gallery intersecting the diagonal passage at the front of the block supplies oil to the left bank cylinder head.

Drilled passages, intersecting longitudinal galleries which parallel the camshafts, lead to the blower and supply oil for the blower drive gears and bearings.

Oil from the right-hand camshaft pocket is directed through a tube to lubricate the water pump drive gear and bearings and the front camshaft gear.

The gear train is lubricated by the overflow of oil from the camshaft pockets spilling into the gear train compartment and by splash from the oil pan. A certain amount of oil also spills into the gear train compartment from both camshaft rear end

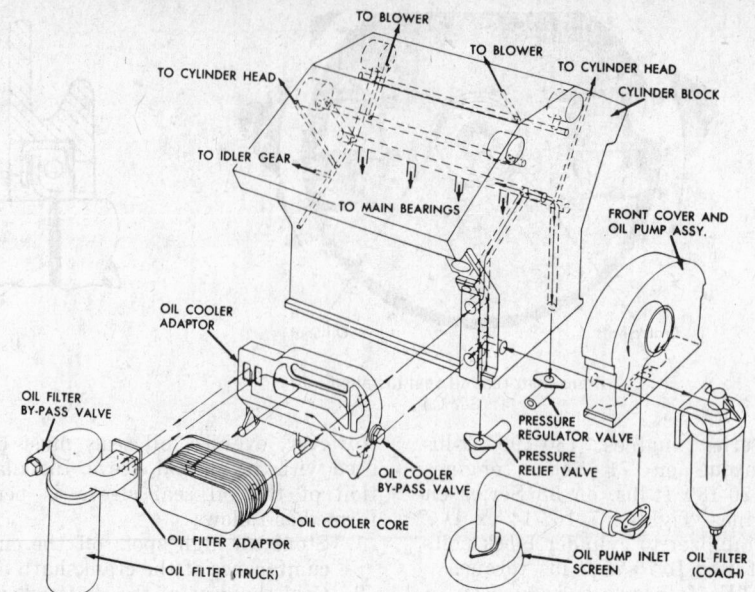

Diagram of engine lubrication system (© G.M.C.)

bearings, the blower drive gear bearing, and the idler gear bearing. The blower drive gear bearing is lubricated through an external pipe from the blower rear end plate to the blower drive support.

The valve and injector operating mechanism is lubricated from a longitudinal oil passage, on the camshaft side of each cylinder head, which connects to the main oil gallery in the cylinder block. Oil from this passage enters the drilled rocker arm shafts through the lower end of the rocker shaft bolts and rocker shaft brackets. Excess oil from the rocker arms lubricates the exhaust valves and cam followers.

Oil Cooler Testing

Pressure Check Core

After the oil cooler core has been cleaned, it may be checked for leaks as follows:
1. Make a suitable plate and attach to flanged side of cooler core. Use a gasket made from rubber to assure a tight seal. The plate

should be drilled and tapped to permit an air hose fitting to be attached at the inlet side of the cooler core.
2. Attach air hose and apply approximately 75 p.s.i. air pressure and submerge cooler core and plate assembly in a container of water. Any leaks will be indicated by air bubbles in the water. If leaks are indicated, replace cooler core.
3. After pressure check is completed, remove plate and air hose from cooler core and dry with compressed air.

By-Pass Spring

Inspect valve and spring for wear and replace if necessary. The by-pass valve spring has a free length of approximately 2-1/64″. A force of $13\frac{3}{8}$ to $14\frac{5}{8}$ pounds is required to compress the spring to a length of 1.793″ when new. When a force of 12 lbs. or less will compress the spring to 1.793″, replace the spring. The spring may be checked in spring tester.

GENERAL SPECIFICATIONS

ENGINE MODEL	BORE AND STROKE	HORSEPOWER @ R.P.M.	TORQUE FT. LBS. @ R.P.M.	COMPRESSION PRESSURE @ 600 R.P.M.	FIRING ORDER	GOVERNED SPEED R.P.M.	IDLE SPEED
D-351-V6	4.56 x 3.58	130 @ 3200	234 @ 2000	500	1-6-5-4-3-2	3200	625-650
D-478-V6	5.125 x 3.86	150 @ 3200	275 @ 2000	500	1-6-5-4-3-2	3200	625-650
DH-478-V6	5.125 x 3.86	170 @ 3200	310 @ 2000	500	1-6-5-4-3-2	3200	625-650
D-637-V8	5.125 x 3.86	195 @ 2600	——	500	1-8-4-3-6-5-7-2	2600	625-650
DH-637-V8	5.125 x 3.86	220 @ 2600	——	500	1-8-4-3-6-5-7-2	2800	625-650

PISTONS, PINS, RINGS, CRANKSHAFT AND BEARINGS

ENGINE MODEL	PISTON CLEARANCE	PISTON RING END GAP MINIMUM COMP.	OIL	PISTON PIN DIAMETER	ROD BEARINGS SHAFT DIAMETER	BEARING CLEARANCE	MAIN BEARING SHAFT DIAMETER	BEARING CLEARANCE	SHAFT END PLAY	THRUST ON BEARING NO.
D-351-VP	.0062-.0068	.015	.013	1.6148-1.6150	2.8112-2.8122	.001-.0031	3.1237-3.1247	.0013-.0039	.003-.008	3
478-V6	.0067-.0073	.017	.017	1.6148-1.6150	2.8112-2.8122	.001-.0031	3.1237-3.1247	.0013-.0039	.003-.008	3
637-V8	.0067-.0073	.017	.017	1.6148-1.6150	2.8112-2.8122	.001-.0031	3.1237-3.1247	.0013-.0039	.003-.008	4

VALVE SPECIFICATIONS

ENGINE MODEL	VALVE LASH (HOT) IN.	EXH.	VALVE FACE ANGLE (DEG.)	VALVE STEM CLEARANCE INTAKE	EXHAUST	VALVE STEM DIAMETER INTAKE	EXHAUST	VALVE SPRINGS FREE LENGTH INNER	OUTER
D351-V6	.010	.018	45	.0015-.003	.002-.0035	.3407-34.12	.3402-.3407	2.46	2.67
478-V6	.010	.018	45	.0015-.003	.0025-.004	.3725-.3730	.4340-.4345	2.46	2.67
637-V8	.010	.018	45	.0015-.003	.0025-.004	.3725-.3730	.4340-.4345	2.46	2.67

ENGINE TORQUE SPECIFICATIONS

ENGINE MODEL	CYLINDER HEAD FT. LBS.	INTAKE MANIFOLD END BOLTS FT. LBS.	EXHAUST MANIFOLD END BOLTS FT. LBS.	MANIFOLD CLAMP BOLTS	CONN. ROD FT. LBS.	MAIN BEARING CAPS FT. LBS.	DAMPER TO CRANKSHAFT BOLT FT. LBS.	FLYWHEEL TO CRANKSHAFT FT. LBS.	CAMSHAFT GEAR BOLT FT. LBS.	OIL PUMP TO BLOCK BOLTS FT. LBS.	OIL PAN BOLTS FT. LBS.
D351-V6	130-135	20-25	15-20	20-25	55-65	——	200-210	100-110	50-60	30-35	10-15
478-V6	130-135	20-25	15-20	20-25	55-65	——	240-260	100-110	50-60	30-35	20-25
637-V8	130-135	20-25	15-20	20-25	55-65	——	240-260	100-110	50-60	30-35	10-15

Trouble Diagnosis

The satisfactory performance of a Diesel engine depends on two items of foremost importance, (1) Sufficiently high compression pressure, and (2) The injection of the proper amount of fuel at the right time.

The first one of these items depends, mainly on pistons, piston rings, valves and valve operating mechanism; the second item depends on injectors and their operating mechanism, and fuel system.

Lack of engine power, uneven running, excessive vibration, and tendency to stall when idling may be caused by either a compression loss or faulty injector operation.

The causes of trouble symptoms may be varied; therefore a hit-and-miss search should be avoided. A proper diagnosis of symptoms is an orderly process of diagnosing the symptoms. An "orderly process" means to check the most probable common cause first; then proceed with the next most probable cause.

Hard Starting
1. Improper oil viscosity.
2. Low battery output.
3. Defective starter or solenoid.
4. Low temperature.
5. Burned or warped valves.
6. Improper valve lash.
7. Worn or broken piston rings.
8. Defective cylinder head gasket.
9. Fuel line shut off valve closed.
10. Engine stop mechanism not released.
11. Low fuel supply.
12. Clogged vent in fuel tank cap.
13. Clogged fuel filters.
14. Broken fuel line allowing air leak into system.
15. Air trapped in system.
16. Water or ice in fuel system.
17. Defective fuel supply pump.
18. Overflow valve stuck or leaking.
19. Low fuel delivery.
20. Improper grade or type fuel.

Abnormal Engine Operation
1. Idle speed set too low.
2. Cylinder misfiring.
3. Burned or sticking valves.
4. Engine temperature too hot or too cold.
5. Fuel system deficiencies.
6. Detonation due to improper injector timing.
7. Defective injection nozzles.
8. Lack of power due to improper valve adjustment.
9. Insufficient fuel.
10. Insufficient air due to clogged air cleaner or intake system.
11. Excessive exhaust back pressure.
12. Low cylinder compression.

GMC Toro-Flow Diesel Engines

Excessive Black or Gray Smoke

1. Insufficient combustion air.
2. Exhaust pipe or muffler clogged.
3. Injection pump improperly timed to engine.
4. Incorrect engine valve timing.
5. Injector nozzles leaky, worn or improper opening pressure.
6. Wrong grade and type of fuel.
7. Engine overheating.
8. Poor cylinder compression.
9. Damaged or worn piston rings and/or pistons.
10. Cylinders overfueled.

High Lubricating Oil Consumption

1. Oil lines or connections leaking.
2. Leaking gaskets.
3. Crankcase oil level too high.
4. Pistons and/or rings worn or damaged.

Excessive White or Blue Smoke

1. Oil level in air cleaner too high.
2. Injection pump improperly timed to engine.
3. Engine crankcase oil level too high.
4. Engine running too cool.
5. Poor cylinder compression.
6. Damaged or worn pistons and/or rings.
7. Wrong grade or type of fuel.
8. Improper or dirty oil in use.

Low Engine Oil Pressure

1. Defective oil gauge or sending unit.
2. Oil viscosity too low.
3. Oil diluted with fuel oil.
4. Suction loss.
5. Weak or broken relief valve spring in oil pump.
6. Worn oil pump.

7. Excessive clearance at crankshaft, camshaft or balance shaft bearings.
8. Overheating.

NOTE: In case of no oil pressure the oil pump drive shaft or drive gear teeth could be worn to the point where oil pump is not driven. Inspection can be made by removing the oil pump and tachometer drive housing and gear at rear of engine.

Engine Overheats

1. Scale deposits in cooling system.
2. Air flow thru radiator restricted or clogged.
3. Loose fan drive belt.
4. Low coolant level.
5. Hoses collapsed.
6. Defective thermostats.
7. Combustion gases in coolant water.
8. Incorrect injection pump to engine timing.

Engine Tune-Up

Description

GMC Toro-Flow Diesel truck engines are four stroke cycle type of 60-degree, V-6 and V-8 design. Fuel for combustion is injected directly into combustion area through hole-type nozzles—one nozzle for each cylinder. Fuel injection pump has a single plunger with distributor type hydraulic head. Pump is mounted between cylinder heads and is driven from drive gear bolted to engine camshaft gear.

Illustrations show D478 engine with typical accessories installed, and typical V-8 engine, model D637.

The descriptions of the various components are applicable to all of the engines in the Toro-Flow Diesel series; namely, D351, D and DH478, and D and DH637 unless otherwise indicated.

Before performing other tune-up procedure, check battery electrolyte

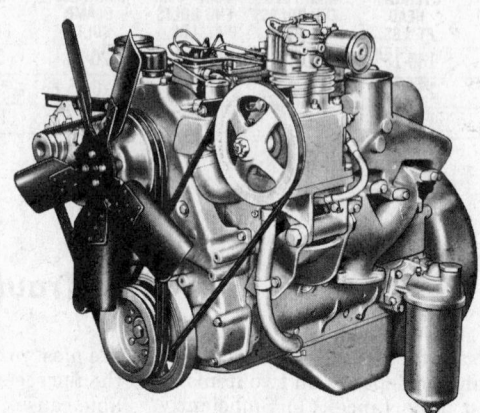

Typical D478 Toro flow diesel engine

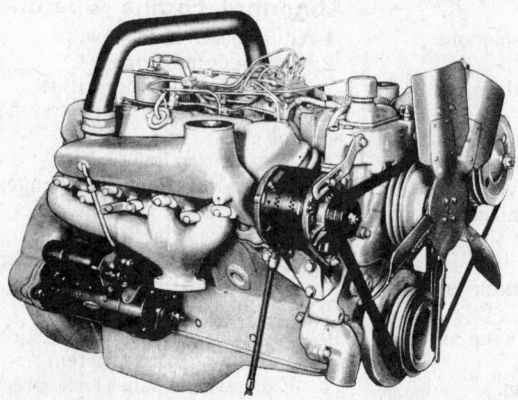

Views of D637 Toro flow diesel engine (© G.M.C.)

specific gravity at each cell to be sure battery is fully charged.

Check Cylinder Compression

1. Start engine and run until normal operation temperature is reached.
2. Stop engine and remove high pressure line clamps. Remove the high pressure line between injection pump and nozzle holder at cylinder to be checked. Remove the nut at each end from the threads at pump and nozzle holder and lift the line away from both units without bending.
3. Disconnect leak-off hoses from tee or elbow on injection nozzle holder. Remove nozzle holder retaining bolts and pull the holder assembly out of well in cylinder head. Clean nozzle holder cavity and gasket seat in cylinder head, then turn engine over with starter to blow loose carbon out of combustion chamber and nozzle holder cavity. Use a nozzle holder copper gasket at bottom of well and install the compression gauge adapter in place of nozzle holder as shown, using the nozzle holder bolts to hold the adapter in place. Connect gauge to the adapter. Be sure all connections are tight.
4. If a rear cylinder is being checked, insert plug in the disconnected leak-off hose. When other cylinders are being checked, join the disconnected

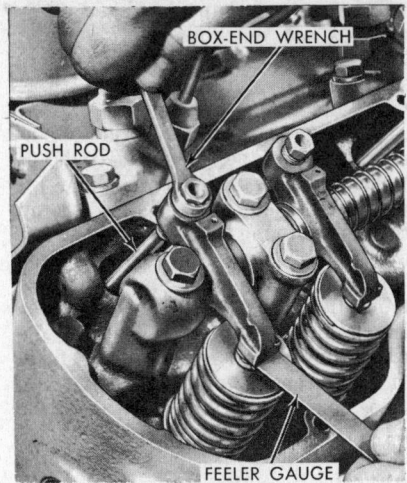

Adjusting valves
(© G.M.C.)

leak-off hoses with a two-way or three-way connector, as required.

5. Install adapter and hose in outlet port in hydraulic head on injection pump and place hose in receptacle to catch fuel oil while running the engine. **CAUTION: Under no condition should the high pressure outlet port on the injection pump be plugged to prevent the escape of fuel.**
6. Start the engine and allow engine to idle at 625 rpm. Observe and record the compression pressure shown on compression gauge. NOTE: Do not attempt to obtain compression pressure by cranking engine with starter.

7. Perform this operation on each cylinder. The compression pressure in any one cylinder should be not less than 500 psi at 625 rpm. Also, there should not be more than a 30 psi difference between cylinders.

Cylinder Head Bolts

Check cylinder head bolts for proper torque. Normally head bolts should not need tightening at tune-up intervals.

Manifold Bolts

Check manifold bolts for proper torque and be sure end bolts have been securely locked.

Valve Clearance

Engine must be at normal operating temperature. Check and set exhaust valves to .018 and intake to .010 inch clearance.

Inspect Fuel System

1. Check all fuel lines for evidence or leaking.
2. Check and service fuel and air filters as directed in "owner's and driver's manual."
3. Check accelerator controls and linkage for binding. Check proper operation of stop control. (manual or solenoid type)
4. Check engine idle speed which should be 625 RPM. To adjust, remove lower acord nut at rear of injection pump. With engine running, loosen lock nut and

Gauge and adapter for checking compression (© G.M.C.)

1 Fuel leak-off hose	5 Adapter	9 Fuel receptacle
2 Injection nozzle bolts and lock washers	6 Fuel injection pump assembly	10 Fuel return line (hose)
3 Compression gauge and hose	7 Adapter and fuel hose assembly	11 Fuel supply pump assembly
4 Leak-off hose connector	8 Overflow valve	12 Fuel secondary filter

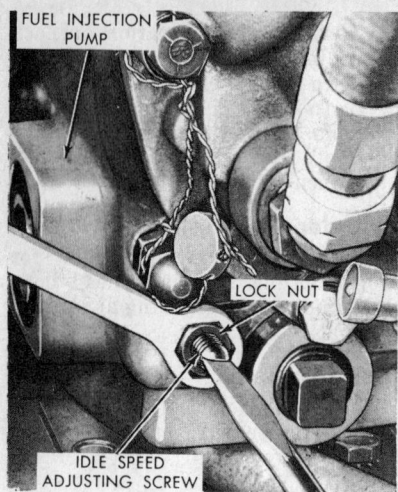

FUEL INJECTION PUMP

LOCK NUT

IDLE SPEED ADJUSTING SCREW

Adjusting engine idle speed
(© G.M.C.)

turn adjusting screw. Retighten lock nut and install acorn nut.

NOTE: In instances where engine still does not perform properly, it may be necessary to remove the fuel injection nozzles and check the nozzle opening pressure and spray pattern.

Fuel System

Description

The fuel system used with Toro-Flow Diesel engines includes the various components schematically illustrated.

Fuel Injection Pump Assembly

The fuel injection pump assembly is of the PSJ series manufactured by American Bosch Arma Corporation.

Injection pump assembly is mounted on machined rear face of engine front cover assembly. Pump assembly is held in place by three studs and nuts.

The high pressure pump is the single plunger type, which distributes equal amounts of fuel to the fuel injection nozzles, at the cylinders, in the proper firing sequence.

Injection pump is gear driven from a pump drive gear bolted to front face of engine camshaft gear. Pump input shaft, on which the pump driven gear is mounted, rotates at engine crankshaft speed. The cam lobes on input shaft actuate the pump plunger to inject fuel to cylinders for each revolution of the engine crankshaft.

Included in the fuel injection pump assembly are the following units or devices:
1. Fuel supply pump.
2. Hydraulic head assembly.
3. Timing advance mechanism.
5. Governor and fuel control unit.
5. Excess fuel device for starting.

An overflow valve assembly is installed on the hydraulic head housing and incorporates a valve assembly which maintains proper fuel pressure within the fuel chamber to assure an adequate supply of fuel to the pump plunger at all times.

Seal wires are installed by the manufacturer and in no case are the seals to be removed or the pump disassembled except by authorized personnel.

Complete replacement pumps are available through G.M.C. Truck dealers.

Fuel Injection Nozzle

Holder Assemblies

The fuel injection nozzle and holder assemblies are utilized to carry the high pressure fuel through the cylinder head to nozzle tip and deliver it to the combustion chamber.

The nozzle holder consists of a forged body which houses the spindle and spring, and has a high pressure fuel duct. The spindle bore and spring chamber are utilized as a passage for leak-off fuel which seeps past and lubricates the nozzle valve.

The nozzle valve and body are a closely matched pair which must always be used together as a unit. *Valves and valve bodies are not interchangable*; however, matched nozzle and valve assemblies are available. Nozzle and valve assembly is located on nozzle holder by dowel pins and the assembly is firmly seated at lapped mating surfaces on body and nozzle holder by the nozzle cap nut. A high pressure duct in nozzle valve body aligns with duct in nozzle holder to carry high pressure fuel to the valve seat area. Pressure spring at upper part of holder seats at a flange on spindle, and spring is retained by the pressure adjusting nut. Spring pressure is transferred through the spindle to the nozzle valve. Spring holds the valve on its seat in nozzle body.

A perfect seal is necessary at valve seat to prevent fuel dribble as well as to prevent any of the combustion gases in cylinder from entering the spray nozzle interior chamber.

The cap nut at upper end of nozzle holder assembly serves to lock the spring adjusting nut and is threaded to accept a leak-off tee of elbow to which is attached the leak-off hose.

A high pressure fuel line from outlet port in injection pump hydraulic head is connected to threaded portion on nozzle body forging.

The fuel injection nozzle holder assemblies are held in place in cylinder heads by two bolts; and a copper gasket is used at bottom of each nozzle well in head.

Fuel Injection Nozzles and High Pressure Lines

Fuel Injection Nozzle Operation

Fuel under pressure is fed from the rotary single plunger injection pump to the appropriate nozzle via the high pressure pipe line and finds its way through the ducts in the nozzle holder and valve body to the lower end of valve. When the pressure reaches 3000 psi, the valve will lift, and fuel is injected through holes in nozzle tip and into combustion chamber. Nozzle valve closes again when the line pressure drops. Any leakback through the leak-off line fitting

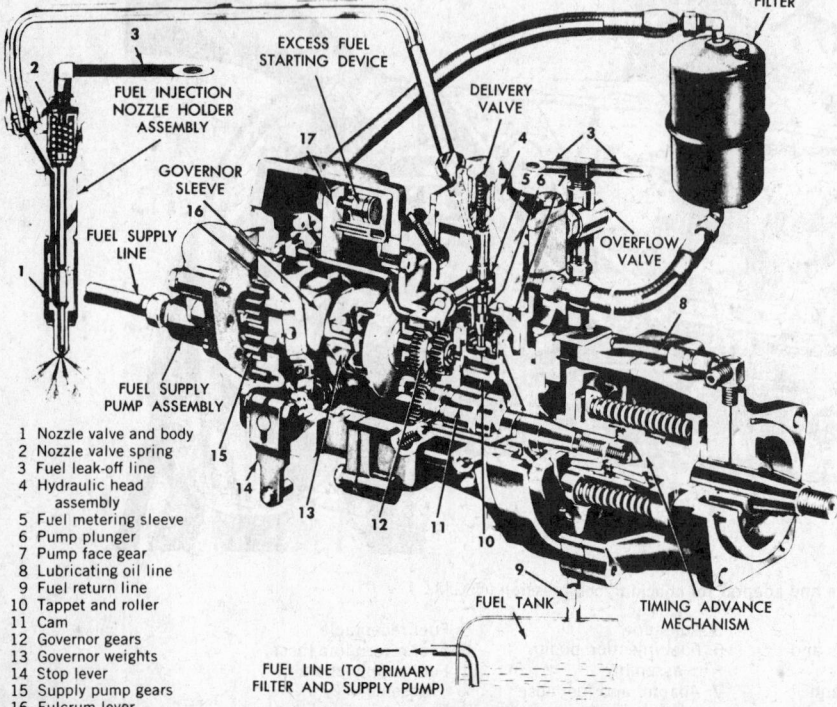

SECONDARY FILTER

EXCESS FUEL STARTING DEVICE

FUEL INJECTION NOZZLE HOLDER ASSEMBLY

DELIVERY VALVE

GOVERNOR SLEEVE

FUEL SUPPLY LINE

OVERFLOW VALVE

FUEL SUPPLY PUMP ASSEMBLY

1 Nozzle valve and body
2 Nozzle valve spring
3 Fuel leak-off line
4 Hydraulic head assembly
5 Fuel metering sleeve
6 Pump plunger
7 Pump face gear
8 Lubricating oil line
9 Fuel return line
10 Tappet and roller
11 Cam
12 Governor gears
13 Governor weights
14 Stop lever
15 Supply pump gears
16 Fulcrum lever
17 Governor stop plate

FUEL TANK

TIMING ADVANCE MECHANISM

FUEL LINE (TO PRIMARY FILTER AND SUPPLY PUMP)

Arrangement of fuel system units (8 cyl. engines) (© G.M.C.)

is passed back to the junction fitting on injection pump overflow valve.

Locating a Misfiring Cylinder

This procedure will not serve to detect a nozzle with incorrect popping pressure within a moderate range. Such a malfunction can only be determined by removal of nozzle assembly from engine and checking on test stand.

1. Connect an accurate tachometer at the tachometer drive fitting on engine.
2. With engine idling, loosen a high pressure line nut at injection nozzle holder. When a high pressure line nut is loosened, the pressure build-up will not be great enough to cause the nozzle to deliver fuel to the cylinder being tested and no power will result. If there is little or no drop in engine rpm when the line nut is loosened, the nozzle may be at fault and it should be removed for cleaning and inspection.

NOTE: Nut at high pressure line must remain loose only long enough to observe the tachometer reading, since fuel will spill from the loosened nut with each discharge from the injection pump port.

Nozzle and Holder Assembly Maintenance

Normally, unless engine performance, or other trouble diagnosis symptoms indicate a malfunction of fuel injection nozzles, the nozzle and holder assemblies should not be removed from engine.

When it becomes necessary to remove nozzle and holder assemblies, follow the procedures given to remove, clean, and test the units.

Removing Nozzle and Holder Assemblies

1. Disconnect the high pressure line nut at the fuel injection nozzle holder and at fuel injection pump hydraulic head.
2. Remove the clamp assembly from line being removed then remove the high pressure line. IMPORTANT: High pressure line must be disconnected at both ends and clamp must be removed so line can be removed without bending. Install dust caps at injection pump outlet port and at nozzle holder when line is removed.
3. Force leak-off hose off fitting in top of nozzle holder. Remove the two nozzle holder bolts and lock washers, then pull nozzle holder assembly out of cylinder head, meanwhile using air hose to blow away any dirt particles which may be loosened as the nozzle holder is lifted out.

1 Spring
2 Cap gasket
3 Pressure adjusting nut
4 Cap nut
5 Leak-off line tee
6 Spindle
7 Nozzle holder
8 Valve
9 Valve body
10 Nozzle cap nut
11 Gasket (copper)

Nozzle holder and valve assembly components (© G.M.C.)

4. Remove copper gasket, and then plug the opening in cylinder head to prevent objects from being accidentally dropped into nozzle well.

Inspection Before Disassembly

NOTE: It is recommended that each nozzle assembly be tested for leakage and valve opening pressure as well as for spray pattern before disassembly. If nozzle assembly is found to be in good condition, it should not be disassembled, but should only be cleaned externally. If the holes in nozzle tip are clogged as indicated by unsatisfactory spray pattern, the components should be disassembled before attempting to clean the holes.

Disassembly

1. With nozzle holder mounted on fixture or held in vise, loosen cap nut and remove nut and gasket from spring pressure adjusting nut.
2. With Allen wrench, remove spring pressure adjusting nut from nozzle holder. Remove spring and spindle. IMPORTANT: Do not proceed with step 3 unless pressure adjusting nut has been loosened to relieve spring pressure.
3. Invert nozzle holder and remove the nozzle cap nut from nozzle holder, then separate valve body from the nozzle holder.

Cleaning and Inspection

The nozzle tip should be cleaned with a brass wire brush before disassembling the valve body from nozzle holder. The nozzle tip orifices should be cleaned using orifice needle and holder. Care must be used when cleaning orifices to avoid breaking orifice cleaning needle in hole. After

disassembly, all parts should be carefully cleaned. If lapped surfaces on holder and mating surface on valve body are scratched, remove dowel pins from nozzle holder and polish the surface on lapping block using special fine lapping compound. Also polish the mating surface on valve body. Install new dowel pins after nozzle holder surface has been lapped.

Assembling Injection Nozzle Parts

1. Dip valve and body in clean fuel and insert valve in body. *NOTE: The valve must be free in the body. By lifting the valve about one-third of its length out of the body, the valve should slide back to its seat without aid when the assembly is held at a 45 degree angle.*
2. Mount nozzle holder in holding fixture, then set valve body in place on nozzle holder dowel pins and screw nozzle cap nut onto nozzle holder. Before tightening cap nut, use centering sleeve over nozzle tip to locate the nozzle valve body in center of opening in cap nut. After nut has been tightened sufficiently to seat parts, remove centering sleeve and tighten cap nut (10) to 50 to 55 foot-pounds.
3. Set nozzle holder upright in holding fixture or vise and insert spindle into holder. Place spring on spindle, then thread spring pressure adjusting nut into threads in nozzle holder. Place cap nut gasket over nut, then install cap nut on threads on adjusting nut.
4. Check nozzle on test fixture for spray pattern and valve opening pressure.
5. Setting Injection Nozzle Valve

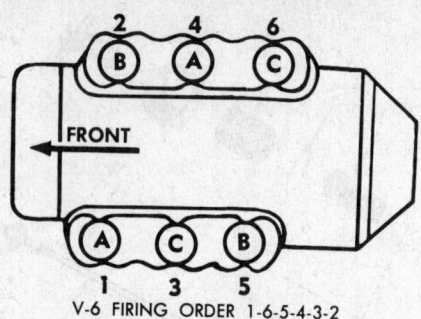

V-6 FIRING ORDER 1-6-5-4-3-2

Firing order and mated cylinders
6 cyl. Toro flow engines
(© G.M.C.)

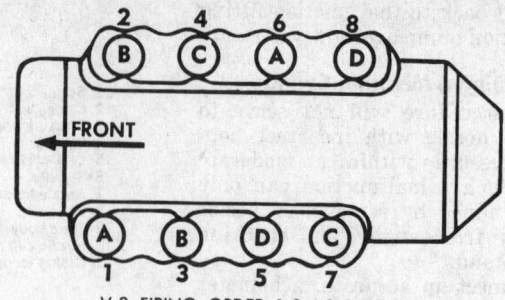

V-8 FIRING ORDER 1-8-4-3-6-5-7-2

Firing order and mated cylinders
8 cyl. Toro flow engines
(© G.M.C.)

Opening Pressure. Using test pump, proceed as instructed below to set the valve opening pressure and test for leaks.

With nozzle holder assembly connected to tester, actuate the test stand handle rapidly (about 25 strokes per minute) to expel air from nozzle and holder and to "settle" the spring and nozzle loading column.

Depress operating handle slowly to raise pressure. Continue to depress handle and note the gauge pressure at which the nozzle valve opens. If opening pressure is not as specified, remove the leak-off tee and loosen cap nut so Allen wrench can be used to adjust spring pressure.

Use Allen wrench to turn pressure adjusting nut and change spring pressure. Turning nut clockwise increases pressure. Repeat valve opening pressure test each time adjusting nut is changed until correct opening pressure is obtained.

NOTE: When new pressure adjusting spring is installed, adjust the opening pressure 10% higher than specification to allow for spring set.

After correct valve opening pressure is obtained, tighten cap nut firmly to lock the adjustment. Recheck for valve-opening pressure after cap nut is tightened.

Operate test pump handle with several fast sharp strokes, meanwhile observing spray pattern. Spray should be in the nature of finely atomized sprays indicating that all holes are open. D351 engines have five orifices in each nozzle tip. Nozzle tips on other engines have only four orifices.

Stroke pump slowly (approx. 3 seconds). Some degree of "chatter" should occur, indicating that nozzle valve is free and component parts are correctly assembled.

Depress the pump handle slowly with a 10-second stroke and hold the pressure slightly (approx. 100 lbs.) below nozzle valve opening pressure to check for valve seat leakage. No fuel droplets should appear. Occasionally a slight wetting of the tip (2) may be noticed. This is permissible since it is caused from hydrauli-

cally balancing the spring and nozzle valve.

Remove the nozzle holder assembly from test stand and if nozzle holder is not to be installed immediately, it should be wrapped in clean paper.

If nozzle does not operate properly, replace valve and body or install rebuilt nozzle holder assembly.

Nozzle Holder Assembly Installation

Use seat cleaner and wrench handle first to clean seat, and the bore in cylinder head through which the injection nozzle valve body extends, then use brush to finish cleaning the well cavity.

1. Use new copper gasket at lower end of injection nozzle assembly and insert the nozzle holder assembly into place in cylinder head.
2. Gradually tighten nozzle holder bolts alternately to avoid any tendency to bend the holder assembly.
3. Final tension on two nozzle holder bolts should be equal to seat the assembly squarely on copper gasket. Correct torque reading is 17 to 20 foot-pounds.
4. Connect leak-off lines at tee or elbow on nozzle holder. Place the high pressure line in place so that fittings on both ends line up squarely with respective threads at both the injection pump and

the nozzle holder. Use torque wrench and adapter to tighten nut at hydraulic head to 16 to 19 ft.-lbs. Tighten nuts at injector nozzle holders to 22 to 26 foot-pounds.

Injection Timing

The fuel injection pump assembly is timed to the engine when pump assembly is installed, and should not require any subsequent alteration if the installation is done properly.

In the event there is reason to suspect that fuel injection is not properly timed, the following procedure is necessary to check the timing.

Same procedure is used both on the 6-cylinder and 8-cylinder engines except as indicated in text.

Checking Injection Timing (With All Fuel Units Installed)

1. Remove the cover and gasket from right side of fuel injection pump to expose the alignment mark in the opening below the control unit lever.
2. On 6-cylinder engines, remove the access cover from top of engine front cover assembly to expose pointer and mark on injection pump hub. On 8-cylinder engines, remove the inspection plate from top of advance mechanism housing to expose the pointer and timing mark.

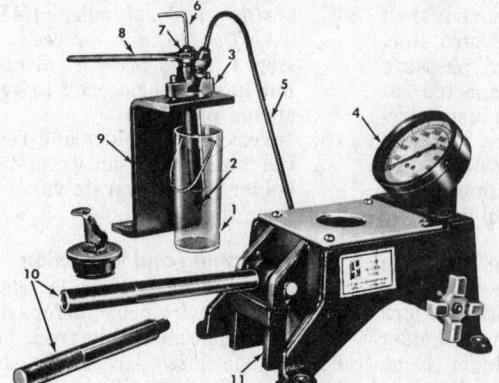

1 Cup
2 Spray tip
3 Nozzle holder assy.
4 Pressure gauge
5 Fuel high pressure line
6 Allen wrench
7 Cap nut
8 Box end wrench
9 Fixture
10 Pump handle
11 Test stand

Test stand for checking injection nozzle operation
(© G.M.C.)

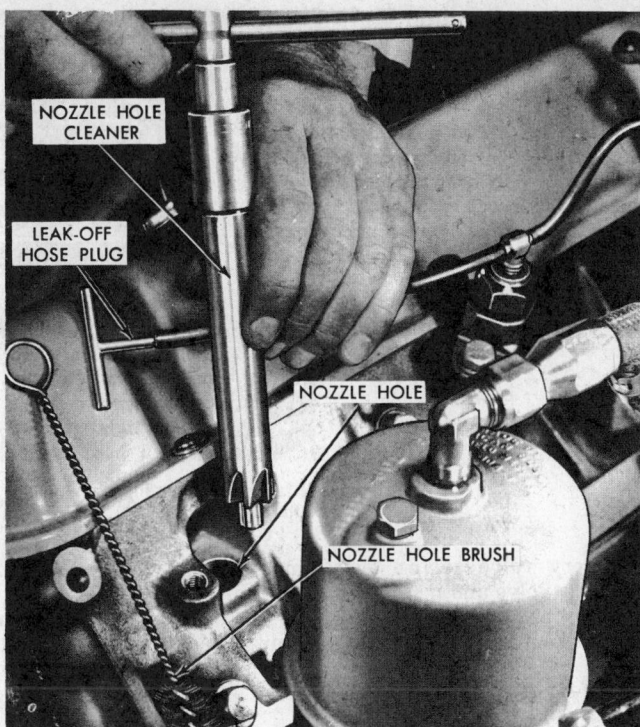

Cylinder head nozzle well cleaning equipment (© G.M.C.)

3. Remove left hand rocker arm cover, then crank engine to place No. 1 piston on compression stroke—(both valves will then be closed and push rods can be turned with fingers). Slowly crank the engine until the "INJ" mark on crankshaft damper is aligned with pointer in engine front cover. At this point pump should be injecting fuel into No. 1 cylinder.

4. If 6-cylinder engine is correctly timed, the mark and pointer will be indexed and mark on face gear viewed through cover opening in pump housing will be in the position shown in the illustration. If 8-cylinder engine is correctly timed, the mark and pointer will be indexed and mark on pump face gear viewed through cover opening in pump housing will be in the position shown in the illustration.

5. If the mark on pump drive hub is not indexed with pointer, or if the mark on face gear is not visible in opening, the injection timing is not correct. However, before proceeding to reset the injection timing, recheck to make sure No. 1 piston is on compression stroke and "INJ" mark on damper is aligned with pointer.

Setting Injection Timing

The elongated holes in injection pump driven gear make possible the setting of injection timing when in-

stalling the fuel injection pump assembly as well as for making correction of timing when necessary.

With engine crankshaft positioned for firing on No. 1 cylinder use special wrench and box end wrench to loosen the driven gear mounting bolts.

Bring engine back to firing position on No. 1 cylinder before loosening the last bolt.

With bolts loose, use special wrench to turn injection pump shaft and hub as necessary to align mark on hub with pointer for 6-cylinder engines or to index and mark pointer when viewed through opening on 8-cylinder engines. Hold the shaft with wrench and tighten driven gear mounting bolts.

Recheck to make sure of exact indexing of marks.

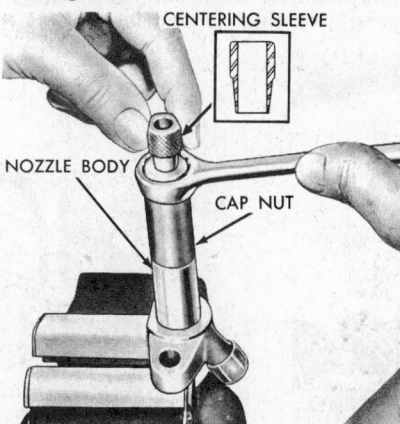

Using centering sleeve to locate valve body while tightening cap nut
(© G.M.C.)

Fuel Injection System

Illustrations show cut-away views of fuel injection pump for 6-cylinder engines. Same construction is used on pumps for 8-cylinder engines.

Service Operations

The service operations contained herein can be accomplished without removing the fuel injection pump assembly from the engine. WARNING: Do not attempt to work on the fuel injection pump, nozzles, or high pressure lines in a dusty area. Extreme precautions must be taken to prevent any dust particles from entering the fuel passages in lines and nozzles. Any dirt or abrasive material permitted to enter the fuel supply port on fuel injection pump could quickly cause extensive damage to, or possi-

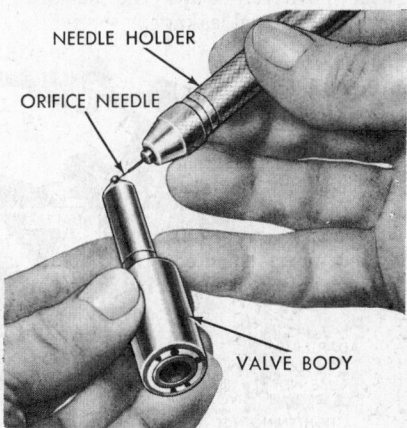

Using special needle to clean spray holes in nozzle valve body
(© G.M.C.)

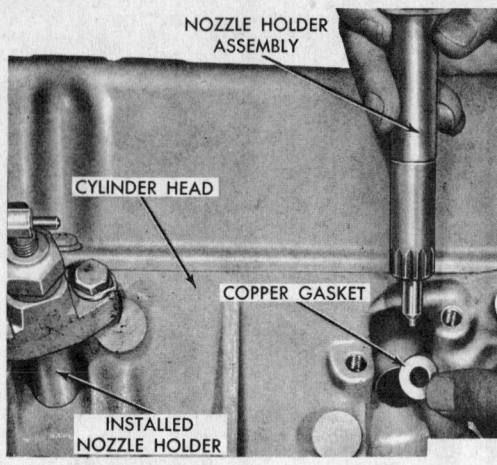

Installing fuel injection nozzle assembly
(© G.M.C.)

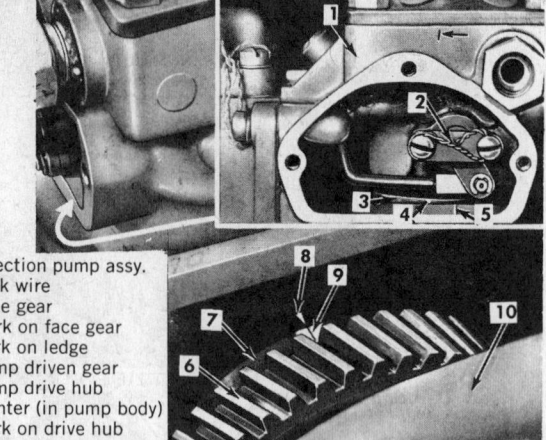

1 Injection pump assy.
2 Lock wire
3 Face gear
4 Mark on face gear
5 Mark on ledge
6 Pump driven gear
7 Pump drive hub
8 Pointer (in pump body)
9 Mark on drive hub
10 Engine front cover

Timing marks at face gear and at injection pump
gear flange (6 cyl. engines)
(© G.M.C.)

ble ruin of the pump hydraulic head mechanism.

Fuel Delivery Valve

Removal

Wash off top of the head assembly and blow dry with air hose.

Remove delivery valve cap, then remove spring from cavity. Remove the valve from its seat.

Using special wrench adapter, remove the delivery valve retainer from head assembly.

Use needle-nose pliers or tweezers to reach into cavity and remove the delivery valve seat.

Installation

CAUTION: Use utmost care to keep delivery valve and retaining parts absolutely clean during installation.

Place fuel delivery valve and seat assembly in position in hydraulic head cavity, then install the retainer. Use adapter and torque wrench to tighten valve retainer to 65 to 70 foot-pounds torque; then loosen the retainer and retorque to 65-70 foot-pounds. Tightening the retainer twice positively seats the parts to prevent any fuel leakage.

Place delivery valve spring over pilot on delivery valve and install cap using torque wrench to tighten cap once to 55 to 60 foot-pounds.

Control Unit Assembly

Removal

Remove the four control unit cover attaching screws, then remove cover and gasket.

Place a clean lint-free cloth in opening so parts cannot fall into pump interior as they are removed.

On early type pumps, remove the retainer ring which secures control rod to lever on control unit. Late production pumps have a different design at control rod and the retainer plate prevents disengagement of rod from lever.

Remove lock wire from plate screws. Remove screws, plate, and the spacers which retain the control unit, then disengage the control rod from lever.

Mark the parts is necessary so the control unit can be installed in original location without changing the pump calibration.

Carefully withdraw the control unit assembly from bore in housing, using two thin screwdrivers diago-

nally across from each other to start the unit out of bore.

Remove O-ring seal from control unit. Hold the control unit level so the pin will not fall out of place.

Installation

Using a clean screwdriver, carefully insert it through the control unit opening in the pump housing and carefully move the metering sleeve downward to its lowest-position on plunger.

Lubricate the control unit O-ring and assemble it to the control unit.

Holding the control unit, pin, and O-ring assembly, with the control unit arm horizontal and pointing toward the rear of the pump and with the pin horizontal and the identification dot up, insert the assembly into the pump housing.

If the pin is not properly positioned, the control unit arm can not be rotated 360 degrees.

Rotate the control unit bushing, which was marked so the screw scallops in the control unit bushing are aligned with the holes for the retainer plate attaching screws and the control unit bushing flange in the same position as before removal.

On early pump, engage the control

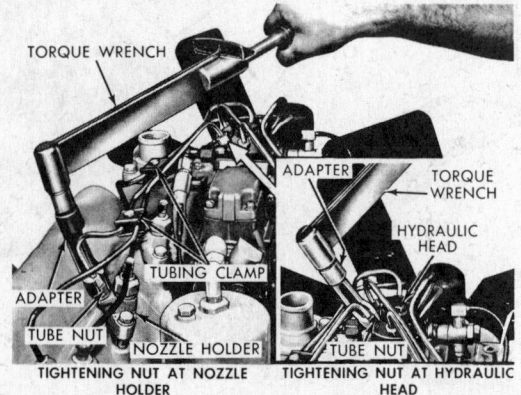

TIGHTENING NUT AT NOZZLE
HOLDER

TIGHTENING NUT AT HYDRAULIC
HEAD

Tightening nuts on high pressure lines
(© G.M.C.)

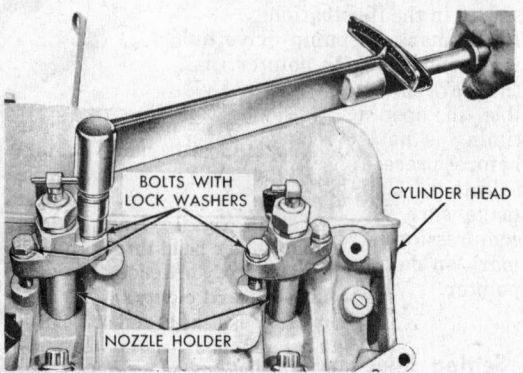

Tightening nozzle holder bolts
with torque wrench
(© G.M.C.)

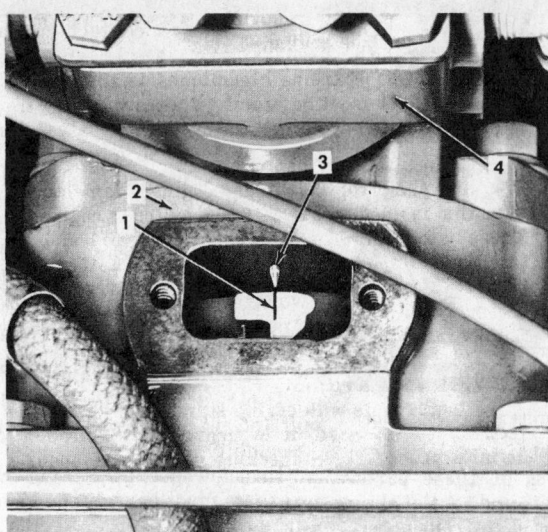

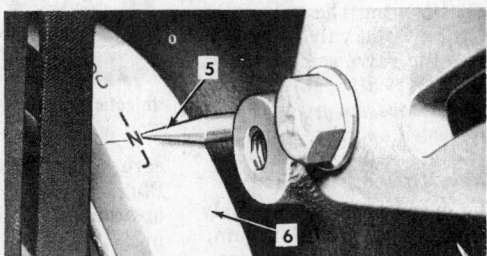

1 "INJ" mark on hub
2 Advance mechanism housing
3 Pointer
4 Fuel injection pump assembly
5 Pointer in front cover
6 Crankshaft damper assembly

POINTER AND MARK AT PUMP ADVANCE COVER OPENING

POINTER AND MARKS AT CRANKSHAFT PULLEY

Timing marks at advance mechanism and crankshaft damper (8 cylinder engines) (© G.M.C.)

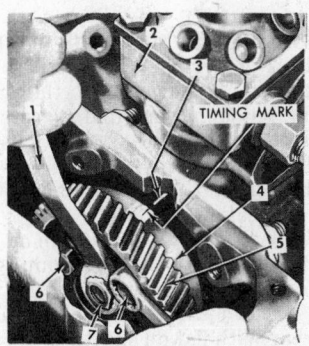

Using special wrench to hold pump shaft while tightening driven gear bolts
(© G.M.C.)

1 Pump shaft holding wrench (special tool)
2 Fuel injection pump assembly
3 Pointer
4 Driven gear mounting hub
5 Injection pump driven gear
6 Driven gear mounting bolts
7 Hub retaining nut

rod with hole in control unit lever, then install new retainer ring which secures rod to lever.

Assemble control unit retaining plate, screws (with lock washers), and spacers then install parts to retain the control unit. Tighten screws to 18 to 33 inch-pounds. Install lock wire through screw heads. On late production pumps, the retaining plate retains the control unit and also prevents control rod from becoming disengaged from lever.

Using new gasket, install control unit cover and secure with screws and lock washers.

Overflow Valve

NOTE: Parts of latest type injection pump overflow valve assembly are shown in illustration. The complete valve assembly may be installed or the valve and/or spring shown may be installed.

Overflow Valve Removal

1. Disconnect leak-off lines from fitting at valve and disconnect fuel return line.
2. Using wrench on hex-shaped portion of valve assembly remove the assembly from injection pump body.

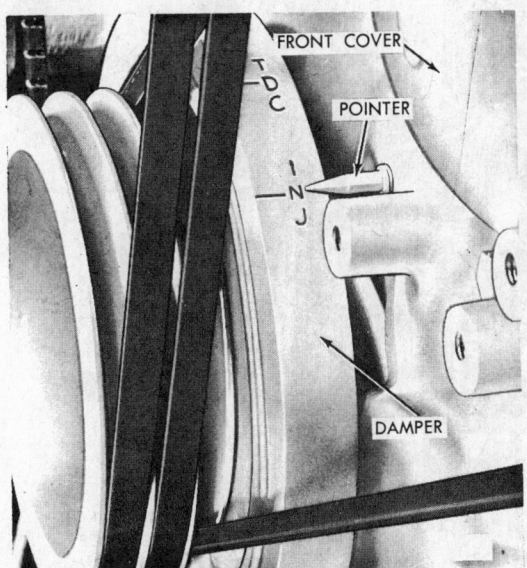

Timing marks (6 cylinder engine)
(© G.M.C.)

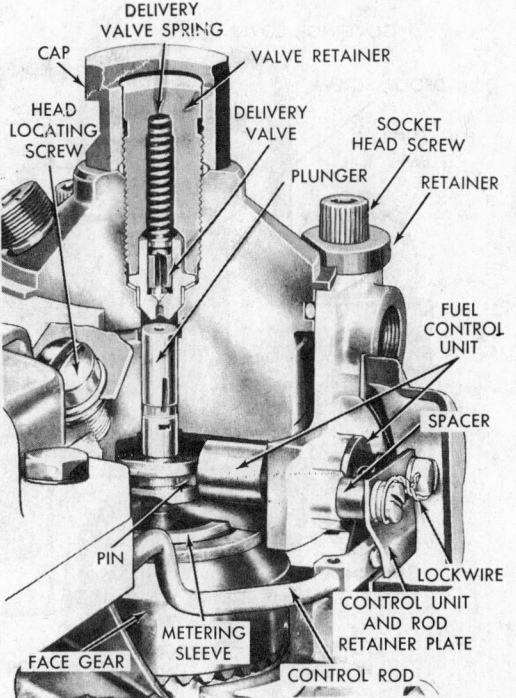

View of hydraulic head showing internal components
(© G.M.C.)

Disassembly

Grip the overflow valve body in vise fitted with soft jaws, then screw the valve seat out of valve body. Remove valve and spring. Some assemblies may have spacer washers which are used in manufacture to provide correct setting.

Assembly

Place spring, (spacers, if removed), and valve in valve body; then after positioning the valve seat over valve, thread the valve seat into body. Tighten the valve seat securely. *NOTE: Overflow valve spring may be shimmed, if necessary, to provide 40 psi at governed speed.*

Overflow Valve Installation

Screw the overflow valve assembly into threaded port in injection pump body, using wrench on the hex-shaped portion of valve.

Install fittings and connect return line and leak-off lines to overflow valve body.

Fuel Supply Pump

Fuel Supply Pump Removal

Disconnect the fuel lines and remove fittings from supply pump.

Remove the three supply pump attaching bolts and lock washers, then remove the supply pump and gasket. CAUTION: As supply pump is removed the governor inner spring guide and spring spacers are free to fall out of place. Exercise necessary

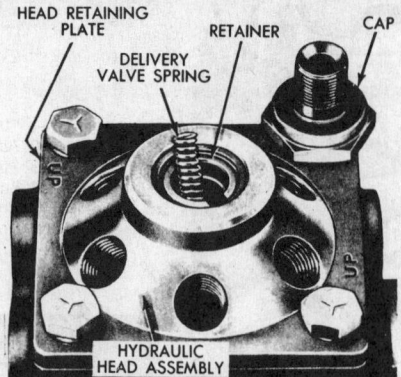

Injection pump cap, spring and valve retainer
(© G.M.C.)

care to prevent loss of these parts. Plug the fuel inlet and outlet line openings to prevent dirt from entering.

Fuel Supply Pump Disassembly

Remove governor outer spring guide from pump assembly.

Remove the five screws and lock washers which attach cover to insert. Tap the insert lightly with plastic hammer to separate insert from cover. Remove gears from insert.

Remove relief valve retainer screw and remove spring and relief valve assembly.

Supply Pump Component Inspection

Inspect for wear in gear cavities in pump insert. Also look for evidence of worn surface or scoring on cover

area at which contact is made with gears.

Note condition of the oil seal in pump insert.

Examine relief valve parts. Use new parts when assembling pump if any of the components are defective. Relief valve should pass fuel through center in only one direction.

Inspect idler gear shaft which is a press fit in bore in insert. If pin is worn, scored, or loose in bore, obtain a new insert and pin assembly.

Assembling Fuel Supply Pump

Dip each of the gears in clean fuel oil and place gears in cavity in insert assembly.

Use new gasket. Screws with lock washers must be tightened evenly and firmly.

Dip relief valve and spring in fuel oil, then install valve with the spring pilot engaged with spring. Install spring retainer cap and tighten firmly.

Fuel Supply Pump Installation

Check governor, inner and outer springs to make certain the inner spring guide and original spacers are in place. Outer spring guide must be in place at pump insert.

Place gasket on cover flange with bolt holes aligned.

Observe through opening in oil seal the position of flat in pump gear bore. A clean screwdriver may be used to turn the gear so as to align the gear with flat on shaft.

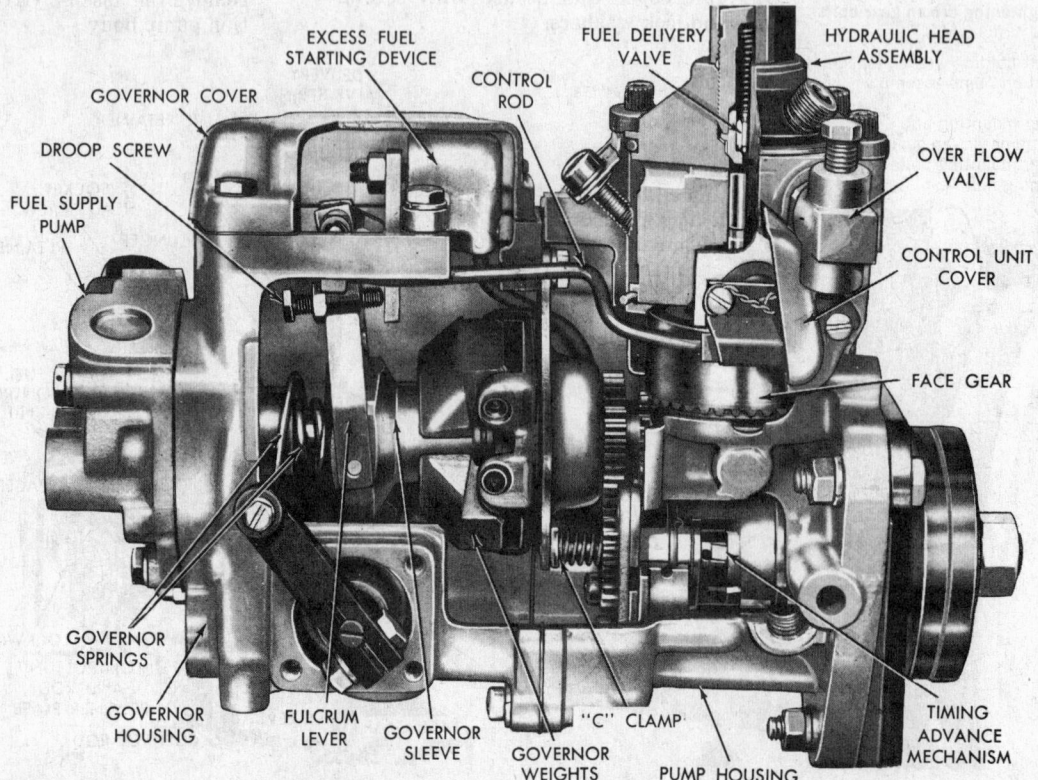

Cut away view of fuel injection pump (6 cyl. engines) (© G.M.C.)

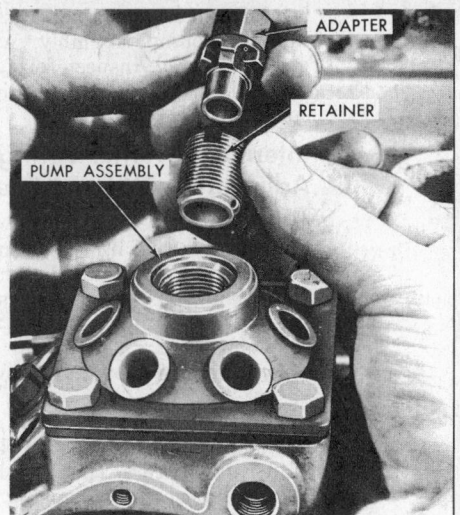 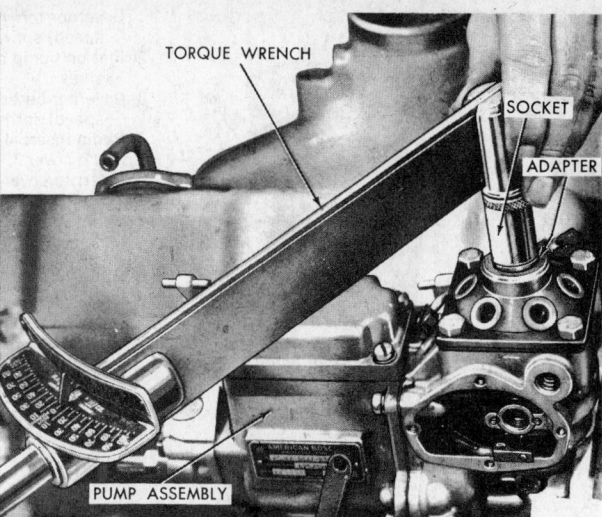

Installing fuel delivery valve retainer with adapter and torque wrench (© G.M.C.)

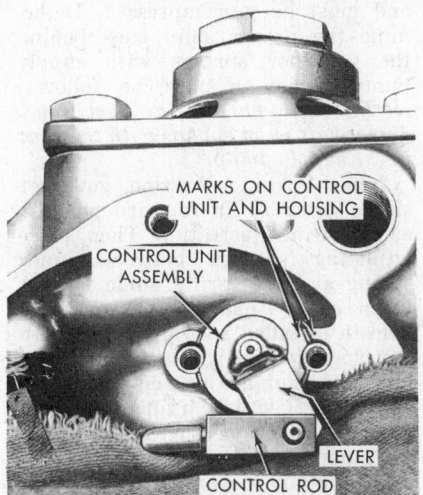

Control unit retaining parts and alignment marks
(© G.M.C.)

Move supply pump assembly into position at injection pump housing using care to engage drive gear with flat on shaft. Install and tighten supply pump attaching bolts.

Connect fuel lines to fittings at supply pump, remove vent plug from top of secondary fuel filter and bleed air from pump and lines.

Replacing Hydraulic Head Assembly

In cases where the hydraulic head is in satisfactory operating condition, but it becomes necessary to remove the head assembly to replace the O-ring seals or other parts of the pump, it is possible to remove and install the hydraulic head without removing the pump assembly from the engine; and recalibrating is not required.

Hydraulic Head Removal

Disconnect and remove fuel high pressure lines, then plug or cap the openings in the hydraulic head and nozzles.

Remove injection pump control unit assembly.

Turn engine crankshaft to place No. 1 piston in firing position with the marking on crankshaft damper aligned with pointer, and the mark on face gear (visible in inspection window in pump housing) indexed with mark on housing. Make sure the No. 1 piston is on compression stroke.

NOTE: Mark on gear face should be indexed with housing mark, even though the damper mark may not be exactly registered with pointer.

Remove the head assembly locating screw and gasket. Remove retaining plate. Tap the hydraulic head assembly lightly with plastic or lead hammer or slightly rotate the assembly

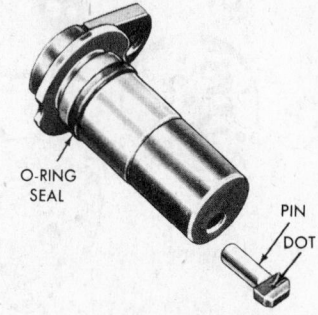

Fuel control unit assembly
(© G.M.C.)

with a wrench on the delivery valve cap to loosen it in housing bore. Raise the hydraulic head assembly up out of the pump housing. Remove upper seal ring from groove in head

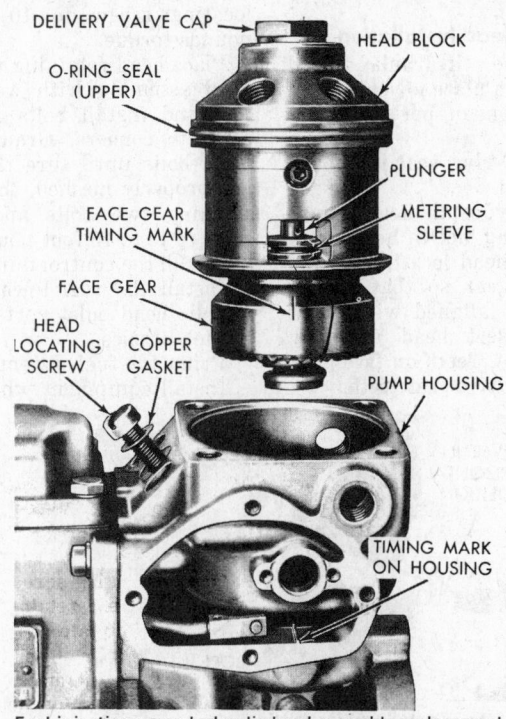

Fuel injection pump hydraulic head assembly replacement
(© G.M.C.)

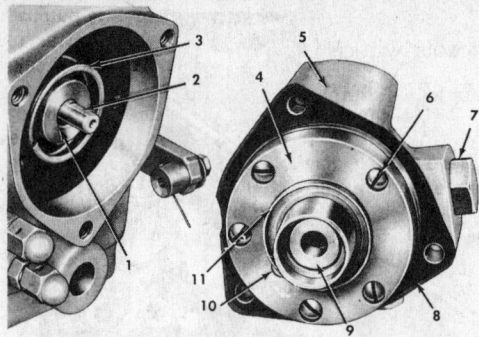

1 Governor inner (high speed) spring guide
2 Flat on pump drive shaft
3 Governor outer (low speed) spring
4 Pump insert
5 Pump cover
6 Insert to cover screws and lock washers
7 Relief valve retainer screw
8 Gasket
9 Oil Seal
10 Pump idler gear pin
11 Outer (low speed) spring guide

Fuel supply pump and injection pump assembly
(© G.M.C.)

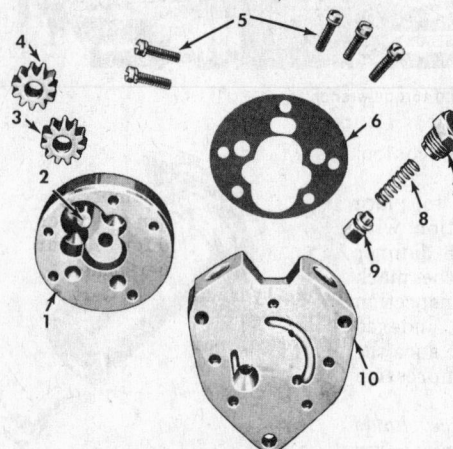

1 Insert assembly
2 Pin (pressed into Item 1)
3 Drive gear
4 Idler gear
5 Screws with lock washers
6 Gasket
7 Valve spring retainer cap
8 Valve spring
9 Relief valve assembly
10 Pump cover

Fuel supply pump components
(© G.M.C.)

assembly. Remove the head lower seal ring from bottom of bore in pump housing. *Do not turn engine crankshaft while hydraulic head is removed from pump.*

Hydraulic Head Installation

Lubricate the hydraulic head O-ring seals, then place lower ring in position at bottom of bore in pump body.

Place upper O-ring seal in groove in hydraulic head.

Position the hydraulic head assembly so the locating slot in head block is aligned with head locating screw; and turn face gear so that timing mark on gear is aligned with mark on housing. Insert head assembly into housing bore. Teeth on face gear must mesh with drive gear teeth.

Install head locating screw with copper washer (gasket) and if necessary, rotate head assembly so the locating screw can be turned freely—indicating no interference. Tighten locating screw to 10 to 11 foot-pounds torque.

Place head retaining plate over the head assembly with word "UP" on top, and install bolts in diagonally opposite corners. Gradually tighten plate bolts until sure the gear teeth are properly meshed, then install remaining two bolts and tighten all bolts to 13 to 15 foot-pounds.

Install the control unit assembly.

Install the fuel lines between hydraulic head oulet ports and fuel injection nozzles.

Prime the fuel system.

Install equipment which may have

been removed to gain access to the injection pump assembly. After engine is started, inspect for fuel leakage at hydraulic head, and high pressure fuel lines.

Internal Governor Spring Pack Adjustment (Without Special Tool)

Break the fuel injection pump seal (if used), back out idle adjusting screw, and remove three bolts attaching the fuel supply pump to the governor housing.

Remove the governor cover and gasket.

Remove fuel supply pump. Remove inner and outer governor springs and adjusting spacers (noting the position of spacers and spring guides. The inner spring is the high speed spring and must have a gap. The outer spring is the low speed spring and must be precompressed. Determine the proper shim pack behind the governor springs with supply pump gasket in place as follows: *NOTE: The flat on the fuel pump drive shaft must engage with the gear in the supply pump.*

Inner Spring: Position governor sleeve assembly in forward position as shown in illustration. Then, place adjusting spacers between inner spring and the spring guide so that when supply pump with gasket is held in installed position (flush with governor housing), inner spring can be turned but has no end play. This can be checked with a finger through governor cover opening. Remove supply pump and remove adjusting spacers to obtain proper gap of 0.089 to 0.108 inch on both the 6-cylinder engines and the 8-cylinder engines.

Outer Spring: Place adjusting spacers between governor sleeve and outer spring so that when supply pump with gasket is held in installed position, the outer spring can be turned but has no end play. Then, remove supply pump and add adjusting spacers with total thickness of 0.059 to 0.079 inch to give proper precompression on outer spring.

NOTE: Do not install outer spring when checking spacer pack for inner spring or install inner spring when checking spacer pack for outer spring.

Reinstall the inner and outer gov-

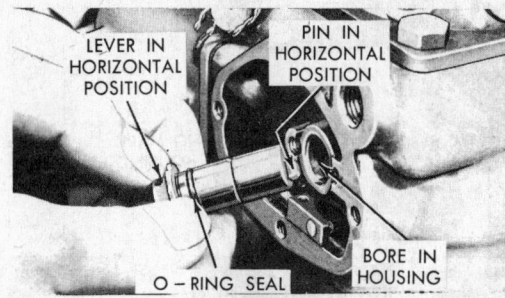

Control unit installation
(© G.M.C.)

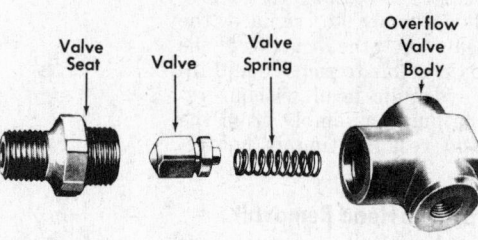

Overflow valve (typical)
(© G.M.C.)

ernor springs with their respective spacers as previously determined.

Reinstall the fuel supply pump and new gasket, taking care that the inner and outer spring guides do not fall off the supply pump insert. The flat on the fuel pump drive shaft must engage with the gear in the supply pump. Exercise care so that seal is not damaged.

Reinstall governor cover. Turn the idle speed adjustment screw until seven or eight threads are visible, then install screw lock nut.

Start engine and adjust idle speed to 625-675 spm with full accessory load. Set high speed at 3440 rpm no-load on 6-cylinder engine, or to 3250 rpm on 8-cylinder engines.

NOTE: Range of idle screw adjustment after setting by above method is approximately one turn. More adjustment than this may allow droop screw to rest against excess fuel piston stop plate and give erratic idle.

Connect the accelerator linkage and stop linkage, making sure that full travel of the accelerator lever can be obtained.

Internal Governor Spring Pack Adjustment (Using Special Tool)

A special tool is available for use to accurately select injection pump governor spring spacers. The tool serves in determining the required spacers, both for the inner (high speed) and the outer (low speed) spring. Refer to illustration for location of spacers. Whenever occasion arises for checking either of the governor springs, the other spring should also be checked. Tool may be used without removing the injection pump assembly from engine.

Assembly and Adjustment After Governor Spring Spacer Selection

Place inner and outer governor springs and their respective spacers in position in governor housing, then install fuel supply pump. The flat on governor shaft must engage drive gear in supply pump. Tighten supply pump mounting bolts to 5 to 6 foot-pounds torque.

Connect fuel lines at fittings in supply pump cover. Install cover on top of governor housing. Connect controls if they have been disconnected; and adjust controls for full travel in both directions.

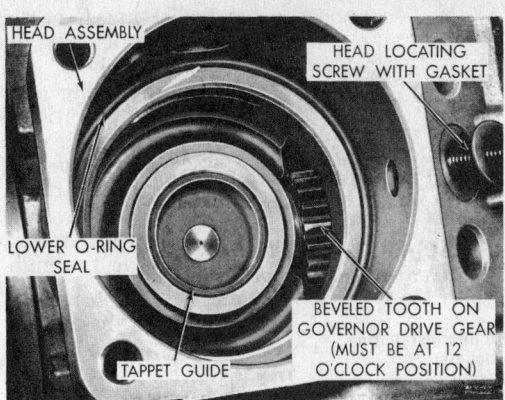

Fuel injection pump with hydraulic head removed
(© G.M.C.)

Check and, if necessary, adjust high speed with no-load. Tighten the adjusting screw lock nuts immediately after making the adjustment.

NOTE: The normal idle and maximum speeds depend almost entirely on the characteristics of the governor springs, hence the speeds can only be varied to a limited degree by the adjusting screws at rear of governor housing.

Checking for Air Leaks in Fuel Supply (Suction) Line

CAUTION: Exercise necessary care to prevent any dirt from entering the fuel system when fuel lines are disconnected in making following tests.

Disconnect outlet line from fuel supply pump and install sight glass, or plastic line so that fuel may be observed while engine is running.

Start engine, then look for air bubbles in the pump discharge line.

If a leak is apparent in the suction lines, filter or fittings, a sight glass or section of plastic line may be installed at inlet fitting on supply pump to determine if air is reaching pump through suction side of system. If no bubbles are observed at inlet connection, but continue to be present at outlet line, a defective supply pump is indicated.

NOTE: Be sure the fuel filter between fuel tank and pump is not clogged, and that filter gaskets are in good condition.

Checking for Inadequate Fuel Pump Delivery and/or Overflow Valve Malfunction

Make necessary inspection to determine that fuel filters are not clogged.

Remove, clean, and inspect the supply pump relief valve assembly and spring. Also, disconnect lines from overflow valve and remove the valve assembly from the fuel injection pump. Disassemble the overflow valve, inspect parts. Replace overflow valve or parts as required.

Check supply pump pressure as follows:

Install pressure gauge in line between supply pump outlet and secondary filter.

Remove overflow valve and install a fuel shut-off valve, connecting the fuel return line to the valve discharge opening.

Open the valve slightly and then start the engine. While running the engine a 2400 rpm momentarily close the fuel shut-off valve. The pressure gauge should show between 55 and 65 psi.

CAUTION: Keeping the shut-off valve closed for any period of time can cause the injection pump plunger to overheat and seize.

If pressures are appreciably lower than specified the supply pump should be overhauled or the unit should be replaced with a new supply pump assembly.

Remove shut-off valve and pressure gauge, and reinstall overflow valve assembly.

Fuel Injection Pump Assembly

NOTE: The fuel injection pump assembly is available for service replacement as a complete assembly. In most instances the required repairs may be made without removing the injection pump assembly from the engine.

Fuel Injection Pump Replacement

Replacement should only be made if it is definitely indicated that the fuel injection pump assembly is at fault and is not repairable without removing the unit from the engine.

Fuel Injection Pump Removal

CAUTION: Do not attempt to work on the fuel injection pump, nozzles, or high pressure lines in a dusty area.

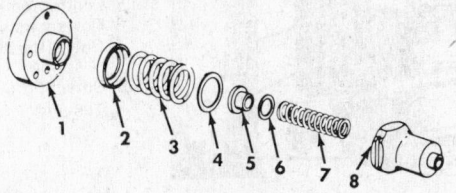

1 Supply pump
2 Outer spring guide
3 Outer spring
4 Outer spring spacer
5 Inner spring guide
6 Inner spring spacer
7 Inner spring
8 Governor sleeve

Governor spring and spacer arrangement
(© G.M.C.)

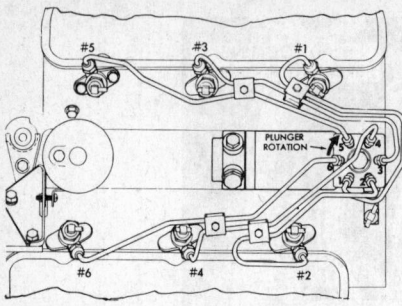

Fuel high pressure line installation
(6 cyl. models)
(© G.M.C.)

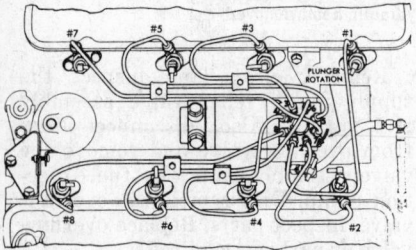

Fuel high pressure line installation
(8 cyl. engines)
(© G.M.C.)

Extreme precautions must be taken to prevent any dust particles from entering the fuel passages in lines and nozzles. Any dirt or abbrasive material permitted to enter the fuel supply port on fuel injection pump could quickly cause extensive damage to, or possible ruin of the pump hydraulic head mechanism.

Preliminary Operations (All Engines)

Depending on type of vehicle and accessory equipment, remove the necessary equipment to provide access to fuel injection pump assembly, high pressure fuel lines and engine controls. Carefully observe the arrangement of accelerator control linkage, and the engine stop mechanism. A stop solenoid or an engine stop air cylinder is mounted on the injection pump assembly. Disconnect wire from stop solenoid terminal (when used). Disconnect stop control cable

and disconnect accelerator linkage from lever at left side of injection pump assembly.

Remove tubing clamps which are shown in illustration. Use the adapters shown and loosen high pressure tubing nuts at fuel injection nozzles and at pump hydraulic head. Use air hose to blow away any loose dirt particles at hydraulic head and at nozzles. Remove the high pressure tubing using care not to bend tubing.

NOTE: To facilitate installation, each line may be tagged with identification number corresponding with cylinder numbers.

Insert plugs in pump ports and cap the fittings at injection nozzles as soon as lines are removed. The importance of cleanliness cannot be over-emphasized.

Illustrations show caps at injection nozzles and plugs in hydraulic head ports.

Disconnect fuel leak-off lines from overflow valve.

Replacing Fuel Injection Pump on 6-Cylinder Engine

Disconnect fuel lines from fittings on supply pump at rear end of injection pump assembly. Also disconnect the fuel line at left side of injection pump housing. Cap or plug all openings to prevent entrance of dust.

At right side of pump assembly, disconnect lubricating oil line and the fuel return line at the overflow valve assembly.

Remove the gear access cover from top of engine front cover to provide access to injection pump driven gear and attaching bolts. The driven gear must be removed from hub on pump shaft before the injection pump assembly can be removed from engine.

CAUTION: To prevent accidental dropping of bolt washers into engine front cover and out of reach, pack clean shop rags into cavity below the gear attaching bolts. Do not crank engine until the rags have been removed.

When removing the last of the three pump driven gear bolts, hold

the pump shaft from turning with special wrench.

Remove the stud nuts and washers which secure the fuel injection pump assembly to engine front cover, then remove the pump assembly. Remove the sealing ring from pump mounting flange.

Remove pump driven gear from opening in front cover.

Installation

Preliminary Operations

If a new fuel injection pump assembly is being installed, it will be necessary to transfer the fuel line and oil line fittings from the removed pump assembly to the new pump. Observe the position of each fitting as it is removed, and install it in same position on new pump assembly.

CAUTION: Carefully clean any accumulated dirt from fittings before installing in replacement pump assembly and install dust caps to exclude any foreign material when the fittings are installed.

Inspect gear teeth on injection pump drive gear and driven gear before installing pump. Be sure the mounting surface on pump flange and the mating surface at engine front cover are clean.

Remove the cover from right-hand side of injection pump so alignment marks on gear face and ledge can be seen.

Fuel Injection Pump Installation

Set fuel injection pump driven gear in place through access opening in top of engine front cover. Driven gear teeth must engage teeth on pump drive gear and flat side of gear must be toward rear.

Check position of rocker arms at No. 1 cylinder to be sure No. 1 piston is on compression stroke. Observe "INJ" mark on crankshaft damper assembly which must be indexed with the pointer on front cover.

Through opening in right-hand side of fuel injection pump assembly, look for mark on edge of gear face. If

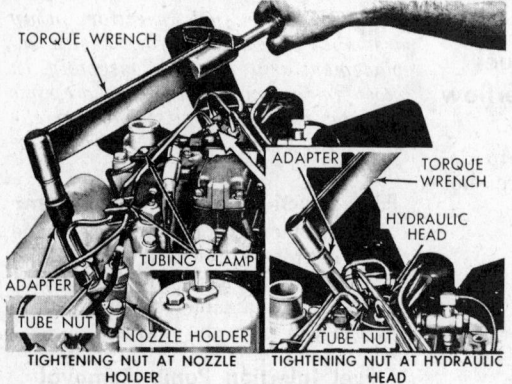

Torque wrench adapter for tightening nuts on high pressure lines
(© G.M.C.)

Governor mechanism
(© G.M.C.)

1 Inner (high speed) spring
2 Fulcrum lever assy.
3 Governor housing
4 Excess fuel starting device
5 Control rod
6 Governor sleeve
7 Outer (low speed) spring
8 Outer spring guide

1 Injection pump assembly
2 Fuel supply line fitting
3 Drive gear
4 Mounting studs
5 Seal ring
6 Overflow valve
7 Lubricating oil fitting
8 Pointer
9 Mark on adapter hub
10 Engine front cover
11 Stop lever

Installing fuel injection pump assembly (6 cyl. engines) (© G.M.C.)

mark is not visible, turn the pump shaft until mark comes into view, then set the mark on hub in alignment with pointer in pump housing.

Install seal ring on pilot at forward side of pump mounting flange, then lift the injection pump assembly into position on mounting studs.

Install plain washer and nut on each of the three pump mounting studs and tighten the nuts evenly to seat pump flange firmly on cover.

Check the position of threaded hole in pump hub in relation to elongated slot in driven gear. If necessary, change the position of gear so threaded hole in hub is approximately centered in slot in gear. The gear teeth can be lifted out of mesh with drive gear to change gear position.

Start one of the driven gear mounting bolts with plain washer through driven gear slot and into mounting hub, then using special wrench hold pump hub timing mark in alignment with pointer and tighten gear mounting bolt firmly.

Crank the engine as necessary to bring other driven gear bolt holes into position for installing remaining two bolts; install bolts and plain washers, then tighten bolts to 35 to 40 foot-pounds.

Install pump side cover and gear access cover connect oil supply line, and fuel return to overflow valve.

Mount a new secondary filter on filter bracket and connect fuel lines.

Connect accelerator control linkage and engine stop mechanism.

CAUTION: The fuel injection nozzles should be inspected and repaired as necessary, or replaced with new nozzles whenever a new or rebuilt fuel injection pump is installed on engine. Attempting to start an engine having defective injection nozzles can damage or ruin a fuel injection pump assembly.

Install high pressure fuel lines.

Prime the fuel system, before attempting to start engine. During priming procedure, the supply line should remain disconnected from fitting on left side of pump assembly until all air has been purged from filters, supply pump, and lines; then continue to flush fuel through open end of line to avoid the possibility of any loose particles entering the fuel chamber surrounding the hydraulic head assembly.

Start engine and inspect all fuel line and lubricating oil line connections for leakage.

Check and adjust engine idle speed if required.

Replacing Fuel Injection Pump on 8-Cylinder Engine

Removal

Disconnect fuel lines from fittings on supply pump at rear end of injection pump assembly. Also disconnect the fuel supply line at left side of injection pump housing. Cap or plug all openings to prevent entrance of dust.

Disconnect and remove lubricating oil lines.

At rear end of injection pump loosen clamps on oil drain hose. Pull hose connector elbow out of grommet

1 injection pump assembly
2 Plugs to seal openings
3 Seal ring
4 Adapter hub
5 Pump mounting studs
6 Drive gear
7 Engine front cover

Installing fuel injection pump assembly (8 cyl. engines) (© G.M.C.)

in valve lifter cover; then remove connector and hose.

Remove two bolts and washers which hold pump support and bracket together; then remove the two bolts and washers at lifter compartment cover. Remove the pump support bracket.

Remove the gear access cover from top of engine front cover to provide access to injection pump driven gear and attaching bolts. The driven gear must be removed from hub on pump shaft before the injection pump assembly can be removed from engine.

When removing the last of the three pump driven gear bolts hold the pump shaft from turning with special wrench.

Remove the two lower mounting bolts and washers, first, then place a suitable wood block under rear of injection pump to support the assembly while two upper bolts and washers are removed from mounting studs.

Remove the fuel injection pump from engine; then remove driven gear from front cover. Remove sealing ring from pilot at pump mounting flange.

Installation

Preliminary Operations

If a new fuel injection pump assembly is being installed, it will be necessary to transfer the fuel line and oil line fittings from the removed pump assembly to the new pump. Observe the position of each fitting as it is removed, and install it in same position on new pump assembly.

CAUTION: Carefully clean any accumulated dirt from fittings before installing in replacement pump assembly and install dust caps to exclude any foreign material when the fittings are installed.

Inspect gear teeth on injection pump drive gear and driven gear before installing pump. Be sure the

Injection pump rear support and lubricating oil drain (8 cyl. engines) (© G.M.C.)

1 Bolt with lock washer and plain washer
2 Bolt with lock washer and plain washer
3 Injection pump support
4 Fuel injection pump assembly
5 Nipple
6 Oil drain hose
7 Hose clamp
8 Pump support bracket
9 Hose connector elbow
10 Grommet
11 Lifter compartment cover

mounting surface on pump flange and the mating surface at engine front cover are clean.

Remove the cover from right-hand side of injection pump so alignment marks on face gear and ledge can be seen.

Fuel Injection Pump Installation

Set fuel injection pump driven gear in place through access opening in top of engine front cover. Driven gear teeth must engage teeth on pump drive gear and flat side of gear must be toward rear.

Check position of rocker arms at No. 1 cylinder to be sure No. 1 piston is on compression stroke. Observe "INJ" mark on crankshaft damper assembly which must be indexed with the pointer on front cover.

Through opening in right-hand side of fuel injection pump assembly, look for mark on edge of face gear. If mark is not visible turn the pump shaft until mark comes into view. Through inspection plate opening in top of advance mechanism

housing check the position of mark in relation to pointer. If necessary turn pump shaft to index mark with pointer.

Install injection pump support on bottom of pump housing using two bolts with lock washers.

Install seal ring on pilot at forward side of pump mounting flange, then lift the injection pump assembly into position on mounting studs. Install plain washer and nut on each of the four pump mounting studs and tighten the nuts evenly to seat pump flange firmly on cover.

NOTE: A suitable block may be inserted between rear of pump and lifter compartment cover to support pump while installing nuts and washers at mounting flange.

Place injection pump support bracket in place on valve lifter cover with holes for bolts aligned with threaded holes in cylinder block and weld nuts on injection pump support. Install two bolts with lock washers and plain washers but tighten these bolts only enough to bring support

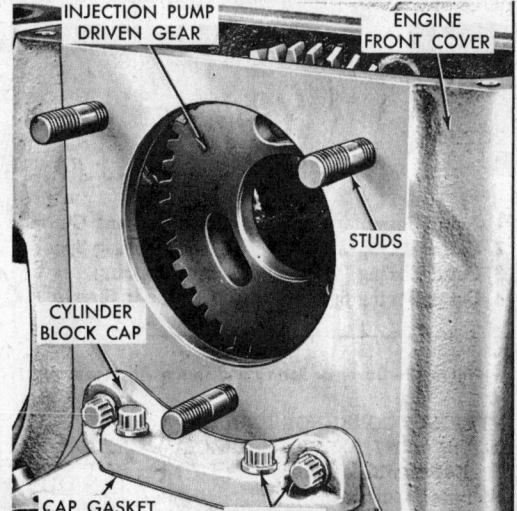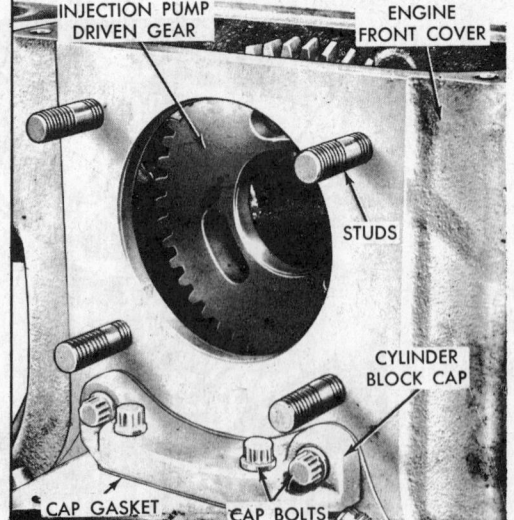

Injection pump driven gear and pump mounting studs (© G.M.C.)

and bracket into contact with each other. Install two bolts with lock washers and plain washers to secure the pump support bracket, then tighten the bolts which clamp the bracket to the pump support. Install nipple in threaded hole in pump housing, insert grommet in hole in lifter cover; then assemble oil drain hose and elbow. Hose must be clamped at nipple and end of elbow must be inserted into grommet.

Check the position of threaded hole in pump hub in relation to elongated slot in driven gear. If necessary, change the position of gear so threaded hole in hub is approximately centered in slow gear. The gear teeth can be lifted out of mesh with drive gear to change gear position.

Start one of the driven gear mounting bolts with plain washer through slot in driven gear and into drive hub on pump shaft. Use special wrench to hold pump shaft so timing mark is indexed with pointer while tightening gear mounting bolt firmly.

Crank the engine as necessary to bring other driven gear bolt holes into position for installing remaining two bolts, then install bolts and plain washers and tighten bolts to 35 to 40 foot-pounds.

Install cover on pump housing. Install inspection plate and gasket on top of advance mechanism housing. Install gear access cover on engine front cover assembly.

Connect lubrication oil supply lines at fittings on housings. Connect the fuel return line to overflow valve assembly.

Thoroughly clean and inspect the fuel lines to be used in connecting the secondary filter. Mount a new secondary fuel filter on filter bracket with the fuel lines connected.

Connect accelerator control linkage and engine stop mechanism.

CAUTION: The fuel injection nozzles should be inspected and/or repaired as necessary, or replaced with new nozzles, whenever a new or rebuilt fuel injection pump is installed on engine. Attempting to start an engine having defective injection nozzles can damage or ruin a fuel injection pump assembly.

Install high pressure fuel lines.

Prime the fuel system before attempting to start engine. During priming procedure the supply line should remain disconnected from fitting in left side of pump assembly until all air has been purged from filters, supply pump, and lines; then continue to flush fuel through open end of line to avoid the possibility of any loose particles entering the chamber surrounding the hydraulic head assembly.

Start engine and inspect all fuel line and lubricating oil line connections for leakage.

Check engine idle speed and, if necessary, adjust to proper setting.

Water Pump

General

The engines covered are equipped with either of two water pumps shown in illustrations.

Type 1 pump is used on all 6-cylinder engines except DH478, unless engine is in vehicle equipped with air conditioning, in which case type 2 pump is used.

Type 2 is the standard water pump used on all 8-cylinder engines and DH478, 6-cylinder engines. Overhaul procedure for each type of pump is given separately under appropriate headings below.

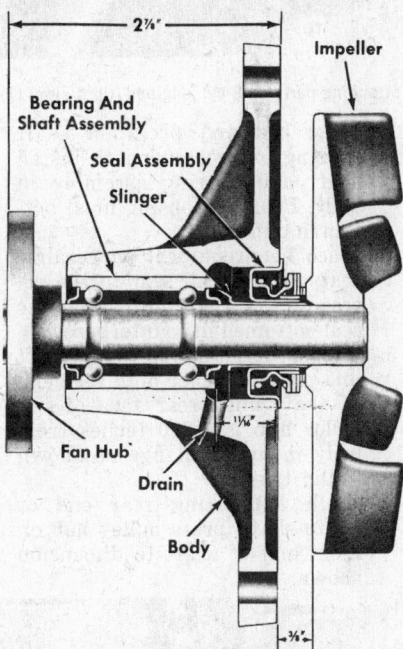

Water pump (6 cyl. engines exc. DH 478)
(© G.M.C.)

Type 1 Water Pump Overhaul
Disassembly
1. Support the hub and press shaft out.
2. Support pump body on press and press on front end of shaft to force shaft bearing, and impeller assembly out of body toward rear.
3. Support impeller as near as possible to hub at forward side, then press on rear end of shaft to remove impeller from shaft. With impeller removed from shaft, remove seal assembly.
4. Shaft and bearing assembly and seal must be replaced whenever pump is disassembled. Surface contacted by seal on impeller hub must be smooth and flat. Late pumps do not have slinger.

Assembly
1. Be sure bore through pump body is clean and free from burrs.
2. Install shaft and bearing assembly in pump body, pressing on bearing outer race. Proper location of bearing and shaft assembly is shown in illustration. Measure from rear face of body to bearing as shown.
3. Coat outer surface of seal which fits into pump body with sealing compound, then press seal assembly into body.
4. Press fan hub on front end of pump shaft to dimension shown in illustration. Hub front face must be located 2-7/8 inches from rear surface of pump body.
5. Install impeller on rear end of pump shaft, using pressure at rear face of impeller hub while supporting front end of shaft. Locate impeller so a space of 3/8 inches exists between impeller and pump body.

Type 2 Water Pump Overhaul

Disassembly
1. Support water pump pulley hub at rear surface and press shaft out of hub.

body and impeller, then while supporting pump body with shaft in vertical position, press the pump shaft out of impeller and force bearing out of body.

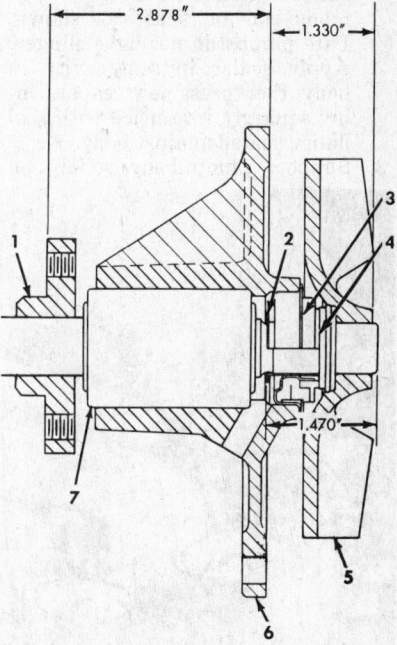

Water pump
(6 cyl. DH 478 and all 8 cyl. engines)
(© G.M.C.)

1 Pulley hub   3 Seal assembly
2 Slinger   4 Seal seat
5 Impeller
6 Pump body
7 Shaft and bearing assembly

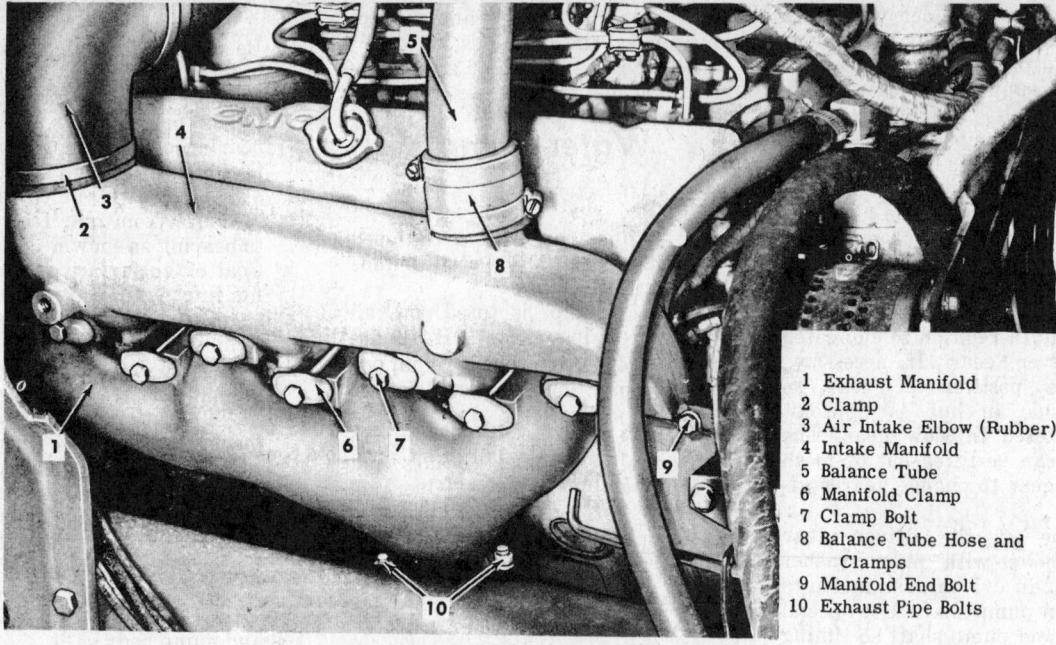

1 Exhaust Manifold
2 Clamp
3 Air Intake Elbow (Rubber)
4 Intake Manifold
5 Balance Tube
6 Manifold Clamp
7 Clamp Bolt
8 Balance Tube Hose and Clamps
9 Manifold End Bolt
10 Exhaust Pipe Bolts

Manifolds and attaching parts on 8 cyl. engines (right side) (© G.M.C.)

3. Remove seal assembly from pump body. If seal seat in impeller is not in good condition, use a thin blade to pry the seal seat out of impeller. Thoroughly clean the counterbore in impeller.

Assembly

1. On early production pumps, the slinger should be 1.470 inches from end of shaft as shown. Late pumps do not have slinger.
2. Apply sealer in seal cavity in body, then press new seal assembly squarely into place with seal flange seated against body.
3. Support pump body solidly on press bed and press on shaft bearing outer housing to install shaft and bearing assembly in body. Bearing housing must bottom in pump body.
4. Place neoprene seal on ceramic seat, then install seat in the recess in impeller with neoprene seal bottomed in counterbore.
5. Support front end of pump shaft and press impeller onto rear end of shaft until rear face of impeller hub is 1.330 inches from body mounting flange as shown in illustration.
6. While supporting rear end of pump shaft, press pulley hub on front end of shaft to dimension shown.

Manifolds

Manifold Removal

NOTE: With engine in vehicle, the exhaust pipes must be disconnected from exhaust manifold flanges before removing exhaust manifolds. Air intake hoses and any brackets bolted to intake manifolds must also be removed. On V-8 engines, the balance tube and connecting hoses must be removed.

Bend lock away from each of the exhaust manifold end bolts, then remove the short bolt attaching intake manifold to cylinder head.

Remove the bolts and clamps hold-

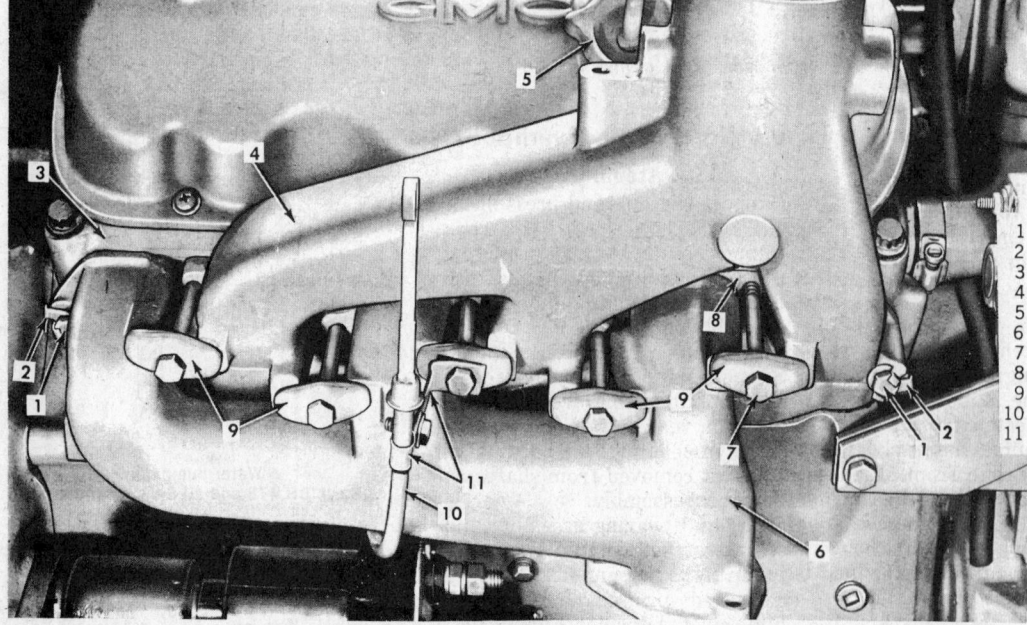

1 End bolt
2 Bolt lock
3 Cylinder head
4 Intake manifold
5 Crankcase ventilation cap
6 Exhaust manifold
7 Clamp bolts
8 Gasket
9 Manifold clamps
10 Dip stick tube
11 Tube clamp and bracket

Manifolds and attaching parts on 6 cyl. engines (right side) (© G.M.C.)

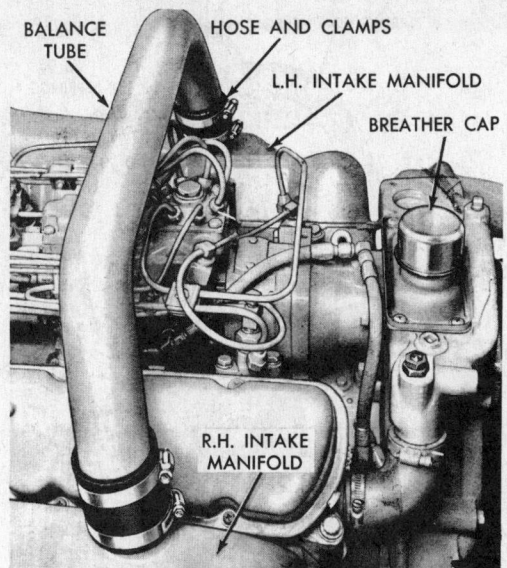

Manifold balance tube installation (8 cyl. engines)
(© G.M.C.)

ing manifolds to cylinder head, then remove the intake manifold.

Remove the short bolt attaching exhaust manifold to cylinder head, and remove the exhaust manifold and gaskets.

Manifold Installation

Manifolds are attached in similar manner at both cylinder heads. In some instances accessory support brackets must be removed to gain access to manifold bolts, if engine is installed in vehicle. Clean the manifold gasket surfaces and cylinder head mating surfaces.

Place bolt lock on exhaust manifold short end bolt, then insert bolt through hole in end of exhaust manifold. Place clamp on clamp bolt and insert clamp bolt through hole in exhaust manifold flange. Place manifold gasket at exhaust manifold so it is held in proper position by bolts.

Hold exhaust manifold and gasket against cylinder head and start the two bolts into tapped holes.

Place lock on long end bolt and insert bolt through hole in intake manifold. Position intake manifold at cylinder head and start end bolt into tapped hole in head. Install a clamp bolt with clamp through hole in intake manifold flange and thread bolt into cylinder head to properly locate intake manifold. Tighten clamp bolts with the respective clamps bridging bosses at intake and exhaust manifolds sufficiently to hold intake and exhaust manifolds in place.

Install remaining clamps and clamp bolts and position the clamps to contact manifold bosses squarely.

Tighten exhaust manifold end bolt to 15 to 20 foot-pounds torque. Tighten intake manifold end bolt to 20 to 25 foot-pounds, and tighten intake and exhaust manifold clamp bolts to 20 to 25 foot-pounds. Bend locks to secure the exhaust manifold end bolts.

Install balance tube between intake manifolds.

Engine Repairs

Cylinder Head Overhaul

General

The two cylinder heads on each engine are identical except for position of water outlet sleeve and core hole plug, and procedure is same to repair either head.

Cylinder heads for D/DH478 and D/DH637 engines are constructed as shown in sectional view of engines. D351 engine cylinder heads are similar to heads on D/DH478 and D/DH637 except that they do not have exhaust valve seat inserts.

Unless otherwise noted, the cylinder head repair procedures apply to all Toro-Flow Diesel engines. Illustration shows cylinder head components and associated parts for 8-cylinder engines.

Cleaning

Flush out water passages, and clean oil drain-back holes and hole at bottom of end rocker arm shaft bracket which supplies oil to valve overhead mechanism.

Disassembling Cylinder Head

1. Using valve spring compressor, relieve pressure of valve springs against retaining parts.
2. Remove valve keys, release the spring compressor, then at intake valve, remove springs and

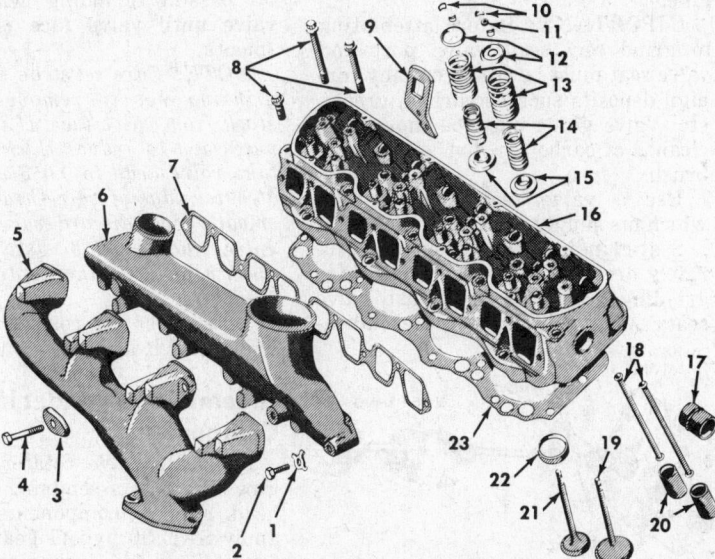

1 Bolt lock
2 End bolt
3 Clamp
4 Clamp bolt
5 Exhaust manifold
6 Intake manifold
7 Manifold gasket
8 Cylinder head bolts
9 Lifting bracket
10 Valve keys
11 Intake valve seal
12 Spring cap
13 Valve outer springs
14 Valve inner springs
15 Valve rotators
16 Cylinder head
17 Water tube
18 Push rods
19 Intake valve
20 Valve lifters
22 Exhaust valve seat
   insert
23 Cylinder head gasket

Cylinder head components and associated parts (8 cyl. engine) (© G.M.C.)

spring cap. After springs and retaining parts are removed from intake valve stem. Remove intake valve from cylinder head.

3. After releasing spring compressor at exhaust valves, remove springs and spring cap, then remove exhaust valve from cylinder head.

4. Remove valve rotators from cylinder head.

Cleaning and Inspection of Cylinder Head Components

1. Clean deposits from valve stems, and ports in cylinder head.
2. Use injection nozzle well cleaning tools to remove any carbon deposits from nozzle wells.
3. Test valve springs for tension with valve spring tester.
4. Check valves for burned or pitted faces and for wear or corrosion at valve stems. NOTE: Valves with oversize stems are available for use when guide bores in cylinder head are excessively worn. Guides are integral with head and must be reamed out to provide proper fit with oversize valve stem.
5. On cylinder heads with replaceable exhaust valve seat rings, inspect rings which must be tight in head and seats must be in good condition.
6. Inspect all plugs for evidence of leakage.
7. Visually inspect valve spring retaining parts. Also examine valve rotators for evidence of wear and damage.

Repair of Cylinder Head Components

Reaming Valve Guides

If clearance between valve stems and guides is excessive, select service valve with oversize stem for use in rebuilding cylinder head. Intake and exhaust valves are available with 0.020, 0.030, and 0.040 inch oversize stems.

Reamers are of proper dimensions to provide correct clearance for valve stem with corresponding oversize.

Replacing Exhaust Valve Seat Rings (All Except D351 Engines)

In cases where exhaust valve seat ring is damaged, the ring may be replaced as follows:

Assemble special exhaust valve seat insert remover in cylinder head as shown. Press tool firmly against ring, then while holding tool with handle, tighten screw to expand tool which will engage inner edge of ring.

Attach slide hammer to threaded end of tool screw, then slide weight sharply to end of rod to remove seat ring.

Clean cylinder head.

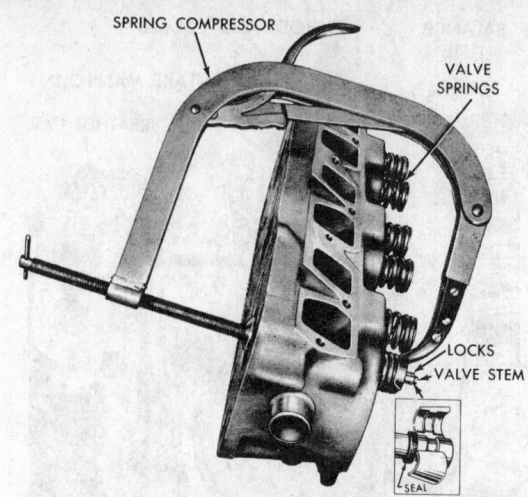

Use of valve spring compressor
(© G.M.C.)

Pack new inserts in dry ice for minimum of 15 minutes.

Preheat cylinder head by immersion in water at a temperature of 180°F., to 200°F. Place cylinder head on bench, blow out each insert counterbore with air, then lay chilled insert in counterbore with valve side toward installer.

Using valve seat installer, insert pilot end of tool into guide, then drive insert down tight into counterbore. This operation must be done quickly while the valve seat inserts are cold. Due to extreme hardness of material, exhaust valve seats must be finished with special grinding equipment.

Reconditioning Valve Seats

Reconditioning of valve seats on high compression engines is very important because seating of valves must be as near perfect as possible in order to obtain the maximum power and performance built into the engine.

IMPORTANT: Before attempting to grind any seat, valve port and valve seat must be free from any foreign deposits such as carbon, grease, etc. Valve guide must be thoroughly cleaned of carbon or dirt with a wire brush.

Use a valve seat grinder pilot which fits snugly in valve guide.

Seat-grinding stone must be accurately dressed to 45-degree angle for grinding intake and exhaust valve seats. After grinding, use dial indica-

EXHAUST VALVE
SEAT RING

SEAT REMOVER   SLIDE HAMMER

CYLINDER HEAD

Removing exhaust valve seat ring
with special tool
(© G.M.C.)

tor to check concentricity of valve seat with pilot hole. Seat should be concentric with pilot within 0.002 inch (Total Indicator Reading).

NOTE: When installing new valve seat inserts and/or new valves in cylinder head, measure distance from cylinder head surface to top of valve head with valve contacting seat. Intake valves must not project more than 0.050 inch above head; and exhaust valves must not project more than 0.045 inch. Grind necessary amount of stock from valve face to obtain proper dimension.

Test valve for perfect contact with seat in cylinder head.

Refacing Valves

Valves that are slightly pitted or burned can be refaced to the proper angle.

Set chuck at a 45 degree angle for both the intake and the exhaust valves.

Take light cut from face of valve by passing grinding wheel across valve until valve face is true and smooth.

NOTE: Care must be taken while grinding not to remove too much stock from valve face. If it should be necessary to reduce thickness of intake valve head to 0.055 inch or less at outer edge of valve head to obtain smooth face, discard valve. Exhaust valves should not be used if head at outer edge is less than 1/16 inch after refacing.

Test valves for contact with seats in cylinder head.

Assembling Cylinder Head Components

Lubricate valve rotators with engine oil before assembling to cylinder head. Keep all components clean and apply SAE 90 hypoid gear lubricant for at least 1½ inches at tip of all valve stems prior to installation.

Set valve rotator in place at cylin-

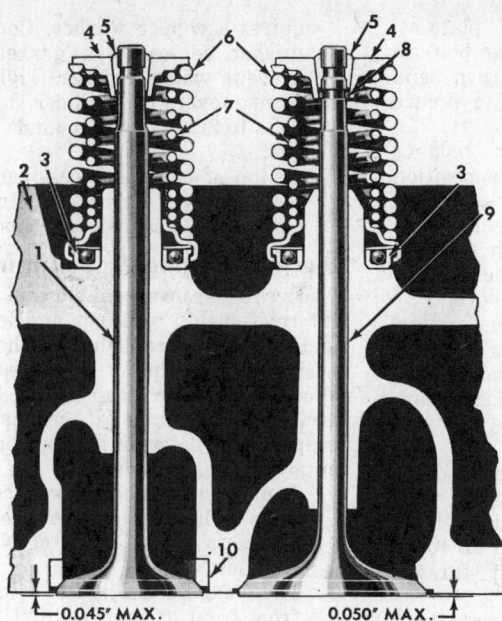

1 Exhaust valve
2 Cylinder head
3 Valve rotator
4 Spring cap
5 Valve keys
6 Outer springs
7 Inner springs
8 Intake valve oil seal
9 Intake valve
10 Exhaust valve seat insert

0.045″ MAX.   0.050″ MAX.

Sectional view through cylinder head and valves
(© G.M.C.)

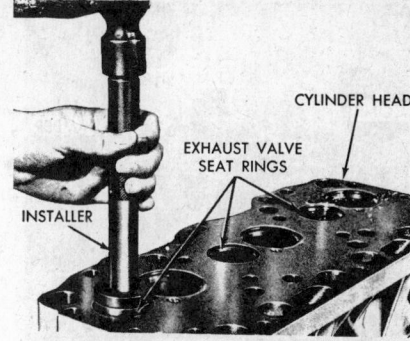

Using special driver to install exhaust valve Seat rings in head (© G.M.C.)

der head and insert valve through guide.

Set inner and outer springs in place on rotator with close-wound coils toward cylinder head and place spring cap (4) on springs.

Use spring compressor to hold valve in place and at same time compress springs. On intake valves install oil seal in lower groove in intake valve stem. Place two valve keys in valve stem groove, then release the spring compressor.

Valve Operating Mechanism

General

Mechanism for operating the intake and exhaust valves consists of mechanical lifters at camshaft, tubular push rods, and rocker arms mounted on hardened steel shaft. Each rocker arm is equipped with a self-locking screw for adjusting valve lash.

On 6-cylinder engines, the rocker arm shafts are each supported on five cast aluminum brackets. Construction on 8-cylinder engines is similar to that on 6-cylinder models but there are two more rocker arms and six brackets support the rocker arm shafts.

Inspection of Rocker Arms and Shafts

1. Wash rocker arms, shaft, and bracket assemblies in solvent. Oil holes in rocker arms must not be clogged.
2. Check torque required to turn adjusting screw in rocker arm. If screw turns with less than 5 foot-pounds, or if screw threads are not in good condition, replace screw.

3. Examine shaft for evidence of wear. Replace worn parts when assembling.

IMPORTANT: When installing new adjusting screws in rocker arms, lubricate screw threads and threads in rocker arm with engine oil prior to assembly.

Damper and Crankshaft Pulley

General

Latest engines have a counterbore in damper hub to accomodate a seal ring which seals against oil leakage. When installing crankshaft damper look for counterbore at rear edge of hub bore and install new seal ring on crankshaft as instructed below. The special installer may not be required on some (late) engines on which the oil seal ring is used but must be used on all other engines. Whenever a new damper assembly is being installed, the late type with counterbore and seal ring should be used.

1. Place the seal ring over end of crankshaft and position the ring squarely on crankshaft at inner end of damper key. *Do not place seal ring against slinger.* The seal must roll into hub counterbore when damper is installed.

2. Apply engine oil on damper hub at seal surface, then start damper onto crankshaft with keyway aligned with key. Assemble installer, screw, and thrust bearing and tighten screw firmly into threaded hole in crankshaft. Turn the nut on screw to force damper into place on crankshaft. Installer is not required on late engines.
3. Assemble cone, retainer and damper retainer bolt. Tighten bolt to specified torque with torque wrench.

Engine Front Cover Removal

NOTE: The thermostat housing and engine water pump assembly may remain on front cover during cover removal.

1. At rear side of engine front cover, remove bolts which attach cap to cylinder block and front cover, then remove cap and gasket.
2. At front side of front cover assembly, remove the bolts and washers which attach cover to cylinder block.
3. Jar the cover loose with a lead hammer, and remove the cover. Cover assembly must be lifted upward sufficiently to clear the injection pump drive gear before it can be moved forward.

Gear Train Removal

1. Remove four bolts attaching injection pump drive gear to camshaft gear, then remove drive gear.
2. Bend lock plates away from plate retaining bolts. Remove bolts and gear retainer plates

Cylinder head tightening sequence (6 cylinder engine)
(© G.M.C.)

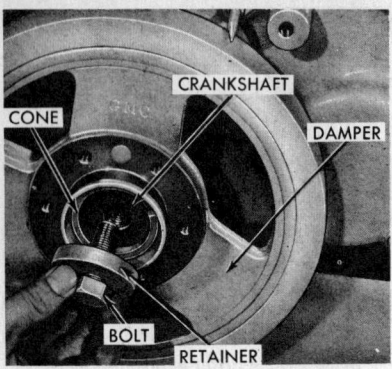

Crankshaft damper retaining parts
(© G.M.C.)

from idler gear shafts. Remove idler gears.

3. Gears are keyed to respective shafts and may remain on shafts until shafts are removed from cylinder block. Crankshaft gear can be removed from crankshaft with special puller, with crankshaft either installed or removed. An arbor press should be used to replace camshaft gear or balance shaft gear. Before the gear can be removed from V-8 engine balance shaft, the balance weight must be removed.

Gear Train Installation

1. Turn crankshaft so "O" mark on gear is at 12 o'clock position. Turn camshaft gear so "X" mark is at 6 o'clock position.
2. Lubricate bushing in camshaft idler gear, then install idler gear on idler gear shaft so "X" marks are indexed and "O" marks are indexed.
3. Using two bolts and lock plate,

install gear retaining plate at idler gear shaft. Tighten bolts to 15 to 20 foot-pounds, then bend corners of lock plate to secure bolts.

4. Lubricate bushing in balance shaft idler gear then position gear so alignment marks are indexed with marks on gears.
5. Retain idler gear with retainer plate and bolts in same manner as described in step 3 above.
6. Using narrow feeler gauge, check gear back lash between gear sets. Make backlash check at several points at each gear; there must be no tight spots in any position.
7. Install injection pump drive gear on camshaft gear with self-locking bolts. Tighten bolts with torque wrench to 35 to 40 foot-pounds with bolt threads oiled.
8. Check torque on camshaft gear retainer bolt. With bolt threads oiled the proper torque is 50 to 60 foot-pounds.

Engine Front Cover Installation

1. Locate front cover gasket at front of cylinder block with bolt holes in gasket aligned with holes in cylinder block.
2. Carefully guide front cover over injection pump drive gear and over end of crankshaft. Oil seal in cover and gear teeth may be damaged if care is not used in positioning cover assembly.
3. Install four short bolts with lock washers, then install long bolts with lock washers on all except the one bolt which passes through water passage which re-

quires a copper washer. Copper washer serves as a gasket to prevent water leakage. Tighten front cover to cylinder block bolts to 22 to 27 foot-pounds torque.

4. At top of cylinder block, behind front cover, install cylinder block cap using new cap gasket.

Checking Engine Valve Timing

Valves must open and close in correct relationship to upper and lower dead center of crankshaft. When a check of valve timing is necessary, following procedure may be used referring to markings at crankshaft damper for determining top dead center on No. 1 cylinder.

1. Remove left-hand rocker cover from cylinder head to provide access to rocker arms at No. 1 cylinder.
2. Turn engine clockwise (viewed from front of engine) to "UDC" mark at crankshaft damper on compression stroke. Both the intake and the exhaust valve on No. 1 cylinder will then be closed.
3. Adjust valve lash on No. 1 exhaust valve (front valve) to exactly 0.054 inch on 6-cylinder engines; or to 0.057 inch on 8-cylinder engines.
4. Turn engine clockwise until No. 1 exhaust valve opens and begins to close, then with fingers try turning push rod of No. 1 exhaust valve as engine is cranked slowly. When push rod rotates with finger pressure, the "UDC" mark on damper should be at pointer. This will be about 1 revolution from starting point. If

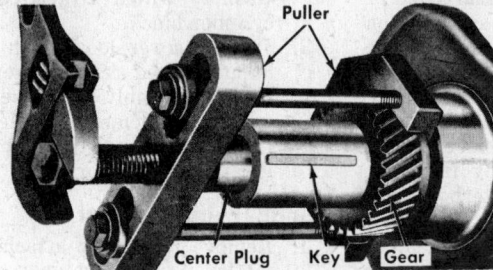

Removing gear from camshaft
(© G.M.C.)

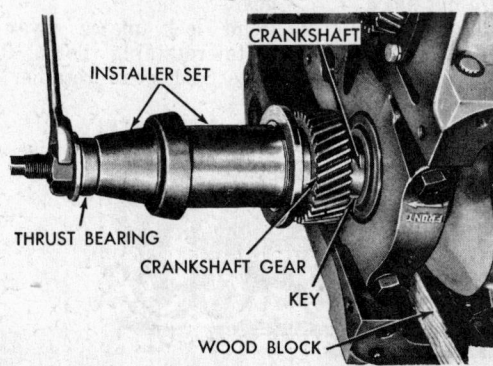

Installing crankshaft gear (using special tools)
(© G.M.C.)

Checking timing gear backlash
(© G.M.C.)

Gear train and timing marks (V6 engine)
(© G.M.C.)

1 Camshaft gear
2 Injection pump drive gear
3 Drive gear bolts
4 Self locking bolt
5 Camshaft gear retainer
6 Lock plates
7 Balance shaft idler gear
8 Balance shaft thrust plate
9 Balance shaft gear
10 Gear retainer plate
11 Plate retaining bolts
12 Crankshaft balancer key
13 Cylinder block
14 Camshaft idler gear

Gear train and timing marks (V8 engine)
(© G.M.C.)

1 Drive gear bolts
2 Camshaft gear
3 Injection pump drive gear
4 Camshaft idler gear
5 Lock plates
6 Plate retaining bolts
7 Gear retainer plate
8 Crankshaft gear
9 Crankshaft
10 Balance shaft gear
11 Balance shaft
12 Hole plug (ball)
13 Balance shaft front weight
14 Thrust plate
15 Balance shaft idler gear

push rod starts to rotate at any point within one-quarter inch either side of "UDC" mark, the valve timing is correct. Be sure to adjust exhaust valve clearance to 0.018 inch after performing the foregoing check.

5. If when checking valve timing, the push rod starts to rotate at a point more than three-quarters inch either side of the "UDC" mark, the camshaft can be considered "out of time." NOTE: When valve timing has been corrected, it will be necessary to retime the fuel injection pump.

Balance Shaft, Gear and Thrust Plate

Balance Shaft Removal

NOTE: Balance shaft cannot be *removed from cylinder block until bolts which secure weights to the balance shaft have been removed. Self-locking bolts secure the weights on V-8 engines. Locks under weight bolt heads are used on V-6 engines. Special wedges are required to expand weights and relieve clamping action so balance shaft can be removed.*

1. Remove balance shaft weight bolts from all weights.
2. Insert wedges in slots in balance shaft weights and tap each wedge lightly to just relieve clamping pressure of weights on shaft.
3. At front of cylinder block, remove the two bolts and washers which secure balance shaft thrust plate to block, then remove balance shaft, by pulling

shaft forward through bearings and removing weights from crankcase as each one drops off shaft.

4. On V-8 engine remove spacer plate from balance shaft and gear assembly.

The balance shaft, gear, and thrust plate may be inspected without removing gear from shaft. Clean **all** parts thoroughly before beginning inspection. Be sure oil passages in shaft are unobstructed and clean.

Inspection

1. Check balance shaft bearing surfaces for evidence of scoring and for wear. Use micrometer to measure shaft diameter at bearing areas. There are four bearing areas on V-6 balance shafts, and five bearing areas on V-8 balance shafts. Diameter of new balance shaft is 1.4705 to 1.4695 inches at bearing areas.
2. On 6-cylinder engines, measure clearance between thrust plate and balance shaft gear hub using feeler gauge. Clearance (endplay) should be 0.003 to 0.005 inch.
3. On 8-Cylinder engines use feeler gauge in manner similar to that shown in figure 45 to measure clearance between thrust plate and gear. Correct clearance between these parts if 0.002 to 0.006 inch.
4. Mount balance shaft in Vee blocks and use dial indicator to measure shaft runout. With shaft supported at ends, runout should not exceed 0.002 inch for 6-cylinder engines or 0.003 inch for 8-cylinder engines (total indicator reading).

Balance Shaft Gear and Thrust Plate Replacement (D351 and D, DH478 Engines)

Removal

Using suitable press plates, support the balance shaft gear while applying pressure on front end of shaft with arbor press. Press shaft out of gear, then remove gear key from slot in shaft and remove thrust plate.

Installation

1. Place thrust plate on balance shaft with the trade mark and part number facing forward.
2. Drive key into place in slot.
3. While supporting rear end of balance shaft, start gear squarely onto shaft with keyway in gear aligned with key in shaft. Extended hub on gear must be toward thrust plate. Timing mark on gear is at front side of gear.
4. Complete the gear installation by using a hollow driver be-

tween press and gear hub. Slowly press gear into position while using feeler gauge to measure the space between gear hub and thrust plate.

Balance Shaft Weight, Gear, and Thrust Plate Replacement (D, DH637 Engines)

General

In the event it should be necessary to replace balance shaft on 8-cylinder engines the balance weight pin hole must be drilled and taper reamed at assembly.

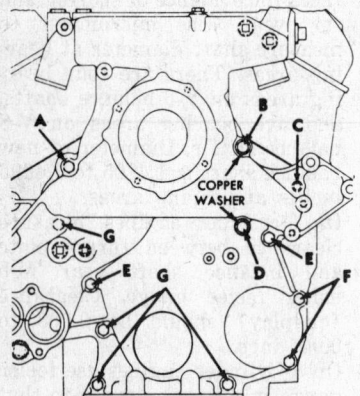

Front cover bolt arrangement
and torquing instructions
(© G.M.C.)

Bolt Location	Length (Inches)	Ft. Lb. Torque
A	4-¼	25-30
B	4-½	10-15
C	2	35-45
D	2-⅞	10-15
E	2-⅞	25-30
F	3-¾	25-30
G	1	25-30

Balance Shaft Weight and Gear Removal

1. Use a suitable punch and hammer to drive tapered pin out of weight and shaft.
2. Support rear side of thrust washer on arbor press plate, then press on front end of balance shaft and force the shaft out of gear and weight. CAUTION: When pressing balance shaft out of gear and weight, press plate used to support the thrust washer, must be positioned so as to allow gear and weight keys to clear the plate without interference. Support the thrust washer close to shaft to avoid bending the thrust washer.

Balance Shaft, Weight, and Gear Assembly, Procedure

General Information

1. The procedure for installing gear and weight on a new balance shaft differs from proce-

dure required when the original balance shaft is used. Each procedure is covered separately under 2. and 3. following:
2. Assembling Parts Using Original Balance Shaft
   a. With keys installed in respective keyways, place thrust washer over keys and into place against shoulder on balance shaft.
   b. Lubricate the front area on balance shaft and the bore through gear and weight with engine oil. Then start balance shaft gear onto front end of shaft with keyway aligned with keys and gear extended hub toward thrust washer. Support rear end of balance shaft and press the gear into place. Hold the thrust plate so it does not catch between gear hub and thrust washer as gear is pressed into place.
   c. Press balance shaft weight onto balance shaft until solidly seated against gear.
   d. Use No. 5 tapered reamer to clean up tapered hole in weight and shaft, then use air hose to blow into drilled oil passage and blow out any cuttings which may be in oil holes. Also flush out oil passages with clean solvent.
   e. Drive in new tapered pin to secure balance shaft front weight to balance shaft referring to illustration for pin installation views. Stake the pin hole in weight to secure tapered pin.
3. Assembling Parts Using New Balance Shaft and Weight
   a. Accomplish the procedure as instructed under steps 2. a., b., and c. to assemble thrust

plate gear and weight on balance shaft.
   b. Using drill press and 0.25 inch, long shank drill piloted in pin hole in balance weight, drill squarely through balance shaft. NOTE: If a previously-used balance weight is used, pilot the drill from the thick side of weight. If a new weight is used a straight pilot hole is provided for use in drilling pin hole in shaft.
   c. After pin hole has been drilled in balance weight shaft, use No. 5 tapered reamer to ream the pin hole in weight and shaft. Ream the pin hole sufficiently to allow new pin to enter freely until only 3/32-inch of pin is above weight surface. When reaming is completed and before pin is driven into place, use air hose and blow all cuttings out of oil hole at center of balance shaft. Also flush out the oil passages with clean solvent.
   d. Using punch and hammer, drive tapered pin into place until large end of pin is just below surface of weight, then use staking tool and stake edge of pin hole to secure the pin.
   e. Plug oil hole in each end of balance weight shaft with 11/32-inch diameter steel ball. Drive each ball in 0.030 inch below surface of shaft.

Balance Shaft Bearings

Inspect balance shaft bearings for scoring. Also use inside micrometer to determine inside diameter at each bearing. If damage to bearings is apparent, or if extensive wear is indi-

TOOL (WEDGE) FOR ASSEMBLING WEIGHTS

A—BALANCE SHAFT AND FRONT WEIGHT IN POSITION

1 Cylinder block
2 Balance weight
3 Bolt with lock
4 Notch in balance shaft
5 Thrust plate
6 Balance shaft gear
7 Thrust plate bolt with lock washer

B—FRONT WEIGHT AND THRUST PLATE SECURED
Balance shaft and weight installation
(© G.M.C.)

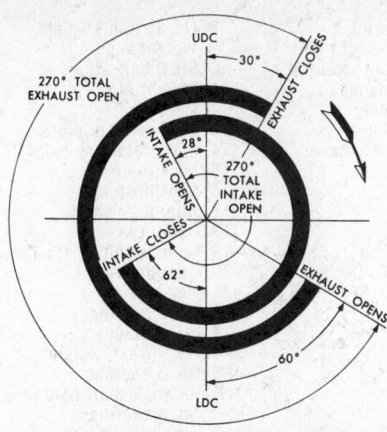

· Valve timing diagram (all Toro flow diesel engines) (Ⓒ G.M.C.)

cated, the balance shaft bearings should be replaced.

Balance Shaft Bearing Replacement

Pre-sized balance shaft bearings are available for service replacement for both the 6-cylinder and 8-cylinder engines. The bearings provide proper fit without reaming. Four bearings are used on 6-cylinder engines and five are used on 8-cylinder engines.

Cylinder block must be stripped and flywheel housing removed to per-mit use of special tools in replacing balance shaft bearings. Crankshaft should also be removed for easy access to balance shaft bearing area.

Since the outside dimension of each bearing is different than other bearings, the balance shaft bearings will fit only their respective bores. Smallest diameter bore is at rear and bore sizes increase progressively toward front. Bearings have a chamfer at outer edge to facilitate installation.

Be sure all oil supply passages in cylinder block are clean before installing new balance shaft bearings. Use suitable tool to provide smooth chamfer at forward edge of each balance shaft bearing bore. Dip bearing assemblies in light engine oil prior to installation.

Notice that two of the bearings have oil holes and oil grooves. When installing these bearings, *an oil hole must be aligned with hole in block.*

IMPORTANT: Be sure to remove from cylinder block any shavings or metal particles which have resulted from the bearing installation procedure. Also recheck each bearing position to see that split in bearing is at top of bore and that oil hole in intermediate bearings registers with oil hole in cylinder block. Plug should be driven in slightly below surface of block.

Piston, Rod, and Bearing Cleaning and Inspection

Cleaning

Immerse all parts in dry-cleaning solvent to loosen and remove accumulated oil, sludge, or other deposits. Remove piston rings and tag rings to identify them with respective piston. If there is a possibility that same rings will be installed at assembly, mark the top of top ring and oil ring so they can be reinstalled with same side upward. Clean ring grooves with groove cleaner. Use suitable tool to clean oil drain holes in oil ring groove.

Checking Piston Ring Fit

After thoroughly cleaning piston rings and ring grooves in pistons, try fit of piston rings in respective cylinder bore.

1. Place piston rings in cylinder one at a time and push ring at least half-way down cylinder bore, using head of piston to position ring squarely in bore. Measure ring gap with feeler gauge. If gap is too small, remove ring and dress off ends of ring with fine cut file or other tool designed for this purpose.

2. If piston ring gap is within specified dimensions, as determined in step 1. above, the rings should

1 Drain plug and gasket	42 Camshaft bearing hole plug
2 Oil pan	43 Camshaft rear (no. 5) bearing
3 Pan gasket	44 Camshaft no. 4 bearing
4 Oil inlet screen and pipe assembly	45 Camshaft no. 3 bearing
5 Pipe flange bolt	46 Camshaft no. 2 bearing
6 Gasket	47 Camshaft no. 1 bearing
7 Engine oil pump	48 Camshaft gear key
8 Pump to block gasket	49 Spacer
9 Plug	50 Camshaft thrust plate
10 Bearing cap rear bolt	51 Camshaft gear
11 Rear bearing cap side bolt	52 Fuel injection pump gear
12 Cap side bolts (except rear)	53 Washer
13 Bearing cap bolt and washer	54 Pump gear bolt
14 No. 1 bearing cap	55 Spacing washer
15 No. 2 bearing cap	56 Fuel injection pump drive gear
16 No. 3 bearing cap	57 Gear retainer washer
17 No. 4 bearing cap	58 Self-locking bolt
18 Rear bearing cap side seal	59 Idler gear shaft
19 Rear bearing cap	60 Balance shaft idler gear
20 Crankshaft no. 1, 2, 3, and 5 lower bearing	61 Gear retainer plate
21 Crankshaft no. 1, 2, 3, and 5 upper bearing	62 Lock plates
22 No. 4 lower bearing	63 Plate retaining bolts
23 Crankshaft	64 Balance shaft front (no. 1) weight
24 No. 4 upper bearing	65 Balance shaft gear
25 Balance shaft	66 Thrust plate
26 Crankshaft rear bearing seal	67 Weight key
27 Flywheel bolts	68 Gear key
28 Bolt lock plates	69 Thrust washer
29 Flywheel assembly	70 Spacer
30 Oil pump drive shaft	71 Camshaft idler gear shaft
31 Balance shaft weight bolt	72 Crankshaft gear key
32 Balance shaft front (No. 1) bearing	73 Crankshaft damper key
33 Balance shaft no. 2 weight	74 Camshaft idler gear
34 Balance shaft no. 2 bearing	75 Gear retainer plate
35 Balance shaft no. 3 bearing	76 Balance shaft weight pin
36 Balance shaft no. 3 weight	77 Crankshaft gear
37 Balance shaft no. 4 bearing	78 Oil slinger
38 Balance shaft no. 4 weight	79 Oil seal assembly
39 Balance shaft rear (no. 5) bearing	80 Oil seal ring
40 Hole plug	81 Damper
41 Camshaft	82 Damper cone
	83 Damper retainer
	84 Retainer bolt
	85 Pulley bolt
	86 Crankshaft pulley

Crankshaft, camshaft, balance shaft and associated parts (8 cyl. engine) Ⓒ G.M.C.

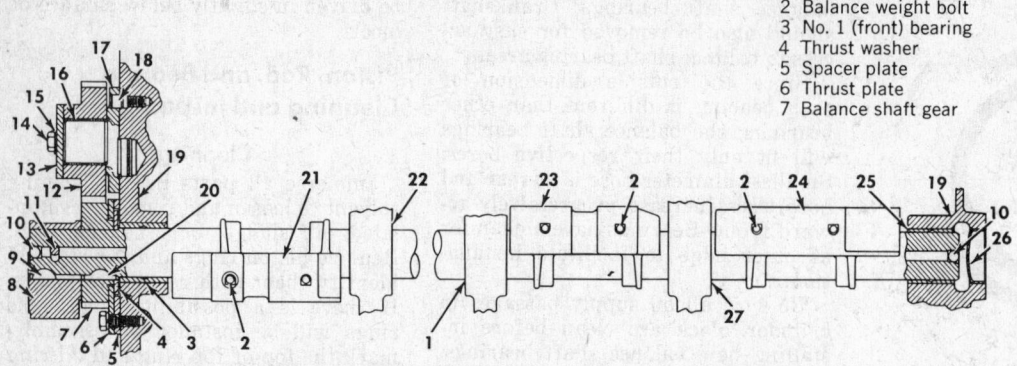

1 No. 3 bearing
2 Balance weight bolt
3 No. 1 (front) bearing
4 Thrust washer
5 Spacer plate
6 Thrust plate
7 Balance shaft gear
8 No. 1 (front) weight
9 Weight key
10 Steel ball (plug)
11 Weight pin
12 Idler gear
13 Gear retainer plate
14 Plate retaining bolts
15 Lock plate
16 Gear bushing
17 Idler gear shaft
18 Dowel pin
19 Cylinder block assembly
20 Gear key
21 No. 2 weight
22 No. 2 bearing
23 No. 3 weight
24 No. 4 (rear) weight
25 No. 5 bearing
26 Balance shaft hole plug
27 No. 4 bearing

Balance shaft, weight and idler gear installed (V8 engines) (© G.M.C.)

be checked for fit in grooves in piston.

a. Using piston ring spreader tool install rings in respective grooves.

b. On all except turbo-charged engines, use feeler gauge to measure clearance between 2nd compression ring and piston ring land.

c. On all engines, use feeler gauge to measure clearance between oil control ring and ring land.

d. At top compression ring and 2nd compresssion ring on turbo-charged engines check fit of each compression ring in groove. Because of normal taper to the keystone and one-half keystone rings, feeler gauges cannot be used to check ring fit.

Clamp piston firmly in horizontal position so piston rings will hang downward with tapered surfaces seated in tapered groove.

Using dial indicator, check position of top ring surface "A" and 2nd compression ring surface "C" in relation to surface on piston ring land "B".

Surface "A" on top compression ring should normally be 0.002 to 0.012 inch below ring land surface, while surface "C" on 2nd compression ring used with turbo-charged engines should normally be 0.000 to 0.014 inch below ring land.

IMPORTANT: When checking position of piston rings, do not use for reference any area on piston other than the surface and do not indicate from surface of steel insert since this is below the surface of piston.

Piston and Pin Inspection

Remove piston pin retainers and push out piston pin.

Measure each piston at 7/8 inch above bottom of skirt at right angle to pin bore and compare dimension with diameter at bottom of cylinder bore in which piston was installed.

Inspect piston ring lands and grooves for wear and other damage. Measure piston pin bores in piston bosses, and compare with diameter of pins. Piston pins should have clearance of 0.005 to 0.007 inch in respective pistons (oversize pins are available).

NOTE: New service pistons are fitted with piston pins.

Connecting Rod and Bearing Inspection

Connecting Rod Inspection

1. With bearing halves removed from rod and cap, install cap and tighten rod bolt nuts to 60

foot-pounds torque. Check bore at lower end for elongation. If bore is out-of-round, the rod should be replaced or reconditioned. Examine surfaces in rod and cap which are contacted by bearing halves. If there is evidence of fretting of metal at these surfaces, the surfaces must be polished to provide a smooth surface for bearing halves to seat against. If bore is round and within specifications, make careful visual inspection of piston pin bushing for evidence of wear. Test fit of piston pin in bushing by measuring pin O.D. and bushing I.D. At 70°F., there should be a clearance of 0.008 to 0.0014 inch.

Connecting Rod Alignment

2. Connecting rod assembly should be placed in suitable aligning fixture to determine if rod is twisted or bent. Piston pin hole must be parallel with axis of crankpin hole in all planes within .002 inch in 7 inches.

Bearings

3. Carefully inspect connecting rod bearing halves for evidence of

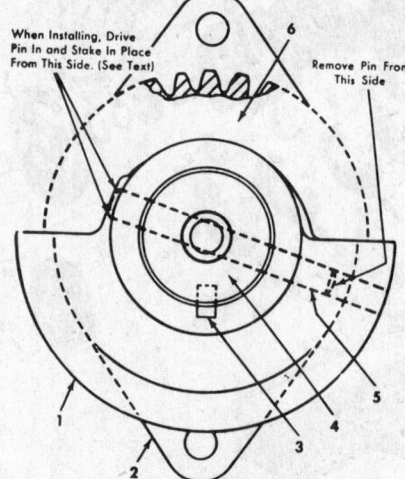

1 No. 1 (front) balance weight
2 Thrust plate
3 Weight key
4 Balance shaft
5 Tapered pin
6 Balance shaft gear

When Installing, Drive Pin In and Stake In Place From This Side. (See Text)

Remove Pin From This Side

Checking clearance between thrust plate and balance shaft gear (V6 engine) (© G.M.C.)

Balance weight shaft weight pin installation (8 cyl. engines) (© G.M.C.)

scores, chipping, or flaking out of bearing metal. If a visual inspection reveals any of these defects, replace with new parts. Measure each bearing half with accurate reading tube micrometer. Take measurement at right angle to split line.

Connecting Rod Repair

If inspection indicates that connecting rod piston pin bushing requires replacement, install new bushing which is furnished for service as follows:

1. Remove old bushing, using driver and arbor press.
2. Inspect bore in upper end of connecting rod. Measure bore I.D. which must be 1.802 to 1.803 inches to assure satisfactory fit of new bushing. Clean oil hole at top of rod.
3. Support connecting rod and press new bushing into place, using care to align oil hole in bushing with hole in top of connecting rod.

4. Diamond bore and hone bushing to provide 0.008 to 0.0014 inch clearance of pin in bushing at normal room temperature of 70°F. Suitable fixture should be used to assure finished hole alignment.
5. Dress bushing edges to remove any portion of bushing extending beyond edge of rod. Be sure oil passage in top of rod is free from cuttings.
6. With connecting rod alignment fixture, check connecting rod for bent or twisted condition.

Piston and Ring Fitting
General

Pistons must be fitted to cylinder bores by measuring to assure correct clearance between piston skirt and cylinder wall. Pistons and piston rings are available in oversize dimensions. The pistons must be assembled on respective connecting rods and installed in original positions in block. Pistons should be stamped with corresponding cylinder number. Number

1 cylinder is at front of left bank and cylinder Number 2 is at front of right bank, thus left bank cylinders are numbered from front to rear as shown for 6-cylinder engine.

Fitting Pistons in New or Reconditioned Cylinder Block

1. When necessary to select replacement pistons for new block or for original bore in block which does not have a ring-travel ridge, refer to selection size number stamped on top of block adjacent cylinder bore. number is used to indicate bore size to permit selection of standard size piston for proper fit. Selection numbers range from 1 through 7 for production. For service use, three standard size pistons are available: Size 1 for use in cylinders marked either 1 or 2; size 3 for cylinders marked 3 or 4; and size 5 for cylinders marked 5 or 6; and size 7 for cylinders marked 7.

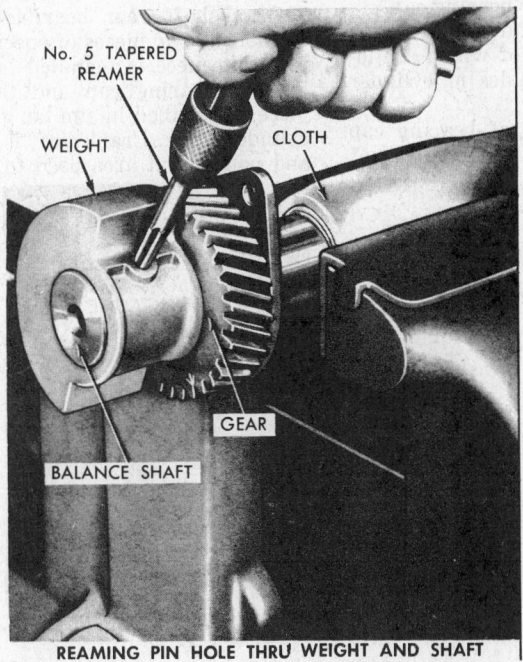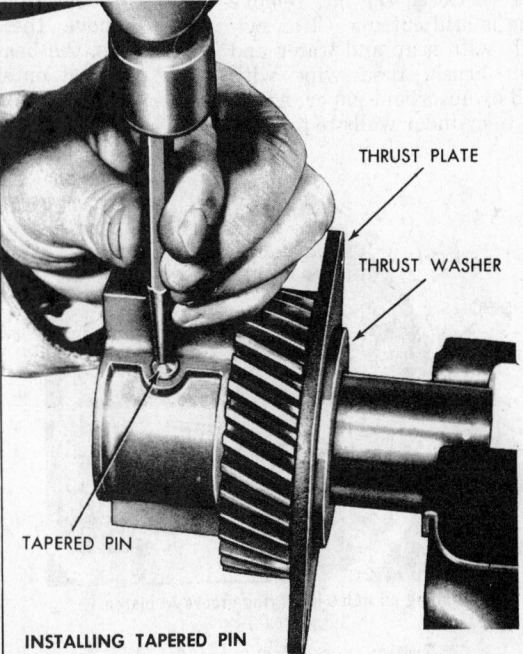

REAMING PIN HOLE THRU WEIGHT AND SHAFT   INSTALLING TAPERED PIN

Installing balance shaft weight pin (8 cyl. engines) (© G.M.C.)

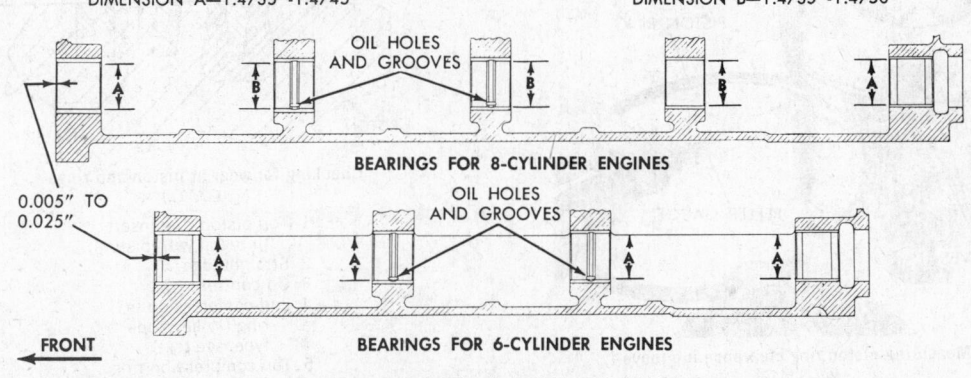

DIMENSION A—1.4735"-1.4745"   DIMENSION B—1.4735"-1.4750"

BEARINGS FOR 8-CYLINDER ENGINES

BEARINGS FOR 6-CYLINDER ENGINES

Balance shaft bearing installation (© G.M.C.)

2. Oversize pistons are available and do not have selection numbers. Cylinders must be rebored and honed to proper dimensions to provide correct fit when installing oversize pistons. CAUTION: Whenever possible, remove crankshaft when reboring or honing cylinder bores. If crankshaft is not removed, it must be covered to keep abrasive away from crankshaft bearings. Wash cylinder bores after honing and wipe dry before checking piston fit.

3. As new piston is fitted to each cylinder, stamp cylinder number on top of piston to assure installation of pistons on corresponding connecting rod and in proper cylinder. Refer to figure for cylinder numbering arrangement.

Preparing Cylinder Bores When Re-Ringing Pistons

Ridge must be removed completely. Use polishing hone to break glaze. At end of honing operation allow hone to work free in bore. Do not remove while hone is still cutting. Clean cylinder walls with soap and water and stiff scrub brush, then wipe with clean cloth or absorbent paper. Apply engine oil to cylinder walls to prevent rusting.

Assembly of Connecting Rod and Piston

1. If original pistons and connecting rods are to be installed, refer to marks (numbers) on connecting rods and assemble rods to pistons from which they were removed. When new pistons are installed, assemble pistons on rods bearing the numbers of cylinders for which pistons were fitted.

2. Apply engine oil on piston pin, rod bushing and pin bosses; then hold rod between pin bosses in piston and push piston pin through piston and connecting rod. Always use a new retainer ring in groove at each end of piston.

3. Check alignment of piston and connecting rod on alignment fixture.

Removing Crankshaft and Bearings

1. Remove plugs which seal rear bolt holes in rear bearing cap.
2. Remove the bearing cap side bolts, the heads of which are located at outer sides of cylinder block.
3. Remove balance of bearing cap bolts and washers. Then at

thrust bearing cap, use bearing cap remover tool and slide hammer to remove the bearing cap. This must be pulled straight out of block to avoid bending or loosening dowel pins as well as to prevent damage to thrust faces on main bearing halves.

4. At other bearing caps, if necessary, use a round bar in bolt hole to work cap out of block; or the tool and slide hammer may be employed at all except rear bearing cap.

5. Lift crankshaft straight out of cylinder block assembly to prevent any binding which could damage thrust faces on thrust bearing.

6. Remove bearing upper halves from cylinder block and lower halves from bearing caps. Tag each pair of bearing halves, so they may be reinstalled in original positions.

Crankshaft Rear Bearing Oil Seal

Crankshaft rear bearing oil seal consists of two pieces of special packing. One piece is installed in groove in rear bearing cap and the other piece is installed in similar groove in cylinder block. Crankshaft is knurled and polished at area used to prevent

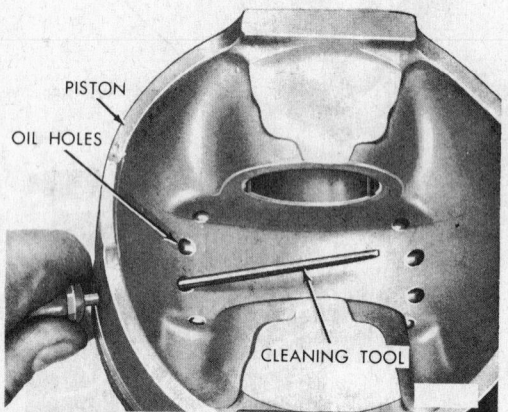

Cleaning oil holes in oil ring groove in piston
(© G.M.C.)

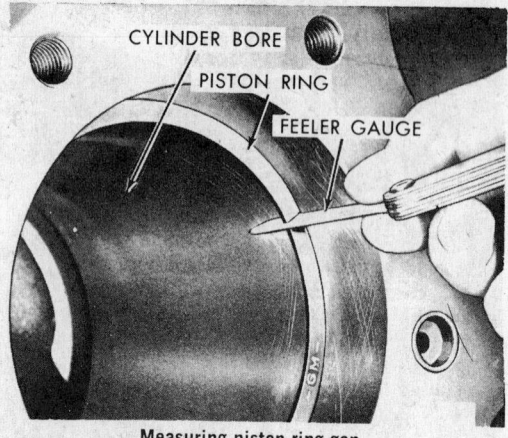

Measuring piston ring gap
(© G.M.C.)

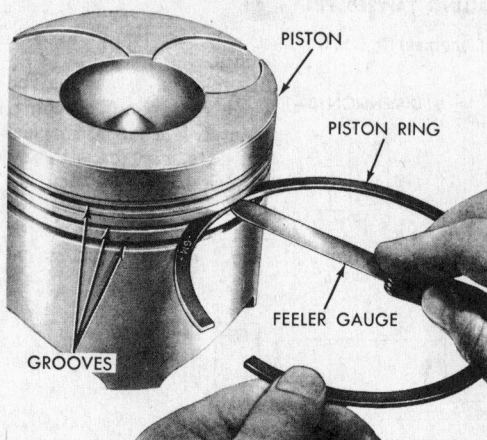

Measuring piston ring clearance in groove
(© G.M.C.)

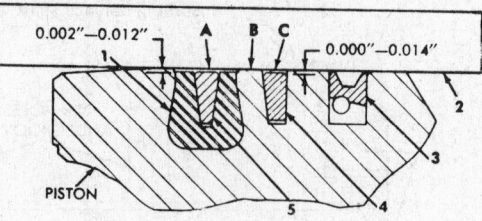

Checking for wear at piston and rings
(© G.M.C.)

1 Top piston ring insert (integral with piston)
2 Straightedge
3 Oil control ring
4 2nd compression ring (½ keystone type, see text)
5 Top compression ring (keystone type)

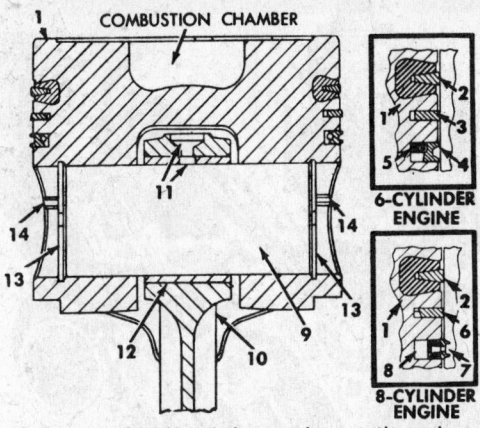

1 COMBUSTION CHAMBER

1 Piston
2 Top compression ring (keystone type)
3 2nd compression ring (tapered face)
4 Oil control ring
5 Expander spring
6 2nd compression ring (taper face, reverse twist)
7 Oil control ring
8 Expander
9 Piston pin
10 Connecting rod
11 Oil hole in rod and bushing
12 Bushing
13 Piston pin retainer
14 Oil groove

6-CYLINDER ENGINE

8-CYLINDER ENGINE

Cross section view of piston and connecting rod
(© G.M.C.)

oil leakage at sides of crankshaft rear bearing cap and plugs are used to seal rear bolt holes.

Seal Installation

1. Apply cement in seal groove in cylinder block and in crankshaft rear bearing cap.
2. Position seal in groove in block and pack seal firmly into place in groove.
3. In similar manner, install seal in bearing cap.
4. Using a sharp knife; cut off both ends of each seal which project out of grooves.

Installing Crankshaft and Bearings

Thoroughly clean to remove all cuttings and abrasive material.
1. Place crankshaft bearing upper halves in respective bearing bores in crankcase, then apply engine oil on each bearing. Apply sealer on cylinder block area contacted by crankshaft rear bearing cap.
2. Lower crankshaft into place in crankcase, and with bearing lower halves in place, position all crankshaft bearing caps. Each cap has the word "FRONT" cast at front edge of cap and caps are numbered from front to rear with numerals 1, 2, 3, etc.
3. Lubricate all bearing cap bolts and bolt washers. Insert the ⅝-11 bolts with washers into holes in bearing caps and start bolts into threads in cylinder block, but do not tighten bolts until side (½-13) bolts have been started into threads in bearing caps.
4. Insert the side bolts from outer sides of crankcase and start each bolt into threads in bearing caps. Insert two ½-13 rear bearing cap (vertical) bolts and start threads into block.
5. Tighten the ⅝-11 bearing cap-to-cylinder block bolts to 170 to 180 foot-pounds.

6. Tighten all ½-13 bolts to 90 to 100 foot-pounds.
7. Apply Diesel fuel oil on the two bearing cap side seals, then insert a seal in groove at each side of rear bearing cap chamfered end first. Chamfer must be toward outer side of crankcase to mate with radius at broached bearing cap seat in block. The narrow edge of side seals must mate with narrow side of grooves in cap. *NOTE: When side seals are fully inserted exposed end will protrude approximately 0.045 inch beyond surface of bearing cap. Do not cut off side seals.*
8. Install two bolt hole plugs in bearing cap to seal the holes and prevent oil leakage.
9. Crankshaft must turn freely with all cap bolts tightened.

Camshaft, Gear and Thrust Plate

Camshaft Removal

Remove thrust plate attaching bolts and lock washers at front end of camshaft. Turn gear as necessary to align holes in gear with bolts.

Carefully pull camshaft assembly forward to remove from cylinder block. Do not damage bearings.
1. Check camshaft for general condition. Bearing journals should not be scored or burred.
2. Support camshaft at end journals, then with dial indicator, check camshaft for straightness.
3. Inspect thrust plate for wear at thrust surfaces. Replace plate and spacer if worn. *NOTE: If the camshaft gear has not been removed from camshaft, the extent of wear on thrust plate may be determined with feeler gauge by measuring space between thrust plate and shoulder on camshaft. Correct clearance is 0.0045 to 0.0085 inch.*
4. Inspect camshaft gear for wear and damage at teeth. If gear teeth are not in good condition: gear must be replaced.

Camshaft Gear Replacement

1. Remove bolt, retainer, and spacer washer from camshaft.
2. Position camshaft and gear assembly in arbor press, and press the camshaft out of gear. A bolt tightened ino tapped hole in camshaft may be used to apply pressure to camshaft.
3. Remove key from slot in camshaft, then remove spacer and thrust plate.
4. Position spacer on camshaft with chamfered side toward camshaft shoulder. Place thrust plate over spacer and install key in keyway.
5. Position camshaft in arbor press with support plate clamped below front journal. Place gear at camshaft with the keyway in gear aligned with key in cam-

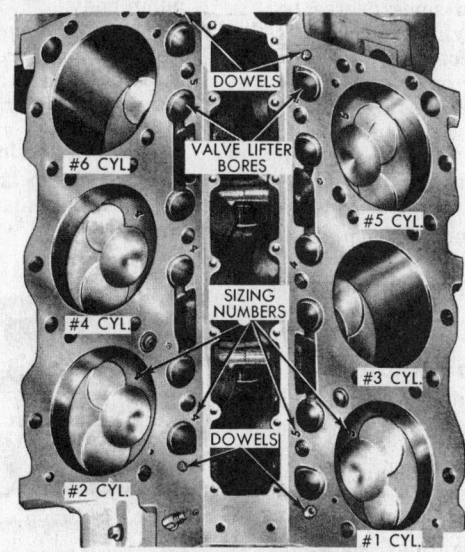

DOWELS
#6 CYL.
VALVE LIFTER BORES
#5 CYL.
#4 CYL.
SIZING NUMBERS
#3 CYL.
DOWELS
#2 CYL.
#1 CYL.

Cylinder numbering arrangement and size selection numbers (6 cylinder engine)
(© G.M.C.)

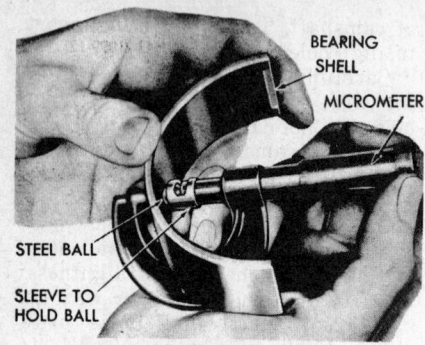

Measuring bearing thickness
(© G.M.C.)

shaft. Use hollow tubing as shown to apply pressure on gear hub, and press gear onto camshaft until hub bottoms solidly at spacer.

6. Install the spacer washer at camshaft gear hub, then install thick retaining washer and self-blocking bolt. Final tightening of bolt may be deferred until camshaft and gear train are installed.

Camshaft Bearings

Inspect each camshaft bearing for scoring, wear, or looseness.

Measure I.D. of camshaft bearings and compare with O.D. of corresponding camshaft journal. If excessive wear is indicated, replace camshaft bearings.

Pre-sized camshaft bearings are available for service replacement for all engines.

Although these bearings provide proper fit without reaming no attempt should be made to replace them without first obtaining the required special tool set.

All camshaft bearings can be pulled forward through the block. It is recommended that all bearings be removed before installing any new camshaft bearings. Replacement procedures vary to some degree between the 6 and 8 cylinder engines.

NOTE: That when installing bearings, split must be on top and chamfered end must be toward the rear. This will properly position oil holes.

IMPORTANT: Make sure all shavings and metal particles which may have resulted from bearing installation, are removed from block. Also recheck each bearing position to insure the split is properly located at top.

Engine Oil Pump and Drive

The engine oil pump assembly is located in the engine crankcase. The pump assembly is bolted to the cylinder block and driven from a gear on the rear portion of the camshaft. Oil intake is through a screen at bottom of oil pan. Illustrations show installed view of oil pump and inlet screen assembly.

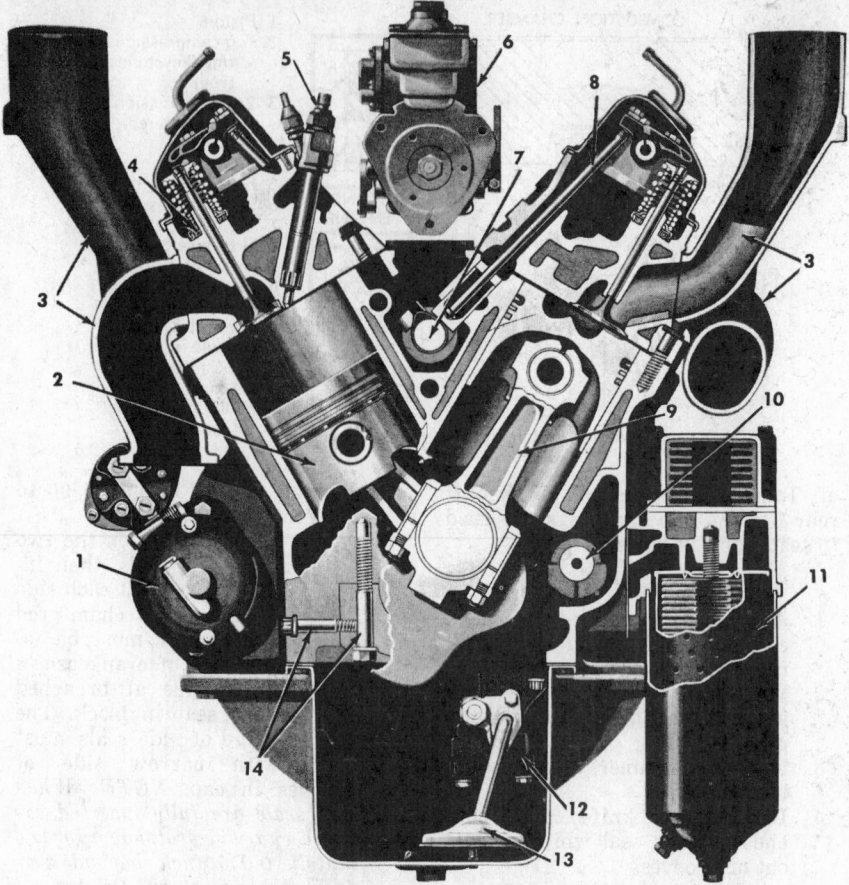

Sectional view of Toro flow diesel engine (© G.M.C.)

1 Starter	9 Connecting rod assy.
2 Piston	10 Balance shaft and weight
3 Manifolds	assy.
4 Valve rotator	11 Oil filter assy.
5 Fuel injection nozzle assy.	12 Engine oil pump assy.
6 Fuel injection pump assy.	13 Oil inlet screen and tube
7 Camshaft	assy.
8 Push rod	14 Main bearing cap bolts

Overhauling Oil Pump

Disassembly

1. Remove spring and valve from pump body.
2. Remove pump cover from body.
3. Remove inner rotor and shaft assembly, then remove outer rotor from cavity in body.

Cleaning and Inspection

1. Be sure screen and suction pipe assembly is clean.

2. Inspect ends of drive shaft which engage drive sockets in oil pump shafts. Replace drive shaft if excessively worn or damaged.
3. Check fit of inner rotor shaft in body. If excessive wear is indicated at either the shaft or body, replace worn parts.
4. Place the outer rotor in body and insert inner rotor and shaft assembly in operating position. Check clearance between outer rotor and body and clearance be-

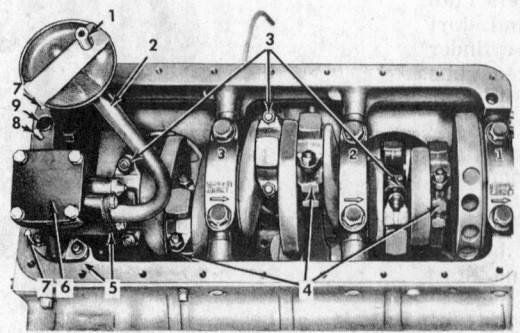

Crankcase ready for Oil pan installation
(© G.M.C.)

1 Damper (in screen support)
2 Oil inlet screen and pipe assy.
3 Rod identification numbers (2, 4 and 6)
4 Rod identification numbers (1, 3 and 5)
5 Gaskets
6 Engine oil pump assy.
7 Bearing cap side seal (2 used)
8 Oil pan stud (1 used)
9 Hole plug (2 used)

tween edge of inner rotor lobe and outer rotor with feeler gauge. If clearances are greater than specified in "Specifications," replace parts as necessary. Inner rotor and shaft are furnished for service as an assembly. With rotors in pump body, place straightedge across rotors and body, then with feeler gauge, measure clearance between rotors and straightedge. Clearance should be 0.0011 to 0.0039 inch.

5. Use straightedge to check rotor side of pump cover for flatness. Also look for grooves and other evidence of wear on pump cover. Replace cover if not in good condition.

6. Inspect relief valve and valve spring. On 6-cylinder engines without turbo-charger spring should have 14.7 to 16.3 pounds pressure when compressed to 1.69 inches. Spring free length should be 2.72 inches.

On turbo-charged 6-cylinder engines and 8-cylinder engines, free length of relief valve spring is 2.72 inches; and spring pressure is 17.8 to 19.4 pounds when compressed to 1.69 inches.

Assembling Oil Pump

1. Coat all oil pump parts with engine oil, then place outer rotor in pump body. Insert inner rotor shaft through bore in oil pump body and move inner rotor into mesh with outer rotor.

2. Install cover on pump body and retain with four cover screws. Try turning pump shaft which should turn freely without drag through six or more revolutions. If shaft cannot be turned or if binding occurs when shaft is turned, disassemble pump and make necessary corrections.

3. Insert relief valve and spring into body, then install spring cap.

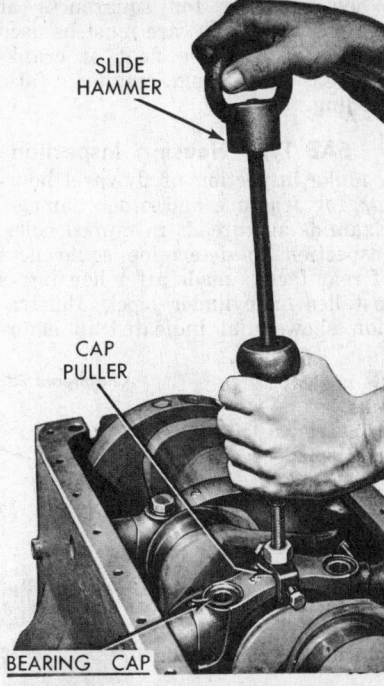

Removing crankshaft No. 3 bearing cap with special tool
(© G.M.C.)

Oil Pump Drive Shaft and Housing

The shaft and housing which drives the engine oil pump are installed at rear of cylinder block. The shaft has an attached gear which meshes with spiral gear teeth on camshaft. When tachometer is used, the cap is omitted and tachometer drive shaft housing is connected to threaded housing.

Inspection

Check oil passages in housing and check clearance between shaft and bushing in housing. If bushing is worn, a new bushing may be installed.

If drive gear or shaft is worn or otherwise damaged, a new shaft and

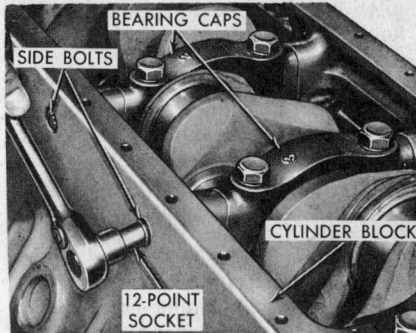

Removing or installing crankshaft rear main bearing side bolts
(© G.M.C.)

gear assembly must be installed at assembly.

Housing Bushing Replacement

1. Drive worn bushing out of housing and discard bushing.
2. Inspect edge of bore in housing and remove any sharp edges.
3. Press new bushing into housing to dimension shown.
4. If necessary to provide free fit of drive gear and shaft assembly in bushing, ream the bushing. Apply engine oil on shaft and bushing and insert shaft into place in housing.

Flywheel Housing and Flywheel Inspection

Cast iron flywheel housing is bolted to rear face of cylinder block and is located by two dowel pins. Some D351 engines have an apron type housing with pressed metal under-pan. These housings have a small pilot opening in which the transmission drive gear retainer fits. Bosses are provided on these housings for attaching engine rear mounting brackets.

Other housings have a large SAE type opening to which the transmission bell housing is attached; inspection to determine squareness of housing rear face and concentricity of

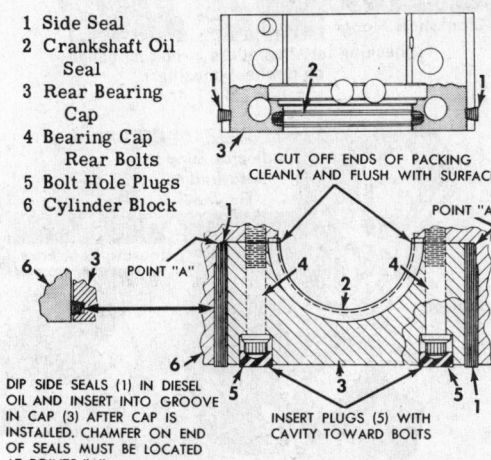

1 Side Seal
2 Crankshaft Oil Seal
3 Rear Bearing Cap
4 Bearing Cap Rear Bolts
5 Bolt Hole Plugs
6 Cylinder Block

CUT OFF ENDS OF PACKING CLEANLY AND FLUSH WITH SURFACE

POINT "A"

DIP SIDE SEALS (1) IN DIESEL OIL AND INSERT INTO GROOVE IN CAP (3) AFTER CAP IS INSTALLED. CHAMFER ON END OF SEALS MUST BE LOCATED AT POINTS "A"

INSERT PLUGS (5) WITH CAVITY TOWARD BOLTS

Cross section at seals and bearing cap bolts
(© G.M.C.)

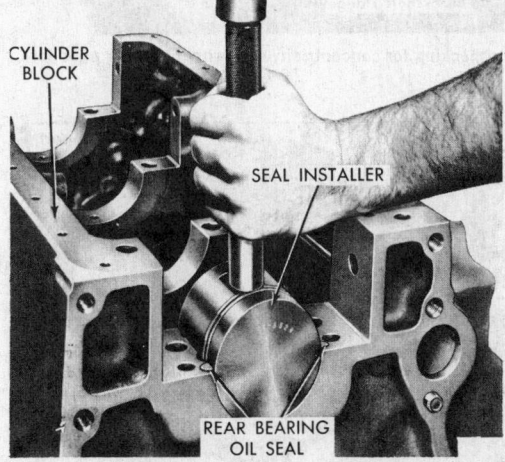

Installing crankshaft rear bearing oil seal with special tool
(© G.M.C.)

pilot opening should be made with housings bolted to cylinder block.

Apron Type Housing Inspection

Check housing for damaged threads, and for cracked or broken flanges. Bolt the housing to cylinderblock and check pilot hole for concentricity and rear face for squareness with dial indicator supported at crankshaft flange or flywheel. Illustrations show typical use of dial indicator for checking pilot hole concentricity and rear face squareness.

When checking for squareness at housing rear face, care must be used to prevent endwise float of crankshaft as this would result in false reading.

SAE Type Housing Inspection

Make inspection of flywheel housing for fracture and other damage. Examine all threads in tapped holes. Inspection to determine squareness of rear face is made after housing is installed on cylinder block. Illustration shows dial indicated to deter-

mine concentricity and squareness at SAE type flywheel housing. Repair of broken or cracked housings is not recommended.

Flywheel Inspection

Inspect flywheel surface which is contacted by clutch facing. Surface must be smooth and should not be grooved or show deep heat checks. On flywheels having notches which drive clutch plate, inspect for wear at notches. Inspect starter ring gear which is shrunk on flywheel.

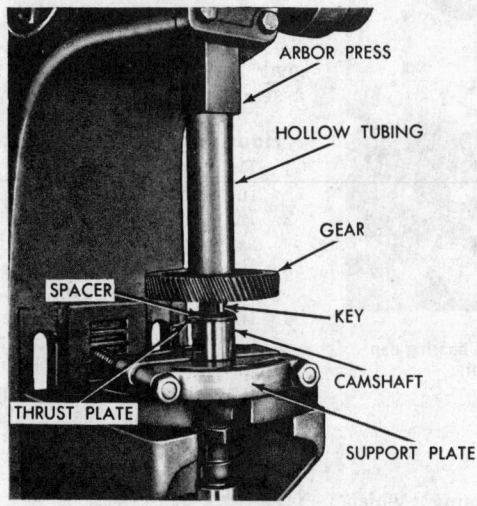

Installing camshaft gear
(© G.M.C.)

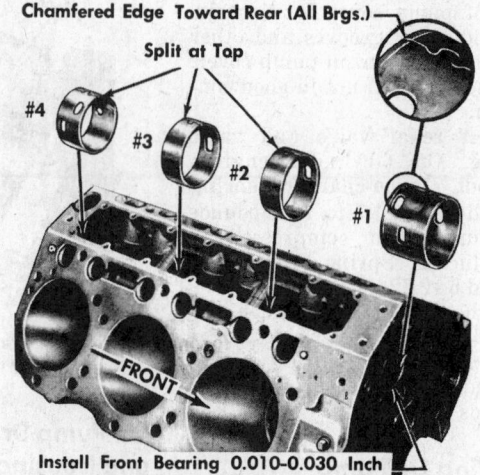

Install Front Bearing 0.010-0.030 Inch Beyond Front Face of Block. Locate Other Bearings Flush With Front of Bore.
Camshaft bearing installation (V6 engine)
(© G.M.C.)

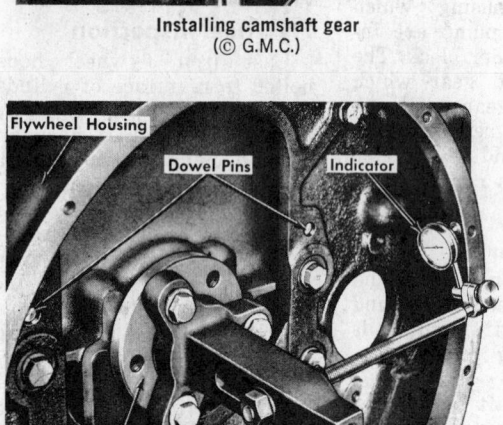

Checking for concentricity of flywheel housing pilot bore
(© G.M.C.)

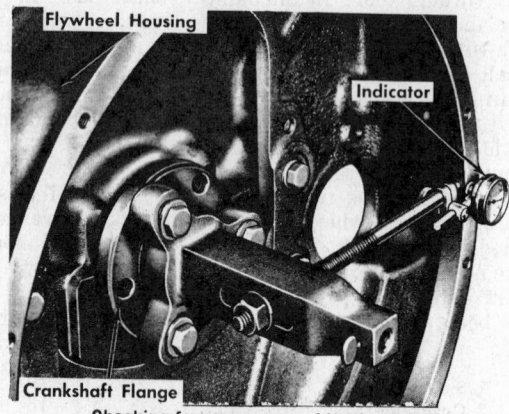

Checking for squareness of bolting flange on flywheel housing
(© G.M.C.)

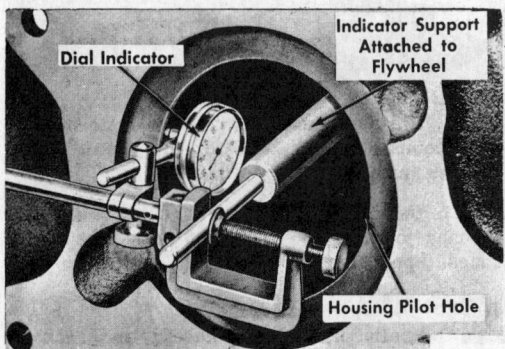

Checking location of pilot hole in apron type housing
(© G.M.C.)

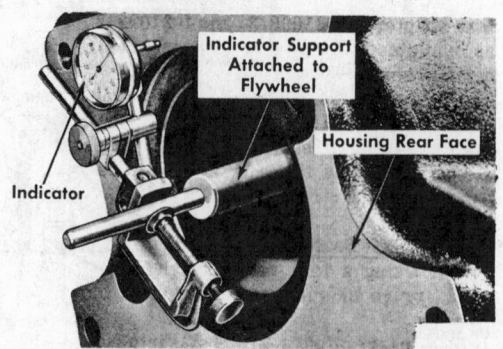

Checking flywheel housing for squareness (apron type)
(© G.M.C.)

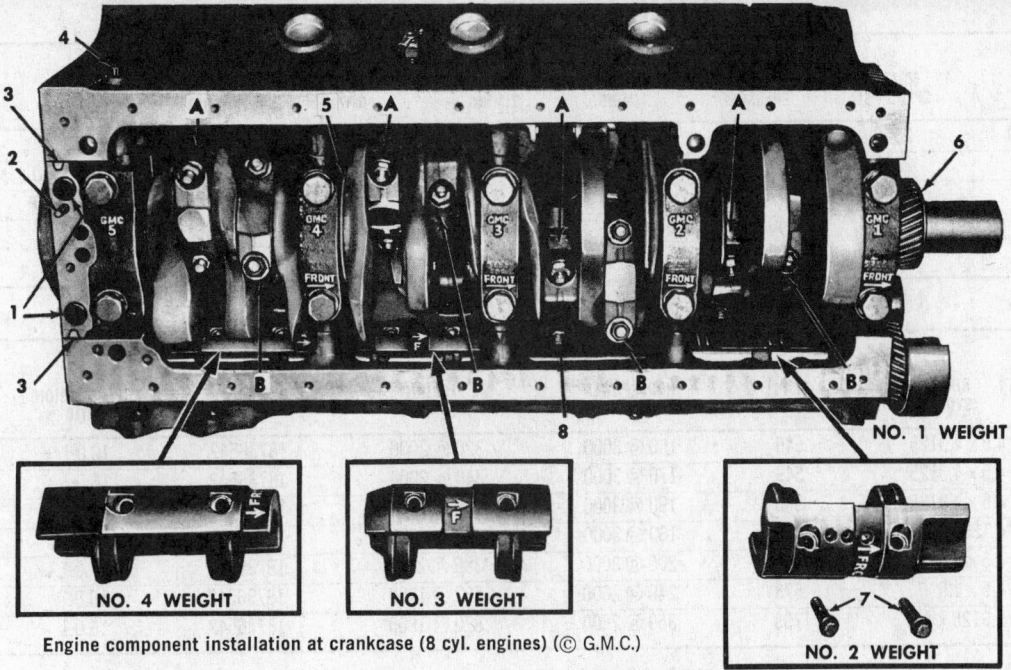

Engine component installation at crankcase (8 cyl. engines) (© G.M.C.)

A - Rod identification numbers (2, 4, 6, 8)
B - Rod identification numbers (1, 3, 5, 7)
1 Hole plug
2 Oil pan stud
3 Rear bearing cap side seal
4 Bearing cap side bolts
5 Crankshaft thrust bearing
6 Oil slinger
7 Weight clamp bolts
8 Balance shaft

Correcting Flywheel Housing Misalignment

If flywheel housing is misaligned as indicated by check, cause of misalignment must be determined if possible and necessary measures must be taken to remedy the condition.

1. Remove housing from cylinder block and look for dirt or burrs on mating surfaces which may prevent housing from seating against cylinder block.
2. If misalignment is due to distortion, or combination of tolerances, try installing another housing from service stock.
3. When a selection of housings is not available, housing bore may be aligned by removing dowel pins from cylinder block, then align housing by shifting hous-

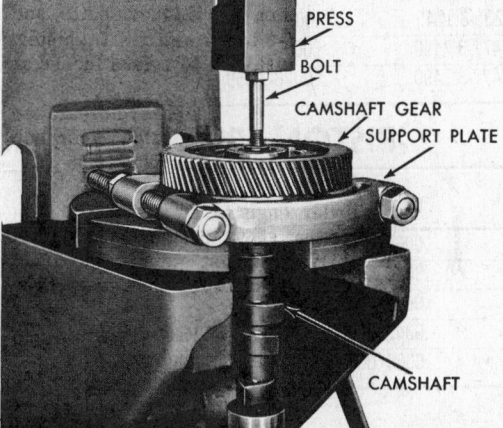

Removing camshaft gear
(© G.M.C.)

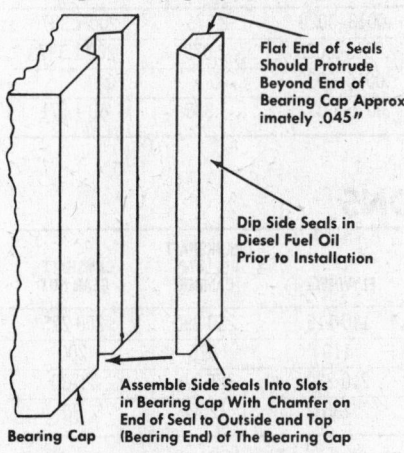

Flat End of Seals Should Protrude Beyond End of Bearing Cap Approximately .045"

Dip Side Seals in Diesel Fuel Oil Prior to Installation

Bearing Cap

Assemble Side Seals Into Slots in Bearing Cap With Chamfer on End of Seal to Outside and Top (Bearing End) of The Bearing Cap

Bearing cap side seal installation
(© G.M.C.)

ing while held loosely in place by mounting bolts. If necessary, enlarge the mounting bolt holes to permit additional movement.

4. When aligned, tighten housing bolts firmly, then ream dowel pin holes to a convenient oversize and install overside dowel pins.

Flywheel Repair

Replacing Starter Gear

1. To remove damaged starter ring gear, heat ring gear with torch, then drive gear off flywheel.
2. Inspect ring gear surface of flywheel. Surface must be free of nicks and burrs. Uniformly heat flywheel ring gear to faint straw color below blue range (400°F.) CAUTION: *Do not heat metal to red heat as metal structure will* *be changed.* Install ring gear on flywheel as soon as possible after ring is heated. Make certain that chamfered edge of ring gear teeth are installed toward engine side of flywheel.

Machining Clutch Surface on Flywheel

When flywheel surface contacted by clutch facing is found to be scored, burned or worn, the flywheel can be restored to serviceable condition by machining to provide a smooth, flat surface. Do not remove more than 1/32-inch of stock in machining. After machining friction surface on recessed type flywheel, remove a corresponding amount of stock from surface to which clutch cover is bolted.

International · Diesel Engines

TUNE-UP SPECIFICATIONS

ENGINE MODEL	NOZZLE OPENS PSI.	COMPRESSION PRESSURE	INJECTION TIMING	VALVE TIMING	VALVE LASH		MAXIMUM SPEED FULL LOAD	LOW IDLE
					INTAKE	EXHAUST		
D-150, 170, 190	2800	375-425	32B	16B	.014C	.016C	3350	600-650
DV-462B	2300	375-425	32B	16B	.014C	.016C	3200-3350	550-600
DV-550B	2300	375-425	34B	16B	.014C	.016C	3200-3350	550-600
DVT-573	3150	400-470	8B	20B	.013C	.025C	2600	575-625
V-800	2300	400-470	22B	30B	.013C	.025C	2600	625-675

GENERAL SPECIFICATIONS

ENGINE MODEL	BORE & STROKE	DISPL. CU. IN.	HORSEPOWER @ RPM	TORQUE @ RPM	FIRING ORDER	COMPRESSION RATIO	OIL CAPACITY W/FILTERS (QTS)
D-150	4.5 x 4.3125	549	150 @ 3000	320 @ 2000	18736542	16.0:1	14
D-170	4.5 x 4.3125	549	170 @ 3000	340 @ 2000	18736542	16.0:1	14
D-190	4.5 x 4.3125	549	190 @ 3000	360 @ 2000	18736542	16.0:1	14
DV-462B	4.125 x 4.3125	461	160 @ 3000	307 @ 2000	18736542	17.0:1	16
DV-550B	4.5 x 4.3125	549	200 @ 3000	389 @ 2000	18736542	17.0:1	16
DVT-573	4.5 x 4.5	573	240 @ 2600	552 @ 1900	18736542	16.0:1	25
V-800	5.3125 x 4.5	798	350 @ 2600	820 @ 1800	18736542	16.0:1	38

CRANKSHAFT AND BEARING SPECIFICATIONS

	MAIN BRG. JRL. DIAMETER	MAIN BRG. CLEARANCE	THRUST ON	SHAFT END PLAY	ROD BRG. DIAMETER	ROD BRG. CLEARANCE
D-150, 170, 190; DV-462B, 550B	3.1235-3.1245	.0013-.0043	Inter.	.004-.010	2.9455-2.9460	.0019-.0044
DVT-573	3.7477-3.7490	.0026-.0059	Rear	.003-.011	3.2000-3.2010	.0018-.0051
V-800	3.7477-3.7490	.0026-.0059	Rear	.003-.011	3.2000-3.2010	.0026-.0056

PISTON, PIN AND RING SPECIFICATIONS

	PISTON CLEARANCE	PISTON PIN DIAMETER	PISTON PIN CLEARANCE	RING END GAP	
				COMPRESSION	OIL
D-150, 170, 190	.0060-.0070	1.4998-1.5000	.0009-.0013	.013-.023	.013-.028
DV-462B	.0055-.0065	1.4998-1.5000	.0005-.0009	.013-.023	.013-.028
DV-550B	.0060-.0070	1.4998-1.5000	.0002-.0006	.013-.023	.013-.028
DVT-573	.0042-.0060	1.6248-1.6250	.0001-.0003	.013-.023	.010-.030
V-800	.0060-.0090	2.0000-2.0002	.0001-.0005	.018-.028	.008-.018

VALVE SPECIFICATIONS

	FACE ANGLE	SEAT ANGLE	STEM DIAMETER		STEM-GUIDE CLEARANCE		VALVE SPRINGS FREE LENGTH (IN.)	PRESSURE LB. @ IN.
			INTAKE	EXHAUST	INTAKE	EXHAUST		
D-150, 170, 190	45	45	.3725-.3730	.3720-.3725	.0008-.0023	.0013-.0028	2.075	200 1.397
DV-462B, 550B	45	45	.3725-.3730	.3720-.3725	.0008-.0023	.0013-.0028	2.075	200 1.397
DVT-573	45	45	.4348-.4355	.4348-.4355	.0015-.0032	.0015-.0032	①	②
V-800	30	45	.3718-.3725	.3718-.3725	.0015-.0032	.0015-.0032	1.835	60 1.571

① Outer: 2.5625; inner: 2.28125   ② Outer: 45 @ 2.089; inner: 25 @ 1.964

ENGINE TORQUE SPECIFICATIONS

	CYL. HEAD	MAIN BRG.	CONN. ROD	NOZZLE CLAMPS	CAMSHAFT FLANGE	FLYWHEEL	CRANKSHAFT PULLEY/DAMPER	CAMSHAFT GEAR NUT
D-150, 170, 190	105-110	110-115①	55②	14-16	40-50	110-115	260-290	200-225
DV-462B, 550B	110	130①	55	15	40	110	325	200
DVT-573	300-320	275-300	100-110	8-12	25-30	210-230	325-375	25-30
V-800	220	390③	35	35	30	235	425④	30

① Tie bolts: 50 ft lb   ③ Cross bolts: 160 ft lb. to be torqued after cap bolts
② Plus 1/6 turn more   ④ Gear nut

The International Harvester DV-8 Series Diesels are all V-8, 90 degree, direct injection, open chamber engines with 5 main bearing crankshafts. All engines use wet cylinder liners, replaceable valve seats, and valve rotators.

A special combustion chamber design is employed in the normally aspirated engines. Virtually the entire air charge is forced into a compact, spherical chamber located in the top of the piston. This raises temperatures and pressures to very high levels before injection, improving combustion efficiency. In all DV-8 engines, a special oil spray nozzle system provides for efficient piston cooling. As long as oil pressure is 55 psi, oil is sprayed onto the bottom surfaces of the pistons. At extremely low rpm, oil pressure is too low to provide piston cooling, thus preventing overcooling at idle.

In all the engines, right and left side cylinder heads are interchangeable. All the engines employ gear driven injector pumps which meter and pressurize the fuel charge. The injectors provide for proper atomization and distribution within the chamber through the configuration of their nozzles and the operation of heavily sprung check valves. An electrically driven pump pulls fuel through the primary fuel filter, and forces it through the secondary filter and to the injector pump.

Engine Fails to Start
1. Tank empty, tank valve closed.
2. Plugged filter or fuel lines.
3. Defective damper valve.
4. Defective transfer pump.
5. Plugged injector line.
6. Defective pump plunger.

Engine Starts Hard
1. Cranking speed too slow (below 250 rpm).
2. Swirl destroyer stuck open (except DVT-573).

3. Accelerator fails to reach full fuel position.
4. Improper fuel.
5. Water in fuel.
6. Improper injection timing.
7. Poor compression.

Erratic Engine Operation
1. Improper fuel.
2. Inadequate transfer pump pressure.
3. Injection lines leaking.
4. Incorrect injector timing.
5. Faulty injector nozzle.
6. Poor compression.

Low Power Without Smoke
1. Accelerator linkage travel restricted.
2. Governor high idle adjustment incorrect.
3. Low transfer pump pressure.
4. Low fuel supply pressure.
5. Improper maximum fuel setting.
6. Injector plungers worn.
7. Exhaust system restricted.

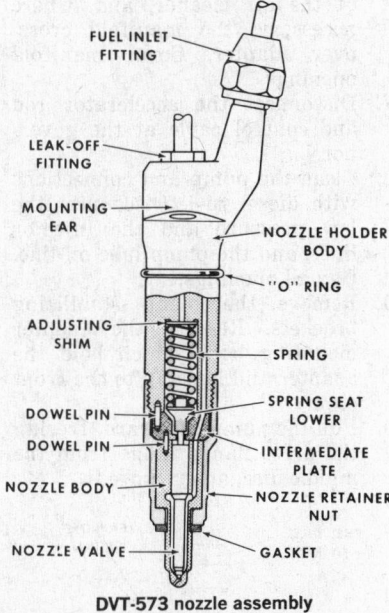

FUEL INLET FITTING
LEAK-OFF FITTING
MOUNTING EAR
NOZZLE HOLDER BODY
"O" RING
ADJUSTING SHIM
SPRING
DOWEL PIN
SPRING SEAT LOWER
DOWEL PIN
INTERMEDIATE PLATE
NOZZLE BODY
NOZZLE RETAINER NUT
NOZZLE VALVE
GASKET

DVT-573 nozzle assembly
(© International Harvester Co.)

8. Swirl destroyer in "on" position (except DVT-573).
9. Air cleaner slightly restricted.
10. Faulty injector nozzles.
11. Improper injection timing.

Engine Smokes, but with No Loss in Power
1. Faulty Nozzles.
2. Faulty maximum fuel setting.

Lube Oil Diluted with Fuel
1. Faulty nozzles.
2. Incorrect delivery valve torque.
3. Faulty pump plunger.
4. Damaged pump barrel seat.
5. Cracked pump housing.

Engine Smokes and Lacks Power
1. Swirl destroyer on (except DVT 573).
2. Air cleaner restricted.
3. Faulty nozzles.
4. Injector pump out of time.
5. Loss of compression in one cylinder.
6. Maximum fuel setting substantially too high.

Engine Tune-up

D-150, 170, 190
1. Position the shut-off control valve in the SHUT-OFF position.
2. Rotate the engine in the direction of normal rotation until #1 cylinder is on the compression stroke. Continue rotating until the 32B mark reaches the pointer.
   NOTE: The engine should be turned manually; if the timing

INDEX POINTER
B.T.D.C. DEGREE SETTING
DAMPER PULLEY

Timing mark and pointer
(© International Harvester Co.)

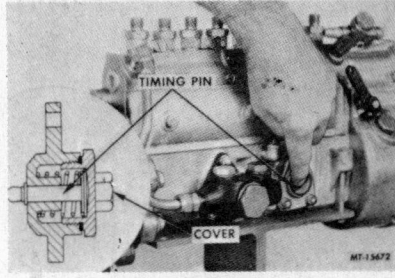

TIMING PIN
COVER

The injection pump timing pin
(© International Harvester Co.)

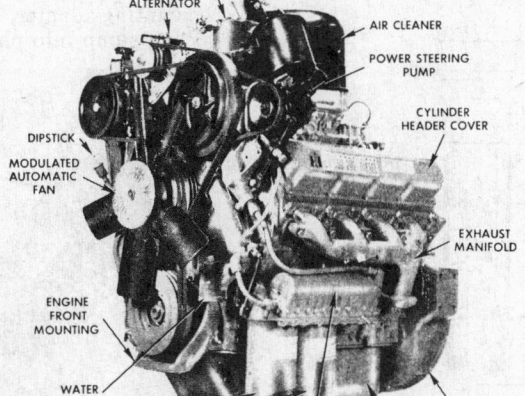

AIR COMPRESSOR
ALTERNATOR
AIR CLEANER
POWER STEERING PUMP
CYLINDER HEADER COVER
DIPSTICK
MODULATED AUTOMATIC FAN
ENGINE FRONT MOUNTING
EXHAUST MANIFOLD
WATER INLET
OIL FILTERS
OIL COOLER
OIL PAN
FLYWHEEL HOUSING

The International DV-8 diesel engine
(© International Harvester Co.)

International • Diesel Engines

mark is passed, back up at least ¼ revolution past the mark and approach it again.

3. Release the shut-off control.
4. Remove the delivery valve, spring and fill piece from the #1 pumping element and install a drip spout. The drip spout can be made from a length of injection pipe and a connector nut.
5. Position the control rack at the load position as follows:
   a. Hold the accelerator lever in the full forward position.
   b. Slowly move the shut-off lever rearward. A distinct click will be heard as the rack moves from the excess fuel to the full load position. On pumps with a torque capsule (D-150, 170) use a spring clip to prevent spring collapse in the torque capsule.
6. Supply fuel to the pump gallery. Fuel should flow from the drip spout at the rate of one drop every three to five seconds. If this rate is not observed, the pump must be removed from the engine and the injection pump drive flange must be repositioned on the engine.

Injection Pump Timing— DV-462B, DV-550B

1. Put engine shutoff control in "shut-off" position.
2. Remove the cap from the timing pin on the left side of the injector pump.
3. Loosen No. 1 injector nozzle in its bore. Rotate the engine. When the nozzle pops up, start watching the timing mark and pointer on the front of the engine. NOTE: Nozzle hold-down bolts must be threaded in at least two turns.
4. Stop rotating the engine when the pointer indicates 34 degrees BTDC on 550B engines, or 32 degrees on 462B engines. If damper is turned beyond the proper mark, turn the engine backwards until it has passed at least ¼ of a turn beyond the mark.

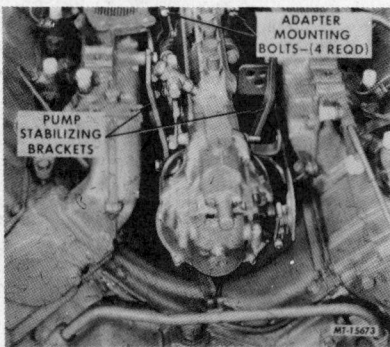

Injection pump with fuel lines removed
(© International Harvester Co.)

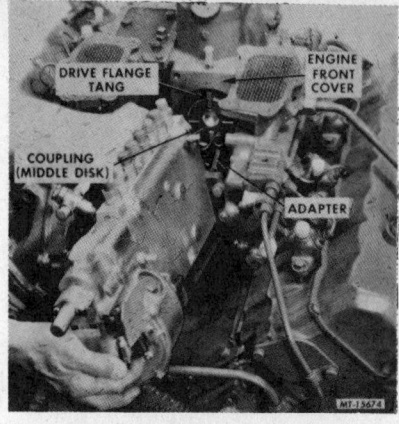

Removing the injection pump
(© International Harvester Co.)

5. Attempt to insert the timing pin into the slot in the pump camshaft. If the pin cannot be inserted, repeat steps 3-5. If the pin can be inserted, adjust timing as described below.
6. Remove upper and lower halves of the air cleaner, and, where necessary, the manifold crossover adapter. Cover manifold openings.
7. Disconnect the accelerator rod and control cable at the governor.
8. Clean the pump and connections with diesel fuel. Disconnect the low pressure and the injector lines, and the pump lube oil line. Cap all openings.
9. Remove the pump stabilizing brackets. Remove the adapter mounting bolts which hold the adapter and housing to the front cover.
10. Pull the pump rearward, freeing the drive flange tangs from the middle disc, and remove it.

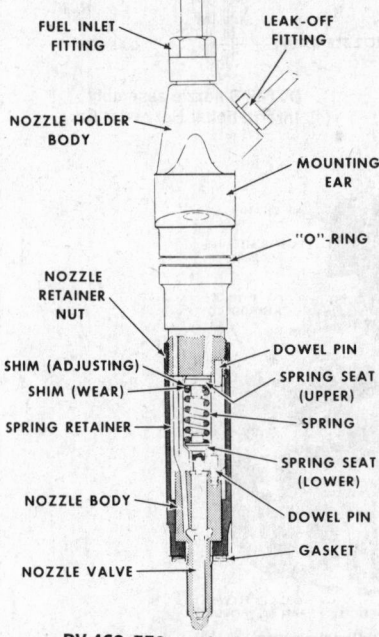

DV-462, 550 nozzle assembly
(© International Harvester Co.)

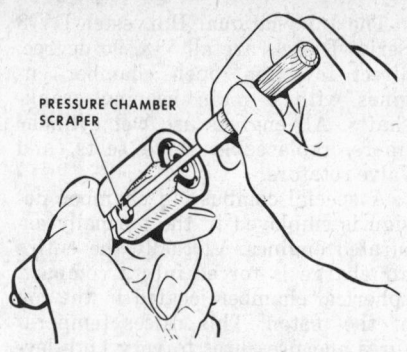

Cleaning the nozzle pressure chamber
(© International Harvester Co.)

11. Install the coupling onto the tangs of the drive shaft, with the blind holes in the center of the coupling facing the pump.
12. Position the drive flange of the pump so that its tangs are horizontal and the timing pin can be engaged with the pump camshaft.
13. Locate the pump on the engine, carefully engaging the drive flange tangs with the coupling. Secure the adapter and pump to the rear of the engine front cover with the mounting bolts.
14. Install the stablilizing brackets.

V-800

1. Remove pump from engine.
2. Remove the cover from the rear of the mounting adapter.
3. Reach through the opening with a deep socket and loosen the pump drive capscrews. Rotate the pump camshaft to bring each of the six capscrews around into view.
4. Remove the timing pin cover.
5. Rotate the pump camshaft while holding in on the notch in the camshaft. Install the timing pin holding tool to keep the pin depressed.
6. Scribe timing marks on gear and align with scribe mark on advance unit.
7. Install and lubricate a new O-ring on the flange of the pump mounting adapter.
8. Push pump into place on engine,

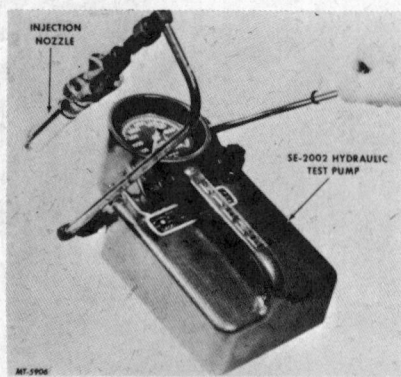

Testing the nozzle
(© International Harvester Co.)

752

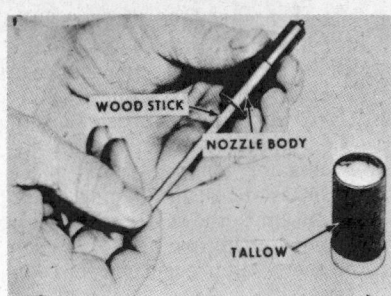

Polishing the valve seat
(© International Harvester Co.)

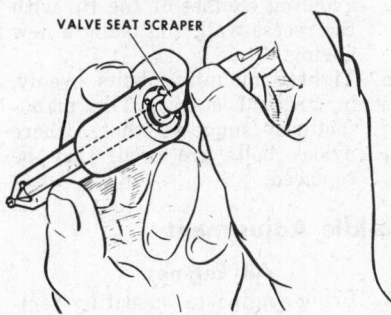

Cleaning the valve seat
(© International Harvester Co.)

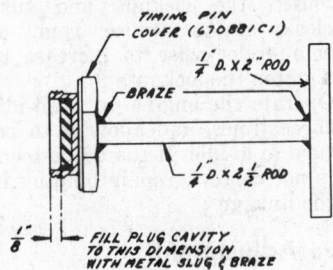

Timing pin tool

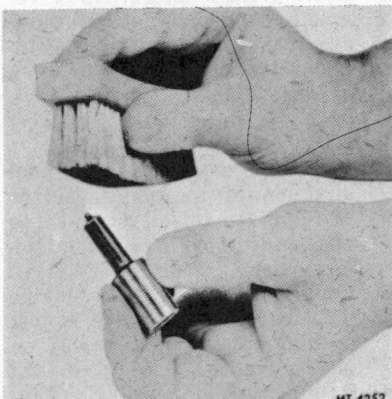

Cleaning the nozzle tip
(© International Harvester Co.)

engaging camshaft gear and compressor idle gear.

9. Visually check, through the adapter opening, that the injection pump gear bolt is near the center of the slotted hole in the advance unit. If not, the pump is out of time one tooth in either direction.

10. Push pump into engagement with the camshaft gear and air

compressor idler gear. Before O-ring enters bore, align the mounting holes in the pump flange with those in the front plate.

NOTE: When the pump is installed, the gear should have rotated far enough clockwise to allow the gear bolts to be positioned in the middle of the slots in the advance unit. If alignment is not correct, remove the pump, rotate the gear one tooth and reinstall.

11. Install all pump mounting bolts.

12. Using a deep socket, reach in through the access hole and tighten the first gear bolt to 50 ft lbs.

13. Remove the timing pin holding tool. Make sure that the timing pin is completely released.

14. Torque the five remaining bolts to 50 ft lbs. each, by rotating the crankshaft through two revolutions, stopping each 120° to tighten the next bolt.

15. Recheck static timing at the vibration damper. It should be 22°BTDC at #1 compression.

16. Install the access cover and gasket and the timing pin cap.

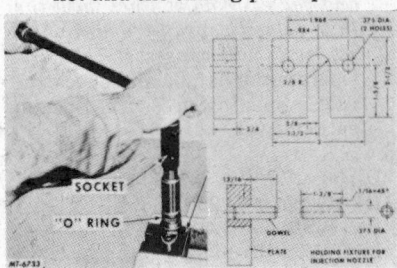

Loosening the nozzle retainer nut
(© International Harvester Co.)

Injection timing marks aligned

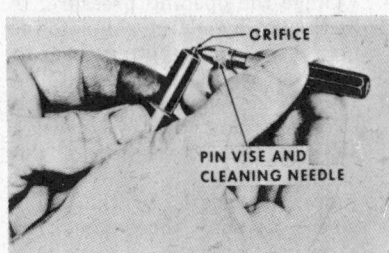

Cleaning the orifice holes
(© International Harvester Co.)

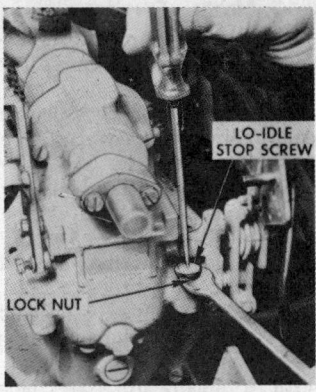

Fast idle adjustment location
(© International Harvester Co.)

Injector Nozzle Service

The injector nozzles direct a metered amount of fuel into the combustion chamber in finely atomized form. The nozzle valve is lifted off its seat by high pressure fuel and is returned to its seat by spring pressure while fuel pressure is still extremely high, thus ensuring the production of a high velocity stream of fine fuel particles of precisely controlled size.

The fuel enters the nozzle holder through the inlet fitting and passes to the pressure chamber just above the nozzle valve seat. When the injector pump plunger has raised the fuel pressure to the specified level, the pressure in the pressure chamber lifts the valve off its seat. The process is reversed as the pump nears the end of its stroke.

Nozzle Troubleshooting

Excessive Leak-off
1. Dirt between pressure face or spring retainer and nozzle holder.
2. Loose retainer nut.
3. Defective nozzle.

Nozzle Blueing (Excessive Heat)
1. Faulty installation or improper torquing of hold-down nuts.
2. Faulty engine cooling system.

Opening Pressure Too High
1. Incorrect shim adjustment.
2. Dirty valve.
3. Sticking or seized valve.
4. Clogged nozzle.

Opening Pressure Too Low
1. Shim adjustment incorrect.
2. Valve spring broken.
3. Worn seat.

Nozzle Drips
1. Carbon deposits.
2. Sticking valve.
3. Faulty nozzle.

Distorted Spray Pattern
1. Carbon deposits on nozzle tip.
2. Partially blocked nozzle hole.
3. Defective nozzle.

Nozzle Service

1. Using kerosene or diesel fuel, clean the fuel line connections very thoroughly.
2. Remove the high pressure and leak-off lines, and cover all openings with caps.
3. Remove the mounting bolts.
4. Pull the nozzles out of the engine, rotating them slightly to break it loose of carbon deposits, if necessary.
5. Cover the openings in the block, and ensure that the nozzle tips are protected.
6. With a clean, lint free cloth, wipe all dirt and carbon from each nozzle assembly.
7. Procure or make a holding fixture similar to that shown in the illustration.
8. Using a 12 point, deep well socket, remove the nozzle retainer nut. Remove the spray nozzle and spring retainer or plate, being careful not to drop any parts. If disassembly is difficult, the assembly may first be soaked in any good carbon solvent.
9. Thoroughly soak all disassembled parts in a pan of solvent.
10. Hold the nozzle valve stem in a revolving chuck, and use mutton tallow on a soft cloth to clean it. An oil soaked piece of wood or a fine wire brush will help in cleaning. *Do not use any hard tools in cleaning.*
11. Clean the inside of the nozzle body, and the valve seat, with special brass scrapers.
12. Using cleaning tallow on a wooden stick, polish the valve seat.
13. Extend the cleaning needle 1/32″ out from the pin vise. Clean the orifice holes. Clean the sac hole (inside tip) with .050″ reamer on 462-550, .047″ on 573-800. For D-150, 170, 190 use Set SE2202.
14. Clean the nozzle tip with a wire brush or soft cloth, but avoid scraping carbon from the surface around the orifices, or damage may result.
15. Rinse all parts of the assembly very thoroughly in clean fuel. Remove all carbon from the retainer nut.
16. Mount the nozzle holder in a fixture and clamp the assembly in a vise with the lapped face of the holder upward.

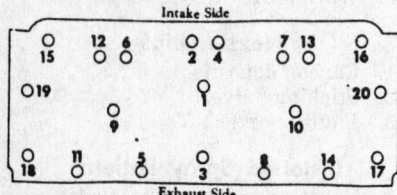

Head bolt tightening sequence DVT-573
(© International Harvester Co.)

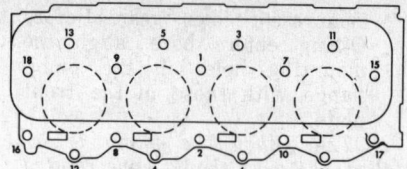

Head bolt tightening sequence for DV-462, 550
(© International Harvester Co.)

17. Install the spring retainer or plate, noting proper location of locating dowels.
18. Install the nozzle retaining nut hand tight, noting that the nozzle is concentric with the holder.
19. Tighten the retainer nut with a 12 point deep socket to 50-60 ft lbs. on 462 and 550 engines, 45-65 ft. lbs. on D-150, 170, 190 engines, 60-70 ft. lbs. on 573 engines, and 35 ft lbs. on 800 engines.
20. Fill a test fixture with fuel oil, and bleed the test pump of air.
21. Connect the nozzle to the test pump and operate the handle several times to remove all air from the whole system. *During this and all procedures involving actual operation of the nozzle and valve, avoid contact with the high velocity spray from the nozzle, as it will penetrate the skin.*
22. Operate the handle and check the nozzle opening pressure. Specifications are:

New
  D-150, 170, 190—2800-2850
  DV-462, 550—2250-2350
  DV-462B, 550B—2225-2375
  DVT-573—3100-3200
Rebuilt
  D-150, 170, 190—2500
  DV-462, 550—2000
  DV-462B, 550B—2150-2375
  DVT-573—2850

The nozzle should operate so as to produce a fine, atomized mist and a high pitched squeak. The atomized streams must not be streaked.

23. Pump up the pressure again to a level just below the pop pressure. Be sure there is no leakage at the the nozzle, spring retainer, and holder. Leakage may be seen at the retainer nut (external) or at the leak-off outlet (internal). If leakage occurs, disassemble and clean again.
24. Change the opening pressure, if necessary, by adding shims to increase it, or subtracting them to decrease it. Place shims at the top of the spring, and ensure that the dark, hardened wear shim is next to the top of the spring.
25. Raise the nozzle to the point where it pops three times to expel air. Wipe the tip to make it perfectly dry. Then, raise pressure to within 500 psi of pop

pressure, and hold it. If the tip remains dry or is not wet enough after five seconds to produce a droplet, it is satisfactory.

26. *Carefully* clean the nozzle bore in the cylinder head, using an appropriately shaped round piece of brass or brush. Pay particular attention to the seating surfaces, as even a tiny particle of carbon on them will prevent proper sealing.
27. Carefully install each nozzle into the bore it was removed from, avoiding contact of the tip with the recess wall, and using a new O-ring.
28. Tighten mounting bolts evenly, in gradual stages. The manufacturer suggests that, where nylock bolts are used, they be replaced.

Lo-Idle Adjustment

All Engines

1. Bring engine to operating temperature. Note tachometer reading. Lo-idle speed is 550-600.
2. If necessary to adjust the lo-idle, loosen the locknut and turn clockwise to reduce rpm, or counterclockwise to increase it. Tighten the locknut.
3. Operate the engine up to hi-idle three times, and allow it to return to lo-idle. If the adjustment is not correct, repair binding in the linkage.

Valve Adjustment

All Engines

1. Remove valve covers. Allow engine to cool down until all parts are at uniform temperature. Intake valves are adjusted to .014″, exhaust to .016″ except on V-800 which is intake .013″, exhaust .025″.
2. Rotate the engine in the normal direction of rotation until the No. 1 intake valve just starts to open. Adjust both valves on No. 6 cylinder.
3. Continue rotating the engine until No. 8 intake valve is just opening, and adjust valves on No. 5 cylinder.
4. Rotate the engine until No. 7 intake valve is just opening, and adjust the valves on No. 4 cylinder.
5. Rotate the engine until No. 3 intake valve is just opening, and adjust the valves on No. 2 cylinder.
6. Rotate the engine until No. 6 intake is just opening, and adjust the valves on No. 1 cylinder.
7. Rotate the engine until No. 5 intake is just opening, and adjust the valves on No. 8 cylinder.
8. Rotate the engine until No. 4 in-

take is just opening, and adjust the valves on No. 7 cylinder.

9. Rotate the engine until No. 2 intake is just opening, and adjust the valves on No. 3 cylinder.

Torquing Cylinder Heads

DV-462B, DV-550B, D-150, 170, 190

1. Torque all bolts in the pattern shown to 30 ft lbs.
2. Torque, in the same pattern, to 60 ft lbs.
3. Torque, in pattern, to 110 ft lbs.

DVT-573

1. Torque all bolts, in the sequence shown, to 100 ft lbs.
2. Torque, in sequence, to 190 ft lbs.
3. Torque, in sequence, to 320 ft lbs.
4. Loosen each bolt one half turn and retorque to 320 ft lbs, one bolt at a time. NOTE: After torquing head bolts, valve adjustment *must* be performed.

V-800

1. Torque bolts in sequence shown to 100 ft lbs.
2. Torque in sequence to 160 ft lbs.
3. Torque in sequence to 185 ft lbs., loosen ¼ turn and torque to 205 ft lbs.
4. After run-in, back off ¼ turn and torque to 220 ft lbs.

V-800 head bolt torque sequence

FUEL SYSTEM

Injector Pump

DV-462B, DV-550B, D-150, 170, 190
Removal

1. Remove upper and lower halves of the air cleaner, and, where necessary, the manifold crossover adapter. Cover manifold openings.
2. Disconnect the accelerator rod and control cable at the governor.
3. Clean the pump and connections with diesel fuel. Disconnect low pressure and injector lines, and the pump lube oil line. Cap all openings.
4. Remove the pump stabilizing brackets. Remove the adapter mounting bolts which hold the adapter and housing to the front cover.
5. Pull the pump rearward, freeing

the drive flange tangs from the middle disc, and remove it.

Inspection

1. Inspect pump mounting flange and the bosses of the rear mounting bracket.
2. Inspect the drive flange for damage, wear, or loose mounting.
3. Torque the securing nut to 75 ft lbs.

Installation

1. Install the coupling onto the tangs of the drive shaft, with the blind hole in the center of the coupling facing the pump.
2. Position the drive flange of the pump so that its tangs are horizontal and the timing pin can be engaged with the pump camshaft.
3. Locate the pump on the engine, carefully engaging the drive flange tangs with the coupling. Secure the adapter and pump to the rear of the engine front cover with the mounting bolts.
4. Install the stabilizer brackets.

DVT-573

Removal

1. Remove air compressor.
2. Remove the front cover plates and gaskets from the heads.
3. Remove all leak-off lines at the injector nozzles. Cap all openings.
4. Remove the injector lines and dampers assembled. Cap all openings.
5. Remove the mounting caps and lockwashers.
6. Lift the injection pump and shaft off the engine. Pull the gasket off the drive housing.

Installation

1. Install a new gasket onto the drive housing.
2. Put the pump in place on the engine.
3. Install the mounting screws and lockwashers.
4. Uncap openings, and install the injector lines and dampers.

5. Uncap and install the leak-off lines.
6. Install the front cover plates and gaskets.
7. Install the air compressor.

V-800

1. Disconnect and remove injection pump oil supply and return lines. Remove oil pressure tube.
2. Remove two accelerator rod-to-pump control bolts. Remove three accelerator rod bracket-to-cylinder head bolts.
3. Remove accelerator rod and bracket.
4. Remove pump bracket-to-engine bolt.
5. Support pump and remove the five attaching bolts.
6. Slide the pump rearward clear of the drive gear.
7. To install, follow instructions under INJECTOR PUMP TIMING.

Fuel System Priming

All Engines

1. Check to make sure that there is plenty of fuel in the tank.
2. Turn the starting switch to "on" position. Turn and hold the primer switch to the "on" position.
3. Open the vent valve on top of the primary filter. When fuel flows without bubbles, close the valve.
4. Open the vent valve on top of the secondary filter. Close the valve after fuel flows out bubble-free.
5. Open the vent valve at the right front of the injection pump. Close the valve when the discharge is free of bubbles.
6. Allow primer switch to return to "off" position.

Electric Fuel Pump Pressure Check

All Engines

1. Install an adapter at the priming vent and run a hose to a special test kit.
2. Operate the engine under load, and read the fuel pressure. It should be 3-5 psi.

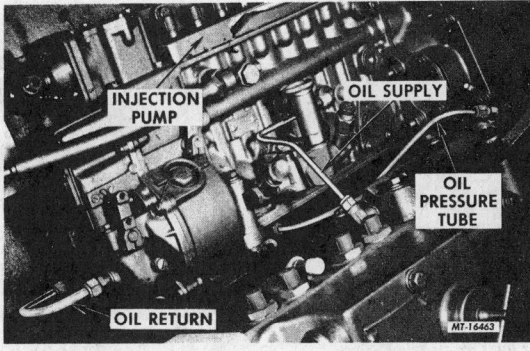

Injection pump oil lines

3. If reading is not to specifications, look for a damaged or plugged supply line or faulty pump.

Transfer Pressure Check

1. Connect an adapter to the priming outlet at the right front of the injector pump. Connect a test gauge, located in the cab, to the adapter.
2. Operate the engine under load, and observe the pressure. It must be 16 psi or more on pumps, and 21 psi or more on C92 pumps.
3. If pressure is low, check for clogged filters. Otherwise, the transfer pump is defective.

ENGINE REMOVAL

DV-462B, DV-550B, D-150, 170, 190

1. Drain the cooling system by opening drain cocks on either side toward the rear of the block, at the rear of the oil cooler, and at the lower left portion of the radiator. Drain the engine oil.
2. Disconnect the battery ground cable. Tape thoroughly the end of the terminal.
3. Remove the secondary fuel filter with a strap wrench. Remove both oil filter assemblies from the cooler.
4. Raise the hood assembly, and block upward with slack in the stop cables.
5. Disconnect the headlight wiring harness from the junction block at right front of the hood. Remove the harness from the clips at the right front fender.
6. Disconnect the hood stop cable yokes at the radiator support.
7. Remove the hood hinge pins, removing the cotter pin and flat washer on either side and then either pin.
8. Remove the hood and fender assembly (which requires two persons). Remove the front bumper.
9. Remove the upper radiator hose. Loosen the lower radiator hose clamp at the engine water inlet and, where heater equipped, remove heater hoses.

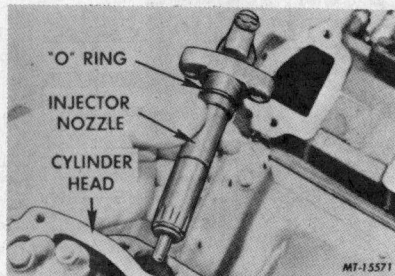

Location of injector nozzle O-ring
(© International Harvester Co.)

Location for checking the fuel transfer pressure
(© International Harvester Co.)

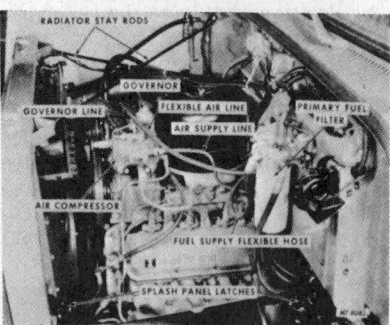

Location of splash panel latches for removal of DV series engines
(© International Harvester Co.)

DVT-573 Cover plates
(© International Harvester Co.)

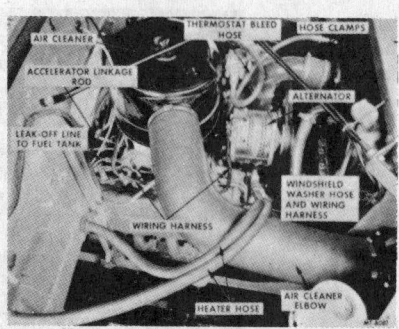

Heater and coolant hose locations on DV series engines
(© International Harvester Co.)

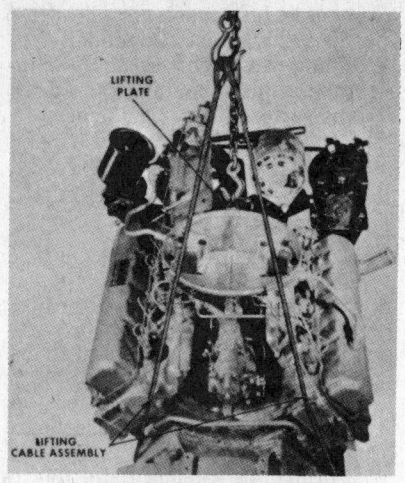

Removal of DV series engines
(© International Harvester Co.)

10. Release both splash panel latches on either side, and remove both panels.
11. Disconnect the shutterstat hose where it connects at the engine water outlet.
12. Disconnect both radiator stay rods at either end.
13. Disconnect the thermostat bleed hose where it enters the upper radiator tank.
14. Disconnect windshield washer hose and wiring harness from the stay rods and radiator support.
15. If the truck is equipped with a shutterstat, disconnect the linkage and spring at the actuating cylinder. Remove the shutter or radiator grille. Remove the radiator shell side panels.
16. Remove the radiator core and tanks.
17. Remove the air cleaner intake elbow and the radiator support assembly.
18. Disconnect the:
    a. fuel supply and return hoses at the rear of the engine.
    b. left and right exhaust pipes at the manifolds (and remove seals).
    c. armature and field wires at the alternator.
    d. temperature indicator wire at the front of the right cylinder head.
    e. oil pressure sending wire.
    f. wiring harness at all clips along the right side of the intake manifold and the rear of the left of the intake manifold.
    g. main air supply line at the compressor
    h. flexible air line at the compressor governor.
    i. throttle and control cable, swirl destroyer cable, and engine shutoff cable at their points of connection on the injection pump.

j. accelerator linkage at the bellcrank and at the injection pump.

k. tachometer cable at the rear of the block.

l. wiring and battery cable at the starter solenoid.

m. alternator mounting strap. Loosen the alternator mounting bolts, remove the belts, and pull the alternator away from the thermostat housing.

n. water line at the thermostat housing and air compressor, and remove it.

19. Remove the:

a. thermostat housing, thermostat, and gasket.

b. governor line from the air compressor and governor.

c. air compressor governor from its mounting bracket.

d. leak-off line and clip from the rear of the right cylinder head.

20. Disconnect both power steering lines at the pump.

21. Install an engine lifting plate to the thermostat mounting surface on the front cover with four ⅜" x 1¼" hex head bolts and flat washers. Bolts should be turned exactly ¼ turn beyond finger tight position.

22. Route a special cable under the flywheel housing and attach both ends to a special lifting hoist. Make sure cable does not contact the timing indicator pin on the flywheel housing.

23. Remove the front mount bolts and self locking nuts. Lower the insulator retainer and the insulator from the mounting.

24. Remove the dust shield from the clutch cross shaft.

25. Disconnect the vertical link on the clutch release linkage.

26. Disconnect the clutch cross shaft linkage and nut on the end of the cross shaft. Remove the bearing retainer caps and pull the shaft outward.

27. Remove the rear engine mounts on either side of the flywheel housing.

28. Remove all (12) bell housing to flywheel housing caps. Make sure the engine ground strap is pulled free when the cap is removed.

29. Pull the engine forward until it clears the transmission main drive gear and clutch driven disc, and then raise it out of the chassis.

DVT-573

1. Drain the cooling system by opening cocks on either side toward the rear of the block, at the rear of the oil cooler, and on the lower radiator tank. Drain the crankcase.

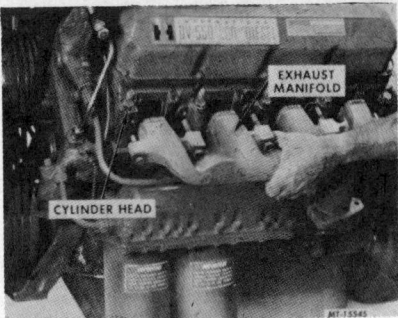

Exhaust manifold removal
(© International Harvester Co.)

2. Disconnect the battery ground cable, and thoroughly tape the end.

3. Disconnect and/or remove the following:

a. the radiator hoses and pipes.

b. all wiring at all connection points on the engine assembly.

c. exhaust pipe at the turbocharger.

d. the air cleaner pipe cleaner pipe at the turbocharger and air cleaner (and remove).

e. throttle linage at the throttle cross-over shaft

f. the throttle return spring at the clip at the left-front of the engine.

g. the fuel line at the final fuel filter.

h. air supply line at the air compressor (and remove).

i. control cable at the other cable unit.

j. heater hoses.

k. engine ground strap at the engine.

l. tachometer cable at the air compressor.

m. power steering lines at the pump.

n. air compressor governor line.

o. fuel leak-off line at the tee at the rear of the front cover plate.

p. transmission remote control shift lever housing.

q. air conditioner compressor lines, *after a complete discharge of the system.*

r. hose at the shutterstat.

s. crankcase breather tube.

t. oil filler tube hose.

4. Securely attach an oversize chain

Location of engine crossover tube clip
(© International Harvester Co.)

at both engine lifting brackets (front and rear).

5. Attach the chain to a hoist and raise just enough to hold engine in position.

6. Remove the front engine mounts and their insulators and retainers.

7. Remove the rear mounts and auxiliary parts.

8. Raise the engine from the chassis, using extreme care to avoid damage to wiring and piping.

V-800

Due to various chassis designs and equipment, disconnect points will vary from vehicle to vehicle.

1. Drain the cooling system and engine oil.

2. Disconnect the battery ground.

3. Remove radiator hoses and pipes.

4. Disconnect all electrical wiring.

5. Disconnect the exhaust pipe at the turbocharger.

6. Remove the air cleaner pipe from the turbocharger and air cleaner.

7. Disconnect the throttle linkage at the throttle shaft and return spring from clip at the left front of the engine.

8. Disconnect the fuel inlet line at the final filter.

9. Remove the air supply line at the compressor.

10. Disconnect the other control cable.

11. Remove the heater hoses.

12. Disconnect the engine ground strap.

13. Disconnect the tachometer cable at the air compressor.

14. Disconnect the power steering line.

15. Remove the air compressor governor line.

16. Disconnect the fuel leak-off line at the front cover plate.

17. Remove the remote control shaft lever housing.

18. Remove the refrigerant compressor lines.

19. Disconnect the shutterstat hose.

20. Remove oil filler tube hose.

21. Rig-up lifting equipment and take up the engine weight.

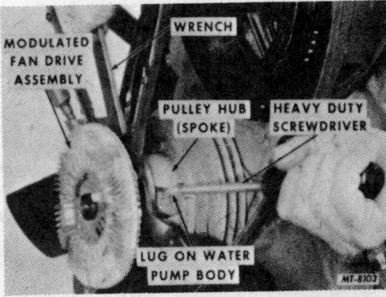

Using a screwdriver as a stop during fan removal
(© International Harvester Co.)

Location of oil cooler lines
(© International Harvester Co.)

22. Remove engine mount bolts, front first, and lift out engine.
NOTE: Use extreme care to avoid damage to pipes and wires.

ENGINE DISASSEMBLY

DV-462B, DV-550B, D-150, 170, 190

NOTE: The disassembly procedures below are intended for disassembly operations using an engine stand. The sequence of disassembly would vary slightly if different equipment were used, although the following serves as a general guide, even where in-chassis work is to be accomplished.

Preliminary Parts Removal

1. Remove both exhaust manifolds.
2. Remove two screws in the oil pan flange, and remove the exhaust manifold heat shield.
3. Remove the starting motor and adapter.
4. Remove the oil cooler return line from the cooler and engine water outlet.
5. Remove the oil cooler water inlet line, disconnecting it at the oil cooler and crossover tube. Remove the water line connecting the crossover tube to the air compressor. Remove the water line clip from the power steering pump mounting bracket bolt.
6. Remove the oil line running from the oil cooler base to the tee at the base of the air compressor mounting. Disconnect the clip

Air cleaner air and oil line locations
(© International Harvester Co.)

from the engine crossover tube clip extension and remove the oil line.
7. Remove the crossover tube and gaskets from the water pump and crankcase.
8. Loosen both fuel filter line nuts at the top of the filter base, loosen the mounting capscrews, and remove the base.
The engine may be mounted on an overhaul stand at this time.

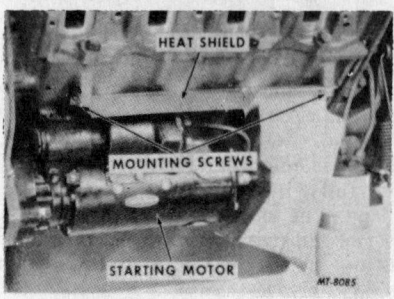

Location of the exhaust manifold heat shield
(© International Harvester Co.)

Major Disassembly

1. Remove the power steering belt, loosening the mounting bracket capscrews and adjusting bolt locknut to permit easy accomplishment of the operation. Remove the pump mounting stud nuts and lockwashers, and remove the pump.
2. Remove the mounting bracket capscrews and lockwashers. Remove the mounting bracket.
3. Loosen the air conditioner compressor mounts, slide the unit inward, and remove the drive belt.
4. Remove the compressor mounting capscrews, lockwashers, and flatwashers, and remove the compressor.
5. Remove the mounting bracket capscrews and lockwashers, and remove the bracket.
6. Remove the oil sending unit from the cooler base.
7. Remove the oil cooler capscrews, and pull the cooler and gasket off the engine.
8. Remove the oil filters from the cooler base.
9. Remove the capscrews from the cooler and filter base, and remove the entire unit and gasket.
10. Remove the alternator mounting bolts, disengage the belts, and remove the alternator from the engine.
11. Using a special wrench and with a heavy duty screwdriver blade against the lug on the water pump body forming a stop against one pulley hub spoke, remove the fan drive assembly. NOTE: Do not permit the screwdriver to become wedged between the water pump body and pulley hub spoke (see illustration).

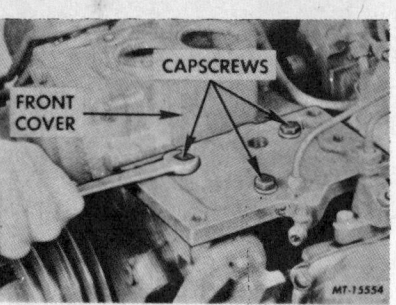

Location of compressor mounting capscrews
(© International Harvester Co.)

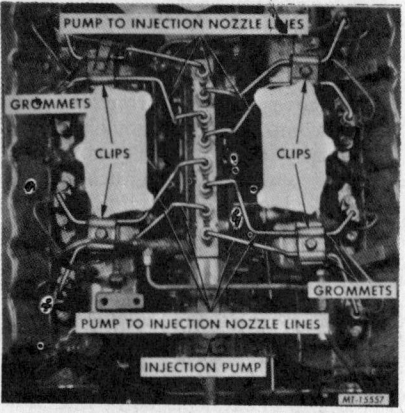

Location of fuel injection line clips
(© International Harvester Co.)

12. Push the idler pulley toward the center of the engine and remove the main and accessory drive belts.
13. Remove the capscrews and washers from the alternator mounting bracket, and remove the unit from the engine.
14. Disconnect the air line from the air compressor to the crossover or air cleaner adapter plate. Disconnect the oil line at the adapter plate.
15. On models equipped with an intake manifold crossover, remove the unit's capscrews and lift it off the manifolds or air cleaner adapter plate.
16. Remove the crossover or adapter mounting gaskets and screens. Tape the intake manifold openings to prevent the entrance of dirt.
17. Remove the air compressor mounting capscrews and their lockwashers. Remove the compressor and gasket from the engine.
18. Remove the mounting capscrews and washers from the compressor adapter plate, and remove the adapter plate. Remove the adapter plate O-ring from the underside of the plate.
19. Remove the ventilator pipe connecting the right cylinder head cover and intake manifold.
20. Remove clips and grommets from the fuel injection lines. Disconnect the lines from the

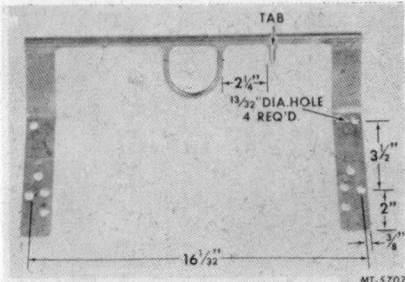

Injector mounting bracket capscrews
(© International Harvester Co.)

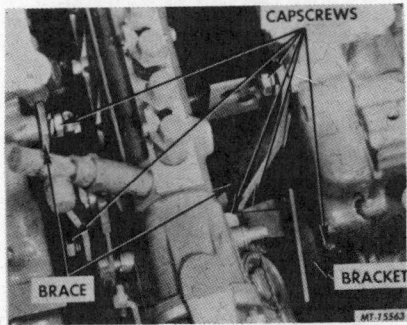

Lifting sling
(© International Harvester Co.)

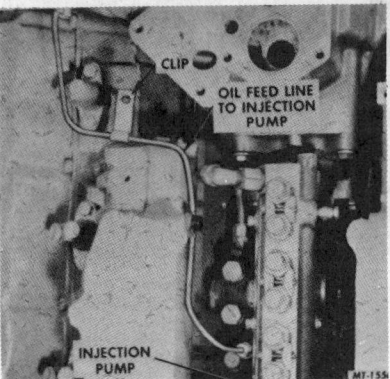

Injector pump oil feed line clip
(© International Harvester Co.)

pump and nozzles, and immediately cap all openings.

21. Disconnect the leak-off lines at the injection pump, at either end of the lines running between manifolds, and between manifold and injection pump. Remove grommets and plug or cap all fittings.

22. Remove the fuel supply line clip from the front of the right valve cover, and disconnect fittings at either end and remove the line.

23. Remove the two clips holding the primary to secondary filter line in place, disconnect fittings at either end, and remove the line. Cap or plug all openings.

24. Remove the clip attaching the injector pump oil feed line to the left intake manifold, disconnect fittings at either end of the line, remove the line, and cap the opening in the pump.

25. Remove the capscrews holding the horizontal bracket to the injector pump, and the pump-to-bracket screw, and remove the bracket.

26. Remove the capscrews securing the brace to the injector pump, and those securing it to the intake manifold. Remove the brace.

27. Remove the injector pump adapter capscrews and lockwashers, and remove the pump and its gasket from the engine. Remove the drive disc from the pump or flange.

28. Unsnap the swirl destroyer linkage ball studs at the manifolds.

29. Remove the intake manifold capscrews and lockwashers, and remove the manifolds and gaskets.

30. Remove the right bank water outlet mounting caps and lockwashers. Remove the dipstick tube, and water outlet and gasket.

31. Repeat the operation on the left bank water outlet (there is no dipstick tube to remove on this side).

32. Remove all cylinder head cover screws and washers, and remove covers and gaskets.

33. Remove the nozzle assemblies from the heads, keeping them in order so each can be reinstalled in the same cylinder at reassembly.

34. Remove the rocker assemblies from the heads.

35. Remove valve tips and pushrods.

36. Install a lifting sling similar to that shown in the illustration, using ⅜"NC × ¾" capscrews. The tab should be made of ⅜" × ¾" cold rolled stock.

37. Remove all head bolts and lift each head off the two dowel sleeves. Remove the gaskets. Proceed carefully and avoid prying, which could damage the gasket surface.

38. Compress the clutch plate with the three capscrews "A" in the illustration. These screws are ⅜"-16NC × 2, and flatwashers should be used under them.

39. Remove the clutch back plate-to-flywheel retaining capscrews, and remove the clutch assembly.

40. Remove the clutch pilot bearing with a slide hammer and puller.

41. Remove the flywheel mounting bolts, and remove the flywheel from the crankshaft flange.

42. Remove the flywheel housing capscrews, and remove the housing, being careful not to damage the roll pins.

43. Remove the capscrews and washers from the camshaft cover plate, and remove the plate and gasket.

44. Disconnect the water crossover pipe at the rear of the crankcase, and remove it.

45. Remove all tappets with a mag-

Locating crankshaft for removal of piston cooling nozzles
(© International Harvester Co.)

netic remover. If necessary, use only a small amount of carburetor solvent to loosen the tappets.

46. Remove oil pan mounting caps. Remove the pan and gasket.

47. Remove the oil pump screen-brace mounting nuts, bolts, and washers, and remove the brace and clamp.

48. Remove the oil pump mounting capscrews, and remove the pump. It may be necessary to turn crankshaft rear counterweight away from the oil pump shaft.

49. Disconnect and remove the idler arm spring from the front cover and idler arm.

50. Remove the idler arm capscrew and the arm and spacer from the bracket.

51. Remove the water pump pulley capscrews, and remove the pulley.

52. Remove the water pump mounting caps, and remove the front portion of the pump and the gasket from the pump body. Remove the remaining pump body screws, and remove the body and lift the body and gasket off the front cover.

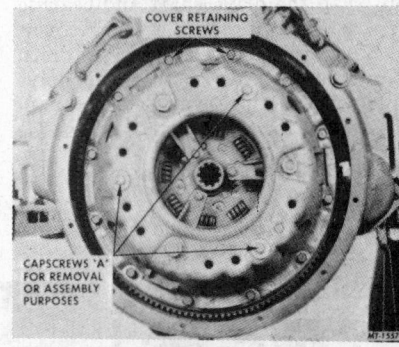

Installation locations for capscrews used in compressing the clutch plate
(© International Harvester Co.)

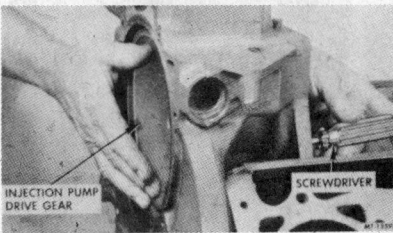

Location of injection pump drive gear
(© International Harvester Co.)

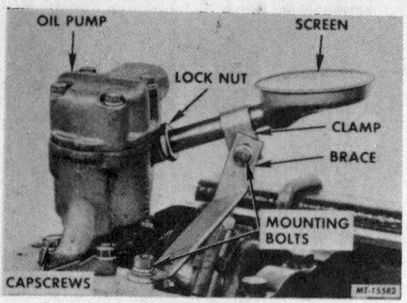

Location of oil pump screenbrace and related parts
(© International Harvester Co.)

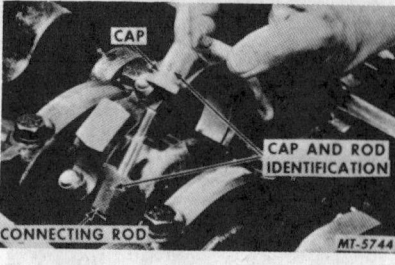

Removing the connecting rod cap
(© International Harvester Co.)

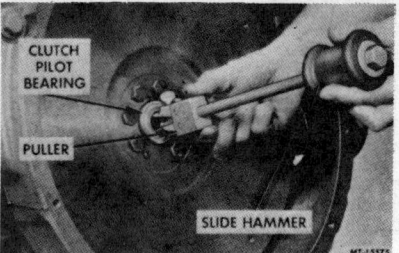

Removing the clutch pilot bearing
(© International Harvester Co.)

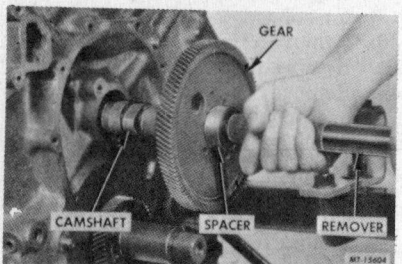

Removing the camshaft, gear, and spacer
(© International Harvester Co.)

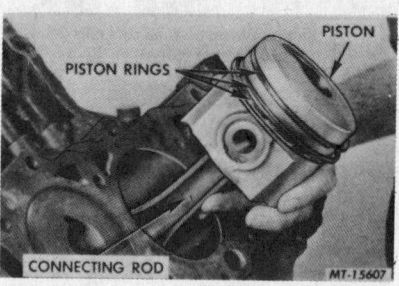
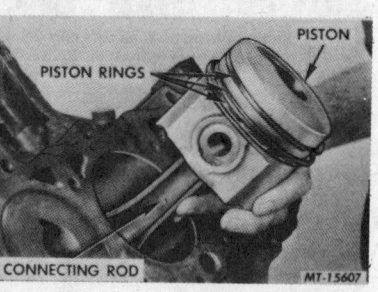
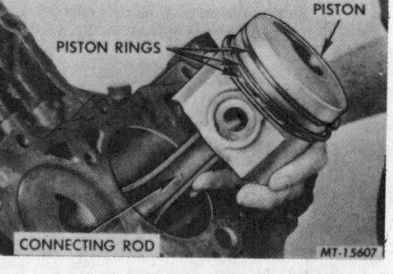

Removing piston and connecting rod assembly
(© International Harvester Co.)

Removing the drive shaft rear bearing
(© International Harvester Co.)

53. Remove the injector pump drive flange nut, and then remove the drive flange with retainer and grip springs using a special puller.

54. Remove the injector pump drive gear mounting caps and washers. Remove the drive gear by inserting a screwdriver into the injection pump oil return hole in back of the gear housing and pushing the drive gear off the shaft.

55. Remove the snap ring from the front bearing of the injector pump drive.

56. Tap the drive shaft and bearing assembly from the front cover, using a hammer and brass drift.

57. Mount the injector pump drive shaft in a press and use an adaptor to force the bearing from the shaft.

58. Remove the drive shaft rear bearing from the front cover with a special remover and handle.

59. Position a wood block between a crankshaft counterweight and the crankcase to prevent shaft rotation. Remove the crankshaft pulley with a special puller.

60. Remove the crankshaft pulley with a special puller.

61. Remove the two Nylock capscrews and lockwashers from the rear of the front cover.

62. Remove all but two of the front cover caps. Support the front cover, remove the remaining caps, and lift the cover and gasket off the dowel pins.

63. Remove the oil seal from the front cover with a hammer and drift.

64. Remove the spring, nuts, and lockwashers from the swirl destroyer mounting plate, and remove the plate and control rod assembly. Remove the flatwashers from the mounting studs.

65. Remove the tappet cover mounting capscrews and washers, and remove the cover from the en-

gine. Pull off the gasket and discard it.

66. Remove the camshaft thrust plate capscrews and camshaft nut, and remove the camshaft, gear and spacer with a special remover.

67. Rotate the crankshaft to the position shown in the illustration.

68. Remove the piston cooling nozzles from both sides of the crankcase.

69. Remove all ridges from upper cylinder bores.

70. For each cylinder, remove the connecting rod cap, push the rod and piston assembly out of the cylinder bore at the top, and replace the cap and bearing inserts on the rods so the numbered sides match.

71. Remove the self locking bolts, tie bolts, and seal washers from each cap.

72. Remove the front four main caps with a remover and slide ham-

Locations of mounting bolts for the oil cooler and accessories
(© International Harvester Co.)

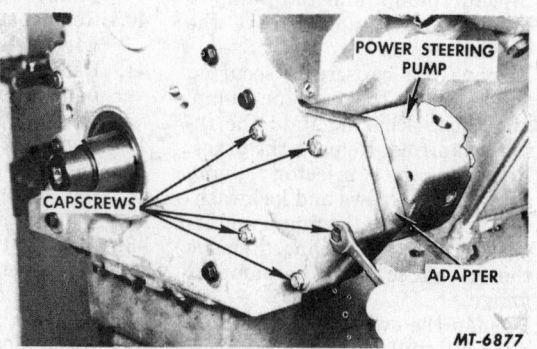

Location of power steering pump mounting capscrews
(© International Harvester Co.)

mer by inserting the hook end of the remover into the horizontal hole toward the center line of the crankshaft.

73. Remove the rear main cap with a special puller and adapter.
74. Remove the rear main cap oil seal.
75. Lift the crankshaft straight upward to remove it from the crankcase.

ENGINE DISASSEMBLY

DVT-573

NOTE: The disassembly procedures below are intended for disassembly operations using an engine stand. The sequence of disassembly would vary slightly if different equipment were used, although the following serves as a general guide, even where in-chassis work is to be accomplished.

Removing rear oil seal using a special remover
(© International Harvester Co.)

Preliminary Parts Removal

1. Remove the front oil filter body and element.
2. Loosen alternator adjusting and mounting bolts, and remove the belt.
3. Loosen the fan pulley and bracket mounting bolts and the bracket adjusting screw, and remove the main drive belts.
4. Loosen and remove hose clamps, and remove the hoses from the water pump.
5. Remove the water pump body, mounting bolts and oil outlet tube mounting bolts. Remove the three bolts mounting the oil filter base, and remove the inlet tube, base, oil cooler, and water pump assembly. Be careful not to lose oil cooler O-rings.
6. Remove the breather pipe hose clamp, clip (at manifold extension), and nut and washer holding the unit to the right cylinder head cover, and remove the unit.

Major Disassembly

1. Remove the starting motor mounting bolts, and remove the motor from the flywheel housing.
2. Remove the alternator adjusting strap screw, flat washer, lockwasher, and all mounting bolts and nuts. Remove the unit from the bracket.

Removing the rear main cap with puller and adapter
(© International Harvester Co.)

3. Remove the fan blade mounting bolts and washers from behind the fan. Remove the blade and adapter.
4. Remove the fan pulley hub mounts and washers, and loosen the adjusting screw. Remove the fan pulley hub and bracket assembly.
5. Loosen the cross-over tube hose clamp at the left thermostat housing. Remove the tube and hose.
6. Remove the capscrews and washers from the front cover access plate, and remove the plate and O-ring.
7. Disconnect water and air lines at the air compressor, and at the left side thermostat housing and cylinder head.
8. Remove both thermostat housings, and remove both thermostats and gaskets.
9. Loosen, but do not fully remove, the crankshaft pulley nut. In this position, pulley will be restrained if it is forced suddenly off the shaft.
10. Install a puller, and pull the crankshaft pulley free of the shaft taper. Then, remove the bolt and pulley.
11. Remove the front mounting bracket.
12. Remove the remaining alternator mounting capscrews, and remove the bracket.
13. Loosen the capscrews which fasten the power steering pump to the front cover, and remove the pump and adapter.

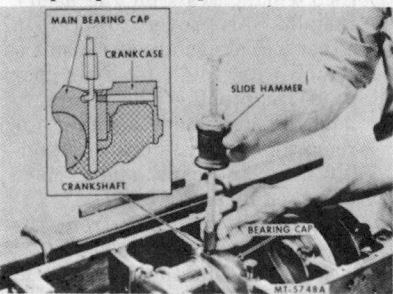

Removing the main caps with a slide hammer
(© International Harvester Co.)

14. Remove capscrews and washers holding the air compressor and the cover plate to the front cover.
15. Remove all remaining front cover capscrews and washers, and remove the front cover and gasket.
16. Remove the front cover oil seal with a drift and hammer.
17. Disconnect the oil level gauge tube clip, and remove the gauge tube.
18. Remove all remaining oil pan capscrews and lockwashers, and remove the oil pan and gasket.
19. Disconnect the throttle rod at the injection pump, and remove the throttle cross shaft, and bracket assembly.
20. Disconnect the oil feed line at the air compressor elbow.
21. Remove the remaining capscrew holding the air compressor to the cover plate, and remove the compressor and gasket.
22. Remove the compressor oil feed line from the crankcase.
23. Disconnect the fuel filter and bypass hoses at the injection pump.
24. Disconnect the leak-off lines at the junction block, and the aneroid control valve hose at the elbow on the right head. Seal off all ends with plastic caps, and plug all openings.
25. Disconnect the hose at the final filter, remove the filter's mounting bolts, and remove the filter from the bracket.
26. Where applicable, remove the oil filler tube and gasket from the flywheel housing.
27. Remove the capscrew and washer from the turbocharger elbow bracket on the flywheel housing. Remove the elbow, bracket, and gasket from the turbocharger, plug or tape the opening.
28. Lift the exhaust sleeve and seal rings out of the turbocharger.
29. Remove the mounting capscrews and washers from the turbocharger exhaust inlet elbow, and remove the elbow and gaskets.
30. Bend back the locking tabs on the rear manifold section rear mounting washer, and remove the capscrews and washers, and manifold and gasket.
31. Bend back the locking tabs on the front manifold section front mounting washer, and remove the capscrews and washers, and manifold and gasket.
32. Remove the capscrews and washers from the center section, and remove the manifold and gasket.
33. Repeat steps 30-32 on the opposite cylinder head.
34. Disconnect the turbocharger oil pressure line at the crankcase

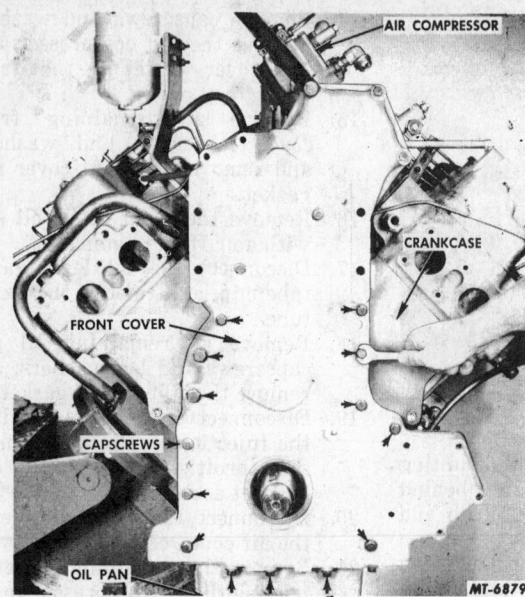

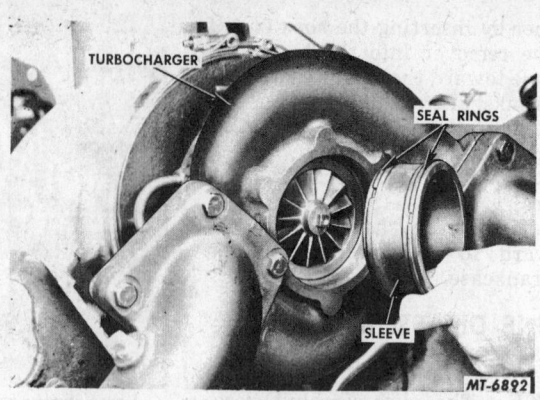

Lifting the exhaust sleeve and seal rings
out of the turbocharger
(© International Harvester Co.)

Location of remaining front cover capscrews
(© International Harvester Co.)

Turbocharger oil pressure and vent lines
(© International Harvester Co.)

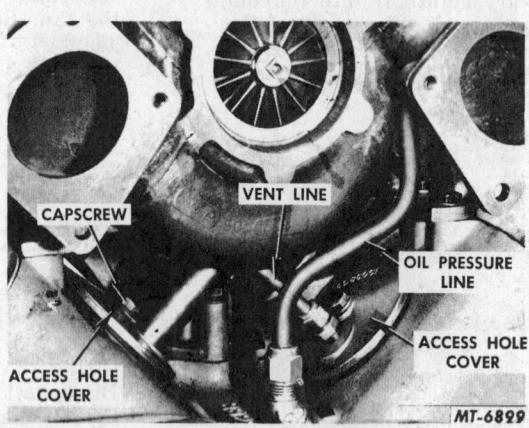

Injector leak-off lines
(© International Harvester Co.)

and the vent line at the access hole cover.

35. Remove the capscrews from the oil pressure line fitting at the turbocharger and remove the oil line with the fitting and gasket.

36. Disconnect the vent line at the turbocharger, and remove it.

37. Remove the turbocharger capscrews and washers, and the capscrew and plastic washer from the left rear access hole cover. Lift the turbocharger and oil return as an assembly off the cylinder heads.

38. Remove the turbocharger adapters and their O-rings, and the access hole cover and adapter gaskets.

39. Remove the oil pressure line, including safety switch tee, from the cylinder head.

40. Remove the cylinder head front cover plates and gaskets.

41. Remove injector leakoff lines from the injectors, and cap or plug all openings.

42. Remove all injector lines and dampers as an assembly, and cap or plug all openings.

43. Remove the injector pump mounting bolts and washers. Lift the pump and drive shaft from the housing. Remove the housing gasket.

44. Remove the oil inlet pipe mounting clip capscrew, disconnect the inlet pipe at the regulator valve housing, and remove the assembly.

45. Remove the oil pressure regulator housing mounting capscrews and washers, and remove the housing, gasket, and valve.

46. Remove the injection nozzle mounting capscrews. Carefully use a prybar to loosen each injector, and then lift each injector with its gasket and O-ring from the head. Place injectors in a stand so as to identify their position in the engine for reassembly.

47. Remove valve cover capscrews and washers, and lift covers and their O-ring seals off the heads.

48. Install a valve cover capscrew in either end of each rocker arm assembly (a total of four per head), tightening just enough to hold the assemblies together for removal. Remove rocker assembly mounting caps, and remove the assemblies.

49. Remove all pushrods from the heads.

50. Remove all but the center head bolt on either side, and collect all loose washers.

51. Install a lifting eye on the front of the head to be removed first. Attach a sling to the lifting eyes, and use a hoist to apply tension to the sling. Remove the center bolt and washer, and lift the head and gasket off the block.

52. Repeat the removal operation for the other head.

53. Remove the capscrews and their washers from the remaining access hole covers, and remove the covers and gaskets.

54. Remove all tappets through the access hole.

55. Place two wood spacer blocks between $1/2'' \times 3/4'' \times 3 1/2''$ between the clutch plate and throwout bearing.

56. Remove the clutch plate mounting capscrews and washers, and remove the plate assembly.

57. Remove the rear disc and facings. Remove the intermediate pressure plate and front disc with facings.

58. Remove flywheel setscrews and intermediate pressure plate locating pins.

59. Remove the clutch pilot bearing with a special puller and slide hammer.

60. Mark the position of the flywheel bore and crankshaft flange, and the position of the lock plates on the flywheel.

Location of spacer blocks required
in clutch plate removal
(© International Harvester Co.)

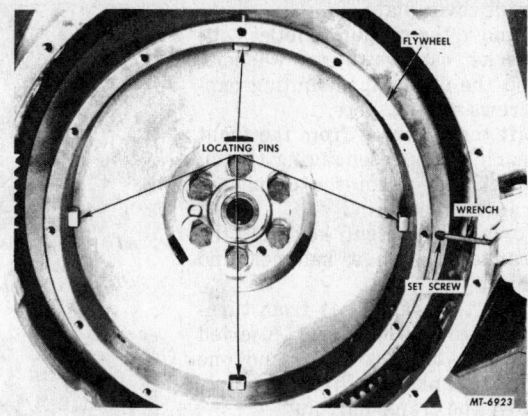

Location of flywheel setscrews and
locating pins
(© International Harvester Co.)

61. Flatten out the tabs on the lock plates, and remove the flywheel mounting capscrews.
62. Install guide pins and puller screws and remove the flywheel. Guide pins may be made by cutting the hex head off $\frac{5}{8}$-18NF $\times$ 4" capscrews. Puller screws are $\frac{1}{2}$-13NC $\times$ 2".
63. Remove the mounting capscrews from the rear oil seal plate, and lift the plate and gasket off the flywheel housing.
64. Remove the crankshaft rear oil

seal by working around it in a clockwise direction with a special remove and slide hammer.
65. Remove the flywheel housing mounting capscrews and washers, and remove the housing. Remove the housing gasket from the rear of the crankcase.
66. Remove the mounting bolts from the injection pump drive gear, and remove the gear. Remove the housing mounting bolts and washers, and remove the drive assembly, including bearings,

shaft spring, and O-ring.
67. Straighten the tabs on the camshaft thrust plate capscrew locks with a hammer and chisel, and remove the capscrews and locks.
68. Put the crankcase in a vertical position with camshaft gear upward. Lift the camshaft, thrust plate, and gear assembly out of the crankcase.
69. Remove the power steering pump idler gear mounting bolt, and remove the idler gear, bearing adapter, and spacer from the

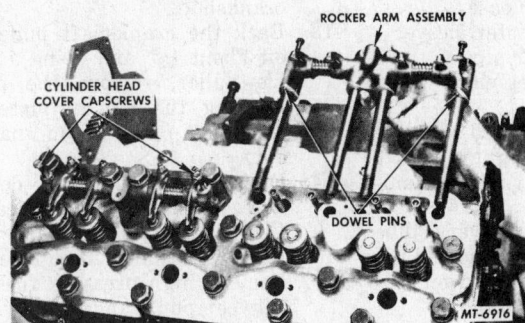

Removing rocker assemblies
(© International Harvester Co.)

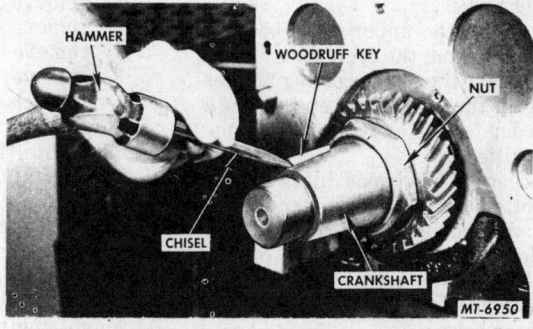

Forcing the Woodruff key off the crankshaft
(© International Harvester Co.)

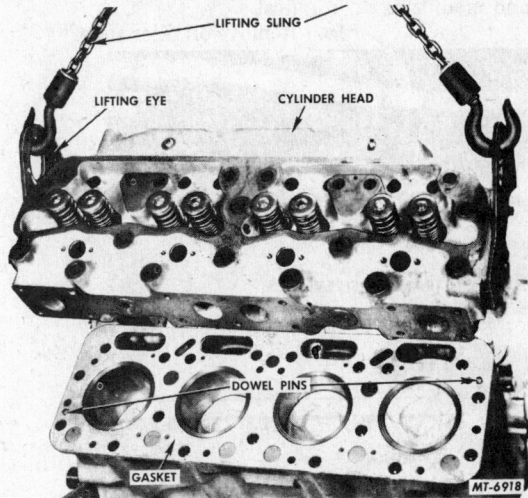

Removing the cylinder head with a lifting sling
(© International Harvester Co.)

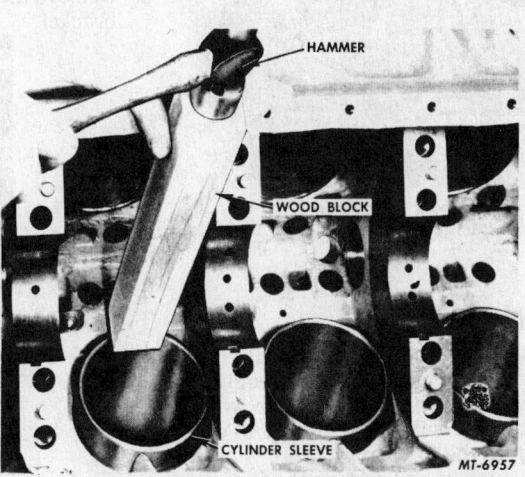

Forcing the cylinder sleeves out of the block
(© International Harvester Co.)

front cover plate.

70. Remove the oil pump outlet tube bracket capscrews and washers, and the oil pump mounting capscrews and washers.

71. Lift the oil pump from the front bearing cap, disengaging the oil outlet tube from the crankcase transfer sleeve.

72. Force the Woodruff key from the crankcase with a hammer and chisel.

73. Block the crankshaft from turning with a wood block inserted between the crankcase and one of the crankshaft throws, and remove the crankshaft nut.

74. Remove the crankshaft gear with a special puller.

75. Unbolt the front cover plate, and remove it and the gasket.

76. Use a ridge reamer to remove ridges in the tops of all cylinder bores.

77. Remove the connecting rod caps, and push each rod and piston assembly from its cylinder bore. Replace caps and bearing inserts into the rods so the numbered sides match.

78. Remove the main cap bolts and washers, and remove the caps from the crankcase.

79. Lift the crankshaft straight up and out of the upper main bearing halves.

80. Remove the upper main bearing halves from the crankcase.

81. Remove the piston cooling nozzles from the crankcase, and place in order for installation in the same positions.

82. Locate a hardwood block on the bottom of each cylinder sleeve, and hammer it to force the sleeve upward and out of the block. Mark each sleeve as it is removed for reinstallation in the same bore.

Aligning timing marks
(© International Harvester Co.)

V-800

1. Remove engine and mount in work stand SE-1962 or equivalent. Before mounting in work stand it will be necessary to remove the following:
   a. unbolt and remove the fan.
   b. remove the two alternator bolts.

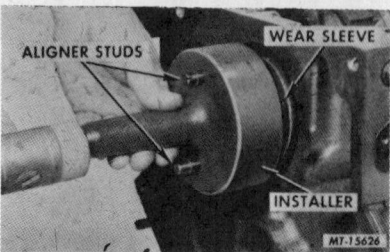

Installing the wear sleeve with a special installer
(© International Harvester Co.)

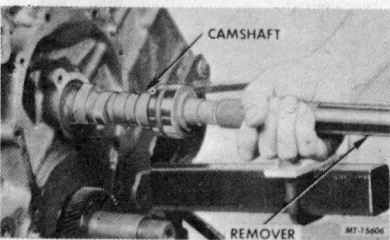

Using a special remover to pull the camshaft
(© International Harvester Co.)

   c. remove the three main drive belts.
   d. unbolt and remove the water pump and O-ring.
   e. remove the volute housing from the oil cooler.
   f. remove oil delivery tube from the filter base.
   g. remove the oil filters.
   h. remove oil filter base.
   i. remove oil transfer sleeve.
   j. remove breather pipe.
   k. remove air inlet elbow from turbocharger.
   l. remove fuel bleed-off line from "T" at right cylinder head and "T" at intake manifold.
   m. disconnect oil delivery tube nut at front of engine and position tube so that it won't interfere with the stand.

2. Mount engine in stand.
3. Check timing before disassembly. Timing is 22 BTDC.
4. Remove starter.
5. Remove alternator and mounting bracket.

6. Remove fan pulley hub and bracket.
7. Remove cross-over tube and hose.

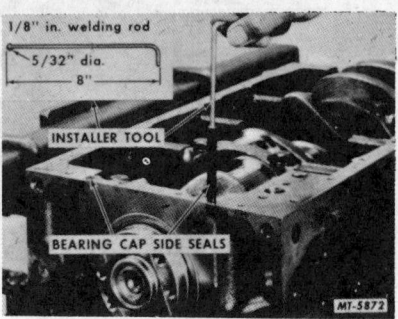

Installing rear main cap side oil seals
(© International Harvester Co.)

8. Remove refrigerant compressor and bracket.
9. Disconnect water hose from left cylinder head thermostat housing and air compressor. Disconnect oil supply line from compressor and injection pump.
10. Unbolt and lift off air compressor. A sling hoist is helpful in that the compressor must be moved straight back, carefully, to disengage gears before lifting.
11. Remove left and right thermostat housings and thermostats.
12. Remove vibration damper from crankshaft.
13. Back the crankshaft pulley nut off about ½" and using a suitable puller, remove the pulley. Leaving the nut in place prevents the pulley from snapping out.
14. Disconnect the fuel delivery hose, supply hose and bleed line. Lift the assembly from the head.
15. Remove the final filter reservoir.
16. Remove high pressure fuel lines between pump and right cylinder head.
17. Remove high pressure fuel lines between pump and left cylinder head. Remove two forward lines first.
18. Remove oil filler tube.

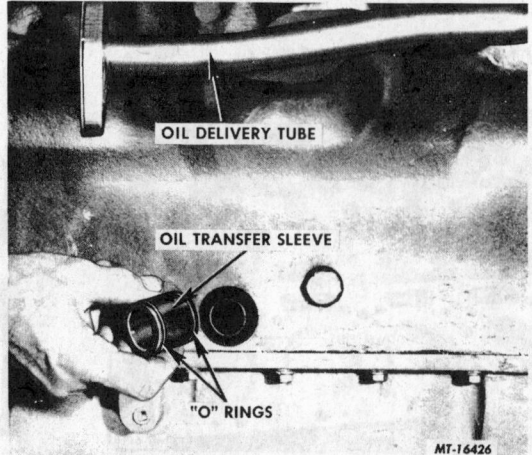

V-800 oil transfer sleeve

19. Remove turbocharger elbow and sleeve.
20. Remove oil return line from hose. Disconnect oil feed line at engine "V."

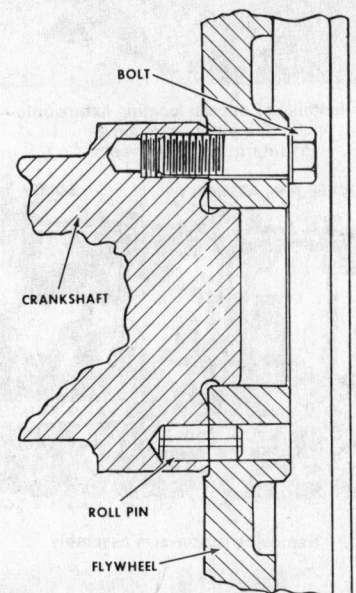

The crankshaft roll pin
(© International Harvester Co.)

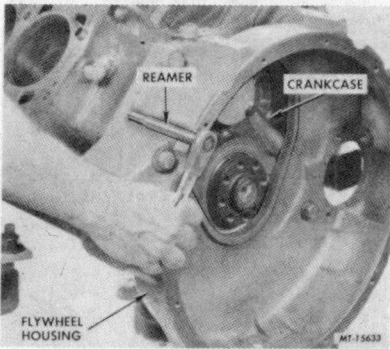

Enlarging the flywheel housing to dowel pin holes
(© International Harvester Co.)

21. Remove oil feed and return lines from turbocharger.
22. Loosen turbocharger inlet and outlet hose clamps.
23. Remove the four turbocharger-to-support manifold lock nuts and lift unit from support manifold.
24. Remove hose between air outlet and intake manifold.
25. Remove modulator-to-intake manifold crossover air line.
26. Remove left cylinder head and injection pump leak-off lines.
27. Unbolt (6) and remove intake manifold crossover.
28. Unbolt and remove the turbocharger support manifold.

Checking rod end clearance
(© International Harvester Co.)

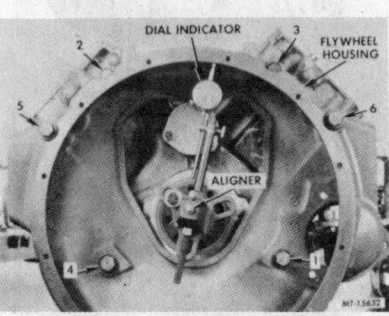

Aligning the flywheel housing with the crankcase
(© International Harvester Co.)

29. Remove the exhaust manifold adapters.
30. Unbolt (3) and remove the oil regulator valve housing.
31. Lift cartridge type valve assembly from engine.
32. Disconnect and remove injection pump oil supply and return lines. Remove oil pressure tube.
33. Unbolt (2) and remove accelerator rod lever from injection pump control. Unbolt (3) and remove accelerator rod bracket from left cylinder head.
34. Remove injection pump bracket-to-engine bolt.

Checking gear backlash
(© International Harvester Co.)

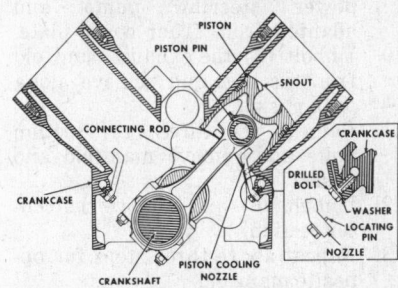

Turn the crankshaft as necessary to insure that the counterweights are positioned away from the oil pump shaft
(© International Harvester Co.)

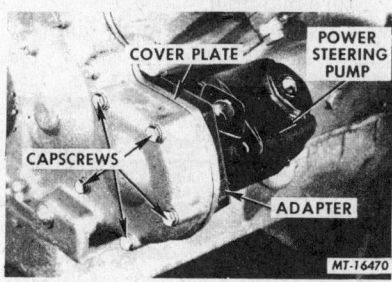

Power steering pump

V-800 injection pump connections

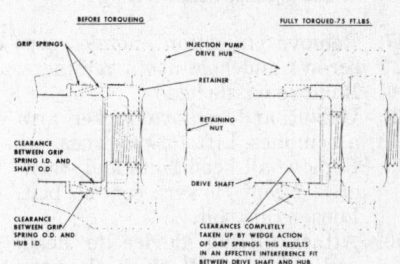

Torquing the injection pump drive hub
(© International Harvester Co.)

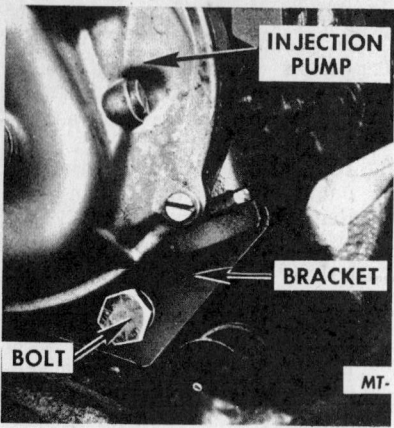

V-800 injection pump lower bracket

35. Support pump in a sling hoist and remove the five retaining bolts.
36. Slide pump rearward to disengage gears.
37. Unbolt and remove fuel filter mounting adapter from right cylinder head.
38. Loosen nut at the base of the oil level gauge tube, unbolt the extension and remove the oil level gauge.
39. Loosen capscrews and remove power steering pump and adapter from front cover plate.
40. Unbolt (4) the exhaust manifold front section and remove along with gasket.
41. Remove the three rear section bolts and remove manifold and gasket.
42. Unbolt (3) and remove the center section.
43. Repeat above three steps for opposite manifold.
44. Unbolt and remove cylinder head covers.
45. Disconnect and remove fuel return manifold.
46. Disconnect fuel supply lines at nozzle and head.

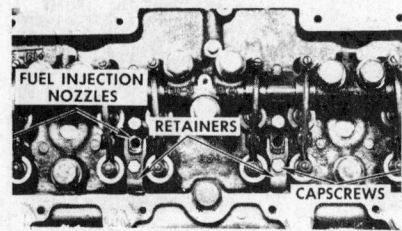

V-800 injection nozzle retainers

47. Remove injection nozzle capscrews and hold-down retainers. Remove nozzle heads.
48. Unbolt and remove rocker arm assemblies. Lift out pushrods.
49. Remove all head bolts and washers except lower center bolt. Loosen this bolt.
50. Attach lifting device to head, and take up all slack. Remove center bolt and lift off head.
51. Remove access hold covers and

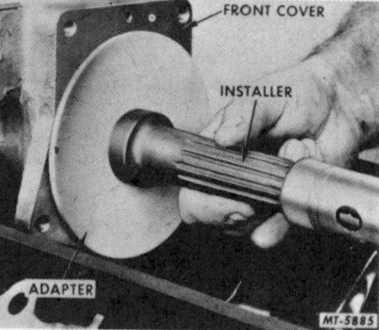

Install the drive shaft rear bearing in the front cover
(© International Harvester Co.)

Positioning the pump drive flange
(© International Harvester Co.)

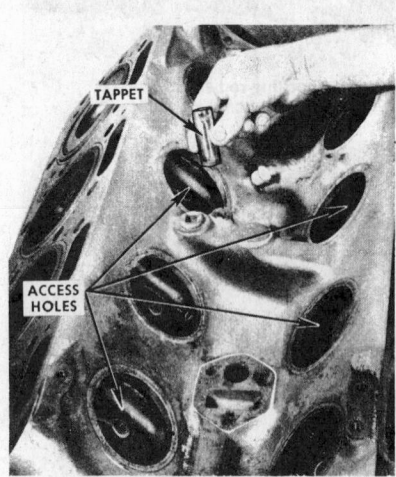

V-800 tappet removal

gaskets, and injection pump support bracket. Remove oil inlet tube.
52. Remove tappets through access holes.
53. Remove clutch assembly from flywheel.
54. Remove six flywheel mounting bolts, install guide pins and puller screws and remove flywheel. Guide pins can be made from $5/8$-18NF $\times$ 4" bolts, by cutting off head.
55. Remove crankshaft rear seal.
56. Remove wear sleeve with a cold chisel.
57. Unbolt and remove oil pan.
58. Unbolt and remove a flywheel housing.
59. Remove engine front cover.
60. At this time, it is a good procedure to check camshaft end play.
61. Straighten the camshaft thrust plate capscrew locks, unbolt and

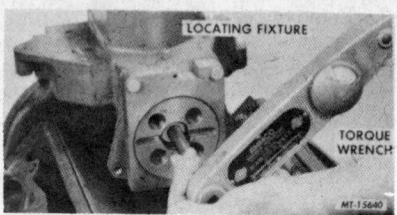

Installing a special locating fixture onto the pump drive flange
(© International Harvester Co.)

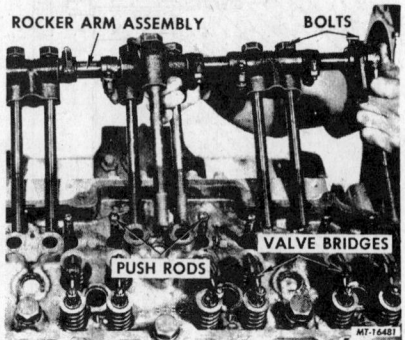

Removing rocker arm assembly

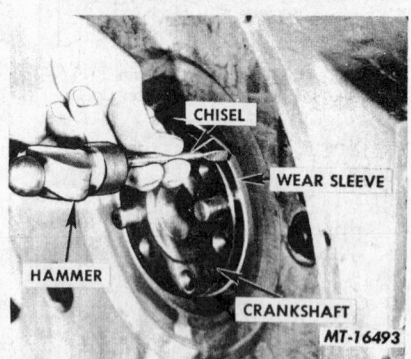

Removing wear sleeve

Removing main bearing caps

remove the camshaft. This is best done with the engine in a vertical position.
62. Remove the air compressor idler gear assembly from the front cover plate.
63. Remove the power steering pump idler gear mounting bolt. Remove idler gear and adapter.
64. Remove the oil pump inlet tube and gasket.
65. Unbolt and remove the oil pump.
66. Remove the crank pulley woodruff key with a chisel.
67. Remove the crankshaft nut.

68. Pull off crankshaft gear.
69. Remove the front cover plate.
70. Remove ridge from each cylinder.
71. Remove and mark connecting rod caps.
72. Remove pistons taking care to avoid damaging spray nozzles.
73. Unbolt and remove main bearing caps. Mark caps for installation.
74. Using a sling hoist, lift crankshaft out of block.
75. Remove upper main halves.
76. Using a wood block and hammer, drive sleeves from block. Mark sleeves for installation.

ENGINE ASSEMBLY

DV-462B, DV-550B, D-150, 170, 190

1. Install the water drain cocks into the block and turn the block upside down.
2. Coat the camshaft lobes, bearing surfaces and bores, and gears with a heavy duty hypoid lubricant.
3. Use a special installer to insert the camshaft into the front of the block.
4. Install the camshaft rear bearing cover plate and gasket with the capscrews and washers.
5. Use a dial indicator, as shown, and check the camshaft end play. It should be .0035″-.0115″. If end play is not to specifications, the thrust flange must be replaced.
6. Wipe and then lubricate with engine oil the cylinder block halves of the main bearings.
7. Place the bearing shell halves in position in the bores of the block, making sure that they are fully seated, that the shell oil holes line up with those in the block, and that the tangs fit snugly into the recesses.
8. Place the lower bearing shell halves in the main bearing caps, following the procedures described in step 7.
9. Relubricate all bearing surfaces, and lift the crankshaft into position, aligning the timing marks.
10. Install the bearing caps with the number toward the right side of the crankcase. Install new self locking bolts and flatwashers and tighten slightly.
11. With a soft mallet, tap each bearing cap to bring its rear machined face flush with the rear machined face on the portion of the block adjacent to it. Bearing must be flush on both left and right sides. Torque bolts to 130 ft lbs.
12. Install the bearing tie bolts and torque to 50 ft lbs.
13. Install the aligner studs of a special oil seal installer into the end of the crankshaft flange.

14. Place the seal over the end of the crankshaft aligning it with the crankcase bore.
15. Drive the seal into the bore until the outer edge is flush with the crankcase. Remove the installer and aligner studs from the crankshaft flange.
16. Insert the rear main cap side oil seals, using an installer made from a piece of ⅛″ welding rod. Such a tool may be made from welding rod by puddling a ball on the end of the rod and filing it to about 5/32″ diameter. If excess material protrudes above the crankcase after installation of the seal, cut the excess off flush with the bottom of the rear main cap mating surface.
17. Rotate the crankshaft to ensure that there is no binding between it and the camshaft. Then, hook up a dial indicator as shown and check the backlash, which should be .0005″-.0045″.
18. Turn the block so the front end faces upward. Rotate the crankshaft so the crankpin for number 1 cylinder is at TDC. Lubricate the cylinder bores, crankshaft journals, pistons, pins, and rings with engine oil. Insert each piston into the ring compressor and install the assembly into the cylinder bore with the word "Top" toward the centerline of the block.

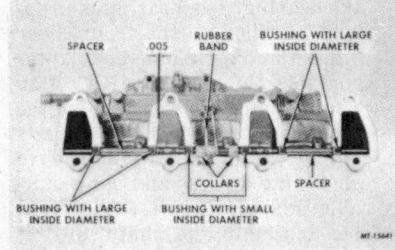

Right side intake manifold
(© International Harvester Co.)

19. Lubricate each connecting rod bearing shell on both sides with engine oil, and install bearings and caps with the number matching and on the numbered side of the rod.
20. Install new bolts, caps, and lockwashers, and torque to 55 ft lbs. Release all torque load, and retorque to 55 ft lbs. Then, turn each nut an additional flat (1/6 turn).
21. Check connecting rod and play as shown, and ensure that it is within .008″-.018″.
22. Rotate the engine so the bottom of the crankcase faces upward.
23. On DV-550 engines, install the piston cooling nozzles with the locating pin engaging the hole in the crankcase. Install bolts and washers finger tight. Lower each

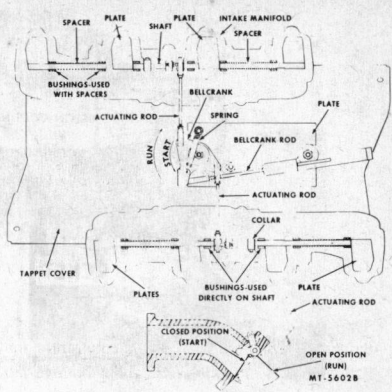

Swirl destroyer linkages
(© International Harvester Co.)

piston in its bore to make sure the nozzle is centered in the slot in the skirt. After this is verified for each nozzle, tighten its mounting bolt to 12 ft lbs.
24. Make sure the crankshaft counterweights are positioned away from the oil pump shaft, rotating the crank as necessary. Install the oil pump, with the capscrews and lockwashers. Make sure the pump is correctly aligned in order to avoid binding.
25. Install the screen to the pump body. Place a straightedge across the pan rails of the pump body, and measure the distance from the bottom of the straightedge to the top of the pump screen at both sides. The screen should be checked and positioned so that both dimensions are 6½″. Then, tighten the pump screen tube locknut to 20 ft lbs.
26. Install the screen clamp and brace, tightening the bolts evenly to avoid disturbing the screen.
27. Place the flywheel housing over the roll pins, and tap into place with a soft hammer. Install cap screws and washers, and tighten in sequence shown to 150 ft lbs. The two lower mounting capscrews require two turns of Teflon thread sealing tape. If new crankcase or flywheel housing is used, check runout as shown. It must not exceed .010″. If necessary, ream roll pin holes with a tapered ½″ reamer and install oversize roll pins. Align the housing with the engine mounted with the rear end upward.
28. Position the flywheel on the crankshaft flange roll pin. Install the self-locking bolts and torque to 110 ft lbs.
29. Coat the outside diameter of the clutch pilot bearing and the inside diameter of the flywheel bore with a locking compound and install the pilot bearing into the flywheel. Keep the compound from getting onto the inside of

Minimal clearance exists between crankshaft
and piston cooling nozzles
(© International Harvester Co.)

Checking crankshaft endplay
(© International Harvester Co.)

the bearing or between the flywheel and crankshaft. Allow two hours curing time.

30. Position the driven disc against the flywheel with the long portion of the hub to the rear. Position the clutch onto the flywheel over the driven disc. Locate the clutch so that the inspection mark (white paint) on the flange of the backing plate or cover is as near as possible to the "L" on the flywheel. Install the two or three mounting capscrews and lockwashers loosely.

31. Insert a clutch aligning arbor or transmission main gear shaft through the hub of the driven disc and into the pilot bearing.

32. Hold the driven disc in position while installing the mounting capscrews and lockwashers in the flange of the clutch backing plate or cover. Tighten the capscrews alternately and evenly until secure. Remove the capscrews installed to keep the clutch in a compressed position.

33. Position the gasket on front, rear, and sides of the tappet cover, and install the tappet cover with the gasket onto the crankcase with bolts and lockwashers.

34. Put a new gasket for the front cover over the dowel pins on the crankcase. Coat the right water outlet gasket with grease and position it to the crankcase, making sure the tab will not block the water passage. Install the front cover and the idler arm mounting bracket. NOTE: The second set of mounting bolts from the bottom of the front cover, and the two on the back side of the cover are Nylock bolts. It is a good practice to replace these bolts, but if a careful inspection indicates they are in good condition, they may be reused.

35. Position the front cover oil seal and a small pilot of a special installer with the wiping lip to-

ward the outer end of the pilot. Make sure the crankshaft pulley key is removed from the end of the crankshaft, and then place the installer over the outer end of the crankshaft and drive the seal into the cover. Make sure the inner shoulder of the tool contacts the machined surface of the cover before removing the tool.

36. Assemble a retaining screw through the bore of the installer and place a $\frac{5}{8}$" diameter adapter over the threaded end of the screw. Install the handle on the retaining screw.

37. Position the injection pump drive shaft rear bearing on the pilot of the installer and install the bearing. The bearing is properly installed on the front cover when the $5\frac{5}{8}$" diameter adapter contacts the machined injection pump mounting surface on the front cover.

38. Mount the injection pump drive shaft into a press and use a suitable adapter to press the front bearing onto the shaft until it bottoms on the shaft flange.

39. Position the shaft and bearing into the front cover. Install the snap ring.

40. Align the pump drive gear teeth with the camshaft gear teeth. Rotate the pump drive shaft

until the holes in the shaft are aligned with the holes in the drive gear. Tap the gear onto the drive shaft with a soft hammer. Install the capscrews and lockwashers and torque.

41. Assemble the injection pump drive hub, grip springs, retainer, and nut, and torque the nut to 75 ft lbs.

42. Install the idler pulley arm to the mounting bracket with the spacer between the arm and the boss on the front cover. Secure the arm with a bolt and lockwasher.

43. Install the woodruff key in the front of the crankshaft and install the pulley with a special installer. Install the nut and washer, and tighten the nut.

44. Rotate crankshaft until No. 1 piston is on compression stroke, and position the shaft so that the timing point aligns with 32 degrees BTDC on 462 engines, or 34 degrees BTDC on 550 engines.

45. Position the pump drive flange so that tangs are vertical with the marked tang up. Install a special locating fixture onto the flange and engage the pin in the roll pin hole at the rear of the front cover. Install the locating fixture mounting bolts.

46. Tighten the drive shaft nut to 75 ft lbs. Remove the injection

Installation of drive plate, washer, and nut
(© International Harvester Co.)

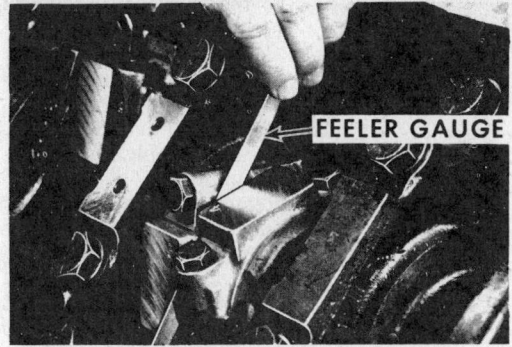

Checking connecting rod end clearance
(© International Harvester Co.)

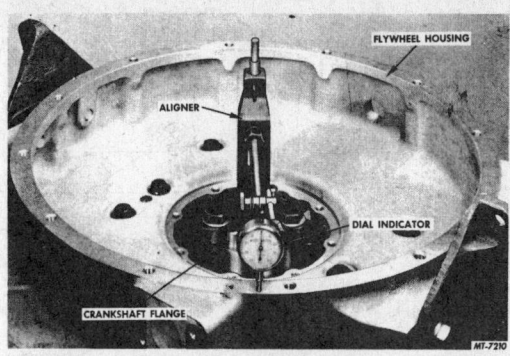

Checking flywheel housing runout
(© International Harvester Co.)

Installation of special adapter and sleeve
(© International Harvester Co.)

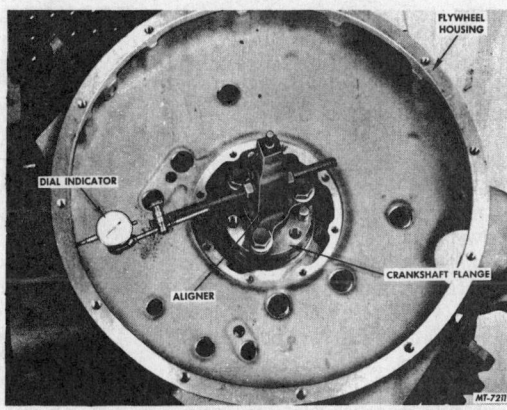

Checking flywheel housing concentricity
(© International Harvester Co.)

pump drive flange locating fixture.

47. Install a new gasket and the water pump body, and secure with screws and lockwashers.

48. Position the O-ring in the water pump housing, and install the pump assembly and flat head screws. Install the water pump housing, using a new Nylock screw in the lower mounting case, leaving the second and hole.

49. Install a new gasket and the oil pan to the bottom of the crankthird bolts on the right pan rail out for later securing of the exhaust manifold heat shield.

50. Using a special tappet installer install each tappet into its original bore.

51. Position new head gaskets over aligning dowel sleeves, making sure all gasket bolt holes are in line with crankcase holes.

52. With a special holding sling, mount each head, aligning it carefully with the crankcase

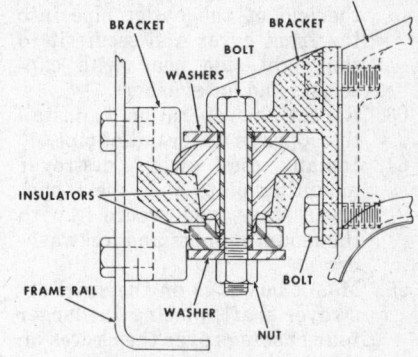

Rear engine mount
(© International Harvester Co.)

dowel sleeves. Install head bolts and washers hand tight.

53. Torque head bolts alternately and evenly, in the sequence shown, to 110 ft lbs.

54. Insert each pushrod in its original position.

55. Install a valve cup on each valve stem.

56. Loosen rocker arm locknuts and back off adjusting screws.

57. Install the rocker arm assemblies, aligning rockers and valves. Install all mounting bolts loosely, check for interference at all rockers and pushrods, and then tighten the mounting bolts.

58. Position each pushrod under its rocker adjusting screw and

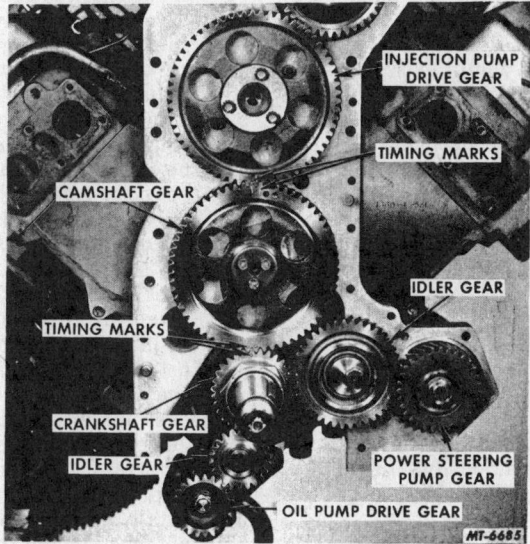

(© International Harvester Co.)

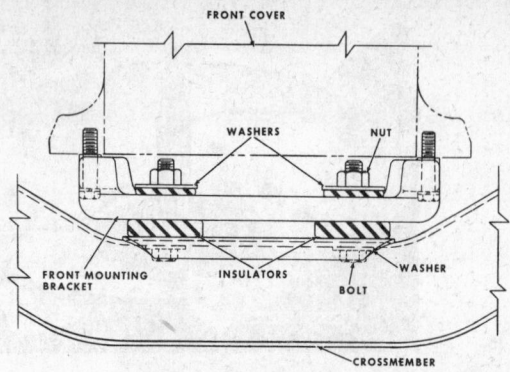

DVT series front mount
(© International Harvester Co.)

tighten adjusting screws until they just contact the pushrods, rotating the engine as necessary.

59. Adjust the valves as described in Engine Tune-Up section.

60. Using new O-rings and gaskets, install each injector nozzle into its original bore, torquing hold-down capscrews to 15 ft lbs. Use a new Nylock screw unless careful inspection reveals the tab is in perfect condition.

61. Install valve covers with new gaskets, and secure with screws and washers.

62. Position new O-rings on the right side water outlet pipe and install it in the water outlet. Position a new O-ring in the water outlet in the front cover. Insert the end of the outlet pipe into the front cover and secure it to the right side head with capscrews and lockwashers.

63. Repeat this operation to install the opposite water outlet pipe.

64. Install the swirl destroyer mounting plate over the tappet cover studs, and secure it with the mounting nuts and lockwashers.

65. Mount bushings on the swirl destroyer shaft, placing the larger four bushings over the sleeves on the shaft.

66. Position the shaft and plates in the manifold ports with the plates in closed position, and slide the shaft over until one of the plates contacts the manifold. Back off .005″ and adjust the collars against the manifold, being careful to prevent binding, and tighten setscrews and locknuts.

67. Install a rubber band as shown in the illustration on swirl destroyer linkage, and install intake manifolds and gaskets. Install capscrews and lockwashers.

68. Connect the actuating rods to swirl destroyer linkages on both sides, adjust rod length so that plates are closed at "Start" position, and tighten locknuts.

69. Turn the injection pump drive coupling until the punch mark on the coupling tang is pointed to-

Installation of capscrews on either side of the cover plate (© International Harvester Co.)

ward the transfer pump mounting flange. Rotate the crankshaft so the injection pump drive shaft flange tang which is punch marked is on top and flanges are vertical. Engage the timing indicator pin with the hole in the flywheel.

70. Install the coupling disc over the tangs of the pump drive flange.

71. Place a gasket on the pump mounting flange, and make sure the swirl destroyer flaps are in an open position.

72. Install the pump, ensuring that the coupling drive tangs engage with the holes in the coupling disc and that the pump adapter housing indexes with the roll pins. Install mounting capscrews and washers.

73. Install the injection pump support braces and bracket with capscrews, lockwashers, and flat washers.

74. Connect the fuel supply line to the secondary fuel filter.

75. Remove all dust caps, and install injection nozzle leak-off lines and grommets, and injector pump to manifold leak-off line. Install the leak-off line at the rear of the intake manifolds.

76. Put O-ring in place, and install the air compressor adapter plate

to the front cover with the capscrews and lockwashers.

77. Connect the oil line located at the compressor adapter plate to the injector pump.

78. Remove all dust caps, and install the fuel supply line to the transfer pump.

79. Remove all dust caps, and install fuel injection lines, and secure them to the manifold with grommets, clips, and capscrews.

80. Remove dust caps, and install the crankcase ventilator pipe between the right cylinder head cover and intake manifold.

81. Install the water crossover pipe to the rear of the crankcase.

82. Install the water pump pulley to the hub and secure it with capscrews and lockwashers.

83. Install the oil dipstick tube.

84. Install the air compressor gasket, and install the air compressor onto the adapter plate.

85. Install the alternator bracket, and install the alternator.

86. Install the air conditioner compressor mounting bracket, and install the compressor.

87. Install the power steering pump mounting bracket, using a class "4" bolt in the lower mount, and install the pump to the mounting bracket.

88. Install accessory drive belts, and adjust each set by moving the compressor or pump on its mounting bracket and tightening when tension is correct.

89. Push the idler pulley inward, and install main drive belts over crankshaft, water pump, and idler pulleys. Then, tension belts and tighten idler pulley mount.

90. Insert two capscrews into either end and the middle of the oil filter and cooler base. Position the gasket against the base and start the threads of the capscrews into the gasket.

91. Place the assembly against the crankcase, and tighten the capscrews alternately to an even torque.

92. Install a new gasket and the oil cooler to the cooler base. Install the oil pressure sending unit into the cooler base.

93. Remove tape from intake manifold openings, and install gaskets and screens with "Air Cleaner Side" marking up.

94. Install the air cleaner adapter to the intake manifold.

95. Install the crankcase ventilator pipe to the right cylinder head cover and intake manifold.

96. Connect the air line from the compressor to the air cleaner adapter.

97. Install the fan to the modulated fan drive. Install the assembly to the pump hub with a special

wrench. Use a heavy screwdriver against the lug on the pump body to stop one of the pulley hub spokes from turning during tightening. Tighten the nut to 40 ft lbs (left hand thread).

98. Install a lifting plate to the thermostat mounting surface and a cable affixed under the flywheel housing, attach suitable hoisting equipment and remove the engine from the overhaul stand. The steps which follow involve installing units to the engine while on the hoist:

   a. Install the crossover tube and its gasket, making connections at the water pump body and crankcase.

   b. Install the water line that runs from the crossover tube to the air compressor.

   c. Install the starting motor and its adapter to the flywheel housing.

   d. Install the exhaust manifold heat shield.

   e. Install the gasket and base for the secondary fuel filter, and connect the fuel supply lines.

   f. Install the exhaust manifolds and gaskets, torquing bolts to 30 ft lbs.

   g. Connect the water inlet line to the oil cooler and engine crossover tube.

   h. Connect the water return line from the oil cooler to the water outlet on the odd-numbered bank.

   i. Connect the oil feed line from the filter base to the tee on the air compressor adapter plate.

ENGINE ASSEMBLY

DVT-573

1. Install the water drain cocks into the block, and turn the block upside down.
2. Apply a non-hardening sealant to one side of the front plate gasket, and position the gasket on the block. Position the front plate over the gasket and dowel pins.
3. Apply a drop of thread locking compound to the threads of each capscrew, and install capscrews and washers, torquing to 30-35 ft lbs.
4. Position each cooling nozzle in its original position and install the capscrew and lockwasher.
5. Coat the lobes, bearing surfaces, gear, and bearing bores of the camshaft, and insert it into the front of the crankcase, taking care not to damage the bearings.
6. Install the capscrews and lock plates for the camshaft thrust flange, and torque the capscrews

to 25-30 ft lbs. Bend lock tabs back against capscrew heads.

7. Install the camshaft gear key, and heat the camshaft gear to 250°F. Then, position the gear, timing marks outward, onto the end of the camshaft. Installation will be facilitated if the key is started before starting the gear onto the camshaft. Start the gear by tapping with a soft hammer and then using the mounting capscrews and flat washers to draw it on the shaft.
8. Remove the mounting capscrews, discard the flatwashers, replace them with external tooth lockwashers, and torque the capscrews to 25-30 ft lbs.
9. Check the camshaft end play with a dial indicator. If it is not .005"-.013", replace the camshaft thrust plate.
10. Wipe the backs of the cylinder block halves of the main bearings, and then lubricate lightly on both sides with engine oil.
11. Position each shell in the cylinder block bore, seating it fully and making sure the oil holes in the bearing shells line up with those in the block. Make sure, also, that the locking tangs on the bearings fit into the recesses.
12. Repeat steps 11 and 12 for the lower shells, and install them into the caps.
13. Lubricate all bearing surfaces with engine oil, and install the crankshaft, aligning the timing marks (two dots) on the camshaft gear with the dot on the crankshaft gear.
14. Install the bearing caps in numbered order, starting at the front of the engine. Use self locking bolts and lockwashers, and torque to 275-300 ft lbs. Check crankshaft end play for conformance to .003"-.011" specifications.
15. Rotate the crankshaft to check that there is no binding of the gears . Check gear backlash with a dial indicator (.007"-.013").
16. Position the block so the front end faces upward. Turn the crank so the No. 1 crankpin is at TDC. Coat the cylinder bore, crankshaft journal, piston, pin and rings for No. 1 cylinder with engine oil. Insert the piston into the ring compressor, and install the assembly into No. 1 cylinder with the "CAMSIDE" marking facing the centerline of the block. Avoid damage to the piston cooling nozzle during this operation, as there is minimal clearance. Coat all connecting rod bearing surfaces with engine oil. Install the cap with the numbered side matching and on the numbered side of the rod. The

large chamfer side of the rod and cap will be to the fillet side of the crankpin. Install new connecting rod cap bolts, and torque to 100-110 ft lbs. Repeat this procedure for all remaining cylinders.

17. Check the conecting rod end play, as shown. End play must be .008"-.020".
18. Wipe off the rear face of the crankcase and flywheel housing, apply a thin film of a non-hardening sealer, and install a new gasket.
19. Put the housing in position over the two dowel pins, install capscrews and lockwashers, and torque evenly.
20. Attach a flywheel housing aligner and a dial indicator to the crankshaft flange, and place the indicator pointer against the flywheel housing flange. Remove the crankshaft end play, record the indicator reading, and then rotate the assembly around the housing, reading it at four points, 90 degrees apart. Keep the end play at zero throughout the operation. Runout must not exceed .008".
21. Position the indicator needle against the inside bore of the flywheel housing (see illustration). Read the indicator, and then rotate the crankshaft slowly and record the greatest variation. It must not exceed .008".
22. Install a new rear oil seal with a special installer (do not drive the seal onto the crankshaft), as follows:

   a. Clean the crankshaft seal surface, and inspect it to ensure that it is free of burrs or nicks.

   b. Inspect the surface of the flywheel housing where the oil seal face plate is installed, remove burrs with a stone, and thoroughly clean the area.

   c. Fasten an adapter to the crankshaft flange with two special bolts. Remove the protective cover from the oil seal and place the seal on the cover, teflon surface outward.

   d. Position the sleeve on the adapter.

   e. Install the drive plate, washer, and nut.

   f. Carefully tighten the nut until the drive plate legs seat on the flywheel housing.

   g. Remove the installing tool.

   h. Lube the seal face of the oil seal face plate with engine oil. Install the face plate and gasket. Torque the capscrews to 18-20 ft lbs.

23. Install the guide pins into the crankshaft flange. Position the flywheel over the pins, lining up

all holes. Draw the flywheel onto the dowel pins using four of the capscrews.

24. Remove the four capscrews and the guide pins. Install the six flywheel capscrews and their locks, torque to 210-230 ft lbs, and bend the locks over the capscrew heads.

25. Install the clutch pilot bearing. Install the front driven disc (the disc whose hub protrudes approximately an equal distance on either side) with the riveted hub flange toward the transmission.

26. Place four square headed driving pins in the flywheel square with the flywheel face. Install two Allen setscrews to secure each drive pin. Install the intermediate plate in the flywheel, and index it with the driving pins.

27. Install the rear driven disc with the longer toward the rear of the transmission. Install the clutch assembly, aligning the splines of the driven discs with the pilot bearing through use of a main drive gear shaft or aligning bar.

28. Tighten the capscrews around the flywheel ring uniformly. Remove the spacer blocks and aligning bar from the clutch assembly.

29. Position the oil outlet pipe in the oil pump cover, and bolt the oil pump to the No. 1 main bearing cap. Bolt the outlet pipe to the No. 2 main cap, making sure that the end is engaged with the opening in the right side of the crankcase. Check the backlash of the gears with a dial indicator, as shown. Limits are .004"-.014".

30. Lubricate the transfer tube O-rings with engine oil and install it in the opening on the right side of the crankcase, checking for proper engagement with the oil outlet tube.

31. Mount the injection pump drive assembly, including bearings, shaft, spring, and O-ring, to the back of the front cover plate with the four capscrews and lockwashers.

32. Slide the injection pump drive gear onto the shaft, aligning its timing mark with the two marks on the camshaft gear, as shown. Install the capscrews and external tooth lockwashers, and torque to 21-24 ft lbs.

33. Insert the power steering pump idler gear spacer into the front cover plate. Place the adapter and bolt in the idler gear bearing, and install the gear to the front of the crankcase. Torque the bolt to 140-160 ft lbs.

34. Install each tappet in its bore through the access openings in the crankcase.

35. Lubricate the cylinder head contacting surfaces, bolts, and lockwashers with engine oil. Check that threaded crankcase holes are free of oil to prevent hydraulic lock.

36. Install one of the cylinder head gaskets, and hoist the head for that side onto the engine, fitting it over the crankcase dowels. Install, but do not torque, the 12 short bolts.

37. Install the pushrods. Install the rocker arm assembly, applying an inward pressure to each bracket in order to hold the assembly until the end brackets locate on the cylinder head dowels.

38. Back off all rocker adjustments, and insert each into its pushrod. Then, install the four long head bolts.

39. Torque the cylinder head bolts, as described under "Torquing Cylinder Heads, DVT-573".

40. Repeat steps 35-39 for the other cylinder head.

41. Clean the nozzle sleeves using a *soft* object, such as a cloth wrapped around a stick.

42. Install a new O-ring on each nozzle assembly, install each in its original bore, and torque the hold-down bolts alternately as follows:
    a. First Bolt: 3-4 ft lbs.
    b. Second Bolt: 3-4 ft lbs.
    c. First Bolt: 5-8 ft lbs.
    d. Second Bolt: 5-8 ft lbs.
    e. First Bolt: 8-12 ft lbs.
    f. Second Bolt: 8-12 ft lbs.
    g. Recheck torque of first bolt.

43. Install all but the left-rear access hole cover, making sure each has a gasket and plastic washers under all capscrews.

44. Install the oil pressure regulator valve and housing, using a new gasket.

45. Rotate the oil pump drive shaft until the blind shaft spline matches that inside the pump drive assembly on the rear of the front cover plate. Install the pump into the pump drive assembly. Install the mounting capscrews, and secure the adapter plate to the flange of the pump drive housing.

46. Install the oil inlet line to pressure regulator valve housing, securing the clip on the line to the front of the right head with capscrew, lockwasher, and flat washer.

47. Apply a non-hardening sealer to one side of the front cover gasket, and install the gasket onto the front cover. Trim the gasket so it is flush with the bottom of the front cover.

48. Lift the front cover onto the locating dowels, and carefully tap it into place. Install the capscrews, lockwashers, and flat washers to the locations arrowed in the illustration. Install one capscrew, and related lockwasher and flat washer, on each side, from the back side of the cover plate.

49. Mount the fan pulley and bracket assembly onto the front cover with the two lower mounting capscrews and their lockwashers and flat washers. Do not torque the capscrews. Install the adjusting screw and its flatwasher.

50. Mount the alternator mounting bracket to the front cover plate with the two upper mounting capscrews, lockwashers, and flatwashers, and nuts. The leak-off line junction block must be on the uppermost capscrew.

51. Install the capscrew (with flatwasher installed) through the fan pulley and bracket assembly, front cover, and alternator bracket. Install the nut and lockwasher.

52. Remove all plastic caps and plugs, and install the fuel injection lines and damper.

53. Install the leak-off lines, connecting to nozzles and junction block.

54. Connect the appropriate hose between the elbow on the right head and the aneroid control valve.

55. Install the leak-off line that runs between the injection pump and the elbow at the junction block.

56. Install the accelerator cross shaft bracket and gasket to the left head with the three capscrews and lockwashers. The final line damper must be under the head of the rear capscrew.

57. Install the final fuel filter bracket, with gasket, to the right head. Install the capscrews and lockwashers, making sure the fuel line damper is installed under the head of the rear screw, and the safety switch tee is under the front screw.

58. Install the oil line to the galley fitting in the V of the block, and to the safety switch tee.

59. Install the final filter to its mounting bracket with four capscrews and lockwashers.

60. Connect the fuel line to the final filter and injection pump.

61. Position thermostats into their housings and install each housing into its position on the appropriate head with capscrews and lockwashers.

62. Mount the air compressor and gasket to the front cover plate, engaging the compressor gear with the injection pump drive gear, with the capscrews, flatwashers, lockwashers, and nuts. The alternator adjusting strap goes under the top right cap-

63. Connect the water inlet and outlet lines between the air compressor and left thermostat housing.

64. Connect the air inlet pipe between the compressor and left cylinder head.

65. Connect the compressor oil feed line between the oil gallery fitting in the V of the block and the compressor.

66. Install the accelerator cross shaft to the bracket, and connect the throttle rod between the cross shaft and injection pump.

67. Install the accelerator mounting bracket to the bracket on the front cover plate.

68. Position the alternator on its bracket, and install the capscrews and nuts, but do not tighten. Attach the adjusting strap with the capscrews, flatwasher, and lockwasher, and tighten finger tight.

69. Apply an anti-seize compound to the slip joint of each exhaust manifold section and, using new gaskets and bolt locks, install the manifold, one section at a time, starting at the front. Apply anti-seize to the bolts, and leave the bolts slightly loose until all sections are in place. Then, torque the bolts to 100-120 ft lbs. Bend the lock tabs.

70. Install the Woodruff key into the crankshaft, and then install the crankshaft pulley and nut, and torque to 325-375 ft lbs.

71. Position the power steering pump gasket onto the front cover plate, and insert the pump into the plate, engaging the drive and idler gears. Install capscrews, flat washers, lockwashers, and nuts.

72. Check that the oil drain line O-ring is in its groove in the left hand rear tappet access cover, and position the access cover on the drain line.

73. Inspect the turbocharger compressor mounting pads on the cylinder head for flatness and complete removal of old gasket material. Position the adapters with O-rings installed onto the turbocharger compressor housing.

74. Position the turbocharger on the mounting pads. Insert the gaskets between the adapters and heads, and install the adapter and bracket mounting capscrews and lockwashers finger tight.

75. Loosen the compressor and turbine clap bolt nuts, and tighten the adapter and bracket mounting capscrews.

76. Coat the joints of the turbocharger inlet exhaust elbows with anti-seize compound, and install elbows and gaskets to the turbine housing. Install mounting capscrews, coating the threads with anti-seize compound. Torque both clamp bolt nuts to 8-12 ft lbs.

77. Insert the exhaust sleeve, with seal rings installed, into the turbine housing. Install the exhaust elbow and bracket, connecting it to the turbocharger and flywheel housing.

78. Slide the access hole cover down on the oil drain line, and mount it onto the crankcase.

79. Force 4-5 oz. of engine oil into the turbocharger bearing housing.

80. Install the turbocharger oil inlet filter screen, gaskets, and a special break-in oil pressure line to the turbocharger. Connect the other end of the oil line to the oil galley. After 5-25 hrs. of operation, the screen and run-in line are replaced with a regular line.

81. Install the vent line, making connections at the bearing housing and the right rear access hole cover.

82. Attach the oil filler elbow and its gasket to the flywheel housing.

83. Install the oil pan and gasket.

84. Install the starting motor to the flywheel housing, and torque capscrews to 105-120 ft lbs.

85. Connect the dipstick tube to the oil pan and clip it to the exhaust manifold.

86. Adjust the valve tappets as described in Engine Tune-Up section.

87. Install the valve covers with new gaskets.

88. Mount the engine front mounting bracket to the front cover. Install the rear mounting brackets to the flywheel housing.

89. Hoist the engine out of the overhaul stand, using a chain fastened to both lifting eyes. Proceed with the steps below with the engine on the hoist.

90. Assemble the oil filter base, cooler, and water pump assembly to the crankcase, using a new gasket between the filter base and crankcase.

91. Install the elbow between the transfer sleeve and oil filter base, using a new O-ring.

92. With a new O-ring in the flange, install the oil inlet line to the oil cooler.

93. Install new oil filters and gaskets, and mount filter cases, torquing mounting bolts to -5-20 ft lbs.

94. Install the hose and clamps between the right thermostat housing and water pump connection.

95. Install the crossover pipe, and its hoses and clamps between the left thermostat housing and the water pump connection.

96. Install the main drive and accessory belts, and tension so that ½" deflection is possible halfway between pulleys.

97. Connect the hose elbow and clamp to the breather on the right rear cylinder head cover.

98. Position the breather pipe in the elbow, and clip the pipe to the right valve cover rear stud with the clip, flat washer, and lockwasher.

V-800

1. Apply Permatex #2 or equivalent to one side of the front plate gasket and install front plate. Torque bolts to 35 ft lbs. Use Loctite "B" on threads.

2. If removed, install piston cooling nozzles in crankcase.

3. Coat camshaft and bore with heavy duty hypoid axle oil and insert camshaft into place.

4. Install and torque thrust plate to 30 ft lbs.

5. If camshaft gear was removed, heat gear to 250°F and install on shaft over key using a soft hammer. As soon as retainer plate clears a few threads, press the gear on with the retainer capscrews. Torque capscrews to 30 ft lbs.

6. Lubricate and install upper main valves. Position crankshaft in block, oil the journals and install the lower halves and caps. Make certain the timing marks align. Use new cap bolts; check bearing

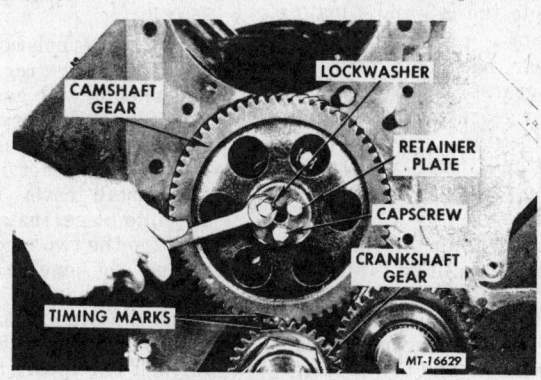

Aligning camshaft timing marks

fit with Plastigage. Torque main caps to 390 ft lbs.

7. Rotate crankshaft and camshaft to check for binding and backlask. Backlash should be .002"-.014".

8. Coat pistons, pins, rings, bores and journals with engine oil. Position respective crank pin on top of travel and insert piston assembly. Number on piston is toward center of engine. Take care to avoid damage to the piston cooling nozzles. There is very little clearance between the bottom of the rod and the nozzle.

9. Coat bearing surfaces with engine oil and install bearing caps. Numbered sides of rod and cap should match. Use new cap bolts and torque to 130 ft lbs.

10. Check connecting rod end play—.008"-.020".

11. Install flywheel housing and gasket. Tighten capscrews evenly.

12. Check flywheel housing runout with a dial indicator adapted off of the crankshaft flange. The indicator pointer should rest on the housing flange. Check reading at four 90° opposed points. Crankshaft endplay should be held at zero. T.I.R. = .008".

13. Check housing concentricity in the same manner with indicator pointer on the inside flange surface. T.I.R. = .008".

14. Install crankshaft rear wear sleeve and oil seal. Do not attempt to install the seal and sleeve without installer tool SE-2205 or its equivalent. Do not drive the seal or sleeve into place. Lip on oil seal faces inward.

15. Install guide pins into crankshaft flange and position flywheel so that attaching bolts can be started. Draw up bolts evenly and torque to 235 ft lbs.

16. Install pilot bearing.

17. Position the oil inlet pipe and gasket on the oil pump cover.

18. Install the oil pump by bolting the pump to the #1 main bearing cap; the outlet pipe to the #2 main bearing cap, and the inlet pipe to the #3 and 4 main bearing caps.

19. Check gear backlash: between pumping gears—.008"-.012"; between idler gears and crankshaft gear: .004"-.014".

20. Lubricate and install the transfer tube in the right side of the block.

21. For injection pump installation, see the section entitled "Injector Pump Timing."

22. Insert power steering pump idler gear spacer in front cover plate. Position adapter and bolt in idler gear bearing. Install gear in

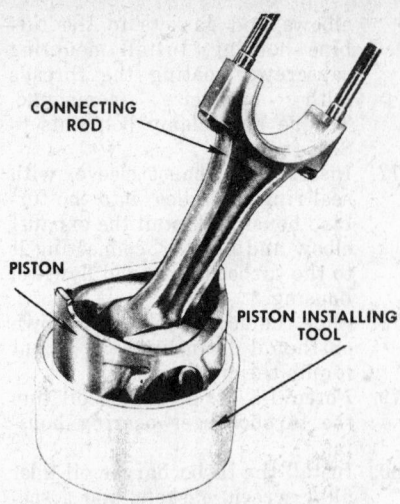
Piston in holding tool

front of block. Torque bolt to 150 ft lbs.

23. Install tappets in their bores as marked.

24. Connect oil delivery tube between oil filters and regulating valve.

25. Lubricate cylinder head mating surfaces, position new head gaskets on the block and install heads with the aid of a hoist.

26. Lubricate the bolt threads and flat washers with Nolykote "G" or equivalent; install and snug the eleven short bolts.

27. Install pushrods.

28. Lubricate and install valve bridges.

29. Install valve lever assembly by applying inward pressure on each end bracket until brackets are located on cylinder head.

30. Back-off adjusting screws and seat lever balls in pushrod sockets and valve lever bridges. Install and snug the nine long head bolts.

31. Torque the head bolts in sequence in the following four steps: 100, 160, 185, back-off ¼ turn then go to 205 ft lbs. After run-in, torque to 220 ft lbs.

32. Clean nozzle sleeves, place new O-rings and gaskets on nozzles, install nozzles in bores with retainers and bolts. Torque bolts to 35 ft lbs.

33. Install access hole covers.

34. Install pressure regulating valve and housing with gasket.

35. Connect oil delivery tube loosely to housing.

36. Install new gasket on front cover and install cover over dowels. Tap into place. Install capscrews.

37. Position the two machined blocks behind the engine front plate (the 3-hole block to the left). Position the bracket and pulley assembly and thread in the adjusting bolt to support the weight of the assembly. Install the four

capscrews but don't tighten until belts are installed.

38. Install the fuel filter mounting adapter and gasket to the right cylinder head.

39. Install the alternator mounting bracket.

40. Install the alternator adjusting strap and lifting eye.

41. Flush and install high pressure fuel lines. Torque connections to 20 ft lbs.

42. Install leak-off manifold to injection nozzles.

43. If front cover oil seal was replaced. install new wear sleeve.

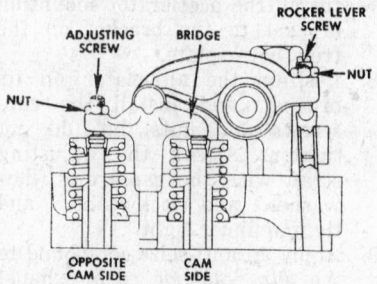
Valve lash adjustment

44. Install crankshaft pulley and torque nut to 425 ft lbs.

45. Install vibration damper and torque nuts evenly.

46. Install oil pan and new gasket. Torque capscrews to:
    ⅜ NC—25-30 ft lbs.
    ½ NC—65-75 ft lbs.

47. Adjust valves cold as follows:
    a. Bring no. 1 to TDC compression.
    b. Adjust intakes 1, 2, 4, 5; exhausts 1, 3, 7, 8.
    c. Turn crankshaft one revolution and bring #6 to TDC compression.
    d. Adjust intakes 3, 6, 7, 8; exhausts 2, 4, 5, 6.

To Perform An Adjustment:
   a. Loosen the bridge adjusting screw nut and back out screw several turns.
   b. Hold bridge firmly against guide to contact cam side valve. There should be no contact between the opposite valve and bridge adjusting screw.
   c. Set the lash on the cam side valve. Leave feeler in place, hold bridge firmly against rocker adjust rocker screw so that feeler is snug and torque nut to 35 ft lbs.
   d. Set lash on opposite side in the same manner. Torque nut to 25 ft lbs.

48. Install head covers and torque capscrews to 25 ft lbs.

49. Install lube oil return line between the injection pump and tappet access cover.

50. Install injection pump linkage and bracket.

51. Install lube line between engine "V" and alternator bracket.

52. Install lube line between rear of alternator bracket and injection pump.

53. Install power steering pump, adapter and gasket into front cover. Install and tighten capscrews and washers.

54. Install intake manifold and gaskets. Tighten capscrews carefully. Install the aneroid sensing tube.

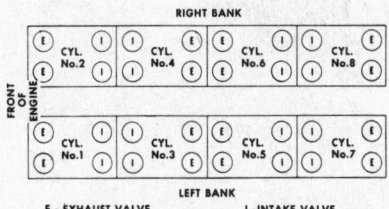

Valve arrangement

55. Connect all fuel return piping at the intake manifold.

56. Using a sling, install air compressor and new gasket. Torque capscrews to 45 ft lbs.

57. Connect the oil supply line between the compressor and the injection pump.

58. Install thermostats and housings.

59. Connect air compressor hoses to the left thermostat housing.

60. Install the refrigerant compressor and brackets to the front cover and left thermostat housing.

61. Cost exhaust couplings with anti-seize compound and insert into manifolds.

62. Place the turbocharger support manifold on the block and loosely install the four bolts.

63. Starting from the front, install the exhaust manifold one section at a time.

64. Torque all bolts *after* all sections are in place. Torque to 85 ft lbs. and retorque after run-up.

65. Lower the turbocharger into place, guiding the oil return pipe and air outlet into their hoses.

66. Carefully align unit and tighten mounting nuts. Tighten hose clamps to 45 ft lbs.

67. Install the turbocharger oil supply line.

68. Install idler pulley bracket and turnbuckle to front cover.

69. Install starter. Torque bolts to 110 ft lbs.

70. Install oil dipstick tube.

71. Install final filter-reservoir assembly.

72. Install the bleed line to the tee on the cylinder head.

73. Connect the hoses from the fuel reservoir to the forward injection pump fitting and from the final filter to the rear of the transfer pump.

74. Install the alternator; do not tighten bolts.

75. Install fan and spacer.

76. Install the front engine mount bracket to the front cover.

77. Install the rear engine mount brackets to the flywheel housing.

78. Install the oil fill tube.

79. Attach hoist and remove engine from stand.

80. Remove the forward filter body and install the oil cooler and filter assembly.

81. Install filter body and drain plugs.

82. Connect oil delivery tube to the oil regulating valve housing and filter assembly.

83. Using a new oiled O-ring install the water pump.

84. Install the by-pass hoses and pipe from the thermostat to the water pump.

85. Install turbocharger air inlet elbow.

86. Install and adjust drive belts.

87. Close all drain plugs and petcocks.

88. Check all hoses and fittings for tightness.

ENGINE INSTALLATION

General

1. Check to be sure that all lines, cables, and similar parts removed from the engine assembly are properly reinstalled. Check all visible connections on units removed.

2. Exercise care throughout reinstallation to avoid pinching the wiring harness.

DV-462B, DV-550B, D-150, 170, 190

1. Lower the engine and align the main drive gear spline of the transmission with the clutch driven disc. Install flywheel housing to bell housing bolts and torque to 150 ft lbs. Install front mount components, and torque locknuts to 95-105 ft lbs. Install the rear mount components, and torque the locknuts to 95-105 ft lbs. Torque mounting bracket to flywheel housing bolts to 150 ft lbs.

2. Connect the wiring, pipes, hoses, and linkages that were disconnected at removal.

3. Remove the lifting plate and cable.

4. Install the thermostats into the front cover. Position the gasket over the thermostats, checking that the air compressor return line hole or cooling system bleeder hole is open, and install the housing. Connect the air compressor water return line to the engine water outlet.

5. Install the compressor governor and governor line.

6. Install the secondary fuel filter as described on the filter body.

7. Install new O-rings on the oil filter base, lubricate them with engine oil, and insert the mounting bolt into the filter body. Place a new element in the body bearing the description "DIESEL LUBE FILTER."

8. Mount the oil filter body to the base, and torque the bolt to 25-35 ft lbs.

9. Repeat 7 and 8 for the other filter.

10. Install the radiator support assembly. Install the radiator into the support. Install the radiator side panels.

11. Install the grille or shutter, and connect the shutter linkage to the cylinder.

12. Install the radiator stay rods.

13. Install the upper and lower rediator hoses and their clamps.

14. Latch both splash panels into position. Install hood and fenders, and attach the hood stop cables to the supports at the radiator. Attach the front bumper to the frame.

15. Connect the battery ground strap and fill the cooling system.

16. If possible, prime the engine lubricating system. DO NOT ROTATE CRANKSHAFT.

17. Fill the crankcase with the proper quantity of oil.

18. Start the engine and operate to determine that all controls are functioning properly. Check for air, water, and oil leaks.

DVT-573

1. Lower the engine into the frame and align front and rear mountings. Install front mount components, and torque locknuts to 80-100 ft lbs. Assemble the rear mounts and note that the insulators are in the carriers. Lower the rear mounting brackets, check that the insulators are still properly located, and install and tighten bolts and nuts.

2. Connect wiring, lines, hoses, and linkage disconnected during removal.

3. Remove the hoisting equipment and lifting chain.

4. Connect transmission and controls.

5. Connect exhaust and air intake lines to the turbocharger.

6. Connect the battery ground strap.

7. Fill the cooling system.

8. Prime the engine lubricating system, if possible. DO NOT ROTATE CRANKSHAFT.

9. Fill the crankcase to the proper level.

10. Start the engine and operate it

to determine that all controls are functioning properly. Check for air, water, and oil leaks.

V-800

1. Install in reverse of removal instructions. Take note of the following:
2. Install transmission and controls *after* removing hoisting equipment.
3. Connect exhaust and air induction systems to turbocharger.
4. Connect battery ground; fill cooling system.
5. Fill crankcase and prime engine. Inspect all hose connections.

Turbocharger AiResearch Units

Pre-Disassembly Checks

1. Check free-spin of wheels.
2. Position a magnetic dial indicator on a vise, clamp and support the unit and measure the axial end play at the shaft end. T.I.R. should be .003-.008".

Position of related parts

3. Remove oil drain cover and position dial indicator pointer on shaft and check radial shaft movement. T.I.R. should be .003-.007".

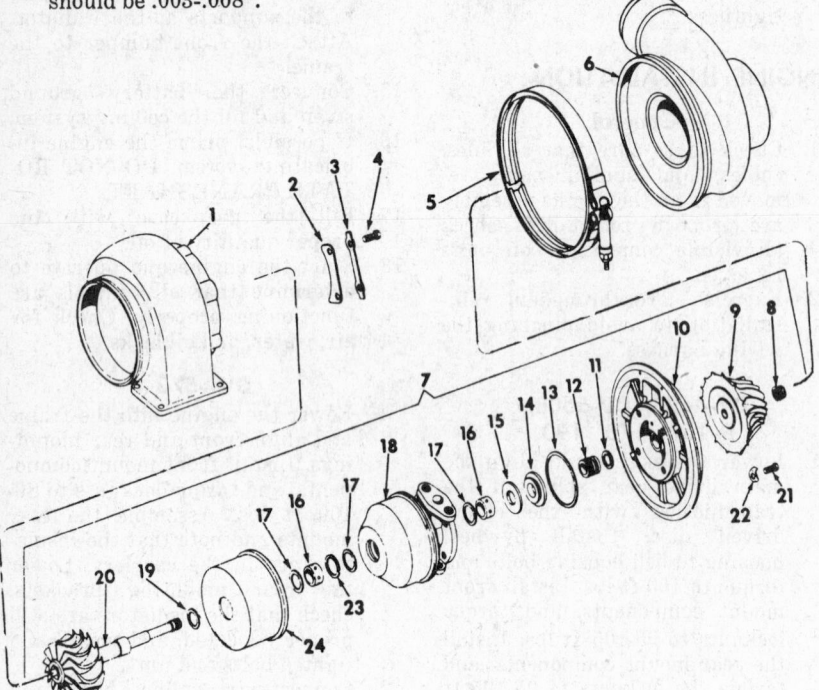

Turbocharger-exploded view

Disassembly

1. Cover intake and exhaust ports; wash unit in safe solvent.
2. Mark all components prior to disassembly.
3. Remove the clamp and lift the turbine housing, with center housing attached, from the compressor housing.
4. Remove the capscrews and lockplates securing the center housing to the turbine housing. Separate housings by tapping with a rubber mallet.
   CAUTION: Never rest the center housing on either the compressor or turbine wheel.
5. Support the center housing in an oil filled container with the compressor wheel submerged. Heat the oil to 350°F for ten minutes.
6. Support the hot until in a press and force the shaft from the wheel.
7. Loosen, but don't remove the bolts securing the center housing to the back plate.
8. Invert the center housing and remove the bolts and lock plates.
9. Remove the spacer, collar and bearing from the center housing.
10. Remove the seal ring from its groove.
11. Remove the bearing and washer from the compressor side.
12. Invert the center housing and remove the retainer ring, bearing and washer from the turbine end.
13. Remove the piston ring from the thrust spacer and shaft.

Assembly

1. Install new inboard retainers into grooves in the center housing bore. Round edge of retainer is to be toward bearing.
2. Install the washers in the center housing bore and install lubricated bearings until they bottom on washers.

Key	Description	Key	Description
1	HOUSING, Turbine	13	RING, Plate Seal
2	CLAMP, Housing (3)	14	COLLAR, Thrust
3	PLATE, Lock (3)	15	BEARING, Inboard Thrust
4	SCREW, Hex Hd Cap (6)	16	BEARING, Shaft (2)
5	CLAMP, V-Band	17	RING, Retaining (3)
6	HOUSING, Compressor	18	HOUSING, Rotating
7	CORE ASSEMBLY	19	RING, Piston Seal
8	NUT, Impeller Lock	20	WHEEL and SHAFT, Turbine
9	WHEEL, Impeller	21	SCREW, Hex Hd Cap (4)
10	PLATE, Back	22	PLATE, Bolt Lock (4)
11	RING, Piston	23	WASHER, Bearing (2)
12	SPACER, Thrust	24	SHROUD, Turbine Wheel

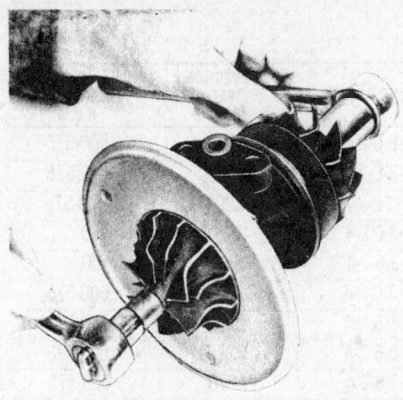

Bearing removal

3. Install new bearing retainer in the turbine end of the center housing.

4. Install a new piston ring into the groove on the shaft.

5. Place the shroud over the turbine end of the center housing. Insert the shaft through the shroud and center housing. A gentle rocking and pushing action should suffice.

6. Slip the bearing over the shaft. The hole and cutout must engage the anti-rotating pins in the center housing.

7. Install a new seal ring onto the spacer. Install the spacer into the bore of the thrust collar.

8. Install the collar, with spacer, over the shaft and on to the bearing.

9. Install a new seal ring in the compressor side.

10. Align the marks and install the back plate over the shaft and thrust spacer. Install capscrews and torque to 80-100 in lbs.

11. Using the same heating process as for disassembly, press the compressor wheel onto the shaft. Tighten the lock nut, using two wrenches to 120 in lbs. hot. When assembly has cooled below 150°F remove the lock nut. Reinstall the locknut and torque to 18-20 in lbs. and 1/3 turn.

12. Check wheel free spin.

13. Align marks and install center housing into turbine housing.

14. Install clamps, lockplates and bolts. Torque bolts to 160-190 in lbs.

15. Align marks and install the compressor housing. Install clamps and torque nut to 100-130 in lbs.

16. Check for free rotation.

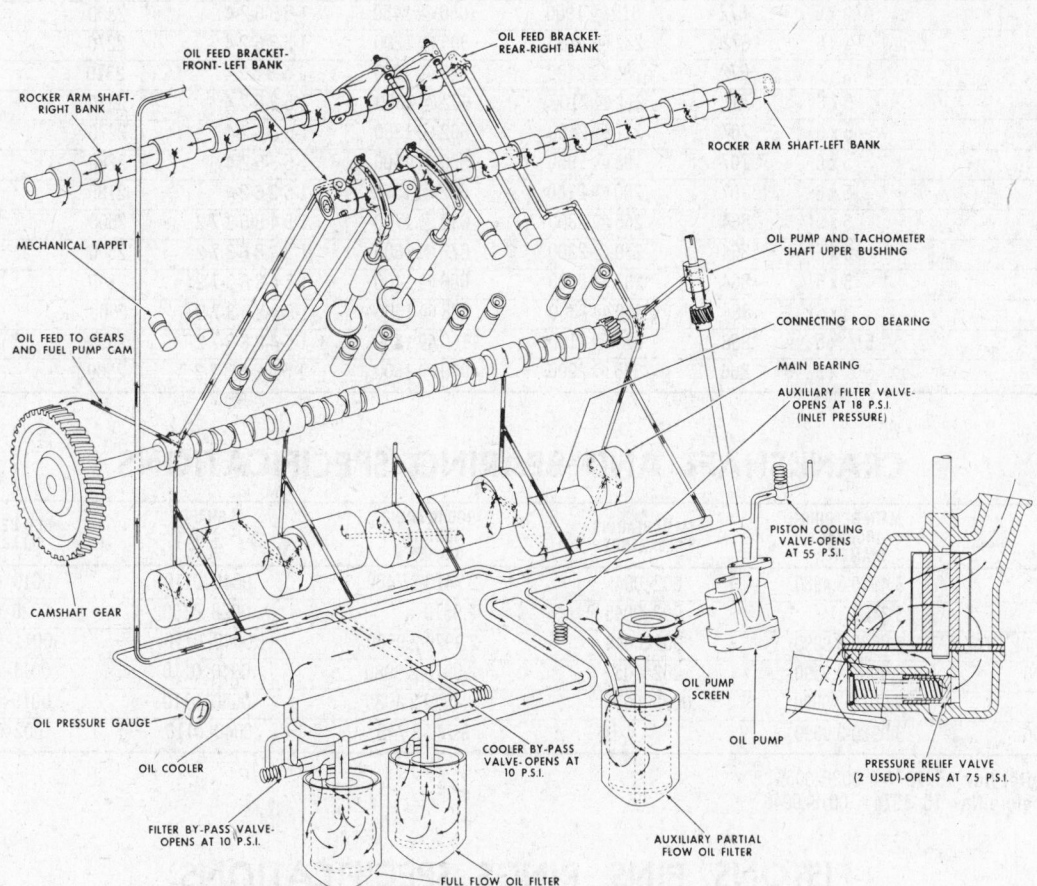

ENGINE LUBRICATION SYSTEM

GENERAL SPECIFICATIONS

ENGINE MODEL	BORE & STROKE	PISTON DISP. CU. IN.	HORSEPOWER @ R.P.M.	TORQUE @ R.P.M.	FIRING ORDER*	GOVERNOR SPEED NO LOAD	IDLE SPEED R.P.M.
END 465	4⁷/₁₆ x 5	464	110 @ 2600	253 @ 1600	1-5-3-6-2-4	2800	500-575
END 465B	4⁷/₁₆ x 5	464	140 @ 2600	325 @ 1600	1-5-3-6-2-4	2800	500-575
END 465C	4⁷/₁₆ x 5	464	155 @ 2600	350 @ 1600	1-5-3-6-2-4	2800	500-575
END 475	4.53 x 4.92	475	140 @ 2400	350 @ 1300	1-5-3-6-2-4	2600	450
ENDT 475	4.53 x 4.92	475	190 @ 2400	470 @ 1500	1-5-3-6-2-4	2600	500
END 673	4⁷/₈ x 6	672	176 @ 2100	496 @ 1200	1-5-3-6-2-4	2270	500-575
END 673B	4⁷/₈ x 6	672	176 @ 2100	496 @ 1200	1-5-3-6-2-4	2270	500-575
END 673A	4⁷/₈ x 6	672	155 @ 2100	426 @ 1400	1-5-3-6-2-4	2270	500-575
END 673C	4⁷/₈ x 6	672	187 @ 2100	527 @ 1400	1-5-3-6-2-4	2270	500-575
END 673P	4⁷/₈ x 6	672	187 @ 2100	527 @ 1400	1-5-3-6-2-4	2270	500-575
END 673E	4⁷/₈ x 6	672	180 @ 2100	540 @ 1400	1-5-3-6-2-4	2270	500-575
ENDT 673	4⁷/₈ x 6	672	225 @ 2100	653 @ 1500	1-5-3-6-2-4	2265	500-575
ENDT 673A	4⁷/₈ x 6	672	225 @ 2100	653 @ 1600	1-5-3-6-2-4	2320	500-575
ENDT 673B	4⁷/₈ x 6	672	250 @ 2100	704 @ 1500	1-5-3-6-2-4	2310	500-575
ENDT 673C	4⁷/₈ x 6	672	250 @ 2100	704 @ 1500	1-5-3-6-2-4	2310	500-575
ETAZ 673A	4⁷/₈ x 6	672	315 @ 1900	1050 @ 1450	1-5-3-6-2-4	2300	500-550
ENDT 675	4⁷/₈ x 6	672	237 @ 1700	906 @ 1200	1-5-3-6-2-4	2270	500-575
ENDT 676ESI	4⁷/₈ x 6	672	285 @ 2100	1080 @ 1200	1-5-3-6-2-4	2310	525-575
END 711	5 x 6	707	211 @ 2100	602 @ 1350	1-5-3-6-2-4	2270	500-575
END 711A	5 x 6	707	211 @ 2100	602 @ 1350	1-5-3-6-2-4	2270	500-575
END 707	5 x 6	707	195 @ 1950	550 @ 1600	1-5-3-6-2-4	2280	500-575
ENDD 711	5 x 6	707	200 @ 2100	602 @ 1350	1-5-3-6-2-4	2185	500-575
END 864	5 x 5½	864	255 @ 2300	639 @ 1700	1-5-4-8-6-3-7-2	2500	500-575
END 864B	5 x 5½	864	270 @ 2300	672 @ 1600	1-5-4-8-6-3-7-2	2500	500-575
ENDD 864	5 x 5½	864	237 @ 2000	686 @ 1400	1-5-4-8-6-3-7-2	2190	550-575
ENDT 864	5 x 5½	864	300 @ 2300	788 @ 1600	1-5-4-8-6-3-7-2	2500	550-575
ENDT 865	5¼ x 5	866	325 @ 2100	1100 @ 1350	1-5-4-8-6-3-7-2	2650	600-650
ENDT 866	5¼ x 5	866	375 @ 2200	1040 @ 1600	1-5-4-8-6-3-7-2	2500	600-650

CRANKSHAFT AND BEARING SPECIFICATIONS

ENGINE SERIES	MAIN BEARING JOURNAL DIAMETER	MAIN BEARING CLEARANCE	ROD BEARING JOURNAL DIAMETER	SHAFT END PLAY	ROD BEARING CLEARANCE
END 465	3.4970-3.4980	.002-.004	2.7470-2.7480	.0040-.0110	.0010-.0035
END 475	3.3465	.002-.0045	2.9510	.0020-.0100	.0020-.0040
END(T) 673, 711, ETAZ 673	3.9980-3.9990	.002-.0045	2.9970-2.9980	.0040-.0110	.0011-.0036
ENDT 675, 676	3.9980-3.9990	.002-.005	2.9970-2.9980	.0040-.0110	.0011-.0036
END(T) 864	3.9980-3.9990	.0025-.0055	3.4970-3.4980	.0040-.0110	.0015-.0045
ENDT 865, 866	3.9980-3.9990	①	3.747-3.748	.0040-.0110	.002-.005

① Engines after Nov. 15, 1974—.0026-.0056
Engines before Nov. 15, 1974—.0016-.0046

PISTONS, PINS, RINGS SPECIFICATIONS

ENGINE SERIES	PISTON CLEARANCE	PISTON PIN DIAMETER	PISTON TO PIN BUSHING CLEARANCE	RING END GAP COMP.	RING END GAP OIL
END 465	.0060-.0070	1.4366	.0001-.0003	.0130-.0250	.0130-.0250①
END 475	.0060	1.6538-1.6540	.0001-.0003	.0200-.0280②	.0120-.0240
END(T) 673, 711, ETAZ 673	.0050-.0070①	1.8742-1.8743⑥	.0001-.0002	.0130-.0250③	.0130-.0250④
ENDT 675, 676	.0050-.0070	1.9988-1.9990	.0001-.0002	.0130-.0250	.0130-.0250
END(T) 864	.0081-.0099	1.8744-1.8743	.0001-.0002	.0150-.0320	.0150-.0320⑤
ENDT 865, 866	.0080	2.1240-2.1241	.0002-.0005	.0150-.0320	.0150-.0320

① Figures apply to 673 engines only. For 711 engines, figures are: .008"-.010".
② Figure refers to #1 ring. #2 and #3 are .0120"-.0240".
③ Figures are for 673 engines. For 711 engines, figures are: .008"-.010".
④ One piece U-shaped ring requires no end gap. Figures are for 673 engines. For 711 engines, figures are: .008"-.010".
⑤ One piece U-shaped ring requires no end gap.
⑥ ENDT 673—1.9988-1.9989

R.H. Side of ENDT475

L.H. Side of ENDT475

CYLINDER HEAD BOLT TIGHTENING SEQUENCE

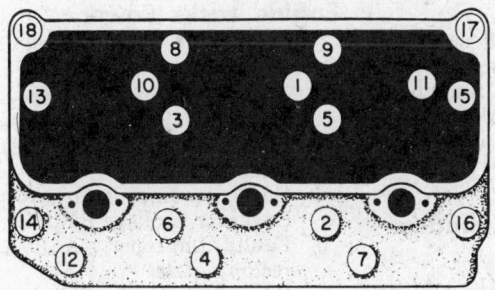

Cylinder Head Nut Torquing
Sequence END465 Series

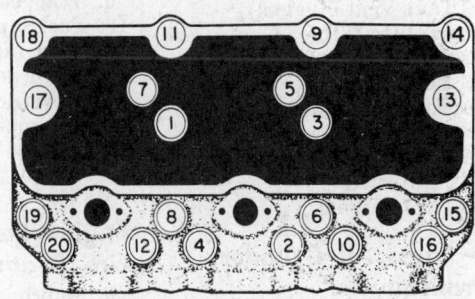

Cylinder Head Nut Torquing
Sequence END, T673 and 711 (5/8 in. Stud)

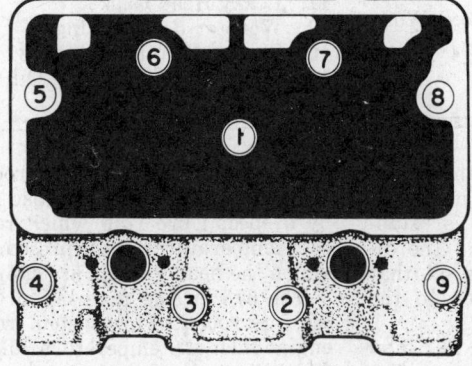

Cylinder Head Nut Torquing
Sequence END475 Series

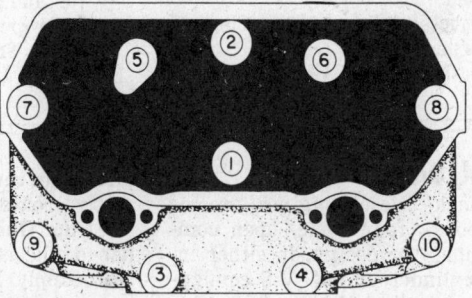

Cylinder Head Nut Torquing
Sequence END, T, 864 Series

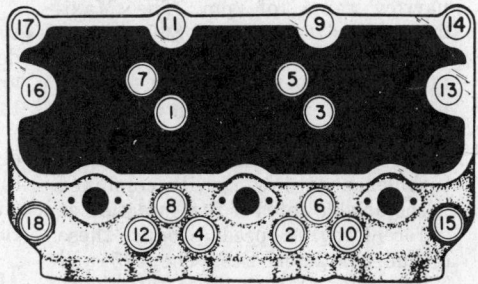

Cylinder Head Nut Torquing
Sequence END673 and 711 (3/4 in. Stud at
15 & 18 Locations)

Mack Diesel Engine

ENGINE TORQUE SPECIFICATIONS

ENGINE SERIES	CYLINDER HEAD FT. LBS.	MAIN BEARING BOLTS FT. LBS.	CONN. ROD BOLTS FT. LBS.	ROCKER BRACKET CAPS FT. LBS.	FLYWHEEL MOUNTING NUTS FT. LBS.	CORE PLUGS FT. LBS.
465	150	150①	150①	40①	150-175②	40-50
475	140①	150	80①	40	149	—
673, 711	150③	150④	150①	55①	150-175	40-50
675, 676	175	200	150①	55①	150-175	40-50
864	175①⑤	240①⑥	150①	55①	150①	40-50
865, 866	225①	350①⑦	⑧	35	150①	40-50

① Oiled
② Nylock—150
③ ⅝" nuts—175, ¾"—300
④ 1 1/16"—200
⑤ 1 1/16"—235 oiled
⑥ Turbocharged—275 oiled
⑦ Buttress Screw—100 oiled
⑧ Caps stamped 3178: 140-150 ftlb., oiled
Caps stamped 3178A: 170-180 ftlb., oiled

Trouble Diagnosis

Engine Cranks, Will Not Start

1. Emergency Shutoff Valve Closed
2. Air Intake Manifold Closed
3. Fuel Tank Vent Plugged
4. Temperature Too Low (Use Starting Aid)
5. Fuel Pump Inoperative, Leaks
6. Injection Pump Rack Stuck
7. Governor Problems
8. Injection Pump Worn, Incorrectly Adjusted
9. Poor Compression

Engine Runs, All Cylinders Fire Erratically

1. Low Fuel Pressure
2. Improper Governor Idle Adjustment
3. Air Intake Manifold Valve Partly Closed
4. Restricted Air Intake, Clogged Air Cleaner
5. Injection Pump Worn, Improperly Adjusted
6. Low Compression
7. Leaky Head Gasket
8. Valve Adjustment Improper
9. Plugged Fuel Tank Vent
10. Fuel Supply Pump Faulty
11. Fuel Lines Leaky or Restricted
12. Governor Shaft Binding
13. Governor Internal Malfunction
14. Improper Governor Adjustments
15. Faulty Injection Pump Overflow Valve

Engine Runs, Some Cylinders Fire Erratically

1. Improper Governor Idle Adjustment
2. Poor Compression (on some cylinders)
3. Injection Pump Condition or Adjustments Improper
4. Improper Valve Clearance, Worn Valve Train
5. Faulty Injector Nozzle
6. Injection Pump Plunger Rack Stuck
7. Improper Valve Clearance, Worn Valve Train

Engine Lacks Power

1. Emergency Shutoff Cock Partly Closed
2. Inadequate Fuel Pressure—Poor Pump, Clogged, Leaky Lines
3. Injection Pump Faulty or Adjustments Improper
4. Governor Improperly Adjusted, Governor Linkage Binding
5. Faulty, Improperly Adjusted Injector Nozzle
6. Poor Compression
7. Improper Valve Clearance
8. Turbocharger Passing Oil
9. Turbocharger Impeller Binding
10. Air Intake Clogged
11. Leaky Head Gasket
12. Worn or Broken Rings
13. Scuffed Pistons or Liners
14. Turbocharger Passing Oil

General Description

All Mack Diesel engines are of water cooled, four cycle, open combustion chamber design, in either in-line six cylinder or 90° V-8 configuration. All in-line engines have dry type cylinder liners, while all V-8 engines use wet type sleeves. All in-line engines employ two cylinder heads which cover three cylinders each. V-8 heads cover 2 cylinders each.

Large in-line engines and V-8 engines employ water-cooled oil coolers, in addition to the sump, to control lubricant temperature. Turbocharged engines employ a special piston cooling system consisting of nozzles located in the crankcase which spray oil over the entire inner surface of each piston. Oil is supplied, in most cases, directly from the oil pump. Nozzles are sized so that oil flow will be eliminated at idle speed to prevent over-cooling.

Some turbocharged engines also employ an anti-lag valve which sup-plies the turbocharger with oil directly routed from the pump at start-up. Once oil pressure is established, the anti-lag closes off the direct supply line, and sends oil that has been cooled and filtered to the turbocharger.

Series 465, 475, 673, 711, 707, and 864 engines are relatively conventional in design, and operate efficiently only within a relatively narrow range of rpm. The Maxi-dyne 675, 676, 865, and 866 engines are different, however, in having a much wider operating range with especially high torque output a relatively low rpm. Combustion chamber and manifold design that produces high turbulence at low speeds combined with variable rates of fuel injection at different engine speeds produce these torque characteristics. The 675 and 676 engines feature taper top piston design and corresponding connecting rod design for distribution of combustion pressure over a greater area of the wrist pin, piston, and rod. The combustion chambers are also sealed by a special fire ring which reduces the temperatures and pressures to which the head gaskets are subjected.

The 676 engine features an air to air intercooler. This system reduces engine inlet air temperature, allowing for the introduction of a greater mass of cooler air to the combustion chamber.

The 865 and 866 engines also features an oversquare or short-stroke design for higher maximum speeds, and pressure regulated oil supply to the piston cooling nozzles.

The 865 engine also features an oversquare or short-stroke design for higher maximum speeds, and pressure regulated oil supply to the piston cooling nozzles.

Troubleshooting—Compression Testing

It should be remembered that Diesels are compression ignition engines which employ only the heat generated during compression of the intake air charge to initiate combustion. While fuel or ignition system problems are most often the cause of actual misfiring in gasoline engines, a Diesel can misfire because of a basic mechanical malfunction such as a burned valve or clogged air intake, as well as fuel system problems. It should also be remembered that engine temperature has such a great effect on compression heating effect that a perfectly maintained engine may fail to fire when cold, requiring use of ether to initiate combustion. Finally, it should be remembered that adequate cranking speed is necessary for full compression temperatures to be generated. If the cranking system is performing even slightly below par, a good engine may not start when cold.

Because low compression is a most common cause of operating problems, compression testing has been included in the Troubleshooting section. The Tune-Up and Fuel System sections which follow should also be consulted in case the engine operates poorly. Because of the importance of efficient atomization in burning the relatively non-volatile Diesel fuel, and because of the importance of proper breathing and compression, these sections are placed adjacent to the Troubleshooting section. Before attempting troubleshooting, it is wise to check for sufficient fuel and oil of the proper types, restricted air intakes, and loose or restricted fuel lines or fittings.

Where misfiring of some cylinders is suspected, the problem cylinder(s) may be identified by loosening the high pressure fitting on top of the fuel injection pump ½ turn for each cylinder, one at a time. Faulty cylinders will produce no additional firing impulses when the fitting is retorqued.

If referral to the chart indicated low compression could be the cause of the malfunction, test the compression of each cylinder as described below:

1. Operate the engine at fast idle until it reaches operating temperature. Stop the engine.
2. Disconnect the fuel line at the nozzle of the cylinder to be checked. Place a container near the open end of the line to catch the fuel.
3. Remove the injection nozzle and holder assembly. Clean carbon from the nozzle tip with a special wire brush.
4. Remove carbon from the cylinder head nozzle hole with a special reamer and wire brush. Use an air gun to remove loose carbon. Crank the engine a few times with the starter to remove any remaining loose material.
5. Install a special adapter onto the

gauge, place a copper gasket on the adapter, and insert the assembly into the nozzle hole. Secure the assembly with the injector hold-down nuts.
6. Start the engine, and set the throttle for 1,000 rpm. Compare the reading with the figures on the chart below.

Engine Series	Minimum	Normal
END 465	486	540
END 475	490	540
ENDT 475	430	470
END 673	500	530
END 711	500	530
ENDT 673	525	575
ENDTL 673	525	575
ENDLT 673B	415	460
END 864	490	525
ENDT 864	450	485
END 707	500	530
END 675, 676	—	460
ENDT 865	—	485
ENDT 866	—	*

* Prior to Eng. #7WO515—485
  After Eng. #7WO515—540

7. Remove the gauge. Reinstall the injection nozzle and holder with a new gasket washer and holder dust seal. *Make sure only one gasket washer is used.*
8. Repeat the test for the remaining cylinders.
9. Compare readings with each other. They should be within 50 psi.
10. Check all nozzle holder and fuel line fittings for proper torque. Start the engine and make sure there are no fuel leaks.

Tune-Up

1. Adjust valve clearance as follows:
   a. Run the engine at fast idle long enough to normalize it. This takes about ½ hour with a cold engine, 15 minutes with an engine which has just come off the road.
   b. Stop the engine, remove rocker covers, and restart the

engine, running it at normal idle.
   c. Using the specified gauge check valve clearance. If necessary, loosen the locknut, rotate the adjusting screw with a screwdriver, and resecure locknut.
   d. Recheck clearance, making sure a light pull will move the gauge between the rocker shoe and valve stem.
   e. Repeat steps C and D for all valves. Shut down engine.
   f. Replace rocker cover gaskets, reinstall rocker covers, and run the engine to check for leaks.
2. Check the injection pump timing as described in the following sections.

Injection Pump Timing to Engine END(T) 465, 673, 711-PSJ-PSM Pump

1. Rotate engine in the normal direction of rotation until no. 1 cylinder is on compression stroke and the flywheel timing mark indicates the timing in degrees specified on the data plate on the pump. Remove the timing gear inspection hole capscrews, cover, and gasket. Remove control lever cover and timing device cover.
2. Rotate the injection pump drive shaft by turning the hub counterclockwise until the place bolts bottom lightly at the ends of the elongated slots. Then, slowly rotate the shaft clockwise until the pointer in the timing device housing and the scribe mark on the timing device spider flange line up. See the illustration. If this cannot be accomplished, shift the position of the driving gear one tooth in the appropriate direction and repeat the procedure.
3. Torque the place bolts on the driven gear to the pump shaft hub. Torque the injection pump drive shaft gear hub nut.

VALVE CLEARANCES

Engine	Cold Static		Hot Idle	
	Inlet	Exhaust	Inlet	Exhaust
END465 series	.017 in.	.023 in.	.016 in.	.022 in.
END475	.014 in.	.018 in.	.012 in.	.016 in.
ENDT475	.014 in.	.028 in.	.012 in.	.026 in.
END673 series	.016 in.	.024 in.	.014 in.	.022 in.
ENDT673	.016 in.	.024 in.	.014 in.	.022 in.
END707 & 711	.016 in.	.024 in.	.014 in.	.022 in.
ENDT675	.016 in.	.024 in.	.014 in.	.022 in.
ENDT676	.016 in.	.024 in.	.014 in.	.022 in.
END864	.016 in.	.024 in.	.014 in.	.022 in.
ENDT864	.016 in.	.024 in.	.014 in.	.022 in.
ENDT865 series	.016 in.	.026 in.	.015 in.	— —
ENDT(B)865 series	.016 in.	.026 in.*	.015 in.	— —
ENDT866	.016 in.	.026 in.	.015 in.	— —

* Press downward with hand on adjustment screw while gauging valve clearance. Cannot be adjusted with engine running. Use Automation Aids Adjusting Tool #MVT-36-3 to prevent damage to parts.

Aligning pointer and scribe mark on type PSJ, PSM pumps

4. Rotate the engine over until no. 1 cylinder is again on the compression stroke and the flywheel timing mark again indicates the specified timing setting.

5. Recheck the timing marks in the control lever window and timing device housing, and on the spider flange and ensure all marks align correctly.

6. Reinstall the control lever cover, timing device cover, and timing gear inspection hole gasket cover. Tighten the capscrews on the timing gear cover evenly to avoid oil leaks.

Injection Pump Timing to Engine END(T) 673, 675, 711-APE and PE Pump

1. Rotate the engine in normal direction of rotation until no. 1 cylinder is on compression stroke and the flywheel timing mark is approaching the setting in degrees which is specified on the valve cover escutcheon plate.

2. Remove no. 1 delivery valve holder and its retaining nut from the injection pump.

3. Install a suitable air line, equipped with a separator and pressure regulator, onto the IN fitting of the injector pump. Failure to equip the line properly may cause serious pump damage because of the admission of moisture.

4. Attach a fixture, fabricated as shown in the illustration, to the delivery valve holder.

5. Turn on the air supply and adjust the regulator so that a steady flow of bubbles exists within the fixture.

6. Slowly rotate the crankshaft, watching the bubbles, just until the flow of bubbles stops.

7. If bubbles do not stop at precisely the right point, loosen the four bolts which secure the hub to the timing gear. Make sure bolts are loose enough to permit the gear to rotate to the fully retarded position.

8. Bar the engine in the normal direction of rotation until no. 1 cylinder is on the compression stroke and the flywhel timing mark indicates the specified timing.

9. Very slowly turn the hub nut with an 11/16" box wrench toward the advanced position. Watch the air bubbles and stop cranking the hub when the bubbles stop. Tighten two capscrews which are opposite each other to secure the hub for the next step.

10. Bar the engine over in the normal direction of rotation, stopping the instant when bubbles stop flowing. If the timing is correct, tighten the other two hub capscrews. Otherwise, repeat steps nine and 10 until timing is correct.

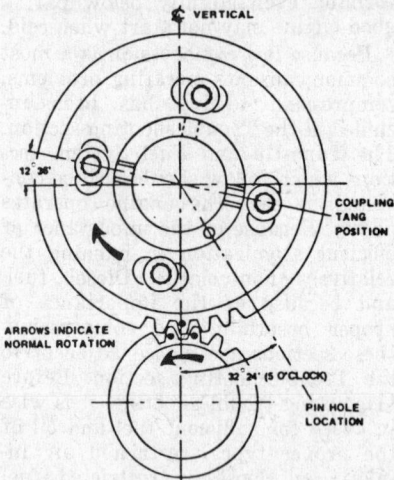

Position of drive coupling tangs and timing marks at No. 1 piston TDC.

11. Shut off the air supply, disconnect the air line, remove the fixture from the injection pump, and then install the delivery valve and spring, torquing the holder to specification.

Injection Pump Timing to Engine END 475

1. Bar the engine in the normal direction of rotation until no. 1 cylinder is on compression stroke and the flywheel timing mark

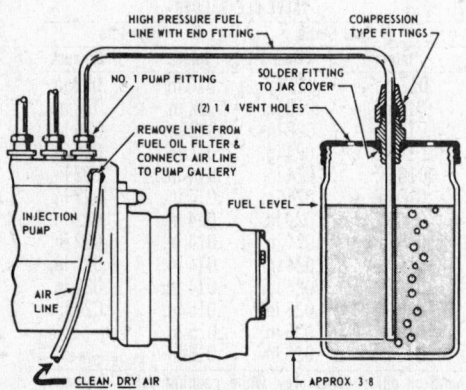

APE injection pump air flow testing device

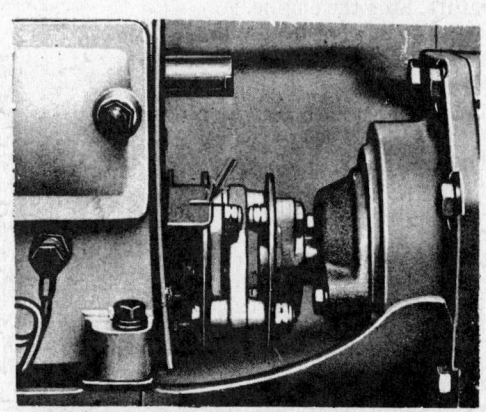

Injection pump closing mark on END475 engine

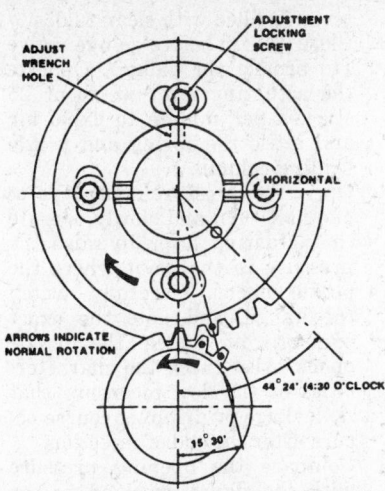

Position of drive coupling tangs and timing marks just prior to pump installation.

corresponds with the recommended setting in degrees shown on the pump data plate.

2. Check the port closing marks on pump and coupling. If they are not aligned, loosen the two coupling adjustment bolts, adjust the coupling until it is aligned, and tighten the bolts.

3. Bar the engine over in normal direction of rotation until crankshaft is again in timing position and check for proper alignment of marks. Repeat the adjustment if necessary.

Injection Pump Timing to Engine END(T) 864, 865, 866

1. Turn the engine in the normal direction until no. 1 cylinder is on compression stroke, and the flywheel timing mark is lined up with the specified mark.

2. Remove the inspection covers on top of the injection pump and auxiliary drive housing.

3. Check to see if the timing mark aligns with the pointer which is visible through the hole on top of the injection pump. If it does not align, leave the coupling screw adjacent to the timing marks tight, but loosen the other two screws, rotating the crankshaft as necessary to accomplish this. Then, repeat step 1.

4. Loosen the third screw as necessary, align the timing marks, and then tighten all three screws. Repeat step 1, check the timing, readjust and recheck as necessary, and reinstall all covers.

Injection Timing

All engines
Exc. Calif.
High Pressure Method:

1. Cap or connect injection lines on all except #1 delivery valve outlet.

2. Remove all fuel return lines at overflow relief valve union fitting, and cap valve port connections.

3. Connect the high pressure line from the portable PC stand to the fuel pump inlet of the pump gallery.

4. Connect return line from the #1 cylinder delivery valve holder to the portable PC stand.

5. Loosen the engine-to-pump coupling bolts and move the injection pump drive counterclockwise (rear view-8 cyl; front view—6 cyl) until pump drive is at end of adjusting slots. Hand tighten bolts.

6. Turn engine crankshaft clockwise (front view) until #1 cylinder is coming up on its compression stroke. Stop at the recommended static port closing degrees BTC for the particular engine involved.

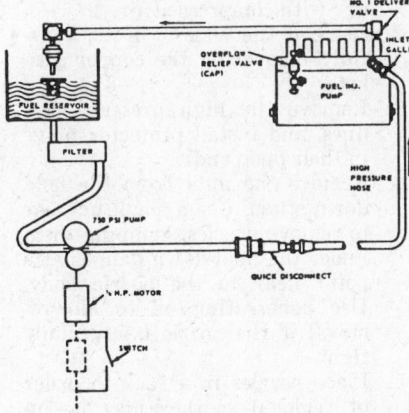

High pressure port closing system diagram.

7. Remove control rack cap plug and insert correct timing plunge gauge—MACK # J-24345-1 for American Bosch and J-24345-2 for Robert Bosch.

NOTE: If pump does not have retard start, a plug gauge is not required. If pump has a torque limiter, a plug gauge is not required when setting static timing; 80-120 psi air pressure must be applied and held to the torque limiter.

8. Secure stop level in running position.

9. Work throttle lever several times and secure in the full load position.

10. Apply fuel pressure to the gallery.

11. Turn pump drive clockwise (rear view—8 cyl; front view—6 cyl) until the fuel flow from #1 delivery outlet changes from a solid stream to drops.

12. Lock the drive gear bolts securely. Do not disturb the relative position of the gear shaft.

13. Turn engine crankshaft clockwise a minimum of 45° followed

by clockwise rotation to check port closing at desired timing on damper or flywheel.

14. Torque the drive gear bolts: 35 ft. lb.—6 cyl, 26 ft. lb.—8 cyl.

15. Remove the timing plug gauge and replace the control rack cap plug.

16. Remove caps from the overflow relief valve union port connections and reconnect all fuel return lines.

California Engines

1. Run engine to normal operating temperature.

2. Run the engine at low idle for one minute.

3. DO NOT ACCELERATE THE ENGINE ABOVE LOW IDLE. Shut the engine off. This makes certain that the extravance timing device will be in the full retard position for pump-to-engine timing.

4. Follow the above procedure for High Pressure Method, steps 1 through 16.

Alternate Low Pressure Method:

1. Remove #1 delivery valve holder from injection pump and remove delivery valve and spring. On Robert Bosch pumps, remove the shim.

2. Install a suitable air line onto the "in" fitting of the pump gallery.

CAUTION: Equip air line with a separator & regulator. Moisture will damage pump parts.

3. Rig up a fixture and attach to the delivery valve holder.

4. Secure stop lever in the running position.

5. Activate throttle lever several times and secure in full load position.

6. Remove control rack cap plug and insert timing plug gauge—MACK # J-24345-1 for American Bosch and J-24345-2 for Robert Bosch.

NOTE: Pumps without retard start and those with torque limiter do not require plus gauge. Those with torque limiter do need 80-120 psi air applied and held to torque limiter.

7. Apply air and just crack regulator so that a steady flow of air bubbles is seen in the fixture jar.

8. Rotate crankshaft slowly in direction of normal rotation. At the instant bubbles stop, stop rotating shaft.

9. Timing indicator must register recommended degrees BTC.

10. Repeat steps 8 & 9 to ensure accuracy.

If timing is not correct:

11. Loosen the engine-to-pump drive coupling bolts and move the pump drive counterclockwise (rear view—8 cyl; front view—

6 cyl (until pump drive is at end of adjusting slots. Snug tighten bolts.

12. Turn engine in direction of normal rotation until #1 cylinder is on compression stroke & timing mark indicates correct degrees BTC.

13. Turn pump drive clockwise (rear view) while checking bubbles in fixture jar. Stop when bubbles cease.

14. Tighten two opposite bolts to secure hub, and rotate crankshaft and check timing.

15. When timing is correct, tighten two remaining bolts.

16. Remove timing plug gauge and replace control rack plug.

17. Replace delivery valve and spring. Torque the delivery valve holder nut.

Engine Auxiliaries

Fuel System

NOTE: Mack fuel injection pumps are factory sealed. No adjustments should be made by other than Mack authorized repair stations, or the engine warranty may be voided. Therefore, these adjustments are not covered here.

Fuel Injector Service

The fuel injectors in Mack diesel engines receive metered fuel at injection pressure from the injector pump. The nozzles are used to provide effective atomization of the fuel through use of a nozzle valve and discharge nozzle, which insure sharp start-up and cutoff of fuel flow and production of four high velocity streams which will effectively penetrate to all parts of the combustion chamber.

The nozzle body at the lower end of the injector consists of the hole type nozzle, the seat for the nozzle valve, and the fuel passage. The nozzle valve spindle extends upward to the top of the nozzle holder, where there is an adjustable spring for maintenance of precise injector pop pressure. Controlled clearance between the nozzle body and the nozzle valve lubricates the mechanism. A leak-off fitting at the top of the injector returns excess fuel to the fuel tank.

Injector nozzles should be removed and checked after 50,000 to 75,000 miles, or if troubleshooting reveals poor performance. *Extreme* care must be taken to ensure absolute cleanliness during nozzle service. It must also be remembered that oil leaving an injector nozzle is moving at extreme speed. Contact with the skin will usually result in penetration. Therefore, testing must be carried out in a manner that will protect the skin from nozzle discharge.

Removal

1. Clean the cylinder head around the nozzles and the tubing connections with solvent. Blow dry with compressed air.

2. Remove the leak-off lines, carefully recovering the copper gaskets.

3. Remove the high pressure fuel lines, and install protector plugs in their open ends.

4. Remove the nuts from the hold down studs. Use a small pry bar to remove nozzles, gripping them under the hold-down flanges at a point near to the nozzle body. Use penetrating oil to aid removal if the nozzle is especially tight.

5. Place nozzles in a rack in order of removal so they may be installed in the engine in original order. Plug the nozzle ports in the cylinder head(s).

Testing

1. Clean carbon from the nozzle with a special wire brush as shown in the illustration.

2. Mount the nozzle holder assembly in a tester. Make sure the tester is filled with clean fuel.

3. Close the pressure gauge valve (to protect the gauge). Operate the actuating lever at about 25 strokes per minute to expel air and settle the spring and nozzle loading column.

4. Open the pressure gauge valve one half turn and slowly operate the actuating lever to raise the pressure to the point where the nozzle opens. Carefully watch the gauge, and note the exact pressure at which the nozzle opens. Also check the characteristics of the flow to ensure that no leakage or dripping course occurs after the end of injection.

5. Compare the opening pressure with the figure specified in the nozzle opening pressure chart and adjust the nozzle to specifications if necessary.

6. Wipe the nozzle tip dry. Operate

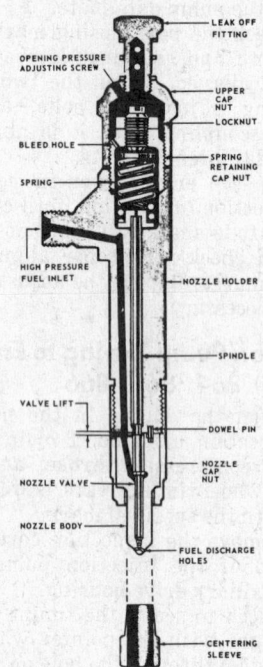

Typical fuel injection nozzle and holder

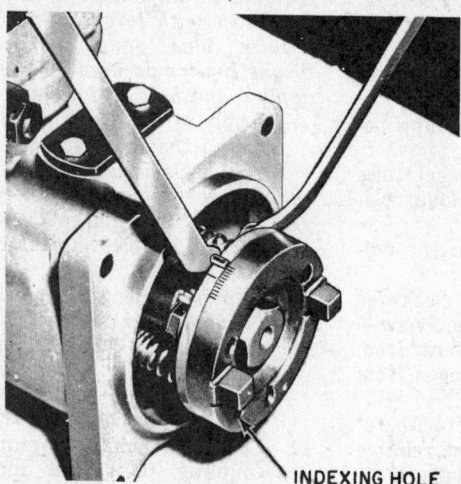

Adjusting injection pump coupling on END864 engine

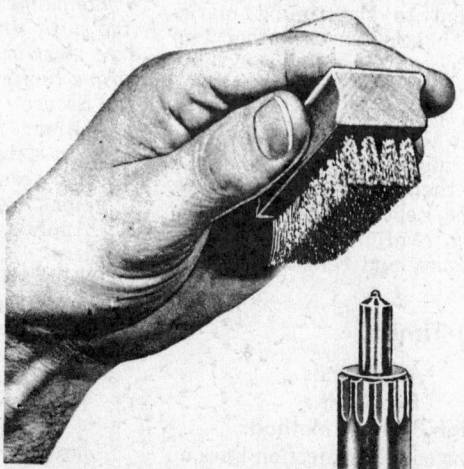

Removing carbon from nozzle tip

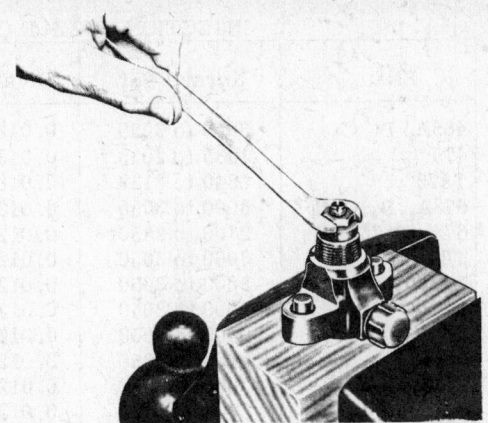

Loosening locknut on pressure adjusting screw

the actuating lever slowly to bring pressure to within 100 psi of opening pressure (20 psi with END 475 nozzles) and maintain the pressure for five seconds. If drops of fuel form or if the nozzle sprays slightly, reject it.

7. Close the gauge valve (to protect the gauge) and operate the lever at about 15 strokes per minute. The spray pattern formed should be sharp, solid, and with uniform quantities and angles between orifices. Use short, rapid strokes on END 475 nozzles to produce a good spray pattern. Reject a nozzle with a poor spray pattern.

8. On all but 475 series nozzles, test the nozzle to make sure it makes a chattering sound. Operate the actuating handle so the stroke takes about two seconds. Close the pressure gauge valve. A distinct and regular chattering sound must be produced, although an occasional variation is acceptable. Reject a nozzle which does not pass the chatter test.

Disassembly

NOTE: The American Bosch AKN 90 SM nozzle holder assembly (21/17MM Injectors used on ENDT 676, 865 and 866 engines) consists of a leak-off adapter, holder body, shims pressure adjusting spring, spring seat, doweled nozzle spacer and a nozzle retaining cap nut. The nozzle and holder should be presoaked thoroughly in a suitable carbon removing solvent prior to disassembly. Only a brass brush should be used on the tip area. Rotate the cap nut, using a box wrench and minimum force. If the nozzle rotates with the cap nut, STOP immediately. The nozzle, in this case, is stuck in the cap nut due to carbon deposits. Any attempt to loosen the cap nut will damage the spacer dowel pins, holder and nozzle. Tool TSE 77105 should be used for safe disassembly in the event of a stuck nozzle.

1. Loosen the opening pressure adjusting screw all the way to re-

live all downward pressure.
2. Position the nozzle assembly on a suitable block in a vise. Remove the upper cap nut.
3. Loosen the locknut, if the nozzle is equipped with one, and loosen the opening pressure adjusting screw all the way to relieve all downward pressure.
4. Use a special wrench to remove the spring retaining capnut. Remove the pressure adjusting spring and spindle assembly.
5. Invert the holder assembly in a soft-jawed vise and remove the nozzle capnut. Remove the nozzle body and valve assembly. Keep these two parts together as they are a mated assembly. Reinstall nozzle cap nut loosely.

Inspection

1. Wash all parts in a safe solvent and inspect for wear as described below:
   a. Check the spring for corrosion and pitting and replace if they are evident.
   b. Check spindle for straightness within .010" T.I.R.
   c. Check the lapped surfaces of the nozzle body and holder for cracks and scratches. Replace parts that are cracked, and lap parts which are slighly scratched. If the holder lapped surface is spalled more than .003" deep in the needle valve area, it must be replaced rather than resurfaced.
   d. Remove the locating dowels from the holder very carefully to prevent damaging the lapped surface. Check the spring retaining cap nut to make sure the bleed hole is open.
   e. Clean the pressure chamber of the nozzle body with a special scraper, as shown in the illustration.
   f. Using a special needle and vise, clean the nozzle body holes, as in the illustration.

Be careful to avoid breaking the needle inside the discharge holes, as fragments may prove to be impossible to remove.
   g. Lift the valve about 1/3 of its length out of the nozzle body, hold the body at 45 degrees from the vertical, and release the valve. It should slide back freely. If the valve is not free, work the valve in the nozzle body using a special polishing tallow. Clean the nozzle valve and body in solvent and blow dry.
   h. Check the nozzle valve lift using a straightedge and dial indicator as shown in the illustration. See the injection Nozzle Opening Pressure Chart for specifications. Place the nozzle in a fixture and run a straightedge across the top of the nozzle body. Mount a dial indicator right

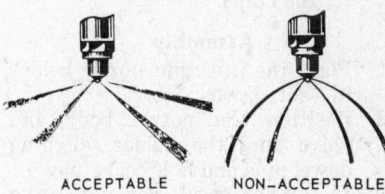

ACCEPTABLE   NON-ACCEPTABLE

Nozzle spray pattern

Removing spring retaining cap nut

Mack Diesel Engine

INJECTION NOZZLE OPENING PRESSURE CHART

END	Normal PSI	Valve Lift	Inj. Pump	Holes	Dia. Inch
465A, B, C	3000 to 3050	0.012 to 0.014	PSJ, PSM	5	0.010
475	1985 to 2035	0.012 to 0.014	CAV	4	0.011
T475	2840 to 3130	0.010 to 0.012	CAV	5	0.011
673A, B, C, D	3000 to 3050	0.012 to 0.014	PSJ, PSM	4	0.0128
673, T673	2300 to 2350	0.012 to 0.014	APE	4	0.0128
T673A, B	3000 to 3050	0.012 to 0.014	PSJ, PSM	4	0.0132
T675	3000 to 3050	0.012 to 0.014	APE	5	0.0126
T673C	3000 to 3050	0.012 to 0.014	APE	5	0.0126
707	3000 to 3050	0.012 to 0.014	APE	4	0.0138
673E	3000 to 3050	0.012 to 0.014	APE	4	0.0138
711A, B	3000 to 3050	0.012 to 0.014	PSJ, PSM	4	0.0138
711	2300 to 2350	0.012 to 0.014	APE	4	0.0128
864	2300 to 2350	0.012 to 0.014	PSJ, PSM	4	0.0138
D864	2300 to 2350	0.012 to 0.014	PSJ, PSM	4	0.0128
T864	3000 to 3050	0.012 to 0.014	PSJ, PSM	4	0.0147
T864	3000 to 3050	0.012 to 0.014	APE	5	0.0126
T865	3000 to 3050	0.012 to 0.014	APE	5	0.0126

IF A NEW SPRING IS USED, OPENING PRESSURE IS TEN PERCENT HIGHER

above the needle valve and zero it. While holding the straightedge, raise the needle valve with a pair of tweezers just until the lower portion of the valve contacts the straightedge. Hold the valve in position while reading the dial indicator. If the valve lift is not to specification, the assembly must be replaced.

i. If the injector uses a filter in the fuel inlet connection, clean it by reverse flushing it with compressed air.

j. Check the ends of the high pressure fuel tubes to make sure they are open. Ream ENDT 864 lines to .085" and all others to .078". Flush out the chips.

Assembly

1. Place the injection nozzle holder in a soft jawed vise.
2. Position the nozzle body and valve onto the holder, aligning dowel pins and holes carefully.
3. Install the nozzle cap nut, using a special centering sleeve during the initial tightening. Remove the centering sleeve and torque the cap nut to specification, using a special adapter.
4. Invert the holder and install the spindle and pressure adjusting spring. Tighten the pressure adjusting screw and lock it in position with the locknut.
5. Install the injector in a test stand and adjust opening pressure as described in the section on service.
6. Install the nozzle holder with a new gasket and install and tighten upper capnuts.

Installation

1. Remove copper nozzle tip gasket

from the hole in the cylinder head. Clean the nozzle cavity with a special reamer and wire brush. Check the gasket seat for trueness and cleanliness. Crank the engine over by hand to blow loose carbon from the cavity.

2. Apply an anti-seize compound to the outside diameter of nozzle and holder, and position the assembly in the head with a new nozzle tip gasket and dust seal O-ring. The nozzle tip gasket

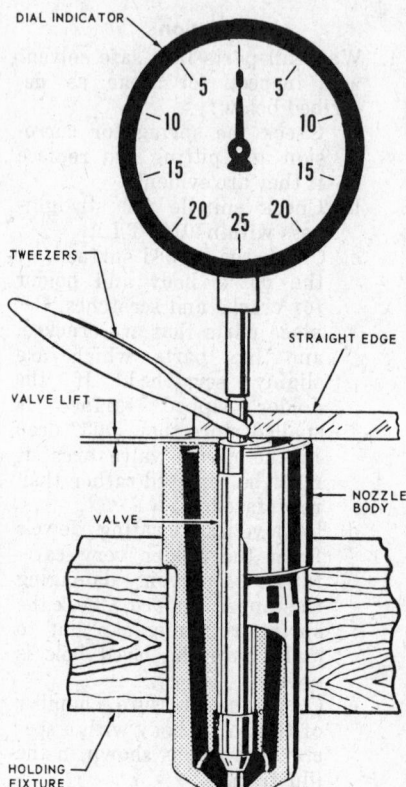

Checking nozzle valve lift

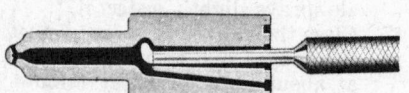

Removing carbon from nozzle body

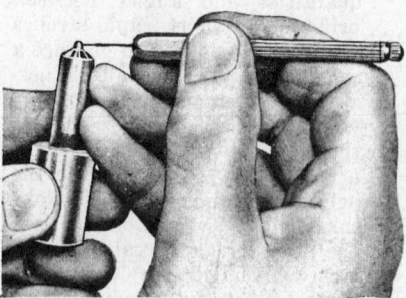

Cleaning nozzle body fuel discharge holes

may be held in place with a small amount of grease.

3. Seat the nozzle squarely and then install and tighten hold down nuts evenly.
4. Install the high pressure fuel lines and torque nuts carefully so that tubing will not be distorted.
5. Install the leak-off lines with their copper gaskets.
6. Operate the engine and check for leaks. Retighten fittings as necessary.

Fuel Filter Service

Because of the close clearances employed in fuel injection systems, extremely efficient filtration of fuel is an absolute necessity. In-line, full flow primary and secondary filters make up the filtration system. The secondary filter incorporates an air bleed valve which returns bleed air and fuel to the vent valve on the fuel tank.

Primary Filter

Both types of primary filters con-

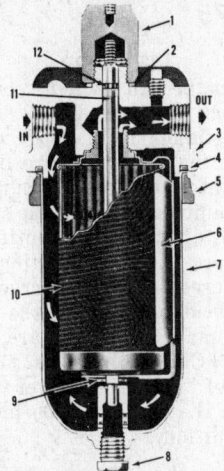

Primary fuel filter

1. Cap nut
2. Cap nut gasket
3. Cover
4. Cover gasket
5. Retaining ring
6. Cleaning blade
7. Housing
8. Drain plug
9. Cotter pin
10. Filter element
11. Shaft
12. O-Ring

tain a wire mesh filter which is cleaned by manually rotating a blade around the element. Every 2,000 miles, the cap nut should be rotated approximately two revolutions. Elements with the high type cap nut require that the cup nut be removed and installed upside-down on the shaft in order for it to be rotated. At this time, the cup nut gasket should be inspected. Replace it, if necessary, before reinstalling the nut. Torque the nut sufficiently to prevent leakage.

Both types of primary filters should be drained every 6,000 miles, or sooner, if fuel is of poor quality. To perform this operation, remove the drain plug until clean fuel appears. Replace the plug carefully, tightening it sufficiently to ensure that there will be no leakage.

On the type of filter with the low cap nut, the shaft o-rings should

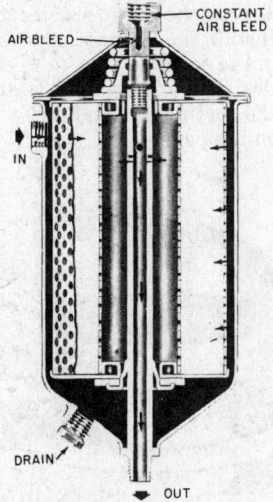

Secondary fuel filter

be replaced every 6,000 miles, or sooner if a leak develops. Replacement may be performed as follows:

1. Remove the shaft circlet and unscrew the shaft nut.
2. Remove the filter housing retaining ring bolts and nuts, and remove the housing.
3. Remove the cotter pin from the end of the shaft, and then withdraw the shaft.

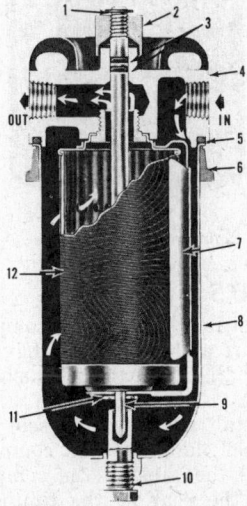

Primary fuel filter

1. Shaft circlet
2. Nut
3. O-ring
4. Cover
5. Gasket
6. Housing retaining ring
7. Cleaning blade
8. Filter housing
9. Shaft
10. Drain plug
11. Cotter pin
12. Filter element

4. Remove the old o-rings and install new ones from the unthreaded end of the shaft.
5. Lubricate the rings and then insert the shaft from the bottom.
6. Install the shaft nut, circlet and cotter pin.
7. Install a new gasket, fill the housing with clean fuel, place the housing in position, and install the retaining ring bolts and nuts.
8. Tighten the nuts evenly, and then prime the filter from the IN side with clean fuel. Operate the engine and check for leaks.

Secondary Filter

The element of the cartridge type fuel filter may be replaced as follows:

1. Remove the filter housing drain plug and allow the filter to drain.
2. On filters with constant air bleed, remove the air bleed line from the filter cover.
3. Remove the housing cover, gasket, and element.
4. Wash the interior of the housing with clean fuel oil. Replace the drain plug.
5. Install a new element. New gaskets should be employed at top and bottom of the new element

in the constant air bleed type of unit. The element in the manually vented type of unit should be kept in the protective plastic bag until right before installation.

6. Fill the housing with clean fuel. Install the cover securely with a new gasket.
7. Install the air bleed line on constant air bleed filters.
8. On manually vented type filters, close the bleed valve by turning it in, and pump the hand priming pump 150-200 strokes. Immediately after pumping stops, open the bleed valve, turning it outward all the way. Then, pump the primer 50 more strokes to pressurize the system.
9. Start the engine, and operate it while checking for leaks.
10. Replacement should again be performed after 6,000 miles on constant air bleed filters, 8,000 miles on manually vented filters. Replacement should also be made if fuel pressure drops below 10 psi with constant bleed filters, or 12 psi with manually vented filters.

Spin On Filters

Spin on type primary and secondary fuel filters are provided on some engines. The filters should be replaced every 16,000 miles, or sooner, if fuel pressure drops below 15 psi with PSJ/PSM pump, or 10 psi with the APE pump.

The filter may be replaced as described below:

1. Clean the adapter, line connections, and adjacent area with sol-

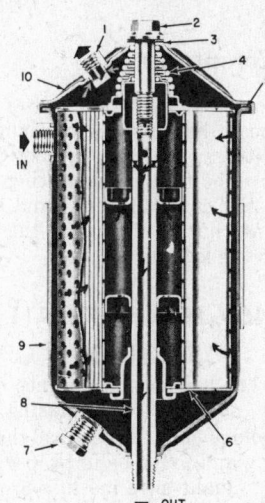

Secondary fuel filter

1. Bleed valve
2. Cover retaining screw
3. Washer
4. Spring
5. Gasket
6. Plastisol end cap
7. Drain plug
8. Center tube
9. Housing
10. Cover

vent spray. Dry with compressed air.

2. Use a special wrench to unscrew the filter from the adapter, remove and save the dust cap, and discard the used filter.
3. Clean the dust cover with solvent and blow dry with compressed air.
4. Install the dust cover on the new filter.
5. Apply a small amount of clean engine oil to the gasket sealing surface.
6. Screw the filter onto the adapter, and turn it by hand one full turn after the gasket contacts the adapter.
7. Loosen the secondary filter to injector pump fuel line at the pump on APE pumps, or the line at the filter bleed shutoff valve (with valve open) on PSJ/PSM pumps.
8. Operate the hand primer through about 200 strokes or until fuel flows, free of bubbles, from the open fitting.
9. Tighten the fuel line fitting on APE type pumps.
10. On APE pumps operate the primer through 20 more strokes, and secure the handle. On PSJ/PSM pumps, operate the primer until bubble-free fuel flows from the filter bleed valve, then close the valve and continue pumping until fuel flows from the filter bleed line. Retighten the bleed line and secure the pump handle.
11. Start the engine and operate it while checking for leaks. Open the filter bleed valve on PSJ/PSM fuel systems.

Air Compressor

Inspection

Inspect the compressor flange for straightness, lack of cracking, and proper condition of mounting holes. Check the coupling teeth and ensure the shaft nut is tight and the tab is bent to the locknut.

END 465, END(T) 673, 711

Installation

Current production engines do not require compressor phasing, but many older models require that the coupling and drive be in proper mesh. Check to make sure mesh is proper if problems occur.

1. Position the adapter ring in the auxiliary housing counterbore.
2. Place the gasket in position on the compressor and position the compressor on the auxiliary housing.
3. Install retaining capscrews with proper torque.

Spin-on secondary filter

END 475

1. Position key in the compressor shaft and install the drive gear.
2. Install the gear lockwasher and nut.
3. Install a new o-ring seal on the inner flange of the compressor, and then install the compressor on the rear of the timing gear housing.
4. Install lockwashers and torque nuts to specification.
5. Check the backlash between compressor and camshaft gear.
6. Install the timing gear inspection cover.

ENDT 864, 865, 866

1. Position the gasket on the compressor flange and position the compressor on the auxiliary drive housing.
2. Install the capscrews loosely. Adjust the position of the compressor until the backlash is correct.
3. Torque capscrews to specification.

Oil Pan

All Engines Except END 475 Series

Inspection

Inspect the pan for dents, cracks or other such damage. Check to make sure the flange is in good condition and that the bolt holes are not elongated. Make sure the baffle is securely mounted and that the drain hole threads are not damaged.

Installation

1. Position the gasket on the pan. Position the pan up against the block with all bolt holes properly lined up.
2. Install the capscrews, and tighten in rotation to the specified torque.

END 475

Inspection

Inspect the pan for dents, cracks or other such damage. Check to make sure the flange is in good condition and that the bolt holes are not elongated. Check the oil pressure relief valve mounting pad to make sure it is flat and that the mounting hole threads are in good condition. Check the oil screen for holes, cracks, and broken welds, and check the oil passages to make sure they are not obstructed. Check that the mounting surface of the screen cover is clean and that the pick up tube hole is in good condition.

Installation

1. Position a spacer on the stud in the center of the pan. Install the screen assembly over the stud, locating the crimped section on the lug. Install the lockwasher and nut.
2. Mount the baffle in the pan with capscrews and lockwashers.
3. Position the new gasket in place on the flange. Position a new o-ring on the pressure relief valve tube. Install a new felt washer on the oil pick up tube.
4. Position the pan on the crankcase and install the capscrews, torquing them evenly.

Oil Sump
END 475

Inspection

Inspect the sump casting and oil screen for dents, cracks or other damage. Make sure the mounting cover is clean and that the mounting surface and holes are not damaged.

Installation

1. Place the space on the stud in the center of the cover.
2. Position the sump over the stud. Install the lockwasher and nut. Torque the nut to specification.
3. Install the magnetic drain plug in the cover.
4. Position a new felt washer on the suction line and a new gasket on the pan.

Oil pan sump, END475 engine

OIL PUMP SPECIFICATIONS IN INCHES

	END 465	END 475	END(T) 673	END(T) 711	END 864	ENDT 675, 676	ENDT 865, 866
Bushings to Inner Rotor Shaft	.0015-.0025	0015-.0025					
Driven Gear to Idler Gear Backlash		.0020-.0040		.0020-.0060	.0020-.0060		.0050-.0070
Idler Gear to Crankshaft Gear Backlash	.0020-.0040	.0020-.0040			.0050-.0070		.004-.009
Gear O.D. to Cavity Side			.0025-.0045	.0035-.0060	.0035-.0060	.0035-.0060	.0035-.0070
Gear to Cover End			.0015-.0035	.0015-.0035	.0015-.0035	.0025-.0050	.0015-.0060①
Gear Backlash In Pump	.0250-.0290		.0250-.0290	.0250-.0290	.0250-.0290	.0295-.0235	.0250-.0290

① with 2" Gears—.0015-.0095"

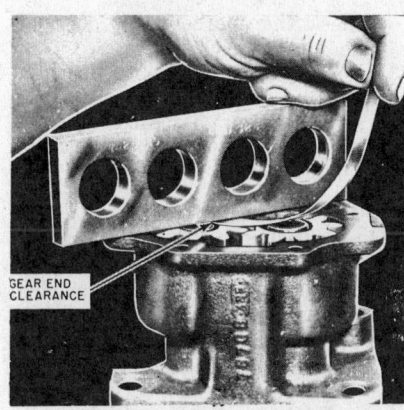

Checking oil pump gear end clearance, END465. 673, T, and 711 engines

5. Install the assembly. Install the capscrews and torque them evenly.

Oil Pump
ENDT 465, 673, 675, 676, 711

Inspection

1. Inspect the oil pump drive gear to make sure the gear teeth are in good condition and worn in a normal pattern. Check to make sure the journal wear is not excessive and that the splines are in good condition. Make sure oil holes are not obstructed.
2. Check the pump housing flanges and gear cavity for scoring or other damage. Make sure the idler shaft is not tight in the housing.
3. Check the oil pressure regulator body and cap to make sure they are clean and that all holes and threads are in good condition. Check regulator spring for proper tension and good condition of the ends.

Assembly

1. Apply a small amount of oil to the drive gear shaft and position the gear in a vise.
2. Install the housing over the shaft and position the drive gear on the shaft splines.
3. On END 465 Series engines, secure the drive gear with the Spirolox retaining ring.
4. Lubricate the idler gear shaft and install the idler gear.
5. Check the end clearance, side clearance, and backlash of the gears as shown in the illustrations. See the Oil Pump Specifications Chart.
6. Mount the oil pump cover in a vise and install the tab lock, pressure relief valve plunger, spring and cap. Torque the cap down and then bend the tab lock.
7. Remove the cover from the vise and position it on the oil pump body. Install the capscrews. Rotate the oil pump shaft to check for free rotation.
8. Lubricate the drive shaft gear and position the assembly on the block. Install the retaining capscrews, and torque and lockwire them.
9. Make sure the oil pick up pipe and shield assembly are clean and unobstructed. Place a gasket on the pump flange and install the oil pick up line. Torque the

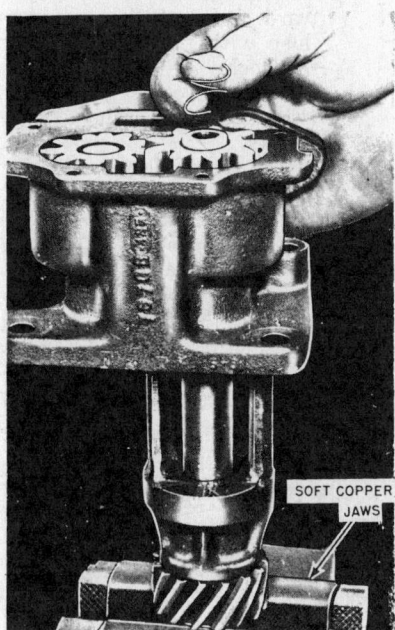

Spirolox retainer ring installation on END465 engine

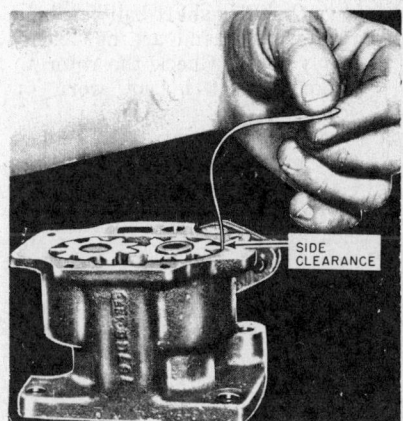

Checking oil pump gear side clearance on END465, 673, T, and 711 engines

screws and lockwire them together.
10. Install the oil screen, support, and inlet pipe.

END 475 Series

Inspection

1. Inspect the housing for cracks, the condition of the holes and flanges (flanges should be flat) and fully open passages.
2. Check the idler and driven gear to make sure they are worn in a regular pattern and that the teeth are in good condition. Make

Checking oil pump gear backlash on END465, 673, T, and 711 engines

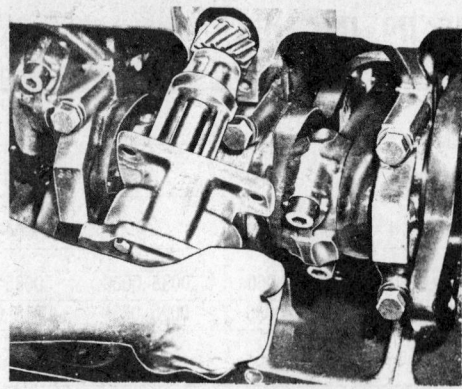

Installing oil pump on END465, 674, T, and 711 engines

Installing oil pump

sure the idler shaft ball bearings rotate easily and are not excessively worn. Check the rotor to make sure it is not worn or scored.

3. Check the bushing for wear. If they must be replaced, use a special tool to install them and ream to provide the required running clearance.

Assembly

1. Install the lock ring in the groove in the idler gear bore of the housing.
2. Install the rear ball bearing and stub shaft into the housing from the rear, as shown in the illustration.
3. Install the spacer washer over the stub shaft and install the front ball bearing.
4. Position the idler with the dish toward the housing, and install the capscrew to retain the idler gear and ball bearings (the capscrew is reverse threaded). Check the idler gear for freedom of rotation.
5. Install the rotor and shaft assembly into the housing. Position the cover on the housing and install the capscrews and lockwashers.
6. Put the key into the shaft, and then position the oil pump drive gear onto the shaft. Install the retaining washer and nut.
7. Check the backlash between idler and driven gear.

Oil pump idler gear END465, 674, T, and 711 engines

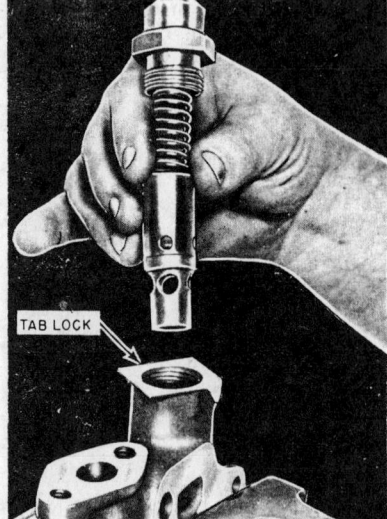

TAB LOCK

Installing oil pump pressure relief valve plunger on END465, 673, T, and 711 engines

Installation

1. Install the new o-ring at the pump outlet and position the assembly on the engine. Install the capscrews and lockwashers.

END 864

Inspection

1. Inspect the housing for cracks, damaged screw holes, bent flanges, or clogged passages.
2. Inspect idler and driven gears for damaged teeth or an improper wear pattern. Check the oil pump pressure and scavenge gears for damaged teeth, an improper wear pattern, or improper fit in the housing.

Assembly

1. Install the idler gear shaft into the housing. Install the pressure pump shaft and gear into the housing and put the driven idler gear onto the shaft.
2. Install the scavenge housing over both shafts and install the scavenge drive and driven gears onto their shafts. Put the scavenge pump cover into position, install and torque the capscrews, and

check the gears for proper end float and side clearance.

3. Position the shield onto the inlet housing and install the capscrews. Push the screen into its proper position in the shield assembly.

Installation

1. Install the locating plate on the pump and position the assembly on the block with the forward face of the mounting legs flush with the block and the guide pin through the locating plate.
2. Check the backlash between idler and drive gears. Tap the housing lightly to obtain the correct backlash, as shown in the illustration. Then, torque the mounting bolts.
3. If the oil pump gears have been replaced or a replacement pump is to be installed, guide pins may have to be realigned in locating plate as follows:
   a. Install the pump with the mounting screws lightly torqued. Make sure both forward faces of the legs are flush with the front face of the block.
   b. Install the locate plate with two lightly torqued capscrews.
   c. Install the guide pin into the

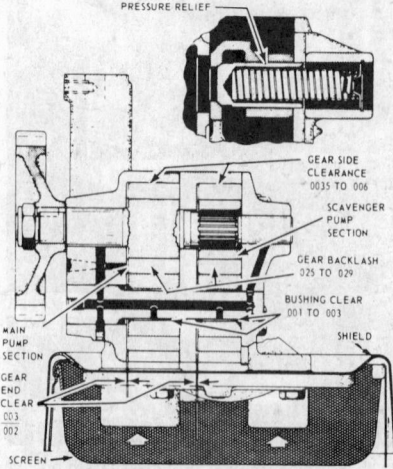

Oil pump END864 engines

Checking oil pump idler gear backlash ENDT864 engine

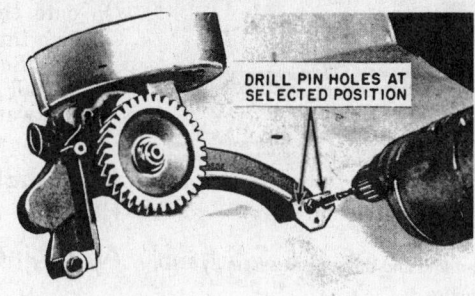

DRILL PIN HOLES AT SELECTED POSITION

Drilling new guide pin holes in oil pump, ENDT864 engine

recess in the block, going through the locating plate.

d. Lightly tap the pump housing to adjust the idler gear to drive gear backlash. When proper lash is established, tighten the locating plate screws.

e. Remove the assembly from the engine without loosening locating plate screws.

f. Drill two holes, .186"-194" in diameter, .188" deep through the locating holes in the locating plate into the housing leg.

g. Install two locating pins into the pump housing. Then, reinstall the pump assembly as previously outlined.

4. Position new gaskets on the block and pump outlet hole. Install the outlet tube with flat washers and nylock capscrews.

5. Install the scavenger tube clamp with the loop toward the centerline of the engine and secure it on the inner side of the bracket toward the center of the engine using a capscrew and elastic stop nut.

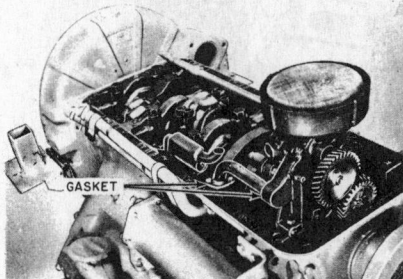

Oil pump outlet tube, END864 engine

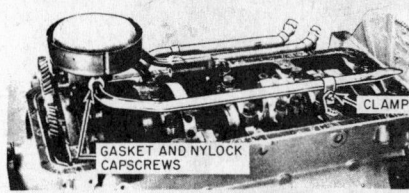

Oil pump scavenge tube, END864 engines

ENDT 865, 866

Inspection

1. Check pump gear for pattern and wear.

865, 866 oil pump idler gear.

2. Check pump shaft for wear.
3. Clear all oil passages.
4. Check housing, flanges and cavity for scoring and cracks.
5. Check idler shaft for scoring and binding.
6. Check safety valve plunger, body and cap for wear and dirt.
7. Check safety valve spring tension.

Installation

1. Align hole in driven gear with driven shaft and press on gear. Install roll pin.
2. Press idler shaft in flush with housing.
3. Lubricate gear and shaft assembly and install in housing.
4. Install pumping gear on splined end of shaft.
5. Press one idler gear bushing from each end into gear.
6. Install idler gear on shaft. Check backlash (.004-.009"), end float (.0015-.0040") and side clearance (.0035-.0070").
7. Install cover on housing. Torque capscrews (30 ft. lb.).
8. Rotate pump shaft to check for binding.
9. Install safety valve plunger, spring, cap, and retainer in cover.
10. Lubricate pump drive shaft gear and install on engine.

Intake Manifold
All Engines

Inspection

1. Inspect the manifold for restrictions.
2. Check the mounting flanges for erosion, cracks, and straightness.

Installation

1. Put new gaskets into position on the block.
2. Put the manifold in position on the block.
3. Install capscrews and tighten to the proper torque.
4. On 864 Series engines, position the O-rings and install the inlet manifold crossover tubes and connecting hose (see illustration).

Exhaust Manifold

Inspection

1. Inspect the manifold for erosion or restrictions.
2. Check mounting flanges for erosion, cracks, and straightness.

Installation

1. Put new gaskets in position on the head.
2. Install and tighten the attaching washers and nuts or the capscrews to the proper torque.

Water Outlet Manifold

Inspection

1. Inspect the manifold for restrictions.
2. Check the mounting flanges for cracks or warping.
3. On END (T) Series engines, check the thermostat seal in the water outlet fitting for looseness and erosion.

Installing thermostat seal END, T, 864 engines

Installation

1. Place new hose and clamps or a new tube and o-ring between front and rear halves of the manifold, as applicable. Put new

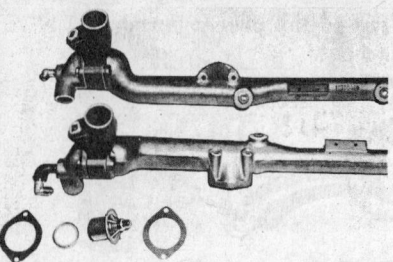
Water manifold and thermostat assembly END, T, 864 engines

gaskets into position on the cylinder head.
2. Install capscrews and tighten to the proper torque. Position the hoses and clamps on engines employing them.
3. Install the thermostat and by-pass fitting and the housing cover as appropriate. See illustrations. On END T 864, install a new thermostat seal, if necessary. See illustration.

Crankshaft Pulley and Vibration Damper

Crankshaft Hub-END 475 Series

Inspection
1. Inspect the hub for scoring, or damage to the threaded holes, or keyway.

Installation
1. Place the hub key in the crankshaft keyway.
2. Align the keyway slot in the hub with the key and start the hub onto the crankshaft. Drive the hub into position with a special tool. See illustration.
3. Install the wedge cone, bolt, and lockwasher. Torque the bolt.

Crankshaft Vibration Damper Hub

All Engines Except END 475

Inspection
1. Inspect the hub for scoring, or damage to the flange, threaded holes, or keyway.

Installation
1. Place the hub key in position in the keyway in the crankshaft.
2. Heat the vibration damper hub to 250° F. in hot oil, unless the damper is an integral part of the hub. In this case, heat the unit in a solution of one to two percent soluble oil and water at 200-225° F. for 13-14 minutes.
3. Wipe the bore with a dry rag, and position the unit onto the crankshaft with the keyway and key aligned. Drive the unit into position.
4. Install the washer and retaining bolt, and torque the bolt to specifications.

NOTE: On ENDT 865, 866 the TDC mark must align with the timing indicator 30 degrees to the left of the engine center line. On 6 cyl engines exc. MB chassis, a spun steel pulley replaces the cast iron original. Spacer 537GC215 must be used.

Combination Pulley and Vibration Damper END 475

Inspection
1. Inspect the v-belt channels, mounting holes, and flanges for excessive wear. Make sure the housing and rubber bond are sound.

Installation
1. Position the pulley and damper assembly on the crankshaft hub. Secure it with hardened washers and capscrews, torquing capscrews to specification. Where pulley and damper parts are separate, torque capscrews which hold them together to specification also.

Engine Disassembly

Mack diesels are easily disassembled with a minimum of special tools. Rather than detail a disassembly procedure, take note of the following helpful hints:
1. Always remove injection nozzles and holders before removing heads.
2. To remove PSJ or PSM pumps from END 465, 673 or 711 series, remove the access plate on timing gear cover. The pump driven gear must be removed before the pump can be removed.
3. Always remove the pump, governor and adapter as an assembly.

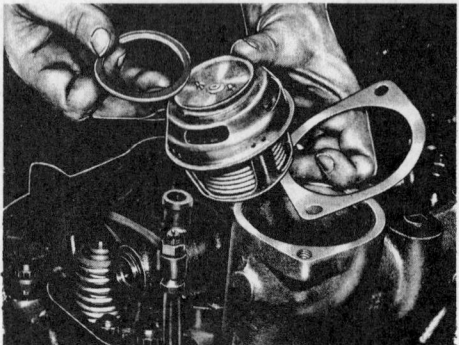
Installing thermostat END465 engines

Installing thermostat END465 engines

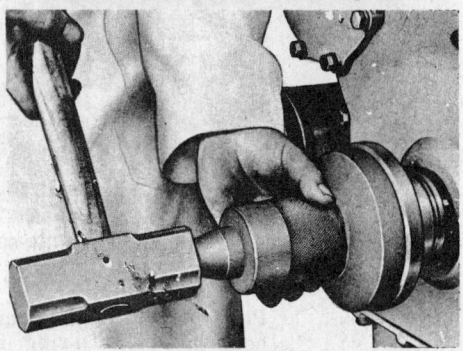
Installing crankshaft hub END475 engines

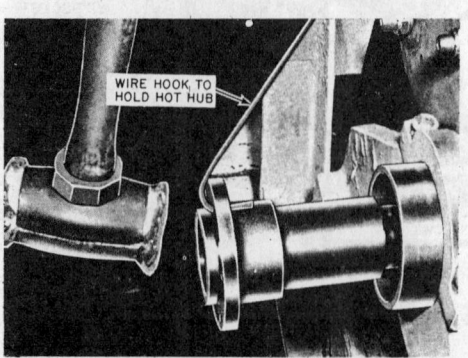
Installing crankshaft hub all engines except END475

4. On ENDT 864, 865, 866 series, be sure to remove buttress screws at main bearings before attempting to remove caps.

5. Use new capscrews on connecting rods at overhaul.

6. When removing piston and rod assemblies on engines with oil-cooled pistons, piston cooling nozzles must be removed before pistons are removed.

Engine Assembly

Cylinder Head

1. If new nozzle holder sleeves are used, use a 7/8-14NF thread, tap at the large end and remove the old sleeve with a puller.

2. To install new nozzle sleeves, heat to 180°F, apply a light coat of sealer to each end of the sleeve, position sleeve, small end first, in hole and press into place.

3. To install new valve guides, chill guides in dry ice and safe solvent, or in freezer for approximately 30 minutes. Install guide from top of head until it extends the required distance.

Engine Series	Guide Protrusion
465	1 9/16
475	63/64
673	1 35/64
711	1 35/64
864	1 7/8
865	1 7/8
866	1 7/8
675	—
676	—

4. Chill new valve seats in dry ice to −112°F and position in head using press. Hold seat in position while it warms up.

5. Grind seats to 30° 30'. END 475 uses a 30° angle.

6. Grind valves to 30° 30' angle; 29° 30' for END 475.

7. Install valves and check valve to piston head clearance.

Engine Series	Gauge
465	J-22149
673	17-T-11324
675	.014" min
676	.014" min
711	17-T-11486
864	J-22099
865	—
866	—

8. Valve spring installation.
   a) END 465
   1. Position inner spring centering washers, inner and outer valve springs, oil shields, and retainer washers (O-ring in intake) over stems.
   2. Install keys.
   b) END 475
   1. Position lower guide washers, inner and outer valve springs and upper guide washers over valve stems.
   2. Install keys.
   3. Install stem wear caps.
   c) END(T) 673, 675, 676, 711
   1. Position inner and outer valve springs over the exhaust and intake valves.
   2. Position the oil shield, O-ring and retainer washers over the intake valve stems and install keys.
   3. Position roto cap washers over the exhaust valve stems and install keys.
   d) END(T) 864, 865, 866
   1. Position lower guide washers, inner and outer valve springs, oil shields, O-rings and retainer washers over the intake valve stems and install keys.
   2. Position the roto coil washers, inner and outer valve springs and retainer washers over the exhaust valve stem and install keys. ENDT engines use roto coil on both intake and exhaust valves.

9. On "T" engines with piston spray coolers, be sure spray fittings are clean and unobstructed. On 673, 675, 677 engines, position the cooling nozzle on the inside crankcase wall. On 864, 865, 866 engines the spray fitting is installed from outside the crankcase and is located by the positioning sleeve.

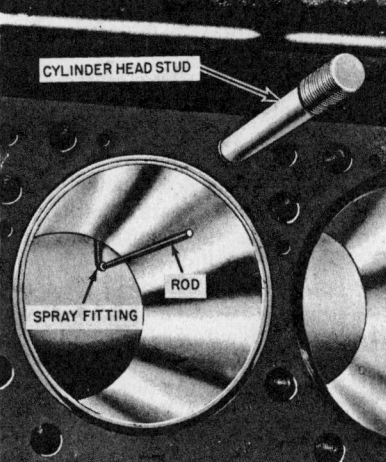

Positioning rod in piston cooling oil nozzle.

10. Cylinder lines—Dry type
   a. Check liner for out-of-round. Take reading in two directions at each of three different levels.
   b. Insert liner in bore. If liner enters bore 1/2 to 2/3 of the way, a light press should force liner into position.
   c. If liner does not enter 1/2 to 2/3 of bore and is within tolerance, shrink liner for 25 min. in dry ice and alcohol.
   d. Check liner shoulder protrusion: .0005 below to .0035 above.

11. Cylinder Sleeves—Wet type
   a. Check sleeves for out of round or taper. Take readings in three directions at each of three levels.
   b. Measure bore at upper and lower locations for sleeve fit.
   c. Position new sleeve in cylinder block, without O-rings to check sleeve-to-deck fit. Shims will adjust this fit.
   d. Check liner-above deck dimensions.
   e. Remove sleeve and install O-rings.
   f. Lubricate sleeve O.D. at ring locations.
   g. Insert sleeve until it contacts the O-rings.
   h. Push down with palms of hands in a quick motion to bottom sleeve. A light tap with a soft mallet is permissable.

NOTE: If, for any reason, the liner must be moved back out, for any distance, the O-rings will be damaged.

END 475 sleeves have a dot at the top which must face forward.

Crankshaft Installation

1. Inspect crankshaft for wear. Check oil passages.
2. Check pattern on crankshaft gear.
3. On END 865, ENDT 865, 866, check condition of wear ring. The wear ring can be removed by splitting it with a cold chisel. To install a new wear ring:
   a. Heat new ring to 400°F.
   b. Position heated ring on crankshaft with arrow on I.D. pointing toward center of shaft.
   c. Press wear ring on shaft to shoulder.
4. Install crankshaft gear aligning key in crankshaft.
5. Heat gear to 400°F (212°F—END 475).
6. Align key way and position gear on shaft with larger chamfer toward engine.
7. Drive gear into place using suitable driver.
8. Insert upper half of main bearing shells in cylinder block. Insert the lower half in the cap.
9. Apply a light coat of oil to the shell surfaces and journals and lower the crankshaft into position.
10. Align dots on timing gears. On END 475 engines, the number of teeth on the idler gear provides

Installing crankshaft wear ring—V8

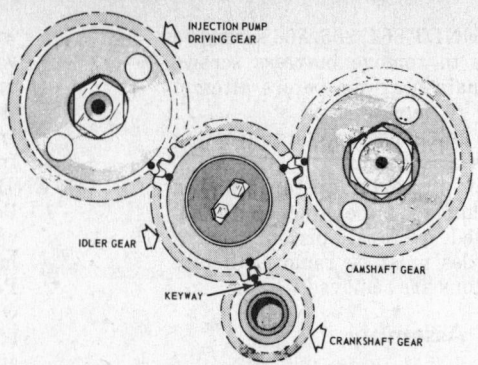

Timing gear alignment CAV pump—END 475

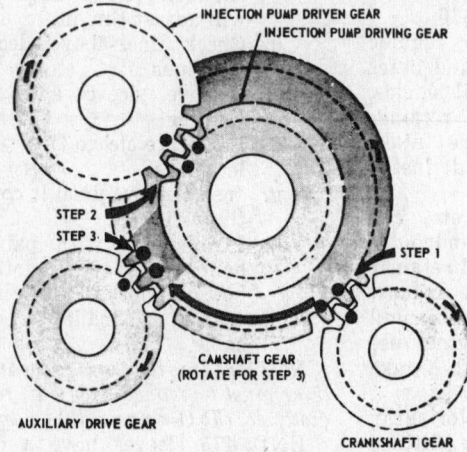

Timing gear alignment APE pump END 673, 711

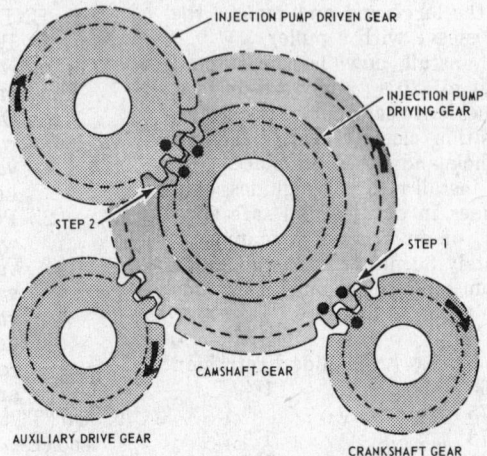

Timing gear alignment PSJ-PSM pump—END 673, 711

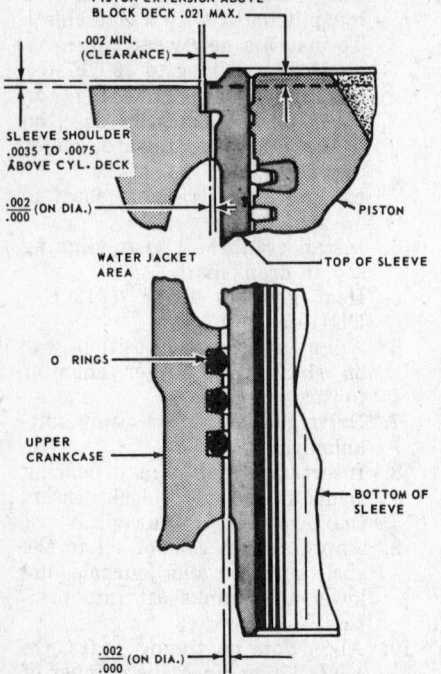

Cylinder sleeve fit END 864

the timing location. When the alignment marks are indexed, the idler gear will align only once every 54 revolutions with the other three gears. The three timing marks on the idler gear are located so that the crankshaft, camshaft and injection pump drive gears are all indexed on #1 cylinder compression.

On 673, 675, 676 and 711 series align the dots on the auxiliary drive gear with those on the injection pump drive gear. On engines without auxiliary drive gear timing marks:

a. Install the gears with the mating numbers outward.
b. Align the dots on the crankshaft and camshaft gears.
c. Align the dots on the injection pump gears.
d. Rotate the camshaft clockwise to align it with the dots on the auxiliary shaft drive gear.

11. Install front and rear thrust washers at center main bearing.
12. Force crankshaft fore and aft to align thrust washers.
13. Check end play.
14. Check main bearing clearance with plastigage.
15. Torque main bearing caps. Check crankshaft free rotation. Use oversize thrust washer or undersize main bearings to correct binding.
16. On 864 engines, buttress screws are used at the 2, 3 and 4 main bearing locations. On 865, 866 engines, buttress screws are used at all main bearing locations. On these engines the short buttress screws are located at the 2 and 4 locations. Torque buttress screws *after* main bearing capscrews are torqued.

Pistons and Connecting Rods exc. V8 Engines

1. Heat piston to 200°F. Position piston on connecting rod and install pin. Arrow on piston crown faces front.
2. Install pin retainers.
3. Install rings on pistons with staggered gaps.
4. Coat piston, rings and liner with light oil.
5. With respective crankpin at bottom center and rod number on auxiliary side, install piston and rod assemblies in their bores with arrows toward front.
6. Guide large end of rod into position. Mack tool #J-21480 is used for this purpose.
CAUTION: If engine is turbocharged, use extreme care to avoid damage to piston spray cooler or rod bearings.
7. Check bearing clearance with plastigage.
8. Coat journals and bearings with oil, install caps and capscrews, check rod and cap side clearance (.007″), and torque capscrews.
9. On turbocharged engines, check

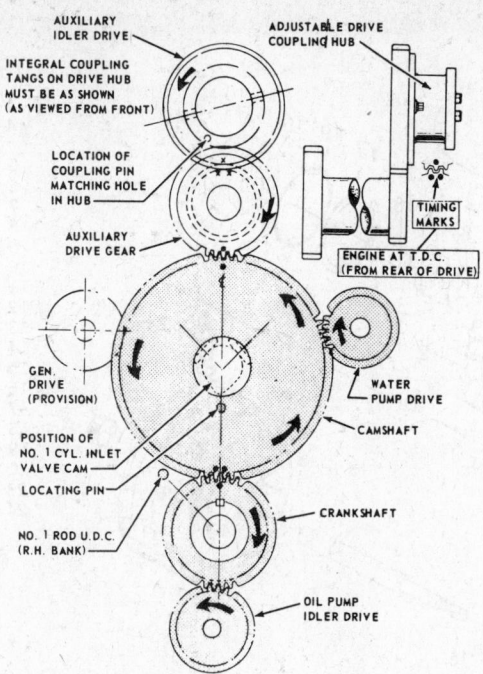

Timing gear alignment BSJ-PSM pump—END 864

Labels in diagram:
AUXILIARY IDLER DRIVE
INTEGRAL COUPLING TANGS ON DRIVE HUB MUST BE AS SHOWN (AS VIEWED FROM FRONT)
ADJUSTABLE DRIVE COUPLING HUB
LOCATION OF COUPLING PIN MATCHING HOLE IN HUB
AUXILIARY DRIVE GEAR
TIMING MARKS
ENGINE AT T.D.C. (FROM REAR OF DRIVE)
GEN. DRIVE (PROVISION)
WATER PUMP DRIVE
CAMSHAFT
POSITION OF NO. 1 CYL. INLET VALVE CAM
LOCATING PIN
CRANKSHAFT
NO. 1 ROD U.D.C. (R.H. BANK)
OIL PUMP IDLER DRIVE

Tip Turbine Fan Cleaning

ENDT 676, ETAZ 673 & 673A

The tip turbine fan should be cleaned every 50,000 miles or sooner if the unit is operated in a dusty environment and the exhaust temperature exceeds 1125°F at full load.

1. Loosen the air bleed hose from the turbocharger crossover and intercooler manifold to the tip turbine fan check valve.
2. Disconnect the oil inlet and loosen the clamp on the oil drain lines. Cap all lines.
3. Disconnect the V-band holding the tip turbine fan to the intercooler manifold. Remove the tip turbine fan.
4. Clean the area around the pipe plug and remove it using a ½" drive.
5. Remove the cooling air inlet housing.

CAUTION: Do not apply axial pressure to the fan wheel or shaft. This could distort the thrust bearing.

6. Secure the fan discharge end of the shaft with a 5/16" 8 point socket mounted to a work bench or vise so that the fan discharge housing is resting on the table or vise with the shaft floating freely in the socket with no axial pressure applied to the shaft. Remove the fan shaft nut.
7. On tip turbine fan part numbers 418GBb3A and 418GB43B, remove the large aluminum disc covering the cupped nose of the fan wheel.
8. On tip turbine fans with the fan wheel pressed onto the shaft, the aluminum fan wheel must be heated so that it can be removed by hand.
9. While still secured to the table or vise, apply heat, up to 700°F, to the fan wheel using an electric hot air heat gun. While the wheel is heating, apply a turning force to the wheel. When it is hot enough it can be turned off the shaft.

---

skirt to spray nozzle clearance (.062-.025"):

a. Position #1 and 6 at BDC.
b. Sight along nozzle. Nozzle should be pointing midway between connecting rod and inside face of piston boss when nozzle is square with engine center line. Bend to adjust.
c. Slide go (.062) and no-go (.125) wires between nozzle tube and skirt.
d. Repeat for other pistons.

Piston and Connecting Rods—V8 Engines

1. Heat pistons to 200°F.
2. Position pistons on crowns with arrows toward right.
3. Install rods 1, 2, 3 and 4 in pistons with the numbers toward you.
4. Install rods 5, 6, 7 and 8 with the numbers away from you.
5. Install pins and retaining rings.
6. Install rings on pistons—stagger gaps.
7. Coat the pistons, rings and liner with light oil.
8. With the crankpin at BDC and rod number toward center of engine, install piston and rod assemblies.

CAUTION: On turbocharged engines, take care that the rod doesn't damage the piston spray nozzle.

9. Check rod bearing clearance with plastigage.
10. Oil the capscrews, install the caps, finger-tighten the capscrews and check side clearance between paired rods (.007-.015").
11. A tapered brass wedge may be used to force rods against crankshaft cheeks to align rod with cap.
12. Torque bearing caps and check crankshaft for free rotation.
13. Check piston-to-cylinder block deck height (.018").

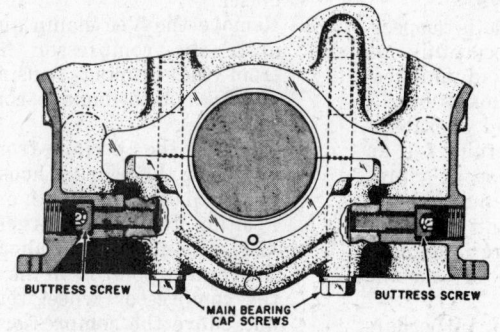

Buttress screws—V8

Labels: BUTTRESS SCREW, MAIN BEARING CAP SCREWS, BUTTRESS SCREW

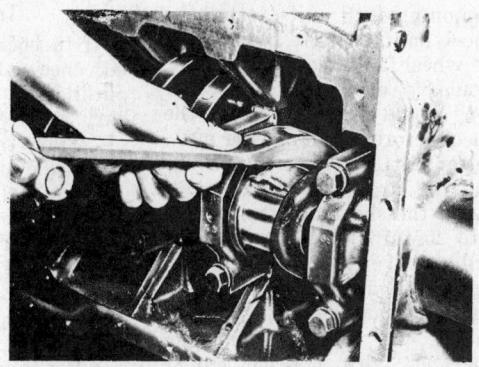

Positioning connecting rod on crankshaft.

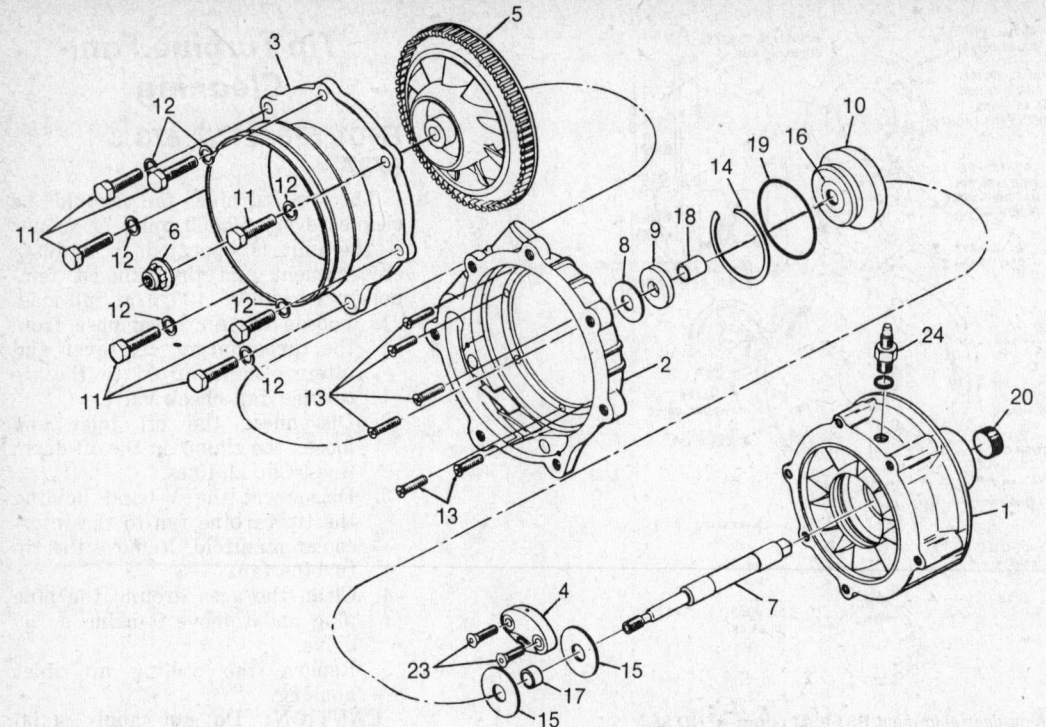

1 Discharge housing assembly
2 Turbine housing nozzle assembly
3 Inlet housing
4 Thrust bearing assembly
5 Wheel
6 Nut
7 Shaft
8 Shim
9 Wheel washer
10 Cover
11 Bolt
12 Lockwasher
13 Pan head screw
14 Retaining ring
15 Thrust plate
16 Seal
17 Spacer
18 Spacer
19 O-ring
20 Plug
23 Screw
24 Connector

Tip turbine fan, exploded view.

NOTE: the 8 shims found behind the wheel are for wheel-to-housing clearance. They must be retained.

10. Replace the pipe plug in the discharge housing to prevent the shaft from falling out.
11. The parts may now be cleaned. Do not let any solvent get into the discharge housing. It will damage the bearings. Do not scrape, sandblast, glass bead, wire brush or file the wheel or housing. The wheel and housing should be cleaned with a non-caustic solvent an a hair bristle brush.
12. Install the 8 shims on the fan shaft.
13. On tip turnine fans with the fan wheel pressed on the shaft, apply a small amount of clean engine oil on the shaft and push the wheel on the shaft.
14. Hold the shaft as described in step 6 above and tighten the wheel nut to press the wheel on the shaft.
15. On non-press fit units, install the wheel on the shaft dry. When the wheel is placed on the shaft, about ½″ of the threaded portion should protrude from the wheel before installing the nut.
16. On fan #418GB43A and 43B, install the large aluminum disc.
17. Install the shaft nut and torque it to 205-215 inlb. The use of a double universal joint and a T-handle will avoid any side loading on the shaft.
18. Check the diesel-to-housing clearance with a feeler gauge. The outer ring of the turbine

fan blades must be within .002″ of the inlet housing mounting surface when the wheel is moved its full travel. Adjust is made with shims.
19. Install the inlet housing. Torque the bolts to 90-110 inlb.
20. Check the wheel for drag or binding.
21. Wrap the pipe plug with two layers of teflon tape or RTV silicone sealant. Install the plug and torque to 400-425 inlb.
22. Install the tip turbine fan to the oil drain, bleed air hose and intercooler manifold. Torque the fan bleed line and oil drain line. V-band to 110-130 inlb.
23. Tighten the hose clamps on the fan bleed line and oil drain line. Secure the oil feed lines.
24. Install and secure the cooling air inlet duct.

Turbocharger
ENDT 673, 675, 676 Series
Troubleshooting

If turbocharger trouble is suspected, check the unit out carefully as outlined in order to avoid unnecessary removal of a functional unit.

1. Start the engine and operate it while listening carefully for an unusual type of noise or vibration from the unit. Unusual noise indicates that the unit is out of balance, or that there is damage to the compressor or turbine wheels.
2. Check for leaks in turbocharger-to-manifold or manifold-to-block plumbing. Make sure

clamps and bolts are tight and that all hoses and gaskets are in good shape.
3. Check the air cleaner and lines supplying the turbocharger for restrictions.
4. Check the exhaust plumbing between block and turbine inlet for loose bolts or connections or blown gaskets. Check for restrictions in the exhaust pipe downstream of the turbine.
5. Check the radial and axial play of the unit. If either is excessive, the unit will have to be removed for overhaul. These are .0230″ and .0040″ respectively.

Disassembly

1. Clean the exterior of the unit with a pressurized spray of cleaning solvent and air blow dry.
2. Mark the alignment of the turbine housing and compressor housing with the bearing housing, using a center punch or small chisel.
3. Remove the Vee clamp and separate the compressor housing from the bearing housing. Be careful not to damage compressor wheel.
4. Remove the o-ring from the groove in the bearing housing, if the unit is equipped with one.
5. Position the turbocharger in a vise. Mount a dial indicator so the contact point is on the side of the compressor wheel retaining nut. Move the compressor wheel away from the indicator while pressing the turbine wheel in the

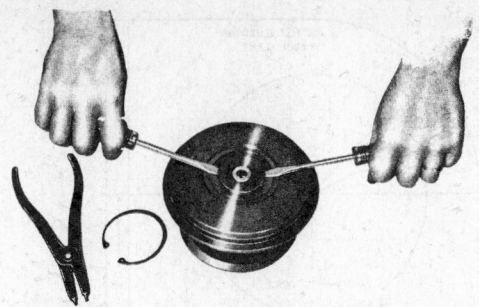

Removing compressor insert

Compressor bearings and snap rings

Loosening locknut on compressor wheel

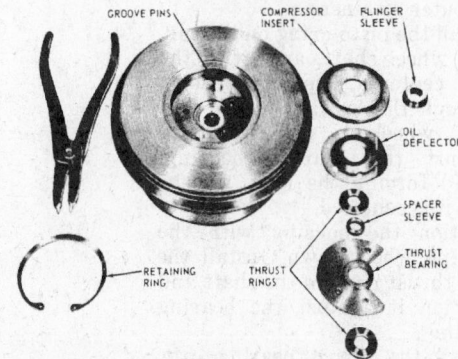

Compressor bearing housing and thrust assembly

opposite direction, and zero the indicator.

6. Press the compressor wheel toward the indicator while pressing the turbine wheel in the opposite direction and read the indicator. If the measurement exceeds the maximum, the turbocharger must be overhauled.

7. Mount the indicator so the end contacts the end of the turbine shaft. Move the shaft toward the indicator and zero it. Then move the shaft to the opposite extreme of end-to-end travel, and record the reading. If this reading is greater than specifications, overhauling is necessary.

8. Remove the vee clamp and separate the turbine housing from the bearing housing. If force is required, use a soft mallet only.

9. Place the bearing housing on a bench and hold the turbine wheel at the hub with a ⅜" eight point or ⅝" 12 point socket. With a ⅝" box wrench, loosen and remove the compressor wheel retaining nut.

10. Slide the compressor wheel off the shaft. Remove the turbine wheel and shaft from the housing. Remove the piston ring from the shaft.

11. Remove the large snap ring from the bearing housing. Use two screwdrivers to carefully pry the compressor insert from the bearing housing. Remove the o-ring from the compressor insert.

12. Remove the finger sleeve, oil deflector, thrust ring, thrust bear-

ing, and spacer from the bearing housing.

13. With tru-arc pliers, remove the snap ring from the compressor end of the bearing housing. Remove the compressor bearing. Then, using tru-arc pliers, remove the rear bearing snap ring. Remove the snap rings and bearings from the turbine end of of the bearing housing in the same way.

14. Remove the oil control sleeve, which is located between the snap ring and the turbine bearing.

15. Soak the turbine wheel and housing (avoid wetting the shaft) in an appropriate carbon removing solvent for 25 minutes. Then, remove carbon with a brush. Clean the remaining parts in a solvent spray and dry with compressed air.

Inspection

1. Inspect the turbine and compressor housing for wear due to contact with rotating parts. Make sure the mounting flange is flat and free from erosion.

2. Check the bearing housing for warped flanges, worn bearing bores or snap ring grooves, and loose groove pins. Make sure all oil passages are clean. Make sure threaded holes and oil inlet and outlet pads are in good condition.

3. Inspect bearings and replace if tin plating is worn off either the inside or outside diameter, or if bearings show signs of imbed-

ment of foreign matter.

4. Make sure the shaft is not scored or worn at bearing locations. Make sure oil seal ring grooves and threads on the ends of the shaft are in good condition.

5. Check turbine and compressor wheels for signs of rubbing or vanes worn to a feather edge. Make sure the compressor wheel bore is not galled.

6. Make sure the thrust bearing is worn evenly, and that the drilled hole is clean.

Preparation for Reassembly

1. Discard all parts that show excessive wear, rubbing, scoring, or galling.

2. Where there is minor surface damage, polish it out using silicon carbide cloth on aluminum parts and crocus cloth on steel parts.

2. Discard piston rings, seal rings, bearing retainers and the compressor retaining nut at reassembly.

Reassembly

Before reassembling each part, check it to ensure it is clean. Lubricate all moving parts with engine oil.

1. Mount the bearing housing on a bench with the compressor end down. Install the tru-arc bearing retainer into the inner snap ring groove. Make sure rounded surface is positioned toward bearing.

2. Install the bearing in the hous-

Mack Diesel Engine

ing bore and seat it against the tru-arc ring.

NOTE: AiResearch turbochargers feature an anti-wear washer in-board of the bearing.

3. Install the oil control sleeve, flat surface toward the bearing, and then install the bearing retainer in the outer groove.

4. Invert the bearing housing so the turbine is down and the compressor up. Install the inner bearing retainer, force the bearing in until it seats, and install the outer retainer.

5. Install the piston ring on the turbine wheel shaft, and, with the ring centered, insert the shaft through the bearing so the turbine wheel is firmly seated against the bearing housing. Avoid forcing the ring, and be sure it is centered.

6. Position the housing with the turbine wheel down. Install the first thrust ring on the shaft and position it against the bearing housing.

7. Install the thrust bearing over the shaft and groove pins, and press it into position against the bearing housing.

8. Install the thrust bearing over the shaft. Install the groove pins and press them into position against the bearing housing.

9. Install the spacer sleeve over the shaft and into the inside diameter of the thrust bearing.

10. Install the second thrust ring. Engage the oil deflector over the groove pins in the same position as the thrust bearing with the baffle downward.

11. Install the piston ring in the groove of the flinger sleeve. Center the ring. Install the flinger sleeve into the compressor insert.

12. Position the o-ring in the groove of the compressor insert and in-

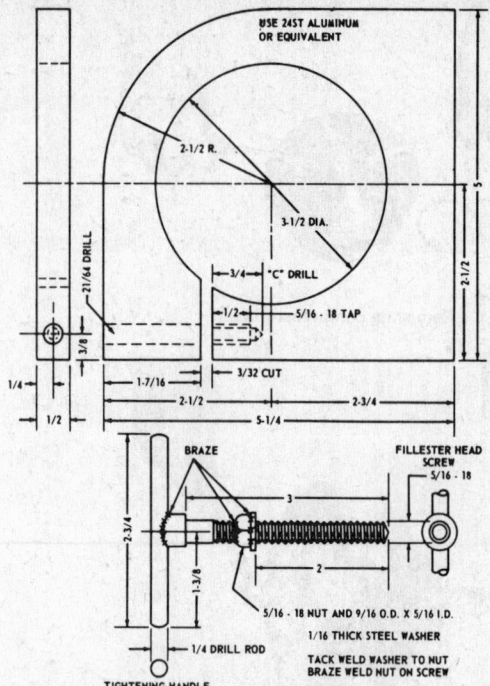

Turbocharger assembly and disassembly device

stall the compressor insert over the shaft and press into position. The insert must be pressed in until it clears the large retaining ring groove.

13. Insert the compressor insert retaining ring with tru-arc pliers. Make sure the ring is properly positioned in the groove.

14. Install the compressor wheel, pushing it down until it bottoms.

15. Coat the threads of the shaft and the face of the washer with graphite grease. Install the locknut onto the shaft, and torque the locknut while holding the turbine wheel with a ⅜" eight point or ⅝" 12 point socket.

16. Spin the compressor wheel to check for absolute freedom from binding.

17. Install the o-ring in the bearing housing compressor flange.

18. Position the compressor and vee clamp onto the bearing housing. Align the marks and torque the vee clamp nut. AiResearch TE06A & TH08A—50 in. lb. others—120 in. lb.

19. Following the same procedure, install the turbine housing.

20. Spin the compressor wheel to ensure freedom from binding.

21. Inspect the air intake system to ensure cleanliness, and make sure no foreign material has found its way into the exhaust manifold.

22. Inspect the oil supply line and drain line to make sure no clogging or leakage has occurred and that any check valves operate freely.

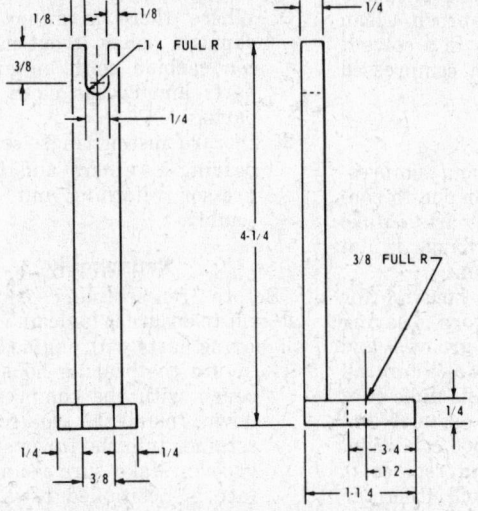

Adaptor for turbocharger dial indicator

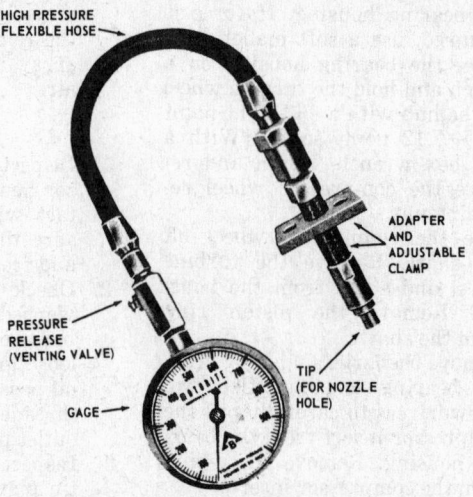

Compression tester

23. Inspect the turbocharger mounting pad to ensure it is not eroded and is flat.

24. Position a new gasket on the exhaust manifold, position the turbocharger, and install the mounting bolts. Torque to 30 ft. lb.

25. Connect the oil supply line, and compressor inlet and outlet piping, torquing all clamps snugly.

26. Pull engine stop button out so engine will not start, and crank the engine until a steady flow of oil comes from the drain line connection. Then, connect the drain line. Make sure the drain line slopes downward along its entire length, with no kinks.

ENDT 864, 865, 866 Series

Preliminary Inspection

1. Remove the compressor housing, force the compressor toward the turbine end of the unit, and spin the compressor. Check for binding, rubbing, or unusual noises.

2. Remove the turbine exhaust outlet and check the end play of the turbine shaft and the radial float of the bearings. Move the shaft with equal pressure on both ends.

3. If there is binding or rubbing, or end play or radial float is excessive, remove the unit for repairs.

Disassembly

1. Clean the exterior of the unit with a spray of cleaing solvent and blow dry.

2. Mount the turbocharger on a bench. Install a locally fabricated dial indicator holder, like that shown in the illustration, to the oil drain mounting pad.

3. Install a plunger type dial indicator, with a one inch travel and three inch extension, to the indicator holder.

4. Zero the indicator. Move the shaft back and forth, applying equal pressure at both ends. The unit must be overhauled if end play exceeds .0070".

5. Check the end float with the indicator mounted to the center housing and its tip on the end of the shaft. Hold the shaft toward the indicator and zero it. Then, move the shaft back and forth and read the total indicator movement. If it exceeds .0070", the unit must be overhauled.

6. Mark the turbine and compressor housings with a center punch to permit correct alignment at reassembly.

7. Remove the vee clamp and pull off the compressor housing, carefully avoiding damage to the compressor.

8. Unlock the tabs and remove the capscrews, which secure the turbine housing. Remove the turbine housing carefully, using only a soft mallet if force is required.

9. Mount the bearing housing in a specially fabricated fixture like that shown in the illustration. Then mount the fixture in a vise so the turbine wheel will be pointed downward. Remove the turbine wheel retaining nut with a socket and t-handle.

10. Heat the bearing housing at 325-375° F. for a maximum of 10 minutes.

11. Remove the bearing housing from the oven. Press the turbine wheel shaft out of the compressor wheel with an arbor press. Keep the shaft centered within the bearing housing until the shaft is clear of the housing.

12. Remove the assembly from the holding fixture, and remove the piston ring from the turbine wheel shaft.

13. Unlock the locking tabs and remove the thrust plate retaining capscrews and the thrust plate. Remove the thrust spacer, and then remove the piston ring from the spacer.

14. Remove the inboard thrust bearing, thrust collar, and seal ring, and bearing from the bearing housing.

15. Use special pliers to remove the two tru-arc rings, and then remove the rear bearing.

Cleaning

1. Soak the turbine housing and wheel in a carbon removing solvent for 25 minutes. Do not immerse the shaft.

2. Remove the assembly from the solvent and use a bristle brush to remove remaining carbon.

3. Clean all parts with a solvent spray and blow dry.

Inspection

1. Inspect turbine and compressor housings for signs of contact with rotating parts. Make sure the mounting flange is flat, and without erosion. Check the condition of the threaded holes and parting surfaces.

2. Check the condition of the bearing housing to ensure the bearing bores are in good condition and that the inboard thust bearing positioning pins are securely in position. Make sure oil holes and threaded holes are in good condition.

3. Check the condition of the inboard thrust bearing and thrust collar, and make sure that thrust collar holes are not obstructed.

4. Check the condition of the thrust plate mating surfaces. Check the condition of the thrust spacer bore.

5. Check the condition of the bearings—especially for the imbedment of foreign matter.

6. Check the turbine wheel shaft for wear or scoring at the journal locations. Check the condition of the oil seal ring groove and threads on the end of the shaft.

7. Inspect the turbine and compressor wheels for signs of rubbing or vanes worn to a feather edge. The compressor wheel bore must not be galled.

8. Check the condition of the turbine wheel shroud. Make sure the mating surface is clean and flat.

9. Reject all parts that show damage. New piston rings, seal ring, bearing retainers, compressor retaining nut, and lock plates should always be used at reassembly.

10. Polish areas showing minor surface damage with silicon carbide cloth on aluminm or crocus cloth on steel.

Reassembly

1. Coat all moving parts with clean engine oil prior to reassembly. Make sure all parts are absolutely clean.
   * AiResearch TE06A & TH08A—50 in. lb.

2. Install the piston ring in the groove in the turbine wheel shaft. Posiion the turbine wheel in the holding fixtue.

3. Install the rear bearing, and then insert the tru-arc retaining ring with the rounded side toward the bearing.

NOTE: AiResearch turbochargers feature an anti-wear washer inboard of the bearing.

4. Install the front bearing tru-arc retaining ring in the groove in the center housing, with the rounded side toward the bearing. Install the bearing.

5. Position the shroud on the rear end of the center housing and install it over the turbine wheel shaft. Gentle rocking while carefully forcing the shroud on will allow the ring to seat and permit the housing to bottom on the shaft. If necessary, a dental pick or similar tool may be used to seat the ring.

6. Make sure the front bearing is in proper location. Then, install the inboard thrust washer over the shaft. Make sure the holes in the washer align with the pins in the center housing, and that the washer is seated flat.

7. Install the thrust collar so it will be positioned next to the inboard thrust washer.

8. Install the square cut oil ring in the groove in the center housing.
9. Position the thrust plate over the shaft and align the oil holes. Install the center housing to the thrust plate retaining capscrews and lock tabs. Torque the capscrews (40 in. lb.) and bend the tab locks.
10. Install the piston ring in the groove in the thrust spacer. Install the spacer over the shaft with the ring up, sliding it down until it is positioned in the thrust plate. Bottom the ring in the groove in the spacer and push the spacer into position.
11. Heat the compressor wheel in an oven at 325-375° F. for 10 minutes or less.
12. Remove the compressor wheel from the oven and install it over the turbine wheel shaft. Install and torque the compressor wheel retaining nut to 120 in-lbs while the compressor is hot, using a t-handle.
13. Allow the compressor to come to room temperature. Remove the compressor wheel retaining nut. Check the surface of the nut which bears against the washer and the compressor wheel for pick-up. They must be smooth and clean.
14. Lightly oil the shaft threads and install the nut, torquing to 18-20 in-lbs. Tighten it about 1/3 of a turn farther. Remove the holding fixture and check the unit for free rotation.
15. Put the turbine housing into position on the bearing housing, aligning the marks made during disassembly. Install the housing clamps, lock plates, and capscrews, coating the threads of the screws with anti-seize compound before installation. Torque the capscrews (120 in. lb.*) and bend over the tabs. Make sure the turbine still rotates freely.
16. Put the compressor housing into position, align marks made during disassembly, and install the vee clamp. Torque the vee clamp bolt nut.
17. Check the assembly for free rotation by pushing the turbine toward the compressor and turn-ing the shaft. Repeat the check while pushing the compressor end of the shaft.
18. Inspect the air intake system to ensure cleanliness, and make sure no foreign material has found its way into the exhaust manifold.
19. Inspect the oil supply line and drain line to make sure no clogging or leakage has occurred and that any check valves operate freely.
20. Inspect the turbocharger mounting pad to ensure it is not eroded and is flat.
21. Position a new gasket on the exhaust manifold, position the turbocharger, and install the mounting bolts. Torque to 35 ft. lb.
22. Connect the oil supply line, and compressor inlet and outlet piping, torquing all clamps snugly.
23. Pull engine stop button out so engine will not start, and crank the engine until a steady flow of oil comes from the drain line connection. Then, connect the drain line. Make sure the drain line slopes downward along its entire length, with no kinks.

Noise Check Procedure

WARNING: Never place hands near the turbocharger intake. The vacuum could draw fingers into the unit and cause serious injury.

Start the engine and check for unusual noise or vibration from the unit. This will indicate out of balance, rubbing or damage to the compressor or turbine wheels. Removal and visual checking will be necessary.

Leakage Checking

1. Start the engine.
2. Check for leaks in the air intake system, from the air cleaner to the intake manifold. Use light oil, squirted from a can, or a soapy water solution, at suspected leak points. Soapy water can be used to detect porous hoses, or a light application of starting fluid can be sprayed on hoses or connections. A rise in rpm will be noticed in the event of a leak.
3. All hoses should be checked for dislocation, chafing, sliding or holes.
4. Check the exhaust manifold for leaks, cracks and missing or blown gaskets.
5. Check turbocharger lubricating oil inlet and outlet lines for chafing, pinching or leaking.
6. Shut down the engine and check all bolts, clamps and connections.
7. If the unit shows signs of oil leakage:
   a. Check the air cleaner and/or compressor inlet line, if the compressor is tied into the air intake system.
   b. Check the back of the compressor wheel and in the crossover tube.

If oil wetting is found, clean the compressor, compressor wheel and crossover tube. Run the engine to normal operating temperature, inspect the unit and if no leakage is evident, check the compressor for oil passing or other points of leakage at the engine other than the turbocharger.

Pressure Checking

1. Install an oil pressure gauge in the turbocharger oil inlet line and a pressure gauge in the intake manifold.
2. Check the oil inlet pressure for a minimum of 10psi with the engine at idle.
3. Check items 2 through 5 under Leakage Checking, above, with the engine running.

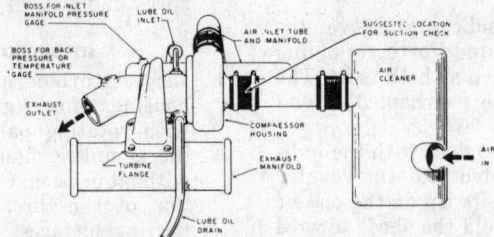

Typical Turbocharger Installation (© Mack Trucks Inc.)

Engine Series	Altitude Feet							
	0	2,000	4,000	6,000	8,000	10,000	12,000	14,000
END673E	530	500	460	430	390	360	340	310
ENDT673C ENDT675, ENDT676	475	445	415	385	355	325	295	275
ENDT(B)865, 866 14.95:1 *	485	455	425	385	355	335	305	285
ENDT(B)865, 866 15.7:1 *	540	510	480	440	410	380	350	330

* Compression Ratio   Minimum Compression Pressure PSI

Inlet Manifold Pressure Chart

Engine	Turbocharger Model	Inlet Manifold	
		Pressure in Hg.	Pressure psi
ENDT475	3LD125	24.0 to 26.0	11.8 to 12.6
ENDT673	4LD556 4LE556	23.0 to 25.0	11.3 to 12.3
ENDT673C	4LE556	27.0 to 29.8	13.3 to 14.6
ENDT675	TE0648 4LE354	30.0 to 35.5	15.0 to 17.4
ENDT676	TV60	36.7 to 50.9	18.0 to 25.0
ENDT(B)865	THO8A or TV70	37.5 to 47.5	18.4 to 23.4
ENDDT(B)865	THO8A or TV70	30.2 to 40.0	14.8 to 19.7
ENDT(B)866	TV70	42.0 to 49.0	20.6 to 24.1

Pressure will vary approximately 1.0 in. Hg. (inches of mercury) with fuel flow change of 2.0 percent at test rpm.

4. Check manifold pressure. This test must be performed with the engine under load and the rack wide open, running at maximum governed speed.

Road Testing

1. Connect a pressure gauge at the manifold, with a long tube so that the gauge can be read in the cab.
2. Increase the engine speed to between 1700 and 1800 rpm. With one foot on the accelerator and one foot on the brakes, apply the brakes at the same time the accelerator is pressed slowly to bring the engine to 100 rpm below governed speed with the throttle wide open and the brakes providing the load.
3. Note the pressure at the instant the speed reaches the specific rpm mark on the tachometer. Do this two or three times to check the results.

Tolerance Check Procedure

Continual, routine oil and/or carbon build-up will affect the amount of mechanical play. The following

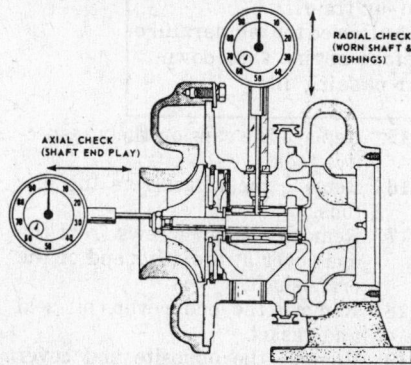

Checking Axial Play (© Mack Trucks Inc.)

should be used as a general guide.
1. Place a dial indicator on the end of the compressor shaft. Move the shaft back and forth by hand and check total indicator movement.
Schwitzer: .004-.006"
AiResearch: .003-.007"
Holset: .004-.006"
2. Check the bearing radial clearance on the AiResearch unit by removing the oil outlet line. Install an adapter plate and dial

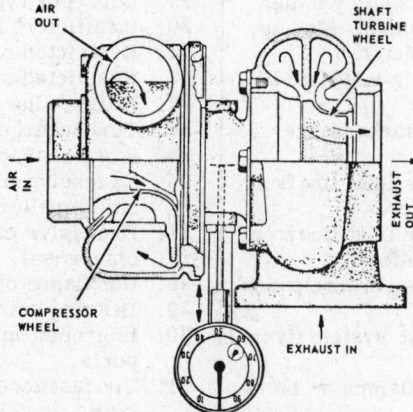

Checking Radial Clearance-AiResearch Unit (© Mack Trucks Inc.)

indicator with extension. Zero the indicator and hold both ends of the compressor shaft. Move the shaft up and down. Exert equal pressure on both ends of the shaft. Bearing radial play should be .003-.007".
3. Check the radial play on Schwitzer and Holset units with a dial indicator.
Position the indicator on the turbocharger shaft and rock the shaft up and down on opposite ends. Bearing radial play, total rocking motion of

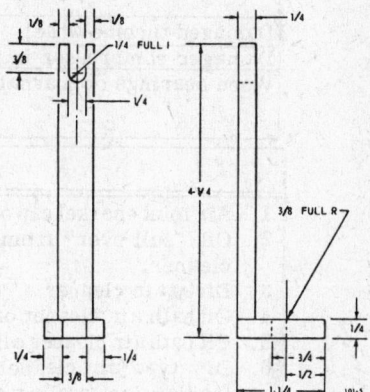

Dial Indicator Adapter (© Mack Trucks Inc.)

the shaft at either end, is .023" maximum.

Air Compressor Bendix-Westinghouse Tu-Flo 400, 500, 1000

Disassembly

1. Clean the unit thoroughly in solvent.
2. Mark the following:
a. block-to-crankcase.
b. end covers-to-crankcase.
c. crankshaft-to-crankcase.
d. head-to-block.
e. oil filter-to-base plate.
3. Remove the capscrews and lift off the head. It may be necessary to tap the head with a soft mallet to loosen it.
4. Remove the inlet valve springs from the head and the inlet valves from the block.
5. Discard the head gasket and thoroughly clean the head and block gasket surfaces.
6. Remove the discharge valve cap

Mack Diesel Engine

Trouble Shooting Chart

Trouble and Symptoms	Probable Causes Check causes given in sequence shown.
Noisy	9, 10, 11, 12, 13, 14, 15, 17, 20, 23, 25, 26, 27, 28, 29, 30
Low turbocharger speed	1, 3, 4, 7, 8, 13, 15, 16, 20, 23, 25, 26, 27, 29, 30, 34
Low turbocharger pressure	1, 3, 5, 7, 14, 11, 12, 13, 15, 16, 23
Cracked turbocharger housings or flanges	1, 3, 5, 7, 30, 11, 12, 13, 15, 16, 19, 20, 23, 26, 27, 28, 31, 32, 33

Engine lacks power
  Black smoke
  Blue smoke
	
	1, 2, 3, 5, 16, 24, 7, 19, 15
	24, 2, 7, 4, 22, 34

Lubrication oil consumption excessive	19, 21, 22, 3, 1, 5, 18, 24, 25, 34
Lubrication oil leakage	21, 19, 18, 22, 24, 25, 34
Insufficient lubrication	19, 21, 17, 31

Oil in Air Inlet System
  Before turbocharger
  After turbocharger
	
	1, 2, 3, 4, 5, 7, 18, 19, 34
	1, 2, 3, 4, 7, 5, 22, 24, 25, 28, 34

Oil in Exhaust System
  Before turbocharger
  After turbocharger
	
	12
	22, 24, 25

Unbalance of rotating assembly	25, 23, 9, 3, 11, 12, 13, 18, 19, 20
Drag or bind in rotating assembly	9, 26, 27, 28, 30, 25, 18, 19, 20, 17, 13, 22, 29, 11, 12, 31, 32, 33, 34
Damaged turbine wheel	12, 13, 15, 25, 27, 28, 30, 32, 33
Damaged compressor impeller	6, 9, 11, 13, 25, 26, 27, 28, 30, 33
Worn bearings or journals	17, 18, 19, 20, 21, 26, 27, 28, 30, 31, 32, 33

Probable Causes

1. Air inlet snorkel cap or screen restricted.
2. Oil "pull over" from oil bath type air cleaner.
3. Dirty air cleaner.
4. Oil bath air cleaner oil viscosity too low.
5. Oil bath air cleaner oil viscosity too high.
6. Dry type air cleaner element missing, leaking, not sealing correctly.
7. Collapsed or restricted air inlet line before turbocharger.
8. Restricted air inlet line turbocharger to engine.
9. Loose, chafed or open air inlet line from air cleaner to turbocharger.
10. Loose, chafed or open air inlet line from turbocharger to inlet manifold.
11. Foreign object in inlet system air cleaner to turbocharger.
12. Foreign object in exhaust system (from engine, check engine).
13. Turbocharger flanges, clamps or bolts loose.
14. Inlet manifold cracked or flange gaskets loose or missing.
15. Exhaust manifold cracked or flange gaskets loose, blown or missing.
16. Restricted exhaust system.
17. Oil lag (delay in oil supply to turbocharger at start up).
18. Lubricating oil contaminated with dirt or other material.
19. Improper type lubricating oil used.
20. Insufficient lubrication.
21. Restricted or leaking oil feed line.
22. Restricted oil drain line.
23. Nozzle ring damaged or restricted.
24. Turbocharger seal leakage.
25. Worn shaft bearings.
26. Excessive dirt build-up behind compressor impeller.
27. Excessive carbon build-up behind turbine wheel.
28. Unbalance of rotating assembly.
29. Bearing seizure.
30. Improper mounting, brackets or supports.
31. Too fast acceleration at initial start (oil lag).
32. Too little warm-up time.
33. Insufficient time allowed for temperature stabilization before engine shut-down.
34. Air compressor passing oil.

nuts and lift out the discharge valve springs and valves.
7. If the discharge valve seats are badly worn or nicked, they may be removed with a cold chisel.
8. Unbolt and remove the baseplate.
9. Remove the oil relief valve set screw and the relief valve.
10. Remove the retaining ring and lift out the oil strainer.
11. If necessary, remove the set screw and lift out the piston bushing and shim.
12. On self-lubricated models, remove the cotter pin from the oil rod cap nuts, remove the nuts and remove the oil pump piston rod and cap.
13. Remove the connecting rod bolts, lock washers and bearing caps.
14. Push the piston and rod out through the top of the block.
15. Replace the caps on their respective rods.
16. Remove the rings from the pistons.
17. Remove the cap screws from the end cover at the drive end of the crankshaft.
18. Remove the end cover, oil seal and gasket.
19. Remove the opposite end cover and gasket.

Oil Level-Self-Lubricated Compressor
(© Mack Trucks Inc.)

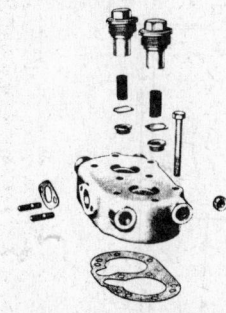

Cylinder Head Exploded View
(© Mack Trucks Inc.)

20. Press the crankshaft and ball bearings from the crankcase. *NOTE: some units have a shoulder for positioning the crankshaft. In these cases the crankshaft must be removed through that one end.*
21. Press the ball bearing from the crankshaft.
22. Remove the inlet strainer, elbow and governor.
23. Remove the capscrews joining the block and crankcase and separate the two parts. Scrape off the old gasket.
24. Remove the unloader spring, spring saddle and spring seat from the block.
25. Remove the unloader guides and plungers, and using compressed air, blow out the unloader pistons from their bores.
26. Remove the intake valve guides.
27. If the intake valve seats are badly worn or nicked, they may be removed with a cold chisel.
28. Unloader bore bushings may be replaced if damaged.
29. Inspect all parts for excessive wear or damage.
30. All parts may be cleaned in solvent.
31. The discharge valves may be cleaned with crocus cloth.
32. Use air pressure to test the cooling system of water-cooled types.
33. Never reuse discharge valve spring cap nuts.
34. Check the fit of the oil ring in the end cover groove. There should be .008-.015" clearance at the gap when the ring is placed in the end bore of the crankshaft.
35. Cylinder bores should not be more than .002" out of round nor more than .003" tapered. Rebor-ing to .030" oversize is allowable.
36. Clearance between cast iron pistons and bores should be .002-.004". Aluminum pistons are cam ground.
37. Piston pins should be a light press fit. Pin-to-rod clearance should not exceed .0015".
38. Crankshaft journals more than .001" out of round should be reground. Regrinding to .030" is allowable. Bearings should be a snug fit.
39. Connecting rod journal-to-bearing clearance must not exceed .0003".
40. Discharge valve travel must be .036-.058" for the 400 and .056-.070" for the 500, 1000.
41. To test for leakage at the discharge valves, apply 100psi air pressure to the discharge ports. Apply a soap suds solution to the valves. If leakage is noticed, leave the air connected and tap the valves off their seats several times with a wooden mallet. This should help the valves seat and reduce leakage. Check for leakage at the discharge valve cap nuts. No leakage is permissable.
42. Intake valves may be dressed with an emery cloth or lapping compound. The distance from the top of the block to the intake valve seat should not exceed .145", nor be less than .101". If anything more than slight damage is present on the valves, they should be replaced.

Assembly

1. Assemble the block and the crankcase, aligning previously made marks and using a new gasket.
2. If the crankshaft is fitted with oil seal rings, install the rings.
3. Position the ball bearings and crankshaft in the crankcase making sure that the crankshaft is installed through the correct end.
4. Using an arbor press, press the crankshaft and bearings into the crankcase.
5. Position a new rear end cover gasket over the rear end of the crankcase making sure that the oil hole in the gasket lines up with the oil hole in the crankcase.
6. Position the end cover, with the oil seal ring, over the crankcase and end cover gasket.
7. Install a new seal in the opposite end cover.
8. Carefully install this end cover and gasket taking care to avoid damage to the seal.
9. If necessary, install a new piston pin bushing. The bushings should be reamed to provide .0001-.0006" clearance.

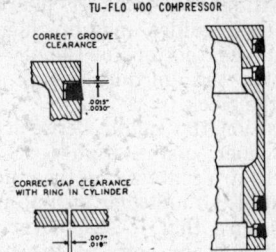

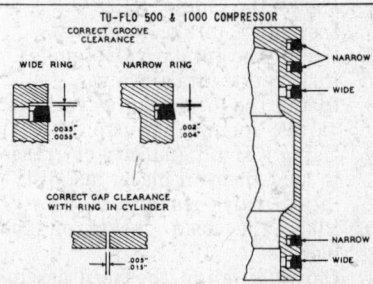

All rings must be located in their proper ring grooves as shown. The rings can be identified by the width and should be installed with the bevel or the pip mark (if any) toward the top of the piston. This applies to Cast Iron Pistons (only as shown above).
Die Cast Pistons use five (5) narrow rings.

Piston Ring Positions-Gaps and Groove Clearance (© Mack Trucks Inc.)

10. Position the connecting rod in the piston and press in the pin so that the lockwire hole in the pin aligns with the same hole in the piston.
11. Install the lockwire, snapping the short end into the piston.
12. Install the piston rings with pipmarks up.
13. Coat the piston, rings, pin and connecting rod bearings with clean engine oil.
14. Remove the rod bearing cap and bolts from the rod.
15. Turn the crankshaft so that one of the rod journals is down.
16. Insert the piston and rod assembly through the top of the block in the bore at which the journal is down.
17. Attach the cap to the rod, making sure that the lock washers are properly positioned. Tighten the nuts evenly and bend the washer.
18. Install the other piston and rod in the same manner.
19. Lubricate the unloader pistons and bores with dimethyl polysiloxane prior to installation.
20. Install the new unloader pistons in their bores taking care to avoid damaging the grommets or distorting the back-up rings.

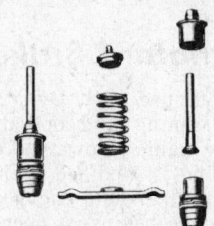

Unloader Mechanism (© Mack Trucks Inc.)

21. Position the unloader plungers in their guides and slip them in and over the tops of the pistons.
22. Install the unloader spring seat in the block.
23. Position the saddle between the unloader piston guides so the forks are centered on the guides.
24. Install the unloader spring making sure that it seats over the spring seats both in the block and on the saddle.
25. Install the intake valve seats.
26. Install the intake valve guides.
27. Install the intake valves.
28. Install the discharge valve seats, valves, springs and cap nuts.
29. Using a small quantity of grease to hold them in place, install the intake valve springs.
30. Place the head gasket on the block.
31. Carefully align the head assembly on the block and install the cap screws and washers.
32. Tighten the capscrews securely and evenly. *NOTE: steps 33-38 are for self-lubricating compressors.*
33. Install the oil pump piston and rod on the crankshaft.
34. Install the oil rod cap nuts and cotter pins.
35. Install the relief valve in the base plate. Test the relief valve at this point. The valve should open between 14 and 24psi. Install the set screw.
36. Place the oil pump screen in the base plate and install the retaining ring.
37. Install the oil filter and cover.
38. Install a new oil seal on the check valve and position a new base plate gasket on the crankcase. The oil pump piston should engage the bushing in the base plate. Install the base plate.
39. Assemble the air strainer and mount it on the block using a new gasket.
40. Install the governor, using a new gasket.
41. Install the compressor on a test rack and check the build-up time. Connect a reservoir with a combined line and tank volume of 1300cuin.
42. Run the compressor at 1700-1750rpm and note build-up time to 100psi.:

400	47 seconds maximum
500	30 seconds maximum
1000	15 seconds maximum

Dynatard Brake

When equipped with Dynatard, inlet valve timing and operation remains the same. However, exhaust valve timing is modified with Dynatard engaged. When engaged, each exhaust valve begins to open before TDC of the compression stroke. To

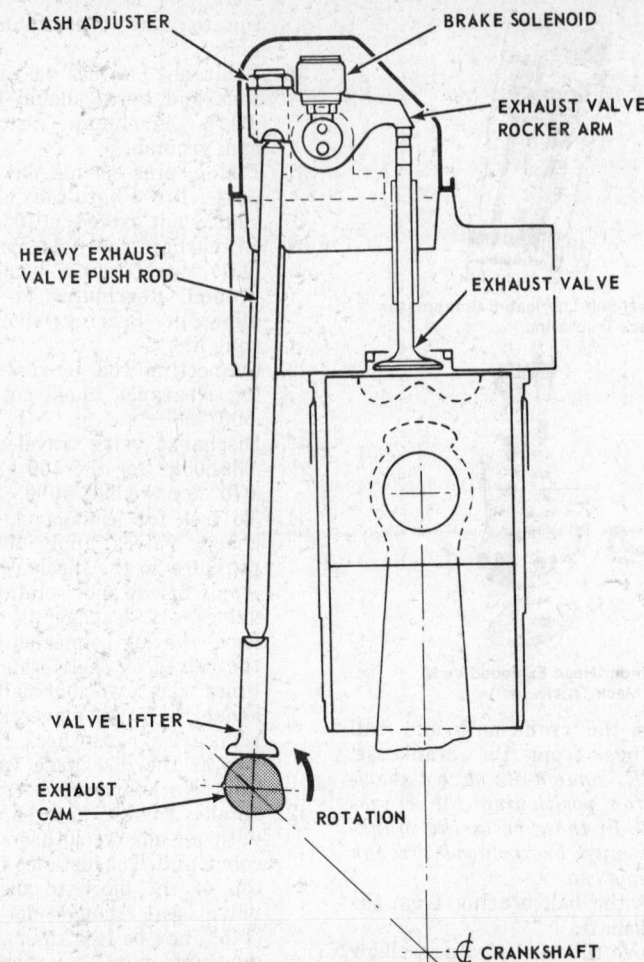

Dynatard Engine Brake Installed (© Mack Trucks Inc.)

accomplish this, valve operation is controlled so that the exhaust valve lifter follows a special contour of the exhaust cam lobe.

When the system is not engaged, valve lash is normal and the lifter will follow the small cam profile. When the system is engaged, the valve lash is reduced to zero and the lifter follows the larger exhaust cam profile causing the valve to open shortly before TDC compression, turning the engine into an air pump. resulting in negative, retarding horsepower.

A two-position, hydraulic lash adjuster is used to control valve timing. When the system is not engaged, oil flows into the lash adjuster through two passages in the exhaust rocker arm. Oil enters the lower portion under normal pressure and the upper portion under high pressure.

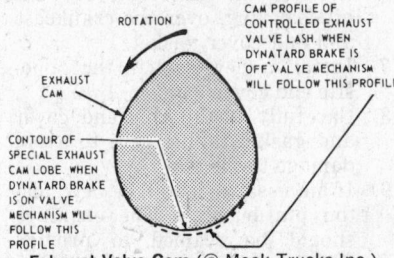

Exhaust Valve Cam (© Mack Trucks Inc.)

The control piston in the upper half is also held down by a spring. In the down position, the piston holds a ball check valve off its seat. This prevents hydraulic locking of the socket piston, allowing the hydraulic assembly to collapse, equalizing pressure in both ends of the lash adjuster. The collapsed assembly provides the necessary lash required for normal operation. With the system actuated, oil flow to the upper portion is shut off by a solenoid mounted on the rocker arm. The oil pressure under the control piston overcomes the spring force and lifts the piston. This allows the ball check valve to seat. Oil is drawn in by the socket piston, the ball check valve blocks its return and the result is a predetermined extension of the overall length of the assembly. The extended socket piston becomes locked, forcing the engine brake cam profile to be transmitted to the lifter.

Solenoid Control Valve

Installation

Solenoid holding fixture J24293 and crow-foot adapter TC-24 are necessary for installation.

1. Place the solenoid in the holding fixture and secure it with a vise.

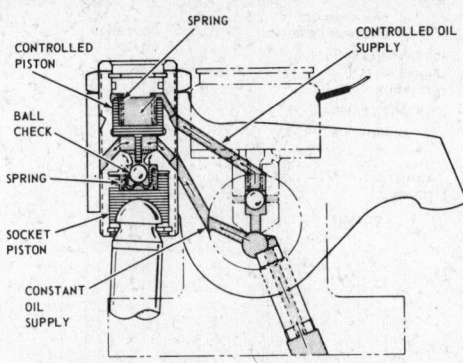

Oil Flow Through Lash Adjuster-Brake Off
(© Mack Trucks Inc.)

Oil Flow Through Lash Adjuster-Brake On
(© Mack Trucks Inc.)

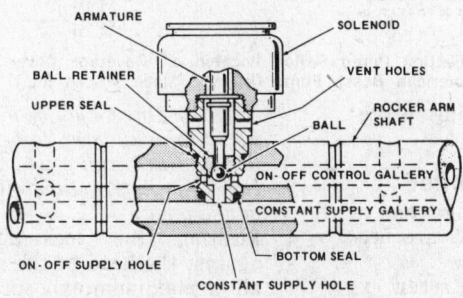

Oil Flow Through Rocker Arm Shaft
(© Mack Trucks Inc.)

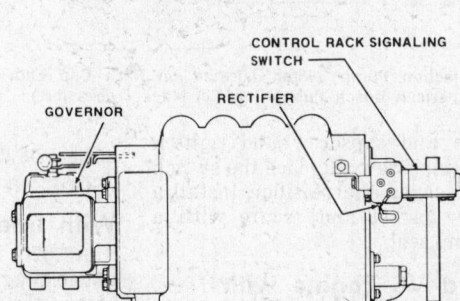

Location Of Fuel Injection Pump Switch on American Bosch V8 Pump (© Mack Trucks Inc.)

Apply Loctite TL-277 to the adapter end and assemble it to the solenoid. Torque it to 7-9 ftlb and remove the holding fixture.

2. Install the solenoid bottom seal in the adapter tapped hole in the rocker arm shaft.

3. Install the upper seal in the top counterbore of the adapter tapped hole in the rocker arm shaft. *NOTE: Current production engines do not use upper seal.*

4. Lubricate the adapter threads with clean 30W engine oil.

5. Install the assembly in the rocker arm shaft and torque it to 18-20 ftlb using crow-foot adapter.

6. Connect the blade end to the solenoid electrical lead.

Injection Pump Switch

V8 Engines with American Bosch V-type Pumps

The switch is located on the rack cap end of the injection pump. It is a micro-type switch. The switch is engaged by the rack extension and activates the system at zero fuel only. The switch is pre-adjusted, and under normal operation needs no further adjustment. If adjustment proves necessary:

1. Measure the height of each of the three electrical terminals on the top cover of the Dynatard switch. Each must extend at least ½ inch above the cover. If

not, remove the cover and adjust the terminal by loosening the locknuts.

2. Cut and remove the seal wire from the switch side cover, remove the side cover, rectifier and gasket. Position the engine brake toggle switch on the dash to the ON position and activate the electrical circuit to the engine brake switch.

3. Depress the microswitch contact button with a screwdriver and check for continuity with a circuit tester.

4. If continuity is good, check the adjustment of the engine brake adjusting screw. Connect the circuit tester from the engine

terminals to ground. Push the control rack extension in to the stop position and the light should go on. If not, turn the adjusting screw until it does.

5. Loosen the cap screw on the end of the rack extension just enough to remove the adjusting screw and bracket. Place a straight edge along the forked end of the adjusting screw bracket and measure the clearance between the straight edge and the screw. Loosen the locknut and adjust the screw to obtain a clearance of .138-.142 inch and tighten the locknut.

6. Slide the adjusting screw bracket between the control rack exten-

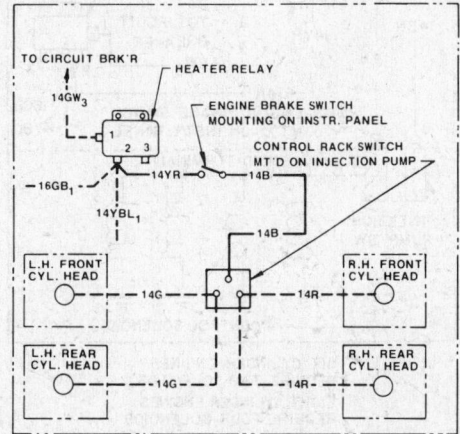

Wiring Arrangement-V8 Engines with American Bosch Vee Type Pump (© Mack Trucks Inc.)

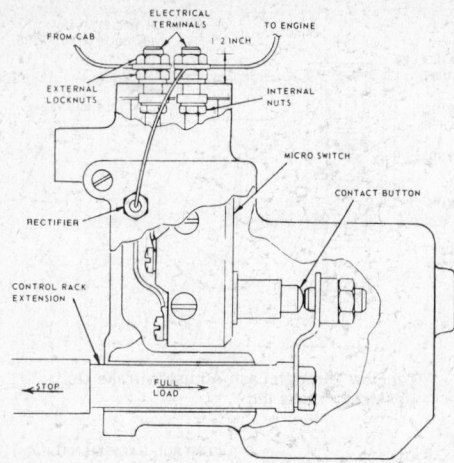

Injection Pump Switch Located on Rack Cap End-
American Bosch Pump Only (© Mack Trucks Inc.)

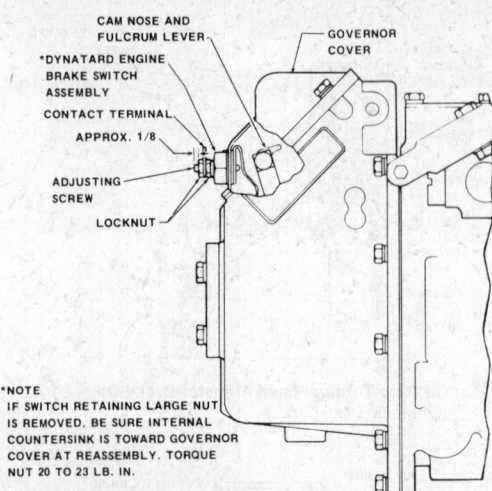

*NOTE
IF SWITCH RETAINING LARGE NUT
IS REMOVED, BE SURE INTERNAL
COUNTERSINK IS TOWARD GOVERNOR
COVER AT REASSEMBLY. TORQUE
NUT 20 TO 23 LB. IN.

Injection Pump Switch Located on Governor Cover-
American Bosch Pump Only (© Mack Trucks Inc.)

sion and capscrew and tighten the capscrew. Replace the switch side cover and rectifier, install a new gasket and secure with a pump seal.

Six and V8 Engine with American Bosch Injection Pump, except V-Type Pump

This switch is a small contact type and is an integral part of the governor housing cover. The switch assembly is activated by the cam nose and fulcrum lever assembly and activates the system at zero fuel only. If adjustment is necessary follow

either of the two following procedures:

With Internal Puff Limiter

A .070 inch feeler gauge and a gauge plug, part # J24659 are necessary for this adjustment.
1. Measure the adjusting screw extension from the first locknut. The screw should extend ⅛ inch. If not, loosen both locknuts and adjust the screw. Tighten the locknuts.
2. Place the pump stop lever in the RUN position.
3. Remove the rack cap plug.

NOTE: On 6-cyl. models, the rack cap may have to be held to prevent its turning when removing the plug or installing the gauge plug.
4. Position the lockring flush against the head of the gauge to allow maximum exposure of the plug gauge screw threads.
5. Screw the special gauge plug into the rack cap until it contacts the rack. Continue to screw it in until it bottoms internally in the injection pump, placing the rack in the full OFF position.
6. Position and lock the lockring by turning it clockwise flush against the rack cap. *NOTE: To keep the lockring fixed in this position,*

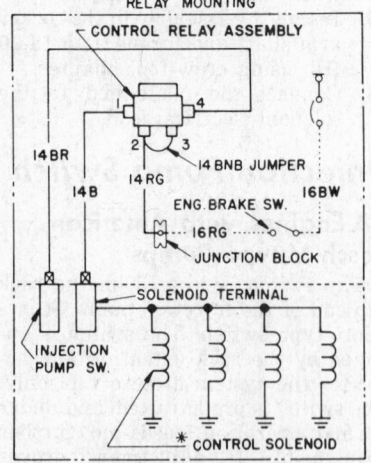

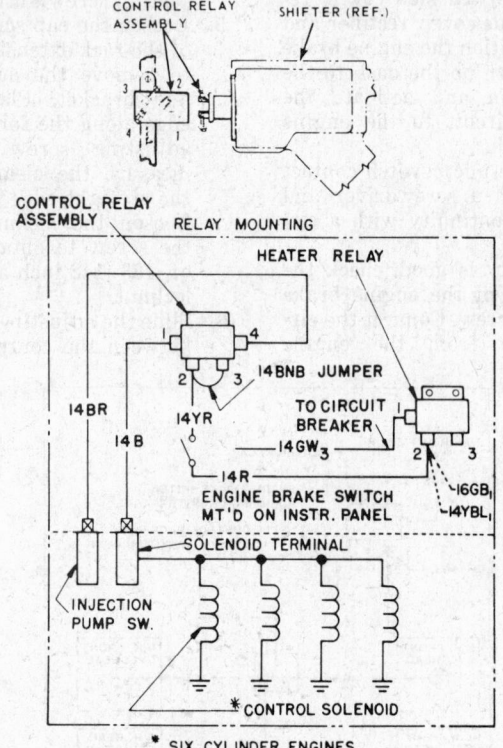

Wiring Arrangement-Dynatard Fuel Switch on American
Bosch Governor Cover-"F" Chassis
(© Mack Trucks Inc.)

*   SIX CYLINDER ENGINES
    REQUIRE TWO SOLENOIDS
    EIGHT CYLINDER ENGINES
    REQUIRE FOUR SOLENOIDS

Wiring Arrangement-Switch on American
Bosch Governor Cover-"R", "RD", "U",
"DM" Chassis (© Mack Trucks Inc.)

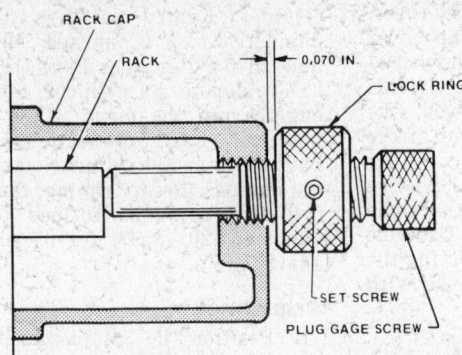

Adjustment-American Bosch Pump with Switch On Governor Housing Cover (© Mack Trucks Inc.)

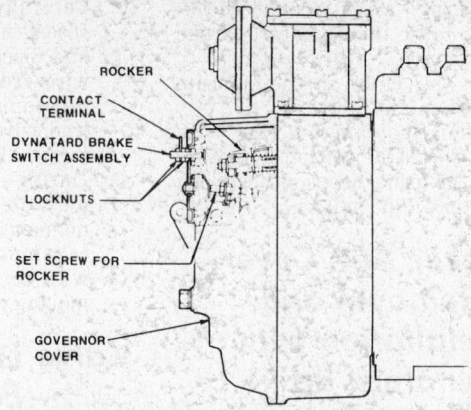

Location Of Switch on Robert Bosch Pumps
(© Mack Trucks Inc.)

screw it in on the set screws placed 180° apart.

7. Turn the special gauge plug counterclockwise until the .070 inch feeler gauge can be inserted between the lockring and the cap.
8. Attach the electrical continuity tester to the contact terminal and ground.
9. Loosen the two adjusting screw locknuts on the engine brake switch adjusting screw.
10. Turn the adjusting screw slowly, counterclockwise until continuity is broken.
11. Turn the screw slowly clockwise until continuity is just achieved. Tighten the inner locknut without disturbing the setting.
12. Place the contact terminal in the upright position and tighten the outer locknut.

With Aneroid Puff Limiter—Robert Bosch Pump

Special gauge plug # J24660 is necessary for this adjustment.

1. Place the injection pump stop lever in the RUN position.
2. Remove the rack cap plug.
3. Screw the plug gauge into the rack cap until it contacts the rack. Continue to screw it in until it bottoms internally in the full OFF position.
4. Move the throttle from idle to wide open several times. Return and securely retain it against the idle stop by using the throttle return spring or its equivalent.
5. Connect the continuity tester between the contact terminal and ground.
6. Loosen the two adjusting screw locknuts on the engine brake switch adjusting screw.
7. Turn the adjusting screw counterclockwise until continuity occurs at the tester.
8. Turn the adjusting screw slowly clockwise until the point at which continuity breaks. Then turn it one full turn more, clockwise.
9. Tighten the inner locknut without disturbing the setting.

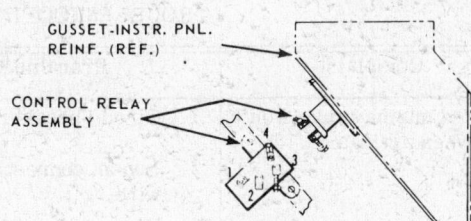

Plug Gauge Screw Installed in a Robert Bosch Pump (© Mack Trucks Inc.)

10. Return the terminal to the upright position and tighten the outer locknut.
11. Remove the plug gauge and install the rack cap securely.

Pumps with Mack Puff Limiter

1. Run the engine to normal operating temperature and establish a smooth idle of 550 rpm.
2. Loosen both nuts on the pump brake switch.
3. On American Bosch Pumps, turn the adjusting screw clockwise; on Robert Bosch Pumps, turn the

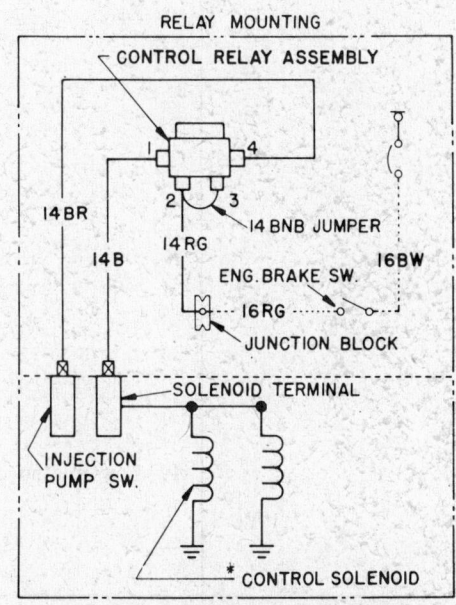

Wiring Arrangement with Robert Bosch Pumps
(© Mack Trucks Inc.)

adjusting screw counterclockwise, until the Dynatard brake comes on.

4. Turn the screw in the opposite direction one full turn.

5. Tighten the jam nut and wire nut. Check the setting by noting that the brake drops out between 1000 and 700 rpm.

Rocker Arm Cover Removal and Installation with Dynatard Brake

Late production engines have a switch, identical in appearance to the pump mounted switch, located on the rocker arm cover. This switch makes removal of the cover simple. However, early production sixes require great care in cover removal.

1. Remove the cover bolts.

2. Carefully raise and tilt the cover. Insert a hand under the cover and disconnect the solenoid lead wire from the terminal. If this procedure is not followed, the lead wire will be torn out of the terminal connection.

3. During installation, the wire must be connected in the same manner it was removed. Extreme care must be taken in routing the wire to avoid interference with moving parts.

Valve Lash Adjustment —Engine with Dynatard Brake

The adjustment must be made cold, i.e. coolant temperature below 100°F. Valves should be adjusted in firing order.

Present production engines have marks identifying TDC of each cylinder, located on the vibration damper. No. 1 is marked. Most V8 and all 6-cyl engines have the mark "1" to denote #1 cylinder. Some V8 engines use the mark "TC" to denote #1. Early production engines have only the #1 cylinder marked on the damper. Before timing, the other cylinders must be scribed on the damper; 120° apart for 6's and 90° apart for 8's.

Adjustment

1. Position the #1 piston at TDC compression.

2. Adjust the intake valve to .016″ 6 or 8 cyl.

3. Adjust the exhaust valve, using special tool MVT-36-6. Press downward with a hand on the adjusting screw while gauging the correct lash. Correct lash is .024 for 6-cyl.; .026 for 8-cyl.

4. Adjust the remaining valves in firing order.

TROUBLESHOOTING CHART

Complaint	Probable Cause	Remedy
Dynatard engine brake won't go on when activated	1. Circuit breaker kicked "out".	1. Check circuit breaker.
	2. Loose connection or broken wire.	2. Check electrical system.
	3. Defective dash switch.	3. Check continuity of dash switch: (a) Start engine (idle O.K.). (b) Place dash switch in "ON" position. (c) Rev engine to governed speed. (d) Remove foot from throttle quickly to permit throttle and governor to close the fuel injection pump switch and energize the engine brake solenoids. (e) If continuity does not exist, solenoids will not activate. Replace defective switch.
	4. Low engine oil pressure.	4. Check engine oil pressure. Oil pressure of 30 psi min. is required to activate hydraulic lash controller.
Dynatard engine brake won't go on when activated. (cont)	5. Excessive oil leakage around and at vent hole of adapters.	5. Check adapter for loose upper seat. If loose, replace with new parts.

TROUBLESHOOTING CHART

Complaint	Probable Cause	Remedy
Dynatard engine brake activates but low braking power exists.	1. Defective brake solenoid.	1. Check brake solenoid operation: (a) Remove cylinder head covers. (b) Turn ignition switch "ON". Do not start engine. (c) Place dash switch "ON" position. (d) Pull engine stop lever to "OFF" position. (e) Check voltage to the solenoids. If low or no voltage is present, replace defective solenoid.
	2. Improper valve adjustment characterized by rough engine idle.	2. Readjust valve clearance.
	3. Restrictions in solenoid control valve and mounting adapter.	3. Locate restricted adapter by manually pushing each solenoid down one at a time. Engine should idle rough and misfire as brake cuts off two cylinders on eight and three cylinders on six cylinder engines. If no change in engine idle, remove solenoid control valve and mounting adapter and clean. Check oil passages to be sure they are not restricted.
Dynatard engine brake goes on when not activated.	1. Restrictions in solenoid control valve and mounting adapter.	1. Remove solenoid control valve and mounting adapters. Clean and be sure internal ball is free and plunger seats ball when solenoid is energized.
	2. Improper lubricating oil.	2. Check lube oil specifications. Thick oil can cause malfunction.
	3. Defective fuel injection pump switch.	3. Hold fuel shut off in the "ON" position and check continuity across input and terminals on the injection pump switch. If continuity exists, replace switch.
	4. Loose lower adapter seat (early production solenoid and adapter only).	4. Replace old style adapter with new parts.

Mitsubishi Diesel Engines

TUNEUP SPECIFICATIONS

INJECTION PRESSURE PSI	PUMP TYPE	INJECTION	IDLE SPEED	VALVE CLEARANCE INTAKE EXHAUST		COMPRESSION PRESSURE PSI RPM	
1705	Bosch PES6A65B	18°B	550-600	.012C	.012C	426	170

GENERAL ENGINE SPECIFICATIONS

MODEL	CYL.-DISPL. CU. IN.	BORE x STROKE	MAX. BHP RPM		MAX. TORQUE FTLB RPM		FIRING ORDER	COMPRESSION RATIO	OIL PRESSURE PSI 2000 RPM
6DR50A	6-243	3.62 x3.94	100	3700	163	2200	153624	20.0:1	42-71

LUBRICATION SYSTEM

PUMP	PAN CAPACITY	FILTER CAPACITY	IDLE PRESSURE	API REQUIREMENT
Trochoid	7 qts.	2 qts.	21 psi	SE/CC

COOLING SYSTEM

CAPACITY	PUMP TYPE	THERMOSTAT	FAN RATIO	CAP PRESSURE
13 qts	Centrifugal	185°F	1.07:1	7 lbs.

ELECTRICAL SYSTEM

BATTERY TYPE	BATTERY AMP	ALTERNATOR	GLOW PLUGS
2-12 volt	70 each	60amp	10.5v; 8.3amp

FUEL SYSTEM

TIMER	CAM LIFT	DELIVERY VALVE DIA.	PLUNGER LEAD	PLUNGER DIA.	GOVERNOR TYPE	FEED PUMP CAM LIFT	NOZZLE DIA.	SPRAY ANGLE	HOLDER TYPE
Centrifugal	.315″	.236″	.591″	.256″	RU	.236″	.039″	4°	KC

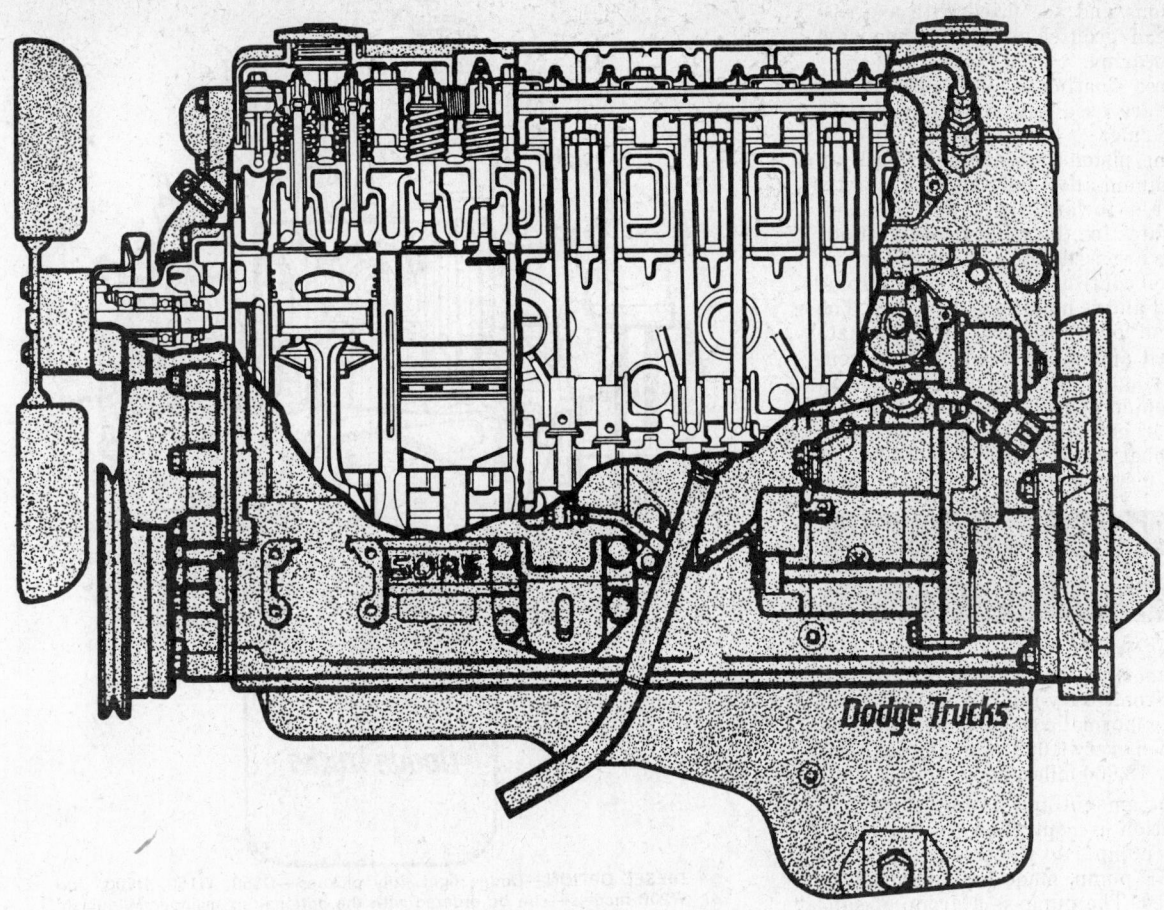

DIESEL OFFERED—An optional engine for Dodge light duty pickup models—D and W150 and D and W200—will be the six cylinder diesel engine from Mitsubishi Motors Corporation. Production is slated for this fall in Dodge pickups.

The Mitsubishi 6DR50A is an in-line 6-cylinder, liquid cooled, naturally aspirated, OHV engine. It will be used by Dodge Truck division in most two and four wheel drive model pick-up trucks in the 6100-8500 lb GVWR class. The engine is not available in California.

The engine is designed to run on either No. 1-D or No. 2-D grade fuel oil with a cetane rating of 45. No. 2-D should be used at all temperatures above 10°F (-12°C).

Lubricating oil must meet service specification SE/CC.

Engine

Cylinder Block

The block is made of cast iron. It extends 2¾" below the centerline of the crankshaft for extra rigidity. Force-fit dry cylinder liners made of cast iron are used. These are removed in the conventional manner, from above with a standard puller. The cylinder bores are jacketed for adequate cooling.

Each crankshaft and camshaft bearing support is directly connected, by drilled passage with the lubricating oil gallery.

Crankshaft

A forged chrome/molybdenum steel crankshaft is used, mounted on seven main bearings. The crankshaft fillets are ground to a .12 inch radius. The main journals and crankpins are cross-drilled for oil passage. Crankshaft bearings are a tri-metal design, steel back with a lead and tin plating. Thrust is taken on the number seven bearing.

Cylinder Head

The head is cast iron with integrally cast intake and exhaust ports. The head is a cross-flow design with the intake ports on the left side; the exhaust ports on the right. Case hardened steel valve inserts are used for both intake and exhaust.

The camshaft is chrome/molybdenum steel and supported by four bearings.

The main coolant guides are located in the left side of the head.

Swirl-type pre-combustion chambers are cast into the head with a steel combustion chamber jet pressed into the head below the nozzle to form the pre-combustion chamber cavity.

A steel core asbestos head gasket is used for positive sealing. The head is secured by twenty five bolts. Steel grommets at the cylinder bores insure positive high pressure gas sealing.

Valve Train

The camshaft is driven off the crankshaft through an idler gear. Cast iron, solid offset lifters provide action to the rocker arms by means of hollow steel, tubular pushrods with steel ball inserts at each end. Intake and exhaust valves are surface-hardened steel and ride in cast iron insert valve guides.

The rocker arms are forged steel and are mounted on a chrome-plated, hollow steel rocker arm shaft. Lead-bronze bushings are fitted in the rocker arms. The rocker shaft is plugged at both ends and serves as an oil gallery for the valve train.

Connecting Rods, Piston Pins and Pistons

The rods are forged steel, I-beam

sections and are fitted with a press-fit, lead/bronze bushing for the piston pin bearing.

Free floating hollow steel piston pins are used and are retained with snap rings.

The pistons are one-piece with an aluminum alloy skirt and are slightly tapered toward the top. This compensates for thermal expansion. Four rings are used; three compression and one oil control. The top ring is chrome faced and is half (top) keystone. The second and third rings are phosphate coated and taper-faced. The oil control ring is chrome plated and has an on-center helical expander.

A ramp from the pre-combustion chamber is machined into the top of each piston to direct the mixture flow.

Fuel System

The intake manifold is cast aluminum and the exhaust manifold is cast iron. A dry type air filter is used. Under normal conditions it should be cleaned every 6,000 miles and changed every 18,000 miles.

The amount and timing of the fuel injection is controlled by the injection pump. A Robert Bosch PES 6A65B pump, made by Nippondenso, is used. The pump is a group of sub-assemblies which consist of a low pressure fuel supply pump, a high pressure injection pump and the engine speed control governor.

The injection pump is driven off the crankshaft by way of an idler gear. The supply pump and the governor mechanism are actuated by the injection pump camshaft. The injection pump camshaft is located in the lower part of the housing supported by two tapered roller bearings and a bushing centered on the shaft.

When the camshaft rotates it lifts tappets, in firing order, which in turn lift plungers in which metered amounts of fuel have been trapped. The fuel is passed through the delivery valve and nozzle at extremely high pressure (1700psi.). The pump contains six independent fuel injection plunger assemblies, in-line, one for each cylinder in the engine.

The governor is attached to the rear of the injection pump housing and is driven off the end of the pump camshaft. The mechanism is designed to control the amount of fuel delivered to the injection nozzles, depending on speed and load conditions.

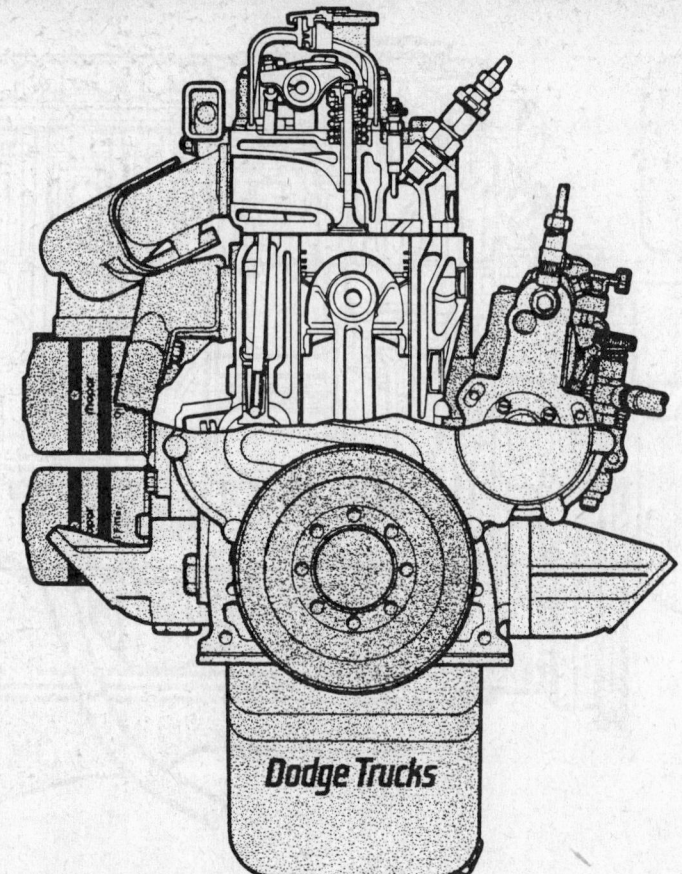

DIESEL OPTION—Dodge light duty pickups—D150, W150, D200, and W200 models—can be ordered with the optional six cylinder Mitsubishi Motors Corporation engine.

Lubricating oil from the engine is delivered through an external pipe to the rear plunger tappet area. The oil drains back to the engine through an overflow allowing an adequate oil level in the pump sump.

The injection pump is fed a low pressure supply of fuel oil by the feed pump, located on the side of the injection pump and driven by the injection pump camshaft. A wire gauze filter located in the feed pump removes large dirt or wax particles from the fuel before it enters the feed pump. After leaving the feed pump, the fuel passes through a paper filter which removes all small particles before reaching the injection pump. The wire gauze filter should be cleaned every 12,000 miles. The paper filter should be inspected every 12,000 miles and replaced every 24,000 miles.

Engine Lubrication

A trochoid type oil pump is driven by the engine camshaft to supply oil pressure. Two passenger car-type, throw-away, full flow oil filters are used. They filter the oil in parallel and each has an internal bypass valve. Normal oil and filter change interval is every 6,000 miles or 6 months, whichever comes first.

Engine oil must conform to service classification SE/CC. Viscosity range is determined by the existing ambient temperature.

A centrifugal water pump delivers coolant to the jackets around the cylinders and then to the head via the coolant guides on the left side. The coolant leaves the head on the right side, through a pipe to the thermostat and radiator. A thermostat bypass system is used to permit a constant coolant flow through the block, even with the thermostat closed.

GENERAL SPECIFICATIONS

ENGINE MODEL	CUBIC IN. DISP.	HORSEPOWER @ R.P.M.—SAE	TORQUE @ R.P.M.—SAE	FIRING ORDER	GOVERNED SPEED	OIL PRESSURE	VALVE CLEARANCE INTAKE	EXHAUST	BORE & STROKE
OM352A	346	144 @ 1800	318 @ 2000	1-5-3-6-2-4	2800	35.5 @ 2800	.007	.011	3.82 x 5.04
OM352	346	120 @ 1800	257 @ 1800	1-5-3-6-2-4	2800	35.5 @ 2800	.007	.011	3.82 x 5.04

CRANKSHAFT AND BEARING SPECIFICATIONS

(INCHES)

ENGINE MODEL	MAIN BEARING JOURNAL DIA.	MAIN BEARING CLEARANCE	SHAFT ENDPLAY	ROD BEARING JOURNAL DIA.	ROD BEARING CLEARANCE
OM352	3.4696	.0019-.0035	.0074	2.3641	.0019-.0035

PISTON ASSEMBLY SPECIFICATIONS

(INCHES)

ENG. MODEL	RING SIDE CLEARANCE 1st	2nd	3rd	4th	5th	PISTON SKIRT DIA.	RING END GAP 1st	2nd	3rd	4th	5th	CYLINDER LINER INSIDE DIA.
OM352	.0017-.0031	.0021-.0032	.0028-.0032	.0013-.0024	.0009-.0020	3.815	.013-.021	.013-.012	.013-.021	.009-.014	.009-.014	3.819

ENGINE TORQUE SPECIFICATIONS

(FT. LB.)

ENGINE MODEL	CYLINDER HEAD	ROCKER SUPPORT	EXHAUST MANIFOLD	CONN. ROD	MAIN BRG. CAP	DAMPER TO CRANKSHAFT	FLYWHEEL	NOZZLE HOLDER	OIL PAN	ROCKER COVER
OM352	①	80	36	72 + 90°	②	362-398	22+90°	50	5	18

① Three Steps 43, 65, 80

② To eng. 100 HP/009386 M14
110 HP/006010 M14 } 86
126 HP/020157 M14
From above engines—101
Type Suffix M15—36+90°

Engine Disassembly

NOTE: Disassembly procedure is done with engine removal from vehicle and mounted in a work stand.

1. Remove all hoses and lines from clips on cylinder head.
2. Unbolt and remove cover and gasket.
3. Remove fuel filter.
4. Remove upper cooling line.
5. Remove injection high pressure lines.
6. Remove rocker arm assemblies.
7. Lift out pushrods and mark order for re-installation.
8. Disconnect pull rod on throttle valve lever.
9. Unbolt exhaust pipe from manifold.
10. Attach lifting device and remove head bolts in reverse order of installation. Lift off head. Place head on its side to avoid damage to nozzles.
11. Remove injection leak-off pipe.
12. Remove nozzle holder pressure screw and remove nozzle holder with a slide hammer or puller.
13. Unscrew nozzle sleeve with slotted socket.
14. Compress valve springs, remove keepers and disassemble valves and springs.
15. Remove flywheel attaching bolts and remove flywheel by tapping with a plastic hammer in a downward direction.
16. Remove alternator and belt.
17. Remove damper bolt and remove damper from crankshaft with puller.
18. Unbolt and remove front cover and timing housing bottom.
19. Remove oil pan.
20. Remove oil pump.
21. Remove cylinder liner ridge.
22. Unbolt and remove rod bearing caps. Mark cap.
23. Push piston and rod assemblies from block.
24. Remove piston cooling nozzles.
25. Unbolt and remove lower main bearing caps.
26. Remove crankshaft from block.
27. Remove upper main halves.
28. Remove oil filler connection from timing housing.
29. Set engine to injection delivery beginning: 18° BTC.
30. Remove fuel lines from injection pump.
31. Remove oil supply line.
32. Unbolt and remove injection pump.
33. Remove pushrod chamber cover.
34. Remove valve tappets.
35. Unbolt and remove air compressor.
36. Unscrew thrust washer from camshaft.

NOTE: Locking plate is only used on engines up to 130 HP/115816.

37. Remove camshaft.
38. Remove starter.
39. Remove electric line from oil pressure transmitter or oil pressure gauge line.
40. Unscrew exchanger from crankcase.
41. Remove upper cooling pipe and thermostat.
42. Remove coolant hoses from water pump.
43. Remove fan and pulley assembly.
44. Unbolt and remove water pump.

Mercedes Benz Diesel Engines

Piston dia.	3.03149″
Stroke	1.1811″
Total displacement	8.54 cid
Delivery gal./min at rpm	46.09/3000
Max. operating pressure	104.5 psi
Filling time from 42.66 to 56.88 psi. using a 9 gal. air tank and 1,500 rpm of air compressor	18① sec

① During repairs max. 23 sec.

Cylinder Liner and Piston
(inches)

CYLINDER LINER	BORE DIA.	PISTON DIA.	PISTON CLEARANCE
Normal	3.0318 / 3.0310	3.0307 / 3.0303	
Normal I	3.0348 / 3.0340	3.03365 / 3.0332	
Rep.-St. I	3.0417 / 3.0409	3.0405 / 3.0401	.00039-.00078 wear limit max. .001181
Rep.-St. II	3.0515 / 3.0507	3.0503 / 3.0499	
Rep.-St. III	3.0614 / 3.0606	3.0602 / 3.0598	

Piston Rings
(inches)

GROOVE	PISTON RING DESIGNATION (PART NO.)	WIDTH OF RING GROOVE	SIDE CLEARANCE	GAP CLEARANCE
I	Angular compression ring (900282 077000) 12 f 77 x 70,2 x 2,5 x KE 54 N 282	.0984	.00039-.00133	.01181-.01771
II	Angular compression ring (900282 077000) 12 f 77 x 70,2 x 2,5 x KE 54 N 282	.0984	.00039-.00133	.01181-.01771
III	Slotted chamfered oil control ring (900282 077000) 42 f 77 x 70,2 x 4 KE 54 N 283	.1574	.00039-.00133	.00984-.01574
IV	Slotted chamfered oil control ring (900278 077000) 30 f 77 x 70,2 x 3 KE 54 N 278	.11811	.00039-.00133	.01181-.01771

Note: The designation of the piston rings refers only to normal stage of piston.

Service Procedures

Pistons

1. Remove rings from pistons.
2. Remove piston pin locks and separate rod, pin, and position.
3. Clean ring grooves and check width:
   Groove 1—.118″
   Groove 2—.118″
   Groove 3—.118″
   Groove 4—.216″
   Groove 5—.216″
4. Assemble rings in grooves.
5. Assemble piston, pin and rod. Install locks.
6. Install bearing halves in rod and cap.

Valve Guides

1. Force guide out of head with mandrel driver.
2. Ream guide bore to next highest step: .00397″, .00787″ or .01968″ over.
3. Coat new guide with graphited oil and drive into head with suitable mandrel driver. Ream guides to .3551″ for intake and .3945″ for exhaust valves.

Injection Nozzles—Testing

1. Remove carbon from nozzle.
2. Install nozzle in tester 000589142700.
3. Push pump lever slowly until reading is 284 psi below opening pressure (2,844 ± 142 psi).
4. The nozzle should show no signs of leakage for at least 10 seconds.
5. Move hand lever at the rate of about one stroke per second and read opening pressure: 2844 psi.
6. Disconnect pressure gauge and operate lever at about 4 to 6 strokes per second. The nozzle should buzz softly. No lateral diffusion of the spray should occur.

Injection Nozzles—Repair

1. Insert nozzle holder into mounting fixture 403489 003100 and remove pressure nut. Disassemble nozzle.

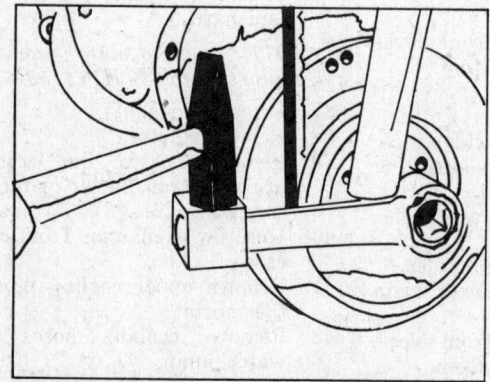

Loosening vibration damper bolt

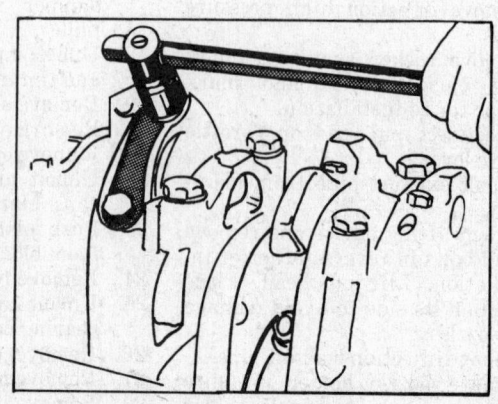

Rod cap removal

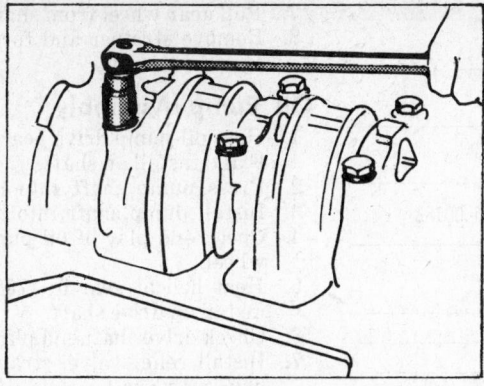

Main cap removal

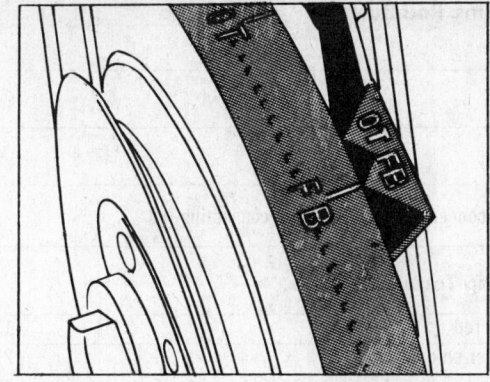

Pointer on timing housing at beginning of delivery

2. Clean nozzle in fuel oil. Injection holes can be cleaned with tool 00589006800 or equivalent.
3. Place nozzle needle and body in diesel fuel and check slidability

by means of a drop test: pull needle ⅓ way out of body and release. Needle should slide back to its seat under its own weight.
4. Assemble nozzle and place in

holder. Torque pressure nut to 57 ft. lbs.

Air Compressor—Disassembly

1. Remove head, unscrew liner and pull off over piston.
2. Unlock and force out piston pin. Remove piston.
3. Unbolt and remove connecting rod, bearing and cap.
4. Unscrew suction valve cap from cylinder head.
5. Remove valve spring, valve disc and suction valve seat.
6. Unscrew pressure valve seat from cylinder head and remove together with valve disc, valve spring and spring housing.
7. Remove piston rings.
8. Clean all parts in alcohol. Check the following specifications:

Piston Pins
(inches)

COLOR CODE	PISTON PIN OD	BORE IN PISTON
yellow	.6304 / .6303	.6316 / .6315
green	.6305 / .6352	.6317 / .6316
Piston pin clearance in connecting rod bushing		.00035-.0009
Piston pin clearance in piston		.00106-.00129

Connecting Rod
(inches)

Basic bore in connecting rod	1.3785 / 1.3779
Basic bore for connecting rod bushing	.7488 / .7480
Distance from center of camshaft lifting pin to center of piston pin bore	3.6220 / 3.6200
Perm. deviation of parallel axle alignment for length of 100 mm	.001181
Width of connecting rod on — connecting rod eye	.8635 / .8615
Width of connecting rod on — piston pin eye	.8779 / .8740
Radial play of crankpins	.00118-.00259
Axial play (end play) of crankpins	.00255-.01248

Air Compressor—Assembly

1. Insert bearing shells into connecting rod and cap.
2. Bolt connecting rod cap to rod and measure bore at 3 places.
3. Remove cap and measure ID of rod bushing.
4. Attach rod assembly to camshaft lifting pin.
5. Install rings on piston and assemble piston and pin onto connecting rod. Install pin locks.
6. Install liner over piston and screw liner into block.
7. Install suction valve components into head.
8. Insert pressure valve components into head and screw on pressure valve seat with socket wrench.
9. Screw cylinder head to liner using new gasket.
10. Install suction and brake hose.

Oil Pump—Disassembly

1. Clamp pump in soft jaw vise.
2. Pull helical gear.
3. Unscrew housing cover.
4. Unscrew oil pressure relief valve plug and remove along with spring and piston.
5. Unscrew valve housing.
6. Remove oil pump gear wheel and drive shaft.

Camshaft Lifting Pin and Connecting Rod Bearing
(inches)

STAGES	LIFTING PIN DIA.	BEARING BORE IN INSTALLED CONDITION	WALL THICKNESS FOR BEARING SHELLS READY FOR INSTALLATION
Normal	1.2598 / 1.2592	1.2624 / 1.2608	.0580-.0585
Normal I	1.2559 / 1.2552	1.2585 / 1.2569	.0600-.06051
Rep.-St. I	1.2499 / 1.2493	1.2525 / 1.2510	.0629-.0634
Rep.-St. II	1.240 / 1.239	1.2427 / 1.2411	.0679-.0683
Rep.-St. III	1.2303 / 1.2296	1.2329 / 1.2313	.0728-.0733

Connecting Rod Bushing (inches)

OD	.7499 / .7490
ID	.6309 / .6312
Overlap of connecting rod bushing in connecting rod	.00055-.00188

Tightening Torques in Ft. lb.

Connecting rod	10.8
Cylinder liner on crankcase	25.3-28.9
Cylinder head on liner	25.3-28.9
Inlet and exhaust valve on cylinder head	86.7-101.2

Oil Pump—Dimensions in M.M.

Radial play of drive shaft	.0006-.0016
Radial play between oil pump gear wheel and oil pump shaft	.0004-.0015
Overlap between oil pump shaft and housing	.00039-.00153
Radial play of oil pump gear wheels between housing and gear wheel	.00118-.00413
End play of oil pump gear heels between housing cover and gear wheel	.00098-.00350
Backlash of oil pump gear wheels	.00590-.00984
Backlash of oil pump drive gears (helical gears)	.00377-.00503
Number of teeth of oil pump gear wheels	7/7
Test torque of oil pump helical gear	50 ft. lb.
Test torque of oil pump gear wheel (driving)	

DIMENSION A	Dia. in housing	.66999 / .66929
	Dia. of drive shaft	.66866 / .66834
Dimension B = Dia. of oil pump shaft		.59208 / .59165
DIMENSION C	Height of gear wheel housing	1.5757 / 1.5748
	Height of oil pump gear wheel	1.5738 / 1.5722
DIMENSION D	Perm. play between driving helical gear and upper edge of pump housing with driving gear at upper edge of pump housing	.00157
Dimension E = Installation height of drive shaft		.01968-.03149

Oil Pump Test

Delivery rate measured with SAE 10 oil Oil temperature 50° C Oil counter pressure 85.32 psi	300 rpm	1.36 gal/min
	1,400 rpm	9.57 gal/min

Springs for oil pressure relief valves

	OD IN.	WIRE DIA. IN.	ADJUSTED TO PSI	UN-LOADED SPRING LENGTH IN.	SPRING PRELOAD		SPRING FINAL LOAD	
					LENGTH IN.	LOAD lb.	LENGTH IN.	LOAD lb.
In oil pump	.366	.066	74	1.94	1.78	9.67	1.425	31.30

7. Pull gear wheel from shaft.
8. Remove strainer and force shaft from housing.

Oil Pump Assembly

1. Heat oil pump drive gear to 175° F and install on shaft.
2. Press pump shaft into housing.
3. Install pump shaft into housing.
4. Check end play of oil pump gear wheel.
5. Heat helical gear to 175° F and install on drive shaft.
6. Check drive shaft end play.
7. Install relief valve, strainer and housing cover.

Water Pump Disassembly

1. Pull hub from pump shaft.
2. Unscrew and remove sealing ring holder.
3. Remove grease cup.
4. Push shaft and bearing from impeller end.
5. Force small bearing and seal from housing.
6. Remove sliding ring seal from housing.
7. Pull bearing from shaft.
8. Remove counter ring from impeller.

Water Pump—Assembly

1. Press radial seal into housing.
2. Press sliding ring seal into housing.
3. Press bearings onto shaft.
4. Fit seal on hub and around bearing.
5. Press hub on shaft.
6. Press shaft into housing.
7. Install seal holder to housing.
8. Install counter ring and seal on impeller.
9. Press impeller onto shaft.
10. Check clearances.

Engine Assembly

1. Bolt water pump and gasket to engine. Torque pump bolts to 35 ft. lbs.
2. Install thermostat and fan. Torque thermostat housing bolts to 21 ft. lbs. and fan-to-hub bolts to 25 ft. lbs.
3. Coat heat exchanger cover with grease and position new gasket on cover.
4. Position sealing discs on holes of oil outlet and inlet duct.
5. Place heat exchanger plates, one on top of another in position. Place sealing rings under bolt heads and hand-tighten bolts.
6. Attach heat exchanger to engine. Torque bolts to 25 ft. lbs. Insert bypass valve.
7. Install starter.
8. Connect oil pressure sending unit line.
9. Carefully slide camshaft into block. Align timing marks with idler gear.
10. Attach thrust washer.

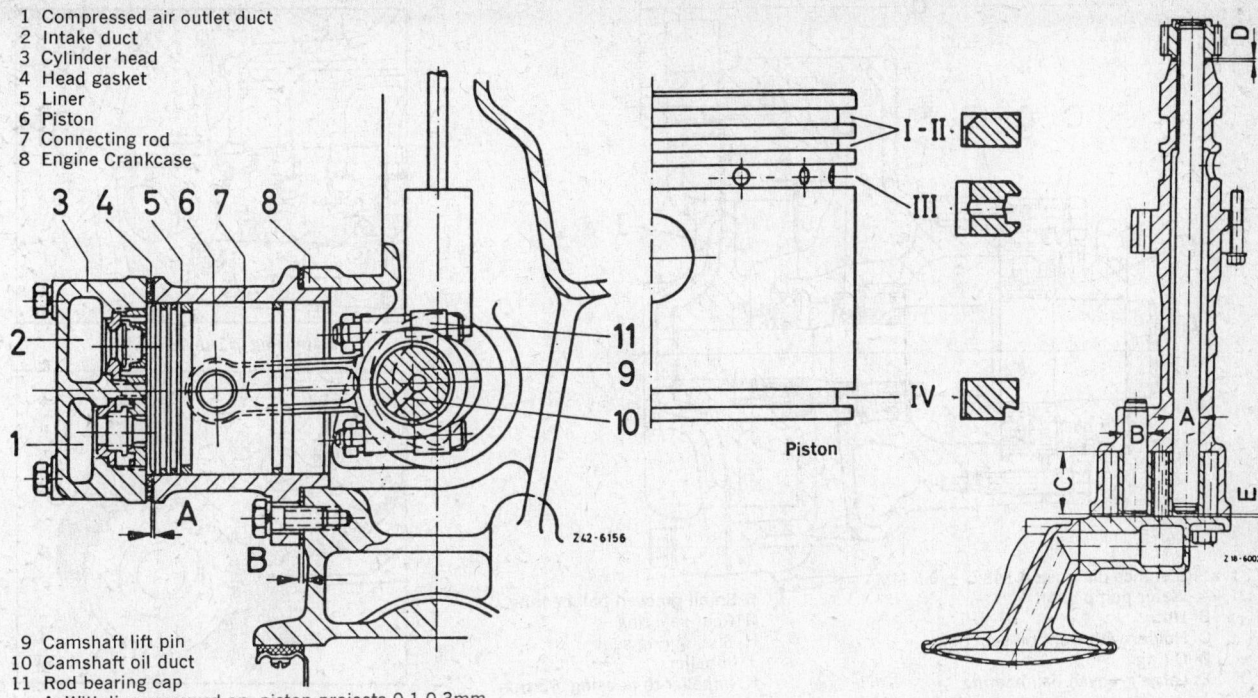

1 Compressed air outlet duct
2 Intake duct
3 Cylinder head
4 Head gasket
5 Liner
6 Piston
7 Connecting rod
8 Engine Crankcase

9 Camshaft lift pin
10 Camshaft oil duct
11 Rod bearing cap
   A With liner screwed on, piston projects 0.1-0,3mm
   B Gasket 0.25 and .05 mm thick

Piston

Tightening Torques in Ft. lb,

Cover to oil pump	25
Strainer to cover	14-18
Oil pressure relief valve to cover	43

Dimensions in M.M.

		on impeller seat	.59208 / .59165
Water pump shaft dia.	Bearing seat	large bearing	1.18145 / 1.18094
		small bearing	.66960 / .66917
Shaft dia. for hub			1.14385 / 1.14334
Bore dia. in hub			1.14255 / 1.14173
Hub dia. for sealing ring front			1.65354 / 1.64724
Shaft dia. for rear sealing ring			.66960 / .66917
Bore dia. in impeller			.59125 / .59055
Press impeller on water pump shaft			Flange surface—impeller flush with housing, flange
Lubrication of water pump			Grease approx. 3 oz.

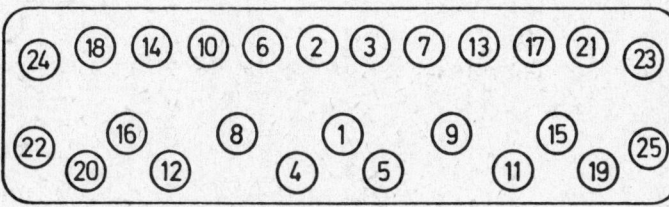

Head bolt torque sequence

11. Insert valve tappets in order as marked.
12. Install pushrod chamber cover.
13. Install injection pump into timing hosing so that mark on pump gear is aligned with arrow in timing housing.
14. Attach fuel supply, oil, and injection lines.
15. Clean all crankshaft bearing surfaces; insert upper main halves in block. Coat with SAE 30. Attach main caps and check bore. Remove caps.
16. Install crankshaft into block. Align timing marks with camshaft.
17. Install caps and torque to specifications.
18. Install piston cooling nozzles.
19. Assemble piston and rod, coat all parts with SAE 30 and install piston with notch forward. Assemble rod caps and torque to specifications.
20. Install oil pump and new gasket.
21. Install oil pan and gasket.
22. Install front cover and timing housing.
23. Heat vibration camper to 175°F and install on crankshaft with key. Install and torque bolt.
24. Install alternator.
25. Set engine to #1 compression at 18° BTDC. Remove oil supply line from injector timing housing. Remove #1 injection line from pump. Remove pipe connection from #1 pump cylinder and remove filler piece, compression spring and valve cone. Reinstall pipe connection and overflow pipe. Unscrew fuel supply line

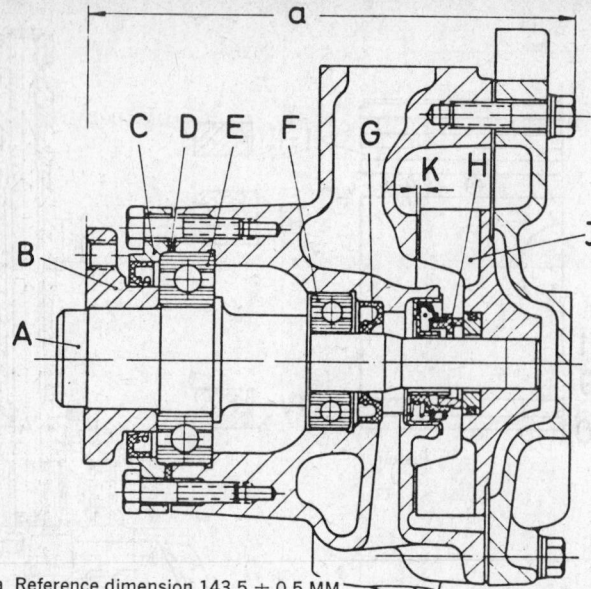

a Reference dimension 143.5 ± 0.5 MM
A Water pump shaft
B Hub
C Holder with front oil seal ring
D O-ring
E Large grooved ball bearing
F Small grooved ball bearing
G Rear seal ring
H Sliding ring seal
J Impeller
K Impeller-to-housing .08 mm

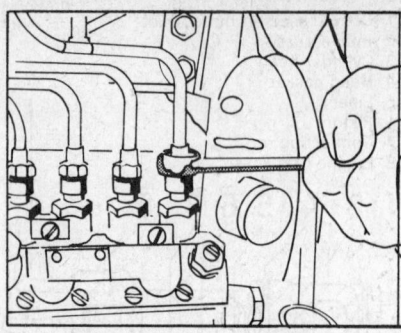

Removing #1 injection line

Timing mark alignment

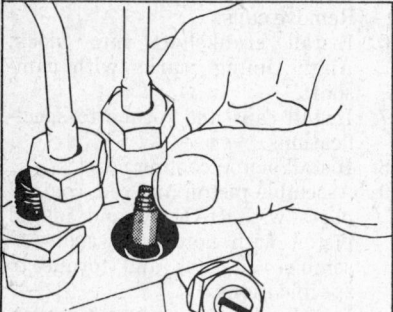

Fuel reservoir in place

Rod cap alignments

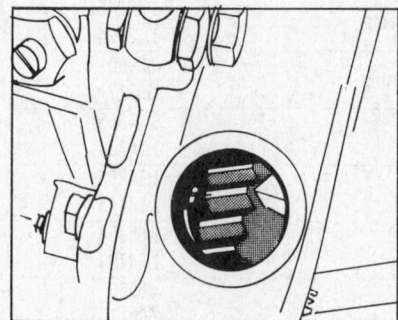

Removing filler from #1 injection port

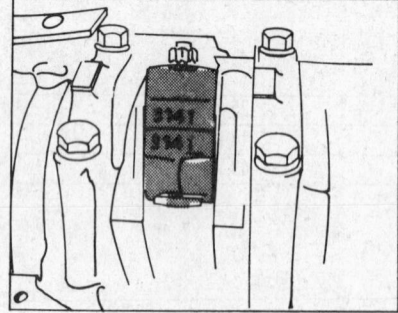

Pointer on timing housing at beginning of delivery

from pump and attach reservoir. Fill reservoir with clean fuel and open shut-off valve. Set control valve on pump to full load. When adjusted ONE DROP of fuel should come out of overflow pipe every 15-20 sec. If flow is greater or less, unbolt pump in swivel until adjustment is correct. Reconnect pump. Reinstall parts.

26. Position flywheel on crankshaft and align bolt holes. Oil bolt threads and tighten in a cross pattern.
27. Assembly valves in head.
28. Install sleeves in head and torque to 43 ft. lbs.
29. Install holder seals in sleeves.
30. Insert holders into sleeves. Tighten pressure nut to 43-50 ft. lbs.
31. Install oil drain pipe and new seals.
32. Place new head gasket on block.
33. Lower head into position, lubricate bolt threads and install bolts. Tighten bolts in pattern shown to 43, then 65 then 80 ft. lbs.
34. Install pushrods as marked.
35. Mount rocker arm assembly and torque to 80 ft. lbs.
36. Install injection lines.
37. Attach exhaust manifold (36 ft. lbs.) and exhaust pipe.
38. Attach and secure pull rod to throttle valve.
39. Install upper cooling line with new seals.
40. Attach fuel filter.
41. Install dipstick.
42. Adjust valves.
43. Install cylinder head cover and attach all lines and hoses clipped to cover.

GENERAL ENGINE SPECIFICATIONS

NO. CYL-CU. IN.	BORE X STROKE	HP @ RPM	TORQUE @ RPM	COMPRESSION RATIO	COMPRESSION PRESS (psi)*	OIL PRESSURE (psi) @ 2000 RPM	WEIGHT DRY (lb.)
6-198	3.27 x 3.94	73 @ 3200	133 @ 1600	22:1	426	50	662

* At 200 rpm

ENGINE TUNE-UP SPECIFICATIONS

INJECTOR OPENING PRESSURE (psi)	LOW IDLE	VALVE CLEARANCE* INTAKE	VALVE CLEARANCE* EXHAUST	INTAKE VALVE OPENS (deg.)	INJECTION TIMING (rpm)	FIRING ORDER
1422	700	.014	.014	28B	20B @ 1000	1-4-2-6-3-5

* Hot or cold

CRANKSHAFT AND CONNECTING ROD

MAIN BEARING JOURNAL DIAMETER (in.)	MAIN BEARING OIL CLEARANCE (in.)	THRUST ON	CONNECTING ROD JOURNAL DIAMETER (in.)	CONNECTING ROD BEARING OIL CLEARANCE (in.)	CONNECTING ROD SIDE CLEARANCE (in.)	CRANKSHAFT END PLAY (in.)
2.7918-2.7988	.0013-.0038	3	2.0840-2.0906	.0013-.0038	.0039-.0079	.0024-.0094

CAMSHAFT SPECIFICATIONS

JOURNAL DIAMETER 1 (in.)	JOURNAL DIAMETER 2 & 3 (in.)	JOURNAL DIAMETER 4 (in.)	JOURNAL-TO-BEARING CLEARANCE 1 (in.)	JOURNAL-TO-BEARING CLEARANCE 2 & 3 (in.)	JOURNAL-TO-BEARING CLEARANCE 4 (in.)	CAM LOBE HEIGHT (in.)	LIFT (in.)	CRANKSHAFT END PLAY (in.)
1.774-1.789	1.714-1.729	1.608-1.624	.0012-.0043	.0015-.0049	.0059.0079	1.468-1.469	.248	.0032-.0102

VALVE SPECIFICATIONS

FACE ANGLE (deg.)	SEAT ANGLE (deg.)	STEM DIAMETER (in.)	STEM-TO-GUIDE CLEARANCE INTAKE	STEM-TO-GUIDE CLEARANCE EXHAUST	SPRING TENSION LB. @ INCHES	SPRING FREE LENGTH (in.)
45	45	.3137-.3143	.0006-.0018	.0016-.0028	33 @ 1.634	1.929

PISTON, RING, AND PIN SPECIFICATIONS

PISTON DIAMETER (in.)	PISTON CLEARANCE (in.)	RING-TO-GROOVE CLEARANCE 1 (in.)	RING-TO-GROOVE CLEARANCE 2 & 3 (in.)	RING-TO-GROOVE CLEARANCE 4 (in.)	RING END GAP (in.)	PISTON PIN CLEARANCE (in.)
3.2617-3.2723	.0047-.0067	.0024-.0039	.0016-.0032	.0008-.0024	.0118-.0197	.00012

TORQUE SPECIFICATIONS

CYLINDER HEAD	MAIN BEARINGS	CONNECTING ROD CAPS	INT. & EXH. MANIFOLDS	CRANKSHAFT DAMPER	FLYWHEEL	INJECTOR NOZZLE HOLDER	INJECTION PUMP
94 large 36 small	109-116	36-40	11-13	217-239	33-36	50-65	15-18

Fuel System
Injection Pump Removal and Installation

NOTE: In some applications, this procedure is best done with the engine removed from the vehicle.

1. Remove the inlet and outlet lines from the oil cooler.
2. Remove the bolts (4) and separate the oil filter and lines from the cooler.
3. Remove the coolant hose between the oil cooler and the head.
4. Remove the bolts (10) and separate the cooler from the block.
5. Disconnect the fuel lines and remove the fuel filter from the bracket.
6. Remove the injection lines from the nozzles and pump. Cover all openings immediately.
7. Remove the fan, spacer and pulley from the water pump.
8. Remove the bypass hose from the pump and thermostat housing.

Oil filter mounting

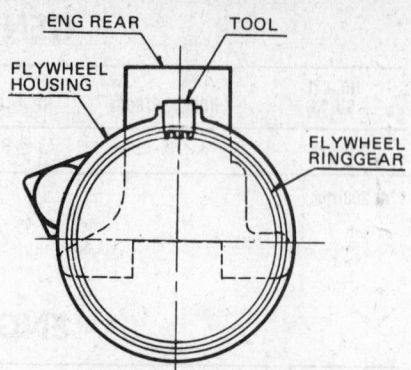

Flywheel locking tool installation

9. Remove the three bolts and lift off the water pump and gasket.
10. Remove the inspection cover and pointer from the flywheel housing and lock the flywheel in place with a locking tool.
11. Flatten the lockwasher and remove the crankshaft pulley nut.
12. Tap evenly around the edge of the pulley using a brass drift, until the cone protrudes from the pulley. Remove the cone.
13. Drive the pulley and damper from the crankshaft with a soft mallet.

14. Remove the inner cover from the timing gear case.
15. Pry out the oil seal.
16. Remove the mounting bolts and tap the case loose with a soft mallet.
17. Remove the tachometer drive support nuts.
18. Remove the timer round nut.
19. Thread the timer extractor, special tool # 57926-581 into the timer weight holder. Remove the timer assembly by tightening the extractor bolt.

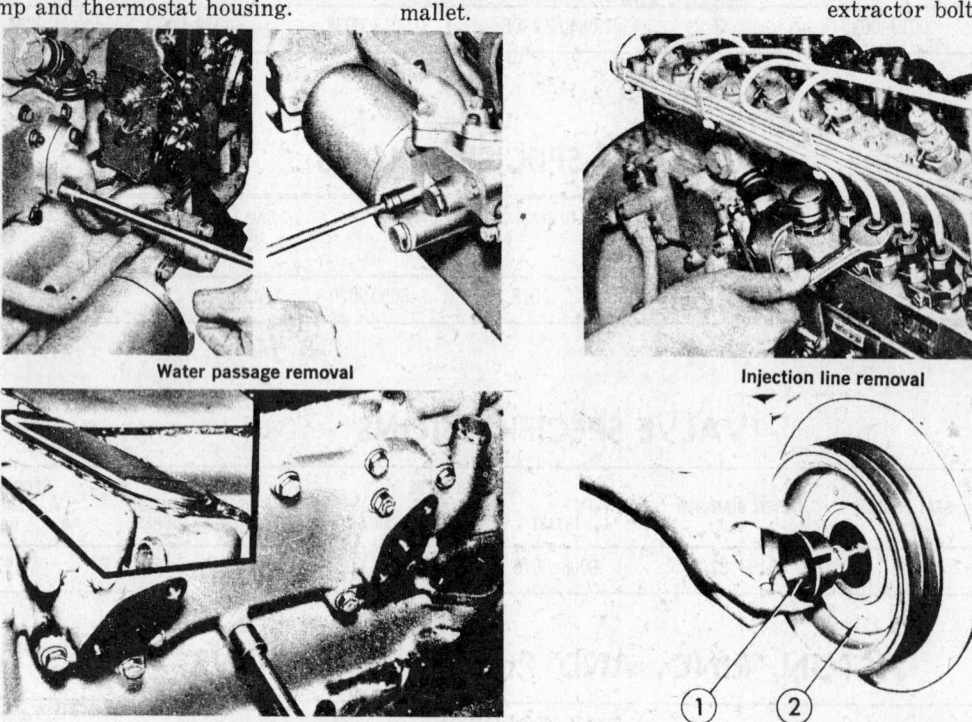

Water passage removal

Injection line removal

Oil cooler removal

Crankshaft cone removal

Timing cover removal

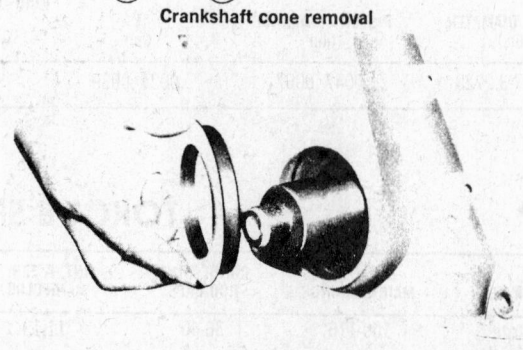

Oil seal removal

Backlash measurement

Timer removal

Tachometer drive removal

1 Injection pump
2 End plate

Timing extractor installation

Injection pump positioning

Tachometer drive coupling installation

Timing mark alignment

20. Unbolt and separate the injection pump from the front end plate.
21. Temporarily install the injection pump and gasket on the front plate.
22. Check the timing marks and bring the #1 piston to TDC.
23. Mesh the injection pump drive gear and idler gear at the timing marks.
24. After aligning the injection pump keyway, install the lockwasher and round nut and torque to 50-58 ftlb.
25. Install the tachometer drive coupling.
26. Check the backlash between the pump drive gear and the idler gear. Backlash should be .0028-.0079″. Adjust if necessary.

Delivery spring removal

27. Remove the #1 cylinder holder clamp, loosen the delivery valve and pull out the delivery spring. Tighten the valve holder to 22-25 ftlb.
28. Connect the fuel supply lines.
29. Bring the #1 piston to 20° BTDC. This can be done by aligning the first mark, in normal rotation, on the crankshaft pulley with the raised line on the gear case.
30. Hand prime the pump. Push the pump in all the way toward the block. Move the pump slowly away from the block until the fuel just stops flowing from the valve holder. Lock the pump in place.
31. Remove the delivery holder and assemble the spring. Torque the holder to 22-25 ftlb.
32. Install remaining parts in reverse order of removal. *NOTE: Oil filter bolt torque is 15-18 ftlb.*

Injection Nozzle Removal, Overhaul and Installation

1. Loosen the injection lines at the pump and nozzles and remove the lines. Cap the openings immediately.
2. Unscrew the injector and holder from the head.
3. Secure the nozzle holder in a vise and remove the lock nut.
4. Remove the nipple.
5. Remove the nozzle holder body from the nozzle nut.
6. Remove the spacer collar and pushrod.
7. Remove the nozzle holder body from the vise and remove the nozzle spring and adjusting shims.
   NOTE: The adjusting shims may be removed with a piece of wire, but great care must be taken to avoid damage to the nozzle tip.
8. Clean fuel oil may be used to clean all parts. Inspect all parts for damage and good fit.
9. Assemble the nozzle in reverse order of disassembly.
10. Install the nozzle in a tester.
11. Operate the tester lever at 1 stroke per second and read the pressure at injection. The pointer will oscillate slightly during injection.
12. Increase or decrease the thickness of the nozzle spring adjusting shims until opening pressure is 1,422.3 psi. A total of 31 different shims are available. A shim thickness of .05mm equals a difference of 85.338 psi.

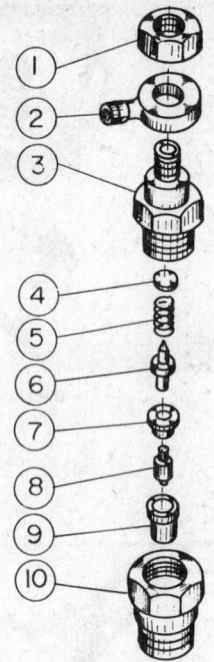

Nozzle and holder
1 Nut   6 Pushrod
2 Nipple   7 Distance piece
3 Holder   8 & 9 Nozzle assembly
4 Shim   10 Nozzle nut
5 Return spring

13. Install the nozzles and lines. Torque the nozzles to 50-65 ftlb.

Governor

RSV Mechanical Type Removal

1. Remove the injection pump and place it in a holding fixture.
2. Install the timer.
3. Drain the cam and governor chambers.
4. Remove the supply pump.

Cover set screw

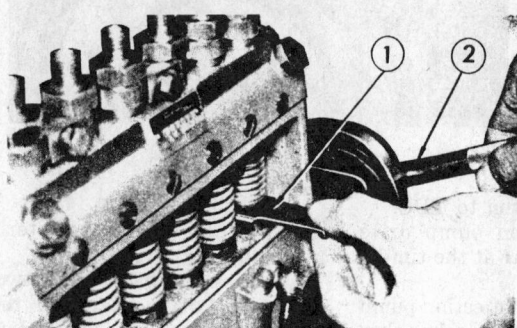

Tappet holder installation

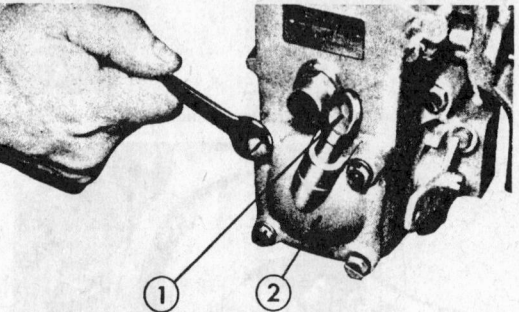

1 Dipstick   2 Rear cover

1 Counterweight   2 57915-010   3 57916-432   4 Round nut

Diaphragm housing

1 Counterweight 2 Special wrench 3 Timer 4 Special wrench

Governor housing removal

1 Delivery valve holder 2 Special wrench

5. Remove the cam cover.
6. Using a special wrench, ST-57916-432, on the timer, turn the camshaft until all the tappets are raised to TDC. Place a tappet holder, 57931-210, between the tappet adjusting bolt and nut for each cylinder.
7. Remove the rear cover and dipstick.
8. Loosen the balance idler spring and the auxiliary idler spring lock nut.
9. Loosen the governor cover lock screw.
10. Unbolt and remove the governor cover from the governor body. Remove the link from the control rack.
11. Remove the start spring from the spring eye.
12. Remove the counterweights from the camshaft.
13. Hold the timer and remove the slotted nuts and lockwashers.
14. Using a puller, ST57926-512, remove the flyweight assembly.
15. Remove the timer.
16. Unbolt and remove the governor body.

MZ Pneumatic Type Removal
1. Follow steps 1 through 6 of RSV Type Removal.
2. Unbolt and remove the diaphragm housing and main spring.
3. Remove the diaphragm ring with a screwdriver.
4. Remove the cotter pin from the connecting rod bolt with a needle-nosed pliers.
5. Remove the diaphragm assembly from the control rack.
6. Remove the five set screws and

remove the governor body by applying force with a screwdriver blade in the slit between the governor and pump housings.
7. Remove the timer.

RSV Type Installation
1. Apply RTV silicone gasket material to the governor body and install the governor on the injection pump.
2. Tighten the upper spring eye screw holding the starting spring.
3. Install the timer.
4. Install the flyweight assembly.
5. Apply RTV silicone gasket material to the governor cover and install the starting spring on the housing side of the spring eye.
6. Install the link leaf spring in the hole in the end of the control rack.
7. Install the cover and set screws.

1 Governor housing 2 Pump housing

MZ Type Installation
1. Apply RTV silicone gasket material to the governor body, position it on the pump body and tap it into position with a plastic mallet. Install the set screws.
2. Install the diaphragm and balance spring on the control rack connecting bolt and lock with a new cotter pin. Apply chassis lube to the diaphragm ring.
3. Insert the main spring and install diaphragm housing with the four bolts.

1 Diaphragm 2 Connecting bolt

Injection Pump Service

Replacing the Delivery Valve
1. Thoroughly clean the area around the nozzle tube and delivery valve.
2. Remove the nozzle tube.
3. Remove the delivery valve holder lock plate.

1 Delivery valve extractor

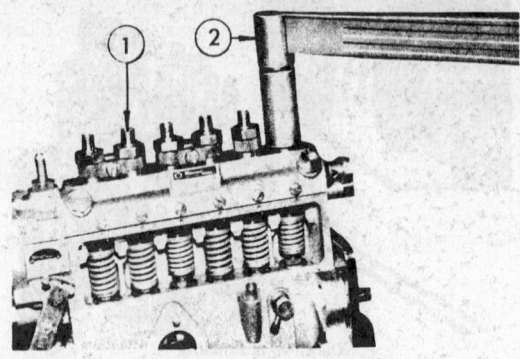

1 Plunger barrel 2 Wire

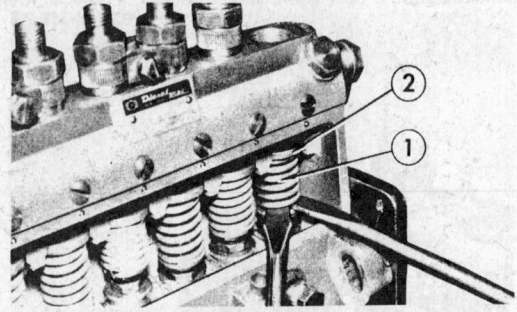

1 Plunger spring 2 Upper spring seat

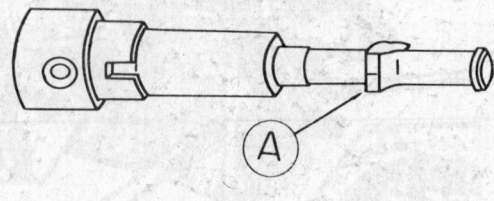

A Match mark

A	Match mark

4. Remove the delivery valve holder and spring.
5. Using ST57930-032, remove the delivery valve.
6. Position the delivery valve in the pump housing making sure no dirt gets between the top of the plunger barrel and the delivery valve.
7. Install a new delivery valve gasket. The gasket is installed with the larger face downward and may be tapped into place through the extractor.
8. Install the delivery valve spring.
9. Install the delivery valve holder and torque it to 22-25 ftlb.
10. Loosen the holder and retorque it.
11. Install the lock plate, nozzle tube and nozzle clamp.

Replacing the Plunger

1. Remove the delivery valve.
2. Push the plunger spring up with two screwdrivers and remove the lower spring seat from the plunger.
3. Insert a hooked wire through the top of the pump housing, down through the plunger opening and hook it on the lead unit of the plunger. Pull up to remove the plunger and barrel.
4. Immerse a new plunger in clean fuel oil and thoroughly wash off the rust preventive.
   NOTE: The plunger was lapped at the factory. Do not hold it by the lapped section.
5. Operate the plunger in clean fuel oil to check its operation.
6. Slowly insert the plunger and barrel into the pump with the barrel groove and plunger notch

facing forward. Make certain the plunger piston pin is properly seated in the groove in the control sleeve.
7. Push the plunger spring up with two screwdrivers and insert the lower spring seat.
8. Install the remaining parts in reverse of disassembly.

Replacing the Tappets

1. Remove the delivery valve, plunger and plunger barrel.
2. Remove the plunger spring.
3. Remove the tappet.
4. Installation is the reverse of removal.

Testing and Adjusting the Fuel Injection Pump

NOTE: It is necessary to inspect and adjust the pump, using a pump tester, whenever it has been disas-

sembled and assembled, when the plunger or plunger barrel have been replaced or when any of the component parts have been replaced. Use nozzle tube 57805-002 and test nozzle 5000-101, starting pressure 1422.3 psi. Clean No. 2 fuel should be used. Rotating direction, from drive side, is clockwise. Sequence is 1-4-2-6-3-5.

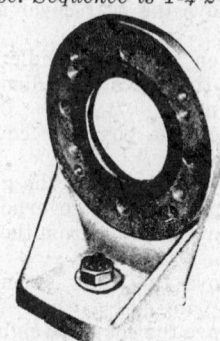

Holding fixture

Installing the injection pump on the tester

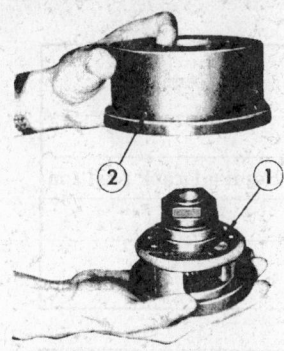

1 Timer 2 Coupling

1. Remove the fuel feed pump and cover plate from the injection pump.
2. Install the injection pump on the tester and holding fixture.
3. Connect the test coupling to the tester drive shaft with the coupling disc.
4. Remove the cap and position the tester dial to measure the camshaft rotating angle.
5. Install a tappet lift gauge on the #1 tappet.
6. Bottom the tappet and set the dial gauge to 0.
7. Bleed the pump at the bleeder screw.
8. Loosen the nozzle holder ball valve.
9. Feed fuel to the pump inlet while slowly turning the pump tester by hand in the normal engine rotation direction. Fuel will flow from the test nozzle. When the fuel stops flowing the injection point has been reached. The tappet, at this precise point, must be .08858-.09251″ above BDC.

10. If the fuel does not stop flowing after .0926″, turn the adjusting bolt counterclockwise to raise the position of the plunger.
11. If the fuel stops flowing before .0886″, turn the adjusting bolt clockwise to lower the plunger position.
12. When adjustment is made, torque the locknut to 43-50 ftlb.
13. With the pump set at initial injection, set the angle scale mark on the tester flywheel at 0 or 180°.
14. If adjustment is correct, fuel should stop flowing at #4 cylinder when the tester has been turned 60° in normal rotation. If fuel does not stop flowing at the correct time, adjust as above.
15. Check and adjust each cylinder in turn.
16. When timing for each cylinder is correct, position the cam at TDC, check the plunger piston pin-to-plunger barrel clearance and make sure that the tappet vertical clearance is at least .0118″ for each tappet.

Standard Fuel Injection Volume Adjustment

1. Determine the zero position of the control by attaching the measuring device to the pump and pushing the index all the way to governor side. Match the scale on the left end of the index and set the 0 (zero) position of the scale at the position at which the measuring device index stops. On RSV mechanical governors,

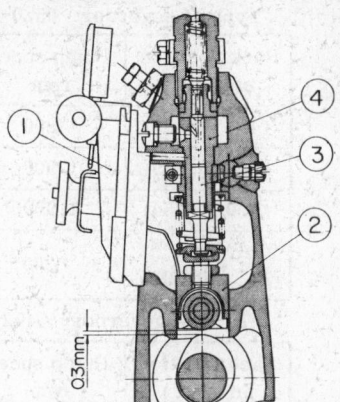

1 Tappet lift gauge 2 Tappet
3 Plunger 4 Plunger barrel

loosen the stop bolt to align this index.
2. Remove the rack guide screw from the rear of the pump housing and apply the lock screw attached to the tester. Secure the control rack in the standard position for adjustment.
NOTE: The lock screw should be tightened by hand; overtightening will bend the rack.
3. Start the tester and run the pump at rated speed.
4. Set the pump feed pressure at 21.3-22.75 psi and measure the injection volume at the rated stroke of the female cylinder.
5. In the same manner, measure the injection volume at rated speed and standard rack position. Compute the rate of unevenness of the injection.
6. If the results show that the mean

Turning the adjusting bolt

Installing the tester lock screw

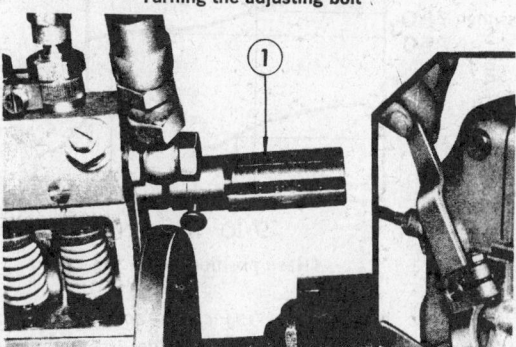

1 Measuring device

Loosening the pinion set screw

Type of governor: 5520-659, 5520-664

Rack position mm (in.)	Pump speed rpm	Mean injection volume cc/1,000 strokes	Allowable deviation (%)	Remarks
14.8 (0.583)	800	41 ± 1.0	± 4.0	
14.2 (0.559)	1,900	43 ± 1.0	± 2.5	Nominal rack position
9.0 (0.354)	1,900	12.8 ± 1.0	± 7.5	
Approx. 10 (0.393)	300	7.5 ± 1.1	± 15.0	

Type of governor: 5412-072

Rack position mm (in.)	Pump speed rpm	Mean injection volume cc/1,000 strokes	Allowable deviation (%)	Remarks
14.5 (0.571)	600	36.4 ± 1	± 2.5	Nominal rack position
13.8 (0.543)	1,500	36.7 ± 1.5	± 4.0	
Approx. 10 (0.393)	300	7.5 ± 1.1	± 15.0	

With RSV (mechanical) type governor

Type of governor: 5410-087, 5410-091

Rack position mm (in.)	Pump speed rpm	Mean injection volume cc/1,000 strokes	Allowable deviation (%)	Remarks
14.0 (0.551)	750	34.4 ± 1	± 2.5	Nominal rack position
12.0 (0.472)	750	21.1 ± 0.9	± 4.0	
Approx. 10 (0.393)	300	7.5 ± 1.1	± 15.0	

RSV (Mechanical) type governor

Type of governor: 5410-087

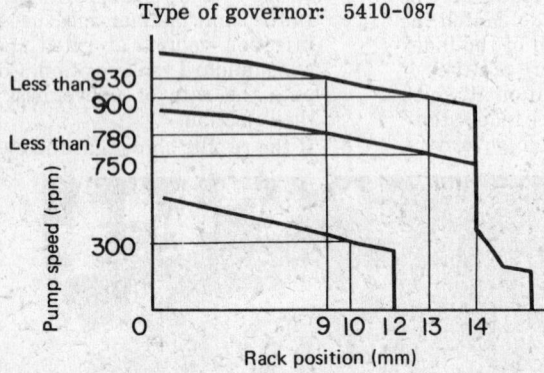

Type of governor: 5412-072

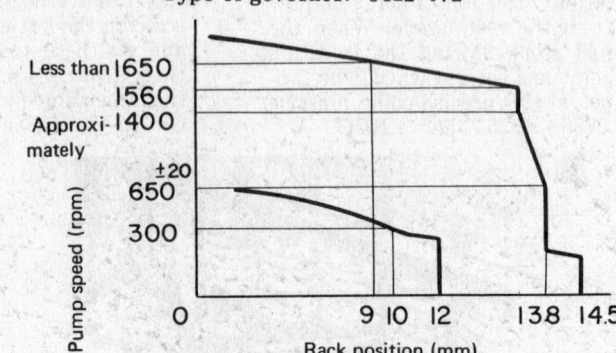

Type of governor: 5410-091

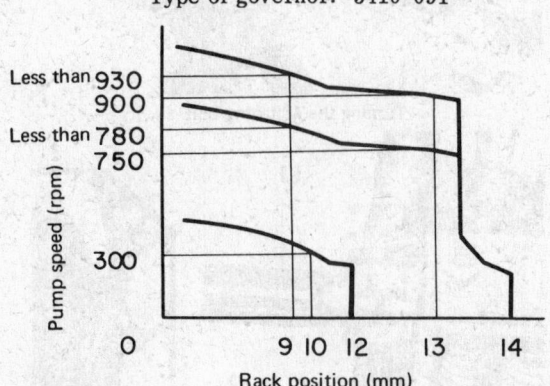

injection volume and rate of unevenness are not within the limits, adjust by changing the relative position of the control pinion and control sleeve. This may be done by:

a. loosen the pinion set screw.
b. place a pin in the hole in the control sleeve and adjust by moving the control sleeve along the control rack a little at a time.
c. when adjustment is completed, secure the pinion set screw.
d. remove the lock screw from the control rack and reinstall the guide screw.

Maximum Fuel Injection Adjustment—MZ Governor

NOTE: See the standard adjustment tables above. The stroke set screw on the bottom of the governor housing is used to adjust the maximum injection volume. To increase the volume, turn the stroke set screw to the left; right to decrease volume.

Testing and Adjusting the Governor

RSV Mechanical Governor

1. Match the adjusting device index to the zero point on the scale and set the control rack to the zero position.
2. Operate the control lever and make certain the full stroke of the rack is .827".
3. Make certain that the rack moves freely in the direction for maximum fuel injection by the spring force of the starting spring.
4. Set the stop bolt to remove any significant load on the governor linkage.
5. Set the stop bolt to give a control rack setting of .0197-.03937".

High Speed Adjustment

6. Remove the governor rear cover.
7. Loosen the full load stop lock nut and adjust the full load stop so that its setting corresponds to an A rack position between rotation and BC. See the accompanying chart.

$$\text{Unevenness} = \frac{\frac{\text{Max. or min. injection volume for each plunger}}{} - \frac{\text{Main injection volume}}{}}{\text{Mean injection volume}} \times 100$$

8. To increase the volume, turn the stop to the right; left to decrease.

Maximum Speed Adjustment

9. Operate the pump at speed G and adjust the maximum speed stop so that the control rack position is G mm.

Speed Fluctuation Adjustment

10. Speed fluctuation can be controlled by varying the spring rate on RSV governors.

Balance Spring Adjustment

11. Set the control lever to the point where it contacts the maximum speed stop and operate the pump at a rate of F, between A and B.
12. Install the balance spring assembly under the tension lever.
13. Tighten the balance spring assembly with tool 57916-212 until the control rack position G is H mm. and secure with the lock nut.
14. Gradually accelerate the engine from speed D and make certain that the control rack is at G mm when the action of the spring ends at speed E.

Idle Speed Adjustment

15. Set the control lever at the stop position so that the control rack is at point B.
16. Set the speed to D and tighten the idle auxiliary spring so that the position of the rack is C mm. Secure with the lock nut.

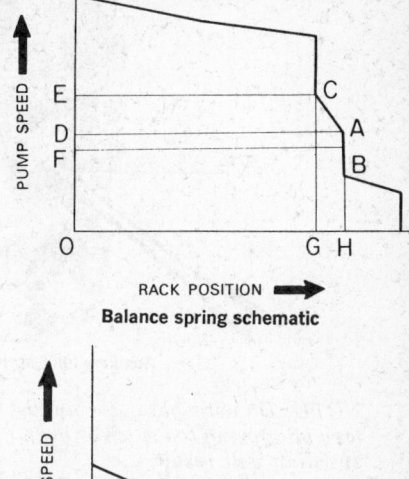
Balance spring schematic

Idle schematic

Timer advance angle performance curve

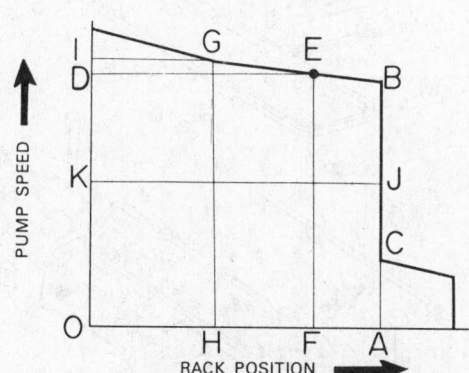
Schematic diagram of the governor

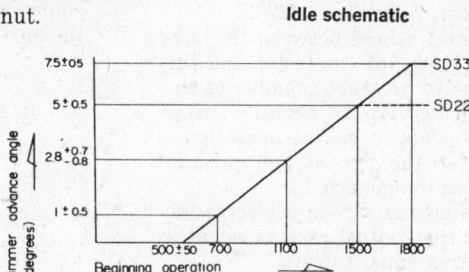
1 Maximum stop bolt 2 Nut

1 Full load stop 2 Nut

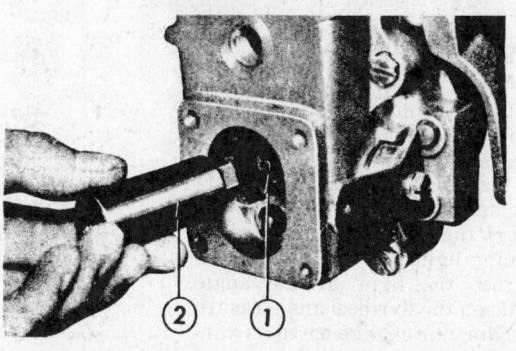
1 Control lever 2 Lever set tool

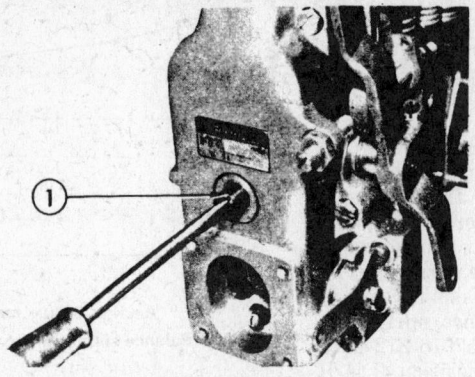

Auxiliary idler spring

1 Synchronizer attachment   2 Coupling   3 Flywheel   4 Strobe light

NOTE: Do not tighten the auxiliary idle spring too much or overspeeding will result.

MZ Pneumatic Governor

1. Set the measuring device index at the zero point on the scale and set the control rack at the zero position.
2. Operate the control lever and make certain the full stroke of the control rack is .827".
3. Connect a hose between the vacuum pump of the tester and the negative pressure chamber of the governor. Apply negative pressure equal to engine operation.
4. Operate the pump at 500 rpm and adjust the governor.
5. Tighten the stroke set screw so that the control rack is set at a position equal to zero.
6. Adjust the stroke of the balance spring to .236".
7. If movement of the control rack deviates considerably from the performance curve, adjust by increasing or decreasing the thickness of the adjusting shims.
8. Operate the tester vacuum pump and check the control rack position and movement in relation to the pressure shown on the gauge. If the movement of the control rack for pressure variations does not conform to the performance curve, the idle speed is either too high or too low. Turn the auxiliary idle spring right to increase and left to decrease speed.

Testing and Adjusting the Timer

1. Install a timing light, such as tool 5783-001 on the tester using the cover plate attaching bolts, so that the synchronizer lever attachment is applied to the tappet.
2. Start the pump and turn on the timing light.
3. Direct the light at the angle scale on the flywheel and measure the angular change based on variations in pump speed.

4. If the tester does not have an angle scale:
   a. Attach an angle scale to the timer coupling and mount a pointer on the tester drive shaft.
   b. Operate the pump and direct the light on the scale.
5. If the angular change is not within limits, disassemble the timer and adjust the spring force by increasing or decreasing the shims, or if necessary, replace the spring.

Engine Repair

NOTE: Disassembly of major components is best done with the engine out of the vehicle and mounted on a rotary work stand.

Cylinder Head Removal, Overhaul and Installation

1. Remove the air cleaner.

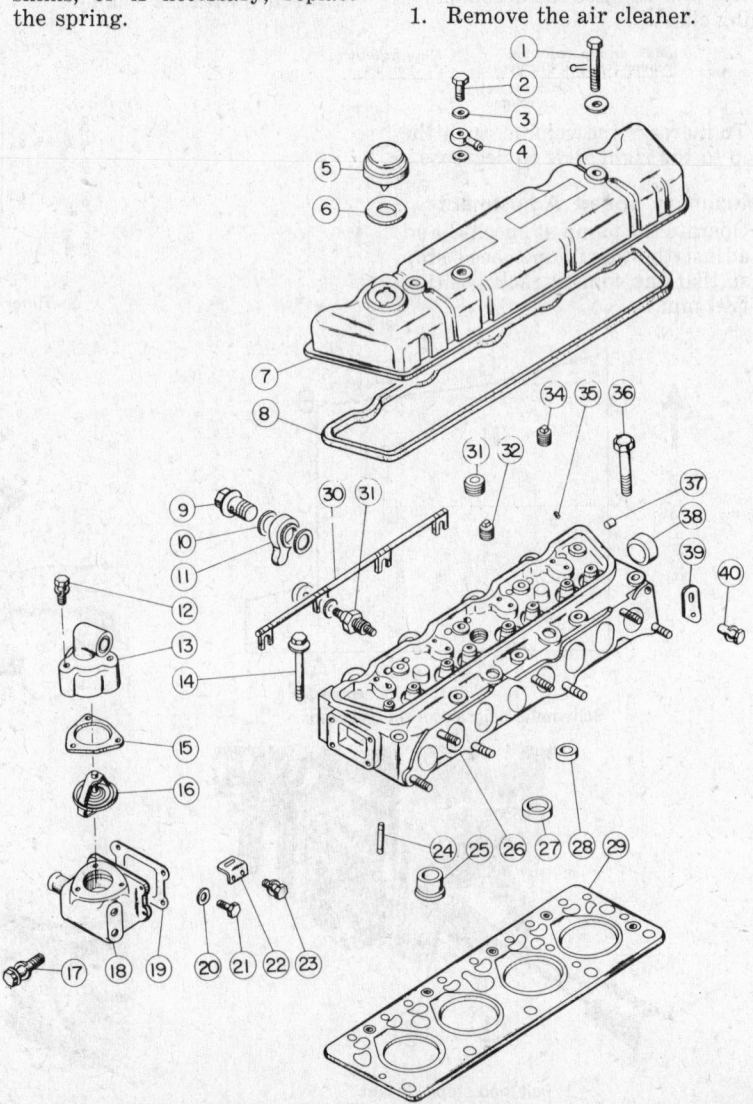

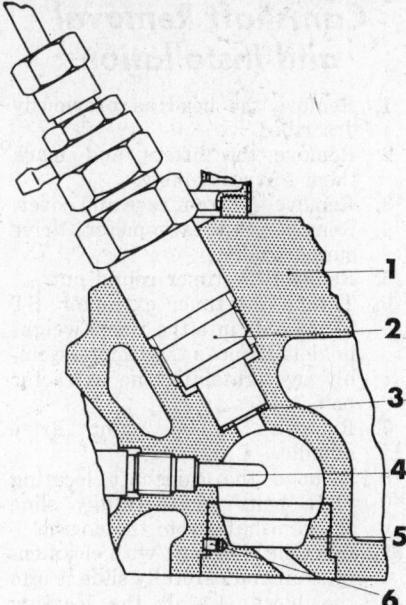

1 Cylinder head   2 Nozzle assembly
3 Gasket   4 Glow plug
5 Precombustion chamber   6 Knock pin

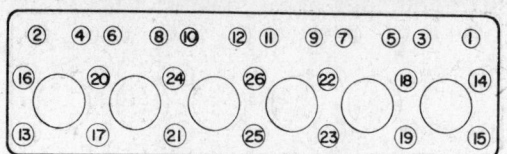

Head bolt removal sequence

2. Remove the crankcase vent hose and remove the intake and exhaust manifolds. These are bolted together.
3. Remove the alternator, bracket and belts.
4. Disconnect the coolant hose between the head and the oil cooler.
5. Remove the fuel filter assembly.
6. Disconnect the injection lines from the pump and the injectors. Cap all openings at once.
7. Remove the bypass hoses between the coolant pump and the thermostat housing.
8. Remove the fan.
9. Remove the rocker arm cover.
10. Remove the rocker arm shaft assembly.
11. Remove the pushrods and keep them in order.

12. Remove the fuel return lines.
13. Remove the nozzles from the head.
14. Remove the cylinder head bolts in the sequence shown.
15. Attach a hoist to the head and lift it clear of the block. On occasion, the precombustion chambers may fall out, especially if the head is bumped or handled roughly. Take care that they are returned to their original positions if this occurs.
16. Remove the head gasket and O-rings.
   NOTE: Before disassembling the head, make all necessary valve train measurements.
17. Place the head in a holding fixture.
18. Disassemble the valves. Mark all parts for assembly in their original positions.
19. Remove the retaining wire and lift off the valve stem seals.
20. Unscrew the glow plugs.
21. Disassemble the rocker arm shaft by removing the cotter pins at either end. Keep all parts in order. If rocker arm brackets prove to be difficult to remove, immerse the assembly in water heated to about 160°F. Immersion for a few minutes will loosen the parts.
22. Clean and inspect all parts.
23. Check the head with a straightedge. Maximum warpage is .0079″. Do not remove more than .011″ from the head.

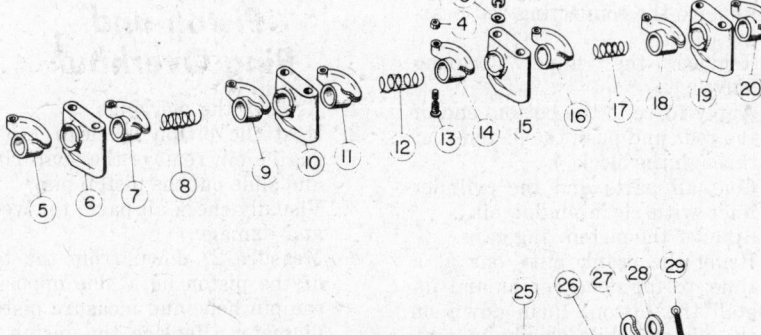

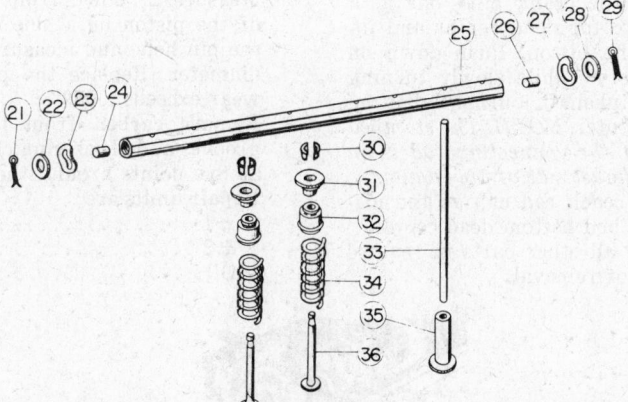

Valve train

1 Bolt	13 Adjusting screw	25 Rocker shaft
2 Lock washer	14 Valve rocker A	26 Plug
3 Flat washer	15 Rocker shaft bracket	27 Outside spring
4 Lock nut	16 Valve rocker B	28 Washer
5 Valve rocker A	17 Inside spring	29 Cotter pin
6 Rocker shaft bracket	18 Valve rocker C	30 Split collar
7 Valve rocker B	19 Rocker shaft bracket	31 Spring seat
8 Inside spring	20 Valve rocker D	32 Valve stem seal
9 Valve rocker C	21 Cotter pin	33 Push rod
10 Rocker shaft bracket	22 Washer	34 Valve spring
11 Valve rocker D	23 Outside spring	35 Valve lifter
12 Inside spring	24 Plug	36 Valve

**39
(1.535)**

Valve installed height-inches in parantheses

24. Check the valve springs for free length and tilt. Free length must not be less than 1.850″ and tilt must not exceed .03937″ (1mm).
25. Valve seats may be removed by cracking with a cold chisel or with a valve seat remover. New valve seats should be cooled in

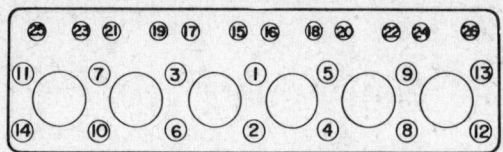

Head bolt tightening sequence

dry ice for five minutes prior to installation, at the same time the head should be immersed in 175°F water.

26. Assemble the head in reverse order of disassembly.
27. Place a new cylinder head gasket on the block with the stainless steel inset side facing up.
28. Install the O-rings around the water and oil passages.
29. Position the head on the block.
30. Coat the head bolts with clean engine oil and torque them in sequence, in stages as follows:
    Large: 43, 94
    Small: 21, 36
31. Install the pushrods, pressing down and turning them to be sure of proper seating.
32. Install the rocker arm shaft assembly, torquing the bolts to 18 ftlb. in sequence from the center to each end.
33. Install the injection nozzles.
34. Install all other parts in reverse order of removal.

Crankshaft Removal and Installation

1. Remove the timing gear case and cover.
2. Remove the flywheel.
3. Remove the flywheel housing.
4. Invert the engine and remove the oil pan.
5. Remove the oil pump.
6. Pull the crankshaft gear from the front end of the shaft.
7. Remove the connecting rod bearing caps.
8. Remove the main bearing caps. *NOTE: Loosen the cap bolts alternately and evenly. A puller, tool #ST16660000 may be necessary to remove the caps.*

.9. Remove the cranksaft with a hoist. *NOTE: The #3 bearing contains a thrust washer. Avoid mix-ups.*
10. Lift out the upper bearing halves. Mark them for reassembly.
11. Coat all parts with clean engine oil. Assembly is the reverse of disassembly. Oil groove on the thrust bearing faces away from the journal. Bearing caps have an F which faces front. Guide tool ST16490000 is available for installation of bearing caps. Rear main seal is replaced at this time.

Piston Removal and Installation

1. Remove the head as described previously.
2. Remove the oil pan.
3. Remove the ridge at the top of each cylinder.
4. Remove the connecting rod bearing caps.
5. Remove the upper bearing halves.
6. Apply force to the bottom end of the rods and push the pistons out through the block.
7. Coat all parts and the cylinder liner with clean engine oil.
8. Stagger the piston ring gaps.
9. Bring the crank pins, one at a time, to top dead center and install the piston. Push down on the piston while slowly turning the crankshaft journal to bottom dead center. *NOTE: The stamped side of the connecting rod is on the exhaust side of the engine.*
10. Install each rod cap as the pisto reached bottom dead center.
11. Install all other parts in reverse order of removal.

Camshaft Removal and Installation

1. Remove the head as previously described.
2. Remove the lifters and mark them for reassembly.
3. Remove the front case and cover.
4. Remove the tachometer drive support nuts.
5. Remove the timer round nut.
6. Thread the timer extractor, ST 57926-581, into the timer weight holder. Remove the timer assembly by tightening the extractor bolt.
7. Remove the oil pump drive spindle.
8. Remove the camshaft locating plate bolts and carefully slide the camshaft from the engine.
9. Coat the camshaft with clean engine oil and carefully slide it into the block. Install the locating plate.
10. Install the oil pump drive spindle by aligning the oil pump drive shaft groove and the camshaft oil pump drive gear with the spindle.
11. Install all other parts in reverse order, following previously given instructions.

Piston and Ring Overhaul

1. Remove the old rings.
2. Heat the piston in clean 140°F engine oil, remove the snap ring and slide out the piston pin.
3. Visually check all parts for wear and damage.
4. Measure 2″ down from the top of the piston on a side opposite the pin hole, and measure piston diameter. Replace the piston if wear exceeds .0059″.
5. Remove carbon from the ring grooves and check ring clearance at five points around the piston. Repair limits are:

#1	.0197″
#2	.0118″
Oil	.0059″

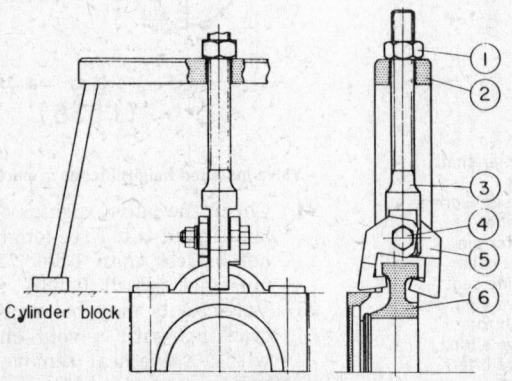

Cylinder block

ST16660000

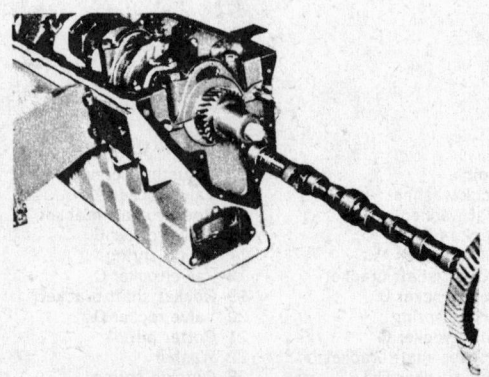

Camshaft removal

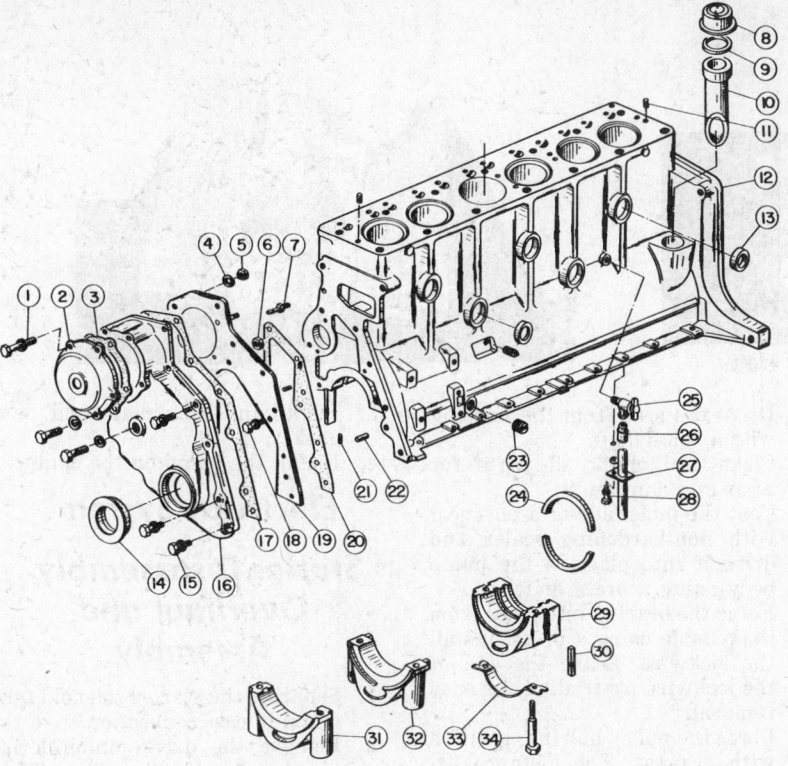

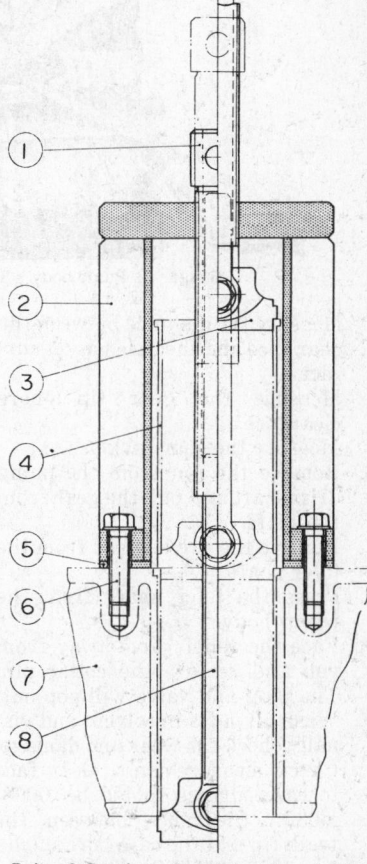

1 Bolt	2 Bracket	3 Patch plate	4 Liner
5 Packing	6 Bolt	7 Block	8 Bolt

1 Bolt	13 Plug
2 Timing gear cover	14 Oil seal
3 Gasket	15 Bolt
4 Lock washer	16 Timing gear case
5 Nut	17 Gasket
6 Nut	18 Bolt
7 Oil jet	19 Front end plate
8 Breather cap	20 Gasket
9 Gasket	21 Plug
10 Breather tube	22 Dowel pin
11 Dowel pin	23 Plug
12 Cylinder block	24 Oil seal

25 Water drain cock
26 Vinyl hose
27 Hose clamp
28 Bolt
29 Bearing cap No. 4
30 Oil seal
31 Bearing cap No. 1
32 Bearing cap No. 3
33 Lock washer
34 Bolt

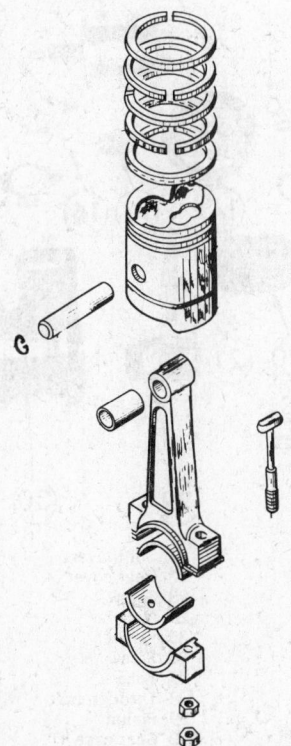

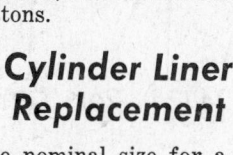

6. Place a ring in the cylinder and square it by pushing it in with a piston. Measure the end gap. Replace the rings if end gap exceeds .0059″.

6. Check piston pin-to-piston clearance. Clearance should not exceed .0039″.

7. Measure bearing-to-crankpin clearance. Clearance should not exceed .0059″.

8. Check connecting rod straightness. Deviation should not exceed .0012″ per 100mm.

9. When all specifications are acceptable, assemble and install the pistons.

Cylinder Liner Replacement

1. The nominal size for a liner is 3.268″. If the liner wear exceeds .0079″, replace the liner.

2. Install a puller such as ST-1030000 over the liner and tighten the extractor bolt, pulling the liner from the block.

3. Carefully clean the new cylinder liner and block surfaces with clean engine oil and install the liner using ST1030000. Make sure the liner is fully seated.

4. Check the following liner dimensions:
   Height above block:
   .00078-.00360
   Block-to-liner: .0004-.0012″
   Nominal bore: 3.268″

Camshaft Bushing Replacement

1. Using a suitable puller, remove the bushings one-by-one starting from the front. The rear bushing, however, must be pulled from the rear.

2. The bushings are driven in, in order, from the rear. Make sure that the beveled side faces the front in all cases. Make certain that oil passages are aligned and the bushings are flush-fit.

Lubrication System

Oil Pump Overhaul

1. Remove the oil pump.
2. Remove the pump cover plate.

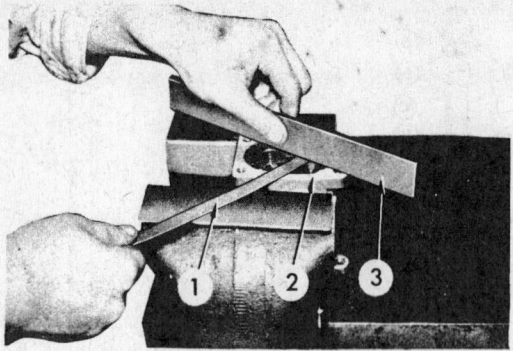
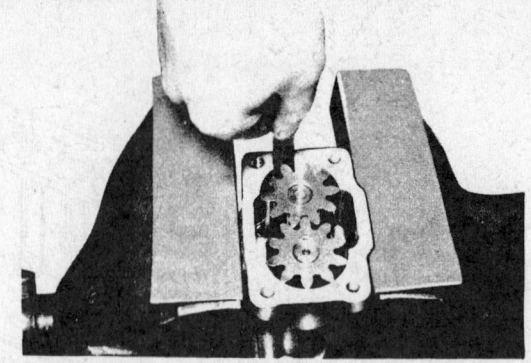

1 Gauge 2 Pump body 3 Straight edge

3. Measure the distance between the gear face and the case gasket surface.
4. Measure the gear tip-to-bore clearance.
5. Measure the gear backlash.
6. Remove the pin from the pump driveshaft and pull the gear from the shaft.
7. Remove the driven gear from the idler shaft.
8. Press the idler shaft from the pump body.
9. Face the relief valve away from you and remove the cotter pin. The plug and valve will pop out.
10. Wash all parts in solvent and visually check for wear and damage.
11. Check measurements. Gear face to case surface should be .0008-.0032". Clearance between the teeth tip and the case bore should be .0029-0059" Backlash should be .0118-.0157". If any measurement exceeds these limits, replace both gears.
12. Measure the shaft diameters. Driveshaft tip end should be .5943-.5950". Driveshaft gear end should be .5148-.5155". Idler shaft should be .5158-.5182". Driveshaft-to-bore clearance should be .00945-.00272".
13. Coat all parts in clean engine oil. Assembly is the reverse of disassembly. *NOTE: Place the beveled side of the gears toward the oil pump body. Check clearances during assembly. Check operation after assembly.*

Cooling System

Water Pump Overhaul

1. Remove the water pump from the engine.
2. Remove the snap ring and, using a suitable puller, remove the hub.
3. Remove the bearing lockwire.
4. Remove the back plate and gasket.
5. Place a tube with a slightly larger diameter than the bearing over the bearing and onto the impeller. Place the pump body, nose down, in a press and press off the impeller and bearing.

6. Drive the seal from the housing with a brass drift.
7. Clean and check all parts for wear and damage.
8. Coat the underside of a new seal with non-hardening sealer and drive it into place in the pump body using a brass drift.
9. Force the bearing into place from the outside using a press. Install the lockwire. Leave the end of the lockwire protruding for easy removal.
10. Force the pulley hub into position with a press. The distance between the front surface of the hub and the back surface of the pump body should be 5.63".
11. Install the snap ring.
12. Place a .028" feeler gauge between the impeller and the pump body and press the impeller into place.

13. Install the back plate and new gasket.
14. Install the pump on the engine.

Electrical System

Starter Disassembly, Overhaul and Assembly

1. Separate the starter solenoid terminal-to-case connection.
2. Remove the drive pinion lever pin.
3. Remove the solenoid.
4. Remove the rear cover.
5. Remove the field coil terminal screws.
6. Using a small piece of bent wire, remove the brushes and springs.
7. Remove the brush holder and thrust washer.

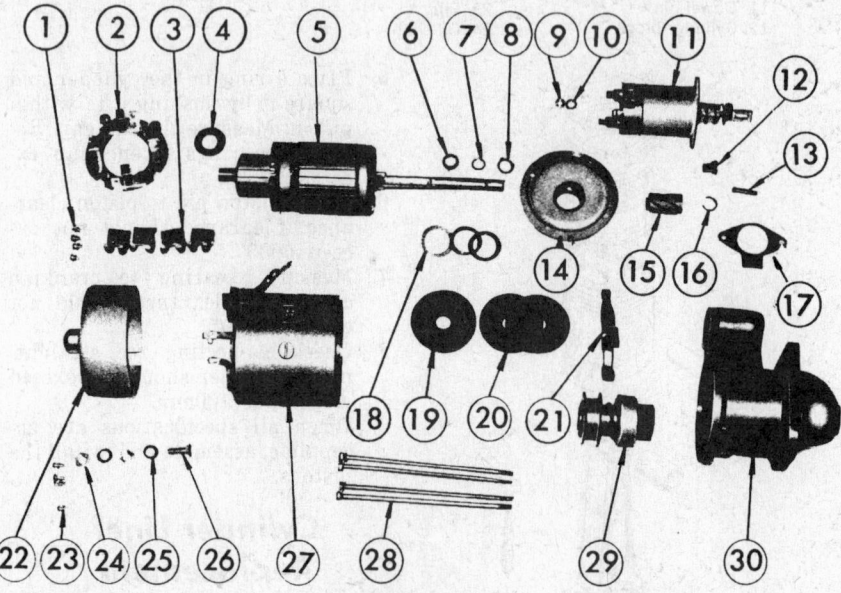

S13-04K Starter

1 Screw	11 Solenoid	21 Shift lever
2 Brush holder	12 Screw	22 Rear cover
3 Brush	13 Shift lever pin	23 Screw
4 Thrust washer	14 Center bearing	24 Nut
5 Armature	15 Advance sleeve	25 Washer
6 Pinion stop	16 Clip	26 Screw
7 Pinion stop clip	17 Dust cover	27 Yoke
8 Pinion stop washer	18 Thrust washer	28 Through bolt
9 Nut	19 Pressure equalizer	29 Pinion
10 Washer	20 Lock washer	30 Gear case

Item	Model		Hitachi LT123-38K Hitachi LT123-64K
Specifications	Polarity		Negative ground
	Method of generation		3-phase, AC, self-rectified type
	Battery voltage	(V)	12
	Nominal output	(W)	300
	Direction of rotation		To the right when viewed from the pulley side
	Diodes	Positive side P-N	S13-00 (Red mark)
		Negative side N-P	S14-00 (Black mark)
Performance	Speed	(rpm)	1,000 - 9,000
	Rated speed	(rpm)	2,500
	No-load voltage	(V/rpm)	13.8 - 14.8 (2,500)
	Output current	(A/V/rpm)	24.5 or more/14/2,500 or less

8. Separate the yoke and gear case.
9. Using a wooden mallet, tap the gear case off the armature.
10. Remove the drive pinion lever.
11. Pry the pinion stop washer off the armature.
12. Remove the pinion stop and pinion assembly.
13. Remove the center bearing from the armature shaft.
14. Remove the thrust washer from the armature shaft.
15. Observe the following specifications:
Commutator undercut: .008"
Brush wear limit: .512"
Brush spring tension limit: 2.07 lb.
16. Assembly is the reverse of disas-

sembly. After assembly, adjust the pinion plunger gap. Gap should be .008-.059" measured between the pinion and pinion stopper. Gap is adjusted by turning the adjusting screw on the end of the plunger.

Alternator Disassembly, Overhaul and Assembly

1. Remove pulley nut, pulley, fan, fan base, key and spacer.
2. Remove brush cover and brushes.
3. Remove the three through-bolts and separate the alternator halves.

4. Remove the bearing retainer screws and front cover.
5. Remove the felt seal.
6. Unsolder the diode connections without allowing the diodes to become excessively hot.
7. Separate the stator and rear cover.
8. Remove the SR holder from the rear cover.
9. Remove the brush holder.
10. Note the followings pecifications:
Rotor coil resistance: 5 ohms
Slip ring deflection limit: .012"
Slip ring O.D.: 1.26"
Brush wear limit: .295"
Brush spring limit: .436 lb.
11. Assembly is the reverse of disassembly.

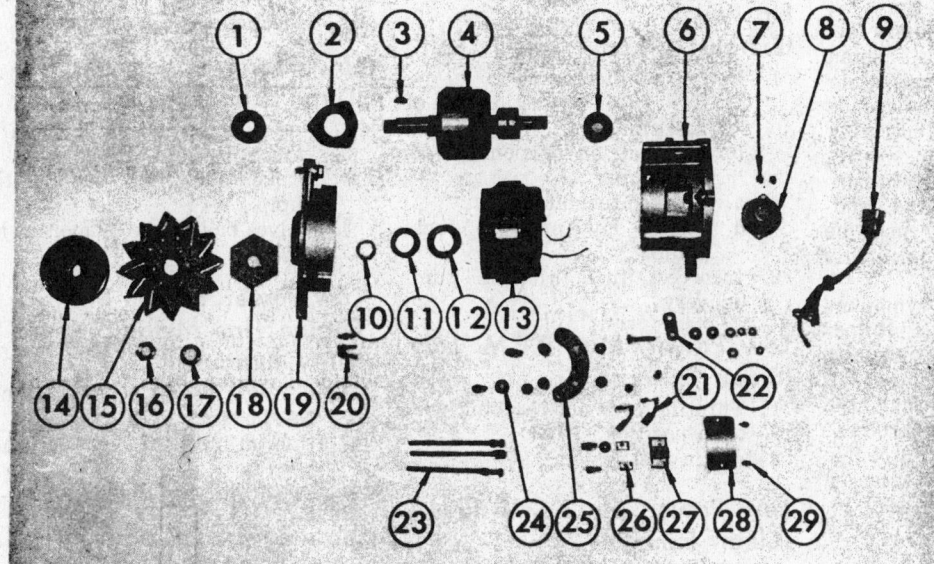

1 Ball bearing
2 Bearing retainer
3 Key
4 Rotor
5 Ball bearing
6 Rear cover
7 Screw
8 Shaft cover
9 Lead
10 Spacer
11 Packing
12 Retainer
13 Stator
14 Pulley
15 Fan
16 Nut
17 Lock washer
18 Fan base
19 Front cover
20 Screw
21 Brush
22 Terminal insulator
23 Through bolt
24 Insulating washer
25 SR holder
26 Terminal plate
27 Brush holder
28 Cover
29 Screw

Alternator

GENERAL SPECIFICATIONS

ENGINE MODEL	BORE AND STROKE	HORSEPOWER @ R.P.M.	TORQUE FT. LBS. @ R.P.M.	COMPRESSION PRESSURE CRANKING SPEED	FIRING ORDER	NO LOAD R.P.M.	GOVERNED SPEED FULL LOAD R.P.M.	IDLE SPEED
354D	3.8785 x 5	120 @ 2800	260 @ 1450	430	153624	3000	2800	525

PISTONS, PINS, RINGS, CRANKSHAFT AND BEARINGS

ENGINE MODEL	PISTON CLEARANCE	PISTON RING END GAP 1st	2nd & 3rd	Oil	PISTON PIN DIAMETER	ROD BEARING SHAFT DIAMETER	BEARING CLEARANCE	MAIN BEARING SHAFT DIAMETER	BEARING CLEARANCE	SHAFT END PLAY	THRUST NO. BEARING ON
354D	.0059	.015-.019	.011-.016	.011-.016	1.375	2.499-2.4995	.0015-.003	2.9985-2.999	.0025-.0045	.002-.014	4

VALVE SPECIFICATIONS

ENGINE MODEL	VALVE LASH (HOT)	VALVE FACE ANGLE	VALVE STEM INTAKE	CLEARANCE EXHAUST	VALVE STEM DIAMETER INTAKE	EXHAUST	FREE LENGTH INNER	OUTER	INSTALLED LENGTH INNER	OUTER	PRESSURE INNER	OUTER
354D	.010	45°	.0015-.0035	.002-.004	.3725-.3735	.3720-.3730	2.0	2.5	1 9/16	1 25/32	33 @ 1 9/64	72 @ 1 23/64

ENGINE TORQUE SPECIFICATIONS

ENGINE MODEL	CYLINDER HEAD FT. LBS.	INTAKE MANIFOLD FT. LBS.	EXHAUST MANIFOLD FT. LBS.	CONN. ROD FT. LBS.	MAIN BEARING CAPS FT. LBS.	CRANKSHAFT END NUT FT. LBS.	FLYWHEEL CRANKSHAFT TO FT. LBS.	CAMSHAFT GEAR BOLT FT. LBS.	FUEL INJECTOR NUTS FT. LBS.	OIL PAN BOLTS FT. LBS.
354D	1	31-36	41-45	65-70	145-150	250-300	75	45-50	10-12	10

1—7/16" studs 55-60 lbs. 1/2" studs 85 lbs.

Trouble Diagnosis

The satisfactory performance of a Diesel engine depends on two items of foremost importance, (1) Sufficiently high compression pressure, and (2) The injection of the proper amount of fuel at the right time.

The first one of these items depends entirely on pistons, piston rings, valves and valve operating mechanism; the second item depends on injectors and their operating mechanism, and fuel system.

Lack of engine power, uneven running, excessive vibration, and tendency to stall when idling may be caused by either a compression loss or faulty injector operation.

The causes of trouble symptoms may be varied; therefore a hit-and-miss search should be avoided. A proper diagnosis of symptoms is an orderly process of diagnosing the symptoms. An "orderly process" means to check the most probable common cause first; then proceed with the next most probable case.

Hard Starting
1. No fuel at injectors.
   a. No fuel in tank.
   b. Stop control in "stop" position.
   c. Air in fuel system.
2. Oil in crankcase too heavy.
3. Wrong type fuel oil.
4. Injection pump timing wrong.
5. Injectors in need of service.
6. Cold starting aid inoperative.
7. Poor compression in one or more cylinders.

No Power (Exhaust Clean)
1. Restricted fuel supply.
   a. Restriction on suction side of fuel pump.
   b. Air leak on suction side of fuel pump.
   a. Blocked fuel filter.
   d. Faulty low pressure fuel pump.
2. Incorrect fuel injection pump delivery.

Low Power, Max R.P.M. Low (Clean Exhaust)
1. Throttle lever not reaching wide open position.
2. Air in fuel system.

Low Power, Engine Misfires (Clean Exhaust)
1. One or more broken or cracked fuel lines.

Low Power, Engine Smooth (Smoky Exhaust)
1. Blocked air cleaner.
2. Wrong injector settings.
3. Incorrect valve lash.
4. Wrong grade fuel oil.
5. Injection pump timing late.

Low Power (Engine and Exhaust Overheat)
1. Injection timing late.

Low Power, Engine Misfires (Dirty Exhaust)
1. One or more injectors faulty.
2. Injection pump timing late.
3. Broken parts in fuel injection pump.
4. Valves bad in one or more cylinders.

Low Power, Engine Misfires, Engine Has Metallic Knock (Dirty Exhaust)
1. Exhaust valves badly carboned.
2. Needle stuck in injector.
3. One or more rocker levers broken.
4. One or more push rods bent.

Blow-By From Breather Tube
1. One or more pistons and/or liners scored.

Power Good, Poor Fuel Mileage (Exhaust Smoky and Black)
1. Fuel pump delivery too high due

Trouble Diagnosis

to unauthorized adjustment.

Fuel Oil in Oil Pan
1. Diaphragm on low pressure pump leaking.
2. "O" ring on fuel injection pump leaking.

3. Bad injectors.

Water in Oil Pan or Oil in Radiator
1. Cracked head or block.
2. Blown head gasket.

Fuel System

General Description
The Perkins-6-354 Diesel engine is a vertical four stroke cycle power unit.

The Perkins-6-354 Diesel engine is a six-cylinder direct injection unit havng a 3⅞" bore and 5" stroke.

The efficient operation of a diesel engine depends on the correct amount of fuel being injected into the cylinder at exactly the right time in the operating cycle. It also depends upon each cylinder being fully charged with air that has been compressed to the correct pressure and temperature in order to provide the necessary oxygen for complete and efficient combustion of the injected fuel oil.

Trouble shooting the fuel system becomes basically a process of elimination of the various components to ascertain which one is not functioning as it should. From this it follows that any malfunction in the combustion cycle will result in inefficient and/or incomplete burning of the fuel and air mixture, therefore, the most common complaint is likely to be "No Power."

—The causes of "No Power" fall into two major categories which are fairly easy to diagnose, the Key being the condition of the exhaust gas.

If the *correct* amount of fuel is being injected but power is low, the exhaust gasses will be heavy and generally black, brown or whitish brown in color due to incomplete combustion in the cylinders.

If too much fuel is being injected the exhaust will be black in color but the power will be good.

If insufficient fuel is being injected and power is low, the exhaust gasses will be hardly visible due to complete burning of the mixture in the cylinders.

Other associated complaints will be in the nature of:
1. Hard starting.
2. Flames in exhaust pipe.
3. Rough idle and missing at light loads.
4. Noisy and vibrating.
5. Overheating (especially the exhaust).
6. Low governed speed.

Fuel Filters
Always be careful to keep dirt of any kind from the fuel system. Within the system itself much has been done to be sure that only clean fuel oil reaches the fuel injection pump and nozzles. Fuel oil filters are provided as follows:

The *primary filter* is a screen unit located in the upper half of the fuel supply pump.

It should be removed for cleaning, every 12,000 miles or oftener as conditions warrant. The primary filter may be removed by unscrewing the bolt which secures the domed cover to the top of the fuel supply pump.

When reassembling after cleaning, make sure that a good joint is made between cover and fuel supply pump body. Any leakage of air here may cause air locks in the fuel system.

The *secondary* or *final filter* contains a paper element, this element being located between top and bottom covers the filter. Do not attempt to clean the paper element, it should be *replaced* every 20,000 miles.

Fuel Filter Replacement
1. Unscrew bolt in center of top cover.
2. Remove filter bottom cover.
3. Remove paper element and discard.
4. Clean filter top and bottom covers.
5. Replace the gaskets and rubber O-ring.
6. After reassembling, prime fuel system.

Fuel Supply Pump
The fuel supply pump is of the diaphragm type similar to the type used on gasoline engines. It is located on the right side of the engine and driven by an eccentric cam on the camshaft. A hand primer is an integral part of the fuel supply pump. Use the hand primer to pump fuel from tank until supply pump, filters, pipes, and injection pump are filled with fuel.

Priming System
Whenever an attempt is made to start a new or rebuilt engine, an engine on which the fuel lines have been disconnected (fuel filter change, service of injection pump, etc.), or an engine which has been standing idle for some time, the fuel system must be primed or bled. The procedure for priming the fuel system is as follows:

1. Loosen air vent screw on top of governor control housing.
2. Loosen vent screw in hydraulic head locking screw on side of fuel pump body.
3. Operate priming lever of fuel supply pump and when fuel, free from bubbles, flows from around each venting point, tighten head locking screw vent first and then vent screw on governor housing. If priming lever does not operate, rotate engine until operation is possible.
4. Loosen pipe union fitting at injection pump fuel inlet and

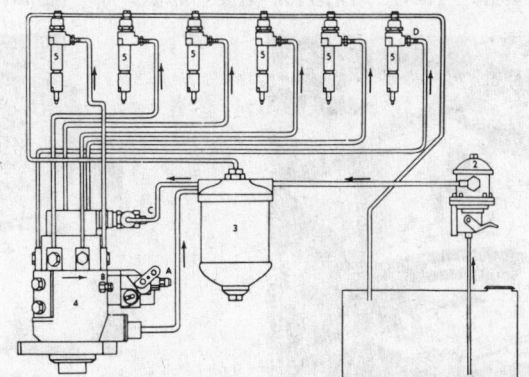

1 Tank
2 Supply pump
3 Filter
4 Injection pump
5 Injection nozzles
A Vent screw—governor housing
B Vent screw—head locking screw
C Fitting—pump fuel inlet
D Unions—Nozzle fuel inlet

Fuel system with C.A.V. hydraulic governed injection pump
(© Chrysler Corp.)

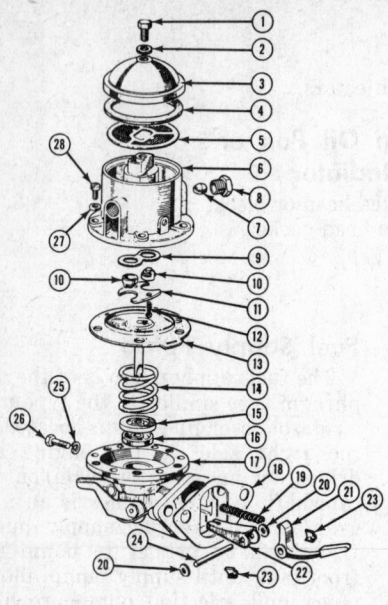

1 Bolt, fuel pump cover
2 Washer
3 Cover, fuel pump
4 Gasket, cover
5 Screen, fuel filter
6 Body, pump, upper half
7 Sleeve
8 Nut
9 Gasket, inlet & outlet valve
10 Valve, inlet & outlet
11 Retainer, valve
12 Screw, retainer
13 Diaphragm
14 Spring, diaphragm
15 Retainer, oil seal
16 Seal, oil
17 Body, pump lower half
18 Gasket, pump mounting
19 Spring, lever
20 Washer, rocker arm pin
21 Arm, rocker
22 Lever
23 Retainer
24 Pin, rocker arm
25 Washer, lock
26 Bolt
27 Washer
28 Screw

Fuel pump
(© Chrysler Corp.)

supply pump. When fuel, free from air bubbles, flows from threads, retighten fitting.

5. Loosen unions at nozzle ends of two of high pressure fuel injection pipes.

6. Set accelerator at full open position and make sure "stop" control is in "run" position.

7. Crank engine until fuel oil, free from air bubbles, flows from both fuel pipes.

8. Retighten unions on fuel injection pipes. The engine is now ready for starting.

Fuel Injection Pump

The C.A.V. Model DPA fuel injection pump is a distributor type pump incorporating an all speed governor and an automatic advance.

Pumping is effected by a single pumping element, having twin opposed plungers, situated transversely in a central rotating member which also acts as a distributor. The rotor revolves in a stationary member, known as the hydraulic head. The pump plungers are operated by contact with cam lobes on a stationary internal cam ring. The fuel is metered before entering the pumping element, and the accurately metered high pressure charges are distributed to the engine cylinders in correct firing order and at the required timing intervals through a system of ports in the rotor and the hydraulic head.

A simple hydraulic type governor is embodied in the pump to give accurate control of engine speed throughout the complete speed range of the engine and under all conditions of engine loading. An automatic advance mechanism is also provided to vary the point of commencement of injection.

Fuel Injection Pump Removal

Disconnect the fuel injector lines at the injectors and at the injector pump. Unbolt the mounting clamp from the cylinder head and remove the six injector lines as an assembly. *Injector lines must be handled* *carefully to avoid bending or damaging. Dirt or chips should not be allowed to enter fuel lines. Do not disassemble injector line assembly unless one or more of the lines are to be replaced. Cover or tape fuel openings on the injection pump to prevent dirt from entering the pump.*

Disconnect the accelerator and stop cable, remove the three injection pump attaching stud nuts and lift out the injection pump.

Fuel Injection Pump Repairs

Disassembly or repair of the fuel injection pump should never be attempted by anyone other than an authorized C.A.V. dealer, since special equipment is needed to properly service and repair this assembly. Satisfactory repair and adjustment are impossible without proper test equipment and procedures.

Fuel Metering

Apart from small losses which occur during the injection stroke, the total volume of fuel introduced into the element is passed on to the nozzle. Metering is effected, therefore, by regulating the volume of fuel which enters the element at each charging stroke. The volume of the charge is governed by two principal factors—the fuel pressure at the inlet port, and the time available for fuel to flow into the element while the inlet ports in the rotor and the hydraulic head are in register. It is by controlling the pressure at the inlet port that accurate metering is achieved.

Fuel Injector Pump Installation and Timing

The correct static injection pump timing (engine stopped) is 24° before top dead center. If the procedure we are covering here is followed carefully, the correct specified timing will result.

The fuel injection pump adapter plate has a slot, approximately ⅛" wide in its inside diameter. Also, on the edge of the adapter plate flange, adjacent to one of the bolt holes a

Removing fuel pump from engine
(© Chrysler Corp.)

Installing fuel filter
(© Chrysler Corp.)

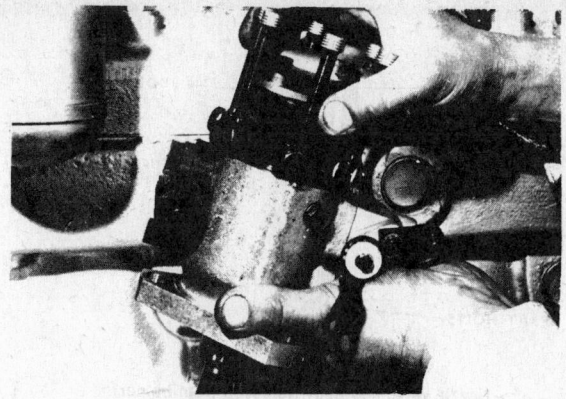

Removing injection pump
(© Chrysler Corp.)

Alignment of timing slots
(© Chrysler Corp.)

scribed line will be found.

Set the adapter plate in position on the three mounting studs with the scribe line next to the stud which is farthest from the center of the engine.

The timing mark for top dead center is located on the front face of flywheel. A pointer is located on flywheel housing.

To check timing, make certain the number one piston is at exact top dead center, and is on the compression stroke as follows: With valve cover removed rotate the crankshaft with the flywheel. When the number six exhaust valve is just seating, align the mark on the flywheel with the pointer. Number one cylinder will be at T.D.C.

A slot approximately 1/8" wide can be seen on the top edge of the fuel injection pump drive hub on the vertical shaft.

Rotate the auxiliary shaft and fuel injection pump drive shaft in the direction of engine rotation, to where the 1/8" slot in injector pump drive hub exactly aligns with the 1/8" slot in inside diameter of adapter plate.

Place the auxiliary drive shaft gear on front of the shaft with chamfered side in, and so when it is against the mounting flange on shaft, the bolt holes in the flange are at center of slotted holes in the gear. If the bolt holes will not center in the slotted holes, turn the gear 120 degrees (one bolt hole). One of the three possible positions will allow the bolt holes to be centered. Install the retaining washer and the three recessed head retaining screws, with lock washers on each screw.

Check to make sure that the 1/8" slot in fuel injection pump drive hub

is exactly aligned with corresponding slot in adapter plate. If some slight correction is necessary, bring the slots into alignment by turning injection pump vertical drive shaft. Then tighten the three gear retaining screws securely.

With the gasket positioned on the studs, align the master spline on the fuel injection pump shaft with the master spline in the vertical shaft, and set injection pump on adapter plate.

Align scribed line on injection pump mounting flange with scribed line on adapter plate, and tighten the three retaining nuts securely.

With Gasket Sealer on each side of gaskets, install camshaft gear cover and auxiliary drive shaft gear cover and tighten the bolts.

Install the vibration damper and

1 Shaft, splined
2 Tube, pilot
3 Seals, oil
4 Ring, snap
5 Lever, control
6 Shaft, pinion
7 Housing, governor
8 Screw, idle
9 Washer
10 Spring, idle
11 Rack
12 Spring, governor
13 Plate, dished (dashpot)
14 Plate, roller to roller
     adjusting
15 Valve, metering
16 Chamber, metering
17 Port, charging
18 Head, hydraulic
19 Rotor, distributing
20 Liner, transfer pump
21 Seal, "O" ring
22 Rotor, transfer pump
23 Plate, end
24 Bolt, end plate
25 Vane, sliding
26 Port, distributor
27 Hole, radial
28 Connector, nozzle pipe
29 Fitting, head locating with
     damper
30 Shoe, roller
31 Screw, cam advance
32 Ring, cam
33 Plunger
34 Plate, roller to roller
     adjusting
35 Plate, drive
36 Roller, cam
37 Housing, pump

C.A.V. fuel injection pump (© Chrysler Corp.)

Removing nozzle assembly
(© Chrysler Corp.)

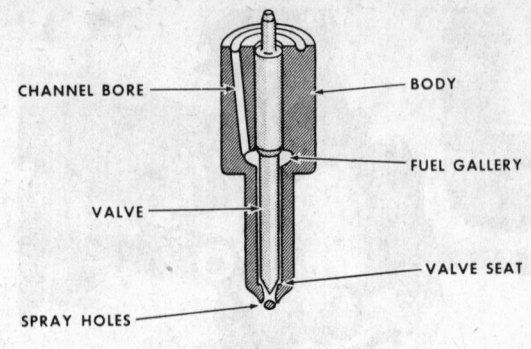

Nozzle valve and body (removed from holder)
(© Chrysler Corp.)

pulley assembly, retaining washer and nut, tightening the nut to 250 foot-pounds.

Injector Nozzles

General Information

The performance of the high speed diesel engine depends largely on the proper function of its fuel injection system. For maximum efficiency it is essential the engine be not only provided with metered amounts of fuel at exactly timed intervals, but also that it receives each charge of fuel in the proper condition so it can be completely consumed without causing smoke in the exhaust. This briefly is the function of the injection nozzle assembly. The complete assembly consists of a nozzle and a nozzle holder.

Nozzle Maintenance

All nozzles should be removed from the engine at regular intervals. Just how long these intervals should be, depends on the different conditions under which the engine will operate. Where ideal conditions of good combustion, adequate cooling, and absolutely clean fuel are realized, the nozzles will need little attention. Since the efficiency of the nozzles vitally affects the performance and economical operation of the diesel engine, no engine should be allowed to run with

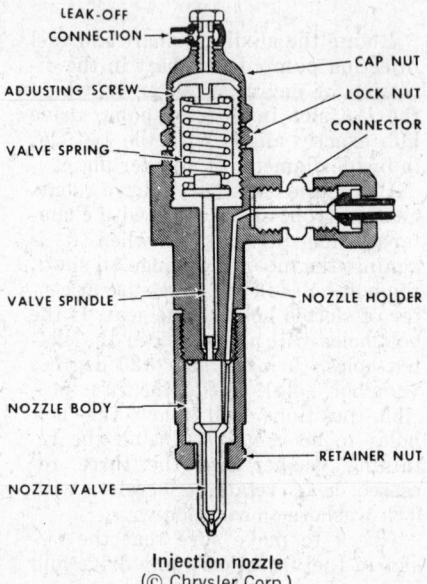

Injection nozzle
(© Chrysler Corp.)

nozzles that are defective. 18,000 miles is a good standard interval between inspections.

Nozzle trouble is usually indicated by one or more of the following symptoms:
1. Cylinder knock.
2. Engine overheating
3. Loss of power
4. Smoky exhaust (black)
5. Increased fuel consumption.
While the above faults may be

caused by defective nozzles, it should also be determined that they are not caused by other engine troubles such as wrong fuel, water in fuel, dirty or damaged filters, incorrect maximum fuel setting, defective engine lubrication, incorrect pump timing, incorrect engine valve timing, or faulty valves.

If everything else is in order and the nozzles are still suspected, the nozzle causing the trouble can often be found by releasing the pipe union nut on each nozzle in turn, with engine running and listening to the idling performance of each of the cylinders.

Nozzle Testing

To test a doubtful nozzle, loosen injection pipe at both ends and remove nozzle holder from cylinder head. *A blast of compressed air around each nozzle before loosening the holder will help prevent dirt or moisture from entering the combustion chamber when nozzle is removed.* Turn the nozzle and holder around the injection pipe so that nozzle points away from engine. Retighten both ends of injection line. Next, loosen pipe unions at the other nozzles to prevent fuel oil from being sprayed into cylinders. Turn engine with starter until the removed nozzle sprays fuel into the air. If spray is "wet," or "streaky," or to one side, or if the

Alignment of scribed line on injector pump and adapter plate
(© Chrysler Corp.)

Timing mark on flywheel and pointer
(© Chrysler Corp.)

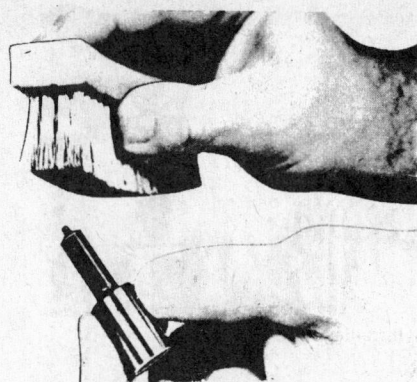

Cleaning nozzle with brass wire brush
(© Chrysler Corp.)

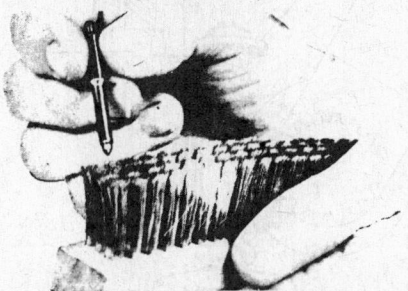

Cleaning valve tip with brass wire brush
(© Chrysler Corp.)

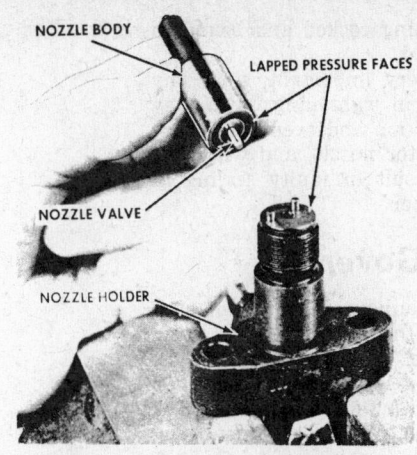

Pressure faces of nozzle and holder
(© Chrysler Corp.)

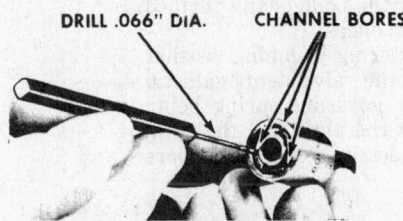

Cleaning nozzle body channel bores
(© Chrysler Corp.)

nozzle "dribbles," remove and install a complete new or reconditioned unit.

WARNING: Keep hands away from spray. The working pressure will cause oil to penetrate the skin.

After making a nozzle replacement, the faulty unit should be sent to a test bench for cleaning and testing.

Nozzle Cleaning, Inspection

Nozzle maintenance like injection pump maintenance should be performed in a clean dust-free place where no other work is performed. A work bench with a holding fixture or small vise with soft jaws is essential for making the disassembly. A hand test pump with a good supply of clean diesel fuel or test oil should also be provided.

1. Attach injecton nozzle to hand test pump and check completely for spray pattern nozzle opening pressure setting, and back leakage.
2. If nozzle is faulty secure nozzle assembly in holding fixture or vise. Release pressure on nozzle valve spring by loosening lock nut and adjusting screw. Remove nozzle retainer nut and nozzle.
3. Inspect nozzle for carbon and note whether valve lifts out freely. Brush all carbon from outside with a brass wire brush. Place body and valve in clean fuel oil or cleaning solvent to soak and soften carbon. *The nozzle should be free from all dam-*

age and not "blued" from overheating. All polished surfaces should be relatively bright without scratches or dull patches. All pressure surfaces must be clean since these must mate to form a high pressure joint between nozzle holder and nozzle.

4. Clean small feed channel bores with drill or wire of .066 inch diameter. These bores are rarely choked and insertion of drill or wire by hand will be sufficent.
5. Insert probe or groove scraper in fuel gallery and press against side of cavity to clear all carbon deposits from this area.
6. Clean all carbon from valve seat area by rotating and pressing probe on to seat.
7. Clear spray holes with appropriate size cleaning wire. Do not break wire in holes since such particles are almost impossible to remove (Hole Dia-.0126 in.).
8. Install nozzle in a flushing tool that can be coupled to a nozzle hand test pump and wash out all carbon particles. Force fuel or test oil through nozzle vigorously since this is best way to remove loosened particles that may still be present both in body cavities and spray holes.
9. Clean nozzle valve tip by brushing away carbon with a brass wire brush. To assemble valve into nozzle, immerse both items in clean fuel oil and fit them together under surface so as to

prevent closely fitting lapped surfaces from being touched by hand. *If nozzle is blued or seat has a dull ring indicating wear or pitting, complete unit should be set aside. No attempt should be made to lap nozzle valve and body unless special training and equipment are available.*

10. Make sure lapped pressure faces on both nozzle and holder are clean and free from metalic particles. If holder has been disassembled, all parts should be thoroughly washed in fuel oil or cleaning solvent.

Injector Installation

Before installing injectors, make certain that all of the old copper sealing washers have been removed from the recesses in the cylinder head and have been discarded. Use only new copper sealing washers of the correct replacement part number. Old copper sealing washers can be removed with a piece of bent wire.

The recess in cylinder head, the faces of the copper washer, and corresponding face on injector must be thoroughly clean to avoid possible combustion leakage.

Place copper washer in position, and then injector. Make sure injector is an easy fit in the recess and on attaching studs, and that it can be placed down on copper washer without any force of any kind.

The nuts on attaching studs should then be tightened down evenly in order to prevent the injector nozzle from becoming cocked and bound in cylinder head.

This is very important, since any unevenness in tightening down may cause distortion and eventual failure of the injector nozzle, and will most certainly result in faulty sealing at copper washer.

Governor

The hydraulic governor is of simple design, the working parts being contained in a small housing. The control lever is carried on a pinion shaft, the pinion meshing with a rack which is free to slide on the metering valve stem.

The metering valve slides in a chamber in the hydraulic head, into which bore the diagonally drilled metering port opens.

A self-centering damping washer is carried on the valve stem against a shoulder, the governor spring being held between the plate and the rack. The plate slides in a cylindrical bore

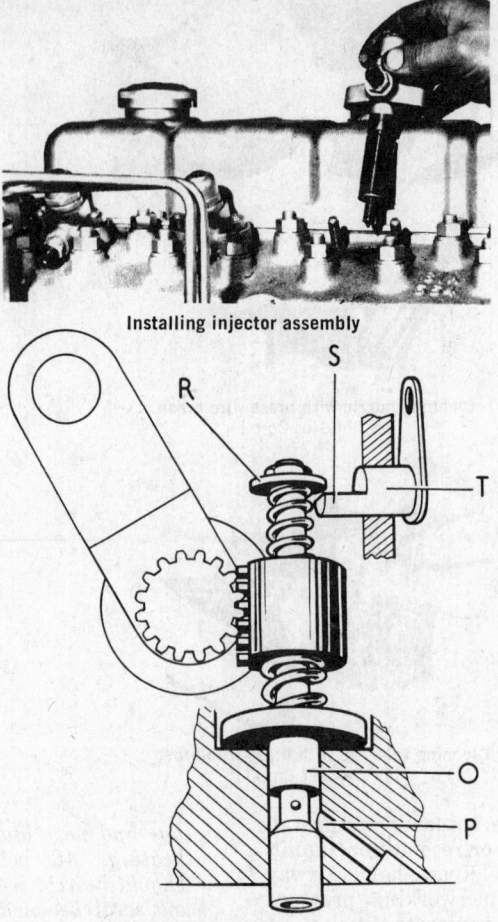

Installing injector assembly

Details of motoring valve and shut-off spindle
of hydraulic governor
(© Chrysler Corp.)

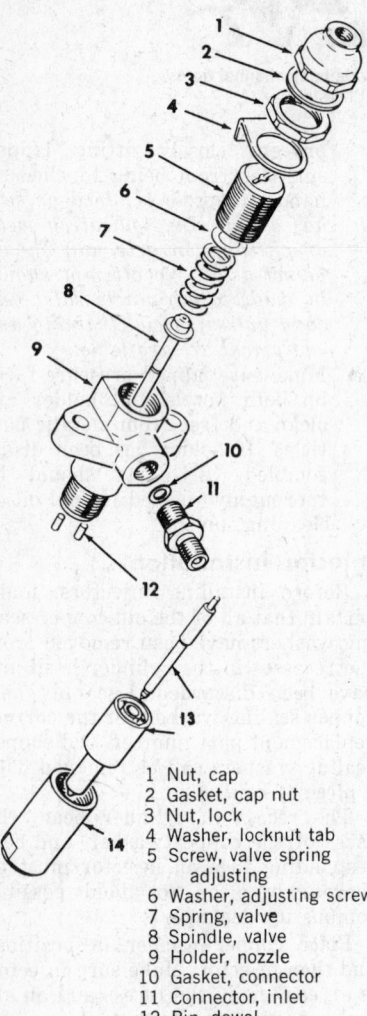

1 Nut, cap
2 Gasket, cap nut
3 Nut, lock
4 Washer, locknut tab
5 Screw, valve spring adjusting
6 Washer, adjusting screw
7 Spring, valve
8 Spindle, valve
9 Holder, nozzle
10 Gasket, connector
11 Connector, inlet
12 Pin, dowel
13 Nozzle, body and valve assy.
14 Nut, nozzle retainer

Nozzle and holder parts
(© Chrysler Corp.)

filled with fuel, and acts as a dashpot to damp out any violent movement of the metering valve.

An idling spring is located between the rack and the washer fitted at the top of the valve spindle and known as the shut-off washer. A screw is placed so as to form an idling stop.

The governor is operated by fuel at transfer pressure which is fed from the annular groove surrounding the pump rotor. The fuel passes through the metering valve, which is hollow, and via transverse holes to an annular space around the valve.

End wise movement of the metering valve varies the area of the metering port which registers with the annulus around the valve; the effective area of the port is that part which is uncovered by the inner edge of the groove or annulus.

The metering valve is loaded by the governor spring and this load may be manually adjusted by moving the control lever.

As the engine control is operated to give increased speed, the valve is pushed to the full open position by the governor spring. When engine speed increases, transfer pressure increases also, and this pressure forces

the valve back against the spring pressure until a balance is reached.

When the engine control is moved towards the idling stop, the idling spring is compressed, and the compression of the main governor spring is reduced. Equilibrium is reached when the forces exerted on the metering valve by the idling spring and fuel at transfer pressure are balanced by compression of the main spring. The latter becomes progressively less as the engine control is moved towards the idling stop, thus enabling the reduced transfer pressure at low speeds to operate the metering valve and perform the governing function throughout the idling range.

Details of the valve and the shut-off mechanism are shown in illustration. The shut-off spindle is carried in the governor housing, and carries a half-round end or "cam," which contacts the underside of the shut-off washer. Rotation of the spindle by the shut-off lever lifts the metering valve to a position where the metering port is blanked off, and thus stops the engine.

Automatic Advance

Under full load conditions, the maximum amount of fuel is introduced into the element, and the plungers and actuating rollers are forced outwards to the limit of their travel. As the rotor turns, the rollers are brought into contact with cam lobes on the cam ring. The point of contact is near the base of the cams.

Under lightly loaded conditions, fueling is decreased and the plunger travel is proportionally reduced. Contact between roller and cam is now made at a point near the cam peak.

The advance device corrects the tendency of the pump to retard timing under varying load conditions at constant speed.

The device illustrated provides progressive advancement of injection timing as engine speed is increased.

The piston is free to slide in a cylinder machined in the body of the device. Movement of this piston is transmitted to the cam ring by the ball-ended lever, causing the cam ring to rotate within the pump housing.

Pressure exerted on the piston by the springs, tends to hold the piston and the cam ring in the fully retarded position.

Fuel oil at transfer pressure enters the device through a fuel passage in the screw which secures the device to the pump housing. Transfer pressure acts upon the piston and tends to move the cam ring towards the fully advanced position.

Transfer pressure increases progressively as the engine speed is raised, and the piston is moved along the cylinder to compress the springs and move the cam ring towards the fully advanced position. When engine speed is decreased, transfer pressure falls, and the piston and cam ring are moved towards the retarded position by spring pressure.

The impact of the actuating rollers on the cam lobes at commencement of injection tends to move the cam ring towards the retarded position. Such movement is prevented by a non-

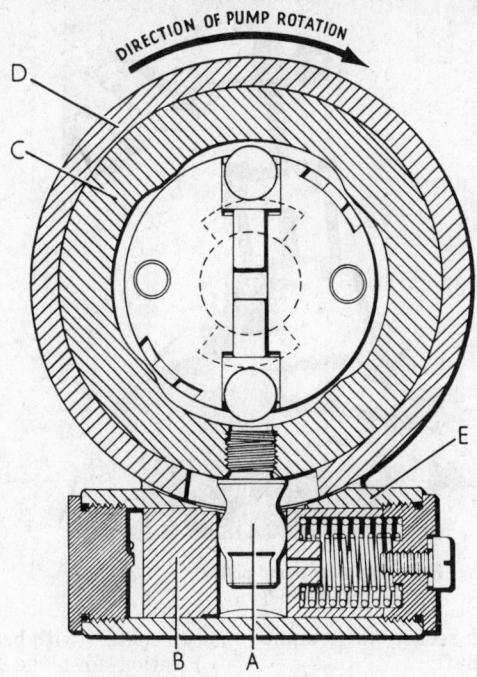

Automatic advance mechanism
(© Chrysler Corp.)

return valve situated in the fuel passage in the screw securing the device to the pump housing. Normal leakage between the piston and the cylinder permits the device to return to the retarded position when the engine speed falls.

Any desired timing advance up to a maximum of 9° (pump) is obtainable, and the engine speed at which this is attained can be varied by fitting stronger or weaker springs. The stronger the spring the higher will be the engine speed at which maximum timing advance is attained. Fine adjustment is made by shims fitted between the springs and the spring cap.

Governor No-Load Speed Setting

1. Warm engine to normal operating temperature.
2. Use accurate tachometer for checking engine speed.

3. Increase engine speed until governor control is apparent; this should be 3600-3700 rpm. If necessary, break seal and readjust by changing tension on governor spring. Increasing tension increases speed 50 rpm per turn of nut.

Water Pump

Removal and Installation

1. Drain cooling system and remove fan belt and fan.
2. Use a suitable puller and remove pump pulley.
3. Remove the pump to block mounting bolts and remove the pump.
4. Install by reversing the removal procedures.

Servicing Water Pump

Remove the self-locking nut and

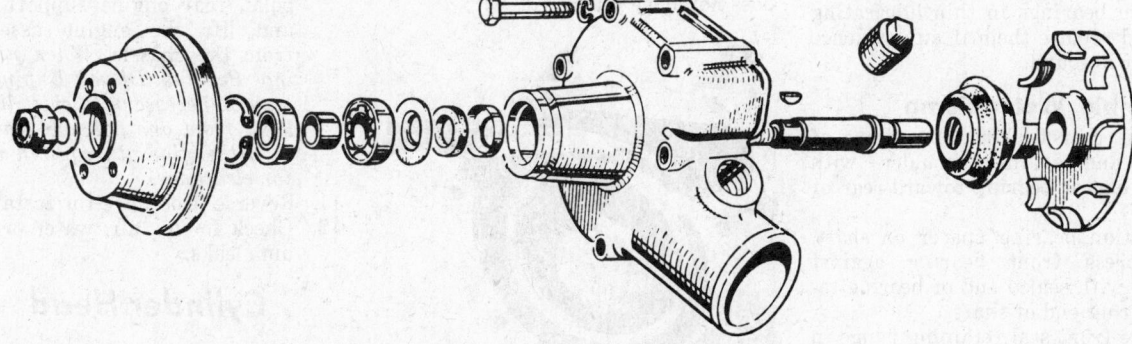

Water pump (disassembled view) (© Chrysler Corp.)

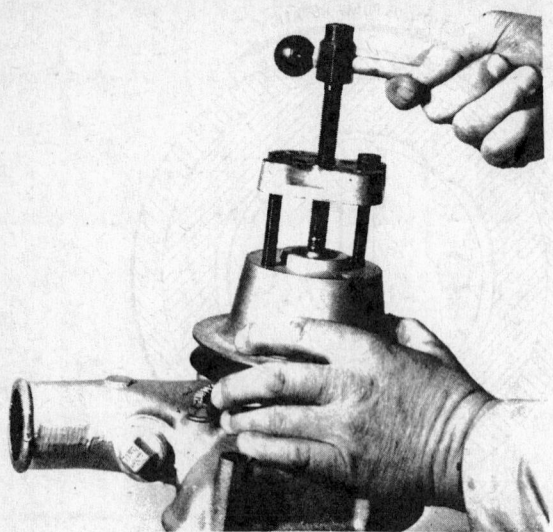

Removing water pump pulley
(© Chrysler Corp.)

Installing bearing retainer snap ring
(© Chrysler Corp.)

flat washer which secure water pump pulley to pump shaft.

Using remover, pull the pulley from pump shaft. Remove key.

Press water pump shaft, complete with impeller, out through the rear of pump body. Press impeller off shaft.

Using snap ring pliers, remove bearing retaining snap ring from front of pump body.

With a suitable tool, press the two bearings and spacer out through front of pump body.

To complete disassembly of water pump, remove front seal, retainer, and flange.

Inspection of Water Pump Parts

Examine pump body for cracks, damage or corrosion.

Examine pump shaft for wear. If pump bearing inner races rotate on shaft, shaft must be replaced.

Clean all rust and scale from impeller, and inspect it for cracks and damage. Examine impeller hub sealing surface for wear or scoring. If impeller shows any damage, cracks or wear at seal surface, it should be replaced.

Wash bearings in thin lubricating oil, and replace them if any evidence of wear is found.

Assemble Water Pump

Press rear bearing onto front of shaft, and against shoulder, with sealed end of bearing toward rear of shaft.

Position bearing spacer on shaft, and press front bearing against spacer with sealed end of bearing toward front end of shaft.

Place front seal retaining flange in position against rear face of rear bearing. This dished flange must be positioned so the center of flange is not in contact with bearing.

Position first the front seal, and then seal retainer against retainer flange.

Half fill the space between the two bearings with short fibre water pump grease, or multi-purpose chassis grease and press the complete assembly into pump body through front end. Then position bearing snap ring firmly in groove, using snap ring pliers.

Place a new rear seal in pump on rear end of pump shaft with carbon face to rear so it will contact the impeller hub. Make sure seal is resting squarely, and is not cocked in any way.

At this point, the shaft should be rotated by hand to be sure it turns without binding.

Some of the attaching bolts must be entered in the holes in the pump body before pressing on the pump pulley, since it is not possible to position them after the pulley is installed

With key in position in shaft and with necessary attaching bolts in their holes, support shaft at its rear end, and press pulley onto shaft, with keyway aligned. Make sure shaft does not move rearward.

Checking impeller to pump body clearance

Install flat washer and self-locking nut on front of pump shaft and tighten nut to 50 to 55 foot-pounds.

Press impeller onto shaft to where clearance between pump body and inner edge of impeller vanes is between .012″ and .019″.

Engine Section

Engine Removal and Installation

1. Disconnect battery.
2. Drain cooling system and engine oil.
3. Disconnect fuel in-line and return line, and exhaust pipe. Remove air cleaner.
4. Disconnect wiring for alternator, gauges, switches, etc.
5. Bleed air booster brake and disconnect line.
6. Disconnect hydraulic clutch line, stop control cable accelerator and hand throttle, coolant and heater hoses.
7. Remove radiator.
8. Remove driveshaft and transmission.
9. Install engine lifting sling.
10. Remove front engine support bolts, rear engine support nuts and lift the engine assembly from the chassis. *When removing Perkins Diesel Engines it may be necessary to shift lifting hook from one notch to another to lift engine at required angle for clearance.*
11. Reverse procedure for installing.
12. Check for oil, air, water or vacuum leaks.

Cylinder Head

Removal

1. Disconnect battery cables, drain radiator and block.

2. Remove air cleaner, and disconnect breather pipe.
3. Remove cylinder head cover from head.
4. Remove rocker arm support stud nuts. The rocker arm assembly complete with lubrication tube and "O" ring can now be removed as an assembly.
5. Carefully lift valve push rods out of engine. CAUTION: It is possible to drop No. 2, 3, 10 and 11 push rods down into oil pan and if this occurs, they cannot be retrieved without removing oil pan.
6. Disconnect fuel injector lines at both ends. Unbolt mounting clamp from cylinder head and remove six injectors lines as an assembly. *Injector lines must be handled carefully to avoid bending or damaging, and dirt or chips should not be allowed to enter fuel lines.*
7. Remove cylinder head to water pump by-pass tube.
8. Remove fuel lines.
9. Remove six injectors and copper sealing washers. *If injectors stick, tap them sideways carefully with a soft hammer to loosen them. Place injectors where tips will not be damaged, do not use a wire brush or other abrasive on injectors.*
10. Remove upper radiator hose and heater hoses.
11. Disconnect exhaust pipe at manifold flange and fuel return line bracket.
12. Remove 32 cylinder head stud nuts.
13. Remove voltage regulator to prevent possible damage. *Avoid damage to cylinder head and block surfaces by careless prying under cylinder block.*
14. Remove No. 4 and 5 rocker shaft studs, install lifting fixture on cylinder head and remove.

Installation

1. Install intake and exhaust manifolds, using new gaskets. Use the short bolts for the third manifold bolt from each end to avoid bolt interference with the push rods.
2. The mating surfaces of cylinder head and cylinder block must be clean. A cracked head, must, of course, be renewed.
3. If the longitudinal bow in the head is more than 0.006", or the transverse boy exceeds 0.003", then machining of the head joint face is necessary. Machining is also necessary if the head has deep scratches, scores, nicks, or fret marks, in particular, across the areas where the gasket bore eyelets seal, or, where they run into water or oil apertures. *No more than 0.010" should ever be removed from the gasket face of the head* and if the head cannot be cleaned by removing 0.010" then it must be renewed. The surface finish of the head should be of a medium ground finish of 60/100 micro inches, or equivalent to the top face of the cylinder block. *Only the minimum amount of metal should ever be removed from the head face* and the amount removed should be stamped on the head face in case further resurfacing becomes necessary. Care should be taken in selecting the area for stamping.
4. If the head bottom face has been machined, check that the valve heads are recessed below the head face by not less than 0.029". If the valve heads are recessed less than 0.029" then the valve seats must be ground until the valve heads are recessed at least 0.029" and not more than 0.049". After machining of the head the atomizer nozzle protrusion must not be greater than 0.2165" and this figure *must not* be achieved by the use of additional atomizer seating washers as these may be inadvertently removed at a later date and result in a loss of power.
5. The cylinder head studs and nuts must be examined for damaged threads or other signs of wear. Studs must fit snugly into the cylinder block. With a light smear of oil on the threads screw the studs into the block tightening to 10 ft/lbs, when they should not be slack or wobble. Any suspect studs should be renewed as should any loose fitting nuts.
6. IMPORTANT: The relief counter-bore between the block and the top of the sleeve *must* be free of carbon deposits. The gasket cannot seal properly unless the counter-bore is perfectly clean.
7. Place gasket on block face, using gasket as supplied, *dry, with no jointing compound.*
8. Fit cylinder head carefully so as not to damage threads on studs, or, gasket. Lightly oil threads and working face of cylinder head nuts and fit nuts.
9. The head must be tightened as follows: Using an accurate torque wrench, with the engine cold and rocker assembly removed from engine:

   a. Pull down to 80 ft/lbs (½" studs) or 55 ft/lbs (7/16" studs) following the sequence shown.
   b. Pull down to 85 ft/lbs (½" studs) or 60 ft/lbs (7/16" studs) following the same sequence.
   c. Pull down to 85 ft/lbs (½" studs) or 60 ft/lbs (7/16" studs) *again*, following same sequence. This final tightening of the nuts should be carried

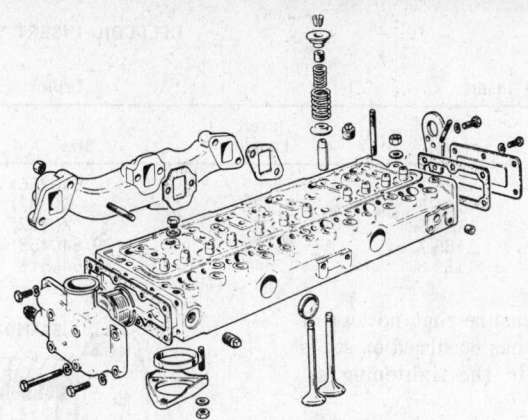

Cylinder head assembly (typical)
(© Chrysler Corp.)

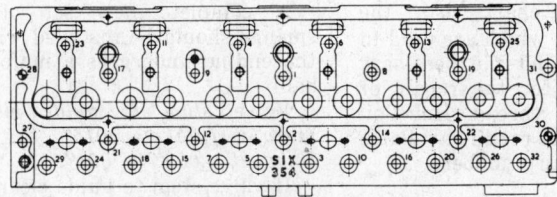

Cylinder head tightening sequence
(© Chrysler Corp.)

Perkins Diesel Engines

HELI-COIL INSERT CHART

Heli-Coil Insert		Insert Length	Drill		Tap	Inserting Tool	Extracting Tool
Thread Size	Part No.		Size		Part No.	Part No.	Part No.
1/2-20	1185-4	3/8"	17/64(.266)		4 CPB	528-4N	1227-6
5/16-18	1185-5	15/32"	Q(.332)		5 CPB	528-5N	1227-6
3/8-16	1185-6	9/16"	X(.397)		6 CPB	528-6N	1227-6
7/16-14	1185-7	21/32"	29/64(.453)		7 CPB	528-7N	1227-16
1/2-13	1185-8	3/4"	33/64(.516)		8 CPB	528-8N	1227-16

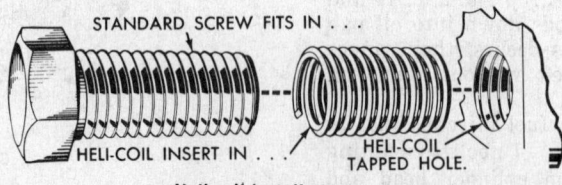

Heli-coil installation
(© Chrysler Corp.)

out to ensure that no loss of tension has occurred on studs earlier in the tightening sequence.

10. Install all push rods, being careful not to drop any of them into the crankcase.

11. Replace rocker assembly. The oil feed pipe to the rockers locates in a hole in the cylinder head and a rubber "O" ring should fit around the stem of the pipe when the rockers have been fitted. In order to ensure the new seal locates correctly, fit it immediately below the lower convolution and as the pipe is inserted into the cylinder head the "O" ring will roll up and over the lower convolution and locate itself correctly between the two convolutions. If, before inserting the pipe in the cylinder head, the "O" ring is fitted *between* the convolutions then it will roll up and over the upper convolution when inserting the pipe and fail to make an effective seal. Always fit a new "O" ring.

12. Rocker assembly shaft pillar securing nuts should be tightened to 29-32 ft/lbs.

13. Adjust valve lash (clearance)— Inlet and Exhaust to 0.012"— engine cold.

14. Using new copper gaskets, install the fuel injectors in cylinder head.

15. Install fuel lines.

16. Install six injector lines as an assembly.

17. Install cylinder head to water pump by-pass tube, upper radiator hose, heater hoses and voltage regulator.

18. Connect exhaust pipe to manifold flange. Install fuel return line bracket.

19. Fill cooling system. Connect battery cables.

20. Start engine and idle (550 r.p.m.) for a few minutes, checking that the valve action is satisfactory and lubricating oil is reaching the rockers.

21. Run engine at 1200/1500 r.p.m. for at least *one hour*, or until coolant temperature is, at least, 180°F or higher. If necessary, a radiator blind should be used to achieve this.

22. Stop engine. Remove rocker assembly. Check head nut tightness quickly before engine loses temperature, following diagram sequence as before. AGAIN check head nut tightness following same sequence.

23. Replace rocker assembly. Check fit of "O" ring as at item 9 and tighten rocker assembly securing nuts to 29-32 ft/lbs. Adjust valve lash (clearance) to 0.010"—engine hot.

24. Restart engine and idle at 550 r.p.m. Check the oil feed to the rockers and check for leak at the "O" ring around the rocker feed pipe. Check that the valve lash (clearance) is correct and that no valve is sticking. IMPORTANT: A retightening of cylinder head nuts must be made approximately 250 miles or at the end of two weeks after the gasket change.

25. Use special adapter wrench to tighten nuts underneath the rocker arms. Readjust tappets to .010" clearance "hot."

Repair of Damaged or Worn Threads

Damaged or worn threads can be repaired by the use of Heli-Coils. Essentially, this repair consists of drilling out worn or damaged threads, tapping the hole with a special Heli-Coil Tap, and installing a Heli-Coil Insert into the tapped hole. This brings the hole back to its original thread size (See Fig. 1).

The following chart lists the threaded hole sizes which are used in the engine block and the necessary tools and inserts for the repair of damaged or worn thread. Heli-Coil tools and inserts are readily available from automotive parts jobbers.

Servicing Cylinder Head

In the Perkins 354 Diesel Engine,

carbon rarely, if ever, forms in the combustion chambers in sufficient quantity to require periodic decarbonizing of the engine. Valve ports require cleaning at every head service.

Due to high thermal efficiency and other characteriestics, the valves in this engine are relatively free from trouble due to overheating.

As noted in the maintenance schedule for this engine, it is recommended that tappet clearances be checked at 18,000 mile intervals. Both intake and exhaust should be adjusted to .010" hot.

Disassembly of Cylinder Head

Remove manifolds.

Note that on this diesel engine no heat riser connection is required between intake and exhaust manifolds.

Clean gasket material from manifolds or cylinder head.

Remove thermostat and housing assembly.

Pry out retaining snap ring, and remove thermostat from housing.

All intake and exhaust valves are numbered consecutively from one to 12, commencing at the front of the engine.

The cylinder head is marked with corresponding numbers opposite the valve seats.

Using a valve spring compressor, remove the valve locks, spring retainers, inner and outer springs, and the intake valve stem seals.

Remove valves from cylinder head, and remove spring seats from top end of valve guides.

A complete new set of valve springs should be installed whenever the engine undergoes a major overhaul.

Valve stem to guide clearance (new): Intake—.0015" to .0035". Exhaust—.002" to .004".

If valve stem to guide clearance is excessive, the worn parts should be replaced. Valve guide bore diameter

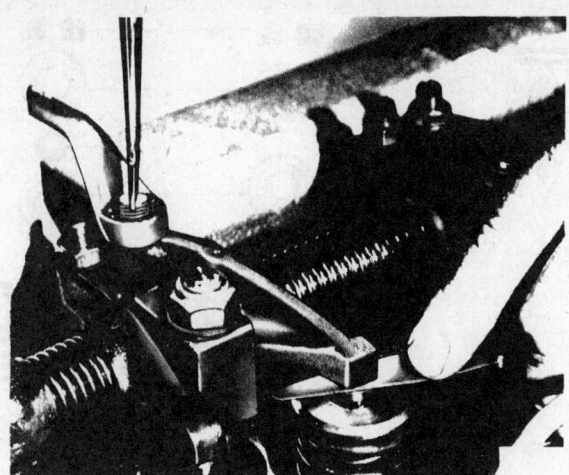

Adjusting tappet clearance
(© Chrysler Corp.)

(new) for intake and exhaust valve guides is .375" to .376".

Stem diameters for new valves are .3725" to .3735" for intake valves and .372" to .373" for exhaust valves.

With a new head and valves, the top surface of all valves should be recessed not less than .029" and not more than .039" below the surface of the cylinder head.

Clean the valves thoroughly, and examine them carefully.

Any new valves used should be stamped with the same number as the discarded valve, which the replaced.

Remove any carbon deposits from the cylinder head combustion chambers and valve ports.

Thoroughly clean water passages in cylinder head. If excessive scale is present, a proper cleaning solution should be used.

Grind Valves

Valves and valve seats should be reconditioned in the conventional way, either using grinding compound or accepted valve and seat refacing equipment. Valve seat and valve face angle are both 45 degrees.

After valves and seats have been reconditioned, the valve head depth must be checked. Use a straight edge and feeler gauge to check this dimension. It must be between the limits of .029" and .060".

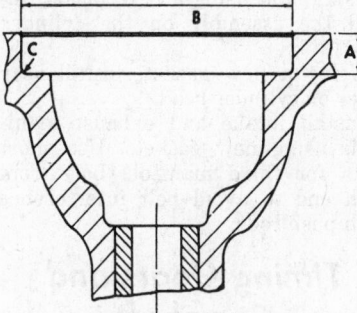

Valve seat counterbore dimensions
(© Chrysler Corp.)

Intake
A .283 to .288 in.
B 2.0165 to 2.0175 in.
C Radius .015 in.
  (max.)
Exhaust
A .375 to .380 in.
B 1.678 to 1.679 in.
C Radius .015 in.
  (max.)

It is essential that valve heights be held within these limits to provide clearance between valve and piston, and to maintain correct combustion chamber volume.

Valve seat inserts are not installed in these engines in production. However, if a seat is worn or damaged to such an extent that reconditioning places the valve depth beyond the limits of .029" to .060", a counterbore can be machined into the cylinder head and a valve seat insert can be installed.

Before you can install a valve seat insert, a new valve guide must be installed.

Valve Guides

Press out old valve guide, and thoroughly clean bore in cylinder head.

Press in a new valve guide, to where the top of the guide protrudes ⅝" above the spring seat machined surface on the cylinder head. Coat valve guides with engine oil to reduce friction during installation.

Using valve guide bore as a pilot, machine a counterbore in the cylinder head to the required dimensions.

Remove all chips and thoroughly clean insert counterbore, removing any burrs which may be present.

Again using valve guide bore as a pilot, press insert home with an installing tool.

Use a steady pressure with either a hand or hydraulic press. The tool will insure squareness. Under no circumstances should the insert be hammered in. Also, do not use lubrication when pressing in the insert. Chill seats with dry ice to reduce interference while installing them.

Visually inspect to make sure that insert has been pressed in squarely, and that it is flush with bottom of counterbore.

Using valve guide bore as a pilot, machine "flare" in insert to dimensions shown. Remove any chips or burrs from seat.

Grind insert seat to a 45° angle (90° included angle), and to where valve head depth below cylinder head face is within limits of .029" to .039".

It is advisable to work as closely as possible to the minimum figure in order to allow for reseating at a later date if required.

An existing insert may also be replaced in the same manner, except that it will not be necessary to machine a counterbore.

The gasket face of cylinder head may be resurfaced to eliminate score marks, warpage, etc. No more than .010" should ever be removed from the gasket face of the cylinder head however.

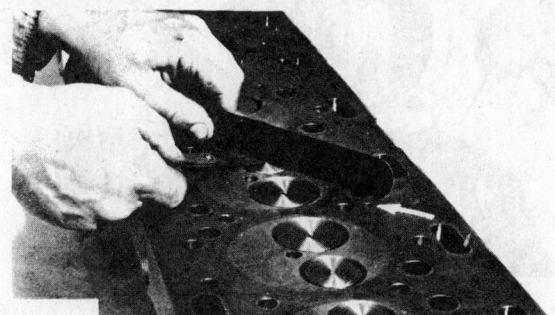

Checking valve head depth
(© Chrysler Corp.)

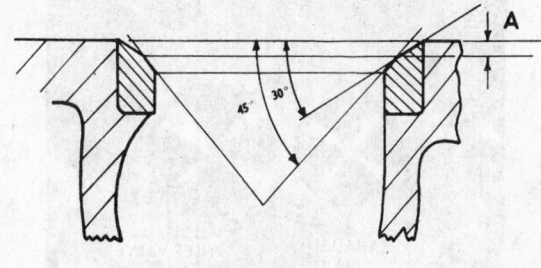

**Relieve seat at 30 degrees
to dimension A .094 to .099 inch**
(© Chrysler Corp.)

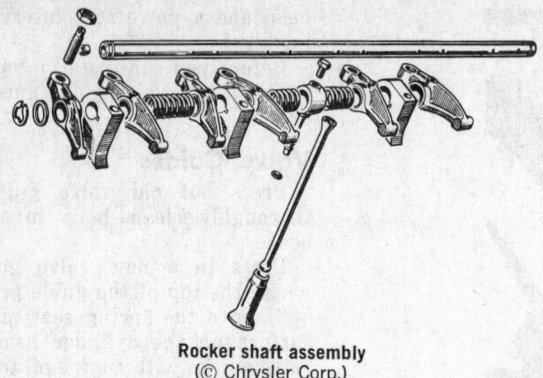

Rocker shaft assembly
(© Chrysler Corp.)

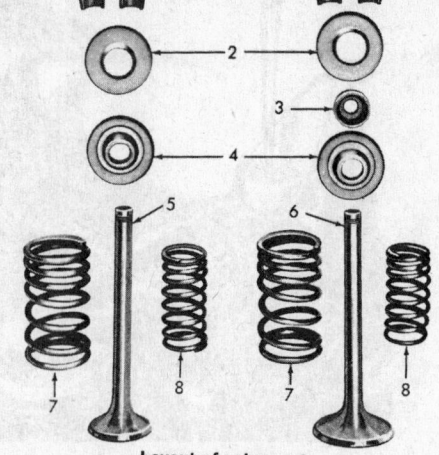

Layout of valve parts
(© Chrysler Corp.)

To replace valve guides, press out old guides. Clean bores in cylinder head, and remove all burrs.

Coat new guides with oil, and press them into where they protrude 5/8" above the cylinder head. Care should be exercised, since the guides are made of cast iron and are, therefore, very brittle.

Rocker Arm Assembly

Remove retaining snap rings and end washers from ends of rocker shaft, and slide off rocker arms and springs.

Remove rocker shaft oil feed pipe locating screw, and slide oil feed pipe off rocker shaft. Remove the small "O" ring from pipe.

Assemble Cylinder Heads and Valves

Install valves in their proper positions as numbered, place valve seat washers in position, and install new valve stem seals (open end to valve guides) on the intake valves.

Position inner and outer valve springs so close coils are next to the cylinder head, and install valve retainers and locks.

All valve springs on this engine incorporate a damper coil (close coil). Care must be taken to assure that this damper coil is replaced nearest to the cylinder head.

Install gasket, thermostat and tapered snap ring in the thermostat housing, and using a new gasket, install the assembly on the cylinder head.

Using a new gasket, install rear cover on cylinder head.

Install intake and exhaust manifolds using new gaskets. Use short bolts for third manifold bolt from each end to avoid bolt interference with push rods.

Timing Gears and Camshaft

Removal

1. Remove vibration damper retaining screw and lock.
2. Slide the vibration damper assembly off the splined end of the crankshaft.
3. Remove camshaft gear cover and auxiliary drive gear cover or tachometer drive adapter from timing gear case.
4. Remove auxiliary shaft drive gear by unscrewing three recessed head retaining screws and the retaining plate.
5. Remove camshaft retaining screw, washer and lock plate.
6. Pull camshaft gear. *Avoid driv-ing the camshaft back against the plug in the rear wall of the cylinder block, since an oil leak could develop if this plug is disturbed.*
7. Remove bolts and remove the timing gear case.
8. Remove two self-locking nuts and remove camshaft idler gear and auxiliary shaft idle gear.

Note that the camshaft idler gear can be identified as having a timing mark (punch mark), while the auxiliary shaft idler gear has no timing mark.

The two idler gear hubs are doweled, and can be readily removed from the hub mounting studs by gripping the end of the rolled pin and tapping hub with a soft hammer.

Camshaft and Auxiliary Drive Shaft

The camshaft is supported by four pressure fed bearings and is driven by helical type gears. The thrust washers used to control end thrust of both the camshaft and auxiliary drive shaft are fitted in machined recesses in the front of the cylinder block and are located by dowel pins. Before the thrust washers, camshaft

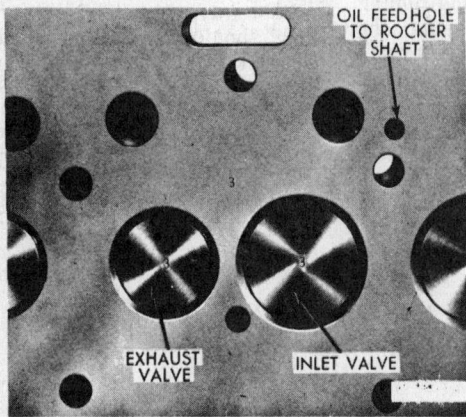

Valve numbering (© Chrysler Corp.)

Removing auxiliary drive gear
(© Chrysler Corp.)

Removing auxiliary drive shaft (© Chrysler Corp.)

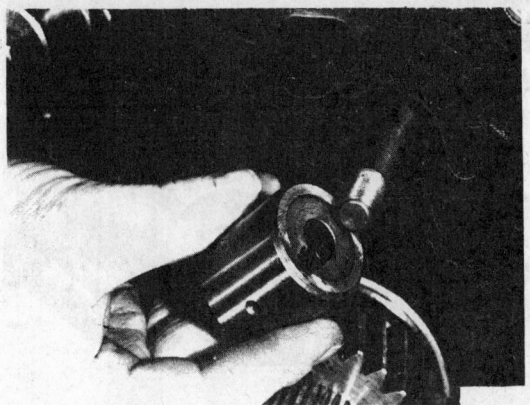

Removing idler gear hub (© Chrysler Corp.)

or auxiliary drive shaft can be removed, the timing gear case must be removed.

Camshaft Removal
(with Gear Case and Gears Removed)

1. Remove the rocker arm cover.
2. Remove the six rocker arm support stud nuts and flat washers. The rocker arm assembly complete with lubrication tube and "O" ring can now be removed as an assembly.
3. Lift the push rods out of the engine. Do this carefully since it is possible to drop No. 2, 3, 10 and 11 push rods down into the crankcase.
4. Disconnect the fuel lines, remove the pump attaching screws and remove the fuel pump from the engine.
5. Remove the alternator.
6. Remove push rod inspection covers.
7. Lift tappets to full travel and hold in full position with clothes pins, carefully remove camshaft and thrust washer.

Auxiliary Drive Shaft Removal
(Gear Case and Gears Removed)

Disconnect the fuel injector lines at the injectors and at the injector pump. Unbolt the mounting clamp from the cylinder head and remove the six injector lines as an assembly.

Injector lines must be handled carefully to avoid bending or damaging. Cover or tape fuel openings on the injection pump to prevent dirt from entering the pump.

1. Disconnect the accelerator and stop cable, remove the three injection pump attaching stud nuts and lift out the injection pump.
2. Remove the three attaching stud nuts and lift the injection pump adapter from the engine.
3. Disconnect the coupling and lines and remove the vacuum pump or air compressor.
4. Loosen the coupling clamp bolt, and remove the coupling from rear end of auxiliary drive shaft.
5. Remove the woodruff key from the rear end of the auxiliary shaft.
6. Pull out the auxiliary drive shaft and the two half moon shaped thrust washers.

If a suitable puller is available the injector and engine oil pump vertical drive shaft assembly can be removed from the top. If tool is not available the following steps are necessary:

Remove the dipstick and dipstick tube.

Remove the attaching screws and remove the oil pump assembly.

Disconnect the oil delivery housing from the crankcase by removing the two attaching bolts.

Loosen the locknut, and unscrew the oil pump locating set screw. This screw is located outside the cylinder block. Now remove the oil pump and oil delivery housing assembly.

The two parts of this assembly are connected by a short pipe which is a push fit in each component.

Inspect the short pipe for wear.

Tap out the injector pump and engine oil pump vertical drive shaft assembly from inside the crankcase.

Remove the seal from the rear end of the auxiliary drive shaft bore in the crankcase. This can be done by prying behind the lip extension of the seal retainer.

The engine oil pump and injector pump vertical drive shaft can be removed with a suitable puller, if the engine oil pan and oil pump are not to be removed.

Injector Pump Vertical Drive Removal

Use a suitable puller with fine jaws to remove the bearing, being careful not to damage the bronze gear.

Remove the four recessed screws, support the gear carefully, and press the shaft out of the gear.

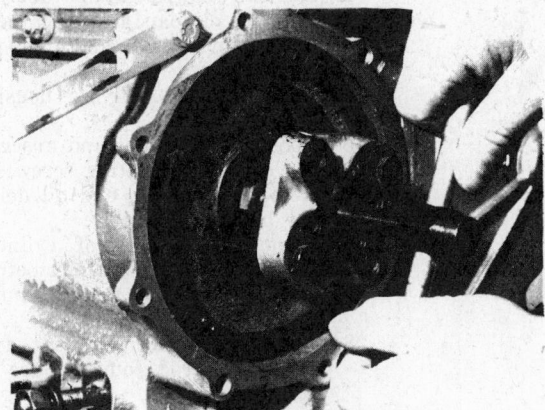

Removing camshaft gear
(© Chrysler Corp.)

Removing idler gear (© Chrysler Corp.)

Removing auxiliary drive gear (© Chrysler Corp.)

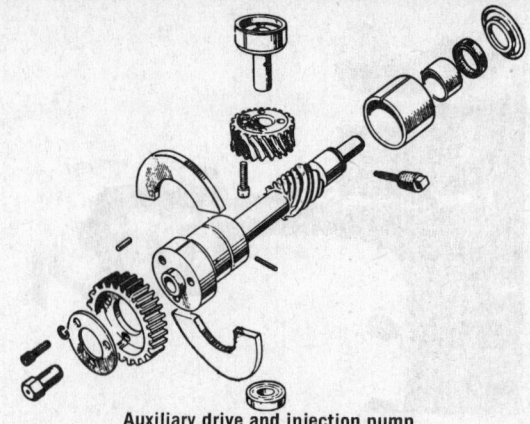

Auxiliary drive and injection pump vertical drive parts
(© Chrysler Corp.)

To reassemble, press the bronze gear onto the injection pump vertical drive shaft with holes in gear and flange aligned, and with counterbored side away from flange and secure it with the retaining screws. Then press a new bearing on the shaft. Use a new gear if necessary.

Insert the vertical shaft, gear and bearing assembly in the cylinder block, and push it down into position.

On a few engines, one or more steel shim .002″ thick is fitted beneath the bearing. Fit the shim (shims) into the block first and then slide complete vertical drive assembly into block on top of shim (shims). Always replace shim (or shims) into block from which they came. *Do not* fit shims into a block which did not have them in the first place.

The vertical shaft and gear must be installed before installing the horizontal positioned auxiliary drive shaft.

Auxiliary Drive Shaft Removal

Place the two thrust washer halves in the groove on the auxiliary drive shaft immediately behind the timing gear mounting flange on the auxiliary drive shaft, and with the chamfer to the rear. Then enter the shaft into the cylinder block bore so the two thrust washer halves straddle the small locating dowel in the recess at the front of the cylinder block. Do this carefully so as not to damage the bronze gear or the worm gear on the auxiliary shaft.

With the key in place in front end of crankshaft, place crankshaft gear on shaft with timing mark on gear toward front of engine, and using a suitable driver, drive gear into place on shaft.

Turn crankshaft until number one and number six pistons are at top dead center. The timing mark on flywheel can be used to accurately determine this position. Also, the line on front face of crankshaft is vertical in respect to engine when number

one and number six are at top dead center.

Install Camshaft and Tappets

Examine camshaft for wear or pitting on cam lobes. If condition is not satisfactory, a new camshaft must be installed.

Oil camshaft lightly, and carefully slide it into place in cylinder block.

Do not drive on front end of camshaft for any reason, since this will dislodge the camshaft plug in the rear of the cylinder block, and an oil leak will result.

Do not install camshaft gear until timing gear case has been installed. The timing case must be installed first.

Place camshaft thrust washer in place, chamfered side in, and so small dowel in cylinder block is entered in hole provided in thrust washer.

Idler Gears

Place the two idler gear hubs on

Vertical drive parts
(© Chrysler Corp.)

their mounting studs at front of cylinder block, so ends with largest chamfer are toward cylinder block and their hollow dowels are aligned with holes provided. Tap hubs into place with a soft hammer.

The hollow dowels in idler gear hubs also serve as oil feed passages for idler gears.

The two idler gears are different. The timing marks are on camshaft idler gear. The auxiliary drive idler gear is plain.

Place camshaft idler gear on its hub, engaging it with crankshaft gear so the two adjacent teeth with timing marks on idler gear straddle the one tooth on crankshaft gear which has a timing mark.

Secure camshaft idler gear with retaining washer and self-locking nut. Tighten nut to 45 to 50 foot-pounds.

Install auxiliary drive idler gear in exactly the same way, except that there are no timing marks to be concerned with on the auxiliary drive idler gear.

Pistons and Rings

Piston and Rod Removal

1. Remove cylinder head assembly as outlined earlier in the engine section.
2. Remove oil dipstick and tube.
3. Remove oil pan attaching bolts and remove oil pan and pump assembly.
4. Disconnect oil delivery housing from crankcase.
5. Loosen the locknut and unscrew the oil pump locating screw.
6. Remove oil pump and oil delivery housing assembly.
7. Remove top ridge of cylinder bores with a ridge reamer before removing pistons from cylinder block.
8. Turn crankshaft so that two connecting rod journals are at bottom center.
9. Remove connecting rod caps and inserts, and remove rod and pis-

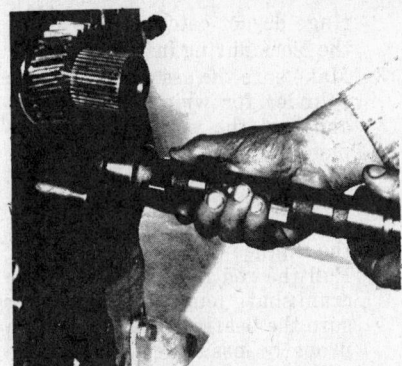

Installing camshaft (© Chrysler Corp.)

Installing camshaft thrust washer
(© Chrysler Corp.)

Toroidal cavity in piston note markings
(© Chrysler Corp.)

ton assemblies from cylinder bores.

10. Keep related parts of each assembly together for convenience when matching for reassembly.

Dissassemble Piston and Rod Assemblies

Remove five piston rings.

Remove the piston pin retaining rings from both ends of pin bore in piston.

Make sure the ends of the pin bores are clean, and remove any burrs around the retainer grooves.

Warm the piston up to about 120°F. This can be done by immersing the pistons in clean liquid of that temperature for a short period of time. Do not use a torch.

After warming the piston to 120°F., it should be possible to remove pin by pushing it out of bore.

If the piston pin bushings are worn, they can be replaced.

Replacing Piston Pin Bushings

Using a suitable pressing tool, press the pin bushing out of the connecting rod.

Piston pins and bushings are supplied in standard size only. Standard pin diameter is 1.375".

Position a new bushing so the oil hole will align with the oil hole at the top of the connecting rod, and press the bushing into the connecting rod bore.

Ream the bushing to provide pin clearance of .0007" to .0017".

Connecting rod alignment should be checked on a reliable fixture. The piston pin should be parallel and in the same plane with the bearing bore centerline within .001".

Cleaning and Inspecting Pistons

Use a ring groove cleaning tool to clean the ring grooves, being careful not to gouge or damage the groove surfaces.

A power wire brush should never be used to clean pistons.

Examine the piston carefully for cracks or other defects.

If a piston ring groove has been damaged by a broken ring, or if ring grooves have been worn excessively wide, due to high mileage, or if piston skirt is scored, the piston should be replaced.

When a new piston is being installed, stamp it for the cylinder position number in which it is to be installed.

All pistons in any one engine must be of same size.

Fitting New Piston Rings

New rings should be checked in the cylinder bore for correct ring gap.

To check ring gap in worn cylinder liners the ring should be positioned at the bottom of the bore, below the ring travel area, where minimum wear has occurred.

With the ring positioned squarely in the cylinder, check the gap with a feeler gauge.

The ring can be positioned squarely in the cylinder bore by using a piston to push it down into the bore.

Ring gap at the top compression ring should be .015" to .019". The second and third compression rings, and also the upper and lower oil rings should have a gap of .011" to .016".

New rings are available in standard size for installation in new liners and worn liners which have not been rebored. They are also available in .030" oversize for use *only* in rebored liners.

New rings must be thoroughly washed before installation.

Assembling Piston and Connecting Rod Assembly

Heat piston to 120°F. before installing piston pin.

Coat the pin with light engine oil.

Position the piston and connecting rod so the word "front" on the piston will be toward the front of the engine when the stamped numbers on the connecting rod are toward the side of the engine opposite the camshaft.

Alignment of timing marks on drive gears
(© Chrysler Corp.)

Connecting rod numbering (© Chrysler Corp.)

Push the piston pin through the piston pin hole and rod bushing, and center the pin between the two retainer ring grooves in the piston.

Make sure the piston is centered between the grooves, and install both retaining rings.

Assure that both pin retaining rings are perfectly seated in their grooves.

When installing the piston pin, *do not* have one of the retaining rings in place, and drive the pin against it to permit installation of the second retaining ring.

This practice can result in damage to the groove or ring, and result in the ring eventually coming out of the groove, permitting the pin to seriously damage the cylinder liner.

Install the two oil rings.

Note that the 2nd and 3rd compression rings are tapered, and are marked "T" for top side.

In current production, the second and third compression rings are internally stepped. These rings must be assembled to the piston so that the step faces the crown of the piston.

The number one (top) compression ring is chrome plated, and can be installed either side up. Install this ring last.

Space the piston ring gaps equally around the piston, and so they are not in line with one another.

Connecting rod bearing upper and lower inserts are both alike. Also note that the parting faces of the rod and cap are serrated to ensure accurate positioning of the cap, and under no circumstances should they be filed.

Connecting rod bolts should be carefully inspected for possible damage, and should be replaced if they are not in excellent condition.

The original self-locking nuts must be replaced with new ones, since they will have lost some of their prevailing torque.

When replacing connecting rod bolts or nuts, use only the correct factory supplied replacement parts.

Connecting rod bearings are available in standard size, and in .010", .020", and .030" undersizes.

Place upper and lower connecting rod bearing inserts in position in the rod bore.

Make sure the piston and rod assembly and the cylinder bore are perfectly clean, and oil them with engine oil.

Piston and Rod Installation

1. Position the crankshaft so the throw for the rod and piston being installed is at bottom center.
2. Install the piston and rod assembly in the cylinder. Use an installing tool, and make sure the

Installing piston (© Chrysler Corp.)

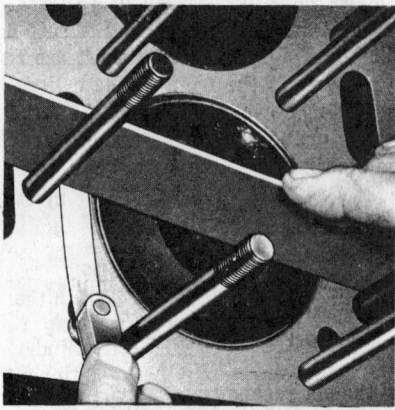
Measuring cylinder liner height (© Chrysler Corp.)

Measuring piston height (© Chrysler Corp.)

Installing connecting rod cap (© Chrysler Corp.)

rings do not catch on the top of the block during installation.

3. Make sure the assembly is in the cylinder for which it is marked and that the word *front* on the piston is toward the front of the engine, and that the numbered side of the connecting rod is opposite from the camshaft side of the engine.
4. Pull the rod into position at the crankshaft journal, and making sure the bearing inserts are both properly positioned, install the connecting rod cap and bearing in place.
5. Install the two connecting rod bolts with new self-locking nuts. Connecting rod bolt heads have two flats. The largest flat must be positioned toward the connecting rod. Tighten to 65 to 70 foot-pounds.

With a properly fitted piston and rod assembly, the piston should protrude above the top of the cylinder block no less than .003, and no more than .0095" when at top dead center. This can be checked by placing a straightedge across the top of the liner and with a feeler gauge, measure the liner protrusion above the top surface of the cylinder block and the gap between the piston top and the straight edge (the smallest height must be to the piston).

The difference between the two dimensions is the height of the piston above the block. If the piston height is excessive, the piston must be removed from the engine and the top machined to bring the height within the specified limits. Should the piston height be too low at top dead center, it must be replaced.

Piston protrusion above the top of the block is most important and should always be checked at TDC. Service pistons should not require machining if they are stamped with the letter "L" on the crown.

Cylinder Liners

Servicing Cylinder Liners

The Dry type cylinder liners are pressed into cylinder block with a .003" to .005" interference fit. Pressing force required is approximately ten tons.

Liners must be bored and honed after they have been pressed into the cylinder block. For a standard piston, the finished honed inside diameter of the liner must be between 3.877" and 3.8785".

If liners are worn or scored, they can be rebored to .030" oversize. Finished inside diameter must be between 3.907" and 3.9085".

All pistons in any one engine must be of the same size. *Liners should be carefully inspected for cracks or other damage before making a decision whether to rebore or replace a liner.*

During reboring, it is most important that true alignment of the bores relative to the crankshaft centerline be maintained.

If cylinder liners are not acceptable for reboring to .030″ oversize, they must be pressed out, and new liners pressed into the cylinder block.

The first step when installing new liners is to remove all cylinder head attaching studs from the top of the cylinder block.

With cylinder block placed in a suitable press, cylinder liners must be pressed out through top of cylinder block, using the proper tool to adapt the press ram to the bottom of the cylinder liner.

Make sure the outside of liners are clean and well oiled before pressing them into the cylinder block.

When pressing in liners, the press load should be released several times during the first inch to allow the liner to centralize itself in the cylinder block parent bore. *Cylinder liners can be removed and installed with engine in truck. Also chilling liners with dry ice may reduce pressing effort required.*

The latest type of cylinder liner (Engine No. 8039047 and later) protrudes .030″ to .035″ above the block face. This liner should be bored and honed to size in the same manner as the earlier recessed-block type liner. *The earlier-type liner must not be pulled up above the top surface of the block.*

After they are pressed in, the liners should be finish bored to 3.877″ to 3.878″.

When work on the cylinder liners has been completed, reinstall cylinder head attaching studs, replacing any which have been weakened by corrosion, or damaged.

Crankshaft and Bearings

Crankshaft Removal

1. Remove oil dipstick and tube.
2. Remove oil pan attaching bolts and remove oil pan and sump assembly.
3. Disconnect oil delivery housing from crankcase.

Location of crankshaft thrust washers
(© Chrysler Corp.)

4. Loosen the locknut and unscrew the oil pump locating screw.
5. Remove oil pump and oil delivery housing assembly.
6. Remove two vertical bolts which attach lower half of rear main bearing oil seal retainer to cylinder block.
7. Remove upper and lower halves of seal retainer.
8. Remove attaching screws from front and rear main bearing bridge pieces and lift both bridge pieces out of cylinder block.
9. Bend down tabs on main bearing bolt locks and remove main bearing bolts.
10. Remove center (no. 4) main bearing cap along with the two lower half crankshaft thrust washers.
11. Remove six remaining main bearing caps and lift out the crankshaft. NOTE: Identification marks on main bearing caps are positioned away from camshaft side.

Servicing Crankshaft and Bearings

Bearing clearance at the seven main bearings is .0025″ to .0045″. Specified clearance at connecting rod bearings is .0015″ to .003″.

Main and connecting rod bearings are both available in standard, .010″, .020″, and .030″ undersize.

Crankshaft end clearance is specified .002″ to .014″, and is controlled by four semi-circular thrust washers, two on each side of the center (No. 4) main bearing. These thrust washers are available in .0075″ oversize to overcome end thrust wear. Lower

thrust washers have locating tabs, while upper thrust washers do not.

.0075″ oversize thrust washers, can be used on both sides of No. 4 main bearing to reduce end play by .015″ when required. If two standard thickness thrust washers are used on one side of No. 4 main bearing, and two oversize washers are used on the other side, crankshaft end play will be reduced by .0075″.

Care must be taken to avoid placing one standard and one oversize thrust washer on the same side of No. 4 main bearing.

Specified end play at connecting rod bearings is .0095″ to .0135″.

Under normal circumstances, by the time the main bearings and thrust washers require replacing, the crankshaft will need to be removed for regrinding. However, if for any reason, one or more of bearings or thrust washers have to be removed for inspection or replacement, this can be done without removing crankshaft from engine.

Main Bearings

To remove a main bearing without removing the crankshaft, take off the cap of the bearing involved, and loosen the remaining main bearing bolts one or two turns.

Before No. 1 or No. 7 main bearing caps can be removed, it is necessary to remove the main bearing bridge piece at the end involved.

To remove front main bearing bridge piece, it is necessary to remove two recessed screws attaching it to cylinder block, as well as the two lower timing case attaching screws.

When removing the rear main bearing bridge, it is necessary to remove the two recessed screws attaching it to the cylinder block, and the two lower rear main bearing oil seal housing attaching screws.

To remove the top half of a main bearing, using a suitable piece of wood or tool, apply pressure to the end of the bearing opposite from the locating tab, and rotate bearing around crankshaft.

Locating tabs are on camshaft side

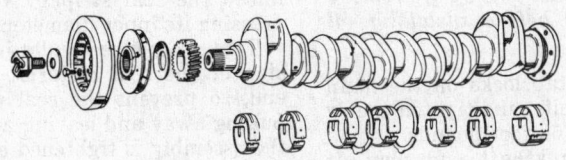

View of crankshaft and bearings
(© Chrysler Corp.)

Removing main bearing upper half
(© Chrysler Corp.)

Aligning bridge piece
(© Chrysler Corp.)

of engine.

When installing a new bearing shell start the upper half, plain end first, and rotate the shell around the crankshaft and into position.

Crankshaft Thrust Washers

To replace crankshaft thrust washers with the crankshaft in place, remove the center (No. 4) main bearing cap, and remove the two bottom half thrust washers from the recesses in the cap.

The two top half thrust washers can now be removed by sliding them around from one side, using a piece of wood or similar tool, and rotating them until they can be removed.

After replacing individual main bearings or thrust washers, make sure all main bearing bolts which have been loosened are tightened to 145 to 150 foot-pounds and secured with new lock tabs.

Main bearing cap bolts now have a steel shim between cap and lock tab to prevent the bolt head from biting into the lock tab.

Crankshaft Installation

Before the crankshaft and bearings are installed, all oil passages in the cylinder block and crankshaft must be thoroughly clean. Inspect the main bearing bolts, and replace as necessary.

CAUTION: In no case should bolts other than the specified main bearing bolts be used, since they are of special heat treated high grade steel.

Make sure the bearing shells and the main bearing bores in the block and caps are free of all foreign matter, and place the main bearing upper halves in position in the cylinder block.

No. 4 main bearing is wider than the others. Also, upper main bearings have large slotted oil holes, which are not found in the lower halves.

1. Oil the upper bearings lightly, and place the crankshaft in position.
2. Lightly lubricate the two upper

thrust washers (no locating tabs) and slide them into places in the recesses provided on each side of No. 4 main bearing in the cylinder block.

3. Place lower main bearings in bearing caps. Place the two lower thrust washers on the No. 4 main bearing cap so their locating tabs are positioned in the slots provided in the cap.
4. The main bearing caps are numbered as to their positions, No. 1 being at the front of the engine, and must be installed accordingly.
5. Install caps with bearings so bearing locating tabs are on the same side of the engine as tabs on upper halves.
6. When caps are properly installed, their stamped serial numbers will be in line on the same side of the engine. Check main bearing clearances, using shim stock or plasti-gage. Clearance at all seven main bearings should be .0025" to .0045".

When using plasti-gage to measure main bearing clearances with engine in truck, support the crankshaft to take up clearance between the upper bearing and the crankshaft journal. This can be done by snugging bearing caps of the adjacent bearings with a strip of .005" to .015" cardboard between lower bearing and journal. When doing this, avoid unnecessary strain on crankshaft and bearing, or false reading may be obtained. Do not rotate crankshaft while plasti-gage is installed. Be sure to remove cardboard before installing oil pan.

7. Use new tab locks on the main bearing bolts, and tighten evenly to 145 to 150 foot-pounds.
8. Check crankshaft end play. It should be .006"–.014". Use oversize thrust washers to correct excessive end play.

9. Bend lock tabs to secure the main bearing bolts. Check crankshaft for free rotation.

Installing Main Bearing Bridge Pieces

Make sure front and rear main bearing bridge pieces are clean, coat attaching surfaces with Gasket Sealer, and place them in position. Place new end seals in grooves provided.

Start recessed attaching screws, and using a straight-edge to align the bridge pieces so they are flush with front and rear faces of cylinder block, tighten screws securely.

Crankshaft Rear Oil Seal

The crankshaft rear oil seal is composed of two rubber cored woven asbestos strips which are inserted in the upper and lower halves of the seal retainer.

The seal surface around the rear of the crankshaft has a shallow spiral oil return groove recessed to a depth of .004" to .008".

The seal stripes come cut to length, and they must never be shortened.

Place one half of seal retainer in a smooth jawed vise with seal groove facing upward.

Imbed about one inch of each end of a seal strip into ends of groove so each end projects .010" to .020" beyond retainer parting face. Allowing middle portion of seal strip to bulge out of groove will enable you to do this.

With thumb or fingers, press the remainder of strip into groove, working from center. Then use any convenient smooth round bar to further imbed the seal strip by rolling and pressing its inner diameter.

This procedure results in compression of the seal material near each end, to prevent the seal ends from pulling away and leaving a gap when the assembly is tightened around the crankshaft.

Install the remaining seal strip in the other half of the seal retainer, using the same methods to position

Removing rear main bearing oil seal
(© Chrysler Corp.)

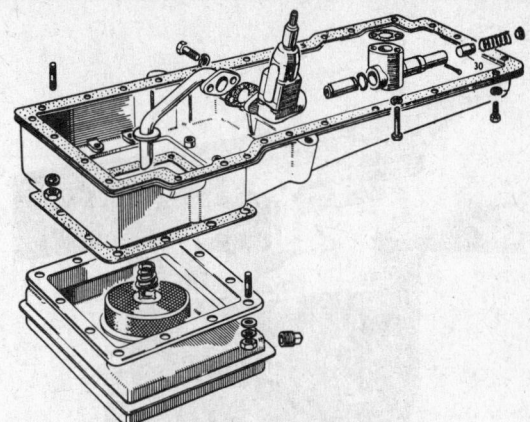

Engine oil pan and sump assembly
(© Chrysler Corp.)

and seat the seal material.

Make sure gasket faces are clean and apply Gasket Sealer, to both sides of new upper and lower retainer gaskets. Also, apply a light coat of sealer to the parting faces between upper and lower seal retainers.

Spread a film of graphited grease over the inside diameter of the upper and lower seals.

Assemble seal retainer halves around crankshaft, drawing the halves together squarely with two bolts.

Make certain upper and lower halves pull completely together, with no ragged edges of the seal ends caught between the parting faces, and are aligned.

Rotate the seal and retainer assembly around crankshaft to make sure it turns freely on the shaft. (This can be done if clutch housing has been removed).

With upper and lower gaskets in place, bolt the seal and retainer assembly to the rear of the engine.

With the retainer to cylinder block attaching bolts pulled up snugly, loosen the two bolts which hold the upper and lower retainer halves together. Make sure the retainer to cylinder block attaching bolts are all tight (Do not back them off). Then retighten the two bolts which attach the upper retainer half to the lower retainer bolt.

Align the index hole in the flywheel with the index hole in the crankshaft flange and install the flywheel. Tighten the bolts to 75 foot-pounds.

Engine Lubrication

Oil Pan Removal

1. Remove dipstick and dipstick tube.
2. Remove oil pan attaching screws and remove oil pan and sump assembly.

Oil Pump Removal

Disconnect oil delivery housing from crankcase by removing two attaching bolts. Loosen the locknut and unscrew the oil pump locating screw. This screw is located outside cylinder block below fuel injection pump.

Remove oil pump and oil delivery housing assembly. The oil pump and oil delivery housing are connected by a short pipe which is a push fit in each component.

Oil Pump and Oil Delivery Housing

The oil pump and oil delivery housing are both a push fit on the connecting pipe. Pull them both loose from the pipe and remove the two "O" rings from their internal grooves in the two assemblies and discard them. Whenever these seals are disturbed, they must be replaced. Inspect the connecting pipe for wear at "O" ring seal surfaces.

Servicing Relief Valve

Remove cotter key, spring retainer, spring and plunger from relief valve bore in oil delivery housing. *Do not* stretch or distort spring.

Wash housing and relief valve parts thoroughly, and replice any worn or damaged parts.

Lubricate bore in delivery housing, and install plunger with small end outward.

Install relief valve spring and retainer, and secure them in place with cotter key.

Servicing Oil Pump

Remove two attaching screws, and remove pump intake tube.

Remove four pump cover to pump body attaching screws and remove cover.

Remove pump shaft and inner rotor and outer rotor.

Thoroughly clean all parts and inspect rotors for cracks and scores.

Install inner and outer rotor in pump body, making sure chamfered edge of outer rotor enters pump body first.

Check clearance between maxi-

Removing engine oil pump
(© Chrysler Corp.)

Checking end clearance of rotors
(© Chrysler Corp.)

Checking clearance between rotors
(© Chrysler Corp.)

mum diameter of inner rotor and minimum diameter of outer rotor at all points, using a feeler gauge. This clearance between inner and outer rotor should be between .004″ and .008″.

The clearance between outer rotor and body must fall between .005″ and .010″. A clearance of .002″ to .005″ must be maintained between top of rotor and end plate.

If pump rotors or body show excessive scoring or other damage, or if the above mentioned clearances are exceeded, the oil pump assembly must be replaced.

The oil pump parts are not available individually. The pump must be replaced as an assembly.

Thoroughly lubricate interior surfaces of pump and rotors. Make sure outer rotor is positioned in pump body chamfered end first, and install pump cover, securing it with four attaching screws.

Clean pump intake tube, and attach it to oil pump, using a new gasket. Tighten the two bolts firmly.

Install new "O" rings in the recesses in the pipe bores in the oil pump and oil delivery housing. Lubricate the "O" rings and connector pipe with engine oil, and place one end of the pipe in the oil pump body and the other end in the oil delivery housing. Be careful not to dislodge the "O" rings from their grooves.

Place oil pump and delivery housing assembly in position in the engine crankcase, while engaging the splines of oil pump shaft with splines in the pump vertical drive shaft.

Install oil pump retaining screw through side of the engine so it engages the oil pump to hold it in posi-

tion. *Do not* tighten screw at this time.

Place a new gasket between the oil delivery housing and engine, and attach housing with two bolts.

Tighten oil pump retaining screw and lock it in place with locknut.

Oil Pan, Sump Well and Screen

Place a new gasket on the oil pan, with Gasket Sealer, on both sides and install the pan on the engine, tightening the bolts firmly. Avoid tightening the bolts too tightly, since this can distort the oil pan sealing surface.

Clean the pump well and screen thoroughly, and using a new gasket and Sealer, attach the sump well to engine pan. Make sure oil pickup tube is properly positioned in the screen as sump well and screen assembly are placed in position.

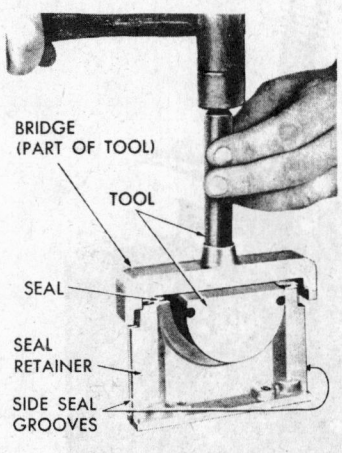

BRIDGE
(PART OF TOOL)

TOOL

SEAL

SEAL
RETAINER

SIDE SEAL
GROOVES

Installing Rear Main Bearing Lower Oil Seal

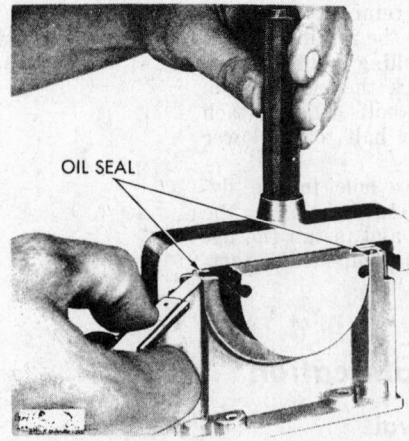

OIL SEAL

Trimming Rear Main Bearing Lower Oil Seal

Trouble and Symptoms	Probable Causes Check causes given in sequence shown.
Noisy	9, 10, 11, 12, 13, 14, 15, 17, 20, 23, 25, 26, 27, 28, 29, 30
Low turbocharger speed	1, 3, 4, 7, 8, 13, 15, 16, 20, 23, 25, 26, 27, 29, 30, 34
Low turbocharger pressure	1, 3, 5, 7, 14, 11, 12, 13, 15, 16, 23
Cracked turbocharger housings or flanges	1, 3, 5, 7, 30, 11, 12, 13, 15, 16, 19, 20, 23, 26, 27, 28, 31, 32, 33

Engine lacks power
   Black smoke
   Blue smoke
	
	1, 2, 3, 5, 16, 24, 7, 19, 15
	24, 2, 7, 4, 22, 34

Lubrication oil consumption excessive	19, 21, 22, 3, 1, 5, 18, 24, 25, 34
Lubrication oil leakage	21, 19, 18, 22, 24, 25, 34
Insufficient lubrication	19, 21, 17, 31

Oil in Air Inlet System
   Before turbocharger
   After turbocharger
	
	1, 2, 3, 4, 5, 7, 18, 19, 34
	1, 2, 3, 4, 7, 5, 22, 24, 25, 28, 34

Oil in Exhaust System
   Before turbocharger
   After turbocharger
	
	12
	22, 24, 25

Unbalance of rotating assembly	25, 23, 9, 3, 11, 12, 13, 18, 19, 20
Drag or bind in rotating assembly	9, 26, 27, 28, 30, 25, 18, 19, 20, 17, 13, 22, 29, 11, 12, 31, 32, 33, 34
Damaged turbine wheel	12, 13, 15, 25, 27, 28, 30, 32, 33
Damaged compressor impeller	6, 9, 11, 13, 25, 26, 27, 28, 30, 33
Worn bearings or journals	17, 18, 19, 20, 21, 26, 27, 28, 30, 31, 32, 33

Probable Causes

1. Air inlet snorkel cap or screen restricted.
2. Oil "pull over" from oil bath type air cleaner.
3. Dirty air cleaner.
4. Oil bath air cleaner oil viscosity too low.
5. Oil bath air cleaner oil viscosity too high.
6. Dry type air cleaner element missing, leaking, not sealing correctly.
7. Collapsed or restricted air inlet line before turbocharger.
8. Restricted air inlet line turbocharger to engine.
9. Loose, chafed or open air inlet line from air cleaner to turbocharger.
10. Loose, chafed or open air inlet line from turbocharger to inlet manifold.
11. Foreign object in inlet system air cleaner to turbocharger.
12. Foreign object in exhaust system (from engine, check engine).
13. Turbocharger flanges, clamps or bolts loose.
14. Inlet manifold cracked or flange gaskets loose or missing.
15. Exhaust manifold cracked or flange gaskets loose, blown or missing.
16. Restricted exhaust system.
17. Oil lag (delay in oil supply to turbocharger at start up).
18. Lubricating oil contaminated with dirt or other material.
19. Improper type lubricating oil used.
20. Insufficient lubrication.
21. Restricted or leaking oil feed line.
22. Restricted oil drain line.
23. Nozzle ring damaged or restricted.
24. Turbocharger seal leakage.
25. Worn shaft bearings.
26. Excessive dirt build-up behind compressor impeller.
27. Excessive carbon build-up behind turbine wheel.
28. Unbalance of rotating assembly.
29. Bearing seizure.
30. Improper mounting, brackets or supports.
31. Too fast acceleration at initial start (oil lag).
32. Too little warm-up time.
33. Insufficient time allowed for temperature stabilization before engine shut-down.
34. Air compressor passing oil.

Volvo Diesel Engines

GENERAL ENGINE SPECIFICATIONS

ENGINE SERIES	NO. CYL. DISPL.	HORSEPOWER RPM (NET)	TORQUE RPM (FT LB)	BORE & STROKE (in.)	COMPRESSION RATIO	OIL PRESSURE (psi)	WEIGHT (LB)
D 60 A	6-334	120 @ 2800	260 @ 1500	3.875 x 4.724	17:1	43-71	1,190
TD 60 A	6-334	180 @ 2800	376 @ 1900	3.875 x 4.724	16:1	43-71	1,235
TD 70 E	6-409	205 @ 2400	492 @ 1400	4.124 x 5.118	14.5:1	40-70	1,433

TUNE-UP SPECIFICATIONS

ENGINE SERIES	INJECTION TIMING (deg.)	INJECTOR OPENING PRESSURE (Psi)	LOW IDLE (RPM)	HIGH IDLE (RPM)	MAXIMUM FULL LOAD SPEED (RPM)	COMPRESSION PRESSURE (psi)	VALVE LASH (in.) INTAKE	VALVE LASH (in.) EXHAUST
60	22B	2840①	625	3000	2800	355②	.016 cold	.018 cold
70	20B	2840①	500	2600	2650	325	.016 cold	.022 cold

① figure given is for in service injectors; for new or rebuilt injectors with new thrust spring: 2958 psi
② with turbocharging: 327 psi

CRANKSHAFT AND CONNECTING ROD SPECIFICATIONS
(inches)

ENGINE SERIES	MAIN BRG. JOURNAL DIA.	MAIN BRG. OIL CLEARANCE	SHAFT END PLAY	CONN. ROD JOURNAL DIA.	ROD BRG. OIL CLEARANCE	ROD BRG. SIDE CLEARANCE
60	2.9980-2.9985	.0026-.0048	.0028-.0106	2.4980-2.4985	.0022-.0040	.006-.014
70	3.2494-3.2500	.0026-.00468	.0027-.0105	2.7495-2.7500	.0027-.0040	.006-.014

VALVE SPECIFICATIONS

ENGINE SERIES	SEAT ANGLE (deg.)	FACE ANGLE (deg.)	SPRING TEST PRESSURE (lb @ in.)	SPRING INSTALLED HEIGHT (in.)	STEM-TO-GUIDE CLEARANCE INTAKE (in.)	STEM-TO-GUIDE CLEARANCE EXHAUST (in.)	STEM DIAMETER INTAKE (in.)	STEM DIAMETER EXHAUST (in.)
60	30①	29.5②	158 @ 1.53	2.24	.0010-.0023	.0020-.0035	.3133-.3140	.3124-.3130
70	30①	29.5②	119 @ 1.417③	1.913④	.0013-.0027	.0025-.0040	.4327-.4331	.4311-.4318

① exhaust: 45   ② exhaust: 44.5   ③ inner: 50 @ 1.26   ④ inner: 1.75

PISTON AND RING SPECIFICATIONS
(inches)

ENGINE SERIES	PISTON-BORE CLEARANCE	RING GAP 1st	RING GAP 2nd	RING GAP 3rd	RING GAP OIL	RING SIDE CLEARANCE 1st	RING SIDE CLEARANCE 2nd	RING SIDE CLEARANCE 3rd	RING SIDE CLEARANCE oil
60	.0043-.0051	.0088-.0236	.0088-.0236	——	.0100-.0236	.0043-.0055	.0028-.0039	——	.0012-.0024
70	.00511	.012-.024	.012-.024	.012-.024	.012-.024	.0035-.0048	.0030-.0047	.0030-.0047	.0016-.0029

TORQUE SPECIFICATIONS
(ft. lb.)

ENGINE SERIES	CYLINDER HEAD BOLTS	CONN. ROD BRG. BOLTS	MAIN BRG. BOLTS	CRANKSHAFT DAMPER BOLT	FLYWHEEL-CRANKSHAFT	INJECTION PUMP FLANGE	INJECTORS
60	123	116	101	188	130	18*	14
70	137 long 100 short	115	100	188	130	50	15

* element fastener

Fuel System

Injector Removal and Installation

Series 60 and 70

1. Remove the rocker arm covers.
2. Remove the fuel delivery lines and cap them.
3. Remove the injector holddown bolts and pull out the injectors. If injectors are difficult to remove, use tool Volvo #2683 and 2991 or equivalent. Clean the copper sleeve contact surfaces.
4. Installation is the reverse of removal. Torque bolts to 14 ftlb.

Injector Removal
(courtesy Volvo of America Corp.)

Injector Sleeve Removal and Installation

Series 60 and 70

NOTE: the cab member or gear lever carrier may have to be removed, depending on which sleeve is to be pulled.

1. Remove the injector as described above.
2. Using extractor, Volvo #2128 or its equivalent, pull the sleeve from the head.
3. Remove the O-ring from the head.
4. Clean the O-ring groove and the sealing surface between the head and sleeve. Install a new O-ring.
5. Manually turn the engine until

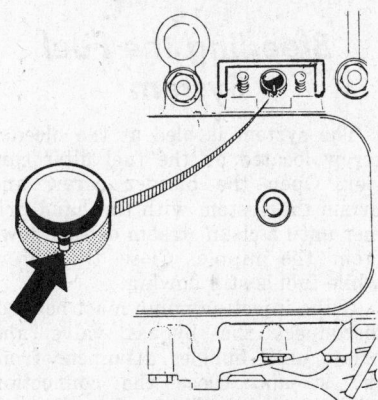

Injector Sleeve Removal
(courtesy Volvo of America Corp.)

the piston corresponding to the sleeve being worked on, is at bottom dead center. This can be determined by removing the inspection cover on the flywheel.

6. Install the sleeve with Volvo tool #6008, or its equivalent, as follows:
   a. Unscrew the tool widening pin and place the sleeve on the tool.
   b. Back off the tool spindle nut.
   c. Screw in the widening pin.
   d. Coat the outside of the sleeve with clean engine oil and push the tool and sleeve into the head. Check that the index mark (recess) for the sleeve points straight upwards.
7. Install the injector holddown nuts and force the widening tool downward until the sleeve bottoms in the head.
8. Hold the widening tool securely and tighten the large nut. The widening pin is pressed through the lower end of the sleeve.
9. Tighten the nut until the tool spindle is free of the sleeve. Pull up on the spindle and remove the rest of the tool from the sleeve.
10. Install the injector and cab member or lever carrier.

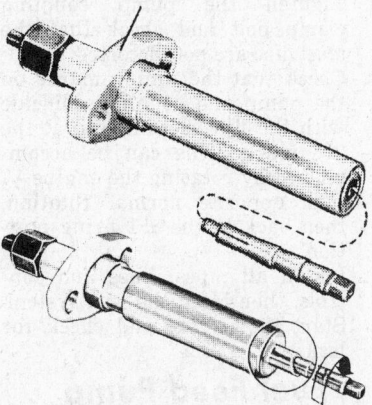

Tool #668 (courtesy Volvo of America Corp.)

Fuel Injection Pump Removal, Installation and Adjustment

Series 60

1. Clean all related parts.
2. Disconnect the fuel delivery pipes at the injectors and pump. Cap all openings.
3. Disconnect all remaining lines, pipes and controls from the pump. Cap all openings.
4. Remove the inspection plate from the flywheel housing and manually turn the engine to #1 TDC, on the compression stroke.
5. Remove the bolts securing the flange and pump drive. Remove

the intermediate section and bolt.
6. Remove the speed sensor.
7. Remove the pump retaining bolts from the timing gear case and lift off the pump.
8. Before installing the pump, make sure that there is 1 pt. of oil in the unit.
9. Turn the engine manually until the timing gradations show 22-23° (D60) or 21-22° (TD60) opposite the pointer.
10. Adjust the pump camshaft until the mark on the end of the shaft inclines about 20° obliquely towards the cylinder block.
11. Apply chassis grease to the sealing ring at the front of the pump.
12. Install the pump on the timing gear case and tighten the bolts. *NOTE: the pump must be positioned so that the stud bolts are opposite the oval holes in the pump.*
13. Install the speed sensor.
14. Connect all pipes, except the delivery pipes, to the pump.
15. Install the gear wheel clamp, lock washer and bolts on the pump drive at the front of the pump. The bolts must be

Adjustment With Wilbar Tube
(courtesy Volvo of America Corp.)

5/16UNCx45mm. Tighten the bolts snugly. DO NOT OVERTIGHTEN.

16. Install a Wilbär tube, or its equivalent, to the delivery pipe for #1 cylinder. Bleed the fuel system.
17. Remove the delivery pipe for #1 cylinder and bleed the discharge valve and Wilbär tube by turning the pump shaft back and forth a few times.
18. Move the throttle control lever to the full throttle position and hold it there with a spring or similar device.
19. Pull the stop arm back as far as possible and then return it to the operating position. This will set the control rod to the full load position.
NOTE: if the stop arm is not pulled all the way back, then returned to the operating position, the control rod will stop at the

cold-start position, giving a faulty adjustment.

20. Turn the pump shaft in the opposite direction of normal rotation and check that the fuel level in the Wilbär tube moves. Open the valve on the Wilbär tube and allow the level to drop to the middle of the sight glass.

21. Turn the pump shaft in the normal direction in small increments until the fuel in the level tube starts to rise. The point at which the fuel just starts to rise is the injection point for #1 cylinder.

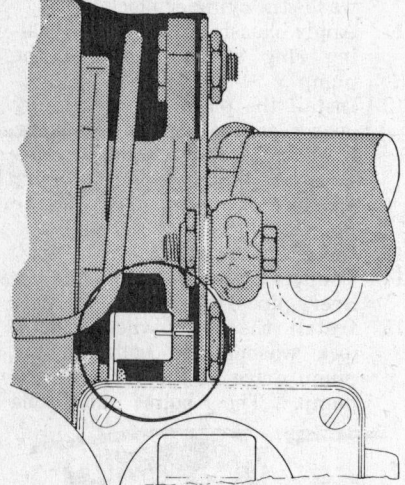

Setting the Pump Coupling—Series 70
(courtesy Volvo of America Corp.)

22. Tighten the bolts at the front of the pump between the flange and the pump gear.

23. Manually turn the engine to check that the flywheel markings coincide with the pump adjustments in steps 20 and 21.

24. When all adjustments are correct, tighten the pump drive bolts and install the flywheel housing inspection plate.

25. Remove the Wilbär tube, connect the delivery pipes and install the controls. Install the cover on the timing gear case.

26. Start the engine and check for leaks.

Series 70

1. Clean all related parts.
2. Disconnect all pipes, lines and controls from the pump. Cap all openings.
3. Manually turn the engine to #1 cylinder TDC compression. Check the rocker arms and timing marks.
4. Mark the position of the pump coupling nuts for exact reassembly.
5. Remove the bolts from the pump coupling. Separate the rear flange from the intermediate section of the pump coupling. *NOTE: the position of the nuts*

must not be changed during removal and installation.

6. Unbolt and remove the pump. Be careful to avoid damage to the steel discs.
7. Before installing the pump, make sure that the unit is correctly filled with oil.
8. While observing the timing marks, manually rotate the engine until the pointer is opposite the 20° mark. Check the mark from a straight-on angle. Viewing from the side can cause an error of several degrees.
9. Loosen the pump coupling clamp bolt and position the pump on the bracket. Push the coupling forward on the shaft.
10. Install the pump coupling rear flange on the shaft. Turn it until the index line on the flange is opposite the index line on the setting plate.
11. Install the intermediate section of the pump coupling on the flange by sliding the coupling on the shaft from the auxiliary drive gear end. Make certain the domed washers are located between the rear flange of the coupling and the steel discs. Tighten the bolts. Make certain that the nuts are in the previously marked positions.
12. Tighten the pump coupling clamp bolt and check that the steel disc are not distorted.
13. Check that the timing marks on the pump and coupling coincide with the flywheel indexed at the 20° mark. This can be accomplished by rotating the engine ½ turn opposite normal rotation, then back to the #1 firing position.
14. Install all pipes, lines and controls, then bleed the fuel system. Start the engine and check for leaks.

Fuel Feed Pump

Series 60 and 70

1. Clean all related parts.
2. Disconnect the fuel lines from the pump.
3. Unbolt and remove the feed pump from the injection pump.
4. Clean and inspect the pump-to-pump mating surfaces and install a new gasket.
5. Install the feed pump.
6. If oil has run out of the injection pump, it should be refilled through the injection pump stroke adjusting hole.

Speed Adjustments

Low Idle

1. Check that the accelerator linkage is operating properly, and that there is no play.

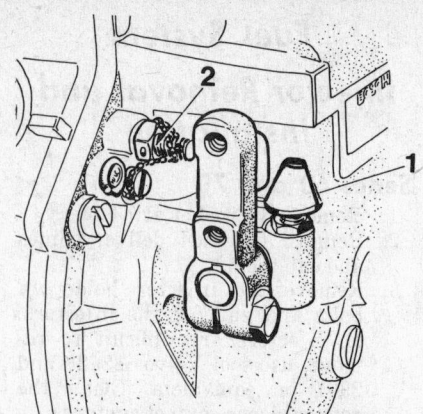

Speed Setting—Series 70
(courtesy Volvo of America Corp.)

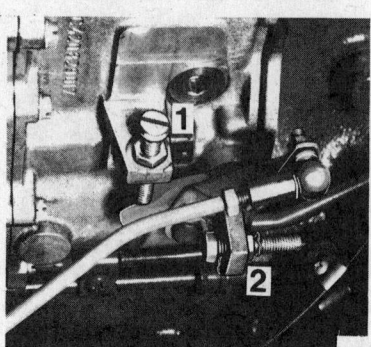

Speed Setting—Series 60
(courtesy Volvo of America Corp.)

2. Run the engine to normal operating temperature.
3. Turn the throttle stop screw to obtain 475-525rpm for the 70 series and 600-650rpm for the 60 series.

High Idle

1. Run the engine to normal operating temperature.
2. Break the lead seal on the speed stop.
3. Run the engine at maximum speed and check that the speed arm on the injection pump touches the maximum speed stop.
4. Adjust the stop to obtain 3000-3100 rpm for the 70 series and 2550-2650rpm for the 60 series. Replace the lead seal.

Bleeding the Fuel System

The system is bled at the bleeder screw located on the fuel filter carrier. Open the bleeder screw and prime the system with the hand primer until a clean stream of fuel flows from the nipple. Close the screw while fuel is still flowing.

If the injection pump must be bled, disconnect the bypass valve and prime until bubbles disappear from the stream. Close the connection while fuel is still flowing. Do not bleed at the pressure equalizer.

Engine Repairs
Engine Removal

Series 60

1. Disconnect the battery and drain the coolant.
2. Remove the radiator.
3. Remove the fan and hub.
4. Disconnect all coolant hoses.
5. Remove the upper cab member.
6. Disconnect the wiring from the oil pressure sender, low pressure indicator and back-up light switch.
7. Disconnect the fuel and brake pipes from the clutch housing.
8. Remove the power steering reservoir bracket.
9. Remove the oil dipstick.
10. Disconnect the shift control lever from the transmission. A puller may be necessary.
11. Disconnect the brake line at the shift lever housing.
12. Disconnect the hand throttle cable and the fuel injection control rod.
13. Disconnect the wiring from the alternator, handbrake switch, temperature sender, tachometer, and lay the wires aside.
14. Remove the alternator cable harness clamp.
15. Disconnect the exhaust pipe.
16. Disconnect the #6 delivery pipe at the injection pump.

17. Disconnect the power steering hoses.
18. Disconnect the exhaust pipe clamp at the clutch housing.
19. Disconnect the ground strap from the clutch housing.
20. Remove the clutch cylinder from the transmission.
21. Disconnect the speedometer cable at the lower end.
22. Disconnect the driveshaft from the transmission.
23. Disconnect the starter cables.
24. Install a lifting device on the engine.
25. Support the engine weight and disconnect the mounts.
26. Remove the engine. The right side cab attachment may have to be removed to provide clearance.
27. Installation is the reverse of removal.
28. Start and test run the engine.

Series 70

1. Disconnect the battery ground and the steering column clamp. Drain the coolant.
2. Disconnect the driveshaft.
3. Disconnect the speedometer cable at the lower end.
4. Remove the high and low ratio control cable.
5. Disconnect the hoses from the exhaust pressure regulator solenoid and the exhaust pressure regulator.
6. Remove the blocking valve hose

from the T-joint in the frame.
7. Disconnect the brake system supply line.
8. Disconnect all piping and wiring at the transmission and disconnect the transmission rear mounts.
9. Remove the six lower bell housing nuts.
10. Remove the exhaust pipe bracket from the bellhousing.
11. Support the transmission with a jack and remove the six upper bellhousing bolts.
12. Remove the transmission.
13. Remove the cab crossmember retaining bolts on the left side.
14. Disconnect the speedometer cable clamp at the crossmember and the fuel line at the bellhousing.
15. Remove the cold start chain and pipe.
16. Disconnect the accelerator arm control and the stop control.
17. Remove the cover plate at the shift lever and remove the shift control.
18. Unbolt and remove the gearshift lever housing.
19. Disconnect the exhaust pipe.
20. Disconnect the wiring from the temperature and tachometer sensors at the alternator.
21. Disconnect the power steering hoses at the left side of the frame.
22. Disconnect all coolant hoses and the air hose at the compressor.
23. Remove the upper exhaust pipe bracket.
24. Remove the turbocharger inlet hose, air cleaner hose and air cleaner.
25. Remove the anti-skid control unit and the hose from the anti-freeze unit. Remove the hose between the solenoid valve and the exhaust pressure governor.
26. Disconnect the wiring from the solenoid valve.
27. Remove the cab crossmember and the air intake.
28. Disconnect the wiring from the starter and the oil pressure sending unit. Disconnect the oil pressure hose at the T-joint.
29. Install a lifting tool and hoist, take up the engine weight and disconnect the mounts.
30. Lift out the engine.
31. Installation is the reverse of removal. Start the engine and test run.

Cylinder Head
Removal and
Installation

Series 60

1. Drain the coolant.
2. Remove the exhaust pipe bracket from the transmission.

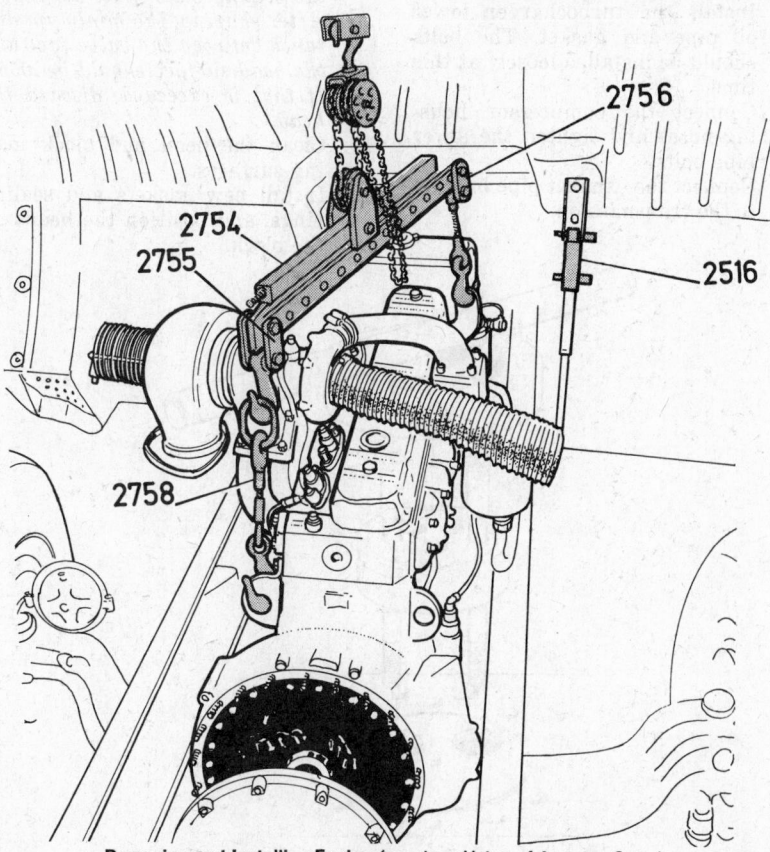

Removing and Installing Engine (courtesy Volvo of America Corp.)

3. Remove the lower oil pipe from the turbocharger on units so equipped.
4. Remove the exhaust elbow from the turbocharger and then lift off the intermediate sleeve on engines without an exhaust brake.
5. Remove the upper oil pipe from the turbocharger.
6. Remove the hoses from the turbocharger compressor housing. Remove the turbocharger from the engine.
7. Remove the exhaust manifold and gaskets.
8. Remove the coolant hose connection between the heads.
9. Remove the injectors.
10. Remove the intake manifold.
11. Remove the right rear lifting eye.
12. Remove the rocker arm covers, rocker arm shaft assemblies and pushrods.
13. Remove the head bolts and lift off the heads.
14. Thoroughly clean the block and head mating surfaces.
15. Clean all related parts.
16. Check the head and block surfaces for warpage, nicks, gouges and cracks. Maximum deviation for the cylinder head face is .0012″.
17. If any machining was done, thoroughly clean head and block mating surfaces. New sealing grooves must be milled.

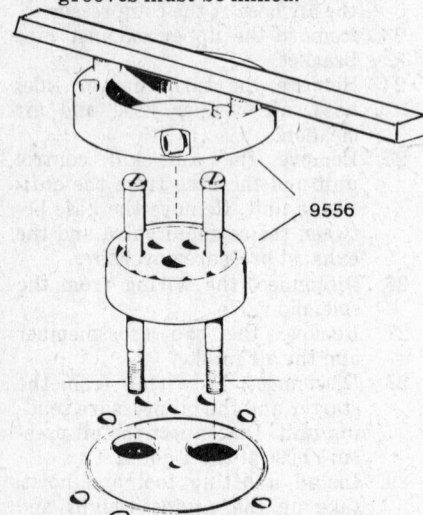

Milling New Sealing Grooves—Series 60 (courtesy Volvo of America Corp.)

18. If removed, install new sealing rings and head gasket.
19. Install the rear head first, then the front head. Torque the head bolts, in the sequence shown, in three stages: 29, 87 and 123 ftlb.
20. Install the pushrods and rocker arm assemblies.
21. Remove the flywheel inspection cover and turn the engine over to #1 TDC compression. Adjust valves 1,2,4,5,7,9. Turn the en-

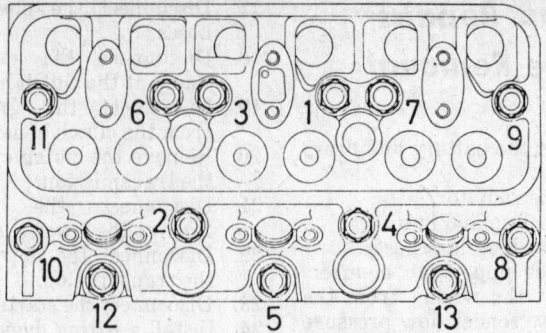

Head Bolt Sequence—Series 60 (courtesy Volvo of America Corp.)

gine in direction of rotation until #6 is at TDC compression (0 on the flywheel). Adjust valves 3,6,8,10,11,12.
22. Install the rocker arm covers and the flywheel inspection cover.
23. Install the rear lift ring and the gearshift carrier.
24. Install the intake manifold.
25. Install the injectors. Torque to 14 ftlb.
26. Install the cylinder head coolant connection.
27. Install the exhaust manifold and gasket.
28. Install the turbocharger and gasket.
29. On vehicles without an exhaust brake, install the spacer sleeve on the exhaust pipe flange.
30. Connect the exhaust pipe.
31. Install the turbocharger upper oil pipe and gasket.
32. Install the turbocharger lower oil pipe and gasket. The bolts should be installed loosely at this time.
33. Connect the compressor housing hoses and tighten the lower pipe bolts.
34. Connect the exhaust pipe bracket to the transmission.

Series 70

1. Drain the cooling system.
2. Remove the rocker arm covers.
3. Remove the cylinder head coolant connectors.
4. Remove the delivery and return lines from the injectors. Cap the lines.
5. Remove the injectors, using tools Volvo #2683 and 2991 if necessary.
6. Remove the intake and exhaust manifolds.
7. Remove the rocker arm assemblies.
8. Unbolt and remove the heads.
9. Remove the head gaskets and rubber seals from the block.
10. Check the head for surface warpage. Maximum deviation is .0012″. *NOTE: if more than .004″ is shaved from the head, the sealing slots must be milled. After shaving, the minimum distance between the valve face and the head surface should be .055″. If that is exceeded, discard the head.*
11. Clean the head and block mating surfaces.
12. Install new gaskets and sealing rings, and position the heads on the block.

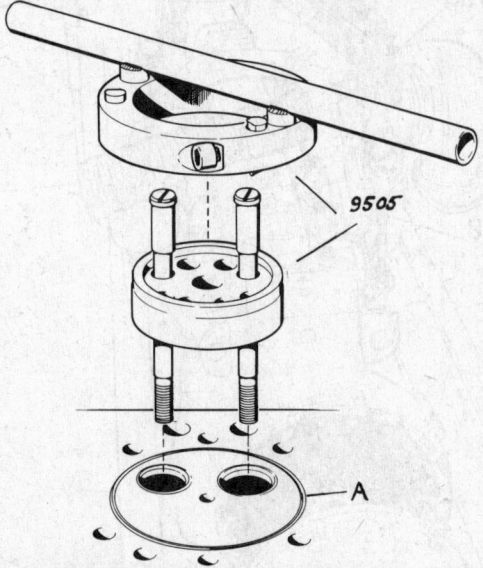

Milling New Sealing Grooves—Series 70 (courtesy Volvo of America Corp.)

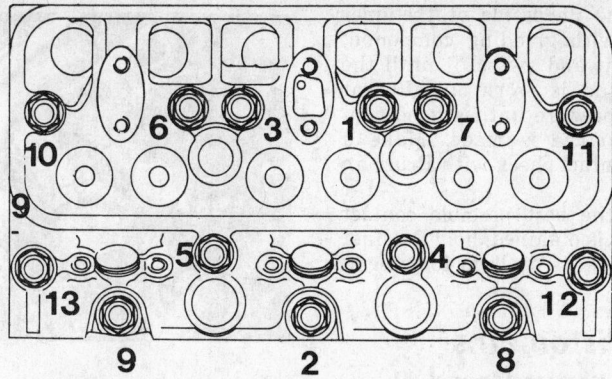

Head Bolt Sequence—Series 70 (courtesy Volvo of America Corp.)

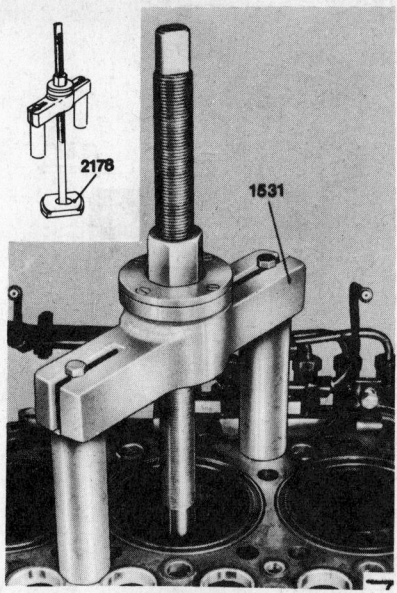

Removing Cylinder Liner
(courtesy Volvo of America Corp.)

13. Coat the head bolts with rust-proofing compound and install them loosely.
14. Install the intake manifold, with the bolts only finger tight.
15. Torque the head bolts, in the sequence shown, in three stages, to 29, 100 and 140 ftlb.
16. Tighten the intake manifold bolts.
17. Install the exhaust manifold and gaskets. Make sure that the sealing ring gaps are 180° to each other. Two of the gaskets are slightly different. These should be located at #3 and 6 cylinders so that the recessed edges point forward. The folded surface of the gaskets face the head.
18. Install the coolant connector pipe.
19. Install the pushrods and rocker arm assemblies.
20. Remove the flywheel inspection cover and turn the engine to #1 TDC compression. Adjust valves 1,2,4,5,7,9. Turn the engine to #6 TDC compression and adjust valves 3,6,8,10,11,12. Clearance is .016″ intake; .022″ exhaust.
21. Install the injectors. Torque the bolts alternately and evenly to 15 ftlb. Install the delivery and return lines.
22. Install the rocker arm covers. Torque to 7.5 ftlb. maximum.

Pistons and Rings Removal and Installation

1. Remove the cylinder head.
2. Remove the oil pan.
3. Turn the crankshaft until the piston to be removed is at BDC.
4. Remove the connecting rod cap.
5. The piston is removed from the top.
6. The rings and piston pin may now be removed.
7. Thoroughly clean the piston and rod, giving special attention to the ring grooves.
8. Install a new piston pin, if neces-

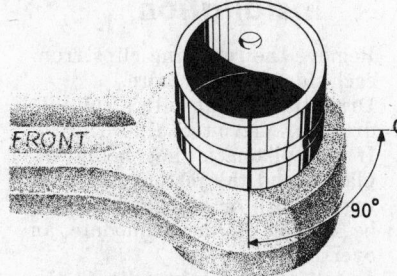

Installing Connecting Rod Bushing
(courtesy Volvo of America Corp.)

sary. Pin-to-hole clearance is .0002″ max.
9. Install new rings. The opening in the oil scraper ring should be placed opposite the oil scraper ring gap.
10. Coat the piston assembly with clean engine oil and install it in the block. The arrow points forward.
11. Make sure the bearings are seated and install the lower cap. Torque it to 113 ftlb. for the series 60 and 115 ftlb for the series 70.

Cylinder Liners Removal and Installation

1. Remove the cylinder head and oil pan.
2. Remove the piston from the liner to be replaced.
3. The liner can be removed with puller Volvo # 1531 and plate #6087 for series 60 or 2178 for series 70, or their equivalent.
4. Remove the sealing rings and thoroughly clean the ring seats and all related surfaces. *NOTE: the upper and lower liner seats should be cleaned with solvent and blown dry. Scraping should never be employed.*
5. Coat the underside of the liner flange with Prussian blue or other marking compound, and insert the liner, without rings, into position.
6. Twist the liner back and forth a few times and remove it. The

marking compound must be evenly distributed around the shoulder. A slight deviation can be corrected with lapping compound. A more serious deviation may be corrected by grinding. Any material removed must be compensated for by steel shims. A special milling tool Volvo #9508, or its equivalent should be used.
7. Check that the stepped edge of the liner is .0094-.0114″ (series 60) or .0114-.0134″ (series 70) above the block face. The difference between liners under the same head must not exceed .0008″. *NOTE: when checking with the sealing rings installed, e.g. before the liner is removed, a pair of clamps such as Volvo #2667 must be installed to hold the linder against the shoulder.*
8. When all measurements have been satisfactorily made, install the lower sealing rings in the block and the upper on the liner. *NOTE: on series 60 engines, the red ring goes at the very bottom. The rings can be coated with a soap solution to prevent twisting.*
9. Install the liners without using force. They should slide in easily.
10. Install a new piston assembly. Install all other parts in reverse order of removal.

Milling and Shimming the Cylinder Liner Seats

Prior to grinding the seat, it should be roughened with grinding paper. This is absolutely necessary if the seat was ground with grinding compound.

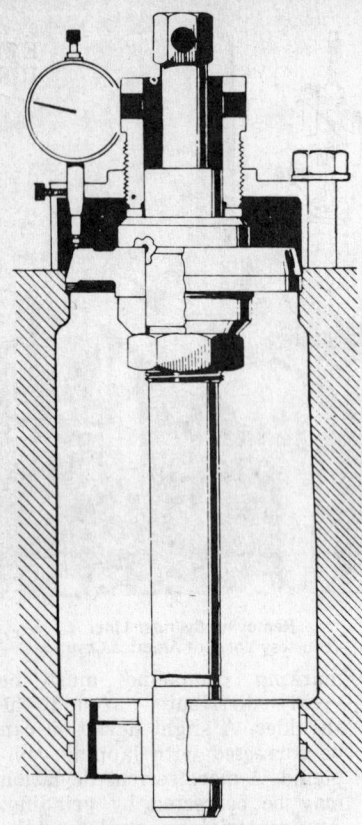

**Reconditioning Cylinder Liner
Seat with Tool 9508**
(courtesy Volvo of America Corp.)

Shims are available in thickness of .008", .012", and .020".

1. Insert the cylinder liner. Correct height above the block should be .0094"-.0114" for the series 60 engines and .0114"-.0134" for the series 70 engines.
2. Figure an additional .0008" for material removed during grinding. From that figure, determine the size and number of shims to be used.
3. Secure a milling tool such as Volvo #9553 for the series 60 engines and 9508 for the series 70 engines.
4. Install the tool in the liner. Use adequate flat washers under the bolt heads. Make sure the feed screw does not press on the tool. Install a dial indicator on the pool.
5. Adjust the feed sleeve to press lightly on the tool, and set the dial indicator at zero.
6. Turn the tool with a 1" socket on a ¾" drive ratchet.
7. Turn the cutter slowly and evenly so that an even feed is maintained.
8. Remove the ratchet and allow the tool to turn several times without feeding to obtain a run-out reading on the dial. Remove the tool when the desired reading is obtained.
9. Check the seat surface and re-check the liner height.

10. Coat the underside of the liner collar with grinding compound, install it and swivel it until the compound is worn out. Repeat this procedure until a good contact surface is made. Make a final contact check with Prussian blue.
11. When the best possible contact surface is obtained, hold the liner and mark its position for assembly.

Piston Pins Removal and Installation

1. Remove the retaining clips from each side of the pin bore.
2. Drive the pin out with a suitable driver or Volvo tool #2071.
3. If the piston is not being replaced, and the piston pin bushing is worn or damaged, it may be reamed to accommodate an oversized pin.
4. Install one retaining clip in the piston.

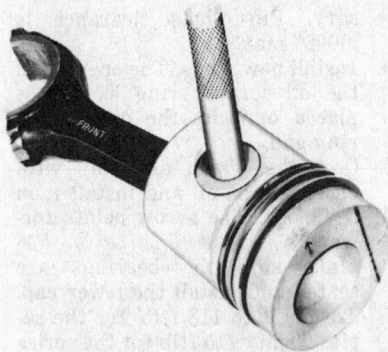

5. Coat the piston pin with clean engine oil.
6. Heat the piston to 212°F and assemble the connecting rod and piston. Insert the piston pin. Tool 2071 may be used to install the pin. The pin should slide in under light pressure. Never force the pin into position.
7. Install the other retaining clip.

Valve Removal, Grinding and Installation

1. Remove the valve stem locks, washers, springs and valves.
2. Grind the valves to give an angle of 29.5° for intakes and 44.5° for exhausts. If, after grinding, the valve face thickness is less than .047" for series 60 intake and exhaust or .060" intake and .039" exhaust on the series 70.
3. Coat the valve face with Prussian blue and check for fit against the seat. If fit is poor, grind the seat to compensate.

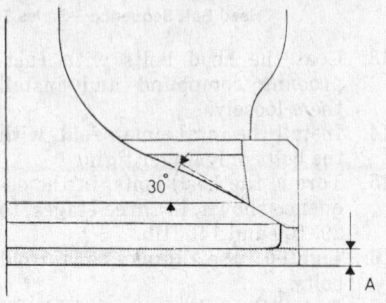

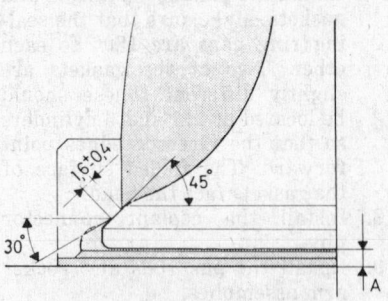

Valve Refacing
(courtesy Volvo of America Corp.)

NOTE: the distance between the valve face and the cylinder head face must not exceed .055".

Replacing the Valve Seats

1. Remove the old valve seat by cracking it with a cold chisel.
2. Thoroughly clean the seat and check for cracks in the head.
3. Measure the valve seat area for a possible oversize fit.
4. Place the new seat in dry ice to a minus 74-94°F.
5. Warm the head with warm water.
6. Drive the seat into position with a brass drift.
7. Machine the new seat to give a valve face-to-head face clearance of .028-.043" for series 60 and .039-.055" for series 70 engines.

Valve Lifter Removal and Installation

1. Remove the rocker arm assemblies and pushrods.

2. Remove the lifter inspection covers from the block.
3. Remove and mark the lifters.
4. Installation is the reverse of removal.

Timing Gears and Cover Removal and Installation

1. Remove all drive belts.
2. Remove the crankshaft pulley and damper, and the fan.
3. Disconnect the battery ground.
4. Remove the radiator.
5. Remove the water pump pipe and hoses.
6. Disconnect the oil pipes from the servo pump.
7. Remove the crankshaft hub center bolts, and using a puller remove the hub.
8. Unbolt and remove the timing gear cover.
9. Manually turn the engine to #1 TDC compression.
10. The timing and drive gears may now be removed. *NOTE: on series 70 engines, the oil pump drive gear has a splash plate in front of it. Pullers will probably be necessary for the removal of most of the gears.*
11. Install all gears making sure that timing marks are aligned. On series 60 engines, the intermediate gear has an axial play of .0020-.0060″.

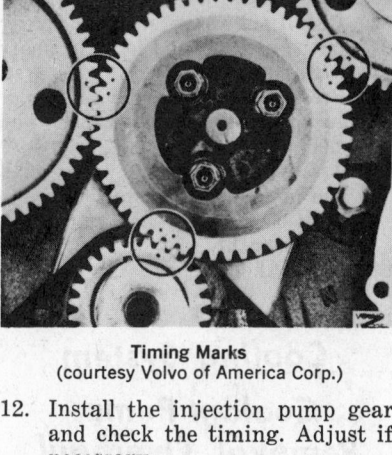

Timing Marks
(courtesy Volvo of America Corp.)

12. Install the injection pump gear and check the timing. Adjust if necessary.
13. Install the oil slinger with the concave side outwards.
14. Thoroughly clean the cover, block and oil pan mating surfaces.
15. Install a new felt ring in the cover.
16. Apply gasket cement to the mating surfaces and install the cover and new gasket.
17. Coat the crankshaft end with moly grease.
18. Heat the crankshaft hub to 212°F, Install a centering tool such as Volvo #2657 on the crankshaft and rapidly knock the hub onto the shaft to .020″ from the gear face.
19. When the hub has cooled, install a washer and tighten the bolt to 188 ftlb.

20. Install the crankshaft pulley and vibration damper. NEVER HAMMER ON THE VIBRATION DAMPER!
21. Finish the job in reverse order of disassembly.

Camshaft Removal and Installation

1. Remove the timing gear cover, rocker arm assemblies, lifters and the camshaft gear and flange.
2. On series 60 engines, remove the narrow plastic grille between the radiator grille and the bumper.
3. Using a puller, carefully remove the camshaft.
4. Maximum out-of-round on the camshaft bearing surfaces is .00275″. Maximum axial runout is .0016″.
   NOTE: it is recommended that the camshaft bearings be replaced only during a complete overhaul.
5. Camshaft installation is the reverse of removal. See timing gear and cover installation.

Crankshaft and Main Bearings Removal and Installation

It is recommended that this job be performed with the engine out.
1. Remove the oil pan, pump, pickup and pressure lines.
2. Remove the timing gear cover.
3. Remove the flywheel.
4. Remove the lower main halves and remove the crankshaft.
5. Carefully check the crankshaft for damage or excessive wear.
6. Maximum permissible out-of-round on the journals is .0030″. Maximum taper is .0020″. When grinding the crankshaft, the fillet radii between the journals and the flange must be .1280-.1378″. Maximum undersize is .050″.
7. Install new bearings in the block and caps. Coat the bearings with clean engine oil.
8. Make a Plastigage check on the bearings and crankshaft. If fit is okay, install the crankshaft and torque the main bolts to 100 ft lb.
9. Check crankshaft end play. Play should be .0027″-.0105″.

Assembly of the remaining parts is the reverse of disassembly. See the appropriate paragraphs in this chapter.

1. Idler gear   2. Camshaft gear   3. Injection pump gear   4. Oil pump idler
5. Crankshaft gear (courtesy Volvo of America Corp.)

Removing Oil Pump Drive Gear
(courtesy Volvo of America Corp.)

Engine Lubrication

Oil Pump
Removal, Overhaul and Installation

1. Remove the oil pan.
2. Unbolt and remove the pump.
3. Remove the circlip and lift off the drive gear.
4. Remove the three bolts and lift off the idler gear.
5. Unbolt and remove the pump housing. If it won't come off readily, two 5/16″ bolts can be used as drivers.
6. Remove the lock ring and press out the shaft and drive gear.
7. Remove the driven gear and press out the journal if it needs replacement.
8. Drive shaft-to-bushing clearance should be .006″. Clearance between the idler gear and journal should be .008″. Gear float clearance in the pump should be .0028″-.0060″. Gear backlash should be .006″-.014″.
9. If the drive gear bushings are replaced, they should be pressed in and reamed to .6305″-.6313″.

10. Press in the driven gear shaft.
11. Install the drive shaft and gear in the bracket and lock the shaft with the circlip.
12. Press the pump drive gear to a gear-to-washer clearance of .008″-.032″.
13. Assemble the pump drive gear and housing and install the idler gear. Check that the pump operates smoothly.
14. Install the pump.

Cooling System

Coolant Pump
Removal, Overhaul and Installation

1. Drain the cooling system, remove the drive belt and disconnect the hoses.
2. Remove the pump.
3. Remove the fan hub attaching bolts.
4. Using a brass drift, drive the pump shaft downwards to free the hub.
5. Place the pump in a holding fixture as shown and using drift Volvo #2268 or equivalent, drive out the impeller, shaft, seal, slinger ring and rear bearings. The slinger rings will be destroyed.
6. Remove the front bearing circlip.
7. Place adapter Volvo #2266 in the pump housing journal and assemble puller Volvo #2265. Remove the pulley and bearing.
8. Drive the bearings from the pulley with a drift.
9. Pack the bearings with EP grease and press them into the pulley using a drift. Install the circlip.
10. Place a holding fixture, Volvo #2269, into the impeller recess. Continue to press on the pulley

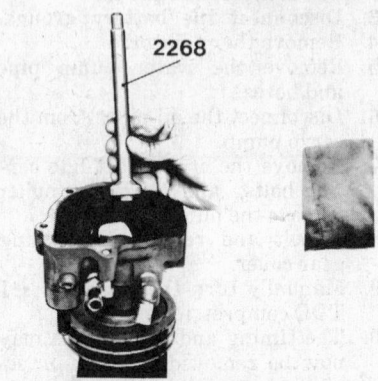

and bearing until they are flush with the impeller.
11. Fill the cavity on the inside of the bearing seat with grease.
12. Press on the rear bearing (housing) using tool #2268, or its equivalent. The sealing washer must be toward the impeller.
13. On series 60 engines: Install the deflector ring with the flange facing away from the bearing and press on the shaft seal.
14. On series 60 engines: install the hub and tighten the bolts. Place the pump hub-side down and oil the impeller shaft. Coat the impeller seal O-ring with soapy water and install it on the impeller. Carefully press the ceramic seal into its seat. Install the shaft through the seal. Press

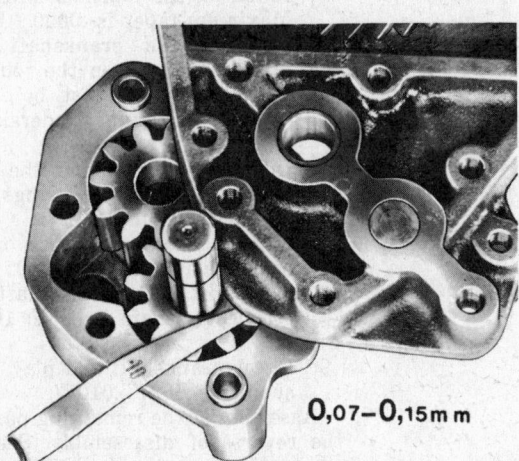

0,07-0,15 m m

Checking End Play (courtesy Volvo of America Corp.)

Checking Backlash (courtesy Volvo of America Corp.)

the shaft and impeller to obtain a clearance between the blades and the housing of .035"-.043".

15. On series 70 engines: Install the pulley and bearing assembly. Install the slinger ring with the flange away from the bearing and press in the shaft seal. Place the pump so that the adapter is against the bearing journal and the pump shaft turns freely. Oil the shaft. Guide the shaft through the seal and press in until there is a clearance of .035"-.043" between the blades and the pump housing.

16. To install the pump on series 60 engines:
    a. clean all mating surfaces.
    b. coat mating surfaces with gasket cement.
    c. position the new gasket and the pump on the block; install the bolts finger tight.
    d. push upwards on the pump with a pry bar.
    e. tighten the bolts and finish all connections.

17. To install the pump on series 70 engines:
    a. clean mating surfaces.
    b. install a new gasket and the pump.
    c. tighten bolts and make all connections.

Measuring Impeller Clearance
(courtesy Volvo of America Corp.)

Turbocharger
AiResearch T-04 B/S-3/1.0 E

TD 60 A Engine

Volvo recommends that the turbocharger be overhauled or replaced every 100,000 miles. Before disassembling unit, clean the outside thoroughly with a non-abrasive solution.

1. Unbolt and separate the compressor housing and turbo housing from the rotor unit. Use of a soft mallet may be necessary for separation.
2. Place the rotor unit in a holding fixture to prevent the turbine wheel from rotating.
3. Remove the compressor wheel

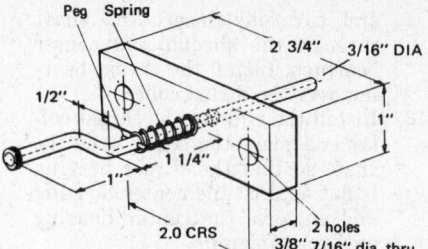

Adaptor for Measuring Radial Bearing Clearance
(courtesy Volvo of America Corp.)

nut with a T-handle to prevent uneven loading of the shaft.

4. Remove the compressor wheel, turbine wheel and shaft from the unit.

CAUTION: The turbine wheel shaft is not locked in the bearing housing and can easily drop out when the compressor wheel is removed.

5. Remove the bolts and lock clamps from the end on the compressor side.

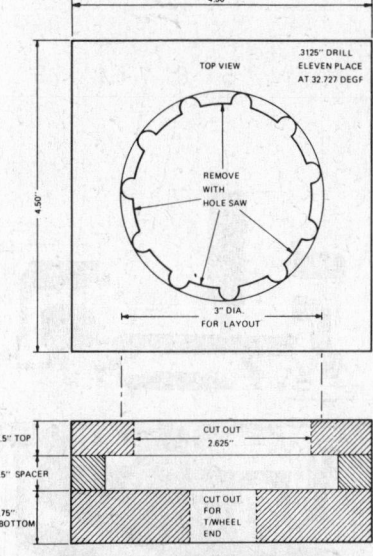

Turbine Wheel Holding Fixture
(courtesy Volvo of America Corp.)

6. Knock off the bearing housing end with a soft mallet.
7. Remove the thrust sleeve and bearing from the bearing housing.
8. Remove the circlips and bearing bushings from the bearing housing. Discard the O-ring.
9. All friction marks, heat marks and other damage should be marked prior to cleaning. All parts should be cleaned in a non-abrasive, decarbonizing solution and scrubbed lightly with a stiff bristle brush. Parts should be dried thoroughly, immediately after cleaning.
10. Before assembling, make sure that no foreign material is present on the parts.
11. Install the bearing bushing inner lock rings.
12. Coat the bushings with clean en-

gine oil and install them along with the outer lockrings.

13. Install the turbine wheel and shaft and place the thrust bearing over the bearing sleeve.
14. Install the thrust sleeve ring and push the thrust sleeve on the shaft so that the sleeve is flush with the bearing housing.
15. Install a new O-ring in the bearing housing groove.
16. Install the thrust spring and align the bearing housing and end bolt holes. Place the bearing housing and end on the shaft and thrust sleeve.

NOTE: the easiest way to fit the bearing housing end is to first install the rings in the end.

17. Install the compressor end bolts and lock washers. Torque to 6-7 ft lb.
18. Install the compressor wheel. The large surface on the lock nut and the leading surface of the compressor wheel must be flush. Apply a little oil to the threads and contact surfaces and torque the nut to 18 in. lb. Use a T-handle to prevent uneven side forces on the shaft. Check axial play (.0012-.0039"). When proper axial play has been reached turn the nut an additional ¼ turn.
19. Place the compressor housing on the rotor unit and install the bolts and lockwashers. Torque the bolts to 8-11 ft lb.
20. Place the turbine housing on the rotor unit. Coat the bolt threads with graphite grease. Install the bolts and lockwashers. Torque the bolts to 8-11 ft lb.
21. Press the turbine wheel in as far as possible and check for free rotation. Press the compressor wheel in and repeat the check.

AiResearch T-0 4B/S-3/1.15 F

TD 70 E Engine

Volvo recommends that the turbocharger be overhauled or replaced every 100,000 miles.

Before disassembling the unit, clean the outside thoroughly with a non-abrasive solution.

1. Remove the compressor and turbine housings from the center housing. A soft mallet may be needed for separation.
2. Place the center housing in a holding fixture to keep the turbine wheel from rotating.
3. Using a T-handle, remove the compressor wheel locknut.
4. Lift the compressor wheel off the shaft and remove the shaft wheel from the center housing.

CAUTION: The turbine wheel shroud is not retained in the center

housing and will fall out when the shaft wheel is removed.

5. Remove the lockplates and bolts from the back plate.
6. Tap the back plate off with a soft mallet.
7. Remove the thrust collar and bearing from the center housing.
8. Remove the bearings and retainers from the center housing, discard the O-ring.
9. Mark all damage prior to cleaning. Clean all parts in a non-abrasive solution and brush with a stiff bristle brush. Dry all parts thoroughly immediately after cleaning.
10. Prior to reassembly, check each part to make sure it is free of foreign particles.
11. Install the inner bearing retainers. Oil the bearings with clean engine oil and install them and the outer retainers.
12. Place the turbine wheel upright and carefully insert the shaft through the shroud and center bearings. Install the thrust bearing over the thrust collar.
13. Install the ring on the thrust collar and place the collar over the shaft so that the thrust bearing is flat against the center housing and engages the center housing anti-rotating pins.
14. Install a new O-ring in the groove in the center housing.
15. Make sure that the thrust plate spring is installed in the back plate. Align the center housing and back plate and place them over the shaft and thrust collar. Back plate is easy to install if the open end of the ring is engaged in the back plate bore first.
16. Install the compressor back plate bolts and lockplates and torque to 75-90 in lb.
17. Install the compressor wheel. The larger face of the lock nut and the front face of the impeller must be smooth and clean. Oil the threads and face of the nut and torque it to 18-20 in lb. Continue to tighten the nut until the shaft protrubence increases by .0055″-.0065″. Tighten the nut with a T-handled wrench to avoid bending the shaft. Axial play should be .001″-.004″.
18. Check for free rotation between the turbine wheel and the shroud. Align the compressor housing and center housing and install the bolts and lockplates. Torque to 100-130 in lb.
19. Align the turbine housing and center housing and coat the bolt threads with graphite grease. Install the bolts and lockplates and torque to 100-130 in lb.
20. Push the rotating assembly as far as possible from the turbine end and check for binding. Repeat the check from the compressor end.

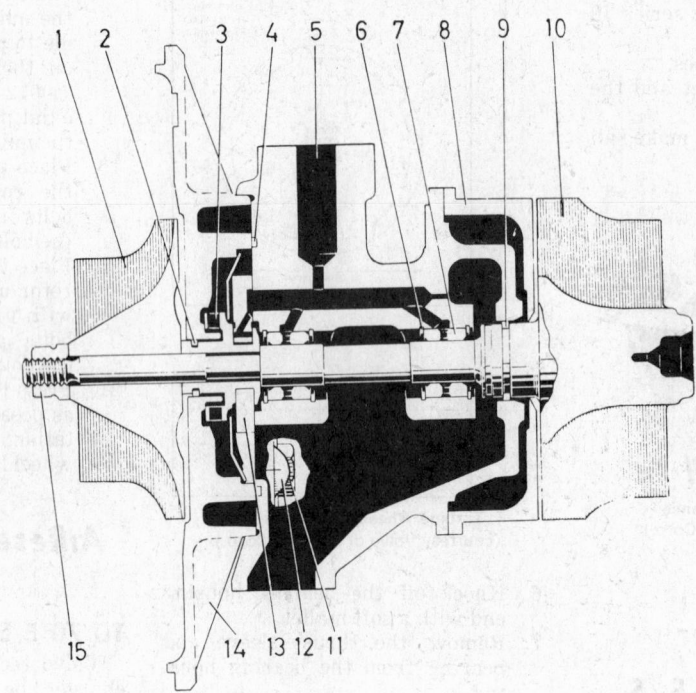

1	Compressor wheel
2	Piston ring seal, compressor side
3	Thrust sleeve
4	O-ring seal
5	Bearing housing
6	Circlip
7	Bearing bushing
8	Protective casing
9	Piston ring seal, turbine side
10	Turbine wheel with shaft
11	Retaining bolt for end
12	Lock washer
13	Axial thrust bearing
14	End
15	Lock nut for compressor wheel

Index

CARTER CARBURETORS

YEAR	MODEL OR TYPE	FLOAT LEVEL (IN.)	FLOAT DROP (IN.)	Pump Travel Setting	CHOKE SETTING Unloader (IN.)	Housing	Idle Screw Turns Open	ON THE TRUCK ADJUSTMENTS Idle Speed (RPM)	Fast Idle Speed (RPM)	Dashpot Plunger Clearance (IN.)
Dodge/Plymouth										
				MODEL BBS						
1971	4658S	¼	—	—	—	—	1-2	550	2600	—
	4659S	¼	—	—	³⁄₁₆	2 R	1-2	550	700	—
	4629S	¼	—	—	—	—	1-2	650	2800	—
	4630S	¼	—	—	³⁄₁₆	2 R	1-2	650	1550	—
	4730S	¼	—	—	—	—	1-2	550	—	—
1972-73	6395S	¼	—	—	.190	—	1-2	800	2000	—
	6396S	¼	—	—	.190	—	1-2	800	1800	—
	6219S	¼	—	—	.190	—	1-2	700	2000	—
	6218S	¼	—	—	.190	—	1-2	700	1800	—
	6220S	¼	—	—	—	—	1-2	700	2800	—
1974	7044S	¼	—	—	³⁄₁₆	—	1-2	800	2000	—
	7045S	¼	—	—	³⁄₁₆	—	1-2	800	1800	—
				MODEL BBD						
1971	4613S	⁵⁄₁₆	—	1.00	¼	2 R	1½	700	1600	—
	4748S	⁵⁄₁₆	—	1.00	¼	2 R	1½	700	1600	—
	4614S	⁵⁄₁₆	—	1.00	¼	2 R	1½	600	1600	—
1972-73	6316SA	¼	—	.280	¼	—	—	750	1700	—
	6317SA	¼	—	.280	¼	—	—	700	1700	—
	6343SA	¼	—	.280	¼	—	—	750	1700	—
	6344SA	¼	—	.280	¼	—	—	700	1700	—
	6221S	¼	—	.310	¼	—	—	750	1600	—
	6222S	¼	—	.310	¼	—	—	750	1900	—
	6363S	¼	—	.310	¼	—	—	750	1600	—
	6364S	¼	—	.310	¼	—	—	750	1700	—
	6365S	¼	—	.340	—	—	—	700	1900	—
	6224S	¼	—	.340	—	—	—	700	1900	—
	6228S	¼	—	.340	—	—	—	600	1600	—
	6368S	¼	—	.310	—	—	—	600, 700 Calif.	1600	—
	6229S	¼	—	.390	—	—	—	600	1600	—

Dodge/Plymouth, Continued

MODEL BBD

YEAR	MODEL OR TYPE	FLOAT LEVEL (IN.)	FLOAT DROP (IN.)	Pump Travel Setting (IN.)	CHOKE SETTING Unloader (IN.)	CHOKE SETTING Housing	ON THE TRUCK ADJUSTMENTS Idle Screw Turns Open	Idle Speed (RPM)	Fast Idle Speed (RPM)	Dashpot Plunger Clearance (IN.)
1974	6610S	¼	—	½	.28	—	—	750	1700	—
	6611S	¼	—	½	.28	—	—	750	1500	—
	8008S	¼	—	½	.28	—	—	750	1700	—
	6613S	¼	—	½	.28	—	—	750	1600	—
	6536S	¼	—	½	.28	—	—	750	1700	—
	6537S	¼	—	½	.28	—	—	750	1500	—
	6585S	¼	—	½	—	—	—	700	1900	—
	6586S	¼	—	½	—	—	—	700	1900	—
1975	8019S	¼	—	½	.28	—	—	800	1500	—
	8020S	¼	—	½	.28	—	—	750	1500	—
	8022S	¼	—	½	.31	—	—	750	1500	—
	8024S	¼	—	½	.28	—	—	750	1500	—
	8025S	¼	—	½	.31	—	—	750	1500	—
	6536S	¼	—	½	.28	—	—	750	1700	—
	6537S	¼	—	½	.31	—	—	750	1500	—
	8013S	¼	—	½	.31	—	—	700	1500	—
	8014S	¼	—	½	.31	—	—	700	1500	—
	6585S	¼	—	½	—	—	—	700	1900	—
	8016S	¼	—	½	.31	—	—	700	1500	—
	8026S	¼	—	½	.31	—	—	700	1500	—
	6586S	¼	—	½	—	—	—	700	1900	—
1976	8081S	¼	—	½	.31	—	—	750	1500	—
	8108S	¼	—	½	.31	—	—	750	1500	—
	8082S	¼	—	½	.28	—	—	750	1500	—
	8085S	¼	—	½	.28	—	—	750	1500	—
	6536S	¼	—	½	.28	—	—	750	1700	—
	6537S	¼	—	½	.28	—	—	750	1500	—
	8013S	¼	—	½	.31	—	—	700	1500	—
	8014S	¼	—	½	.31	—	—	700	1500	—
1977-78	8081S	¼	—	½	.31	—	—	750	1500	—
	8082S	¼	—	½	.28	—	—	750	1500	—
	8085S	¼	—	½	.28	—	—	750	1500	—
	8147S	¼	—	½	.28	—	—	750	1700	—
	8146S	¼	—	½	.31	—	—	750	1500	—
	8113S	¼	—	½	.31	—	—	750	1500	—
	8110S	¼	—	½	.28	—	—	700	1500	—

MODEL TQ

YEAR	MODEL OR TYPE	FLOAT LEVEL (IN.)	FLOAT DROP (IN.)	Pump Travel Setting (IN.)	CHOKE SETTING Unloader (IN.)	CHOKE SETTING Housing	ON THE TRUCK ADJUSTMENTS Idle Screw Turns Open	Idle Speed (RPM)	Fast Idle Speed (RPM)	Dashpot Plunger Clearance (IN.)
1973	6446	1.0	—	31/64	.19	—	—	700	1700	—
	6518	1.0	—	31/64	.19	—	—	700	1700	—
1974	9022S	1.0[1]	—	35/64	.31	—	—	750	1800	—
	9025S	1.0[1]	—	31/64	.31	—	—	700	1700	—
	9017S	1.0[1]	—	15/32	.31	—	—	700	1700	—
	6545S	1.0[1]	—	31/64	.31	—	—	700	1700	—
1975	9034S	29/32	—	½	.31	—	—	700	1700	—
	9035A	29/32	—	½	.31	—	—	700	1700	—
	6545S	29/32	—	½	.31	—	—	700	1700	—
	9036S	29/32	—	½	.31	—	—	700	1700	—
1976	6545S	29/32	—	31/64	.31	—	—	700	1700	—
	9036S	29/32	—	31/64	.31	—	—	700	1700	—
1977-78	6545S	29/32	—	½	.31	—	—	700	1700	—
	9096S	29/32	—	½	.31	—	—	700	1700	—

Note[1]—Brass float, 29/32 with plastic float

Ford

YEAR	MODEL OR TYPE	FLOAT LEVEL (IN.)	FLOAT DROP (IN.)	Pump Travel Setting (IN.)	CHOKE SETTING Unloader (IN.)	Housing	Idle Screw Turns Open	ON THE TRUCK ADJUSTMENTS Idle Speed (RPM)	Fast Idle Speed (RPM)	Dashpot Plunger Clearance (IN.)
					MODEL YF					
1971	D1TF-VA	3/8	1¼	—	.28	Index	—	—	—	Solenoid
	D1TF-XA	3/8	1¼	—	.28	Index	—	—	—	—
	D1TF-YA	3/8	1¼	—	.28	Index	—	—	—	7/64
	D1TF-ABA	7/32	1¼	—	.28	1 L	—	—	—	7/64
	D1TF-ZA	7/32	1¼	—	.28	1 L	—	—	—	Solenoid
	D1TF-AAA	7/32	1¼	—	.28	Index	—	—	—	Solenoid
1972	D2TF-MA	3/8	1¼	—	.28	Index	—	—	—	Solenoid
	D2TF-LA	3/8	1¼	—	.28	1 L	—	—	—	Solenoid
	D2TF-FA	3/8	1¼	—	.28	1 L	—	—	—	Solenoid
	D2TF-VA	3/8	1¼	—	.28	1 L	—	—	—	Solenoid
	D2TF-SA	3/8	1¼	—	.28	1 L	—	—	—	7/64
1973	D3TF-TA	3/8	1¼	—	.28	Index	—	—	—	Solenoid
	D3TF-KA	3/8	1¼	—	.28	1 L	—	—	—	Solenoid
	D3TF-AAA	3/8	1¼	—	.28	1 L	—	—	—	Solenoid
	D3TF-ABA	3/8	1¼	—	.28	Index	—	—	—	Solenoid
	D3TF-LA	3/8	1¼	—	.28	1 R	—	—	—	.1
1974	D4TE-ZA	3/8	1¼	—	.28	1 L	—	—	—	Solenoid
	D4TE-YC	3/8	1¼	—	.28	Index	—	—	—	.1
1975	D5TE-ALA	3/8	1¼	—	.28	Index	—	—	—	—
	D5TE-APA	3/8	1¼	—	.28	Index	—	—	—	—
	D5TE-ADA	3/8	1¼	—	.28	Index	—	—	—	—
	D5TE-AKA	3/8	1¼	—	.28	Index	—	—	—	—
	D5TE-AMA	3/8	1¼	—	.28	1 R	—	—	—	—
	D5TE-AGA	3/8	1¼	—	.28	1 R	—	—	—	—
	D5TE-ANA	3/8	1¼	—	.28	1 R	—	—	—	—
1976-77	D5TE-ADA	3/8	1¼	—	.28	Index	—	—	—	—
	D5TE-ADB	23/32	1¼	—	.28	Index	—	—	—	—
	D5TE-AKA	3/8	1¼	—	.28	Index	—	—	—	—
	D5TE-AKB	23/32	1¼	—	.28	Index	—	—	—	—
	D5UE-EA	3/8	1¼	—	.28	Index	—	—	—	—
	D5TE-BSB	23/32	1¼	—	.28	Index	—	—	—	—
	D5TE-CBA	3/8	1¼	—	.28	Index	—	—	—	—
	D5UE-FA	3/8	1¼	—	.28	Index	—	—	—	—
	D5TE-APB	23/32	1¼	—	.28	Index	—	—	—	—
	D5TE-CAB	23/32	1¼	—	.28	Index	—	—	—	—
	D6TE-HA	23/32	1¼	—	.28	Index	—	—	—	—
	D5UE-AAB	23/32	1¼	—	.28	Index	—	—	—	—
	D5UE-HB	23/32	1¼	—	.28	Index	—	—	—	—
	D6TF-DA	23/32	1¼	--	.28	Index	—	—	—	—
	D5UE-GB	23/32	1¼	—	.28	Index	—	—	—	—
	D5TE-AGA	3/8	1¼	—	.28	1 R	—	—	—	—
	D5TE-AGB	3/8	1¼	—	.28	1 R	—	—	—	—
	D5TE-AFA	3/8	1¼	—	—	—	—	—	—	—
	D5TE-AFB	3/8	1¼	—	—	—	—	—	—	—
	D5TE-AJA	3/8	1¼	—	—	—	—	—	—	—
	D5TE-AJB	3/8	1¼	—	—	—	—	—	—	—

Jeep

YEAR	MODEL OR TYPE	FLOAT LEVEL (IN.)	FLOAT DROP (IN.)	Pump Travel Setting (IN.)	CHOKE SETTING Unloader (IN.)	Housing	Idle Screw Turns Open	ON THE TRUCK ADJUSTMENTS Idle Speed (RPM)	Fast Idle Speed (RPM)	Dashpot Plunger Clearance (IN.)
					MODEL BBD					
1977	8107	¼	—	—	.28	2 R	—	—	1700	—
1978	8107	¼	—	.44	.28	2 R	—	—	1700	—

Carter Carburetors

YEAR	MODEL OR TYPE	FLOAT LEVEL (IN.)	FLOAT DROP (IN.)	Pump Travel Setting (IN.)	CHOKE SETTING Unloader (IN.)	Housing	Idle Screw Turns Open	ON THE TRUCK ADJUSTMENTS Idle Speed (RPM)	Fast Idle Speed (RPM)	Dashpot Plunger Clearance (IN.)
Jeep					**MODEL YF**					
1972	6287S	$^{29}/_{64}$	$1\frac{1}{4}$	—	$^{19}/_{64}$	Index	—	—	1600	$^{3}/_{32}$
	6288S	$^{29}/_{64}$	$1\frac{1}{4}$	—	$^{19}/_{64}$	Index	—	—	1600	$^{3}/_{32}$
1973	6299S	$^{29}/_{64}$	$1\frac{1}{4}$	—	$^{9}/_{32}$	1 R	—	—	1600	$^{3}/_{32}$
	6300S	$^{29}/_{64}$	$1\frac{1}{4}$	—	$^{9}/_{32}$	1 R	—	—	1600	$^{3}/_{32}$
	6401S	$^{29}/_{64}$	$1\frac{1}{4}$	—	$^{9}/_{32}$	1 R	—	—	1600	—
1973	64013	$^{29}/_{64}$	$1\frac{1}{4}$	—	$^{9}/_{32}$	1 R	—	—	1600	$^{3}/_{32}$
1974	6431	.48	$1\frac{3}{8}$	—	.28	1 R	—	—	1600	.095
	6511	.48	$1\frac{3}{8}$	—	.28	1 R	—	—	1600	.095
	7001	.48	$1\frac{3}{8}$	—	.28	1 R	—	—	1600	—
	7029	.48	$1\frac{3}{8}$	—	.28	1 R	—	—	1600	.095
1975	7043	.48	$1\frac{3}{8}$	—	.28	1 R	—	—	1600	
	7041	.48	$1\frac{3}{8}$	—	.28	1 R	—	—	1600	.075
	7040	.48	$1\frac{3}{8}$	—	.28	1 R	—	—	1600	.075
1976	7088	.48	$1\frac{3}{8}$	—	.28	1 R	—	—	1600	—
	7084	.48	$1\frac{3}{8}$	—	.28	1 R	—	—	1600	.075
	7109	.48	$1\frac{3}{8}$	—	.28	2 R	—	—	1600	.075
	7083	.48	$1\frac{3}{8}$	—	.28	2 R	—	—	1600	.075
	7085	.48	$1\frac{3}{8}$	—	.28	1 R	—	—	1600	.075
1977	7154	.48	$1\frac{3}{8}$	—	.28	1 R	—	—	1600	—
	7151	.48	$1\frac{3}{8}$	—	.28	1 R	—	—	1600	—
	7153	.48	$1\frac{3}{8}$	—	.28	Index	—	—	1600	—
	7110	.48	$1\frac{3}{8}$	—	.28	2 R	—	—	1600	—
	7111	.48	$1\frac{3}{8}$	—	.28	2 R	—	—	1800	—
1978	7201	.48	$1\frac{3}{8}$	—	.28	Index	—	—	1600	—
	7228	.48	$1\frac{3}{8}$	—	.28	1 R	—	—	1600	—
	7230	.48	$1\frac{3}{8}$	—	.28	1 R	—	—	1500	—
	7231	.48	$1\frac{3}{8}$	—	.28	2 R	—	—	1500	—

International

					MODEL TQ					
	6591S, 6550S	1.06	—	.34/.14[1]	.28-.32	1 R	—	800	1600	.06/.08[2]
	6590S, 6592S, 6551S	1.06	—	.34/.14[1]	.28-.32	1 R	—	700	1600	.06/.08[2]

Note [1]—Primary/secondary.
Note [2]—Manual/automatic.

SINGLE BARREL—YF TYPE

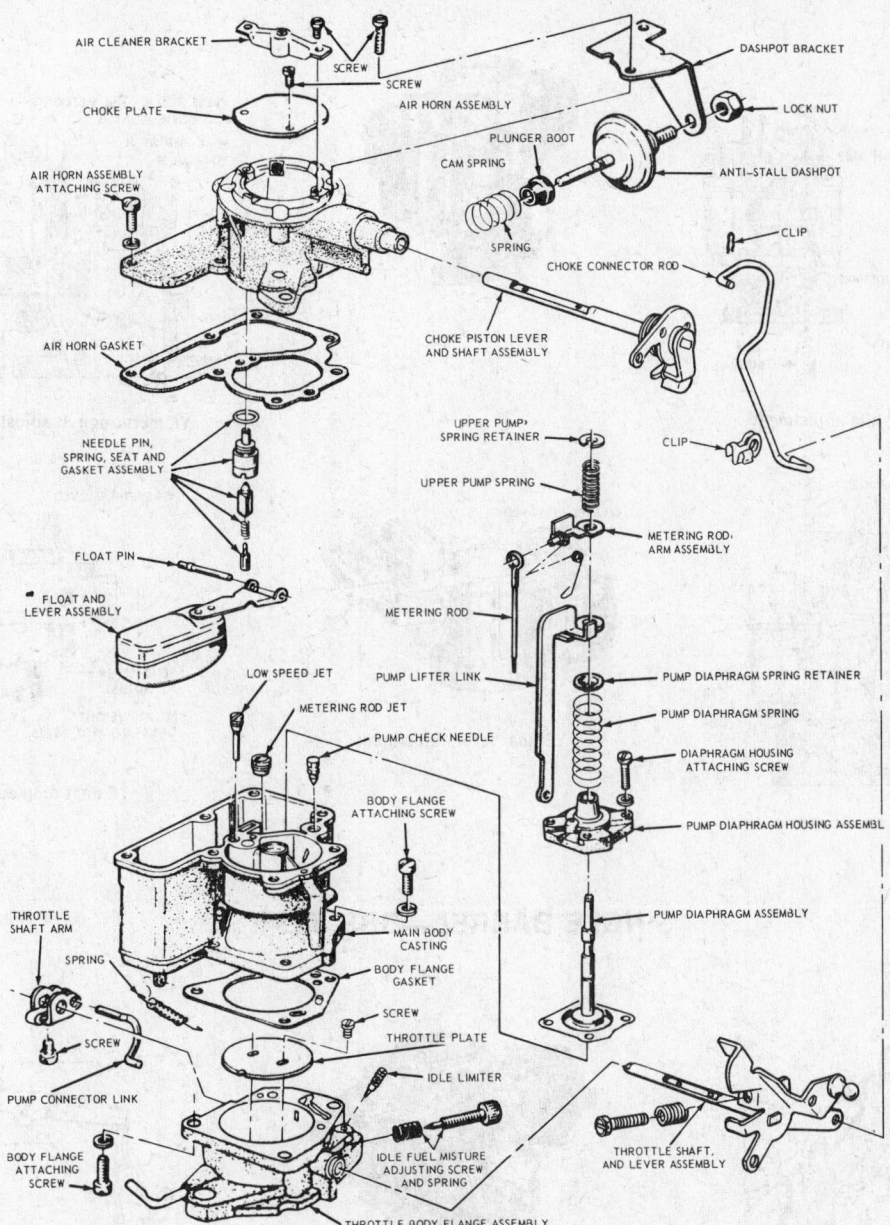

AIR CLEANER BRACKET

SCREW

SCREW

CHOKE PLATE

AIR HORN ASSEMBLY

DASHPOT BRACKET

LOCK NUT

PLUNGER BOOT

CAM SPRING

AIR HORN ASSEMBLY
ATTACHING SCREW

ANTI-STALL DASHPOT

SPRING

CLIP

CHOKE CONNECTOR ROD

AIR HORN GASKET

CHOKE PISTON LEVER
AND SHAFT ASSEMBLY

UPPER PUMP
SPRING RETAINER

CLIP

NEEDLE PIN,
SPRING, SEAT AND
GASKET ASSEMBLY

UPPER PUMP SPRING

METERING ROD,
ARM ASSEMBLY

FLOAT PIN

METERING ROD

FLOAT AND
LEVER ASSEMBLY

LOW SPEED JET

PUMP LIFTER LINK

PUMP DIAPHRAGM SPRING RETAINER

METERING ROD JET

PUMP DIAPHRAGM SPRING

PUMP CHECK NEEDLE

DIAPHRAGM HOUSING
ATTACHING SCREW

BODY FLANGE
ATTACHING SCREW

PUMP DIAPHRAGM HOUSING ASSEMBL

THROTTLE
SHAFT ARM

MAIN BODY
CASTING

PUMP DIAPHRAGM ASSEMBLY

SPRING

BODY FLANGE
GASKET

SCREW

THROTTLE PLATE

SCREW

PUMP CONNECTOR LINK

IDLE LIMITER

BODY FLANGE
ATTACHING
SCREW

THROTTLE SHAFT,
AND LEVER ASSEMBLY

IDLE FUEL MISTURE
ADJUSTING SCREW
AND SPRING

THROTTLE BODY FLANGE ASSEMBLY

Single barrel YF carburetor—exploded View

Carter Carburetors

SINGLE BARREL—YF TYPE

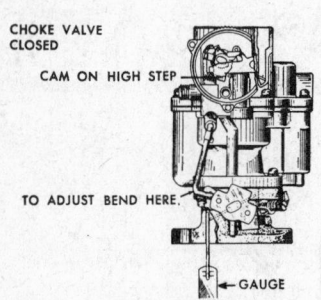

CHOKE VALVE CLOSED

CAM ON HIGH STEP

TO ADJUST BEND HERE

GAUGE

YF fast idle adjustment

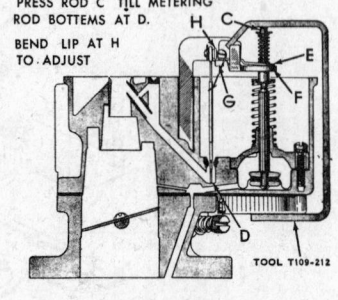

PRESS ROD C TILL METERING ROD BOTTEMS AT D.

BEND LIP AT H TO ADJUST

H C
E
G F
D

TOOL T109-212

YF metering rod adjustment

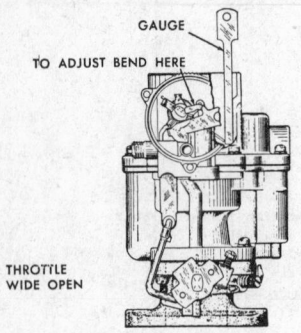

GAUGE

TO ADJUST BEND HERE

THROTTLE WIDE OPEN

YF unloader adjustment

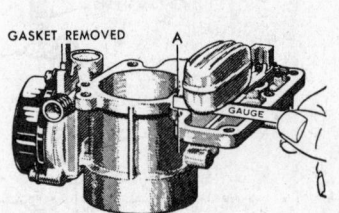

GASKET REMOVED
A
GAUGE

YF float level adjustment

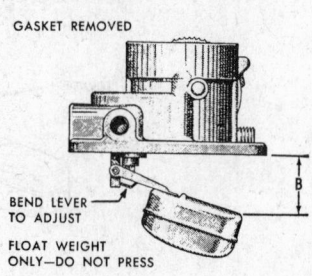

GASKET REMOVED

BEND LEVER TO ADJUST

FLOAT WEIGHT ONLY—DO NOT PRESS

B

YF float drop adjustment

SINGLE BARREL—BBS TYPE

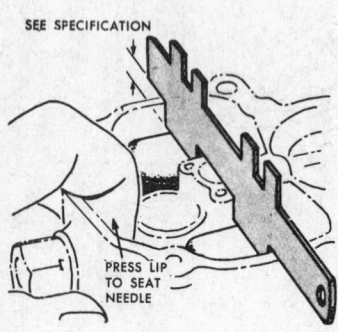

SEE SPECIFICATION

PRESS LIP TO SEAT NEEDLE

BBS float adjustment

BBS pump adjustment

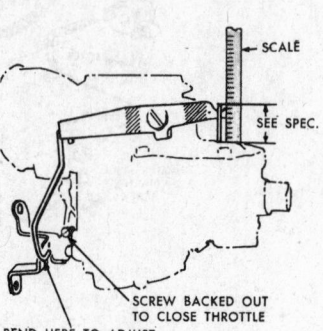

SCALE

SEE SPEC.

SCREW BACKED OUT TO CLOSE THROTTLE

BEND HERE TO ADJUST

BBS pump adjustment

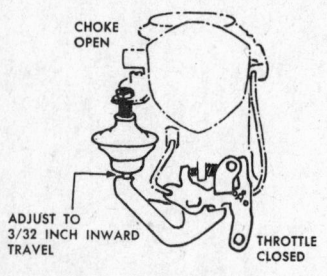

CHOKE OPEN

ADJUST TO 3/32 INCH INWARD TRAVEL

THROTTLE CLOSED

BBS dashpot adjustment

INSERT GAUGE

BEND HERE TO ADJUST

THROTTLE CLOSED

CHOKE VALVE CLOSED

BBS fast idle adjustment

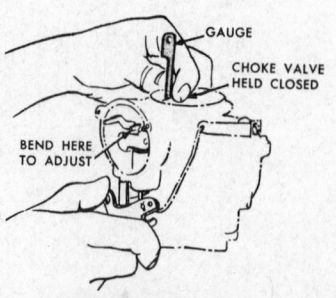

GAUGE

CHOKE VALVE HELD CLOSED

BEND HERE TO ADJUST

BBS unloader adjustment

872

SINGLE BARREL—BBS TYPE

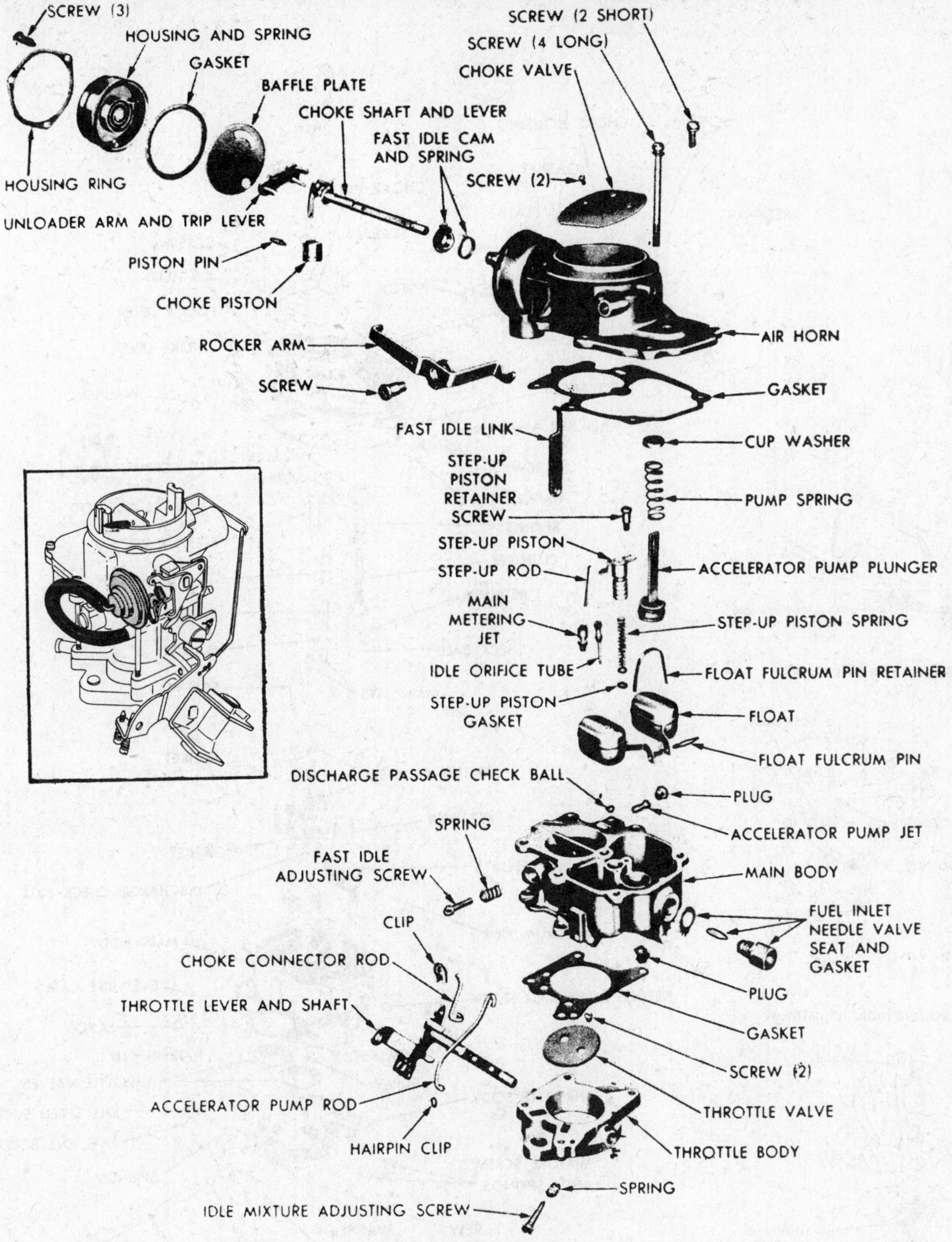

SCREW (3)

HOUSING AND SPRING

GASKET

BAFFLE PLATE

CHOKE SHAFT AND LEVER

FAST IDLE CAM AND SPRING

SCREW (2 SHORT)

SCREW (4 LONG)

CHOKE VALVE

SCREW (2)

HOUSING RING

UNLOADER ARM AND TRIP LEVER

PISTON PIN

CHOKE PISTON

AIR HORN

ROCKER ARM

SCREW

GASKET

FAST IDLE LINK

CUP WASHER

STEP-UP PISTON RETAINER SCREW

PUMP SPRING

STEP-UP PISTON

STEP-UP ROD

ACCELERATOR PUMP PLUNGER

MAIN METERING JET

STEP-UP PISTON SPRING

IDLE ORIFICE TUBE

FLOAT FULCRUM PIN RETAINER

STEP-UP PISTON GASKET

FLOAT

FLOAT FULCRUM PIN

DISCHARGE PASSAGE CHECK BALL

PLUG

SPRING

ACCELERATOR PUMP JET

FAST IDLE ADJUSTING SCREW

MAIN BODY

CLIP

FUEL INLET NEEDLE VALVE SEAT AND GASKET

CHOKE CONNECTOR ROD

THROTTLE LEVER AND SHAFT

PLUG

GASKET

ACCELERATOR PUMP ROD

SCREW (2)

HAIRPIN CLIP

THROTTLE VALVE

THROTTLE BODY

SPRING

IDLE MIXTURE ADJUSTING SCREW

Carter BBS single barrel—typical

TWO BARREL—BBD TYPE

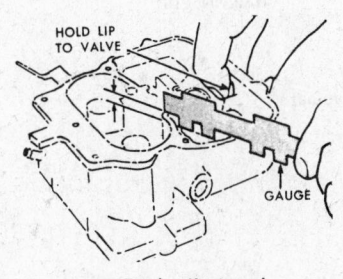

HOLD LIP TO VALVE

GAUGE

BBD float adjustment

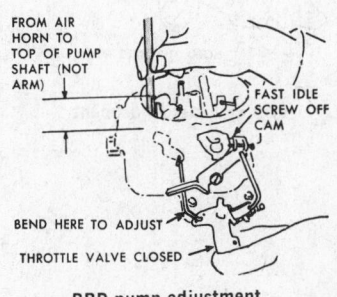

FROM AIR HORN TO TOP OF PUMP SHAFT (NOT ARM)

FAST IDLE SCREW OFF CAM

BEND HERE TO ADJUST

THROTTLE VALVE CLOSED

BBD pump adjustment

TWO BARREL—BBD TYPE continued

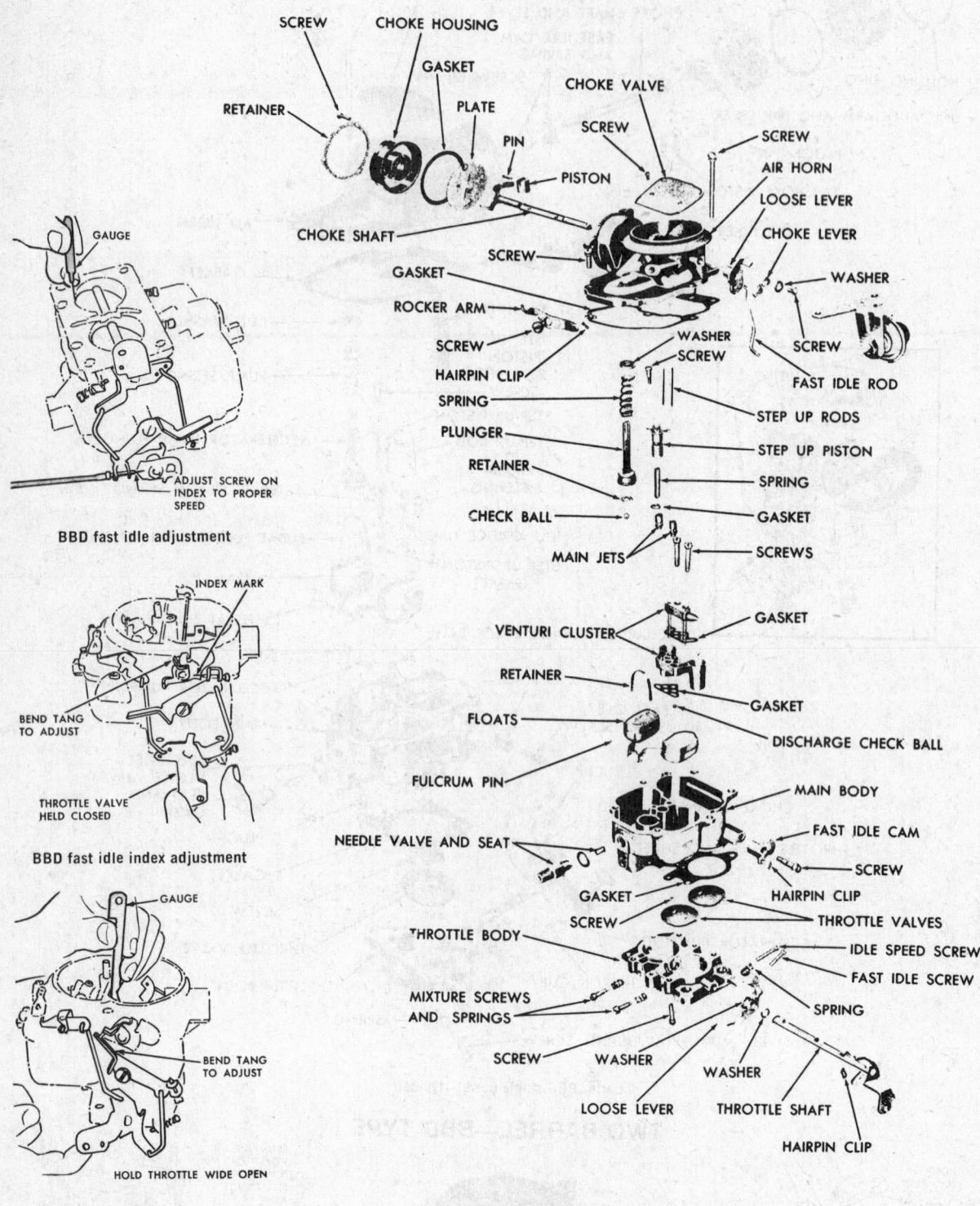

BBD fast idle adjustment

ADJUST SCREW ON INDEX TO PROPER SPEED

GAUGE

INDEX MARK

BEND TANG TO ADJUST

THROTTLE VALVE HELD CLOSED

BBD fast idle index adjustment

GAUGE

BEND TANG TO ADJUST

HOLD THROTTLE WIDE OPEN

BBD unloader adjustment

SCREW
CHOKE HOUSING
GASKET
PLATE
RETAINER
PIN
CHOKE VALVE
SCREW
PISTON
SCREW
AIR HORN
CHOKE SHAFT
LOOSE LEVER
CHOKE LEVER
SCREW
GASKET
WASHER
ROCKER ARM
WASHER
SCREW
SCREW
SCREW
HAIRPIN CLIP
FAST IDLE ROD
SPRING
STEP UP RODS
PLUNGER
STEP UP PISTON
RETAINER
SPRING
CHECK BALL
GASKET
MAIN JETS
SCREWS

VENTURI CLUSTER
GASKET
RETAINER
GASKET
FLOATS
DISCHARGE CHECK BALL
FULCRUM PIN
MAIN BODY
NEEDLE VALVE AND SEAT
FAST IDLE CAM
SCREW
GASKET
HAIRPIN CLIP
SCREW
THROTTLE VALVES
THROTTLE BODY
IDLE SPEED SCREW
FAST IDLE SCREW
MIXTURE SCREWS AND SPRINGS
SPRING
SCREW
WASHER
WASHER
THROTTLE SHAFT
LOOSE LEVER
HAIRPIN CLIP

Carter—BBD two barrel—typical

FOUR BARREL—TQ

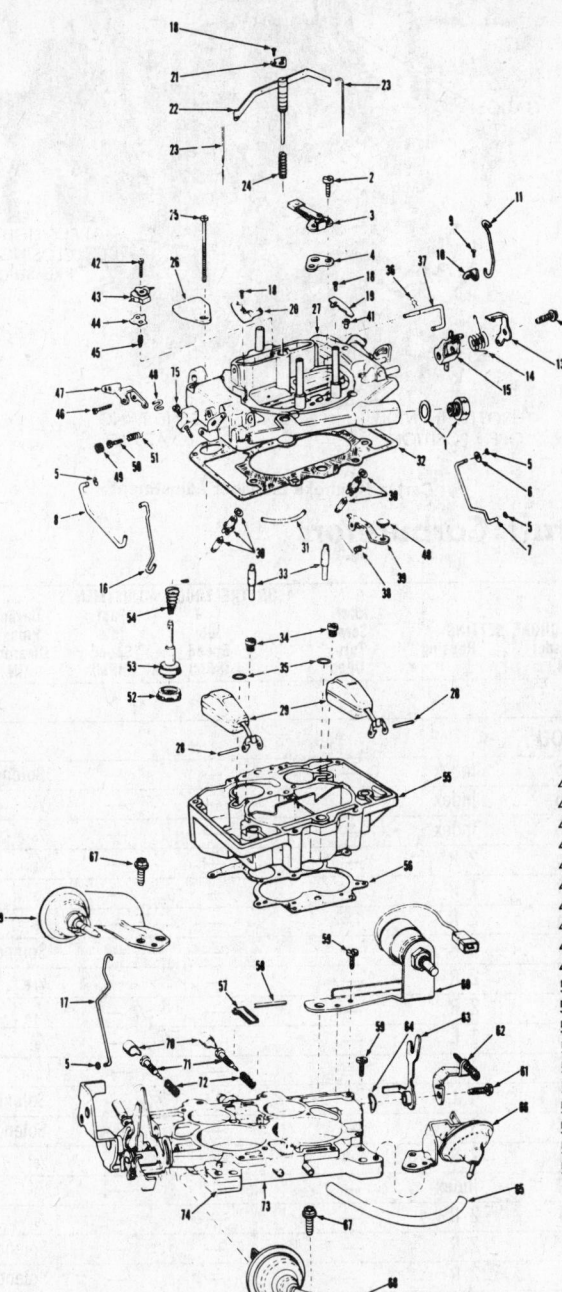

1. Fuel Inlet Nut and Gasket
2. Idle Compensator Screw
3. Idle Compensator
4. Idle Compensator Gasket
5. "E" Retainer
6. Primary Diaphragm Choke Pull-Off Rod Washer
7. Primary Diaphragm Choke Pull-Off Rod
8. Auxiliary Diaphragm Choke Pull-Off Rod (if equipped)
9. Choke Lever Screw
10. Choke Lever
11. Choke Connector Rod
12. Countershaft Lever Screw
13. Countershaft, Lever, Outer
14. Countershaft Lever Spring
15. Countershaft Lever, Inner
16. Fast Idle Cam Rod
17. Throttle Connector Rod
18. Cover Plate Screw
19. Metering Rod Cover Plate (opposite pump)
20. Metering Rod Cover Plate (pump side)
21. Step-Up Piston Cover Plate
22. Step-Up Piston & Hanger Assembly
23. Metering Rod
24. Step-Up Piston Spring
25. Bowl Cover Screw
26. IH Part Number Location
27. Bowl Cover Assembly
28. Float Pin
29. Float Assembly
30. Needle, Seat, and Gasket
31. Pump Passage Tube
32. Bowl Cover Gasket
33. Secondary Metering Jet
34. Primary Metering Jet
35. Quad Rings
36. Pin Spring Retainer
37. Bowl Vent Valve Lever, Upper
38. Bowl Vent Valve Lever Spring
39. Bowl Vent Valve Arm

40. Bowl Vent Valve Grommet
41. Rivet Plug
42. Pump Housing Screw
43. Pump Housing
44. Pump Housing Gasket
45. Discharge Check Needle
46. Pump Arm Screw
47. Pump Arm
48. Pump "S" Link
49. Air Valve Lock Plug
50. Air Valve Adjustment Plug
51. Air Valve Spring
52. Pump Intake Check Assembly
53. Plunger Assembly
54. Plunger Spring
55. Main Body
56. Main Body Gasket
57. Step-Up Piston Lifter
58. Step-Up Piston Lifter Lever Pin
59. Solenoid & Diaphragm Choke Pull-Off

Bracket Screw
60. Solenoid
61. Solenoid Operating Lever Screw
62. Curb Idle Speed Screw & Lever
63. Bowl Vent Lever, Lower
64. Throttle Shaft Washer
65. Hose
66. Primary Diaphragm Choke Pull-Off Bracket
67. Auxiliary Choke Pull-Off and Dashpot
68. Auxiliary Choke Pull-Off and Bracket (if equipped)
69. Dashpot and Bracket
70. Limiter Cap
71. Idle Mixture Screw
72. Idle Mixture Screw Spring
73. Throttle Body Assembly
74. Carter Part Number Location
75. Low Idle Speed Screw

Exploded view of typical Carter Thermo-Quad®

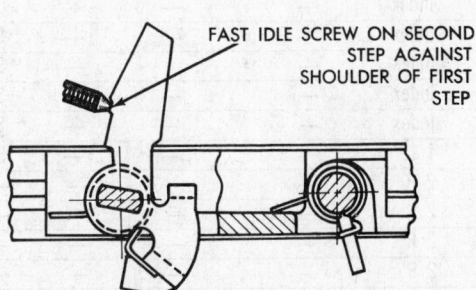

FAST IDLE SCREW ON SECOND STEP AGAINST SHOULDER OF FIRST STEP

Carter TQ fast idle speed adjustment cam position

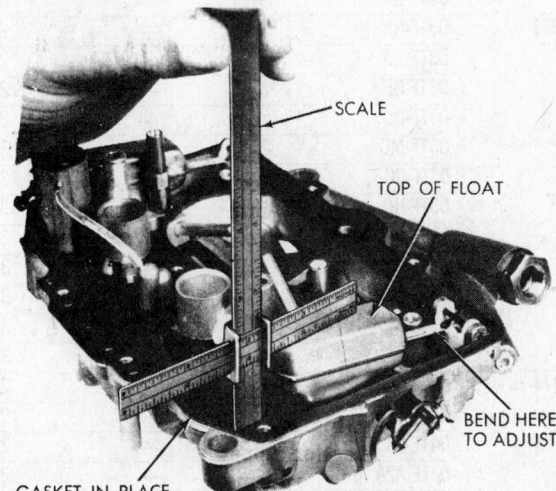

SCALE

TOP OF FLOAT

BEND HERE TO ADJUST

GASKET IN PLACE

Carter TQ float height measurement

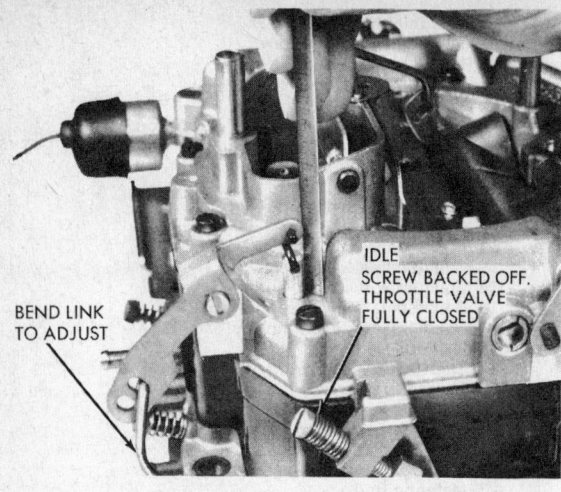

Carter TQ accelerator pump stroke adjustment

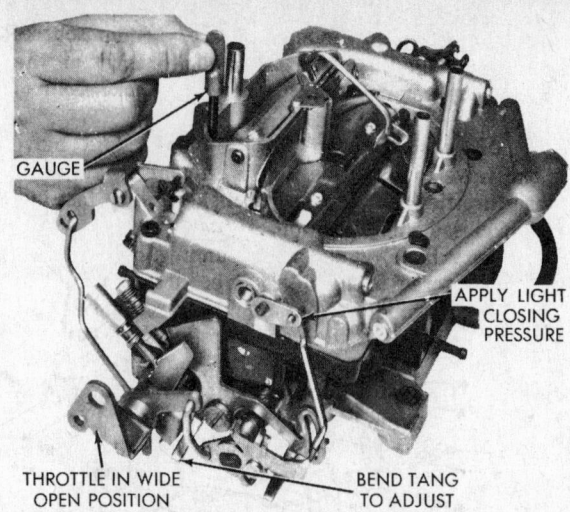

Carter TQ choke unloader adjustment

Ford/Autolite/Motorcraft Carburetors

MODEL 2100[1]

YEAR	MODEL OR TYPE	FLOAT LEVEL (IN.)	FLOAT DROP (IN.)	Pump Travel Setting (IN.)	CHOKE SETTING Unloader (IN.)	CHOKE SETTING Housing	ON THE TRUCK ADJUSTMENTS Idle Screw Turns Open	ON THE TRUCK ADJUSTMENTS Idle Speed (RPM)	ON THE TRUCK ADJUSTMENTS Fast Idle Speed (RPM)	Dashpot Plunger Clearance (IN.)
Ford										
1971	D1TF-AMA	$13/16$	—	3	.06	Index	—	—	—	Solenoid
	D1TF-EA	$13/16$	—	3	.06	Index	—	—	—	1/8
	D1TF-HA	7/8	—	3	.06	Index	—	—	—	1/8
	D1TF-ALA	7/8	—	3	.06	2 R	—	—	—	1/8
	D1TF-NA	$13/16$	—	3	.06	1 R	—	—	—	1/8
	D1TF-AKA	$13/16$	—	3	.06	2 R	—	—	—	1/8
1972	D2TF-EA	$13/16$	—	2	.06	2 R	—	—	—	Solenoid
	D2TF-CA	$13/16$	—	3	.06	2 R	—	—	—	1/16
	D2TF-DB	$13/16$	—	3	.12	2 R	—	—	—	1/16
	D2TF-AA	7/8	—	3	.16	1 L	—	—	—	1/8
	D2TF-AB	7/8	—	3	.16	1 L	—	—	—	1/8
	D2TF-FC	7/8	—	2	.14	1 L	—	—	—	Solenoid
	D2TF-AAA	7/8	—	2	.14	1 L	—	—	—	Solenoid
	D2TF-JA	7/8	—	3	.14	2 R	—	—	—	1/8
	D2TF-GA	$13/16$	—	3	.14	Index	—	—	—	1/8
	D2TF-KA	$13/16$	—	3	.14	2 R	—	—	—	1/8
1973	D3TF-NR	$13/16$	—	2	—	2 R	—	—	—	Solenoid
	D3TF-NA	$13/16$	—	2	—	2 R	—	—	—	Solenoid
	D3TF-BE	$29/32$	—	2	—	2 R	—	—	—	Solenoid
	D3TF-DD	7/8	—	3	—	Index	—	—	—	Solenoid
	D3TF-MC	$13/16$	—	3	—	Index	—	—	—	Solenoid
	D3TF-GC	7/8	—	3	—	Index	—	—	—	—
	D3TF-HC	$13/16$	—	3	—	Index	—	—	—	.065
	D3TF-DE	7/8	—	3	—	Index	—	—	—	Solenoid
	D3TF-UA	$13/16$	—	3	—	Index	—	—	—	Solenoid
	D3TF-MC	$13/16$	—	3	—	Index	—	—	—	Solenoid
	D3TF-VA	7/8	—	3	—	Index	—	—	—	Solenoid
	D3TF-XA	$13/16$	—	3	—	1 L	—	—	—	Solenoid
1974-75	D4TE-EA	$13/16$	—	2	—	2 R	—	—	—	—
	D4TE-CAA	$13/16$	—	2	—	2 R	—	—	—	—
	D4TE-LA	$13/16$	—	2	—	2 R	—	—	—	—
	D4TE-AZA	$13/16$	—	2	—	2 R	—	—	—	—
	D4TE-GA	$13/16$	—	3	—	2 R	—	—	—	—
	D4TE-TA	7/8	—	3	—	Index	—	—	—	—
	D4TE-BAA	7/8	—	3	—	1 R	—	—	—	—

Ford/Autolite/Motorcraft Carburetors

YEAR	MODEL OR TYPE	FLOAT LEVEL (IN.)	FLOAT DROP (IN.)	Pump Travel Setting (IN.)	CHOKE SETTING Unloader (IN.)	CHOKE SETTING Housing	Idle Screw Turns Open	ON THE TRUCK ADJUSTMENTS Idle Speed (RPM)	Fast Idle Speed (RPM)	Dashpot Plunger Clearance (IN.)
Ford, Continued										
1976-77	D5TE-ABA	.81	$7/16^2$	4	—	—	—	—	—	—
	D5TE-ASA	.88	$31/64^2$	4	—	Index	—	—	—	—
	D5TE-ATA	.88	$31/64^2$	4	—	Index	—	—	—	—
	D5TE-BYA	$7/8$	$31/64^2$	3	—	2 R	—	—	—	—
	D5TE-AAF	$7/8$	$31/64^2$	3	—	3 R	—	—	—	—
	D5TE-YF	$7/8$	$31/64^2$	3	—	2 R	—	—	—	—
	D5TE-ZB	$7/8$	$31/64^2$	3	—	2 R	—	—	—	—
MODEL 2150[1]										
1975	D5TE-BHA	$7/8$	—	hole 2	—	Index	—	—	—	—
	D5TE-LA	$7/8$	—	2	—	Index	—	—	—	—
	D5TE-PA	$7/8$	—	2	—	3 R	—	—	—	—
	D5TE-BCA	$7/8$	—	3	—	2 R	—	—	—	—
	D5TE-BCB	$7/8$	—	3	—	2 R	—	—	—	—
	D5TE-BFA	$7/8$	—	3	—	2 R	—	—	—	—
	D5TE-BFB	$7/8$	—	3	—	2 R	—	—	—	—
	D5TE-VA	$7/8$	—	.3	—	2 R	—	—	—	—
	D5TE-AUB	$7/8$	—	3	—	2 R	—	—	—	—
	D5TE-BGA	$7/8$	—	3	—	3 R	—	—	—	—
	D5TE-ZA	$7/8$	—	3	—	3 R	—	—	—	—
	D5TE-YD	$7/8$	—	3	—	3 R	—	—	—	—
	D5TE-AAD	$7/8$	—	3	—	3 R	—	—	—	—
	D5TE-BDA	$7/8$	—	3	—	2 R	—	—	—	—
	D5TE-BDB	$7/8$	—	3	—	2 R	—	—	—	—
	D5TE-BEA	$7/8$	—	3	—	2 R	—	—	—	—
1976-77	D5TE-BEB	$7/8$	$31/64^2$	3	—	2 R	—	—	—	—
MODEL 4300										
1974	D4TE-BB	$3/4$	—	hole 1	—	Index	—	—	$.20^3$	—
MODEL 4350										
1975	D5TE-ARC	$15/16$	—	hole 1	.30	Index	—	—	$.17^3$	—
	D5TE-ARD	1.0	—	1	.30	Index	—	—	$.17^3$	—
	D5TE-BBA	$15/16$	—	1	.30	Index	—	—	$.17^3$	—
	D5TE-BBB	$59/64$	—	1	.30	Index	—	—	$.17^3$	—
	D5TE-BBC	1.0	—	1	.30	Index	—	—	$.17^3$	—
1976-77	D6UE-KA	1.0	—	3	.30	Index	—	—	$.17^3$	—
	D6UE-LA	1.0	—	3	.30	Index	—	—	$.17^3$	—
	D6TE-NA	1.0	—	3	.30	Index	—	—	$.17^3$	—
	D6TE-UA	1.0	—	3	.30	Index	—	—	$.17^3$	—

Note [1]—Fuel level in bowl given for float level.
Note [2]—Dry float setting.
Note [3]—Fast idle cam setting.

MODEL 2100										
1972	2DM2	$3/8$	—	—	$13/64$	1 R	—	—	1600	$1/8$
1973	3DM2	—	$3/4$	—	$1/4$	1 R	—	—	1600	$1/8$
1974	4DMJ2	—	$25/32$	—	$1/4$	2 R	—	—	1600	—
	4DM2	—	$25/32$	—	$1/4$	2 R	—	—	1600	$9/64$
	4RA2	—	$25/32$	—	$1/4$	1 R	—	—	1600	—
	4RHD2	—	$3/4$	—	$1/4$	2 R	—	—	1600	—
1975	5DM2J	—	$25/32$	—	$1/4$	2 R	—	—	1600	$1/8$
	5DM2	—	$25/32$	—	$1/4$	2 R	—	—	1600	$1/8$
1976	6DM2J	—	$15/16$	—	$1/4$	1 R	—	—	1600	$1/8$
	6DM2	—	$15/16$	—	$1/4$	2 R	—	—	1600	$1/8$
	6DA2J	—	$15/16$	—	$1/4$	1 R	—	—	1600	$1/8$
	6RHM2	—	.93	—	$1/4$	2 R	—	—	1600	—
	6RHA2	—	.93	—	$1/4$	2 R	—	—	1600	—

Ford Carburetors

YEAR	MODEL OR TYPE	FLOAT LEVEL (IN.)	FLOAT DROP (IN.)	Pump Travel Setting (IN.)	CHOKE SETTING Unloader (IN.)	Housing	Idle Screw Turns Open	ON THE TRUCK ADJUSTMENTS Idle Speed (RPM)	Fast Idle Speed (RPM)	Dashpot Plunger Clearance (IN.)
Ford, Continued										
				MODEL 2100[1]						
1977	6RHM2	.56	.93	.115	—	2 R	—	—	1600	—
	6RHA2	.56	.93	.115	—	2 R	—	—	1600	—
	6DM2	.56	.93	.120	—	2 R	—	—	1600	.09
	6DA2J	.56	.93	.126	—	1 R	—	—	1600	—
	6DM2J	.56	.93	.120	—	1 R	—	—	1600	.09
1978	8DM2	.56	.93	.120	—	2 R	—	—	1500	—
	8DM2C	.56	.93	.120	—	1 R	—	—	1500	—
	8DA2J	.56	.93	.126	—	1 R	—	—	1600	—
	8DA2JC	.56	.93	.126	—	1 R	—	—	1600	—
	6RHA2	.56	.93	.115	—	2 R	—	—	1600	—
	6RHM2	.56	.93	.115	—	2 R	—	—	1600	—
				MODEL 2150						
1977	7DM2A	.56	.93	.089	—	2 R	—	—	1600	—
	7DA2A	.56	.93	.089	—	1 R	—	—	1600	—
1978	8DA2A	.56	.93	.078	—	2 R	—	—	1600	—
	8DM2A	.56	.93	.078	—	2 R	—	—	1600	—
				MODEL 4300						
1971-73	—	$13/16$	—	—	$9/32$	—	—	—	1600	$9/64$
1974	4TA4	.82	—	—	.33	2 R	—	—	1600	$9/64$
	4THD4	.82	—	—	.33	2 R	—	—	1600	—
				MODEL 4350						
1975	5THA4	.90	—	—	.33	2 R	—	—	1600	—
	5THM4	.90	—	—	.33	2 R	—	—	1600	—
1976	6THA4	.90	—	—	.33	2 R	—	—	1600	—
	6THM4	.90	—	—	.33	2 R	—	—	1600	—
	6THA4C	.90	—	—	.33	2 R	—	—	1600	—
1977	6THA4	.90	—	.14	.31	2 R	—	—	1600	—
	6THM4	.90	—	.14	.31	2 R	—	—	1600	—
	6THA4C	.90	—	.14	.31	2 R	—	—	1600	—
1978	6THA4	.90	—	.14	.33	2 R	—	—	1600	—
	6THM4	.90	—	.14	.33	2 R	—	—	1600	—

Note [1]—Dry.
Note [2]—Wet fuel level, measured on vehicle.
Note [3]—Parallel with float bowl floor (inverted).

TWO BARREL

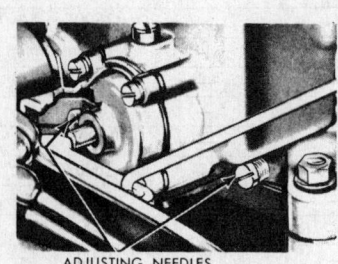

ADJUSTING NEEDLES

Idle speed adjustment

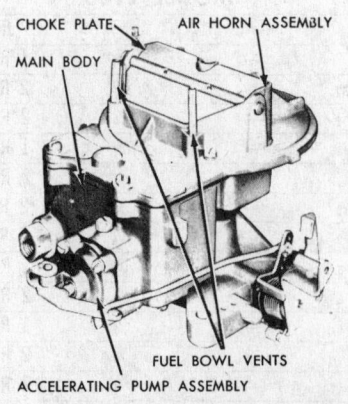

CHOKE PLATE AIR HORN ASSEMBLY
MAIN BODY
FUEL BOWL VENTS
ACCELERATING PUMP ASSEMBLY

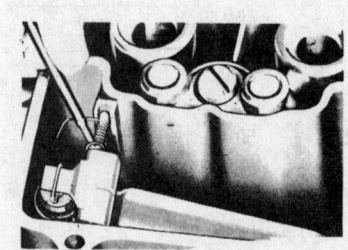

Float shaft retainer removal

TWO BARREL

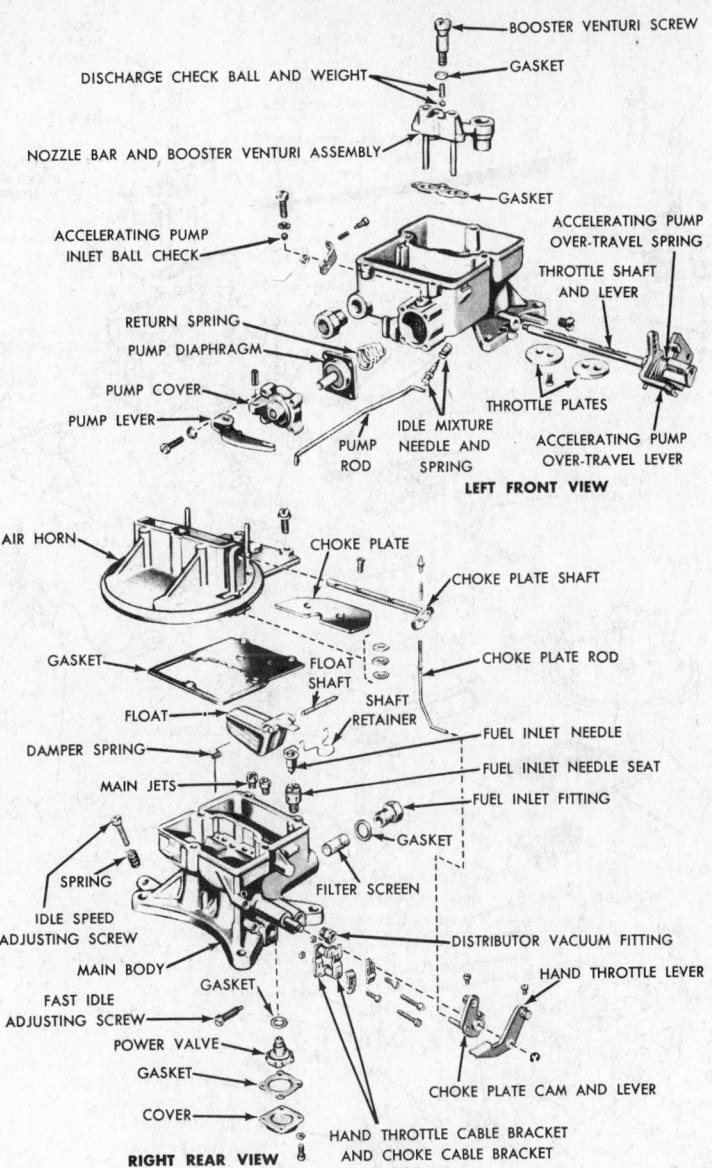

BOOSTER VENTURI SCREW

GASKET

DISCHARGE CHECK BALL AND WEIGHT

NOZZLE BAR AND BOOSTER VENTURI ASSEMBLY

GASKET

ACCELERATING PUMP INLET BALL CHECK

ACCELERATING PUMP OVER-TRAVEL SPRING

THROTTLE SHAFT AND LEVER

RETURN SPRING

PUMP DIAPHRAGM

PUMP COVER

PUMP LEVER

PUMP ROD

IDLE MIXTURE NEEDLE AND SPRING

THROTTLE PLATES

ACCELERATING PUMP OVER-TRAVEL LEVER

LEFT FRONT VIEW

AIR HORN

CHOKE PLATE

CHOKE PLATE SHAFT

GASKET

FLOAT SHAFT

CHOKE PLATE ROD

FLOAT

SHAFT RETAINER

DAMPER SPRING

FUEL INLET NEEDLE

MAIN JETS

FUEL INLET NEEDLE SEAT

FUEL INLET FITTING

GASKET

SPRING

FILTER SCREEN

IDLE SPEED ADJUSTING SCREW

MAIN BODY

DISTRIBUTOR VACUUM FITTING

HAND THROTTLE LEVER

FAST IDLE ADJUSTING SCREW

GASKET

POWER VALVE

GASKET

COVER

CHOKE PLATE CAM AND LEVER

HAND THROTTLE CABLE BRACKET AND CHOKE CABLE BRACKET

RIGHT REAR VIEW

Ford two barrel—disassembled view

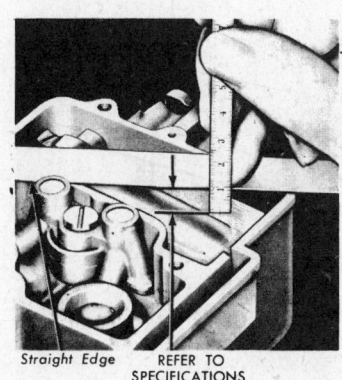

Straight Edge   REFER TO SPECIFICATIONS

Fuel level adjustment

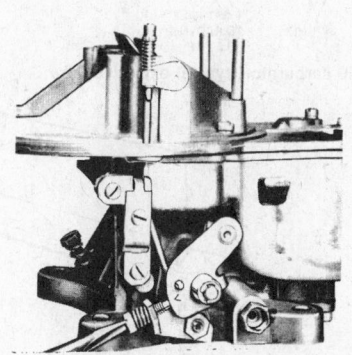

Fast idle adjustment

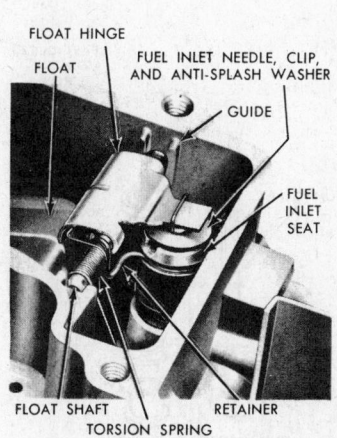

FLOAT HINGE

FLOAT

FUEL INLET NEEDLE, CLIP, AND ANTI-SPLASH WASHER

GUIDE

FUEL INLET SEAT

FLOAT SHAFT

TORSION SPRING

RETAINER

Float assembly installed

FOUR BARREL

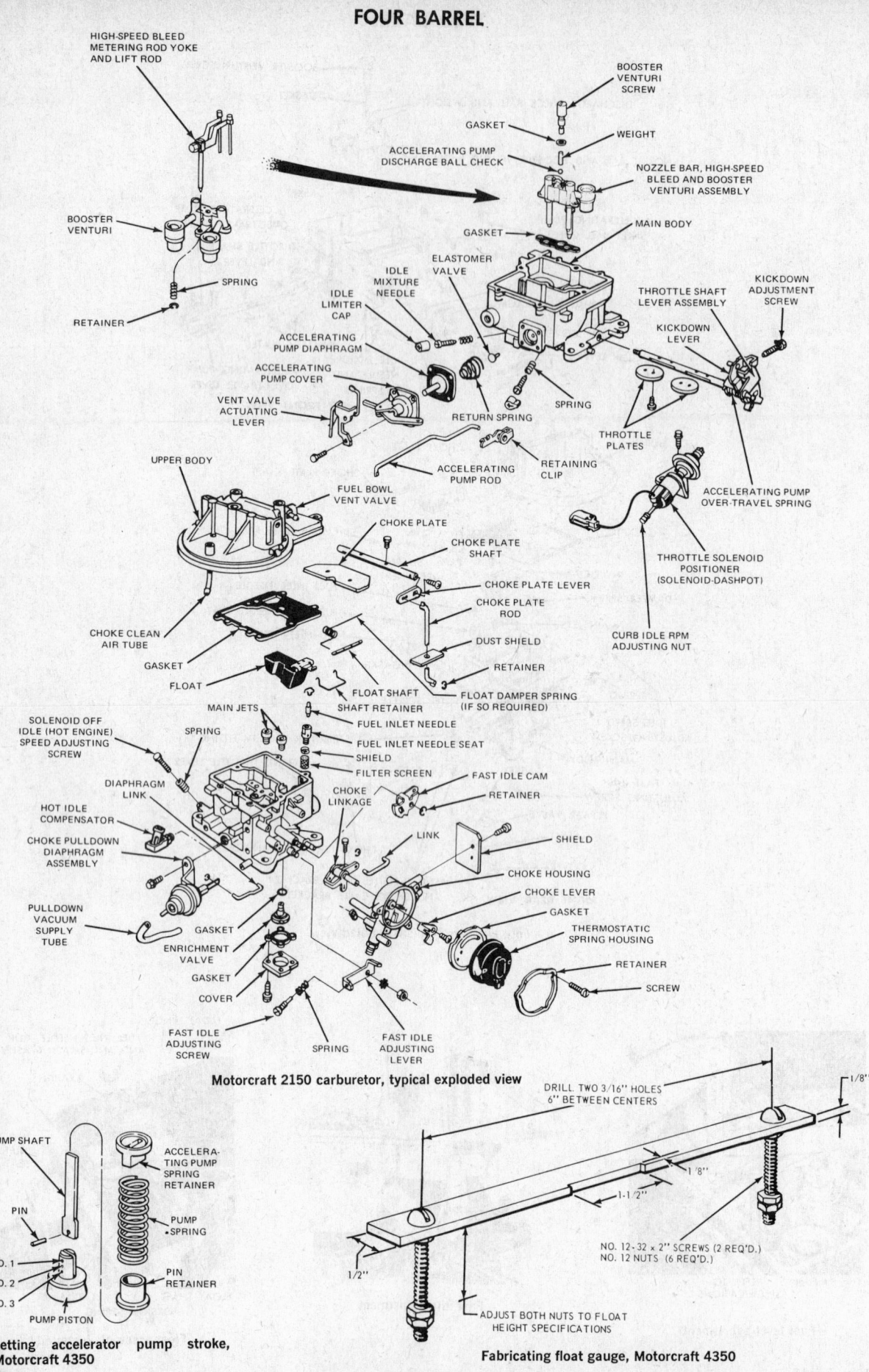

HIGH-SPEED BLEED
METERING ROD YOKE
AND LIFT ROD

BOOSTER
VENTURI
SCREW

GASKET

WEIGHT

ACCELERATING PUMP
DISCHARGE BALL CHECK

NOZZLE BAR, HIGH-SPEED
BLEED AND BOOSTER
VENTURI ASSEMBLY

BOOSTER
VENTURI

GASKET

MAIN BODY

SPRING

ELASTOMER
VALVE

THROTTLE SHAFT
LEVER ASSEMBLY

KICKDOWN
ADJUSTMENT
SCREW

RETAINER

IDLE
MIXTURE
NEEDLE

IDLE
LIMITER
CAP

KICKDOWN
LEVER

ACCELERATING
PUMP DIAPHRAGM

ACCELERATING
PUMP COVER

VENT VALVE
ACTUATING
LEVER

RETURN SPRING

SPRING

THROTTLE
PLATES

ACCELERATING PUMP
OVER-TRAVEL SPRING

UPPER BODY

FUEL BOWL
VENT VALVE

ACCELERATING
PUMP ROD

RETAINING
CLIP

THROTTLE SOLENOID
POSITIONER
(SOLENOID-DASHPOT)

CHOKE PLATE

CHOKE PLATE
SHAFT

CHOKE PLATE LEVER

CHOKE PLATE
ROD

CHOKE CLEAN
AIR TUBE

GASKET

FLOAT

DUST SHIELD

RETAINER

CURB IDLE RPM
ADJUSTING NUT

MAIN JETS

FLOAT SHAFT

SHAFT RETAINER

FLOAT DAMPER SPRING
(IF SO REQUIRED)

SOLENOID OFF
IDLE (HOT ENGINE)
SPEED ADJUSTING
SCREW

SPRING

FUEL INLET NEEDLE

FUEL INLET NEEDLE SEAT

SHIELD

FILTER SCREEN

FAST IDLE CAM

RETAINER

DIAPHRAGM
LINK

HOT IDLE
COMPENSATOR

CHOKE
LINKAGE

LINK

SHIELD

CHOKE PULLDOWN
DIAPHRAGM
ASSEMBLY

CHOKE HOUSING

CHOKE LEVER

PULLDOWN
VACUUM
SUPPLY
TUBE

GASKET

ENRICHMENT
VALVE

GASKET

COVER

FAST IDLE
ADJUSTING
SCREW

SPRING

FAST IDLE
ADJUSTING
LEVER

GASKET

THERMOSTATIC
SPRING HOUSING

RETAINER

SCREW

Motorcraft 2150 carburetor, typical exploded view

PUMP SHAFT

PIN

ACCELERA-
TING PUMP
SPRING
RETAINER

PUMP
SPRING

NO. 1
NO. 2
NO. 3

PIN
RETAINER

PUMP PISTON

**Setting accelerator pump stroke,
Motorcraft 4350**

DRILL TWO 3/16'' HOLES
6'' BETWEEN CENTERS

1/8''

1 8''

1-1/2''

NO. 12-32 x 2'' SCREWS (2 REQ'D.)
NO. 12 NUTS (6 REQ'D.)

1/2''

ADJUST BOTH NUTS TO FLOAT
HEIGHT SPECIFICATIONS

Fabricating float gauge, Motorcraft 4350

FOUR BARREL

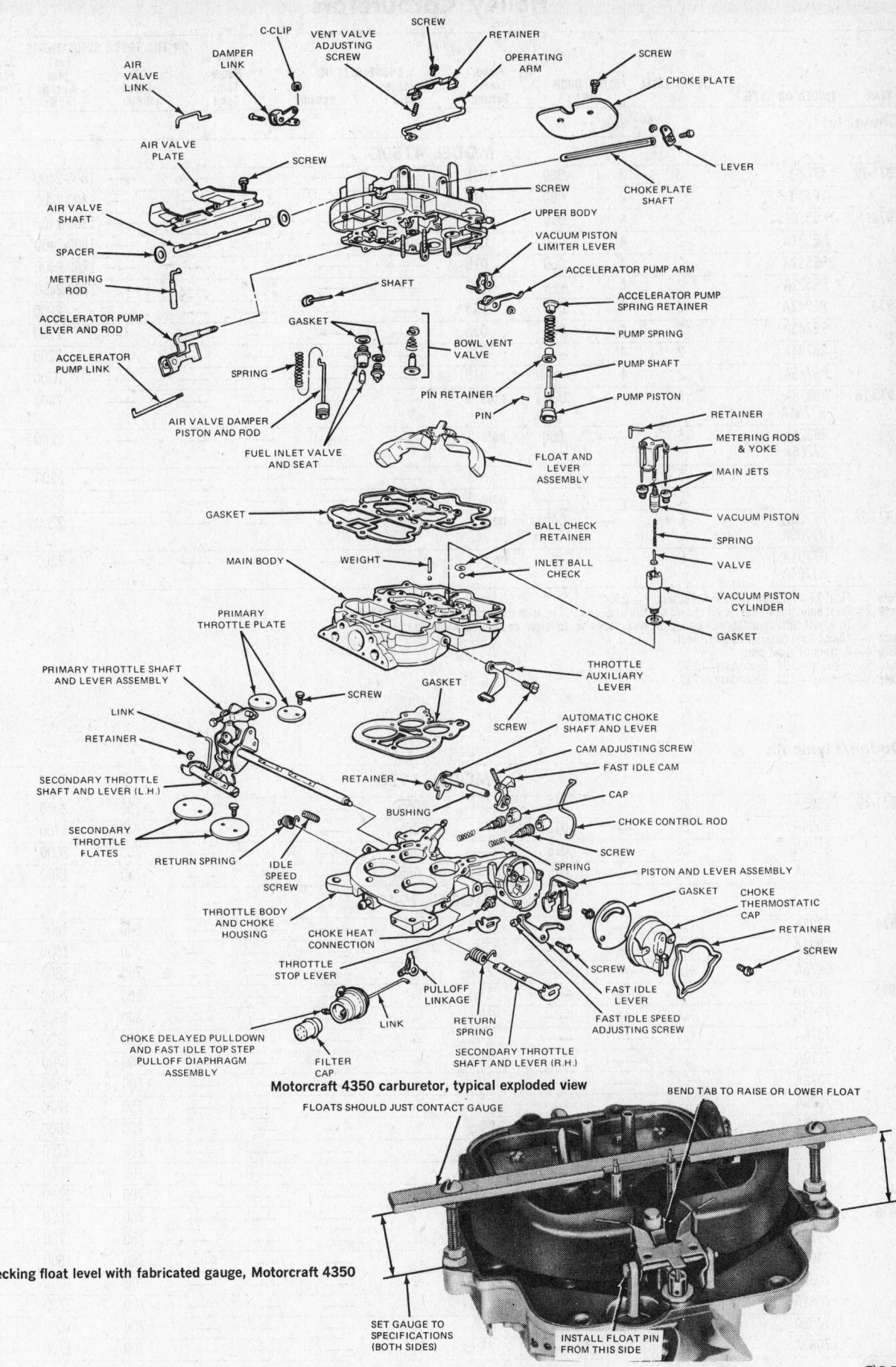

Motorcraft 4350 carburetor, typical exploded view

AIR VALVE LINK

C-CLIP

DAMPER LINK

VENT VALVE ADJUSTING SCREW

SCREW

RETAINER

OPERATING ARM

SCREW

CHOKE PLATE

AIR VALVE PLATE

SCREW

LEVER

SCREW

CHOKE PLATE SHAFT

AIR VALVE SHAFT

UPPER BODY

SPACER

VACUUM PISTON LIMITER LEVER

METERING ROD

SHAFT

ACCELERATOR PUMP ARM

ACCELERATOR PUMP LEVER AND ROD

GASKET

ACCELERATOR PUMP SPRING RETAINER

ACCELERATOR PUMP LINK

SPRING

BOWL VENT VALVE

PUMP SPRING

PUMP SHAFT

PIN RETAINER

PUMP PISTON

AIR VALVE DAMPER PISTON AND ROD

PIN

FUEL INLET VALVE AND SEAT

FLOAT AND LEVER ASSEMBLY

RETAINER

METERING RODS & YOKE

MAIN JETS

GASKET

VACUUM PISTON

BALL CHECK RETAINER

SPRING

MAIN BODY

WEIGHT

INLET BALL CHECK

VALVE

VACUUM PISTON CYLINDER

PRIMARY THROTTLE PLATE

GASKET

PRIMARY THROTTLE SHAFT AND LEVER ASSEMBLY

SCREW

THROTTLE AUXILIARY LEVER

LINK

GASKET

AUTOMATIC CHOKE SHAFT AND LEVER

RETAINER

SCREW

CAM ADJUSTING SCREW

SECONDARY THROTTLE SHAFT AND LEVER (L.H.)

RETAINER

FAST IDLE CAM

CAP

SECONDARY THROTTLE FLATES

BUSHING

CHOKE CONTROL ROD

RETURN SPRING

IDLE SPEED SCREW

SCREW

PISTON AND LEVER ASSEMBLY

THROTTLE BODY AND CHOKE HOUSING

SPRING

GASKET

CHOKE THERMOSTATIC CAP

CHOKE HEAT CONNECTION

RETAINER

THROTTLE STOP LEVER

SCREW

SCREW

PULLOFF LINKAGE

FAST IDLE LEVER

CHOKE DELAYED PULLDOWN AND FAST IDLE TOP STEP PULLOFF DIAPHRAGM ASSEMBLY

LINK

RETURN SPRING

FAST IDLE SPEED ADJUSTING SCREW

FILTER CAP

SECONDARY THROTTLE SHAFT AND LEVER (R.H.)

FLOATS SHOULD JUST CONTACT GAUGE

BEND TAB TO RAISE OR LOWER FLOAT

Checking float level with fabricated gauge, Motorcraft 4350

SET GAUGE TO SPECIFICATIONS (BOTH SIDES)

INSTALL FLOAT PIN FROM THIS SIDE

Holley Carburetors

Chevrolet

MODEL 4150G

YEAR	MODEL OR TYPE	FLOAT LEVEL (IN.)	FLOAT DROP (IN.)		Pump Travel Setting	CHOKE SETTING		Idle Screw Turns Open	ON THE TRUCK ADJUSTMENTS		
						Unloader (IN.)	Housing		Idle Speed (RPM)	Fast Idle Speed (RPM)	Dashpot Plunger Clearance (IN.)
1971-72	4772-1	3	2	.060	.015	—	—	3	—	1800-2400	—
	4773-1	3	2	.060	.015	—	—	3	—	1800-2400	—
1973	R6510A	2	4	.060	.015	—	—	3	—	1800-2400	—
	R6521A	2	4	—	.015	—	—	3	—	1800-2400	—
	R6511A	2	4	.060	.015	—	—	3	—	1800-2400	—
	R6522A	2	4	—	.015	—	—	3	—	1800-2400	—
1974	R6742A	2	4	.060	.015	—	—	—	—	2200	—
	R6743A	2	4	.060	.015	—	—	—	—	2200	—
	R6744A	2	4	—	.015	—	—	—	—	2200	—
	R6745A	2	4	—	.015	—	—	—	—	2200	—
1975-76	R6928A, R7264A	5	—	.060	hole 1	—	—	—	—	2200	—
	R6929A, R7266A	5	—	.060	hole 1	—	—	—	—	2200	—
	R6930A	5	—	—	hole 1	—	—	—	—	2200	—
	R6931A	5	—	—	hole 1	—	—	—	—	2200	—
1977-78	R7700A, R7703A	6	—	.045/.075	hole 1	—	—	—	—	2200	—
	R7701A, R7704A	6	—	—	hole 1	—	—	—	—	2200	—

Note 1—Float Level. Primary—.195". Secondary—.250".
Note 2—Float adjustment: Fuel level should be plus or minus 1/32 in. with threads at bottom of sight holes.
  To adjust turn adjusting nut on top of bowl clockwise, to lower, counterclockwise to raise.
Note 3—Adjust float parallel to bowl floor.
Note 4—Bottom of sight plug.
Note 5—Primary—.197, Secondary—.166.
Note 6—Primary—.194, Secondary—.213.

Dodge/Plymouth

MODEL 1920

YEAR	MODEL OR TYPE	FLOAT LEVEL (IN.)	FLOAT DROP (IN.)		Pump Travel Setting	CHOKE SETTING		Idle Screw Turns Open	ON THE TRUCK ADJUSTMENTS		
						Unloader (IN.)	Housing		Idle Speed (RPM)	Fast Idle Speed (RPM)	Dashpot Plunger Clearance (IN.)
1971-73	6593A	.26	—	.015	—	.065	—	1½	750	2000	—
	6594A	.26	—	.015	—	.065	—	1½	750	1700	—
	6595A	.26	—	.015	—	.065	—	1½	750	2000	—
	6596A	.26	—	.015	—	.065	—	1½	750	1700	—

MODEL 1945

YEAR	MODEL OR TYPE	FLOAT LEVEL (IN.)	FLOAT DROP (IN.)		Pump Travel Setting	CHOKE SETTING		Idle Screw Turns Open	ON THE TRUCK ADJUSTMENTS		
						Unloader (IN.)	Housing		Idle Speed (RPM)	Fast Idle Speed (RPM)	Dashpot Plunger Clearance (IN.)
1974	6725A	3/64	—	—	2 7/32	¼	—	—	800	1600	—
	6921A	3/64	—	—	2 7/32	¼	—	—	800	1600	—
	6875A	3/64	—	—	2 11/32	¼	—	—	750	1800	—
1975	7074A	3/64	—	—	2 7/32	¼	—	—	800	1600	—
	7209A	3/64	—	—	2 7/32	¼	—	—	800	1600	—
	7076A	3/64	—	—	2 21/64	¼	—	—	750	1700	—
	7210A	3/64	—	—	2 21/64	¼	—	—	750	1700	—
	7078A	3/64	—	—	2 7/32	¼	—	—	800	1600	—
	7079A	3/64	—	—	2 21/64	¼	—	—	750	1700	—
	7080A	3/64	—	—	2 7/32	¼	—	—	700	1600	—
	7081A	3/64	—	—	2 21/64	¼	—	—	700	1700	—
	7082A	3/64	—	—	2 7/32	¼	—	—	700	1600	—
	7083A	3/64	—	—	2 21/64	¼	—	—	700	1700	—
1976	7428A	2	—	—	2 7/32	¼	—	—	750	1600	—
	7429A	2	—	—	2 21/64	¼	—	—	750	1700	—
	7401A	2	—	—	2 7/32	¼	—	—	750	1600	—
	7080A	2	—	—	2 7/32	¼	—	—	700	1600	—
	7081A	2	—	—	2 21/64	¼	—	—	700	1700	—
	7082A	2	—	—	2 7/32	¼	—	—	700	1600	—
	7083A	2	—	—	2 21/64	¼	—	—	700	1700	—

Year	MODEL OR TYPE	Float Level (IN.)	Fuel Level (IN.)	Bowl Vent Valve (IN.)	Pump Travel Setting (IN.)	CHOKE SETTING Unloader (IN.)	Housing	Idle Screw Turns Out	ON THE TRUCK ADJUSTMENTS Air Bypass Turns Open	Idle Speed (RPM)	Fast Idle Speed (RPM)	Dashpot Plunger Clearance (IN.)
Dodge/Plymouth, Continued												
					MODEL 1945							
1977-78	7847A	2	—	—	$2\frac{7}{32}$	$\frac{1}{4}$	—	—	—	750	1600	—
	7848A	2	—	—	$2\frac{21}{64}$	$\frac{1}{4}$	—	—	—	750	1700	—
	7815A	2	—	—	$2\frac{7}{32}$	$\frac{1}{4}$	—	—	—	750	1600	—
	7816A	2	—	—	$2\frac{21}{64}$	$\frac{1}{4}$	—	—	—	750	1700	—
					MODEL 2210							
1971-73	6484A	.18	—	.015	$\frac{1}{4}$	.25	—	—	—	750	1900	—
	6484A	.18	—	.015	$\frac{1}{4}$	.17	—	—	—	700	1900	—
	6575A	.18	—	.015	$\frac{1}{4}$	.17	—	—	—	750	1900	—
	6485	.18	—	.015	$\frac{1}{4}$	.17	—	—	—	750	1900	—
	6486A	.18	—	.015	$\frac{1}{4}$	.17	—	—	—	700, 750 Calif.	1800	—
	6454A	.18	—	.015	$\frac{1}{4}$	.17	—	—	—	700	1800	—
	6472A	.18	—	.015	$\frac{1}{4}$	.17	—	—	—	700	1800	—
	6487A	.18	—	.015	$\frac{1}{4}$	.17	—	—	—	700	1800	—
	6488A	.18	—	.015	$\frac{1}{4}$	.17	—	—	—	700	1800	—
1974	6764A	.18	—	—	.27	.17	—	—	—	750	1700	—
	6765A	.18	—	—	$\frac{1}{4}$	.17	—	—	—	750	1800	—
	6886A	.18	—	—	$\frac{1}{4}$	.17	—	—	—	750	1800	—
	7052A	.18	—	—	$\frac{1}{4}$	.17	—	—	—	750	1800	—
1975	6764A	.18	—	—	.26/.31[1]	.17	—	—	—	750	1700	—
	6765A	.18	—	—	.26/.31[1]	.17	—	—	—	750	1800	—
1976	6764A	.18	—	—	.27/.32[1]	.17	—	—	—	750	1700	—
	6765A	.18	—	—	.26/.31[1]	.17	—	—	—	750	1800	—
	68861A	.18	—	—	.26/.31[1]	.17	—	—	—	700	1400	—
1977-78	6764A	.18	—	—	.27/.32[1]	.17	—	—	—	750	1700	—
	7870A	.18	—	—	.26/.31[1]	.17	—	—	—	750	1800	—
	6886-1A	.18	—	—	.26/.31[1]	.17	—	—	—	700	1400	—
					MODEL 2245							
1974	6762-1A	.18	—	.015	.26	.22	—	—	—	750	1700	—
	6860A	.18	—	.015	.26	.17	—	—	—	750	1800	—
	6990A	.18	—	.015	.26	.17	—	—	—	750	1600	—
1975	7187A	.18	—	—	.26/.31[1]	.17	—	—	—	700	1600	—
	7188A	.18	—	—	.26/.31[1]	.17	—	—	—	700	1600	—
	7103A	.18	—	—	.26/.31[1]	.17	—	—	—	750	1600	—
	7088A	.18	—	—	.26/.31[1]	.17	—	—	—	750	1600	—
	7089A	.18	—	—	.26/.31[1]	.17	—	—	—	750	1600	—
	7090A	.18	—	—	.26/.31[1]	.17	—	—	—	700	1600	—
	7091A	.18	—	—	.26/.31[1]	.17	—	—	—	750	1600	—
	7092A	.18	—	—	.26/.31[1]	.17	—	—	—	750	1600	—
1976	7403A	.18	—	—	.26/.31[1]	.17	—	—	—	750	1600	—
	7188A	.18	—	—	.26/.31[1]	.17	—	—	—	700	1600	—
1977-78	7871A	.25	—	—	.26/.31[1]	.17	—	—	—	750	1700	—
	8036A, 7403A	.25	—	—	.26/.31[1]	.17	—	—	—	700	1600	—
	7697A	.25	—	—	.26/.31[1]	.17	—	—	—	700	1600	—
	8182A	.25	—	—	.26/.31[1]	.17	—	—	—	800	1600	—

Note 1—from curb idle/from closed throttle
Note 2—Flush with top of bowl cover gasket

Year	MODEL OR TYPE	Float Level (IN.)	Fuel Level (IN.)	Bowl Vent Valve (IN.)	Pump Travel Setting (IN.)	CHOKE SETTING Unloader (IN.)	Housing	Idle Screw Turns Out	Air Bypass Turns Open	Idle Speed (RPM)	Fast Idle Speed (RPM)	Dashpot Plunger Clearance (IN.)
					MODEL 2300G							
1971	2951107	1	1	—	.015-.063	—	—	1.0	—	500	—	—
1972-73	6277A	1	1	—	.015-.063	—	—	1.0	—	700	—	—
	6278A	1	1	—	.015-.063	—	—	1.0	—	600, 700 Calif.	—	—

Holley Carburetors

Dodge/Plymouth

MODEL 2300G

Year	MODEL OR TYPE	Float Level (IN.)	Fuel Level (IN.)	Bowl Vent Valve (IN.)	Pump Travel Setting (IN.)	CHOKE SETTING Unloader (IN.)	Housing	Idle Screw Turns Out	ON THE TRUCK ADJUSTMENTS Air Bypass Turns Open	Idle Speed (RPM)	Fast Idle Speed (RPM)	Dashpot Plunger Clearance (IN.)
1974	6769-1A	1	1	—	.015²	—	—	1.0	—	700	—	—
	6770A	1	1	—	.015²	—	—	1.0	—	700	—	—
1975-76	6719-1A	1	1	—	.015²	—	—	—	—	700	—	—
	7137A	1	1	—	.015²	—	—	—	—	700	—	—

MODEL 4150G

Year	MODEL OR TYPE	Float Level (IN.)	Fuel Level (IN.)	Bowl Vent Valve (IN.)	Pump Travel Setting (IN.)	CHOKE SETTING Unloader (IN.)	Housing	Idle Screw Turns Out	ON THE TRUCK ADJUSTMENTS Air Bypass Turns Open	Idle Speed (RPM)	Fast Idle Speed (RPM)	Dashpot Plunger Clearance (IN.)
1971-72	2951109	1	1	—	.015-.063	—	—	1.0	—	500	—	—
1975-76	6771A	1	1	—	.015²	—	—	—	—	700	—	—
	7138A	1	1	—	.015²	—	—	—	—	700	—	—

Note 1—Float setting. (Dry) Float parting line parallel with floor of bowl.
(Wet) Fuel level with bottom of sight plug hole.
Note 2—Setting at full throttle.

Ford

MODEL 4150

YEAR	CARB. PART NO.	MODEL OR TYPE	Float Level	Fuel Level	Bowl Vent Valve	Pump Travel Setting	CHOKE SETTING Unloader	Housing	Idle Screw Turns Out	ON THE TRUCK ADJUSTMENTS Air Bypass Turns Open	Idle Speed	Fast Idle Speed	Dashpot Plunger Clearance
1976-77	D5TE-CB	3	1	—	hole 2	—	—	—	—	—	—	—
	D4TE-EAA	3	1	—	2	—	—	—	—	—	—	—
	D4HE-BA	3	1	—	1	—	—	—	—	—	—	—
	D4HE-CA	3	1	—	1	—	—	—	—	—	—	—
	D5HE-CA	3	1	—	1	—	—	—	—	—	—	—
	D4HE-DA	3	1	—	1	—	—	—	—	—	—	—
	D5HE-BA	3	1	—	1	—	—	—	—	—	—	—

MODEL 4160

YEAR	CARB. PART NO.	MODEL OR TYPE	Float Level	Fuel Level	Bowl Vent Valve	Pump Travel Setting	CHOKE SETTING Unloader	Housing	Idle Screw Turns Out	ON THE TRUCK ADJUSTMENTS Air Bypass Turns Open	Idle Speed	Fast Idle Speed	Dashpot Plunger Clearance
1975	D5TE-DA	3	1	—	—	.32	Index	—	—	—	top step	—
	D5TE-EA	3	1	—	—	.32	Index	—	—	—	—	—
	D5TE-FA	3	1	—	—	.32	Index	—	—	—	—	—
	D5TE-GA	3	1	—	—	.32	Index	—	—	—	—	—

GMC

MODEL 4150G

YEAR	MODEL OR TYPE	Float Level	Fuel Level	Bowl Vent Valve	Pump Travel Setting	CHOKE SETTING Unloader	Housing	Idle Screw Turns Out	ON THE TRUCK ADJUSTMENTS Air Bypass Turns Open	Idle Speed	Fast Idle Speed	Dashpot Plunger Clearance
1972	R-6292A	1	2	—	.015	—	—	—	—	500	—	—
	R-6293A	1	2	—	.015	—	—	—	—	500	—	—
1973	R6510A	2	4	.060	.015	—	—	3	—	—	1800-2400	—
	R6521A	2	4	—	.015	—	—	3	—	—	1800-2400	—
	R6511A	2	4	.060	.015	—	—	3	—	—	1800-2400	—
	R6522A	2	4	—	.015	—	—	3	—	—	1800-2400	—
1974	R6742A	2	4	.060	.015	—	—	—	—	—	2200	—
	R6743A	2	4	.060	.015	—	—	—	—	—	2200	—
	R6744A	2	4	—	.015	—	—	—	—	—	2200	—
	R6745A	2	4	—	.015	—	—	—	—	—	2200	—
1975-76	R-6928A, R7264A	3	—	.060	hole 1	—	—	—	—	—	2200	—
	R-6929A, R7266A	3	—	.060	hole 1	—	—	—	—	—	2200	—
	R-6930A	3	—	—	hole 1	—	—	—	—	—	2200	—
	R-6931A	3	—	—	hole 1	—	—	—	—	—	2200	—

Holley Carburetors

Year	MODEL OR TYPE	Float Level (IN.)	Fuel Level (IN.)	Bowl Vent Valve (IN.)	Pump Travel Setting (IN.)	CHOKE SETTING Unloader (IN.)	Housing	Idle Screw Turns Out	ON THE TRUCK ADJUSTMENTS Air Bypass Turns Open	Idle Speed (RPM)	Fast Idle Speed (RPM)	Dashpot Plunger Clearance (IN.)
GMC												
				MODEL 4150G								
1977-78	R7700A, R7703A	6	—	.045/.075	hole 1	—	—	—	—	—	—	2200	—
	R7701A, R7704A	6	—	—	hole 1	—	—	—	—	—	—	2200	—

Note 1—Float level. A—Primary—.195. B—Secondary—.250.
Note 2—Lower edge of sight plug hole.
Note 3—Primary—.197, Secondary—.166.
Note 4—Bottom of sight plug.

CARB. MODEL	CARB. PART NO.	Float Level	Pump Link Position	Fuel Level	Idle Mix. Screw Turn Out	Fast Idle Setting	Dashpot Clearance	Choke Setting
International								
1920	4405	1	hole 2	$27/32$	3.0	½-1 turn	.11-.14	—
	4542-1	1	hole 2	$27/32$	3.0	½-1 turn	—	—
	6286	1	hole 2	$27/32$	2.0	½-1 turn	.05-.08	—
	6266	1	hole 2	$27/32$	2.0	½-1 turn	.11-.14	—
	3993-1	1	hole 2	$27/32$	3.0	½-1 turn	.13-.16	—
	4487	1	hole 2	$27/32$	3.0	½-1 turn	.13-.16	—
	4591	1	hole 2	$27/32$	3.0	½-1 turn	.08-.11	—
1940C	All	2	$25/32$	$1 1/16$	—	2000 rpm	.07-.09	1 R
2110G	4326	$1 11/32$	—	¾	1½-2	—	—	—
2140G	4334, 4335, 4344, 864-1, 977	¼	hole 2	½	⅞	—	—	—
	4337, 4338, 4339, 4636, 4635, 4340, 4341	¼	hole 2	½	1½-2	—	—	—
2140SG	All	¼	hole 2	½	1½-2	—	—	—
2210C	6443	.2	.7	½	—	2200 rpm	.07-.09	Index/2 R[a]
	6443-1	.2	.7	½	—	2200 rpm	.07-.09	Index/2 R[a]
	6776	.2	.7	½	—	2200 rpm	.07-.09	Index
	6620	.2	.7	½	—	2200 rpm	—	1 R
	6674-1	.2	.7	½	—	2200 rpm	.07-.09	—
2300	2997-2, 2976-2	—	—	⅜	3.0	.00	.09-.12	—
	3991-1, 4079-1	—	—	⅜	3.0	.00	.09-.12	—
	4306, 4308	—	—	⅜	3.0	.00	.09-.12	—
	4574	—	—	⅜	3.0	.00	—	—
	4594	—	—	⅜	3.0	.00	.09-.12	—
	4081-1, 4083-1	—	—	⅜	3.0	.00	.09-.12	—
	6379	—	—	⅜	3.0	.00	.08-.11	—
	6391, 6391-1	—	—	⅜	1½-2	.11	—	—
2300G	2979	—	—	⅜	3.0	.02	—	—
	2980, 2975	—	—	⅜	3.0	.02	—	—
	4310-1, 4311-1	—	—	⅜	1½-2	.02	—	—
	4310-2	—	—	⅜	1½-2	2000 rpm	—	—
	6623, 6624-1	—	—	⅜	1½-2	.019	—	—
	6801-1, 6802	—	—	⅜	—	.019	—	—
	7213	—	—	⅜	—	2400 rpm	—	—
	6908	—	—	⅜	—	2000 rpm	—	—
	6899	—	—	⅜	—	2000 rpm	—	—
	7216	—	—	⅜	—	2400 rpm	—	—
	7198	—	—	⅜	—	2000 rpm	—	—
	7656	—	—	⅜	—	2000 rpm	—	—

Holley Carburetors

CARB. MODEL	CARB. PART NO.	Float Level	Pump Link Position	Fuel Level	Idle Mix. Screw Turn Out	Fast Idle Setting	Dashpot Clearance	Choke Setting
International								
2300C	3679-3, 2520-3	—	—	3/8	3.0	—	.09-.12	1 L
	3937-2, 3936-2	—	—	3/8	3.0	—	.09-.12	Index
	4078-1, 4080-1	—	—	3/8	3.0	—	.09-.12	1 L
	4082-1, 4084-1	—	—	3/8	3.0	—	.09-.12	1 L
	4307, 4309	—	—	3/8	3.0	—	.09-.12	1 L
	4593	—	—	3/8	3.0	—	—	Index
	4595	—	—	3/8	3.0	—	.09-.12	Index
	6393	—	—	3/8	1½-2	2000 rpm	—	1 L
	6394, 6393-1, 6394-1	—	—	3/8	1½-2	2000 rpm	.09-.12	1 L
	6380	—	—	3/8	3.0	2000 rpm	.08-.11	4 L/1 L[a]
	6386	—	—	3/8	3.0	2000 rpm	.08-.11	4 L/1 L[a]
4150	4318	—	—	½ 5/8[4]	3.0	.025	—	—
	4237	—	—	½ 5/8[4]	3.0	.025	—	—
	4572	—	—	½ 5/8[4]	3.0	.025	—	—
	4320	—	—	½ 5/8[4]	3.0	.025	.06-.09	—
	4264, 4624-1	—	—	½ 5/8[4]	3.0	.025	.06-.09	—
	4601, 4601-1, 4601-2	—	—	½ 5/8[4]	3.0	.025	.09-.12	—
4150C	4319	—	—	½ 5/8[4]	3.0	.025	—	3 L
	4312	—	—	½ 5/8[4]	3.0	.025	—	2 L
	4599	—	—	½ 5/8[4]	3.0	.025	—	Index
	4313	—	—	½ 5/8[4]	3.0	.025	.06-.09	2 L
	4602	—	—	½ 5/8[4]	3.0	.025	.09-.12	Index
	4321	—	—	½ 5/8[4]	3.0	.025	.06-.09	3 L
4150G	4323	—	—	3/8 5/8[4]	3.0	.025	—	—
	4324	—	—	3/8 5/8[4]	3.0	.025	—	—
	6803, 6803-1	—	—	3/8 5/8[4]	3	2000 rpm	—	—
	6803-2	—	—	3/8 5/8[4]	—	2000 rpm	—	—
	7215	—	—	3/8 5/8[4]	—	2400 rpm	—	—
	7251	—	—	3/8 5/8[4]	—	2400 rpm	—	—
	6911	—	—	3/8 5/8[4]	—	2000 rpm	—	—
	6974	—	—	3/8 5/8[4]	—	2000 rpm	—	—
	7028	—	—	3/8 5/8[4]	—	.018	—	—
	7218	—	—	3/8 5/8[4]	—	2400 rpm	—	—
	7219-1	—	—	3/8 5/8[4]	—	2400 rpm	—	—
	7029	—	—	3/8 5/8[4]	—	2400 rpm	—	—
	7029-1	—	—	3/8 5/8[4]	—	2400 rpm	—	—
852-FFG	4329, 4330	1¼	—	5/8	1½-2	—	—	—
	6398, 6398-1, 6438	1¼	—	5/8	1½-2	—	—	—
	4438, 4679, 4378	1¼	—	5/8	1½-2	—	—	—
	4331, 4338, 4377, 4332	1¼	—	5/8	1½-2	—	—	—
885-FFG	4327, 4328, 6397, 6397-1, 6437	7/32	—	½	1½-2	—	—	—
	6625	7/32	—	½	—	—	—	—

Note 1—Use special gauge.
Note 2—Flush with top of bowl.
Note 3—Summer/winter.
Note 4—Primary/secondary.

HOLLEY MODEL 1920

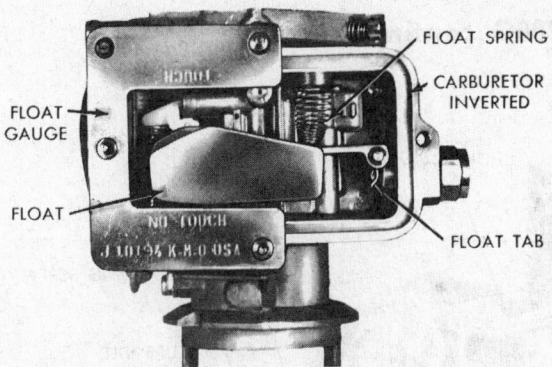

Checking float setting

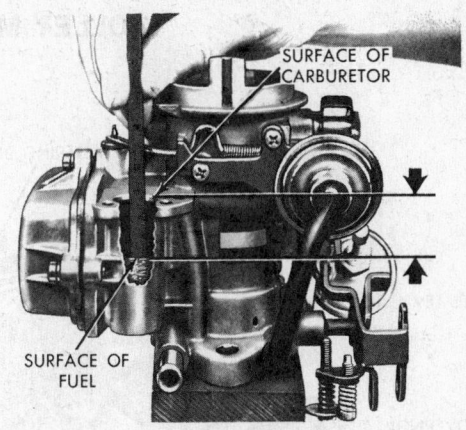

Measuring wet fuel level

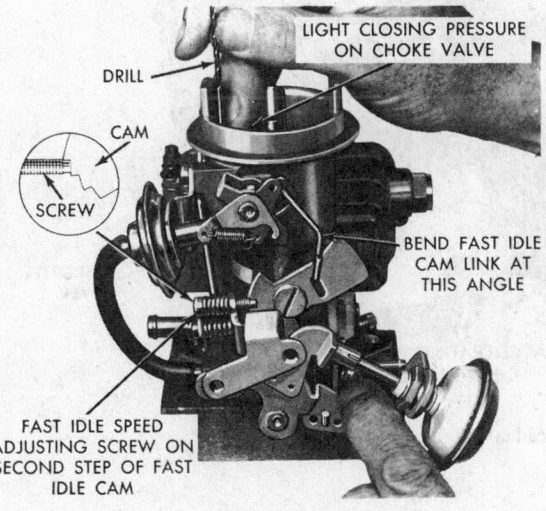

Fast idle cam adjustment

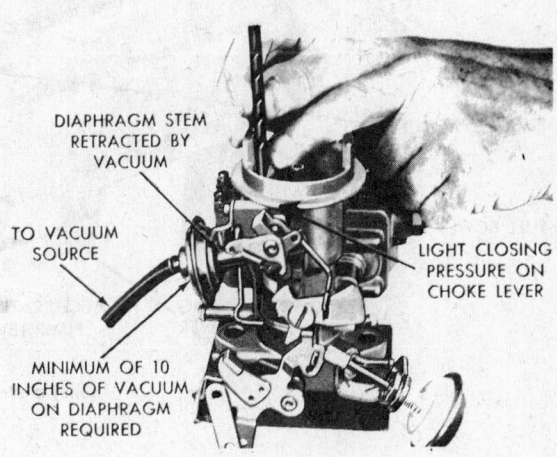

Choke vacuum kick setting

HOLLEY MODEL 2300G

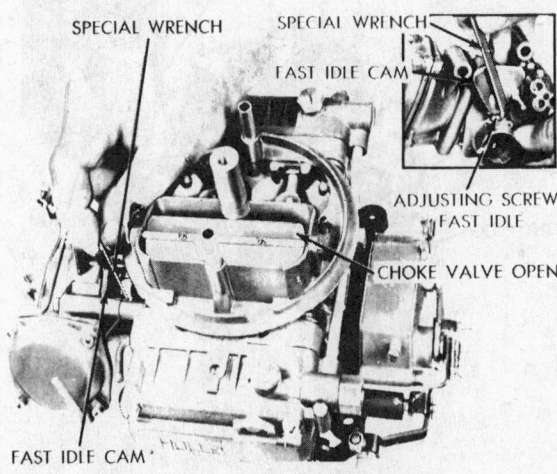

Fast idle speed adjustment

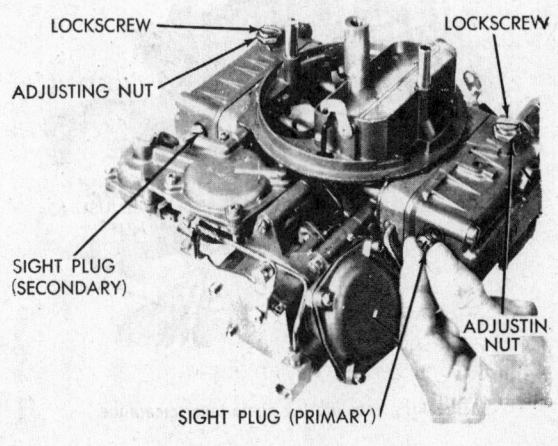

Fuel level sight plug location

HOLLEY MODEL 2300G continued

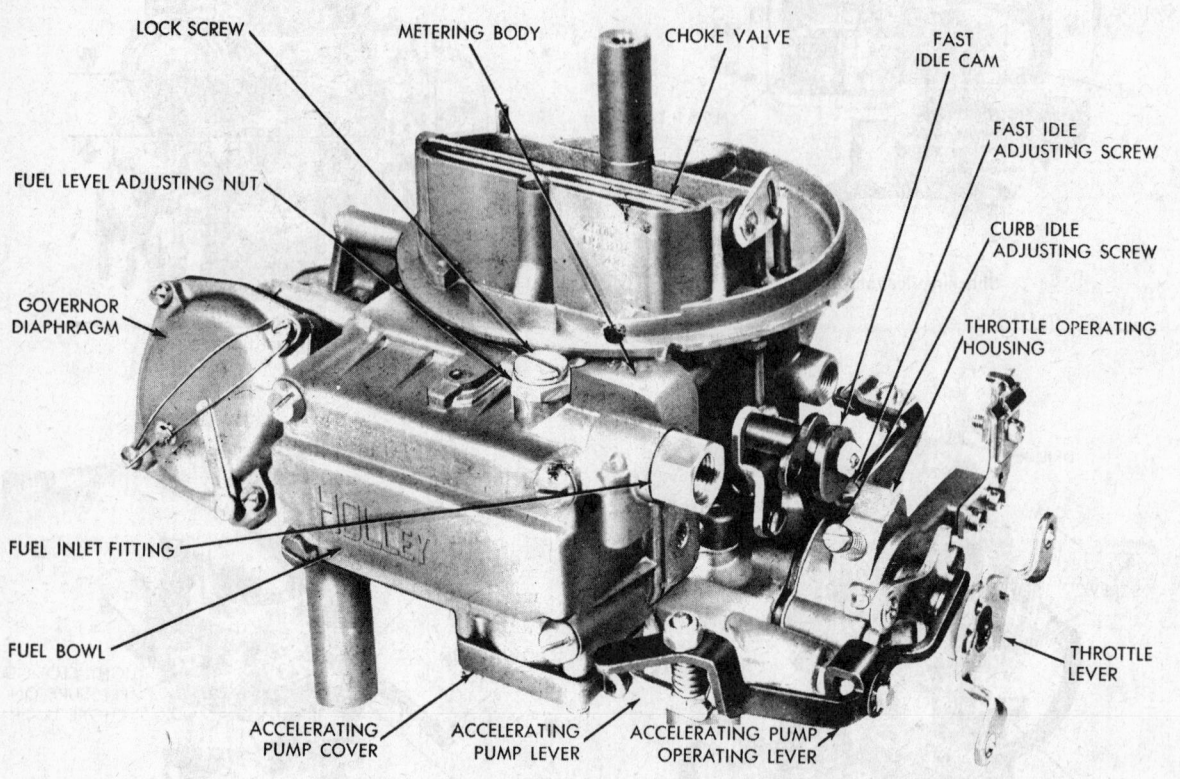

LOCK SCREW

METERING BODY

CHOKE VALVE

FAST IDLE CAM

FUEL LEVEL ADJUSTING NUT

FAST IDLE ADJUSTING SCREW

CURB IDLE ADJUSTING SCREW

GOVERNOR DIAPHRAGM

THROTTLE OPERATING HOUSING

FUEL INLET FITTING

THROTTLE LEVER

FUEL BOWL

ACCELERATING PUMP COVER

ACCELERATING PUMP LEVER

ACCELERATING PUMP OPERATING LEVER

Holley model 2300G carburetor

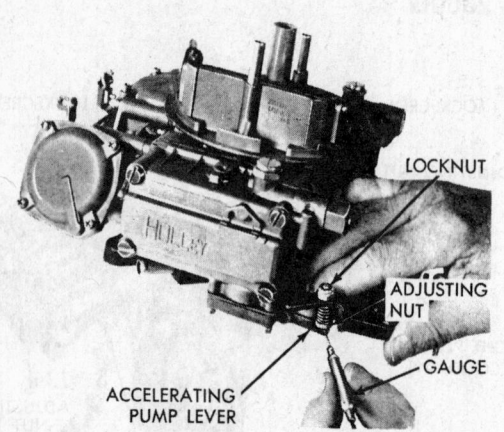

LOCKNUT

ADJUSTING NUT

GAUGE

ACCELERATING PUMP LEVER

Checking accelerating pump lever clearance

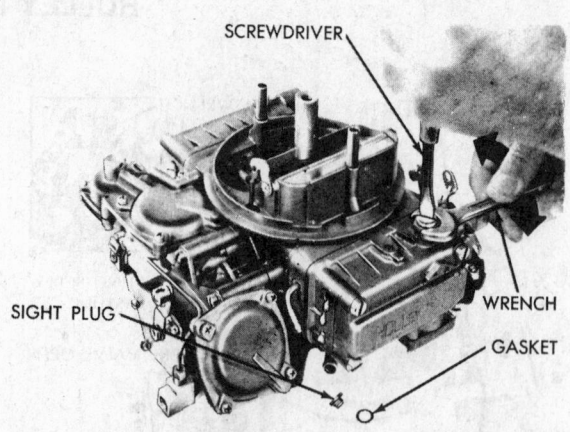

SCREWDRIVER

SIGHT PLUG

WRENCH

GASKET

Adjusting fuel level

HOLLEY MODEL 4150G

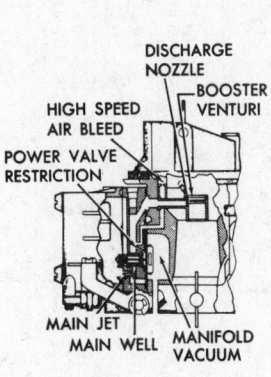

Power system

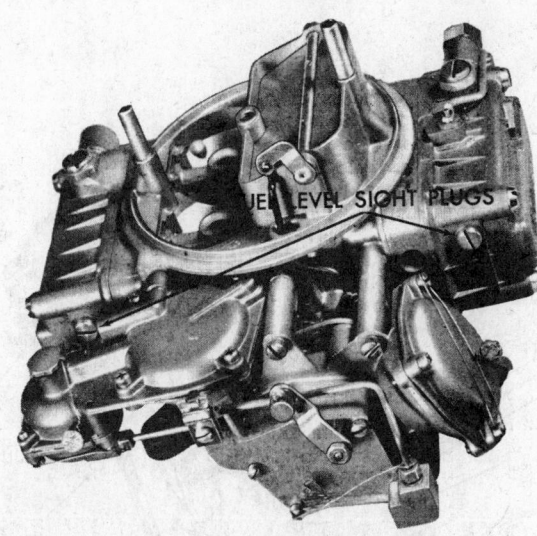

Model 4150G four barrel

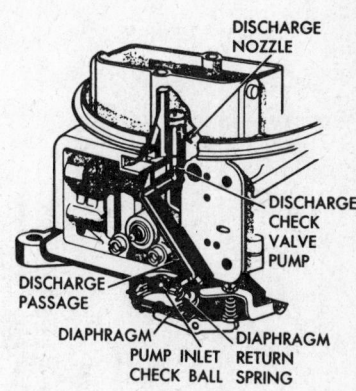

Pump system

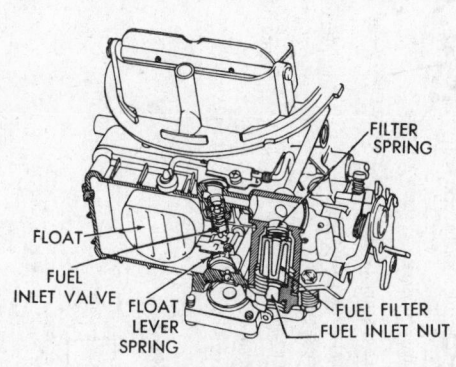

Fuel inlet system

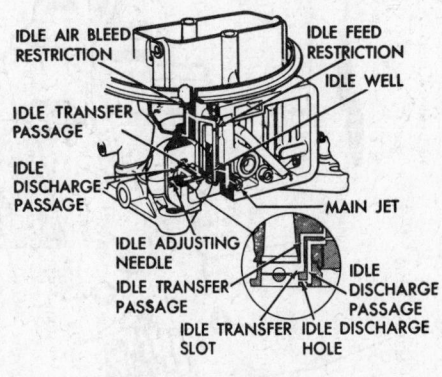

Idle system

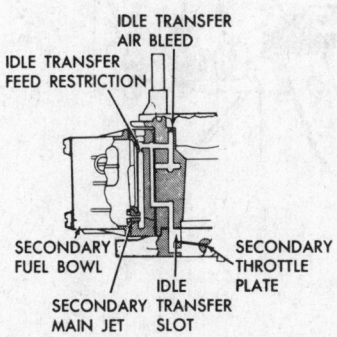

Secondary idle system (idle and low speed)

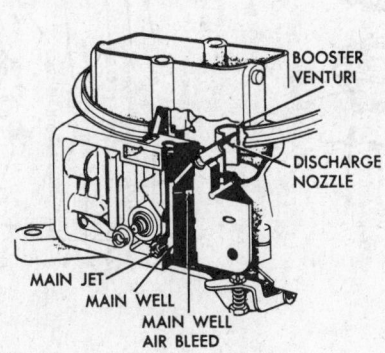

Main metering system

Secondary throttle operating system

HOLLEY TWO BARREL

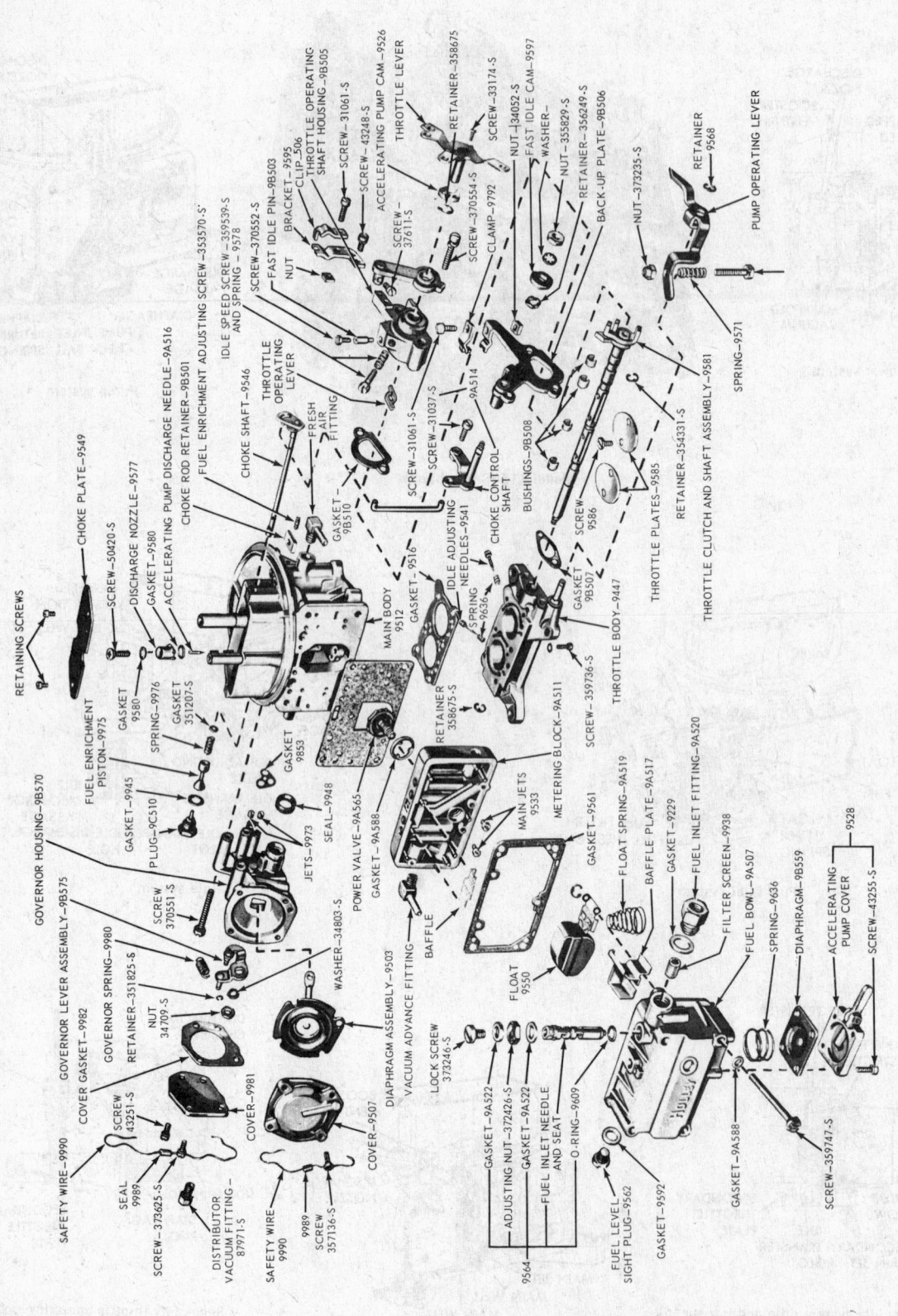

Holley two barrel—typical

HOLLEY FOUR BARREL

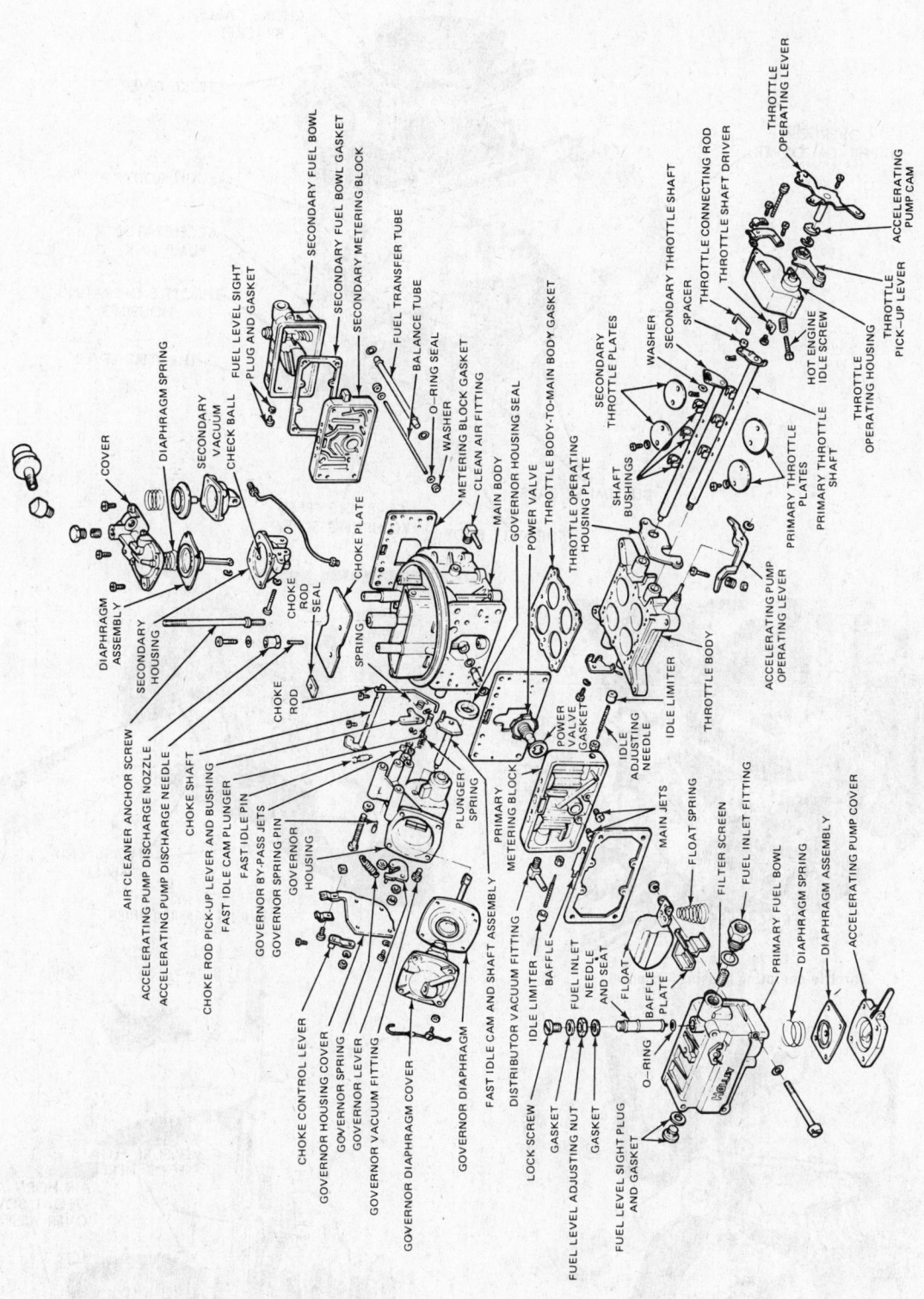

Holley four barrel—typical

HOLLEY MODEL 852 FFG

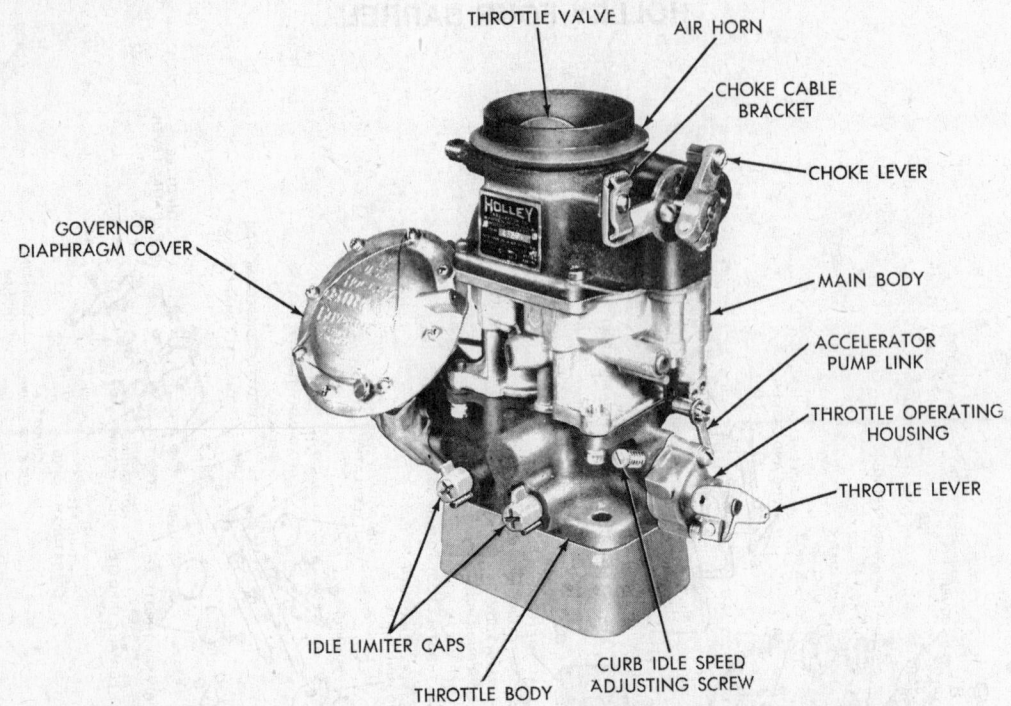

THROTTLE VALVE

AIR HORN

CHOKE CABLE BRACKET

CHOKE LEVER

GOVERNOR DIAPHRAGM COVER

MAIN BODY

ACCELERATOR PUMP LINK

THROTTLE OPERATING HOUSING

THROTTLE LEVER

IDLE LIMITER CAPS

THROTTLE BODY

CURB IDLE SPEED ADJUSTING SCREW

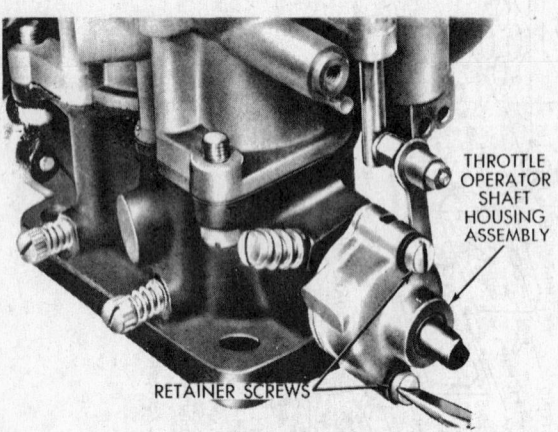

THROTTLE OPERATOR SHAFT HOUSING ASSEMBLY

RETAINER SCREWS

Throttle operating housing—removal

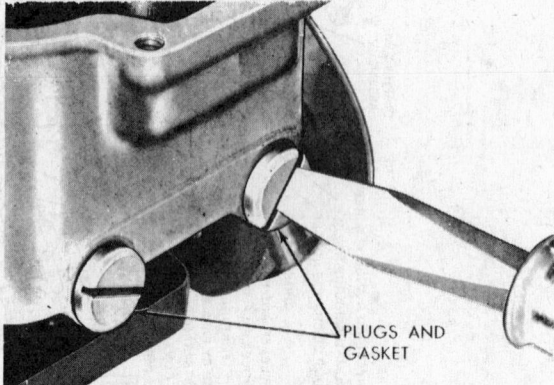

PLUGS AND GASKET

Main jet passage plugs

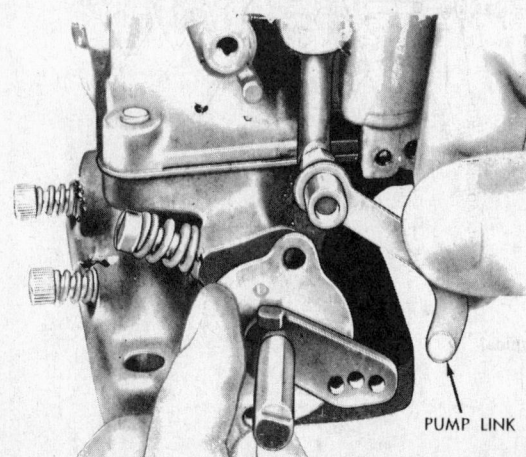

PUMP LINK

Accelerator pump link

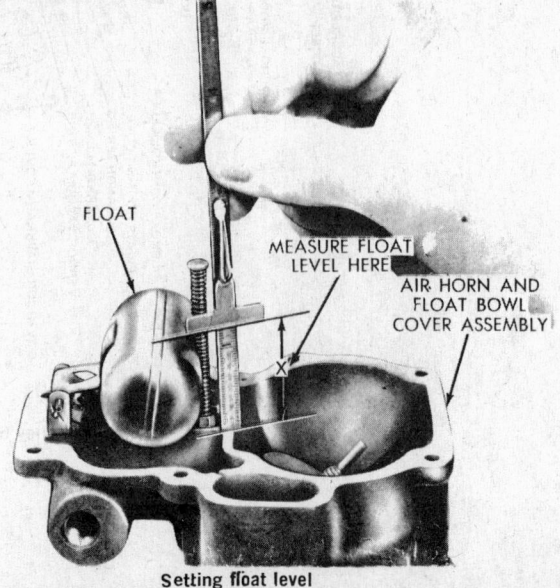

FLOAT

MEASURE FLOAT LEVEL HERE

AIR HORN AND FLOAT BOWL COVER ASSEMBLY

X

Setting float level

HOLLEY MODEL 2110 AND 2110G

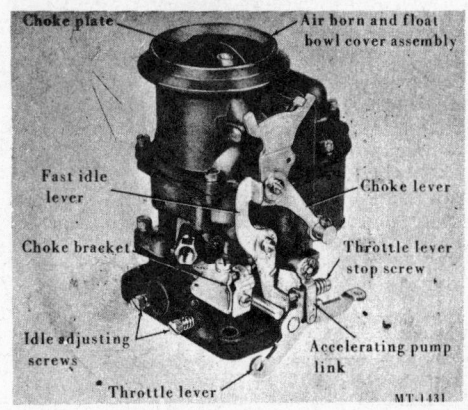

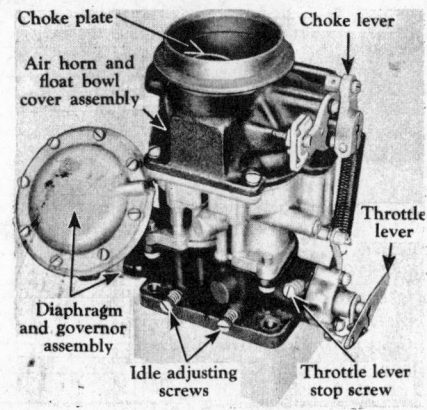

HOLLEY MODEL 2300

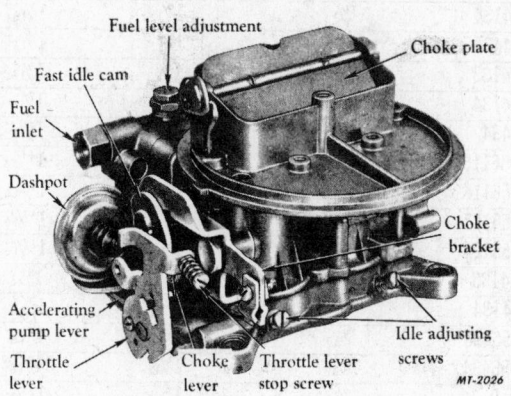

Rochester Carburetors

YEAR	MODEL OR TYPE	FLOAT LEVEL (IN.)	FLOAT DROP (IN.)	PUMP ROD	IDLE VENT
Chevrolet					
	MODEL 2G, 2GC, 2GV				
1971	7041105	$2^{1}/_{32}$	$1^{3}/_{4}$	$1^{3}/_{64}$	——
	7041111	$2^{1}/_{32}$	$1^{3}/_{4}$	$1^{3}/_{64}$	——
	7041125	$2^{1}/_{32}$	$1^{3}/_{4}$	$1^{3}/_{64}$	——
	7041126	$2^{1}/_{32}$	$1^{3}/_{4}$	$1^{3}/_{64}$	——
	7041138	$2^{3}/_{32}$	$1^{1}/_{4}$	$1^{5}/_{32}$	——
	7041139	$2^{3}/_{32}$	$1^{1}/_{4}$	$1^{5}/_{32}$	——
1972	7042102	$2^{1}/_{32}$	$1^{9}/_{32}$	$1^{5}/_{16}$	——
	7042103	$2^{1}/_{32}$	$1^{9}/_{32}$	$1^{5}/_{16}$	——
	7042104	$2^{1}/_{32}$	$1^{9}/_{32}$	$1^{5}/_{16}$	——
	7042105	$2^{1}/_{32}$	$1^{9}/_{32}$	$1^{5}/_{16}$	——
	7042108	$2^{5}/_{32}$	$1^{9}/_{32}$	$1^{1}/_{2}$	——
	7042109	$2^{5}/_{32}$	$1^{9}/_{32}$	$1^{1}/_{2}$	——
	7042123	$2^{3}/_{32}$	$1^{9}/_{32}$	$1^{1}/_{2}$	——
	7042124	$2^{3}/_{32}$	$1^{9}/_{32}$	$1^{1}/_{2}$	——
	7042822	$2^{1}/_{32}$	$1^{9}/_{32}$	$1^{5}/_{16}$	——
	7042823	$2^{1}/_{32}$	$1^{9}/_{32}$	$1^{5}/_{16}$	——
	7042824	$2^{1}/_{32}$	$1^{9}/_{32}$	$1^{5}/_{16}$	——
	7042825	$2^{1}/_{32}$	$1^{9}/_{32}$	$1^{5}/_{16}$	——

Rochester Carburetors

YEAR	MODEL OR TYPE	FLOAT LEVEL (IN.)	FLOAT DROP (IN.)	PUMP ROD	IDLE VENT

Chevrolet Truck, Continued

MODEL—2G Cont.

YEAR	MODEL OR TYPE	FLOAT LEVEL (IN.)	FLOAT DROP (IN.)	PUMP ROD	IDLE VENT
1973	7043103	$2\frac{1}{32}$	$1\frac{9}{32}$	$1\frac{5}{16}$	—
	7043105	$2\frac{1}{32}$	$1\frac{9}{32}$	$1\frac{5}{16}$	—
	7043108	$2\frac{1}{32}$	$1\frac{9}{32}$	$1\frac{7}{16}$	—
	7043123	$2\frac{3}{32}$	$1\frac{9}{32}$	$1\frac{7}{16}$	—
	7043124	$2\frac{3}{32}$	$1\frac{9}{32}$	$1\frac{7}{16}$	—
	7043424	$2\frac{3}{32}$	$1\frac{9}{32}$	$1\frac{7}{16}$	—
	7047796	$2\frac{3}{32}$	$1\frac{9}{32}$	$1\frac{7}{16}$	—
1974	7044113	$1\frac{9}{32}$	$1\frac{9}{32}$	$1\frac{9}{32}$	—
	7044114	$1\frac{9}{32}$	$1\frac{9}{32}$	$1\frac{3}{16}$	—
	7044123	$1\frac{9}{32}$	$1\frac{9}{32}$	$1\frac{9}{32}$	—
	7044124	$1\frac{9}{32}$	$1\frac{9}{32}$	$1\frac{3}{16}$	—
	7044133	$1\frac{9}{32}$	$1\frac{9}{32}$	$1\frac{9}{16}$	—
	7044134	$1\frac{9}{32}$	$1\frac{9}{32}$	$1\frac{7}{16}$	—
	7044434	$1\frac{9}{32}$	$1\frac{9}{32}$	$1\frac{7}{16}$	—
1975	7045115	$2\frac{1}{32}$	$3\frac{1}{32}$	$1\frac{5}{8}$	—
	7045116	$2\frac{1}{32}$	$3\frac{1}{32}$	$1\frac{5}{8}$	—
	7045123	$2\frac{1}{32}$	$3\frac{1}{32}$	$1\frac{5}{8}$	—
	7045124	$2\frac{1}{32}$	$3\frac{1}{32}$	$1\frac{5}{8}$	—
	7044133	$1\frac{9}{32}$	$1\frac{9}{32}$	$1\frac{9}{16}$	—
	7044134	$1\frac{9}{32}$	$1\frac{9}{32}$	$1\frac{7}{16}$	—
	7044434	$1\frac{9}{32}$	$1\frac{9}{32}$	$1\frac{7}{16}$	—
1976	7044133	$1\frac{9}{32}$	$1\frac{9}{32}$	$1\frac{9}{16}$	—
	7044134	$1\frac{9}{32}$	$1\frac{9}{32}$	$1\frac{7}{16}$	—
	704434	$1\frac{9}{32}$	$1\frac{9}{32}$	$1\frac{7}{16}$	—
	17056115	$2\frac{1}{32}$	$1\frac{9}{32}$	$1\frac{11}{16}$	—
	17056116	$2\frac{1}{32}$	$1\frac{9}{32}$	$1\frac{11}{16}$	—
	17056123	$2\frac{1}{32}$	$1\frac{9}{32}$	$1\frac{11}{16}$	—
	17056124	$2\frac{1}{32}$	$1\frac{9}{32}$	$1\frac{11}{16}$	—
1977-78	7044133	$1\frac{9}{32}$	$1\frac{9}{32}$	$1\frac{9}{16}$	—
	7044134	$1\frac{9}{32}$	$1\frac{9}{32}$	$1\frac{7}{16}$	—
	17056433	$1\frac{9}{32}$	$1\frac{9}{32}$	$1\frac{9}{16}$	—
	17056434	$1\frac{9}{32}$	$1\frac{9}{32}$	$1\frac{7}{16}$	—
	All L.D.	$1\frac{9}{32}$	$1\frac{9}{32}$	$1\frac{21}{32}$	—

L.D.—Light Duty (10, 20, 30 series)

MONOJET ONE BARREL

YEAR	MODEL OR TYPE	Float Level (IN.)	Idle Speed Screw Setting	Metering Rod	Choke Rod	Fast Idle (Running)	Idle Vent
1971	7041021	$\frac{1}{4}$	1	.080	.180	2400	—
	7041022	$\frac{1}{4}$	1	.070	.300	2400	—
	7041023	$\frac{1}{16}$	1	—	.120	2400	—
	7041024	$\frac{1}{16}$	1	—	.080	2400	—
	7041025	$\frac{1}{4}$	1	.070	.180	2400	—
	7041026	$\frac{1}{4}$	1	.070	.275	2400	—
1972	7042011	$\frac{1}{4}$	1	.070	.150	2400	—
	7042012	$\frac{1}{4}$	1	.070	.150	2400	—
	7042021	$\frac{1}{4}$	1	.078	.150	2400	—
	7042022	$\frac{1}{4}$	1	.079	.125	2400	—
	7042025	$\frac{1}{4}$	1	.070	.180	2400	—
	7042026	$\frac{1}{4}$	1	.070	.275	2400	—
	7042991	$\frac{1}{4}$	1	.076	.150	2400	—
	7042992	$\frac{1}{4}$	1	.078	.125	2400	—
1973	7043021	$\frac{1}{4}$	1	.080	.275	2400	—
	7043022	$\frac{1}{4}$	1	.080	.245	2400	—
	7043025	$\frac{1}{4}$	1	.070	.350	2400	—
	7043026	$\frac{1}{4}$	1	.070	.375	2400	—
	7043026	$\frac{1}{4}$	1	.070	.375	2400	—

Rochester Carburetors

MONOJET ONE BARREL

YEAR	MODEL OR TYPE	Float Level (IN.)	Idle Speed Screw Setting	Metering Rod	Choke Rod	Fast Idle (Running)	Idle Vent
	7043009	¼	1	.070	——	1800-2400	——
	7043012	¼	1	.070	——	1800-2400	——
	7043312	¼	1	.070	——	1800-2400	——
1974	7044021	.295	——	.080	.275	1800	——
	7044022	.295	——	.080	.245	1800	——
	7044321	.295	——	.080	.300	1800	——
	7044025	¼	——	.070	.245	2400	——
	7044026	¼	——	.070	.275	2400	——
	7044011	¼	——	.070	.150	——	——
	7044012	¼	——	.070	.150	——	——
1975	7045002	¹¹/₃₂	——	.080	.260	1800	——
	7045003	¹¹/₃₂	——	.080	.275	1800	——
	7045004	¹¹/₃₂	——	.080	.245	1800	——
	7045005	¹¹/₃₂	——	.080	.275	1800	——
	7045302	¹¹/₃₂	——	.080	.245	1800	——
	7045303	¹¹/₃₂	——	.080	.275	1800	——
	7045304	¹¹/₃₂	——	.080	.245	1800	——
	7045305	¹¹/₃₂	——	.080	.275	1800	——
	7044012	¼	——	.070	.150	——	——
1976	7044012	¼	——	.070	.150	——	——
	17056002	¹¹/₃₂	——	.080	.130	——	——
	17056003	¹¹/₃₂	——	.080	.145	——	——
	17056004	¹¹/₃₂	——	.080	.130	——	——
	17056006	¼	——	.080	.130	——	——
	17056007	¼	——	.070	.130	——	——
	17056008	¼	——	.070	.150	——	——
	17056009	¹¹/₃₂	——	.070	.150	——	——
	17056302	¹¹/₃₂	——	.080	.155	——	——
	17056303	¼	——	.080	.180	——	——
	17056308	¼	——	.070	.150	——	——
	17056309	¼	——	.070	.150	——	——
1977-78	17057011	¼	——	.065	.150	2400	——
	17057001	⅜	——	.080	.125	——	——
	17057002	⅜	——	.080	.110	——	——
	17057004	⅜	——	.080	.110	——	——
	17057005	⅜	——	.080	.125	——	——
	17057010	⅜	——	.080	.110	——	——
	17057302	⅜	——	.080	.110	——	——
	17057303	⅜	——	.090	.125	——	——
	17057006	⁵/₁₆	——	.070	.150	——	——
	17057007	⁵/₁₆	——	.070	.150	——	——
	17057008	⁵/₁₆	——	.065	.150	——	——
	17057009	⁵/₁₆	——	.065	.150	——	——
	17057308	⁵/₁₆	——	.065	.150	——	——
	17057309	⁵/₁₆	——	.065	.150	——	——

Note 1—Adjust idle speed screw until it just contacts the lever tang. For automatic transmission add another 1¼ turns and 2 turns for standard transmission.

Rochester Carburetors

QUADRAJET FOUR BARREL

YEAR	MODEL OR TYPE	Float Level (IN.)	Air Valve Dashpot	Secondary Metering Rod	PUMP ROD Adj.	PUMP ROD Hole	Idle Vent	Fast Idle (Bench)	Choke Rod	Vacuum Break	Choke Unloader	Air Valve Spring Wind-up
1971	7041206	¼	.020	—	—	—	—	—	.100	.260	.450	—
	7041208	¼	.020	—	—	—	—	—	.100	.260	.450	—
	7041209	¹¹/₃₂	.020	—	—	—	—	—	.100	.260	.450	—
	7041211	¼	.020	—	—	—	—	—	.100	.260	.450	—
1972	7042206	⁷/₃₂	.020	—	—	—	—	—	.100	.250	.450	—
	7042207	⁵/₁₆	.020	—	¹³/₃₂	—	—	—	.100	.250	.450	—
	7042208	³/₁₆	.020	—	¹³/₃₂	—	—	—	.100	.260	.450	—
	7042210	³/₁₆	.020	—	³/₈	—	—	—	.100	.215	.450	—
	7042211	³/₁₆	.020	—	³/₈	—	—	—	.100	.215	.450	—
	7042218	¼	.020	—	³/₈	—	—	—	.100	.250	.450	—
	7042219	⁵/₁₆	.020	—	³/₈	—	—	—	.100	.260	.450	—
	7042910	³/₁₆	.020	—	³/₈	—	—	—	.100	.215	.450	—
	7042911	³/₁₆	.020	—	³/₈	—	—	—	.100	.215	.450	—
1973	7043202	⁷/₃₂	—	—	¹³/₃₂	—	—	—	—	.215	.450	½
	7043203	⁷/₃₂	—	—	¹³/₃₂	—	—	—	—	.215	.450	½
	7043210	⁷/₃₂	—	—	¹³/₃₂	—	—	—	—	.215	.450	½
	7043211	⁷/₃₂	—	—	¹³/₃₂	—	—	—	—	.215	.450	½
	7043208	⁵/₁₆	—	—	¹³/₃₂	—	—	—	—	.215	.450	½
	7043215	⁵/₁₆	—	—	¹³/₃₂	—	—	—	—	.215	.450	½
	7043200	¼	—	—	¹³/₃₂	—	—	—	—	.250	.450	¹¹/₁₆
	7043216	¼	—	—	¹³/₃₂	—	—	—	—	.250	.450	¹¹/₁₆
	7043207	¼	—	—	¹³/₃₂	—	—	—	—	.250	.450	¹¹/₁₆
	7043507	¼	—	—	¹³/₃₂	—	—	—	—	.275	.450	¹¹/₁₆
1974	7044202, 7044502	¼	—	—	¹³/₃₂	—	—	—	.430	.230	.450	⅞
	7044203, 7044503	¼	—	—	¹³/₃₂	—	—	—	.430	.230	.450	⅞
	7044218, 7044518	¼	—	—	¹³/₃₂	—	—	—	.430	.215	.450	⅞
	7044219, 7044519	¼	—	—	¹³/₃₂	—	—	—	.430	.215	.450	⅞
	7044213, 7044513	¹¹/₃₂	—	—	¹³/₃₂	—	—	—	.430	.215	.450	⅞
	7044223, 7044227	.675	—	—	¹³/₃₂	—	—	—	.430	.220	.450	⁷/₁₆
	7044212, 7044217	.675	—	—	¹³/₃₂	—	—	—	.430	.230	.450	⁷/₁₆
	7044512, 7044517	.675	—	—	¹³/₃₂	—	—	—	.430	.230	.450	⁷/₁₆
	7044500, 7044520	.675	—	—	¹³/₃₂	—	—	—	.430	.250	.450	⁷/₁₆
	7044224	¹¹/₃₂	—	—	¹³/₃₂	—	—	—	.430	.215	.450	⅞
	7044214, 7044514	¹¹/₃₂	—	—	¹³/₃₂	—	—	—	.430	.215	.450	⅞
	7044215, 7044515	¹¹/₃₂	—	—	¹³/₃₂	—	—	—	.430	.215	.450	⅞
	7044216, 7044516	¹¹/₃₂	—	—	¹³/₃₂	—	—	—	.430	.215	.450	⅞
1975	7045212	³/₈	.015	—	.275	Inner	—	—	.430	.225	.450	⁷/₁₆
	7045213	¹¹/₃₂	.015	—	.275	Inner	—	—	.430	.210	.450	⅞
	7045214	¹¹/₃₂	.015	—	.275	Inner	—	—	.430	.215	.450	⅞
	7045215	¹¹/₃₂	.015	—	.275	Inner	—	—	.430	.215	.450	⅞
	7045216	¹¹/₃₂	.015	—	.275	Inner	—	—	.430	.210	.450	⅞
	7045217	³/₈	.015	—	.275	Inner	—	—	.430	.225	.450	⁷/₁₆
	7045225	¹¹/₃₂	.015	—	.275	Inner	—	—	.430	.200	.450	¾
	7045229	¹⁵/₃₂	.015	—	.275	Inner	—	—	.430	.200	.450	¾
	7045583	¹¹/₃₂	.015	—	.275	Inner	—	—	.430	.230	.450	⅞
	7045584	¹¹/₃₂	.015	—	.275	Inner	—	—	.430	.230	.450	⅞
	7045585	¹¹/₃₂	.015	—	.275	Inner	—	—	.430	.230	.450	⅞
	7045586	¹¹/₃₂	.015	—	.275	Inner	—	—	.430	.230	.450	⅞
	7045588	¹¹/₃₂	.015	—	.275	Inner	—	—	.430	.230	.450	¾

QUADRAJET FOUR BARREL

YEAR	MODEL OR TYPE	Float Level (IN.)	Air Valve Dashpot	Secondary Metering Rod	PUMP ROD Adj.	Hole	Idle Vent	Fast Idle (Bench)	Choke Rod	Vacuum Break	Choke Unloader	Air Valve Spring Wind-up
1975	7045589	$1\frac{1}{32}$	.015	——	.275	Inner	——	——	.430	.230	.450	$\frac{3}{4}$
	7045202	$1\frac{5}{32}$	.015	——	.275	Inner	——	——	.300	.180/.170[a]	.325	$\frac{7}{8}$
	7045203	$1\frac{5}{32}$	.015	——	.275	Inner	——	——	.300	.180/.170[a]	.325	$\frac{7}{8}$
	7045218	$1\frac{5}{32}$	.015	——	.275	Inner	——	——	.325	.180/.170[a]	.325	$\frac{3}{4}$
	7045219	$1\frac{5}{32}$	.015	——	.275	Inner	——	——	.325	.180/.170[a]	.325	$\frac{3}{4}$
	7045220	$1\frac{7}{32}$	.015	——	.275	Inner	——	——	.300	.200/.550[a]	.325	$\frac{9}{16}$
	7045512	$1\frac{7}{32}$	.015	——	.275	Inner	——	——	.300	.180/.550[a]	.325	$\frac{9}{16}$
	7045517	$1\frac{7}{32}$	.015	——	.275	Inner	——	——	.300	.180/.550[a]	.325	$\frac{9}{16}$
1976	7045213	$1\frac{1}{32}$	.015	——	$\frac{9}{32}$	Inner	——	——	.290	.145	.295	$\frac{7}{8}$
	7045214	$1\frac{1}{32}$	.015	——	$\frac{9}{32}$	Inner	——	——	.290	.145	.295	$\frac{7}{8}$
	7045215	$1\frac{1}{32}$	.015	——	$\frac{9}{32}$	Inner	——	——	.290	.145	.295	$\frac{7}{8}$
	7045216	$1\frac{1}{32}$	.015	——	$\frac{9}{32}$	Inner	——	——	.290	.145	.295	$\frac{7}{8}$
	7045225	$1\frac{1}{32}$	.015	——	$\frac{9}{32}$	Inner	——	——	.290	.138	.295	$\frac{3}{4}$
	7045229	$1\frac{1}{32}$	.015	——	$\frac{9}{32}$	Inner	——	——	.290	.138	.295	$\frac{3}{4}$
	7045583	$1\frac{1}{32}$	.015	——	$\frac{9}{32}$	Inner	——	——	.290	.155	.295	$\frac{7}{8}$
	7045584	$1\frac{1}{32}$	.015	——	$\frac{9}{32}$	Inner	——	——	.290	.155	.295	$\frac{7}{8}$
	7045585	$1\frac{1}{32}$	.015	——	$\frac{9}{32}$	Inner	——	——	.290	.155	.295	$\frac{7}{8}$
	7045586	$1\frac{1}{32}$	.015	——	$\frac{9}{32}$	Inner	——	——	.290	.155	.295	$\frac{7}{8}$
	7045588	$1\frac{1}{32}$	.015	——	$\frac{9}{32}$	Inner	——	——	.290	.155	.295	$\frac{3}{4}$
	7045589	$1\frac{1}{32}$	.015	——	$\frac{9}{32}$	Inner	——	——	.290	.155	.295	$\frac{3}{4}$
	17056212	$\frac{3}{8}$	.015	——	$\frac{9}{32}$	Inner	——	——	.290	.155	.295	$\frac{7}{16}$
	17056217	$\frac{3}{8}$	.015	——	$\frac{9}{32}$	Inner	——	——	.290	.155	.295	$\frac{7}{16}$
	17056208	3	.015	——	$\frac{9}{32}$	Inner	——	——	.325	.185	.325	$\frac{7}{8}$
	17056209	3	.015	——	$\frac{9}{32}$	Inner	——	——	.325	.185	.325	$\frac{7}{8}$
	17056218	$\frac{5}{16}$	.015	——	$\frac{9}{32}$	Inner	——	——	.325	.185	.325	$\frac{7}{8}$
	17056219	$\frac{5}{16}$	.015	——	$\frac{9}{32}$	Inner	——	——	.325	.185	.325	$\frac{7}{8}$
	17056508	3	.015	——	$\frac{9}{32}$	Inner	——	——	.325	.185	.325	$\frac{7}{8}$
	17056509	3	.015	——	$\frac{9}{32}$	Inner	——	——	.325	.185	.325	$\frac{7}{8}$
	17056512	$\frac{7}{16}$	.015	——	$\frac{9}{32}$	Inner	——	——	.325	.185	.275	$\frac{7}{8}$
	17056517	$\frac{7}{16}$	.015	——	$\frac{9}{32}$	Inner	——	——	.325	.185	.275	$\frac{7}{8}$
	17056518	$\frac{5}{16}$	.015	——	$\frac{9}{32}$	Inner	——	——	.325	.185	.325	$\frac{7}{8}$
	17056519	$\frac{5}{16}$	.015	——	$\frac{9}{32}$	Inner	——	——	.325	.185	.325	$\frac{7}{8}$
1977-78	17057202	$1\frac{5}{32}$	.015	——	$\frac{9}{32}$	Inner	——	——	.325	.160	.280	$\frac{7}{8}$
	17057204	$1\frac{5}{32}$	.015	——	$\frac{9}{32}$	Inner	——	——	.325	.160	.280	$\frac{7}{8}$
	17057502	$1\frac{5}{32}$	.015	——	$\frac{9}{32}$	Inner	——	——	.325	.165	.280	$\frac{7}{8}$
	17057582	$1\frac{5}{32}$	.015	——	$\frac{3}{8}$	Outer	——	——	.325	.182	.280	$\frac{7}{8}$
	17057584	$1\frac{5}{32}$	.015	——	$\frac{3}{8}$	Outer	——	——	.325	.180	.280	$\frac{7}{8}$
	17057503	$1\frac{5}{32}$	.015	——	$\frac{9}{32}$	Inner	——	——	.325	.165	.280	$\frac{7}{8}$
	17057504	$1\frac{5}{32}$	.015	——	$\frac{9}{32}$	Inner	——	——	.325	.165	.280	$\frac{7}{8}$
	17057209	$\frac{7}{16}$	.015	——	$\frac{9}{32}$	Inner	——	——	.325	——	.325	$\frac{7}{8}$
	17057218	$\frac{7}{16}$	.015	——	$\frac{9}{32}$	Inner	——	——	.325	.160	.280	$\frac{7}{8}$
	17057222	$\frac{7}{16}$	.015	——	$\frac{9}{32}$	Inner	——	——	.325	.160	.280	$\frac{7}{8}$
	17057518	$\frac{7}{16}$	.015	——	$\frac{9}{32}$	Inner	——	——	.325	.165	.280	$\frac{7}{8}$
	17057522	$\frac{7}{16}$	.015	——	$\frac{9}{32}$	Inner	——	——	.325	.165	.280	$\frac{7}{8}$
	17057586	$\frac{7}{16}$	.015	——	$\frac{3}{8}$	Outer	——	——	.325	.180	.295	$\frac{7}{8}$
	17057588	$\frac{7}{16}$	.015	——	$\frac{3}{8}$	Outer	——	——	.325	.180	.280	$\frac{7}{8}$
	17057219	$\frac{7}{16}$	.015	——	$\frac{9}{32}$	Inner	——	——	.325	.165	.280	$\frac{7}{8}$
	17057519	$\frac{7}{16}$	.015	——	$\frac{9}{32}$	Inner	——	——	.325	.165	.280	$\frac{7}{8}$
	17057512	$\frac{7}{16}$	.015	——	$\frac{9}{32}$	Inner	——	——	.325	.165	.240	$\frac{7}{8}$
	17057517	$\frac{7}{16}$	.015	——	$\frac{9}{32}$	Inner	——	——	.325	.165	.240	$\frac{7}{16}$
	17056212	$\frac{3}{8}$	.015	——	$\frac{9}{32}$	Inner	——	——	.290	.120	.295	$\frac{7}{16}$
	17057221	$\frac{3}{8}$	.015	——	$\frac{9}{32}$	Inner	——	——	.325	——/.160	.325	$\frac{7}{8}$
	17056217	$\frac{3}{8}$	.015	——	$\frac{9}{32}$	Inner	——	——	.290	.120	.295	$\frac{7}{16}$
	17057213	$1\frac{1}{32}$	.015	——	$\frac{9}{32}$	Inner	——	——	.285	.115	.205	$\frac{7}{8}$
	17057215	$1\frac{1}{32}$	.015	——	$\frac{9}{32}$	Inner	——	——	.285	.115	.205	$\frac{7}{8}$
	17057216	$1\frac{1}{32}$	.015	——	$\frac{9}{32}$	Inner	——	——	.285	.115	.205	$\frac{7}{8}$
	17057525	$1\frac{1}{32}$	.015	——	$\frac{9}{32}$	Inner	——	——	.285	.120	.225	$\frac{3}{4}$
	17057514	$1\frac{1}{32}$	.015	——	$\frac{9}{32}$	Inner	——	——	.285	.120	.280	$\frac{7}{8}$

QUADRAJET FOUR BARREL

YEAR	MODEL OR TYPE	Float Level (IN.)	Air Valve Dashpot	Secondary Metering Rod	PUMP ROD Adj.	PUMP ROD Hole	Idle Vent	Fast Idle (Bench)	Choke Rod	Vacuum Break	Choke Unloader	Air Valve Spring Wind-up
1977-78	17057529	$1^{1}/_{32}$	.015	——	$^{9}/_{32}$	Inner	——	——	.285	.110	.205	$^{7}/_{8}$
	17057229	$1^{1}/_{32}$	.015	——	$^{9}/_{32}$	Inner	——	——	.285	.110	.205	$^{7}/_{8}$
	7045583	$1^{1}/_{32}$	.015	——	$^{9}/_{32}$	Inner	——	——	.285	.120	.295	$^{7}/_{8}$
	7045585	$1^{1}/_{32}$	.015	——	$^{9}/_{32}$	Inner	——	——	.285	.120	.295	$^{7}/_{8}$
	7045586	$1^{1}/_{32}$	.015	——	$^{3}/_{8}$	Outer	——	——	.285	.120	.295	$^{7}/_{8}$

Note 1—Vacuum Break. A—Auto. Trans.—.245. Stand. Trans.—.275.
Note 2—Front/Rear
Note 3—Needle seat with groove at upper edge—5/16; without groove—7/16.

YEAR	Carb. Model	Fuel Level (Wet)	Float Level (Dry)	Float Drop	Pump Setting	Idle Vent Setting	Hot Idle Speed (RPM)	Idle Mix Screw Setting	Choke Setting
GMC									
			MODEL 2G						
1971	7041123	——	$^{23}/_{32}$	$1^{3}/_{4}$	$1^{5}/_{32}$	.025	——	——	
	7041124	——	$^{23}/_{32}$	$1^{3}/_{4}$	$1^{5}/_{32}$	.025	——	——	
1972	7042123	——	$^{23}/_{32}$	$1^{9}/_{32}$	$1^{1}/_{2}$	——	——	——	
	7042124	——	$^{23}/_{32}$	$1^{9}/_{32}$	$1^{1}/_{2}$	——	——	——	
1973	7043123	——	$^{23}/_{32}$	$1^{9}/_{32}$	$1^{7}/_{16}$	——	——	——	
	7043124	——	$^{23}/_{32}$	$1^{9}/_{32}$	$1^{7}/_{16}$	——	——	——	
	7043424	——	$^{23}/_{32}$	$1^{9}/_{32}$	$1^{7}/_{16}$	——	——	——	
	7047796	——	$^{23}/_{32}$	$1^{9}/_{32}$	$1^{7}/_{16}$	——	——	——	
1974-76	7044133	——	$1^{9}/_{32}$	$1^{9}/_{32}$	$1^{9}/_{16}$	——	——	——	
	7044134	——	$1^{9}/_{32}$	$1^{9}/_{32}$	$1^{7}/_{16}$	——	——	——	
	7044434	——	$1^{9}/_{32}$	$1^{9}/_{32}$	$1^{7}/_{16}$	——	——	——	
1977-78	7044133	——	$1^{9}/_{32}$	$1^{9}/_{32}$	$1^{9}/_{16}$	——	——	——	
	7044134	——	$1^{9}/_{32}$	$1^{9}/_{32}$	$1^{7}/_{16}$	——	——	——	
	17056433	——	$1^{9}/_{32}$	$1^{9}/_{32}$	$1^{9}/_{16}$	——	——	——	
	17056434	——	$1^{9}/_{32}$	$1^{9}/_{32}$	$1^{7}/_{16}$	——	——	——	
			MODEL 2GV, 2GC						
1971	7041105	——	$2^{1}/_{32}$	$1^{3}/_{4}$	——	——	——	——	.040
	7041125	——	$2^{1}/_{32}$	$1^{3}/_{4}$	——	——	——	——	.040
	7041138	——	$2^{3}/_{32}$	$1^{1}/_{4}$	——	——	——	——	.100
	7041139	——	$2^{3}/_{32}$	$1^{1}/_{4}$	——	——	——	——	.100
1972	7042102	——	$2^{1}/_{32}$	$1^{9}/_{32}$	$1^{5}/_{16}$	——	——	——	.080
	7042104	——	$2^{1}/_{32}$	$1^{9}/_{32}$	$1^{5}/_{16}$	——	——	——	.080
	7042822	——	$2^{1}/_{32}$	$1^{9}/_{32}$	$1^{5}/_{16}$	——	——	——	.080
	7042824	——	$2^{1}/_{32}$	$1^{9}/_{32}$	$1^{5}/_{16}$	——	——	——	.080
	7042103	——	$2^{1}/_{32}$	$1^{9}/_{32}$	$1^{5}/_{16}$	——	——	——	.110
	7042105	——	$2^{1}/_{32}$	$1^{9}/_{32}$	$1^{5}/_{16}$	——	——	——	.110
	7042823	——	$2^{1}/_{32}$	$1^{9}/_{32}$	$1^{5}/_{16}$	——	——	——	.110
	7042825	——	$2^{1}/_{32}$	$1^{9}/_{32}$	$1^{5}/_{16}$	——	——	——	.110
	7042108	——	$2^{5}/_{32}$	$1^{9}/_{32}$	$1^{1}/_{2}$	——	——	——	.170
	7042113	——	$2^{3}/_{32}$	$1^{9}/_{32}$	$1^{1}/_{2}$	——	——	——	.100
	7042833	——	$2^{3}/_{32}$	$1^{9}/_{32}$	$1^{1}/_{2}$	——	——	——	.100
	7042114	——	$2^{3}/_{32}$	$1^{9}/_{32}$	$1^{1}/_{2}$	——	——	——	.100
	7042834	——	$2^{3}/_{32}$	$1^{9}/_{32}$	$1^{1}/_{2}$	——	——	——	.100
	7042100	——	$2^{5}/_{32}$	$1^{31}/_{32}$	$1^{5}/_{16}$	——	——	——	.040
	7042820	——	$2^{5}/_{32}$	$1^{31}/_{32}$	$1^{5}/_{16}$	——	——	——	.040
	7042101	——	$2^{5}/_{32}$	$1^{31}/_{32}$	$1^{5}/_{16}$	——	——	——	.075
	7042821	——	$2^{5}/_{32}$	$1^{31}/_{32}$	$1^{5}/_{16}$	——	——	——	.075
1973	7043105	——	$2^{1}/_{32}$	$1^{9}/_{32}$	$1^{5}/_{16}$	——	——	——	.150
	7043103	——	$2^{1}/_{32}$	$1^{9}/_{32}$	$1^{5}/_{16}$	——	——	——	.150
	7043108	——	$2^{1}/_{32}$	$1^{9}/_{32}$	$1^{5}/_{16}$	——	——	——	.200
1974	7044113	——	$1^{9}/_{32}$	$1^{9}/_{32}$	$1^{9}/_{32}$	——	——	——	.200
	7044114	——	$1^{9}/_{32}$	$1^{9}/_{32}$	$1^{3}/_{16}$	——	——	——	.245
	7044123	——	$1^{9}/_{32}$	$1^{9}/_{32}$	$1^{9}/_{32}$	——	——	——	.200
	7044124	——	$1^{9}/_{32}$	$1^{9}/_{32}$	$1^{3}/_{16}$	——	——	——	.245

YEAR	Carb. Model	Fuel Level (Wet)	Float Level (Dry)	Float Drop	Pump Setting	Idle Vent Setting	Hot Idle Speed (RPM)	Idle Mix Screw Setting	Choke Setting
GMC, Continued									
			MODEL—2 GV, 2 GC						
1975	7045115	——	$2\frac{1}{32}$	$1\frac{9}{32}$	$1\frac{5}{8}$	——	——	——	.400
	7045116	——	$2\frac{1}{32}$	$1\frac{9}{32}$	$1\frac{5}{8}$	——	——	——	.400
	7045123	——	$2\frac{1}{32}$	$1\frac{9}{32}$	$1\frac{5}{8}$	——	——	——	.400
	7045124	——	$2\frac{1}{32}$	$1\frac{9}{32}$	$1\frac{5}{8}$	——	——	——	.400
1976	17056115	——	$2\frac{1}{32}$	$1\frac{9}{32}$	$1\frac{11}{16}$	——	——	——	.260
	17056116	——	$2\frac{1}{32}$	$1\frac{9}{32}$	$1\frac{11}{16}$	——	——	——	.260
	17056123	——	$2\frac{1}{32}$	$1\frac{9}{32}$	$1\frac{11}{16}$	——	——	——	.260
	17056124	——	$2\frac{1}{32}$	$1\frac{9}{32}$	$1\frac{11}{16}$	——	——	——	.260
1977-78	All	——	$1\frac{9}{32}$	$1\frac{9}{32}$	$1\frac{21}{32}$	——	——	——	.260
			MONOJET ONE BARREL						
1971 (M)	7040011	——	$\frac{1}{4}$	——	——	——	——	——	.100
	7040012	——	$\frac{1}{4}$	——	——	——	——	——	.100
(MV)	7041021	——	$\frac{1}{4}$	——	——	——	——	——	.180
	7041022	——	$\frac{1}{4}$	——	——	——	——	——	.300
	7041025	——	$\frac{1}{4}$	——	——	——	——	——	.180
	7041026	——	$\frac{1}{4}$	——	——	——	——	——	.275
1972 (MV)	7042011	——	$\frac{1}{4}$	——	——	——	——	——	.150
	7042012	——	$\frac{1}{4}$	——	——	——	——	——	.150
	7042014	——	$\frac{1}{4}$	——	——	——	——	——	.125
	7042021	——	$\frac{1}{4}$	——	——	——	——	——	.150
	7042022	——	$\frac{1}{4}$	——	——	——	——	——	.125
	7042025	——	$\frac{1}{4}$	——	——	——	——	——	.180
	7042026	——	$\frac{1}{4}$	——	——	——	——	——	.275
	7042991	——	$\frac{1}{4}$	——	——	——	——	——	.150
	7042992	——	$\frac{1}{4}$	——	——	——	——	——	.125
	7042984	——	$\frac{1}{4}$	——	——	——	——	——	.125
	7042987	——	$\frac{1}{4}$	——	——	——	——	——	.150
1973 (M)	7043009	——	$\frac{1}{4}$	——	——	——	——	——	.100
	7043012	——	$\frac{1}{4}$	——	——	——	——	——	.100
	7043312	——	$\frac{1}{4}$	——	——	——	——	——	.100
(MV)	7043021	——	$\frac{1}{4}$	——	——	——	——	——	.275
	7043022	——	$\frac{1}{4}$	——	——	——	——	——	.245
	7043025	——	$\frac{1}{4}$	——	——	——	——	——	.350
	7043026	——	$\frac{1}{4}$	——	——	——	——	——	.375
1974 (MV)	7044021	——	.295	——	——	——	——	——	.275
	7044022	——	.295	——	——	——	——	——	.245
	7044321	——	.295	——	——	——	——	——	.300
	7044025	——	$\frac{1}{4}$	——	——	——	——	——	.245
	7044026	——	$\frac{1}{4}$	——	——	——	——	——	.275
(M)	7044011	——	$\frac{1}{4}$	——	——	——	——	——	.150
	7044012	——	$\frac{1}{4}$	——	——	——	——	——	.150
1975 (MV)	7045002	——	$\frac{11}{32}$	——	——	——	——	——	.260
	7045003	——	$\frac{11}{32}$	——	——	——	——	——	.275
	7045004	——	$\frac{11}{32}$	——	——	——	——	——	.245
	7045005	——	$\frac{11}{32}$	——	——	——	——	——	.275
	7045302	——	$\frac{11}{32}$	——	——	——	——	——	.245
	7045303	——	$\frac{11}{32}$	——	——	——	——	——	.275
	7045304	——	$\frac{11}{32}$	——	——	——	——	——	.245
	7045305	——	$\frac{11}{32}$	——	——	——	——	——	.275
(M)	7044012	——	$\frac{1}{4}$	——	——	——	——	——	.150
1976 (M)	7044012	——	$\frac{1}{4}$	——	——	——	——	——	.150
(1MV)	17056002	——	$\frac{11}{32}$	——	——	——	——	——	.130
	17056003	——	$\frac{11}{32}$	——	——	——	——	——	.145
	17056004	——	$\frac{11}{32}$	——	——	——	——	——	.130
	17056006	——	$\frac{1}{4}$	——	——	——	——	——	.130

Rochester Carburetors

YEAR	Carb. Model	Fuel Level (Wet)	Float Level (Dry)	Float Drop	Pump Setting	Idle Vent Setting	Hot Idle Speed (RPM)	Idle Mix Screw Setting	Choke Rod
GMC, Continued									
			MONOJET ONE BARREL						
1976	17056007	—	1/4	—	—	—	—	—	.130
	17056008	—	1/4	—	—	—	—	—	.150
	17056009	—	1/4	—	—	—	—	—	.150
	17056302	—	11/32	—	—	—	—	—	.155
	17056303	—	11/32	—	—	—	—	—	.180
	17056308	—	1/4	—	—	—	—	—	.150
	17056309	—	1/4	—	—	—	—	—	.150
1977-78 (M)	17057011	—	1/4	—	—	—	—	—	.150
(1ME)	17057001	—	3/8	—	—	—	—	—	.125
	17057002	—	3/8	—	—	—	—	—	.110
	17057004	—	3/8	—	—	—	—	—	.110
	17057005	—	3/8	—	—	—	—	—	.125
	17057010	—	3/8	—	—	—	—	—	.110
	17057302	—	3/8	—	—	—	—	—	.110
	17057303	—	3/8	—	—	—	—	—	.125
	17057006	—	5/16	—	—	—	—	—	.150
	17057007	—	5/16	—	—	—	—	—	.150
	17057008	—	5/16	—	—	—	—	—	.150
	17057009	—	5/16	—	—	—	—	—	.150
	17057308	—	5/16	—	—	—	—	—	.150
	17057309	—	5/16	—	—	—	—	—	.150
			QUADRAJET FOUR BARREL						
1971 (4MV)	7041208	—	1/4	—	—	—	—	—	.100
	7041211	—	1/4	—	—	—	—	—	.100
	7041206	—	1/4	—	—	—	—	—	.100
	7041209	—	11/32	—	—	—	—	—	.100
1972 (4MV)	7042202	—	1/4	—	3/8	—	—	—	.100
	7042203	—	1/4	—	3/8	—	—	—	.100
	7042206	—	1/4	—	13/32	—	—	—	.100
	7042207	—	11/32	—	13/32	—	—	—	.100
	7042208	—	3/16	—	3/8	—	—	—	.100
	7042210	—	3/16	—	3/8	—	—	—	.100
	7042211	—	3/16	—	3/8	—	—	—	.100
	7042215	—	1/4	—	3/8	—	—	—	.100
	7042218	—	1/4	—	3/8	—	—	—	.100
	7042219	—	11/32	—	3/8	—	—	—	.100
	7042220	—	1/4	—	3/8	—	—	—	.100
	7042902	—	1/4	—	3/8	—	—	—	.100
	7042903	—	1/4	—	3/8	—	—	—	.100
	7042910	—	3/16	—	3/8	—	—	—	.100
	7042911	—	3/16	—	3/8	—	—	—	.100
1973 (4QJ)	7043200	—	1/4	—	13/32 [6]	—	—	—	.430
	7043202	—	7/32	—	13/32 [6]	—	—	—	.430
	7043203	—	7/32	—	13/32 [6]	—	—	—	.430
	7043207	—	1/4	—	13/32 [6]	—	—	—	.430
	7043208	—	5/16	—	13/32 [6]	—	—	—	.430
	7043210	—	7/32	—	13/32 [6]	—	—	—	.430
	7043211	—	7/32	—	13/32 [6]	—	—	—	.430
	7043215	—	5/16	—	13/32 [6]	—	—	—	.430
	7043216	—	1/4	—	13/32 [6]	—	—	—	.430
	7043507	—	1/4	—	13/32 [6]	—	—	—	.430
1974 (4QJ)	7044202, 7044502	—	1/4	—	13/32	—	—	—	.430
	7044203, 7044503	—	1/4	—	13/32	—	—	—	.430

QUADRAJET FOUR BARREL

YEAR	Carb. Model	Fuel Level (Wet)	Float Level (Dry)	Float Drop	Pump Setting	Idle Vent Setting	Hot Idle Speed (RPM)	Idle Mix Screw Setting	Choke Rod
1974	7044218, 7044518	——	1/4	——	13/32	——	——	——	.430
	7044219, 7044519	——	1/4	——	13/32	——	——	——	.430
	7044213, 7044513	——	11/32	——	13/32	——	——	——	.430
	7044223, 7044227	——	.675	——	13/32	——	——	——	.430
	7044212, 7044217	——	.675	——	13/32	——	——	——	.430
	7044512, 7044517	——	.675	——	13/32	——	——	——	.430
	7044500, 7044520	——	.675	——	13/32	——	——	——	.430
	7044224	——	11/32	——	13/32	——	——	——	.430
	7044214, 7044514	——	11/32	——	13/32	——	——	——	.430
	7044215, 7044515	——	11/32	——	13/32	——	——	——	.430
	7044216, 7044516	——	11/32	——	13/32	——	——	——	.430
1975 (4MV)	7045212	——	3/8	——	.275	——	——	——	.430
	7045213	——	11/32	——	.275	——	——	——	.430
	7045214	——	11/32	——	.275	——	——	——	.430
	7045215	——	11/32	——	.275	——	——	——	.430
	7045216	——	11/32	——	.275	——	——	——	.430
	7045217	——	3/8	——	.275	——	——	——	.430
	7045225	——	11/32	——	.275	——	——	——	.430
	7045229	——	15/32	——	.275	——	——	——	.430
	7045583	——	11/32	——	.275	——	——	——	.430
	7045584	——	11/32	——	.275	——	——	——	.430
	7045585	——	11/32	——	.275	——	——	——	.430
	7045586	——	11/32	——	.275	——	——	——	.430
	7045588	——	11/32	——	.275	——	——	——	.430
	7045589	——	11/32	——	.275	——	——	——	.430
(M4MC)	7045202	——	15/32	——	.275	——	——	——	.300
	7045203	——	15/32	——	.275	——	——	——	.300
	7045218	——	15/32	——	.275	——	——	——	.325
	7045219	——	15/32	——	.275	——	——	——	.325
	7045220	——	17/32	——	.275	——	——	——	.300
(M4MCA)	7045512	——	17/32	——	.275	——	——	——	.300
	7045517	——	17/32	——	.275	——	——	——	.300
1976 (4MV)	7045213	——	11/32	——	9/32	——	——	——	.290
	7045214	——	11/32	——	9/32	——	——	——	.290
	7045215	——	11/32	——	9/32	——	——	——	.290
	7045216	——	11/32	——	9/32	——	——	——	.290
	7045225	——	11/32	——	9/32	——	——	——	.290
	7045229	——	11/32	——	9/32	——	——	——	.290
	7045583	——	11/32	——	9/32	——	——	——	.290
	7045584	——	11/32	——	9/32	——	——	——	.290
	7045585	——	11/32	——	9/32	——	——	——	.290
	7045586	——	11/32	——	9/32	——	——	——	.290
	7045588	——	11/32	——	9/32	——	——	——	.290
	7045589	——	11/32	——	9/32	——	——	——	.290
	17056212	——	3/8	——	9/32	——	——	——	.290
	17056217	——	3/8	——	9/32	——	——	——	.290
(M4MC)	17056208	——	5	——	9/32	——	——	——	.325
	17056209	——	5	——	9/32	——	——	——	.325
	17056218	——	5/16	——	9/32	——	——	——	.325
	17056219	——	5/16	——	9/32	——	——	——	.325
	17056508	——	5	——	9/32	——	——	——	.325

Rochester Carburetors

QUADRAJET FOUR BARREL

YEAR	Carb. Model	Fuel Level (Wet)	Float Level (Dry)	Float Drop	Pump Setting	Idle Vent Setting	Hot Idle Speed (RPM)	Idle Mix Screw Setting	Choke Rod
1976	17056509	——	5	——	9/32	——	——	——	.325
	17056512	——	7/16	——	9/32	——	——	——	.325
	17056517	——	7/16	——	9/32	——	——	——	.325
	17056518	——	5/16	——	9/32	——	——	——	.325
	17056519	——	5/16	——	9/32	——	——	——	.325
(M4ME)	17056221	——	7/16	——	9/32	——	——	——	.300
1977-78 (M4MC)	17057202	——	15/32	——	9/32	——	——	——	.325
	17057204	——	15/32	——	9/32	——	——	——	.325
	17057502	——	15/32	——	9/32	——	——	——	.325
	17057582	——	15/32	——	3/8	——	——	——	.325
	17057584	——	15/32	——	3/8	——	——	——	.325
	17057503	——	15/32	——	9/32	——	——	——	.325
	17057504	——	15/32	——	9/32	——	——	——	.325
	17057209	——	7/16	——	9/32	——	——	——	.325
	17057218	——	7/16	——	9/32	——	——	——	.325
	17057222	——	7/16	——	9/32	——	——	——	.325
	17057518	——	7/16	——	9/32	——	——	——	.325
	17057522	——	7/16	——	9/32	——	——	——	.325
	17057586	——	7/16	——	3/8	——	——	——	.325
	17057588	——	7/16	——	3/8	——	——	——	.325
	17057219	——	7/16	——	9/32	——	——	——	.325
	17057519	——	7/16	——	9/32	——	——	——	.325
	17057229	——	11/32	——	9/32	——	——	——	.285
	17057525	——	11/32	——	9/32	——	——	——	.285
	17057529	——	11/32	——	9/32	——	——	——	.285
(M4ME)	17057221	——	3/8	——	9/32	——	——	——	.325
(4MV, M4MC)	17057512	——	7/16	——	9/32	——	——	——	.325
	17057517	——	7/16	——	9/32	——	——	——	.325
	17056212	——	3/8	——	9/32	——	——	——	.290
	17056217	——	3/8	——	9/32	——	——	——	.290
	17057213	——	11/32	——	9/32	——	——	——	.285
	17057215	——	11/32	——	9/32	——	——	——	.285
	17057216	——	11/32	——	9/32	——	——	——	.285
	17057525	——	11/32	——	9/32	——	——	——	.285
	17057514	——	11/32	——	9/32	——	——	——	.285
	17057529	——	11/32	——	9/32	——	——	——	.285
	17057229	——	11/32	——	9/32	——	——	——	.285
	7045583	——	11/32	——	9/32	——	——	——	.285
	7045585	——	11/32	——	9/32	——	——	——	.285
	7045586	——	11/32	——	9/32	——	——	——	.285

Note 6—Pump rod location: Inner.
Note 5—Needle seat with groove at upper edge—5/16; without groove—7/16.

ROCHESTER MODEL 2G

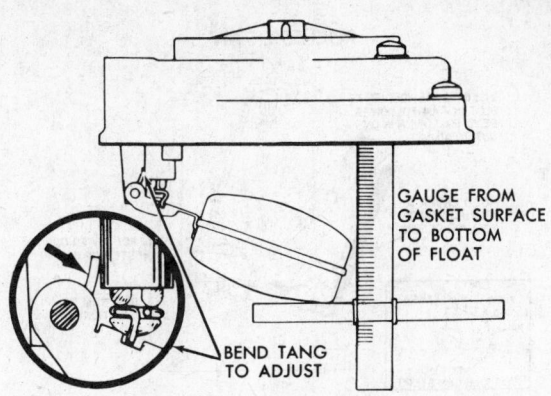

Float drop adjustment

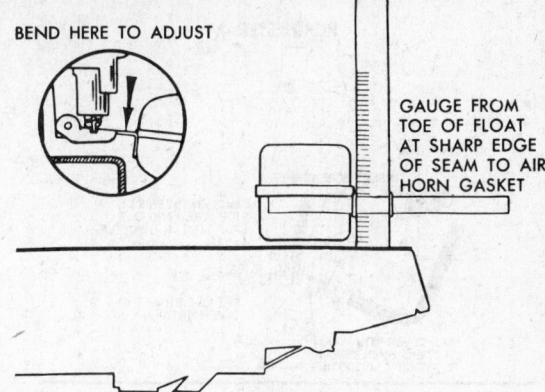

Float level adjustment

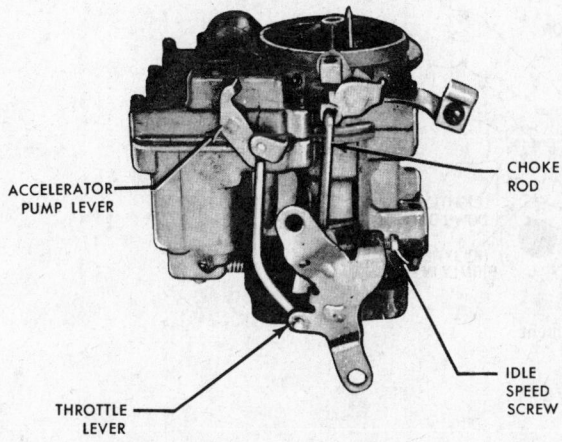

ROCHESTER model 2G carburetor

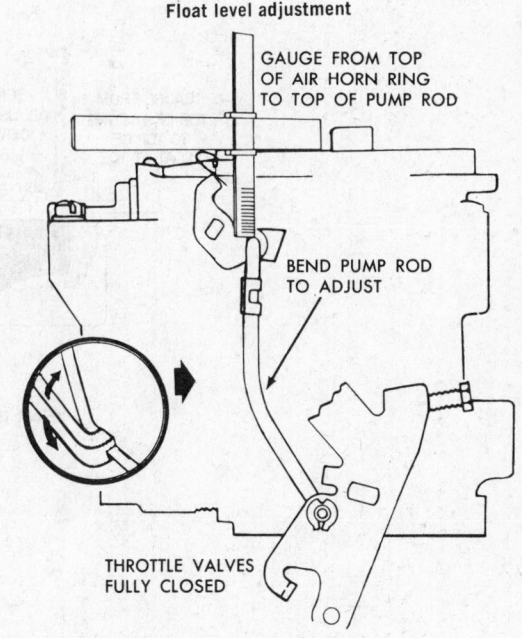

Accelerator pump rod adjustment

ROCHESTER MODELS M AND MV

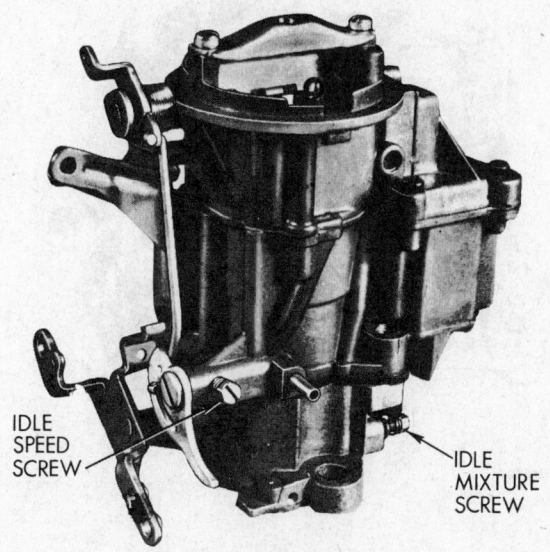

Rochester monojet carburetor

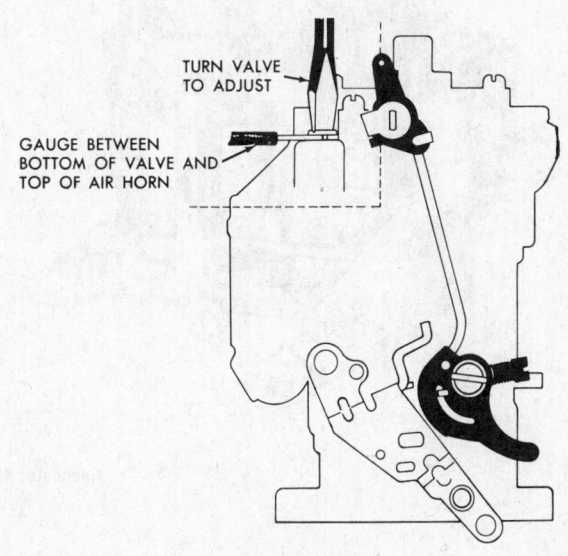

Idle vent adjustment

ROCHESTER MODELS M AND MV

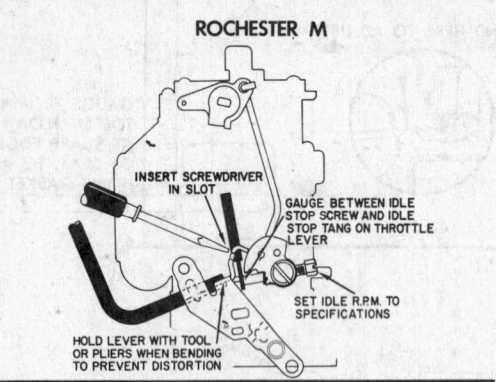

ROCHESTER M

INSERT SCREWDRIVER IN SLOT

GAUGE BETWEEN IDLE STOP SCREW AND IDLE STOP TANG ON THROTTLE LEVER

SET IDLE R.P.M. TO SPECIFICATIONS

HOLD LEVER WITH TOOL OR PLIERS WHEN BENDING TO PREVENT DISTORTION

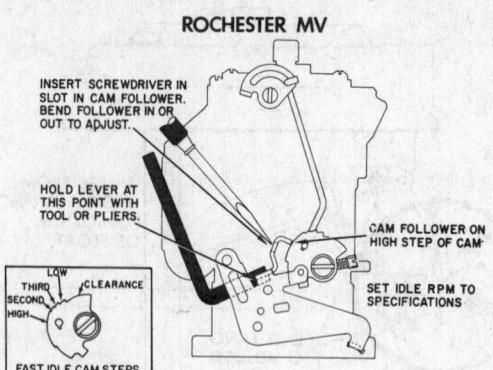

ROCHESTER MV

INSERT SCREWDRIVER IN SLOT IN CAM FOLLOWER. BEND FOLLOWER IN OR OUT TO ADJUST.

HOLD LEVER AT THIS POINT WITH TOOL OR PLIERS.

CAM FOLLOWER ON HIGH STEP OF CAM

SET IDLE RPM TO SPECIFICATIONS

LOW
THIRD   CLEARANCE
SECOND
HIGH

FAST IDLE CAM STEPS

Fast idle adjustment

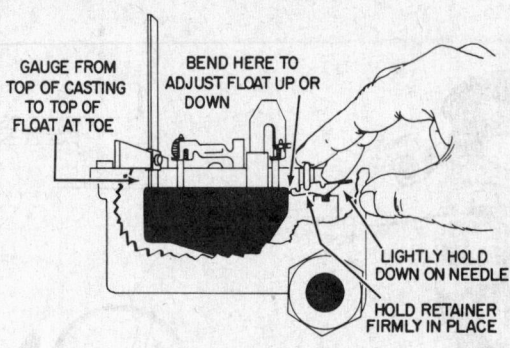

GAUGE FROM TOP OF CASTING TO TOP OF FLOAT AT TOE

BEND HERE TO ADJUST FLOAT UP OR DOWN

LIGHTLY HOLD DOWN ON NEEDLE

HOLD RETAINER FIRMLY IN PLACE

Float level adjustment

ROCHESTER MODEL 4MV

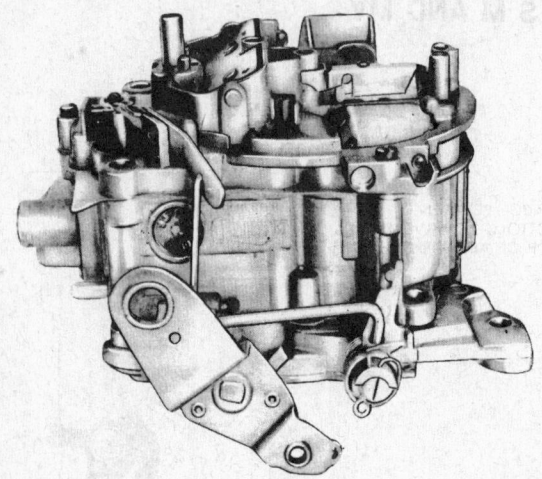

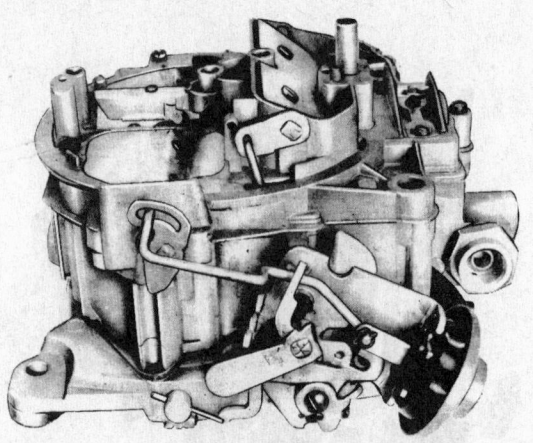

Rochester 4MV carburetor

ROCHESTER MODEL 4MV continued

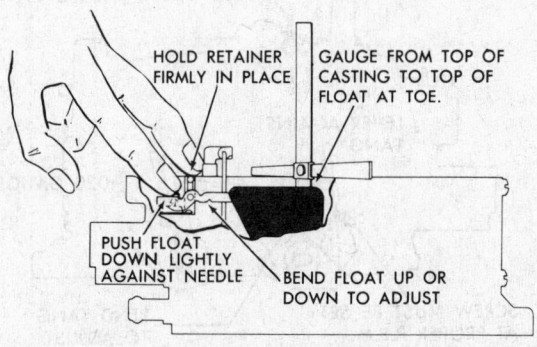

HOLD RETAINER FIRMLY IN PLACE

GAUGE FROM TOP OF CASTING TO TOP OF FLOAT AT TOE.

PUSH FLOAT DOWN LIGHTLY AGAINST NEEDLE

BEND FLOAT UP OR DOWN TO ADJUST

Float level adjustment

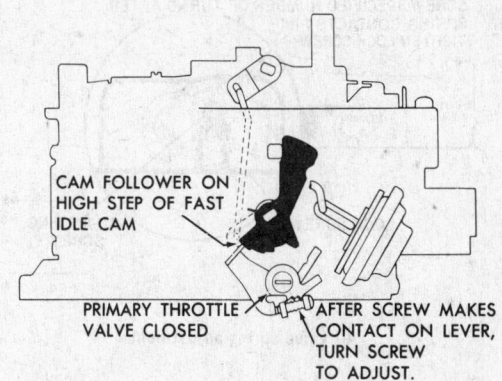

CAM FOLLOWER ON HIGH STEP OF FAST IDLE CAM

PRIMARY THROTTLE VALVE CLOSED

AFTER SCREW MAKES CONTACT ON LEVER, TURN SCREW TO ADJUST.

Fast idle adjustment

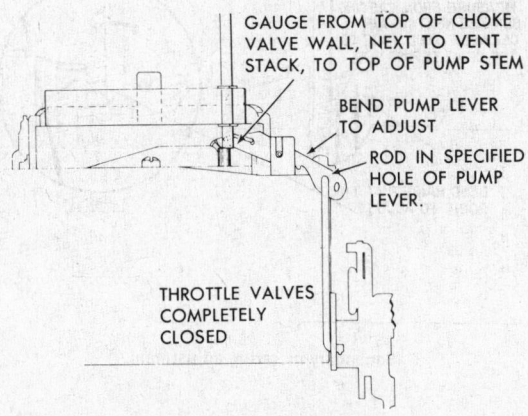

GAUGE FROM TOP OF CHOKE VALVE WALL, NEXT TO VENT STACK, TO TOP OF PUMP STEM

BEND PUMP LEVER TO ADJUST

ROD IN SPECIFIED HOLE OF PUMP LEVER.

THROTTLE VALVES COMPLETELY CLOSED

Pump rod adjustment

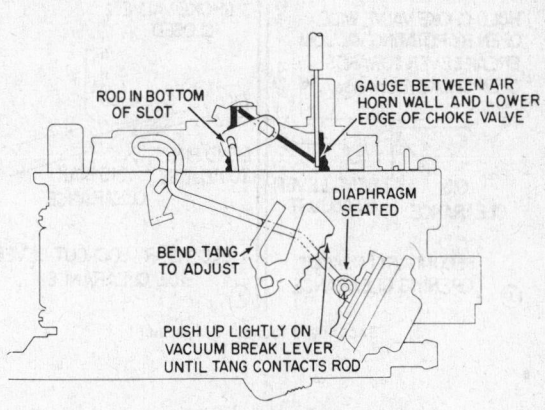

ROD IN BOTTOM OF SLOT

GAUGE BETWEEN AIR HORN WALL AND LOWER EDGE OF CHOKE VALVE

DIAPHRAGM SEATED

BEND TANG TO ADJUST

PUSH UP LIGHTLY ON VACUUM BREAK LEVER UNTIL TANG CONTACTS ROD

Vacuum break adjustment

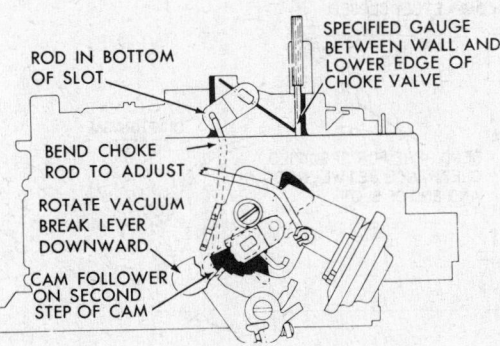

ROD IN BOTTOM OF SLOT

SPECIFIED GAUGE BETWEEN WALL AND LOWER EDGE OF CHOKE VALVE

BEND CHOKE ROD TO ADJUST

ROTATE VACUUM BREAK LEVER DOWNWARD

CAM FOLLOWER ON SECOND STEP OF CAM

Choke rod adjustment

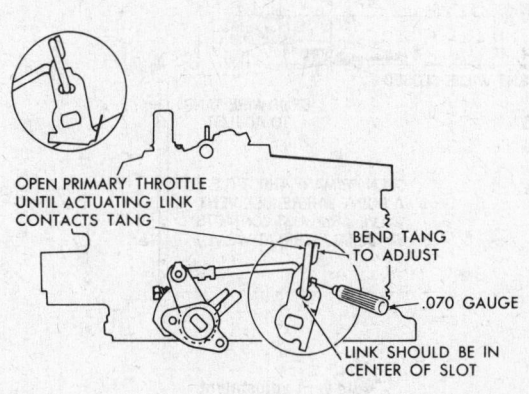

OPEN PRIMARY THROTTLE UNTIL ACTUATING LINK CONTACTS TANG

BEND TANG TO ADJUST

.070 GAUGE

LINK SHOULD BE IN CENTER OF SLOT

Secondary opening adjustment

ROCHESTER MODEL 4MV continued

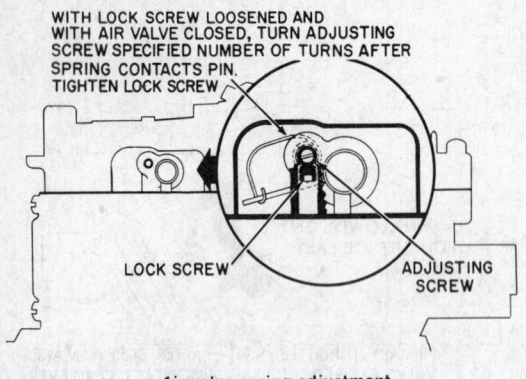

WITH LOCK SCREW LOOSENED AND WITH AIR VALVE CLOSED, TURN ADJUSTING SCREW SPECIFIED NUMBER OF TURNS AFTER SPRING CONTACTS PIN. TIGHTEN LOCK SCREW.

LOCK SCREW

ADJUSTING SCREW

Air valve spring adjustment

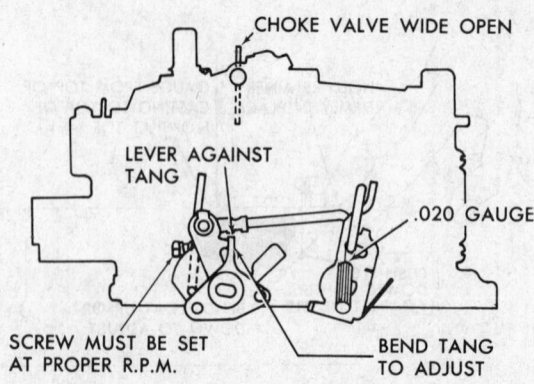

CHOKE VALVE WIDE OPEN

LEVER AGAINST TANG

.020 GAUGE

SCREW MUST BE SET AT PROPER R.P.M.

BEND TANG TO ADJUST

Secondary closing adjustment

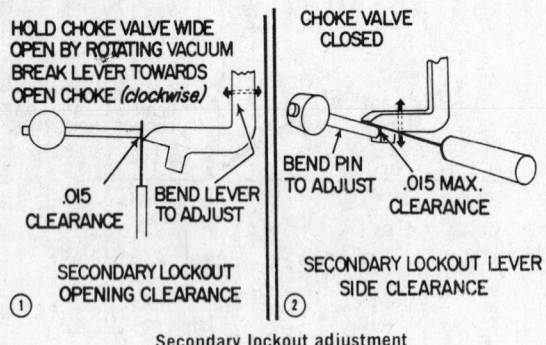

HOLD CHOKE VALVE WIDE OPEN BY ROTATING VACUUM BREAK LEVER TOWARDS OPEN CHOKE (clockwise)

CHOKE VALVE CLOSED

.015 CLEARANCE

BEND LEVER TO ADJUST

BEND PIN TO ADJUST

.015 MAX. CLEARANCE

① SECONDARY LOCKOUT OPENING CLEARANCE

② SECONDARY LOCKOUT LEVER SIDE CLEARANCE

Secondary lockout adjustment

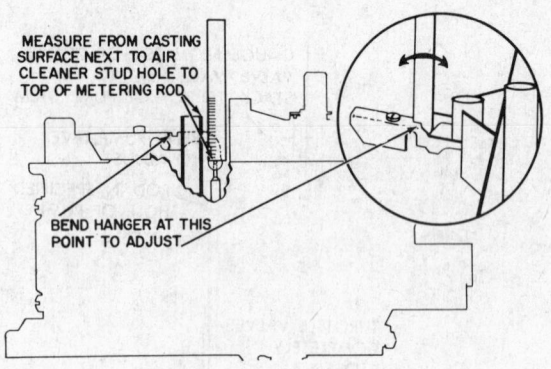

MEASURE FROM CASTING SURFACE NEXT TO AIR CLEANER STUD HOLE TO TOP OF METERING ROD

BEND HANGER AT THIS POINT TO ADJUST

Secondary metering adjustment

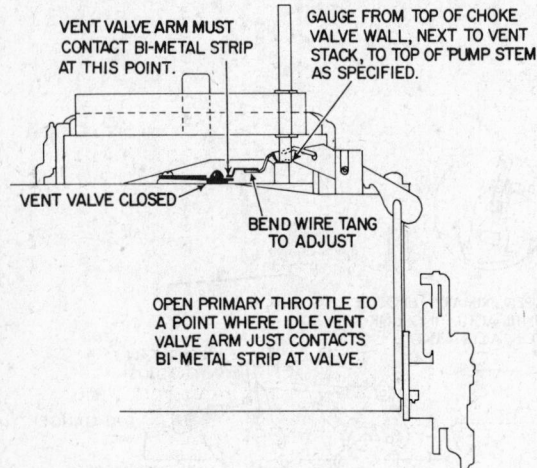

VENT VALVE ARM MUST CONTACT BI-METAL STRIP AT THIS POINT.

GAUGE FROM TOP OF CHOKE VALVE WALL, NEXT TO VENT STACK, TO TOP OF PUMP STEM AS SPECIFIED.

VENT VALVE CLOSED

BEND WIRE TANG TO ADJUST

OPEN PRIMARY THROTTLE TO A POINT WHERE IDLE VENT VALVE ARM JUST CONTACTS BI-METAL STRIP AT VALVE.

Idle vent adjustment

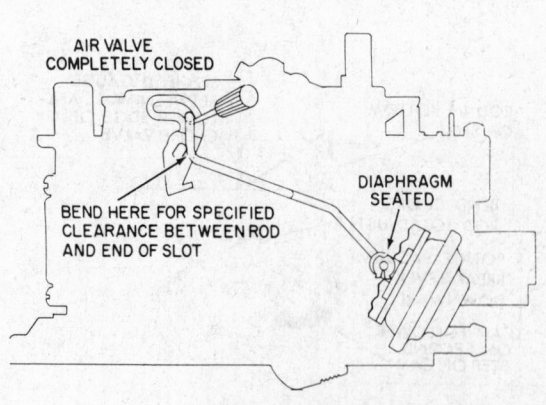

AIR VALVE COMPLETELY CLOSED

DIAPHRAGM SEATED

BEND HERE FOR SPECIFIED CLEARANCE BETWEEN ROD AND END OF SLOT

Air valve dashpot adjustment

Stromberg Carburetors

Dodge/Plymouth

MODEL WW3

YEAR	Model or Type	Float Level	Fast Idle Cam Position	Vacuum Kick	Choke Setting	Bowl Vent	Choke Unloader	Fast Idle Speed	Idle Mix. Screw Adjust
1971-72	310	7/32	—	—	—	—	—	—	1 1/8
	308	7/32	—	—	—	—	—	—	1 7/8
	283	7/32	—	—	—	.060	—	—	1 1/4
	287	7/32	.089	.221	2 R	.060	15/64	700	1 1/4
	298	7/32	.089	.094	2 R	.060	15/64	700	1 1/4
	300	7/32	—	—	—	.050	—	—	1 1/2
	302	7/32	—	—	—	.050	—	—	1 1/2
	303, 304	7/32	—	—	—	—	5/16	1400	1 1/4-1 3/8 1 1/2
	299	7/32	.161	—	Index	.050	5/16	1400	1 1/2
	301	7/32	.161	—	Index	.050	5/16	1400	1 1/2

Chevrolet

MODEL WW, WWC

YEAR	Carb. Model	Fuel Level (Wet)	Float Level (Dry)	Float Drop	Hot Idle Speed (RPM)	Pump Setting	Idle Mix. Screw Setting	Idle Vent Setting	Choke Setting
1973	23-257	—	3/16	—	—	.97	—	—	—
	23-258	—	3/16	—	—	.97	—	—	—
	23-259	—	7/32	—	—	.28	—	—	—
	23-260	—	7/32	—	—	.28	—	—	—
1974	23-264	—	.22	—	—	.97	—	—	—
	23-265	—	.22	—	—	.68	—	—	—

GMC

MODEL WW

YEAR	Carb. Model	Fuel Level (Wet)	Float Level (Dry)	Float Drop	Hot Idle Speed (RPM)	Pump Setting	Idle Mix. Screw Setting	Idle Vent Setting	Choke Setting
1971	671829	—	.190	—	—	.420-.450	—	—	—
1972	23-244	—	.190	—	—	.420-.450	—	—	—
	23-248	—	.190	—	—	.420-.450	—	—	—
1973	23-257	—	3/16	—	—	.97	—	—	—
	23-258	—	3/16	—	—	.97	—	—	—
1974	23-264	—	.22	—	—	.97	—	—	—

MODEL WWC

YEAR	Carb. Model	Fuel Level (Wet)	Float Level (Dry)	Float Drop	Hot Idle Speed (RPM)	Pump Setting	Idle Mix. Screw Setting	Idle Vent Setting	Choke Setting
1973	23-259	—	7/32	—	—	.28	—	—	—
	23-260	—	7/32	—	—	.28	—	—	—
1974	23-265	—	.22	—	—	.68	—	—	—

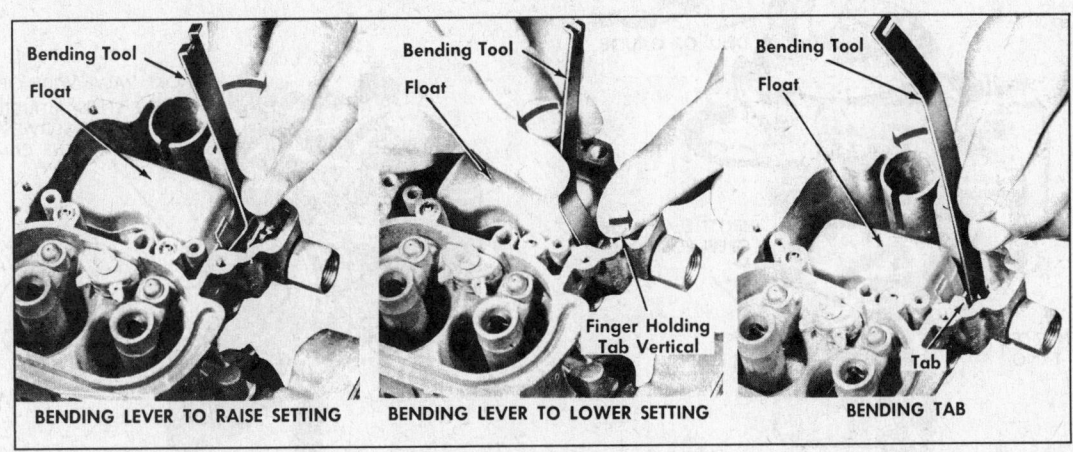

Float level adjustment Type WW-3 carburetor

STROMBERG TYPE—WW3

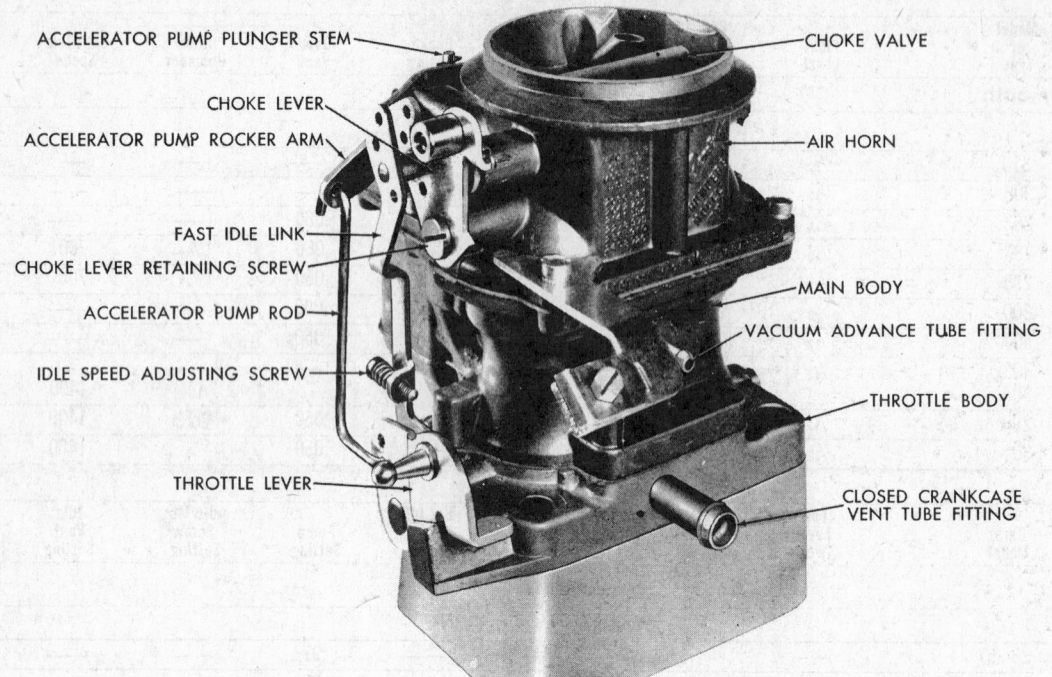

ACCELERATOR PUMP PLUNGER STEM

CHOKE LEVER

ACCELERATOR PUMP ROCKER ARM

FAST IDLE LINK

CHOKE LEVER RETAINING SCREW

ACCELERATOR PUMP ROD

IDLE SPEED ADJUSTING SCREW

THROTTLE LEVER

CHOKE VALVE

AIR HORN

MAIN BODY

VACUUM ADVANCE TUBE FITTING

THROTTLE BODY

CLOSED CRANKCASE VENT TUBE FITTING

Type WW-3 carburetor

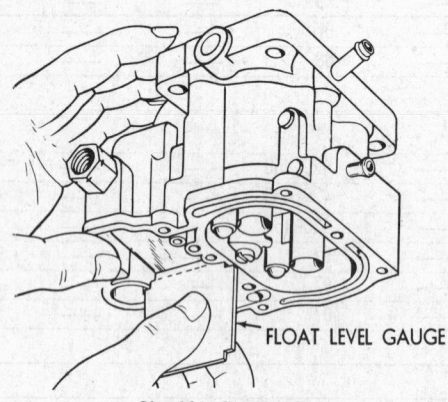

FLOAT LEVEL GAUGE

Checking float setting

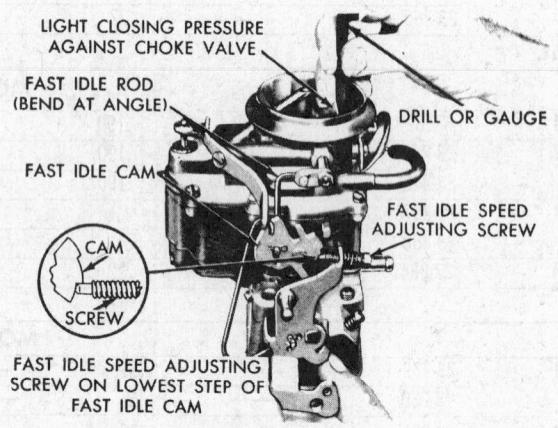

LIGHT CLOSING PRESSURE AGAINST CHOKE VALVE

FAST IDLE ROD (BEND AT ANGLE)

FAST IDLE CAM

DRILL OR GAUGE

FAST IDLE SPEED ADJUSTING SCREW

CAM

SCREW

FAST IDLE SPEED ADJUSTING SCREW ON LOWEST STEP OF FAST IDLE CAM

Fast idle cam adjustment

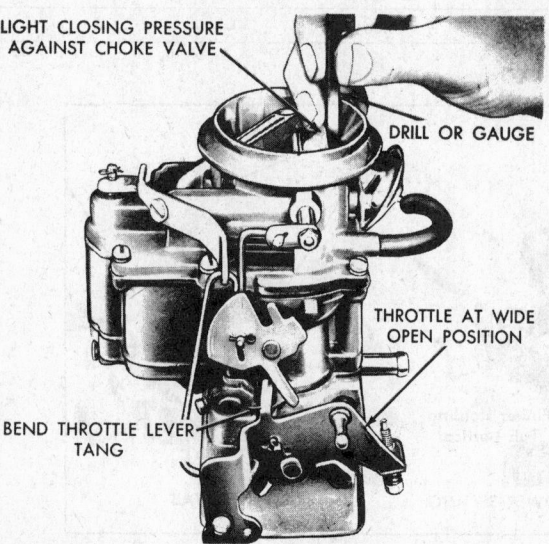

LIGHT CLOSING PRESSURE AGAINST CHOKE VALVE

DRILL OR GAUGE

THROTTLE AT WIDE OPEN POSITION

BEND THROTTLE LEVER TANG

Unloader adjustment (wide open kick)

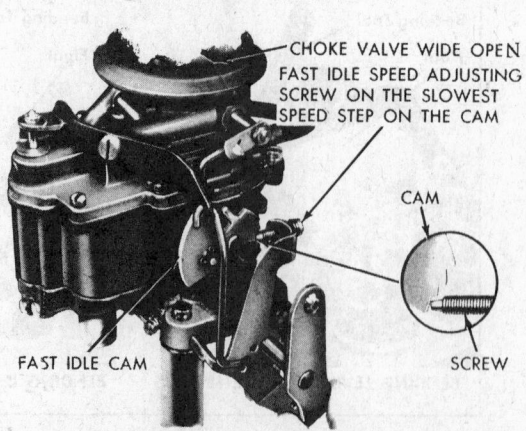

CHOKE VALVE WIDE OPEN
FAST IDLE SPEED ADJUSTING SCREW ON THE SLOWEST SPEED STEP ON THE CAM

CAM

SCREW

FAST IDLE CAM

Fast idle speed adjustment (on vehicle)

Index

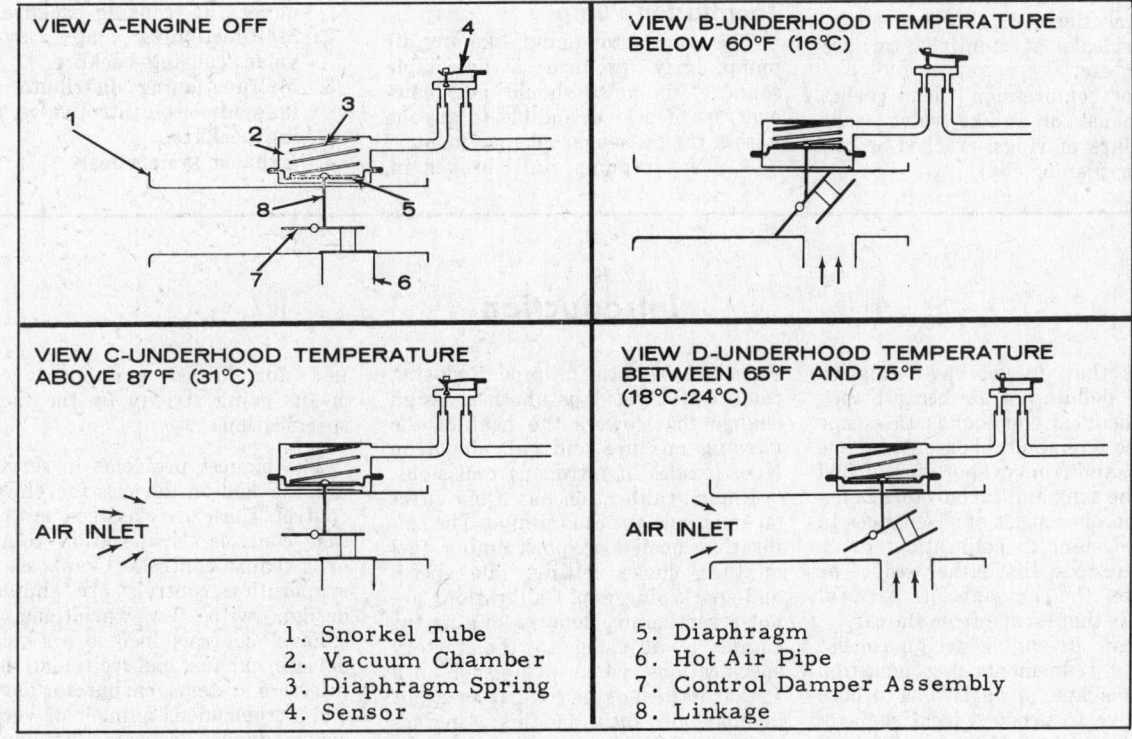

VIEW A-ENGINE OFF

VIEW B-UNDERHOOD TEMPERATURE BELOW 60°F (16°C)

VIEW C-UNDERHOOD TEMPERATURE ABOVE 87°F (31°C)

VIEW D-UNDERHOOD TEMPERATURE BETWEEN 65°F AND 75°F (18°C-24°C)

AIR INLET

1. Snorkel Tube
2. Vacuum Chamber
3. Diaphragm Spring
4. Sensor
5. Diaphragm
6. Hot Air Pipe
7. Control Damper Assembly
8. Linkage

Thermostatically Controlled Air Cleaner Operation

Diagnosis Guide

Symptoms—Rough Idle
Fuel System

1. Leak in vacuum system (lines, connections, diaphragms, etc.).
2. Engine idle too low.
3. Idle fuel mixture incorrect.
4. Air leaks at manifold, carburetor, etc.
5. Power valve leaking fuel.
6. Idle fuel system air bleeds or fuel passages restricted.
7. Secondary throttle not closing (4-V carburetor).
8. Float setting incorrect.
9. Hot and cold air intake system stuck in Heat On position.

Ignition System

1. Defective or poorly adjusted points.
2. Poorly functioning spark plugs.
3. Incorrect ignition timing.
4. Insufficient secondary voltage at plug wires.

Exhaust System

1. Exhaust control valve inoperative (if so equipped).

Engine

1. Leak in vacuum system (lines, connections, diaphragms, etc.).
2. Air leaks at manifold, carburetor, etc.
3. Poor compression (head gasket, exhaust or intake valve leaks, failure of rings, cracked or broken piston, etc.).

4. Inoperative crankcase ventilator (PCV) valve, or restricted tubing.
5. Improperly adjusted valve tappets.
6. Worn camshaft lobes.
7. Incorrect valve timing.

Overheating at Idle

With engine operating, check warning light, or temperature gauge. Trouble could develop in the distributor vacuum control valve (temperature sensing valve).

Engine Stalls
Fuel System

1. Idle speed too low.
2. Idle mixture out of adjustment.
3. Carburetor fast idle too low.
4. Choke out of calibration or adjustment.
5. Choke mechanism binding.
6. Float level too high.
7. Interference with fuel flow.
8. Bad fuel pump.
9. Obstructed fuel lines or tank vent.
10. Carburetor icing.
11. Vapor lock.
12. Inoperative carburetor dashpot.

Engine Noises
Thermactor Pump

The thermactor pump, like any air pump, may produce a detectable sound. This noise should not, however, be of a level audible to anyone within the passenger compartment.

A new air pump, until broken in, may produce a slight squealing or chirping sound.

If pump noise is objectionable, check the following:
1. Drive belt alignment and tension.
2. Loose mounting.
3. Hoses disconnected or leaking.
4. Interference of hoses and body.
5. Defective centrifugal filter fan.
6. Defective relief valve.
7. Improper pressure-setting plug, broken spring, or plug missing. If, after checking the above noise factors, the trouble still persists, replace or repair the pump.

Ignition

1. Ignition, initial timing, too far advanced.
2. Poor engine mounting.
3. Leaking cylinder head gasket.
4. Crankcase ventilator circuit inoperative (stuck open or closed).
5. Poor compression, valves, tappet clearance, or piston and ring assemblies.
6. Worn camshaft lobes.
7. Incorrect valve timing.

Exhaust System

1. Air by-pass valve vacuum line collapsed, plugged, disconnected or leaking, causing backfire.
2. Malfunctioning air by-pass valve, causing backfire.
3. Malfunctioning distributor vacuum advance control valve, causing backfire.
4. Exhaust system leak.

Introduction

Vehicles that do not have emission controls pollute the air because they allow chemical compounds to escape from the engine crankcase, from the exhaust, and from evaporation of fuel out of the tank and carburetor. Emission controls consist of: 1 changes in engine design. 2. calibration, or 3. add-on devices, that either reduce or eliminate the amount of harmful chemicals that escape from the car.

Changes in engine design consist mostly of refinements in combustion chamber shape, or variations in bore and stroke to produce ideal surface-to-volume ratios. If the amount of surface in the combustion chamber is kept to a minimum, the emissions will be reduced because there is less chance for gasoline to cling to the surface without burning. The unburned gasoline is swept out the exhaust and causes high hydrocarbon

emissions from the tailpipe. Reducing compression ratios is another design change that lowers the heat of the burning mixture and cuts down on NOx (oxides of nitrogen) emissions.

Engine calibration has a big effect on emissions out the tailpipe. The calibration consists of spark timing, fuel mixture, choke setting, idle speed, and spark plug gap. Calibrations are not a service problem as long as the engine is adjusted to the factory specifications, which are either on a sticker in the engine compartment, or in manuals such as this. Engines must be adjusted to these factory specifications, or emissions will be high. It is common knowledge that adjustments outside of the specifications will make many (but not all) engines run better. The days when we could adjust engines for best performance are gone. Now we must ad-

just for lowest emissions, which means going strictly by the factory specifications.

The biggest problems in servicing are the add-on devices for emission control. These are classified as Crankcase controls, Evaporation controls, or Exhaust controls. Crankcase and evaporation controls are simple in design, with few variations. But exhaust controls include air cleaner devices, exhaust gas recirculation, air injection systems, carburetor devices, and a tremendous number of vacuum spark advance devices. Following is a description of each group of controls and how they work to reduce emissions.

NOTE: The emission control devices required on light trucks are determined by the weight classification.

Light duty models use the same emission controls as passenger cars; these are models with a Gross Vehicle Weight (GVW) of 6000 lbs or less. In certain years, passenger carrying vehicles such as window vans with a greater GVW were also considered to be light duty models. Heavy duty models operate under less stringent rules and use a few less emission control devices; these have a GVW over 6000 lbs. These general rules do not necessarily apply to vehicles sold in California; this state has stricter emission control regulations.

Crankcase Controls

The first emission control was the positive crankcase ventilation (PCV) system, which appeared in the early 1960s. Ventilation of a crankcase is necessary because of the compression blow-by past the piston rings. This blowby is mostly unburned gasoline. If allowed to stay in the crankcase, it dilutes the oil and increases engine wear. Before PCV systems, the crankcase was vented through a road draft tube. The suction of airflow past the end of the tube drew out the crankcase fumes and fresh air entered through the oil breather cap. When the vehicle was moving, there was a continuous flow of fresh air through the crankcase.

The PCV system accomplishes the same thing, but it uses engine vacuum instead of the road draft to draw out the crankcase fumes. The crankcase or the rocker arm cover is connected by a hose to engine vacuum at the intake manifold or carburetor. When the engine is running, the crankcase fumes are drawn into the engine and burned in the combustion chamber. Fresh air enters the crankcase through the oil filler cap on the open system. When the oil filler cap is connected to the air cleaner, it is known as a closed system.

At wide open throttle, there is little vacuum in the engine, so the PCV system doesn't pull any fumes out of the crankcase. On the open system the fumes go out through the oil filler cap into the atmosphere at wide open throttle. On the closed system the fumes go into the air cleaner, where they are drawn into the engine by the rush of air through the cleaner, so they end up being burned in the engine anyway.

Because the hose connection from the crankcase to the intake manifold acts like a vacuum leak, there has to be some kind of control to limit the air flow. The PCV valve is the control. It can be an actual valve, with an internal plunger, or a simple orifice without any moving parts. In the plunger types, a spring moves the

plunger against engine vacuum, allowing less flow at high vacuum and more flow at low vacuum. If there is an intake manifold cough back or spit back, the plunger moves to close the PCV valve and prevent a crankcase explosion.

Originally all the PCV systems used a simple hose from the rocker cover to the intake manifold or carburetor, with the PCV valve mounted at one end of the hose. Fresh air always entered through the oil filler cap whether it connected to the air cleaner or not. On later models, the plumbing is not as simple, but the principle is still the same. Fresh air enters the air cleaner and goes through a hose to the crankcase or rocker cover. The fumes exit the crankcase and enter the intake manifold, either through a hose or some other type of connection, usually with a PCV valve controlling the flow.

Most systems use some kind of PCV filter, usually mounted at the end of the hose in the air cleaner. The filter keeps dust from entering the crankcase, and also prevents oil fumes from ruining the air cleaner element.

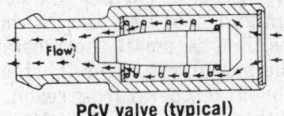

PCV valve (typical)

Testing Crankcase Controls

Checking crankcase vacuum is the most effective way to test any PCV system. If there is vacuum in the crankcase, then the major part of the system has to be working.

Inspect the system to find out where the fresh air enters the engine. This is usually through a hose attached to the air cleaner, but it may be through the oil filler cap on some models. If the fresh air entry is separate from the oil filler cap, remove the hose and plug it so fresh air cannot enter the crankcase. If the fresh

air entry is through the oil filler cap, simply remove the cap.

On all models, use a piece of paper or a PCV tester to measure the crankcase vacuum at the oil filler cap, with the cap removed, and the engine idling in Park or Neutral. It may take a few seconds for the vacuum to build up enough to suck the piece of paper against the oil filler hole. If the vacuum does not build up, check to be sure you have plugged the fresh air entry. An alternate method on some models is to use the piece of paper or PCV tester on the end of the fresh air entry hose. When you do it that way, the oil filler cap must be the solid type and you must leave it in place.

If there is no crankcase vacuum, pull the PCV valve from the crankcase and hold your finger over the end of it. You should feel full manifold vacuum with the engine idling. If not, the valve is plugged or there is an obstruction in a hose or passageway. On some designs the valve may be screwed into its mounting, with a hose leading to the rocker cover or crankcase. If the valve has good suction, but there is no crankcase vacuum, check the hose to be sure it is open. PCV valves that are restricted or plugged must be replaced, unless they are the type that will come apart for cleaning. Lack of crankcase vacuum can also be caused by vacuum leaks at rocker cover, oil pan, or other engine gaskets. Usually, tightening the bolts will stop the leak.

In some extreme cases, usually on high mileage engines, the PCV system is in good shape, but the blowby past the rings is so much that the system can't handle it, and the engine will blow smoke out the oil filler hole. Switching to a PCV valve with a higher flow may temporarily correct the problem, but the only good solution is to do a ring job on the engine.

After checking crankcase vacuum, always check the condition of the fresh air filter and hose, to be sure they are clean and not clogged.

Evaporation Controls

Most evaporation fuel losses come from the fuel tank. On an uncontrolled vehicle the vapors go out through the tank vent, which may be in several places at the top of the

tank, or in the cap. There are also some losses through the bowl vent on the carburetor, but these are minor compared to the tank.

Evaporation controls are made up

of hoses which allow the tank and carburetor vapors to go either to the engine crankcase or to a canister filled with charcoal. When the engine is running, vacuum from the PCV

Emission Control Systems

system cleans the vapors out of the crankcase. If a charcoal canister is used, a hose to the intake manifold or carburetor base allows engine vacuum to pull fresh air through the canister, drawing the vapors into the engine where they are burned. Fresh air enters the canister through a filter, which keeps the charcoal clean.

When the engine is running, air must enter the tank to replace the fuel that is used up and prevent a vacuum. On Chrysler crankcase storage systems, this air enters through a breather cap on the engine. AMC crankcase storage models allow the air to enter through the pressure-vacuum fuel tank cap. Ford crankcase storage models have a 3-way valve mounted forward of the fuel tank on the frame. It opens under vacuum and allows air to enter.

On all makes of canister storage models, air enters the tank through the filter in the canister, but air can also enter the tank through the pressure-vacuum tank cap.

All evaporation control systems use some sort of vapor separator at the fuel tank to prevent liquid fuel from traveling along the vent line to the crankcase or the canister. The early

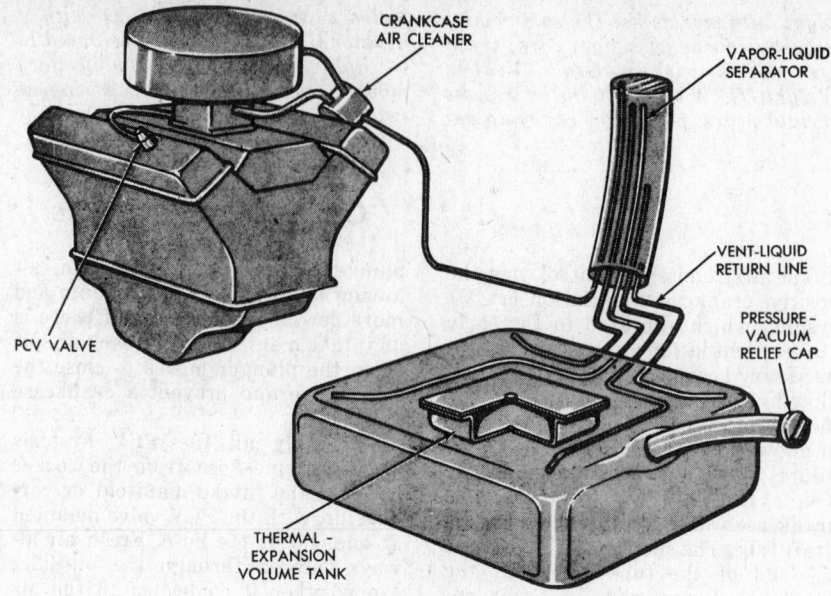

Vapor emission control system components

models had very elaborate separators mounted separately from the tank, but now they are simpler and usually attached to the top of the tank. The

only periodic servicing required on evaporation controls is replacement of the canister filter on those models on which it is replaceable.

Exhaust Controls

Exhaust controls vary considerably in design. There are almost 60 different systems or devices used on the domestic makes to control exhaust emissions. Following are basic descriptions of the common systems.

Thermostatic Air Cleaner

Fresh air supplied to the air cleaner comes either from the normal snorkel, or from a tube connected to an exhaust manifold stove. A door in the snorkel regulates the source of incoming air so that a warm engine always takes in warm air, approximately 100°F. The door may be controlled by a thermostatic spring or expansion bulb, or it may be vacuum operated. The vacuum operated designs use a thermostatic bimetal switch inside the air cleaner that bleeds off vacuum as the engine warms up, and regulates the position of the air door. On all late models, the snorkel is connected to a long tube so it takes in cooler air from outside the engine compartment. In hot climates the cool air tube is necessary because underhood air can easily reach 200°F.

Vacuum operated air doors are all designed so that the air cleaner takes in cold air when there is no vacuum. This means that an air door in the hot air position will switch to the cold position at wide open throttle because of the loss of manifold vacuum. The

sudden switching of the door from hot to cold may cause a stumble or misfire in the engine, so some designs include a modulator valve mounted on the side of the air cleaner to block the vacuum and hold the door in the hot air position. A small thermostat inside the modulator opens it when the underhood temperatures reach normal. Other designs use a delay valve that allows the air door to move to the cold position slowly, to prevent stumble.

Testing Air Cleaners, Non-Vacuum Type

To test the non-vacuum type of heated air cleaner found on some Ford Motor Co. and American Motors Corp. engines, start with an engine that is cold enough to have the air door in the hot air position. Remove the top of the air cleaner and put a thermometer inside the cleaner, then replace the cover without the nuts. Start the engine and watch the air door through the end of the air

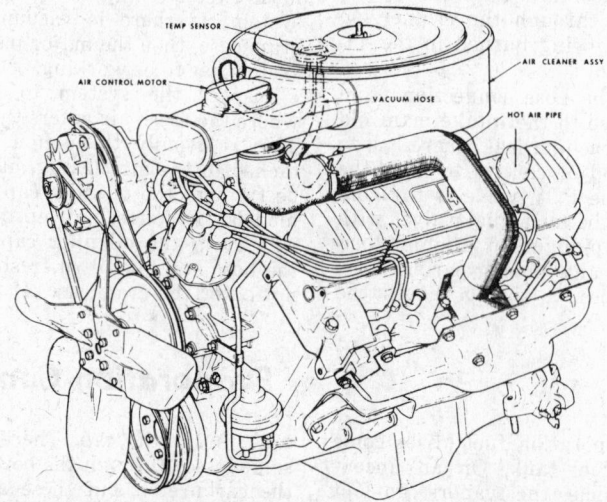

A typical heated air cleaner system, with the hot air pipe connected to the left exhaust manifold (© G.M. Corp.)

cleaner. You may have to remove some air ducting to be able to see the air door. As soon as the air door starts to move from the hot air position, lift the top off the air cleaner and read the temperature. If the temperature is between 130 and 150°F. the thermostat is working correctly. If not, replace the thermostat.

CAUTION: Do not replace the thermostat if the temperature is off by only a few degrees. It must be considerably out of specification, or perhaps not opening at all, to affect the running of the vehicle.

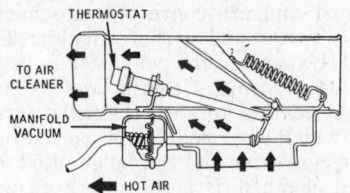

Duct and valve assembly in "heat on" position — warm-up

Testing Air Cleaners, Vacuum Type

To test the vacuum type of heated air cleaner, inspect the air door with the engine off. It should be in the cold air position. Start the engine. If the engine is cold, the air door should move to the hot air position. As the engine warms up, the air door should move to a mid position, depending on the outside air temperature.

If the outside air is extremely cold, the air door may stay in the hot air position indefinitely. On a warm day, after the engine warms up the air door should move to the cold air

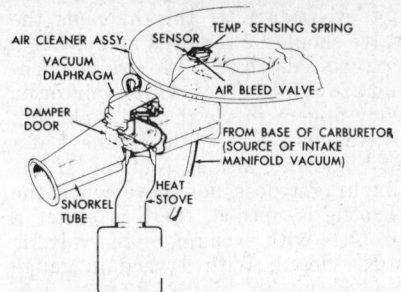

Components of the thermostatically controlled air cleaner (© G.M. Corp)

position. If it doesn't, the temperature sensor inside the air cleaner might be faulty, or the air door itself might be hanging up. Check the air door by running a hose from manifold vacuum to the vacuum motor. Connect and disconnect the hose to see if the air door moves freely. If the air door is free, check out the hoses for leaks or blockage. If the hoses are okay, the trouble must be in the temperature sensor, and it should be replaced.

Both General Motors and Ford use a modulator in the air cleaner vacuum line on some engines. The modulator mounts on the side of the air cleaner and has two hose connections, one to the air cleaner temperature sensor, and the other to the vacuum motor. Below 50-80°F. the modulator is a one-way check valve, which allows vacuum to move the air door to the hot air position, but traps the vacuum so the door will not jump back to the cold air position during acceleration. This prevents a stumble.

After the modulator warms up, the check valve unseats so that the vacuum can pass freely in either

direction, and the air door then operates normally. The connections for the modulator are important. The connection in the center goes to the vacuum motor, and the connection on the edge goes to the vacuum source, which is the temperature sensor.

To test the modulator on a cold engine, apply enough vacuum to the edge port to move the air door to the hot position. Then remove the hose from the port, and the air door should stay in the hot position. Make the same test when the engine is warmed up, and the air door should move to the cold position when you pull off the hose.

Exhaust Gas Recirculation

NOx (oxides of nitrogen) is a tailpipe emission caused by the oxidation of nitrogen in the combustion chamber. When the peak combustion temperatures go over 2500°F. NOx is formed in excessive amounts. To keep the combustion temperatures down, exhaust gas is recirculated on some 1972 and later models. Recirculation is accomplished by allowing intake manifold vacuum to draw exhaust gas into the intake manifold. Usually, an EGR (exhaust gas recirculation) valve is used to control the flow of exhaust gas. All EGR valves look alike, and are operated by vacuum. When the vacuum is off, the valve is closed. Several different types of controls are used to turn the vacuum to the EGR valve on and off. Most of them have to do with engine temperature, as described later.

Ported vacuum EGR systems are the simplest. When the EGR valve hose is connected to the base of the

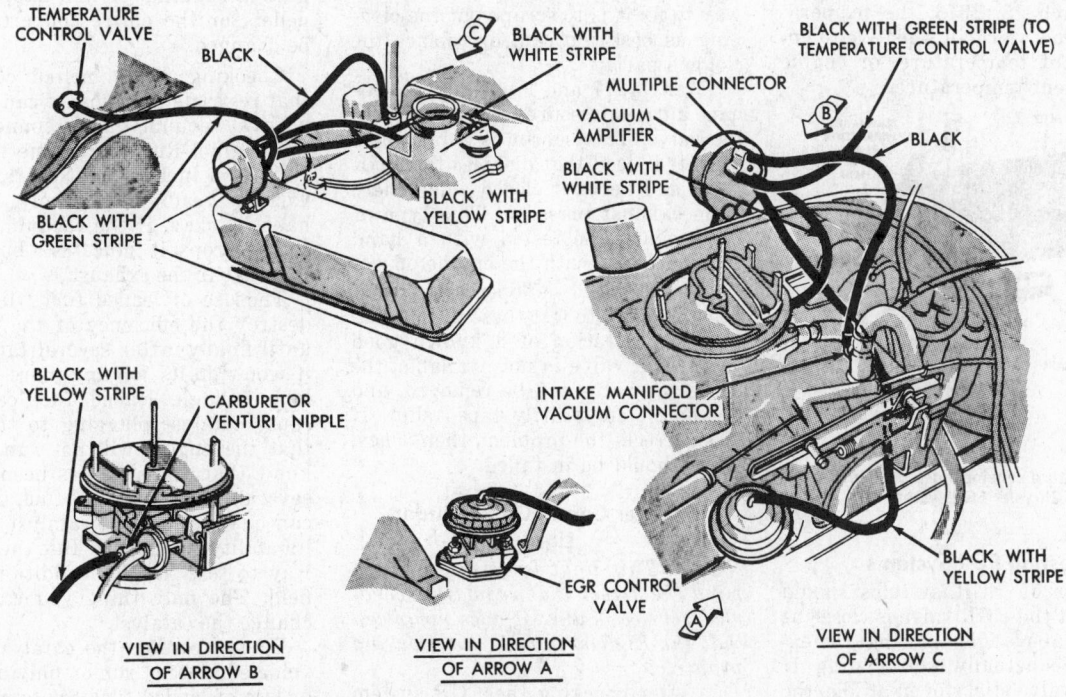

Typical hose routing and component layout of a Chrysler Proportional EGR system

Emission Control Systems

carburetor, without a separate amplifier, the system is operated by ported vacuum. The hose may not run directly from the EGR valve to the carburetor, but may go through a temperature control valve of some sort. In a ported vacuum system, the vacuum to operate the EGR valve is taken from a port that is above the throttle plate at idle, and thus not subject to vacuum. Because there is no vacuum, the spring in the EGR valve closes it, and the exhaust gas does not recirculate. As the throttle is opened, the port is exposed to vacuum, and the EGR valve opens.

Venturi vacuum systems, with an amplifier, are the most complicated, because of the number of hoses. Manifold vacuum is connected to the amplifier by a hose, and then connects to the EGR valve. The amplifier also connects to venturi vacuum. At idle there is no venturi vacuum, but above idle the air moves through the carburetor venturi fast enough to create a vacuum. This slight amount of vacuum opens the amplifier, which then allows manifold vacuum to open the EGR valve.

Temperature controls for EGR systems come in many different designs. They are all made so that the EGR valve stays closed when the engine or the outside air is cold. After the engine or the outside air warms up, the temperature control allows the EGR valve to operate normally. Before March 15, 1973, many EGR systems used a temperature control that was sensitive to outside air temperature. Even with a fully warmed up engine, the EGR system would stay off if the outside temperature was cold enough. On vehicles made after March 15, 1973, the temperature controls were all sensitive to engine coolant temperature, or engine compartment temperature.

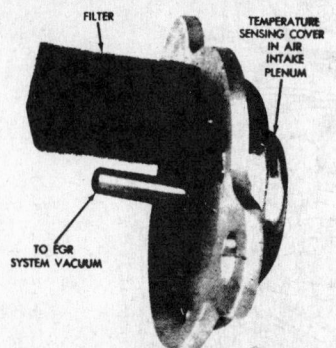

Temperature sensing valve used on the Chrysler EGR system

Testing EGR Systems

Testing of EGR systems should verify that the EGR valve is closed at idle, open above idle, and that the exhaust gas is actually recirculating. If the EGR valve sticks open at idle, the engine will run very rough, or may

not even start. If this happens the valve should be removed and cleaned, or replaced. To check for valve opening above idle, check with a mirror or your fingers to see if the diaphragm or stem moves when the engine is at a fast idle in Park or Neutral. If the diaphragm does not move when the throttle is opened, there is either a problem with vacuum, or the valve is stuck closed. With a vacuum gauge hooked up to the EGR port, you should see vacuum on the gauge when the throttle is opened. EGR valves should not leak when tested with a hand vacuum pump. If they do they must be replaced.

To find out if the exhaust gas is actually recirculating, use a hand vacuum pump or mouth suction through a hose to open the EGR valve with the engine idling. If the engine runs rough or dies, you know the exhaust gas is recirculating. If the engine does not run rough, make a second test at 2500 rpm. Opening the EGR valve at that rpm should cause a change in engine speed. If it does, you know the exhaust gas is recirculating. To make the 2500 rpm test, remove and plug the hose from the EGR port. Attach your suction hose to the EGR valve before running the engine at 2500 rpm. Simply pulling off the EGR hose at 2500 rpm is not a valid test, because the extra air entering the engine through the hose could cause a speed change all by itself. On most engines you won't have to go this far, because opening the EGR valve at idle will prove that the exhaust is recirculating.

If the exhaust is not recirculating, it means that a passageway or the valve itself is clogged up. The only way to fix it is to scrape out the clogging as best you can, or replace the clogged part.

Many 1977 and later EGR valves have a back-pressure sensor built into the valve. This sensor is a pressure-operated bleed that disables the EGR valve and keeps it closed when there is no exhaust pressure. This type of valve cannot be tested with a hand vacuum pump with the engine off because the bleed is open. The only practical way to test these new valves is by substitution of a known good valve. If a valve is not available, the suspected valve can be removed, and the holes temporarily taped shut. If this corrects the problem, then a new valve should be installed.

Chrysler Corp. EGR Reminder Light

NOTE: This light is designed to remind the driver that regularly scheduled service is due; it does not mean that the EGR system is not working properly.

1. After checking the EGR system for proper operation, slide the

rubber boot on the EGR reminder odometer on the speedometer cable up, out of the way.
2. Reset the odometer with a small screwdriver.
3. Slide the boot back down over the odometer. The light will come on again when the next 15,000 mile check-up is due.

Catalytic Converter

A catalytic converter is a chamber in the exhaust system that contains a catalyst. When hydrocarbons or carbon monoxide pass over the catalyst they react with the oxygen in the exhaust and are converted into harmless water and carbon dioxide. The catalyst inside the converter is made in two forms. General Motors and American Motors use the pellet form, in which loose pellets are packed into the converter and can be emptied out and changed, if necessary. Ford and Chrysler use the honeycomb catalyst, which is built into the converter shell and is not replaceable. On Ford and Chrysler products the entire converter must be replaced if it goes bad.

There is no way to test a converter in the field to see if it is actually working. Tailpipe readings may be used to set carburetor idle mixtures, when the car maker requires it, but taking a tailpipe reading to determine if the converter is working is not possible.

The one field check that is recommended in all cases is to inspect for mechanical damage. If a converter gets overheated, the catalyst can melt and block the exhaust. Pellets or pieces of the catalyst may even come flying out the tailpipe while the engine is running. If this happens, the pellets or the entire converter must be changed.

Checking for a melted converter that restricts the exhaust can be done with a vacuum gauge connected to the engine. Run the engine at about 2500 rpm in Park or Neutral. If the vacuum reading is steady, the exhaust is okay. If the vacuum reading slowly drops, it indicates a buildup of pressure in the exhaust.

The use of leaded fuel will slowly destroy the efficiency of the catalyst until finally, after several tanks full, it won't do its job any more. If used long enough, leaded fuel can even cause catalyst plugging to the point that the engine will not run. If you know that a vehicle has been run on several tanks of leaded fuel, then you can be sure that the catalyst has lost its ability to convert. But there is no way to test for this condition in the field. The only thing you can do is change the catalyst.

Do not change the catalyst if the vehicle has been run on only one tank or less of leaded fuel. Switching back to lead free fuel will allow the cata-

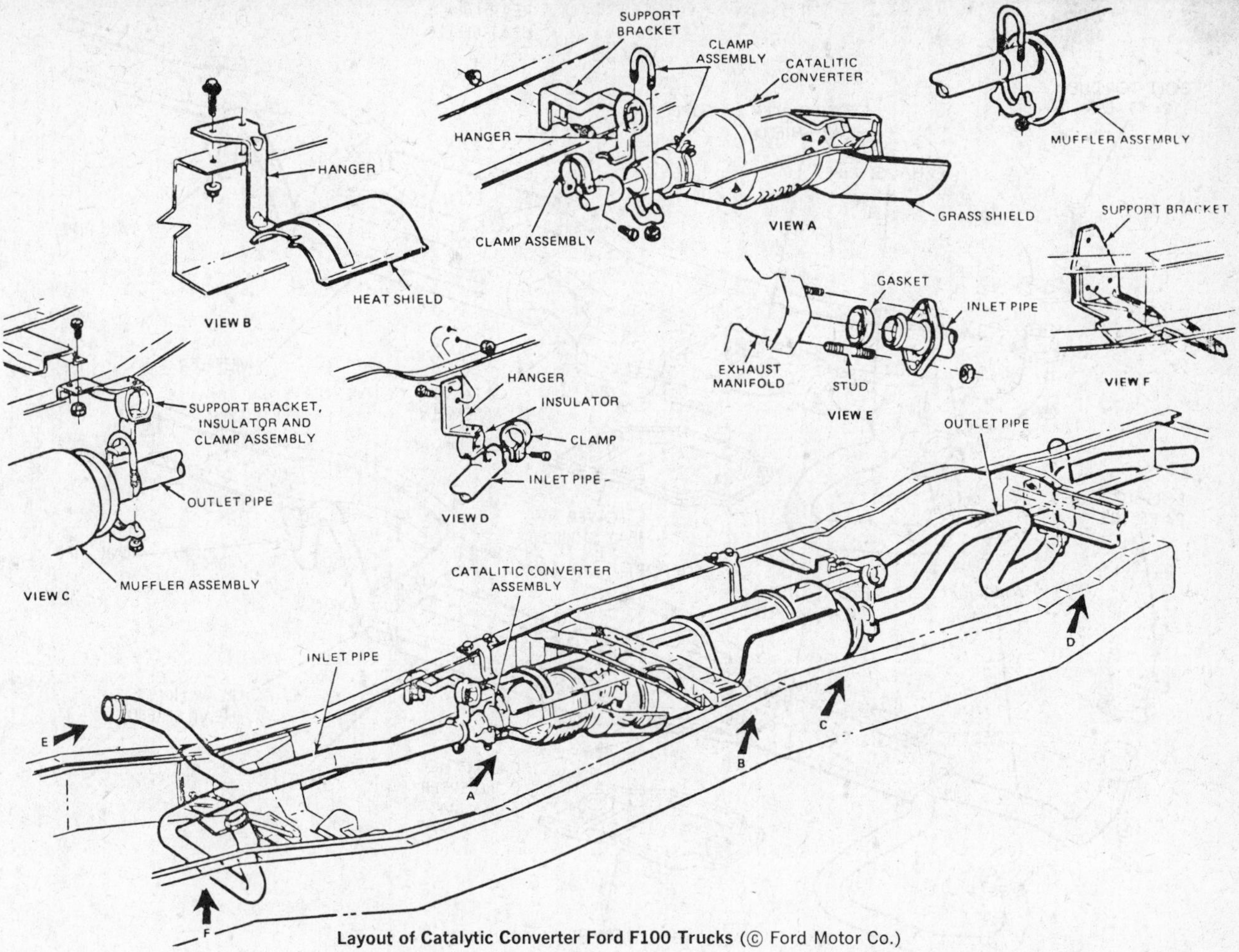

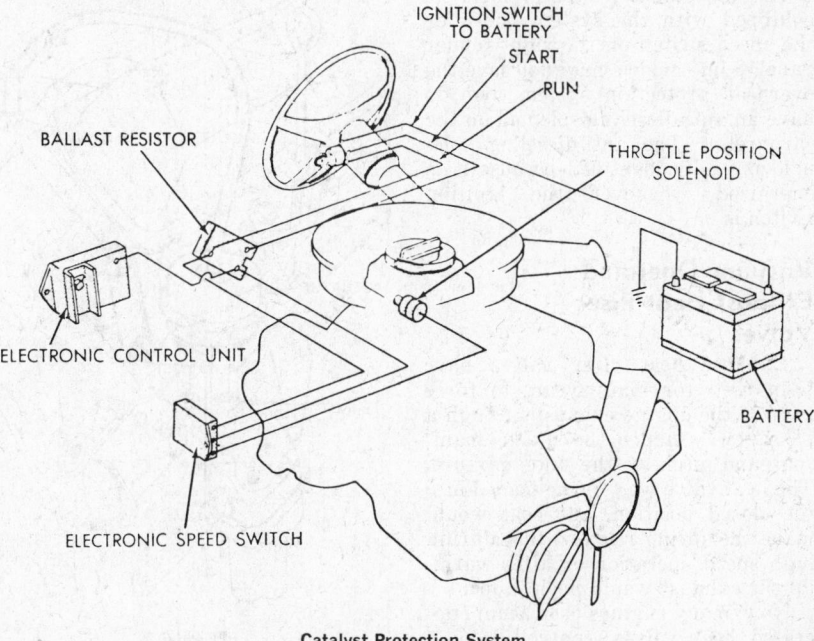

Layout of Catalytic Converter Ford F100 Trucks (© Ford Motor Co.)

lyst to recover and be almost as efficient as it was.

Converter Overheat Protection

Some models have overheat protection systems for the converter. Ford Motor Co. sometimes uses a heat sensitive switch mounted in the floorpan above the converter. The switch turns a vacuum solenoid on and off to control the vacuum to the air pump bypass valve. When the vacuum is shut off the bypass valve dumps the pump air into the atmosphere so that it doesn't pump into the exhaust any more. Without the air in the exhaust, the converter can't convert, and it cools down.

Chrysler Corporation uses an overheat protection system that holds the throttle open to prevent high speed closed throttle deceleration. Any engine decelerating on closed throttle is usually running rich, because the high vacuum pulls so much fuel out of the carburetor bowl through the idle circuit.

To prevent this, Chrysler uses a solenoid on the carburetor that is identical to an anti-dieseling solenoid. The solenoid is controlled by an electronic speed switch so that it only comes on when the engine speed is above 2000

rpm. When the solenoid is on, its stem extends to the equivalent of a 1500 rpm fast idle setting. If the driver takes his foot off the throttle, the throttle does not close, but rests against the extended solenoid stem.

The solenoid goes off below 2000 rpm so that the engine doesn't run away in traffic.

To test the system put the transmission in Park or Neutral and operate the throttle from under the hood.

Catalyst Protection System

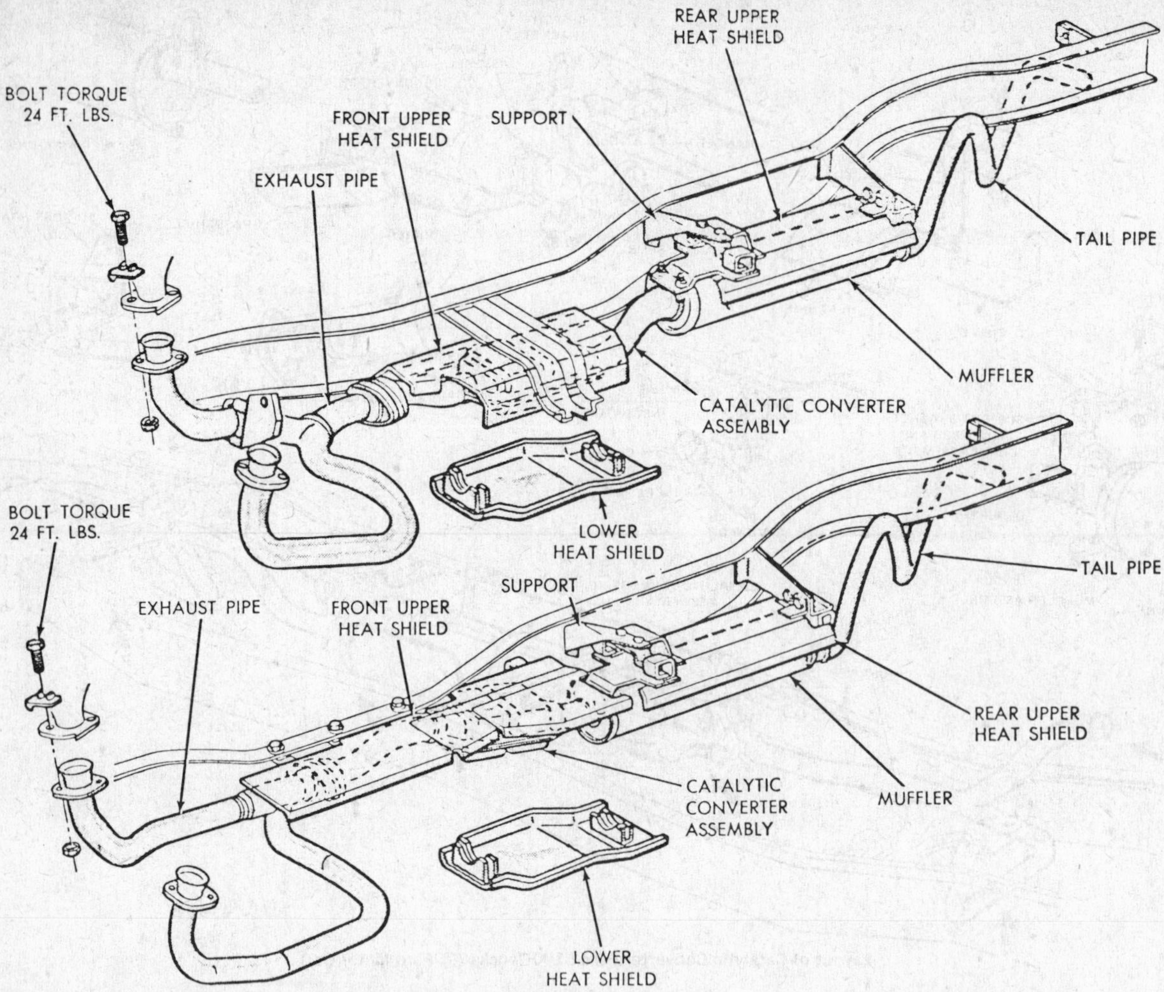

BOLT TORQUE
24 FT. LBS.

REAR UPPER
HEAT SHIELD

FRONT UPPER
HEAT SHIELD

SUPPORT

EXHAUST PIPE

TAIL PIPE

MUFFLER

CATALYTIC CONVERTER
ASSEMBLY

BOLT TORQUE
24 FT. LBS.

LOWER
HEAT SHIELD

SUPPORT

EXHAUST PIPE

FRONT UPPER
HEAT SHIELD

TAIL PIPE

REAR UPPER
HEAT SHIELD

CATALYTIC
CONVERTER
ASSEMBLY

MUFFLER

LOWER
HEAT SHIELD

Layout of Catalytic Converter Ramcharger/Trailduster—2 w/Drive top, 4 w/Drive Bottom (© Chrysler Corp.)

Slowly increase the engine speed until it is above 2000 rpm. The solenoid stem should extend. As the speed drops below 2000 rpm, the stem should retract.

To determine if the vehicle is equipped with the system, look for the speed switch on the right fender panel. Some models may not have the overheat protection system, but do have an anti-dieseling solenoid on the carburetor. The anti-dieseling solenoid is easily identified because it is energized whenever the ignition switch is on.

Vacuum Operated Exhaust Heat Riser Valves

Exhaust heat riser valves have been used for many years to force part of the engine exhaust through a passageway under the intake manifold and preheat the fuel mixture. The heat valve was spring loaded into the closed position, but heat would make the spring relax so that during high speed operation or after warm-up the exhaust would push it open.

Now, many engines use vacuum operated heat valves, controlled by a

vacuum switch that is sensitive to engine temperature. Ford calls their system simply a vacuum operated exhaust heat valve. General Motors refers to theirs as Early Fuel Evaporation, and Chrysler calls theirs a Power Heat Control Valve.

On all these systems, manifold vacuum is used to close the valve, and force the exhaust gases through the crossover passage in the intake manifold. All the systems have some kind of temperature valve that shuts the vacuum off when the engine warms up.

Both Chrysler and Ford products use a simple coolant temperature-sensitive vacuum switch mounted on the intake manifold coolant passage. The Chrysler switch has two hose connections. It actually does triple duty because it also controls the vacuum supply to the idle enrichment system and the air switching valve.

Ford's vacuum switch has three hose connections, but one of them is a vent with a filter to keep the dirt out.

General Motors uses either a coolant vacuum switch, or a vacuum solenoid connected to an oil temperature switch. The coolant vacuum switch has two hose connections and a vent when it controls the heat valve only. When it is tied into other emission control systems, it can have as many as five hose connections, and a

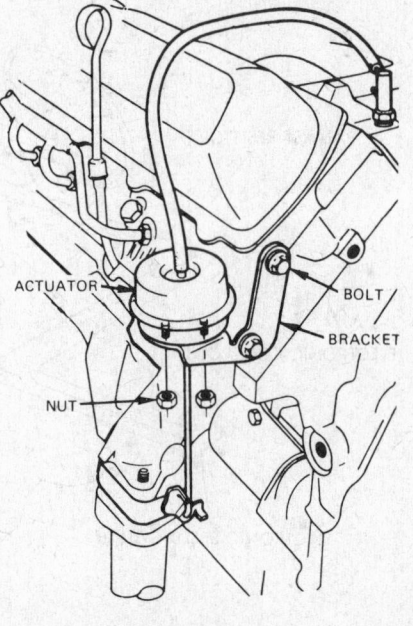

ACTUATOR

BOLT

BRACKET

NUT

EFE valve

vent. Many General Motors models also have a check valve in the hose so that vacuum will be trapped in the heat valve actuator when the engine is accelerated. This keeps the heat valve in the closed position and prevents a rattle.

Testing Vacuum Operated Exhaust Heat Riser Valves

Testing the vacuum operated heat riser valve is a matter of making sure it closes and opens freely. You can move it by hand to see if it works, on a warm engine. On a cold engine, the valve should be closed, and disconnecting the hose should allow it to open. On a cold engine, there should be vacuum at the vacuum actuator, and on a warm engine the vacuum should be shut off.

Air Aspirator System

1977 and later Chrysler Corporation models which use this system have done away with the air pump. The complete air aspirator system consists of a hose from the clean side of the air cleaner, the aspirator valve mounted on top of the engine, and a tube connecting the valve with the exhaust manifold. The suction in the exhaust draws in air through the air cleaner and this extra air helps the catalytic converter burn up the pollutants. The aspirator valve is similar to the check valve used with all air pump systems. It keeps the exhaust from flowing back into the air cleaner, but allows clean air to go into the exhaust.

Testing The Air Aspirator System

Testing the air aspirator valve is done by disconnecting the hose from the air cleaner and checking for slight suction at idle with a piece of paper over the end of the valve. Speeding the engine up slightly will show if the valve is leaking. Exhaust should not come out of the valve. Vibration of the valve diaphragm is normal, due to exhaust impulses.

Air Injection Systems

A belt-driven air pump supplies air to small tubes positioned in the exhaust port near each exhaust valve. The air mixes with unburned hy-

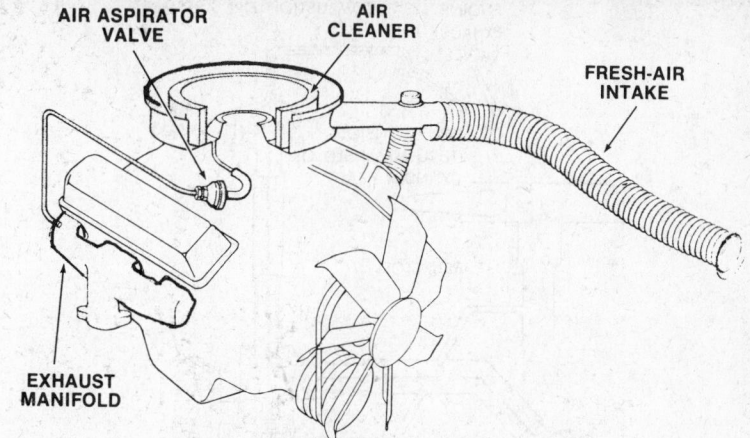

Chrysler Air Aspirator system (© Chrysler Corp.)

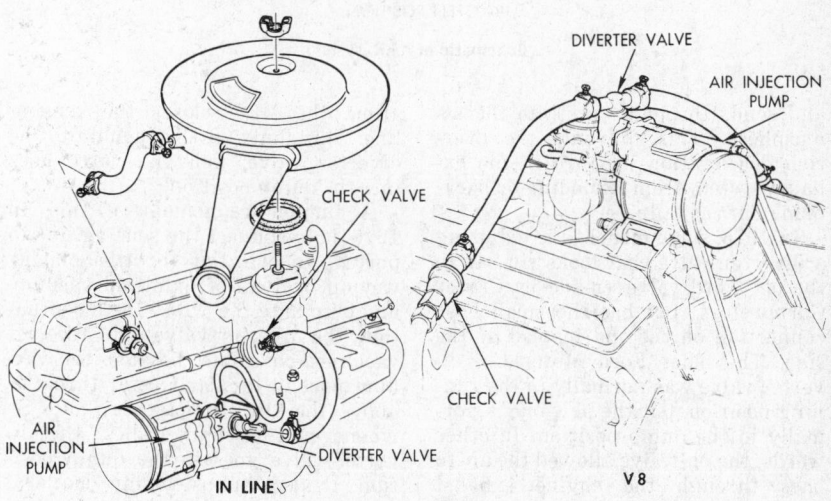

Air injection reactor system

drocarbons in the exhaust and the hydrocarbons actually burn up in the exhaust system. On late model engines, air may not be pumped to every exhaust port, and some engines have only a single air injection fitting on the exhaust pipe near its connection to the exhaust manifold. Air injection systems are frequently used on engines with catalytic converters, so that the converter gets enough air to keep the reaction going.

Plumbing on air injection systems varies considerably. At first, all the plumbing was external, with individual tubes inserted into each exhaust port either through the cylinder head

or the exhaust manifold. Now most engines have internal passageways to duct the air to the exhaust port.

A check valve is used between the pump and the exhaust port nozzle to keep hot exhaust gases from traveling up the plumbing and destroying the pump. V8s use two check valves.

An anti-backfire valve, also called bypass valve or diverter valve, is used between the pump and the check valve. Usually, the diverter valve is mounted on the pump or near it. A small sensing hose connects the diverter valve to intake manifold vacuum. When the vacuum rises during deceleration, the diverter valve opens,

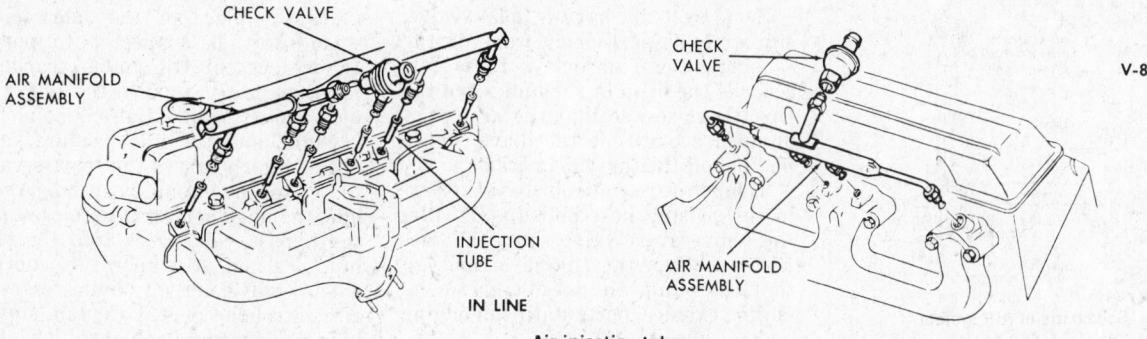

Air injection tubes

Emission Control Systems

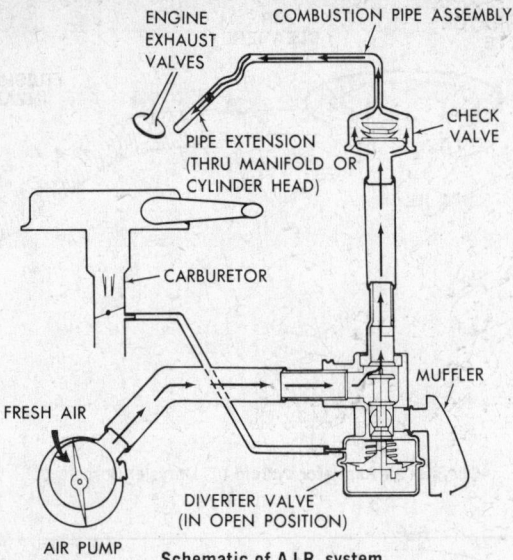

Schematic of A.I.R. system

and sends the pump air into the atmosphere. This prevents the overrich deceleration mixture in the exhaust system from exploding or backfiring out the tailpipe.

In 1975, some models started using a diverter valve that looks similar to the old Ford valve (made by Carter Carburetor), but has the small hose connection on the end instead of the side. The older Ford Motor Co. diverter valve was normally in the running position, but the new one is normally in the dump position. In other words, the old valve allowed the air to pass through the engine exhaust ports regardless of whether the small sensing line was hooked up. The new valve, being normally in the dump position, must have the small sensing line hooked up to manifold vacuum, which pulls the valve mechanism from the dump position into the normal running position.

Unfortunately, the new style valve will not go into the dump position automatically during deceleration. To get the valve to dump, a vacuum differential valve (VDV) is connected in the sensing line. Manifold vacuum goes through the VDV and then to the diverter valve. When the manifold vacuum increases during deceleration, the VDV closes the sensing line. This shuts off the vacuum to the diverter valve, and the valve goes into the dump position.

A further refinement of this, in 1976, is to connect the sensing line to ported (above the throttle plates) vacuum instead of manifold vacuum, and eliminate the VDV. In this situation, the diverter valve only receives vacuum above idle, because the vacuum port in the carburetor throat is above the throttle plate at idle. So whenever the engine idles, the diverter valve goes to the dump position. It also dumps during deceleration, because the throttle at that time is in the idle position.

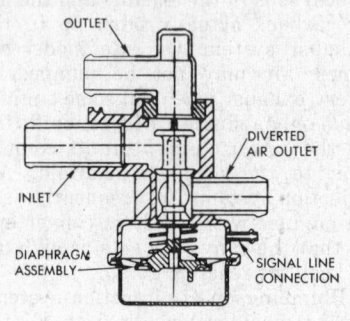

VALVE IN OPEN POSITION

Diverter valve

Some systems have a delay valve, similar to a spark delay valve, in the sensing hose. This delays for a few seconds the drop in vacuum when the throttle closes, so that the air is not dumped every time the driver takes his foot off the throttle in traffic.

Temperature controls are also used in the sensing hose hookup. Usually, the temperature valve shuts the vacuum off when the engine is cold, so that the pump air doesn't go to the engine exhaust ports until the engine warms up.

Some models have a temperature sensor mounted above the catalytic converter. If the converter overheats, the sensor turns off a solenoid which shuts off the air to the diverter valve. The diverter valve then goes to the dump position, shutting off the air to the exhaust and, hopefully, keeping the converter from melting or burning up.

Some 1976 and later Ford Motor Company inline 6 engines use a unique air bypass valve, with two small sensing hoses connected to it. Each of the hoses connects to one side of a diaphragm in the valve. The hose on the body of the valve connects to manifold vacuum, and the hose closer to the end connects to a separate on-off valve.

The diaphragm has a small hole so that the vacuum or pressure on each side will equalize. As long as the end chamber is sealed by the separate valve being closed, nothing happens, and the air flows through the bypass valve on the way to the exhaust ports. But if the separate valve is opened, it admits atmospheric pressure to one side of the diaphragm, and the vacuum on the other side moves the bypass valve to the dump position, exhausting the pump air into the atmosphere.

Two types of separate valves are used, one of them an electric solenoid operated valve, and the other a vacuum-operated valve. The electric solenoid is controlled by a Thermo Actuated Valve (TAV) in the air cleaner. When the engine is cold, the TAV closes, which energizes the solenoid. Atmospheric pressure then enters the upper chamber on the bypass valve and it goes to the dump position. When the engine warms up, the TAV opens, shuts off the solenoid, and the bypass valve goes into the normal running position.

On some California engines the solenoid is connected so that manifold vacuum passes through the solenoid to get to the bypass valve. A small filter-vent is placed over the end of the nozzle on the end cap of the bypass valve. With the same electrical hookup, this setup has the same action as that described earlier.

The vacuum operated valve, which takes the place of the solenoid on some sixes, is connected to ported (above the throttle plates) carburetor vacuum. It is called the Idle Vacuum Valve. At idle, there is no ported vacuum, and the idle vacuum valve opens, which causes the bypass valve to go to the dump position. Above idle, the idle vacuum valve closes, and the bypass valve goes into the running position. A temperature control, a delay valve, and a vacuum reservoir all control the ported vacuum supply to the idle vacuum valve.

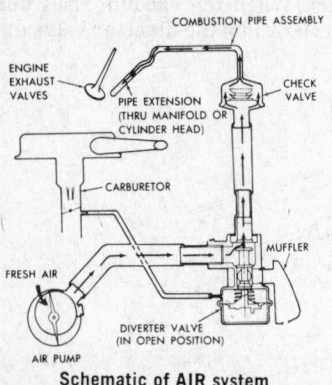

Schematic of AIR system
(© G.M. Corp.)

Air Pump Tests

CAUTION: Do not hammer on, pry or bend the pump housing while tightening the drive belt or testing the pump.

Before proceeding with the tests, check the pump drive belt tension.

If the belt squeals when the engine is running, the pump may be dragging or seized. Remove the belt and turn the pump by hand to check for seizure. Disregard any chirping, squealing, or rolling sounds from inside the pump when turning it by hand, as these are normal.

Check the hoses and connections for leaks. Hissing or a blast of air is indicative of a leak. Soapy water, applied lightly around the area in question, is a good method for detecting leaks.

To test air output, disconnect the air hose from the pump wherever it is convenient. If you disconnect it from one check valve on a V8, the other hose should also be disconnected and plugged for the test. Run the engine at idle and feel the blast of air from the hose with your hand. Increase the engine speed to 1500 rpm and feel the blast of air again. If the blast increases, and is steady, the pump is okay.

Pump Noise Diagnosis

The air pump is normally noisy; as engine speed increases, the noise of the pump will rise in pitch. The rolling sound the pump bearings make is normal. However, if this sound becomes objectionable at certain speeds, the pump is defective and will have to be replaced.

A continual hissing sound from the air pump pressure relief valve at idle indicates a defective valve. Replace the relief valve.

If the pump rear bearing fails, a continual knocking sound will be heard. Since the rear bearing is not separately replaceable, the pump will have to be replaced as an assembly.

Anti-backfire Valve Tests

Detach the hose, which runs from the bypass valve to the check valve.

Connect a tachometer to the engine. With the engine running at normal idle speed, check to see that air is flowing from the bypass valve hose connection.

Speed the engine up, so that it is running at 1,500-2,000 rpm. Allow the throttle to snap shut. The flow of air from the bypass valve at the check valve hose connection should stop momentarily and air should then flow from the exhaust port on the valve body or the silencer assembly.

Let the throttle snap shut several times. If the flow of air is not diverted into the atmosphere from the valve exhaust port or if it fails to

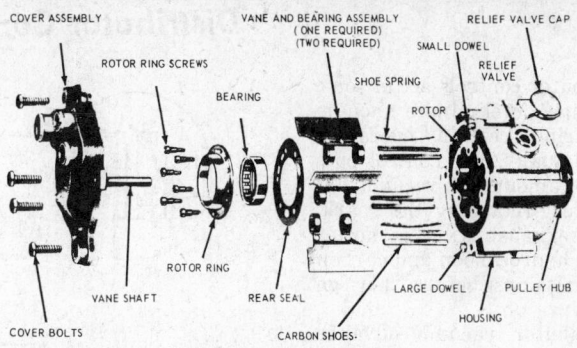

Air injection reactor pump

stop flowing from the hose connection, check the vacuum lines and connections. If these are tight, either the bypass valve or one of the accessory valves in the small sensing hose is defective and must be replaced.

A leaking diaphragm will cause the air to flow out both the hose connection and the exhaust port at the same valve.

Late model, systems should stop flowing at idle, as described earlier. If not, the bypass valve or accessory valve is defective.

Check Valve Test

Remove the hose from the check valve. With the engine running at 1,500 rpm in Park or Neutral, hold the back of your hand near the check valve to test for exhaust gas leakage. If the valve leaks, it must be replaced.
NOTE: Vibration and flutter of the valve at idle is a normal condition caused by exhaust pulsations. It does not mean that the valve is defective.

Vacuum Differential Valve Test

Disconnect the small sensing hose at the bypass valve and connect a vacuum gauge to the hose. With the engine idling in Park or Neutral, the gauge should read full manifold vacuum.

Run the engine at a steady 2500 rpm in Park or Neutral, and release the throttle. As the engine decelerates, the vacuum gauge should drop close to zero, then return to full manifold vacuum as the engine speed drops to idle. If not, the VDV is defective and must be replaced.
NOTE: The small hose nozzle should be connected to manifold vacuum.

Solenoid Vacuum Valve Tests (Ford Products)

Solenoid vacuum valves used with the air injection system on 1975 and later Ford products are of two types, normally closed and normally open. On the normally closed type, applying electric current to the terminals will open the vacuum valve. On the normally open type, applying current will close the valve. The closed valve has both hose connections at the bottom end, and the manifold vacuum connects to the bottom nozzle, furthest from the electrical connector. The open valve has the connection separated, with one at the top and the other at the bottom. Manifold vacuum connects to the top nozzle, nearest the electric connector.

Type I (Normally Closed)

With the engine idling in Park or Neutral, detach the vacuum supply hose from the solenoid bottom nozzle. Vacuum should be felt at the end of the hose with your finger. If not, check the hose and source of vacuum. When vacuum is good at the hose, reconnect it to the solenoid bottom nozzle.

Disconnect the other hose from the solenoid and connect a vacuum gauge to the solenoid. Disconnect the electricity from the solenoid. With the engine idling, there should be no reading on the gauge. Connect one terminal of the solenoid to the battery positive post, and the other terminal to ground. The vacuum gauge should read full manifold vacuum. Disconnect the battery hookup. The vacuum gauge should drop to zero. If the solenoid does not operate correctly, replace it.

Type II (Normally Open)

With the engine idling in Park or Neutral, detach the vacuum supply hose from the solenoid upper nozzle. Vacuum should be felt at the end of the hose with your finger. If not, check the hose and source of vacuum. When vacuum is good at the hose, reconnect it to the solenoid bottom nozzle.

Disconnect the other hose from the solenoid and connect a vacuum gauge to the solenoid. Disconnect the electricity from the solenoid. With the engine idling, full manifold vacuum should appear on the gauge. Connect one terminal of the solenoid to the battery positive post and the other terminal to ground. The vacuum gauge should drop to zero. Disconnect the battery hookup, and full manifold vacuum should appear on the gauge.

If the solenoid does not operate correctly, replace it.

Distributor Controls

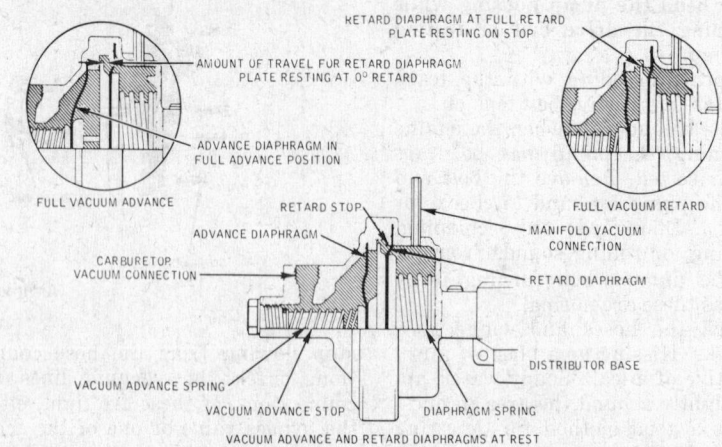

Dual diaphragm vacuum advance mechanism

All distributor controls act in some way to change or eliminate vacuum advance during certain operating conditions. Usually, the control cuts down on the amount of vacuum advance, in effect retarding the spark, so that the exhaust will get hotter and burn up hydrocarbon and carbon monoxide emissions before they go out the tailpipe.

The distributor vacuum advance unit might be connected, according to factory design, to either manifold vacuum or ported (above the throttle plates) carboretor vacuum. Either way, the vacuum spark advance curve is approximately the same for all running conditions above idle. At idle, however, the manifold vacuum hookup results in full advance, while the ported hookup gives zero advance. If the hoses are hooked up the wrong way, the addition or lack of advance will affect idle speed, requiring a readjustment of the throttle position to bring the idle speed back to specifications. When this is done, emissions will usually be high, so it is important to keep the hoses hooked up correctly.

Dual Diaphragm Distributors

These distributors have two hose connections, one in the normal position, and the other closer to the distributor body. The hose fitting next to the body is for the retard diaphragm, and is connected to manifold vacuum. The retard diaphragm affects the spark only at idle, when there is no vacuum on the advance diaphragm. In effect, the retard diaphragm provides a movable resting place for the advance diaphragm. When ported vacuum is not acting on the advance diaphragm, it returns to the neutral or no-advance position against the retard diaphragm. At idle, manifold vacuum pulls the retard diaphragm to the retard position, and the advance diaphragm follows along to retard the spark.

the distributor and plug the hose. With the engine running, increase the speed to a fast idle and watch the timing marks. The timing should advance. If not, either the vacuum unit is faulty, the vacuum port is plugged, or there is a temperature control device that is shutting off the vacuum. Apply hand pump or mouth suction vacuum to the advance diaphragm and the timing should advance. If not, the distributor must be disassembled and repaired. Failure to advance could be caused by a faulty diaphragm or a sticking advance plate.

Remove the advance hose from the vacuum unit and read the timing at normal idle speed. Remove the plug that was inserted in the retard hose, and check for full manifold vacuum at the end of it. If there is no vacuum, temperature controls may be shutting it off.

Connect the hose to the retard diaphragm, or apply vacuum from another source. The timing should immediately retard several degrees. If not, the diaphragm is not working, and the unit must be replaced. Reconnect all hoses as they were originally.

Distributor Vacuum Deceleration Valve

First used on Chrysler Corporation engines as part of the original Clean Air Package, this valve was later used on AMC and Ford engines. It was commonly known as a spark valve. Its purpose is to advance the spark during deceleration, by sending full manifold vacuum to the vacuum advance unit. At all other times the vacuum advance unit receives ported (above the throttle plates) carburetor vacuum.

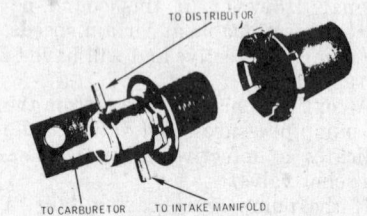

Distributor vacuum advance control valve

Three checks should be made on the valve: the amount of vacuum at the distributor, any valve leaks, and the adjustment. To check the amount

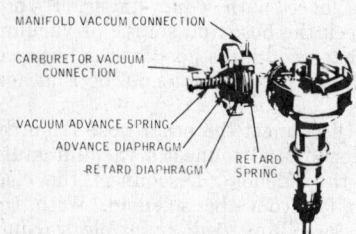

Distributor with dual-diaphragm vacuum advance

Testing Dual Diaphragm Distributors

To test a dual diaphragm distributor, connect a timing light to the engine. Remove the retard hose from

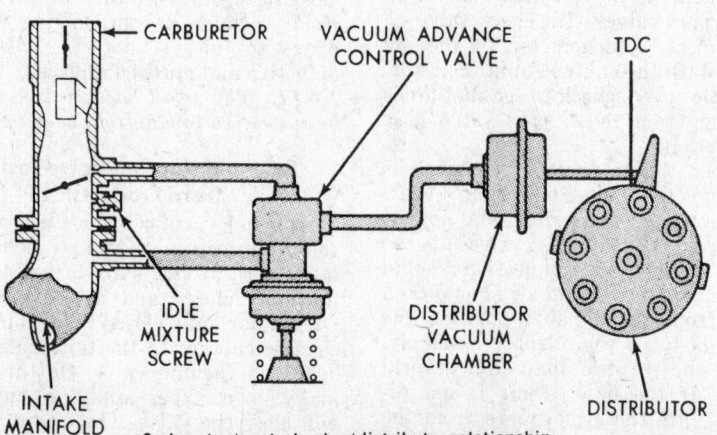

Carburetor/control valve/distributor relationship

of vacuum at the distributor, use a T-fitting and a short length of vacuum hose to connect a vacuum gauge into the distributor vacuum line near the distributor. At idle, with the engine fully warmed up, the vacuum on the gauge should be less than 1 Hg. If the gauge shows more than 1 Hg. the idle speed is too fast, or the valve is leaking. To check for a leak, remove the large manifold vacuum hose on the side of the valve. If the vacuum drops, the valve is leaking and must be replaced. If the vacuum stays high, reduce the engine idle speed so that the port in the carburetor is covered.

To check the valve adjustment, connect the manifold vacuum hose and run the engine at 2000 rpm for 5 seconds. Then release the throttle. The distributor vacuum should go over 16 in. Hg. and stay there for about one second. Within about three seconds after you release the throttle, the distributor vacuum should drop to below 6 in. Hg. If the carburetor is equipped with a dashpot to make the throttle close slowly, the time may be about one second longer. If the time is too long, remove the cover on the valve and turn the screw clockwise to reduce the time. To increase the time, turn the screw counterclockwise. If the valve will not adjust properly, it must be replaced, and the new valve adjusted to specifications.

Spark Delay Valve

This small valve is connected between the carburetor and the distributor vacuum advance, so that the ported (above the throttle plates) vacuum to the distributor must pass through the valve. A restriction in the valve delays the vacuum applied

Testing the Ford Spark Delay Valve —the black side should be connected to the vacuum source.

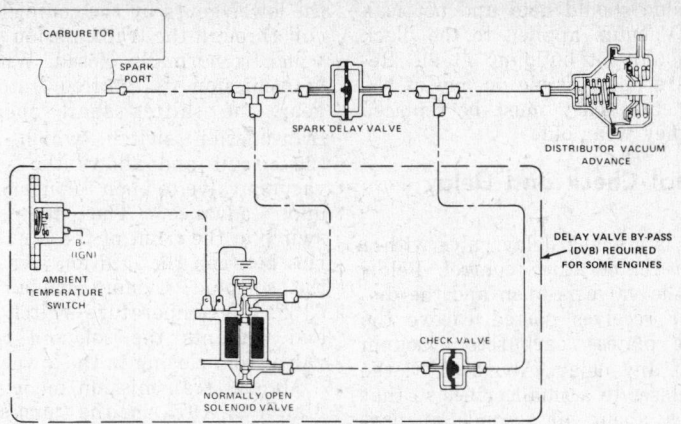

Ford Spark delay valve circuit
(© Ford Motor Co.)

to the vacuum advance unit so that the advance comes in slowly. When there is no vacuum at the carburetor port, as during idle or wide open throttle a check valve inside the spark delay valve opens and dumps the vacuum so that the vacuum advance unit returns to the no-advance position without any delay.

Ford Products use spark delay valves with one side black and the other colored. The colored side indicates the amount of delay, which can be from one to 28 seconds. The valve should always be installed with the black side toward the source of vacuum, and the colored side toward the distributor.

General Motors spark delay valves are a different shape than Ford, and are marked on both sides with the names of the components they connect to. Usually, they are marked CARB on one side, and either TVS or DIST on the other. Of course, the CARB side must be connected to the carburetor port.

Spark delay valves can be tested for correct operation and leaks with a source of vacuum such as a hand vacuum pump or a running engine, and a vacuum gauge. Connect the vacuum gauge to the distributor side of the valve, and the vacuum source to the other side. The gauge should rise slowly until it reads the amount of vacuum available. The time to rise to the maximum reading should be from one to 28 seconds. If the vacuum gauge does not read anything, the valve is plugged. If the vacuum reads instantly, without any delay, the valve is open. In either case, the spark delay valve must be replaced. To test the check valve part of the spark delay valve, remove the vacuum source and the vacuum gauge should drop instantly to zero without any delay. If there is any delay, the spark delay valve is defective and must be replaced.

Delay Vacuum Bypass

This system, used on Ford Products

in 1973 only, bypasses the spark delay valve below an ambient temperature of 49-50°F. so that the vacuum advance unit receives ported carburetor vacuum without any delay. Above 49-65-F. the ported vacuum must pass through the spark delay valve.

The system consists of an ambient temperature switch mounted in the front door hinge post (either side), a vacuum solenoid mounted on the engine near the distributor, a check valve, a spark delay valve, and the connecting hoses and wires.

To test for correct operation, disconnect the hose from the distributor vacuum advance and set the throttle so the engine runs at approximately 1500 rpm with the transmission in Park or Neutral. Connect a vacuum gauge to the hose and see if the vacuum rises slowly to about 5 in. Hg. If the vacuum gauge shows a few inches of vacuum immediately, without a slow rise, it means the system is in the by-pass mode, or the system is not working right. To be sure the system is not in the bypass mode, warm the temperature switch in the door hinge post with a hot cloth, and make the test again. The bypass system will not affect the maximum amount of vacuum on the gauge, only the rate of rise. Each time you test it you must disconnect and reconnect the vacuum gauge to check the rate of rise.

If the system doesn't work, test each component. The temperature switch should be electrically open below 49-65°F. and closed above that figure. The vacuum solenoid is normally open, and closes when current is applied. Use only a continuous source of vacuum, such as a running engine, to test the solenoid. A hand vacuum pump cannot be used because of the internal bleed. The black nozzle on the solenoid connects to the vacuum source. The black sides of both the check valve and the spark delay valve should be connected to the vacuum source. When testing the check valve, vacuum applied to the white side should hold and not leak down. Vacuum applied to the black side should not build up at all. Repairs are not possible on any of the components. They must be replaced when they wear out.

Thermal Check and Delay Valve

This is a spark delay valve with a built-in temperature control. Below 50°F. the valve is open and the distributor receives ported (above the throttle plates) carburetor vacuum without any delay. Above 50°F. the valve closes to a small orifice so that it takes about 40 seconds at part throttle before the distributor gets all of the ported vacuum.

To test the valve, connect a hand vacuum pump to the CARB nozzle and a vacuum gauge to the TVS nozzle. Be sure the valve is at room temperature (68°F.). Work the pump rapidly to create a vacuum of about 20 in. Hg. on the pump gauge. The vacuum gauge should lag behind. When you stop pumping, the pump gauge should drop slightly, and in a few seconds should read the same as the vacuum gauge. If not, the valve is defective and must be replaced. When the valve is cold, it is open, and vacuum should pass freely through so that both gauges register the same with no lag.

Transmission Controlled Spark

This system is used widely on General Motors vehicles but variations of it are also found on American Motors, Chrysler Corporation, and Ford Motor Company products. The object of transmission controlled spark is to eliminate vacuum spark advance in the lower gears. Once the transmission gets into high gear, vacuum spark advances is allowed for better gas mileage and part throttle response. Because each of the maker's systems are different, we will describe them separately.

American Motors/Jeep

Manual transmission vehicles produced before March 15, 1973 eliminate vacuum advance in the lower gears when the air temperature at the front of the vehicle is over 63°F. If the temperature is above 63°F. vacuum advance is allowed in all gears.

The vacuum supply to the distributor vacuum advance unit is controlled by a solenoid vacuum valve mounted on the top of the engine. This valve receives current whenever the ignition switch is on, and is grounded to complete the circuit through a solenoid control switch on the transmission. The solenoid vacuum value is normally open, but is held closed in

the lower gears by the completed circuit through the transmission switch, which is normally closed. When the transmission is shifted into high gear, the shifter shaft opens the transmission switch, which breaks the circuit and allows the solenoid vacuum valve to open for normal vacuum advance. The temperature switch at the front of the car is in series between the ignition switch and the solenoid vacuum valve. Below 63°F. the temperature switch opens, and prevents the solenoid vacuum valve from closing in the lower gears.

Manual transmission models after March 15, 1973 use the same system, except that the temperature switch is not used.

Automatic transmission models built before March 15, 1973 eliminate vacuum advance below 34 mph when the air temperature at the front of the vehicle is over 63°F. If the temperature is above 63°F. vacuum advance is allowed at all speeds. The same solenoid vacuum valve is used, as on the manual transmission models, but it is connected to a solenoid control switch on the transmission where the speedometer cable connects. The operation of the solenoid vacuum valve and solenoid control switch are the same as on the manual transmission models, except that the solenoid control switch is sensitive to speed instead of gear position.

After March 15, 1973, the solenoid control switch on automatic transmission models was moved up to the top of the engine and made sensitive to governor hydraulic pressure only. A hydraulic line from the transmission conducts governor pressure to the switch. In the top center of the switch is a small Allen screw that is used to adjust the switching point to exactly 34 mph. The temperature switch is not used after March 15, 1973.

To test the system, connect a vacuum gauge to the distributor vacuum hose, using enough additional hose to come out from under the hood and through the side window so that the vacuum gauge can be seen while driving. Then drive to test the system. On a manual transmission you should see vacuum on the gauge in high gear only. On an automatic, you should see vacuum above approximately 34 mph

only. On models with temperature control, the temperature must be above 63°F. Because the distributor runs on ported vacuum, you must have the throttle open a little to get vacuum. Also, you must slow down to approximately 25 mph before the solenoid vacuum valve will close. This means that once you have gone above 34 mph, you will continue to see vacuum on the gauge when the throttle is open, as long as the vehicle does not go below the speed that closes the solenoid. If the system does not work correctly, check out the individual units or the hose connections.

Chevrolet/GMC

GM used transmission controlled spark (TCS) through 1974. In several years there were important changes in the design from the previous year. The end result was the same, in that vacuum spark advance was eliminated in the lower gears, but the design changes from year to year mean that the units operate differently.

In all years, the basic system uses a grounding switch at the transmission, and a vacuum solenoid that opens and closes the distributor vacuum hose. On some models, the vacuum solenoid is mounted on the carburetor and is called a Combined Emission Control (CEC) valve. The CEC valve opens or closes the vacuum passage the same as a vacuum solenoid but it also does something else, as explained later under the years it is used.

1971 6-Cylinder and V8

The vacuum solenoid is mounted on the carburetor, and is called a CEC valve. It is easily recognizable because it has vacuum hoses connected to it. The CEC valve is normally closed. When it is energized, it opens, allows vacuum advance, and also opens the throttle slightly. The transmission switch is also normally closed, but a relay mounted on the firewall reverses its action. There is also a time relay that energizes the CEC valve for about 15 seconds whenever the ignition switch is turned on. This feature gives a faster idle just after a hot restart, with less chance of the engine dying.

Vacuum advance is allowed in all gears below 82°F. coolant tempera-

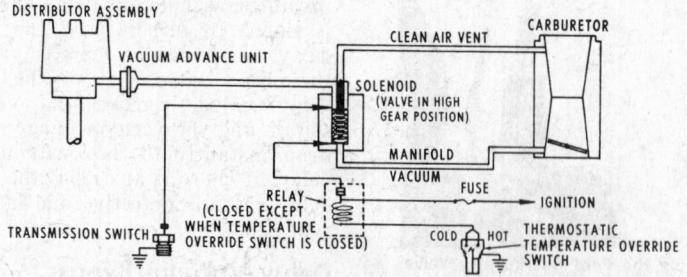

A typical transmission controlled spark system (© General Motors Corp)

ture. Some models have a hot override that allows vacuum advance in all gears above 230°F.

In operation the system works like this. On a cold start the cold override cancels the system and allows vacuum advance in all gears until the coolant gets up to 82°F. When the engine is warm, and the transmission is shifted into High gear, the transmission switch opens. This interrupts the current going through the reversing relay, allowing the spring in the relay, allowing the spring in the relay to close the points and energize the CEC valve. When the CEC valve is energized, it opens the vacuum advance hose and also extends its stem to slightly open the throttle. The slightly open throttle prevents closed throttle deceleration in high gear for lowered emissions.

If the engine is restarted warm, the cold override will not operate, but the time relay takes over and completes the ground circuit through the CEC valve so that vacuum advance is allowed and the throttle is opened slightly for about 15 seconds.

1972-73 6-Cylinder

The system allows vacuum advance only in high gear on all transmissions. The transmission switch is a normally open design, which eliminates the need for a reversing relay. The transmission switch is open in the lower gears. It closes when in high gear, and turns on the CEC valve. The time relay turns on the CEC valve for 20 seconds after the ignition switch is turned on. The relay has a different part number from earlier models, and is wired differently.

Cold override temperature control switching temperature is 82°F. for

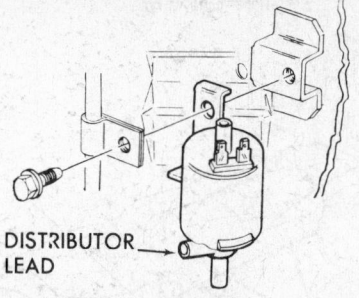

Chevrolet vacuum advance solenoid — other GM cars similar (© G.M. Corp)

1972, and 93°F. for 1973. Six-cylinder engines also have an idle stop solenoid mounted next to the CEC valve. It is important that you don't confuse the two. The CEC valve has vacuum hoses connected to it.

1974 6-Cylinder

The CEC valve is discontinued for 1974, and a vacuum solenoid is used instead. This is a normally closed solenoid, mounted on the side of the engine at the coil mounting bolt. Other than the CEC valve action, operation of the 1974 6-cylinder system is the same as in 1973.

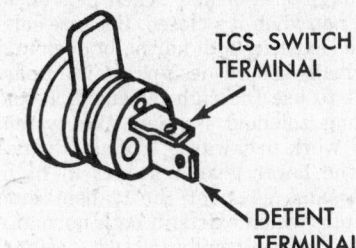

The terminals on the Turbo-Hydramatic 400 Transmission Switch (© G.M. Corp.)

1972 V8

The CEC valve is not used. Vacuum advance is turned on and off by

a vacuum solenoid with normally closed design. New for 1972 is a delay relay, that keeps the vacuum solenoid off for 20 seconds after shifting into high gear. The delay relay is used only on 307, 350, and 400 V8s. The cold override switch turns on the vacuum solenoid below 82°F. coolant temperature. The cold override switch is built into the sending unit that operates the red HOT light on the dash.

1973-74 V8

There is no CEC valve. The delay relay is eliminated, and a time relay similar to that used in 1971 is used. The time relay turns on the vacuum solenoid to give vacuum advance for about 20 seconds every time the ignition switch is turned on. Because all 1973-74 V8 engines with TCS operate the distributor vacuum advance on full manifold vacuum, turning on the solenoid gives a faster idle. The vacuum solenoid is a normally closed design, used with a normally open transmission switch.

Ford Motor Company

The 1972 Ford Motor Company system is called Transmission Regulated Spark. Vacuum advance is allowed only in high gear. The vacuum is controlled by a solenoid vacuum valve, sometimes called a Distributor Modulator Valve. The vacuum valve receives current from the ignition switch, but this current passes through a temperature switch in the front door post of the car. Below 49-65°F. the temperature switch is open, blocking the current and preventing the vacuum valve from being energized. The vacuum valve is a normally open design. It is grounded through a normally closed transmission switch, which stays closed in the lower gears. When the transmission is in high gear, the transmission switch opens, breaking the circuit and opening the vacuum valve, which allows vacuum advance.

The 1973 Ford Motor Company system is called TRS+1. As far as the vacuum advance is concerned, the TRS+1 does exactly the same thing that the 1972 TRS system does. The difference is that the temperature switch and transmission switch also operate a three-way vacuum solenoid that operates the exhaust gas recirculation valve on carburetor spark port vacuum in first and second gear, and EGR port vacuum in high gear. After March 15, 1973 the door post temperature switch was discontinued.

1974 and later Ford Motor Company vehicles do not use Transmission Regulated Spark.

Testing TCS Systems

Testing the system is done by connecting a vacuum gauge to the distributor vacuum line with a long hose

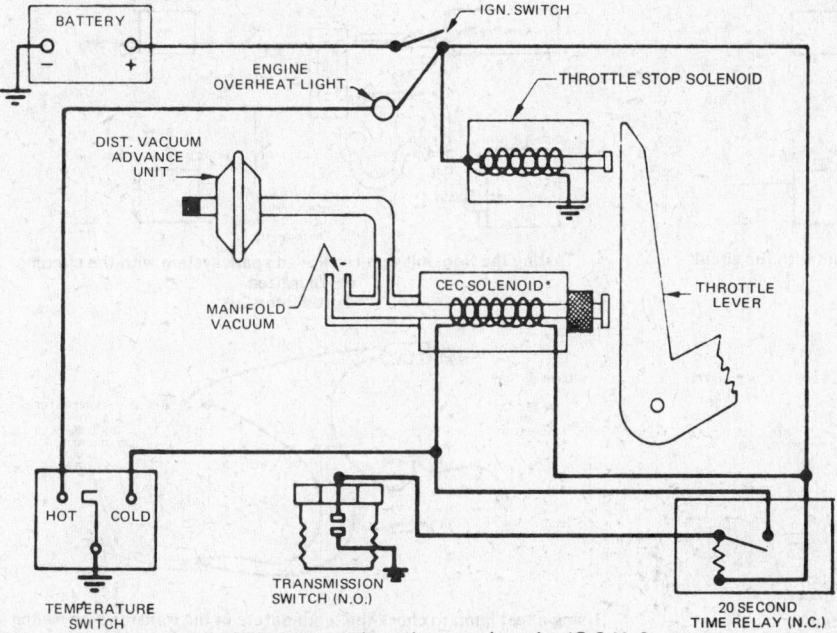

1972-73 CEC system without the reversing relay (© G.M. Corp)

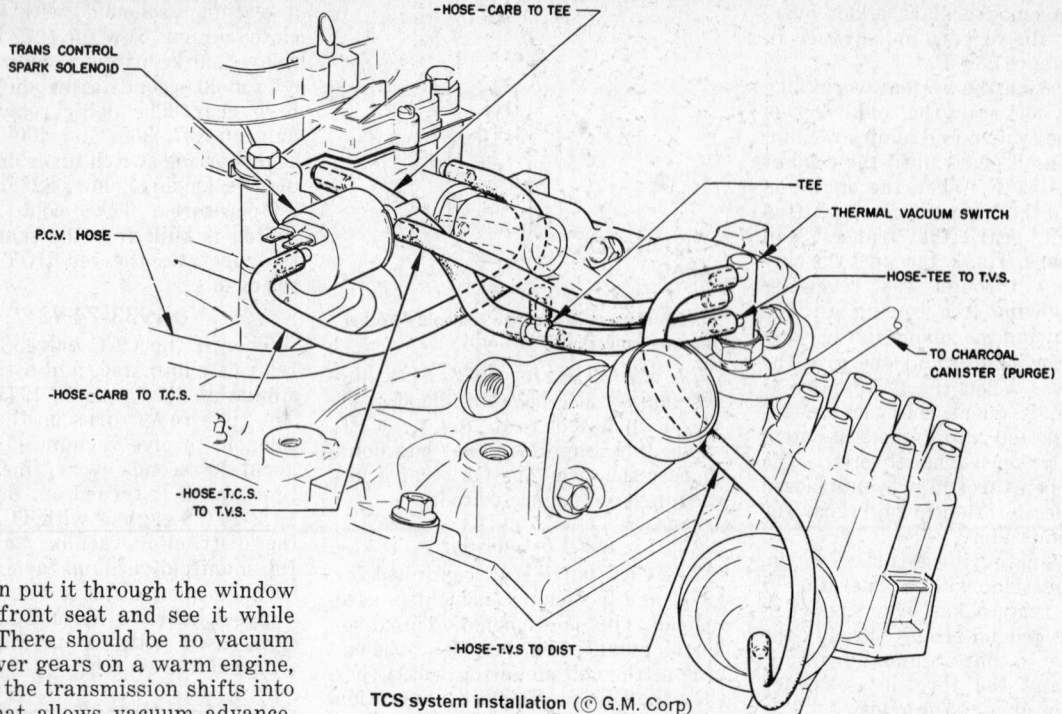

TCS system installation (© G.M. Corp)

so you can put it through the window into the front seat and see it while driving. There should be no vacuum in the lower gears on a warm engine, but after the transmission shifts into a gear that allows vacuum advance, you should see vacuum on the gauge. Engines that run their distributors on manifold vacuum will show vacuum at all times when in the proper gear. Engines that use ported (above the throttle plates) vacuum will show vacuum in the proper gear only when the throttle is open. If you don't get vacuum when you should, test the individual units in the system.

Vacuum solenoids can be tested by disconnecting all wiring and connecting hot and ground wires to the solenoid terminals, to make it open or close. You should be able to blow

through the solenoid when it is open, but not when it's closed. Because solenoids exist in both normally open and normally closed designs, it is important to use the right solenoid. If the wrong solenoid is used, the system will work backwards, giving advance in the lower gears but not in high. The same goes for the transmission switch, which exists in both normally open and normally closed designs. The term "normally open" means that the solenoid or switch is open when it is not energized or activated.

In the case of a vacuum solenoid, normally open means that if you were holding the solenoid in your hand without any wires connected to it, the vacuum passages would be open, allowing vacuum to pass. In the case of a transmission switch, the term "normally open" refers to the electrical path, which is "open" or "off" so that it will not conduct electricity. Normally closed, of course, means that the electric contacts are closed so that the current can pass. But normally closed on a vacuum solenoid means

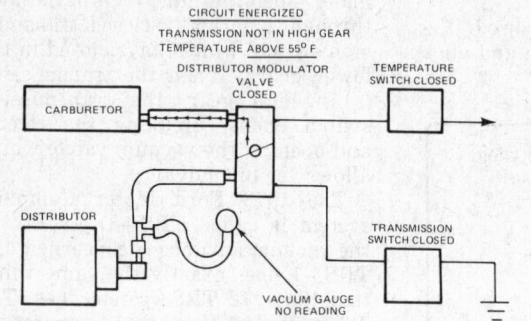

Testing the transmission-controlled spark system with the circuit energized
(© Ford Motor Co)

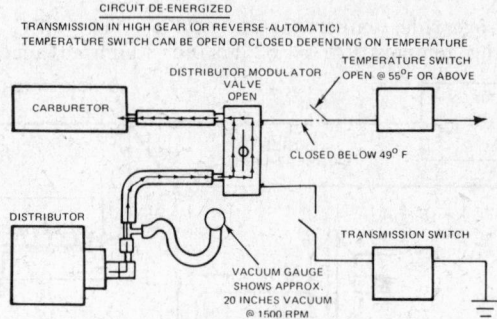

Testing the transmission-controlled spark system with the circuit de-energized
(© Ford Motor Co)

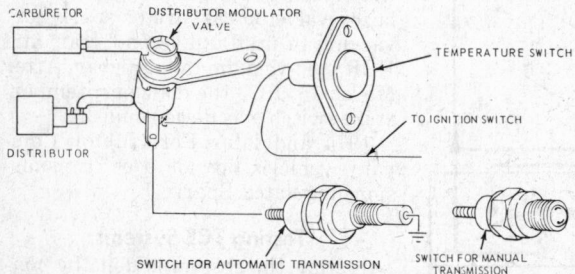

Ford transmission controlled spark system
(© Ford Motor Co)

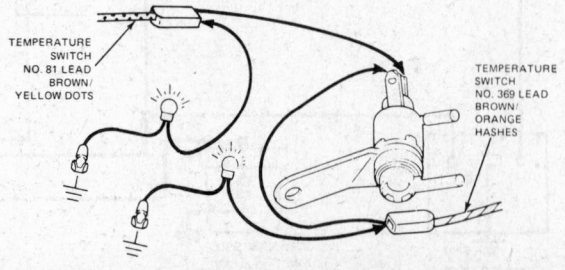

Using a test lamp to check the temperature of the transmission switch
(© Ford Motor Co)

that the vacuum passage is blocked so the vacuum can't get through.

All electrical switches should be tested with a penlight-powered test light. Testing with a car battery and a light bulb is dangerous, because you might put so much current through the switch that it burns up.

Unless you are familiar with handling small electrical probes, relays should be tested by elimination. Test everything else first, and if the system still doesn't work, it must be the relay. It's not that relays can't be tested, but they are expensive, and one small slip with a hot wire can burn up $20 in parts.

Temperature Activated Vacuum (TAV) and Cold Temperature Activated Vacuum (CTAV) Systems

This system, used only on Ford 6-cylinder engines, switches the vacuum source back and forth between the carburetor spark port and EGR port, according to the air temperature. A 3-nozzle vacuum solenoid is used, connected to a temperature switch located in the front door post on models built before March 15, 1973. Below approximately 55°F. outside air temperature, the temperature switch is open, and the solenoid is not energized. In this position, the solenoid connects the spark port to the vacuum advance unit. Above 55°F. the temperature switch closes, and energizes the solenoid. In this position, the solenoid connects the EGR port to the vacuum advance unit.

Models built after March 15, 1973 use the same system, but the temperature switch is located in the air cleaner, and a latching relay is added, on the firewall. Once the temperature switch has closed, the relay latches so that any sudden rush of cold air through the air cleaner will not cycle the solenoid on and off. The latching relay keeps the solenoid energized as long as the ignition switch is on. When the ignition switch is turned off, the relay unlatches and the system is ready for the next start, whether the air temperature is hot or cold. If the air at the temperature switch is over 55°F. the latching relay will come on when the ignition switch is turned on.

Test the system with a vacuum gauge connected to the vacuum advance hose at the distributor. With the temperature above 65°F. (to be sure the temperature switch has closed) you should be getting vacuum from the EGR port. If you disconnect the EGR port hose and the vacuum drops, you know the system is working. When making a cold test, the vacuum should come from the spark port hose, so disconnecting that hose should make the vacuum drop. Because both ports are above the throttle plate, the throttle must be opened slightly to get vacuum at the hose.

Identifying the spark port and EGR ports on the carburetor is easy if they are marked. If there is no marking on the carburetor, connect two vacuum gauges, one to each port. At idle you should not have any vacuum. If you do see vacuum, it usually

means the engine is idling too fast. Close the throttle slightly to slow down the idle and the vacuum should drop to almost zero.

When you open the throttle, you will see vacuum on one gauge before the other. The gauge that gets vacuum first is connected to the spark port.

Orifice Spark Advance Control (OSAC)

This is strictly a Chrysler Corporation system, used on several years and models. In effect, it is simply a mechanism that delays the application of vacuum to the distributor vacuum advance unit. When the throttle is opened, the carburetor port is exposed to vacuum. This vacuum goes through a hose to the OSAC valve, and then to the distributor vacuum advance. The OSAC valve is sometimes mounted on the firewall, and sometimes on the air cleaner. Inside the OSAC valve is a calibrated orifice that delays the vacuum as much as 27 seconds, depending on the calibration of the valve.

Some OSAC valves have temperature control that senses the temperature inside the air cleaner or inside the plenum chamber behind the firewall, depending on where the valve is mounted. If the valve contains temperature control, it will be wide open below 60°F. bypassing the orifice and allowing vacuum advance without any delay. Above 60°F. the bypass closes and the delay takes over.

To test the valve, just connect a

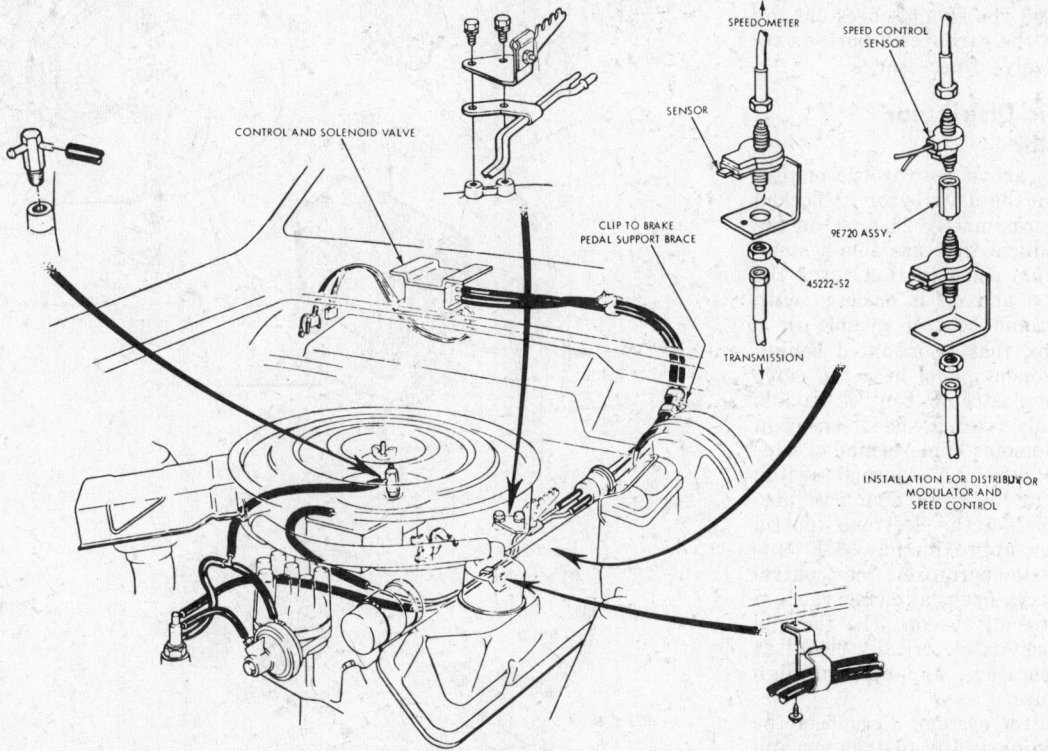

Typical distributor modulator installation (© Ford Motor Company)

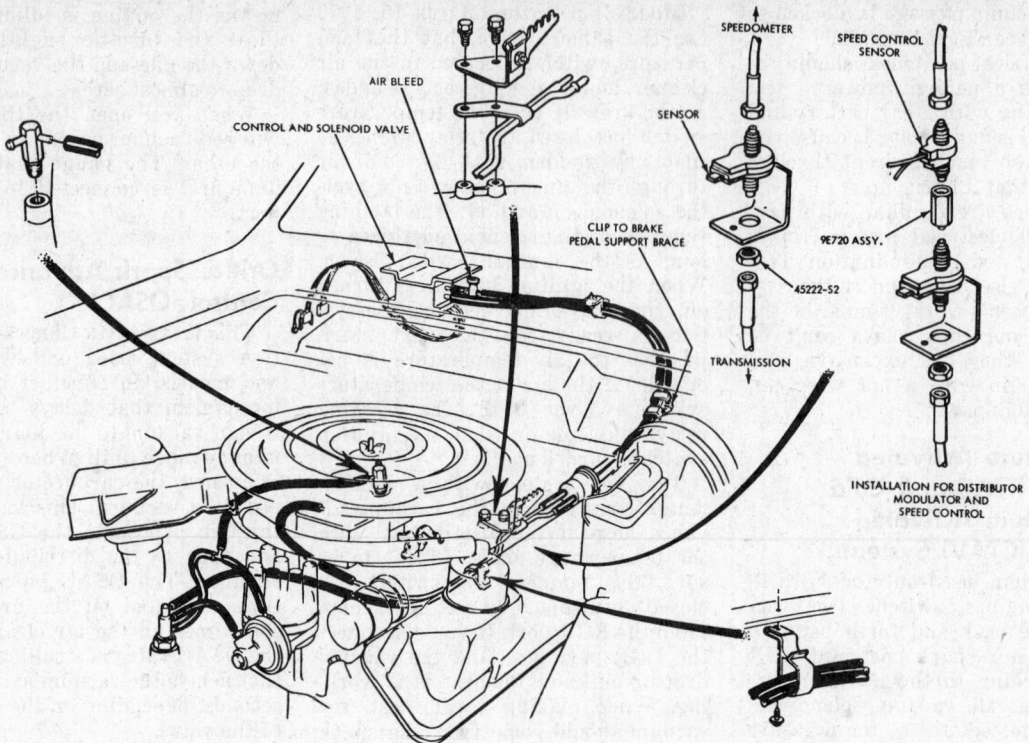

Typical distributor modulator installation (© Ford Motor Co)

vacuum gauge to the DIST connection on the valve. With the engine idling, you should have no reading on the gauge. If there is a reading, the engine is idling too fast. With the engine idling, open the throttle to a fast idle, and hold it steady. The vacuum on the gauge will rise slowly until it reaches a maximum reading. If not, there is something wrong with the system, and you should check out the hoses and the carburetor port, or replace the valve if necessary.

Electronic Distributor Modulator

Ported (above the throttle plates) vacuum to the distributor is blocked below approximately 25 mph, on the Ford products that use this system. The vacuum solenoid that turns the vacuum on and off is enclosed with the electronic control module in a plastic box that is mounted behind the instrument panel near the glove box. The electronic control module gets signals from a speed sensor in the speedometer cable behind the instrument panel. A thermal switch mounted in the front door post also sends signals to the electronic control unit. Below approximately 68°F. outside air temperature, the control module stays open, allowing vacuum advance at all speeds. The thermal switch itself is electrically closed at low temperature, and open at high temperature.

To test the system, disconnect the vacuum hose at the distributor and connect a vacuum gauge to the hose.

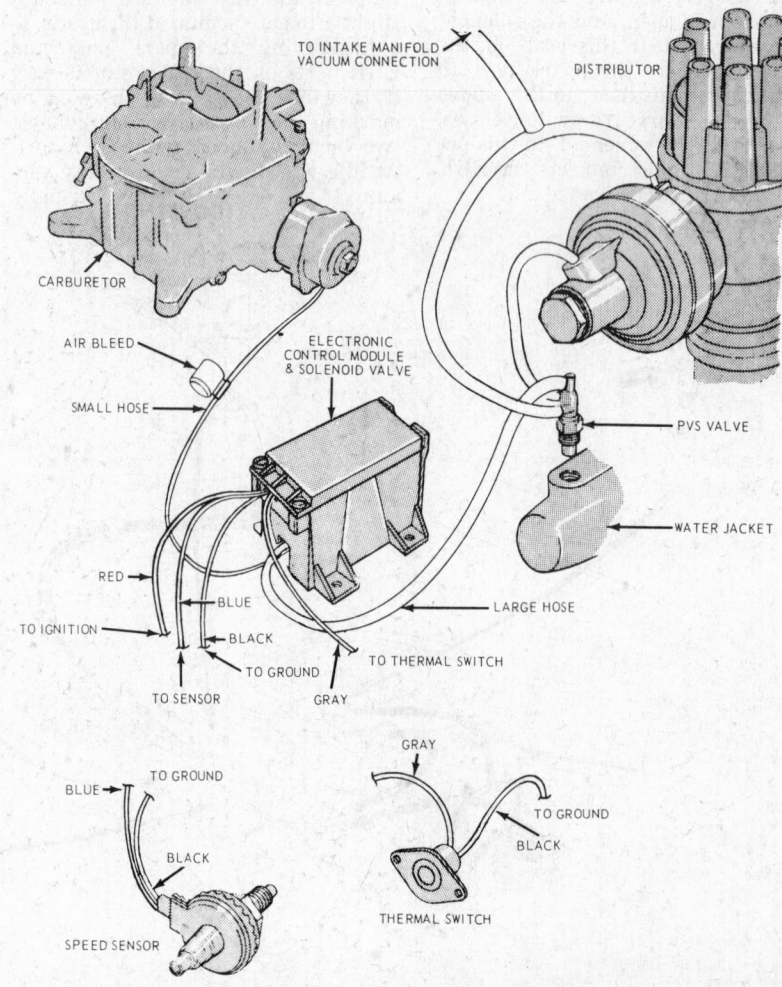

Distributor modulator details (© Ford Motor Company)

Position the gauge so you can see it while driving the car. Above approximately 25 mph you should get vacuum on the gauge. When decelerating, the vacuum is shut off at approximately 18 mph. Check this by opening the throttle slightly when you are below 18 mph. You should not get any vacuum.

The speed sensor can be checked with an alternating current voltmeter. If you get some indication of voltage when the sensor is turned, it is probably okay. There is no reliable test for the control module. If the speed sensor and the thermal switch check okay, the trouble is probably in the control module.

Electronic Spark Control

Electronic spark control is a refined version of the electronic distributor modulator system. The vacuum solenoid is separate from the control module, and is called the distributor modulator valve. The control module is much smaller, and is now called the amplifier. Vacuum to operate the distributor vacuum advance runs through the modulator valve. The modulator valve is closed at low speeds, and opens at 25-40 mph, depending on the car. The speed at which vacuum is allowed depends on the amplifier, which is available in several settings, each a different color. The amplifier is mounted behind the instrument panel near the glove box.

A speed sensor in the speedometer cable behind the instrument panel sends signals to the amplifier. The amplifier receives current from the ignition switch, but this current must first pass through the temperature switch in the front door post. Thus, the temperature switch is in series between the ignition switch and the amplifier. This is different from the electronic distributor modulator system, in which the temperature switch provides a ground.

Testing of the electronic spark control system is the same as for the electronic distributor modulator described earlier, using a vacuum gauge while driving the car.

Vacuum Reducer Valve

Inserted between the manifold vacuum source and the distributor, this valve reduces the vacuum acting on the advance diaphragm by about 3 in. Hg. This valve is always used on a system that includes a distributor thermal vacuum switch. The vacuum advance unit operates on ported (above the throttle plates) vacuum, except when the engine overheats above 225°F. This opens the thermal vacuum switch and sends full manifold vacuum through the vacuum reducer valve to the advance unit. Thus, the vacuum reducer valve is only operating when the engine is overheated.

To test the valve, connect a vacuum gauge to the TVS nozzle, and a hand vacuum pump to the MAN nozzle. When you pump up 15 in. Hg. vacuum on the hand pump, the vacuum on the separate gauge should be 3 to 4 in. Hg. lower. Both gauges should hold the vacuum without leakdown. If not, the valve is defective and must be replaced.

Retard Delay Valve

When the throttle is suddenly opened, engine vacuum drops immediately, and this causes the vacuum advance to move quickly from the advance position to the neutral or no-advance position. A retard delay valve is a restriction with a one-way check valve. It allows the vacuum to act on the vacuum advance unit normally, but when the vacuum drops,

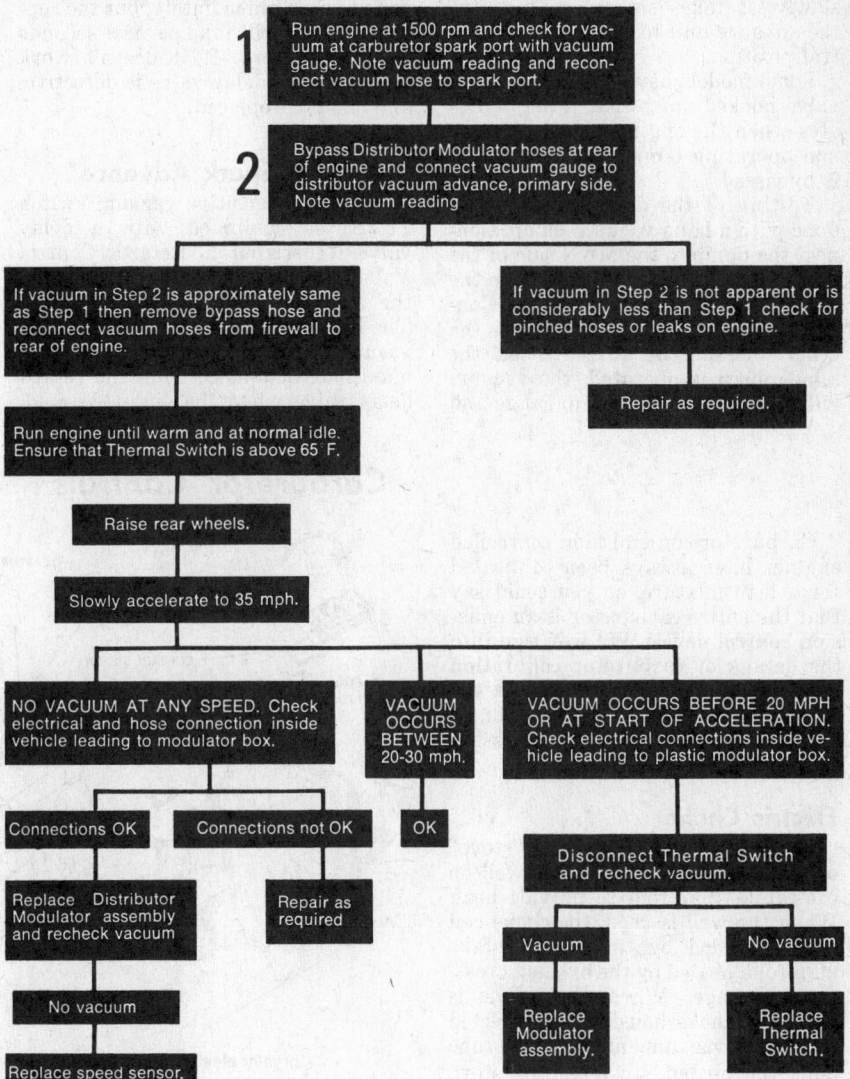

DISTRIBUTOR MODULATOR SYSTEM TROUBLESHOOTING CHART

1 Run engine at 1500 rpm and check for vacuum at carburetor spark port with vacuum gauge. Note vacuum reading and reconnect vacuum hose to spark port.

2 Bypass Distributor Modulator hoses at rear of engine and connect vacuum gauge to distributor vacuum advance, primary side. Note vacuum reading

If vacuum in Step 2 is approximately same as Step 1 then remove bypass hose and reconnect vacuum hoses from firewall to rear of engine.

If vacuum in Step 2 is not apparent or is considerably less than Step 1 check for pinched hoses or leaks on engine.

Repair as required.

Run engine until warm and at normal idle. Ensure that Thermal Switch is above 65 F.

Raise rear wheels.

Slowly accelerate to 35 mph.

NO VACUUM AT ANY SPEED. Check electrical and hose connection inside vehicle leading to modulator box.

VACUUM OCCURS BETWEEN 20-30 mph.

VACUUM OCCURS BEFORE 20 MPH OR AT START OF ACCELERATION. Check electrical connections inside vehicle leading to plastic modulator box.

Connections OK

Connections not OK

OK

Disconnect Thermal Switch and recheck vacuum.

Replace Distributor Modulator assembly and recheck vacuum

Repair as required

Vacuum

No vacuum

Replace Modulator assembly.

Replace Thermal Switch.

No vacuum

Replace speed sensor.

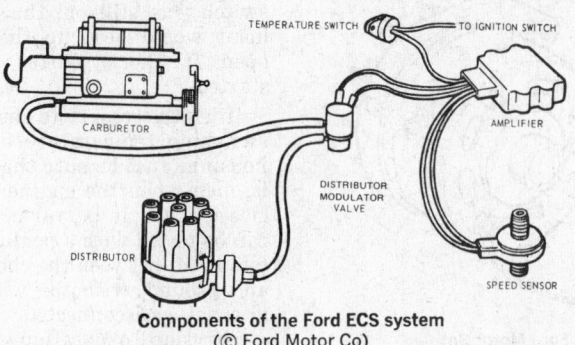

Components of the Ford ECS system
(© Ford Motor Co)

the delay valve traps the vacuum in the advance unit and lets it out slowly. It takes several seconds for the advance unit to return to the neutral position.

Some models have the retard delay valve hooked up so that it only operates when the engine is cold. At normal operating temperature the delay is bypassed.

Testing of the delay valve can be done with a hand vacuum pump. Connect the pump to the MAN side of the valve, or the side that connects to the vacuum source on the engine. Connect a separate vacuum gauge to the other side of the valve. When the hand pump is operated, the vacuum will rise on both the pump gauge and

the separate gauge equally. When the release is pulled, the pump gauge will drop to zero immediately, but the separate gauge will take several seconds to drop to zero. If it doesn't work that way, the delay valve is defective, and must be replaced.

Cold Start Spark Advance

A coolant sensitive vacuum switch (PVS) is combined with a delay valve (Distributor Retard Control Valve) to provide retard delay when the engine coolant is below 128°F. The hose routing is set up so that the vacuum advance unit operates on manifold vacuum through the retard delay valve when the engine is cold,

and on ported vacuum through a spark delay valve when the engine is warm. The system also has an overheat PVS that switches the vacuum advance over to manifold vacuum (through the spark delay valve) when the engine coolant gets over 235°F.

Testing the spark delay valve is covered in this section under Spark Delay Valve. Testing for the Distributor Retard Control Valve is the same as for the Retard Delay Valve in this section.

When the 128° PVS is cold, connection No. 2 is blocked and D and 1 are connected. When it is over 128°F No. 1 is blocked and D and 2 are connected.

Carburetor Controls

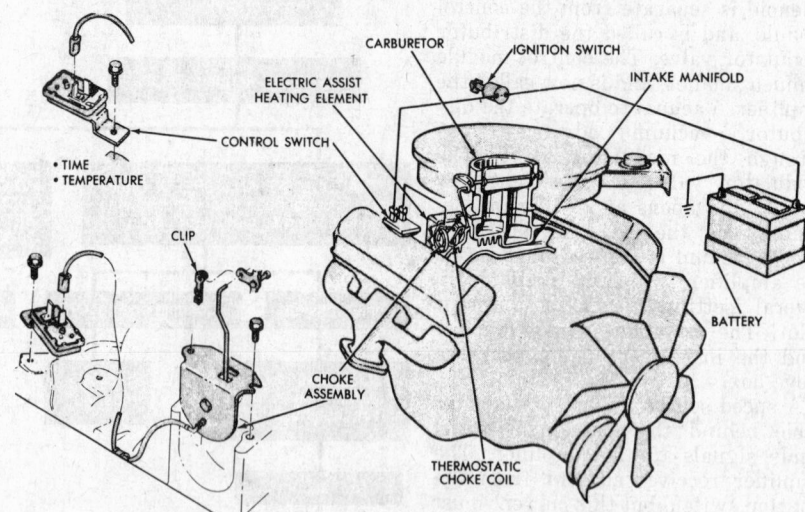

Carburetors on emission controlled engines have always been calibrated for a lean mixture, so you could say that the entire carburetor is an emission control device. We won't go into the details of carburetor calibration here. What we want to cover is the devices, both on and off the carburetor, that work with it for emission control.

Electric Choke

A non-electric choke uses a "stove" on the exhaust manifold or a well on the intake manifold to provide heat. When the well is used, the choke coil is surrounded by the warm intake manifold, heated by the exhaust crossover passage. When the stove is used, the choke housing is connected to engine vacuum, and a long tube pulls the heated air from the stove into the choke housing to heat up the choke coil and cause the choke to open as the engine warms up. When an electric choke is used, it can be in addition to all the above, or it can be the only source of choke heat, depending on the design.

The electric choke has a small heater next to the choke coil. This heater receives its current from different sources, depending on the car maker.

Ford Motor Company and Ameri-

can Motors chokes are powered from the alternator "center tap," which produces about 7 volts. As the alternator is only putting out voltage when the engine is running, the electric choke is automatically shut off when the engine is off. It is important that the choke is connected only to the special "center tap" provided on the alternator. The description "center tap" refers to the construction of the alternator wiring, and not to the location of the connection.

Chrysler electrically assisted choke system components © Chrysler Corp

Inside the Ford 4-bbl. choke cover is a thermostatic switch that turns on the heating element at approximately 80°F. Above that, the element stays on as long as the engine is running. The 80°F. figure was selected because the engine is warm enough at that temperature to keep running without the choke. When the heater comes on, the choke opens very quickly. When the engine is shut off and cools down, the choke switch may stay on to as low as 65°F. at the choke housing. On a warm restart, where the choke switch was still on, the heating element would heat up the choke and open it shortly after the engine started.

It isn't necessary to check the exact switching temperature of the choke housing. Just be sure that the switch is open when the engine is cold, and closed when it is warm. The switch can be tested with a penlight-powered test light, between the choke terminal and ground, with the wire from the alternator disconnected.

Chrysler Corporation vehicles with

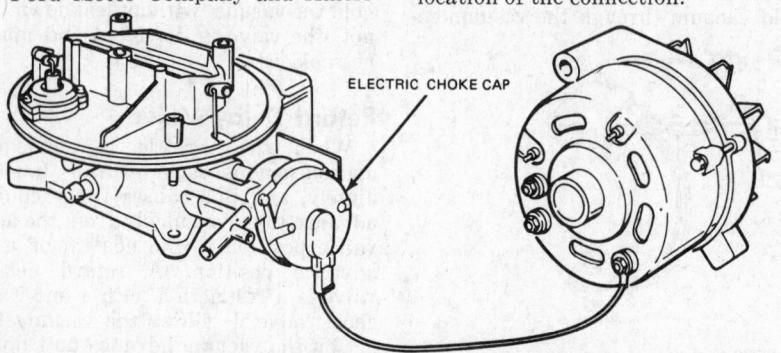

Ford and AMC electrically assisted choke hook-up (© Ford Motor Co)

an electric choke use a well type choke, which receives heat both from the intake manifold and the electric choke heater. A separate choke control unit is mounted on top of the intake manifold and connected to the heater with a wire. This wire disconnects at the choke control unit only, not at the heater.

Choke control units may be single and double stage. The double stage is recognized by the external resistor alongside the unit. The single stage unit turns on the choke heater at approximately 60°F. and off at 110°F. The double stage unit keeps the heater on below 60°F. but the current runs through the resistor. At approximately 60°F. the resistor is taken out of the circuit and the heater gets full current. At 110°F. the control unit turns the heater off.

Testing can be done with a non-powered test light on the choke terminal to find out if the heater is on or off. The ignition switch must be on. If the light glows, you know the control unit is on. On two-stage units, the light will glow dimly when the resistor is in the circuit, and brightly when the resistor is out. The current to the control unit comes from the ignition switch, and there is no fuse.

Chevrolet and GMC use an electric choke that is mounted on the carburetor. The choke has a dual element behind the coil spring. Whenever the engine is running, the choke heater is in operation. Below 50-70°F. a bimetal snap disc in the choke cover turns off the large section of the heating element so that only the small section gives off heat. Above 50-70°F. the disc switches on the large heating element for faster choke opening.

Current to the choke is controlled by a three-terminal oil pressure switch. One of the terminals is a ground for the red oil pressure light on the instrument panel. The other two terminals are a switch in series between the ignition switch and the choke heater. Oil pressure operates the switch so that the choke gets current only when the engine is running. The circuit is fused through the backup light or transmission fuse in the fuse block.

NOTE: Failure of the choke heater circuit will cause the oil pressure light to go on.

Staged Choke Pulldown

Ford Motor Company 2-bbl. and 4-bbl. V8 carburetors have a vacuum diaphragm housing and bracket mounted on the carburetor base on the choke housing side of the carburetor. The housing is connected by a hose to intake manifold vacuum. On the opposite end of the housing a pull rod connects to the choke linkage. Inside the housing are two chambers, one for vacuum and one filled with silicone fluid. The two chambers are separated by an orifice. When the engine starts, manifold vacuum in the vacuum chamber pulls the fluid through the orifice and slowly creates a vacuum in the second chamber. The vacuum in the second chamber slowly pulls the choke open. The length of time to pull the choke open is controlled by the size of the orifice, and varies from approximately 15 to 60 seconds.

Below 60°F. a bimetal valve in the vacuum end of the housing shuts off the vacuum so the unit does not work. On both 2-bbl. and 4-bbl. carburetors, the staged pulldown is in addition to the normal pulldown that opens the choke partway as soon as the engine starts. The staged choke pulldown was used only in 1972, on all V8 carburetors.

To test the pulldown, have the engine running at idle, and disconnect the hose. Plug the hose with your finger so the engine doesn't die, and wait for the pull rod to extend all the way. Then hook up the hose and see how long it takes for the rod to pull into the housing. If the rod doesn't pull in at all, or pulls in immediately, without any delay, the unit should be replaced.

Delayed Choke Pulldown

Some 1974 Ford Motor Company 460 V8s used a Carter Thermo-Quad 4-bbl. carburetor with delayed choke pulldown. The normal choke diaphragm is on the same side of the carburetor as the choke housing. It opens the choke part way as soon as the engine starts. On the opposite side of the carburetor from the choke housing is the delayed pulldown diaphragm. It is connected to manifold vacuum, but a restriction inside the diaphragm housing delays the movement of the pulldown for three to ten seconds. The delayed pulldown not only opens the choke further than the initial pulldown, but also pulls the throttle off the top step of the fast idle cam onto the second step.

The pulldown diaphragm can be checked by connecting it to vacuum of a running engine, or by disconnecting and reconnecting its hose on a running engine to see how long it takes to make a full stroke. It should take approximately three to ten seconds to make a stroke. If not, it should be replaced.

1975 and later Ford-Motorcraft 4300 4-bbl. carburetors also use a delayed pulldown system. When the engine starts, a vacuum piston inside the choke housing opens the choke partway. About 6 to 18 seconds later, the delayed pulldown located on the carburetor in front of the choke housing pulls the choke open further, and also pulls the fast idle cam to a lower step. This gives more precise choking, and slows the engine down to prevent damage to the catalytic converter from overly long fast idle.

The pulldown diaphragm housing has an internal restriction in the vacuum passage. It receives full manifold vacuum when the engine starts, but the restriction delays the stroke of the pull rod, giving the engine a few seconds to warm up before it opens the choke and slows down the fast idle.

The action of the pulldown diaphragm can be checked on a running engine by disconnecting the hose, waiting until the pull rod extends, and then connecting the hose again. The pull rod should take several seconds to stroke back into the housing. If not, the unit should be replaced.

Fast Idle Pulloff

1976 Chevrolet and GMC 454 V8s use an electric choke 4-bbl. carburetor with an extra vacuum diaphragm on the front. The diaphragm is connected by a hose to the same thermal vaccum switch that controls the Early Fuel Evaporation actuator. Below 150°F. coolant temperature, the vacuum is shut off. Above that temperature the TVS opens the passage and allows vacuum to operate the diaphragm, which then pulls the throttle off the high step of the fast idle cam onto the next lower step. Reducing the idle this way prevents damage to the catalytic converter if the engine is left to warm up unattended.

Testing the vacuum diaphragm can be done with a hand vacuum pump. The pull rod should make a full stroke, and the diaphragm should hold vacuum without leaking.

Temperature Controlled Choke Vacuum Break

1975 and later General Motors vehicles use this system on 6-cylinder inline engines. The system uses an extra vacuum break diaphragm or electric solenoid to open the choke as the engine warms up. There are three different, carburetors used on these two engines, and each carburetor uses a slightly different system.

The normal vacuum break unit is on the choke coil side of the carburetor. It opens the choke partway as soon as the engine starts. The temperature controlled vacuum break is on the throttle lever side of the carburetor. It opens the choke to an almost wide open position whenever the engine is running, and the coolant temperature is above 80°F. Manifold vacuum comes through a hose from a thermal vacuum switch on the right front of the cylinder head. Above the switch is a manifold vacuum fitting screwed into the intake manifold part of the head. (This system is used only on the engine with the integral head and manifold.) Below 80°F. coolant temperature, the thermal vacuum switch is closed. Above 80°F. it is open, and supplies vacuum to the vacuum break unit at all times during engine operation.

Testing of the vacuum break unit can be done by applying vacuum to see that it moves through a full stroke, and does not leak. The thermal vacuum switch can be tested by blowing through it to see that it is open above approximately 80°F.

Idle Enrichment System

Some Chrysler Corporation automatic transmission models, 1975 and later, have an idle enrichment valve built into the carburetor. The valve opens or closes a passageway that admits extra air to the idle system. When the valve is open, the idle mixture is lean from the excess air. When the valve is closed, the idle mixture is rich, because the air is shut off. The valve is turned on and off by manifold vacuum, connected by a hose.

Some models have a coolant temperature control valve called a Coolant Control Idle Enrichment valve. This is a mechanical valve, mounted on a coolant passage and connected by hoses between the manifold vacuum source and the idle enrichment valve. When the engine is cold, the valve is open, allowing vacuum to operate the idle enrichment valve and richen the idle. When the engine warms up, the valve closes, and stops the idle enrichment.

Some engines also have a vacuum solenoid connected to the same timer that provides EGR delay. On most engines the solenoid has three hose connections, one to manifold vacuum, one to the idle enrichment valve, and the third to the EGR amplifier. When the solenoid is not energized, it allows vacuum to go to the EGR valve, but blocks the vacuum to the idle enrichment valve. When energized, the vacuum to the EGR amplifier is blocked, but the vacuum passes to the idle enrichment valve. The timer energizes the solenoid during the first 35 or 60 seconds, depending on the model. This means that when the engine is cold, the idle is enriched during the first 35 or 60 seconds of engine operation.

One engine, the 49-State 318 V8 with catalytic converter, uses two separate solenoids, one for EGR and one for idle enrichment, but the working of the system is the same.

Next to the idle enrichment valve, inserted in the hose, is a small air bleed. It lets a constant small supply of air into the hose to keep it purged of fuel vapor.

Testing the system can be done on a cold engine by disconnecting the hose at the carburetor and connecting a vacuum gauge to the hose. Start the engine and note the length of time that vacuum appears on the gauge. At the end of the timed period, the gauge should drop to zero. Allow the engine to warm up to operating temperature and make the test again. This time you should not see any vacuum on the gauge, because the CCIE valve should be closed. If there is no timer, you will see vacuum for several minutes after a cold start, until the engine warms up.

To check the effect of the idle enrichment use a hand vacuum pump on the idle enrichment valve on the carburetor. With vacuum applied, the valve will be closed, richening the idle, and changing the idle speed. Release the vacuum and the speed should go back where it was. If there is no speed change, either the valve is not working, or a carburetor passage is blocked with dirt. The valve should also hold vacuum without leaking down.

Altitude Compensation

Air at high altitude is much thinner than at sea level, so engines run rich. To keep the mixture correct, and prevent rich running that causes high emissions, many 1976 and later 4-bbl. carburetors have an altitude compensation system.

The heart of an altitude compensation system is a sealed bellows chamber, called an aneroid. The aneroid is sealed at sea level, and expands at high altitude. This expansion is used to open or close a passageway and lean out the mixture. The Carter Thermo-Quad, used on Chrysler Corporation products, and the Ford-Motorcraft 4300 4-bbl. use an aneroid that opens an air passage to lean the mixture. The Thermo-Quad bleeds this air into the main metering system, while the Ford 4300 bleeds the air into the main venturi.

The General Motors Rochester 4-bbl. uses an aneroid that works with the fuel metering adjustable part throttle feature. The adjustable part throttle fuel feed is adjusted at the factory to give the right fuel mixture at sea level. When the aneroid expands at high altitude, it shuts off the

adjustable fuel passage to lean the mixture.

Idle Speedup Solenoid

Many 1976 and later vehicles use an idle speedup solenoid on air conditioned models. The solenoid looks just like the old anti-dieseling or idle stop solenoid. The difference is that the idle speedup solenoid is connected to the air conditioning system and only comes on when the air conditioning is turned on. Its only purpose is to speed up the idle so the engine won't die.

On carburetors equipped with the solenoid, curb idle speed adjustments are made with the throttle screw, not the solenoid screw. There is a specification for the engine speed with the solenoid energized, but it is higher than the normal curb idle.

Automatic Idle Speed Diaphragm

1976 and later Chrysler Corporation 6-cylinder engines with a 1-bbl. carburetor use a special diaphragm or dashpot on some models. The stem of the dashpot touches the throttle lever. The dashpot is connected to manifold vacuum, which compresses a spring inside the dashpot housing. If the load on the engine is changed by turning the air conditioner on or shifting into DRIVE, the vacuum will drop, and the dashpot spring will open the throttle to bring the speed back up to what it was. Theoretically, whatever load is put on the engine will be balanced by the dashpot, and the idle speed will remain constant.

On 1976 models, the diaphragm was not adjustable, but the throttle speed screw actually rested against the diaphragm stem. In 1977 models this was changed so that the throttle speed screw is separate, and the diaphragm stem pushes against the throttle lever. The 1977 dashpot has a threaded housing and locknut.

To adjust the 1977 dashpot, start the engine in Neutral and position the throttle lever so the actuating tab on the lever is touching the stem of the dashpot, but not depressing it. Wait 30 seconds to allow the engine to settle down, but keep the throttle so it is just touching the stem. In that position, the engine speed should be 2500 rpm. If not, move the throttle so the speed is 2500, and adjust the dashpot by loosening the locknut and turning the housing so the stem just touches the throttle lever.

Index

MANUAL STEERING

POWER STEERING

STEERING TROUBLE DIAGNOSIS

Condition	Possible Cause	Correction
	Steering Gear	
Excessive Play or Looseness in the Steering	(1) Steering gear shaft adjusted too loose or shaft and/or bushing badly worn.	(1) Replace worn parts and adjust according to instructions.
	(2) Excessive steering gear worm end play due to bearing adjustment.	(2) Adjust according to instructions.
	(3) Steering linkage loose or worn.	(3) Replace worn parts.
	(4) Front wheel bearings improperly adjusted.	(4) Adjust wheel bearings.
	(5) Steering arm loose on steering gear shaft.	(5) Inspect for damage to the gear shaft and steering arm, replace parts as necessary.
	(6) Steering gear housing attaching bolts loose.	(6) Tighten the attaching bolts to specifications.
	(7) Steering arms loose at steering knuckles.	(7) Tighten according to specifications.
	(8) Worn king pins or bushings.	(8) Replace king pins and bushings.
	(9) Loose spring shackles.	(9) Adjust or replace parts as necessary.
Hard Steering	(1) Low or uneven tire pressure.	(1) Inflate the tires to recommended pressures.
	(2) Insufficient lubricant in the steering gear housing or in steering linkage.	(2) Lubricate as necessary.
	(3) Steering gear shaft adjusted too tight.	(3) Adjust according to instructions.
	(4) Improper caster or toe-in.	(4) Align the wheels.
	(5) Steering column misaligned.	(5) See "Steering Gear Alignment."
	(6) Loose, worn or broken pump belt.	(6) Adjust or replace belt.
	(7) Air in system.	(7) Bleed air from system.
	(8) Low fluid level in the pump reservoir.	(8) Fill to correct level.
	(9) Pump output pressure low.	(9) See "Pressure Test."
	(10) Leakage at power cylinder piston rings. (Linkage type).	(10) Replace piston rings and repair as required.
	(11) Binding or bent cylinder linkage. (Linkage type).	(11) Replace or repair as required.
	(12) Valve spool and/or sleeve sticking. (Linkage type).	(12) Free-up or replace as required.
Pull to One Side (Tendency of the Vehicle to Veer in one Direction Only)	(1) Incorrect tire pressure or tire sizes not uniform.	(1) Check tire sizes and inflate the tires to recommended pressures.
	(2) Wheel bearings improperly adjusted.	(2) Adjust wheel bearings.
	(3) Dragging brakes.	(3) Inspect for weak, or broken brake shoe spring, binding pedal.
	(4) Improper caster, camber or toe-in.	(4) Adjust to specifications.
	(5) Grease, dirt, oil or brake fluid on brake linings.	(5) Inspect, replace and adjust as necessary.
	(6) Broken or sagging rear springs.	(6) Replace the rear springs.
	(7) Bent front axle, linkage or steering knuckle.	(7) Replace the parts as necessary.
	(8) Worn or tight king pin bushings.	(8) Lubricate or replace as necessary.
Wander or Weave	(1) Improper caster, camber or toe-in.	(1) Adjust to specifications.
	(2) Worn king pin and bushings.	(2) Replace parts as required.
	(3) Worn or improperly adjusted front wheel bearings.	(3) Adjust or replace parts as necessary.
	(4) Loose spring shackles.	(4) Adjust or replace parts as necessary.
	(5) Incorrect tire pressure or tire sizes not uniform.	(5) Check tire sizes and inflate tires to recommended pressure.
	(6) Loose steering gear mounting bolts.	(6) Tighten to specifications.
	(7) Tight king pin bushings.	(7) Lubricate or ream to proper fit.
	(8) Tight king pin thrust bearings.	(8) Adjust to .001 to .005 inch clearance.
Wheel Tramp (Excessive Vertical Motion of Wheels)	(1) Incorrect tire pressure.	(1) Inflate the tires to recommended pressures.
	(2) Improper balance of wheels, tires and brake drums.	(2) Balance as necessary.
	(3) Loose tie rod ends or steering connections.	(3) Inspect and repair as necessary.
	(4) Worn or inoperative shock absorbers.	(4) Replace the shock absorbers.
	(5) Excessive run-out of brake drums, wheels or tires.	(5) Repair or replace as required.
Shimmy	(1) Badly worn and/or unevenly worn tires.	(1) Rotate tires or replace if necessary.
	(2) Wheels and tires out of balance.	(2) Balance wheel and tire assemblies.
	(3) Worn or loose steering linkage parts.	(3) Replace parts as required.
	(4) Worn king pins and bushings.	(4) Replace king pins and bushings.
	(5) Loose steering gear adjustments.	(5) Adjust steering gear as necessary.
	(6) Loose wheel bearings.	(6) Adjust wheel bearings.
	(7) Improper caster setting.	(7) Adjust caster to specifications.
	(8) Weak or broken springs.	(8) Replace as required.
	(9) Incorrect tire pressure or tire sizes not uniform.	(9) Check tire sizes and inflate tires to recommended pressure.
	(10) Faulty shock absorbers.	(10) Replace as necessary.
	(11) Power cylinder loose at bracket.	(11) Tighten or replace bushings as necessary.
	(12) Power cylinder not aligned in same plane as tie rod.	(12) Correct the alignment.
	(13) Improper valve spool nut adjustment allowing control valve end play.	(13) Adjust according to instructions.
	(14) Loose steering gear mounting bolts.	(14) Tighten to specifications.

STEERING TROUBLE DIAGNOSIS

Condition	Possible Cause	Correction
Intermittent or No Power Assist	(1) Belt slipping and/or low fluid level.	(1) Adjust or replace belt. Add fluid as necessary.
	(2) Piston or rod binding in power cylinder. (Linkage type).	(2) Repair or replace piston and rod.
	(3) Sliding sleeve stuck in control valve. (Linkage type).	(3) Free-up or replace sleeve.
	(4) Improper pump operation.	(4) Refer to "Power Steering Pump."
Poor or No Recovery from Turns	(1) Improper caster setting.	(1) Adjust to specifications.
	(2) Steering gear adjustments too tight.	(2) Adjust according to instructions.
	(3) Improper spool nut adjustment. (Linkage type).	(3) Adjust according to instructions.
	(4) Valve spool installed backwards. (Linkage Type).	(4) Install valve spool correctly.
	(5) Low tire pressure.	(5) Inflate tires to recommended pressure.
	(6) Tight steering linkage.	(6) Lubricate as necessary.
	(7) King pins frozen.	(7) Lubricate as necessary.

POWER STEERING PUMP

Condition	Possible Cause	Correction
Intermittant Assist	(1) Flow control valve sticking.	(1) Pressure test pump and service as necessary.
	(2) Slipping belt.	(2) Adjust belt.
	(3) Low fluid level.	(3) Inspect and correct fluid level.
	(4) Low pump efficiency.	(4) Pressure test pump and service as necessary.
No Assist	(1) Pump seizure.	(1) Replace pump.
	(2) Broken slipper spring(s).	(2) Recondition pump or replace as necessary.
	(3) Flow control bore plug ring not in place.	(3) Replace snap ring. Inspect groove for depth.
	(4) Flow control valve sticking.	(4) Pressure test pump and service as necessary.
No Assist When Parking Only	(1) Wrong pressure relief valve.	(1) Install proper relief valve.
	(2) Broken "O" ring on flow control bore plug.	(2) Replace "O" ring.
	(3) Loose pressure relief valve.	(3) Tighten valve. DO NOT ADJUST.
	(4) Low pump efficiency.	(4) Pressure test pump and service as necessary.
Noisy Pump	(1) Low fluid level.	(1) Inspect and correct fluid level.
	(2) Belt noise.	(2) Inspect for pulley alignment, paint or grease on pulley and correct. Adjust belt.
	(3) Foreign material blocking pump housing oil inlet hole.	(3) Remove reservoir, visually check inlet oil hole and service as necessary.
Pump Vibration	(1) Pump hose interference with sheet metal or brake lines.	(1) Reroute hoses.
	(2) Belt loose.	(2) Adjust belt.
	(3) Pulley loose or out of round.	(3) Replace pulley.
	(4) Crankshaft pulley loose or damaged.	(4) Replace crankshaft pulley.
	(5) Bracket pivot bolts loose.	(5) If unable to tighten, replace bracket.
Pump Leaks	(1) Cap or filler neck leaks.	(1) Correct fluid level.
	(2) Reservoir solder joints leak.	(2) Resolder or replace reservoir as necessary.
	(3) Reservoir "O" ring leaking.	(3) Inspect sealing area of reservoir. Replace "O" ring or reservoir as necessary.
	(4) Shaft seal leaking.	(4) Replace seal.
	(5) Loose rear bracket bolts.	(5) Tighten bolts.
	(6) Loose or faulty high pressure ferrule.	(6) Tighten fitting to 24 foot-pounds or replace as necessary.
	(7) Rear bolt holes stripped or casting cracked.	(7) Repair, if possible, or replace pump.

Manual Steering Gear

Manual Steering Gear Service

Steering Gear Alignment

Before any steering gear adjustments are made, it is recommended that the front end of the truck be raised and a thorough inspection be made for stiffness or lost motion in the steering gear, steering linkage and front suspension. Worn or damaged parts should be replaced, since a satisfactory adjustment of the steering gear cannot be obtained if bent or badly worn parts exist.

It is also very important that the steering gear be properly aligned in the truck. Misalignment of the gear places a stress on the steering worm shaft, therefore a proper adjustment is impossible. To align the steering gear, loosen the steering gear-to-frame mounting bolts to permit the gear to align itself. Check the steering gear to frame mounting seat, if there is a gap at any of the mounting bolts, proper alignment may be obtained by placing shims where excessive gap appears. Tighten the steering gear-to-frame bolts. Alignment of the gear in the truck is very important and should be done carefully so that a satisfactory, trouble-free gear adjustment may be obtained.

Gemmer Worm and Double Roller Tooth Type

With Screw Adjusted Mesh

The steering gear is of the worm and roller type with a 24 to 1 gear ratio. The cross shaft is straddle mounted with a bearing surface at the top and bottom points of the shaft mounting areas. The three tooth cross shaft roller is mounted in ball bearings. The proper lubricant used in the gear box is S.A.E. 90 Extreme Pressure Lubricant.

The external adjustments given below will remove all play from the steering gear.

Worm Bearing Adjustment

1. Turn the steering wheel about one full turn from straight ahead and secure it so it doesn't move.
2. Determine if there is any worm gear end-play by shaking the front wheel sideways and noting if there is any end movement that may be felt between the steering wheel hub and the steering jacket tube. (Be sure any movement noted is not looseness in the steering jacket tube.)
3. If end play is present, adjust the worm bearings by loosening the four cover cap screws about 1/8". Separate the top shim, using a knife blade, and remove it. Do not damage the remaining shims or gaskets.
4. Replace the cover and recheck the end-play again. If necessary, repeat steps 2 and 3 until the end-play movement is as small as possible without tightening the steering gear too much.

NOTE: Adjustment may be done with the Pitman arm disconnected. With the steering wheel turned about one full turn from straight ahead,

WORM BEARING PRELOAD SHIMS

WORM AND ROLLER
MESH ADJUSTMENT

Steering gear adjustments
(© Ford Motor Co.)

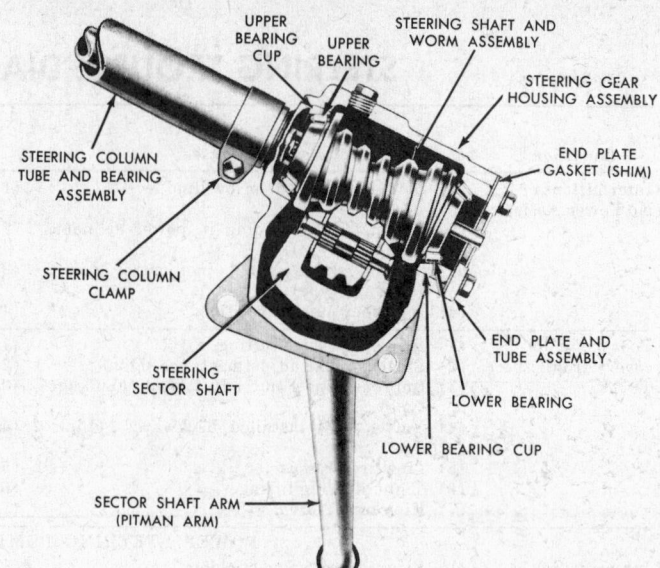

UPPER BEARING CUP
UPPER BEARING
STEERING SHAFT AND WORM ASSEMBLY
STEERING GEAR HOUSING ASSEMBLY
STEERING COLUMN TUBE AND BEARING ASSEMBLY
END PLATE GASKET (SHIM)
STEERING COLUMN CLAMP
END PLATE AND TUBE ASSEMBLY
STEERING SECTOR SHAFT
LOWER BEARING
LOWER BEARING CUP
SECTOR SHAFT ARM (PITMAN ARM)

Worm and roller type steering gear
(© Ford Motor Co.)

using a spring scale tool, adjust with the shims as given above until the spring scale pull is between 1/4 and 5/8 ft. lbs.

Cross Shaft Roller and Worm Mesh Adjustment

1. Turn the steering wheel to the middle of its turning limits with the Pitman arm disconnected. The steering gear roller should be on the worm high spot.
2. Shake the Pitman arm sideways to determine the amount of clearance between the worm cross shaft roller. Movement of more than 1/32" indicates that the roller and worm mesh must be adjusted.
3. Loosen the adjusting screw lock nut and tighten the external cross shaft adjusting screw a small amount. Recheck the clearance by shaking the Pitman arm. Repeat until the clearance is correct. (Do not overtighten.) *NOTE: The cross shaft roller and worm mesh adjustment may be done, using a spring scale tool, by measuring the amount of wheel pull as the external cross shaft adjusting screw is tightened. When the spring scale pull is between 7/8 and 1 1/8 ft. lbs., the adjustment is correct.*
4. Tighten the Pitman arm attaching nut to 100-125 ft. lbs. The steering wheel nut (if loosened) should be tightened to 15-20 ft. lbs. torque.

Disassembly

1. Remove steering gear oil seal, using a suitable puller.
2. Remove cross shaft, using an arbor to prevent bearings from dropping out.

3. Remove cover, shims and cover gasket.
4. Remove worm gear, thrust bearings and bearing cups.

Assembly

1. Clean and inspect all parts, replace as necessary. *NOTE: If either thrust bearing is excessively worn, replace them both.*
2. Reassemble steering gear, using new oil seal.
3. Perform worm bearing and cross shaft roller and worm mesh adjustments.
4. Lubricate to specifications.

Ford Steering Gear— Recirculating Ball Type

Steering Worm and Sector Gear Adjustments

The ball nut assembly and the sector gear must be adjusted properly to maintain a minimum amount of steering shaft end play and a minimum amount of backlash between the sector gear and the ball nut. There are only two adjustments that may

Inch-Pounds Torque Wrench

Checking steering gear preload
(© Ford Motor Co.)

be done on this steering gear and they should be done as given below:

1. Remove the steering gear from the vehicle.
2. Loosen the locknut on the sector shaft adjustment screw and turn the adjusting screw counterclockwise about three turns.
3. Measure the worm bearing preload by attaching an in. lbs. torque wrench to the input shaft. Note the reading required to rotate input shaft about 1½ turns either side of center. If the torque reading is not about 4-5 in. lbs., adjust the gear as given in the next step.
4. Loosen the steering shaft bearing adjuster lock nut and tighten or back off the bearing adjusting screw until the preload is within the specified limits.
5. Tighten the steering shaft bearing adjuster lock nut, and recheck the preload torque.
6. Turn the input shaft slowly to either stop. Turn gently against the stop to avoid possible damage to the ball return guides. Then rotate the shaft three turns to center the ball nut.
7. Turn the sector adjusting screw clockwise until the proper torque (9-10 in. lbs.) is obtained that is necessary to rotate the worm gear past its center (high spot).
8. With the input shaft centered, hold the sector shaft and check the lash between the ball nuts, balls, and worm shaft by applying 15 lbs. torque to the steering input shaft in both right and left turn directions. The total travel of the wrench should not exceed 1¼".
9. Tighten the sector adjusting screw locknut, and recheck the backlash. Install the steering gear.

Disassembly

1. Rotate the steering shaft three turns from either stop.
2. Remove the sector shaft adjusting screw locknut and loosen the screw one turn. Remove the steering shaft bearing adjuster, and the housing cover bolts and remove the sector shaft. Remove the shaft by turning the screw clockwise. Keep the shim with the screw.
3. Remove the sector shaft from the housing.
4. Carefully pull the steering shaft and ball nut from the housing, and remove the steering shaft lower bearing. Do not run the ball nut to either end of the worm gear to prevent damaging the ball return guides. Disassemble the ball nut only if there are signs of binding or tightness.

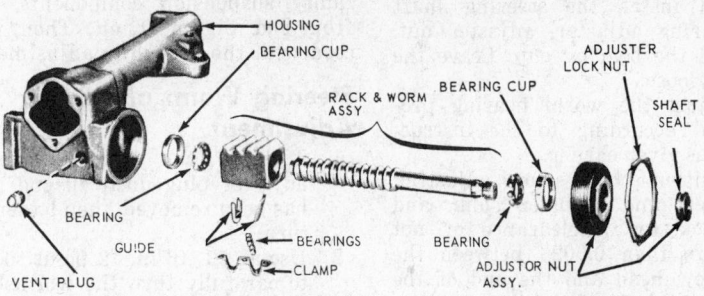

Steering shaft and related parts—Ford recirculating ball gear
(© Ford Motor Co.)

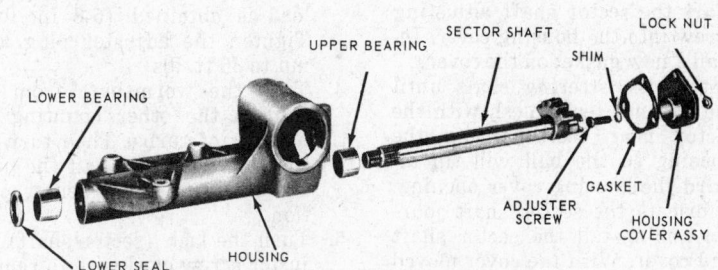

Sector shaft and housing—Ford recirculating ball gear
(© Ford Motor Co.)

5. To disassemble the ball nut, remove the ball return guide clamp and the ball return guides from the ball nut. Keep ball nut clamp side up until ready to remove the ball bearings.
6. Turn the ball nut over and rotate the worm shaft from side to side until all 50 balls have dropped out into a clean pan. With all balls removed, the nut will slide off the wormshaft.
7. Remove the upper bearing cup from the bearing adjuster and the lower cup from the housing. It may be necessary to tap the housing or the adjuster on a wooden block to jar the bearing cups loose.

Inspection

1. Carefully clean and inspect all parts. If the inspection shows bearing damage, the sector shaft bearing and the oil seal should be pressed out.
2. If the sector shaft bearing and oil seals were removed, press new bearings and oil seals into the housing. Do not clean, wash, or soak seals in cleaning solvent.
3. Apply the recommended steering gear lubricant to the housing and seals, filling the pocket between sector shaft bearings.

Assembly

1. Install the bearing cup in the lower end of the housing and a bearing cup in the adjuster nut. Install a new seal in the bearing adjuster if the old seal was removed.

2. Apply gear lube to the outside of the worm shaft and the inside of the ball nut. Lay the steering shaft down, and position the ball nut on the shaft with the guide holes upward and the shallow end of the teeth to the left of the steering wheel position. Align the grooves in worm and ball nut by sighting through the guide holes.
3. Insert the ball guides into the holes in the ball nut, lightly tapping them, if necessary, to seat them.
4. Insert 25 balls into the hole in the top of each ball guide. If necessary, rotate the shaft slightly to distribute the balls evenly in the circuit.
5. Install the ball guide clamp, tightening the screws to the proper torque. Check that the worm shaft rotates freely.
6. Coat the threads of the steering shaft bearing adjuster, the housing cover bolts, and the sector adjusting screw with a suitable oil-resistant sealing compound. Do not apply sealer to female threads and do not get sealer on the steering shaft bearings.
7. Coat the worm bearings, sector shaft bearings, and gear teeth with steering gear lubricant.
8. Clamp the housing in a vise, with the sector shaft axis horizontal, and place the steering shaft lower bearing in its cup. Place the steering shaft and ball nut assemblies in the housing.
9. Position the steering shaft upper bearing on top of the worm gear

and install the steering shaft bearing adjuster, adjuster nut, and the bearing cup. Leave the nut loose.

10. Adjust the worm bearing preload according to the instructions given earlier.

11. Position the sector adjusting screw and adjuster shim, and check for a clearance of not more than 0.002″ between the screw head and the end of the sector shaft. If the clearance exceeds 0.002″, add enough shims to reduce the clearance to under 0.002″ clearance.

12. Start the sector shaft adjusting screw into the housing cover. Install a new gasket on the cover.

13. Rotate the steering shaft until the ball nut teeth mesh with the sector gear teeth, tilting the housing so the ball will tip toward the housing cover opening.

14. Lubricate the sector shaft journal and install the sector shaft and cover. With the cover moved to one side, fill the gear with lubricant (about 0.97 lb.). Push the cover and the sector shaft into place, and install the two top housing bolts. Do not tighten the bolts until checking to see that there is some lash between the ball nut and the sector gear teeth. Hold or push the cover away from the ball nut and tighten the bolts to the proper torque (30-40 ft. lbs.).

15. Loosely install the sector shaft adjusting screw lock nut and adjust the sector shaft mesh load as given earlier. Tighten the adjusting screw lock nut.

Saginaw Recirculating Ball Type

The steering gear is of the recirculating ball nut type. The ball nut, mounted on the worm gear, is driven by means of steel balls which circulate in helical grooves in both the worm and nut. Ball return guides attached to the nut serve to recirculate the two sets of balls in the grooves. As the steering wheel is turned to the right, the ball nut moves upward. When the wheel is turned to the left, the ball nut moves downward.

The sector teeth on the pinion shaft and the ball nut are designed so that they fit the tightest when the steering wheel is straight ahead. This mesh action is adjusted by an adjusting screw which moves the pinion shaft endwise until the teeth mesh properly. The worm bearing adjuster provides proper preloading of the upper and lower bearings.

Before doing the adjustment procedures given below, ensure that the steering problem is not caused by

faulty suspension components, bad front end alignment, etc. Then, proceed with the following adjustments.

Steering Worm and Sector Adjustment

1. Tighten the worm bearing adjuster plug until all end play has been removed, then loosen ¼ turn.

2. Use an 11/16 in. 12 point socket to carefully turn the wormshaft all the way into the right corner then turn back about ½ turn.

3. Tighten the adjuster plug until the proper thrust bearing preload is obtained (5-8 in. lbs.). Tighten the adjuster plug locknut to 85 ft. lbs.

4. Turn the wormshaft from one stop to the other counting the number of turns. Then turn the shaft back exactly half the number of turns to the center position.

5. Turn the lash (sector shaft) adjuster screw clockwise to remove all lash between the ball nut and sector teeth. Tighten the locknut to 25 ft. lbs.

6. Using an 11/16 in. 12 point socket and an in. lb. torque wrench, observe the highest reading while the gear is turned through the center position. It should be 16 in. lbs. or less.

7. If necessary repeat steps 5 and 6.

Disassembly

1. Place the steering gear in a vise, clamping onto one of the mounting tabs. The wormshaft should be in a horizontal position.

2. Rotate the wormshaft from stop to stop and count the total number of turns. Turn back exactly halfway, placing the gear on center.

3. Remove the three self locking bolts which attach the sector cover to the housing.

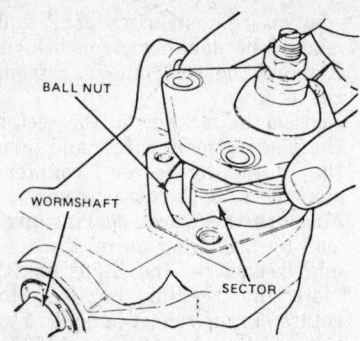

Removing sector shaft assembly—Saginaw

4. Using a plastic hammer, tap lightly on the end of the sector shaft and lift the sector cover and sector and sector shaft assembly from the gear housing.

   NOTE: It may be necessary to turn the wormshaft by hand until the sector will pass through the opening in the housing.

5. Remove the locknut from the adjuster plug and remove the adjuster plug assembly.

6. Pull the wormshaft and ball nut assembly from the housing.

   NOTE: Damage may be done to the ends of the ball guides if the ball nut is allowed to rotate to the end of the worm.

7. Remove the worm shaft upper bearing from inside the gear housing.

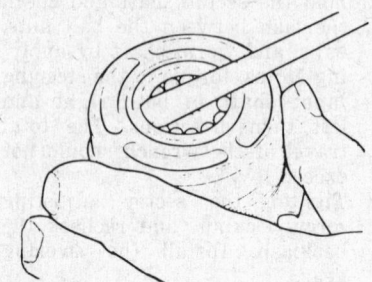

Removing bearing retainer from adjuster— Saginaw recirculating ball (© Ford Motor Co)

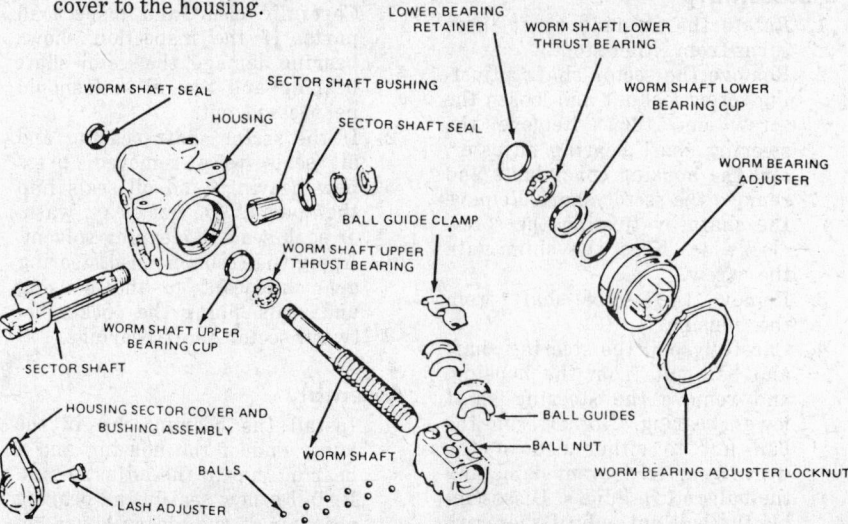

Steering gear assembly—Saginaw recirculating (© Ford Motor Company)

8. Pry the wormshaft lower bearing retainer from the adjuster plug housing and remove the bearing.

9. Remove the locknut from the lash adjuster screw in the sector cover. Turn the lash adjuster screw clockwise and remove it from the sector cover. Slide the adjuster screw and shim out of the slot in the end of the sector shaft.

10. Pry out and discard both the sector shaft and wormshaft seals.

Inspection

1. Wash all parts in cleaning solvent and blow dry with an air hose.

2. Use a magnifying glass and inspect the bearings and bearing caps for signs of indentation, or chipping. Replace any parts that show signs of damage.

3. Check the fit of the sector shaft in the bushings in the sector cover and housing. If these bushings are worn, a new sector cover and bushing assembly or housing bushing should be installed.

4. Check steering gear wormshaft assembly for being bent or damaged.

Shaft Seal Replacement

1. Using a screwdriver, pry out the old seal.

2. Install the new seal by pressing the outer diameter of the seal with a suitable size socket.

NOTE: Make sure the socket is large enough to avoid damaging the external lip of the seal.

Sector Shaft Bushing Replacement

1. Place the steering gear housing in an arbor press.

2. Press the sector shaft bushing from the housing.

NOTE: Service bushings are bored to size and require no further reaming.

Sector Cover Bushing Replacement

1. The sector cover bushing is not

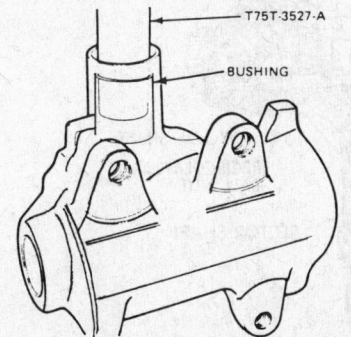

Removing sector shaft bushing—Saginaw recirculating ball (© Ford Motor Company)

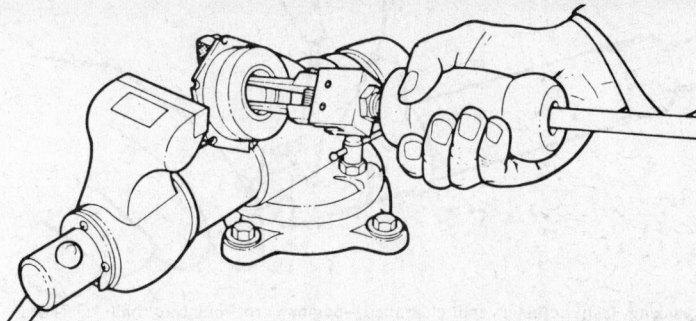

Removing wormshaft lower bearing cup from the adjuster plug—Saginaw recirculating ball (© Ford Motor Company)

serviced separately. The entire sector cover assembly including the bushing must be replaced as a unit.

Ball Nut Service

If there is any indication of binding or tightness when the ball nut is rotated on the worm the unit should be disassembled, cleaned and inspected as follows:

Ball Nut Disassembly

1. Remove the screws and clamp retaining the ball guides in the ball nut. Pull the guides out of the ball nut.

2. Turn the ball nut upside down and rotate the wormshaft back and forth until all the balls have dropped out of the ball nut. The ball nut can now be pulled endwise off the worm.

3. Wash all parts in solvent and dry them with air. Use a magnifying glass and inspect the worm and nut grooves and the surface of all balls for signs of indentation. Check all ball guides for damage at the ends. Replace any damaged parts.

Ball Nut Assembly

1. Slip the ball nut over the worm with the ball guide holes up and the shallow end of the ball nut teeth to the left from the steering wheel position. Sight through the ball guide to align the grooves in the worm.

2. Place two ball guide halves together and insert them in the upper circuit in the ball nut. Place the two remaining guides together and insert them in the lower circuit.

3. Count out 25 balls and place them in a suitable container. This is the proper number of balls for one circuit.

4. Load the 25 balls into one of the guide holes while turning the wormshaft gradually away from that hole.

5. Fill the remaining ball circuit in the same manner.

6. Assemble the ball guide clamp to the ball nut and tighten the screws to 18-24 in. lbs.

7. Check the assembly by rotating the ball nut on the worm to see that it moves freely.

NOTE: Do not rotate the ball nut to the end of the worm threads as this may damage the ball guides.

Assembly

1. Coat the threads of the adjuster plug, sector cover bolts and lash adjuster with a non-drying oil resistant sealing compound.

NOTE: Do not apply compound to the female threads. Use extreme care when applying compound to the bearing adjuster so that it does not come in contact with the wormshaft bearing.

2. Place the steering gear housing in a vise with the wormshaft bore horizontal and the sector cover opening up.

3. Make sure that all seals, bushings and bearing cups are installed in the gear housing and that the ball nut is installed on the wormshaft.

4. Slip the wormshaft upper bearing assembly over the wormshaft and insert the wormshaft and ball nut assembly into the housing, feeding the end of the shaft

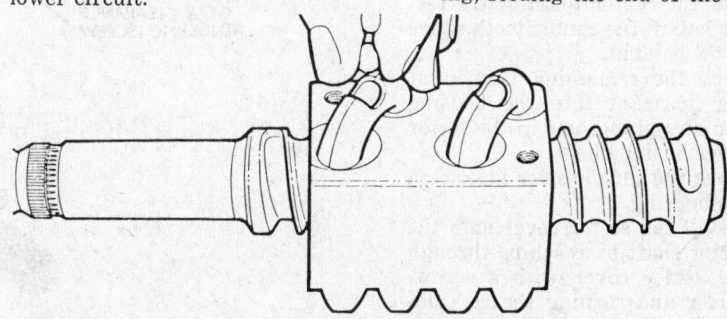

Filling the ball circuits—Saginaw recirculating ball (© Ford Motor Company)

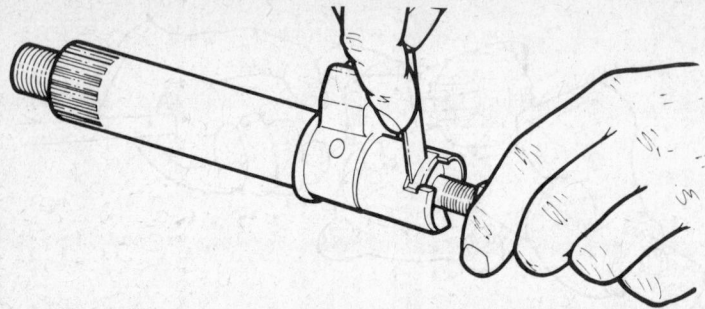

Checking lash adjuster end clearance—Saginaw recirculating ball (© Ford Motor Company)

through the upper ball bearing cup and seal.

5. Place the wormshaft lower bearing assembly in the adjuster plug bearing cup and press the stamped retainer into place with a suitable size socket.

6. Install the adjuster plug and locknut into the lower end of the housing while carefully guiding the end of the wormshaft into the bearing until nearly all end play has been removed from the wormshaft.

7. Position the lash adjuster including the shim in the slotted end of the sector shaft.

NOTE: End clearance should not be greater than .002. If the end clearance is greater than .002 a shim package is available with thicknesses of .063, .065, .067, .069.

8. Lubricate the steering gear with 11 oz. of steering gear grease. Rotate the wormshaft until the ball nut is at the other end of its travel and then pack as much new lubricant into the housing as possible without losing out the sector shaft opening. Rotate the wormshaft until the ball nut is at the other end of its travel and pack as much lubricant into the opposite end as possible.

9. Rotate the wormshaft until the ball nut is in the center of travel. This is to make sure that the sector shaft and ball nut will engage properly with the center tooth of the sector entering the center tooth space in the ball nut.

10. Insert the sector shaft assembly including lash adjuster screw and shim into the housing so that the center tooth of the sector enters the center tooth space in the ball nut.

11. Pack the remaining portion of the lubricant into the housing and also place some in the sector cover bushing hole.

12. Place the sector cover gasket on the housing.

13. Install the sector cover onto the sector shaft by reaching through the sector cover with a screwdriver and turning the lash adjuster screw counterclockwise

until the screw bottoms, then back the screw off one-half turn. Loosely install a new lock nut onto the adjuster screw.

14. Install and tighten the sector cover bolt to 30 ft. lbs.

Chrysler Recirculating Ball Type

The steering gear is of the recirculating ball nut type. The ball nut, mounted on the worm gear, is driven by means of steel balls which circulate in helical grooves in both the worm and nut. Ball return guides attached to the nut serve to recirculate the two sets of balls in the grooves. As the steering wheel is turned to the right, the ball nut moves upward. When the wheel is turned to the left, the ball nut moves downward.

The sector teeth on the pinion shaft and the ball nut are designed so that they fit the tightest when the steering wheel is straight ahead. This mesh action is adjusted by an adjusting screw which moves the pinion shaft endwise until the teeth mesh properly. The worm bearing adjuster provides proper preloading of the upper and lower bearings.

Worm Bearing Pre-load Adjustment

1. Remove the steering gear arm and lockwasher from the sector shaft, using a suitable gear puller.

2. Remove the horn button or horn ring.

3. Loosen the cross-shaft adjusting screw locknut, and back out the adjusting screw about two turns.

4. Turn the steering wheel two complete turns from the straight ahead position, and place an in. lb torque wrench on the steering shaft nut.

5. Rotate the steering shaft at least one turn toward the straight ahead position while measuring the torque on the torque wrench. The torque should be between $1\frac{1}{8}$ and $4\frac{1}{2}$ in. lbs. to move the steering wheel. If torque is not within these limits, loosen the worm shaft bearing adjuster locknut and turn the adjuster clockwise to increase the preload or counterclockwise to decrease the preload. When the preload is correct, hold the adjuster screw steady and tighten the locknut. Recheck preload.

Ball Nut Rack and Sector Mesh Adjustment

NOTE: this adjustment can be accurately made only after proper preloading of worm bearing.

1. Turn steering wheel gently from one stop to the other, counting the number of turns. Turn the steering wheel back exactly half way, to the center position.

2. Turn the cross-shaft adjusting screw clockwise to remove all lash between ball nut rack and the sector gear teeth, then tighten adjusting screw locknut to 35 ft. lbs.

3. Turn the steering wheel about $\frac{1}{4}$ turn away from the center or high spot position. With the torque wrench on the steering wheel nut measure the torque required to turn the steering wheel through the high spot at the center position. The reading should

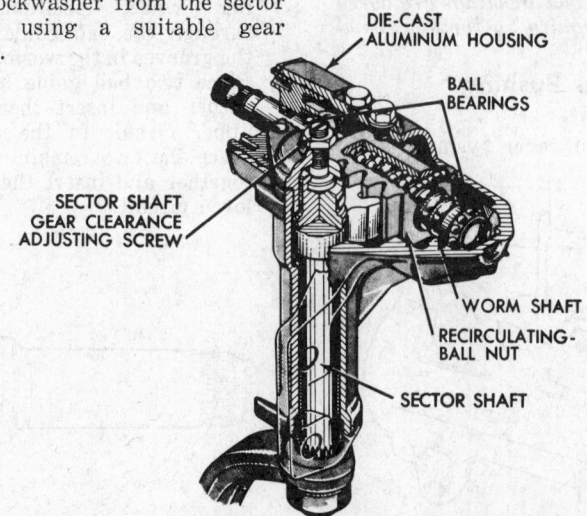

Chrysler recirculating ball type steering gear (© Chrysler Corp.)

DIE-CAST ALUMINUM HOUSING

BALL BEARINGS

SECTOR SHAFT GEAR CLEARANCE ADJUSTING SCREW

WORM SHAFT

RECIRCULATING-BALL NUT

SECTOR SHAFT

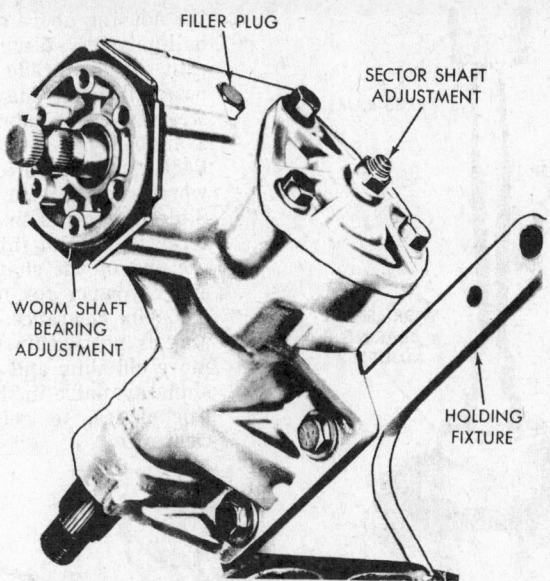

FILLER PLUG

SECTOR SHAFT
ADJUSTMENT

WORM SHAFT
BEARING
ADJUSTMENT

HOLDING
FIXTURE

Steering gear adjustment locations (© Chrysler Corp.)

be between 8 and 11 in. lbs. This is the total of the worm shaft bearing preload and the ball nut rack and sector gear mesh load. Readjust the cross-shaft adjustment screw if necessary to obtain a correct torque reading.

4. After completing the adjustments, place the front wheels in a straight ahead position, and with the steering wheel and steering gear centered, install the steering arm on cross-shaft. Tighten the steering arm retaining nut to 180 ft. lbs.

Steering Gear Disassembly and Assembly

1. Attach the steering gear assembly to a holding fixture and put the holding fixture in a bench vise. Thoroughly clean the outside surface before disassembly.
2. Loosen the cross-shaft adjusting screw locknut, and back out the adjusting screw about two turns to relieve the mesh load between the ball nut rack and the sector gear teeth. Remove the cross-shaft seal as given in the procedure for cross-shaft seal replacement.
3. Position the steering gear worm

shaft in a straight ahead position.

4. Remove the attaching bolts from the cross-shaft cover and slowly remove the cross-shaft while sliding an arbor tool into the housing. Remove the locknut from the adjusting screw and remove the screw from the cover by turning it clockwise. Slide the adjustment screw and its shim out of the slot in the end of the

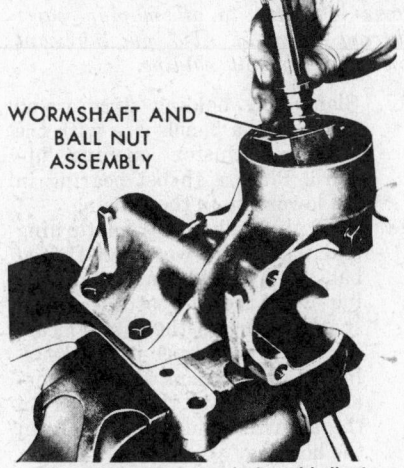

WORMSHAFT AND
BALL NUT
ASSEMBLY

Removing the wormshaft and ballnut assembly (© Chrysler Corp.)

cross-shaft.

5. Loosen the worm shaft bearing adjuster locknut with a brass drift (punch) and remove the locknut. Hold the worm shaft steady while unscrewing the adjuster. Slide the worm adjuster off the shaft.

CAUTION: Handle the adjuster carefully to avoid damaging the aluminum threads. Also, do not run the ball nut down to either end of the worm shaft to avoid damaging the ball guides.

6. Carefully remove the worm and ball nut assembly. This assembly is serviced as a complete assembly only and is not to be disassembled or the ball return guides removed or disturbed.
7. Remove the cross-shaft needle bearing by placing the gear housing in an arbor press; insert a tool in the lower end of the housing and press both bearings through the housing.

The cross-shaft cover assembly, including a needle bearing or bushing, is serviced as an assembly.

8. Remove the worm shaft oil seal from the worm shaft bearing adjuster by inserting a blunt punch behind the seal and tapping alternately on each side of the seal until it is driven out of the adjuster.
9. Remove the worm shaft in the same manner as that given in step 8. *Be careful not to cock the bearing cup and distort the adjuster counter bore.*
10. Remove the lower cup if necessary. Pull the bearing cup out.
11. Wash all parts in clean solvent and dry thoroughly. Inspect all parts for wear, scoring, pitting, etc. Test operation of the worm shaft and ball nut assembly. If ball nut does not travel smoothly and freely on the worm shaft or if there is binding, replace the assembly.

NOTE: extreme care must be taken when handling the aluminum worm bearing adjuster to avoid thread damage. Also, be careful not to damage the threads in the gear housing. Always lubricate the worm

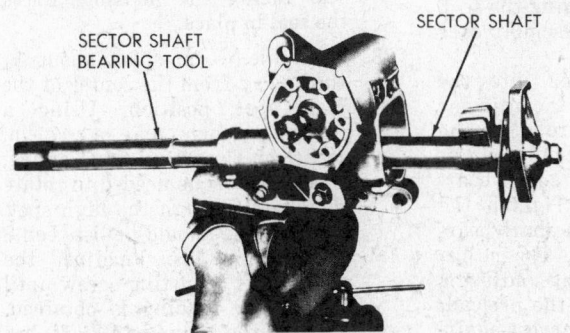

SECTOR SHAFT
BEARING TOOL

SECTOR SHAFT

Removing the cross shaft (© Chrysler Corp.)

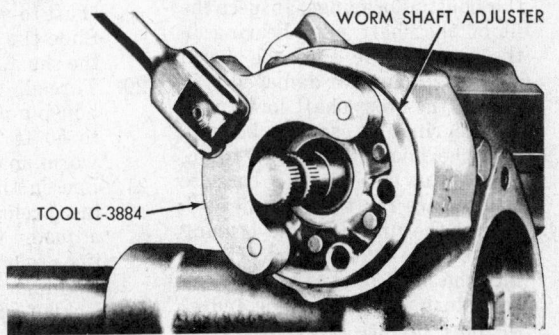

WORM SHAFT ADJUSTER

TOOL C-3884

Removing the wormshaft adjuster (© Chrysler Corp.)

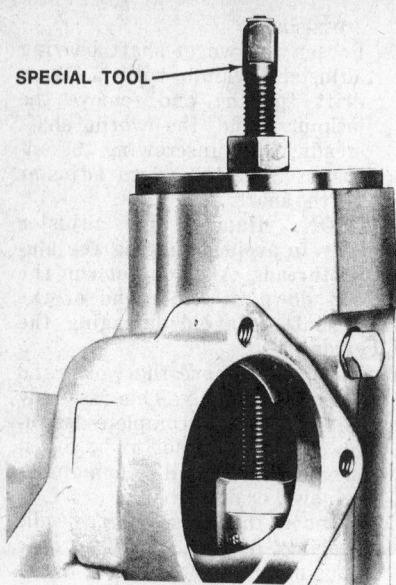

Removing the lower bearing cup
(© Chrysler Corp.)

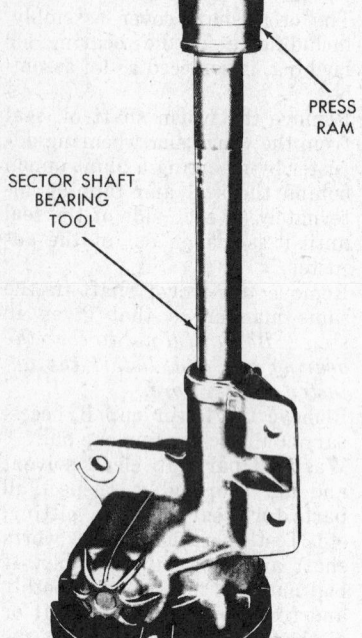

Removing the cross shaft inner and outer
bearings (© Chrysler Corp.)

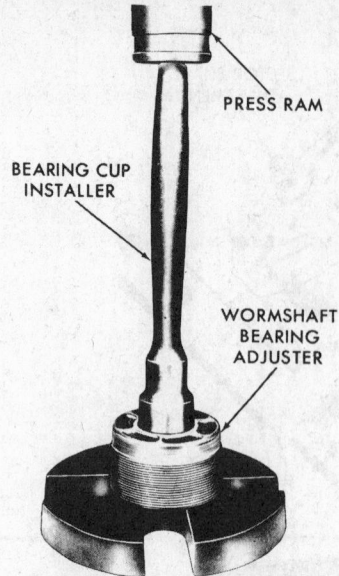

Installing the wormshaft upper bearing
cup (© Chrysler Corp.)

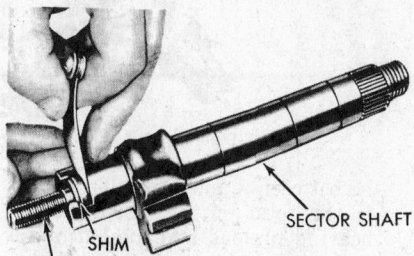

Measuring the cross shaft adjusting screw
and clearance (© Chrysler Corp.)

*bearing adjuster before screwing it
into the housing.*

12. Inspect the cross-shaft for wear and check the fit of the shaft in the housing bearings. Inspect the fit of the shaft pilot bearing in the housing. Be sure the worm shaft is not bent or damaged.

13. Install the cross-shaft lower needle bearing. Press the bearing into the housing about 7/16 in. below the end of the bore to leave space for the new oil seal.

14. Install the upper needle bearing in the same manner and press it into the inside end of the housing bore flush with the inside end of the bore surface.

15. Install the worm shaft bearing

cups (upper and lower) by placing them and their spacers in the adjuster nut and press them into place.

16. Install the worm shaft oil seal by placing the seal in the worm shaft adjuster with the metal seal retainer up. Drive the seal into place with a suitable sleeve until it is just below the end of the bore in the adjuster.

NOTE: apply a coating of steering gear lubricant to all moving parts during assembly. Also, put lubricant on and around oil seal lips.

17. Clamp the holding fixture and housing in a bench vise with the bearing adjuster opening upward. Place a thrust bearing in the lower cup in the housing.

18. Hold the ball nut from turning and insert the worm shaft and ball nut assembly into the housing with the end of the worm shaft resting in the thrust bearing. Place the upper thrust bearing on the worm shaft. Thoroughly lubricate the threads on the adjuster and the threads in the housing.

19. Place a protective sleeve of tape over the splines on the worm shaft to avoid damaging the seal. Slide the adjuster assembly over the shaft.

20. Thread the adjuster into the housing and tighten the adjuster to 50 ft. lbs. while rotating the worm shaft to seat the bearings.

21. Loosen the adjuster so no bearing preload exists. Tighten the adjuster for a worm shaft bearing preload of 1⅛ to 4½ in. lbs. Tighten the bearing adjuster locknut and recheck the preload.

22. Before installing the cross-shaft, pack the worm shaft cavities in

the housing above and below the ball nut with steering gear lubricant. A good grade of multi-purpose lubricant may be used if steering gear lubricant is not available. *Do not use gear oil.* Pack enough lubricant into the worm cavities to cover the worm.

23. Slide the cross-shaft adjusting screw and shim into the slot in the end of the shaft. Check the end clearance for no more than 0.004 in. clearance. If the clearance is not within the limit, remove old shim and install a new shim, available in three different thicknesses, to get the proper clearance.

24. Start the cross-shaft and adjuster screw into the bearing in the housing cover. Using a screwdriver through the hole in the cover, turn the screw counterclockwise to pull the shaft into the cover. Install the adjusting screw locknut, but do not tighten at this time.

25. Rotate the worm shaft to center the ball nut.

26. Place a new gasket on the housing cover and install the cross-shaft and cover assembly into the steering gear housing. *Be sure to coat the cross-shaft and sector teeth with steering gear lubricant before installing the cross-shaft in the housing.* Allow some lash between the cross-shaft sector teeth and the ball nut rack. Install and tighten the cover bolts to 25 ft. lbs.

27. Place the cross-shaft seal on the cross-shaft with the lip of the seal facing the housing. Press the seal in place.

28. Turn the worm shaft about ¼ turn away from the center of the high spot position. Using a torque wrench and a ¾ in. socket on the worm shaft spline, check the torque needed to rotate the shaft through the high spot. The reading should be between 8 and 11 in. lbs. Readjust the cross-shaft adjusting screw until the proper reading is obtained. Tighten the locknut to 35 ft. lbs. and recheck cross-shaft torque.

Cross-Shaft Oil Seal Replacement

1. Remove the steering gear arm retaining nut and lockwasher.
2. Remove seal with a seal puller or other appropriate tool.
3. Place a new oil seal onto the splines of the cross-shaft with the lip of the seal facing the housing.
4. Remove the tool, and install the steering gear arm, lockwasher, and retaining nut. Tighten the nut to 180 ft. lbs. torque.

Ross Cam and Twin Lever Type

The cam and lever steering gear consists of a spiral cam and a cross shaft and lever assembly with two lever studs. When the steering wheel is turned, the cam moves the studs, causing the cross shaft to rotate and the steering arm to move. There are two adjustments that must be done: end play of the steering shaft, and backlash adjustment of the lever studs (tapered pins).

End-Play Adjustment

The steering gear is adjusted for the cam groove. Loosen the cover enough to remove one or more shims as required. As each shim is removed, retighten the cover and check the end-play. Adjustment should leave a slight drag in the steering gear but still allow it to turn easily. Replacement shims are available in three thicknesses: 0.002", 0.003", and 0.010". (0.0508, 0.0762, and 0.254 mm).

Backlash Adjustment

The steering gear lever studs (tapered pins) backlash adjustment is done by turning the adjusting screw in until a slight drag is felt through the mid-position when turning the steering wheel slowly from one stop to the other. This adjustment is done when the steering wheel is centered in the straight ahead driving position. The cam groove is cut shallow in this area to provide a close adjustment of the pins. Do not adjust the screw in any other position of the steering gear.

Steering Gear

Disassembly

NOTE: When disassembling early production models, the line across the face of the steering arm and the end of the shaft should match. On later production models, blind splines on the lever shaft and in the steering gear arm engage, allowing correct positioning of the arm.

1. Remove the steering gear arm.
2. Loosen the adjusting screw lock nut and unscrew the adjusting screw two turns.
3. Remove the side cover screws and washers. Remove the side cover and gasket.
4. Remove the lever shaft.
5. Remove the upper cover plate screws, the upper cover, and gasket. Remove the cam, wheel tube, and bearing assembly.

Assembly

1. Inspect all bearings, seals, cam grooves, etc. for wear, chipping, scoring and other damage. Replace all worn parts.
2. Assemble all parts to the wheel tube in the reverse order of disassembly. Assemble the cam, wheel tube, and bearing assembly and install in the housing, seating the lower bearing ball cup correctly in the housing.

NOTE: New plastic retainer type cam bearings are available for Ross steering gears that re-place, and are interchangeable with, the lock ring type cam bearings on gears equipped with early type cams.

3. With the adjusting shims installed in the housing, install the upper cover and adjust the cam bearings.
4. Install the lever shaft, gasket, and side cover on the housing. Set the adjusting screw for a minimum backlash of the studs in the cam groove, with the steering gear at the center point of travel.
5. Assemble the upper bearing spring and spring seat in the jacket tube, being sure the spring seat is positioned correctly. Install it with the long flange down against the bearing and not up inside the spring coil.
6. Install the steering gear assembly in the truck, reversing the removal procedure.
7. After installing the steering gear assembly in the truck, install the steering wheel temporarily and position the steering gear in the midposition (straight ahead). Do this by slowly turning the steering wheel from one stop to the other, then turning it to the middle of its range.
8. With the steering wheel held steady and the front wheels facing straight ahead, install the steering arm on the lever shaft, with the ball end down. If properly installed, the line across the face of the arm and the end of the shaft should match.

Ross Worm and Roller Steering Gear

Bearing Preload Adjustment

1. Loosen the four capscrews which fasten the end cover to the steering gear housing.

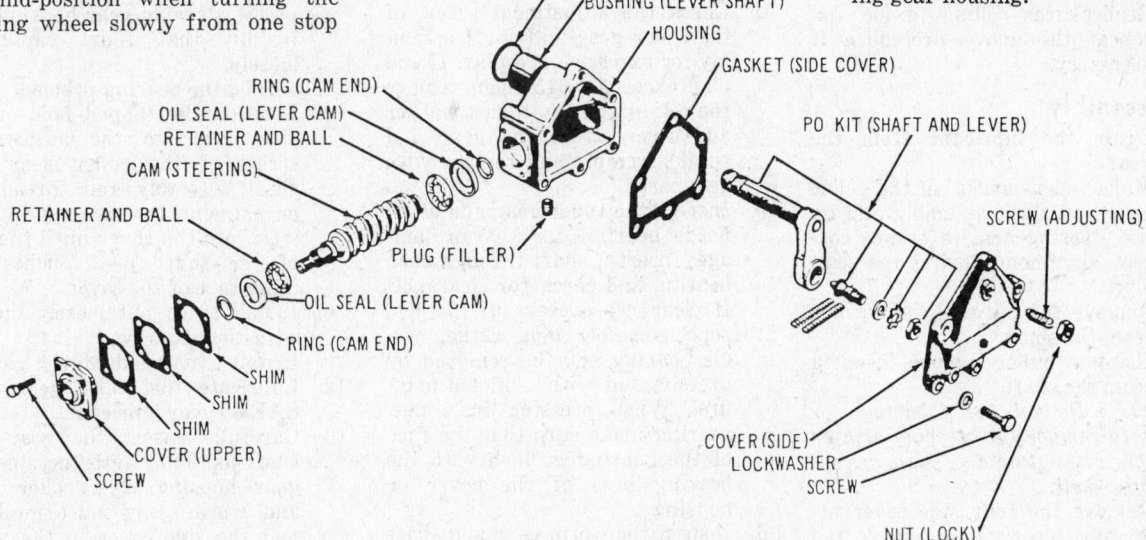

OIL SEAL (LEVER SHAFT)
BUSHING (LEVER SHAFT)
HOUSING
GASKET (SIDE COVER)
RING (CAM END)
OIL SEAL (LEVER CAM)
RETAINER AND BALL
CAM (STEERING)
RETAINER AND BALL
PLUG (FILLER)
OIL SEAL (LEVER CAM)
RING (CAM END)
SHIM
SHIM
SHIM
COVER (UPPER)
SCREW
PO KIT (SHAFT AND LEVER)
SCREW (ADJUSTING)
COVER (SIDE)
LOCKWASHER
SCREW
NUT (LOCK)

Ross cam & lever steering gear (© Chrysler Corporation)

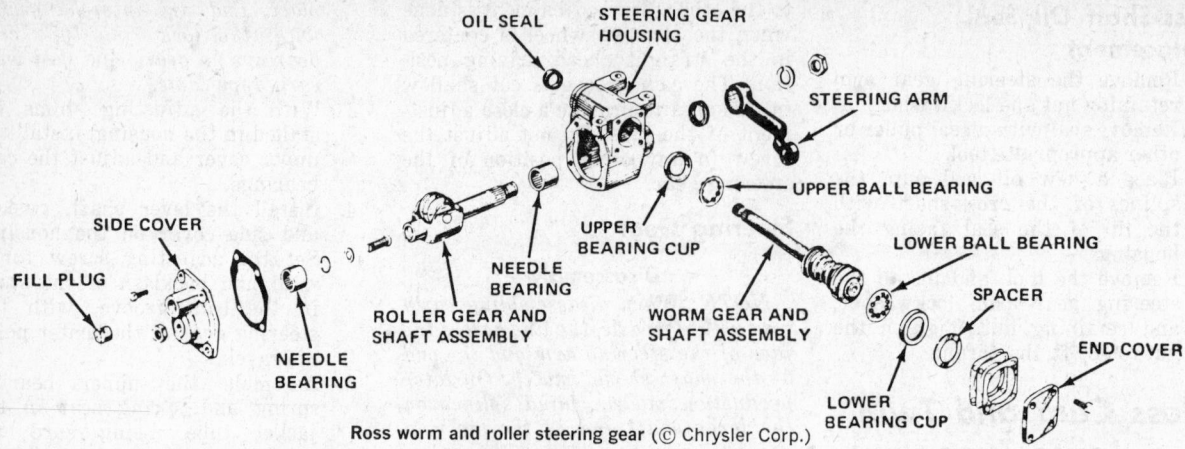

OIL SEAL · STEERING GEAR HOUSING · STEERING ARM · UPPER BALL BEARING · SIDE COVER · UPPER BEARING CUP · LOWER BALL BEARING · FILL PLUG · NEEDLE BEARING · SPACER · ROLLER GEAR AND SHAFT ASSEMBLY · WORM GEAR AND SHAFT ASSEMBLY · END COVER · NEEDLE BEARING · LOWER BEARING CUP

Ross worm and roller steering gear (© Chrysler Corp.)

2. Alternately tighten the cap-screws evenly and rotate the worm gear shaft. Tighten the screws to 18 to 22 ft. lbs.
3. If necessary, remove the end cover and either add or subtract from the number of shims, and repeat steps 1 and 2 to obtain the correct bearing preload.

Worm Gear and Roller Gear Backlash Adjustment

1. Loosen the locknut and turn the adjustment at the side cover counterclockwise until the worm gear shaft turns freely through its entire range of travel.
2. Count the number of turns necessary to rotate the worm gear shaft through its entire range of travel.
3. Turn the shaft back exactly half the number of turns to the center position.
4. Turn the shaft back and forth through its center of travel and tighten the adjustment screw to obtain a rolling torque requirement of 7 to 12 inch lbs.
5. Hold the adjustment screw in position and torque the locknut to 16-20 ft. lbs.
6. Recheck the rolling torque and repeat the above procedure if necessary.

Disassembly

1. Drain the lubricant from the gear.
2. Make index marks on the roller gear and shaft assembly and on the steering arm to assure correct alignment during reassembly.
3. Remove the nut and lockwasher from the shaft.
4. Using a puller remove the arm from the shaft.
   NOTE: Do not use a hammer or wedge to remove the steering arm or damage to the gear and shaft assembly may result.
5. Remove the four side cover attaching screws and remove the cover and roller gear and shaft

assembly as a unit.
6. Remove the locknut from the adjustment screw and turn the screw clockwise until it is completely unthreaded from the side cover, then remove the roller gear and shaft assembly from the cover.
7. Remove the four end cover attaching screws and remove the cover from the housing.
8. Withdraw the worm gear and shaft assembly from the housing.
9. Remove the lower and upper bearing cups and ball bearings from the shaft.
10. Remove and discard both the worm gear shaft and roller gear shaft housing oil seals.

Inspection

1. Clean all parts with a suitable cleaning solvent and blow dry with an air hose.
2. Check the steering gear housing for cracks, leaks or breaks and replace if damaged.
3. Examine the roller gear to assure that it has proper freedom of movement and does not have excessive lash or roughness. Replace if necessary.
4. Check the adjustment screw of the roller gear and shaft assembly for excessive end play. If end play exceeds 0.015 inch, remove the retaining ring, thrust washer and screw from the gear and shaft assembly and replace with new parts.
5. Inspect the roller gear and shaft needle bearings for wear or damage. Insert a shaft through each bearing and check for clearance. If clearance exceeds 0.010 inch, replace the bearings. Either needle bearing may be removed by pressing out with a piloted mandrel. When pressing in a new bearing make sure that the face of the bearing is flush with the bearing boss of the cover or housing.
6. Inspect the worm gear and shaft assembly for wear, scoring or

pitting. Polish the assembly with a fine abrasive cloth or replace if necessary.
7. Check the upper and lower ball bearings and cups of the worm gear and shaft assembly for wear and damage. Replace the ball bearings as a full set if worn or damaged.

Assembly

1. Press new oil seals into the worm gear shaft and roller gear shaft oil seal bores of the housing with the longer lip of each seal facing into the housing.
2. Lubricate the worm gear and shaft assembly and upper ball bearing and cup with SAE 80 gear lubricant.
3. Install the bearing and cup on the shaft.
4. Carefully install the shaft assembly into the steering gear housing.
5. Lubricate the lower end of the worm gear and shaft assembly and lower ball bearing and cup with SAE 80 gear lubricant.
6. Install the bearing, cups and spacer on the shaft.
7. Position the shims and end cover on the steering gear housing and install the four capscrews loosely.
8. Adjust the bearing preload.
9. Position the tapped hole of the side cover to the adjustment screw of the roller gear and shaft assembly and thread the adjustment screw counterclockwise into the cover until the end of the shaft just touches the inner face of the cover.
10. Install a locknut loosely on the adjustment screw.
11. Install a new side cover gasket.
12. Lubricate the roller gear with SAE 80 gear lubricant.
13. Carefully insert the gear and shaft assembly into the steering gear housing. The roller gear and worm gear must mesh to seat the side cover to the housing.

14. Tighten the side cover capscrews to 18-22 ft. lbs.
15. Make a worm gear and roller gear backlash adjustment.
16. Clamp the exposed section of the roller gear and shaft assembly firmly into a soft jaw vise.
17. Align the index marks made during disassembly and position the steering arm to the splined end of the shaft.
18. Install the lockwasher and nut on the shaft threads and tighten the nut to draw the arm into position on the splines.
19. Fill the steering gear housing to the required level with SAE 80 gear lubricant.

Power Steering Gear

General Information

The procedures for maintaining, adjusting, and repairing the power steering systems and components discussed in this chapter are to be done only after determining that the steering linkages and front suspension systems are correctly aligned and in good condition. All worn or damaged parts should be replaced before attempting to service the power steering system. After correcting any condition that could affect the power steering, do the preliminary tests of the steering system components.

Preliminary Tests

Lubrication

Proper lubrication of the steering linkage and the front suspension components is very important for the proper operation of the steering systems of trucks equipped with power steering. Most all power steering systems use the same lubricant in the steering gear box as in the power steering pump reservoir, and the fluid level is maintaned at the pump reservoir.

With power cylinder-assist power steering, the steering gear is of the standard mechanical type and the lubricating oil is self contained within the gear box and the level is maintained by the removal of a filler plug on the gear box housing. The control valve assembly is mounted on the gear box and is lubricated by power steering oil from the power steering pump reservoir, where the level is maintained.

Air Bleeding

Air bubbles in the power steering system must be removed from the fluid. Be sure the reservoir is filled to the proper level and the fluid is warmed up to operating temperature. Then, turn the steering wheel through its full travel three or four times until all the air bubbles are removed. Do not hold the steering wheel against its stops. Recheck the fluid level.

Fluid Level Check

1. Run the engine until the fluid is at the normal operating temperature. Then, turn the steering wheel through its full travel three or four times, and shut off the engine.
2. Check the fluid level in the steering reservoir. If the fluid level is low, add enough fluid to raise the level to the Full mark on the dipstick or filler tube.

Pump Belt Check

1. Inspect the pump belt for cracks, glazing, or worn places. Using a belt tension gauge, check the belt tension for the proper range of adjustment. The amount of tension varies with the make of truck and the condition of the belt. New belts (those belts used less than 15 minutes) require a higher figure. The belt deflection method of adjustment may be used only if a belt tension gauge is not available. The belt should be adjusted for a deflection of ⅜" to ½".

Fluid Leaks

Check all possible leakage points (hoses, power steering pump, or steering gear) for loss of fluid. Turn engine on and rotate the steering wheel from stop to stop several times. Tighten all loose fittings and replace any defective lines or valve seats.

Turning Effort

Check the turning effort required to turn the steering wheel after aligning the front wheels and inflating the tires to the proper pressure.
1. With the vehicle on dry pavement and the front wheel straight ahead, set the parking brake and turn the engine on.
2. After a short warm-up period for the engine, turn the steering wheel back and forth several times to warm the steering fluid.
3. Attach a spring scale to the steering wheel rim and measure the pull required to turn the steering wheel one complete revolution in each direction. The effort needed to turn the steering wheel should not exceed the limits specified.

NOTE: This test may be done with the steering wheel removed and a torque wrench applied on the steering wheel nut.

Power Steering Pump Flow

Since the power steering pump provides all the power assist in a power steering system, the pump must operate properly at all times for the system to work. After performing all the checks given above, the power steering pump may be tested for proper flow by the following procedure:
1. Disconnect the pressure and return lines at the power steering pump and connect the test pressure and return lines. The test lines are connected to a pressure gauge and two manual valves.
2. Open the two manual valves, connect a tachometer to the engine, and start the engine. Run the engine at idle speed until the reservoir fluid temperature reaches about 165-175 degrees Fahrenheit. This temperature must be maintained during the test. Manual valve B may be partially opened to create a back pressure of no more than 350 psi to aid the temperature rise. Reservoir fluid must be at the proper level.
3. After the engine and the reservoir fluid are sufficiently

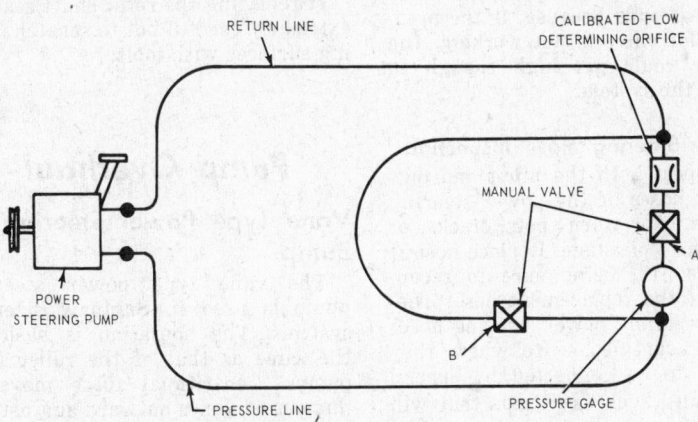

Power steering pump test circuit diagram (© Ford Motor Co.)

warmed up, close the manual valve B. Note the pressure gauge reading. It must be a minimum of 620 psi.

4. If the pressure reading is below the minimum acceptable pressure, the pump is defective and must be repaired. If the pressure reading is at or above the minimum value, the pump is normal. Open manual valve B and proceed to the pump fluid pressure test.

Power Steering Pump Fluid Pressure Test

1. Keep the lines and pressure gauge connected as in the Pump Flow Test.
2. With manual valve A and B opened fully, run the engine at the proper idle speed. Then, close manual valve A and manual valve B, in that order. **CAUTION: Do not keep both valves closed for more than 5 seconds since the fluid temperature will increase abnormally and cause unnecessary wear to the pump.**
3. With both manual valves closed, the pressure reading should be as given in the specifications. If the pressure is below the minimum reading, the pump is defective and must be repaired. If the pressure reading is at or above the minimum reading, the pump is normal and the power steering gear or power assist control valve must be checked.

Checking the Oil Flow and Pressure Relief Valve in the Pump Assembly

When the wheels are turned hard right, or hard left, against the stops, the oil flow and pressure relief valves come into action. If these valves are working and are not stuck there should be a slight buzzing noise.

CAUTION: Do not hold the wheels in the extreme position for over three or four seconds because, if the pressure relief valve is not working, the pressure could get high enough to damage the system.

Power Steering Hose Inspection

1. Inspect both the input and output hoses of the power steering pump for worn spots, cracks, or signs of leakage. Replace hose if defective, being sure to reconnect the replacement hose properly. Many power steering hoses are identified as to where they are to be connected by special means, such as fittings that will only fit on the correct pump fitting, or hoses of special lengths.

Test Driving Truck to Check the Power Steering

When test driving to check power steering, drive at a speed between 15 and 20 mph. Make several turns in each direction. When a turn is completed, the front wheels should return to the straight ahead position with very little help from the driver.

If the front wheels fail to return as they should and yet the steering linkage is free, well oiled and properly adjusted, the trouble is probably due to misalignment of the power cylinder or improper adjustment of the spool valve.

The power steering pump supplies all the power assist used in power steering systems of all designs. There are various designs of pumps used by the truck manufacturers but all pumps supply power to operate the steering systems with the least effort. All power steering pumps have a reservoir tank built onto the oil pump. These pumps are driven by belts turned by pulleys on either the engine, the rear of the generator, or the front of the crankshaft.

During operation of the engine at idle speed, there is provision for the power steering pump to supply more fluid pressure. During driving speeds or when the truck is moving straight ahead, less pressure is needed and the excess is relieved through a pressure relief and flow control valve. The pressure relief part of the valve is inside the flow control and is basically the same for all pumps. The flow control valve regulates, or controls, the constant flow of fluid from the pump as it varies with the demands of the steering gear. The pressure relief valve limits the hydraulic pressure built up when the steering gear is turned against its stops.

During pump disassembly, make sure all work is done on a clean surface. Clean the outside of the pump thoroughly and do not allow dirt of any kind to get inside. Do not immerse the shaft oil seal in solvent.

If replacing the rotor shaft seal, be extremely careful not to scratch sealing surfaces with tools.

Pump Overhaul

Vane Type Power Steering Pump

The vane type power steering pump is used in Saginaw steering systems. The operation is basically the same as that of the roller type pumps. Centrifugal force moves a number of vanes outward against the pump ring, causing a pumping action of the fluid to the control valve.

Removal

1. Disconnect hoses at the pump, securing them in a raised position to prevent oil drainage. Cap or cover the ends of the hoses to keep dirt out.
2. Install two caps on the pump fittings to prevent oil drainage.
3. Loosen the bracket-to-pump mounting nuts, move pump toward engine slightly, and remove the pump drive belt.
4. Remove the bracket-to-pump bolts and remove the pump from the truck.
5. While holding the drive pulley steady, loosen and remove the pulley attaching nut. Slide the pulley off the shaft. *NOTE: Do not hammer the pulley off the shaft.*

Installation

1. To install the pump on the truck, reverse the removal procedure. Always use a new pulley nut, tightening it to 25-45 ft. lbs. torque.
2. After reconnecting the hoses to the pump, fill the reservoir with fluid and bleed the pump of air by turning the drive pulley counterclockwise (as viewed from the front) until air bubbles do not appear.
3. Install the pump drive belt over the pulley, move the pump against the belt until tight enough, then tighten the mounting bolts and nuts.
4. Bleed the air from the system.

Disassembly

1. Clean the outside of the pump in a non-toxic solvent before disassembling.
2. Mount the pump in a vise, being careful not to squeeze the front hub too tight.
3. Remove the union and seal.
4. Remove the reservoir retaining studs and separate the reservoir from the housing.
5. Remove the mounting bolt and union O-rings.
6. Remove the filter and filter cage; discard the element.
7. Remove the end plate retaining ring by compressing the retaining ring and then prying it out with a screwdriver. The retaining ring may be compressed by inserting a small punch in the 1/8" diameter hole in the housing and pushing in until the ring clears the groove.
8. Remove the end plate. The end plate is spring-loaded and should rise above the housing level. If it is stuck inside the housing, a slight rocking or gentle tapping should free the plate.
9. Remove the shaft woodruff key and tap the end of the shaft gen-

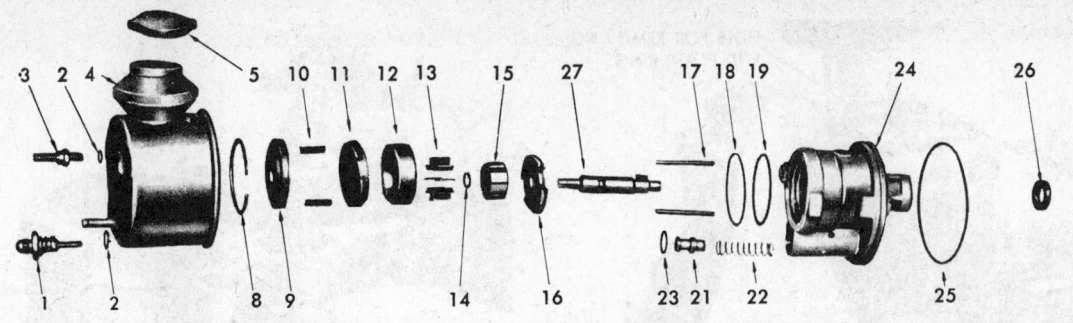

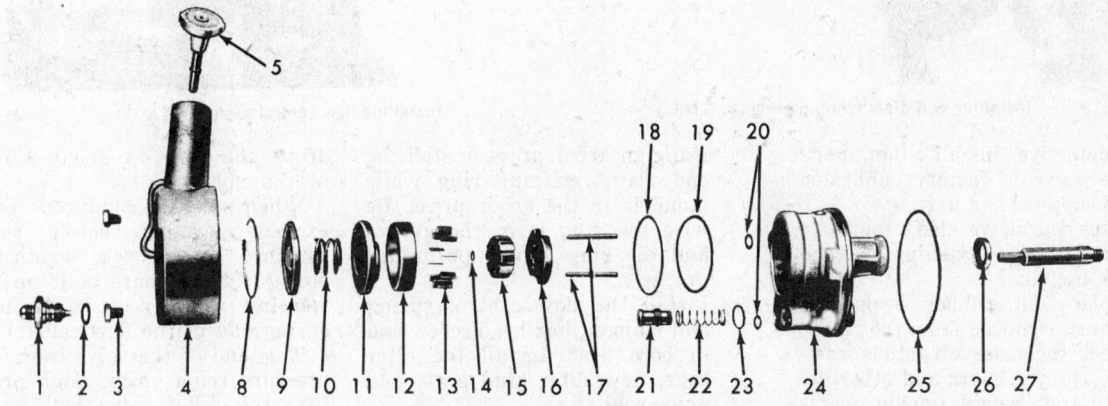

Vane type power steering pumps (© G.M.C.)

1 Union
2 Union "O" ring seal
3 Mounting studs
4 Reservoir
5 Dip stick and cover
8 End plate retaining ring
9 End plate
10 Spring
11 Pressure plate

12 Pump ring
13 Vanes
14 Drive shaft retaining ring
15 Rotor
16 Thrust plate
17 Dowel pins
18 End plate "O" ring
19 Pressure plate "O" ring
20 Mounting stud square ring

21 Flow control valve
22 Flow control valve spring
23 Flow control valve square ring seal
24 Pump housing
25 Reservoir "O" ring seal
26 Shaft seal
27 Shaft

thrust plate. Remove these parts as one unit.

10. Remove the end plate O-ring. Separate the pressure plate, pump ring, rotor assembly, and thrust plate.

Inspection

Clean all metal parts in a non-toxic solvent and inspect them as given below:

1. Check the flow control valve for free movement in the housing bore. If the valve is sticking, see if there is dirt or a rough spot in the bore.
2. Check the cap screw in the end of the flow control valve for looseness. Tighten if necessary being careful not to damage the machined surfaces.
3. Inspect the pressure plate and

the pump plate surfaces for flatness and check that there are no cracks or scores in the parts. Do not mistake the normal wear marks for scoring.

4. Check the vanes in the rotor assembly for free movement and that they were installed with the radiused edge toward the pump ring.
5. If the flow control valve plunger

Removing end plate ring (© G.M.C.)

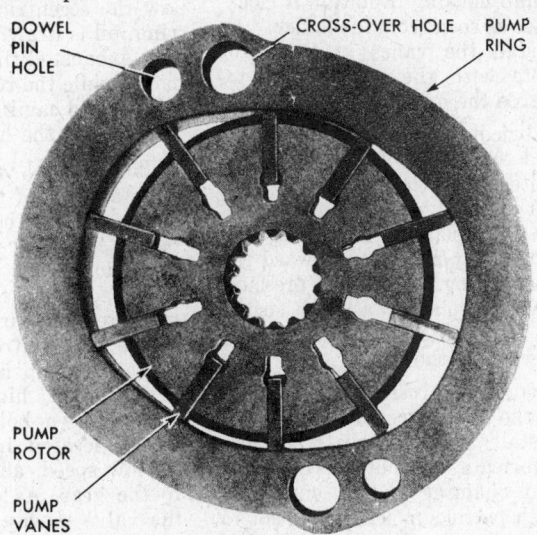

Correct vane assembly (© G.M.C.)

945

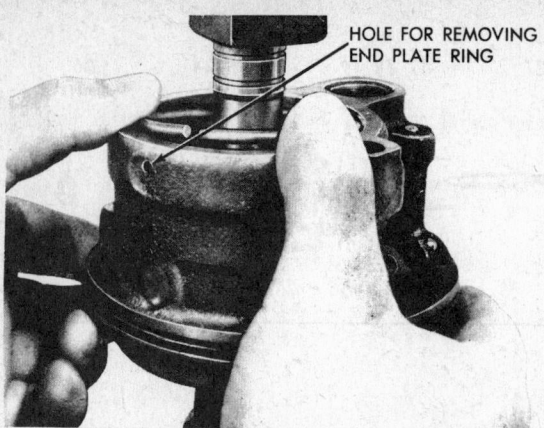
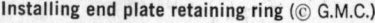

Installing end plate retaining ring (© G.M.C.)

Installing flow control valve (© G.M.C.)

is defective, install a new part. The valve is factory calibrated and supplied as a unit.

6. Check the drive shaft for worn splines, breaks, bushing material pick-up, etc.
7. Replace all rubber seals and O-rings removed from the pump.
8. Check the reservoir, studs, casting, etc. for burrs and other defects that would impair operation.

Assembly

1. Install a new shaft seal in the housing and insert the shaft at the hub end of housing, splined end entering mounting face side.
2. Install the thrust plate on the dowel pins with the ported side facing the rear of the pump housing.
3. Install the rotor on the pump shaft over the splined end. Be sure the rotor moves freely on the splines. Countersunk side must be toward the shaft.
4. Install the shaft retaining ring. Install the pump ring on the dowel pins with the rotation arrow toward the rear of the pump housing. Rotation is clockwise as seen from the pulley.
5. Install the vanes in the rotor slots with the radius edge towards the outside.
6. Lubricate the outside diameter and chamfer of the pressure plate with petroleum jelly so as not to damage the O-ring and install the plate on the dowel pins with the ported face toward the pump ring. Seat the pressure plate by placing a large socket on top of the plate and pushing down with the hand.
7. Install the pressure plate spring in the center groove of the plate.
8. Install the end plate O-ring. Lubricate the outside diameter and chamfer of the end plate with petroleum jelly so as not to damage the O-ring and install the end plate in the housing,

using an arbor press. Install the end plate retaining ring while pump is in the arbor press. Be sure the ring is in the groove and the ring gap is positioned properly.

9. Install the flow control spring and plunger, hex head screw end in bore first. Install the filter cage, new filter stud seals and union seal.
10. Place the reservoir in the normal position and press down until the reservoir seats on the housing. Check the position of the stud seals and the union seal.
11. Install the studs, union, and drive shaft woodruff key. Support the shaft on the opposite side of the key when tapping the key into place.

Roller Type Power Steering Pump

The roller type power steering pump is designed similar to other constant flow centrifugal force pumps. A star-shaped rotor forces 12 steel rollers against the inside surface of a cam ring. As the rollers follow the eccentric pattern of the cam ring, oil is drawn into the inlet ports and exhausted through the discharge ports while the rollers are moved into vee shaped cavities of the rotor, forcing oil into the high pressure circuit.

A flow control valve permits a regulated amount of fluid to return to the intake side of the pump when excess output is produced during high speed operation. This reduces the power needs to drive the pump and minimizes temperature build-up.

The flow control valve used in one make of pump is a two-stage valve. Fluid under high pressure passes through two holes into a metering circuit located in a sealed passage. At low speed, about 2.7 gpm. passes to the gear. As speed increases and the valve moves, excess fluid is bypassed to the inlet and the valve blocks flow through one hole. This

drops the flow to about 1.6 gpm. at high speeds.

When steering conditions produce excessive pressure needs (such as turning the wheels against the stops), the pressure built up in the steering gear exerts force on the spring end of the flow control valve.

This end of the valve contains the pressure relief valve. High pressure lifts the relief valve ball from its seat, allowing fluid to flow through a trigger orifice located in the front land of the flow control valve. This reduces pressure on the spring end of the valve which then opens and allows the fluid to return to the intake side of the pump. This action limits the maximum pressure output of the pump to a safe level. Normally, the pressure needs of the pump are below the maximum limits, causing the pressure relief ball and the flow control valve to remain closed.

Removal

1. Loosen the pump mounting and locking bolts and remove the belt.
2. Disconnect both hoses at the pump. Cap and tie the hoses out of the way. Cap the hose fittings on the pump.
3. Remove the mounting and locking bolts, the pump and brackets from the truck.

Installation

1. Position the pump and brackets on the engine and install the mounting and locking bolts.
2. Install the drive belt and adjust for the proper tension.
3. Connect the pressure and return hoses, using a new pressure hose O-ring
4. Fill the pump reservoir to the top of the filler neck with power steering fluid.
5. Start the engine and turn the steering wheel several times from stop to stop to bleed the pump of air. Check the level and add fluid if necessary.

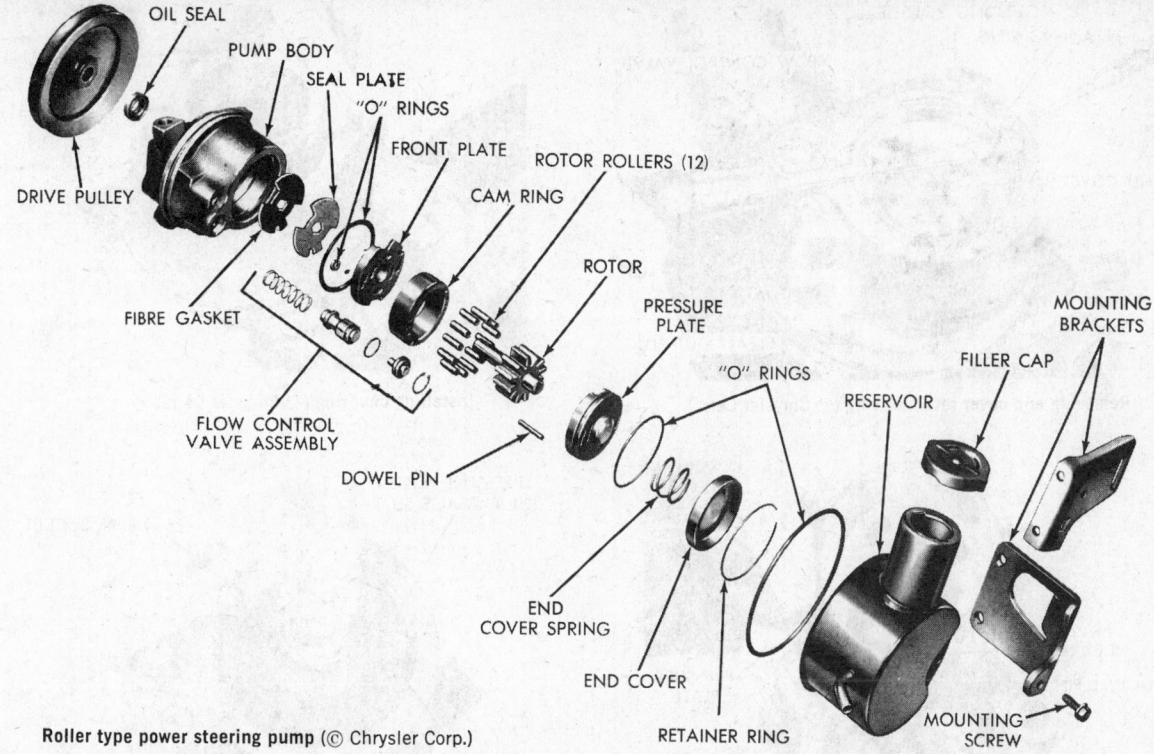

OIL SEAL
PUMP BODY
SEAL PLATE
"O" RINGS
FRONT PLATE
ROTOR ROLLERS (12)
CAM RING
DRIVE PULLEY
ROTOR
FIBRE GASKET
PRESSURE PLATE
MOUNTING BRACKETS
FILLER CAP
"O" RINGS
RESERVOIR
FLOW CONTROL VALVE ASSEMBLY
DOWEL PIN
END COVER SPRING
END COVER
MOUNTING SCREW
RETAINER RING

Roller type power steering pump (© Chrysler Corp.)

NOTE: *When checking the level, see that the level is as follows: engine cold-bottom of filler tube; engine hot-half way up filler tube.*

Disassembly

1. Remove pump from engine, drain reservoir, and clean outside of pump. Clamp the pump in a vise at the mounting bracket.
2. Remove the drive pulley.
3. Remove the shaft seal by installing the seal remover adapter over the end of the drive shaft with the large end toward the pump. Place the seal remover tool over the shaft and through the adapter. Then, screw the tapered thread well into the metal portion of the seal. Tighten the large drive nut and remove the seal.
4. Remove the pump from the vise and remove the bracket mount-

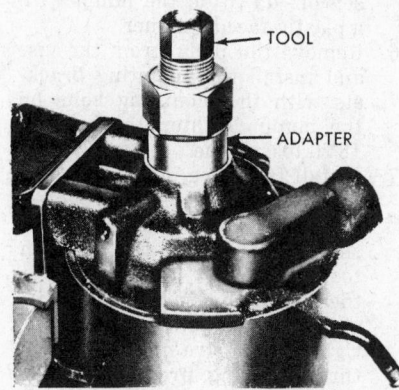

TOOL

ADAPTER

Removing shaft seal (© Chrysler Corp.)

ing bolts. Remove the bracket.
5. Remove the reservoir and place the pump in a soft-faced vise with the shaft down. Discard the mounting bolt and the reservoir O-rings.
6. Move the end cover retaining ring around until one end of the ring lines up with the hole in the pump body. Insert a small punch in the hole and push it in far enough to bend the ring so a screwdriver can be inserted between the ring and the housing. Remove the ring.
7. Remove the end cover and spring from the housing. It may be necessary to tap the cover gently to loosen it in the housing.
8. Remove the pump from the vise and turn the pump over so the rotating group may come out of the housing. Tap the end of the drive shaft to loosen these parts. Lift the pump body off the rotating group. Check that the seal plate is removed from the bottom of the housing bore.
9. Discard the O-rings from the pressure plate and end cover.
10. Remove the snap-ring, bore plug, flow control valve and spring from the housing. Discard the O-ring. If necessary to dismantle the flow control valve for cleaning, see the procedure for disassembly.

Inspection

1. Remove the clean out plug with an Allen wrench.

2. Wash all metal parts in clean, non-toxic solvent. Blow out all passages with compressed air and air dry all cleaned parts.
3. Inspect the drive shaft for excessive wear and the seal area for nicks or scoring. Replace if necessary.
4. Inspect the end plates, rollers, rotor and cam ring for nicks, burrs, or scratches. If any of the components are damaged enough to cause poor operation of the pump, all the interior parts may have to be replaced to prevent later failures.
5. Inspect the pump body drive shaft bushing for excessive wear. Replace the pump body and bushing as one assembly.

Assembly

1. Install the 1/8" pipe clean out plug, tightening it to 80 in. lbs. torque.
2. Place the pump body on a clean flat surface and install a new shaft seal into the bore.
3. Install a new end cover O-ring into the groove in the pump bore. Be sure to lubricate the O-ring with power steering fluid before installing it.
4. Lubricate and install a new O-ring in the groove on the pump body where the reservoir fits snugly.
5. Install the brass seal plate to the bottom of the housing bore. Align the notch in the seal plate with the dowel pin hole in the housing.
6. Carefully install the front plate

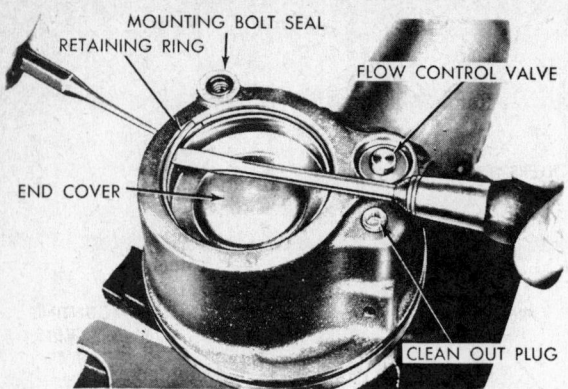

Removing end cover retaining ring (© Chrysler Corp.)

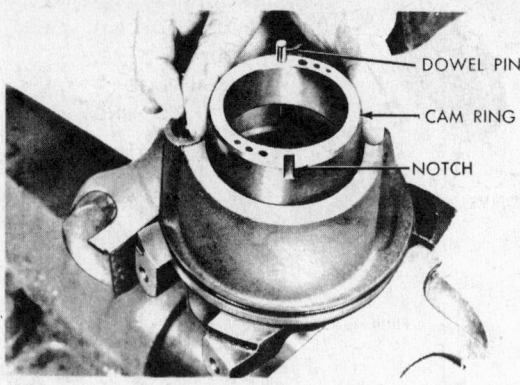

Installing cam ring (© Chrysler Corp.)

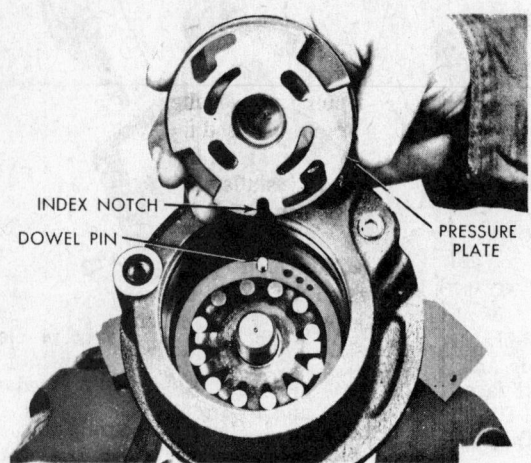

Installing pressure plate (© Chrysler Corp.)

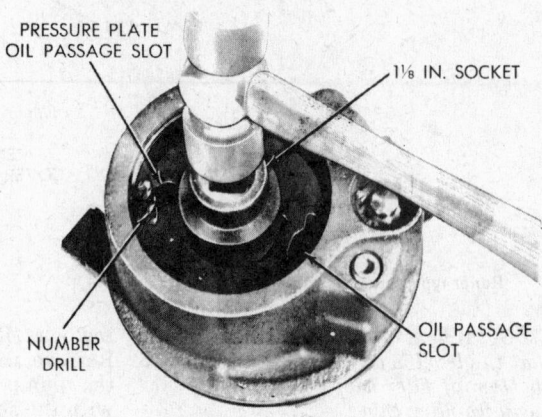

Seating pressure plate (© Chrysler Corp.)

with the chamfered edge down in the pump bore. Align the index notch in the plate with the dowel pin hole in the housing. **CAUTION: Be extremely careful to align the dowel pin hole properly. Pump can be completely assembled with the dowel pin not seated properly in the hole.**

7. Place the dowel pin in the cam ring and position the cam ring inside the pump bore. Notch in the cam ring must be facing up (away from the pulley end of pump housing). If the cam ring has two notches, one machined and one cast, install the cam ring with the machined notch up. Check the amount of dowel pin extending above the cam ring surface. If more than 3/16″ is showing, the dowel pin is not seated in the index hole in the housing.

8. Install the rotor and shaft in the cam ring and carefully install the 12 steel rollers in the cavities of the rotor. Lubricate the rotor, rollers, and the inside surface of the cam ring with power steering fluid. Rotate the shaft by hand to be sure all the rollers are seated parallel with the

shaft and are not sticking or binding.

9. Position the pressure plate by carefully aligning the index notch on the plate with the dowel pin and inserting a clean drill (number 13 to 16) in the cam ring oil hole next to the dowel pin notch until it bottoms on the housing floor.

10. Lubricate and install a new O-ring on the pressure plate. Position the pressure plate in the pump bore so that the dowel pin is in the index notch on the plate and the drill extends through the oil passage in the pressure plate. Seat the pressure plate on the cam ring using a clean 1⅛″ socket and a soft-faced hammer to tap it gently. Remove the drill and inspect the plate at both oil passage slots to be sure that the plate is squarely seated on the cam ring.

11. Place the large coil spring over the raised portion of the installed pressure plate.

12. Place the end cover, lip edge facing up, over the spring. Press the end cover down below the retaining ring groove. Install the retaining ring in the groove. Be sure the end cover chamfer is squarely seated against the

snap-ring.

13. Replace the reservoir mounting bolt seal.

14. Lubricate the flow control valve assembly with power steering fluid and insert the valve spring and valve in the bore. Install a new O-ring on the bore plug, lubricate with fluid, and carefully install in the bore. Install the snap-ring with the sharp edge up. Do not depress the bore plug more than 1/16″ below the snap-ring groove.

15. Place the reservoir on the pump body and visually align the mounting bolt hole. Tap the reservoir down on the pump with a plastic-faced hammer.

16. Remove the pump from the vise and install the mounting brackets with the mounting bolts on the pump. Tighten the bolts to 18 ft. lbs. torque.

17. Install the drive pulley by using the installer tool as follows: place the pulley on the end of the shaft and thread the installer tool into the ⅜″ threaded hole in the end of the shaft. Put the installer shaft in a vise and tighten the drive nut against the thrust bearing, pressing the pulley on the shaft until it is flush. Do not try to press the pulley on

the shaft without the special installer tool since the pump interior will be damaged by any other installation procedure. A small amount of drive shaft end play will be seen when the pulley is installed. This end play is necessary and will be minimized by a thin coat of oil between the rotor and the end plates when the pump is operating.

18. Install the pump assembly on the engine, install the drive belt and hoses (use new O-ring on pressure hose), and check for leaks.

Flow Control Valve Disassembly

1. After removing the pump from the engine and the reservoir from the pump, remove the snap-ring and plug from the flow bore. Discard the O-ring.
2. Depress the control valve against the spring pressure and allow the valve to spring out of the bore. If the valve is stuck in the bore or it did not come out of the bore far enough, it may be necessary to tap the housing lightly to remove it.
3. If the valve has dirt or foreign particles on it or in its bore, the rest of the pump needs cleaning. The hoses should be flushed and the steering gear valve body reconditioned. If the valve bore is badly scored, replace the pump body and the flow control valve.
4. Remove any nicks or burrs by gently rubbing the valve with crocus cloth. Clamp the valve land in a vise with soft-jaws and remove the hex head ball seat and shims. Note the number and gauge (thickness) of the shims on the ball seat. They must be re-installed for the same shim thickness to keep the same value of relief pressure.
5. Remove the valve from the vise and remove the pressure relief ball, guide, and spring.

Flow Control Valve Assembly

1. Insert the spring, guide and pressure relief ball in the end of the flow control valve.
2. Install the hex head plug using the exact number and thickness shims that were removed. Tighten the plug to 80 in. lbs. torque.

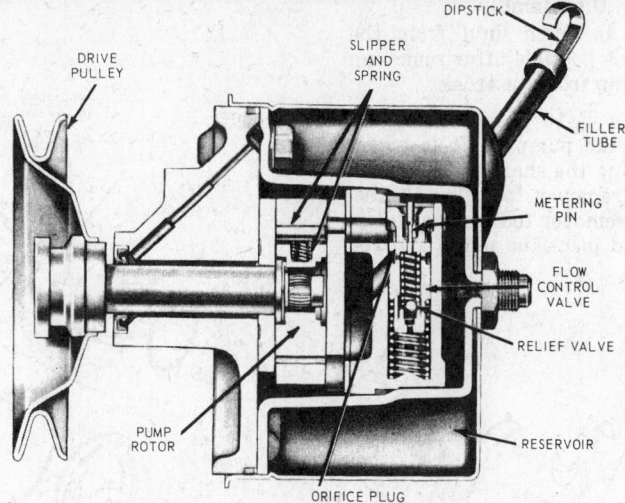

Ford Thompson power steering pump, sectional view
(© Ford Motor Co.)

3. Lubricate the valve with power steering fluid and insert the flow control valve spring and valve in the housing bore. Install a new O-ring on the bore plug, lubricate with fluid and carefully install into the bore. Install the snap-ring. Do not depress the bore plug more than 1/16" beyond the snap-ring groove.

Slipper Type Power Steering Pump

The slipper type power steering pump is a belt-driven constant displacement assembly that uses a number of spring-loaded slippers in the pump rotor to force fluid from the inlet side to the flow control valve. Openings in the metering pin allow a flow of about two gpm. of fluid to the steering gear before the flow control valve directs the excess fluid to the inlet side of the pump again. Maximum pressure in the pump is limited by the pressure relief valve which opens when the pressure exceeds the maximum limits.

The slipper type power steering pump discussed in this section is used on Ford trucks and is called the Ford-Thompson power steering pump.

Removal

1. Drain the fluid from the pump reservoir by disconnecting the

fluid return hose at the pump. Then, disconnect the pressure hose from the pump.
2. Remove the mounting bolts from the front of the pump. On eight cylinder engines, there is a nut on the rear of the pump that must be removed. After removing all the mounting bolts and nuts from the pump, move the unit inward to loosen the belt tension and remove the belt from the pulley. Then remove the pump from the truck.

Installation

1. Position the pump on the mounting bracket and loosely install the mounting bolts and nuts. Put the drive belt over the pulley and move the pump outward against the belt until the proper belt tension is obtained. Measure the belt tension with a gauge for the proper adjustment. Only in cases where a belt tension gauge is not available should the belt deflection method be used. If the belt deflection method is used, be sure to check the belt with a tension gauge at the earliest time since the deflection method is not accurate.
2. Tighten the mounting bolts and nuts to the specified torque limits.
5. Tighten the pressure hose fitting hex nut to the proper torque. Then connect the pressure hose to the pump and tighten the hose nut to the proper torque.
4. Connect the fluid return hose to the pump and tighten the clamp.
5. Fill the pump reservoir with power steering fluid and bleed the air bubbles from the system.
6. Check for leaks and recheck the fluid level. If necessary, add fluid to raise the level properly.

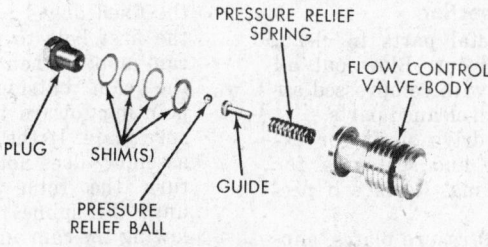

Flow control valve (© Chrysler Corp.)

Power Steering Gears

Disassembly

1. Drain as much fluid from the pump as possible after removing the pump from the truck.

2. Install a ⅜-16″ capscrew in the end of the pump shaft to avoid damaging the shaft end with the pulley remover tool. Install the pulley remover tool on the pulley hub and place the pump and re-

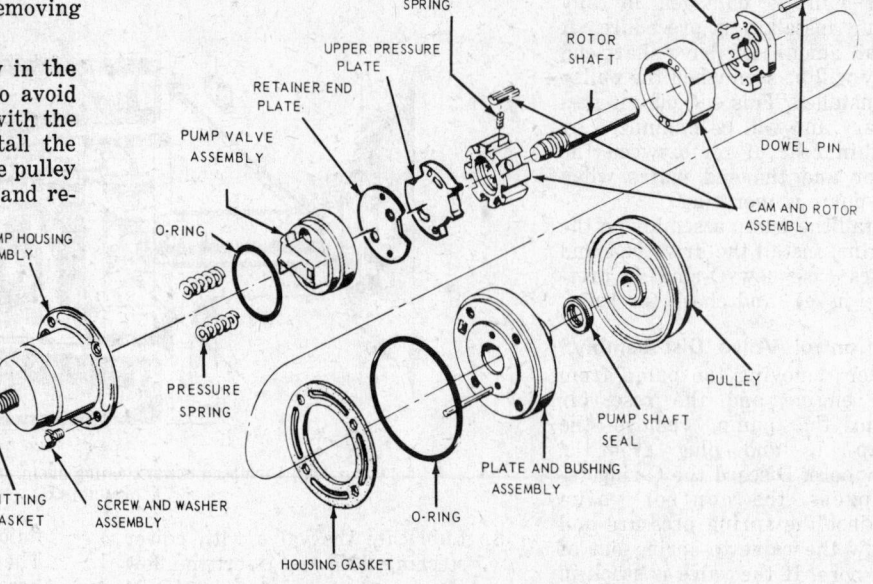

Ford Thompson power steering pump disassembled (© Ford Motor Co.)

mover tool in a vise. Hold the pump steady and turn the tool nut counterclockwise to draw the pulley off the shaft. The pulley must be removed without in and out pressure on the pump shaft to avoid damaging the internal thrust washers.

3. Remove the pump reservoir by installing the pump in a holding fixture with an adapter plate in a vise with the reservoir facing up.

4. Remove the outlet fitting hex nut and any other attaching parts from the reservoir case.

5. Invert the pump so the reservoir is now facing down. Using a wooden block, remove the reservoir by tapping around the flange until the reservoir is loose. Remove the reservoir O-ring seal and the outlet fitting gasket from the pump.

6. Again invert the pump assembly in the vise, remove the pump housing holding bolts and the pump housing.

7. Remove the housing cover, the O-ring seal and the pressure springs from inside the pump housing. Remove the pump cover gasket and discard it.

8. Remove the retainer end plate and upper pressure plate. In some pumps, the end plate and the upper pressure plate are made as one unit.

9. Remove the loose fitting dowel pin. Be careful not to bend the fixed dowel pin which remains in the housing plate assembly.

10. Remove the rotor assembly being careful not to let the slippers and springs fall out of the rotor. It may not be necessary to

disassemble the rotor assembly unless the lower pressure plate, housing plate, rotor shaft and/or seal is to be replaced. However, the rotor assembly may be disassembled by removing the slippers and springs from the cam ring.

11. Remove any rust, dirt, burrs, or scoring from the pulley end of the rotor shaft before removing the shaft from the housing plate. The shaft must come out without restrictions to avoid scoring or damaging the bushing. Remove the pump rotor shaft.

12. Remove the lower pressure plate.

13. Remove the rotor shaft seal after first wrapping a piece of 0.005″ shim stock around the shaft and pushing it into the inside of the seal until it touches the bushing. With a sharp tool, pierce the seal body and pry the seal out. Do not damage the bushing, housing, or the shaft. Install a new seal using a soft-faced hammer.

14. If the pump has a flow control valve, disassemble according to instructions given in the section on the roller type power steering pump.

Inspection

1. Wash all metal parts in clean, non-toxic solvent. Blow out all oil passages with compressed air and air dry all cleaned parts.

2. Inspect the drive shaft for excessive wear and seal area for nicks or scoring. Replace if necessary.

3. Inspect the pressure plates, slippers, rotor, and cam ring for

nicks, burrs, or scratches. If any of the parts are damaged enough to cause poor operation or binding of the pump, replace the defective part.

4. Inspect the pump body drive shaft bushing for excessive wear. Replace if necessary.

Assembly

1. With the pump assembly positioned on the adapter plate in the holding fixture, install the lower pressure plate on the anchor pin with the chamfered slots at the center hole facing up.

2. Lubricate the rotor shaft with power steering fluid and insert the shaft into the lower pressure and housing plates.

3. Assemble the rotor, slippers, and springs by wrapping a piece of wire around the rotor, installing the springs, and sliding a slipper in each groove of the rotor over the springs. Then, insert the assembly into the cam ring. Be sure the flat side of the slippers are toward the left side. Be sure that the springs are installed straight and are not cocked to one side under the slippers.

4. Install the cam ring and rotor assembly on the drive shaft with the fixed dowel passing through the first hole to the left of the cam notch when the arrow on the cam outside diameter is pointing toward the lower pressure plate. If the cam and rotor assembly does not seat properly, turn the rotor shaft slightly until the spline teeth mesh, allowing the cam and rotor to drop into position.

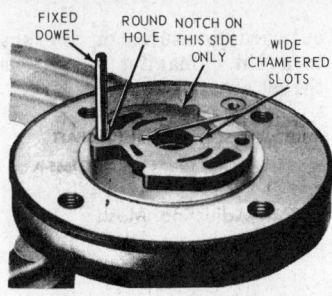

Lower pressure plate installed
(© Ford Motor Co.)

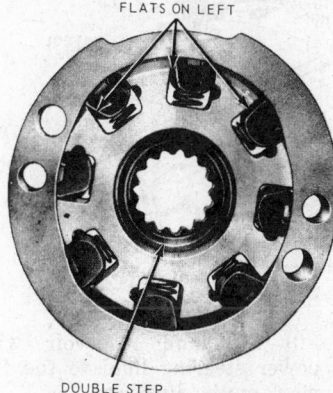

FLATS ON LEFT

DOUBLE STEP

Correct slipper installation—Ford
(© Ford Motor Co.)

5. Insert the looser fitting dowel through the cam insert and lower pressure plate into the hole in the housing plate assembly. When both dowels are installed properly, they will be the same height.

6. Install the upper pressure plate so the tapered notch is facing down against the cam insert. The fixed dowel should pass through the round dowel hole and the loose dowel through the long hole. The slot between the ears on the outside of the pressure plate should match the notch

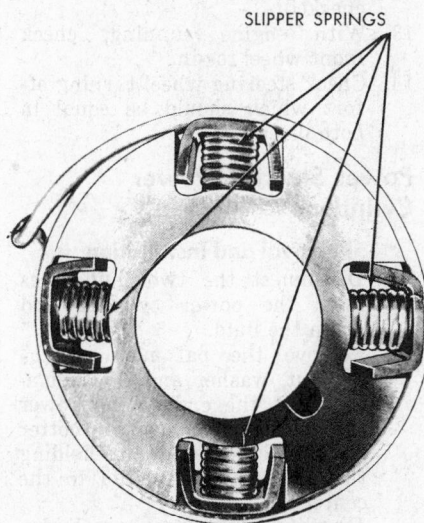

SLIPPER SPRINGS

Correct slipper installation—Dodge
(© Chrysler Corp.)

on the cam insert.

7. Install the retainer end plate so the slot on the end plate matches the notches on the upper pressure plate and the cam insert.

8. Install the pump valve assembly O-ring seal on the pump valve assembly. Do not twist the seal.

9. Place the pump valve assembly on top of the retainer end plate with the large exhaust slot on the pump valve in line with the outside notches of the cam, upper pressure plate, and retainer end plate. All parts must be fully seated. If correctly installed, the relief valve stem will be in line with the lube return hole in the pump housing plate.

10. Put small amounts of vaseline on the pump housing plate to hold the cover gasket in place. Install the cover gasket in place.

11. Insert the pressure plate springs into the pockets in the pump valve assembly.

12. Plug the intake hole in the housing.

13. Lubricate the inside of the housing and the housing cover seal with power steering fluid. Install two studs for use as positioning guides, one in the bolt hole nearest the drain hole and the other in the bolt hole on the opposite side of the housing plate.

14. Align the small lube hole in the housing rim and the lube hole in the housing plate. Install the housing, using a steady, even, downward pressure. Do not jar the pressure spring out of position. Remove the guide studs and loosely install the housing retaining bolts finger tight.

FIXED DOWEL   DOUBLE STEP

ARROW POINTING DOWN

Cam and rotor installation
(© Ford Motor Co.)

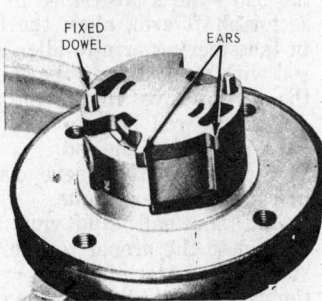

FIXED DOWEL   EARS

Upper pressure plate installation
(© Ford Motor Co.)

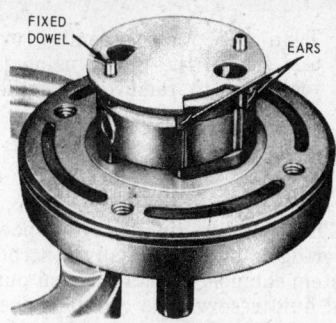

FIXED DOWEL   EARS

Retainer end plate installation
(© Ford Motor Co.)

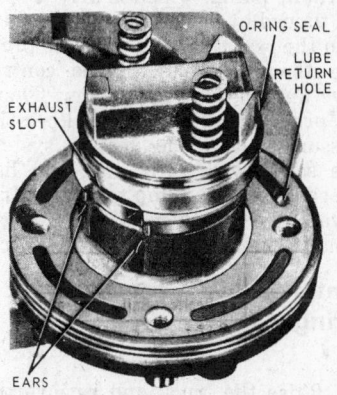

O-RING SEAL
LUBE RETURN HOLE
EXHAUST SLOT
EARS

Valve and pressure spring installation
(© Ford Motor Co.)

15. Tighten the retaining bolts evenly to 28-32 ft. lbs. until the housing flange contacts the gasket.

16. Install a 3/8-16 hex head screw into the end of the rotor shaft and put a torque wrench on it. Check the amount of torque needed to rotate the rotor shaft. If the torque is more than 15 in. lbs., loosen the retaining bolts slightly and rotate the rotor shaft. Then, retighten the retaining bolts evenly. Do not use the pump if the shaft torque exceeds 15 in. lbs.

17. Release the pin in the bench holding fixture and shake the pump assembly back and forth. If there is a rattle, the pressure springs have fallen out of their seats and must be reinstalled.

18. Install the reservoir O-ring seal on the housing plate without twisting it. Lubricate the seal and install the reservoir, aligning the notch in the reservoir flange with the notch in the outside edge of the pump housing plate and bushing assembly. Using only a soft-faced hammer, tap at the rear outer corners of the reservoir. Inspect the assembly to be sure the reservoir is fully seated on the housing plate.

19. Install the identification tag (if one was removed) on the outlet valve fitting. Install the outlet valve fitting nut and tighten to 43-45 ft. lbs. torque.

20. Turn the pump assembly over and install the pulley using the tool used to remove the pulley.

Turn the tool nut clockwise to draw the pulley on the shaft until it is flush with the shaft end. Do not exert inward and outward pressures on the shaft to avoid damaging the internal thrust areas. Remove the tool.

Bendix Linkage—Type Power Steering System

The Bendix linkage-type power steering is a hydraulically controlled system composed of an integral pump and fluid reservoir, a control valve, a power cylinder, connecting fluid lines, and the steering linkage. The hydraulic pump, which is driven by a belt turned by the engine, draws fluid from the reservoir and provides pressure through hoses to the control valve and the power cylinder. There is a pressure relief valve to limit the pressures within the steering system to a safe level. After the fluid has passed from the pump to the control valve and the power cylinder, it returns to the reservoir.

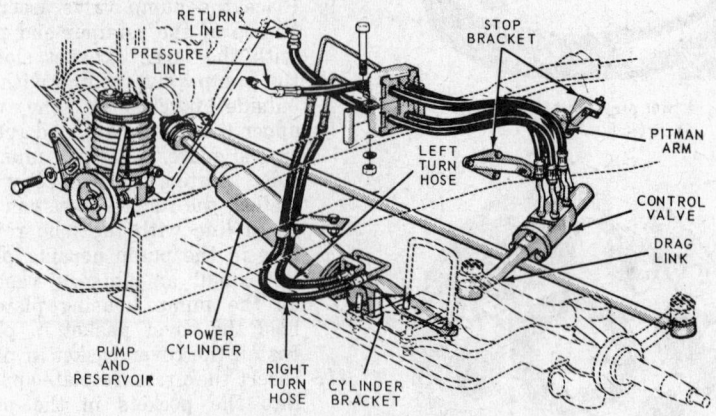

Power steering installation—typical
(© Ford Motor Co.)

Control Valve Centering Spring

Adjustment

1. Raise the truck and remove the spring cap attaching screws and remove the spring cap. CAUTION: Be very careful not to position the hoist adapters of two post hoists under the suspension and/or steering components. Place the hoist adapters under the front suspension lower arms.
2. Tighten the adjusting nut snug (about 90-100 in. lbs.) ; then, loosen the nut ¼ turn (90 degrees). Do not turn the adjusting nut too tight.
3. Place the spring cap on the valve housing. Lubricate and install the attaching screws and washers. Tighten the screws to 72-100 in. lbs. torque.
4. Lower the truck and start the engine. Check the steering effort using a spring scale attached to the steering wheel rim for a pull of no more than 12 lbs.

Power Steering Control Valve

Removal

1. Raise the truck on a hoist. If a two post hoist is used, be sure to place the hoist adapters under the front suspension steering arms. Do not allow the hoist adapters to contact the steering linkage.
2. Disconnect the four fluid line fittings at the control valve and drain the fluid from the lines. Turn the front wheels back and forth to force all the fluid from the system.
3. Loosen the clamping nut and

bolt at the right end of the sleeve.
4. Remove the roll pin from the steering arm-to-idler arm rod through the slot in the sleeve.
5. Remove the control valve ball stud nut.
6. Remove the ball stud from the sector shaft arm.
7. After turning the front wheels fully to the left, unthread the control valve from the center link steering arm-to-idler arm rod.

Installation

1. Thread the valve on the center link until about four threads are still visible.
2. Position the ball stud in the sector shaft arm.
3. Measure the distance between the grease plug in the sleeve and the stud at the inner end of the left spindle connecting rod. If the distance is not correct, disconnect the ball stud from the sector shaft arm and turn the valve on the center link until the correct distance is obtained.
4. When the distance is correct and the ball stud is positioned in the sector shaft arm, align the hole in the steering arm-to-idler arm rod with the slot near the end of the valve sleeve. Install the roll pin in the rod hole to lock the valve in place on the rod.
5. Tighten the valve sleeve clamp bolt to the proper torque.
6. Install the ball stud nut and tighten to the proper torque. Install a new cotter pin.
7. Connect all fluid lines to the control valve and tighten all fittings securely. Do not over-tighten.

8. Fill the fluid reservoir with power steering fluid to the full mark on the dipstick.
9. Start the engine and run it for a few minutes to warm the fluid in the power steering system. Turn the steering wheel back and forth to the stops and check the system for leaks.
10. Increase the engine idle speed to about 1000 rpm. Turn the steering wheel back and forth several times, then stop the engine. Check the control valve and hose connections for leaks.
11. Recheck the fluid level and add fluid if necessary.
12. Start the engine again, and check the position of the steering wheel when the front wheels are straight ahead. Do not make any adjustments until toe-in is checked.
13. With engine running, check front wheel toe-in.
14. Check steering wheel turning effort which should be equal in both directions.

Power Steering Power Cylinder

Removal and Installation

1. Disconnect the two fluid lines from the power cylinder and drain the fluid.
2. Remove the pal nut, attaching nut, washer and the insulator from the end of the power cylinder rod. Remove the cotter pin and castellated nut holding the power cylinder stud to the center link.
3. Disconnect the power cylinder stud from the center link.
4. Remove the insulator sleeve and

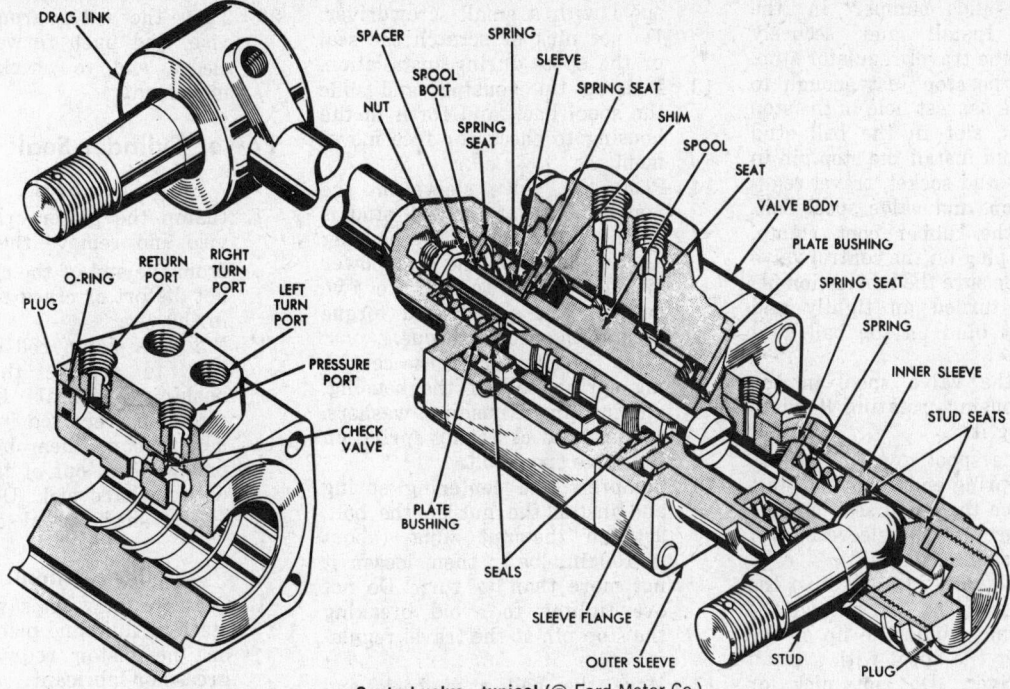

DRAG LINK

SPACER SPRING

SPOOL SLEEVE
BOLT

NUT SPRING SEAT

SPRING SHIM
SEAT SPOOL

SEAT

VALVE BODY

PLATE BUSHING

SPRING SEAT

SPRING

INNER SLEEVE

STUD SEATS

O-RING RETURN RIGHT
PORT TURN
PORT LEFT
TURN
PLUG PORT

PRESSURE
PORT

CHECK
VALVE

PLATE
BUSHING

SEALS

SLEEVE FLANGE

OUTER SLEEVE STUD PLUG

Control valve—typical (© Ford Motor Co.)

washer from the end of the power cylinder.

5. Inspect the tube fittings and seats in the power cylinder for nicks, burrs, or other damage. Replace the seats or tubes if damaged.

6. Install the washer, sleeve and the insulator on the end of the power cylinder rod.

7. While extending the rod as far as possible, insert the rod in the bracket on the frame and then, compress the rod so the stud may be inserted in the center link. Secure the stud with the castellated nut and a new cotter pin.

8. Install the insulater, washer, nut, and a pal nut on the power cylinder rod.

9. Connect the two fluid lines to their proper ports on the power cylinder.

10. Fill the reservoir with power steering fluid to the full mark on the dipstick. Start the engine and run for a few minutes to warm the fluid. Turn the steering wheel back and forth to the stops to fill the system. Stop the engine.

11. Recheck the fluid level and add fluid if necessary. Check for fluid leaks.

12. Start the engine again, turn the steering wheel back and forth, and check for leaks while the engine is running.

Control Valve

Disassembly

1. Clean the outside of the control valve of dirt and fluid.

2. Remove the centering spring cap from the valve housing. The control valve should be put in a soft-faced bench vise during disassembly. Clamp the control valve around the sleeve flange only to avoid damaging the valve housing, spool, or sleeve.

3. Remove the nut from the end of the valve spool bolt. Remove the washers, spacer, centering spring, adapter, and the bushing from the bolt and valve housing.

4. Remove the two bolts holding the valve housing and the sleeve together. Separate the valve housing and the sleeve.

5. Remove the plug from the sleeve. Push the valve spool out of the centering spring end of the valve housing, and remove the seal from the spool.

6. Remove the spacer, bushing and valve housing.

7. Drive the pin out of the travel regulator stop with a punch and hammer. Pull the head of the valve spool bolt tightly against the travel regulator stop before driving the pin out of the stop.

8. Turn the travel regulator stop counterclockwise in the valve sleeve to remove the stop from the sleeve.

9. Remove the valve spool bolt, spacer, and rubber washer from the travel regulator stop.

10. Remove the rubber boot and clamp from the valve sleeve. Slide the bumper, spring, and ball stud seat out of the valve sleeve and remove the ball stud socket from the sleeve.

11. Remove the return port hose seat and the return port relief valve.

12. Remove the spring plug and O-ring. Then remove the reaction limiting valve.

13. Replace all worn or damaged hose seats by using an Easy-Out screw extractor or a bolt of proper size as a puller. Tap the existing hole in the hose seat, using a starting tap of the correct size. Remove all metal chips from the hose seat after tapping. Place a nut and washer on a bolt of the same size as the tapped hole. The washer must be large enough to cover the hose seat port. Insert the bolt in the tapped hole and remove the hose seat by turning the nut clockwise and drawing the bolt out. Install a new hose seal in the port, and thread a bolt of the correct size in the port. Tighten the bolt enough to bottom the seal in the port.

Assembly

1. Coat all parts of the control valve assembly with power steering fluid. Seals should be coated with lubricant before installation.

2. Install the reaction limiting valve, spring and plug. Install the return port relief valve and the hose seat.

3. Insert one of the ball stud seats (flat end first) into the ball stud socket, and insert the threaded end of the ball stud into the socket.

4. Place the socket in the control valve sleeve so that the threaded end of the ball stud can be pulled out through the slot.

5. Place the other ball stud seat,

spring, and bumper in the socket. Install and securely tighten the travel regulator stop.

6. Loosen the stop just enough to align the nearest hole in the stop with the slot in the ball stud socket and install the stop pin in the ball stud socket, travel regulator stop, and valve spool bolt.

7. Install the rubber boot, clamp, and the plug on the control valve sleeve. Be sure the lubrication fitting is turned on tightly and does not bind on the ball stud socket.

8. Insert the valve spool in the valve housing, rotating it while installing it.

9. Move the spool toward the centering spring end of the housing, and place the small seal bushing and spacer in the sleeve end of the housing.

10. Press the valve spool against the inner lip of the seal and, at the same time, guide the lip of the seal over the spool with a small screwdriver. Do not nick or scratch the seal or the spool during installation.

11. Place the sleeve end of the housing on a flat surface so that the seal, bushing and spacer are at the bottom end; then push down the valve spool until it stops.

12. Carefully install the spool seal and bushing in the centering spring end of the housing. Press the seal against the end of the spool, guiding the seal over the

spool with a small screwdriver. Do not nick or scratch the seal or the spool during installation.

13. Pick up the housing, and slide the spool back and forth in the housing to check for free movement.

14. Place the valve sleeve on the housing so that the ball stud is on the same side of the housing as the ports for the two power cylinder lines. Install the two bolts in the sleeve, and torque them to the proper torque.

15. Place the adapter on the centering spring end of the housing, and install the bushing, washers, spacers and centering spring on the valve spool bolt.

16. Compress the centering spring and install the nut on the bolt. Tighten the nut snug (about 90-100 in. lbs.); then, loosen it not more than $1/4$ turn. Do not over-tighten to avoid breaking the stop pin at the travel regulator stop.

17. Move the ball stud back and forth to check for free movement.

18. Lubricate the two cap attaching bolts. Install the centering spring cap on the valve housing, and tighten the two cap bolts to the proper torque.

19. Install the nut on the ball stud so that the valve can be put in a vise. Then, push forward on the cap end of the valve to check the valve spool for free movement.

20. Turn the valve around in the vise, and push forward on the sleeve end to check for free movement.

Power Cylinder Seal

Removal

1. Clamp the power cylinder in a vise and remove the snap-ring from the end of the cylinder. Do not distort or crack the cylinder in the vise.

2. Pull the piston rod out all the way to remove the scraper, bushing, and seals. If the seals cannot be removed in this manner, remove them by carefully prying them out of the cylinder with a sharp pick. Do not damage the shaft or seal seat.

Installation

1. Coat the new seals with power steering fluid and place the parts on the piston rod. Coat with grease or lubricant.

2. Push the rod in all the way, and install the parts in the cylinder with a deep socket slightly smaller than the cylinder opening.

Power Steering Pump

Removal and Replacement

To remove or install the power steering pump, see the section on the slipper type pump.

Saginaw Linkage—Type Power Steering System

Control Valve

Removal

1. Raise front of vehicle and place on stands.

2. Remove relay rod to control valve clamp bolt.

3. Disconnect two pump to control valve hose connections and allow fluid to drain into a container, then disconnect the valve to power cylinder hoses.

4. Remove ball stud to pitman arm retaining nut and disconnect control valve.

5. Turn steering gear so that pitman arm is away from valve, to allow working room, and unscrew control valve from relay rod.

6. Remove control valve from vehicle.

Disassembly

1. Place valve assembly in vise with dust cap end up, then remove dust cap.

2. Remove adjusting nut.

3. Remove valve to adapter bolts then remove housing and spool

from adapter.

4. Remove spool from housing.

5. Remove spring, reaction spool, washer, reaction spring and seal. O-ring may now be removed from reaction spool.

6. Remove annulus spacer, valve shaft washer and plug-to-sleeve key.

7. Carefully turn adjuster plug out of sleeve. Use care not to nick the top surface.

8. If necessary to replace a connector seat, tap threads in center hole using a 5/16-18 tap. Thread a bolt with a nut and a flat washer into the tapped hole so the washer is against the face of the port boss and the nut is against the washer. Hold the bolt from turning while backing the nut off the bolt. This will force the washer against the port boss face and back out the bolt, drawing the connector seat from the top cover housing. Discard the old connector seat and clean the housing out thoroughly to remove any metal tapping chips. Drive a new connector seat against the housing seat,

being careful not to damage either seat.

9. Remove adapter from vise and turn over to allow spring and one of the two ball seats to drop out.

10. Remove ball stud with other ball seat and allow sleeve to fall free.

Inspection

1. Wash all parts in non-toxic solvent and blow dry with air.

2. Inspect all parts for scratches, burrs, distortion, excessive wear and replace all worn or damaged parts.

3. Replace all seals and gaskets.

Assembly

1. Install sleeve and ball seat in adapter, then the ball stud and then the other ball seat and spring (small end down).

2. Place adapter in vise. Put the shaft through the seat in the adjuster plug and screw adjuster plug into sleeve.

3. Turn plug in until tight, then back off until slot lines up with notches in sleeve.

4. Insert key. Be sure small tangs

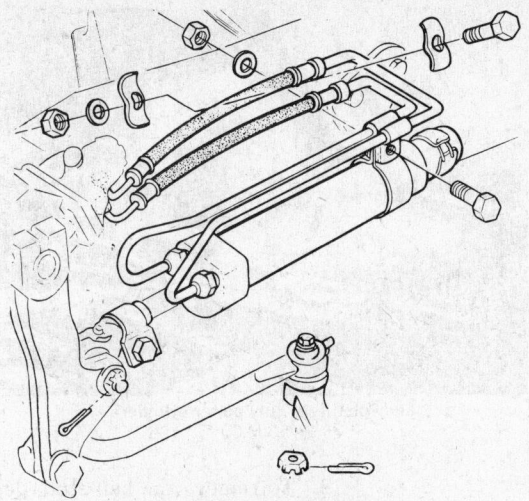

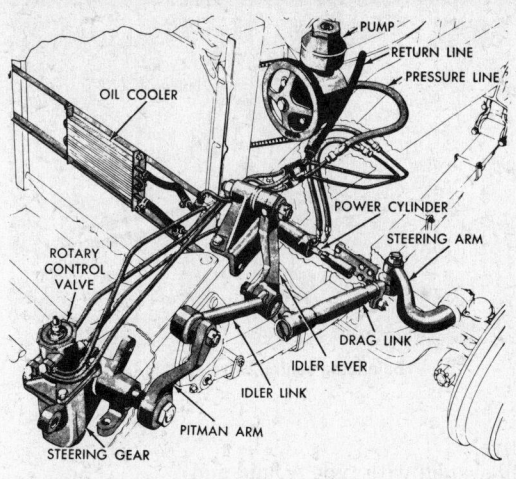

Chevrolet side mounted power cylinder
(© G.M.C.)

G.M.C. power steering system—tilt cab—typical
(© G.M.C.)

on end of key fit into notches in sleeve.

5. Install valve shaft washer, annulus spacer, reaction seal (lip up), spring retainer, reaction spring and spool, then washer and adjustment spring. Install O-ring seal on reaction spool before installing spool on shaft. Install washer with chamfer up.

6. Install seal on valve spool with lip down. Then install spool, being careful not to jam spool in housing.

7. Install housing with spool onto adapter. The side ports should be on the same side as the ball stud. Bolt the housing to the adapter.

8. Depress the valve spool and turn the locknut into the shaft about four turns. Use a clean wrench or socket.

NOTE: Always use a new nut.

Installation and Balancing

1. Install the control valve on the relay rod so that control valve bottoms, then back off enough

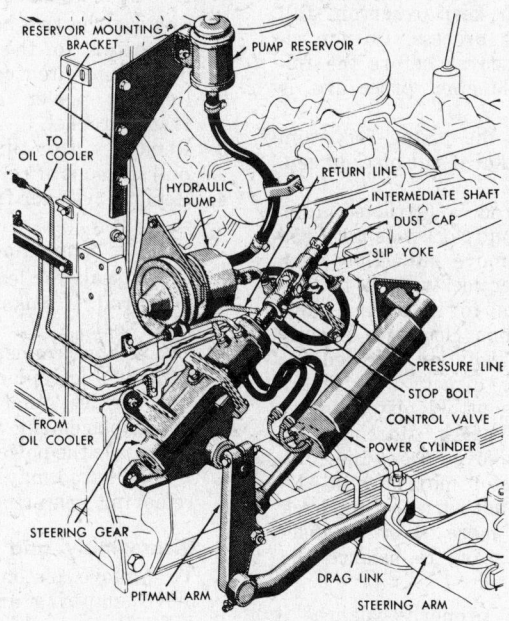

G.M.C. power steering system—conventional cab—typical
(© G.M.C.)

(if necessary) to install the clamp bolt. Do not back off more than two turns. There should be approximately 1/16-1/8″ gap.

2. Tighten control valve clamping-bolt and install ball stud to pitman arm.

3. Reconnect the four hoses to the

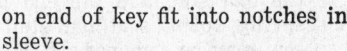

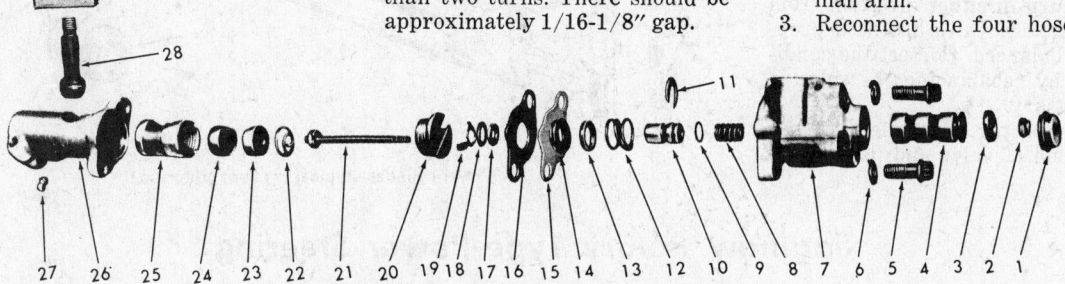

Chevrolet power steering control valve (© G.M.C.)

1 Dust cover	spring	15 Annulus spacer	22 Ball seat spring
2 Adjusting nut	9 "O" ring seal	16 Gasket	23 Ball seat
3 Vee block seal	10 Valve reaction	17 Valve shaft	24 Ball seat
4 Valve spool	spool	washer	25 Sleeve bearing
5 Valve mounting	11 Spring thrust	18 "O" ring seal	26 Adapter housing
bolts	washer	19 Plug to sleeve	27 Lubrication
6 Lock washer	12 Valve spring	key	fitting
7 Valve housing	13 Spring retainer	20 Ball adjuster nut	28 Ball stud
8 Valve adjustment	14 Annulus seal	21 Valve shaft	29 Cover

Power Steering Gears

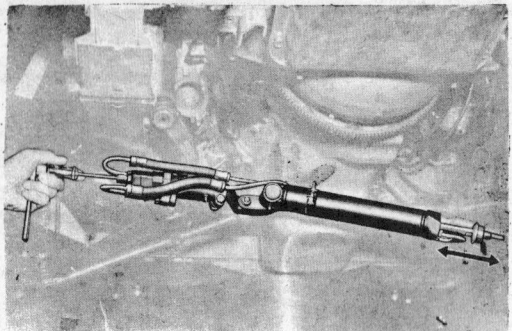

Balancing control valve (© G.M.C.)

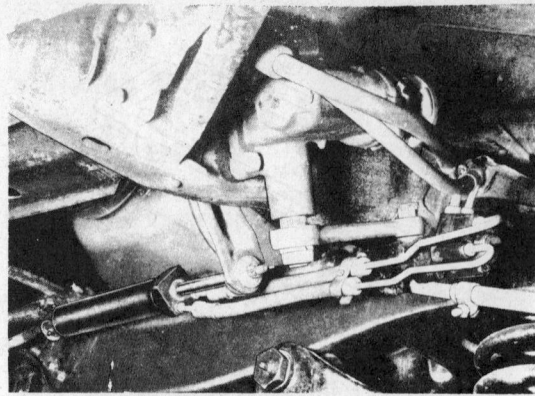

Chevrolet light duty power cylinder
(© G.M.C.)

valve.

4. Fill system with type A fluid and bleed air by running engine, then slowly turning wheels from lock to lock with engine idling. Be sure to keep reservoir full during this process. Do not replace dust cover before the following balancing procedure is completed.

5. Disconnect the piston rod from frame bracket if not already separated.

6. If piston rod is retracted, turn adjusting nut clockwise until rod begins to move out. Then turn nut counterclockwise until rod just begins to move in. Now, turn the nut clockwise exactly half the rotation needed to change the direction of piston rod movement. If piston rod is extended before starting, reverse the above to get the midpoint in piston movement. **CAUTION: Do not turn the nut back and forth more than is absolutely necessary to balance the valve.**

7. With valve properly balanced it should be possible to move the rod in and out manually.

8. Shut off engine and connect piston rod to frame bracket.

9. Restart engine with front wheels still off ground. If the wheels do not turn in either direction from center, the valve has been properly balanced. Correct the condition by rebalancing the valve if necessary.

10. After proper adjustment, grease the end of valve and install dust cap.

Power Cylinder

Removal

1. Remove the two hoses which are connected to the cylinder and drain fluid into a container.
2. Remove power cylinder from frame bracket.
3. Remove cotter pin and nut and pull stud out of relay rod.
4. Remove cylinder from vehicle.

Inspection

1. Check seals for leaks around cylinder rod. If leaks are found, replace seals.
2. Check hose connection seats for damage and replace if necessary.
3. For service other than seat or seal replacement, it is necessary to replace the power cylinder.
4. The ball stud may be replaced by removing snap-ring.

Disassembly and Assembly

1. To remove piston rod seal, remove snap-ring and pull out on rod. Remove back-up washer, piston rod scraper and piston rod seal from rod.

2. To remove the ball stud, depress the end plug and remove the snap-ring. Push on the end of the ball stud and the end plug, spring, spring seat, ball stud and seal may be removed. If the ball seat is to be replaced, it must be pressed out.

3. Reverse disassembly procedure. Be sure snap-ring is properly seated.

Installation

1. Install power cylinder on vehicle in reverse of removal procedure.
2. Reconnect the hydraulic lines, fill system and bleed out air as described in the installation and balancing section of control valve servicing.

Power Steering Hoses

Carefully inspect the hoses. When installing, be sure to place in such a position as to avoid all chafing or other abuse when making sharp turns.

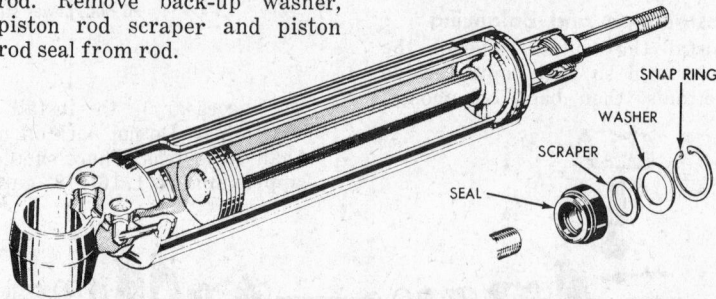

Power cylinder—typical (© Ford Motor Co.)

Saginaw Rotary Type Power Steering

The rotary type power steering gear is designed with all components in one housing.

The power cylinder is an integral part of the gear housing. A double-acting type piston allows oil pressure to be applied to either side of the piston. The one-piece piston and power rack is meshed to the sector shaft.

The hydraulic control valve is composed of a sleeve and valve spool. The spool is held in the neutral position by the torsion bar and spool actuator. Twisting of the torsion bar moves the valve spool, allowing oil pressure to be directed to either side of the power piston, depending upon the directional rotation of the steering

wheel, to give power assist.

On many trucks of the General Motors Corporation, a modified version of the rotary valve power steering system provides variable ratio steering to assist the driver to steer the truck easier and safer. The steering gear ratio will vary from a high ratio of about 16:1 while steering straight ahead to a lower gear ratio of about 12.1:1 while making a full turn to either side.

Roller Pump

Removal

Remove the reservoir cover and use a suction gun to empty the reservoir. Disconnect the hoses from the pump and tie them in a raised position to prevent oil drainage. Loosen the pump adjusting screw and remove the pump belt, then take out the retaining bolts and remove the pump and reservoir.

Installation

Position the pump assembly and install the retaining bolts. Be sure there is clearance between the pump bracket and the engine front support bracket. Install the hoses and place the pump belt on the pulley. Adjust the belt to $\frac{1}{2}''$ deflection, then tighten the adjusting screw.

Connect the hoses to the pump assembly.

Fill the reservoir to within $\frac{1}{2}''$ of the top with automatic transmission fluid type A.

Start the engine and rotate the steering wheel several times to the right and left to expel air from the system, then recheck the oil level and install the reservoir cover.

Power Steering Unit

Fluid Used

This unit uses automatic transmission fluid type A. The fluid capacity is $4\frac{1}{2}$ pints.

Bleeding the System

Fill the pump reservoir to within

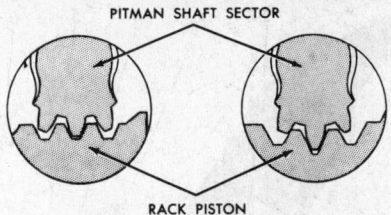

PITMAN SHAFT SECTOR

RACK PISTON

CONSTANT RATIO · VARIABLE RATIO

Comparison of constant ratio and variable ratio steering sector shaft teeth (© G.M.C.)

$\frac{1}{2}''$ of the top. Start and run the engine to attain normal operating temperatures. Now, turn the steering wheel through its entire travel three or four times to expel air from the system, then recheck the fluid level.

Checking Steering Effort

Run the engine to attain normal operating temperatures. With the wheels on a dry floor, hook a pull scale to the spoke of the steering wheel at the outer edge. The effort required to turn the steering wheel should be $3\frac{1}{2}$-5 lbs. If the pull is not within these limits, check the hydraulic pressure.

Pressure Test

To check the hydraulic pressure, disconnect the pressure hose from the gear. Now connect the pressure gauge between the pressure hose from the pump and the steering gear housing. Run the engine to attain normal operating temperatures, then turn the wheel to a full right and a full left turn to the wheel stops.

Hold the wheel in this position only long enough to obtain an accurate reading.

The pressure gauge reading should be within the limits specified. If the pressure reading is less than the minimum needed for proper operation, close the valve at the gauge and see if the reading increases. If the pressure is still low, the pump is defective and needs repair. If the pressure reading is at or near the minimum reading, the pump is normal and needs only an adjustment of the power steering gear or power assist control valve.

Worm Bearing Preload and Sector Mesh Adjustments

Disconnect the pitman arm from the sector shaft, then back off on the sector shaft adjusting screw on the sector shaft cover.

Center the steering on the high point, then attach a pull scale to the spoke of the steering wheel at the outer edge. The pull required to keep the wheel moving for one complete turn should be $\frac{1}{2}$-$\frac{2}{3}$ lbs.

If the pull is not within these limits, loosen the thrust bearing locknut and tighten or back off on the valve sleeve adjuster locknut to bring the preload within limits. Tighten the thrust bearing locknut and recheck the preload.

Slowly rotate the steering wheel several times, then center the steering on the high point. Now, turn the sector shaft adjusting screw until a steering wheel pull of 1-$1\frac{1}{2}$ lbs. is required to move the worm through the center point. Tighten the sector shaft adjusting screw locknut and recheck the sector mesh adjustment.

Install the pitman arm and draw the arm in position with the nut.

1 Locknut	28 Side cover	42 "O" ring
2 Retaining ring	29 "O" ring	43 Plug
3 Dust seal	30 Adjuster retainer	44 "O" ring
4 Oil seal	31 Shim	45 Housing end cover
5 Bearing	32 Adjuster screw	46 Retainer ring
6 Adjuster plug		
7 "O" ring		
8 Thrust washer (large)		
9 Thrust bearing		
10 Thrust washer (small)		
11 Spacer		
12 Retainer		
13 Spool valve spring		
14 "O" ring		
15 Spool valve		
16 Teflon oil rings		
17 "O" rings		
18 Valve body		
19 Stub shaft	33 Thrust washer	
20 "O" ring	34 Spring	
21 Wormshaft	35 Pitman shaft	47 Needle bearing
22 Thrust washer	36 Screws and lock washers	48 Oil seal
23 Thrust bearing	37 Clamp	49 Back up washer
24 Thrust washer	38 Ball return guide	50 Oil seal
25 Housing	39 Balls	51 Back up washer
26 Locknut	40 Rack-piston	52 Retaining ring
27 Attaching bolts and washers	41 Teflon oil seal	

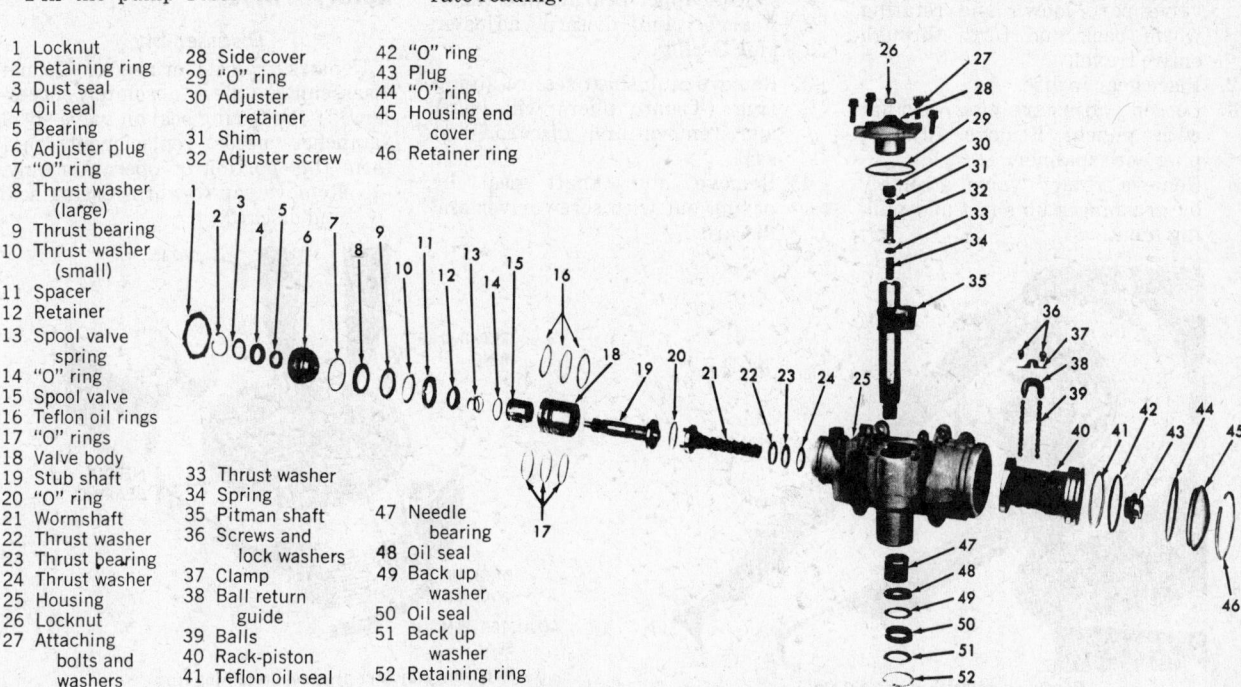

Power steering gear—light duty trucks (© G.M.C.)

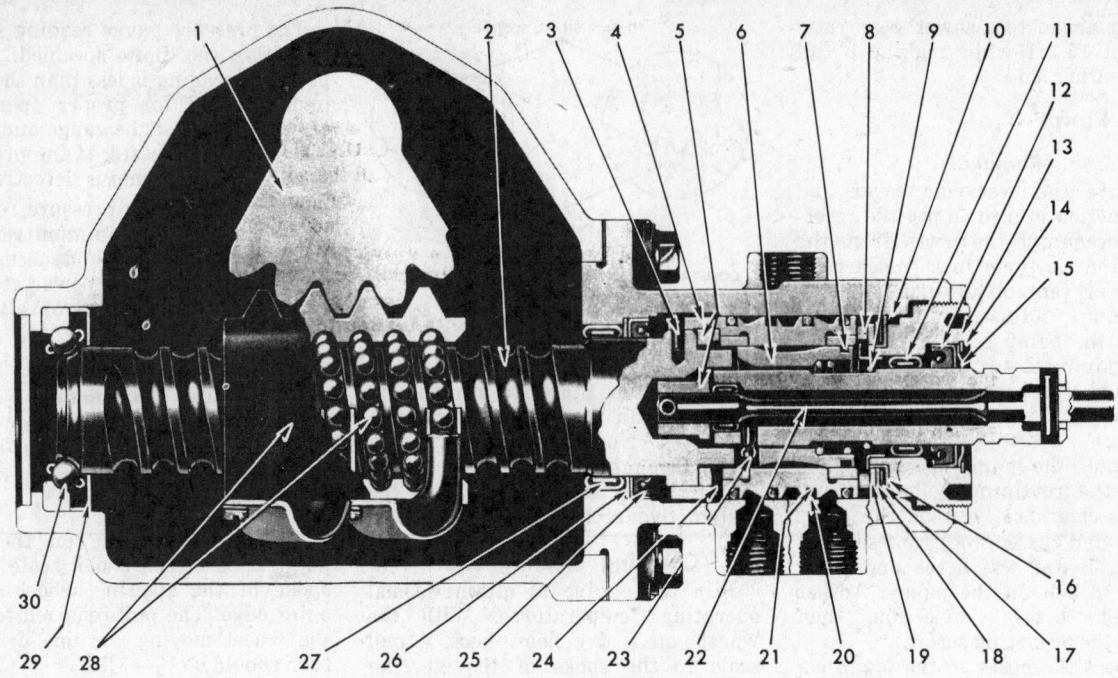

Power steering gear—medium duty trucks (© G.M.C.)

1 Sector
2 Wormshaft
3 Body drive pin
4 Valve body
5 Cap assembly
6 Valve spool
7 Spool dampener "O" ring
8 Thrust bearing spacer
9 Spool spring
10 Adjuster plug "O" ring seal

11 Adjuster plug needle bearing
12 Adjuster plug shaft seal
13 Adjuster plug
14 Adjuster plug snap ring
15 Adjuster plug dust seal
16 Bearing race
17 Upper thrust bearing
18 Bearing race
19 Spacer
20 Valve body ring

21 Ring back-up seal
22 Torsion bar
23 Spool valve pin
24 Valve body pin
25 Top cover (valve) seal
26 Back-up washer
27 Top cover (valve) bearing
28 Ball nut and balls
29 Bearing retainer
30 Lower thrust bearing

Service Operations
Adjuster Plug and Rotary Valve

Removal

1. Thoroughly clean exterior of gear assembly. Drain by holding valve ports down and rotating worm back and forth through entire travel.
2. Place gear in vise.
3. Loosen adjuster plug locknut with punch. Remove adjuster plug with spanner.
4. Remove rotary valve assembly by grasping stub shaft and pulling it out.

Adjuster Plug

Disassembly & Assembly

1. Remove upper thrust bearing retainer with screwdriver. Be careful not to damage bearing bore. Discard retainer. Remove spacer, upper bearing and races.
2. Remove and discard adjuster plug O-ring.
3. Remove stub shaft seal retaining ring (Truarc pliers will help) and remove and discard dust seal.
4. Remove stub shaft seal by prying out with screwdriver and discard.

5. Examine needle bearing and, if required, remove same by pressing from thrust bearing end.
6. Inspect thrust bearing spacer, bearing rollers and races.
7. Reassemble in reverse of above.

Rotary Valve

Disassembly

Repairs are seldom needed. Do not disassemble unless absolutely necessary. If the O-ring seal on valve spool dampener needs replacement, perform this portion of operation only.

1. Remove cap-to-worm O-ring seal and discard.

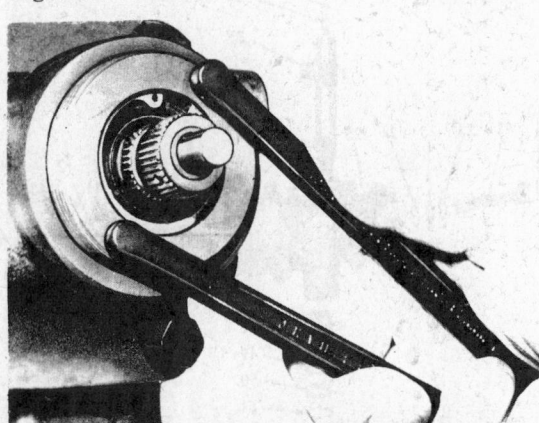

Removing adjuster plug (© G.M.C.)

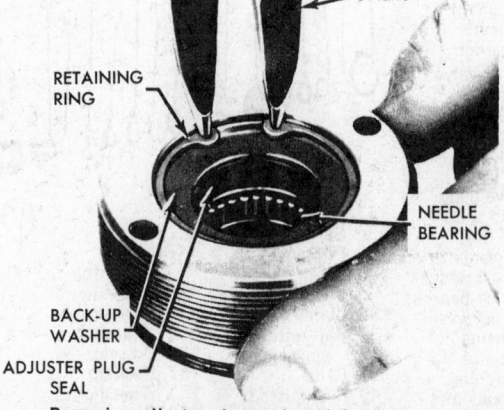

Removing adjuster plug seal retaining ring
(© G.M.C.)

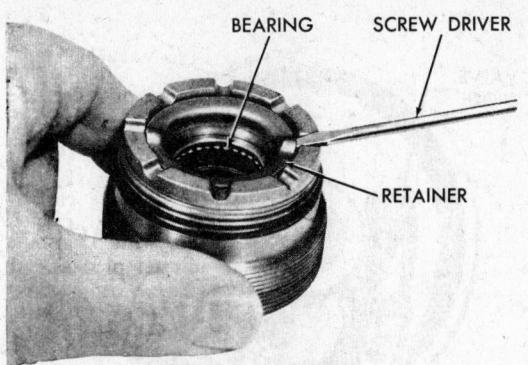

Removing thrust bearing retainer
(© G.M.C.)

1 Retaining ring
2 Dust seal
3 Oil seal   8 Thrust bearing
4 Needle bearing   9 Thrust washer (small)
5 Adjuster plug   10 Spacer
6 "O" ring   11 Retainer
7 Thrust washer (large)

Adjuster plug (© G.M.C.)

2. Remove valve spool spring by prying on small coil with small screwdriver to work spring onto bearing surface of stub shaft. Slide spring off shaft. Be careful not to damage shaft surface.
3. Remove valve spool by holding the valve assembly in one hand with the stub shaft pointing down. Insert the end of pencil or wood rod through opening in valve body cap and push spool until it is out far enough to be removed. In this procedure, rotate to prevent jamming. If spool becomes jammed it may be necessary to remove stub shaft, torsion bar and cap assembly.

Assembly

CAUTION: All parts must be free and clear of dirt, chips, etc., before assembly and must be protected after.
1. Lubricate three new back-up O-ring seals with automatic transmission oil and reassemble in the ring grooves of valve body. Assemble three new valve body rings in the grooves over the O-ring seals by carefully slipping over the valve body. NOTE: If the valve body rings seem loose or twisted in the grooves, the heat of the oil during operation will cause them to straighten.
2. Lubricate a new dampener O-ring with automatic transmission oil and install in valve spool groove.
3. Assemble stub shaft torsion bar and cap assembly in the valve body, aligning the groove in the valve cap with the pin in the valve body. Tap lightly with soft remainder of assembly. Valve body pin must be in the cap groove. Hold parts together during the remainder of assembly.
4. Lubricate spool. With notch in spool toward valve body, slide the spool over the stub shaft. Align the notch on the spool with the spool drive pin on stub shaft and carefully engage spool in valve body bore. Push spool

evenly and with slight rotating motion until it reaches the drive pin. Rotate slowly, with some pressure, until notch engages pin. Be sure dampener O-ring seal is evenly distributed in the spool groove. CAUTION: Use extreme care because spool to valve body clearance is very small. Damage is easily caused.
5. With seal protector over stub shaft, slide valve spool spring over shaft, with small diameter of spring going over shaft last. Work spring onto shaft until small coil is located in stub shaft groove.
6. Lubricate a new cap to O-ring seal and install in valve body.

Adjuster Plug and Rotary Valve

Installation

1. Align narrow pin slot on valve body with valve body drive pin on the worm. Insert the valve assembly into gear housing by pressing against valve body with finger tips. Do not press on stub shaft or torsion bar. The return hole in the gear housing should be fully visible when properly assembled. CAUTION: Do not press on stub shaft as this may cause shaft and cap to pull out of valve body, allowing the spool dampener O-ring seal to slip into valve body oil grooves.
2. With seal protector over end of stub shaft, install adjuster plug assembly into gear housing snugly with spanner, then back plug off approximately one-eighth turn. Install plug locknut but do not tighten. Adjust preload as described in the adjustment section.
3. After adjustment, tighten locknut.

Pitman Shaft

Removal and Installation

1. Completely drain the gear assembly and thoroughly clean the outside.

2. Place gear in vise.
3. Rotate stub shaft until pitman shaft gear is in center position. Remove side cover retaining bolts.
4. Tap end of pitman shaft with soft hammer and slide shaft out of housing.
5. Remove and discard side cover O-ring seal.
6. The seals, washers, retainers and bearings may now be removed and examined.
7. Examine all parts for wear or damage and replace as required.
8. Install in reverse of above. Make proper adjustment as described in adjustment section.

Rack-Piston Nut and Worm Assembly

Removal

1. Completely drain the gear assembly and thoroughly clean the outside.
2. Remove pitman shaft assembly as previously described.
3. Rotate housing end plug retaining ring so that one end of ring is over hole in gear housing. Spring one end of ring so screwdriver can be inserted to lift it out.
4. Rotate stub shaft to full left turn position to force end plug out of housing.
5. Remove and discard housing end plug O-ring seal.
6. Remove rack-piston nut end plug with 1/2" square drive.
7. Insert special tool in end of worm. Turn stub shaft so that rack-piston nut will go into tool and then remove rack-piston nut from gear housing.
8. Remove adjuster plug and rotary valve assemblies as previously described.
9. Remove worm and lower thrust bearing and races.
10. Remove cap O-ring seal and discard.

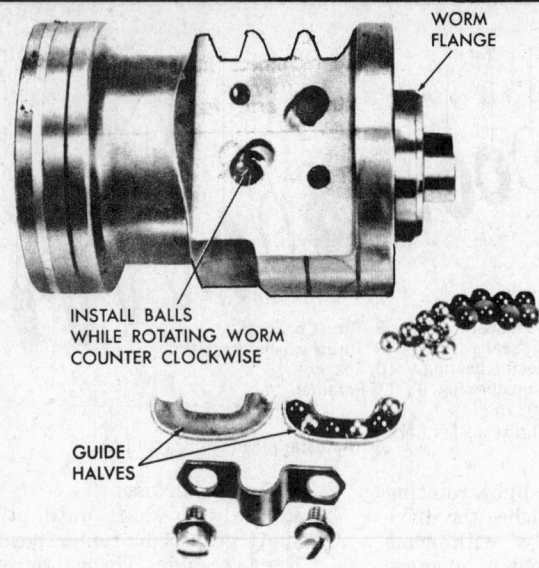

INSTALL BALLS WHILE ROTATING WORM COUNTER CLOCKWISE

WORM FLANGE

GUIDE HALVES

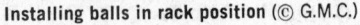

Installing balls in rack position (© G.M.C.)

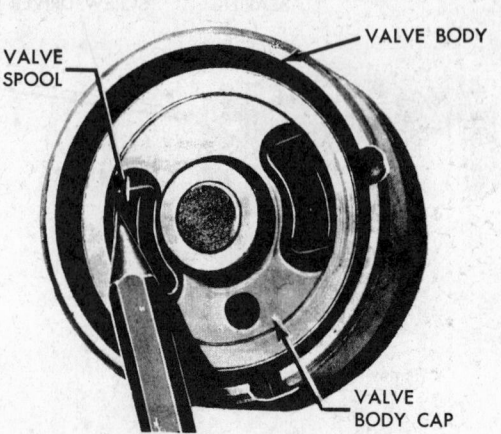

VALVE SPOOL

VALVE BODY

VALVE BODY CAP

Separating valve spool from valve body
(© G.M.C.)

Rack-Piston Nut and Worm

Disassembly and Assembly

1. Remove and discard piston ring and back-up O-ring on rack piston nut.
2. Remove ball guide clamp and return guide.
3. Place nut on clean cloth and remove ball retaining tool. Make sure all balls are removed.
4. Inspect all parts for wear, nicks, scoring or burrs. If worm or rack-pinion nut need replacing, both must be replaced as a matched pair.
5. In reassembling reverse the above.

NOTE: When assembling, alternate black and white balls, and install guide and clamp. Packing with grease helps in holding during assembly.

When new balls are used, various sizes are available and a selection must be made to secure proper torque when making the high point adjustment.

Rack-Piston Nut and Worm Assembly

Installation

1. Install in reverse of removal procedure.
2. In all cases use new O-ring seals.
3. Make adjustments as previously described.

Saginaw Model 170,170-D Integral Power Steering Gear

The model 170, 170-D power steering gear unit is used in conjunction with the heavier pitman and steering arms, and eliminates the need for power cylinder assist units attached to the axle and to the steering linkage.

The unit uses a remote mounted, belt driven, vane type hydraulic pump for fluid pressure and directs the fluid to and from the gear unit by the use of pressure and return hoses.

As the vehicle operator turns the steering wheel, the control valve is moved within the gear housing, and closes the pressure relief port and directs fluid pressure to the opposite ends of the primary and secondary pistons. The pressure assists the movement of the pistons as they rotate the pitman shaft, which in turn, moves the steering linkage to turn the wheels. The greater the turning effort, the more pressure is applied to the piston ends, therefore assuring the operator a smooth hydraulic assist in turning at all times.

As the steering effort is stopped, the control valve is returned to its neutral position, the fluid pressure to the piston ends are equalized on both sides, the pressure is directed to the re-

lief port and returned to the pump reservoir, and the steering gear is returned to the neutral or straight ahead position.

Adjustments

There are no on-the-vehicle adjustments of the integral type steering gear.

Removal

1. Center the steering gear and remove the pitman arm bolt.
2. Spread the pitman arm clamp boss slightly to remove the arm. Do not spread the arm clamp boss over .004 inch.
3. Remove the pot joint to stub shaft clamp bolt, loosen the steering column assembly and pull upward until the shaft coupling clears the stub shaft.
4. Disconnect the hydraulic lines and plug them. Remove the steering gear attaching bolts and with the aid of an assistant, turn the gear in a vertical position and lower the gear between the frame and the inner fender panel.

Installation

1. Install adapter plate to the gear assembly, if removed. (Install

the lower forward bolt through the adapter plate before attaching it to the gear housing.)
2. With the gear in a vertical position, (stub shaft up), move the gear upward between the fender panel and the frame. Loosely install the bolts.
3. Unplug the hydraulic lines and install them into the fittings of the gear housing.
4. Tighten the gear to frame bolts and torque to specifications.
5. With the aid of one or more assistants, center and push the steering shaft over the stub shaft until the coupling lines up with the cross groove in the stub shaft.
6. Install the clamp bolt in the cross groove clamp and tighten. Tighten the steering column assembly.
7. Install the pitman arm, install the bolt and torque to specifications.
8. Fill the reservoir and bleed the system as outlined previously.

Gear Unit

Disassembly

1. Place the steering gear box in a holding fixture or a vise. With a

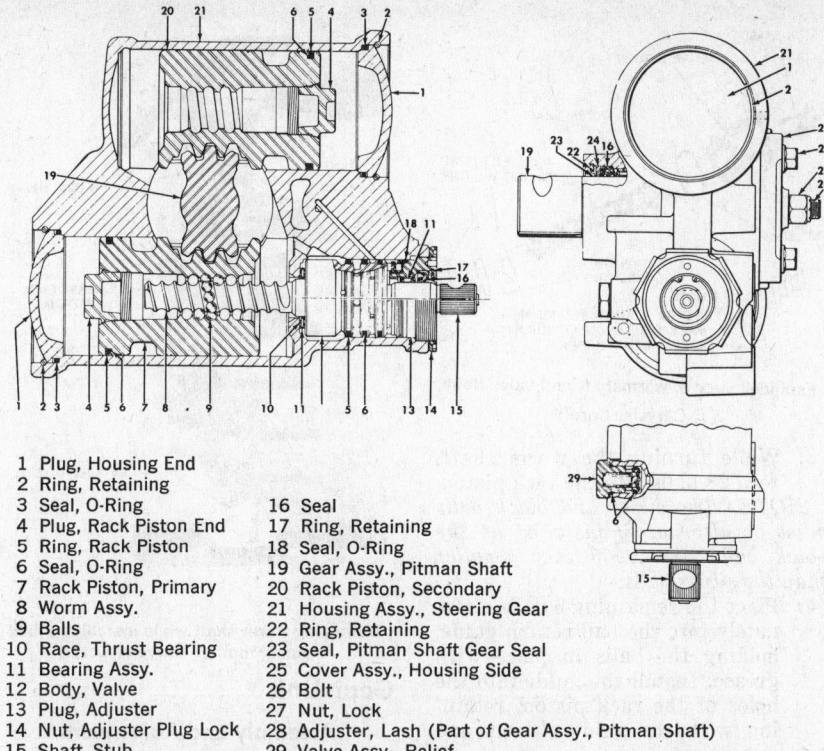

1 Plug, Housing End
2 Ring, Retaining
3 Seal, O-Ring
4 Plug, Rack Piston End
5 Ring, Rack Piston
6 Seal, O-Ring
7 Rack Piston, Primary
8 Worm Assy.
9 Balls
10 Race, Thrust Bearing
11 Bearing Assy.
12 Body, Valve
13 Plug, Adjuster
14 Nut, Adjuster Plug Lock
15 Shaft, Stub

16 Seal
17 Ring, Retaining
18 Seal, O-Ring
19 Gear Assy., Pitman Shaft
20 Rack Piston, Secondary
21 Housing Assy., Steering Gear
22 Ring, Retaining
23 Seal, Pitman Shaft Gear Seal
25 Cover Assy., Housing Side
26 Bolt
27 Nut, Lock
28 Adjuster, Lash (Part of Gear Assy., Pitman Shaft)
29 Valve Assy., Relief

Integral power steering gear and control unit—Model 170-170D (© G.M.C.)

small pin punch, dislodge the end cover retaining rings from their grooves in the primary and secondary piston housings and pry them out with the aid of a screwdriver.

2. Turn the stub shaft counterclockwise to force the cover from the primary cylinder. Remove the cover and "O" ring seal.

3. Remove the rack piston end plug, the sector preload adjuster nut, and the four side cover bolts.

4. Using a ¼ in. Allen wrench, turn the preload adjusting nut clockwise until the side cover separates from the sector shaft and remove the cover.

5. Turn the stub shaft counterclockwise until the sector shaft teeth are out of engagement with the teeth of the rack piston. *NOTE: If the secondary piston end cover is stuck, turn the stub shaft counterclockwise until the rack piston bottoms in the housing, then engage the sector end tooth in the center tooth spacing on the primary rack piston. Turn the stub shaft counterclockwise until the secondary rack piston forces the end cover from the housing.*

6. Remove the secondary rack piston from the bore in the gear housing. Do not remove the end plug unless it is to be replaced.

7. Rotate the stub shaft clockwise until the teeth of the sector and the rack piston clear each other and the rack piston can move freely.

8. Insert a ball retainer tool or its equivalent into the bore of the rack piston. Turn the stub shaft counterclockwise while holding the tool firmly against the worm, forcing the rack piston over the tool and retaining the recirculating balls in place. Remove the rack piston from the housing.

1. Locknut
2. Retaining Ring
3. Back-Up Washer
4. Stub Shaft Seal
5. Needle Bearing
6. Adjuster Plug
7. "O" Ring
8. Thrust Race (Upper)
9. Thrust Bearing
10. Thrust Race
11. Spacer
12. Retainer
13. Dampener "O" Ring
14. Valve Spool
15. Teflon "O" Rings

9. Rotate the sector shaft teeth to clear the housing and remove the shaft from the gear housing.

10. Remove the adjuster plug lock nut and with the aid of a spanner wrench, remove the adjuster plug from the stub shaft end of the gear housing.

11. Remove the valve and worm as an assembly with the thrust bearings and races and separate the worm from the valve assembly.

Adjuster Plug

Disassembly and Assembly

1. Reinstall the adjuster plug into the gear housing and snug it finger tight. Remove the snap retaining ring and the back-up washer.

2. Remove the seal from the plug by prying the seal outward, being careful not to damage the bore.

3. Pry the thrust bearing retainer from the bore with the aid of a small screwdriver. Remove the spacer, washer, bearing and second washer.

4. The needle bearing assembly can be removed from the plug by driving it out.

5. The assembly of the plug can be accomplished by the reversal of the disassembly procedure. Install new "O" rings, seals, and bearings as needed and lubricate the parts with power steering fluid.

Valve and Stub Shaft

Disassembly and Assembly

1. Hold the valve assembly by hand

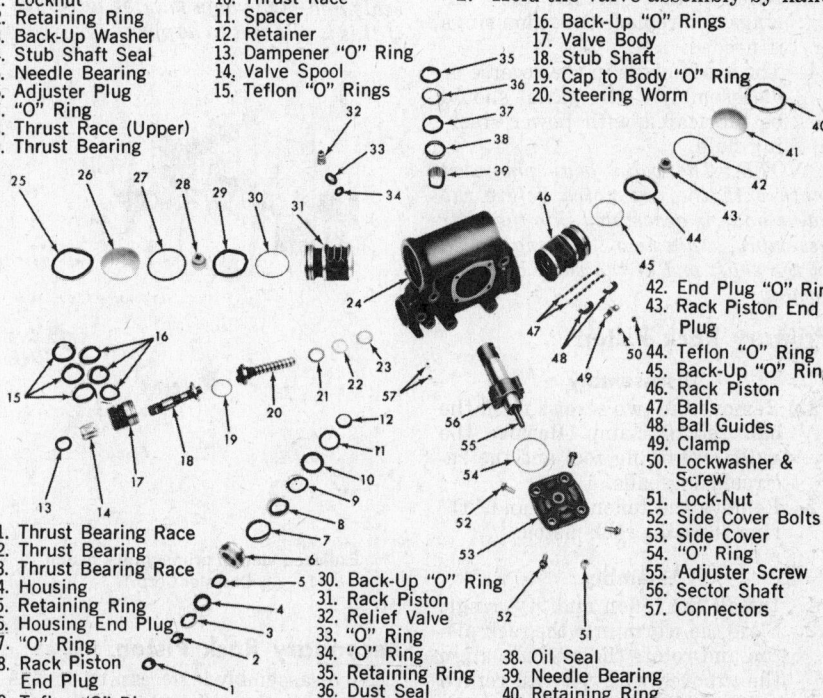

16. Back-Up "O" Rings
17. Valve Body
18. Stub Shaft
19. Cap to Body "O" Ring
20. Steering Worm

42. End Plug "O" Ring
43. Rack Piston End Plug
44. Teflon "O" Ring
45. Back-Up "O" Ring
46. Rack Piston
47. Balls
48. Ball Guides
49. Clamp
50. Lockwasher & Screw
51. Lock-Nut
52. Side Cover Bolts
53. Side Cover
54. "O" Ring
55. Adjuster Screw
56. Sector Shaft
57. Connectors

21. Thrust Bearing Race
22. Thrust Bearing
23. Thrust Bearing Race
24. Housing
25. Retaining Ring
26. Housing End Plug
27. "O" Ring
28. Rack Piston End Plug
29. Teflon "O" Ring

30. Back-Up "O" Ring
31. Rack Piston
32. Relief Valve
33. "O" Ring
34. "O" Ring
35. Retaining Ring
36. Dust Seal
37. Back-Up Washer

38. Oil Seal
39. Needle Bearing
40. Retaining Ring
41. Housing End Plug

Exploded view—Model 170-170D steering gear (© Chrysler Corp.)

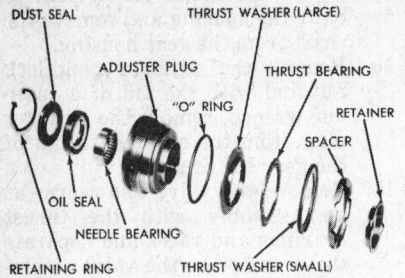

Exploded view of adjuster plug (© Chrysler Corp.)

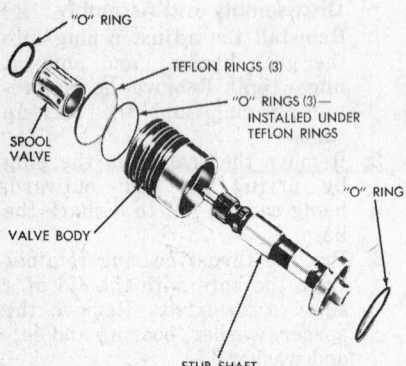

Exploded view of valve body and stub shaft
(© Chrysler Corp.)

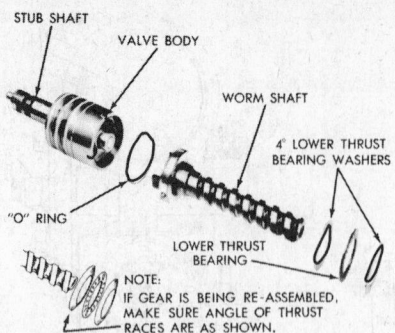

Exploded view of wormshaft and valve body
(© Chrysler Corp.)

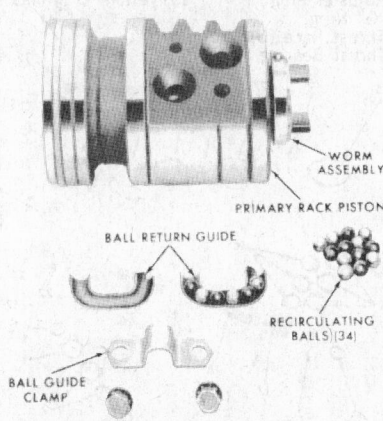

Exploded view of primary rack piston
(© Chrysler Corp.)

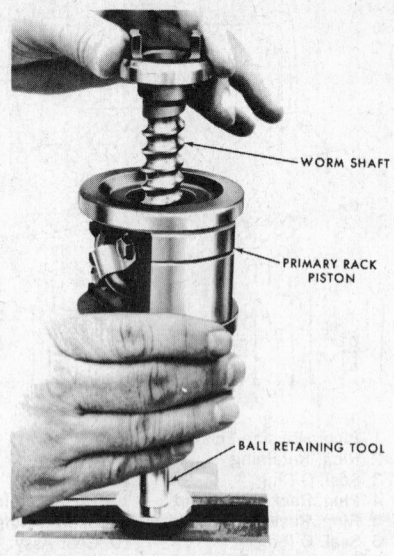

Removing worm shaft while installing a ball retaining tool (© Chrysler Corp.)

with the stub shaft down. Lightly tap the stub shaft against a wood block until the cap is raised from the valve body approximately ¼ inch.

2. Remove the shaft assembly from the spool by disengaging the shaft pin, and remove the spool from the valve body by rotating it.

3. Remove and discard the "O" rings and replace the teflon rings if needed.

4. The assembly is in the reverse of disassembly. All parts should be lubricated with power steering fluid.

NOTE: The valve body pin must mate with the cap notch before the valve body is assembled into the gear assembly, and a new "O" ring placed in the shaft end of the valve body assembly.

Primary Rack Piston

Disassembly

1. Remove the two screws from the ball return clamp. Remove the guide, retaining tool and the re-circulating balls.

2. Remove the teflon ring and "O" ring from the rack piston.

Assembly

1. Install the teflon and "O" rings.

2. Slide the worm into the rack piston and rotate the worm to align the grooves with the ball return guide hole nearest the piston ring.

3. While turning the worm shaft, feed 28 balls into the rack piston.

NOTE: The silver and black balls must be alternately installed as the black balls are .0005 inch smaller than the silver balls.

4. Place the remaining 6 balls alternately into the ball return guide, holding the balls in place with grease. Install the guide into the holes of the rack piston, retaining with the guide clamps and screws.

5. Install the ball retaining tool in place of the worm shaft, being careful not to allow any balls to drop.

NOTE: When installing the teflon rings, looseness will be noticed. The teflon rings will heat-shrink when the gear is operated. Therefore, care must be exercised when assembling the internal parts into the housing, to insure that all parts are lubricated with power steering fluid and not forced during the reassembly. The seals and "O" rings may be damaged if this is allowed to happen.

Secondary Rack Piston

No disassembly is necessary on this unit unless the teflon and "O" rings are to be replaced.

Gear Unit

Assembly and Adjustment

1. Install the thrust washer, bearing and second thrust washer over the end of the worm and lubricate with power steering fluid.

NOTE: The tapered surfaces of the washers should be parallel to each other and the cupped side towards the stub shaft.

2. Install the "O" ring in the valve body so that it is seated against the lower shaft cap and lubricate the valve body, rings and seals with power steering fluid.

3. Align the narrow notch in the valve body with the pin in the worm and install the unit into the gear housing, by exerting pressure on the valve body and not the stub shaft.

4. The return hole in the gear housing should be fully uncovered when the valve body is fully seated. Screw in the adjuster plug assembly and seat it against the valve body.

5. Adjust the thrust bearing preload by torquing the adjuster plug to 20 ft. lbs. to seat the thrust bearings.

6. Mark the steering housing in line with one of the tool hole locations on the adjuster plug. Measure counterclockwise 3/16 to ¼ inch and remark the housing.

7. Loosen the adjuster until the tool hole is in line with the second mark on the steering housing and install the lock nut and tighten while maintaining the alignment of the adjuster tool hole with the mark on the housing.

8. With the aid of a torque wrench, turn the stub shaft evenly and observe the torque reading. The reading should be from 4 to 10 inch pounds.

9. Continue the adjustment as necessary to obtain the specified torque reading.

10. With the ball retaining tool in position, lubricate and install the primary rack piston into the gear housing until the retaining tool bottoms against the center of the worm.

11. Turn the stub shaft clockwise to thread the rack piston onto the worm. Keep the retaining tool tight against the worm while turning the stub shaft.

12. Remove the ball retainer tool when the rack piston is completely threaded onto the worm.

Center the rack teeth in the sector shaft opening.

13. Install the secondary rack piston in the gear housing and line up the center tooth space with the teeth of the primary rack piston.

14. Slide the sector shaft into the gear housing with the tapered teeth engaging the primary rack piston.

15. Install a new "O" ring on the side cover and push the cover into the housing until contact is made with the preload adjuster screw. With the aid of a ¼ inch Allen wrench inserted through the cover, turn the adjusting screw counterclockwise until the cover bottoms on the housing.

16. Install the side cover bolts and torque to 45 ft. lbs. Install the rack piston plug and torque to 75 ft. lbs.

17. Install the primary and secondary end covers, "O" rings, and install the retainer rings.

18. With the steering gear on center, tighten the sector adjusting screw. Install and tighten the lock nut and check the overcenter torque while rotating the stub shaft through an arc of 180 degrees, with a torque wrench. Adjust the sector shaft accordingly until the correct torque is obtained.
New gears—4 to 8 in. lbs., but not over 18 in. lbs. combined torque.
Used gears—4 to 5 in. lbs., but not over 14 in. lbs. combined torque.

NOTE: Combined torque includes the thrust bearing adjustment reading, over-center and internal friction.

Chrysler Full-Time Power Steering (Constant Control Type)

The Chrysler Corporation Constant Control Type Power Steering Gear System consists of a hydraulic pressure pump, a power steering gear and connecting hoses.

The power steering gear housing contains a gear shaft and sector gear, a power piston with gear teeth milled into the side of the piston which is in constant mesh with the gear shaft sector teeth, a worm shaft which connects the steering wheel to the power piston through a coupling. The worm shaft is geared to the piston through recirculating ball contact.

A pivot lever is fitted into the spool valve at the upper end and into a drilled hole in the center thrust bearing race at the lower end. The center thrust bearing race is held firmly against the shoulder of the worm shaft by two thrust bearings, bearing races and an adjusting nut. The pivot lever pivots in the spacer which is held in place by the pressure plate.

When the steering wheel is turned to the left the worm shaft moves out of the power piston a few thousandths of an inch, the center thrust bearing race moves the same distance since it is clamped to the worm shaft. The race thus tips the pivot lever and moves the spool valve down, allowing oil under pressure to flow into the left-turn power chamber and force the power piston down. As the power piston moves, it rotates the cross-shaft sector gear and, through the steering linkage, turns the front wheels.

On a right turn the worm shaft moves into the power piston, the center thrust bearing race thus tips the pivot lever and moves the spool valve up, allowing oil under pressure to flow into the right power chamber and force the power piston up.

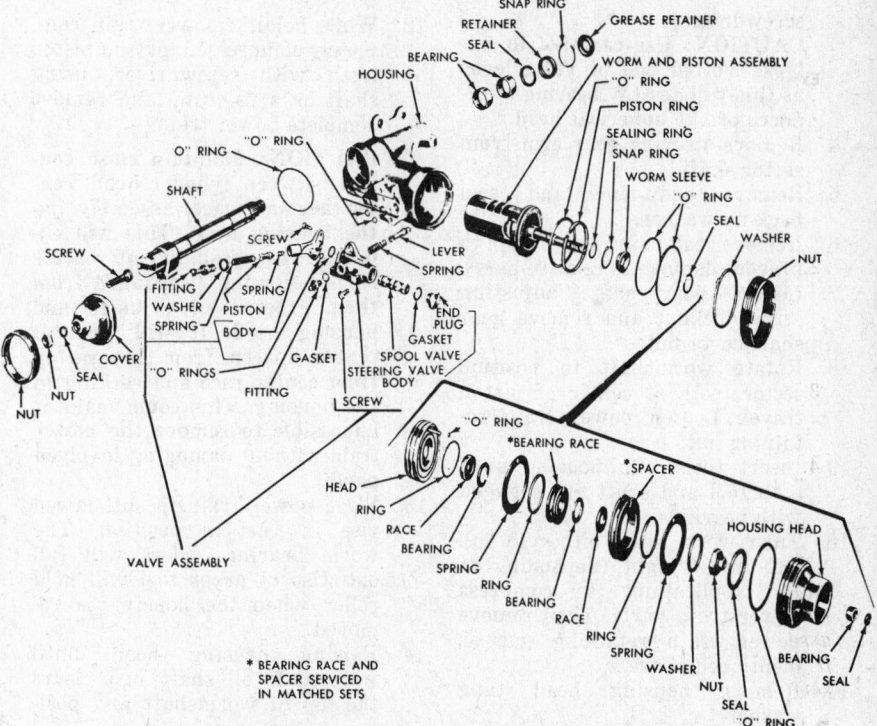

Chrysler power steering disassembled view (© Chrysler Corp.)

Pressure Test

Connect the pressure test hoses with the pressure gauge installed between the pump and steering gear.

Now, fill the reservoir to the level mark, then start the engine and bleed the system. Allow the engine to idle until the fluid in the reservoir is between 150° F. and 170° F. Now turn the steering wheel to the extreme right and check the pressure reading, then turn to the extreme left and check the reading again. The gauge reading should be equal in each direction. If not, it indicates excessive in-

ternal leakage in the unit.

The pressure should agree with the specifications in Pump Section for satisfactory power steering operation.

Reconditioning

1. Drain gear by turning worm shaft from limit to limit with oil connections held downward. Thoroughly clean outside.

2. Remove valve body attaching screws, body and three O-rings.

3. Remove pivot lever and spring. Pry under spherical head with a

Power Steering Gears

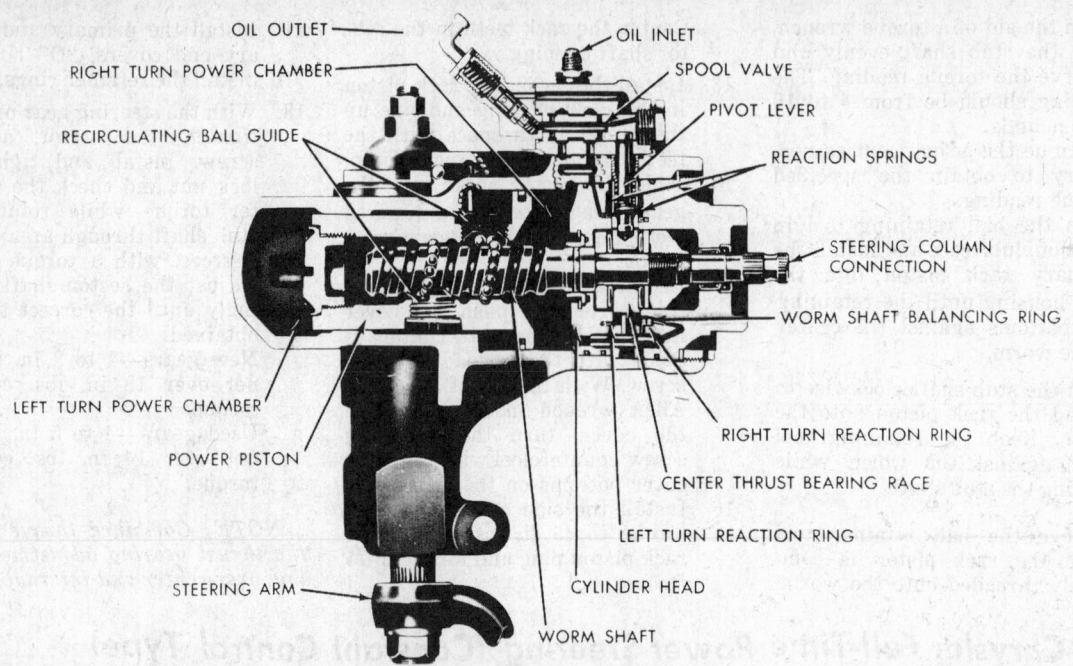

Chrysler power steering gear (© Chrysler Corp.)

screwdriver.

CAUTION: Use care not to collapse slotted end of valve lever as this will destroy bearing tolerances of the spherical head.

4. Remove steering gear arm from sector shaft.
5. Remove snap-ring and seal back-up washer.
6. Remove seal, using proper tool to prevent damage to relative parts.
7. Loosen gear shaft adjusting screw locknut and remove gear shaft cover nut.
8. Rotate wormshaft to position sector teeth at center of piston travel. Loosen power train retaining nut.
9. Insert tools into housing until both tool and shaft are engaged with bearings.
10. Turn worm shaft either to full left or full right (depending on car application) to compress power train parts. Then remove power train retaining nut as mentioned above.
11. Remove housing head tang washer.

12. While holding power train completely compressed, pry on piston teeth with screwdriver, using shaft as a fulcrum, and remove complete power train.

CAUTION: Maintain close contact between cylinder head, center race and spacer assembly and the housing head. This will eliminate the possibility of reactor rings becoming disengaged from their grooves in cylinder and housing head. It will prohibit center spacer from separating from center race and cocking in the housing. This could make it impossible to remove the power train without damaging involved parts.

13. Place power train in soft-jawed vise in vertical position. The worm bearing rollers will fall out. Use of arbor tool will hold roller when the housing is removed.
14. Raising housing head until wormshaft oil shaft just clears the top of wormshaft and posi-

tion arbor tool on top of shaft and into seal. With arbor in position, pull up on housing head until arbor is positioned in bearing. Remove when the housing is removed.
15. Remove large O-ring from housing head groove.
16. Remove reaction seal from groove in face of head with air pressure directed into ferrule chamber.
17. Remove reactor spring, reactor ring, worm balancing ring and spacer.
18. While holding wormshaft from turning, turn nut with enough force to release staked portions from knurled section and remove nut.

NOTE: Pay strict attention to cleanliness.

19. Remove upper thrust bearing race (thin) and upper thrust bearing.
20. Remove center bearing race.
21. Remove lower thrust bearing and lower thrust bearing race (thick).

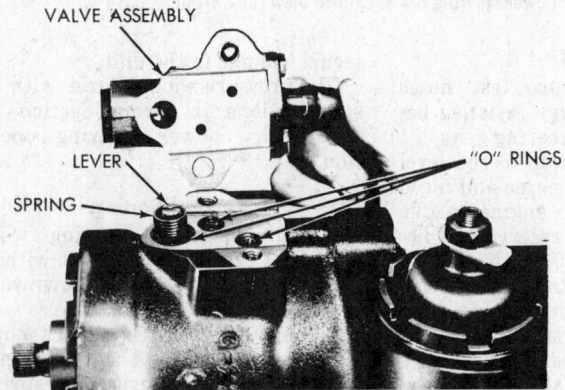

Removing valve body assembly (© Chrysler Corp.)

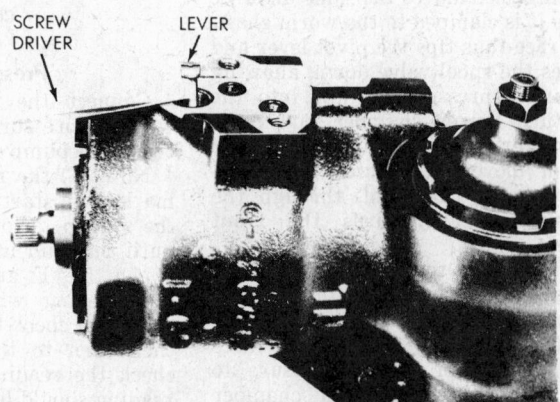

Removing pilot lever (© Chrysler Corp.)

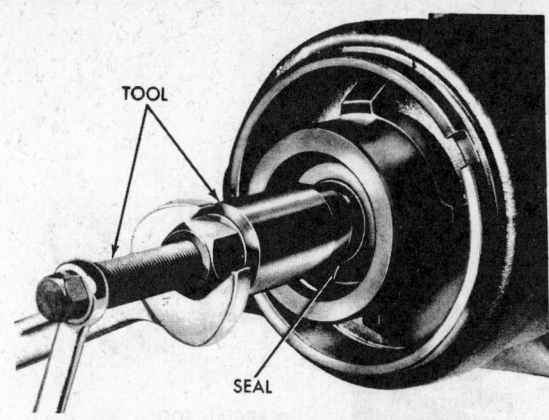

Removing worm shaft oil seal (© Chrysler Corp.)

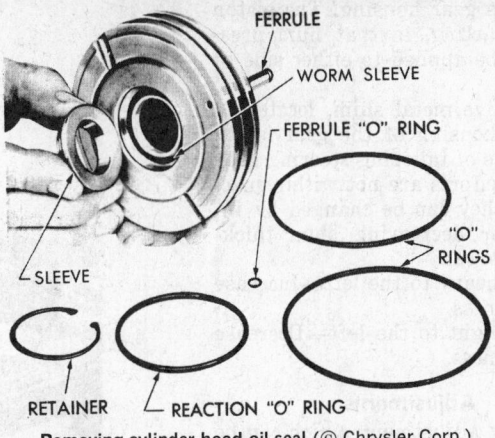

Removing cylinder head oil seal (© Chrysler Corp.)

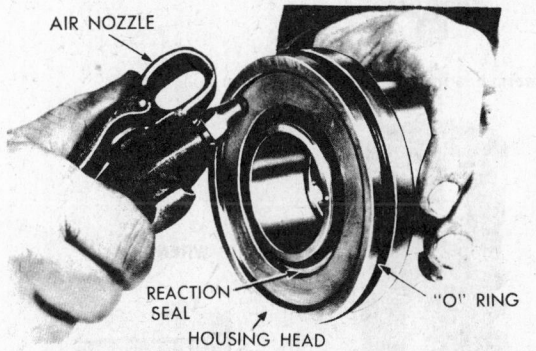

Removing reaction seal from wormshaft support (© Chrysler Corp.)

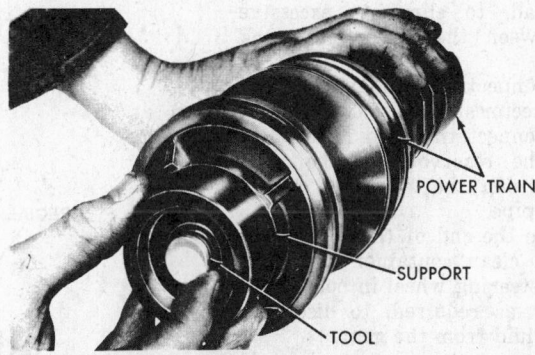

Retaining bearing rollers with arbor tool (© Chrysler Corp.)

22. Remove lower reaction ring and reaction spring.
23. Remove cylinder head assembly.
24. Remove O-rings from outer grooves in head.
25. Remove reaction O-ring from groove in face of cylinder head. Use air pressure in oil hole located between O-ring grooves.
26. Remove snap-ring, sleeve and rectangular oil seal from cylinder head counterbore.
27. Test wormshaft operation. Not more than 2 in. lbs. should be required to turn it through its entire travel, and with a 15 ft. lb. side load.

NOTE: The worm and piston is serviced as a complete assembly and should not be disassembled.

28. Shaft side play should not exceed 0.008 in. under light pull applied 2 5/16 in. from piston flange.
29. Assemble in reverse of above, noting proper adjustments and preload requirements following.

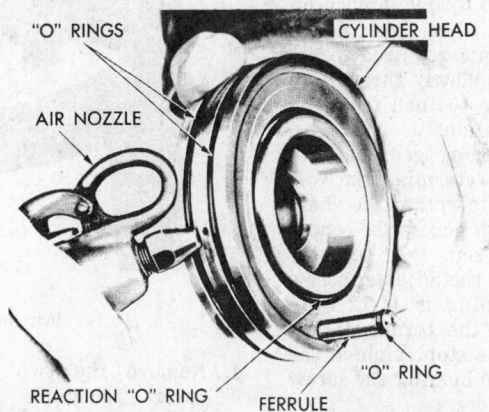

Removing reaction seal from cylinder head (© Chrysler Corp.)

30. When cover nut is installed, tighten to 20 ft-lbs. torque.
31. Valve mounting screws should be tightened to 200 in.-lbs. torque.
32. With hoses connected, system bled, and engine idling roughly, center valve unit until not self-steering. Tap on head of valve body attaching screws to move valve body up, and tap on end plug to move valve body down.
33. With steering gear on center, tighten gear shaft adjusting screw until lash just disappears.
34. Continue to tighten 3/8 to 1/2 turn and tighten locknut to 50 ft. lbs.

Ford Integral Power Steering Gear

The Ford integral power steering unit is a torsion-bar type.

The torsion bar power steering unit includes a worm and one-piece rack piston, which is meshed to the gear teeth on the steering sector shaft. The unit also includes a hydraulic valve, valve actuator, input shaft and torsion bar assembly which are mounted on the end of the worm shaft and operated by the twisting action of the torsion bar.

The torsion-bar type of power steering gear is designed with the one piece rack-piston, worm and sector shaft in one housing and the valve spool in an attaching housing. This makes possible internal fluid passages between the valve and cylinder, thus eliminating all external lines and hoses, except the pressure and return hoses between the pump and gear assembly.

The power cylinder is an integral

part of the gear housing. The piston is double acting, in that fluid pressure may be applied to either side of the piston.

A selective metal shim, located in the valve housing of the gear is for the purpose of tailoring steering gear efforts. If efforts are not within specifications they can be changed by increasing or decreasing shim thickness as follows:

Efforts heavy to the left—Increase shim thickness.

Efforts light to the left—Decrease shim thickness.

Adjustments

The only adjustment which can be performed is the total over center position load, to eliminate excessive lash between the sector and rack teeth.

1. Disconnect the Pitman arm from the sector shaft.
2. Disconnect the fluid return line at the reservoir, at the same time cap the reservoir return line pipe.
3. Place the end of the return line in a clean container and cycle the steering wheel in both directions as required, to discharge the fluid from the gear.
4. Turn the steering wheel to 45 degrees from the left stop.
5. Using an in-lb torque wrench on the steering wheel nut, determine the torque required to rotate the shaft slowly through an approximately ⅛ turn from the 45 degree position.
6. Turn the steering gear back to center, then determine the torque required to rotate the shaft back and forth across the center position. Loosen the adjuster nut, and turn the adjuster screw until the reading is 11-12 in-lb greater than the torque 45 degrees from the stop. Tighten the lock nut while holding the screw in place.
7. Recheck the readings and replace the Pitman arm and the steering wheel hub cover.
8. Correct the fluid return line to the reservoir and fill the reservoir with specified lubricant to the proper level.

Valve Centering Shim

Removal and Installation

1. Hold the steering gear over a drain pan in an inverted position and cycle the input shaft several times to drain the remaining fluid from the gear.
2. Mount the gear in a soft-jawed vise.
3. Turn the input shaft to either stop then, turn it back approximately 1¾ turns to center the gear.

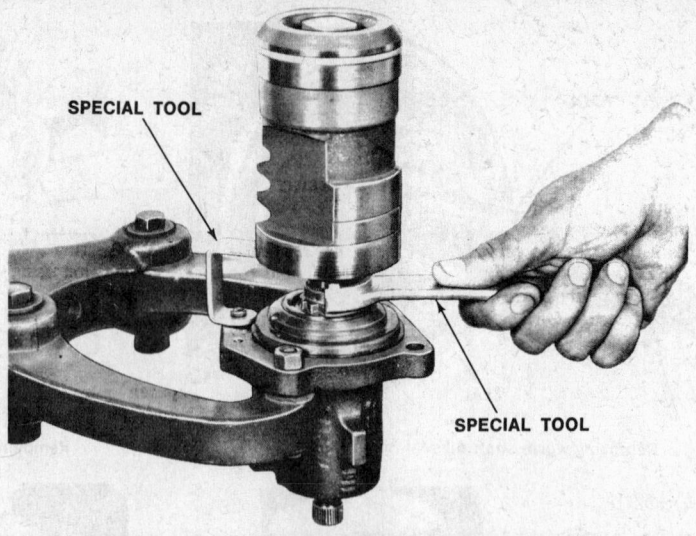

Removing worm bearing race nut (© Ford Motor Co.)

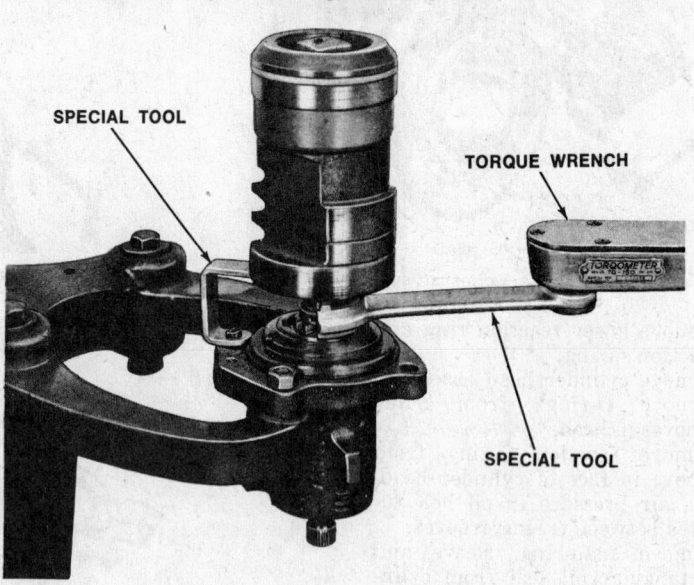

Installing worm bearing race nut (© Ford Motor Co.)

4. Remove the two sector shaft cover attaching screws, the brake line bracket and the identification tag.
5. Tap the lower end of the sector shaft with a soft-faced hammer to loosen it, then lift the cover and shaft from the housing as an assembly. Discard the O-ring.
6. Remove the four valve housing attaching bolts. Lift the valve housing from the steering gear housing while holding the piston to prevent it from rotating off the worm shaft.
7. Remove the valve housing and the lube passage O-rings and discard them.
8. Place the valve housing, worm and piston assembly in the bench mounted holding fixture with the piston on the top.
9. Rotate the piston upward (back off) 3½ turns.

10. Insert Tool T66P-3553-C (with the arm facing away from the piston) into a bolt hole in the valve housing. Rotate the arm into position under the piston.
11. Loosen the Allen head race nut set screw from the valve housing.
12. Using Tool T66P-3553-B, loosen the worm bearing race nut.
13. Lift the piston-worm assembly from the valve housing. During removal hold the piston to prevent it from spinning off at the shaft.
14. Change the power steering valve centering shim.
15. Install the piston-worm assembly into the valve housing. Hold the piston worm to prevent it from spinning off of the shaft.
16. Install the worm bearing race nut and torque to 2-8 in-lbs using Tool T66P-3553-B.

17. Install the race nut set screw (Allen head) through the valve housing.
18. Rotate the piston upward (back off) ½ turn and remove Tool T66P-3553-C.
19. Remove the valve housing, worm, and piston assembly from the holding fixture.
20. Position a new lube passage O-ring in the counterbore of the gear housing.
21. Apply vaseline to the teflon seal on the piston.
22. Place a new O-ring on the valve housing.
23. Slide the piston and valve into the gear housing being careful not to damage the teflon seal.
24. Align the lube passage in the valve housing with the one in the gear housing, and install but do not tighten the attaching bolts.
25. Rotate the ball nut so that the teeth are in the same place as the sector teeth. Tighten the four valve housing attaching bolts to 35-45 ft-lbs.
26. Position the sector shaft cover O-ring in the steering gear housing. Turn the input shaft as required to center the piston.
27. Apply vaseline to the sector shaft journal; then, position the sector shaft and cover assembly in the gear housing. Install the brake line bracket, steering gear identification tag and the two sector shaft cover attaching studs.
28. Position an in-lb torque wrench on the gear input shaft and adjust the meshload to approximately 4 in-lbs. Then, torque the sector shaft cover attaching studs to 55-70 ft-lbs.
29. After the cover attaching bolts have been tightened to specification, adjust the mesh load to 17 in-lbs with an in-lb torque wrench.

Steering Gear Disassembly

1. Hold the steering gear over a drain pan in an inverted position and cycle the input shaft several times to drain the remaining fluid from the gear.
2. Mount the gear in a soft-jawed vise.
3. Remove the lock nut from the adjusting screw.
4. Turn the input shaft to either stop then, turn it back approximately 1¾ turns to center the gear.
5. Remove the two sector shaft cover attaching studs, the brake line bracket and the identification tag.
6. Tap the lower end of the sector shaft with a soft-hammer to loosen it, then lift the cover and shaft from the housing as an assembly. Discard the O-ring.
7. Turn the sector shaft cover counterclockwise off the adjuster screw.
8. Remove the four valve housing attaching bolts. Lift the valve housing from the steering gear housing while holding the piston to prevent it from rotating off the worm shaft. Remove the valve housing and the lube passage O-rings and discard them.
9. Stand the valve body and piston on end with the piston end down. Rotate the input shaft counterclockwise out of the piston allowing the ball bearings to drop into the piston.
10. Place a cloth over the open end of the piston and turn it upside down to remove the balls.
11. Remove the two screws that attach the ball guide clamp to the ball nut and remove the clamp and the guides.
12. Install the valve body assembly in the holding fixture (do not clamp in a vise) and loosen the race nut screw (Allen head) from the valve housing and remove the worm bearing race nut.
13. Carefully slide the input shaft, worm and valve assembly out of the valve housing. Due to the close diametrical clearance between the spool and housing, the slightest cocking of the spool may cause it to jam in the housing.
14. Remove the shim from the valve housing bore.

Valve Housing

1. Remove the dust seal from the rear of the valve housing and discard the seal.
2. Remove the snap-ring from the valve housing.
3. Turn the fixture to place the valve housing in an inverted position.
4. Insert special tool in the valve body assembly opposite the seal end and gently tap the bearing and seal out of the housing. Discard the seal. Caution must be exercised when inserting and removing the tool to prevent damage to the valve bore in the housing.
5. Remove the fluid inlet and outlet tube seats with an EZ-out if they are damaged.
6. Coat the fluid inlet and outlet tube seats with vaseline and position them in the housing. Install and tighten the tube nuts to press the seats to the proper location.
7. Coat the bearing and seal surface of the housing with a film of vaseline.

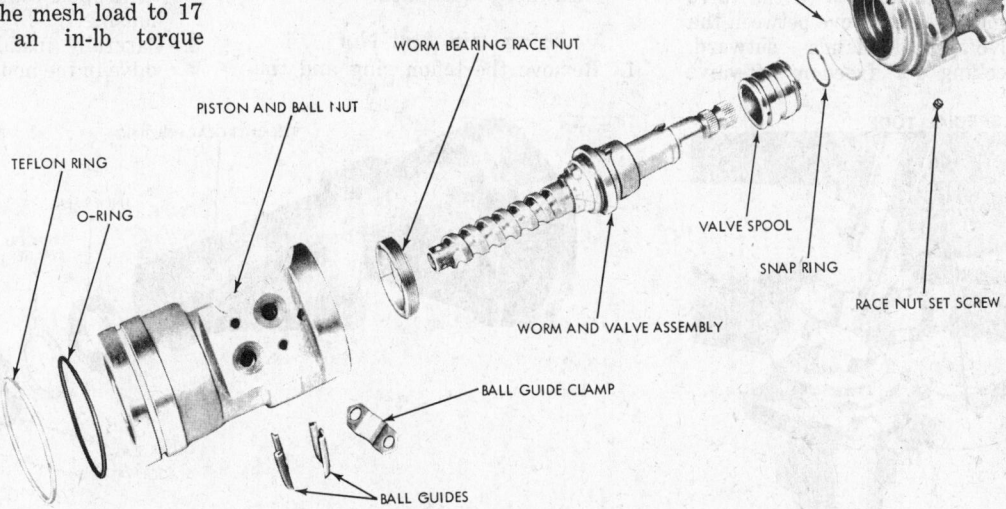

Ball nut and valve housing (© Ford Motor Co.)

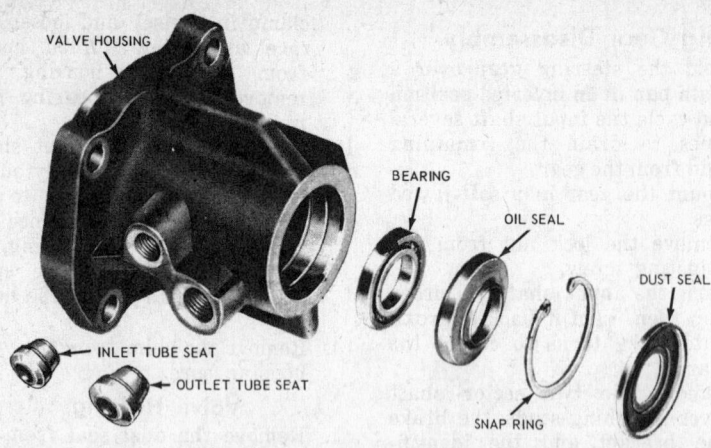

Valve housing disassembled (© Ford Motor Co.)

8. Seat the bearing in the valve housing. Make sure that the bearing is free to rotate.
9. Dip the new oil seal in gear lubricant; then, place it in the housing with the metal side of the seal facing outward. Drive the seal into the housing until the outer edge of seal does not quite clear the snap-ring groove.
10. Place the snap-ring in the housing; then, drive on the ring until the snap-ring seats in its groove to properly locate the seal.
11. Place the dust seal in the housing with the dished side (rubber side) facing out. Drive the dust seal into place. The seal must be located behind the undercut in the input shaft when it is installed.

Worm and Valve

1. Remove the snap-ring from the end of the actuator.
2. Slide the control valve spool off the actuator.
3. Install the valve spool evenly and slowly with a slight oscillating motion into the flanged end of valve housing with the valve identification groove between the valve spool lands outward, checking for freedom of valve movement within the housing working area. The valve spool should enter the housing bore freely and fall by its own weight.
4. If the valve spool is not free, check for burrs at the outward edges of the working lands in the housing and remove with a hard stone.
5. Check the valve for burrs and if burrs are found, stone the valve in a radial direction only. Check for freedom of the valve again.
6. Remove the valve spool from the housing.
7. Slide the spool onto the actuator making sure that the groove in the spool annulus is toward the worm.
8. Install the snap-ring to retain the spool. The beveled ID of the snap-ring must be assembled toward the spool.
9. Check the clearance between the spool and the snap-ring. The clearance should be between .0005-.035 inch. If the clearance is not within these limits, select a snap-ring that will allow a clearance of .002 inch.

Piston and Ball Nut

1. Remove the teflon ring and the O-ring from the piston and ball nut.
2. Dip a new O-ring in gear lubricant and install it on the piston and ball nut.
3. Install a new teflon ring on the piston and ball nut being careful not to stretch it any more than necessary.

Steering Gear Housing

1. Remove the snap-ring and the spacer washer from the lower end of the steering gear housing.
2. Remove the lower seal from the housing. Lift the spacer washer from the housing.
3. Remove the upper seal in the same manner as the lower seal. Some housings require only one seal and one spacer.
4. Dip both sector shaft seals in gear lubricant.
5. Apply lubricant to the sector shaft seal bore of the housing and position the sector shaft inner seal into the housing with the lip facing inward. Press the seal into place. Place a spacer washer (0.090 inch) on top of the seal and apply more lubricant to the housing bore.
6. Place the outer seal in the housing with the lip facing inward and press it into place. Then, place a 0.090 inch spacer washer on top of the seal.
7. Position the snap-ring in the housing. Press the snap-ring into the housing to properly locate the seals and engage the snap-ring in the groove.

Steering Gear Assembly

Do not clean, wash, or soak seals in cleaning solvent.

1. Mount the valve housing in the holding fixture with the flanged end up.
2. Place the required thickness valve spool centering shim in the housing.
3. Carefully install the worm and valve in the housing.

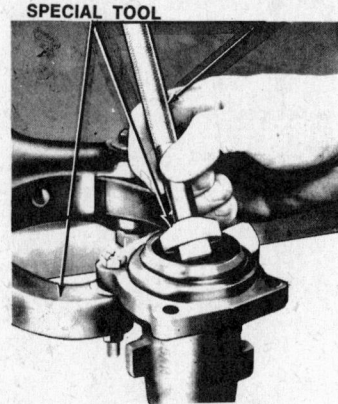

Removing bearing and oil seal
(© Ford Motor Co.)

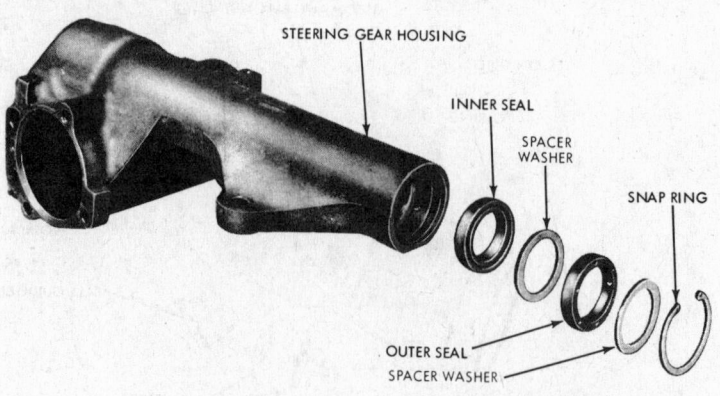

Steering gear housing (© Ford Motor Co.)

4. Install the race nut in the housing and torque it to 42 ft-lbs.

5. Install the race nut set screw (Allen head) through the valve housing and torque to 20-25 in-lbs.

6. Place the piston on the bench with the ball guide holes facing up. Insert the worm shaft into the piston so that the first groove is in alignment with the hole nearest to the center of the piston.

7. Place the ball guide in the piston. Place the 27 to 29 balls, depending on the piston design, in the ball guide turning the worm in a clockwise direction as viewed from the input end of the shaft. If all of the balls have not been fed into the guide upon reaching the right stop, rotate the input shaft in one direction and then in the other while installing the balls. After the balls have been installed, do not rotate the input shaft or the piston more than 3½ turns off the right stop to prevent the balls from falling out of the circuit.

8. Secure the guides in the ball nut with the clamp.

9. Position a new lube passage O-ring in the counterbore of the gear housing.

10. Apply petroleum jelly to the teflon seal on the piston.

11. Place a new O-ring on the valve housing.

12. Slide the piston and valve into the gear housing being careful not to damage the teflon seal.

13. Align the lube passage in the valve housing with the one in the gear housing and install but do not tighten the attaching bolts.

14. Rotate the ball nut so that the teeth are in the same plane as the sector teeth. Tighten the four valve housing attaching bolts to 35-45 ft-lbs.

15. Position the sector shaft cover O-ring in the steering gear housing. Turn the input shaft as required to center the piston.

16. Apply vaseline to the sector shaft journal then position the sector shaft and cover assembly in the gear housing. Install the brake line bracket, the steering identification tag and two sector shaft cover attaching bolts. Torque the bolts to 55-70 ft-lbs.

17. Attach an in-lb torque wrench to the input shaft. Adjust the mesh load to 17 in-lbs.

Ross HF-54 and HF-64 Integral Power Steering Gear

Adjustments

Unloader Valve Adjustment

This unloader valve adjustment is for right turn only on HF-54 Model gears and for both turns on HF-64 Model gears. Prior to performing the following procedure, obtain the vehicle's straight ahead position by driving the vehicle with hands off the steering wheel thus allowing the unit to find its own center. Now mark the steering column to steering wheel with chalk or masking tape.

1. Check the front wheel turning angles and adjust as required with the wheels off the ground.

2. Position the wheels straight ahead and lower the vehicle.

3. RIGHT TURN—HF-54 OR HF-64 GEARS: With the engine at idle, the vehicle standing still and the fluid at normal operating temperature, rotate the steering wheel to the right the prescribed number of turns.

```
H-54 ..................... 1¾
H-64 - 2 port ............. 1¼
H-64 - 4 port ............. 1½
```

Hold in this position.

4. Loosen the locknut and turn the unloader valve pressure adjusting screw until an audible hiss is heard. Tighten the locknut.

5. Return the wheel to a straight ahead position while the vehicle is moving. With the vehicle standing still, again rotate the steering wheel the prescribed turns; then, check for the audible hiss. Readjust if necessary as in Step 4 and check once more as in this step. It is important to remember that the HF-64 gear has a pitman arm stop for the right turn cast on the gear housing. The pitman arm must not contact this stop prior to contacting the unloader valve. When the hiss is heard during the adjustment, the clearance between the pitman arm and the cast stop should be 1/16 to ⅛-inch minimum.

6. *Left turn—HF-64 gear only:* Repeat Steps 3, 4, and 5 while rotating the steering wheel to the left 1¾ turns.

Sector Shaft Adjustment

1. Disconnect the drag link from the Pitman arm.

2. Center the steering wheel. Grasp the Pitman arm and check it for free movement (lash) between the sector shaft and the rack piston.

3. If free movement is noted (lash), remove the steering gear from the vehicle.

4. Loosen the sector shaft adjustment screw locknut on the side cover.

5. After rotating the input shaft through its full travel for a minimum of five cycles, adjust the sector shaft adjusting screw to provide 15-20 in-lb torque as the

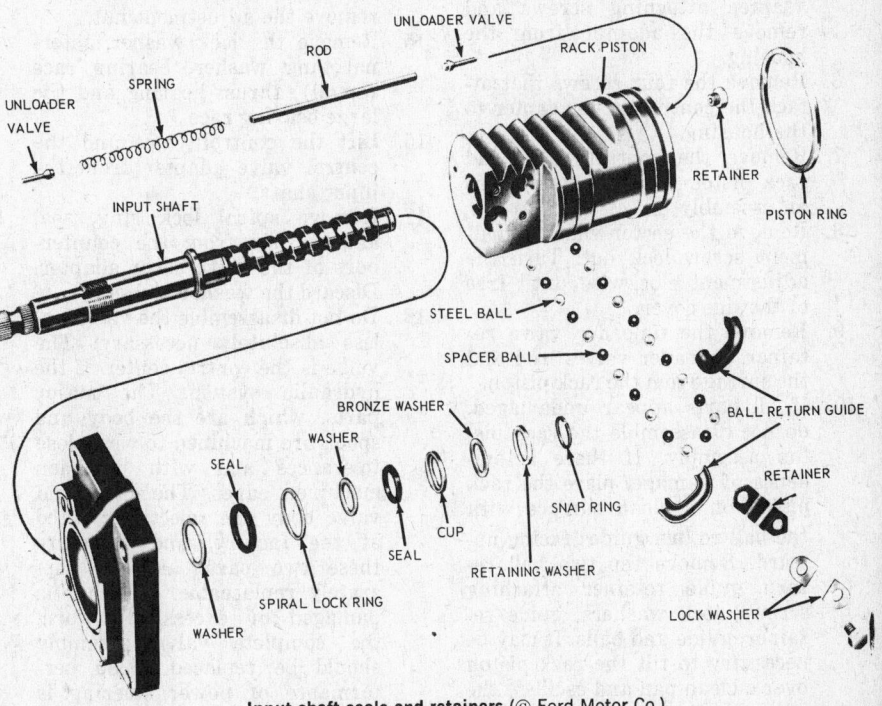

Input shaft seals and retainers (© Ford Motor Co.)

input shaft is rotated 90 degrees each side of center.

6. Back out the adjusting screw one turn and note the torque required to move the input shaft 90 degrees each side of the center position. Move the adjusting screw in to provide an increase in torque of 2-4 in-lb at a point within 45 degrees each side of center after the adjusting screw jam nut is first tightened snug. Now torque to a final 20-25 ft-lb. The input torque of the completely assembled gear, minus hydraulic oil, should not exceed 15 in-lb for the full travel of the output shaft.

7. Install the steering gear in the vehicle.

8. Connect the drag link to the Pitman arm.

9. Connect the pump lines and refill the system with the specified fluid.

Disassembly

1. Rotate the input shaft so the index mark on the end of the sector shaft is perpendicular to the center-line of the gear (straight-ahead position).

2. Remove the side cover attaching screws and washers.

3. Tap lightly on the end of the sector shaft with a soft hammer to disengage the side cover seal and allow the housing to drain.

4. Remove all nicks, burrs, rust and paint before removing the shaft. Lift the side cover and sector shaft from the housing as an assembly.

5. Remove the sector shaft seal adapter attaching screws and remove the adapter from the housing.

6. Remove the four screws that attach the control valve adapter to the housing.

7. Remove the control valve and rack piston from the housing as an assembly.

8. Remove the sector shaft adjustment screw lock nut. Turn the adjustment clockwise until free of the side cover.

9. Remove the unloader valve retainer, unloader valves, rod and the spring from the rack piston.

10. If all parts appear undamaged, do not disassemble the rack piston assembly. If there is evidence of damage, place the rack piston on a clean surface with the ball return guides facing upward. Remove the two ball return guide retainer attaching screws, lock washers, guide retainer, guide and balls. It may be necessary to tilt the rack piston over a clean pan and oscillate the worm shaft to empty the rack of

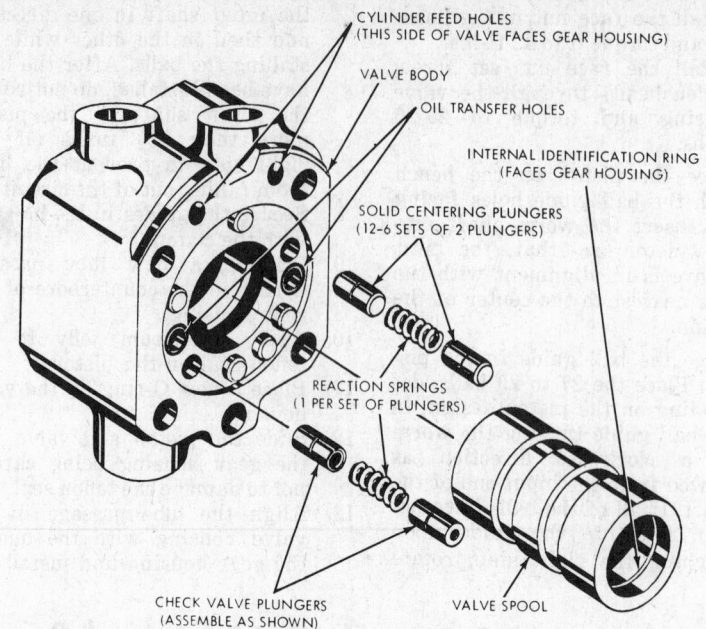

all the balls. Lift the worm shaft from the rack after all the balls have been removed.

11. Carefully hold the input shaft in a vise equipped with soft jaws. Remove the snap-ring, washer, bronze washer, cup, seal and washer. It may be necessary to cut the teflon cup off the shaft.

12. Remove the valve cover dirt and water seal.

13. Remove the four valve cover attaching screws. Lift the cover from the control valve.

14. Unstake the thrust bearing adjustment nut lock washer and remove the adjustment nut.

15. Remove the lock washer, internal tang washer, bearing race (small), thrust bearing and the large bearing race.

16. Lift the control valve and the control valve adapter from the input shaft.

17. Remove spiral lock ring, seal and washer from the counterbore of the valve cover adapter. Discard the washer and seal.

18. Do not disassemble the valve unless absolutely necessary. The valve is the control center of the hydraulic system. The major parts, which are the body and spool, are machined to very close tolerances and with precision machined edges. The spool and valve body are selectively fitted at the factory, and therefore these two parts are not separately replaceable. If either is damaged or excessively worn, the complete valve assembly should be replaced. Good performance of power steering is not assured if a mismatched

valve spool and body are used. Care should be exercised in the handling of these parts to prevent damage. Sealing edges of the valve bore and the spool should not be broken. This will result in excessive leakage and reduced hydraulic power. If valve parts should drop out during gear disassembly, reassemble the valve as follows:

a. Clean all parts with a clean petroleum base solvent and blow dry with clean, dry air.

b. Insert the valve spool in the control valve making certain that the machined identification groove in the ID of one end of the spool is toward the gear housing.

c. There are 7 sets of plungers, each set having one reaction spring. Insert the 6 solid centering plunger sets first along with one spring per set. The remaining plunger set should be inserted in the valve body with the small hole on each plunger facing outboard.

Cleaning and Inspection

1. All parts should be cleaned in a clean petroleum base solvent and blown dry with clean dry air. Avoid wiping parts with a cloth, since lint may cause binding and sticking of closely fitted components.

2. Inspect the worm grooves in the rack piston and on the input shaft for wear scores. Inspect the OD of the rack piston and the ring or teeth for wear or scores. On the HF-64 gear, the ball nut and input shaft are serviced as a

Hydraulic valve control assembly (© Ford Motor Co.)

matched assembly. Therefore, both must be replaced if either are worn or damaged. The rack piston is not matched.

3. Inspect the inside ends of the ball return guides for wear or damage.
4. Inspect the housing bore for wear or scores or being cracked and replace as required.
5. Inspect the sector shaft teeth for wear or the bearing surfaces for wear or scores.
6. Replace the sector shaft bearings if worn or damaged. Note the sector shaft bearing in the

side cover is replaced as part of the side cover assembly.

7. Replace all seals at time of disassembly.

Assembly

1. Lubricate all rubber parts prior to assembly.
2. If the sector shaft bearing was removed from the steering gear housing, install the snap-ring in the outboard side of the housing (HF-54 gear only).
3. Place the steering gear housing in a press with the side cover area on a wood block to prevent

damage to the machined area.

4. Position the bearing on the housing with the numbered end facing up. Carefully press the bearing into the housing until the outer surface is flush. Use a tool that pilots in the ID of the bearing and contacts the bearing end surface.
5. Coat the unloader valve pressure adjusting screw O-ring liberally with clean grease or oil. Carefully slide it into the groove on the non-threaded end of the adjusting screw.
6. Thread the adjusting screw into the lower end of the housing leaving 7/8 inch of the screw exposed. Install the lock nut on the adjusting screw and tighten it securely.
7. Carefully secure the input shaft in a vise equipped with soft jaws to permit access to both ends of the shaft.

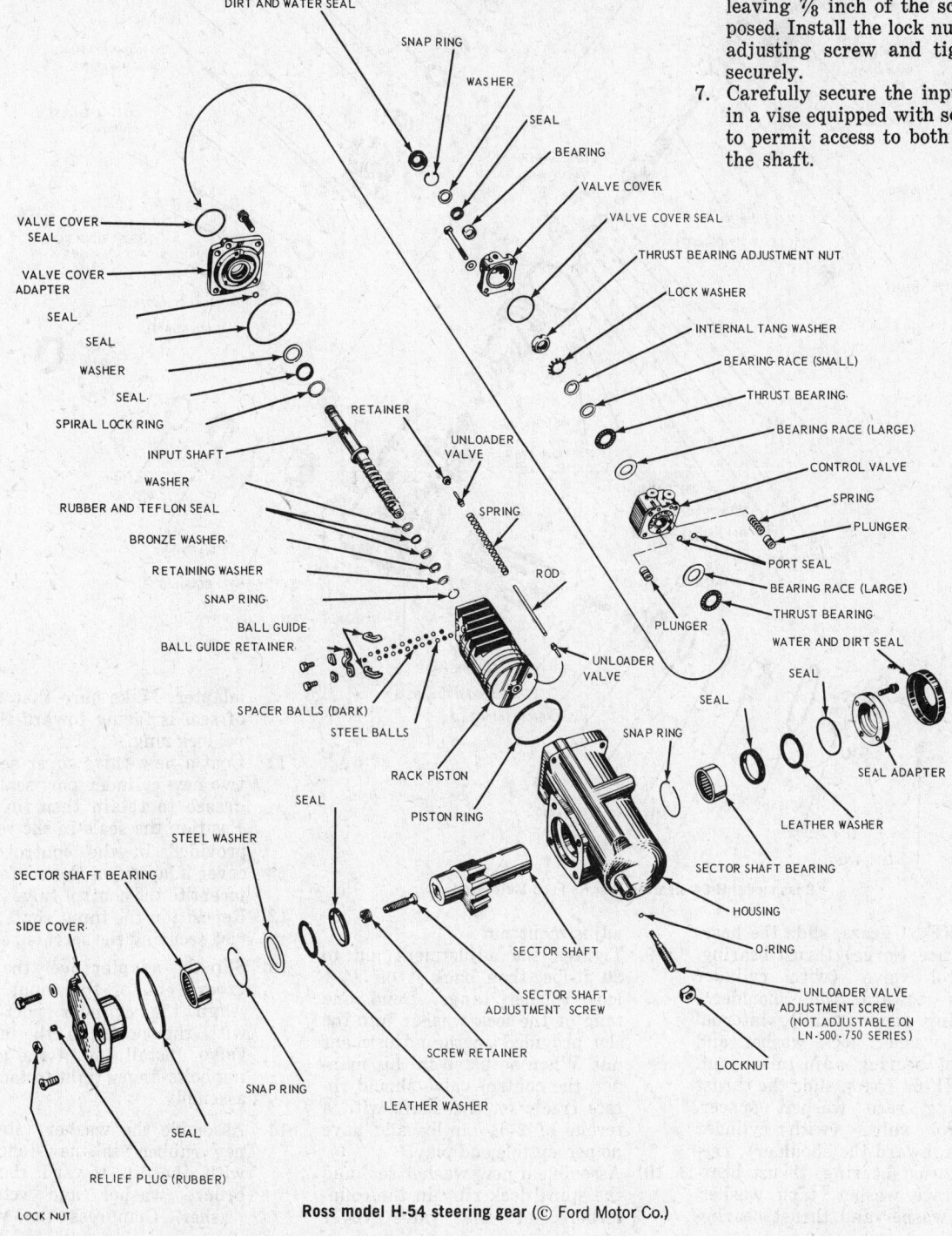

Ross model H-54 steering gear (© Ford Motor Co.)

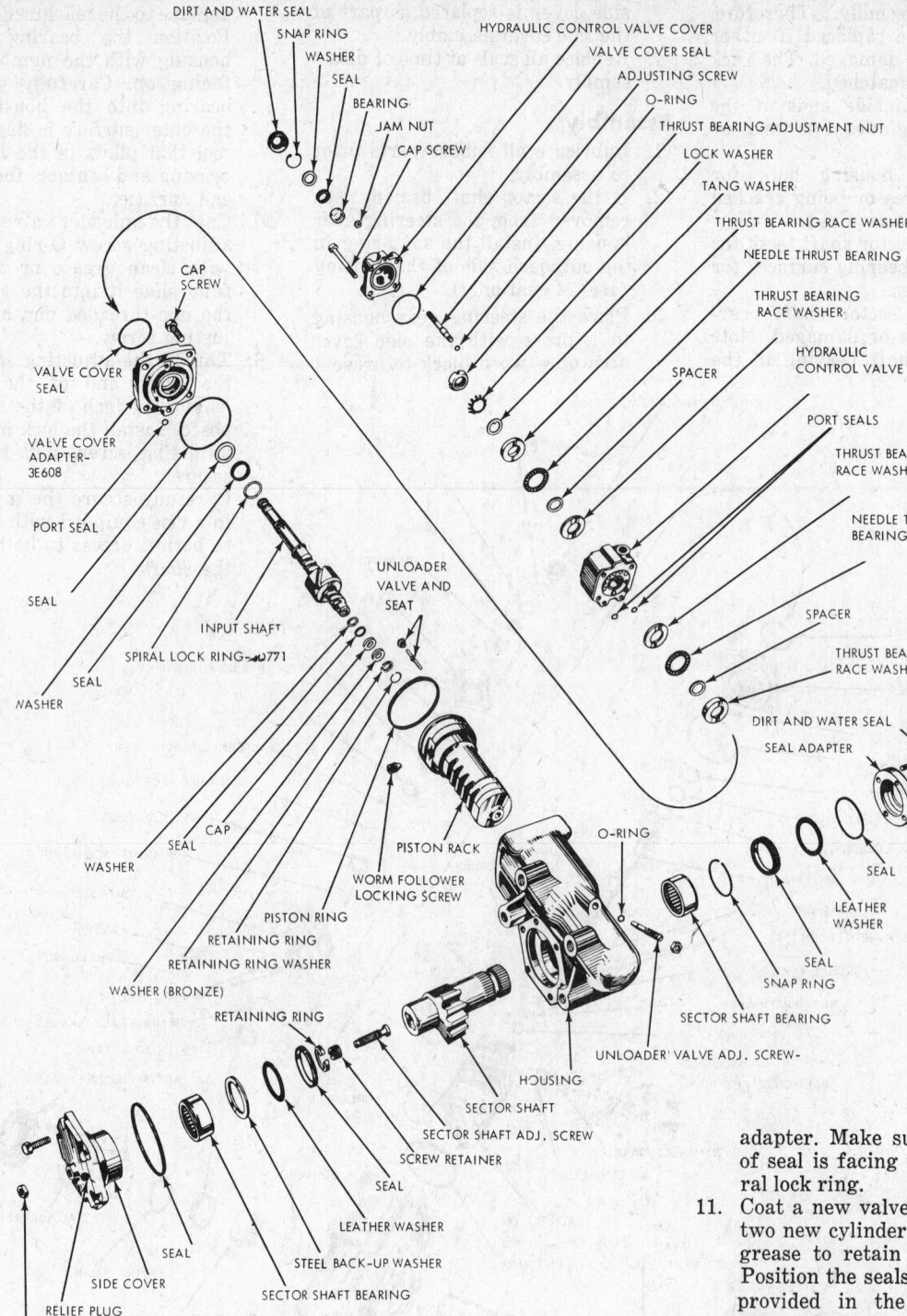

DIRT AND WATER SEAL
SNAP RING
WASHER
SEAL
BEARING
JAM NUT
CAP SCREW

HYDRAULIC CONTROL VALVE COVER
VALVE COVER SEAL
ADJUSTING SCREW
O-RING
THRUST BEARING ADJUSTMENT NUT
LOCK WASHER
TANG WASHER
THRUST BEARING RACE WASHER
NEEDLE THRUST BEARING
THRUST BEARING RACE WASHER
HYDRAULIC CONTROL VALVE
PORT SEALS
THRUST BEARING RACE WASHER
NEEDLE THRUST BEARING
SPACER
THRUST BEARING RACE WASHER
DIRT AND WATER SEAL
SEAL ADAPTER
SEAL
LEATHER WASHER
SEAL
SNAP RING
SECTOR SHAFT BEARING
UNLOADER VALVE ADJ. SCREW

CAP SCREW
VALVE COVER SEAL
VALVE COVER ADAPTER-3E608
PORT SEAL
SEAL
SEAL
WASHER
INPUT SHAFT
SPIRAL LOCK RING—40771
UNLOADER VALVE AND SEAT
SPACER

CAP
SEAL
WASHER
PISTON RACK
WORM FOLLOWER LOCKING SCREW
PISTON RING
RETAINING RING
RETAINING RING WASHER
WASHER (BRONZE)
RETAINING RING
O-RING
HOUSING
SECTOR SHAFT
SECTOR SHAFT ADJ. SCREW
SCREW RETAINER
SEAL
LEATHER WASHER
STEEL BACK-UP WASHER
SECTOR SHAFT BEARING
SEAL
SIDE COVER
RELIEF PLUG
LOCK NUT

Ross model H-64 steering gear (© Ford Motor Co.)

8. On HF-54 gears, slide the bearing race (large) thrust bearing, control valve (with cylinder ports toward the shoulder), bearing race (small), internal tang washer, lock washer and thrust bearing adjustment nut. On HF-64 gears, slide the thrust bearing race washer, spacer, control valve (with cylinder ports toward the shoulder), needle thrust bearing, thrust bearing race washer, tang washer, lock washer and thrust bearing adjustment nut.

9. Tighten the adjustment nut to 20 ft-lbs, then back it off ½-1 lock washer tangs. Bend one tang of the lock washer into the slot provided on the adjustment nut. When adjusted in this manner, the control valve should rotate freely on the shaft with a torque of 2-3½ in-lbs and have no perceptible end play.

10. Assemble a new washer seal and the spiral lock ring in the counterbore of the valve cover

adapter. Make sure that the lip of seal is facing toward the spiral lock ring.

11. Coat a new valve cover seal and two new cylinder port seals with grease to retain them in place. Position the seals in the recesses provided in the control valve cover adapter on the surface adjacent to the control valve.

12. Reposition the input shaft in the vise securing the serrated end.

13. Slip the adapter over the worn groove end of the input shaft. Align the cylinder port seals with the ports in the control valve. Install one of the attaching bolts finger tight to facilitate assembly.

14. Assemble the washer (steel), a new rubber seal, new teflon cup with the lip toward the seal, bronze washer and retaining washer. Compress the washer

and seal, then install the snap-ring on the end of the input shaft. Make sure that the snap-ring is fully seated in the groove and the recessed area of the retaining washer.

15. Secure the rack piston in a soft-jawed vise with the ball guide holes facing upward.

16. Carefully expand the piston ring and install it in the piston groove.

17. Place the two unloader valves, spring, and rod in the rack piston. Apply a drop of sealer to the threads of the retainer. Install and torque the retainer to 25 ft-lbs.

18. Coat the input shaft seal at the end of worm with grease and place it in the rack piston bore.

19. Assemble sixteen balls while rotating the input shaft counterclockwise. The black spacer balls and the polished steel balls must be installed alternately. Coat the ball return guides with grease to retain the balls, then install the six remaining balls in the guides making sure that the balls in the guide alternate with the last balls installed in the rack piston. If a ball is lost, no more than three black spacer balls may be used for replacement. Secure the guide retaining clip to the rack piston with two screws and washers. Torque the screws to 30-35 ft-lbs and bend the tab of the locking washer against the flat.

20. Grip the serrated end of the sector shaft in a soft-jawed vise.

21. Coat the head of the sector shaft adjusting screw with grease. Position the head of adjusting screw into the slot in the end of sector shaft.

22. Install a new sector shaft adjustment screw retainer in the end of shaft. Tighten the retainer to permit free rotation of screw without perceptible end play. Stake the retainer in the two slots provided and recheck the rotation effort.

23. If the pressure relief plug has been removed or ruptured, press a new one into the side cover until it is flush with the surface.

24. Assemble the snap-ring, steel washer (with taper toward the snap-ring), leather washer and the two piece seal into the side cover. The seal has Oil Side molded into one side and must be visible after installation.

25. Coat the end of sector shaft with lubricant. Rotate the sector shaft adjusting screw counterclockwise to thread it into the side cover. Rotate the screw until a firm stop is reached. Make sure that the shaft seal has not fallen out of position.

26. Place the outer seal in the seal adapter. Then install the leather washer and inner seal making sure that the side having the mold "Oil Side" is visible after installation.

27. If the input shaft needle bearing has been removed from the control valve cover it must be installed with a tool that will pilot in the bearing and have clearance in the cover bore. The bearing must be pressed from the part number end, and to a depth of $1\frac{1}{8}$ inches from the face of the valve cover. After installation of the bearing, make sure that all rollers rotate freely.

28. Install the seal on the control valve cover with the lip facing toward the needle bearing. Coat the washer with grease and install it on the cover. Install the snap-ring to secure the seal and washer.

29. Pack the new dirt and water seal with grease and install it on the control valve cover.

30. Secure the steering gear housing in a vise equipped with soft jaws.

31. Lubricate the steering gear housing bore. Start the rack piston into the bore, then compress the ring and move the piston into position so that the teeth are visible through the side cover opening. Install the four adapter - to - housing attaching bolts. Remove the one bolt that was previously installed.

32. Lubricate a new valve cover seal with grease and position it in the recess of the valve cover.

33. Slide the valve cover onto the input shaft and install the four cover-to-control valve attaching bolts.

34. Rotate the input shaft as required to align the center tooth of the rack piston (marked tooth) with the side cover opening.

35. Lubricate a new side cover O-ring and position it on the side cover.

36. Position the sector shaft and side cover to the steering gear housing making sure that the center tooth (marked tooth) engages the center space (marked space).

37. Install the four side cover attaching bolts and lock washers. Torque the bolts to 45-55 ft-lbs.

38. Adjust sector shaft adjustment screw as outlined in the Sector Shaft Adjustment.

39. Cover the sector shaft serrations with a layer of scotch tape to prevent damage to the seal in the adapter.

40. Position the adapter over the sector shaft and on the housing. Install and tighten the attaching bolts.

41. Pack the seal adapter outer seal with grease, then install it on the adapter to prevent water entry.

Final Checks

1. After rotating the input shaft through its full travel for a minimum of five cycles, recheck the sector shaft adjustment. No rotational lash or bind of the sector shaft in center position is permissible.

2. If the gear is properly assembled and adjusted, the input torque of the empty gear should not exceed 15 in-lbs over full travel of 95 degrees at the output shaft.

3. Reverse-torque applied to output shaft for full gear travel should not exceed 50 ft-lbs.

International—Semi Integral Power Steering Gear

This gear is a semi-integral hydraulic steering gear which incorporates a hydraulic control valve on a single stud cam and lever mechanical steering gear. When the steering wheel is turned it actuates the control valve. The valve then directs fluid from a pump to a power cylinder located in the linkage.

Adjustments

The following are the two principal adjustments on this type gear:

1. Adjustment of the needle thrust bearings on the cam shaft on each side of the centering washer assembly.

2. Adjustment of the tapered stud in the cam groove for backlash.

If adjustments are made with the steering gear mounted in the vehicle, free the steering gear of all load by disconnecting the drag link from the steering gear arm. Disconnect the coupling if the gear is the stub shaft type.

Thrust Bearing Adjustment

If the gear is mounted in the vehicle it may be necessary to remove any parts restricting the removal of the actuator housing.

1. Remove the four screws retaining the valve to the actuator housing and remove the valve.

2. Pull out the actuator lever.

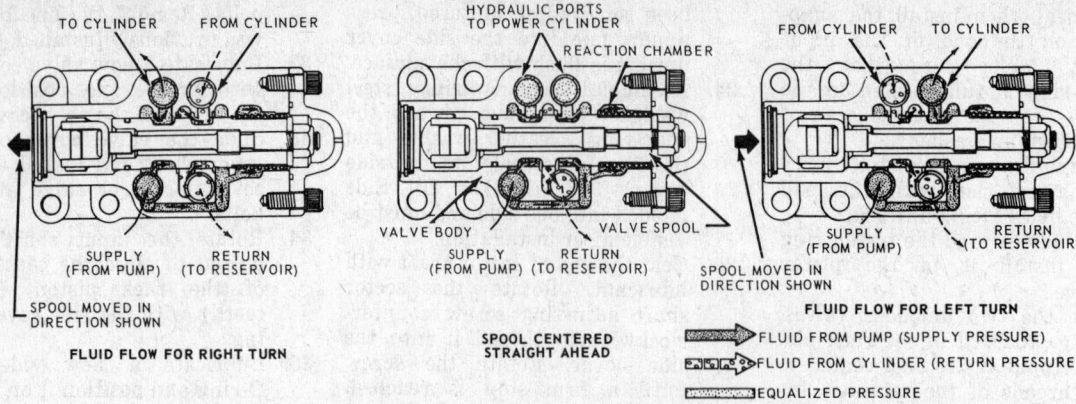

Semi-Integral Power Steering Gear Valve Power Flow

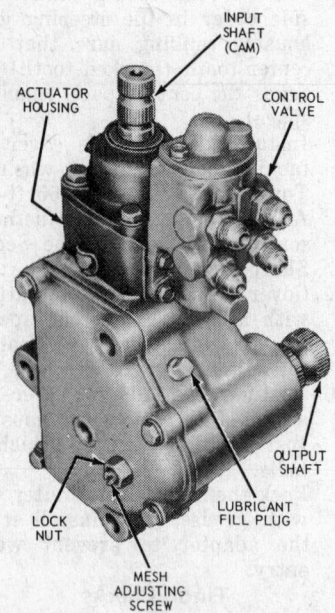

Semi integral power steering gear assembly
(© Ford Motor Co.)

3. Remove the four actuator housing mounting screws and remove the actuator housing.
4. Remove the adjusting nut, tongued washer, upper thrust washer and needle bearing. Wash all parts in solvent and coat with a light oil.
5. Reassemble the thrust bearing parts, tongued washer and adjusting nut.
6. Turn the steering gear off its center position to free the levershaft stud in the cam groove.
7. Torque the nut to 10 ft. lbs. Back the nut off 10°-20° and restake the lip of the adjusting nut in the slot of the cam shaft.
8. Position the actuator housing to the gear housing. Make sure that the pin engages in the hole of the actuator housing.
9. Position the actuator lever in the actuator housing making sure that the slotted end of the lever straddles the centering washer assembly and position on the slot in the other end of the lever so that the pin in the clevis of the valve fixture rod will fit freely into it when mounting the valve.
10. Mount the valve and gasket on the actuator housing making sure the valve spool clevis pin fits freely into the slot of the actuator lever. Install all four mounting screws and tighten lightly in rotation, then torque to 10-15 ft. lbs.

NOTE: Careless tightening may cause the valve spool to be pulled off center by actuator lever interference with the clevis pin.

Backlash adjustment (stud in cam groove)

1. Turn the adjusting screw clockwise in the side cover until a very slight drag is felt when turning the gear through the mid position.
2. Lock the adjusting screw with the locknut and turn the gear from the extreme left to the extreme right and back again to check adjustment.

Steering Gear

Disassembly

1. Unscrew the locknut and adjusting screw.
2. Slide the sub-assembly of the side cover and the levershaft from the housing and remove the gasket.
3. Remove the four screws retaining the valve to the actuator housing and remove the valve.
4. Pull out the actuator lever.
5. Remove the four actuator housing mounting screws and remove the actuator housing.
6. Remove the cam and wheel tube and centering washer assembly as a unit from the housing.
7. Remove the adjusting nut then remove the tongued spacer washer and upper thrust washer and needle bearing.
8. Remove the centering washer assembly and bearing and without losing any springs remove the spring retainer.

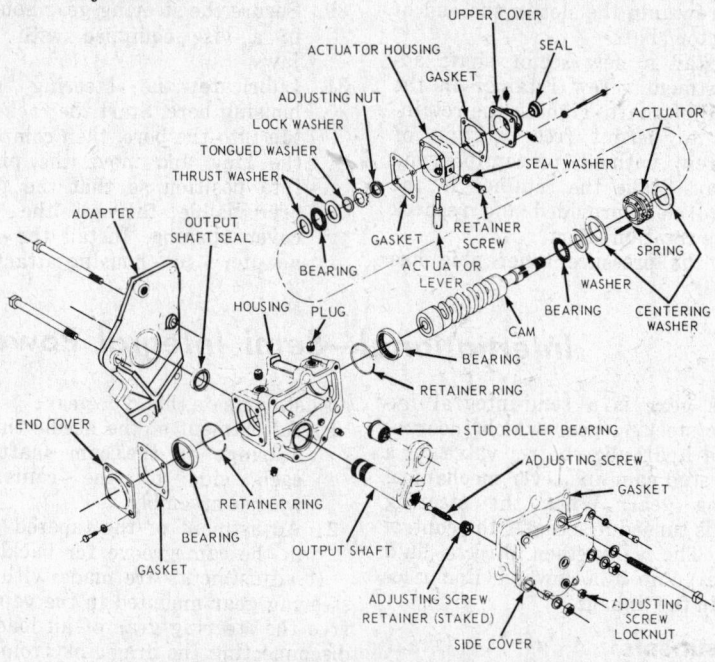

Exploded view—semi integral power steering gear (typical) (© Ford Motor Co.)

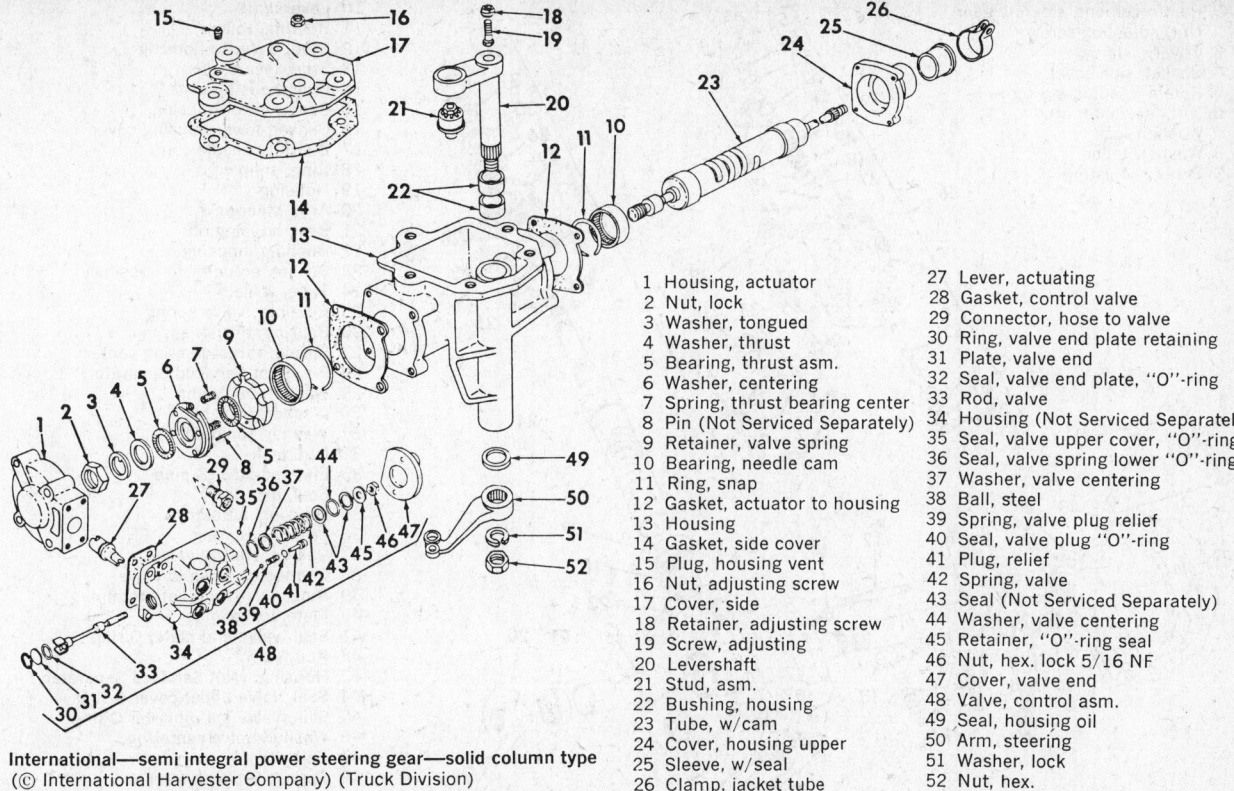

International—semi integral power steering gear—solid column type
(© International Harvester Company) (Truck Division)

1 Housing, actuator
2 Nut, lock
3 Washer, tongued
4 Washer, thrust
5 Bearing, thrust asm.
6 Washer, centering
7 Spring, thrust bearing center
8 Pin (Not Serviced Separately)
9 Retainer, valve spring
10 Bearing, needle cam
11 Ring, snap
12 Gasket, actuator to housing
13 Housing
14 Gasket, side cover
15 Plug, housing vent
16 Nut, adjusting screw
17 Cover, side
18 Retainer, adjusting screw
19 Screw, adjusting
20 Levershaft
21 Stud, asm.
22 Bushing, housing
23 Tube, w/cam
24 Cover, housing upper
25 Sleeve, w/seal
26 Clamp, jacket tube

27 Lever, actuating
28 Gasket, control valve
29 Connector, hose to valve
30 Ring, valve end plate retaining
31 Plate, valve end
32 Seal, valve end plate, "O"-ring
33 Rod, valve
34 Housing (Not Serviced Separately)
35 Seal, valve upper cover, "O"-ring
36 Seal, valve spring lower "O"-ring
37 Washer, valve centering
38 Ball, steel
39 Spring, valve plug relief
40 Seal, valve plug "O"-ring
41 Plug, relief
42 Spring, valve
43 Seal (Not Serviced Separately)
44 Washer, valve centering
45 Retainer, "O"-ring seal
46 Nut, hex. lock 5/16 NF
47 Cover, valve end
48 Valve, control asm.
49 Seal, housing oil
50 Arm, steering
51 Washer, lock
52 Nut, hex.

Inspection

1. Clean all parts with solvent and blow dry with compressed air.
2. Examine all parts for chips, wear or pitting.
NOTE: The cam is copper plated. The operation of the stud in the cam groove will wear away the copper plating. This is a normal condition.

Assembly

1. Use new gaskets and seals.
2. Replace any needle bearings in the ends of the housing that were removed.
3. Install the valve spring retainer over the wheel tube and then assemble in the following order: Springs into retainer, bearing, centering washer assembly, bearing, thrust washer, tongued washer, adjusting nut.
4. Adjust the bearings (See "Thrust Bearing Adjustment").
5. Assemble the cam in the housing making sure it rotates freely.
6. Use a new gasket and install the actuator housing and pin to the gear housing. Make sure the pin engages in the hole of the actuator housing.
7. Assemble the retainer, screw and seal washer to the actuator housing. Screw the retainer in far enough to eliminate play between the adjusting screw and levershaft but the screw must be free to rotate.
8. Assemble the side cover to the adjusting screw. Position the gasket and install the side cover

and levershaft in the housing.
9. Install the locknut to the adjusting screw and adjust the screw properly (see "Backlash Adjustment").
NOTE: On stub shaft type gears, assemble the lubricated oil seal with the lip out, into the counterbore of the actuator housing.

Control Valve

Disassembly

1. Remove the retainer ring, cover plate and O-ring seal.
2. Remove the end cover from the valve body.

3. Remove the elastic stop nut and washer from the end of the fixture rod and pull the fixture rod out of the spool.
4. Push the spool out in the same direction to permit removal of the centering washers, O-ring and centering spring from the valve body.
5. Remove the O-ring from the spool.
6. Remove the by-pass valve parts, plug assembly and spring and ball.

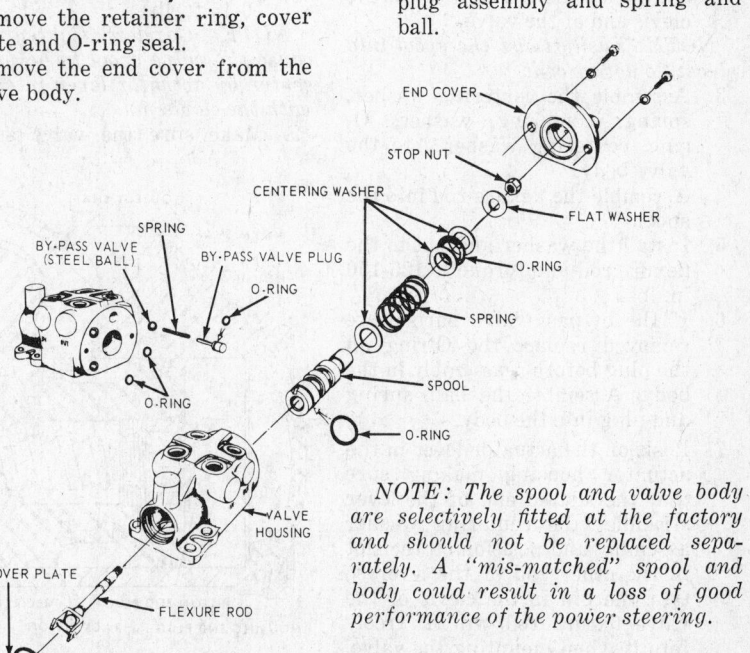

NOTE: The spool and valve body are selectively fitted at the factory and should not be replaced separately. A "mis-matched" spool and body could result in a loss of good performance of the power steering.

Exploded view—semi integral power steering gear control valve (typical) (© Ford Motor Co.)

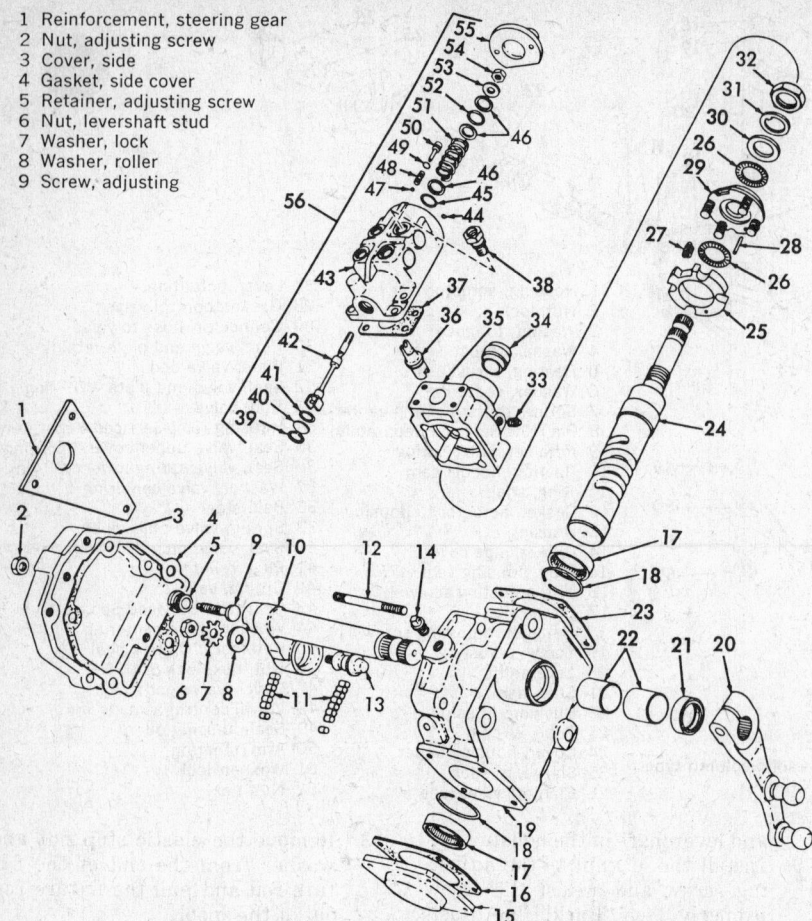

1 Reinforcement, steering gear
2 Nut, adjusting screw
3 Cover, side
4 Gasket, side cover
5 Retainer, adjusting screw
6 Nut, levershaft stud
7 Washer, lock
8 Washer, roller
9 Screw, adjusting
10 Levershaft
11 Bearing, roller
12 Stud, cover to housing
13 Stud, levershaft
14 Plug, housing vent
15 Cover, lower housing
16 Gasket, lower housing cover
17 Bearing, needle cam
18 Ring, snap
19 Housing
20 Arm, steering
21 Seal, housing oil
22 Bushing, housing
23 Gasket, actuator to housing
24 Tube, w/cam
25 Retainer, valve spring
26 Bearing, thrust, assy.
27 Spring, thrust bearing center
28 Pin, (Not Serviced Separately)
29 Washer, centering
30 Washer, thrust
31 Washer, tongued
32 Nut, lock
33 Plug, sq-hd 1/8 pipe
34 Seal, oil
35 Housing, actuator
36 Lever, actuating
37 Gasket, control valve
38 Connector, hose to valve
39 Ring, valve end plate retaining
40 Plate, valve end
41 Seal, valve end plate, O-ring
42 Rod, valve
43 Housing, (Not Serviced Separately)
44 Seal, valve upper cover O-ring
45 Seal, valve spring lower O-ring
46 Washer, valve centering
47 Ball, steel
48 Spring, valve plug relief
49 Seal, valve plug O-ring
50 Plug, relief
51 Spring, valve
52 Seal, (Not Serviced Separately)
53 Retainer, O-ring seal
54 Nut, hex, lock 5/16 NF
55 Cover, valve end
56 Valve, control, assy.

International—semi integral power steering gear—jointed column type
(© International Harvester Company) (Truck Division)

Assembly

1. Install the O-ring on the spool.
2. Apply a light coat of lubricating oil to the spool and O-rings and position the spool in the valve body with the O-ring toward the clevis end of the valve.

NOTE: Easily twist the spool into place: do not force it.

3. Assemble the centering washer, spring, centering washer, O-ring, centering washer into the valve body.
4. Assemble the flexure rod into the spool.
5. Install the washer and nut to the flexure rod and torque to 125-150 in. lbs.
6. If the by-pass valve parts were removed replace the O-ring on the plug before reassembly in the body. Assemble the ball, spring and plug into the body.
7. Position the actuator lever in the actuator housing making sure that the slotted end of the lever straddles the centering washer assembly and position on the slot in the other end of the lever so that the pin in the clevis of the valve fixture rod will fit freely into it when mounting the valve.
8. Mount the valve and gasket on the actuator housing making sure the valve spool clevis pin fits freely into the slot of the actuator lever. Install all four mounting screws and tighten lightly in rotation then torque to 10-15 ft. lbs.

NOTE: Careless tightening may cause the valve spool to be pulled off center by actuator lever interference with the clevis pin.

9. Make sure the valve spool actuates before assembling the end covers.

NOTE: Valve spool travel should be a minimum of .065 in. each direction for full flow.

10. Position the O-rings on the end of the valve body and install the end cover.
11. Install the O-ring, cover plate, and retaining ring to the clevis end of the valve.

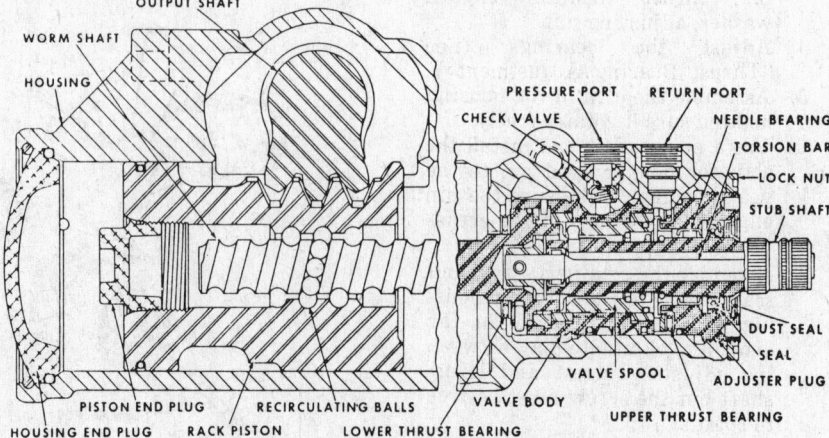

Rotary Valve Power Steering Gear (Cross-Section View)
(©International Harvester Company) (Truck Division)

Truck Front End Trouble Diagnosis

Steering Wheel Spoke Position Not Centered

1. Start with steering gear set on high-spot.
2. Check for proper toe-in.
3. Check for proper relation between lengths of each tie rod.

Front End Rides Hard

1. Improper tires.
2. Improper air pressure.
3. Shock absorbers too severe or malfunctioning.

Rides Too Soft

1. Improper tire pressure.
2. Loss of spring load-rate, (weak springs).
3. Weak or leaking shock absorbers.

Truck Steers to One Side At All Times

1. Incorrect caster angle.
2. Incorrect camber angle.
3. Incorrect kingpin inclination or wheel support angle.
4. Unequal air pressure or unequal tread.
5. Unequal or one-side brake drag.
6. Unequal shock absorber control.
7. Bent or damaged steering suspension components.
8. Uneven or weak spring condition, front or rear, causing truck to sit unevenly.
9. Improper tracking.

Truck Steers to One Side Only When Brakes Are Applied

1. Improper brake adjustment, damaged or worn shoes or anchors.
2. Grease or foreign substance on brake lining.
3. Excessive wear or bent condition in suspension components.

Truck Steers Down Off Crowned Road But Normally on Flat Road

1. Excessive positive camber at one or both sides.

2. Weak or uneven shock absorber action.
3. Excessive or unequal wear in suspension components.

Truck Wanders—Steers Erratically

1. Incorrect caster.
2. Improper tire pressure or unequal treads.
3. Excessively worn or damaged suspension components.
4. Power steering gear damaged, causing power assist to function abnormally.

Truck Steers Hard

1. Binding steering or suspension parts.
2. Improper lubrication.
3. Improper (too large) tires.
4. Low tire pressure.

Tires Cup on Outside Edge With Ripple Wear Pattern

1. Generally incorrect camber toe-in.

Tires Wear Unevenly in Center and Faster Than Outer Edges

1. Generally too much tire pressure.

Tires Wear and Scuff on Both Outer Edges, Not in Center

1. Generally low tire pressure.

Uneven Outer Wear—Center and Inner Edge Wear Normal

1. With adjustments normal this is usually caused by driving into turns at too high speed. Do not confuse with outer edge cupping.

Unequal Tire Wear Between Front Wheels

1. Unequal tire pressure.
2. Unequal tire quality or size.
3. Bent or worn steering suspension components.
4. Improper tracking.

Tire Squeal on Turns

1. Low tire pressure.
2. Driving into turn too fast.
3. Damaged or misaligned parts causing improper front wheel toe-out steering radius.
4. Improper camber adjustment.

Wheel Bounce

1. Unequal tire pressure.
2. Unbalanced wheels or tires.
3. Excessive wheel or tire run-out.
4. Weak or broken front springs.
5. Inoperative shock absorber.
6. Loose or damaged wheel bearings.

Noisy Front End

1. Lack of, or improper, lubrication.
2. Loose tie rod ends.
3. Worn spring shackle.
4. Inoperative shock absorber.
5. Loose U-bolts or clips.
6. Broken spring.
7. Worn universal (FWD).
8. Worn differential (FWD).

Lubricant Leaking Into Drums

1. Too much differential lubricant (FWD).
2. Clogged axle housing vent (FWD).
3. Damaged or worn universal driveshaft oil seal (FWD).
4. Loose steering knuckle flange bearings (FWD).
5. Defective outer oil seal.
6. Rough spindle bearing surface.
7. Wheel bearings over-packed or using wrong type lubricant.
8. Clogged wheel bearing oil slinger drain.
9. Cracked steering knuckle outer flange.

	RAPID WEAR AT SHOULDERS	RAPID WEAR AT CENTER	CRACKED TREADS	WEAR ON ONE SIDE	FEATHERED EDGE	BALD SPOTS
CONDITION						
CAUSE	UNDER INFLATION	OVER INFLATION	UNDER-INFLATION OR EXCESSIVE SPEED	IMPROPER CAMBER	INCORRECT TOE	WHEEL UNBALANCED
CORRECTION	ADJUST PRESSURE TO SPECIFICATIONS WHEN TIRES ARE COOL			ADJUST CAMBER TO SPECIFICATIONS	ADJUST FOR TOE-IN 1/8 INCH	DYNAMIC OR STATIC BALANCE WHEELS

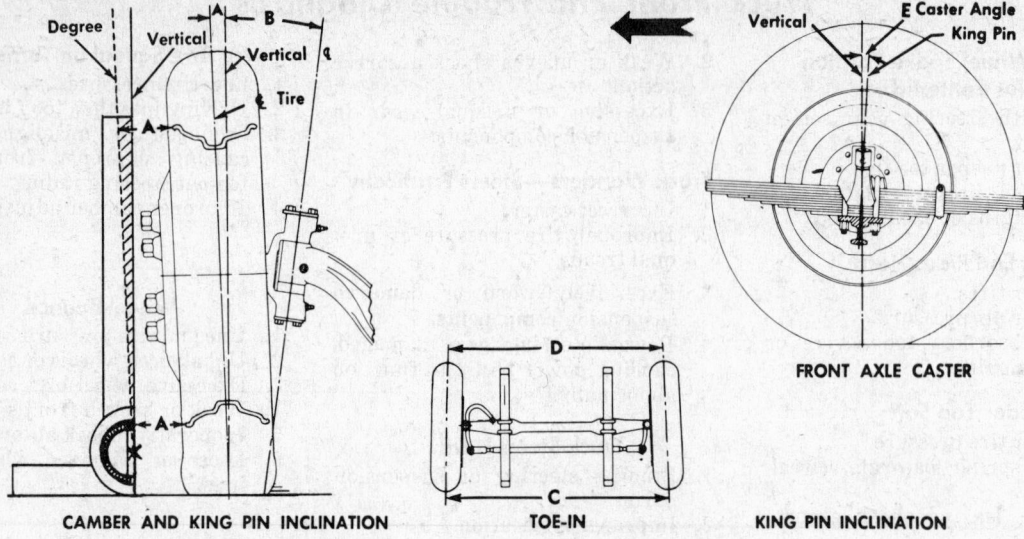

*A Camber (degrees positive)
*B King pin inclination (degrees)
*C Minus D Toe-in (inches)

*E Caster (degrees positive)
*Refer to "Specifications" for correct dimensions

CAMBER AND KING PIN INCLINATION   TOE-IN   KING PIN INCLINATION

FRONT AXLE CASTER

Front end alignment chart (© G.M.C.)

Wheel Alignment

For a truck to have safe steering control with a minimum of tire wear, certain established rules must be followed. These rules fix the values of planes, angles and radii relative to each other and to truck and tire dimensions. Some factors are built in, with no provision for adjustment; others are adjustable within limits. The entire system depends upon all value factors, separately and combined. It is therefore difficult to change some of the established settings without influencing others.

This system is called steering geometry or wheel alignment and requires a complete check of all the factors involved. Definitions of these factors and the effect each one has on the truck are given in the following paragraphs.

For adjustment data relative to each separate truck and year, refer to the individual truck sections.

Steering Wheel Position

Always check steering wheel alignment in conjunction with and at the same time as, toe-in. In fact, the steering wheel spoke position, with the truck on a straight section of highway, may be the first indication of front end misalignment.

If the truck has been wrecked, or indicates any evidence of steering gear or linkage disturbance, the pitman arm should be disconnected from the sector shaft. The steering wheel (or gear) should be turned from extreme right to extreme left to determine the halfway point in its turning scope. This will be the spot on the gear that is in action during straight ahead driving and in which position

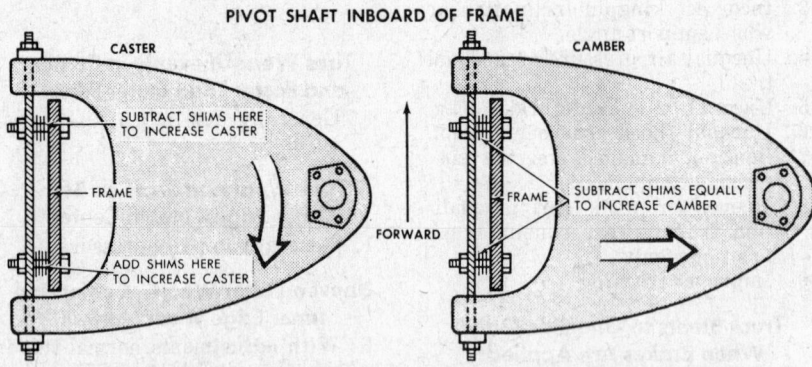

PIVOT SHAFT INBOARD OF FRAME

CASTER

SUBTRACT SHIMS HERE TO INCREASE CASTER

FRAME

ADD SHIMS HERE TO INCREASE CASTER

FORWARD

CAMBER

FRAME

SUBTRACT SHIMS EQUALLY TO INCREASE CAMBER

Caster-camber adjustment on upper arm equipped front suspension (© G.M.C.)

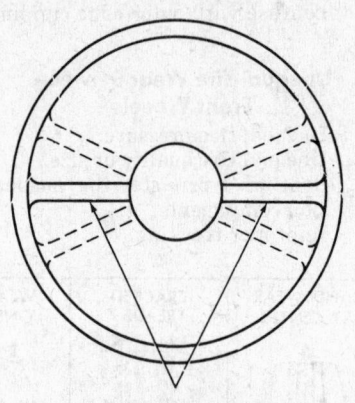

Steering wheel position

the steering gear should be adjusted. With the steering wheel in the straight-ahead position and the steering gear adjusted to zero lash status, reconnect the pitman arm.

Steering Geometry

Camber Angle

Camber is the amount that the front wheels are inclined outward at the top. Camber is spoken of, and measured, in degrees from the perpendicular.

The purpose of the camber angle is to take some of the load off the spindle outboard bearing.

Caster Angle

Caster is the amount that the king-pin (or in the case of trucks without king-pins, the knuckle support pivots) is tilted towards the back of the truck. Caster is usually spoken of, and measured, in degrees. Positive caster means that the top of the king-pin is tilted toward the back of the truck. Positive caster is indicated by the sign +.

Negative caster is exactly the opposite; the top of the kingpin is tilted toward the front of the truck. This is generally indicated by the sign —.

Negative caster is sometimes referred to as reverse caster.

The effect of positive caster is to cause the truck to steer in the direction in which it tends to go. Positive caster in the front wheels may cause the truck to steer down off a crowned road or steer in the direction of a

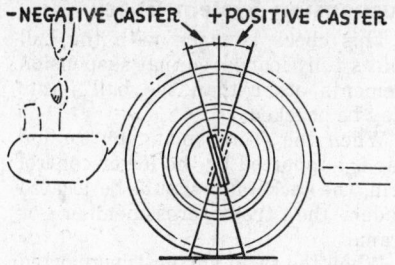

Caster angle, note that if the pivot tilts forward, caster is negative, tilted back, caster is positive

cross wind. For this reason, many of our modern trucks are arranged with negative caster so that the opposite is true; the truck tends to steer up a crowned road and into a cross wind.

Angle of Kingpin Inclination

In addition to the caster angle, the kingpins (or knuckle support pivots) are also inclined toward each other at the top. This angle is known as kingpin inclination and is usually spoken of, and measured, in degrees.

The effect of kingpin inclination is to cause the wheels to steer in a straight line, regardless of outside forces such as crowned roads, cross winds, etc., which may tend to make it steer at a tangent. As the spindle is moved from extreme right to extreme left it apparently rises and falls. Notice that it reaches its highest position when the wheels are in the straight-ahead position. In actual operation, the spindle cannot rise and fall because the wheel is in constant contact with the ground.

Therefore, the truck itself will rise at the extreme right turn and come to its lowest point at the straight-ahead position, and again rise for an extreme left turn. The weight of the truck will tend to cause the wheels to come to the straight-ahead position, which is the lowest position of the truck itself.

Included Angle

Included angle is the name given to that angle which includes kingpin inclination and camber. It is the relationship between the centerline of the wheel and the centerline of the kingpin (or the knuckle support pivots). This angle is built into the knuckle (spindle) forging and will remain

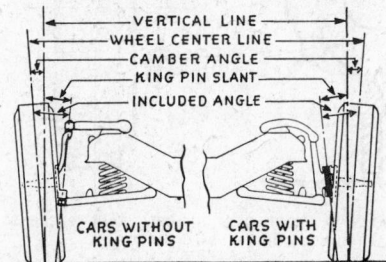

Camber, king pin slant and included angle

constant throughout the life of the truck, unless the spindle itself is damaged.

When checking a truck on the front end stand, always measure kingpin inclination as well as camber unless some provision is made on the stand for checking condition of the spindle. Where no such provision is made, add the kingpin inclination to the camber for each side of the truck. These totals should be exactly the same, regardless of how far from the norm the readings may be.

For example the left side of the truck checks $5\frac{1}{2}°$ kingpin inclination and $1°$ positive camber—total $6\frac{1}{2}°$. Since both sides check exactly the same for the included angle, it is unlikely that both spindles, in this instance, are bent. Adjusting to correct for camber will automatically set correct kingpin inclination.

A bent spindle would show up like this: left side of the truck has $\frac{3}{4}°$ positive camber, $5\frac{1}{4}°$ kingpin inclination—total $6°$ included angle. Right side of truck has $1\frac{1}{4}°$ positive camber, $6°$ kingpin inclination—total $7\frac{1}{4}°$ included angle. One of these spindles is bent and if adjustments are made to correct camber, the kingpin inclination will be incorrect due to the bent spindle.

Since the most common cause of a bent spindle is striking the curb when parking, which causes the spindle to bend upward, the side having the greater included angle usually has the bent spindle. It will be found impossible to achieve good alignment and minimum tire wear unless the bent spindle is replaced.

Toe-in

Toe-in is the amount that the front wheels are closer together at the front than they are at the back. This dimension is usually spoken of, and measured, in inches or fractions of inches.

Generally speaking, the wheels are toed-in because they are cambered. When a truck operates with $0°$ camber it will be found to operate with zero toe-in. As the required camber increases, so does the toe-in. The reason for this is that the cambered wheel tends to steer in the direction in which it is cambered. Therefore it is necessary to overcome this tendency of the wheel by compensating very slightly in the direction opposite to that in which it tends to roll. Caster and camber both have an effect on toe-in. Therefore toe-in is the last thing on the front end which should be corrected.

Toe-out Steering Radius

When a truck is steered into a turn, the outside wheel of the vehicle scribes a much larger circle than the inside wheel. Therefore, the outside

wheel must be steered to a somewhat less angle than the inside wheel. This difference in angle is often called toe-out.

The change in angle from toe-in in the straight-ahead position to toe-out in the turn is caused by the relative position of the steering arms to the kingpin and to each other.

If a line were drawn from the center of the kingpin through the center of the steering arm-tie rod attaching hole at each wheel, these lines would be found to cross almost exactly in the center of the rear axle.

If the front end angles, including toe-in, are set correctly, and the toe-out is found to be incorrect, one or both of the steering arms are bent.

Tracking

While tracking is more a function of the rear axle and frame, it is difficult to align the front suspension when the truck does not track straight. Tracking means that the centerline of the rear axle follows exactly the path of the centerline of the front axle when the truck is moving in a straight line.

On trucks that have equal tread, front and rear, the rear tires will follow in exactly the tread of the front tires, when moving in a straight line. However, there are many trucks whose rear tread is wider than the front tread. On such trucks, the rear axle tread will straddle the front axle tread an equal amount on both sides, when moving in a straight line.

Perhaps the easiest way to check a truck for tracking is to stand directly in back of it and watch it move in a straight line down the street. If the observer will stand as near to the center of the truck as possible, he can readily observe, even with the difference in perspective between the front and rear wheels, whether or not they are tracking properly. If the truck is found to track incorrectly, the diffi-

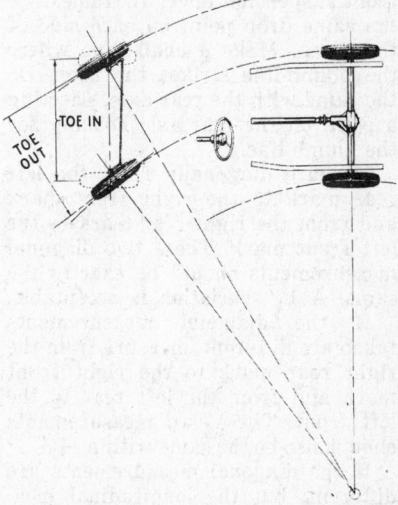

Steering geometry on turns (© Chrysler Corp.)

Front Axle and Suspension

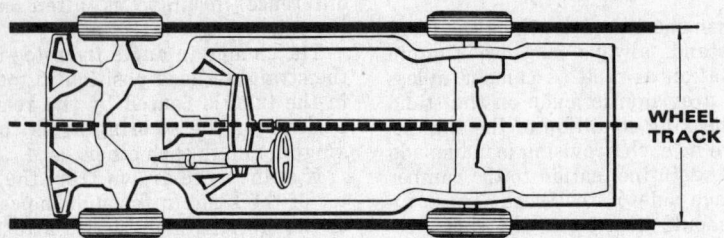

Typical parallel wheel track (© Hunter Engineering Co.)

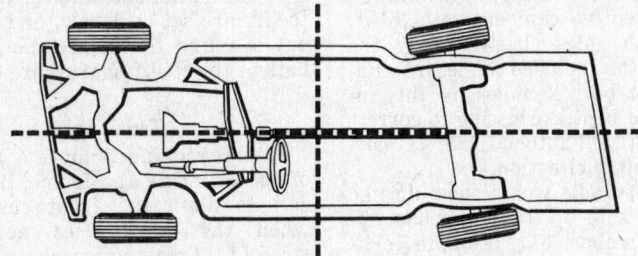

Bent Frame, diamond shaped (© Hunter Engineering Co.)

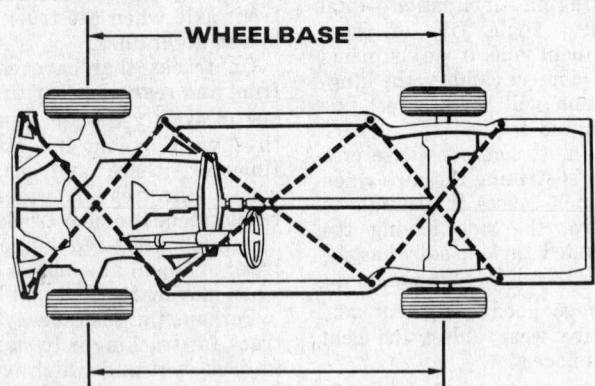

Measuring corresponding points of frame (© Hunter Engineering Co.)

culty will be found in either the frame or in the rear axle alignment.

Another more accurate method to check tracking is to park the truck on a level floor and drop a plumb-line from the extreme outer edge of the front suspension lower A-frame. Use the same drop point on each side of the truck. Make a chalk line where the plumb-line strikes the floor. Do the same with the rear axle, selecting a point on the rear axle housing for the plumb-line.

Measure diagonally from the left rear mark to the right front mark and from the right rear mark to the left front mark. These two diagonal measurements should be exactly the same. A 1/4" variation is acceptable.

If the diagonal measurements taken are different, measure from the right rear mark to the right front mark and from the left rear to the left front. These two measurements should also be the same within 1/4"

If the diagonal measurements are different, but the longitudinal measurements are the same, the frame is

swayed (diamond shaped).

However, in the event that the diagonal measurements are unqeual and the longitudinal measurements are also unequal, and the truck is tracking incorrectly, the rear axle is misaligned.

If the diagonal and longitudinal measurements are both unequal, but the truck appears to track correctly on the street, a kneeback is indicated.

NOTE: A kneeback means that one complete side of the front suspension is bent back. This is often caused by crimping the front wheels against the curb when parking the vehicle, then starting up without straightening the wheels out.

Suspension and Ball Joint Checks

When checking the suspension and ball joints, it is advisable to follow the manufacturer's recommendations. For all practical purposes, however, the following general procedures are applicable.

Suspension System Check

This check is made with the ball joints fully loaded, so that suspension elements other than the ball joints may be checked.

When the front spring or torsion bar is supported by the lower control arm, the jackstand should be located under the front crossmember or frame.

When the front spring is supported by the upper control arm, the jackstand should be located under the lower control arm.

Vertical or horizontal movement at the road wheel should not exceed the following:

Up to and including 16 inches	1/4"
16 to 18 8nches	1/3"
More than 18 inches	1/2"

Ball Joint Check

When checking the ball joints for any wear, they must be free of any load.

When the front spring or torsion bar is supported by the lower control arm, the jackstand should be positioned under the lower control arm.

When the front spring is supported

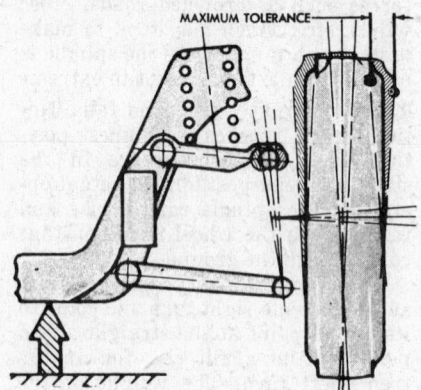

Rock tire top and bottom
Reject if movement at tire sidewall exceeds maximum tolerance, but do not confuse wheel bearing looseness with ball joint wear

Check ball joint radial (side) play

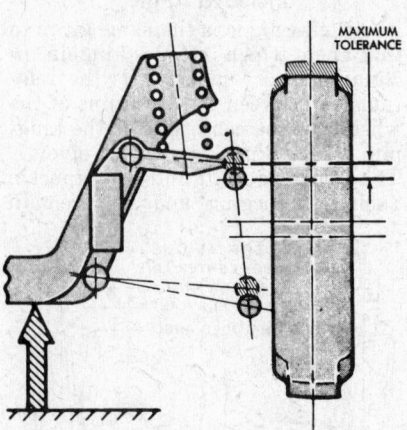

Reject if axial play in ball joint exceeds maximum tolerance

Check ball joint axial (up and down) play

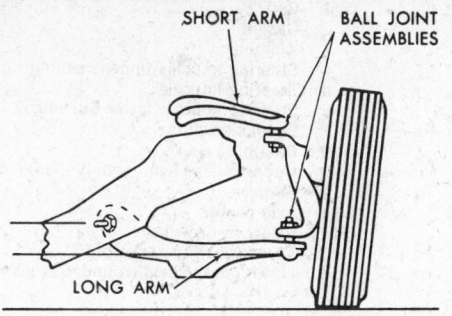

SHORT ARM
BALL JOINT ASSEMBLIES
LONG ARM

Use of control arms and ball joints for independent suspension (© G.M.C.)

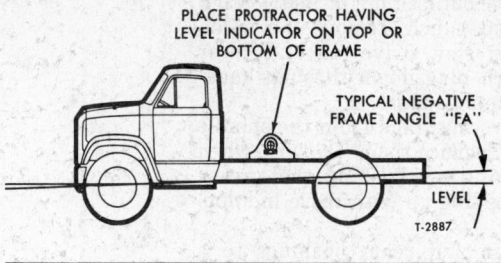

PLACE PROTRACTOR HAVING LEVEL INDICATOR ON TOP OR BOTTOM OF FRAME

TYPICAL NEGATIVE FRAME ANGLE "FA"

LEVEL

T-2887

Method of checking frame angle
(© G.M.C.)

by the upper control arm, the jack-stand should be located under the front crossmember or frame.

Replace the upper ball joint if any noticeable play is present in the joint when the spring is supported by the upper control arm.

Replace the load carrying ball joint if the sideplay (horizontal motion) of the wheel, when rocked, exceeds specifications; or if the up and down (vertical motion) exceeds specifications.

Wheel Bearing and Seal Replacement

Place jack under lower suspension arm. Remove hub cover and grease cap. Remove spindle nut, keyed washer and outer bearing. Slide off hub and drum.

NOTE: In some cases, drum removal may require loosening of brake adjustment.

At this point, brakes and drums should be inspected for their condition.

With hub and drum on bench, remove seal and inner bearing. Thoroughly clean all parts. Drive out inner and outer races of roller type. Use care not to mar the bearing surfaces.

Pack bearings with approved lubricant. When replacing cups, use a bearing race driver if possible. If a punch is used, make sure it is blunt and then drive parts in very carefully to avoid cocking the bearings.

Install new grease seal in hub. Assemble hub and drum on spindle and replace the outer bearing, key washer and nut.

A common method of adjustment is to tighten to zero clearance and then back off to first cotter pin castellation. Some manufacturers recommend tightening to approximately 10 to 12 ft. lbs., then backing off nut 1/6 turn. If cotter pin hole does not line up, loosen slightly.

Readjust brake if necessary and install grease cap and hub cover. Remove jack.

Kingpin and Bushing Replacement

Kingpins and bushings can be placed in two general classes; (A) with bushings in knuckle, (B) with bushings in spindle.

Jack up the truck and remove the hub as described in the wheel bearing section.

1. Remove backing plate to knuckle bolts and lift assembly, with brakes, from the knuckle. Suspend it with a piece of wire to prevent damage to brake hose.

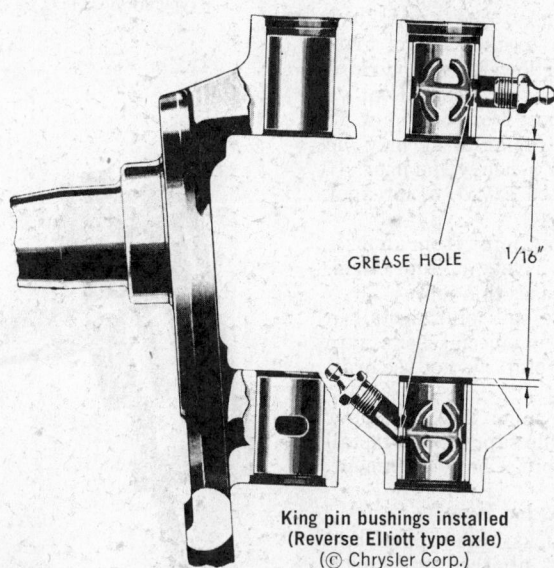

GREASE HOLE   1/16"

King pin bushings installed
(Reverse Elliott type axle)
(© Chrysler Corp.)

1. HAND SPIN WHEEL

3. BACK OFF NUT UNTIL JUST LOOSE (1/4—1/2 TURN)

2. "SNUG-UP" THE NUT TO FULLY SEAT BEARINGS—THIS OVERCOMES ANY BURRS ON THREADS.

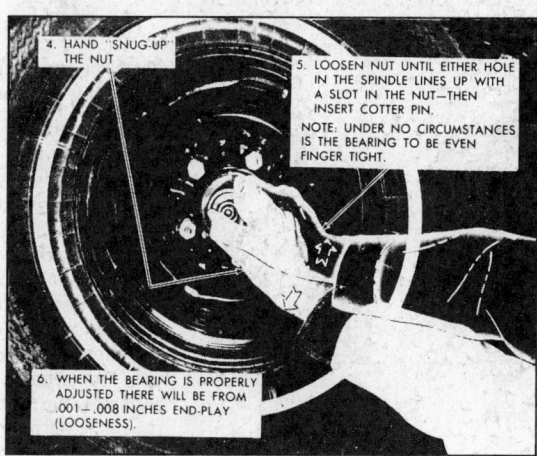

4. HAND "SNUG-UP" THE NUT

5. LOOSEN NUT UNTIL EITHER HOLE IN THE SPINDLE LINES UP WITH A SLOT IN THE NUT—THEN INSERT COTTER PIN

NOTE: UNDER NO CIRCUMSTANCES IS THE BEARING TO BE EVEN FINGER TIGHT.

6. WHEN THE BEARING IS PROPERLY ADJUSTED THERE WILL BE FROM .001—.008 INCHES END-PLAY (LOOSENESS).

Wheel bearing adjustment procedure (© G.M.C.)

Front Axle and Suspension

2. Drive out lock pin or bolt. With a sharp punch, remove top welch plug. Now drive pin and lower welch plug down through knuckle and support.

3. Drive bushings from the spindle and replace them. Be sure, when driving new bushing, that grease holes line up with those in knuckle.

4. Align and ream bushings to a snug running fit for the new kingpin.

5. Insert the kingpin through the top of the spindle, support, thrust bearing (with shims to control vertical play) and into the spindle bottom. Keep the kingpin in proper rotation so that the lockpin can be inserted. Install lockpin or bolt. Install upper and lower welch plugs.

6. Install backing plate with steering arms and lubricate properly.

7. Install hubs, drums and wheels, then remove jack.

Ball Joint Replacement

Upper Ball Joint R & R
Riveted Type

On some trucks, the upper ball joint is riveted to the control arm. Place jack under lower arm and raise wheel clear of the floor. Remove wheel. Remove nut from ball joint. If joint is being replaced, it may be driven out with a heavy hammer. If threads are to be saved, a spreader tool should be used.

After removing joint from knuckle support, cut off rivets at upper arm. Drilling rivets eases this job.

To replace the ball joint: install in upper arm, using special bolts supplied with new joint. Do not use ordinary bolts.

Next, set the taper into the upper end of the knuckle support and install nut and cotter pin. Check alignment.

Threaded Type

On some trucks, the upper ball joint is threaded into the control arm.

Place jack under lower control arm and relieve load on torsion bar. Raise wheel clear of floor. Remove wheel.

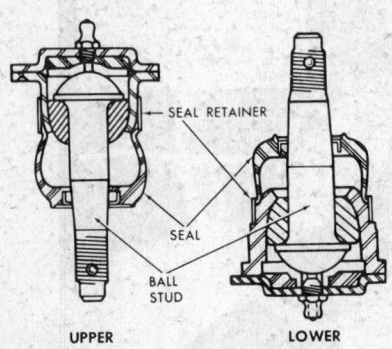

Typical ball joint assemblies (© G.M.C.)

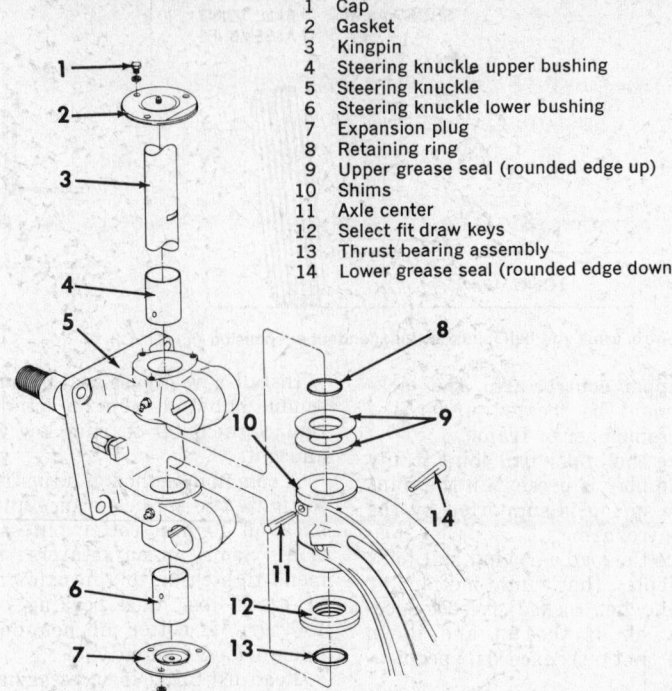

```
1   Cap
2   Gasket
3   Kingpin
4   Steering knuckle upper bushing
5   Steering knuckle
6   Steering knuckle lower bushing
7   Expansion plug
8   Retaining ring
9   Upper grease seal (rounded edge up)
10  Shims
11  Axle center
12  Select fit draw keys
13  Thrust bearing assembly
14  Lower grease seal (rounded edge down)
```

Exploded view of spindle bolt and bushing attachment of a steering knuckle (Typical) (© G.M.C.)

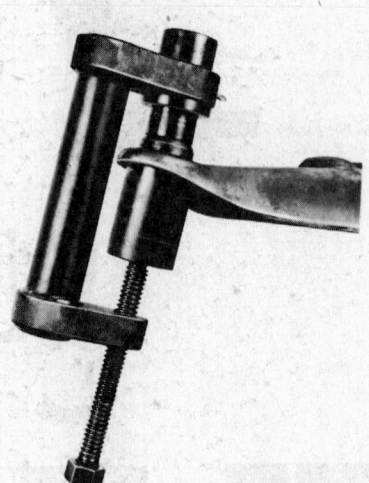

Installing a pressed-in lower ball joint with a "C" clamp type tool   (© Chrysler Corp.)

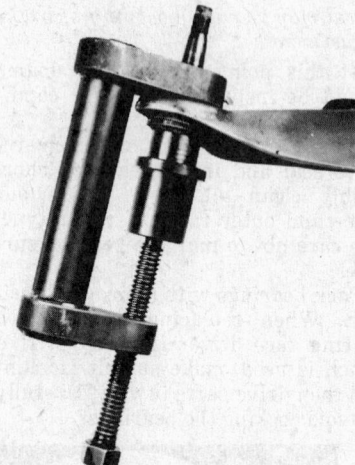

Removing a pressed-in lower ball joint with a "C" clamp type tool   (© Chrysler Corp.)

Remove nut from ball joint. If ball joint is being replaced, it may be driven out with a heavy hammer. If threads are to be saved, use a spreader tool.

After removing from knuckle support, the ball joint can be unscrewed from the support arm. Special tools are recommended for this operation.

When replacing the ball joint, be sure to engage the threads into the control arm squarely. Torque to 125 ft. lbs. If this torque cannot be obtained, check for bad threads in arm or on joint. Install new balloon seal.

Place joint in knuckle and install nut. Reload torsion bar (if so equipped) and reset height.

Lower Ball Joint R & R

Pressed Type

These ball joints are pressed into support arms. To replace pressed-in units, it is necessary to remove the front spring and support arm.

After removing wheel and drum, loosen nut slightly at ball joint taper and hammer lightly around area to loosen. If new ball joints are being in-

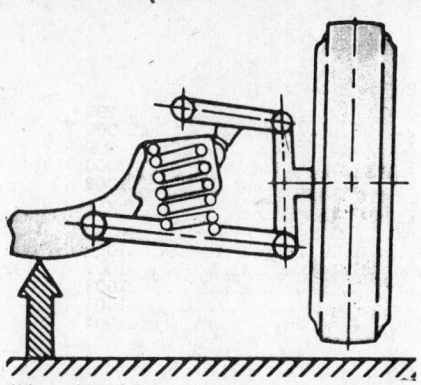

When the spring is supported on the lower control arm, vehicle must be jacked from the frame or cross member

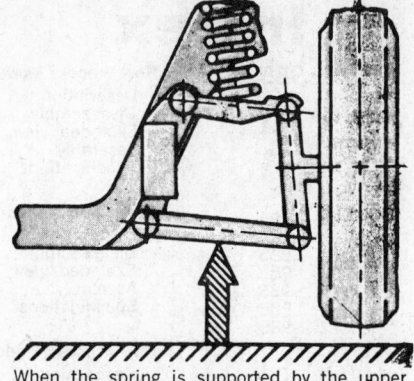

When the spring is supported by the upper control arm, the vehicle must be jacked at the lower control arm

Steering and suspension jacking procedure

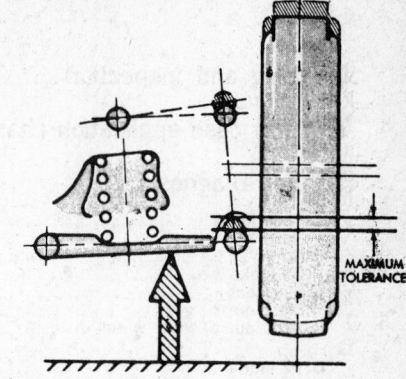

Reject if axial play in ball joint exceeds maximum tolerance

Check ball joint axial (up and down) play

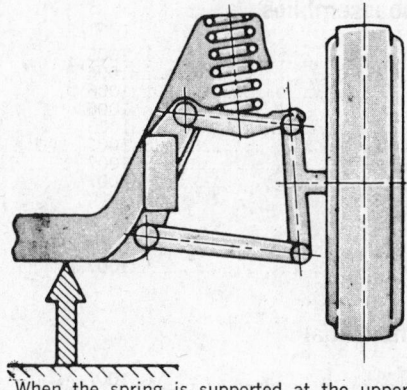

When the spring is supported at the upper control arm, the truck must be hoisted at the frame

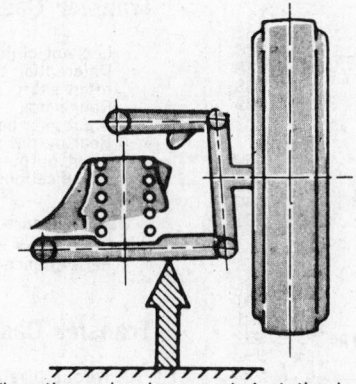

When the spring is supported at the lower control arm, truck must be hoisted at the arm. Reject if upper joint is perceptibly loose

Ball joint inspection jacking procedure

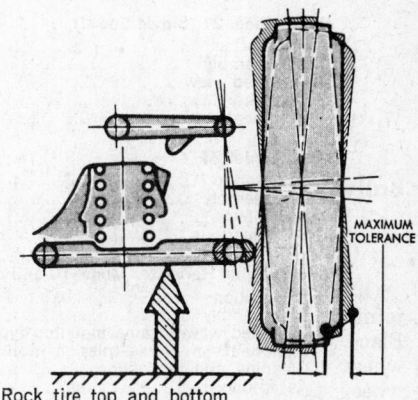

Rock tire top and bottom
Reject if movement at tire sidewall exceeds maximum tolerance, but do not confuse wheel bearing looseness with ball joint wear

Check ball joint radial (side) play

stalled, it is not necessary to protect the threads.

Place support arm in an arbor press with a suitable tool and press ball joint from the arm.

Install ball joint by reversing the pressing procedure.

NOTE: Special tools of the C-clamp type are available and can be used on some trucks to avoid removal of front spring and support arm.

Integral Type

On some trucks, the lower ball joint is integral with the steering arm and is not serviced separately.

To service this unit: Remove the upper arm bumper. Raise truck so that front suspension is under no load. If jacks are used, a support must be placed between the jack and K-member.

1. Remove the wheel and drum assembly. Remove the two lower bolts holding the steering arm to the backing plate.

2. Disconnect tie-rod end from the steering arm. Do not damage seal.

3. Remove the ball joint stud from the lower control arm. A spreading tool will aid in this operation.

4. Install new seal on ball joint. Bolt the steering arm to the backing plate. Insert the ball joint into control arm and torque nut.

5. Connect the tie-rod end. Install drum and wheel.

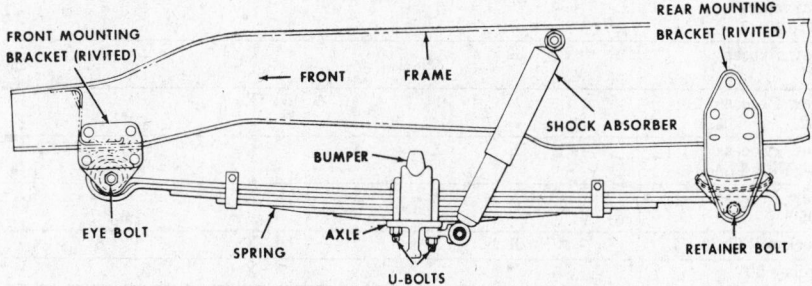

Typical front leaf spring installation (© G.M.C.)

FRONT MOUNTING BRACKET (RIVITED)
REAR MOUNTING BRACKET (RIVITED)
FRAME
← FRONT
SHOCK ABSORBER
BUMPER
EYE BOLT
SPRING
AXLE
U-BOLTS
RETAINER BOLT

Index

TRANSFER CASE APPLICATION CHART

	Chevrolet	Dodge	Ford	GMC	International Harvester	Jeep	Plymouth
Dana 20	X		①	X	X		
Dana 21			②				
Dana 24			③				
International Harvester TC-143					④		
New Process 201		X					
New Process 202					X		
New Process 203 Full Time	X	X	X	X			X
New Process 205	X	X	X	X	X		
Rockwell T223		X	X		X		
Spicer 20					X	X	
Warner Quadra-Trac						X	

① Bronco only
② F-100 only
③ F-250 only through 1973
④ Scout only

Transfer Case Trouble Analysis

Slips Out of Gear (High-Low)
1. Shifting poppet spring weak.
2. Bearing broken or worn.
3. Shifting fork bent.
4. Improper control rod adjustment.

Slips Out of Front Wheel Drive
1. Shifting poppet spring weak or broken.
2. Bearing worn or broken.
3. Excessive shaft end-play.
4. Shifting fork bent.

Hard Shifting
1. Lack of lubricant.

2. Shift lever binding on shaft.
3. Shifting poppet ball scored.
4. Shifting fork bent.
5. Low tire pressure.

Backlash
1. Companion yoke loose.
2. Transfer case loose on mounts.
3. Internal parts excessively worn.

Noisy
1. Low lubricant level.
2. Bearings improperly adjusted or excessively worn.
3. Gears worn or damaged.

4. Improper alignment of driveshafts or U-joints.

Oil Leakage
1. Excessive amount of lubricant in case.
2. Vent clogged.
3. Gaskets or seals leaking.
4. Bearings loose or damaged.
5. Driveshaft yoke mating surfaces scored.

Overheating
1. Excessive or insufficient amount of lubricant.
2. Bearing adjustment too tight.

Cleaning and Inspection

Cleaning
During overhaul, all components of the transfer case (except bearing assemblies) should be thoroughly cleaned with solvent and dried with air pressure prior to inspection and reassembly.
1. Clean the bearing assemblies as follows.

NOTE: Proper cleaning of bearings is of utmost importance. Bearings should always be cleaned separately from other parts.

a. Soak all bearing assemblies in CLEAN solvent, gasoline, or fuel oil. Bearings should never be cleaned in a hot solution tank.
b. Slush bearings in solvent until all old lubricant is loosened. Hold races so that bearings will not rotate; then clean bearings with a soft bristled brush until all dirt has been removed. Remove loose particles of dirt by tapping bearing flat against a block of wood.
c. Rinse bearings in clean solvent; then blow bearings dry with air pressure.

CAUTION: Do not spin bearings while drying.
d. After drying, rotate each bearing slowly while examining balls or rollers for roughness, damage, or excessive wear. Replace all bearings that are not in first class condition.

NOTE: After cleaning and inspecting bearings, lubricate generously with recommended lubricant, then wrap each bearing in clean paper until ready for reassembly.
2. Remove all portions of old gaskets from parts, using a stiff brush or scraper.

Inspection
1. Inspect all parts for discoloration or warpage.
2. Examine all gears and splines for chipped, worn, broken or nicked teeth. Small nicks or burrs may be removed with a fine abrasive stone.
3. Inspect the breather assembly to make sure that it is open and not damaged.
4. Check all threaded parts for damaged, stripped, or crossed threads.

5. Replace all gaskets, oil seals and snap-rings.
6. Inspect housings, retainers and covers for cracks or other damage. Replace the damaged parts.
7. Inspect keys and keyways for condition and fit.
8. Inspect shift forks for wear, distortion or any other damage.
9. Check detent ball springs for free length, compressed length, distortion or collapsed coils.
10. Check bearing fit on their respective shafts and in their bores or cups. Inspect bearings, shafts and cups for wear.

NOTE: If either bearings or cups are worn or damaged, it is advisable to replace both parts.

11. Inspect all bearing rollers or balls for pitting or galling.
12. Examine detent balls for corrosion or brinelling. If shift bar detents show wear, replace them.
13. Replace all worn or damaged parts. When assembling the transfer case, coat all moving parts with recommended lubricant.

Dana Model 20

Description
The Dana Model 20 is a two-speed gearbox that controls the power from the transmission to the front and rear driving axles. Positions of the transfer case are: four-wheel-drive low (4L), neutral (N), two-wheel-drive high (2H) and four-wheel-drive high (4H).

Disassembly

Transfer Case
1. Clean any dirt from the transfer case and remove the bottom cover plate.

2. Remove the retaining plug, flat washer, detent spring and ball which engages the front drive shift rail detent rod. Then, remove plug from front drive detent rod access hole.
3. Remove the retaining plug, detent spring and ball which engages the rear drive shift rail detent rod.
4. Remove the idler shaft lockplate.
5. Using a hammer and soft drift, drive the idler shaft rearward and out of the case; then lift out

the thrust washers and idler gear.
NOTE: When removing the idler gear, do not lose any of the rollers.
6. Remove the flange retaining nuts from the front and rear output shafts.
7. Remove the flange from the front and rear output shafts. Discard the O-ring.
8. Remove the bolts securing the adapter housing to the case; then remove the adapter as an assembly.
9. Remove the bolts which attach

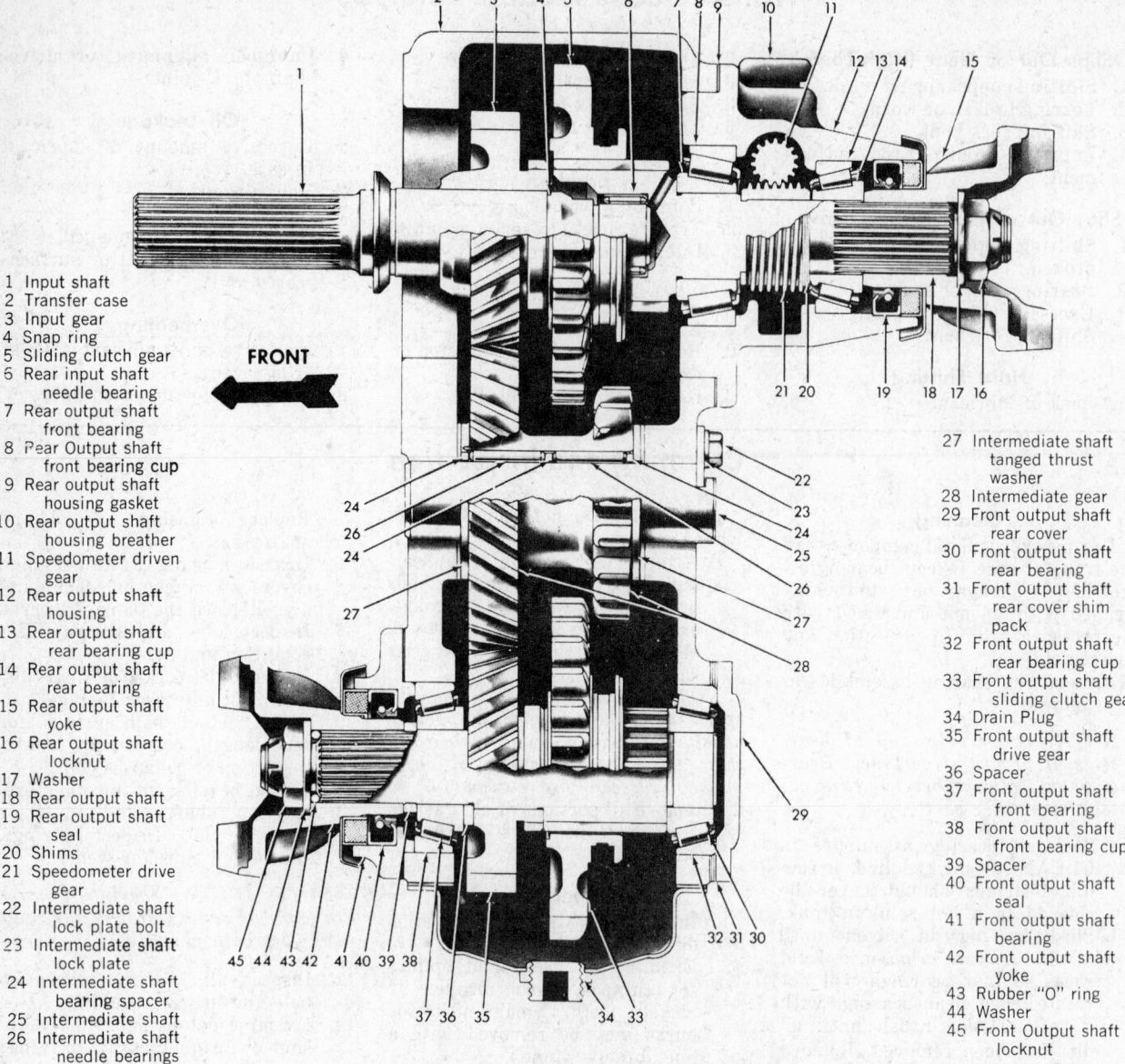

1 Input shaft
2 Transfer case
3 Input gear
4 Snap ring
5 Sliding clutch gear
6 Rear input shaft needle bearing
7 Rear output shaft front bearing
8 Rear Output shaft front bearing cup
9 Rear output shaft housing gasket
10 Rear output shaft housing breather
11 Speedometer driven gear
12 Rear output shaft housing
13 Rear output shaft rear bearing cup
14 Rear output shaft rear bearing
15 Rear output shaft yoke
16 Rear output shaft locknut
17 Washer
18 Rear output shaft
19 Rear output shaft seal
20 Shims
21 Speedometer drive gear
22 Intermediate shaft lock plate bolt
23 Intermediate shaft lock plate
24 Intermediate shaft bearing spacer
25 Intermediate gear
26 Intermediate shaft needle bearings

27 Intermediate shaft tanged thrust washer
28 Intermediate gear
29 Front output shaft rear cover
30 Front output shaft rear bearing
31 Front output shaft rear cover shim pack
32 Front output shaft rear bearing cup
33 Front output shaft sliding clutch gear
34 Drain Plug
35 Front output shaft drive gear
36 Spacer
37 Front output shaft front bearing
38 Front output shaft front bearing cup
39 Spacer
40 Front output shaft seal
41 Front output shaft bearing
42 Front output shaft yoke
43 Rubber "O" ring
44 Washer
45 Front Output shaft locknut

FRONT

Cutaway view of the Dana 20 transfer case (© G.M.C.)

the rear output shaft bearing retainer to the case; then remove the retainer and output shaft as an assembly.

NOTE: Be sure not to lose any of the rollers.

10. Disconnect the shift rail link from the two shift rails.
11. Lift out the rear output shaft sliding gear.
12. Remove the setscrew securing the rear fork to the shift rail; then remove the rear drive shift rail and fork.
13. Remove the front output shaft rear cover and shims. Fasten the shims together.
14. Remove the front output shaft bearing retainer and gasket.
15. Tap the threaded end of the front output shaft; then remove the rear cup.
16. Angle the front output shaft

front bearing away from the main drive gear to allow removal of the snap-ring; then tap the shaft and rear bearing out of the case.
17. Lift out the sliding gear, main drive gear, front bearing, spacer and snap-ring.
18. Remove the front cup.
19. Remove the setscrew securing the front shift fork to the shift rail; then remove the rail and fork.
20. Remove the detent rods.
21. Remove shift rail oil seal.

Input Shaft

1. Remove the snap-ring from the front of the shaft.
2. Place the adapter housing and input shaft on a press and force the shaft out of the main drive gear and housing.

3. Remove the bearing retaining snap-ring; then remove bearing.
4. Remove the seal in the adapter housing.

Rear Output Shaft

1. Remove needle bearings from bore of shaft.
2. Remove speedometer driven gear.
3. Place bearing retainer and shaft assembly in a press; then force shaft out of retainer.
4. Lift off speedometer drive gear and shims. Tag shims for reassembly.
5. Press out the outer cup, bearing and seal.
6. Remove the inner cup.
7. Remove the inner bearing.

Front Output Shaft

Using the sliding gear as a base, press rear bearing off shaft.

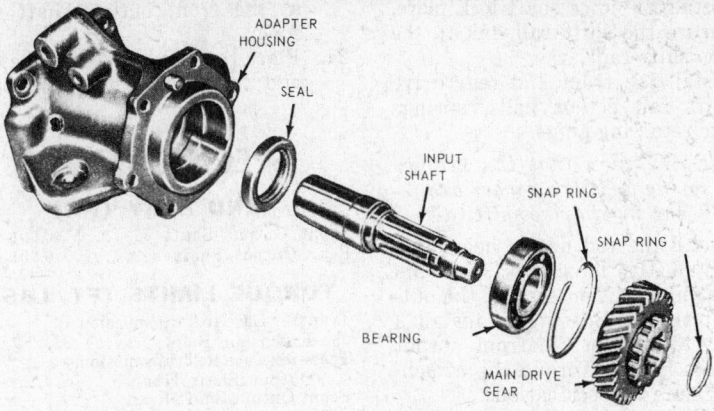

Exploded view of the input shaft on a Dana 20 transfer case
(© Ford Motor Company)

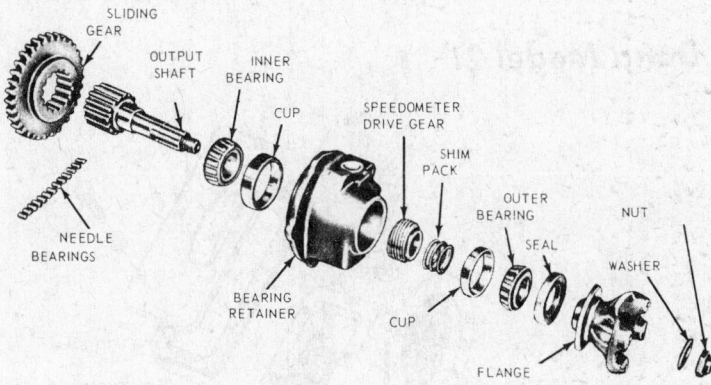

Exploded view of the rear output shaft on a Dana 20 transfer case
(© Ford Motor Company)

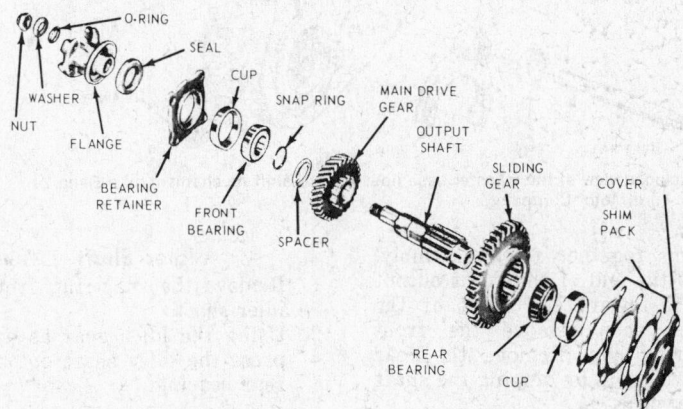

Exploded view of the front output shaft on a Dana 20 transfer case
(© Ford Motor Company)

Assembly

Input Shaft

1. Install a new seal in the adapter housing.
2. Install bearing in the housing and secure with snap-ring.
3. Using the main drive gear as a base, force the input shaft through the housing, seal, bearings and main drive gear. Secure with snap-ring on front of shaft.

Rear Output Shaft

1. Press the shaft into the inner bearing.
2. Install outer cup in the bearing retainer.
3. Install the inner cup.
4. Position the outer bearing in the retainer; then place the shims and speedometer drive gear on the shaft. Install shaft in the bearing retainer housing.
5. Place the bearing retainer and shaft in a vise. Install the out-

put shaft flange and torque the retaining nut to specifications.
6. With a dial indicator on the flange end of the shaft, measure end-play. Adjust shim pack between the speedometer drive gear and outer bearing to achieve correct clearance.
7. After setting correct end-play, remove flange and press bearing retainer seal into housing.
8. Install the flange, washer and nut. Tighten the nut to specifications.

Front Output Shaft

Using a press, force front output rear bearing on shaft.

Shift Rail Oil Seals

Using the tool shown in illustration, install the two shift rail oil seals.

Transfer Case

1. Install the front detent rod in the case.
2. Slide the front drive shift rail all the way into the case and place the shift fork on the rail as it enters the case. Secure the fork to the rail with the setscrew.
3. Position the front output shaft sliding gear in the shift fork.
4. Install the rear detent rod.
5. Slide the rear drive shift rail into the case and position the shift fork on the rail as the rail enters the case. Secure the fork to the rail with the setscrew.

NOTE: The shift rails should be inserted so that the detents are positioned as shown in illustration.

6. While holding the sliding gear and main drive gear in position, install the front output shaft and rear bearing assembly through the two gears.
7. Install the main drive gear spacer and secure with the snapring.
8. Install the front output shaft rear bearing cup.
9. Place the front output shaft rear cover and shims on the case and install the attaching bolts.
10. Install the front output shaft front bearing on the shaft. Install the front bearing cup.
11. If the front bearing retainer oil seal was removed, install a new seal. Position the bearing retainer and gasket to the case and install the attaching bolts.
12. Place the rear output shaft rear bearing retainer on a work bench and install 13 needle bearings in the splined hub of the output shaft, using vaseline or grease.
13. Position the rear output shaft rear bearing retainer assembly

to the case and install the attaching bolts.

14. Install the rear output shaft sliding gear in the shifting fork and on the splines of the output shaft.
15. Position the adapter housing assembly on the rear output shaft and case. Install the attaching bolts.
16. Install the roller bearings in the bore of the idler shaft gear with vaseline or grease.
17. Position the idler gear and thrust washers in the case; then drive the idler shaft into the rear of the case through the idler gear and thrust washers.

NOTE: After installing the idler shaft, tap the sides of the case to relieve any possible binding.

Description

The Dana Model 21 is a single-speed gearbox that transmits power to the front driving axle. There are two positions of the transfer case; front drive axle engaged and front drive axle disengaged.

Disassembly

Transfer Case

1. Clean all dirt from transfer case and drain lubricant.
2. Remove bolts that attach the cover to the top of the case; then remove the cover.
3. Remove the setscrew securing the shift fork to the rail. Tap the shift rail rearward; then remove the rail cap from the rear of the case.
4. Remove the shift rail and fork.
5. Remove the detent spring and ball which engages the front drive shift rail.
6. Remove the flange attaching nuts, flat washer and O-ring from the front and rear output shafts. Discard the O-rings.
7. Remove the flange from the front and rear output shafts.
8. Remove the bolts that attach the rear output shaft bearing retainer to the case; then remove the retainer and output shaft as an assembly.
9. Remove the front and rear idler shaft covers.
10. Using a hammer and soft drift, drive the idler shaft and rear idler bearing rearward out of the case; then lift out the front bearing and idler gear.
11. Remove the front output shaft bearing retainer and gasket. Remove the retainer seal if it is worn or damaged.
12. Remove the front output shaft rear cover and shims. Tie the

18. Install the idler shaft lock plate.
19. Secure the shift rail link to the two shift rails.
20. Install the front and rear drive shift rail detent balls, springs and retaining plugs.

NOTE: Be sure that the heavier loaded spring and flat washer are installed in the front drive shift rail.

21. Install the rod access hole plug.
22. Install the flange, washer and retaining nut on each of the output shafts. Be sure to install a new O-ring in the front output shaft flange. Torque the attaching nuts to specifications.
23. With a dial indicator on the front drive output shaft, check the end-play. If not within specifications, adjust the shim pack

shims together for reassembly.
13. Tap the end of the front output shaft toward the front of the case; then remove the front bearing cup. Remove the rear bearing cup by tapping the shaft rearward.
14. Angle the front output shaft front bearing away from the main drive gear to remove the snap-ring from its groove in the shaft. Drive the output shaft and rear bearing out of the case.
15. Remove the sliding gear, main drive gear, front bearing, thrust washer and snap-ring from the case.
16. Remove the shift rail seal.

Front Output Shaft Bearing

To remove the front output shaft rear bearing, use the sliding gear as a base and press off the bearing.

at the front output shaft rear cover.
24. Place the cover plate on the case and install the attaching bolts.

SPECIFICATIONS

END PLAY (IN.)

Front Output Shaft	0.001-0.005
Rear Output Shaft	0.001-0.005

TORQUE LIMITS (FT. LBS.)

Transfer Case to Transmission Extension Bolts	20-30
Transfer Case to Transmission Output Shaft Nut	60-80
Front Output Shaft Rear Cover Bolts	25-32
Front Output Shaft Bearing Retainer Bolts	25-32
Idler Shaft Cover Bolts	25-32
Front and Rear Output Flanges	80-85

Dana Model 21

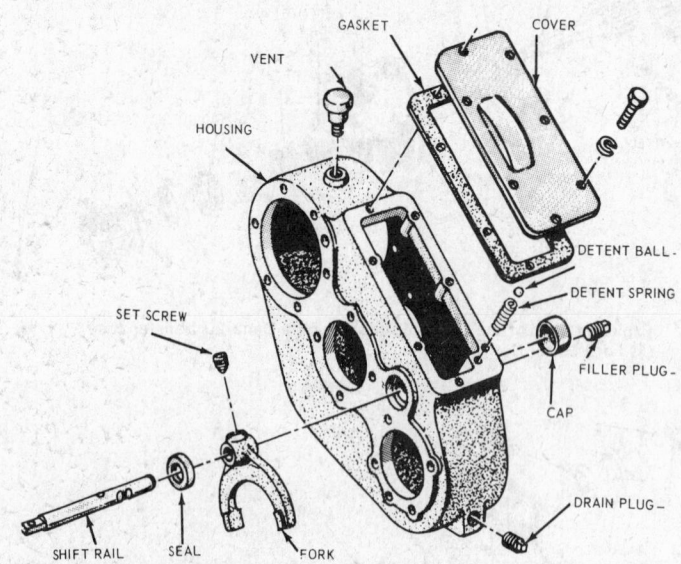

Exploded view of the transfer case housing and shift mechanism on a Dana 21
(© Ford Motor Company)

Idler Shaft

1. Remove the snap-ring from the idler shaft.
2. Using the idler gear as a base, press the idler shaft out of the rear bearing.

Rear Output Shaft

1. To remove the output shaft from the bearing retainer tap shaft rearward. Remove the shims and spacer.
2. Remove the inner bearing from the output shaft.
3. Place the bearing retainer on a press and force out the outer cup, outer bearing and oil seal.
4. Using a soft drift, drive out the inner bearing cup.

Assembly

Front Output Shaft

Using an arbor press, force rear bearing on front output shaft.

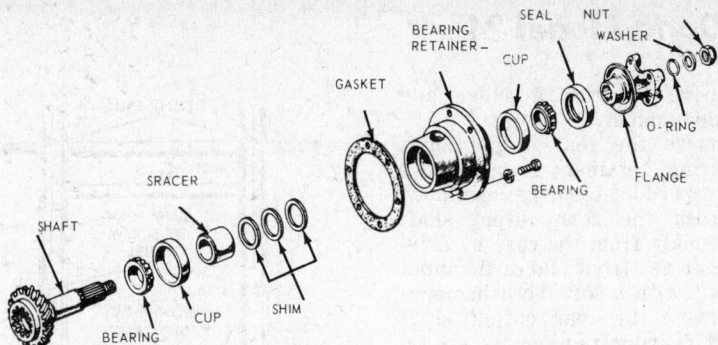

Exploded view of the rear output shaft on a Dana 21 (© Ford Motor Company)

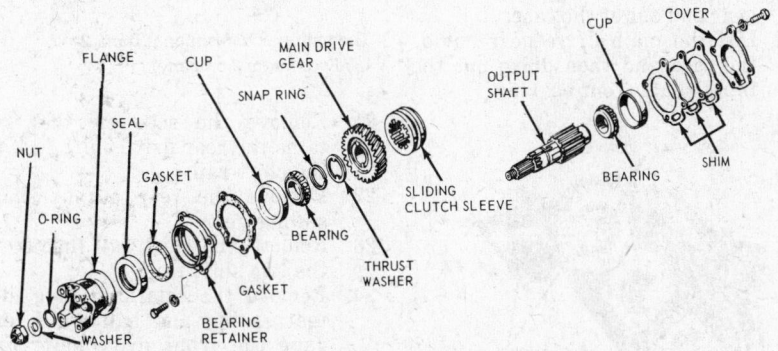

Exploded view of the front output shaft on a Dana 21 (© Ford Motor Company)

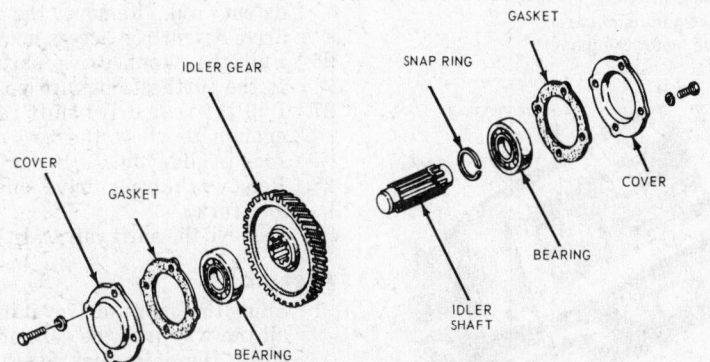

Exploded view of the idler shaft on a Dana 21 (© Ford Motor Company)

Idler Shaft

1. Using a press, install rear idler bearing on the shaft.
2. Install snap-ring.

Rear Output Shaft

1. Press the inner bearing on the output shaft.
2. Using a soft-faced hammer, tap the inner bearing cup into the retainer.
3. Install the outer cup.
4. Place spacer and shims on the output shaft; then install the shaft in the bearing retainer housing.
5. Install the outer bearing on the shaft.
6. Place the bearing retainer and output shaft in a vise. Measure end-play with a dial indicator on end of the shaft. If not within specifications, adjust shim pack between the spacer and the front and rear bearing cones.
7. After setting correct end-play, install the bearing retainer seal.

Transfer Case

1. While holding the drive gear, sliding gear and thrust washer in the case, install the front output shaft, from the rear, through the gears and washer. Install the snap-ring.
2. Install the front output shaft rear bearing cup.

3. Place the front output shaft rear cover and shims on the case. After removing old sealant from all mating surfaces with thinner, apply gasket sealer to the attaching bolts and torque to specifications. With the cover installed, apply sealer to the outside edge of the adjusting shims, case and cover joints.
4. Install the front output shaft rear bearing on the shaft. Install the front bearing.
5. If the front bearing retainer oil seal was removed, install a new seal. Position the bearing retainer and gasket to the case and install attaching bolts.
6. Install the flange, new O-ring, washer and attaching nut on the front output shaft.
7. With a dial indicator on the front drive output shaft, check the end-play. If not within specified limits, increase or decrease the shim pack thickness at the front output shaft rear cover.
8. Place the idler gear in the case; then install the idler shaft through the gear. Install the front bearing.
9. Place the front and rear idler covers and gaskets on the case; then install attaching bolts.
10. Install a new shift rail seal.
11. Install the shift rail detent ball and spring in the top of the case.
12. Slide the shift rail into the case and position the fork on the rail as the rail enters the case. Depressing the detent ball and spring will allow the rail to pass. Secure the fork to the rail with the setscrew. Install the shift rail cap.
13. Position the rear output shaft and bearing retainer assembly to the case, then install the attaching bolts.
14. Install the flange, new O-ring, washer and attaching nut on the rear output shaft.
15. Place the top cover and gasket on the case, then install attaching bolts.
16. Fill the transfer case to the proper level with the recommended lubricant.

SPECIFICATIONS

END PLAY (IN.)

Front Output Shaft 0.001-0.005
Rear Output Shaft 0.001-0.005

TORQUE LIMITS (FT. LBS.)

Transfer Case to Transmission
Extension Bolts 20-30
Transfer Case to Transmission
Output Shaft Nut 125-150
Front Output Shaft Rear
Cover Bolts 25-32
Front Output Shaft Bearing
Retainer Bolts 25-32
Idler Shaft Cover Bolts 25-32

Dana Model 24

Description

The Dana Model 24 is a two-speed gearbox that is manually controlled by a shift lever in the cab. The transfer case positions are: four-wheel-drive low (4L), neutral (N), two-wheel-drive high (2H) and four-wheel-drive high (4H).

Disassembly

1. Clean any dirt from the transfer case and remove the power take-off cover plate.
2. Remove both idler shaft bearing retainers.
3. Using a soft-faced hammer, tap the idler shaft and bearing to the rear until the bearing is free of the case.
4. Remove the idler shaft, two gears and spacer.
5. Remove the idler shaft front bearing.
6. Remove the flange retaining nuts from the front output shaft, the input shaft and the rear output shaft.
7. Remove the flanges and washers.
8. Remove the front output shaft front and rear bearing retainers.
9. Tap the front output shaft and rear bearing through the gears and case. Remove the high speed gear.
10. Remove the front output shaft front bearing and washer.

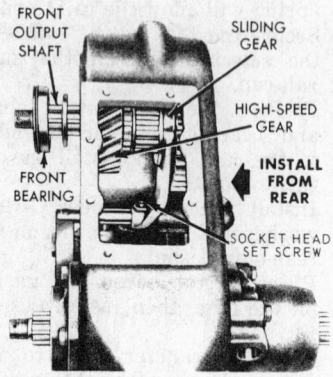

Removing/installing the front output shaft on a Dana 24 transfer case
(© Ford Motor Company)

11. Remove the setscrew that retains the front drive shaft fork to the shift rail.
12. Remove the front output shaft sliding gear.
13. If the input shaft oil seal is to be replaced, remove it with a four-jaw puller and slide hammer.
14. Remove the input shaft bearing retainer.
15. If the output shaft bearing retainer oil seal is to be replaced,

remove it with a puller and slide hammer.
16. Remove the rear output shaft bearing retainer; then remove the speedometer drive assembly.
17. Loosen the rear output shaft assembly from the case by driving on the front end of the input shaft with a soft-faced hammer.
18. Remove the rear output shaft and bearing retainer as an assembly.
19. Tap the input shaft through the front bearing, through the main drive gear, through the sliding gear and out of the case.
20. Lift the main drive gear out of the case and then drive out the input shaft front bearing.

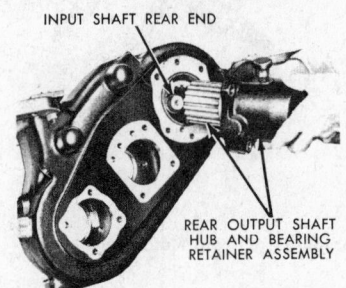

Removing/installing the rear output shaft on a Dana 24 transfer case
(© Ford Motor Company)

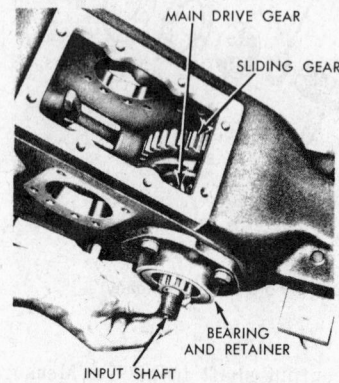

Removing/installing the input shaft on a Dana 24 transfer case
(© Ford Motor Company)

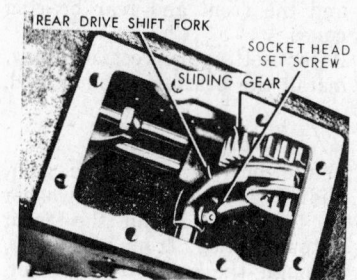

The rear drive shift fork and sliding gear on a Dana 24 transfer case
(© Ford Motor Company)

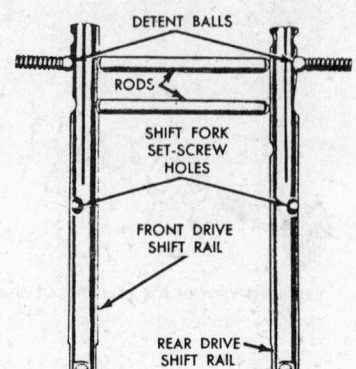

The shift mechanism on a Dana 24
(© Ford Motor Company)

21. Remove the setscrew that retains the rear drive shift fork to the shift rail.
22. Remove the rear output shaft sliding gear.
23. Remove the shift rail link from the two shift rails.
24. Remove the retaining plug, detent spring and ball which engage the front drive shift rail detent rod.
25. Remove the retaining plug detent spring and ball which engages the rear drive shift rail detent rod. Remove the front drive detent rod access hole plug.
26. Pull the front drive shift rail to the furthest outward position.
27. Pull the rear drive shift rail far enough to allow the two detent rods to slide out.
28. Remove the rear drive shift rail and fork.
29. Remove the shift rail seals.

Assembly

1. Slide the front drive shift rail all the way into the case and position the shift fork on the rail as the rail enters the case.
2. Install the two detent rods in the case.
3. Install the rear drive shift fork and hold the detent rods and the fork in place as the rear drive shift rail is pushed in as far as possible.
   NOTE: In steps 2 and 3, the shift rails should be inserted so that the detents are positioned as shown in illustration.
4. Pull the front drive shift rail out to its next detent. This will permit the rear drive shift rail to be pushed in to the full extent of its travel. After pushing the rear drive shift rail all the way in, push the front drive shift rail back to its extreme inward position.
5. Install the rear drive shift detent ball, spring and retaining

plug; then install the access hole plug.

6. Install the front drive shift rail detent ball, spring and retaining plug.

7. Secure the shift rail link in the two shift rails.

8. Place the rear output sliding gear in the shift fork and secure the fork to the rear drive shift rail with the setscrew.

9. Install the input shaft front bearing and retainer assembly. Coat retainer and bolts with sealer.

10. Place the main drive gear in the case; then slide the input shaft into the rear of the case through the main drive gear and through the front bearing and retainer.

11. Install the roller bearings in the splined hub of the rear output shaft assembly with petrolatum or grease. Then install the shaft and bearing retainer assembly, making sure that the output shaft is aligned correctly with the input shaft. Coat the case,

bearing retainer and bolts with sealer.

12. Position the front output sliding gear in the shift fork and secure the folk to the front drive shift rail with the setscrew.

13. While holding the sliding gear and high speed gear in position, install the front output shaft and rear bearing assembly through the two gears from the rear of the case.

14. After coating with sealer, install the front output shaft rear bearing retainer and gasket.

15. Install the washer and bearing over the front output shaft at the front of the case and then install the front bearing retainer and gasket. Coat retainer with sealer.

16. Install the flange, washer, flange retaining nut and cotter key on each of the three shafts. Torque to specifications.

17. Place the idler shaft gears in the case and install the shaft and rear bearing assembly from the rear. After applying sealer

to the plate and bolts, install the rear bearing retainer.

18. Position the spacer on the front end of the idler shaft and install the front bearing. Tap the bearing lightly with a mallet or soft-faced hammer.

19. Install the washer, retaining nut and cotter pin on the front end of the idler shaft.

20. After applying sealer to the plates and bolts, install the idler shaft front bearing retainer and the power take-off cover plate.

SPECIFICATIONS

END PLAY (IN.)
Front Output Shaft 0.003-0.007
Rear Output Shaft 0.003-0.007

TORQUE LIMITS (FT. LBS.)
Transfer Case to Transmission
 Extension Bolts 20-30
Transfer Case to Transmission
 Output Shaft Nut 60-80
Front Output Shaft Rear
 Cover Bolts 25-32
Front Output Shaft Bearing
 Retainer Bolts 25-32
Idler Shaft Cover Bolts 25-32

International Harvester Model TC-143

The IH Model TC-143 "Silent Drive" transfer case is a chain driven single speed unit. Unlike conventional gear driven transfer cases, this unit has a high-strength link-belt type loop of chain driving two broad-faced sprockets. There is no neutral position. There are two variations of this transfer case; one is frame mounted with a short intermediate drive shaft between the input shaft

of the transfer case and the transmission output shaft and the other type is mounted directly to the rear of the transmission.

Disassembly

1. After removing the transfer case from the vehicle and draining all of the lubricant out, clean the outside of the case.

2. Remove the shift cover.

3. Unscrew and remove the indicator light switch from the case.

4. With the rear output shaft flange clamped in a soft jawed vise, remove the flange retaining nut. Remove the flange from the vise and remove the flange from the rear output shaft. Use a puller, if necessary.

5. Turn the case so it rests on the flanges and remove the bolts securing the two halves of the case. Lift the top (rear) half of the case from the assembly and discard the gasket.

6. If present, remove the short spacer from the rear side of the input shaft. Also, if the thrust washer did not stay with the cover when removed, remove it now from the front output shaft.

7. With the case again secured in a soft jawed vise, place your thumbs on the ends of the shafts with your fingers under the sprockets. Pull the sprockets together with the chain off the shafts and out of the case as an assembly.

8. Unhook and remove the shift spring.

9. Remove the shift assembly mounting bolt and spring stud from the case.

10. Pull the shift cranks from the bosses inside the case.

11. If present, remove the long spacer from the input shaft and the washer from the front output shaft.

1. Input shaft end nut	12. Chain
2. Washer	13. Input shaft
3. Seal	14. Short spacer
4. Flange bolt	15. Dowel ring
5. Frame-mounted type case	16. Cover
6. Gasket	17. Flange nut
7. Shift cover	
8. Bearing snap ring	
9. Ball bearing	
10. Long spacer	
11. Upper sprocket (Old style)	

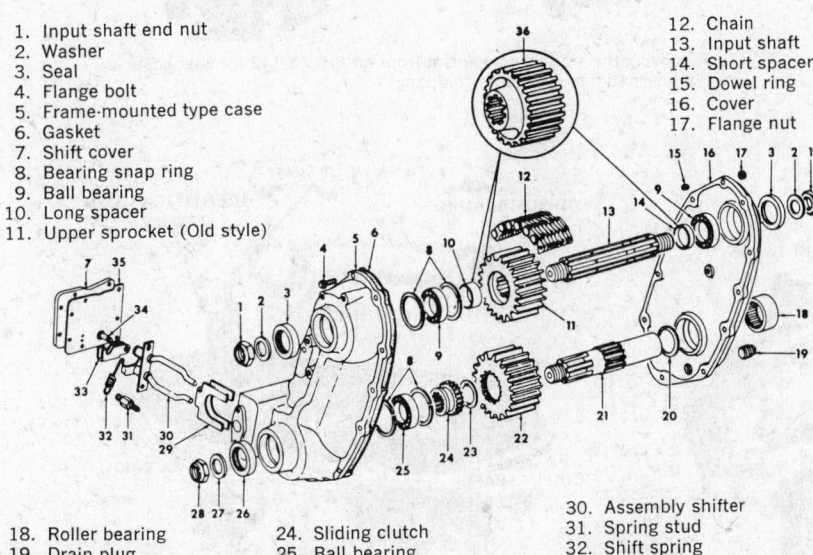

18. Roller bearing	24. Sliding clutch
19. Drain plug	25. Ball bearing
20. Thrust washer	26. Seal
21. Output shaft	27. Washer
22. Lower sprocket	28. Output shaft end nut
23. Thrust washer	29. Shift shoe

30. Assembly shifter	
31. Spring stud	
32. Shift spring	
33. Shift clevis	
34. Clevis pin	
35. Shift cover gasket	
36. Upper sprocket (new style)	

Exploded view of a frame-mounted type International Harvester TC-143 transfer case (© International Harvester Company)

12. Lift the sliding clutch and its shift shoe from the front output shaft.

13. The two shafts are removed from the case with a press or by tapping them out with a soft hammer.

NOTE: If the transfer case is a transmission mounted unit, a snap ring on the input shaft must be removed before the shaft can be removed from the case.

14. After the shafts are removed, remove the oil seals and the bearing snap rings, and press or tap out the two ball bearings.

15. Pry and remove the thrust washer from the boss for the front output shaft roller bearing in the other half of the case.

16. Press the roller bearing cage from the inside and out of the cover.

17. Press the ball bearing on the outer race from the outside and out of the cover.

Cleaning and Inspection

1. Clean all parts in solvent, removing all traces of old gaskets, sealants and lubricants, and dry the parts with compressed air.

2. Examine all the ball and roller bearings for wear or damage and replace as necessary.

3. Inspect the sprocket teeth and bores for damage and wear. Check the internal splines and clutch teeth for chipped surfaces. Small nicks or burrs can be removed with a file.

4. Check the smooth and splined surfaces of the shafts for wear or damage. The sliding clutch must move freely on the output shaft, but excessive clearance is remedied by replacement of parts.

5. Examine the chain for bent or broken links. If either condition exists, replace the chain.

Assembly

1. Install the two snap rings in the

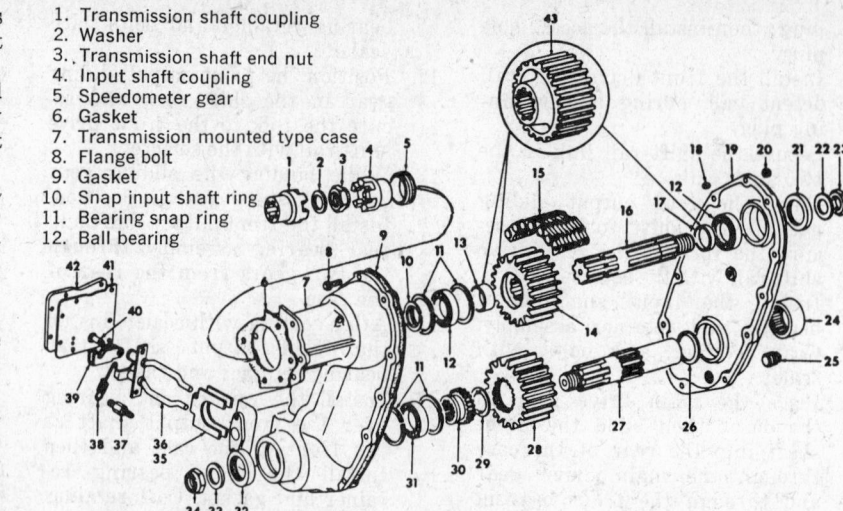

1. Transmission shaft coupling
2. Washer
3. Transmission shaft end nut
4. Input shaft coupling
5. Speedometer gear
6. Gasket
7. Transmission mounted type case
8. Flange bolt
9. Gasket
10. Snap input shaft ring
11. Bearing snap ring
12. Ball bearing

Exploded view of a transmission-mounted type International Harvester TC-143 transfer case (© International Harvester Company)

13. Long spacer	23. End nut	33. Washer
14. Upper sprocket (Old style)	24. Roller bearing	34. Output shaft end nut
15. Chain	25. Drain plug	35. Shift shoe
16. Input shaft	26. Thrust washer	36. Assembly shifter
17. Short spacer	27. Output shaft	37. Spring stud
18. Dowel ring	28. Lower sprocket	38. Shift spring
19. Cover	29. Thrust washer	39. Shift clevis
20. Flange nut	30. Sliding clutch	40. Clevis pin
21. Seal	31. Ball bearing	41. Shift cover gasket
22. Washer	32. Seal	42. Shift cover
		43. Upper sprocket (new style)

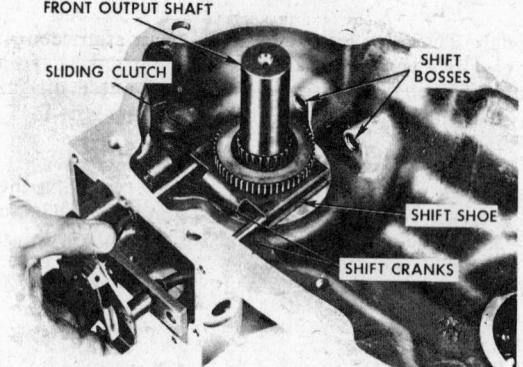

Removing the shifter mechanism from an I.H. TC-143 transfer case (© International Harvester Company)

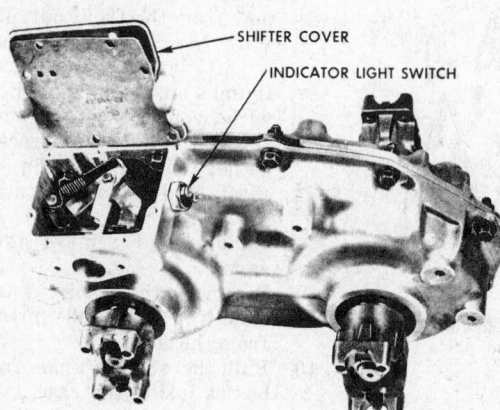

Removing the shifter cover on an I.H. TC-143 transfer case (© International Harvester Company)

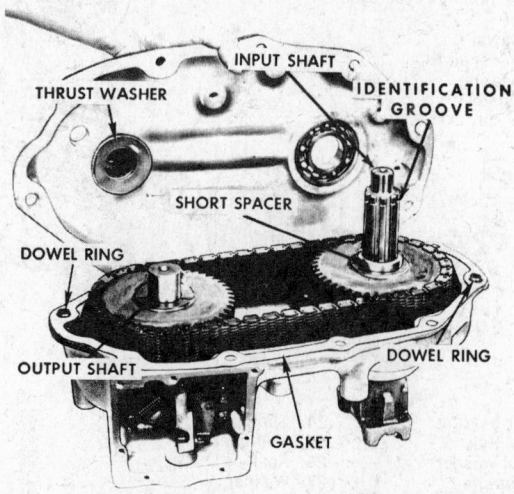

Removing the rear half of the case (cover) from an I.H. TC-143 transfer case (© International Harvester Company)

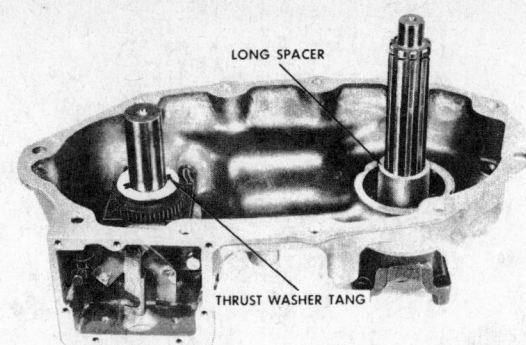

Installation of the thrust washer and long spacer, if so equipped, on an I.H. TC-143 transfer case (© International Harvester Company)

outer grooves of both bearing bores in the front half of the case.

2. Coat both bearing bores with lubricant and press or tap with a soft hammer both ball bearing assemblies into the case.

3. Install the two snap rings in the inner grooves of the two bearing bores in the front case half.

4. Install the shaft oil seals from the outside of the front case half. They are best installed with a press.

5. On a frame mounted transfer case, position the lightly lubricated shafts in the bearings, then, pull the shafts into the bearings by tightening the flange attaching nuts with the flanges installed. The input shaft is installed with the identification groove toward the rear of the transfer case. Tighten the flange attaching nut on the front output shaft to 200-250 ft. lbs., and the flange attaching nut on the input shaft until it bottoms, then back off two turns and retighten to 140-150 ft. lbs.

6. The output shaft in transmission mounted transfer cases is installed in the same manner as outlined for frame mounted units in step 5.

7. The input shaft in transmission mounted transfer cases has a snap ring to be installed on the front end of the shaft prior to installation in the bearing. The shaft is then pressed or tapped into the bearing with a soft

hammer until the snap ring is bottomed against the bearing.

8. Assemble the shift shoe to the sliding clutch and install the sliding clutch to the front output shaft.

9. Insert the shifter assembly into the transfer case so the shift cranks of the shifter pass through the shift shoe before being guided into the shift bosses. Make sure the shifter operates the sliding clutch and then secure the assembly with the bolt and spring stud.

NOTE: The flange bolt must be installed before the spring stud because the position of the spring stud when installed prevents the installation of the flange bolt.

10. Install the thrust washer to the front output shaft and the long spacer to the input shaft, if so equipped with a long spacer.

NOTE: Be sure the thrust washer tangs fit down into the splines on the shaft.

11. Lightly lubricate the bores of both sprockets and secure the case in a soft jawed vise by the end of the input shaft.

12. Install the upper sprocket to the outer end of the input shaft. Do not slide the sprocket completely into the case.

13. Position the chain over the sprocket and place the lower sprocket inside the chain.

14. Pull down on the lower sprocket enough to slide the lower sprocket onto the output shaft.

15. Slide both sprockets and chain onto the sprockets as far as they will go.

16. Install the short spacer on the input shaft (not required on later models).

17. Install the shift spring between the spring post and the shifter assembly.

18. Press the ball bearing into the input shaft bore of the rear cover from the inside surface.

19. Press the roller bearing into the output shaft bore of the rear cover from the outside surface until it is flush with the bearing bore.

20. Press the seal for the rear of the input shaft (rear output) into the cover until it is flush.

21. Place the cover on a bench, inside facing up. Coat the back side of the thrust washer with a thin coat of sealant and position it on the roller bearing boss with the tang on the washer mated with the oil passage in the boss.

NOTE: Use the sealant sparingly and make sure none of it enters the bearing or blocks the two oil passages.

22. Position a new gasket and two dowel rings on the mating surface of the case.

23. Position the cover onto the case, guiding the two shafts into their respective bearings. Secure the cover to the case with the attaching nuts and bolts, tightening them to 29-38 ft. lbs.

24. Install the indicator light switch to the case and check its operation with a test light.

25. Install the shift cover gasket and cover on the case and secure it with bolts and lockwashers tightened to 4-6 ft. lbs. Do not overtighten.

26. Install the rear output flange on the rear of the input shaft so both the input flange and the output flange are on the same plane. The flanges must be assembled in this manner to prevent vibration.

27. Install the washer and a nylon insert locknut on the shaft tightened to 140-150 ft. lbs. to secure the flange.

New Process Model T201

Description

The New Process Model 201 transfer case is a 2-speed gearbox which provides speed reduction and couples power to the front and rear driving axles.

CAUTION: Do not engage front driving axle when operating truck on hard surfaced roads at high speeds.

De-Clutch and Shift Rod Adjustment

NOTE: All adjustments must be made with the front axle engaged and the transfer case in low range.

1. Disconnect de-clutch and shift rods at shift levers.

2. Adjust de-clutch rod length until lever clears rear end of slot

in cab underbody by 5/8 in. Secure adjusting yoke with locknut.

3. Adjust shift rod length until distance between protusions on shift and de-clutch levers is 1/4 in. Secure adjusting yoke with locknut.

4. Connect de-clutch and shift rods at shift levers and then road test vehicle.

1 Ball, shift bar poppet
2 Spring, shift bar poppet ball
3 Bar, front output shaft
4 Fork, shift
5 Screw, set
6 Clutch, front output sliding
7 Fork, shift
8 Gear, front output drive
9 Bearing, roller (70 req'd)
10 Spacer, bearing
11 Shaft, front output front gear
12 Washer, thrust
13 Bearing, ball
14 Retainer, lock
15 Nut, bearing retaining
16 Shaft, idler gear
17 Cone, idler gear bearing
18 Cup, idler gear bearing
19 Shim, set
20 Spacer, bearing
21 Shaft input
22 Ring, snap
23 Bearing, roller (15 req'd)
24 Gear, rear output drive w/shaft
25 Gear, speedometer drive
26 Spacer, speedometer drive gear
27 Clutch, two-speed sliding
28 Plug, pipe
29 Gasket, set
30 Bolt, hex head, w/lock washer
31 Retainer, front output shaft rear bearing
32 Bolt, hex head, w/lock washer
33 Washer, lock
34 Nut, hex
35 Breather
36 Cup, rear output shaft outer bearing
37 Cone, rear output shaft outer bearing
38 Support, hand brake
39 Seal, rear output shaft outer bearing oil
40 Stud
41 Retainer, rear output drive gear bearing
42 Plate, idler shaft cover
43 Cup, rear output shaft inner bearing
44 Cone, rear output shaft inner bearing
45 Case, transfer
46 Bolt w/washer
47 Bolt, hex head
48 Washer, lock
49 Cover, power take-off opening

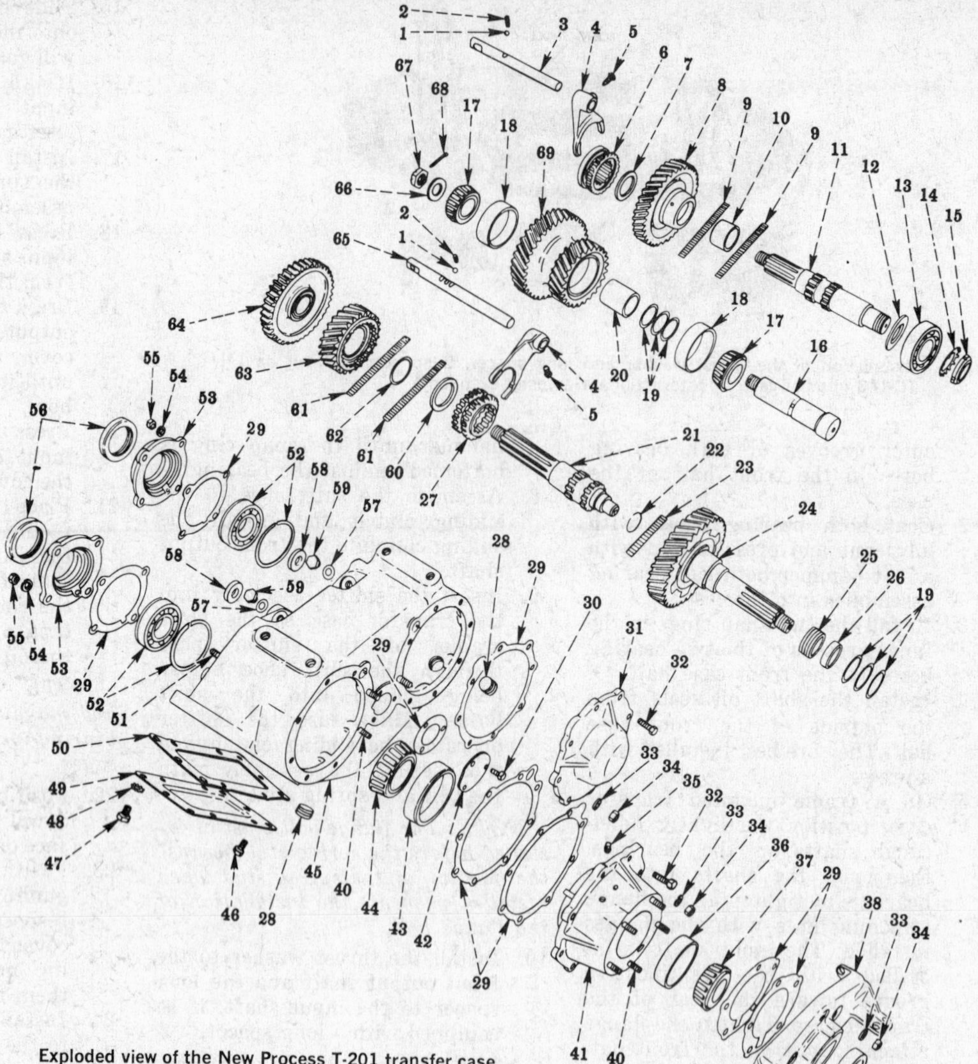

50 Gasket, power take-off cover
51 Stud
52 Bearing, input and front output ball
53 retainer, input and front output bearing
54 Washer, lock
55 Nut, hex
56 Seal, input and front output bearing retainer oil
57 Gasket
58 Seal, front and rear output shift bar oil
59 Screw, poppet ball retainer
60 Washer, thrust
61 Bearing, roller (70 req'd)

62 Spacer, bearing
63 Gear, input drive
64 Gear, power take-off
65 Bar, two-speed shift
66 Washer
67 Nut, hex
68 Pin, cotter
69 Gear, idler

Exploded view of the New Process T-201 transfer case
(© Chrysler Corporation)

Disassembly

Rear Output Shaft

1. Remove cotter pin and flange nut at rear output shaft.
2. Remove attaching bolts and nuts, then remove bearing retainer and gasket. Discard gasket.
3. Remove output shaft assembly from case.
4. Remove nut and washer at rear of shaft.
5. Remove brake drum by tapping on it lightly (if necessary).
6. Remove shaft and gear from bearing retainer.
7. Remove shims, spacer and speedometer drive gear. Tie shims together for reassembly.

8. Remove inner bearing cone, using a suitable puller.
9. Remove snap-ring and roller bearings from shaft gear bore.
   NOTE: The rear output shaft and gear is serviced as an assembly. Do not remove gear from shaft.
10. Remove attaching nuts, brake support, bearing and oil seal. Discard seal.

Front Output Shaft

1. Remove attaching bolts and nuts, then remove rear bearing retainer and gasket. Discard gasket.
2. Remove cotter pin and nut at front output shaft.

3. Remove attaching nut, washer and companion flange.
4. Remove the front output shaft assembly through the rear of the case.
5. Remove sliding clutch gear.
6. Position front output shaft assembly in a soft-jawed vise.
7. Remove lock retainer and bearing nut.
8. Using a suitable puller, remove bearing.
9. Remove front output shaft drive gear by lifting upward and holding the thrust washers against the hub to retain bearing rollers.
10. Carefully set aside thrust washers, then remove the two rows of bearing rollers (70 rollers)

and the spacer separating them.

11. Remove attaching nuts and bearing retainer with gaskets. Note number of gaskets used.

Idler Gear, Shift Bar and Fork

1. Remove safety wire and set-screws securing shift forks to shift bars.
2. Remove poppet ball retaining plugs, gaskets and poppet ball springs.
3. Pull shift bars out of the case, then remove shift forks and poppet balls.
4. Remove cotter pin, nut and washer from front end of idler shaft.
5. Remove attaching bolts, gasket and idler gear shaft cover plate.
6. Remove idler gear assembly.

NOTE: When removing the idler gear assembly, use an arbor as shown in illustration and perform the following:

a. Install arbor on threaded end of shaft so that it seats against the shaft shoulder.
b. Drive arbor until shaft is free at rear of case.
c. Separate shaft and arbor.
d. Remove gear with arbor through front output shaft opening.
e. Remove arbor, shims, spacer and bearing cones. Tie shims together for reassembly.

Input Shaft

1. Remove cotter pin and input shaft flange nut.
2. Remove nut, flat washer and companion flange.
3. Remove input shaft assembly through rear output shaft opening in case.

NOTE: Make certain to retain power take-off and drive gear on the shaft.

4. Mount input shaft in a soft-jawed vise.
5. Remove power take-off gear.
6. Remove input drive gear while holding thrust washers tightly against hub.
7. Carefully remove the two rows of bearing rollers (70 rollers) and the separating spacer.
8. Remove attaching nuts, bearing retainer, gaskets and oil seal. Note the number of gaskets used and discard the oil seal.

Assembly

The transfer case is assembled in the reverse order of disassembly. The following procedures concerning preloading bearings and bearing adjustment however, must all be completed before (or in some cases, during) installation of shaft and gear assemblies.

Input Shaft

1. The input shaft drive gear bearing consists of 70 rollers divided into two rows by a spacer.
2. Coat gear bore with grease, then position rollers and spacer.
3. Hold thrust washers against hub to retain rollers, then install gear on shaft.
4. Input shaft end-play is controlled by gasket thickness between case and bearing retainer.
5. Position retainer on case, then measure clearance with a feeler gauge.
6. Select gasket(s) with a thickness .005 in. more than measured clearance.
7. Remove retainer and install selected gasket(s), then reinstall retainer.

Rear Output Shaft

1. Rear output shaft bearing preload is set by shim set size selection.
2. Using an arbor press and suitable sleeve, install inner bearing cone on shaft. The cone should be firmly seated against gear.
3. Install speedometer drive gear, spacer and original bearing shim set.
4. Install rear output shaft in bearing retainer, then position outer bearing cone on shaft.
5. Install parking brake drum, flat washer and slotted nut. Torque nut to 125 ft. lbs.
6. Mount shaft and retainer assem-

Removing/installing the rear output shaft on a New Process T-201 transfer case (© Chrysler Corporation)

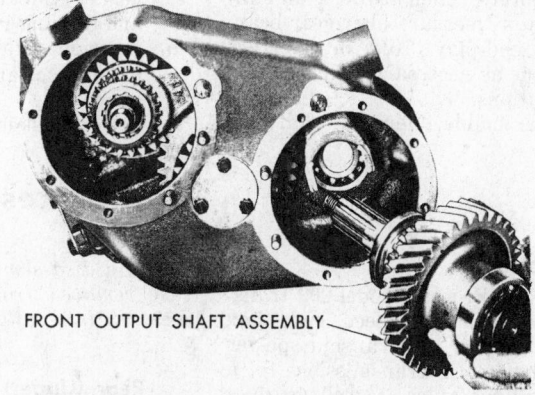

FRONT OUTPUT SHAFT ASSEMBLY

Removing/installing the front output shaft on a New Process T-201 transfer case (© Chrysler Corporation)

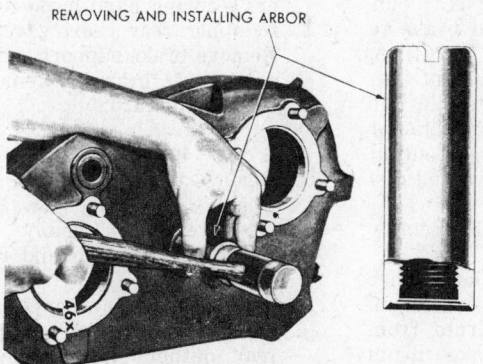

REMOVING AND INSTALLING ARBOR

Removing/installing the idler gear shaft on a New Process T-201 transfer case (© Chrysler Corporation)

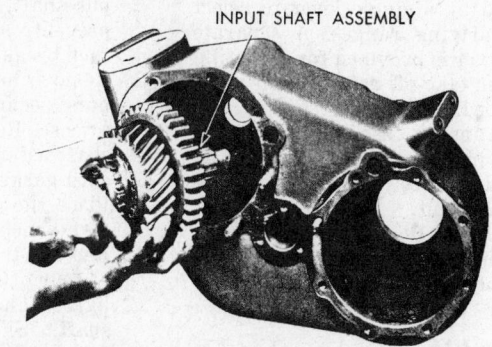

INPUT SHAFT ASSEMBLY

Removing/installing the input shaft assembly on a New Process T-201 transfer case (© Chrysler Corporation)

bly in a vise so that shaft is free to rotate.

7. Turn shaft until bearing rolls smoothly. Then, using an inch-pound torque wrench, measure bearing preload with the wrench in motion. Subtract or add shims as necessary to meet specifications.

NOTE: Shims for adjusting bearing preload are available in four thicknesses.

8. After the final adjustment, remove slotted nut, washer and brake drum.
9. Install parking brake support and new support oil seal.
10. Reinstall brake drum, washer and slotted nut. Torque nut to specfications.

Idler Gear

1. Idler gear bearing end-play is controlled by shim set size selection.
2. Clamp the idler gear shaft (large end) in a soft-jawed vise.
3. Install bearing cone against shoulder on shaft.
4. Install bearing spacer and original shim set.
5. Install idler gear (small end down) on shaft, then install other bearing cone.
6. Position spacer tool as shown in illustration, then install flat washer and nut. Torque nut to specifications.
7. Rotate gear until bearing rolls smoothly. Then using a dial indicator, measure idler gear bearing end-play. Add or subtract shims as necessary to meet specifications.
8. Disassemble idler gear shaft as-

sembly. Tie newly selected shims together.

9. To install gear assembly in the case, position bearing cone in in the large end of the gear, then place gear (small end up) on a bench.
10. Install the same arbor used to remove idler gear shaft, a spacer, new shims and other bearing cone in the idler gear.
11. Hold the idler gear assembly in the case with the small diameter gear facing rearward, then insert the shaft through the case (from rear) and thread it into the previously installed arbor.
12. Tap on the shaft until arbor extends through opposite side of case.
13. Remove arbor.

Two-Speed Clutch

When installing the two-speed clutch on the input shaft, make certain that the recessed side of the gear is toward the front of the case.

Front Output Shaft

1. The front output shaft drive gear bearing consists of 70 bearing rollers divided into two rows by a spacer.
2. Coat gear bore with grease, then position rollers and spacer.
3. Hold thrust washers against each hub to retain rollers, then install gear on shaft.
4. Shaft end-play is controlled by gasket thickness between case and bearing retainer.
5. Position retainer on case, then measure clearance with a feeler gauge.
6. Select gasket(s) with a thick-

ness .005 in. more than measured clearance.

7. Remove retainer and install selected gasket(s), then reinstall retainer.
8. When installing the front output shaft assembly in the case, locate front output shift bar as far toward rear of case as possible. This will make it possible to position shift fork in clutch gear collar and to align clutch gear splines on front output shaft.

SPECIFICATIONS
END PLAY (IN.)
Input Shaft .... Approx. 0.005
Front Output Shaft ...... Approx. 0.005
Idler Gear 0.000-0.002

BEARING PRELOAD (IN. LBS.)
Rear Output Shaft Bearings 15-30

TORQUE LIMITS (FT. LBS.)
Front Retainer Nut 35-55
Poppet Screw 15-25
Idler Cover Screw 15-25
Drain and Filler Plugs 25-45
Top Cover 38-42
All Bearing Caps 38-42
Flange and Idler Shaft Nut 125
Input Shaft 300-400
Front Output Shaft 300-400
Rear Output Shaft 300-400
Idler Shaft Nut 140-160
Front Output Shaft Flange Nut .. 140-160
Input Shaft Flange Nut 140-160
Driveshaft Mating Flange Nuts 35
Breather 8-12
Brake Support Nut 25-45
Brake Retaining Screw 20-40
Brake Retainer Nut 25-45
Brake Drum 60-66
Brake Mounting 60-66
Brake Drum Nut 140-160
P.T.O. Screw 8-12
P.T.O. to Case 38-42
P.T.O. Bearing Cap 38-42
P.T.O. Top Cover 38-42
P.T.O. Shaft 300-400

New Process Model 202

Description

The New Process Model 202 transfer case is a two-speed gear box which is used to transmit power from the main transmission to a front driving axle, as well as to a conventional rear axle. Sliding clutch gears in the transfer case are controlled by a single lever to select various driving ranges. A separate control lever is provided for operating the power take-off assembly on units so equipped.

The transfer case should not be operated in four-wheel high range on dry, hard surface roads, as rapid tire wear will result.

CAUTION: Do not operate transfer case in neutral range for extended periods of time when power take-off assembly is disengaged.

Disassembly

NOTE: The following procedure

covers a disassembly of the transfer case removed from the chassis and mounted in a suitable stand.

Rear (Upper) Output Shaft

1. Remove cotter pin from flange nut located at end of rear output shaft. Apply hand brake to prevent shaft from rotating; then break the torque on nut.
2. Remove hex head bolts and two nuts securing bearing retainer to case. Remove the rear output shaft assembly. Remove and discard gasket.
3. Place the output shaft assembly on a bench.
4. Remove the nut and flat washer at rear of shaft. Tap brake drum lightly and remove drum from shaft. Slide off brake drum flange spacer.
5. Remove output shaft with drive

gear from upper rear bearing retainer. Slide off speedometer drive gear, spacer and drive gear inner bearing. Remove snap-ring and fifteen roller bearings.

6. Remove hex nuts and lock washers securing hand brake support to upper rear bearing retainer. Remove brake support and drive gear outer bearing. Remove oil seal and discard.

Input Shaft

1. Remove ten capscrews and lock washers attaching power take-off cover or assembly (if so equipped). Remove and discard gasket.
2. Remove shift bar poppet screw, spring and ball. Through the rear output shaft opening in case, remove roll pin securing two-speed shift fork on bar. Use

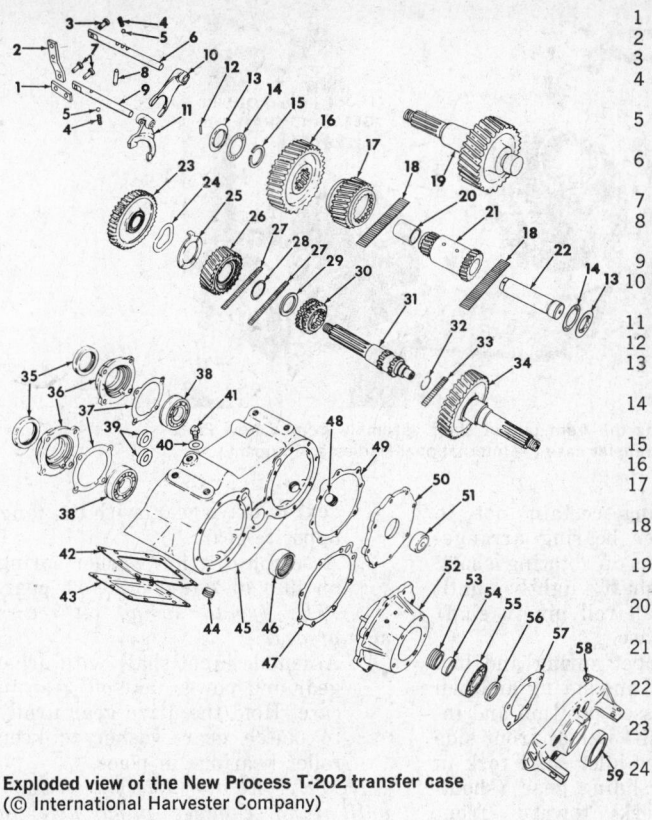

Exploded view of the New Process T-202 transfer case
(© International Harvester Company)

1 Link, shift lever
2 Lever, shift
3 Pin, rod
4 Spring, shift bar poppet
5 Ball, shift bar poppet
6 Bar, shift, two-speed
7 Pin, rod
8 Plunger, shift bar interlock
9 Bar, shift idler
10 Fork, shift, two-speed
11 Fork, shift, idler
12 Pin, idler shaft
13 Bearing, thrust, race
14 Bearing, thrust, needle
15 Ring, snap, idler
16 Gear, idler drive
17 Gear, sliding, idler
18 Bearing, roller, idler, shaft
19 Gear, lower shaft (front output)
20 Spacer, rotating shaft
21 Shaft, idler gear, rotating
22 Shaft, idler gear, stationary
23 Gear, PTO, input shaft, upper
24 Spring, friction washer, input shaft

25 Washer, friction, input shaft
26 Gear, drive, input shaft, upper
27 Bearing, roller, input shaft, front
28 Spacer, input shaft drive gear bearing
29 Washer, input shaft sliding clutch
30 Gear, clutch, input shaft, upper
31 Shaft, input
32 Ring, snap
33 Bearing, roller, drive gear
34 Gear, shaft, output drive
35 Seal, oil, input and output shaft, front
36 Retainer, bearing, input and output shaft, front
37 Gasket, bearing retainer
38 Bearing, input and output shaft, front
39 Seal, oil, shift bar
40 Gasket, shift bar poppet screw
41 Screw, shift bar poppet
42 Gasket, PTO cover
43 Cover, PTO opening

44 Plug, pipe
45 Case, transfer
46 Bearing, gear, output shaft, inner
47 Gasket, rear bearing retainer, upper
48 Plug, pipe
49 Gasket, rear bearing retainer, lower
50 Retainer, lower rear bearing (front output shaft)
51 Bearing, lower shaft, rear
52 Retainer, upper rear (output shaft) bearing
53 Gear, speedometer drive gear
54 Spacer, speedometer drive gear
55 Bearing, output shaft drive gear, outer
56 Spacer, brake drum flange
57 Gasket, brake drum support
58 Support, hand brake
59 Seal, oil, output shaft bearing, rear

a 5/32 in. diameter steel rod to drive out roll pin.

NOTE: Where case interference prevents complete removal of roll pin, drive the pin until shift bar can be removed. Shift fork can then be moved toward input shaft clutch gear to provide clearance for removal of roll pin.

3. Position idler shift bar in four-wheel high drive position (shift bar to extreme rear in case). Simultaneously remove input shaft clutch gear and two-speed shift fork.

4. Remove cotter pin from nut at outer end of input shaft. Use a suitable flange holder to prevent shaft from turning; then break torque on nut. Remove the nut,

Removing/installing the rear output shaft from a New Process T-202 transfer case
(© International Harvester Company)

flat washer and companion flange.

5. Move the input shaft and gear assembly toward rear of case by lightly tapping on the outer end of shaft. This will free shaft from bearing and permit shaft and power take-off gear to be removed.

CAUTION: Hold input drive gear firmly against shoulder on shaft to retain roller bearings in bore of gear.

6. Place input shaft on a bench and remove drive gear, roller bearings, spacer and sliding clutch gear washer. Remove friction washer and washer spring from hub of power take-off gear.

7. Remove input shaft bearing retainer, gasket and bearing. If necessary, remove and discard oil seal from bearing retainer.

Front (Lower) Output Shaft

1. Remove cotter pin from nut at outer end of front output shaft. Use a suitable flange holder to prevent shaft from turning; then break torque on nut. Remove nut, flat washer and flange.

2. Remove hex head bolts and nuts attaching input shaft rear bearing retainer to case. Lightly tap on end of shaft, moving it toward rear of case to displace shoulder on rear bearing re-

tainer. Remove drive gear and shaft.

3. Press drive gear rear bearing from retainer. Remove front bearing retainer and bearing. If lubricant leakage is evident at the retainer, remove oil seal and discard.

Idler Shift Bar and Gear

1. Remove roll pin securing fork to idler shift bar. Use a 5/32" diameter steel rod to drive out pin. Withdraw shift bar and fork.

NOTE: Spring tension on poppet ball will displace ball from case upon removing shift bar.

2. Remove shift bar interlock plunger and spring.

3. Remove roll pin from outer end of idler gear stationary shaft. Using a brass drift, drive the idler shaft toward rear of case. Remove shaft.

4. Remove idler gear rotating shaft assembly through the opening from which the rear output shaft assembly was removed. Place the gear assembly on a bench.

5. Remove thrust bearing and thrust bearing race at each end of rotating shaft. Remove individual roller bearing and spacer.

6. Remove snap-ring; then separate idler drive gear and shaft. Remove sliding clutch gear.

Transfer Cases

Assembly

Lubricate all bearings, bushings, spline shafts and shift bar forks at their contact surfaces during assembly. This will provide initial lubrication and avoid possible damage when the transfer case is first operated.

Idler Shift Bar and Gear

1. Place the idler gear rotating shaft in a vertical position with the large splined gear end resting on bench.
2. Position sliding clutch gear on shaft with the groove which receives idler shift fork, facing upward.
3. Position drive gear on shaft so that gear internal splines engage splines on shaft.

CAUTION: The long shoulder side of drive gear must face sliding clutch gear to provide clearance with the inside face of case.

4. Install snap-ring and check for proper side clearance with drive gear. This is accomplished by inserting a feeler gauge between the ring and gear. Three different thicknesses of snap-rings are available for obtaining proper clearance. Select and install proper size. Refer to Specifications for recommended snap-ring side clearance.

NOTE: Snap-ring side clearance must be maintained to prevent possible ring breakage.

5. Coat the bore of idler gear rotating shaft with Lubriplate No. 110; then assemble roller bearings and spacer. Apply Lubriplate to thrust bearing and bearing race; then assemble at each end of idler rotating shaft. Carefully place rotating shaft assembly in case. Align shaft with opening in front and rear of case for installing the stationary shaft. Install stationary

Removing the front output shaft assembly from a New Process T-202 transfer case (© International Harvester Company)

shaft, making certain not to disturb roller bearing arrangement in bore of rotating shaft. Tap the shaft lightly until seated. Install roll pin in shaft at front of case.

6. Position poppet spring and ball in case. Hold the spring and ball in a compressed position and insert idler shift bar at front side of case. Place idler shift fork in position on sliding gear (shoulder on fork toward front of case) and install on shift bar. Secure fork to shift bar with roll pin.

NOTE: Side of shift bar employing only one detent must be facing upward.

Input Shaft

1. Place the input drive gear on a bench and assemble roller bearings and spacer. Coat the gear bore with Lubriplate No. 110 to retain bearing rollers.
2. Install sliding clutch washer; then carefully position drive gear with bearings on shaft. Place friction washer on shaft

next to drive gear, with the tang opposite gear.
3. Assemble friction washer spring on hub of power take-off gear.

NOTE: Install spring on recess side of gear.

4. Assemble input shaft with drive gear and power take-off gear in case. Hold the drive gear firmly to clutch gear washer to keep roller bearings in place.

NOTE: Rotate friction washer until tang engages blind hole in power take-off gear.

5. With the idler shift bar in four-wheel high drive position (shift bar to extreme rear of case), insert interlock plunger in case.
6. Install two-speed shift bar in case so that side of bar with the single detent is facing the idler shift bar. Install sliding clutch gear and two-speed shift bar fork. Install roll pin to secure fork to shift bar.
7. Install input shaft bearing.

CAUTION: Movement of input shaft toward rear of case while installing bearing will permit rollers to become displaced from bore of drive gear.

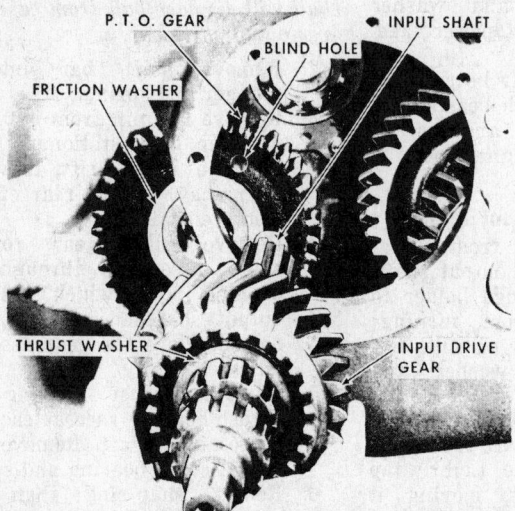

Removing the input shaft and P.T.O. gear from a New Process T-202 transfer case (© International Harvester Company)

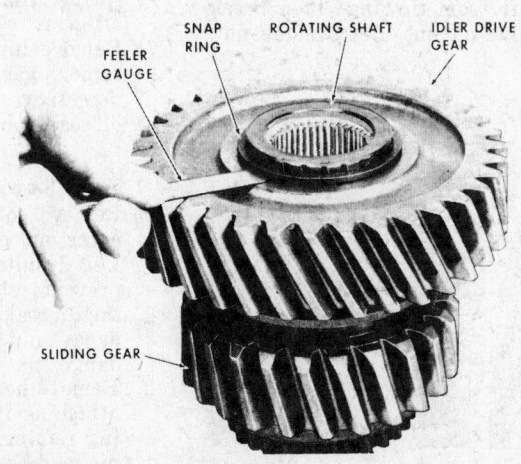

Checking the side clearance between the idler drive gear and retaining snap ring; New Process T-202 transfer case (© International Harvester Company)

8. Position bearing retainer to case and install hex head bolts finger tight. Do not torque bolts at this time since gasket thickness must be determined following installation of rear output shaft assembly.

Front (Lower) Output Shaft

1. Install front output shaft bearing in face of case.
2. Install front output shaft and drive gear.
3. Press drive gear rear bearing in retainer. Install bearing with stamped end of bearing resting on press. Position gasket and retainer at rear face of case. Install hex head bolts with lock washers and tighten securely.
4. Install new oil seal in bearing retainer. Assemble retainer to front face of case and install hex head bolts finger tight. Do not torque bolts at this time.

Rear (Upper) Output Shaft

1. Support rear output shaft bearing retainer in a press; then install shaft inner bearing flush with outer surface of retainer.
2. Place bearing retainer on a bench and assemble drive gear and shaft. Position speedometer drive gear and spacer on shaft. Assemble outer bearing on shaft and install in upper rear bearing retainer. Install hand brake drum spacer.
3. Install new oil seal in hand brake support. Install gasket and brake support to upper rear bearing retainer.
4. Position gasket to bearing retainer and install the rear output assembly on case.

Input and Front Output Shaft Bearing Adjustment

The fit of the input and front output shaft bearing retainer to their respective bearings is controlled by gasket thickness. To determine the correct thickness or number of gaskets required, measure the clearance between retainer and case with a feeler gauge. Select gasket(s) which will give a bearing retainer fit within specified limits. See Specifications. Remove the retainer; then install selected gasket(s) and retainer.

SPECIFICATIONS

BEARING CLEARANCE

Input and Front Output Shaft Bearings	0.003-0.006 in.
Gasket Thickness Available	0.009-0.011 in.
	0.0135-0.0165 in.

SNAP RING

Side Clearance	0.007-0.013 in.
Thickness Available	0.092-0.094 in.
	0.095-0.097 in.
	0.098-0.100 in.

POPPET BALL SPRING

Free Length	1.0 in.
Pressure @ 21/32"	30 Lbs.

TORQUE LIMITS

Flange Nut	Minimum 125 ft lbs.

New Process Model 205

Description

The New Process Model 205 transfer case is a two-speed gearbox mounted between the main transmission and the rear axle. The gearbox transmits power from the transmission and engine to the front and rear driving axles.

Disassembly

Transfer Case

1. Clean the exterior of the case.
2. Remove the nuts from the universal joint flanges.
3. Remove the front output shaft rear bearing retainer, front bearing retainer and drive flange.
4. Tap the front output shaft assembly from the case with a soft hammer. Remove the sliding clutch, front output high gear, washer and bearing from the case.
5. Remove the rear output shaft housing attaching bolts and remove the housing, output shaft, bearing retainer and speedometer gear.
6. Slide the rear output shaft from the housing.

NOTE: Be careful not to lose the 15 needle bearings that will be loose when the rear output shaft is removed.

7. Drive the two ¼ in. shift rail pin acess hole plugs into the transfer case with a punch and hammer.
8. Remove the two shift rail detent nuts and springs from the case. Use a magnet to remove the two detent balls.
9. Position both shift rails in neutral and remove the shift fork retaining roll pins with a long punch.

1. Shift lever link	
2. Bar	30. Gasket
3. Bar	31. Retainer
4. Plunger	32. Cone
5. Seal	33. Cup
6. Screw	34. Shim set
7. Gasket	35. Gear
8. Spring	36. Spacer
9. Ball	37. Shaft
10. Plug	38. Gasket
11. Nut	39. Cover
12. Washer	40. Bearing
13. Seal	41. Shaft
14. Retainer	42. Ring
15. Gasket	43. Washer
16. Bearing	44. Bearing
17. Washer	45. Gear
18. Gear	46. Washer
19. Shaft	47. Bearing
20. Pin	48. Gear
21. Clutch	49. Spacer
22. Fork	50. Retainer
23. Pin	51. Breather
24. Bearing	52. Gasket
25. Spacer	53. Retainer
26. Gear	54. Seal
27. Washer	55. Case
28. Ring	56. Gasket
29. Bearing	57. Cover

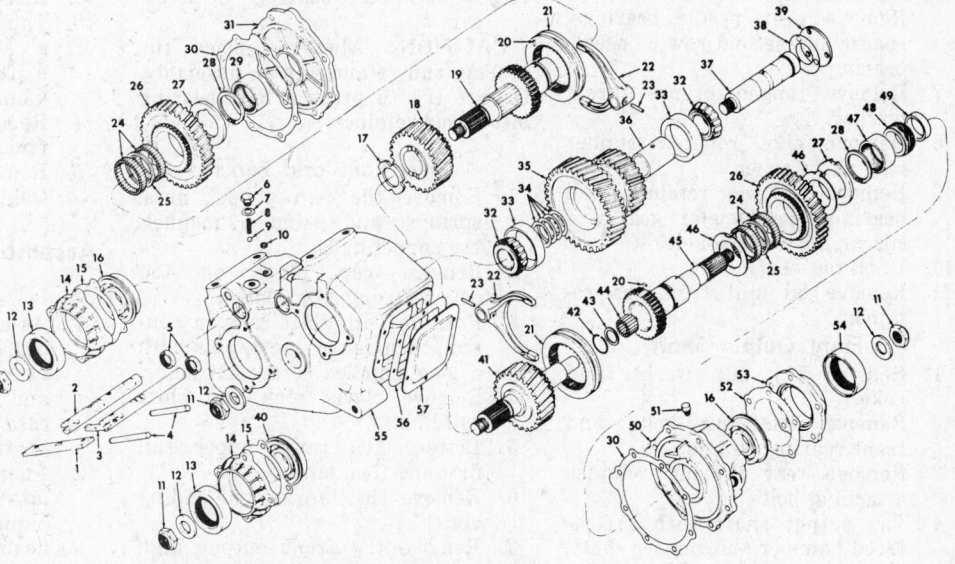

Exploded view of a New Process 205 transfer case
(© International Harvester Company)

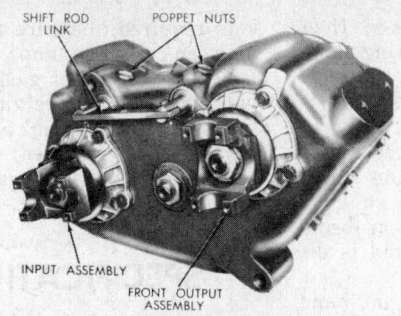

Front view of a New **Process** 205 transfer case. This unit would have a short driveshaft connected to the input shaft. Others are bolted directly to the transmission (© International Harvester Company)

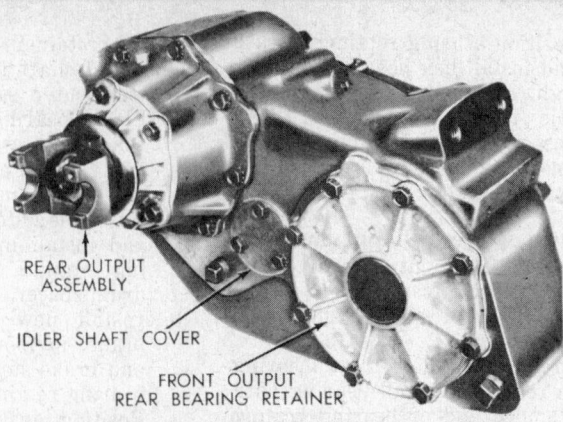

Rear view of a New Process 205 transfer case (© G.M.C.)

10. Remove the clevis pin from one shift rail and rail link.
11. Remove the range shift rail first, then the 4WD shift rail.
12. Remove the shift forks and and sliding clutch from the case. Remove the input shaft bearing retainer, bearing and shaft.
13. Remove the cup plugs and rail pins, if they were driven out, from the case.
14. Remove the locknut from the idler gear shaft.
15. Remove the idler gear shaft rear cover.
16. Remove the idler gear shaft, using a soft hammer and a drift.
17. Roll the idler gear assembly to the front output shaft hole and remove the assembly from the case.

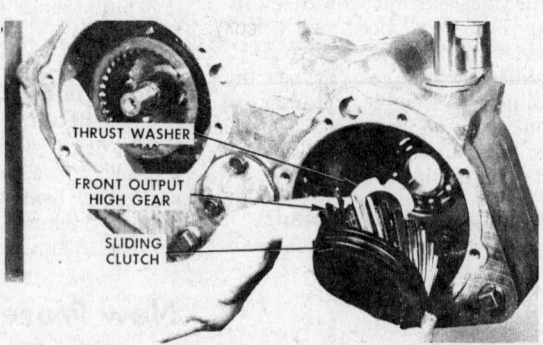

Removing the sliding clutch, front output high gear, washer and bearing— New Process 205 (© International Harvester Company)

Rear Output Shaft and Yoke
1. Loosen rear output shaft yoke nut.
2. Remove shaft housing bolts, then remove the housing and retainer assembly.
3. Remove retaining nut and yoke from the shaft, then remove the shaft assembly.
4. Remove and discard snap ring.
5. Remove thrust washer and pin.
6. Remove tanged bronze washer. Remove gear needle bearings, spacer and second row of needle bearings.
7. Remove tanged bronze thrust washer.
8. Remove pilot rollers, retainer ring and washer.
9. Remove oil seal retainer, ball bearing, speedometer gear and spacer. Discard gaskets.
10. Press out bearing.
11. Remove oil seal from the retainer.

Front Output Shaft
1. Remove lock nut, washer and yoke.
2. Remove attaching bolts and front bearing retainer.
3. Remove rear bearing retainer attaching bolts.
4. Tap output shaft with a soft-faced hammer and remove shaft, gear assembly and rear bearing retainer.

5. Remove sliding clutch, gear, washer and bearing from output high gear.
6. Remove sliding clutch from the high output gear; then remove gear, washer and bearing.
7. Remove gear retaining snap-ring from the shaft, using large snap-ring picks. Discard ring.
8. Remove thrust washer and pin.
9. Remove gear, needle bearings and spacer.
10. Replace rear bearing, if necessary.

CAUTION: Always replace the bearing and retainer as an assembly. Do not try to press a new bearing into an old retainer.

Shift Rails and Forks
1. Remove the two poppet nuts, springs, and using a magnet, the poppet balls.
2. Remove cup plugs on top of case, using a ¼" punch.
3. Position both shift rails in neutral, then remove fork pins with a long handled screw extractor.
4. Remove clevis pins and shift rail link.
5. Lower shift rails; upper rail first and then lower.
6. Remove shift forks and sliding clutch.
7. Remove the front output high gear, washer and bearing. Remove the shift rail cup plugs.

Input Shaft
1. Remove snap-ring in front of bearing. Tap shaft out rear of case and bearing out front of case, using a soft-faced hammer or mallet.
2. Tilt case up on power take-off and remove the two interlock pins from inside.

Idler Gear
1. Remove idler gear shaft nut.
2. Remove rear cover.
3. Tap out idler gear shaft, using a soft-faced hammer and a drift approximately the same diameter as the shaft.
4. Remove idler gear through the front output shaft hole.
5. Remove two bearing cups from the idler gear.

Assembly

Transfer Case
1. Assemble the idler shaft gears, bearings, spacer and shims, and bearings on a dummy shaft tool and install the assembly into the case through the front output shaft bore, large end first.
2. Install the idler shaft from the large bore side, using a soft hammer to drive it through the bearings, spacer, gears, and shims.
3. Install a washer and new lock-

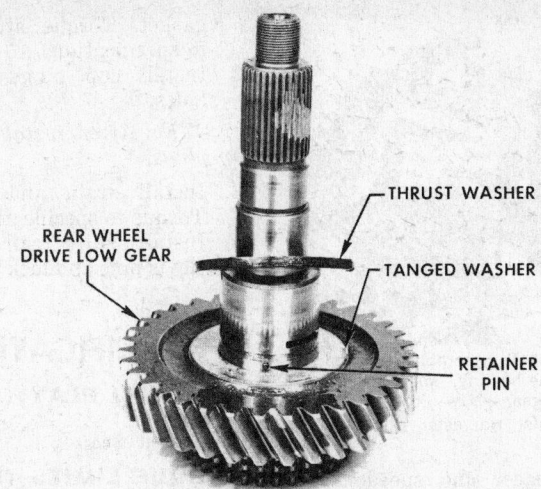

REAR WHEEL
DRIVE LOW GEAR

THRUST WASHER

TANGED WASHER

RETAINER
PIN

Removal of thrust washer, retainer pin and tanged bronze washer from the rear output shaft assembly on a New Process 205 (© International Harvester Company)

nut on the end of the idler shaft. Check to make sure the idler gear rotates freely. Tighten the locknut to specification.

4. Install the idler shaft cover with a new gasket so the flat side faces the rear bearing retainer of the front output shaft. Install and tighten the two retaining screws to the proper torque.

5. Install the interlock pins into the interlock bore through the front of the output shaft opening.

6. Start the 4WD shift rail into the front of the case, solid end of the rail first, with the detent notches facing up.

7. Position the shift fork onto the shift rail with the long end facing inward. Push the rail through the fork and into the Neutral position.

8. Position the input shaft and bearing in the case.

9. Start the range shift rail into the case from the front, with the detent notches facing up.

10. Position the sliding clutch to the shift fork. Place the sliding clutch on the input shaft and align the fork with the shift rail. Push the rail through the fork into the Neutral position.

11. Install the roll pins that lock the shift forks to the shift rails with a long punch.

12. Position the front wheel drive high gear and its thrust washer in the case. Position the sliding clutch in the shift fork. Shift the rail and fork into the front wheel drive (4WD-Hi) position ,while at the same time, meshing the clutch with the mating teeth on the front wheel drive high gear.

13. Align the thrust washer, high gear and sliding clutch with the bearing bore in the case and insert the front output shaft and

low gear into the high gear assembly.

14. Install a new seal in the front bearing retainer of the front output shaft, and install the bearing and retainer and new gasket in the case. Tighten the bearing retainer cap screws to the proper torque.

15. Lubricate the roller bearing in the front output shaft rear bearing retainer, which is the aluminum cover, and install it over the front output shaft and to the case. Install and tighten the retaining screws to the proper torque.

16. Move the range shift rail to the High position and install the rear output shaft and retainer assembly to the housing and input shaft. Use one or two new gaskets, as required, to adjust the clearance on the input shaft pilot. Install the rear output shaft housing retaining bolts and tighten them to specification.

17. Using a punch and sealing compound, install the shift rail pin access plugs.

18. Install the fill and drain plugs and the cross-link clevis pin.

Idler Gear

1. Press the two bearing cups in the idler gear.

2. Assemble the two bearing cones, spacer, shims and idler gear on a dummy shaft, with bore facing up. Check end-play.

3. Install idler gear assembly (with dummy shaft) into the case, large end first, through the front output shaft bore.

4. Install idler shaft from large bore side, driving it through with a soft-faced hammer or mallet.

5. Install washer and new locknut. Check for free rotation and

measure end-play. Torque locknut to specifications.

6. Install idler shaft cover and new gasket. Torque cover bolts to specifications.
NOTE: Flat side of cover must be positioned towards front output shaft rear cover.

Shift Rails and Forks

1. Press the two rail seals into the case.
NOTE: Install seals with metal lip outward.

2. Install interlock pins from inside case.

3. Insert slotted end of front output drive shift rail (with poppet notches up) into back of case.

4. While pushing rail through to neutral position, install shift fork (long end inward).

5. Install input shaft and bearing into case.

6. Install end of range rail (with poppet notches up) into front of case.

7. Install sliding clutch on fork, then place over input shaft in case.

8. Push range rail, while engaging sliding clutch and fork, through to neutral position.

9. Drive new lockpins into forks through holes at top of case.
NOTE: Tilt case on power take-off opening to install range rail lockpin.

Front Output Shaft and Gear

1. Install two rows of needle bearings in the front low output gear and retain with grease.
NOTE: Each row consists of 32 needle bearings and the two rows are separated by a spacer.

2. Position front output shaft in a soft-jaw vise, with spline end down. Place front low gear over shaft with clutch gear facing down; then install thrust washer pin, thrust washer and new snap-ring.
NOTE: Position snap ring gap opposite the thrust washer pin.

3. Place front drive high gear and washer in case. Install sliding clutch in the shift fork, then put fork and rail into 4-High position, meshing front drive high gear and clutch teeth.

4. Align washer, high gear and sliding clutch and bearing bore. Insert front output shaft and low gear assembly through the high gear assembly.

5. Install front output bearing and retainer with a new seal in the case.

6. Clean and grease rollers in front output rear bearing retainer. Install on case with one gasket and bolts coated with sealant. Torque bolts to specifications.

7. Install front output yoke, washer and locknut. Torque locknut to specifications.

Rear Output Shaft

1. Install two rows of needle bearings into the output low gear, retaining them with grease.

NOTE: Each row consists of 32 needle bearings and the two rows are separated by a spacer.

2. Install thrust washer (with tang down in clutch gear groove) onto the rear output shaft.
3. Install output low gear onto shaft with clutch teeth facing downward.
4. Install thrust washer over gear with tab pointing up and away. Install washer pin.
5. Install large thrust washer over shaft and pin. Turn washer until tab fits into slot located approximately 90° away from pin.
6. Install snap-ring and measure shaft end-play.
7. Grease pilot bore and install needle bearings.

NOTE: There are 15 pilot needle bearings.

8. Install thrust washer and new snap-ring in pilot bore.
9. Press new bearing into retainer housing.
10. Install housing on output shaft assembly.

Rear output shaft assembly removed; removal of the bearing, spacer, and speedometer gear—New Process 205
© International Harvester Company)

11. Install spacer and speedometer gear. Install rear bearing.
12. Install rear bearing retainer seal.
13. Install bearing retainer assembly on housing, using one or two gaskets to achieve specified clearance. Torque attaching bolts to specifications.
14. Install yoke, washer and locknut on output shaft.
15. Position range rail in high, then install output shaft and retainer assembly on case. Torque housing bolts to specifications.

Case

1. Install power take-off cover and gasket. Torque attaching bolts to specifications.
2. Install cup plugs at rail pin holes.

NOTE: After installing, seal the cup plugs.

3. Install drain and filler plugs. Torque to specifications.
4. Install shift rail cross link, clevis pins and lock pins.

SPECIFICATIONS

END PLAY (IN.)

Idler Gear	0.000-0.002 in.
Rear Output Shaft	0.002-0.027 in.

TORQUE LIMITS (FT. LBS.)

Idler Shaft Locknut	150
Idler Shaft Cover	20
Front Output Shaft Front Bearing Retainer	30-35
Front Output Shaft Yoke Locknut	130-150
Rear Output Shaft Bearing Retainer and Housing	30-35
Rear Output Shaft Yoke Locknut	130-150
P.T.O. Cover	15
Front Output Shaft Rear Bearing Retainer	30-35
Filler and Drain Plugs	30
Case to Frame	130
Case to Adapter	25
Adapter Mount	75
Case Bracket to Frame	
Upper	30
Lower	65
Adapter to Transmission	
Manual Transmission	30-35
Automatic Transmission	30-35

Dana/Spicer Model 20

Description

The Spicer Model 20 transfer case is essentially a two-speed gear box located at the rear of the standard transmission, which provides low and direct gear ranges. It also provides a means of connecting the power to the front axle.

Disassembly

1. Remove cover and gasket from bottom of transfer case.
2. Remove the front drive shift bar poppet spring access hole plugs from front output shaft bearing retainer. If necessary to remove shift bar lock plungers, remove the two expansion plugs from the housing.
3. Remove the idler gear shaft lock plate bolt and lock plate from rear of transfer case.
4. Using a hammer and brass drift, remove idler gear shaft, driving from the front end of the transfer case (opposite slotted end of shaft). As shaft passes thrust washers, remove washers. Remove idler gear with bearings and spacers as a sub-assembly. Remove bearings and spacer from bore of idler gear.
5. Remove end yoke retaining lock

nuts from the front and rear output shafts. Using a suitable puller, remove the end yokes from both shafts.

6. Remove steady springs from shift levers. Remove setscrew securing shift lever pivot pin. Remove pin and lift out each lever as it is freed.
7. Remove socket-head setscrew from underdrive shift fork.
8. Move underdrive shift fork and rear output shaft sliding gear forward far enough for gear to clear splines on rear output shaft. Swing fork and gear toward cover opening and lift out gear.
9. Remove rear output shaft bearing retainer and output shaft as an assembly.
10. Remove front output shaft rear cover and shims. Tag shims for reassembly.
11. Remove socket-head setscrew from front wheel drive shift fork.
12. Remove front output shaft bearing retainer with shift bars and gasket.
13. Pull forward on front output shaft as far as possible to permit removal of bearing cup. Remove bearing cone from front

output shaft, using suitable puller.

14. Remove front output shaft bearing snap-ring. Using a soft-faced hammer, tap on front end of output shaft and remove it from the rear side of housing. Take out driven gear and sliding gear with shift fork as shaft is withdrawn.
15. Remove shift bars from retainer.

CAUTION: Be sure the poppet balls and springs are secured when withdrawing shift bars. This will prevent possible personal injury.

16. Remove shift bar oil seals.
17. Remove speedometer driven gear.
18. Support inner face of bearing retainer on an arbor press and press output shaft from retainer.
19. Remove oil seal from retainer bore.
20. Remove tapered bearing cone and cup from rear bore of retainer. Remove tapered bearing cup from front bore. Remove O-ring seal.
21. Using a suitable puller, remove bearing cone from output shaft. Remove shims from shaft and tie together before laying aside. Remove speedometer drive gear.

22. Using a puller, remove bearing cone from front output shaft.

Assembly

1. Press bearing cone on rear end of front output shaft.
2. Install speedometer drive gear, shims and bearing cone on output shaft.
3. Position new O-ring seal in bore of rear output shaft bearing retainer. Install tapered bearing cup in front bore of retainer. Install other tapered bearing cup in rear bore of bearing retainer.
4. Place rear output shaft into bearing retainer and support shaft in an arbor press. Position bearing cone over end of shaft and using an adapter slightly larger in diameter than the shaft, press bearing cone on shaft. Locate rear output shaft bearing retainer assembly in a fixed position. Set up a dial indicator and check shaft endplay. Add or remove shims between speedometer drive gear and rear bearing cone to bring end-play within specifications.

NOTE: It is very important to set end-play correctly, since it controls the seating of the tapered bearings. Incorrect endplay will shorten the life of bearings.

5. Install new oil seal in rear bore of bearing retainer, using proper adapter.
6. Install speedometer driven gear.
7. Using an adapter of correct size, install shift bar oil seals in front output shaft bearing retainer.
8. Install poppet springs and balls into bearing retainer.

NOTE: make certain that the heavier loaded spring is installed for the front drive shift bar.

9. Depress poppet balls and insert shift bars into front output shaft bearing retainer.

NOTE: Be sure when installing shift bars that the seal lips are not damaged.

10. Start front output shaft into transfer case from rear. As shaft emerges inside of case, sliding gear with shift fork and front output driven gear can be installed. Hold shaft in position and install bearing cup in rear bore of case. Install front output shaft bearing snap-ring.

11. Support rear end of front output shaft and using an adapter of correct diameter, tap bearing cone onto front end of shaft. With shaft held in position, install bearing cup in front bore of case.

12. Using a new gasket, install front output shaft bearing retainer with shift bars on front of transfer case.

13. Install socket-head setscrew in front wheel drive shaft fork.

14. Install shims and front output shaft rear cover on case. Secure with bolts and torque to specifications. Set up a dial indicator at front end of front output shaft and check end-play. Add or remove shims under rear cover to meet specifications.

NOTE: It is essential to set end-play within specifications, since it controls seating of the tapered bearings. Improper end-play can shorten life of bearings.

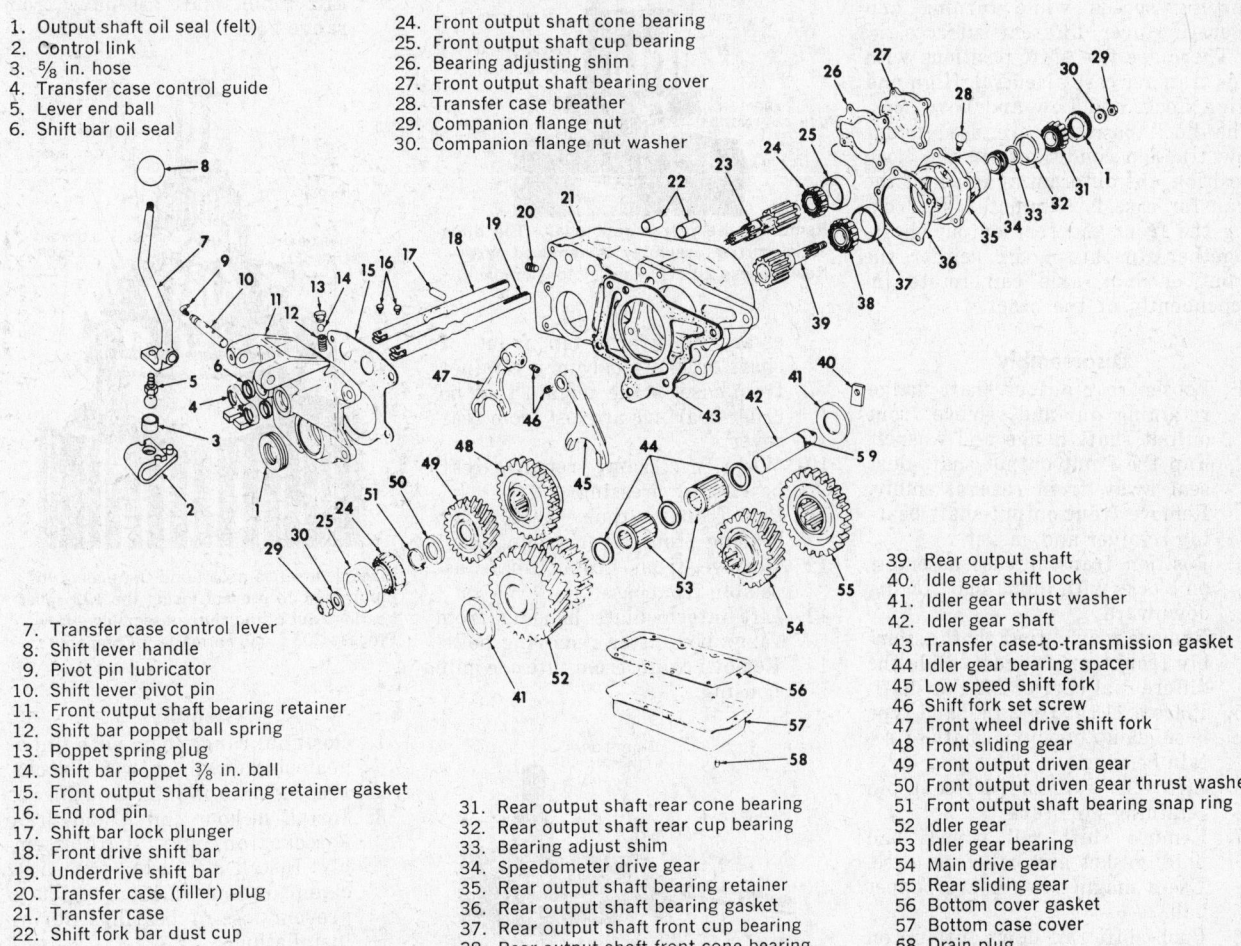

1. Output shaft oil seal (felt)
2. Control link
3. ⅝ in. hose
4. Transfer case control guide
5. Lever end ball
6. Shift bar oil seal
7. Transfer case control lever
8. Shift lever handle
9. Pivot pin lubricator
10. Shift lever pivot pin
11. Front output shaft bearing retainer
12. Shift bar poppet ball spring
13. Poppet spring plug
14. Shift bar poppet ⅜ in. ball
15. Front output shaft bearing retainer gasket
16. End rod pin
17. Shift bar lock plunger
18. Front drive shift bar
19. Underdrive shift bar
20. Transfer case (filler) plug
21. Transfer case
22. Shift fork bar dust cup
23. Front output shaft
24. Front output shaft cone bearing
25. Front output shaft cup bearing
26. Bearing adjusting shim
27. Front output shaft bearing cover
28. Transfer case breather
29. Companion flange nut
30. Companion flange nut washer
31. Rear output shaft rear cone bearing
32. Rear output shaft rear cup bearing
33. Bearing adjust shim
34. Speedometer drive gear
35. Rear output shaft bearing retainer
36. Rear output shaft bearing retainer gasket
37. Rear output shaft front cup bearing
38. Rear output shaft front cone bearing
39. Rear output shaft
40. Idle gear shift lock
41. Idler gear thrust washer
42. Idler gear shaft
43. Transfer case-to-transmission gasket
44. Idler gear bearing spacer
45. Low speed shift fork
46. Shift fork set screw
47. Front wheel drive shift fork
48. Front sliding gear
49. Front output driven gear
50. Front output driven gear thrust washer
51. Front output shaft bearing snap ring
52. Idler gear
53. Idler gear bearing
54. Main drive gear
55. Rear sliding gear
56. Bottom cover gasket
57. Bottom case cover
58. Drain plug

Exploded view of a Spicer 20 transfer case. This model has single lever control. Some models in I.H. trucks have dual lever control. Spicer 20 models installed in Jeep vehicles have a slightly different single lever control mechanism.
ⓒ International Harvester Company

15. Install rear output shaft bearing retainer and output shaft as an assembly. Secure with bolts and tighten to specifications.

16. Place rear output shaft sliding gear in underdrive shift fork. Swing fork and gear into case until gear can be positioned on rear output shaft.

17. Install socket-head setscrew in underdrive shift fork.

18. Position shift levers in front output shaft bearing retainer and install shift lever pivot pin. Install pivot pin retaining bolt. Install steady springs on shift levers.

19. Install end yokes on front and rear output shafts and secure with retaining lock nuts.

20. Install spacers and bearings in bore of idler gear. Position idler gear and thrust washers in transfer case. Insert idler shaft into rear of case until slot is flush with case face. Install lock plate and secure with retaining bolt. Tighten to specifications.

21. Install shift bar lock plungers and expansion plugs. Install shift bar poppet spring access hole plugs.

22. Position new gasket on bottom of transfer case and install cover. Secure with bolts and tighten to specifications.

SPECIFICATIONS
END PLAY

Rear Output Shaft	0.003-0.005 in.
Front Output Shaft	0.000-0.001 in.

BALL POPPET SPRING

Free Length

Red	1.028 in.
Yellow	1.078 in.

Pressure @ Test Length

Red @ 59/64 in.	26 Lbs.
Yellow @ 49/64 in.	12 Lbs.

TORQUE LIMITS (FT. LBS.)

Rear Output Shaft Bearing Retainer	25-32
Front Output Shaft Rear Bearing Cover	25-32
Front Output Shaft Front Bearing Retainer	25-32
Bottom Cover	10-15
Idler Shaft Lock Bolt	12-17
Companion Flange Nut	225-300

New Process Model 203 (Full Time Four Wheel Drive)

Description

The New Process Model 203 transfer case is a full-time 4WD unit that operates in 4WD at all times. The unit incorporates a differential similar to axle differentials; compensating for different speeds of the front and rear axles resulting from varying speeds while turning and operating over different surfaces.

There are five shift positions with this transfer case; Neutral, High and High Lock, and Low and Low Lock. The Lock positions are used under low traction conditions. In the Lock position, the differential action of the transfer case is eliminated, by locking the front and rear output shafts together. In this mode, neither the front or rear axle can rotate independently of the other.

Disassembly

1. Loosen rear output shaft flange retaining nut and remove front output shaft flange and washer.
2. Tap the front output shaft dust seal away from case assembly. Remove front output shaft bearing retainer and gasket.
3. Position transfer case assembly on blocks with input shaft facing downward.
4. Remove rear output shaft assembly from transfer case. Slide the differential carrier off the shaft.
5. Place a 1½ in. to 2 in. band type hose clamp on input shaft to retain bearings.
6. Lift shift rail and driveout pin retaining shift fork.
7. Remove shift rail poppet ball plug, gasket and ball from case. Use a magnet to remove poppet ball.
8. Push shift rail down, lift up on lockout clutch and remove shift fork from clutch assembly.
9. Remove front output shaft rear bearing retainer. It may be nec-

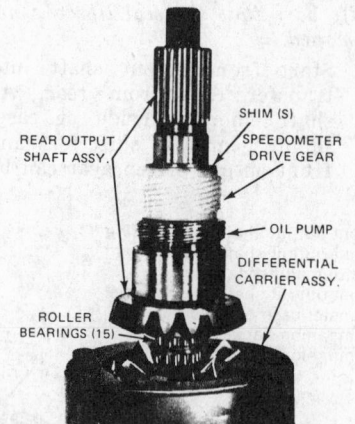

Removing the rear output shaft. The differential carrier assembly is removed next—New Process 203 (© Ford Motor Company)

essary to gently tap front of shaft or cautiously pry retainer from case. Make certain that no roller bearings are lost from rear cover.

10. When necessary, remove rear bearing by pressing.
11. Pry front output shaft front bearing from lower side of case.
12. Remove front otuput shaft assembly from case.
13. Lift intermediate housing from range box, after removing bolts.
14. Remove chain from intermediate housing.

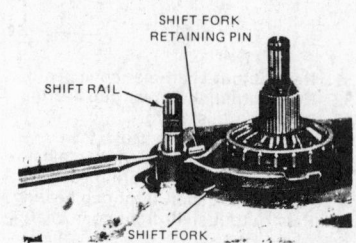

Driving the pin retaining the shift fork to the shift rail in a New Process 203 (© Ford Motor Company)

15. Remove lockout clutch, drive gear and input shaft from range box.
16. Install a 1½" to 2" band type hose clamp on end of input shaft to retain roller bearings.
17. Pull up on shift rail and remove rail from link.
18. Lift input shaft assembly from range box.

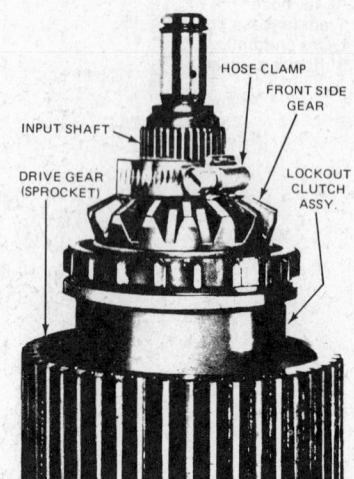

Place a hose clamp around the end of the input shaft to prevent losing the 123 roller bearings out of the clutch assembly—New Process 203 (© Ford Motor Company)

Assembly

1. Position range box with input gear side down, on wood blocks.
2. Place gasket on input housing.
3. Install lockout clutch and drive sprocket on input shaft assembly. Install a 2" band type hose clamp on end of input shaft to prevent loss of bearings during installation.
4. Place input shaft, lockout clutch and drive sprocket in range box. Align tab on bearing retainer with notch in gasket.

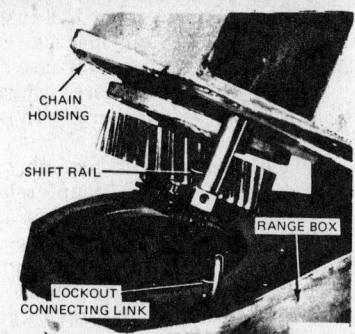

Disengaging the shift rail from the lockout connecting link in a New Process 203 (© Ford Motor Company)

5. Engage lockout clutch shift rail to the connector link. Position rail in housing bore and turn shifter shaft lowering rail into the housing. This will prevent the link and rail from becoming disconnected.

6. Place the drive chain in housing with the chain around the outer wall.

7. Secure the chain housing to the range box. Be sure that the shift rail engages the channel of the housing. Place the chain on the input drive sprocket.

8. Place the front output sprocket in transfer case. Turn the clutch drive gear to assist in positioning chain on sprocket.

9. Position the shift fork and rail on the clutch assembly. Install the clutch assembly completely into the drive sprocket. Insert retaining pin in shift fork and rail.

10. Install front output bearing, gasket, retainer, bolts, flange, gasket, seal, washer and retaining nut.

11. If rear bearing was removed from front output shaft, press a new bearing into the outside face of cover until bearing is flush with opening.

12. Install front output shaft, rear bearing retainer, gasket and bolts.

13. Slip differential carrier assembly on the input shaft. Bolts on carrier must face rear of shaft.

14. Load bearings in pinion shaft, install rear output housing assembly, gasket and bolts.

15. Install a dial indicator on the rear housing. The indicator must contact the end of the output shaft. While holding the rear flange, rotate the front output shaft and find the highest point of gear hop. Reset indicator and with rear output shaft at high point, pull up on the end of the shaft to determine end play. Remove indicator and install shim pack to control end play to between 0 and .005″. The shim pack is positioned on the shaft in front of the rear bearing. Check for binding of rear output shaft.

16. Insert lockout clutch shift rail poppet ball, spring and screw plug in transfer case.

17. Install poppet plate spring, gasket and plug, if they were not previously installed.

18. Install shift levers on the range box, if these were not left on vehicle.

19. Torque all bolts, locknuts and plugs to specifications.

20. Fill transfer case with specified lubricant until the proper level is reached. Secure filler plug.

Transfer Case Subassemblies Overhaul

Lockout Clutch Assembly

Disassembly

1. Remove front side gear from input shaft assembly.

2. Remove thrust washer, roller bearings and spacers from front side gear bore. The position of the spacers must be noted.

3. Remove the snap ring which holds the drive sprocket to clutch assembly. Slip the drive sprocket from the front side gear.

4. Remove the lower snap ring.

5. Remove sliding gear, spring and spring cup washer from the front side gear.

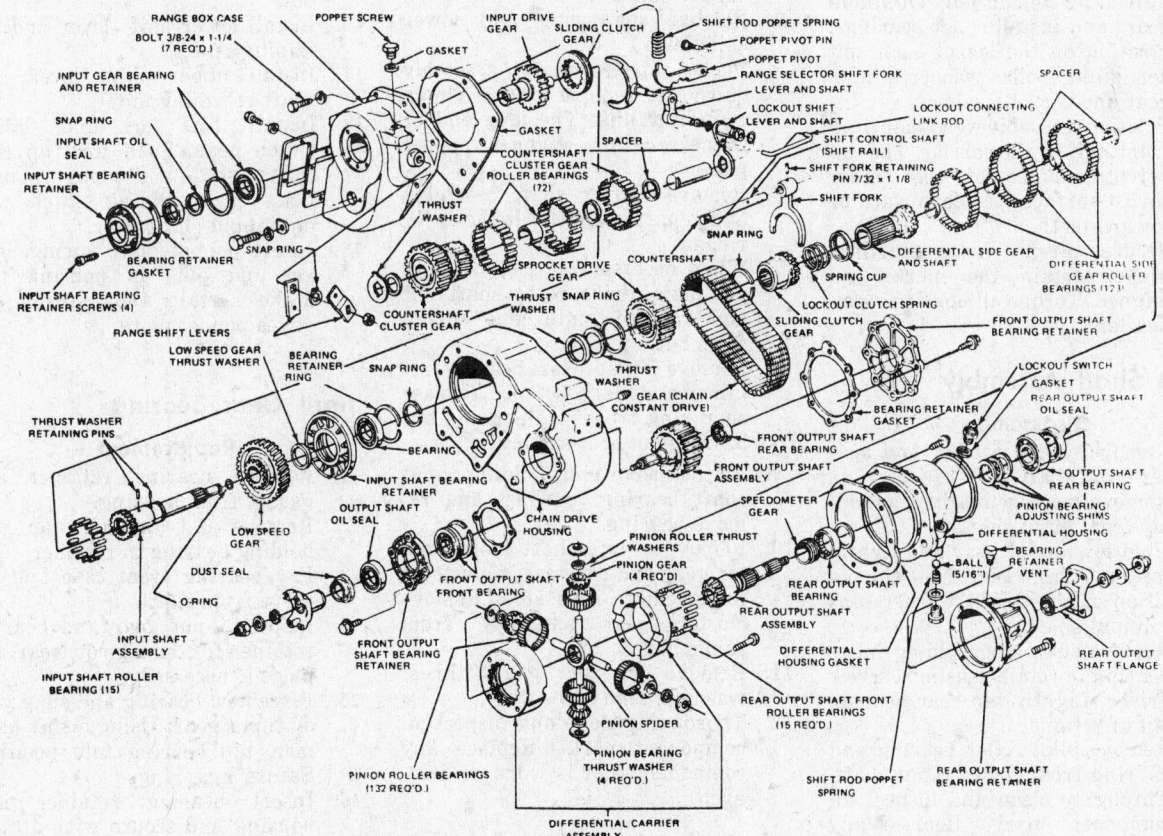

Exploded view of the New Process 203 full-time 4WD transfer case (© Ford Motor Company)

6. Throughly clean and inspect all component parts. Replace any component that is worn or defective.

Assembly

1. Place spring cup washer, spring and sliding clutch gear on front side gear.
2. Secure sliding clutch to front side gear with a snap ring.
3. Spread petroleum jelly on front side gear and install roller bearings and spacers.
4. Place thrust washer in gear end of front side gear.
5. Slide drive sprocket on clutch splines and secure with snap ring.

Differential Carrier Assembly

Disassembly

1. Separate differential carrier sections and lift out pinion gear and spider assembly.
2. Note that undercut side of pinion gear spider faces toward front of side gear.
3. Remove pinion thrust washers, pinion roller washer gears and roller bearings from spider unit.
4. Throughly clean and inspect all component parts. Replace any component that is worn or damaged.

Assembly

1. Spread petroleum jelly on pinion gears and install roller bearings.
2. Position on the leg of each spider, pinion roller washer, pinion gear and thrust washer.
3. Position the spider assembly in front half of the carrier. The undercut surface of the spider thrust surface face downward or toward teeth.
4. Secure carrier halfs together. Make certain the marks are aligned. Torque all bolts to specifications.

Input Shaft Assembly

Disassembly

1. Remove thrust washer and spacer from shaft.
2. Remove bearing retainer assembly from input shaft.
3. Hold low speed gear and lightly tap shaft from gear. Note the position of the thrust washer pins in input shaft.
4. Remove snap ring holding input bearing in retainer using a screw driver. Lightly tap rear bearing out of retainer.
5. Remove pilot roller bearing and 'O' ring from end of input shaft.
6. Throughly clean and inspect all component parts. Replace any component that is worn or damaged.

Assembly

1. Tap or press input bearing into retainer. Be sure that ball loading slots are toward concave side of retainer. Install securing snap ring. Make certain that selective snap ring of proper thickness is used to provide tightest fit.
2. Position low speed gear on shaft, clutch end facing gear end of input shaft.
3. Place thrust washers on input shaft, align slot in washer with pin in shaft. Slide or tap washers into position.
4. Place input bearing retainer on shaft and secure with snap ring. Snap rings are selective. Use snap ring that provides tightest fit.
5. Slip spacer and thrust washer on shaft and align with locating pin.
6. Spread heavy grease on end of shaft and install roller bearings.
7. Install rubber 'O' ring at end of shaft.

Range Box

Disassembly

1. Remove poppet plate spring, plug and gasket.
2. Remove clutch fork and sliding gear by disengaging sliding clutch gear from input gear.
3. Remove upper shift lever from shifter shaft.
4. Remove snap ring and lower shift lever.
5. Push shifter shaft assembly down and remove lockout clutch connector link. The long end of connector link engages poppet plate.
6. Remove shifter shaft assembly and separate shafts. Remove 'O' rings.
7. When necessary to remove poppet plate, drive pivot shaft out and remove plate and spring from bottom of case.
8. Remove input gear bearing retainer and seal assembly. Release snap ring from retainer and tap bearing out of assembly.
9. Release snap ring holding input shaft bearing to shaft and remove bearing.
10. Remove countershaft from cluster gear and case assembly from intermediate case side. Remove cluster gear assembly from range box.
11. Remove cluster gear thrust washers from case.
12. Thoroughly clean and inspect all component parts. Replace any component that is worn or damaged.

Assembly

1. Spread heavy grease in cluster bore and using proper tool install roller bearings and spacers.
2. Spread heavy grease on case and install thrust washers. Engage tab on thrust washers with slot in case.
3. Place cluster gear assembly in case and install countershaft through front of range box and into gear assembly. Flat face of countershaft must be aligned with case gasket.
4. Place bearing on input gear shaft with snap ring groove facing out, install a new retaining ring. Insert input gear and bearing in housing. The retaining ring used in this operation is a select fit. Use ring that provides the tightest fit.
5. Secure input gear and bearing with a snap ring.
6. Match up oil slot in retainer with drain hole in case and insert input gear bearing retainer and gaskets. Install bolts and torque to specifications.
7. Spread sealant on pin and install poppet pin and pivot pin in housing.
8. Lubricate and install new 'O' rings on inner and outer shifter shafts.
9. Insert shifter shafts in housing and engage long end of lockout clutch connector link with outer shifter shaft. Complete this operation before assembly bottoms out.
10. Install lower shift lever and retaining ring.
11. Install upper shift lever and shaft retaining nut.
12. Install shift fork and sliding clutch gear. Push fork up into shifter shaft and engage poppet plate. Move sliding clutch gear onto input shaft gear.
13. Insert poppet plate spring, gasket and plug in housing top. Make certain that spring engages poppet plate.

Input Gear Bearing

Replacement

1. Remove bearing retainer and gasket from housing.
2. Remove and discard snap ring holding bearing in retainer.
3. Pry bearing from case and remove from shaft.
4. Inspect input gear and bearing retainer for damage or wear. Replace if necessary.
5. Place new bearing and snap ring on input gear. Using a soft hammer, tap bearing into position. Secure snap ring.
6. Insert bearing retainer into housing and secure with attaching bolts. Tighten bolts to specifications.

Input Gear Retainer Seal

Replacement

1. Remove bearing retainer from housing.
2. Remove seal from retainer by prying.
3. Place new seal on retainer and install with proper seal driver.
4. Install bearing retainer in housing and secure with attaching bolts. Tighten bolts to specifications.

Rear Output Shaft Housing Assembly

Disassembly

1. Remove speedometer driven gear from housing.
2. Remove rear output flange and washer, if they have not been removed previously.
3. Using a soft hammer tap on flange end of pinion and remove the pinion. If speedometer drive gear does not come off with pinion reach into case and remove.
4. Remove old seal from bore with suitable prying tool.
5. Remove snap ring retaining rear output rear bearing.
6. Tap bearing out of housing.
7. Install a long drift into rear opening of housing and drive out front output bearing. Remove seal and discard.

Assembly

1. Spread grease on front bearing seal and place in bore. Place bearing in bore and press until it bottoms in housing.
2. Using a soft hammer tap rear bearing into place. Secure with proper snap ring. Snap rings are selective, use the one that provides the tightest fit.
3. Place rear seal in bore and drive into position with suitable tool. When seal is in position it should be approximately 1/8" to 3/16" below housing face.
4. Place speedometer drive gear on output shaft with shims of approximately .050" thickness. Insert output shaft into carrier through housing front opening.
5. Install flange and washer on output shaft. Leave retaining nut loose until shim requirements are known.
6. Install speedometer driven gear.

Front Output Shaft Bearing Retainer Seal

Replacement

1. Remove old seal from retainer bore.
2. Inspect and clean retainer.
3. Spread sealer on outer edge of new seal.
4. Place new seal in retainer bore and drive into position with proper tool.

Front Output Bearing

Replacement

1. Remove rear cover from case assembly and discard gasket.
2. Press old bearing from cover.

3. Place new bearing on outside face of cover. Cover bearing with a wood block and press into cover until bearing is flush with opening.
4. Place gasket on transfer case and tap cover into position using a soft hammer. Secure cover with attaching bolts and tighten to specifications.

SPECIFICATIONS

TORQUE (FT. LBS.)

Adapter to Transfer Case Bolts	38
Adapter to Transmission Bolts	40
Transfer Case to Frame Nuts (Upper)	50
Transfer Case to Frame Nuts (Lower)	65
Shift Lever Attaching Nuts	25
Shift Lever Rod Swivel Locknuts	50
Shift Lever Locking Arm Nut	150 in. lbs.
Skid Plate Bolt Retaining Nuts	45
Crossmember Bolt Retaining Nuts	45
Adapter Mount Bolts	25
Intermediate Case to Range Box Bolts	30
Front Output Bearing Retainer Bolts	30
Output Shaft Yoke Nuts	150
Front Output Rear Bearing Retainer Bolts	30
Differential Assembly Screws	45
Rear Output Shaft Housing	30
Poppet Ball Retainer Nut	15
PTO Cover Bolts	15
Front Input Bearing Retainer Bolts	20
Filler Plug	25

Warner Quadra-Trac

Description

The Quadra-Trac transfer case provides full-time, four-wheel drive under all driving conditions. The front and rear driveshafts are driven by a limited slip differential in the transfer case. The limited slip differential is connected to the input shaft by a link-belt type chain. In operation, if the rear axle loses traction, then the engine torque will be transferred through the transfer case differential to the front axle.

The transfer case contains a manually actuated lockout system that locks the front and rear driveshafts together, cancelling the differential action. This feature is used under extreme marginal traction situations.

NOTE: In order to spare the transfer case differential side gears and brake cones from excessive, and possibly, damaging wear, do not spin the wheels excessively when the vehicle is stuck or bogged down.

An optional gear reduction unit mounted at the rear of the input shaft is available for the Quadra-Trac unit, making it a two-speed transfer case.

Performance Checks
Transfer Case Differential Torque Bias Check

1. With the lock-out feature *not* engaged and the transmission in Park, raise the vehicle until the front wheels are free of the ground.
2. Disconnect the rear driveshaft from the transfer case.
3. Turn the rear yoke retaining nut with a torque wrench and socket, taking note of how much torque is required to force the cone clutches to slip. They should slip when 110 to 270 ft. lbs. are applied.

Slippage below 110 ft. lbs. indicates replacement of the differential is needed. If no slippage occurs at 270 ft. lbs., improper lubrication is indicated. Drain and refill the transfer case and reduction unit, if so equipped, with the proper lubricant mixture.

Drive Chain Tension Check

1. Drain the lubricant from the transfer case.
2. Remove the chain inspection plug and insert a steel rule into the hole.
3. A new chain will be 1.575 in. from the outer edge of the plug hole. When the slack in the chain reaches 1/2-3/4 in., the chain should be replaced. No adjustment is possible.
4. Reinstall the drain and chain inspection plugs, and refill the unit with the proper lubricant mixture.

Rear Case Cover

Removal

Most Quadra-Trac components can be serviced without removing the complete unit from the vehicle. To gain access to the rear output shaft, drive sprocket and thrust washer, chain, differential and needle bearing, or the diaphragm control system, just the rear cover has to be removed.

1. Lift and support the vehicle.
2. If the vehicle is equipped with a reduction unit, continue on to the next step for the reduction unit removal procedure. If the vehicle is not equipped with a reduction unit, proceed to Step 7.
3. Loosen all the bolts that attach the reduction unit to the transfer case cover.
4. Move the reduction unit backward just enough to allow the oil to drain from the unit.
5. Loosen the cable retaining bolt at the shift control lever. Loosen the cable clamp bolt and remove the control cable from the clamp bracket and control lever.
6. When the oil has drained, remove the bolts which hold the reduction unit to the transfer case cover. Move the reduction unit rearward to clear the transmis-

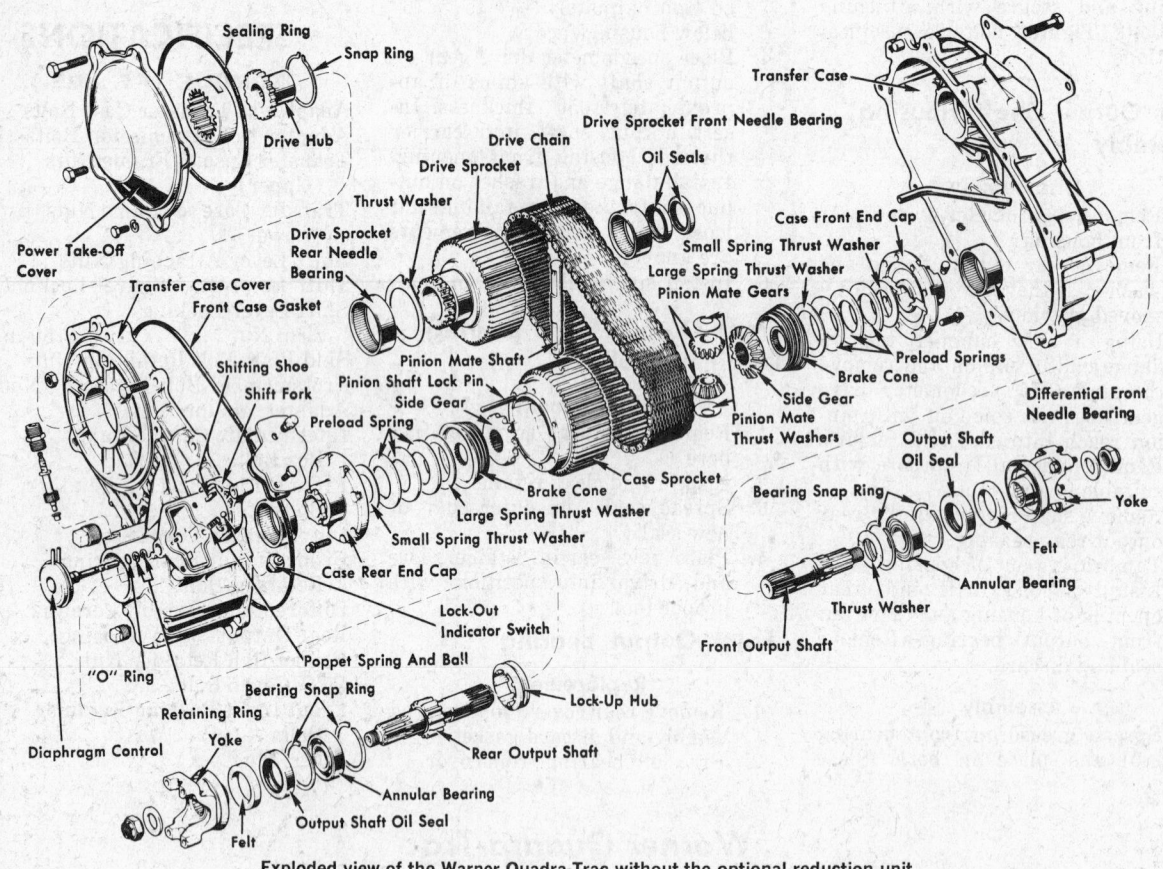

Exploded view of the Warner Quadra-Trac without the optional reduction unit (© Jeep Corporation)

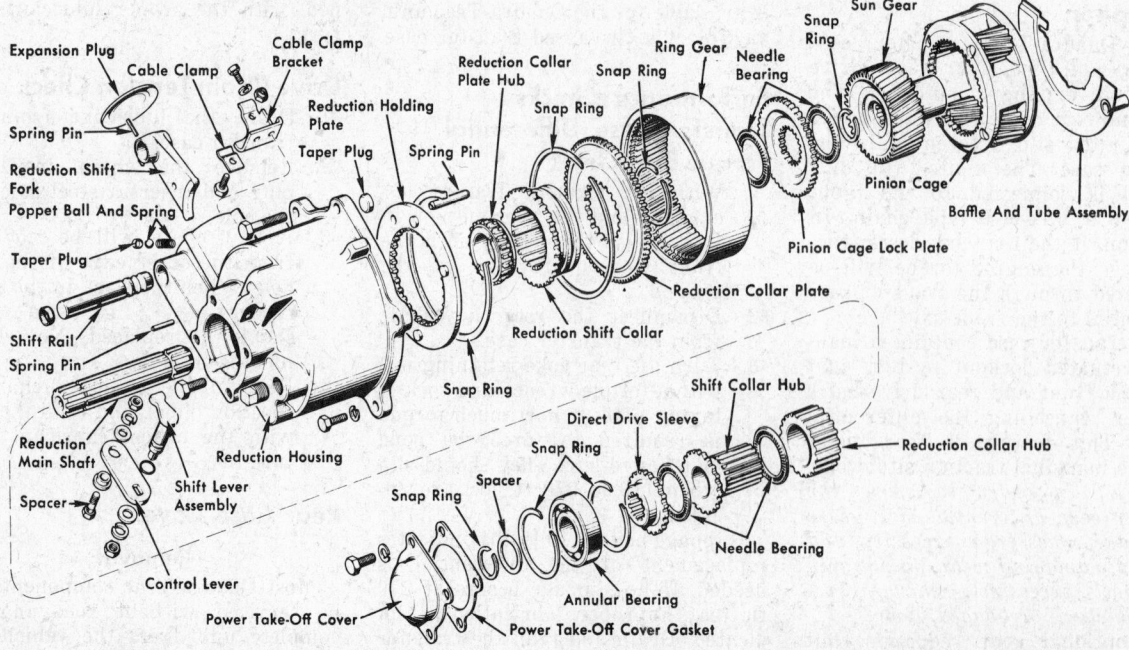

Exploded view of the optional reduction unit for a Warner Quadra-Trac transfer case (© Jeep Corporation)

The differential and drive sprocket positioned for installation of the drive chain —Warner Quadra-Trac (© Jeep Corporation)

The Quadra-Trac transfer case positioned for disassembly (© Jeep Corporation)

sion output shaft and pinion cage which is attached to the transfer case drive sprocket. The pinion cage will remain with the transfer case assembly.

NOTE: The pinion cage should not be removed if the transfer case cover assembly is to be removed, but may be removed for inspection or replacement if the transfer case cover assembly is to remain in the vehicle. Removal of the pinion cage involves only removing the snap-ring which holds the cage to the sprocket and sliding the cage backward.

7. Remove the transfer case drain plug and allow the unit to drain.
8. Mark the rear output shaft yoke and universal joint to provide an alignment reference during reassembly. Disconnect the rear drive shaft front universal joint from the transfer case rear yoke.
9. Mark the diaphragm control vacuum hoses for identification during reassembly, then disconnect them. Also remove the lock-up indicator switch wire and the speedometer cable. Remove the indicator switch.
10. Disconnect the parking brake cable guide from the pivot on the right frame side.
11. Remove the bolts which attach the case cover assembly to the case (front housing). Carefully

slide the cover assembly backward off the front output shaft and the transmission output shaft.

Disassembly

1. To disassemble the unit, remove the rear output shaft yoke.
2. If the unit is *not* equipped with a reduction unit, remove the power take-off cover from the rear of the transfer case cover. Remove the sealing ring from the transfer case cover.
3. Using a piece of wood 2 in. $\times$ 4 in. and 6 in. long, support the cover and drive sprocket.
4. If *not* equipped with a reduction unit, remove the drive hub and sleeve from the drive sprocket rear splines by expanding the internal snap-ring. The ring expanding tabs are accessible through a slot in the outside edge of the drive sleeve.
5. If equipped with a reduction unit, remove the pinion cage snap ring and carrier.
6. Lift the case cover from the drive sprocket and differential. The cover, rear output shaft, bearings and seal, drive sprocket rear needle bearing, and lock-up hub can now be serviced without any further disassembly of other components.

7. Slide the drive sprocket toward the differential unit and remove the chain. The differential unit may now be serviced without any further disassembly of other components.

Assembly

1. Position the drive sprocket on a block of wood 2 in. $\times$ 4 in. and 6 in. long.
2. Place the differential assembly about 2 in. from the drive sprocket and with the front end of the differential on the bench.
3. Position the drive chain around the drive sprocket and the differential assembly. Be sure that the chain is properly engaged with the sprocket and differential teeth and that the slack is removed from the chain.
4. Insert the rear output shaft into the differential.
5. Shift the lock-up hub rearward in the case cover. Lubricate the drive sprocket thrust washer and insert it in position on the case cover.
6. Carefully align the case cover and position it onto the drive sprocket and differential. The output shaft may have to be slightly rotated to align it with the lock-up hub. Be sure that the drive sprocket thrust washer stays positioned correctly.
7. If equipped with a reduction unit, install the pinion cage onto the drive sprocket rear splines. Install the snap-ring. Be sure that the snap-ring seats properly in the groove.
8. If the vehicle is *not* equipped with a reduction unit, assemble the drive hub, drive sleeve, and snap-ring, then install them onto the drive sprocket rear splines. Be sure the snap-ring seats properly.
9. Turn the drive sleeve or pinion cage to make sure the drive sprocket thrust washer did not come out of position. No binding should be present.
10. If *not* equipped with a reduction unit, install the power take-off sealing ring and cover and tighten the attaching screws.
11. Install the speedometer gear on the rear output shaft.
12. Install the rear output shaft oil seal and the rear yoke and nut. Tighten the nut to specification.

Installation

13. Clean the groove which the front oil seal gasket fits into and install the seal.
14. Install two $\frac{3}{8}$ in. 16×2 in. long pilot studs into the transfer case front cover housing.
15. On 1973-74 models only, insert the oil tube into the case bore

Transfer Cases

Matched Front Set   Matched Rear Set

Front End Cap — Small Thrust Washer — Preload Springs — Large Thrust Washer — Brake Cone — Side Gear — Case Sprocket — Side Gear — Brake Cone — Large Thrust Washer — Preload Springs — Small Thrust Washer — Rear End Cap

Exploded view of the differential assembly—Warner Quadra-Trac
(© Jeep Corporation)

at the front output shaft bearing boss. Insert a 6 in. length of 5/16 in. rod into the tube. The rod will be used as a pilot to align the tube with the case cover. Lift the cover assembly and align the tube pilot with the hole in the cover. Move the assembly forward over the pilot studs.

16. Move the cover assembly forward to mesh with the front output shaft and transmission output shaft. It may be necessary to rotate the rear output shaft slightly to allow the two sets of splines to engage.

17. After the cover assembly has been moved forward and is evenly touching the front half of the case, remove the pilot studs and install the rear cover attaching bolts. Tighten the bolts alternately and evenly to specifications.

18. Install the lock out indicator switch and connect the lock out switch wire, diaphragm control vacuum hoses, and the speedometer cable.

19. Install the rear drive shaft.

20. Install the parking brake cable guide to the pivot on the right frame side.

21. Install the reduction unit, if so equipped as follows:

22. Position the reduction unit to the transfer case and mesh the caged pinions with the sun gear and ring gear, and align the sun gear inner splines with the transmission output shaft splines.

23. Move the reduction unit forward until it touches the sealing ring.

24. Install the attaching screws loosely, then tighten them alternately to specification.

25. Connect the shift control cable and adjust it by first removing the swivel block from the control lever. Move the control lever to the most forward position. Thread the swivel block in or out on the cable end to obtain the correct length to fit the swivel block in the control lever.

26. Install the proper type and amount of lubricant and lower the vehicle.

NOTE: Use 8oz. of Jeep Lubricant

Concentrate Part No. 8123004 or 5356068 or Lubrizol® 762 (there is no substitute) mixed with SAE 30 non-detergent motor oil (Valvoline preferred). 3.5 pints of the mixture is required to fill the transfer case without a reduction unit, 4.5 pints with a reduction unit.

Differential Assembly

Disassembly

1. Mark end caps and case sprocket with paint. Marks must be used to identify front end cap, rear end cap and proper orientation of caps to case sprocket.

2. Remove front end cap. If necessary, tap gently with a soft hammer.

3. Remove thrust washer, preload springs, brake cone and side gears from case sprocket. Care must be taken to keep the various pieces together as they must be installed as a unit.

4. Invert the case sprocket and remove rear cap. If necessary, tap gently with a soft hammer.

5. Remove the thrust washers, preload springs, brake cone and side gears. Care must be taken to keep the various pieces together as they must be installed as a unit.

6. Raise the case sprocket. The pinion shaft lock pin should fall out. If the pin does not fall, drive the pin out with a 1/4 in. pin punch.

7. Drive the pinion mate shaft out of case sprocket, using a brass drift and hammer. Care must be taken to avoid damaging the pinion mate thrust washers.

8. Thoroughly clean and inspect all component parts. Replace any damaged or worn parts with a complete matched set.

Assembly

Prelubricate all bearings and thrust surfaces with Jeep Lubricant Concentrate Part No. 8123004 or 5356068 or Lubrizol 762 (there is no substitute) prior to installation.

1. Slide the pinion mate shaft in the case sprocket three inches.

2. Install the pinion mate thrust

Mark the end caps and case sprocket of the differential assembly before disassembling the differential—Warner Quadra-Trac
(© Jeep Corporation)

washers and gears on shaft in the proper order.

3. Align the pinion mate shaft lockpin hole with hole in case sprocket. Lightly drive the pinion mate shaft into case sprocket until lockpin holes are exactly aligned.

4. Move the pinion mate gears apart until the gears are pressing the washers against the case sprocket.

5. Engage the pinion mate gear with the front side gears. Insert the brake cone over the gear and in case sprocket. Install the large thrust washer and preload springs, concave side of springs facing toward brake cone.

6. Lubricate the small thrust washer and place it on the front end cap. Install the front end cap, secure with attaching screws and alternately tighten to proper torque. Make certain that alignment marks are in order.

7. Invert the case sprocket and end cap and install the pinion shaft lock pin.

8. Mesh remaining side gear with pinion mate gears.

9. Insert the remaining brake cone over the side gear. Install the large thrust washer and preload spring, concave side of springs facing toward brake cone.

10. Lubricate the small thrust

washer and place it on the rear end cap. Install the rear cap, tighten attaching screws finger tight.

11. Insert the front and rear output shafts in the differential and rotate until both shafts have aligned with the splines on the brake cones and side gears. Alternately tighten the retaining screws to proper torque.

SPECIFICATIONS

TORQUE (FT. LBS.)

Transfer Case
Breather 6-10
Chain Measuring Access
   Hole Plug 6-14
Drain Plug 15-25
Fill Plug 15-25
Lock-Up Cover to
   Transfer Case 8-10
Lock-Out Indicator Switch . . .10-15
Output Shaft Nut 90-150
PTO Cover to Transfer Case
   bolts: 3/8 in.-16 15-25
          5/16 in.-18 10-20
Speedometer Adapter 20-30
Transfer Case Cover to
   Transfer Case 15-25
Transfer Case to Transmission
   Extension Bolt 30-50
Reduction Unit
Cable Housing Clamp Nut 7-12
Fill Plug 15-25
Shift Lever Cable
   Clamp Nut 10-20
Shift Lever to Shift Nut 15-25
Reduction PTO Cover
   to Case 15-25
Reduction Unit to Transfer
   Case Bolts: 3/8 in.-16 15-25
           5/16 in.-18 8-10

Diaphragm Control, Shift Fork and Lock-Up Hub

Disassembly

1. Remove the vent cover and seal ring.
2. Remove the retaining rings positioning the shift fork on diaphragm. Carefully pry the shift fork forward to gain access to the retaining rings. Remove the spring with a magnet.
3. Caution must be exercised in removal of the diaphragm control

Removing the diaphragm control rod—
Warner Quadra-Trac (© Jeep Corporation)

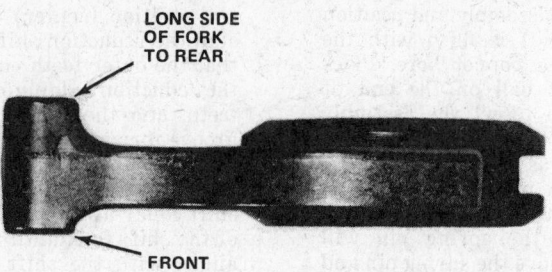

LONG SIDE OF FORK TO REAR

FRONT

Install the lock-up hub assembly shift fork with the long side of the fork facing toward the rear—Warner Quadra-Trac (© Jeep Corporation)

rod as it is retained by a spring loaded detent ball. Insert a magnet into the hole to hold the detent ball. Slip the diaphragm control rod out of case. Remove detent ball and spring.
4. Remove shifting fork, plastic shifting shoes and lock-up hub.

Assembly

1. Lubricate the shifting shoes and place them in the shift fork.
2. Install the shift fork and lock-up hub assembly in the case cover, long end of shift fork first (toward rear). Make certain that the shift fork does not separate from the lock-up hub by reaching through the needle bearings.
3. Insert the diaphragm control rod in the case and shifting fork, stopping before the detent ball hole is reached.
4. Install the detent ball and spring. Depress the detent ball with a 1/4 in. pin punch and slide the diaphragm control rod into place.
5. Install the shift fork retaining pin, (clips) and the diaphragm retaining spring, the spring should be below the surface of the bore.
6. Install the vent cover and seal ring.

Reduction Unit

Disassembly

1. Remove PTO cover and gasket.
2. Remove snap ring from reduction main shaft rear end, slide the reduction main shaft and sun gear assembly forward and out. Remove needle bearings.
3. Remove as an assembly, the ring gear, reduction collar plate, pinion cage lock plate, shift collar hub and reduction collar hub. Using a soft hammer, remove the shift collar hub from the pinion cage lock plate.
4. Remove the pinion cage lock plate, needle bearing, ring gear, reduction collar plate and shift collar hub. Separate the reduction collar hub and needle bearing from shift collar hub.

5. When necessary, separate the reduction collar plate and ring gear by removing retaining snap ring.
6. Remove needle bearing and direct drive sleeve from the reduction shift collar.
7. Shift the reduction shift collar to the neutral (center) detent with the control lever. Disengage the shift fork. Place the shift in the direct drive detent position (rear), align the collar outer teeth with the inner teeth in the reduction holding plate. Place the fork and collar in the reduction position (front) detent, and remove the reduction shift collar.
8. Remove the annular bearing rear snap ring and bearing.
9. Remove the shift fork locating spring pin, large expansion plug, shift rail taper plug, and control lever.
10. Drive the spring pin out of the shift fork and rail with a 3/16 in. pin punch. Slide the shift rail forward out of the shift fork, and remove the shift fork. Remove the spring shift fork poppet ball. Drive the poppet taper plug into the shift rail bore and remove the plug and spring.
11. Remove the shift lever retaining pin and the shift lever assembly.
12. Remove the reduction holding plate snap ring and reduction holding plate.

Assembly

1. Align the shift fork locating spring holes in the reduction holding plate and housing. Install the reduction holding plate. Locating pins should index the plate in the case. Secure the plate with a snap ring, tabs facing forward.
2. Install the shift lever assembly into the housing, lever towards the rear. Position seal ring on groove in shift lever shaft.
3. Move the shift lever assembly inward and install taper pin.
4. Install the shift rail in the shift rail rear bore, grooved end first. Position the shift rail with flat side towards the poppet spring.

Engage the shift rail with the shift lever assembly and position the rail so it is flush with the edge of the poppet bore. Place the poppet ball on the end of spring and insert the assembly in the poppet bore, using a spring pin as an installation tool. Depress the poppet ball and slide the shift rail over the poppet ball as far as the spring pin will allow. Remove the spring pin and place the shift rail in the first detent position.

5. Position the shift rail so the flat side is facing the shift lever assembly and the spring pin bore is aligned with the spring pin bore in the shift fork. Once the spring pin holes are aligned, install the spring pin so that it is flush with the outside surface of the shift fork.

6. Install the shift rail taper plug, poppet bore taper plug, shift rail cover expansion plug, shift fork spring locating pin and the control lever.

7. Place the shift fork in the neutral position (center) detent. Install the reduction shift collar so that the outer teeth engage with the reduction holding plate inner teeth, and the shift collar fork groove forward of the shift fork. Place the shift fork in the direct drive (rear) detent. Move the shift collar away and to the rear of the shift fork until the groove aligns with the shift fork. Engage the collar groove with the shift fork.

8. Place the direct drive sleeve in the reduction shift collar, needle bearing surface facing toward the front. Lubricate and install the needle bearing, against the direct drive sleeve.

9. Assemble the reduction collar plate hub and ring gear. Make certain that snap rings are seated in their grooves.

10. Install the needle bearing and reduction collar hub on the shift collar hub.

11. Install the ring gear, reduction collar plate and hub on the shift collar hub.

12. Place a needle bearing on the shift collar hub and the reduction collar hub.

13. Tap the pinion cage lock into place on the shift collar hub with a soft hammer and install assembly in housing. Place needle bearings on the shift collar hub and the pinion cage lock plate.

14. Install the reduction main shaft and sun gear into shift collar hub and through the direct drive sleeve and annular bearing. With a brass drift gently tap the assembly as far to the rear as possible. Place the rear spacer on the main shaft and secure with the selective snap ring which gives the tightest fit, between .004 in. and .009 in. clearance. Snap rings are available in thicknesses ranging from 0.089 in. to 0.105 in.

15. Install PTO cover and gasket, tighten attaching bolts to proper torque.

Rockwell T-223

Description

The Rockwell T-223 transfer case is a two speed unit used to transmit power to both the front and rear driving axles. The transfer case is mounted behind the transmission and is driven by a short drive shaft. A parking brake drum is mounted to the case on the opposite side of the front output shaft.

Disassembly

Transfer Case

1. Clean the exterior of the case and remove the three driving yokes from the input and output shafts, using a slide hammer.
2. Remove the retaining screws and lift the top cover and gasket. Discard the gasket and clean the mating surfaces.

Shift Components

1. Remove the detent springs and remove the detent balls with a magnet.
2. Pull both of the shift shafts out through the front of the case. If the shafts cannot be removed this way, drive the shafts and expansion plugs out through the rear of the case.
3. Remove the shift shaft oil seals from the front of the case.
4. Lift the range and declutch shift forks from the case.

Front Output Shaft

1. Remove the front output shaft

bearing cap and gasket and pull the shaft and declutch collar from the case.
2. Slide a long bar through the front output shaft gear and tap out the small and large expansion plugs in the brake hub.

Parking Brake

1. Cut the wire lock and remove the retaining screws from the brake drum, and remove the brake drum from the hub.
2. Remove the brake hub retaining snap ring and remove the brake hub from the front output shaft, using a large gear puller.
3. Disconnect the brake shoe return springs and remove the brake shoes and lever.
4. Remove the brake backing plate mounting bolts and remove the backing plate, washer, deflector plate and washer.

NOTE: Do not try to remove the front output shaft at this time. The idler shaft must be removed first, due to the interference of the gears.

Idler Shaft

1. Remove the idler shaft front and rear bearing caps. Wire the forward bearing cap shims together for reassembly.
2. Remove the screws and retainer plate from the front end of the idler shaft and press the idler shaft out through the rear of the case.
3. Lift the Low gear, spacer and front bearing from the case.
4. Remove the High gear and tap the idler shaft front bearing cup from the case.
5. Remove the retaining snap ring and press the bearing from the idler shaft.

Front Output Gear

1. After removing the idler shaft assembly, remove the front output shaft rear bearing cap and gasket.
2. Remove the bearing retaining snap ring and tap the front output gear and bearing into the case. Reach through the cover

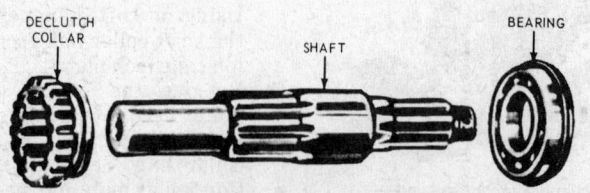

Font output shaft assembly—Rockwell T-223 (© Ford Motor Company)

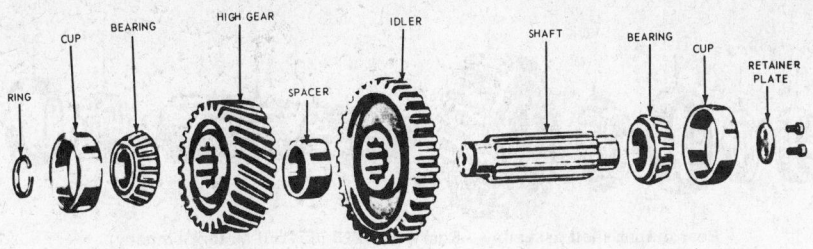

Idler shaft assembly—Rockwell T-223 (© Ford Motor Company)

opening and lift the gear and bearing out of the case.

3. The bearing has to be pressed off the front output shaft.

Rear Output Shaft

1. Remove the retaining screws and lift the rear output shaft front and rear bearing caps off the case. Wire the shims together for reassembly.

2. Remove the speedometer drive gear and spacer.

3. Block the rear output gear with a piece of wood and press the rear output shaft and front bearing out of the case. The rear output gear and rear bearing can be lifted out through the cover opening.

Input Shaft

1. Remove the input shaft front and rear bearing covers and wire the shims together for reassembly.

2. Place a block of wood between the sliding gear and the case and press out the input shaft and front bearing.

NOTE: If the gear should cock and become bound on the shaft, try to block it evenly with an additional piece of wood. Do not pound on the shaft.

3. Lift the sliding gear, drive gear, and spacer from the case.

4. Remove the front bearing from the input shaft.

Assembly

Input Shaft

1. Slide the input shaft front bearing onto the input shaft with the shielded side against the shaft shoulder.

2. Start the input shaft into the case. Mount the sliding gear, spacer and drive gear with its bushing onto the input shaft, working through the cover opening.

3. Tap the input shaft into position using a suitable sleeve placed against the inner race of the front bearing and a soft hammer.

4. Position the input shaft front cover and the original shim pack plus an adidtional 0.010 in. shim on the case. Tighten the retaining screws to the proper torque.

5. Install the rear thrust washer on the shaft and install the rear bearing with the shielded side toward the inside of the case.

6. Position a new gasket on the rear bearing cover and install the cover. Tighten the retaining screws to the proper torque.

7. Mount a dial indicator to the

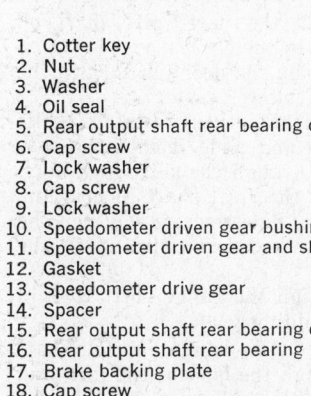

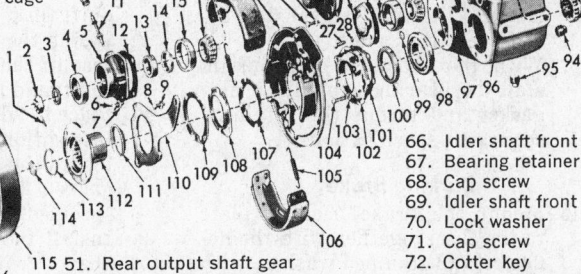

1. Cotter key
2. Nut
3. Washer
4. Oil seal
5. Rear output shaft rear bearing cap
6. Cap screw
7. Lock washer
8. Cap screw
9. Lock washer
10. Speedometer driven gear bushing
11. Speedometer driven gear and shaft
12. Gasket
13. Speedometer drive gear
14. Spacer
15. Rear output shaft rear bearing cup
16. Rear output shaft rear bearing
17. Brake backing plate
18. Cap screw
19. Lock washer
20. Input shaft rear bearing cage
21. Gasket
22. Idler shaft rear bearing
23. Input shaft rear bearing
24. Gasket
25. Spacer

26. Shift shaft hole plug
27. Cap screw
28. Lock washer
29. Snap ring
30. Idler shaft rear bearing cup
31. Idler shaft rear bearing
32. Front output shaft
33. Idler shaft hi gear
34. Gear spacer
35. Direct drive gear
36. Direct drive gear bushing
37. Low speed sliding gear

38. Range shift fork
39. Detent ball
40. Detent spring
41. Beather
42. Cap screw
43. Washer
44. Housing cover
45. Detent spring
46. Detent ball
47. Shift fork set screw
48. Shift fork set screw
49. Front axle declutch shift fork
50. Idler shaft lo gear

51. Rear output shaft gear
52. Shift shaft oil seal
53. Direct drive gear spacing washer
54. Range shift shaft
55. Input shaft
56. Input shaft front bearing
57. Shims
58. Input shaft front bearing cap
59. Lock washer
60. Screw
61. Oil seal
62. Washer
63. Nut
64. Idler shaft
65. Idler shaft front bearing

66. Idler shaft front bearing cup
67. Bearing retainer plate
68. Cap screw
69. Idler shaft front bearing cap
70. Lock washer
71. Cap screw
72. Cotter key
73. Declutch shift shaft
74. Rear output shaft
75. Rear output shaft front bearing
76. Rear output shaft front bearing cup
77. Shims
78. Rear output shaft front bearing cap
79. Cap screw
80. Locker washer
81. Nut
82. Cotter key
83. Washer
84. Oil seal
85. Cap screw
86. Lock washer

87. Front output shaft front bearing cap
88. Gasket
89. Shims
90. Front output shaft front bearing (with snap ring)
91. Front output shaft
92. Declutch collar
93. Shift shaft oil seal
94. Oil filler plug (some units employ elbow arrangement)

95. Drain plug
96. Cover to housing gasket
97. Transfer case housing
98. Front output shaft rear bearing (with snap ring)
99. Shift shaft hole plug
100. Oil seal
101. Gasket
102. Front output shaft rear bearing cap
103. Washer
104. Cap screw
105. Brake shoe return spring
106. Brake shoe
107. Gasket
108. Oil deflector
109. Washer
110. Brake lever
111. Oil baffle
112. Brake hub
113. Snap ring
114. Small expansion plug
115. Brake drum
116. Lock washer
117. Large expansion plug
118. Cap screw
119. Lock wire

Exploded view of the Rockwell T-223 transfer case (© Chrysler Corporation)

transfer case with the stem against the front end of the input shaft and check the amount of end play present. Remove enough shims from under the front bearing cover to provide 0.003 to 0.005 in. end play. Tighten the front bearing cover retaining screws to specification.

Rear Output Shaft

1. Press the front bearing onto the rear output shaft.
2. Hold the rear output gear in position inside the case and slide the shaft through it.
3. Install the front bearing cup and the original shim pack plus an additional 0.010 in. shim, together with the front bearing cover. Install the retaining screws and tighten them to specifications.
4. Press the rear bearing onto the rear output shaft and tap the bearing cup into place.
5. Install the speedometer drive gear and spacer on the shaft.
6. Position a new gasket and the bearing cap and oil seal over the shaft. Install the retaining screws and tighten them to the proper torque.
7. Rotate the shaft to seat the bearings, then install a dial indicator against the rear end of the shaft to check the end play. Remove enough shims from under the front bearing cap to provide zero end play and zero preload.

Front Output Gear

NOTE: The front output gear must be installed in the case before the idler shaft assembly is installed.
1. Install the ball bearing on the front output gear hub.
2. Position the gear and bearing in the case and install the retaining snap ring on the bearing.
3. Install the rear bearing cap over a new gasket and tighten the retaining screws to the proper torque.

Idler Shaft

1. Press the rear bearing onto the idler shaft and install the retaining snap ring.
2. Hold the high gear in position inside the case and tap the idler shaft through it with a soft hammer. The short hub side of the gear faces toward the rear of the case.
3. Install the gear spacer on the idler shaft, then install the low gear with the long hub toward the front of the case.
4. Install the rear bearing cup in the case and install the rear bearing cap with a new gasket, tightening the retaining screws to the proper torque.

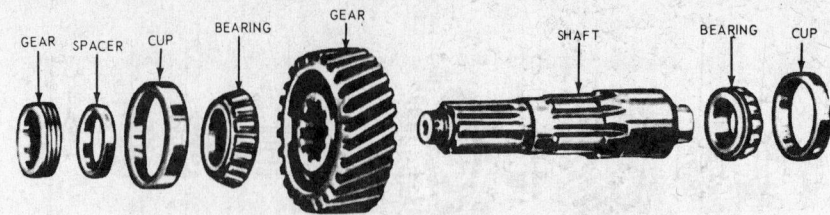

Rear output shaft assembly—Rockwell T-223 (© Ford Motor Company)

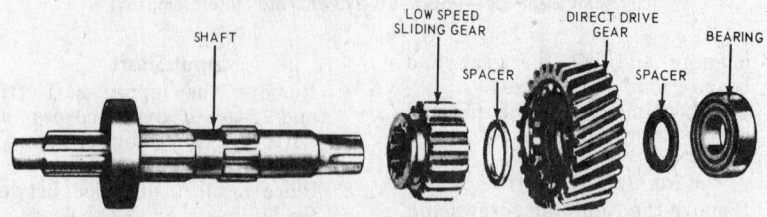

Input shaft assembly—Rockwell T-223 (© Ford Motor Company)

5. Drive the front bearing onto the idler shaft while holding the shaft rigid to avoid damaging the rear bearing and cup. Install the bearing retainer plate, torque the retaining screws and lock wire them.
6. Tap the front bearing cup into place in the case.
7. Install the front bearing cap with the original shim pack and some extras in order to set up the shaft end play. Torque the front bearing cap retaining screw to specification.
8. Mount a dial indicator on the case with the stem set against the inside face of the lower gear. Check the amount of end play by working the assembly back and forth with a pry bar. Remove enough shims to arrive at 0.003 to 0.005 in. end play.

Front Output Shaft

1. Install the ball bearing onto the front output shaft with the retaining snap ring toward the front.
2. Install the sliding declutch collar onto the shaft and install the shaft into the case.
3. With the shaft in position, install the bearing cap with a new gasket and torque the retaining screws to specification.

Parking Brake

1. Mount the brake backing plate to the case, together with the deflector and stamped washer, with a new gasket between the backing plate and the case. Tighten the retaining screws with star washers to the proper torque.
2. Position the brake lever on the backing plate.
3. Position the brake shoes on the backing plate with the actuating pawl in the web slot.
4. Install the brake shoe return springs.
5. Slide the brake hub onto the splines of the front output shaft

and install the retaining snap ring.
6. Install the brake drum with the attaching lockwashers and screws. Tighten the screws to specification and insert lock wires.
7. Install the expansion plugs. The smaller plug goes in the bore of the front output gear.

Shift Components

1. Install the new shift shaft oil seals in the case.
2. Position the declutch fork in the shift collar.
3. Lubricate the declutch shift shaft and slide it into the case and through the shift fork bore.
4. With the shift shaft in position, insert the set screw. Tighten the screw and lock wire it to the fork.
5. Position the range shift fork in the sliding gear.
6. Slide the range shift shaft through the bore of the fork and install the set screw. Tighten the set screw and lock wire it to the shift fork.
7. Install the expansion plugs at the rear of the case and flatten them to expand them.
8. Place the detent balls and springs in position in the case bores and install the case cover with a new gasket, tightening them to the specified torque.
9. Install the driving yokes on the shafts with their lockwashers and retaining nuts and tighten them to the proper torque.

SPECIFICATIONS

TORQUE (FT. LBS.)

Top cover-to-case bolts	38-49
Bearing cap retaining bolts	38-49
Brake drum-to-hub bolts	60-77
Brake mounting plate bolts	60-77
Input shaft yoke nut	300-400
Output shaft yoke nuts	300-400

Index

Manual Transmissions

Diagnosis

Jumping out of High Gear

1. Misalignment of transmission case or clutch housing.
2. Worn pilot bearing in crankshaft.
3. Bent transmission shaft.
4. Worn high speed sliding gear.
5. Worn teeth in clutch shaft.
6. Insufficient spring tension on shifter rail plunger.
7. Bent or loose shifter fork.
8. End-play in clutch shaft.
9. Gears not engaging completely.
10. Loose or worn bearings on clutch shaft or mainshaft.

Sticking in High Gear

1. Clutch not releasing fully.
2. Burred or battered teeth on clutch shaft.
3. Burred or battered transmission main-shaft.
4. Frozen synchronizing clutch.
5. Stuck shifter rail plunger.
6. Gearshift lever twisting and binding shifter rail.
7. Battered teeth on high speed sliding gear or on sleeve.
8. Lack of lubrication.
9. Improper lubrication.
10. Corroded transmission parts.
11. Defective mainshaft pilot bearing.

Jumping out of Second Gear

1. Insufficient spring tension on shifter rail plunger.
2. Bent or loose shifter fork.
3. Gears not engaging completely.
4. End-play in transmission mainshaft.
5. Loose transmission gear bearing.
6. Defective mainshaft pilot bearing.
7. Bent transmission shaft.
8. Worn teeth on second speed sliding gear or sleeve.
9. Loose or worn bearings on transmission mainshaft.
10. End-play in countershaft.

Sticking in Second Gear

1. Clutch not releasing fully.
2. Burred or battered teeth on sliding sleeve.
3. Burred or battered transmission main-shaft.
4. Frozen synchronizing clutch.
5. Stuck shifter rail plunger.
6. Gearshift lever twisting and binding shifter rail.
7. Lack of lubrication.
8. Second speed transmission gear bearings locked will give same effect as gears stuck in second.
9. Improper lubrication.
10. Corroded transmission parts.

Jumping out of Low Gear

1. Gears not engaging completely.
2. Bent or loose shifter fork.
3. End-play in transmission mainshaft.
4. End-play in countershaft.
5. Loose or worn bearings on transmission mainshaft.
6. Loose or worn bearings in countershaft.
7. Defective mainshaft pilot bearing.

Sticking in Low Gear

1. Clutch not releasing fully.
2. Burred or battered transmission main-shaft.
3. Stuck shifter rail plunger.
4. Gearshift lever twisting and binding shifter rail.
5. Lack of lubrication.
6. Improper lubrication.
7. Corroded transmission parts.

Jumping out of Reverse Gear

1. Insufficient spring tension on shifter rail plunger.
2. Bent or loose shifter fork.
3. Badly worn gear teeth.
4. Gears not engaging completely.
5. End-play in transmission mainshaft.

6. Idler gear bushings loose or worn.
7. Loose or worn bearings on transmission mainshaft.
8. Defective mainshaft pilot bearing.

Sticking in Reverse Gear

1. Clutch not releasing fully.
2. Burred or battered transmission main-shaft.
3. Stuck shifter rail plunger.
4. Gearshift lever twisting and binding shifter rail.
5. Lack of lubrication.
6. Improper lubrication.
7. Corroded transmission parts.

Failure of Gears to Synchronize

1. Binding pilot bearing on mainshaft, will synchronize in high gear only.
2. Clutch not releasing fully.
3. Detent springs weak or broken.
4. Weak or broken springs under balls in sliding gear sleeve.
5. Binding bearing on clutch shaft.
6. Binding countershaft.
7. Binding pilot bearing in crankshaft.
8. Badly worn gear teeth.
9. Scored or worn cones.
10. Improper lubrication.
11. Constant mesh gear not turning freely on transmission mainshaft. Will synchronize in that gear only.

Gears Spinning When Shifting into Gear from Neutral

1. Clutch not releasing fully.
2. In some cases an extremely light lubricant in transmission will cause gears to continue to spin for a short time after clutch is released.
3. Binding pilot bearing in crankshaft.

Cleaning of Transmission Components

Cleanliness of parts, tools, and work area is of the utmost importance. All transmission components (except bearing assemblies) should be cleaned in cleaning solvent and dried with compressed air before any inspection or work is begun. Great care should be taken when cleaning bearings. Bearings should always be cleaned separately from other parts in clean cleaning solvent and not gasoline. They must never be cleaned in a hot solution tank. It is advisable that they be soaked in cleaning fluid and then tapped against a block of

wood in order to free any solidified lubricant that may be trapped inside.

Rinse bearings thoroughly in clean solvent and then dry them with moisture-free compressed air being careful not to spin the bearings with the air stream. Rotate each bearing slowly and inspect rollers or balls for any signs of excessive wear, roughness, or damage. Those bearings not in excellent condition must be replaced. If they pass this inspection, they should be dipped in clean oil and wrapped in clean lintless cloth to protect them until installation.

Inspection of Transmission Components

All parts must be completely and carefully inspected and replaced for any signs of wear, stress, discoloration or warpage due to excessive heat. Whenever available, the magna flux process should be used on all parts except roller and ball bearings, to detect small cracks unseen by the eye. Inspect the breather assembly to see that it is not clogged or damaged and check all threaded parts for stripped or cross threads. Oil passages must be cleared of obstructions by the use of air pressure or brass rods and all gaskets, oil seals, lock wires, cotter pins, and snap rings are to be replaced. Small nicks or burrs in gears or splines can be removed with a fine abrasive stone. It is important that any housings or covers having cracks or other damage should be replaced and not welded. Synchronizers, not in excellent condition, must be replaced. The bronze synchronizer cone should be checked for wear or for any steel chips that may have become imbedded in it. Springs must be inspected for free length, compressed length, distortion, or collapsed coils.

IMPORTANT: The splines on many clutch gears, mainshafts, etc., are equipped with a machined relief called a "hopping guard". With the clutch gear engaged, the mating gear is free to slip into this notch, preventing the two gears from separat- ing or "walking out of gear" under various load conditions. This is not a worn or chipped gear. Do not grind or discard the gear.

Check all shafts for spline wear or damage. If the mainshaft 1st and reverse sliding gear or clutch hub have worn into the sides of the splines, the shaft should be replaced. Shift forks, shift rods, interlock balls and pins must be replaced if scored, worn, distorted or damaged.

Clark 280V Series-Five Speed Transmissions

Disassembly of Transmission

1. Remove the remote control or shift tower from the control cover. Remove the control cover capscrews and lock washers. Remove the control cover assembly from the transmission. Remove the back-up light switch.
2. Remove the universal joint assembly and the drive shaft from the parking brake drum.
3. Remove the parking brake drum. Disconnect the parking brake actuating lever from the linkage.
4. Remove the transmission spline flange. Remove the bolts holding the carrier plate to the transmission housing. Slide the plate with the brake shoes and retaining springs off the transmission.
5. Lock the transmission in two gears and remove the brake drum retaining nut. Remove the brake drum.
6. Remove the output shaft rear bearing retainer and the speedometer drive gear.
7. Remove the countershaft rear bearing retainer. Remove the bearing snap ring.
8. Remove the input shaft bearing retainer and pull the input shaft out of the case. *Be careful not to drop the output shaft pilot bearing rollers into the transmission case.*
9. Move the output shaft rearward until the rear bearing is exposed. bearing is exposed. Remove the rear bearing with a suitable bearing puller.
10. Lift the output shaft out of the transmission case.
11. Remove the reverse idler shaft using the special tool. Lift the reverse idler gear, bearing and thrust washer out of the case.
12. Move the countershaft rearward until the rear bearing is exposed. Remove the bearing and the oil slinger with a suitable puller.
13. Lift the countershaft out of the case.
14. If the countershaft front bearing or the pilot bearing is to be replaced, remove the clutch housing from the transmission.
15. Press the pilot bearing out of the transmission case. *Do not ham-* mer or drive the bearing out of the as damage to the bearing bore may result.

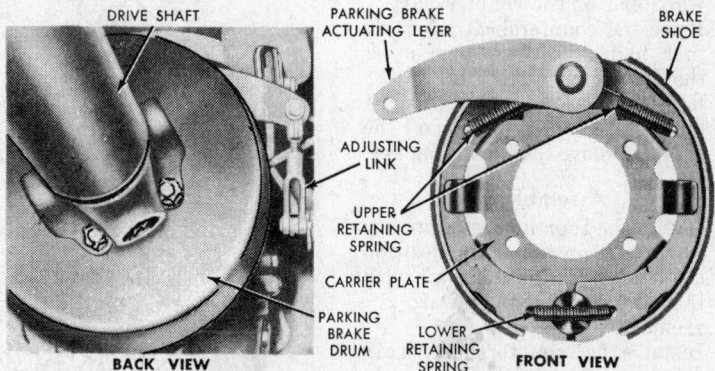

Transmission parking brake—Clark 280V (© Ford Motor Co.)

Sub-Assemblies

Output Shaft Disassembly

1. Remove the first reverse gear from the output shaft.
2. Clamp the output shaft, front end facing up, in a soft-jawed vise. Remove the fourth-fifth speed synchronizer assembly from the shaft. Remove the snap ring retaining the fourth gear and remove the shift hub sleeve and the fourth gear.
3. Remove the third gear snap ring and lift the locating washer and the gear off the shaft.
4. Lift the second-third speed synchronizer off the shaft.
5. Remove the snap ring retaining the second-third speed gear shift hub sleeve and lift the sleeve off the shaft.
6. Remove the second gear snap ring and remove the locating washer and the gear.

Assembly

1. Install the output shaft, forward end up, in a soft-jawed vise.
2. Install the second gear lower snap ring and the locating washer on the output shaft. Install the second gear on the output shaft with the clutching teeth facing upward. Install the upper snap ring.
3. Install the second-third shift hub sleeve and snap ring. *While* pressing downward on the second gear to compress the lower helical snap ring, check the gap between the second gear and the upper snap ring. This gap must be minimum of 0.006 inch.

4. Install the second-third speed synchronizer assembly on the shaft.
5. Install the third gear on the shaft with clutching teeth facing downward. Install the third gear locating washer and snap ring.
6. Install the fourth gear with the clutching teeth facing upward.
7. Install the bottom cone of the fourth-fifth synchronizer over the clutching teeth of the fourth gear.
8. Install the fourth-fifth shift sleeve hub. *Be sure the chamfered side is facing down.* Install the snap ring.
9. Install the fourth-fifth synchronizer on the shift hub sleeve.
10. Reverse the output shaft and install the first-reverse gear with the shift fork facing downward. *A minimum of 0.006 inch end play must be maintained on all mainshaft gears. Synchronizer end play on both synchronizers must be maintained at 0.060 inch minimum and 0.160 inch maximum.*

Input Shaft Disassembly

1. Remove the bearing retaining snap ring from the input shaft. Remove the bearing, using the special tool illustrated.

Assembly

1. Install the oil slinger and press the bearing on the input shaft. Be sure the snap ring groove is facing the forward end of the shaft. *To prevent damage to the bearing, apply pressure on the inner race of the bearing only.*
2. Install the bearing snap ring.

Countershaft

Disassembly

1. Remove the snap ring at the forward end of the countershaft.
2. Place the countershaft assembly in a hydraulic press, and press the drive gear off the shaft.
3. Remove the key from the shaft and press the fourth gear off the shaft. Remove the remaining key.

Assembly

1. Install the fourth speed gear key in the slot on the shaft. Position the fourth gear on the shaft with the long hub facing toward the front.
2. Install the countershaft main drive gear key in the keyway in the shaft. Install the countershaft drive gear on the shaft with the long hub facing toward the rear. Install the snap ring on the countershaft.

Gear Shift Housing

Disassembly

NOTE: It is not necessary to disassemble the gear shift housing assembly to determine if the parts are worn. Note the condition of the shifter fork shafts and forks by visual inspection. If the forks are excessively worn or if they bind when shifted, disassemble the gear shift housing assembly to make repairs. Check the interlocking system.

1. With the control cover in neutral, pry the fourth-fifth shift fork to the fourth speed position (toward the rear of the cover). Remove the front rail support capscrews and remove the front rail support.
2. Remove the interlock tapered

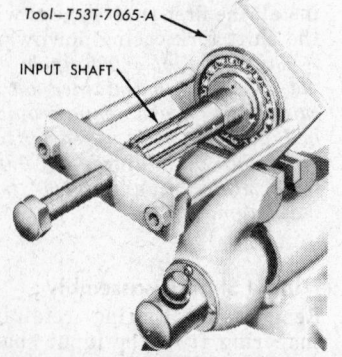

Tool—T53T-7065-A

INPUT SHAFT

Removing the input shaft bearing—Clark 280V
(© Ford Motor Co.)

1018

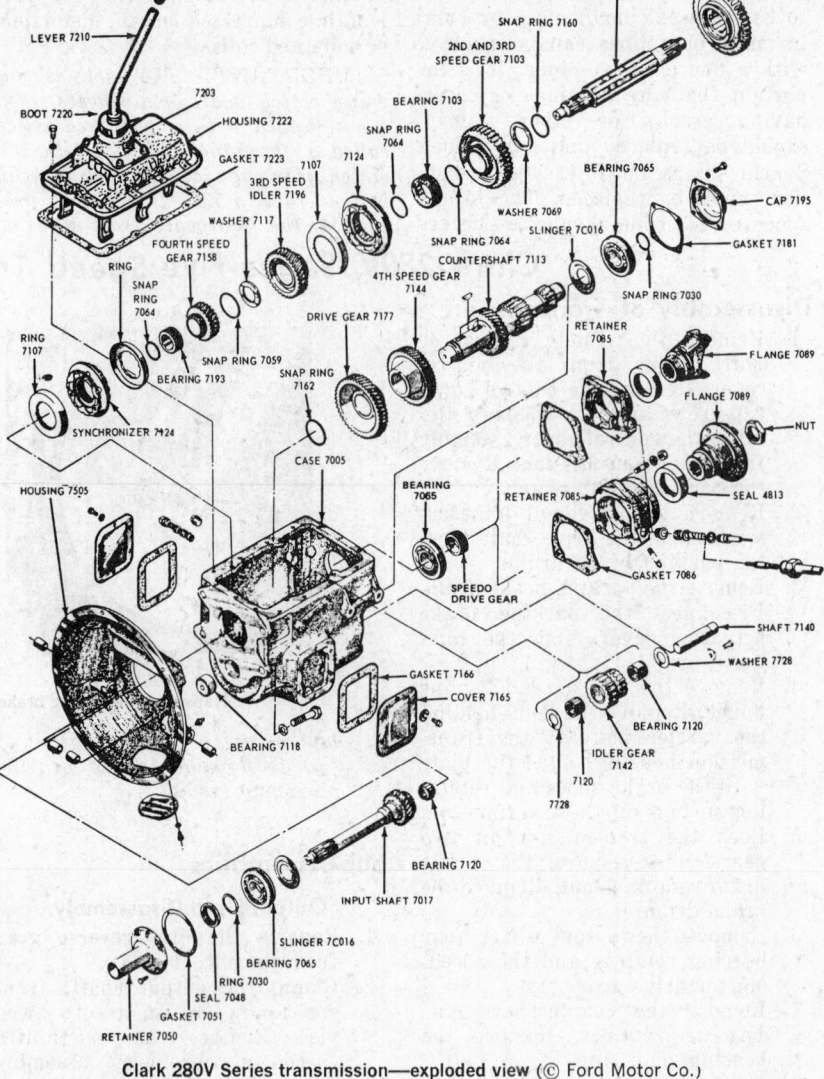

Clark 280V Series transmission—exploded view (© Ford Motor Co.)

pin supports. Note the position of the interlock tapered pins for reassembly.

3. Remove the rear rail support capscrews and remove the rear rail support.
4. Remove the first-reverse shift fork and rail assembly. Remove the fourth, fifth, second, and third shift fork and rail assembly. Use caution so as not to lose the interlock cross pin, interlock tapered pins, or the mesh lock poppet balls.
5. Remove the first-reverse shift rail. Remove the four mesh lock poppet balls. Remove the four poppet springs.
6. Remove the first-reverse rocker arm. Remove the reverse latch plunger spring retaining plug and the reverse latch plunger spring and plunger.
7. If the fork bushings are worn, secure the fork in a soft-jawed vise and remove the worn bushings with a drift. Install new bushings in the fork. Turn the

fork over on the anvil of the vise and secure the bushings in the fork, using a prick punch and upsetting the bushing metal on the outside of the fork.

Assembly

1. Position the first-reverse rocker arm on the pivot pin. Install the reverse latch plunger, spring and retaining plug. Tighten the plug securely.
2. Install the four poppet springs and the four mesh lock poppet balls. Note the first-reverse shift fork rail poppet ball in the pocket.
3. Align one tapered interlock cross pin with the hole in the first-reverse shift rail. Position the rail on the poppet ball, in neutral position with the interlock pin aligned with the first interlock tapered pin. Install the second interlock tapered pin. Align the pin with the interlock cross pin hole.
5. Position the fourth-fifth shift

fork and rail assembly on the poppet ball in neutral position. Slightly raise the rear of the rail and align the second interlock tapered pin with the cross hole in the rail. Note the positions of the tapered interlock pins and the shift rails.

6. Install the first-reverse shift fork and rail assembly on the poppet ball in a neutral position. Align the first-reverse rocker arm in the notch at the rear of the rail. Position the rear rail support.

7. Install the rail support capscrews and washers. Tighten the capscrews slightly. Install the interlock tapered pin supports. Tap the fourth-fifth shift fork to the rear (fourth speed position).

8. Position the front rail support and install the capscrews and washers. Tighten the front and rear support capscrews to 20-25 ft-lbs.

9. Tap the fourth-fifth shift fork and rail assembly forward to a neutral position.

Assembly of Transmission

1. Coat the countershaft needle bearings with grease and install them in the front countershaft bore.

2. Tip the rear of the countershaft down and lower it into the case. Push the countershaft forward and insert the shaft through the bearing.

3. Position the countershaft rear bearing oil slinger and bearing on the shaft. Drive the bearing in to the bearing bore in the case. The countershaft drive gear should be supported while driving the shaft to prevent damage to the front bearing. Install the rear bearing retainer.

4. Coat the reverse idler thrust washers with grease and position them in the case.

5. Insert the two roller bearings in the reverse idler bear bore. Place the gear assembly in position in the transmission case, with the small gear toward the rear of the case.

6. Insert the reverse idler shaft through the hole in the case, through the reverse idler gear, and into the forward support boss. Drive the reverse idler shaft into the case until the slot in the shaft is lined up with the lock bolt hole. Install the retainer in the slot and secure the retainer with the lock bolt. Tighten the bolt to specifications.

7. Tilt the rear of the output shaft assembly downward and insert the end of the shaft through the output shaft bore in the case.

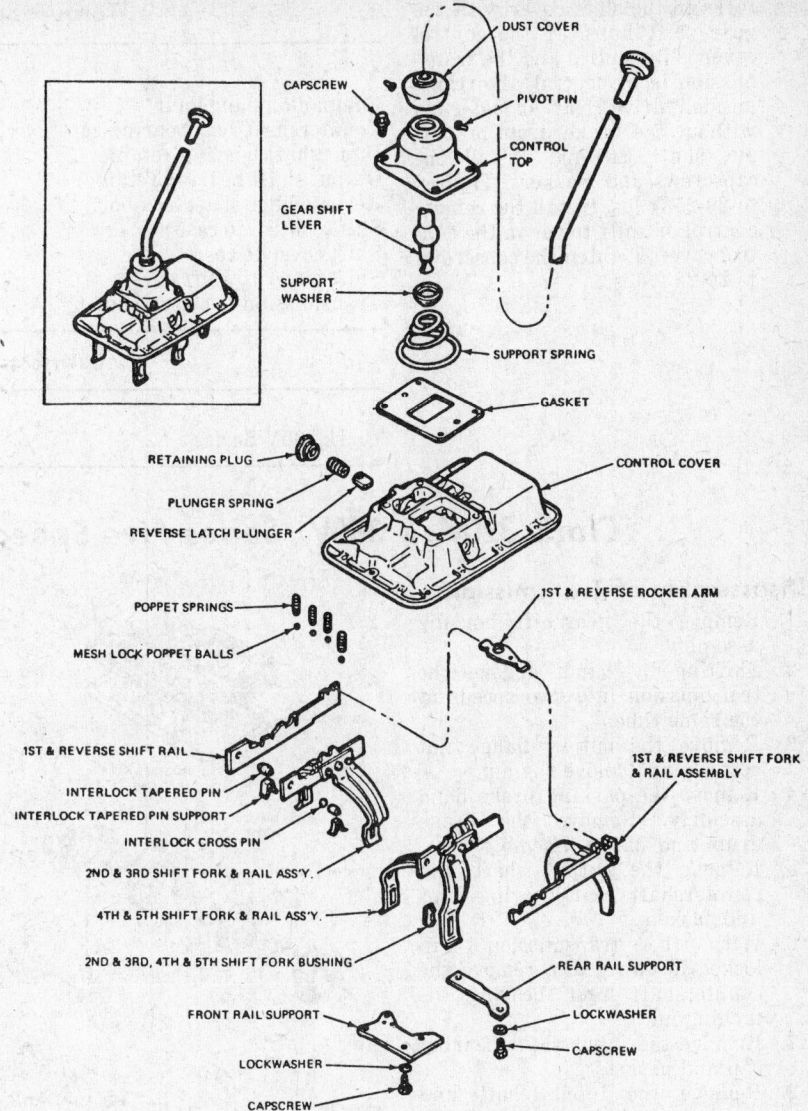

Gear shift housing—exploded view—Clark 280V (© Ford Motor Co.)

Lower the front end of the output shaft until it is in line with the pilot bearing opening. Move the assembly forward into position.

8. Insert the pilot bearing in the input shaft bore.

9. Position the input shaft and bearing assembly in the forward end of the transmission case. Tap the front end of the shaft with a soft faced hammer until the snap ring is seated against the case. Be sure the clutching teeth of the input shaft gear mesh with the fifth speed synchronizer without binding.

10. Position a new gasket on the input shaft bearing retainer. *Be sure that the oil return holes in the retainer and gasket are aligned with the holes in the transmission case.*

11. Install the lock washers and bolts on the input shaft bearing retainer and tighten the bolts to specifications.

12. Position the output shaft rear bearing on the shaft and drive the bearing into the bore until the snap ring is seated against the case.

13. Position the countershaft rear bearing cap and gasket on the case and install the lock washers and bolts. Tighten the bolts to specification.

14. Install the speedometer drive gear on the output shaft and position a new oil seal in the output shaft bearing retainer. Be sure the seal is correctly installed. Position a new gasket and the bearing retainer on the case and install the cock washers and bolts. Tighten the bolts to specification.

15. Install the parking brake drum and companion flange assembly. Tighten the yoke retaining nut to specification.

16. With the transmission in neutral, position the control cover over the gears, aligning the shift

forks in the shift cover with the gear shift hubs. If the control cover is in neutral and the transmission is in neutral, the transmission drive gear should turn without the brake drum or output shaft turning. Install the capscrews and washers. Tighten to 20-25 ft lbs. Install the remote control or shift tower in the control cover. Tighten the capscrews to 20-25 ft-lbs.

Torque Specifications

	Foot-Pounds
Companion flange nut	350-420
Countershaft rear bearing cap	20-25
Input shaft bearing retainer	20-25
Output shaft bearing retainer	20-25
Reverse idler shaft lock bolt	20-25
Shift control to case	20-25
Shift cover to case	20-25
Shift tower to shift cover	20-25
Transmission to clutch housing	60-80

Lubricant Capacity

	Pints
Clark 280V Series	8

Clark 320V & 380V Series-Five Speed Transmissions

Disassembly of Transmission

1. Remove the gearshift housing assembly.
2. Shifting by hand, engage the transmission into two speeds at the same time.
3. Remove the output flange nut cotter pin. Remove the nut.
4. Remove the parking brake band assembly. Remove the brake drum and flange assembly.
5. Remove the output shaft and countershaft rear bearing caps and gaskets.
6. With the transmission still locked in two gears, remove the countershaft rear bearing attaching nut.
7. Remove the input shaft bearing cap and gasket.
8. Remove the input shaft and bearing assembly from the front of the case.

CAUTION: Do not drop the pilot bearing from the pocket of the drive gear if it remains with the gear. If the bearing is not in the pocket of the gear, remove it from the end of the output shaft pilot.

9. Remove the speedometer drive gear and spacer from the output shaft. Move the output shaft assembly toward the rear of the case so that the rear bearing clears the case. Working with a suitable puller, remove the bearing from the end of the shaft.
10. Tilt the forward end of the output shaft upward to clear the front of the case and remove the entire assembly.
11. Remove the screw and lock washer holding the reverse idler shift lock plate at the rear of the case and remove the lock.
12. Working with a suitable puller in the slot in the end of the reverse idler shaft, pull the shaft from the case.

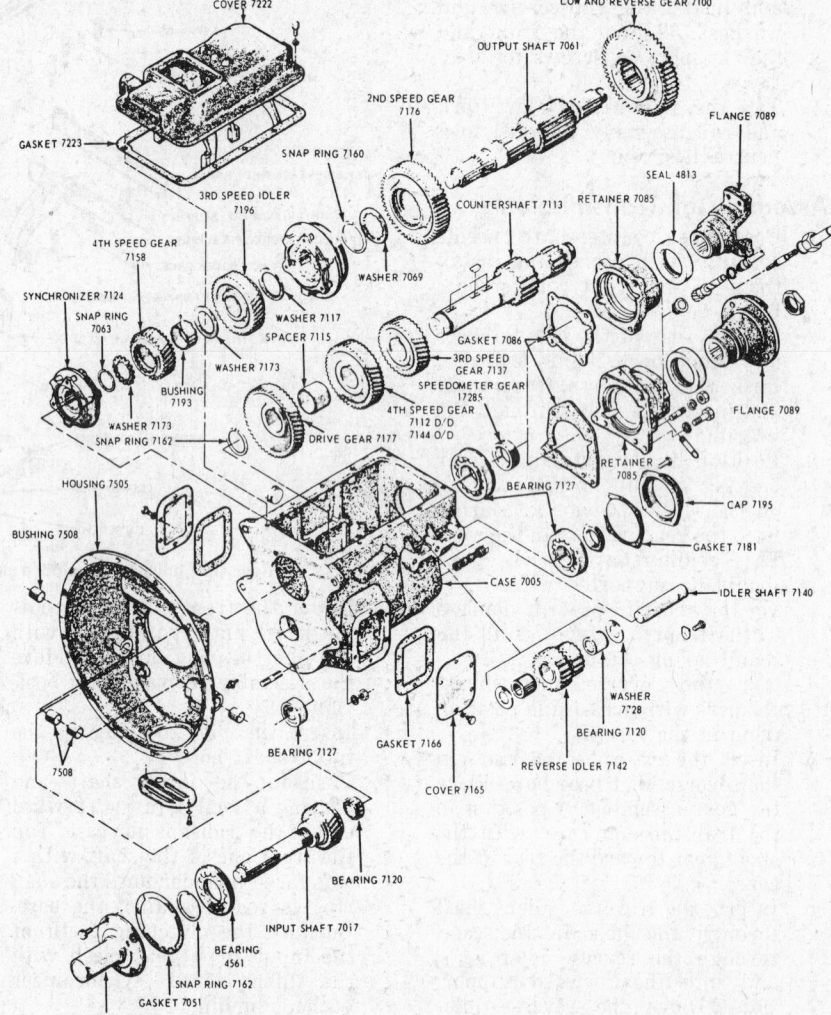

Clark 320V or 380 V Series Transmission—exploded view (© Ford Motor Co.)

13. Lift the reverse idler gear and two thrust washers from the case.
14. Push the countershaft toward the rear of the transmission case, far enough that the rear countershaft bearing can be removed from the countershaft.
15. Tilt the forward end of the countershaft upward to clear the front of the case and remove the entire assembly.
16. If the countershaft front bearing or pilot bearing is to be replaced, remove the clutch housing from the transmission.

17. Press the pilot bearing out of the case. *Do not hammer or drive out the bearing as damage to the bearing bore in the case will result.*

Sub-Assemblies

Output Shaft Disassembly

1. Remove the first-reverse sliding gear and the fourth-fifth synchronizer from the output shaft.
2. Place the remainder of the output shaft assembly in a soft-jawed vise, with the front end facing upward.
3. Remove the snap ring holding the fourth gear. Remove the gear and washer from the shaft.
4. Remove the assembly from the vise and with the front end facing downward, tap the assembly lightly on a block of wood, allowing the weight of the third gear to force the fourth gear bushing off the output shaft. Once the bushing has loosened, remove it from the shaft along with the fourth gear locating washer. When the bushing is removed, the locating pin will come off with it. Be sure that this pin is not lost.
5. Remove the third gear from the front end of the output shaft.
6. Remove the second-third synchronizer assembly.
7. Remove the second gear retaining ring and washer. Remove the second gear.
8. Remove the second gear bushing and locating washer from the output shaft.

Assembly

1. Clamp the output shaft in a soft-jawed vise with the front end facing upward.
2. Install the second gear locating washer on the output shaft.
3. Position the lock pin in the second gear bushing. Install the bushing on the output shaft with the lock pin toward the upper end.
4. *To be sure that the second gear locating washer, bushing sleeve, retaining washer, and snap ring are properly tight, use the following procedure.* Before the second gear is installed on the bushing sleeve, place the retaining washer on the shaft and hold it down tightly against the bushing sleeve. Hold the snap ring next to the retaining washer, and slip the snap ring in the groove in the shaft. Select a washer that will give a 0.004 inch maximum snap ring clearance. Remove the washer and install the gear over the bushing sleeve. Place the washer in position and install the snap ring on the shaft so that it is locked

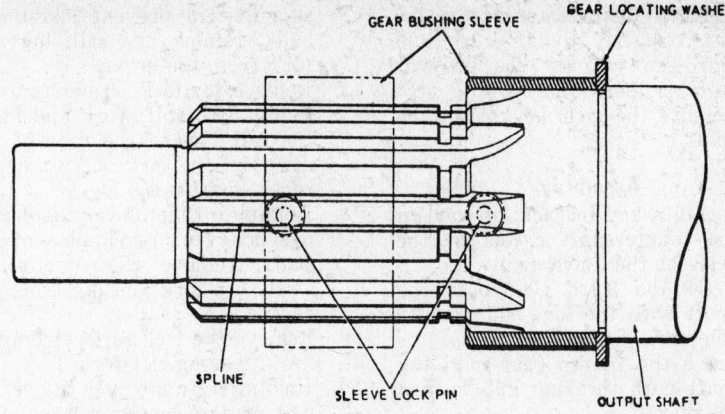

Installing the bushing sleeve—Clark 320V & 380V (© Ford Motor Co.)

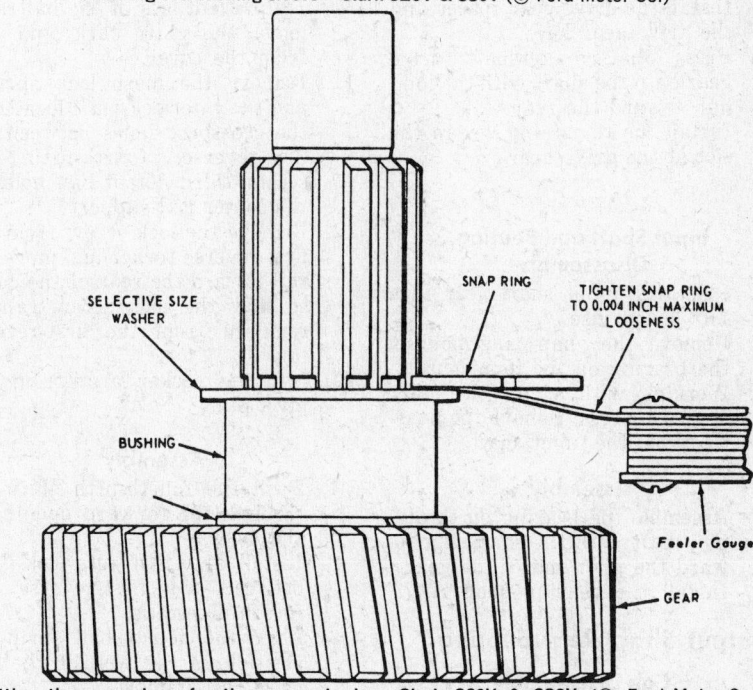

Setting the second or fourth gear end play—Clark 320V & 380V (© Ford Motor Co.)

securely in the groove. End play of the second gear must be a minimum of 0.006 inch.

5. Install the second-third synchronizer assembly on the splines of the shaft and drop it into position.
6. Install the third gear thrust washer.
7. Install the third gear on the shaft with the toothed hub facing downward.
8. Install the fourth gear locating washer.
9. Position the lock pin in the fourth gear bushing. Install the bushing on the shaft with the lock pin toward the upper end.
10. *To be sure that the fourth gear locating washer, bushing sleeve, retaining washer, and snap ring are properly tight, use the following procedure.* Before the fourth gear is installed on the bushing sleeve, place the retaining washer on the shaft and hold it down tightly against the bushing sleeve. Hold the snap ring

next to the retaining washer, and slip the snap ring in the groove on the shaft. Select a washer that will give a 0.004 inch maximum clearance. Remove the washer and install the gear over the bushing sleeve. Place the washer in position and install the snap ring on the shaft so that it is locked securely in the groove. Output shaft gears must maintain a minimum gear end play as follows; Fourth-0.006 inch, Third-0.009 inch, Second-0.006 inch.

11. Remove the assembly from the vise and install the sliding gears on the shaft as follows; fourth-fifth synchronizer hub with the large end facing toward the rear, and the first-reverse gear with the fork slot toward the front.

Countershaft Disassembly

1. Remove the snap ring from the front end of the countershaft.

2. Remove the countershaft drive gear, spacer, fourth gear and third gear with an arbor press or suitable gear puller.
3. Remove the gear keys from the shaft.

Assembly

1. Install a key in each keyway on the countershaft except at the drive at the drive gear.
2. Press the third gear onto the shaft with the long hub toward the rear.
3. Press the fourth gear onto the shaft with the long hub toward the rear.
4. Install the drive gear spacer and the drive gear key.
5. Press the countershaft drive gear onto the shaft with the long hub toward the rear.
6. Install the remaining key in the slot at the drive gear.

Input Shaft and Bearing Disassembly

1. Place the input shaft gear in a soft-jawed vise.
2. Remove the snap ring holding the bearing on the input shaft.
3. Working with a suitable puller or arbor press, remove the bearing from the input shaft.

Assembly

1. Assemble the bearing on the input shaft with the snap ring toward the pilot end of the gear.
2. Install the bearing snap ring.

Output Shaft Rear Bearing

Cap Disassembly

Press or drive the oil seal out. *This should only be done if the seal is in need of replacement.*

Assembly

Press a new oil seal in the bearing cap with the lip of the seal toward the transmission.

Gearshift Housing Disassembly

1. Place the cover on a bench in an inverted position with the extension plug opening facing upward.
2. Remove the expansion plugs from the forward openings in the cover.
3. Remove the lock wires from the shifter lugs and forks.
4. Shift all rails to their neutral position.
5. Remove the lock screws from the second-third shift fork and move the rail toward the rear of the cover so that the lock screw can be removed.
6. Holding a thumb over the detent ball hole in the center rail support (to prevent losing the ball), remove the rail, lug and fork from the cover.
7. Remove the lock screw from the fourth-fifth shift fork and move the rail toward the rear of the cover so the lock screw can be removed from the lug.
8. Holding a thumb over the detent ball hole (to prevent loss of the ball), remove the fourth-fifth shift rail, fork and lug from the cover.
9. Remove the lock screw from the first-reverse shift fork.
10. Holding a thumb over the detent ball in the center rail support (to prevent loss of the ball), remove the shift fork and rail from the cover.
11. Remove the mesh lock springs and the interlock balls located in the crossbore holes between the first-reverse, fourth-fifth and second-third detent ball holes of the center rail support.
12. Remove the lock screw from the first-reverse rocker lug, move the rail toward the rear of the cover so that the lock screw can be removed from the first-reverse lug.
13. Lift the rocker arm from the pivot pin.

Assembly

1. Push the fourth-fifth shift rail through the forward opening in the cover.
2. Install the two interlock balls between the first-reverse and fourth-fifth rails.
3. Place the detent ball spring in position in the well of the forward rail support.
4. Place the detent ball in position on the spring.
5. Working with a suitable tool, compress the spring and ball so they will pass between the guide fingers of the first-reverse rocker lug if the fingers are aligned.
6. Slide the fourth-fifth shift fork on the rail with the offset for the lock screw toward the rear of the cover.
7. Install the lock screw in the fork, and wire it securely.
8. Shift the assembly into neutral.
9. Push the second-third rail through the forward opening in the cover.
10. Install the two interlock balls between the second-third and fourth-fifth detent ball holes on the forward rail support.
11. Place the detent ball spring in position in the well of the rail support.
12. Using a suitable tool, compress the spring and ball to allow the rail to pass through the support.
13. Slide the second-third shift fork onto the rail with the offset for

the lock screw toward the front of the cover.
14. Install the lock screw in the fork and wire securely.
15. Shift the cover into neutral position.
16. Install the expansion plugs in the forward opening.
17. Install the detent ball lock spring and ball in the well of the first-reverse shift fork.
18. Using a suitable tool, compress the ball and spring so that the rail will pass through the fork.
19. Push the first-reverse rail in the rear opening of the cover.
20. Holding the fork in position on the rocker arm, and with the off-set to the rear of the cover, push the rail through the fork and into the rail support.
21. Install the lock screw in the rail support and rail and wire it securely.
22. Check the assembly of the cover by shifting in out of all speed ranges. Shift into the fifth gear position and attempt to shift into first, reverse, second, or third at the same time. This provides a check for the proper assembly of the interlock balls and pin. The shift lever should not shift into more than one speed at a time.
23. Return the shift rail to neutral.

Gear Shift Lever and Tower Disassembly

1. Remove the gearshift knob and dust cover.
2. Remove the shift lever, support spring, and washer.
3. Remove the shift lever from the shift tower.

Assembly

1. Position the shift lever in the shift tower.
2. Place the shift lever on the transmission case. Replace the washer and spring support.
3. Replace the dust cover and the gearshift knob.

Reverse Lug Assembly Disassembly

1. Remove the cotter key from the end of the plunger.
2. Remove the nut from the end of the plunger.
3. Remove the spring and plunger from the well in the lug.

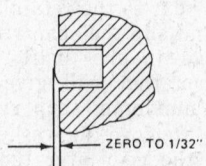

ZERO TO 1/32"

Reverse lug plunger adjustment tolerance —Clark 320V & 380V (© Ford Motor Co.)

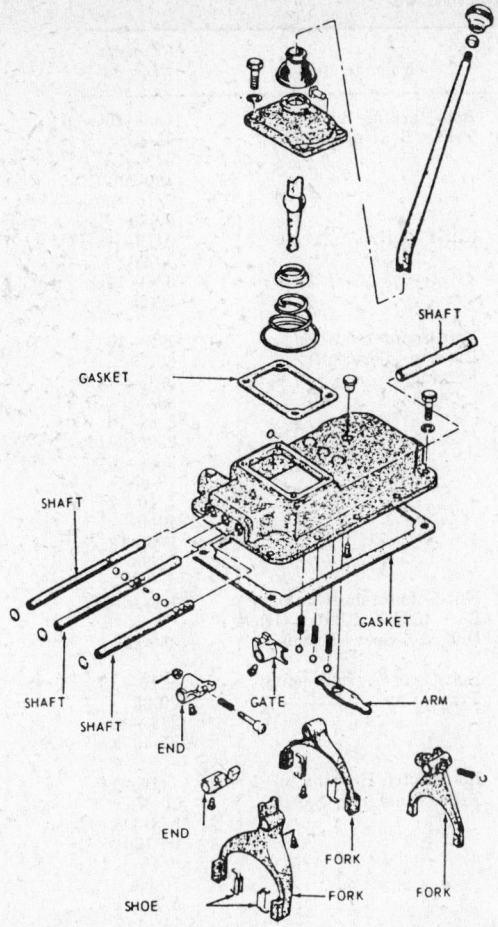

Gear shift housing—exploded view—Clark 320V & 380V (© Ford Motor Co.)

Assembly

1. Install the plunger and spring in the well in the lug.
2. Install the nut on the end of the plunger. Adjust the plunger flush to 1/32 inch above the face of the lug.
3. Install the cotter key.

Assembly of Transmission

1. If the countershaft front bearing is to be replaced, coat the outer diameter of the pilot bearing with a suitable sealer.
2. From the clutch housing end of the case with the open end of the bearing toward the rear of the transmission, press the bearing 0.007 inch below the front face of the case.
3. Tip the rear of the countershaft down and lower it into the case, running it through the countershaft opening, in the rear of the case, far enough so that the front of the countershaft may be lowered into position. Push the countershaft into position.
4. Position the countershaft rear bearing oil slinger and bearing on the shaft. Drive the bearing into the bore. The countershaft drive gear should be supported while driving the shaft to prevent damage to the front bear-

ing. Install the rear bearing retainer ring.
5. Coat the reverse idler thrust washers with grease and position them in the case.
6. Insert the two roller bearings in the reverse idler gear bore. Place the gear assembly in position in the case, with the small gear toward the rear of the case.
7. Insert the reverse idler shaft through the hole in the case, through the reverse idler gear, and into the forward support boss. Drive the reverse idler shaft into the case until the slot in the shaft is lined up with the lock bolt hole. Install the retainer in the slot and secure the retainer with the lock bolt.

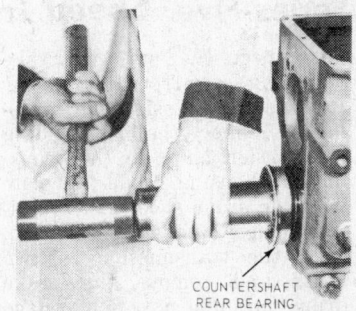

Installing the rear countershaft Bearing—Clark 320V & 380V (© Ford Motor Co.)

Tighten the bolt to specification.

8. Tilt the rear end of the output shaft down and lower it into, and through, the opening in the rear of the transmission case. Lower the front end, in line with the pilot bearing opening, and move the output shaft assembly forward into position. Let the shaft rest in this position until assembly is completed.
9. Assemble the output shaft rear bearing over the end of the output shaft with the snap ring in the outer race facing toward the rear.
10. Install the pilot bearing on the output shaft.
11. Install the input shaft and bearing assembly in the case. Be sure the input shaft gear engages and meshes with the countershaft gear, and the output shaft and pilot bearing are seated in the input shaft bore. If a new pilot bearing is being used, it will be in a plastic sleeve, stand the drive gear on end. Set the bearing and sleeve over the bearing pocket in the drive gear. Slide the bearing rollers and cage from the plastic sleeve into the bearing pocket.

NOTE: When a new bearing is used, always install a new bearing race on the output shaft. If the old pilot bearing is used, set the rollers in the bearing cage and hole them in place with a rubber band. Slide the bearing rollers and the cage from the rubber band into the bearing pocket.

12. Install the output shaft bearing cap with a new gasket. Be sure that the oil hole in the cap is lined up with the oil return hole in the case. This passage must be open and clear.
13. Lock the gears in two gears at one time, and install the countershaft rear bearing attaching nut. Tighten the nut to specifications and stake it in position.
14. Install the countershaft rear bearing cap and gasket.
15. Install the speedometer drive gear spacer and the speedometer drive gear on the output shaft.
16. Install the output shaft rear bearing cap assembly and gasket.
17. Install the parking brake drum and the companion flange assembly.
18. Install the companion flange nut. Tighten the nut to specification.
19. Move the gears to neutral and assemble the gearshift housing and gasket to the transmission case. Make sure the shift forks engage the slots on the gears.
20. Assemble the gasket, gearshaft lever, and tower assembly to the shift bar housing.

SPECIFICATIONS

Nomenclature	Torque Limits	Nomenclature	Torque Limits
Bolt-Reverse Lockout Plunger Retainer	11/16—16 90-100	Bolt-Parking Brake to Trans. Bearing Retainer	1/2—20C 45-55 5/8—18 180-220 7/16—20 50-70
Bolt-Clutch Housing to Engine Block	7/16—14 24-32 33-45	Nut-Bellcrank to Trans.	9/16—18 70-90 3/8—16C 20-25
Bolt-Output Shaft Bearing Retainer to Trans. Case	3/8—16 20-25 7/16—14 30-38 1/2—13 50-62 5/8—11 96-120	Bolt-Countershaft & Reverse Idler Shaft Retainer	1/2—13 60-70 5/16—18 25-30 3/8—16 25-37 3/8—16 18-25 7/16—14 40-45 1/2—13 8C-85
Bolt-Gear Shift Lever Tower to Gearshift Housing	3/8—16 20-25 5/16—18 11-16 7/16—14 30-45		
Nut-Drum Parking Brake to Companion Flange	3/8—24 37-50 7/16—20 40-55 1/2—20 74-86	Nut-Countershaft Bearing Lock (5-Speed Extra Heavy Duty & 5-Speed Exclusive)	1 1/4—18 350-450
		Bolt-Gear Shaft Housing Trans. Case	5/16—18 20-25 3/8—16 35-40
Bolt—Lever Assy. to Trans.	3/8—16 20-25	Bolt-Clutch Housing to Trans. Case	7/16—14 30-38 5/8—11 96-120 9/16—12 70-90 5/8—18C 115-140 9/16—18C 81-102
Nut-Handbrake Anchor Bar to Trans. Case (except 5-Speed Extra-Heavy Duty)	9/16—18 100-110		
Nut Handbrake Anchor Bar to Trans. Case (5-Speed Extra-Heavy Duty only) Nut-U-Joint Flange to Trans. Output Shaft	9/16—18 120-130 1.00—20 90-125 1 1/4—18 225-275 1 1/2—18 275-350		
Nut-Input Shaft Yoke to Aux. Trans.	1 1/4—18 275-350	Bolt—Gear Shift Housing Trans. Case	7/16—14 45-50
Bolt-Countershaft Rear Bearing Retainer	5/16—18 25-30 3/8—16 35-40 7/16—14 45-55 1/2—13 60-70	Nut-Countershaft Bearing Lock (5-Speed Exclusive Heavy Duty)	1 1/2—18 350-450
		Bolt-Countershaft Front Bearing Retainer	5/16—18 25-30 3/8—16 25-35 7/16—14 50-55
Jam Nut—Adjusting Height Main Trans.	150-200		

Lubricant Capacity			Pints
Clark 320V & 380V (SAE 50 Engine Oil)			13

Clark 390V Series-Five Speed Transmission

Disassembly of Transmission

1. Remove the remote-control or shift tower from the control cover. Remove the control cover capscrews and lockwashers. Remove the control cover assembly from the transmission. Remove the back-up light switch.
2. Remove the universal joint assembly and the drive shaft from the parking brake drum.
3. Remove the parking brake drum. Disconnect the parking brake actuating lever from the linkage.
4. Remove the transmission spline flange. Remove the bolts holding the carrier plate to the transmission housing. Slide the plate with the brake shoes and retaining springs off the transmission.
5. Lock the transmission in two gears and remove the brake drum retaining nut. Remove the brake drum from the output shaft.
6. Remove the output shaft rear bearing retainer and the speedometer drive gear.
7. Remove the countershaft rear bearing retainer, then remove the bearing snap ring.
8. Remove the input shaft bearing retainer and pull the input shaft out of the case. Be careful not to drop the output shaft pilot bearing rollers into the transmission case.
9. Force the output shaft rearwar

until the rear bearing is exposed. Working with a bearing puller, remove the rear bearing from the bore.

10. Lift the output shaft out of the case.

11. Remove the reverse idler shaft using special tool. Lift the reverse idler gear, bearing and thrust washer out of the transmission case.

12. Move the countershaft rearward until the rear bearing is exposed. Remove the bearing and oil slinger with a suitable gear puller.

13. Lift the countershaft from the case.

14. If the countershaft front bearing or pilot bearing is to be replaced, remove the clutch housing from the transmission.

15. Press the pilot bearing out of the case. *Do not hammer or drive the bearing out as damage or distortion to the bearing bore will result.*

Sub-Assemblies

Output Shaft Disassembly

1. Clamp the output shaft, front end facing upward, in a soft-jawed vise. Remove the fifth gear synchroninizer and the fourth and fifth gear synchronizer assembly.

2. Remove the shift hub thrust bearing and race.

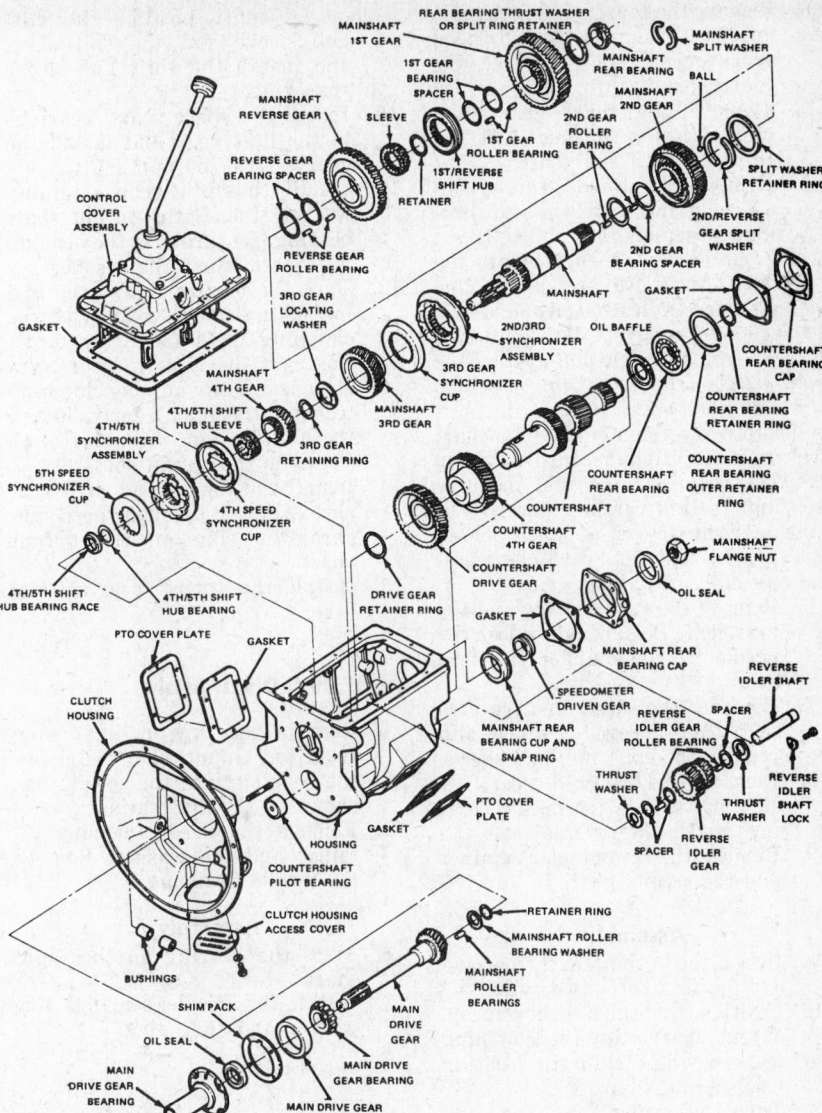

Clark 390V Series transmission—exploded view (© Ford Motor Co.)

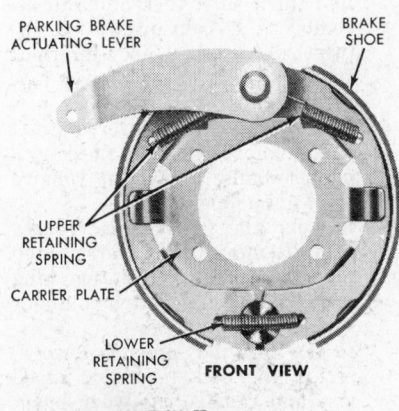

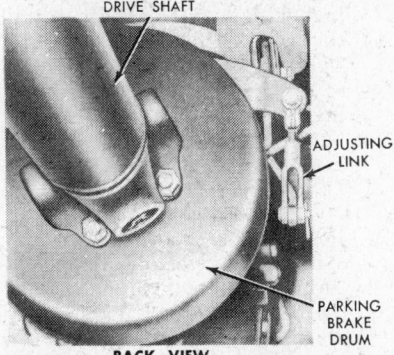

Transmission parking brake—Clark 390V
(© Ford Motor Co.)

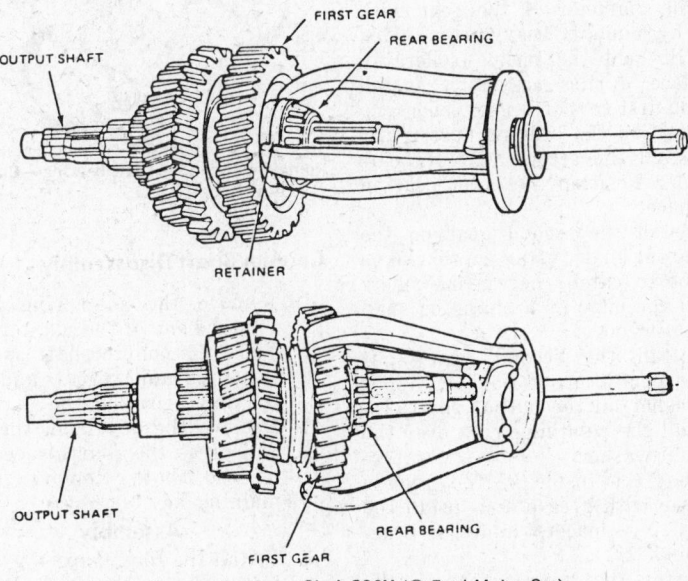

First gear retention—Clark 390V (© Ford Motor Co.)

3. Remove the fourth and fifth shift hub sleeve and the fourth speed synchronizer ring. Remove the fourth gear.
4. Remove the third gear snap ring, locating washer. Lift the third gear off the shaft.
5. Remove the third gear synchronizer ring and the synchronizer assembly.

NOTE: *There is a variation in the mainshaft first gear retention. One version has a split washer and a split washer retainer ring. Working with a three-legged puller, pull against the split washer retainer ring to remove the rear bearing.*

The other version of the mainshaft first gear retention is a thrust washer between the first gear and the rear bearing. In this version there is not enough room to get a puller behind the thrust washer and it will be necessary to pull on the first gear.

6. Remove the rear bearing and the first gear. Be sure not to lose the needle bearings under the first gear.
7. Remove the first and reverse shift hub sleeve retainer and remove the shift hub and sleeve.
8. Remove the reverse gear. Be sure not to lose the needle bearings under the reverse gear.
9. Remove the second gear retainer and the second gear.

Assembly

1. Install the output shaft, forward end up, in a soft-jawed vise.
2. Position the third gear on the output shaft with the clutching teeth down. Install the locating washer and snap ring.
3. Turn the output shaft over and install the third gear synchronizer ring and the second and third synchronizer assembly.
4. With the clutching teeth of the second gear down, coat the inside diameter of the gear with a high quality heavy grease. This will hold the needle rollers in place during assembly. Install the first row of needle rollers.
5. Position the bearing spacer and install the second row of bearings. Position the outer bearing spacer.
6. Install the second gear on the output shaft, using caution so as not to catch the needle rollers on the edge of a spline or snap ring groove.
7. Install the second gear split washer locating ball and split washer on the output shaft. Install the retaining ring over the split washer.
8. Coat the inside of the reverse gear with grease and install the bearing spacers and the bearings.
9. Install the reverse gear on the

output shaft, position the shift hub sleeve and the shift hub, and install the shift hub sleeve snap ring.
10. Install the spacers and bearings in the first gear and install the gear on the output shaft.
11. Install the first gear retaining washer. Install the output shaft bearing. Be sure that the bearing is tight against the washer.
12. Turn the output shaft over and install the fourth gear with the clutching teeth facing upward.
13. Position the fourth gear synchronizer ring on the clutching teeth of the fourth gear. Install the fourth and fifth shift hub sleeve on the output shaft.
14. Install the fourth and fifth synchronizer and the fifth speed synchronizer ring on the output shaft.
15. Install the thrust bearing and race.

Input Shaft Disassembly

1. Remove the bearing snap ring from the input shaft. Remove the bearing with the special tool shown in the illustration.
2. Remove the needle bearing retainer and the washer. Remove the needle bearings.

Assembly

1. Press the bearing on the input shaft.
2. Install the needle bearings, the washer and snap ring.

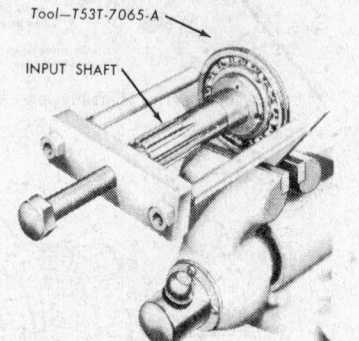

Tool—T53T-7065-A

INPUT SHAFT

Removing the input shaft bearing—Clark 390V
(© Ford Motor Co.)

Counter Shaft Disassembly

1. Remove the snap ring at the forward end of the counter shaft.
2. Place the countershaft assembly in a hydraulic press, and press the drive gear off the shaft.
3. Remove the key from the shaft and press the fourth-speed gear off the shaft. Remove the remaining key.

Assembly

1. Install the fourth-speed gear key in the slot on the shaft. Position

the fourth-speed gear on the shaft with the long hub facing toward the front.
2. Install the countershaft main drive gear key in the keyway on the shaft. Install the countershaft drive gear on the shaft with the long hub facing toward the rear. Install the snap ring on the countershaft.

Gear Shift Housing Disassembly

NOTE: *It is not necessary to disassemble the gear shift housing assembly to determine if the parts are worn. Note the condition of the shifter fork shafts and the forks by visual inspection. If the forks are excessively worn, or if they bind when shifted into the various positions, disassemble the gear shift housing. Check the interlocking system.*

1. With the control cover in neutral, pry the fourth-fifth speed shift fork to the fourth speed position (toward the rear of the cover). Remove the front rail support capscrews and remove the front rail support.
2. Remove the interlock tapered pin supports. Note the position on the interlock tapered pins for reassembly.
3. Remove the rear rail support capscrews and remove the rear rail support.
4. Remove the first-and-reverse shift fork and rail assembly, remove the fourth, fifth, second and third shift fork and rail assembly. Use caution so that the interlock cross pin, interlock tapered pins, or the mesh lock poppet balls are not lost.
5. Remove the first-reverse shift rail. Remove the four mesh lock poppet balls, and then remove the four poppet springs.
6. Remove the first-reverse rocker arm. Remove the reverse latch plunger spring retaining plug and the reverse latch plunger spring and plunger.
7. If the fork bushings are worn, secure the fork in a soft-jawed vise and remove the worn bushings with a suitable drift. Install the new bushings in the fork. Turn the fork over on the anvil of the vise and secure the bushing in the fork, using a pick punch and upsetting the bushing metal on the outside of the fork.

Assembly

1. Position the first-reverse rocker arm on the pivot pin. Install the reverse latch plunger, spring and the retaining plug. Tighten the plug securely.
2. Install the four poppet springs and the four mesh lock poppet balls.

NOTE: the first-reverse poppet ball in the pocket.

3. Align one tapered interlock cross pin with the hole in the first-reverse shift rail. Position the rail on the poppet ball, with the rail in the neutral position. Note the position of the tapered interlock pin in relation to the rail.

4. Install the interlock cross pin in the second-third shift rail. Position the rail on the poppet ball, in neutral position with the interlock pin aligned with the first interlock pin aligned with the first tapered pin. Install the second interlock tapered pin. Align the pin with the interlock cross pin hole.

5. Position the fourth-fifth shift fork rail assembly on the poppet ball in the neutral position. Slightly raise the rear of the rail and align the second interlock tapered pin with the cross hole in the rail. Note the positions of the tapered interlock pins in the shift rails.

6. Install the first-reverse shift fork and rail assembly on the poppet ball in a neutral position. Align the first-reverse rocker arm in the rear of the rail.

7. Install the rail support capscrews and washers. Tighten the capscrews slightly. Install the interlock tapered pin supports. Tap the fourth-fifth shift fork to the rear (fourth-speed position).

8. Position the front rail support and install the capscrews and washers. Tighten the front and rear support capscrews to 20-25 ft-lbs.

9. Tap the fourth-fifth shift fork and rail assembly forward to a neutral position.

Assembly of Transmission

1. Coat the front counter shaft needle bearings with grease and install them in the front countershaft bore.

2. Tap the rear of the countershaft down and lower into the case. Push the countershaft forward and insert the shaft into the front bearing.

3. Position the countershaft rear bearing oil slinger and bearing on the shaft. Drive the bearing into the bore. The countershaft drive gear should be supported while driving the shaft to prevent damage to the front bearing. Install the rear bearing retainer.

4. Coat the reverse idler thrust washer with grease and position them in the transmission case.

5. Insert the bearings in the reverse idler bore. Place the gear assembly in position in the case,

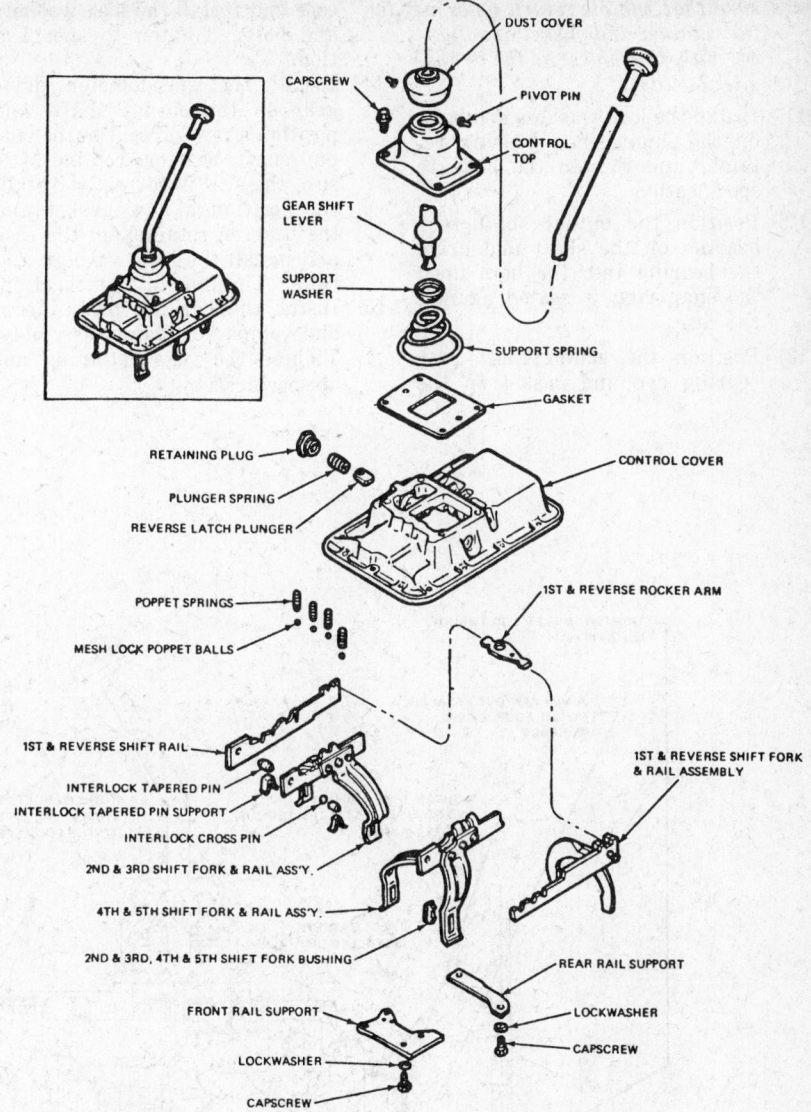

Gear shift housing—exploded view—Clark 390V (© Ford Motor Co.)

Installing the rear countershaft bearing—Clark 390V (© Ford Motor Co.)

with the small gear toward the rear of the case.

6. Insert the reverse idler shaft through the hole in the case, through the reverse idler gear, and into the forward support boss. Drive the reverse idler shaft into the case until the slot in the shaft is lined up with the

lock bolt hole. Install the retainer in the slot and secure the retainer with the lock bolt. Tighten the bolt to specification.

7. Tilt the rear of the output shaft assembly downward and insert the end of the shaft through the output shaft in the bore in the case. Lower the front end of the output shaft until it is in line with the pilot bearing opening. Move the assembly forward into position.

8. Insert the pilot bearing in the input shaft bore.

9. Position the input shaft and bearings in the forward end of the transmission case. Tap the front end of the shaft with a soft faced hammer until the snap ring is seated against the case. Be sure the clutching teeth of the input shaft gear mesh with the fifth-speed synchronizer without binding.

10. Position a new gasket on the input shaft bearing retainer. *Be*

sure that the oil return holes in the retainer and gasket are aligned with the holes in the transmission case.

11. Install the lock washers and bolts on the input shaft bearing retainer and tighten the bolts to specification.

12. Position the output shaft rear bearing on the shaft and drive the bearing into the bore until the snap ring is seated against the case.

13. Position the countershaft rear bearing cap and gasket on the case and install the lock washers and bolts. Tighten to specification.

14. Install the speedometer drive gear on the output shaft, and position a new oil seal in the output shaft bearing retainer. Be sure the seal is correctly installed. Position a new gasket and the bearing retainer on the case and install the lock washers and bolts. Tighten to specification.

15. Install the parking brake drum and companion flange assembly. Tighten the yoke retaining nut to specification.

16. With the transmission in neutral, position the control cover over the gears, aligning the shift forks in the shift cover with the gear shift hubs. If the control cover is in neutral and the transmission is in neutral, the transmission drive gear should turn without the brake drum or output shaft turning. Install the capscrews and washers, and tighten to 20-25 ft-lbs. Install the remote control or shift tower in the control cover. Tighten the capscrews to 20-25 ft-lbs.

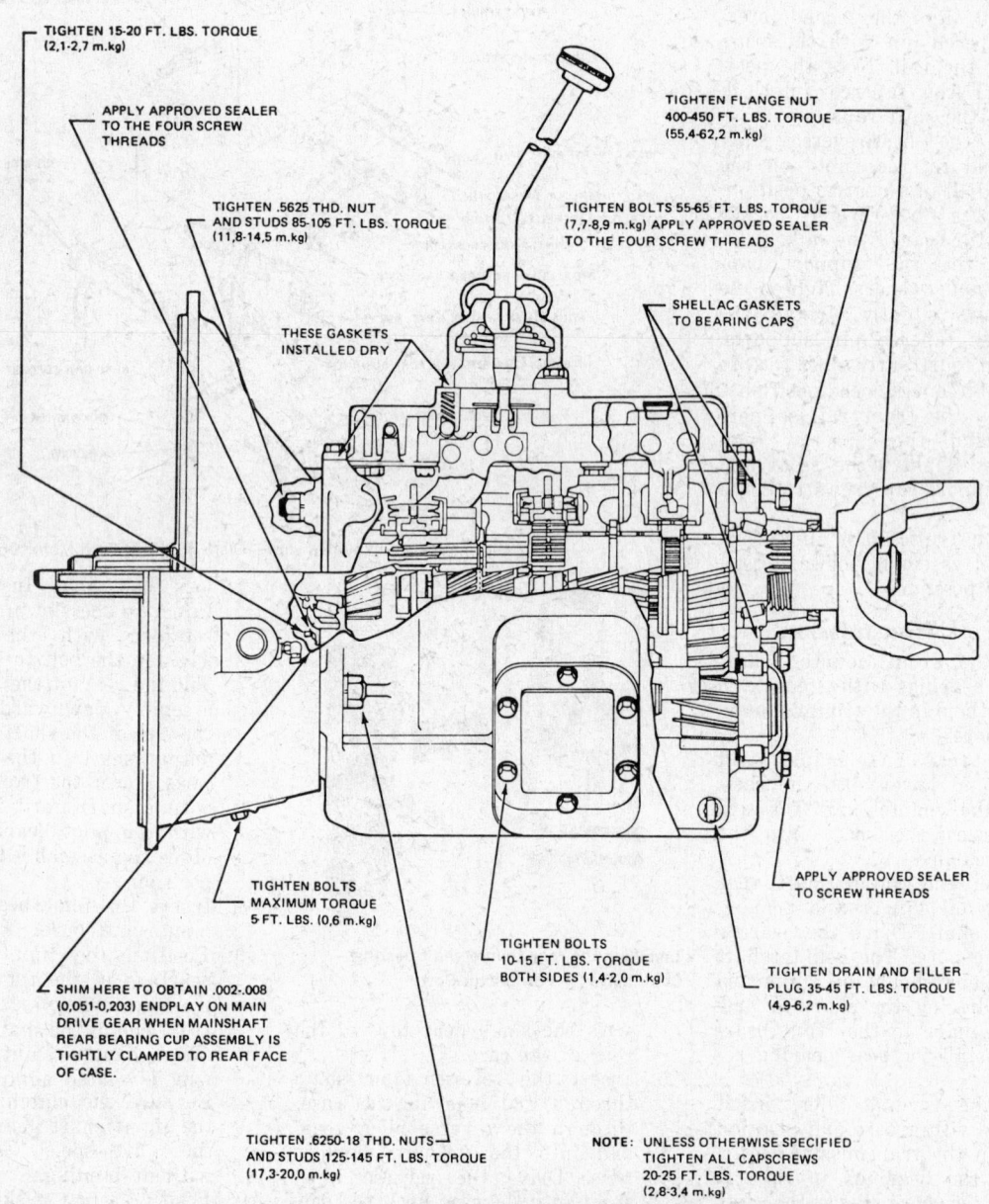

TIGHTEN 15-20 FT. LBS. TORQUE (2,1-2,7 m.kg)

APPLY APPROVED SEALER TO THE FOUR SCREW THREADS

TIGHTEN .5625 THD. NUT AND STUDS 85-105 FT. LBS. TORQUE (11,8-14,5 m.kg)

THESE GASKETS INSTALLED DRY

TIGHTEN FLANGE NUT 400-450 FT. LBS. TORQUE (55,4-62,2 m.kg)

TIGHTEN BOLTS 55-65 FT. LBS. TORQUE (7,7-8,9 m.kg) APPLY APPROVED SEALER TO THE FOUR SCREW THREADS

SHELLAC GASKETS TO BEARING CAPS

APPLY APPROVED SEALER TO SCREW THREADS

TIGHTEN BOLTS MAXIMUM TORQUE 5 FT. LBS. (0,6 m.kg)

SHIM HERE TO OBTAIN .002-.008 (0,051-0,203) ENDPLAY ON MAIN DRIVE GEAR WHEN MAINSHAFT REAR BEARING CUP ASSEMBLY IS TIGHTLY CLAMPED TO REAR FACE OF CASE.

TIGHTEN BOLTS 10-15 FT. LBS. TORQUE BOTH SIDES (1,4-2,0 m.kg)

TIGHTEN DRAIN AND FILLER PLUG 35-45 FT. LBS. TORQUE (4,9-6,2 m.kg)

TIGHTEN .6250-18 THD. NUTS AND STUDS 125-145 FT. LBS. TORQUE (17,3-20,0 m.kg)

NOTE: UNLESS OTHERWISE SPECIFIED TIGHTEN ALL CAPSCREWS 20-25 FT. LBS. TORQUE (2,8-3,4 m.kg)

Torque specifications—Clark 390V (© Ford Motor Co.)

Dodge/Plymouth A-230 3 Speed

Description

The Dodge A-230 is a three-speed transmission equipped with two synchonizer units to assist in the engagement of all forward gears. Lubricant capacity is 5 pints.

Disassembly of Transmission

Shift Housing and Mechanism

1. Shift to second gear.
2. Remove side cover. If shaft O-ring seals need replacement:
a. Pull shift-forks out of shafts.
b. Remove nuts and operating levers from shafts.
c. Deburr shafts. Remove shafts.

Drive Pinion Retainer and Extension Housing

1. Remove pinion bearing retainer from front of transmission case.

Pry off retainer oil seal. For clearance:
a. With a brass drift, tap drive pinion as far forward as possible. Rotate cut away part of second gear next to countershaft gear. Shift second-third synchronizer sleeve forward.
b. Remove speedometer pinion adapter retainer. Work adapter and pinion out of extension housing.
c. Unbolt extension housing. Break housing loose with plastic hammer and carefully remove.

Idler Gear and Mainshaft

1. Insert dummy shaft in case to push reverse idler shaft and key out of case.

2. Remove dummy shaft and idler rollers.
3. Remove both tanged idler gear thrust washers.
4. Remove mainshaft assembly through rear of case.

Countershaft Gear and Drive Pinion

1. Using a mallet and dummy shaft, tap the countershaft rearward enough to remove key. Drive countershaft out of case, being careful not to drop the washers.
2. Lower countershaft gear to bottom of case.
3. Remove snap-ring from pinion bearing outer race (outside front of case).

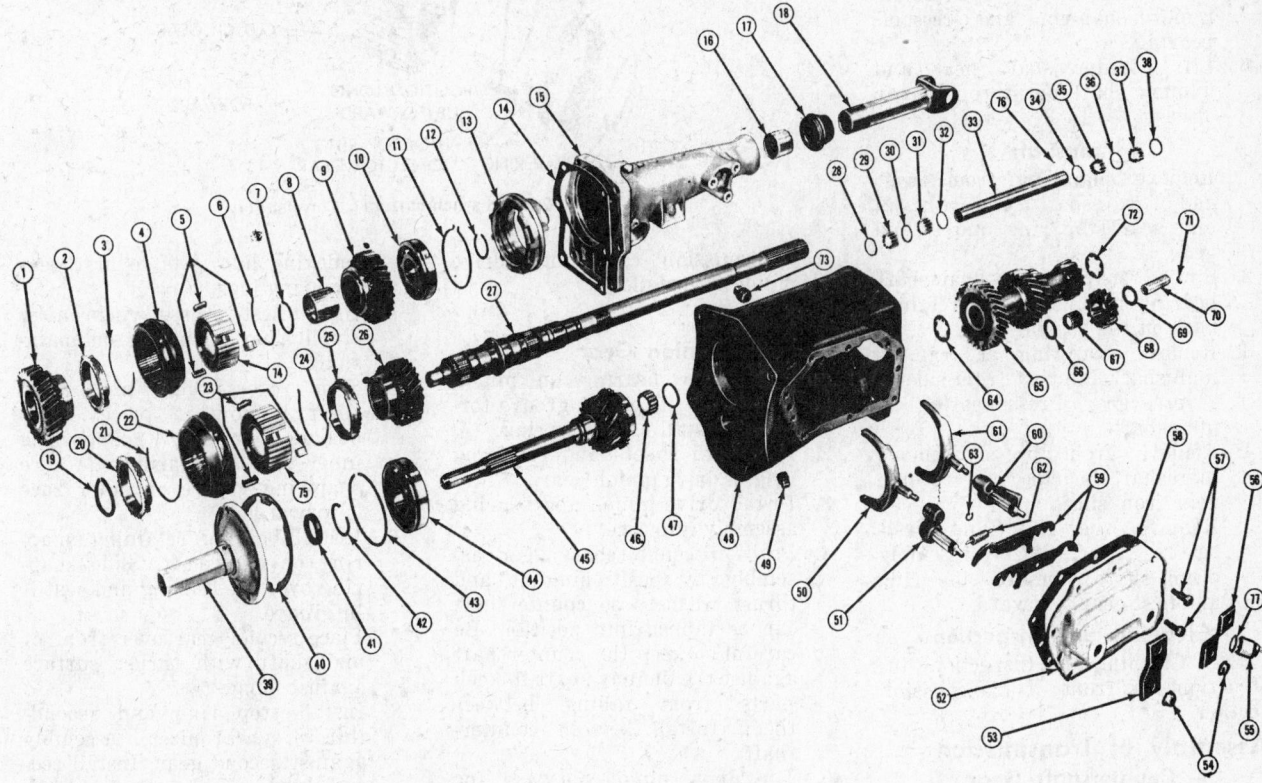

1 Gear first	20 ring	39 Retainer
2 Ring	21 Spring	40 Gasket
3 Spring	22 Sleeve	41 Seal
4 Sleeve	23 Struts (3)	42 Snap ring
5 Struts (3)	24 Spring	43 Snap Ring
6 Spring	25 Ring	44 Bearing
7 Snap ring	26 Gear, second	45 Pinion, drive
8 Bushing	27 Shaft, output	46 Roller
9 Gear, reverse	28 Washer	47 Snap ring
10 Bearing	29 Roller	48 Case
11 Snap ring	30 Washer	49 Plug, Drain
12 Snap ring	31 Roller	50 Fork
13 Retainer	32 Washer	51 Lever
14 Gasket	33 Countershaft	52 Housing
15 Extension	34 Washer	53 Lever
16 Bushing	35 Roller	54 Nut, locking
17 Seal	36 Washer	55 Switch
18 Yoke	37 Roller	56 Lever
19 Snap ring	38 Washer	57 Bolt
		58 Gasket

59 Lever, interlock
60 Lever
61 Fork
62 Spring
63 Snap ring
64 Washer
65 Gear, countershaft
66 Washer
67 Roller
68 Gear, idler
69 Washer
70 Shaft
71 Key
72 Washer
73 Plug, filler
74 Gear, Clutch
75 Gear, clutch
76 Key
77 Gasket

A-230 3 speed transmission (© Chrysler Corp.)

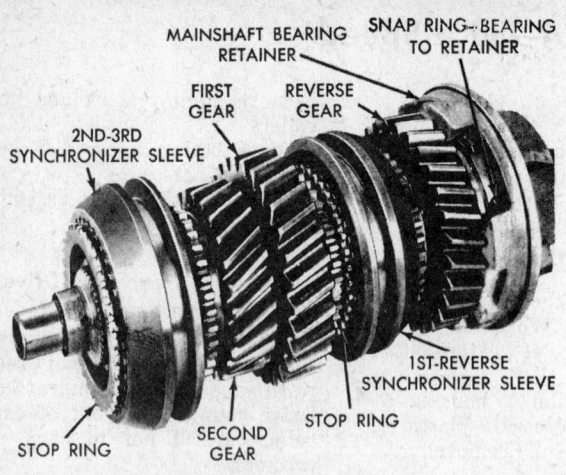

Mainshaft assembly (© Chrysler Corp.)

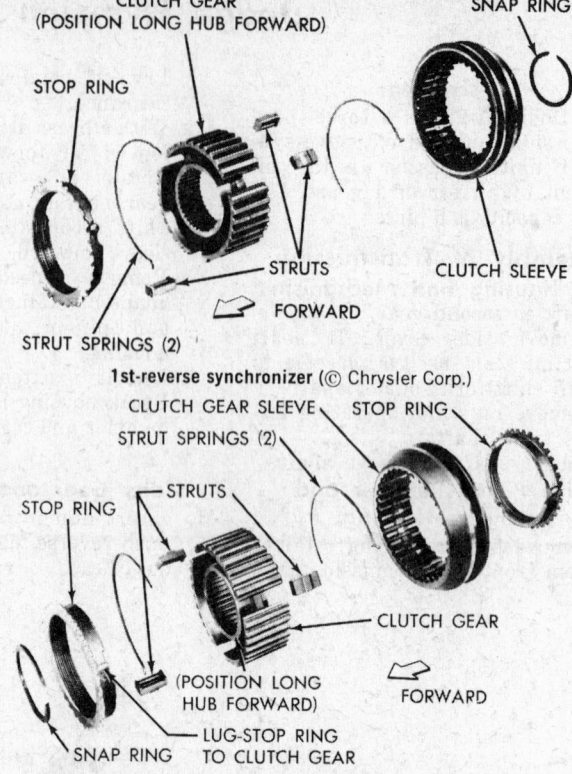

1st-reverse synchronizer (© Chrysler Corp.)

2nd-3rd synchronizer (© Chrysler Corp.)

4. Drive pinion shaft into case with plastic hammer. Remove assembly through rear of case.
5. If bearing is to be replaced, remove snap-ring and press off bearing.
6. Lift counter shaft gear and dummy shaft out through rear of case.

Mainshaft

1. Remove snap-ring from front end of mainshaft along with second gear stop ring and second gear.
2. Spread snap-ring in mainshaft bearing retainer. Slide retainer back off the bearing race.
3. Remove snap-ring at rear of mainshaft. Support front side of reverse gear. Press bearing off mainshaft.
4. Remove from press. Remove mainshaft bearing and reverse gear from shaft.
5. Remove snap-ring and first-reverse synchronizer assembly from shaft. Remove stop ring and first gear rearward.

Cleaning and Inspection

See "Cleaning and Inspection" instructions at front of transmission section.

Assembly of Transmission
Countershaft Gear

1. Slide dummy shaft into counter-shaft gear.
2. Slide one roller thrust washer over dummy shaft and into gear, followed by 22 greased rollers.
3. Repeat Step 2, adding one roller thrust washer on end.
4. Repeat steps 2 and 3 at other end of countershaft gear. There is a total of 88 rollers and 6 thrust washers.
5. Place greased front thrust washer on dummy shaft against gear with tangs forward.
6. Grease rear thrust washer and stick it in place in the case, with tangs rearward. Place counter-shaft gear assembly in bottom of

transmission case until drive pinion is installed.

Pinion Gear

1. Press new bearing on pinion shaft with snap-ring groove forward. Install new snap-ring.
2. Install 15 rollers and retaining ring in drive pinion gear.
3. Install drive pinion and bearing assembly into case.
4. Position countershaft gear assembly by positioning it and thrust washers so countershaft can be tapped into position. Be careful to keep the countershaft against the dummy shaft to keep parts from falling between them. Install key in countershaft.
5. Tap drive pinion forward for clearance.

Mainshaft

1. Place a stop ring flat on the bench. Place a clutch gear and a sleeve on top. Drop the struts in their slots and insert a strut spring with the tang inside on strut. Turn the assembly over and install second strut spring, tang in a different strut.
2. Slide first gear and stop ring over rear of mainshaft and against thrust flange between assembly over rear of mainshaft, first and second gears on shaft.
3. Slide first-reverse synchronizer

indexing hub slots to first gear stop ring lugs.
4. Install first-reverse synchronizer clutch gear snap-ring on mainshaft.
5. Slide reverse gear and mainshaft bearing in place. Press bearing on shaft, supporting inner race of bearing. Be sure snap-ring groove on outer race is forward.
6. Install bearing retaining snap-ring on mainshaft. Slide snapring over the bearing and seat it in groove.
7. Place second gear over front of mainshaft with thrust surface against flange.
8. Install stop ring and second-third synchronizer assembly against second gear. Install second-third synchronizer clutch gear snap-ring on shaft.
9. Move second-third synchronizer sleeve forward as far as possible. Install front stop ring inside sleeve with lugs indexed to struts.
10. Rotate cut out on second gear toward countershaft gear for clearance.
11. Insert mainshaft assembly into case. Tilt assembly to clear cluster gears and insert pilot rollers in drive pinion gear. If assembly is correct, the bearing retainer will bottom to the case without force. If not, check for a misplaced strut, pinion roller, or stop ring.

Reverse Idler Gear

1. Place dummy shaft into idler gear. Insert 22 greased rollers.
2. Position reverse idler thrust washers in case with grease.
3. Position idler gear and dummy shaft in case. Install idler shaft and key.

Extension Housing

1. Remove extension housing yoke seal. Drive bushing out from inside housing.
2. Align oil hole in bushing with oil slot in housing. Drive bushing into place. Drive new seal into housing.
3. Install extension housing and gasket to hold mainshaft and bearing retainer in place.

Drive Pinion Bearing Retainer

1. Install outer snap-ring on drive pinion bearing. Tap assembly back until snap-ring contacts case.
2. Using seal installer tool or equivalent, install a new seal in retainer bore.
3. Position main drive pinion bearing retainer and gasket on front of case. Coat threads with sealing compound, install bolts, torque to 30 ft. lbs.

Gearshift Mechanism and Housing

1. If removed, place two interlock levers in pivot pin with spring hangers offset toward each other, so that spring installs in a straight line. Place E-clip on pivot pin.

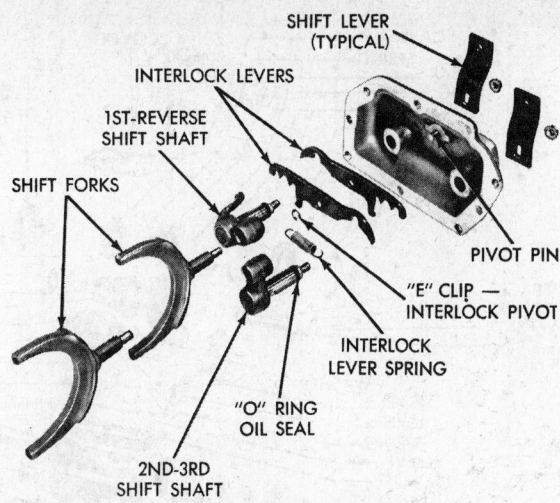

Gearshift mechanism and housing
(© Chrysler Corp.)

2. Grease and install new O-ring seals on both shift shafts. Grease housing bores and insert shafts.
3. Install spring on interlock lever hangers.
4. Rotate each shift shaft fork bore to straight up position. Install shift forks through bores and under both interlock levers.
5. Position second-third synchronizer sleeve to rear, in second gear position. Position first-reverse synchronizer sleeve to middle of travel, in neutral position. Place shift forks in the same positions.
6. Install gasket and gearshift mechanism. The bolt with the extra long shoulder must be installed at the center rear of the case. Torque bolts to 15 ft. lbs.

7. Install speedometer drive pinion gear and adapter. Range number on adapter, which represents the number of teeth on the gear, should be in 6 o'clock position.

TORQUE CHART

Manual A-203 3-Speed

	Foot-Pounds
Back up light switch	15
Extension housing bolts	50
Drive pinion bearing retainer bolts	30
Gearshift operating lever nuts	18
Transmission to clutch housing bolts	50
Transmission cover retaining bolts	12
Transmission drain plug	25

Dodge/Plymouth A-250 3 Speed

Description

The Dodge A-250 is a three speed transmission equipped with a synchronizer between second and third gears. Lubricant capacity is 4½ pints.

Disassembly of Transmission

1. Remove case cover and gasket.
2. Measure the synchronizer "float" with a pair of feeler gauges. Measurement is made between the synchronizer outer ring pin and the opposite synchronizer outer ring. This measurement must be made on two pins 180 degrees apart with equal gap on both ends for "float" determination. The measurement should be between 0.060-0.117 inch. A snug fit should be maintained between feeler gauge and pins.
3. Remove the bolt and retainer holding the speedometer pinion adapter in the extension housing. Carefully work the adapter

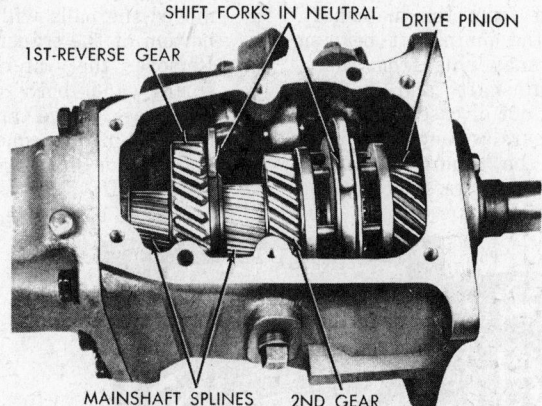

Cover removal—Dodge A-250

and pinion out of the extension housing.
4. Remove extension housing bolts and extension housing.
5. Remove the bolts that attach the drive pinion bearing retainer to case, then slide the retainer off the pinion. Pry the seal out of

retainer using a screwdriver. Be cautious not to nick or scratch the bore.
6. Rotate the drive pinion so that the blank clutch tooth area is opposite the countershaft for removal clearance.
7. Slide drive pinion assembly

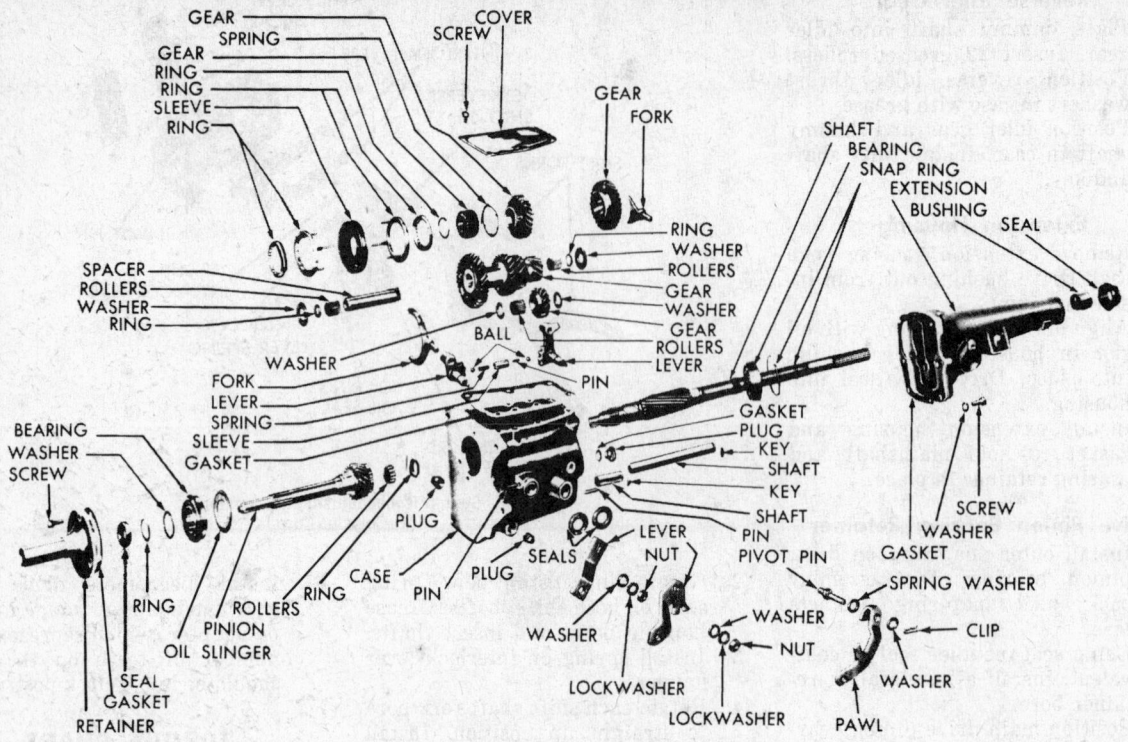

GEAR
GEAR
SPRING
RING
RING
SLEEVE
RING
COVER
SCREW
GEAR
FORK
SHAFT
BEARING
SNAP RING
EXTENSION
BUSHING
SEAL
SPACER
ROLLERS
WASHER
RING
RING
WASHER
ROLLERS
GEAR
WASHER
GEAR
ROLLERS
LEVER
WASHER
FORK
LEVER
SPRING
SLEEVE
GASKET
BALL
PIN
GASKET
PLUG
KEY
SHAFT
KEY
SHAFT
PIN
PIVOT PIN
SCREW
WASHER
GASKET
BEARING
WASHER
SCREW
PLUG
LEVER
NUT
SPRING WASHER
SEALS
PLUG
PIN
CASE
PIN
RING
RING
ROLLERS
PINION
OIL SLINGER
WASHER
WASHER
NUT
WASHER
CLIP
WASHER
SEAL
GASKET
RETAINER
LOCKWASHER
LOCKWASHER
PAWL

Dodge A-250 transmission—exploded view

slightly out of case. Move the synchronizer front inner stop ring from the short splines on the pinion shaft. Slowly remove drive pinion assembly.

8. Remove snap ring that holds bearing on pinion shaft. Remove pinion bearing washer. Using an arbor press, press pinion shaft out of bearing. Remove oil slinger.

9. Remove snap ring and bearing rollers from the end of the drive pinion.

10. Remove clutch gear retaining snap ring from the mainshaft.

11. Remove the mainshaft bearing securing snap ring from case.

12. Slide mainshaft and bearing rearward out of case while holding the gears as they drop free.

13. Remove the snap ring from mainshaft and press the bearing off of mainshaft.

14. Remove the synchronizer components, second gear, first-reverse gear and shift forks from case.

NOTE: Steps 15 thru 18 need only be performed if gear shift lever seals are leaking.

15. Remove the shift levers from the shift shafts.

16. Drive out the tapered retaining pin from the first-reverse shift shaft. Remove the shift shaft from inside the case.
As the detent balls are spring loaded, when the shafts are removed the balls will drop to the bottom of the transmission case.

17. Remove the interlock sleeve, spring and both detent balls from case. Drive tapered retaining pin out of second-third shaft and remove shaft from case.

18. Drive shift shaft seals out of case with a suitable drift.

19. Check end play of countershaft gear with a feeler gauge. The end play should be between 0.005--0.022 inch. This measurement is used to determine if a new thrust washer is necessary during re-assembly.

20. Using a countershaft bearing arbor, drive the countershaft towards the rear of the case until the small key can be removed from the countershaft.

21. Drive the countershaft the rest of the way out of the case, keeping the arbor tight against the end of the countershaft. This will prevent loss of roller bearings.

22. Remove the countershaft gear, front thrust washer and rear thrust washer from the case.

23. Remove the bearing rollers, spacer ring and center spacer from the countershaft gear.

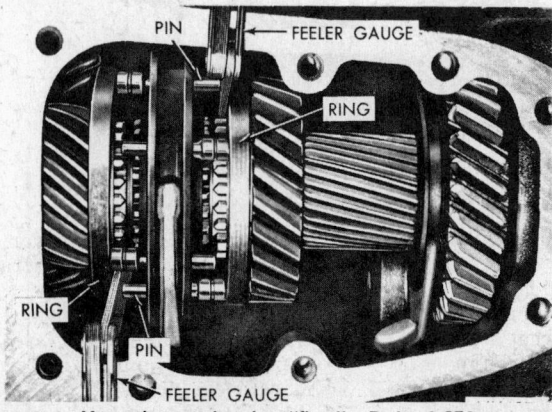

PIN
FEELER GAUGE
RING
RING
PIN
FEELER GAUGE

Measuring synchronizer "float"—Dodge A-250

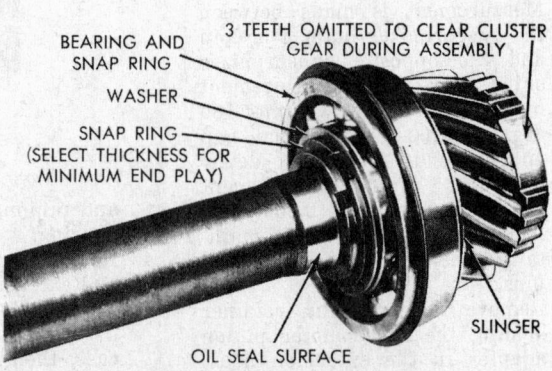

BEARING AND
SNAP RING
3 TEETH OMITTED TO CLEAR CLUSTER
GEAR DURING ASSEMBLY
WASHER
SNAP RING
(SELECT THICKNESS FOR
MINIMUM END PLAY)
SLINGER
OIL SEAL SURFACE

Drive pinion assembly—Dodge A-250

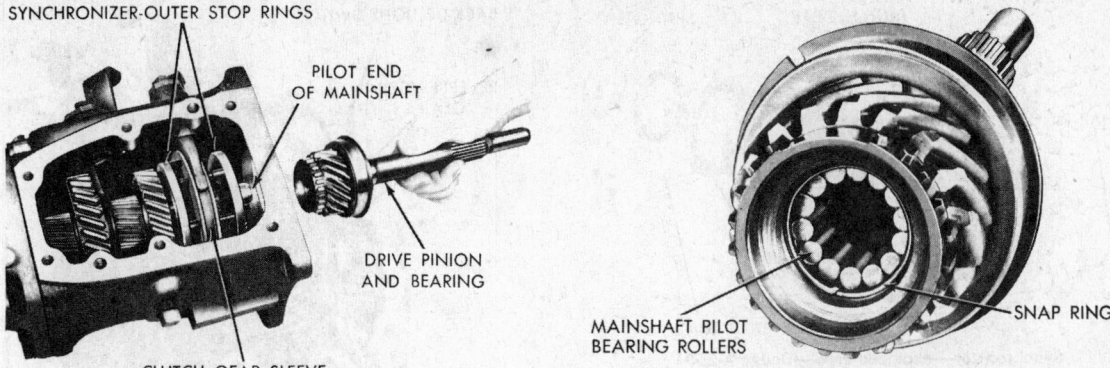

SYNCHRONIZER-OUTER STOP RINGS

PILOT END OF MAINSHAFT

DRIVE PINION AND BEARING

CLUTCH GEAR SLEEVE

Removing or Installing drive pinion—Dodge A-250

MAINSHAFT PILOT BEARING ROLLERS

SNAP RING

Mainshaft pilot bearings in the end of the drive pinion—Dodge A-250

CLUTCH GEAR SNAP RING ON MAINSHAFT

SPRING-SYNCHRONIZER STOP RING

Removing or installing clutch gear snap ring—Dodge A-250

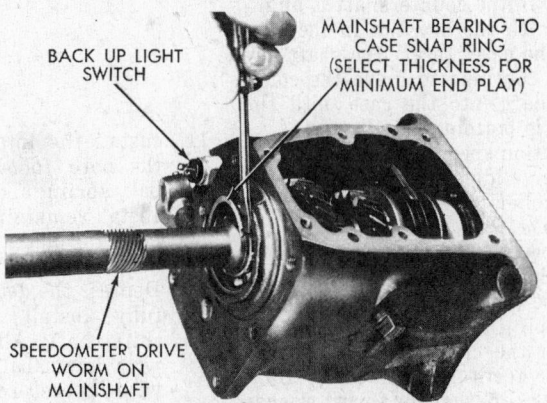

BACK UP LIGHT SWITCH

MAINSHAFT BEARING TO CASE SNAP RING (SELECT THICKNESS FOR MINIMUM END PLAY)

SPEEDOMETER DRIVE WORM ON MAINSHAFT

Removing mainshaft-to-case snap ring

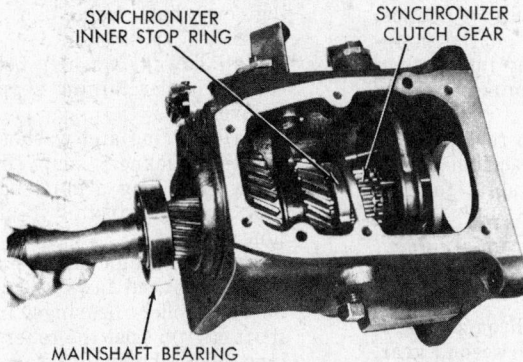

SYNCHRONIZER INNER STOP RING

SYNCHRONIZER CLUTCH GEAR

MAINSHAFT BEARING

Removing mainshaft from case and gear—Dodge A-250

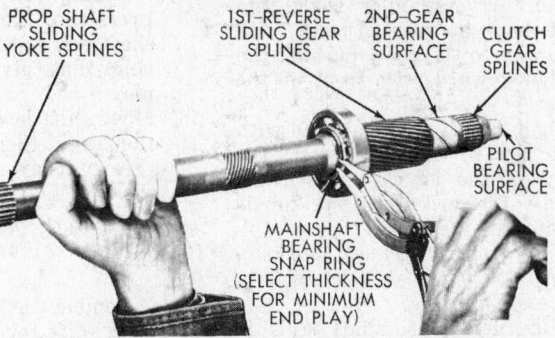

PROP SHAFT SLIDING YOKE SPLINES

1ST-REVERSE SLIDING GEAR SPLINES

2ND-GEAR BEARING SURFACE

CLUTCH GEAR SPLINES

PILOT BEARING SURFACE

MAINSHAFT BEARING SNAP RING (SELECT THICKNESS FOR MINIMUM END PLAY)

Mainshaft and bearing assembly—Dodge A-250

24. Drive the reverse idler gear shaft out of the transmission case using a suitable drift. Remove the woodruff key from the end of the reverse idler shaft.
25. Remove the reverse idler gear and thrust washers out of the case. Remove the bearing rollers from the gear.

Cleaning and Inspection

See "Cleaning and Inspection" instructions at the front of the transmission section.

Assembly of Transmission

1. Slide the countershaft gear bearing roller spacer over arbor tool. Coat the bore of gear with lubricant and slide tool and spacer into gear bore.
2. Lubricate the bearing rollers with heavy grease and install two rows of 22 rollers each in

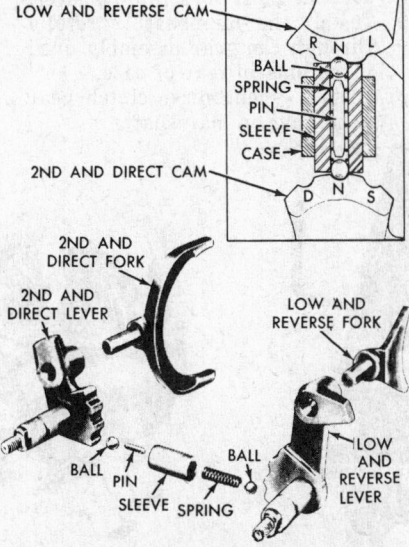

LOW AND REVERSE CAM

BALL
SPRING
PIN
SLEEVE
CASE

R N L

D N S

2ND AND DIRECT CAM

2ND AND DIRECT FORK

2ND AND DIRECT LEVER

LOW AND REVERSE FORK

LOW AND REVERSE LEVER

BALL PIN
SLEEVE SPRING
BALL

Shift forks and levers—Dodge A-250

both ends of gear in area around arbor. Cover with heavy grease and install bearing spacer rings in each end of gear and between roller rows.

3. If countershaft gear end play was found to be excessive during disassembly, install new thrust washers. Cover with heavy grease and install thrust washer and thrust needle bearing and cap at each end of countershaft gear and over arbor. Install gear and arbor in the case, and make sure that tabs on rear thrust washer slide into grooves in the case.
4. Drive the arbor forward out of the countershaft gear and through the bore in the front of the case using the countershaft and a soft faced hammer. When the countershaft is almost

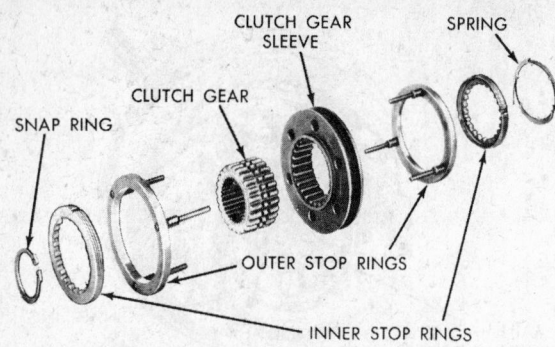

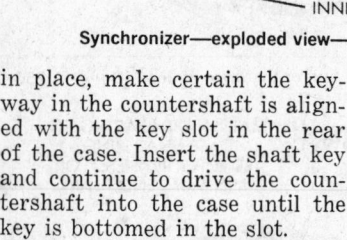

Synchronizer—exploded view—Dodge A-250

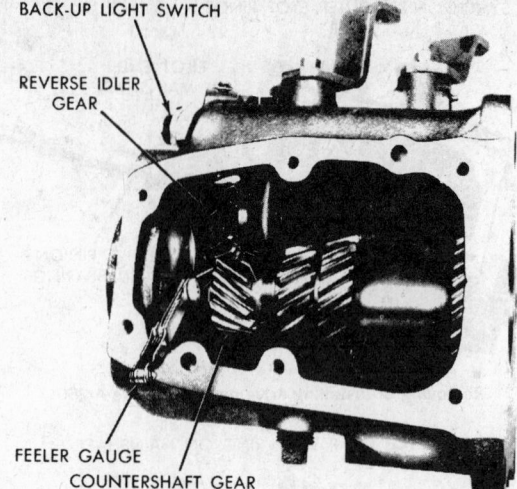

Checking countershaft gear end play—Dodge A-250

in place, make certain the keyway in the countershaft is aligned with the key slot in the rear of the case. Insert the shaft key and continue to drive the countershaft into the case until the key is bottomed in the slot.

5. Position special arbor tool in the reverse idler gear and install the 22 roller bearings using a heavy grease.

6. Place the front and rear thrust washers at each end of the reverse idler gear. Position the assembly in the transmission case with the chamfered end of the gear teeth towards the front. Make sure that the thrust washer tabs engage the slots in case.

7. Insert reverse idler shaft into the bore at rear of case with keyway to the rear, pushing the arbor towards the front of the case.

8. When the keyway is aligned with the slot in the case, insert the key in the keyway. Drive the shaft forward until the key is seated in the recess.

NOTE: Steps 9 thru 14 need only be performed if the shift levers have been disassembled.

9. Place new shift shaft seals in the case and drive it into position with suitable drift.

10. Carefully slide the first-reverse shift shaft into the case and lock into place with a tapered retaining pin. Position the lever so that the center detent is aligned with the interlock bore.

11. Install the interlock sleeve into the bore followed by a detent ball, spring and pin.

12. Install remaining detent ball and hold in place with detent ball holding tool.

13. Depress the detent ball and carefully install the second-third shift shaft. Align center detent with detent ball and secure lever with tapered retaining pin.

14. Install shift levers and tighten retaining nuts to 18 foot-pounds.

15. Press the bearing on the mainshaft and select and install snap ring that gives minimum end play.

16. Move shift lever to reverse position, and then place the first-reverse gear and shift fork in the case.

NOTE: Both shift forks are offset toward the rear of the transmission case.

17. Assemble the synchronizer parts with shift fork and second gear.

18. Place the second gear assembly in the transmission case and insert the shift fork into its lever.

19. Install the mainshaft carefully through the gear assembly until it bottoms in rear of case.

20. Install synchronizer clutch gear snap ring on mainshaft.

21. Select and install mainshaft bearing snap ring in case.

22. If "float" measurement was found to be outside specifications, install or remove shims to place "float" within range.

23. Install oil slinger on drive pinion shaft and slide against the gear.

24. Slide the bearing over the pinion shaft with snap ring groove away from gear, then seat bearing on shaft using an arbor press.

25. Install keyed washer between bearing and retaining snap ring groove.

26. Secure bearing and washer with selected thickness snap ring. If large snap ring around bearing was removed, install it at this time.

27. Place drive pinion shaft in a vise with soft faced jaws and install the 14 roller bearings in the shaft cavity. Coat the roller bearings with a heavy grease and install retaining ring in groove.

28. Rotate the drive pinion so that the blank clutch tooth area is next to the countershaft. Guide the drive pinion through the front of case and engage the inner stop ring with the clutch teeth. Then seat pinion bearing.

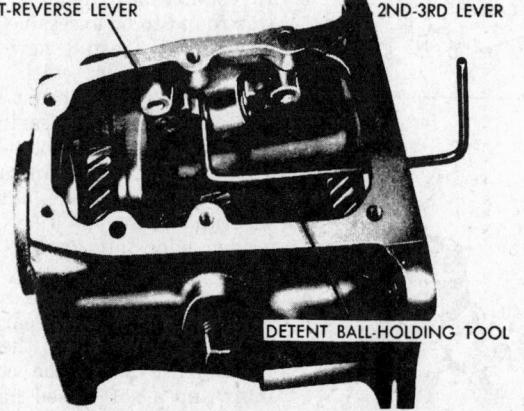

Installing shift levers and detent—Dodge A-250

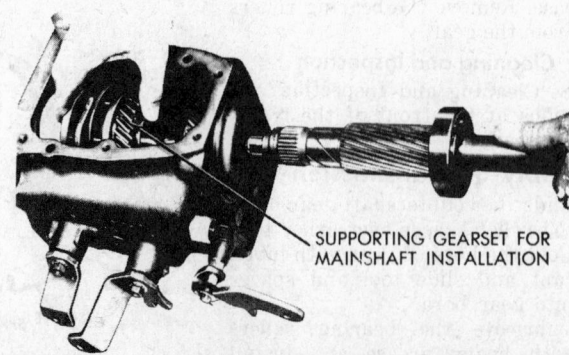

Installing mainshaft into the gears and case—Dodge A-250

Drive pinion seal replacement—Dodge A-250

Synchronizer shim location—Dodge A-250

The pinion shaft is fully seated when the snap ring is in full contact with the case.

29. Install a new seal in the pinion bearing retainer.

30. Position retainer assembly and new gasket on the case. Use sealing compound on bolts and tighten to 30 foot-pounds.

31. Slide the extension housing and a new gasket over mainshaft. Guide shaft through bushing and oil seal. Use sealing compound on the bolt used in the hole tapped through the transmission case. Install remaining bolts and tighten all to 50 foot-pounds.

32. Install the transmission cover and gasket and tighten cover bolts to 12 foot-pounds.

33. Rotate the speedometer pinion gear and adapter assembly so that the number on the adapter corresponding to the number of teeth on the gear is in the 6 o'clock position as the assembly is installed.

34. Fill the transmission with the proper lubricant and install the drain plug and tighten to 25 ft.-lbs. Install the back-up light switch and tighten to 15 ft.-lbs.

35. Rotate the drive pinion shaft and check operation of transmission by running the transmission through all gear ranges.

Dodge/Plymouth A-390 Three Speed

Description

The A-390 is a three speed synchromesh transmission. Lubricant capacity is 4½ pints.

Disassembly of Transmission

1. Remove the bolts that attach the cover to the case. Remove the cover and gasket.

2. Remove the long spring that retains the detent plug in the case. Remove the detent plug with a small magnet.

3. Remove the bolt and retainer securing the speedometer pinion adapter to the transmission case. Carefully work the adapter and pinion out of the extension housing.

4. Remove the bolts that attach the extension housing to the transmission case. Slide the extension housing off the output shaft.

5. Remove the bolts that attach the input shaft bearing retainer to the case. Slide the retainer off the shaft. Using a suitable tool, pry the seal out of the retainer. Be careful not to nick or scratch the bore in which the seal is pressed of the surface on which the seal is bottomed.

6. Remove the lubricant fill plug from the right side of the case. Working through the fill plug opening, drive the roll pin out of the countershaft with a ¼ inch punch.

7. Working with the countershaft bearing arbor and a soft faced hammer, tap the countershaft toward the front of the case with the arbor tool to remove the expansion plug from the countershaft bore at the front of the case. The countershaft is a loose fit in the case and will slide easily.

8. Insert the arbor tool through the front of the case and push the countershaft out of the rear of the case so the roll pin hole in the countershaft does not travel through the roller bearings. The countershaft gear will drop to the bottom of the case. Remove the countershaft from the rear of the case.

9. Place both shift levers in neutral (center) position.

10. Remove the input shaft assembly and stop ring from the front of the case.

11. Remove the set screw that secures the first-reverse shift fork to the shift rail. Slide the

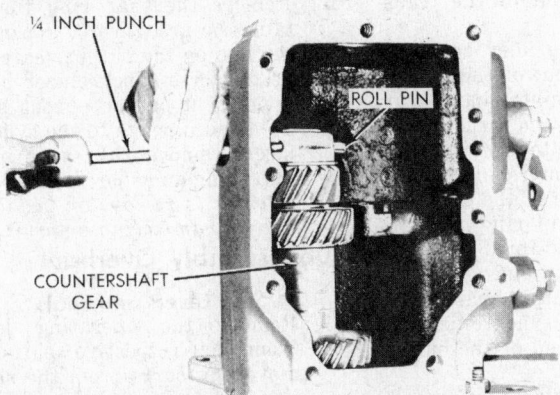

Removing the roll pin from the countershaft—A-390 (© Chrysler Corp.)

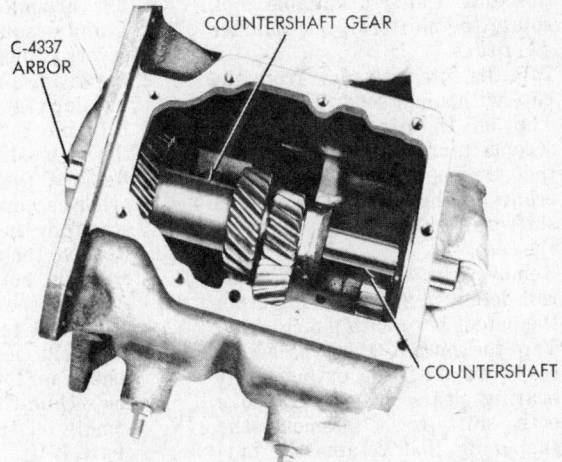

Removing the countershaft—A-390 (© Chrysler Corp.)

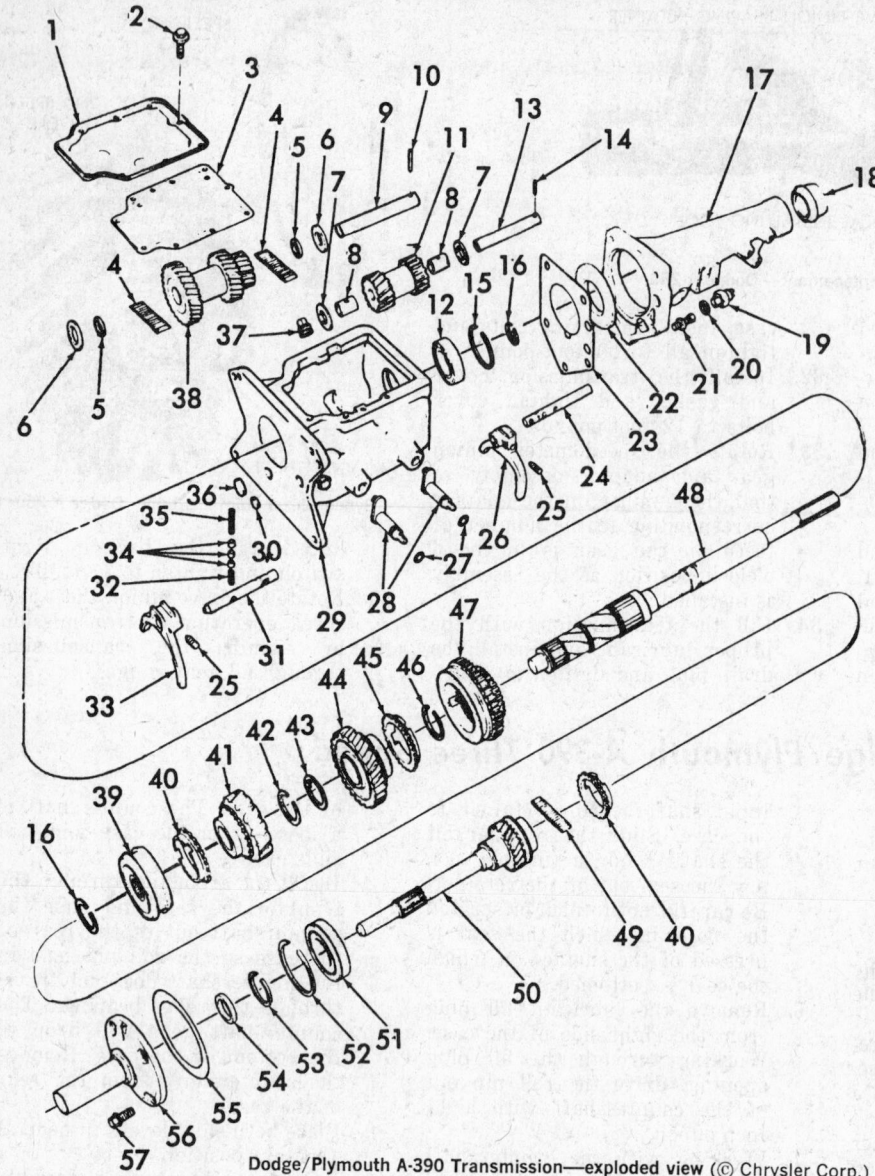

Dodge/Plymouth A-390 Transmission—exploded view (© Chrysler Corp.)

1 Transmission case cover
2 Transmission case cover screw
3 Transmission case cover gasket
4 Countershaft roller bearing
5 Countershaft bearing washer
6 Countershaft thrust washer
7 Reverse idler thrust washer
8 Reverse idler bushing
9 Countershaft
10 Countershaft roll pin
11 Reverse idler gear
12 Output shaft bearing
13 Reverse idler shaft
14 Reverse idler stop pin
15 Output shaft bearing outer snap ring
16 Output shaft inner snap ring
17 Extension
18 Extension seal
19 Back-up light switch
20 Back-up lamp switch gasket
21 Extension screw and lockwasher
22 Output shaft bearing retainer
23 Extension gasket
24 First-reverse shift rail
25 Shift fork set screw
26 First-reverse shift fork
27 Shift lever oil seal
28 Gearshift lever
29 Transmission case
30 Plug
31 Second-third shift rail
32 Shift detent spring pin
33 Second-third shift fork
34 Shift detent pin
35 Shift detent pin spring
36 Plug
37 Transmission case filler plug
38 Countershaft gear
39 Second-third synchronizer assembly
40 Second-third synchronizer stop ring
41 Second gear
42 Low speed gear thrust washer snap ring
43 Low speed gear thrust washer
44 Low speed gear
45 First-reverse synchronizer stop ring
46 First-reverse synchronizer clutch gear snap ring
47 First-reverse synchronizer assembly
48 Output shaft
49 Output shaft pilot roller bearing
50 Input shaft
51 Input shaft bearing
52 Bearing outer snap ring
53 Bearing inner snap ring
54 Bearing retainer oil seal
55 Bearing retainer snap ring
56 Bearing retainer
57 Bearing retainer screw

first-reverse shift rail out through the rear of the case.

12. Move the second-third shift fork rearward for access to the set screw. Remove the setscrew from the fork. Using a suitable tool, rotate the shift rail one quarter (¼) turn.

13. Lift the interlock plug from the case with a magnet.

14. Tap on the inner end of the second-third shift rail to remove the expansion plug from the front of the case. Remove the shift rail through the front of the case.

15. Remove the second-third shift rail detent plug and spring from the detent bore with a magnet.

16. Tap the output shaft assembly rearward until the output shaft bearing clears the case. Remove both shift forks. Remove the snap ring that retains the output shaft bearing to the output shaft.

17. Assemble the output shaft bearing removal tool over the output shaft and bearing. Remove the output shaft bearing.

18. Remove the output shaft assembly through top of the case.

19. Using a suitable drift, drive the reverse idler gear shaft toward the rear, and out of the transmission case.

20. Lift the reverse idler gear and thrust washer out of the case.

21. Remove the countershaft gear, arbor assembly, and thrust washers from the bottom of the case.

22. Remove the countershaft roll pin from the bottom of the case.

23. Remove the snap ring that retains the second-third synchronizer clutch gear and sleeve assembly on the output shaft. Slide the second-third synchronizer assembly off the end of the output shaft.

NOTE: Do not separate the second-third synchronizer clutch gear, sleeve,

struts, or spring unless inspection reveals that a replacement is necessary.

24. Slide the second gear and stop ring off the output shaft.

25. Remove the snap ring and thrust washer retaining the first gear. Slide the first gear and stop ring off the output shaft.

26. Remove the snap ring that retains the first-reverse synchronizer hub on the output shaft. The first-reverse synchronizer hub is a press fit on the output shaft. To avoid damage to the synchronizer, remove the synchronizer hub using an arbor press. Do not attempt to remove or install the hub by hammering or prying.

Sub-Assembly Overhaul

Shift Levers and Seals

1. Remove the operating levers from their respective shafts. Remove any burrs from the shafts to avoid damage to the case.

2. Push the shift levers out of the

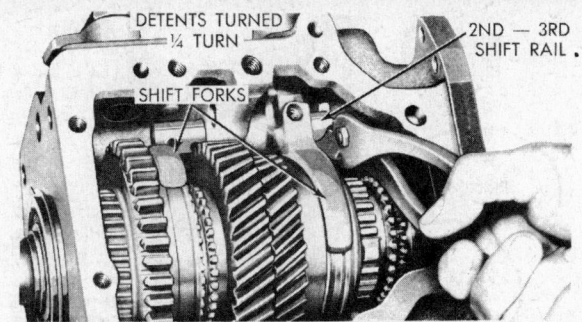

Rotating the second-third shift rail ¼ turn—A-390 (© Chrysler Corp.)

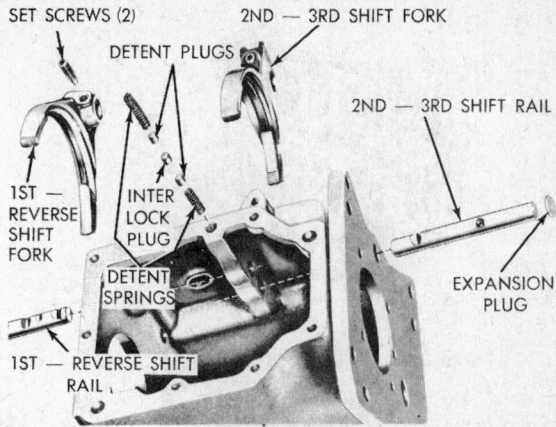

Shift rails and forks—exploded view—A-390 (© Chrysler Corp.)

transmission case. Remove and discard the "O" ring seal from each shaft.

3. Lubricate the new seals with transmission oil and install them on the shafts.

4. Install the shift levers in the case.

5. Install the operating levers and tighten the retaining nuts to 18 ft-lbs.

Input Shaft Bearing and Rollers

1. Remove the snap ring securing the bearing on the input shaft. Carefully press the input shaft out of the bearing with an arbor press.

2. Remove the fifteen bearing rollers from the cavity in the end of the input shaft.

3. Install the 15 bearing rollers in the cavity of the input shaft. Coat the rollers with a thin film of grease to retain them during installation.

4. Slide the input shaft bearing over the input shaft, snap ring groove away from the gear end. Seat the bearing assembly on the input shaft with an arbor press.

5. Secure the bearing with the snap ring. Be sure the snap ring is properly seated. If a large snap ring around the bearing was removed, be sure to install it at this time.

Synchronizers

NOTE: If either synchronizer is to be disassembled, mark all parts so that they will be reassembled in the same position. Do not mix parts from the two synchronizers.

1. Push the synchronizer hub off each synchronizer sleeve.

2. Separate the struts and springs from the hubs.

3. Install the spring on the front side of the first-reverse synchronizer hub, making sure that all three strut slots are fully covered. Hang the three struts on the spring and in the slots with the wide end of the strut inside the hub.

4. With the alignment marks on the hub and sleeve aligned, push the

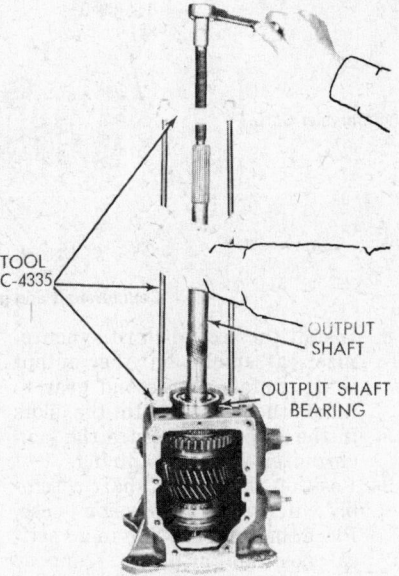

Removing the output shaft bearing—A-390 (© Chrysler Corp.)

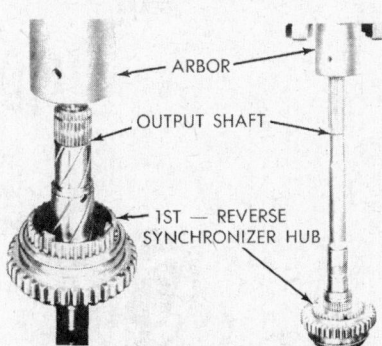

Removing and installing the first-reverse synchronizer hub—A-390 (© Chrysler Corp.)

sleeve down on the hub until the struts are in the neutral detent. Place the stop ring on top of the synchronizer assembly.

5. With the alignment marks on the second-third synchronizer sleeve and hub aligned, slide the sleeve on the hub. Drop in the three struts in the strut slots. Install the spring with the hump in the center, into the hollow of the strut. Turn the assembly over and install the other spring so

that the hump in the center of the spring is inserted in the same strut. Place the stop ring on each end of the synchronizer assembly.

Countershaft Gear and Bearing

1. Remove the countershaft bearing arbor, the roller bearings and the two bearing retainers from the countershaft gear.

2. Coat the bore in each end of the countershaft gear with grease.

3. Insert the countershaft arbor and install twenty five roller bearings and the retainer washer in each end of the countershaft gear.

4. Position the countershaft gear and arbor assembly in the transmission case. Align the gear bore and the thrust washers with the bores in the case and install the countershaft.

5. Using a feeler gauge, check the countershaft gear end play. The end play should be within 0.004-0.018 inch. If the clearance is not within limits, replace the thrust washers.

6. After establishing the correct end play, install the arbor tool in the countershaft gear and lower the gear and tool out of the bottom of the transmission case.

Assembly of Transmission

1. Coat the countershaft gear thrust surfaces in the case with a thin film of grease and position the two thrust washers in place. Place the countershaft gear and arbor assembly in the proper position in the bottom of the transmission case. The countershaft gear will remain in the bottom of the case until the output and input shafts are installed.

2. Coat the reverse idler gear thrust surfaces in the case with a thin film of grease and position the two thrust washers in place. Install the reverse idler gear in the case and align the gear bore with

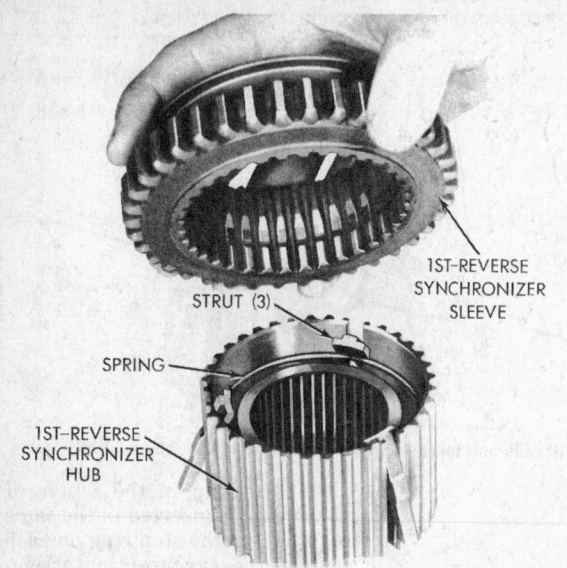

Assembling the first-reverse synchronizer—A-390 (© Chrysler Corp.)

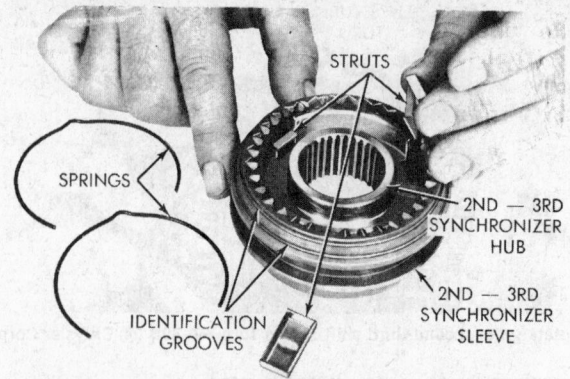

Assembling the second-third synchronizer—A-390 (© Chrysler Corp.)

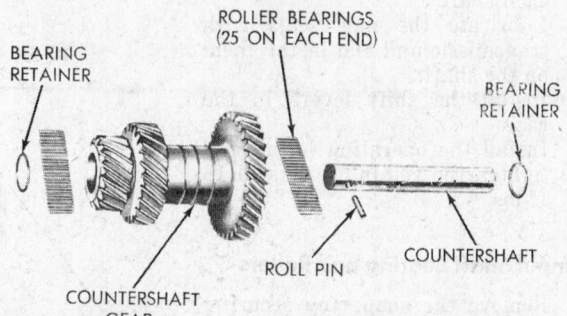

Countershaft and gear—exploded view—A-390 (© Chrysler Corp.)

the thrust washers in the case bore. Install the reverse idler shaft.

3. Measure the reverse idler gear end play with a feeler gauge. End play should be 0.004-0.018 inch. If the clearance is not within limits, replace the thrust washers. If the end play is correct, leave the reverse idler gear in place.

4. Lubricate the output shaft splines and the machined surfaces with transmission oil.

5. Slide the first reverse synchronizer onto the output shaft with the fork groove toward the front. The first-reverse synchronizer hub is a press fit on the output shaft. To eliminate the possibility of damage to the hub, install the hub using an arbor press. *Do not attempt to install the hub by hammering or driving.* Secure the hub on the output shaft with the snap ring.

6. Slide the first gear and stop ring onto the output shaft, aligning the slots in the stop ring with the struts. Install the thrust washer and snap ring.

7. Slide the second gear and stop ring on the output shaft.

8. Install the second-third synchronizer assembly on the output shaft. Rotate the second gear to index the struts with the slots in the stop ring. Secure the synchronizer with a snap ring.

9. Position the output shaft assembly in the transmission case. Place the transmission in a vertical position with the front of the case flat on the work bench.

Place a 1¼ inch block of wood under the end of the output shaft. The block of wood will hold the output shaft assembly up during installation of the output shaft bearing.

10. Install the large snap ring on the output shaft bearing. Place the bearing on the output shaft with the large snap ring up. Drive the bearing on the shaft until it

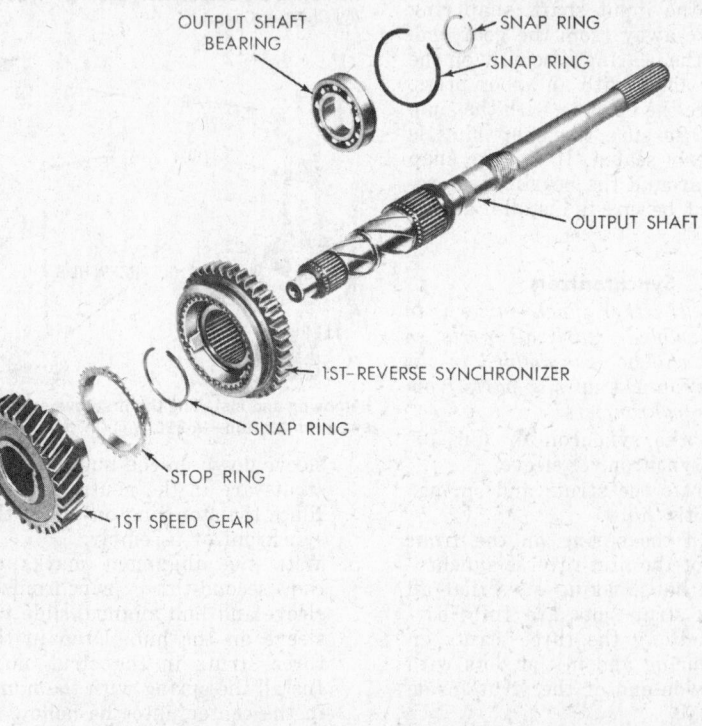

Output shaft assembly—exploded view—A-390 (© Chrysler Corp.)

is seated on the shaft. Secure the bearing on the output shaft with the snap ring. Return the transmission to a horizontal position.

11. Insert both shift forks in the case and in their proper sleeves. Push the output shaft assembly into position and tap it forward until the output shaft bearing is seated in the transmission case.

12. Install the *shortest* detent spring followed by a detent plug into the case. Place the second-third synchronizer assembly in the second gear position.

13. Align the second-third shift fork and install the second-third shift rail. The second-third shift rail is the shortest of the two shift rails. It will be necessary to depress the detent plug to enter the shift rail in the bore. Move the rail inward until the detent plug engages the forward notch (second gear position).

14. Secure the fork to the rail with the set screw. Move the synchronizer to the neutral position.

15. Install a new expansion plug in the transmission case.

16. Install the interlock plug in the transmission case with a magnet. If the second-third shift rail is in the neutral position, the top of the interlock plug will be slightly lower than the surface of the first-reverse shift rail bore.

17. Align the first reverse fork and install the first-reverse shift rail. Move the rail inward until the center notch (neutral) is aligned with the detent bore. Secure the fork to the rail with the set screw.

18. Using a suitable tool, install a new oil seal in the input shaft bearing retainer bore.

19. Coat the bore of the input shaft gear with a thin film of grease. *A thick, heavy grease will plug the lubricant holes and prevent lubrication of the roller bearings.* Install the fifteen roller bearings in the bore.

20. Place the stop ring, slots aligned with the struts, into the second-third synchronizer. Tap the input shaft assembly into place in the case while holding the output shaft to prevent the roller bearings from dropping.

21. Roll the transmission over so that it rests on both the top edge and the shift levers. The countershaft gear will drop into place. Using a screw driver, align the countershaft gear and thrust washers with the bore in the transmission case.

22. Working from the rear of the case, slide the countershaft into position being careful to keep the countershaft in contact with the arbor to avoid dropping parts out of position. Be sure that the roll pin hole in the countershaft aligns with the roll pin hole in the case.

23. Install the roll pin. Install a new expansion plug in the counter-

shaft bore at the front of the case. Install the plug flush of below the face of the case to prevent interference with the clutch housing.

24. Slide the extension housing, with a new gasket, over the output shaft and against the case. Coat the attaching bolt threads with a sealing compound. Install and tighten the attaching bolts to 50 ft-lbs.

25. Install the input shaft bearing retainer and a new gasket. Make sure that the oil return slot is at the bottom. Coat the threads with a sealing compound, install the attaching bolts and tighten to 30 ft-lbs.

26. Install the remaining detent plug into the case followed by the detent spring.

27. Install the filler plug and the back-up light switch. Pour lubricant over the entire gear train while rotating the input shaft and the output shaft.

28. Place the cover and a new gasket on the transmission. Coat the attaching screw threads with a sealing compound. Install and tighten the attaching screws to 22 ft-lbs.

TORQUE SPECIFICATIONS

	Foot-Pounds
Cover to case screws	22
Back-up light switch	15
Extension housing to case bolts	50
Extension housing to cross member bolts	50
Gearshift lever nuts	18
Input shaft bearing retainer bolts	30
Shift fork to shift rail set screw	10
Transmission to clutch housing bolts	50
Transmission drain plug	25
Transmission filler plug	15

Dodge/Plymouth A-745 3 Speed

Description

The Dodge A-745 is a three-speed synchromesh transmission having helical type gears. Lubricant capacity is 3¼ pints.

Disassembly of Transmission

1. Remove output flange nut, then the drum and flange assembly, if so equipped. Remove parking brake assembly, if so equipped.

2. Remove case cover. Measure synchronizer float with a feeler gauge between the end of a synchronizer pin and the opposite synchronizer outer ring. A measurement from .050-.090 in. is acceptable for 1964-67 models. The measurement should be .060-.117 in. for 1968 and up models.

3. Remove the extension housing from the case.

4. Remove the mainshaft rear bearing, if it did not come off with the extension housing.

5. Remove transmission case cover and gasket.

6. Remove the drive pinion bearing retainer.

7. When removing the drive pinion and bearing assembly from the transmission case, slide the front inner stop ring from the short splines on the pinion as the assembly is being removed from the case.

8. Remove the main drive pinion bearing snap-ring.

9. Press bearing off pinion shaft and remove oil slinger.

10. Remove mainshaft pilot bearing snap-ring from the cavity of the pinion gear.

11. Remove the 15 pilot roller bearings.

12. Remove seal from pinion retainer.

13. Remove mainshaft rear bearing snap-ring from groove in mainshaft rear bearing bore in the case.

14. Remove the rear bearing snapring from groove in mainshaft rear bearing bore in the case.

14. Remove the rear bearing from the case by moving the mainshaft and rear bearing assembly to the rear.

15. Remove synchronizer assembly from case.

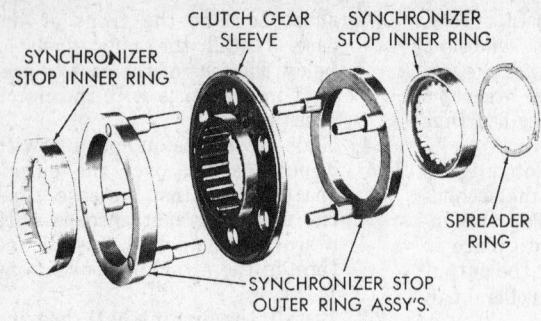

SYNCHRONIZER STOP INNER RING · CLUTCH GEAR SLEEVE · SYNCHRONIZER STOP INNER RING · SPREADER RING · SYNCHRONIZER STOP OUTER RING ASSY'S.

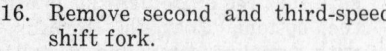

Synchronizer components (© Chrysler Corp.)

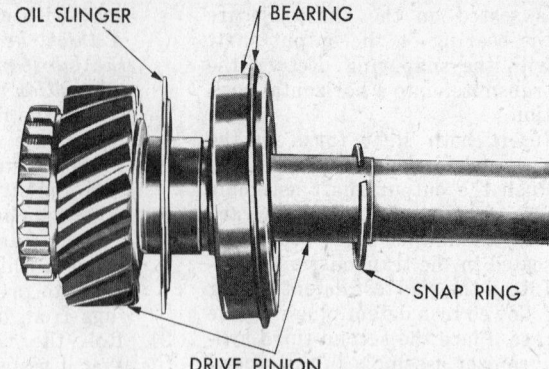

OIL SLINGER · BEARING · SNAP RING · DRIVE PINION

Drive pinion assembly (© Chrysler Corp.)

16. Remove second and third-speed shift fork.
17. Remove synchronizer clutch gear, snap-ring, clutch gear, second-speed gear, and first and reverse sliding gear from the mainshaft.
18. Withdraw mainshaft and bearing out through the rear of the case.
19. Remove the synchronizer clutch gear, second-speed gear, low and reverse sliding gear, and low and reverse shift fork from the case.
20. With a dummy shaft, drive the countershaft toward the rear of the case until the small key can be removed from the countershaft.
21. Remove the countershaft from the case.

22. Lift the cluster gear, the thrust washers and the dummy shaft assembly out of the case.
23. Dismantle the cluster gear, (88 rollers, four spacer rings and the center spacer from the cluster).
24. With a blunt drift, drive the reverse idler shaft toward the rear of the case far enough to remove the key from the shaft.
25. Completely remove the shaft from the case, then remove the idler gear.
26. Remove the thrust washers and 22 rollers.
27. Wtih a small punch, remove low and reverse gear lever shaft tapered lock pin by driving it toward the top of the transmission case.

28. Remove the second and third gear lever shaft in the same manner.
29. Remove the lever shafts from the transmission case being careful not to lose the spring-loaded detent balls.
30. Remove the interlock sleeve, spring pin and detent balls.
31. Remove both lever shaft seals and discard same.

Assembly of Transmission

1. Place oil slinger on the main drive pinion with the offset outer portion next to the drive pinion teeth.
2. Place the main drive pinion bearing on the pinion shaft with

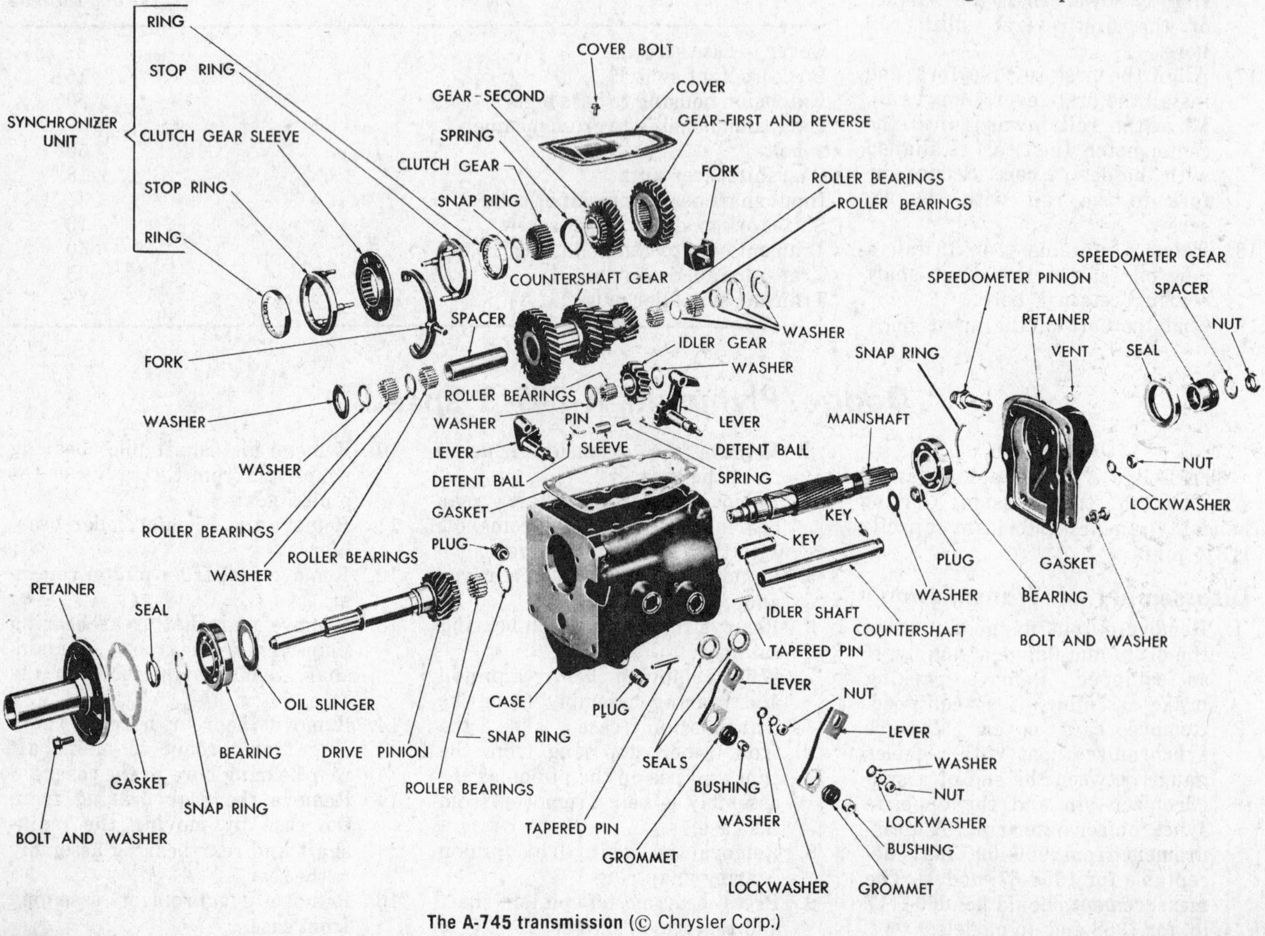

The A-745 transmission (© Chrysler Corp.)

the outer snap-ring away from the pinion gear.

3. Press the bearing into position so it is seated firmly against the oil slinger and pinion gear.

4. Install the bearing retaining snap-ring in its groove on the pinion shaft.

5. Heavily grease the 15 pilot bearing rollers and install them in the cavity at the rear of the main drive pinion.

6. Install the snap-ring.

7. Place the bearing spacer in the center of the bore in the cluster gear and use the dummy shaft to assist in assembling the roller bearings.

8. Install a row of 22 rollers next to one end of the spacer, using heavy grease to hold them.

9. Place one of the four bearing spacer rings next to the row of rollers, and install another row of 22 rollers next to the spacer ring.

10. Install another spacer ring at the outside end of the second row of bearing rollers.

11. At the opposite end of the cluster gear bore, install the remaining spacer rings and bearing rollers in the same sequence as listed in Steps 8, 9, and 10.

12. With a small amount of grease, install the front thrust washer on the dummy shaft at the front end of the cluster gear, with the tabs outward.

13. Install the tabbed rear thrust washer onto the dummy shaft against the rear of the cluster gear with the tabs inserted in the cluster gear grooves.

14. Install the remaining rear thrust washer plate onto the rear of the gear and dummy shaft with the step in the washer facing upward, as viewed from the rear.

15. Align tabs of the front thrust washer vertically to index with notches in the transmission case and with the step in the rear thrust washer positioned upward. Insert the cluster gear and dummy shaft in the transmission case.

16. Using the countershaft, drive the dummy shaft forward, out of the case. Countershaft end play should be .0045-.028 in.

17. Position a dummy shaft in the reverse idler gear and, using heavy grease, install the 22 roller bearings into the gear.

18. Place the thrust washers at each end of the reverse idler gear, then place the assembly in the case with the chamfered end of the gear teeth toward the front.

19. Insert the reverse idler shaft into the rear case bore with the keyway to the rear, pushing the dummy shaft toward and out of the front of the transmission.

20. With the keyway in proper alignment, insert the key and continue driving the shaft forward until the key seats in the recess.

21. Install two new lever shaft seals in the transmission case.

22. Lubricate and install second and third-speed lever shaft in the bores of the case.

23. Install the second and third speed lever shaft lock pin in the hole in the case.

24. Place interlock parts in the case in the following order: ball, sleeve, spring, pin and ball.

25. Enter low and reverse lever shaft in the case bore, depress the detent ball against spring tension and push the lever shaft firmly into position, in order to prevent the ball from escaping.

26. Install low and reverse lever shaft lock pin in the case.

27. Place low and reverse fork in the lever shaft, with the offset facing the rear.

28. While holding the low and reverse sliding gear in position in the fork, with the hub extension to the rear, insert the mainshaft with the rear bearing through the rear of the case and into the sliding gear.

29. Place synchronizer stop ring spring, then the rear stop ring, on the synchronizer splines of the second-speed gear. Install the second-speed gear onto the mainshaft. Synchronizer shims must be added if synchronizer float is more than the maximum in Step 2, disassembly. If float was less than minimum, the six pins must be shortened. gear on the mainshaft with the

30. Install the synchronizer clutch shoulder to the front.

31. Select the thickest synchronizer clutch gear snap-ring that can be used, and install it in the mainshaft groove.

32. Check to see that clearance between clutch gear and second-speed gear is .004-.014 in.

33. Hold the synchronizer clutch gear sleeve and two outer rings together with pins properly entered into the holes in the clutch gear sleeve and with the clutch gear sleeve engaged in the groove of the second- and third-speed shift fork, position the fork in the second- and third-speed lever shaft.

34. While holding the synchronizer parts and fork in position, slide the mainshaft forward, entering the synchronizer clutch gear into the clutch gear sleeve and simultaneously entering the mainshaft rear bearing in the case bore.

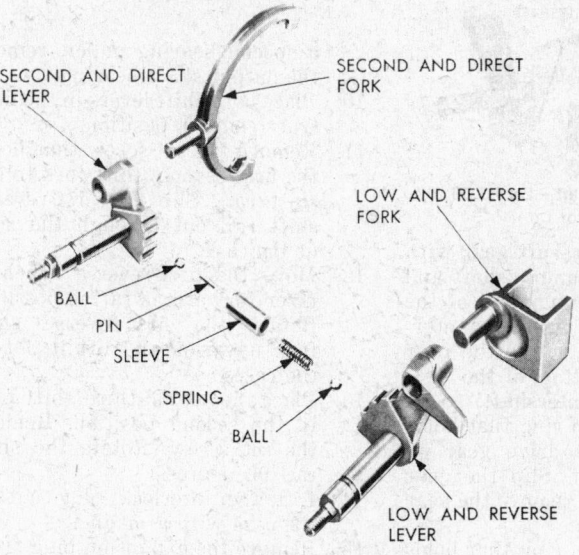

Shift fork assembly (© Chrysler Corp.)

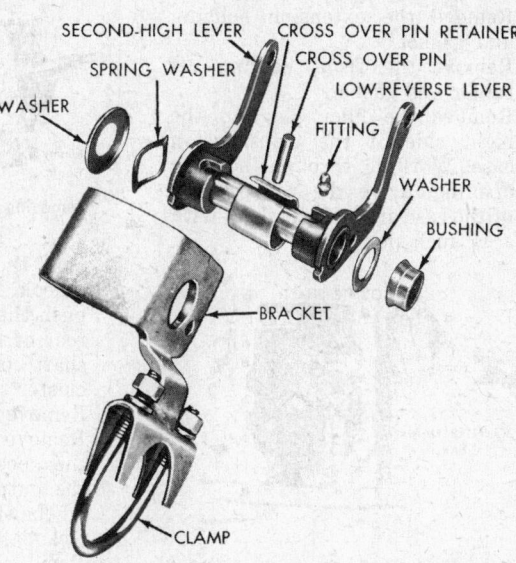

Gearshift controls (© Chrysler Corp.)

35. While still holding the synchronizer parts in position, tap the mainshaft forward until the rear bearing bottoms in the case bore.
36. Install the mainshaft rear bearing snap-ring into place in the case bore.
37. Install a new drive pinion retainer seal.
38. Place the synchronizer front inner ring in position in the front outer ring, and enter the main drive pinion through the case bore.
39. Engage the splines on the rear of the pinion with the inner stop ring, and tap the drive pinion into the case until the outer snap-ring on the pinion bearing is against the transmission case.
40. Place the drive pinion bearing retainer over the pinion shaft and against the transmission case. While holding the retainer against the transmission case, measure the clearance between the retainer and case and choose a gasket .003-.005 in. thicker than this reading.
41. Torque the front bearing retainer bolts to 30 ft. lbs.
42. Install a new extension housing seal.
43. Install extension housing and torque the bolts to 50 ft. lbs.
44. Install the parking brake assembly, on vehicles so equipped.
45. Install the parking brake drum (if so equipped) and flange assembly and torque to 175 ft. lbs.
46. Install the drain plug in the transmission case.
47. Install the gearshift operating levers, and torque to 12 ft. lbs.
48. Install the back-up light switch.
49. Install the speedometer cable and drive gear. Bring transmission to lubricant level.

TORQUE CHART

Gearshift rod and swivel assembly nuts	5/16-24	70 (in.-lbs.)
Backup light switch hole plug	9/16-18	15
Case drain plug	3/8	20
Case filler plug	1/2	30
Cast to clutch housing bolt	7/16-18	50
Drive pinion bearing retainer bolt	5/16-18	15
Extension bolt	3/8-16	30
Gearshift fork lock bolt	3/8-24	30
Gearshift housing lower bolt	5/16-18	15
Gearshift housing upper bolt	5/16-18	20
Gearshift operating lever nut	3/8-24	35
Gearshift selector ball spring bolt	1/2-20	25
Gearshift selector lever washer nut	5/16-24	20
Manual remote gearshift lever shaft bolt	1/4-20	10
Shaft flange nut	3/4-16	175

Ford 3.03 3 Speed

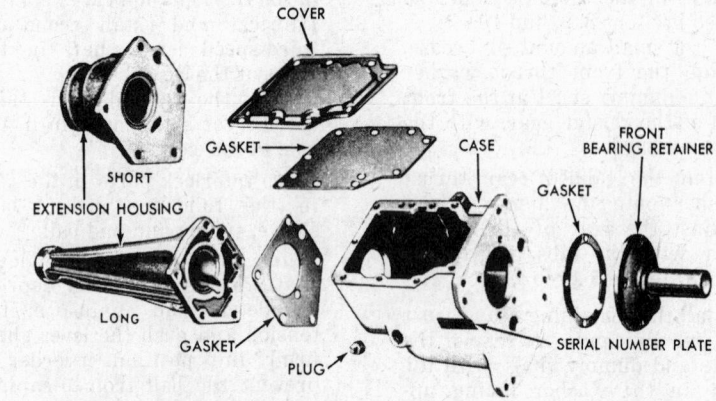

Transmission case and exterior parts—Ford 3.03 (© Ford Motor Co.)

Description

The Ford 3.03 is a fully synchronized three speed transmission. All gears except reverse are in constant mesh. Forward speed gear changes are accomplished with synchronizer sleeves.

Disassembly of Transmission

1. Drain the lubricant by removing the lower extension housing bolt.
2. Remove the case cover and gasket.
3. Remove the long spring that holds the detent plug in the case and remove the detent plug with a small magnet.
4. Remove the extension housing and gasket.
5. Remove the front bearing retainer and gasket.
6. Remove the filler plug on the right side of the transmission case. Working through the plug opening, drive the roll pin out of the case and countershaft with a 1/4 inch punch.

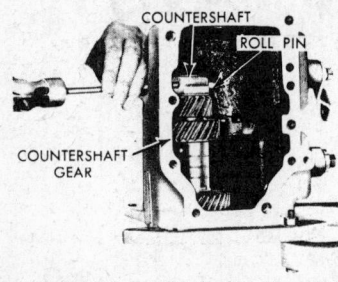

Removing countershaft roll pin—Ford 3.03
(© Ford Motor Co.)

Removing countershaft—Ford 3.03
(© Ford Motor Co.)

7. Hold the countershaft gear with a hook. Install dummy shaft and push the countershaft out of the rear of the case. As the countershaft comes out, lower the gear cluster to the bottom of the case. Remove the countershaft.
8. Remove the snap ring that holds the speedometer drive gear on the output shaft. Slip the gear off the shaft and remove the gear lock ball.
9. Remove the snap ring that holds the output shaft bearing. Using a special bearing puller, remove the output shaft bearing.
10. Place both shift levers in the neutral (center) position.
11. Remove the set screw that holds the first-reverse shift fork to the shift rail. Slip the first-reverse shift rail out through the rear of the case.
12. Move the first-reverse synchronizer forward as far as possible. Rotate the first-reverse shift fork upwards and lift it out of the case.
13. Place the second-third shift fork in the second position. Remove the set screw. Rotate the shift rail 90 degrees.
14. Lift the interlock plug out of the case with a magnet.
15. Remove the expansion plug from the second-third shift rail by

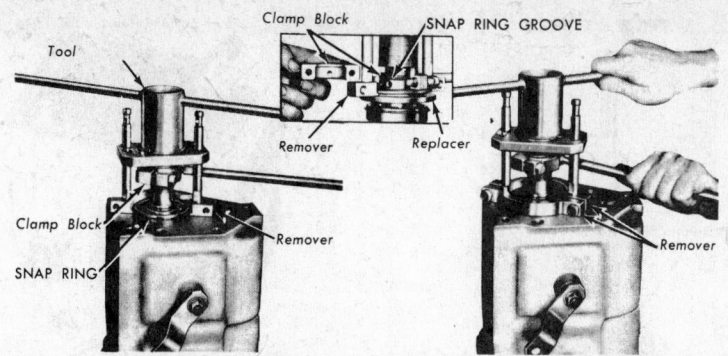

Removing output shaft bearing—Ford 3.03 (© Ford Motor Co.)

lightly tapping the end of the rail. Remove the second-third shift rail.

16. Remove the second-third shift rail detent plug and spring from detent bore.

17. Remove the input gear and shaft from the case.

18. Rotate the second-third shift fork upwards and remove from case.

19. Using caution, lift the output shaft assembly out through top of case.

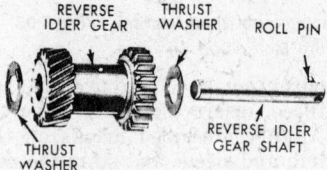

Reverse idler shaft—exploded view—Ford 3.03 (© Ford Motor Co.)

20. Lift the reverse idler gear and thrust washers out of case. Remove the countershaft gear, thrust washer and dummy shaft from case.

21. Remove the snap ring from the front of the output shaft. Slip the synchronizer and second gear off shaft.

22. Remove the second snap ring from output shaft and remove the thrust washer, first gear and blocking ring.

23. Remove the third snap ring from the output shaft. The first reverse synchronizer hub is a press fit on the output shaft. Remove the synchronizer hub with an arbor press.

CAUTION: Do not attempt to remove or install the synchronizer hub by prying or hammering.

Disassembly and Assembly of Sub-Assemblies

Shift Levers and Seals

1. Remove shift levers from the shafts. Slip the levers out of case. Discard shaft sealing O-rings.

2. Lubricate and install new O-rings on shift shafts.

3. Install the shift shafts in the case and secure shift levers.

Input Shaft Bearings

1. Remove the snap ring securing the input shaft bearing. Using an arbor press, remove the bearing.

2. Press the input shaft bearing onto shaft using correct tool.

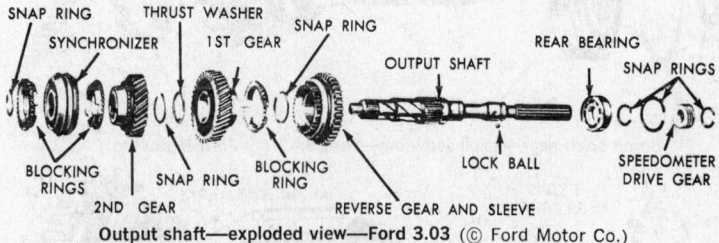

Output shaft—exploded view—Ford 3.03 (© Ford Motor Co.)

Rotating second-third shift rail—Ford 3.03 (© Ford Motor Co.)

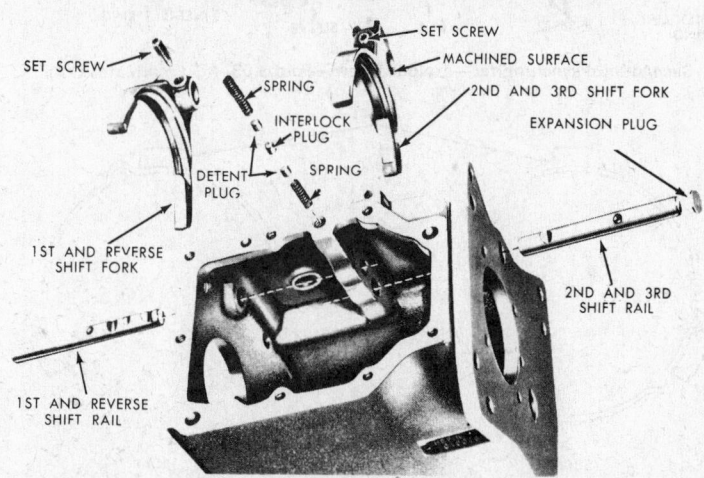

Shift rails and forks—exploded view—Ford 3.03 (© Ford Motor Co.)

Synchronizers

1. Scribe alignment marks on synchronizer hubs before disassembly. Remove each synchronizer hub from the synchronizer sleeves.

2. Separate the inserts and insert springs from the hubs.

CAUTION: Do not mix parts from the seperate synchronizer assemblies.

3. Install the insert spring in the hub of the first-reverse synchronizer. Be sure that the spring covers all the insert grooves. Start the hub on the sleeve making certain that the scribed marks are properly aligned. Place the three inserts in the hub, small ends on the inside. Slide the sleeve and reverse gear onto hub.

4. Install one insert spring into a groove on the second-third syn-

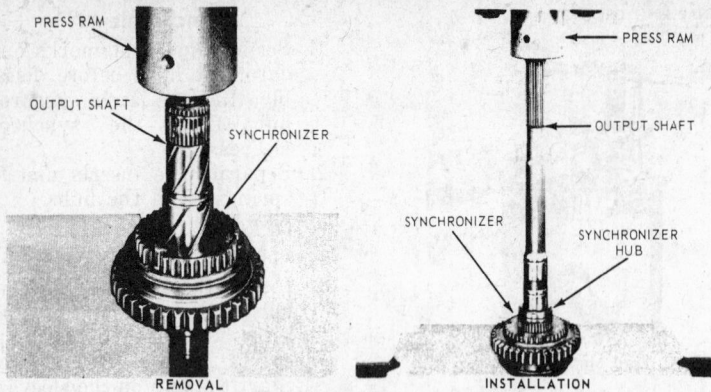

Removing and installing first-reverse synchronizer—Ford 3.03 (© Ford Motor Co.)

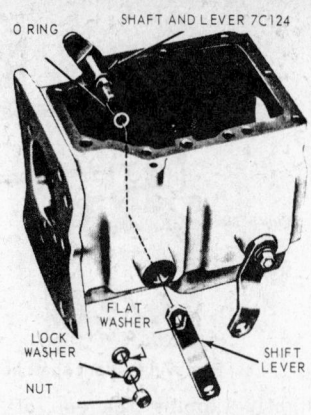

Shift lever and shaft—exploded view— Ford 3.03 (© Ford Motor Co.)

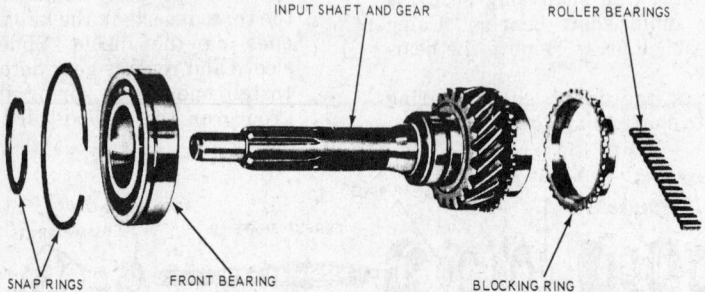

Input shaft gear—exploded view—Ford 3.03 (© Ford Motor Co.)

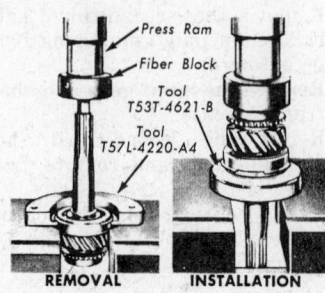

Replacing input shaft bearing—Ford 3.03 (© Ford Motor Co.)

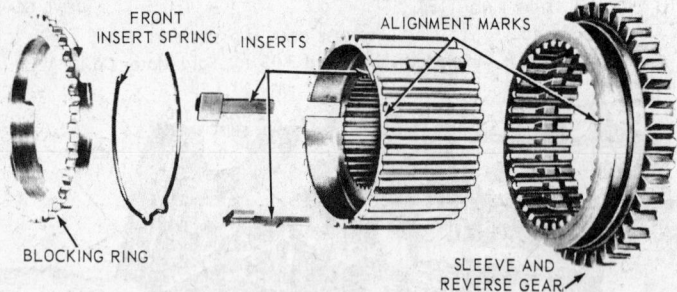

First-reverse synchronizer—exploded view—Ford 3.03 (© Ford Motor Co.)

chronizer hub. Be sure that all three insert slots are covered. Align the scribed marks on the hub and sleeve and start the hub into the sleeve. Position the three inserts on the top of the retaining spring and push the assembly together. Install the remaining retainer spring so that the spring ends cover the same slots as the first spring. Do not stagger the springs. Place a synchronizer blocking ring on the ends of the synchronizer sleeve.

Countershaft Gear Bearings

1. Remove the dummy shaft, needle bearings and bearing retainers from the countershaft gear.
2. Coat the bore in each end of the countershaft gear with grease.
3. Hold the dummy shaft in the gear and install the needle bearings in the case.
4. Place the countershaft gear, dummy shaft, and needle bearings in the case.
5. Place the case in a vertical position. Align the gear bore and the thrust washers with the bores in the case and install the countershaft.
6. Place the case in a horizontal position. Check the countershaft gear end play with a feeler gauge. Clearance should be between 0.004-0.018 inch. If clearance does not come within specifications, replace the thrust washers.
7. Install the dummy shaft in the countershaft gear and leave the

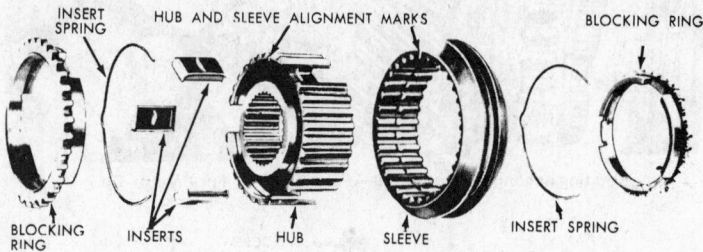

Second-third synchronizer—exploded view—Ford 3.03 (© Ford Motor Co.)

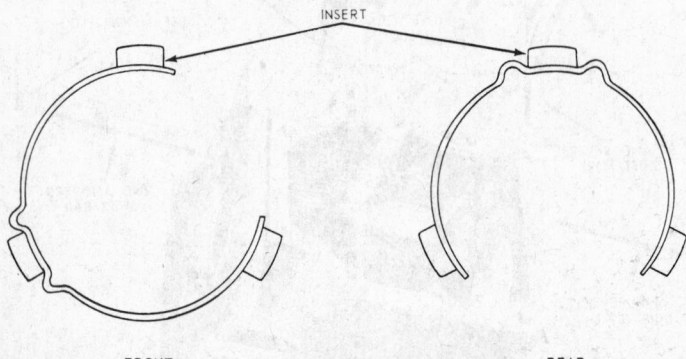

First-reverse synchronizer insert spring installation—Ford 3.03 (© Ford Motor Co.)

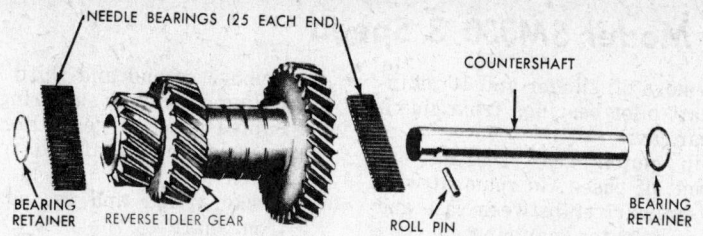

NEEDLE BEARINGS (25 EACH END)

COUNTERSHAFT

BEARING RETAINER

REVERSE IDLER GEAR

ROLL PIN

BEARING RETAINER

Countershaft gear—exploded view—Ford 3.03 (© Ford Motor Co.)

gear at the bottom of the transmission case.

Assembly of Transmission

1. Cover the reverse idler gear thrust surfaces in the case with a thin film of lubricant, and install the two thrust washers in the case.
2. Install the reverse idler gear and shaft in the case. Align the case bore and thrust washers with gear bore and install the reverse idler shaft.
3. Measure the reverse idler gear end play with a feeler gauge; clearance should be between 0.004-0.018 inch. If end play is not within specifications, replace the thrust washers. If clearance is correct, leave the reverse idler gear in case.
4. Lubricate the output shaft splines and machined surfaces with transmission oil.
5. The first-reverse synchronizer hub is a press fit on the output shaft. Hub must be installed in an arbor press. Install the synchronizer hub with the teeth-end of the gear facing towards the rear of the shaft.
CAUTION: Do not attempt to install the first-reverse synchronizer with a hammer.
6. Place the blocking ring on the tapered surface of the first gear.
7. Slide the first gear on the output shaft with the blocking ring toward the rear of the shaft. Rotate the gear as necessary to engage the three notches in the blocking ring with the synchronizer inserts. Install the thrust washer and snap ring.
8. Slide the blocking ring onto the tapered surface of the second gear. Slide the second gear with blocking ring and the second-third synchronizer on the mainshaft. Be sure that the tapered surface of second gear is facing the front of the shaft and that the notches in the blocking ring engage the synchronizer inserts. Install the snap ring and secure assembly.
9. Cover the core of the input shaft with a thin coat of grease.
CAUTION: A thick film of grease will plug lubricant holes and cause damage to bearings.
   Install bearings.

10. Install the input shaft through the front of the case and insert snap ring in the bearing groove.
11. Install the output shaft assembly in the case. Position the second-third shift fork on the second-third synchronizer.
12. Place a detent plug spring and a plug in the case. Place the second-third synchronizer in the second gear position (toward the rear of the case). Align the fork and install the second-third shift rail. It will be necessary to depress the detent plug to install the shift rail in the bore. Move the rail forward until the detent plug enters the forward notch (second gear).
13. Secure the fork to the shift rail with a set screw and place the synchronizer in neutral.
14. Install the interlock plug in the case.
15. Place the first-reverse synchronizer in the first gear position (towards the front of the case). Place the shift fork in the groove of the synchronizer. Rotate the fork into position and install the shift rail. Move the shift rail inward until the center notch (neutral) is aligned with the detent bore. Secure shift fork with set screw.
16. Install a new shift rail expansion plug in the front of the case.
17. Hold the input shaft and blocking ring in position and move the

output shaft forward to seat the pilot in the roller bearings on the input gear.
18. Tap the input gear bearing into place while holding the output shaft. Install the front bearing retainer and gasket. Torque attaching bolts to specifications.
19. Install the large snap ring on the rear bearing. Place the bearing on the output shaft with the snap ring end toward the rear of the shaft. Press the bearing into place using a special tool. Secure the bearing to the shaft with the snap ring.
20. Hold the speedometer drive gear lock ball in the detent and slide the speedometer drive gear into position. Secure with snap ring.
21. Place the transmission in the vertical position. Working with a screwdriver through the drain hole in the bottom of the case, align the bore of the countershaft gear and the thrust washer with the bore in the case.
22. Working from the rear of the case, push the dummy shaft out of the countershaft gear with the countershaft. Align the roll pin hole in the countershaft with the matching hole in the case. Drive the shaft into place and install the roll pin.
23. Position the new extension housing gasket on the case with sealer. Install the extension housing and torque to specification.
24. Place the transmission in gear and pour gear oil over entire gear train while rotating the input shaft.
25. Install the remaining detent plug and long spring in case.
26. Position cover gasket on case with sealer and install cover. Torque cover bolts to specifications.
27. Check operation of transmission in all gear positions.

TORQUE SPECIFICATIONS

	Foot-Lb.
Input shaft gear bearing retainer to transmission case	30-36
Transmission to flywheel housing	37-42
Transmission cover to transmission case	14-19
Speedometer cable retainer to transmission extension	3-4.5
Transmission extension to transmission case	42-50
Flywheel housing to engine	40-50
Gear shift lever to cam and shaft assembly lock nuts	18-23
U-Joint flange to output shaft	60-80
Filler plug	10-20
Shifter fork set screws	10-18
T.R.S. switch to case	15-20

Lubricant Refill Capacity (Pints)	U.S.	Imp.
Ford Type 3.03	3.5	3.0

Muncie Model SM330 3 Speed

Description

The G.M. Corporation Model SM 330 (Muncie) is a three-speed transmission using helical constant mesh gears. The engagement of all gears except reverse is assisted by synchronizers.

General Data

Type	3-Speed
Synchromesh gears	
	1st, 2nd, and 3rd
Model	SM330

Gear Ratios	
1st Speed	3.03:1
2nd Speed	1.75:1
3rd Speed	1.00:1
Reverse	3.02:1

Transmission Disassembly

1. Remove side cover and shift forks.
2. Unbolt extension and rotate to line up groove in extension flange with reverse idler shaft. Drive reverse idler shaft and key out of case with a brass drift.
3. Move second-third synchronizer sleeve forward. Remove extension housing and mainshaft assembly.
4. Remove reverse idler gear from case.
5. Remove third speed blocker ring from clutch gear.
6. Expand snap-ring which holds mainshaft rear bearing. Tap gently on end of mainshaft to remove extension.
7. Remove clutch gear bearing retainer and gasket.
8. Remove snap-ring. Remove clutch gear from inside case by gently tapping on end of clutch gear.

9. Remove oil slinger and 16 mainshaft pilot bearings from clutch gear cavity.
10. Slip clutch gear bearing out front of case. Aid removal with a screwdriver between case and bearing outer snap-ring.
11. Drive countershaft and key out to rear.
12. Remove countergear and two tanged thrust washers.

Mainshaft Disassembly

1. Remove speedometer drive gear. Some speedometer drive gears, made of metal, must be pulled off.
2. Remove rear bearing snap-ring.
3. Support reverse gear. Press on rear of mainshaft to remove reverse gear, thrust washer, and rear bearing. Be careful not to cock the bearing on the shaft.
4. Remove first and reverse sliding clutch hub snap-ring.
5. Support first gear. Press on rear of mainshaft to remove clutch assembly, blocker ring, and first gear.

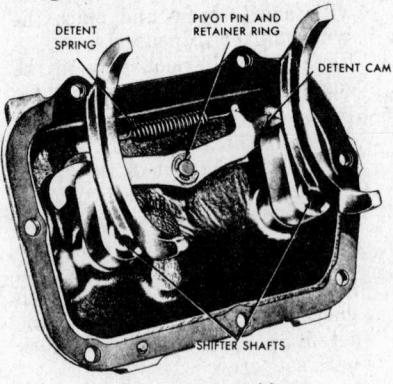

Side cover assembly
(© G.M.C.)

6. Remove second and third speed sliding clutch hub snap-ring.
7. Support second gear. Press on front of mainshaft to remove clutch assembly, second speed blocker ring, and second gear from shaft.

Cleaning and Inspection

For more detailed information, see the "Cleaning and Inspection" instructions at front of transmission section.
1. Wash all parts in solvent.
2. Air dry.

Clutch Keys and Springs

Keys and sprnigs may be replaced if worn or broken, but the hubs and sleeves must be kept together as originally assembled.
1. Mark hub and sleeve for reassembly.
2. Push hub from sleeve. Remove keys and springs.
3. Place three keys and two springs, one on each side of hub, so all three keys are engaged by both springs. The tanged end of the springs should not be installed into the same key.
4. Slide the sleeve onto the hub, aligning the marks.

Extension Oil Seal and Bushing

1. Remove seal.
2. Using bushing remover and installer, or other suitable tool, drive bushing into extension housing.
3. Drive new bushing in from rear. Lubricate inside of bushing and seal. Install new oil seal with extension seal installer or suitable tool.

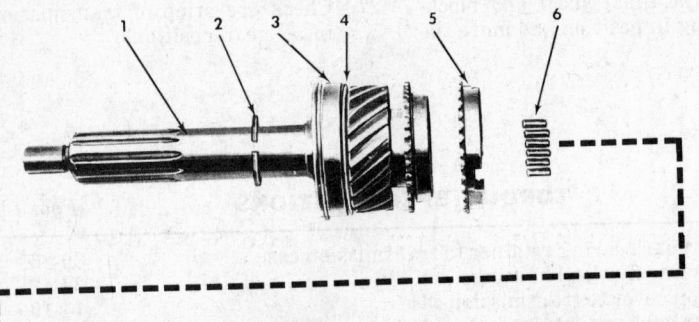

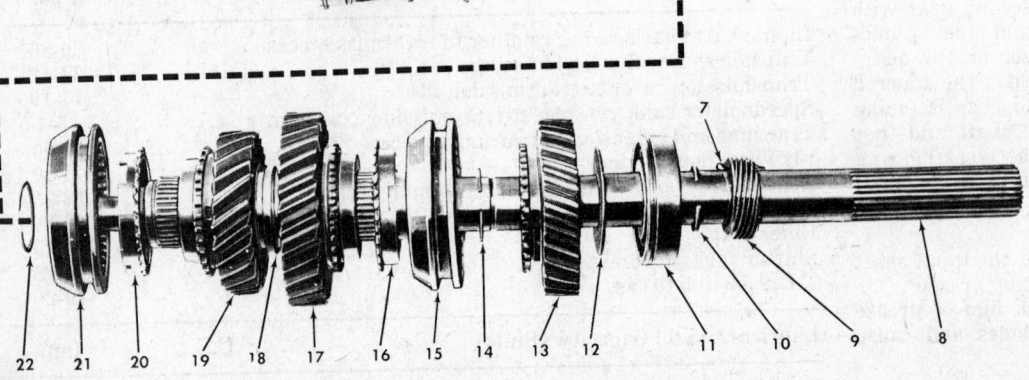

1 Main drive gear	6 Mainshaft pilot bearings
2 Snap ring	7 Speedometer retainer clip
3 Main drive gear bearings	8 Mainshaft
4 Oil slinger	9 Speedometer drive gear
5 3rd speed blocker ring	10 Snap ring
	11 Rear bearing
	12 Reverse gear thrust washer
	13 Reverse gear
	14 Snap ring
	15 1st & reverse synchronizer assembly
	16 First speed blocker ring
	17 First speed gear
	18 Shoulder (part of mainshaft)
	19 Second speed gear
	20 Second speed blocker ring
	21 2nd and 3rd synchronizer assembly
	22 Snap ring

Main drive gear and mainshaft assembly (© G.M.C.)

Clutch Bearing Retainer Oil Seal

1. Pry old seal out.
2. Install new seal using seal installer or suitable tool. Seat seal in bore.

Mainshaft Assembly

1. Lift front of mainshaft.
2. Install second gear with clutching teeth up; the rear face of the gear butts against the mainshaft flange.
3. Install a blocking ring with clutching teeth downward. All three blocking rings are the same.
4. Install second and third synchronizer assembly with fork slot down. Press it onto mainshaft splines. Both synchronizer assemblies are identical but are assembled differently. The second-third speed hub and sleeve is assembled with the sleeve fork slot toward the thrust face of the hub; the first-reverse hub and sleeve, with the fork slot opposite the thrust face. Be sure that the blocker ring notches align with the synchronizer assembly keys.
5. Install synchronizer snap-ring. Both synchronizer snap-rings are the same.
6. Turn rear of shaft up.
7. Install first gear with clutching teeth upward; the front face of the gear butts against the flange on the mainshaft.
8. Install a blocker ring with clutching teeth down.
9. Install first and reverse synchronizer assembly with fork slot down. Press it onto mainshaft splines. Be sure blocker ring notches align with synchronizer assembly keys and synchronizer sleeves face front of mainshaft.
10. Install snap-ring.
11. Install reverse gear with clutching teeth down.
12. Install steel reverse gear thrust washer with flats aligned.
13. Press rear ball bearing onto shaft with snap-ring slot down.
14. Install snap-ring.
15. Install speedometer drive gear and retaining clip.

Transmission Assembly

1. Place a row of 29 roller bearings, a bearing washer, a second row of 29 bearings, and a second bearing washer at each end of

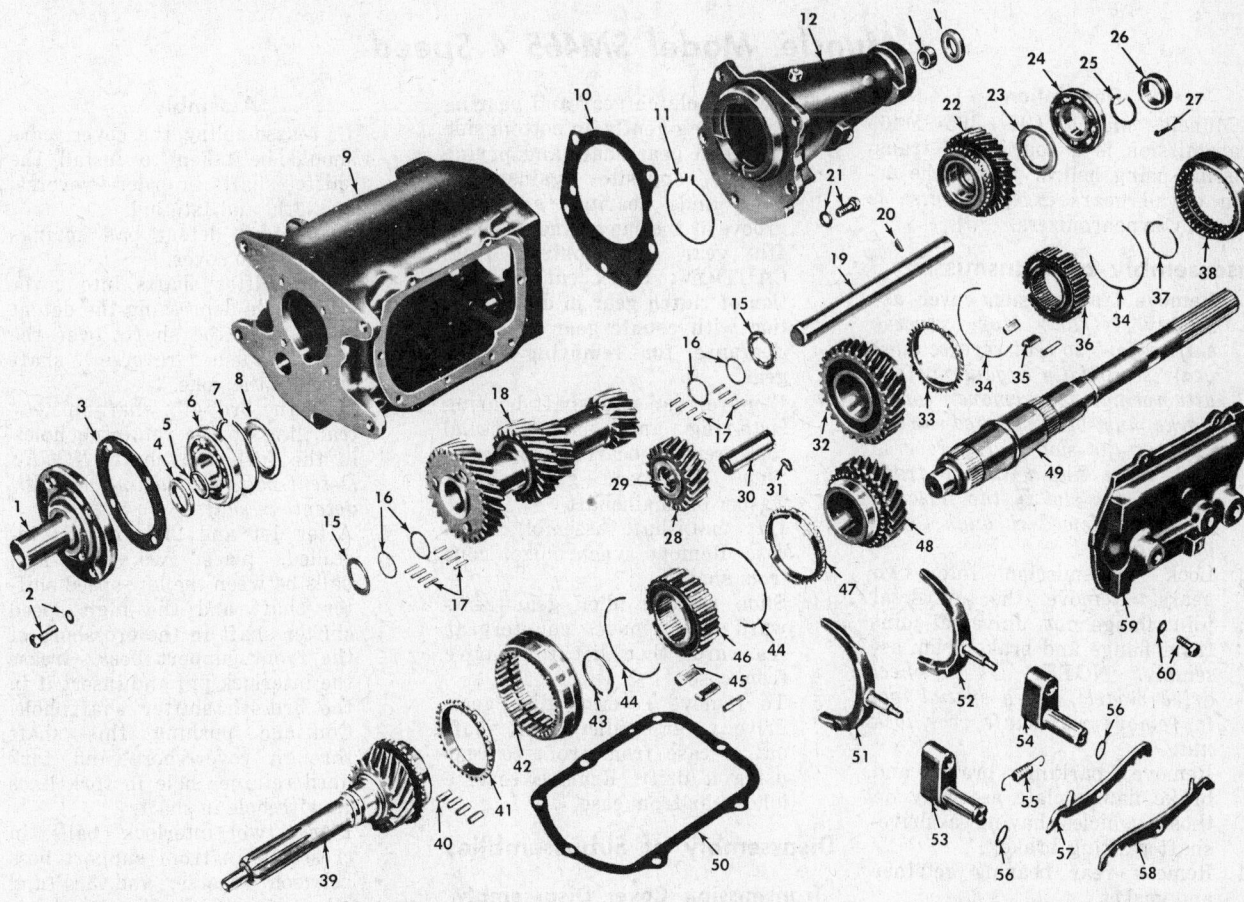

1 Bearing retainer
2 Bolt and lock washer
3 Gasket
4 Oil seal
5 Snap ring (bearing-to-main drive gear)
6 Main drive gear bearing
7 Snap ring bearing
8 Oil slinger
9 Case
10 Gasket
11 Snap ring (rear bearing-to-extension)
12 Extension
13 Extension bushing
14 Oil seal
15 Thrust washer
16 Bearing washer
17 Needle bearings
18 Countergear
19 Countershaft
20 Woodruff key
21 Bolt (extension-to-case)
22 Reverse gear
23 Thrust washer
24 Rear bearing
25 Snap ring
26 Speedometer drive gear
27 Retainer clip
28 Reverse idler gear
29 Reverse idler bushing
30 Reverse idler shaft
31 Woodruff key
32 1st speed gear
33 1st speed blocker ring
34 Synchronizer key spring
35 Synchronizer keys
36 1st and reverse synchronizer hub assembly
37 Snap ring
38 1st and reverse synchronizer collar
39 Main drive gear
40 Pilot bearings
41 3rd speed blocker ring
42 2nd and 3rd synchronizer collar
43 Snap ring
44 Synchronizer key spring
45 Synchronizer keys
46 2nd and 3rd synchronizer hub
47 2nd speed blocker ring
48 2nd speed gear
49 Mainshaft
50 Gasket
51 2nd and 3rd shifter fork
52 1st and reverse shifter fork
53 2-3 shifter shaft assembly
54 1st and reverse shifter shaft assembly
55 Spring
56 O-ring seal
57 1st and reverse detent cam
58 2nd and 3rd detent cam
59 Side cover
60 Bolt and lock washer

SM330 transmission components (© G.M.C.)

the countergear. Hold in place with grease.

2. Place countergear assembly through rear case opening with a tanged thrust washer, tang away from gear, at each end. Install countershaft and key from rear of case. Be sure that thrust washer tangs are aligned with notches in case.

3. Place reverse idler gear in case. Do not install reverse idler shaft yet. *NOTE: The reverse idler gear bushing may not be replaced separately—only as a unit.*

4. Expand snap-ring in extension. Assemble extension over mainshaft and onto rear bearing. Seat snap-ring.

5. Load 16 mainshaft pilot bearings into clutch gear cavity. Assemble third speed blocker ring onto clutch gear clutching surface with teeth toward gear.

6. Place clutch gear assembly, without front bearing, over front of mainshaft. Make sure that blocker ring notches align with keys in second-third synchronizer assembly.

7. Stick gasket onto extension housing with grease. Assemble clutch gear, mainshaft, and extension to case together. Make sure that clutch gear teeth engage teeth of countergear antilash plate.

8. Rotate extension housing. Install reverse idler shaft and key.

9. Torque extension bolts to 45 ft. lbs.

10. Install oil slinger with inner lip facing forward. Install front bearing outer snap-ring and slide bearing into case bore.

11. Install snap-ring to clutch gear stem. Install bearing retainer and gasket and torque to 20 ft. lbs. Retainer oil return hole must be at 6 o'clock.

12. Shift both synchronizer sleeves to neutral positions. Install side cover, inserting shifter forks in synchronizer sleeve grooves.

13. Torque side cover bolts to 20 ft. lbs.

TORQUE SPECIFICATIONS
Foot-Pounds

Extension to case attaching	45
Drain plug	30
Filler plug	15
Side cover attaching bolts	22
Main drive gear retainer bolts	22
Transmission case to clutch Housing bolts	45

Muncie Model SM465 4 Speed

Description
Muncie model CH 465-SM465 transmission is a four speed transmission using helical gears. The action of all gears except reverse is aided by synchronizers.

Disassembly of Transmission

1. Remove transmission cover assembly. *NOTE: Move reverse shifter fork so that reverse idler gear is partially engaged before attempting to remove cover. Forks must be positioned so rear edge of the slot in the reverse fork is in line with the front edge of the slot in the forward forks as viewed through tower opening.*

2. Lock transmission into two gears. Remove the universal joint flange nut, universal joint front flange and brake drum assembly. *NOTE: On 4-wheel drive models, use a special tool to remove mainshaft rear lock nut.*

3. Remove parking brake and brake flange plate assembly on those vehicles having a driveshaft parking brake.

4. Remove rear bearing retainer and gasket.

5. Slide speedometer drive gear off mainshaft.

6. Remove clutch gear bearing retainers and gasket.

7. Remove countergear front bearing cap and gasket.

8. Using a screwdriver, pry off countershaft front bearing.

9. Remove countergear rear bearing snap-rings from shaft and bearing. Using special tool, remove countergear rear bearings.

10. Remove clutch gear bearing outer race to case retaining ring.

11. Remove clutch gear and bearing by tapping gently on bottom side of clutch gear shaft and prying directly opposite against the case and bearing snap-ring groove at the same time. Remove 4th gear synchronizer ring. **CAUTION: Index cut out section of clutch gear in down position with countergear to obtain clearance for removing clutch gear.**

12. Remove rear mainshaft bearing snap-ring and, using special tools, remove bearing from case. Slide 1st speed gear thrust washer off mainshaft.

13. Lift mainshaft assembly from case. Remove synchronizer cone from shaft.

14. Slide reverse idler gear rearward and move countergear rearward, then lift to remove from case.

15. To remove reverse idler gear, drive reverse idler gear shaft out of case from front to rear using a drift. Remove reverse idler gear from case.

Disassembly of Subassemblies

Transmission Cover Disassembly

1. Remove shifter fork retaining pins and drive out expansion plugs. NOTE: The third and fourth shifter fork must be removed before the reverse shifter head pin can be removed.

2. With shifter shafts in neutral position, remove shafts. CAUTION: Care should be taken when removing the detent balls and springs since removal of the shifter shafts will cause these parts to be forcibly ejected.

3. Remove retaining pin and drive out reverse shifter shaft.

Assembly

1. In reassembling the cover, care should be taken to install the shifter shafts in order—reverse, 3rd.-4th, and 1st.-2nd.

2. Place fork detent ball springs and balls in cover.

3. Start shifter shafts into cover and, while depressing the detent balls, push the shafts over the balls. Push reverse shaft through the yoke.

4. With the 3rd.-4th. shaft in neutral, line up the retaining holes in the fork and shaft. *NOTE: Detent balls should line up with detents in shaft.*

5. After 1st and 2nd fork is installed, place two inner-lock balls between the low speed shifter shaft and the high speed shifter shaft in the crossbore of the front support boss. Grease the interlock pin and insert it in the 3rd-4th shifter shaft hole. Continue pushing this shaft through cover bore and fork until retainer hole in fork lines up with hole in shaft.

6. Place two interlock balls in crossbore in front support boss between reverse, and 3rd and 4th shifter shaft. Then push remaining shaft through fork and cover bore, keeping both balls in position between shafts until retaining holes line up in fork and shaft. Install retaining pin.

7. Install 1st/2nd fork and reverse fork retaining pins. Install new shifter shaft hole expansion plugs.

Clutch Gear and Shaft Disassembly

1. Remove mainshaft pilot bearing rollers from clutch gear if not already removed, and remove

roller retainer. Do not remove snap-ring on inside of clutch gear.

2. Remove snap-ring securing bearing on stem of clutch gear.

3. To remove bearing, position a special tool to the bearing and, with an arbor press, press gear and shaft out of bearing.

Assembly

1. Press bearing and new oil slinger onto clutch gear shaft using a special tool. Slinger should be located flush with bearing shoulder on clutch gear. CAUTION: Be careful not to distort oil slinger.

2. Install bearing snap-ring on clutch gear shaft.

3. Install bearing retainer ring in groove on O.D. of bearing. **CAUTION: The bearing must turn freely on the shaft.**

4. Install snap-ring on I.D. of mainshaft pilot bearing bore in clutch gear.

5. Lightly grease bearing surface in shaft recess, install transmis-

1 Transmission cover
2 Interlock balls
3 3rd-4th shifter shaft
4 Reverse shifter shaft
5 Fork retaining pin
6 Detent ball
7 Detent spring
8 3rd-4th shifter fork
9 "C" ring lock clip
10 Reverse shifter fork
11 Shifter shaft hole plugs
12 1st-2nd shifter fork
13 Interlock plunger spring
14 Reverse interlock plunger
15 1st-2nd shifter shaft
16 Interlock pin
17 Cover gasket

Shift cover assembly components (© G.M.C.)

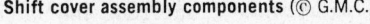
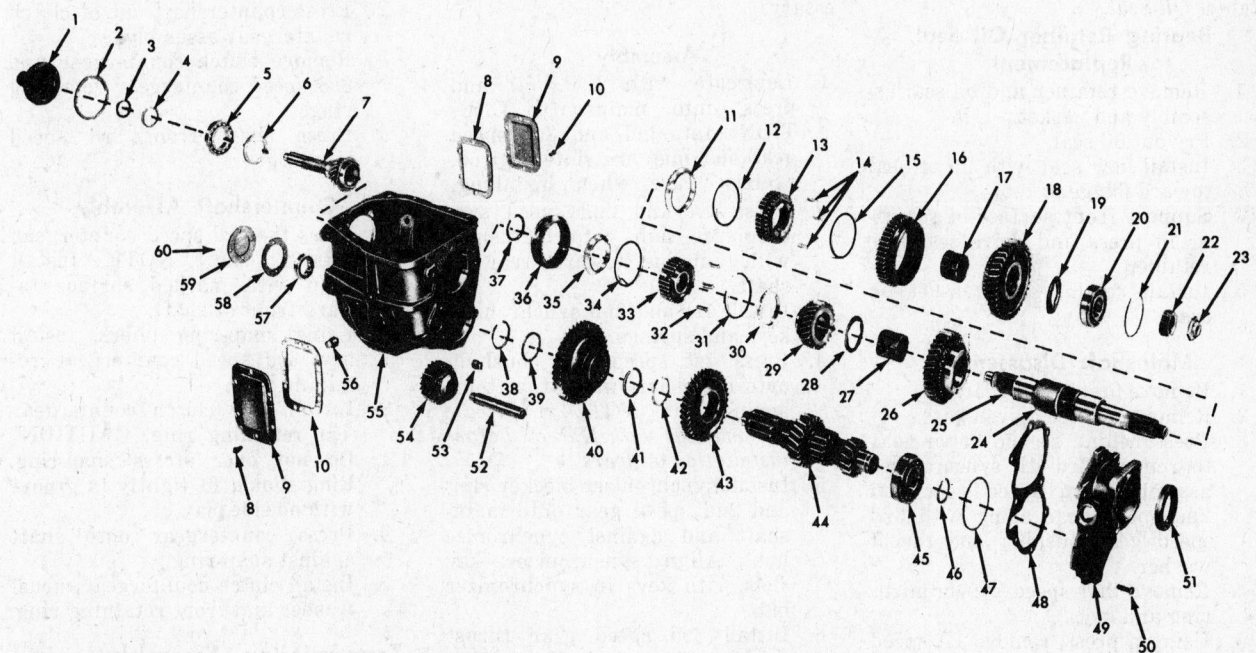

1 Clutch gear bearing retainer	15 Synchronizer spring	31 Synchronizer spring
2 Retainer gasket	16 Reverse driven gear	32 Synchronizer keys
3 Lip seal	17 1st gear bushing	33 3rd-4th synchronizer hub
4 Snap ring	18 1st gear	34 Synchronizer spring
5 Clutch gear bearing	19 Thrust washer	35 3rd-4th speed blocker ring
6 Oil slinger	20 Rear main bearing	36 3rd-4th speed synchronizer sleeve
7 Clutch gear and pilot bearings	21 Bearing snap ring	37 Snap ring
8 Power take-off cover gasket	22 Speedometer gear	38 Snap ring
9 Power take-off cover	23 Rear mainshaft lock nut	39 Thrust washer
10 Retaining screws	24 2nd speed bushing (on shaft)	40 Clutch countergear
11 1st-2nd speed blocker ring	25 Mainshaft	41 Snap ring
12 Synchronizer spring	26 2nd speed gear	42 Snap ring
13 1st-2nd speed synchronizer hub	27 3rd speed bushing	43 3rd speed countergear
14 Synchronizer keys	28 Thrust washer	44 Countergear shaft
	29 3rd speed gear	45 Countergear rear bearing
	30 3rd speed blocker ring	46 Snap ring
		47 Bearing outer snap ring
		48 Rear retainer gasket
		49 Rear retainer
		50 Retainer bolts
		51 Retainer lip seal
		52 Reverse idler shaft
		53 Drain plug
		54 Reverse idler gear
		55 Case
		56 Fill plug
		57 Countergear front bearing
		58 Gasket
		59 Front cover
		60 Cover screws

Transmission components (© G.M.C.)

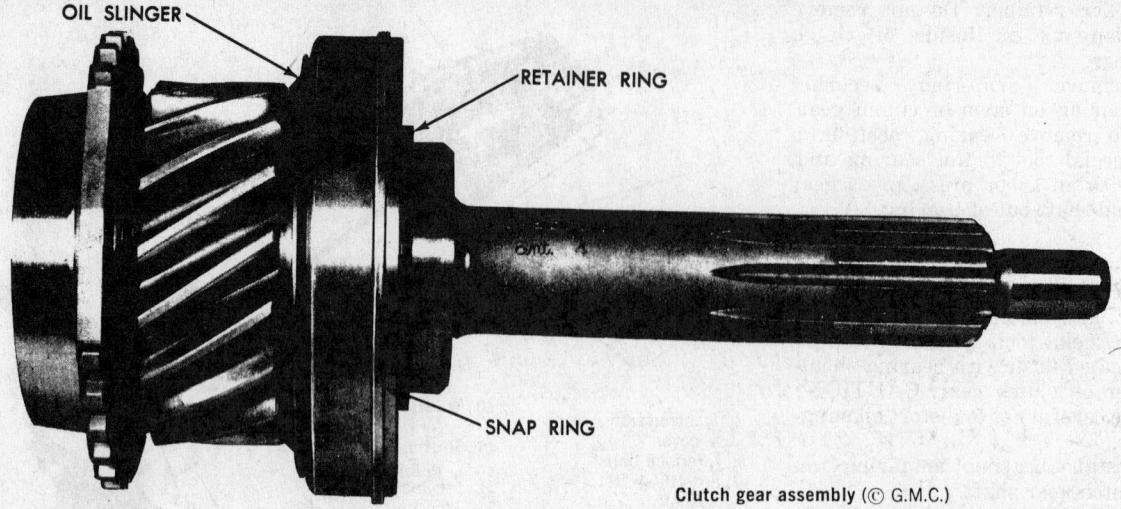

OIL SLINGER

RETAINER RING

SNAP RING

Clutch gear assembly (© G.M.C.)

sion mainshaft pilot roller bearings and install roller bearing retainer.

NOTE: This roller bearing retainer holds bearings in position, and, in final transmission assembly, is pushed forward into recess by mainshaft pilot. Clutch Gear Bearing Retainer Oil Seal.

Bearing Retainer Oil Seal Replacement

1. Remove retainer and oil seal assembly and gasket.
2. Pry out oil seal.
3. Install new seal with lip of seal toward flange of tool.
4. Support front surface of retainer in press and drive seal into retainer.
5. Install retainer and gasket on case.

Mainshaft Disassembly

1. Remove first speed gear.
2. Remove reverse driven gear.
3. Press behind second speed gear to remove 3rd-4th synchronizer assembly, 3rd speed gear and 2nd speed gear along with 3rd speed gear bushing and thrust washer.
4. Remove 2nd speed synchronizer ring and keys.
5. Using a press, remove 1st speed gear bushing and 2nd speed synchronizer hub.
6. Without damaging the mainshaft, chisel out the 2nd speed gear bushing.

Inspection

Wash all parts in cleaning solvent and inspect them for excessive wear or scoring.

NOTE: Third and fourth speed clutch sleeve should slide freely on clutch hub but clutch hub should fit snugly on shaft splines.

Third speed gear must be running fit on mainshaft bushing and mainshaft bushing should be press fit on shaft.

First and reverse sliding gear must be sliding fit on synchronizer hub and must not have excessive radial or circumferential play. If sliding gear is not free on hub, inspect for burrs which may have rolled up on front end of half-tooth internal splines and remove by honing as necessary.

Assembly

1. Lubricate with E.P. oil and press onto mainshaft. **CAUTION: 1st, 2nd and 3rd speed gear bushings are sintered iron, exercise care when installing.**
2. Press 1st and 2nd speed synchronizer hub onto mainshaft with annulus toward rear of shaft.
3. Install 1st and 2nd synchronizer keys and springs.
4. Press 1st speed gear bushing onto mainshaft until it bottoms against hub. *NOTE: Lubricate all bushings with E.P. oil before installation of gears.*
5. Install synchronizer blocker ring and 2nd speed gear onto mainshaft and against synchronize hub. Align synchronizer key slots with keys in synchronizer hub.
6. Install 3rd speed gear thrust washer onto mainshaft inserting washer tang in slotted shaft. Then press 3rd speed gear bushing onto mainshaft against thrust washer.
7. Install 3rd speed gear and synchronizer blocker ring against 3rd speed gear thrust washer.
8. Align synchronizer key ring slots with synchronizer assembly keys and drive 3rd and 4th synchronizer assembly onto mainshaft. Secure assembly with snap-ring.
9. Install reverse driven gear with fork groove toward rear.
10. Install 1st speed gear against 1st and 2nd synchronizer hub. In-

stall 1st speed gear thrust washer.

Countershaft Disassembly

1. Remove front countergear retaining ring and thrust washer. Do not re-use this snap-ring or any others.
2. Press countershaft out of clutch countergear assembly.
3. Remove clutch countergear and 3rd speed countergear retaining rings.
4. Press shaft from 3rd speed countergear.

Countershaft Assembly

1. Press the 3rd speed countergear onto the shaft. NOTE: Install gear with marked surface toward front of shaft.
2. Using snap-ring pliers, install new 3rd speed countergear retaining ring.
3. Install new clutch countergear rear retaining ring. **CAUTION: Do not over stress snap-ring. Ring should fit tightly in groove with no side play.**
4. Press countergear onto shaft against snap-ring.
5. Install clutch countergear thrust washer and front retaining ring.

Transmission Assembly

1. Lower the countergear into the case.
2. Place reverse idler gear in transmission case with gear teeth toward the front. Install idler gear shaft from rear to front, being careful to have slot in end of shaft facing down and flush with case.
3. Install mainshaft assembly into case with rear of shaft protruding out rear bearing hole in case. Rotate case onto front end. *NOTE: Install 1st speed gear thrust washer on shaft, if not previously installed.*
4. Install snap-ring on bearing

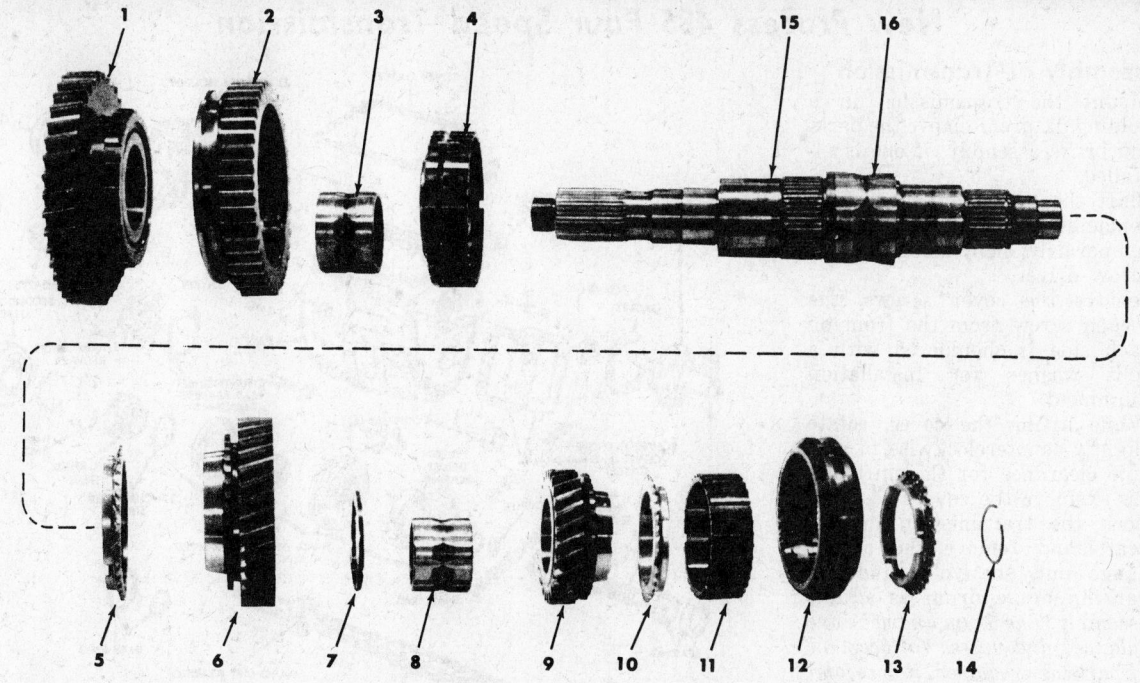

1 1st speed gear
2 Reverse driven gear
3 1st gear bushing
4 1st-2nd gear synchronizer
   hub assembly

5 2nd speed blocker ring
6 2nd speed gear
7 Thrust washer
8 3rd speed bushing
9 3rd speed gear

10 3rd speed blocker ring
11 3rd-4th speed synchronizer
    hub assembly
12 3rd-4th speed synchronizer
    sleeve

13 4th speed blocker ring
14 Snap ring
15 Mainshaft
16 2nd speed gear bushing

Mainshaft assembly (© G.M.C.)

O.D. and place rear mainshaft bearing on shaft. Drive bearing onto shaft and into case.

5. Install synchronizer cone on mainshaft and slide rearward to clutch hub. **CAUTION: Make sure three cut-out sections of 4th speed synchronizer cone align with three clutch keys in clutch assembly.**

6. Install snap-ring on clutch gear bearing O.D. Index cut out portion of clutch gear teeth to obtain clearance over countershaft drive gear teeth, and install into case.

7. Install clutch gear bearing retainer and gasket and torque to 15-18 ft. lbs.

8. Rotate case onto front end.

9. Install snap-ring on countergear rear bearing O.D., and drive bearing into place. Install snap-ring on countershaft at rear bearing.

10. Tap countergear front bearing assembly into case.

11. Install countergear front bear-

ing cap and new gasket and torque to 20-30 in. lbs.

12. Slide speedometer drive gear over mainshaft to bearing.

13. Install rear bearing retainer with new gasket. Be sure snap-ring ends are in lube slot and cut out in bearing retainer. Install bolts and tighten to 15-18 ft. lbs. Install brake backing plate assembly on those models having driveshaft brake. *NOTE: On models equipped with 4-wheel drive, install rear lock nut and washer and torque to 120 ft. lbs. and bend washer tangs to fit slots in nut.*

14. Install parking brake drum and/or universal joint flange. *NOTE: Lightly oil seal surface.*

15. Lock transmission in two gears at once. Install universal joint flange locknut and tighten to 90-120 ft. lbs.

16. Move all transmission gears to neutral except the reverse idler gear which should be engaged approximately ⅜ of an inch

(leading edge of reverse idler gear taper lines up with the front edge of the 1st speed gear). Install cover assembly and gasket. Shifting forks must slide into their proper positions on clutch sleeves and reverse idler gear. Forks must be positioned as in removal.

17. Install cover attaching bolts and gearshift lever and check operation of transmission.

TORQUE CHART

	Foot-Pounds
Rear bearing retainer	18
Cover bolts	25
Filler plug	35
Drain plug	35
Clutch gear bearing retainer bolts	18
Universal joint front flange nut	95
Power take off cover bolts	18
Parking brake	22
Countergear front cover screws	25
Rear mainshaft lock nut (4 wheel drive models)	95

New Process 435 Four Speed Transmission

Disassembly of Transmission

1. Mount the transmission in a holding fixture. Remove the parking brake assembly, if one is installed.

2. Shift the gears into neutral by replacing the gear shift lever temporarily, or by using a bar or screw driver.

3. Remove the cover screws, the second screw from the front on each side is shouldered with a split washer for installation alignment.

4. While lifting the cover, rotate slightly counterclockwise to provide clearance for the shift levers. Remove the cover.

5. Lock the transmission in two gears and remove the output flange nut, the yoke, and the parking brake drum as a unit assembly. *The drum and yoke are balanced and unless replacement of parts are required, it is recommended that the drum and yoke be removed as a assembly.*

6. Remove the speedometer drive gear pinion and the mainshaft rear bearing retainer.

7. Before removal and disassembly of the drive pinion and mainshaft, measure the end play between the synchronizer stop ring and the third gear. *Record this reading for reference during assembly.* Clearance should be within 0.050-0.070 inch. If necessary, add corrective shims during assembly.

8. Remove the drive pinion bearing retainer.

9. Rotate the drive pinion gear to align the space in the pinion gear clutch teeth with the countershaft drive gear teeth. Remove the drive pinion gear and the tapered roller bearing from the transmission by pulling on the pinion shaft, and rapping the face of the case lightly with a brass hammer.

10. Remove the snap ring, washer, and the pilot roller bearings from the recess in the drive pinion gear.

11. Place a brass drift in the front center of the mainshaft and drive the shaft rearward.

12. When the mainshaft rear bearing has cleared the case, remove the rear bearing and the speedometer drive gear with a suitable gear puller.

13. Move the mainshaft assembly to the rear of the case and tilt the front of the mainshaft upward.

14. Remove the roller type thrust washer.

15. Remove the synchronizer and stop rings separately.

16. Remove the mainshaft assembly.

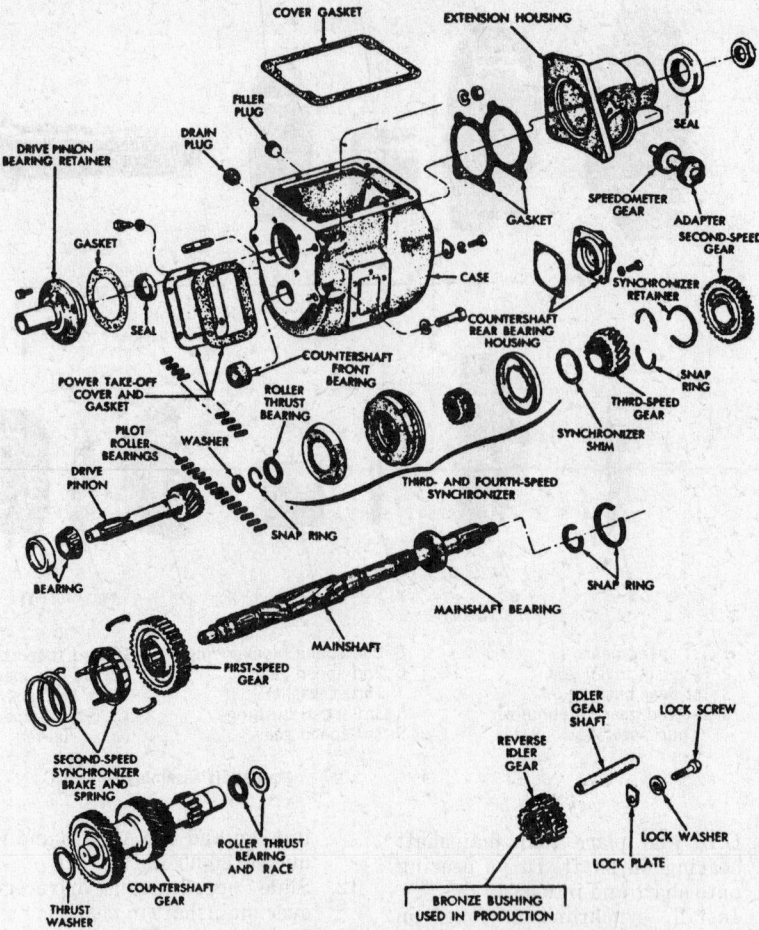

New Process 435 Transmission—exploded view (© Chrysler Corp.)

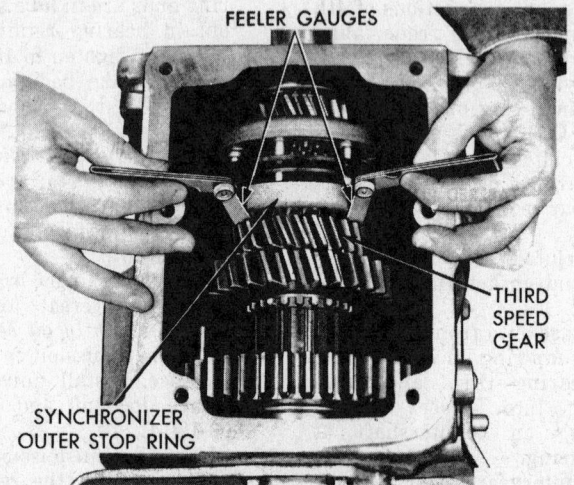

Measuring the synchronizer end-play—New Process 435 (© Chrysler Corp.)

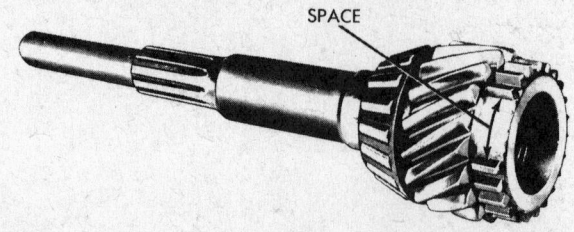

Drive pinion gear showing the teeth removed—New Process 435 (© Chrysler Corp.)

17. Remove the reverse idler lock screw and lock plate.
18. Using a brass drift held at an angle, drive the idler shaft to the rear while pulling.
19. Lift the reverse idler gear out of the case.

NOTE: If the countershaft gear does not show signs of excessive side play or end play and the teeth are not badly worn or chipped, it may not be necessary to replace the countershaft gear.

20. Remove the bearing retainer at the rear end of the countershaft. The bearing assembly will remain with the retainer.
21. Tilt the cluster gear assembly and work it out of the transmission case.
22. Remove the front bearings from the case with a suitable driver.

Overhaul of Sub-Assemblies

Mainshaft
Disassembly

1. Remove the clutch gear snap ring.
2. Remove the clutch gear, the synchronizer outer stop ring to third gear shim, and the third gear.
3. Remove the special split lock ring with two screw drivers. Remove the second gear and synchronizer.

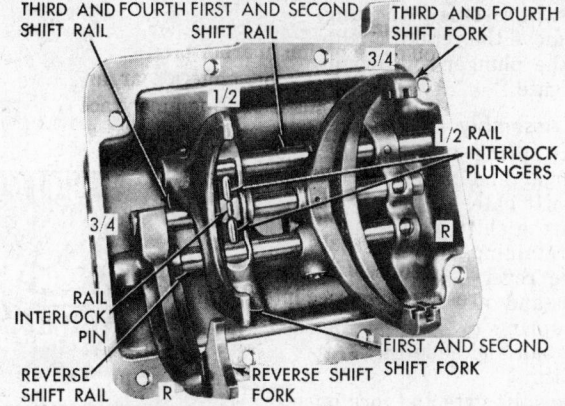

Cover and shift fork assembly—New Process 435 (© Chrysler Corp.)

4. Remove the first-reverse sliding gear.
5. Drive the old seal out of the bearing retainer.

Assembly

1. Place the mainshaft in a soft-jawed vise with the rear end up.
2. Install the first-reverse gear. Be sure the two spline springs, if used, are in place inside the gear as the gear is installed on the shaft.
3. Place the mainshaft in a soft-jawed vise with the front end up.

4. Assemble the second speed synchronizer spring and synchronizer brake on the second gear. Secure the brake with a snap ring making sure that the snap ring tangs are away from the gear.
5. Slide the second gear on the front of the mainshaft. Make sure that the synchronizer brake is toward the rear. Secure the gear to the shaft with the two piece lock ring. Install the third gear.
6. Install the shim between the third gear and the third-fourth synchronizer stop ring. Refer to the measurements of end play made during disassembly to determine if additional shims are needed.

NOTE: The exact determination of end-play must be made after the complete assembly of the mainshaft and the main drive pinion is installed in the transmission case.

Reverse Idler Gear

DO NOT disassemble the reverse idler gear. If it is no longer serviceable, replace the assembly complete with the integral bearings.

Cover and Shift Fork Assembly

NOTE: The cover and shift fork assembly should be disassembled ONLY if inspection shows worn or damaged parts, or if the assembly is not working properly.

Disassembly

1. Remove the roll pin from the first-second shift fork and the shift gate with an "easy out".

NOTE: A square type or a closely wound spiral "easy out" mounted in a tap is preferable for this operation.

2. Move the first-second shift rail forward and force the expansion plug out of the cover. Cover the detent ball access hole in the cover with a cloth to prevent it from flying out. Remove the rail, fork, and gate from the cover.
3. Remove the third-fourth shift rail, then the reverse rail in the manner outlined in steps 1 and 2 above.

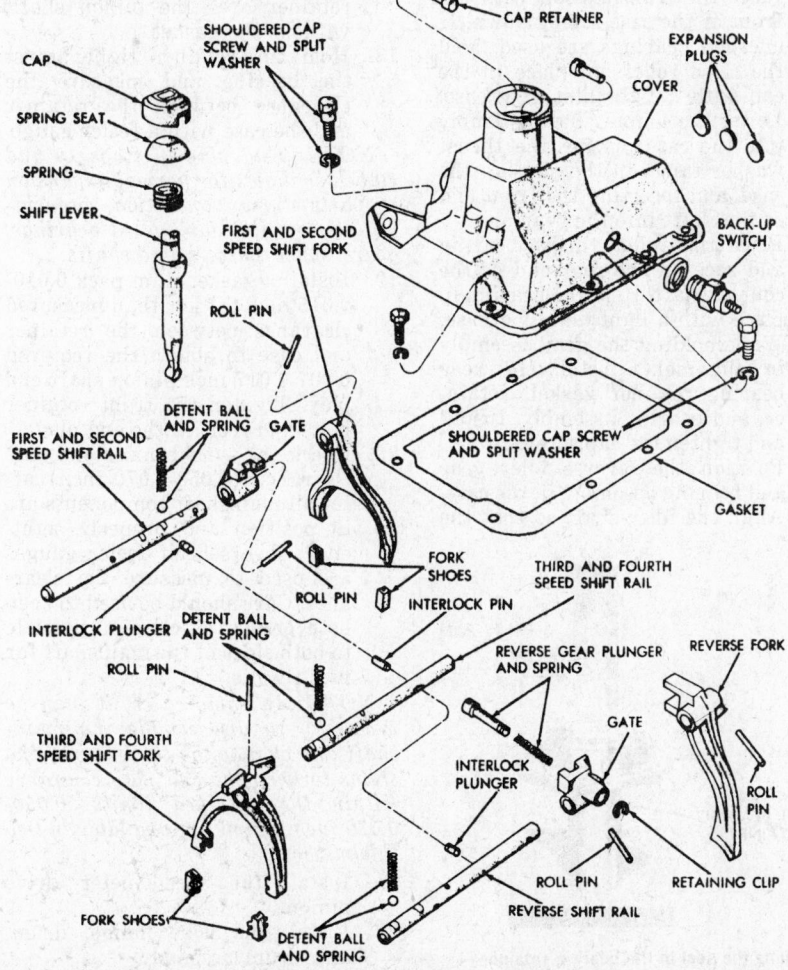

Cover and shift fork assembly—exploded view—New Process 435 (© Chrysler Corp.)

4. Compress the reverse gear plunger and remove the retaining clip. Remove the plunger and spring from the gate.

Assembly

1. Install the spring on the reverse gear plunger and hold it in the reverse shift gate. Compress the spring in the shift gate and install the retaining clip.
2. Insert the reverse shift rail in the cover and place the detent ball and spring in position. Depress the ball and slide the shift rail over it.
3. Install the shift gate and fork on the reverse shift rail. Install a new roll pin in the gate and the fork.
4. Place the reverse fork in the neutral position.
5. Install the two interlock plungers in their bores.
6. Insert the interlock pin in the third-fourth shift rail. Install the shift rail in the same manner as the reverse shift rail.
7. Install the first-second shift rail in the same manner as outlined above. Make sure the interlock plunger is in place.
8. Check the interlocks by shifting the reverse shift rail into the "Reverse" position. It should be impossible to shift the other rails with the reverse rail in this position.
9. If the shift lever is to be installed at this point, lubricate the spherical ball seat and place the cap in place.
10. Install the back-up light switch.
11. Install new expansion plugs in the bores of the shift rail holes in the cover. Install the rail interlock hole plug.

Drive Pinion and Bearing Retainer
Disassembly

1. Remove the tapered roller bearing from the pinion shaft with a suitable tool.
2. Remove the snap ring, washer, and the pilot rollers from the gear bore, if they have not been previously removed.
3. Pull the bearing race from the front bearing retainer with a suitable puller.
4. Remove the pinion shaft seal with a suitable tool.

Assembly

1. Position the drive pinion in an arbor press.
2. Place a wood block on the pinion gear and press it into the bearing until it contacts the bearing inner race.
3. Coat the roller bearings with a light film of grease to hold the bearings in place, and insert them in the pocket of the drive pinion gear.

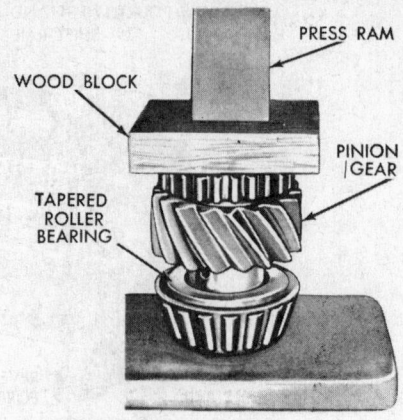

Installing the pinion bearing—New Process 435 (© Chrysler Corp.)

4. Install the washer and snap ring.
5. Press a new seal into the bearing retainer. Make sure that the lip of the seal is toward the mounting surface.
6. Press the bearing race into the retainer.

Assembly of Transmission

1. Press the front countershaft roller bearings into the case until the cage is flush with the front of the transmission case. Coat the bearings with a light film of grease.
2. Place the transmission with the front of the case facing down. If uncaged bearings are used, hold the loose rollers in place in the cap with a light film of grease.
3. Lower the countershaft assembly into the case placing the thrust washer tangs in the slots in the case, and inserting the front end of the shaft into the bearing.
4. Place the roller thrust bearing and race on the rear end of the countershaft. Hold the bearing in place with a light film of grease.
5. While holding the gear assembly in alignment, install the rear bearing retainer gasket, retainer, and bearing assembly. Install and tighten the cap screws.
6. Position the reverse idler gear and bearing assembly in the case.
7. Align the idler shaft so that the

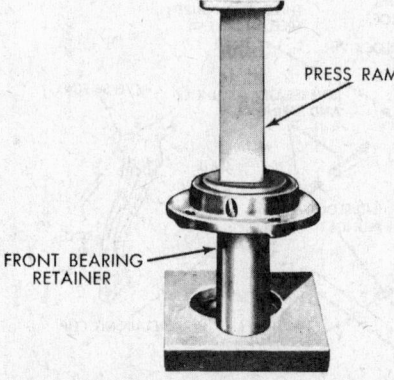

Installing the seal in the bearing retainer—New Process 435 (© Chrysler Corp.)

lock plate groove in the shaft is in position to install the lock plate.

8. Install the lock plate, washer, and cap screw.
9. Make sure the reverse idler gear turns freely.
10. Lower the rear end of the mainshaft assembly into the case, holding the first gear on the shaft. Maneuver the shaft through the rear bearing opening.
    NOTE: With the mainshaft assembly moved to the rear of the case, be sure the third-fourth synchronizer and shims remain in position.
11. Install the roller type thrust bearing.
12. Place a wood block in-between the front of the case and the front of the mainshaft.
13. Install the rear bearing on the mainshaft by carefully driving the bearing onto the shaft and into the case, snap ring flush against the case.
14. Install the drive pinion shaft and bearing assembly. Make sure that the pilot rollers remain in place.
15. Install the spacer and speedometer drive gear.
16. Install the rear bearing retainer and gasket.
17. Place the drive pinion bearing retainer over the pinion shaft, without the gasket.
18. Hold the retainer tight aginst the bearing and measure the clearance between the retainer and the case with a feeler gauge.
    NOTE: End play in steps 19 and 20 below allows for normal expansion of parts during operation, preventing seizure and damage to bearings, gears, synchronizers, and shafts.
19. Install a gasket shim pack 0.010-0.015 inch thicker than measured clearance between the retainer and case to obtain the required 0.007-0.017 inch pinion shaft end play. Tighten the front retainer bolts and recheck the end play.
20. Check the synchronizer end play clearance (0.050-0.070 inch) after all mainshaft components are in position and properly tightened. Two sets of feeler gauges are used to measure the clearance. Care should be used to keep both gauges as close as possible to both sides of the mainshaft for best results.
    NOTE: In some cases, it may be necessary to disassemble the mainshaft and change the thickness of the shims to keep the end play clearance within the specified limits, 0.050-0.070 inch. Shims are available in two thicknesses.
21. Install the speedometer drive pinion.
22. Install the yoke flange, drum, and drum assembly.
23. Place the transmission in two

gears at once, and tighten the yoke flange nut.

24. Shift the gears and/or synchronizers into all gear positions and and check for free rotation.

25. Cover all transmissions components with a film of transmission oil to prevent damage during start up after initial lubricant fill-up.

26. Move the gears to the neutral position.

27. Place a new cover gasket on the transmission case, and lower the cover over the transmission.

28. Carefully, engage the shift forks into their poper gears. Align the cover.

29. Install a shouldered alignment screw with split washer in the screw hole second from the front of the cover. Try out gear operation by shifting through all ranges. Make sure everything moves freely.

30. Install the remaining cover screws.

LUBRICANT CAPACITY

New Process 435	7 pt.

TORQUE SPECIFICATIONS

	Foot-Pounds
Cover screws	20-40
Drive gear retaining screw	15-25
Front countershaft retainer screw	15-25
Front countershaft bearing washer screw	12-22
Flange nut	125
Mainshaft rear retainer screw	15-25
Rear countershaft retainer screw	15-25
PTO cover screws	8-12
Filler and drain plugs	25-45
Reverse idler shaft lock screw	20-40
Brake link shoulder screw	20-40

New Process Model 445 Four Speed Transmission

Disassembly of Transmission

1. Place the transmission in a holding fixture and drain the lubricant.

2. Shift the transmission gears into neutral. Remove the gearshift cover attaching bolts. Note that the two bolts opposite the tower are shouldered to properly position the cover. Lift the cover straight up and remove.

3. Lock the transmission in two gears at once and remove the mainshaft nut and yoke.

4. Loosen and remove the extension housing bolts. Remove the mainshaft extension housing and the speedometer drive pinion.

5. Remove the bolts from the drive pinion front bearing retainer and pull the bearing retainer and gasket off.

6. Rotate the drive pinion gear to align the pinion gear flat with the countershaft drive gear teeth. Remove the drive pinion gear and the tapered roller bearing from the transmission.

7. Remove the mainshaft thrust bearing.

8. Push the mainshaft assembly to the rear of the transmission and tilt the front of the mainshaft up.

9. Remove the mainshaft assembly from the transmission case.

10. Remove the reverse idler lock screw and lock plate.

11. Using a suitable size brass drift, carefully drive the reverse idler shaft out the REAR of the case.

NOTE: DO NOT ATTEMPT TO DRIVE THE REVERSE IDLER

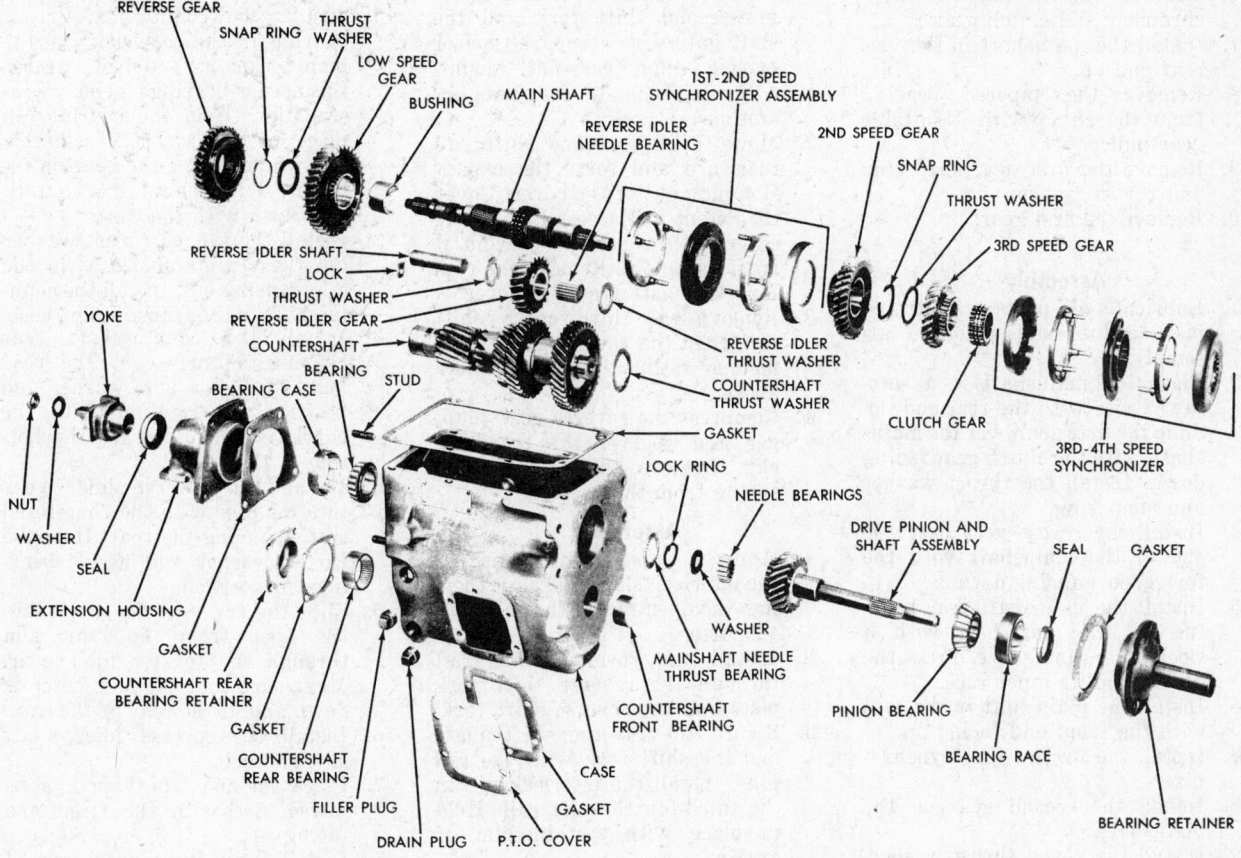

New Process 445 transmission—exploded view (© Chrysler Corp.)

Drive pinion gear—New Process 445 (© Chrysler Corp.)

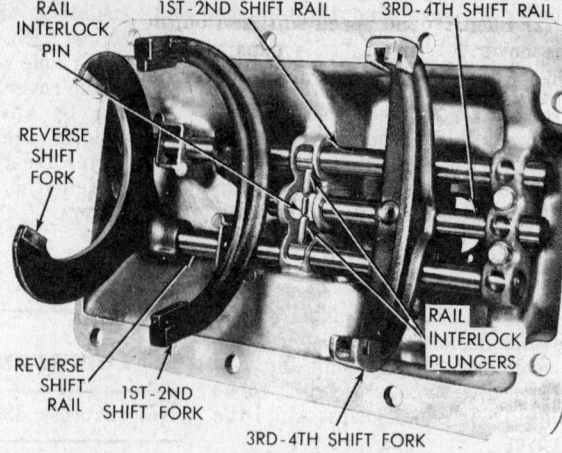

Cover and shift fork assembly—New Process 445 (© Chrysler Corp.)

SHAFT FORWARD! This will damage the transmission case and the reverse idler shaft.

12. Remove the countershaft rear bearing retainer.
13. Slide the countershaft to the rear, then up and out of the case.
14. Drive the countershaft forward, out of the bearing and the case.

Overhaul of Sub-Assemblies

Mainshaft
Disassembly

1. Place the mainshaft in a soft-jawed vise with the front end up.
2. Lift the third-fourth synchronizer and high speed clutch off the mainshaft.
3. Remove the third gear.
4. Remove the second gear snap ring. Lift off the thrust washer.
5. Remove the second gear.
6. Remove the first-reverse synchronizer and clutch gear.
7. Install the mainshaft in the vise rear end up.
8. Remove the tapered bearing from the shaft with a suitable gear puller.
9. Remove the first gear snap ring and thrust washer.
10. Remove the first gear.

Assembly

1. Lubricate all parts with transmission lubricant prior to assembly.
2. Place the mainshaft in a soft-jawed vise with the rear end up.
3. Slide the first gear over the mainshaft, with the clutch gear facing down. Install the thrust washer and snap ring.
4. Install the revese gear over the end of the mainshaft with the fork groove facing down.
5. Install the mainshaft rear bearing on the mainshaft with a sleeve of suitable size. Press the bearing on its inner race.
6. Install the mainshaft in the vise with the front end facing up.
7. Install the first-reverse synchronizer.
8. Install the second gear on the mainshaft.
9. Install the keyed thrust washer, ground side toward the second

gear and secure with the snap ring.
10. Install the third gear and one shim on the mainshaft.
11. Install the third fourth synchronizer over the mainshaft. Make sure that the slotted end of the clutch gear is positioned toward the third gear.

Cover and Shift Fork Assembly
Disassembly

NOTE: The cover and shift fork assembly should be disassembled ONLY if inspection shows worn or damaged parts, or if the assembly is not working properly.

1. Remove the roll pin from the first-second shift fork and the shift gate. Use a square-type or spiral wound "easy-out" mounted in a tap handle for these operations.
2. Move the first-second shift rail rearward and force the expansion plug out of the cover. Cover the detent ball access hole in the cover with a cloth to prevent it from flying out. Remove the rail, fork, and gate from the cover.
3. Remove the third-fourth shift rail, then the reverse rail in the manner outlined in steps 1 and 2 above.
4. Compress the reverse gear plunger and remove the retaining clip. Remove the plunger and spring from the gate.

Assembly

1. Apply a thin film of grease on the interlock slugs and slide them into the openings in the shift rail supports.
2. Install the reverse shift rail through the reverse shift fork plate and the reverse shift fork.
3. Secure the reverse shift plate and the shift fork with the roll pins. Install the interlock pin in the third-fourth shift rail. Hold in place with a thin film of grease.
4. Slide the third-fourth shift rail

into the rail support from the rear of the cover. Slide the rail through the third-fourth shift fork and poppet ball and spring. Secure the third-fourth shift fork with the roll pin.
5. Install the interlock pin in the first-second shift rail and secure with a light coat of grease. Slide the first-second shift rail into the case, through the shift fork and shift gate. Hold the poppet ball and spring down until the shaft rail passes.
6. Secure the first-second shift rail and gate with the roll pins.

Assembly of Transmission

1. Install the countershaft front bearing in the case using a 1⅜ inch socket as a driver. Grease the needle bearings prior to installation. Hold the bearings in place with a socket of suitable size while seating the bearing retainer. Drive the retainer in until it is flush with the case.
2. Install the tanged thrust washer on the countershaft with the tangs facing out. Install the countershaft in the transmission case.
3. Install the countershaft rear bearing retainer over the rear bearing. Use a new washer and position the retainer with the curved segment toward the bottom of the case.
4. Install the reverse idler gear into the case with the chamfered section facing the rear. Hold the thrust washer and needle bearings in position.
5. Slide the reverse idler shaft into the case, from the rear, and through the reverse idler gear. Make sure that the lock notch is down and at the rear of the case.
6. Install the reverse idler shaft lock and bolt.
7. Place the mainshaft in a soft-jawed vise with the front end facing up.
8. Install the drive gear on top of the mainshaft.

9. Measure the clearance between the high-speed synchronizer and the drive gear with two feeler gauges.
   If the clearance is greater than 0.043-0.053 inch, install synchronizer shims between the third gear and the synchronizer brake drum. After the required shims have been installed, remove the drive gear from the mainshaft.

10. Install the mainshaft into the transmission case. Place the thrust washer over the pilot end of the mainshaft.

11. Position the drive gear so that the cutaway portion of the gear is facing down. Slide the drive gear into the front of the case and engage the mainshaft pilot in the pocket of the drive gear.

12. Slip the drive gear front bearing retainer over the shaft on gasket, and do not secure with bolts.

13. Install the mainshaft rear bearing retainer. Tighten the screws to specifications.

14. Hold the retainer against the front of the transmission case and measure the clearance between the front bearing retainer and the front of the case with a feeler gauge. Record the measurement and remove the bearing retainer.

15. Install a gasket pack on the front bearing retainer which is 0.010-0.015 inch thicker than the clearance measured in step 14. Install the front bearing retainer and torque attaching screws to specification.

16. The end play float of the front synchronizer must be checked before installation of the transmission cover assembly. Measure the end play "float" by inserting two feeler gauges opposite one another between the third gear and the synchronizer stop ring. Accurate measurement can be made only after all mainshaft parts are in place and torqued to specification.

17. If the front synchronizer end play "float" does not fall between 0.050-0.070 inch, shims should be added or removed as required, from between the third gear and the synchronizer stop ring.

18. Install the yoke retaining nut on the rear of the mainshaft. Shift the transmission into two gears at the same time and torque the yoke nut to 125 ft.-lbs.

19. Shift the transmission into neutral.

20. Install the cover gasket.

21. Shift the transmission into second gear. Shift the cover into second.

22. Carefully lower the cover into position. It may be necessary to position the reverse gear to permit the fork to engage its groove.

23. Install the cover aligning screws (shouldered) and tighten with fingers only.

24. Install the remaining cover screws and tighten to specifications.

TORQUE SPECIFICATIONS

	Foot-Pounds
Cover screws	20-40
Drive gear retaining screw	15-25
Front countershaft retaining screw	15-25
Front countershaft bearing washer screw	12-22
Flange nut	125
Mainshaft rear retainer screw	15-25
Rear countershaft retainer screw	15-25
PTO cover screws	8-12
Filler and drain plugs	25-45
Reverse idler shaft lock screw	20-40
Brake link shoulder screw	20-40

LUBRICANT CAPACITY

New Process 445	7½ pts.

New Process 540 & 542 Series Five Speed

Disassembly of Transmission

1. Place the transmission on a stand or a bench.

2. Shift the transmission into the 2nd speed for the 540 and 3rd speed for the 542.

3. Remove the screws and remove the transmission cover by lifting upward and carefully rotating the housing counterclockwise. Note the location of the alignment screws. The alignment screws use split type lockwashers.
   NOTE: On the 542 transmission, it may be necessary to move the 1st speed gear back slightly allowing the offset curve in the shift fork to clear the rim of the gear.

4. Lock the transmission in two gears and remove the output flange nut, with the yoke and parking brake drum as an assembly.
   NOTE: The drum and yoke are balanced and unless replaced it is recommended that the drum and yoke be removed as a unit assembly.

5. Before removal and disassembly of the drive pinion and mainshaft measure the end play between the synchronizer outer stop ring and the 4th speed gear. Record the reading for reference during reassembly and shim as

necessary to obtain the ideal end play of .050-.070 inch.

6. Remove the drive pinion bearing retainer.

7. Remove the drive pinion assembly from the case while pulling on the shaft and tapping with a small hammer.

8. Remove the mainshaft rear bearing retainer and speedometer drive gear.

9. Using a brass hammer, tap the front of the mainshaft rearward to drive the rear bearing from its bore then using a puller remove the bearing from the mainshaft.

10. Remove the mainshaft assembly from the case by lifting the front end upward and forward until the 1st speed can pass through the notch areas in the case.

11. Remove the reverse idler lock screw and lock plate.

12. With a brass drift held at an angle drive the idler shaft to the rear and pull the shaft.

13. Lift the reverse idler gear from the shaft.

14. On the 542 transmission, push the bearing retainer including the needle bearings and radial thrust bearing out the back of the case.
    NOTE: On the 542 transmission,

the countershaft must be laid in the bottom of the case to make easier the reverse idler gear removal.

15. Remove the reverse gear shaft with the integral gear, sliding gear and thrust washer.

16. Push the caged front needle bearing out of the case.

17. Remove the countershaft front bearing cover and gasket.

18. To prevent the countershaft from turning, insert a hammer handle between the gear set and the case.

19. Remove the spiral roll pin, screw, retaining washer and C-pin.

20. After removing the gear bearing retainer cap screws, drive against the front end of the countershaft with a brass drift, driving through the front bearing toward the rear, until the countershaft rear bearing and retainer cap comes out of the case.
    NOTE: On the 542 transmission, the idler gear must be removed before the countershaft can be removed from the case.

21. Remove the bearing and cap from the countershaft and lift the countershaft from the case.

22. Remove the countershaft front bearing from the bore in the case by tapping the outer bearing race from inside the case.

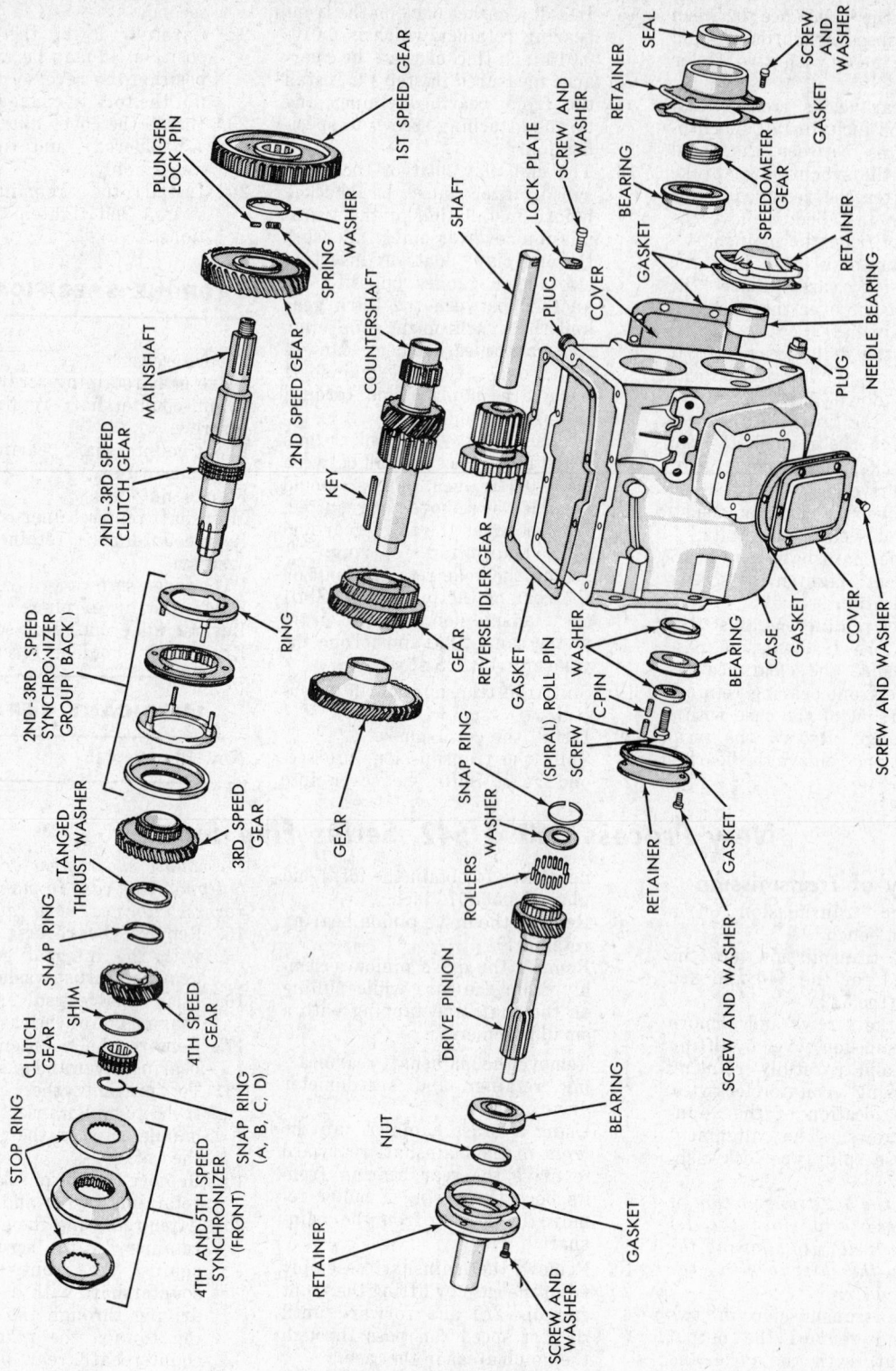

Labels in figure:

PLUNGER LOCK PIN
1ST SPEED GEAR
SHAFT
LOCK PLATE
SCREW AND WASHER
SEAL
RETAINER
SCREW AND WASHER
GASKET
BEARING
SPEEDOMETER GEAR
RETAINER
NEEDLE BEARING
PLUG
PLUG
COVER
GASKET
SPRING
WASHER
MAINSHAFT
2ND SPEED GEAR
2ND-3RD SPEED CLUTCH GEAR
COUNTERSHAFT
KEY
RING
2ND-3RD SPEED SYNCHRONIZER GROUP (BACK)
TANGED THRUST WASHER
SNAP RING
3RD SPEED GEAR
GEAR
REVERSE IDLER GEAR
GEAR
GASKET
(SPIRAL) ROLL PIN
WASHER
SCREW
C-PIN
WASHER
BEARING
CASE
GASKET
COVER
SCREW AND WASHER
RETAINER
GASKET
SCREW AND WASHER
CLUTCH GEAR
SHIM
4TH SPEED GEAR
SNAP RING (A, B, C, D)
STOP RING
4TH AND 5TH SPEED SYNCHRONIZER (FRONT)
SNAP RING
ROLLERS
WASHER
DRIVE PINION
NUT
BEARING
RETAINER
SCREW AND WASHER
GASKET

New Process 540 transmission

Disassembly of Sub-Assemblies

Mainshaft Disassembly

1. Remove the 1st speed gear from the mainshaft.
2. Remove the 2nd speed gear by depressing the plunger lock, and rotating the splined thrust washer. On the 542 remove the snap ring and thrust washer.

3. Remove the 2nd-3rd speed synchronizer unit.
4. Clamp the mainshaft in a soft jawed vise and remove the 4th and 5th speed synchronizer assembly, clutch gear snap ring and clutch gear.
5. Remove the 4th speed gear and shim.
6. Remove the 3rd speed gear snap ring and tanged washer. Remove the 3rd speed gear.

Assembly

1. Place the mainshaft with the forward end up in a soft jawed vise.
2. Place the 3rd speed gear on the shaft with the clutching teeth facing down. Install the one piece snap ring and thrust washer.
3. Place the 4th speed gear on the shaft with the clutching teeth

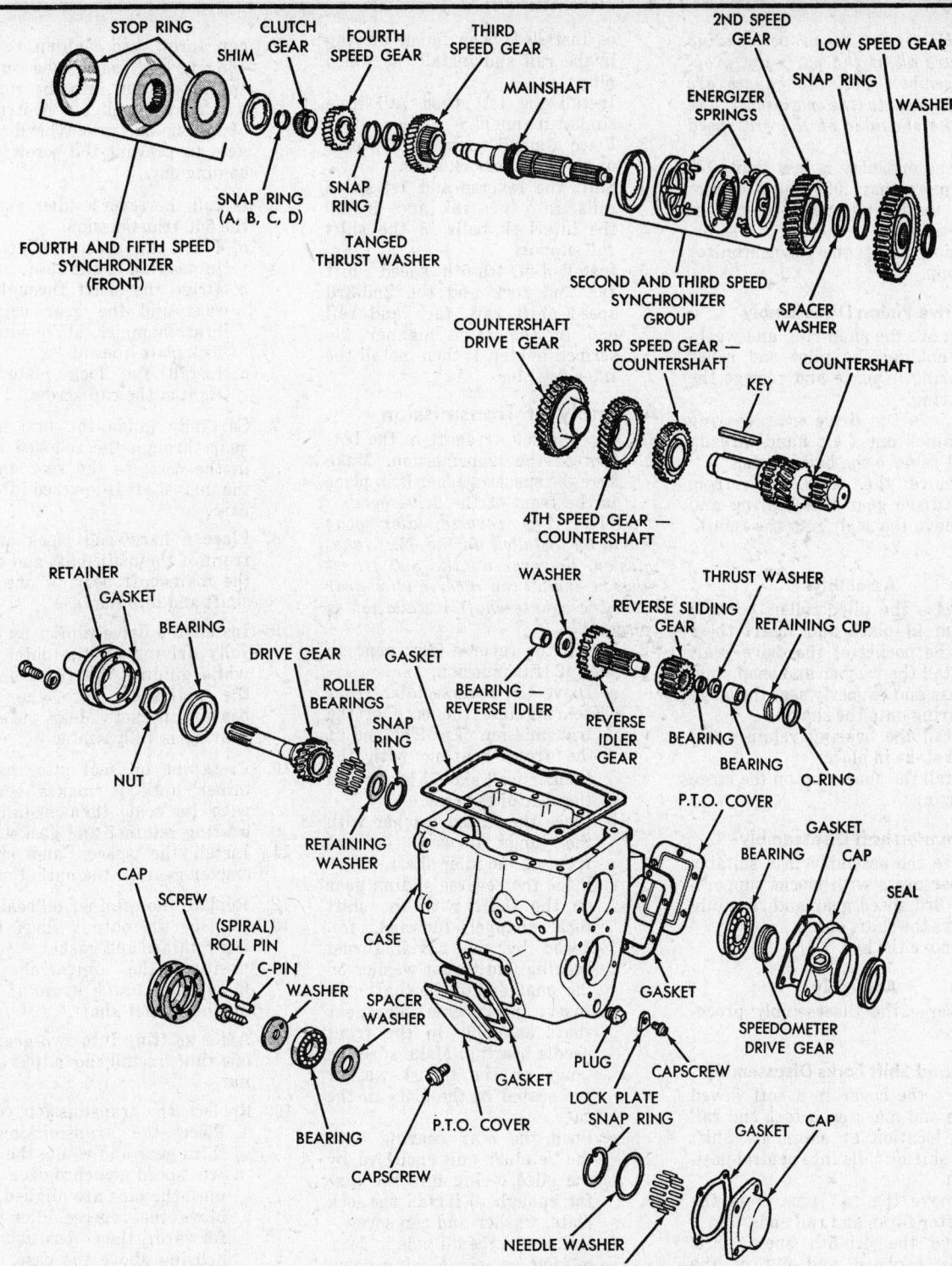

New Process 542 transmission

up. Refer to the end play dimension recorded earlier and select shims to provide .050-.070 inch end play between the gear and the front synchronizer.

4. Place the 4th-5th speed synchronizer clutch gear with the oil slots down on the mainshaft. Select a snap ring of the greatest possible thickness to eliminate all end play of the clutch gear.

5. Remove the mainshaft from the

vise and install the 2nd-3rd speed synchronizer group.

NOTE: the synchronizer sleeve is marked "FRONT" for proper installation.

6. Place the 2nd speed gear on the shaft. On the 540 lock in place by installing the plunger spring, plunger and splined washer. Push in the washer and lock by rotating until the splines are aligned. On the 542 install the thrust washer and snap ring.

Place the 1st speed gear on the shaft with fork groove facing the front end of the shaft.

7. Checking the end play float (.050-.090) at the rear synchronizer (2nd-3rd speeds) is mandatory and be performed during the mainshaft. This can be done by using two equal size feeler gauges diametrically opposite each other between the 3rd speed outer stop ring and the 3rd speed gear.

NOTE: To get the proper reading make sure all of the parts are properly assembled and the gauges are inserted close to the mainshaft and up on the shoulder of the 3rd speed gear.

8. If the end play is less than .070 or more than .090 shims cannot be used and new component parts must be used for the assembly of the synchronizer group.

Drive Pinion Disassembly

1. Remove the snap ring and washer holding the pilot and roller bearing in place and remove the bearing.
2. Remove the drive gear bearing retainer nut (left hand thread) and remove the ball bearing.
3. Remove the snap ring from the drive gear ball bearing and remove the seal from the retainer.

Assembly

1. Grease the pilot rollers to hold them in place and insert them in the pocket of the drive gear. Install the washer and snap ring.
2. Press and properly seat the large bearing onto the shaft.
3. Install the bearing retainer nut and stake in place.
4. Install the snap ring on the large bearing.

Countershaft Disassembly

1. Place the assembly in a suitable arbor press with blocks supporting 3rd speed gear and carefully press the shaft out.
2. Remove the key.

Assembly

1. Reverse the disassembly procedure.

Cover and Shift Forks Disassembly

1. Place the cover in a soft jawed vise and mark each fork and rail for location at assembly. Shift the shifter rails into neutral position.
2. Remove the roll pins from the shifter forks and rail ends.
3. Drive the 4th-5th speed shift rails forward and out of the cover, then, the remaining center (2nd-3rd) rail.
4. Drive out the reverse and 1st speed rails.
5. Remove the six interlock balls and two interlock pins from the shift rail support.

Assembly

1. Drive the reverse rail into the housing only far enough to install the reverse gate, poppet ball and spring. Continue to drive the rail through the support until the reverse fork can

be installed, then finish driving in the rail and install the welch plug.
2. Install the 1st speed rail in a similar manner.
3. Place a small quantity of grease on the six interlock balls.
4. Shift the reverse and 1st speed rails into neutral and install the interlock balls in the shift rail support.
5. Install the 4th-5th speed shift rail and fork and the 2nd-3rd speed shift rail, fork and rail end in the same manner described in step 1, then install the interlock pins.

Assembly of Transmission

1. Lay the countershaft in the bottom of the transmission. Make sure the spacer washer is in place in the front of the drive gear.

NOTE: The reverse idler gear should be installed on the 542 transmission, however on the 540 transmission, install the reverse idler gear after the countershaft installation is complete.

2. Install the reverse idler gear on the 542 transmission.
   a. Drive the reverse idler front bearing into the bore of the transmission case. The end of the front bearing with the thicker wall should be toward the rear of the case.
   b. Place the thrust washer with the tangs forward on the front of the idler shaft.
   c. Place the reverse sliding gear on the shaft with the shift fork channel forward, followed by the radial thrust bearing and thrust washer on the small end of the shaft.
   d. Insert the reverse idler gear shaft assembly in the front needle bearing. Make sure the tangs on the thrust washer are seated in the slots in the case.
   e. Push the rear bearing with the retainer cup encircled by the oiled o-ring into the case far enough to install the lock plate, washer and cap screw.
   f. Make sure the oil hole is fully in view as seen looking down into the case, and the lock plate is flat against the case.
3. With the countershaft front bearing journal protruding through the front bearing bore, install the front bearing.
4. Install the countershaft rear needle bearing on the 540 and the roller bearing on the 542. Install the gasket and retainer.
5. Install the front bearing retainer washer into position, with the large roll pin through it and in the corresponding hole in the countershaft. Install a $\frac{5}{8}$ inch

cap screw and tighten to 100-135 ft. lbs. Install the smaller spiral lock pin into the roll pin leaving the lock pin protruding about one-half screw head thickness to prevent the screw from coming out.
6. Install the reverse idler gear on the 540 transmission.
   a. Place the reverse idler gear in position in the case.
   b. Drive the shaft through the case and the gear using a brass hammer. Make sure the lock plate lines up.
   c. Install the lock plate and tighten the cap screw.
7. Carefully guide the first speed gear through the relieved areas in the case, as the rear end of the mainshaft is lowered into the case.
8. Place a hardwood block at the front of the mainshaft and drive the mainshaft bearing onto the shaft and into the case.
9. Install the drive pinion by carefully driving on the outer race while guiding the front end of the mainshaft into the pilot bearing pocket. Make sure the bearing is fully seated.
10. Press the oil seal into the retainer until it makes contact with its seat, then install the bearing retainer and gasket.
11. Install the spacer and speedometer gear on the output shaft.
12. Replace the retainer oil seal.
13. Replace the output shaft bearing retainer and gasket.
14. Position the universal joint flange and brake drum if used on the output shaft.
15. After shifting into two gears at one time, install the output shaft nut.
16. Replace the transmission cover.
    a. Place the transmission in third gear and rotate the 2nd-3rd speed synchronizer unit until the pins are aligned.
    b. Move the reverse idler gear forward, then, position the housing above the case.
    c. Lower the cover into position while guiding the reverse fork through the case and pass the synchronizer pins. Move the first speed gear slightly forward as necessary to engage the fork in the groove in the gear.
    d. Install the shouldered aligning screws and split lockwashers in the second hole from the front and tighten thumb tight.
    e. Install the remaining cover screws.

Torque Chart

Nomenclature	Nuts and/or Bolts and Torque Limits		Nomenclature	Nuts and/or Bolts and Torque Limits		Nomenclature	Nuts and/or Bolts and Torque Limits	
Bolt—gear shift lever tower to gearshift housing	3/8-16 20-25	7/16-14 30-35	Bolt—bellcrank to trans.	3/8-16 20-25		Nut—countershaft bearing lock (5-speed extra H.D. & 5-speed exclusive)	1¼-18 350-450	
Bolt—clutch housing to trans. case	7/16-14 30-38 5/8-11 96-120	9/16-12 70-90	Bolt—reverse lockout plunger retainer	11/16-16 80-100		Nut—countershaft bearing lock (5-speed exclusive H.D.)	1½-18 350-450	
Nut—U-joint flange to trans. output shaft	1.00-20 90-125	1½-18 275-350 1¼-18 225-275	Bolt—countershaft rear bearing retainer	5/16-18 25-30 3/8-16 35-40	7/16-14 45-55 1/2-13 60-70	Bolt—input shaft bearing retainer to trans. case	5/16-18 25-30 3/8-16 25-30	7/16-14 50-55
Nut—drum parking brake to companion flange	3/8-24 35-45	7/16-20 50-70	Bolt—countershaft & reverse idler shaft retainer	5/16-18 25-30 3/8-16 25-37 3/8-16† 18-25†	7/16-14 40-45 1/2-13 80-85	Bolt—countershaft front bearing retainer	5/16-18 25-30 3/8-16 25-35	7/16-14 40-45
Nut—bellcrank to trans.	9/16-18 70-90		Bolt—gear shift housing to trans. case	5/16-18 20-25 3/8-16 35-40	3/8-16†† 30-35†† 7/16-14 45-50			
Bolt—lever assy. to trans.	3/8-16 20-25		Bolt—power take off cover to trans. case	3/8-16 20-30				
Nut—handbrake anchor bar to trans. case (5-speed extra-heavy duty only)	9/16-18 120-130							

Lubricant Capacity

Model	Pints
540	9.5
542	9

New Process 7550 & 7590 Five Speed Transmissions

Disassembly of Transmission

1. Mount the transmission in a housing fixture and remove the parking brake assembly.
2. Shift the gears into neutral.
3. Remove the shift cover screws, the second screw from the front on each side is shouldered with split washers for installation alignment.
4. Lift the cover and turn slightly either way to clear the shift forks, remove the cover.
5. Lock the transmission in two gears at once and remove the out-put flange nut.
6. Remove the brake drum and yoke assembly by tapping with a brass hammer. *The drum and yoke assembly are balanced and unless replacement of parts is required, it is recommended that the drum and yoke be removed as a single unit.*
7. Remove the brake band assembly bracket and support bolts and lock washers. Remove the brake band assembly as a complete unit.
8. Before removal and disassembly of the drive pinion and mainshaft, measure the end-play between the synchronizer stop ring and the fourth gear. *Record this measurement for reference during assembly.* The correct end-play is 0.050-0.070 inch. Note any difference from the limits so that corrective shims can be added or removed as required.
9. Remove the drive pinion bearing retainer.
10. Remove the drive pinion gear and ball bearing from the trans-

mission by pulling on the shaft and rapping the face of the case with a brass hammer.
11. Remove the speedometer drive pinion and the mainshaft rear bearing retainer.
12. Place a brass drift in the front center of the mainshaft and drive the mainshaft to the rear.
13. Pull the rear bearing from the mainshaft with a suitable gear puller.
14. Once the mainshaft rear bearing has cleared the case, remove the rear bearing and the speedometer gear with a suitable puller.
15. Move the mainshaft assembly to the rear and tilt the front of the mainshaft up.
16. Hold the first-reverse gear and the fourth-fifth synchronizer to keep them from sliding off the shaft, remove the mainshaft from the transmission case.
17. Remove the reverse idler lock screw and lock plate.
18. Using a brass drift, held at an angle, drive the idler shaft to the rear while pulling.

19. Lift the reverse idler gear and thrust washers out of the case.
NOTE: Loose needle bearings are usually replaced. Never mix old and new bearings.
20. Remove the countershaft rear bearing retainer, gasket and bearing.
21. Tip the countershaft upward and remove it from the case. Remove the thrust washer from the front end of the countershaft.
22. Remove the countershaft front needle bearing from the case bore by tapping on the bearing cage from inside the case, with a suitable driver.

Overhaul of Sub-Assemblies

Drive Pinion Disassembly

1. Remove the snap ring and washer holding the pilot needle bearings in place and remove the bearings.
2. Relieve the staked area, remove the drive pinion ball bearing retainer nut and remove the ball

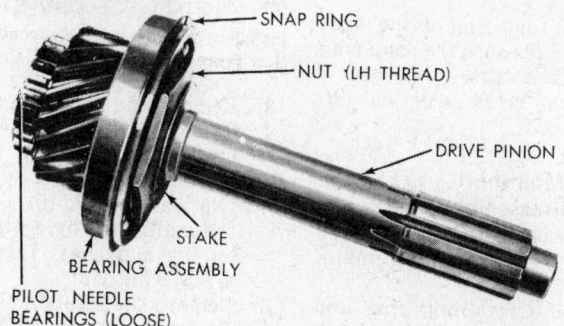

SNAP RING — NUT (LH THREAD) — DRIVE PINION — STAKE — BEARING ASSEMBLY — PILOT NEEDLE BEARINGS (LOOSE)

Drive pinion shaft assembly—New Process 7550-7590 (© Chrysler Corp.)

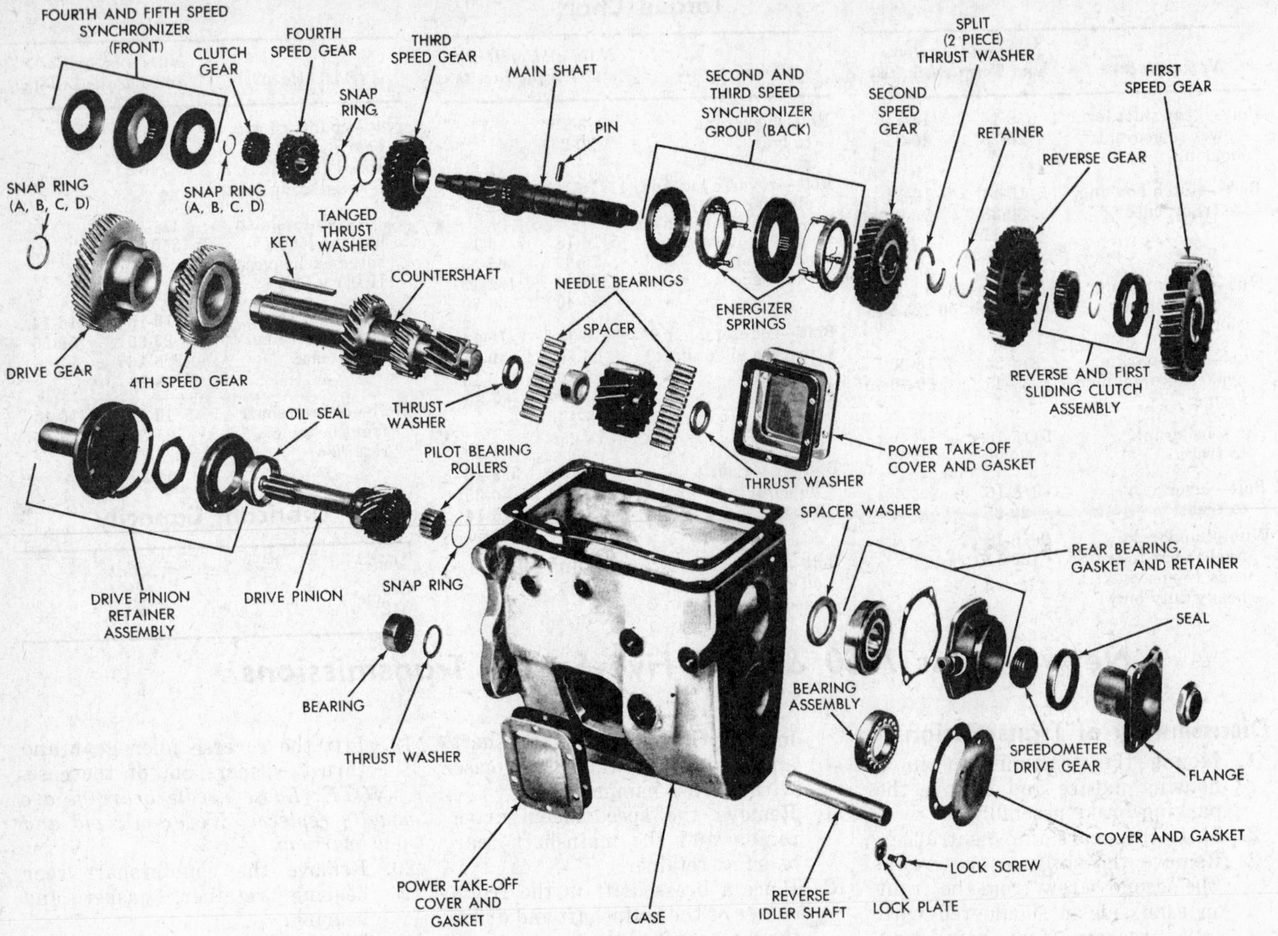

FOURTH AND FIFTH SPEED SYNCHRONIZER (FRONT)

CLUTCH GEAR

FOURTH SPEED GEAR

THIRD SPEED GEAR

MAIN SHAFT

SNAP RING

PIN

SECOND AND THIRD SPEED SYNCHRONIZER GROUP (BACK)

SPLIT (2 PIECE) THRUST WASHER

SECOND SPEED GEAR

RETAINER

FIRST SPEED GEAR

REVERSE GEAR

SNAP RING (A, B, C, D)

SNAP RING (A, B, C, D)

TANGED THRUST WASHER

KEY

COUNTERSHAFT

NEEDLE BEARINGS

SPACER

ENERGIZER SPRINGS

DRIVE GEAR

4TH SPEED GEAR

OIL SEAL

THRUST WASHER

PILOT BEARING ROLLERS

THRUST WASHER

POWER TAKE-OFF COVER AND GASKET

REVERSE AND FIRST SLIDING CLUTCH ASSEMBLY

SPACER WASHER

REAR BEARING, GASKET AND RETAINER

SEAL

DRIVE PINION RETAINER ASSEMBLY

DRIVE PINION

SNAP RING

BEARING

THRUST WASHER

BEARING ASSEMBLY

SPEEDOMETER DRIVE GEAR

FLANGE

COVER AND GASKET

POWER TAKE-OFF COVER AND GASKET

CASE

REVERSE IDLER SHAFT

LOCK PLATE

LOCK SCREW

New Process 7550 & 7590 transmission—exploded view (© Chrysler Corp.)

bearing. *The ball bearing retainer nut has left hand threads.*

3. Remove the snap ring from the drive pinion ball bearing.
4. Remove the seal from the drive pinion bearing retainer.

Assembly

1. Grease the loose pilot bearings to hold them in place and insert them into the pocket of the drive gear. Install the washer and snap ring.
2. Press the large bearing on the pinion shaft. Make sure the bearing is properly seated.
3. Install the bearing retainer nut and tighten securely. Stake in place.
4. Install the snap ring on the large bearing. Make sure the snap ring is properly seated.
   The bearing retainer nut has left hand threads.

Mainshaft Disassembly

1. Remove the spacer washer and the first gear from the mainshaft.
2. Remove the retaining ring and the first-reverse clutch and clutch gear assembly.

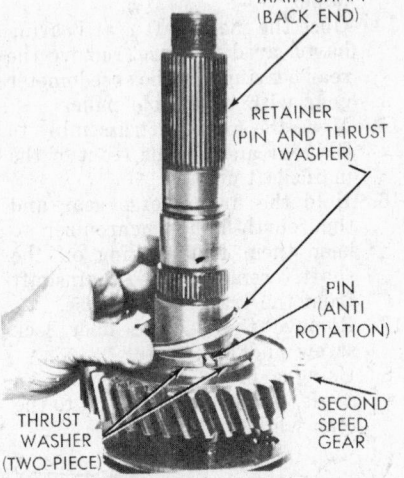

MAIN SHAFT (BACK END)

RETAINER (PIN AND THRUST WASHER)

PIN (ANTI ROTATION)

SECOND SPEED GEAR

THRUST WASHER (TWO-PIECE)

Installing or removing the second gear— New Process 7550-7590 (© Chrysler Corp.)

3. Remove the reverse gear.
4. Remove the second gear retaining thrust washer. This two-piece split washer consists of two halves held in position on the mainshaft by a pin in a hole on the mainshaft together with a retaining ring.
5. Remove the second gear.
6. Remove the second-third synchronizer assembly. The second-

third clutch gear is integral with the mainshaft.

7. Remove the snap ring and the four-fifth synchronizer assembly and clutch gear.
8. Remove the fourth gear, retain the shims for assembly.
9. Remove the retaining snap ring, tanged thrust washer and the third gear.

Assembly

1. Place the mainshaft, front end up, in a soft-jawed vise.
2. Place the third gear on the shaft with the clutching teeth facing down. Install the tanged thrust washer and the one piece snap ring.
3. Place the fourth gear on the shaft with the clutching teeth up.
4. Check the end-play measurements recorded during disassembly and select shims to provide 0.050-0.070 inch end play between the fourth gear and the fourth-fifth synchronizer.
5. Place the fourth-fifth synchronizer clutch gear, with the oil slots down, on the mainshaft. Select a snap ring of the greatest possible thickness to eliminate all end-play of the clutch gear.

6. Remove the mainshaft from the vise and install the second-third synchronizer group. The synchronizer sleeve is marked "Front" for proper installation.
7. Place the second speed gear on the mainshaft.
8. Place the thrust washer retaining pin in the hole in the mainshaft and position the two thrust washer halves. Install the thrust washer retaining ring with the large diameter contacting the second gear.
9. Install the reverse gear.
10. Position the reverse-first clutch gear on the shaft. Install the retaining snap ring. Select a snap ring with the greatest possible thickness to eliminate all end-play of the clutch gear.
11. Position the sliding clutch on the clutch gear. Install the first gear on the mainshaft.
12. Place the spacer washer on the mainshaft.

Countershaft

Note: It is only necessary to disassemble the countershaft if inspection shows signs of damage or malfunction.

Disassembly

1. Remove the snap ring.
2. Place the assembly in an arbor press with a block supporting the drive gear. Carefully press the shaft out.
3. Support the fourth gear with wood blocks and carefully drive the shaft out.
4. Remove the key.

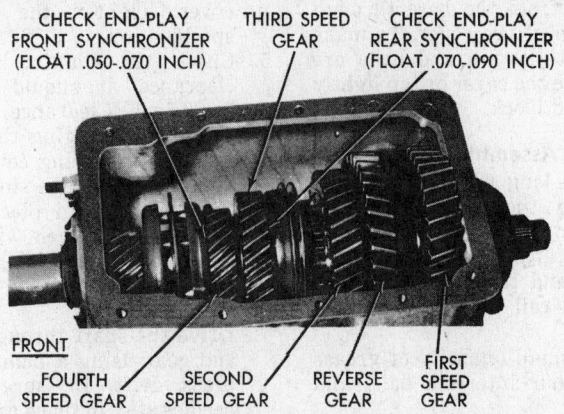

CHECK END-PLAY FRONT SYNCHRONIZER (FLOAT .050-.070 INCH) THIRD SPEED GEAR CHECK END-PLAY REAR SYNCHRONIZER (FLOAT .070-.090 INCH)

FRONT

FOURTH SPEED GEAR   SECOND SPEED GEAR   REVERSE GEAR   FIRST SPEED GEAR

Synchronizer end-play "float" check points—New Process 7550-7590 (© Chrysler Corp.)

Assembly

1. Place the key in position on the countershaft.
2. Press the gears on the countershaft until properly seated.

Note: Make sure that the key does not move out of position as the gears are being pressed on the shaft.

3. Install the snap ring on the countershaft. Select a snap ring with the greatest possible thickness to eliminate all possible end-play.
4. Install the washer on the countershaft drive gear.

Cover and Shift Fork Assembly

Note: The cover and shift fork assembly should be disassembled ONLY if inspection shows worn or damaged parts, or if the assembly is not working properly.

Disassembly

1. Mount the cover assembly in a soft-jawed vise. Mark each fork and rail for location during assembly.
2. Place the shift forks in neutral. Remove the spiral roll pins from the shift forks and lugs. The roll pins may be removed by working with a "easy-out" mounted in a tap handle.
3. Drive the short first-reverse shift rail toward the rear and out of the shift cover. Remove the fork.
4. Remove the first-reverse shift rail pivot bolt and nut. Remove the crossover level.
5. Drive the fourth-fifth and the second-third shift rail forward and out of the shift cover. Remove the long first-reverse shift rail in the same manner.

Note: Place a cloth over the shift rails while driving the shift rails out of the cover to prevent the poppet balls and springs from flying out.

6. Remove the four interlock balls

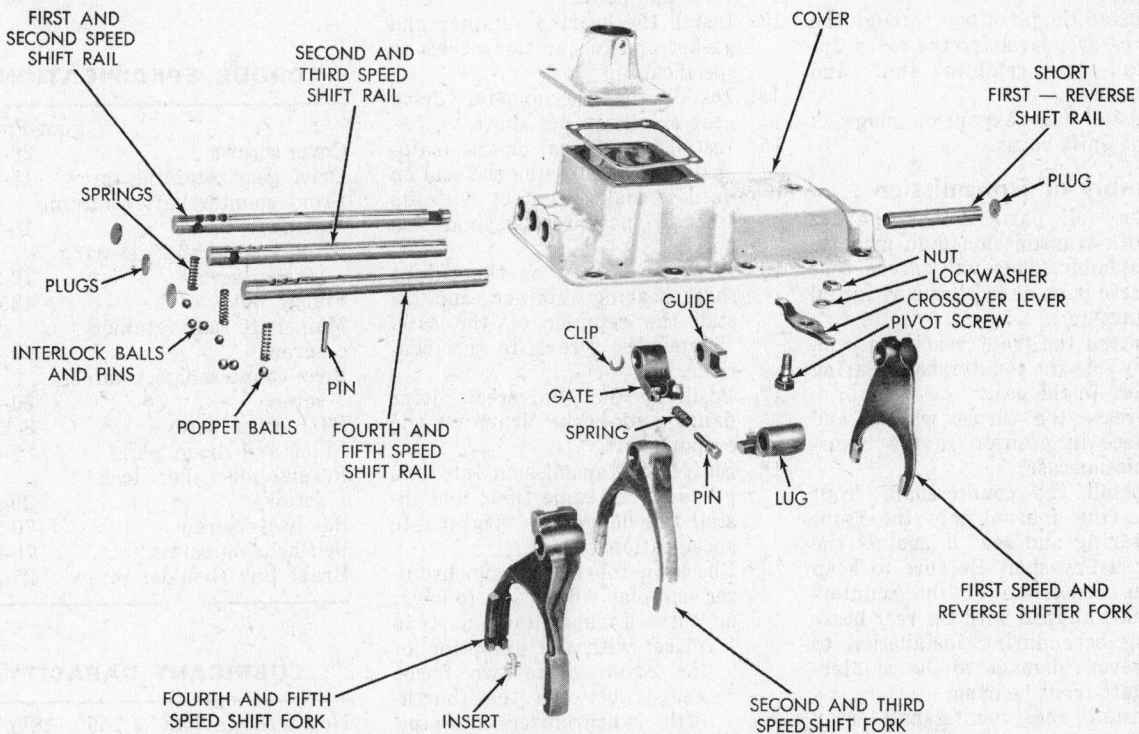

FIRST AND SECOND SPEED SHIFT RAIL

SECOND AND THIRD SPEED SHIFT RAIL

COVER

SHORT FIRST — REVERSE SHIFT RAIL

SPRINGS

PLUGS

INTERLOCK BALLS AND PINS

POPPET BALLS   FOURTH AND FIFTH SPEED SHIFT RAIL

PIN

PLUG

NUT
LOCKWASHER
CROSSOVER LEVER
PIVOT SCREW

GUIDE

CLIP

GATE

SPRING

PIN   LUG

FIRST SPEED AND REVERSE SHIFTER FORK

FOURTH AND FIFTH SPEED SHIFT FORK   INSERT   SECOND AND THIRD SPEED SHIFT FORK

Cover and shift fork assembly—exploded view—New Process 7550-7590 (© Chrysler Corp.)

and pin from the bore through the width of the cover. To make sure that the pin and balls are out, shake the cover or tap lightly on a wood block.

Assembly

1. Push the long first-reverse shift rail into the cover bore far enough to permit the installation of the gates, poppet ball and springs and the roll pins. Move the shift rail into neutral position.
2. Place a small quantity of grease on the four interlock balls and pin.
3. Place two balls in the interlock bore. Move both of the balls toward the shift rail to seat the ball in the neutral notch.
4. Grease the interlock pin and place it in the hole located in the second-third shift rail. Install the second-third shift rail, gate, and fork as outlined in step 1. Move the rail into neutral.
5. Install the remaining balls into the interlock bore.
6. Push the fourth-fifth shift rail into the cover bore and install the shift fork as outlined in step 1. Move the shift rail into neutral.
7. Place the crossover lever in position in the cover in such a way that the short first-reverse shift rail, fork, and roll pin can be installed. Install the short first-reverse shift rail and parts.
8. Reposition the crossover level to mate in the notches in both the long and short first-reverse shift rails.
9. Install the pivot bolt through the crossover level and the cover. Install the retaining nut and washer.
10. Install new expansion plugs in the shift cover.

Assembly of Transmission

1. Coat all parts and assemblies with transmission fluid prior to assembly. This will insure that there is no damage during initial start-up.
2. Install the front bearing assembly into the countershaft bearing bore in the case.
3. Grease the thrust washer and place in position in the transmission case.
4. Install the countershaft front bearing journal into the front bearing and seat it against the thrust washer. Be sure to keep the centerline of the countershaft aligned with the rear bearing bore during installation to prevent damage to the countershaft front bearing.
4a. Install the countershaft rear bearing assembly, gasket, and

cover. Tighten the screws to specification.
5. Check the counter shaft end clearance, it should be 0.008-0.020 inch. Clearance can be adjusted by changing the countershaft rear bearing cover gasket.
6. With the reverse idler needle bearings held in place on each side of the spacer with grease, place the reverse idler gear and thrust washer in position in the case.
7. Drive the shaft through the case and gear using a hammer and a brass drift. Be sure that the needles stay in place and that the lock strap slot in the shaft will line up so that the lock strap, cap, and cap screw can be installed.
8. Install the lock strap on the shaft and tighten the cap screw securely.
9. Carefully lower the rear end of the mainshaft into the case while holding the first gear and washer on the shaft.
10. Place a hardwood block at the front of the mainshaft and drive the mainshaft bearing onto the shaft and into the case.
11. Install the drive pinion by carefully driving on the bearing outer race, forcing it into the case while guiding the front end of the mainshaft into the pilot bearing pocket. Make sure the bearing is fully seated.
12. Replace the retainer oil seal, pressing the seal into the retainer until the seal makes contact with its seat. Do not press beyond this point.
13. Install the bearing retainer and gasket and torque the screws to specification.
14. Install the speedometer drive gear on the output shaft.
15. Install the oil seal on the mainshaft flange, pressing the seal on until it makes contact with its seat. Do not press beyond this point.
16. Place the gasket on the output shaft bearing retainer, and install the retainer on the case. Tighten the screws to specifications.
17. Position the universal joint flange and brake drum on the output shaft.
18. Shift the transmission into two gears at the same time, and install the flange nut. Tighten to specifications.
19. Check the fourth-fifth synchronizer end-play "float" as follows:
    a. With all transmission parts in place, with the exception of the cover, place two feeler gauges between the fourth-fifth synchronizer and the stop ring. End-play should

be 0.050-0.070 inch.
    b. If end-play is not within limits, shims should be added or removed, as required, from between the fourth-fifth synchronizer and the stop ring.
    c. Reassemble and recheck the end-play.
20. Check the second-third synchronizer end-play "float" as follows:
    a. With all transmission parts in place, with the exception of the cover, place two feeler gauges between the second-third synchronizer and the outer stop ring. End-play should be 0.070-0.090 inch.
    b. If the end-play is not within limits, install new parts as required. *Shims cannot be used at this point.*
    c. Reassemble and recheck the end-play.
21. Place the transmission gears and the shift cover in neutral.
22. Position the cover gasket on the transmission.
23. Carefully lower the cover into position on the transmission. Make sure that all shift forks engage their grooves correctly.
24. Install the shouldered aligning screws and split washers, in the second hole from the front, and tighten finger tight.
25. Install the remaining screws and tighten all screws to specification.
26. Shift the transmission through all gear ranges to be sure that the transmission is working properly.

TORQUE SPECIFICATIONS

	Foot-Pounds
Cover screws	20-40
Drive gear retaining screw	15-25
Front countershaft retaining screw	15-25
Front countershaft bearing washer screw	12-22
Flange nut	125-175
Mainshaft rear retainer screw	20-40
Rear countershaft retainer screw	20-40
PTO cover screws	8-12
Filler and drain plugs	25-45
Reverse idler shaft lock screw	20-40
Bar brake screw	70-110
Bell housing screw	70-110
Brake link shoulder screw	25-45

LUBRICANT CAPACITY

New Process 7550 & 7590   18¼ pts.

Saginaw Three Speed (GM-SM326)

Description

The G.M. Corporation Model SM326 (Saginaw) is a synchromesh three-speed transmission using helical constant mesh gears. The engagement of all gears except reverse is assisted by synchronizers.

General Data

Type   3-Speed

Synchromesh Gears 1st, 2nd, and 3rd

Model SM326 and SM326 w/Overdrive

Gear Ratios

1st Speed	2.85:1
2nd Speed	1.68:1
3rd Speed	1.00:1
Reverse	2.95:1

Transmission Disassembly

1. Remove side cover assembly and shift forks.
2. Remove clutch gear bearing retainer.
3. Remove clutch gear bearing to gear stem snap-ring. Pull clutch gear outward until a screwdriver can be inserted between bearing and case. Remove clutch gear bearing.
4. Remove speedometer driven gear and extension bolts.
5. Remove reverse idler shaft snap-ring. Slide reverse idler gear forward on shaft.
6. Remove mainshaft and extension assembly.
7. Remove clutch gear and third-speed blocker ring from inside case. Remove 14 roller bearings from clutch gear.
8. Expand the snap-ring which retains the mainshaft rear bearing. Remove the extension.
9. Usnig a dummy shaft, drive the countershaft and key out the rear of the case. Remove the gear, two tanged thrust washers, and dummy shaft. Remove bearing washer and 27 roller bearings from each end of countergear.
10. Use a long drift to drive the reverse idler shaft and key through the rear of the case.
11. Remove reverse idler gear and tanged steel thrust washer.

Mainshaft Disassembly

1. Remove second and third speed sliding clutch hug snap-ring from mainshaft. Remove clutch assembly, second speed blocker ring, and second speed gear from front of mainshaft.
2. Depress speedometer drive gear retaining clip. Remove gear. Some units have a metal speedometer drive gear which must be pulled off.
3. Remove rear bearing snap-ring.
4. Support reverse gear. Press on rear of mainshaft. Remove reverse gear, thrust washer, spring washer, rear bearing, and snap-ring. When pressing off the rear bearing, be careful not to cock the bearing on the shaft.
5. Remove first and reverse sliding clutch hub snap-ring. Remove clutch assembly, first speed blocker ring, and first gear.

Cleaning and Inspection

See "Cleaning and Inspection" instructions at front of transmission section.

Clutch Keys and Springs

Keys and springs may be replaced if worn or broken, but the hubs and sleeves are matched pairs and must be kept together.

1. Mark hub and sleeve for reassembly.
2. Push hub from sleeve. Remove keys and springs.
3. Place three keys and two springs, one on each side of hub, in position, so all three keys are engaged by both springs. The tanged end of the springs should not be installed into the same key.
4. Slide the sleeve onto the hub, aligning the marks.

NOTE: A groove around the outside of the synchronizer hub marks the end that must be opposite the fork slot in the sleeve when assembled.

Extension Oil Seal and Bushing

1. Remove seal.
2. Using bushing remover and installer tool, or other suitable tool, drive bushing into extension housing.
3. Drive new bushing in from the rear. Lubricate inside of bushing and seal. Install new oil seal with extension seal installer tool or other suitable tool.

Clutch Bearing Retainer Oil Seal

1. Pry old seal out.
2. Install new seal using seal installer or suitable tool. Seat seal in bore.

Mainshaft Assembly

1. Turn front of mainshaft up.
2. Install second gear with clutching teeth up; the rear face of the gear butts against the flange on the mainshaft.
3. Install a blocker ring with clutching teeth down. All three blocker rings are the same.
4. Install second and third speed synchronizer assembly with fork slot down. Press it onto mainshaft splines. Both synchronizer assemblies are the same. Be sure that blocker ring notches align with synchronizer assembly keys.
5. Install synchronizer snap-ring. Both synchronizer snap-rings are the same.
6. Turn rear of shaft up.
7. Install first gear with clutching teeth up; the front face of the gear butts against the flange on the mainshaft.
8. Install a blocker ring with clutching teeth down.
9. Install first and reverse synchronizer assembly with fork slot down. Press it onto mainshaft splines. Be sure blocker ring notches align with synchronizer assembly keys.
10. Install snap-ring.
11. Install reverse gear with clutching teeth down.
12. Install steel reverse gear thrust washer and spring washer.
13. Press rear ball bearing onto shaft with snap-ring slot down.
14. Install snap-ring.
15. Install speedometer drive gear and retaining clip. Press on metal speedometer drive gear.

Transmission Assembly

1. Using dummy shaft load a row of 27 roller bearings and a thrust washer at each end of countergear. Hold in place with grease.
2. Place countergear assembly into case through rear. Place a tanged thrust washer, tang away from gear at each end. Install countershaft and key, making sure that tangs align with notches in case.
3. Install reverse idler gear thrust washer, gear, and shaft with key from rear of case. Be sure thrust washer is between gear and rear of case with tang toward notch in case. *NOTE: The reverse idler gear bushing may not be replaced separately—only as a unit with the gear.*
4. Expand snap-ring in extension. Assemble extension over rear of mainshaft and onto rear bearing. Seat snap-ring in rear bearing groove.
5. Install 14 mainshaft pilot bearings into clutch gear cavity. Assemble third speed blocker ring onto clutch gear clutching surface with teeth toward gear.
6. Place clutch gear, pilot bearings, and third speed blocker ring assembly over front of mainshaft assembly. Be sure blocker rings

1. Thrust washer—front
2. Bearing washer
3. Needle bearings
4. Countergear
5. Needle bearings
6. Bearing washer
7. Thrust washer—rear
8. Countershaft
9. Woodruff key
10. Bearing retainer
11. Gasket
12. Oil seal
13. Snap ring—bearing to case
14. Snap ring—bearing to gear
15. Clutch gear bearing
16. Case
17. Clutch gear
18. Pilot bearings
19. 3rd speed blocker ring
20. Retainer E-Ring
21. Reverse idler gear
22. Reverse idler gear bushing (not serviced separately)
23. Reverse idler shaft
24. Woodruff key
25. Snap ring—hub to shaft
26. 2-3 synchronizer sleeve
27. Synchronizer key spring
28. 2-3 Synchronizer hub assy.
29. 2nd speed blocker ring
30. 2nd speed gear
31. Mainshaft
32. 1st speed gear
33. 1st speed blocker ring
34. 1st and reverse synchronizer hub assembly
35. 1st and reverse synchronizer sleeve
36. Snap ring—hub to shaft
37. Reverse gear assy.
38. Thrust washer
39. Thrust washer
40. Rear bearing
41. Snap ring—bearing to shaft
42. Speedometer drive gear
43. Gasket
44. Snap ring—rear bearing to extension
45. Extension

46. Oil seal
47. Gasket
48. 2-3 shift fork
49. 1st and reverse shift fork
50. 2-3 shifter shaft assembly
51. 1st and reverse shifter shaft assembly

52. O-ring seal
53. Detent cam retainer ring
54. Spring
55. 2nd and 3rd detent cam
56. 1st and reverse detent cam
57. Side cover

SM326 transmission components (© G.M.C.)

align with keys in second-third synchronizer assembly.

7. Stick extension gasket to case with grease. Install clutch gear, mainshaft, and extension together. Be sure clutch gear engages teeth of countergear anti-lash plate. Torque extension bolts to 45 ft. lbs.

8. Place bearing over stem of clutch gear and into front case bore. Install front bearing to

clutch gear snap-ring.

9. Install clutch gear bearing retainer and gasket. The retainer oil return hole must be at the bottom. Torque to 10 ft. lbs.

10. Install reverse idler gear shaft E-ring.

11. Shift synchronizer sleeves to neutral positions. Install cover, gasket, and forks, aligning forks with synchronizer sleeve grooves. Torque side cover bolts

to 10 ft. lbs.

12. Install speedometer driven gear.

TORQUE SPECIFICATIONS
Foot-Pounds

Extension to case attaching bolts	35-55
Drain and filler plugs	10-15
Side cover attaching bolts	18-24
Clutch gear retainer bolts	18-24

Saginaw Four Speed

Disassembly of Transmission

1. Remove the side cover bolts and the side cover assembly.
2. Remove the drive gear bearing retainer.
3. Remove the drive gear to gear stem snap ring. Remove the drive gear bearing by pulling outward on the gear until a screwdriver can be inserted between the bearing and the large snap ring. The drive gear bearing is a slip fit on the gear and into the case bore. This provides clearance between

the case bore and the shaft for removal of the drive gear and the mainshaft assembly.

4. Remove the extension housing bolts.
5. Remove the drive gear, mainshaft and the extension housing assembly through the rear case opening. Remove the drive gear and bearing from the mainshaft.
6. Using snap ring pliers, expand the snap ring in the extension housing which holds the mainshaft rear bearing and remove

the housing.

7. Using a suitable size drift at the front of the countershaft, drive the countershaft and the woodruff key out the rear of the case. The drift will hold the countergear bearings in place. Remove the gear and thrust washers.
8. Remove the reverse idler gear stop ring. Use a long drift or punch through the front bearing case bore and drive the reverse idler shaft and woodruff key out of the rear of the case.

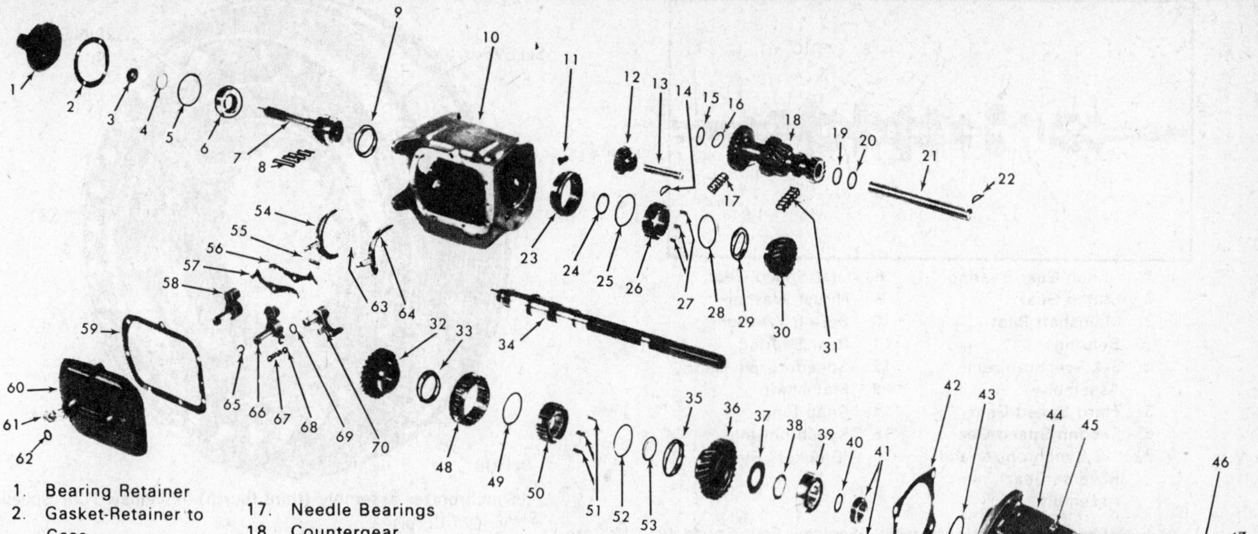

Saginaw Four Speed Transmission—exploded view (© GM Corp.)

1. Bearing Retainer
2. Gasket-Retainer to Case
3. Oil Seal
4. Snap Ring-Bearing to Shaft
5. Snap Ring-Bearing to Case
6. Drive Gear Bearing
7. Drive Gear
8. Mainshaft Pilot Bearings
9. 4th Speed Blocker Ring
10. Case
11. Filler Plug
12. Reverse Idler Gear
13. Reverse Idler Shaft
14. Woodruff Key
15. Thrust Washer-Front Gear
16. Needle Retainer Washer
17. Needle Bearings
18. Countergear
19. Needle Retainer Washer
20. Thrust Washer-Rear Gear
21. Countershaft
22. Woodruff Key
23. Synchronizer Sleeve
24. Snap Ring-Hub to Shaft
25. Key Retainer
26. 3-4 Synchronizer Hub
27. Clutch Keys
28. Key Retainer
29. 3rd Speed Blocker Ring
30. 3rd Speed Gear
31. Needle Bearings
32. Second Speed Gear
33. 2nd Speed Blocker Ring
34. Mainshaft
35. 1st Speed Blocker Ring
36. First Speed Gear
37. Thrust Washer
38. Wave Washer
39. Rear Bearing
40. Snap Ring-Bearing to Shaft
41. Speedo Drive Gear and Clip
42. Gasket-Extension to Case
43. Snap Ring-Extension to Rear Bearing
44. Extension
45. Vent
46. Bushing
47. Oil Seal
48. 1-2 Synchronizer Sleeve and Reverse Gear
49. Key Retainer
50. 1-2 Synchronizer Hub
51. Clutch Keys
52. Key Retainer
53. Snap Ring-Hub to Shaft
54. 3-4 Shift Fork
55. Detent Spring
56. 3-4 Detent Cam
57. 1-2 Detent Cam
58. 3-4 Shifter Shaft
59. Gasket-Cover to Case
60. Cover
61. TCS Switch and Gasket
62. Lipseal
63. Detent Cam Retainer
64. 1-2 Shift Fork
65. "O" Ring
66. 1-2 Shift Shaft
67. Spring
68. Ball
69. "O" Ring
70. Reverse Shifter Shaft and Fork

Overhaul of Sub-Assemblies

Cleaning and Inspection
Transmission Case

1. Wash the transmission case thoroughly with clean solvent. Inspect the case for cracks.
2. Check the front and rear faces for burrs. If burrs are present, dress them off with a mill file.

Front and Rear Bearings

1. Wash the front and rear bearings with clean solvent.
2. Blow the bearings out with compressed air.

NOTE: Do not allow the bearings to spin. Turn them slowly by hand. Spinning bearings will damage the race and balls.

3. Make sure that the bearings are clean. Lubricate the bearings with light engine oil and check them for roughness by turning the race slowly by hand.

Bearing Rollers

All clutch gear and countergear bearing rollers should be closely inspected and replaced if they show signs of wear. Inspect the countershaft and the reverse idler shaft, re-

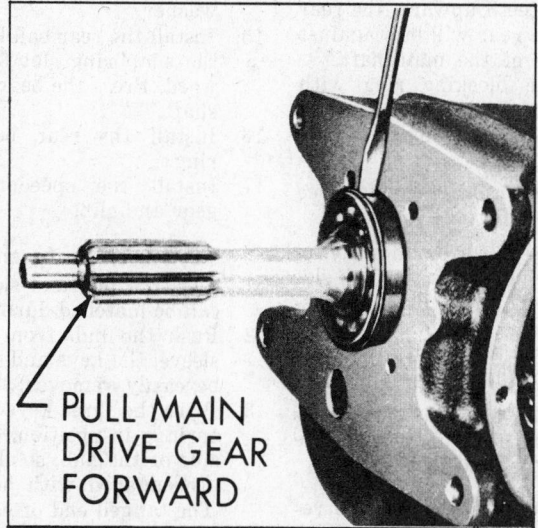

Removing the drive gear bearing—Saginaw Four Speed (© GM Corp.)

place if damaged. Replace all worn washers.

Gears

1. Inspect all gears for excessive wear, chips, or cracks. Replace as necessary.
2. Check both clutch sleeves to see that they slide freely on their hubs.

Mainshaft
Disassembly

1. Using snap ring pliers, remove the third-fourth sliding clutch hub snap ring from the mainshaft. Remove the clutch assembly, third gear blocking ring, and the third gear from the mainshaft.

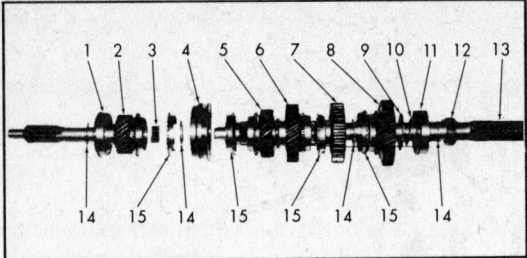

1. Clutch Gear Bearing
2. Clutch Gear
3. Mainshaft Pilot Bearings
4. 3-4 Synchronizer Assembly
5. Third Speed Gear
6. Second Speed Gear
7. 1-2 Synchronizer and Reverse Gear Assembly
8. First Speed Gear
9. Thrust Washer
10. Spring Washer
11. Rear Bearing
12. Speedo Drive Gear
13. Mainshaft
14. Snap Ring
15. Synchronizing "Blocker" Ring

Mainshaft and clutch gear—exploded view—Saginaw Four Speed (© GM Corp.)

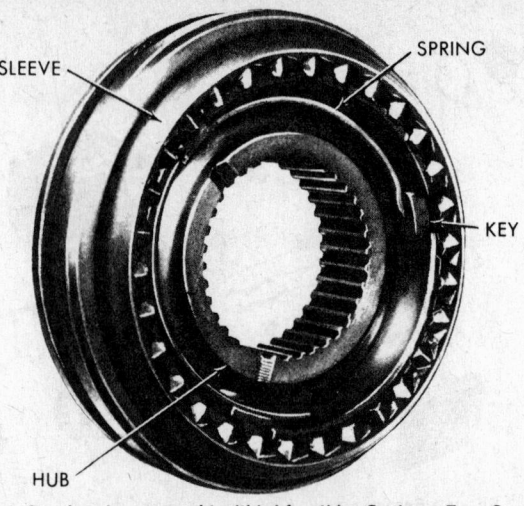

Synchronizer assembly (third-fourth)—Saginaw Four Speed (© GM Corp.)

2. Depress the speedometer retaining clip and slide the gear from the shaft.
3. Remove the rear bearing snap ring from the shaft groove.
4. Support the first gear with press plates and press on the shaft to remove the first gear, thrust washer, and rear bearing.
5. Remove the first-second sliding clutch hub snap ring from the shaft and remove the clutch assembly, the second gear blocking ring, and the second gear.

Assembly

1. Position the mainshaft with the front end facing upward.
2. Install the third gear with the clutching teeth upward, the rear face of the gear will butt against the flange of the mainshaft.
3. Install the blocking ring with the clutching teeth downward over the synchronizing surface of the third gear.

NOTE: All four blocking rings used in the transmission are identical.

4. Install the third-fourth synchronizer assembly with the fork slot facing downward. Push the assembly onto the splines on the mainshaft until it bottoms against the flange.

CAUTION: Be sure that the notches of the blocking ring align with the keys on the synchronizer assembly.

5. Install the synchronizer hub retaining snap ring on the mainshaft.
6. Position the mainshaft with the rear end facing uward.
7. Install the second gear with the clutching teeth facing upward. The front face of the gear will butt against the flange of the mainshaft.
8. Install the blocker ring with the clutching teeth downward over

the synchronizing surface of the second speed gear.
9. Install the first-second synchronizer assembly with the fork slot facing downward.

NOTE: Be sure the notches of the blocking ring align with the keys of the synchronizer assembly.

10. Install the synchronizer retaining snap ring on the mainshaft.
11. Install a blocker ring with the notches downward so that they align with the keys on the first-second synchronizer assembly.
12. Install the first gear with the clutching teeth facing downward.
13. Install the first gear steel thrust washer.
14. Install the first gear spring washer.
15. Install the rear ball bearing with the snap ring slot facing downward. Press the bearing onto the shaft.
16. Install the rear bearing snap ring.
17. Install the speedometer drive gear and clip.

Synchronizer Assemblies

1. Mark the hub and sleeve so they can be matched during assembly.
2. Push the hub from the sliding sleeve, the keys and springs may be easily removed.
3. Place the three keys and the two springs in position, one on each side of the hub, so all three keys are engaged with both springs. The tanged end of each synchronizer spring should be installed into different key cavities on either side. Slide the sleeve onto the hub, aligning the marks made before disassembly.

Note: A groove around the outside of the synchronizer hub identifies the end that must be opposite the fork slot in the sleeve when assembled. This groove indicates the end of the hub with the greatest recess depth.

Assembly of Transmission

1. Using a suitable size drift, load a row of roller bearings (27) and a bearing washer at each end of the countergear. Use heavy grease to hold the bearings in place.
2. Install the countergear ssembly through the rear opening in the case, along with a tanged thrust washer, tang away from the gear, at each end of the gear. Install the countershaft and woodruff key from the rear of the case.

CAUTION: Be sure that the countershaft picks up both thrust washers and that the tangs are aligned with the notches in the case.

3. Install the reverse idler gear and shaft with its woodruff key from the rear of the case.
4. Using snap ring pliers, expand the snap ring in the extension housing and install the housing over the rear of the mainshaft and onto the rear bearing. Seat the snap ring in the rear bearing groove.
5. Load the mainshaft pilot bearings (14) into the drive gear cavity and assemble the fourth gear blocker ring onto the drive gear clutching surface, with the clutching teeth toward the gear.
6. Install the drive gear, pilot bearings and fourth gear blocker ring assembly over the front of the mainshaft.

NOTE: Be sure that the notches in the blocker ring align with the keys in the third-fourth synchronizer assembly.

7. Place the extension housing gasket on the rear of the case, hold in place with grease. From the rear of the case, install the mainshaft and extension housing assembly.

8. Install the extension housing retaining bolts. Use a sealing compound on the bottom bolt only. Tighten to specification.

9. Install the front bearing outer snap ring and position the bearing over the stem of the drive gear and into the front of the case.

10. Install the snap ring on the drive gear stem. Install the bearing retainer and gasket on the case. *Make sure that the bearing retainer oil hole is at the bottom.*

11. Shift the synchronizer sleeves to the neutral position and install the cover, gasket and fork assembly on the transmission. Make sure that the forks align with the synchronizer sleeve grooves.

12. Tighten all bolts to specification.

TORQUE SPECIFICATIONS

	Foot-Pounds
Clutch gear retainer to case	15
Side cover to case	15
Extension to case	45
Shift lever to shaft bolts	25
Filler plug	18
Case to clutch housing	75
Crossmember to frame	25
Crossmember to mount—	
mount to extension	40
Mount to transmission	32

Spicer 5000 Series 5 Speed

Description

The extra heavy-duty five-speed transmission (Spicer,) is a manually shifted, synchromesh, helical gear type. Fifth forward speed is direct drive. A power take-off is located on the right and left side of the transmission case.

Disassembly of Transmission

1. Remove the gear shift housing from the transmission case. Remove the detent balls from the housing and shift the transmission into two gears.
2. Remove the brake drum and spline flange.
3. Remove the brake shoe assembly, the output shaft bearing retainer, speedometer driving gear, and spacer.
4. Remove the countershaft rear bearing retainer, gasket, and countershaft nut.
5. Remove the left-side power take-off cover.
6. Remove the input shaft bearing retainer and gasket from the case. Using a soft drift, drive the input shaft and front bearing from the case. Remove the pilot rollers from the drive gear.
7. With a hardwood or fiber block placed against the front side of the second-speed gear, drive the input shaft assembly rearward until the output shaft bearing clears the case. Be careful not to hit the second-speed gear against the countershaft reverse gear. Remove the bearing from the output shaft.
8. When removing the output shaft from the case, slide off the first-speed gear.
9. Using a puller, remove the countershaft rear bearing.
10. Lift the countershaft assembly out of the case.
11. Remove the countershaft front bearing retainer, gasket and bearing from the front of the case.
12. Remove the reverse idler gear, and bearings from the gear bores.

13. Remove any of the 14 output shaft pilot rollers which may have dropped into the case.

Disassembly of Sub-Assemblies

Output Shaft

1. Remove the fourth and fifth speed synchronizer assembly snapping and thrust washer at the front of the fourth-speed gear. Remove the gear.
2. Remove the snap-ring at the front of the third-speed gear sleeve. The second and third-speed synchronizer can then be removed by bouncing the front of the output shaft on a block of wood.
3. Press the output shaft out of the second and third-speed syn-

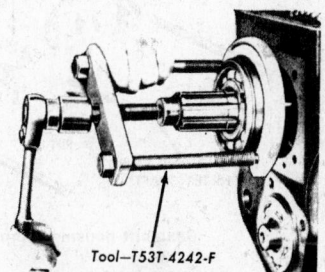

Tool—T53T-4242-F

Removing output shaft bearing using special puller
(© Dana Corp., Spicer Div.)

chronizer clutch gear and second-speed gear.

Countershaft

When replacing the countershaft or countershaft gear, press off one gear at a time. To remove the second-speed gear use special tool.

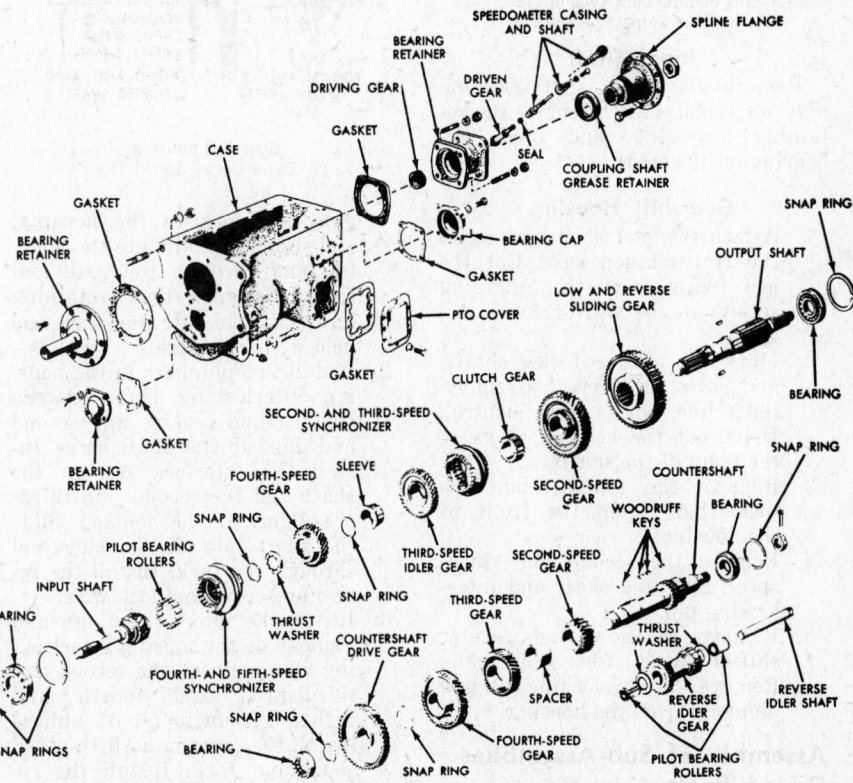

Spicer 5 speed transmission (© Dana Corp., Spicer Div.)

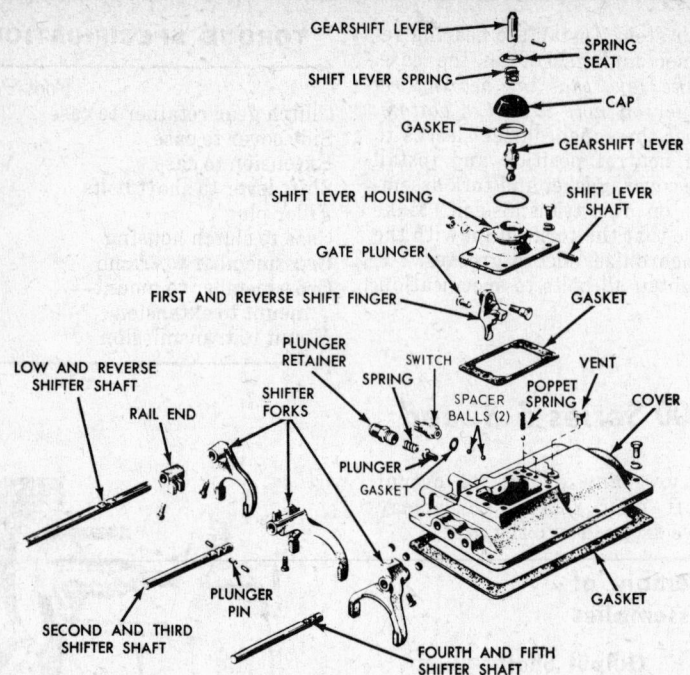

GEARSHIFT LEVER
SPRING SEAT
SHIFT LEVER SPRING
CAP
GASKET
GEARSHIFT LEVER
SHIFT LEVER HOUSING
SHIFT LEVER SHAFT
GATE PLUNGER
FIRST AND REVERSE SHIFT FINGER
GASKET
PLUNGER RETAINER
SWITCH
LOW AND REVERSE SHIFTER SHAFT
SPRING
VENT
SHIFTER FORKS
SPACER BALLS (2)
POPPET SPRING
COVER
RAIL END
PLUNGER
GASKET
PLUNGER PIN
SECOND AND THIRD SHIFTER SHAFT
FOURTH AND FIFTH SHIFTER SHAFT
GASKET

Gearshift housing components (© Dana Corp., Spicer Div.)

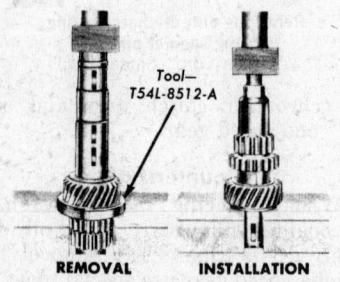

Tool—T54L-8512-A

REMOVAL   INSTALLATION

Replacing countershaft second speed gear (© Dana Corp., Spicer Div.)

Input Shaft

Remove the input shaft bearing only for replacement. Remove the retaining snap-ring, and press the bearing off the shaft.

Gearshift Housing

1. Attach the gear shift housing to the transmission case. Cut the lock wire from the retaining screws in the shifter forks and gates.
2. Mark the shifter forks, shafts, and gates for correct assembly. Shift the shafts into neutral. Drive out the housing plugs at the front of the shafts.
3. Remove the fourth and fifth speed-shaft from the front of the housing.
4. Remove the second and third-speed fork and shaft and interlocking pin.
5. Remove the low and reverse shifter shaft, fork, and gate. Remove the interlocking pin and plungers from the housing.

Assembly of Sub-Assemblies
Gearshift Housing

1. Place the low and reverse shifter

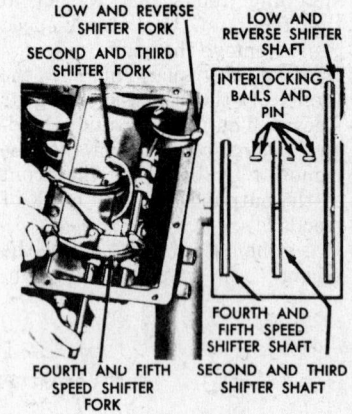

LOW AND REVERSE SHIFTER FORK
LOW AND REVERSE SHIFTER SHAFT
SECOND AND THIRD SHIFTER FORK
INTERLOCKING BALLS AND PIN
FOURTH AND FIFTH SPEED SHIFTER FORK
FOURTH AND FIFTH SPEED SHIFTER SHAFT
SECOND AND THIRD SHIFTER SHAFT

Gearshift housing
(© Dana Corp., Spicer Div.)

gate and fork in the housing, and slide the shaft into the housing and through the gate and fork. Install the retaining screws in the gate and fork and hold with lock wire.

2. Install two plungers in the housing interlocking bore between the low and reverse and second and third shifter shaft bores. Install the interlock pin in the shaft and the second and third-speed fork in the housing. Slide the shaft into the housing and through the fork. Install the retaining screw, and lock wire.

3. Install the interlocking pin and plunger in the housing interlocking bore between the second and third-speed and fourth and fifth-speed shifter shaft guides. Install the fourth and fifth-speed shaft and fork. Install the retaining screw and lock wire.

4. Check the interlocking system

for correct operation and, using sealer, install the housing plugs. Remove the housing from the transmission case.

Output Shaft

1. Place the second-speed gear onto the output shaft, with the clutch teeth facing forward.
2. Insert the two Woodruff keys in the output shaft and install the second and third-speed synchronizer clutch gear.
3. Place the second and third-speed synchronizer, and third speed gear and sleeve on the output shaft. Press the sleeve onto the shaft until it bottoms on the synchronizer clutch gear. The third-speed gear sleeve slots must line up with the Woodruff keys in the output shaft.
4. Remove the assembly from the press and install the snap-ring at the front of the third-speed gear sleeve.
5. Install the fourth-speed gear, thrust washer, and snap-ring on the output shaft. Install the fourth and fifth-speed synchronizer on the output shaft.

Countershaft

Install gears and spacer onto shaft and hold with the snap-rings. Each gear takes a specific Woodruff key, so install them one at a time.

Input Shaft

Press the input shaft bearing onto the shaft using special tool.

Assembly of Transmission

NOTE: As a protection against scoring, coat all parts with transmission lubricant.

1. Tap the countershaft front bearing into the case and install retainer and new gasket. Line up the oil return holes in the retainer, gasket and case and torque the 7/16″ retainer bolts at 50-55 ft. lbs. (30 ft. lbs. for smaller bolts).
2. Place the assembled countershaft in the transmission case into the front bearing.
3. Drive the countershaft rear bearing onto the countershaft and into the case.
4. Install the idler gear bearings, and gear in the case. Drive the idler gear shaft into position, and install the power takeoff cover.
5. Tap the input shaft and bearing into the case. Place the pilot bearing rollers in the input shaft.
6. Install the input shaft bearing retainer without a gasket and tighten the bolts. With a feeler gauge, check the clearance between the bearing retainer and

the case to determine gasket size.

7. Install the bearing retainer and gasket, making certain that the oil drain-hole is in line with the gasket and case holes. Torque retainer bolts to 30 ft. lbs., 40 ft. lbs. for 7/16" bolt.

8. Install the low and reverse gear on the output shaft and place the assembly in the case. Drive the output shaft bearing into position.

9. Shift the transmission into two gears. Install the countershaft nut and torque at 350-450 ft. lbs. Install the countershaft rear bearing retainer and torque the 7/16" bolts to 45 ft. lbs.

10. Install a new oil seal in the output shaft bearing retainer. Place the spacer and speedometer driving gear on the output shaft, and install the bearing retainer. Torque the bolts to specification.

11. Install the parking brake shoe assembly.

12. Install the brake drum and the spline flange. Torque the output shaft nut to specification.

13. Shift the transmission and gear shift housing into neutral, and install the gear shift housing and, using 7/16" bolts, torque to 45 ft. lbs.

TORQUE CHART

Nomenclature	Nuts and/or Bolts and Torque Limits		Nomenclature	Nuts and/or Bolts and Torque Limits	
Bolt-gear shift lever tower to gearshift housing	3/8-16 20-25	7/16-14 30-35	Bolt—countershaft & reverse idler shaft retainer	5/16-18 25-30 3/8-16 25-37 3/8-16† 18-25†	7/16-14 40-45 1/2-13 80-85
Bolt-clutch housing to trans. case	7/16-14 30-38 5/8-11 96-120	9/16-12 70-90	Bolt—gear shift housing to trans. case	5/16-18 20-25 3/8-16 35-40	3/8-16†† 30-35†† 7/16-14 45-50
Nut-U-joint flange to trans. output shaft	1.00-20 90-125	1 1/2-18 275-350 1 1/4-18 225-275	Bolt—power take off cover to trans. case	3/8-16 20-30	
Nut—drum parking brake to companion flange	3/8-24 35-45	7/16 20 50-70	Nut—countershaft bearing lock (5-speed extra h.d. & 5-speed exclusive)	1 1/4-18 350-450	
Nut—bellcrank to trans.	9/16-18 70-90		Nut—countershaft bearing lock (5-speed exclusive h.d.)		
Bolt—lever assy. to trans.	3/8-16 20-25				
Nut—handbrake anchor bar to trans. case (5-speed extra-heavy duty only)			Bolt—input shaft bearing retainer to trans. case	5/16-18 25-30 3/8-16 25-30	7/16-14 40-45
Bolt—bellcrank to trans.					
Bolt—reverse lock-out plunger retainer	11/16-16 80-100				
Bolt—countershaft rear bearing retainer	5/16-18 25-30 3/8-16 35-40	7/16-14 45-55 1/2-13 67-70	Bolt—countershaft front bearing retainer	5/16-18 25-30 3/8-16 25-35	7/16-14 50-55

Spicer 6000 Series 5 Speed

Description

This transmission is a 5-speed synchromesh helical gear design with direct drive in 5th speed on all models, except the 6853C, which has overdrive in 5th speed.

Engagement of all gears, except first and reverse, is aided by sleeve type synchronizers. All gears are of helical design, with the exception of first and reverse gears. Lubricant capacity is 17 pints.

General Data

Make	Spicer
Type	5-Speed Synchromesh
Models	6852S, 6852K, 6852G, and 6853C
Clutch housing	S.A.E. #2

Gear ratios—all ratios are (to 1)

Transmission Model	6852G	6852K
1st	6.70	6.70
2nd	3.52	4.02
3rd	1.97	2.49
4th	1.17	1.57
5th	1.00	1.00
Reverse	6.72	6.72
Transmission Model	6852S	6853C
1st	5.71	5.71
2nd	3.20	3.00
3rd	1.89	1.78
4th	1.15	1.00
5th	1.00	0.85
Reverse	5.73	5.73

Disassembly of Transmission
Removal of Mainshaft

1. Remove clutch housing and clutch release mechanism as a unit.

2. Engage 2nd and 3rd synchronizer with mainshaft 2nd speed gear, and engage 4th and 5th synchronizer with mainshaft 4th speed or overdrive gear to lock transmission in two gears.

3. Remove companion flange or yoke retaining nut. Use puller to remove flange or yoke.

4. Remove speedometer driven gear and adapter (if used) from mainshaft rear bearing cap.

5. Remove mainshaft and countershaft rear bearing caps and gaskets.

6. Remove cotter pin and countershaft rear bearing nut.

7. Remove speedometer drive gear or spacer from rear end of mainshaft.

8. Remove mainshaft rear bearing snap-ring.

9. Using bearing puller, remove mainshaft rear bearing.

10. Remove mainshaft and gear assembly from the transmission case by clearing it from main drive gear and sliding assembly up and forward out of rear bearing bore.

Disassembly of Mainshaft

1. Remove the 1st and reverse sliding gear from mainshaft.

2. Remove 4th and 5th synchronizer.

3. Remove snap-ring and thrust washer.

4. Slide mainshaft 4th speed or ov-

Using puller to remove mainshaft rear bearing
(© Dana Corp., Spicer Div.)

Mainshaft components (© Dana Corp., Spicer Div.)

1. 4th and 5th speed synchronizer
2. Snap ring
3. Thrust washer
4. 4th speed or overdrive gear
5. Sleeve pin
6. 4th speed or over-drive gear sleeve
7. 3rd speed gear
8. 2nd and 3rd speed synchronizer
9. Snap ring
10. 2nd and 3rd clutch gear
11. 2nd speed gear
12. Mainshaft
13. 1st and reverse sliding gear

Main drive gear components
(© Dana Corp., Spicer Div.)

1. Main drive gear bearing retainer ring
2. Main drive gear bearing
3. Main drive gear
4. Pilot bearing rollers

erdrive gear from end of mainshaft.

5. Remove mainshaft 3rd speed gear and mainshaft 4th speed or overdrive gear sleeve. *NOTE: If necessary to press off gear and sleeve, shift 2nd and 3rd synchronizer into engagement with mainshaft 2nd speed gear and support under 3rd speed gear. Be sure to remove sleeve pin from the inside of 4th speed or overdrive gear sleeve.*

6. Slide 2nd and 3rd synchronizer from mainshaft.

7. Remove snap-ring from 2nd and 3rd speed clutch gear.

8. Support mainshaft 2nd speed gear under arbor press and press mainshaft out of 2nd and 3rd speed clutch gear and 2nd speed gear.

Removal of Maindrive Gear

1. Remove 14 mainshaft pilot bearing rollers which may have remained in cavity of main drive gear.
2. Remove main bearing cap and gasket.
3. Remove snap-ring from main drive gear bearing.
4. Remove main drive gear and bearing assembly from transmission case.

Disassembly of Main Drive Gear

1. Remove main drive gear bearing retaining snap-ring.
2. Using bearing remover plates with an arbor press, remove bearing from main drive gear.

Reverse Idler Gear Removal

IMPORTANT: When removing the reverse idler gear shaft, support the gear to prevent it from being damaged.

1. Using remover tool, remove reverse idler gear shaft, gear and bearings.

Removal of Countershaft

1. Using puller, remove countershaft rear bearing.

1. Main drive gear
2. Main drive gear bearing cap
3. Mainshaft pilot bearing rollers
4. Snap ring
5. Clutch housing
6. 4th and 5th shift rod
7. 4th and 5th shift fork
8. 4th and 5th synchronizer
9. Poppet ball
10. Poppet spring
11. Snap ring
12. Thrust washer
13. Mainshaft 4th speed or overdrive gear
14. Mainshaft 4th speed or overdrive gear sleeve
15. Mainshaft 3rd speed gear
15. Mainshaft 3rd speed gear
16. 2nd and 3rd shift fork
17. Snap ring
18. Mainshaft 2nd and 3rd speed clutch gear
19. 2nd and 3rd synchronizer
20. Mainshaft 2nd speed gear
21. 1st and reverse shift rod
22. 1st and reverse shift fork
23. Mainshaft 1st and reverse sliding gear
24. Mainshaft
25. Shifter housing
26. Mainshaft rear bearing
27. Snap ring
28. Speedometer driven gear
29. Mainshaft rear bearing cap oil seal
30. Companion flange
31. Companion flange nut
32. Mainshaft rear bearing cap
33. Speedometer drive gear
34. Countershaft rear bearing
35. Countershaft rear bearing cap
36. Countershaft rear bearing nut

NOTE
Reverse Idler Gear Shaft (Item 38) Is Intentionally Shown Out Of Normal Position

37. Snap ring
38. Reverse idler gear shaft
39. Reverse idler gear bearings
40. Countershaft 1st gear teeth
41. Countershaft 1st gear teeth
42. Countershaft reverse gear
43. Countershaft 2nd speed gear
44. Countershaft 3rd speed gear
45. Countershaft 4th speed or overdrive gear
46. Countershaft
47. Countershaft drive gear
48. Transmission case
49. Countershaft front bearing
50. Main drive gear bearing
51. Snap ring
52. Main drive gear bearing cap oil seal

Spicer 6000 Series 5 speed transmission (© Dana Corp., Spicer Div.)

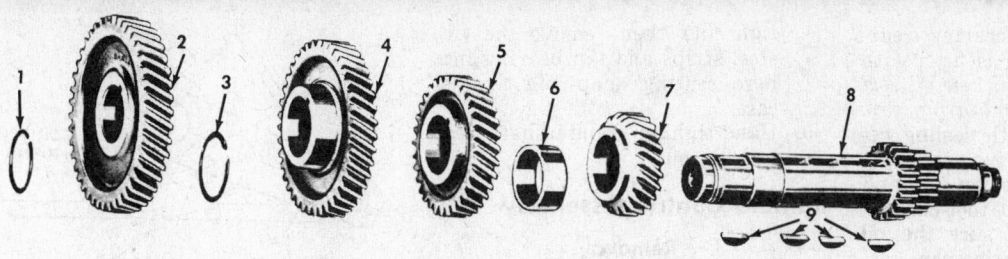

1 Snap ring
2 Countershaft drive gear
3 Snap ring
4 4th speed or overdrive gear
5 3rd speed gear
6 2nd and 3rd gear spacer
7 2nd speed gear
8 Countershaft
9 Countershaft gear keys

Countershaft components (© Dana Corp., Spicer Div.)

2. Lift the countershaft assembly out of transmission case.
3. Press or drive countershaft front bearing from bore of transmission case.

Disassembly of Countershaft

1. Support countershaft drive gear with parallel bars under hub and press countershaft free of gear.
2. Remove exposed countershaft gear key and snap-ring. Support 4th speed or overdrive gear and press countershaft free of gear.
3. Follow the same procedure and remove 2nd and 3rd speed gear.
4. Remove the remaining countershaft gear key.

Assembly of Subassemblies

Cleanliness is of the utmost importance. The transmission should be rebuilt in a clean working area. All parts, except those actually being worked on should be covered with clean lint-free paper. Avoid nicking, marring, or burring all surfaces.

IMPORTANT: Coat all thrust washers, splines of shafts, and bores of all gears with lubricant to provide initial lubrication thus preventing scoring or galling.

Assembly of Mainshaft

1. Position mainshaft in a soft-jawed vise, front end up. NOTE: Fit of new parts may require the use of an arbor press. If so, set up vertically and follow same procedure.
2. Position 2nd speed gear on mainshaft with clutch teeth and synchronizer cone facing up.
3. Using a suitable sleeve, press or drive 2nd and 3rd speed clutch gear on mainshaft. Install snap-ring in mainshaft groove. NOTE: Minimum end clearance between 2nd and 3rd speed clutch gear and 2nd speed gear should be 0.004". Correct accordingly.
4. Slide 2nd and 3rd speed synchronizer on mainshaft until engaged with 2nd and 3rd speed clutch gear. IMPORTANT: The 2nd and 3rd speed synchronizer is often assembled backward on the mainshaft. Make sure that the long hub on synchronizer clutch gear faces the 2nd speed gear.

5. Place 3rd speed gear on mainshaft with clutch teeth and synchronizer cone facing downward.
6. Assemble sleeve pin to 4th speed or overdrive gear sleeve with head of pin inside sleeve, with flanged end of sleeve facing the 3rd speed gear, align sleeve pin with splines and press on the mainshaft.
7. Place 4th speed or overdrive gear on sleeve with clutch hub facing up and secure with thrust washer and snap-ring.

Assembly of Main Drive Gear

1. Press or drive main drive gear bearing onto main drive gear shaft and install snap-ring.

Assembly of Countershaft

1. Position first countershaft gear key in slot of countershaft. Press 2nd speed gear and, 2nd and 3rd gear spacer onto countershaft.
2. Install the remaining countershaft gear keys. It may be necessary to dress the keys with a file.
3. Press 3rd speed gear on the countershaft, followed by 4th speed gear, a snap-ring and drive gear secured by a snap-ring.

Assembly of Transmission

Installation of Mainshaft

1. Place mainshaft 1st and reverse sliding gear at an angle in the rear of the transmission case with shift fork collar facing toward front of case.
2. Place 4th and 5th synchronizer on mainshaft. Shift synchronizer clutch collar into engagement with mainshaft 4th speed or overdrive gear to help lock synchronizer in place during installation in case.
3. Lower rear of mainshaft into case, through 1st and reverse sliding gear and out mainshaft rear bearing bore. NOTE: There must be fourteen pilot bearing rollers in the main drive gear pocket.
4. Lower front of mainshaft to mesh with countershaft gears. Slide rear of mainshaft into pocket of main drive gear.

5. Slide rear bearing onto mainshaft with snap-ring facing the rear.
6. Tap bearing into case bore with its snap-ring flush with rear of case.
7. Press speedometer drive gear or spacer onto the mainshaft against bearing. NOTE: Install new oil seal in mainshaft rear bearing cap.
8. Install gasket with sealing cement and mainshaft rear bearing cap and torque cap screws to 35 to 40 foot-pounds. NOTE: Install speedometer driven gear (when used) through opening in mainshaft rear bearing cap.
9. Engage 2nd and 3rd synchronizer with mainshaft 2nd speed gear, and 4th and 5th synchronizer with mainshaft 4th speed or overdrive gear to lock transmission in two gears.
10. Install companion flange or yoke on mainshaft and secure with a washer and nut at a torque of 320 to 350 foot-pounds.
11. Remove rear bearing cap from countershaft and torque bearing nut to 320 to 350 foot-pounds and secure with cotter pin.
12. Install gasket and countershaft rear bearing (with sealing cement) and torque cap screws to 35 to 40 foot-pounds. NOTE: The projection on the countershaft rear bearing cap locks the reverse idler gear shaft into proper position.
13. Rotate main drive gear to check for free rotation of all gears and shafts.
14. Use pressure type oil can, filled with transmission lubricant (S.A.E. 50 engine oil of good quality) to force oil through holes and end slots of all mainshaft gears to open oil passageways. NOTE: Using the pressure type oil can, spray gear teeth with transmission lubricant to provide initial lubrication and to prevent corrosion.

Installation of Main Drive Gear

1. Install main drive gear and bearing from inside the transmission case, by tapping bearing through case bore.
2. Install snap-ring on outer race

of bearing. Tap bearing rearward, so snap-ring is flush with case. *NOTE: Install new oil seal in main drive gear bearing cap.*

3. Install gasket (with sealing cement) and main drive gear bearing cap to transmission case and torque to 60 to 80 foot-pounds. **IMPORTANT: Be sure the oil passages in the bearing cap, gasket and transmission case are all aligned.**

4. Coat the pocket of main drive gear with Light Weight Ball and Roller Bearing Grease.

5. Place the fourteen pilot bearing rollers in main drive gear pocket.

Installation of Reverse Idler Gear

1. Install reverse idler gear bearings into gear and install in transmission case, with large gear on idler gear toward front of case. Mesh idler gear with countershaft and align bore of bearings with hole in transmission case.

2. Insert reverse idler gear shaft in rear of case, noting that "flat" on shaft is squared toward countershaft so that it can be locked by rear bearing cap.

3. Set countershaft rear bearing cap in place to check lock of reverse idler shaft. Finger-tighten bearing cap screws to prevent countershaft from moving during installation of mainshaft.

Installation of Countershaft

1. Press countershaft front bearing on front of countershaft.

2. Lower countershaft assembly into case, guiding rear end of shaft out through rear of case.

3. Guide countershaft front bearing into countershaft front bearing bore.

4. Place two strips of flat steel stock, approximately 3/8-inch thick, between countershaft drive gear and wall of case.

5. Seat bearing on shaft (should be

tight fit) then remove the two steel strips and tap bearing into bore seating snap-ring against case.

6. Hand-tighten countershaft rear bearing nut.

Remote Control Assembly

Removal

NOTE: The Spicer 6000 transmission uses two different remote control assemblies: Prop shaft type and rod type.

1. Remove retainer, plunger pin spring, and plunger.

2. Remove remote control assembly from transmission as described below: **IMPORTANT: Tilt the remote control assembly slightly to the left during removal to prevent the plunger from falling into the transmission.**

3. Remove the plunger from the 1st and reverse shift finger.

Installation

1. Coat remote control assembly position on shifter housing.

2. Position plunger in 1st and reverse shift finger; then carefully install remote control assembly as shown, keeping the assembly slightly tilted to the left to prevent the plunger from falling into transmission.

3. Install plunger pin, plunger pin spring, and retainer.

4. Install remote control assembly-to-shifter housing attaching parts. Tighten cap screws firmly.

Control Tower

Installation

Install control tower as described below only when the transmission is going into storage.

1. Coat gasket with sealing cement and position on shifter housing.

2. Position plunger in 1st and reverse shift finger, then place the assembly and new gasket (with ce-

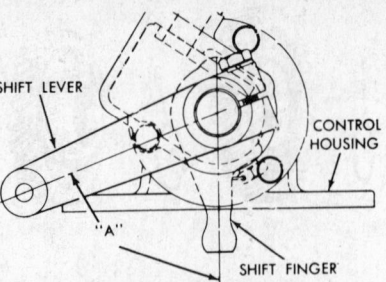

Rod type remote control adjustment
(© Dana Corp., Spicer Div.)

ment) on the shifter housing.

3. Install plunger, plunger spring, and plunger retainer and tighten down the tower with washers and capscrews.

Shifter Housing

Removal

NOTE: The Spicer 6000 transmission uses a forward control shifter housing and a center control shifter housing.

1. Remove shifter housing-to-transmission case attaching

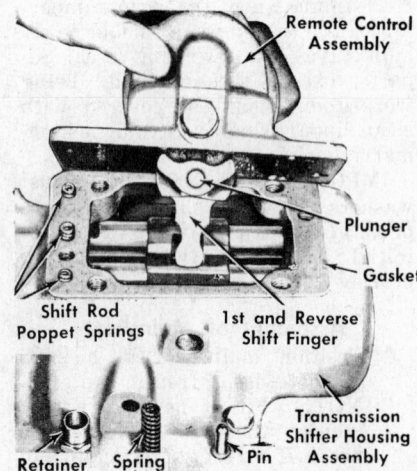

Replacing remote control assembly
(© Dana Corp., Spicer Div.)

1 1st and reverse shift rod bracket
2 Poppet ball
3 Poppet spring
4 1st and reverse shift rod
5 1st and reverse shift fork
6 Breather
7 Forward control shifter housing
8 Expansion plug
9 2nd and 3rd shift fork
10 Setscrew
11 4th and 5th shift rod
12 4th and 5th shift fork
13 Interlock
14 Interlock pin
15 2nd and 3rd shift rod
16 Setscrew
17 Expansion plug
18 2nd and 3rd shift rod bracket
19 Setscrew
20 Shift rod thimble

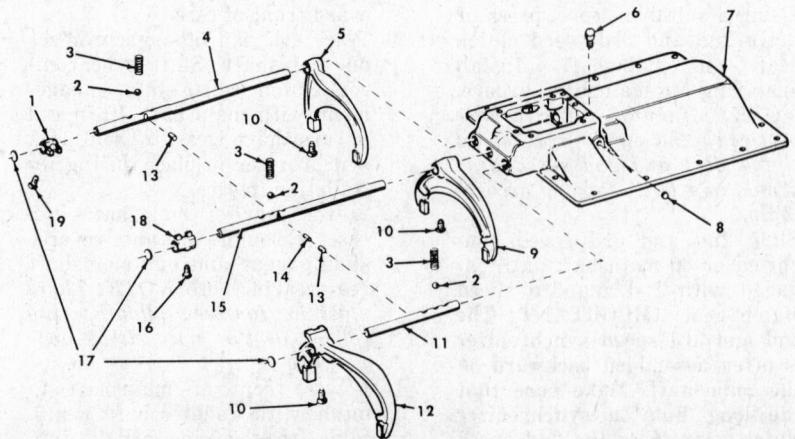

Shifter housing components (© Dana Corp., Spicer Div.)

parts. *NOTE: Carefully remove forward control shifter housing to prevent loss of the three poppet balls and springs.*

2. Carefully remove shifter housing from the case.

Installation

NOTE: Make certain that shift forks and transmission clutch collars are in neutral position.

1. Carefully position shifter housing and gasket (use cement) on transmission case and make sure all three shift forks are in their corresponding shift collar and tighten down.

2. Use large screwdriver or small pry bar and check movement of each shift rod for proper shift action. Return shift rods to neutral position.

IMPORTANT: On transmissions using the forward control shifter housing, place the three poppet balls and springs in the shifter housing.

TORQUE SPECIFICATIONS

Location	Foot-Pounds
Shift fork set screws	45-50
Main drive gear bearing cap retaining cap screws	60-80
Mainshaft rear bearing cap retaining cap screws	35-40
Mainshaft flange or yoke retaining nut	320-350
Countershaft rear bearing retaining nut	320-350
Countershaft rear bearing cap retaining cap screws	35-40
Clutch housing retaining cap screws	90-95

Warner T-10 4 Speed

Description

The Warner T-10 is a fully synchronized four-speed transmission having a floor mounted shift lever. Lubricant capacity is 2½ pints.

Disassembly of Transmission

1. Drain transmission, mount in stand and remove side cover and shift controls.
2. Remove front bearing retainer and gasket.
3. Remove output shaft companion flange.
4. Drive out lock pin and pull reverse shift shaft out about ⅛ in. to disengage shifter fork from reverse gear.
5. Remove bolts and tap the case extension (with soft hammer) rearward. When idler gear shaft is out as far as it will go, move extension to the left so the reverse fork clears the reverse gear. Remove extension and gasket.
6. Remove rear bearing snap-ring from mainshaft.
7. Remove case extension oil seal.
8. Using a puller, remove speedometer drive gear.
9. Remove the reverse gear, reverse idler gear and tanged thrust washer.
10. Remove self-locking bolt holding the rear bearing retainer to transmission case.
11. Remove the entire mainshaft assembly.
12. Unload bearing rollers from main drive gear and remove fourth-speed synchronizer blocking ring.
13. Lift the front half of reverse idler gear and its thrust washer from the case.
14. Remove the main drive gear snap-ring and spacer washer.
15. With soft hammer, tap main drive gear out of front bearing.
16. From inside the case, tap out front bearing and snap-ring.
17. From the front of the case, tap out the countershaft with a dummy shaft.
18. Then lift out the countergear assembly with both tanged washers.
19. Dismantle the countergear, consisting of 80 rollers, six .050 in. spacers and a roller tubular spacer.
20. Remove mainshaft front snapring and slide third and fourth-speed clutch assembly, third-speed gear and synchronizer ring, second and third-speed gear thrust bearing, second-speed gear and second-speed synchronizer ring from front of mainshaft.
21. Press mainshaft out of retainer.
22. Remove the mainshaft rear snap-ring.
23. Support first and second-speed clutch assembly and press on rear of mainshaft to remove shaft from rear bearing, first-speed gear, and synchromesh ring, first and second-speed clutch sliding sleeve and first-speed gear bushing.

Assembly of Transmission

Mainshaft

1. From the rear of the mainshaft, assemble first and second-speed clutch assembly to mainshaft (sliding clutch sleeve taper toward the rear, hub to the front) and press the first-speed gear bushing onto the shaft.
2. Install first speed gear synchronizing ring aligning notches in ring with keys in hub.
3. Install first-speed gear (hub toward front) and thrust washer with the washer grooves facing first-speed gear.
4. Press on the rear bearing, with the snap-ring groove toward the front of the transmission. Be sure the bearing is firmly seated against the shoulder on the mainshaft.
5. Install the selective fit snap-ring onto the mainshaft behind the rear bearing. Use the thickest ring that will fit between the rear face of the bearing and the front face of the snap-ring.
6. From the front of the mainshaft, install the second-speed gear synchronizing ring so that the ring notches correspond with the hub keys.
7. Install the second-speed gear (hub toward the back) and the second and third-speed gear thrust bearing.
8. Install third-speed gear (hub to front) and third-speed gear synchronizing ring (notches front).
9. Install third and fourth-speed gear clutch assembly (hub and sliding sleeve) with taper front, being sure keys in the hub correspond with notches in third-speed gear synchronizing ring.
10. Install snap-ring (.086-.088 in. thickness) into mainshaft groove in front of the third and fourth-speed clutch assembly.
11. Install rear bearing retainer plate. Spread the snap-ring on the plate to allow the snap-ring to drop around the rear bearing and press on the end of the mainshaft until the snap-ring engages the groove in the rear bearing.
12. Install reverse gear (shift collar to the rear).
13. Press speedometer drive gear onto the mainshaft so that there is a measurement of 4½ in. from the center of the gear to the flat surface of the rear bearing retainer.
14. Install special snap-ring into the groove at the rear of the mainshaft.

Countergear

1. Install countergear dummy and tubular roller bearing spacer into the countergear.
2. Using heavy grease to hold the rollers, install 20 bearing rollers in either end of the countergear, two spacers, 20 more rollers, then one spacer. Install the same combination of rollers and spac-

ers in the other end of the countergear.

3. Set the countergear assembly in the bottom of the transmission case making sure the tanged thrust washers are in their proper position.

Main Drive Gear

1. Press bearing (snap-ring groove front) onto main drive gear until the bearing fully seats against the shoulder on the gear.

2. Install spacer washer and selective fit snap-ring in the groove in the main drive gear shaft.

NOTE: Variable thickness snap-rings are available to obtain a prescribed clearance of .000-.005 in. between the rear face of the snap-ring and the front face of the spacer washer.

Assembly of Transmission

1. Install main drive gear and bearing assembly through the side cover opening and into position in the transmission front bore. After assembly is in place, install front bearing snap-ring.

2. Lift countergear and thrust washers into place. Install

Woodruff key into end of countershaft, then from the rear of the case, press the countershaft in until flush with rear of case and the dummy shaft is displaced. Maximum countergear end play is .025 in.

3. Install the 14 bearing rollers into the grease-coated end of the main drive gear.

4. Using heavy grease, position gasket on front face of rear bearing retainer. Install the fourth-speed synchronizing ring onto main drive gear with clutch key notches toward rear of transmission.

1. Bearing retainer
2. Gasket
3. Selective fit snap ring
4. Spacer washer
5. Bearing snap ring
6. Main drive gear bearing
7. Transmission case
8. Rear bearing retainer gasket
9. Main drive gear
10. Bearing rollers (14)
11. Snap ring (.086" to .088")
12. Fourth speed gear synchronizing ring
13. Third and fourth speed clutch sliding sleeve
14. Third speed synchronizing ring
15. Third speed gear
16. Second and third speed gear thrust washer (needle roller bearing)
17. Second speed gear
18. Second speed gear synchronizing ring
19. Mainshaft
20. First and second speed clutch assembly
21. Clutch key spring
22. Clutch keys
23. Clutch hub
24. Clutch key spring
25. First and second speed clutch sliding sleeve
26. First speed gear synchronizing ring
27. First speed gear
28. First speed gear bushing
29. First speed gear thrust washer
30. Rear bearing snap ring
31. Rear bearing
32. Rear bearing retainer
33. Selective fit snap ring
34. Reverse gear
35. Speedometer drive gear
36. Rear bearing retainer to case extension gasket

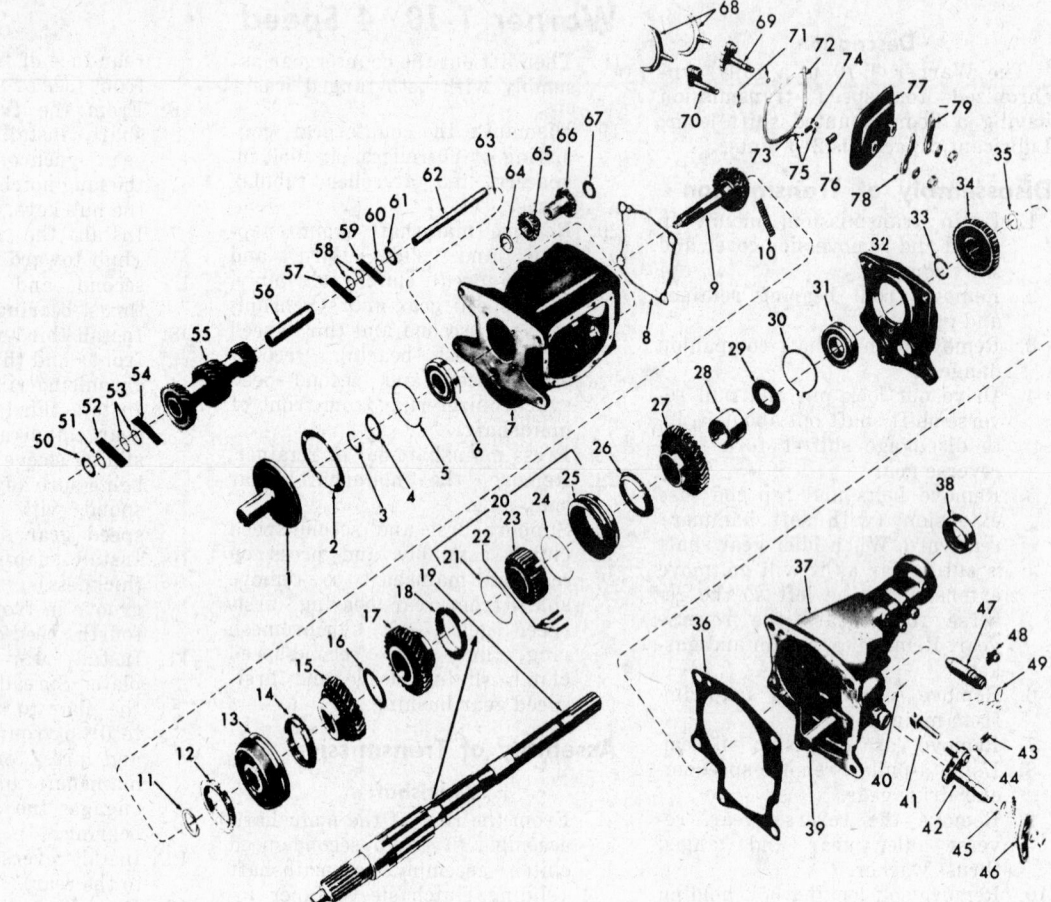

37. Case extension
38. Rear oil seal
39. Reverse idler shaft
40. Reverse shifter shaft lock pin
41. Reverse shift fork
42. Reverse shifter shaft and detent plate
43. Reverse shifter shaft ball detent spring
44. Reverse shifter shaft detent ball
45. Reverse shifter shaft "O" ring seal

46. Reverse shifter lever
47. Speedometer driven gear and fitting
48. Retainer and bolt
49. "O" ring seal
50. Tanged washer
51. Spacer (.050")
52. Bearing rollers (20)
53. Spacer (2—.050")
54. Bearing rollers (20)
55. Countergear
56. Countergear roller spacer
57. Bearing rollers (20)
58. Spacers (2—.050")

59. Bearing rollers (20)
60. Spacer (.050")
61. Tanged washer
62. Countershaft
63. Countershaft woodruff key
64. Reverse idler front thrust washer (flat)
65. Reverse idler gear (front)
66. Reverse idler gear (rear)
67. Tanged thrust washer
68. Forward speed shift forks
69. First and second speed gear shifter shaft

and detent plate
70. Third and fourth speed gear shifter shaft and detent plate
71. "O" ring seals
72. Gasket
73. Interlock pin
74. Poppet spring
75. Detent balls
76. Interlock sleeve
77. Transmission side cover
78. Third and fourth speed shifter lever
79. First and second speed shifter lever

The Warner T-10 4-speed transmission—typical (© Borg Warner Corp.)

5. Position the reverse idler gear thrust washer on the machined face of the ear cast in the case for the reverse idler shaft. Position the front reverse idler gear on top of the thrust washer, hub facing toward rear of case.

6. Lower the mainshaft assembly into the case, with the fourth-speed synchronizing ring notches aligning with the keys in the clutch assembly.

7. Install self-locking bolt, attach the rear bearing retainer to the transmission case and torque the bolt to 20-30 ft. lbs.

8. From the rear of the case, insert the rear reverse idler gear, engaging the splines with the portion of the gear within the case.

9. Place a greased gasket on the rear face of the rear bearing retainer.

10. Install remaining tanged thrust washer into place on reverse idler shaft, being sure the tang on the thrust washer is in the notch in the idler thrust face of the extension.

11. Place the two clutches in neutral position.

12. Pull reverse shifter shaft to left side of extension and rotate shaft to bring reverse fork to extreme forward position in the extension. Align forward and reverse idler gears.

13. Position the extension onto the transmission case by uniting the reverse idler shaft with the idler gears and engaging the shift fork with the reverse shift collar by turning the shifter shaft, the reverse gear will move rearward thus enabling installation of the extension.

14. Install three extension and retainer to case attaching bolts at 35-45 ft. lbs. and two extension to retainer attaching bolts (20-30 ft. lbs.). Use sealer on the lower, right bolt.

15. Align groove in reverse shift shaft with hole in boss and drive in lock pin.

16. Install the main drive gear bearing retainer and gasket aligning the oil well with the oil outlet hole and torque the sealer-coated bolts to 15-20 ft. lbs.

17. Install a shift fork into each clutch sleeve.

18. With both clutches in neutral, install side cover and gasket and torque to 10-20 ft. lbs. Use sealer on the lower right bolt.

19. Install first and second, and third and fourth shift levers.

TORQUE CHART

Part	Location	Thread Size	Torque Ft. Lbs.
Bolt	Front bearing retainer to transmission case	5/16-18	15-20
Bolt	Side cover bolts	5/16-18	15-20
Nut	Shift lever to shaft	5/16-18	12-18
Bolt	Transmission to flywheel housing	1/2-13	45-60
Bolt	Flywheel housing to engine	3/8-16	30-35

Warner T-14A, T-15A 3 Speed

Description

The Warner T-14A, T-15A are fully synchronized three-speed transmissions having helical drive gears throughout. Lubricant capacity is 2½ pints.

Disassembly of Transmission

1. Separate transfer case from transmission by removing five capscrews.
2. Remove gearshift housing and disassembly by removing shift rails, poppet balls, springs, and shift forks.
3. Remove nut, flat washer, transfer case drive gear, adapter, and spacer.
4. Remove main drive gear bearing retainer gasket.
5. Remove main drive gear and mainshaft bearing snap-rings and bearings.
6. Remove main drive gear and mainshaft assembly. NOTE: The T-15A transmission must be shifted into second gear to allow removal of the mainshaft and gear assembly.
7. On remote shift models, remove roll pins from lever shafts and housing. From inside case, slide levers and interlock assembly out. Remove forks and lever assemblies.
8. Remove lock plate from reverse idler shaft and countershaft.
9. Drive countershaft out to rear with dummy shaft. Remove countergear and two thrust washers. Remove spacer washers, rollers, and spacer from gear.
10. Drive reverse idler shaft out to rear. Remove gear, washers, and roller bearings.
11. Remove clutch hub snap-ring and second-third synchronizer assembly.
12. Remove second and reverse gears.
13. Remove clutch hub snap-ring and low synchronizer assembly.
14. Remove low gear.

Synchronizer

Disassembly and Assembly

1. Remove springs. Low synchronizer has only one spring; second-third, two.
2. Mark sleeve and hub before separating.
3. Remove hub.
4. Remove three shifter plates from hub.
5. Inspect all parts for wear.
6. Assembly in reverse order of disassembly. On second-third unit, make sure that spring openings are 120 degrees from each other, with spring tension opposed.
NOTE: If a synchronized assembly is replaced on a floor shift unit, the shift fork operating the synchronizer being replaced must have the letter A just under the shaft hole on the side opposite the pin.

Inspection

1. Wash all parts in solvent.
2. Air dry but do not spin bearings with air pressure.
3. Check case bearing and shaft bores for cracks or burrs.
4. Check all gears and bronze blocking rings for cracks, and chipped, worn, or cracked teeth. If any gears are replaced, also replace the meshing gears.
5. Check all bearings and bushings for wear or damage.
6. Check that synchronizer sleeves slide freely on clutch hubs.

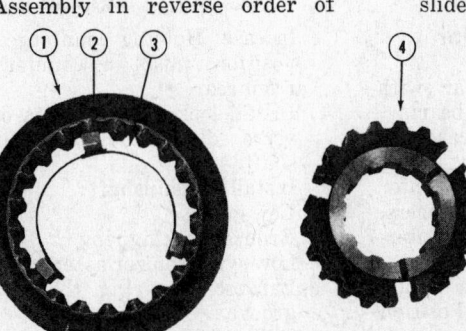

1 Clutch hub
2 Shifter plate
3 Synchronizer spring (1)
4 Clutch sleeve

Low synchronizer assembly (© Borg Warner Corp.)

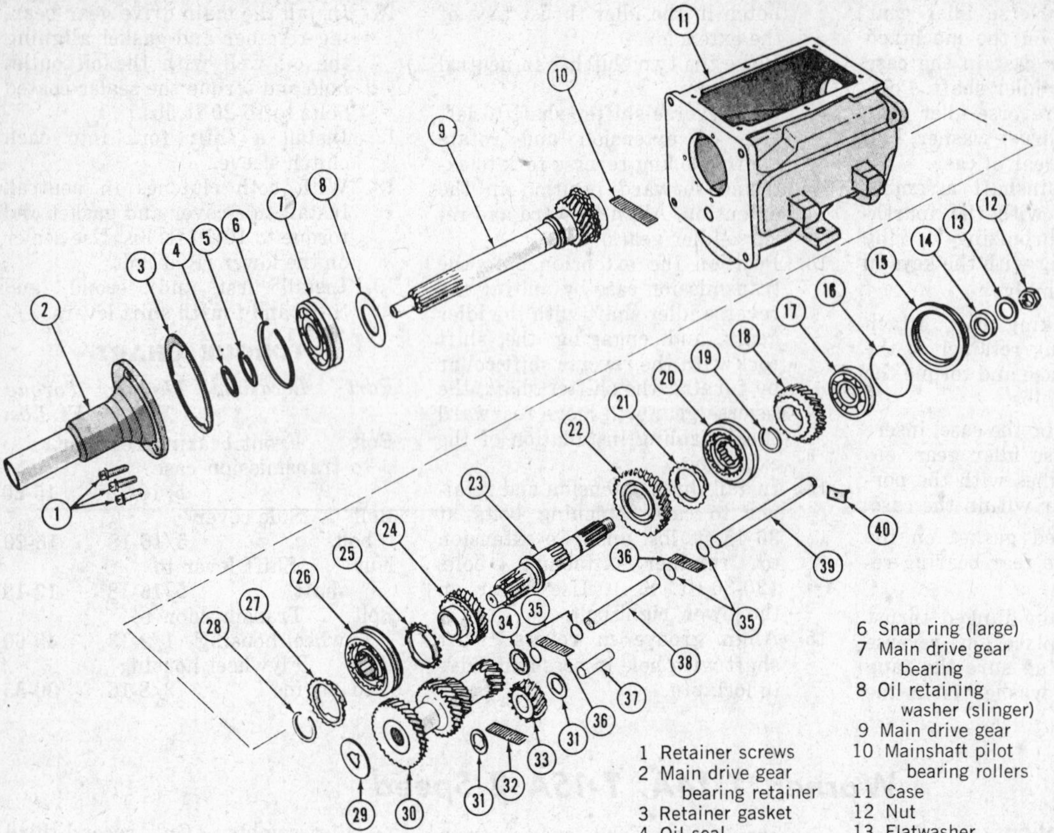

15 Bearing adapter
16 Snap ring
17 Mainshaft bearing
18 Reverse gear
19 Snap ring
20 Low synchronizer assembly
21 Synchronizer blocking ring
22 Low gear
23 Mainshaft
24 Second gear
25 Synchronizer blocking ring
26 Second-third synchronizer assembly
27 Synchronizer blocking ring
28 Snap ring
29 Countershaft front thrust washer (large)
30 Countershaft gear
31 Reverse idler gear bearing washer
32 Reverse idler gear roller bearings
33 Reverse idler gear
34 Countershaft rear thrust washer (small)
35 Countershaft bearing spacer washer
36 Countershaft roller bearings
37 Reverse idler shaft
38 Spacer
39 Countershaft
40 Lockplate

6 Snap ring (large)
7 Main drive gear bearing
8 Oil retaining washer (slinger)
9 Main drive gear
10 Mainshaft pilot bearing rollers
11 Case
12 Nut
13 Flatwasher
14 Spacer

1 Retainer screws
2 Main drive gear bearing retainer
3 Retainer gasket
4 Oil seal
5 Snap ring (small)

T-14A, T-15A three-speed transmission (© Borg Warner Corp.)

1 Reverse gear
2 Low synchronizer assembly
3 Low gear
4 Second gear
5 Second-Third synchronizer assembly
6 Main drive gear

Mainshaft assembly
(© Borg Warner Corp.)

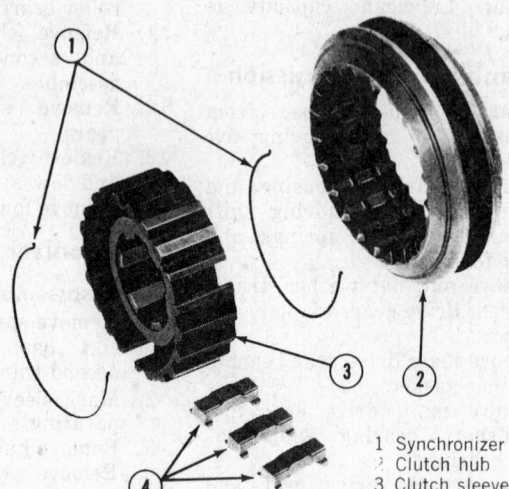

1 Synchronizer spring (2)
2 Clutch hub
3 Clutch sleeve
4 Shifter plate

Second—third synchronizer assembly (© Borg Warner Corp.)

Assembly of Transmission

1. Place reverse idler gear with dummy shaft, roller bearing, and thrust washers in case. Install reverse idler shaft.
2. Assembly countershaft center spacer, four bearing spacers, and bearing rollers in countershaft gear.
3. Install large countergear thrust washer in front of case. Position small thrust washer on countergear hub with lip facing groove in case. Holding countergear in position, push in countershaft from rear.
4. Install lock plate in slots of reverse idler shaft and countershaft.
5. Install to mainshaft:
   a. Low gear.
   b. Bronze blocking ring.
   c. Low synchronizer assembly.
   d. Largest snap-ring that fits in groove.
   e. Second gear.
   f. Bronze blocking ring.
   g. Second-third synchronizer assembly.
   h. Largest snap-ring that fits in groove.
   i. Reverse gear.
6. Install mainshaft assembly through top of case.
7. Install bronze blocking ring to second-third synchronizer assembly.
8. On remote shift units, install shifter shafts, with new O-rings, into case. NOTE: T-15 interlock levers are marked as to location.

1 Low-Reverse shift fork
2 Screwdriver
3 Second-Third interlock lever
4 Second-Third shift fork

Installing shifter forks
(© Borg Warner Corp.)

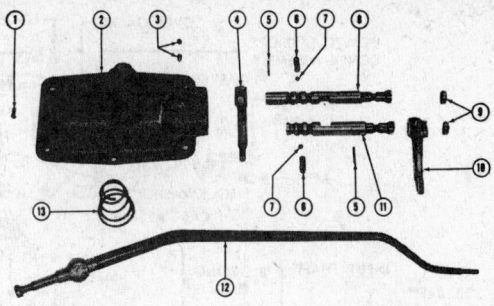

1 Control lever housing pin
2 Control housing
3 Interlock plunger and plug
4 Second-third shift fork
5 Shift fork pin
6 Poppet spring
7 Poppet ball
8 Second-third shift rail
9 Shift rail caps
10 Low-Reverse shift fork
11 Low-reverse shift rail
12 Shift lever
13 Shift lever support spring

Shift control assembly components
(© Borg Warner Corp.)

T-14 levers have no marks and are interchangeable.

9. Depress interlock lever while installing shift fork into shift lever and synchronizer clutch sleeve. Install poppet spring. Install tapered pins securing shafts in case.

10. Install main drive gear roller bearings.

11. Install main drive gear and oil slinger into case with cutaway portion of gear toward countergear. Install main drive gear to mainshaft.

12. Using bearing installer and thrust yoke tool, install main drive gear and mainshaft bearings and drive into position. The thrust yoke is needed to prevent damage to the synchronizer clutch.

13. Install main drive gear and mainshaft bearing snap-rings. The mainshaft bearing snap-ring is .010 thicker than main drive gear bearing snap-ring.

14. Install mainshaft rear bearing adapter, spacer, transfer case drive gear, flat washer, and nut. Torque nut to 130-170 ft. lbs.

15. Install main drive gear bearing retainer (with new oil seal) and gasket. Align oil drain holes in retainer and gasket.

16. Install case cover gasket. On remote shift units, install cover gasket with vent holes to left side.

17. Position gear train and floor-shift assembly in neutral. Insert shifter forks into clutch sleeves and torque to 8-15 ft. lbs.

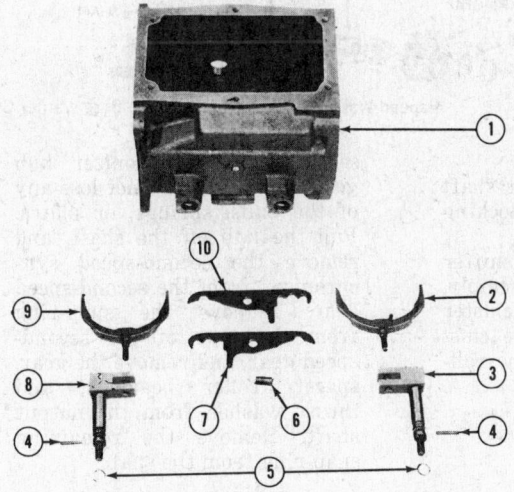

1 Case
2 Low-Reverse shift fork
3 Low-Reverse shift lever shaft
4 Tapered pin
5 O-ring
6 Poppet spring
7 Second-Third interlock lever
8 Second-Third shift lever shaft
9 Second-Third shift fork
10 Low-Reverse interlock lever

Shift bar housing components—remote control (© Borg Warner Corp.)

GENERAL INFORMATION

Model	T14A
Make	Warner
Ratios :	
Low	3.100:1
Second	1.612:1
High	1.000:1
Reverse	3.100:1

Warner T-18, T-19 Series 4 Speed

Description

The Warner T-18, T-19 are four-speed fully synchronized transmissions having a floor-mounted shift lever and a power take-off opening on the right side of the case. The T-18 first and reverse gears are spur gears while the others are helical. The T-19 has all helical gears.

Disassembly of Transmission

1. After draining the transmission and removing the parking break drum (or shoe assembly), lock the transmission in two gears and remove the U-joint flange, oil seal, speedometer driven gear and bearing assembly. Lubricant capacity is 6½ pints.

2. Remove the output shaft bearing retainer and the speedometer drive gear and spacer.

3. Remove the output shaft bearing snap-ring, and remove the bearing.

4. Remove the countershaft and idler shaft retainer and the power take-off cover.

5. After removing the input shaft bearing retainer, remove the snap-rings from the bearing and the shaft.

6. Remove the input shaft bearing and oil baffle.

7. Drive out the countershaft (from the front). Keep the dummy shaft in contact with the countershaft to avoid dropping

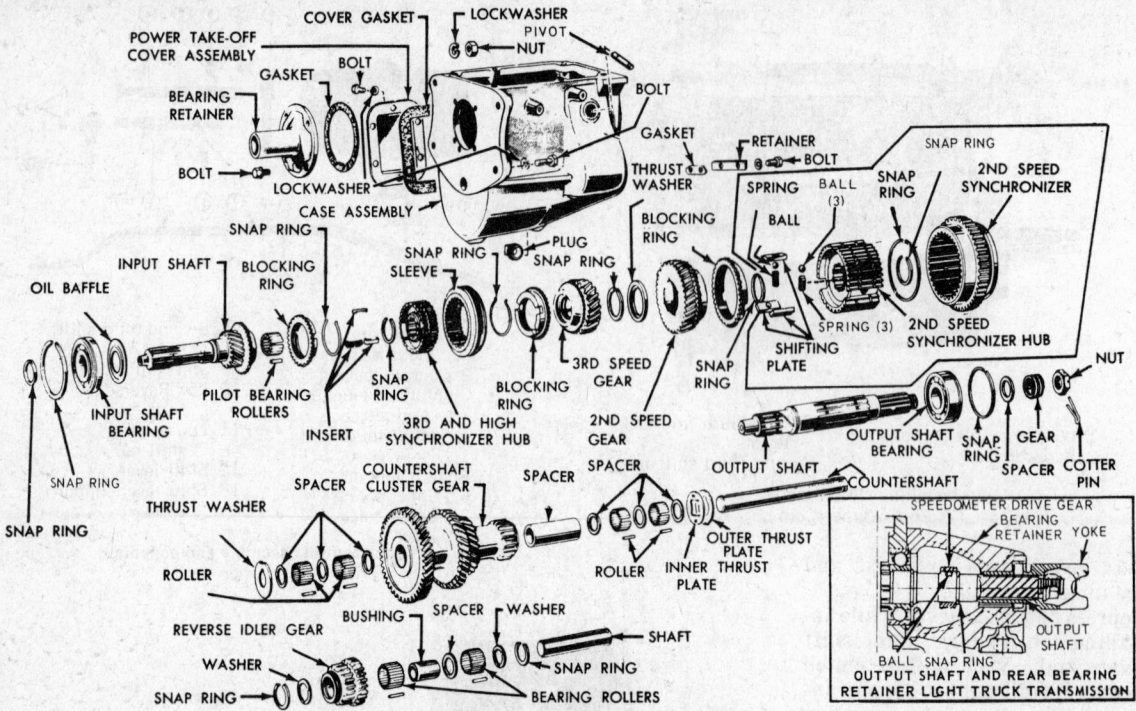

4-speed Warner T-18 transmission (© Borg Warner Corp.)

any rollers.

8. After removing the input shaft and the synchronizer blocking ring, pull the idler shaft.

9. Remove the reverse gear shifter arm, the output shaft assembly, the idler gear, and the cluster gear. When removing the cluster, do not lose any of the rollers.

Disassembly of Sub-Assemblies

Output Shaft

1. Remove the third- and high-speed synchronizer hub snap-ring from the output shaft, and slide the third- and high-speed synchronizer assembly and the third-speed gear off the shaft. Remove the synchronizer sleeve and the inserts from the hub. Before removing the two snap-rings from the ends of the hub, check the end play of the second-speed gear (0.005-0.024 inch).

2. Remove the second-speed synchronizer snap-ring. Slide the

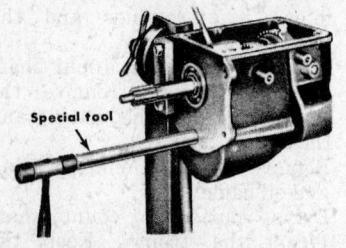

Special tool

Removing countershaft
(© Borg Warner Corp.)

second-speed synchronizer hub gear off the hub. Do not lose any of the balls, springs, or plates. Pull the hub off the shaft, and remove the second-speed synchronizer from the second-speed gear. Remove the snap-ring from the rear of the second-speed gear, and remove the gear, spacer, roller bearings, and thrust washer from the output shaft. Remove the remaining snap-ring from the shaft.

Cluster Gear

Remove the dummy shaft, pilot bearing rollers, bearing spacers, and center spacer from the cluster gear.

Reverse Idler Gear

Rotate the reverse idler gear on the shaft, and if it turns freely and smoothly, disassembly of the unit is not necessary. If any roughness is noticed, disassemble the unit.

Gear Shift Housing

1. Remove the housing cap and lever. Be sure all shafts are in neutral before disassembly.

2. Tap the shifter shafts out of the housing while holding one hand over the holes in the housing to prevent loss of the springs and balls. Remove the two shaft lock plungers from the housing.

Assembly of Sub-Assemblies

Cluster Gear Assembly

Slide the long bearing spacer into the cluster gear bore, and insert the dummy shaft in the spacer. Hold the

cluster gear in a vertical position, and install one of the bearing spacers. Position the 22 pilot bearing rollers in the cluster gear bore. Place a spacer on the rollers, and install 22 more rollers and another spacer. Hold a large thrust washer against the end of cluster gear and turn the assembly over. Install the rollers and spacers in the other end of the gear.

Reverse Idler Gear Assembly

1. Install a snap-ring in one end of the idler gear, and set the gear on end, wtih the snap-ring at the bottom.

2. Position a thrust washer in the gear on top of the snap-ring. Install the bushing on top of the washer, insert the 37 bearing rollers, and then a spacer followed by 37 more rollers. Place the remaining thrust washer on the rollers, and install the other snap-ring.

Output Shaft Assembly

1. Install the second speed gear thrust washer and snap-ring on the output shaft. Hold the shaft vertically, and slide on the second-speed gear. Insert the bearing rollers in the second-speed gear, and slide the spacer into the gear. (The T-18 model does not contain second speed gear rollers or spacer). Install the snap-ring on the output shaft at the rear of the second-speed gear. Position the blocking ring on the second-speed gear. Do not invert the shaft because the bearing rollers will slide out of the gear.

2. Press the second-speed synchronizer hub onto the shaft, and install the snap-ring. Position the shaft vertically in a soft-jawed vise. Position the springs and plates in the second-speed synchronizer hub, and place the hub gear on the hub.

3. With the T-19 model, press the first and second speed synchronizer onto the shaft and install the snap-ring. Install the first speed gear and snap-ring on the shaft and press on the reverse gear. For the T-19, ignore steps 2 and 4.

4. Hold the gear above the hub spring and ball holes, and position one ball at a time in the hub, and slide the hub gear downward to hold the ball in place. Push the plate upward, and insert a small block to hold the plate in position, thereby holding the ball in the hub. Follow these procedures for the remaining balls.

5. Install the third speed gear and synchronizer blocking ring on the shaft.

6. Install the snap-rings at both ends of the third and high-speed synchronizer hub. Stagger the openings of the snap-rings so that they are not aligned. Place the inserts in the synchronizer sleeve, and position the sleeve on the hub.

7. Slide the synchronizer assembly onto the output shaft. The slots in the blocking ring must be in line with the synchronizer in-

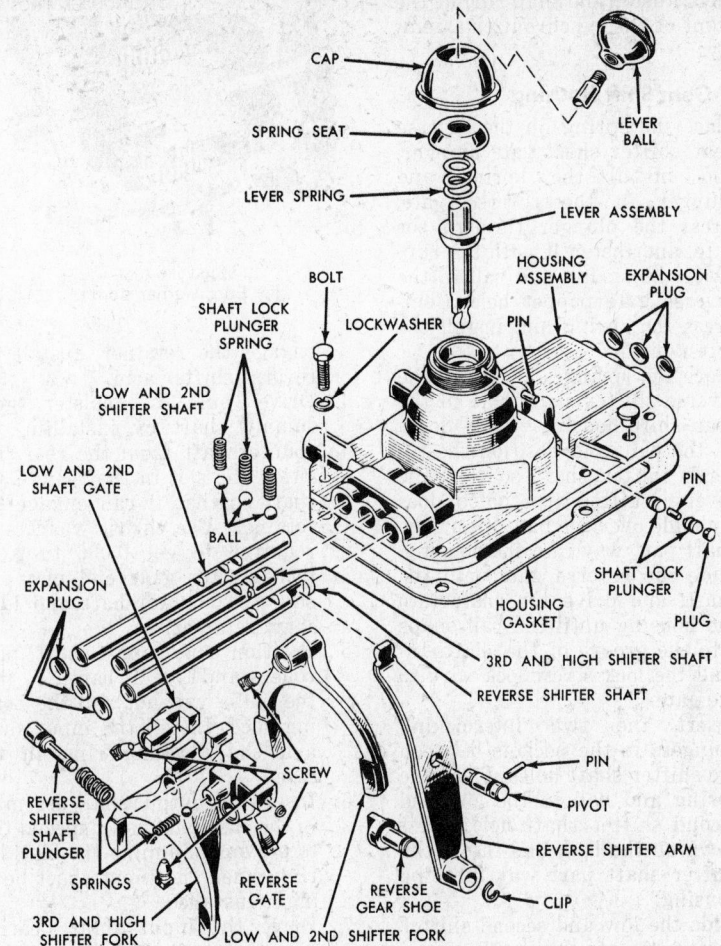

Conventional cab shift linkage (© Borg Warner Corp.)

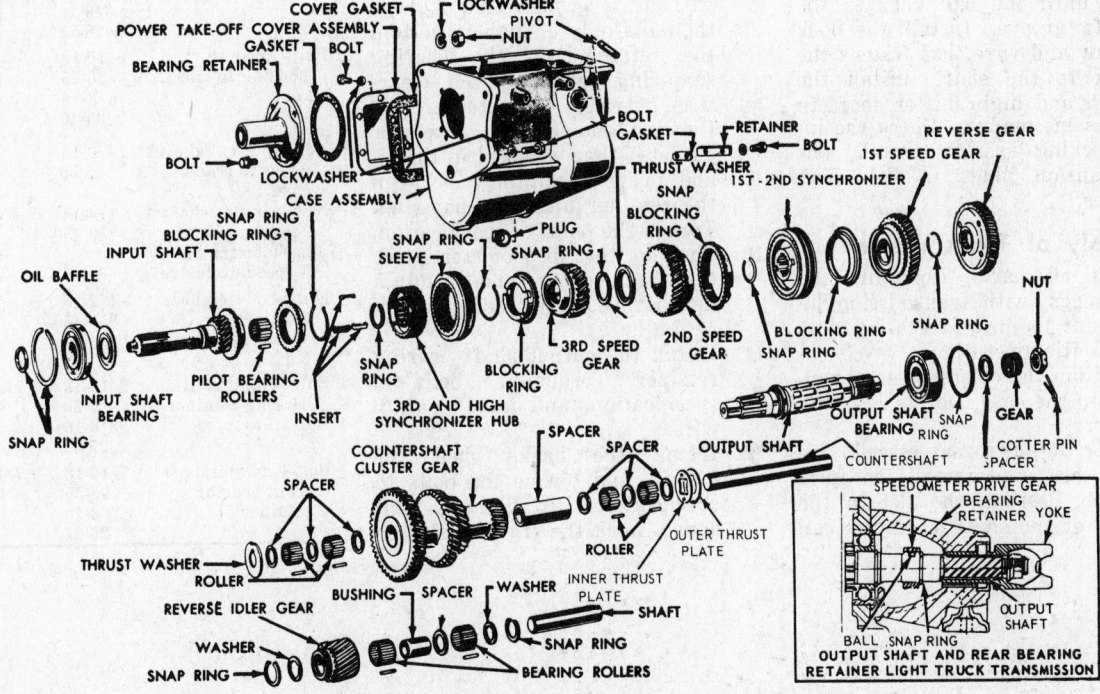

4-speed Warner T-19 transmission (© Borg Warner Corp.)

serts. Install the snap-ring at the front of the synchronizer assembly.

Gear Shift Housing

1. Place the spring on the reverse gear shifter shaft gate plunger, and install the spring and plunger in the reverse gate. Press the plunger through the gate, and fasten it with the clip. Place the spring and ball in the reverse gate poppet hole. Compress the spring and install the cotter pin.
2. Place the spring and ball in the reverse shifter shaft hole in the gear shift housing. Press down on the ball, and position the reverse shifter shaft so that the reverse shifter arm notch does not slide over the ball. Insert the shaft part way into the housing.
3. Slide the reverse gate onto the shaft, and drive the shaft into the housing until the ball snaps into the groove of the shaft. Install the lock screw lock wire to the gate.
4. Insert the two interlocking plungers in the pockets between the shifter shaft holes. Place the spring and ball in the low and second shifter shaft hole. Press down on the ball, and insert the shifter shaft part way into the housing.
5. Slide the low and second shifter shaft gate onto the shaft, and install the corresponding shifter fork on the shaft so that the offset of the fork is toward the rear of the housing. Push the shaft all the way into the housing until the ball engages the shaft groove. Install the lock screw and wire that fastens the fork to the shaft. Install the third and high shifter shaft in the same manner. Check the interlocking system. Install new expansion plugs in the shaft bores.

Assembly of Transmission

1. Coat all parts, especially the bearings, with transmission lubricant to prevent scoring during initial operation.
2. Position the cluster gear assembly in the case. Do not lose any rollers.
3. Place the idler gear assembly in the case, and install the idler shaft. Position the slot in the rear of the shaft so that it can

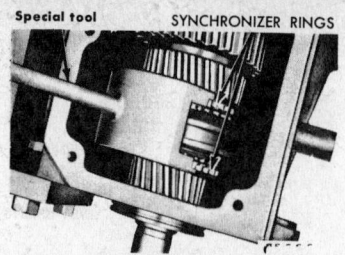

Special tool · SYNCHRONIZER RINGS

Stop yoke tool
(© Borg Warner Corp.)

engage the retainer. Install the reverse shifter arm.
4. Drive out the cluster gear dummy shaft by installing the countershaft from the rear. Position the slot in the rear of the shaft so that it can engage the retainer. Use thrust washers as required to get 0.006 to 0.020 inch cluster gear end play. Install the countershaft and idler shaft retainer.
5. Position the input shaft pilot rollers and the oil baffle, so that the baffle will not rub the bearing race. Install the input shaft and the blocking ring in the case.
6. Install the output shaft assembly in the case, and use a special tool to prevent jamming the blocking ring when the input shaft bearing is installed.
7. Drive the input shaft bearing onto the shaft. Install the thickest select-fit snap-ring that will fit on the bearing. Install the input shaft snap-ring.
8. Install the output shaft bearing.
9. Install the input shaft bearing without a gasket, and tighten the bolts only enough to bottom the retainer on the bearing snap-ring. Measure the clearance between the retainer and the case, and select a gasket (or gaskets) that will seal in the oil and prevent end play between the retainer and the snap-ring. Torque the bolts to specification.
10. Position the speedometer drive gear and spacer, and install a new output shaft bearing retainer seal.
11. Install the output shaft bearing retainer. Torque the bolts to specification, and install safety wire.
12. Install the brake shoe (or drum), and torque the bolts to specification. Install the U-joint flange. Lock the transmission in

two gears and torque the nut to specification.
13. Install the power take-off cover plates with new gaskets. Fill the transmission according to specifications.

TORQUE CHART

Nomenclature	Nuts and/or Bolts and Torque Limits		
Bolt—gear shift lever tower to gearshift housing	3/8-16 20-25	7/16-14 30-35	
Bolt—clutch housing to trans. case	7/16-14 30-38 5/8-11 96-120	9/16-12 70-90	
Nut—U-joint flange to trans. output shaft	1.00-20 90-125	1 1/2-18 275-350	1 1/4-18 225-275
Nut—drum parking brake to companion flange	3/8-24 35-45	7/16-20 50-70	
Nut—bellcrank to trans.	9/16-18 70-90		
Bolt—lever assy. to trans.	3/8-16 20-25		
Nut—handbrake anchor bar to trans. case (5-speed extra-heavy duty only)	9/16-18 120-130		
Bolt—belcrank to trans.	3/8-16 20-25		
Bolt—reverse lock-out plunger retainer	11/16-16 80-100		
Bolt—countershaft rear bearing retainer	5/16-18 25-30 3/8-16 35-40	7/16-14 45-55 1/2-13 60-70	
Bolt—countershaft & reverse idler shaft retainer	5/16-18 25-30 3/8-16 25-37 3/8-16† 18-25†	7/16-14 40-45 1/2-13 80-85	
Bolt—gear shift housing to trans. case	5/16-18 20-25 3/8-16 35-40	3/8-16†† 30-35†† 7/16-14 45-50	
Bolt—power take off cover to trans. case	3/8-16 20-30		
Nut—countershaft bearing lock (5-speed extra h.d. & 5-speed exclusive)	1 1/4-18 350-450		
Nut—countershaft bearing lock (5-speed exclusive h.d.)	1 1/2-18 350-450		
Bolt—input shaft bearing retainer to trans. case	5/16-18 25-30 3/8-16 25-30	7/16-14 40-45	
Bolt—countershaft front bearing retainer	5/16-18 25-30 3/8-16 25-35	7/16-14 50-55	

Warner T-85N 3 Speed W/O.D.

Description

The Warner T-85N overdrive is a synchronized 3 speed transmission with a 2 speed planetary gear transmission attached to the rear of the housing. An electrical control system automatically controls the overdrive shifts. Lubricant capacity is 4 pints.

Disassembly of Transmission

1. Mount the transmission in a vise.
2. Remove the gearshift housing assembly and gasket from the transmission.
3. Remove the shaft levers from the camshafts. Pull the shifter forks and cams out of the gearshift housing. With the cams removed, the interlock balls, sleeve, and spring will fall out of the housing.
4. Pull the shifter forks out of the cam and shaft assembly. Remove the oil seals from the camshafts.
5. Pull the solenoid body about ¼ turn, and remove it. Remove the governor.
6. With a sharp punch, pierce the snap-ring hole cover and remove the cover.
7. Remove the overdrive housing bolts, and the overdrive control shaft and lever pin.

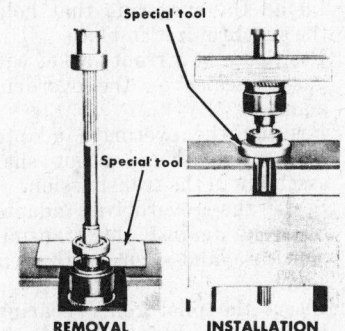

Mainshaft bearing removal or installation
(© Borg Warner Corp.)

8. Pull out the manual control shaft and lever as far as possible. Spread the snap-ring that retains the overdrive main shaft front bearing, and then remove the overdrive housing. It may be necessary to tap the overdrive main shaft to free the main shaft bearing from the housing.
9. Remove the overdrive main shaft from the assembly and the free-wheel unit rollers.
10. Remove the speedometer gear and drive ball. To install the main shaft bearing, press on the new bearing, and then install the thickest snap-ring that will fit. Snap-rings are available in the following sizes: 0.086-0.088,

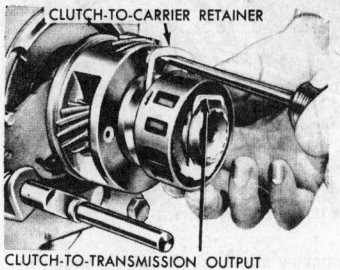

Removing free wheel unit retainer
(© Borg Warner Corp.)

0.0890-0.091, 0.092-0.094, and 0.095-0.097 inch.

11. Remove the free-wheel unit retainers. The free-wheel unit, planetary gear, sun gear, and shaft rail can now be removed.
12. Remove the snap-ring from the adapter, and then remove the plate and trough, balk ring and gear, and pawl.
13. Remove the input shaft bearing retainer and gasket and replace seal if necessary.
14. Rotate the overdrive adapter to expose the countershaft lock, and remove the lock.
15. With a drift, drive the countershaft toward the rear until it just clears the front case bore, and then push the countershaft out the rear with the tool shown.
16. Tap the input shaft and bearing out of the front of the case.
17. If the input shaft bearing is to

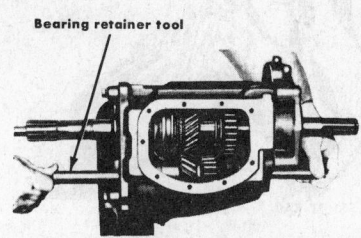

Removing countershaft—typical
(© Borg Warner Corp.)

be replaced, press off the bearing. The bearing baffle acts as a slinger, and must be installed so that it does not rub the bearing outer race as the shaft turns.

18. Remove the overdrive adapter and transmission output shaft as an assembly.
19. Remove the snap-ring at the front of the transmission output shaft, and then slide the synchronizer assembly, intermediate gear, and the sliding low and reverse gear off the shaft.
20. Disassemble the synchronizer unit by sliding the intermediate and high sleeve off the hub. Remove the three inserts and two springs from the hub.
21. Remove the snap-ring that holds the output shaft in the adapter and tap the bearing and shaft out of the adapter. Remove the baffle from the adapter.

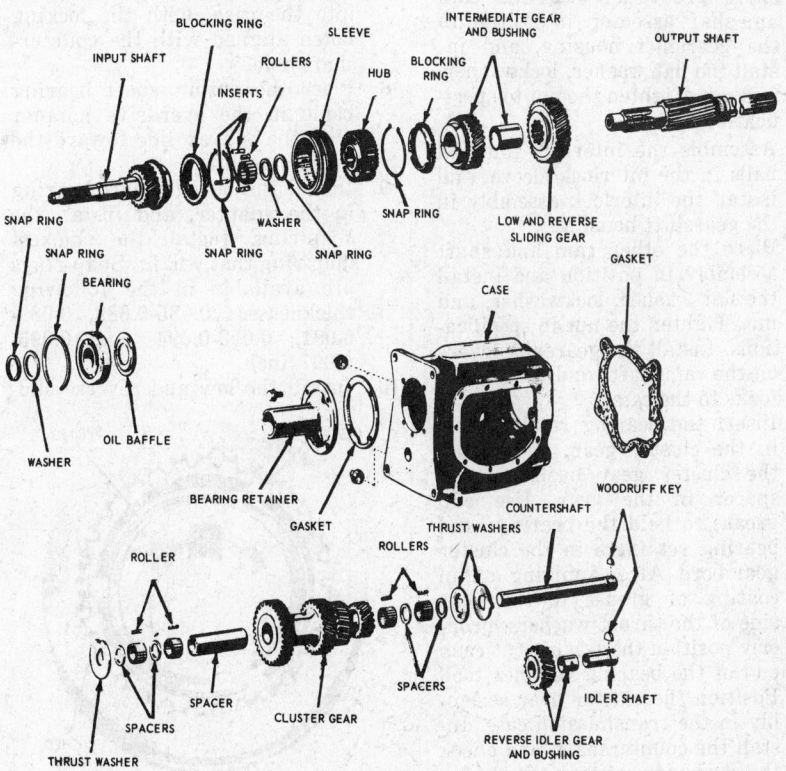

Warner 3-speed transmission
(© Borg Warner Corp.)

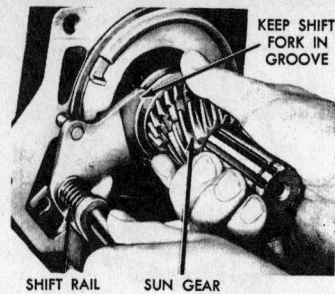

KEEP SHIFT FORK IN GROOVE

SHIFT RAIL   SUN GEAR

Synchronizer assembly
(© Borg Warner Corp.)

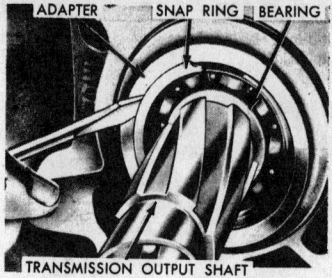

ADAPTER   SNAP RING   BEARING

TRANSMISSION OUTPUT SHAFT

Removing output shaft bearing snap-ring
(© Borg Warner Corp.)

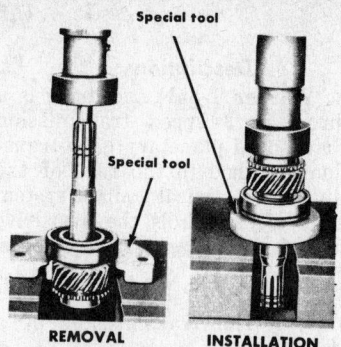

Special tool

Special tool

REMOVAL   INSTALLATION

Input shaft bearing replacement
(© Borg Warner Corp.)

22. If the output shaft or bearing is to be replaced, press the old bearing off and the new bearing on and install the thickest snap-ring that will fit. Snap-rings are available in the following thicknesses: 0.0890-0.091, 0.092-0.094, 0.095-0.097, and 0.100-0.0102 inch.

23. While holding the washers and bearing retainer tool at the small end of the cluster gear to prevent the roller bearings and bearing retainer tool from falling out, lift the cluster gear assembly from the case.

24. Drive the reverse idler gear shaft out of the rear of the case.

Assembly of Transmission

NOTE: Always use new gaskets and gasket cement during assembly. Put a thin coating of lubricant on all parts before installation.

1. Install new oil seals in the camshaft grooves. Place one cam and shaft assembly in position in the gearshift housing, and install the flat washer, lockwasher, and nut. Tighten the nut to specifications.

2. Assemble the interlock pin and balls in the interlock sleeve, and install the interlock assembly in the gearshift housing.

3. Place the other cam and shaft assembly in position and install the flat washer, lockwasher, and nut. Tighten the nut to specifications. Install the gearshift levers on the camshafts and the shifter forks to the cams.

4. Insert the bearing retainer tool in the cluster gear, and install the cluster gear bushings and spacer in the gear. Use cup grease to hold the bearings and bearing retainers in the cluster gear bore. After applying a thin coating of grease to the gear side of the thrust washers, properly position the washers at each end of the bearing retainer tool.

5. Position the cluster gear assembly in the transmission case. Install the countershaft, and check the clearance between the cluster gear front thrust washer and the case (0.004-0.018 inch).

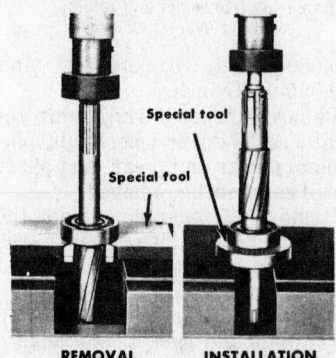

Special tool

Special tool

REMOVAL   INSTALLATION

Removing adapter and output shaft
(© Borg Warner Corp.)

Remove the countershaft by installing the bearing retaining tool.

6. Position the reverse idler gear in the case, with the chamfered gear teeth ends toward the front.

7. Drive the reverse idler shaft into the case, with the locking notch aligned with the countershaft hole.

8. Place the output shaft bearing baffle in the overdrive adapter with the convex side toward the overdrive unit.

9. Install the output shaft bearing in the adapter, and install the snap-ring. Install the thickest snap-ring that will fit. Snap-rings are available in the following thicknesses: 0.086-0.088, 0.089-0.091, 0.092-0.094 and 0.095-0.097 inch.

10. Install the low and reverse slid-

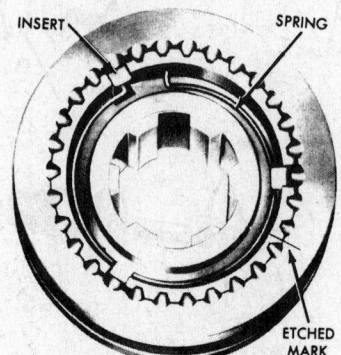

INSERT   SPRING

ETCHED MARK

Installing sun gear and shift rail
(© Borg Warner Corp.)

ing gear on the output shaft, with the shifter fork groove toward the front.

11. Install the intermediate gear on the output shaft, with the clutch teeth toward the front.

12. Assemble the synchronizer unit by installing the two springs on the hub, and placing the three inserts in the hub. Hook one spring end in an insert.

13. After lining up the etched marks on the sleeve and hub splines, slide the sleeve onto the hub.

14. Place a blocking ring on the intermediate gear, and install the synchronizer assembly on the transmission output shaft, with the hub thrust surface toward the rear.

15. Install the snap-ring that holds the synchronizer in place.

16. Position a new front gasket with gasket sealer on the overdrive adapter.

17. Position the overdrive adapter and transmission output shaft assembly in the transmission.

18. Seat the overdrive adapter squarely against the transmission case, and secure with a cap screw.

19. Insert the pilot roller bearings in the input shaft and hold with grease.

20. Place the blocking ring on the input shaft gear.

21. Tap the input shaft and bearing into the case while lining up the slots in the blocking ring with the synchronizer inserts.

22. Position the input shaft bearing retainer without a gasket, and tighten the bolts to bottom the retainer on the bearing outer race snap-ring. Check the clearance between the retainer gasket surfaces and the case.

23. Select a gasket (or gaskets) which will seal the oil, and at the same time prevent end play between the bearing outer race snap-ring and the retainer. Gaskets are available in 0.010, 0.015, 0.020 and 0.025 inch thicknesses.

24. Install the input shaft bearing retainer on the transmission

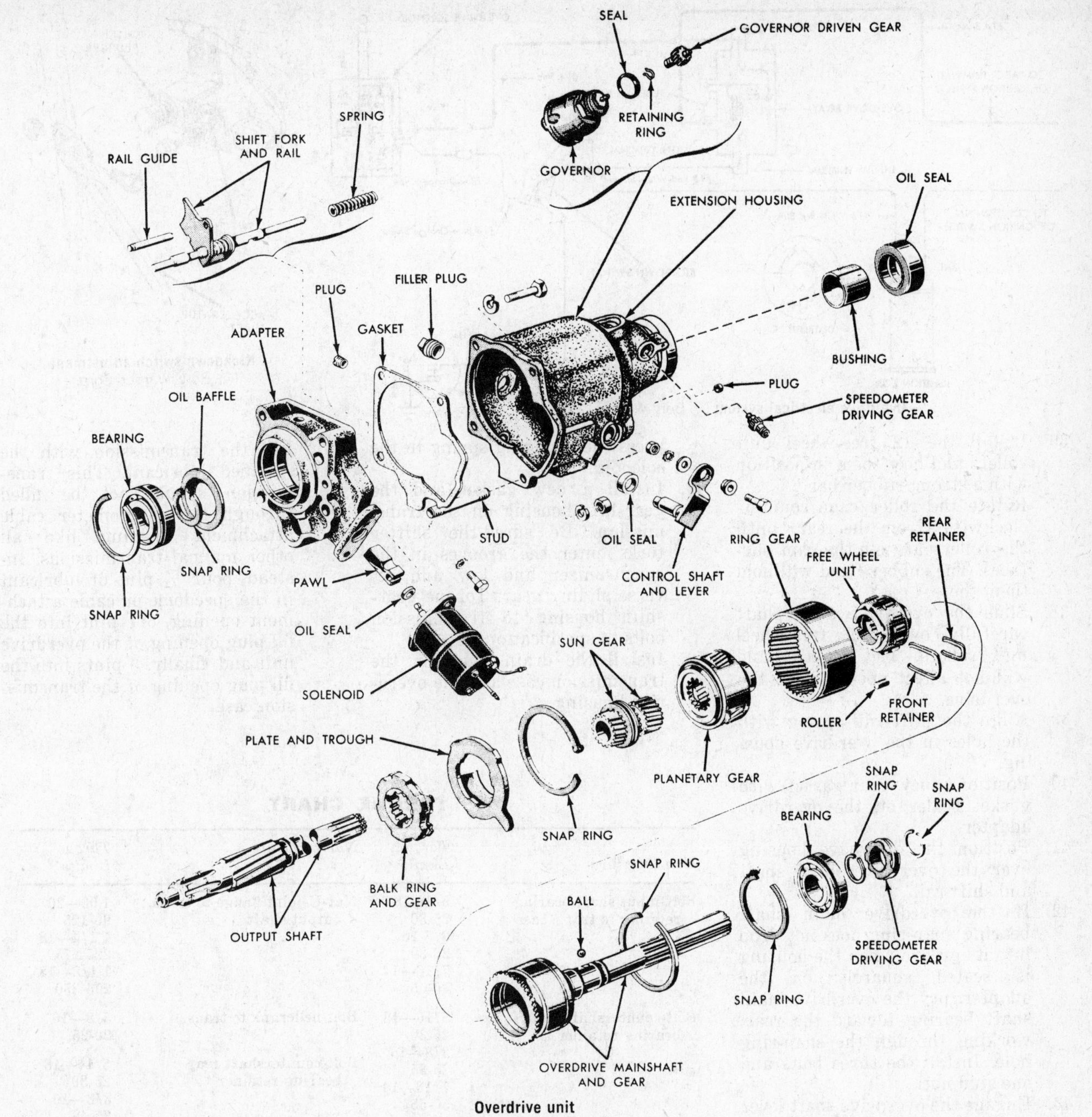

SEAL

GOVERNOR DRIVEN GEAR

RETAINING RING

GOVERNOR

EXTENSION HOUSING

OIL SEAL

RAIL GUIDE

SHIFT FORK AND RAIL

SPRING

BUSHING

PLUG

SPEEDOMETER DRIVING GEAR

PLUG

FILLER PLUG

ADAPTER

GASKET

OIL BAFFLE

BEARING

SNAP RING

SNAP RING

PAWL

OIL SEAL

SOLENOID

PLATE AND TROUGH

STUD

OIL SEAL

CONTROL SHAFT AND LEVER

SUN GEAR

RING GEAR

FREE WHEEL UNIT

REAR RETAINER

ROLLER

FRONT RETAINER

PLANETARY GEAR

SNAP RING

SNAP RING

BEARING

SNAP RING

SNAP RING

OUTPUT SHAFT

BALK RING AND GEAR

SNAP RING

BALL

SPEEDOMETER DRIVING GEAR

SNAP RING

OVERDRIVE MAINSHAFT AND GEAR

Overdrive unit
(© Borg Warner Corp.)

case. Torque the bolts to specification.

25. Pull out the adapter about ¼ inch, and rotate it to expose the countershaft hole.

26. Work the cluster gear into normal position by rotating the input and output shafts.

27. Push the countershaft into the case from the rear.

28. Align the slot in the countershaft with the slot in the reverse idler shaft, and install the lock plate.

29. Rotate the adapter to its normal position and seat it squarely against the transmission case. Check the block rings to make sure the slots are aligned with the synchronizer inserts.

30. Move the transmission so that the input shaft is pointing down.

31. Install the stud in the transmission case.

PAWL   BALK RING GEAR

INSTALL WITH MACHINED
RECESS IN THIS POSITION

Installing plate and trough
(© Borg Warner Corp.)

32. Place the balk ring gear assembly and pawl in the adapter.

33. Place the plate and trough assembly in the adapter and install its snap-ring.

34. Install the sun gear and shift rail.

35. Install the planet carrier and the clutch cam. Install the retainers.

Installing Mainshaft assembly
(© Borg Warner Corp.)

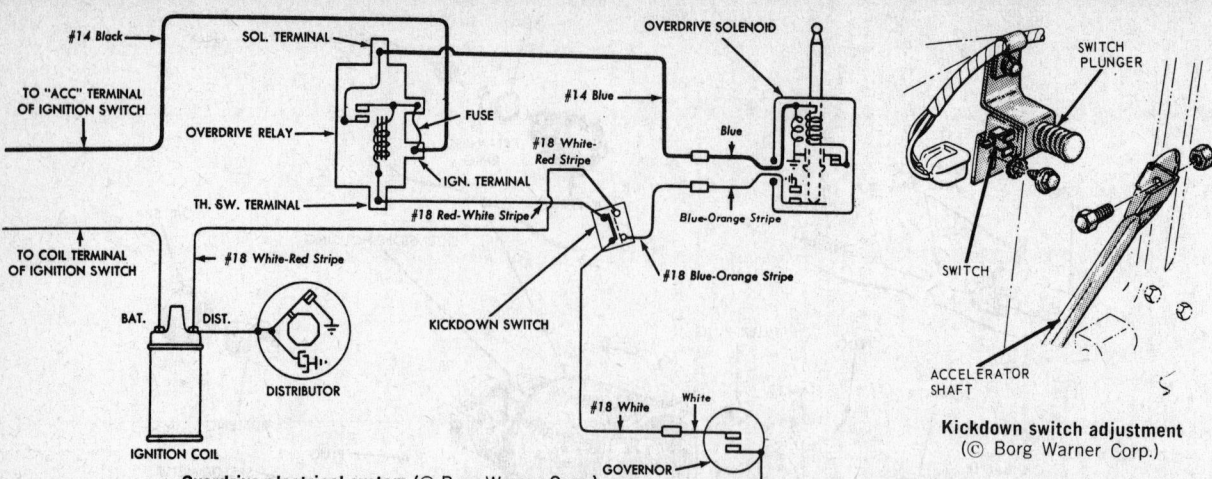

Overdrive electrical system (© Borg Warner Corp.)

Kickdown switch adjustment
(© Borg Warner Corp.)

36. Install the 12 free-wheel unit rollers and hold them in position with a strong rubber band.

37. Rotate the roller cage counter-clockwise (from the rear) until the rollers are off the cam surfaces. The rubber band will hold them there.

38. Slide the overdrive main shaft carefully over the free-wheel unit rollers. The rubber band will not affect operation of the overdrive.

39. Align the shift rail spring with the holes in the overdrive housing.

40. Position a new rear gasket with gasket sealer on the overdrive adapter.

41. Position the overdrive housing over the overdrive main shaft and shift rail.

42. If the overdrive main shaft bearing snap-ring does not drop into its groove when the housing is seated squarely on the adapter, pry the overdrive main shaft bearing toward the rear, working through the snap-ring hole. Install the three bolts and one stud nut.

43. Engage the overdrive shaft lever by pushing it inward. The lever is correctly engaged when a spring load is apparent as the lever is pushed.

44. Install the retaining pin in the overdrive housing to hold the control shaft in place.

45. If necessary, replace the overdrive housing bushing.

46. Thread the governor into the overdrive housing.

47. With the cap drain hole at the bottom, rotate the solenoid 1/4 turn from normal position, so that the half ball on the solenoid stem can engage the pawl. Install the two solenoid cap screws.

48. If the solenoid stem is properly engaged, the solenoid cannot be removed from the overdrive in its normal position. Any attempt to pull it out will merely compress the engaging spring in the solenoid.

49. Install a new gasket and the gearshift housing on the transmission. Be sure the shifter forks enter the grooves in the synchronizer and low and reverse sliding gear. Torque gearshift housing to transmission bolts to specifications.

50. Install the drain plugs in the transmission case and the overdrive housing.

51. Fill the transmission with the specified lubricant. This transmission should not be filled through the speedometer cable attachment opening like all other manual transmissions. Instead, pour 1/2 pint of lubricant in the speedometer cable attachment opening, one pint into the fill plug opening of the overdrive unit and, finally, 3 pints into the fill plug opening of the transmission case.

TORQUE CHART

Nomenclature	Torque Limits	Nomenclature	Torque Limits
Bolt-input shaft bearing retainer to trans. case	5/16-18 25-30 3/8-16 25-30 7/16-14 40-45	Nut-U-joint flange to trans. output shaft	1.00-20 90-125 1 1/4-18 225-275 1 1/2-18 275-350
Bolt-countershaft front bearing retainer	5/16-18 25-30 3/8-16 25-35 7/16-14 50-55	Bolt-bellcrank to trans.	3/8-16 20-25
		Bolt-countershaft rear bearing retainer	5/16-18 25-30 3/8-16 35-40 7/16-14 45-55 1/2-13 60-70
Bolt-reverse lockout plunger retainer	11/16-16 90-100		
Bolt-clutch housing to engine block	7/16-14 24-32 33-45	Nut-bellcrank to trans.	9/16-18 70-90 3/8-16C 20-25
Nut-drum parking brake to companion flange	3/8-24 37-50 7/16-20 40-55 1/2-20 74-86	Bolt-countershaft & reverse idler shaft retainer	1/2-13 60-70 5/16-18 25-30 3/8-16 25-37 3/8-16 18-25 7/16-14 40-45 1/2-13 80-85
Bolt-lever assy. to trans.	3/8-16 20-25		
Bolt-clutch housing to trans. case	7/16-14 30-38 5/8-11 96-120 9/16-12 70-90 5/8-18C 115-140 9/16-18C 81-102	Bolt-gear shaft housing trans. case	5/16-18 20-25 3/8-16 35-40

Warner T86 Three Speed

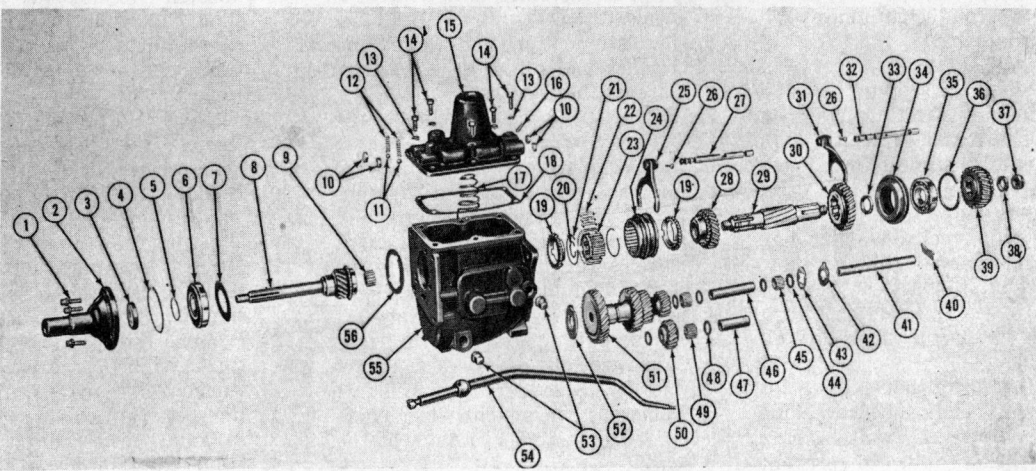

1 Bearing retainer screws
2 Main drive gear bearing retainer
3 Bearing retainer oil seal
4 Bearing snap ring
5 Main drive gear snap ring
6 Main drive gear bearing
7 Front bearing oil retaining washer
8 Main drive gear
9 Pilot roller bearing
10 Shift rail cap
11 Poppet ball
12 Poppet spring
13 Lock washer
14 Shift housing bolt

15 Control housing
16 Interlock plunger
17 Shift lever spring
18 Shift tower gasket
19 Blocking ring
20 Clutch hub snap ring
21 Synchronizer spring
22 Synchronizer plate
23 Clutch hub
24 Clutch sleeve
25 High and intermediate clutch fork
26 Shift fork pin
27 High and intermediate shift rail
28 Second speed gear
29 Main shaft
30 Low and reverse sliding gear

31 Low and reverse shift fork
32 Low and reverse shift rail
33 Bearing spacer
34 Rear bearing adapter
35 Rear bearing
36 Rear bearing snap ring
37 Nut
38 Washer
39 Transfer case drive gear
40 Lock plate
41 Countershaft
42 Rear Countershaft thrust washer (steel)
43 Rear countershaft

thrust washer (bronze)
44 Countershaft bearing washer
45 Countershaft bearing
46 Countershaft center bearing spacer
47 Reverse idler gear shaft
48 Reverse idler gear bearing washer
49 Reverse idler gear roller bearings
50 Reverse idler gear
51 Countershaft gear
52 Countershaft front thrust washer
53 Plug
54 Shift lever
55 Transmission case
56 Retainer gasket

T-86 three-speed transmission—floor shift (© Borg Warner Corp.)

Description

The Warner T-86 is a synchronized three-speed transmission having either a floor-mounted or a column shift lever.

Disassembly of Transmission

Except Jeep Models

1. Remove cover.
2. Remove front bearing cap, clutch shaft snap-ring and bearing lock-ring.
3. Remove the front bearing, using a bearing puller and a thrust yoke to prevent damaging synchronizer clutches.
4. Remove extension housing. Drive out seal from inside housing with oil seal remover and installer tool. Use bushing remover and installer tool to replace bushing.
5. Move mainshaft assembly back about ¾ in. Lower front end of clutch shaft, move mainshaft assembly over countergear and out of shift forks. Remove clutch shaft from front of case.
6. Check clutch shaft roller bearings for wear or damage.
7. Remove snap-ring speedometer drive gear, and key.

8. Remove snap-ring, synchro-clutch assembly, second gear, friction ring, and low-reverse sliding gear. Press off rear bearing.
9. Remove shifter forks.
10. Remove shaft lock plate from rear of case.
11. Drive countershaft out to rear, using dummy shaft. Lower countergear to bottom of case.
12. Drive out reverse idler shaft. Remove reverse idler gear and countergear.
13. Remove outer shift levers and shifter shaft lock pins. Remove shifter shafts, two interlock ball bearings, sleeve and spring. Remove shaft oil seals from case.

Assembly of Transmission

1. Install new shift shaft oil seals.
2. Install low-reverse shift shaft, interlock sleeve, ball bearing, pin, and spring. Install second-high shift shaft and second ball bearing.
3. Shift mechanism into any gear position with one end of interlock sleeve against shift shaft quadrant, clearance between opposite end of sleeve and quadrant on other shaft should be

.001-.007 in. Interlock sleeves are available in several sizes for adjustment.
4. Install lock pins and shift levers.
5. Install dummy shaft and bearings in countergear. Install thrust washers; the bronze front washer must index with the case.
6. Place countergear and dummy shaft in bottom of case.
7. Install reverse idler gear with chamfered side of teeth to front of case. Drive in shaft from rear.
8. Align slots in countershaft and reverse idler shaft. Position countergear and drive in countershaft. Install lock plate.
9. Install shifter forks.
10. Press rear bearing onto mainshaft. Install key, speedometer drive gear, and snap-ring.
11. Install low-reverse sliding gear on mainshaft with sliding collar to front. Gear should slide easily. Install second gear with tapered cone to front.
12. Install rear bearing snap-ring—the thickest possible.
13. Install synchro-clutch assembly and front snap-ring.
14. When synchro-clutch hub is

pressed against snap-ring, there should be .003-.010 in. clearance between second gear and shoulder on mainshaft.

15. Install 14 greased rollers in clutch shaft.
16. Install front friction ring on clutch shaft and insert through top of case. Install mainshaft assembly through rear of case, moving to right to engage shifter forks in synchro-clutch collar and low-reverse sliding gear. Guide clutch shaft onto main shaft.
17. Install extension housing oil seal.
18. Install extension housing.
19. Install oil slinger, concave side to rear. Drive in front bearing using thrust yoke to prevent synchronizer damage.
20. Install thickest clutch shaft snap-ring that will fit in groove.
21. Install front bearing cap with new gasket. Choose thickness of gasket to give zero clutch shaft end play.
22. Clearance of friction rings should be .056-145 in.
23. Install cover end gasket.

Disassembly of Transmission

Jeep Models

1. Drain lubricant and flush out the case.
2. If a transfer case is involved, remove its rear cover.

3. If a power take-off is involved, remove the shift unit which replaces the cover.
4. Remove cotter pin, nut and washer and remove the transfer case main drive gear.
5. Remove the transmission shift cover.
6. Loop a piece of wire around the mainshaft just back of second-speed gear. Twist the wire and attach one end to the right front cover screw, the other end to the left cover screw. Tighten the wire to prevent the mainshaft from pulling out of the case when the transfer case is removed. Should the mainshaft come out, the synchronizer parts will drop into the bottom of the case.
7. Remove transfer case screws then tap lightly on the end of the transmission mainshaft to sep-

Transfer case driving gear
(© Borg Warner Corp.)

arate the two units. The transmission mainshaft bearing should slide out of the transfer case and stay with the transmission.
8. Remove front main drive gear bearing retainer and gasket.
9. Remove the oil collector hollowhead screws.
10. Remove lock plate from the reverse idler shaft and countershaft, at the rear of the case.
11. Drive the countershaft out the rear of the case with a dummy shaft and a brass drift.
12. Remove the mainshaft assembly through the case rear opening. Remove main drive gear.
13. Remove the countershaft gear set and three thrust washers from the bottom of the case, then dismantle the countershaft gear assembly.
14. Remove the reverse idler shaft and gear using a brass drift.
15. On column shift models, check clearance between ends of interlock sleeve and notched surface of each shift lever. The correct clearance is .001-.007 in. Several sizes of interlock sleeves are available for adjustment.

Assembly of Transmission

To assemble, reverse the disassembly procedures, giving the following points particular attention:

1. The countershaft gear set, when assembled in the case, should

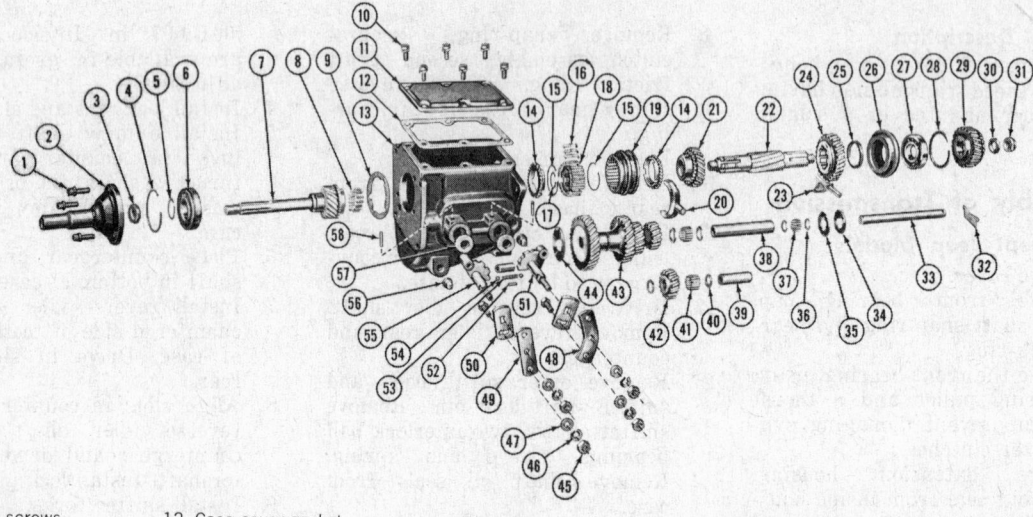

1 Retainer screws	12 Case cover gasket	27 Bearing
2 Main drive gear retainer	13 Case	28 Snap ring
3 Oil seal	14 Blocking ring	29 Transfer case drive gear
4 Snap ring (large)	15 Synchronizer spring	30 Washer
5 Snap ring (small)	16 Shifting plate	31 Nut
6 Main drive gear bearing	17 Snap ring	32 Locking plate
7 Main drive gear	18 Clutch hub	33 Countershaft
8 Main shaft bearing rollers	19 Clutch sleeve	34 Washer (steel)
9 Retainer gasket	20 Shift fork	35 Washer (bronze)
10 Cover bolt and lock washer	21 Second speed gear	36 Spacer
11 Case cover	22 Main shaft	37 Needle bearings
	23 Shifting shoe	38 Spacer washer
	24 Sliding gear	39 Reverse idler gear shaft
	25 Seal	
	26 Bearing adapter	

40 Spacer washer	49 Intermediate and high control lever (outer)
41 Reverse idler gear roller bearings	50 Control lever (inner)
42 Reverse idler gear	51 Low and reverse shift lever
43 Countershaft gear	52 Spacer
44 Countershaft front thrust washer	53 Spring
45 Nut	54 Ball
46 Lock washer	55 Intermediate and high shift lever
47 Control lever washer	56 Plug
48 Low and Reverse control lever (outer)	57 Oil seal
	58 Taper pin

T-86 three-speed transmission—remote control (© Borg Warner Corp.)

have from .012-.018 in. end-play controlled by the thickness of the rear steel thrust washer.

2. Assemble the large bronze with the lip entered in the slot in the case.

3. The bronze-faced steel washer is placed next to the gear at the rear end, and the steel washer next to the case.

4. To assemble the countershaft bearing rollers, use a dummy shaft. Use grease and a loading sleeve to facilitate reassembly of the countershaft gear components.

5. In assembling the mainshaft gears, low and reverse gear is installed with the shift shoe groove toward the front.

6. In assembling the synchronizer unit, install the two springs in the high and intermediate clutch hub with spring tension opposed.

Place the right lipped end of a spring in the hub slot and place the spring in the hub. Turn the hub around and make the same installation with the other spring, starting with the same slot. Install the three synchronizer shifting plates into the three slots in the hub, wih the smooth sides of the plates out. Hold the plate in position and slip the second and direct clutch sleeve over the hub, with the long beveled edge toward the long part of the clutch hub. Install the completed assembly onto the mainshaft with the beveled edge of the clutch sleeve toward the front end of the shaft.

7. When installing the mainshaft, be sure the bearing rollers are in place in the pilot bore of the clutch gear.

8. Be sure that the countershaft

and reverse idler shaft lock plate are in position and completely recessed into the indents of the transfer case.

Interlock sleeve clearance
(© Borg Warner Corp.)

Warner T-89 Series 3 Speed

Description

The Warner T-89 is a fully synchronized three-speed transmission controlled by a column mounted shift lever.

Disassembly of Transmission

1. Install the transmission in a work stand.

2. Remove the gear shift housing.

3. Shift the transmission into two gears. Remove the output shaft nut and remove the parking brake assembly.

4. Remove the input shaft bearing retainer. Tap the input shaft and bearing toward the front as far as it will go.

5. Remove the output shaft bearing retainer bolts. Turn the retainer to expose the end of the countershaft.

6. Drive the countershaft toward the rear until it just clears the hole at the front of the case. Push the counter shaft out of the rear of the case. Remove the Woodruff key from the countershaft.

7. Remove the output shaft assembly. Remove the pilot rollers from the rear of the input shaft.

8. Remove the bearing snap-ring

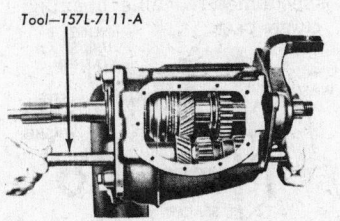

Tool—T57L-7111-A

Removing countershaft
(© Borg Warner Corp.)

and spacer washer from the input shaft. Drive the input shaft through the bearing. Remove the input shaft and bearing from the case.

9. Drive the reverse idler shaft out of the rear of the case.

10. Remove the cluster gear roller bearings and washers.

11. Remove the snap-ring at the front of the output shaft, and slide the synchronizer, intermediate gear, and low and reverse sliding gear off the shaft.

12. Remove the snap-ring that holds the outer race, and tap the output shaft bearing and shaft out of the bearing retainer casting and replace the bearing if worn.

Parts Repair or Replacement

Synchronizer

1. Slide the clutch sleeve off the hub and remove the three inserts and springs.

2. To assemble the synchronizer, align the index mark on the hub with the one on the sleeve.

3. Position the insert on the hub and slide the sleeve into place.

4. Install the insert springs.

Shift Cover

Check the condition of the shift levers and forks. If there is any binding or clashing of gears when the lever is operated, disassemble the cover.

Assembly

1. Install new seal rings in the shifter cam grooves.

2. Position one cam assembly in the gear shift cover.

3. Assemble the interlock spring and balls in the interlock, install the interlock assembly in the gear shift cover, and position the other cam assembly.

4. Install the gear shift levers and forks on the cams.

Assembly of Transmission

1. Position the cluster gear bearings and washers.

2. Position the cluster gear assembly in the case. The tang on the front thrust washer must be fitted into the slot in the case thrust surface. The tab on the rear steel washer must be up. Install the countershaft, and check the cluster gear end play (0.006-0.020 inch).

3. Place the reverse idler gear in the case with the chamfered teeth ends toward the front. Drive the reverse idler shaft through the gear from the rear of the case. When the end of the shaft is nearly flush with the case install the Woodruff key.

4. Tap the input shaft bearing into the case until the snap-ring bottoms on the case. Install the input shaft bearing retainer without a gasket. Install the retainer and barely tighten the cap screws.

5. Position the baffle on the input shaft and hold it there with a light coat of grease. It must be installed at the rear of the bearing with the dished side away from the bearing so that the baffle does not rub the bearing outer race.

6. Place the input shaft (with baffle) in the case and drive the

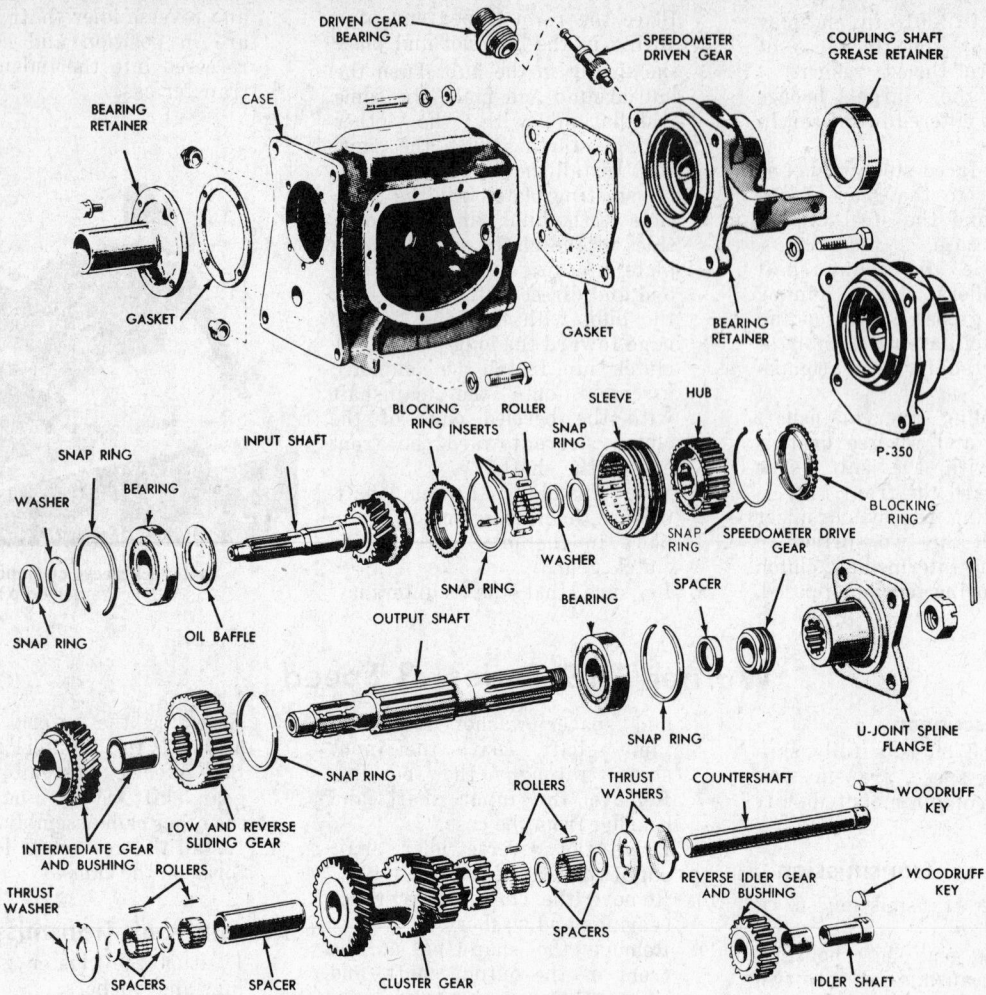

DRIVEN GEAR BEARING · SPEEDOMETER DRIVEN GEAR · COUPLING SHAFT GREASE RETAINER · BEARING RETAINER · CASE · GASKET · GASKET · BEARING RETAINER · P-350 · BLOCKING RING INSERTS · ROLLER · SLEEVE · HUB · SNAP RING · BLOCKING RING · SNAP RING · WASHER · BEARING · INPUT SHAFT · SNAP RING · WASHER · SNAP RING · SPACER · SPEEDOMETER DRIVE GEAR · SNAP RING · OIL BAFFLE · OUTPUT SHAFT · BEARING · SNAP RING · U-JOINT SPLINE FLANGE · SNAP RING · ROLLERS · THRUST WASHERS · COUNTERSHAFT · WOODRUFF KEY · INTERMEDIATE GEAR AND BUSHING · LOW AND REVERSE SLIDING GEAR · ROLLERS · SPACERS · REVERSE IDLER GEAR AND BUSHING · WOODRUFF KEY · THRUST WASHER · SPACERS · SPACER · CLUSTER GEAR · SPACERS · IDLER SHAFT

Warner T-89 series transmission—typical (© Borg Warner Corp.)

shaft into the bearing with a hardwood block.

7. Remove the input shaft bearing retainer, and install the spacer washer and snap-ring on the shaft. Install the thickest snap-ring that will fit. Snap-rings vary from 0.086-0.103 inch. Tap the input shaft and bearing as far to the front as they will go. Place the pilot bearing rollers in the input shaft and hold them with grease.

8. Install the output shaft and bearing in the bearing retainer casting. Install the thickest snap-ring that will fit. Assemble the low and reverse gear, intermediate gear and synchronizer assembly on the output

Removing or installing output shaft
(© Borg Warner Corp.)

shaft. Install the snap-ring at the front of the output shaft, and place the pilot bearing flat washer on the shaft.

9. Place a new gasket on the output shaft bearing retainer casting and install assembly in the case. Position the bearing retainer casting so that the countershaft hole is exposed.

10. Tap the input shaft bearing toward the rear until the bearing outer race snap-ring hits the case.

11. Raise the cluster gear assembly and install the countershaft. When the end of the countershaft is nearly flush with the case, install the Woodruff key.

12. Turn the output shaft bearing retainer to a normal position, and torque the retainer bolts to specification.

13. Install a new grease retainer, if necessary, and apply lubricant to the seal lips and cavity.

14. Install the input shaft bearing retainer and gasket. Torque the bolts to specification. Select a gasket that will seal and prevent end play. Gaskets are available

in the following thicknesses: 0.010, 0.015, 0.020, 0.025 inch.

15. Install the spacer and speedometer driving gear on the output shaft. Install the parking brake assembly.

16. Shift the transmission into two gears and install the output shaft nut.

17. Pour the specified lubricant over the entire gear train to prevent scoring when the transmission is initially operated.

18. Shift the transmission into neutral. Install the shift cover and gasket and torque the bolts to specification.

19. Fill the transmission with the proper lubricant through the speedometer cable opening.

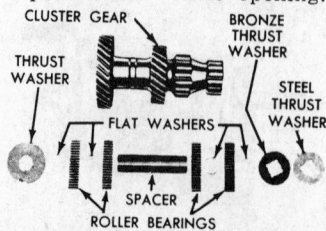

CLUSTER GEAR · BRONZE THRUST WASHER · THRUST WASHER · STEEL THRUST WASHER · FLAT WASHERS · SPACER · ROLLER BEARINGS

Cluster gear and components
(© Borg Warner Corp.)

TORQUE CHART

Nomenclature	Nuts and/or Bolts and Torque Limits	
Bolt—gear shift lever tower to gearshift housing	3/8-16 20-25	7/16-14 30-35
Bolt—clutch housing to trans. case	7/16-14 30-38 5/8-11 96-120	9/16-12 70-90
Nut-U-joint flange to trans. output shaft	1.00-20 1 1/2-18 90-125 275-350	1 1/4-18 225-275
Nut—drum parking brake to companion flange	3/8 24 35-45	7/16-20 50-70
Nut—bellcrank to trans.	9/16-18 70-90	
Bolt—lever assy. to trans.	3/8-16 20-25	

Nomenclature	Nuts and/or Bolts and Torque Limits	
Nut—handbrake anchor bar to trans. case (5-speed extra-heavy duty only)	9/16-18 120-130	
Bolt—bellcrank to trans.	3/8-16 20-25	
Bolt—reverse lockout plunger retainer	11/16-16 80-100	
Bolt—countershaft rear bearing retainer	5/16-18 25-30 3/8-16 35-40	7/16-14 45-55 1/2-13 60-70
Bolt—countershaft & reverse idler shaft retainer	5/16-18 25-30 3/8-16 25-37 3/8-16† 18-25†	7/16-14 40-45 1/2-13 80-85

Nomenclature	Nuts and/or Bolts and Torque Limits	
Bolt—gear shift housing to trans. case	5/16-18 20-25 3/8-16 35-40	3/8-16†† 30-35†† 7/16-14 45-50
Bolt—power take off cover to trans. case	3/8-16 20-30	
Nut—countershaft bearing lock (5-speed extra H.D. & 5-speed exclusive)	1 1/4-18 350-450	
Nut—countershaft bearing lock (5-speed exclusive H.D.)	1 1/2-18 350-450	
Bolt—input shaft bearing retainer to trans. case	5/16-18 25-30 3/8-16 25-30	7/16-14 40-45
Bolt—countershaft front bearing retainer	5/16-18 25-30 3/8-16 25-35	7/16-14 50-55

Tremec T-150 Three Speed Transmission

Disassembly of Transmission

1. Remove the bolts securing the transfer case to the transmission. Remove the transfer case.
2. Remove the transfer case drive gear locknut, flat washer, and drive gear. Remove the large fiber washer from the rear bearing adapter. Move the second-third clutch sleeve forward and the first-reverse sleeve to the rear before removing the locknut.
3. Remove the transmission oil plug and drive the countershaft out of the case with a suitable size drift. *Do not lose the countershaft access plug when removing the countershaft.* With the countershaft removed the countershaft gear will lie at the bottom of the case, leave it there until the mainshaft is removed.
4. Punch alignment marks in the front bearing cap and the transmission case for assembly reference.
5. Remove the front bearing cap and gasket.
6. Remove the large lock ring from the front bearing.
7. Remove the clutch shaft, front bearing, and the second-third synchronizer assembly. *A special tool is required for this operation, see illustration.*
8. Remove the rear bearing and adapter assembly with a brass drift and hammer. Drive the adapter out the rear of the case with light blows from the hammer.
9. Remove the mainshaft assembly. Tilt the spline end of the shaft downward and lift the front end

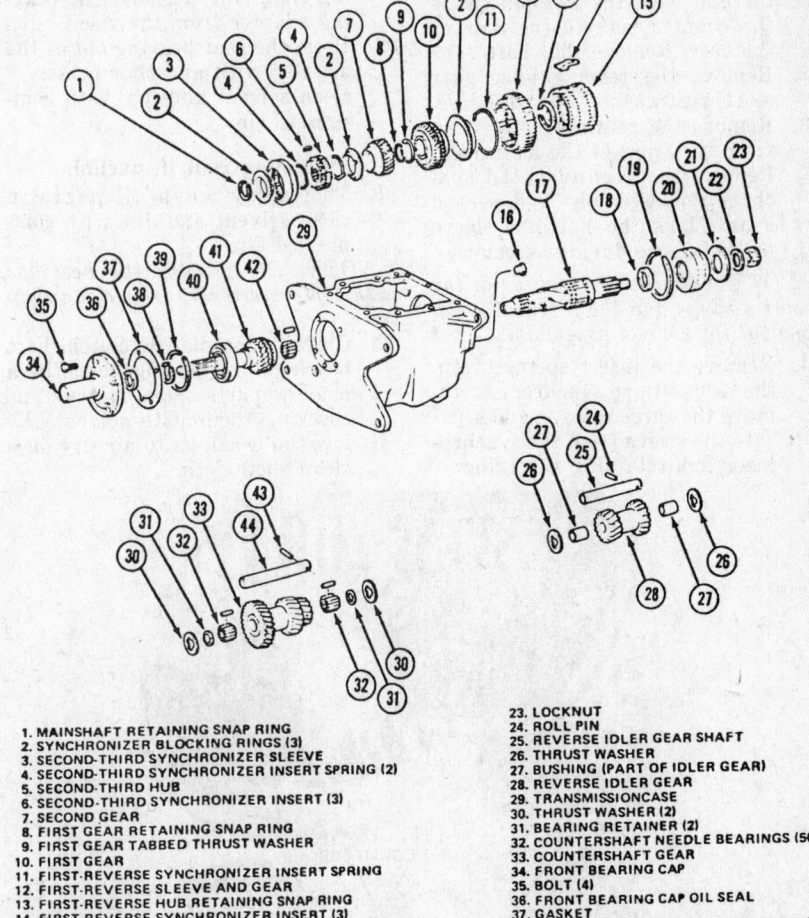

1. MAINSHAFT RETAINING SNAP RING
2. SYNCHRONIZER BLOCKING RINGS (3)
3. SECOND-THIRD SYNCHRONIZER SLEEVE
4. SECOND-THIRD SYNCHRONIZER INSERT SPRING (2)
5. SECOND-THIRD HUB
6. SECOND-THIRD SYNCHRONIZER INSERT (3)
7. SECOND GEAR
8. FIRST GEAR RETAINING SNAP RING
9. FIRST GEAR TABBED THRUST WASHER
10. FIRST GEAR
11. FIRST-REVERSE SYNCHRONIZER INSERT SPRING
12. FIRST-REVERSE SLEEVE AND GEAR
13. FIRST-REVERSE HUB RETAINING SNAP RING
14. FIRST-REVERSE SYNCHRONIZER INSERT (3)
15. FIRST-REVERSE HUB
16. COUNTERSHAFT ACCESS PLUG
17. MAINSHAFT
18. MAINSHAFT SPACER
19. REAR BEARING ADAPTER LOCK RING
20. REAR BEARING AND ADAPTER ASSEMBLY
21. FIBER WASHER
22. FLAT WASHER
23. LOCKNUT
24. ROLL PIN
25. REVERSE IDLER GEAR SHAFT
26. THRUST WASHER
27. BUSHING (PART OF IDLER GEAR)
28. REVERSE IDLER GEAR
29. TRANSMISSION CASE
30. THRUST WASHER (2)
31. BEARING RETAINER (2)
32. COUNTERSHAFT NEEDLE BEARINGS (50)
33. COUNTERSHAFT GEAR
34. FRONT BEARING CAP
35. BOLT (4)
36. FRONT BEARING CAP OIL SEAL
37. GASKET
38. FRONT BEARING RETAINER SNAP RING
39. FRONT BEARING LOCKRING
40. FRONT BEARING
41. CLUTCH SHAFT
42. MAINSHAFT PILOT ROLLER BEARINGS (3)
43. ROLL PIN
44. COUNTERSHAFT

T-150 Transmission—exploded view (© American Motors)

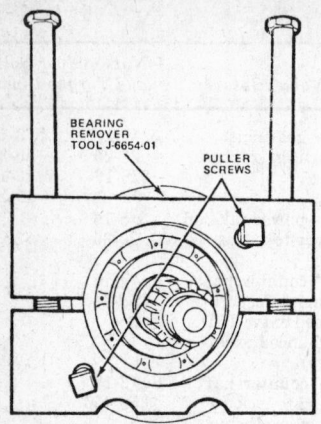

Removing the clutch shaft—T-150
(© American Motors)

up and out of the case.

10. Remove the countershaft tool and arbor as an assembly. Remove the countershaft thrust washers, countershaft roll pin, and any pilot roller bearings that may have fallen into the case.

11. Remove the reverse idler shaft. Insert a brass drift through the clutch shaft bore in the front of the case and tap the shaft until the end with the roll pin clears the counter bore in the rear of the case. Remove the shaft.

12. Remove the reverse idler gear and thrust washers from the case.

13. Remove the retaining snap ring from the front of the mainshaft. Remove the second-third synchronizer assembly and second gear. Mark the hub and sleeve for reference during assembly.

NOTE: Observe the position of the insert springs and the inserts during removal for correct assembly.

14. Remove the insert springs from the second-third synchronizer, remove the three inserts, and separate the sleeve from the synchronizer hub retaining snap ring.

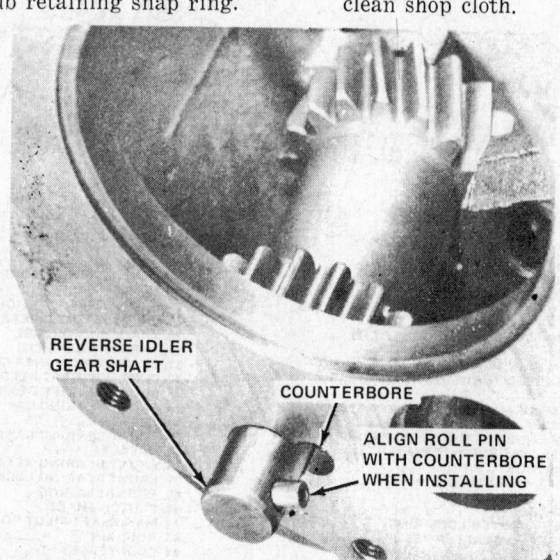

REVERSE IDLER GEAR SHAFT

COUNTERBORE

ALIGN ROLL PIN WITH COUNTERBORE WHEN INSTALLING

Removing or installing the reverse idler shaft– T-150 (© American Motors)

15. Remove the snap ring and the tabbed thrust washer from the mainshaft and remove the first gear blocking ring.

16. Remove the first-reverse synchronizer hub snap ring.

NOTE: Observe the position of the insert springs and the inserts during removal for correct assembly.

17. Remove the first-reverse sleeve, insert spring and the three insert from the hub. Remove the spacer from the rear of the mainshaft.

CAUTION: Do not attempt to remove the press fit hub by hammering. Hammer blows will damage the hub and mainshaft.

18. Remove the front bearing retaining snap ring and any remaining roller bearings from the clutch shaft.

19. Press the front bearing off the clutch shaft with an arbor press.

CAUTION: Do not attempt to remove the bearing by hammering. Hammer blows will damage the bearing and the clutch shaft.

20. Clamp the rear bearing adapter in a soft-jawed vise. *Do not over-tighten.*

21. Remove the rear bearing retaining snap ring. Remove the bearing adapter from the vise.

22. Press the rear bearing out of the adapter with an arbor press. clean solvent and dry with compressed air.

Cleaning and Inspection

1. Thoroughly wash all parts in clean solvent and dry with compressed air.

NOTE: Do not dry the bearings with compressed air, use a clean shop cloth.

2. Clean the needle and clutch shaft bearings by placing them in a shallow parts cleaning tray and covering them with solvent. Allow the bearings to air dry on a clean shop cloth.

3. Check the case for the following:
   a. Cracks in the bores, bosses, or bolt holes.
   b. Stripped threads in bolt holes.
   c. Nicks, burrs, rough surfaces in the shaft bores or on the gasket surfaces.

4. Check the gear and synchronizer assemblies for the following:
   a. Broken, chipped, or worn gear teeth.
   b. Damaged splines on the synchronizer hubs or sleeves.
   c. Bent or damaged inserts.
   d. Damaged needle bearings or bearing bores in the countershaft gear.
   e. Broken or worn teeth or excessive wear of the blocking rings.
   f. Wear of galling of the countershaft, clutch shaft, or reverse idler shaft.
   g. Worn thrust washers.
   h. Nicked, broken, or worn mainshaft or clutch shaft splines.
   i. Bent, distorted, or weak snap rings.
   j. Worn bushings in the reverse idler gear. Replace the gear if the bushings are worn.
   k. Rough, galled, or broken front or rear bearings.

Assembly of Transmission

1. Lubricate the reverse idler shaft bore and bushings with transmission oil.

2. Coat the transmission case reverse idler gear thrust washer surfaces with petroleum jelly and install the thrust washers in the case.

NOTE: Make sure the locating tangs on the thrust washers are aligned in the slots in the case.

3. Install the reverse idler gear. Align the gear bore, thrust washers, and case bore. Install the reverse idler shaft from the rear of the transmission case. Be sure to align and seat the roll pin in the shaft into the counter bore in the rear of the case.

4. Measure the reverse idler gear end-play by inserting a feeler gauge between the thrust washer and the gear. End-play should be 0.004-0.018 inch. If end play exceeds 0.018 inch, remove the reverse idler gear and replace the thrust washers.

5. Coat the needle bearing bores in the countershaft gear with petroleum jelly. Insert the arbor tool in the bore of the gear and install the twenty-five (25) needle bearings and the retainer washers at each end of the countershaft gear.

6. Coat the countershaft gear thrust washer surface with petroleum jelly and position the thrust washers in the case.

NOTE: Make sure the locating tangs on the thrust washers are aligned in the slots in the case.

7. Insert the countershaft into the bore at the rear of the case just far enough to hold the thrust washer in place.

8. Install the countershaft gear in the case. *Do not install the roll pin at this time.* Align the gear bore, thrust washers, the bores in the case, and install the countershaft.

NOTE: Do not remove the arbor tool completely.

9. Measure the countershaft gear end-play by inserting a feeler gauge between the washer and the countershaft gear. End-play should be 0.004-0.018 inch. If the end-play exceeds 0.018 inch, remove the gear and replace the thrust washer.

10. When the correct countershaft gear end-play has been obtained, install the countershaft arbor and remove the countershaft. Allow the countershaft gear to remain at the bottom of the case, leave the countershaft in the case enough to hold the thrust washer in place.

11. Coat the splines and machined surfaces on the mainshaft with transmission oil. Install the first-reverse synchronizer on the output shaft spines by hand. The end of the hub with the slots should face the front of the shaft. Use an arbor press to complete the hub installation. Install the retaining snap ring in the groove farthest to the rear.

CAUTION: Do not attempt to drive the hub on the shaft with a hammer.

12. Coat the splines of the first-reverse hub with transmission oil and install the first reverse sleeve and gear halfway onto the hub, with the gear end of the sleeve facing the rear of the shaft. Align the marks made during disassembly.

13. Install the insert spring in the first-reverse hub. Make sure the spring bottoms in the hub and covers all three insert slots. Position the three "T" shaped inserts in the hub with the small ends in the hub slots and the large ends inside the hub. Push the inserts fully into the hub so they seat on the insert spring, slide the first-reverse sleeve and gear over the inserts until the inserts engage in the sleeve.

14. Coat the bore and the blocking ring surface of first gear with transmission oil and place blocking ring on the tapered surface of the gear.

15. Install the first gear on the output shaft. Rotate the gear until the notches in the blocking ring

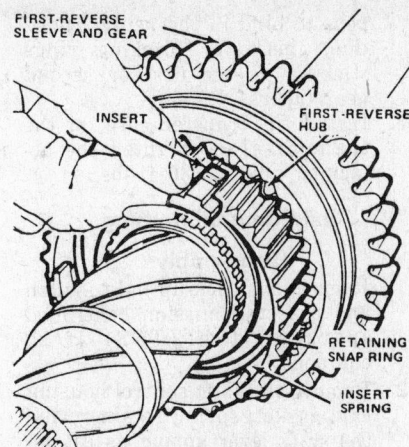
Installing the inserts in the first-reverse synchronizer hub–T-150 (© American Motors)

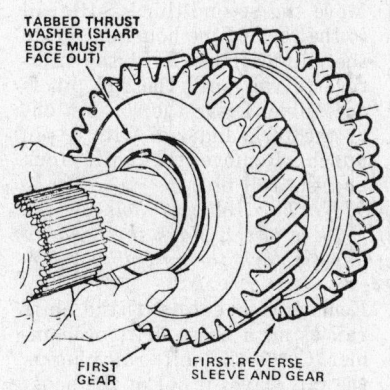
Installing the first gear thrust washer on the mainshaft— T-150 (© American Motors)

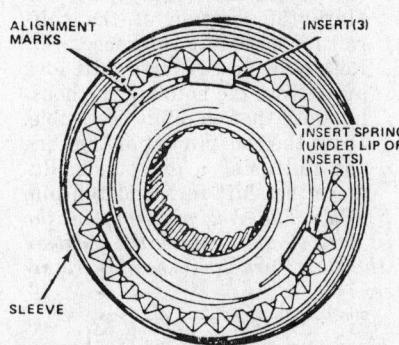
Installing the second gear on the mainshaft— T-150 (© American Motors)

Measuring the mainshaft end-play—T-150 (© American Motors)

engage the inserts in the first-reverse synchronizer assembly. Install the tanged thrust washer, sharp end facing out, and retaining snap ring on the mainshaft.

16. Coat the bore and blocking ring surface of the second gear with transmission oil. Place the second gear blocking ring on the tapered surface of second gear.

17. Install the second gear on the output shaft with the tapered surface of the gear facing the front of the mainshaft.

18. Install one insert spring into the second-third synchronizer hub. Be sure that the spring covers all three insert slots in the hub. Align the second-third sleeve with the hub using the marks made during disassembly. Start the sleeve onto the hub.

19. Place the three inserts into the hub slots and on top of the insert spring. Push the sleeve fully onto the hub to engage the inserts in the sleeve. Install the remaining insert spring in the exact position as the first spring. The ends of both springs must cover the same slot in the hub and not be staggered.

NOTE: The inserts have a small lip on each end. When they are correctly installed, this lip will fit over the insert spring.

20. Install the second-third synchronizer assembly on the mainshaft. Rotate the second gear until the notches in the blocking ring engage the inserts in the second-third synchronizer assembly.

21. Install the retaining snap ring on the mainshaft and measure the end-play between the snap ring and the second-third synchronizer hub. The end-play should be 0.040-0.014 inch. If the end-play exceeds the limit, replace the thrust washer and all the snap rings on the mainshaft assembly. Install the spacer on the rear of the mainshaft.

22. Install the mainshaft assembly in the case. Be sure that the first-reverse sleeve and gear is in the neutral (centered) position.

23. Press the rear bearing into the rear bearing adapter with an arbor press. Install the rear bearing retaining ring and the bearing adapter lockring.

24. Support the mainshaft assembly and install the rear bearing and adapter assembly in the case. Use a soft faced hammer to seat the adapter in the case.

25. Install the large fiber washer in the rear bearing adapter. Install the transfer drive gear, flat washer, and locknut. Tighten the locknut to 150 ft-lbs. torque.

26. Press the front bearing onto the clutch shaft. Install the bearing

retaining snap ring on the clutch shaft and the lockring into its groove.

27. Coat the bore of the clutch shaft assembly with petroleum jelly and install the fifteen (15) roller bearings in the clutch shaft bore.

CAUTION: Do not use chassis grease or a similar heavy grease in the clutch shaft bore. Heavy grease will plug the lubricant holes in the shaft and prevent proper lubrication of the roller bearings.

28. Coat the blocking ring surface of the clutch shaft with transmission oil. Position the blocking ring on the clutch shaft.

29. Support the mainshaft assembly and insert the clutch shaft through the front bearing bore in the case. Seat the mainshaft pilot in the clutch shaft roller bearings. Tap the bearings into place with a shoft faced hammer.

30. Apply a thin film of sealer to the front bearing cap gasket and position the gasket on the case. Be sure the cutout in the gasket is aligned with the oil return hole in the case.

31. Remove the front bearing cap oil seal with a suitable tool. Install a new seal with a suitable driver.

32. Install the front bearing cap and tighten the bolts to 33 ft-lbs. Be sure that the marks on the cap and the transmission case are aligned and the oil return slot in the cap lines up with the oil return hole in the case.

33. Make a wire loop about 18-20 inches long and pass the wire under the countershaft gear assembly. The wire loop should raise and support the countershaft gear assembly when it is pulled upward.

34. Raise the countershaft gear with the wire. Align the bore in the countershaft gear with the front thrust washer and the countershaft. Start the countershaft into the gear with a soft faced hammer.

35. Align the roll pin hole in the countershaft with the roll pin holes in the case and complete the installation of the countershaft. Install the countershaft access plug in the rear of the case and seat with a soft faced hammer.

36. Install the countershaft roll pin in the case. Use a magnet or needle nose pliers to insert and start the pin in the case. Use a ½ inch punch to seat the pin. Install the transmission filler plug.

37. Shift the synchronizer sleeves through all gear ranges and check their operation. If the clutch shaft and mainshaft appear to bind in the neutral position, check for blocking rings sticking on the first or second gear tapers.

38. Install the transfer case on the transmission. Tighten the attaching bolts to 30 ft-lbs.

Shift Control Housing

Disassembly

1. Remove the back-up light switch and the transmission controlled spark switch (TCS) if so equipped.

2. Remove the shift control housing cap, gasket, spring retainer, and the shift lever spring as an assembly.

3. Invert the housing and mount in a soft-jawed vise.

4. Move the second-third shift rail to the rear of the housing, rotate the shift fork toward the first-reverse rail until the roll pin is accessible. Drive the roll pin out of the fork and rail with a pin punch. Remove the shift fork and the roll pin.

NOTE: The roll pin hole in the shift fork is offset. Mark the position of the shift fork for assembly reference.

5. Remove the second-third shift rail using a brass drift or hammer. Catch the shift rail plug as the rail drives it out of the housing. Cover the shift and poppet ball holes in the cover to prevent the poppet ball from flying out. Mark the location of the shift rail for assembly reference.

6. Rotate the first-reverse shift fork away from the notch in the housing until the roll pin is accessible. Drive the role pin out of the fork and rail using a pin punch. Remove the shift fork and roll pin.

NOTE: The roll pin hole in the shift fork is offset. Mark the position of the shift fork for assembly reference.

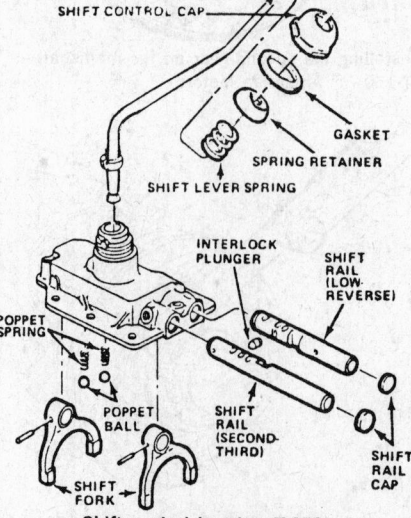

Shift control housing—T-150
(© American Motors)

7. Remove the first-reverse shift rail using a brass drift or hammer. Catch the shift rail plug as the rail drives it out of the housing. Cover the shift and poppet ball holes in the cover to prevent the poppet ball from flying out. Mark the location of the shift rail for assembly reference.

8. Remove the poppet balls, springs, and the interlock plunger from the housing.

Assembly

1. Install the poppet springs and the detent plug in the housing.

2. Insert the first-reverse shift rail into the housing, and install the shift fork on the shift rail.

3. Install the poppet ball on the top of the spring in the first-reverse rail.

4. Using a punch or wooden dowel, push the poppet ball and spring downward into the housing bore and install the first-reverse shift rail.

5. Align the roll pin holes in the first-reverse shift fork and install the roll pin. Move the shift rail to the neutral (center) detent.

6. Insert the second-third shift rail into the housing and install the poppet ball on top of the spring in the shift rail bore.

7. Using a punch or wooden dowel, push the poppet ball and spring downward into the housing bore and install the second-third shift rail.

8. Align the roll pin holes in the second-third shift rail and the shift fork and install the roll pin. Move the shift rail to the neutral (center) position.

9. Install the shift rail plugs in the housing, and remove the shift control cover from the vise.

10. Install the shift lever, shift lever spring, spring retainer, gasket, and the shift control housing cap as an assembly. Tighten the cap securely.

11. Install the back-up light switch and the TCS switch if so equipped.

TORQUE SPECIFICATIONS

	Foot-Pounds
Back-up light switch	15-20
Fill and drain plugs	10-20
Front bearing cap bolt	30-36
Shift control housing bolts	20-25
Transfer case drive gear locknut	150
Transfer case to transmission bolts	30
TCS switch	18

LUBRICANT CAPACITY

SAE 80-90 Gear Lube	3 pts.

Index

Allison AT Series

TROUBLE SYMPTOMS	IN TRUCK	OUT OF TRUCK
		ITEMS TO CHECK
Shifts At Too High a Speed	KL	
Shifts At Too Low a Speed With Full Throttle	L	
Low Main Pressure In All Ranges	BEFG	a
Low Main Pressure In First Gear	F	c
Intermittent Buzzing Noise	ABEG	
Excessive Creep In First And Reverse Gears	H	
Low Lubrication Pressure I	BDFI	
Oil Leaking From Converter Housing		abc
Transmission Overheating In All Ranges	BJI	b
No Response To Movement Of Shift Lever	CF	
High Stall Speed	BF	cde
Low Stall Speed	M	b
Rough Shifting	CFHKL	
Engine Over Speeds On Full Throttle Upshift	L	c
Excessive Slippage And Clutch Chatter In One Range	F	defgh
Dirty Oil	AEI	defgh
Slippage In All Forward Ranges	BDFG	ad
Slippage In Fourth And Reverse Only		ch
Slippage In First and Reverse Only		ce
Vehicle Moves Forward In Neutral	C	d
Vehicle Moves Backwards In Neutral		h
Throws Oil From Filler Tube	BN	

KEY TO CHECKS

A. Oil	F. Valve Body	K. Modulator	a. Oil Pump	f. Second Clutch
B. Oil Level	G. Oil Intake Pipe	L. Governor	b. Converter	g. Third Clutch
C. Linkage	H. Idle Speed (engine)	M. Engine	c. Seals	h. Fourth Clutch
D. Oil Pressure Check	I. Cooler Lines	N. Breather	d. Forward Clutch	
E. Oil Filter	J. Cooling System (engine)		e. First-Reverse Clutch	

General Information

This transmission is a hydraulically operated automatic transmission with four forward speeds and one reverse.

The transmission consists basically of a torque converter, a planetary gear train and a hydraulic control system for shifting gears.

Power is transmitted from engine to transmission gearing through a three element torque converter. The torque converter operates as a torque multiplier and a fluid coupling.

The forward and reverse speeds are controlled by the planetary gear ratios. The planetary gear train is operated by hydraulic clutches. All gearing is in constant mesh.

Adjustments

Range Selector Lever Linkage Adjustment

Chevrolet and GMC

1. Place selector in N (neutral) position.
2. Adjust cable so .38 inch dimension is attained (end of cable to

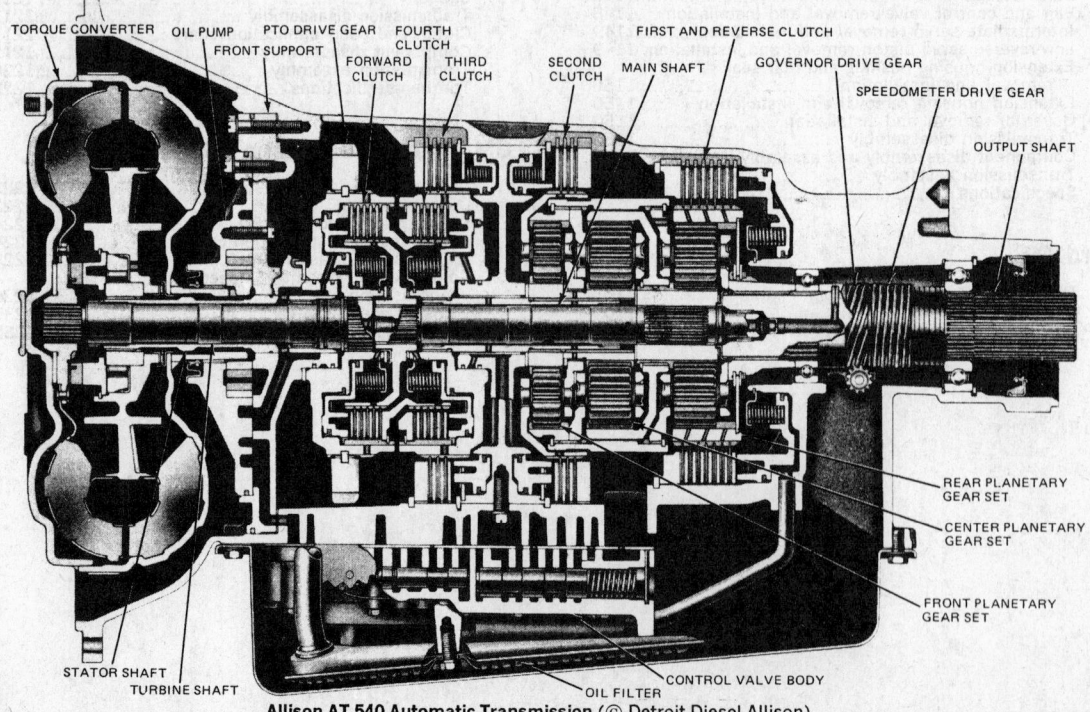

Allison AT 540 Automatic Transmission (© Detroit Diesel Allison)

center of fitting on lever assembly). Secure cable to base of selector bracket.

3. Disconnect clevis from transmission manual shift lever. Secure cable to transmission with existing clamp.

4. Place transmission manual shift lever in N (neutral) position. Neutral position is found by moving lever completely forward, then back one notch.

5. Align clevis and lever for free-entry of pin. Next, shorten cable by moving clevis two turns clockwise. Tighten jam nut against clevis, install clevis pin and secure with cotter pin.

Ford

1. Place selector lever in R (reverse) position.
2. Disconnect cable at transmission manual shift lever and move lever all the way forward.
3. Adjust sleeve so it enters manual shift lever freely. Install flat washers, spring washer and secure with cotter pin.
4. Check operation of shift linkage.

Neutral Safety and Back Up Light Switch Adjustment

Chevrolet and GMC

NOTE: Selector level adjustment should be made prior to this operation.

1. Remove secondary wire from distributor cap. Block front wheels and set parking brake.
2. Place selector lever in N (neutral) position. Loosen jam nuts and adjust push rod to .79 inch. Tighten jam nut.
3. Check each range to insure that starter only operates in N (neutral) position. Place selector in R (reverse) and check back up lamps).

4. Reconnect secondary wire.

Ford

1. Loosen neutral switch retaining screws.
2. Position switch so it will only function in N (neutral) position on selector.
3. Check operation of switch.

Maintenance

Checking Oil Level

The oil level should be checked at least every 1000 miles, if necessary add oil to attain proper level on dipstick. Avoid overfilling as this causes transmission overheating.

1. Run transmission until normal operating temperature is reached (150° to 200°F)
2. Position vehicle on level ground. Set parking and service brakes.
3. Move transmission through all drive ranges so that all clutch cavities and oil passages are filled.
4. Place selector in N (neutral)

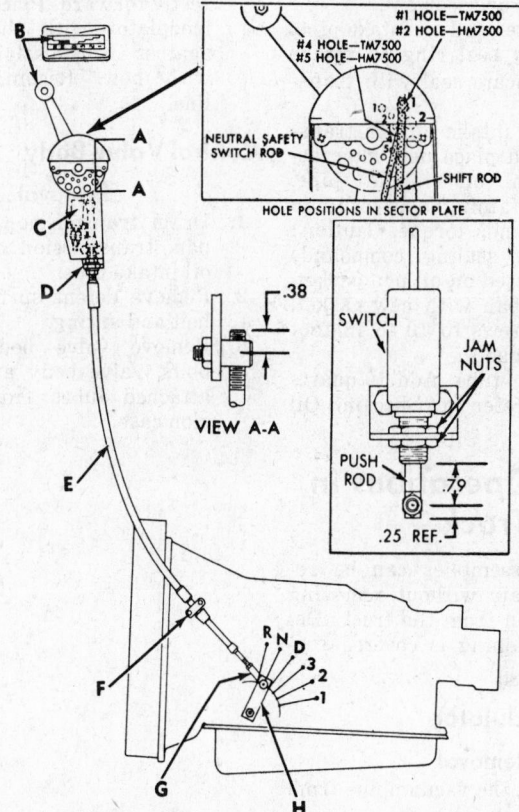

Chevrolet and GMC control linkage (© Chevrolet)

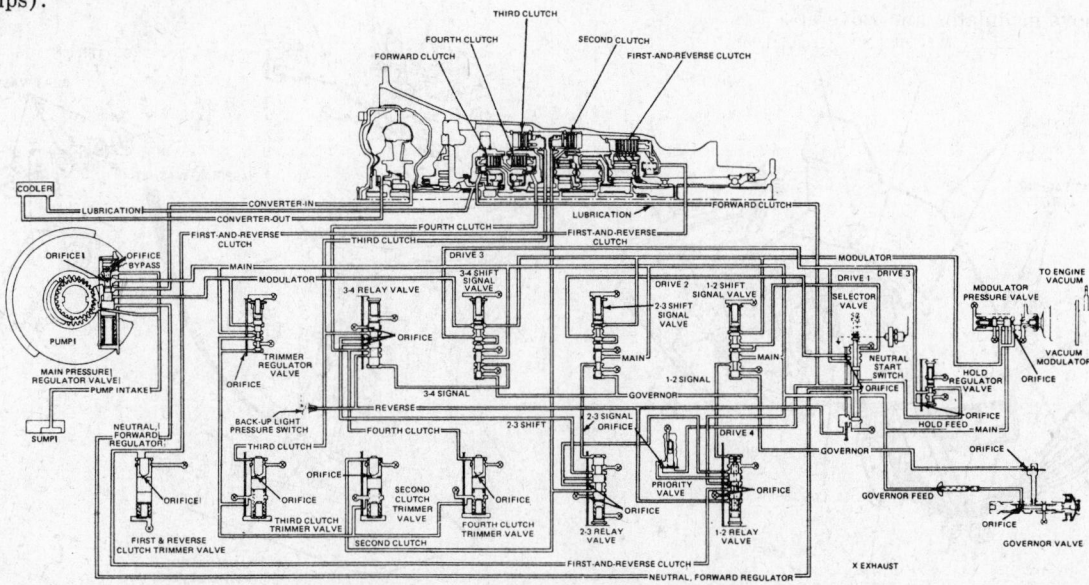

Hydraulic control system (© Detroit Diesel Allison)

and operate engine at idle.

5. Check oil level on dipstick.

Changing Oil and Filter

Transmission oil should be changed every 25,000 miles or 12 months. A new filter is also required at this interval. Use only Dexron® type automatic transmission fluid.

1. Run transmission to normal operating temperature.
2. Remove filler tube and drain oil.
3. Remove transmission oil pan. Discard the pan gasket and clean pan with proper solvent.
4. Remove filter and oil intake pipe.
5. Install new seal ring on intake pipe. Lubricate seal with transmission oil.
6. Replace oil intake pipe in transmission and place new filter assembly on oil intake pipe. Tighten retaining bolt to 10 to 13 foot pounds torque. Caution: No gasket sealing compounds should be used on oil pan gasket.
7. Install oil pan with new gasket. Tighten screws to 10 to 13 foot pounds torque.
8. Install filler tube. Add 10 quarts of oil and refer to "Checking Oil Level"

Service Operations in Truck

Some sub assemblies can be removed for repair without removing the transmission from the truck. Detailed reconditioning is covered further in the text.

Vacuum Modulator

Removal

1. Disconnect the vacuum line from modulator.
2. Remove retaining bolt and retainer.
3. Remove modulator and valve actuating rod from transmission case. Remove the seal ring from the modulator.

Installation

1. Install the valve actuating rod in case. Larger diameter end inserted first.
2. Place a new seal ring on modulator. Lubricate the seal ring with oil-soluble grease and install modulator in case.
3. Position the modulator so that vacuum connector pipe faces directly forward. Place retainer on modulator with the bent tabs against transmission and secure with bolt. Reconnect vacuum line.

Control Valve Body

Removal

1. Drain transmission. Remove oil pan, transmission oil filter and oil intake pipe.
2. Remove detent spring retaining bolt and spring.
3. Remove valve body attaching bolts, valve body and the three attached tubes from transmission case.

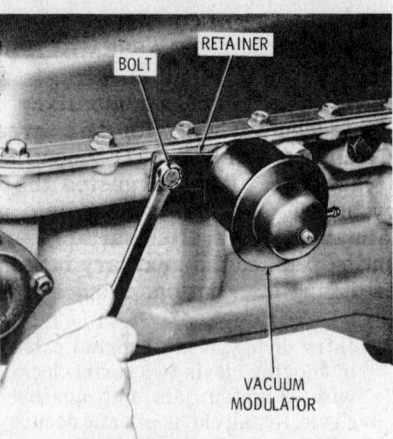

Removing vacuum modulator
(© Detroit Diesel Allison)

CAUTION: Do not drop selector valve out of valve body.

Installation

1. Place the governor feed tube, first-reverse tube and pressure tube in valve body.
2. Align tubes with holes in case and install valve body. Install attaching bolts, make certain that long bolt is placed near center of

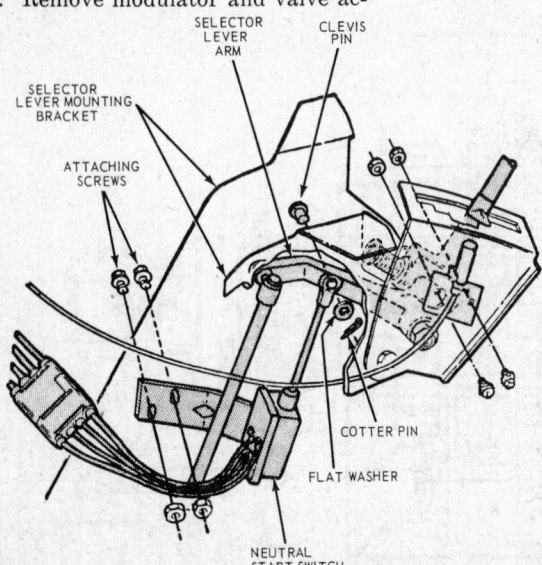

Ford neutral start switch (© Ford)

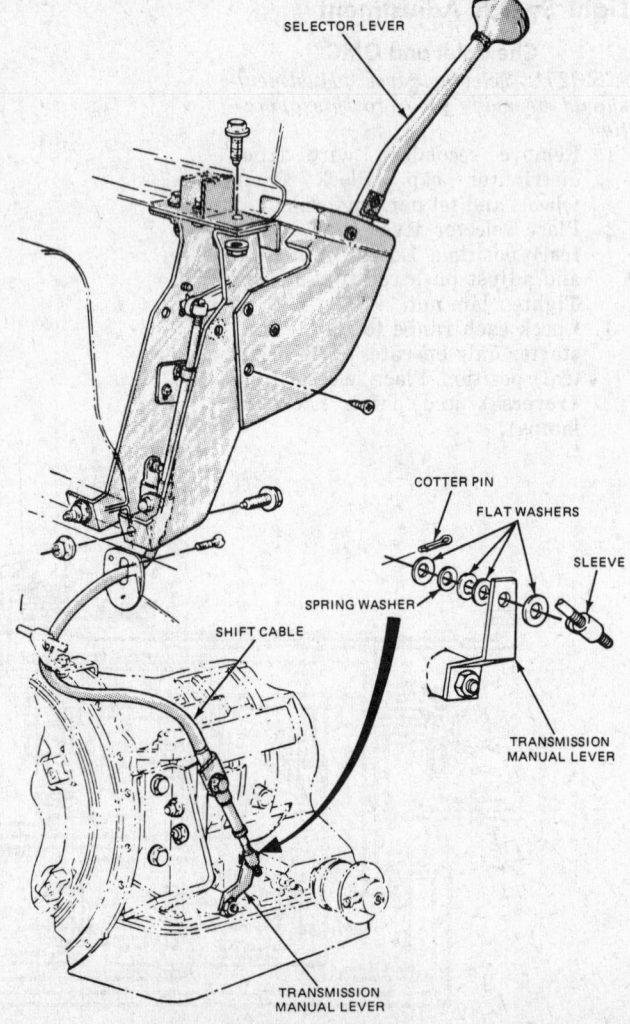

Ford control linkage (© Ford)

valve body. Torque bolts to 8-12 foot pounds.

3. Place the detent spring on valve body with spring roller in center of selector lever. Install attaching bolt.

4. Install new seal ring on intake pipe and place pipe in transmission. Install filter and oil pan with new gasket.

Governor

Removal

1. Remove governor cover and gasket from case. Rotate the governor clockwise to disengage drive gears. Remove governor.

Installation

1. Install governor in case, replace gasket and cover.

Extension Housing Rear Oil Seal

Removal

1. Disconnect drive shaft. Remove parking brake linkage.
2. Remove parking brake drum.
3. Remove seal with proper tool.

Installation

1. Lubricate the oil seal lip with a high temperature grease. Coat the outer edge of oil seal with a non-hardening sealer.
2. Place the oil seal, lip first, squarely into rear of transmission case. Using a seal installer, drive the oil seal into case until driver contacts against case. Note: The oil seal should be .51 to .55 inch forward of brake mounting surface on case.
3. Install output flange components, tighten retaining bolts to 83 to 100 foot pounds torque. Stake the tab washer into the flange washer and bend the tab washer

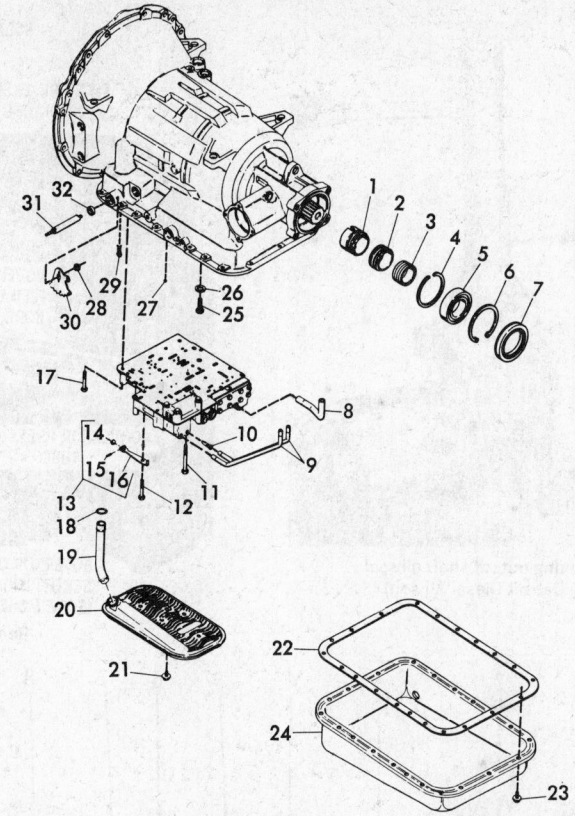

1	Governor Drive Gear	12	Bolt	22	Oil pan gasket
2	Speedometer Drive Gear	13	Detent Roller and	23	Washer-head screw
3	Spacer		Spring Assembly	24	Oil pan
4	Snap Ring	14	Detent Roller	25	Bolt
5	Ball Bearing	15	Retainer Pin	26	Plain washer
6	Snap Ring	16	Detent Spring	27	Steel Ball
7	Oil Seal	17	Bolt	28	Locknut
8	First-Reverse Tube	18	Seal Ring	29	Selector shaft retainer pin
9	Governor Tube	19	Intake Pipe	30	Selector lever
10	Governor Oil Screen	20	Oil Filter	31	Selector shaft
11	Bolt	21	Washer-head screw	32	Seal

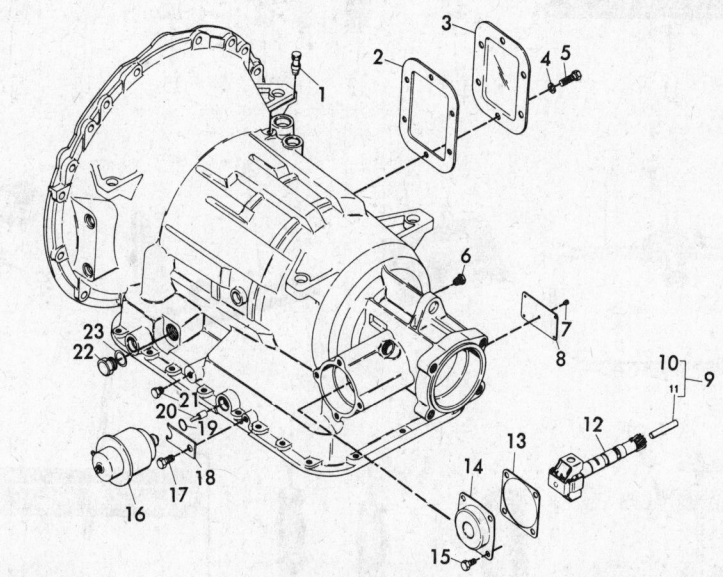

1	Breather	9	Case and Pin Assembly	17	Bolt
2	Gasket	10	Transmission Case	18	Modulator Retainer
3	PTO Cover	11	Governor Drive Pin	19	Modulator Seal Ring
4	Lockwasher	12	Governor Assembly	20	Modulator Valve
5	Bolt	13	Gasket		Actuating Rod
6	Pipe Plug	14	Governor Cover	21	Pipe Plug
7	Drive Screw	15	Bolt	22	Neutral Start Plug
8	Name Plate	16	Vacuum Modulator	23	Gasket

Transmission external components (© Detroit Diesel Allison)

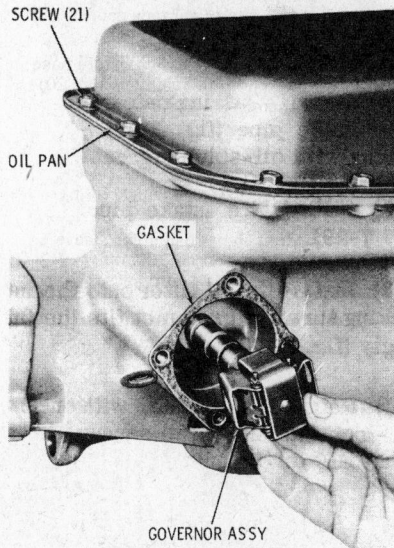

SCREW (21)

OIL PAN

GASKET

GOVERNOR ASSY

Installing governor assembly
(© Detroit Diesel Allison)

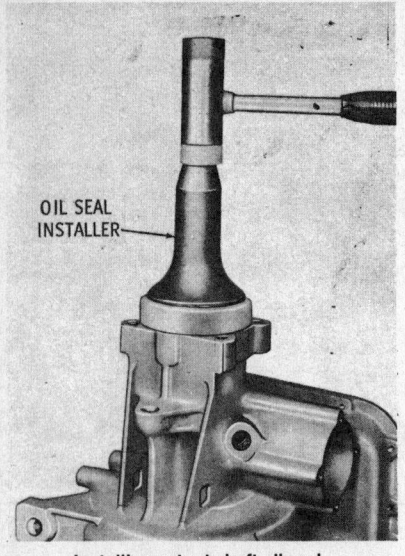

Installing output shaft oil seal
(© Detroit Diesel Allison)

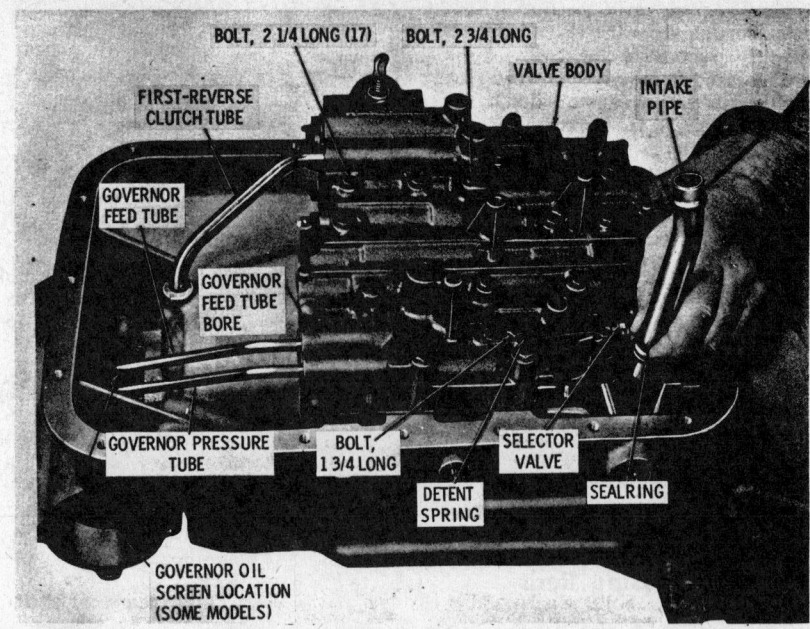

BOLT, 2 1/4 LONG (17)   BOLT, 2 3/4 LONG   VALVE BODY   INTAKE PIPE
FIRST-REVERSE CLUTCH TUBE
GOVERNOR FEED TUBE
GOVERNOR FEED TUBE BORE
GOVERNOR PRESSURE TUBE   BOLT, 1 3/4 LONG   DETENT SPRING   SELECTOR VALVE   SEALRING
GOVERNOR OIL SCREEN LOCATION (SOME MODELS)

Removing oil intake pipe (© Detroit Diesel Allison)

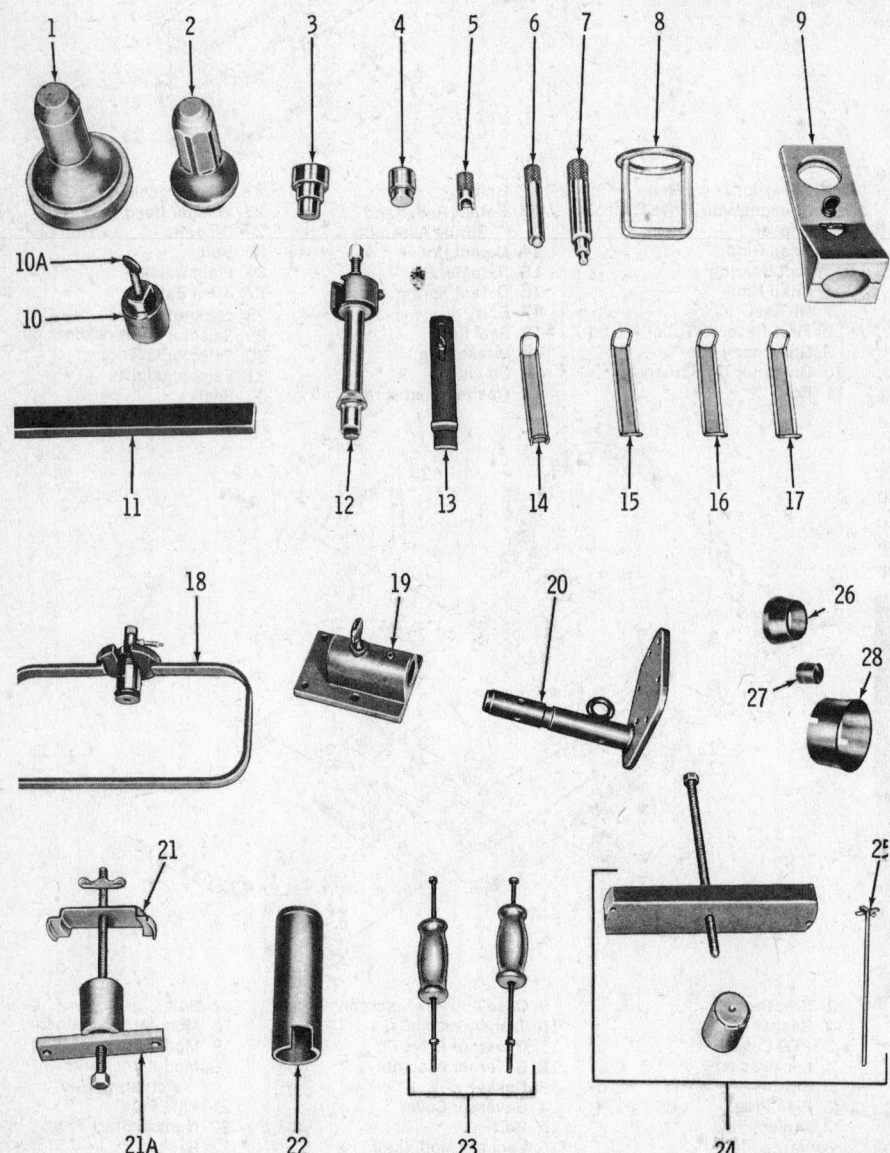

1 Output shaft seal installer
2 Oil pump seal installer
3 Stator shaft rear bearing installer
4 Stator shaft front bushing installer (also sun gear shaft bushings)
5 Shift valve adjusting ring tool
6 Selector shaft seal installer
7 Output shaft bushing installer
8 Forward and fourth-clutch spring compressor
9 Center support lifting bracket
10 Sun gear shaft retainer
10A Thumb screw for item 10
11 Thrust washer selection gauge bar
12 Converter end play checking fixture
13 Spacer selection gauge
14 Forward clutch clearance gauge
15 Fourth-clutch clearance gauge
16 First-and-reverse clutch clearance gauge
17 Third clutch clearance gauge
18 Converter leak test fixture
19 Transmission holding fixture base (used with holding fixture 20)
20 Transmission holding fixture (used with base 19)
21 First-and-reverse clutch spring compressor
21A First-and-reverse clutch spring compressor base
22 Output shaft positioning sleeve
23 Front support slide hammer (2)
24 Center support compressor assembly
25 Snapring gauge
26 Forward clutch inner seal protector
27 First-and-reverse clutch inner seal protector
28 Forward and fourth clutch outer seal protector

Special tools   (© Detroit Diesel Allison)

against a flat of the bolt head.

4. Install the parking brake drum, parking brake linkage and drive shaft.

Transmission Disassembly and Assembly

Disassembly

1. Remove converter.

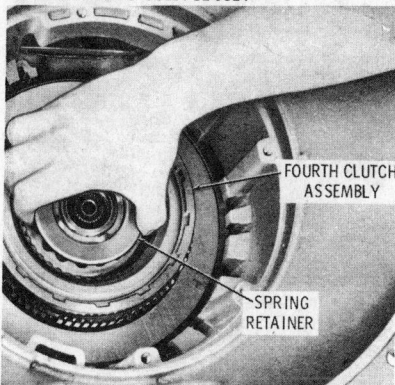

Removing fourth clutch assembly
(© Detroit Diesel Allison)

FOURTH CLUTCH ASSEMBLY
SPRING RETAINER

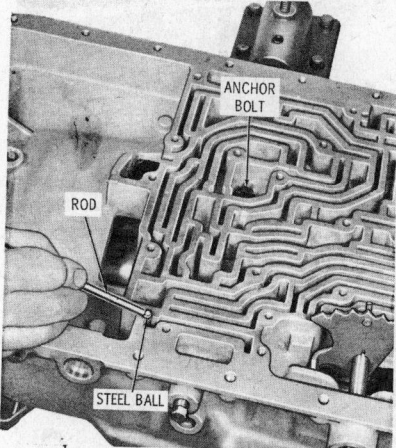

Removing governor check valve ball
(© Detroit Diesel Allison)

ANCHOR BOLT
ROD
STEEL BALL

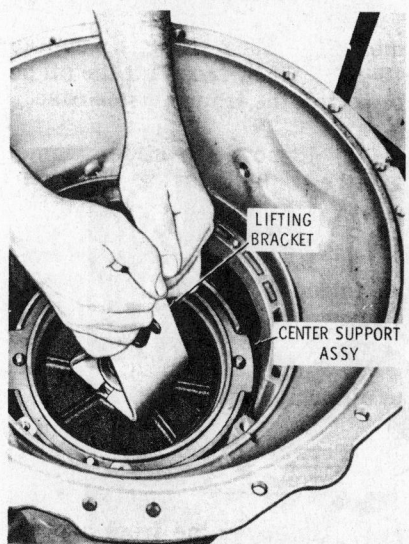

Removing center support assembly
(© Detroit Diesel Allison)

LIFTING BRACKET
CENTER SUPPORT ASSY

2. Remove vacuum modulator, governor, oil pan, oil filter and intake pipe.
3. Remove detent spring and valve body with attached oil tubes. CAUTION: Do not drop selector valve out of valve body.
4. Remove the three oil tubes from the valve body and the oil screen at the governor feed tube bore.
5. Remove governor check valve spring ball from channel in case with magnet. Remove center support anchor bolt and flat washer.
6. Remove bolts and lockwashers from front support.
7. Remove two bolts and washers from opposite holes in the oil pump body. Install slide hammers in holes.
8. Hammer upward with slide hammers to free oil pump assembly. Remove oil pump assembly from case.
9. Remove slide hammers and replace bolts, do not tighten.
10. Remove hook type seal rings and thrust washers from oil pump assembly.
11. Remove front support gasket. Lift the turbine shaft and forward clutch assembly out of case. Remove nylon thrust washer from rear of clutch assembly.
12. Remove the fourth clutch assembly by holding the spring retainer and lifting out of case.
13. Remove sun gear shaft.
14. Remove snap ring from third clutch back plate and back plate.
15. Remove third clutch plates.
16. Remove the snap ring that holds the center support assembly.
17. Install center support lifting tool into recess between the hook type seal rings on center support hub.
18. Remove center support carefully, lifting straight upward on tool. CAUTION: The center support is fitted to the case with very little clearance. If the case is cold the assembly may bind. Heat the case slightly with a sun lamp or warm air. DO NOT use a torch. Once assembly is moved upward and binds, tap lightly down to straighten and lift again.
19. Remove thrust washers from planetary sun gear.
20. Remove rear output flange.
21. Hold the main shaft and lift the planetary gears and shaft as a unit. NOTE: If assembly binds, tap lightly upward on output shaft with a soft faced mallet while lifting.
22. Remove governor and speedometer drive gears if they did not come out with planetary gear pack.
23. Remove snap ring that holds the

second clutch. Remove second clutch plates and backs.
24. Remove snap ring that holds the first-reverse clutch. Remove first-reverse clutch plates and backs.
25. Remove ring gear and remaining clutch plates.
26. Install the first-reverse spring compressor and tighten until snap ring is cleared.

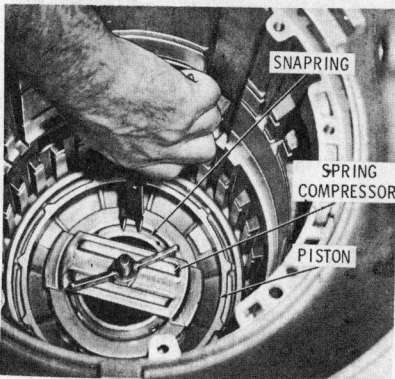

Removing first-and-reverse clutch spring retainer spring
(© Detroit Diesel Allison)

SNAPRING
SPRING COMPRESSOR
PISTON

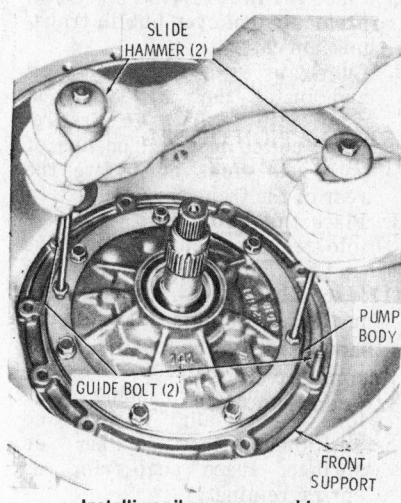

Installing oil pump assembly
(© Detroit Diesel Allison)

SLIDE HAMMER (2)
PUMP BODY
GUIDE BOLT (2)
FRONT SUPPORT

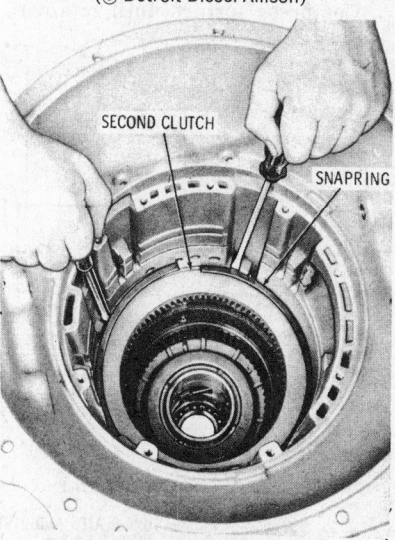

Removing second clutch snapring
(© Detroit Diesel Allison)

SECOND CLUTCH
SNAPRING

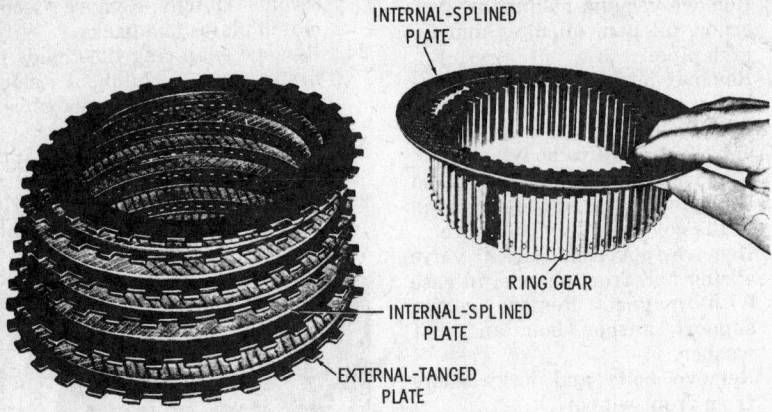

Installing first-and-reverse clutch plates onto rear ring gear (© Detroit Diesel Allison)

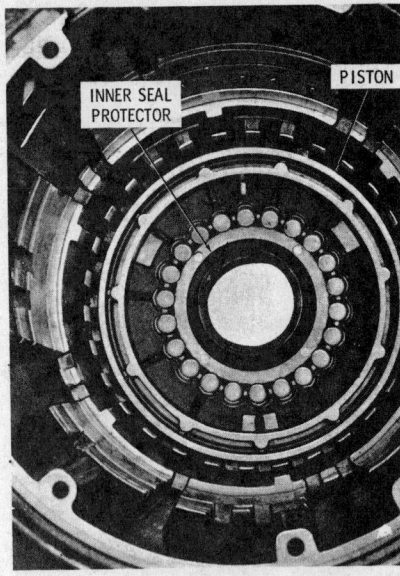

Installing first-and-reverse clutch piston
(© Detroit Diesel Allison)

27. Remove snap ring, spring compressor tool, spring retainer, piston return spring and piston.
28. Remove the two lip type seal rings from piston.
29. Remove output shaft seal. Remove the rear bearing snap ring and bearing.

Assembly

1. Place the first-reverse inner seal protector tool over hub in transmission case.
2. Lubricate with transmission oil and install new lip type seal rings into grooves in first-reverse clutch piston. The lips of both seal rings must face the rear of the transmission.
3. Place the piston and seal rings into transmission, making sure that piston tang engages slot in transmission case. Check and make certain that seal ring on the outside of piston is not distorted. Remove protective tool.
4. Place piston return spring in piston recesses, align springs and position spring retainer on springs. Place snap ring on spring retainer.
5. Install spring compressor tool. Compress springs until retainer

clears snap ring groove. Engage snap ring and remove tool.
6. Place the first-reverse ring gear on a flat surface, front side down. Starting with a internal splined plate, alternately install six internal splined plates and six external tanged plates onto the rear of the ring gear.
7. Align the external tangs on the clutch plates. Install the ring gear and clutch plates as a unit. Extended teeth on the ring gear must remain at top of gear after installation. Install the two remaining plates, steel plate first. Install back plate and secure with snap ring.
8. Check clearance between the snap ring and back plate. The clearance range for first-reverse clutch is .0405 to .1005 inch. If clearance is excessive new plates should be installed. If clearance is insufficient a thinner back plate is required. Back plates are stamped with numbers to identify thickness.

#1 .... .683 to .693 inch
#2 .... .647 to .657 inch
#3 .... .611 to .621 inch

9. Holding the main shaft, lower the unit into the transmission.

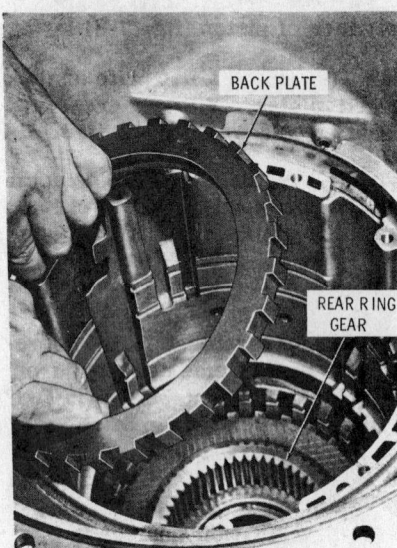

Installing first-and-reverse clutch back plate
(© Detroit Diesel Allison)

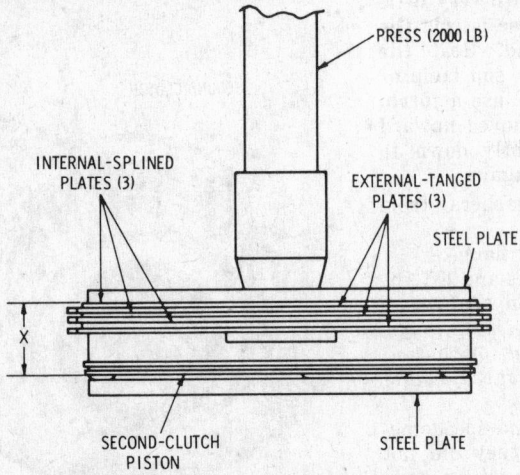

MEASURE PLATES AND PISTON AT X
Checking second-clutch plate running clearance
(© Detroit Diesel Allison)

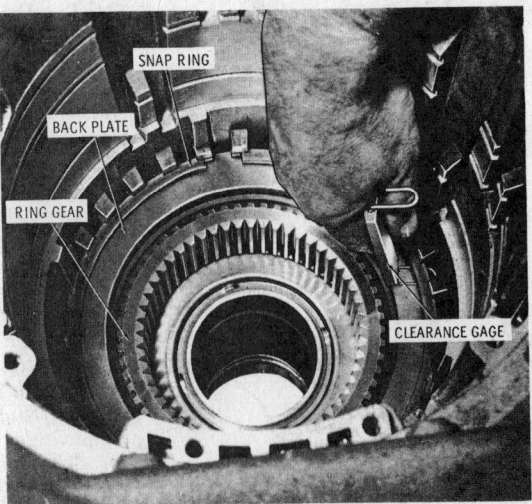

Checking first-and-reverse clutch running clearance
(© Detroit Diesel Allison)

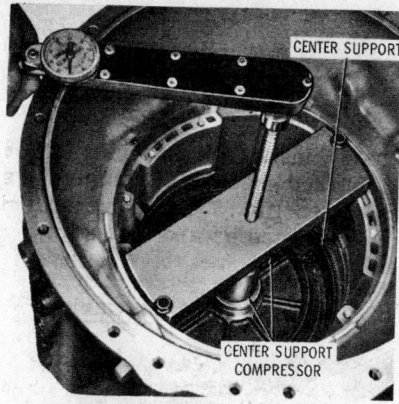

Compressing center support for snapring measurement
(© Detroit Diesel Allison)

Measuring for selection of rear bearing spacer
(© Detroit Diesel Allison)

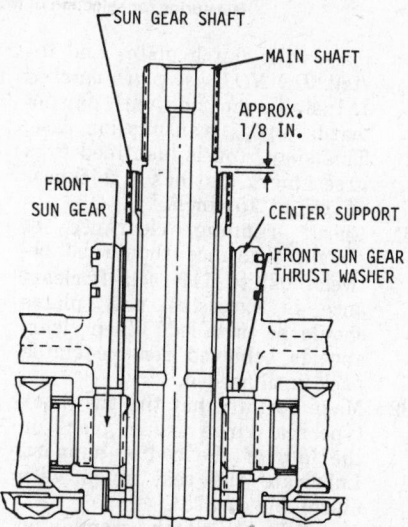

Checking end play of output shaft
(© Detroit Diesel Allison)

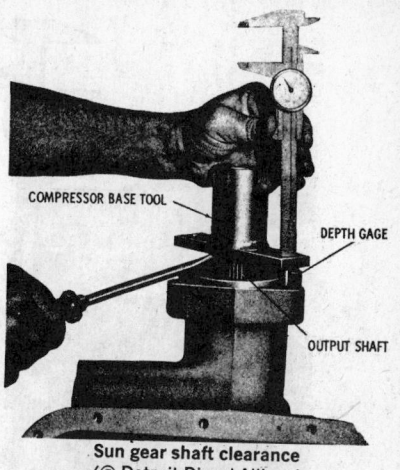

Sun gear shaft clearance
(© Detroit Diesel Allison)

Positioning components for front thrust washer measurements
(© Detroit Diesel Allison)

Mesh the internal teeth with the previously installed ring gear with the rear planetary carrier pinion teeth. Be sure that assembly bottoms.

10. Install sun gear shaft and front sun gear thrust washer.

11. Place the second clutch back plate in case. Install the clutch plates alternately starting with the internal splined plate. The single tangs of the clutch plate must align with the single tang on the back plate. Install snap ring, marked green (.155 to .157 inch), with gap located towards the top of case.

12. Remove the third clutch piston and its attaching parts from center support.

13. Install the center support assembly with the lifting tool. Make certain that tapped hole in support is aligned with the bolt hole in the bottom of case.

14. Remove the lifting tool. Install the plain washer and new center support anchor bolt. **CAUTION: Do not tighten bolt at this time.**

15. Install center support compressor tool. Use two oil pump retaining bolts to secure tool. Tighten the compressor bolt to 8-10 foot pounds.

16. Measure snap ring groove clearance and select proper size from chart below.

.148 to .150 inch	Blue
.152 to .154 inch	Yellow
.155 to .157 inch	Green
.158 to .160 inch	Red

17. Remove center support compressor and install lifting tool. Remove center support assembly. Clean assembly and remove any foreign matter from piston bore.

18. Inspect the third clutch piston seal rings for damage. Lubricate and install seal rings. Install third clutch piston and its attaching parts. Engage the small lug on the third clutch spring retainer with the slot in the center support. Make certain that the lips of the seal ring face the bottom of the piston cavity. Caution: Both the second and third clutch pistons must bottom in their bores. Failure to do this will result in poor transmission performance.

19. Install center support lifting tool and carefully lower assembly into transmission. Alignment of anchor bolt hole is important and should be checked prior to installation. Install the anchor bolt with plain washer. *DO NOT tighten.*

20. Install previously selected snap ring. Make certain that snap ring gap is located towards top of case. It may be necessary to use center support compressor to install snap ring.

21. Install sun gear shaft retainer on main shaft, retainer sleeve must be seated on shaft while tightening thumb screw.

22. Place the transmission with its rear end upward. Install the governor drive gear, engaging slot on governor with pin on output shaft. Install speedometer drive gear and tap with a soft drift to seat all internal components.

23. Place the spacer selector gauge against the output shaft and push the straight section against the speedometer drive gear. Position the lipped section against the rear bearing front snap ring. Tighten thumbscrew and remove gauge. Measure the distance between the straight section and lip section with a depth micrometer. Use this measurement to select proper spacer from following chart.

Mirometer Dimension	Identification
1.0003 to 1.0138	1 Groove
1.0138 to 1.0273	2 Grooves
1.0273 to 1.0408	3 Grooves
1.0408 to 1.0543	4 Grooves
1.0543 to 1.0678	5 Grooves
1.0678 to 1.0813	6 Grooves

24. Install the selected spacer, rear bearing and snap ring. Snap ring should be installed flat side down.

25. Position the first-reverse clutch spring compressor base, flange

Automatic Transmissions

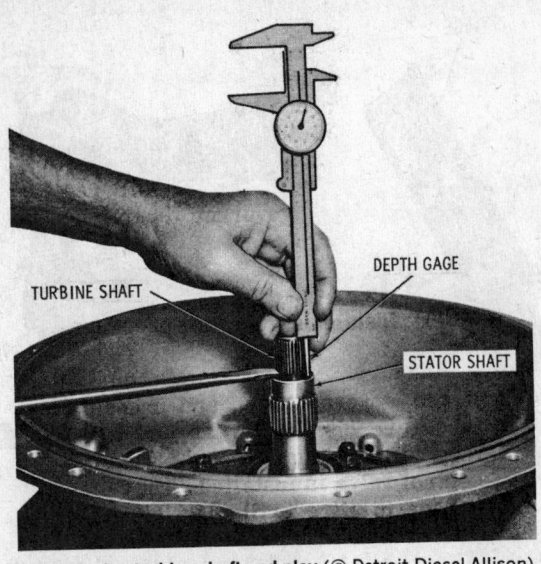

Measuring turbine shaft end play (© Detroit Diesel Allison)

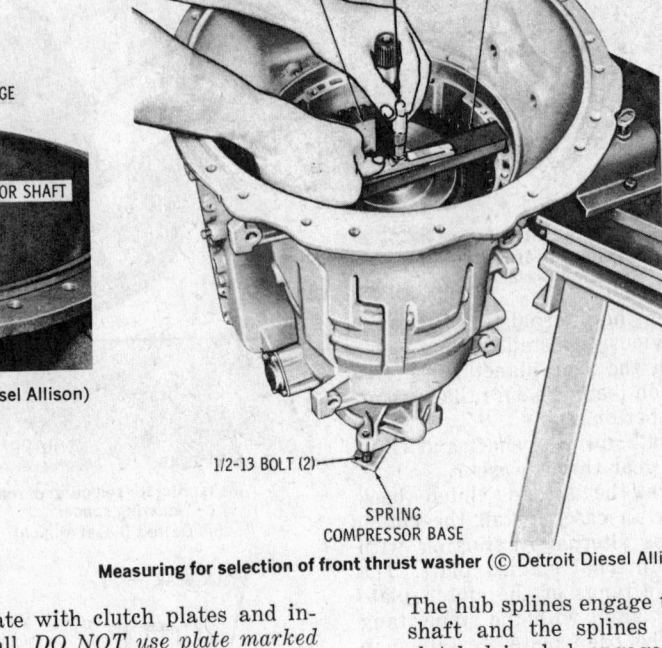

Measuring for selection of front thrust washer (© Detroit Diesel Allison)

down, on output shaft. Secure the tool to the shaft. Lifting the output shaft with a screwdriver measure the distance from top of flange to rear of transmission. Release output shaft and repeat the operation. A measurement of .003 ot .042 inch is acceptable.

26. Remove spring compressor, invert and secure to transmission case at parking brake flange. Tighten securing bolts to 5 to 8 foot pounds torque and center bolt to 5 pounds torque. All components are now properly aligned for selection of thrust washer.

27. Place the transmission with front upward. Cautiously remove sun gear shaft retainer without movement of shaft. **CAUTION: Do not remove the sun gear shaft assembly before measuring clearances.**

28. There should be a 1/8 inch clearance from the end of the sun gear shaft to the shoulder of the main shaft. If this clearance is not present the sun gear shaft is not properly seated. To seat shaft, first, slowly rotate shaft with slight up and down motion. If this is not successful, note sun gear shaft position relative to main shaft, remove shaft and center the front sun gear thrust washer so sun gear shaft will pass. The sun gear shaft assembly must be seated properly so accurate clearance between forward clutch housing, front support and bearing assembly is maintained.

29. Starting with external tanged clutch plate, alternately install three external tanged and three internal splined plates. The single tang on the plates must engage the single slot in the transmission case.

30. Align tangs on third clutch back

plate with clutch plates and install. *DO NOT use plate marked 1.* Install snap ring with gap towards top of transmission case. This snap ring is identified by a green mark and has a thickness of .155 to .157 inch.

31. Check running clearance of clutch, clearance should be between .029 to .119 inch. If clearance is excessive new plates should be installed. When clearance is achieved remove clutch pack until needed.

32. Make certain that the two hook type seal rings are in place on the hub of the center support. Lubricate the seal rings with oil-soluble grease.

33. Install fourth clutch assembly on splines of sun gear. Install third clutch pack (previously removed, refer to step 29 for procedure).

34. Lubricate with oil-soluble grease and install thrust washer on hub of forward clutch assembly.

35. Place the forward clutch and turbine shaft assembly in case.

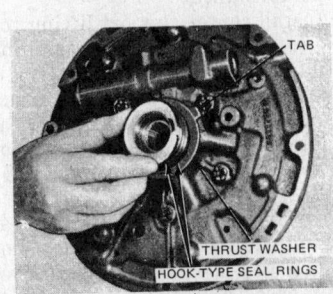

Removing or installing front support seal rings and thrust washer (© Detroit Diesel Allison)

The hub splines engage the main shaft and the splines of forth clutch drive hub engage internal splined plates of forth clutch pack. Pushing forward, rotate the forward clutch assembly one or two times.

36. Place the thrust washer selection gauge bar in transmission. Position a micrometer so that the stem passes through hole in center of gauge bar. Align the gauge bar so that micrometer stem is above the thrust washer surface of forward clutch housing. Measure distance from top of bar to thrust washer surface. Subtract the thickness of the bar and note the difference. Select the proper washer from the following chart.

Dimension	Identification
.7329 to .7493	0
.7493 to .7656	1
.7656 to .7820	2
.7820 to .7983	3
.7983 to .8714	4
.8147 to .8311	5

37. Lubricate selected thrust washer and install on oil pump assembly. Tab on washer must engage recess in front support. Install two hook type seal rings on hub of front support.

38. Position front support gasket in transmission so that holes in gasket are aligned with bolt holes in case.

39. Install slide hammers in two opposite holes on front support. Install two headless guide bolts in two opposite holes on transmission case.

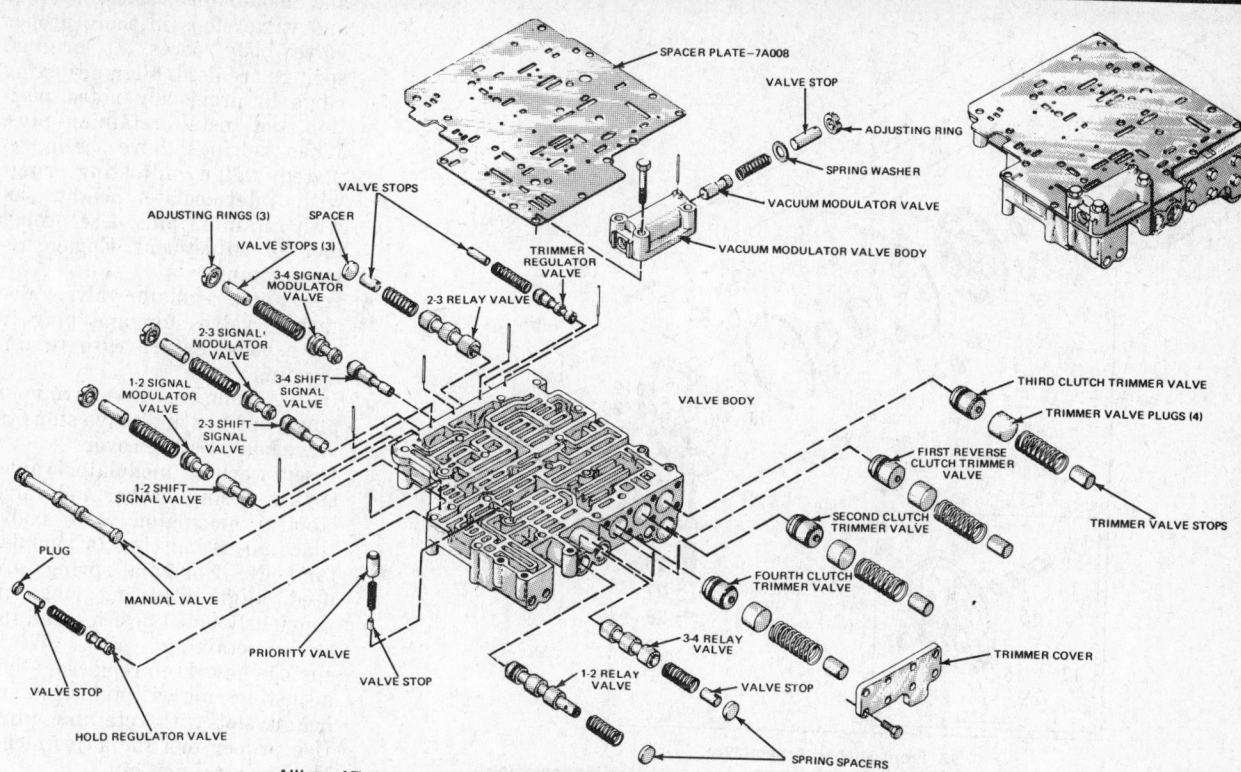

Allison AT series automatic transmission valve body (© Detroit Diesel Allison)

40. Install front support and oil pump assembly in case. Make certain that assembly bottoms. Remove slide hammers and guide bolts. Install self locking bolts with new rubber coated washers, into front support and transmission case. Torque bolts to 15 to 20 foot pounds.

41. Place a Vernier dial caliper on turbine shaft. Raise the shaft and bring the gauge to bear on sector shaft, note the dial reading. Release the shaft and note dial reading. If the dial reading does not drop within the end play range (.0053 minimum to .0337 maximum), the thrust washer selected in step 36 must be recalculated.

42. Remove the spring compressor

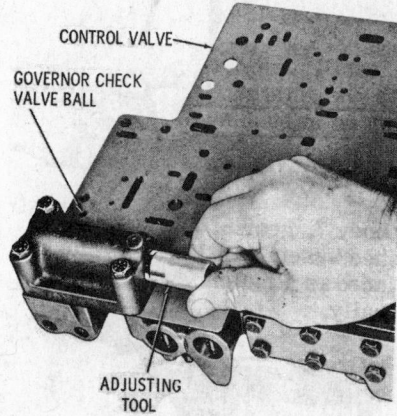

CONTROL VALVE

GOVERNOR CHECK VALVE BALL

ADJUSTING TOOL

Adjusting Vacuum modulator valve ring
(© Detroit Diesel Allison)

base from output shaft and flange. Tighten center support anchor bolt to 39 to 46 foot pounds torque.

43. Install governor, vacuum modulator, extension housing oil seal, control valve body, oil intake pipe, oil filter and oil pan, as outlined in previous section.

44. Install torque converter, engaging turbine shaft with converter turbine, sector shaft with sector and hub with oil pump drive gear. Secure torque converter for handling by attaching strap between drive lug and transmission case.

Transmission Subassemblies Overhaul

Control Valve Body

Disassembly

1. Remove manual selector valve, vacuum modulator body and spacer plate.
2. Remove priority valve, spring and valve stop. *NOTE: Prior to removing retainer pin from modulator valve body, note slot in adjusting ring where pin is engaged.*
3. Keeping pressure on adjusting ring, remove retainer pin from modulator valve body.
4. From the vacuum modulator valve body, remove adjusting ring, stop, spring washer, valve

spring and vacuum modulator valve. *NOTE: Hold trimmer cover while removing, as it is spring loaded.*

5. Remove trimmer cover, trimmer springs, valve stops, trimmer plugs, trimmer valves from valve body. *NOTE: Relay valve spring spacers are spring loaded and must be held during removal.*
6. Remove retainer pins, spring spacers, valve stops, relay valve springs and relay valves. *NOTE: Valve plug on hold regulator valve is spring loaded and must be held during removal.*
7. Remove retainer pin, valve plug, valve stop, valve spring and hold regulator valve. *NOTE: Prior to removing retainer pins from shift signal valves, note slot in adjusting rings and pins are engaged. The adjusting rings are spring loaded and must be held during removal.*
8. Remove retaining pins, adjusting rings, valve stop springs, modulator valves and signal valves. *NOTE: Trimmer regulator valve and 3-4 relay valve spacers and stops are spring loaded and must be held during removal.*
9. Remove retainer pins, spring spacers, valve stops, valve springs, 3-4 relay valve and trimmer valve.

Cleaning and Inspection

1. Carefully clean all parts with volatile mineral spirits. Clean oil passages by running a pipe

Automatic Transmissions

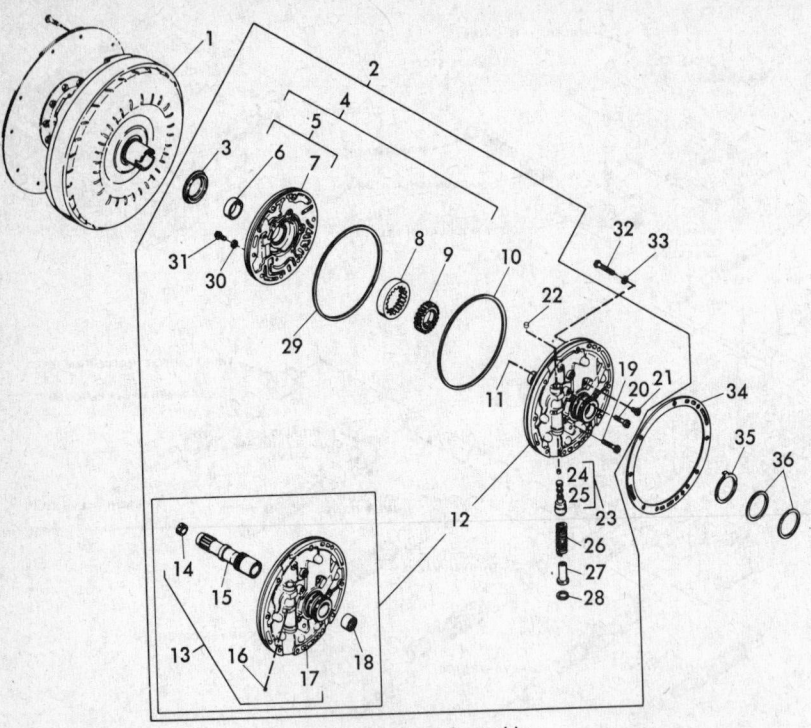

1 Torque Converter
2 Oil Pump
3 Oil Seal
4 Pump Body & Gear
5 Pump Body Assembly
6 Bushing
7 Pump Body
8 Pump Driven Gear
9 Pump Drive Gear
10 Seal Ring
11 Pin
12 Front Support & Bearing

13 Front Support Assembly
14 Stator Shaft Bushing
15 Stator Shaft
16 Plug
17 Front Support
18 Roller Bearing
19 Self-Locking Bolt
20 Self-Locking Bolt
21 Self-Locking Bolt
22 Valve Plug
23 Main Pressure Regulator & Plug Assembly
24 Orifice Plug

25 Main Pressure Valve
26 Valve Spring
27 Spring Stop
28 Retainer Ring
29 Seal Ring
30 Rubber Coat Washer
31 Self-Locking Bolt
32 Self-Locking Bolt
33 Rubber Coat Washer
34 Gasket
35 Thrust Washer
36 Hook Type Seal Ring (2)

Torque converter and oil pump assemblies—exploded view (© Detroit Diesel Allison)

cleaner through and flushing with mineral spirits. Dry all parts with compressed air. Make certain that all passages are free of foreign matter and obstructions.

CAUTION: Never use abrasives or scraping tools in cleaning valve body components.

2. Thoroughly check all parts for dirt, cracks, wear, breaks and burrs. Burrs may be removed with a soft abrasive stone. Do not damage the sharp edges of valves. Replace all parts that are in question.
3. Check all springs free length and condition.
4. Prior to assembly check all parts for free movement in bores, never force.
5. Cover all parts until assembly to exclude dirt.

Assembly

CAUTION: Proper assembly of control valve body is critical! Before making any installation be certain that the position of all parts, the identity of all springs and the configuration of all valves are thoroughly checked.

1. Insert the trimmer regulator valve, 3-4 relay valve, springs and stops in valve body. Align valve stop pin hole with hole in body, place pressure on stop or spacer and insert pin.
2. Insert 1-2, 2-3 relay valves, springs, valve stops and spring spacers in valve body. Secure with retaining pin.
3. Insert 1-2, 2-3, 3-4 shift signal

and modulator valves, springs and valve stops in there proper valve body bores. If original springs are used, align adjusting rings in previously noted position and install retaining pins. When springs have been replaced, align adjusting rings with intermediate height slot with retaining pins. Use proper tool for adjustment. Engage retaining pins.

4. Insert hold regulator valve, valve spring, valve stop and plug in valve body. Secure with retaining pin.
5. Insert trimmer valves, trimmer plugs, springs and valve stops in valve body. Secure cover.
6. Insert vacuum modulator valve, spring, spring washer and valve stop in modulator valve body. Place adjusting ring in modulator body, if original springs are used, align adjusting rings in previously noted position and install retaining pins. When springs have been replaced, align adjusting ring with intermediate height slot with retaining pins. Use proper tool for adjustment. Engage retaining pin.
7. Insert the priority valve stop, spring and valve in valve body. Depress the priority valve and install the spacer plate.
8. Install the vacuum modulator valve body.
9. Install the manual selector valve.

Oil Pump

Disassembly

1. Remove large seal ring from outer edge of forward support.
2. Remove bolts and seals from oil pump body and front support. Separate front support and oil pump.
3. Remove drive and driven gears from pump body.
4. Remove seal ring from outer edge of pump body. Drive oil seal out of pump body.

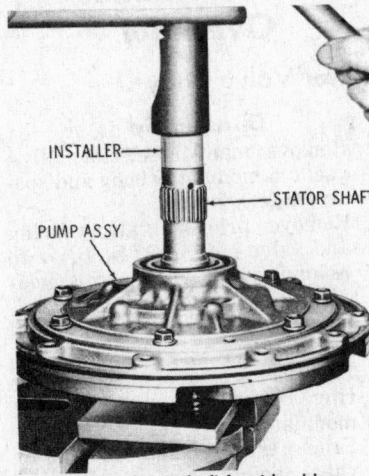

Installing stator shaft front bushing
(© Detroit Diesel Allison)

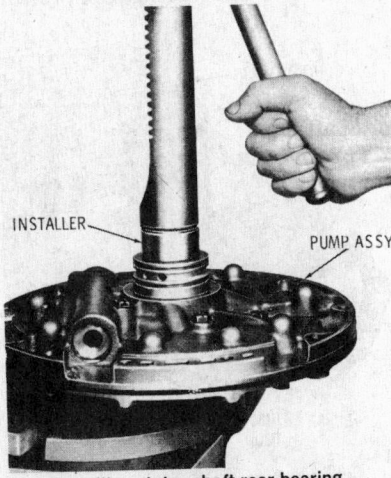

Installing stator shaft rear bearing
(© Detroit Diesel Allison)

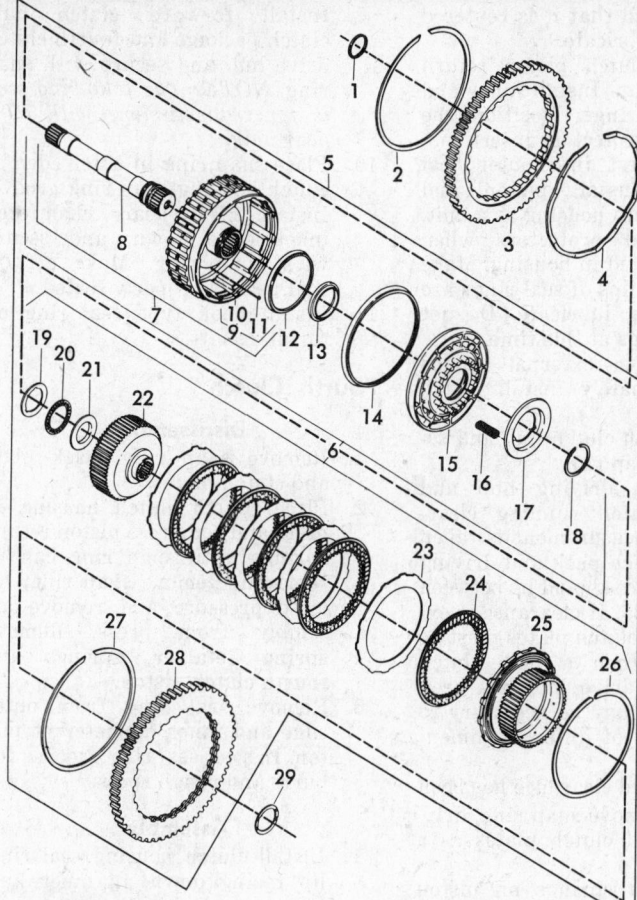

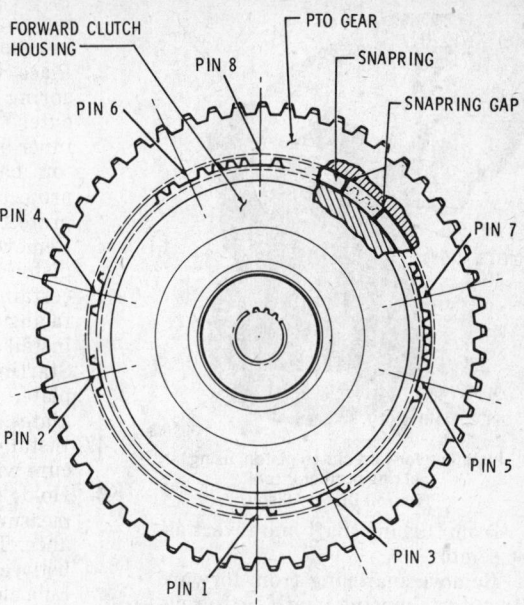

Removing PTO gear from forward clutch housing
(© Detroit Diesel Allison)

1 Hook Type Seal Ring	15 Forward Clutch Piston
2 Snapring (Early)	16 Clutch Spring (16)
3 PTO Gear (Early)	17 Spring Retainer
4 Wave Type Snap Ring (Early)	18 Snap Ring
5 Forward Clutch and Turbine	19 Thrust Bearing Race
Shaft Assembly (w/PTO)	20 Thrust Needle Bearing
6 Forward Clutch Assembly	21 Thrust Bearing Race
7 Housing & Shaft Assy.	22 Forward Clutch Hub
8 Turbine Shaft	23 External Tanged Clutch (5)
9 Forward Clutch Housing Assy.	24 Internal Splined Clutch (5)
10 Forward Clutch Housing	25 Fourth Clutch Driving Hub
11 Ball	26 Snap Ring
12 Clutch Housing Seal Ring	27 Snap Ring (Late)
13 Piston Inner Seal Ring	28 PTO Gear (Late)
14 Piston Outer Seal Ring	29 Thrust Washer

Forward clutch and turbine shaft assembly (© Detroit Diesel Allison)

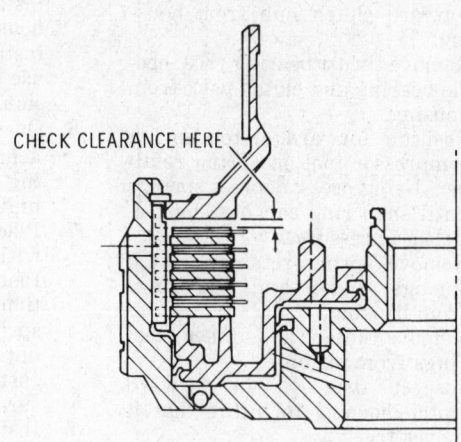

Forward clutch clearance point (© Detroit Diesel Allison)

edge of front support assembly.

Forward Clutch and Turbine Shaft

Disassembly

NOTE: On transmissions not equipped with PTO gear proceed to step 5. Early models with PTO gear use step 4.

1. Place forward clutch housing on flat surface, shaft side up. Using 5/32 drill rod, prepair eight, five inch sections with pointed ends.

2. Locate snap ring gap. Place one pin, point end first, in flat area opposite snap ring gap. Push pin between snap ring and PTO gear, insert remaining pins in same manner alternately from first.

3. Tap PTO gear downward and remove.

4. On early models, place forward clutch assembly shaft up. Using a screwdriver remove snap ring and PTO gear.

5. Remove hook type seal rings

---

5. If parts are being replaced, remove bushing from sector shaft. Be cautious not to damage bore. Remove roller bearing from sector shaft.

   CAUTION: Do not remove sector shaft or bushing from oil pump body. If bushing is defective replace oil pump body.

6. Remove main pressure regulator valve assembly by depressing spring stop and removing retaining ring.

7. Invert front support and remove retaining pin from small end of main pressure regulator bore, remove plug.

Assembly

1. Insert valve plug in oil pump side of front support and secure with retaining pin.

2. Invert front support, insert main

---

pressure regulator valve (stem first), valve spring and spring stop. Depress spring stop and install retainer ring.

3. Press a new bearing (lettered end first) on sector shaft until it is .595 to .615 inch from end of shaft. Press new bushing on opposite end of shaft until it is .005 inch from end of shaft.

4. Coat oil seal bore in pump body with a non-hardening sealer. Drive oil seal, lip first, into oil pump body.

5. Position a new seal ring on outer edge of pump body.

6. Place oil pump body and front support together, aligning bolt holes. Install all bolts, with new rubber coated washers where required. Torque bolts to 15 to 20 foot pounds.

7. Install new seal ring on outer

Installing forward clutch piston, using seal ring protector tool
(© Detroit Diesel Allison)

from turbine shaft and invert assembly.

6. Remove snap ring from forward clutch housing with a screwdriver. Remove fourth clutch driving hub, clutch plates and forward clutch hub from housing.

7. Remove thrust bearing race, needle bearing and clutch pack from housing.

8. Position forward clutch spring compressor tool on spring retainer. Using press, depress springs until snap ring can be removed. Release pressure.

9. Remove spring retainer, clutch springs, piston and seal rings from housing.

10. Remove outer and inner seal rings from piston.

11. Inspect the ball in forward clutch housing to insure that it moves freely.

Assembly

1. Place the forward clutch housing on flat surface with turbine shaft down. Install seal ring, be sure that lip of seal ring is fac-

ing down and that it is centered in groove, lubricate.

2. Place the clutch piston return spring side up. Install inner and outer seal rings. Position the inner seal protector, insert piston back first into outer seal protector. Install piston and protectors into housing as a unit. Remove seal protectors when piston is seated in housing. Make certain that lips of seal rings are facing down, lubricate. Do not install springs at this time.

3. Starting with external tanged plate, alternately install clutch plates.

4. Install fourth clutch hub and secure with snap ring.

5. Hold clutch driving hub and measure clutch running clearance. The measurement is taken between clutch pack and driving hub, clearance should be between .0765 to .1125. If clearance is excessive new clutch plates must be installed. When a new clutch housing or drive hub has been installed, it may be necessary to use a piston of different dimension.

6. Once specified clearance has been achieved remove snap ring, driving hub and clutch plates as a unit.

7. Place clutch springs on piston with spring retainer on top. Place assembly in press and position spring compressor tool on spring retainer. Apply pressure until snap ring groove is clear. Install snap ring. Release pressure and remove from press.

8. Install thrust bearing race and thrust needle bearing on hub of forward clutch housing. Make certain that lips on race cover bearing. Hold the bearing assembly in position with oil-soluble grease.

9. Install forward clutch hub, clutch package and fourth clutch drive hub and secure with snap ring. *NOTE: The following step is for transmissions with PTO gear only.*

10. Place snap ring in outer edge of clutch housing snap ring groove. Install PTO gear, chamfered inner edge down, and secure with snap ring. Make certain that gear is properly seated.

11. Install hook type seal ring on turbine shaft.

Fourth Clutch

Disassembly

1. Remove snap ring, back plate and clutch plates.

2. Place fourth clutch housing in press and compress piston return springs until snap ring can be removed. Remove snap ring, release pressure and remove assembly from press. Remove spring retainer, springs and fourth clutch piston.

3. Remove seal ring from outer edge and inner diameter of piston. Inspect seal ring grooves for burrs and rough spots.

Assembly

1. Install clutch housing seal ring, lip facing down, in inner hub. Center seal ring in groove and lubricate.

2. Install seal ring on outer edge of piston, lip facing down, lubricate. No seal ring is used on inner diameter of this piston. Place seal ring protector on outside of piston and install as a unit, be certain that piston bottoms. Remove seal protector.

3. Starting with external tanged clutch plate, install alternately clutch plates. Position clutch back plate, flat side first, and se-

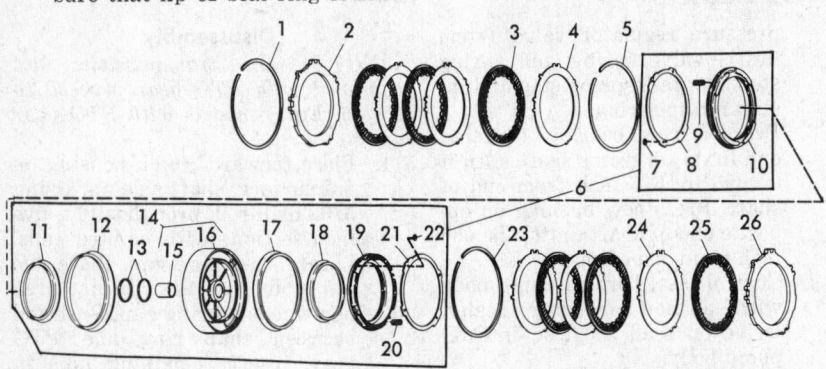

1 Snap Ring	10 Third Clutch Piston	19 Second Clutch Piston
2 Third Clutch Back Plate	11 Piston Inner Seal Ring	20 Piston Return Spring
3 Internal Splined Clutch (3)	12 Piston Outer Seal Ring	21 Spring Retainer
4 External Tanged Clutch (3)	13 Hook Type Seal Ring	22 Self-Locking Washer (4)
5 Snap Ring	14 Support & Bushing Assy.	23 Snap Ring
6 Center Support Assembly	15 Bushing	24 External Tanged Clutch (3)
7 Self Locking Retainer Washer (4)	16 Center Support	25 Internal Splined Clutch (3)
8 Spring Retainer	17 Piston Inner Seal Ring	26 Second Clutch Back Plate
9 Piston Return Spring (12)	18 Piston Outer Seal Ring	

Second clutch, third clutch and center support assembly

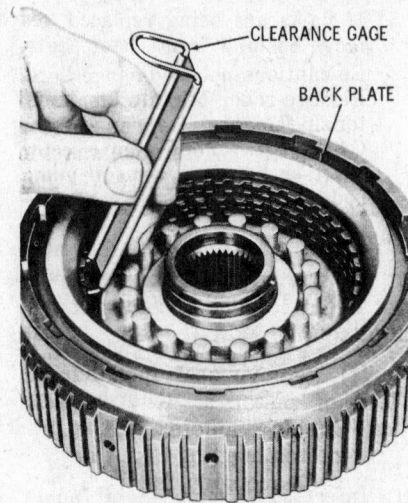

Checking fourth clutch clearance
(© Detroit Diesel Allison)

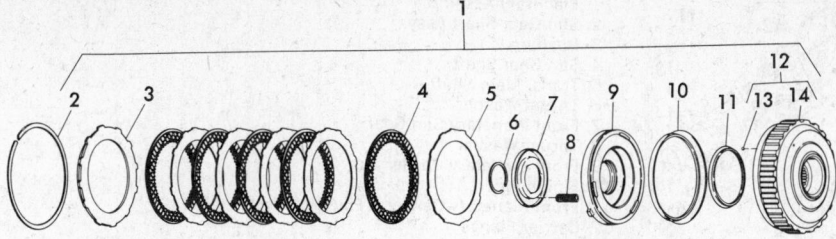

1. Fourth Clutch Assembly
2. Snapring
3. Clutch Back Plate
4. Internal Splined Clutch (5)
5. External Tanged Clutch (5)
6. Snap Ring
7. Spring Retainer
8. Clutch Return Spring
9. Fourth Clutch Piston
10. Piston Outer Seal Ring
11. Clutch Housing Seal Ring
12. Fourth Clutch Housing Assy.
13. Ball
14. Fourth Clutch Housing

Fourth clutch (© Detroit Diesel Allison)

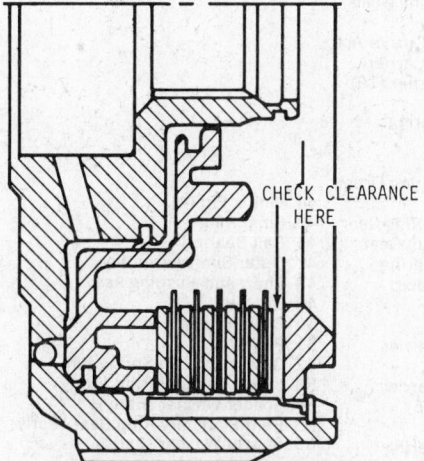

Fourth clutch clearance check point
(© Detroit Diesel Allison)

cure with snap ring.

4. Hold back plate firmly and measure clutch running clearance. Measurement is made between clutch pack and back plate and should range from .0625 to .1125 inch. If clearance is not within range install new clutch plates. *NOTE: If clutch housing or back plate has been replaced it may be necessary to change piston to bring running clearance into proper range.*

5. Place springs on piston with retainer on top. Position assembly on press. Depress spring retainer until snap ring groove is cleared, install snap ring and release pressure. Remove clutch assembly from press.

Center Support Assembly

Disassembly

1. Remove from support assembly, spring retainers and attached parts.
2. Remove retainer washers, separate parts which are freed.
3. Remove inner and outer seal rings from pistons.
4. Remove two hook type seal rings from support assembly.
5. When parts replacement is necessary, place support assembly in press, seal ring side up, and

press out bushing. Be careful not to damage bore.

Assembly

1. Place center support assembly in a press, seal ring side up, and install flush with surface of bore. It is important that notch on bushing is in its specified place, refer to illustration for location.
2. Place return springs on pistons with spring retainers on top. Align holes in retainers with ejector pins. Depress return springs and start self-locking washers on ejector pin bosses. Insert piston assembly in support. Pistons should bottom before washers touch side of support. Check alignment of tangs in slots of support assembly. Remove assembly. Secure washers to seat springs.
3. Lubricate and install inner and outer seal rings on pistons. Seal ring lips must face down. Carefully insert pistons in to center support assembly, engaging tangs of spring retainers with slots.
4. Install two hook type seal rings on center support hub.

Planetary Gears and Shafts

Disassembly

1. Remove sun gear shaft, thrust washer, front planetary sun gear, front planetary carrier assembly.
2. Remove thrust washers from front planetary carrier assembly and center planetary carrier assembly.
3. Remove snap ring retaining front planetary ring gear and ring gear and center planetary carrier assembly.
4. Remove center planetary sun gear and thrust washer.
5. Remove snap ring retaining rear planetary carrier assembly and output shaft with attaching parts.
6. Remove needle bearing and race between output shaft and rear sun gear.

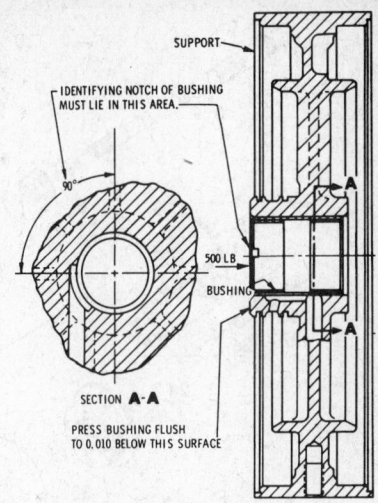

Center support assembly
(© Detroit Diesel Allison)

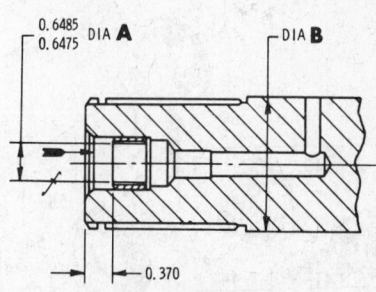

WHEN MOUNTED ON DIA **A** TOTAL RUNOUT OF DIA **B** TO BE WITHIN 0.002

BUSHING MUST WITHSTAND APPROX. 350 LB LOAD IN DIRECTION SHOWN

Output shaft assembly
(© Detroit Diesel Allison)

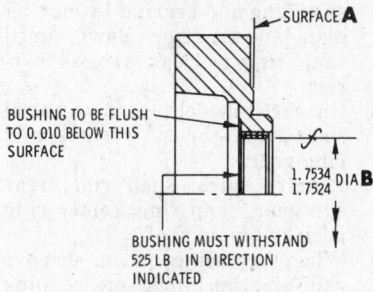

BUSHING MUST WITHSTAND 525 LB IN DIRECTION INDICATED

DIA **B** MUST BE SQUARE WITH SURFACE **A** WITHIN 0.001 TOTAL

Front planetary carrier bushing
(© Detroit Diesel Allison)

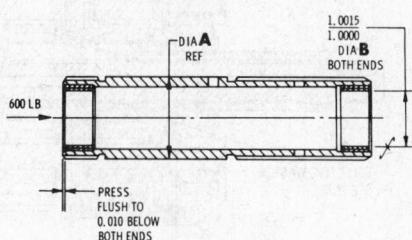

DIA **B** MUST BE CONCENTRIC WITH DIA **A** WITHIN 0.002 TIR

EACH BUSHING MUST WITHSTAND THE SPECIFIED LOAD IN THE INDICATED DIRECTION

Sun gear and shaft bushings
(© Detroit Diesel Allison)

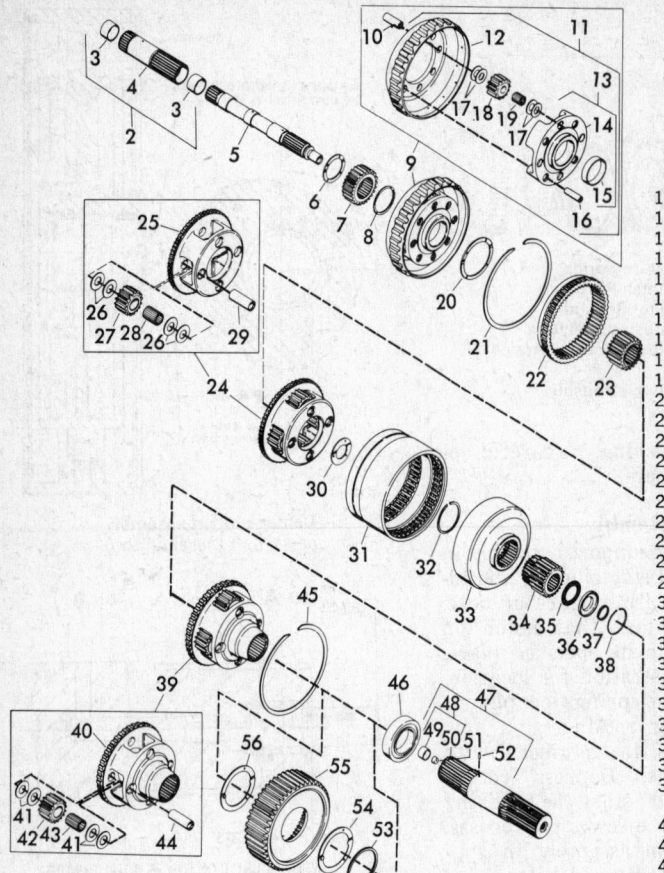

1 Planetary Assembly
2 Sun Gear Shaft Assy.
3 Bushing (2)
4 Sun Gear Shaft
5 Trans. Main Shaft
6 Thrust Washer
7 Front Planetary Sun Gear
8 Thrust Washer
9 Front Planetary Carrier Assy.
10 Pinion Pin
11 Front Planetary Carrier & Flange Assy.
12 Carrier Flange
13 Carrier & Bushing Assy.
14 Planetary Carrier
15 Bushing
16 Carrier Rivet Pin
17 Pinion Thrust Washer
18 Pinion
19 Needle Roller Bearing
20 Thrust Washer
21 Snap Ring
22 Front Planetary Ring Gear
23 Center Sun Gear
24 Center Planetary Carrier Assy.
25 Center Planetary Carrier
26 Pinion Thrust Washer (16)
27 Pinion
28 Needle Roller Bearing
29 Pinion Pin
30 Thrust Washer
31 Planetary Connecting Drum
32 Snap Ring
33 Center Planetary Ring Gear
34 Rear Planetary Sun Gear
35 Needle Thrust Bearing
36 Thrust Bearing Race
37 Spiral Snap Ring
38 Snap Ring
39 Rear Planetary Carrier Assembly
40 Rear Planetary Carrier
41 Thrust Washer (16)
42 Pinion (4)
43 Needle Roller Bearing
44 Pinion Pin (4)
45 Snap Ring
46 Ball Bearing
47 Outer Shaft Assembly
48 Shaft and Bushing Assy.
49 Bushing
50 Output Shaft
51 Plug
52 Governor Drive Spring Pin
53 Spiral Retainer Ring (Early)
54 Spacer Washer (Early)
55 Rear Planetary Ring Gear (Early)
56 Spacer Washer (Early)

Planetary gears and shafts assembly   (© Detroit Diesel Allison)

7. Position output shaft with rear planetary carrier up. Remove snap ring and bearing by moving planetary carrier down until snap ring is clear, remove carrier.
8. On early models, remove spiral snap ring, thrust washers and ring gear.
9. Remove spiral snap ring, rear sun gear, snap ring, center ring gear and main shaft.
10. When replacing parts, remove roll pin, bushings, orifice plug from output shaft and bushings from sun gear shaft.

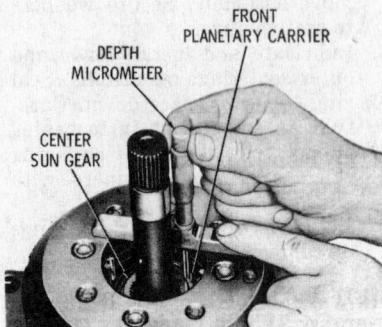

Measuring depth of front planetary carrier
(© Detroit Diesel Allison)

Assembly

1. Install new bushings on output shaft and sun gear shaft, if they were removed.
2. Install orifice plug and roll pin on output shaft.
3. Install rear planetary sun gear, small end first, spiral snap ring, center ring gear and snap ring.
4. Install bearing assembly over snap ring groove on output shaft, output shaft with snap ring groove first and snap ring.
5. On early models, install rear ring gear.
6. Install spacer washer on hub of rear planetary carrier, rear planet ring gear on carrier hub, spacer washer and spiral snap ring.
7. Place positioning tool over output shaft and tighten bolt to 30 foot pounds. Install rear planetary carrier assembly in carrier connecting drum and secure with snap ring. Broad groove in carrier connecting drum is positioned away from rear carrier.
8. Lubricate and install bearing assembly over output shaft next to rear sun gear.
9. Insert small end of main shaft, with attached parts, in output shaft.
10. Install thrust washer on front of rear sun gear.

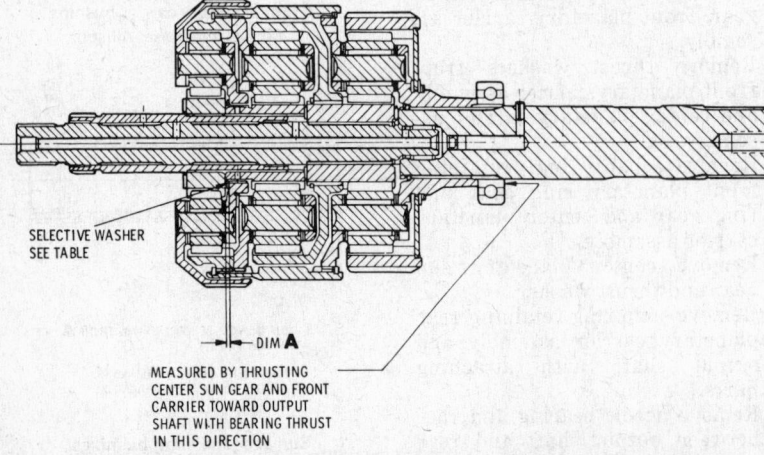

SELECTIVE WASHER SEE TABLE

DIM **A**

MEASURED BY THRUSTING
CENTER SUN GEAR AND FRONT
CARRIER TOWARD OUTPUT
SHAFT WITH BEARING THRUST
IN THIS DIRECTION

Selection of thrust washer for planetary gears assembly (© Detroit Diesel Allison)

11. Install center planetary carrier small end first, in planetary connecting drum. Engage carrier splines with splines in drum.

12. Install center sun gear, large end first against thrust washer.

13. Install front planetary ring gear, large end first in planetary connecting drum, attach snap ring.

13. Lubricate and install thrust washer on front planetary carrier hub. Install front carrier assembly on center sun gear.

14. With a depth micrometer, measure the depth of the center planetary sun gear and front planetary carrier. Difference in measurements determines thrust washer thickness. Select washer from following chart.

Thickness	Identification
.071 to .090	#1
.090 to .109	#2
.109 to .128	#3

Remove positioning tool from output shaft. Lubricate and install selected thrust washer. Reinstall.

15. Install front planetary sun gear, spline chamfer first, in planetary carrier.

NOTE: Thrust washer and sun gear shaft after planetary gears and shafts are positioned in transmission.

Transmission Case

Disassembly

1. Remove retainer pin and lock nut from manual shift lever.
2. Pull the selector shaft carefully through oil seal out of case. Remove seal.
3. Do not remove snap ring from rear of output flange bearing unless replacement is necessary.
4. Remove governor support pin if damaged.
5. Remove and inspect test plugs, replace if necessary.

Assembly

1. Install output bearing rear snap ring if it was removed. Inspect for damage to bore.
2. Install test plugs.
3. Lubricate and install selector shaft oil seal, lip first. Slide selector shaft through seal and into case. Attach locknut and retainer pin to shaft. Tighten locknut to 15 to 20 foot pounds.
4. Install governor support pin if it was removed. Press to dimensions shown in illustration. Alignment of pin must be correct or damage to governor will occur.

AUTOMATIC TRANSMISSION SPECIFICATIONS

GENERAL SPECIFICATIONS

Model AT-540—4 speed	
Dry Weight	275 lbs.
Oil Capacity	15 Qts.
Temperature	
Sump	250°F Max.
To cooler —converter out	300°F Max.
Normal operation	160°-220°F
Minimum	100°F
Clutches	Oil wet, hydraulic-actuated, spring released, self compensating for wear.
Gearing	Planetary, straight-cut spur, constant mesh.
Power take-off— (If equipped)	Converter driven, located on right side, viewed from rear.
Oil filter	Integral (in sump)
Oil type	Dexron®
Drive range and sequences	Reverse, N, 1-2-3-4 1-2-3 1-2 1
Shift control—	
External	Mechanical
Internal	Hydraulic
Shift Modulation	Vacuum or mechanical

TORQUE LIMITS FOOT POUNDS

Oil Pump to Support Bolts	15 to 20
Output Flange Bolts	83 to 100
Front Support to Case Bolts	15 to 20
PTO Cover Bolts	15 to 20
Pressure Tap Plugs	10 to 15
Governor Cover Bolts	15 to 20
Vacuum Modulator Retaining Bolts	15 to 20
Modulator Valve Body to Control Valve Body	8 to 12
Trimmer Cover to Valve Body Bolts	8 to 12
Detent Spring to Valve Body Bolt	8 to 12
Control Valve Body to Case Bolts	8 to 12
Oil Filter Bolt	10 to 15
Oil Pan Bolts	10 to 13
Center Support Anchor Bolt	39 to 46
Selector Shaft Nut	15 to 20
Converter to Flywheel Bolts	34 to 40

MAIN PRESSURE CHECKS

Forward and Neutral	150 psi
Reverse	250 psi

CLUTCHES ENGAGED FOR THE DRIVE RANGES AND THE RESULTING GEAR RATIOS

RANGE	CLUTCHES ENGAGED	RATIO
Neutral	First and reverse	0
First	Forward and first-and-reverse	3.45:1
Second	Forward and second	2.25:1
Third	Forward and third	1.41:1
Fourth	Forward and fourth	1.00:1
Reverse	Fourth and first-and-reverse	5.02:1

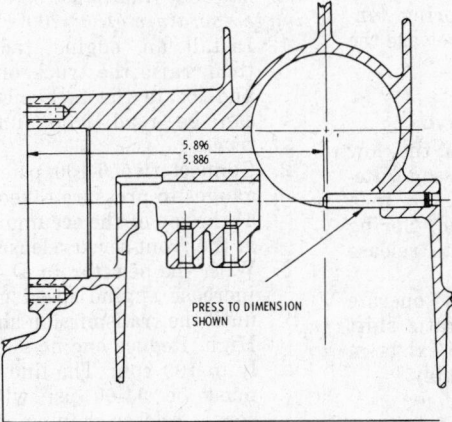

Governor support pin location (© Detroit Diesel Allison)

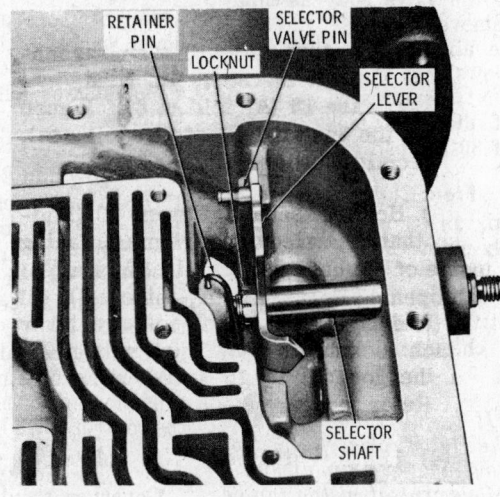

Selector shaft components (© Detroit Diesel Allison)

Dodge/Plymouth Loadflite

DIAGNOSIS

This guide covers the most common symptoms and is an aid to careful diagnosis. The items to check are listed in sequence to be followed for quickest results. Follow the checks in the order given for the particular transmission type.

TROUBLE SYMPTOMS	ITEMS TO CHECK	
	IN TRUCK	OUT OF TRUCK
Harsh N to D or N to R shift	CDEFGIJ	ab
Delayed Shift—N to D	ACIHKJ	ca
Runaway on upshift and 3-2 kickdown	ABCDLIHK	b
Harsh upshift and 3-2 kickdown	BCDLIHJ	b
No upshift	ABCDLIHKM	b
No kickdown on normal downshift	ABNCDLIHKM	d
Erratic shifts	ABNCFOPQ IKMJ	c
Slips in forward drive positions	ACIHK	abd
Slips in reverse only	CEGIK	
Slips in all positions	ACOIK	e
No drive in any position	ACOQIKJ	ca
No drive in forward positions	CDOLIHK	abd
No drive in reverse	CEGKMI	b
Drives in neutral	NIJ	a
Drags or locks	DELG	abfd
Noises	AORPMIJ	a
Hard to fill or blows out	AORSTQI	
Transmission overheats	ADEORTMI	abe

Key to Checks

A. Oil level	H. Accumulator
B. Control linkage	I. Valve body assembly
C. Oil pressure check	J. Manual valve lever
D. Kickdown band	K. Air pressure check
E. Low-reverse band	L. K-D servo link
F. Improper engine idle	M. Governor
G. Servo linkage	N. Gear shift cable

O. Regulator valve and/or spring	a. Front-kickdown clutch
P. Output shaft bushing	b. Rear clutch
Q. Strainer	c. Front pump and/or sleeve
R. Converter control valve	d. Overrunning clutch
S. Breather clogged	e. Converter
T. Cooler or lines	f. Planetary

Tests

Air Pressure Tests

The front clutch, rear clutch, kickdown servo and low and reverse servo may be checked with air pressure, after the valve body assembly has been removed.

To make air pressure tests, proceed as follows:

CAUTION: Compressed air must be free of dirt and moisture. Use pressure of 30-100 psi.

Front Clutch

Apply air pressure to the front clutch apply passage and listen for a dull thud. This will indicate operation of the front clutch. Hold the air pressure at this point for a few seconds and check for excessive oil leaks.

NOTE: If a dull thud cannot be heard in the clutch, place finger tips on clutch housing and again apply air pressure. Movement of piston can be felt as clutch is applied.

Rear Clutch

Apply air pressure to the rear clutch apply passage and proceed in an identical manner as that described in the previous paragraph.

Kickdown Servo

Air pressure applied to the kickdown servo apply passage should tighten the front band. Spring tension should be sufficient to release the band.

Low and Reverse Servo

Direct air pressure into the low and reverse servo apply passage. Response of the servo will result in a tightening of the rear band. Spring tension should be enough to release the band.

If clutches and servos operate properly, no upshift or erratic shift conditions existing, trouble exists in the control valve body assembly.

Governor

Governor troubles can usually be found during a road or pressure test.

Hydraulic Control Pressure Checks

Line Pressure and Front Servo Release Pressure

NOTE: These pressure checks must be made in the D position with the rear wheels free to turn. The transmission fluid must be at operating temperature (150°-200°F).

1. Install an engine tachometer, then, raise the truck on a hoist and locate the tachometer so it can be read from under the truck.
2. Connect two 0-100 psi pressure gauges to pressure takeoff points at the top of the accumulator and at the front servo release.
3. With the selector in D position, increase engine speed gradually until the transmission shifts into High. Reduce engine speed slowly to 100 rpm. The line pressure must be 54-60 psi with front servo release having no more than a 3 psi drop.
4. Disconnect throttle linkage from

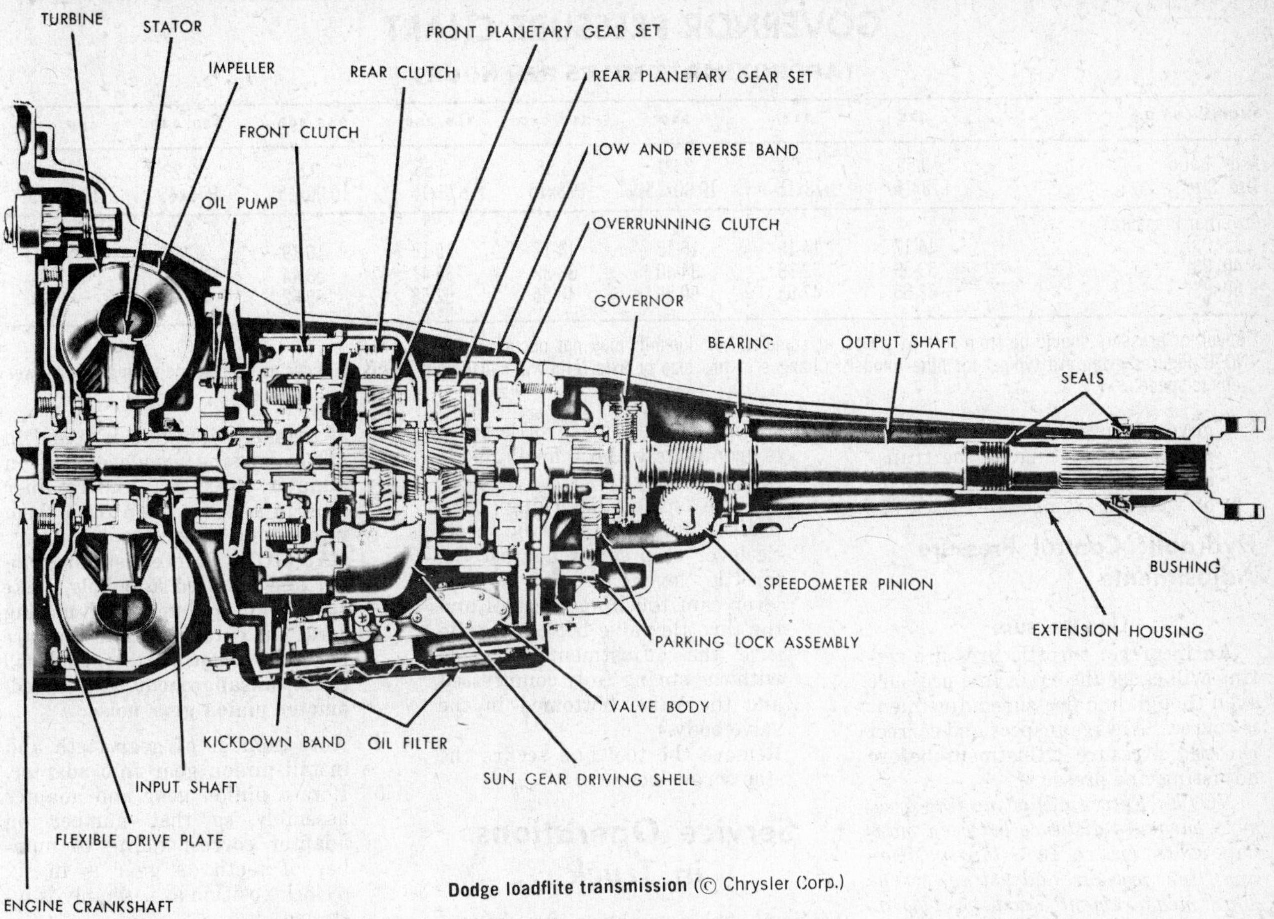

TURBINE   STATOR   FRONT PLANETARY GEAR SET

IMPELLER   REAR CLUTCH   REAR PLANETARY GEAR SET

FRONT CLUTCH   LOW AND REVERSE BAND

OIL PUMP   OVERRUNNING CLUTCH

GOVERNOR   BEARING   OUTPUT SHAFT

SEALS

SPEEDOMETER PINION   BUSHING

PARKING LOCK ASSEMBLY   EXTENSION HOUSING

VALVE BODY

KICKDOWN BAND   OIL FILTER   SUN GEAR DRIVING SHELL

INPUT SHAFT

FLEXIBLE DRIVE PLATE

ENGINE CRANKSHAFT

Dodge loadflite transmission (© Chrysler Corp.)

transmission throttle lever and more throttle lever gradually to full throttle position. Line pressure must rise to maximum of 90-96 psi just before or at kickdown into low gear. Front servo pressure must follow line pressure up to kickdown point and should not be more than 3 psi below line pressure. If pressure is not 54-60 psi at 1000 rpm, adjust line pressure.

If line pressure is not as above, adjust the pressure as outlined under the heading: Hydraulic Control Pressure Adjustments—"Line Pressure."

If front servo release pressures are less than specified, and line pressures are within limits, there is excessive leakage in the front clutch and/or front servo circuits.

Lubrication Pressures

A lubrication pressure check should be made when line pressure and front servo release pressures are checked.

1. Install a T fitting between the cooler return line fitting and the fitting hole in the transmission case at the rear left side of the transmission. Connect a 0-100

psi pressure gauge to the T-fitting.
2. At 1000 engine rpm, with throttle closed and transmission in High, lubrication pressure should be 5-15 psi. Lubrication pressure will approximately double as throttle is opened to maximum line pressure.

Rear Servo Apply Pressure

1. Connect a 0-300 psi pressure gauge, to the apply pressure take off point at the rear servo.
2. With the control in the R position, and the engine running at 1600 rpm, the reverse servo apply pressure should be 230-300 psi.

Governor Pressure

1. Connect a 0-100 psi gauge (same as the one used for line pressure and front servo release pressure) to the governor pressure takeoff point. This location is at the lower left rear corner of the extension mounting flange.

2. The governor pressure should respond smoothly during the changes in speed and should return to 0 to 1½ PSI, when the vehicle is stopped. A pressure reading above 2 PSI, at standstill, will prevent the transmission from downshifting. If the

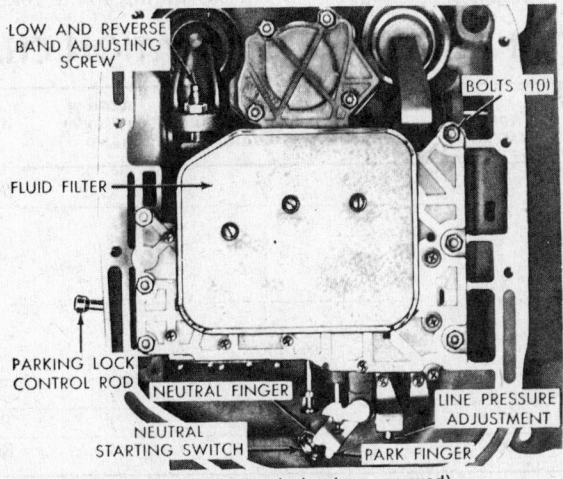

LOW AND REVERSE BAND ADJUSTING SCREW

BOLTS (10)

FLUID FILTER

PARKING LOCK CONTROL ROD

NEUTRAL FINGER   LINE PRESSURE ADJUSTMENT

NEUTRAL STARTING SWITCH   PARK FINGER

Bottom view of transmission (pan removed)
(© Chrysler Corp.)

GOVERNOR PRESSURE CHART

(APPROXIMATE MILES PER HOUR)

ENGINE (C.I.D.)	225	318	318	318/360	318/360	318-360	400/440	400/440
Axle Ratio	3.55	3.90	3.90	3.55	3.55	3.55	3.23	3.23
Tire Size	E78x15	H78x15	10.00x15	E78x15	H78x15	10.00x15	H78x15	10.00x15
Governor Pressure*								
15 PSI	14-17	14-16	15-17	14-17	15-18	16-19	17-20	18-21
40 PSI	33-39	32-38	34-40	33-39	36-41	38-44	39-45	42-48
60 PSI	48-55	47-53	50-56	48-55	52-58	55-62	57-64	60-68

* Governor pressure should be from zero to 1.5 psi at stand-still or donshift may not occur.
NOTE: Figures given are typical for other models. Changes in tire size or axle ratio will cause shift points to occur at corresponding higher or lower vehicle speeds.

governor pressure is erratic during the upshifting of the transmission, the governor valve and/or weights are probably sticking.

Hydraulic Control Pressure Adjustments

Line Pressure

An incorrect throttle pressure setting will cause incorrect line pressure even though line pressure adjustment is correct. Always inspect and correct throttle pressure adjustment before adjusting line pressure.

NOTE: Before adjusting line pressure, measure distance between manual valve (valve in 1-low position and line pressure adjusting screw. This measurement must be 1 7/8 in. Correct by loosening spring retainer screws and repositioning spring retainer. The regulator valve may cock and hang up in its bore if spring retainer is out of position.

If line pressure is not correct remove valve body assembly to adjust. The correct adjustment is 1-5/16 in. measured from valve body to inner edge of adjusting nut. Vary adjustment slightly to obtain specified line pressure.

One complete turn of the adjusting screw (Allen head) changes closed throttle line pressure about 1.66 psi. Turning the screw counterclockwise increases pressure, clockwise decreases pressure.

Throttle Pressure

Because throttle pressures cannot be checked, exact adjustments should be checked and made correct whenever the valve body is disturbed.

1. Remove the valve body assembly, as outlined in a succeeding coverage entitled, Valve Body Assembly and Accumulator Piston.
2. Loosen throttle lever stop screw locknut. and back off the screw about five turns.
3. Insert gauge pin between the throttle lever cam and the kickdown valve.
4. Push on the tool and compress the kickdown valve against its

spring, so that the throttle valve is completely bottomed inside the valve body.
5. As the spring is being compressed, finger tighten the throttle lever stop screw against the throttle lever tang, with the lever cam touching the tool and the throttle valve bottomed. (Be sure the adjustment is made with the spring fully compressed and the valve bottomed in the valve body.)
6. Remove the tool and secure the stop screw locknut.

Service Operations in Truck

Some sub-assemblies can be removed for repairs without removing the transmission from the truck. Detailed reconditioning of sub-assemblies is covered further in the text.

Speedometer Pinion

Removal and Installation

Rear axle gear ratio and tire size determine pinion gear size.
1. Remove bolt and retainer securing speedometer pinion adapter in extension housing.
2. With cable housing connected, carefully work adapter and pinion out of extension housing.

3. If transmission fluid is found in cable housing, replace seal in adapter. Start seal and retainer ring in adapter, then push them into adapter until tool bottoms. CAUTION: Before installing pinion and adapted assembly make sure adapter flange and mating area on extension housing are perfectly clean. Dirt or sand will cause misalignment and speedometer pinion gear noise.
4. Note number of gear teeth and install pinion gear into adapter.
5. Rotate pinion gear and adapter assembly so that number on adapter corresponding to number of teeth on gear is in six o'clock position as assembly is installed.
6. Install retainer and bolt with retainer tangs in adapter positioning slots. Tap adapter firmly into extension housing and tighten retainer bolt to 100 in. lbs.

Output Shaft Oil Seal

Replacement

1. Mark parts for reassembly. Disconnect driveshaft at rear universal joint. Carefully pull shaft yoke out of transmission extension housing. Be careful not to scratch or nick ground surface of sliding spline yoke.

CLUTCH AND BAND APPLICATION CHART

LEVER POSITION DRIVE-RATIO	FRONT CLUTCH	REAR CLUTCH	FRONT (KICKDOWN) BAND	REAR (LOW-REV) BAND	OVERRUNNING CLUTCH
N-NEUTRAL					NO MOVEMENT
D-DRIVE (Breakaway)		X			HOLDS
(Second)		X	X		OVERRUNS
(Direct)	X	X			OVERRUNS
KICKDOWN (To Second)		X	X		OVERRUNS
(To Low)		X			HOLDS
2-Second		X	X		OVERRUNS
1-Low		X		X	PARTIAL HOLD
R-REVERSE	X			X	NO MOVEMENT

X=Applied

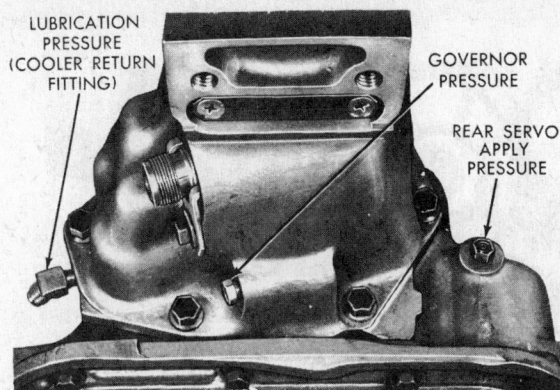

Pressure test locations (right side of case)
(© Chrysler Corp.)

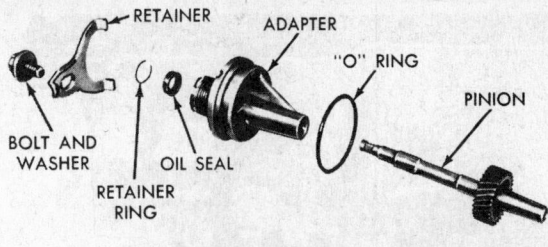

Speedometer drive (© Chrysler Corp.)

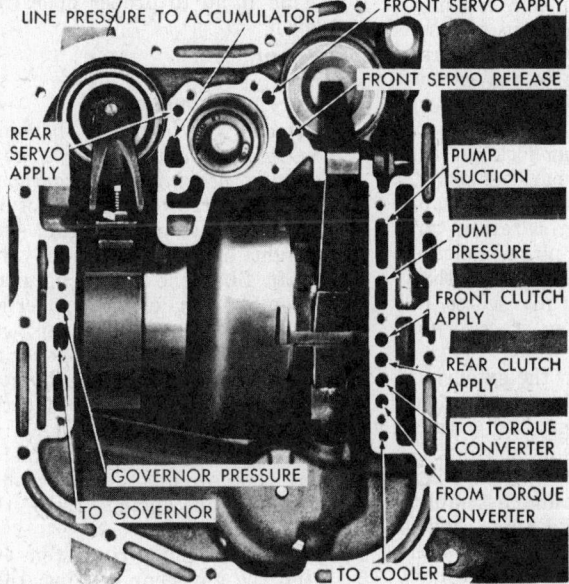

Air pressure test points (© Chrysler Corp.)

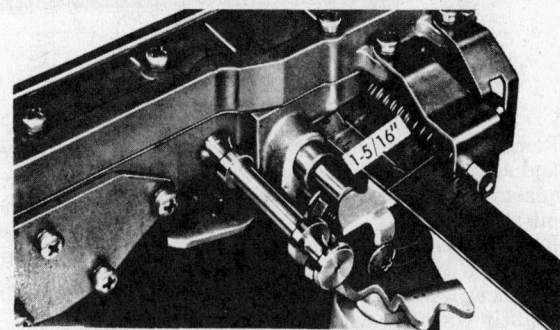

Line pressure adjustments (© Chrysler Corp.)

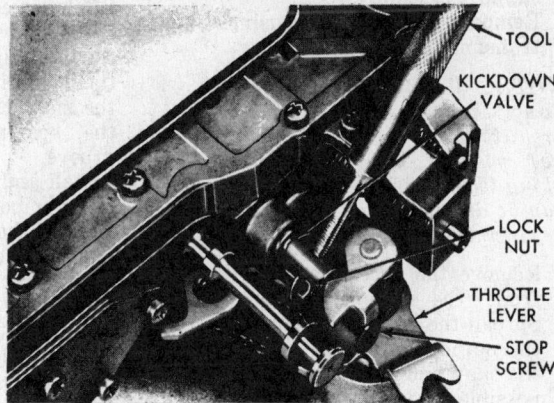

Throttle pressure adjustment (© Chrysler Corp.)

2. Remove extension housing yoke seal by gently tapping out around circumference of seal with slide hammer.
3. To install new seal, place seal in opening of extension housing and drive it into housing with suitable drift.
4. Carefully guide front universal joint yoke into extension housing and onto the mainshaft splines. Align marks made at removal and connect driveshaft to pinion shaft yoke.

Short Extension Housing

Removal

1. Mark parts for reassembly. Disconnect driveshaft at rear universal joint. Carefully pull shaft out of extension housing.
2. Remove speedometer pinion and adapted assembly. Drain approximately two quarts of fluid from transmission.
3. Remove bolts securing extension housing to crossmember. Raise transmission slightly with serv-

ice jack and remove center crossmember and support assembly.
4. Remove extension housing to transmission bolts. On console shifts, remove two bolts securing gearshift torque shaft lower bracket to extension housing. Swing bracket out of way.

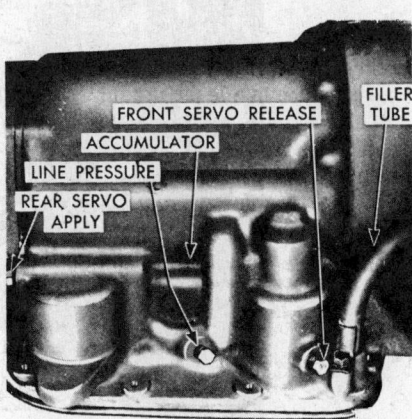

Pressure test locations (right side of case)
(© Chrysler Corp.)

NOTE: Gearshift lever must be in 1-low position so that parking lock control rod can be engaged or disengaged with parking lock sprag.

5. Remove two screws, plate, and gasket from bottom of extension housing mounting pad.
6. Spread large snap ring from output shaft bearing.
7. With snap ring spread as far as possible tap extension housing gently off output shaft bearing.
8. Carefully pull extension housing rearward to bring parking lock control rod knob past parking sprag and remove housing.

Long Extension Housing Bushing and Output Shaft Bearing

Removal

1. Mark drive shaft components for reassembly then disconnect drive shaft at rear universal joint. Carefully, pull the drive shaft assembly out of the extension housing.

Measuring spring retainer location (© Chrysler Corp.)

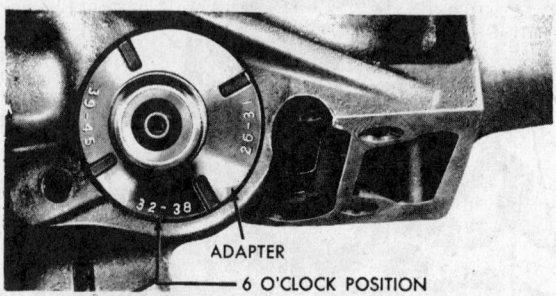

ADAPTER

6 O'CLOCK POSITION

Speedometer pinion and adapter installed
(© Chrysler Corp.)

2. Remove the speedometer pinion and adapter assembly. Drain two quarts of fluid from the transmission.
3. Remove the bolts securing the extension housing to the crossmember. Raise the transmission slightly with a service jack, then remove the crossmember and support assembly.
4. Remove the extension housing to transmission bolts.

NOTE: When removing or installing the extension housing the gearshift lever must be in the 'l' (low) position. This positions the parking lock control rod rearward so it can be disengaged or engaged with the parking lock sprag.

5. Remove the access plate gasket from the extension housing. Spread the large snap ring from the output shaft bearing. When the snap ring is spread as far as possible, carefully tap the extension housing off the output shaft bearing. Carefully pull the extension housing rearward to remove parking lock rod control rod knob past the parking sprag, then remove the extension housing.

Bushing Replacement

1. Remove the oil seal with special Chrysler tool.
2. Press or drive out the bushing.
3. Slide a new bushing on installation tool. Align the oil hole in the bushing with the oil slot in the housing, then press or drive the bushing into place.
4. Position a new seal in the opening of the extension housing and drive it into the housing.

Bearing Replacement

1. Using a heavy duty snap ring pliers, remove the output shaft bearing rear snap ring and remove the bearing from the shaft.
2. Install a new bearing on the shaft with the outer race ring groove toward the front. Install the rear snap ring.

Installation

1. Place a new extension housing gasket on the transmission case. Position output shaft bearing retaining snap ring in the extension housing. Slide the extension housing on the output shaft, guiding the parking lock control rod knob past the parking sprag. While spreading the large snap ring in the housing, carefully tap the housing into place. Release the snap ring. Make sure that the snap ring is fully seated in the bearing outer race ring groove.
2. Install and tighten the extension housing bolts to 24 foot-pounds.
3. Install the gasket and access plate on the extension housing.
4. Install the center crossmember and the rear mount assembly. Tighten the retaining bolts. Lower the transmission, install the extension housing and torque bolts to 40 foot-pounds.
5. Install the speedometer pinion and adapter.
6. Carefully guide the front universal joint yoke into the extension housing and on the output shaft splines. Align the marks made during removal and connect the drive shaft to the rear axle pinion shaft yoke.

7. Add fluid to the transmission to bring it up to proper operating level.

Governor

Removal

1. Remove extension housing. *NOTE: Remove output shaft, support bearing if so equipped.*
2. With a screwdriver, carefully pry the snaping from the weight end of governor valve shaft. Slide the valve and shaft assembly out of the governor housing.
3. Remove the large snap-ring from the weight end of the governor housing and lift out the governor weight assembly.
4. Remove snap-ring from inside governor weight, remove inner weight and spring from the outer weight.
5. Remove the snap-ring from behind the governor housing, then slide the governor housing and support assembly from the output shaft. If necessary, remove the four screws and separate the governor housing from the support.

Cleaning and Inspection

The primary cause of governor operating trouble is sticking of the

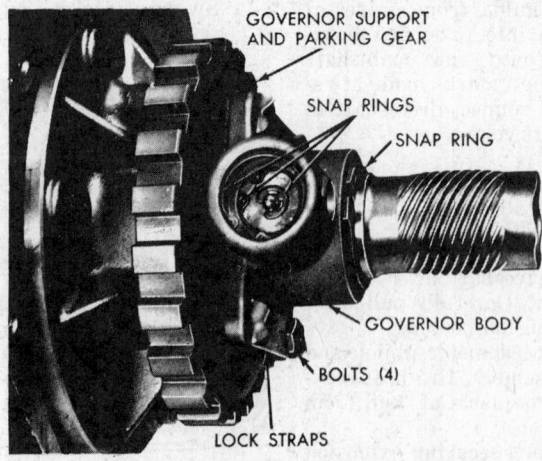

GOVERNOR SUPPORT AND PARKING GEAR

SNAP RINGS

SNAP RING

GOVERNOR BODY

BOLTS (4)

LOCK STRAPS

Governor shaft and weight snap-rings
(© Chrysler Corp.)

valve or weights. This is brought about by dirt or rough surfaces. Thoroughly clean and blow dry all of the governor parts, crocus cloth any burrs or rough bearing surfaces and clean again. If all moving parts are clean and operating freely, the governor may be reassembled.

Installation

1. Assemble governor body and screen to the support, if disassembled, and tighten the bolts finger tight. Make sure the oil passages of the governor body aligns with the passage in the support.
2. Position the support and the governor on the output shaft. Align the assembly so that the valve shaft hole in the governor body alighs with the hole in the output shaft, then slide the assembly in place. Torque the body support bolts to 100 inch-pounds. Bend the ends of the lock straps over the bolt heads.

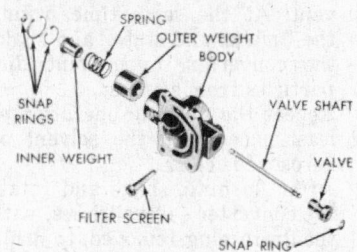

Exploded View—Governor Assembly
(© Chrysler Corp.)

3. Assemble the governor weights and spring, and secure with the snap ring inside of the large governor weight. Place the weight assembly in the governor body and install the snap ring.
4. Place the governor valve on the valve shaft, insert the assembly into the body and through the governor weights. Install the valve shaft retaining snap ring. Inspect the valve and weight assembly for free-movement after installation.
5. Install the output shaft bearing and the extension housing.

Parking Lock Components

Removal

1. Remove the extension housing.
2. Slide the shaft out of the extension housing to remove the parking sprag and spring. Remove the snap ring and slide the reaction plug and pin assembly out of the housing.
3. To replace the parking lock rod, refer to "Valve Body-Removal and Installation".

Assembly

1. Install the reaction plug and pin

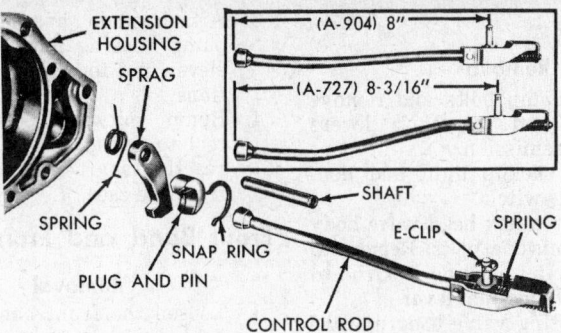

Parking lock components (© Chrysler Corp.)

assembly in the housing and secure with a snap ring.
2. Position the sprag and spring in the housing and insert the shaft. Make sure that the square lug on the sprag is toward the parking gear, and the spring is positioned so it moves the sprag away from the gear.
3. Install the extension housing.

Removal

1. Raise vehicle on hoist and loosen oil pan bolts, tap pan to break it loose, allowing fluid to drain.
2. Remove pan and gasket.
3. Disconnect throttle and gear shift linkage from levers on transmission. Loosen clamp bolts and remove levers.
4. Remove E clip securing parking lock rod to valve body manual lever.
5. Remove backup light and neutral start switch.
6. Place drain pan under transmission, remove ten hex head valve body to transmission case bolts. Hold valve body in position while removing bolts.
7. While lowering valve body down out of transmission case, disconnect parking lock rod from lever.

To remove rod pull if forward out of case. If necessary rotate driveshaft to align parking gear and sprag to permit knob on end of control rod to pass sprag.
8. Withdraw accumulator piston from transmission case. Inspect piston for scoring, and rings for wear or breakage.
9. If valve body manual lever shaft seal requires replacement, drive it out of case with punch.
10. Drive new seal into case with 15/16 in. socket and hammer.

Installation

1. If parking lock rod was removed, insert it through opening in rear of case with knob positioned against plug and sprag. Move front end of rod toward center of transmission while exerting rearward pressure on rod to force it past sprag. Rotate driveshaft if necessary.

2. Install accumulator piston in transmission case.
3. Place accumulator spring on valve body.
4. Place valve body manual lever in low position. Lift valve body into its approximate position, connect parking lock rod to manual lever and secure with E clip. Place valve body in case, install retaining bolts finger tight.
5. With neutral start switch installed, place manual lever in neutral position. Shift valve body if necessary to center neutral finger over neutral switch plunger. Snug bolts down evenly. Torque to 100 in. lbs.
6. Install gearshift lever and tighten clamp bolt. Check lever shaft for binding by moving lever through all detends. If lever binds, loosen valve body bolts and re-align.
7. Make sure throttle shaft seal is in place, install flat washer and lever and tighten clamp bolt. Connect throttle and gearshift linkage and adjust as required.
8. Install oil pan using new gasket. Add transmission fluid to proper level.

Detailed Unit Reconditioning

The following reconditioning data covers the removal, disassembly, inspection, repair, assembly and installation procedures for each sub-assembly in detail.

NOTE: In the event that any part has failed in the transmission, the converter should be thoroughly flushed to insure the removal of fine particles that may cause damage to the reconditioned transmission.

Oil Pan

Removal

1. Secure transmission in a repair stand.
2. Unscrew attaching bolts and remove oil pan and gasket.

Valve Body

Removal

1. Loosen clamp bolts and remove throttle and gearshift levers from transmission.
2. Remove backup light and neutral start switch.
3. Remove ten hex head valve body to transmission bolts. Remove E clip securing parking lock rod to valve body manual lever.
4. While lifting valve body upward out of transmission case, disconnect parking lock rod from lever.

Accumulator Piston and Spring

Removal

Lift the spring from the accumulator piston and withdraw the piston from the case.

Checking Drive Train End-Play

1. Attach a dial indicator to the extension housing and seat the plunger on the end of the output shaft.
2. Pry the output shaft out and tap it in to register the extreme shaft end-play.
3. Record this reading for possible future use. Correct end play is found at front of section.

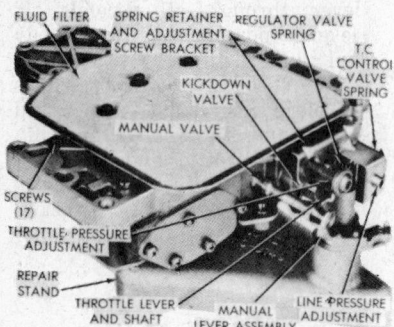

Control valve body assembly (© Chrysler Corp.)

Governor and Support

Removal

1. Remove the snap-ring from the weight end of the governor valve shaft. Slide the valve and shaft assembly from the governor housing.
2. Remove the snap-ring from behind the governor housing, then slide the governor housing and support from the output shaft.

Oil Pump and Reaction Shaft Support

Shaft Support Removal

1. Tighten the front band on the front clutch retainer.
2. Remove front pump housing retaining bolts.

3. Attach slide hammers to the pump housing flange, using the eleven and four o'clock hole locations.
4. Bump outward, evenly, with tool to withdraw oil pump and reaction shaft support assembly from the case.

Front Band and Front Clutch

Removal

1. Loosen the front band adjuster, remove the head strut and slide the band from the case.
2. Slide the front clutch assembly from the case.

Input Shaft and Rear Clutch

Removal

Grasp the input shaft and slide the shaft and rear clutch assembly out of the case.

NOTE: Don't lose the thrust washer located between the rear end of the input shaft and the front end of the output shaft.

Planetary Gear Assemblies, Sun Gear, Driving Shell, Low and Reverse Drum

Removal

While hand-supporting the output shaft and driving shell, carefully slide the assembly forward and out of the case.

Rear Band and Low-Reverse Drum

Removal

Remove low-reverse drum, loosen rear band adjuster, remove band strut and link, and remove band from case.

Overrunning Clutch

Removal

1. Notice the established position of the overrunning clutch rollers and springs before disassembly.
2. Slide out the clutch hub and remove rollers and springs.

Kickdown Servo

Removal

1. Compress kickdown servo spring using engine valve spring compressor. Then remove snap ring.
2. Remove the rod guide, spring and piston rod from the case. Don't damage the piston rod or guide during removal.
3. Withdraw piston from the transmission case.
4. If so equipped, disassemble the "controlled load" servo piston assembly by removing small snapring from servo piston then remove the washer, spring and piston rod from servo piston.

Low and Reverse Servo

Removal

1. Using a suitable tool, depress the piston spring retainer and remove the snap ring.
2. Remove the spring retainer, spring, servo piston and plug assembly from the case.

Flushing the Torque Converter

1. The torque converter must be removed in order to flush it.
2. Place the converter in a horizontal position and pour two quarts of new clean solvent into the converter through the impeller hub.
3. Turn and shake the converter, swirling the solvent through the internal parts. Turn the turbine and stator with the input and reaction shafts to dislodge foriegn material.
4. Position the converter in its normal position with the drain plug at its lowest point. Drain the solvent. At the same time, rotate the turbine and stator and shake the converter, to prevent dirt particles from settling.
5. Repeat the flushing operation at least once, until the solvent or kerosene is clear.
6. After flushing, shake and rotate the converter several times, with the drain plug removed, to drain out any residual solvent or dirt.
7. Flush any remaining solvent or dirt with two quarts of new transmission fluid. Tighten the drain plug to 110 in. lbs.
8. Flush and blow out the oil cooler and lines.

Before removing any of the transmission sub-assemblies, thoroughly clean the exterior of the unit, preferably by steam. When disassembling, each part should be washed in a suitable solvent, and either set aside to drain or dried with compressed air. Do not wipe with shop towels. All of the transmission parts require extremely careful handling to avoid nicks and other damage to the accurately machined surfaces.

Sub-Assembly Reconditioning

The following procedures cover the disassembly, inspection, repair and assembly of each sub-assembly as removed from the transmission.

The use of crocus cloth is permissible but not encouraged as extreme care must be used to avoid rounding off sharp edges of valves. The edge portion of valve body and valves is very important to proper functioning.

NOTE: Use all new seals and gaskets, and coat each part with auto-

matic transmission fluid, type A, suffix A, during assembly.

Valve Body

Disassembly

NOTE: This area is extremely critical, and sensitive to distortion. Never clamp any portion of the valve body or transfer plate in a vise. Clean with new solvent and dry with compressed air. Start all valves into their respective chambers with a twisting motion, seeing that they are well lubricated with automatic transmission fluid.

▶ Disassembly 1971-74

1. Place the valve body on a clean repair stand. Never place any part of the valve body or transfer plate in a vise, since distortion can cause sticking valves and excessive leakage.
2. Remove the three screws from the fluid filter and remove the filter.
3. Remove the transfer plate retaining screws and 2 of the spring retainer mounting screws.
4. Lift off the transfer plate assembly. Separate the stiffener and separator plate for cleaning.
5. Remove the seven balls and spring from the valve body. Tag all springs as they are removed, for identification.
6. Turn the valve body over and remove the shuttle valve cover plate.
7. Remove the governor plug end plate and slide out the shuttle valve, throttle valve and spring, 1-2 shift valve governor plug and the 2-3 shift valve governor plug.
8. Remove the shuttle valve E-clip and remove the shuttle valve. If equipped, also remove the secondary spring and guides which are held by the clip.
9. Hold the spring retainer firmly against the spring and remove the last screw from the valve body.
10. Remove the spring retainer, line pressure adjusting screw (do not disturb the setting), and the line pressure and torque converter regulator springs.
11. Slide the torque converter and line pressure valves out of their bores.
12. Remove the E-clip and washer from the throttle lever shaft. Remove any burrs from the shaft, and, while holding the manual lever detent ball and spring in their bore, slide the manual lever from the throttle shaft. Remove detent ball and spring.
13. Slide the manual valve from its bore.
14. Remove the throttle lever stop

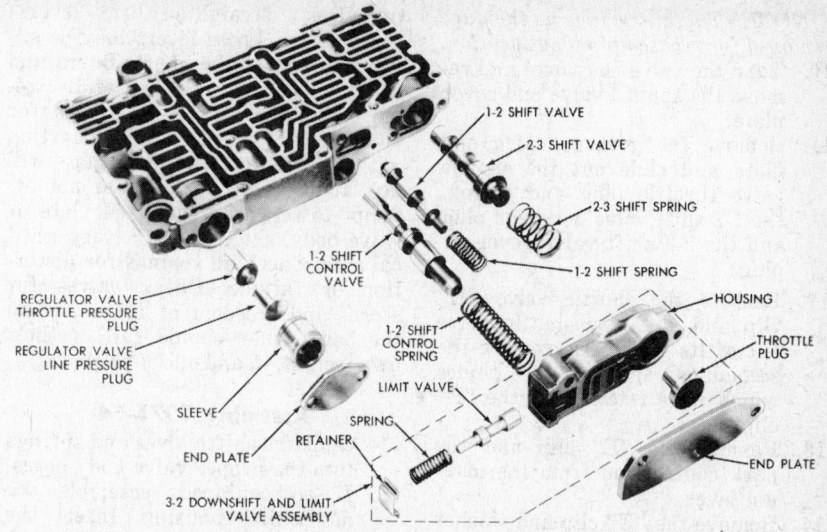

Exploded view—shift valves and pressure regulator valves (© Chrysler Corp.)

screw assembly from the valve body and slide out the kickdown detent, kickdown valve, throttle valve spring and throttle valve.

15. Remove the line pressure regulator valve end plate and slide out the regulator valve sleeve, line pressure plug and throttle pressure plug. If equipped, remove end plate and downshift housing assembly.
16. Remove the throttle plug housing and slide the throttle plug out. If equipped, remove retainer. limit valve, and spring.
17. Remove the shift valve springs and slide both shift valves from their bores.

Disassembly 1975 and Later

1. Place the valve body on a repair stand. Remove the three screws from the fluid filter and lift off the filter.
2. Remove the top and bottom screws from the spring retainer and adjustment screw basket.

3. Hold the spring retainer firmly against the spring force while removing the last screw from the side of the valve body.
4. Remove the spring retainer, with the line and throttle pressure adjusting screws (do not disturb the setting). Remove the line pressure and torque converter regulator springs.
5. Slide the torque converter and line pressure valves out of their bores.
6. Remove the transfer plate retaining screws and lift off the transfer plate and separator plate assembly.
7. Remove the screws from the stiffener and separate parts for cleaning.
8. Remove the rear clutch ball check valve from the transfer plate and regulator valve screen from the separator plate for cleaning.
9. Remove the seven balls and spring from valve body.

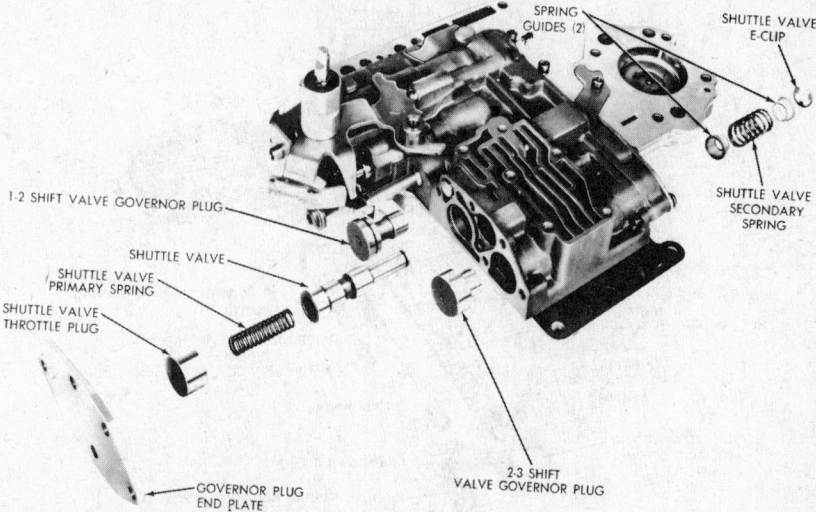

Exploded view—Shuttle valve and governor plugs (© Chrysler Corp.)

Automatic Transmissions

NOTE: Tag all springs as they are removed for reassembly identification.

10. Turn the valve body over and remove the shuttle valve and cover plate.
11. Remove the governor plug end plate and slide out the shuttle valve throttle plug and spring, the 1-2 shift valve governor plug and the 2-3 shift valve governor plug.
12. Remove the shuttle valve "E" clip and slide the shuttle valve out of its bore. Also remove the secondary spring and guides which were retained by the "E" clip.
13. Remove the "E" clip and the park control rod from the manual lever.
14. Remove the "E" clip and washer from the throttle lever shaft. Remove any burrs from the shaft, then while holding the manual their bore, slide the manual lever off the throttle shaft. Remove the detent ball and spring.
15. Slide the manual valve out of its bore.
16. Slide out the kickdown detent, kickdown valve, throttle valve spring and throttle valve.
17. Remove the line pressure regulator valve end plate and slide out the regulator valve sleeve, line pressure plug, and the throttle pressure plug.
18. Remove the end plate and the downshift housing assembly.
19. Remove the throttle plug from the housing.
20. Slide the retainer plug from the housing and remove the limit valve and spring.
21. Remove the three springs and shift valves from the valve body.

Cleaning and Inspection

Inspect all components for scores, loose or bent levers, burrs and warping. Don't straighten bent levers; renew them. Loose levers may be silver soldered at the shaft. Burrs and minor nicks may be carefully removed with crocus cloth. Check for valve body warpage or distortion with a surface plate (plate glass will do) and a feeler gauge. Do not attempt to service a distorted plate or valve body, since this is a very critical area. Check all springs for distortion or fatigue. Check valves for scores and freedom of movement in the bores, they should fall of their own weight, in and out of the bore.

Assembly 1971-74

1. Slide the shift valves and springs into the proper valve body bores.
2. If so equipped, assemble the downshifts housing. Insert the limit valve and spring into the housing. Slide the spring retainer into the groove.
3. Insert the throttle plug into the housing bore and install the housing on the valve body. Torque the screws to 28 in. lbs.
4. Install the throttle pressure plug, line pressure plug and sleeve, fastening the end plate. Torque to 28 in. lbs.
5. Install the throttle valve, throttle valve spring, kickdown valve, kickdown detent and throttle lever stop screw with the locknut (do not adjust yet).
6. Slide the manual valve into its bore.
7. Install the throttle lever and shaft on the valve body. Insert the detent ball and spring into its bore in the valve body. Depress the ball and spring and slide the manual lever over the throttle shaft. Be sure that it engages the manual valve and detent ball. Install the seal, retaining washer and E-clip on the throttle shaft.

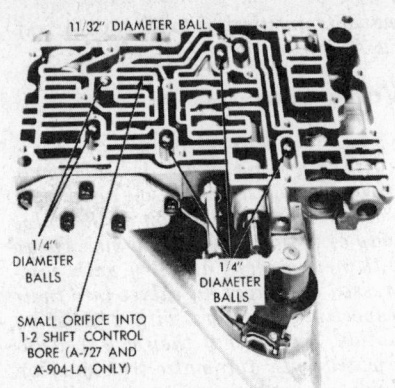

Valve body steel ball locations (© Chrysler Corp.)

8. Insert the torque converter control valve and spring into the valve body.
9. Insert the line pressure regulator and spring into the valve body.
10. Install the line pressure adjusting screw assembly and spring retainer on the springs temporarily with one screw.
11. Place the 1-2 and 2-3 shift valve governor plugs in their proper bores.
12. Install the shuttle valve, spring, and shuttle valve throttle plug. If so equipped, install the secondary spring with two guides and clip on the other end.
13. Install the governor plug end plate and torque the screws to 28 in. lbs.
14. On those valve bodies not having a secondary spring, install the E-clip on the end of the shuttle valve.
15. Install the shuttle valve cover plate and torque screws to 28 in. lbs.
16. Install the spring and seven balls in the valve body. The seven include: five 1/4 in. balls, one 3/8 in. diameter ball in the corner and one 11/32 in. diameter ball in the large chamber.
17. Place the separator plate on the transfer plate. Make sure all bolt holes are aligned and torque the two transfer plate screws and two stiffener plate screws to 28 in. lbs.
18. Place the transfer plate assembly on the valve body. Align the spring loaded ball as the 17 shorter screws are installed. Start at the center and work outward, tightening the screws to 35 in. lbs.
19. Install the oil filter and torque to 35 in. lbs.
20. Check spring engagement with the tang and adjusting nut. Install the remaining spring retainer screws. Check alignment and torque to 28 in. lbs.
21. After valve body has been serviced and completely assembled,

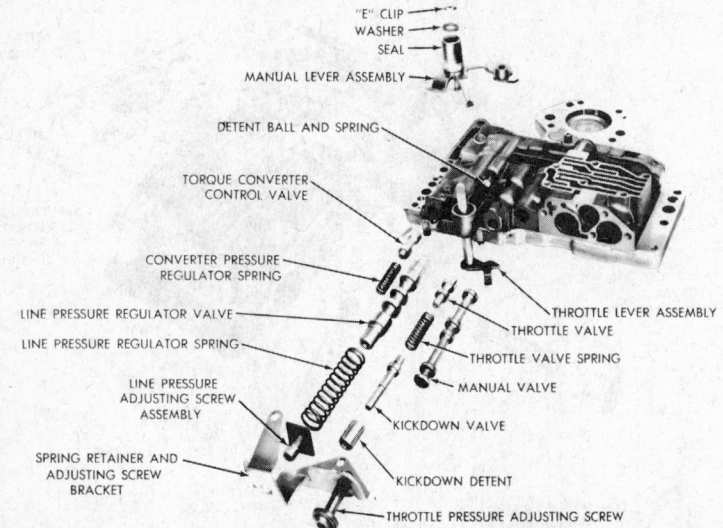

Exploded view—pressure regulator and manual control (© Chrysler Corp.)

adjust throttle and line pressures. If pressures were satisfactory prior to disassembly, use the original settings.

Assembly 1975 and Later

1. Slide the shift valves and springs into their proper valve body bores.
2. Assemble the downshift housing as follows:
   a. Insert the limit valve and spring into the housing.
   b. Slide the spring retainer into the groove in the housing.
   c. Insert the throttle plug in the housing bore. Position the assembly against the shift valve springs.
3. Install the end plate and tighten the screws to 28 inch-pounds.
4. Install the throttle pressure plug, line pressure plug and sleeve, then fasten the end plate to the valve body. Tighten to 28-inch pounds torque.
5. Install the throttle valve, throttle valve spring, kickdown valve, and the kickdown detent.
6. Slide the manual valve into its bore.
7. Install the throttle lever and shaft on the valve body. Insert the detent spring and ball in its bore in the valve body. Depress the ball and spring and slide the manual lever over the throttle shaft so that it engages the manual valve and detent ball. Install the seal, retaining washer, and "E" clip on the throttle shaft.
8. Install the torque converter control valve and spring into the valve body.
9. Insert the line pressure regulator valve and spring into the valve body.
10. Install the pressure adjusting screw and bracket assembly on the springs and fasten it with the screw which goes into the side of the valve body. Start the screws on the top and bottom, do not tighten, and then tighten the screw on the side of the valve body.

11. Place the 1-2 and 2-3 shift valve governor plugs in their respective bores.
12. Install the shuttle valve and hold it in the bore with your index finger, while installing on the other end of it the secondary spring with the guides and retaining "E" clip.
13. Install the primary shuttle valve spring and the throttle plug.
14. Install the governor plug end plate and tighten the five retaining screws to 28 inch-pounds.
15. Install the shuttle valve cover plate and tighten the six retaining screws to 28 inch-pounds.
16. Install the spring and seven balls in the valve body in their respective locations.
17. Place the separator plate on the transfer plate. Install the stiffener plate and retaining screws in their original positions.
18. Make sure that all the bolt holes are aligned, then tighten two transfer plate screws and two stiffener plate screws to 28 inch pounds.
19. Place the transfer plate assembly on the valve body. Be careful to align the spring loaded ball as the 17 shorter screws are installed, the three longer srcews are for the oil filter.
20. Starting in the center and working outward tighten the screws to 35 inch-pounds.
21. Check the spring engagement and install the remaining top and bottom screws in the adjusting screw bracket. Tighten the side screw first to 28 inch pounds, then tighten the top and bottom screws.
22. Install the oil filter and tighten to 35 inch-pounds.
23. After the valve body has been serviced and completely assembled, adjust the throttle and line pressures. If pressures were satisfactory prior to disassembly, use oirginal settings.

Accumulator Piston and Spring

Inspect both seal rings for wear and freedom in the piston grooves. Check the piston for scores, burrs, nicks and wear. Check the piston bore for corresponding damage and check piston spring for distortion and fatigue. Replace parts as required.

Governor

Disassembly

1. Carefully remove the snap-ring from weight end of governor valve shaft and pull out valve and shaft. Remove the large snap-ring from the weight end of governor housing and lift out the governor weight assembly.

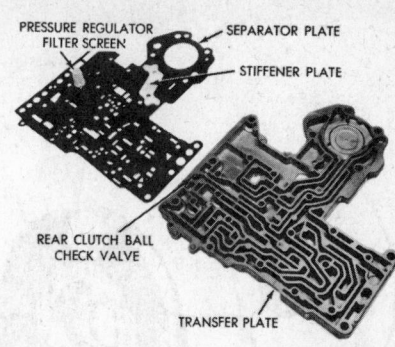

Transfer and separator plates. (© Chrysler Corp.)

2. Remove the snap-ring from inside the governor weight, remove the inner weight and spring from the outer weight.

NOTE: throughly clean all parts in a suitable and clean solvent. Check for damage and free movement before assembly.

3. If lugs on support gear are damaged, remove four bolts and separate support from governor body.

Assembly

1. If support was separated from governor body, assemble and tighten bolts finger-tight. Make sure the oil passage of the body aligns with passage in the support. Position support and governor on output shaft so that the valve shaft hole in the governor aligns with the hole in the output shaft. Install a snap-ring behind the governor body and tighten the bolts to 100 in. lbs.
2. Assemble the governor weights and spring, then secure with snap-ring inside large governor weight.
3. Place the weight assembly in the governor housing and install snap-ring.

Oil Pump and Reaction Shaft Support

Disassembly

1. Remove the bolts from the rear side of the reaction shaft support and lift the support off the pump.
2. Remove the rubber seal ring from the pump body flange.
3. Drive the oil seal out with a blunt punch.

Inspection

1. Inspect the interlocking steel rings on the reaction shaft for wear or broken locks, make sure they turn freely in the grooves.
2. Inspect the machined surfaces on the pump body and the reaction shaft support for nicks and burrs.
3. Inspect the pump body and reaction shaft support bushings for wear or scores.

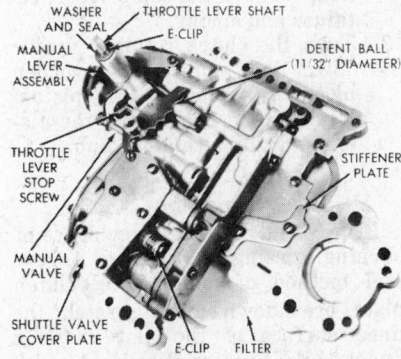

Valve body and control assembly
(© Chrysler Corp.)

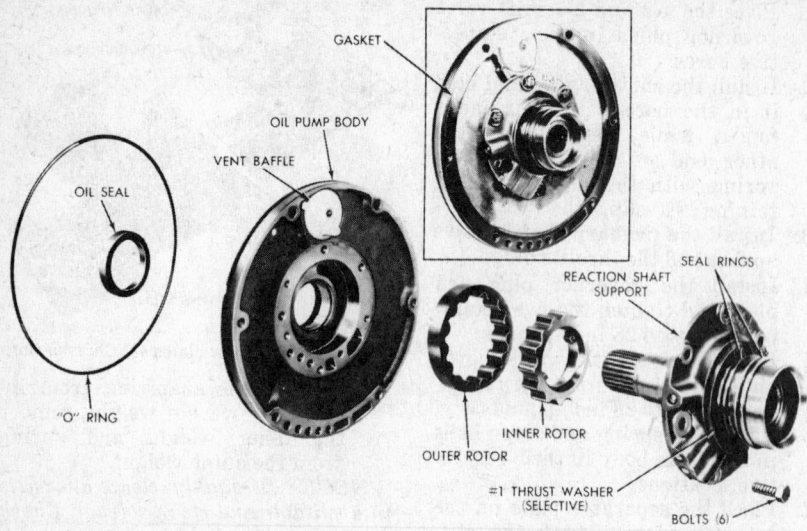

Exploded view—Oil pump and reaction shaft support (© Chrysler Corp.)

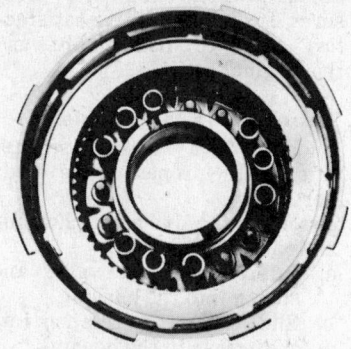

Location of front clutch springs—two types——9 and 13 springs (© Chrysler Corp.)

4. Inspect the pump rotors for scoring or pitting.

5. With the rotors cleaned and installed in the pump body, place a straight edge across the face of the rotors and pump body. Use a feeler gauge to measure the clearance between the straight edge and the face of the rotors. Clearance limits are 0.0015-0.003 inch. Also measure the rotor tip clearance between the inner and outer teeth. Clearance limits are 0.005-0.010 inch. Clearance between outer rotor and its bore in the oil pump body should be from 0.004-0.008 inch.

Oil Pump Bushing

Replacement

1. Place the pump housing on a clean smooth surface with the rotor cavity down.

2. Place the removing head of the special bushing tool in the bushing, and install the tool handle.

3. Drive the bushing straight down and out of the bore. Be careful not to cock the tool in the bore.

4. Place a new bushing on the installation tool.

5. With the pump housing on a smooth clean surface, hub end down, start the bushing and installation head in the bushing bore. Install the tool handle in the installation head.

6. Drive the bushing into the housing until the tool bottoms in the pump cavity. Be careful not to cock the tool during installation.

7. Stake the bushing in place using a blunt punch or similar tool. A gentle tap at each stake slot location will suffice.

8. Using a narrow bladed knife, remove high points of burrs around the staked area. Do not use a file that will remove more

metal than is necessary.

9. Thoroughly clean the pump housing before installation.

Reaction Shaft Bushing

Replacement

CAUTION: Do not clamp any part of the reaction shaft or support in a vise.

1. Assemble the special bushing removal tool which consists of: cup, hex nut and removal tool.

2. With the cup held firmly against the reaction shaft, thread the remover into the bushing as far as possible by hand.

3. Using a wrench to screw remover into the bushing 3 to 4 additional turns to firmly engage the threads in the bushing.

4. Turn the hex nut down against the cup to pull the bushing out of the reaction shaft. Throughly clean the reaction shaft to remove chips made by the remover threads.

5. Lightly grip the bushing in a vise or with pliers and back the tool out of the bushing. Be careful not to damage the threads on the bushing remover.

6. Slide a new bushing (chamfered end first) on the installing head of the special tool and start them in the bore of the reaction shaft.

7. Support the reaction shaft upright on a clean smooth surface and install the installation tool handle. Drive the bushing into the shaft until the tool bottoms.

8. Thoroughly clean the reaction shaft support assembly before installation.

Assembly

1. Assemble the pup rotors and 'O' ring in the pump housing.

2. Install the reaction shaft support. Install the retaining bolts

and torque to 160 inch-pounds.

3. Place a new oil seal in the opening of the pump housing, lip of the seal facing inward, and drive the seal into the housing until it bottoms.

Front Clutch

Disassembly

Exploded view of front clutch assembly is illustrated.

1. With screwdriver or pick, remove large snap-ring, which holds the pressure plate in the clutch piston retainer. Lift pressure plate and clutch plates out of the retainer.

2. Install spring compresison tool or similar tool, over piston spring retainer. Compress spring and remove snap-ring, then, slowly release tool until the spring retainer is free of the hub. Remove the compressor, retainer and spring.

3. Turn the clutch retainer upside down and bump on a wooden block to remove the piston. Remove seal rings from the piston and clutch retainer hub.

Inspection

Inspect clutch discs for evidence of burning, glazing and flaking. A general method of determining clutch plate breakdown is to scratch the lined surface of the plate with a finger nail. If material collects under the nail, replace all driving discs. Check driving splines for wear or burrs. Inspect steel plates and pressure plate surfaces for discoloration, scuffing or damaged driving lugs. Replace if necessary.

Check steel plate lug grooves in clutch retainer for smooth surfaces. Plate travel must be free. Inspect band contacting surface of clutch retainer, being sure the ball moves freely. Check seal ring surfaces in clutch retainer for scratches or nicks, light annular scratches will not interfere with the sealing of neoprene rings.

Inspect inside bore of piston for score marks. If light marks exist, polish with crocus cloth. Check seal ring grooves for nicks and burrs. Inspect neoprene seal rings for deterioration, wear and hardness. Check piston spring, retainer, and snap-ring for distortion and fatigue.

Front Clutch Retainer Bushing

Replacement

1. Lay clutch retainer (open end down) on a clean smooth surface and place removing head in the bushing. Install the bushing removal tool handle.
2. Drive the bushing straight down and out of clutch retainer bore. Be careful not to cock the tool in the bore.
3. Lay the clutch retainer (open end up) on a clean smooth surface. Slide a new bushing on the installation head tool, and start them in the clutch retainer bore.
4. Install the bushing installation tool handle and drive the bushing into the clutch retainer until the tool bottoms.
5. Thoroughly clean the clutch retainer before assembly and installation.

Assembly

1. Lubricate and install inner seal ring onto hub of clutch retainer.

Removing front clutch retainer snap-ring
(© Chrysler Corp.)

Be sure that lip of seal faces down and is properly seated in the groove.
2. Lubricate and install outer seal ring onto clutch piston, with lip of seal toward the bottom of the clutch retainer. Place piston assembly in retainer and, with a twisting motion, seat the piston in the bottom of the retainer.
3. Place spring on the piston hub and position spring retainer and snap-ring on spring. Compress spring with tool, or suitable ring compressor, and seat snap-ring in the hub groove. Remove compressor.
4. Lubricate all clutch plates, then, install a steel plate, followed by a lined plate, until all plates are installed. Install the pressure plate and snap-ring. Be sure the snap-ring is correctly seated.
5. With front clutch assembled, insert a feeler gauge between the pressure plate and snap-ring.

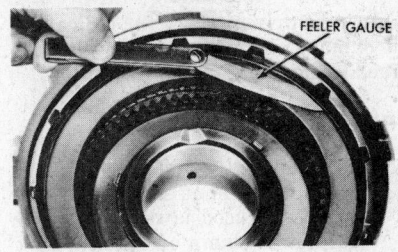

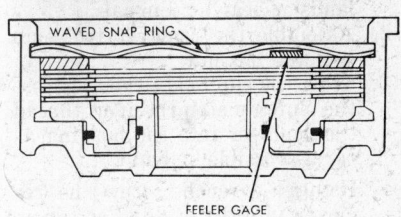

Measuring front clutch plate clearance
(© Chrysler Corp.)

The clearance should be to specification. If not, install a snap-ring of proper thickness.

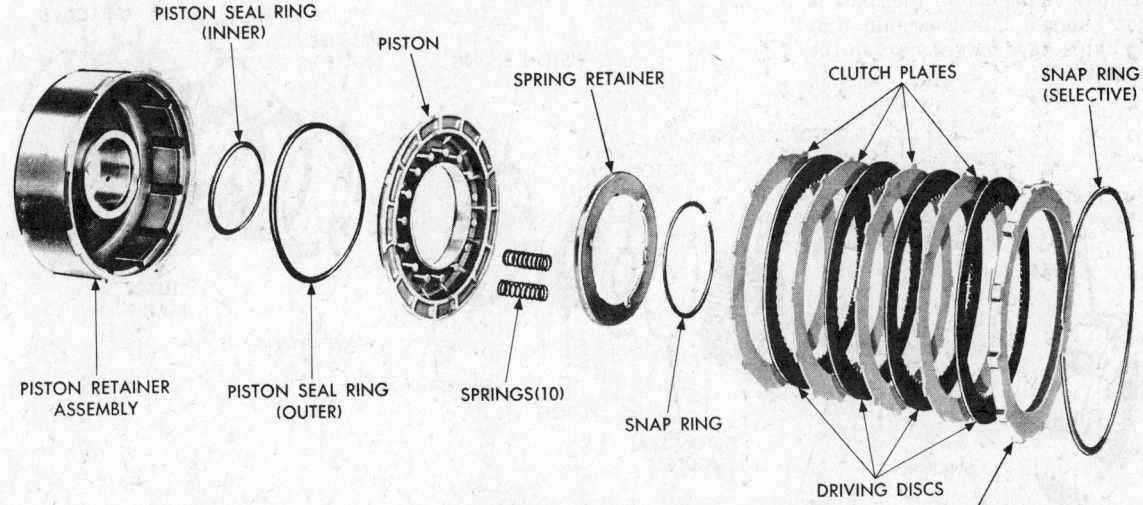

Front clutch assembly (© Chrysler Corp.)

Automatic Transmissions

Rear Clutch

Disassembly

1. With a small screwdriver or pick, remove the large snap-ring that secures the pressure plate in the clutch piston retainer. Lift the pressure plate, clutch plates, and inner pressure plate from the retainer.
2. Carefully pry one end of wave spring out of its groove in clutch retainer and remove wave spring, spacer ring and clutch piston spring.
3. Turn clutch retainer assembly upside down and bump on a wood block to remove the piston. Remove seal rings from the piston.
4. If necessary, remove snap-ring and press the input shaft from the clutch piston retainer.

Inspection

Inspect driving discs for indication of damage; handle as previously outlined under front clutch inspection.

Input Shaft Bushing

Replacement

1. Clamp the input shaft in a vise with soft faced jaws, being careful not to clamp on the seal ring lands or bearing journals.
2. Assemble the remover tool, cup tool and hex nut
3. With the cup held firmly against the clutch piston retainer, thread the remover into the bushing as far as possible by hand.
4. Using a wrench, screw the remover into the bushing as far as possible by hand.
5. Turn the hex nut down against the cup to pull the bushing from the input shaft.
6. Thoroughly clean the input shaft to remove the chips made by the remover threads. Make sure that the small lubrication hole next to the ball in the end of the shaft is not plugged. Make certain that no chips have lodged next to the steel ball.

7. Slide a new bushing on the installation head of the bushing tool and start them in the bore of the input shaft.
8. Stand the input shaft upright on a clean surface and position the handle on the installation tool. Drive the bushing into the shaft until the tool bottoms.
9. Thoroughly clean the input shaft and clutch piston retainer before assembly and installation.

Assembly

1. If removed, press input shaft into the piston retainer and install snap ring .
2. Lubricate, then install inner and outer seal rings onto the clutch piston. Be sure the seal lips face toward the head of the clutch retainer and seals are properly seated in the piston grooves.
3. Place piston assembly in retainer and, with a twisting motion, seat piston in bottom of retainer.
4. Place spring over piston with outer edge of spring positioned below snap ring groove. Start one end of snap ring in groove. Make sure spring is exactly centered on piston. Progressively tap snap ring into groove. Be sure snap ring is fully seated in groove.
5. Install inner pressure plate into clutch retainer, with raised portion of plate resting on the spring.
6. Lubricate all clutch plates, then install one lined plate, followed by a steel plate, until all plates are installed. Install outer pressure plate and snap-ring.
7. With rear clutch completely assembled, insert a feeler gauge between the pressure plate and snap-ring. The clearance should be to specification. If not, install snap-ring of proper thickness to obtain the required clearance.

NOTE: *Rear clutch plate clearance is very important to obtaining satisfactory clutch performance. Clearance is influenced by the use of various thickness outer snap-rings.*

Planetary Gear Train End Play

NOTE: *Before removal of the planetary gear assemblies, sun gear and drive shell parts from the output shaft, measure the end play as follows:*

1. With the assembly in an upright position, push the rear annulus gear support downward on the output shaft.
2. Insert a feeler gauge between the rear annulus gear support hub and sholder on the output shaft.
3. The clearance should be 0.01-0.025 inch. If the clearance is not within limits, replace thrust washer and/or necessary parts.

Disassembly

Refer to illustrations for aid in the assembly and disassembly of these units.

1. Remove thrust washer from forward end of output shaft.
2. Remove snap-ring from forward end of output shaft, then, slide front planetary assembly from the shaft.
3. Slide front annulus gear off planetary gear set. Remove thrust washer from rear side of planetary gear set.
4. Slide the sun gear, driving shell, and rear planetary assembly, from the output shaft.
5. Remove sun gear and driving shell from the rear planetary assembly, remove thrust washer from inside driving shell, remove snap-ring and steel washer from sun gear (rear side of driving shell). Slide sun gear out of driving shell, then remove snap-ring

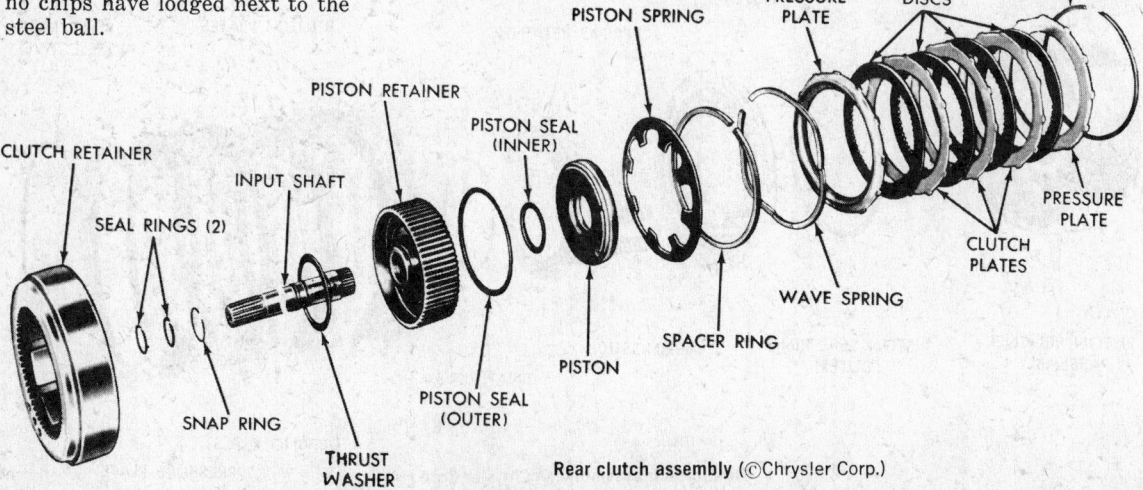

Rear clutch assembly (©Chrysler Corp.)

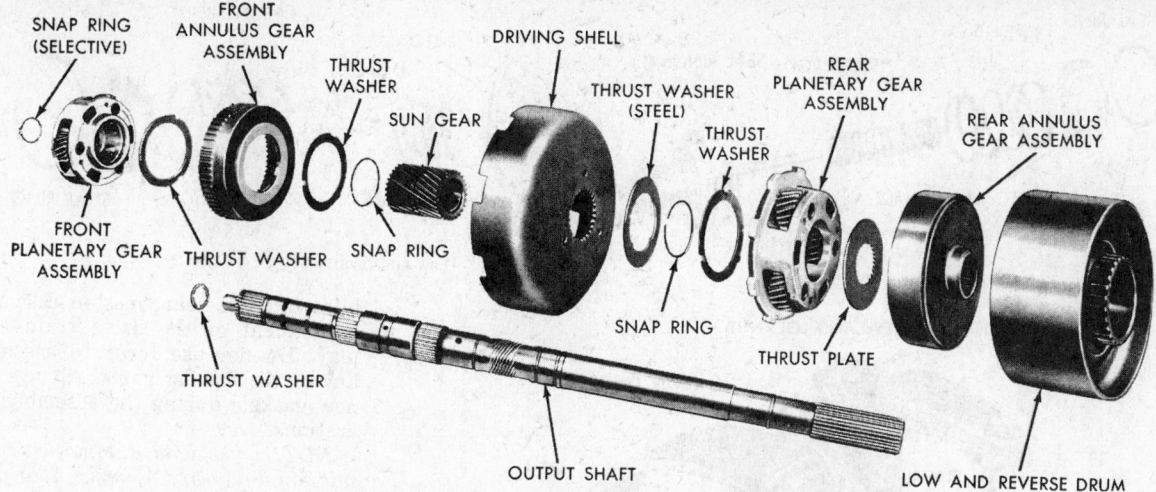

SNAP RING (SELECTIVE)

FRONT ANNULUS GEAR ASSEMBLY

THRUST WASHER

SUN GEAR

DRIVING SHELL

THRUST WASHER (STEEL)

THRUST WASHER

REAR PLANETARY GEAR ASSEMBLY

REAR ANNULUS GEAR ASSEMBLY

FRONT PLANETARY GEAR ASSEMBLY

THRUST WASHER

SNAP RING

SNAP RING

THRUST PLATE

THRUST WASHER

OUTPUT SHAFT

LOW AND REVERSE DRUM

Planetary gear train and output shaft assembly (© Chrysler Corp.)

and steel washer from opposite end of sun gear, if necessary.

6. Remove thrust washer from forward side of rear planetary assembly. Remove the planetary gear set and thrust plate from the rear annulus gear.

Inspection

Inspect output shaft bearing surfaces for burrs or other damage. Light scratches or burrs may be polished out with crocus cloth or a fine stone. Check speedometer drive gear for damage, and make sure all oil passages are clear.

Check bushings in the sun gear for wear or scores. Replace sun gear assembly if bushings show wear or other damage. Inspect all thrust washers for wear and scores. Replace if necessary. Check lockrings for distortion and fatigue. Inspect annulus gear and driving gear teeth for damage. Inspect planetary gear carrier for cracks and the pinions for broken or worn gear teeth.

Assembly

1. Install rear annulus gear on output shaft. Apply thin coat of grease on thrust plate, place it on shaft, and in annulus gear making sure teeth are over shaft splines.
2. Position rear planetary gear assembly in rear annulus gear. Place thrust washer on front side of planetary gear assembly.
3. Install snap-ring in front groove of sun gear (long end of gear). Insert sun gear through front side of driving shell. Install rear steel washer and snap-ring.
4. Carefully slide driving shell and sun gear assembly on output shaft, engaging sun gear teeth with rear planetary pinion teeth. Place thrust washer inside front driving shell.
5. Place thrust washer on rear hub of front planetary gear set. Slide

assembly into front annulus gear.

6. Carefully work front planetary and annulus gear assembly on output shaft, meshing planetary pinions with sun gear teeth.
7. With all components properly positioned, install selective snap ring on front end of output shaft. Measure end-play of assembly. Adjust end-play with selective snap rings.

Overrunning Clutch

Inspection

Inspect clutch rollers for smooth round surfaces, they must be free of flat spots, chipped edges and flaking. Inspect roller contacting surfaces on both cam and race for pock marks and roller wear-marks. Check springs for distortion and fatigue and inspect low and reverse drum thrust. Inspect cam set screw for tightness. If loose, tighten and restake the case.

Overrunning Clutch Cam

Replacement

If the overrunning clutch cam or roller spring retainer are found to be defective, replace the cam and spring in the following manner:

1. Remove the set screw from the case below the clutch cam.
2. Remove the four bolts securing the output shaft support to the rear transmission case. Insert a punch through the bolt holes and drive the cam from the case. Alternately, punch from one bolt hole to another so the cam will be driven evenly from the case.

NOTE: The output shaft support must be in the case to install the overrunning clutch cam. If the support requires replacement, drive it rearward out of the case with a wood block and hammer. To install, screw two pilot studs into the case. Chill the support with dry ice. Quickly posi-

tion the support over the pilot studs, and drive it firmly into the case with a wood block and hammer.

3. Clean all burrs and chips from the cam area in the case.
4. Place the spring retainer on the cam, making sure the retainer lugs snap firmly into the notches on the cam.
5. Position the cam in the case with the cam serrations aligned with those in the case. Tap the cam evenly into the case as far as possible with a soft faced mallet.
6. Position the cam installation tool on the cam and tighten the hex nut on the tool to seat the cam in the case. Make sure that the cam is firmly seated. Install the cam retaining set screw and stake the case around the set screw to prevent it coming loose.
7. Remove the cam installation tool. Install the support retaining screws and tighten to 140 inch-pounds torque. Stake the case around the cam in twelve places with a blunt chisel.

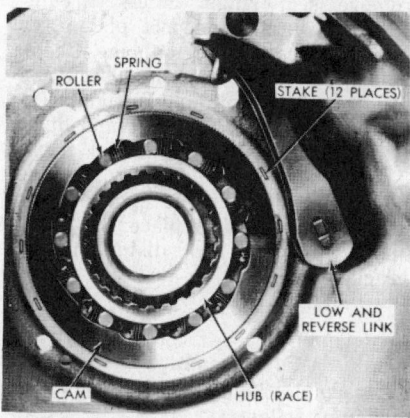

SPRING

ROLLER

STAKE (12 PLACES)

LOW AND REVERSE LINK

CAM

HUB (RACE)

Overrunning clutch installation—with low-reverse band link in position (© Chrysler Corp.)

Kickdown Servo and Band

Disassembly

Disassemble the controlled load

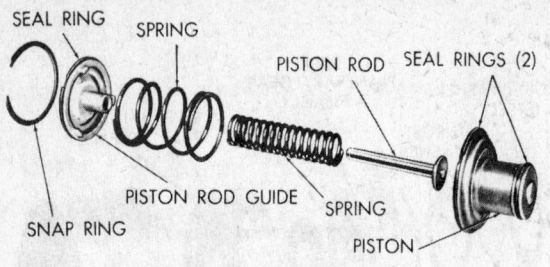

Kickdown servo (© Chrysler Corp.)

Low and reverse servo (© Chrysler Corp.)

Low and reverse band and linkage
(© Chrysler Corp.)

servo piston by removing the small snap ring from the servo piston. Remove the washer, spring and piston rod from the servo piston.

Inspection

Inspect piston and guide seal rings for wear, and be sure of their freedom in grooves. It is not necessary to remove seal rings, unless circumstances warrant. Inspect piston for scores, burrs or other damage. Check fit of guide on piston rod. Check piston for distortion and fatigue. Inspect band lining for wear and fit of lining material to the metal band. This lining is grooved; if grooves are not still visible at the ends or any part of the band, replace the band. Inspect band for distortion or cracked ends.

Assembly

Assemble the controlled load servo piston as follows:
1. Grease the "O" ring and install on the piston rod.
2. Install the piston rod into the servo piston.
3. Install the spring, flat trasher and snap ring to complete the assembly.

Low and Reverse Servo and Band

Disassembly

Remove snap-ring from piston and remove the piston plug and spring.

Inspection

Inspect neoprene seal ring for damage, rot, or hardness. Check piston and piston plug for nicks, burrs, scores and wear. The piston plug must operate freely in the piston. Check the piston bore in the case for scores or other damage. Examine springs for distortion and fatigue.

Check band lining for wear and the fit of the lining to the metal band. This lining has a grooved surface; if the grooves are worn away at the ends or at any part of the band, replace the band. Inspect the band for distortion or cracked ends.

Assembly

Lubricate and insert the piston plug and spring into the piston, and secure with the snap-ring.

Sub-Assemblies
Installation

The following assembly procedures include the installation of sub-assem-

blies into the transmission case and adjustment of the drive train endplay. Do not use force to assemble any of the mating parts. Always use new gaskets during the assembly operations.

NOTE: use only automatic transmission fluid, type A, suffix A, or fluid of equivalent chemical structure, to lubricate automatic transmission parts during, or after, assembly.

Overrunning Clutch

With transmission case in upright position, insert clutch race inside cam. Install overrunning clutch rollers and springs as shown in figure.

Low and Reverse Servo and Band

1. Carefully work servo piston assembly into the case with a twisting motion. Place spring, retainer and snap-ring over the piston.
2. Using a valve spring compressor, compress the spring and install the snap-ring.
3. Position rear band in the case, install the short strut, then connect the long lever and strut to the band. Screw in band adjuster just enough to hold struts in place. Install low-reverse drum. Be sure long link and anchor assembly is installed to provide running clearance for low-reverse drum.

Kickdown Servo

1. If equipped with a controlled load servo piston, sub-assemble the unit as follows: grease the O ring and install on the piston rod; install the piston rod into the servo piston; install the spring, flat washer and snapring.
2. Carefully insert servo piston into case bore. Install piston rod, two springs and guide.
3. Compress the kickdown servo springs by using a engine valve spring compressor. Install the snapring.

Planetary Gear Assemblies, Sun Gear, Driving Shell, Low and Reverse Drum

1. While supporting the assembly in the case, insert the output shaft through the rear support. Carefully work the assembly rearward, engaging the carrier

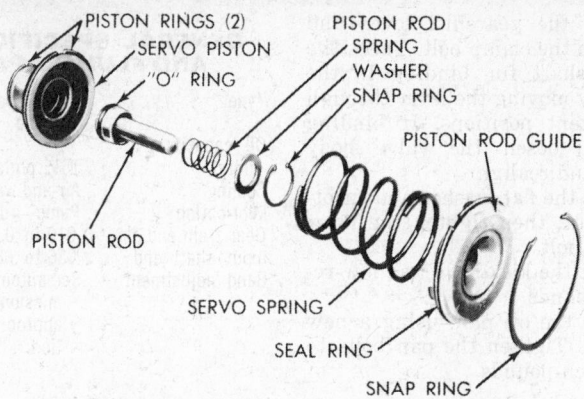

PISTON RINGS (2)
SERVO PISTON
"O" RING
PISTON ROD
PISTON ROD
SPRING
WASHER
SNAP RING
PISTON ROD GUIDE
SERVO SPRING
SEAL RING
SNAP RING

Exploded view—Kick down servo (© Chrysler Corp.)

lugs with low-reverse drum slots.

CAUTION:' Be careful not to damage the ground surfaces of the output shaft during installation.

Front and Rear Clutch

1. The following method may be used to support the transmission; cut a 3½" hole in a bench, small drum or box, strong enough to support the transmission; file notches at the edge so the output shaft support will lie flat; insert the output shaft into the hole and support the transmission upright.
2. Apply a coat of grease to the input shaft to output shaft thrust washer and install the thrust washer on the front end of the output shaft.
3. Align the front clutch plate inner splines, and place the assembly in position on the rear clutch. Be sure the clutch plate splines are fully engaged.
4. Align the rear clutch plate inner splines and lower the two clutch assemblies in to the case.
5. Carefully, work the clutch assemblies in a circular motion to engage the rear clutch splines over the front annulus gear splines. Make sure the front clutch drive lugs are fully engaged in the driving shell.

Front Band

1. Slide the band over the front clutch assembly.
2. Install band strut, screw in the adjuster just enough to hold the band in place.

Oil Pump and Reaction Shaft Support

1. If drive train end-play was not within specifications, replace the thrust washer on the reaction shaft support hub, with one of proper thickness (see specifications).
2. Screw (two) pilot studs into front pump opening in the case.
3. Place a new rubber seal ring in groove on outer flange of pump. Be sure the seal ring is not twisted.
4. Install the assembly into the case, tap lightly with a soft mallet if necessary. Install four bolts, remove pilot studs, install remaining bolts and pull down evenly.
5. Rotate the input and output shafts to see if any binding exists. Tighten bolts to 175 inch-pounds. Check shafts again for free roattion. Adjust both bands.

Governor and Support

1. Place the governor and support on the output shaft. Position it so that the governor valve shaft hole aligns with the hole in the output shaft, then slide the assembly into place. Install snapring behind the governor housing. Torque housing-to-support screws to specification.
2. Place the governor valve on the valve shaft, insert the assembly into the housing and through the governor weights. Install the valve shaft retaining snap-ring.

Extension Housing-Short

1. Position a new gasket on the extension housing. Carefully slide the extension housing into place. Install the remaining bolts and washers and tighten to 24 foot-pounds torque.
2. Install the transmission yoke. Install the nut with its three projections toward the washer. Hold the yoke and tighten to 175 foot-pounds.
3. Install the speedometer pinion and adapter assembly.

Extension Housing-Long

1. Install the snap ring in the front groove on the output shaft. Install the bearing on the shaft with its outer race ring groove toward the front. Press or tap the bearing tight against the front snap ring. Install the rear snap ring.
2. Position a new gasket on the transmission case. Place the output shaft bearing retaining snap ring in the extension housing.

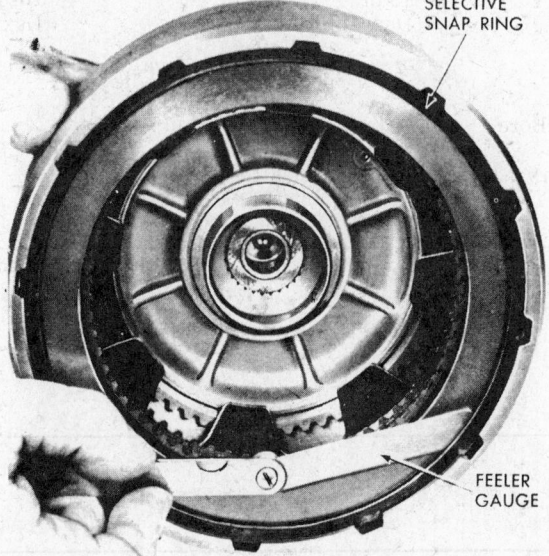

SELECTIVE
SNAP RING

FEELER
GAUGE

Measuring rear clutch plate clearance

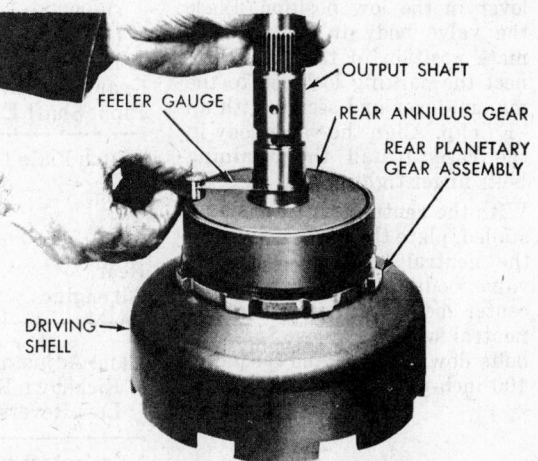

FEELER GAUGE
OUTPUT SHAFT
REAR ANNULUS GEAR
REAR PLANETARY GEAR ASSEMBLY
DRIVING SHELL

Measuring end play of planetary gear assembly

Spread the snap ring as far as possible. Carefully, tap the extension housing into place. Make sure that the snap ring is fully seated in the bearing groove.

3. Install and tighten the extension housing bolts to 24 foot-pounds.
4. Install the access plate and gasket on the side of bottom of the extension housing mounting pad.
5. Install the speedometer pinion and adapter assembly.
6. Measure the input shaft end play, correct if necessary.

Valve Body Assembly and Accumulator Piston

1. Clean the mating surfaces and inspect for burrs on both the transmission case and the valve body steel plate.
2. Install the accumulator piston in the transmission case and place the piston spring on the accumulator piston.

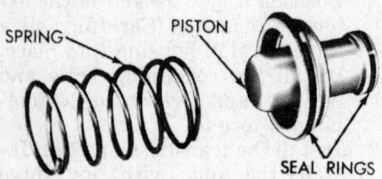

SPRING   PISTON

SEAL RINGS

Accumulator piston and spring
(© Chrysler Corp.)

NOTE: There is no spring used with the 440 cu. in. engine.

3. Make sure that the back-up light and neutral start switches have been removed.
4. Insert the parking lock rod through the opening in the rear of the case with the knob positioned against the reaction plug and sprag. Move the front end of the rod toward the center of the transmission while exerting rearward pressure on the rod, to force it past the sprag, rotate the output shaft if necessary.
5. Place the valve body manual lever in the low position. Place the valve body in its approximate position in the case. Connect the parking lock rod to the manual lever and secure with an "E" clip. Align the valve body in the case, install the retaining bolts finger tight.
6. With the neutral start switch installed, place the manual valve in the neutral position. Shift the valve body if necessary to center neutral finger over the neutral switch plunger. Snug the bolts down evenly, and torque to 100 inch-pounds.

7. Install the gearshift lever and tighten the clamp bolt. Check the lever shaft for binding in the case by moving the lever through all detent positions. If binding exists, loosen the valve body bolts and realign.
8. Install the flat washer and throttle lever, then tighten the lever clamp bolt.
9. Adjust the kickdown and low-reverse bands.
10. Install the oil pan, using a new gasket. Tighten the pan bolts to 150 inch-pounds.

GENERAL SPECIFICATIONS AND FLUID CAPACITY

Type	Three speed—fully automatic
Oil capacity (dry)	16½ pints "Dexron"
Cooling	Air and water cooled
Lubrication	Pump—rotor type
Gear train end play	.010 to .025 inch
Input shaft end	.036 to .082 inch
Band adjustment	See automatic transmission chapter under appropriate truck section

Torque Reference

	Foot-Pounds	Inch-Pounds
Kickdown band adjusting screw locknut (eight cylinder cars)	29	
Kickdown lever shaft plug	25	150
Reverse band adjusting screw locknut (eight cylinder cars)	35	
Reverse band adjusting locknut (six cylinder cars)	20	110
Cooler line fitting		
Converter drive plate to crankshaft bolt	55	
Converter drive plate to torque converter bolt		270
Extension housing to transmission case bolt		
Extension housing to insulator mounting bolt	40	
Extension housing—crossmember to frame bolt	75	—
Oil pump housing to transmission case bolt		175
Governor body to parking sprag bolt		100
Neutral starter switch	25-30	
Oil pan bolt		150
Overrunning clutch cam set screw		40
Reaction shaft support to front oil pump bolt		160
Transmission to engine bolt	25-30	
Valve body screw		35
Valve body to transmission case bolt		100

Specifications—Dodge Loadflite

Torque Converter Diameter	11¾ in	10¾ in
Oil Capacity	19 pt	16¼ pt
Oil Type	Dexron	Dexron
Cooling	Water	
Lubrication	Pump (Rotor Type)	
Pump Clearances		
Outer Rotor to Case Bore	0.004-0.008 in	
Outer to Inner Tip	0.005-0.010 in	
End Clearance-Rotors	0.0015-0.003 in	
Gear Train End Play	0.010-0.025 in	
Input Shaft End Play	0.036-0.082 in	

Clutch Plate Data	No. of Discs	Clearance (in)	No. of Springs
Front			
Six	3	0.076-0.123	13
V-8	4	0.088-0.145	9
Rear			
All engines	4	0.025-0.045	1

Band Adjustments		
Kickdown Band-Turns*	2½	1
Low-Reverse-Turns*	2	

* Backed off from 72 in-lbs.
1 440 engine 2 (two) turns.

New Process A-345

DIAGNOSIS

This guide covers the most common symptoms and is an aid to careful diagnosis. The items to check are listed in sequence to be followed for quickest results. Follow the checks in the order given for the particular transmission type.

TROUBLE SYMPTOMS	ITEMS TO CHECK IN TRUCK	OUT OF TRUCK
Harsh N to D or N to R shift	CDEFGIJ	ab
Delayed Shift—N to D	ACIHKJ	ca
Runaway on upshift and 3-2 kickdown	ABCDLIHK	b
Harsh upshift and 3-2 kickdown	BCDLIHJ	b
No upshift	ABCDLIHKM	b
No kickdown on normal downshift	ABNCDLIHKM	d
Erratic shifts	ABNCFOPQ IKMJ	c
Slips in forward drive positions	ACIHK	abd
Slips in reverse only	CEGIK	
Slips in all positions	ACOIK	e
No drive in any position	ACOQIKJ	ca
No drive in forward positions	CDOLIHK	abd
No drive in reverse	CEGKMI	b
Drives in neutral	NIJ	a
Drags or locks	DELG	abfd
Noises	AORPMIJ	a
Hard to fill or blows out	AORSTQI	
Transmission overheats	ADEORTMI	abe

Key to Checks

A. Oil level
B. Control linkage
C. Oil pressure check
D. Kickdown band
E. Low-reverse band
F. Improper engine idle
G. Servo linkage

H. Accumulator
I. Valve body assembly
J. Manual valve lever
K. Air pressure check
L. K-D servo link
M. Governor
N. Gear shift cable

O. Regulator valve and/or spring
P. Output shaft bushing
Q. Strainer
R. Converter control valve
S. Breather clogged
T. Cooler or lines

a. Front-kickdown clutch
b. Rear clutch
c. Front pump and/or sleeve
d. Overrunning clutch
e. Converter
f. Planetary

General Specifications and Fluid Capacity

Type ...... 4 speed fully automatic
Oil Capacity (Dry) .....28½ pints
Cooling ....Water
Oil Filter Type
.........Bottom suction screen
Clutches 4
Overrunning Clutches 2
Bands & Servos 2
Planetary Gear Sets 3

Band Adjustment

See automatic transmission chapter under appropriate truck section.

Sub-Assembly Removal

NOTE: Before removing any transmission sub-assemblies, plug all openings and thoroughly clean the exterior of the unit with steam. Cleanliness during the entire disassembly and assembly cannot be over-emphasized. When disassembling, each part should be washed in a suitable solvent, then dried by compressed air. Do not wipe parts with shop towels. All mating surfaces in the transmission are accurately machined; therefore, careful handling of parts must be exercised to avoid damage.

Input Shaft End Play

Measurement of the input shaft end play before disassembly will usually indicate when a thrust washer should be changed. The thrust washer is located between the reaction shaft support and the front clutch retainer.

1. Attach a dial indicator to the transmission bell housing with its plunger seated against the end of the input shaft. Move the input shaft in and out to obtain the end play reading. End play should be 0.018-0.062 inch.

2. Record the indicator reading for reference during reassembly.

Oil Pan

1. Place the transmission assembly in repair stand.
2. Remove the oil pan bolts and remove the oil pan. Discard the gasket.

Valve Body

1. Loosen the clamp bolts and remove the throttle and gearshift levers.
2. Remove the back-up light and neutral safety switches.
3. Remove the bolts securing the valve body to the transmission case, remove the valve body.

Accumulator Piston and Spring

1. Remove the accumulator piston from the case and lift the spring from the case.

Transmission Compounder

Disassembly

NOTE: Before removing the com-

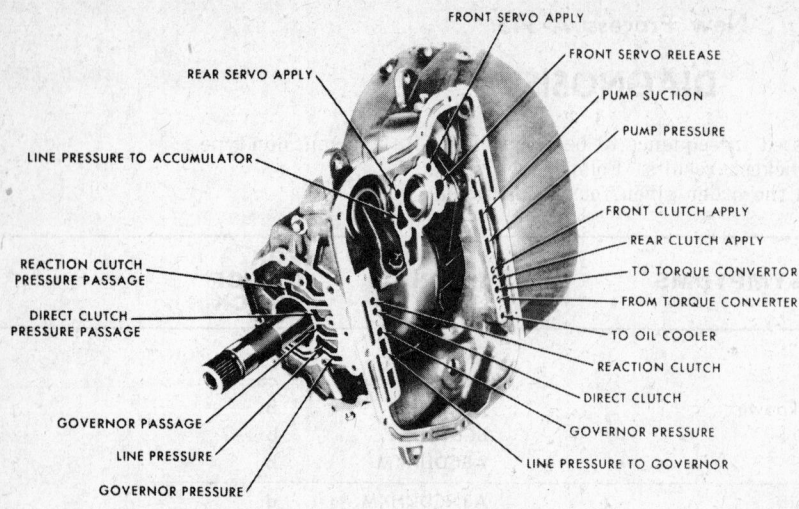

FRONT SERVO APPLY
FRONT SERVO RELEASE
REAR SERVO APPLY
PUMP SUCTION
PUMP PRESSURE
LINE PRESSURE TO ACCUMULATOR
FRONT CLUTCH APPLY
REAR CLUTCH APPLY
TO TORQUE CONVERTOR
REACTION CLUTCH PRESSURE PASSAGE
FROM TORQUE CONVERTER
DIRECT CLUTCH PRESSURE PASSAGE
TO OIL COOLER
REACTION CLUTCH
DIRECT CLUTCH
GOVERNOR PASSAGE
GOVERNOR PRESSURE
LINE PRESSURE
GOVERNOR PRESSURE
LINE PRESSURE TO GOVERNOR

Passages for air pressure tests (© Chrysler Corp.)

pounder, reduce the input shaft end play to zero by inserting and wedging a flat screw driver between the low-reverse drum and the transmission case. When the low-reverse drum is as far forward as possible, tighten the rear band until the band is tight on the low-reverse drum. Remove the screw driver. This prevents the #2 thrust washer or one of the clutches from coming out of position.

1. Remove the propeller shaft companion flange and yoke retaining nut and washer from the output shaft and remove the drum and yoke. Carefully pull the shaft yoke out of the transmission extension housing. Remove the parking brake assembly.

2. Remove the speedometer pinion and governor cover and gasket.
3. Remove the governor assembly from the extension housing.
4. Remove the compounder extension to compounder adapter screws and remove the compounder extension for further disassembly on work bench.
5. Remove the snap ring retaining the compounder annulus gear to intermediate shaft and remove the annulus gear and fiber washer and direct clutch.
6. Remove the two seal rings from the adapter. Unlock the seal ring hooks by compressing the rings.
7. Spring the seal rings just enough so that they can be re-

moved without scratching the adapter hub.

NOTE: The two seal rings for the direct clutch are different metal than the two seal rings for the reaction clutch. Do not mix these seal rings.

8. Remove the snap ring retaining the reaction clutch to the adapter clutch hub using snap ring tool.
9. Remove the reaction clutch assembly.
10. *DO NOT remove the reaction clutch seal rings unelss they are worn or damaged.*
11. Remove the snap ring retaining the compounder planetary gear assembly to the output shaft.
12. Remove the planetary gear assembly.
13. Remove the driving shell and the sun gear assembly.
14. Remove the overrunning clutch hub rollers and springs. Note the position of the overrunning clutch rollers and springs before disassembly to assist during assembly.
15. Remove the output shaft oil seal with a suitable hook type slide hammer. *Place the puller hooks between the output shaft and under the side of the seal so that the hooks do not damage the housing bore.*
16. Remove the output shaft bearing snap ring with special tool.
17. Remove the output shaft from the rear of the compounder extension.

NOTE: The output shaft bearing is a sliding fit in the housing bore and an interference fit on the output

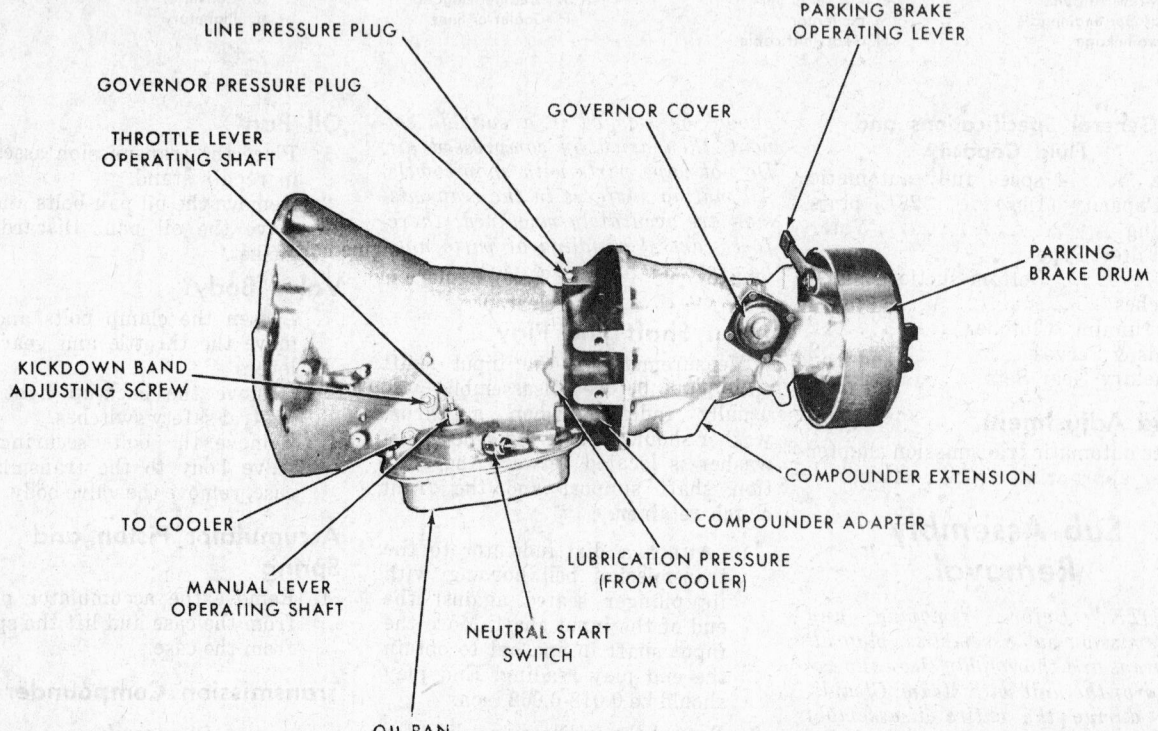

LINE PRESSURE PLUG
GOVERNOR PRESSURE PLUG
THROTTLE LEVER OPERATING SHAFT
GOVERNOR COVER
PARKING BRAKE OPERATING LEVER
PARKING BRAKE DRUM
KICKDOWN BAND ADJUSTING SCREW
TO COOLER
MANUAL LEVER OPERATING SHAFT
NEUTRAL START SWITCH
LUBRICATION PRESSURE (FROM COOLER)
COMPOUNDER ADAPTER
COMPOUNDER EXTENSION
OIL PAN

New Process Model A-345 Automatic Transmission (© Chrysler Corp.)

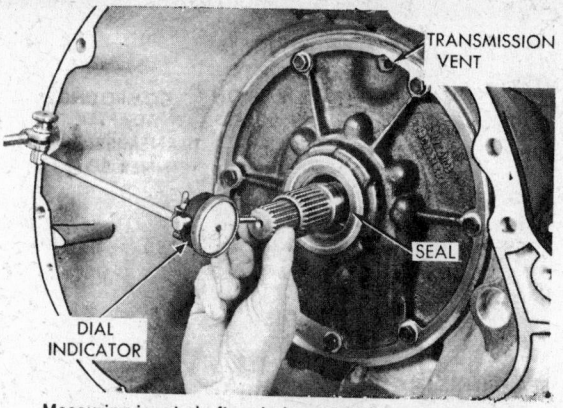

Measuring input shaft end play—A-345 (© Chrysler Corp.)

Removing or installing direct clutch seal rings—A-345 (© Chrysler Corp.)

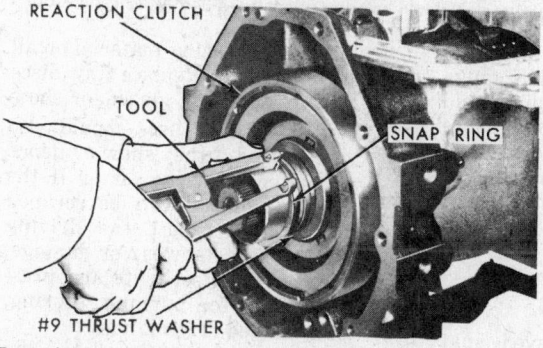

Removing or installing the reaction clutch snap ring (© Chrysler Corp.)

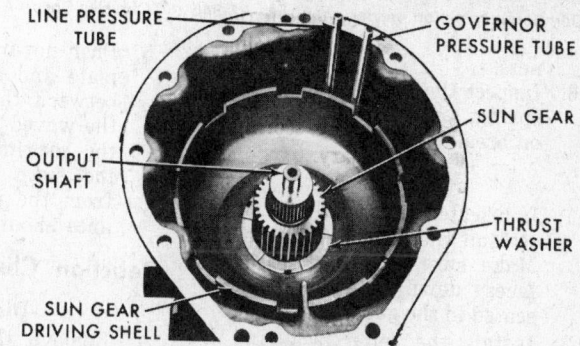

Removing or installing the rear driving shell—A-345 (© Chrysler Corp.)

shaft. *The overrunning clutch cam and spring retainer cannot be removed from the extension housing. If the clutch cam or roller spring retainer is damaged, it will be necessary to replace the extension housing with the cam and retainer as an assembly.*

18. Remove the output shaft bearing by tapping the threaded end of the output shaft on a hard wood block or press. Note the position, then remove the snap ring, governor drive gear, the drive ball, and the speedometer drive gear from the output shaft.
19. Remove the line pressure plug and remove the governor screen retainer clip and the screen.
20. Thoroughly clean the inside of the compounder extension housing with a suitable solvent. Blow out all passages and blow dry with compressed air. Reinstall or replace the governor screen, retainer and line pressure plug.

Direct Clutch

Disassembly

1. Remove the large waved snap ring which secures the pressure plate in the clutch piston retainer. Lift the pressure plate and clutch plates out of the retainer.
2. Install the compressor over the spring retainer. Compress the springs and remove the snap ring. Slowly release the pressure on the springs until the spring retainer is free of the hub. Remove the compressor tool, retainer, and springs.

3. Invert the clutch retainer assembly and bump it on a wood block to remove the piston. Remove the seals from the piston and the clutch retainer hub.

Inspection

1. Inspect the clutch plates and discs for flatness. They must not be warped or cone shaped.
2. Inspect the facing mateiral on driving discs. Replace the discs that are charred, glazed or heavily pitted. Discs should also be replaced if they show evidence of material flaking off or if the facing material can be scraped off easily. Inspect the driving disc

splines for wear or damage. Inspect the steel plate and pressure plate surfaces for burning, scoring of damaged driving lugs. Replace if necessary.
3. Inspect the steel plate lug grooves in the clutch retainer for smooth surfaecs, the plates must travel freely in the grooves. Note the ball check in the retainer, make sure the ball moves freely.
4. Inspect the seal surfaces in the clutch retainer for nicks of deep scratches, light scratches will not interfere with the neopreme seals.
5. Inspect the neopreme seals for deterioration, wear, and hard-

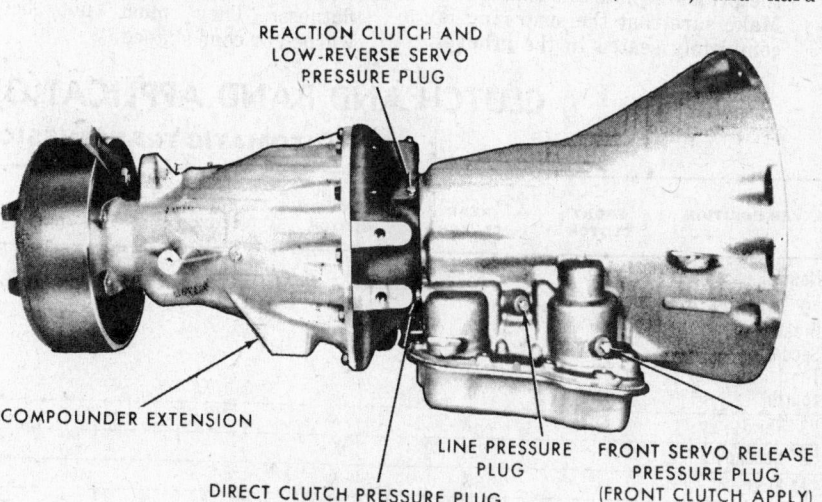

Pressure test locations (right side of case) (© Chrysler Corp.)

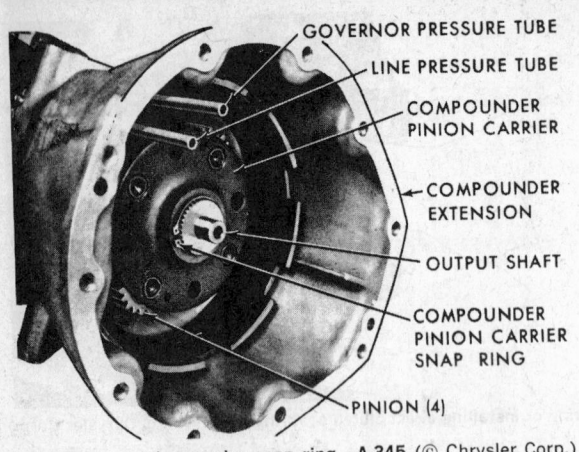

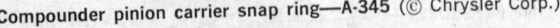

Compounder pinion carrier snap ring—A-345 (© Chrysler Corp.)

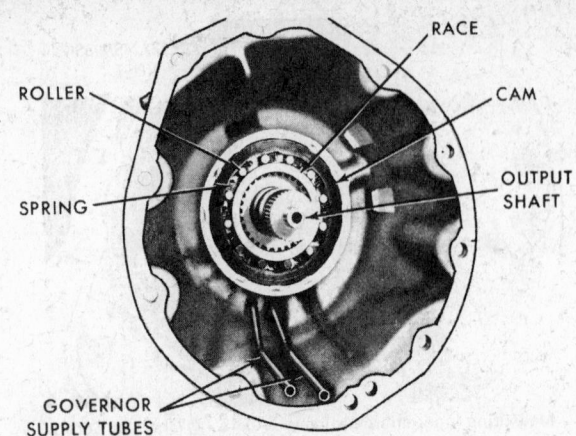

Rear overrunning clutch and output shaft—A-345 (© Chrysler Corp.)

ness.

6. Inspect the piston springs, wave spring, and spacer for distortion or breakage.

Assembly

1. Lubricate and install the inner seal on the clutch retainer hub. Make sure the lip of the seal faces down and is properly seated in the groove.

2. Install the outer seal in the clutch piston and with the lip of the seal toward the bottom of the clutch retainer, apply a coating of wax type lubricant (Door Ease) to the outer edge of the seals. Place the piston assembly in the retainer and carefully seat the piston in the retainer.

3. Install the eight springs on the piston. Position the spring retainer and snap ring over the springs. Compress the springs with special tool and seat the snap ring in the hub groove. Remove the compressor tool.

4. Lubricate all the clutch plates. Install one steel plate by a friction plate (wave type) until the four discs are installed. Install the pressure plate and snap ring. Make sure that the snap ring is completely seated in the groove.

5. Push downward on the pressure plate and insert a feeler gauge between the pressure plate and the waved snap ring to measure the maximum clearance where the snap ring is waved way from the pressure plate. Clearance should be 0.088-0.145 inch.

Reaction Clutch

Disassembly

1. Remove the large waved snap ring that secures the pressure palte in the clutch retainer. Lift the pressure plate and the clutch plates out of the retainer.

2. Install the spring compressor tool over the piston spring retainer. Compress the springs and remove the snap ring. Slowly release the tool until the spring retainer is free of the hub. Remove the tool, the retainer, and springs.

3. Invert the clutch retainer assembly and bump it on a wood block to remove the piston. Remove the seals from the piston and the clutch retainer hub.

Inspection

1. Inspect the plates and discs for flatness. They must not be warped or cone shaped.

2. Inspect the facing material on all driving discs. Replace any discs that are charred, glazed, or show heavy pitting. Discs should also be replaced if they show evidence of material flaking off or if the facing material can be scraped off easily. Inspect the driving disc splines for wear or damage. Inspect the steel plate and pressure plate for burning, scoring, or damaged driving lugs.

3. Inspect the steel plate lug grooves in the clutch retainer for smooth surfaces, plates must travel freely in the grooves. Inspect the seal surfaces in the clutch retainer for nicks or deep scratches. Light scratches will not interfere with the neoprene seals. Inspect the neoprene seals for deterioration, wear and hardness.

4. Inspect the pitson spring, wave spring and spacer for distortion or breakage.

Assembly

1. Lubricate and install the inner seal on the hub of the clutch retainer. Make sure that the lip faces down and is properly seated in the groove.

2. Install the outer seal on the

CLUTCH AND BAND APPLICATION CHART

A-345 AUTOMATIC TRANSMISSION

LEVER POSITION	FRONT CLUTCH	REAR CLUTCH	REACTION CLUTCH	DIRECT CLUTCH	FRONT (KICKDOWN) BAND	REAR (LOW-REV.) BAND	FRONT OVERRUNNING CLUTCH	REAR OVERRUNNING CLUTCH
Neutral							NO MOVEMENT	
"D" Drive							HOLDS	HOLDS
Breakaway		X			X		OVERRUNS	HOLDS
Second		X					OVERRUNS	HOLDS
Third	X	X					OVERRUNS	OVERRUNS
Fourth	X	X		X			OVERRUNS	OVERRUNS
"3" Third	X		X				OVERRUNS	PARTIAL HOLD
"2" Second		X	X			X	OVERRUNS	PARTIAL HOLD
"1" First		X	X			X	PARTIAL HOLD	PARTIAL HOLD
Reverse	X		X			X	NO MOVEMENT	

X=Applied

clutch piston with the lip of the seal toward the bottom of the clutch retainer. Apply a coat of a wax type lubricant (Door Ease) to the outer edge of the seals. Place the piston assembly in the retainer and carefully seat the piston.

3. Install the eight springs on the piston. Position the spring retainer and snap ring over the springs. Install the spring compressor tool and compress the piston spring until the snap ring is seated in the groove on the hub. Remove the compressor tool.

4. Lubricate all the clutch plates. Install one steel plate by a friction plate (plain gray) until four discs are installed. Install the pressure plate and snap ring. Make sure that the snap ring is properly seated.

5. Push downward on the pressure plate and insert a feeler gauge between the pressure plate and the waved snap ring. Clearance should be 0.088-0.145 inch.

Governor Bushing

Installation

1. Loosely bolt the reamer fixture to the compounder extension.
2. Place the alignment arbor into the reamer fixture and down into the governor bore.
3. Righten the screws on the reamer fixture to 8-12 foot pounds.

CAUTION: Do not over-torque and strip the threads.

NOTE: Be sure the alignment arbor rotates freely after the screws are properly tightened.

4. Remove the alignment arbor.
5. Working with the proper reamer and drive ratchet, hand ream the governor bore in the following manner:
   a. Oil the reamer, reamer fixture, and governor bore.
   b. Using a 5-10 pound feeding force on the reamer, ream until the reamer bottoms in the extension and then con-

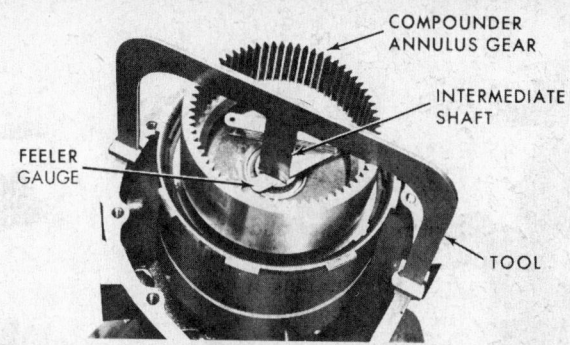

Selecting the proper thrust washer thickness with special tool and feeler gauge—A-345 (© Chrysler Corp.)

tinue to rotate the reamer 10 (ten) complete revolutions.
   c. Remove the reamer using a *clockwise rotation* and 5-10 pounds force upward.

CAUTION: Pulling the reamer out without using a rotating motion may score the governor bore and cause a leak between the extension and the bushing.

6. Remove the reamer fixture from the extension.
7. Thoroughly clean the chips from the extension. Visually check the governor feed holes to insure that they are free of chips.
8. Install the bushing using the following operation:
   a. Note the position of the two (2) notches at one end of the bushing.
   b. Position the notches so that one notch is at the 11 o'clock position and the other is at the 2 o'clock position, viewing the extension in its normal installed, parallel position.
   c. Use the alignment arbor and the bushing installation tool to drive the bushing into the extension. A brass hammer should be used to strike the hardened steel bushing installation tool.
   d. Drive the bushing until it is flush with the top of the bore.
9. Oil a new governor and insert it into the installed bushing .The

governor should spin freely. If slight honing on the bushing is required, use crocus cloth around your ringer and rotate the cloth within the new bushing.

Transmission Compounder

Assembly

NOTE: Be sure the #2 thrust washer, front clutch discs, and rear clutch discs are in their proper position.

1. Push the intermediate shaft forward to reduce the input shaft end play to zero.
2. Insert and wedge a flat screw driver between the low-reverse drum and the transmission case, moving the low-reverse drum as far forward as possible.
3. Tighten the rear (low-reverse) band adjusting screw until the band is tight on the low-reverse drum. Remove the screw driver.
4. Install the two seal rings on the forward grooves on the adapter hub.
5. Inspect the seal rings for wear or broken locks. Be sure that the rings move freely in the grooves.
6. Install the reaction clutch assembly on the adapter hub.
7. Install the fiber thrust washer. Install the snap ring.
8. Install the two seal rings on the rear grooves of the adapter hub. Inspect for wear or broken hock locks. Be sure that the rings turn freely in the groves.
9. Install the direct clutch assembly on the adapter with the direct clutch drum gear teeth engaging all the clutch discs in the reaction clutch.
10. Use the thrust washer selection tool and a feeler gauge to select the proper thrust washer thickness.
11. Assemble the compounder annulus gear to the annulus gear support and retain with a snap ring.
12. Install the selective fiber thrus washer on the compounder annulus gear support.
13. Install the annulus gear assembly into the direct clutch, engaging all the clutch discs in the di-

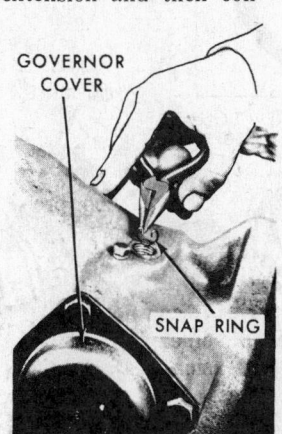

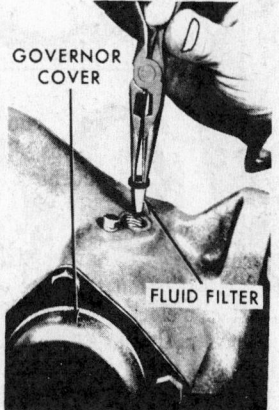

Removing or installing the governor fluid filter—A-345 (© Chrysler Corp.)

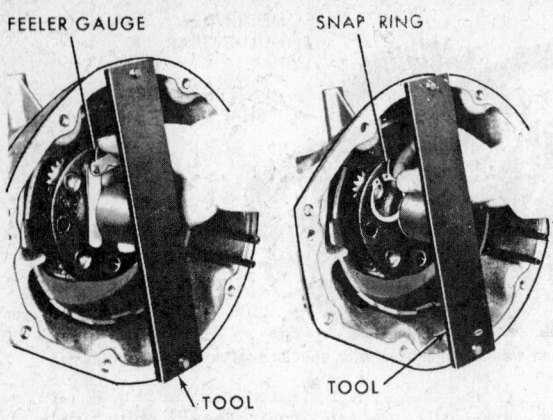

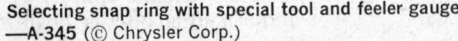

Selecting snap ring with special tool and feeler gauge
—A-345 (© Chrysler Corp.)

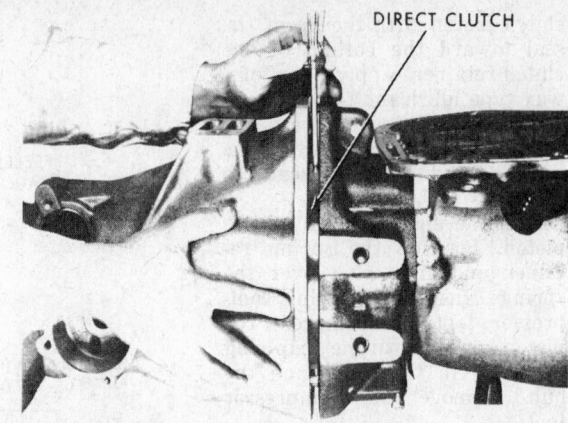

Aligning the direct clutch with the driving shell, and the annulus gear with the compounder planetary pinions—A-345 (© Chrysler Corp.)

rect clutch. Install the snap ring in the groove on the intermediate shaft.

14. Install the speedometer drive gear, drive ball, governor drive gear, and snap ring on the output shaft.

15. Assemble the bearing to the output shaft. Install the output shaft and bearing into the extension housing.

16. Install the bearing retainer snap ring. Avoid scoring the housing bore.

17. Install a new output shaft oil seal.

18. Place the extension housing in an upright position, insert the overrunning clutch hub (race) inside the clutch cam.

19. Install the overrunning clutch rollers and spring.

20. Install the governor pressure and line pressure tubes in the extension.

21. Assemble the sun gear and the overrunning clutch roller retainer to the driving shell. Install the snap ring.

22. Install the driving shell and sun gear assembly on the output shaft.

23. Install the bronze thrust washer in the driving shell. Install the planetary carrier assembly and install the selective snap ring.

NOTE: Use the snap ring selector tool and a feeler gauge to measure the clearance between the pinion carrier and selector tool hub. Position the feeler gauge in the output shaft snap ring groove. Do not rest the extension housing assembly on the output shaft.

24. Install the speedometer gear and the governor assembly in the extension housing. Install the governor cover with a new gasket. Tighten the cover screws to 12 foot-pounds torque.

25. Install the gasket and position the extension housing assembly on the compounder adapter

using pilot studs. Carefully align the pressure tubes with the matching holes in the adapter.

With light pressure against the extension housing, use a 3/16 inch diameter phillips head screw driver in the hole in the direct clutch drum, rotate the drum to align the tangs of the drum with the slots in the driving shell and the compounder annulus gear with the compounder planetarp gear pinions.

26. Install the compounder extension housing to the compounder adapter. Tighten the retaining bolts to 30 foot-pounds.

27. Install the parking brake assembly, drum and yoke on the compounder extension housing. Tighten the output shaft nut to 175 foot-pounds.

Oil pump and the reaction shaft support—A-345 (© Chrysler Corp.)

Transmission Disassembly

Oil Pump and Reaction Shaft Support

1. Tighten the front band adjusting screw until the band is tight on the front clutch retainer. This prevents the clutch retainer from coming out with the oil pump which might cause damage to the clutches.
2. Remove the oil pump housing retaining bolts.
3. Install the two slide hammers on the pump housing flange.
4. Move the slide hammer weights outward evenly to remove the pump and reaction shaft from the case.

Front Band and Front Clutch

1. Loosen the front band adjuster remove the band strut and slide the band out of the transmission case.
2. Slide the front clutch assembly out of the transmission case.

Input Shaft and Rear Clutch

1. Grasp the input shaft and slide the input shaft and rear clutch assembly out of the transmission acse. *Be careful not to lose the thrust washer located between and the forward end of the intermediage shaft.*

Planetary Gear Assemblies, Sun Gear and Driving Shell

1. While supporting the intermediate shaft and drive shell, carefully slide the assembly forward and out through the case.

Rear Band and Low-Reverse Drum

1. Loosen the rear band adjuster, remove the low-reverse drum, remove the band strut and link. Remove the band from the case.

Front Overrunning Clutch

1. Note the position of the overrun-

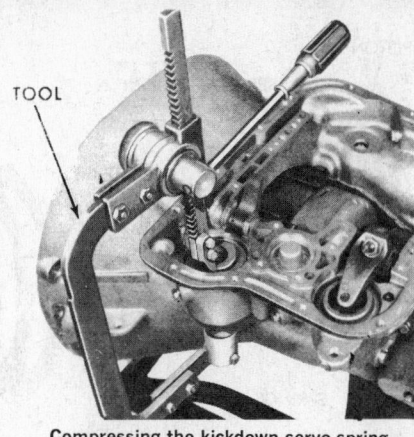

Compressing the kickdown servo spring—A-345 (© Chrysler Corp.)

ning clutch rollers and springs before disassembly to assist during reassembly.

2. Carefully slide the clutch hub out of the case and remove the rollers and springs. *If the overrunning clutch cam and roller spring retainer are found to be damaged or worn, refer to replacement procedures later in this section.*

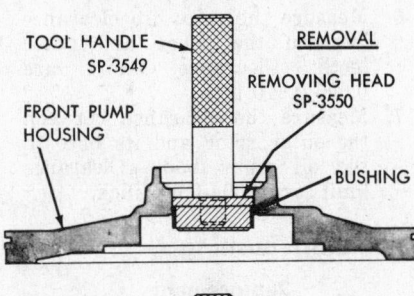

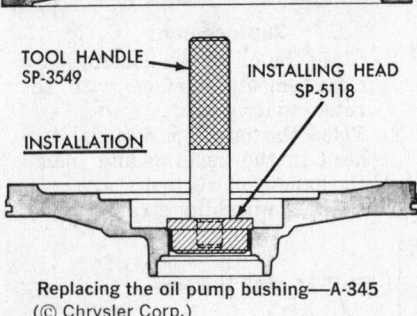

Replacing the oil pump bushing—A-345 (© Chrysler Corp.)

Kickdown Servo (Front)

1. Compress the kickdown servo spring with an engine valve spring compressor and remove the snap ring.
2. Remove the rod guide, springs and piston rod from the case. Be careful not to damage the piston rod or guide during removal.
3. Withdraw the piston from the transmission case.

Low and Reverse Servo (Rear)

1. Compress the low and reverse servo piston spring with an engine valve spring compressor and remove the snap ring.
2. Remove the spring retainer, spring, and servo piston and plug asembly from the transmission case.

Compounder Adapter

1. Remove the ten compounder to transmission case bolts. Remove the adapter and discard the gasket.,
2. Inspect the intermediate shaft bushings and replace if necessary.

Sub-Assemblies Reconditioning

Accumulator Piston and Springs

Inspection

1. Inspect the seal rings for wear and make sure they turn freely in the piston grooves. It is not necessary to remove the rings unless conditions warrant.
2. Inspect the piston for nicks, burrs, scores, and wear.
3. Inspect the piston bore in the case for scores or other damage.
4. Inspect the piston spring for distortion.
5. Replace parts as required.

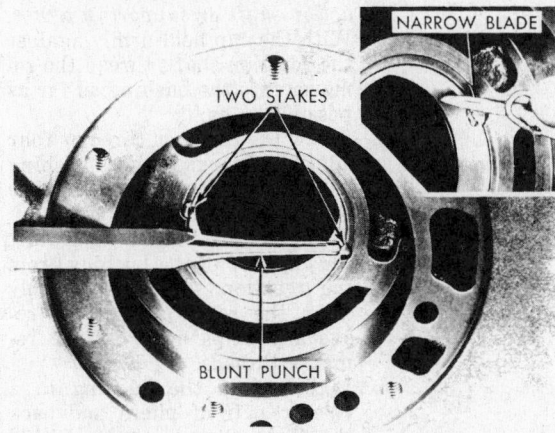

Staking the oil pump bushing—A-345 (© Chrysler Corp.)

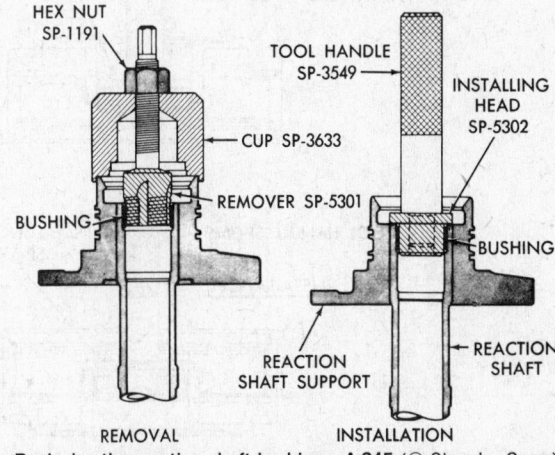

Replacing the reaction shaft bushing—A-345 (© Chrysler Corp.)

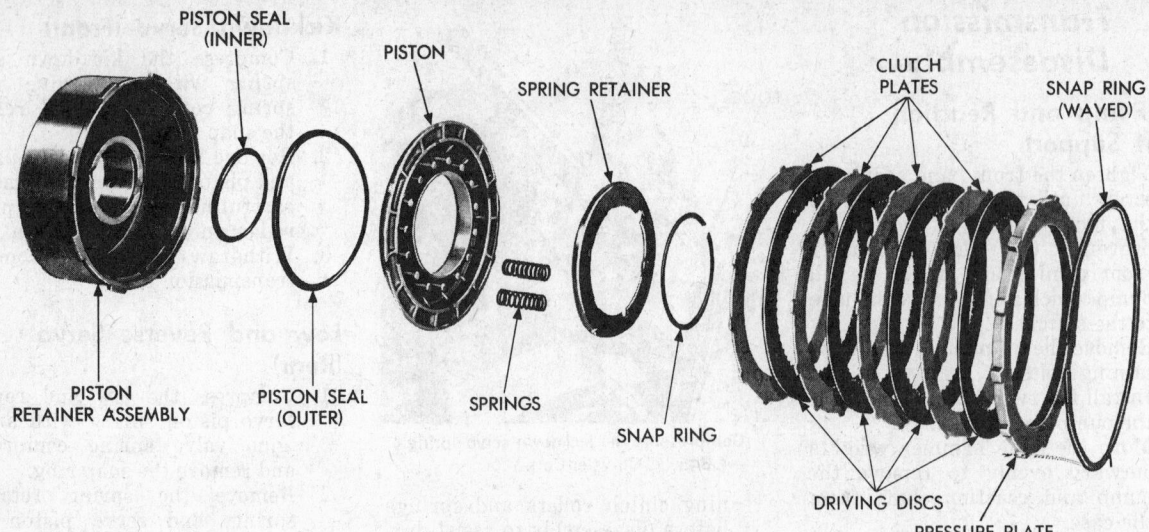

Front clutch assembly (© Chrysler Corp.)

Oil Pump and Reaction Shaft Support

Disassembly

1. Remove the bolts from the rear side of the reaction shaft support, and lift the support off the pump.
2. Remove the rubber seal ring from the pump body flange.
3. Drive the oil seal out with a blunt punch.

Inspection

1. Inspect the interlocking steel rings on the reaction shaft support for wear or broken locks. Make sure the rings turn freely in the grooves.
2. Inspect the machined surfaces on the pump body and reaction shaft support for nicks and burrs.
3. Inspect the pump body and reaction shaft support bushings for wear or scoring.
4. Inspect the pump rotors for scoring or pitting.

5. Clean the rotors and install them in the pump body. Place a straight edge across the face of the rotors and pup body. Use a feeler gauge to measure the clearance between the straight edge and the face of the rotors. Clearance limits are 0.0015 0.003 inch.
6. Measure the rotor tip clearance between the inner and outer teeth. Clearance limits are 0.005-0.010 inch.
7. Measure the clearance between the outer rotor and its bore in the oil pump body. Clearance limits are 0.004-0.008 inch.

Oil Pump Bushing

Replacement

1. Place the oil pump housing on a clean smooth surface with the rotor cavity down.
2. Place the bushing removal tool head in the bushing and install the handle on the tool.
3. Drive the bushing straight down

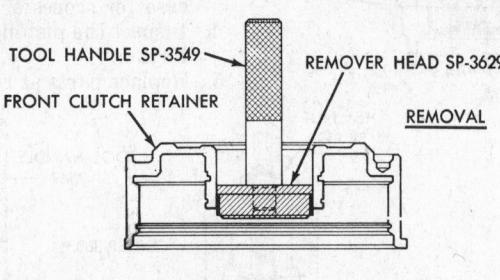

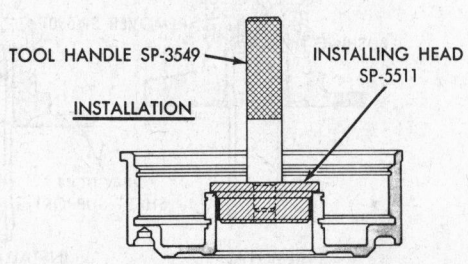

Replacing the front clutch retainer bushing—A-345 (© Chrysler Corp.)

and out of the bore. Be careful not to cock the tool in the bore.
4. Place a new bushing on the installation tool head.
5. With the pump housing on a clean smooth surface, hub end down, start the bushing and installation head in the bushing bore. Install the tool handle.
6. Drive the bushing into the housing until the tool bottoms in the pump cavity. Be careful not to cock the tool during installation. Remove the installation tool.
7. Stake the bushing in place by using a blunt punch. A gentle tap at each stake slot should be sufficient.
8. Working with a narrow-bladed knife, remove the high points or burrs around the staked area. Do not use a file that will remove more metal than necessary.
9. Thoroughly clean the pump housing before installation.

Reaction Shaft Bushing

Replacement

1. Assemble the removal tool complete with the cup and hex nuts. *Do not clamp any part of the reaction shaft or support in a vise.*
2. With the cup held firmly against the reaction shaft, thread the remover into the bushing as far as possible by hand.
3. Screw the remover three or four additional turns into the bushing with a wrench to firmly engage the threads in the bushing.
4. Turn the hex nut down against the cup to pull the bushing from the reaction shaft. Thoroughly clean the reaction shaft to remove any chips made by the remover threads.
5. Lightly grip the bushing in a vise or pair of pliers and back the removal tool out of the old bushing. Be careful not to dam-

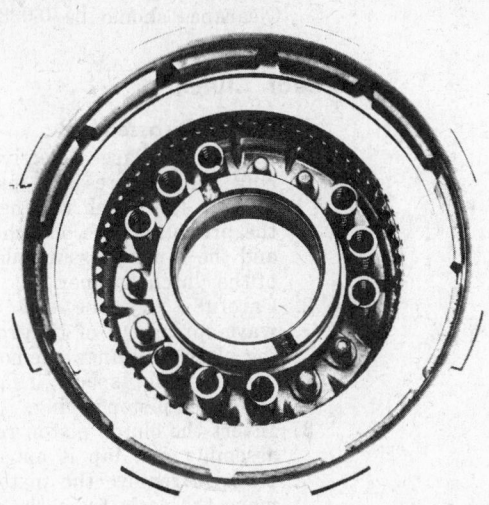

Front spring location—9 springs—A-345 (© Chrysler Corp.)

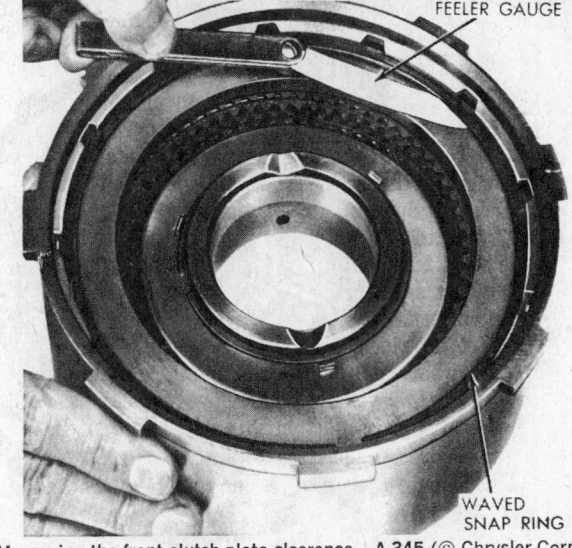

FEELER GAUGE

WAVED SNAP RING

Measuring the front clutch plate clearance—A-345 (© Chrysler Corp.)

age the threads on the bushing remover.

6. Slide a new bushing (chamfered end first) on the installation head of bushing tool and start the bushing and tool into the reaction shaft bore.

7. Support the reaction shaft in an upright position on a clean smooth surface and install the installation tool handle. Drive the bushing into the shaft until the installation head bottoms. Remove the bushing installation tool.

8. Thoroughly clean the reaction shaft support assembly before installation.

Assembly

1. Assemble the pump rotor and "O" ring in the pump housing.

2. Install the reaction shaft support. Install the retaining bolts and tighten to 160 inch-pounds.

3. Place a new oil seal in the opening of the pump housing (lip of the seal facing inward) using a suitable drift.

Front Clutch

Dissassembly

1. Remove the large waved snap

ring that secures the pressure plate in the clutch piston retainer. Lift the clutch plates out of the retainer.

2. Install the piston spring compressor over the piston spring retainer. Compress the springs and remove the snap ring. Slowly release the tool until the spring retainer is free of the hub. Remove the tool, the spring retainer and the springs.

3. Invert the clutch retainer assembly and tap it on a wood block to remove the piston. Remove the seals from the piston and the clutch retainer hub.

Inspection

1. Inspect the facing material on all the driving plates. Replace any plates that are charred, glazed or heavily pitted. Plates should also be replaced if they show evidence of material flaking off or if the facing material can be scraped off easily.

2. Inspect the driving plate splines for wear or other damage.

3. Inspect the steel plate and pressure plate surfaces for burring, scoring or damaged driving lugs.

4. Inspect the steel plate lug grooves in the clutch retainer for smooth surfaces, plates must travel freely in the grooves.

5. Inspect the band contacting surface on the clutch retainer for scores.

6. Note the ball check in the clutch retainer, make sure that the ball moves freely.

7. Inspect the seal surfaces in the clutch retainer for nicks or deep scratches, light scratches will not interfere with the neoprene seals.

8. Inspect the clutch retainer bushing for wear or scoring.

9. Inspect the inside bore of the piston for score marks, if light remove with crocus cloth.

10. Inspect the neoprene seals for deterioration, wear, and hardness.

11. Inspect the piston springs, retainer and snap ring for distortion.

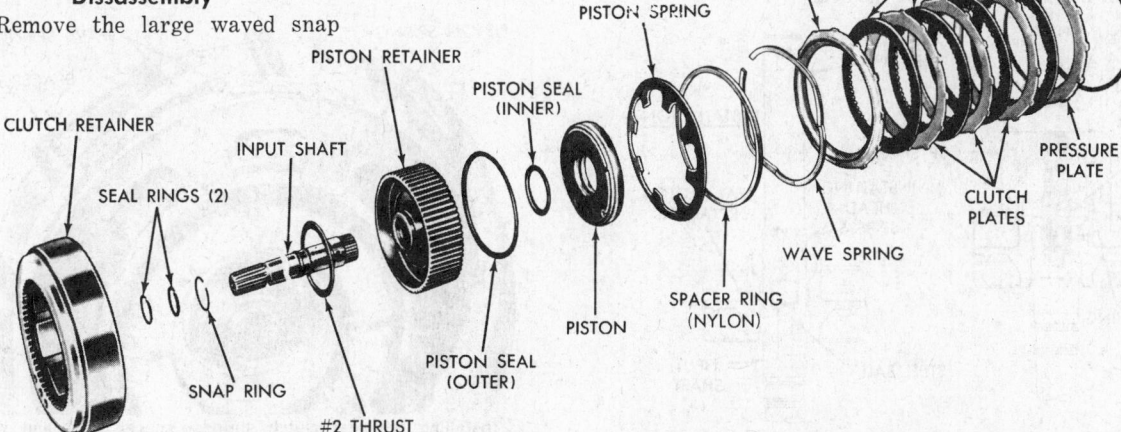

SNAP RING (SELECTIVE)

DRIVING DISCS

PRESSURE PLATE

PISTON SPRING

PISTON RETAINER

PISTON SEAL (INNER)

CLUTCH RETAINER

INPUT SHAFT

SEAL RINGS (2)

PRESSURE PLATE

CLUTCH PLATES

WAVE SPRING

SPACER RING (NYLON)

PISTON

PISTON SEAL (OUTER)

SNAP RING

#2 THRUST WASHER

Rear clutch—exploded view—A-345 (© Chrysler Corp.)

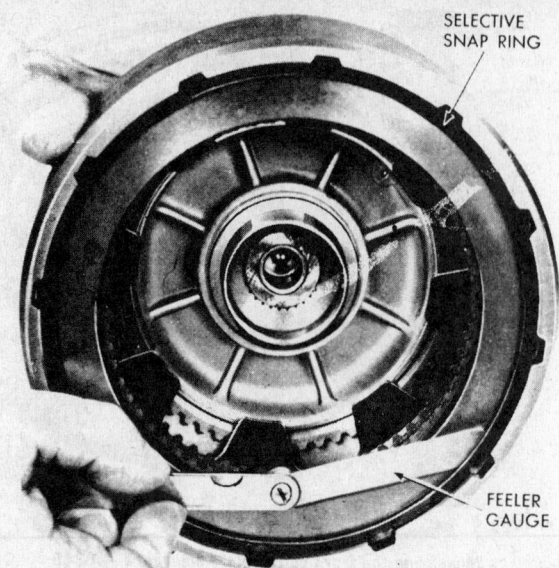

Measuring the rear clutch plate clearance—A-345 (© Chrysler Corp.)

Front Clutch Retainer Bushing

Replacement

1. Place the clutch retainer (open end down) on a clean smooth surface. Install the removing head of bushing tool in the bushing. Install the bushing tool handle.
2. Drive the bushing straight down and out of the clutch retainer bore. Be careful not to cock the tool in the bore.
3. Lay the clutch retainer (open end up) on a clean smooth surface. Slide a new bushing on the installation head of the bushing tool. Start the tool and bushing in the clutch retainer bore.
4. Install the bushing tool handle. Drive the bushing into the clutch retainer until the tool bottoms.
5. Thoroughly clean the clutch retainer before assembly and installation.

Assembly

1. Lubricate and install the inner seal on the clutch retainer hub. Make sure that the lip of the seal faces down and is properly seated in the groove.
2. Install the outer seal on the clutch piston, with the lip toward the bottom of the clutch retainer. Apply a coating of wax type lubricant (Door Ease) to outer edges of the seals. Place the piston assembly in the retainer and carefully seat the piston in the bottom of the retainer.
3. Install the clutch piston springs on the piston exactly as they were removed.
4. Place the spring retainer and snap ring over the springs. Install the spring compressor tool. Compress the springs and seat the snap ring in its groove. Remove the compressor tool.
5. Lubricate all clutch plates. Install one steel plate followed by a lined (driving) plate until all the plates are installed. Install the pressure plate and snap ring. Make sure that the snap ring is correctly seated in the groove.
6. When the front clutch is completely assembled, push downward on the pressure plate and insert a feeler gauge between the pressure plate and the snap ring. Clearance should be 0.088-0.145 inch.

Rear Clutch

Disassembly

1. Remove the large selective snap ring that secures the pressure plate in the clutch retainer. Lift the pressure plate, clutch plates and the inner pressure plate out of the clutch retainer.
2. Carefully pry one end of the wave spring out of its groove in the clutch retainer. Remove the wave spring, spacer ring, and the clutch piston spring.
3. Invert the clutch piston retainer assembly and tap it on a wood block to remove the piston. Remove the seals from the piston.
4. If necessary, remove the snap ring and press the input shaft from the clutch retainer.

Inspection

1. Inspect the facing material on all the driving plates. Replace any plates that are charred, glazed or heavily pitted. Plates should also be replaced if they show evidence of material flaking off or if the facing material can be scraped off easily.
2. Inspect the driving plate splines for wear or other damage.
3. Inspect the steel plate and the pressure plates surfaces for burring, scoring or damaged driving lugs.
4. Inspect the steel plate lug grooves in the clutch retainer for smooth surfaces, plates must travel freely in the grooves.
5. Inspect the band contacting surface on the clutch retainer for scores.
6. Note the ball check in the clutch retainer, make sure that the ball moves freely.
7. Inspect the seal surfaces in the clutch retainer for nicks or deep scratches, light scratches will not interfere with the neoprene seals.

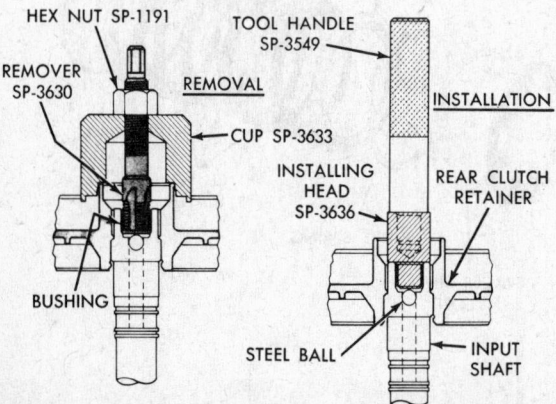

Replacing the input shaft bushing—A-345 (© Chrysler Corp.)

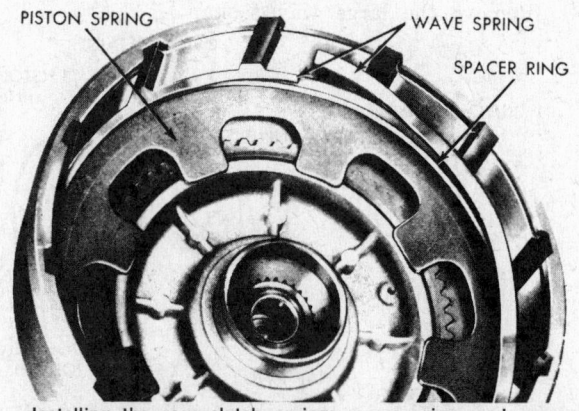

Installing the rear clutch springs, spacer ring, and wave spring—A-345 (© Chrysler Corp)

8. Inspect the neoprene seals for deterioration, wear, and hardness.

9. Inspect the piston springs, wave spring, and spacer for distortion or breakage.

10. Inspect the interlocking seal rings on the input shaft for wear or broken locks. Make sure that they turn freely in the grooves. Do not remove the seal rings unless necessary.

11. Inspect the bushing in the input shaft for wear or scores.

12. Inspect the rear clutch to front clutch thrust washer for wear. Washes thickness should be 0.061-0.063 inch, replace if necessary.

Input Shaft Bushing

Replacement

NOTE: Perform this operation only if necessary.

1. Clamp the input shaft in a vise with soft faced jaws. Be careful not to clamp on the seal ring lands or the bearing journals.

2. Assemble the removal tool, cup and hex nut.

3. Hold the cup firmly against the clutch piston retainer and thread the remover into the bushing as far as possible by hand.

4. Working with a wrench, screw the remover into the bushing three or four additional turns to firmly engage the threads in the bushing.

5. Turn the hex nut down against the cup to pull the bushing out of the input shaft.

6. Thoroughly clean the input shaft to remove any chips made by the remover threads. Make sure that the smapp lubrication hole next to the ball in the end of the shaft is not plugged. Be sure no chips have lodged next to the steel ball.

7. Place a new bushing on the installation head of the bushing tool and start them into the input shaft bore.

8. Stand the input shaft on a clean smooth surface and install the installation tool handle. Drive the bushing into the shaft until the tool bottoms.

9. Thoroughly clean the input shaft and the clutch piston retainer before installation.

Assembly

1. If removed, press the input shaft into the clutch piston retainer, and install the snap ring.

2. Lubricate and install the inner and outer seal rings on the clutch piston. Be sure that the lip of the seals face toward the head of the clutch retainer, and that they are properly seated in the piston grooves.

3. Place the piston assembly in the retainer and with a twisting action, seat the piston in the bottom of the retainer.

4. Place the clutch retainer over the piston retainer splines and support the assembly so that the clutch retainer remains in position.

5. Place the clutch piston spring and the spacer ring on the top of the piston in the clutch retainer. Make sure that the spring and the spacer ring are positioned in the retainer recess. Start one end of the waved spring in the retainer groove, then progressively push or tap the spring into place. Make sure that the waved spring is fully seated in its groove.

6. Install the inner pressure plate in the clutch retainer with the raised position of the plate resting on the spring.

7. Lubricate all the clutch plates, install one lined plate followed by a steel plate until all the plates are installed. Install the outer pressure plate and secure with the selective snap ring.

8. Measure the rear clutch plate clearance by having an assistant press downward firmly on the outer pressure plate, then insert a feeler gauge between the plate and the snap ring. Clearance should be 0.025-0.045 inch. If clearance is not within limits, install a snap ring of the proper thickness to obtain the proper

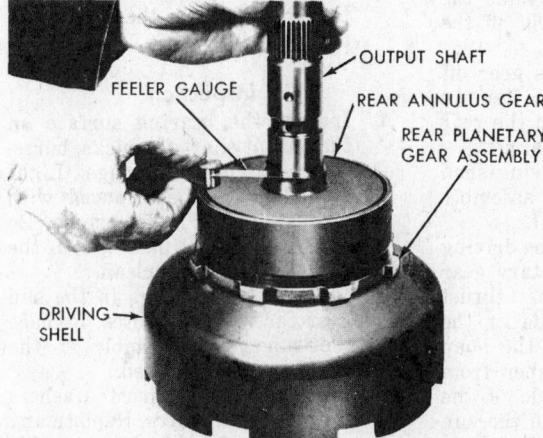

Measuring the end play of the planetary gear train—A-345 (© Chrysler Corp.)

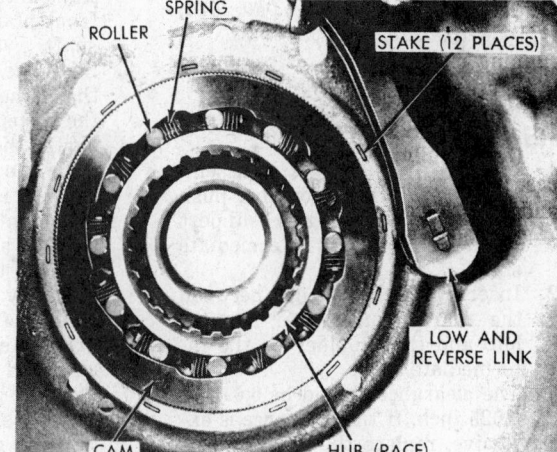

Overrunning clutch and the low-reverse band link—A-345 (© Chrysler Corp.)

Removing the front overrunning clutch cam—A-345 (© Chrysler Corp.)

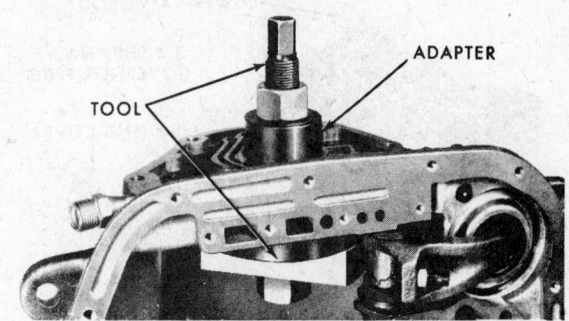

Installing the front overrunning clutch cam—A-345 (© Chrysler Corp.)

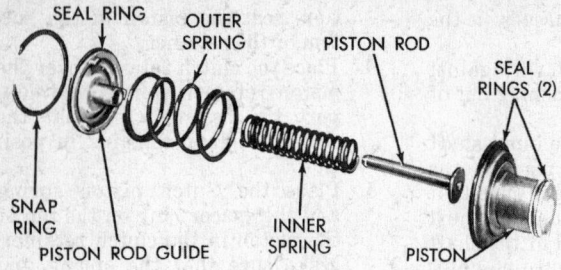

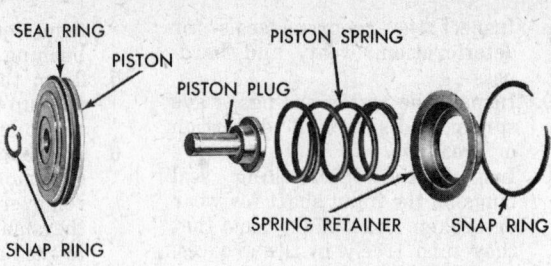

Kickdown servo-exploded view—A-345 (© Chrysler Corp.)

Low-reverse servo-exploded view—A-345 (© Chrysler Corp.)

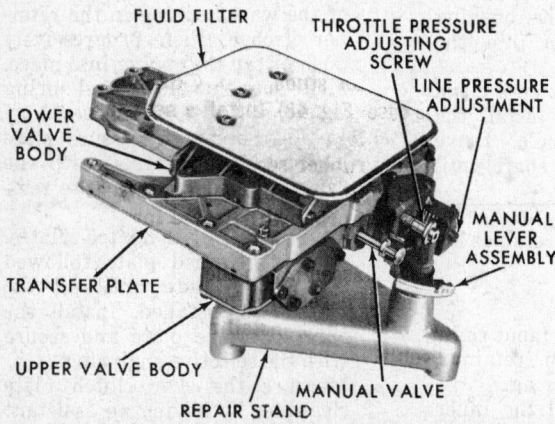

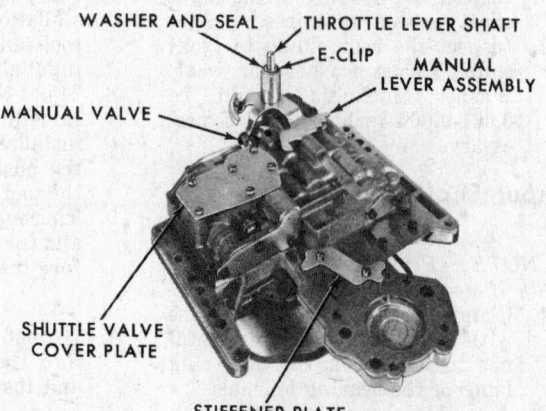

Valve body assembly mounted in the repair stand—A-345 (© Chrysler Corp.)

Valve body controls—A-345 (© Chrysler Corp.)

clearance. Low limit clearance is desirable. *Rear clutch plate clearance is very important in obtaining the proper clutch operation. The clearance can be adjusted by the use of various thickness outer snap rings. Snap rings are available in .060, .074, .088, and .106 inch thickness.*

Planetary Gear Train End Play

1. Place the planetary gear assembly in an upright position, push the rear annulus gear support downward on the intermediate shaft.
2. Insert a feeler gauge between the rear annulus gear support hub and the shoulder on the intermediate shaft.
3. The clearance should be 0.010 to 0.025 inch. If the clearance is excessive, replace the thrust washers and/or necessary parts.

Disassembly

1. Remove the thrust washer from the forward end of the intermediate shaft.
2. Remove the selective snap ring from the forward end of the intermediate shaft, then slide the front planetary assembly off the shaft.
3. Slide the front annulus gear off the planetary gear set. Remove the thrust washer from the rear side of the planetary gear set.
4. Slide the sun gear, driving shell and the rear planetary assembly off the intermdiate shaft.
5. Lift the sun gear and the driving shell off the rear planetary gear assembly. Remove the thrust washer from the inside of the driving shell. Remove the snap ring and the steel washer from the sun gear (rear side of the driving shell) and slide the sun gear out of the shell. Remove the front snap ring from the sun

gear if necessary. Note that the front end of the sun gear is longer than the rear.
6. Remove the thrust washer from the forward side of the rear planetary gear assembly. Remove the planetary gear set and the thrust plate from the rear annulus gear.

Inspection

1. Inspect the bearing surface on the output shaft for nicks, burrs, scores or other damage. Light srcatches can be removed with crocus cloth or a fine stone. Be sure that all oil passages in the shaft are open and clean.
2. Inspect the bushings in the sun gear for wear or scores. Replace the sun gear assembly if the bushings are damaged.
3. Inspect all the thrust washers for wear and scores. Replace any washer that is in questionable condition.

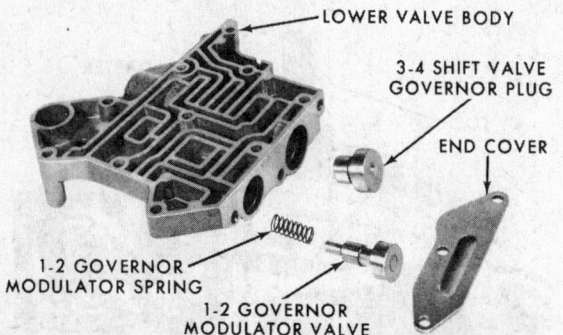

Governor modulator and the 3-4 governor plug—A-345 (© Chrysler Corp.)

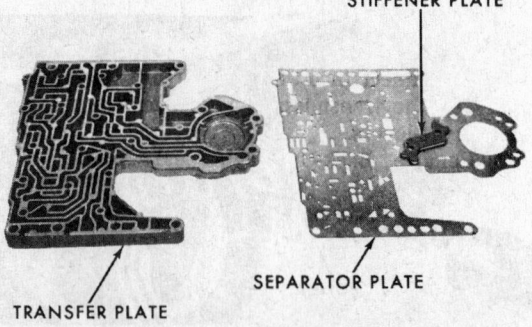

Transfer and separator plates—A-345 (© Chrysler Corp.)

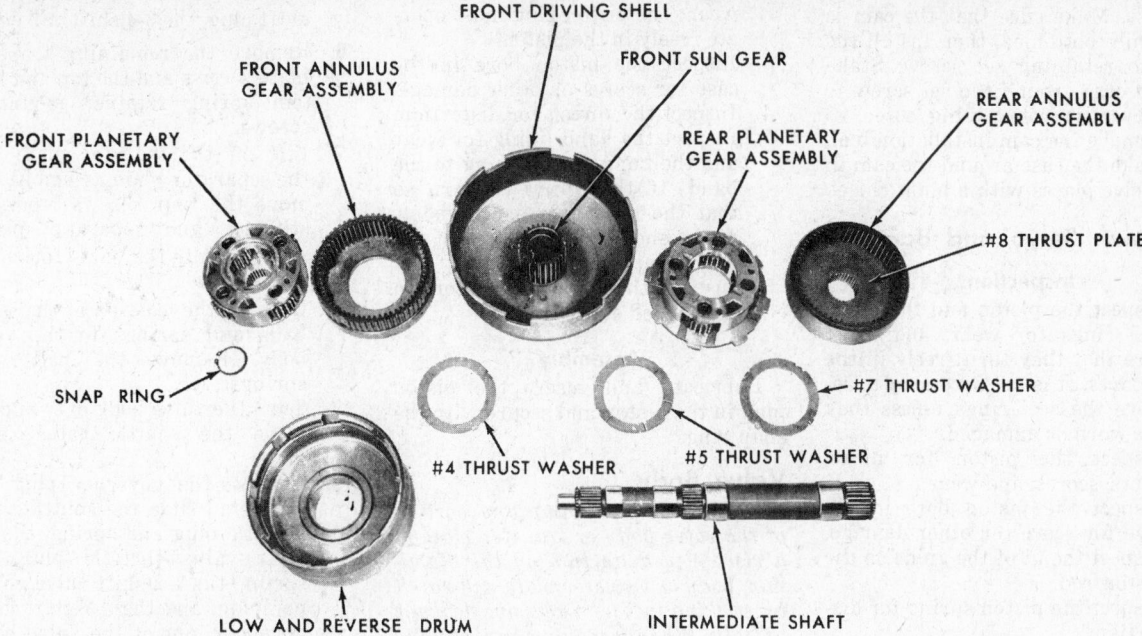

FRONT DRIVING SHELL

FRONT SUN GEAR

FRONT ANNULUS GEAR ASSEMBLY

REAR ANNULUS GEAR ASSEMBLY

FRONT PLANETARY GEAR ASSEMBLY

REAR PLANETARY GEAR ASSEMBLY

#8 THRUST PLATE

SNAP RING

#7 THRUST WASHER

#4 THRUST WASHER

#5 THRUST WASHER

LOW AND REVERSE DRUM

INTERMEDIATE SHAFT

Front planetary gear train and the intermediate shaft-exploded view—A-345 (© Chrysler Corp.)

4. Inspect the thrust faces of the planetary gear carriers for wear, scoring or other damage. Replace as necessary.
5. Inspect the planetary gear carriers for cracks and pinions with broken teeth or ware. Check the pinion shaft for broken lock pins.
6. Inspect the annulus gear and the driving gear teeth for damage. Replace distorted lock rings.

Assembly

1. Install the rear annulus gear on the intermediate shaft. Apply a thin coat of petroleum jelly on the thrust plate, place it on the shaft and in the annulus gear making sure that the teeth are over the shaft splines.
2. Position the rear planetary gear assembly in the rear annulus gear. Place the thrust washer on the front side of the planetary gear assembly.
3. Install the snap ring in the front groove of the sun gear (long end of the gear). Insert the sun gear through the front side of the driving shell, install the rear steel washer and snap ring.
4. Carefully slide the driving shell and the sun gear assembly on the intermediate shaft, engaging the sun gear teeth with the rear planetary pinion teeth. Place the thrust washer inside the front driving shell.
5. Place the thrust washer on the rear hub on the front planetary gear set and slide the assembly into the front annulus gear.
6. Carefully work the front planetary and annulus gear assembly on the output shaft, meshing the

planetary pinions with the sun gear teeth.
7. With all components properly positioned, install the selective snap ring on the front end of the intermediate shaft. Remeasure the end play of the assembly.

The clearance can be adjusted by use of various thickness snap rings. Snap rings are available in .048, .055, and .062 inch thickness.

Front Overrunning Clutch

Inspection

1. Inspect the clutch rollers for smooth round surfaces, they must be free of flat spots and chipped edges.
2. Inspect the roller contacting surfaces in the cam and race for brinnelling.
3. Inspect the roller springs for distortion, wear or other damage.
4. Inspect the cam set screw for tightness. If loose, tighten and restake the case around the screw.

Front Overrunning Clutch Cam

Replacement

NOTE: If the overrunning clutch cam or the roller spring retainer are damaged, replace the cam and spring retainer in the following manner:
1. Remove the set screw from the case below the clutch cam.
2. Insert a blunt punch through the bolt holes and drive the cam from the case. Alternate the punch from one hole to another so that the cam will be driven evenly from the case.
3. Clean all burrs and chips from the cam area in the case.
4. Place the spring retainer on the cam, making sure that the retainer lugs snap firmly into the notches on the cam.
5. Position the cam in the case with the cam serrations aligned with those in the case. Tap the cam *evenly* into the case as far as possible, use a soft faced mallet.
6. Install the cam replacement tool and adapter, tighten the nut on the tool to seat the cam in the

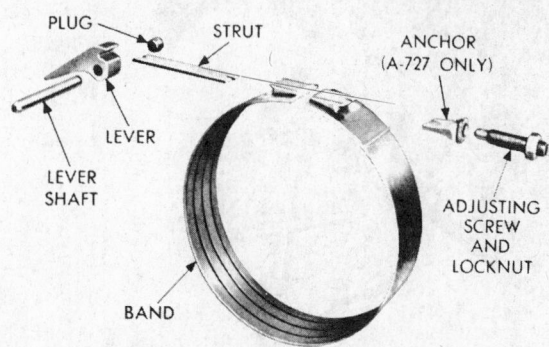

PLUG

STRUT

ANCHOR (A-727 ONLY)

LEVER

LEVER SHAFT

ADJUSTING SCREW AND LOCKNUT

BAND

Kickdown band and linkage—A-345 (© Chrysler Corp.)

case. Make sure that the cam is firmly bottomed, then install the cam retaining set screw. Stake the case around the set screw to prevent it from coming loose.

7. Remove the cam installation tool. Stake the case around the cam in twelve places with a blunt chisel.

Kickdown Servo and Band

Inspection

1. Inspect the piston and the guide seal rings for wear, and make sure that they turn freely in the grooves. It is not necessary to remove the seal rings unless they are worn or damaged.
2. Inspect the piston for nicks, burrs, scores, and wear.
3. Inspect the piston bore in the case for scores or other damage.
4. Inspect the fit of the guide on the piston rod.
5. Inspect the piston spring for distortion.
6. Inspect the band lining for wear and the bond of the lining to the band. Check the lining for burn marks, glazing, non-uniform wear patterns and flaking. If the lining is worn so that the grooves are not visible at the ends or any portion of the bands, replace the band.
7. Inspect the band for distortion or cracked ends.

Low-Reverse Servo and Band

Disassembly

Remove the snap ring from the piston and remove the piston plug.

Inspection

1. Inspect the seal for deterioration, wear and hardness.
2. Inspect the piston and the piston plug for nicks, burrs, scores and

wear; the piston plug must operate freely in the piston.

3. Inspect the piston bore in the case for scores or other damage.
4. Inspect the spring for distortion.
5. Inspect the band lining for wear and the bond of the lining to the band. If the lining is worn so that the grooves are not visible at the ends or any portion of the bands, replace the band.
6. Inspect the band for distortion or cracked ends.

Assembly

Lubricate and insert the piston plug in the piston and secure with the snap ring.

Valve Body

NOTE: Do not clamp any portion of the valve body or transfer plate in a vise. Any distortion on the aluminum body or transfer plate will result in sticking valves excessive leakage or both. When removing or installing the valves or plugs, slide them in or out carefully. Do not use force.

Disassembly

1. Place the valve body assembly on the special repair stand. Remove the three screws from the fluid filter and lift off the filter.
2. Remove the lower valve body and the steel plate from the transfer plate. Observe the two steel balls in the transfer plate for proper location.

NOTE: Tag all springs as they are removed for assembly identification.

3. Remove the flat end plate, 3-4 shaft valve governor plug, 1-2 governor modulator valve spring.
4. Remove the end cover from the opposite side of the lower valve body.
5. Remove the part throttle down-

shift plug, the 3-4 shift valve.

6. Remove the remaining transfer plate screws and the top and bottom spring retainer mounting screws.
7. Lift off the transfer plate and the separator plate assembly. Remove the four screws from the stiffener and separator plates and separate the parts for cleaning.
8. Observe the location of the seven balls and springs in the valve body. Remove the balls and springs.
9. Turn the valve body over and remove the shuttle valve cover plate.
10. Remove the governor plug end plate and slide the shuttle valve throttle plug and spring, the 1-2 shift valve thrtotle plug and spring, the 1-2 shift valve governor plug, and the 2-3 shift valve governor, out of the valve body.
11. Remove the shuttle valve "E" clip and slide the shuttle valve out of its bore. Remove the secondary guides and spring which are retained by an "E" clip.
12. Hold the spring retainer firmly against the spring force while removing the last retaining screw from the side of the valve body.
13. Remove the spring retainer, with the line and the throttle pressure adjusting screws (do not disturb the settings) and the line pressure and torque converter regulator springs.
14. Slide the torque converter and line pressure valves out of their bores.
15. Remove the "E" clip and washer from the throttle lever shaft. Remove any burrs from the shaft. While holding the manual lever detent ball and spring in their

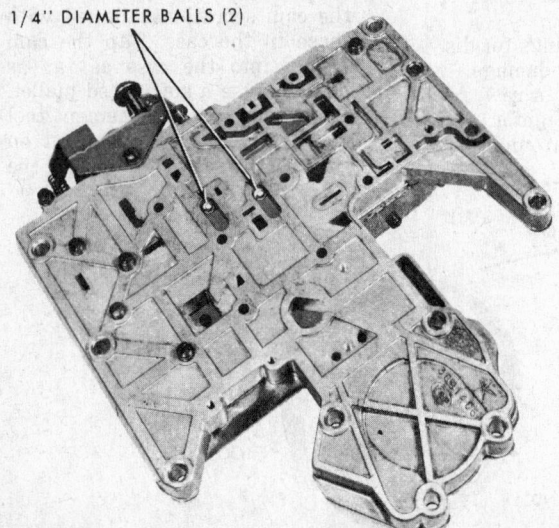

1/4" DIAMETER BALLS (2)

Stell ball locations (two)—A-345 (© Chrysler Corp.)

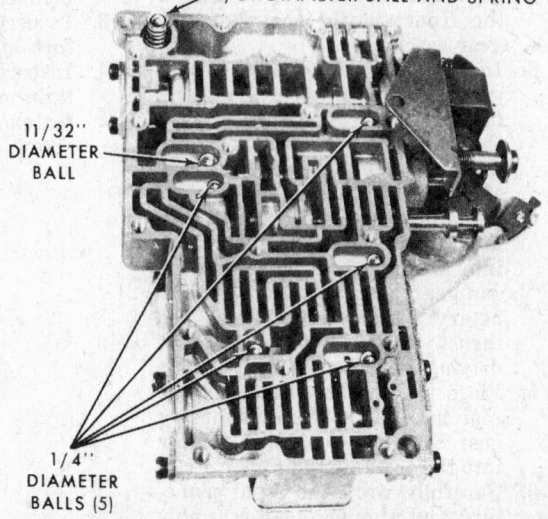

3/8" DIAMETER BALL AND SPRING

11/32" DIAMETER BALL

1/4" DIAMETER BALLS (5)

Steel ball locations (seven)—A-345 (© Chrysler Corp.)

bore with a suitable tool, slide the manual lever off of the throttle shaft. Remove the detent ball and spring.

16. Slide the manual valve out of its bore.
17. Slide the kickdown detent, kickdown valve, throttle valve spring and the throttle valve out of their bores in the valve body.
18. Remove the line pressure regulator valve end plate. Slide the pressure regulator valve sleeve, line perssure plug, and the throttle pressure plug out of the valve body.
19. Remove the shift valve end plate.
20. Remove the three springs, two shift valves and the 1-2 shift control valve from the valve body.

Cleaning and Inspection

1. Soak all parts in a suitable solvent for a few minutes. Wash thoroughly and blow dry with filtered compressed air. Make sure that all passages are clean and free of obstructions.
2. Inspect the manual valve operating levers and shafts for distortion or wear. If a lever is loose on its shaft, it may be *silver soldered* only, or replace the shaft assembly. *Do not attempt to straighten a bent lever.*
3. Inspect all mating surfaces for burrs, nicks and scratches. Minor defects can be removed with crocus colth, *use very light pressure.*
4. Inspect all mating surfaces for warpage or distortion with a straight edge. Slight distortion can be corrected with a surface plane.

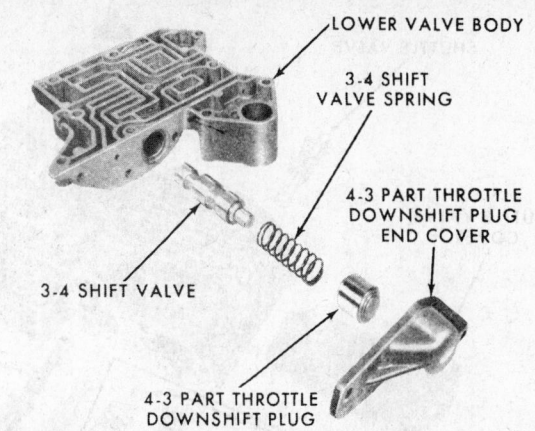

The 3-4 shift valve and the downshift plug—A-345 (© Chrysler Corp.)

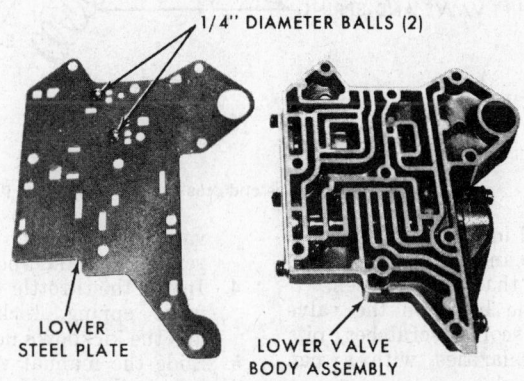

Lower valve body and steel plate—A-345 (© Chrysler Corp.)

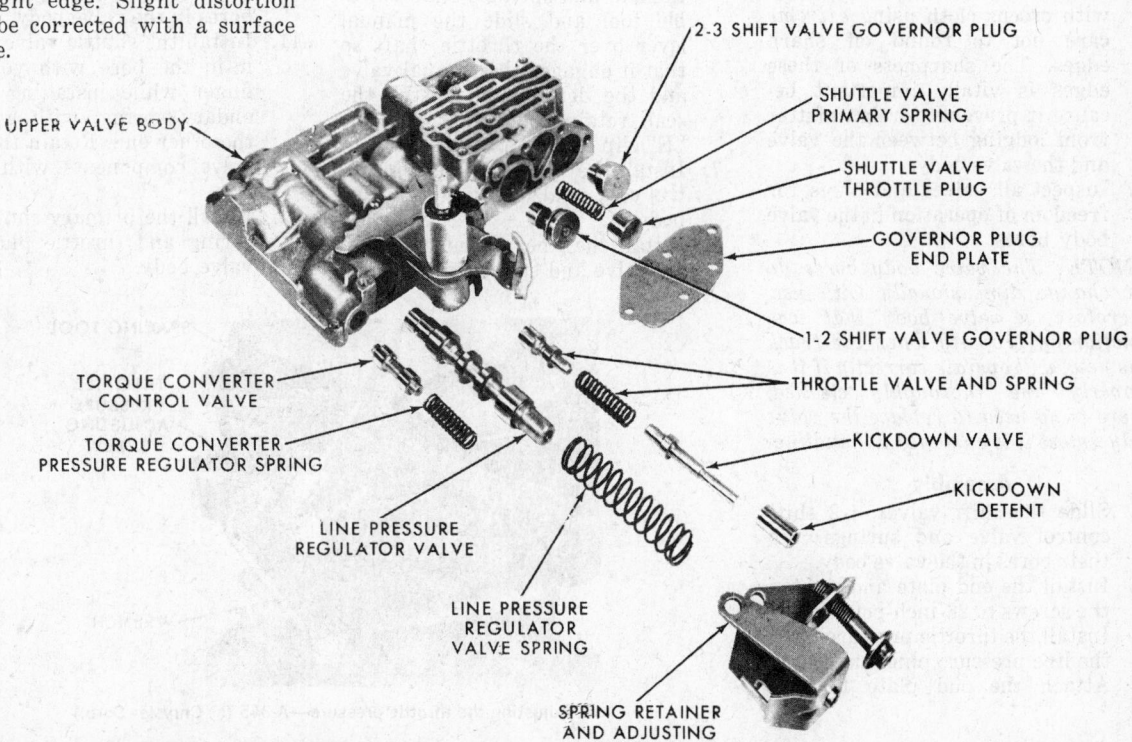

Pressure regulators and governor plug—A-345 (© Chrysler Corp.)

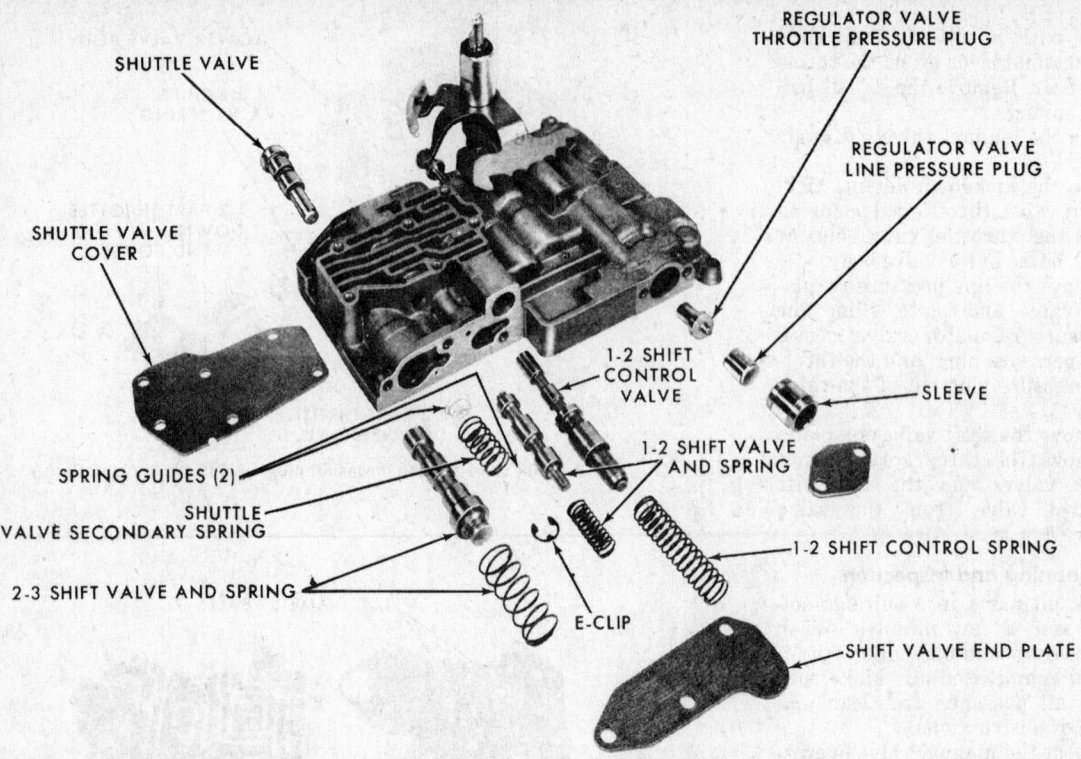

SHUTTLE VALVE

REGULATOR VALVE THROTTLE PRESSURE PLUG

SHUTTLE VALVE COVER

REGULATOR VALVE LINE PRESSURE PLUG

1-2 SHIFT CONTROL VALVE

SLEEVE

1-2 SHIFT VALVE AND SPRING

SPRING GUIDES (2)

SHUTTLE VALVE SECONDARY SPRING

1-2 SHIFT CONTROL SPRING

2-3 SHIFT VALVE AND SPRING

E-CLIP

1-2 SHIFT CONTROL SPRING

SHIFT VALVE END PLATE

Shift valves and pressure regulator valve plugs—A-345 (© Chrysler Corp.)

5. Inspect all metering holes in the steel plate and the valve body to make sure that they are open.
6. Inspect the bores in the valve body for scores, scratches, pits and irregularities with a pen light.
7. Inspect all valve springs for distortion and collapsed coils.
8. Inspect all valves and plugs for burrs, nicks or scoring. Small nicks and scores may be removed with crocus cloth using *extreme* care not to round off sharp edges. The sharpness of these edges is vitally important because it prevents foreign matter from lodging between the valve and the valve body.
9. Inspect all valves and plugs for freedom of operation in the valve body bores.

NOTE: The valve body bores do not change dimensionally with use. Therefore, a valve body that was functioning properly when the truck was new, will operate correctly if it is properly and thoroughly cleaned. There is no need to replace the valve body unless it is damaged in handling.

Assembly

1. Slide the shift valves, 1-2 shift control valve and springs into their bores in the valve body.
2. Install the end plate and tighten the screws to 28-inch-pounds.
3. Install the throttle pressure plug, the line pressure plug and sleeve. Attach the end plate for the

valve body and tighten the screws to 28 inch-pounds.
4. Install the throttle valve, throttle valve spring, kickdown valve, and the kickdown detent.
5. Slide the manual valve into the bore in the valve body.
6. Install the throttle lever and shaft on the vavle body. Insert the detent spring and ball in the bore of the valve body. Depress the ball and spring with a suitable tool and slide the manual lever over the throttle shaft so that it engages the manual valve and the detent ball. Install the seal, retaining washer and the "E" clip on the throttle shaft.
7. Install the torque converter control valve and spring in the valve body.
8. Install the line pressure regulator valve and spring in the valve

body.
8. Install the line pressure regulator valve and spring in the valve body.
9. Install the pressure adjusting screw and bracket assembly on the springs and fasten, finger tight, with the screw which goes on the side of the valve body.
10. Install the 1-2 and the 2-3 shift valve governor plugs into their bores in the valve body.
11. Install the shuttle valve and hold it in the bore with your index finger while installilng the secondary spring, with guides, on the other end. Retain the shuttle valve components with an "E" clip.
12. Install the primary shuttle valve spring and throttle plug in the valve body.

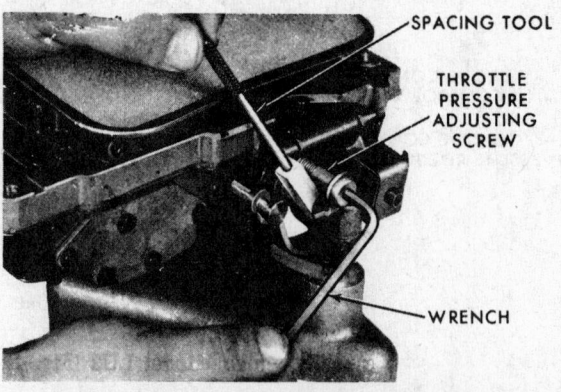

SPACING TOOL

THROTTLE PRESSURE ADJUSTING SCREW

WRENCH

Adjusting the throttle pressure—A-345 (© Chrysler Corp.)

Ford C-4
DIAGNOSIS

This diagnosis guide covers the most common symptoms and is an aid to careful diagnosis. The items to check are listed in the sequence to be followed for quickest results. Thus, follow the checks in the order given for the particular transmission type.

TROUBLE SYMPTOMS	ITEMS TO CHECK	
	IN TRUCK	OUT OF TRUCK
Rough initial engagement in D1 or D2	KBUFEG	a
1-2 or 2-3 shift points incorrect or erratic	ABLCDUER	
Rough 1-2 shifts	BJGUEF	
Rough 2-3 shifts	BJUFGER	bl
Dragged out 1-2 shift	ABJUGEFR	c
Engine overspeeds on 2-3 shift	CABJUEFG	bl
No 1-2 or 2-3 upshift	CLBDUEGJ	bc
No 3-1 shift in D1 or 3-2 shift in D2	DE	
No forced downshift	LEB	
Runaway engine on forced downshift	UJGFEB	c
Rough 3-2 or 3-1 shift at closed throttle	KBJEF	
Shifts 1-3 in D1 and D2	GJBEDR	
No engine braking in first gear—manual low	CHIEDR	
Creeps excessively	KW	
Slips or chatters in first gear, D1	ABUFE	acg
Slips or chatters in second gear	ABJGUFER	ac
Slips or chatters in R	ABHUIFER	bcl
No drive in D1	ACUER	g
No drive in D2	ACUJER	cg
No drive in L	ACUEIR	cg
No drive in R	ACHUIER	bcl
No drive in any selector position	ACUFER	cd
Lockup in D1		bec
Lockup in D2	HI	becg
Lockup in L	GJ	bec
Lockup in R	GJ	aec
Parking lock binds or does not hold	C	e
Transmission overheats	OFBU	i
Maximum speed too low, poor acceleration	VW	i
Transmission noisy in N and P	AF	df
Transmission noisy in any drive position	AF	fadg
Fluid leaks	AMNOPQ STBIJX	hik
Car moves forward in N	C	a

Key to Checks

A. Fluid level
B. Vacuum diaphragm unit or tube
C. Manual linkage
D. Governor
E. Valve body
F. Pressure regulator
G. Intermediate band
H. Reverse band
I. Reverse servo
J. Intermediate servo
K. Engine idle speed
L. Downshift linkage—inner lever position
M. Converter drain plug

N. Oil pan and/or filler tube gaskets/seals
O. Oil cooler and/or connections
P. Manual or downshift lever shaft seal
Q. Pipe plug, side of case
R. Perform air pressure checks
S. Extension housing-to-case gasket or washers
T. Extension housing rear oil seal
U. Make control pressure test
V. Engine performance
W. Vehicle brakes
X. Speedometer driven gear adaptor seal

a. Forward clutch
b. Reverse—high clutch
c. Hydraulic system leakage
d. Front pump
e. Parking brake linkage
f. Planetary assembly
g. Planetary one-way clutch
h. Engine rear oil seal
i. Front oil pump seal
j. Converter oneway clutch
k. Front pump-to-case seal or gasket
l. Reverse—high clutch piston air bleed valve

Transmission Checks

Transmission Fluid Leakage Checks

Make the following checks if a leakage is suspected from the transmission case:

1. Clean all dirt and grease from the transmission case.

2. Inspect the speedometer cable connection at the extension housing of the transmission. If fluid is leaking here, disconnect the cable and replace the rubber seal.

3. Inspect the oil pan gasket and attaching bolts for leaks. Tighten any bolts that appear loose to the proper torque (10-13 ft-lbs) Recheck for signs of leakage. If necessary, remove the pan attaching bolts and old pan gasket and install new gasket and reinstall the pan and its attaching bolts.

4. Check filler tube connection at the transmission for signs of

leakage. If tube is leaking, tighten the connection to stop the leak. If necessary, disconnect the filler tube, replace the O-ring, and reinstall the filler tube.

5. Inspect all fluid lines between the transmission and the cooler core in the lower radiator tank. Replace any lines or fittings that appear to be worn or damaged. Tighten all fittings to the proper torque.

6. Inspect the engine coolant for signs of transmission fluid in the radiator. If there is transmission fluid in the engine coolant, the oil cooler core is probably leaking. The oil cooler core may be tested further by disconnecting all lines to it and applying 50-75 psi air pressure through the fittings. Remove the radiator cap to relieve any pressure buildup outside the cooler core. If air bubbles appear in the coolant or if the cooler core will not hold pressure, the oil cooler core is leaking and must be replaced. Oil cooler core repair and replacement is discussed in the section on Cooling Systems in this manual.

7. Inspect the openings in the case where the downshift control lever shaft and the manual lever shaft are located for leaks. If necessary, replace the defective seal.

8. Inspect all plugs or cable connections in the transmission for signs of leakage. Tighten any loose plugs or connectors to the proper torque according to the specifications.

CLUTCH AND BAND APPLICATION CHART— C-4 AUTOMATIC TRANSMISSION

GEAR	FORWARD CLUTCH	REVERSE HIGH CLUTCH	INTER-MEDIATE BAND	LOW REVERSE BAND	ONE-WAY CLUTCH
1-st*	on	off	off	off	holding
2-nd*	on	off	on	off	over-running
3-rd*	on	on	off	off	over-running
Low(1)	on	off	off	on	holding
Reverse	off	on	off	on	not affected

* Transmission selector in "D" position

9. Remove the lower cover from the front of the bellhousing and inspect the converter drainplugs for signs of leakage. If there is a leak around the drainplugs, loosen the plug and coat the threads with a sealing compound and tighten the plug to the proper torque.

NOTE: Fluid leaks from around the converter area may be caused by the converter drain plug, oil pump seal, and/or gaskets, engine oil leakage past the rear main bearing seal, oil gallery plugs loose, valve cover gaskets, or the power steering system. To determine the exact cause of the leak before beginning repair procedures, an oil-soluble Aniline or flourescent dye may be added to the leak detection process. When using the dye, a black light must be used to detect the dye within the oil.

If further converter checks are necessary, remove the transmission from the truck and the converter from the transmission. The converter cannot be disassembled for cleaning or repair. If the converter is leaking, it must be replaced with a new unit. To further check the converter for leaks, assemble and install the converter leak checking tool shown and fill the converter with 20 psi air pressure. Then, place the converter in a tank of water and watch for air bubbles. If no air bubbles are seen, the converter is not leaking.

Control Pressure Check for Automatic Transmissions (C4)

When the vacuum diaphragm unit operates properly and the downshift linkage is adjusted correctly, all transmission shifts (automatic and kickdown) should occur within the specified road speed limits. If these shifts do not occur within the limits or if the transmission slips during a shift point, perform the following procedure to locate the problem:

1. Connect the Automatic Transmission Tester (see illustration) as follows:

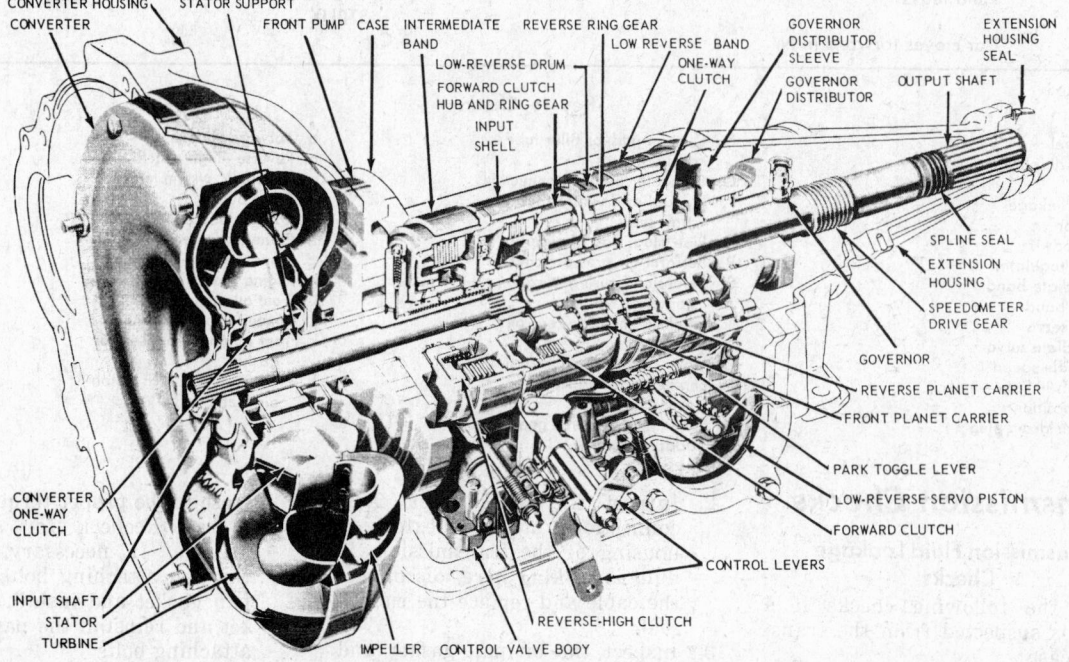

C-4 automatic transmission (© Ford Motor Co.)

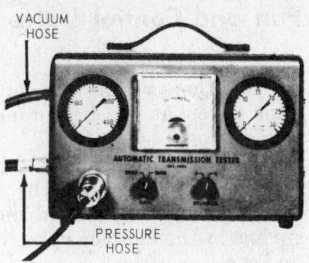

Automatic transmission tester
(© Ford Motor Co.)

a. Tachometer cable to engine
b. Vacuum gauge hose to the transmission vacuum diaphragm unit (see illustration)

2. Apply the parking brake and start the engine. On a truck equipped with a vacuum brake release, disconnect the vacuum line or use the service brake since the parking brake will release automatically when the transmission is put in any Drive position.

3. Check engine idle speed and throttle and downshift linkage for correct operation. Check the transmission diaphragm unit for leaks.

Vacuum Diaphragm Check (Off Truck)

With the use of a vacuum pump, set the vacuum to read 18 inches, with the end of the hose blocked off. On single area diaphragms, connect the vacuum hose to the manifold vacuum hose port. On the dual area diaphragms, connect the vacuum hose to the EGR port, leaving the manifold vacuum port open to the atmosphere. If the gauge holds at 18 inches, the vacuum unit diaphragm is not leaking. A second check can be made by holding the control rod end by a finger, with the other end inserted into the unit. Install the vacuum hose from the vacuum pump onto the diaphragm port. The control rod should move inward and when the vacuum supply is cut off by the removal of the hose, the rod should move outward.

Air Pressure Checks

If the truck will not move in one or more ranges, or, if it shifts erratically, the items at fault can be determined by using air pressure at the indicated passages.

Drain the transmission and remove the oil pan and the control valve assembly.

NOTE: Oil will spray profusely during this operation.

Front Clutch

Apply sufficient air pressure to the front clutch input passage. (See illustration.) A dull thud can be heard when the clutch piston moves. Check also, for leaks.

Governor

Remove the governor inspection cover from the extension housing. Apply air to the front clutch input passage. (See illustration). Listen for a sharp click and watch to see if the governor valve snaps inward as it should.

Rear Clutch

Apply air to the rear clutch passage (See illustration) and listen for the dull thud that will indicate that the rear clutch piston has moved. Listen also for leaks.

Front Servo

Apply air pressure to the front servo apply tube (See illustration.) and note if front band tightens. Shift the air to the front servo release tube, which is next to the apply tube, and watch band release.

Rear Servo

Apply air pressure to the rear servo apply passage. The rear band should tighten around the drum.

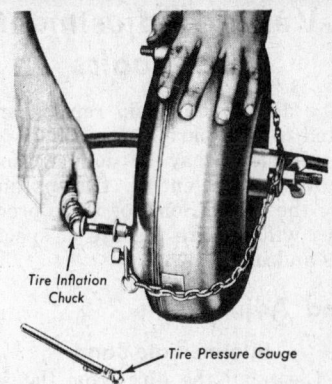

Converter leak checking tool installed
(© Ford Motor Co.)

Conclusions

If the operation of the servos and clutches is normal with air pressure, the no-drive condition is due to the control valve and pressure regulator valve assemblies, which should be disassembled, cleaned and inspected.

If operation of the clutches is not normal; that is, if both clutches apply from one passage or if one fails to move, the aluminum sleeve (bushing) in the output shaft is out of position or badly worn. (See illustration.)

Use air pressure to check the passages in the sleeve and shaft, and also check the passages in the primary sun gear shaft.

If the passages in the two shafts and the sleeve are clean, remove the clutch assmblies, clean and inspect the parts.

Erratic operation can also be caused by loose valve body screws. When reinstalling the valve body be careful to tighten the control valve body screws as specified in the Torque Limits table.

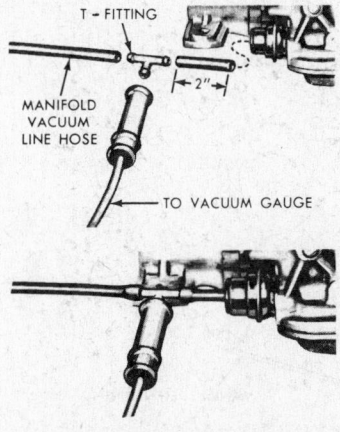

Typical vacuum test line connections
(© Ford Motor Co.)

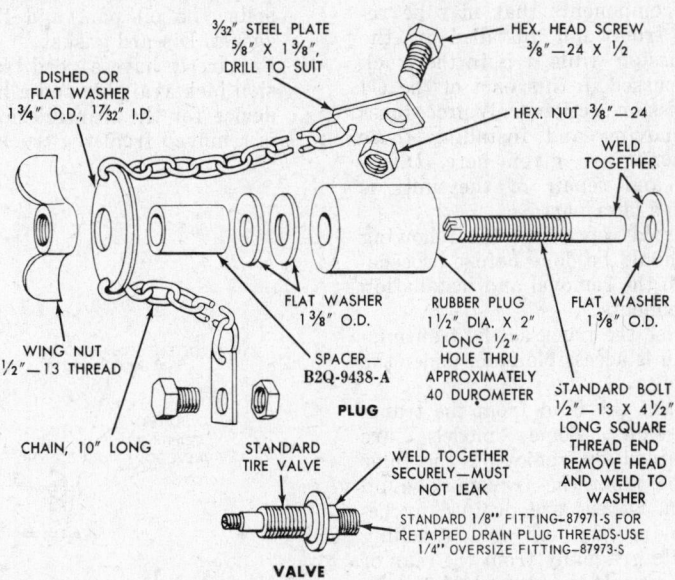

Converter leak checking tool (© Ford Motor Co.)

Automatic Transmissions

In-Vehicle Adjustments and Repairs

The adjustments and repairs presented in this part of the section on transmissions may be done without removing the entire transmission from the truck. Some of these procedures will require the use of special tools and instruments.

Band Adjustments

Intermediate Band

1. Clean all the dirt from the adjusting screw and remove and discard the locknut.
2. Install a new locknut on the adjusting screw. Using the tool shown in the illustration, tighten the adjusting screw until the wrench clicks and breaks at 10 ft-lbs. torque.
3. Back off the adjusting screw *Exactly* 1¾ turns.
4. Hold the adjusting screw steady and tighten the locknut to the proper torque.

Low-Reverse Band

1. Clean all dirt from around the band adjusting screw, and remove and discard the locknut.
2. Install a new locknut on the adjusting screw. Using the tool shown in the illustration, tighten the adjusting screw until the wrench clicks and breaks at 10 ft-lbs torque.
3. Back off the adjusting screw *Exactly 3 full turns*.
4. Hold the adjusting screw steady and tighten the locknut to the proper torque.

Transmission Component

Removal and Installation

The components that may be removed from and installed in the transmission while it is in the truck are discussed in this part of the C4 transmission section. Only procedures for removing and installing these components are given here. Disassembly and repair of the units is given in a later part.

To avoid repetition, the following tasks should be done before proceeding with the removal and installation of components.

1. Raise the truck so the transmission is accessible from under the truck.
2. Drain the fluid from the transmission. Some models are drained by removing the filler tube from the transmission oil pan. Others are drained by removing the oil pan attaching bolts gradually from the rear of the pan. If the same fluid is to be reused, filter it through a 100

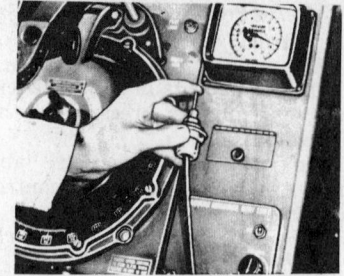

Testing transmission vacuum for leakage
(© Ford Motor Co.)

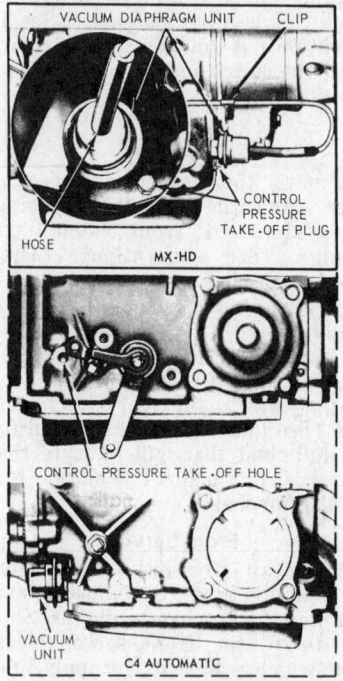

Typical vacuum diaphragm and control pressure connecting point
(© Ford Motor Co.)

mesh screen. Reuse the fluid only if it is in good condition.

3. Remove the oil pan attaching bolts, the oil pan, and the old gasket. Discard gasket.
4. Be sure to have a good transmission jack available and a holding device for the transmission if it is removed from the truck later.

Oil Pan and Control Valve

Removal

1. Do all the preliminary operations given at the beginning of this section.
2. Shift the transmission to Park position and remove the two bolts holding the manual detent spring to the control valve body and case.
3. Remove all the valve body-to-case attaching bolts. Hold the manual valve in place and remove the valve body from the case.
   CAUTION: If the manual valve is not held in place, it could be bent or damaged.
4. Refer to the Component Disassembly and repair section for control valve body repair procedures.
5. Thoroughly clean the old gasket material from the case and remove the nylon shipping plug from the oil filler tube hold. This nylon plug is installed before shipment and should be discarded when the transmission oil pan is removed.

Installation

1. Be sure the transmission is in the Park position (manual detent lever is in P detent position). Install the valve body in the case. Position the inner downshift lever between the downshift lever stop and the downshift valve. Be sure the two lands on the end of the manual valve engage the actuating pin on the manual detent lever.
2. Install seven valve body attaching bolts but do not tighten them.
3. Place the detent spring on the lower valve body and install the spring-to-case bolt finger tight.
4. While holding the detent spring roller in the center of the manual detent lever, install the detent spring-to-lower valve body

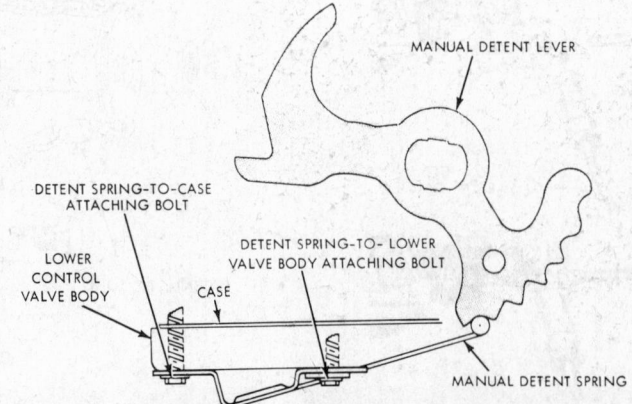

Control valve body detent spring installed (© Ford Motor Co.)

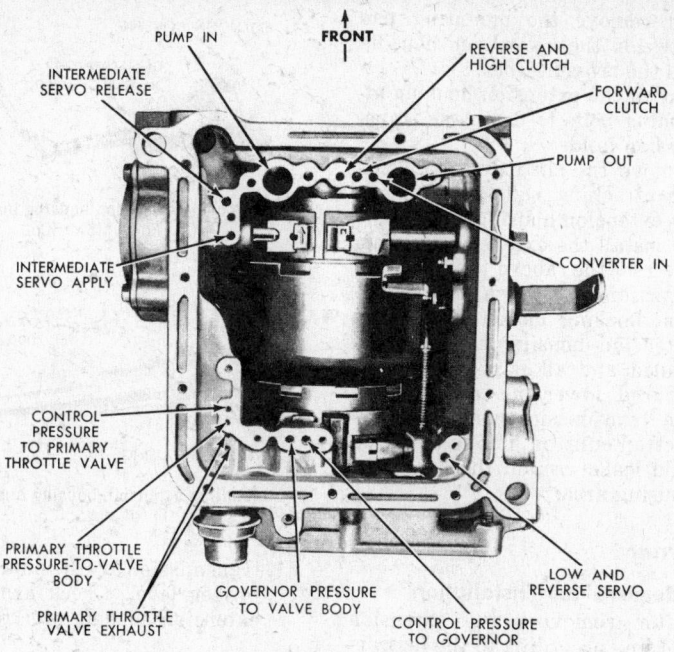

Case fluid passage hole identification (© Ford Motor Co.)

Labels: PUMP IN, FRONT, REVERSE AND HIGH CLUTCH, INTERMEDIATE SERVO RELEASE, FORWARD CLUTCH, PUMP OUT, CONVERTER IN, INTERMEDIATE SERVO APPLY, CONTROL PRESSURE TO PRIMARY THROTTLE VALVE, PRIMARY THROTTLE PRESSURE-TO-VALVE BODY, PRIMARY THROTTLE VALVE EXHAUST, GOVERNOR PRESSURE TO VALVE BODY, CONTROL PRESSURE TO GOVERNOR, LOW AND REVERSE SERVO

Adjusting intermediate band
(© Ford Motor Co.)

Adjusting low-reverse band
(© Ford Motor Co.)

bolt and tighten it to 80-120 in-lbs. torque.

5. Tighten the remainder of the control valve body attaching bolts to 80-120 in-lbs. torque.
6. Put a new gasket on the oil pan, install pan in place, and install and tighten all the pan attaching bolts to the proper torque.
7. If the filler tube was removed, reinstall it and tighten securely. If necessary, replace the oil seal around the filler tube to prevent leakage.
8. Lower the car and fill the transmission with enough fluid to bring the level up to the FULL mark on the dipstick. Check for fluid leaks at this time.

Intermediate Servo

Removal and Installation

1. Raise the truck and remove the four servo cover attaching bolts (right-hand side of case). Remove the cover and identification tag (*do not lose tag*).
2. Remove the gasket, piston, and piston return spring.
3. Install the piston return spring in the case. Place a new gasket on the cover. Install the piston and cover in the transmission case, using two 51/6-18 x 1¼ bolts 180 degrees apart to align the cover against the case.
4. Install the transmission identification tag and two attaching bolts. Remove the two 1¼ bolts and install the other two cover attaching bolts. Tighten all cover attaching bolts to the proper torque.
5. Adjust the intermediate band.

Lower the truck and fill the transmission with enough fluid to raise the fluid level to the FULL mark on the dipstick.
6. If the intermediate band cannot be adjusted correctly, remove the oil pan and control valve body and see if the struts are installed correctly. Adjust the struts and reinstall the control valve body and oil pan with a new gasket. Refill the transmission with fluid.

Low-Reverse Servo Piston

Removal and Installation

1. Raise the truck on a hoist.
2. Loosen the reverse band adjusting screw locknut and tighten

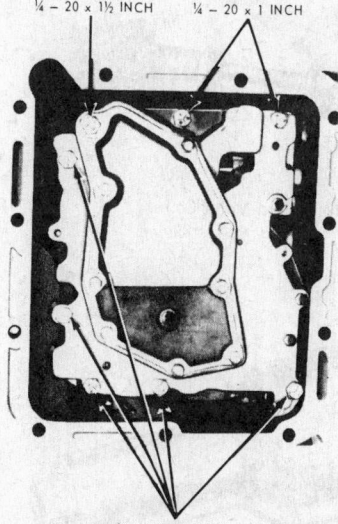

¼ – 20 x 1½ INCH   ¼ – 20 x 1 INCH

¼ – 20 x 1 INCH

Control valve body attaching bolts
(© Ford Motor Co.)

the adjusting screw to 10 ft-lbs torque. This operation will hold the band strut against the case and prevent it from falling when the reverse servo piston is removed.
3. Remove the four servo cover attaching bolts and remove the servo cover and seal from the case.
4. Remove the servo piston from the case. *The piston and piston seal are bonded together and must be replaced together.*
5. Install the servo piston assembly into the case. Place a new cover seal on the cover and position them by installing two 51/6-18 bolts, 1¼ in. long, at 180 degrees apart on the case. Install two cover attaching bolts with the identification tag.
6. Remove the two positioning bolts and install the other two cover bolts. Tighten all the cover attaching bolts to the proper torque.
7. Adjust the low-reverse band. Lower the truck and fill the transmission with enough fluid to raise the fluid level to the FULL mark on the dipstick.
8. If the low-reverse band cannot be adjusted properly, the transmission must be drained and the oil pan and valve body removed. Check the alignment of the band struts. Reinstall the valve body and the oil pan with a new gasket, and refill the transmission with fluid.

Extension Housing Bushing and Rear Seal

Removal and Installation

1. Disconnect the drive shaft from the transmission.
2. If only the rear seal needs replacing, carefully remove it with a tapered chisel or use the tools shown in the illustration. Remove the bushing as shown. Be careful not to damage the spline seal with the bushing remover.
3. Install the new bushing, using the special tool shown.
4. Before installing a new rear seal, inspect the sealing surface of the universal joint yoke for scores. If the universal joint yoke is scored, replace the yoke.
5. Inspect the housing counterbore for burrs and remove them with crocus cloth if necessary.
6. Install the new rear seal into the housing, using the tool shown in the illustration. The seal should be firmly seated in the housing. Coat the inside diameter of the fiber portion of the seal with lubricant.
7. Coat the front universal joint spline with lubricant and install the drive shaft.

Extension Housing

Removal and Installation

1. Raise the truck on a hoist.
2. Remove the drive shaft. Place a transmission jack under the transmission for support.
3. Remove the speedometer cable from the extension housing.
4. Remove the extension housing-to-crossmember mount attaching bolts. Raise the transmission and remove the mounting pad between the extension housing and the crossmember.
5. Loosen the extension housing attaching bolts to drain the transmission fluid.
6. Remove the six extension housing attaching bolts and remove the extension housing.
7. To install the extension housing, reverse the above removal instructions. Install a new extension housing gasket. When the extension housing has been installed and all parts have been secured, lower the truck and fill the transmission with the correct amount of fluid. Check for fluid leaks around the extension housing area.

Governor

Removal and Installation

1. After removing the extension housing according to the instructions above, remove the governor housing-to-governor distributor attaching bolts. Remove the governor housing from the distributor.
2. Refer to the Component Disassembly and Repair section for instructions on repairing the governor assembly.
3. Install the governor housing on the governor distributor and tighten the attaching bolts to the proper torque.
4. Install the extension housing with a new gasket according to the instructions above.
5. When the extension housing has been installed and all bolts have been tightened to the proper torque, lower the truck and fill the

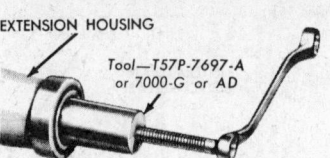

Removing extension housing bushing
(© Ford Motor Co.)

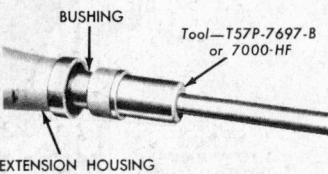

Installing extension housing bushing
(© Ford Motor Co.)

transmission with fluid to the proper level. Check around the extension housing area for leaks.

Transmission Overhaul Procedures

The transmission overhaul procedures presented here are the checks and repairs that must be done with the transmission out of the truck. Disassembly of each transmission subassembly is illustrated by exploded views of the subassembly showing how the individual parts fit together. Reassembly of the subassembly is often the reverse of the disassembly procedure except for alignment, special tolerances, etc.

Procedures for removing the transmission from the truck and reinstalling it back in the truck are given in the Truck Section.

During the transmission disassembly and reassembly operations, ten thrust washers that are installed between the subassemblies of the gear train must be removed and reinstalled correctly. Since it is very important that these thrust washers be installed correctly, they are shown in their positions and they are numbered for further identification. The No. 1 thrust washer is located at the front pump, and the No. 10 thrust washer is located at the packing pawl ring gear.

During all repairs to the transmission subassemblies, the following instructions must be followed:

1. Be sure that no dirt or grease gets in the transmission. All parts must be clean. *Remember—a little dirt can disable a transmission completely if it gets in a fluid passage.*
2. Handle all transmission parts carefully to avoid burring or nicking bearing or mating surfaces.
3. Lubricate all internal parts of

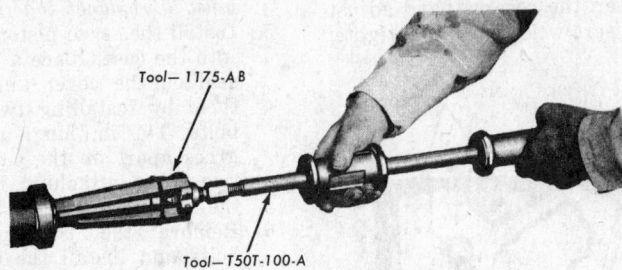

Removing extension housing seal (© Ford Motor Co.)

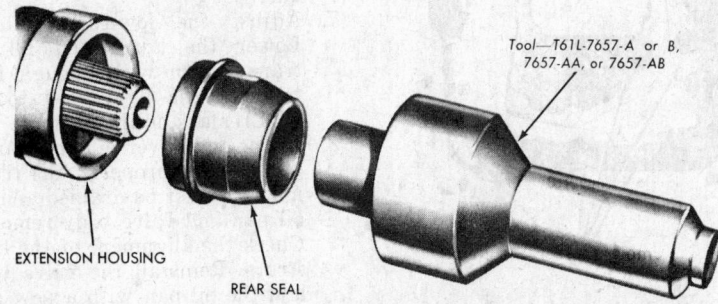

Installing extension housing seal (© Ford Motor Co.)

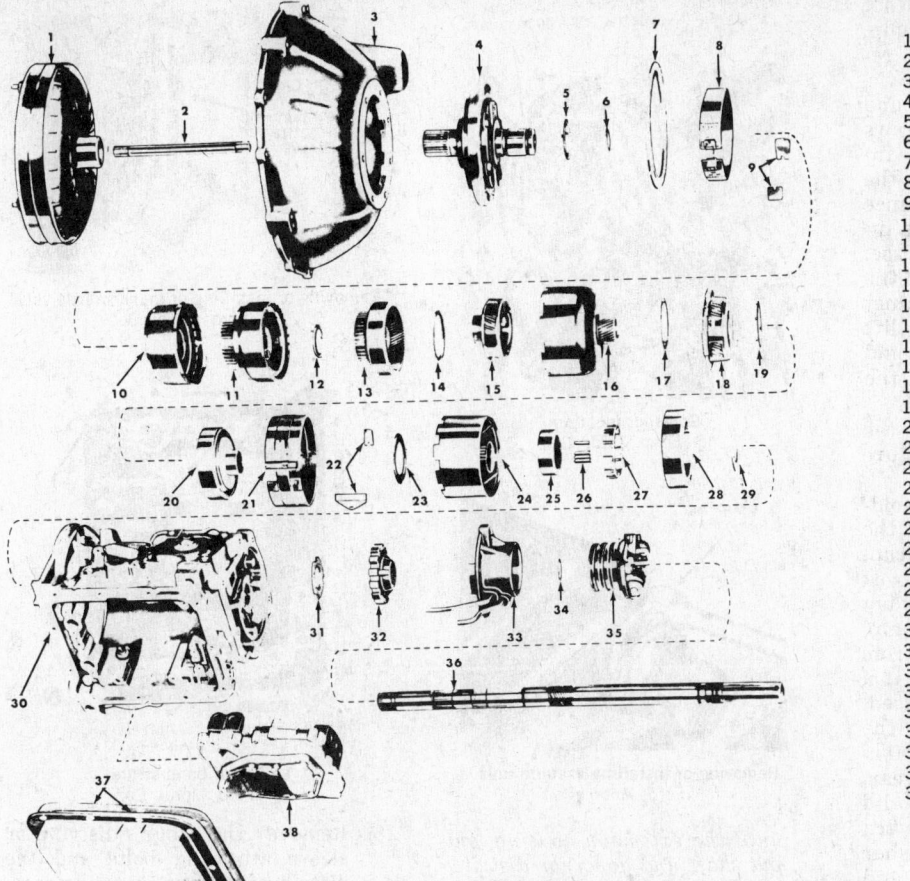

1 Converter
2 Input shaft
3 Converter housing
4 Front pump
5 Thrust washer no. 1
6 Thrust washer no. 2
7 Front pump gasket
8 Intermediate band
9 Band struts
10 Reverse and high clutch drum
11 Forward clutch and cylinder
12 Thrust washer no. 3
13 Forward clutch hub and ring gear
14 Thrust washer no. 4
15 Front planet carrier
16 Input shell and thrust washer no. 5
17 Thrust washer no. 6
18 Reverse planet carrier
19 Thrust washer no. 7
20 Reverse ring gear and hub
21 Low and reverse bands
22 Band struts
23 Thrust washer no. 8
24 Low and reverse drum
25 One-way clutch inner race
26 Roller and spring (12 each)
27 Spring and roller cage
28 One-way clutch outer race
29 Thrust washer no. 9
30 Case
31 Thrust washer no. 10
32 Parking gear
33 Governor distributor sleeve
34 Snap ring
35 Governor and distributor assy.
36 Output shaft
37 Oil pan and gasket
38 Control valve body

Transmission subassemblies—typical—C-4 (© Ford Motor Co.)

the transmission with clean transmission fluid before assembling. *Do not use any other lubricants except on gaskets or thrust washers which may be coated lightly with vaseline to ease assembly.*

4. Always use new gaskets when assembling the parts of the transmission.

5. Tighten all bolts and screws to the recommended torque limits using a torque wrench.

Transmission Disassembly

Disassemble the transmission by following the procedures below:

1. Thoroughly clean the outside of the transmission to prevent dirt or grease from getting inside the mechanism. *Do this before removing any subassembly.*

2. Place the transmission in the transmission holder. See illustration.

3. Remove the converter from the transmission front pump and converter housing.

4. On a C4 automatic transmission, remove the transmission vacuum unit with the tool shown in the illustration. Remove the vacuum unit gasket and the control rod.

5. On a C4 automatic transmission, remove the primary throttle valve from the opening at the

rear of the case.

6. Remove the transmission pan attaching bolts, oil pan, and gasket.

7. Remove the control valve body attaching bolts and then lift the control valve body from the case.

8. Loosen the intermediate band adjusting screw and remove the intermediate band struts from the case. Loosen the low-reverse band adjusting screw and remove the low-reverse band struts.

Transmission End-Play Check

1. Remove one of the converter housing attaching bolts and mount the dial indicator support tool in the hole. Mount a dial indicator on the support so that a contact rests on the end of the input shaft. See illustration.

2. Install the extension housing seal replacer tool on the output shaft to provide support and alignment for the shaft.

3. Using a screwdriver, move the input shaft and the gear train to the rear of the case as far as possible. Set the dial indicator at zero while holding a slight pressure on the screwdriver.

4. Remove the screwdriver and insert it behind the input shell. Move the input shell and the

front part of the gear train forward.

5. Record the dial reading for later reference during transmission reassembly. The end play reading should be from 0.008 to 0.042 in. If the end play reading is not within this range, selective thrust washers must be used to obtain the proper reading. The selective thrust washers to be used are listed in the table shown.

6. Remove the dial indicator, its support bar, and the extension housing seal replacer tool.

Removal of Case and Extension Parts

1. Rotate the transmission in the holding fixture until it is in a vertical position with the converter housing up.

2. Remove the five converter housing attaching bolts and remove the converter housing from the transmission case.

3. Remove the seven front pump attaching bolts. Remove the front pump by inserting a screwdriver behind the input shell and pushing it forward until the front pump seal is above the edge of the case. Remove the front pump and gasket from the case. If the selec-

tive thrust washer No. 1 did not come out with the front pump, lift it from the top of the reverse-high clutch.

4. Remove the intermediate and low-reverse adjusting screws from the case. Rotate the intermediate band to align the band ends with the clearance hole in the case. Remove the intermediate band from the case.

5. Using a screwdriver between the input shell and the rear planet carrier (see illustration), lift the input shell upward and remove the forward part of the gear train as an assembly.

6. Place the forward part of the gear train in the holding fixture shown.

7. With the gear train in the holding fixture, remove the reverse-high clutch and drum from the forward clutch. If thrust washer No. 2 did not come out with the front pump, remove the thrust washer from the forward clutch cylinder. If a selective spacer was used, remove the spacer. Remove the forward clutch from the forward clutch hub and ring gear.

8. If the thrust washer No. 3 did not come out with the forward clutch, remove the thrust washer from the forward clutch hub and lift the forward clutch hub and ring gear from the front planet carrier.

9. Remove thrust washer No. 4 and the front planet carrier from the input shell.

10. Remove the input shell, sun gear and thrust washer No. 5 from the holding fixture.

11. From inside the transmission case, remove thrust washer No. 6 from the top of the reverse planet carrier. Remove the reverse planet carrier and thrust washer No. 7 from the reverse ring gear and hub.

12. Move the output shaft forward and, with the tool shown in the illustration, remove the reverse ring gear and hub from the output shaft. Remove the thrust washer No. 8 from the low and reverse drum.

13. Remove the low-reverse band from the case. Remove the one-way clutch inner race by rotating the race clockwise as it is removed.

14. Remove the 12 one-way clutch rollers, springs and the spring retainer from the outer race. *Do not lose or damage any of the 12 springs or rollers. The outer race of the one-way clutch cannot be removed from the case*

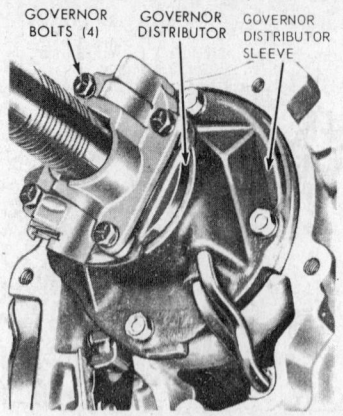

Governor location
(© Ford Motor Co.)

GOVERNOR BOLTS (4)   GOVERNOR DISTRIBUTOR   GOVERNOR DISTRIBUTOR SLEEVE

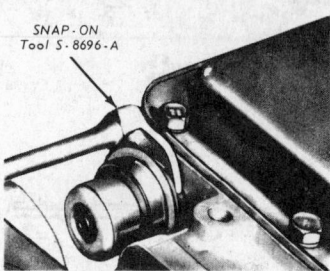

SNAP-ON Tool S-8696-A

Removing or installing vacuum unit
(© Ford Motor Co.)

until the extension housing, output shaft and governor distributor sleeve are removed.

15. Remove the transmission from the holding fixture. Place the transmission on the bench in a vertical position with the extension housing up. Remove the four extension housing attaching bolts, the extension housing, and the gasket from the case.

16. Pull outward on the output shaft and remove the output shaft and governor distributor assembly (if so equipped) from the governor distributor sleeve.

17. Remove the governor distributor lock ring from the output shaft. Remove the governor distributor from the output shaft.

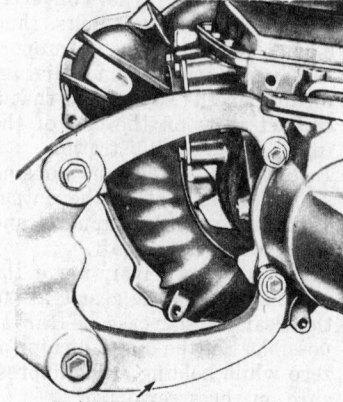

Tool—T57L-500-A or 6005-M or 6005-MS

Transmission mounted in holding fixture
(© Ford Motor Co.)

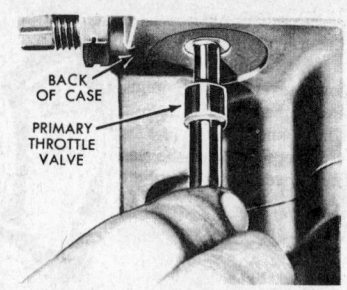

BACK OF CASE

PRIMARY THROTTLE VALVE

Removing or installing primary throttle valve
(© Ford Motor Co.)

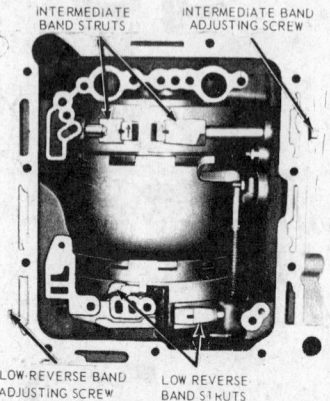

INTERMEDIATE BAND STRUTS   INTERMEDIATE BAND ADJUSTING SCREW

LOW-REVERSE BAND ADJUSTING SCREW   LOW REVERSE BAND STRUTS

Layout of band struts
(© Ford Motor Co.)

18. Remove the four distributor sleeve attaching bolts and the distributor sleeve from the case. *Do not bend or distort the fluid tubes as the tubes are removed from the case with the distributor sleeve.*

19. Remove the parking pawl return spring, pawl, and pawl retaining pin from the case.

20. Remove the parking gear and thrust washer No. 10 from the case.

21. Remove the six one-way clutch outer race attaching bolts with the tool shown. As the bolts are removed, hold the outer race that is located inside the case in position. Then, remove the outer race and thrust washer No. 9 from the case.

Component Disassembly and Assembly

Downshift and Manual Linkage

Removal

1. Loosen the outer downshift lever nut with penetrating oil. Remove the nut and the inner and outer downshift levers.
   From inside the case, remove the upper retaining ring and the flat washer from the manual lever link. Remove the upper end of the lever link from the case retaining pin.

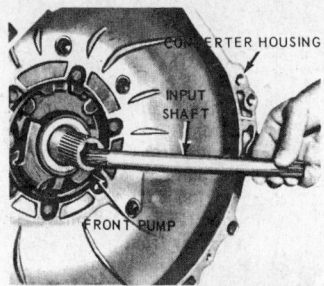

Removing or installing input shaft
(© Ford Motor Co.)

GEAR TRAIN
END PLAY
LIMITS
0.008-0.042
INCH

Tool—4201-C

Checking gear train end play
(© Ford Motor Co.)

PRY
INPUT SHELL
FORWARD

Removing front pump
(© Ford Motor Co.)

2. From the back of the transmission case, remove the upper retaining clip and flat washer from the parking pawl link. Remove the pawl link and spacer from the case retaining pin, and remove the parking pawl link, toggle rod, and the manual lever link as an assembly.

3. Remove the inner manual lever retaining nut and lever. Remove the outer manual lever from the case. Remove manual lever seal, and drive a new seal into the case with an appropriate driver. Install outer manual lever into the case. Install inner manual lever and retaining nut. Torque nut to specifications.

4. From back of the transmission case, install parking toggle rod

and link into the case. Install parking pawl link spacer onto the case retaining pin. Dimpled side of the spacer should be facing the link.

5. Install the parking pawl link onto the case retaining pin. Install flat washer and retaining ring.

6. Position inner manual lever behind the manual lever link, with the cam on the lever contacting the lower link pin.

7. Install the upper end of the manual lever link onto the case retaining pin. Install flat washer and retaining ring.

8. Operate the manual lever and check for correct linkage operation.

Control Valve Body

Removal

1. Remove the screws attaching the oil screen to the valve body and remove the oil screen. Be careful not to lose the throttle pressure limit valve and spring when separating the oil screen from the valve body.

2. Remove the attaching screws from the lower valve body and separate the upper and lower valve bodies, the gasket, separator plate and the hold-down plate. Be careful not to lose the upper valve body shuttle valve and check valve when separating the upper and lower valve bodies.

3. Slide the manual valve out of the body.

Control Valve Body

Disassembly

1. Remove the screws attaching the oil screen to the valve body and remove the oil screen. Be careful not to lose the throttle pressure limit valve and spring when separating the oil screen from the valve body.

2. Remove the attaching screws from the lower valve body and separate the upper and lower valve bodies, the gasket, separator plate and the hold-down plate. Be careful not to lose the upper valve body shuttle valve and check valve when separating the upper and lower valve bodies.

3. Slide the manual valve out of the body.

4. Pry the low servo modulator valve retainer from the body and remove the retainer plug, spring, and valve from the valve body. While working in the same bore, pry the retainer, spring, and the downshift valve from the valve body (see illustration).

INNER DOWNSHIFT
LEVER

INPUT SHELL

Lifting input shell and gear train
(© Ford Motor Co.)

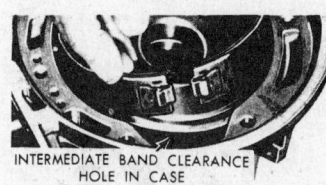

INTERMEDIATE BAND CLEARANCE
HOLE IN CASE

Position of intermediate band for removal or installation
(© Ford Motor Co.)

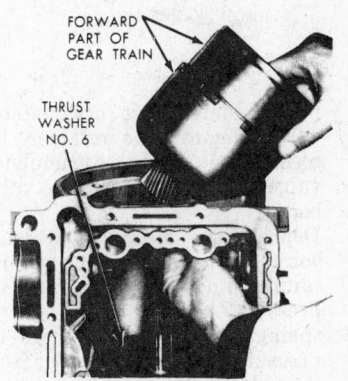

FORWARD
PART OF
GEAR TRAIN

THRUST
WASHER
NO. 6

Removing or installing forward part of gear train
(© Ford Motor Co.)

5. Depress the throttle booster plug and remove the retaining pin. Then, remove the plug, valve, and spring from the valve body.

6. Remove the cover over the cut-back valve and the transition valve from the valve body.

7. Remove the cut-back valve and the transition valve spring, transition valve, 2-3 back-out valve, and spring from the valve body.

8. Remove the cover from over the 1-2 shift valve and the 2-3 shift valve. Remove the 2-3 shift valve, spring, and the throttle modulator valve from the valve body. Remove the 1-2 shift valve, D-2 valve, and spring from the valve body.

9. Remove the retaining pin from the retainer after depressing the

1153

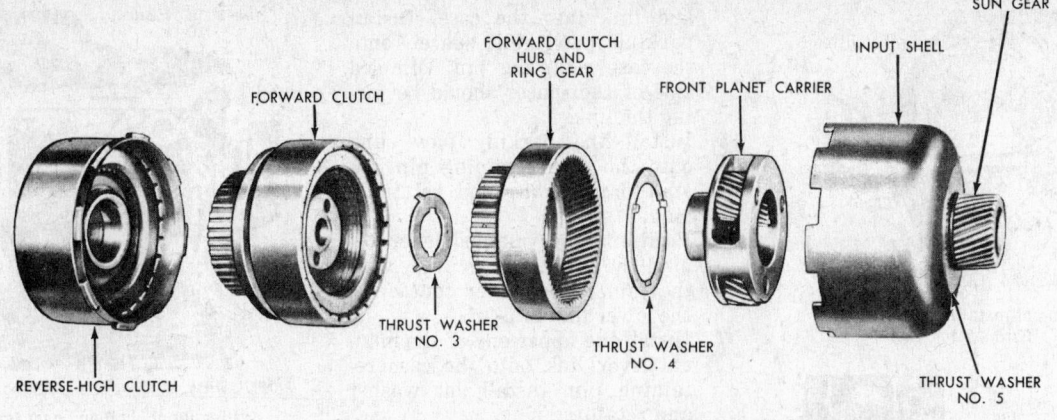

Forward part of gear train disassembled (© Ford Motor Co.)

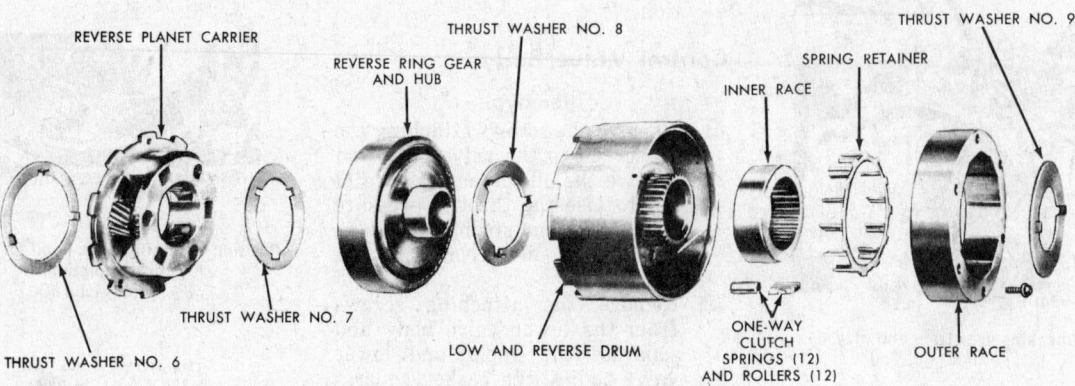

Lower part of gear train disassembled (© Ford Motor Co.)

intermediate servo accumulator valve. Remove the retainer, intermediate servo accumulator valve, and spring from the valve body.

10. Depress the main oil pressure booster valve and remove the retaining pin. Remove the main oil pressure booster valve, sleeve, springs, retainer, and main oil pressure regulator valve from the valve body.

11. Remove the line coasting boost valve retaining clip, the spring, and line coasting boost valve from the valve body.

Front Pump

Disassembly

1. Remove the four seal rings from the stator support.

2. Remove the five bolts that secure the stator support to the front pump housing. Remove stator support from pump housing.

3. Replace the stator bushings if they are worn or damaged. Use a cape chisel to cut the bushing through. Then, pry up the loose ends of the bushing with an awl and remove the bushing. Press a new bushing into the stator support.

NOTE: Ensure that the oil hole in the bushing lines up with the hole in the stator support.

4. Remove drive and driven gears from the front pump housing.

5. Replace the bushing in the pump housing with the tools shown, making sure the slot and groove are toward the rear of the body and 60° below the horizontal centerline of the pump.

6. Install the drive and driven gears in the pump housing. The chamfered side of each gear has an identification mark that must be positioned downward, against the face of the pump housing.

7. Place stator support in pump housing. Install and torque the five retaining bolts.

FORWARD GEAR TRAIN ASSEMBLY

Holding Fixture Tool—77530-A

Forward part of gear train in holding fixture (© Ford Motor Co.)

8. Install four seal rings onto the stator support. The two large oil rings are assembled first, in the oil ring grooves toward the front of the stator support. Install the O-ring seal onto the pump housing.

9. Check pump gears for free rotation by placing pump on the converter drive hub and turning pump housing.

10. If the front pump seal must be replaced, mount the pump in the transmission case and remove the seal with a seal removing tool.

Reverse-High Clutch

Disassembly and Assembly

1. Remove pressure plate retaining snap-ring.

2. Remove the pressure plate, and the drive and driven clutch plates.

CAUTION: Use no detergent or other cleaning solution on the lined clutch plates. Wipe the plates with a lint-free cloth.

3. Remove piston spring retainer by applying pressure to the clutch hub. Compress piston return springs, and remove the retainer.

4. Remove the piston return spring.

5. Remove the piston by applying

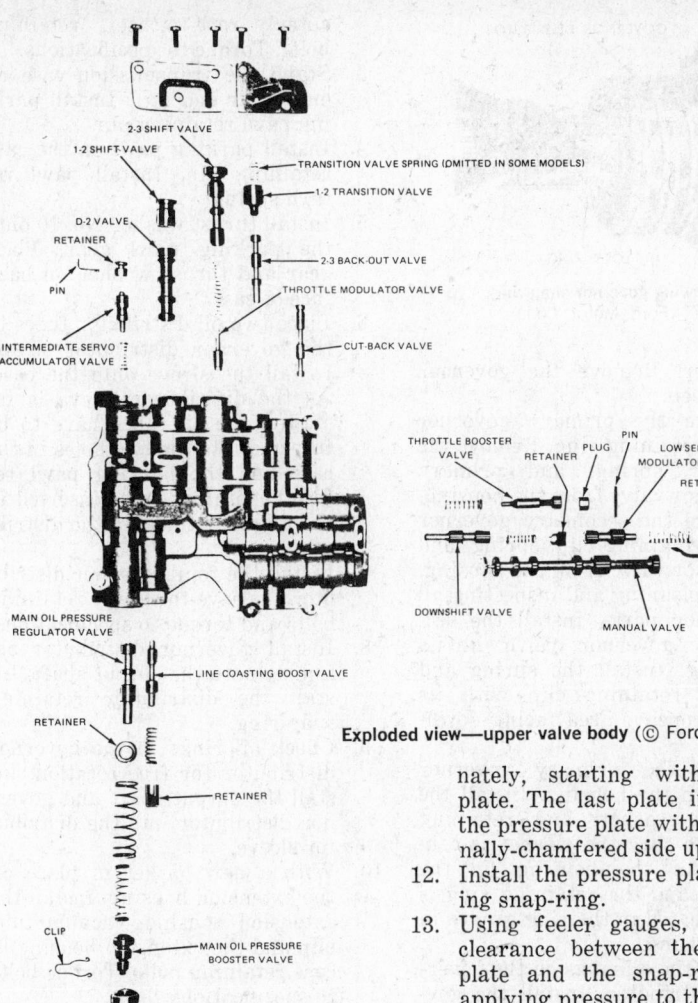

2-3 SHIFT VALVE
1-2 SHIFT VALVE
TRANSITION VALVE SPRING (OMITTED IN SOME MODELS)
1-2 TRANSITION VALVE
D-2 VALVE
RETAINER
PIN
2-3 BACK-OUT VALVE
THROTTLE MODULATOR VALVE
INTERMEDIATE SERVO
ACCUMULATOR VALVE
CUT-BACK VALVE
MAIN OIL PRESSURE
REGULATOR VALVE
LINE COASTING BOOST VALVE
RETAINER
RETAINER
CLIP
MAIN OIL PRESSURE
BOOSTER VALVE
SLEEVE

THROTTLE BOOSTER VALVE
RETAINER
PLUG
PIN
LOW SERVO MODULATOR VALVE
RETAINER PLUG
RETAINER
DOWNSHIFT VALVE
MANUAL VALVE

Exploded view—upper valve body (© Ford Motor Co.)

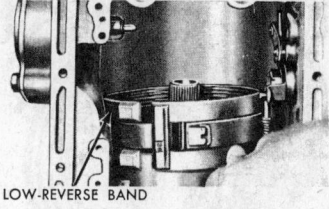

LOW-REVERSE BAND

Removing low-reverse band
(© Ford Motor Co.)

air pressure to the piston apply hole of the clutch hub.

6. Remove piston outer seal from the piston and the piston inner seal from the clutch drum.

7. Remove the drum bushing if it is worn or damaged. Use a cape chisel to cut the bushing seam until it is broken. Pry up the loose ends of the bushing with an awl and remove the bushing. *To prevent leakage at the stator support O-rings, do not nick or damage the hub surface with the chisel.*

8. Position the drum and a new bushing in a press and install the bushing with the tool shown.

9. Install a new inner seal into the clutch drum and a new outer seal onto the clutch piston. Lubricate and install the piston into the clutch drum.

10. Locate the piston return spring on the piston. Place retainer on top of the springs. Compress the assembly with a press, and install the retainer snap-ring.

11. Soak new composition plates in transmission fluid before installation. Install clutch plates, alter-

nately, starting with a steel plate. The last plate installed is the pressure plate with the internally-chamfered side up.

12. Install the pressure plate retaining snap-ring.

13. Using feeler gauges, check the clearance between the pressure plate and the snap-ring while applying pressure to the plate. If clearance is not within specifications, selective thickness snap-rings are available.

Forward Clutch

Disassembly and Assembly

1. Remove the clutch pressure plate retaining snap ring.

2. Remove the pressure plate, drive and driven plates from the hub.

3. Remove the disc spring retaining snap-ring.

4. Apply air pressure to the clutch piston pressure hole to remove the piston from the hub.

5. Remove piston outer seal and the inner seal from the clutch hub.

6. Install new piston seals onto the clutch piston and drum.

7. Lubricate and insert the piston into the clutch hub. Install the disc spring and retaining snap-ring.

8. Install the lower pressure plate, with the flat side up and the radiused side downward. Install one composition clutch plate and alternately install the drive and driven plates. Install the pressure plate with the internally chamfered side up.

9. Install pressure plate retaining snap-ring.

10. With a feeler gauge, check clearance between the snap ring and the pressure plate. Downward pressure on the plate should be used when making this check. If clearance is not within specifications, selective snap-rings are available.

Forward Clutch Hub and Ring Gear

Disassembly and Assembly

1. Remove forward clutch hub retaining snap-ring.

2. Separate clutch hub from ring gear.

3. Press the bushing from the clutch hub.

4. Install a new bushing into the clutch hub as shown.

5. Install clutch hub into ring gear.

6. Install hub retaining snap-ring.

Input Shell and Sun Gear

Disassembly and Assembly

1. Remove external snap-ring from sun gear.

Tool—
T65P-7B456-B

Removing one way clutch outer race retaining bolts
(© Ford Motor Co.)

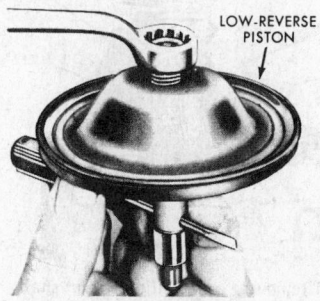

LOW-REVERSE PISTON

Low-reverse servo piston-removal
(© Ford Motor Co.)

REVERSE RING GEAR
HUB RETAINING RING

Reverse ring gear retaining ring-removal
(© Ford Motor Co.)

2. Remove thrust washer No. 5 from input shell and sun gear.
2. From inside the shell, remove the sun gear. Remove internal snap-ring from sun gear.
4. If the sun gear bushings are to be replaced use the tool shown in the illustration and press both bushings through the gear.
5. Press a new bushing into each end of the sun gear.
6. Install internal snap-ring onto sun gear. Install sun gear into the input shell.
7. Install thrust washer No. 5 onto sun gear and input shell.
8. Install external snap-ring onto sun gear.

Governor and Oil Distributor

Disassembly

1. Remove the oil rings from the governor oil distributor.
2. Remove the governor housing attaching bolts and remove the governor assembly from the dis-

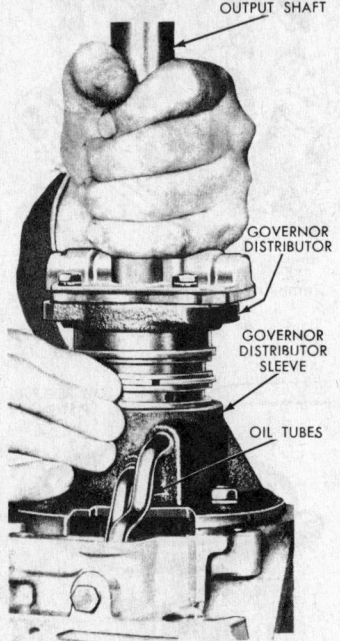

OUTPUT SHAFT

GOVERNOR DISTRIBUTOR

GOVERNOR DISTRIBUTOR SLEEVE

OIL TUBES

Removing or installing output shaft and governor distributor
(© Ford Motor Co.)

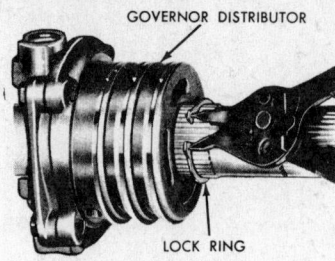

GOVERNOR DISTRIBUTOR

LOCK RING

Removing governor snap-ring
(© Ford Motor Co.)

tributor. Remove the governor oil screen.
3. Remove the primary governor valve retaining ring. Remove the washer, spring, and primary governor valve from the housing.
4. Remove the secondary governor valve retaining clip, spring, and governor valve from the housing.
5. After cleaning and inspecting all governor parts, install the secondary governor valve in its housing. Install the spring and spring retaining clip with its small concave area facing downward.
6. Install the primary governor valve in the housing. Install the spring, washer, and retaining clip. Be sure the washer is centered in the housing on the spring and the retaining ring is fully seated in the ring groove in the housing.
7. Install the oil rings on the governor distributor. Install the governor oil screen and mount the governor assembly on the distributor, tightening the attaching bolts to the proper torque.

Transmission Assembly

1. Install thrust washer No. 9 inside the transmission case.
2. Place the one-way clutch outer race inside the case. From the rear of the case, install the six

outer race-to-case retaining bolts. Torque to specifications.
3. Stand the transmission case on end (rear end up). Install parking pawl retaining pin.
4. Install parking pawl on the case retaining pin. Install pawl return spring.
5. Install thrust washer No. 10 onto the parking pawl gear. Place gear and thrust washer on back face of case.
6. Place two oil distributor tubes in the governor distributor sleeve. Install the sleeve onto the case. As the distributor sleeve is installed, the oil tubes have to be inserted into the two holes in the case and the parking pawl retaining pin has to be inserted in the alignment hole in the distributor.
7. Install the four governor distributor sleeve-to-case retaining bolts and torque to specifications.
8. Install governor distributor assembly onto the output shaft. Install the distributor retaining snap-ring.
9. Check oil rings in the governor distributor for free rotation. Install the output shaft and governor distributor into the distributor sleeve.
10. With a new gasket in place on the extension housing, install the extension housing, vacuum tube clip and the extension housing to case retaining bolts. Torque bolts to specifications.
11. Rotate transmission case so that front end is up, making sure that thrust washer No. 9 is in position at the bottom of the case.
12. Install the 12 one-way clutch springs onto the spring retainer.
13. Place the one-way clutch spring retainer, with springs installed, into the outer race, located inside the transmission.
14. Install the inner race inside the

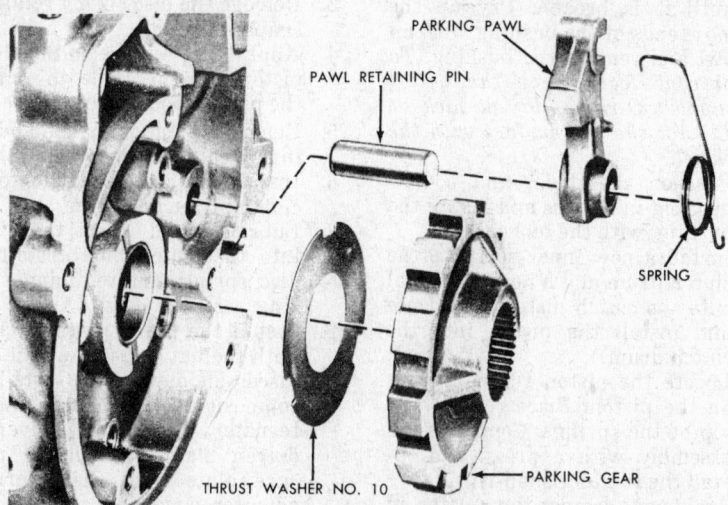

PARKING PAWL

PAWL RETAINING PIN

SPRING

THRUST WASHER NO. 10

PARKING GEAR

Parking pawl, disassembled (© Ford Motor Co.)

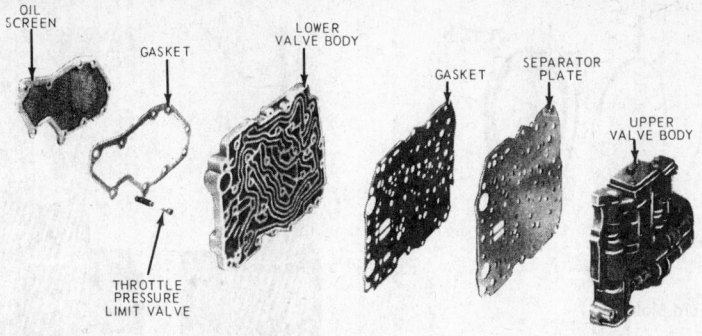

Upper & lower valve bodies C-4 (© Ford Motor Co.)

OIL SCREEN — GASKET — LOWER VALVE BODY — GASKET — SEPARATOR PLATE — UPPER VALVE BODY — THROTTLE PRESSURE LIMIT VALVE

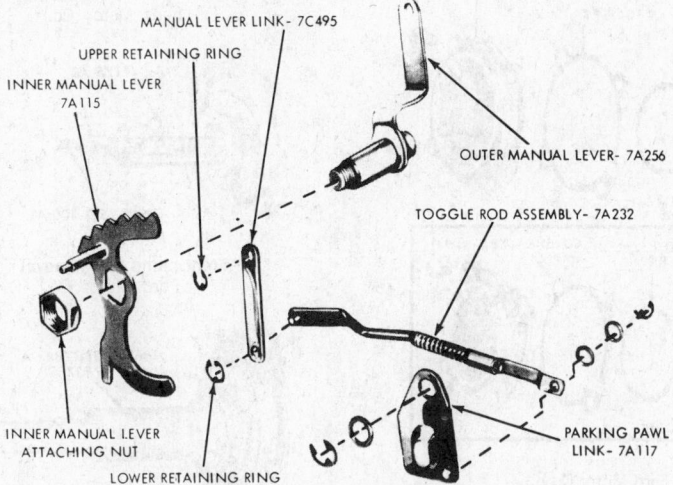

Transmission case internal linkage (© Ford Motor Co.)

MANUAL LEVER LINK- 7C495 — UPPER RETAINING RING — INNER MANUAL LEVER 7A115 — OUTER MANUAL LEVER- 7A256 — TOGGLE ROD ASSEMBLY- 7A232 — INNER MANUAL LEVER ATTACHING NUT — LOWER RETAINING RING — PARKING PAWL LINK- 7A117

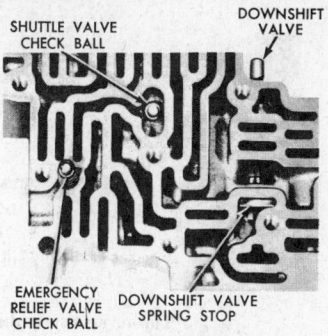

Upper valve body check ball and spring location (© Ford Motor Co.)

SHUTTLE VALVE CHECK BALL — DOWNSHIFT VALVE — EMERGENCY RELIEF VALVE CHECK BALL — DOWNSHIFT VALVE SPRING STOP

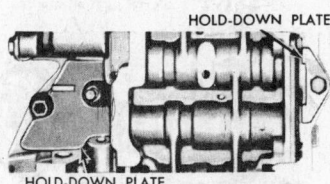

Hold down plate location (© Ford Motor Co.)

HOLD-DOWN PLATE — HOLD-DOWN PLATE

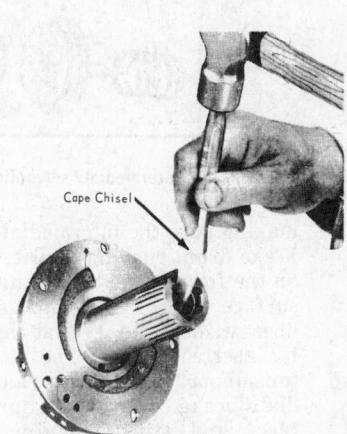

Removing stator support bushings (© Ford Motor Co.)

Cape Chisel

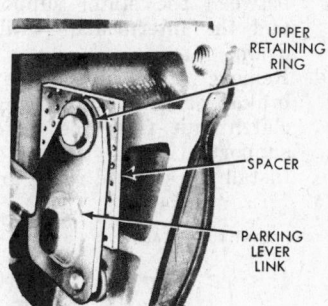

Parking lever pawl link and spacer (© Ford Motor Co.)

UPPER RETAINING RING — SPACER — PARKING LEVER LINK

spring retainer and 12 springs.

15. Starting at the back of the one-way clutch outer race, install the 12 clutch rollers.

16. After the clutch has been assembled, rotate the inner race clockwise to center the rollers and springs. Install low and reverse drum. The splines of the drum must engage the splines of the one-way clutch inner race. Check the clutch operation by rotating the low and reverse drum. The drum should rotate clockwise but not counterclockwise.

17. Install thrust washer No. 8 on top of the low and reverse drum. Install the low and reverse band into the case, with the small strut end facing the low-reverse servo.

18. Install the reverse ring gear and hub onto the output shaft.

19. Move the output shaft forward and install the reverse ring gear hub-to-output shaft retaining ring.

20. Place thrust washers Nos. 6 and 7 on the reverse planet carrier. Install planet carrier into the reverse ring gear and engage the tabs of the carrier with the slots in the low-reverse drum.

21. From inside the transmission case, install the inner downshift lever.

22. Install the forward clutch into the reverse-high clutch by meshing the reverse-high clutch plates with the splines of the forward clutch. Using the end-play check reading obtained during the transmission disassembly, determine which No. 2 thrust washer is necessary to get the proper end-play reading and proceed as follows:

a. Place the stator support vertically on the bench and install the correct No. 2 thrust washer or washer and spacer as required to bring the end-play within the correct range.

b. Install the reverse-high clutch and the forward clutch on the stator support.

c. Invert the complete assembly

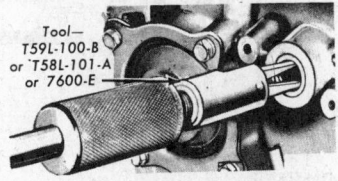

Removing manual lever seal (© Ford Motor Co.)

Tool— T59L-100-B or T58L-101-A or 7600-E

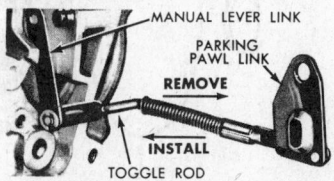

Parking pawl toggle rod (© Ford Motor Co.)

MANUAL LEVER LINK — PARKING PAWL LINK — REMOVE — INSTALL — TOGGLE ROD

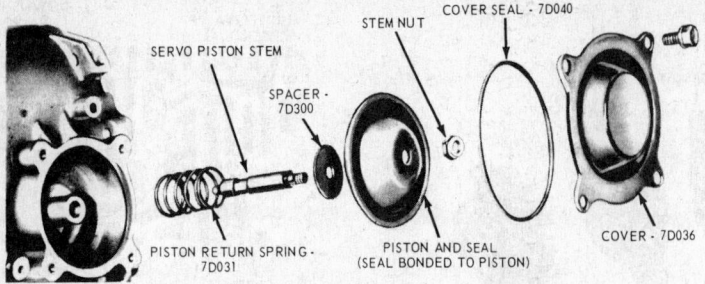

Low-reverse servo disassembled (© Ford Motor Co.)

V-8 ENGINE

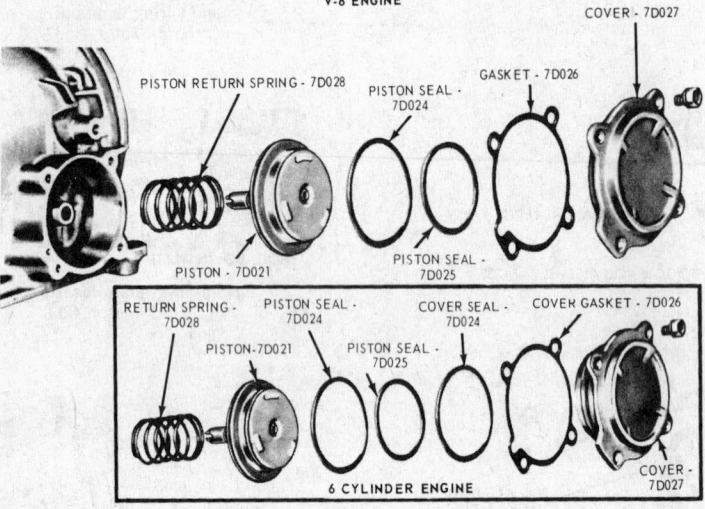

6 CYLINDER ENGINE

Intermediate servo-disassembled (© Ford Motor Co.)

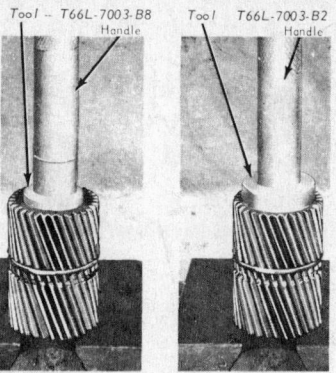

REMOVAL INSTALLATION

Replacing sun gear bushing
(© Ford Motor Co.)

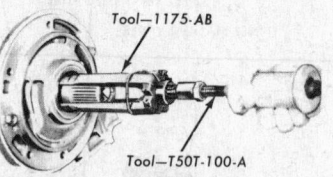

Front pump seal removal
(© Ford Motor Co.)

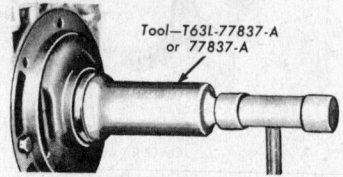

Installing front pump seal
(© Ford Motor Co.)

making sure the intermediate brake drum bushing is seated on the forward clutch mating surface. Select the thickest fiber washer (No. 1) that can be inserted between the stator support and the intermediate brake drum thrust surfaces and still maintain a slight clearance. Do not select a washer that must be forced between the stator support and the intermediate brake drum.

d. Remove the intermediate brake drum and forward clutch unit from the stator support.

e. Install the selected No. 1 and

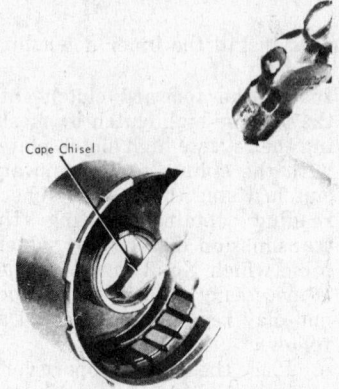

Cape Chisel

Removing reverse-high clutch bushing
(© Ford Motor Co.)

No. 2 thrust washers on the front pump stator support using vaseline to hold the thrust washers in place while installing the front pump.

23. Install thrust washer No. 3 onto the forward clutch.

24. Install forward clutch hub and ring in the forward clutch by rotating the units to mesh the forward clutch plates with the splines on the forward clutch hub.

25. Install thrust washer No. 4 on the front planet carrier, and install the planet carrier into the

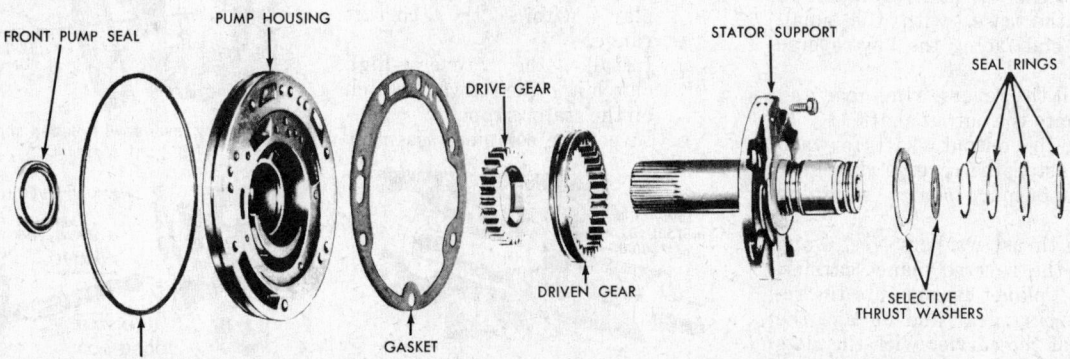

Typical front pump and stator support (© Ford Motor Co.)

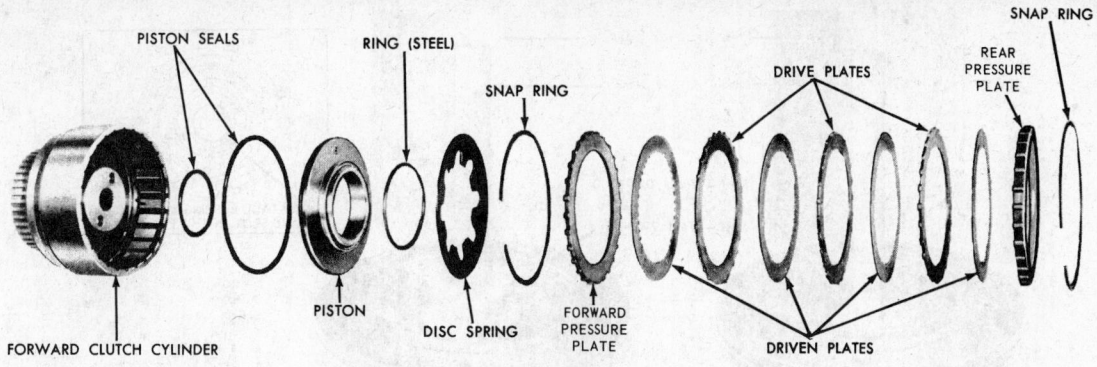

Forward clutch-disassembled (© Ford Motor Co.)

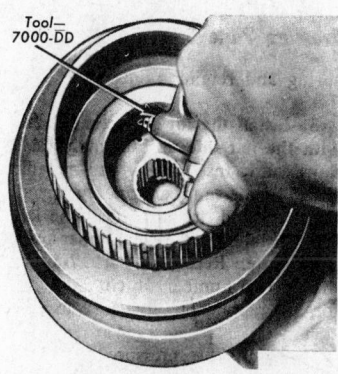

Removing forward clutch piston
(© Ford Motor Co.)

Removing or installing clutch piston retainer snap-ring (© Ford Motor Co.)

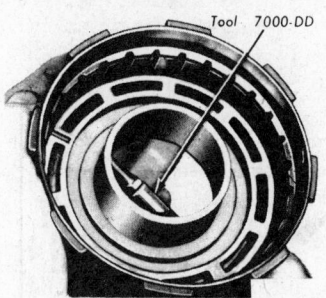

Removing reverse-high clutch piston
(© Ford Motor Co.)

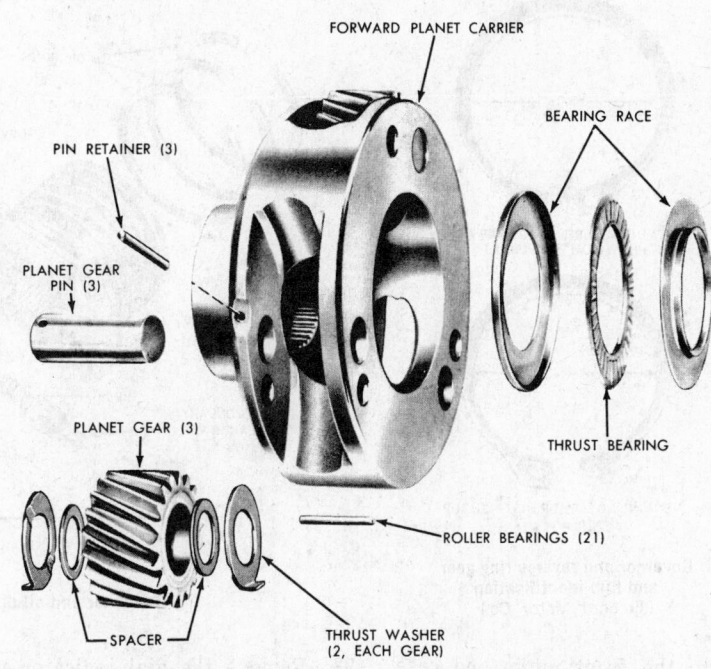

Forward planet carrier-disassembled (© Ford Motor Co.)

forward clutch hub and ring gear.

26. Install input shell and sun gear onto the gear train. Rotate the input shell to engage the drive lugs of the reverse-high clutch. If the drive lugs will not engage, the outer race inside the forward planet carrier is not centered in the end of the sun gear inside the input shell. Center the thrust bearing race and install the input shell.

27. Hold the gear train together and install the forward part of the gear train assembly into the case. The input shell sun gear must mesh with the reverse pinion gears. The front planet carrier internal splines must mesh with the splines of the output shaft.

28. Install intermediate band

through front of case. The side of the band with the anchor tabs faces the back of the transmission. If using a new band, soak it in transmission fluid prior to installation.

29. Install a new front pump gasket onto the case. Install selective thrust washers No. 1 and 2 onto the front pump stator support.

Use vaseline to hold the washers in place.

30. Lubricate the front pump O-ring with transmission fluid, and install the front pump stator support into the reverse-high clutch. Align the pump to the case and install the front pump to case retaining bolts.

31. Install the converter housing

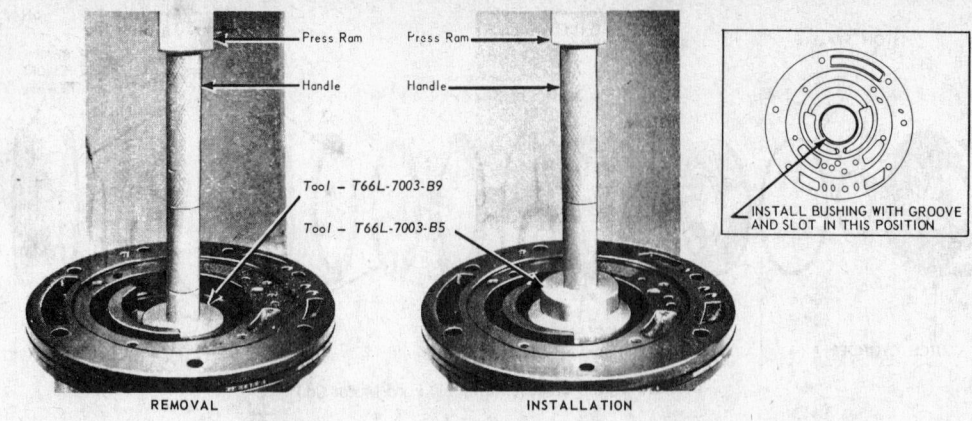

Removing front pump housing bushing (© Ford Motor Co.)

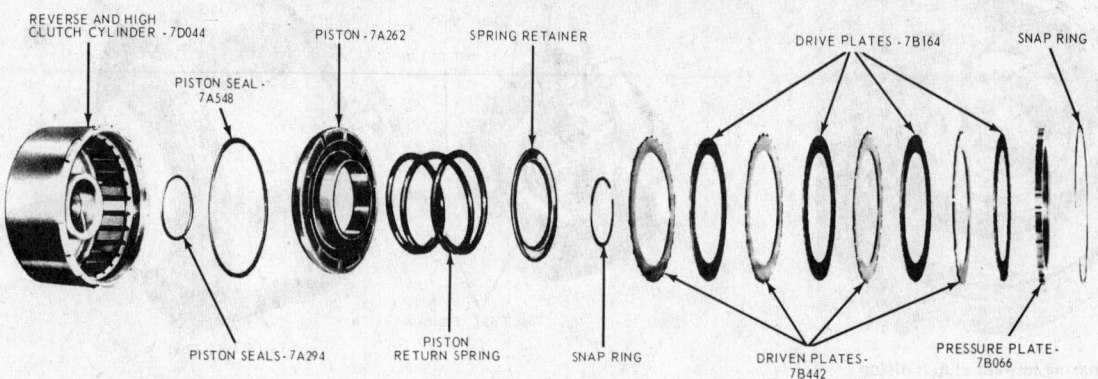

Reverse-high clutch   C-4 (© Ford Motor Co.)

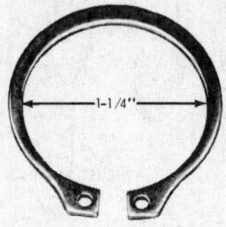

RETAINS REVERSE RING GEAR AND
HUB TO OUTPUT SHAFT

RETAINS GOVERNOR DISTRIBUTOR
TO OUTPUT SHAFT

**Governor and reverse ring gear
and hub identification
(© Ford Motor Co.)**

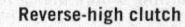

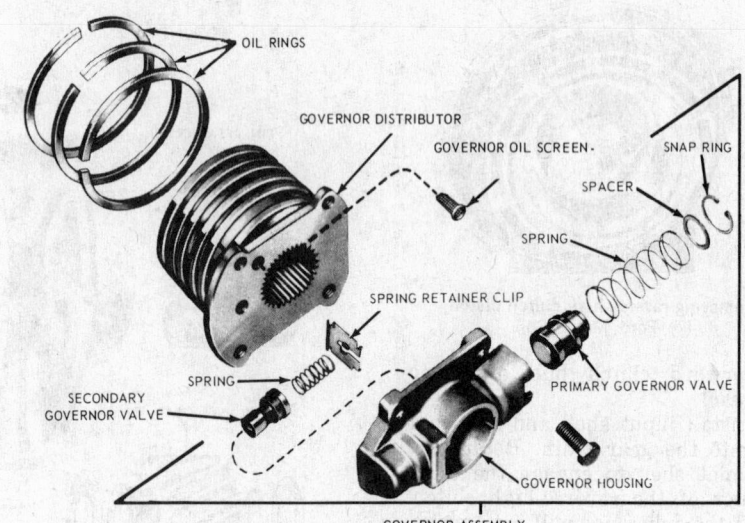

Governor and oil distributor (© Ford Motor Co.)

onto the front pump and case. Install the six converter housing to case retaining bolts. Torque bolts to specifications.

32. Install input shaft. Place transmission in horizontal position, then check transmission end play. If end-play is not within limits, either the wrong selective thrust washers were used, or one of the 10 thrust washers is improperly positioned.

33. Remove the dial indicator used for checking end play and install the one converter housing to case retaining bolt. Torque the bolt to specifications.

34. Install the intermediate and low-reverse band adjusting screws into the case. Install the struts for each band.

35. Adjust intermediate and low-reverse band.

36. Install a universal joint yoke

onto the output shaft. Rotate the input and output shafts in both directions to check for free rotation of the gear train.

37. Install control valve body. As the valve body is installed, engage the manual and down-shift valves with the inner control levers. Torque the eight control valve body-to-case bolts to specifications.

38. With a new oil pan gasket in place, install the oil pan and

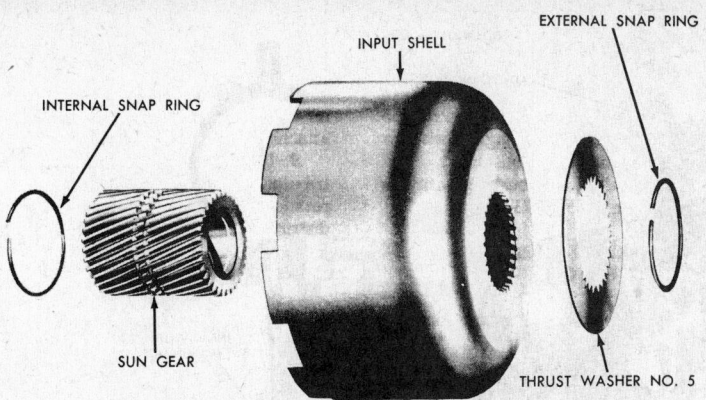

Input shell and sun gear disassembly (© Ford Motor Co)

INTERNAL SNAP RING
INPUT SHELL
EXTERNAL SNAP RING
SUN GEAR
THRUST WASHER NO. 5

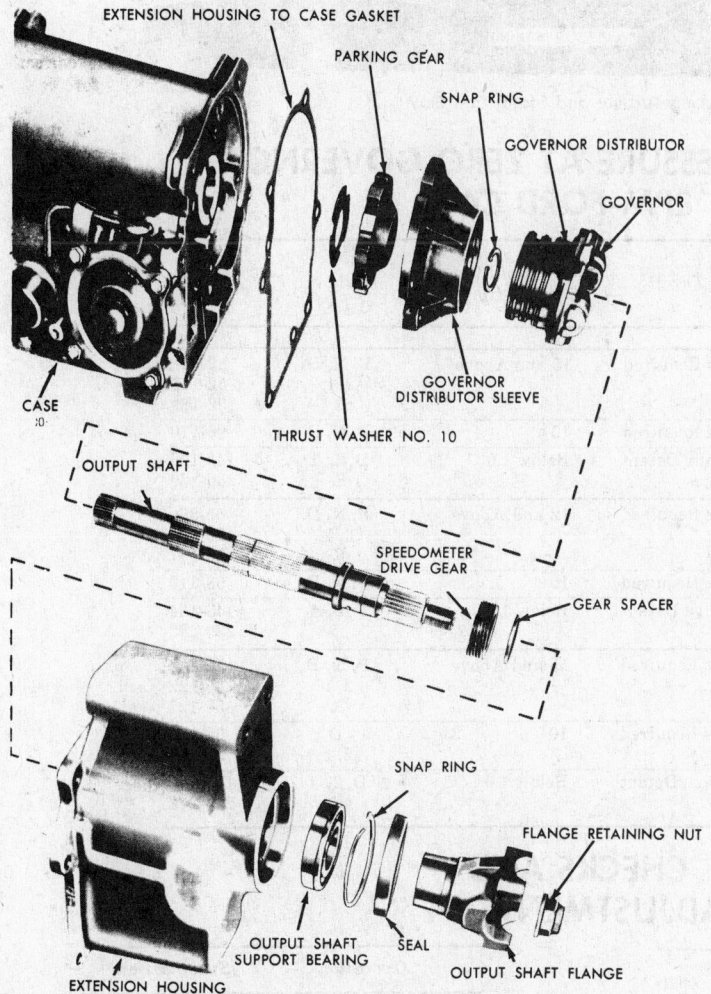

Extension housing and output shaft related parts (© Ford Motor Co)

EXTENSION HOUSING TO CASE GASKET
PARKING GEAR
SNAP RING
GOVERNOR DISTRIBUTOR
GOVERNOR
GOVERNOR DISTRIBUTOR SLEEVE
CASE
THRUST WASHER NO. 10
OUTPUT SHAFT
SPEEDOMETER DRIVE GEAR
GEAR SPACER
SNAP RING
FLANGE RETAINING NUT
OUTPUT SHAFT SUPPORT BEARING
SEAL
OUTPUT SHAFT FLANGE
EXTENSION HOUSING

torque the bolts to specifications.

39. Install primary throttle valve into the transmission case.

40. Install vacuum unit, gasket and control rod into the case.

41. Make sure the input shaft is properly installed in the front pump, stator support and gear train. Install the converter into the front pump and the converter housing.

FORD C4 TORQUE SPECIFICATIONS

	Foot Pounds
Converter to flywheel	20-30
Converter housing to transmission case	28-40
Front pump to transmission case	28-40
Overrunning clutch face to case	13-20
Oil pan to case	12-16
Rear servo cover to case	12-20
Stator support to pump	12-20
Converter cover to converter housing	12-16
Intermediate servo cover to case	16-22
Diaphragm assembly to case	15-23
Extension assembly to transmission case	28-40
Engine to transmission	23-33
Starter attaching bolts	20-30
Engine support-to-crossmember	40-60
Pressure gauge tap	6-12
Band adjusting screw locknut to case	35-45
Yoke to output shaft	60-120
Reverse servo piston to rod	12-20
Converter drain plug	20-30
Manual valve inner lever to shaft	30-40
Downshift lever to shaft	12-16
Filler tube to engine	23-33
Filler tube to pan	32-42
Transmission to engine	40-50
Distributor sleeve to case	12-20
T.R.S. switch to case	4-8
Transfer case to transmission	21-30
Crossmember attaching bolts	14-24

	Inch Pounds
Lower to upper valve body	40-55
Reinforcement plate to body	40-55
Screen and lower to upper valve body	40-55
Neutral switch to case (Econoline)	55-75
Neutral Switch to column	20
Seperator plate to lower valve body	40-55
Control Assembly to case	80-120
Governor assembly to collector body	80-120
Cooler line fittings	80-120
Detent spring to lower valve body	80-120
Upper valve body to lower valve body	80-120
Oil tube connector	80-120
End plates to body	20-35

SELECTIVE THRUST WASHERS

Thrust washer No. 1			Thrust Washer No. 2
Nylon Thrust Washer W/Tangs	Color of Washer	No. Stamped On Washer	Metal Thrust Washer
.053-.0575	Red	1	.041-.043
.070-.074	Green	2	.056-.058
.087-.091	Natural (White)	3	.073-.075

Thrust washer No. 1		Thrust Washer No. 2	
.104-.108	Black	Spacer	.032-.036①
.121-.125	Yellow		

① This is a selective spacer. The spacer must be installed next to the stator support to obtain correct end play.

Checking Turbine and Stator End Play

CONTROL PRESSURE AT ZERO GOVERNOR RPM-FORD C4

Year	Engine Speed	Throttle	Manifold Vac. Ins. Hg.	Range	P.S.I.
1971-72	As Required	As Required	15 and Above	P, N, D 2, 1 R	52-85 52-115 52-180
	As Required	As Required	10	P, N, D	96-110
	Stall	Thur Detent	Below 1.0	D, 2, 1 R	143-160 230-260
1973	As Required	As Required	12 and Above	P, N, D 2, 1 R	55-86 55-122 55-197
	As Required	As Required	10	P; N, D	98-110
	Stall	Thru Detent	Below 1.0	D, 2, 1 R	143-164 239-272
1974-78	Idle	As Required	15 and Above	P, N, D 2, 1 R	55-86 55-122 55-197
	As Required	As Required	10	D 2, 1	98-110 90-115
	Stall	Thru Detent	Below 1.0	D, 2, 1 R	143-164 239-272

CHECKS AND ADJUSTMENTS C4

Operation	Specification
Transmission end play	0.008-0.042 inch (selective thrust washers available)
Turbine and stator end play	Model PEE, PEA, PEF—New or rebuilt 0.023 max. Used 0.C40 max.
Intermediate band adjustment	Remove and discard lock nut. Adjust screw to 10 ft-lbs torque, then back off 1¾ turns. Install new lock nut and torque to specification.

Operation	Specification
Low-reverse band adjustment	Remove and discard lock nut. Adjust screw to 10 ft-lbs torque, then back off 3 turns. Install new lock nut and torque to specification.
Forward clutch pressure plate to snap ring clearance	0.025-0.050 Selective snap ring thicknesses 0.082-0.078, 0.068-0.064, 0.054-0.050
Reverse-high clutch pressure plate to snap ring clearance	0.050-0.071 Selective snap ring thicknesses 0.096-0.092, 0.082-0.078, 0.068-0.064, 0.054-0.050

Ford C-6
DIAGNOSIS

This diagnosis guide covers the most common symptoms and is an aid to careful diagnosis. The items to check are listed in the sequence to be followed for quickest results. Thus, follow the checks in the order given for the particular transmission type

TROUBLE SYMPTOMS	ITEMS TO CHECK	
	IN TRUCK	OUT OF TRUCK
No drive in D, 2 and 1	CWER	ac
Rough initial engagement in D or 2	KBWFE	a
1-2 or 2-3 shift points incorrect or erratic	ABLCDWER	
Rough 1-2 upshifts	BJGWEF	
Rough 2-3 shifts	BJWFGER	br
Dragged out 1-2 shift	ABJWGEFR	c
Engine overspeeds on 2-3 shift	CABJWEFG	br
No 1-2 or 2-3 shift	CLBDWEGJ	bc
No 3-1 shift in D	DE	
No forced downshifts	LEB	
Runaway engine on forced 3-2 downshift	WJGFEB	c
Rough 3-2 or 3-1 shift at closed throttle	KBJEF	
Shifts 1-3 in D	GJBEDR	
No engine braking in first gear—1 range	CHEDR	
Creeps excessively	K	
Slips or chatters in first gear, D	ABWFE	aci
Slips or chatters in second gear	ABJGWFER	ac
Slips or chatters in R	ABHWFER	bcr
No drive in D only	CWE	i
No drive in 2 only	ACWJER	c
No drive in 1 only	ACWER	c
No drive in R only	ACHWER	bcr
No drive in any selector lever position	ACWFER	cd
Lockup in D only		gc
Lockup in 2 only	H	bgci
Lockup in 1 only		gc
Lockup in R only		agc
Parking lock binds or does not hold	C	g
Transmission overheats	OFBW	ns
Maximum speed too low, poor acceleration	YZ	n
Transmission noisy in N and P	AF	d
Transmission noisy in first, second, third or reverse gear	AF	hadi
Fluid leak	AMNOPQS UXBJ	jmp
Car moves forward in N	C	a

Key to Checks

Fluid level	N.	Oil pan gasket, filler tube or seal		Reverse-high clutch	F.
Vacuum diaphragm unit or tubes restricted—leaking—adjustment	O.		b.	Leakage in hydraulic system	
		Oil cooler and connections	c.	Front pump	L.
Manual linkage	P.	Manual or downshift lever shaft seal	d.	Parking linkage	B.
Governor	Q.		g.	Planetary assembly	G.
Valve body	R.	⅛ inch pipe plugs in case	h.	Planetary one-way clutch	K.
Pressure regulator	S.	Perform air pressure check	i.	Engine rear oil seal	C.
Intermediate band	U.	Extension housing-to-case gasket	j.	Front pump oil seal	H.
Low-reverse clutch	W.	Extension housing rear oil seal	m.	Converter one-way clutch	M.
Intermediate servo	X.	Perform control pressure check	n.	Front pump to case gasket or seal	E.
Engine idle speed	Y.	Speedometer driven gear adapter seal	p.		A.
Downshift linkage—including inner level position	Z.	Engine performance	r.	Reverse-high clutch piston air bleed valve	
Converter drain plugs	a.	Vehicle brakes	s.	Converter pressure check valves	J.
		Forward clutch			D.

General Information

The C6 automatic transmission is very similar to the C4 automatic transmission, and most of the maintenance and overhaul procedures given for the C4 transmission will apply to the C6 transmission. One important difference between the C6 and the C4 transmissions is that the C6 transmission uses a low-reverse clutch in place of the low-reverse band. Otherwise, the gear trains are the same as are the clutch combinations. The hydraulic control systems are very similar, except for minor differences in design. All components which are different from the C4 components are illustrated and the procedures to repair them are given.

Transmission Checks

Vacuum Diaphragm Check (Off Truck)

With the use of a vacuum pump, set the vacuum to read 18 inches, with the end of the hose blocked off. On single area diaphragms, connect the vacuum hose to the manifold vacuum hose port. On the dual area diaphragms, connect the vacuum hose to the EGR port, leaving the manifold vacuum port open to the atmosphere. If the gauge holds at 18 inches, the vacuum unit diaphragm is not leaking. A second check can be made by holding the control rod end by a finger, with the other end inserted into the unit. Install the vacuum hose from the vacuum pump onto the diaphragm port. The control rod should move inward and when the vacuum supply is cut off by the removal of the hose, the rod should move outward.

Vacuum Diaphragm

A screw is provided in the inlet tube of the vacuum diaphragm assembly, to permit small adjustments in control pressure. If control pressure is uniformly high or low, in all ranges, it may be brought within specifications by turning this screw. Control pressure may also be varied to alter shift feel, but in no case should it go beyond the specified minimum or maximum.

Control pressure is increased by turning the adjusting screw clockwise, and reduced by turning counterclockwise. One full turn will change control pressure approximately 2-3 psi.

Adjustments

Intermediate Band

1. Raise the truck on a hoist or place on jack stands.
2. Clean threads of the intermediate band adjusting screw.
3. Loosen adjustment screw locknut.
4. Tighten the adjusting screw to 10 ft. lbs., and back the screw off exactly 1½ turns. Tighten the adjusting screw locknut.

Transmission Disassembly

1. Thoroughly clean outside of transmission.

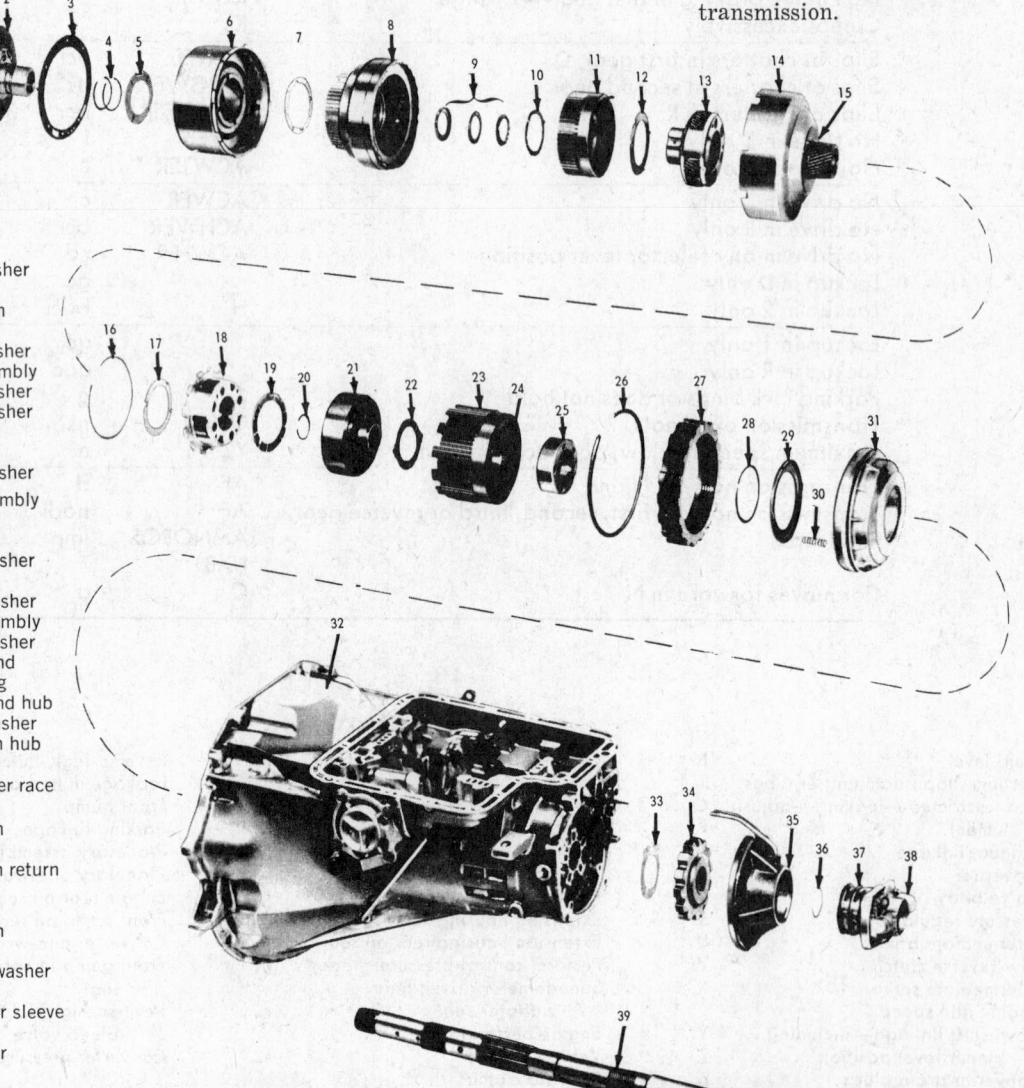

1 Front pump seal ring
2 Front pump
3 Gasket
4 Seal
5 Number 1 thrust washer (selective)
6 Reverse—high clutch assembly
7 Number 2 thrust washer
8 Forward clutch assembly
9 Number 3 thrust washer
10 Number 4 thrust washer
11 Forward clutch hub assembly
12 Number 5 thrust washer
13 Forward planet assembly
14 Input shell and sun gear assembly
15 Number 6 thrust washer
16 Snap ring
17 Number 7 thrust washer
18 Reverse planet assembly
19 Number 8 thrust washer
20 Reverse ring gear and hub retaining ring
21 Reverse ring gear and hub
22 Number 9 thrust washer
23 Low—reverse clutch hub
24 One-way clutch
25 One-way clutch inner race
26 Snap ring
27 Low—reverse clutch
28 Snap ring
29 Low—reverse piston return spring retainer
30 Return spring
31 Low—reverse piston
32 Case
33 Number 10 thrust washer
34 Parking gear
35 Governor distributor sleeve
36 Snap ring
37 Governor distributor
38 Governor
39 Output shaft

Drive train disassembled (© Ford Motor Co.)

2. Secure the unit in a repair stand, and drain the oil.
3. Remove converter from the unit.
4. Unbolt and remove the valve body from the case.
5. Check and record gear train end-play.
6. Slip the input shaft out of the front pump. Remove the vacuum diaphragm, rod and primary throttle valve from the case.
7. Remove the front pump attaching bolts. Pry the gear train forward to remove the pump.
8. Loosen the band adjusting screw and remove the two struts. Rotate the band 90° counterclockwise, to align the ends with the slot in the case. Slide the band off the reverse-high clutch drum.
9. Remove the forward part of the gear train as an assembly. Remove the large snap-ring that holds the reverse planet carrier in the low-reverse clutch hub. Lift the planet carrier from the drum.
10. Remove the snap-ring that holds the reverse ring gear and hub on the output shaft. Slide the ring gear and hub from the shaft.
11. Rotate the low-reverse clutch hub clockwise, and withdraw it from the case.
12. Remove the low-reverse snap-ring from the case, then remove the clutch discs, plates and pressure plate from the case.
13. Remove the extension housing bolts and vent tube from the case. Remove the extension housing and gasket.
14. Slide the output shaft assembly from the transmission case.
15. Remove the distributor sleeve attaching bolts and remove the sleeve, parking gear, and the thrust washer.
16. Compress the low-reverse clutch piston release spring. Remove the snap-ring. Remove the tool and spring retainer.
17. Remove the one-way clutch inner race attaching bolts from the rear of the case. Remove the

inner race from the inside of the case.
18. Remove the low-reverse clutch piston from the case.

Component Disassembly and Assembly

NOTE: For most component services, see Component Disassembly and Assembly, in the C4 Section. The exceptions are as follows.

One-Way Clutch

1. Remove the snap-ring and rear bushing from the rear of the low-reverse clutch hub.
2. Remove the springs and rollers from the spring retainer and lift the spring retainer from the hub.
3. Remove the remaining bushing and snap-ring from the hub.
4. Install the snap-ring in the forward snap-ring groove of the low-reverse clutch hub.
5. Place the forward clutch bushing against the snap-ring with the flat side up. Install the one-way clutch spring retainer on top of

the bushing. Install the retainer in the hub so that the springs load the rollers in a counterclockwise direction.
6. Install a spring and roller into each of the spring retainer compartments by slightly compressing each spring and placing the roller between the spring and the spring retainer.
7. Install the rear bushing on top of the retainer with the flat side down.
8. Install the remaining snap-ring at the rear of the low-reverse clutch hub.

Servo

1. Apply air pressure to the port in the servo cover in order to remove the piston and stem.
2. Remove the seals from the piston.
3. Dip the new seals in transmission fluid.
4. Install the new seals onto the piston and cover.
5. Dip the piston in transmission

CLUTCH AND BAND APPLICATION CHART— C-6 AUTOMATIC TRANSMISSION

GEAR	INTER-MEDIATE BAND	DIRECT CLUTCH	FORWARD CLUTCH	REVERSE CLUTCH	ONE-WAY CLUTCH
1-st*	off	off	on	off	holding
2-nd*	on	off	on	off	overrunning
3-rd*	off	on	on	off	overrunning
Low(1)	off	off	on	on	holding
Reverse	off	on	off	on	not affected

* Transmission selector in "D" position

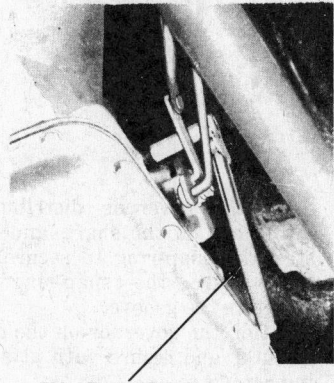

TOOL - T 59 P - 77370 - B

Adjusting intermediate band
(© Ford Motor Co.)

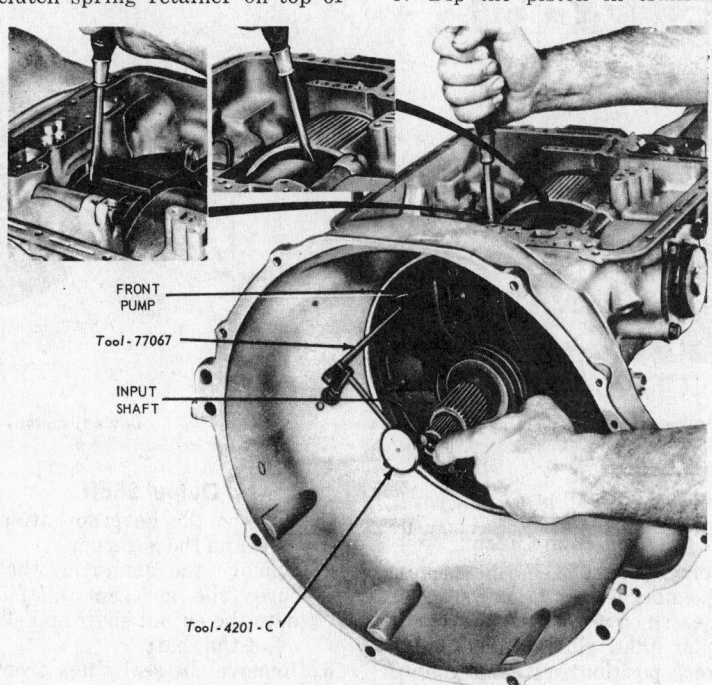

FRONT PUMP

Tool - 77067

INPUT SHAFT

Tool - 4201 - C

Checking gear train end play (© Ford Motor Co.)

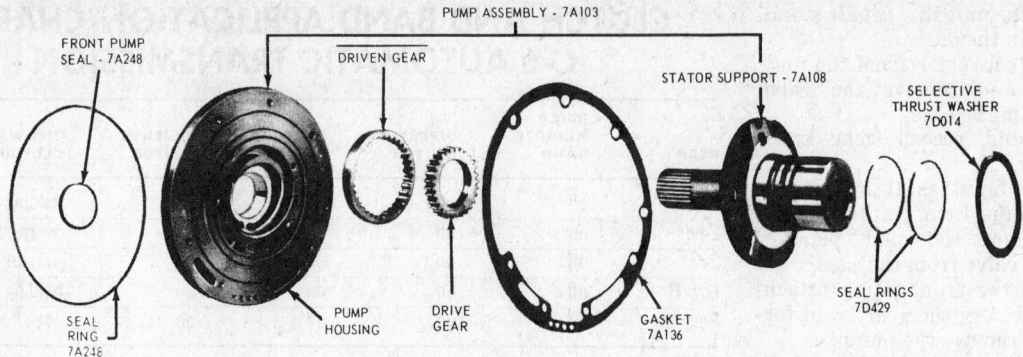

FRONT PUMP SEAL - 7A248

PUMP ASSEMBLY - 7A103

DRIVEN GEAR

STATOR SUPPORT - 7A108

SELECTIVE THRUST WASHER 7D014

SEAL RING 7A248

PUMP HOUSING

DRIVE GEAR

GASKET 7A136

SEAL RINGS 7D429

Front pump disassembled (© Ford Motor Co.)

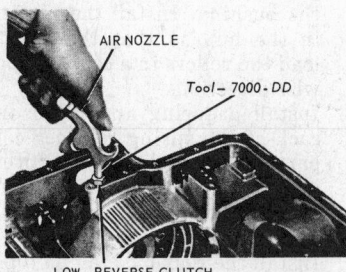

AIR NOZZLE

Tool - 7000 - DD

LOW - REVERSE CLUTCH APPLY PASSAGE

Removing reverse clutch piston (© Ford Motor Co.)

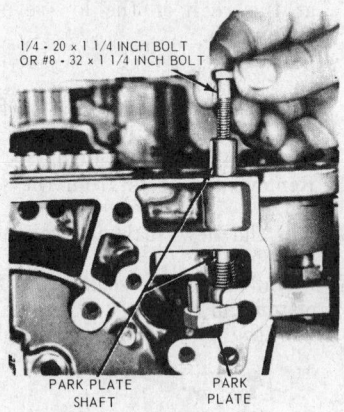

1/4 - 20 x 1 1/4 INCH BOLT OR #8 - 32 x 1 1/4 INCH BOLT

PARK PLATE SHAFT

PARK PLATE

Removing park plate shaft (© Ford Motor Co.)

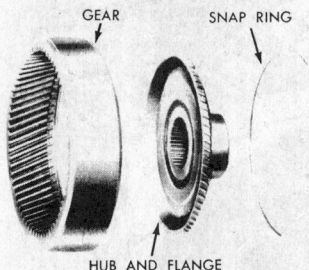

GEAR

SNAP RING

HUB AND FLANGE

Output shaft hub and ring gear (© Ford Motor Co.)

GUIDE PLATE   PIN   SPRING   CUPPED PLUG

ROD

THRUST WASHER

PARKING GEAR

SHAFT

SPRING   PARKING PAWL   PARK PLATE

Parking pawl mechanism (© Ford Motor Co.)

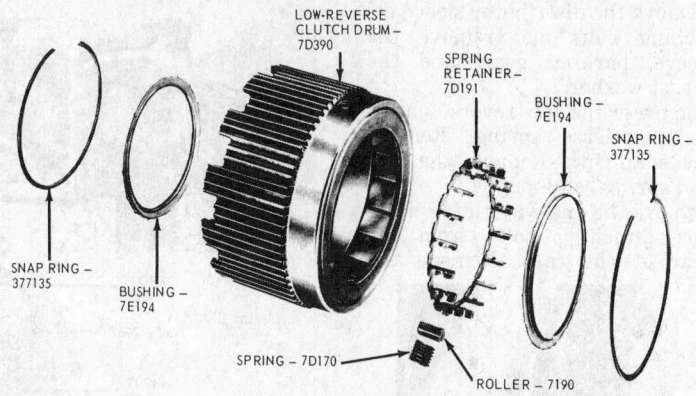

LOW-REVERSE CLUTCH DRUM - 7D390

SPRING RETAINER - 7D191

BUSHING - 7E194

SNAP RING - 377135

SNAP RING - 377135

BUSHING - 7E194

SPRING - 7D170

ROLLER - 7190

One way clutch—disassembled (© Ford Motor Co.)

fluid and install it into the cover.

Reverse-High Clutch

Reverse-high clutch disassembly and assembly is similar to the procedure described for the C4, with the exception that, during disassembly the *exact* position of the piston return spring must be noted, and that they must be installed in the same position.

Output Shaft

1. Remove the governor attaching bolts and the governor.
2. Remove the snap-ring that secures the governor distributor onto the output shaft and slide it off of the shaft.
3. Remove the seal rings from the distributor.
4. Carefully install new seal rings onto the distributor.

5. Slide the governor distributor into place on the shaft, and install the snap-ring to secure it. Make sure the snap-ring is seated in the groove.
6. Position the governor on the distributor and secure with attaching screws.

Parking Pawl Linkage

1. Unbolt and remove the parking

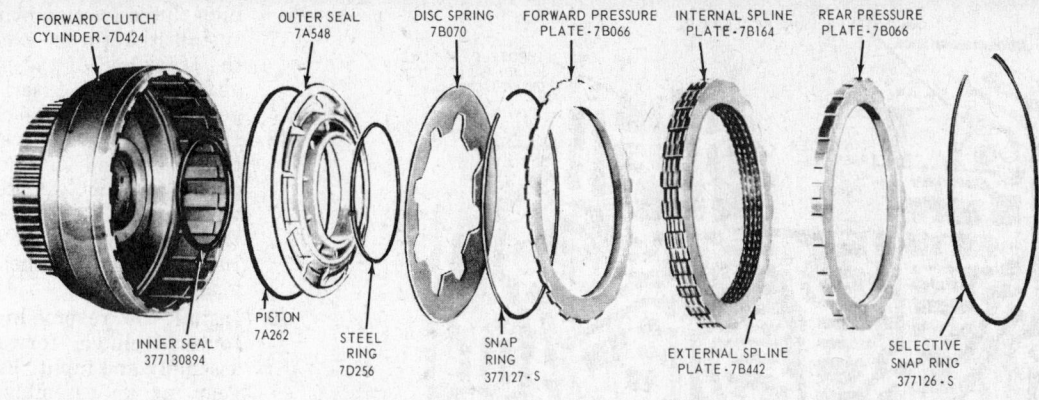

Forward clutch disassembled (© Ford Motor Co.)

Labels on figure: FORWARD CLUTCH CYLINDER-7D424, OUTER SEAL 7A548, DISC SPRING 7B070, FORWARD PRESSURE PLATE-7B066, INTERNAL SPLINE PLATE-7B164, REAR PRESSURE PLATE-7B066, INNER SEAL 377130894, PISTON 7A262, STEEL RING 7D256, SNAP RING 377127-S, EXTERNAL SPLINE PLATE-7B442, SELECTIVE SNAP RING 377126-S

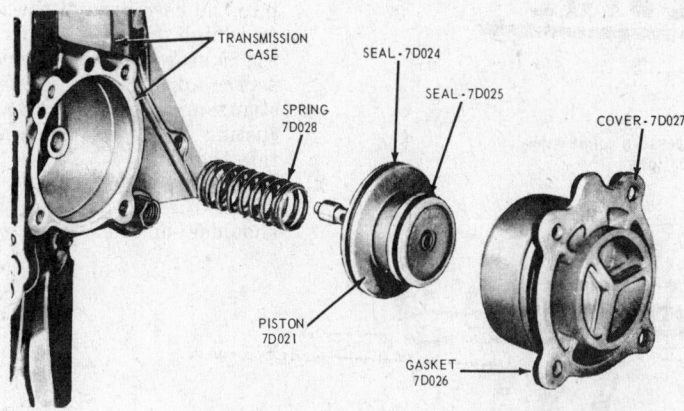

Servo—disassembled (© Ford Motor Co.)

Labels on figure: TRANSMISSION CASE, SEAL-7D024, SEAL-7D025, COVER-7D027, SPRING 7D028, PISTON 7D021, GASKET 7D026

pawl guide plate from the case.

2. Remove the spring, parking pawl, and shaft from the case.

3. Drill a 1/8 in. hole in the park plate shaft retainer plug, and pull the plug out with a wire hook. Unhook the spring from the park plate.

4. Thread a bolt into the park plate shaft, and pull the shaft from the case. Remove the park plate and spring.

5. Position the spring and park plate in the case and install the shaft. Hook the spring on the park plate.

6. Install a new retainer plug.

7. Install the parking pawl shaft in

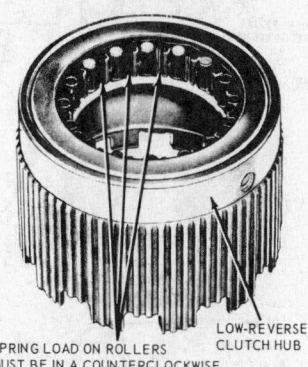

Labels on figure: SPRING LOAD ON ROLLERS MUST BE IN A COUNTERCLOCKWISE DIRECTION FOR INSTALLATION, LOW-REVERSE CLUTCH HUB

Installing one-way clutch (© Ford Motor Co.)

the case, and slip the parking pawls and spring onto the shaft.

8. Bolt the guide plate to the case, making sure that the actuating rod is seated in the slot of the plate.

Transmission Assembly

1. Place transmission case in a holding fixture.

2. Tap the reverse clutch piston into place with a rubber hammer.

3. Hold the one-way clutch inner race in position, install and torque attaching bolts.

4. Install a low-reverse clutch return spring into each pocket of the reverse clutch piston. Press the springs firmly into the piston to prevent their falling out.

5. Position the spring retainer over the springs and position the retainer snap-ring in place on the one-way inner race.

6. Install the compressing tool and compress the springs just enough to install the low-reverse clutch piston retainer snap-ring.

7. Place the transmission case on the bench, bellhousing facing downward.

8. Position the parking gear thrust washer and gear on the case.

9. Position the oil distributor sleeve and tubes in place on the rear of the case. Install and torque the

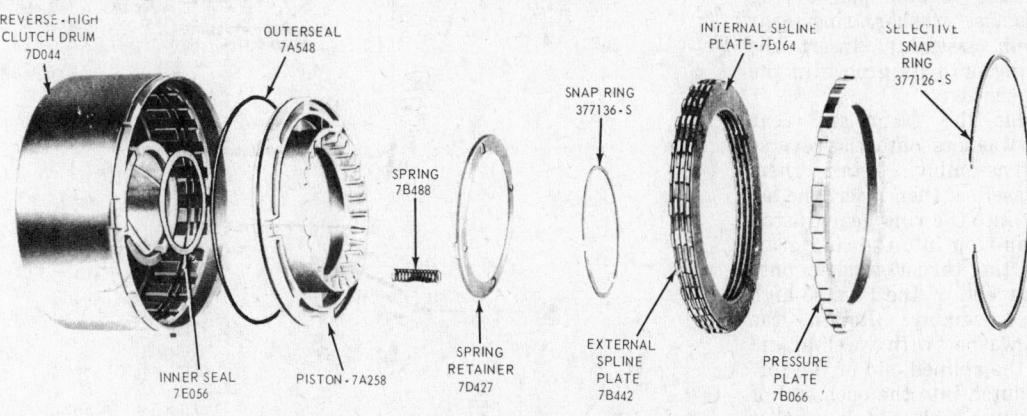

Reverse high clutch disassembled (© Ford Motor Co.)

Labels on figure: REVERSE-HIGH CLUTCH DRUM 7D044, OUTER SEAL 7A548, SPRING 7B488, SNAP RING 377136-S, INTERNAL SPLINE PLATE-7B164, SELECTIVE SNAP RING 377126-S, INNER SEAL 7E056, PISTON-7A258, SPRING RETAINER 7D427, EXTERNAL SPLINE PLATE 7B442, PRESSURE PLATE 7B066

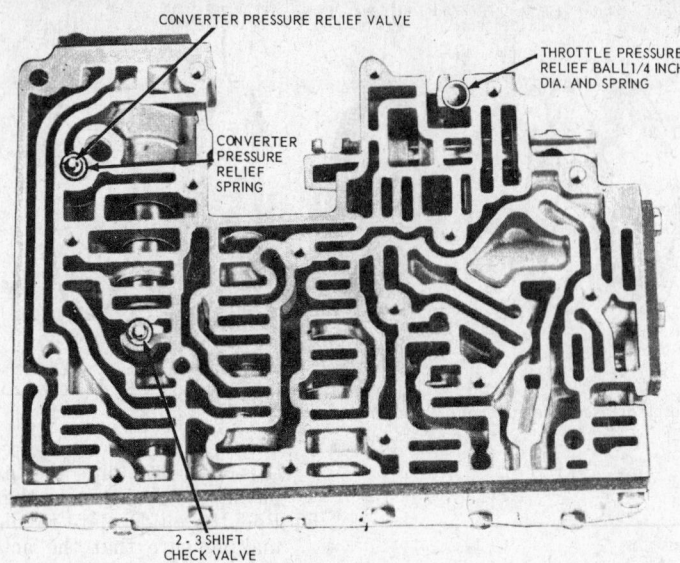

CONVERTER PRESSURE RELIEF VALVE

THROTTLE PRESSURE RELIEF BALL 1/4 INCH DIA. AND SPRING

CONVERTER PRESSURE RELIEF SPRING

2 - 3 SHIFT CHECK VALVE

Converter pressure relief valve, throttle pressure relief valve and 2-3 shift check valve locations
(© Ford Motor Co.)

attaching bolts.

10. Install the output shaft and as an assembly.

11. Place a new gasket on the rear of the transmission case. Position the extension housing on the case and install attaching bolts. Return the case to the holding fixture.

12. Align the low-reverse clutch hub and one-way clutch with the inner race at the rear of the case. Rotate the low-reverse clutch hub clockwise, while applying pressure to seat it.

13. Install the low-reverse clutch plates. Start with a steel plate and follow with friction and steel plates, alternately. If new composition (friction) plates are being used, soak them in transmission fluid for 15 minutes before installation. Install the pressure plate and snap-ring. Test the operation of the low-reverse clutch by applying air pressure at the clutch pressure apply hole in the case.

14. Install the reverse planet ring gear thrust washer, ring gear and hub assembly. Insert the snap-ring onto its groove in the output shaft.

15. Assemble the front and rear thrust washers onto the reverse planet assembly. Retain them with vaseline, then insert the assembly into the ring gear. Install the snap-ring into the ring gear.

16. Install the thrust washer onto the rear end of the reverse-high clutch assembly. Retain the thrust washer with vaseline and insert the splined end of the forward clutch into the open end of the reverse-high clutch so that the splines engage the reverse-

high clutch friction plates.

17. Install the thrust washer onto the front end of the planet ring gear and hub. Insert the ring gear into the forward clutch.

18. Install the thrust washer onto the front end of the forward planet assembly. Retain the washer with vaseline and insert the assembly into the ring gear. Install the input shell and sun gear assembly.

19. Install the reverse-high clutch, forward clutch, forward planet assembly and input shell and sun gear, as an assembly, into the transmission case.

20. Insert the intermediate band into the case around the reverse and high clutch cylinder, with the narrow band end facing the servo apply lever. Install the struts and tighten the band adjusting screw just enough to retain the band.

21. Place a selective thickness bronze thrust washer on the rear shoulder of the stator support

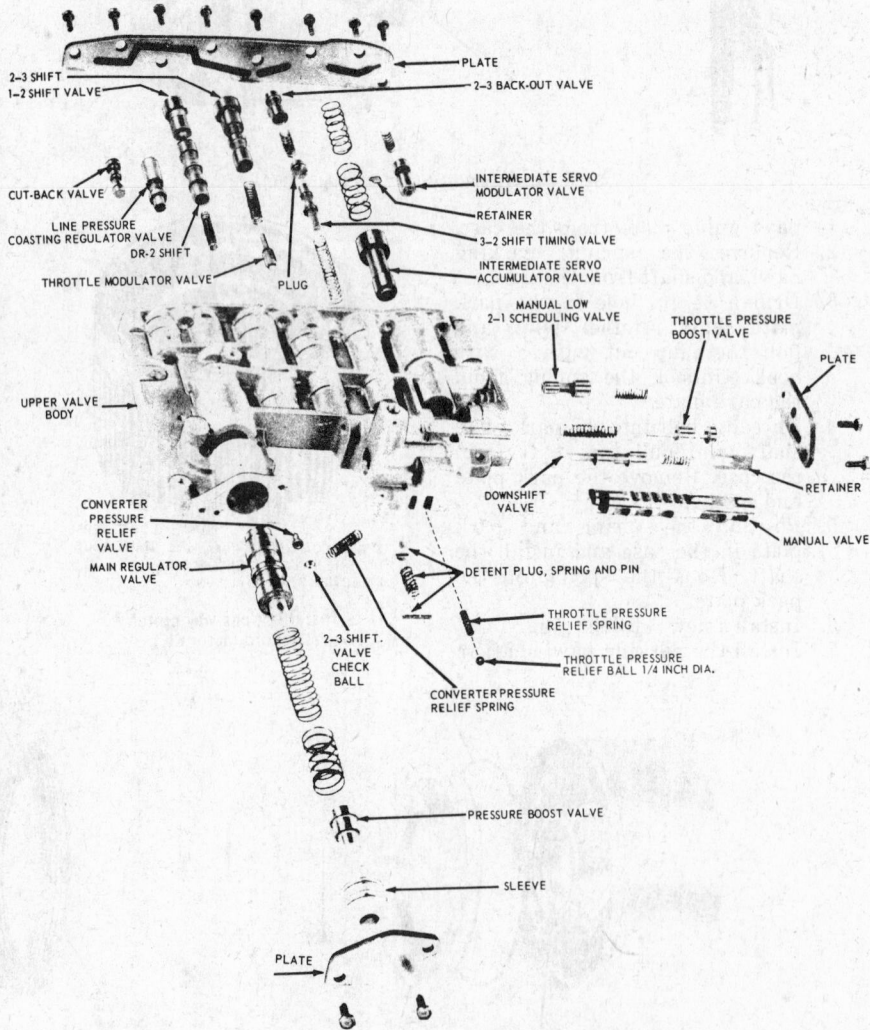

PLATE
2-3 SHIFT
1-2 SHIFT VALVE
2-3 BACK-OUT VALVE
CUT-BACK VALVE
INTERMEDIATE SERVO MODULATOR VALVE
RETAINER
LINE PRESSURE COASTING REGULATOR VALVE
3-2 SHIFT TIMING VALVE
DR-2 SHIFT
INTERMEDIATE SERVO ACCUMULATOR VALVE
THROTTLE MODULATOR VALVE
PLUG
MANUAL LOW 2-1 SCHEDULING VALVE
THROTTLE PRESSURE BOOST VALVE
PLATE
UPPER VALVE BODY
DOWNSHIFT VALVE
RETAINER
CONVERTER PRESSURE RELIEF VALVE
MANUAL VALVE
MAIN REGULATOR VALVE
DETENT PLUG, SPRING AND PIN
2-3 SHIFT VALVE CHECK BALL
THROTTLE PRESSURE RELIEF SPRING
THROTTLE PRESSURE RELIEF BALL 1/4 INCH DIA.
CONVERTER PRESSURE RELIEF SPRING
PRESSURE BOOST VALVE
SLEEVE
PLATE

Upper control valve body –disassembled (© Ford Motor Co.)

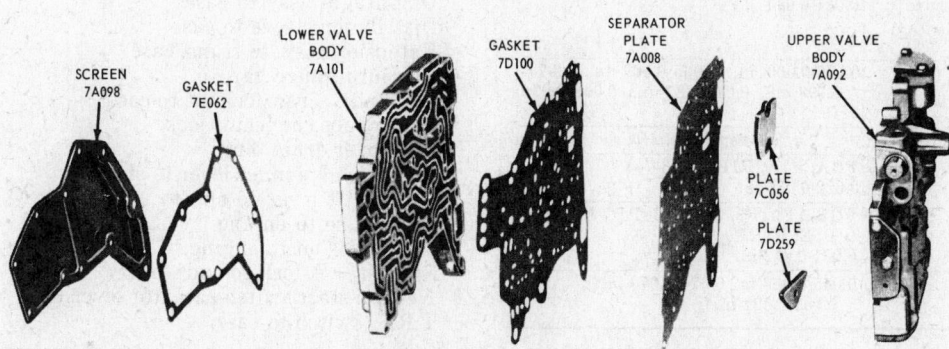

Output shaft disassembled (© Ford Motor Co.)

Upper and lower valve bodies—disassembled (© Ford Motor Co.)

and retain it with vaseline. Lay a new gasket on the rear mounting face of the pump and position it on the case, being careful not to damage the O-ring. Install six of the seven mounting bolts. Adjust the intermediate band as previously described. Then, install the input shaft.

22. Install a dial indicator stand in place of the seventh mounting bolt, and check the transmission end-play. Remove the tool, and install the remaining bolt.

23. Install the control valve into the case, making sure that the levers engage the valves properly.

24. Install the primary throttle valve, rod, and the vacuum diaphragm into the case.

25. Install a new oil pan gasket, and the oil pan.

26. Install the converter assembly.

CONTROL PRESSURE AT ZERO GOVERNOR RPM

Year	Engine Speed	Throttle	Manifold Vac. In. Hg.	Range	Psi
1971	Idle	As Required	Above 15	P, N, D, 2, 1 R	56-62 71-86
	As Required	As Required	10	D, 2, 1	100-115
	Stall	Thru Detent	Below 1.0	D, 2, 1	160-190
1972	Idle	As Required	Above 15	D, 2, 1 R	56-94 78-132
	As Required	As Required	10	D, 2, 1	100-125
	Stall	Thru Detent	Below 1.0	D, 2, 1 R	160-190 240-300
1973-78	Idle	As Required		D, 2, 1, P, N R	56-112 78-161
	As Required	As Required	10	D, 2, 1	100-125
	Stall	Thru Detent	Below 1.0	D, 2, 1 R	160-190 240-300

① It may not be possible to obtain 18 inches of engine vacuum at idle. For idle vacuum of less than 18 inches, the following chart provides idle speed pressure specifications in the "D" range:

Vac.	Psi	Vac.	Psi
17	56-69	13	56-98
16	56-75	12	56-105
15	56-84	11	56-111
14	56-92		

Automatic Transmissions

Checks and Adjustments—

Operation	Specification
Transmission end play	0.008-0.044 (selective thrust washers available)
Turbine and stator end play	New or rebuilt 0.021 in .max. Used 0.030 in. max. ①
Intermediate band adjustment	Remove and discard locknut. Adjust screw to 10 ft-lbs torque, then back off 1 turn ,install new lock nut and tighten locknut to specification. ②
Forward clutch pressure plate-to-snap ring clearance	0.031-0.044
Selective snap ring thicknesses	0.056-0.060 in., 0.065-0.069 in., 0.074-0.078 in., 0.083-0.87 in., 0.092-0.096 in.
Reverse-high clutch pressure plate-to-snap ring clearance	*Transmission Models* PGA, PJA, PJB (1973-74) 0.022-0.036 in. PGB-AF2, F3, G3, PJB, PJC-A, B, E, F, PJD ③ 0.027-0.043 in.
Selective snap ring thicknesses	0.065-0.069 in., 0.074-0.078 in, 0.083-0.087 in. ④

① To check end play, exert force on checking tool to compress turbine to cover thrust washer wear plate. Set indicator at zero.
② 1971-74 models: back off adjusting screw 1½ turns.
③ PJB only to 1972.
④ For 1973-4 add 0.056-0.060; 0.092-0.096.

Selective Thrust Washers—

Identification No.	Thrust Washer Thickness— Inch	Identification No.	Thrust Washer Thickness— Inch
1	0.056-0.058	4	0.103-0.105
2	0.073-0.075	5	0.118-0.120
3	0.088-0.090		

Torque Limits—

	Ft-Lbs
Converter to flywheel	20-30
Front pump to trans. case	12-20
Overrunning clutch race to case	18-25
Oil pan to case	12-16
Stator support to pump	12-16
Converter cover to converter hsg.	12-16
Guide plate to case	12-16
Intermediate servo cover to case	10-14
Diaphragm assy to case	15-23
Distributor sleeve to case	12-16
Extension assy. to trans. case	25-30
Pressure gauge tap	9-15
Band adj. screw locknut to case	35-45
Cooler tube connector lock	25-35②
Converter drain plug	14-28
Manual valve inner lever to shaft	30-40
Downshift lever to shaft	12-16
Filler tube to engine	20-25
Transmission to engine	40-50
Steering col. lock rod adj. nut	10-20
Neutral start switch actuator lever bolt	6-10
T.R.S. switch-to-case	4-8

	In-Lbs
End plates to body	20-30
Inner downshift lever stop	20-30
Reinforcement plate to body	20-30
Screen and lower to upper valve body	50-60
Neutral switch to case	55-75
Neutral switch-to column	20
Control assy. to case	90-125
Gov. body to collector body	80-120
Oil tube connector	80-145①

① 1972 torque limits—80-120 in. lbs.;
 1973-75 torque limits—60-120 in. lbs.
② 1972-75 torque limits—20-35 in. lbs.

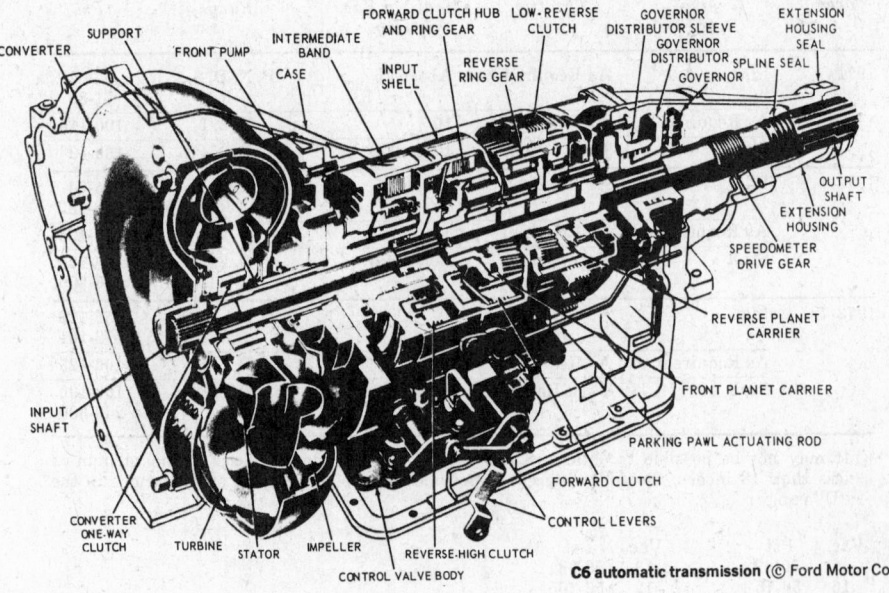

C6 automatic transmission (© Ford Motor Co)

Ford FMX
DIAGNOSIS

This diagnosis guide covers the most common symptoms and is an aid to careful diagnosis. The items to check are listed in the sequence to be followed for quickest results. Thus, follow the checks in the order given for the particular transmission type.

TROUBLE SYMPTOMS	ITEMS TO CHECK	
	IN TRUCK	OUT OF TRUCK
Rough initial engagement in D1 or D2	KBWFEG	
1-2 or 2-3 shift points incorrect or erratic	ABCDWEL	
Rough 2-3 shifts	BGFEJ	
Engine overspeeds on 2-3 shift	BGEF	m
No 12 or 2-3 upshift	DECJG	bcf
No 3-1 shift	KBE	
No forced downshift	LWE	
Runaway engine on forced downshift	GFEJB	c
Rough 3-2 or 3-1 shift at closed throttle	KBE	
Creeps excessively	KZ	
Slips or chatters in first gear, D1	ABWFE	acfi
Slips or chatters in second gear	ABGWFEJ	ac
Slips or chatters in R	AHWFEIB	bcf
No drive in D1	CE	i
No drive in D2	ERC	acf
No drive in L	CER	acf
No drive in R	HIERC	bef
No drive in any selector position	ACWFER	cd
Lockup in D1	CIJ	bgc
Lockup in D2	CHI	bgci
Lockup in L	GJE	bjc
Lockup in R	GJ	agc
Parking lock binds or does not hold	C	g
Transmission overheats	OFG	l
Engine will not push-start	ACFE	ec
Maximum speed to low, poor acceleration	Y	l
Transmisison noisy in N and P	F	ad
Noisy transmission during coast 30-20 mph with engine stopped		e
Transmission noisy in any drive position	F	hbad
Fluid leaks	MNOPQSTUX	jkl

Key to Checks

A. Fluid level
B. Vacuum diaphragm unit or tubes
C. Manual linkage
D. Governor
E. Valve body
F. Pressure regulator
G. Front band
H. Rear band
I. Rear servo
J. Front servo
K. Engine idle speed
L. Downshift linkage
M. Converter drain plug
N. Oil pan, filler tube and/or seals

O. Oil cooler and/or connections
P. Manual or throttle shaft seals
Q. Pipe plug, side of case
R. Perform air pressure checks
S. Extension housing-to-case gasket or washer
T. Center support bolt lock washer
U. Extension housing rear oil seal
W. Make control pressure check
X. Speedometer drive gear adaptor seal
Y. Engine performance
Z. Vehicle brakes

a. Front clutch
b. Rear clutch
c. Hydraulic system leakage
d. Front pump
e. Rear pump
f. Fluid distributor sleeve— output shaft
g. Parking linkage
h. Planetary assembly
i. Planetary oneway clutch
j. Engine rear oil seal
k. Front oil pump seal
l. Front pump-to-case seal or gasket
m. Rear clutch piston air bleed valve

General Information

This section provides procedures for testing, inspection, adjustment, and repair of the FMX 3-speed automatic transmission. Where there are differences in procedures or specifica-tions for various model changes, these differences will be outlined.

The Ford FMX 3-speed automatic transmission is a three-speed unit that provides automatic upshifts and downshifts through three forward gear ratios and also provides manual selection of first and second gears. The transmission consists of a torque converter, planetary gear assembly, two multiple disc clutches, and a hydraulic control system.

The FMX transmission cools its transmission fluid through a cooler

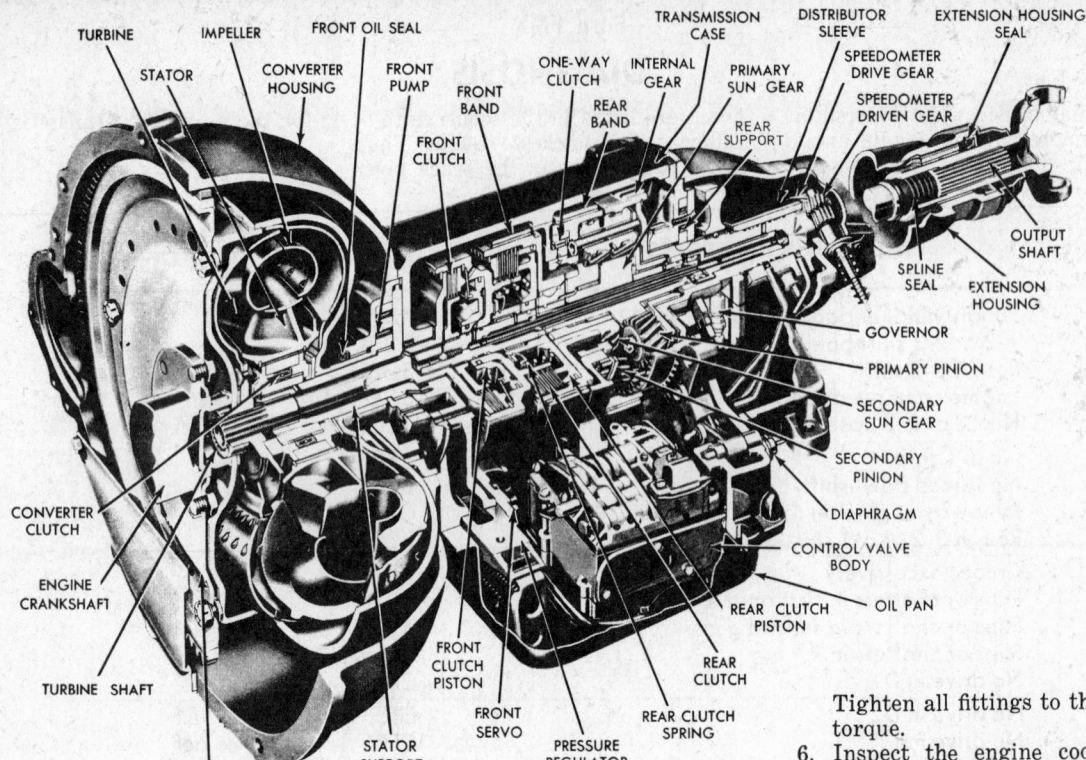

Typical MX—H.D. transmission (© Ford Motor Co.)

Labels on diagram: TURBINE, IMPELLER, FRONT OIL SEAL, TRANSMISSION CASE, DISTRIBUTOR SLEEVE, EXTENSION HOUSING SEAL, STATOR, CONVERTER HOUSING, FRONT PUMP, ONE-WAY CLUTCH, INTERNAL GEAR, PRIMARY SUN GEAR, SPEEDOMETER DRIVE GEAR, FRONT BAND, FRONT CLUTCH, REAR BAND, PRIMARY PINION, SPEEDOMETER DRIVEN GEAR, REAR SUPPORT, OUTPUT SHAFT, SPLINE SEAL, EXTENSION HOUSING, GOVERNOR, PRIMARY PINION, SECONDARY SUN GEAR, SECONDARY PINION, DIAPHRAGM, CONTROL VALVE BODY, OIL PAN, REAR CLUTCH PISTON, CONVERTER CLUTCH, ENGINE CRANKSHAFT, REAR CLUTCH, REAR CLUTCH SPRING, REAR CLUTCH, FRONT CLUTCH PISTON, FRONT SERVO, PRESSURE REGULATOR BODY, STATOR SUPPORT, TURBINE SHAFT, FLYWHEEL

core in the radiator lower tank when a steel converter is used. If an aluminum converter is used, the transmission fluid is air-cooled.

Transmission Checks

Prior to performing any of the tests or adjustments described below, engine idle speed, manual linkage adjustment, and transmission fluid level must be checked.

If fluid level is excessively low, check for leakage as follows.

Transmission Fluid Leakage Checks

Make the following checks if leakage is suspected from the transmission:

1. Clean all dirt and grease from the transmission case.
2. Inspect the speedometer cable connection at the extension housing of the transmission. If fluid is leaking here, disconnect the cable and replace the rubber seal.
3. Inspect the oil pan gasket and attaching bolts for leaks. Tighten any bolts that appear loose to the proper torque (10-13 ft. lbs.). Recheck for signs of leakage. If necessary, remove the pan attaching bolts and old pan gasket and install new gasket. Reinstall the pan and its attaching bolts.
4. Check filler tube connection at the transmission for signs of leakage. If tube is leaking, tighten the connection to stop

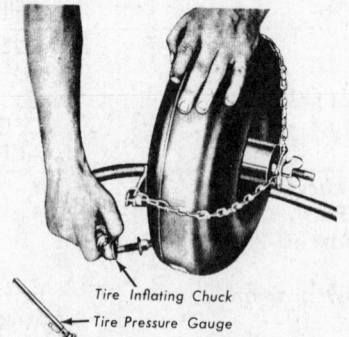

Typical converter leak checking tool
(© Ford Motor Co.)

Labels on image: Tire Inflating Chuck, Tire Pressure Gauge

the leak. If necessary, disconnect the filler tube, replace the O-ring, and reinstall the filler tube.
5. Inspect all fluid lines between the transmission and the cooler core in the lower radiator tank. Replace any lines or fittings that appear to be worn or damaged.

Tighten all fittings to the proper torque.
6. Inspect the engine coolant for signs of transmission fluid in the radiator. If there is transmission fluid in the engine coolant, the oil cooler core is probably leaking. *NOTE: The oil cooler core may be tested further by disconnecting all lines to it and applying 50-75 psi air pressure through the fittings. Remove the radiator cap to relieve any pressure build-up outside the cooler core. If air bubbles appear in the coolant or if the cooler core will not hold pressure, the oil cooler core is leaking and must be replaced.*
7. Inspect the openings in the case where the downshift control lever shaft and the manual lever shaft are located. If necessary, replace the defective seal.
8. Inspect all plugs or cable connections in the transmission for signs of leakage. Tighten any loose plugs or connectors to the proper torque (see torque chart at end of this section).
9. Remove the lower cover from the front of the bellhousing and in-

CLUTCH AND BAND APPLICATION CHART— FMX AUTOMATIC TRANSMISSION

GEAR	FORWARD CLUTCH	REAR CLUTCH	FRONT BAND	REAR BAND	OVERRUNNING CLUTCH
1st*	on	off	off	off	holding
2nd*	on	off	on	off	overrunning
3rd*	on	on	off	off	overrunning
Low(1)	on	off	off	on	overrunning
Reverse	off	on	off	on	not affected

* Transmission selector lever in "D" position

spect the converter drainplugs for signs of leakage. If there is a leak around the drainplugs, loosen the plug and coat the threads with sealing compound. Tighten the plug to the proper torque.

NOTE: Fluid leaks from around the converter drainplug may be caused by engine oil leaking past the rear main bearing or from the oil gallery plugs. To determine the exact cause of the leak before beginning repair procedures, an oil-soluble aniline or fluorescent dye may be added to the transmission fluid to find the source of the leak and whether the transmission is leaking. If a fluorescent dye is used, a black light must be used to detect the dye.

If further converter checks are necessary, remove the transmission from the truck and the converter from the transmission. The converter cannot be disassembled for cleaning or repair. If the converter is leaking, it must be replaced with a new unit. To check the converter for leaks, assemble and install the converter leak checking tool and fill the converter with 20 psi air pressure. Then, place the converter in a tank of water and watch for air bubbles. If no air bubbles are seen, the converter is not leaking.

Control Pressure Check

When the vacuum diaphragm unit operates properly and the downshift linkage is adjusted correctly, all transmission shifts (automatic and kickdown) should occur within the specified road speed limits. If these shifts do not occur within the limits or if the transmission slips during a shift point, perform the following procedure to locate the problem:

1. Connect the Automatic Transmission Tester (see illustration) as follows:
   a. Tachometer cable to engine.
   b. Vacuum gauge hose with a T-fitting between the vacuum hose and vacuum diaphragm (see illustration).
   NOTE: On vehicles equipped with a dual area diaphragm (DAD), check the control pressure at 10 in. of vacuum by removing the exhaust gas recirculation (EGR) control hose from the diaphragm and plugging the hose. Do not plug the EGR port in the diaphragm; this port must be left open to atmospheric pressure. When checking the control pressure at stall and idle, keep the hose connected.
   c. Pressure gauge to the control pressure outlet on the transmission (see illustration).
2. Apply the parking brake and start the engine. On a truck equipped with a vacuum brake

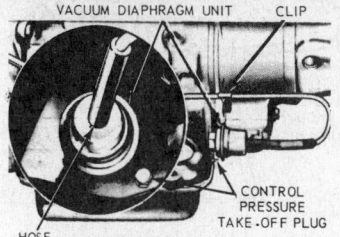

Typical control pressure connecting points
(© Ford Motor Co.)

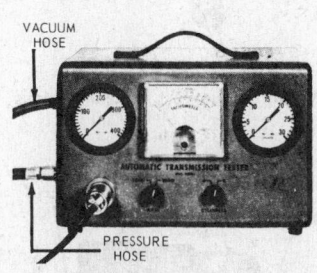

**Rotunda are—
29-05 automatic transmission tester**
(© Ford Motor Co.)

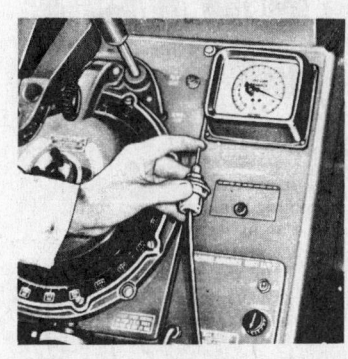

Testing transmission vacuum unit for leakage
(© Ford Motor Co.)

release, use the service brakes since the parking brake will release automatically when the transmission is put in any Drive position.

3. Check the transmission diaphragm unit for leaks (see below).
4. Check control pressure in all selector lever positions at specified manifold vacuum (see specifications). Record readings and compare to specifications.

Vacuum Diaphragm Unit Check

1. Remove the vacuum diaphragm unit from the transmission using crowfoot wrench, after disconnecting the vacuum hose (see illustration).
2. Adjust a vacuum pump until the vacuum gauge shows 18 in/Hg. with the vacuum hose blocked.
3. Connect the vacuum hose to the vacuum diaphragm unit and note the reading on the vacuum gauge. If the reading is 18 inches of vacuum, the vacuum diaphragm unit is good. While removing the vacuum hose from the vacuum diaphragm unit, hold a finger over the end of the control rod. As the vacuum is released, the internal spring of the vacuum diaphragm unit will push the control rod out.

Air Pressure Checks

If the truck will not move in one or more ranges, or, if it shifts erratically, the items at fault can be determined by using air pressure at the indicated passages.

Drain the transmission and remove the oil pan and the control valve assembly.

NOTE: Oil will spray profusely during this operation.

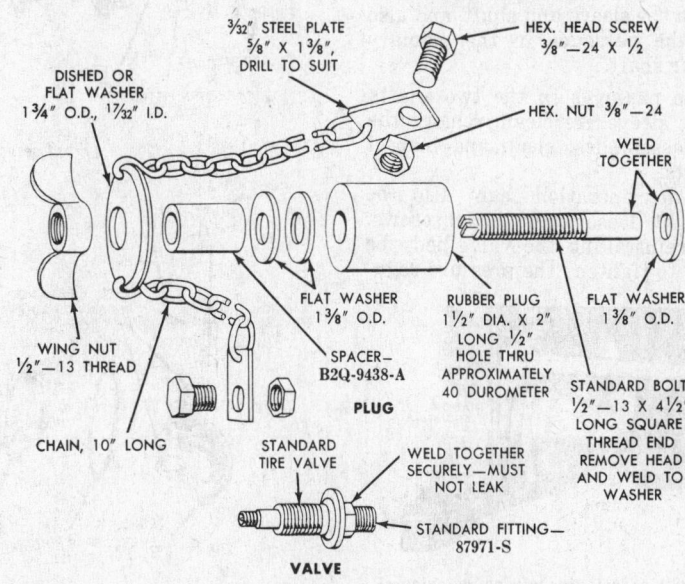

Converter leak checking tool (© Ford Motor Co.)

Front Clutch

Apply sufficient air pressure to the front clutch input passage. A dull thud can be heard when the clutch piston moves. Check also, for leaks.

Governor

Remove the governor inspection cover from the extension housing. Apply air to the front clutch input passage. (See illustration). Listen for a sharp click and watch to see if the governor valve snaps inward as it should.

Rear Clutch

Apply air to the rear clutch passage (See illustration) and listen for the dull thud that will indicate that the rear clutch piston has moved. Listen also for leaks.

Front Servo

Apply air pressure to the front servo apply tube and note if front band tightens. Shift the air to the front servo release tube, which is next to the apply tube, and watch band release.

Rear Servo

Apply air pressure to the rear servo apply passage. The rear band should tighten around the drum.

Conclusions

If the operation of the servos and clutch is normal with air pressure, the no-drive condition is due to the control valve and pressure regulator valve assemblies, which should be disassembled, cleaned and inspected.

If operation of the clutches is not normal; that is, if both clutches apply from one passage or if one fails to move, the aluminum sleeve (bushing) in the output shaft is out of position or badly worn. (See illustration.)

Use air pressure to check the passages in the sleeve and shaft, and also check the passages in the primary sun gear shaft.

If the passages in the two shafts and the sleeve are clean, remove the clutch assemblies, clean and inspect the parts.

Erratic operation can also be caused by loose valve body screws. When reinstalling the valve body be careful to tighten: the pressure regu-

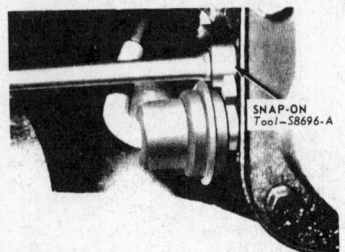

Removing or installing vacuum diaphragm
(© Ford Motor Co.)

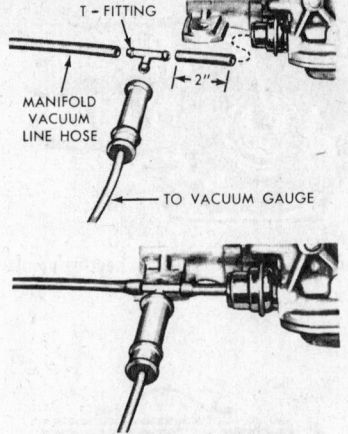

Typical vacuum test line connections
(© Ford Motor Co.)

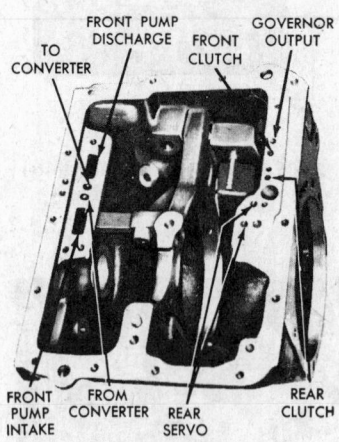

Case fluid hole identification
(© Ford Motor Co.)

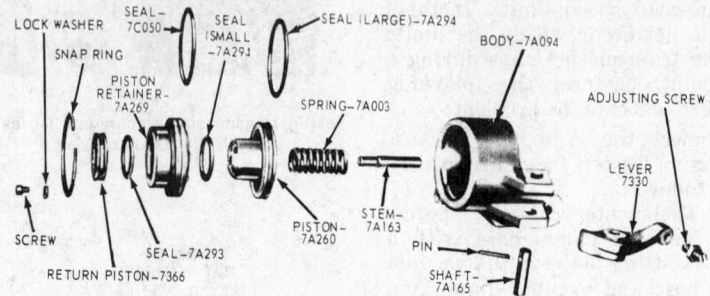

Front servo disassembled (© Ford Motor Co.)

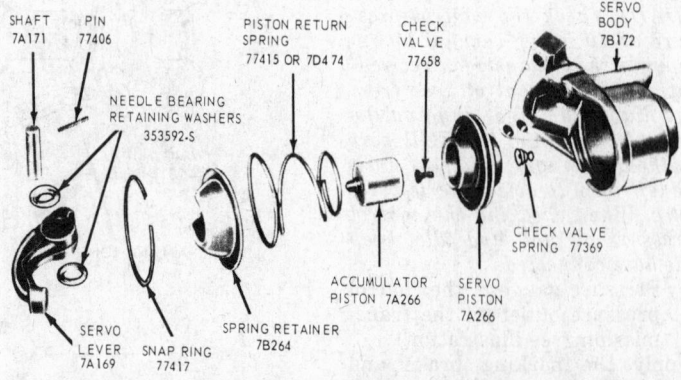

Rear servo disassembled (© Ford Motor Co.)

lator valve to case bolts to 17-22 ft. lbs., the pressure regulator valve cover screws to 20-30 in. lbs., the control valve body screws to 20-30 in. lbs., the 1/4-20 capscrew (lower to upper valve body) to 4-6 ft. lbs., and the control valve body to case bolts to 8-10 ft. lbs.

Shift Point Checks for Automatic Transmissions

To determine if the transmission is shifting at the proper road speeds, use the following procedure:

1. Check the minimum throttle upshifts by placing the transmission selector lever in the Drive position and noting the road speeds at which the transmission shifts from first gear to second gear to third gear. All shifts should occur within the specified limits.

2. While driving in third gear, depress the accelerator pedal past the detent (to the floor). Depending on vehicle speed, the transmission should downshift from third gear to second gear or from second gear to first gear.

3. Check the closed-throttle downshift from third gear to first gear by coasting down from about 30 mph in third gear. This downshift should occur at the specified road speed.

4. With the transmission in third gear and the truck moving at a road speed of 35 mph, the transmission should downshift to second gear when the selector lever

is moved from D to 2 to 1. This check will determine if the governor pressure and shift control valves are operating properly. If the transmission does not shift within the specified limits or certain gears cannot be obtained, refer to the Trouble Diagnosis chart at the beginning of this section.

In-Vehicle Adjustments and Repairs

The following adjustments and repairs may be performed without removing the transmission.

Band Adjustments

Front Band Adjustment

When it is necessary to adjust the front band of the transmission, perform the following procedure:

1. Drain the transmission fluid and remove the oil pan, fluid filter screen, and clip. The same transmission fluid may be reused if it is filtered before being installed. Only transmission fluid in good condition should be used.
2. Clean the pan and filter screen and remove the old gasket.
3. Loosen the front servo adjusting screw locknut.
   NOTE: Special band adjusting wrenches are recommended to do this operation correctly and quickly.
4. Pull back the actuating rod and insert a ¼ in. spacer bar between the adjusting screw and the servo piston stem. Tighten the adjusting screw to 10 in. lbs. torque. Remove the spacer bar and tighten the adjusting screw an additional ¾ turn. Hold the adjusting scew and tighten the locknut securely (20-25 ft. lbs.).
5. Install the transmission fluid filter screen and clip. Install the pan with a new pan gasket.
6. Refill the transmission to the FULL mark on the Dipstick. Start the engine, run for a few minutes, shift the selector lever through all positions, and place it in Park. Recheck the fluid level again and add fluid to proper level is necessary.

Rear Band Adjustments

The rear band of the FMX transmission may be adjusted by any of the methods given below. On most trucks the basic external band adjustment is satisfactory. The internal adjustment may be performed in cases where the adjustment required is outside the range of the external adjustment.

Rear Band External Adjustment

The procedure for adjusting the rear band externally is as follows:

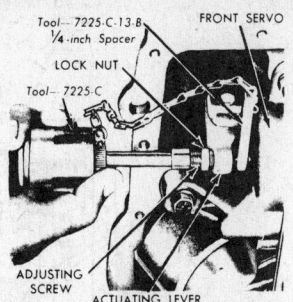

Front band adjustment—typical
(© Ford Motor Co.)

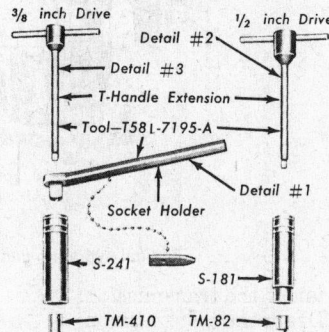

Front and rear band adjusting tools
(© Ford Motor Co.)

1. Locate the external rear band adjusting screw on the transmission case, clean all dirt from the threads, and coat the threads with light oil.
   NOTE: The adjusting screw is located on the upper right side of the transmission case. Access is often through a hole in the front floor to the right of center under the carpet.
2. Loosen the locknut on the rear band external adjusting screw.
3. Using the special preset torque

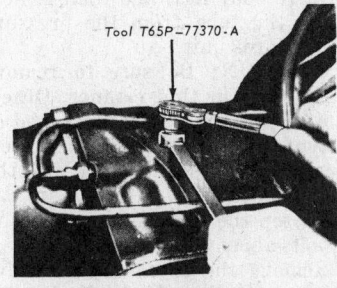

Rear band adjustment
(© Ford Motor Co.)

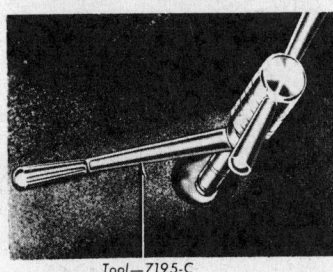

Adjusting rear band
(© Ford Motor Co.)

wrench shown, tighten the adjusting screw until the handle clicks at 10 ft. lbs. torque. If the adjusting screw is tighter than 10 ft. lbs. torque, loosen the adjusting screw and retighten to the proper torque.
4. Back off the adjusting screw 1½ turns. Hold the adjusting screw steady while tightening the locknut to the proper torque (35-40 ft. lbs.).
   CAUTION: Severe damage may result if adjusting screw is not backed off exactly 1½ turns.

Rear Band Internal Adjustment

The rear band is adjusted internally as follows:

1. Drain the transmission fluid. If it is to be resued, filter it as it drains from the transmission. Reuse the transmission fluid only if it is in good condition.
2. Remove and clean the pan, fluid filter, and clip.
3. Loosen the rear servo adjusting locknut.
4. Pull the adjusting screw end of the actuating lever away from the servo body and insert the spacer tool (see illustration) between the servo accumulator piston and the adjusting screw. Be sure the flat surfaces of the tool are placed squarely between the adjusting screw and the accumulator piston. Tool must not touch servo piston and the handle must not touch the servo piston spring retainer.
5. Using a torque wrench with an allen head socket adapter tighten the adjusting screw to 24 in. lbs.
6. Back off the adjusting screw exactly 1½ turns. Hold adjusting screw steady and tighten the locknut securely. Remove the spacer tool.
7. Install the fluid filter, clip, and pan with a new gasket.
8. Fill the transmission with the correct amount of fluid.

Transmission Component

Removal and Installation

Various components of the FMX transmission may be removed while the transmission is in the truck. Installation is often the reverse of the removal instructions except for adjustment and alignment. Repair of the individual components is given in the overhaul section.

Governor Assembly

Removal

1. Raise the truck so that the transmission extension housing is accessible.

2. Remove the governor inspection cover from the extension housing.
3. Rotate the drive shaft until the governor is in line with the inspection hole.
4. Remove the governor valve body from the counterweight.
CAUTION: Do not drop the attaching bolts of the valve parts into the extension housing.

Installation

1. Lubricate the new governor valve parts with clean transmission fluid.
   NOTE: The valve must move freely in the valve body bore.
2. Install the governor body on the counterweight so that the valve body cover is facing the rear. Tighten the attaching bolts.
3. Install the governor inspection cover with a new gasket on the extenson housing. Tighten the attaching bolts to proper torque specifications. (Torque specifications are at the end of this section).

Extension Housing Bushing and Rear Seal

Removal

1. Raise the truck and remove the drive shaft. Remove the parking brake drum if so equipped. Remove the transmission output shaft and attaching nut with special tool.
2. Working with a sharp chisel, remove the seal from the extension housing.
   CAUTION: Do not allow any metal chips to enter the output shaft bearing.
3. To remove the output shaft bearing, remove the snap ring securing the bearing to the extension housing.
4. Remove the bearing from the extension housing.

Installation

1. Position a new bearing in the extension housing and secure with a snap ring.
2. Install a new extension housing seal with the proper seal replacement tool. Install the output shaft yoke and retaining nut. Torque the attaching nut to specifications. Install the parking brake drum, if so equipped.
3. Install the drive shaft and torque bolts to specifications.
4. Lower the truck and check the transmission fluid level.

Control Valve Body and Oil Pan

Removal

1. Raise the truck on a hoist or jackstands and place a drain pan

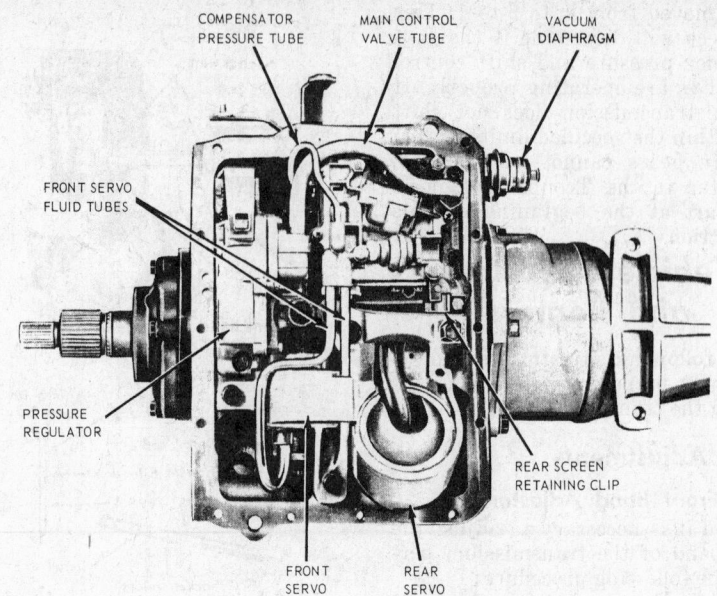

Typical hydraulic control system (© Ford Motor Co.)

under the transmission.
2. Drain the transmission.
3. Remove the oil pan, fluid filter screen, and clip and clean them thoroughly. Discard the old pan gasket.
4. Remove the vacuum diaphragm assembly using a crowfoot wrench. *Do not use pliers, pipe wrenches, etc. to remove the vacuum diaphragm unit. Do not let any solvents enter the vacuum diaphragm unit.* Remove the push rod, the fluid screen and its retaining clip.
5. Remove the small compensator pressure tube.
6. Disconnect the main pressure oil tube by carefully loosening the end connected to the control valve body first and then removing the tube from the pressure regulator unit.
   CAUTION: Be sure to remove the tube in this manner. Otherwise, the tube could be kinked or bent causing improper fluid pressures and possible damage to the transmission.
7. Loosen the front servo attaching bolts about three turns.
8. Remove the three control valve body attaching bolts and care-

fully lower the valve body, sliding it off the front servo tubes. *Do not damage the valve body or the tubes.*

Installation

1. When installing the control valve body, align the front servo tubes with the holes in the valve body. Shift the manual lever to the 1 detent and place the inner downshift lever between the downshift lever stop and the downshift valve. Be sure the manual lever engages the actuating pin in the manual detent lever.
2. Loosely install the control valve body attaching bolts and move the control valve body toward the center of the transmission case until there is a clearance of 0.050 in. between the manual valve and the actuating pin on the manual detent lever.
3. Tighten the attaching bolts to 8-10 ft. lbs. torque. Ensure that the rear fluid filter retaining clip is installed under the valve body.
4. Install the main pressure oil tube, connecting the end to the pressure regulator unit first and then connecting the other end to the main control valve assembly

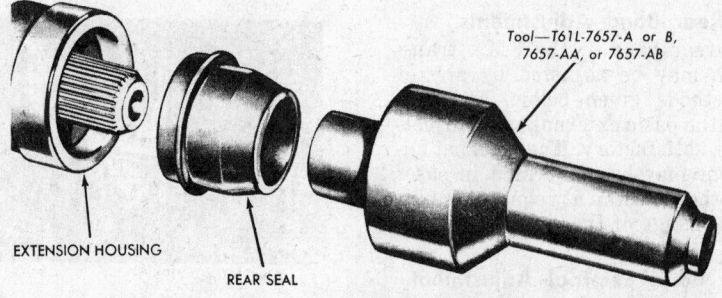

Installing extension housing seal (© Ford Motor Co.)

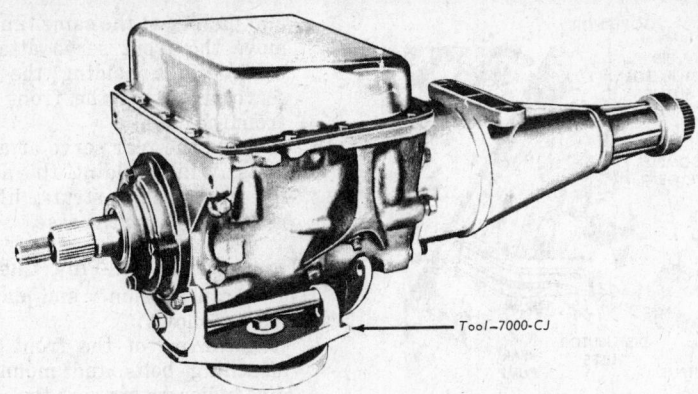

Transmission mounted in holding fixture (© Ford Motor Co.)

by gently tapping it with a soft-faced hammer.

5. Install the compensator pressure tube on the pressure regulator and control valve body.

6. Check the manual lever for free motion in each detent position by rotating it one full turn. If the manual lever binds in any detent position, loosen the valve body attaching bolts and move the valve body away from the center of the transmission case until the binding is relieved. Retighten the attaching bolts according to step 3.

7. Place the pushrod in the bore of the vacuum diaphragm unit and install the vacuum diaphragm unit.

8. Tighten the front servo attaching bolts.

9. Adjust the front band.

10. Install the fluid filter and its retaining clip.

11. Adjust the rear band.

12. Install the oil pan with a new, pan gasket.

13. Fill transmission with fluid. Start and run engine for a few minutes and check the fluid level after shifting the transmission through all positions. *Do not overfill the transmission.*

14. Check the adjustment of the transmission control linkage.

Pressure Regulator

Removal

1. Drain the transmission of fluid and remove the pan, fluid filter screen, and its retaining clip. Discard the used pan gasket.

2. Remove the compensator pressure tube from between the control valve body and the pressure regulator.

3. Remove the main pressure oil tube by gently prying off the end connected to the control valve first and then disconnect the other end from the pressure regulator. *Be sure to remove the tube in this order to prevent kinking or bending it.*

4. Loosen the spring retainer clip and carefully release the spring tension on the pressure springs. Remove the valve springs, retainer and valve stop, and the valves from the pressure regulator body.

5. Remove the pressure regulator attaching bolts and washers and take the regulator body out of the transmission case.

Installation

1. Place the replacement regulator body on the transmission case and install the two securing bolts, tighten the bolts to specifications.

2. Check the converter pressure and the control pressure valves to be sure that the valves are operating freely in the bores.

3. Install the valve springs, spacer and retainer.

4. Install the main pressure oil tube.

NOTE: Be sure to install the end of the tube that connects to the pressure regulator assembly first. Then, install the other end of the tube into the main control assembly by tapping it gently with a soft faced hammer.

5. Install the small compensator pressure tube.

6. Install the fluid screen and the oil pan. Fill the transmission to the correct level with the specified fluid.

Front Servo

Removal

1. Drain the transmission fluid from the transmission and remove the pan, fluid filter screen, and its retaining clip.

2. Remove the vacuum diaphragm unit.

3. Loosen the three control valve body attaching bolts.

4. Remove the front servo attaching bolts, hold the band strut steady, and remove the front servo unit.

Installation

1. After repairing the front servo unit, install it by first positioning the front band forward in the transmission case with the end of the band facing downward. Be sure the front servo anchor pin is placed in the case web. Align the large end of the servo strut with the servo actuating lever, and align the small end with the band end.

2. Rotate the band, strut, and servo to align the anchor end of the band with the anchor in the case. Push the servo unit onto the control valve body tubes.

3. Install the attaching bolts and tighten them to 30-35 ft. lbs. torque.

4. Tighten the control valve body attaching bolts to 8-10 ft. lbs. torque. Check the clearance (0.050 in.) between the manual valve and the manual lever actuating pin.

5. Adjust the front band.

6. Install the vacuum diaphragm unit and its pushrod.

7. Install the fluid filter screen, its retaining clip, and the pan with a new pan gasket.

8. Fill the transmission with fluid.

9. Adjust the downshift and manual shift linkage.

Rear Servo

Removal

1. Drain the transmission fluid from the transmission, and remove the pan, fluid filter screen, and its retaining clip.

2. Remove the vacuum diaphragm unit.

3. Remove the control valve body and the two front servo tubes.

4. Remove the rear servo attaching bolts, hold the actuating and anchor struts, and remove the rear servo unit.

Installation

1. Before installing the rear servo unit, position the servo anchor strut on the servo band and rotate the band to engage the strut.

2. While holding the servo anchor strut in place, position the actuating lever strut and install the rear servo unit in place.

3. Loosely install the rear servo attaching bolts, with the longer bolt in the inner bolt hole.

4. Move the rear servo unit toward the center of the transmission case against the attaching bolts. While holding the servo in this position, tighten the attaching bolts.

5. Install the two front servo tubes and the control valve body. Check for proper clearance (0.050 in.) between the manual

valve and the manual actuating pin.

6. Adjust the rear band.
7. Install the fluid filter screen, its retaining clip, and the oil pan with a new gasket. Fill the transmission with fluid.

Transmission Overhaul Procedures

The transmission overhaul procedures presented here are the checks and repairs that must be done with the transmission out of the truck. Each transmission subassembly is illustrated by exploded views of the subassembly showing how the individual parts fit together. Assembly is often the reverse of the disassembly procedure except for alignment, special tolerances, etc.

During all repairs to the transmission subassemblies, the following instructions must be followed:

1. Be sure that no dirt or grease gets in the transmission. All parts must be clean. *Remember —a little dirt can disable a transmission completely if it gets in a fluid passage.*
2. Handle all transmission parts carefully to avoid burring or nicking bearing or mating surfaces.
3. Lubricate all internal parts of the transmission with clean transmission fluid before assembling. *Do not use any other lubricants except on gaskets or thrust washers which may be coated lightly with vaseline to ease assembly.*
4. Always use new gaskets when assembling the transmission.
5. Tighten all bolts and screws to the recommended torque limits using a torque wrench.

Transmission Disassembly

1. Thoroughly clean the outside of the transmission to prevent dirt or grease from getting inside. *Do this before removing any subassembly.*
2. Place the transmission in the transmission holder or stand.
3. Remove the transmission oil pan, gasket, and fluid filter retaining clip.
4. Lift the fluid filter screen off the forward tube, and then off the rear tube.
5. Remove the spring seat from the pressure regulator. *Maintain constant pressure on the spring seat and release slowly to prevent spring distortion and personal injury.*
6. Remove the pressure regulator springs and pilots, but do not remove the valves yet.
7. Loosen but do not remove the

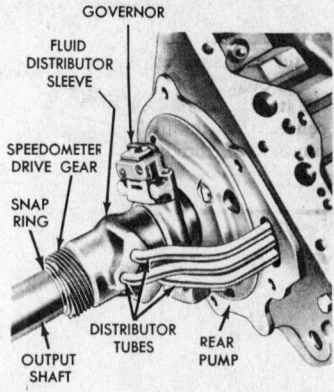

Output shaft and rear pump installed
(© Ford Motor Co.)

pressure regulator attaching bolts.

8. Remove the small compensator pressure tube from the pressure regulator and the control valve body.
9. Remove the main pressure oil tube from the pressure regulator and the main control valve body assembly. Gently pry off the end connected to the main control valve body first and then remove the tube from the pressure regulator. *Failure to do this may kink or bend the tube causing damage to the transmission.*
10. Loosen the front and rear band adjusting screws five turns. Loosen the front servo attaching bolts three turns.
11. Remove the vacuum diaphragm unit and pushrod.
12. Remove the control valve body attaching bolts and align the levers to allow removal of the valve body. Lift the valve body up and pull it off the servo tubes. Place the valve body on a clean surface.
13. Remove the pressure regulator from the case. Keep the control pressure valve and the converter pressure valve in the regulator body to avoid damaging the valves.
14. Remove the front servo supply and release tubes by twisting

and pulling at the same time. Remove the front servo attaching bolts. While holding the front servo strut, lift the front servo from the case.

15. Remove the rear servo attaching bolts. While holding the actuating and anchor struts, lift the rear servo from the case.

Transmission End-Play Check

The transmission end-play is checked as follows:

1. Remove one of the front pump attaching bolts and mount the dial indicator support tool in the hole. Mount a dial indicator on the support so that the stem rests on the end of the turbine shaft.
2. Install the extension housing seal replacer on the output shaft to provide support for the shaft.
3. Using a screwdriver, move the front clutch cylinder to the rear of the transmission case as far as possible. Set the dial indicator to zero while holding a slight pressure on the screwdriver.
4. Remove the screwdriver. Insert it between the large internal gear and the case and move the front clutch cylinder to the front of the case.
5. Record the indicator reading for later use during transmission reassembly. The end-play reading should be between 0.010-0.029 in. If the reading is not within these limits, a new selective thrust washer must be used when reassembling the transmission.

Extension Housing Parts and Case

Removal

1. Remove the front pump attaching bolts. Remove the front pump and gasket. It may be necessary to tap the screw bosses with a soft faced hammer to loosen the pump from the case. *NOTE: If the parking brake drum is mounted on the transmission, remove the drum and yoke from the output shaft.*

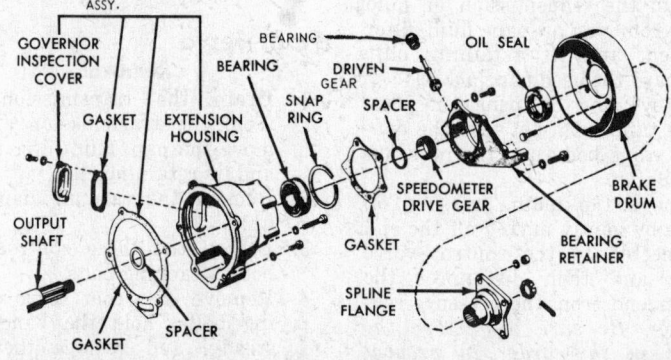

Parking brake drum and extension housing components—Ford FMX (© Ford Motor Co.)

2. Remove the extension housing seal. Remove the lubrication tube from the case. Remove the transmission-to-extension housing bolts and the extension housing. These bolts also secure the rear support to the case.

3. Remove the output shaft assembly. Be sure not to bend the pressure tubes between the distributor sleeve and case, as the tubes are being removed from the case.

4. Place the output shaft assembly on the bench and remove the oil distributor tubes from the sleeve.

5. Slide the speedometer drive gear ball off the shaft.

6. If the drive gear ball does not fall out as the speedometer gear is removed, remove the ball from the seat in the output shaft.

7. Remove the distributor sleeve. Remove the four seal rings from the output, be careful not to break the rings.

8. Remove the governor snap ring from the output shaft. Slide the governor assembly off the output shaft. Remove the governor drive ball. Remove the extension housing and pump gaskets.

9. Remove the rear pump drive key from the output shaft. Remove the bronze thrust washer from the output shaft.

10. Remove the selective thrust washer from the rear pinion carrier.

11. Remove the two seal rings from the primary gear shaft. Remove the pinion carrier.

12. Remove the primary sun gear rear thrust bearing and race from the pinion carrier.
NOTE: Record the rear band position for reference during assembly.

13. The end of the band next to the adjusting screw has a depression in the center of the boss. Squeeze the ends of the rear band together, tilt the band to the rear, and remove the rear band from the case.

14. Remove the two center support outer bolts from the transmission case.

15. Hold the clutch units together on the input shaft. Remove the center support and the front and rear clutches as a unit.

16. Place the clutch assemblies in a holding fixture.

17. Remove the thrust washers from the front of the input shaft.

18. Position the front band ends between the case webbing and tilt the bottom of the band rearward. Squeeze the ends of the band together and remove the front band from the rear of the case.

19. Lift the front clutch assembly from the primary sun gear shaft.

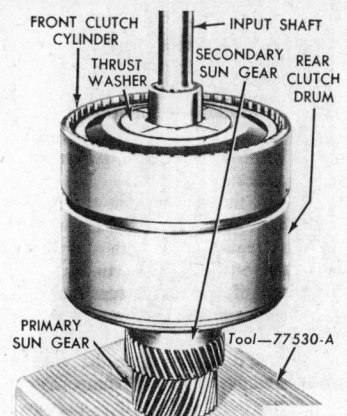

Input shaft and clutch mounted in holding fixture
(© Ford Motor Co.)

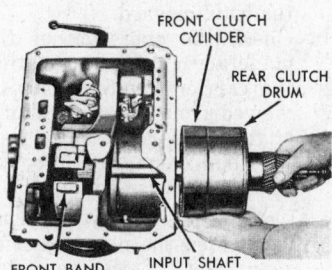

Removing input shaft and clutch
(© Ford Motor Co.)

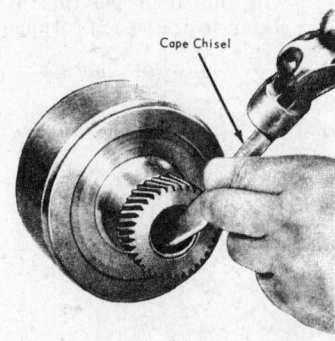

Removing rear clutch sun gear bushing
(© Ford Motor Co.)

20. Remove the bronze and the steel thrust washers from the rear clutch assembly, wire the thrust washers together to be sure of correct assembly.

21. Remove the front clutch seal rings from the primary sun gear shaft. Lift the rear clutch assembly from the primary sun gear shaft and remove the rear clutch seal rings from the primary sun gear shaft.
CAUTION: Do not break the rear clutch seal rings.
Remove the primary sun gear front thrust washer.
NOTE: If the transmission case bushing is worn or damaged, replace the bushing.

22. If the rear brake drum support bushing is worn or damaged, replace it.

23. Remove the output shaft bushing if it is worn or damaged. Use a chisel and cut along the bushing seam until the chisel breaks through the bushing wall. Pry the loose ends of the bushing up with an awl and remove the bushing.

24. Install a new bushing with its proper installation tool. The bushing must be pressed into position.

Assembly

NOTE: Do not use force to assemble mating parts. If the parts do not go together easily, examine them for the cause of the problem. Always use new gaskets and seals during assembly operations.

1. Install the front band in the transmission case with the anchor end aligned with the anchor in the case.

2. Make sure the thrust washer is in place on the input shaft. Lift the clutch assemblies out of the holding fixture.
CAUTION: Do not allow the clutches to separate.

3. Install the clutch sub-assemblies

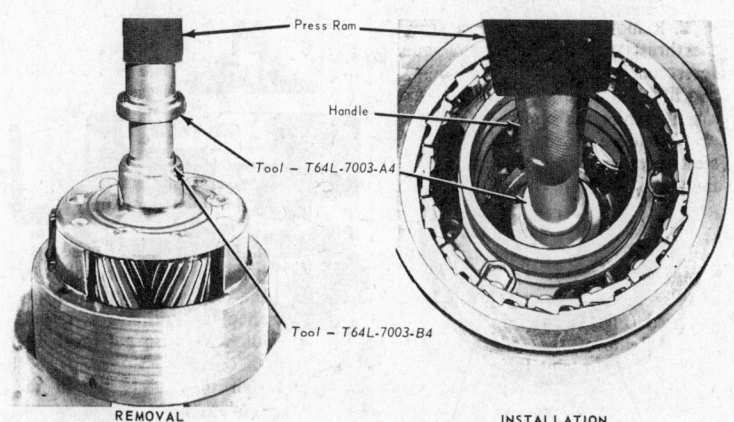

Replacing rear brake drum support bushing (© Ford Motor Co.)

in the transmission case while positioning the servo band on the drum. Hold the units together while installing them.

4. Assemble and install the center support assembly as follows:
   a. Install the center support and the rear band in the case.
   b. Install the one piece needle bearing and race assembly in the planet carrier, be sure that the black oxide coated race is facing the front of the transmission.
   c. Lubricate the bearing surface on the center support, the rollers of the planetary clutch, and the cam race in the carrier with petroleum jelly.
   d. Install the planetary clutch in the carrier.
   e. Carefully, position the planet carrier on the center support. Move the carrier forward until the clutch rollers are felt to contact the bearing surface on the center support.
   f. While applying forward pressure on the planet carrier, rotate it counterclockwise, as viewed from the rear. The clutch rollers will roll towards the large opening end of the cams in the race, compressing the spring slightly and the rollers will ride up

Installing output shaft bushing
(© Ford Motor Co.)

the chamber on the planetary support onto the inner race.
   g. Push the planet carrier all the way forward.
   h. Check the operation of the planetary clutch by rotating the carrier counterclockwise, viewed from the rear, with a slight drag, and should lock up when attempting to rotate in the clockwise direction.
   i. Install the selective thrust washer on the pinion carrier rear pilot. If the end play was not within specifications when checked prior to disassembly, replace the washer with one of proper thickness. Refer to the specifications at

the end of this section.
   j. Install the output shaft, carefully meshing the internal gear with the pinions.

5. With the center support properly assembled, position the rear pump drive key in the keyway on the output shaft.
6. Position the new front and rear gaskets on the pump body. Retain the gaskets with clean transmission fluid. Then install the rear pump. Be sure that the key is aligned with the keyway in the pump drive gear.
7. Position the governor drive ball in the pocket in the output shaft. Retain the ball with clean transmission fluid.
8. Install the governor assembly aligning the groove with the ball in the output shaft.
9. Install the governor with the governor body plate facing towards the rear of the transmission. Install the governor snap ring.
10. Place the four seal rings in the distributor sleeves, and check the ring gap.
11. Check the fit of the grooves in the output shaft. The rings should rotate freely. Install the rings in the grooves of the output shaft.
12. Install the three tubes in the distributor sleeve.
13. Install the distributor sleeve on the output shaft, with the chamfer facing forward. Lubricate the parts with clean transmission fluid to facilitate assembly. Slide the sleeve over the four rings and at the same time start the tubes into the case. The distributor sleeve is located between the governor snap ring and the speedometer drive gear.
14. Make sure that the speedometer drive gear lock ball is in place, then install the speedometer drive gear.
15. Install the output shaft ball bearing front spacer and the ball bearing with the snap ring toward the rear.
16. Install the output shaft rear spacer and the speedometer drive gear.
17. Install the rear bearing retainer gasket and the retainer. Torque the bolts to specification.
18. Install the parking brake drum, if so equipped, and/or the u-joint yoke. Torque the nut to specification. Tighten the nut to the nearest cotter pin hole, and install a new cotter pin.
19. Install the front pump gasket in the counter bore of the transmission case.
20. Install the front pump, aligning the pump bolt holes with the holes in the case. Install the

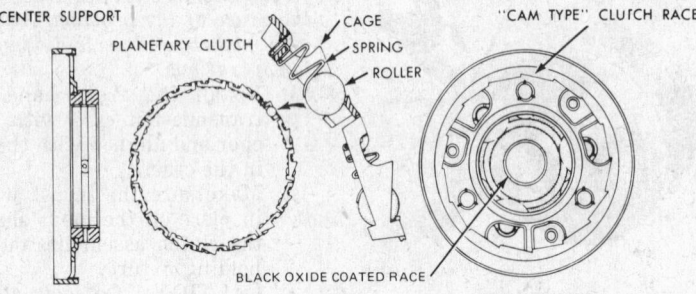

Planetary clutch, planet carrier and center support—Ford FMX (© Ford Motor Co.)

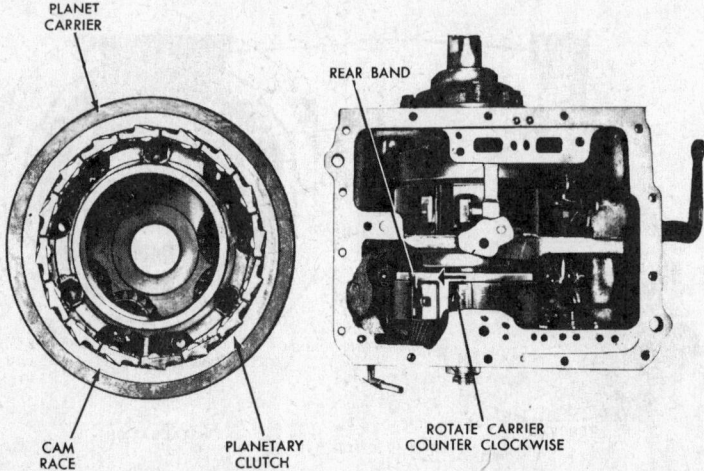

Installation of planetary clutch in carrier-Ford FMX (© Ford Motor Co.)

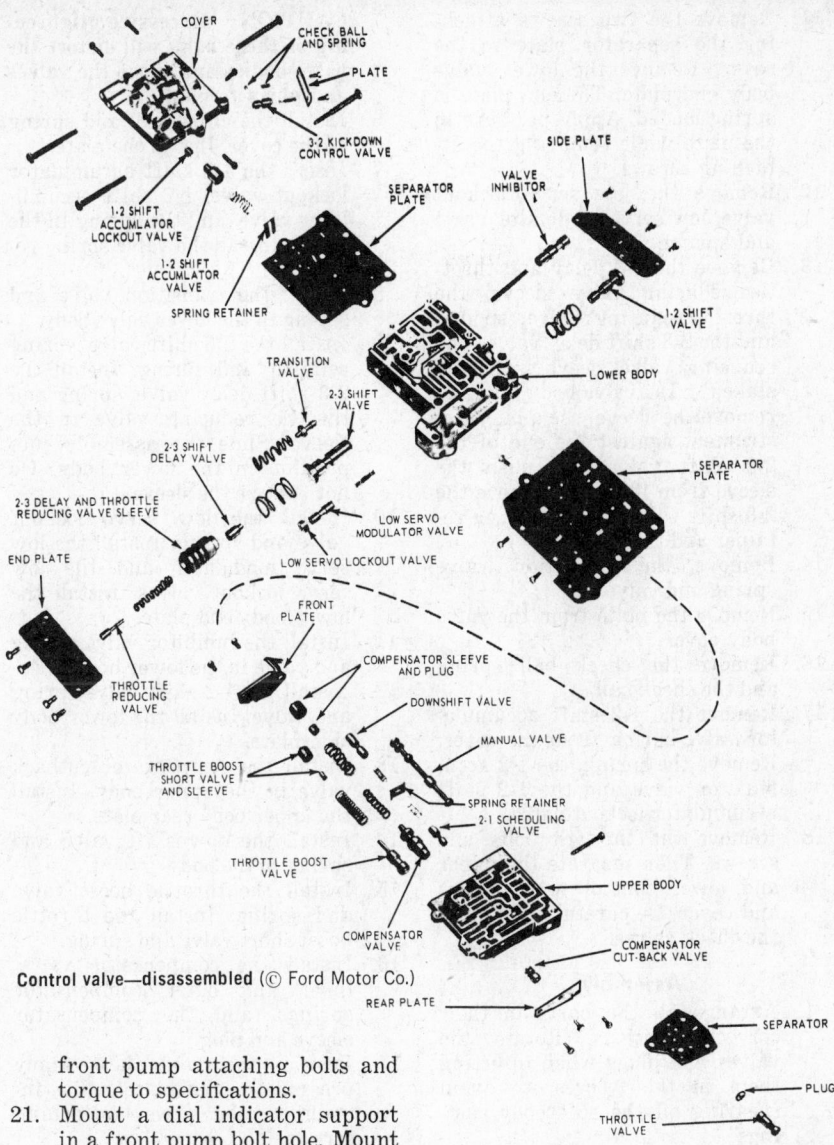

COVER

CHECK BALL AND SPRING

PLATE

3-2 KICKDOWN CONTROL VALVE

1-2 SHIFT ACCUMLATOR LOCKOUT VALVE

1-2 SHIFT ACCUMLATOR VALVE

SPRING RETAINER

SIDE PLATE

VALVE INHIBITOR

SEPARATOR PLATE

1-2 SHIFT VALVE

TRANSITION VALVE

2-3 SHIFT VALVE

LOWER BODY

2-3 SHIFT DELAY VALVE

2-3 DELAY AND THROTTLE REDUCING VALVE SLEEVE

SEPARATOR PLATE

END PLATE

LOW SERVO MODULATOR VALVE

LOW SERVO LOCKOUT VALVE

FRONT PLATE

THROTTLE REDUCING VALVE

COMPENSATOR SLEEVE AND PLUG

DOWNSHIFT VALVE

MANUAL VALVE

THROTTLE BOOST SHORT VALVE AND SLEEVE

SPRING RETAINER

2-1 SCHEDULING VALVE

THROTTLE BOOST VALVE

UPPER BODY

COMPENSATOR VALVE

COMPENSATOR CUT-BACK VALVE

REAR PLATE

SEPARATOR

PLUG

THROTTLE VALVE

THROTTLE VALVE BODY

Control valve—disassembled (© Ford Motor Co.)

front pump attaching bolts and torque to specifications.

21. Mount a dial indicator support in a front pump bolt hole. Mount a dial indicator on the support so that the contact rests on the end of the turbine shaft.

22. Pry the output shaft all the way forward by using a screw driver between the large internal gear and the case.

23. Lightly block the output shaft in the forward position to eliminate all output shaft end play.

24. Inserting a screwdriver between the planet carrier and the large internal gear move the pinion carrier all the way forward. Maintain slight forward pressure, and set the dial indicator at zero.

25. Measure and record the end play between the front of the case and the large internal gear by prying between the front clutch cylinder and the case. This end play should be 0.010-0.029 inch. Total end play including the output shaft must not exceed 0.044 inch. If the end play is not within limits, a new selective thrust washer must be used.

26. Remove the dial indicator and install the remaining pump retaining bolt, tighten to specifications.

27. Position the front band forward in the case with the band end up.

28. Position the servo strut with the slotted end aligned with the servo actuating lever, and the small end aligned with the band end. Rotate the band, strut and servo into position, engaging the anchor end of the band with the anchor pin in the case.

29. Locate the servo on the case, and install the attaching bolts. Tighten the attaching bolts only 2 or 3 threads.

30. Install the servo tubes.

31. Position the servo anchor strut, and rotate the rear band to engage the strut.

32. Position the servo actuating lever strut with a finger, and

then install the servo and attaching bolts. Move the servo toward the centerline of the case, against the attaching bolts. While holding the servo in this position, torque the attaching bolts to specification.

33. Install the pressure regulator body and attaching bolts. Torque the bolts to specifications.

34. Install the control and converter valve guides and springs. Install the spring retainer.

35. Install the control valve assembly, carefully alining the servo tubes with the control valve. Align the inner downshaft lever between the stop and the downshift valve. Shift the manual lever to the #1 position. Align the manual valve with the actuating pin in the manual detent lever. Do not tighten the attaching bolts.

36. Install the main pressure oil tube. Be sure to install the end of the tube that connects to the pressure regulator assembly first. Then, install the other end of the tube into the main control assembly, by tapping gently with a soft faced hammer.

37. Install the small control pressure compensator tube in the valve body and regulator.

38. Move the control valve body toward the center of the case, until the clearance is less than 0.050 inch between the manual valve and the actuating pin on the manual detent lever.

39. Torque the attaching bolts to specifications. Be sure that the rear fluid screen retaining clip is installed under the valve body bolt.

40. Turn the manual valve one full turn in each manual lever detent position. If the manual valve binds against the actuating pin in any detent position, loosen the valve body attaching bolts and move the body away from the center of the case. Move the body only enough to relieve the binding. Torque the attaching bolts to specifications. Recheck the manual valve for binding.

41. Torque the front servo attaching bolts to specifications.

42. Adjust the front and rear bands as outlined in the "Adjustment" section.

43. Position the control rod in the bore of the vacuum diaphragm unit and install the diaphragm unit. Make sure that the control rod enters the throttle valve as the vacuum unit is installed, torque the retaining bolts to specifications.

44. Position the fluid screen under the rear clip and over the front pump inlet tube. Press the screen

down firmly. Install the screen retaining clip.

45. Place a new gasket on the transmission case and install the pan. Install the attaching bolts and lock washers. Tighten the bolts to specifications.

46. If the converter and converter housing were removed from the transmission, install these components. Position the transmission assembly on a transmission jack and install in the vehicle.

Control Valve Body

Disassembly

NOTE: During disassembly of the control valve assembly, avoid damage to the valve parts and keep the valve parts clean. Place the valve assembly on a clean shop towel while performing the disassembly operations. Do not separate the upper and lower valve bodies and cover until after the valves have been removed.

1. Remove the manual valve.
2. Remove the throttle valve body and the separator plate. Be careful not to lose the check valve when removing the separator plate from the valve body. Remove the throttle valve and plug.
3. Remove one screw attaching the separator plate to the lower valve body. Remove the upper body front plate. The plate is spring-loaded. Apply pressure to the plate while removing the attaching screws.
4. Remove the compensator sleeve and plug, remove the compensator valve springs. Remove the compensator valve.
5. Remove the throttle boost short valve and sleeve. Remove the throttle boost spring and valve.
6. Remove the downshift valve and spring.
7. Remove the upper valve body rear plate.
8. Remove the compensator cut back valve.
9. Remove the lower body side plate. The plate is spring-loaded. Apply pressure to the plate while removing the attaching screws.
10. Remove the 1-2 shift valve and spring. Remove the inhibitor valve and spring.

11. Remove the two screws attaching the separator plate to the cover. Remove the lower valve body end plate. The end plate is spring-loaded. Apply pressure to the plate while removing the attaching screws.
12. Remove the low servo lockout valve, low servo modulator valve and spring.
13. Remove the 2-3 delay and throttle reducing valve sleeve, the throttle reducing valve, spring, and the 2-3 shift delay valve. The reducing valve sleeve is lightly staked in the valve body bore. To remove the sleeve, use a blunt instrument against the end of the 2-3 shift valve and push the sleeve from its bore. Remove the 2-3 shift valve spring, spring retainer and the valve.
14. Remove the transition valve spring and valve.
15. Remove the plate from the valve body cover.
16. Remove the check ball spring and the check ball.
17. Remove the 1-2 shift accumulator valve spring from the cover. Remove the spring, the 1-2 accumulator valve and the 1-2 shift accumulator lockout valve.
18. Remove the through bolts and screws. Then separate the upper and lower control valve bodies and cover. Be careful not to lose the check valves.

Assembly

1. Arrange all the parts in their correct positions. Rotate the valves and plugs when inserting them in their bores, to avoid shearing off the soft body castings.
2. Place the check valve in the upper body, then position the separator plate on the body.
3. Position the lower body on the upper body, and start, but do not tighten the attaching bolts.
4. Position the cover and separator plate on the lower body. Start the four through bolts.
5. Align the separator with the upper and lower valve body attaching bolt holes. Install and torque the four valve body bolts to specifications.

CAUTION: Excessive tightening of these bolts will distort the valve bodies and cause the valves and plugs to bind.

6. Install the check ball and spring in the cover. Install the plate.
7. Insert the 1-2 shift accumulator lockout valve, 1-2 shift accumulator valve, and the spring in the cover. Install the valve spring retainer.
8. Install the transition valve and spring in the lower valve body.
9. Install the 2-3 shift valve, spring retainer and spring. Install the 2-3 shift delay valve, spring and throttle reducing valve in the sleeve. Slide the assembly into position in the lower body. Do not restake the sleeve.
10. Install the low servo lockout valve and spring. Install the low servo modulator and the low servo lockout valves. Install the lower body end plate.
11. Install the inhibitor valve spring and valve in the lower body.
12. Install the 1-2 shift valve spring and valve. Install the lower body side plate.
13. Install the compensator cutback valve in the upper body. Install the upper body rear plate.
14. Install the downshift valve and spring in the body.
15. Install the throttle boost valve and spring. Install the throttle boost short valve and spring.
16. Install the compensator valve, inner and outer compensator springs, and the compensator sleeve and plug.
17. Position the front plate. Apply pressure to the plate while installing the two attaching screws.
18. Install the throttle valve, plug, and check valve in the throttle valve body. Position the separator on the upper body and install the throttle valve body. Install the three attaching screws.
19. Install the four screws attaching the cover to the lower body, two screws attaching the separator plate to the lower body. Torque the cover and body screws to specifications.
20. Install the manual valve.

Front Servo

Disassembly

1. Remove the servo piston retainer snap ring. The servo piston is spring loaded. Apply pressure to the piston when removing the snap ring.
2. Remove the servo piston retainer, and the servo piston from the servo body. If necessary, tap the piston stem lightly with a soft-faced hammer to separate the piston retainer from the servo body.

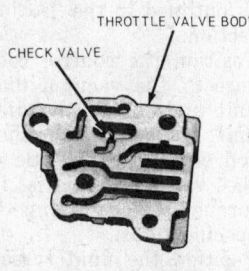
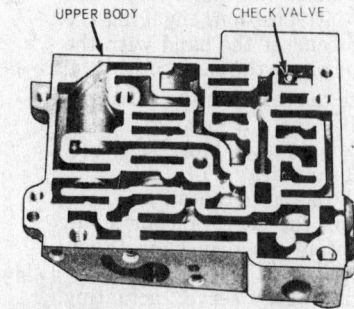

Check valve locations—Ford FMX (© Ford Motor Co.)

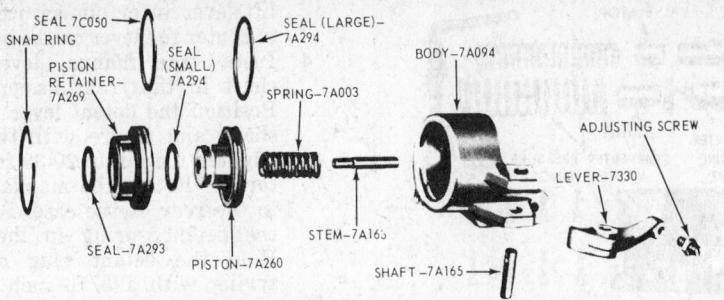

Front servo disassembled—Ford FMX (© Ford Motor Co.)

3. Remove all the seal rings, and remove the spring from the servo body.
4. Inspect the servo body for cracks, and the piston bore and the servo piston stem for scores. Check the fluid passages for obstructions.
5. Check the actuating lever for free movement and inspect it for wear. To replace the actuating lever shaft, it will be necessary to press the shaft out of the bracket. The shaft is retained in the body by serrations on one end of the shaft. These serrations cause a press fit at the end. To remove the shaft, press on the end opposite the serrations. Inspect the adjusting screw threads and the threads in the lever.
6. Check the servo spring and the servo band strut for distortion.
7. Inspect the servo band lining for excessive wear and bonding of the metal. The band should be replaced if worn to a point where the grooves are not clearly evident.
8. Inspect the band ends for cracks and check for band distortion.

Assembly

1. Lubricate all parts of the front servo with clean transmission fluid before starting assembly.
2. Install the inner and outer 'O' rings on the piston retainer. Install a new 'O' ring on the return piston and on the servo piston.
3. Position the servo piston release spring in the servo body. Install the servo piston, retainer, and return piston in the servo body. Compress the assembly into the body and secure it with the snap ring. Make sure that the snap ring is fully seated in the groove.
4. Install the adjusting screw and lock nut in the actuating lever if they were removed.

Rear Servo

Disassembly

1. Remove the servo actuating lever shaft retaining pin with a 1/8 inch punch.
2. Press down on the servo spring retainer, and remove the snap ring. Release the pressure on the retainer slowly to prevent the spring from flying out.
3. Remove the retainer and servo spring.
4. Force the piston out of the servo body with air pressure. Hold one hand over the piston to prevent damage.
5. Remove the piston seal ring.

Assembly

1. Install a new piston seal ring on the servo piston.
2. Install the piston in the servo body. Lubrication of the parts with clean transmission fluid will facilitate assembly. Install the servo spring with the small coiled end against the servo piston.
3. Install the spring retainer. Compress the spring with a C-clamp. Then, install the snap ring. The snap ring must be fully seated in the groove.
4. Install the actuating lever with

the socket in the lever bearing on the piston stem. Install the actuating lever shaft, aligning the retaining pin holes, and install the pin.
5. Check the actuating lever for free movement.

Governor

Disassembly

1. Remove the governor valve body cover.
2. Remove the valve body from the counterweight.
3. Remove the plug, sleeve, and the valve and spring from the governor body.
4. Remove the screen from its bore in the valve body.

Assembly

1. Install the governor valve and spring assembly in the bore of the valve body. Install the sleeve and plug.
2. Install the screen.
3. Install the body counterweight. Make sure that the fluid passages in the body and counterweight are aligned.
4. Position the valve body cover on the body, and install the screws.

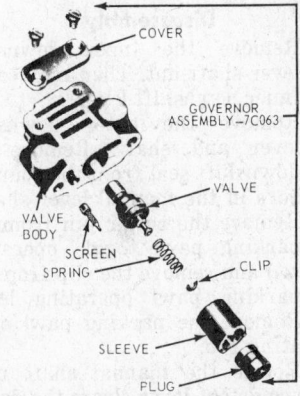

Governor disassembled—Ford FMX (© Ford Motor Co.)

Pressure Regulator

Disassembly

1. Remove the valves from the regulator body.
2. Remove the regulator body cover attaching screws, and remove the cover.
3. Remove the separator plate.
4. Wash all parts thoroughly in a clean solvent and blow dry with moisture-free compressed air.
5. Inspect the regulator body and cover mating surfaces for burrs.
6. Check all fluid passages for obstructions.
7. Inspect the control pressure converter pressure valves and bores for burrs and scoring. Remove all burrs with crocus cloth.
8. Check the free movement of the valve in their bores. Each valve

Rear servo disassembled—Ford FMX (© Ford Motor Co.)

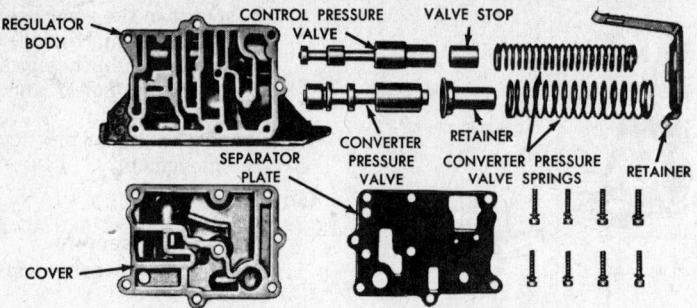

REGULATOR BODY — CONTROL PRESSURE VALVE — VALVE STOP — RETAINER

SEPARATOR PLATE — CONVERTER PRESSURE VALVE — CONVERTER PRESSURE VALVE SPRINGS — RETAINER

COVER

Pressure regulator assembly (© Ford Motor Co.)

should fall freely into its bore when both the valve and bore are dry.

9. Inspect the valve springs for distortion.

Assembly

1. Position the separator plate on the regulator cover.
2. Position the regulator cover and separator plate on the regulator body and install the attaching screws. Tighten to specifications.
3. Insert the valves in the pressure regulator body.

Downshift and Manual Linkage

Disassembly

1. Remove the inner downshift lever shaft nut. Then remove the inner downshift lever.
2. Remove the outer downshift lever and shaft. Remove the downshift seal from the counterbore in the manual lever shaft.
3. Remove the cotter pin from the parking pawl toggle operating rod and remove the clip from the parking pawl operating lever. Remove the parking pawl operating rod.
4. Rotate the manual shaft until the detent lever clears the detent plunger. Then, remove the detent plunger and spring. Do not allow the detent plunger to fly out of the case.
5. Remove the manual lever shaft nut and remove the detent lever. Remove the outer manual lever and shaft from the transmission case.
6. Tap the toggle lever sharply toward the rear of the case to remove the plug and pin.
7. Remove the pawl pin by working the pawl back and forth. Remove the pawl and toggle lever assembly and then disassemble.
8. Remove the manual shaft seal and the case vent tube.

Assembly

1. Coat the outer diameter of a new manual shaft seal with sealer and install the seal in the transmission case with a driver.

2. Install the vent tube in the transmission case.
3. Assemble the link to the pawl with the pawl link pin, washer, and pawl return spring. Assemble the toggle lever to the link with the toggle link pin. Position the pawl return spring over the toggle link pin and secure in place with the washer and small retainer clip. Position the pawl assembly in the transmission case and install the pawl pin, and the toggle lever pin. Press the retaining plug tightly against the toggle lever pin. Install the torsion lever assembly. Position the spring on the torsion lever with a screwdriver. Make sure that the short side of the toggle lift lever does not extend beyond the largest diameter of the ball on the toggle link. Tap the toggle

lift lever in or out as necessary to center the lever on the ball.

4. Install the manual lever and shaft in the transmission case. Position the detent lever on the shaft, and secure with the nut. Tighten the nut to 20-30 ft-lbs of torque. Rotate the manual lever to the rear of the case. Position the detent spring in the case. Hold the detent plug on the spring with a 3/16 inch socket wrench, then depress the spring until the plug is flush with the case. Carefully rotate the manual lever to the front of the case to secure the plug. A piece of thin walled tubing may be used to depress the plug if a small socket wrench is not available.
5. Position the ends of the parking pawl operating rod in the detent lever and toggle lift lever, secure with the two small retaining pins.
6. Install a new seal on the downshift lever shaft, then install the lever and shaft in the transmission case. Position the inner end of the shaft with the mark "O" racing toward the center of the case. Install the lock washer and nut, then tighten the nut to 17-20 ft-lbs torque.
7. Check the operation of the linkage. The linkage should operate freely without binding.

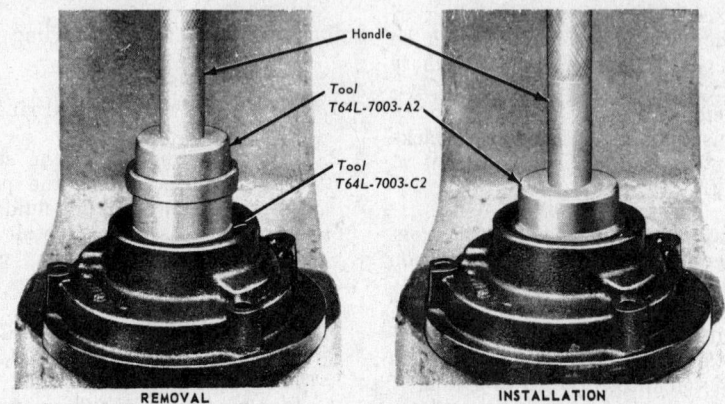

Handle — Tool T64L-7003-A2 — Tool T64L-7003-C2

REMOVAL — INSTALLATION

Replacing front pump housing bushing (© Ford Motor Co.)

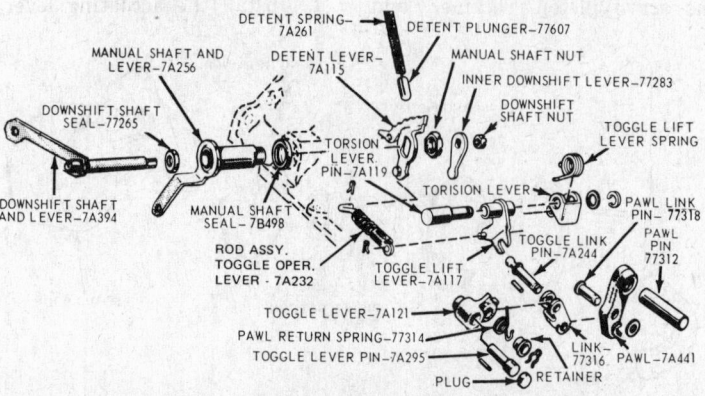

DETENT SPRING—7A261 — DETENT PLUNGER—77607
MANUAL SHAFT AND LEVER—7A256 — DETENT LEVER—7A115 — MANUAL SHAFT NUT
INNER DOWNSHIFT LEVER—77283
DOWNSHIFT SHAFT SEAL—77265 — DOWNSHIFT SHAFT NUT
TORSION LEVER PIN—7A119 — TOGGLE LIFT LEVER SPRING
DOWNSHIFT SHAFT AND LEVER—7A394 — TORSION LEVER — PAWL LINK PIN—77318
MANUAL SHAFT SEAL—7B498 — PAWL PIN 77312
ROD ASSY. TOGGLE OPER. LEVER—7A232 — TOGGLE LINK PIN—7A244
TOGGLE LIFT LEVER—7A117
TOGGLE LEVER—7A121 — LINK 77316 — PAWL—7A441
PAWL RETURN SPRING—77314
TOGGLE LEVER PIN—7A295 — PLUG — RETAINER

Transmission case control linkage—Ford FMX (© Ford Motor Co.)

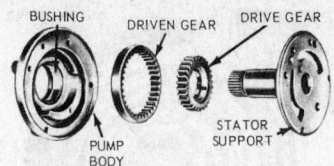

Front pump disassembled
(© Ford Motor Co.)

Front Pump

Disassembly

1. Remove the stator support attaching screws and remove the stator support. Mark the top surface of the pump driven gear with Prussian blue to assure correct assembly. Do not scratch the pump gears.
2. Remove the drive and driven gears from the pump body.
3. Inspect the pump body housing, gear pockets and crescent for signs of wear or damage.
4. If the pump housing bushing is worn or damaged, replace the bushing with the proper tool.
5. If any parts other than the stator support, bushings of oil seal are found to be defective, replace the pump as a unit. Minor burrs and scores may be removed with crocus cloth. The stator support is serviced separately.
6. If the oil seal requires replacement, bolt the front pump to the transmission case. Install the seal removing tool and pull the seal from the body.
7. Clean the pump body counterbore. Inspect the bore for rough spots. Smooth up the counterbore with crocus cloth.
8. Remove the pump body from the transmission case.

Assembly

1. If the oil seal was removed, coat the outer diameter with sealing compound. Position the seal in the pump body and drive the new seal into the pump body with the proper tool. Be sure that the seal is firmly seated.
2. Place the pump driven gear in the pump body with the mark on the gear or tooth gear chamfer facing down. Install the drive gear in the pump body with the chamfered side of the flats facing down.
3. Install the stator support and the attaching screws. Check the pump gears for free rotation.

Rear Pump

Disassembly

1. Remove the screws and lock washers securing the pump cover to the pump body and remove the cover.
2. Remove the drive gear from the pump body.
3. If the pump housing bushing is worn or damaged, replace the bushing with the proper installation and removal tool.

Assembly

1. Install the drive gear in the pump body.
2. Install the pump cover, attaching screws, and lock washers. Torque the screws to specifications.

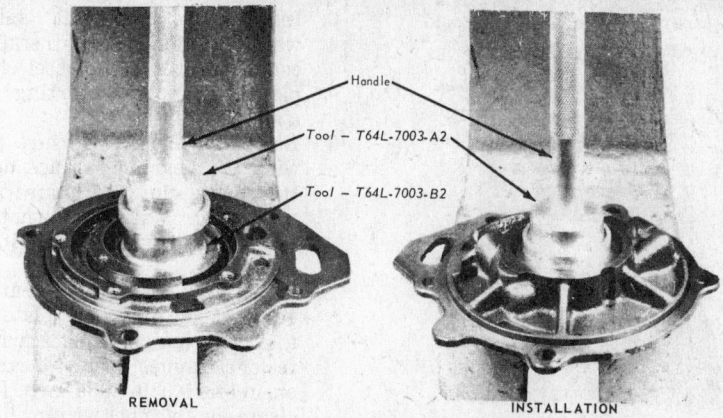

Replacing rear support housing bushing (© Ford Motor Co.)

Rear Clutch

Disassembly

1. Remove the clutch pressure plate snap ring and remove the plate from the drum. Remove the waved cushion spring. Remove the composition and steel plates.
2. Compress the spring with the special spring compression tool, and remove the snap ring.
3. Guide the spring retainer while releasing the pressure, this will prevent retainer from locking in the snap ring grooves.
4. Position the primary sun gear shaft in the rear clutch. Place an air hose nozzle in one of the holes in the shaft, and place one finger over the other hole. Force the clutch piston out of the clutch drum with air pressure. Hold one hand over the piston to prevent damage to the piston during removal.
5. Remove the inner and outer seal rings from the clutch piston.
6. Remove the rear clutch sun gear bushing if it is worn or damaged. Use a cape chisel and cut along the bushing seam until the chisel breaks through the bushing wall. Pry the loose ends of

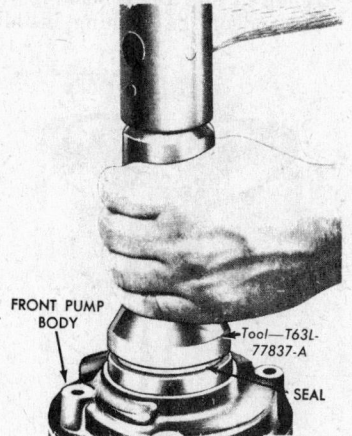

FRONT PUMP BODY

Tool—T63L-77837-A

SEAL

Installing front pump seal
(© Ford Motor Co.)

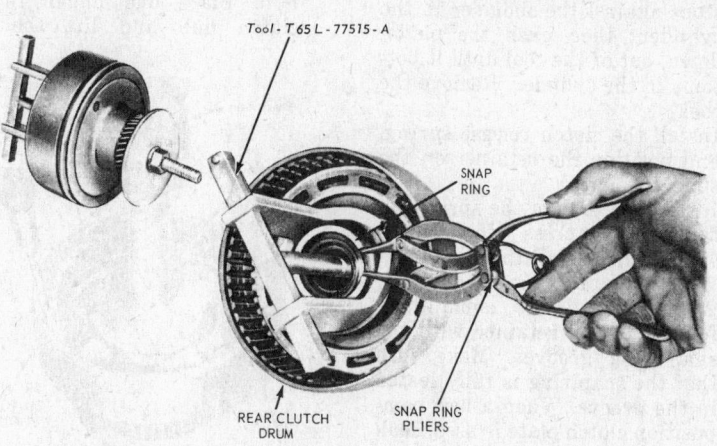

Removing rear clutch spring snap-ring (© Ford Motor Co.)

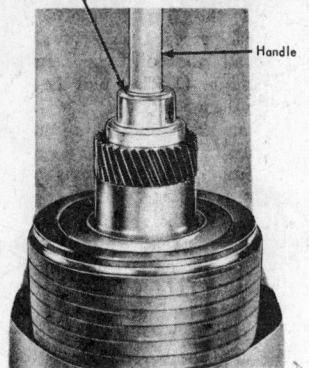

Tool - T64L-7003-A3 OR Tool - T64L-7003-A4

Handle

Installing rear clutch sun gear bushing
(© Ford Motor Co.)

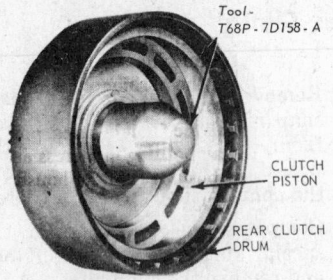

Tool - T68P - 7D158 - A

CLUTCH PISTON

REAR CLUTCH DRUM

Installing rear clutch piston
(© Ford Motor Co.)

the bushing up with an awl and remove the bushing.

Assembly

1. If the rear sun gear bushing was removed, press a new bushing into the rear clutch sun gear.
2. Install a new inner and outer seal ring on the piston.
3. Lubricate the piston seals and tools with clean transmission fluid. Push the small fixture over the cylinder hub. Insert the piston into the large fixture with the seal toward the thin walled end. Hold the piston and large fixture and insert as unit into the cylinder. Push down over the small fixture until the large tool stops against the shoulder in the cylinder; then push the piston down, out of the tool until it bottoms in the cylinder. Remove the tool.
4. Install the clutch release spring, and position the retainer on the spring.
5. Install the tool on the spring retainer. Compress the clutch spring and install the snap ring. While compressing the spring, guide the retainer to avoid interference of the retainer with the snap ring grooves. Make sure that the snap ring is fully seated in the groove. When a new composition clutch plate is used, soak the plates in clean transmission fluid for 15 minutes before assembly.

6. Install the external tabbed waved cushion spring. Install the composition and the steel clutch plates alternately, starting with a steel plate.
7. Install the clutch pressure plate with the bearing surface down. Install the clutch pressure plate snap ring. Make sure that the snap ring is fully seated in the groove.
8. Check the free pack clearance between the pressure plate and the first internal plate with a feeler gauge. The clearance should be 0.030-0.055 inch. If the clearance is not within limits, selective snap rings are available in the following thicknesses: 0.060-0.064, 0.074-0.078, 0.088-0.092 and 0.102-0.106 inch. Insert the correct size snap ring and recheck the clearance.
9. Install the thrust washer on the primary sun gear shaft. Be sure the thrust washer is installed with the tabs of the washer away from the sun gear thrust face. Lubricate all parts with clean transmission fluid or with petroleum jelly. Install the two center seal rings.
10. Install the rear clutch on the primary sun gear shaft. Be sure all of the needles are in the hub if the unit is equipped with loose needles. Install the two seal rings in the front grooves.
11. Install the steel and the bronze thrust washers on the front of the secondary sun gear assembly. If the steel washer is chamfered, place the chamfered side down.

Front Clutch

Disassembly

1. Remove the clutch cover snap rings with a screwdriver, and remove the input shaft from the front clutch drum.
2. Remove the thrust washer from the thrust surface on the clutch hub. Place one finger in the clutch hub and lift the hub

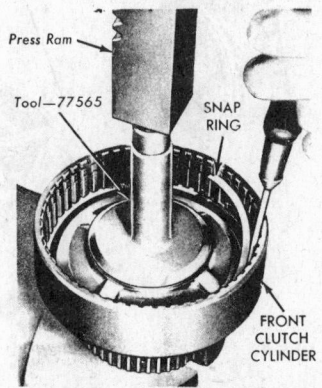

Press Ram

Tool—77565

SNAP RING

FRONT CLUTCH CYLINDER

Removing front clutch snap-ring
(© Ford Motor Co.)

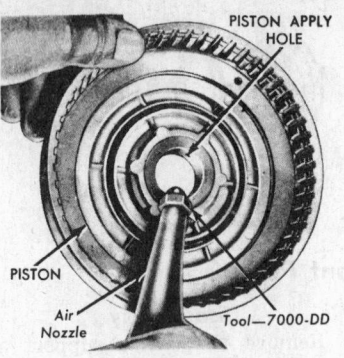

PISTON APPLY HOLE

PISTON

Air Nozzle

Tool—7000-DD

Removing front clutch piston
(© Ford Motor Co.)

straight up to remove the hub from the clutch drum.
3. Remove the composition and the steel pressure plates, and then remove the pressure plate from the clutch drum.
4. Place the front clutch spring compressor on the release spring, position the clutch drum on the bed of an arbor press, and compress the release spring with the arbor press until the release spring snap ring can be removed.
5. Remove the clutch release spring from the clutch drum.
6. Install the special air nozzle on an air line. Place the nozzle against the clutch apply hole in the front clutch housing and force the piston out of the housing.
7. Remove the piston inner seal from the clutch housing. Remove the piston outer seal from the groove in the piston.
8. Remove the input shaft bushing if it is worn or damaged. Use a cape chisel and cut along the bushing seam until the chisel breaks through the bushing wall. Pry the loose ends of the bushing up with an awl and remove the bushing.

Assembly

1. If the input shaft bushing was removed, slip a new bushing over the end of the bushing installa-

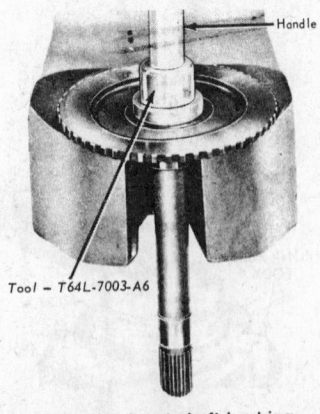

Handle

Tool - T64L-7003-A6

Installing input shaft bushing
(© Ford Motor Co.)

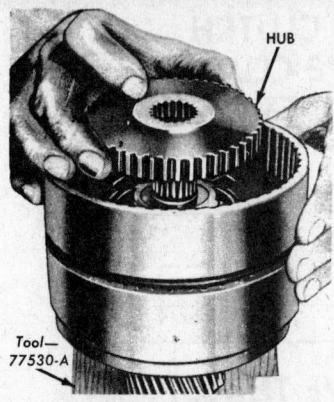

Installing front clutch hub
(© Ford Motor Co.)

Tool—77530-A

Installing clutch plates
(© Ford Motor Co.)

tion tool and place the tool and bushing on the bushing hole. Press the bushing into the input shaft.

2. Lubricate all parts with clean transmission fluid. Install a new piston inner seal ring in the clutch cylinder. Install a new piston outer seal in the groove in the piston.

3. Install the piston in the clutch housing. Make sure that the steel bearing ring is in place on the piston.

4. Position the release spring in the clutch cylinder with the concave side up. Place the release spring compressor on the spring and compress the spring with an arbor press. Install the snap ring. Make sure that the snap ring is firmly seated in the groove.

5. Install the front clutch housing on the primary sun gear shaft. Rotate the clutch units to mesh the rear clutch plates with the serrations on the clutch hub. Do not break the seal rings.

6. Install the clutch hub in the clutch cylinder with the deep counterbore down. Install the thrust washer on the clutch hub.

7. Install the pressure plate in the clutch cylinder with the bearing surface up. Install the composi-

tion and steel clutch plates alternately, starting with a composition plate. When new composition clutch plates are used, soak them in clean transmission fluid for 15 minutes before assembly.

8. The finial friction plate to be installed is selective. Install the thickest plate what will be a minimum of 0.010 inch below the

input shaft shoulder in the cylinder. For all other plates, use the thinnest available (see specifications at end of section).

9. Install the turbine shaft in the clutch cylinder, and secure with the snap ring. Make sure that the snap ring is firmly seated in the groove.

10. Install the thrust washer on the turbine shaft.

Primary Sun Gear Shaft

1. Position the primary sun gear shaft in the clutch bench fixture.

2. Check the fit of the seal rings in their bores. A clearance of 0.002-0.009 inch should exist between the ends of the rings.

3. Replace the seal rings, and check for free movement in the grooves.

TORQUE LIMITS FOR FMX AUTOMATIC TRANSMISSION

	Ft. Lbs.
Converter to flywheel	23-28
Converter hsg. to trans. case	40-50
Front pump to trans. case	17-22
Front servo to trans. case	30-35
Rear servo to trans. case	40-45
Upper valve body to lower valve body	4-6
Oil pan to case	10-13
Converter cover to converter hsg.	12-16
Regulator to case	17-22
Planetary support to trans. case	20-25
Control valve body to trans. case	8-10
Diaphragm assy. to case	20-30
Cooler return check valve	9-12
Extension assy. to trans. case	30-40
Pressure gauge tap	7-15
Converter drain plug	15-28
Rear band adjusting screw to case	35-40
Front band adjusting screw locknut	20-25
Manual valve inner lever to shaft	20-30
Downshift lever to shaft	17-20
Filler tube to engine	20-25
Transmission to engine	40-50
Neutral start switch actuator lever bolt	6-10
Steering col. lock rod adj. nut	10-20
T.R.S. switch to case	4-8

	In Lbs.
Governor to counterweight	50-60
Governor valve body cover screws	20-30
Pressure regulator cover screws	20-30
Pressure regulator cover screws	20-30
Control valve body screws (10-24)	20-30
Front servo release piston	20-30
End plates to body	20-30
Stator support to pump	25-35
Lower body and cover plate to valve body	20-30
T.V. body to valve body	20-30
Lower valve body cover and plate to valve body	48-72

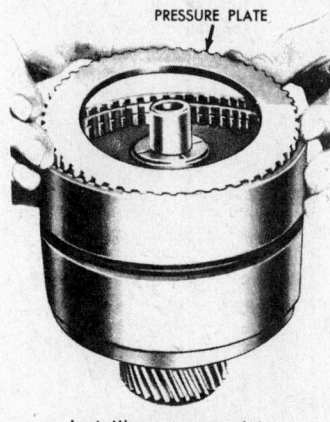

PRESSURE PLATE

Installing pressure plate
(© Ford Motor Co.)

SPECIFICATIONS

Checks and Adjustments Operation	Specification
Transmission end play	0.010-0.029 plus end play between output shaft ball bearing and retainer. Not to exceed a total of .044 inch.
Turbine and stator end play check	New or rebuilt 0.023 maximum. Used 0.040 maximum.
Front band adjustment (Use ¼ inch spacer between adjustment screw and servo stop piston stem)	Adjust screw to 10 in-lbs torque. Remove spacer, then tighten screw an additional ¾ turn. Hold screw and tighten lock nut.
Rear band adjustment	Adjust screw to 10 ft-lbs torque, and back off 1½ turns. Hold screw and tighten lock nut.
Primary sun gear shaft ring end gap check	0.002-0.009

FRONT CLUTCH SELECTIVE CLUTCH PLATE-FMX

Thickness (in.)	Identification
0.0565-0.0605	No Stripe
0.0705-0.0745	One Stripe
0.0845-0.0885	Two Stripes
0.0985-0.1025	Three Stripes

SELECTIVE THRUST WASHERS-FMX

Thickness (in.)	Identification
0.061-0.063	by thickness
0.067-0.069	by thickness
0.074-0.076	by thickness
0.081-0.083	by thickness
0.092-0.094	by thickness
0.105-0.107	by thickness

CAPACITY

Model	Refill (qt.)
All ①	11

① F-100 with 302 engine-9¼ qts.

CONTROL PRESSURE AT ZERO GOVERNOR RPM

Year	Range	Control Pressure (psi) Manifold Vacuum (in. Hg)		
		① 18 and Above	10	Below 1.0
		15 and Above		
1971-1972	P, N, D, 2, 1	56-81	82-112	—
	R	60-110	—	185-221
	D, 2, 1		—	146-175
		12 and Above		
1973	P, N, D, 2, 1	61-107	75-120	—
	R	90-156	—	185-225
	D, 2, 1		—	154-188
1974-78	D, 2, 1	58-80	75-120	143-188
	R	64-124	—	185-225
	P, N	56-80	—	—

① 15 and above-1971-72 and 1974-76; 12 and above 1973

DIAGNOSIS

This diagnosis guide covers the most common symptoms and is an aid to careful diagnosis. The items to check are listed in the sequence to be followed for quickest results. Thus, follow the checks in the order given for the particular transmission type

TROUBLE SYMPTOMS	ITEMS TO CHECK	
	IN TRUCK	OUT OF TRUCK
Car will not move in any selector position	ABC	cab
Engine speed flares, as slipping clutch	ABDFE	def
Engine speed flares on upshift	ADBG	ghi
Transmission will not upshift	HIJ	klm
Harsh upshift	JDGK	
Harsh deceleration downshift	DLMNG	
No downshift	MHI	
Clutch failure, burnt plates	DAHO	hi
Excessive creep in drive	JL	
Car creeps in neutral	J	hd
No drive in reverse	J	jng
Improper shift points	JIH	l
Unable to push-start car		m
Oil leaks	PQG	oa
Oil forced out at filler tube	AQ	p

Key to Checks

A. Oil level
B. Oil screen
C. Pressure regulator valve
D. Band adjustment
E. Servo seal
F. Servo blocked
G. Vacuum modulator or line
H. Governor
I. Throttle valve

J. Throttle linkage
K. Hydraulic modulator valve
L. Too high idle speed
M. Valves malfunctioning
N. Make pressure tests
O. Driving too fast in low
P. Oil leaks at external points
Q. Oil cooler or lines

a. Front pump
b. Input shaft
c. Front pump priming valve
d. Low band
e. Low band linkage
f. Converter stator
g. Clutch feed blocked
h. High clutch

i. Front clutch relief valve
j. Reverse clutch relief valve
k. Low clutch valve stuck
l. Rear pump priming valve
m. Rear pump or drive
n. Low clutch
o. Front pump attaching bolts
p. Pump circuit leakage

General Description

The Powerglide transmission is a two speed unit with a one piece aluminum case and an aluminum case extension.

Driving ranges are low, high and reverse, with a throttle controlled downshift to low range available for sudden acceleration.

The oil pump is the gear type; the pump housing is used as the front bulkhead of the transmission. The torque converter is a three element welded unit bolted to the engine flywheel; it drives through a two speed planetary gearset. Low range uses a band clutch; drive and reverse ranges use disc clutches. The valve body assembly is bolted to the bottom of the transmission case; the modulator valve bore is in the upper part of the valve body assembly. The governor is mounted to the output shaft, inside the case extension.

Transmission removal and installation, shift linkage adjustment, low band adjustment in the truck, neutral safety switches, and transmis-

sion downshift linkages are covered in the truck section.

Fluid Change Schedule

The manufacturer recommends draining the transmission sump every 24,000 miles; 2 quarts to all models. The fluid level should then be rechecked. This should be done with the engine idling, the selector lever in Neutral, and the transmission at operating temperature.

Transmission Disassembly

Extension, Governor and Rear Oil Pump

Removal

1. Place transmission in a holding fixture, if possible.
2. Remove converter holding tool, then lift off the converter.
3. If replacement is necessary, remove speedometer driven gear. Loosen cap screw and retainer clip and remove gear from extension.

4. Remove transmission extension by removing five attaching bolts. Note seal ring on rear pump body.

5. Remove speedometer drive gear from output shaft.

6. Remove C-clip from governor shaft of the weight side of governor, then remove the shaft and governor valve from the opposite side of the governor assembly and the two belleville springs.

7. Loosen the governor drive screw and slide the governor over the end of the output shaft.

8. Remove four bolts holding the rear oil pump to the transmission case and remove the pump body, drain back baffle, extension seal ring, drive and driven gears.

9. Remove oil pump drive pin. (*This is of extreme importance.*)

10. Remove the rear pump wear plate.

Transmission Internal Components

Removal

11. Rotate holding fixture, or turn the transmission, until the front end is pointing up. Then remove the seven front oil pump bolts. (The bolt holes are of unequal spacing to prevent incorrect location upon installation.)
12. Remove the front oil pump and stator shaft assembly and the selective fit thrust washer using an inertia puller or substitute.
13. Release tension on the low band adjustment, then with transmission horizontal, grasp the transmission input shaft and carefully work it and the clutch drum out of the case. Be careful not to lose the low sun gear bushing from the input shaft. The low sun gear thrust washer will probably remain in the planet carrier.
14. The low brake band and struts may now be removed.
15. Remove the planet carrier and the output shaft thrust caged bearing from the front of the transmission.
16. Remove reverse ring gear if it did not come out with the planet carrier.
17. With a large screwdriver, remove the reverse clutch pack retainer ring and lift out the re-

1 Reverse ring gear
2 Reverse clutch pack snap ring
3 Reverse clutch
   pressure plate
4 Reverse clutch
   reaction plates
5 Reverse clutch drive plates
6 Reverse clutch cushion spring
7 Reverse clutch piston return
   spring retainer snap ring
8 Reverse clutch piston return
   spring retainer
9 Reverse clutch piston return
   springs
10 Reverse clutch piston inner
   seal
11 Reverse clutch piston
12 Reverse clutch piston outer
   seal

CLUTCH AND BAND APPLICATION —POWERGLIDE AUTOMATIC TRANSMISSION

GEAR	HIGH CLUTCH	LOW BAND	REVERSE CLUTCH
1st*	off	on	off
2nd*	on	on	off
Low	off	on	off
Reverse	off	off	on

* Shift lever in "D" position

verse clutch plates and the cushion spring.

18. Install reverse piston spring compressor through rear bore of the case, with the flat plate on the rear face of the case, and turn down wing nut to compress the rear piston spring retainer and springs. Then remove the snap ring. A spring compressor may be made up from a suitable length bolt and large flat washers.
19. Remove the compression tool, the reverse pistol spring retainer, and the 17 piston return springs.
20. Remove the rear piston by applying air pressure to the reverse port in the rear of the transmission case. Remove inner and outer seals.

21. Remove the three servo cover bolts, servo cover, piston and spring.

Oil Pan and Valve Body

Removal

NOTE: the oil pan and valve body may be serviced without removing the extension, and internal components, covered in the preceding steps.

22. Rotate the transmission until the unit is upside down (oil pan on top). Remove oil pan attaching bolts, oil pan, and gasket.
23. Remove vacuum modulator and gasket, and vacuum modulator plunger, dampening spring, and valve.
24. Remove two bolts holding the detent guide plate to the valve body and the transmission case. Remove the guide plate and the range selector detent roller spring.

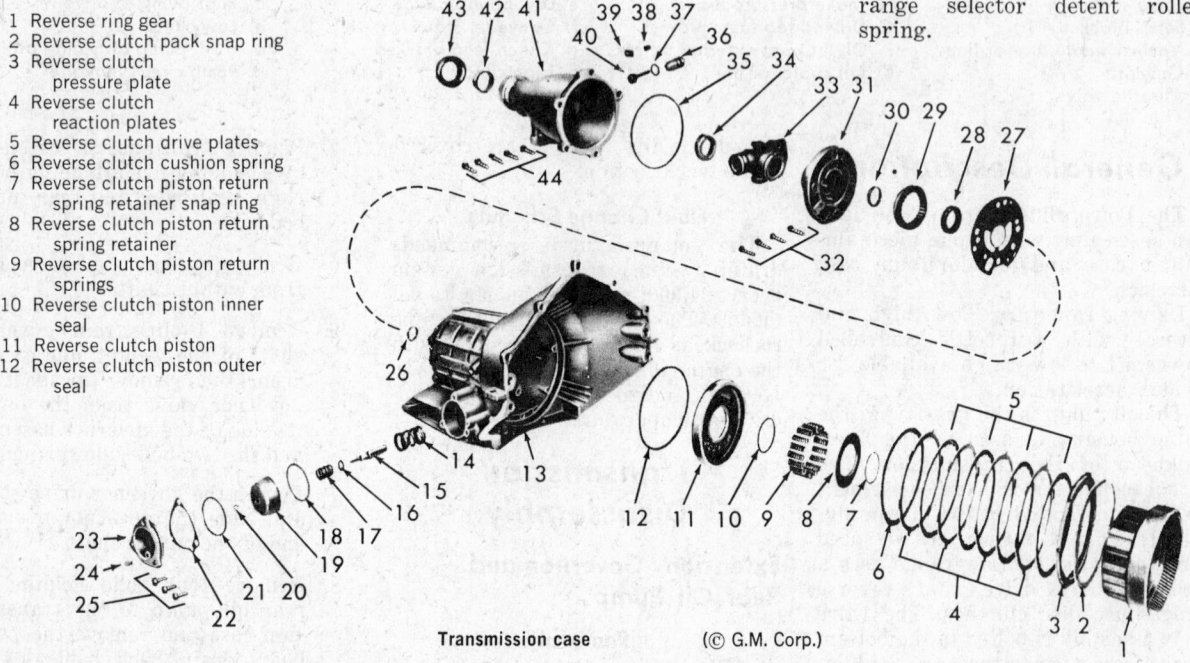

Transmission case   (© G.M. Corp.)

13 Transmission case
13A Transmission case screen
14 Servo piston return spring
15 Servo piston rod
16 Servo piston apply
   spring seat
17 Servo piston apply spring
18 Servo piston seal ring
19 Servo piston

20 Servo piston rod spring
   retainer
21 Servo cover seal
22 Servo cover gasket
23 Servo cover
24 Servo cover plug
25 Servo cover bolts
26 Transmission case bushing
27 Gasket
30 Governor support bushing

31 Governor support
32 Governor support to case
   attaching bolts
33 Governor assembly
34 Speedometer drive gear
   and clip
35 Seal
36 Speedometer shaft fitting
37 Speedometer shaft fitting

   oil seal
38 Lock plate attaching screw
39 Lock plate
40 Speedometer driven gear
41 Transmission extension
42 Extension bushing
43 Extension oil seal
44 Extension to case
   attaching screws

25. Remove the remaining valve body-to-transmission case attaching bolts and lift out the valve body and gasket. Disengage the servo apply tube from the transmission case as the valve body is removed.

26. If necessary, the throttle valve, shift and parking actuator levers and the parking pawl and bracket may be removed.

Unit Assembly Overhaul

Converter and Stator

The converter is a welded assembly and no internal repairs are possible. Check the seams for stress or breaks and replace the converter if necessary.

Front Pump

Seal Replacement

If the front pump seal requires replacement, remove the pump from the transmission, pry out and replace

REAR PUMP DRIVE PIN

Removing rear oil pump drive pin (© G.M. Corp.)

the seal. Drive new seal into place. Then, if no further work is needed on the front pump, reinstall it into the case. (The outer edge of the seal should be coated with non-hardening sealer before installation.)

Disassembly

1. Remove pump cover-to-body attaching bolts and the cover.
2. Remove pump gears from body.
3. Remove rubber seal from pump body.

Inspection

1. Wash all parts in solvent. Blow out all oil passages.
2. Inspect pump gears for nicks or damage.
3. Inspect body and cover faces for nicks or scoring. Inspect cover hub outside diameter for nicks or burrs.
4. Check for free operation of priming valve. Replace if necessary.
5. Inspect body bushing for galling or scoring. Check clearance between body bushing and converter pump hub. Maximum clearance is .005 in. If the bushing is damaged, replace the pump body.
6. Inspect converter housing hub outside diameter for nicks or burrs. Repair or replace.
7. If oil seal is damaged or leaking, pry out and drive in a new seal.
8. Check condition of oil cooler bypass valve. Replace if leaking. An "Easy-Out" type remover may be used to remove the valve. Tap new valve seat into place with a soft hammer or brass drift so that it is flush or up to .010 in. below the surface.
9. With all parts clean and dry, install pump gears and check:
a. clearance between outside diameter of driven gear and body should be .0035-.0065 in.

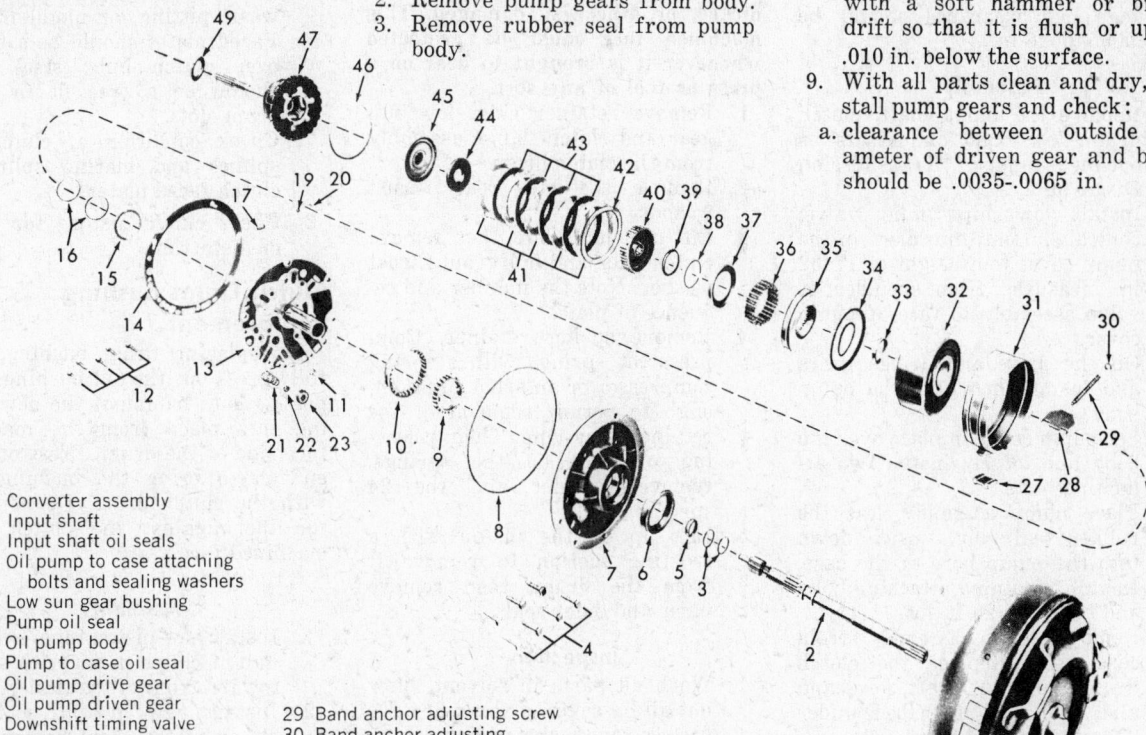

1 Converter assembly
2 Input shaft
3 Input shaft oil seals
4 Oil pump to case attaching bolts and sealing washers
5 Low sun gear bushing
6 Pump oil seal
7 Oil pump body
8 Pump to case oil seal
9 Oil pump drive gear
10 Oil pump driven gear
11 Downshift timing valve
12 Oil pump cover to pump body attaching screws
13 Oil pump cover and converter stator shaft
14 Oil pump gasket
15 Clutch drum thrust washer (selective fit)
16 High clutch seal rings
17 Pump priming valve
18 Pump priming valve spring
20 Pump priming valve spring retaining pin
'21 Oil cooler by-pass valve spring
*22 Oil cooler by-pass valve
*23 Oil cooler by-pass valve seat
27 Band apply strut
28 Band anchor strut

29 Band anchor adjusting screw
30 Band anchor adjusting screw nut
31 Low brake band
32 Clutch drum bushing
33 Clutch drum bushing
34 Clutch piston outer and inner seals
35 Clutch piston
36 Clutch return springs
37 Clutch spring retainer
38 Clutch spring retainer snap ring
39 Clutch hub front thrust washer
40 Clutch hub
41 Clutch driven plates (flat)
42 Clutch cushion spring (waved)

Internal Mechanism (© G.M. Corp.)

43 Clutch drive plates (waved)
44 Clutch hub rear thrust washer
45 Low sun gear and clutch flange assembly
46 Clutch flange retainer ring
47 Planet carrier and output shaft assembly
49 Output shaft thrust bearing

*Except air cooled and 11" converter models

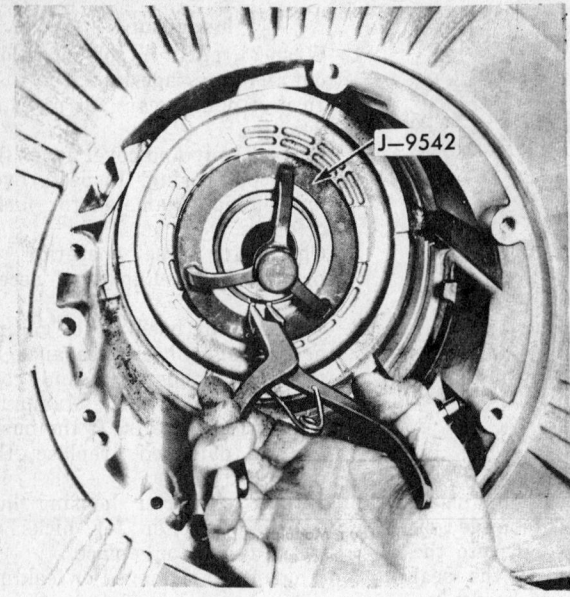

Removing rear piston spring retainer snap-ring
(© G.M. Corp.)

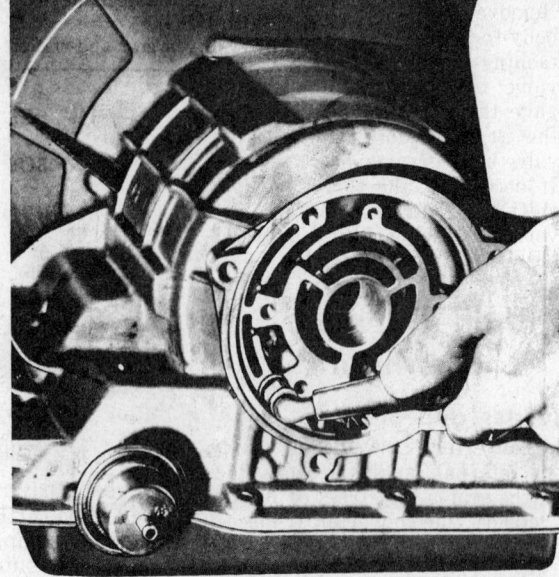

Applying air pressure to remove rear piston
(© G.M. Corp.)

b. clearance between inside diameter of driven gear and crescent should be .003-.009 in.

c. gear end clearance should be .0005-.0015 in.

Assembly

1. Remove the input shaft, clutch drum, low band and struts as outlined under "Transmission Disassembly."

2. Install downship timing valve, conical end out, into place in the pump cover to a height of 17/32 in. measured from shoulder of valve assembly to face of pump cover.

3. Oil the drive and driven gears and install them into the pump body.

4. Set pump cover in place over the body and loosely install two attaching bolts.

5. Place pump assembly, less the rubber seal ring, upside down into the pump bore of the case. Install remaining attaching bolts and torque to 20 ft. lbs.

6. Remove pump assembly from case bore. Replace the clutch drum and input shaft, low band and struts as described under "Transmission Assembly."

7. Renew rubber seal ring in its groove in the pump body and install the pump assembly in place in the case bore, using a new gasket. Be sure that the selective fit thrust washer is in place.

8. Install attaching bolts. (Use new bolt sealing washers if necessary.)

Clutch Drum

Disassembly

CAUTION: When working with the clutch drum, use extreme care that the machined face on the front of the drum not be scratched, scored, nicked, or otherwise damaged. This machined face must be protected whenever it is brought to bear on a press or tool of any sort.

1. Remove retainer ring, low sun gear and clutch flange assembly from the clutch drum.

2. Remove the hub rear thrust washer.

3. Lift out clutch hub, then remove clutch pack and hub front thrust washer. Note the number and sequence of plates.

4. Remove spring retainer. Compress the springs with a spring compressor or an arbor press enough to permit removal of the retainer snap-ring. Then, releasing pressure on the springs, remove retainer and the 24 springs.

5. Lift up on the piston with a twisting motion to remove it from the drum, then remove inner and outer seals.

Inspection

1. Wash all parts in solvent, blow out all passages, and air dry. Do not use rags to dry parts.

2. Check drum bushing for scoring or excessive wear.

3. Check steel ball relief valve in clutch drum. Be sure that it is free to move and that orifice in front face of drum is open. If ball is loose enough to come out, or not loose enough to rattle, replace drum. Do not attempt replacement or restaking of ball.

4. Check fit of low sun gear and clutch flange assembly in clutch drum slots. There should be no appreciable radial play.

5. Check low sun gear for nicks or burrs. Check gear bore for wear.

6. Check clutch plates for burring, wear, pitting, or metal pick-up. Faced plates should be a free fit over clutch hub; steel plates should be a free fit in clutch drum slots.

7. Check condition of clutch hub splines and mating splines of clutch faced plates.

8. Check clutch pistons for cracks or distortion.

Clutch Drum Bushing Replacement

If replacing drum bushing, carefully press out the old bushing. Then press (don't hammer) the new bushing into place from the machined face side of the drum. Press only far enough to bring the bushing flush with the clutch drum. Do not force the tool against the clutch drum machined face.

Assembly

1. Install new piston inner seal into hub of clutch drum with seal lip toward front of transmission.

2. Install new piston seal into clutch piston. Seal lips must be pointed toward the clutch drum, (front of transmission). Lubricate the seals and install piston into clutch drum with a twisting motion.

3. Place 24 springs in position on the piston, then place the retainer on the springs.

4. Depress the retainer plate and springs far enough to permit installation of the spring retainer snap-ring into its groove on the clutch drum hub.

5. Install the hub front washer

with its lip toward the clutch drum, then install the clutch hub.

6. Install cushion spring if used. Install the steel reaction plates and drive (faced) plates alternately, beginning with a steel reaction plate.

NOTE: the number and sequence of plates varies with the power and torque requirements of the truck model involved. On some models, the first driven plate is a selective fit. See the clutch assemblies chart for details.

Clutch Assemblies

Engines	307 V8; 250-L-6 292 L-6	350 V-8
Drive plate	4	5
Driven plate	5	6
Cushion Spring	1	None

350 V-8 Selective Driven Plate Chart

Plate Stack Height (Less Selective Plate)	Plate Part Number	Color Code	Plate Thickness
.903-.872	3883903	Orange	.060±.0025
.872-.798	3883904	Blue	.090±.0025

7. Install the rear hub thrust washer with its flange toward the low sun gear, then install the low sun gear and flange assembly and secure with retaining ring. When installed, the openings in the retainer ring should be adjacent to one of the lands

4. Check input sun gear for tooth

8. Check assembly by turning the clutch hub. If free, assembly is OK.

Low Band

Due to band design and transmission characteristics, this band should require very little attention. However, while the transmission is disassembled, the band should be thoroughly cleaned, then replaced if any trace of scoring, burning, cracks, or excessive wear or damage is found.

Planet Assembly and Input Shaft

Inspection

1. Wash planet carrier and input shaft in cleaning solvent, blow out all passages, and air dry. Do not use rags to dry parts.
2. Inspect planet pinions for nicks or other tooth damage.
3. Check end clearance of planet gears. The clearance should be .006-.030 in. of the clutch drum. damage. Check thrust washer for damage.
5. Inspect output shaft bearing surface and input pilot bushing for nicks or scoring.

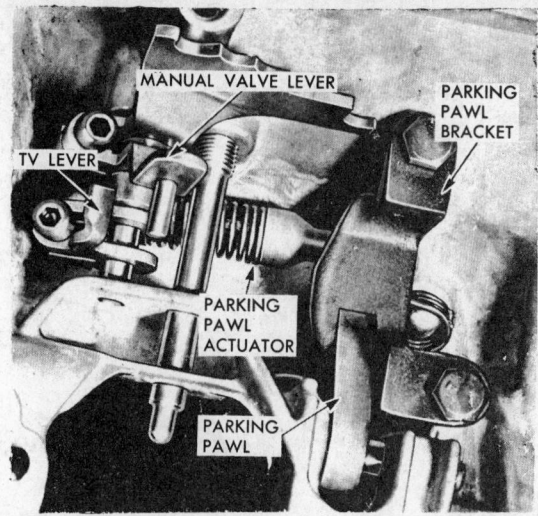

Inner control levers, parking pawl and bracket
(© G.M. Corp.)

Checking planet gear end clearance
(© G.M. Corp.)

6. Inspect input shaft splines for nicks or damage. Check fit in clutch hub, input sun gear, and turbine hub.
7. Check oil seal rings for damage; rings must be free in input shaft ring grooves. Remove rings and insert in stator support bore. Check to see that hooked ring ends have clearance. Replace rings on shaft.

Repairs

NOTE: some large planet carrier assemblies have the pinion shafts flared at each end for retention in the carrier. No overhaul of this type of carrier assembly should be attempted. If inspection shows excessive wear or damage, replace the entire carrier assembly.

1. Place the planet carrier assembly in a padded vise so that the front (parking lock gear end) of the assembly is up.
2. Using a prick punch, mark each pinion shaft and the carrier assembly so that, when reassembling, each shaft will be returned to its original location.
3. Remove pinion shaft lock plate screws and rotate plate counterclockwise far enough to remove it.
4. Starting with a short planet pinion, drive the lower end of the

pinion shaft up until the shaft is above the press fit area of the output shaft flange. Feed a dummy shaft into the short planet pinion from the lower end, pushing the planet pinion shaft ahead of it until the tool is centered in the pinion and the pinion shaft is removed.

5. Remove short planet pinion.
6. Remove dummy shaft, needle and bearing spacers from short pinion. *NOTE: twenty needle bearings are used in each end of each gear and are separated by a bearing spacer in the center.*
7. By following Steps 4, 5, and 6, remove the adjacent long planet pinion that was paired, by thrust washers, to the short pinion now removed.
8. Remove upper and lower thrust washers.
9. Remove and disassemble remaining planet pinions, in pairs, as above.
10. Remove low sun gear needle thrust bearing, input sun gear, and thrust washer.
11. Wash all parts in solvent and blow dry, then inspect.
12. Inspect input shaft bushing in base of output shaft. If damaged, it may be removed by using a slide hammer. New bearing can be installed by using pilot end of input shaft as press tool.
13. Using dummy shaft, assemble needle bearings and spacer (20 rollers in each end) in one of the long planet pinions. Use petroleum jelly to aid in holding the rollers in position.
14. Position long planet gear, with dummy shaft centered in the pinion and with thrust washers at each end, in the planet carrier. Oil grooves on thrust washers must be toward the gears. *NOTE: long pinions are located opposite the closed portions of*

1. Park lock and range selector outer lever and shaft
2. Throttle valve control shaft oil seal
3. Throttle valve control shaft washer
4. Throttle valve control lever and shaft
5. Throttle valve control inner lever to control shaft attaching screw and nut
6. Throttle valve control inner lever
7. Park lock and range selector inner lever
8. Park lock and range selector inner lever attaching screw and nut
9. Park lock pawl disengaging spring
10. Range selector detent roller spring
11. Park lock actuator assembly
12. Range selector detent roller spring retainer
13. Park lock pawl shaft
14. Park lock pawl
15. Park lock pawl shaft retaining ring
16. Park lock pawl reaction bracket
17. Park lock pawl reaction bracket attaching bolts
18. Park lock actuator to park lock and range selector inner lever retaining clip

Manual levers—typical (© G.M. Corp.)

the carrier and short pinions are located in the openings.

15. Feed a second dummy shaft in from the top, picking up the upper thrust washer and the pinion and pushing the already installed dummy shaft out the lower end. As the first dummy is pushed down, be sure that it picks up the lower thrust washer.
16. Select the correct pinion shaft, as marked in Step 2, lubricate the shaft and install it from the top, pushing the assembling tools (dummy) ahead of it.
17. Turn the pinion shaft so that the slot or groove at the upper end faces the center of the assembly.
18. With a brass drift, drive the shaft in until the lower end is flush with the lower face of the planet carrier.
19. Following the same procedure as outlined in Steps 13 through 18, assemble and install a short planet pinion into the planet carrier adjacent to the long pinion now installed. *NOTE: the thrust washers, already installed with the long planet pinion, also serve for this short planet pinion, because the the two pinions are paired together on one set of thrust washers.*
20. Install the input sun gear thrust washer, input sun gear, and low sun gear needle thrust bearing.
21. Assemble and install the remaining planet pinions, in pairs, as previously explained.
22. Check end clearance of planet gears. This clearance should be .006-.030 in.
23. Place the shaft lock plate in position. Then, with the extended portions of the lock plate aligned with slots in the planet pinion shafts, rotate the lock plate clockwise until the three attaching screw holes are accessible.
24. Install lock plate attaching screws and torque to 2½ ft. lbs.

Governor

The governor assembly is a factory balanced unit. If body replacement is needed, the two sections must be replaced as a unit.

Disassembly

NOTE: the governor valve and shaft were removed in Step 6 of "Transmission Disassembly" procedures.

1. Remove the outer weight by sliding toward center of body.
2. Remove smaller inner weight retaining snap-ring and remove inner weight and spring.
3. Remove the four body assembly bolts and separate the body, hub and gasket. Remove the two seal rings.

Inspection

1. Clean all parts in solvent and air dry.
2. Check all parts. Replace all bent, scored, or otherwise damaged parts. Body and hub must be replaced as a unit.

Assembly

1. Reassemble governor weights and install into body bore. Replace seal ring on hub.
2. Slide hub into place on output shaft and lock into place with drive screw. Install gasket and governor body over output shaft, install governor shaft, line up properly with output shaft and install body attaching bolts. Torque bolts to 6-8 ft. lbs.
3. Check governor weight for free fit in body after the four attaching bolts are torqued. If the weight sticks or binds, loosen the bolts and retorque.

Valve Body

Removal

Remove valve body, as described under "Transmission Disassembly". If performing the operation on the truck, the vacuum modulator valve, oil pan and gasket, guide detent plate and range selector detent roller spring must be removed in order to remove the valve from the transmission.

Disassembly

1. Remove manual valve, suction screen and gasket.
2. Remove cover bolts, then remove lower valve body and transfer plate from upper valve body. Discard gaskets.
3. Remove the front and rear pump check valves and springs.
4. From the upper valve body, remove the throttle valve and detent valve and the downshift timing valve as follows:
A. Throttle Valve and Detent Valve—Remove the retaining pin

Checking gear end play (© G.M. Corp.)

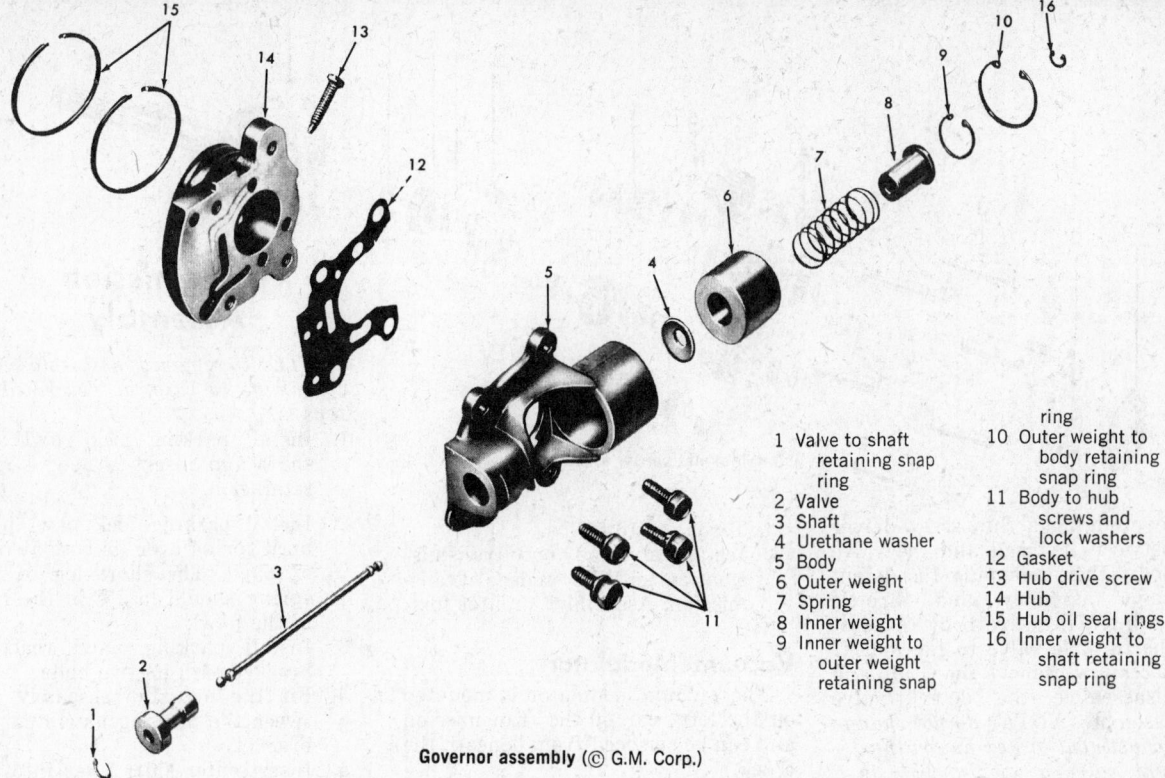

1 Valve to shaft
  retaining snap
  ring
2 Valve
3 Shaft
4 Urethane washer
5 Body
6 Outer weight
7 Spring
8 Inner weight
9 Inner weight to
  outer weight
  retaining snap
  ring
10 Outer weight to
   body retaining
   snap ring
11 Body to hub
   screws and
   lock washers
12 Gasket
13 Hub drive screw
14 Hub
15 Hub oil seal rings
16 Inner weight to
   shaft retaining
   snap ring

Governor assembly (© G.M. Corp.)

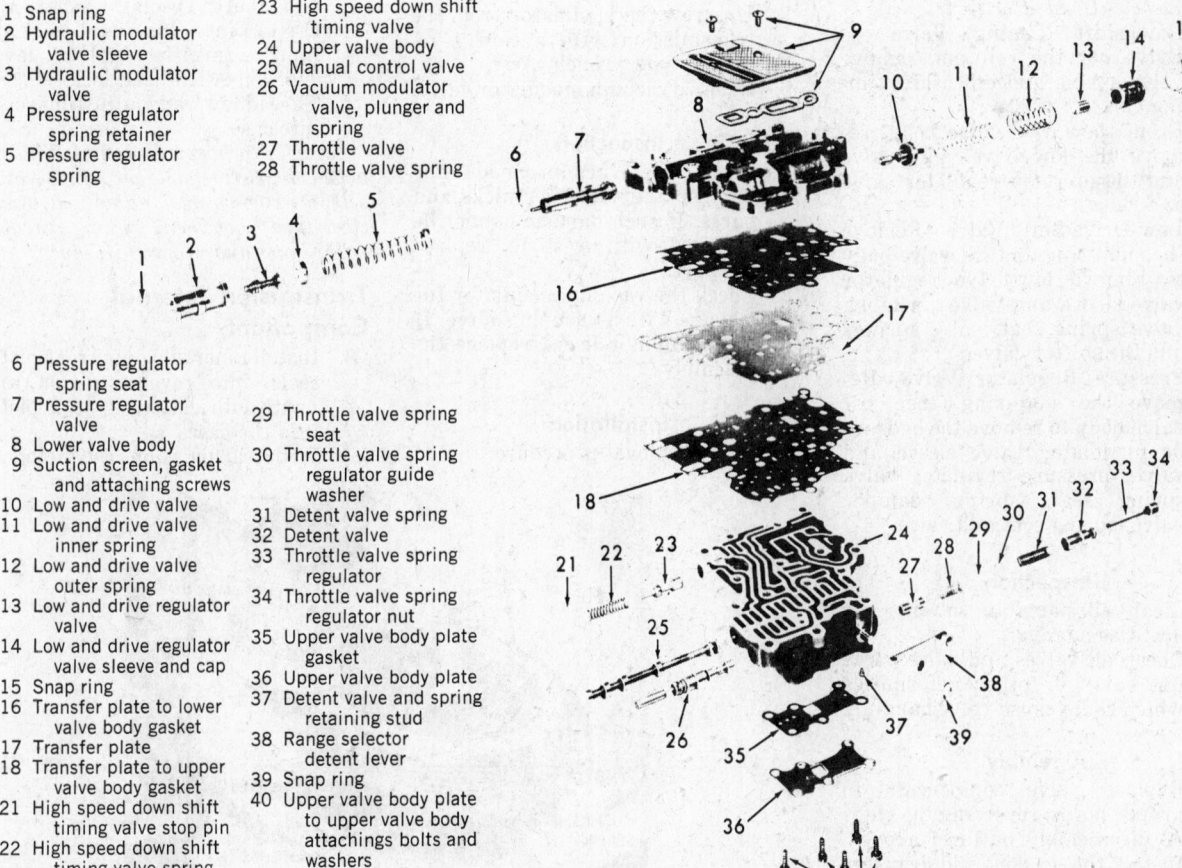

1 Snap ring
2 Hydraulic modulator
  valve sleeve
3 Hydraulic modulator
  valve
4 Pressure regulator
  spring retainer
5 Pressure regulator
  spring

6 Pressure regulator
  spring seat
7 Pressure regulator
  valve
8 Lower valve body
9 Suction screen, gasket
  and attaching screws
10 Low and drive valve
11 Low and drive valve
   inner spring
12 Low and drive valve
   outer spring
13 Low and drive regulator
   valve
14 Low and drive regulator
   valve sleeve and cap
15 Snap ring
16 Transfer plate to lower
   valve body gasket
17 Transfer plate
18 Transfer plate to upper
   valve body gasket
21 High speed down shift
   timing valve stop pin
22 High speed down shift
   timing valve spring

23 High speed down shift
   timing valve
24 Upper valve body
25 Manual control valve
26 Vacuum modulator
   valve, plunger and
   spring
27 Throttle valve
28 Throttle valve spring

29 Throttle valve spring
   seat
30 Throttle valve spring
   regulator guide
   washer
31 Detent valve spring
32 Detent valve
33 Throttle valve spring
   regulator
34 Throttle valve spring
   regulator nut
35 Upper valve body plate
   gasket
36 Upper valve body plate
37 Detent valve and spring
   retaining stud
38 Range selector
   detent lever
39 Snap ring
40 Upper valve body plate
   to upper valve body
   attachings bolts and
   washers

Valve body assembly sequence (© G.M. Corp.)

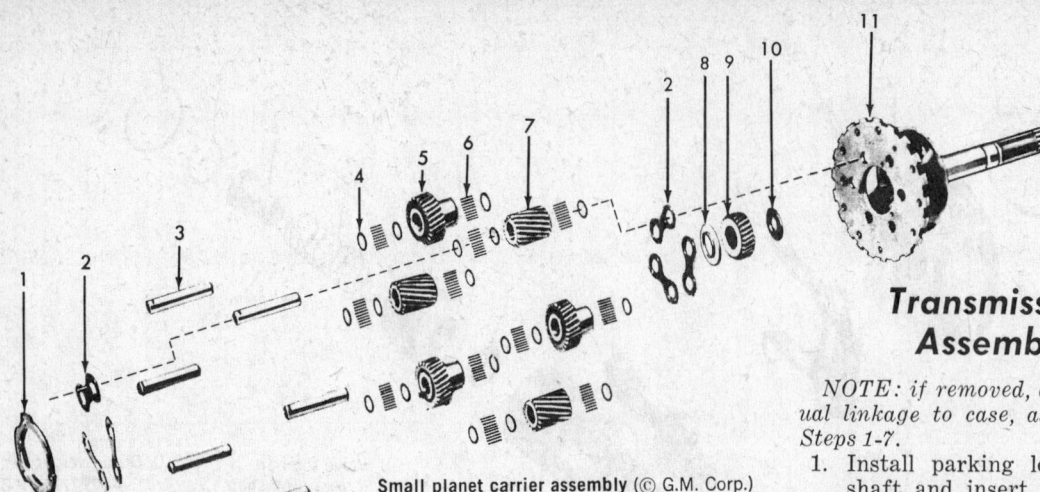

Small planet carrier assembly (© G.M. Corp.)

by wedging a thin screw-driver between its head and the valve body, then removing the detent valve assembly and throttle spring. Tilt valve body to allow the throttle valve to fall out. If necessary, remove the C clip and disassemble the detent valve assembly. *NOTE: do not change adjustment of hex nut on the detent valve assembly. This is a factory setting and should not normally be changed. However, some adjustment is possible if desired. See "Throttle Valve Adjustment," in later text.*

B. Downshift Timing Valve — Drive out the roll pin, remove valve spring and downshift timing valve.

5. From the lower valve body, remove the low-drive shift valve and the pressure regulator valve as follows:

A. Low-Drive Shift Valve—Remove the snap-ring and tilt valve body to remove low-drive regulator valve sleeve and valve assembly, valve spring seat, valve springs and the shifter valve.

B. Pressure Regulator Valve—Remove the snap-ring, then tilt valve body to remove the hydraulic modulator valve sleeve and valve, pressure regulator valve spring seat, spring, damper valve, spring seat and valve.

Inspection

1. Clean all parts in solvent; air dry. Use no rags.
2. Check all valves and valve bores for burrs or other deformities which could cause valve hang-up.

Assembly

1. Replace valve components in proper bores, reversing the steps of disassembly outlined above.
2. Install the gasket and transfer plate.
3. Install lower valve body and gasket and install attaching bolts.

Torque to 15 ft. lbs.
4. Install valve body onto transmission, as outlined under "Transmission, Assembly" in later text.

Vacuum Modulator

The vacuum modulator is mounted on the left rear of the transmission and can be serviced from beneath the truck.

Removal

1. Remove vacuum line at the modulator.
2. Unscrew the modulator from the transmission with a thin 1 in. tappet-type wrench.
3. Remove vacuum modulator valve.

Inspection

1. Check the vacuum modulator plunger and valve for nicks and burrs. If such damage cannot be repaired with a stone, replace the part.
2. Check the vacuum modulator for leakage with a vacuum source. If the modulator leaks, replace the assembly.

Installation

Reverse removal procedure.

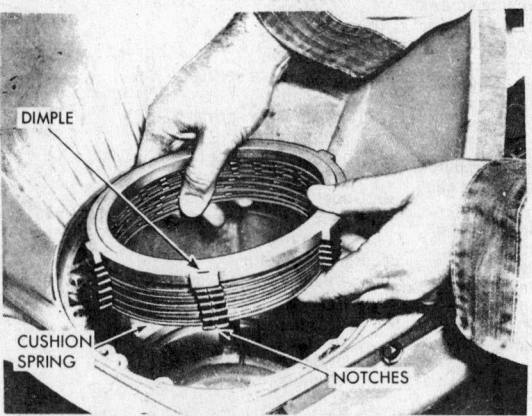

Installing clutch plates—typical
(© G.M. Corp.)

Transmission Assembly

NOTE: if removed, assemble manual linkage to case, as described in Steps 1-7.

1. Install parking lock pawl and shaft and insert a new E-ring retainer.
2. Install parking lock pawl pull-back spring over its boss at rear of pawl. The short leg of the spring should locate in the hole in the pawl.
3. Install parking pawl reaction bracket with its two bolts.
4. Fit the actuator assembly between the parking pawl and the bracket.
5. Insert outer shift lever into the case. Pick up inner shift lever and parking lock assembly. Tighten Allen-head lock.
6. Insert outer throttle valve lever and shaft, special washer, and O-ring into case and pick up inner throttle valve lever. Tighten Allen-head lock.
7. Thread low band adjusting screw into case.

NOTE: to prevent possible binding between throttle lever and range selector controls, allow .010-.020 clearance between inner throttle valve lever and inner shift lever.

Transmission Internal Components

8. Install inner and outer rear piston seals onto reverse piston and, (with lubrication) install piston into the case.
9. With transmission case facing up,

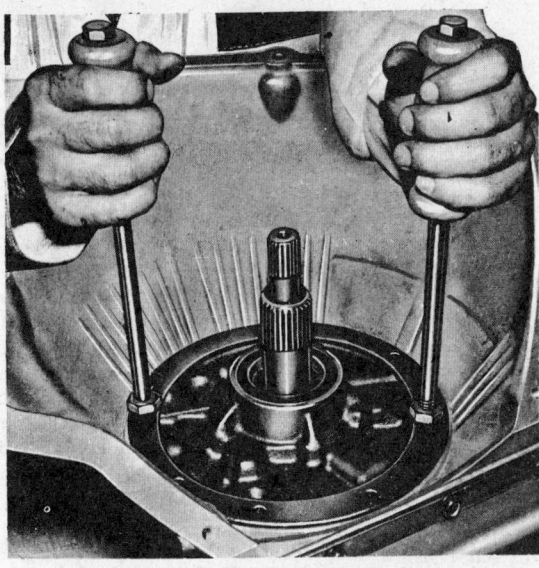

Removing Oil Pump

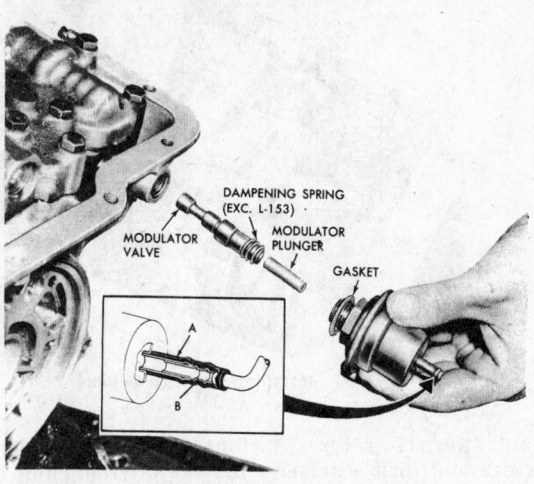

DAMPENING SPRING
(EXC. L-153)

MODULATOR
VALVE

MODULATOR
PLUNGER

GASKET

A

B

Vacuum Modulator, Dampening Spring,
Plunger and Valve

LOW SUN GEAR
NEEDLE THRUST
BEARING

LOW SUN GEAR
BUSHING

Removing Clutch Drum and Input Shaft

Checking Pump Body Bushing to
Converter Pump Hub Clearance

Checking Driven Gear to Pump Body Clearance

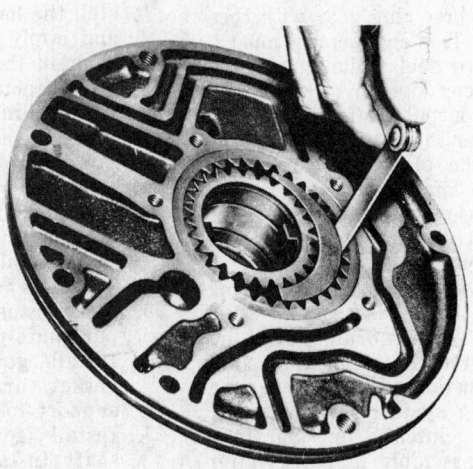

Driven Gear to Crescent Clearance

Installing detent guide plate
(© G.M. Corp.)

install the 17 reverse piston springs and their retainer ring.

10. Install spring compressing tool. Compress the return springs, allowing the retaining ring snap-ring to be installed. Remove the compressor.
11. Install the cushion spring.
12. Lubricate and install reverse clutch pack, beginning with a reaction spacer plate and alternating with the faced plates until all plates are installed.

NOTE: the number and sequence of plates varies with the power and torque requirements of the truck model involved.

The notched lug on each reaction plate is installed in the groove at the seven o'clock position in the case. Then, install the thick pressure plate which has a dimple in one lug to align with the same slot in the case as the notched lugs on the other reaction plates.

13. Install clutch plate retainer ring.
14. Turn rear of transmission case down.
15. Align the internal lands and grooves of the reverse clutch pack faced plates, then engage the reverse ring gear with these plates. This engagement must be made by feel while turning the ring gear.
16. Place output shaft thrust bearing over the output shaft and install the planetary carrier and output shaft into the transmission case.
17. Move transmission to horizontal position.
18. The two input shaft seal rings should be in place on the shaft. Install clutch drum (machined face first) onto the input shaft and install the low sun gear bushing against shoulder.
19. Install clutch drum and input shaft assembly into case, aligning thrust washer on input shaft and indexing low sun gear with the short pinions on the planet carrier.

20. Remove rubber seal ring from the front pump body and install front pump, gasket and selective fit thrust washer into case. Install pump-to-case bolts.
21. To check for correct thickness of the selective fit thrust washer, move transmission so that output shaft points down and proceed as follows:
A. Mount a dial indicator so that the indicator plunger is resting on the end of the input shaft. Zero the indicator.
B. Push up on the output shaft and watch the total dial movement.
C. The indicator should read .028-.059 in. If reading is not within specifications, remove front pump change to a thicker or thinner selective thrust washer. Repeat above checking procedure. *NOTE: washers are available in thicknesses of .061, .078, .092 in. and .106 in.*
22. Install servo piston, piston ring, and spring into the servo bore. Then, using a new gasket and O-ring, install the servo cover.
23. Remove front pump and selective fit washer from the case, and install the low brake band, anchor and apply struts into the case. Tighten the low band adjusting screw enough to prevent struts from falling out of case.
24. Place the seal ring in the groove around front pump body and the two seal rings on the pump cover extension. Install the pump, gasket and thrust washer into the case. Install all pump bolts. Torque bolts to 15 ft. lbs.
25. Turn transmission so that output shaft points upward.
26. Install governor support and gasket, drain back baffle, and support to case attaching bolts.
27. Install governor over output shaft. Install governor shaft and valve, urethane washer, and retaining C clips. Center shaft in output shaft bore and tighten governor hub drive screw.

28. Install speedometer gear to output shaft.
29. Place extension seal ring over governor support. Install transmission extension and five retaining bolts.
30. Replace speedometer driven gear.

Oil Pan and Valve Body

31. With transmission upside down, the manual linkage and the selector lever detent roller installed, install the valve body with a new gasket. (Carefully guide the servo apply line into its boss in the case as the valve body is set in place.) Position the manual valve actuating lever fully forward to more easily pick up the manual valve. Install six mounting bolts and the range selector detent roller spring. Install new gasket and suction screen to valve body.
32. Install the guide plate. Install attaching bolts.
33. Install vacuum modulator valve, the vacuum modulator and the gasket.
34. Install oil pan, using a new gasket, then the pan attaching bolts.
35. Install the converter and safety holding strap.

Low Band Adjustment

Tighten the low servo adjusting screw to 70 in. lbs. The input and output shaft must be rotated simultaneously to properly center the low band on the clutch drum. Then, back off four complete turns for a band which has been in use for 6,000 miles or more, or three turns for one in use less than 6,000 miles, and tighten the locknut.

CAUTION: The amount of backoff is very critical. Back off exactly three or four turns.

Throttle Valve TV Adjustment

No provision is made for checking TV pressures. However, if operation of the transmission is such that some adjustment of the TV is indicated, pressures may be raised or lowered by adjusting the position of the jam nut on the throttle valve assembly.

To raise TV pressure 3 psi, back off the jam nut one full turn.

Conversely, tightening the jam nut one full turn, lowers TV pressure 3 psi. A difference of 3 psi in TV pressure will cause a change of about 2-3 mph in the wide open throttle upshift point. The end of the TV adjusting screw has an Allen head so the screw may be held stationary while the jam nut is locked.

NOTE: Use care in changing this adjustment, as no pressure tap is provided to check TV pressure.

GM Turbo Hydra-Matic 250

DIAGNOSIS

This diagnosis guide covers the most common symptoms and is an aid to careful diagnosis. The items to check are listed in the sequence to be followed for quickest results. Thus, follow the checks in the order given for the particular transmission type.

TROUBLE SYMPTOMS	ITEMS TO CHECK	
	IN TRUCK	OUT OF TRUCK
All ranges—slips	ACDFGHOP	abc
Drive slips—no first gear	ACDFGHOP	abcdei
Line pressure—all low	ADEFGOP	abcd
Line pressure—all high	BCEGL	a
1-2 intermediate pressure low	ADFGHINP	abcdg
2-3 direct clutch pressure high	BG	a
2-3 direct clutch pressure low	ADFGJ	abcf
No 1-2 upshift	BEFGIP	acg
1-2 upshift—early/late	BEFHI	a
1-2 upshift—with wide open throttle	BIL	a
Slips—1-2 upshifts	ACFGHINP	abcdg
Rough—1-2 upshifts	BCGP	a
No 2-3 upshift	FK	acf
2-3 upshift—early/late	BCEFHJL	a
Slips—2-3 upshift	ACFGHJ	abcf
Rough—2-3 upshift	BCGJN	a
No part throttle downshift	JL	—
No full throttle downshift	L	—
No wide open throttle 1-2 upshift	M	a
2-3 upshift—wide open throttle only	L	—
Harsh downshift	H	—
L-1 range—no engine brake	GIKO	aceh
L-2 range—no engine brake	GO	acde
Neutral—drives in neutral	—	e
Reverse—no reverse	GHO	acefh
Slips in reverse	ACDFGHJO	acfh
Spews fluid out of breather	A	j
Hunts between 2 and 3, 3 and 2 shifts	L	—

A. Low fluid level/water in fluid
B. Vacuum leak
C. Modulator and/or valve
D. Strainer and/or gasket
E. Governor—valve or screen
F. Valve body—gasket or plate
G. Pressure regulator and/or boost valve
H. Valve body check balls
I. 1-2 shift valve
J. 2-3 shift valve

K. Manual low control valve
L. Detent valve and linkage
M. Detent regulator valve
N. 2-3 accumulator
O. Manual valve and linkage
P. L-2 accumulator

a. Porosity check-case, passageways
b. Pumps—gears
c. Clutch seal rings

d. Intermediate servo
e. Forward clutch assembly
f. Forward clutch assembly
g. Intermediate band assembly
h. Low and reverse clutch assembly
i. Low and reverse roller clutch assembly
j. Converter assembly

CAPACITY AND GENERAL SPECIFICATIONS

Oil capacity (dry) .... 20 pints
Cooling .... Water
Oil filter type .... Suction screen
Clutch units .... Three
Roller clutch .... One
Band (adjustable) ..... One
Planetary gear sets ... Two

Description

The Turbo Hydra-Matic 250 transmission is a fully automatic unit, consisting of a three element torque converter, two planetary gear sets, three multiple-disc clutches, one roller clutch, and an adjustable intermediate band. A radiator cooler is used to assist in the cooling of the transmission oil.

Transmission Disassembly

Converter and Modulator Valve

Removal

1. Clean the outside of the transmission case throughly, to prevent dirt from entering the unit during disassembly.

2. Install the transmission in a holding fixture or other suitable tool arrangement, and remove the converter assembly.

3. Remove the vacuum modulator assembly retainer and bolt, and remove the valve unit frm the transmission case. Discard the "O" ring from the modulator sleeve.

Rear Extension Housing and Components, Oil Pan and Screen

Removal

1. Remove four extension housing bolts, remove the housing from the case, and remove the square cut "O" ring seal from the extension housing.

2. With the aid of a puller, remove the speedometer drive gear and retaining clip.

3. Remove the governor cover, retained by a press fit into the case, by gently tapping along the cover lip, with the use of a screwdriver blade and light hammer.

4. Pull the governor assembly from the case, and inspect the case bore and governor sleeve for

scoring.

5. Remove the oil pan attaching screws, oil pan, and gasket.
6. Remove the oil pump suction screen attaching screws and remove the screen and gasket.

Valve Body and Linkage

Removal

1. Remove the detent spring and roller assembly from the valve body, and remove the body to case attaching bolts.
2. Remove the control wire by removing the actuator pin from the detent actuator valve lever.
3. Remove the manual control link and remove the valve body from the case.
4. Remove the intermediate servo return spring.
5. Remove the transfer support plate and bolts. Remove the upper gasket, transfer plate, and lower gasket.
6. Locate, mark, and remove the four check balls from the case.
7. Remove the oil pressure screen, and the governor feed screen from the holes in the case.
8. Remove the retainer for the manual control shaft with a screwdriver blade.
9. Loosen and remove the nut holding the range selector inner lever to the manual shaft.
10. Remove the range selector lever and parking pawl actuator rod from the case. Disengage the actuator rod from the inner lever.
11. Remove the parking lock, and lock bracket from the case.
12. Remove the parking pawl disengaging spring, parking pawl shaft retaining plug, shaft, and pawl.

CLUTCH AND BAND APPLICATION—TURBO HYDRA-MATIC 250 TRANSMISSION

GEAR	DIRECT	FORWARD	LOW AND REVERSE CLUTCH	LOW AND REVERSE ROLLER CLUTCH	INTERMEDIATE BAND
1st*	off	on	off	locked	off
2nd*	off	on	off	freewheeling	on
3rd*	on	on	off	freewheeling	off
low	off	on	on	freewheeling	off
reverse	on	off	on	freewheeling	off

* Transmission lever in "D" position

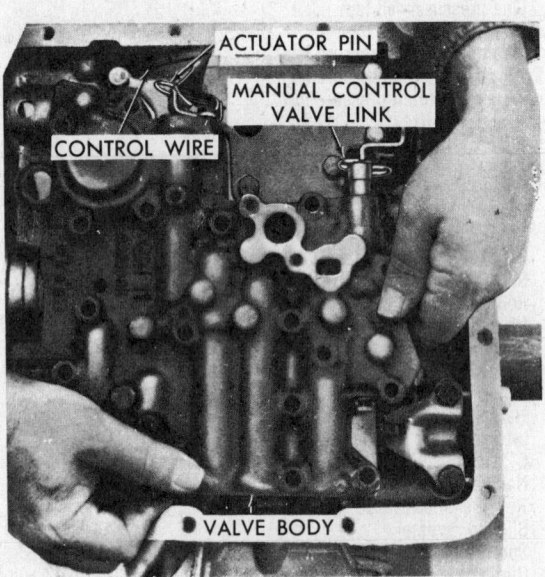

Removing manual control valve link, valve body, and detent actuating lever (© G.M.C.)

Internal Case Components and Oil Pump

Removal

1. Remove the front oil pump retaining bolts and washers.
2. By using two slide hammers, threaded into the pump body, remove the pump from the case.
3. Loosen the intermediate band anchor bolt and remove the band.
4. Remove the intermediate servo from the case, and disassembly can be accomplished by the re-

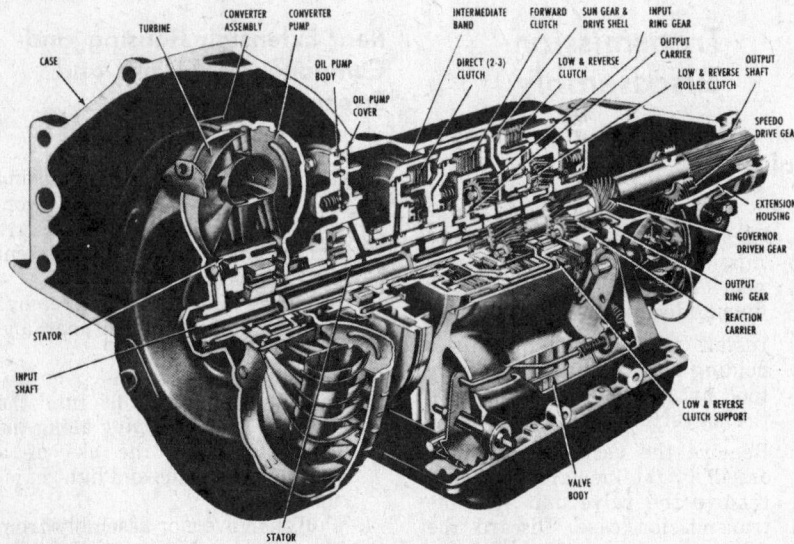

Sectional view—250 Turbo Hydra-Matic (© G.M.C.)

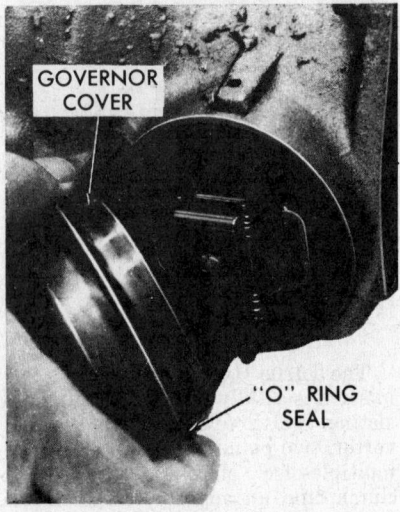

Governor cover removal (© G.M.C.)

Location of detent roller and spring (© G.M.C.)

Check ball location in case pockets (© G.M.C.)

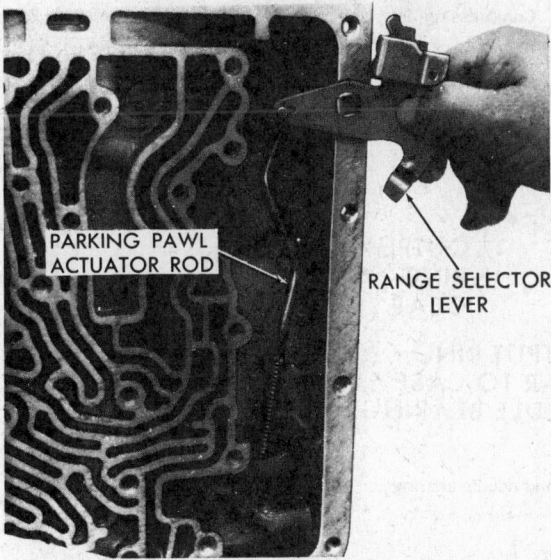

Removal of parking pawl actuator rod and range selector lever from case (© G.M.C.)

Parking pawl, shaft, and retaining plug (© G.M.C.)

Using slide hammers to remove front oil pump assembly (© G.M.C.)

Intermediate band and anchor bolt (© G.M.C.)

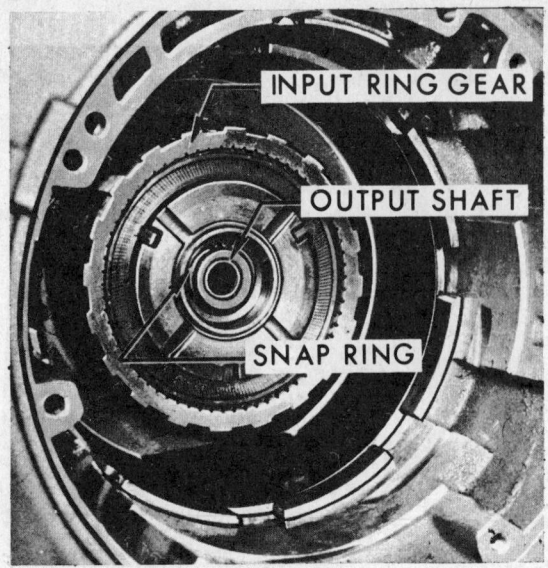

Removal of snap ring—retaining input ring gear to output shaft (© G.M.C.)

Compressing low and reverse clutch spring retainer with special tool usage (© G.M.C.)

moval of the snap rings at each end of the apply rod.

5. Remove the direct and forward clutch assemblies from the case.
6. Remove the input ring gear front thrust washer, noting the positions of the three tangs on the washer.
7. Remove the output shaft snap ring and remove the input ring gear.
8. Remove the input ring gear rear thrust washer and remove the output carrier assembly and the sungear drive shell assembly.
9. Remove the retaining snap ring from the case that holds the low and reverse roller clutch support in place, and remove the support assembly, the race assembly, and the anticlunk spring.
10. Remove the low and reverse clutch pack. (Steel and faced plates).
11. Separate the reaction carrier assembly from the output ring gear and shaft assembly.
12. Remove the tanged thrust washer and the output ring gear to case needle bearing assembly.
13. Compress the low and reverse clutch piston spring retainer and remove the nap ring and retainer. Remove the seventeen piston springs from the piston.
14. With air pressure applied to the low and reverse clutch piston apply passage, remove the piston from the case.
15. Remove the seals from the piston and inspect piston and bore for scoring.

Accumulator Valve

Removal

1. With the aid of a suitable tool, compress the accumulator piston cover inward, and remove the

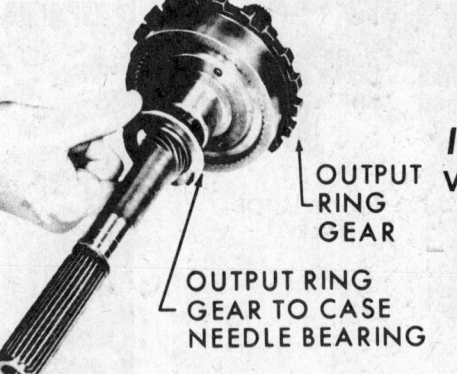

Output shaft, ring gear, and needle bearing assembly (© G.M.C.)

OUTPUT RING GEAR

OUTPUT RING GEAR TO CASE NEEDLE BEARING

snap ring retainer.
2. Remove the cover, and piston spring.
3. Remove the 1-2 accumulator piston with the inner and outer hook type seals.

APPLY AIR PRESSURE HERE TO REMOVE LOW AND REVERSE CLUTCH PISTON

Location of apply passageway for low and reverse clutch piston (© G.M.C.)

Disassembly, Inspection, and Reassembly of Individual Components

Valve Body

Disassembly

1. Position the valve body with the core face up and the direct clutch accumulator piston pocket positioned to the upper left. *NOTE: The use of a multi-channelled steel or wood block assists in keeping the valves and springs in their proper order during the disassembly and assembly of the valve body.*
2. Remove the manual control valve, lower left.
3. From the lower right hand bore, remove the pressure regulator valve train, consisting of the retaining pin, boost valve sleeve, intermediate boost valve, reverse and modulator boost valve, pressure regulator valve and spring.
4. From the second right hand bore from the bottom, remove the 2-3 shift valve retaining pin, sleeve, control valve spring, 2-3 shift control valve, shift valve spring, and the 2-3 shift valve.
5. From the third bore, remove the retaining pin, sleeve, shift control valve spring, 1-2 control valve, and the 1-2 shift valve.
6. From the fourth bore, remove the retaining pin, plug, manual low control valve spring, and the manual low control valve.
7. From the fifth bore, remove the retaining pin, spring, seal, and detent regulator valve.
8. From the upper left bore, remove the detent actuating lever bracket bolt, bracket stop, spring

Exploded view—1-2 accumulator (© G.M.C.)

Removal and replacement of 1-2 accumulator piston cover retaining ring with special tool usage (© G.M.C.)

retainer, seat, outer and inner spring washers, and the detent valve.

Inspection

1. Inspect all the valves for scoring, cracks, distortion, and free movement in their respective bores.
2. Inspect all springs for distortion of the coils or being collapsed in length.
3. Inspect the valve body for cracks, scores in the bores, interconnected passageways, and flatness of the body.

Assembly

1. Reverse the disassembly procedures for the assembly of the valve body.
2. Lubricate the bores and valves with automatic transmission fluid during the assembly.

Oil Pump

Disassembly

1. Remove the pump cover to body attaching bolts.
2. Remove the two forward clutch oil seal rings and the three direct clutch oil seal rings from the pump hub.

3. Remove the selective thrust washer from the pump cover.
4. Separate the pump cover and stator shaft assembly from the pump body.
5. Remove the oil pump gears and "O" ring seal.

Inspection

1. Inspect the gears for nicks and abnormal wear patterns.
2. Inspect the pump body and cover for nicks and scoring.
3. Remove the outer seal and in-

spect the bushing for galling or scoring.
4. Install the pump gears into the body and measure the pump body to gear face clearance. The clearance should be .0005 to .0015 inch.
5. Check the oil passages in the body to insure of no restrictions.

Assembly

1. Replace the hub seal, using a non-harding sealer on the outer diameter of the seal, and seat it

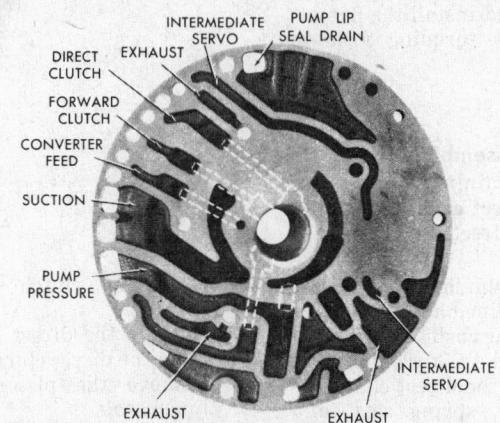

Oil passage locations in oil pump cover (© G.M.C.)

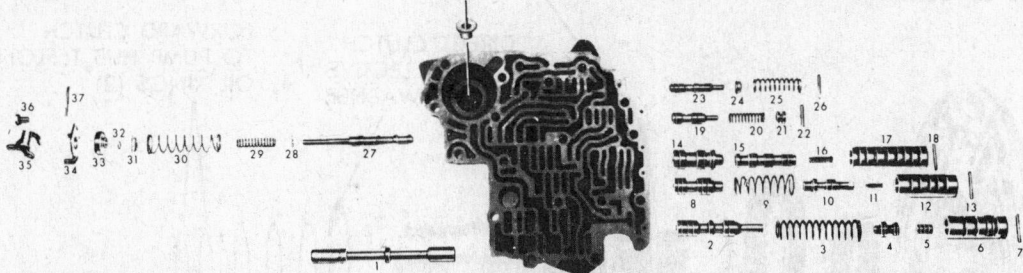

Exploded view—Valve body—typical (© G.M.C.)

1 Manual valve	14 1-2 shift valve
2 Pressure regulator valve	15 1-2 shift control valve
3 Pressure regulator valve spring	16 1-2 Shift control valve spring
4 Reverse and modulator boost valve	17 1-2 shift control valve sleeve
5 Intermediate boost valve	18 Retaining pin
6 Boost valve sleeve	19 Manual low control valve
7 Retaining pin	20 Manual low control valve spring
8 2-3 shift valve	21 Plug
9 2-3 shift valve spring	22 Retaining pin
10 2-3 shift control valve	23 Detent regulator valve
11 2-3 shift control valve spring	24 Detent regulator valve spring seat
12 2-3 shift control valve sleeve	25 Detent regulator valve tspring
13 Retaining pin	26 Retaining pin

27 Detent valve
28 Washer
29 Detent valve inner spring
30 Detent valve outer spring
31 Detent valve outer spring seat
32 Detent valve spring retainer
33 Detent valve spring stop
34 Detent valve actuating lever bracket
35 Detent valve actuating lever
36 Retaining bolt
37 Retaining pin
38 Cap

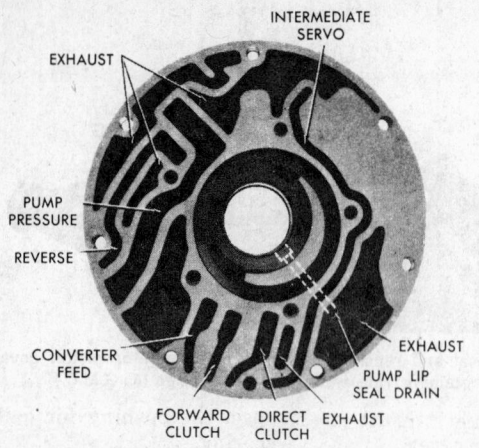

Oil passage locations in oil pump body (© G.M.C.)

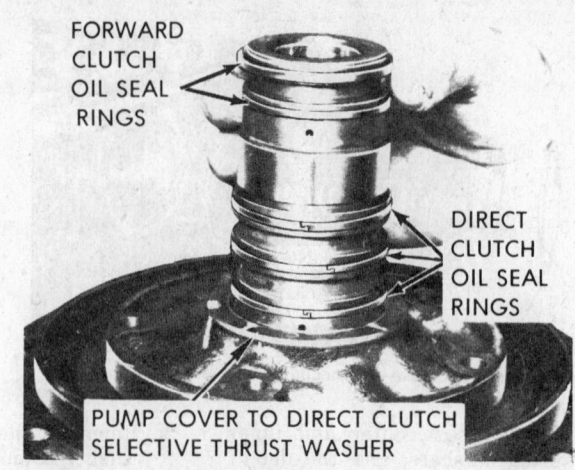

Location of hub oil seal rings and selective thrust washer (© G.M.C.)

firmly into the counter bore.

2. Install the drive and driven gears, and align the marks on the gears.
3. Install the pump cover thrust washer, the three direct clutch oil seal rings, and the two forward clutch oil seal ring.
4. Install the pump outside diameter "O" ring, align the body to the cover, and install the five attaching bolts, torquing them to 18 ft. lbs.

Direct Clutch

Disassembly

1. Remove retaining snap ring from the direct clutch drum and remove the direct clutch pressure plate.
2. Remove the clutch pack, consisting of three lined and three steel plates, and the cushion spring.
3. With the aid of a compression tool, remove the direct clutch retaining ring, spring seat, and

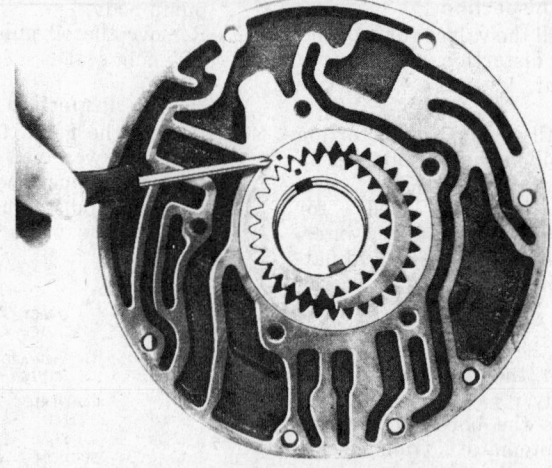

Alignment of oil pump gears (© G.M.C.)

seventeen clutch return coil springs.

4. Remove the direct clutch piston from the direct clutch drum.
5. Remove the piston outer and inner seals.

Inspection

1. Inspect the drive and driven clutch plates for signs of overheating, scoring, or wear.
2. Inspect the seventeen springs for being collapsed or distorted.

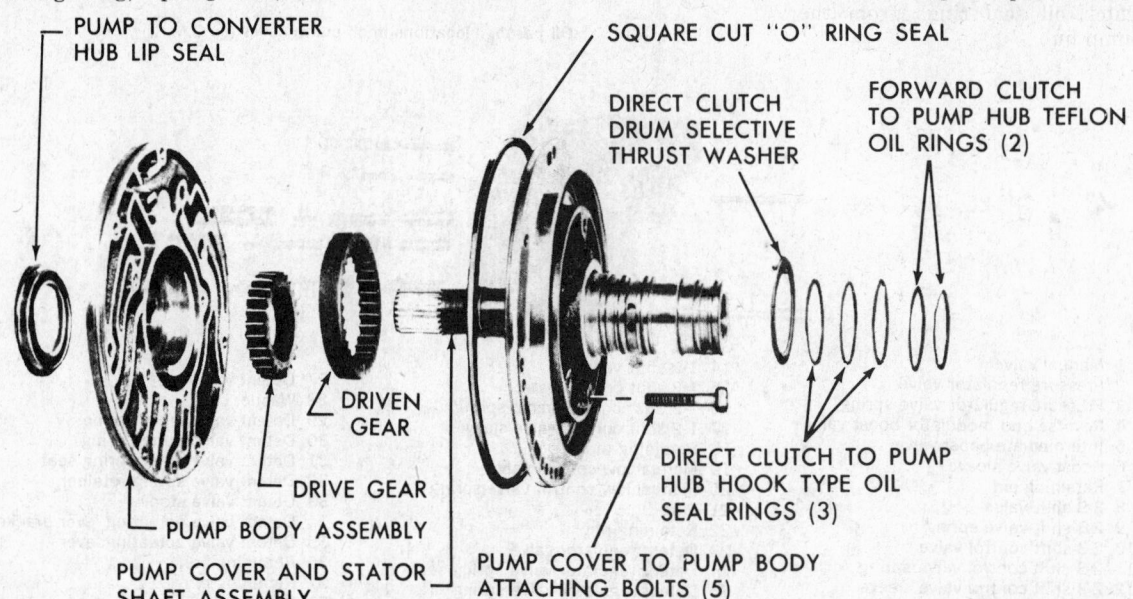

Exploded view—oil pump assembly (© G.M.C.)

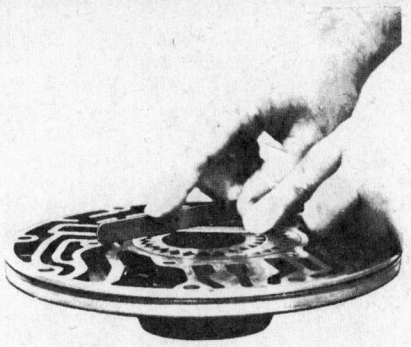

Checking gear face to pump body clearance
(© G.M.C.)

3. Inspect piston for cracks or scoring, ad the direct clutch drum piston seal mating surfaces for scores, wear, oil passages open, and free operation of the check ball.

Assembly

1. Install new inner and outer seals on the direct clutch piston and install it into the drum housing. *NOTE: Lubricate the seals with automatic transmission fluid and aid in the installation of the seals with the use of a feeler gauge blade or a .020 inch piano wire, crimped into a copper tubing.*

.020" PIANO WIRE CRIMPED INTO COPPER TUBING

PISTON

DIRECT CLUTCH HOUSING

Use of piano wire to assist in the installation of piston seals—typical (© G.M.C.)

2. Install the seventeen clutch return springs.

3. Install the return spring seat, compress with a compression type tool, and install the retaining snap ring.

4. Install the clutch pack, beginning with the cushion spring, and alternating with the steel plates and the lined plates.

5. Install the direct clutch pressure plate and retaining rings.

Forward Clutch

Disassembly

1. Remove the needle roller bearing from the forward clutch housing.
2. Remove the retaining ring and pressure plate from the forward clutch drum.
3. Remove the clutch pack from the drum.
4. With the aid of a compression tool, remove the retaining ring and the piston return seat.
5. Remove the twenty-one clutch return springs.

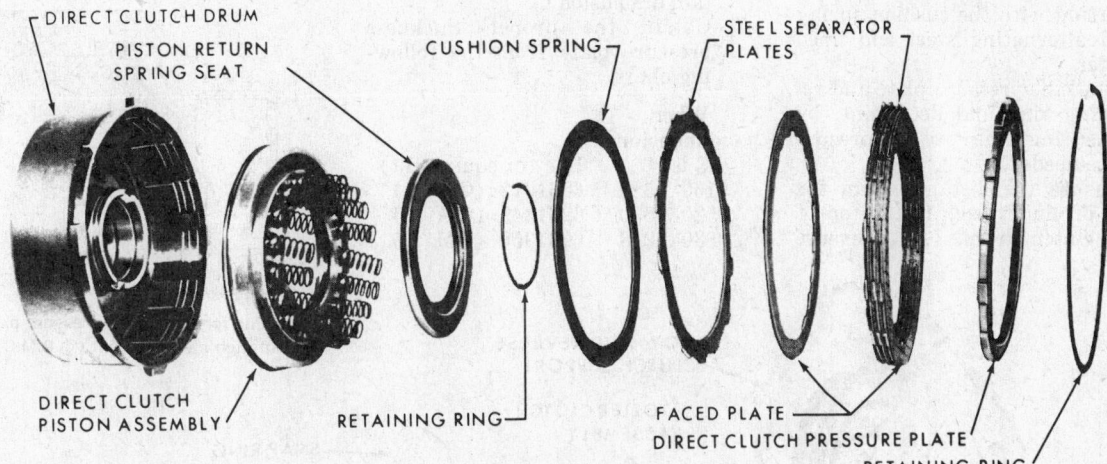

DIRECT CLUTCH DRUM

PISTON RETURN SPRING SEAT

CUSHION SPRING

STEEL SEPARATOR PLATES

DIRECT CLUTCH PISTON ASSEMBLY

RETAINING RING

FACED PLATE

DIRECT CLUTCH PRESSURE PLATE

RETAINING RING

Exploded view—direct clutch assembly (© G.M.C.)

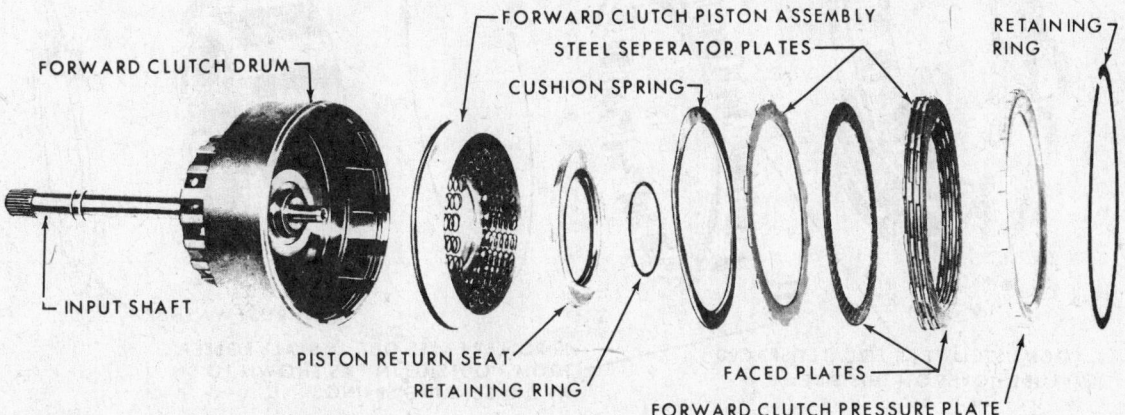

FORWARD CLUTCH PISTON ASSEMBLY

STEEL SEPERATOR PLATES

RETAINING RING

FORWARD CLUTCH DRUM

CUSHION SPRING

INPUT SHAFT

PISTON RETURN SEAT

RETAINING RING

FACED PLATES

FORWARD CLUTCH PRESSURE PLATE

Exploded view—forward clutch assembly (© G.M.C.)

6. Remove the forward clutch piston assembly and remove the inner and outer piston seal.

Inspection

1. Inspect the clutch pack for signs of wear, scoring, or overheating.
2. Inspect the twenty-one springs for distortion, or having collapsed coils.
3. Inspect the piston for cracked or scored surfaces, and the forward clutch drum piston seal mating surface for scores, wear, oil passages open, and free operation of the ball check.

Assembly

1. Install new inner and outer seals on the piston and install it into the forward clutch drum housing.
   NOTE: Lubricate the seals with automatic transmission fluid and aid in the installation of the seals with the use of a feeler gauge blade or a .020 inch piano wire, crimped into a copper tubing.
2. Install the twenty-one clutch return springs.
3. Install the spring retainer and compress, and install the snap ring.
4. Install the cushion spring face plate and steel separator plates, starting with the cushion spring and alternating steel and lined plates.
5. Install the pressure plate and retaining ring and determine, by measuring, the proper presure plate needed.
6. Measure the distance from the top of clutch pack to the top of the clutch drum, (A). Measure

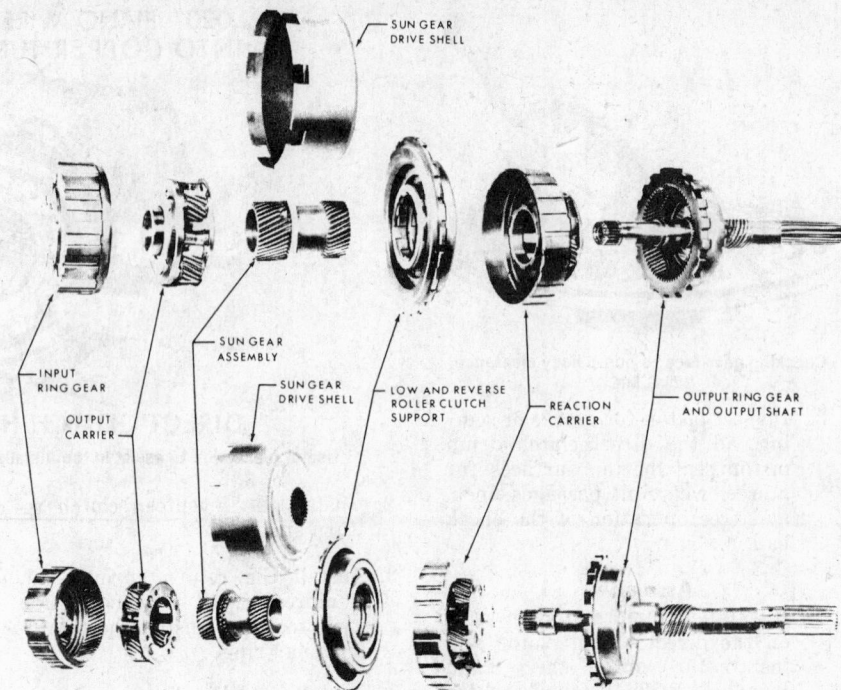

Exploded view—planetary gear train (© G.M.C.)

the distance from the lower edge of the notch on the inner surface of the drum end of the drum (B). Subtract B from A to obtain dimension C.

7. Obtain the proper thickness pressure plate from the following chart.

When dimension C is:

C is:	Use (or equivalent)
.0160-.0520	6261072 (GM #)
.0520-.0830	6261349 (GM #)
.0830-.1218	6261350 (GM #)

WHEN DIM. C IS	USE (OR EQUIVALENT)
.0160— .0520	6261072
.0520— .0830	6261349
.0830— .1218	6261350

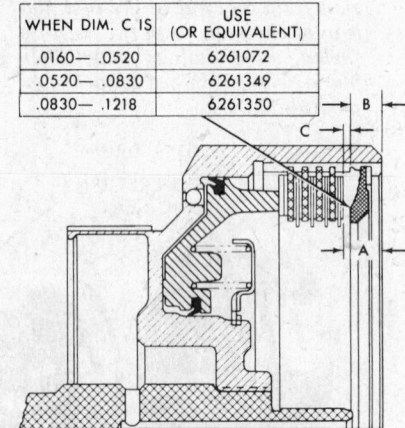

Selecting forward clutch pressure plate— —through measurement (© G.M.C.)

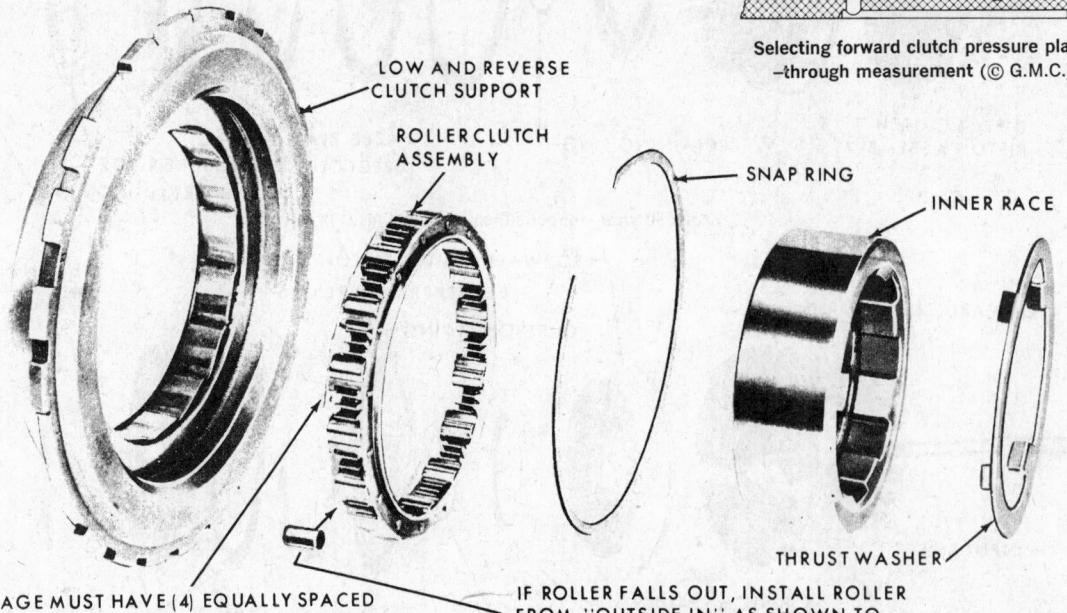

LOW AND REVERSE CLUTCH SUPPORT

ROLLER CLUTCH ASSEMBLY

SNAP RING

INNER RACE

THRUST WASHER

CAGE MUST HAVE (4) EQUALLY SPACED .091 LUBE HOLES ON THIS SIDE

IF ROLLER FALLS OUT, INSTALL ROLLER FROM "OUTSIDE IN" AS SHOWN TO AVOID BENDING SPRINGS

Exploded view—low and reverse clutch assembly (© G.M.C.)

Sun Gear and Sun Gear Drive Shaft

Disassembly

1. Remove the retaining snap ring from the sun gear hub, and remove the flat thrust washer.
2. Remove the sun gear assembly from the drive shell.
3. Remove the sun gear to drive shell front retaining snap ring and discard.

Inspection

1. Inspect the shell and sun gear for wear or damage.

Assembly

1. Install new sun gear to drive shell front retaining snap ring.
2. Install the sun gear assembly into the drive shell.
3. Install the flat steel thrust washer and retaining snap ring on the sun gear hub.

Low and Reverse Roller Clutch Support

Disassembly

1. Remove the thrust washer from the inner race and remove the race from the support.
2. Remove the low and reverse clutch roller retaining ring.
3. Remove the roller clutch from the support.

Inspection

1. Inspect the rollers, races, for scratches, indentations, wear and distortion of the roller springs.

Assembly

1. Install the roller clutch assembly into the inner race with the oil holes towards the rear of the transmission.
2. Install the outer race and the retaining snap ring into the groove on the clutch support. *NOTE: The low and reverse overrun clutch inner race should free wheel in the clockwise direction only.*

Governor Assembly

NOTE: The components of the governor assembly, except the driven gear, are a select fit and each assembly is calibrated and is serviced as a complete assembly. The driven gear can be serviced separately, but the governor assembly must be disassembled to replace it. A repair kit is available, which consists of a gear, gear retainer pin, and two governor weight retaining pins.

The following procedure involves the replacement of the driven gear.

Disassembly

1. Cut off one end of each governor

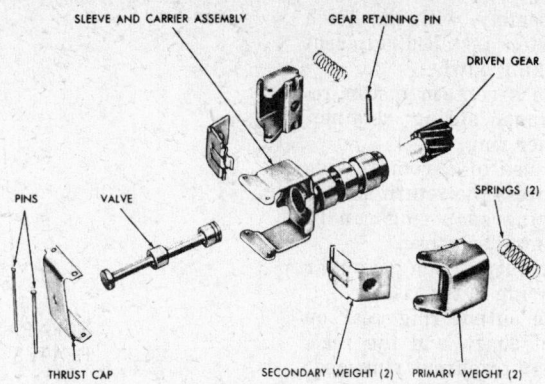

Exploded view—governor assembly—typical (© G.M.C.)

weight pin and remove the pins.
2. Remove the governor valve from the sleeve and carrier assembly.
3. Drive out the driven gear retaining split pin.
4. With the aid of a press, force the driven gear out of the sleeve.
5. Press the new gear into the sleeve and carrier assembly, cleaning any chips from the gear hub before bottoming it on the sleeve shoulder.
6. Locate a new pin hole, 90 degrees from the existing hole, and drill through the sleeve and gear, using a 1/8 inch drill bit.
7. Install the new split pin and clean the governor of any chips that may have collected.

Inspection

While the governor is out of its bore and disassembled, it should be inspected for the following defects:

1. Inspect the sleeve and carrier assembly, and the governor valve for nicks, burrs, scoring, or galling.
2. Inspect the governor valve for free operation in the bore of the governor sleeve.
3. Inspect the governor weights and springs for freedom of operation, distortion, and looseness.

Assembly

1. Install the governor valve into the bore of the sleeve and carrier assembly.
2. Install the governor weights and springs, along with the thrust caps, on the sleeve assembly.
3. Align the holes of the thrust caps, governor sleeve, and weight assemblies, and install the new pins. Crimp the ends of the pins to prevent them from coming out.
4. Check the weights and governor valve for free movement.

Bushing Replacement

Most internal bushings are replaceable, and the use of the proper removal and installation tools are most important when replacement is needed, to insure proper alignment, and to avoid installation damage to the bushings by using improper tools. It is advisable to match the installed bushing to its operating surface before the transmission is assembled to insure proper bushing fit.

Transmission Assembly

Transmission Internal Components

Assembly

1. Install the low and reverse clutch

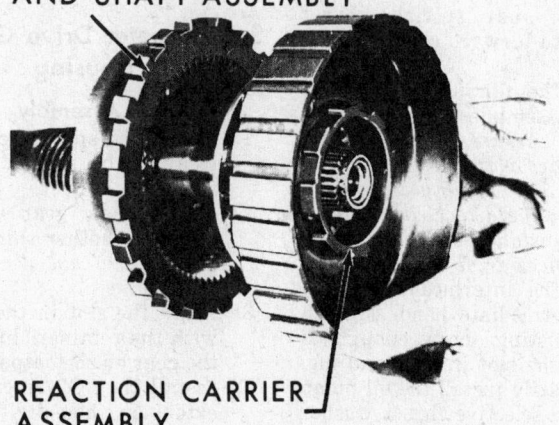

Reaction carrier assembly (© G.M.C.)

piston assembly, with the notch in the piston installed adjacent to the parking pawl.

2. Install the seventeen piston return springs, spring retainer, and retainer ring.

3. With the use of a compressing tool, compress the return seat so the retaining snap ring can be installed into its groove.

4. Install the output ring gear rear thrust bearing in the case.

5. Install the output ring gear on the output shaft, and the reaction carrier to output ring gear front thrust washer, (three tanged), into the output ring gear support.

6. Install the output shaft into the case and install the reaction carrier assembly into the output ring gear and shaft assembly.

7. Starting with a steel plate, and alternating with a lined plate, install the low and reverse clutch pack. Install the clutch support retainer, (anti-clunk), spring. *NOTE: Notch in the steel separator plates should be placed towards the bottom of the case.*

8. Align the splines on the inner race of the roller clutch with the splines on the reaction carrier, and install the low and reverse support assembly with the notch aligned with the support retainer (anti-clunk) spring.

9. Install the thrust washer on the low and reverse roller clutch inner race.

10. Install the retainer snap ring for the low and reverse clutch support into the case groove, with the anti-clunk spring between the gap.

11. Install the rear thrust washer and the sun gear drive shell assembly, and install the output carrier assembly.

12. Install the input ring gear rear thrust washer and install the input ring gear and retain it with a snap ring.

13. Install the input gear front thrust washer, direct clutch assembly, and special thrust washer to forward clutch assembly.

14. Install the forward and direct clutch assemblies into the case. *NOTE: Assure that forward clutch face plates are positioned over the input ring gear and the tangs on the direct clutch housing are installed into the slots on the sun gear drive shell.*

15. Install the intermediate servo and intermediate band. Tighten the adjusting screw enough to engage the slot in the band lug.

16. Temporarily install the oil pump with the selective thrust washer and gasket in place, using two guide bolts to position pump.

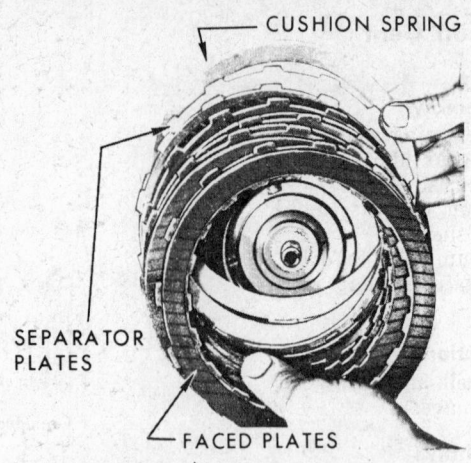

CUSHION SPRING

SEPARATOR PLATES

FACED PLATES

Typical clutch pack (© G.M.C.)

17. Install two pump to case bolts and tighten snugly.

18. Position the transmission so that the input shaft points down, and install a dial indicator to contact the input shaft and zero the indicator.

19. Lift upward on the input shaft and note the total indicator movement. The proper limits are from .032 to .064 inch. If the readings are not correct, the pump must be removed and thicker or thinner selective washers must be obtained *NOTE: Selective fit Thrust washers are available in the following thicknesses.*
.065-.067
.082-.084
.099-.101

21. Install the pump assembly with a new gasket and "O" ring.

CAUTION: If the input shaft cannot be rotated as the pump is placed into the case, the direct and forward clutch housing and clutches have not been properly installed. Correct the condition before continuing.

22. Adjust the intermediate band by adjusting the screw to 30 in. lbs. and backing off three complete turns, and tightening the lock nut to 15 ft lbs.

Speedometer Drive Gear and Extension Housing

Assembly

1. Install the speedometer gear retaining clip on to the output shaft.

2. Heat a new gear with a heat lamp or another suitable source. *NOTE: Do not use a torch or open flame.*

3. Align the slot in the drive gear with the retainer clip and install the gear on the output shaft.

4. Install the "O" ring seal on the extension housing, install the housing to the case, and torque the attaching bolts to 25 ft lbs.

5. Install a new extension housing seal.

Manual Linkage

Assembly

1. Install the parking pawl, tooth towards the inside of the case, install the pawl shaft into the case and through the parking pawl, and install the shaft retainer plug, staking in three places to retain it in the case.

2. Install the parking pawl disengaging spring, square end hooked on the pawl.

3. Install the park lock bracket and torque the bolts to 29 ft lbs.

4. Install the range selector inner lever to the parking pawl actuator rod, and position it under the park lock bracket and parking pawl.

5. Install the manual shaft through the case and range selector inner lever, and install the retaining nut on the manual shaft.

6. Install the manual shaft to case spacer clip.

Valve Body, Oil Pan, and Gasket Assembly

Assembly

1. Install the governor feed screen and oil pump pressure screen.

2. Install the four check balls into their respective case pockets.

3. Install the lower gasket and transfer plate, followed by the upper valve gasket.

4. Install the intermediate servo return spring.

5. Install the valve body and connect the manual control valve link to the range selector inner lever. Torque the bolts to 130 in. lbs.

6. Install the transfer support plate and torque the bolts to 130 in. lbs.

7. Connect the detent control valve wire to the detent valve actuating lever, and attach the lever to the valve body.

Location of oil pump pressure screen (© G.M.C.)

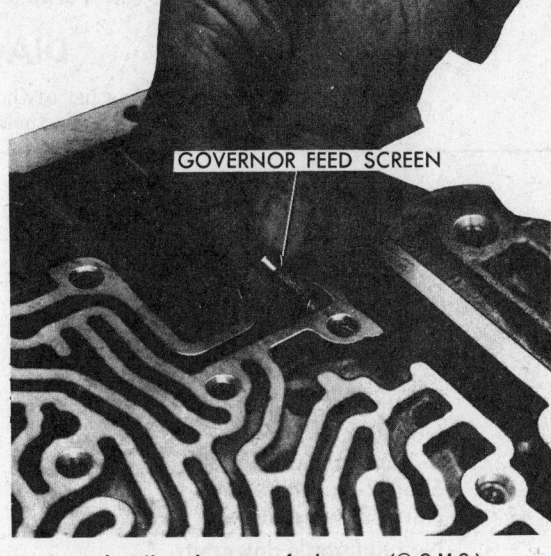

Location of governor feed screen (© G.M.C.)

8. Install the detent roller and spring assembly to the valve body.
9. Install the suction screen and gasket.
10. Install the oil pan and gasket, and torque the bolts to 130 in. lbs.

Governor and Vacuum Modulator

Assembly

1. Install the governor assembly, cover and seal. Tap the cover into place on the case.
2. Install the vacuum modulator valve, modulator and retainer clip. Torque bolt to 130 in. lbs.

Accumulator

Assembly

1. Install the 1-2 accumulator piston assembly and spring.
2. Install new "O" ring in the case groove and install the cover.

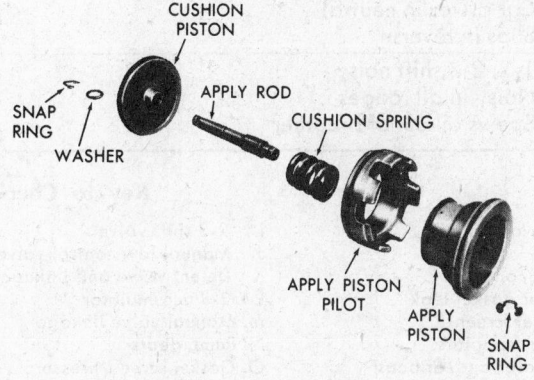

Exploded view—intermediate servo (© G.M.C.)

3. Compress the cover and install the retaining snap ring.

Converter

Assembly

1. Engage the converter to the splines of the input shaft and the stator shaft assembly.

2. Install a holding bracket to the converter from the case, so as not to lose the converter during the installation into the vehicle.

Removing 1-2 Accumulator Piston Cover Retaining Ring

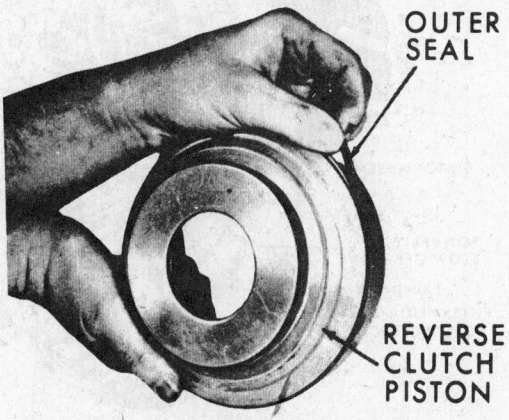

Removing Low and Reverse Clutch Piston Outer Seal

GM Turbo Hydra-Matic 350

DIAGNOSIS

This diagnosis guide is a list of the most common troubles and their causes. The items to check are listed in the sequence to be followed.

TROUBLE SYMPTOMS	Items to Check	
	In Truck	Out of Truck
Slips in all ranges	ACDFGMO	N
Drive slips—no 1st gear	ACDFGMO	NQV
No 1-2 upshift	BEFGH	SU
1-2 upshift, early or late	BEFH	
Slips, 1-2 upshift	ACFGHL	NSU
Harsh 1-2 upshift	BCG	
No 2-3 upshift	FI	R
2-3 upshift, early/late	BCEFIK	
Slips, 2-3 upshift	ACFGI	NR
Harsh 2-3 upshift	BCGIL	
No full throttle downshift	K	
2-3 upshift, wide open throttle only	BK	
L₁ gear—no engine braking	GHJM	PQU
Car drives in neutral	K	Q
Slips in reverse	ACDFGHMO	RT
1-2, 2-3 shift noisy	A	RSY
Noisy in all ranges	ADFO	NXYW
Spews oil out of breather	AD	

Key to Checks

A. Low oil level/water in oil
B. Vacuum leak
C. Modulator and/or valve
D. Strainer and/or gasket leak
E. Governor valve/screen
F. Valve body gasket/plate
G. Pressure regulator and/or boost valve
H. 1-2 shift valve

I. 2-3 shift valve
J. Manual low control valve
K. Detent valve and linkage
L. 2-3 accumulator
M. Manual valve linkage
N. Pump gears
O. Gasket screen-pressure
P. Band—intermediate overrun roller clutch

Q. Forward clutch assembly
R. Direct clutch assembly
S. Intermediate clutch assembly
T. Low & reverse clutch assembly
U. Intermediate roller clutch
V. Low & reverse roller clutch
W. Parking pawl/linkage
X. Converter assembly
Y. Gear set and bearings

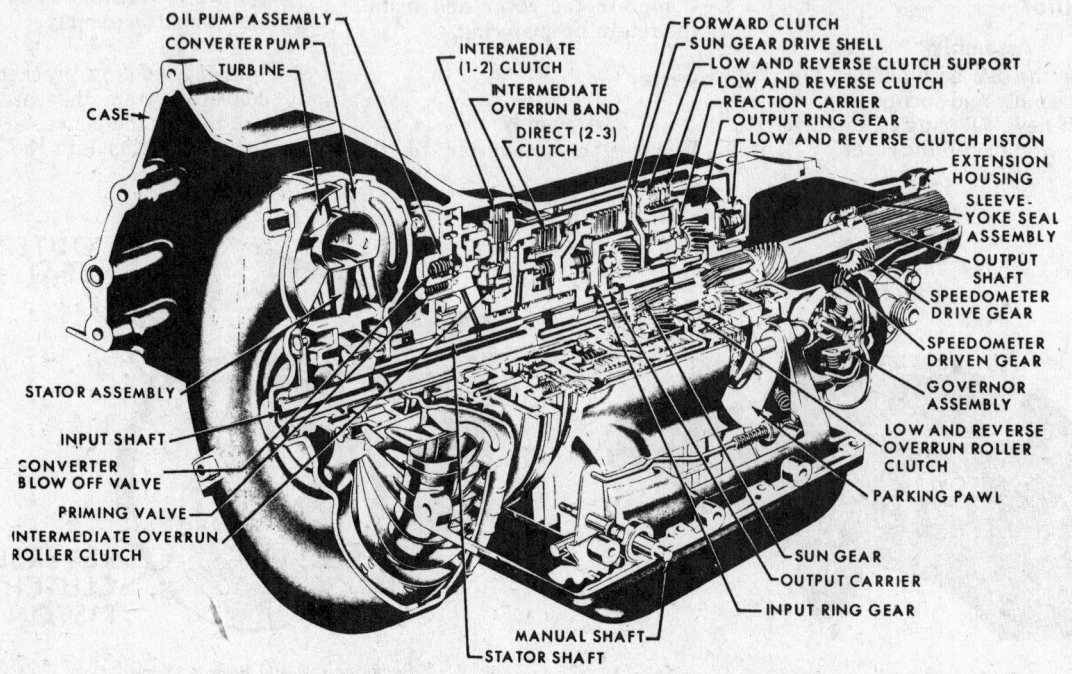

Sectional view of the General Motors Turbo Hydra-Matic 350 (© General Motors)

General Description

The 350 Turbo Hydra-Matic transmission is a fully automatic unit, consisting of a three element torque converter, two planetary gear sets, four multiple-disc clutches, two roller clutches, and an non-adjustable intermediate band. A radiator cooler is used to assist in the cooling of the transmission oil.

CAPACITY AND GENERAL SPECIFICATIONS

Oil Capacity (dry) ....20 Pints
Cooling ....Water
Oil Filter Type
.........Bottom Suction Screen
Clutches 4
Roller Clutches 2
Band (non-adjustable) 1

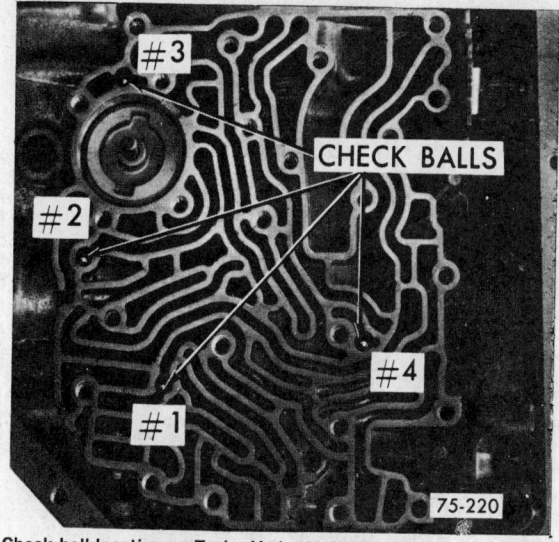

Check ball locations—Turbo Hydra-Matic 350 (© General Motors)

Transmission Disassembly

Clean outside of transmission thoroughly to prevent dirt from entering the unit.

1. With transmission in a holding fixture, lift off torque converter assembly.
2. Remove vacuum modulator assembly attaching bolt and retainer.
3. Remove vacuum modulator assembly, O-ring seal and modulator valve from the case.
4. Remove four extension housing-to-case attaching bolts.
5. Remove extension housing and the square cut O-ring seal.
6. Remove extension housing lip seal from output end of housing, using a screwdriver.
7. Remove extension housing bushing, using chisel to collapse bushing.
8. Drive in new extension housing bushing.
9. Install extension housing lip seal.
10. Depress speedometer drive gear retaining clip, then slide speedometer drive gear off output shaft.
11. Remove governor cover retainer wire with a screwdriver.
12. Remove governor cover and O-ring seal from case, then remove O-ring seal from governor cover.
13. Remove governor assembly from case.
    NOTE: Check governor bore and sleeve for scoring.
14. Remove oil pan attaching screws, pan, and gasket.
15. Remove two oil pump suction screen (strainer) to valve body attaching screws.
16. Remove oil pump screen (strainer) and gasket from valve body.
17. Remove detent roller and spring assembly from valve body. Remove valve body-to-case attaching bolts.
18. Remove manual control valve link from range selector inner lever. Remove valve body. Remove detent control valve link from detent actuating lever.
    NOTE: Refer to later text for valve body disassembly and passage identification.
    NOTE: At this time, when handling valve body assembly, do not touch sleeves, because retaining pins
    may fall into the transmission.
19. Remove valve body-to-spacer plate gasket.
20. Remove spacer support plate gasket. Remove spacer support plate.
21. Remove valve body spacer plate and plate-to-case gasket.
22. Remove four check balls from passages in case face.
    NOTE: If, during assembly, any of the balls are omitted or installed in the wrong locations, transmission failure will result.
23. Remove oil pump pressure screen from oil pump pressure hole in the case.
24. Remove governor feed screen from governor feed passage in case. Remove the manual control valve link from the range selector inner lever.
25. Remove manual shaft to case retainer with a screwdriver.
26. Remove jam nut holding range selector inner lever to manual shaft. Remove manual shaft.
27. Disconnect parking pawl actuating rod from range selector inner lever. Remove both from case.
28. Remove manual shaft to case lip oil seal, using a screwdriver.

CLUTCH AND BAND APPLICATION— 350 TURBO HYDRA-MATIC TRANSMISSION

GEAR	INTER-MEDIATE CLUTCH	DIRECT CLUTCH	FORWARD CLUTCH	LOW AND REVERSE CLUTCH	INTER-MEDIATE OVERRUN BAND	LOW AND REVERSE ROLLER CLUTCH	INTER-MEDIATE OVERRUN ROLLER CLUTCH
1st*	off	off	on	off	off	Locked	Locked
2nd*	on	off	on	off	off	Freewheel	Locked
3rd*	on	on	on	off	off	Freewheel	Freewheel
Low	off	off	on	on	off	Locked	Locked
Reverse	off	on	off	on	off	Freewheel	Freewheel

* Gear shift lever in "D" position

MANUAL SHAFT TO CASE RETAINER ←SCREWDRIVER

OIL PUMP PRESSURE SCREEN

OIL PUMP PRESSURE HOLE

Removing the manual shift to case retainer—Turbo Hydra-Matic 350
(© General Motors)

Removing the oil pump pressure screen-Turbo Hydra-Matic 350
(© General Motors)

29. Remove parking lock bracket.
30. Disconnect and remove parking pawl disengaging spring.
31. Remove parking pawl shaft retaining plug with a bolt extractor. Cock parking pawl on shaft. Using drift and hammer, tap on drift to force the parking pawl shaft from the case. Remove the parking pawl.
32. Remove intermediate servo piston and metal oil seal ring. Remove washer, spring seat, and apply pin.
33. Remove eight pump attaching bolts with washer-type seals. Discard seals.
34. Install two threaded slide hammers into threaded holes in pump body. Tighten jam nuts and remove pump assembly from the case.
35. Remove pump assembly to case gasket and discard.
36. Remove intermediate clutch cushion spring.
37. Remove intermediate clutch faced plates and steel separator plates. Inspect lined plates for pitting, flaking, wear, glazing, cracking and chips or metal particles imbedded in lining. Replace any lined plates showing any of these conditions. Inspect steel plates for heat spot discoloration or surface scuffing. If the surface is smooth and has an even color smear, the steel plates may be re-used.
38. Remove intermediate clutch pressure plate.
39. Remove intermediate overrun brake band.
40. Remove direct and forward clutch assemblies.
41. Remove forward clutch housing-to-input ring gear front thrust washer.

NOTE: Washer has three tangs.
42. Remove output carrier-to-output shaft snap-ring.
43. Remove input ring gear.
44. Remove input ring gear to output carrier thrust washer.
45. Remove output carrier assembly.
46. Remove sun gear drive shell assembly.
47. Remove low and reverse roller clutch support to case retaining ring.
48. Remove low and reverse roller clutch, support assembly, and retaining spring.
49. Remove low and reverse clutch faced plates and steel separator plates.
50. Remove reaction carrier assembly from output ring gear and shaft assembly.
51. Remove output ring gear and shaft assembly from case.
52. Remove reaction carrier to output ring gear tanged thrust washer.
53. Remove output ring gear-to-case needle bearing assembly.
54. Compress low and reverse clutch piston spring retainer and remove piston retaining ring and spring retainer.
55. Remove 17 piston return coil springs from piston.
56. Remove low and reverse clutch piston by application of air pressure.
57. Remove low and reverse clutch piston outer seal.
58. Remove low and reverse clutch piston center and inner seal.
59. Install suitable tool to compress intermediate clutch accumulator cover and remove retaining ring.
60. Remove intermediate clutch accumulator piston cover. Remove cover O-ring seal from case.

61. Remove intermediate clutch accumulator piston spring.
62. Remove intermediate clutch accumulator piston assembly. Remove inner and outer hook-type oil seal rings if required.

Component Disassembly

Valve Body

Disassembly

1. Position valve body assembly with cored face up and direct clutch accumulator piston pocket located as illustrated in valve body assembly illustration.
2. Remove manual valve from lower left-hand bore, A.
3. From lower right-hand bore, B, remove the pressure regulator valve train retaining pin, boost valve sleeve, intermediate boost valve, reverse and modulator boost valve, pressure regulator valve spring, and the pressure regulator valve.
4. From the next bore, C, remove the second-third shift valve train retaining pin, sleeve, control valve spring, second-third shift control valve, shift valve spring, and the second-third shift valve.
5. From the next bore, D, remove the first-second shift valve train retaining pin, sleeve, shift control valve spring, first-second shift control valve, and the first-second shift valve.
6. From the next bore, E, remove retaining pin, plug, manual low control valve spring, and the manual low control valve.
7. From the next bore, F, remove retaining pin, spring seat and the detent regulator valve.

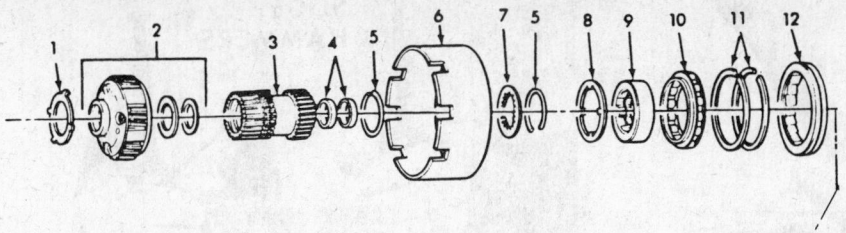

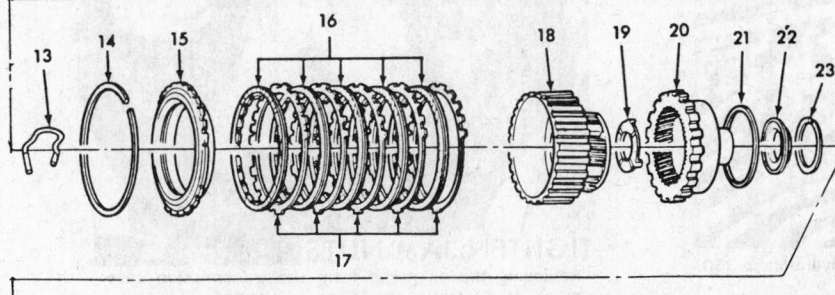

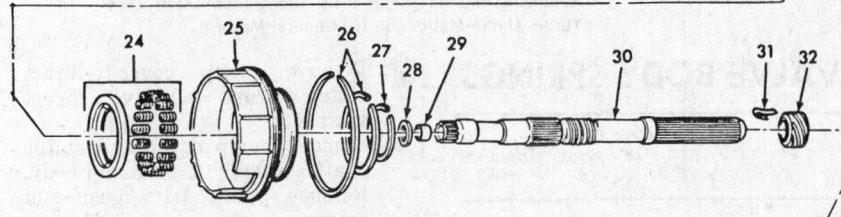

1 Input ring gear thrust washer
2 Output planet carrier assembly
3 Sun gear assembly
4 Sun gear bushings
5 Sun gear drive shell retaining ring
6 Sun gear drive shell
7 Sun gear thrust washer
8 Sun gear rear thrust washer
9. Low and reverse clutch race
10 Low and reverse clutch assembly
11 Low and recerse clutch ring
12 Low and reverse clutch cam (part of support assembly)
13 Low and reverse clutch support spring
14 Low and reverse clutch support spring retainer
15 Low and reverse clutch support assembly
16 Low and reverse reaction plate
17 Low and reverse clutch drive plate ass'y
18 Reaction planet carrier assembly
19 Output ring gear front thrust washer
20 Output ring gear
21 Output ring
22 Output ring gear rear thrust bearing
23 Output ring gear support to output shaft ring
24 Low and reverse clutch seat assembly with springs
25 Low and reverse clutch piston
26 Low and reverse clutch piston seal unit
27 Low and reverse clutch piston return spring seat retainer
28 Reaction planet carrier bushing
29 Output shaft bushing
30 Output shaft assembly
31 Drive gear retaining clip
32 Speedometer drive gear
33 Intermediate clutch accumulator piston retainer
34 Intermediate clutch accumulator piston cover
35 Intermediate clutch piston accumulator cover seal
36 Intermediate clutch accumulator piston spring
37 Intermediate clutch accumulator piston assembly with seals
38 Case vent assembly
39 Vacuum modulator valve assembly
40 Vacuum modulator to transmission case seal
41 Vacuum modulator retainer
42 Vacuum modulator to transmission bolt
43 Vacuum modulator
44 Transmission case bushing
45 Extension housing to case seal
46 Extension housing to transmission case bolt
47 Extension housing assembly
48 Extension housing bushing
49 Extension housing oil seal assembly
50 Speedometer driven gear
51 Speedometer drive fitting assy with seal
52 Speedometer drive fitting retainer
53 Speedometer drive fitting retainer to case bolt
54 Speedometer drive fitting seal
55 Governor cover retainer
56 Governor cover
57 Governor cover seal
58 Governor assembly
59 Shaft assembly with parking pawl plug
60 Parking lock pawl
61 Parking lock spring
62 Parking lock reaction bracket
63 Parking lock bracket bolt
64 Parking lock and range selector shaft retaining ring
65 Parking lock and range selector shaft (outer)
66 Manual shift seal assembly
67 Manual valve control link
68 Detent spring assembly
69 Parking lock actuator assembly
70 Parking lock and range selector lever (inner)
71 Selector shaft to lever nut
72· Transmission case assembly

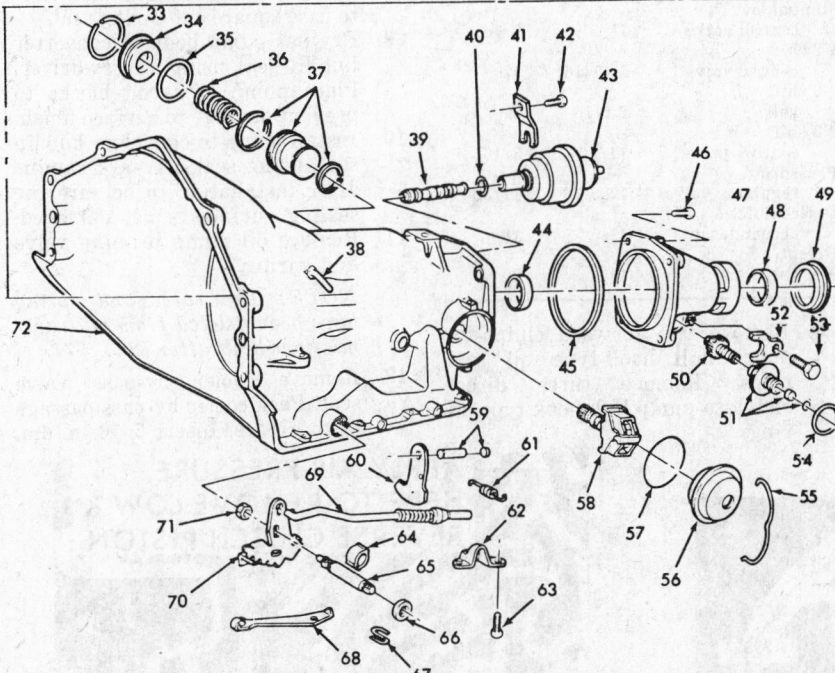

General Motors Turbo Hydra-Matic 350—exploded view—transmission case, extension housing and planet carrier parts (© General Motors)

8. Install spring compressor onto direct clutch accumulator piston and remove retaining E-ring at H.

9. At location H, remove direct clutch accumulator piston, then metal oil seal ring and spring.

10. From the upper left hand bore, G, remove the detent actuating lever bracket bolt, bracket, actuating lever and retaining pin, stop, spring retainer, seat, outer spring, and the detent valve.

Inspection

1. Wash all parts in solvent. Air dry. Blow out all passages.

2. Inspect all valves for scoring, cracks, and free movement in their bores.

3. Inspect sleeves for cracks,

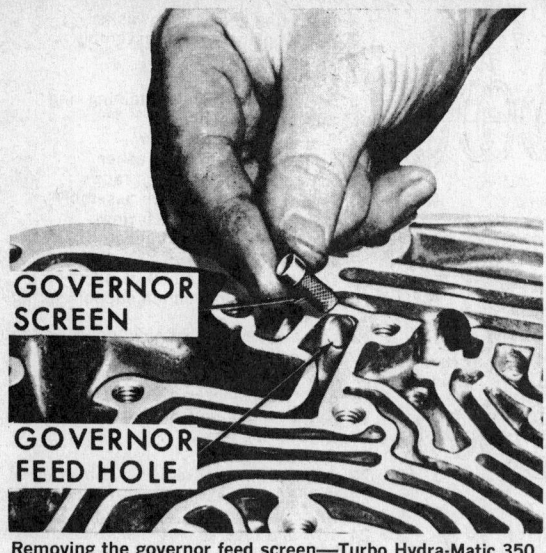

GOVERNOR SCREEN

GOVERNOR FEED HOLE

Removing the governor feed screen—Turbo Hydra-Matic 350 (© General Motors)

SLIDE HAMMERS

TIGHTEN JAM NUTS

Removing the oil pump from the transmission case— Turbo Hydra-Matic 350 (© General Motors)

scratches, or distortion.
4. Inspect valve body for cracks, scored bores, interconnected oil passages, and flatness of mounting face.
5. Check all springs for distortion or collapsed coils.

Assembly

1. Reverse disassembly procedures for assembly. Refer to Valve Body Springs chart for identification of springs.

Oil Pump

Disassembly

1. Remove five pump cover-to-body attaching bolts. Remove spring seat retainer.
2. Remove the intermediate clutch spring seat retainer, intermediate clutch return springs, and the intermediate clutch piston assembly.
3. Remove intermediate clutch piston inner and outer seals.

VALVE BODY SPRINGS

Valve	Free Length of Spring (in.)	Diameter (in.)
Detent regulator	1⅞	9/16
Manual low control valve	1½	7/16
1-2 shift control valve	1 15/16	¼
2-3 shift valve	2 1/16	⅞
2-3 shift control valve	11/16	3/16
Pressure regulator valve	1 11/16	17/32
Direct clutch accumulator	1¾	1½
Detent valve	1⅞	¾

4. Remove two forward clutch-to-pump hub hook-type oil seal rings. Remove three direct clutch-to-pump hub hook type oil rings.

5. Remove pump cover-to-direct clutch drum selective thrust washer.
6. Remove pump cover and stator shaft assembly from pump body.
7. Remove pump drive gear and driven gear from pump body.
8. Remove pump outside diameter-to-case square cut O-ring seal.
9. Pry out pump body to converter hub lip seal, using a screwdriver. Place pump on wood blocks to prevent damage to surface finish.
10. Install pump-to-converter hub lip seal, using seal driver. Examine after installation to be sure the sealing surface is not damaged.
11. Remove oil pump priming valve and spring.

NOTE: The valve and spring have been deleted from all transmissions built after Feb. 1971.

12. Remove cooler by-pass valve seat. Pack cooler by-pass passage with grease. Insert 5/16 in. dia.

J-23327

Compressing the low and reverse clutch piston spring retainer —Turbo Hydra-Matic 350 (© General Motors)

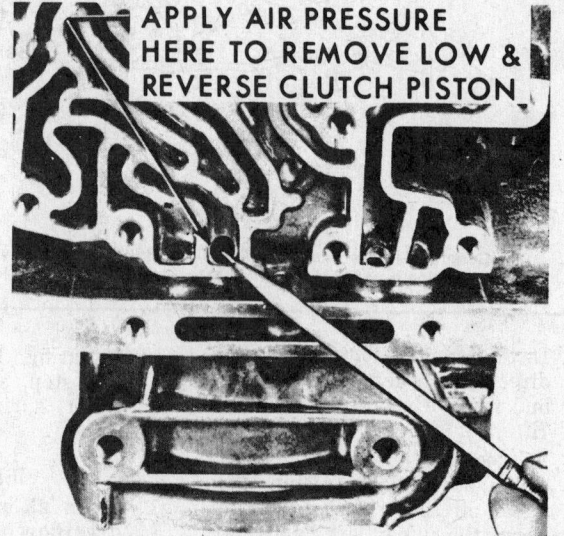

APPLY AIR PRESSURE HERE TO REMOVE LOW & REVERSE CLUTCH PISTON

Removing low and reverse clutch piston—Turbo Hydra-Matic 350 (© General Motors)

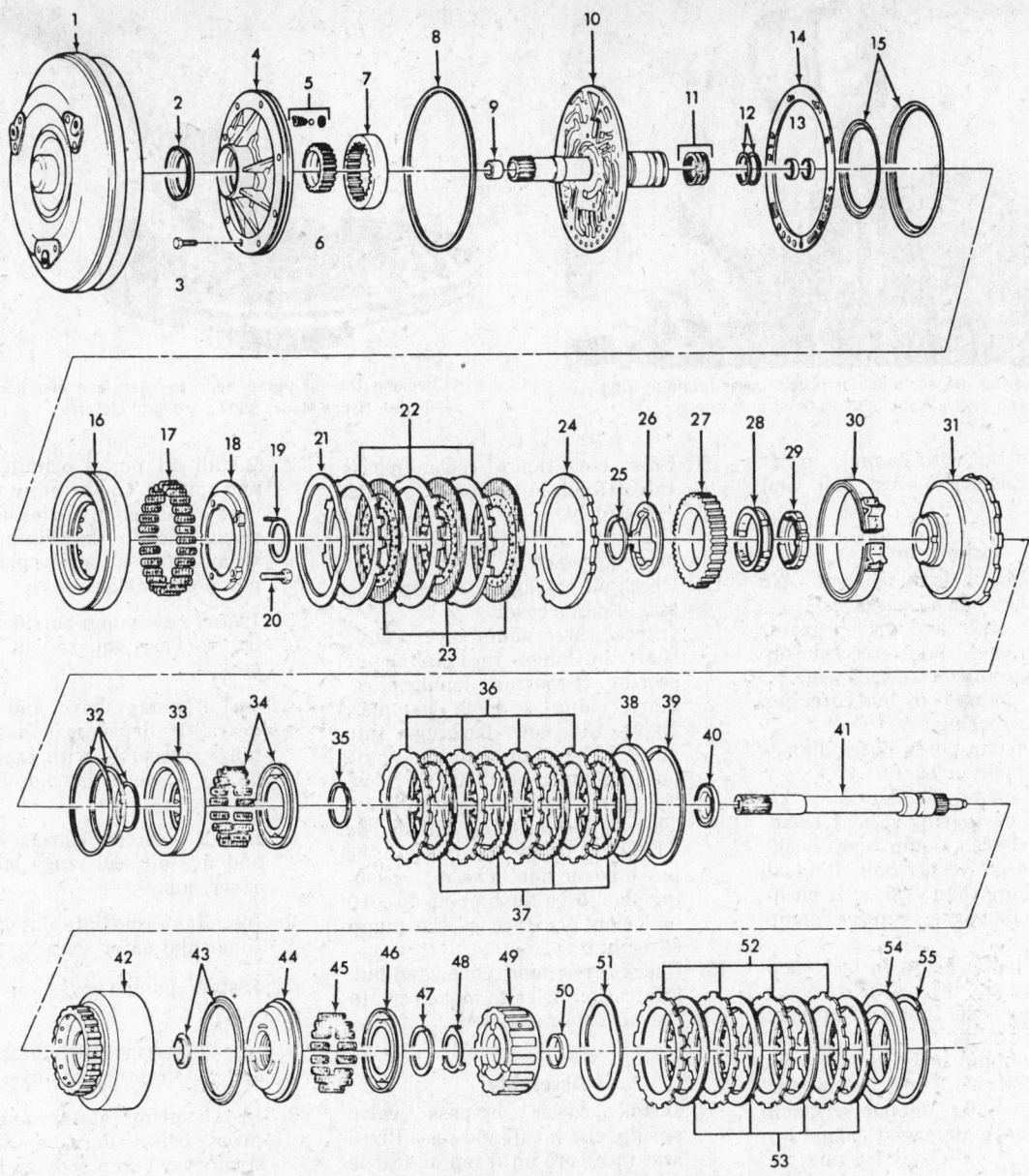

General Motors Turbo Hydra-Matic 350—exploded view—pump, central support, forward, direct and intermediate clutchs (© General Motors)

1 Torque converter assembly
2 Front oil pump seal
3 Oil pump to case screw
4 Oil pump body assembly
5 Oil cooler bypass valve
6 Oil pump drive gear
7 Oil pump driven gear
8 Oil pump to case seal
9 Startor shaft front bushing
10 Oil pump cover assembly
11 Direct clutch drum oil seal
12 Forward clutch housing oil seal
13 Stator shaft rear bushing
14 Oil pump to case gasket
15 Intermediate clutch piston seal
16 Intermediate clutch piston
17 Intermediate clutch piston return spring
18 Intermediate clutch piston return spring seat
19 Direct clutch drum front thrust washer
20 Oil pump to cover bolt
21 Intermediate clutch piston cushion spring
22 Intermediate clutch reaction plate
23 Intermediate clutch drive plate assembly
24 Intermediate clutch pressure plate assembly
25 Intermediate overrunning clutch retainer
26 Intermediate overrunning clutch outer race
27 Intermediate overrunning clutch inner race
28 Intermediate overrunning clutch assembly

29 Intermediate overrunning brake band assembly
30 Intermediate overrunning brake and band assembly
31 Direct clutch drum assembly
32 Direct clutch piston seal
33 Direct clutch piston
34 Direct clutch piston return spring and seat assembly
35 Direct clutch piston return spring seat retainer ring
36 Direct clutch drive plate
37 Direct clutch drive plate assembly
38 Direct clutch pressure plate assembly
39 Direct clutch pressure plate retaining ring
40 Direct clutch drum rear thrust bearing
41 Imput shaft
42 Forward clutch housing assembly
43 Forward clutch piston seal unit (inner and outer)
44 Forward clutch piston
45 Forward clutch piston return spring
46 Forward clutch piston return spring seat
47 Forward clutch piston return spring seat retaining ring
48 Input ring gear front thrust washer
49 Input ring gear assembly with support
50 Input ring gear bushing
51 Forward clutch piston cushion spring
52 Forward clutch driven plate
53 Forward clutch drive plate assembly
54 Forward clutch pressure plate
55 Forward clutch pressure plate retaining ring

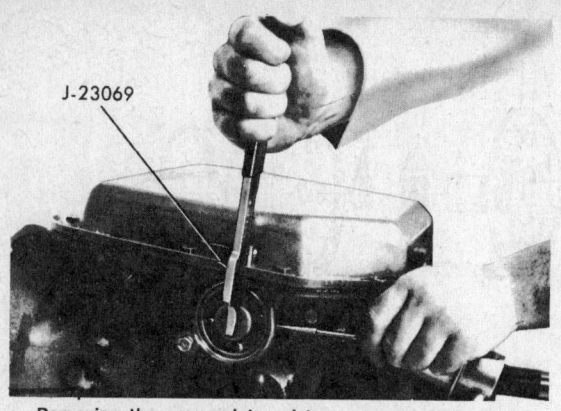

Removing the accumulator piston cover retaining ring
—Turbo Hydra-Matic 350 (© General Motors)

Checking the oil pump body to gear face clearance
—Turbo Hydra-Matic 350 (© General Motors)

rod and tap with hammer to lift seat. Remove check ball and spring.

Inspection

1. Wash all parts in solvent. Air dry. Blow out all passages.
2. Inspect drive and driven gears, gear pocket, and crescent for nicks, galling or other damage.
3. Inspect pump body and cover for nicks or scoring.
4. Check pump cover outer diameter for nicks or burrs.
5. Inspect pump body bushing for galling or scoring. Check clearance between pump body bushing and converter hub. It must be no more than .005 in. If bushing is damaged, replace pump body.
6. Install pump gears in body and check pump body face to gear face clearance. It should be from .0005-.0015 in.
7. Inspect pump body to converter hub lip oil seal. Inspect converter hub for nicks or burrs which might have damaged pump lip oil seal or pump body bushing.
8. Check priming valve for free operation. Replace if necessary.

9. Check condition of cooler bypass valve. Replace valve if it leaks excessively.
10. Check all springs for distortion or collapsed coils.
11. Check oil passages in pump body and in pump cover.
12. Inspect three pump cover stator shaft bushings for galling or scoring. If they are damaged, remove using a slide hammer. Drive the new bushings into place. The front stator shaft bushing must be .250 in. below the front face of the pump body. The center bushing should be 11/32 in. below the face of the pump cover hub. The rear bushing should be flush or up to .010 in. below the face of the pump cover hub.
13. Check three pump cover and hub lubrication holes to make certain they are not restricted.

Assembly

1. Install cooler by-pass valve spring, check ball and seat. Press seat into bore until top of seat is flush to .010 in. with face of pump body.

2. Install oil pump priming valve and spring. The priming valve is used on all pump bodies having a reamed hole in the priming valve area and on all replacement pump assemblies.
3. Install new pump outside diameter to case square cut O-ring seal.
4. Install pump drive and driven gears. If drive gear has offset tangs, assemble with tangs face up to prevent damage by converter.
5. Install selective thrust washer and five oil seal rings on pump cover hub.
6. Install intermediate clutch piston inner and outer seals.
7. Install pump cover on pump body.
8. Install intermediate clutch piston and clutch return springs.
9. Install spring retainer and pump cover bolts. Position aligning strap over pump body and cover. Tighten strap. Torque pump bolts to 18 ft. lbs.

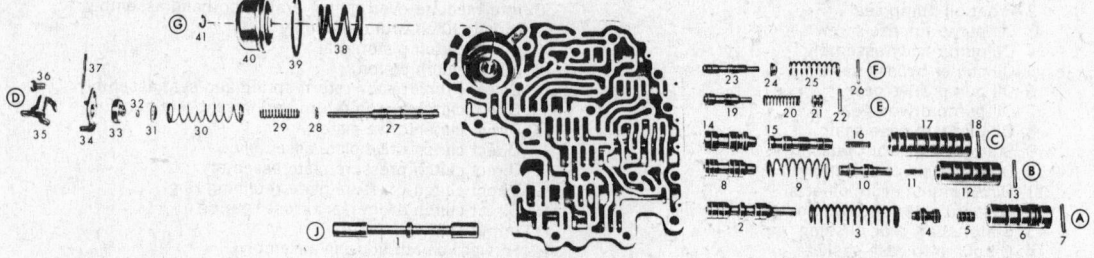

J–1	MANUAL VALVE		14	1-2 SHIFT VALVE		27	DETENT VALVE
2	PRESSURE REGULATOR VALVE		15	1-2 SHIFT CONTROL VALVE		28	WASHER
3	PRESSURE REGULATOR VALVE SPRING	C	16	1-2 SHIFT CONTROL VALVE SPRING		29	DETENT VALVE INNER SPRING
4	REVERSE AND MODULATOR BOOST VALVE		17	1-2 SHIFT CONTROL VALVE SLEEVE		30	DETENT VALVE OUTER SPRING
A 5	INTERMEDIATE BOOST VALVE		18	RETAINING PIN		31	DETENT VALVE OUTER SPRING SEAT
6	BOOST VALVE SLEEVE		19	MANUAL LOW CONTROL VALVE	D	32	DETENT VALVE SPRING RETAINER
7	RETAINING PIN	E	20	MANUAL LOW CONTROL VALVE SPRING		33	DETENT VALVE STOP
8	2-3 SHIFT VALVE		21	PLUG		34	DETENT VALVE ACTUATING LEVER BRACKET
9	2-3 SHIFT VALVE SPRING		22	RETAINING PIN		35	DETENT VALVE ACTUATING LEVER
B 10	2-3 SHIFT CONTROL VALVE		23	DETENT REGULATOR VALVE		36	RETAINING BOLT
11	2-3 SHIFT CONTROL VALVE SPRING		24	DETENT REGULATOR VALVE SPRING SEAT		37	RETAINING PIN
12	2-3 SHIFT CONTROL VALVE SLEEVE	F	25	DETENT REGULATOR VALVE SPRING		38	DIRECT CLUTCH ACCUMULATOR SPRING
13	RETAINING PIN		26	RETAINING PIN		39	OIL SEAL RING
					G	40	DIRECT CLUTCH ACCUMULATOR PISTON
						41	RETAINER RING

Exploded view of the valve body—Turbo Hydra-Matic 350 (© General Motors)

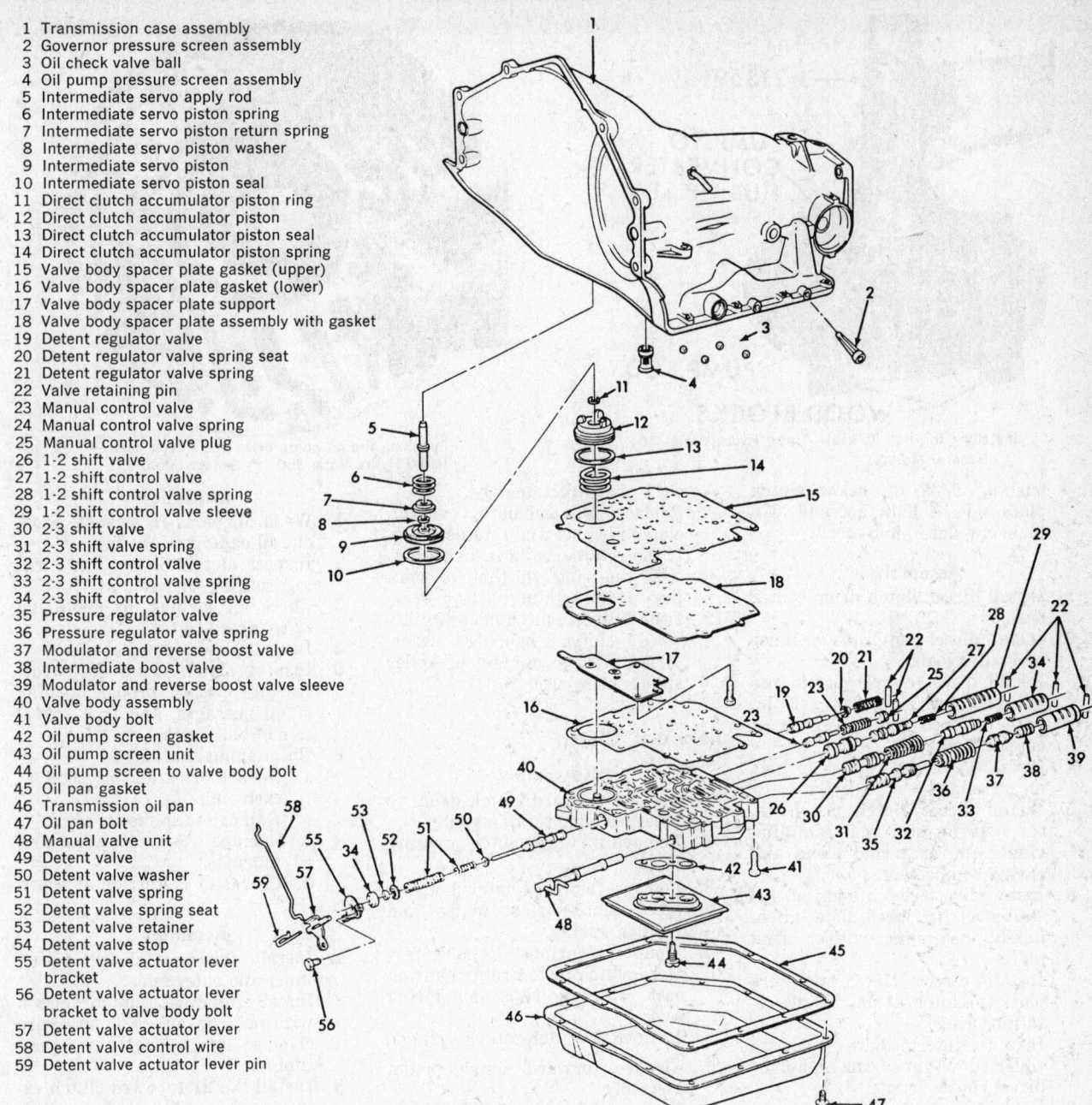

1 Transmission case assembly
2 Governor pressure screen assembly
3 Oil check valve ball
4 Oil pump pressure screen assembly
5 Intermediate servo apply rod
6 Intermediate servo piston spring
7 Intermediate servo piston return spring
8 Intermediate servo piston washer
9 Intermediate servo piston
10 Intermediate servo piston seal
11 Direct clutch accumulator piston ring
12 Direct clutch accumulator piston
13 Direct clutch accumulator piston seal
14 Direct clutch accumulator piston spring
15 Valve body spacer plate gasket (upper)
16 Valve body spacer plate gasket (lower)
17 Valve body spacer plate support
18 Valve body spacer plate assembly with gasket
19 Detent regulator valve
20 Detent regulator valve spring seat
21 Detent regulator valve spring
22 Valve retaining pin
23 Manual control valve
24 Manual control valve spring
25 Manual control valve plug
26 1-2 shift valve
27 1-2 shift control valve
28 1-2 shift control valve spring
29 1-2 shift control valve sleeve
30 2-3 shift valve
31 2-3 shift valve spring
32 2-3 shift control valve
33 2-3 shift control valve spring
34 2-3 shift control valve sleeve
35 Pressure regulator valve
36 Pressure regulator valve spring
37 Modulator and reverse boost valve
38 Intermediate boost valve
39 Modulator and reverse boost valve sleeve
40 Valve body assembly
41 Valve body bolt
42 Oil pump screen gasket
43 Oil pump screen unit
44 Oil pump screen to valve body bolt
45 Oil pan gasket
46 Transmission oil pan
47 Oil pan bolt
48 Manual valve unit
49 Detent valve
50 Detent valve washer
51 Detent valve spring
52 Detent valve spring seat
53 Detent valve retainer
54 Detent valve stop
55 Detent valve actuator lever bracket
56 Detent valve actuator lever bracket to valve body bolt
57 Detent valve actuator lever
58 Detent valve control wire
59 Detent valve actuator lever pin

General Motors Turbo Hydra-Matic 350—exploded view—front and rear servos, and control valve body (© General Motors)

Direct Clutch and Intermediate Overrun Roller Clutch

Disassembly

1. Remove intermediate overrun clutch front retainer ring and retainer.
2. Remove intermediate overrun clutch outer race.
   NOTE: Before removal, check for correct assembly. The outer race should free wheel counter-clockwise only.
3. Remove intermediate overrun roller clutch assembly.
4. Remove intermediate overrun roller clutch cam.
5. Remove direct clutch drum to forward clutch housing special needle bearing.
6. Remove direct clutch pressure plate to clutch drum retaining ring and pressure plate.
7. Remove direct clutch lined and steel plates.
8. Remove direct clutch piston return spring seat retaining ring and spring seat.
9. Remove 17 clutch return coil springs and piston.
10. Remove direct clutch piston inner and outer seals.
11. Remove direct clutch piston center seal.

Inspection

1. Wash all parts in solvent. Blow out all passages. Air dry.
2. Inspect clutch plates for burning, scoring, or wear.
3. Check all springs for collapsed coils or distortion.
4. Inspect piston for cracks and free operation of ball check. Ball should be loose enough to rattle but not to fall out.
5. Check overrun clutch inner cam and outer race for scratches, wear, or indentations.
6. Inspect overrun roller clutch assembly rollers for wear. Check springs for distortion.
7. Inspect clutch drum for wear, scoring, cracks, proper opening of oil passages, wear on clutch plate drive lugs, and free operation of ball check.
8. Check direct clutch drum bushing for galling or scoring. When replacing bushing, drive the new

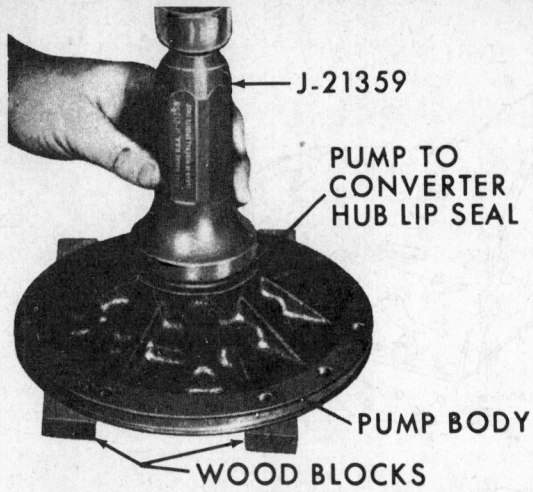

J-21359

PUMP TO
CONVERTER
HUB LIP SEAL

PUMP BODY

WOOD BLOCKS

Installing the hub lip seal—Turbo Hydra-Matic 350
(© General Motors)

Installing the oil pump drive and driven gears—
Turbo Hydra-Matic 350 (© General Motors)

bushing 9/32 in. below clutch plate side of hub face and .010 in. below slot in hub face.

Assembly

1. Install direct clutch drum center seal.
2. Install direct clutch piston inner and outer seals.
3. Install direct clutch piston into housing with a loop of .020 in. wire crimped into a length of copper tubing.
4. Install 17 clutch return springs on piston.
5. Install direct clutch piston return spring seat and retaining ring using snap-ring pliers and spring compressor.
6. Install direct clutch housing. Install steel and faced plates alternately, beginning with a steel plate.
7. Install direct clutch pressure plate in clutch drum. Install retaining ring.
8. Install intermediate overrun roller clutch inner cam on hub of direct clutch drum.
9. Replace intermediate overrun roller clutch assembly.
10. Install intermediate overrun clutch outer race. The outer race must free-wheel in the counter-clockwise direction only.
11. Replace intermediate overrun clutch retainer and retainer ring. If the retainer ring is dished, install the ring so that it compresses the retainer.
12. Install direct clutch drum to forward clutch housing needle thrust washer on hub of roller clutch inner race.

Forward Clutch

Disassembly

1. Remove forward clutch drum to pressure plate retaining ring.
2. Remove forward clutch pressure plate.
3. Remove forward clutch housing faced plates, steel plates and cushion spring.
4. Compress springs. Remove forward clutch piston return spring seat retaining ring and spring seat.
5. Remove 21 clutch return springs.
6. Remove forward clutch piston assembly.
7. Remove forward clutch piston inner and outer seals.
8. Be sure the ball check exhaust in the drum is operable and free of dirt.

Inspection

1. Wash all parts in solvent. Blow out all passages. Air dry.
2. Inspect clutch plates for burning, scoring, or wear.
3. Check all springs for distortion or collapsed coils.
4. Inspect piston for cracks.
5. Inspect clutch drum for wear, scoring, cracks, proper opening of oil passages, and free operation of ball check.
6. Check input shaft for:
   a. Open lubrication passages at each end.
   b. Damage to splines or shaft.
   c. Damage to ground bushing journals.
   d. Cracks or distortion of shaft.

Assembly

1. Install forward clutch piston inner and outer seals.
2. Install forward clutch piston with a loop of .020 in. wire crimped into a length of copper tubing.
3. Install the 21 forward clutch return springs. These springs are identical to those used in the direct clutch. Install spring seat.
4. Compress springs and replace spring seat retaining ring.

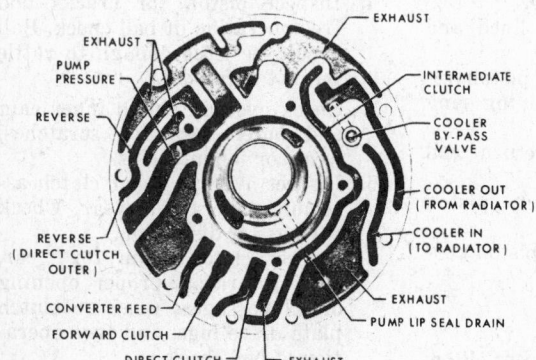

EXHAUST

PUMP
PRESSURE

REVERSE

REVERSE
(DIRECT CLUTCH
OUTER)

CONVERTER FEED

FORWARD CLUTCH

DIRECT CLUTCH

EXHAUST

INTERMEDIATE
CLUTCH

COOLER
BY-PASS
VALVE

COOLER OUT
(FROM RADIATOR)

COOLER IN
(TO RADIATOR)

EXHAUST

PUMP LIP SEAL DRAIN

PUMP BODY

Oil pump oil passages—Turbo Hydra-Matic 350 (© General Motors)

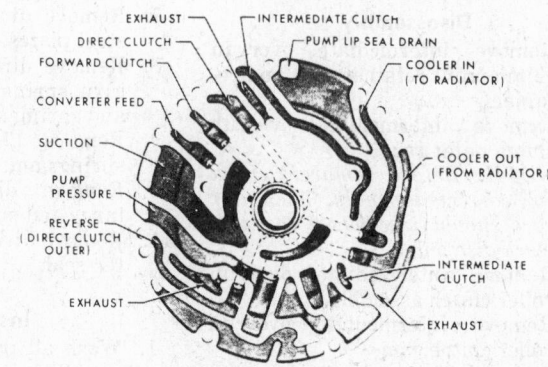

EXHAUST

DIRECT CLUTCH

FORWARD CLUTCH

CONVERTER FEED

SUCTION

PUMP
PRESSURE

REVERSE
(DIRECT CLUTCH
OUTER)

EXHAUST

INTERMEDIATE CLUTCH

PUMP LIP SEAL DRAIN

COOLER IN
(TO RADIATOR)

COOLER OUT
(FROM RADIATOR)

INTERMEDIATE
CLUTCH

EXHAUST

PUMP COVER AND STATOR SHAFT ASSEMBLY

Oil pump cover oil passages—Turbo Hydra-Matic 350 (© General Motors)

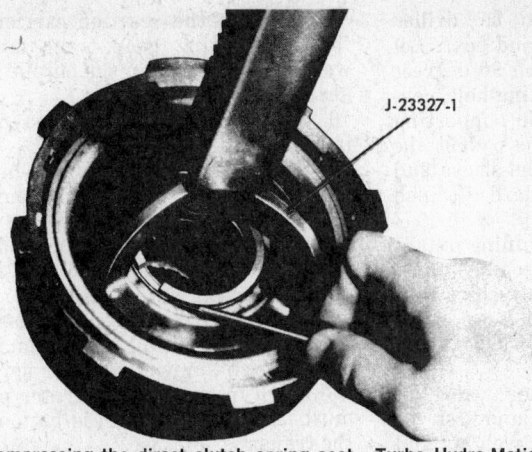

Compressing the direct clutch spring seat—Turbo Hydra-Matic 350
(© General Motors)

.020" PIANO WIRE CRIMPED INTO COPPER TUBING

PISTON

DIRECT CLUTCH HOUSING

Install the direct clutch piston—Turbo Hydra-Matic 350
(© General Motors)

5. Replace forward clutch housing cushion spring. Replace steel and faced plates alternately, starting with a steel plate.
6. Install forward clutch pressure plate and retaining ring.
7. Use a feeler gauge to check the clearance between the top of the clutch pack to the top of the clutch; dimension A, (use the chart that follows to select the correct pressure plate). Measure the distance between the lower edge of the notch on the inner surface of the drum and the end of the drum; dimension B. Subtract dimension B from dimenison A to get dimension C.

Subtract dim. B from C
to obtain dim. A

When dim. A is:	use plate number
.016-.052 in.	6261072
.052-.083 in.	6261349
.083-.122 in.	6261350

8. Install direct clutch drum on input shaft and align faced plates with splines on forward clutch housing.

Sun Gear and Sun Gear Drive Shell

Disassembly

1. Remove sun gear to sun gear drive shell rear retaining ring.
2. Remove sun gear to drive shell flat rear thrust washer.
3. Remove sun gear and front retaining ring from drive shell.
4. Remove front retaining ring from sun gear.

Inspection

1. Wash all parts in solvent. Air dry.
2. Inspect sun gear and sun gear drive shell for wear or damage.
3. Inspect sun gear bushings for galling or scoring. Drive out damaged bushings. Install new bushings flush to .010 in. below surface of counterbore.

Assembly

1. Install new front retaining ring on sun gear. Be careful not to over-stress this ring.
2. Install sun gear and retaining ring in drive shell.
3. Install sun gear to drive shell flat rear thrust washer.
4. Install new sun gear to sun gear drive shell retaining ring.

Low and Reverse Roller Clutch Support

Disassembly

1. Remove low and reverse clutch to sun gear shell thrust washer.
2. Remove low and reverse overrun clutch inner race.
3. Remove low and reverse roller clutch retaining ring.
4. Remove low and reverse roller clutch assembly.

Inspection

1. Wash all parts in solvent. Air dry.
2. Inspect roller clutch inner and outer races for scratches, wear, or indentations.
3. Inspect roller clutch assembly rollers for wear and roller springs for distortion. If rollers are removed from assembly, install rollers from outside in to avoid bending springs.

Assembly

1. Install low and reverse roller clutch assembly.
2. Install low and reverse roller clutch retaining ring.

3. Install low and reverse overrun clutch inner race. Inner race must free-wheel in the clockwise direction only.

Governor

NOTE: All components of the governor, with the exception of the driven gear, are a select fit and each assembly is calibrated. The governor, including the driven gear, is serviced as a complete assembly. It is necessary to disassemble the governor to replace the driven gear. Disassembly, may also be necessary due to improper operation of the governor.

Disassembly & Inspection

1. Cut off one end of each governor weight pin and remove pins, governor thrust cap, governor weights and springs. Governor weights are interchangeable from side to side and need not be identified.
2. Remove the governor valve from the governor sleeve. Be careful not to damage the valve.
3. Wash all parts in clean solvent, air dry and blow out all passages with filtered compressed air.
4. Inspect the governor sleeve for nicks, burrs, scoring or galling.
5. Check the governor sleeve for free operation in the bore of the transmission case.
6. Inspect the governor valve for nicks, burrs or galling.
7. Check governor valve for free operation in the bore of the governor sleeve.
8. Inspect the governor driven gear for nicks, burrs or damage.

CLUTCH PLATES

Year	Model	Intermediate Clutch		Direct Clutch		Forward Clutch		Low-Reverse Clutch	
		Driven Plates	Drive Plates	Driven Plates	Drive Plates	Driven Plates	Drive Plates	Driven Plates	Drive Plates
1971-78	All	3	3	4	4	5	5	5	5

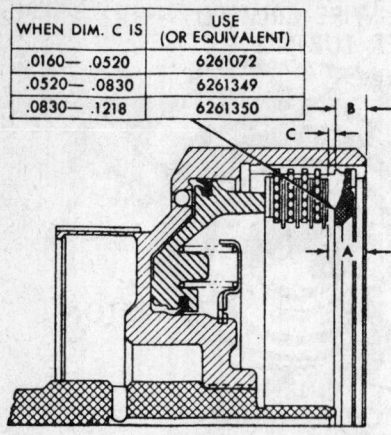

WHEN DIM. C IS	USE (OR EQUIVALENT)
.0160— .0520	6261072
.0520— .0830	6261349
.0830— .1218	6261350

Determining the selective fit for the forward clutch pressure plate—Turbo Hydra-Matic 350 (© General Motors)

9. Check the driven gear for excessive play on the governor sleeve.
10. Inspect the governor weight springs for distortion or damage.
11. Check governor weights for free operation in their retainers.
12. Check the valve opening at the entry and exhaust, 0.020 inch minimum.

Governor Driven Gear

Replacement

1. Using a small punch, drive out the governor gear retaining split pin.
2. Support the governor on 3/16 inch plates installed in the exhaust slots of the governor sleeve, place in an arbor press, and with a long punch, press the driven gear out of the sleeve.
3. Carefully, clean the governor sleeve of chips that remain from original rear installation.
4. Support the governor on 3/16 inch plates installed in the exhaust slots of the sleeve, position the new gear in the sleeve and, with a suitable socket, press the gear into the sleeve until nearly seated. Carefully, remove any chips that may have shaved off the gear hub, and press the gear in until it bottoms on the shoulder.

5. A new pin must be drilled through the sleeve and gear. Locate the hole position 90 degrees away from the existing hole, center punch and while supporting the governor in press, drill the new hole through the sleeve and gear using a standard 1/8 inch drill.
6. Install new split retaining pin.
7. Wash governor assembly thoroughly to remove any chips that may have collected.

Assembly

1. Install the governor valve in bore of sleeve, large land first.
2. Install the governor weights and springs, and the thrust cap on the governor sleeve.
3. Align the holes in the thrust cap, governor weight assemblies and governor sleeve, install new pins. Crimp both ends of the pins to prevent them from falling out.
4. Check the operation of the governor weight assemblies. Pins should operate freely.
5. Check the governor valve for free movement in the governor sleeve.

Transmission Assembly

NOTE: During assembly of transmission, use only clean transmission fluid or petroleum jelly to lubricate or retain parts. Lubricate all bearings, seals rings, and clutch plates before assembly.

Internal Components

1. Install the low and reverse clutch piston assembly with the piston installed next to the parking pawl.
2. Install the piston return springs, there are seventeen.
3. Install the spring retainer and retaining ring. Use the special spring compressing tool. Compress the return seat so the spring retainer retaining ring can be installed. Install the output ring gear thrust bearing in the transmission case.
4. Install the output ring gear on the output shaft.

5. Position the reaction carrier on the output ring gear front tanged washer, and install in the output ring gear support.
6. Install the output shaft assembly in the transmission case.
7. Place the reaction carrier in the output ring gear and shaft assembly.
8. Lubricate (with clean transmission fluid) and install the low and reverse clutch steel reaction and face plates. Starting with a steel plate and alternating with a face plate, assemble the clutch. Install the low and reverse clutch support retainer spring, the notch in the steel separator plate must be placed toward the bottom of the transmission case.
9. Install the low and reverse clutch support assembly. The notch must be aligned so that the retainer spring is in proper contact.
10. Install the low and reverse roller clutch inner race to the sun gear shell thrust washer.
11. Install the low and reverse clutch support to case snap ring with retainer spring between gap.
12. Install thrust rear thrust washer and sun gear drive shell assembly.
13. Install output carrier assembly.
14. Install the input ring gear rear thrust washer.
15. Install the ring gear.
16. Install the new input ring gear

CAUTION: Do not over stress the snap ring.

17. Install the input gear front thrust washer.
18. Install the direct clutch assembly, and the special thrust washer to the forward clutch assembly.
19. Install the clutch assemblies in the transmission case.

CAUTION: Be sure the forward clutch face plates are positioned over the input ring gear. Make certain that the tangs on the direct clutch housing are installed in the slots on the sun gear drive shell.

20. Install the intermediate clutch overrun brake band.
21. Install the intermediate clutch pressure plate.
22. Lubricate (with clean transmis-

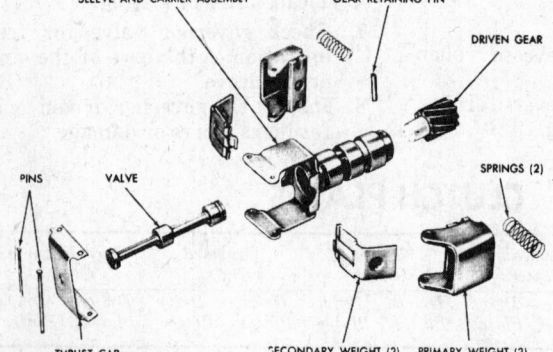

Typical governor assembly, exploded view—Turbo Hydra-Matic 350 (© General Motors)

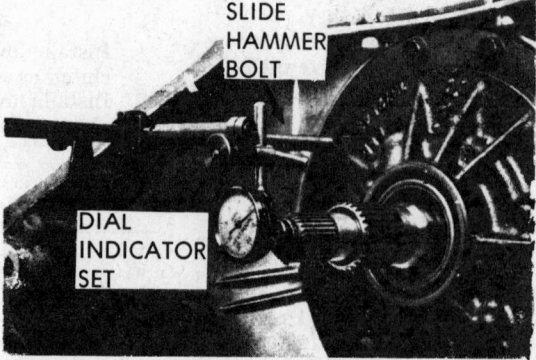

Checking the end play for proper thrust washer selection—Turbo Hydra-Matic 350 (© General Motors)

sion fluid) and install the steel and face plates of the intermediate clutch. Starting with a face plate and alternating with a steel plate, assemble the clutch pack. The steel reaction plates are installed toward the selector lever inner bracket.

23. Install the intermediate clutch cushion spring.

24. Use the following procedure to check for the proper thickness of the selective thrust washer used between the oil pump cover and the direct clutch assembly.

   a. Install the selective fit thrust washer, oil pump gasket. Install the oil pump with two slide hammer bolts and two pump to case bolts.
   b. Move the transmission so that the output shaft points down. Mount a dial indicator on one of the slide hammer bolts so that the indicator plunger is resting on the end of the input shaft. Zero the indicator.
   c. Push the transmission output shaft upward and observe the total indicator movement.
   d. The indicator should read .032-.064 inch. If the reading is within these limits, the proper washer is being used. If the reading is not within the limits, it will be necessary to remove the pump and change the selective fit thrust washer to obtain the proper clearance. Selective fit thrust washers are available in thickness of; 0.066, 0.083, and 0.100 inch.

25. Install a new pump to case gasket.

26. Install a new pump to case square cut oil seal ring.

27. Install the guide pins in the case.

28. Install the pump assembly in the transmission case. Install attaching bolts and new washer type seals.

NOTE: If the input shaft can not be rotated as the pump is being pulled into place, the direct and forward clutch housings have not been properly installed to index the faced plates with their respective parts. This condition must be corrected before the pump is pulled into place.

Extension and Speedometer Drive Gear

Installation

1. Place the speedometer drive gear retaining clip in the hole on the output shaft.
2. Heat a new speedometer drive gear using a suitable heat source.
3. Align the slot in the speedometer drive gear with the retaining clip and install the gear.

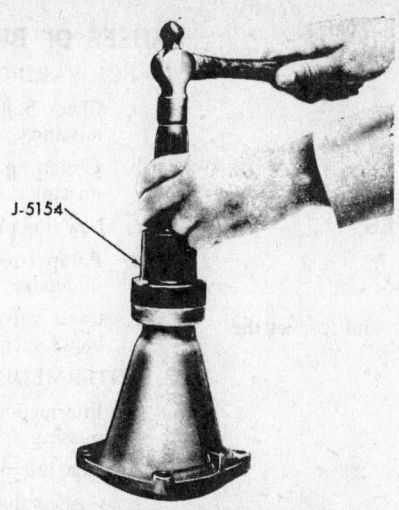

J-5154

Installing the extension housing seal— Turbo Hydra-Matic 350 (© General Motors)

4. Install the extension housing to case square cut 'O' ring.
5. Install the extension housing on the transmission case with the attaching bolts and tighten to 25 foot pounds torque.
6. Install a new extension housing rear seal, if necessary, with the proper size drift.

Manual Linkage Assembly

Installation

1. Install a new manual shift shaft seal using ¾ inch rod, be sure the seal seats in the trassmission case.
2. Install the parking pawl, tooth toward the inside, in the transmission case.
3. Install the parking pawl shaft in the case through the parking pawl.
4. Install the parking pawl shaft retaining plug. Drive the retaining plug into the transmission case with a ⅜ inch rod, until the retaining plug is 0.1330-0.170 inch below the face of the transmission case. Stake the retaining plug in three places.
5. Install the parking pawl disengaging spring, with the square end hocked on the parking pawl.
6. Install the park lock bracket and tighten the bolts to 29 foot pounds torque.

CAUTION: These bolts are type 290 M and have six marks on their heads.

7. Install the actuating rods between the range selector inner lever and parking pawl.
8. Install the actuating rod under the park lock bracket and parking pawl.
9. Install the manual shift shaft through the transmission case and the range selector inner lever.
10. Install the retaining nut on the manual shift shaft and tighten to 30 foot pounds of torque.
11. Install the manual shift shaft to transmission case spacer clip.

Intermediate Servo Piston, Valve Body, Oil Pan, and Gasket

Installation

1. Install the intermediate servo piston, apply pin, spring, and spring seat.
2. Install the intermediate servo piston and the metal oil seal ring.
3. Install the four check balls into the proper transmission case pockets.
4. Install the oil pump pressure screen and governor feed screen.
5. Install the valve body transfer plate and the gasket.
6. Install the valve body to the transfer plate.
7. Install the valve body. Connect the manual control valve link to the range selector inner lever. Torque the retaining bolts in a random sequence to 130 inch pounds.
8. Install the spacer support plate, tighten the bolts to 130 inch pounds torque.
9. Connect the detent control valve wire to the detent valve actuating lever, attach the lever to the valve body.
10. Attach the roller and spring assembly to the valve body.
11. Align the lube holes in the strainer with the holes in the valve body and install the strainer assembly gasket and strainer.
12. Install the oil pan with a new gasket. Tighten the bolts to 130 inch pounds torque.

Governor and Vacuum Modulator

Installation

1. Install the governor assembly, cover and retainer wire.
2. Install the vacuum modulator valve.
3. Install the vacuum modulator retaining clip and tighten the retaining bolts to 130 inch pounds.

NOTE: Position the retainer with the tangs pointing toward the modulator.

Intermediate Clutch Accumulator

Installation

1. Install the intermediate clutch accumulator piston assembly.
2. Install the intermediate clutch accumulator spring.

3. Install a new 'O' ring in the groove of the transmission case before installing cover.

4. Install the intermediate clutch accumulator cover and retaining ring.

TRANSMISSION CLUTCH PLATES DIAGNOSIS

1. Lined Drive Plates.

    a. Dry plates with compressed air and inspect the lined surface for:

        1. pitting and flaking

        2. wear

        3. glazing

        4. cracking

        5. charring

        6. chips or metal particles imbedded in lining.

        If a lined drive plate exhibits any of the above conditions, replacement is required. Do not diagnose drive plates by color.

2. Steel Driven Plates

    Wipe plates dry and check for heat discoloration. If the surface is smooth and an even color smear is indicated, the plate should be reused. If severe heat spot discoloration or surface scuffing is indicated, the plate must be replaced.

3. Clutch Release Springs

    Evidence of extreme heat or burning in the area of the clutch may have caused the springs to take a heat set and would justify replacement of the springs.

CAUSES OF BURNED CLUTCH PLATES

1. FORWARD CLUTCH

    a. Check ball in clutch housing damaged, stuck or missing.

    b. Clutch piston cracked, seals damaged or missing.

    c. Low line pressure.

    d. Pump cover oil seal rings missing, broken or undersize; ring groove oversize.

    e. Case valve body face not flat or porosity between channels.

2. INTERMEDIATE CLUTCH

    a. Intermediate clutch piston seals damaged or missing.

    b. Low line pressure.

    c. Case valve body face not flat or porosity between channels.

3. DIRECT CLUTCH

    a. Restricted orifice in vacuum line to modulator (poor vacuum response).

    b. Check ball in direct clutch piston damaged, stuck or missing.

    c. Defective modulator bellows.

    d. Clutch piston seals damaged or missing.

    e. Case valve body face not flat or porosity between channels.

    f. Clutch installed backwards.

NOTE: Burned clutch plates can be caused by incorrect usage of clutch plates. Also, antifreeze in transmission fluid can cause severe damage, such as large pieces of composition clutch plate material peeling off.

OIL PRESSURE CHECK

Year	Engine	Oil Pressure (psi) @ Altitude (± 5 psi) ①								
		Sea Level			2,000 ft.			8,000 ft.		
		D, P, N	L1, L2	R	D, P, N	L1, L2	R	D, P, N	L1, L2	R
1971	L6 & V8	168	166	254	158	159	240	133	141	202
1972-73	L6 & V8	150	150	244	150	150	233	126	150	194
1974-78	L6	173	166	262	163	159	248	138	141	210
	V8	167	166	254	158	159	240	133	140	202

① Pressures are at 0 (zero) output speed, 1,200 engine rpm and vacuum line disconnected and plugged.

TORQUE SPECIFICATIONS

	Foot pounds	Inch Pounds
Pump cover to pump body	17	—
Pump assembly to case	18½	—
Valve body and support plate	—	130
Parking lock bracket	29	—
Oil suction screen	—	40
Oil pan to case	—	130
Extension to case	25	—
Modulator retainer to case	—	130
Inner selector lever to shaft	25	—
Detent valve actuating bracket	—	52

	Foot pounds	Inch Pounds
Converter to flywheel bolts	32	—
Upper pan to transmission case	—	110
Transmission case to engine	35	—
Oil cooler pipe connection to transmission case or radiator	—	125
Oil cooler pipe to connections	10	—
Detent cable to transmission	—	75
Detent cable to carburetor	—	112

GM Turbo Hydra-Matic 400/475

DIAGNOSIS

This diagnosis guide covers the most common symptoms and is an aid to careful diagnosis. The items to check are listed in the sequence to be followed for quickest results. Thus, follow the checks in the order given for the particular transmission type.

TROUBLE SYMPTOMS	Items to Check	
	In Truck	Out of Truck
No drive in D range	ABCD	abc
No drive in R or slips in reverse	ABCEGHIJL	efa
Drive in neutral	B	a
First speed only—no 1-2 shift	KG	g
1-2 shift at full throttle only	MNG	
First and second speeds only—no 2-3 shift	MNG	f
Slips in all ranges	ACEFG	dfahc
Slips—1-2 shift	ACELGO	gh
Rough 1-2 shift	CELGJOGJ	hi
Slips 2-3 shift	ACELG	fh
Rough 2-3 shift	CELO	
Shifts occur—too high or too low car speed	CKEMGL	
No detent downshifts	NMG	
No part throttle downshift (heavy-duty model "OE" only)	CEG	
No engine braking—super range—second speed	O	i
No engine braking—low range—first speed	GJ	ik
Park will not hold	BPQ	
Poor performance or rough idle—stator not functioning	RS	lo
Noisy transmission	TF	dafgn

Key to Checks

A. Oil level
B. Manual linkage (external)
C. Check oil pressure
D. Manual control disconnected inside
E. Modulator and/or lines
F. Clogged strainer or intake leaks
G. Valves, body and/or leaks
H. Reverse feed passages

I. Valve check balls
J. Rear servo and accumulator
K. Governor and/or feed line seals
L. Pump regulator and boost valve
M. Detent solenoid
N. Detent switch
O. Front servo and accumulator
P. Internal linkage

Q. Parking pawl and/or link
R. Stator switch
S. Valve body-stator section
T. Cooler or lines
a. Front clutch
b. Clutch feed seals and gaskets
c. Low sprags
d. Front pump
e. Rear band
f. Direct clutch

g. Intermediate clutch
h. Pump-to-case gasket
i. Intermediate check valve ball in case
j. Front band
k. Rear band
l. Turbine shaft
m. Converter assembly
n. Planetary assembly

General Information

The Turbo Hydra-Matic Transmission is a fully automatic unit, consisting of a three stage torque converter, three multiple-disc clutches, one gear unit, two roller clutches, and two non-adjustable bands. An electrical operated detent valve is used for forced 3-2 downshifts. While the design and repair procedures are similar, components vary with truck applications and should not be considered interchangeable.

Transmission Disassembly

Clean outside of the unit thoroughly to prevent dirt from entering the unit.

1. With transmission in a work cradle or on a clean bench, lift the converter straight off the transmission input shaft.
2. With transmission bottom-up, remove modulator assembly attaching screw and retainer, then remove the modulator assembly and O-ring seal.

Governor, Speedometer Driven Gear, Oil Pan, Strainer and Intake Pipe

Removal

1. Remove attaching screws, governor cover and gasket, then withdraw governor from the case.
2. Remove speedometer driven gear attaching screw and retainer, then withdraw the driven gear assembly.
3. Remove oil pan attaching screws, then the oil pan. Discard gasket.
4. Remove pump intake pipe and strainer assembly, then the pipe-to-case O-ring seal.

Control Valve Assembly, Governor Pipes and Detent Spring Assembly

Removal

1. Remove the control valve body attaching screws and detent roller and spring assembly.
2. Remove the control valve assembly and the governor pipes. NOTE: Do not remove the solenoid attaching screws. CAUTION: The front servo parts may fall from the transmission when the unit is still in the ve-

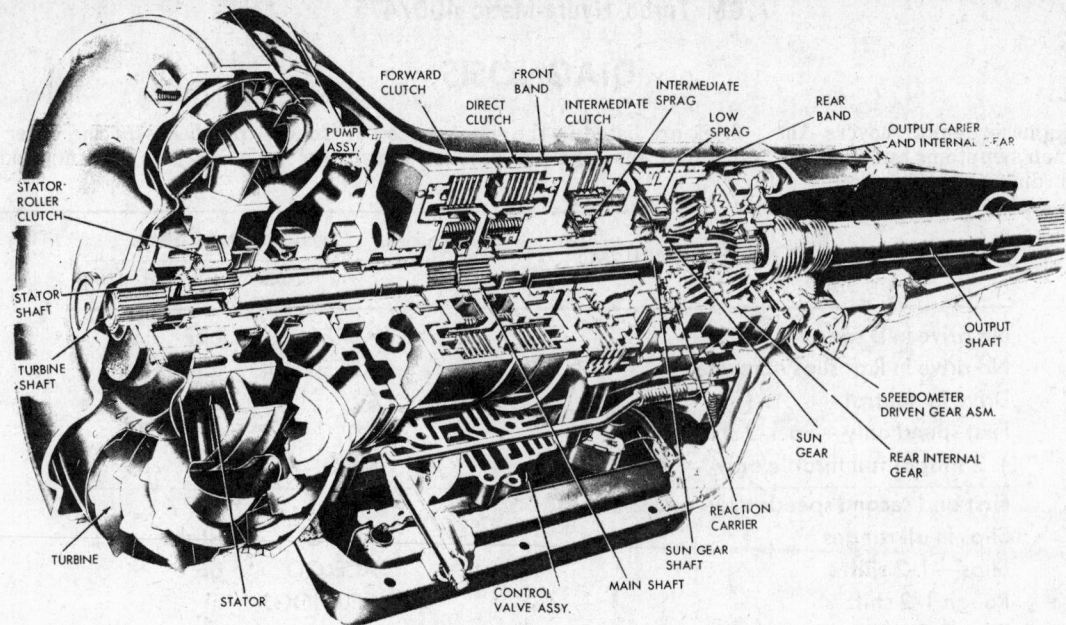

General Motors type turbo hydra matic transmission (© G.M. Corp.)

hicle. Do not drop the manual valve from the valve body.

3. Remove the governor screen from the governor feed pipe hole in the case or from the end of the governor feed pipe.

4. Remove the governor pipes from the control valve assembly.

5. Disconnect the lead wire of the solenoid from the connector terminal.

Rear Servo, Solenoid, Valve Body Spacer, Front Servo, Manual Detent and Parking Linkage

Removal

1. Remove rear servo cover attaching screws, the cover and gasket, then the rear servo assembly from the case.

2. Remove servo accumulator springs.

3. Disconnect the solenoid leads from connector terminal. Withdraw connector and O-ring seal.

4. Make the band apply pin selection check to determine the possible cause of malfunction, proceed as follows:
   a. Attach the band apply pin selection gauge to the transmission case with attaching screws. Make sure that the gauge pin does not bind in the servo pin hole.

FLUID SPECIFICATIONS

Oil capacity, transmission and converter
    Approx. 22 pints
Capacity between marks on dip stick
    1 pint
Type of oil
    Automatic transmission fluid
    Type A
Drain and refill
    24,000

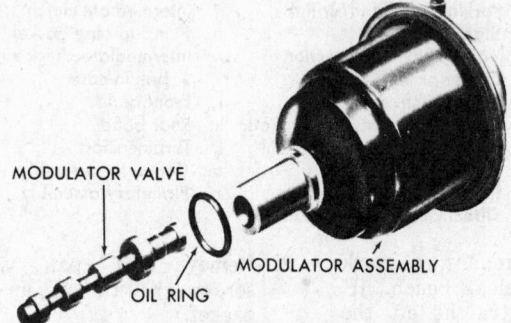

Modulator unit, seal, and valve (© G.M.C.)

BAND AND CLUTCH APPLICATION— 400/475 TURBO HYDRA-MATIC TRANSMISSION

GEAR	FORWARD CLUTCH	DIRECT CLUTCH	FRONT BAND	REAR BAND	INTER-MEDIATE CLUTCH	INTER-MEDIATE ROLLER CLUTCH	LOW ROLLER CLUTCH
1st*	on	off	off	off	off	off	locked
2nd*	on	off	off	off	on	locked	off
3rd*	on	on	off	off	on	off	off
Low(L1)	on	off	off	on	off	off	on
L2	on	off	on	off	on	on	off
Reverse	off	on	off	on	off	off	off

* Shift lever in "D" position

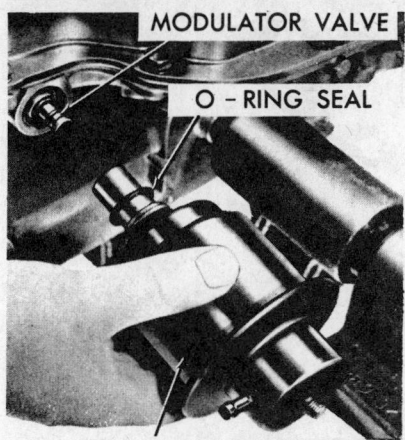

MODULATOR VALVE

O - RING SEAL

VACUUM MODULATOR

Removing vacuum modulator and valve
(© G.M. Corp.)

b. Apply 25 foot pounds torque and select the proper pin to be used during reassembly of the transmission.

Selection of the proper length pin is equivalent to adjusting the band. The band lug end of each selection apply pin bears an identification in the form of one, two or three rings.

c. If both steps of the gauge lever are below the gauge surface, the long pin, identified by three rings must be used.

d. If the gauge surface is between the steps, the medium pin, identified by 2 rings, should be used.

e. If both steps are above the gauge surface, the short pin identified by 1 ring, should be used.

NOTE: If the transmission is in the vehicle, be careful when the detent solenoid is removed as it prevents the spacer plate, gasket, and check balls from dropping down.

5. Remove the detent solenoid attaching screws, detent solenoid and gasket.

6. Remove the electrical connector and 'O' ring seal.
7. Remove valve body spacer plate and gasket.
8. Remove six check balls from cored passages in transmission case.
9. Remove the front servo assembly.
10. Loosen the jam nut which holds the detent lever to the manual shaft. Then remove detent lever from manual shaft and remove manual shaft.
11. Remove the parking actuator rod and detent lever assembly. Then remove detent lever E-ring and the detent lever.
12. Remove attaching screws and park bracket; then the parking pawl return spring.
13. Remove parking pawl shaft retainer, then the parking pawl shaft, O-ring and pawl.

Rear Oil Seal and Extension Housing

Removal

1. Pry rear oil seal from extension housing.
2. Remove housing attaching bolts, then remove extension housing and housing-to-case oil seal.
3. Check the front unit end play as follows:
   a. Remove one front pump bolt and install a ⅜"-16 threaded slide hammer bolt.
   b. Mount a dial indicator on the slide hammer bolt and index the indicator to register with the end of the turbine shaft.
   c. Push the turbine shaft rearward and the output shaft forward.
   d. Set the dial indicator to Zero.
   e. Pull the turbine shaft forward and read the resulting travel or end play. Indicator readings should be .003-.024 inch.

The selective thrust washer

J 8763-01

Transmission in holding fixture
(© G.M. Corp.)

that controls the end play is located between the pump cover and the forward clutch housing. If end play is not within specifications, select a washer thickness from the following chart for use during reassembly.

Thickness (inch)	Color
0.060-0.064	Yellow
0.071-0.075	Blue
0.082-0.086	Red
0.093-0.097	Brown
0.104-0.108	Green
0.115-0.119	Black
0.126-0.130	Purple

NOTE: An oil soaked washer may tend to discolor, so it may be necessary to measure the washer for its actual thickness.

Oil Pump, Forward Clutch and Gear Unit

Removal

1. Pry front seal from the pump. Then remove pump attaching bolts.
2. With slide hammers attached, remove pump from transmission

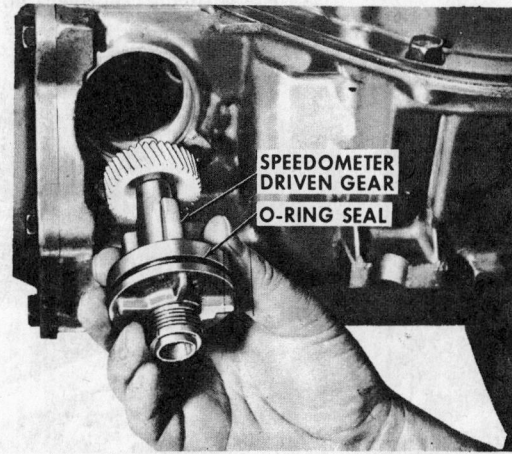

SPEEDOMETER DRIVEN GEAR O-RING SEAL

Removing speedometer drive gear and sleeve
(© G.M. Corp.)

Removing rear servo
(© G.M. Corp.)

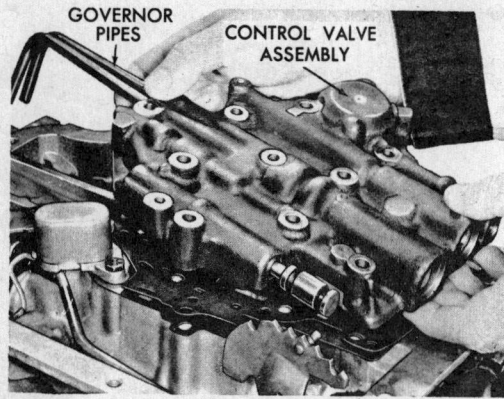

GOVERNOR PIPES
CONTROL VALVE ASSEMBLY

Removing control valve assembly and governor pipes from case
(© G.M. Corp.)

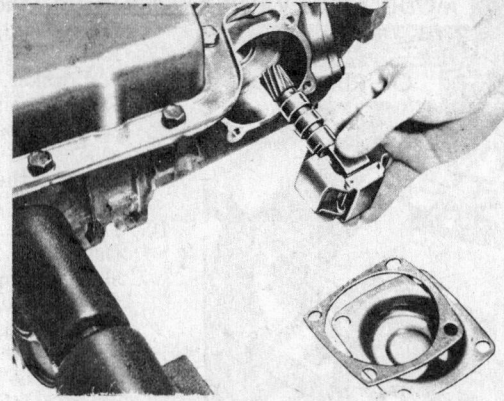

Removing governor
(© G.M. Corp.)

case. Discard pump-to-case seal ring. Discard pump-to-case gasket.

3. Remove turbine shaft from transmission.

4. Remove forward clutch assembly. Be sure that the bronze thrust washer came out with the clutch housing assembly.

5. Remove the direct clutch assembly. Remove front band and sun gear shaft.

6. Check the rear end play as follows:

   a. Install a ⅜" threaded bolt into one of the extension housing bolt holes. Mount a dial indicator on the bolt end and index with the end of the shaft.

   b. Move the output shaft in and out to read the end play. End play should be between 0.007-0.019 inch. The selective thrust washer controlling this end play is composed of steel having three lugs and is located between the output shaft thrust washer and the rear face of the transmission case.

If the end play readings are not within limits, select one from the following chart and install during reassembly.

Thickness (inch)	Identification Notches	Numeral
0.074-0.078	None	1
0.082-0.086	1-tab side	2
0.090-0.094	2-tab side	3
0.098-0.102	1-tab O.D.	4
0.106-0.110	2 tabs O.D.	5
0.114-0.118	3-tabs O.D.	6

7. Remove the case center support-to-case bolt using a ⅜" 12-point thin wall deep socket.

8. Remove intermediate clutch backing plate-to-case snap-ring. Then remove the backing plate, three composition, and three steel clutch plates.

9. Remove the center support-to-case retaining snap-ring.

10. Remove the entire gear unit assembly.

11. Remove the output shaft-to-case thrust washer from the rear of the output shaft, or from inside the case.

12. Remove rear unit selective washer from transmission case.

13. Remove rear band assembly.

14. Remove support to case spacer from inside case.

15. Remove rear band assembly.

Component Disassembly, Inspection, and Assembly

Governor

All components of the governor, except the driven gear, are a select fit and so calibrated. Therefore, service this unit as an assembly.

Clean and inspect all parts for wear or other damage. Check valve opening at feed port with a feeler gauge, holding the governor with the weights extended completely outward. Check valve opening at exhaust port, holding governor with weights completely inward. If either opening is less than .020 in., replace governor assembly.

If a new governor drive gear is installed, a new pin hole must be drilled 90 degrees from the original hole.

Front Servo Inspection

1. Inspect servo pin for damage.

2. Inspect servo piston for damaged oil ring groove, cracks, or porosity. Check freedom of oil seal ring in groove.

3. Check fit of servo pin in piston.

Rear Servo

Disassembly

1. Remove E-ring which holds the servo piston to band apply pin.

2. Remove servo piston and seal from band apply pin. Remove second washer from band apply pin.

3. Remove washer, spring, and retainer.

Inspection

1. Check freedom of accumulator rings in piston.

2. Check fit of band apply pin in servo piston. Inspect pin for scores or cracks.

3. Inspect accumulator and servo pistons for cracks or porosity.

Assembly

1. Install spring retainer, spring,

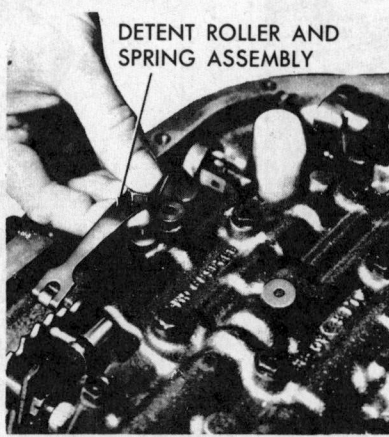

DETENT ROLLER AND SPRING ASSEMBLY

Removing detent roller and spring assembly
(© G.M. Corp.)

GASKET

Removing detent solenoid and gasket
(© G.M. Corp.)

and washer on band apply pin.

2. Install band apply pin, retainer, spring, and washer into bore of servo piston and secure with E-ring.
3. Install oil seal on servo piston. Install outer and inner oil seal rings on accumulator piston and assembly into bore of servo piston.

Control Valve Body

Disassembly

When disassembling control valve body, be careful to identify springs so that they can be replaced in their proper locations.

1. Position valve assembly with cored face up and accumulator pocket on bottom.
2. Remove manual valve from upper bore.
3. Compress accumulator piston spring and remove E-ring retainer. Remove accumulator piston and spring.
4. Press out retaining pin from upper right bore. Remove 1-2 modulator bushing, 1-2 regulator valve and spring, 1-2 detent valve, and 1-2 shift valve. 1-2 regulator valve and spring may be inside of 1-2 modulator bushing.
5. Press out retaining pin from center right bore. Remove 2-3 modulator bushing, 2-3 shift valve spring, 2-3 modulator valve, 3-2 intermediate spring, and 2-3 shift valve. 2-3 modulator valve will be inside of 2-3 modulator bushing.
6. Press out retaining pin from lower right bore. Hold hand over

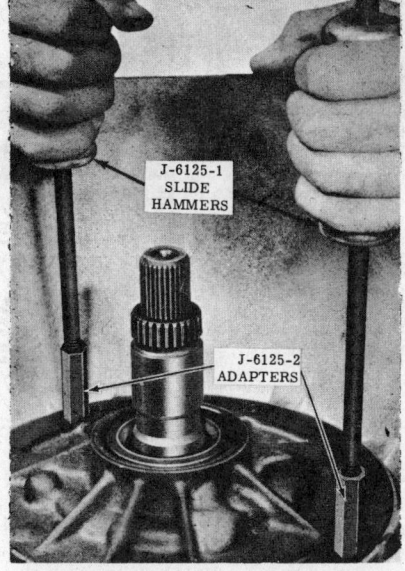

Removing front pump using slide hammers
(© G.M. Corp.)

bore, as plug may pop out. Remove plug, 3-2 valve spring, spacer, and 3-2 valve.
7. Holding hand over bore, press out retainer pin from upper left bore. Remove bore plug, detent valve, detent regulator valve, spacer, and detent regulator valve spring.
8. Pry out grooved retainer ring from lower left bore with long nose pliers. Remove bore plug, 1-2 accumulator bushing, 1-2 accumulator valve, secondary spring, primary 1-2 accumulator valve, and spring.
9. Remove governor oil feed screen from oil feed hole in valve body.

Inspection

1. Wash control valve body, valves and other parts in solvent. Do not allow valves to bump together. Air dry parts and blow out all passages.
2. Inspect all valves and bushings carefully. Burrs may be removed with a fine stone or fine crocus cloth and light oil. Be careful not to round off shoulders of valves.
3. Test all valves and bushings for free movement in their bores. All valves should fall freely of their own weight.
4. The manual valve is the only valve that can be serviced separately. If any of the other valves are defective or damaged, install a new control valve assembly.
5. Inspect body for cracks or scored bores. Check all springs for distortion or collapsed coils.

Assembly

1. Replace front accumulator spring and piston into valve body. Compress spring and piston, assuring that piston pin is correctly aligned with hole in piston and that oil seal ring does not catch on lip of bore when installing piston. Secure piston and spring with E-ring retainer.
2. Install 1-2 accumulator primary spring, 1-2 primary valve, and 1-2 accumulator bushing into lower left bore. Place 1-2 accumulator secondary valve, stem end out, into the 1-2 accumulator bushing. Place 1-2 accumulator secondary spring over stem end of valve. Replace bore plug and retaining pin.

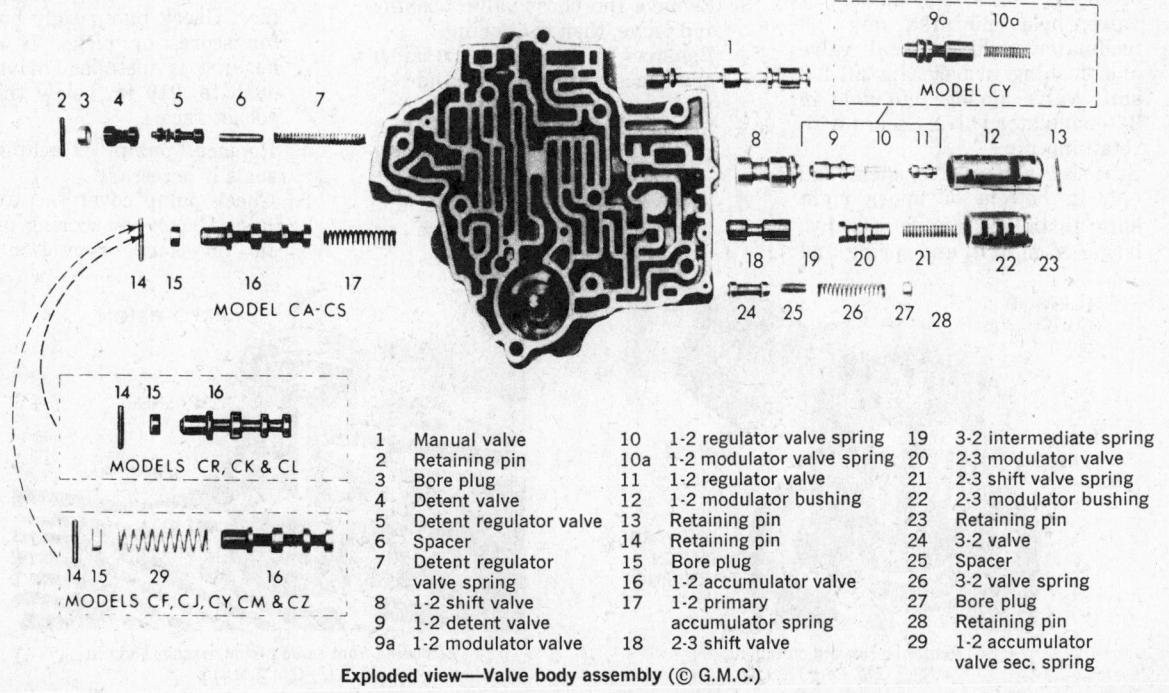

1	Manual valve	10	1-2 regulator valve spring	19	3-2 intermediate spring
2	Retaining pin	10a	1-2 modulator valve spring	20	2-3 modulator valve
3	Bore plug	11	1-2 regulator valve	21	2-3 shift valve spring
4	Detent valve	12	1-2 modulator bushing	22	2-3 modulator bushing
5	Detent regulator valve	13	Retaining pin	23	Retaining pin
6	Spacer	14	Retaining pin	24	3-2 valve
7	Detent regulator valve spring	15	Bore plug	25	Spacer
8	1-2 shift valve	16	1-2 accumulator valve	26	3-2 valve spring
9	1-2 detent valve	17	1-2 primary accumulator spring	27	Bore plug
9a	1-2 modulator valve	18	2-3 shift valve	28	Retaining pin
				29	1-2 accumulator valve sec. spring

Exploded view—Valve body assembly (© G.M.C.)

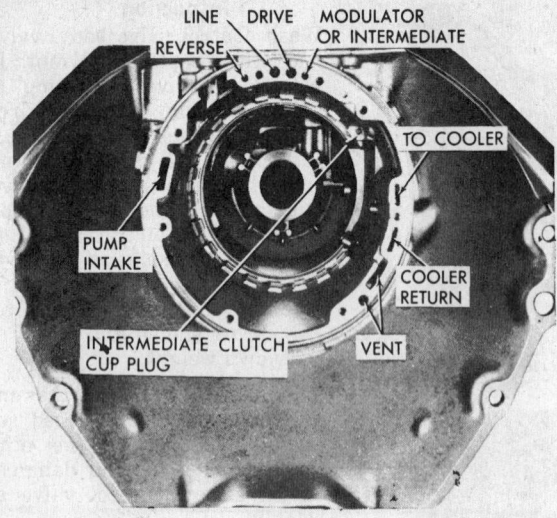

Transmission front view—oil passage identification (© G.M.C.)

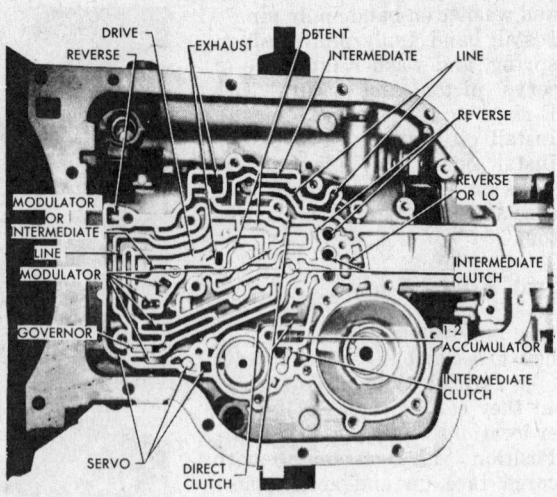

Transmission oil passage identification (© G.M.C.)

3. Install detent regulator valve spring and spacer into upper left bore, making certain that spring seats in bottom of bore. Compress spring and hold with a small screwdriver between end of spring and wall on cored side of valve body. Insert detent regulator valve, stem end out, and detent valve, small land first. Insert bore plug, press inward, remove screwdriver, and install retaining pin.

4. Insert 3-2 valve in bottom right bore. Place spacer inside 3-2 valve spring and insert spacer and spring in bore. Install bore plug and retaining pin.

5. Install 3-2 intermediate spring on stem end of 2-3 shift valve. Install valve and spring, valve first, into center right bore. Be sure that valve seats in bottom of bore. Place 2-3 modulator valve, hole end first, into 2-3 modulator bushing. Install valve and bushing in bore. Install 2-3 shift valve spring into hole in 2-3 modulator valve. Secure with retaining pin.

6. Seat 1-2 shift valve, stem end out, in bottom of upper right bore. Install 1-2 regulator valve, larger stem first, and spring and 1-2 detent valve, hole end first, into 1-2 modulator bushing. Align spring in bore of 1-2 detent valve. Install assembly into upper right bore of control valve body. Install retaining pin.

7. Replace governor oil feed screen assembly in governor oil feed hole.

8. Install manual valve with detent pin groove to the right.

Oil Pump

Disassembly

1. Place the pump over a hole in the bench, shaft down, cover up.

2. Compress regulator boost valve bushing against the pressure regulator spring and remove the snap-ring.
   CAUTION: Presure regulator spring is under extreme pressure.

3. Remove the boost valve bushing and valve, then the spring.

4. Remove valve spring retainer and spacer/s, if present, and regulator valve.

5. Remove pump cover to body attaching bolts, then remove the cover.

6. Remove the retaining pin and bore plug from the pressure regulator bore.

7. Remove the two hook-type oil rings from the pump cover.

8. Remove pump to forward clutch housing selective washer.

9. Mark drive and driven gears for reassembly, then remove the gears.

Inspection

1. Inspect drive gear, driven gear, gear pocket, and crescent for scoring, galling, or other damage.

2. Replace pump gears in pump and check pump body face to face gear clearance. The clearance should be .0008-.0035 in.

3. Check pump body face for scoring or nicks. Check oil passages in pump body for roughness or obstructions. Check condition of cover bolt attaching threads. Check for flatness of pump body face. Check pump body bushing, for scores or nicks. If a new bushing is installed, drive it in flush to .010 in. below the gear pocket face.

4. Replace pump attaching bolt seals if necessary.

5. Check pump cover face for flatness. Check for scoring or chips in pressure regulator bore.

Removing forward clutch
(© G.M. Corp.)

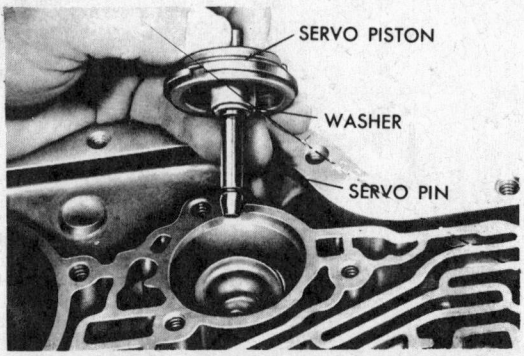

Removing front servo piston, washer and pin
(© G.M. Corp.)

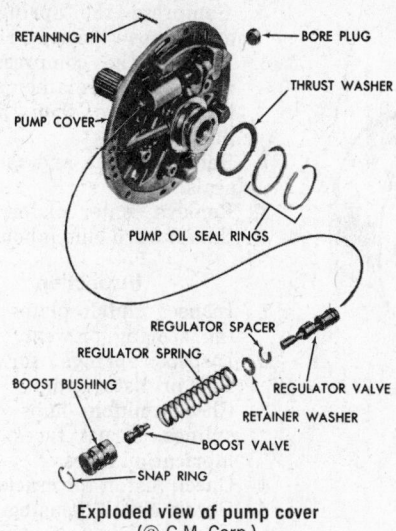

Exploded view of pump cover
(© G.M. Corp.)

Installing pump drive gear
(© G.M. Corp.)

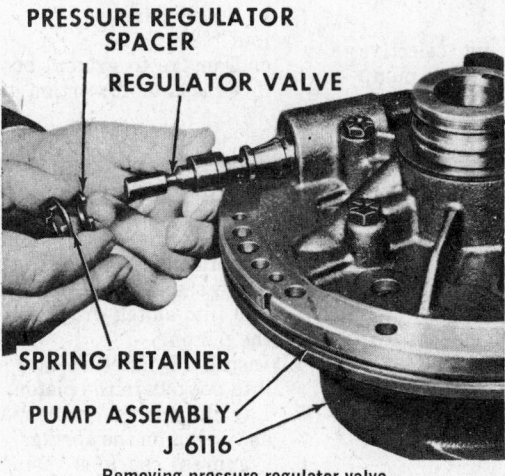

Removing pressure regulator valve
(© G.M. Corp.)

Check pump body face to gear clearance
(© G.M. Corp.)

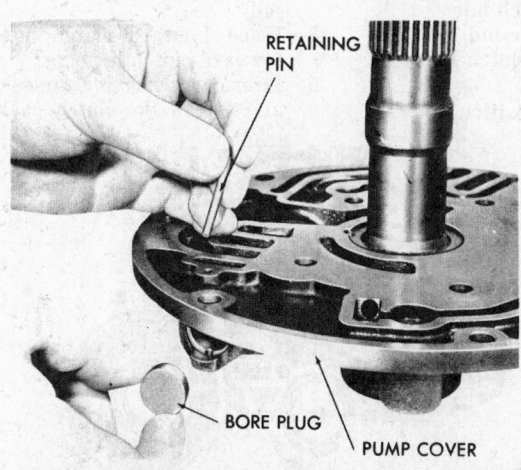

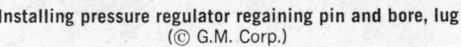

Installing pressure regulator regaining pin and bore, lug
(© G.M. Corp.)

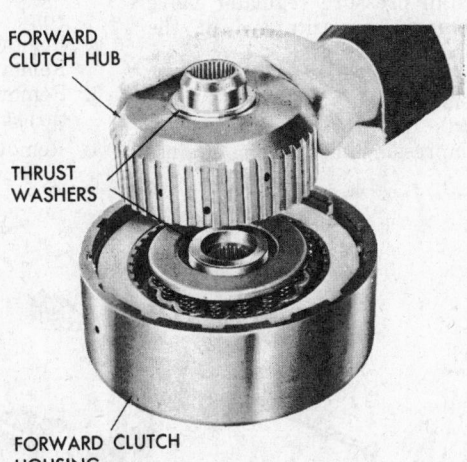

Removing forward clutch hub and thrust washers
(© G.M. Corp.)

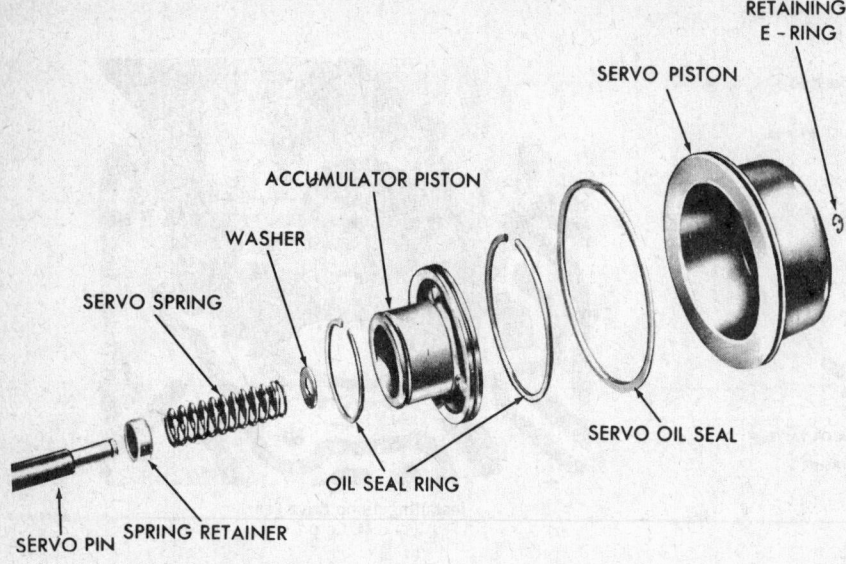

RETAINING
E - RING

SERVO PISTON

ACCUMULATOR PISTON

WASHER

SERVO SPRING

SERVO OIL SEAL

OIL SEAL RING

SERVO PIN   SPRING RETAINER

Exploded view of rear servo and accumulator (© G.M. Corp.)

Check that all passages are unobstructed. Check for scoring or damage at pump gear face. Check that breather hole in pump cover is open.

6. Check condition of stator shaft splines and bushings.

7. Check oil ring grooves for damage or wear. Check selective thrust washer for wear.

8. Make sure that pressure regulator valve and boost valve operate freely.

Assembly

1. Install drive and driven gears into the pump body, alignment marks up and in proper index, (drive gear with drive tangs up).

2. Install pressure regulator spring spacer/s if required, retainer and spring into the pressure regulator bore.

3. Install pressure regulator valve from the opposite end of the bore, stem end first.

4. Install boost valve into bushing, stem end out, and install both parts into the pump cover by compressing the bushing against

the spring. Install the snap-ring.

5. Install the regulator valve bore plug and retaining pin into opposite end of bore.

6. Install the front unit selective thrust washer over the pump cover delivery sleeve.

7. Install two hook-type oil seal rings.

8. Assemble pump cover to pump body with attaching bolts. (Leave bolts one turn loose at this time.)

9. Place pump aligning strap over pump body and cover, then tighten tool.

10. Install pump cover bolts. pump-to-case O-ring seal and gasket.

Forward Clutch

Disassembly

1. Remove forward clutch housing-to-direct clutch hub snap-ring.

2. Remove direct clutch hub.
Remove clamp and install

3. Remove forward clutch hub and thrust washers.

4. Remove five composition and five

steel clutch plates. Press out the turbine shaft.

5. Compress the spring retainer and remove the snap-ring.

6. Remove the compressor, snapring, spring retainer and 16 release springs, then lift out the piston.

7. Remove inner and outer piston seals.

8. Remove center piston seal from the forward clutch housing.

Inspection

1. Inspect clutch plates for burning, scoring, or wear.

2. Inspect springs for collapsed coils or distortion.

3. Check clutch hubs for worn splines, thrust faces, and open lubrication holes.

4. Check piston for cracks.

5. Check clutch housing for wear, scoring, open oil passages, and free operation of ball check.

6. Inspect turbine shaft for:
a. Open lubrication passages at each end.
b. Spline damage.
journals.
c. Damage to ground bushing
d. Cracks or distortion of shaft.

Assembly

1. Place new inner and outer oil seals on clutch piston, lips away from spring pockets.

2. Place a new center seal on the clutch housing, lip faces up.

3. Place a seal protector (thimble) tool into clutch drum and install the piston.

4. Install 16 clutch release springs into pockets in the piston.

5. Lay the spring retainer and the snap-ring on the springs.

6. Compress springs using compressor, and install snap-ring. Press in turbine shaft.

7. Install the direct clutch hub washers; retain with petroleum jelly.

8. Place forward clutch hub into forward clutch housing.

9. Lubricate with transmission oil and install the clutch pack, five

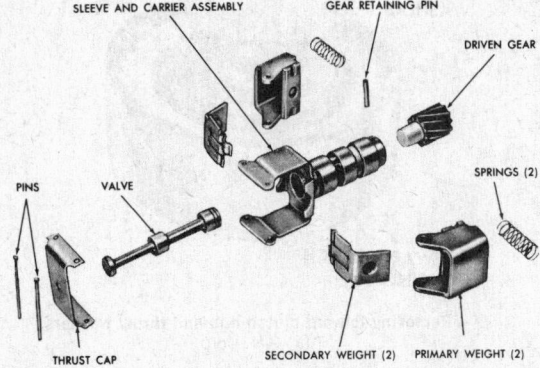

SLEEVE AND CARRIER ASSEMBLY   GEAR RETAINING PIN

DRIVEN GEAR

SPRINGS (2)

PINS

VALVE

THRUST CAP   SECONDARY WEIGHT (2)   PRIMARY WEIGHT (2)

Governor assembly
(© G.M. Corp.)

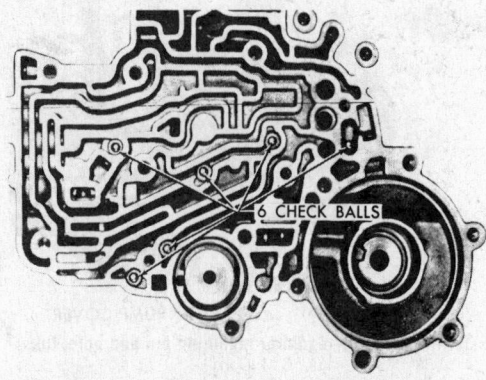

6 CHECK BALLS

Location of check balls
(© G.M. Corp.)

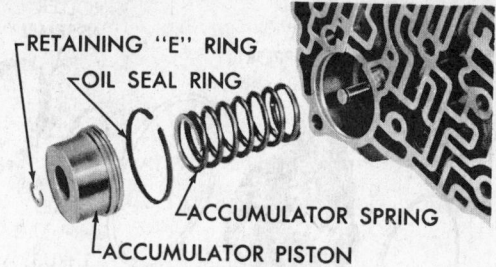

RETAINING "E" RING
OIL SEAL RING
ACCUMULATOR SPRING
ACCUMULATOR PISTON

Removing front accumulator piston and spring
(© G.M. Corp.)

composition, and five steel plates, starting with steel and alternating steel and composition.

10. Install clutch hub and retaining snap-ring.

11. Place forward clutch housing on pump delivery sleeve and air-check clutch operation.

Direct Clutch and Intermediate Sprag

Disassembly

1. Remove intermediate clutch retainer snap-ring, then the retainer.
2. Remove clutch outer race, bushings and sprag assembly.
3. Invert the unit and remove backing plate to clutch housing snap-ring.
4. Remove direct clutch backing plate, five composition, and five steel plates.
5. Using clutch compressor tool compress the spring retainer and remove the snap-ring.
6. Remove retainer and 16 piston release springs.
7. Remove direct clutch piston, then remove the outer and inner seals from the piston.
8. Remove center piston seal from the direct clutch housing.

Inspection

1. Check for popped or loose sprags.

2. Check sprag bushings for distortion or wear.
3. Inspect inner and outer races for scratches or wear.
4. Check clutch housing for cracks, wear, proper opening of oil passages, and wear on clutch plate drive lugs.
5. Check clutch plates for wear or burning.
6. Check backing plate for scratches or damage.
7. Check clutch piston for cracks and free operation of ball check.

Assembly

1. Install a new clutch piston seal onto the piston, lips facing away from spring pockets. Apply transmission fluid to oil seals.
2. Install a new outer piston seal, and a new center seal onto the clutch housing, lip of seal facing up.
3. Place seal protectors over hub and clutch housing, then install piston.
4. Install 16 springs into piston,

place retainer and snap-ring on retainer.

5. With clutch compressor, compress the clutch and install snap-ring.
6. Install five composition and five steel clutch plates, starting with steel and alternating composition and steel.
7. Install clutch backing plate, then install the backing plate snap-ring.
8. Invert the unit and install one sprag bushing, cup side up, over the inner race.
9. Install sprag assembly into outer race.
10. With ridge on inner cage facing down, start sprag and outer race over inner race with clockwise turning motion. NOTE: outer race should not turn counter-clockwise.
11. Install sprag bushing over sprag cup side down.
12. Install clutch retainer and snap-ring.
13. Place direct clutch assembly over center support and air check operation of direct clutch.

NOTE: it is normal for air applied to reverse passage to escape from direct clutch passage. Air applied to direct clutch passage should move direct clutch.

Center Support

Disassembly

1. Remove four hook-type oil rings from center support.

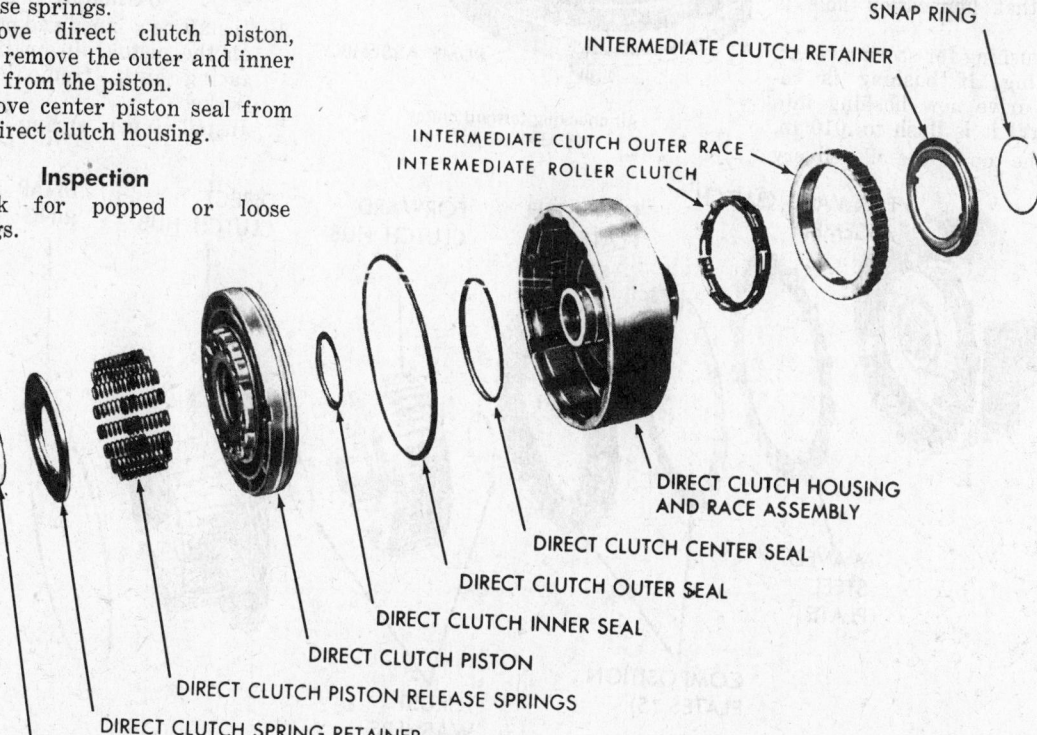

SNAP RING
INTERMEDIATE CLUTCH RETAINER
INTERMEDIATE CLUTCH OUTER RACE
INTERMEDIATE ROLLER CLUTCH
DIRECT CLUTCH HOUSING AND RACE ASSEMBLY
DIRECT CLUTCH CENTER SEAL
DIRECT CLUTCH OUTER SEAL
DIRECT CLUTCH INNER SEAL
DIRECT CLUTCH PISTON
DIRECT CLUTCH PISTON RELEASE SPRINGS
DIRECT CLUTCH SPRING RETAINER
SNAP RING

Exploded view—direct clutch and intermediate roller assembly (© G.M.C.)

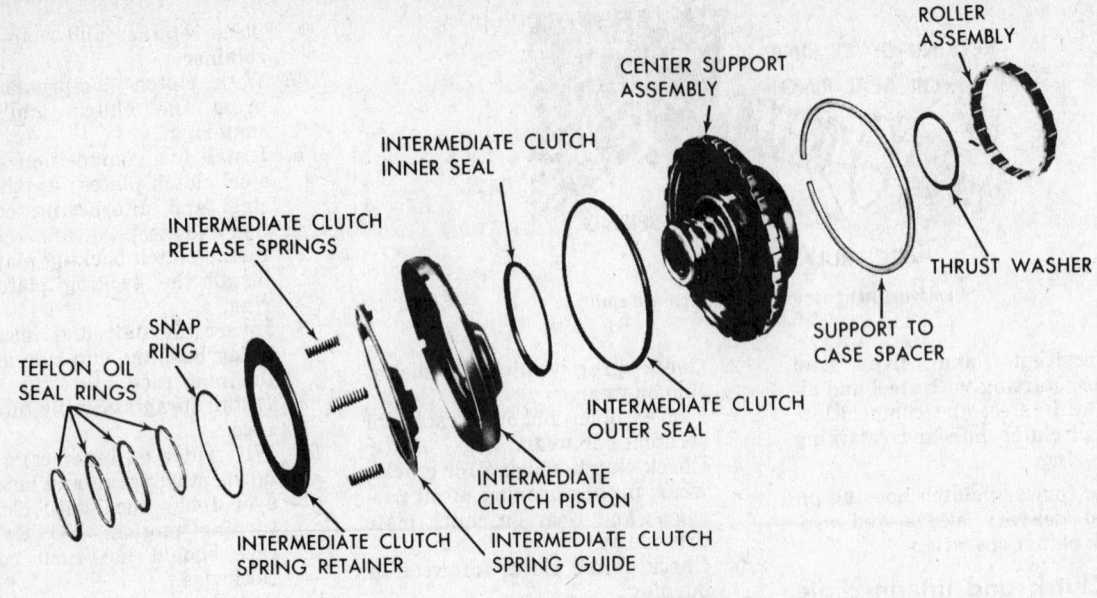

Exploded view—center support assembly (© G.M.C.)

2. Using clutch fingers, compress the spring retainer and remove the snap-ring.
3. Remove spring retainer and 12 clutch release springs.
4. Remove intermediate clutch piston.
5. Remove inner piston seal.
   NOTE: do not remove the three screws holding the roller clutch inner race to the center support.
6. Remove outer piston seal.

Inspection

1. Inspect roller clutch inner race for scratches or identations. Check that lubrication hole is open.
2. Check bushing for scoring, wear, or galling. If bushing is replaced, drive new bushing into bore until it is flush to .010 in. below the top of the oil delivery

Air checking forward clutch
(© G.M. Corp.)

sleeve.

3. Check oil seal rings and ring grooves in the center support tower for damage.
4. Make air pressure check of oil passages to be sure they are not interconnected.
5. Inspect piston sealing surfaces for scratches. Inspect piston seal grooves for damage. Check piston for cracks or porosity.
6. Check release springs for distortion or collapsed coils.
7. Check support to case spacer for burrs or raised edges. Repair with a stone or fine sand paper.

Assembly

1. Install new inner and outer seals on the piston, lip on inner seal facing away from the spring pocket.
2. Install inner spring protector

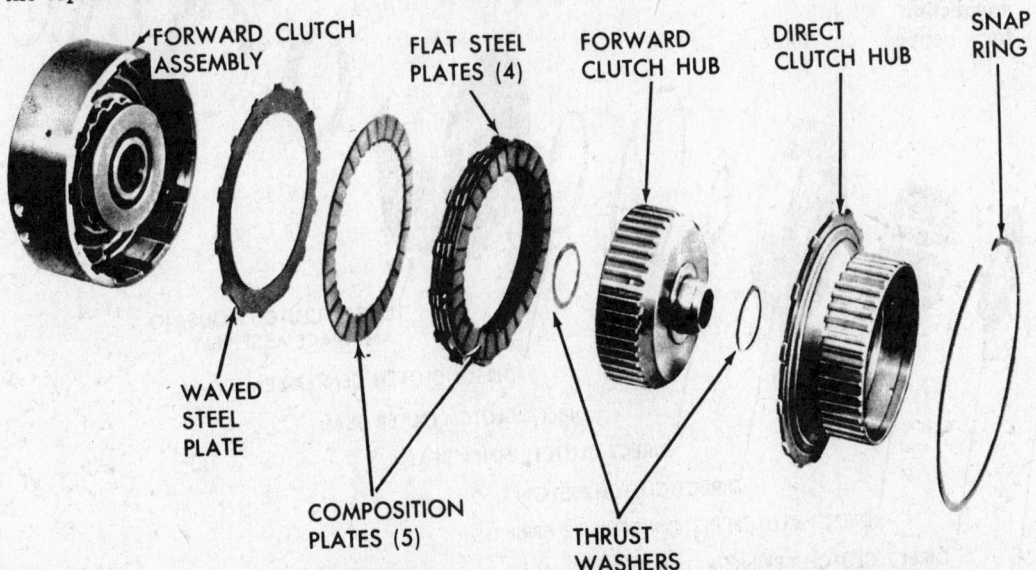

Exploded view of forward clutch (© G.M. Corp.)

Removing forward clutch center seal
(© G.M. Corp.)

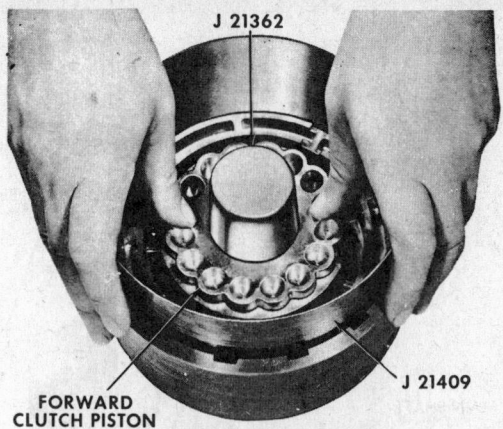

Installing forward clutch piston
(© G.M. Corp.)

tool onto the center support hub; lubricate the seal and install the piston.

3. Install 12 release springs into the piston.
4. Place spring retainer and snap-ring over the springs.
5. Using the clutch spring compressor, compress the springs and install the snap-ring.
6. Install four hook-type oil rings.
7. Air check the operation of intermediate clutch piston.

Torque Converter Inspection

The torque converter is a welded assembly and must be serviced as a unit. If converter output shaft has more than .050 in. end-play, renew the unit.

Check for leaks as follows:

A. Install converter leak test fixture and tighten.
B. Fill converter with air, 80 psi.
C. Submerge in water and check for bubbles.

Planetary Gear Unit

Disassembly

1. Remove center support assembly.
2. Remove center support to reaction carrier thrust washer.
3. Remove center support to sun gear races and thrust bearing. One race may already have been removed with the center support.
4. Remove reaction carrier and roller clutch assembly.
5. Remove front internal gear ring from output carrier assembly.
6. Remove sun gear.
7. Take off reaction carrier to output carrier thrust washer.
8. Turn carrier assembly over. Remove output shaft to output carrier snap ring.
9. Output shaft may now be removed.
10. Measure to determine speedometer drive gear location with rela-

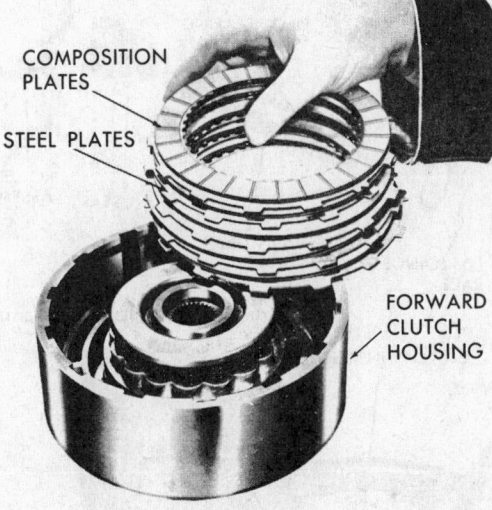

Installing forward clutch composition and steel plates
(© G.M. Corp.)

tion to the end of the shaft for reassembly. Remove nylon speedometer drive gear by depressing retaining clip and sliding gear off output shaft. Remove steel speedometer drive gear with a suitable puller.

11. Remove output shaft to rear internal gear thrust bearing and two races.
12. Remove rear internal gear and mainshaft. Remove rear internal gear to sun gear thrust bearing and two races. Remove rear internal gear to mainshaft snap ring. Remove mainsahft.

Inspection of Reaction Carrier, Roller Clutch, and Output Carrier Assembly

1. Insert band surface reaction carrier for burring or scoring.
2. Check roller clutch outer race for scoring or wear. Check thrust washer for wear.
3. Check bushing for damage. If bushing is damaged, replace reaction carrier.
4. Check reaction carrier pinions

for damage, rough bearings, or excessive tilt. Check pinion end play. It should be .009-.024 in. Pinions may be replaced if necessary.

5. Check roller clutch for damaged members. Check roller clutch cage for damage.
6. Inspect front internal gear (out-put carrier) for damaged teeth. Inspect output carrier pinions, for damage, rough bearings, and excessive tilt. Check pinion end play. It should be .009-.024 in.
7. Inspect parking pawl lugs for cracks or damage. Inspect output locating splines for damage.
8. Check front internal gear ring for flaking.

Assembly

1. Install rear internal gear onto end of mainshaft, then install the snap-ring.
2. Install sun gear to internal gear thrust races and bearings against inner face of rear internal gear as follows:

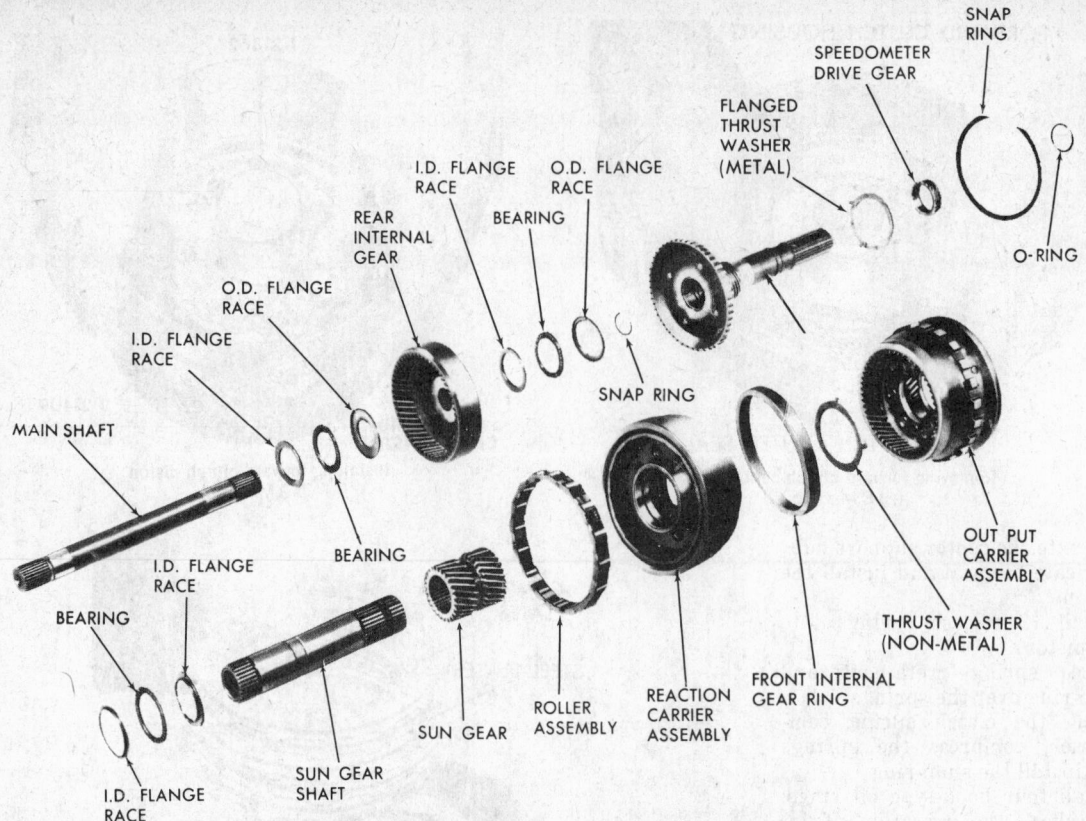

I.D. FLANGE RACE

I.D. FLANGE RACE

O.D. FLANGE RACE

REAR INTERNAL GEAR

I.D. FLANGE RACE

BEARING

O.D. FLANGE RACE

FLANGED THRUST WASHER (METAL)

SPEEDOMETER DRIVE GEAR

SNAP RING

O-RING

MAIN SHAFT

BEARING

I.D. FLANGE RACE

BEARING

I.D. FLANGE RACE

SUN GEAR SHAFT

SUN GEAR

ROLLER ASSEMBLY

REACTION CARRIER ASSEMBLY

SNAP RING

FRONT INTERNAL GEAR RING

THRUST WASHER (NON-METAL)

OUT PUT CARRIER ASSEMBLY

I.D. FLANGE RACE

Exploded view—planetary gear unit assembly (© G.M.C.)

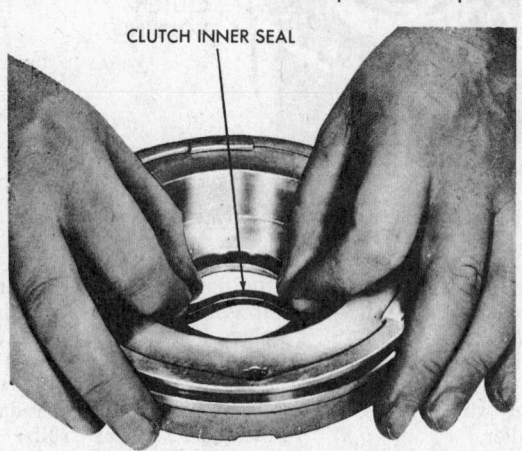

CLUTCH INNER SEAL

Installing direct clutch inner seal
(© G.M. Corp.)

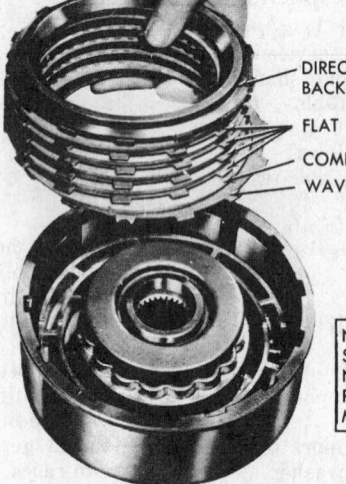

DIRECT CLUTCH BACKING PLATE

FLAT STEEL PLATES (4)

COMPOSITION PLATES (5)

WAVED STEEL PLATE (1)

NOTE: 5 FLAT STEEL PLATES— NO WAVED STEEL PLATE IN PQ & PS MODELS

Installing direct clutch backing plate, and clutches
(© G.M. Corp.)

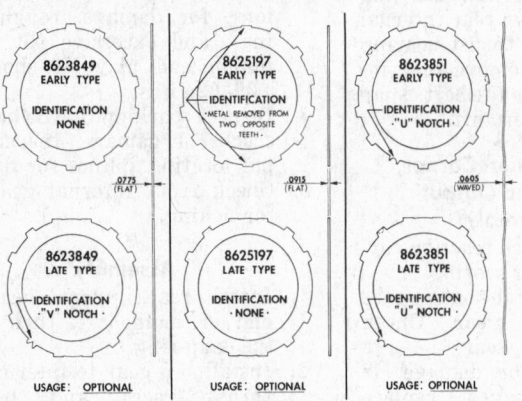

8623849 EARLY TYPE IDENTIFICATION NONE	8625197 EARLY TYPE IDENTIFICATION (METAL REMOVED FROM TWO OPPOSITE TEETH)	8623851 EARLY TYPE IDENTIFICATION "U" NOTCH
.0775 (FLAT)	.0915 (FLAT)	.0605 (WAVED)
8623849 LATE TYPE IDENTIFICATION "V" NOTCH	8625197 LATE TYPE IDENTIFICATION NONE	8623851 LATE TYPE IDENTIFICATION "U" NOTCH
USAGE: OPTIONAL	USAGE: OPTIONAL	USAGE: OPTIONAL

Clutch plate identification
(© G.M. Corp.)

Removing direct clutch assembly
(© G.M. Corp.)

FORWARD CLUTCH PISTON

CLUTCH OUTER SEAL

Removing or installing forward clutch piston outer seal
(© G.M. Corp.)

a. Place large race against the internal gear, with flange facnig up.

b. Place thrust bearing against race.

c. Place small race against the bearing, with inner flange facing the bearing, or down.

3. Install the output carrier over the mainshaft so that the pinions mesh with the rear internal gear.

4. Place the above portion of the assembly through a hole in the bench so that the mainshaft hangs downward.

5. Install the rear internal gear to output shaft thrust races and bearings as follows:

a. Small diamete race against internal gear, center flange facing up.

b. Bearing onto the race.

c. Second race onto the bearing, outer flange cupped over the bearing.

6. Install output shaft into the output carrier assembly.

7. Install output shaft to output carrier snap-ring.

8. Turn assembly over and support it so that the output shaft hangs downward.

9. Install the reaction carrier to output carrier thrust washer, tabs facing down and in their pockets.

10. Install sun gear, splines, chamfer down. Install gear ring over output carrier.

11. Install sun gear shaft, then the reaction carrier.

12. Install the center support to sun gear thrust races and bearing as follows:

A. Install the large race, center flange up over the sun gear shaft.

B. Install thrust bearing.

C. Install the second race, center flange up.

13. Install roller clutch to reaction carrier outer race. Install the center support to reaction carrier thrust washer into the recess in the support. Retain with petroleum jelly.

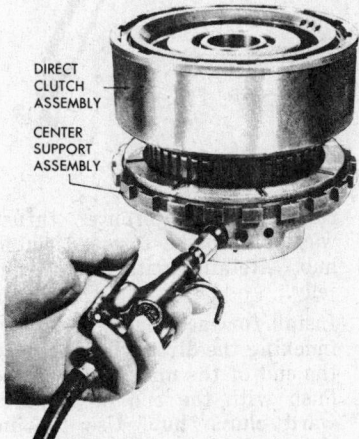

DIRECT CLUTCH ASSEMBLY

CENTER SUPPORT ASSEMBLY

Air checks direct clutch assembly
(© G.M. Corp.)

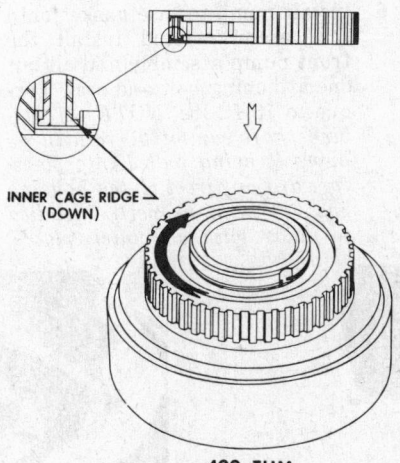

INNER CAGE RIDGE (DOWN)

400 THM

Correct sprag rotation
(© G.M. Corp.)

14. Install center support into reaction carrier and roller clutch assembly. With reaction carrier held, center support should only turn counterclockwise.

15. Install a gear unit assembly holding tool to hold units in place. Install output shaft to case thrust washer tabs in pockets and retain with petroleum jelly.

Transmission Assembly

Parking Mechanism, Rear Band and the Complete Gear Assembly

Installation

1. Install O-ring seal onto parking pawl shaft, then install parking pawl, tooth toward inside of case.

2. Install the pawl shaft retaining clip and the return spring, square-end hooked on the pawl.

3. Install parking brake bracket guides over pawl, using two attaching bolts.

4. Install rear band assembly so that the two lugs index with the two anchor pins. Install support to case spacer with ring gap adjacent to band anchor pin.

5. Install rear selective washer into slots provided inside rear of transmission case. Dip washer in transmission fluid.

6. Install the complete gear unit assembly into the case.

7. Lubricate and install center support to case snap-ring. Install bevel side up. *NOTE: The support to case spacer is .040 in. thick and is flat on both sides. The center support to case snap-ring has one side beveled. The intermediate clutch backing plate to case snap ring is .093 in. thick and is flat on both sides.*

8. Lubricate and install case to center support bolt and torque to 22 ft. lbs. Remove the locating screw.

9. Install three steel and three composition clutch plates. Alternate the plates, starting with steel.

10. Install the backing plate, ridge up, then the snap-ring. Locate snap-ring gap opposite band anchor pin.

11. Check rear end-play as follows:

A. Install a threaded rod or a long bolt into an extension hous-attaching bolt hole.

B. Mount dial indicator on the rod and index it with the end of the output shaft.

c. Move the output shaft in and out. Read the end-play. End-play should be .007-.019 in. The selective washer controlling the end-play is the steel washer, having three lugs, that is located between the thrust washer and the rear face of the transmission case.

If a different washer thickness is required to obtain proper end-play, it can be selected from the following chart.

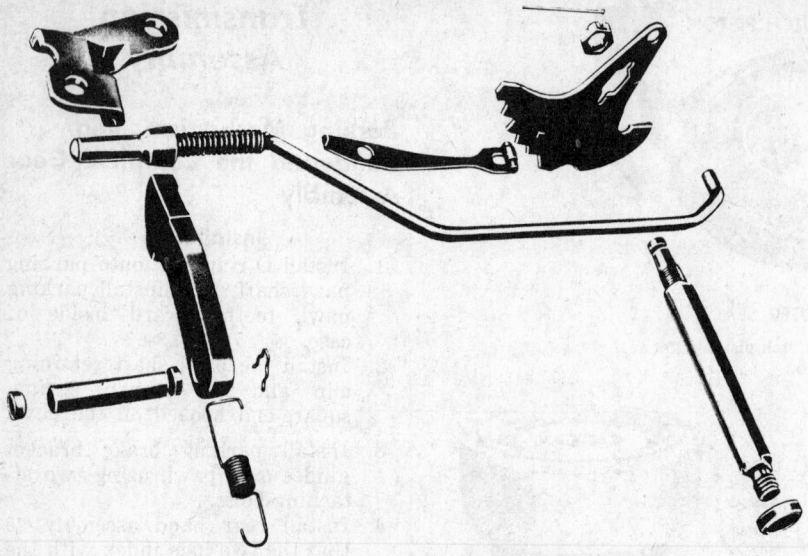

Exploded view—manual and parking linkage (© G.M.C.)

Thickness	Notches and/or Numeral	
.074—.078 in.	None	1
.082—.086 in.	1 Tab Side	2
.090—.094 in.	2 Tabs Side	3
.098—.102 in.	1 Tab O.D.	4
.106—.110 in.	2 Tabs O.D.	5
.114—.118 in.	3 Tabs O.D.	6

Front Band, Direct Clutch and Forward Clutch

Installation

1. Install front band, with band anchor hole placed over the band anchor pin, and apply lug facing the servo hole.
2. Install the direct clutch and intermediate sprag assembly. (Removal of direct clutch plates may help.)
3. Install forward clutch hub to direct housing bronze thrust washer onto the forward clutch hub. Retain with petroleum jelly.
4. Install forward clutch assembly, indexing the direct clutch hub so the end of the mainshaft will be flush with the end of the forward clutch hub. Use turbine shaft as a tool.
5. Install turbine shaft; end with the short spline goes into the forward clutch housing.
6. Install pump-to-case gasket onto the case face and install the front pump assembly and all but one attaching bolt and seal. Torque to 18 ft. lbs. *NOTE: If turbine shaft cannot be rotated as pump is being pulled into place, forward or direct clutch housing has not been properly installed to index with all clutch plates.*

This condition must be corrected before pump is fully pulled into place.

7. Drive in a new front seal.
8. Check front unit end-play as follows:
   A. Install one rod of slide hammer, with 5/16-18 in. thread, into the empty pump assembly attaching bolt hole.
   B. Mount dial indicator on the rod and adjust the indicator probe to contact end of turbine shaft.
   C. Hold output shaft forward while pushing turbine shaft rearward to its stop.
   D. Set dial indicator to zero.
   E. Pull turbine
   F. Read end-play, as registered on dial. The reading should be .003-.024 in.

The selective washer controlling this end-play is located between the pump cover and the forward clutch housing. If more, or less, washer thickness is required to bring end-play within specifications, make selection from the thickness-color chart.

Thickness	Color
.060-.064 in.	Yellow
.071-.075 in.	Blue
.082-.086 in.	Red
.093-.097 in.	Brown
.104-.108 in.	Green
.115-.119 in.	Black
.126-.130 in.	Purple

9. Remove dial indicator and install the remaining front pump attaching bolt and seal. Torque to 18 ft. lbs.

Rear Extension Housing Assembly

Installation

1. Install extension housing-to-case

Installing output shaft
(© G.M. Corp.)

Installing reaction carrier and roller assembly
(© G.M. Corp.)

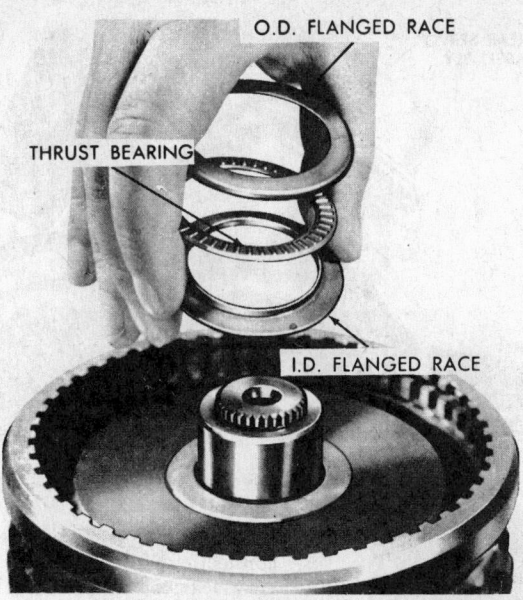

O.D. FLANGED RACE

THRUST BEARING

I.D. FLANGED RACE

Installing rear internal gear to output shaft bearing and races
(© G.M. Corp.)

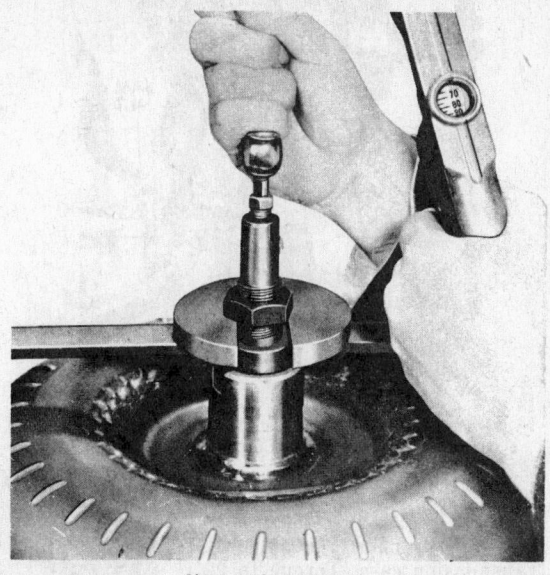

Air checking converter
(© G.M. Corp.)

Installing direct clutch assembly
(© G.M. Corp.)

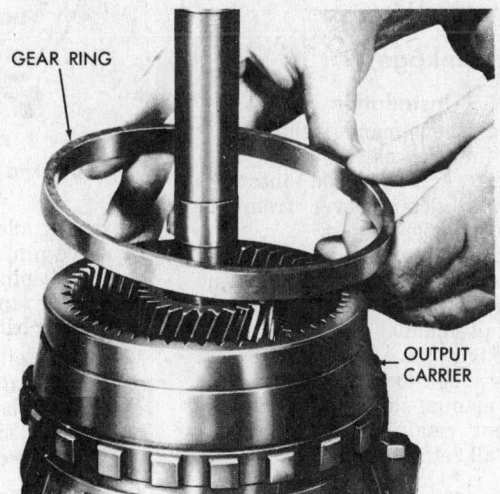

GEAR RING

OUTPUT CARRIER

Installing front internal gear ring to output carrier
(© G.M. Corp.)

FRONT BAND

Installing front band
(© G.M. Corp.)

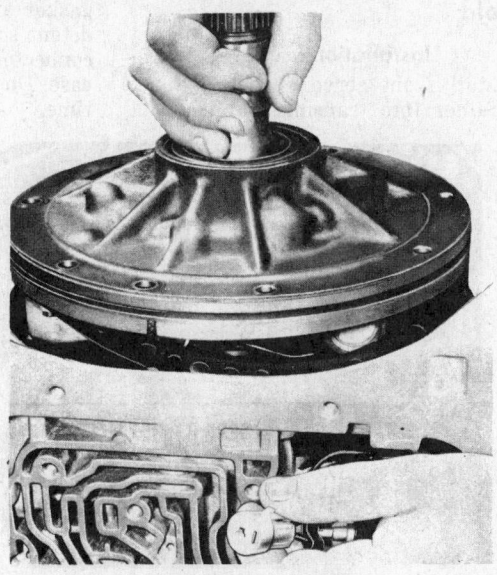

Installing pump assembly
(© G.M. Corp.)

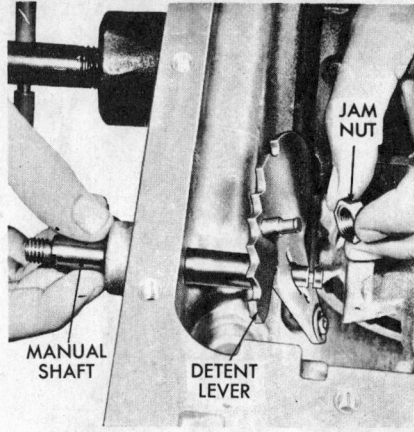

Installing dent lever and jam nut to manual shaft
(© G.M. Corp.)

REAR SERVO ASSEMBLY

Installing Rear Servo Assembly

O-ring seal onto extension housing.

2. Attach extension housing to transmission case. Torque to 22 ft. lbs.
3. Drive in a new extension housing rear seal.

Manual Linkage

Installation

1. Install new manual shift shaft seal into the case.
2. Insert actuator rod into the manual detent lever from the side opposite the pin.
3. Install actuator rod plunger under the parking bracket and over the pawl.
4. Install manual lever and shaft through the case and detent lever, then lock with hex nut on the manual shift shaft. (Be sure detent retaining nut is tight.) Install retaining pin.

Check Balls, Front Servo Gaskets, Spacer and Solenoid

Installation

1. Install front servo spring and retainer into transmission case.

J-21601

J-8001

Checking front unit end play
(© G.M. Corp.)

2. Install flat washer on front servo pin, on end opposite taper. Install pin and washer into case so that tapered end of pin is contacting band.
3. Install oil seal ring on front servo piston, and install on apply pin so that identification numbers on shoulders are exposed. Check freeness of piston in bore.
4. Install six check balls into transmission case pockets.
5. Install valve body spacer to case gasket and spacer plate. Install detent solenoid and gasket, with connector facing outer edge of case. Do not tighten bolts at this time.

6. Install O-ring seal on electrical connector. Lubricate and install electrical connector with locator tab in notch on side of case. Connect detent wire and lead wire to electrical connector. Be sure to install electrical wire clip.

Rear Servo

Installation

1. Before installing the rear servo assembly, check the band apply pin, using rear band apply fixture as follows:
   A. Attach band apply pin selection gauge, to the transmission case with attaching screws.
   B. Apply 25 ft. lbs. torque and select proper servo pin to be used from scale on the tool.
   C. Remove tool and make note of proper pin to be used during assembly.
   There are three selective pins:
   The identification consists of a ring located on the band lug end of the pin. Selecting the proper pin is equivalent to adjusting the band.
2. Install rear accumulator spring.
3. Install servo assembly, then the gasket and cover. Torque bolts to 18 ft. lbs.

Installing forward clutch assembly
(© G.M. Corp.)

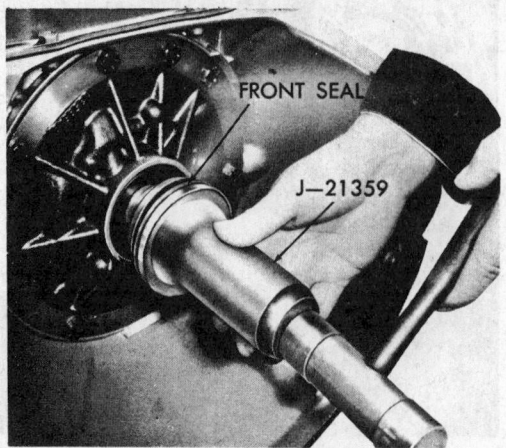

FRONT SEAL

J—21359

Installing front pump oil seal
(© G.M. Corp.)

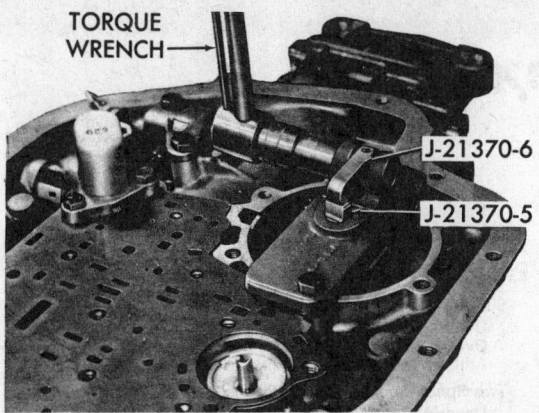

Checking Rear Band Pin

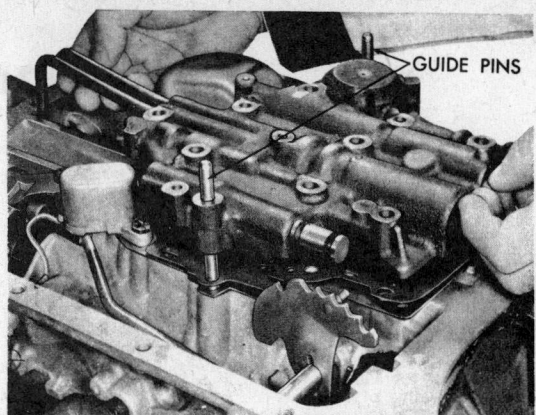

Installing Control Valve Assembly and Governor Pipes

Control Valve and Governor Pipe

Installation

1. Install control valve-to-spacer gasket, then install governor pipes into the control valve body assembly.
2. Install two guide pins, then install the control valve body and governor pipe assembly into the transmission. *NOTE: be sure the manual valve is properly indexed with the pin on the manual detent lever.*
3. Remove guide pins and install valve assembly attaching bolts and manual detent and roller assembly.
4. Tighten detent solenoid and control valve attaching bolts to 8 ft. lbs. torque.

Strainer and Intake Pipe

Installation

1. Install case-to-intake pipe O-ring onto strainer and intake pipe assembly.
2. Install strainer and pipe assembly. Install new filter on models so equipped.
3. Install new pan gasket, then install the pan. Torque to 12 ft. lbs.
4. Install modulator shield and all pan attaching screws. Torque pan attaching screws.

Modulator Assembly

Installation

1. Install modulator valve into the case, stem end out.
2. Install O-ring seal onto the vacuum modulator, then install assembly into the case.
3. Install modulator retainer and attaching bolt. Torque to 18 ft. lbs.

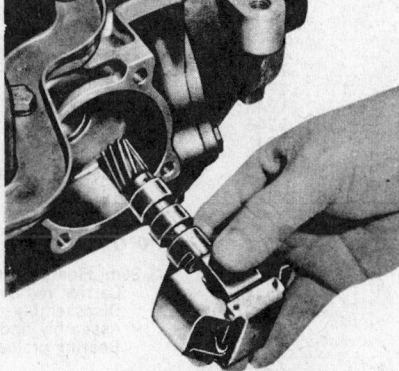

Installing governor assembly
(© G.M. Corp.)

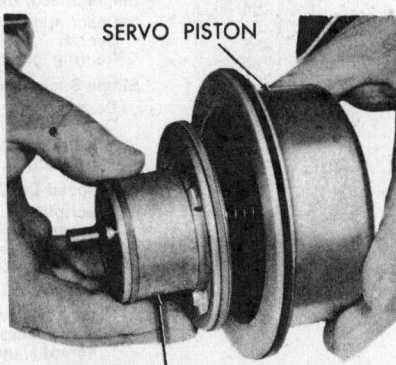

SERVO PISTON

ACCUMULATOR PISTON

Removing Servo Accumulator Piston
(© GM Corp.)

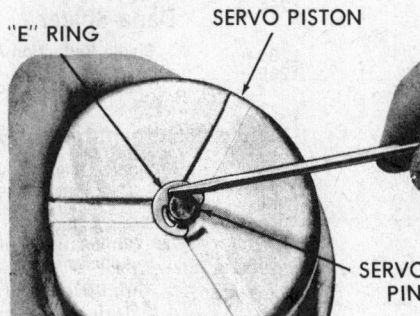

"E" RING   SERVO PISTON

SERVO PIN

Removing Servo "E" Clip
(© GM Corp.)

Governor and Speedometer Driven Gear

Installation

1. Install governor assembly.
2. Attach governor cover and gasket with four bolts. Torque to 18 ft. lbs.
3. Install speedometer driven gear assembly. Install retainer and attaching bolt.

Converter

Installation

1. Place transmission in cradle or portable jack.
2. Install converter assembly to pump assembly, making certain that the converter hub drive slots are fully engaged with the pump drive gear tangs and that the converter is installed all the way toward the rear of the transmission.

TORQUE SPECIFICATIONS
Foot-Pounds

Pump cover bolts	18
Parking pawl bracket bolts	18
Center support bolt	22
Pump-to-case attaching bolts	18
Extension-to-case attaching bolts	22
Rear servo cover bolts	18
Detent solenoid bolts	8
Control valve body bolts	8
Bottom pan attaching screws	12
Modulator retainer bolt	18
Governor cover bolts	18
Manual lever-to-manual shaft nut	20
Linkage swivel clamp screw	20
Transmission-to-engine mounting bolts	40
Rear mount-to-transmission bolts	40
Oil cooler line	16
Filter retainer bolt	10
Pressure switch assembly	8

Index

DRIVE AXLE SERVICE DIAGNOSIS

Condition	Possible Cause	Correction
Rear Wheel Noise	(1) Wheel loose.	(1) Tighten loose nuts.
	(2) Faulty, brinelled wheel bearing.	(2) Faulty or brinelled bearings must be replaced. Check rear axle shaft end play.
	(3) Excessive axle shaft end play.	(3) Readjust axle shaft end play.
Rear Axle Drive Shaft Noise	(1) Misaligned axle housing.	(1) Inspect rear axle housing, alignment. Correct as necessary.
	(2) Bent or sprung axle shaft.	(2) Replace bent or sprung axle shaft.
	(3) End play in drive pinion bearings.	(3) Refer to Pinion Bearing Pre-load.
	(4) Excessive gear lash between ring gear and pinion.	(4) Check adjustment of ring gear and pinion. Correct as necessary.
	(5) Improper adjustment of drive pinion shaft bearings.	(5) Adjust pinion bearings.
	(6) Loose drive pinion companion flange nut.	(6) Tighten drive pinion flange nut to torque specified.
	(7) Improper wheel bearing adjustment.	(7) Check axle shaft end play. Readjust as necessary.
	(8) Scuffed gear-tooth contact surfaces.	(8) If necessary, replace scuffed gears.
Rear Axle Drive Shaft Breakage	(1) Improperly adjusted wheel bearings.	(1) Replace broken shaft and readjust end play.
	(2) Misaligned axle housing.	(2) Replace broken shaft after correcting rear axle housing alignment.
	(3) Vehicle overloaded.	(3) Replace broken shaft. Avoid excessive weight on vehicle.
	(4) Abnormal clutch operation.	(4) Replace broken shaft, after checking for other possible causes. Avoid erratic use of clutch.
	(5) Grabbing clutch.	(5) Replace broken shaft. Inspect clutch and make necessary repairs or adjustments.
	(6) Normal fatigue.	(6) Replace broken shaft. Inspect to determine causes or damage.
Differential Case Breakage	(1) Improper adjustment of differential bearings.	(1) Replace broken case; examine gears and bearings for possible damage. At reassembly, adjust differential bearings.
	(2) Excessive ring gear clearance.	(2) Replace broken case; examine gears and bearings for possible damage. At reassembly, adjust ring gear and pinion backlash.
	(3) Vehicle overloaded.	(3) Replace broken case; examine gears and bearings for possible damage. Avoid excessive weight on vehicle.
	(4) Erratic clutch operation.	(4) Replace broken case. After checking for other possible causes, examine gears and bearings for possible damage. Avoid erratic use of clutch.
Differential Side Gear Broken at Hub	(1) Excessive axle housing deflection.	(1) Replace damaged gears. Examine other gears and bearings for possible damage. Check rear axle housing alignment.
	(2) Misaligned or bent axle shaft.	(2) Replace damaged gears. Check axle shafts or alignment. Examine other gears and bearings for possible damage.
	(3) Worn thrust washers.	(3) Replace damaged gears. Examine other gears and bearings for possible damage. Replace thrust washers that are badly worn.
Scoring of Differential Gears	(1) Insufficient lubrication.	(1) Replace scored gears. Scoring marks on the pressure face of gear teeth or in the bore are caused by instantaneous fusing of the mating surfaces. Scored gears should be replaced. Fill rear axle to required capacity with proper lubricant.
	(2) Improper grade of lubricant.	(2) Replace scored gears. Inspect all gears and bearings for possible damage. Clean out and refill axle to required capacity with proper lubricant.
	(3) Excessive spinning of one wheel.	(3) Replace scored gears. Inspect all gears, pinion bores and shaft for scoring, or bearings for possible damage. Service as necessary.
Tooth Breakage (Ring Gear and Pinion)	(1) Overloading.	(1) Replace gears. Examine other gears and bearings for possible damage. Replace parts as needed. Avoid overloading of vehicle.
	(2) Erratic clutch operation.	(2) Replace gears, and examine remaining parts for possible damage. Avoid erratic clutch operation.
	(3) Ice-spotted pavements.	(3) Replace gears. Examine remaining parts for possible damage. Replace parts as required.
	(4) Normal fatigue.	(4) Replace gears. Examine broken parts to determine cause of normal fatigue.
	(5) Improper adjustment.	(5) Replace other parts for possible damage. Make sure ring gear and pinion backlash is correct.

DRIVE AXLE SERVICE DIAGNOSIS

Condition	Possible Cause	Correction
Rear Axle Noise	(1) Insufficient lubricant.	(1) Refill rear axle with correct amount of the proper lubricant. Also check for leaks and correct as necessary.
	(2) Improper ring gear and pinion adjustment.	(2) Check ring gear and pinion tooth contact.
	(3) Unmatched ring gear and pinion.	(3) Remove unmatched ring gear and pinion. Replace with a new matched gear and pinion set.
	(4) Worn teeth on ring gear or pinion.	(4) Check teeth on ring gear and pinion for contact. If necessary, replace with new matched set.
	(5) Loose drive pinion bearings.	(5) Adjust drive pinion bearings.
	(6) Loose differential gear bearings.	(6) Adjust differential gear bearings.
	(7) Misaligned or sprung ring gear.	(7) Check ring gear for runout.
	(8) Loose carrier housing bolts.	(8) Tighten carrier housing nuts to Specifications. Also, check for oil leaks and correct as necessary.
Loss of Lubricant	(1) Lubricant level too high.	(1) Drain excess lubricant by removing filler plug and allow lubricant to level at lower edge of filler plug hole.
	(2) Worn axle shaft oil seals.	(2) Replace worn oil seals with new ones. Prepare new seals before replacement.
	(3) Cracked rear axle housing.	(3) Repair or replace housing as required.
	(4) Worn drive pinion oil seal.	(4) Replace worn drive pinion oil seal with a new one.
	(5) Scored and worn companion flange.	(5) Replace worn or scored companion flange and oil seal.
Overheating of Unit	(1) Lubricant level too low.	(1) Refill rear axle.
	(2) Incorrect grade of lubricant.	(2) Drain, flush and refill rear axle with correct amount of the proper lubricant.
	(3) Bearings adjusted too tightly.	(3) Readjust bearings.
	(4) Excessive wear in gears.	(4) Check gears for excessive wear or scoring. Replace as necessary.
	(5) Insufficient ring gear to pinion clearance.	(5) Readjust ring gear and pinion backlash and check gears for possible scoring.

Differential Component Failure Diagnosis

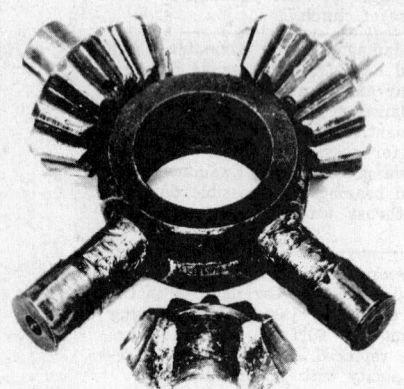  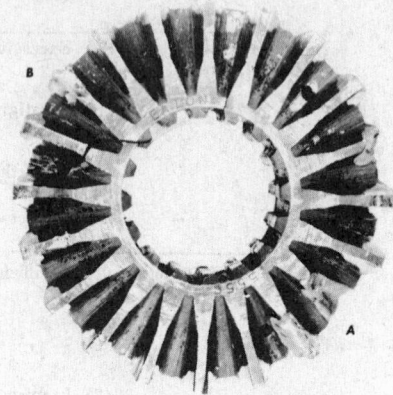

Scoring and seizure of spider and pinion gears

The spider arms and pinion gears were badly discolored by heat, caused by the unit operating for a long time after the initial scoring took place. The most probable cause of this type of failure is excessive wheelspin, particularly in off-road or icy road conditions. Other possible causes are inadequate lubrication or overstress. Friction causes the hardened areas to overheat, score, and eventually to seize. The best way to prevent this problem is to avoid wheelspin and overloading under rough terrain or poor traction conditions.

Shock fracture

These differential pinion and side gears show a grainy structure which indicates a shock fracture. This type of damage occurs instantaneously. The usual cause is a sudden excessive load, as might be caused by sudden clutch engagement at high engine speed. Another cause is a rapidly spinning wheel suddenly reaching a good traction area. This failure can be prevented by proper clutch operation, and by avoiding wheelspin and overloading under rough terrain or poor traction conditions.

Fatigue fracture of the differential side gears

This damage occurs in stages. An initial stress caused a crack, and repeated stresses caused complete failure. Some of the gear teeth were broken off in the later stages. The failures can be seen best at points A and B. All differential gears should be checked when this type of failure is found, very often the other gears will be in the initial stages of failure and must be replaced. This is most often caused by abuse such as sudden clutch engagement or incorrect two-speed axle operation, combined with overloading.

Differential Component Failure Diagnosis

Scored and Scuffed Gear Teeth

This wear pattern is a result of the gears running without enough lubrication between the tooth surfaces. Either poor quality gear lube or low lubricant level can cause this condition. Excessive torque input to the rear can also cause this wear since it will break down even the best of gear lubes. Changing gear lube at regular recommended intervals and keeping excessive torque input to a minimum will usually prevent this problem.

Overheated Gear Set

This problem can be caused by one of three, or any combination of the following circumstances. The causes are low gear lubricant level; improper gear lubricant; or infrequent lubricant change. When one or more of these conditions is present in the rear, it causes the lubricant to break down and allows the gear surface to build up heat because of increased friction. In the failure shown, the gears became so hot the pinion bearing fused to the pinion gear and the pinion gear teeth became distorted. To prevent this problem a good quality gear lubricant must be used in the rear to prevent the breakdown of lubricant under a heavy load.

Fatigue Fractured Pinion Gear

This type of fracture develops over a period of time. The fracture works through the gear tooth until the tooth is not strong enough to support the load applied. Failure happens and a section of the tooth breaks away. Continued use of pitted gears is the usual cause of this type of gear failure. As the gear pits, the support area is reduced and must carry the entire load of the gear tooth. As this continues the gear tooth fatigues and the final result is failure of the gear. To prevent this problem the ring and pinion must be replaced if there is any pitting on the gears.

MT-2306

Fractured Gear Teeth

This problem is caused by improper gear adjustment. The picture on the left shows the result of excessive backlash between the ring and pinion gear. Such backlash allows overloading of the heel section of the gear; gear fracture will follow. The picture on the right shows the result of too little backlash thus allowing the toe section of the gear to overload and become fractured. The best way to eliminate this problem is to correctly adjust the ring and pinion gears, when necessary, according to specifications.

Misalignment Fatigue Fracture

This problem comes from misalignment in the axle shafts. This kind of failure can also happen when the axle shaft breaks. If twisted, bent or sprung axle shafts are not replaced after they are damaged, this kind of failure to the side gears can occur. Bent axle housing can also cause this to happen. In most cases, this type of failure is not instantaneous. It tends to happen over a period of time. The usual cause of this type of failure is abusive operation of the vehicle and severe overloading.

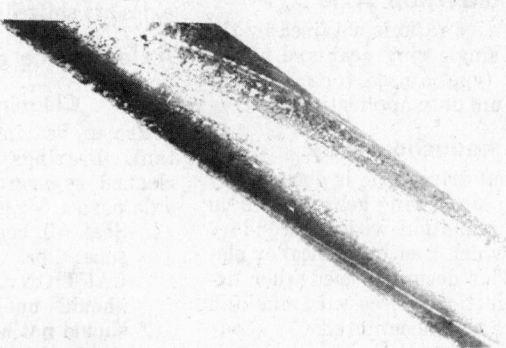

Twisted Axle Shaft

This problem with the axle comes from abusive and/or extremely severe operation of the vehicle. This is only the first stage of failure where the axle shaft has only twisted, but has not yet started to crack. At this stage the shaft should be replaced. If it is not, the shaft will continue to twist and eventually will break. When this happens it will almost certainly damage other axle parts. To eliminate this problem, the shaft should be replaced if found to be twisted. The driver of the vehicle should be informed to adopt better driving procedures.

Differential Component Failure Diagnosis

Pitted Pinion Teeth

This problem is the result of extremely high pressure on the gear teeth due to severe use. The pitting located at the heel end of the pinion gear teeth happens when overloading of the pinion moves the pinion out of its proper position relative to the ring gear. The result is a concentrated area of contact on the heel part of the gear teeth which will break down the oil film, and thus allow the pinion teeth to pit. Sometimes the ring gear will appear to be undamaged. This is because ring gear damage might not be visible to the naked eye; but the contour of the gear teeth will have changed. The ring and pinion gears must be replaced as a pair, or early failure will occur. The best way to eliminate this problem is to use good quality gear lube. The more severely the vehicle is used the better quality the gear lube should be.

Scuffed Gear Teeth on the Coast Side Only

This wear can be caused by two different things. The first is worn pinion bearings which allow excessive end play in the pinion gear. The result is incorrect contact between the ring and pinion gear teeth on the coast side. This allows excessive pressure to build up on the gear teeth and will break down the oil film, resulting in scoring of the teeth. The second cause is hard, abusive driving in vehicles equipped with a manual transmission. This usually happens when going down a steep grade at high speed and slowing the vehicle by using the clutch to brake the speed. The best way to eliminate this problem is to replace the pinion bearing if worn and recommend good driving procedures.

General Axle Service Section

Types of Drive Axles

Full Floating Axles

Support of the vehicle and the payload weight is by the axle housing. The wheels are driven by splined shafts which "float" within the axle housing.

Semi-Floating Axle

This axle design provides for the support of the payload and vehicle weight to be carried by the axle shaft through the wheel bearings to the axle housing.

Single Reduction Axle

Final drive ratio is obtained by the use of a single ring gear and pinion set. This type is used for most light and medium duty applications.

Double Reduction Axle

The final drive ratio is obtained by the use of single ring gear and pinion set in combination with a secondary gear set which is either helical or planetary. This design is used when extreme reduction is necessary and high speeds are not encountered.

2-Speed Axles

Final drive ratios are obtained by the use of a single ring gear and pinion set in combination with a secondary gear set, as in the double reduction axles, except the 2-speed axles have the facility to shift from fast ratios to slow ratios. This design is usually found on the medium duty vehicles.

3-Speed Axles

Final drive ratios of the 3-speed axles are obtained as follows: 2-speed axles in tandems are operated in either low, intermediate or high range. In low range, the range selector of both axles is shifted to low range. In intermediate range, the range selector of the forward axle is shifted into high and the selector for the rear axle is shifted into low. The power divider unit of the forward axle splits the difference of the ranges and therefore affect the intermediate range. In high range, the range selector for both axles are shifted to high.

Axle Service and Inspection

Cleaning Bearings

Proper bearing cleaning is important. Bearings should always be cleaned separately from other rear axle parts.
1. Soak all bearings in clean kerosene or Diesel fuel oil. **CAUTION: Ordinary gasoline should not be used. Bearings should not be cleaned in hot solution tank.**
2. Slush bearings in cleaning solution until all oil lubricant is loosened. Brush bearings with soft bristled brush until ALL dirt has been removed. Remove loose particles of dirt by striking flat against a wood block.
3. Rinse bearings in clean fluid.

While holding races to prevent rotation, blow dry with compressed air. **CAUTION: Do not spin bearings while drying.**
4. After bearings have been inspected, lubricate thoroughly with regular axle lubricant; then wrap each bearing in clean cloth until ready to use.

Cleaning Parts

Immerse all parts in suitable cleaning fluid and clean thoroughly. Use a stiff bristle brush as required to remove foreign deposits. Clean all lubricant passages or channels in pinion cage, carrier, caps and retainers. Make certain that interior of housing is thoroughly cleaned. Clean vent plugs and breathers.

Small parts such as cap screws, bolts, studs, nuts etc., should be cleaned thoroughly.

Inspection

Magna Flux all steel parts, except ball and roller bearings, to detect presence of wear and cracks.

Bearings

Rotate each bearing and check to see if the rollers are worn, chipped, rough or in any other way damaged. Check the cage to see if it is in any way damaged. If either the bearing rollers or the cage are damaged the bearing must be replaced.

Gears

Examine drive gear and drive pinion, differential pinions and differential side gears carefully, for damaged teeth, worn spots in surface hardening, distortion and where drive gear is attached to differential case with rivets, inspect rivets for looseness, replace loose rivets. Check radial clearances between differential side gears and differential case. Check fit of differential pinions on spider.

Differential Case

Inspect case for cracks, distortion or damage, if in good condition, thoroughly clean case and cover; then assemble case with bolts and mount in lathe centers of "V" block stand. If lathe is not available, install differential side bearings and mount case in differential carrier. Install dial indicator and check differential case run out.

Differential case with drive gear installed is checked in the same manner, except that dial indicator reading must be taken at gear instead of at case flange.

Whenever run-out exceeds limits, it may be corrected as later described under "Repair" in this section. However, the support case used in the 2-speed axle cannot be repaired and should be replaced with new case.

Checking drive gear run-out
(© G.M.C.)

Axle Shafts

Examine splined end of axle shaft for twisted or cracked splines, twisted shaft, and worn dowel holes in flange. Install new shafts if necessary.

Install axle shaft assembly in lathe centers and check shaft run-out with dial indicator, if run-out exceeds limits, replace shaft. Place dial indicator so that indicator shaft end contacts inner surface of flange near outer edge of flange and check flange run-out.

Shims

Carefully inspect shims for uniform thickness. Where various thickness of shims are used in a pack, it is recommended that the thickest shims be used between the thin shims.

Thrust Washers

Replace all thrust washers.

Spider

Carefully inspect spider arms for wear or defects.

Differential Pinion Bushings

Examine bushings (when used) for excessive wear, looseness, or damage. Check fit of gears on spider for excessive clearance. See "repair" paragraph for directions on bushing replacement.

Axle Housing Sleeves

Sleeves showing damaged threads, wear, or other damage should be replaced if hydraulic press is available, otherwise replace housing.

Housing Check

Before Removal

A check for bent axle housing can be made with unit in vehicle; however, conventional alignment instruments can be used if available.

1. Raise rear axle with a jack until wheels clear floor. Block up axle under each spring seat.
2. Check wheel bearing adjustment and adjust if necessary, then check wheels for looseness and tighten wheel nuts if necessary.
3. Place a chalk mark on outer side wall of tires at bottom. Measure across tires at chalk marks with a toe-in gauge.

4. Turn wheels half-way around so that chalk marks are positioned at top of wheel. Measure across tires again. If measurement at top is 1/8" or more, smaller than measurement at bottom of wheels, axle housing has sagged and is bent. If measurement at top exceeds bottom dimension by 1/8" or more, axle housing is bent at ends.
5. Turn chalk marks on both wheels so that marks are level with axle and at rear of vehicle. Take measurement with toe-in gauge at chalk marks; then turn both chalk marks to front and level with axle and take another measurement. If measurement at front exceeds rear dimension by 1/8" or more, axle is bent to the rear. If the measurement condition is the reverse, the axle is bent forward.

After Removal

Place two straightedges across the housing flanges and measure the distance between the ends of the straightedges at a point 11 inches from the tube center. Relocate the straightedge 180 degrees and remeasure. If the straightedges are parallel in both measurements within 3/32 inch, the housing is serviceable.

General Repair

Oil Seal Contact Surfaces

Surface of parts, contacted by oil seals must be free of corrosion, pits and grooves. When abrasive cleaning fails to clean up the seal contact surface and restore smooth finish, a new part must be installed.

Oil Seal

Removal

Oil seals can be removed with a drift pin. When removing a seal, be careful that it does not become cocked and result in damage to the retainer. Clean surface of retainer carefully, so that seal will seat properly in retainer.

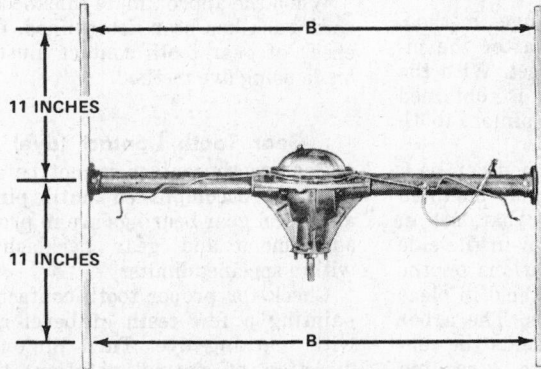

Checking housing alignment with straight edge bars (© American Motors Corp.)

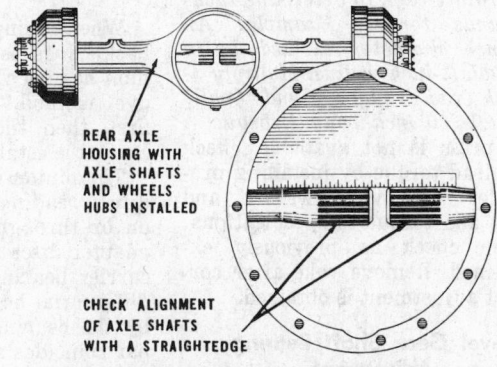

REAR AXLE HOUSING WITH AXLE SHAFTS AND WHEELS HUBS INSTALLED

CHECK ALIGNMENT OF AXLE SHAFTS WITH A STRAIGHTEDGE

Method of checking axle housing alignment— with full floating axles (© International Harvester Co.)

Drive Axles

Installation

Coat outer surface of seal retainer with a light coat of sealer, to prevent lubricant leaks. Carefully start seal in retainer. Cutting, scratching, or curling of lip of seal seriously impairs its efficiency and usually results in premature replacement. Lip of seal should be coated with a high temperature grease containing zinc oxide to help prevent scoring and damage to parts during installation.

Seals must always be installed so that seal lip is toward the lubricant.

Pinion Bearing Adjustments (Pre-Load)

Pinion bearing must be adjusted for pre-load before assembly is installed in carrier.

Do not install oil seal until after adjustment is made—installation of seal would produce false rotating torque.

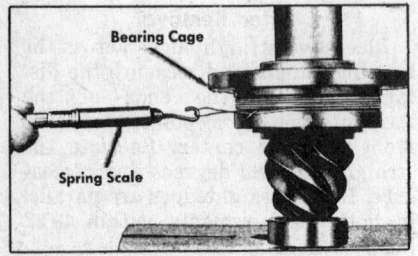

Checking pinion bearing pre-load
(© G.M.C.)

Cage Type

1. With pinion bearings, and adjusting spacers (or shims) installed in cage, check bearing contact by rotating cage.
2. Using a press, apply pressure (approx. 20,000 lbs.) to outer bearing.
3. Wrap soft wire around cage and pull on horizontal line with spring scale. Rotating (not starting) torque should be within limits recommended by manufacturer. *NOTE: Method of determining inch-pounds torque with scale is to determine radius of cage. Multiply radius in inches by pounds pull required to rotate cage to determine inch-pounds torque. Example: An 8-inch diameter divided by 2 equals 4-inch radius. Multiply 4-inch (radius) by 5 pounds (pull) equals 20 inch pounds torque.*
4. If press is not available, check preload torque by installing propeller shaft yoke, washer, and nut and torque to specifications; then check as previously explained. Remove yoke after correct adjustment is obtained.

Bevel Gear Shaft Bearing Adjustment

Bevel gear shaft bearings must be

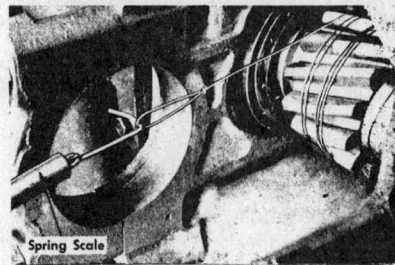

Checking pre-load on bevel gear cross shaft
(© G.M.C.)

adjusted for pre-load before pinion and cage assembly and differential assembly are installed in carrier.

1. Wrap several turns of soft wire around gear teeth on cross shaft, then pull on a horizontal line with spring scale. Rotating (not starting) torque should be used. *NOTE: Method of determining inch-pounds torque with scale is to determine radius. Multiply radius in inches by pounds pull required to rotate shaft to determine inch-pounds torque. Example: An 8-inch diameter divided by 2 equals 4-inch radius times 5 pounds (pull) equals 20 inch-pounds torque.*
2. Remove or add shims from under cage or cap opposite bevel gear to obtain specified bearing pre-load.
3. When making bevel gear and pinion tooth contact or backlash adjustments it is sometimes necessary to remove or add shims from one side. *Always remove or add an equal thickness to the opposite side so to maintain correct pre-load.*

Gear Tooth Contact and Backlash

Pinion Depth Measurement Methods

Methods of adjusting pinions to obtain the proper depths will vary with the axle type and the manufactures recommendations. Pinion depth settings and gear teeth contact may be determined by the use of pinion setting gauges or by the use of marking dye on the gear teeth.

When using the gauge method, backlash is established after the pinion has been properly set. With the dye method, backlash is obtained first, then the proper pinion tooth contact is established.

The pinion gauge method can be a direct reading micrometer, mounted on or through an arbor bar, set in adapter discs and located in the side carrier bearing cup locations on the differential housing and held in place by the bearing cup caps. The arbor bar coincides and represents the center line of the axle shafts. A reading is taken by the mounted micrometer,

from the arbor bar to the head of the pinion to determine the need to add to or remove shims from the shim pack total, to adjust the pinion to the proper nominal assembly dimension or standard pinion depth.

Another method using the arbor bar and discs, is the use of a gauge block with a spring loaded plunger and a thumb screw to lock the plunger upon expansion. A micrometer is used to measure the gauge block after the plunger has been allowed to expand between the arbor bar and the pinion head. As in the mounted micrometer procedure, the shim pack thickness is determined by the reading obtained.

A third method is the use of a gauge block tool, installed in the housing in place of the pinion gear, and a large arbor bar placed in the axle housing differential bearing seats and tightened securely. A measurement is taken between the arbor bar and the pinion tool by either a feeler gauge or the use of individual shims from the shim pack. This measurement represents the shim pack needed for a zero marked pinion.

Setting New Pinion (Without Gauge)

Whenever a pinion setting gauge is not available, the approximate thickness of the pinion shim pack at the rear pinion bearing cup, change the sign of the marking (individual variation distance) on the *new* pinion (plus to minus or minus to plus), then add the variation of the old pinion (sign unchanged) which will determine the amount the original shim pack must be changed when installing a new pinion.

On those types of axles where the shims are located between the pinion cage and differential carrier, change the sign of the marking (individual variation distance) on the *old* pionion (plus to minus or minus to plus), then add variation of the new pinion (sign unchanged) which will determine how much the original shim pack must be altered when installing a new pinion.

When the approximate thickness of shim pack has been determined, final check of gear tooth contact must be made using dye method.

Gear Tooth Contact (dye)

Gear tooth contact cannot be successfully accomplished until pinion and bevel gear bearings are in proper adjustment and gear backlash is within specified limits.

Check for proper tooth contact by painting a few teeth of bevel gear with marking dye. Turn pinion in direction of normal rotation, then check tooth impression on bevel gear.

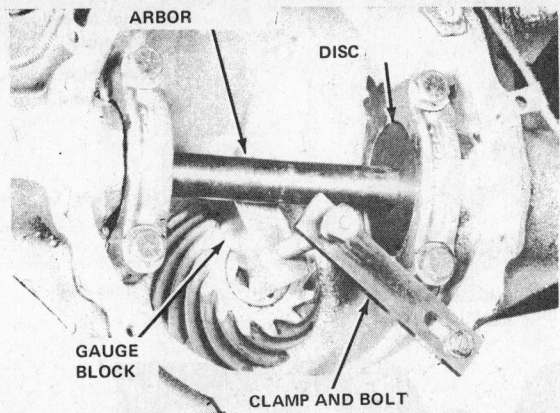

ARBOR
DISC
GAUGE BLOCK
CLAMP AND BOLT

Placement of arbor and gauge block for pinion depth measurement
(© Jeep Corp.)

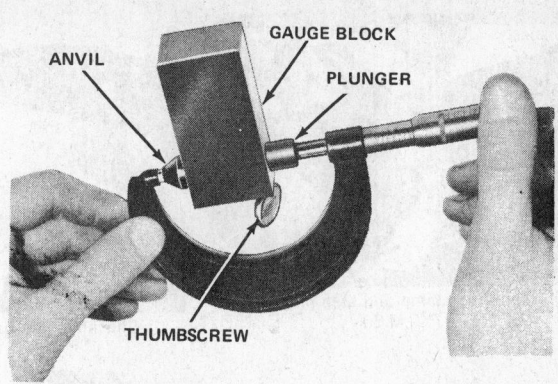

ANVIL
GAUGE BLOCK
PLUNGER
THUMBSCREW

Method of measurement of the gauge block
(© Jeep Corp.)

Gear Backlash

Gears that have been in extended service, form running contacts due to wear of teeth; therefore the original shim pack (between pinion cage and carrier) should be maintained when checking backlash. If backlash exceeds maximum tolerance, reduce backlash only in the amount that will avoid overlap of worn tooth section. Smoothness and roughness can be noted by rotating bevel gear.

If a slight overlap is present at worn tooth section, rotation will be rough.

If new gears are installed, check backlash with dial indicator.

Backlash is increased by moving bevel gear away from pinion, and may be decreased by moving bevel gear toward pinion.

When the drive gear is attached to the differential, backlash is accomplished at differential bearing adjusting rings. It should be remembered that when one ring is tightened, the opposite ring must be loosened an equal amount to maintain previously established bearing adjustment.

On axles where the bevel gear is supported by cross shaft, backlash is

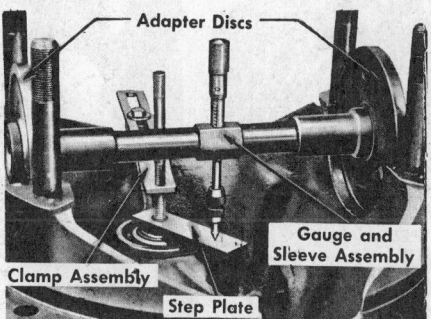

Adapter Discs
Gauge and Sleeve Assembly
Clamp Assembly
Step Plate

Position of pinion setting gauge
(© G.M.C.)

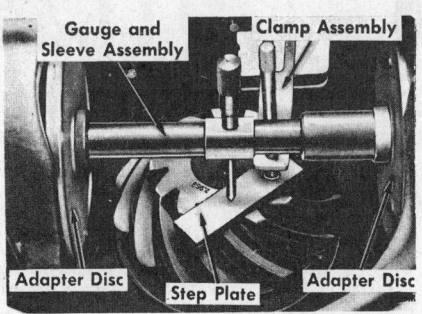

Gauge and Sleeve Assembly
Clamp Assembly
Adapter Disc
Step Plate
Adapter Disc

Installation of pinion gauge (Timken 2-speed)
(© G.M.C.)

accomplished by adding or removing shims under bearing cages.

Terms Used

Certain dimensions must be determined when using the pinion setting gauge:

1. *Nominal Assembly Dimension.* (Standard Pinion depth) This dimension (varying with axle model) is the distance between the center line of the drive gear (or differential carrier bore) and the end of the drive pinion. This dimension may be marked on the pinion or listed on the "Nominal Assembly Dimension and Adapter Disc" chart.

2. *Individual Variation Distance,* (Pinion depth variance) This dimension is a plus or minus variation of the *Nominal Assembly Dimension* on each individual pinion which may be caused by manufacturing variations.

3. *Corrected Nominal Dimension* (Desired Pinion depth) This dimension is the *Nominal Assembly Dimension* plus or minus the *Individual Variation Distance.*

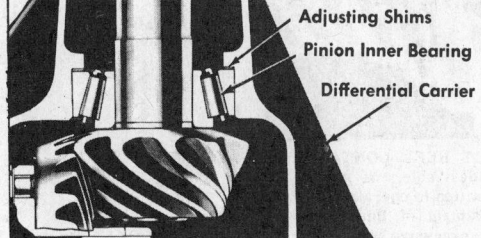

Adjusting Shims
Pinion Inner Bearing
Differential Carrier

Pinion Markings		Difference
Old Pinion	New Pinion	Between Markings
+5	+8	−3 Remove .003" Shim
+8	+5	+3 Add .003" Shim
−3	−5	+2 Add .002" Shim
−5	−3	−2 Remove .002" Shim
−3	+4	−7 Remove .007" Shim
+2	−4	+6 Add .006" Shim

The sign of the new pinion is changed and then added algebraically to the old pinion sign.

Determining pinion shim pack thickness
(if shim pack is located at rear pinion bearing cup)
(© G.M.C.)

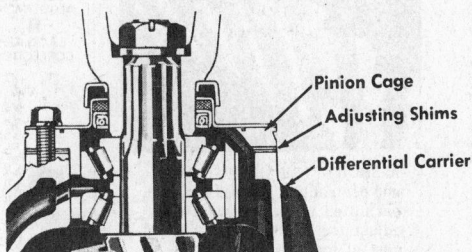

Pinion Cage
Adjusting Shims
Differential Carrier

Pinion Markings		Difference
Old Pinion	New Pinion	Between Markings
+8	+6	−2 Remove .002" Shims
+6	−2	−8 Remove .008" Shims
−4	+4	+8 Add .008" Shims
+2	+6	+4 Add .004" Shims
−7	−4	+3 Add .003" Shims
−2	−6	−4 Remove .004" Shims

The sign of the old pinion is changed and then added algebraically to the new pinion.

Determining pinion shim pack thickness
(if shims are located between pinion cage and differential carrier)
(© G.M.C.)

Drive Axles

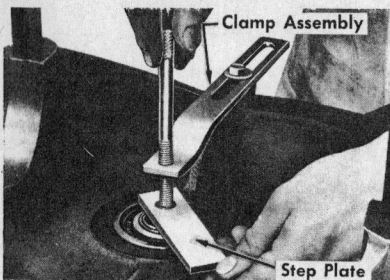

Installing clamp and step plate
(© G.M.C.)

Checking gear backlash (bevel gear)
(© G.M.C.)

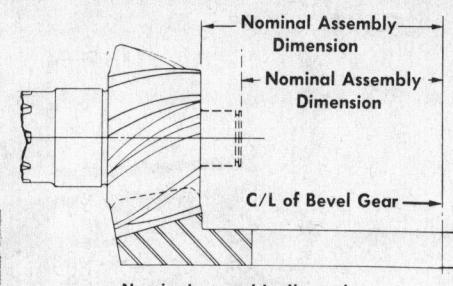

Nominal assembly dimension
(© G.M.C.)

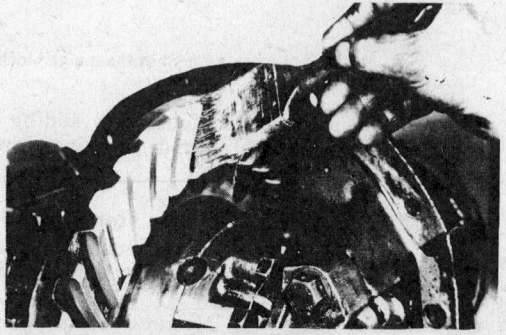

PAINTING GEAR TEETH

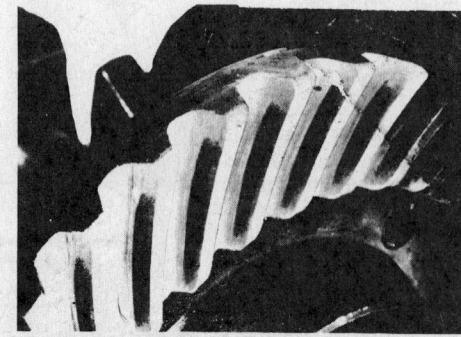

CORRECT TYPE TOOTH CONTACT

A HIGH NARROW CONTACT is not desirable. If gears are permitted to operate with an adjustment of this kind, noise, galling and rolling over of top edge of teeth will result. To obtain correct contact, move pinion toward bevel gear. This lowers contact area to proper location. This adjustment will decrease the backlash which may be corrected by moving bevel gear away from pinion.

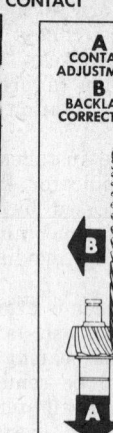

A LOW NARROW CONTACT is not desirable. If gears are permitted to operate with an adjustment of this type, galling, noise and grooving of teeth will result. To obtain correct contact, move pinion away from drive gear. This will raise contact area to proper location. A correct backlash is obtained by moving bevel gear toward pinion.

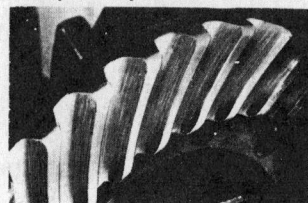

A SHORT TOE CONTACT is not desirable. If gears are permitted to operate with an adjustment of this type, chipping at tooth edges and excessive wear due to small contact area will result. To obtain correct contact, move drive gear from pinion. This will increase the lengthwise contact and move contact toward heel of tooth. Correct backlash is obtained by moving pinion toward bevel gear.

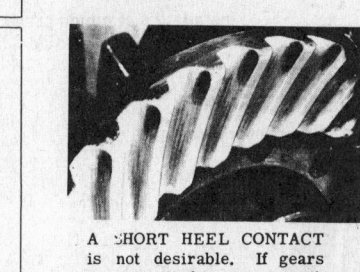

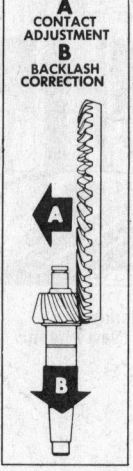

A SHORT HEEL CONTACT is not desirable. If gears are permitted to operate with an adjustment of this type, chipping, excessive wear and noise will result. To obtain correct contact, move drive gear toward pinion to increase lengthwise contact and move contact toward toe. A correct backlash is obtained by moving pinion away from drive gear.

Gear tooth contact chart (© G.M.C.)

4. *Corrected Micrometer Distance* is the *Corrected Nominal Dimension* less the thickness of the gauge set step plate (0.400'')

mounted on end of pinion.

5. *Initial Micrometer Reading* is the dimension taken by micrometer to the gauge step plate.

6. *Shim Pack Correction* is determined by the difference between the *Corrected Micrometer Distance* and the *Initial Micrometer*

Reading, and represents the amount of shim pack to be added or removed as later explained.

7. *Measured Pinion Depth.* This measurement is the distance between the axle center line and the top of the pinion gear. If a step plate or other type gauge tool is used, this measurement is included in the total.

Markings on the Pinion and Drive Gears

Drive gears and pinions are tested at the time of manufacture to detect machining variances and to obtain desirable tooth contact and quietness. When the correct setting is achieved, the gears are considered matched and a set of numbers, along with other identifying marks are etched on the gear set.

A + (plus) or − (minus) sign is used, followed by a digit to represent the factory setting where the tooth contact and quietness were the best. This is called the *Pinion Depth Variance* or *Individual Variation Distance.*

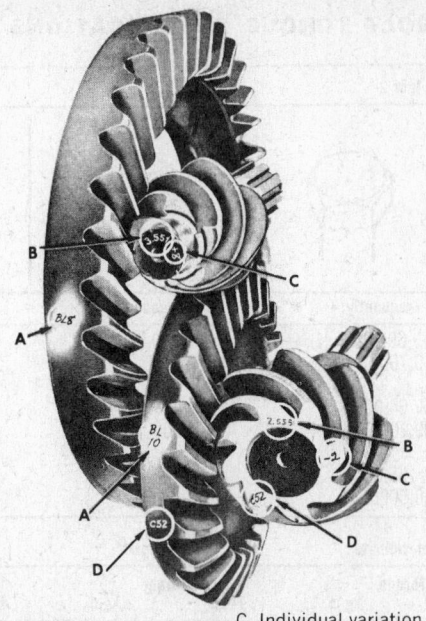

A Backlash
B Nominal assembly dimension
C Individual variation distance
D Gear and pinion matching number

Pinion and bevel gear markings (typical)
(© G.M.C.)

If the pinion is marked + 5 for example, this means the distance from the pinion gear rear face to the axle shaft center line is .005 in. more then the standard setting, and if the pinion gear is marked − 5, this means that the distance is .005 in. less than the standard setting. To move the pinion to the standard setting, compensating for the variation, shims must be either added to or subtracted from the total shim pack, located under the rear pinion bearing cup, between the pinion cage and the differential carrier, or under the rear pinion bearing, depending upon the differential model being serviced.

The procedures to follow in the adjustment of the pinion and drive gears are outlined in the respective differential model disassembly and assembly chapters.

As a rule of thumb on the addition or removal of shims for the pinion depth adjustment, draw a diagram as shown and determine which way the pinion must be moved to obtain the desired pinion depth.

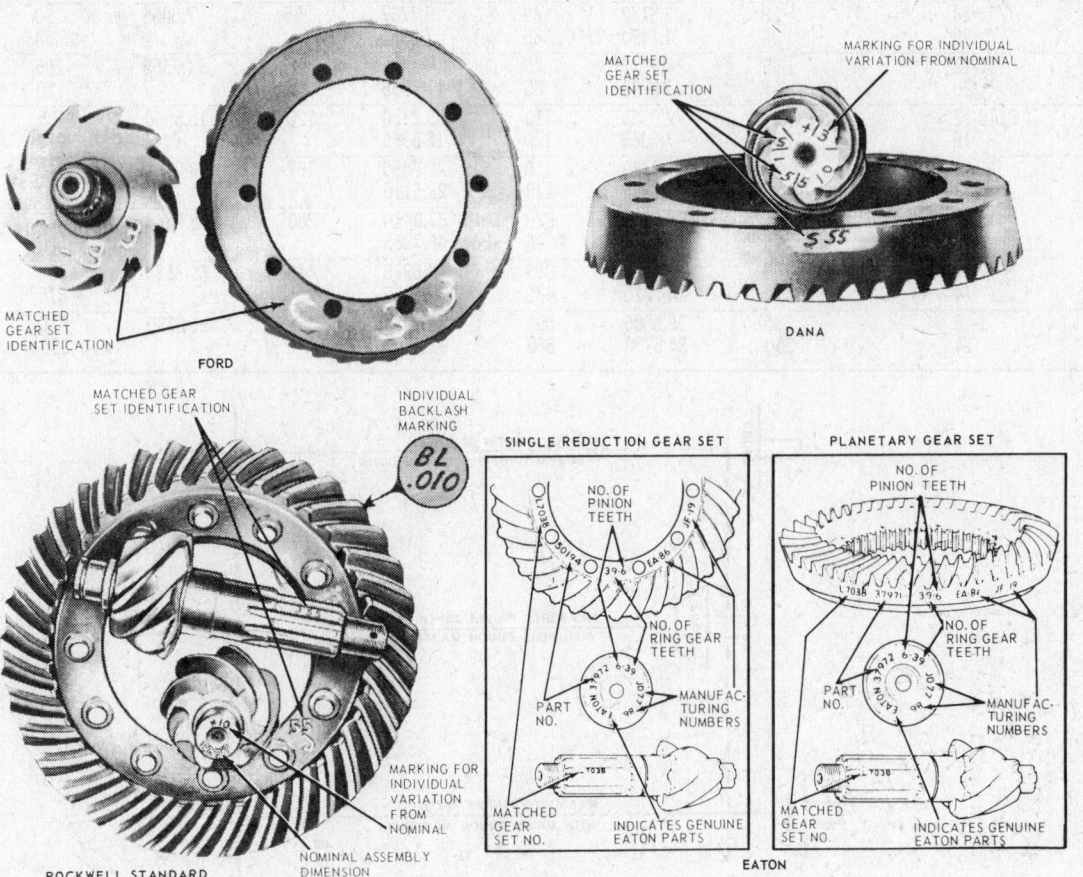

Typical gear set marking codes (© Ford Cotor Co.)

Standard Torque Specifications and Capscrew Markings

Because of the varied bolt sizes used in the many models of differentials, the torque specifications are not always available to the technician for a specific bolt. By determining the grade of bolt, size, and thread, the proper torque limit can be found in the following chart.

BOLT TORQUE SPECIFICATIONS

SAE Grade Number	1 or 2	5	6 or 7	8
Capscrew Head Markings Manufacturer's marks may vary. Three-line markings on heads shown below, for example, indicate SAE Grade 5. 				
Usage	Used Frequently	Used Frequently	Used at Times	Used at Times
Capscrew Diameter and Minimum Tensile Strength psi (Kg/sq cm)	To ½-69,000 (4850.7000) To ¾-64,000 (4499.2000) To 1 -55,000 (3866.5000)	To ¾-120,000 (8436.0000) To 1 -115,000 (8084.5000)	To ⅝-140,000 (9842.0000) To ¾-133,000 (9349.9000)	150,000 (10545.0000)

Quality of Material	Indeterminate		Minimum Commercial		Medium Commercial		Best Commercial	
Capscrew Body Size (Inches) — (Thread)	Torque Ft-Lb	kg m	Torque Ft-Lb	kg m	Torque Ft-Lb	kg m	Torque Ft-Lb	kg m
¼-20	5	0.6915	8	1.1064	10	1.3830	12	1.6596
-28	6	0.8298	10	1.3830			14	1.9362
5/16-18	11	1.5213	17	2.3511	19	2.6277	24	3.3192
-24	13	1.7979	19	2.6277			27	3.7341
⅜-16	18	2.4894	31	4.2873	34	4.7022	44	6.0852
-24	20	2.7660	35	4.8405			49	6.7767
7/16-14	28	3.8132	49	6.7767	55	7.6065	70	9.6810
-20	30	4.1490	55	7.6065			78	10.7874
½-13	39	5.3937	75	10.3725	85	11.7555	105	14.5215
-20	41	5.6703	85	11.7555			120	16.5960
9/16-12	51	7.0533	110	15.2130	120	16.5960	155	21.4365
-18	55	7.6065	120	16.5960			170	23.5110
⅝-11	83	11.4789	150	20.7450	167	23.0961	210	29.0430
-18	95	13.1385	170	23.5110			240	33.1920
¾-10	105	14.5215	270	37.3410	280	38.7240	375	51.8625
-16	115	15.9045	295	40.7985			420	58.0860
⅞- 9	160	22.1280	395	54.6285	440	60.8520	605	83.6715
-14	175	24.2025	435	60.1605			675	93.3525
1- 8	235	32.5005	590	81.5970	660	91.2780	910	125.8530
-14	250	34.5750	660	91.2780			990	136.9170

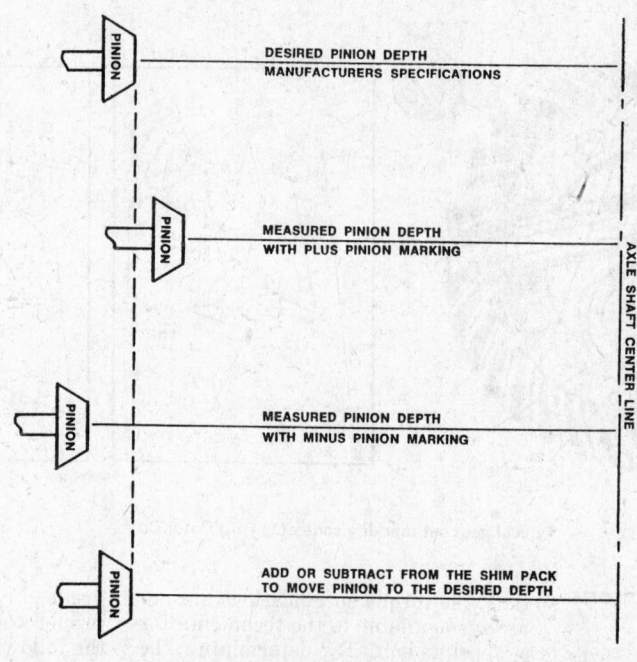

DESIRED PINION DEPTH
MANUFACTURERS SPECIFICATIONS

MEASURED PINION DEPTH
WITH PLUS PINION MARKING

AXLE SHAFT CENTER LINE

MEASURED PINION DEPTH
WITH MINUS PINION MARKING

ADD OR SUBTRACT FROM THE SHIM PACK
TO MOVE PINION TO THE DESIRED DEPTH

Movement of pinion to obtain desired pinion depth

Chevrolet Full Floating Single Speed

This is a full floating axle design using special hypoid type drive and pinion gears. The pinion gear is supported by three bearings, two in front of the pinion gears and one behind. The differential assembly has either two or four pinions depending upon the application.

Rear Axle Conversion

8½, 8⅞ in.—3300 to 3600 lbs.
9¾, 10½ in.—5500 lbs. (Dana)
10½ in.—5200 to 7200 lbs.
12¼ in.—11,000 lbs.

Differential Carrier Removal

1. With vehicle raised and supported securely, drain the lubricant.
2. Remove the axle shafts from the axle.
3. Remove the two drive shaft "U" bolts from the rear yoke and separate the rear universal joint from the yoke. *NOTE: The bearing can be left in place in the trunnion and can be secured with tape.*
4. Push the drive shaft to one side and secure it in place to the frame side rail.
5. Place a jack under the carrier assembly and remove the bolts and lock washers that hold the carrier to the axle housing and supporting the carrier with the jack roll it from under the truck.

Differential Carrier

Disassembly

1. Mount the carrier assembly in a bench vise or holding fixture.
2. On axles so equipped, loosen the ring gear thrust pad lock nut and remove the thrust pad.
3. Remove the differential adjusting nut locks and bearing cap bolts and lockwashers.
4. Mark the bearing caps and carrier for reassembly in the same position. Remove the bearing caps and adjusting nuts by tapping on the bosses of the caps until free from the dowels.
5. Remove the differential and ring gear assembly from the carrier.
6. Remove the bolts which attach the pinion bearing retainer to the carrier.
7. Remove the pinion and bearing assembly from the carrier. *NOTE: It may be necessary to drive this unit from the carrier. Use a brass drift against the pilot end of the pinion.*
8. On all axles except the 11,000 lb. remove shims from inside the carrier housing making note of the number and total thickness of shims removed.
9. On all axles except the 11,000

1 Universal joint yoke
2 Pinion bearing retainer and oil seal
3 Oil seal packing
4 Oil seal
5 Front pinion bearing
6 Drive pinion
7 Pinion bearing shim
8 Rear pinion bearing
9 Ring gear
10 Differential spider

11	Differential pinion (spider) gear	13	Differential case—left half		adjusting nut
12	Differential side gear	14	Axle shaft	17	Adjusting nut lock
		15	Differential bearing	18	Differential case—right half
		16	Differential bearing		

5200 and 7200 lb. axle—cross section (© Chevrolet Div., G.M. Corp.)

lb., the pinion rear bearing outer race and roller assembly is pressed into the carrier. Remove the assembly by driving it from its seat using a soft drift or punch.

Pinion and/or Bearing

Replacement

1. Clamp the pinion drive flange in a bench vise.
2. Remove the cotter pin, nut and washer from the end of the pinion.
3. Remove the drive flange and bearing retainer assembly.
4. Drive the oil seal, and packing if present, from the retainer.
5. On the 11,000 lb. axle, remove the pinion rear bearing snap ring and press the bearing from the pinion.
6. On the 13,500, 15,000 and 17,000 lb. axles press the inner race of the rear bearing from the pinion.
7. On the 5200 and 7200 lb. axles press the pinion from the bearing.
8. On the 11,000 lb. axle, press the front bearing from the pinion.

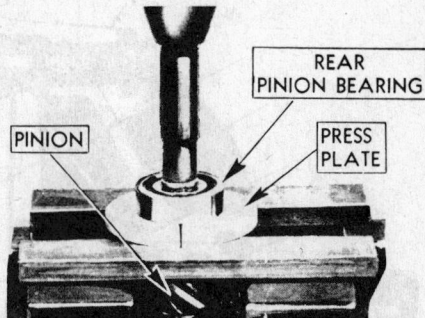

Pinion rear bearing removal (11,000 lb. axle)
(© Chevrolet Div., G.M. Corp.)

9. On the 13,500, 15,000 and 17,000 lb. axles press the front bearing from the pinion. Wash all the parts in solvent and inspect the pinion gear for signs of wear, chipping, pitting or scoring. Check the splines on the pinion shaft for signs of wear or distortion. Check the bearings for signs of wear, roughness or defects.
10. Soak the new oil seal (and packing) in engine oil. Install the felt packing, if so equipped, in the bottom of the retainer. Press the oil seal into the retainer.

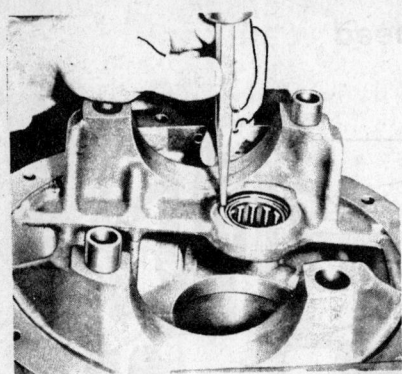

Pinion rear bearing removal—exc. 11,000 lb. axle
(© Chevrolet Div., G.M. Corp.)

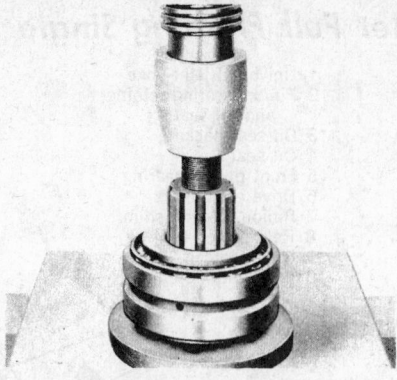

Drive pinion front bearing removal—typical
(© Chevrolet Div., G.M. Corp.)

Pinion inner bearing inner race removal
(13,500-15,000 and 17,000 lb. axles)
(© Chevrolet Div., G.M. Corp.)

11. On the 5200, 7200, 13,500, 15,000 and 17,000 lb. axles, lubricate the pinion rear bearing and press it into the carrier. Then install the inner race on the 13,500, 15,000 and 17,000 lb. axle pinion shaft.

12. On the 11,000 lb. axle, install the pinion rear bearing on the pinion shaft making sure that the chamfered side of the inner race seats against the shoulder on the

Pinion rear bearing inner race installation
(13,500 15,000 and 17,000 lb. axles)
(© Chevrolet Div., G.M. Corp.)

1 Universal joint yoke
2 Pinion bearing and oil seal retainer
3 Oil seal
4 Gasket
5 Front pinion bearing
6 Drive pinion
7 Pinion Bearing shim
8 Rear pinion bearing
9 Ring gear thrust pad
10 Ring gear
11 Differential spider
12 Differential pinion (spider) gear

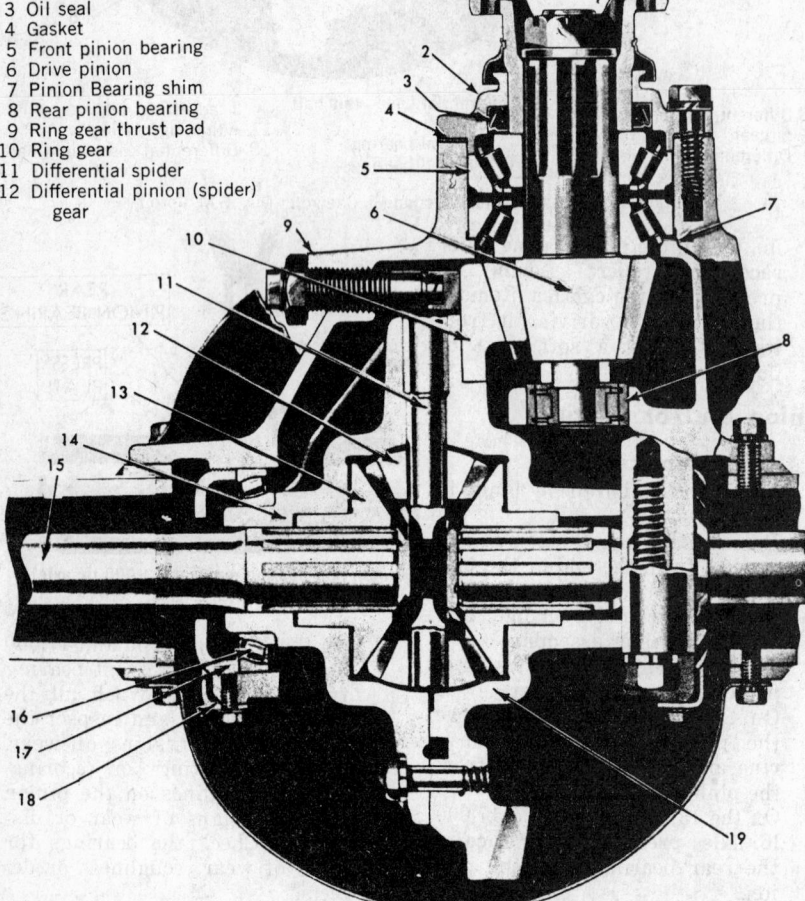

pinion shaft. Then install the pinon bearing lock ring.

13. On the 11,000 lb. axle, position the one piece double row ball bearing on the pinion shaft so that the extended portion of the inner race is toward the pinion head. Press the bearing onto the shaft until it seats against the pinion head.

14. On the 5200, 7200, 13,500, 15,000 and 17,000 lb. axles, place one cone and roller assembly on the pinion shaft so that the large end of bearing is toward the pinion. Then position the outer race, spacer and cone and roller assembly on the pinion shaft. Press the bearing until it seats against the pinion head.

15. Slide the oil seal retainer on the pinion shaft, then tap the flange onto the splines.

16. Clamp the drive flange in a bench vise and install the flange washer and nut. Torque the nut to 220 ft. lbs. and install the cotter pin without backing off on the nut.

Differential

Disassembly

1. Check the differential case to make sure that the two halves are marked so they may be reassembled in the same relation.

2. Remove the bolts holding the case and cover together.

3. Separate the cover from the case and remove the differential side gears and thrust washers, pinion gears with thrust washers and differential spider.

4. Remove the ring gear from the case by tapping the back of the gear with a soft faced hammer.

Assembly

1. Install two guide pins (made from cap screws with the heads cut off and ends slotted) to the new gear opposite each other.

2. Start the guide pins through the case flange and tap the ring gear on the case.

13 Differential side gear
14 Differential case—left half
15 Axle shaft
16 Differential bearing
17 Differential bearing
18 Adjusting nut lock
19 Differential case—right half
adjusting nut

5200 and 7200 lb. axle—with adjusting screw (© Chevrolet Div., G.M. Corp.)

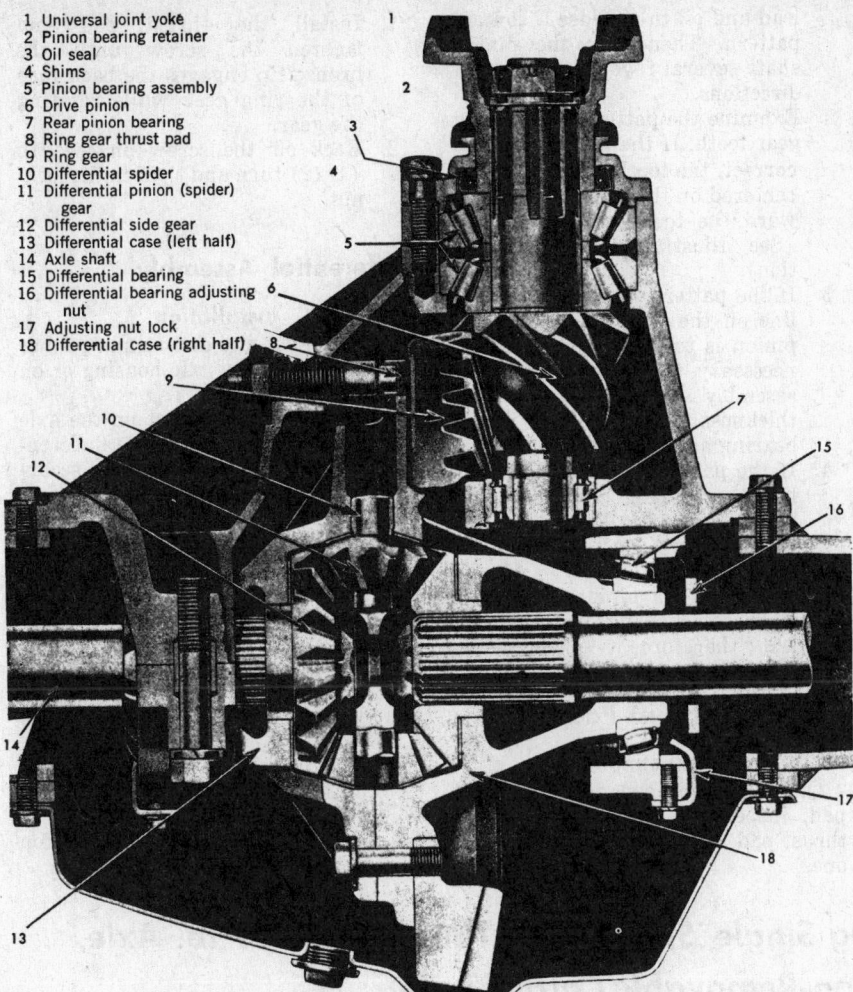

1 Universal joint yoke
2 Pinion bearing retainer
3 Oil seal
4 Shims
5 Pinion bearing assembly
6 Drive pinion
7 Rear pinion bearing
8 Ring gear thrust pad
9 Ring gear
10 Differential spider
11 Differential pinion (spider) gear
12 Differential side gear
13 Differential case (left half)
14 Axle shaft
15 Differential bearing
16 Differential bearing adjusting nut
17 Adjusting nut lock
18 Differential case (right half)

13,500, 15,000 and 17,000 lb. axle—cross section (© Chevrolet Div., G.M. Corp.)

3. Lubricate the differential side gears, pinions and thrust washers.
4. Place the differential pinions and thrust washers on the spider.
5. Assemble the side gears and pinions and thrust washers to the left half of the case.

6. Assemble the case halves being sure to line up the marks on the two halves.
7. Install the differential to ring gear bolts and lock washers and tighten evenly until the ring gear is flush with the case flange.
8. Remove the two guide pins and install the remaining two bolts.
9. Torque all bolts to specifications:
   Except below—110 ft. lbs.
   15,000 lb. and 17,000 lb. axles—160 ft. lbs.

Differential Bearing

Replacement

1. Using a bearing puller, remove the bearings from the case.
2. Place the new bearing on the hub of the case, thick side of inner race toward case. Using a bearing driver, drive the bearing into place.

Differential Carrier

Assembly

To facilitate adjusting the pinion depth in the ring gear, there are five

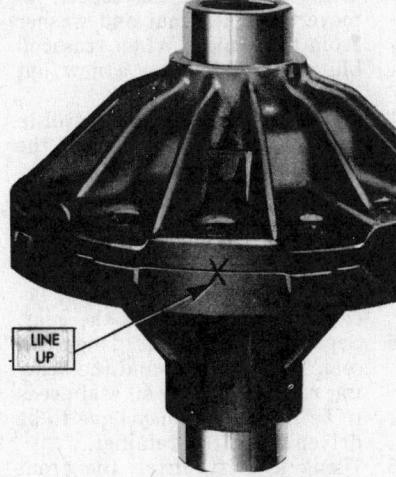

Differential case line up marks
(© Chevrolet Div., G.M. Corp.)

shims available for service use. They are .012″, .015″, .018″, .021″ and .024″.

NOTE: Pinion depth adjustment shims are not required for the 11,000 lb. axle.

If the original ring gear and pinion are to be used it is advisable to replace the same thickness of shims in the carrier counter bore that were removed.

If a new ring gear and pinion are used, one .021″ shim should be used as a standard starting set up.

1. Place the shim in the bore in the carrier.
2. Place a new pinion bearing retainer gasket on the retainer and install the pinion assembly in the carrier.
3. Install and tighten the pinion bearing retainer bolts and lock ing rollers with engine oil and washers.
4. Lubricate the differential bearplace the outer races over them.
5. Install the differential assembly in the carrier and install the adjusting nuts.
6. Install the differential bearing caps making sure the marks on the caps line up with the marks on the carrier.
7. Install the bearing cap bolts and lock washers and tighten until the lock washers just flatten out.

Ring Gear and Pinion

Adjustment

1. With the differential bearing cap bolts loosened just enough to permit turning the bearing adjusting nuts, remove all lash between the ring gear and pinion.
2. Back off the left hand adjusting nut one or two notches to a locking position.
3. Tighten the right hand adjusting nut firmly to force the differential in solid contact with the left hand adjusting nut.

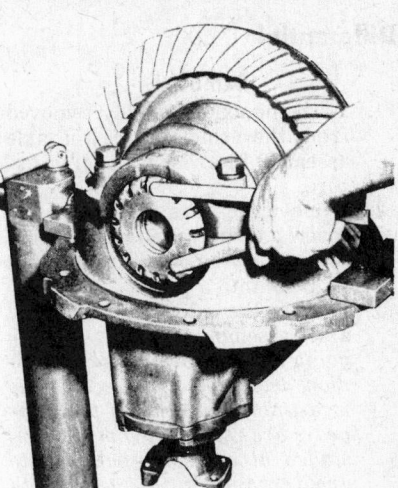

Ring gear and pinion adjustment
(© Chevrolet Div., G.M. Corp.)

4. Back off the right hand adjusting nut and again tighten snugly against the bearing.
5. Tighten the right hand adjusting nut from one to two additional notches to a locking position. *NOTE: This method of adjustment provides for proper preload of the bearings.*
6. Mount a dial indicator on the carrier and check the backlash between the ring gear and pinion Backlash should be .003" to .012" (.005" to .008" preferred). If the backlash is more than .012" loosen the right hand adjusting nut one notch and tighten the left hand adjusting nut one notch. If the backlash is less than .003" loosen the left hand adjusting nut one notch and tighten the right hand adjusting nut one notch.
7. Tighten the bearing cap bolts to specifications.
Except below—205 ft. lbs.
5200 and 7200 lb. axles—
100 ft. lbs.
8. Install the side bearing adjusting nut locks and torque to 15 ft. lbs.

Checking Pinion Depth

1. Coat the ring gear teeth lightly and evenly with a mixture of red lead and oil to produce a contact pattern. Then turn the pinion shaft several revolutions in both directions.
2. Examine the pattern on the ring gear teeth. If the pinion depth is correct, the tooth pattern will be centered on the pitch line and toward the toe of the ring gear. (See Illustration in this section).
3. If the pattern is below the pitch line on the ring gear teeth, the pinion is too deep and it will be necessary to remove the pinion assembly and increase the shim thickness between the pinion bearing and the carrier.
4. If the pattern is above the pitch line on the ring gear teeth, the pinion is too shallow and it will be necessary to decrease the shim thickness.
5. Changing the pinion depth will make some change in the backlash; therefore, it will be necessary to readjust the backlash.

Ring Gear Thrust Pad

Adjustment

On axles equipped with a thrust pad, inspect the bronze tip of the thrust pad and if worn install a new one.

1. Install the thrust pad and tighten the screw until the bronze tip engages the back face of the ring gear while rotating the gear.
2. Back off the screw one-twelfth (1/12) turn and tighten the locknut.

Differential Assembly

Installation

1. Clean out any dirt or sludge that may be in the axle housing or on the cover.
2. Install a new gasket on the axle housing and install the differential carrier to the housing securing it in place with the lockwashers and nuts. Torque the nuts to 70 ft. lbs.
3. Install the housing cover, if it was removed, using a new gasket.
4. Reconnect the drive shaft to the yoke on the differential and torque the "U" bolts to 20 ft. lbs.
5. Install the axle shafts in the axle.
6. Fill the axle with gear lubricant until the level is even with the bottom of the filler hole.
7. Road test the vehicle to check for noise and proper operation.

Chevrolet-Full Floating Single Speed 5200 Through 17,000 lb. Axle, Non-Removable Carrier Type

This axle is a full floating type that uses special Hypoid type drive and pinion gears. The pinion gear is supported by three bearings, two in front of the pinion gear and one behind. The differential assembly has either two or four pinions depending on the application of the axle. This axle assembly must be removed from the vehicle to remove and service the differential.

Differential

Removal

1. With the axle assembly removed from the vehicle, place the axle assembly in a vise or holding fixture.
2. Remove the bolts that retain the cover assembly and remove the cover, allowing the gear lubricant to drain into a pan.
3. Remove the axle shafts from the axle assembly. *NOTE: Before going any further, check the pinion backlash and record the measurement so that if the same gears are reused they may be installed at the same backlash to avoid changing the gear tooth pattern.*
4. From the bearing caps, remove the adjusting nut lock retainers.
5. Mark the bearing caps so they may be reinstalled in the same position and remove the bearing caps.
6. Loosen the side bearing adjusting nut and remove the differential carrier from the axle housing.

Pinion Assembly

Removal

1. Remove the differential assembly from the axle.
2. Check the pinion bearing for the proper preload. The force required to turn the pinion should be 25-35 in. lbs. for new bearings and 5-15 in. lbs. for used bearings. If there is no reading, shake the companion flange to check for any looseness in the bearing. If there is any looseness present the bearing should be replaced.
3. Remove the retaining bolts for the pinion bearing from the axle housing.
4. Remove the bearing retainer and pinion assembly from the axle housing. It may be necessary to tap the pilot end of the pinion shaft to help remove the pinion assembly from the carrier.
5. Record the thickness of the shims that are removed from between the carrier assembly and the bearing retainer assembly.

Drive Pinion

Disassembly

1. With the pinion assembly clamped in a vise, install a holder assembly on the flange.
2. Using the proper size socket, remove the pinion nut and washer from the pinion. When reassembling the pinion use a new nut and washer assembly.
3. With the holder assembly still in place, use a puller to remove the flange from the pinion.
4. With the bearing retainer supported in a press, press the pinion out of the retainer assembly. Be careful not to allow the pinion gear to fall onto the floor because this can damage the gear.
5. Separate the pinion flange, oil seal, front bearing and the bearing retainer. If the oil seal needs to be replaced it may have to be driven from the retainer.
6. Using a drift, drive the front and rear bearing cups from the bearing retainer.

7. Support the pinion assembly in a press, with the bearing supported, Press the bearing from the pinion gear.

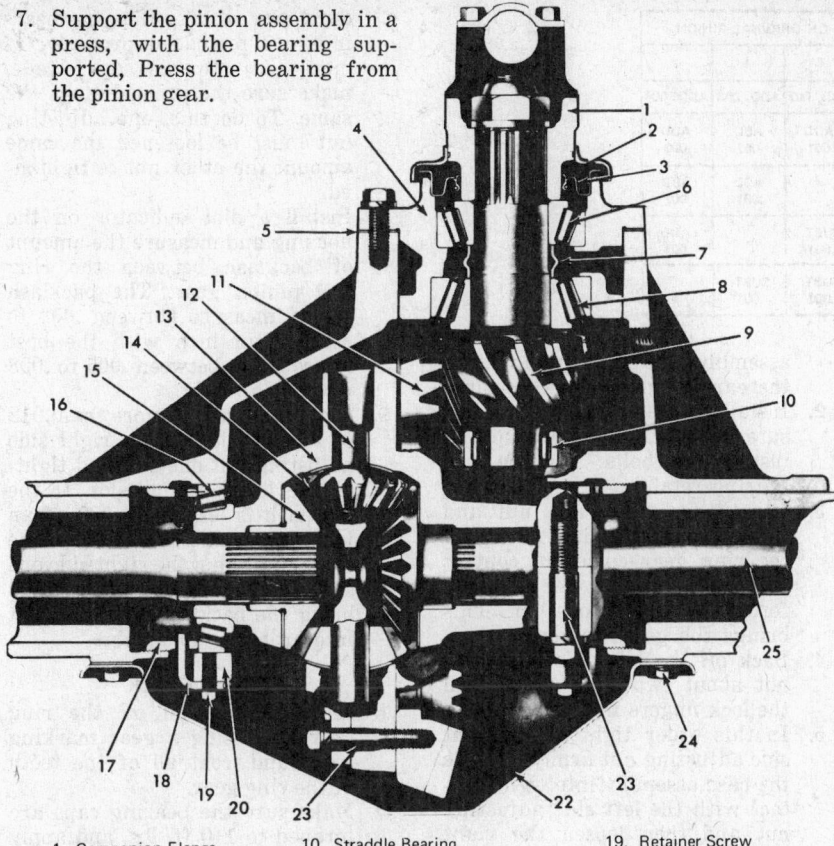

1. Companion Flange
2. Oil Deflector
3. Oil Seal
4. Bearing Retainer
5. Shim
6. Pinion Front Bearing
7. Collapsible Spacer
8. Pinion Rear Bearing
9. Drive Pinion
10. Straddle Bearing
11. Ring Gear
12. Differential Spider
13. Differential Case
14. Differential Pinion
15. Differential Side Gear
16. Side Bearing
17. Side Bearing Adjusting Nut
18. Adjusting Nut Retainer
19. Retainer Screw
20. Bearing Cap
21. Case-to-Ring Gear Bolt
22. Differential Cover
23. Bearing Cap Bolt
24. Cover Screw
25. Axle Shaft

Chevrolet 5,200-17,000 lb. axle cross-section (© Chevrolet Div., G.M. Corp.)

8. Using a drift, drive the straddle bearing from the carrier assembly.

Cleaning and Inspection

1. Clean off all the parts in solvent and blow dry.
2. Check the pinion gear for signs of wear, chips, cracks or any other imperfections. Check the splines for signs of wear or distortion.
3. Check the bearings for signs of wear or pitting on the rollers and races and check the bearing cage for dents and bends. Check the bearing retainer for any cracks, pits, grooves or corrosion.
4. Check the pinion flange splines for any signs of wear or distortion.
5. Replace parts that show any of the signs mentioned above.

Differential Case

Disassembly

1. Scribe a line across the two halves of the differential case so they may be reassembled in the same position, and with the ring gear removed, separate the two halves. To remove the ring gear, remove the ring gear bolts and washers, and using a soft hammer tap the ring gear from the case.
2. Remove the internal parts from the inside of the case and set them aside in order that they may be reassembled in the same position.

Cleaning and Inspection

1. Check the differential gears, pinions, thrust washers and spider for any signs of unusual wear, chips, cracks or pitting.
2. Check all mating surfaces for signs of wear.
3. Replace parts that show any of the signs mentioned above.

Differential Case

Assembly

1. Using a good quality gear lubricant coat all of the parts.
2. Assemble the differential pinions and thrust washers onto the spider and install the assembly into the differential case.
3. Line up the scribe marks on the two halves of the differential case

and install the ring gear. Install the ring gear washers and bolts and torque the bolts to approx. 110 ft. lbs.

Side Bearing

Replacement

1. Install a bearing puller on the bearing and remove the bearing assembly from the differential case.
2. Check the bearings for any signs of wear on distortion.
3. Install the new bearing by setting it in place on the differential case and, using a bearing driver, drive the bearing onto the case assembly until it seats against the shoulder on the case.

Drive Pinion

Assembly and Adjustment

1. Coat all of the parts with a good quality gear lubricant.
2. With the pinion gear in a press, press the rear bearings onto the pinion assembly.
3. In the bearing retainer, install the front and rear bearing cups using a driver of the proper size.
4. In the axle housing, install the straddle bearing assembly using the proper size driver.
5. Install the bearing retainer with the bearing cups in place on the pinion gear and install a new collapsible spacer.
6. Press the front bearing onto the pinion gear.
7. Lubricate the oil seal with a good quality high pressure grease and install the seal into the retainer bore. Be sure to press the seal down until it rests against the internal shoulder.
8. Install the pinion flange and oil deflector onto the splines of the pinion gear and install a new lock washer and pinion nut.
9. With the pinion flange clamped in a vise and a holder assembly installed on the flange, tighten the nut to obtain the proper preload. Measure the amount of torque required to turn the pinion gear. For a new bearing the torque required is 25-35 in. lbs. and for an old bearing it is 5-15 in. lbs. To preload the bearing, tighten the pinion nut to approx. 350 ft. lbs. and take a reading of the torque required to turn the pinion. Continue tightening the nut until the proper preload is obtained.

CAUTION: Do not tighten the nut too tightly because it will collapse the spacer too much. This will make replacement necessary.

		CODE NUMBER ON ORIGINAL PINION				
		+2	+1	0	-1	-2
CODE NUMBER ON SERVICE PINION	+2	—	ADD .001	ADD .002	ADD .003	ADD .004
	+1	SUBT. .001	—	ADD .001	ADD .002	ADD .003
	0	SUBT. .002	SUBT. .001	—	ADD .001	ADD .002
	-1	SUBT. .003	SUBT. .002	SUBT. .001	—	ADD .001
	-2	SUBT. .004	SUBT. .003	SUBT. .002	SUBT. .001	—

Drive Pinion Assembly

Installation

1. If installing a new pinion gear, check the top of the new gear for the depth code number.
2. Compare the new number with the old number on top of the old pinion and check the pinion depth chart for preliminary setting of the pinion depth.
3. Check the thickness of the original shims removed from the pinion and either add or subtract from the shims according to the chart.
4. Place the shim on the carrier assembly and line the holes up with those in the axle housing. Make sure the surfaces are clean of all dirt and grease.
5. Install the retainer and pinion assembly in the housing making sure the holes line up and install the retaining bolts. Torque the bolts to approx. 45 ft. lbs.

Differential Case

Installation and Adjustments

1. Place the bearing cups over the side bearings on the differential assembly and place the unit into the carrier in the axle housing.
2. Install the bearing caps making sure the marks are lined up and install the bolts. Tighten the bearing retaining bolts.
3. Loosen the right side nut and tighten the left side nut until the ring gear comes in contact with the pinion gear. Do not force the gears together. This brings the gears to zero lash.
4. Back off the left side adjusting nut about two slots and install the lock fingers into the nut.
5. In this order tighten the right side adjusting nut firmly to force the case assembly into tight contact with the left side adjusting nut and then loosen the right side nut until it is free from the bearing.
6. Again retighten the right side adjusting nut until it comes in contact with the bearing. Tighten the right adjusting nut about two slots if it is an old bearing or three slots if it is a new bearing.
7. Install the lock retainers into the slots and torque the bearing cap bolts to 100 ft.-lbs. This procedure now insures that the bearings are preloaded properly. If more adjustments are made, make sure the preload stays the same. To do this, one adjusting nut must be loosened the same amount the other nut is tightened.
8. Install a dial indicator on the housing and measure the amount of backlash between the ring and pinion gear. The backlash should measure between .003 to .012 of an inch with the best figure being between .005 to .008 of an inch.
9. If the backlash is more than .012 of an inch, loosen the right side adjusting nut one slot and tighten the left side one slot. If the backlash is less than .003 of an inch, loosen the left side nut one slot and tighten the right side one slot. These adjustments should bring the backlash measurement into an acceptable range.

Pattern Check

1. Clean all the oil off the ring gear and using a gear marking compound, coat all of the teeth of the ring gear.
2. Make sure the bearing caps are torqued to 110 ft. lbs. and apply load to the gears while rotating the pinion. Rotate the ring gear one full turn in both directions. *NOTE: Load must be applied to the assembly while rotating or the pattern will not show completely.*
3. Check the pattern on the ring gear and following the chart, adjust the assembly to get the contact pattern located centrally on the face of the ring gear teeth.

Chevrolet-Semifloating Single Speed, Non-Removable Carrier Type 2400 to 3500 lb. Axle

This axle assembly is the semifloating type with Hypoid type drive pinion and ring gears. The drive pinion gear is supported by two bearings. The differential case contains two pinion gears. The carrier assembly is not removable since it is part of the axle assembly but the design allows for the differential assembly to be serviced while the axle is still in the vehicle. The ring gear is bolted to a one piece differential case that is supported by two preloaded roller bearings.

Differential Case

Removal

1. Remove the inspection cover from the axle housing and drain the gear lubricant into a pan.
2. Remove the screw or pin that holds the pinion shaft in place and remove the shaft.
3. Push the axle shafts in a little and remove the "C" locks from the ends of the shafts. Remove the axle shafts from the housing.
4. Before going any further, the backlash should be measured and recorded. This will allow the old gears to be reassembled at the same amount of lash to avoid changing the gear tooth pattern. It also helps to indicate if there is gear or bearing wear, and if there is any error in the original backlash setting.
5. Roll the differential pinions and thrust washers out of the case and also remove the side gears and thrust washers. Make sure to mark the pinions and side gears so they can be reassembled in their original position.
6. Mark the bearing caps and housing and loosen the retaining bolts. Tap the caps lightly to loosen them. When the caps are loose, take the bolts all the way out and then reinstall the bolts just a few turns. This will keep the case from falling out of the housing when it is pried loose.
7. With a pry bar, very carefully pry the case assembly loose. Be careful not to damage the gasket surface on the housing when prying. The case assembly may suddenly come free if the bearings were preloaded, so pry very slowly.
8. When the case assembly is loose, remove the bolts for the bearing caps and remove the caps. Place the caps so they may be

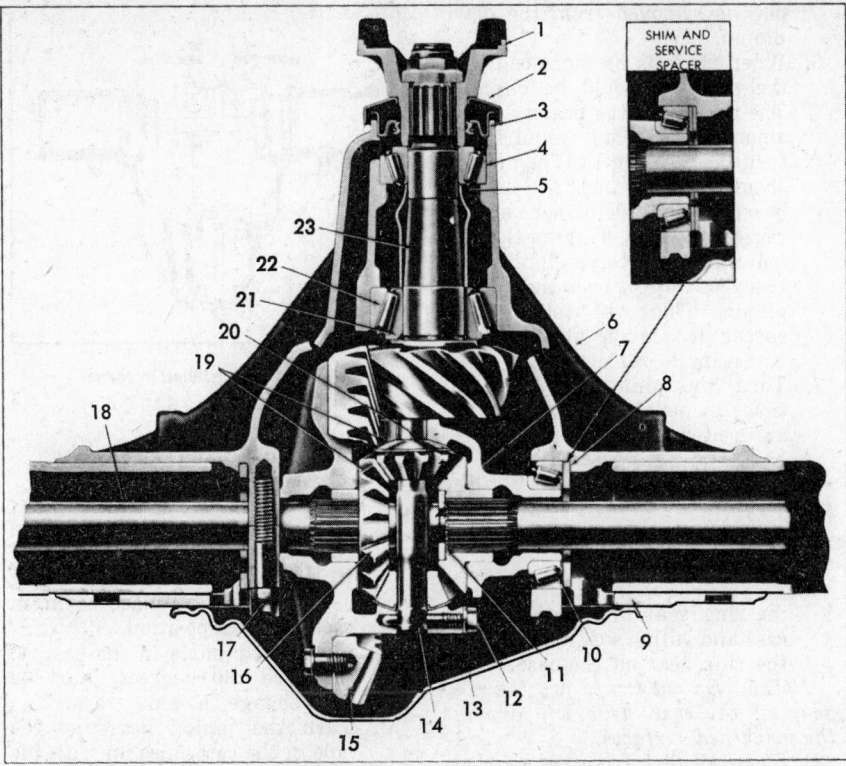

Chevrolet 2,400-3,600 lb. axle cross section (© Chevrolet Div., G.M. Corp.)

1. Companion Flange
2. Deflector
3. Pinion Oil Seal
4. Pinion Front Bearing
5. Pinion Bearing Spacer
6. Differential Carrier
7. Differential Case
8. Shim (A) with Service Shim
9. Gasket
10. Differential Bearing
11. "C" Lock
12. Pinion Shaft Lock Bolt
13. Cover
14. Pinion Shaft
15. Ring Gear
16. Side Gear
17. Bearing Cap
18. Axle Shaft
19. Thrust Washer
20. Differential Pinion
21. Shim
22. Pinion Rear Bearing
23. Drive Pinion

reinstalled in the same position. Place any shims that are removed with the cap they were removed from.

Drive Pinion

Removal

1. With the differential removed, check the pinion preload. Do this by checking the amount of torque needed to turn the pinion gear. For a new bearing, it should be 20-25 in. lbs., and for a used bearing it should be 10-15 in. lbs. If there is no preload reading check the pinion for looseness. If there is any looseness the bearing should be replaced.
2. With a holder assembly installed on the flange, use a socket of the proper size and remove the flange nut and washer.
3. Remove the flange by using a puller assembly and drawing the flange off the pinion splines.
4. Thread the pinion nut a few turns onto the pinion shaft. Using a brass drift and hammer, lightly tap the end of the pinion shaft to remove the pinion from the carrier. Be careful not to allow the pinion to fall out of the carrier after it breaks loose.
5. With the pinion removed from

the carrier, discard the old seal, pinion nut and collapsible spacer and install new ones when reassembling.

Cleaning and Inspection

1. Clean all parts in solvent and blow dry.
2. Check all of the parts for any signs of wear, chips, cracks or distortion. Replace any parts that are defective.
3. Check the fit of the differential side gears in the case and the fit of the side gear and axle shaft splines.

Differential Bearing

Replacement

1. With a bearing puller attached to the bearing, pull the bearing from the case.
2. Place the new bearing on the case hub with the thick side of the inner race toward the case. Using a bearing driver, drive the bearing onto the case until it seats against the shoulder on the case.

Drive Pinion Bearing

Replacement and Adjustment

1. Depending on the bearing that is being replaced, remove the front

or rear bearing cup from the carrier assembly.
2. With the pinion gear mounted in a press, press the rear bearing from the pinion shaft. Be sure to record the thickness of the shims that are removed from between the bearing and the gear.
3. Using a bearing driver of the proper size, install a new bearing cup for each one that was removed. Make sure the cups are seated fully against the shoulder in the housing.
4. The pinion depth must now be checked to determine the nominal setting. This allows for machining variations in the housing and enables you to select the proper shim so that the pinion depth can be set for the best gear tooth contact.
5. Clean the housing and carrier assemblies to insure accurate measurement of the pinion depth.
6. Lubricate the front and rear pinion bearings with gear lubricant and install them in their races in the carrier assembly.
7. Using a Pinion Setting Gauge, select the proper clover leaf plate, and install it on the preload stud.
8. Insert the stud through the rear bearing, with the proper size pilot on the stud, and through the front bearing using the proper pilot. Install the hex nut and tighten it until it is just snug.
9. Holding the preload stud with a wrench, tighten the hex nut until 20 in. lbs. of torque are required to rotate the bearings.
10. Install the side bearing discs on the ends of the arbor assembly, using the step of the disc that fits the bore of the carrier.
11. Install the arbor and plunger assembly into the carrier. Make sure the side bearing discs fit properly.
12. Install the bearing caps in the carrier assembly finger tight to make sure the discs do not move.
13. Mount a dial indicator on the mounting post of the arbor. Have the contact button resting on the

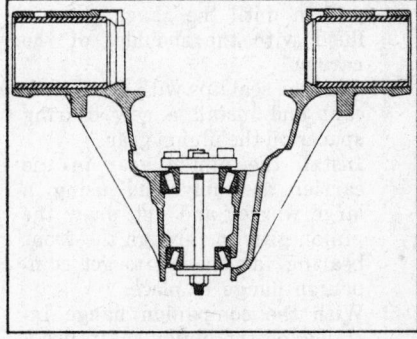

Gauge plate installed

(© Chevrolet Div., G.M. Corp.)

top surface of the plunger.

14. Preload the dial indicator by turning it one-half revolution and tightening it in this position.

15. Use the button on the gauge plate that corresponds to the ring gear size and turn the plate so the plunger rests on top of it.

16. Rock the plunger rod back and forth across the top of the button until the dial indicator reads the greatest amount of variation. Set the dial indicator to zero at the point of most variation. Repeat the rocking of the plunger several times to check the setting.

17. Turn the plunger until it is removed from the gauging plate button. The dial indicator will now read the pinion shim thickness required to set the "nominal pinion depth." Make a note of the reading.

18. Check for the pinion code number on the rear face of the pinion gear being used. This number will indicate the necessary change to the pinion shim thickness. If the pinion is marked with a plus (+) and a number, add that much to the reading you got from the dial indicator. If the pinion has no mark, use the reading from the dial indicator as the correct shim thickness. If the pinion is marked with a minus (−) and a number, subtract that much from the reading on the dial indicator.

19. Remove the depth gauging tools from the carrier assembly and install the proper size shim on the pinion gear.

20. Lubricate the bearing with gear lubricant and using a press, press the bearing into place on the pinion shaft.

Pinion Gear

Installation and Adjustment

1. Lubricate the front bearing with gear lubricant and install it in the front cup.

2. Install the pinion seal in the bore. Using a seal driver and the proper size gauge plate, drive the seal in until the gauge plate is flush with the shoulder of the carrier.

3. Coat the seal lips with gear lubricant and install a new bearing spacer on the pinion gear.

4. Install the pinion gear in the carrier assembly and using a large washer and nut, draw the pinion gear in through the front bearing far enough to get companion flange in place.

5. With the companion flange installed on the pinion shaft, use a holder assembly and tighten the pinion nut until all of the end

play is removed from the drive pinion.

6. When there is no more end play the preload should be checked. The preload of the bearing is the amount of torque required to turn the pinion gear. The preload should be 20-25 inch. lbs. on new bearings and 10-15 inch. lbs. on reused bearings. Tighten the pinion nut until these figures are reached. *Do Not* over tighten the pinion. This will collapse the spacer too much and make it necessary to replace it.

7. Turn the pinion gear several times to make sure the bearings are seated and recheck the preload.

Ring Gear

Replacement

1. Remove all of the bolts that hold the ring gear to the differential case and with a soft hammer, tap the ring gear off the case.

NOTE: Do not try to pry the ring gear off the case. This will damage the machined surfaces.

2. Clean all dirt from the case assembly and lubricate the case with gear lube. Align the ring gear bolt holes with the holes in the carrier and lightly press the ring gear onto the case assembly. Install all of the bolts and tighten them all evenly, using a criss-cross pattern to avoid cocking the ring gear.

3. When the ring gear is firmly seated against the case, tighten the bolts to 60 ft. lbs.

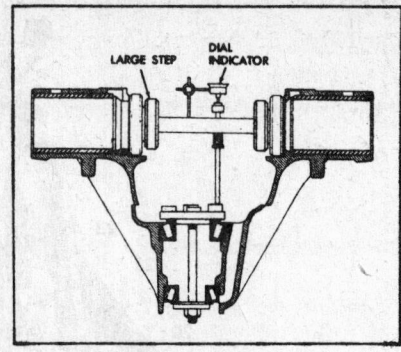

Gauge tools installed in carrier
(© Chevrolet Div., G.M. Corp.)

Differential Case Assembly

Installation and Adjustment

1. Install the thrust washers and side gears into the case assembly. If the original parts are being used, be sure to place them in their original position.

2. Place the pinions in the case so they are 180 degrees apart as they engage the side gears.

3. Turn the pinion gears so the hole in the case lines up with the holes in the gears. When the holes are aligned, install the pinion shaft and lock screw. Do not tighten the lock screw too tightly at this time.

4. Check the bearings, bearing cups, cup seat and carrier caps to make sure they are in good condition.

5. Lubricate the bearings with gear lube. Install the cups on the proper bearings and install the differential assembly in the car-

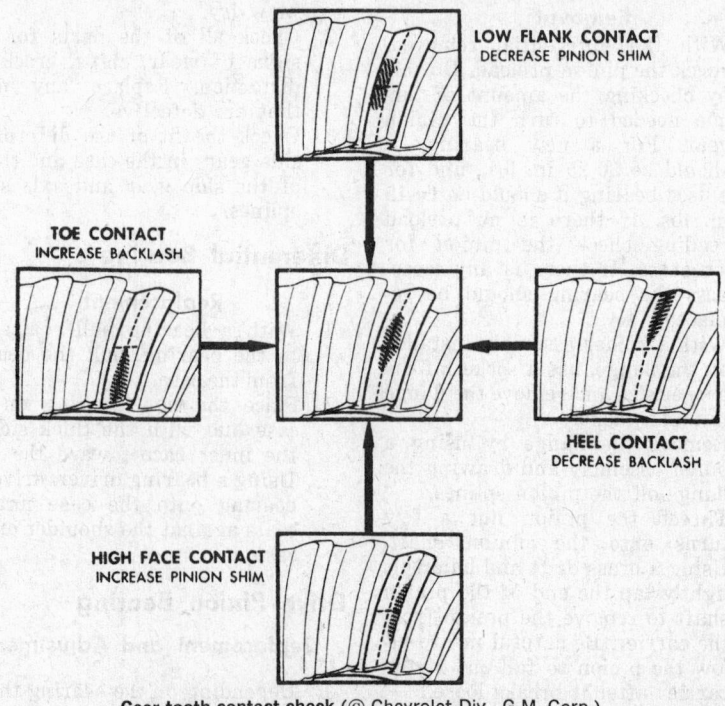

Gear tooth contact check (© Chevrolet Div., G.M. Corp.)

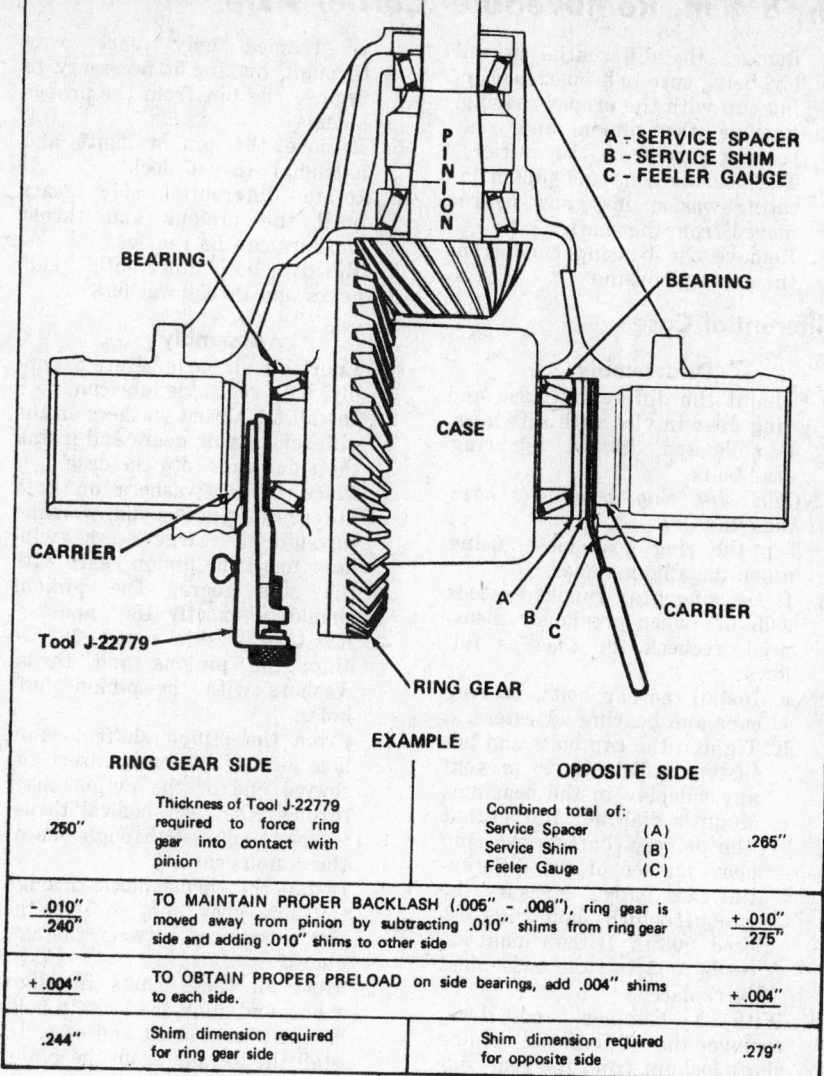

EXAMPLE

RING GEAR SIDE		OPPOSITE SIDE	
.250"	Thickness of Tool J-22779 required to force ring gear into contact with pinion	Combined total of: Service Spacer (A) Service Shim (B) Feeler Gauge (C)	.265"
- .010" / .240"	TO MAINTAIN PROPER BACKLASH (.005" - .008"), ring gear is moved away from pinion by subtracting .010" shims from ring gear side and adding .010" shims to other side		+ .010" / .275"
+ .004"	TO OBTAIN PROPER PRELOAD on side bearings, add .004" shims to each side.		+ .004"
.244"	Shim dimension required for ring gear side	Shim dimension required for opposite side	.279"

Shim pack selection chart (© Chevrolet Div., G.M. Corp.)

rier. Support the carrier assembly to keep it from falling.

6. Install a support strap on the left side bearing and tighten the bearing bolts to an even, snug fit.

7. With the ring gear tight against the pinion gear, insert a gauging tool between the left side bearing cup and the carrier housing.

8. While lightly shaking the tool back and forth, turn the adjusting wheel until a slight drag is felt. Tighten the lock nut.

9. Between the right side bearing and carrier, install a service spacer, .170 of an inch thick, a service shim and a feeler gauge. The feeler gauge must be thick enough so a light drag is felt when it is moved between the carrier and the shim.

10. Add the total of the service spacer, service shim and the feeler gauge. Remove the gauging tool from the left side of the carrier and using a micrometer, measure the thickness in at least

three places. Average the readings and record the result.

11. Refer to the chart to determine the proper thickness of the shim packs.

12. Install the left side shim first, then install the right side shim between the bearing cup and spacer. Position the shim so the chamfered side is outward or next to the spacer. If there is not enough chamfer around the outside of the shim, file or grind the chamfer a little to allow for easy installation.

13. If there is difficulty in installing the shim, partially remove the case from the carrier and slide both the shim and case back into place.

14. Install the bearing caps and torque them to 60 ft. lbs. Tighten the pinion shaft lock screw.

NOTE: *The differential side bearings are now preloaded. If any adjustments are made in later procedures, make sure not to change the*

preload. *Do Not change the total thickness of the shim packs.*

15. Mount a dial indicator on the carrier assembly with the indicator button perpendicular to the tooth angle and in line with the gear rotation.

16. Measure the amount of backlash between the ring and pinion gears. The backlash should be between .005-.008 of an inch. Take readings at four different spots on the gear. There should not be variations greater than .002 of an inch.

17. If there are variations greater than .002 of an inch between the readings, check the runout between the case and ring gear. The gear runout should not be greater than .003 in. If the runout does exceed .003 in. check the case and ring gear for deformation or dirt between the case and gear.

18. If the gear backlash exceeds .008 in., increase the thickness of the shims on the ring gear side and decrease the thickness of the shims on the opposite side, an equal amount.

19. If the backlash is less than .005 in., decrease the shim thickness on the ring gear side and increase the shim thickness on the opposite side an equal amount.

Gear Pattern Check

Before final assembly of the differential, a pattern check of the gear teeth must be made. This determines if the teeth of the ring and pinion gears are meshing properly, for low noise level and long life of the gear teeth. The most important thing to note is if the pattern is located centrally up and down on the face of the ring gear.

1. Wipe any oil out of the carrier and wipe all dirt and oil from the teeth of the ring gear.

2. Coat the teeth of the ring gear with a gear marking compound.

3. With the bearing caps torqued to 55 ft.-lbs., expand the brake shoes until it takes 20-30 ft. lbs. of torque to turn the pinion gear.

4. Turn the companion flange so the ring gear makes one full rotation in one direction, then turn it one full rotation in the opposite direction.

5. Check the pattern on the teeth and refer to the chart for any adjustments necessary.

6. With the gear tooth pattern checked and properly adjusted, install the axle housing cover gasket and cover and tighten securely. Fill the axle with gear lube to the correct level.

7. Road test the vehicle to check for any noise and proper operation of the rear.

Dodge/Plymouth 8¾ in. Removeable Carrier Axle

See the Dodge/Plymouth truck section for external identification and axle shaft service.

Differential

Removal

1. Raise and support vehicle under rear axle housing.
2. Remove wheels, drums and rear axle shafts, as described in truck section. Drain lubricant.
3. Disconnect rear universal joint and place out of way.
4. Remove attaching bolts and remove carrier assembly to bench.

Reconditioning

NOTE: Before disassembling the differential, check and record the side bearing play and ring gear run-out. Also, make a gear tooth contact pattern test, and measure pinion bearing preload.

1. Mount carrier in holding tool, with pinion flange up.
2. Remove pinion shaft nut and washer. Remove flange with puller.
3. With puller screwed into oil seal, pull seal from housing.
4. Rotate assembly with holding tool to allow oil slinger, shim pack and spacer (where used), to drop from the carrier.
5. Matchmark the differential pedestals, adjusting nuts, and bearing caps for reassembly identification.
6. Remove the bearing lockscrews and locks.
7. Remove the cap bolts, caps and back off bearing adjusting nuts slightly.

8. Remove the differential assembly, being sure to keep each bearing cup with the proper bearing.
9. Remove the pinion and rear bearing cone from the carrier.
10. The bearing cone and pinion locating washer may now be removed from the shaft.
11. Remove the bearing cups from the carrier housing.

Differential Case

Disassembly

1. Mount the differential case and ring gear in vise with soft jaws. Remove and discard the ring gear bolts.
   NOTE: the ring gear bolts have left-hand threads.
2. Tap the ring gear loose, using a non-metallic hammer.
3. If the ring gear runout exceeds .005 in., when previously measured, recheck the case as follows:
   a. Install the cap bolts, bearing caps and bearing adjusters.
   b. Tighten the cap bolts and adjusters sufficiently to prevent any sideplay in the bearings.
   c. Mount a dial indicator so that the pointer contacts the ring gear surface of the differential case flange. Measure the runout, which should not exceed .003 in. If the runout exceeds .003 in., the case must be replaced.
4. With a hammer and drift, remove the differential pinion shaft lockpin from the rear side of the ring gear flange. The hole

is reamed only part way through, making it necessary to remove the pin from the proper side.
5. Remove the pinion shaft and axle shaft thrust block.
6. Rotate differential side gears until the pinions and thrust washers can be removed.
7. Remove both differential side gears and thrust washers.

Assembly

1. Lubricate all parts before assembly, with rear axle lubricant.
2. Install the thrust washers on the differential side gears and install the side gears into the case.
3. Place thrust washers on both differential pinions and, working through the large access window, mesh the pinion gears with the side gears. The pinions should be exactly 180° apart.
4. Rotate the side gears 90° to align the pinions and thrust washers with the pinion shaft holes.
5. From the pinion shaft lockpin hole side of the case, insert the slotted end of the pinion shaft through the case, conical thrust washer and just through one of the pinion gears.
6. Install the thrust block through the side gear hub, so that the slot is centered between the side gears.
7. Hold all these parts in alignment, and align the lockpin holes in the pinion shaft and case. Install the lockpin from the pinion shaft side of the ring gear

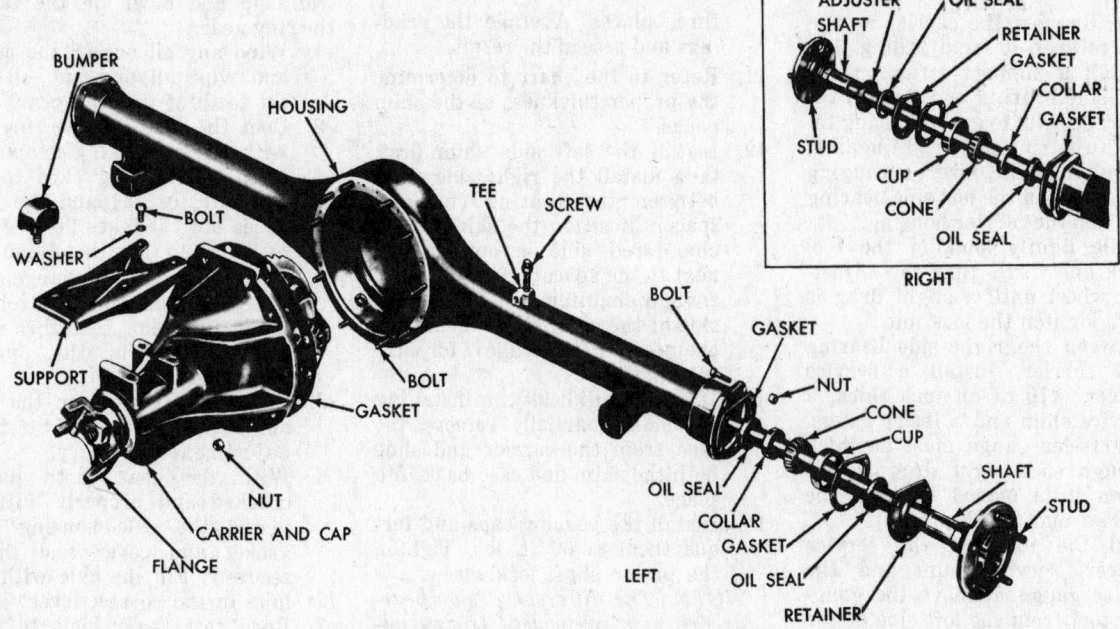

8¾ in. Dodge/Plymouth rear axle assembly (© Chrysler Corp.)

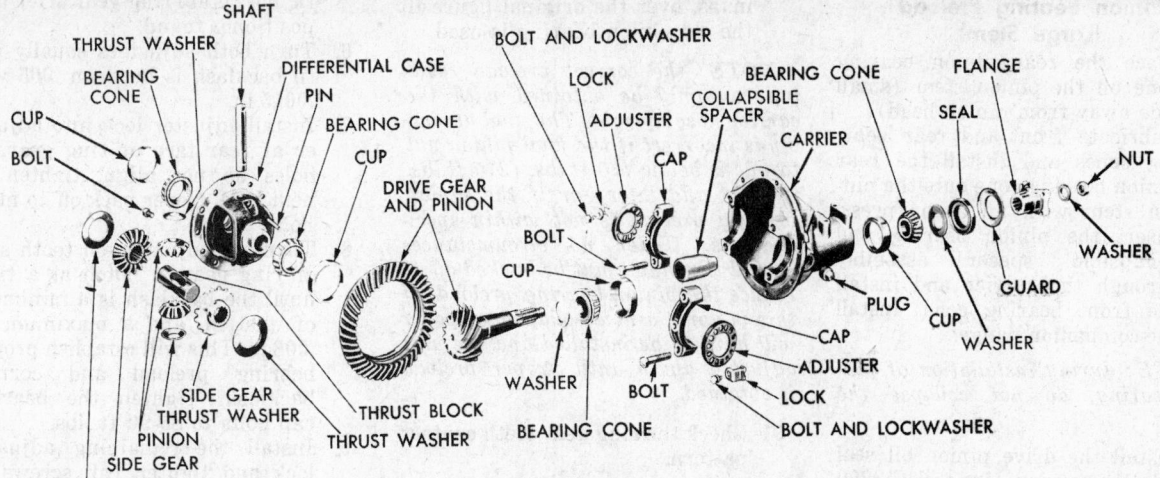

CUP
BOLT
THRUST WASHER
BEARING CONE
SHAFT
DIFFERENTIAL CASE
PIN
BEARING CONE
CUP
DRIVE GEAR AND PINION
BOLT
CUP
WASHER
BOLT
SIDE GEAR
THRUST WASHER
PINION
SIDE GEAR
THRUST WASHER
THRUST WASHER
THRUST BLOCK
THRUST WASHER
BEARING CONE

BOLT AND LOCKWASHER
LOCK
ADJUSTER
CAP
COLLAPSIBLE SPACER
BEARING CONE
CARRIER
FLANGE
SEAL
NUT
WASHER
GUARD
WASHER
CUP
PLUG
CAP
ADJUSTER
LOCK
BOLT AND LOCKWASHER

8¾ in. Dodge/Plymouth differential carrier assembly (© Chrysler Corp.)

flange.

8. With a stone, relieve the edge of the chamfer on the inside diameter of the ring gear.
9. Heat the ring gear (fluid bath or heat lamp) to a temperature not exceeding 300° F.

NOTE: do not heat ring gear with a torch.

10. Align the ring gear with the case. Insert new ring gear screws through the case flange and into the ring gear.
11. Alternately tighten each cap screw to 55 ft lbs.
12. Position each differential bearing cone on the hub of the differential case (taper away from ring gear) and install the bearing cones with an arbor press.
13. Install the pinion bearing cups squarely in the bores of the carrier.

Differential Assembly, Pinion Bearing Preload and Depth of Mesh

This type axle uses two types of pinions. Pinion depth of mesh and bearing preload are determined in the same manner for small and large stem pinions; only the sequence of adjustments varies. Small stem pinions require bearing preload adjustment first, while large stem pinions require pinion depth of mesh adjustment first. The position of the drive pinion, relative to the position of the ring gear (depth of mesh) is determined by the location of the bearing cup shoulders in the carrier and by the portion of the pinion behind the rear bearing.

NOTE: factory service procedures recommend the use of very specialized tools to assure proper adjustment of the rear axle without duplicating labor. However, the special tools are not readily available. The following procedures are substituted in place of special factory tools.

Assembly

1. Assembly procedures, excluding adjustments and selection of shims, are the reverse of disassembly.

NOTE: some rear axles of this type use a collapsible spacer. Once this spacer has been collapsed or the pinion bearing preload exceeded, the collapsible spacer must be replaced with a new one. Before assembly, gather together shims of several sizes and several spacers in case the assembly is not properly adjusted the first time.

If the differential assembly was satisfactory when disassembled, the drive pinion may be assembled with the original components, excepting the pinion front bearing and collapsible spacer (if used). If any replacement parts are installed, a complete adjustment will be necessary. Ring gears and pinions are available in matched sets only, and the adjustment position for best tooth contact is marked on the end of the pinion head.

Proper pinion setting, relative to the ring gear is determined by a shim, selected before the pinion is installed in the carrier. Pinion bearing shims are available in .002 in. increments (small stem) and in .001 in. increments (large stem with collapsible spacer).

Depth of Mesh

The head of the pinion is marked with a (+) or (−) followed by a number ranging from 0 to 4. If the old and new pinion have the same marking and the old bearing is being installed, use a shim of the original thickness. But, if the old pinion is marked (0), for example, and the new pinion is marked +2, try a .002 in. thinner shim. If the new pinion is marked −2, try a .002 in. thicker shim. The exact size of the pinion shim cannot be determined exactly

without special equipment. This method provides a starting point. When the unit is assembled, check the ring gear teeth contact pattern and adjust the shim size accordingly. The entire unit must be disassembled to change the shim size.

Pinion Bearing Preload (Small Stem)

1. If the bearings are being replaced, place the bearings in the carrier and drive into place with a drift.
2. Assemble the pinion shim (chamfered side toward the gear) onto the pinion stem. Install the tubular spacer if equipped, and the preload shims on the pinion stem.
3. Insert the pinion assembly into the carrier.
4. Install the front pinion bearing cone, U joint flange, Belleville washer (convex side up) and nut. Do not install the oil seal.
5. Tighten the flange nut to 240 ft lbs. (210 ft.lbs. in 1973 and later).
   Rotate the pinion to properly seat the bearing rollers. The preload torque required to turn the pinion with the bearings oiled, should be 20-30 in. lbs. with new bearings and 0-15 in. lbs. for used bearings. Use a thinner shim pack to increase preload and a thicker shim pack to decrease preload.
6. After the correct pinion depth has been established and the preload obtained, remove the drive pinion flange.
7. Lubricate and install the drive pinion oil seal.
8. Install the pinion flange, washer and nut. Torque the nut to 240 ft.lbs. (210 ft.lbs. in 1973 and later).

Pinion Bearing Preload (Large Stem)

1. Place the rear pinion bearing cone on the pinion stem (small side away from pinion head).
2. Lubricate front and rear bearing cones and install the rear pinion bearing cone onto the pinion stem with an arbor press.
3. Insert the pinion bearing and collapsible spacer assembly through the carrier and install the front bearing cone. Install the companion flange.

NOTE: during installation of pinion bearing, do not collapse the spacer.

4. Install the drive pinion oil seal into the carrier .Use tool C-3890 to insure proper seal depth.
5. Support the pinion in the carrier and install the anti-rattle washer.
6. Install the Belleville washer (convex side up) and pinion nut.
7. Hold the companion flange and tighten the pinion nut to remove all end-play, while rotating the pinion to insure proper bearing seating. Remove the tools and rotate the pinion several revolutions.
8. Torque the pinion nut to 170 ft. lbs. (210 ft.lbs. in 1973 and later). With an in. lb. torque wrench, measure the pinion bearing preload, which should be 20-30 in.lbs. for new bearings or 10 in.lbs. over the original figure if the old pinion bearing is used.

NOTE: the correct preload reading can only be obtained with the carrier nose upright. The final assembly is incorrect if the final pinion nut torque is below 170 ft.lbs. (210 ft.lbs. in 1973 and later) or if the pinion bearing preload is not within specifications. Under no circumstances should the pinion nut be backed off to reduce the pinion bearing preload. If this is done, a new collapsible spacer will have to be installed and the unit adjusted again until proper preload is obtained.

9. Check the ring gear teeth contact pattern.

Carrier Assembly

1. Install the differential bearing cups onto their respective bearings and position assembly in the carrier.
2. Install the caps and bolts and tighten bolts finger tight. Be sure all identification markings are properly aligned and positioned.
3. Tighten the adjusters enough to square the bearing cups with the bearings and eliminate end-play. Allow some backlash to remain.
4. Tighten one corresponding bolt on each cap to 85-90 ft lbs.
5. Set dial gauge to contact outer end of ring gear tooth and take readings at 90° intervals to find the spot with the least clearance. Do not rotate ring gear after this position is found.
6. Turn both adjusters equally until backlash is between .005 and .0015 in.
7. Install adjuster lock into adjuster at rear face of ring gear. If holes do not align, tighten to next hole. Never back off to meet hole.
8. Turn the adjuster, on tooth side of ring gear, a notch at a time until the backlash is a minimum of .006 in. and a maximum of .008 in. This will establish proper bearing preload and correct backlash. Tighten the bearing cap bolts to 85-90 ft. lbs.
9. Install the remaining adjuster lock and tighten cap screws to 15-20 ft.lbs.

Final Installation

1. Thoroughly clean the gasket surfaces of the carrier and housing.
2. Install the carrier on the housing using a new gasket. Tighten the nuts to 45 ft.lbs.
3. Install the axle shafts and adjust the axle shaft end-play if necessary.
4. Install the driveshaft and torque the screws to 15 ft.lbs.
5. Install the backing plates, hubs and drums and bleed and adjust the brakes (if necessary).
6. Fill the differential with SAE 90 lubricant to the bottom of the filler plug opening. Lower and road-test the vehicle.

Dodge/Plymouth 8⅜ and 9¼ in. Integral Carrier Axle

See the Dodge/Plymouth truck section for external identification and axle shaft service.

Differential

Removal

1. Keep a pencil and paper handy to record data.
2. Jack up the rear of the vehicle and remove the wheels, drums, and axle shafts.
3. Loosen the housing cover screws and drain the lubricant from the rear axle.
4. Remove the cover.
5. Clean the inside of the differential case with solvent and blow dry with compressed air.
6. Check for differential side-play by inserting a large screwdriver between the left side of the axle housing and the differential case flange. Using a prying motion, determine whether side-play exists. There should be no side-play.
7. Paint the ring gear teeth and make a gear tooth contact pattern. Determine if proper depth of mesh can be obtained.
8. If side-play was found in step six, proceed to step nine. If no side-play was found in step six, check the drive gear run-out. Mount a dial indicator and index the indicator stem at right angles to the rear face of the ring gear. Rotate the ring gear and mark the ring gear and case at the point of greatest run-out. Total indicator reading should not exceed 0.005 in. If it does, the possibility exists that the case must be replaced.
9. Measure and record the pinion bearing preload. Use an in. lb. torque wrench to measure the preload.
10. Remove the pinion nut, washer and pinion flange.
11. Remove and discard the pinion oil seal.
12. Match-mark the axle housing and the differential bearing caps.
13. Remove the threaded adjusters and the differential bearing caps. There is a special wrench to do this through the axle tube on late models.
14. Remove the differential case from the housing. The differential bearing cups and threaded adjusters must be kept together so they can be installed in their original position.

Disassembly

1. To remove the drive pinion or front bearing cone, drive the pinion rearward out of the bearing. This will result in damage to the bearing and cup. The bearing cone and cup must be replaced with new parts. Discard the collapsible spacer.
2. Drive the front and rear bearing cups from the housing with a brass drift. Remove the shim from behind the rear bearing cup and record the thickness. Discard the shim.
3. Remove the rear bearing cone from the pinion stem with a puller.
4. Clamp the differential case and ring gear in a vise with soft jaws.
5. Remove the ring gear bolts (left-hand thread). Tap the ring

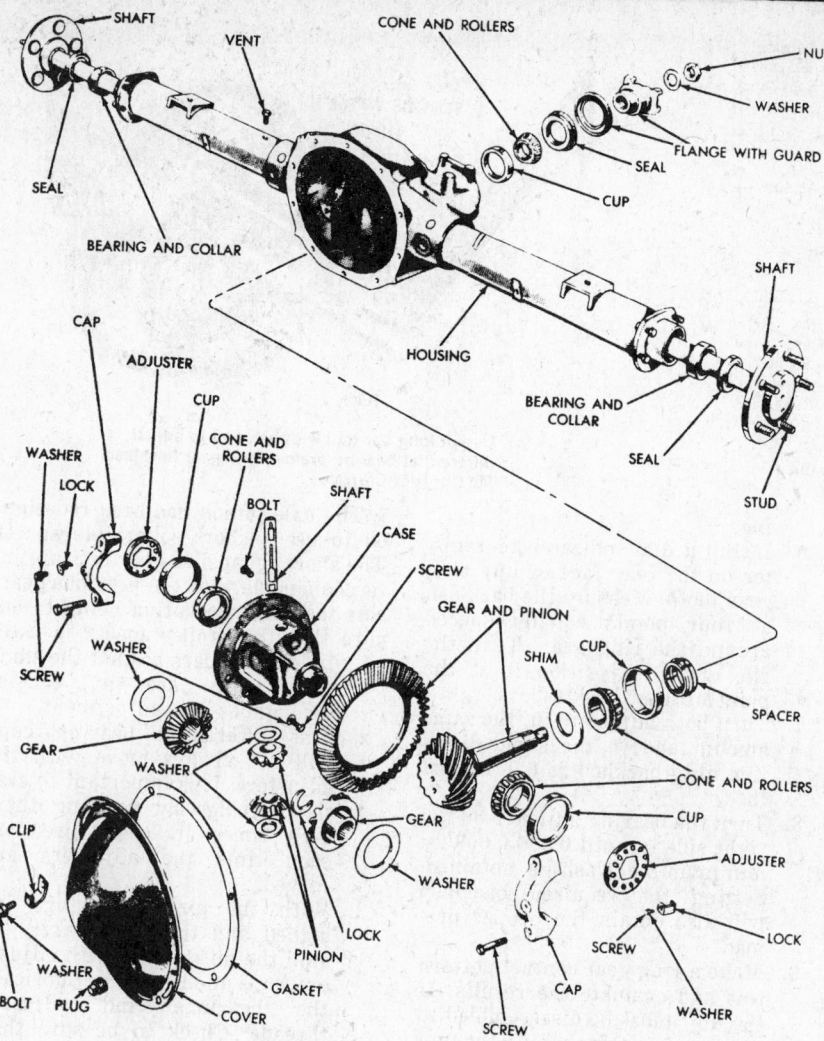

9¼ in. integral carrier Dodge/Plymouth axle (© Chrysler Corp.)

shaft side of the ring gear flange.

8. With a stone, relieve the edge of the chamfer on the inside diameter of the ring gear.

9. Heat the ring gear (fluid bath or heat lamp) to a temperature not exceeding 300° F.

NOTE: do not heat ring gear with a torch.

10. Align the ring gear with the case. Insert the ring gear screws through the case flange and into the ring gear.

11. Alternately tighten each cap screw to 55 ft. lbs.

12. Position each differential bearing cone on the hub of the differential case (taper away from ring gear) and install the bearing cones. An arbor press may be helpful.

Pinion Depth of Mesh

1. The proper pinion setting (relative to the ring gear) is determined by a shim which has been selected before the pinion is to be installed in the carrier. Pinion bearing shims are available in 0.001 in. increments.

2. The head of the pinion is marked with a "plus" (+) or a "minus" (−) mark that is followed by a number ranging from zero to four. If the old and new pinions have the same marking and the old bearing is being installed, use a shim of the original thickness. If the old pinion is marked zero (0), however, and the new pinion is marked plus two (+2), try a shim that is 0.002 in. thinner. If the new pinion is marked axle housing cup bore and install minus two (−2), try a shim that is 0.002 in. thicker.

3. Position the selected shim in the the rear bearing cup.

4. Place the rear pinion bearing cone on the pinion stem (small side away from pinion head).

5. Lubricate the front and rear bearing cones and install the rear pinion bearing cone onto the pinion stem with an arbor press.

6. Insert the pinion bearing and collapsible spacer assembly through the carrier and install the front bearing cone. Install the companion flange.

NOTE: during installation of the pinion bearing do not collapse the spacer.

7. Install the drive pinion oil seal into the carrier. Be sure to properly seat the seal.

8. Support the pinion in the carrier and install the anti-clang washer.

9. Install the Belleville washer (convex side up) and pinion nut.

10. Hold the companion flange and tighten the pinion nut to remove all end-play, while rotating the

gear loose with a soft-faced mallet.

6. If the ring gear run-out exceeded 0.005 in., recheck the case as follows. Install the differential case, cups, caps, and adjusters in the housing. Turn the adjusters to eliminate all side-play and tighten the differential cap bolts snugly. Measure the run-out at the ring gear flange face. Total indicator reading should not exceed 0.003 in. It is often possible to reduce run-out by removing the ring gear and remounting 180° from its original position. Remove the differential case from the housing.

7. Remove the pinion shaft lockscrew and remove the pinion shaft.

8. Rotate the differential side gears until the differential pinion shafts can be removed through the opening in the case.

9. Remove the differential side gears and thrust washers.

10. Using a puller or a press and press plates, remove the differential side bearings.

Assembly

1. Lubricate all parts, before assembly, with rear axle lubricant.

2. Install the thrust washers on the differential side gears and install the side gears into the case.

3. Place thrust washers on both differential pinions and, working through the large access window mesh the pinion gears with the side gears. The pinions should be exactly 180° apart.

4. Rotate the side gears 90° to align the pinions and thrust washers with the pinion shaft holes.

5. From the pinion shaft lockpin hole side of the case, insert the slotted end of the pinion shaft through the case, conical thrust washer and just through one of the pinion gears.

6. Install a thrust block through the side gear hub, so that the slot is centered between the side gears.

7. Hold all these parts in alignment, and align the lockpin holes in the pinion shaft and case. Install the lockpin from the pinion

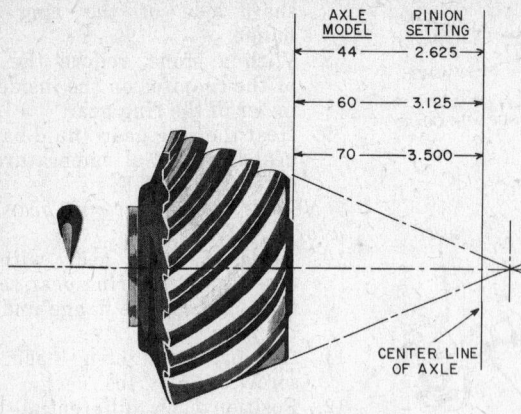

Pinion setting depth measurement
(© Chrysler Corp.)

AXLE MODEL	PINION SETTING
44	2.625
60	3.125
70	3.500

CENTER LINE OF AXLE

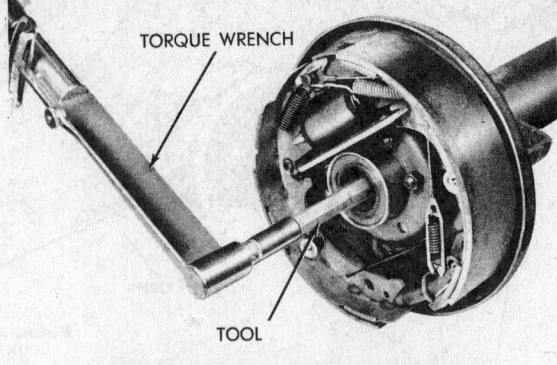

Use of long bar tool with hex end to adjust
differential bearing preload and gear backlash
(© Chrysler Corp.)

TORQUE WRENCH

TOOL

pinion to ensure proper bearing seating. Remove the tools and rotate the pinion several revolutions.

11. Torque the pinion nut to 170 ft. lbs. (210 from 1972 on). With an in. lbs torque wrench, measure the pinion bearing preload, which should be 20-30 in. lbs. for new bearings or 10 in. lbs. over the original figure if the old pinion bearing is used.

NOTE: *The correct preload reading can only be obtained with the carrier nose upright. The final assembly is incorrect if the final pinion nut torque is below 170 ft.lbs. (210 from 1972 on) or if the pinion bearing preload is not within specifications. Under no circumstances should the pinion nut be backed off to reduce the pinion bearing preload; if this is done, a new collapsible spacer will have to be installed and the unit adjusted again until proper preload is obtained.*

Differential Bearing Preload and Ring Gear-to-Pinion Backlash— First Type

NOTE: *for later axles see Side Bearing Preload Adjustment following.*

1. With the pinion bearings installed and preload set, install the differential case and ring gear in the housing with the respective bearing caps and adjusters. The adjusters should be flush with the caps. Tighten the bolts on each cap to 10 in. lbs.
2. Turn the right adjuster until the assembly has approximately 0.005 in. spread, measured with a dial indicator.
3. Tighten all four cap bolts to 60 ft.lbs.
4. Turn the adjuster on the right until all spread is removed.
5. Turn the adjusters in until all bearing free-play is eliminated and there is some backlash between the ring and pinion gear. Turn the ring gear and pinion several times to seat the bearings.

6. Install a dial indicator to register on the rear face of any ring gear tooth. Measure the backlash at four points equally spaced around the ring gear. Turn the ring gear to position it at the point of least backlash.
7. Turn both adjusters in, the same amount and in the same direction until backlash is 0.001-0.002 in.
8. Turn the bearing adjuster on the right side in until 0.006-0.008 in. minimum backlash is obtained. Setting the required backlash will also obtain the proper preload.
9. Make a ring gear contact pattern test and evaluate the results. If the unit must be disassembled to put in a new shim behind the pinion, the entire assembly operation must be repeated.
10. Install the axle shafts, drums, and wheels.
11. Install a new cover gasket and the differential cover. Refill the unit with the specified lubricant.
12. Lower and road-test the car.

Side Bearing Preload— Second Type

A number of changes have been incorporated into these axles. The new type of axles still use threaded adjusters to control side bearing preload. The threaded adjuster uses a hex drive hole, and requires special tool C-4164 to adjust the side bearing preload through the axle tube. An adjuster lock with two pointed teeth which engage in the exposed adjuster thread when the lock is tightened is provided. Previously, spacers were used from 0.084-0.100 in. The new shims will range from 0.020-0.038 in. and will be equipped with internal centering tabs. The new shims, marked with a number which represents its thickness in thousandths of an inch, can be installed with either side against the pinion head. To accommodate the new adjuster locks, a new axle cover is used.

The axle pinion has been redesigned to use a short collapsible spacer The short collapsible gear sets are interchangeable with the previous gear-sets the only precaution being to ensure that the proper spacer is used.

1. Index the gears so that the same gear teeth are in contact throughout the adjustment.
2. The differential bearing cups will not always move with the adjusters. It is important to seat the bearings by rotating them 5-10 times in each direction, each time the adjusters are moved.
3. With the pinion bearings installed and the preload set, install the differential with adjusters, caps and bearings. Lubricate the bearings and adjuster threads. Check to be sure that there are no crossed threads. Tighten the cap screws on the right and left to 10 ft. lbs. Tighten the bottom cap screws finger-tight until the head is just seated on the bearing cap.
4. Using the tool, check to be sure that the adjuster rotates freely. Turn both adjusters in until bearing play is eliminated with some drive gear backlash (0.010 in.). Seat the bearing rollers.
5. Install and register a dial indicator against the drive side of a gear tooth. Check the backlash at four positions to find the point of minimum backlash. Rotate the gear to the position of least backlash and mark the tooth so that all readings will be taken at the same point.
6. Loosen the right adjuster and turn the right adjuster until the backlash is 0.003-0.004 in. with each adjuster tightened to 10 ft. lbs. Seat the bearings rollers.
7. Tighten the differential bearing cap screws to 100 ft. lbs.
8. Tighten the right adjuster to 70 ft. lbs. and seat the rollers, until the torque remains constant at 70 ft. lbs. Measure the backlash. If the backlash is not 0.006-0.008

in. increase the torque on the right adjusters and seat the rollers until the correct backlash is obtained. Tighten the left adjuster to 70 ft. lbs. and seat the bearings until the torque remains constant.

9. If the assembly is properly done, the initial reading on the left ad-juster will be approximately 70 ft. lbs. If it is substantially less, the entire procedure should be repeated.

10. After adjustments are complete, install the adjuster locks. Be sure the teeth are engaged in the adjuster threads. Torque the lockscrews to 90 in. lbs.

Final Assembly

1. Install the axle shafts.
2. Install the cover on the differential housing, using a new gasket.
3. Refill the rear axle housing with SAE 90 lubricant. On 8⅜ in. axles, fill to ¼ below the plug opening through 1971, and ½ in. below from 1972 on. Fill 9¼ in. axles to the bottom of the plug opening.

Dodge RA115 12½ in. Axle

The RA115 is a full-floating rear axle with hypoid drive gear and pinion and uses a two-pinion differential assembly. The drive pinion shaft is supported by three tapered roller bearings (two front and one rear).

IMPORTANT: Before proceeding with disassembly, consult the "General Axle Service" section for instructions on important pre-assembly inspections and checks. After removing differential assembly from the carrier, remove the drive gear thrust block from the drive gear side of carrier housing. Break weld and remove thrust block adjusting screw.

Disassembly

1. Unscrew nut from drive pinion shaft and remove companion yoke.
2. Remove drive pinion oil seal.
3. Loosen drive pinion front bearing retainer lock screw and remove bearing.
4. Remove drive pinion shaft assembly and front bearing forward cone from carrier.
5. Remove front bearing rear cone and pre-load spacer from pinion shaft, then pull rear bearing from shaft.
6. Complete disassembly by driving cups out of carrier.

Assembly

1. Install pinion bearing cups in carrier, making sure they are seated evenly. The front bearing cup is designed so that front bearing cones are seated in taper of cup from opposite sides. The cup is marked with the letter "Y", which must face toward the front (engine side). The front bearing cone that operates in this cup is also marked "Y". The cone must be assembled with corresponding marks mated together.

2. Press rear bearing on pinion shaft. Install front bearing rear cone on shaft.

3. Lubricate bearing cones and cups with axle lubricant and insert shaft assembly into carrier. Adjustment of pinion bearing pre-load is required whenever a change is made in pinion or bearings. Adjustment is made with spacers available in .020 inch graduations from .1790 to .1930 inch.

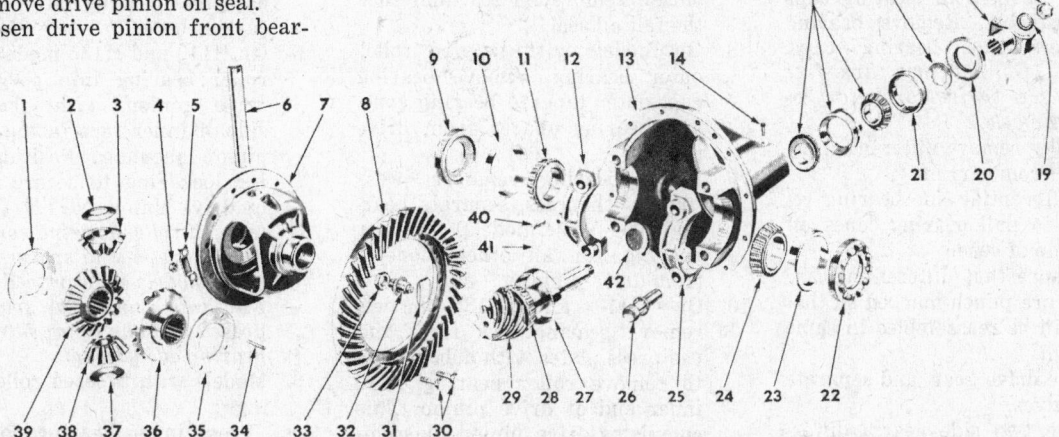

RA115 Dodge differential assembly (© Chrysler Corp.)

1. Differential side gear thrust washer	22. Differential bearing adjuster
2. Differential pinion thrust washer	23. Differential bearing cup
3. Differential pinion	24. Differential bearing cone and rollers
4. Rear axle drive gear bolt nut	25. Differential bearing cap
5. Rear axle drive gear bolt nut lock	26. Differential bearing cap screw and lock washer
6. Differential pinion shaft	27. Rear axle drive pinion rear bearing cup
7. Differential case	28. Rear axle drive pinion rear bearing cone and rollers
8. Rear axle drive gear	29. Rear axle drive pinion
9. Differential bearing adjuster	30. Rear axle gear bolts
10. Differential bearing cup	31. Rear axle drive gear thrust pad
11. Differential bearing cone and rollers	32. Rear axle gear thrust screw
12. Differential bearing cup	33. Rear axle drive gear thrust screw nut and lock washer
13. Differential carrier	34. Differential pinion shaft lock pin
14. Rear axle drive pinion front bearing retainer nut lock screw	35. Differential side gear thrust washer
15. Rear axle drive pinion front bearing assembly	36. Differential side gear
16. Rear axle drive pinion front bearing washer	37. Differential pinion thrust washer
17. Rear axle drive pinion companion yoke	38. Differential pinion
18. Rear axle drive pinion companion yoke nut	39. Differential side gear
19. Rear axle drive pinion companion yoke nut washer	40. Differential bearing adjuster lock pin
20. Rear axle drive pinion bearing oil seal	41. Differential bearing adjuster lock pin cotter pin
21. Rear axle drive pinion front bearing retainer nut	42. Rear axle drive pinion bearing spacer

4. Select and install a spacer between the two front bearings. This spacer will separate front bearing rear cone from cup.

5. Install and tighten bearing retainer nut. Rotate pinion shaft to seat bearings.

6. Rotate pinion shaft with an inch-pound torque wrench. Torque should be from 35 to 70 inch-pounds while shaft is turning (40 inch-pounds preferred). It may be necessary to change spacer to arrive at specified torque.

7. Remove spacer from between two front bearings and install it in its proper location. After tightening bearing retainer nut, recheck bearing pre-load. Torque should be from 60 to 95 inch-pounds while rotating pinion (70 to 80 inch-pounds preferred). Tighten front bearing retainer nut lock screw.

8. Place front bearing washer on pinion shaft and install new oil seal.

9. Install companion yoke, washer and nut and tighten nut to 325-foot-pounds. Install cotter key.

Install drive gear thrust block before installing differential assembly in carrier. Thread thrust block adjusting screw through carrier with one thread inside carrier. Hold thrust block on screw with heavy grease.

Drive Gear Thrust Block Adjustment

Turn in thrust block screw approximately four turns to force block against back face of drive gear. Install lock nut on screw, back out screw 1/8 turn to provide .006-.008 inch clearance, then tighten lock nut securely. Spot-weld lock nut to carrier and screw.

Torque Chart

Axle drive shaft nut (cotter key type)	175-250
Axle drive shaft flange nut	30-35
Axle drive shaft bearing retainer bolt	30-35
Carrier to housing screw	40
Differential bearing cap bolt	85-90
Differential bearing adjuster lock bolt	20
Differential case bolt	45
Drive gear bolts to differential case (anti-spin)	40-45
Drive pinion yoke nut (two bearing pinion)	180-200
Drive pinion yoke nut (three bearing pinion)	325

GMC Single Speed Axle Service
H052, H072, H110, H130, H135, H150

Disassembly of Subassemblies

Differential Disassembly

1. Remove lock nut, adjusting screw, and thrust block.

2. Remove two adjuster lock cap screws and locks.

3. Punch-mark bearing caps and carrier to help in locating caps for assembly. Remove bearing adjusters and bearing caps. *NOTE: Do not pry caps free with a screwdriver or distort locating dowels.*

4. Carefully remove differential assembly from carrier.

5. Use differential side bearing remover to pull bearing cones off each side of case.

6. Make sure that differential case halves are punch-marked so that they can be reassembled in same position.

7. Remove drive gear, and separate case halves.

8. Remove two side gears; differential spider, and four differential pinions.

9. On H110, H135, and H150 models remove pinion and side gear thrust washers to complete differential disassembly.

Drive Pinion Disassembly

1. Remove seal retainer and gasket from carrier.

2. Use brass drift against inner end of pinion to drive out pinion and bearings assembly.

3. Remove shim pack from carrier from those models having tapered roller outer bearings.

4. According to the model, it may be necessary to use a drift to remove the pinion rear bearing.

5. Clamp yoke in soft-jawed vise. Remove yoke nut and washer and separate drive pinion from yoke.

6. Separate yoke from oil seal retainer.

7. Place retainer in a soft-jawed vise and, using a hammer and chisel, remove oil seal and then the felt oil seal.

8. On models with tapered roller outer bearing, remove bearing cup, outer tapered bearing cone, and bearing spacer from drive pinion.

9. Using bearing remover press plate with press, separate bearing cone (some models) or roller bearing (on all other models) from drive pinion.

10. On H110 and H135 models remove bearing lock ring, and use press plates with arbor press to remove roller bearing from inner end of drive pinion. This completes drive pinion disassembly.

Assembly of Subassemblies

NOTE: Thoroughly clean and lubricate all components with axle lubricant before reassembling.

Drive Pinion Assembly

1. Clean counterbore of oil seal retainer. Saturate felt seal in oil and install evenly in retainer. Soak oil seal in light engine oil for about one hour before installing. Coat outer surface of seal lightly with sealing compound to prevent oil leaks between seal and retainer.

2. Install oil seal into retainer with lip of seal toward inner side of retainer. Using a seal installer, press oil seal into retainer with face of seal flush with retainer face.

3. Retainer surface must be clean and smooth to prevent oil leaks between retainer and carrier.

4. On H052, H072, and H150 models, press bearing into place into carrier bore.

5. On H110 and H135 models, press roller bearing into position on drive pinion with chamfered side of inner race facing toward pinion shoulder. Position bearing lock ring to secure bearing on drive pinion. *NOTE: Opposed tapered roller bearing cones, two bearing cups, and spacer used on some models are serviced and replaced as a unit. The spacer is a preselected one to provide proper bearing adjustment.*

6. Models with tapered roller bearings:

a. Press inner bearing cone into place with largest side of cone facing pinion gear end.

b. Install original shim pack in carrier. If original ring gear and pinion are reinstalled, use shims that were removed. Shims are available in five thicknesses -0.012, 0.015, 0.018, 0.021, and 0.024 inch. When using new gears, start with one 0.021 inch shim and refer to "General Axle Service" section for details on checking pinion depth.

c. Insert pinion assembly into carrier (on H150 model), align roller bearing with carrier boss. Install bearing spacer, bearing cup and bearing cone with wide side facing pinion splines.

7. Models with double-row ball bearing: Using a 2-inch pipe or tubing, drive bearing unit into proper seating position.
8. With pinion assembly properly positioned in carrier, install new gasket. Install seal retainer onto yoke, and assemble yoke and retainer assembly onto splined end of drive pinion.
9. Secure retainer to carrier with lock washers and cap screws and torque to specifications.
10. Secure pinion assembly with yoke washer and nut and torque to 220 ft. lbs. This completes drive pinion assembly.

Differential Assembly

1. To facilitate installation of drive gear, install two guide pins (cut ½"-20x2" bolts) in gear. Start guide pins through case flange holes and tap drive gear onto case. If one differential gear is bad, the complete set should be replaced.
2. Lubricate differential case inner walls and all component parts with axle lubricant. Place differential pinions and thrust washers (thrust washers are used only on H110, H135, and H150 axles) on spider.
3. Assemble side gears, pinions and

side gear and pinion thrust washers to left half of differential.
4. Assemble drive gear half (right half of differential, being sure to line up marks on the two halves.
5. Install differential-to-drive gear cap screw and lock washers and tighten evenly until drive gear is flush with case flange. Remove guide pins and install cap screws and torque to specifications.
6. Differential side bearing cones can be installed with special installer tool.

Installation of Subassemblies

Differential Installation

1. Install bearing cap locating dowels in caps. Lubricate side bearings and place bearing cups on bearings.
2. Install differential assembly into carrier. Carefully install bearing adjusters into carrier.
3. Install bearing caps, aligning punch marks previously made. Be sure that bearing adjuster threads are engaged with carrier and caps. Tighten adjusters alternately and evenly. Tighten bearing cap screws until lock

washers are flat.

Drive Gear and Pinion Adjustment

1. Loosen bearing cap screws just enough to loosen right-hand bearing adjuster (pinion side) and tighten left-hand bearing adjuster (opposite pinion side.) Using adjuster, remove all backlash between drive gear and pinion.
2. Back off left-hand bearing adjuster about two notches to point where notch in adjuster is aligned with lock. Tighten right-hand bearing adjuster solidly to seat bearing. Again loosen right-hand adjuster enough to free bearing; then retighten snugly against bearing. Draw up right-hand adjuster one or two more notches until adjuster notch aligns with lock.
3. With dial indicator on carrier adjuster, slowly oscillate drive gear and take backlash reading. Backlash should be 0.005 to 0.008-inch.
4. If backlash exceeds 0.008-inch, loosen right-hand adjuster one notch; then tighten left-hand adjuster one notch. If less than 0.005 inch, loosen left-hand adjuster one notch and tighten right-hand adjuster one notch.

SPECIFICATIONS

GMC H-052, H-072, H-110, H-135, H-150

Make	GMC	Pinion gear	0.058"-0.062"
Type	Hypoid	Spider	
Adjustment and Clearances		Diameter of arms—H-052, H-072	0.808".0.809"
Backlash—gear to pinion	0.005"-0.008"	Diameter of arms—H-110, H-135	0.874"-0.875"
Adjustment method	See text	Diameter of arms—H-150	0.9365"-0.9375"
Pinion backlash adjustment		Thrust Block	
Models with tapered bearings	Shims	Thickness	0.1845"-0.1885"
Models with ball bearing	None	Clearance—block to gear	0.005"-0.007"
Shim pack thickness	0.021" Initial	Axle Shafts	
Shims available	0.012"-0.015"-0.018"-0.021"-0.024"	Diameter of splines—H-052, H-072	1.5275"-1.5325"
Differential bearing adjustment method	Threaded rings	Number of splines—H-052, H-072	17
Drive Gear		Diameter of splines—H-110, H-135	1.724"-1.732"
Backlast—gear to pinion	0.005"-0.008"	Number of splines—H-110, H-135	27
Adjustment method	See text	Diameter of Splines—H-150	1.848"-1.856"
Runout (mounted to case)	0.006"	Number of Splines—H-150	29
Drive Pinion		Torque Specifications (Ft. Lbs.)	
Backlash—pinion to drive gear	0.005"-0.008"	Drive gear bolts	
Adjustment method	See text	H-052, H-072	85-95
Differential Case		H-110, H-135	100-110
Runout at flange (max.)	0.002"	H-150	150-170
Diameter at side gear		Differential side bearing cap bolts	
H-052, H-072	1.927"-1.929"	H-052, H-072	95-105
H-110, H-135	2.193"-2.195"	H-110, H-135, H-150	190-220
H-150	2.409"-2.411"	Pinion bearing retainer bolts	
Side Gear		H-052, H-072	90-100
Backlash—side gear to pinion gear	0.004"-0.007"	H-110, H-135, H-150	160-170
Hub diameter—H-052, H-072	1.923"-1.925"	Pinion flange nut	160-280
Hub diameter—H-110, H-135	2.189"-2.191"	Diff. bearing adj. nut lock	10-20
Hub diameter—H-150	2.405"-2.407"	Axle shaft flange to hub	
Pinion Gear		H-052, H-072	85-95
Inside diameter—H-052 H-072	0.814"-0.815"	H-110, H-135, H-150	10-20
Inside diameter—H-110, H-135	0.880"-0.881"	Carrier to housing	
Inside diameter—H-150	0.9435"-0.9445"	H-052, H-072	40-50
Thrust Washer Thickness (H-110, H-135, H-150)		H-110, H-1.5, H-150	75-90
Side gear	0.058"-0.062"	Axle shaft nuts	
		All models	80-100

5. After backlash has been adjusted, again tighten bearing cap screws until their respective lock washers flatten out.
6. Check drive gear run-out.
7. Install side bearing adjusting nut lock and secure with cap screws and lock washers.

Checking Pinion Depth (Models with Tapered Roller Bearings Only)

NOTE: Refer to tooth contact chart in the "General Axle Service" section.

1. Coat drive gear with red lead.

The GMC T150 is a 2-speed axle (planetary type) which can be shifted by the driver when desired. Operating the shift control button activates an electric or vacuum shift unit (attached to differential carrier) which, in turn, changes axle speed from low to high or vice-versa.

CAUTION: There is no "Neutral" position. There are just "High" and "Low" ratios. To attempt to coast with axle in a supposedly "Neutral" position could cause serious axle damage.

Differential Carrier

Removal

1. Remove all but two cap screws holding carrier to axle housing. Loosen the two remaining cap screws and, with a soft hammer, break carrier loose from housing. **CAUTION: Do not use pinch bar. This might cause damage to carrier or housing.**
2. Place a roller jack under carrier, remove the two cap screws and remove carrier and differential assembly from housing.
3. Place carrier and differential assembly in a suitable repair stand.
4. Remove shift unit.

Differential and Planetary

Removal

1. Remove anchor from end of sun gear.
2. Drive shift lever out of carrier.
3. Remove shift yoke and sleeve assembly. Separate yoke from sleeve.
4. Remove adjusting nut lock.
5. Remove oil trough.
6. Tap bearing caps from carrier.
7. Remove differential and planetary assembly from carrier assembly.

CAUTION: DO NOT drop bearing cones and cups.

Turn pinion shaft several revolutions in both directions while applying considerable drag on drive gear.

2. Pinion depth is determined by shim pack selection. Shim packs are available in thicknesses of: 0.012, 0.015, 0.018, 0.021, and 0.024- inch.
3. Changing pinion depth will again require adjusting backlash. After pinion depth and backlash have been adjusted, torque bearing caps to specifications.

Model T150

Drive Pinion and Retainer

Removal

1. With carrier still in repair stand, use holding bar to hold propeller shaft yoke while removing nut and washer.
2. Using a soft faced hammer or a blunt chisel, work retainer cap and pinion assembly from carrier.
3. Remove pinion adjusting shim pack and record or tag thickness for reassembly information.
4. If necessary, use hammer and drift to remove drive pinion rear bearing from carrier.

Drive Pinion and Retainer

Disassembly

1. Carefully pry pinion oil seal from pinion and bearing retainer.
2. Using a press, press downward on end of pinion shaft to separate pinion assembly from retainer. Remove yoke and deflector.
3. Remove pinion front bearing cone from retainer. Pull bearing preload spacer off drive pinion shaft.
4. When necessary, use bearing cup remover to force bearing cups out of retainer cap.
5. Slide pinion bearing remover up over pinion head, and press drive pinion out of intermediate bearing cone.

Differential and Planetary

Disassembly

1. Punch-mark drive gear, differential support case, and differential case cover to facilitate proper alignment during reassembly.
2. Remove drive gear and case cover.
3. Tag sun gear and four planetary gears so they may be correctly

Thrust Block Installation

1. Install thrust block and lock nut to adjusting screw. Thread screw and block into carrier until block contacts drive gear.

Rotate gear and note change of drag. Adjust these parts until point of greatest drag is reached. Back screw off about a 30 degree turn to provide 0.005 to 0.007-inch clearance between block and gear. Make certain screw does not turn at all when tightening lock nut to 135 foot-pounds torque.

reassembled. Remove planet gears.

4. Lift differential and planet support assembly away from support case.
5. Punch-mark planet support cover and support and separate.
6. Again mark or tag all pinion and side gears for reassembly to same positions, and remove them along with their respective thrust washers. Slide four pinion gears off differential spider.
7. Remove differential side bearing cups. Insert brass drift through holes in case and cover and drive left and right side bearing cones off hubs.

NOTE: Alternately tap drift from hole on one side to hole on other side to prevent race from binding on hub.

Axle Assembly

Before axle is assembled, make sure that all parts have been thoroughly cleaned, especially oil passages. Lightly lubricate all parts while assembling.

When assembling axle unit, it is recommended that new lock washers and gaskets be installed wherever possible. See "General Axle Service" section for details on cleaning and inspection.

Drive Pinion and Retainer

Assembly

Drive pinion and drive gear must be replaced as a matching set.

1. If it was necessary to remove pinion rear bearing, use a driver handle with bearing installer to bottom bearing into position in carrier.
2. Position pinion intermediate bearing cone and cup in pinion bearing retainer with thicker end of cup facing inward. Use driver handle with bearing cup installer to firmly seat cup in place.
3. Follow step 1 procedure to in-

stall pinion front bearing and cup to pinion bearing retainer. Seat cup with handle and cup installer.

4. Use arbor press with a 2 inch inner diameter pipe of suitable length to position pinion intermediate bearing cone with thicker end up against shoulder of drive pinion.

5. Place new pinion bearing spacer on pinion up against bearing cone.

6. Position drive pinion assembly in retainer cap assembly, and install pinion front bearing cone in place. Thicker end of cone should face outward.

7. Again use arbor press with a 2 inch inner diameter pipe of suitable length to press inner race of cone up tightly against spacer.

8. Install pinion oil seal in retainer. See "oil seal replacement" in the "General Axle Service" section.

Differential and Planetary

Assembly

1. Lubricate thrust washers, pinion gears, spider, and side gears. Assemble pinion gears to arms of differential spider and install all above components into planetary support (match up punch-marks made earlier).

2. Align and secure support cover with new lock washers and cap screws. Tighten cap screws to specifications.

3. Install left and right differential bearing cones to support case and support case cover hubs. Use bearing installer for left and right bearings.

4. Install planet gears on journal arms of support. Position sun gear in the planet gears.

5. Align punch-marks and install drive gear to support case cover. Improvised guide pins can be made from drive gear retaining cap screws.

6. Turn differential assembly over so drive gear teeth face downward. Position assembled planet and differential assembly in cover, making sure thrust washer at case bottom is properly positioned.

7. Using lock washers and cap screws, join support case, case cover and drive gear together. Remove guide pins and install rest of lock washers and cap screws. Tighten all cap screws in a cross-wise pattern to specified torque.

Differential and Planetary

Installation

1. Position left and right side bearing cups over their respective cones. Position differential and planetary assembly into carrier.

2. Install differential adjusting nuts at both sides.

3. Position bearing caps on carrier making certain adjusting nuts properly engage threads in caps.

If bearing caps seat flush in place while adjusting nuts turn freely, caps are properly installed. Use soft hammer to seat caps properly. Tap lightly **CAUTION: Do not force bearing caps on nuts.**

4. Tighten cap screws evenly until adjusting nuts turn freely in threads.

Drive Pinion and Retainer

Installation

1. Install original shim pack to carrier, making sure shim holes are aligned with carrier cap screw holes.

2. Properly install pinion and retainer assembly to carrier bore. Make sure pilot end of pinion properly seats in pinion rear bearing.

3. Assemble retainer to carrier with new lock washers and cap screws and torque to specifications.

4. Tap yoke onto pinion shaft splined end, making sure proper spline engagement is made.

5. Install and torque pinion washer and nut to end of drive pinion.

6. Continue tightening nut in small degrees until torque required to rotate drive pinion in carrier is from 25 to 35 inch-pounds (when new pinion bearings are installed). Only 5 to 15 in. lbs. torque is required if same bearings are reinstalled.

Drive Gear and Pinion

Adjustment

1. Loosen right differential adjusting nut. Tighten left adjusting nut. Use differential nut wrench on right side nut and a drift hammer to the left side nut.

2. Tighten left adjusting nut until drive gear contacts pinion and zero lash is obtained; however, do not force gears into contact so as to bind them.

3. Back off left adjusting nut about two notches to a point where nut lock and nut are aligned. Install nut lock and torque to specifications.

4. Tighten right adjusting nut firmly to force the differential and planet assembly into solid contact with left adjusting nut. Loosen right adjusting nut until

it no longer contacts its bearing, then retighten until nut contacts bearing. Tighten right adjusting nut from one to two notches more if old bearings are used, and two to three notches if new bearings are used, to a position where nut and nut lock are aligned. Install nut lock and torque to specifications. *NOTE: At this point the differential bearings are properly preloaded. If any additional adjustments are required, make sure that this proper preload remains. If one adjusting nut is loosened, the other nut must be tightened an equal amount to maintain this preload.*

5. Mount a dial indicator on the housing and measure the backlash between the drive gear and pinion. Backlash should be from 0.003 to 0.012 inch with 0.005 to 0.008 inch preferred.

NOTE: If backlash is more than 0.012 inch, loosen the right adjusting nut one notch and tighten left adjusting nut one notch. If backlash is less than 0.003 inch, loosen the left adjusting nut one notch and tighten the right adjusting nut one notch.

Checking Pinion Depth

1. Thoroughly clean drive and pinion gear teeth.

2. Paint drive gear teeth lightly and evenly with a mixture of powdered red lead and oil.

3. Rotate pinion through several revolutions in both directions until a definite contact pattern is developed on drive gear. Apply pressure to drive gear while turning pinion. This will create load on gears to simulate a driving pattern.

4. Examine the pattern on drive gear teeth. If the pinion depth is correct, the tooth pattern will be centered on the pitch line and toward the toe of drive gear.

5. If the pattern is below the pitch line on drive gear teeth, the pinion is too deep and it will be necessary to remove the pinion assembly and increase the shim thickness between the pinion bearing retainer and the carrier.

6. If the pattern is above the pitch line on drive gear teeth, the pinion is too shallow and it will be necessary to remove the pinion assembly and decrease the shim thickness between the pinion bearing retainer and the carrier.

7. Changing the pinion depth will cause some change in backlash. Therefore adjust backlash to maintain correct specifications.

8. Torque bearing cap screws to specifications and recheck drive

gear to pinion backlash. Install adjusting nut locks and torque to specifications.

9. Install oil trough to differential carrier, adjust trough to drive gear so that clearance is 0.03 to 0.09-inch. Torque retaining cap screws.

Shifter Components

Installation

1. Position shift yoke in shift sleeve groove.

2. Slide yoke and sleeve assembly over sun gear. Splines of sleeve and sun gear should properly mesh.

3. Install shift lever through carrier housing into yoke mating splines together. Tap end of lever lightly to seat it in yoke.

4. Position the shifter unit and gasket over mounting pad, aligning lever in carrier with shift rod in shift unit. *NOTE: On axles equipped with electrical shift units, shift electrical leads must be connected to shift unit before installing unit to carrier.*

5. Install and torque bolts.

6. Install differential assembly into axle housing.

SPECIFICATIONS

Make	G. M. Corp.
Type	2-Speed planetary
Ratios	5.83:1-7.95:1
Adjustments and Clearances	
Backlash—gear to pinion	0.005"-0.008"
Backlash adjustment method	See text
Pinion backlash adjustment	Shims
Shim pack thickness (starting)	0.021"
Shims Available	0.006"-0.009"-0.012"-0.015"-0.018"-0.021"-0.024"
Pinion bearing adjustment method	Pre-selected spacer
Differential bearing adjustment method	Threaded adjusters
Drive Gear	
Backlash—gear to pinion	0.005"-0.008"
Adjustment method	See text
Gear teeth	
5.83:1-7.95:1	35
Run out-max.	0.005"
Drive Pinion	
Backlash—gear to pinion	0.005"-0.008"
Adjustment method	See text
Bearing pre-load (max. in. lbs.)	
New bearing	15-25
Used bearing	5-15
Gear teeth	
5.83:1-7.95:1	6
Differential Spider	
Diameter of arms	0.9365"-0.9375"
Differential Side Gear	
Backlash	0.007"-0.009"
Hub diameter	2.405"-2.407"
Differential Pinion	
Backlash	0.007"-0.009"
Inside diameter	0.9435"-0.9445"

Differential Case	
Diameter of planet gear studs	1.3122"-1.3132"
Distance across studs	6.583"-6.593"
Run-out at flange (max.)	0.002"
Thrust Washer Thickness	
Side gear	0.058"-0.062"
Pinion gear	0.058"-0.062"
Differential case to case cover	0.058"-0.062"
Planet Gear	
Bushing inside diameter	1.3158"-1.3164"
Sun Gear	
Diameter over retainer	3.339"-3.400"
Backlash	0.004"-0.008"
Shifter Yoke	
Yoke width	0.395"-0.400"
Shifter sleeve	
Backlash	0.004"-0.008"
Width at yoke groove	0.445"-0.457"
Axle Shaft	
Diameter at splines	1.848"-1.856"
Number of splines	29
Torque Specifications (Ft. Lbs.)	
Differential case cap screws	100-110
Planetary support case & cover to bevel gear cap screws	100-110
Differential bearing cap screws	190-220
Differential carrier to housing cap screws	75-90
Universal joint yoke nut	160-280
Anchor plate to differential bearing cap screws	40-45
Differential bearing lock cap screws	10-15
Shift housing to carrier cap screws	25-30
Axle shaft cover cap screws	10-20

GMC Single Reduction, Integral Housing Axles

Models H024, H029, H033, H035, H036

IMPORTANT: Before disassembling unit, perform the following:

1. Inspect axle housing for lubricant leaks before cleaning; then clean thoroughly.

2. Check and record drive gear and pinion backlash.

Differential Case and Drive Pinion

Differential Case

Removal

1. Remove wheels and brake drums.

2. Drain lubricant and remove cover.

3. Remove pinion shaft from differential.

4. Remove "C" locks from axle shafts and pull shafts from housing.

5. Extract the differential pinions, side gears and thrust washers and tag them for reassembly.

6. Punch-mark differential bearing caps and housing for proper reassembly and remove caps. *NOTE: Do not attempt to pry caps free as this may damage machined face of caps.* **CAUTION: When prying out differential case do not damage gasket surface on carrier housing.**

7. Using a pry bar, pry differential case and ring gear assembly out of carrier. If the differential bearings are under preload, the case will fall free at a certain point; therefore, prevent case from falling.

8. Place both the left and right bearing cups and shims with their respective bearing caps so they will be re-installed in their original positions.

Drive Pinion

Removal

1. Using a suitable tool, remove and discard flange nut and washer and pull off flange.

2. Using a soft hammer, remove drive pinion. Remove pinion front bearing cone and pinion

bearing spacer. Replace spacer when assembling.

Pinion Oil Seal

Replacement

1. Pry oil seal out of carrier.
2. Pack the cavity between lips of new oil seal with lithium base lubricant. Place seal in carrier bore and place gauge plate over seal between seal flange and carrier.
3. On H024 axle, use installers and press seal into carrier. On all others, drive seal into carrier.
4. Press or drive seal into carrier bore until gauge plate is snug between seal flange and carrier. Rotate gauge plate 180 degrees from original position; seal must be square in carrier bore.
5. Using gauge plates will create ⅛" gap between the carrier and seal flange. Applying unnecessary pressure after seal installation may distort the seal.
6. If necessary to replace oil deflector on companion flange, clean up stake points on flange, then install the deflector. Stake in place at three new equally spaced positions.

Pinion and/or Bearing

Replacement

1. Tap out front and rear pinion bearing cups from carrier. Remove carefully to avoid cocking the cups in carrier.
2. Use installers to install front and rear bearing cups against carrier shoulders. Do not cock the cups.
3. Remove pinion rear bearing cone and roller assembly, using suitable tools:
   H024 Axle - use press plate holder with remover plates.
   All Other Axles - use a discarded ring gear as a press plate holder for remover plates. Remove shim and record thickness.
4. If the original ring gear and drive pinion and the rear bearing assembly are re-used, then the original shim can also be used.
5. If any one of the three components; the ring gear, pinion, or rear bearing assembly require replacement, attain correct shim thickness by the following methods:
   a. Lubricate pinion bearing cone and roller assemblies and place them in proper cups.
   b. Position pinion depth setting gauge plate in pinion rear bearing.
      (1) H024 Axle - position gauge plate with lower surface of

gauge (X stamped near low side) toward top of carrier.
      (2) All Other Axles - position gauge plate with higher surface toward top of carrier.
   c. Insert clamp screw through gauge plate and both of the pinion bearings.
   d. Position plug on threaded end of clamp screw and index it in front bearing. Install hex nut and rotate bearings several times to seat them.
   e. Tighten hex nut until 20 inch-pounds of torque is required to rotate bearings.
   f. After establishing torque, check position of gauge plate in carrier (see step b).
   g. Position dial indicator on gauge post of barrel adapter so that indicator button rests on top of tool plunger.
   h. Position tool in carrier. Swing tool body so that plunger does not touch gauge plate, and set indicator dial at zero. NOTE: Pinion setting gauge sleeves must be used when gauging all except H024 axles. Adapter must be seated in bearing bores of carrier before measuring depth.
   i. Slowly swing inner end of tool plunger across gauge plate until highest indicator reading is obtained. Record and recheck this measurement.
   j. All service pinion gears are stamped with a code number on the threaded end of the pinion. The numerical difference between the code number and the gauge reading obtained in step i determines the required thickness of the pinion locating shim.

Example:

Pinion code number	45
Dial indicator reading	16
Difference	29
Proper shim thickness	0.029-in.

NOTE: Shims are available in thicknesses from 0.021 to 0.037-inch. Each shim is etched with its thickness.

6. Remove pinion setting gauge and pinion bearing cone and roller assemblies from carrier.
7. Position the selected shim against pinion head on shaft.
8. Install new cone and roller assembly, using a suitable installer. Press cone tight against shim.

Differential Bearing

Replacement

1. Install the differential bearing remover making sure puller legs are fitted securely in notches of case and against bearing cone.
2. Tighten puller screw to remove

bearing.
3. Place new bearing on one hub with thick side of inner race against case and drive into place using suitable tool.
4. Before installing bearing on opposite hub, support differential case on adapter plug to allow case to rest on adapter instead of bearing cage. Install remaining bearing as instructed in step 3.

Ring Gear or Differential Case

Replacement

1. Remove ring gear. *NOTE: Do not attempt to pry ring gear from case; to do so may cause damage.*
2. Install guide pins, made from ⅜"-24 x 1½" long cap screws to ring gear.
3. Position ring gear over pilot diameter of the case. Install every other ring gear bolt and lock washer and tighten so that gear face is flush with face of case.
4. Remove guide pins and install remaining bolts and torque to 40-60 ft.lbs.
5. Install thrust washers and side gears in case; if same parts are used, replace in original sides. Position pinions and thrust washers through loading holes in case 180 degrees apart so that they engage side gears. Rotate gears until pinion shaft holes are aligned and install shaft and lock screw.

Axle Assembly

Drive Pinion and Bearing

1. Position pinion and rear bearing cone and roller assembly in carrier, install a new pinion bearing spacer on shaft, and install pinion front bearing cone and roller assembly.
2. Install flange on pinion shaft using companion flange installer and tool.
3. Pack the cavity between end of pinion splines and pinion flange with a non-hardening sealer before installing washer and nut on pinion.
4. Install washer and a new self-locking nut on pinion shaft. Tighten nut to remove end play. Continue tightening, in small increments, and checking preload with torque wrench and adapter until torque required to rotate pinion is 20 to 30 inch-pounds for new bearings and seal or 5 to 15 inch-pounds when used parts are reinstalled.

Differential Bearing Preload and Ring Gear Adjustment

1. Check bearing cups and cup seat for nicks and burrs and replace if necessary.
2. Lubricate bearings, position cups to proper bearings then install differential assembly in carrier. Install right bearing cap and torque to 55 foot-pounds. Check to be sure cup is seated in carrier bore.
3. Service spacers are available in one thickness only for each axle: 0.170 ± .001-inch for model H024 and 0.160 ± .001-inch for all other axles. Steel service shims are used with the spacer. See shim chart below for shim availability and identification.

DIFFERENTIAL BEARING SHIM IDENTIFICATION

| Thickness* | Identification Notches | | | |
| | I.D. | | O.D. | |
	H024	All Others	H024	All Others
.064	0	0	0	2
.066	0	0	1	3
.068	0	0	2	4
.070	0	1	3	0
.072	1	1	0	1
.074	1	1	1	2
.076	1	1	2	3
.078	1	1	3	4
.080	2	2	0	0
.082	2	2	1	1
.084	2	2	2	2
.086	2	2	3	3
.088	3	2	0	4
.090	3	3	1	0
.092	3	3	2	1
.094	3	3	3	2
.096		3		3
.098		3		4
.100		4		0

*Additional shims are available for the H-024 axle.

These shims, with a thickness of 0.052-inch through 0.062-inch, have the shim thickness stamped on the shim. *NOTE: Do not reuse original bearing preload shims. They are made of cast iron and may be broken easily.*

4. With ring gear tight against the pinion gear (0.000 to 0.001-inch backlash) and with gauging tool inserted between left bearing cup and carrier housing, turn tool adjusting nut clockwise until a noticeable drag is produced. *NOTE: Do not apply pressure on bearing so as to cause preload.*
5. Tighten locking bolt on the side of the gauge and remove gauge. Using a micrometer, measure the thickness of the two gauging plates. By subtracting the thickness of the service spacer to be used from the thickness of the plates, the thickness of the service shim may be determined.

Example:

Gauge thickness	0.254"
Service spacer to be used	0.070"
Service shim size	0.084"

6. Install the selected shim between the service spacer and bearing. Install the left bearing cap and check backlash for a 0.000 to 0.001-inch reading.
7. Remove right bearing cap and insert tool between right bearing cap and carrier housing. Turn tool adjusting nut clockwise until a noticeable drag is produced. Tighten locking bolt on the side of the tool and remove gauge.
8. Using a micrometer, measure the thickness of the gauge plate. Subtract the service spacer thickness from the total thickness and add 0.008-inch to this figure. This total will be the size of the service shim to be used on the right side. *NOTE: Service*

Example:

Gauge thickness	0.224"
Service spacer	0.170"
	0.054"
Preload	0.008"
Total shim thickness (right side)	0.062"

shims are available in 0.002-inch graduations. If the shim measurement falls between the available shims, select the shim that is 0.001 inch thinner.

9. Using a soft hammer, tap the selected shim between the service spacer and bearing.
10. Install right bearing cap and torque both caps to 55 foot-pounds.
11. Mount a dial indicator on the carrier and check backlash between ring gear and pinion. Backlash should range within 0.003 to 0.010-inch with a reading of 0.005 to 0.008-inch preferred. Check reading at four equally spaced positions around the ring gear. Variation in reading should not exceed 0.002-inch. *NOTE: The dial indicator button should be perpendicular to tooth angle and in line with gear rotation.*
12. If variation in backlash exceeds 0.002-inch, measure ring gear and case runout. Gear runout should be a maximum of 0.002-inch; should it be more, check ring gear and case for deformation and/or foreign matter between case and gear.
13. If gear lash is not within limits, correct by decreasing shim thickness on one side and increasing thickness of the other shim the same amount. Total shim thickness must be maintained to maintain the proper preload. (By decreasing shim on right side 0.002-inch and increasing shim thickness on left side by 0.002-inch backlash will decrease by 0.001-inch.)

SPECIFICATIONS

Make	GMC
Type	Hypoid
Models	H-024, H-029, H-033, H-035, and H-036
Adjustment & Clearances	
Backlash—gear to pinion	0.005"-0.008"
Adjustment method	See text
Pinion bearing pre-load (in.-lbs.)	20-35
Pinion backlash adjustment shims	
Available (0.001" increments)	0.021" to 0.037"
Differential bearing and backlash adjustment method	Selective spacer
Thrust Washer Thickness	
Pinion gear	0.028"-0.032"
Side gear	0.029"-0.031"
Differential pinion shaft	
Diameter	0.808"-0.809"
Differential pinion gear	
Inside diameter	0.813"-0.816"
Torque Specifications (Ft. Lbs.)	

TORQUE SPECIFICATIONS FOOT-POUNDS

Carrier cover bolts	
H-024, H-029	15-25
H-033, H-035, H-036	20-25
Differential bearing cap bolts	55
Differential pinion lock bolt	15-25
Filler plug	20
Ring gear to case bolts	50

Single-Speed, Single Reduction Eaton

This is a full-floating axle having spiral bevel drive gear and pinion, and using a four-pinion differential assembly. The pinion has three roller bearings—two in front of the pinion teeth and one behind.

Disassembly

1. Remove thrust block screw.
2. Punch mark carrier leg, bearing cap and bearing adjusting nut to aid in reassembly.
3. Remove bearing caps and adjusting nuts.
4. Lift assembly out of carrier housing.
5. Punch mark differential case halves for correct alignment. Separate case halves.
6. Remove pinion shaft pinions, side gears, thrust washers and bearing cones.
7. Separate drive gear from cage.
8. Remove pinion shaft yoke.
9. Using a drift, drive pinion and cage assembly from carrier. **CAUTION: Support bearing cage when driving assembly from carrier.**
10. Wire bearing cage shim pack together to facilitate reassembly.
11. Lift bearing cage from pinion and remove cork seal. Remove pinion drive yoke washer and front bearing cone from cage.
12. Remove pinion bearing spacer washer, spacer, inner bearing cone and pilot bearing from pinion shaft.
13. Remove pinion cage bearing cups.
14. Clean all parts thoroughly in a suitable solvent and blow dry with compressed air. The bearings should be immersed in clean solvent and rotated by hand until clean. After cleaning, blow dry with compressed air. Do not spin bearings with air pressure, as they might score due to absence of any lubrication. Inspect all parts for wear or imperfections. The thrust shoulders must be flat, so that bearing cups will seat properly. Install a new case if cracked or distorted. (For further details, see "cleaning" and "inspection" in the "General Axle Service" section).

Assembly

1. Press drive pinion inner bearing cone and rear pilot bearing firmly against pinion shoulder.
2. Stake pilot bearing in at least four places. Install pinion cage bearing cups.
3. Place bearing spacer and washer on pinion shaft. Lubricate bearing cone with SAE 90 oil and position bearing cage on bearing. Lubricate and press outer bearing on pinion shaft. Rotate cage to assure normal bearing contact. If a press is not available, install pinion yoke washer and nut and torque to specifications.
4. While the assembly is in the press under pressure or pinion nut torqued to specifications, check pinion bearing pre-load. The correct pressures and torques for preload are:

Pinion Shaft Thread Size and Number of Threads Per Inch	Required Pressure to Obtain Correct Pre-Load lbs.	Required Torque to Obtain Correct Pre-Load ft. lbs.
7/8 in. x 16	10,000	200-325
1 1/8 in. x 18	12,000	325-450
1 1/4 in. x 12	15,000	400-600

5. Wrap a soft wire around cage and pull on horizontal line with pound scale. Measure diameter of pinion cage. Assuming cage diameter is 6 inches, pulling radius would be 3 inches; therefore, 6 pounds pull on scale would equal 18 inch-pounds preload.

1 Lockwire
2 Cap screw
3 Cotter pin
4 Adjuster lock
5 Differential carrier and bearing caps (matched parts) (conical type)
6 Differential bearing adjust (RH)
7 Differential bearing cup (RH)
8 Differential bearing cone (RH)
9 Lockwire
10 Cap screw
11 Nut

12 Differential case (Plain Half)
13 Side gear thrust washer (RH)
14 Side gear (RH)
15 Side pinion thrust washer
16 Side pinion
17 Spider
18 Side gear (LH)
19 Side gear thrust washer (LH)
20 Bolt and nut

21 Ring gear and drive pinion (Matched Set)
22 Differential case (Flanged Half)
23 Bolt
24 Differential bearing cone (LH)
25 Differential bearing cup (LH)
26 Differential bearing Adjuster (LH)
27 Carrier gasket
28 Pinion pilot bearing
29 Pinion bearing cone (Inner)
30 Pinion bearing spacer
31 Pinion bearing spacer washer
32 Pinion bearing cup (Inner)

33 Pinion bearing cage shims
34 Pinion bearing cage
35 Lock washer
36 Cap screw
37 Pinion bearing cup (Outer)
38 Pinion bearing cone (Outer)
39 Companion flange flat washer
40 Strip sealer (Seal Retainer)
41 Oil seal
42 Oil seal retainer (Pressed-in-Type)
43 Companion flange
44 Cotter pin
45 Pinion nut
46 Flat washer

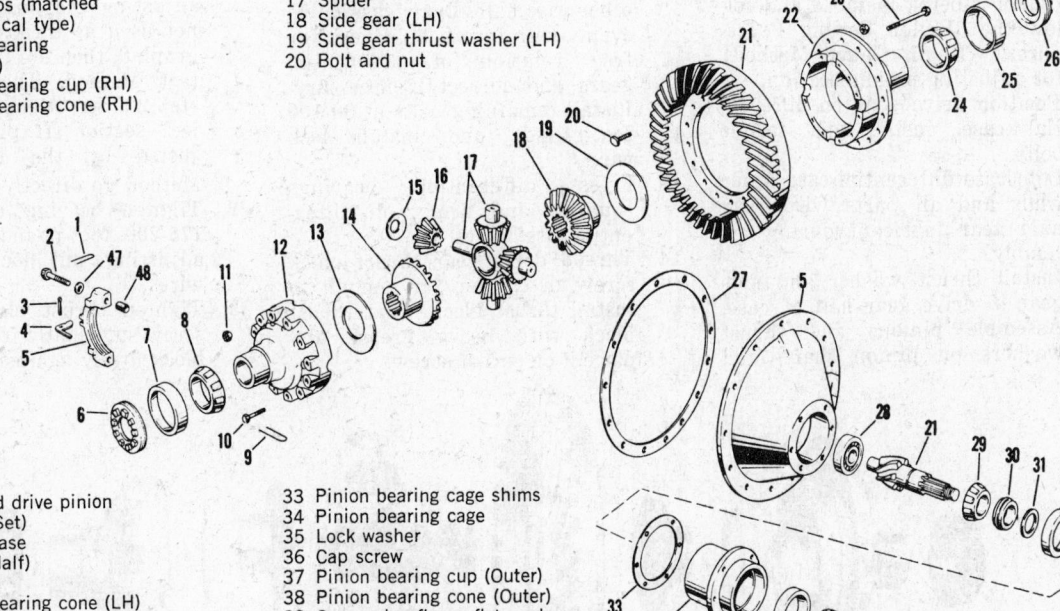

Eaton single-speed, single reduction rear axle (© Eaton Yale & Towne Inc.)

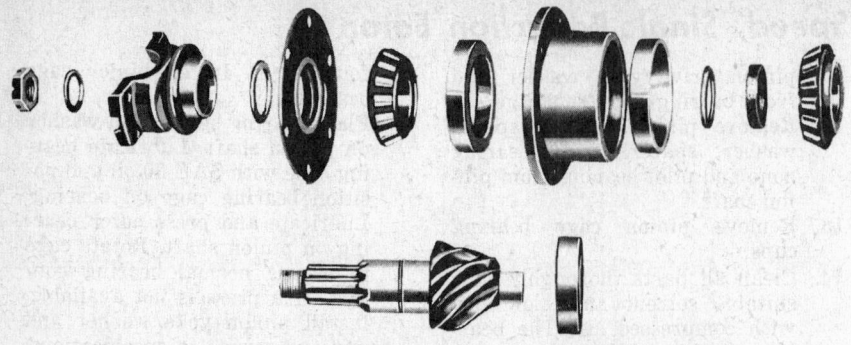

Pinion and cage assembly components (© Eaton Yale & Towne Inc.)

6. Use rotating torque, not starting torque. If rotating torque is not within 15 to 35 inch-pounds, use thinner spacer to increase preload or thicker spacer to decrease pre-load. Torque must be near low limit inch-pounds with original pinion bearings and near high limit when using new bearings.

7. Upon obtaining correct preload, install a new pinion oil seal assembly. Install pinion yoke spacer, yoke and nut, and torque nut to specifications. Install a new cork seal on pinion cage.

8. Place original shim pack on carrier, tap pinion and cage assembly into place with a soft mallet. Make sure oil passages in cage and carrier align. Install lock washers and bolts at 100-125 foot-pounds. See the "General Axle Service" section if a new pinion is being installed and follow the "Pinion Setting Procedures" (Depth Gauge Method) for adjusting pinion location.

9. Position drive gear on differential case, install and torque bolts.

10. Lubricate differential case inner walls and all parts with rear axle gear lubricant during assembly.

11. Install thrust washer and side gear in drive gear half of case. Assemble pinions and thrust washers on pinion shaft, and

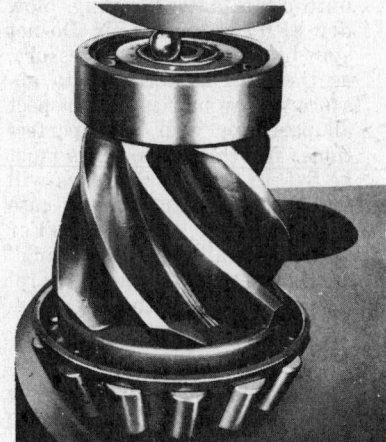

Staking pinion shaft with bearing ball
(© Eaton Yale & Towne Inc.)

place in case. Place other side and thrust washer on four pinions.

12. Align mating marks and install other case half. Draw case down evenly with four bolts. Check for free rotation of differential gears and correct if necessary. Install remaining bolts at 80-100 foot-pounds and install lock wire.

13. Press differential bearings squarely and firmly on differential case halves.

14. Thread drive gear thrust block screw in carrier far enough to install thrust block. Coat thrust block with heavy grease and place it on end of screw.

15. Lubricate differential bearings and cups with rear axle lubricant. Place cups over bearing and position assembly in carrier housing. Hand-tighten adjusting nuts against bearing cups.

16. Install bearing caps in correct location and tap lightly into position, making sure they are not cross-threaded. Make certain that some backlash is present between drive gear and pinion. Install bearing cap bolts snugly with allowance for turning adjusting nuts.

17. Attach a dial indicator to the carrier with the pointer resting against back face of drive gear. Eliminate all end play by turning right adjusting nut clockwise.

18. Tighten adjusting nuts, one notch each, to pre-load differential bearings. Align notches for installation of adjusting nut locks.

19. Position dial indicator so that the pointer rests against face of one of the drive gear teeth. Check backlash between drive gear and pinion at 90 degree intervals of rotation.

20. Adjust backlash to .008-.015 inch (.008 preferred, especially on new gears.) When adjusting backlash, back off one adjusting nut and advance opposite nut the same amount to maintain bearing pre-load.

21. If the Depth Gauge Method for adjusting the pinion depth was not used as mentioned in paragraph 8, then use the Tooth Contact Method. Then proceed to step 22. See "General Axle Service" section. If pinion was adjusted by the Depth Gauge Method go directly to step 22.

22. Tighten bearing cap bolts, to 175-200 foot-pounds. Install the adjusting nut locks and lock wires.

23. Tighten thrust block adjusting screw sufficiently to locate thrust block firmly against back face of

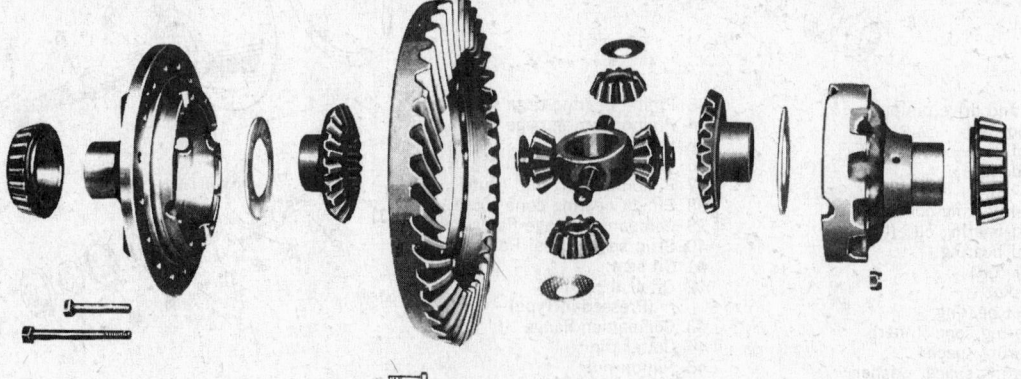

Differential assembly components (© Eaton Yale & Towne Inc.)

gear. Back off adjusting screw ¼ turn to provide .010-.015 inch clearance and lock securely with nut.

24. Place oil distributor in carrier with taper side of scoop down, install spring and retaining plug.

TORQUE CHART

Carrier to housing bolt or stud nut	7/16-14	50-70	Differential bearing adjuster lock bolt	9/16-12	100-125
	1/2-13	80-100		5/8-11	150-175
	5/8-11	150-175	Electric shift unit		
Differential case bolt	7/16-14	50-70	(mounting)	7/16-20	50-70
	1 2-13	80-100	Pinion cage bolt	9/16-12	100-125
Differential bearing cap bolt			Pinion shaft yoke nut	7/8-16	200-325
	5/8-11	150-175		1-1/8-18	325-450
	11/16-11	175-200		1-1/4-12	400-600
	13/16-10	225-250			

EATON AXLE DRIVE PINION ADJUSTMENT SPECIFICATIONS

		Specifications for Drive Pinion Bearing Preload Adjustment				Drive Pinion Position Adjustment	
	Axle Models	Pinion Bearing Spacer Thickness	Drive Pinion Nut Torque Setting (ft.-lbs.)	Arbor Press Preload Pressure Setting (tons)	Spring Scale Reading (pounds) To obtain 15 to 35 in.-lbs. Torque	Pinion Bearing Cage Shim Pack	Depth Gauge Setting
Single	17101, 17121	0.639	400-600	7-8	6-14	0.030 + 0.125 Spacer	4.2187
Reduction	18101, 18121	0.639	400-600	7-8	6-14	0.030 + 0.125 Spacer	4.2187
Axles	19121	0.185	550-850	10-11	5-11	0.024	4.600

Eaton Two-Speed and Planetary Double Reduction Axles

Description

The two-speed axle is a full-floating type with bevel drive gear and pinion, and uses a four-pinion differential assembly. The pinion has three roller bearings—two tapered bearings in front of the pinion teeth and one straight roller bearing be-

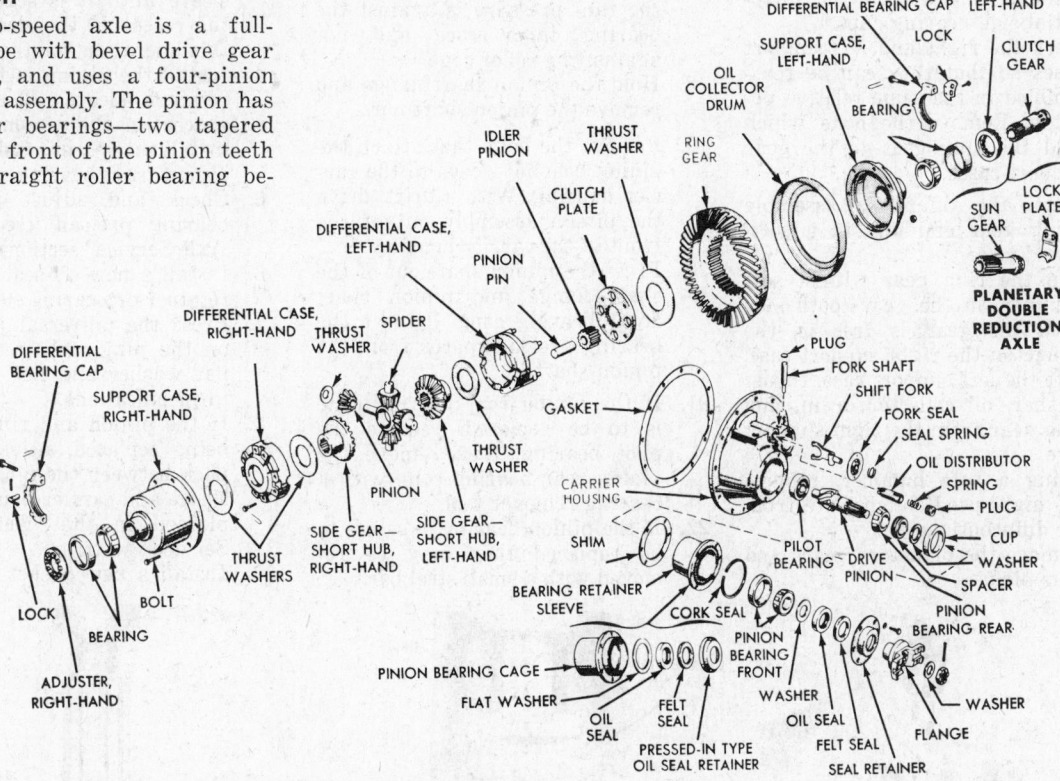

Eaton planetary two-speed or double reduction axle (© Eaton Yale & Towne Inc.)

hind the teeth. An oiling system provides lubrication to the pinion bearings, differential bearings and planetary unit. The planetary double reduction axle combines the ring and pinion and the planetary unit in one axle housing. Procedures for disassembly are the same as the two-speed

except the double reduction axle has no shift mechanism and, instead of a sliding clutch gear, it uses a sun gear.

Differential Carrier

Disassembly

1. Inspect the carrier before disas-

sembly. (See "cleaning" and "inspection" paragraph in the "General Axle Service" section).

2. After removing the oil distributor plug, pull the oil distributor and spring out of the differential carrier.

3. Remove the shift fork shaft

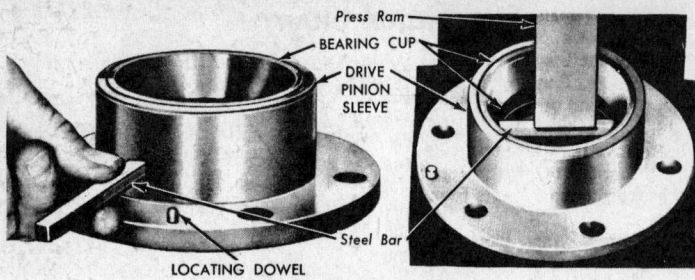

Removal or installation of pinion bearing cup
(© Eaton Yale & Towne Inc.)

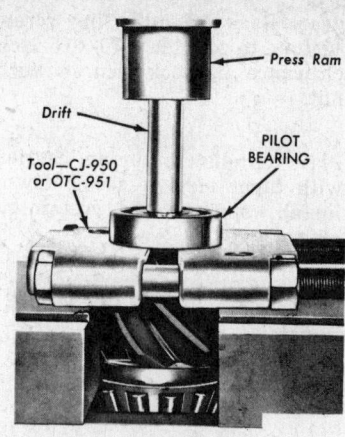

Removing the pinion pilot bearing
(© Eaton Yale & Towne, Inc.)

plugs, and shaft.

4. Remove the shift fork seal and the shift fork from the housing. Pull the sliding clutch gear out of the differential carrier.

5. Punch the right and left bearing adjusters to aid in relocating the adjustment when reassembling. Remove the differential bearing cap bolts. Back off one bearing adjusting nut to relieve the bearing preload.

6. Remove the right and left bearing caps, bearing adjusters, and locks.

7. Lift the differential and planetary assembly out of the differential side bearing cups.

8. Mark the right and left support cases so that they can be reassembled in the same relative position. Remove the bolts which hold the ring gear to the gear support cases.

9. Clamp the differential assembly (ring gear end up) in a soft-jawed vise.

10. Tap the ring gear with a soft hammer on the gear tooth side until the gear is free of the flange on the right support case.

11. Lift the left support case, thrust washer, oil collector drum, and ring gear from the right support case.

12. Using a soft hammer, remove the high-speed clutch plate from the differential case.

13. Remove the planetary gears and gear pins.

14. Lift the differential assembly and the thrust washer out of the right support case. Mark the two pieces of the differential case for proper assembly.

15. Separate the right and left differential cases. Remove the differential side gears and thrust washers.

16. Remove the differential spider gears, thrust washers, and the differential spider.

17. If the differential support case bearings are to be replaced, they may be removed with a drift. Press on the new bearing so that the ram pressure is against the bearing inner race and not against the roller cage.

18. Hold the pinion shaft flange and remove the pinion shaft nut.

19. Remove the bolts that attach the pinion bearing sleeve to the carrier housing. With a drift, drive the pinion assembly out of the front of the carrier housing.

20. Press the pinion shaft out of the shaft flange and pinion shaft front bearing cone. Remove the bearing preload spacer from the pinion shaft.

21. If the pinion rear bearing cone is to be replaced, remove the pilot bearing, then remove the pinion rear bearing cone with a bearing remover tool.

22. If the pinion bearing cups are to be replaced they may be removed with a small steel bar.

Assembly

1. Press the pilot bearing on the pinion shaft and stake the pinion shaft end at 4 points.

2. Press the pinion rear bearing onto the shaft until seated.

3. To install a new bearing cup, place the pinion bearing retainer sleeve in a press and, using a steel bar, press the cup in the sleeve until it is seated against the recess in the sleeve. Repeat this operation for the other cup.

4. Install the pinion in the bearing retainer sleeve. Place the old spacer on the pinion shaft and install the spacer washer and the pinion front bearing.

5. Check and adjust the pinion bearing preload (see "General Axle Service" section).

6. Install a new oil seal in the seal retainer or bearing sleeve.

7. Press the universal joint flange on the pinion shaft. Install the flat washer and nut and torque to specifications.

8. If the pinion and ring gear are being replaced, adjust the shim pack between the pinion bearing sleeve and carrier housing as explained in the "General Axle Service" section.

9. Install a new gasket in the pin-

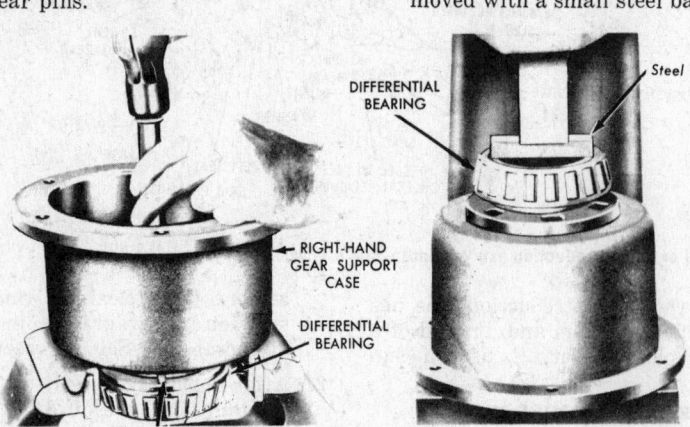

Removing or installing the differential bearing (© Eaton Yale & Towne, Inc.)

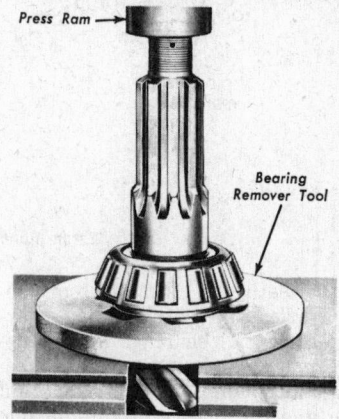

Removing the pinion rear bearing cone
(© Eaton Yale & Towne, Inc.)

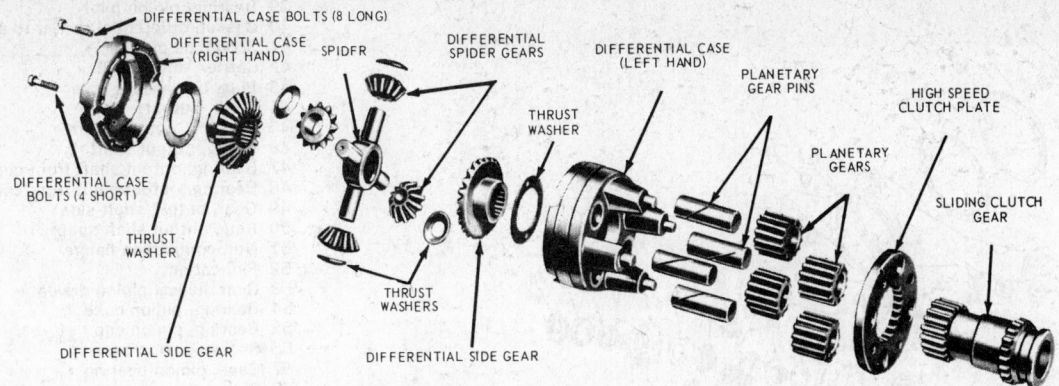

Planetary and differential assembly (© Eaton Yale & Towne Inc.)

ion bearing sleeve, if so equipped.

10. Place the shim pack on the bearing sleeve and install the pinion and sleeve assembly in the carrier housing. Install the sleeve to housing bolts and lock washers and torque to specifications.

11. Place the left half of the differential support case in a vise with the planetary side facing downward. Place one of the differential side gear thrust washers in position in the case. Place the differential left-hand side gear (with the short hub) in position in the case.

12. Install the four differential pinions and thrust washers on the differential spider, then place the spider in the differential case. Place the right side gear and thrust washer on the differential pinions.

13. Place the right half of the differential case in position over the differential spider, then install and torque the case bolts.

14. Insert the four planetary gear pins in the holes in the differential case.

15. Place the four planetary gears on the planetary gear pins. Position the high speed clutch plate on the planetary gear pins with the chamfered teeth facing the planetary gears. Tap the plate in place with a brass hammer.

16. Position the gear support thrust washer in the right gear support case.

17. Place the differential assembly in the gear support case with the planetary end facing upward. Place the ring gear (teeth facing down) over the differential assembly, and engage the planetary gears with the internal teeth on the ring gear.

18. Align the ring gear bolt holes with those in the gear support case. Place the oil collector drum on the ring gear (open side facing gear) with the notches between the bolt holes in the ring gear.

19. Apply oil to each side of the high-speed clutch plate thrust washer and position on the high-speed clutch plate. Place the left gear support case on the ring gear, and line up the bolt holes in the ring gear and both gear support cases.

20. Install the 6 bolts through the gear support cases with the bolt heads against the flange of the right gear support case. Install and torque the nuts and secure with lock wire.

21. Install the differential assembly in the carrier housing and adjust backlash and preload (see "General Axle Service" section).

22. Check and adjust the gear tooth contact pattern.

23. Install the sliding clutch gear and attach the shift fork to it. Install the shift fork shaft through the hole in the differential carrier and into the shift fork. Install the shift fork shaft plugs. Place the shift fork seal and retainer on the shift fork. Shift the axle into the low range.

24. Place the oil distributor and spring in the carrier. Install the oil distributor plug and gasket.

Eaton Thru-Drive Type Tandem Axles

Description

The Thru-Drive tandem axles are of the single reduction, double reduction, two speed and three speed types.

The three speed axle type is basically a two speed axle unit, electrically controlled by the operator to operate the air or vacuum lock-out mechanism for the power-divider and the speeds are obtained as follows:

High Speed—Both axles in high range and in single reduction.
Low Speed—Both axles in low range and in double reduction.
Intermediate Speed—Forward axle in high range and single reduction and the rear axle in low range and in double reduction,

and the inter-axle differential, or power divider, equalizing the differences in RPMs, between the two axles to give a speed midway between the high and low range speeds by the movement of the differential gears in the power divider.

A lock-out is used to disengage the power divider and to provide maximum traction and positive drive to both axles during the high and low ranges.

NOTE: This service procedure will cover the forward rear axle of the tandem unit as the rear axle is of the conventional single reduction, two speed or double reduction design and is covered in detail in another chapter.

Differential Carrier and Power Divider

Disassembly

1. Place the differential carrier and power divider assembly in a repair stand and inspect before disassembly. Refer to the "cleaning" and "inspection" paragraphs in the "General Axle Service" section.

2. Remove the carrier cover to differential carrier capscrews and lock washers, and remove the carrier cover assembly.

3. Remove the inner-axle differential off the output shaft side gear.

4. Punch mark the inner-axle differential case before disassembly. Remove the lock nuts and bolts. Separate the cases and remove the thrust washers, side pinions, bushings and spider.

Removal of the Output Shaft

1. Remove the shaft holdout springs and thrust washers if used, and discard. Lift out the output shaft assembly from the carrier assembly by tapping at the companion flange end of the shaft, and remove the bearing cup from the carrier.

2. To disassemble the output shaft,

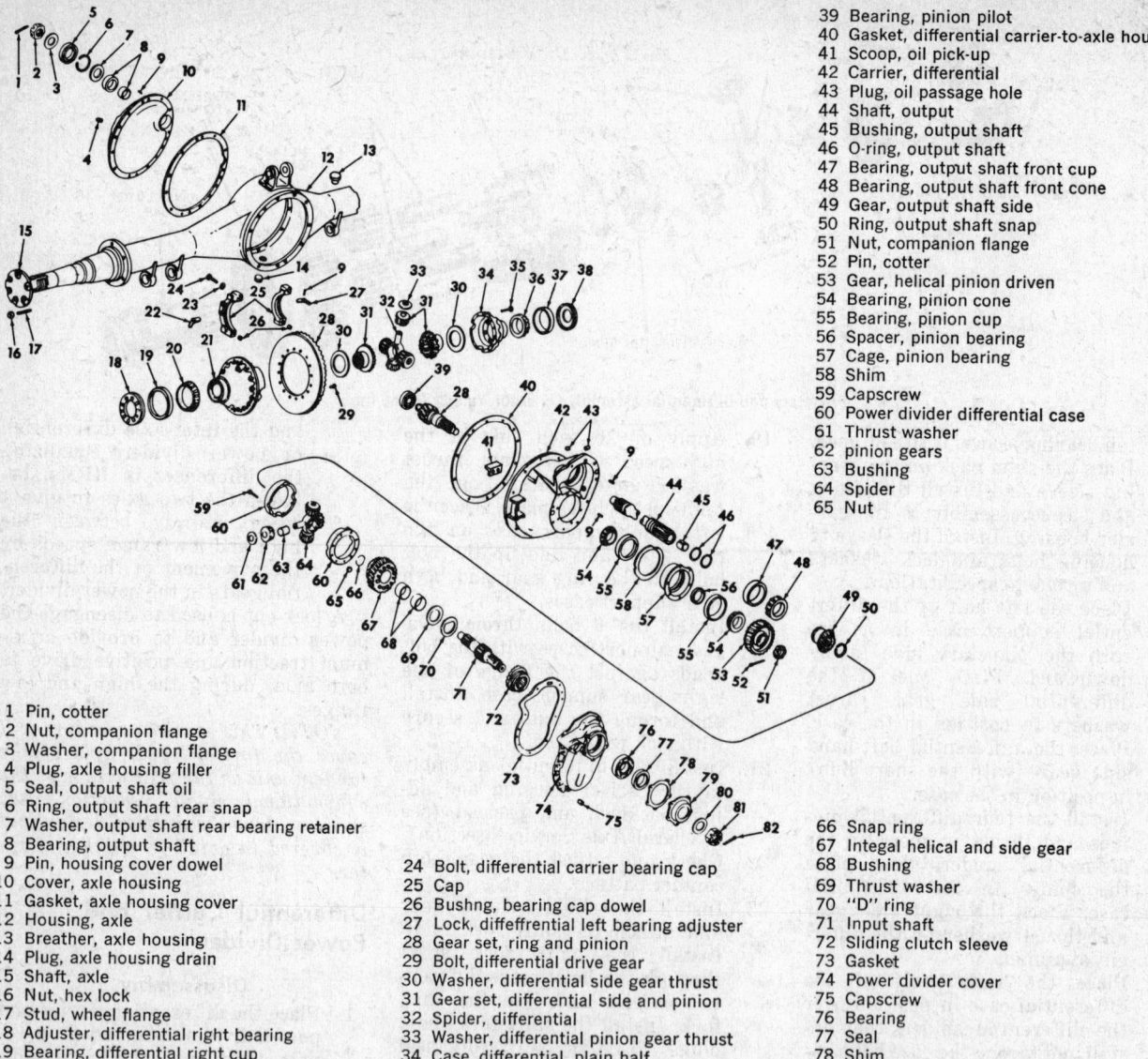

39 Bearing, pinion pilot
40 Gasket, differential carrier-to-axle housing
41 Scoop, oil pick-up
42 Carrier, differential
43 Plug, oil passage hole
44 Shaft, output
45 Bushing, output shaft
46 O-ring, output shaft
47 Bearing, output shaft front cup
48 Bearing, output shaft front cone
49 Gear, output shaft side
50 Ring, output shaft snap
51 Nut, companion flange
52 Pin, cotter
53 Gear, helical pinion driven
54 Bearing, pinion cone
55 Bearing, pinion cup
56 Spacer, pinion bearing
57 Cage, pinion bearing
58 Shim
59 Capscrew
60 Power divider differential case
61 Thrust washer
62 pinion gears
63 Bushing
64 Spider
65 Nut

1 Pin, cotter
2 Nut, companion flange
3 Washer, companion flange
4 Plug, axle housing filler
5 Seal, output shaft oil
6 Ring, output shaft rear snap
7 Washer, output shaft rear bearing retainer
8 Bearing, output shaft
9 Pin, housing cover dowel
10 Cover, axle housing
11 Gasket, axle housing cover
12 Housing, axle
13 Breather, axle housing
14 Plug, axle housing drain
15 Shaft, axle
16 Nut, hex lock
17 Stud, wheel flange
18 Adjuster, differential right bearing
19 Bearing, differential right cup
20 Bearing, differential right cone
21 Case, differential, flanged half
22 Lock, differential right bearing adjuster
23 Washer, differential carrier bearing cap bolt

24 Bolt, differential carrier bearing cap
25 Cap
26 Bushng, bearing cap dowel
27 Lock, diffeffrential left bearing adjuster
28 Gear set, ring and pinion
29 Bolt, differential drive gear
30 Washer, differential side gear thrust
31 Gear set, differential side and pinion
32 Spider, differential
33 Washer, differential pinion gear thrust
34 Case, differential, plain half
35 Bolt, differential case
36 Bearing, differential left cone
37 Bearing, differential left cup
38 Adjuster, differential left bearing

66 Snap ring
67 Integal helical and side gear
68 Bushing
69 Thrust washer
70 "D" ring
71 Input shaft
72 Sliding clutch sleeve
73 Gasket
74 Power divider cover
75 Capscrew
76 Bearing
77 Seal
78 Shim
79 Bearing retainer
80 Washer
81 Nut, input shaft flange
82 Pin, cotter

Exploded view—Eaton single reduction forward rear axle—typical (© International Harvester Co.)

remove the snap ring and lift off the side gear assembly. Discard the output shaft O-rings.

3. Remove the bushing from the bore of the output shaft and replace it, if necessary.

4. Remove the output shaft front bearing from the side gear, with the use of a press.

Removal of the Input Shaft

1. Remove the snap ring from the input shaft and lift off the helical side gear, helical side thrust washer and the "D" washer.

2. If replacement is necessary, press the bushing from the bore of the helical side gear.

3. Remove the cotter pin, nut and input companion flange from the input shaft.

4. Remove the input shaft bearing

retainer-to-differential carrier cover capscrews and lift the bearing cover and shims off. *NOTE: Observe the size and quantity of shims removed to ease the reassembly adjustment.*

5. Remove the oil seal and felt seal from the bearing cover.

6. Drive the input shaft out of the differential carrier cover.

7. Remove the bearing from the input shaft by the use of a press and pressing the shaft out of the bearing.

Differential and Gear Assembly Single Reduction Carrier

Removal

1. Remove the lock wires, if used. Mark the differential carrier legs and the bearing caps for proper

reassembly. Remove the adjusting nut locks.

2. Remove the bearing cap stud nuts or capscrews, the bearing caps and the adjusting nuts.

3. Loosen the jam nut and back off the thrust block adjusting screw.

4. Lift out the differential and gear assembly.

5. Remove the thrust block from inside the carrier housing.

6. Mark the differential case halves with a punch to aid in the reassembly.

7. Remove the lock wire (if equipped), the bolts and separate the case halves.

8. Remove the spider, pinions, side gears and thrust washers.

9. Remove the rivets or bolts and remove the ring gear from the case. *NOTE: Do not chisel the*

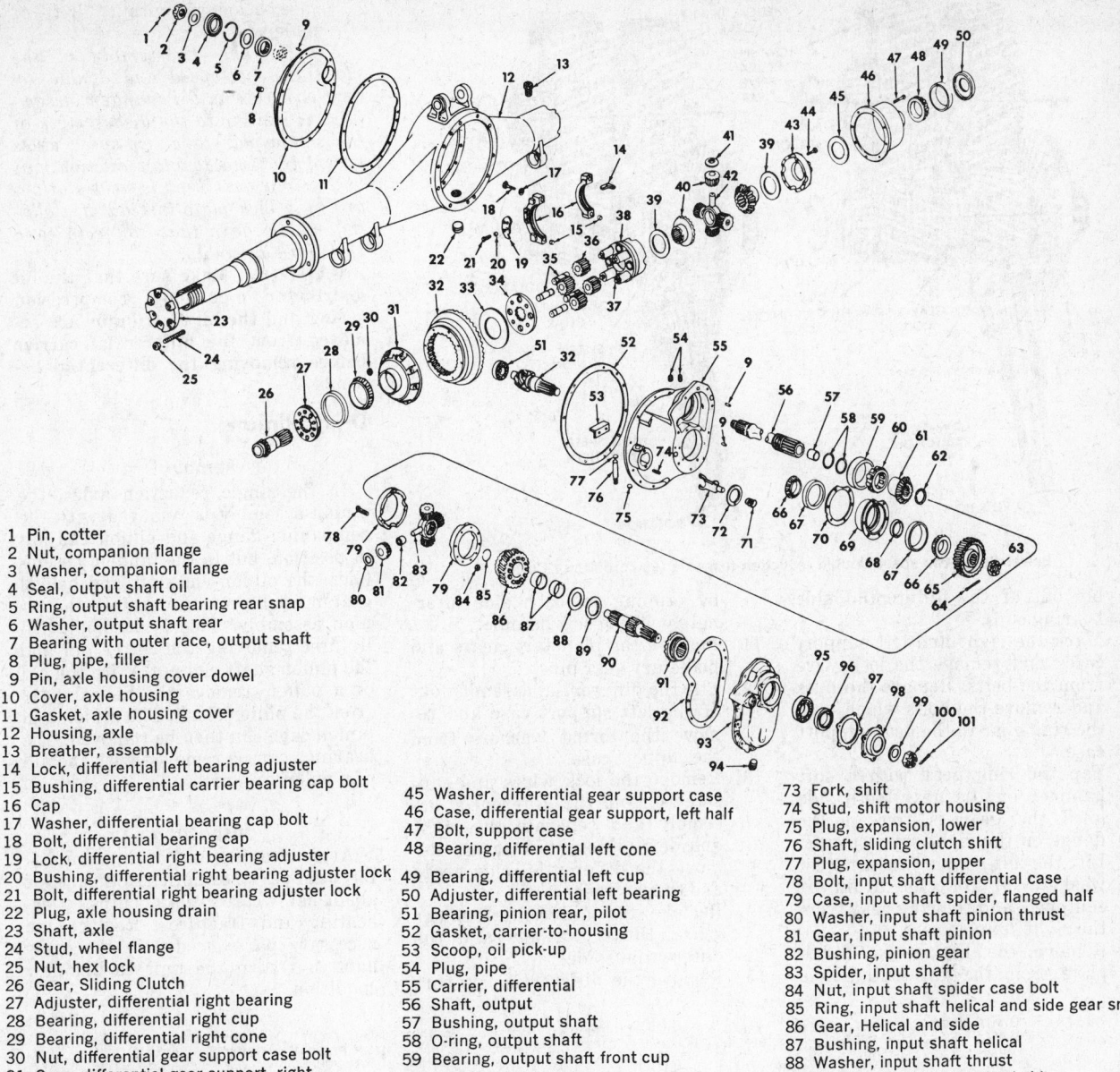

1 Pin, cotter
2 Nut, companion flange
3 Washer, companion flange
4 Seal, output shaft oil
5 Ring, output shaft bearing rear snap
6 Washer, output shaft rear
7 Bearing with outer race, output shaft
8 Plug, pipe, filler
9 Pin, axle housing cover dowel
10 Cover, axle housing
11 Gasket, axle housing cover
12 Housing, axle
13 Breather, assembly
14 Lock, differential left bearing adjuster
15 Bushing, differential carrier bearing cap bolt
16 Cap
17 Washer, differential bearing cap bolt
18 Bolt, differential bearing cap
19 Lock, differential right bearing adjuster
20 Bushing, differential right bearing adjuster lock
21 Bolt, differential right bearing adjuster lock
22 Plug, axle housing drain
23 Shaft, axle
24 Stud, wheel flange
25 Nut, hex lock
26 Gear, Sliding Clutch
27 Adjuster, differential right bearing
28 Bearing, differential right cup
29 Bearing, differential right cone
30 Nut, differential gear support case bolt
31 Case, differential gear support, right
32 Gear set, ring and pinion
33 Washer, clutch plate thrust
34 Plate, high speed clutch
35 Shaft, differential Idler gear
36 Gear, differential idler
37 Pin, high speed clutch plate
38 Case, differential, right half
39 Washer, side gear thrust
40 Gear set, pinion and side
41 Washer, pinion gear thrust
42 Spider, differential
43 Case, differential, left half
44 Bolt, differential case

45 Washer, differential gear support case
46 Case, differential gear support, left half
47 Bolt, support case
48 Bearing, differential left cone
49 Bearing, differential left cup
50 Adjuster, differential left bearing
51 Bearing, pinion rear, pilot
52 Gasket, carrier-to-housing
53 Scoop, oil pick-up
54 Plug, pipe
55 Carrier, differential
56 Shaft, output
57 Bushing, output shaft
58 O-ring, output shaft
59 Bearing, output shaft front cup
60 Bearing, output shaft front cone
61 Gear, output shaft side
62 Ring, output shaft snap
63 Nut, companion flange
64 Pin, cotter
65 Gear, Helical pinion driven
66 Bearing, pinion cone
67 Bearing, pinion cup
68 Spacer, pinion bearing
69 Cage, pinion bearing
70 Shim, pinion bearing cage
71 Spring, shift fork
72 Seal, shift motor

73 Fork, shift
74 Stud, shift motor housing
75 Plug, expansion, lower
76 Shaft, sliding clutch shift
77 Plug, expansion, upper
78 Bolt, input shaft differential case
79 Case, input shaft spider, flanged half
80 Washer, input shaft pinion thrust
81 Gear, input shaft pinion
82 Bushing, pinion gear
83 Spider, input shaft
84 Nut, input shaft spider case bolt
85 Ring, input shaft helical and side gear snap
86 Gear, Helical and side
87 Bushing, input shaft helical
88 Washer, input shaft thrust
89 Washer, input shaft helical side
90 Shaft, input drive
91 Gear, input shaft clutch sliding
92 Gasket, differential carrier cover
93 Cover, differential carrier
94 Plug, pipe, drain
95 Bearing, input shaft
96 Seal, input shaft oil
97 Shim, oil seal retainer
98 Retainer, input shaft bearing
99 Washer, Companion flange
100 Nut, companion flange
101 Pin, cotter

Exploded view—Eaton two and three speed forward rear tandem axle assembly— typical (© International Harvester Co.)

rivets from the ring gear and case. Damage to the case can result. Always drill through the rivet and press the rivet from the gear and case.

10. Remove the differential bearings with the use of a suitable puller.

Two Speed and Double Reduction Carriers
Removal

1. Remove the lock wires. Mark the right and left bearing adjusters to aid in the reassembly. Mark the bearing caps and the differential carrier legs for proper reassembly.

2. Remove the oil distributor plug, if equipped, and pull the oil distributor and spring out of the differential carrier.

3. Remove the shift fork shaft plugs and remove the shaft.

4. Remove the shift fork seal and the shift fork from the housing. Remove the sliding clutch gear from the differential carrier.

5. Remove the differential bearing cap bolts and back off the bearing adjusting nuts to relieve the bearing preload.

6. Remove the bearing caps, adjusters, and locks and remove the differential and planetary assem-

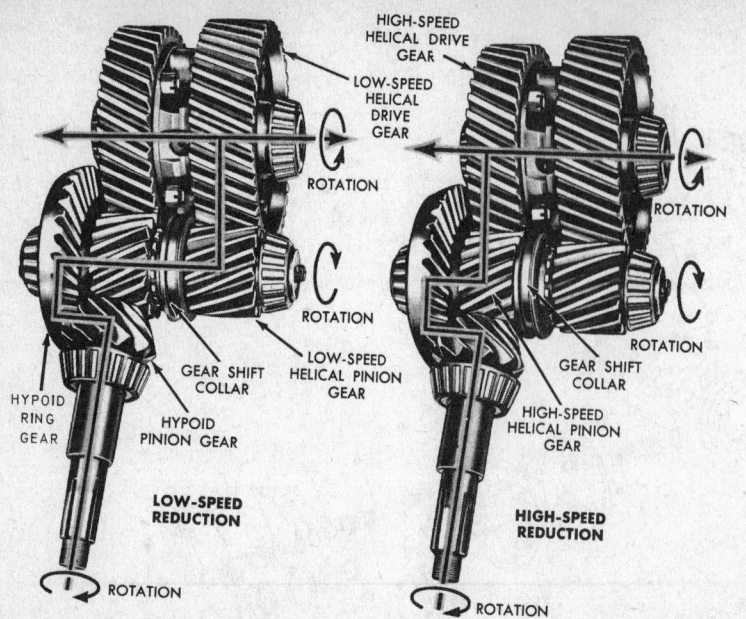

HIGH-SPEED HELICAL DRIVE GEAR

LOW-SPEED HELICAL DRIVE GEAR

ROTATION

ROTATION

ROTATION

ROTATION

GEAR SHIFT COLLAR

LOW-SPEED HELICAL PINION GEAR

HIGH-SPEED HELICAL PINION GEAR

GEAR SHIFT COLLAR

HYPOID RING GEAR

HYPOID PINION GEAR

LOW-SPEED REDUCTION

HIGH-SPEED REDUCTION

ROTATION

ROTATION

Power flow—two speed double reduction rear axle (© Ford Motor Co.)

bly out of the differential side bearing cups.

7. Mark the right and left support cases and remove the lock wire from the bolts. Remove the nuts and remove the bolts which hold the ring gear to the gear support case.

8. Tap the ring gear with a soft hammer on the gear tooth side until the gear is free of the flange on the right support case.

9. Lift the left support case, thrust washer, oil collector drum, if equipped, and the ring gear from the right support case.

10. Remove the high-speed clutch plate from the differential case

by tapping on each side alternately with a soft hammer.

11. Remove the planetary gears and planetary gear pins.

12. Lift the differential assembly out of the left support case and remove the thrust washers from the support case.

13. Remove the lock wires and capscrews from the differential assembly and separate the right and left differential cases. Remove the differential side gears and thrust washers.

14. Remove the differential spider gears, thrust washers and the differential spider.

15. Remove the differential support

case bearings with a drift or puller.

NOTE: The disassembly of the planetary two-speed and double reduction axles used in tandem assemblies is similar to the disassembly of the single planetary two-speed axles except that the double reduction unit uses a sun gear that is held stationary by a lock plate instead of a sliding clutch gear that connects to a shift fork assembly.

CAUTION: Make sure that the oil distributor pipe plug, compression spring and the oil distributor are removed from the differential carrier before removing the differential assembly.

Drive Pinion

Removal

In the single reduction axles, the pinion assembly is removed with the companion flange and pinion end nut in position, but in the tandem assemblies, the pinion end nuts and helical gear must be removed before the pinion assembly. This is accomplished by first removing the cotter pin and the pinion shaft end nut. With the aid of a puller, remove the helical gear from the pinion shaft. The pinion and pinion cage can then be removed. The bearings can be removed with the aid of a puller or a press.

Inspection

After cleaning and drying the differential parts, examine each part for abnormal wear, cracks, chips, overheating and fractures. Replace the necessary parts needed in the overhaul to return the unit to like-new condition.

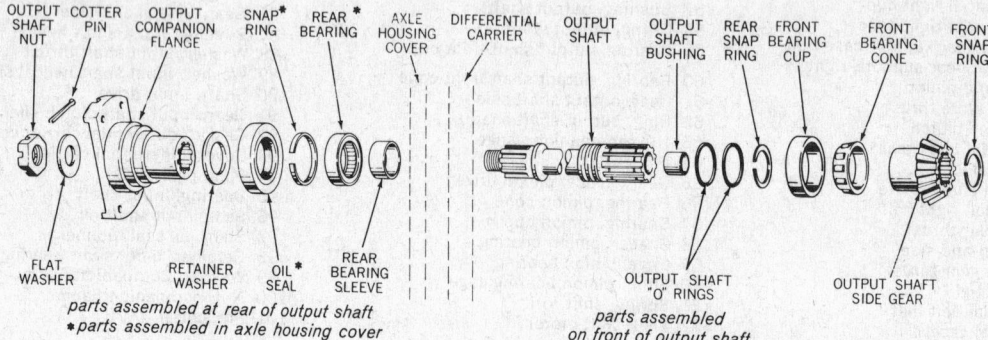

OUTPUT SHAFT NUT | COTTER PIN | OUTPUT COMPANION FLANGE | SNAP* RING | REAR* BEARING | AXLE HOUSING COVER | DIFFERENTIAL CARRIER | OUTPUT SHAFT | OUTPUT SHAFT BUSHING | REAR SNAP RING | FRONT BEARING CUP | FRONT BEARING CONE | FRONT SNAP RING

FLAT WASHER | RETAINER WASHER | OIL* SEAL | REAR BEARING SLEEVE | OUTPUT SHAFT "O" RINGS | OUTPUT SHAFT SIDE GEAR

parts assembled at rear of output shaft
*parts assembled in axle housing cover

parts assembled on front of output shaft

Exploded view—output shaft components—Eaton Axles (© International Harvester Co.)

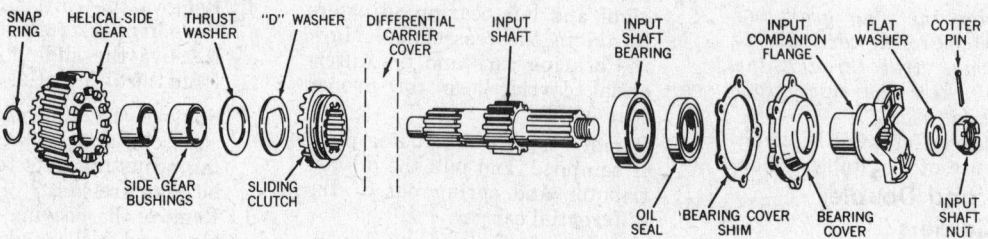

SNAP RING | HELICAL-SIDE GEAR | THRUST WASHER | "D" WASHER | DIFFERENTIAL CARRIER COVER | INPUT SHAFT | INPUT SHAFT BEARING | INPUT COMPANION FLANGE | FLAT WASHER | COTTER PIN

SIDE GEAR BUSHINGS | SLIDING CLUTCH | OIL SEAL | BEARING COVER SHIM | BEARING COVER | INPUT SHAFT NUT

Exploded view—input shaft—Eaton axle (© International Harvester Co.)

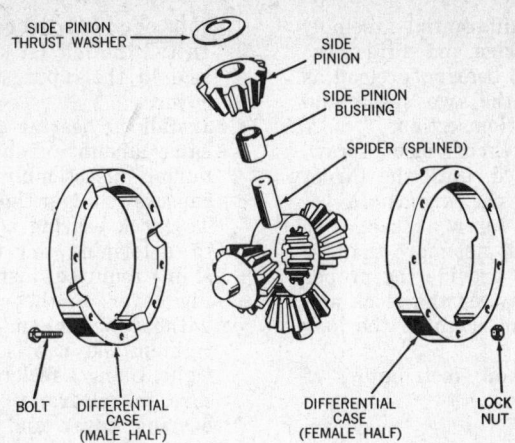

SIDE PINION THRUST WASHER

SIDE PINION

SIDE PINION BUSHING

SPIDER (SPLINED)

BOLT

DIFFERENTIAL CASE (MALE HALF)

DIFFERENTIAL CASE (FEMALE HALF)

LOCK NUT

Exploded view—Inter-axle differential (© International Harvester Co.)

Drive Pinion

Installation

1. Install the necessary bearings and cups on the pinion shaft and the pinion bearing cage.
2. Measure the preload of the pinion bearings by assembling the bearing and pinion shaft without the oil seal in place.
3. Torque the pinion nut to the specified torque and place the universal joint flange in a vise and wrap a strong cord around the pinion bearing retainer.
4. Attach a pound pull scale to the cord and note the pull required to keep the bearing retainer moving. *NOTE: The reading on the pull scale multiplied by one-half the diameter (radius) of the bearing retainer at the point the cord is wrapped (in inches) is the in. lb. torque.*
5. The bearing preload is adjusted by the installation of spacers of different lengths. *NOTE: A shorter spacer increases the preload and a longer spacer decreases the preload.*
6. If the pinion and ring gear are being replaced, adjust the shim pack accordingly between the pinion bearing sleeve and the carrier housing, as per the shim chart.
7. Install a new gasket in the pinion bearing sleeve, is used.
8. Install the shim pack on the bearing sleeve and install the pinion and sleeve assembly into the carrier housing. Install the bolts and torque to specified tension.

Two Speed and Double Reduction Carriers

Installation

1. Place the left half of the differential support case in a vise with the planetary side facing downward.
2. Place one of the side gear thrust washers in position in the case. Position the left hand side gear in the case. (Gear with the short Hub).
3. Install the four differential pinions and thrust washers on the differential spider and place the spider in the differential case. Install the right side gear and thrust washer on the differential pinions.
4. Place the right half of the differential case over the differential spider and install the case bolts. Tighten to specified torque.
5. Install the four planetary gear pins in the holes in the differential case.
6. Place the four planetary gears on the four planetary gear pins. Position the high speed clutch plate on the planetary gear pins with the chamfered teeth facing the planetary gears. Tap the plate in place with a soft hammer.
7. Place the right gear support case in a vise with the bearing side facing downward. Position the

gear support thrust washer in the case.
8. Position the differential assembly in the gear support case with the planetary end facing upward. Place the ring gear over the differential assembly with the teeth facing downward and mesh the planetary gears with the internal teeth on the ring gear.
9. Align the ring gear bolt holes with the holes in the gear support case. Place the oil collector drum on the ring gear with the open side facing the gear, if equipped.
10. Lubricate the high speed clutch plate thrust washer and position it on the high speed clutch plate. Place the left gear support case on the ring gear and align the holes in the ring gear and the gear support cases.
11. Install the bolts through the gear support cases, install the nuts and tighten to specifications.
12. Install the differential into the carrier housing and adjust the backlash and preload by meshing the ring gear with the pinion so that a small amount of backlash exists.
13. Set the differential bearing adjusters in the carrier bearing pocket threads so that they just contact the bearing cups.
14. Position the bearing caps on the carrier and match the marks made upon disassembly. Install the bolts or nuts and tighten to specifications, while moving the adjuster to assure freedom of movement. *NOTE: The bolts or nuts of the bearing caps will have to be loosened to allow the adjusters to move properly upon adjustment of the backlash of the gears.*
15. Alternately tighten the bearing adjusters until there is some backlash between the ring gear and pinion, with some preload on the bearings. *NOTE: While the bearings are preloaded, rotate the ring gear several times in each direction to seat the bearing rollers in the cups. This bearing roller seating is most important to assure the bearing rollers of longer life.*

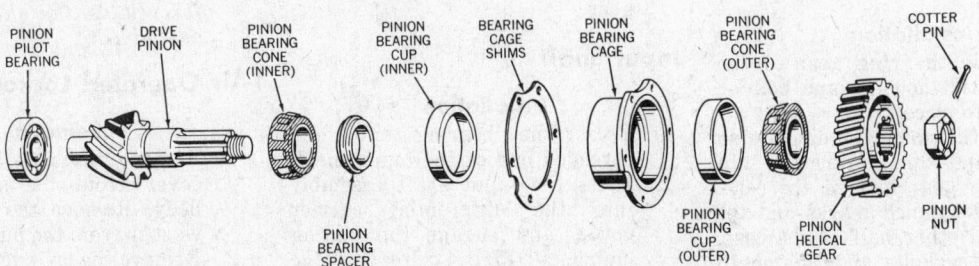

PINION PILOT BEARING

DRIVE PINION

PINION BEARING CONE (INNER)

PINION BEARING CUP (INNER)

BEARING CAGE SHIMS

PINION BEARING CAGE

PINION BEARING CONE (OUTER)

COTTER PIN

PINION BEARING SPACER

PINION BEARING CUP (OUTER)

PINION HELICAL GEAR

PINION NUT

Exploded view—drive pinion—Eaton axle (© International Harvester Co.)

16. Loosen the bearing adjusters until they are clear of the bearing cups and then tighten them until they just touch the bearing cups.
17. Install a dial indicator on the carrier housing with the button against the back face of the ring gear.
18. Adjust the differential end play to zero, without preloading the bearings, and then tighten each adjuster nut one notch to preload the bearings.
19. Check the ring gear to pinion backlash at four equally spaced points on the ring gear. Refer to the specifications for the proper back lash clearance.
20. To reduce the back lash, loosen the pinion side adjuster and tighten the ring gear side adjuster the same number of notches to maintain the bearing preload. To increase the backlash, preform the same operation in reverse. *NOTE: When moving the adjusters, the final movement should always be made in the tightening direction. For example, if one adjuster is to be loosened one notch, loosen it two notches, then retighten it one notch. This procedure will assure the adjuster is in contact with the bearing cup and the cup will not shift after being put into service.*
21. Adjust the backlash to the best tooth contact pattern within the specification limits.
22. Torque all bolts and nuts to specifications and install the locking wire and secure.
23. Install the sliding clutch gear and position the shift fork on the sliding clutch gear.
24. Install the shift fork shaft through the hole in the differential carrier and into the shift fork. Install the shift fork shaft plugs. Install the shift fork shaft seal and retainer on the shift fork and shift the axle into low range.
25. Install the oil distributor and spring in position in the carrier and install the plug and gasket, if so equipped.

Single Reduction Carrier

Installation

1. Assemble the ring gear to the differential housing and bolt or rivet into place.
2. Install the spider, pinions, side gears and thrust washers into the ring gear half of the case. Align the punch marks and install the other half of the case. Install the bolts and tighten to specifications. Install the lock wires.

3. Install the differential assembly into the carrier and adjust the backlash and bearing preload as outlined in the two speed and double reduction section.
4. Tighten the thrust block screw, if so equipped, until the thrust block touches the ring gear.
5. Back off the screw at least 1/16 of a turn, but not more than ¼ of a turn to provide the proper clearance between the block and the ring gear. Tighten the lock nut.
6. Install the oil distributor, if equipped.

Output Shaft

Installation

1. If removed, install the front bearing cup in the differential carrier and bushing in the bore of the output shaft.
2. Press the front bearing cone on the output shaft side gears.
3. Install the rear snap ring on the output shaft and place the side gear assembly on the output shaft and install the front snap ring.
4. Install the two O rings in the grooves of the output shaft.
5. Lubricate the components and install the output shaft assembly in the differential carrier by placing the front bearing cone on the bearing cup in the bore of the carrier.

Differential Carrier and Power Divider

Installation

1. Lubricate the internal parts of the inter-axle differential and install the bushings, side pinions and thrust washers on the journals of the differential spider.
2. Place the spider assembly into the male differential housing and mate the opposite housing, while aligning the punch marks.
3. Secure the assembly with bolts and nuts and tighten to specifications.
4. Install the inter-axle differential onto the output shaft. Position the differential with the nuts away from the output shaft side gear.

Input Shaft

Installation

1. Press the bearing onto the threaded end of the input shaft.
2. Press the input shaft assembly into the differential carrier cover, and engage the sliding clutch. *NOTE: Assure that the oil drain back holes are properly aligned in the carrier cover,*

shims and bearing cover.
3. Install the felt (if used), and oil seal in the input shaft bearing cover.
4. Install the bearing cover and the same amount of shims removed during disassembly. Install the capscrews finger tight.
5. To check bearing adjustment or to determine the thickness of shims required, install the bearing cover on the carrier cover without shims and install and tighten the cap screws finger tight. Using a feeler gauge, measure the clearance between the bearing cover and the carrier cover. This clearance plus 0.001 inch would equal the thickness of the shims required for the correct bearing clearance.
6. Install the companion flange on the input shaft, along with the nut and washer. Tighten the nut to specifications and install the cotter pin.
7. Tighten the bearing cover capscrews to the correct torque.
8. If removed, press the two bushings in the bore of the helical side gear. Place the "D" washer, thrust washer and helical side gear assembly on the input shaft. Install the snap ring on the input shaft to secure these parts.

Differential Carrier Cover

Installation

1. Install a new gasket on the differential carrier and install the cover assembly onto the carrier, aligning the dowel pins with the holes in the cover.
   NOTE: Rotate the input shaft slowly to engage the input shaft splines with the splines in the bore of the inter-axle differential spider.
2. Install two cover screws while observing for any binding between the cover and the carrier. If no binding exists, install the remaining cap screws and tighten to specifications.
   NOTE: After the assembly, check for the inter-axle differential for operation by rotating the input campanion flange, while holding the output shaft stationary. If the assembly differentiates, the unit is correctly assembled.

Air Operated Lockout

Removal

1. Remove the nuts, washers and cover from the shift cylinder body. Remove the nut and flat washer from the push rod.
2. Remove the cylinder body assembly and compression spring from the carrier cover. *NOTE: Check*

DRIVE PINION ADJUSTMENT SPECIFICATIONS

Axle Models		Specifications for Drive Pinion Bearing Preload Adjustment				Drive Pinion Position Adjustment	
		Pinion Bearing Spacer Thickness	Drive Pinion Nut Torque Setting (ft.-lbs.)	Arbor Press Preload Pressure Setting (tons)	Spring Scale Reading (pounds) To obtain 15 to 35 in.-lbs. Torque	Pinion Bearing Cage Shim Pack	Depth Gauge Setting
2-speed Axles	13802, 15201	0.530	225-350	4-5	7-16	0.040 + 0.250	3.6244
	16802, 16244	0.528	325-450	5-6	7-16	0.043	4.125
	17201, 17221	0.638	400-600	7-8	6-14	0.037	4.062
	18201, 18221	0.638	400-600	7-8	6-14	0.037	4.4062
	19221	0.188	600-800	10-11	5-11	0.025	4.7812
Planetary Double Reduction Axles	18301	0.638	400-600	7-8	6-14	0.037	4.4062

TORQUE CHART

Description		Ft.-Lbs.
Differential bearing adjuster lock	Cap screw 9/16-12 (Grade 5)	115-125
	Cap screw 5/8-11 (Grade 5)	160-175
	Cap screw 3/4-10 (Grade 5)	275-300
Differential bearing cap to carrier	Cap screw 5/8-11 (Grade 7)	150-170
	Cap screw 11/16-11 (Grade 7)	210-250
	Cap screw 13/16-10 (Grade 7)	350-425
	Cap screw 7/8-9 (Grade 7)	425-500
Differential carrier to axle housing	Cap screw 7/16-14 (Grade 5)	45-55
	Cap screw 1/2-13 (Grade 5)	75-85
	Cap screw 5/8-11 (Grade 5)	160-175
	Stud nut 1/2-20 (Grade 8)	110-130
	Stud nut 5/8-18 (Grade 8)	220-240
Differential case (axle)	Cap screw 7/16-14 (Grade 7)	55-65
	Cap screw 1/2-13 (Grade 7)	90-105
	Cap screw 9/16-12 (Grade 7)	135-155
Ring gear support case	Bolt/nut 7/16-20 (Grade 7)	60-80
	Bolt/nut 1/2-20 (Grade 7)	90-100
	Bolt/nut 9/16-18 (Grade 7)	130-150
	Bolt/nut 5/8-18 (Grade 7)	190-210
	Locknut 5/8-18	165-190

Description		Ft.-Lbs.
Drive pinion	Nut 1-20	225-350
	Nut 1-1/8-18	325-450
	Nut 1-1/4-12	400-600
	Nut 1-1/2-18	500-700
	Nut 1-3/4-12	600-800
Drive pinion bearing cage to carrier	Cap screw 9/16-12 (Grade 5)	115-125
	Cap screw 9/16-12 (Grade 7)	135-155
	Cap screw 5/8-11 (Grade 5)	160-175
	Cap screw 5/8-11 (Grade 7)	170-190
Inspection cover to axle housing	Cap screw 7/16-14 (Grade 3)	35-50
	Cap screw 7/16-14 (Grade 5)	45-55
Shift units to carrier (2-speed)	Stud nut 7/16-20 (Grade 5)	35-45
	Stud nut 1/2-20 (Grade 5)	85-95
Cover plate to carrier (pdr) (At shift unit opening)	Cap screw 7/16-14 (Grade 5)	45-55
	Cap screw 1/2-13 (Grade 5)	75-85
Oil pickup through to carrier	Nylock screw 1/4-20 (Grade 5)	35-50 in.-lbs.

strainer plate vent screen and felt for a plugged condition. Replace if necessary.

3. Remove the shift fork and push rod assemblies and the input shaft sliding clutch from the carrier cover.

Installation

1. Install the push rod, seal retainer and felt seal on the carrier cover.
2. Engage the shift fork with the groove in the input shaft sliding clutch and place the assembly into the carrier cover.
3. Install the shift cylinder body, compression spring, and gasket over the push rod and on the carrier cover. Secure the body with capscrews and lockwashers. Tighten to specified torque.
4. Install the felt oilers and grommets on the piston and insert the piston assembly into the body and onto the push rod. Install the flat washer and nut, and tighten to 20 to 26 ft. lbs. torque.
5. Install the grommets on the cylinder body cover and install the cover, lockwashers and nuts.

Vacuum Operated Lockout

Removal

1. Remove the boot from the shift lever and remove the clevis pin. Remove the shift lever.
2. Remove the capscrews, lockwashers and shift lever bracket from the carrier cover.
3. Compress the compression spring and remove the pin, retaining washer and spring from the push rod.
4. Remove the shift fork and push rod assembly and input shaft sliding clutch from the carrier cover.

Installation

1. Install the push rod, seal retainer and felt seal on the carrier cover. Engage the lockout shift fork with the groove in the input shaft sliding clutch. Place this assembly into the carrier cover.
2. Place the compression spring and retainer washer on the push rod, depress the spring and install the retainer washer pin.
3. Install a new gasket and shift lever bracket on the carrier cover and install the capscrews and lock washers. Tighten the screws to specifications.
4. Place the shift lever into the bracket, engage the slot in the push rod, and install the clevis pin and cotter key.
5. Place the boot over the shift lever and secure the boot to the bracket.

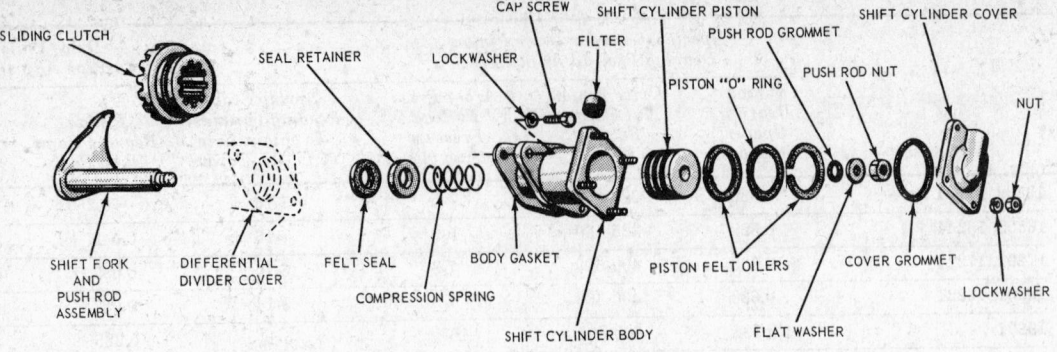

Exploded view—air operated lockout—Eaton axle (© International Harvester Co.)

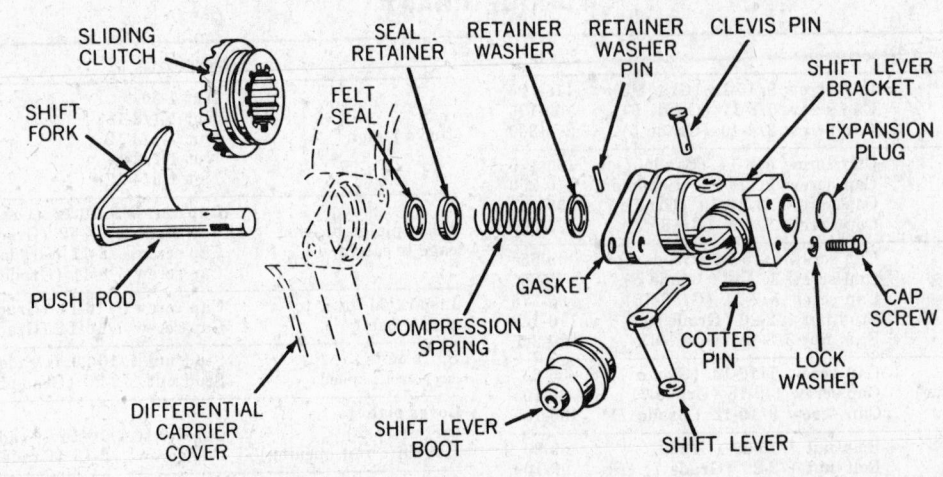

Exploded view—vacuum operated lockout—Eaton axle (© International Harvester Co.)

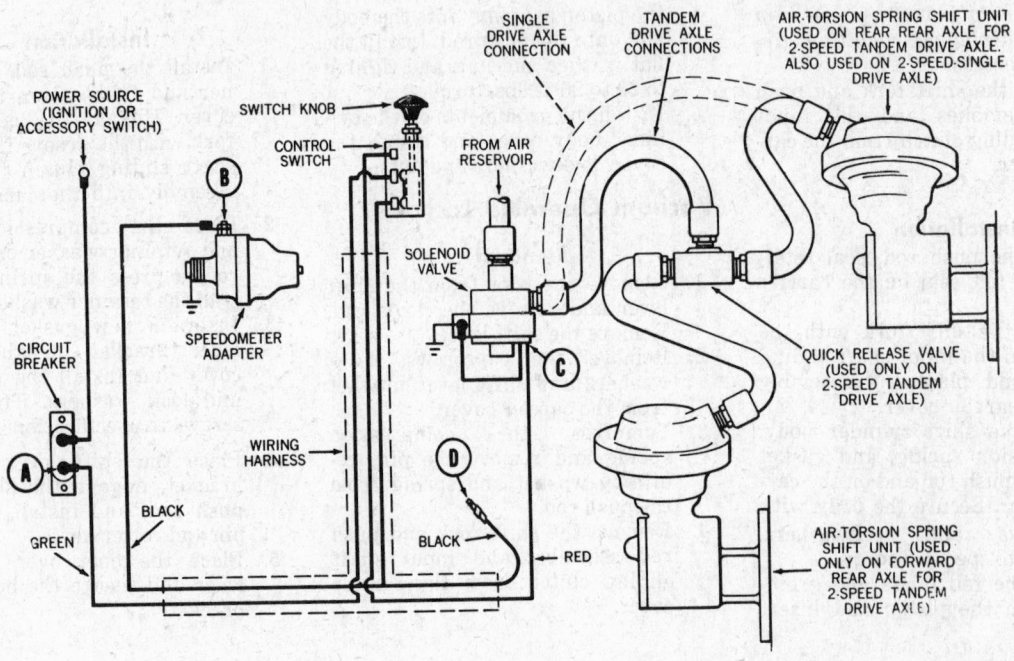

Two speed air shift control—electric over air—single and tandem axles
(© International Harvester Co.)

Ford Semi-Floating Single Speed Axle, Removable Carrier Type

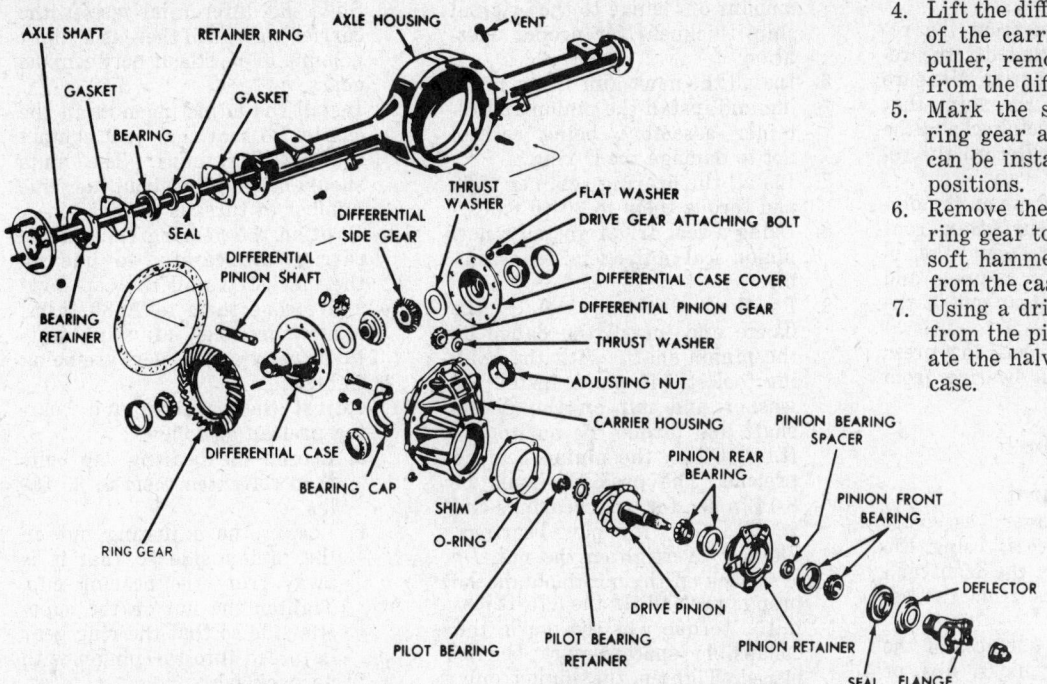

4. Lift the differential assembly out of the carrier. Using a bearing puller, remove the side bearings from the differential case.
5. Mark the side of the case, the ring gear and the cover so they can be installed in their original positions.
6. Remove the bolts that retain the ring gear to the case and using a soft hammer, tap the ring gear from the case.
7. Using a drift, drive the lock pin from the pinion shaft and seperate the halves of the differential case.

This is a conventional type axle used on light duty Ford trucks. The axle design uses a removable carrier with the assembly bolted to the axle housing. The axle uses hypoid type gears and has the pinion gear mounted below the center line on the ring gear. The pinion gear is supported by two bearings in front of the gear and one behind. It is important to refer to the tag showing the axle and model number which is secured to the housing to obtain proper replacement parts.

Carrier Assembly

Removal

1. With the vehicle raised on a lift, remove the axle shafts from the housing.
2. Remove the drive shaft from the carrier assembly.
3. With a drain pan under the axle, remove the retaining bolts from the carrier and drain the gear lube.
4. Remove the carrier assembly from the axle.

Installation

1. Clean the surfaces of the carrier and the axle housing. Install a new gasket.
2. Position the carrier assembly on the studs in the housing and install the retaining nuts. Torque the nuts to 30-40 ft. lbs.
3. Install the drive shaft and torque the bolts to 13-17 ft. lbs.
4. Install the axles in the housing and secure.
5. Fill the axle housing to the pro-

per level with gear lube and road test for proper operation.

Differential Case

Removal and Disassembly

1. Remove the carrier assembly from the axle housing and mount the carrier in a holding fixture.
2. Mark the bearing caps and adjusting nuts so they may be installed in their original positions when assembling.
3. Remove the adjusting nut locks, bearing caps and adjusting nuts.

8. Drive the pinion shaft out of the case using a brass drift and remove the thrust washers and gears.

Drive Pinion and Bearing Retainer

Removal and Disassembly

1. With a holding fixture installed on the flange, remove the pinion nut and washer. Leave the holding fixture on the flange and using a puller, remove the flange

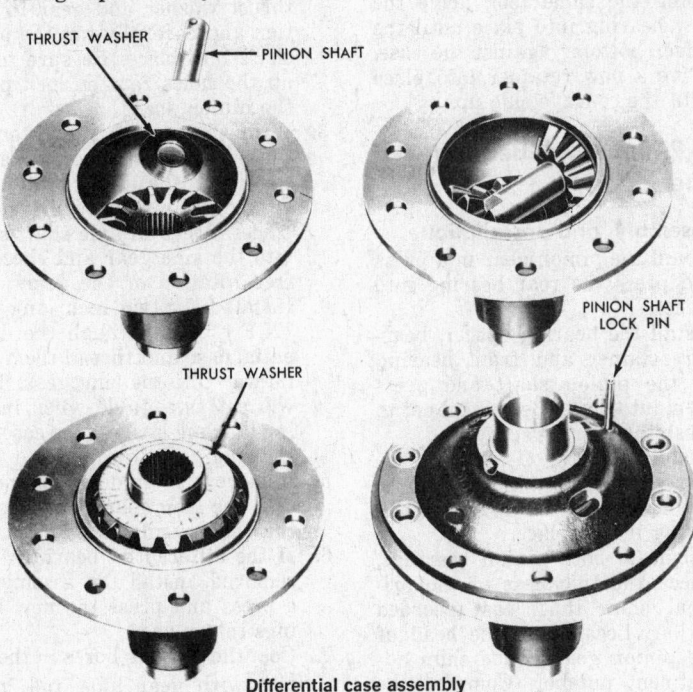

Differential case assembly

from the pinion shaft.

2. Using a seal puller, remove the pinion seal from the retainer assembly.

3. Remove the bolts from the retainer assembly and lift the retainer from the carrier. Measure the thickness of the shim that was between the retainer and the carrier assembly. Record the result.

4. Install a piece of hose on the pinion pilot bearing surface in front of the pinion gear. Mount the retainer assembly in a press and press the pinion gear out of the retainer.

5. Mount the pinion shaft in a press and press the rear bearing from the pinion shaft.

Pinion Bearing Cup

Replacement

1. With the retainer assembly mounted in a press, using the proper tool, press the front and rear bearing cups from the assembly.

2. Check the inside surfaces of the retainer for any nicks, dirt or distortion.

3. Install the new cups by pressing them into place with the proper tool. When the cups are installed, make sure they are seated in the retainer by trying to fit a .0015 in. feeler gauge, between the cup and the bottom of the bore.

Pilot Bearing

Replacement

1. Using a bearing driver, drive the bearing and retainer out of the carrier assembly.

2. Using the same tool, drive the new bearing into place until the driver bottoms against the case.

3. Drive a new retainer into place with the concave side up.

Drive Pinion and Bearing Retainer

Assembly and Installation

1. Mount the pinion gear in a press and press the rear bearing into place.

2. Install the bearing spacer, bearing retainer and front bearing on the pinion shaft and press them into place. Be careful not to crush the bearing spacer.

3. Install a new O-ring in the groove in the retainer assembly. Do not twist the O-ring when fitting it into place.

4. Lubricate both pinion bearings.

5. Check the thickness of the original shim that was recorded earlier. Located on the head of the pinion gear is the shim adjustment number. Compare the number on the old pinion with the one on the new pinion. Refer to the table which indicates the amount of change to the original shim thickness for proper operation.

6. Install the new shim on the housing and install the pinion and retainer assembly, being careful not to damage the O-ring.

7. Install the bearing retainer bolts and torque them to 30-40 ft.lbs.

8. Using a seal driver, install a new pinion seal in the retainer assembly.

9. Position a holding tool on the flange and install the flange on the pinion shaft. With the holding tool still in place, install the washer and nut on the pinion shaft and torque the nut to 175 ft.lbs. Check the pinion bearing preload. The preload should be 8-14 in.lbs. for used bearings and 22-32 in.lbs. for new bearings. *Do Not* overtighten the nut. *Do Not* back off the nut to obtain the proper preload. If the 175 ft.lbs. initial torque was too much, the collapsible spacer must be replaced. Tighten the pinion only enough to obtain the right preload torque.

Differential Case

Assembly and Installation

1. Lubricate all of the differential parts with gear lube before assembling.

2. Install a side gear and thrust washer in the case bore. Using a soft hammer, drive the pinion shaft into the case far enough to hold a pinion thrust washer and gear. Place the second pinion thrust washer and gear in position and carefully tap the pinion shaft into place. Be sure to line up the holes for the lock pin in the pinion shaft.

3. With the second side gear and thrust washer in place, install the cover on the differential case. Drive the pinion lock pin into place. Insert an axle shaft spline into the side gear and check for free rotation of the gears.

4. Install two, two inch long 7/16 (N.F.) bolts through the differential case and thread them a little way into the ring gear. These will act as a guide when installing the ring gear on the case. Tap the ring gear into place.

5. Remove the guide pins and install the ring gear bolts. Tighten the bolts evenly to 65-85 ft.lbs.

6. If the differential bearings were removed, install the assembly in a press and press the new bearings into place.

7. Coat the bearing bores in the carrier with gear lube and install the bearing cups on the bearings. Place the differential assembly in the carrier.

8. Slide the differential case in the carrier bore until there is a slight amount of backlash between the gears.

9. Install the adjusting nuts in the carrier so that they just contact the bearing cups. The nuts should be engaged about the same number of threads on each side.

10. Position the bearing caps in the carrier. Be careful to line up the marks. Install the cap bolts and torque them to 70-80 ft.lbs. Make sure the adjusting nuts turn freely as the bolts are being tightened.

11. Adjust the backlash and bearing preload as follows;

    a. Loosen the bearing cap bolts then retighten them to 35 ft.-lbs.

    b. Loosen the adjusting nut on the pinion side so that it is away from the bearing cup. Tighten the nut on the opposite side so that the ring gear is forced into the pinion with no backlash.

    c. Recheck the nut on the pinion side to make sure it is still loose. Now tighten this nut until it contacts the bearing cup. After is contacts the cup, turn it two more notches.

    d. Rotate the ring gear several times in each direction. This helps to seat the bearings in the cups.

    e. Again loosen the nut on the pinion side. If there is any backlash between the gears, tighten the nut on the ring gear side until the backlash is removed.

    f. Install a dial indicator on the carrier assembly. Tighten the nut on the pinion side until it just contacts the cup. With the dial indicator set at zero, tighten the pinion side nut until the case is spread .008-.012 in. with new bearings and .005-.008 in. with old bearings. As this preload is applied the ring gear is forced away from the pinion and usually results in the correct backlash.

    g. Mount the dial indicator on the ring gear and check the gear for backlash. Make sure the bearing caps are torqued to 75-85 ft.lbs.

    h. The backlash should be between .008-.012 in. If the backlash is not correct, loosen one nut and tighten the other an equal amount to move the ring gear in or out to correct the measurement. When making final adjustments, always move the adjusting nuts in a tightening direction. To do this, if a nut had to be loosened one notch, loosen it two notches and tighten it one. This makes certain the nut is in contact with the cup and

will not shift when the vehicle is in operation.

i. Coat the ring gear teeth with a marking compound and check the tooth pattern. If the

pattern is not correct make the necessary changes to bring it into adjustment.

12. Install the carrier assembly in the vehicle and road test for proper operation.

Single-Speed, Single Reduction Rockwell-Standard

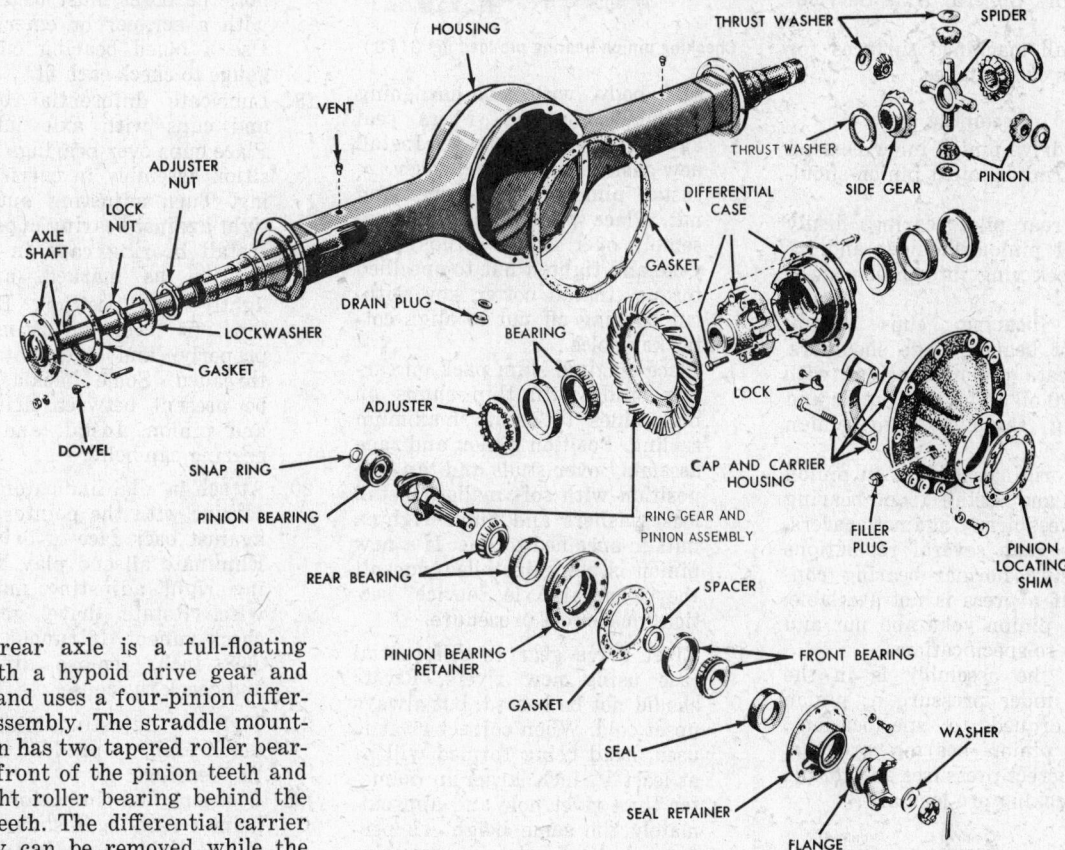

single-speed, single reduction rear axle (© Ford Motor Co.)

This rear axle is a full-floating type with a hypoid drive gear and pinion, and uses a four-pinion differential assembly. The straddle mounted pinion has two tapered roller bearings in front of the pinion teeth and a straight roller bearing behind the pinion teeth. The differential carrier assembly can be removed while the axle remains in the truck.

Differential Carrier Assembly

Removal

1. Remove axle shafts.
2. Drain lubricant. Disconnect propeller shaft at pinion shaft yoke.
3. Remove carrier from axle housing and clean thoroughly.

Disassembly

1. Punch mark carrier leg, bearing cap, and bearing adjusting nut to assist in reassembly.
2. Remove screws, adjusting nut locks, bearing caps and adjusting nuts.
3. Loosen lock nut and back off drive gear thrust block adjusting screw.
4. Lift differential out of carrier and remove thrust block from end of adjusting screw inside of carrier.
5. Punch mark differential case halves for correct reassembly alignment and separate case halves.
6. Remove pinion shaft, pinions,

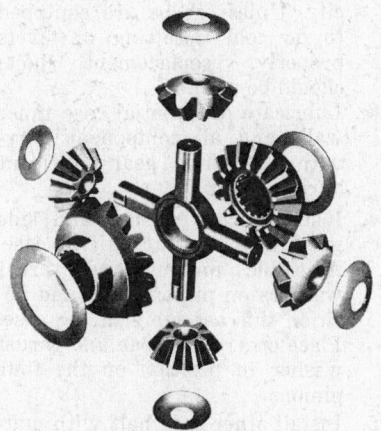

Exploded view of side gear and pinion assembly
(© Chrysler Corp.)

side gears, thrust washers and differential bearing cones.

7. To remove drive gear, carefully center punch each rivet in center of rivet head. Use a drill 1/32 inch smaller than the body of

rivet to drill through the rivet head. Press out rivets.

8. Remove pinion shaft nut, washer and yoke. Driving yoke off will cause runout.

9. Remove pinion bearing cover and oil seal assembly and, using puller screws, remove bearing cage. Using a pinch bar to remove cage will damage shims. Driving pinion from inner end with a drift will damage bearing lock ring groove.

10. Wire bearing cage shim pack together to facilitate adjustment when reassembling.

11. Tap pinion shaft out of cage with soft mallet or press shaft from cage. Remove bearing from cage.

12. Remove spacers and inner bearing from shaft.

13. Remove pinion shaft rear pilot bearing lock ring; and then bearing.

14. Remove oil seal assembly from bearing cover.

15. Clean all parts thoroughly in a suitable solvent and blow dry with compressed air. Do not spin bearings with air pressure as they might score due to absence of any lubrication.

Inspect all parts for wear or roughness and replace if necessary. (For details, see "cleaning" and "inspection" in "General Axle Service" section).

Inspect all machined surfaces for nicks, burrs or scratches.

Assembly

1. Press drive pinion inner bearing cone firmly against pinion shoulder.
2. Press rear pilot bearings firmly against pinion shoulder and install lock ring into pinion shaft groove.
3. Press bearing cups firmly against bearing cage shoulders.
4. Lubricate pinion bearings with SAE 90 oil and insert pinion and bearing assembly into pinion cage.
5. Place original spacers on pinion shaft, and install front bearing and press firmly against spacers. Rotate cage several revolutions to assure normal bearing contact. If a press is not available, install pinion yoke and nut and torque to specifications.
6. While the assembly is in the press under pressure or pinion nut torqued to specifications, check pinion bearing pre-load. The correct pressures and torque for checking pre-load are:

Pinion Shaft Thread Size and Number of Threads Per Inch	Required Pressure to Obtain Correct Pre-Load lbs.	Required Torque to Obtain Correct Pre-Load ft. lbs.
1 in. x 20	12,000	300- 400
1¼ in. x 18	22,000	700- 900
1½ in. x 12	28,000	800-1100
1½ in. x 18	28,000	800-1100
1¾ in. x 12	28,000	800-1100

7. Wrap a soft wire around cage and pull on horizontal line with pound scale when determining pinion bearing pre-load, first measure diameter of the pinion cage. Assuming cage diameter is 6 inches, the pulling radius would be 3 inches; therefore, 5 pounds pull on the scale would equal 15 inch-pounds pre-load.
8. Use rotating torque, not starting torque. If rotating torque is not within 5 to 15 inch-pounds, use thinner spacer to increase pre-load or thicker spacer to decrease pre-load. Torque must be near low limit inch-pounds with original pinion bearings and near high limit when using new bearings. Remove yoke and install new oil seal.
9. Lubricate pinion shaft oil seal and lightly coat outer edge of

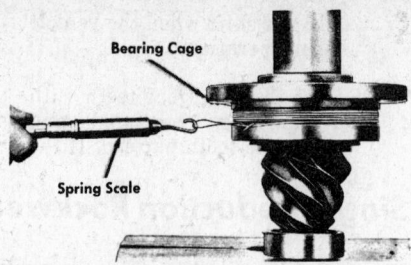

Checking pinion bearing pre-load (© G.M.C.)

seal body with non-hardening sealing compound. Press seal against cover shoulder. Install new gasket and bearing cover.
10. Install pinion yoke, washer and nut. Place pinion and cage assembly over carrier studs, hold yoke and tighten nut to specified torque. Install cotter key without backing off nut to align cotter key holes.
11. Place original shim pack on carrier studs with thin shims on both sides to create maximum sealing. Position pinion and cage assembly over studs and tap into position with soft mallet. Install lock washers and nuts. Tighten nuts to specified torque. If a new pinion is being installed, consult the "General Axle Service" section for correct procedure.
12. Rivet drive gear to differential case using new rivets. Rivets should not be heated, but always upset cold. When correct rivet is used, head being formed will be at least ⅛ inch larger in diameter than rivet hole and approximately the same height as performed head. Avoid excessive pressure as it might distort case holes and cause gear eccentricity. Unless shops are equipped to do cold upsetting of rivets properly, replacement bolts should be used.
13. Lubricate differential case inner walls and all component parts with rear axle gear lubricant during assembly.
14. Install thrust washer and side gear in drive gear half of case. Assemble pinions and thrust washers on pinion shaft and position this assembly in the case. Place other side gear and thrust washer in position on the four pinions.
15. Install other case half with mating marks aligned. Draw case down evenly with four bolts. Check for free rotation of differential gears and correct if necessary. Install and torque remaining bolts and then lock wire.
16. Press differential bearings squarely and firmly on differential case halves. Differential

bearing cup fit in the pedestal bores should be checked before installing assembly in carrier.
17. Temporarily install bearing cups, threaded adjusting nuts or split ring and bearing caps. Tighten cap bolts to specified torque. Bearing cups must be of a hand push fit in the bores; if not, the bores must be enlarged with a scraper or emery cloth. Use a blued bearing cup as a gauge to check each fit.
18. Lubricate differential bearings and cups with axle lubricant. Place cups over bearings and position assembly in carrier housing. Turn adjusting nuts hand tight against bearing cups.
19. Install bearing caps in correct location as marked, and tap lightly into position. Be sure caps fit over adjusting nuts properly and are not crossthreaded. Some backlash must be present between drive gear and pinion. Install and torque bearing cap bolts.
20. Attach a dial indicator to the carrier with the pointer resting against back face of drive gear. Eliminate all end play by turning right adjusting nut clockwise. Rotate drive gear and check runout. If runout exceeds .008 inch, remove differential and check the cause.
21. Tighten adjusting nuts, one notch each, to pre-load differential bearings.
22. Position dial indicator so the pointer rests against face of one of the drive gear teeth. Check backlash between drive gear and pinion at 90 degree intervals of rotation.
23. Adjust backlash to .006-.012 inch (.006 preferred, especially on new gears). When adjusting backlash, back off one adjusting nut and advance opposite nut the same amount to maintain bearing pre-load.

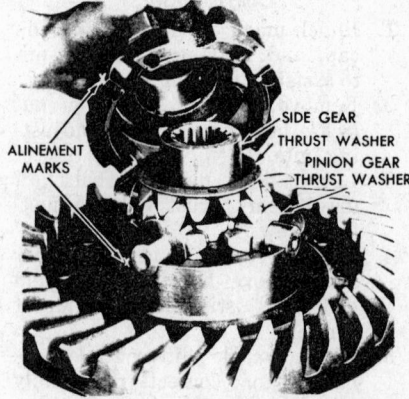

Differential case alignment marks (© G.M.C.)

24. If the "Pinion Setting Procedures" (Depth Gauge Method) for adjusting pinion depth (see "General Axle Service" section) was not used, adjust gears using the Tooth Contact Method. Then proceed to step 25. If Depth Gauge Method was used, go directly to step 25.

25. Torque bearing cap bolts and install and torque adjusting nut locks and cap screws and lock wire.

26. Hold drive gear thrust block on rear face of gear with heavy grease, rotate gear until hole in thrust block aligns with adjusting screw hole in carrier. Install adjusting screw and lock nut, tighten screw enough to locate thrust block firmly against back face of gear. Back off adjusting screw ¼ turn to create .010-.015 inch clearance and lock securely with nut. Recheck to assure minimum clearance of .010 inch during full rotation of drive gear.

Installation

1. Install a new gasket on axle housing flange. Start carrier into clean housing and hold in place with four equally spaced washers and nuts. Tighten nuts alternately to draw carrier evenly into housing. Install and torque carrier flange lock washers and nuts.

2. Install axle drive shafts and connect universal joint at pinion flange.

3. Fill axle housing to proper level and road test vehicle.

TORQUE CHART

Carrier to housing screw or stud nut	1/2-20	94-102
	5/8-18	186-205
	3/4-16	325-360
Differential case bolt	1/2-20	94-102
	9/16-18	132-145
	5/8-18	186-205
	3/4-16	325-360
Differential bearing cap bolt	5/8-11-18	127-140
	3/4-10-16	230-250
	7/8-9	345-370
	7/8-14	375-415
	1-12	555-615
Differential bearing adjuster lock bolt	5/16-18	15-17
	1/2-13	85-91
	9/16-12	120-129
	5/8-11	168-180
Pinion shaft yoke nut	1-20	300-400
	1-1/4-18	700-900
	1-1/2-18	800-1100
	1-1/2-12	800-1100
	1-3/4-12	800-1100
Shaft flange stud	7/16-20	52-58

ROCKWELL-STANDARD BEARING PRELOAD

Axle Model	Pinion Shaft Nut-Thread Size and Torque Limits (Foot-Pounds)		Press Ram Pressure for Preload Check (Tons)	Pinion and Cross Shaft Bearing Preload (Inch-Pounds)	Backlash Limits (Inches)	Differential Bearing Preload Adjusting Nut Notches Tighten from Zero End Play (each Adjusting Nut)
Single-speed	1-20	300-400	6	5-15	0.005-0.015	1
Single reduction	1¼ x 18	700-900	11	5-15	0.005-0.015	1
	1¼ x 18	800-1100	14	5-15	0.005-0.015	1
	1½ x 12	800-1100	14	5-15	0.005-0.015	1
	1½ x 18	800-1100	14	5-15	0.005-0.015	1

Rockwell-Standard Single-Speed Double Reduction Axles

This axle includes a hypoid helical drive and a two-step gear reduction. The first reduction is through the hypoid pinion and ring gears with the second reduction through the helical gears. The helical pinion gear engages with the helical drive gear. The axle housings are one-piece with full-floating axle shafts.

Disassembly of Rear Axle

Differential Carrier

1. Before disassembling, inspect the unit as described in the "cleaning" and "inspection" paragraph of the "General Axle Service" section.

2. Punch-mark one differential bearing cap and the corresponding carrier leg. This will be an aid to proper reassembly.

3. Remove the locking wires and bolts from both bearing caps and remove both bearing adjuster nuts.

4. Back off both bearing adjusters one full turn to reduce bearing pre-load. Remove both bearing caps and adjusters and wire together.

5. Lift the differential out of the carrier.

6. Remove the U-joint flange nut and then remove flange with a rawhide mallet. If this is unsuccessful, the flange can be pressed off when the hypoid pinion cage is being disassembled.

7. Using two 3 inch 38—16 puller screws, remove the cage and pinion gear from the carrier.

8. Remove the pinion adjusting shims from the carrier and re-use if the same hypoid gear set is to be used again when the carrier is assembled.

9. Place a wooden block under the helical pinion gear to support the cross shaft. Then remove the bolts from the cross shaft bearing cap mounted on the differential carrier—opposite the hypoid ring gear.

10. Force the bearing cap out of the carrier about ¼ inch by prying the hypoid ring gear away from the side of the carrier.

11. Place metal strips under both puller screw holes in the bearing cap and, using puller screws, hold them in place against the adjusting shims. Do not tighten the screws directly onto the adjusting shims.

12. Tighten the puller screws evenly against the metal strips, and remove the bearing cap and the

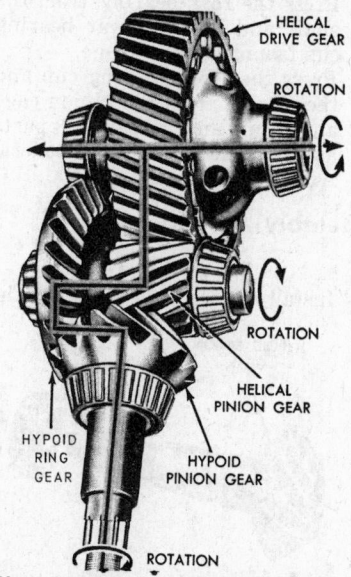

HELICAL DRIVE GEAR
ROTATION
ROTATION
HELICAL PINION GEAR
HYPOID RING GEAR
HYPOID PINION GEAR
ROTATION

Single-speed, double reduction power flow
(© Rockwell Stand. Div.)

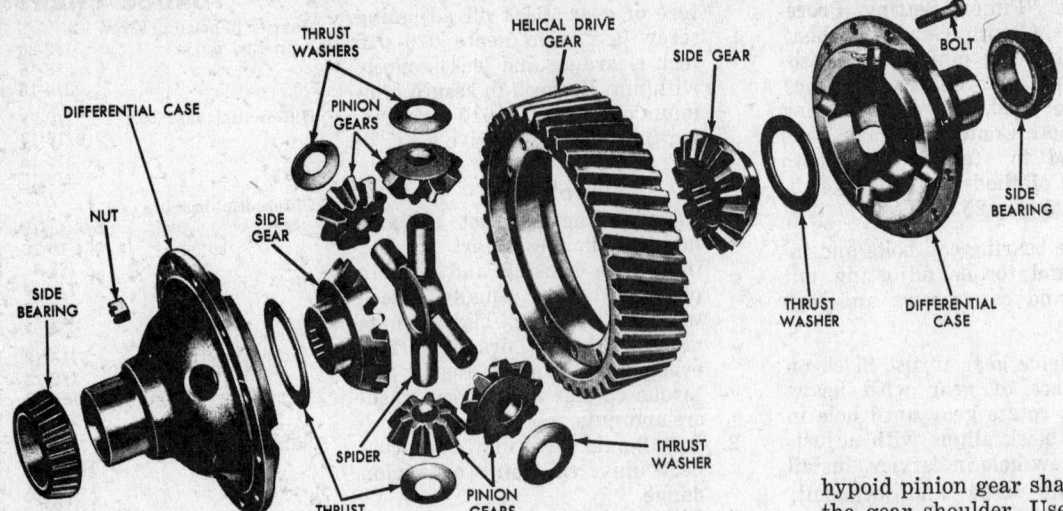

Differential gear and case components
(© Rockwell Stand. Div.)

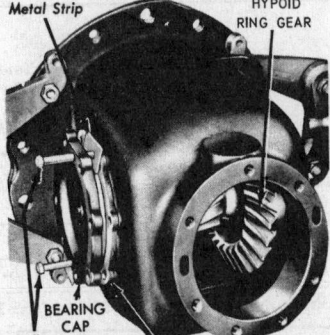

Removing cross shaft bearing cap
and positioning metal strips
(© Rockwell Stand. Div.)

adjusting shims from the carrier.

13. Carefully remove the hypoid ring gear and cross shaft from the carrier.

14. Replace either cross shaft bearing cup if necessary. Remove the damaged or worn cup from the bearing cap with a suitable puller. To operate the cup in the bearing cap, tap the cap out of the carrier about ¼ inch with a soft drift and remove from the carrier.

Cross Shaft

1. Remove the locking wire, two screws, and the bearing retaining plate from the hypoid ring gear end of the cross shaft.

2. Press the cross shaft out of the ring gear and the bearing next to the gear. Remove the Woodruff key that holds the gear on the shaft.

3. Press the remaining bearing from the cross shaft.

Differential Gear and Case

1. Mount the differential in a soft-jawed vise. Then, if necessary, remove both differential side bearings.

2. Punch-mark each case half to aid proper assembly.

3. Separate the helical drive gear and the two case halves. If rivets are binding these parts, center—punch the head of each rivet followed by a ⅛ inch pilot hole at each punch-mark, and then a ½ inch hole drilled ½ inch deep and complete by pressing out all the rivets with a ½ inch punch. **CAUTION: Do not attempt to remove the rivets with a hammer and chisel.**

4. Remove the spider, pinion gears, thrust washers, and side gears.

Pinion Cage

1. Press the hypoid pinion gear shaft out of the pinion cage, and then remove the bearing preload spacer from the shaft.

2. Press the rear bearing from the shaft and pull the rear bearing cup from the pinion cage.

3. Press the front bearing cup and the oil seal from the pinion cage and clean and inspect all parts as explained in the "General Axle Service" section.

Assembly of Rear Axle

Pinion Cage

1. Install the rear bearing on the

hypoid pinion gear shaft against the gear shoulder. Use a 3 inch sleeve about one inch long under the bearing race so that the press ram won't damage roller cage.

2. Press the rear bearing cup *firmly against the shoulder in the pinion cage.*

3. Install the front bearing cup in the pinion cage making sure that the cup seats firmly against the shoulder in the cage.

4. Coat the pinion rear bearing with axle lubricant, and position the pinion gear in the cage.

5. Position the original bearing preload spacer on the pinion gear shaft and press the pinion front bearing onto the shaft against the spacer. Use a 3 inch sleeve about 5 inches long over the bearing race so that the press ram doesn't damage the bearing roller cage.

6. Rotate the pinion cage several times to be sure that the bearings are properly seated. Then check the pinion bearing preload as explained in the "General Axle Service" section. If the preload is too low, install a thinner spacer under the front bearing. To decrease the preload, install a thicker spacer under the bearing.

7. When the preload is correct, install the oil seal in the pinion cage.

Differential Gear and Case

1. Check the mating surfaces of the

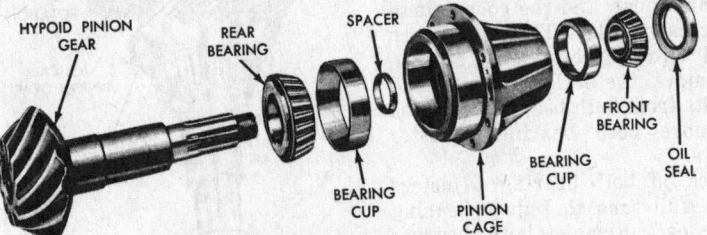

Hypoid pinion gear and cage (exploded view)
(© Rockwell Stand. Div.)

differential case halves and the helical drive gear to be sure they are clean and free of burrs. Then coat all differential parts and the inner walls of the case, with axle lubricant.

2. Install the differential side bearings, making sure they are properly seated. Use a 3¼ inch sleeve about one inch long over the bearing race so that the force of the press ram doesn't damage the bearing roller cage.

3. Position a thrust washer and a side gear in one of the differential case halves. Then place the spider, the pinion gears, and their thrust washers in position. Install the remaining side gear and thrust washer in the case.

4. Position the helical drive gear and install the other case half, aligning the marks on both halves.

5. Install 6 bolts in alternate holes around the case on the side of the gear having the smaller offset. Tighten the bolts and nuts enough to draw the gear and both case halves together. Then check the rotation of the side gears and pinion gears.

6. Install the remaining 6 bolts and nuts, and torque them evenly to specifications and lock wire.

Cross Shaft

1. Place the Woodruff key (tapered end toward pinion gear) in the cross shaft. Then press the hypoid ring gear on the shaft—seating it against the shaft shoulder.

2. Carefully press the shaft into the bearing, seating the bearing against the shoulder.

3. Install the remaining bearing on the other end of the shaft. Use a 3 inch sleeve about one inch long over the bearing race so that the press ram won't damage the bearing roller cage.

4. Install the bearing retaining plate and torque and wire the screws.

5. Press the cross shaft bearing cups into the bearing caps. Be sure that the cups are properly-seated against the shoulders in the cap bores.

Differential Carrier

1. Coat all parts with axle lubricant before assembly.

2. Position the original shim pack on the right cross shaft bearing cap and install and torque the cap on the differential carrier.

3. Install the cross shaft in the carrier so that the end of the shaft opposite the hypoid ring gear enters the bore of the left bearing cap. The right bearing should enter and seat in its bearing cap. Place a wooden block under the helical pinion gear to support the cross shaft.

4. Position the original shim pack on the left cross shaft bearing cap and install and torque the cap on the differential carrier.

5. Rotate the cross shaft to test for normal bearing contact. Then check the cross shaft bearing preload. *NOTE: Do not read the pull required to start the cross shaft rotating. Read only the steady rotating pull on the scale.* To change the scale reading into in-lbs. torque, multiply the scale reading by one-half the diameter of the helical pinion gear at the point where the cord is wound around the gear.

6. If the bearing preload torque is not within 5-15 in-lbs., add shims to decrease the preload, or remove shims to increase the preload. *NOTE: To prevent changing the hypoid ring gear backlash setting, change shims only at the bearing cap opposite the ring gear.*

7. Install the pinion cage-filler hole at top. Install and torque the bolts and lock washers. Use the original shim pack between the pinion cage and the differential

carrier if the original hypoid gears (ring gear and pinion set) are used. If new hypoid gears are installed, follow the shim selection procedure in the "General Axle Service" section.

8. Install the U-joint flange, torque the nut and install cotter pin.

9. The backlash can be adjusted by transferring shims from one cross shaft bearing cap to the other. To move the hypoid ring gear away from the hypoid pinion gear, transfer shims from the right cap to the left cap. To move the ring gear closer to the pinion gear, transfer shims from left to right. For each 0.010 inch movement of the ring gear, the backlash changes about 0.008 inch. Adjust backlash to 0.010 inch and check the gear tooth contact pattern. After obtaining a good pattern at 0.010 backlash, increase backlash to 0.020-0.026 inch regardless of the backlash marking on the ring gear.

10. Check and adjust the tooth contact pattern (see the "General Axle Service" section).

11. Check the fit of the differential side bearing cups and the bearing adjusters in the bearing caps. The adjusters should thread freely into the caps, and they should move the bearing cups into the bores with the cap bolts tightened to normal torque. If the cups do not move when the adjusters are hand tightened, remove the caps and clean the bearing surfaces.

12. Position a bearing cup on each differential side bearing. Then place the differential on the carrier so that the bearing cups rest in the carrier legs centering the drive gear on the helical pinion gear.

13. Place the bearing adjusters on the threads in the carrier legs, and position the bearing caps on the carrier leg with the matching marks aligned. Check to see that the threads in the

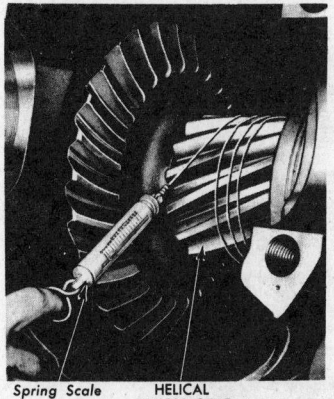

Checking cross shaft bearing preload
(© Rockwell Stand. Div.)

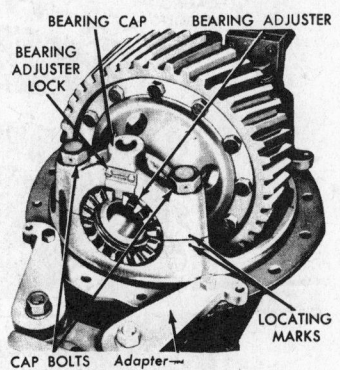

Installation of bearing caps
(© Rockwell Stand. Div.)

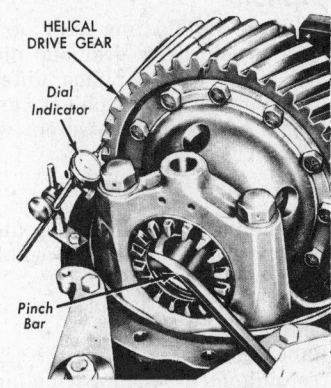

Determining differential end play
(© Rockwell Stand. Div.)

Drive Axles

BEARING PRELOAD

Axle Model	Pinion Shaft Nut-Thread Size and Torque Limits (Foot-Pounds)		Press Ram Pressure for Preload Check (Tons)	Pinion and Cross Shaft Bearing Preload (Inch-Pounds)	Backlash Limits (Inches)	Differential Bearing Preload Adjusting Nut Notches Tighten from Zero End Play (each Adjusting Nut)
Single-Speed	1¼ x 18	700-900	11	5-15	0.020-0.026	1
Double Reduction	1½ x 12	800-1100	14	5-15	0.020-0.026	1
	1½ x 18	800-1100	14	5-15	0.020-0.026	1

TORQUE CHART

Location	Cap Screws or Stud Nuts		
	Diameter	Threads per. in.	Torque-Lb. Ft. Min.-Max.
These torques are	3/8	24	38-49
given according	3/8	16	33-43
to diameter and	7/16	14	53-77
threads per inch.	7/16	20	53-67
The torque will	1/2	13	81-104
be the same for	1/2	20	81-104
a specific size	9/16	12	116-149
no matter where	9/16	18	116-149
the bolt or cap	5/8	11	160-205
screw is used	5/8	18	160-205
on the axle	3/4	10	290-370
except for those	3/4	16	290-370
listed below.	7/8	9	470-595
	7/8	14	510-655
	1	14	580-745
Adjusting nut lock	5/16	18	16-20
	1/2	13	75-96
Inspection cover	3/8	16	.27-35
Shift unit (mounting)	3/8	16	27-35
Shift unit lock nut, set	3/8	16	30-33
screw and clamp screw	7/16	14	30-33
Shift unit travel limiting	1/2	13	40-45
screws	5/8	11	30-33

Torques given apply to parts coated with machine oil; for dry (as received) parts increase torques 10%; for parts coated with multi-purpose gear oil decrease torques 10%. Nuts on studs to use same torque as for driving the stud.

caps and legs are matched—if so, install and torque the cap bolts.

14. Hand tighten the bearing adjusters until they just touch the bearing cups. Then install a dial indicator and check the differential end play using a pinch bar to move the differential away from indicator. Then tighten the bearing adjuster opposite the dial indicator until there is no side-to-side movement of the differential, and the dial indicator shows no end play.

15. After obtaining the correct end play, adjust the differential bearing preload by tightening the bearing adjusters an additional ¾-2½ notches (total for both adjusters).

16. Install the bearing adjuster locks and cap screws.

17. Install locking wires on the lock cap screws and bearing cap bolts.

Rockwell-Standard Two-Speed Double Reduction Axles

Description

This is a full-floating rear axle having a hypoid drive gear and pinion and using a four-pinion differential assembly. The pinion and cross shaft are each mounted on two tapered roller bearings.

Disassembly

1. Punch-mark one carrier leg and bearing cap for reassembly reference.
2. Remove differential and gear assembly.
3. Punch-mark case halves for correct alignment when assembling.
4. Separate differential case halves. Remove side gears, thrust wash-

Differential carrier on stand
(© Rockwell Stand. Div.)

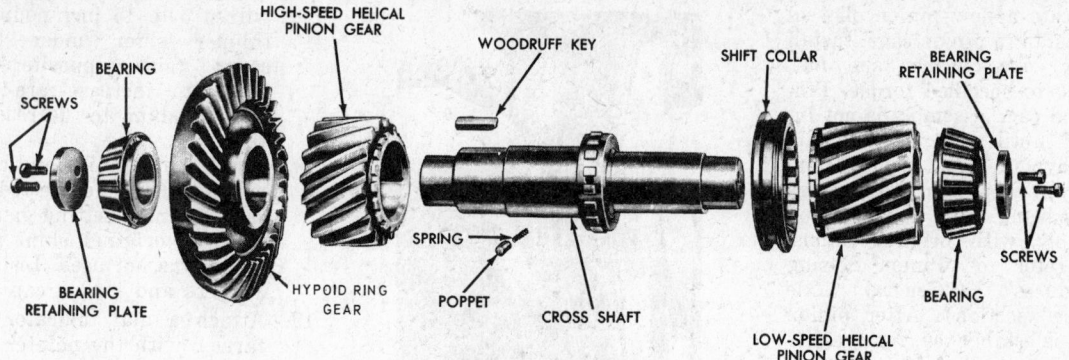

Cross shaft components
(© Rockwell Stand. Div.)

ers, pinions and pinion shaft.

5. Remove differential side bearing.

6. Separate gears from case. Center-punch rivets in center of head. Use drill 1/32 inch smaller than body of rivet to drill through head, and then press out rivets.

7. Remove pinion shaft yoke.

8. Remove pinion bearing cage assembly. If cage is not free, tap loose, or use puller screws in holes provided.

9. Wire bearing cage shim pack together to facilitate adjustment when reassembling.

10. Press pinion shaft out of cage and remove adjusting spacers from shaft. Remove rear bearing from pinion shaft.

11. Press front bearing and pinion shaft oil seal from pinion cage.

12. Remove two bearing cups from pinion cage.

13. Remove shift fork shaft and shift fork.

14. Tap sleeve from carrier *with a soft mallet.* Wire shim pack together to facilitate reassembly.

15. Remove screws and force out bearing cage by using a small pinch bar between back of hypoid gear and carrier housing. Wire shim pack together to facilitate reassembly.

16. Remove cross shaft and hypoid gear assembly.

17. Remove cap screws and press cross shaft from bearing and hy-

poid gear (ring gear). Slide off high speed pinion and remove shift collar, plungers and springs. Precautions should be taken when removing shift collar as the plungers and springs will fly out. Reinstall shift collar and press shaft through low speed pinion and bearing.

18. Remove cross shaft bearing cups. If cross shaft bearing cover is removed (hypoid gear side), be sure to wire shim pack together.

19. Clean all parts thoroughly in a suitable solvent and blow dry with compressed air. Remove any foreign material from carrier housing. The bearings should be submerged in clean solvent and rotated by hand until clean. After cleaning, blow dry with compressed air. **CAUTION: Do not spin bearings with air pressure, as they might score due to absence of any lubrication.** See the "General Axle Service" section for details on cleaning and inspection.

Assembly

1. Press drive pinion rear bearing cone firmly against pinion shoulder.

2. Press pinion cage bearing cups squarely and firmly against cage shoulder.

3. Place bearing spacer on pinion shaft with bevel side toward front bearing. Lubricate bearing

cone with axle lubricant and position bearing cage on shaft. Lubricate and press front bearing on pinion shaft. Rotate cage to assure normal bearing contact. If a press is not available, install pinion yoke and nut and torque to specifications.

4. While assembly is in the press under pressure or pinion nut torqued to specifications, check pinion bearing pre-load. The correct pressures and torques for checking pre-load are:

Pinion Shaft Thread Size and Number of Threads Per Inch	Required Pressure to Obtain Correct Pre-Load lbs.	Required Torque to Obtain Correct Pre-Load ft. lbs.
1 in. x 20	12,000	300- 400
1¼ in. x 18	22,000	700- 900
1½ in. x 12	28,000	800-1100

5. Wrap a soft wire around cage and pull on horizontal line with pound scale. When determining pinion bearing pre-load, first measure diameter of pinion cage. Assuming cage diameter is 6 inches, pulling radius would be 3 inches; therefore, 4 pounds pull on scale would equal 12 inch-pounds pre-load.

6. Use rotating torque, not starting torque. If rotating torque is not within 5 to 15 inch-pounds, use thinner spacer to increase preload or thicker spacer to decrease pre-load. Torque must be near low limit inch-pounds with original pinion bearings and near high limit when using new bearings.

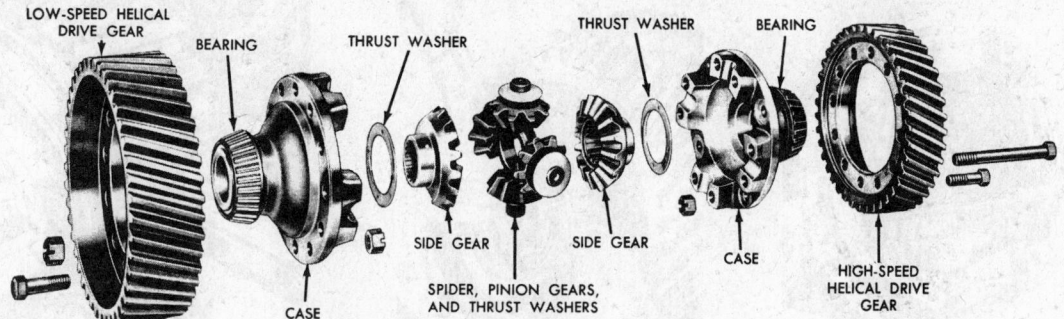

Differential gears and case (exploded view)
(© Rockwell Stand. Div.)

7. Lubricate a new pinion oil seal and install in pinion cage. Install pinion yoke, washer and nut, tighten to specified torque. Pinion and cage assembly is not installed until after cross shaft and hypoid gear installation. However, if pinion depth gauge tool is being used, install pinion assembly with original shims, then refer to "Pinion Setting Procedures" ("General Axle Service" section). After pinion depth is established, remove pinion and cage assembly.

8. Lubricate inner bearing surfaces on low and high speed pinions with axle lubricant. On some axles (H341, L345), the low speed (small) pinion is located next to the hypoid gear. On the others (Q345, RT340) the high speed (large) pinion is next to the hypoid gear.

9. Position correct pinion on hypoid gear end of cross shaft, with splined row of teeth toward center of cross shaft. Install key in shaft and start cross shaft into hypoid gear in line with keyway, then press shaft into hypoid gear. Insert feeler gauge between end of helical pinion and thrust surface on cross shaft must be .010-.025 inch.

10. Coat the three plungers and springs with axle lubricant, and install in cross shaft. Align the three tapered splines in shift collar with the three plungers, slide collar over plungers with side of collar marked LOW SIDE toward low speed (small) pinion.

11. Install remaining pinion on cross shaft with splined row of teeth toward center of shaft. Press cross shaft bearing firmly against cross shaft shoulder. Pinion end play must be .010 to .026 inch.

12. Install bearing retaining washers and torque cap screws to 42-45 foot-pounds and install lock wire.

13. Press bearing cups firmly against shoulder in the bearing covers.

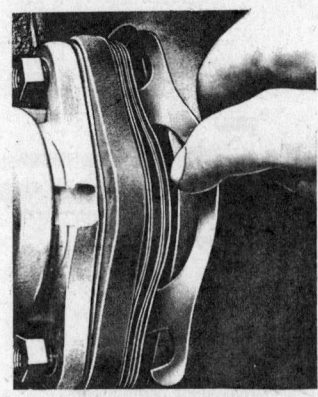

Shift unit adjustment
(© Rockwell Stand. Div.)

14. Install cross shaft bearing cover (hypoid gear side) in carrier housing with original shim pack. Torque cap screws.

15. Lubricate bearing cones with axle lubricant, then position cross shaft assembly in bearing cup in carrier. If a front mounted shift unit is used, position shift fork and install shift shaft through its bearings in the carrier and fork. Align hole in fork with detent in shaft. Torque lock screw and install lockwire. Start the other bearing cover and original shim pack in carrier housing (side opposite hypoid gear). Tap cover into position and torque down cap screws. Rotate cross shaft and gear assembly several times to assure normal bearing contact.

16. Lock low speed pinion and cross shaft with shift collar. Wrap a soft wire around the pinion and pull horizontally with a pound scale. To calculate cross shaft bearing pre-load, first measure diameter of pinion gear. Assuming pinion diameter is 4 inches, pulling radius would be 2 inches; therefore, 7 pounds pull on scale would equal 14 inch-pound pre-load.

17. Use rotating torque, not starting torque. If rotating torque is not

within 5 to 15 inch-pounds, use thinner shim under bearing cover (side opposite hypoid gear) to increase pre-load or thicker shim to decrease preload.

18. Install drive pinion and cage assembly using corrected shim pack if pinion setting gauge was used, or original shim pack if gauge was not used. Install lock washers and torque cap screws.

19. Attach a dial indicator to the carrier with the pointer resting against face of one of the drive gear teeth and adjust backlash temporarily to .010 inch.

20. To increase backlash, remove sufficient shims from under cross shaft bearing cover (opposite hypoid gear), and insert shims of equal thickness under opposite bearing cover. To decrease backlash, reverse procedure. Following this method cross shaft bearing pre-load will be retained.

21. With backlash temporarily set at .010 inch, check gear tooth pattern. Refer to "Gear Tooth Contact" in the "General Axle Service" section.

22. After correct tooth contacts have been made, readjust backlash to measure .020 to .026 inch.

23. If axle has a side mounted shift unit, tap shaft unit sleeve into carrier housing with original shim pack. Install lock washers and tighten nuts to 26-29 foot-pounds.

24. If electric shift unit is used, hold shift fork in the collar, lubricate and slide shift shaft through sleeve and into collar. Install and tighten lock screw. Tighten lock nut to 30-35 foot-pounds and install lock wire. Install shift unit.

25. Check clearance of shift fork pads in shift collar. The clearance should be .010 minimum on each side of fork in both high and low speed positions. Add or remove shims to achieve correct adjustment.

26. If air shift control is used, hold

Determining cross shaft preload
(© Rockwell Stand. Div.)

Split ring installation
(© Rockwell Stand. Div.)

Determining shift fork clearance
(© Rockwell Stand. Div.)

BEARING PRELOAD

Axle Model	Pinion Shaft Nut-Thread Size and Torque Limits (Foot-Pounds)		Press Ram Pressure for Preload Check (Tons)	Pinion and Cross Shaft Bearing Preload (Inch-Pounds)	Backlash Limits (Inches)	Differential Bearing Preload Adjusting Nut Notches Tighten from Zero End Play (each Adjusting Nut)
2-Speed	1¾ x 12	800-1100	14	5-15	0.020-0.026	1
Double Reduction	1 x 20	300-400	6	5-15	0.020-0.026	1
	1¼ x 18	700-900	11	5-15	0.020-0.026	1
	1½ x 12	800-1100	14	5-15	0.020-0.026	1

shift fork in position in the collar, lubricate and slide air shift unit shaft through the sleeve and into the fork. Install lock screw. Tighten lock nut to 30-35 foot-pounds and install lockwire.

27. On those axles with a front mounted air shift unit, shift both shift unit and shift fork in HI position using no gaskets. Position the shaft unit and bellcrank assembly in slot of shift shaft. Install lock washers and nut and torque to specifications.

28. With collar and fork shifted to engage helical pinion next to hypoid gear, adjust Allen screw and nut in top of carrier to center the fork in the collar within .005".

29. With collar and fork shifted to engage the helical pinion away from hypoid gear, adjust hex head bolt and lock nut in cross shaft cage to center the fork in the collar within .005". *NOTE: When checking shift fork clearance, make certain collar is flush with end face of pinion being engaged.*

30. Install high and low speed helical gears on their respective differential case halves. If rivets are used, install them cold. Instead of rivets, bolts can be used but must be tightened to specified torque.

31. Lubricate case inner walls and all parts with axle lubricant.

32. Install thrust washer and side gear in one of case halves. Assemble pinions and thrust washers on pinion shaft (spider) and install assembly. Engage the other side gear and thrust washer with pinions.

33. Aligning mating marks, unite case halves with four of the long bolts equally spaced. Check assembly for free rotation of gears and correct if necessary. Torque remaining bolts and install lock wire.

34. Press differential side bearings squarely and firmly on case halves.

35. Temporarily install bearing cups, threaded adjusting nuts or split ring and bearing caps. Torque cap bolts. Bearing cups

must be hand pushed into the bores; or enlarged with a scraper or emery cloth until a hand push fit is obtained. Use a blued bearing cup as a gauge to check fit. If split rings cannot be turned by hand, reduce their O.D. slightly with a fine mill file.

36. Lubricate side bearings with axle lubricant. Place bearing cups in the cones, then position differential assembly between grooves in carrier legs.

37. Insert thin split rings in carrier leg grooves, making certain there is clearance betwen bearing cup faces and the rings.

38. Attach a dial indicator to the carrier housing with its pointer resting against side surface of one of the drive gears. With a pair of small pinch bars, manipulate assembly back and forth between the split rings and measure the end play.

39. Remove and measure split ring thickness. To the total thickness of the two thin rings, add the end play figure, plus another .017 to .022 inch to obtain total thickness of two thicker rings required to obtain proper bearing pre-load. For example: If necessary thin rings were .290 inch each for a total of .580 inch, and the end play is .005 inch, then .580 inch plus .005 inch equals .585 inch. Adding an additional .020 inch for a total of .605 inch thickness for the two split rings would provide .020 inch pre-load on the bearings. The .605 inch may be divided between the two rings such as .300 and .305 inch.

40. Insert one split ring in carrier leg groove and move differential assembly tightly against ring. Install opposite split ring (with gap upward) by tapping it into the groove with a blunt drift.

41. Tap bearing caps into correct location. Torque cap bolts and install lock wire.

TORQUE CHART

Carrier to housing bolt or stud nut	7/16-14-20	54-58
	1/2-13-20	85-91
	5/8-11-18	168-180
Cross shaft bearing cage and cover screw	1/2-13-20	85-91
	9/16-12-18	120-129
	5/8-11	168-180
Cross shaft bearing lock screw	7/6-14	42-45
	9/16-12	92-101
Differential case bolts	3/8-16	34-37
	7/16-14	54-58
	1/2-20	94-102
	9/16-18	132-145
	5/8-18	186-205
	3/4-16	325-360
Differential bearing cap bolt	5/8-11-18	127-140
	3/4-10-16	230-250
	7/8-9-14	345-360
	7/8-14	375-415
	1-14	375-415
Pinion cage bolt	3/8-16	34-37
	7/16-14-20	54-58
	1/2-13-20	85-91
	9/16-12-18	120-129
	5/8-11	168-180
Pinion shaft yoke nut	7/8-20	175-250
	1-20	300-400
	1-1/4-18	700-900
	1-1/2-12-18	800-1100
	1-3/4-12	800-1100

Rockwell Standard Tandem Axle

Through-drive Type—Three Gear Transfer Train With Inter-Axle Differential

IMPORTANT: Before the carrier assembly can be removed from the housing, the through-shaft must be removed from the housing. The removal of the carrier assembly is accomplished in the conventional manner, except for the removal of the through-shaft. The procedure is as follows:

1. Remove the shift shaft housing cap screws and lock washers. Remove the shift shaft housing assembly.

2. Disassemble and remove the shift lever attaching nut, button, lever, cup, and spring. The body fit bolt should not be removed.

3. Remove the through-shaft cage cap screws and lock washers, and remove the through-shaft, cage, and yoke assembly. *NOTE: To free the through-shaft cage*

Drive Axles

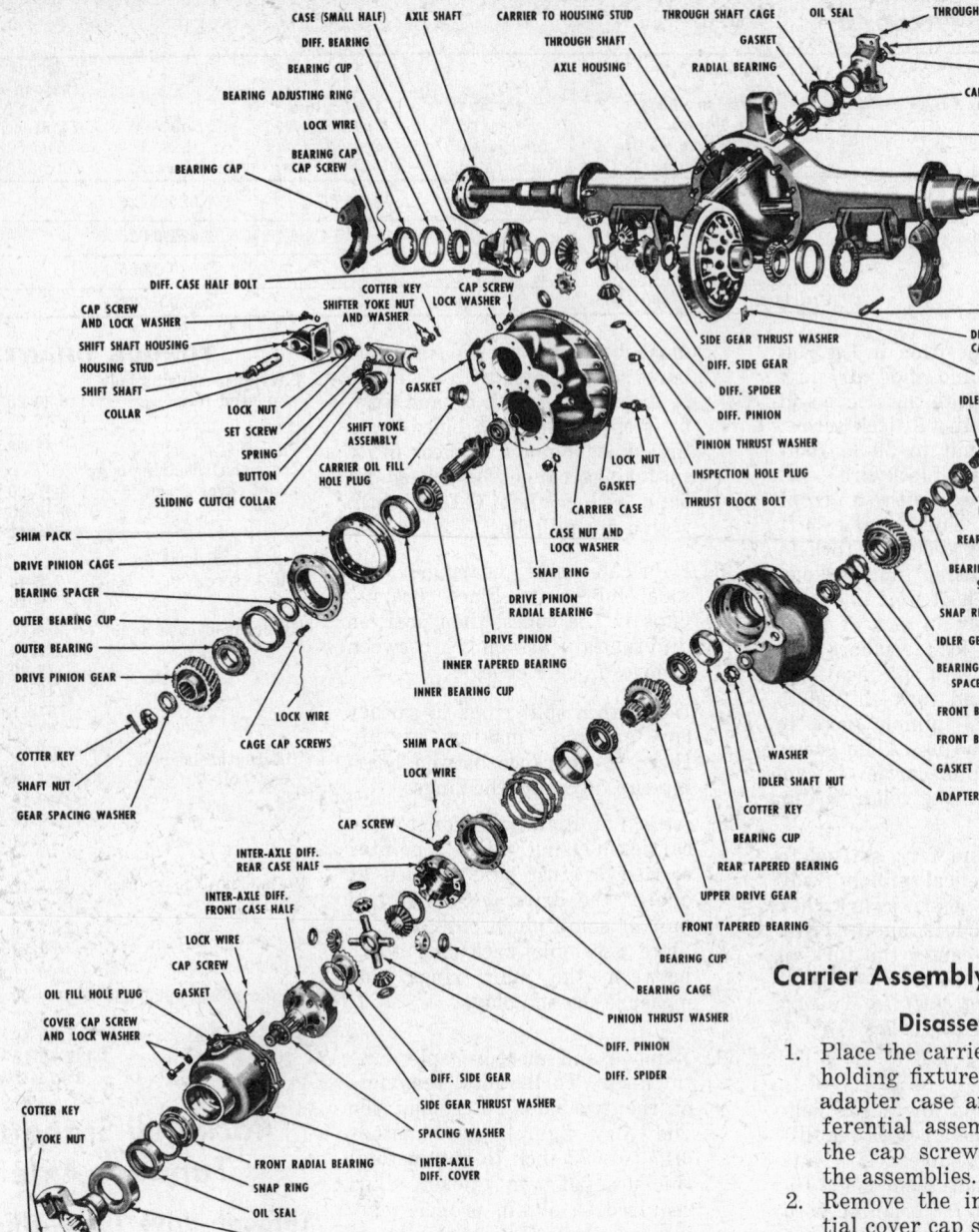

Exploded view—Rockwell Standard tandem axle—through-drive type
(© International Harvester Co.)

from its case bore, it may be necessary to tap the yoke with a soft hammer.

4. Thread the through-shaft assembly from the housing. The sliding clutch must be eased along the shaft at the shift lever opening and when the through-shaft clears the opening, the sliding clutch may be lifted out.

5. Complete the conventional removal of the carrier assembly.

Through-shaft Assembly

Disassembly

1. Remove the through-shaft yoke cotter pin and the nut. Remove the yoke from the shaft.

2. Press the through-shaft from the

cage assembly, using a suitable tool against the inner bearing race. Remove the cage snap ring, if necessary.

3. Tap the radial bearing out of the bore of the cage at the seal end. Discard the seal.

Assembly

1. Install a new bearing into the cage bore and install a new seal.

2. Press the through-shaft into the bearing with the use of suitable tools.

3. Install the yoke, nut, and tighten to specifications, and install the cotter pin.

Carrier Assembly

Disassembly

1. Place the carrier assembly into a holding fixture, and remove the adapter case and inter-axle differential assembly by removing the cap screws and separating the assemblies.

2. Remove the inter-axle differential cover cap screws and lift the assembly from the adapter case.

3. Remove the input shaft cotter pin, nut and washer.

4. Press the inter-axle differential assembly from the cover, using suitable press tools.

5. Mark the differential case halves with a punch to assist in the assembly.

6. Remove the lock wire, bolts and separate the case halves and remove the spider, spider pinions, side gears and thrust washers. *NOTE: Do not remove the radial bearing from the case unless replacement is necessary. If the bearing is replaced, install a new oil seal.*

7. Remove the lock wire from the bolts in the adapter case assembly and remove the bolts from the bearing cage.

8. Insert two cap screws in the puller holes of the cage and

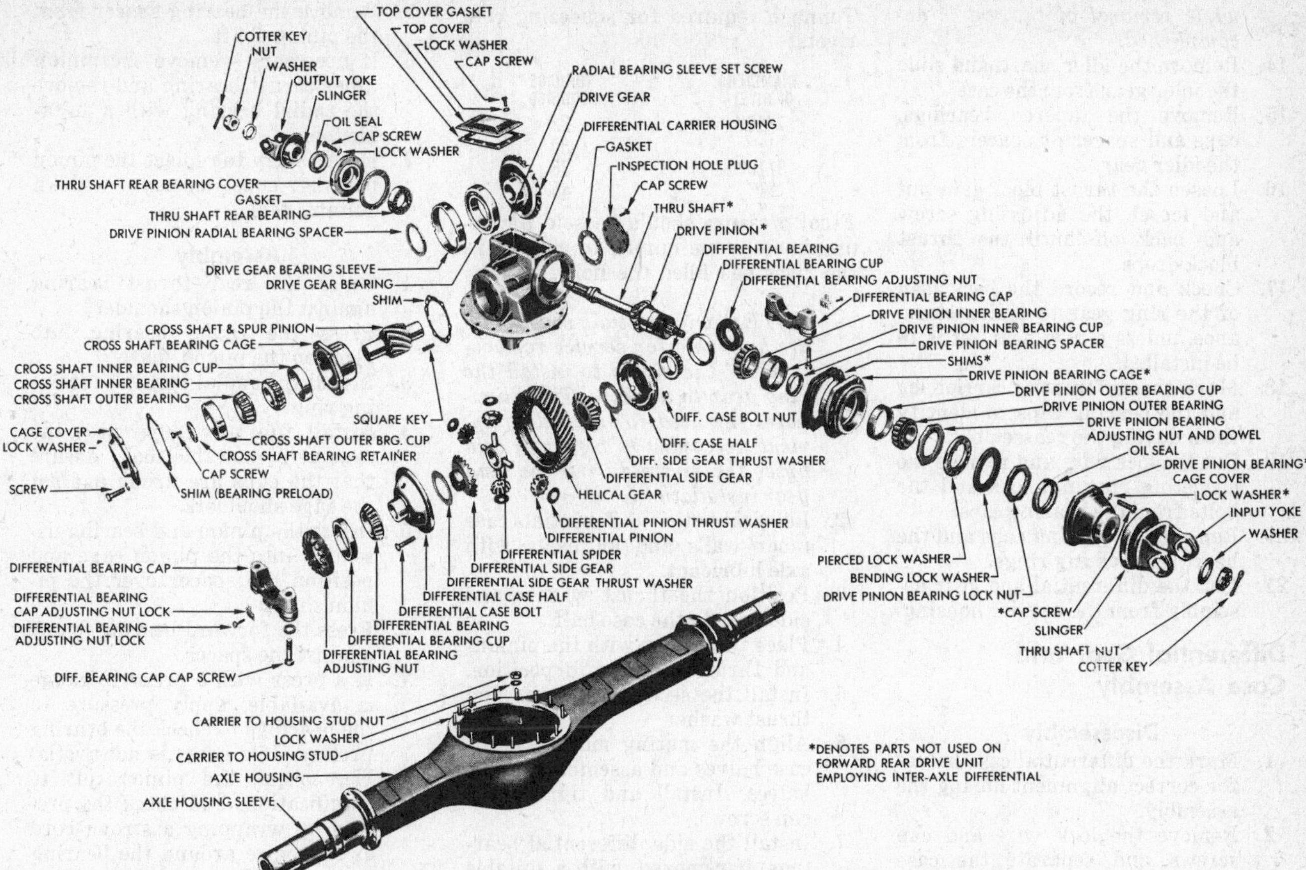

Exploded view—Rockwell Standard—SUD, SUDD, STD, STDD, series—typical rear axles (© International Harvester Co.)

Carrier assembly in holding fixture
(© International Harvester Co.)

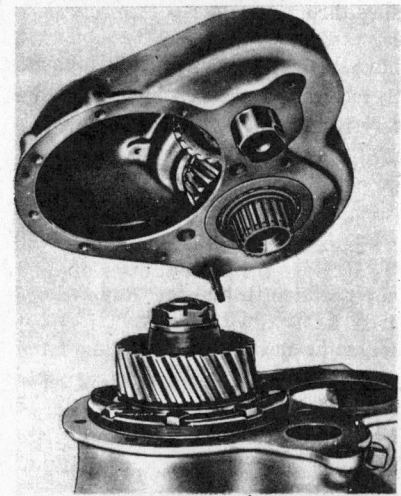

Removing or installing adapter case assembly
(© International Harvester Co.)

tighten to remove the cage from the adapter case.

9. Lift off the cage and the shim pack. Be sure to keep the shim pack intact for the reassembly.

10. From the rear of the adapter case, tap the helical gear assembly with a soft hammer and remove the assembly from the front of the case.

11. If replacement of the bearing cups are necessary, remove them from the cage and adapter case.

12. Remove the tapered bearings from the gear with a suitable puller, if replacement is necessary.

13. Remove the idler gear shaft cotter pin, nut and spacing washer. *NOTE: Flat areas are provided on the idler shaft for holding*

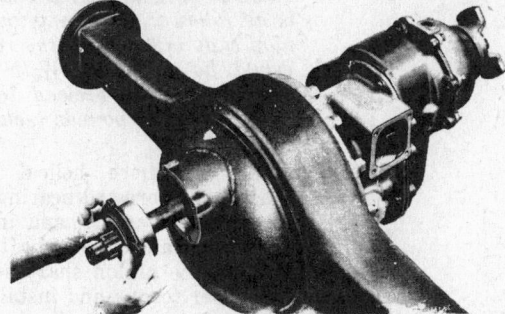

Removing or installing through-shaft assembly
(© International Harvester Co.)

Removing or installing shift shaft housing
(© International Harvester Co.)

while removal of the nut is accomplished.

14. Remove the idler shaft, and slide the idler gear from the case.
15. Remove the tapered bearings, cups and spacer or spacers from the idler gear.
16. Loosen the thrust block jam nut and loosen the adjusting screw and back off until the thrust block drops.
17. Check and record the back lash of the ring gear to pinion clearance, unless a new gear set is to be installed.
18. Mark the differential carrier leg and the bearing caps to identify them during the reassembly.
19. Cut the lock wire and remove the adjusting ring nut lock and the bolts from the bearing caps.
20. Remove the bearing caps and the bearing adjusting rings.
21. Lift the differential and gear assembly from the carrier housing.

Differential Gear and Case Assembly

Disassembly

1. Mark the differential case halves for correct alignment during the assembly.
2. Remove the lock wire and cap screws and separate the case halves.
3. Remove the spider, spider pinions, side gears and the thrust washers.
4. If the ring gear is to be replaced, remove the rivets by drilling and press out the rivets. **CAUTION: Never chisel the head of the rivet to remove the ring gear. Damage to the case holes can result.**
5. The differential bearings can be removed with the aid of a puller.

Assembly

1. Rivet the ring gear to the case half with new rivets.

Tonnage required for squeezing cold rivets:

DIAMETER OF RIVET	TONNAGE REQUIRED
7/16″	22
½″	30
9/16″	36
⅝″	45

Final pressure should be held for approximately one minute to make sure the rivet has filled the hole.

NOTE: Differential case bolts are available for service replacement of the rivets to install the ring gear to the case. This eliminates the need for special equipment necessary to correctly cold upset the rivets during the ring gear installation.

2. Lubricate the differential case inner walls and all parts with axle lubricant.
3. Position the thrust washer and side gear in the case half.
4. Place the spider with the pinions and thrust washers in position.
5. Install the second side gear and thrust washer.
6. Align the mating marks on the case halves and assemble the two halves. Install and tighten the cap screws.
7. Install the side differential bearings, if removed, with a suitable sleeve and press.

Pinion and Cage Assembly

Disassembly

1. Remove the lock wire and remove the pinion cage cap screws.
2. Remove the pinion cage by tapping the end of the pinion shaft with a soft hammer and brass bar.
3. Remove the pinion cotter pin and nut from the shaft. Remove the gear and spacer.
4. Remove the outer bearing from the cage.

Differential case pinion and side gear assembly (© International Harvester Co.)

5. Remove the bearing spacer from the pinion shaft.
6. If necessary, remove the pinion inner thrust bearing and remove the radial bearing with a suitable tool.
7. If necessary to replace the pinion bearing cups, remove with a suitable tool.

Assembly

1. Press the rear thrust bearing against the pinion shoulder.
2. Press the radial bearing into place on the pinion shaft.
3. Install the radial bearing retaining ring.
4. Install the new cups into the cage using suitable tools. Assure that the cups are firmly against the cage shoulders.
5. Insert the pinion and bearing assembly into the pinion cage and position the spacer over the pinion shaft.
6. Press the forward bearing firmly against the spacer.
7. If a press with a pressure gauge is available, apply pressure to the bearings to check the bearing preload. If a press is not available, torque the pinion nut to specifications and check the preload by wrapping a strong cord or soft wire around the bearing cage and pull on a horizontal line with a pound scale. The preload should be within 5-15 inch pounds. Use a thinner spacer to increase or a thicker spacer to decrease the preload torque.

The correct pressures and nut torques for checking the pinion bearing preload are as follows.

PINION SHAFT THREAD SIZE	REQUIRED NUT TORQUE TO OBTAIN CORRECT PRE-LOAD	REQUIRED PRESSURE TO OBTAIN CORRECT PRE-LOAD
1″ x 20	300-400 lb. ft.	6 tons
1¼″ x 18	700-900 lb. ft.	11 tons
1½″ x 12	800-1100 lb. ft.	14 tons
1½″ x 18	800-1100 lb. ft.	14 tons
1¾″ x 12	800-1100 lb. ft.	14 tons
2″ x 16	800-1100 lb. ft.	14 tons

Use rotating torque, not starting torque.

NOTE: To find the correct bearing preload, the cage diameter must be measured. An example is as follows: Assuming the pinion cage is 6 inches, the radius would be 3 inches. With a 5 pound pull, the preload torque would equal 15 pounds inches.

8. Press the drive helical gear against the forward bearing (or spacer where used), and install the washer and pinion shaft nut.
9. Tighten the pinion shaft nut to its proper torque and install the cotter pin.
10. Recheck the pinion bearing preload torque and if not within

Obtaining zero preload—differential bearings
(© International Harvester Co.)

Checking ring gear—pinion backlash
(© International Harvester Co.)

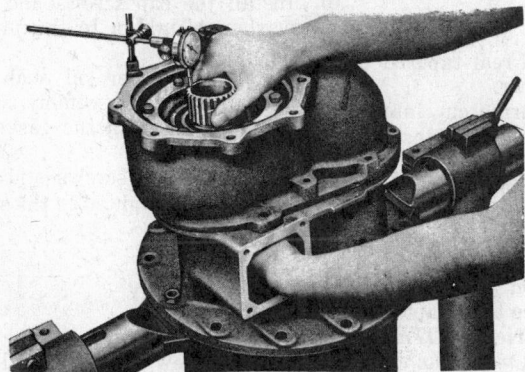

Adjusting tapered bearing end play
(© International Harvester Co.)

Adapter case bearing cage and shim pack
(© International Harvester Co.)

specifications, repeat the procedure to adjust the bearing preload.

11. If the original gears are to be used, install the original shim pack. If the gears have been replaced, alter the shim pack as follows: Record the variation (etched) and the nominal assembly dimension (stamped) on the head of the new and the old pinion gear. Increase or reduce the shim pack in regards to the change in the variation from the old to the new pinion. After changing the sign of the old variation, plus to minus or minus to plus, add to the new variation, sign unchanged, and the result will be the shim pack increase or

decrease in thousands of an inch.

12. Position the pinion and cage assembly in the carrier pinion cage bore and lightly tap into place with a soft hammer.

13. Install the pinion cage screws and torque to specifications.

NOTE: Do not install the lock wire until the final carrier adjustments are made.

Carrier Assembly

Assembly

1. Lubricate the differential bearings and cups with axle lubricant.

2. Place the cups over the bearings and position the carrier assembly over the legs of the carrier

housing and lower into place.

3. Install the bearing adjusting nuts and turn hand tight against the bearing cups.

4. Install the bearing caps on the carrier legs in relation to the previously made marks. **CAUTION: If the bearing caps do not position properly, the adjusting nuts may be crossthreaded. Remove the bearing caps and reposition the adjusting nuts. Forcing the caps into position may result in irreparable damage to the carrier housing or to the bearing caps.**

5. Install the carrier leg cap screw and tighten to specifications. *NOTE: Do not install the adjusting nut locks, cotter keys and lock wire until the final adjustments are made.*

6. Install a dial indicator so that the button is mounted on the backface of the gear.

7. Loosen the bearing adjustment nut on the side opposite the ring gear, sufficient to notice end play on the dial indicator.

8. Tighten the same adjusting nut enough to obtain zero end play.

9. Check gear for runout. If the runout exceeds .008 inch, remove the differential assembly and check for the cause.

10. If the runout is within specifications, tighten the adjusting nuts

Inter-axle differential case assembly—marked for disassembly (© International Harvester Co.)

one notch each from the zero end play, to preload the bearings.

11. If new gears were not used, the established backlash recorded before disassembly should be used. For new gears, the new backlash should be set to .010 inch. Adjust the backlash by moving the ring gear only. This is done by backing off on one adjusting nut and tightening the opposite adjusting nut the same amount.

12. Check the tooth contact between the ring gear and the pinion as outlined in the beginning of this section.

13. Install the adjuster nut locking cotter pins.

14. Install the lock wire in the appropriate bolts.

15. Install the thrust block in the carrier assembly by firmly tightening the adjusting screw until the thrust block is against the back face of the ring gear.

16. Loosen the adjusting screw ¼ turn and lock securely with the lock nut.

Adapter Case

Assembly

1. Install the snap ring and press the idler gear inner bearing cup squarely against the snap ring.

2. Install the idler gear cup spacing sleeve against the opposite side of the snap ring.

3. Press the idler gear outer bearing cup squarely against the spacing sleeve.

4. Position the idler gear inner and outer bearings into the cups with the spacer or spacers between them.

5. Slide the assembly through the adapter case drive pinion opening and position the assembly so that the bearings are aligned with the adapter case shaft hole.

6. Tap the idler shaft through the idler gear assembly so that the inner bearing is against the idler shaft shoulder.

7. Install the washer and nut and torque to 350-400 ft. lbs.

8. Measure the idler shaft bearing end play by the use of a dial indicator mounted to the adapter case with the stem set against the idler gear face. The correct end play is .001-.005.

9. If the bearing end play does not measure within these limits, use a thinner or thicker spacer or combination of the two spacers as required.

10. After the end play has been established, remove the nut and washer and insert the O-ring and reinstall the nut and washer. Torque to 350-400 ft. lbs. and install the cotter pin.

Adjusting differential bearing preload
(© Rockwell Stand. Div.)

Upper Drive Gear

Assembly

1. Press the front and rear tapered bearings into the gear.

2. Press the front bearing cup into the bearing cage, if previously removed.

3. Tap the rear bearing into the adapter case and slide the gear assembly into the case.

4. Mount the carrier assembly in the upright position and install the gasket and place the adapter case over the carrier. *NOTE: The idler shaft flat must line up with the corresponding flat in the carrier.*

5. Lower the adapter case into position on the carrier and install the lock washers and cap screws. Torque to the proper specifications.

6. Using the original shim pack, install the bearing cage over the upper helical gear with the "TOP" mark up.

7. Install the lock washers and cap screws and tighten to the specified torque.

8. Set a dial indicator button against the end of the gear and check the end play. Adjust the tapered bearing end play to .001-.005 inch by adding or subtracting the shims from the shim pack.

9. Install the lock wire in the respective bolts.

Inter-Axle Differential

Assembly

1. Lubricate the differential case walls and all the component parts with lubricant.

2. Position the thrust washer and rear side gear into the case rear half.

3. Place the spider with the pinions and thrust washers in position and install the forward side gear and thrust washer.

4. Align the mating marks and position the forward case half.

5. Install the case cap screws and tighten to the correct torque. Install the lock wire.

6. If the cover assembly was disassembled, install the forward radial bearing and snap ring.

7. Install the spacer on the input shaft.

8. Position the cover over the input shaft and tap the cover down until the bearing seats against the spacer.

9. Install the gasket and install the cover and differential assembly over the upper drive gear hub. Tap the assembly into position. *NOTE: It will be necessary to line up the splines of the drive gear with those of the side gear.*

10. Install the cap screws and lock washers. Tighten to the proper torque.

11. Install the cover oil seal and mount the cover assembly on the adapter case with the gasket in position.

12. Install the cap screws and lock washers and tighten to the specified torque.

Carrier Assembly

Installation

1. Using a new gasket, install the carrier assembly onto the housing and secure with the nuts and lock washers. Tighten the nuts evenly to pull the carrier assembly squarely into the housing.

2. Install the through-shaft rear radial bearing into the cage and lock in place with the snap ring.

Torque Chart

Carrier to housing bolt or stud nut		
	7/16-14-20	54-58
	1/2-13-20	85-91
	5/8-11-18	168-180
Cross shaft bearing cage and cover screw		
	1/2-13-20	85-91
	9/16-12-18	120-129
	5/8-11	168-180
Cross shaft bearing lock screw		
	7/6-14	42-45
	9/16-12	92-101
Differential case bolts		
	3/8-16	34-37
	7/16-14	54-58
	1/2-20	94-102
	9/16-18	132-145
	5/8-18	186-205
	3/4-16	325-360
Differential bearing cap bolt		
	5/8-11-18	127-140
	3/4-10-16	230-250
	7/8-9-14	345-370
	7/8-14	375-415
	1-14	375-415
Pinion cage bolt		
	3/8-16	34-37
	7/16-14-20	54-58
	1/2-13-20	85-91
	9/16-12-18	120-129
	5/8-11	168-180
Pinion shaft yoke nut	7/8-20	175-250
	1-20	300-400
	1-1/4-18	700-900
	1-1/2-12-18	800-1100
	1-3/4-12	800-1100

3. Press the cage and bearing assembly on the splined end of the through-shaft with a suitable sleeve.
4. Install the through-shaft cage oil seal.
5. Enter the through-shaft and cage assembly with a new cage gasket into the cage bore in the rear of the axle housing until the forward end of the shaft is even with the shift lever opening.
6. Install the sliding shift collar over the forward end of the shaft through the shift housing opening. Ease the shaft into the forward side gear of the inter-axle differential, while at the same time, passing the shift col-

lar onto the collar splines.
7. Install the through-shaft cage cap screws and lock washers and tighten to the proper torque.
8. Install over the shift lever bolt, the shift lever spring, cup and lever. Locate the lever inner yoke in the collar groove at this time.
9. Install the shift lever button and nut. Tighten the nut securely and install the cotter pin.
10. Position the shift housing and a new gasket on the carrier assembly.
11. Install the cap screws and lock washers and tighten to the specified torque.
12. Install the through-shaft yoke on the shaft and tighten the nut to

the proper torque and install the cotter pin.

Shift Shaft Adjustment

1. With the shift shaft moved back to its full travel, locking the inter-axle differential, turn the adjusting screw in until the end of the screw touches the end of the shift shaft.
2. Move the adjusting screw 1 to 1¼ turn more and lock the adjusting screw with the jam nut. This will allow approximately .012 inch clearance between the yoke and the groove of the collar and will thus eliminate yoke or collar wear.

Dana-Spicer Single Reduction Models 27, 30, 44, 44-1, 53, 60, 60-35, 70

Differential

Removal

1. Drain lubricant.
2. Remove cover and gasket. *NOTE: Attached to a cover bolt is a metal tag which shows the number of teeth on pinion and ring (drive) gear.*
3. Remove bearing cap screws. Note the matching marks on cap and carrier and make sure caps are reassembled to correct markings.
4. Using a spreader tool, spread carrier a maximum of 0.020 inch and measure amount of spread with a dial indicator. *IMPORTANT: Carrier may be permanently damaged if spread more than 0.020 inch. Do not attempt differential removal without using a spreader.*
5. Carefully lift differential assembly out of carrier.
6. Remove the spreader assembly after removing the differential assembly from the housing.

Drive Pinion

Disassembly

1. Pull flange (yoke) from shaft splines of drive pinion.
2. Using a press or soft hammer, drive pinion and inner bearing cone assembly out of carrier.

Bearing Preload

Mark locations on carrier and caps
(© Dana Corp., Spicer Div.)

Installation of spreader and dial indicator
(© Dana Corp., Spicer Div.)

3. Remove and tag shim pack from splined end of pinion. *NOTE: If either ring (drive) gear or pinion are to be replaced, write down markings (+), (−), or (0) located at face end of pinion*

for reassembly reference.
4. Remove oil seal assembly from carrier bore. This frees oil seal gasket, oil slinger, and bearing cone.
5. If replacement of the pinion tapered bearings is necessary, the bearing cups should be removed from carrier as follows:
   a. Use remover with a driver or slide hammer to remove inner bearing cup from carrier. This frees shim pack. Remove and tag shims for reassembly.
   b. Remove outer bearing cup.
6. Use remover set to separate bearing cone from drive pinion.
7. Separate oil slinger from pinion. *NOTE: This oil slinger is only found on some axle models.*

Differential

Disassembly

1. Remove and label the two bearing cups.
2. Use a suitable type puller to remove the bearing cones. Remove and label adjusting shims.
3. Drive out pinion shaft lock pin. *NOTE: On the Spicer Model 70 rear axle, punch-mark the differential case halves (for reassembly reference) and separate. Remove the differential spider, pinion gears, side gears and thrust washers.*
4. Separate ring gear from case.

Axle Model	Pinion Shaft Nut-Thread Size and Torque Limits (Foot-Pounds)		Press Ram Pressure for Preload Check (Tons)	Pinion and Cross Shaft Bearing Preload (Inch-Pounds)	Backlash Limits (Inches)	Differential Bearing Preload Adjusting Nut Notches Tighten from Zero End Play (each Adjusting Nut)
2-Speed	1¾ x 12	800-1100	14	5-15	0.020-0.026	1
Double Reduction	1 x 20	300-400	6	5-15	0.020-0.026	1
	1¼ x 18	700-900	11	5-15	0.020-0.026	1
	1½ x 12	800-1100	14	5-15	0.020-0.026	1

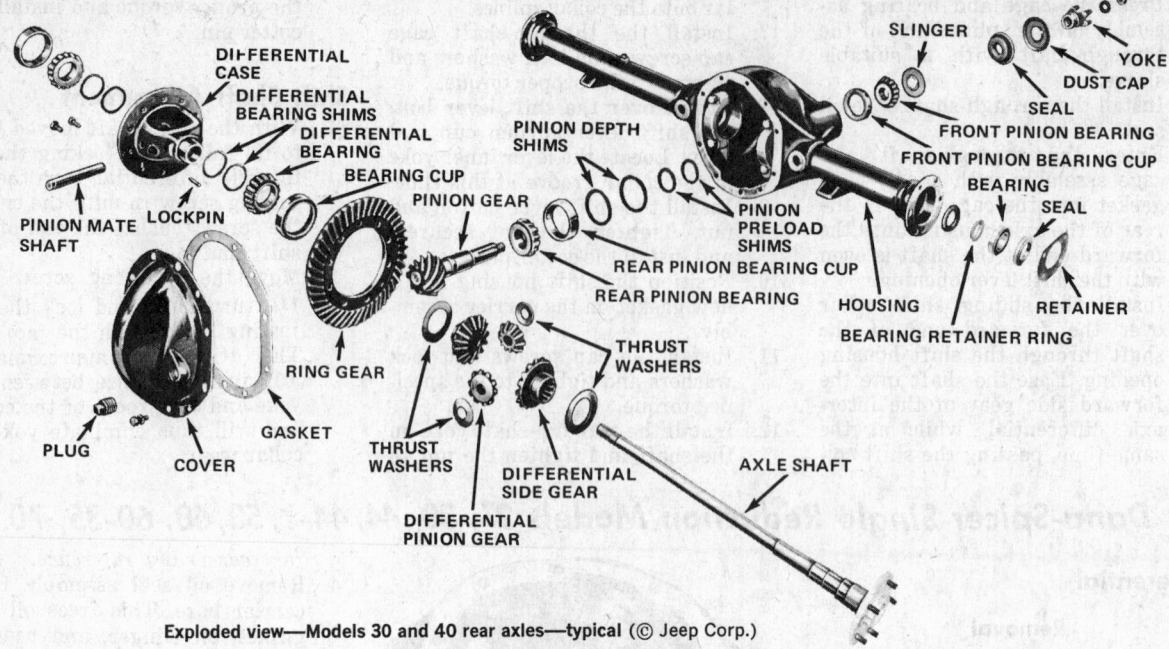

Exploded view—Models 30 and 40 rear axles—typical (© Jeep Corp.)

5. Remove pinion shaft, two pinions, two side gears, and four thrust washers from case.

Assembly

1. Place side gears with new thrust washers in position inside case.
2. Place pinions and thrust washers in position in case.
3. Install the differential pinion shaft in position in case between two pinions. Align shaft lock pin hole with lock pin hole in case and install pinion shaft lock pin. Peen hole to prevent pin from falling out. *NOTE: On the Spicer Model 70 rear axle, install the differential spider along with the differential pinion gears, side gears and thrust washers into the differential case halves. Bolt the two halves together making sure the punch-marks line up.*
4. Place ring (drive) gear in proper position against flange of case and bolt ring gear to case. Alternately tighten these bolts until all bolts are tightened to proper torque.

NOTE: Do not install differential cones or shim packs until pinion depth and bearing preload have been checked out. Differential bearing adjustment is a part of axle assembly procedure.

Differential Bearing

Adjustment

1. Press fit bearing cones tightly against shoulders on case. *IMPORTANT: Do not install shims at this time.*
2. Install bearing cups.
3. Install spreader tool and dial indicator, and spread carrier as described in "Differential Removal."
4. Place differential assembly into carrier.
5. Install bearing caps using their respective cap screws. Make sure caps are assembled to their correct markings. Hand tighten

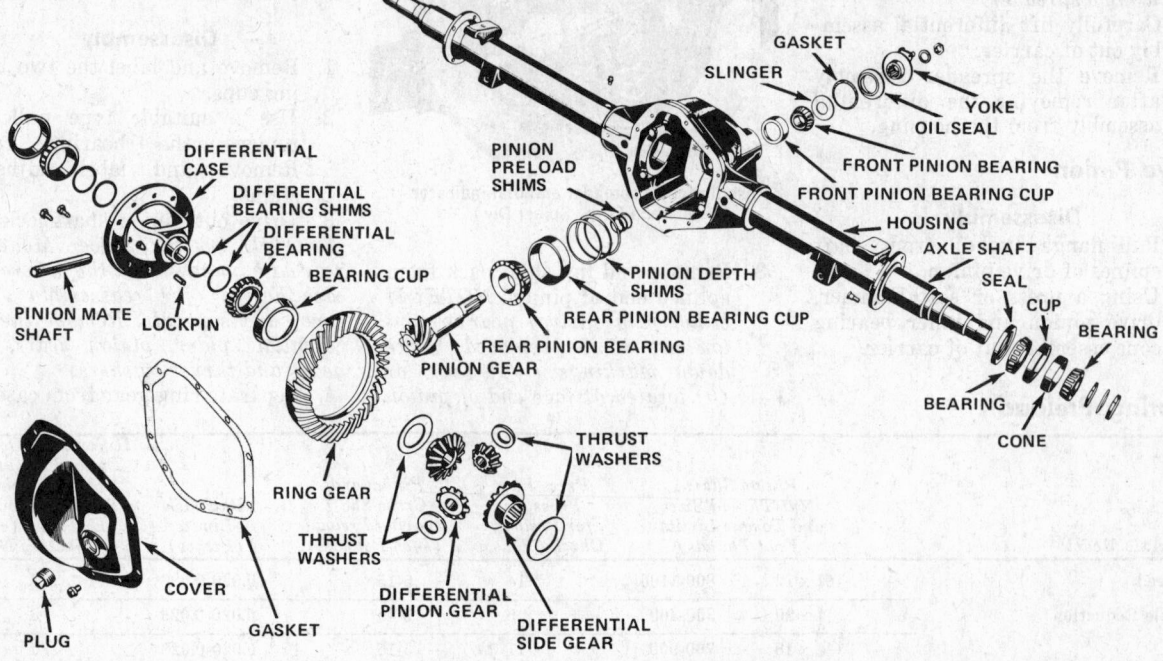

Exploded view—Model 60 rear axle—typical (© Jeep Corp.)

bearing cap screws.

6. Install dial indicator at carrier with indicator button contacting back of ring (drive) gear. Rotate ring gear and check run-out.

7. If run-out exceeds 0.002-inch, remove the differential assembly and remove ring gear from the case.

8. Reinstall differential assembly without ring gear and check run-out of differential case flange. If run-out still exceeds 0.002-inch, the defect is probably due to bearings or case, and should be corrected before proceeding.

9. Remove differential from carrier.

NOTE: Do not install shims behind the bearings until final installation.

Drive Pinion

Installation

1. If either drive pinion or ring (drive) gear must be replaced, they must be installed as a set. (These parts are matched and lapped at time of manufacture to obtain the correct gear tooth contact.)

2. Whenever it is necessary to install a new drive pinion, the plus (+) or minus (−) marking on face of rear end of pinion must be considered. Select a new pinion and ring gear set with markings as near as possible to those on old pinion. If marking on both old and new pinion is the same, do not change thickness of shim pack.

3. The approximate difference between markings on old and new drive pinion is the adjustment that will have to be made in the shim packs.

4. In the first listing below note that the new pinion is a plus eight (+8) while the old pinion is a plus five (+5). Making a difference of plus three (+3). This means that the thickness of each shim pack must be decreased by 0.003-inch. Other examples are:

Pinions New Pinion	Old Pinion	Difference Between Markings	Amount To Change Each Shim Pack (in.)
+8	+5	+3	Dec. 0.003
+5	+8	−3	Inc. 0.003
−5	−3	−2	Inc. 0.002
−3	−5	+2	Dec. 0.002
+5	−3	+8	Dec. 0.008
−4	+2	−6	Inc. 0.006

5. Once proper adjustment in shim packs has been made, place oil slinger, if so equipped, over pinion shaft. Install pinion inner bearing cone over shaft, and use bearing installer and an arbor press to press bearing onto pinion shaft. Bearing must be seated tightly against shoulder or oil slinger.

6. Use pinion front bearing cup installer to install outer bearing cup into carrier bore.

7. Install the selected inner shim pack in carrier. Then use pinion rear bearing cup installer to install inner bearing cup.

8. Insert pinion, oil slinger (when used) and inner bearing cone assembly into carrier and place the selected shim pack into position on outer end of pinion shaft.

9. Place outer bearing cone over pinion shaft, then use installer to seat bearing tight against shim pack.

10. Install pinion flange (yoke), washer and nut. Hold flange while tightening nut to proper torque.

NOTE: Install oil slinger and oil seal only after pinion depth and pinion bearing preload have been checked out.

Checking Pinion Depth Adjustment

1. A pinion depth gauge and correct adapter, which gives a micrometer reading, should be used to determine pinion depth. The actual pinion depth setting can be determined by adding gauge reading to thickness of step plate and comparing result with the nominal dimension of 2.625-inch (models 44/60-35) or 3.125-inch (model 60).

2. If the pinion setting is within minus (−) 0.001-inch to plus (+) 0.003-inch of this nominal dimension, the pinion position can be considered satisfactory.

3. If pinion setting exceeds these limits, it must be corrected by adjusting thickness of shim pack behind the pinion inner bearing cup.

Pinion Bearing Preload Adjustment

1. Use a torque wrench to check pinion bearing preload.

2. Rotating torque of pinion should be from 15 to 30 inch-pounds.

3. Add or remove shims from pack just behind outer bearing cone to bring preload within these torque limits.

Differential

Installation

1. Use dial indicator and spreader tool as described in "Differential Removal," to spread carrier a maximum of 0.020-inch.

2. Install bearing cups and place differential assembly in carrier. Rotate differential and, with a soft hammer, tap ring (drive) gear to assure a proper bearing seating.

3. Reinstall bearing caps in their proper locations as indicated by marks made during the removal procedure. Finger tighten cap screws. Relieve the spreader tool pressure, and tighten cap screws to 70-90 foot-pounds.

4. Move differential assembly tightly against drive pinion.

5. Install dial indicator securely to carrier, then set button at zero and against back of drive gear.

6. Move the differential toward the dial indicator and note the reading. For accuracy, repeat this operation several times.

7. Remove the differential assembly from carrier. Install a shim pack behind differential bearing cone at drive gear side, equal to the dimension indicated by dial indicator.

8. Subtract the indicator reading from the reading previously obtained in paragraph "Differential Bearing Adjustment."

9. To the above result should be added 0.015 to 0.020-inch in shims to provide bearing preload.

10. Install the above shim pack behind differential bearing cone at side opposite to drive gear.

11. Spread differential carrier, using spreader tool.

12. Install differential bearing cups then locate differential assembly in carrier.

13. Rotate differential assembly, tapping gear to seat bearings.

14. Install differential bearing caps in their correct location as indicated by marks made upon disassembly. Finger tighten cap screws.

15. Remove differential carrier spreader tool. Tighten differential bearing cap screws to proper torque.

16. Install dial indicator and check drive gear to drive pinion backlash at four equally spaced points around the drive gear. Backlash must be held to 0.003 to 0.006-inch and must not vary more than 0.002-inch between positions checked.

17. Whenever backlash is not within limits, differential bearing shim pack should be corrected.

NOMINAL ASSEMBLY DIMENSION
AND ADAPTER DISC CHART

Model	Dimension	Tool Number	Model	Dimension	Tool Number
1618-9	4.125"	SE-1065-9-G			SE-1065-9-I
		SE-1065-9-I	17800-1	4.4062"	SE-1065-9-M
1790A-1A	4.2190"	SE-1065-9-CC			SE-1065-9-G
		SE-1065-9-E	G361	2.625"	SE-1065-9-PP
G161	3.551"	SE-1065-9-E			SE-1065-9-O
H140	3.551"	SE-1065-9-E	H340	2.625"	SE-1065-9-PP
H162	3.719"	SE-1065-9-F			SE-1065-9-O
13800	3.6244"	SE-1065-9-G	H362	2.937"	SE-1065-9-R
		SE-1065-9-E			SE-1065-9-PP
16802-3	4.125"	SE-1065-9-G	28M	3.4725"	SE-1065-9-D

Axle Model	Nominal Assembly Dimension	Adapter Disc Tool Number
44	2.625"	SE-1065-9-SS
60*	3.125"	SE-1065-9-Y
70*	3.500"	SE-1065-9-Y

*Model 70—Use a 0.375 shim under dial pointer

International Harvester Co. Single Speed, Single Reduction Light and Heavy Duty Rear Axles

The rear axles may vary as to the design and the construction, but the components of the axles perform similarly regardless of the type. The components of the rear axle that the serviceman will be concerned with are the drive gears, the differential assembly and the axle housing. The removal and installation of the carrier assembly is accomplished in the conventional manner, regardless of a single or tandem axle.

Carrier Assembly
Disassembly

1. Mount the carrier assembly in a suitable fixture.
2. Remove the cotter pins from the bearing adjuster locks and remove the locks from the bearing caps.
3. Match mark the carrier legs to the bearing caps to identify properly upon reassembly.
4. Remove the ring gear thrust block and adjusting screw from the carrier housing.
5. Cut and remove the lock wire. Remove the bearing cap stud nuts or cap screws. Remove the bearing cap stud nots or cap screws. Remove the bearing caps and adjusting nuts. *NOTE: Bearing cap pilot rings may be used on some axle models. Do not lose or damage.*

6. Tip the differential assembly away from the pinion and lift the assembly from the housing.
NOTE: Due to the weight of the differential assembly, a lifting device may be used to assist in the removal.

Differential Case and Gear Aseembly
Disassembly

1. Match mark the differential case halves for the proper reassembly.
2. Cut the lock wire and remove the cap screws or stud nuts and separate the case halves.
3. Remove the spider, pinions, side gears and thrust washers from the case halves.
4. Remove the ring gear rivets by center punching each rivet head and using a drill, 1/32 inch smaller than the rivet body, drill through the rivet head. Use a punch to press out the remaining part of the rivet.
CAUTION: Never use a chisel to cut off the head of the rivets or damage to the differential case can result.

Pinion and Cage Assembly
Removal

1. Remove the pinion cage cap screws and remove the pinion cage assembly from the differential carrier. *NOTE: Puller screw holes are provided on some pinion cages, to assist in the removal of the cage from the housing. If no puller screw holes are present, a brass drift can be used on the inner end of the pinion to force the pinion and cage assembly from the carrier housing.*

CAUTION: Do not use a drift on pinion shafts that have the straddle bearing retained by a snap ring. The snap ring groove may collapse.

2. Retain the shim pack for use during the reassembly.
3. Remove the companion flange from the pinion shaft, after the removal of the cotter pin and nut. *NOTE: The companion flange may have to be tapped of the pinion shaft with a soft hammer.*
4. Remove the outer bearing from the cage by holding the cage in a vise and tapping on the pinion shaft end, and forcing the shaft through the cage. **CAUTION: Do not allow the component parts to fall.**
5. Remove the spacer or spacer combination from the pinion shaft.
6. Remove the rear tapered thrust

Differential spider
Spider gears (4)
Differential case halves
Thrust washers (4)
Thrust washer (2)
Side gears (2)

Exploded view—differential assembly—typical
(© International Harvester Co.)

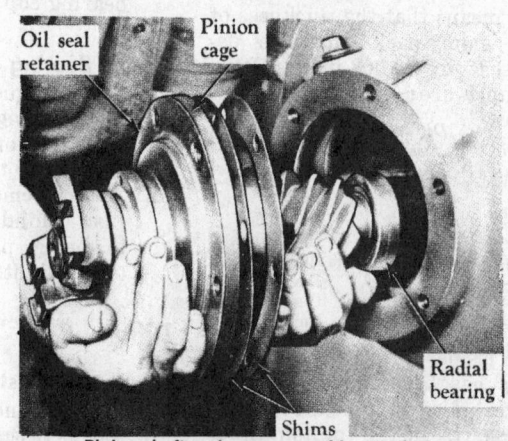

Oil seal retainer
Pinion cage
Radial bearing
Shims

Pinion shaft and cage assembly—typical
(© International Harvester Co.)

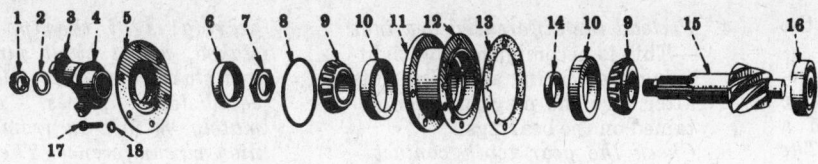

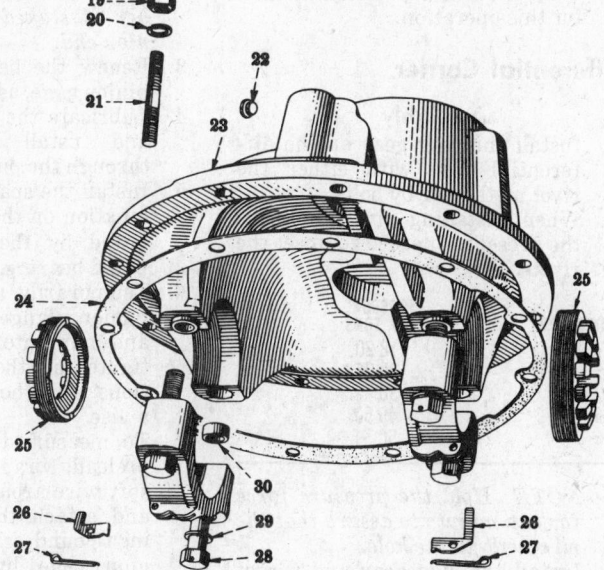

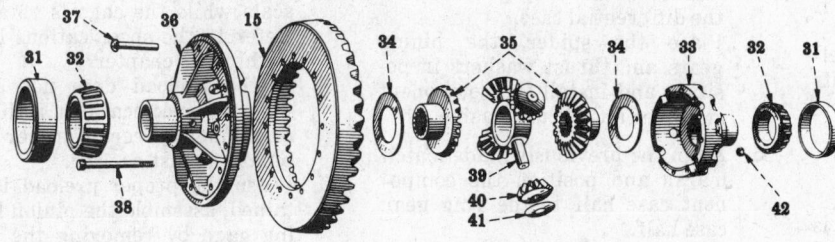

1 Nut, pinion end
2 Washer
3 Flange
4 Slinger
5 Retainer, pinion oil seal
6 Seal
7 No longer used
8 Seal, O-ring type
9 Bearing, pinion thrust
10 Cup, bearing
11 Cage, pinion bearing
12 Not used.
13 Shim pack
14 Spacer, pinion bearing
15 Gear set, drive and pinion
16 Bearing, radial
17 Bolt, hex head
18 Washer, lock
19 Nut, stud
20 Washer, lock
21 Stud
22 Plug, pipe
23 Carrier, with caps, assembly
24 Gasket, carrier to axle housing
25 Adjuster, bearing
26 Lock, adjuster
27 Pin, cotter.
28 Bolt, bearing cap
29 Washer
30 Bushing, pilot ring
31 Cup, bearing
32 Bearing, differential
33 Case, differential, plain half
34 Washer, thrust
35 Gear, differential side
36 Case, differential, flanged half
37 Rivet, drive gear to case
38 Bolt, differential case
39 Spider, differential
40 Gear, spider
41 Washer, thrust
42 Nut, hex
43 Breather, axle housing vent
44 Housing, axle assembly
45 Shaft, axle
46 Stud, wheel flange
47 Nut, hex
48 Plug, pipe

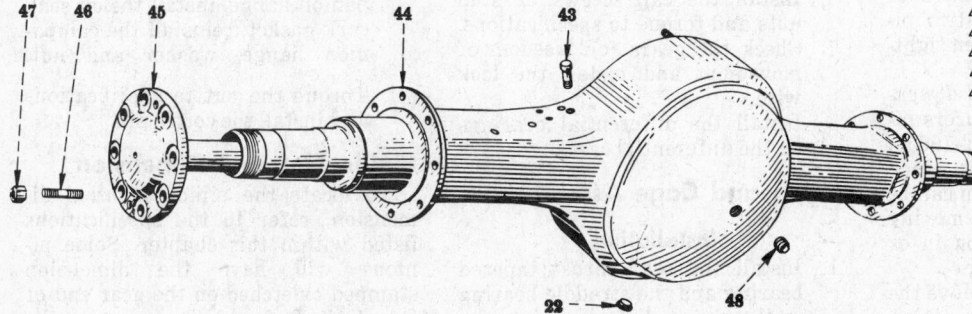

Exploded view—heavy duty single reduction rear axle—International Harvester Co.
(© International Harvester Co.)

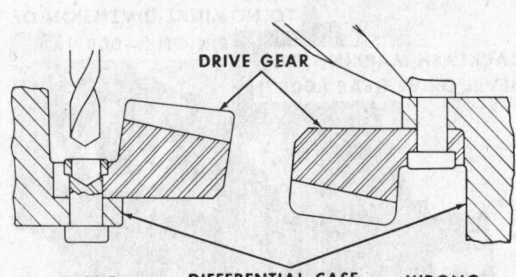

Right and wrong way to remove ring gear rivets
(© International Harvester Co.)

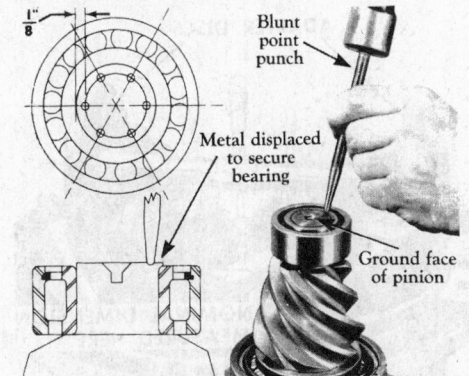

Staking straddle bearing to pinion shaft
(© International Harvester Co.)

bearing from the pinion with the aid of a suitable puller.

7. Remove the straddle bearing retainer, if equipped, and remove the bearing with the aid of a suitable puller. *NOTE: The straddle bearing may be retained by staking of the pinion shaft end, by a snap ring, or by a cap screw and washer.*

8. Remove the cork seal, and oil seal from the pinion cage.

Differential Carrier Assembly

Precautions to be Observed During the Reassembly

1. Before assembly, lubricate the bearings and cups and rewrap to maintain cleanliness.

2. Use correct rivet pressure when installing the ring gear to the differential case, or if bolts are available, be assured that the proper torque is applied in tightening.

3. Be sure that the bearing caps and adjusting nuts are correctly aligned and that the bearing cups fit properly. Irreparable damage can result to the differential carrier or bearing caps if the alignment is off.

4. Observe the proper torque settings when tightening any nuts or bolts.

Five Steps in the Reassembly of the Differential Assembly

1. *Pinion bearing preload*—This is determined by the thickness of the spacer between the two pinion thrust bearings, when tightened in the pinion cage.

2. *Establish pinion nominal dimension*—Use the manufacturers pinion setting gauge (SE-1065), or use an equivalent tool. Changes to this dimension can be made by adding or removing shims to move the pinion in or out of the carrier housing.

3. *Set the ring gear lash*—Move the ring gear to or from the pinion by means of the differential bearing adjusters.

4. *Preload the differential bearings*—This is accomplished by tightening the bearing adjusting nuts after zero end play has been obtained on the bearings.

5. *Check the gear tooth contact*—Use the paint impression method for this operation.

Differential Carrier

Assembly

1. Install the ring gear on the differential case with either the rivet method or by bolts.

2. When installing rivets, observe the pressures needed to upset the rivets.

RIVET SIZE	PRESSURE U.S. TONS
7/16	18-20
1/2	20-25
9/16	36
5/8	45-50
3/4	50

NOTE: Hold the pressure force for one minute to assure that the rivet will fill the hole.

3. Install the side gear and thrust washer in the ring gear half of the differential case.

4. Place the spider, the pinion gears and thrust washers in position and install the component side gear and thrust washer.

5. Align the previously made match marks and position the component case half to the ring gear case half.

6. Install the cap screws or stud nuts and torque to specifications.

7. Check the gears for freedom of movement and install the lock wire.

8. Install the differential bearings on the differential case.

Pinion and Cage Assembly

Installation

1. Install the rear thrust tapered bearing and the straddle bearing on the pinion shaft.

2. Install the straddle bearing retainer. *NOTE: If the straddle*

bearing is of the type to be staked, use a blunt pin punch and stake in at least four to six equidistant places, approximately 1/8 inch in from the pinion circumference. The size of the pinion will dictate the number of staked points on the pinion end.

3. Renew the bearing cups in the pinion cage, as necessary.

4. Lubricate the bearings and cups and install the pinion shaft through the pinion cage.

5. Install the spacer or spacer combination on the pinion shaft, followed by the outer pinion tapered bearing.

6. Temporarily assemble the companion flange and the washer and nut onto the pinion shaft, tightening the nut to specifications while holding the flange in a vise.

7. To measure the pinion bearing preload, wrap a strong cord or soft wire around the pinion cage and attach the other end to a inch-pound scale. Rotate the pinion cage by pulling on the spring scale and reading the scale while the cage is rotating. Refer to the specifications listed within this chapter.

8. If the preload does not agree with the specifications, a thicker or thinner spacer or spacer combination must be used.

9. When the proper preload is obtained, assemble the pinion bearing cage by removing the companion flange, install the oil seal, cork gasket, reinstall the campanion flange, washer and nut.

Torque the nut to specifications and install the cotter pin.

Pinion Nominal Dimension

To locate the pinion nominal dimension, refer to the specifications listed within this chapter. Some pinions will have the dimension stamped or etched on the gear end of the shaft. Refer and compare to the specifications. The pinion variation, noted in thousands of an inch, will be

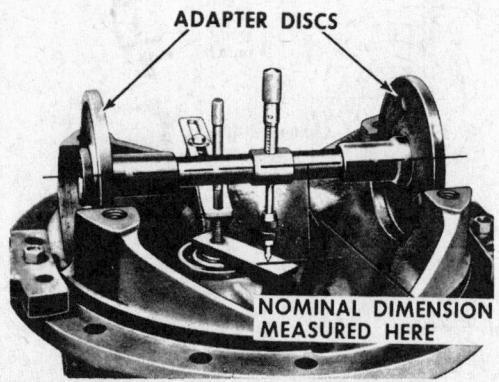

Use of the pinion setting gauge—typical
(© International Harvester Co.)

Location of pinion setting markings—typical
(© International Harvester Co.)

etched on the gear end of the pinion shaft. This figure will be used in determining the amount of shims needed to locate the pinion gear in the proper relationship to the ring gear centerline. *NOTE: Refer to the beginning of the Rear Axe Drive Section for the procedure to follow in the use of the pinion setting gauge tool. If the pinion setting gauge tools are not available, the pinion depth will have to be adjusted by assembling the carrier assembly, installing the pinion cage assembly and the differential assembly into the carrier housing, and observing the tooth contact pattern on the ring gear. This is a trial and error method and very time consuming.*

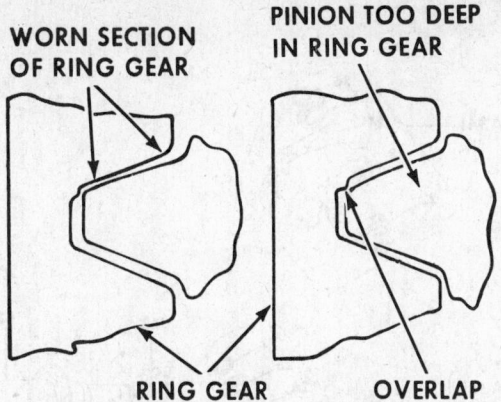

CORRECT   **INCORRECT**

Correct and incorrect gear lash—using worn gears
(© International Harvester Co.)

Differetial Assembly

Installation, Preload, Backlash

1. Install the differential assembly with the bearing cups on the differential bearings into the legs of the carrier housing.
2. Install the bearing adjusting nuts and the bearing caps. Install the cap screws or stud nuts and turn the adjusting nuts while tightening the bearing caps to assure freedom of movement of the adjusting nuts.

   CAUTION: If the bearing caps are not positioned properly, the adjusting nuts may be cross-threaded, and irreparable damage to the carrier housing or to the bearing cups may result.

3. With the side bearing caps loosened to permit the bearing cup movement, loosen the adjusting nuts only enough to notice end play on a dial indicator, mounted on the carrier assembly with the button contacting the back side of the ring gear.
4. Tighten the adjusting nuts to obtain zero end play on the indicator.
5. Move the dial indicator to the coast side of the ring gear teeth, and determine the amount of back lash present between the pinion and the ring gear.
6. To adjust the back lash, move the ring gear towards or away from the pinion by means of the differential bearing adjusting nuts. Move the adjusting nuts the same distance, either in or out to maintain the differential bearing zero end play.
7. When the correct backlash clearance is established, tighten each adjusting nut one or two notches (depending upon the axle model), to preload the differential bearings. Tighten the bearing cap screws or stud nuts to the proper torque and recheck the gear backlash. Install the adjusting nut locks and cotter pins.

8. Coat approximately twelve teeth of the ring gear with oiled red lead paint and rotate the pinion in its normal rotation and check the drive side of the ring gear teeth for the tooth contact impression. *NOTE: A sharper tooth contact impression may be obtained by applying a small amount of resistance to the gear with a flat steel bar and using a wrench to turn the pinion.*
9. If the area of contact starts near the toe end of the ring gear and extends about 2/3 of the tooth length, the tooth contact is satisfactory.
10. Install the ring gear thrust block, if equipped. Adjust the block firmly against the back face of the ring gear and back off the screw 1/4 turn and lock the jam nut.
11. Install the carrier assembly into the housing, following the reverse procedure of the removal operation.

International Harvester Co. Thru-Drive Single Reduction Rear Axle

The forward rear axle used in the tandem arrangement, has a integral power divider and interaxle differential. The differential parts, ring gear and pinion sets, bearings and the axle shafts are common between the forward and rear axle units. The power divider may be removed individually with out removing the differential carrier. The overhaul of the carrier assembly is basically the same as the light and heavy duty models.

Thru-Shaft Assembly

Removal

1. Disconnect the inter-axle control linkage from the lever at the lock control housing. *NOTE: If equipped with air or vacuum controls, disconnect the air or vacuum lines at the cylinder.*
2. Disconnect the drive shaft from the input and thru-shaft flanges.
3. Remove the thru-shaft bearing retainer bolts and loosen the bearing retainer from the housing.
4. Withdraw the bearing cage and the thru-shaft from the axle housing.

Power Divider Assembly

Removal

1. Remove the bolts from the inter-axle lock control housing and remove the housing.
2. Remove the inter-axle differential cover bolts and lift the inter-axle assembly from the power divider intermediate case.
3. Remove the intermediate case to differential carrier bolts and lift the intermediate case from the differential carrier.

Disassembly

1. Lift the inter lock clutch ring from the power divider input gear.
2. Remove and disassemble the input gear.

Axle Model RA 341

a. Free the bearing snap ring and remove by pressing the input shaft to the rear.
b. Remove the input shaft gear and the bearing by pressing forward to free the bearing from the bore.
c. Remove the bearing retainer nut.
d. Pull the bearing from the input gear with a puller tool.

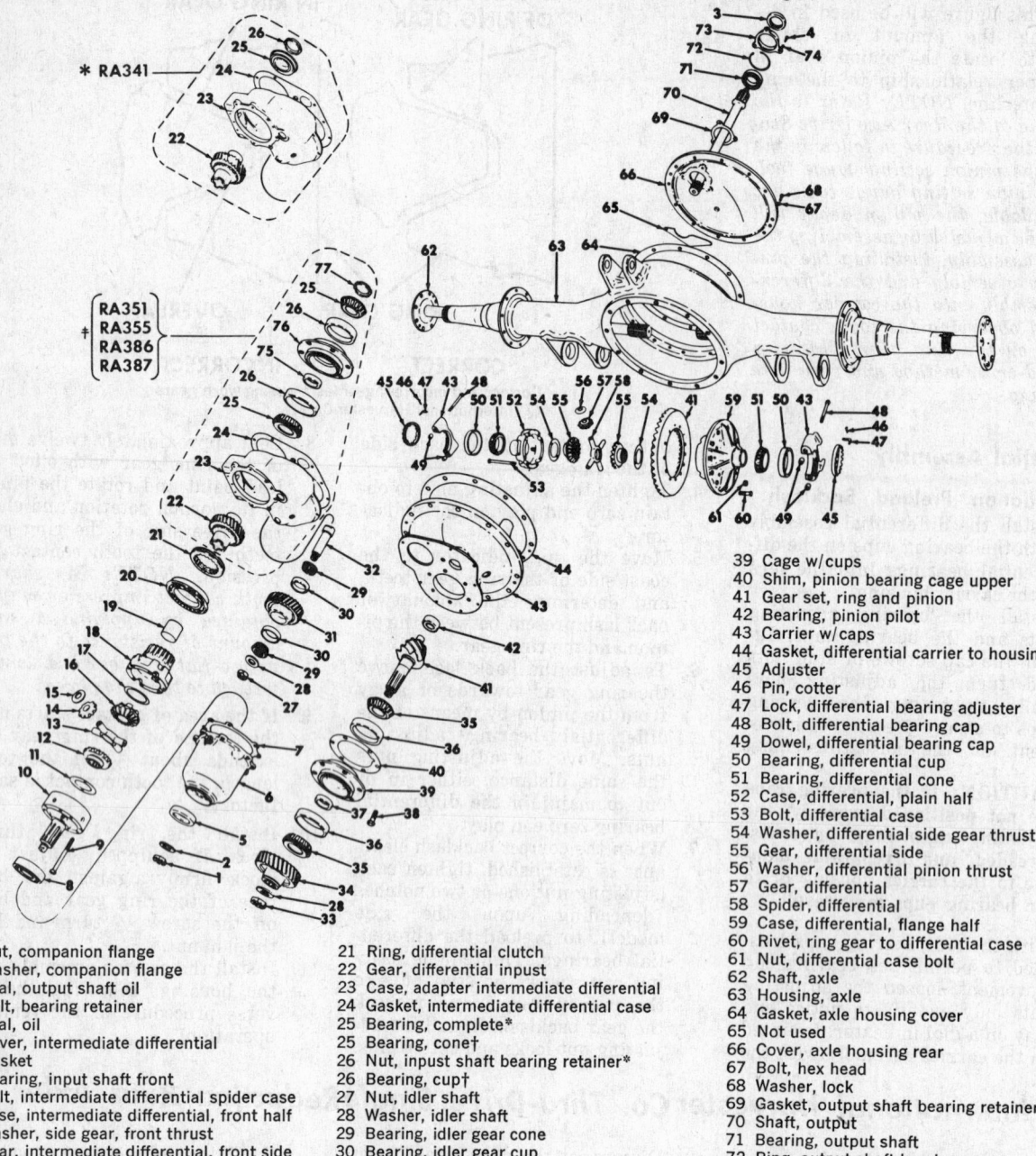

* RA341

RA351
‡ RA355
RA386
RA387

39 Cage w/cups
40 Shim, pinion bearing cage upper
41 Gear set, ring and pinion
42 Bearing, pinion pilot
43 Carrier w/caps
44 Gasket, differential carrier to housing
45 Adjuster
46 Pin, cotter
47 Lock, differential bearing adjuster
48 Bolt, differential bearing cap
49 Dowel, differential bearing cap
50 Bearing, differential cup
51 Bearing, differential cone
52 Case, differential, plain half
53 Bolt, differential case
54 Washer, differential side gear thrust
55 Washer, differential side
56 Washer, differential pinion thrust
57 Gear, differential
58 Spider, differential
59 Case, differential, flange half
60 Rivet, ring gear to differential case
61 Nut, differential case bolt
62 Shaft, axle
63 Housing, axle
64 Gasket, axle housing cover
65 Not used
66 Cover, axle housing rear
67 Bolt, hex head
68 Washer, lock
69 Gasket, output shaft bearing retainer
70 Shaft, output
71 Bearing, output shaft
72 Ring, output shaft bearing snap
73 Retainer, output shaft bearing
74 Washer, lock
75 Washer, rear input gear bearing†
76 Retainer, rear input gear bearing†
77 Nut, input gear bearing retainer†

1 Nut, companion flange
2 Washer, companion flange
3 Seal, output shaft oil
4 Bolt, bearing retainer
5 Seal, oil
6 Cover, intermediate differential
7 Gasket
8 Bearing, input shaft front
9 Bolt, intermediate differential spider case
10 Case, intermediate differential, front half
11 Washer, side gear, front thrust
12 Gear, intermediate differential, front side
13 Spider, intermediate differential
14 Gear, intermediate differential pinion
15 Washer, pinion gear front thrust
16 Gear, intermediate differential rear side
17 Washer, rear side gear thrust
18 Case w/bushing, rear half
19 Bushing, intermediate differential side gear
20 Bearing, case rear half

21 Ring, differential clutch
22 Gear, differential inpust
23 Case, adapter intermediate differential
24 Gasket, intermediate differential case
25 Bearing, complete*
25 Bearing, cone†
26 Nut, inpust shaft bearing retainer*
26 Bearing, cup†
27 Nut, idler shaft
28 Washer, idler shaft
29 Bearing, idler gear cone
30 Bearing, idler gear cup
31 Gear w/bearing
32 Shaft, idler gear
33 Nut, pinion shaft
34 Gear, pinion shaft
35 Bearing, pinion cone
36 Bearing, pinion cup
37 Spacer, pinion bearing
38 Bolt hex head lock

* RA-341 only
† RA-351, RA-355, RA-386, and RA-387

Exploded view—forward tandem axle—International Harvester Co.
(© International Harvester Co.)

Axle Models RA 351, 355, 386, and 387

a. Remove the bearing retainer mounting bolts from the rear of the intermediate case and tap the input gear, bearings and retainer assembly out towards the front.

b. Remove the bearing retainer nut and disassemble the retainer and bearing from the

input gear. *NOTE: The inner bearing is a press fit and should only be removed if it is to be replaced.* **CAUTION: Clamp the input gear in a soft jawed vise only, when removing the bearing retainer nut.**

3. Loosen and remove the idler shaft end nut while holding the shaft on the flats provided.

4. Support the rear face of the intermediate cover and press the idler shaft from the case.

5. Slide the idler gear and bearings from the cover.

6. Separate the bearings, bearing races, and bearing spacers from the idler gear. *NOTE: Axle models RA 351, 355, 386, and 387 use two bearing spacers.*

Assembly

1. Assemble the idler shaft bearings, bearing races and the bearing spacer (s) to the idler gear. **CAUTION: Axle models RA 351, 355, 386, and 387, position the bearing spacers correctly to index the oil groove in the rear spacer with the oil passage in the idler shaft.**

2. Position the idler shaft in the intermediate case so that the flats on the rear shaft will index with the flats on the differential carrier.

3. Support the front of the intermidiate case and press the idler shaft into position.

4. Install the idler shaft end nut and washer and torque to specifications.

5. Using a dial indicator, measure the end play of the idler gear. See specifications for allowable limits.

6. If the end play is not within the allowable limits, increase or decrease the end play by installing a thicker or thinner bearing spacer.

7. Assemble the input gear and install.

Axle Model RA 341

a. Press the bearing on the input shaft gear.

b. Install the bearing retainer nut, and torque to specifications.

Axle Models RA 351, 355, 386, and 387

a. Use the original bearing spacer as a trial assembly. Install the bearing and cage on the input shaft and clamp the assembly in a holding fixture.

b. Install the bearing retainer nut and tighten to the specified torque.

c. Measure the input gear bearing end play with a dial indicator. Tolerance should be .001-.003 inch.

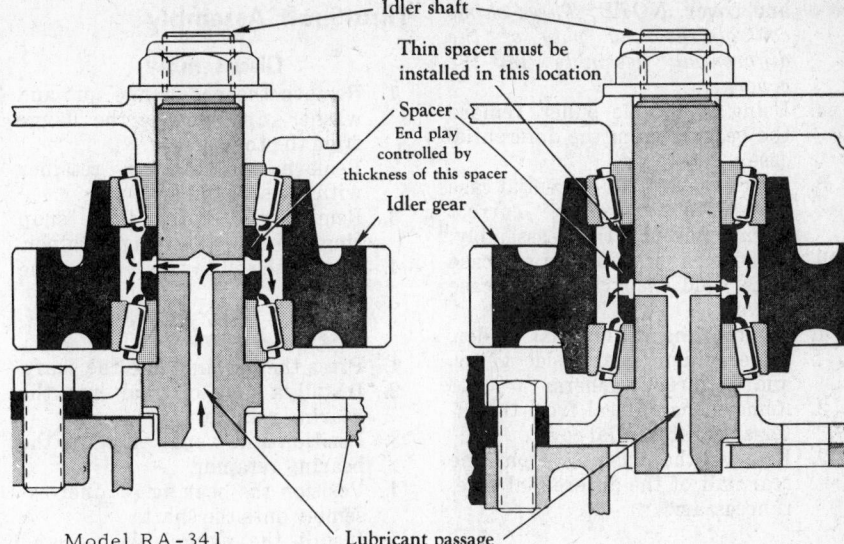

Model RA-341   Model RA-351, 355, 386 and 387

Cross section of idler shaft installation (© International Harvester Co.)

d. Use bearing spacers as necessary to obtain the correct end play.

e. Stake the retainer nut at the milled slot in the input gear, after the proper end play has been obtained.

f. Install the input gear bearings and cage assembly into the intermediate case and secure with the six hex-headed bolts.

Inter-Axle Differential Assembly

Removal

1. Remove the bolts from the in-ter-axle lock control housing and remove the housing.

2. Remove the inter-axle differential cover bolts and lift the inter-axle assembly from the power divider intermediate case.

Disassembly

1. Secure the input flange to a holding tool and remove the flange nut.

2. Using a press, force the inter-axle differential assembly from the cover and flange. **CAUTION: Do not allow the differential assembly to fall to the floor.**

3. Note and retain the shims located between the front bearing

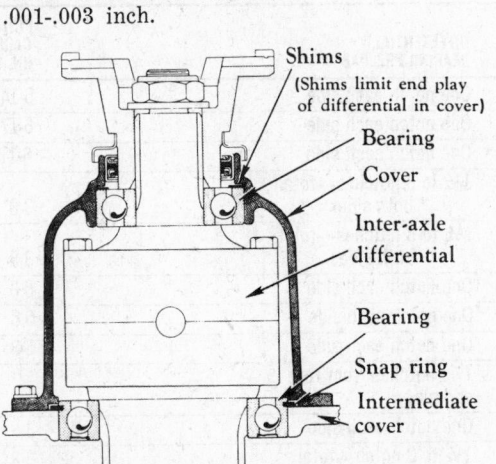

Cross section of Inter-axle differential assembly (© International Harvester Co.)

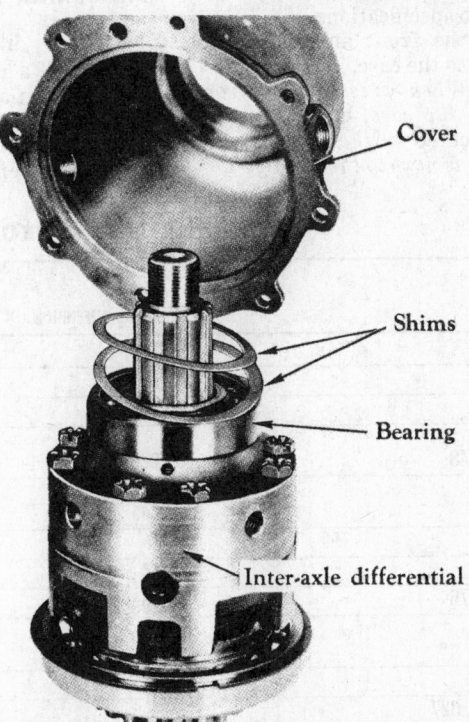

Cover removal—Inter-axle differential (© International Harvester Co.)

Drive Axles

and cover. *NOTE: These shims control the end play of the differential assembly in the cover.*

4. Using a suitable puller, remove the bearing from the differential case.
5. Match mark the differential case halves to insure the correct alignment upon the reassembly.
6. Remove the differential case bolts and separate the case halves.
7. Remove the differential spider, spider pinion gears, side gears, and the thrust washers.
8. Remove the oil seal from the inter-axle differential cover.
9. Remove the bushing from the rear half of the differential case, if necessary.

Assembly

1. Install the bushing, if removed, in the rear half of the differential case.
2. Lubricate all the differential component parts.
3. Position the thrust washer and the rear side gear in the rear half of the differential case.
4. Position the spider, spider pinion gears and the thrust washers in the rear half of the differential case.
5. Position the thrust washer and front side gear in the front half of the differential case.
6. Align the match marks and assemble the differential case. Install the bolts and tighten in stages. Check the gears for freedom of rotation and torque the bolts to specifications.
7. Press the front and rear bearings into the case.

*NOTE: When pressing the front bearing on the case, **be certain** that the large radius on the inner race of the bearing is next to the case.*

Thru-Shaft Assembly

Disassembly

1. Remove the end flange nut and washer, and remove the flange from the thru-shaft.
2. Remove the bearing retainer with the seal and snap ring.
3. Remove the seal and the snap ring from the bearing retainer.
4. Press the bearing from the shaft.

Assembly

1. Press the bearing onto the shaft.
2. Install a new oil seal into the bearing retainer.
3. Position a new snap ring into the bearing retainer.
4. Position the bearing retainer assembly onto the shaft.
5. Install the flange, washer and nut. Torque the nut to specifications.

Forward and Rear Rear Axle Carrier Assemblies

Disassembly and Assembly

Because of the similarity of the drive axle assemblies, refer to the light and heavy duty rear axle section for the procedures necessary for the replacement and adjustment of the ring and pinion gears, the overhaul of the differential gear assembly, and the preloading of the pinion and differential bearings. Tolerance and torque specifications will be found in the accompanying charts.

Power Divider and Inter-Axle Differential Assemblies

Installation

1. Install a new case gasket on the differential cover.
2. Position the intermediate case assembly on the differential carrier. Align the flats on the idler shaft to the flats within the carrier.
3. Install the intermediate case bolts and tighten to the specified torque.
4. Position the inter-axle lock clutch ring on the input gear.
5. Place a new gasket in position on the intermediate case.
6. Install the inter-axle differential assembly on the intermediate case, lining up the teeth to mesh properly of the input gear and the gear within the differential case.
7. Install the differential bolts and torque to specifications.
8. Install a new gasket on the lock control housing opening on the intermediate case, and install the lock control housing.
9. Install the lock control housing bolts and torque to specifications.

Thru-Shaft Assembly

Installation

1. Place the thru-shaft bearing retainer gasket on the axle housing and insert the thru-shaft into position. Rotate the thru-shaft to mesh the shaft splines with the splines in the inter-axle differential side gears.
2. Install the bearing retainer bolts and tighten to the specified torque.
3. Install the drive shaft to the input and thru-shaft flanges.
4. Connect the inter-axle lock control linkage to the lever at the lock control housing.

NOTE: If equipped with air or vacuum controls, reconnect the air or vacuum lines.

SPECIFICATIONS

AXLE MODEL	PINION NOMINAL DIMENSION (IN.)	DIFFERENTIAL BEARING PRELOAD	PINION CAGE PRELOAD (IN. LBS.)
RA 15, 20	2.9830	One notch each side	5-14
RA 25, 30	3.2530	One notch each side	6-17
RA 29, 39	3.4725	One notch each side	6-17
RA 43, 44, 47, 48	3.7695	1½ to 3 notches—total both sides	3-8
RA 57	4.2845	1½ to 3 notches—total both sides	3-8
RA 70	①	One notch each side	5-8
RA 71	4.344	One notch each side	5-8
RA 72, 73, 74, 76	4.613	One notch each side	5-8
RA 75, 78	4.600	Two notches from zero end play	
RA 341	3.4725	One notch each side	3-8
RA 351, 3, 386, 387	3.7695	1½ to 3 notch—total	3-8

① Pinion straddle bearing secured with a snap ring—4.344
Pinion straddle bearing secured with a cap screw—4.596

Index

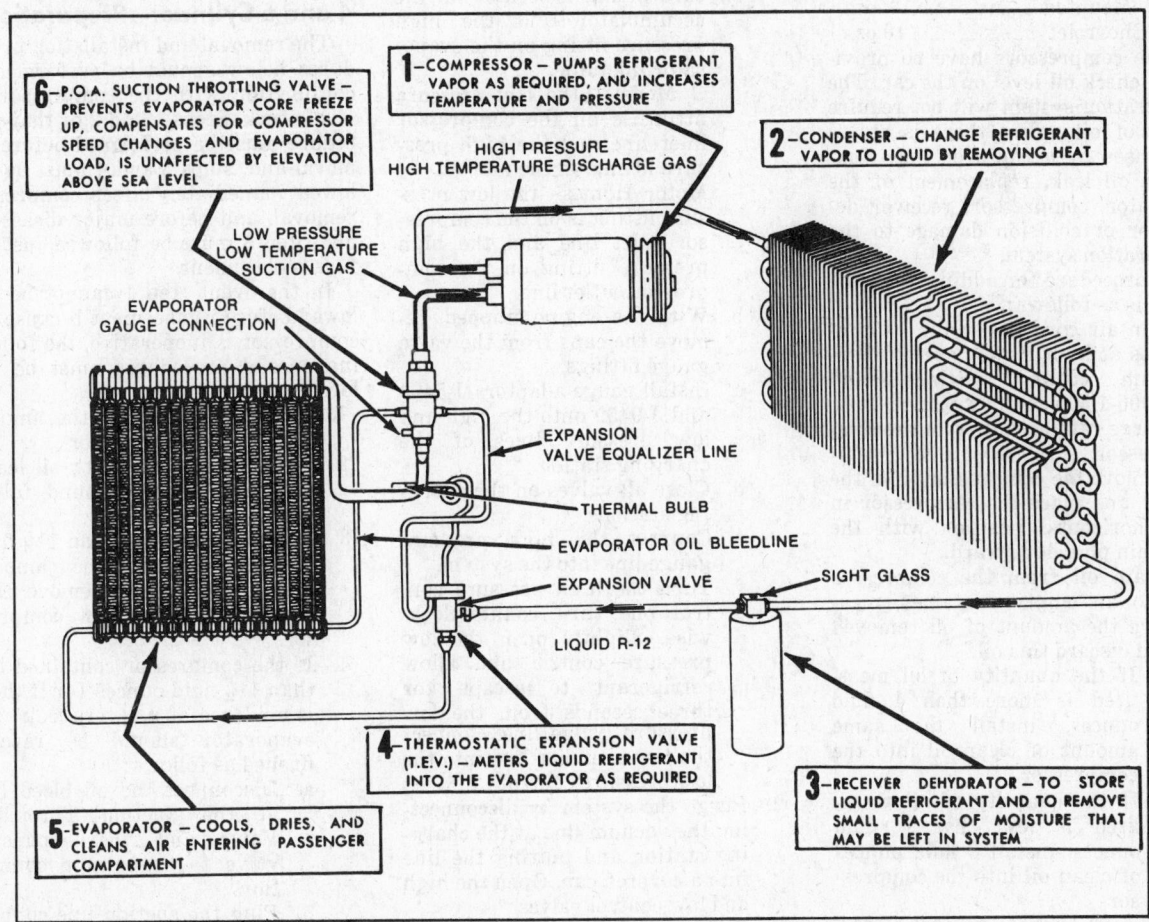

1-COMPRESSOR – PUMPS REFRIGERANT VAPOR AS REQUIRED, AND INCREASES TEMPERATURE AND PRESSURE

2-CONDENSER – CHANGE REFRIGERANT VAPOR TO LIQUID BY REMOVING HEAT

6-P.O.A. SUCTION THROTTLING VALVE – PREVENTS EVAPORATOR CORE FREEZE UP, COMPENSATES FOR COMPRESSOR SPEED CHANGES AND EVAPORATOR LOAD, IS UNAFFECTED BY ELEVATION ABOVE SEA LEVEL

HIGH PRESSURE
HIGH TEMPERATURE DISCHARGE GAS

LOW PRESSURE
LOW TEMPERATURE
SUCTION GAS

EVAPORATOR
GAUGE CONNECTION

EXPANSION
VALVE EQUALIZER LINE

THERMAL BULB

EVAPORATOR OIL BLEEDLINE

EXPANSION VALVE

SIGHT GLASS

LIQUID R-12

4-THERMOSTATIC EXPANSION VALVE (T.E.V.) – METERS LIQUID REFRIGERANT INTO THE EVAPORATOR AS REQUIRED

3-RECEIVER DEHYDRATOR – TO STORE LIQUID REFRIGERANT AND TO REMOVE SMALL TRACES OF MOISTURE THAT MAY BE LEFT IN SYSTEM

5-EVAPORATOR – COOLS, DRIES, AND CLEANS AIR ENTERING PASSENGER COMPARTMENT

Typical Refrigerant Cycle Diagram

Chevrolet

Compressor Oil Level Check

4-Cylinder Radial Compressors

The radial 4-cylinder compressor carries a normal charge of 5.5-6.5 ounces of 525 viscosity refrigerant oil. During normal operation a small amount of oil will circulate with the refrigerant. If a component is replaced, a corresponding amount of oil will have to be added to the system to replace that lost in the removed component.

To check the oil level in the event that a leak is suspected, it is necessary to drain the system by removing the compressor drain plug. The oil should be drained into a measured container. The compressor should be refilled with the prescribed amount of new, clean refrigerant oil.

6-Cylinder Axial Compressors—Charging Station Not Available

Factory installed compressors originally contain the following amounts of 525 viscosity refrigerant oil.

Four Season ....10 oz.
Roof Mounted ....13 oz.
GM Chevrolet ....10 oz.

These compressors have no provision to check oil level on the car. The refrigeration system will not require adding of oil unless there is an oil loss caused by a ruptured line, compressor oil leak, replacement of the evaporator, compressor, receiver dehydrator or collision damage to the refrigeration system.

The procedure for adding oil to the system is as follows:

1. Run air conditioner for 10 minutes at high blower speed, maximum cooling settings, and 1,000-1,500 engine r.p.m.
2. Purge refrigerant from the compressor.
3. Remove the compressor from the car and place the compressor in a horizontal position with the drain plug downward.
4. Drain oil from the compressor into an empty container, measure the amount of oil removed and discard this oil.
5. a. If the quantity of oil measured is more than 4 fluid ounces, install the same amount of clean oil into the compressor.
   b. If the quantity of oil measured is less than 4 fluid ounces, install 6 fluid ounces of clean oil into the compressor.
6. If a component is being replaced, the following amounts of oil

should be poured directly into the component.

Evaporator ....3 oz.
Condenser ....1 oz.
Receiver ....1 oz.

If an evaporator is installed, pour oil into the inlet pipe with the pipe held vertically so oil will drain into the core. If any other components such as valves or hoses are replaced, no additional oil is necessary.

7. Replace the compressor and system components.
8. Evacuate, charge and perform operational test.

6-Cylinder Axial Compressor—Charging Station Available

The Charging Station is a portable assembly consisting of a vacuum pump, refrigerant supply, gauges, valves and a five pound metering refrigerant charging cylinder. This unit combines all evacuation and charging equipment.

Refrigerant oil may be added as follows:

1. Installation of charging station.
   a. C&K Models—the low pressure fitting is located on the accumulator and the high pressure fitting on the evaporator inlet line.
   G Models—the low pressure fitting is on the compressor inlet line and the high pressure fitting on the muffler.
   Motor Homes—the low pressure fitting is on the compressor inlet line and the high pressure fitting on the compressor outlet line.
   b. With the engine stopped, remove the caps from the valve gauge fittings.
   c. Install gauge adaptors J-5420 and J-9459 onto the high and low pressure lines of the charging station.
   d. Close all valves on the charging stations.
   e. Connect the high pressure gauge line into the system.
   f. Turn the high pressure control one turn counterclockwise. Slightly open the low pressure control and allow refrigerant to escape for three seconds from the low pressure gauge line. Connect the low pressure line into the low pressure control.
2. Purge the system by disconnecting the vacuum line at the charging station and putting the line into a covered can. Open the high and low control valves.
3. Connect the vacuum line to the pump.

4. Disconnect the low pressure line and insert tool J-24095 into fitting. Insert pickup tube into graduated container of clean oil.
5. Turn on the vacuum pump and open the vacuum control valve. *NOTE: tool holds ½ ounce.*
6. Note level of oil in container. Open valve on oil adding tool until container level is equal to that lost during discharge plus ½ ounce.
7. Disconnect and cap tool.

Clutch Air Gap

NOTE: The clutch air gap is not adjustable. However, these figures may be used to check the general condition of the clutch. Replace if not to specifications. Chevrolet clutch air gaps should be .022-.057 for 6 cylinder compressors, and .020-.040 for 4 cylinder compressors.

NOTE: For overhaul procedures see the compressor unit section.

Compressor Removal and Installation

4 and 6 Cylinder—Preparation

The removal and installation procedures below should be performed in conjunction with the compressor oil level check. Steps 1 and 2 of that procedure must be performed before removal and, steps 3 and 4 must be followed immediately after compressor removal, and before major disassembly. Step 5 must be followed just before replacement.

In the event step 1 cannot be followed before replacement because the compressor is inoperative, the following special procedures must be followed:

1. Drain and measure the amount of oil in the compressor.
2. Inspect the system for oil leaks. If a sizable leak is found, follow step 4.
3. If there was more than 1½ fluid ounces, subtract the amount from 11 oz. Then remove that amount from the new compressor.
4. If the compressor contained less than 1½ fluid ounces (or if there is evidence of a severe leak) the evaporator should be reverse flushed as follows:
   a. Disconnect the oil bleed line at the Suction Throttling Valve and the expansion valve to evaporator connection.
   b. Plug the suction line connection at the suction throttling valve outlet.

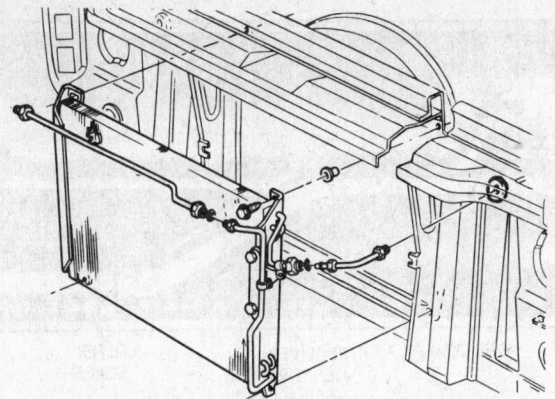

Condenser Installation, C and K models (© G.M. Corp.)

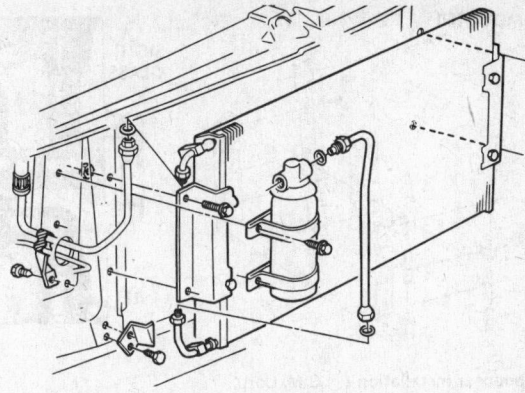

Condenser Installation, G models (© G.M. Corp.)

c. Place a clean pan under the evaporator inlet fitting.

d. Apply pressure with a cylinder of refrigerant to the oil bleed fitting on the STV valve to force any oil out of the evaporator inlet fitting.

e. Inspect the oil for chips, moisture, etc., and clean the system and replace the receiver-drier if oil is dirty. The compressor may then be installed with the full 11 oz. charge new compressors contain.

(Except 1974-78 G and Motor Home)

NOTE: On 1972-76 models, thermal limiter electrical connector must be disconnected before discharging the system.

1. Purge the system of refrigerant.
2. Remove the bolt holding the suction and discharge connector to the compressor. Remove the connector and cover the openings in the compressor and lines.
3. Disconnect the magnetic clutch wire.
4. Remove the drive belt.
5. Remove the compressor mounting bolts and lift out the compressor.
6. Set the compressor in position.
7. Install the compressor mounting bolts.
8. Install the compressor drive belt.
9. Connect the magnetic clutch wire.
10. Install new "O" ring seals and position the suction and discharge connector to the compressor.
11. Evacuate, leak test and charge the system.

G and Motor Home Chassis Models 1974-78

1. Connect battery ground cable.
2. Disconnect compressor clutch connector.
3. Purge system of refrigerant.
4. Remove the drive belt.
5. G models—remove the two bolts

and two clamps that hold the engine cover and remove cover.
6. Remove the air cleaner.
7. Remove the fitting and muffler assembly and cap or plug all open connections.
8. Remove the nuts and bolts holding the compressor to the bracket.
9. Remove the engine oil tube support bracket bolt and nut from the compressor. Remove the clutch ground lead.
10. Refill oil level in compressor.
11. Position compressor on bracket and secure with nuts and bolts. Install ground lead.
12. Install the connector assembly to the compressor rear head. Use new O-ring coated with clean refrigerant oil.
13. Connect the electrical lead to the coil. Install the belt.
14. Evacuate, charge and leak test the system.
15. Replace air cleaner and G model engine cover.
16. Connect battery.

Condenser Removal and Installation

Typical—All Models Through 1973

1. Disconnect the battery ground, and discharge the refrigerant system through the center manifold hose.
2. Remove the hood catch from the radiator support.
3. Remove radiator upper brackets.
4. Disconnect the two condenser lines. Cap open ends immediately.
5. Remove the radiator shroud mounting screws to give clearance so the radiator may move rearward. This will give access to the condenser mounting screws.
6. Remove the four condenser mounting screws. The condenser may then be pulled upward to remove it.
7. Add one fluid ounce of clean re-

frigerant oil to a replacement condenser.
8. Position the condenser, and install the four mounting screws.
9. Reposition the radiator, and replace the radiator shroud screws.
10. Uncap the line and condenser openings, and reconnect the refrigerant lines, using new "O" rings coated with clean refrigerant oil.
11. Replace the radiator upper brackets and hood catch. Reconnect the battery ground.
12. Evacuate, leak test, and recharge the system.

Typical—All Models, 1974

1. Disconnect battery.
2. Purge refrigerant.
3. Remove grille.
4. Remove radiator grille center support.
5. Remove the left grille support to upper fender support screws.
6. Disconnect the condenser and cap all openings.
7. Remove the condenser to radiator support screws.
8. Bend the left grille support outboard to gain access to condenser.
9. Pull condenser forward and lower it from vehicle.
10. Reverse the above procedure adding one ounce of clean refrigerant oil to a new condenser. Always use new O-rings coated with clean refrigerant oil.
11. Evacuate, charge and leak test the system.

FOUR SEASON & GM CHEVROLET SYSTEMS

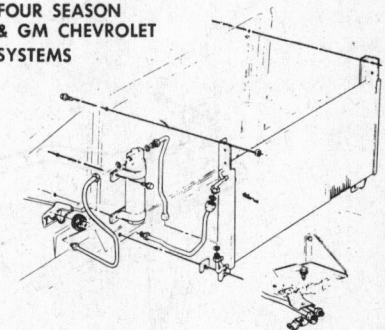

Condenser installation (© G.M. Corp.)

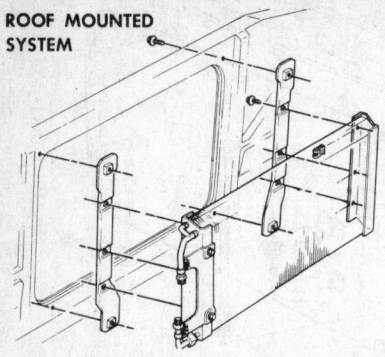

ROOF MOUNTED SYSTEM

Condenser installation (© G.M. Corp.)

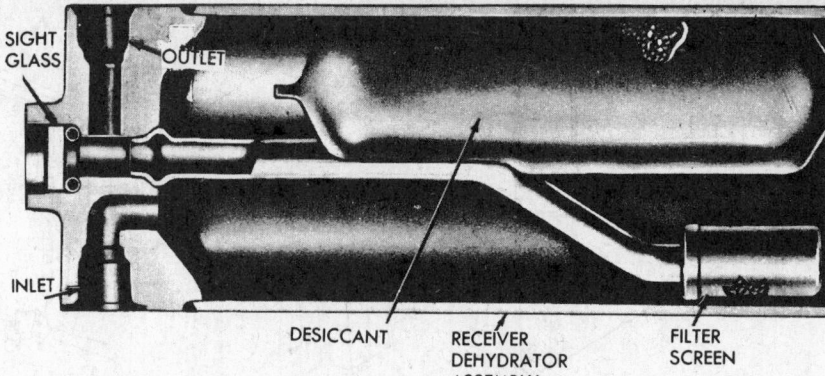

SIGHT GLASS
OUTLET
INLET
DESICCANT
RECEIVER DEHYDRATOR ASSEMBLY
FILTER SCREEN

Receiver dehydrator—inside view (© G.M. Corp.)

Typical—All Models, 1975-78

1. Disconnect the battery ground.
2. Discharge the system.
3. Remove the grille assembly.
4. Remove the radiator center support.
5. Remove the left grille support bracket.
6. Disconnect the refrigerant lines from the condenser. Cap all openings.
7. Remove the condenser retaining screws.
8. Pull condenser forward and down out of vehicle.
9. For installation, reverse the above using new O-rings coated with clean refrigerant oil.
10. Add one ounce of clean refrigerant oil to the system.
11. Evacuate, charge and leak test the system.

Accumulator Removal and Installation

Typical, 1974-78

1. Disconnect the battery and remove the compressor clutch connector.
2. Purge the system.
3. Disconnect the accumulator inlet and outlet lines and cap all openings.
4. Remove the accumulator bracket bolt and lift out unit.
5. Reverse the above procedure using new O-rings soaked in clean refrigerant oil.

6. Add one ounce of new oil to a new accumulator.
7. Refill to level equal to amount lost during removal.
8. Evacuate, charge and leak test the system.

Receiver-Drier Removal and Installation

Typical—All Models Through 1973

1. Disconnect the battery ground cable.
2. Discharge all refrigerant through center hose of gauge manifold.
3. Disconnect inlet and outlet connections, bracing the top of the dehydrator against twisting forces by applying pressure to the oblong sections at the top so as to oppose the torque on the lines.
4. Remove the mounting bolts, and then remove the unit. Cap all openings, unless a new dehydrator is to be installed immediately. *NOTE: Leave all caps in place until the last possible moment.*
5. Position the receiver-dehydrator so all mounting holes are properly lined up, and replace the mounting bolts.

6. Uncap all capped connections, and position new "O" rings coated with refrigerant oil in the two openings in the dehydrator. Then, reconnect the inlet and outlet lines.
7. Evacuate, leak test, and charge the system.

G Models, 1974-78

1. Disconnect the battery ground cable and the compressor clutch connector.
2. Purge the system.
3. Remove the left and right head lamp bezel and parking lamp lens. Cover the upper surface of the front bumper with protective tape. Remove the grille bolts and grille.
4. Disconnect the receiver-dehydrator inlet and outlet lines and cap all openings.
5. Remove the receiver-dehydrator bracket attaching screws.
6. Remove the receiver-dehydrator from the vehicle.
7. With a new unit, add one ounce of refrigerant oil.
8. Connect the inlet and outlet lines using new O-rings coated in clean refrigerant oil.
9. Install the receiver-dehydrator by reversing the removal procedure.
10. Evacuate, recharge and leak test the system.

Motor Home Chassis 1974-78

1. Disconnect the battery ground cable.
2. Purge the system.
3. Disconnect the inlet and outlet lines at the receiver-dehydrator, and cap all openings at once.
4. Remove the receiver-dehydrator bracket attaching screws, and remove the bracket and unit.
5. To install a new receiver-dehydrator, reverse the above procedure adding one ounce of new oil to a new unit. Always use new O-rings coated with clean refrigerant oil.

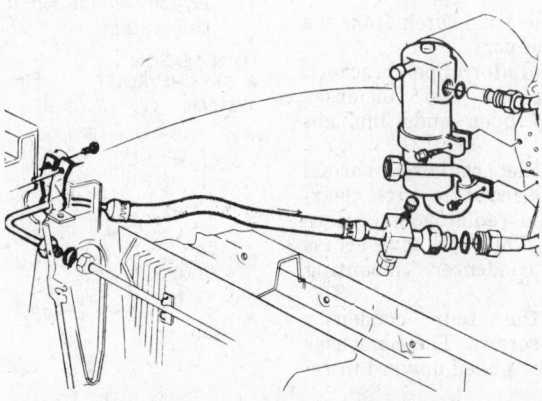

Accumulator installation, C and K models (© G.M. Corp.)

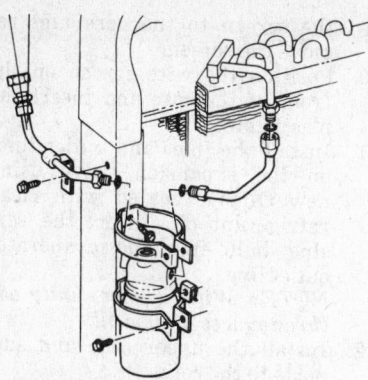

Receiver-dehydrator installation, motor home chassis (© G.M. Corp.)

6. Evacuate, charge and leak test the system.

Blower Motor Removal and Installation

Typical, All Models Through 1973

1. Disconnect the battery ground cable.
2. Support the hood in the fully raised position to ensure safe accomplishment of the next step.
3. Accurately scribe the locations on hood and fender of the right side hood hinge. Then remove the hinge.
4. Disconnect the blower feed wire at the blower flange terminal. Note the flange position in relation to the blower case.
5. Disconnect the motor cooling tube.
6. Remove the screws that hold the blower assembly in place.
7. Carefully remove the assembly, prying gently to pull the flange from the case if the sealer adheres to the case.
8. Remove the nut that holds the blower wheel to the motor shaft, and remove the wheel from the shaft.
9. Install the blower on to the motor shaft. The open end of the blower wheel goes away from the motor.
10. Place the assembly in the proper position on the blower case.

11. Connect the ground strap, cooling tube, and blower feed wire.
12. Reposition the hood hings, lining it up precisely with marks scribed onto the hood and fender before tightening bolts.
13. Check hood alignment, and correct if necessary. Reconnect battery ground cable.

C, K Models 1974-78

1. Disconnect the battery ground cable.
2. Disconnect the blower motor lead and ground wires.
3. Disconnect the blower motor cooling tube.
4. Remove the screws attaching the blower to the case and remove the blower assembly.
5. Remove the nut attaching the blower wheel to the motor shaft. Separate the assemblies.
6. For installation, reverse the above.

C, K Models with Overhead System 1974-78

1. Disconnect the battery ground cable.
2. Remove the rear duct by disconnecting the drain tube and removing the screws securing the duct to the roof panel and rear header brackets.
3. Disconnect the blower motor ground strap.
4. Disconnect the blower motor lead wire.
5. Remove the lower to upper blower-evaporator case screws and lower the lower case and motor assembly.
6. Remove the motor retaining strap and remove the motor and wheels. Remove the wheels from the motor shaft.
7. Place the blower wheels onto the motor shaft, making sure the wheel tension springs are installed on the hubs.
8. Install the blower motor retaining strap and foam.
9. Place the blower motor and wheel assembly into the lower case. Align the blower wheels so that they don't contact the case.
10. Place the lower case and blower motor assembly in position and

install the lower to upper case screws.
11. Install the center ground wire and connect the blower lead wire.
12. Install the rear duct assembly as described previously.
13. Connect the battery ground cable.

G Models 1974-78

1. Purge the system.
2. Disconnect the battery ground.
3. Remove the blower-evaporator shield. Remove the instrument panel reinforcement screws.
4. Remove the heater intermediate duct.
5. Remove the mounting bolts and remove the engine cover.
6. Loosen one and remove one instrument panel steering column reinforcement screw.
7. Remove the left foot cooler bracket to instrument panel screws, disconnect the outlet from the duct and remove the outlet and bracket assembly.
8. Disconnect the speedometer cable.
9. Remove the instrument panel-to-lower reinforcement screw Rotate the instrument panel up and out of the way.
10. Disconnect the evaporator inlet and outlet lines. Cap all openings.
11. Remove the blower-evaporator support bracket-to-door pillar and forward engine housing screws and lower the blower-evaporator assembly. Disconnect electrical connections, pull drain hose through the dash panel and remove the blower-evaporator assembly.
12. Separate the front and rear case halves.
13. Remove the blower motor retaining strap and insulator strip and remove the motor and blower wheels.
14. Mark the blower wheel locations and remove the wheels from the shaft.
15. For installation, reverse the above using new O-rings coated with clean refrigerant oil.
16. Evacuate, charge and leak test the system.

G Models with Overhead System 1974-78

1. Disconnect the battery ground cable and compressor clutch connector.
2. Remove the blower-evaporator shroud by disconnecting the drain tubes and removing the attaching screws.
3. Remove the blower motor ground straps at the center connector between the motors.

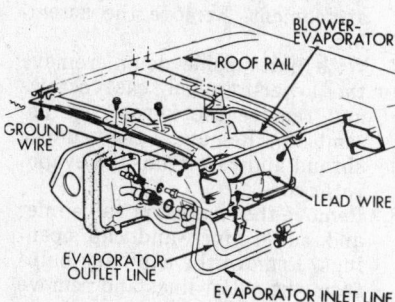

Blower-evaporator installation, C and K models, overhead system (© G.M. Corp.)

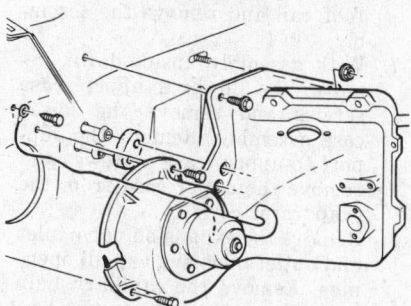

Blower-evaporator installation, C and K models (© G.M. Corp.)

4. Disconnect the blower motor lead wires.
5. Support the lower case and remove the lower to upper blower-evaporator case screws and lower case and motor assemblies.
6. Remove the motor retaining strap and remove the motor and wheels. Remove wheels from shaft.
7. Place wheels on shaft.
8. Install the blower motor retaining strap and foam strip.
9. Place the two blower motor assemblies into the lower case. Make sure wheels do not touch case.
10. Place the lower case into position in the vehicle and install the lower-to-upper case screws.
11. Install the center ground wires and connect the blower lead wires.
12. Install the blower-evaporator shroud assembly.
13. Connect the battery ground cable and compressor clutch connector.

Evaporator Removal and Installation

Typical, All Models Through 1973

1. Disconnect the battery ground. Discharge refrigerant from the system through the center manifold hose.
2. Remove the rear duct as described above.
3. Disconnect the blower motor lead and ground wire, and the resistor connections.
4. Disconnect the refrigerant lines at the rear of the assembly. Cap all open connections.
5. Remove the nuts and washers which hold the evaporator case to the roof panel. Remove the assembly and place it on a workbench upside down.
6. Remove the screws which fasten the case halves to one another, and pull the lower case assembly off the upper case half. Then, remove the evaporator core.
7. Remove the expansion valve connections. Immediately cap the open ends.
8. Remove the capillary bulb from the suction line, noting its position, and remove the expansion valve.
9. Remove the screen by pulling off the plastic pins that hold it to the core and pulling it away.
10. Replace the core screen and retaining pins.
11. Reconnect the expansion valve inlet and outlet lines using new "O" rings coated with clean refrigerant oil.
12. Replace the expansion valve capillary bulb in its former position,

clamping it firmly for good contact with the suction line.
13. Add 3 oz. of refrigerant oil if installing a new core.
14. Position the two halves of the case around the core, and install the screws which fasten the case halves together.
15. Position the assembly on the roof panel, and reinstall the attaching washers and nuts.
16. Remove the caps, and reinstall the refrigerant lines, using new "O" rings coated with clean refrigerant oil.
17. Connect the resistor harness, blower leads, and ground strap.
18. Install the rear duct as described above.
19. Connect battery ground. Then evacuate, leak test, and charge the system.

C, K Models 1974-78

1. Disconnect the battery ground.
2. Purge the system.
3. Remove the nuts from the selector duct studs at the dash panel.
4. Remove the cover-to-dash and cover-to-case screws and remove the evaporator case cover.
5. Disconnect the evaporator core inlet and outlet lines and cap all openings.
6. Remove the thermostatic switch and expansion tube assemblies.
7. Remove the evaporator core.
8. For installation, reverse the above. Add three ounces of clean oil to a new evaporator core.
   NOTE: Make sure that the thermostatic switch capillary is properly installed in the hole provided in the core.
9. Evacuate, charge and leak test the system.

C, K Models with Overhead System 1974-78

1. Disconnect the battery ground.
2. Purge the system.
3. Remove the rear duct.
4. Disconnect the blower motor lead and ground wires.
5. Disconnect the refrigerant lines at the rear of the blower-evaporator and cap all openings.
6. Remove the screws from the blower-evaporator support-to-roof rail and remove the assembly.
7. With assembly upside-down, remove the lower-to-upper case screws and remove the lower case assembly. Remove the support-to-upper case screws and remove the upper case from the evaporator core.
8. Remove the expansion valve inlet and outlet lines and cap all openings. Remove the capillary bulb from the evaporator outlet line and remove the valve.
9. Remove the plastic pins holding

the screen to the core and remove the screen.
10. Position the wire screen on the front of the core and insert the plastic pins.
11. Install the inlet and outlet lines on the expansion valve using new O-rings coated with clean refrigerant oil. Insert the sensing bulb into the evaporator outlet line.
    NOTE: With a new unit, add three ounces of new oil.
12. Install the upper case and supports to the core.
13. Install the lower case and blower assembly.
14. Install the blower-evaporator assembly to the roof and tighten the screws.

G Models 1974-78

1. Purge the system.
2. Follow the procedure for removal of the blower assembly steps 2-9.
3. Disconnect the evaporator inlet and outlet lines. Cap all openings.
4. Remove the blower-evaporator support bracket-to-door pillar and forward engine housing screws, and remove the unit.
5. Remove the left floor outlet duct screw and remove the assembly.
6. Disconnect the vacuum hoses and electrical harness tabs. Remove the air distributor duct screws and duct. Remove the air distributor duct screws and lift out the duct.
7. To install, reverse the above.
8. Evacuate, charge and leak test the system.

G Models with Overhead System 1974-78

1. Disconnect the battery ground and compressor clutch connector.
2. Purge the system.
3. Remove the blower-evaporator shroud.
4. Disconnect the blower motor leads and ground wire.
5. Disconnect blower-evaporator refrigerant lines and cap all openings.
6. Support the case and remove blower-evaporator-to-roof panel attachment. Remove the assembly.
7. With unit upside down, remove the lower-to-upper case screws and remove the lower case assembly. Remove the upper shroud and case from the evaporator core.
8. Remove the expansion valve inlet and outlet lines and cap openings. Remove the capillary bulbs from the outlet lines and remove the valves.
9. Remove the plastic pins and lift off screen from core.

10. Install the wire screen to the front of the new core and install the pins.
11. Attach the inlet and outlet lines to the expansion valve using new O-rings coated in fresh refrigerant oil. Install the sensing bulbs to the evaporator outlet line. Add three ounces of new oil with a new core.
12. Install the upper case and upper shroud to the core.
13. Install the lower core case and blower assemblies.
14. Install the blower-evaporator to the roof panel.
15. Connect the refrigerant lines to the blower-evaporator using new O-rings coated with clean refrigerant oil.
16. Connect the blower lead wires and ground straps.
17. Install the blower-evaporator shroud.
18. Connect the battery ground cable and the compressor clutch connector.
19. Evacuate, charge and leak test the system.

Motor Home Chassis 1974-78

1. Remove the cover plate and separate the upper and lower case halves.
2. Remove the inlet and outlet lines from the xpansion valve. Remove the sensing bulb from the evaporator outlet manifold. Remove the expansion valve and cap openings.
3. Remove the evaporator core retaining screws, and remove the core.
4. Reverse the above using 3 ounces of oil for a new unit.

Suction Throttling Valve Removal and Installation

All Models

1. Disconnect the battery ground. Discharge the refrigerant from the system through the center manifold hose.
2. Remove the expansion valve equalizer line and evaporator oil bleed line from the POA valve (VIR on 1974-76 G and motor home). Cap connections as soon as they are separated to minimize the entry of dirt and moisture.
3. Remove the screw for the clamp which holds the valve in place. Remove the valve outlet pipe clamp mounting screw.
4. Disconnect inlet and outlet fittings, and then remove the valve. Cap all openings immediately.
5. Remove the clamp from the valve.

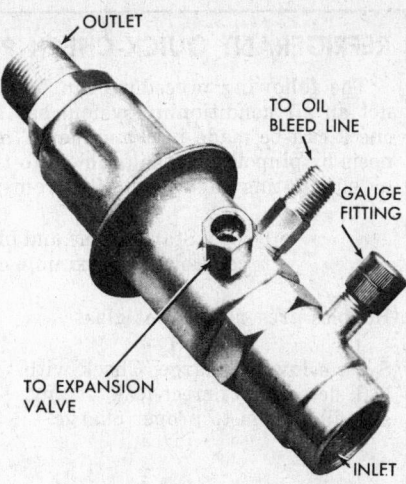

P.O.A. valve (© G.M. Corp.)

6. Reposition the mounting clamp on the valve. Remove caps and reconnect inlet and outlet lines, using new O-rings coated with clean refrigerant oil.
7. Reposition and retighten valve and outlet line clamps.
8. Reconnect oil bleed and equalizer lines, using new seals coated with clean refrigerant oil.
9. Connect battery ground. Evacuate, leak test and charge the system.

Expansion Valve Removal and Installation

Typical, All Models Through 1973

1. Disconnect the battery ground. Discharge the refrigerant from the system through the center manifold hose.
2. Loosen the high pressure line clamp located on the bracket next to the expansion valve.
3. Disconnect the capillary bulb from the suction line, noting its location. Disconnect the equaliezr line at the suction throttling valve, capping the opening on the valve.

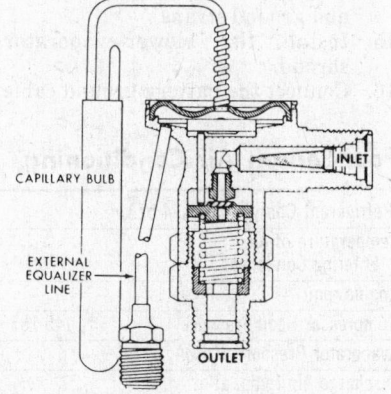

Expansion valve (© G.M. Corp.)

4. Disconnect the inlet and outlet connections and cap the openings.
5. Remove the screw that holds the expansion valve on the mounting bracket, and remove the valve.
6. Reposition the expansion valve on the mounting bracket, and replace the mounting screw.
7. Remove caps, and reconnect inlet and outlet fittings. Tighten the inlet pipe clamp.
8. Connect the equalizer line, using a new "O" ring coated with clean refrigerant oil.
9. Reposition the capillary bulb in its former position, tightening the clamp firmly and repositioning the insulation so the bulb is completely covered.
10. Reconnect the battery ground. Evacuate, leak test, and charge the system.

C, K Models 1974-78

NOTE: C, K Models use an expansion tube in place of an expansion valve.

1. Purge the system.
2. Disconnect the condenser to evaporator line at the evaporator inlet. Cap openings.
3. With needle-nose pliers, remove the expansion tube from the evaporator core inlet line. Remove the O-ring.
4. For installation, reverse the above. Use a new O-ring coated with clean refrigerant oil.
5. Evacuate, charge and leak test the system.

C, K Models with Overhead System 1974-78

1. Disconnect the battery ground.
2. Purge the system.
3. Remove the rear duct.
4. Disconnect the blower motor lead and ground wires.
5. Remove the lower-to-upper evaporator case screws. Remove the lower case and motor.
6. Remove the expansion tube sensing bulb clamps.
7. Disconnect the inlet and outlet lines from the valve and remove the valve unit. Cap all openings.
8. Install the valve and connect the lines using new O-rings coated in fresh oil.
9. Place the sensing bulb in the core outlet line.
10. Install the lower case and blower motor assemblies. Connect the blower motor lead and ground wires.
11. Install the rear duct.
12. Connect the battery ground.
13. Evacuate, charge and leak test the system.

G Models 1974-78

1. Purge the system.

2. Follow steps 2-9 of "Blower Removal."
3. Loosen the expansion valve sensing bulb clamps and remove the bulb.
4. Disconnect the inlet and outlet lines and remove the expansion valve. Cap all openings.
5. For installation, reverse the above. Use new O-rings coated with clean refrigerant oil. Evacuate, charge and leak test the system.

G Models with Overhead System 1974-78

1. Disconnect the battery ground and compressor clutch connector.
2. Purge the system.
3. Support the case and remove the blower-evaporator shroud.
4. Disconnect the center ground wire and lead wires. Remove the lower-to-upper case screws and remove the lower case and motor.
5. Disconnect the sensing bulb from the core outlet line.
6. Disconnect the core inlet line and outlet line and remove the valve. Cap all openings.
7. Install new valve and connect lines using new O-rings coated with fresh refrigerant oil.
8. Install the sensing bulb in the core outlet line.
9. Install the lower case and blower motor assemblies.
10. Install the blower-evaporator shroud.
11. Connect the battery ground and compressor clutch connector.
12. Evacuate, charge and leak test the system.

Outer Valve

1. Disconnect the battery ground and compressor clutch connector.
2. Purge the system.
3. Remove the blower-evaporator shroud.
4. Disconnect the blower wires and leads.
5. Disconnect the refrigerant lines at the rear of the blower-evaporator. Cap all openings.
6. Remove the blower-evaporator-to-roof panel attachments and lower the blower-evaporator, and remove from vehicle.
7. With unit upside down, remove the lower to upper case screws and remove the lower case assembly. Remove the upper shroud and remove the upper case from the core.
8. Remove the expansion valve bulb from the evaporator outlet line, and cap all openings. Remove valve.
9. Remove the cap and install and connect the new valve. Install sensing bulb.

10. Install the upper case to core making sure the sealing strips are positioned correctly. Install the upper shroud on the case.
11. Install the lower case and blower assemblies.
12. Install the blower-evaporator to the roof panel.
13. Connect the refrigerant lines at the rear of the blower-evaporator using new O-rings coated with clean refrigerant oil.
14. Connect the blower lead wires and ground straps.
15. Install the blower-evaporator shroud.
16. Connect the battery ground cable and the compressor clutch connector.
17. Evacuate, charge and leak test the system.

Motor Home Chassis 1974-78

1. Remove the cover plate and separate upper and lower case halves.
2. Remove inlet and outlet lines from expansion valve. Remove sensing bulb and remove the valve. Cap all openings.
3. Reverse the above for installation.
4. Evacuate, charge and leak test the system.

REFRIGERANT QUICK-CHECK PROCEDURE

The following procedure can be used to quickly determine whether or not an air conditioning system has a proper charge of refrigerant. This check can be made in a manner of minutes thus facilitating system diagnosis by pinpointing the problem to the amount of charge in the system or by eliminating this possibility from the overall checkout.

Start engine and place on fast idle. Set controls for maximum cold with blower on high.

Bubbles present in sight glass.

System low on charge. Check with leak detector. Correct leak, if any, and fill system to proper charge.

No appreciable temperature differential noted at compressor.

System empty or nearly empty. Turn off engine and connect charging station. Induce 1/2# of refrigerant in system (if system will not accept charge, start engine and draw 1/2# in through low pressure side). Check system with leak detector.

If refrigerant in sight glass remains clear for more than 45 seconds (before foaming and then settling away from sight glass) an overcharge is indicated. Verify with a performance check.

No bubbles. Sight glass clear.

System is either fully charged or empty. Feel high and low pressure pipes at compressor. High pressure pipe should be warm; low pressure pipe should be cold.

Temperature differential noted at compressor.

Even though a differential is noted, there exists a possibility of overcharge. An overfilled system will result in poor cooling during low speed operation (as a result of excessive head pressure). An overfill is easily checked by disconnecting the compressor clutch connector while observing the sight glass.

If refrigerant foams and then settles away from sight glass in less than 45 seconds, it can be assumed that there is a proper charge of refrigerant in system. Continue checking out system using performance checks outlined previously.

Four-Season Air Conditioning

(Refrigerant Charge—3 lbs.-4 oz.)

Temperature of Air Entering Condenser	70°	80°	90°	100°	110°	120°
Engine rpm			2000			
Compressor Head Pressure	145-155	185-195	200-210	215-225	255-265	285-295
Evaporator Pressure at POA			Dependent Upon Altitude			
Discharge Air Temp. at Right Hand Outlet	38-41	38-41	40-43	41-44	43-46	48-51

GM Chevrolet Air Conditioning

(Refrigerant Charge—3 lbs.-4 oz.)

Temperature of Air Entering Condenser	70°	80°	90°	100°	110°	120°
Engine rpm			2000			
Compressor Head Pressure	110-120	135-145	160-170	190-200	220-230	260-270
Suction Pressure psi*	6	7	9	10	10	13
Discharge Air Temperature*	40-45	41-46	41-46	42-47	44-49	44-49

Roof Mounted Air Conditioning

(Refrigerant Charge—5 lbs.-8 oz.)

Temperature of Air Entering Condenser	70°	80°	90°	100°	110°	120°
Engine rpm			2000			
Compressor Head Pressure*	200-210	210-220	225-235	255-265	285-295	330-340
Suction Pressure psi*	28	29.5	29-31	34-36	39-41	43-45
Front Discharge Air Temperature*	34-37	38-43	41-46	47-52	52-57	56-61
Rear Discharge Air Temperature*	41-46	48-53	51-56	58-63	65-70	68-73

Disassembly

1. Place the compressor in a holding fixture and clamp the fixture in a vise.
2. Hold the clutch hub and remove and discard the shaft nut.
3. Thread the clutch plate and hub remover into the hub, hold the tool body with a wrench and turn the center screw into the remover body.
4. Remove the shaft key.
5. Remove the seal retainer ring and thoroughly clean all related areas.
6. Install a seal seat remover over the shaft and into the recessed area of the seal seat. Tighten the tool clockwise to engage the tangs with the seat. With a twist-pull motion, remove and discard the seat.
7. Insert a seal remover over the shaft and press downward to overcome the spring and engage the seal. Pull and discard the seal.
8. Remove the O-ring from the compressor neck.
9. Remove the rotor and bearing assembly retaining ring and mark the position of the clutch coil terminals.
10. Install a rotor and bearing puller guide over the end of the shaft and seat it on the front head.
11. Install the puller down into the rotor until the puller arms engage the recessed edge of the rotor hub. Hold the puller and arms in place and tighten the puller screw against the guide to remove the clutch rotor and bearing assembly.
12. Place the rotor and bearing assembly on blocks. Drive the bearing out of the rotor hub.
13. Remove the pulley rim mounting screws and discard the lock washers.
14. Slide the pulley rim off the rotor and hub.
15. Unbolt and remove the front head assembly; remove and discard the O-ring.
16. Remove the two thrust and one belleville washer from the shaft. Note the assembled positions of these parts.
17. Place the front head on two blocks and drive out the main bearing.
18. Pry the shell retaining strap away from the cylinder and place the unit in a holding fixture with the step block protrusions engaging the shell. Install

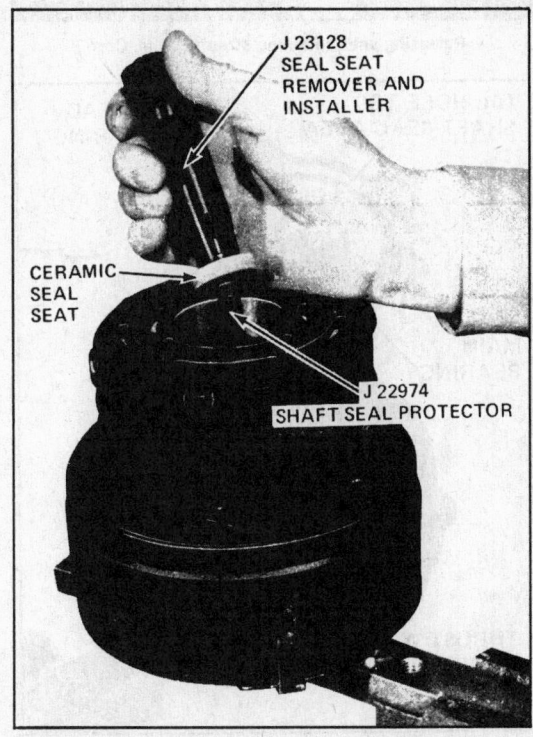

Removing and installing seal seat (© G.M. Corp.)

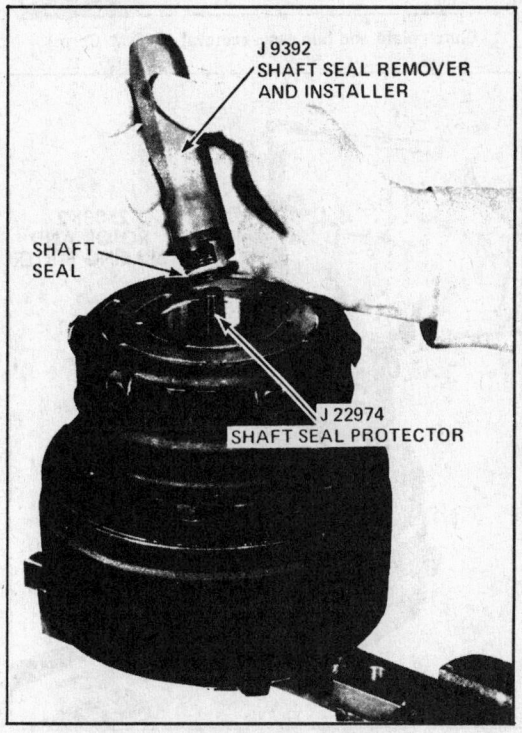

Removing and installing shaft seal (© G.M. Corp.)

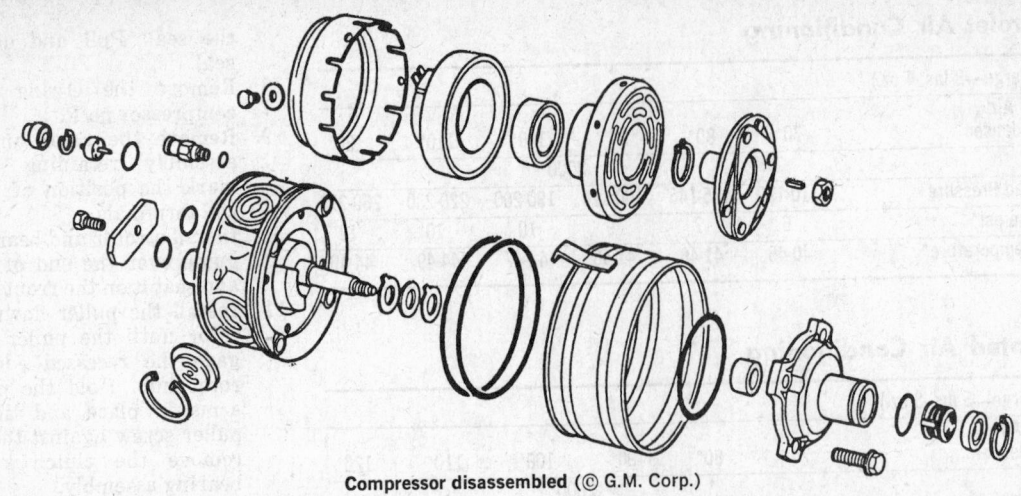

Compressor disassembled (© G.M. Corp.)

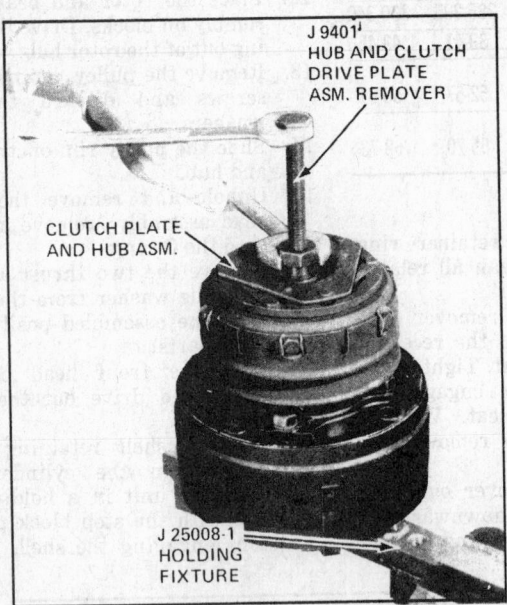

J 9401
HUB AND CLUTCH
DRIVE PLATE
ASM. REMOVER

CLUTCH PLATE
AND HUB ASM.

J 25008-1
HOLDING
FIXTURE

Clutch plate and hub assy. removal (© G.M. Corp.)

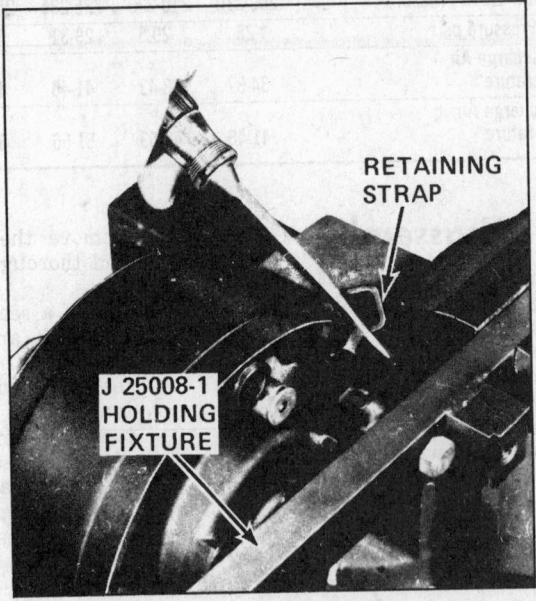

RETAINING
STRAP

J 25008-1
HOLDING
FIXTURE

Releasing shell retaining strap (© G.M. Corp.)

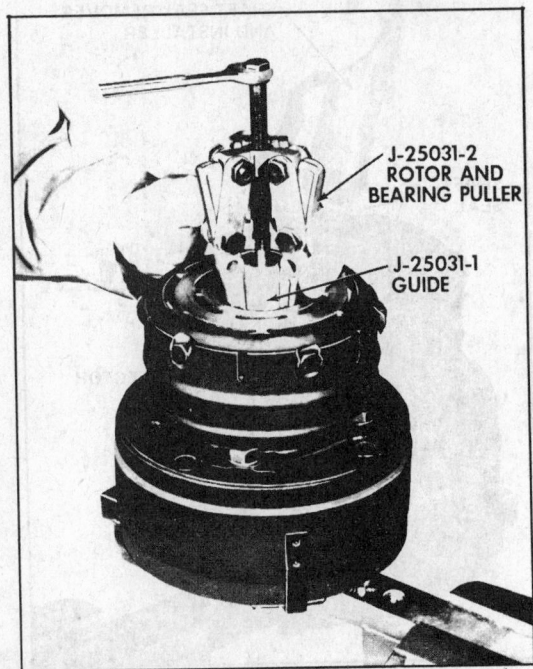

J-25031-2
ROTOR AND
BEARING PULLER

J-25031-1
GUIDE

Removing clutch rotor assy. (© G.M. Corp.)

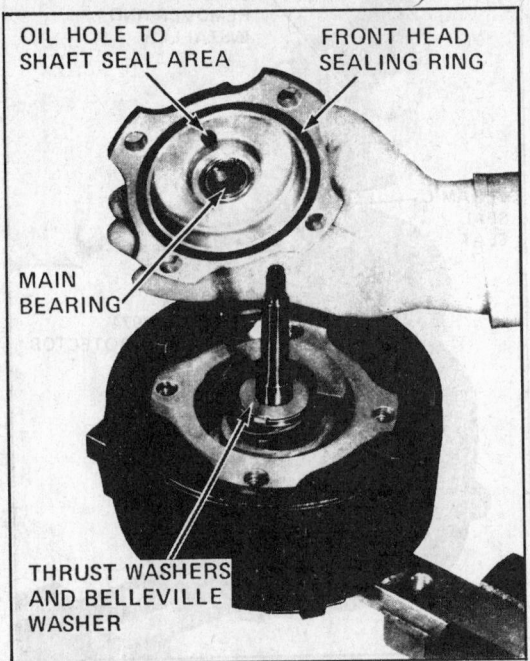

OIL HOLE TO
SHAFT SEAL AREA

FRONT HEAD
SEALING RING

MAIN
BEARING

THRUST WASHERS
AND BELLEVILLE
WASHER

Removing front head assy. (© G.M. Corp.)

Assembling clutch coil, pulley rim, rotor and bearing
(© G.M. Corp.)

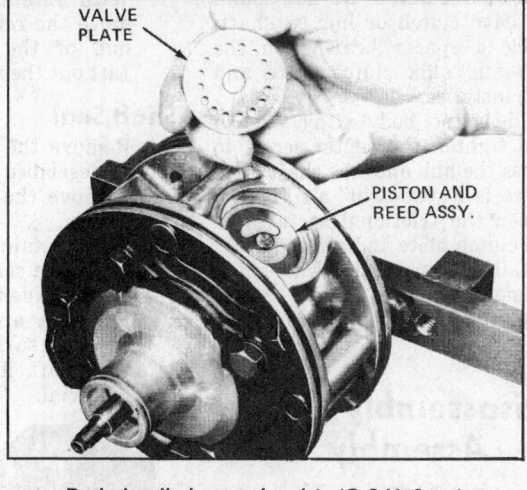

Replacing discharge valve plate (© G.M. Corp.)

PULLEY RIM

CLUTCH COIL
AND HOUSING
ASM.

VALVE
PLATE

PISTON AND
REED ASSY.

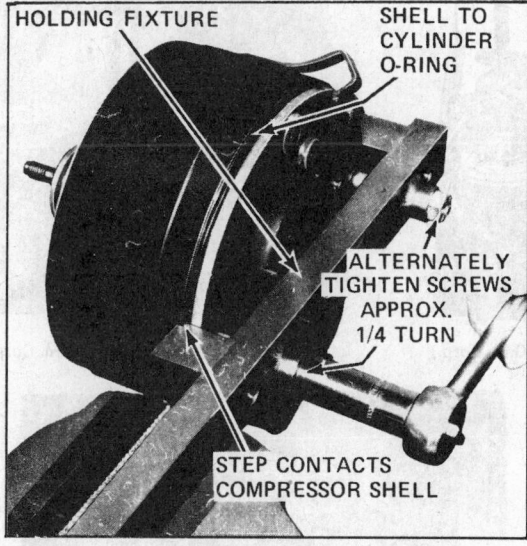

Compressor shell removal (© G.M. Corp.)

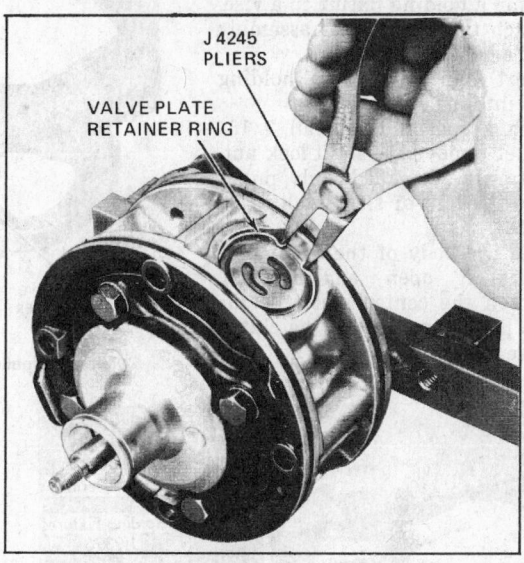

Replacing valve plate retainer (© G.M. Corp.)

HOLDING FIXTURE

SHELL TO
CYLINDER
O-RING

ALTERNATELY
TIGHTEN SCREWS
APPROX.
1/4 TURN

STEP CONTACTS
COMPRESSOR SHELL

J 4245
PLIERS

VALVE PLATE
RETAINER RING

medium length bolts through the holding fixture and thread them into the cylinder until the steps of the fixture protrusions contact the compressor shell, finger tight, on both sides.

19. Using a wrench, alternately tighten each bolt approximately 1/4 turn until the shell is free of the O-rings. Cylinder O-rings should be discarded.
20. Remove the valve plate retainer rings. Remove the compressor discharge valve plates.
21. Remove the high pressure relief valve and O-ring from the cylinder.
22. Remove the retainer ring and lift out the superheat switch and O-ring.
23. Remove cylinder and shaft assembly from holding fixture for replacement.

Assembly

1. Attach cylinder and shaft assem-

bly to holding fixture clamped in vise.
2. Coat new superheat switch O-ring with clean refrigerant oil and install switch and O-ring.
3. Coat new pressure relief valve O-ring with clean refrigerant oil and install.
4. Install discharge valve plates and retainers.
5. Check the parts to make sure they are free of lint, dirt, etc.
6. Coat the new cylinder-to-shell O-rings with clean refrigerant oil and install in the cylinder grooves.
7. Coat the inner surface of the shell with clean refrigerant oil. Place the shell on the cylinder and rotate the retaining strap to its original location.
8. Attach a shell installing tool to the holding fixture using the long bolts and plate washers of the tool set.
9. Align the step projections of the installing fixture to contact the

compressor shell evenly on both sides.
10. Push the shell, evenly, as close to the O-ring as possible and tighten the screws finger tight.
11. Using a wrench, alternately tighten each bolt 1/4 turn to install shell.
12. When shell is seated against the stops, bend the retaining strap down into place by gently tapping with a hammer.
13. Install clutch rotor and bearing assembly on front head and install head and O-ring. Be certain that the clutch coil terminals are in proper location and that the three protrusions on the rear of the coil align with the locator holes in the front head.
14. Install the rotor and bearing retainer ring.
15. Install the shaft seal and seat.
16. Install the shaft key. Allow the key to project about 3/16" out of the keyway.
17. Assemble the clutch plate and

hub on the shaft. Do not pound or drive clutch or hub on shaft.

18. Place a spacer bearing on the hub and affix clutch plate and hub installer.

19. Hold the tool body with a wrench and tighten the center screw to press the hub onto the shaft until there is a .020-.040" air gap between the frictional surfaces of the clutch plate and rotor.

20. Install new shaft nut with small diameter of boss against crankshaft shoulder. Tighten nut to 8-12 ft lb.

Disassembly and Assembly

Hub and Drive Plate

1. Clamp a holding fixture in a vise.
2. Attach the compressor assembly to the holding fixture.
3. Insert the drive plate holding tool into the drive plate.
4. With a special thin wall 7/16" socket remove the shaft lock nut.
5. Install a threaded hub puller onto the hub of the clutch drive plate.
6. Hold the body of the hub puller with an open end wrench, tighten the center screw of the hub puller and lift off the clutch drive plate and woodruff key.

7. With retaining ring pliers, remove the retainer ring from the hub of the clutch drive plate. Lift out the spacer.

Shaft Seal

1. Remove the hub and drive plate as described in above paragraph. Remove the sleeve retainer and sleeve.
2. With retaining ring pliers, remove the seal seat retainer ring from inside the front head. Clean the area around the shaft, the exposed surface of the seal, and the shaft itself of any foreign material.

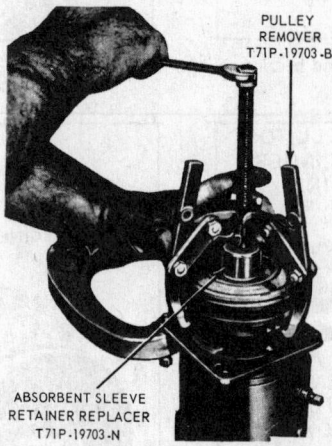

Removing pulley (© G.M. Corp.)

3. Using a seal seat remover and installer, disassemble shaft seal seat. Grasp the flange of the shaft seal seat with the tool and pull straight out.
4. Using seal remover and installer insert the tool into the hub of the front head, press downward and twist clockwise to engage the tabs of the shaft seal, and gently, but firmly, pull straight out.
5. Remove the seal seat "O" ring from the inside hub of the front head. A piece of wire bent at the end is an effective tool for the job.

Pulley and Bearing

1. Remove the clutch drive plate. See Hub and Drive Plate Removal.

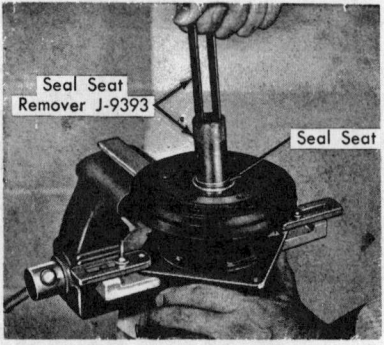

Removing seal seat (© G.M. Corp.)

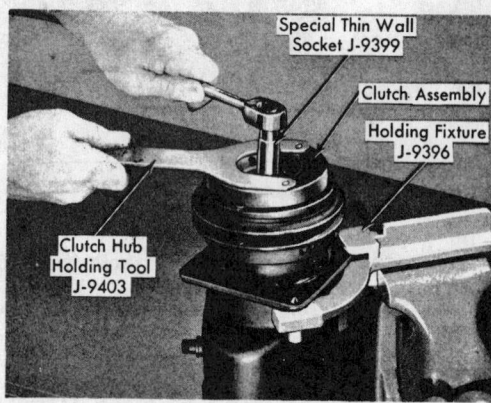

Removing lock nut (© G.M. Corp.)

Removing pulley bearing retainer ring (© G.M. Corp.)

Removing hub and drive plate (© G.M. Corp.)

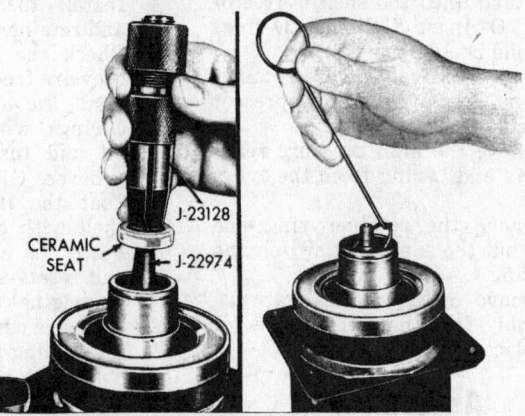

Removing seal seat and "O-ring" (© G.M. Corp.)

2. Using retainer ring pliers, take out the bearing-to-head retainer ring. Remove the retainer, and pry out the absorbent sleeve.
3. Put puller pilot on the hub of the front head and take off the pulley assembly, using a pulley puller.
4. With a small screwdriver, withdraw the bearing-to-pulley retaining ring.
5. Using puller pilot and handle, drive out the bearing.

Inspection

If, upon inspection, the friction surface of the pulley shows evidence of excessive wear due to slippage, the clutch hub, pulley and drive plate assembly should be replaced.

Clutch Coil and Housing

1. Remove the hub and drive plate, and pulley and bearing. See previous paragraphs.
2. Note and mark the position of the electrical terminals on the coil housing to ensure correct location of the terminals when the coil is reinstalled.
3. Using retaining ring pliers, take out the coil housing retaining ring.
4. Lift off the coil assembly.

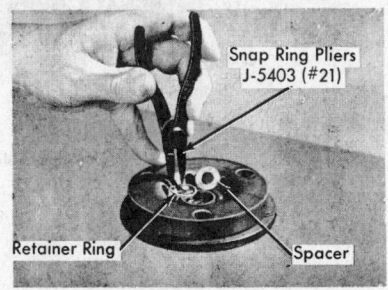

Removing hub retainer ring & spacer (© G.M. Corp.)

Rear Head, Oil Pump, Rear Discharge Valve Plate and Rear Suction Valve Reed Disc

NOTE: Keep the compressor and parts clean.

If further disassembly is intended, first remove the oil drain plug, drain and measure oil, and remove the clutch drive plate assembly, clutch pulley assembly, clutch coil and hous-

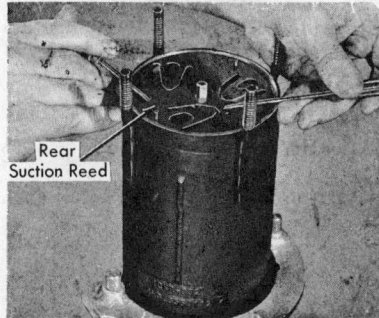

Removing rear suction reed (© G.M. Corp.)

 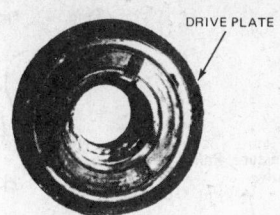

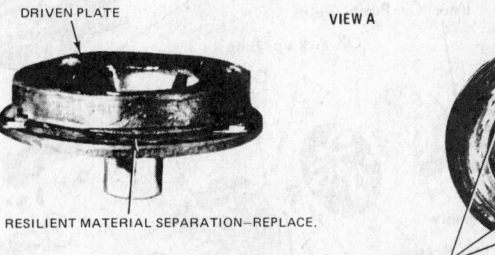

SCORING OF DRIVE PLATE AND DRIVEN PLATE IS NORMAL—DO NOT REPLACE FOR THIS CONDITION.

VIEW A

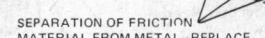 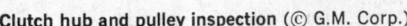

RESILIENT MATERIAL SEPARATION—REPLACE.

VIEW B

SEPARATION OF FRICTION MATERIAL FROM METAL—REPLACE. VIEW C

Clutch hub and pulley inspection (© G.M. Corp.)

ing and shaft seal assembly, as described in previous paragraphs.

1. Attach the compressor to a holding fixture and mount the holding fixture in a vise, with the front of the compressor down.
2. Remove the four lock nuts from the rear of the compressor, and lift off the rear head by tapping it with a mallet. Remove suction screen.
3. Mark the top side of both pump gears and lift out. Pump gears must be reinstalled in precisely the same position.
4. Remove and discard the shell-to-head "O" ring.
5. Pry out the rear discharge valve plate and rear suction valve reed

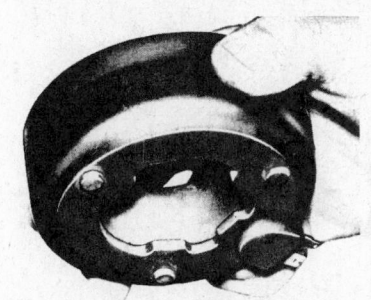

Removing coil housing (© G.M. Corp.)

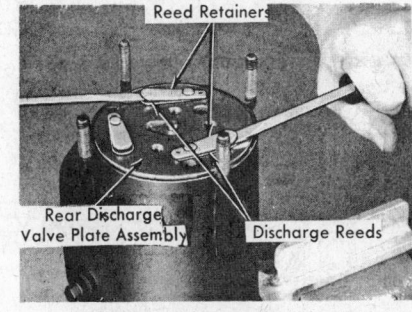

Removing rear discharge valve plate (© G.M. Corp.)

disc with screwdrivers. Place the screwdrivers under the reed retainers to lift discharge valve plate off. Avoid placing the screwdrivers between the reeds and their seats when removing the valve plate. When removing the suction reed disc, avoid lifting on the suction reeds.
6. Remove the oil inlet tube and its O-ring, using oil pick up tube remover.

Compressor Front Head and Cylinder Unit

1. Place an internal assembly support block over the oil pump shaft. Invert the compressor while holding support block in place, and rest the assembly on the bench on top of the support block.
2. Lift front head and shell assembly away from internal mechanism. If necessary, tap on front head with a soft hammer.
3. Remove the compressor front head, using a rubber mallet or wood block to unseat the head from the shell. Protect the Teflon surfaces.
4. Remove and discard the front head-to-shell "O" ring seal.

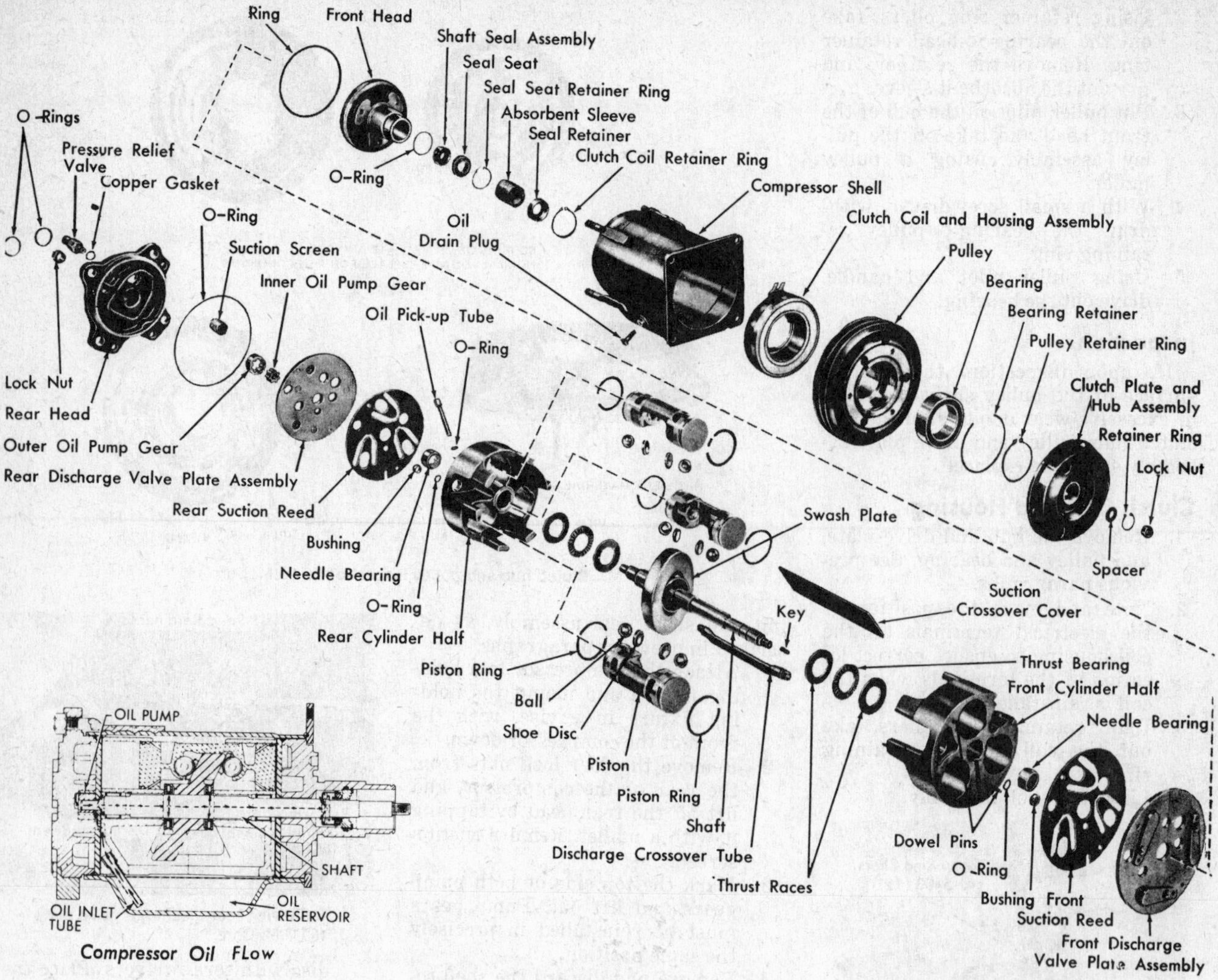

Ring

Front Head

Shaft Seal Assembly

Seal Seat

Seal Seat Retainer Ring

Absorbent Sleeve

Seal Retainer

O–Rings

Pressure Relief Valve

Copper Gasket

O–Ring

O–Ring

Clutch Coil Retainer Ring

Compressor Shell

Clutch Coil and Housing Assembly

Pulley

Bearing

Bearing Retainer

Pulley Retainer Ring

Suction Screen

Oil Drain Plug

Inner Oil Pump Gear

Oil Pick-up Tube

O–Ring

Lock Nut

Rear Head

Outer Oil Pump Gear

Rear Discharge Valve Plate Assembly

Rear Suction Reed

Bushing

Needle Bearing

O–Ring

Rear Cylinder Half

Swash Plate

Clutch Plate and Hub Assembly

Retainer Ring

Lock Nut

Spacer

Suction Crossover Cover

Thrust Bearing

Front Cylinder Half

Needle Bearing

Piston Ring

Ball

Shoe Disc

Piston

Piston Ring

Shaft

Key

Discharge Crossover Tube

Thrust Races

Dowel Pins

O–Ring

Bushing Front

Suction Reed

Front Discharge Valve Plate Assembly

OIL PUMP

SHAFT

OIL INLET TUBE

OIL RESERVOIR

Compressor Oil Flow

Compressor—exploded view (© G.M. Corp.)

DOWEL PIN HOLES

FRONT HEAD

FRONT DISCHARGE VALVE PLATE ASSEMBLY

DOWEL PINS

CYLINDER AND SHAFT ASSEMBLY

VIEW A

SUCTION REED

RECESS

LARGE HOLE

REAR DISCHARGE VALVE PLATE

FRONT DISCHARGE VALVE PLATE

VIEW B

Front head assembly (© G.M. Corp.)

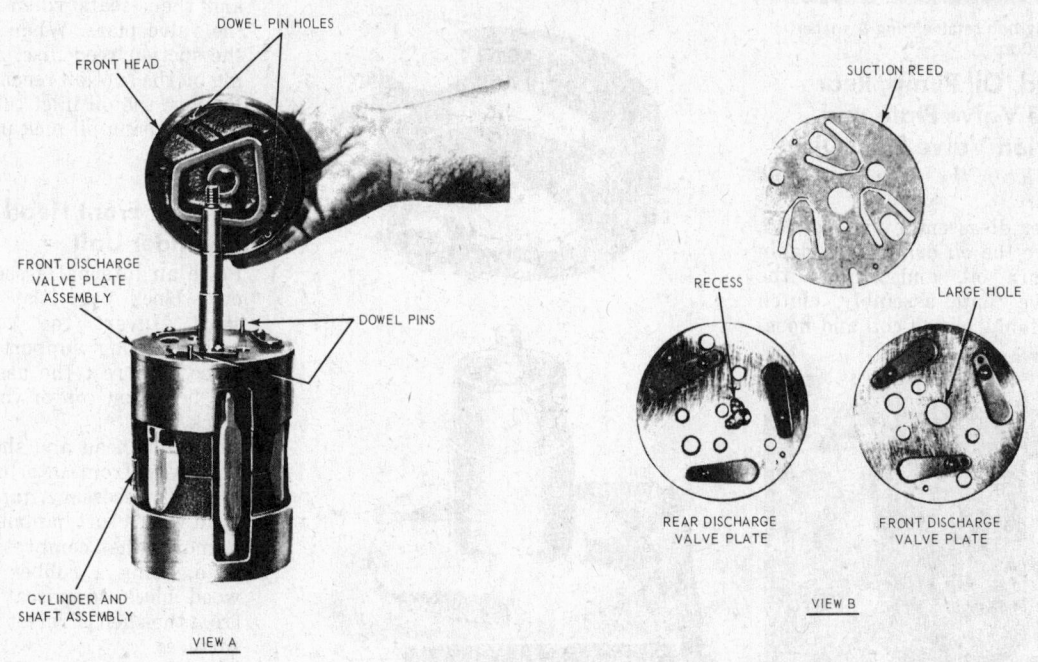

Cylinder Unit Disassembly

1. Remove the front discharge valve plate, and the front suction reed plate. Remove the suction crossover with a screwdriver and discard the seal.
2. Number the pistons and cylinders to facilitate reassembly. Separate cylinder halves using a rubber mallet and wood block. Make sure discharge cross-over tube does not contact axial plate during separation. Do not strike either end of compressor shaft.

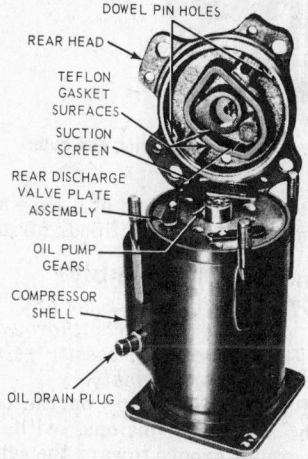

Rear head removed (© G.M. Corp.)

3. Disassemble the rear cylinder half and discharge tube from the front cylinder assembly.
4. Carefully remove from the cylinder assembly and lay out the following parts: 1, 2, and 3 pistons, piston drive balls, and piston rings. To disassemble, rotate the swash plate until the piston is at its highest point, raise the swash plate approximately ½″ and lift out the piston and related parts one at a time. Discard the shoe discs and rear needle thrust bearings and races.
5. Lift out the shaft and swash plate assembly and front needle thrust bearing races. Discard the front needle thrust bearing and races.
6. Wash all salvaged parts of the cylinder assembly in trichlor-

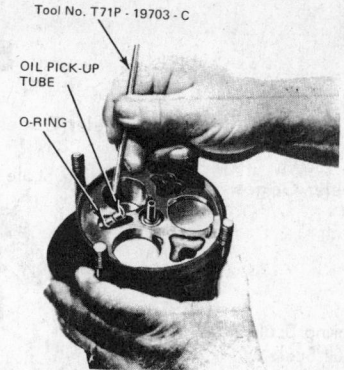

Removing oil pick-up (© G.M. Corp.)

ethylene, alcohol, or similar solvent and dry the parts with filtered, dry compressed air.

NOTE: If internal parts of a cylinder assembly are to be replaced, it will be necessary to adjust shaft end play and piston shoe disc clearance prior to assembly of the cylinder assembly.

NOTE: Compressors on 1976 and later trucks have pistons equipped with Teflon piston rings. Special care is required when replacing these rings. The following procedure applies to the replacement of Teflon piston rings only.

Teflon Piston Ring Replacement

1. Remove the old piston rings with great care, by slicing through the ring with a knife or heavy razor. Hold the blade almost flat with the piston surface to avoid damage to the piston surface or ring groove.
2. Clean the piston and ring grooves with a good solvent and blow dry.
3. With the piston on a flat, clean surface, install the ring installer guide J-24608-2, or its equivalent, on the end of the piston.
4. Install a Teflon ring on the guide with the dished or dull side down and the glossy side up.
5. Push the ring installer J-24608-5, or its equivalent, down over the guide to install the ring in the groove. If the ring becomes

slightly crooked in the groove, it can be positioned with a fingernail or other small, blunt instrument.
6. Lubricate the piston ring area with clean refrigerant oil and rotate the piston and ring assembly into the ring sizer J-24608-6, or its equivalent, until the assembly rotates relatively freely in the ring sizer.
CAUTION: The assembly must be rotated at a slight angle to avoid damage to the end of the piston caused by the needle bearings.
7. Remove the piston and ring assembly, wipe with a clean cloth and push the assembly, wipe with a clean cloth and push the assembly into the ring gauge J-24608-1, or its equivalent. The piston should go through the gauge with a 6 lb. force or less without lubrication.
8. Repeat procedure for other end of piston.
9. Lubricate both ends of the piston before insertion into cylinder bore.

Compressor Assembly, Adjusting Compressor Shaft End Play, and Piston Shoe Disc Clearance

1. Install a compressing fixture on the holding fixture, and mount the front cylinder half on the compressing fixture, with the flat side down. Obtain four ZERO thrust races, three ZERO shoe discs, and two new thrust bearings.
2. Apply clear petroleum jelly to a ZERO thrust race. Place these parts in order on the front end of the compressor main shaft.
3. Set the front half of the cylinder on a holding fixture. Insert the threaded end of the shaft (with the front thrust bearing assembly) through the front cylinder half and allow the thrust race assembly to rest of the hub of the cylinder.
4. Place a ZERO thrust race, a new thrust bearing and a second

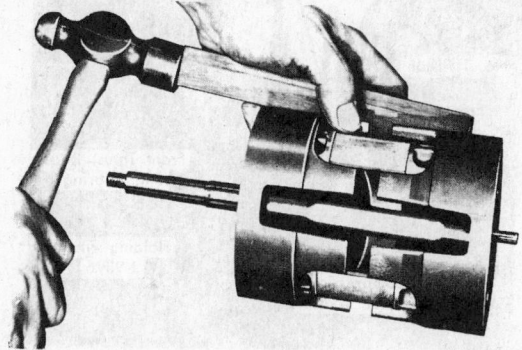

Separating cylinder halves (© G.M. Corp.)

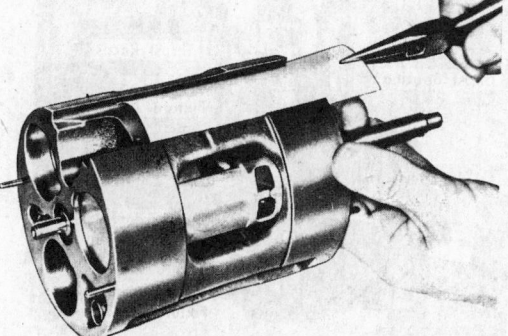

Removing suction crossover cover (© G.M. Corp.)

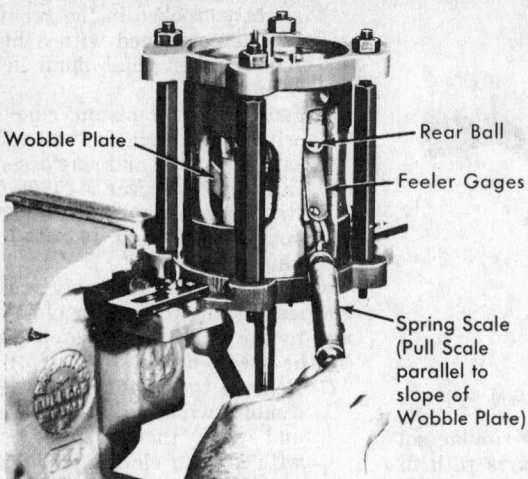

Checking clearance between rear ball and wobble plate (© G.M. Corp.)

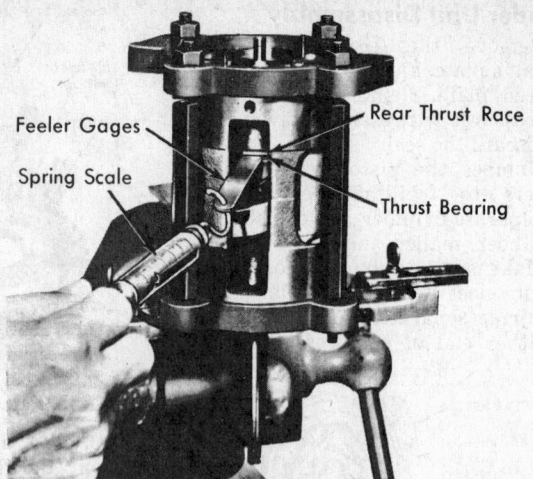

Checking clearance between rear thrust bearing and outer race (© G.M. Corp.)

ZERO thrust race on the rear of the compressor main shaft so that it rests on the hub of the swash plate.

5. Apply a light smear of clear petroleum jelly to the ball pockets of each of the three pistons.
6. Put balls in the ball pockets.
7. Apply petroleum jelly to the cavities of each of the three new ZERO shoe discs.
8. Lay a ZERO shoe disc over each ball in the front end of the piston. Do not put shoes on the piston rear balls.
9. Rotate the shaft and swash plate until the high point of the swash plate is above piston cylinder bore #1. Insert the front end of #1 piston (notched end) and at the same time, put the front ball and shoe and the rear ball over the swash plate.
10. Repeat for #2 and #3 pistons.
11. Align and tap into place the rear cylinder assembly in a compressing fixture with the front end of the compressor shaft pointing up and with the discharge tube opening between fixture bolts. Place the fixture head ring and nuts on the cage. Torque the nuts to 15 foot pounds.

13. Measure the clearance between the rear ball and the swash plate on each piston. Use a feeler gauge, with a spring balance to measure gauge tension. A guage must be found that will require 4-8 oz. tension.
14. Take a second gauge reading of the same ball and plate after rotating the shaft 120 degrees.
15. Take a third reading of the same ball and plate after rotating the shaft another 120 degrees.
16. Take the lowest reading of the three, and select a shoe of the same number as the thickness in thousandths of inches.
17. Identify the shoe by the piston number for which it was selected and lay it out for use later.
18. Repeat steps 13 to 17 for the other two pistons.
19. With a feeler guage measure the clearance between the rear thrust bearing and the rear outer thrust race.
20. Take a thrust race from stock which has a part number the last three digits of which are the same as the measured clearance in thousandths of an inch, and lay it out for use as the rear outer race.

21. Remove one piston at a time and lay them out, identified, for use.

Cylinder Unit Assembly

NOTE: Use all new seals and "O" rings and wet all parts thoroughly with Frigidaire 525 viscosity refrigerant oil during assembly.

1. Place a piston ring on the back of the three pistons, with the scraper groove toward the center of the pistons.
2. Assemble piston drive balls and shoe discs onto #1 piston and apply clear petroleum jelly to piston pockets and shoe discs so that the balls and discs stick to the pistons.
3. Place the front and rear thrust bearing assemblies on the shaft. Install the shaft in the front cylinder half. Orient the swash plate so that the high point of the swash plate is above the cylinder bore #1. Assemble piston #1, with the ball and ZERO shoe on the front end, and a ball on the selected numbered shoe on the rear end, over the swash plate. Position the piston rings such that their gaps are toward the center of the cylinder.
4. Compress a piston ring and in-

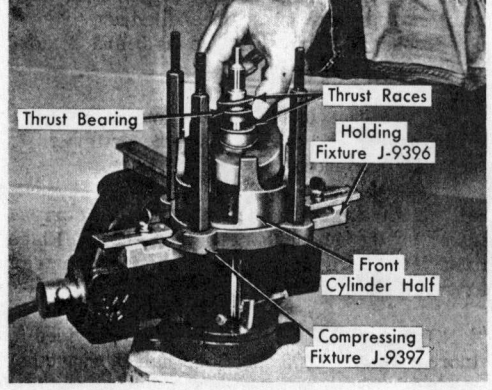

Installing rear thrust races & bearings (© G.M. Corp.)

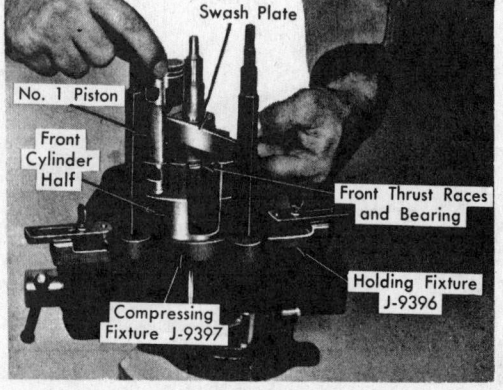

Installing piston (© G.M. Corp.)

sert this into the front half cylinder. Repeat for cylinders 2 and 3.

5. Insert one end of the service discharge crossover tube into the hole in the front cylinder.

6. Orient the shaft so that the pistons are arranged in "stair step" fashion. Orient piston ring gaps toward the center of the cylinder. Place the rear half cylinder over the shaft and start the pistons into the cylinder bores.

7. Compress the piston ring on each piston to permit entry into the cylinder.

8. After all pistons have been inserted into the cylinders, align the end of the discharge crossover tube with the hole in the rear half cylinder, making sure the flattened portion of this tube faces the inside of the compressor for swash plate clearance.

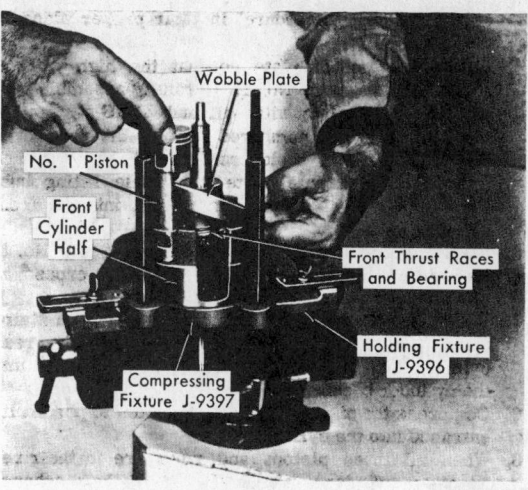

Installing piston, front ball and seat and ball (© G.M. Corp.)

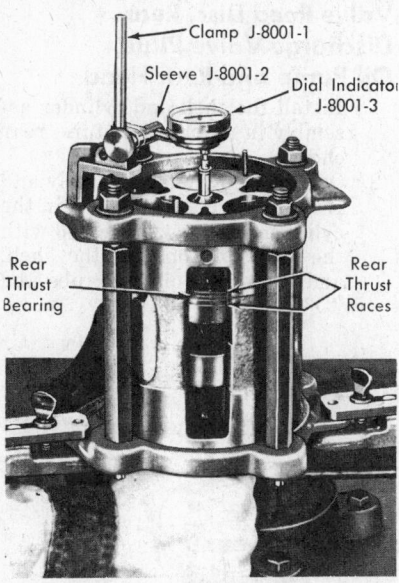

Gaging rear thrust race (© G.M. Corp.)

9. After all parts are aligned, seal the rear cylinder half by tapping it with a fiber block and mallet.

10. Thoroughly lubricate all moving parts with Frigidaire 525 viscosity refrigerator oil. Check the mechanism for free movement.

11. Select a new rectangular suction crossover cover seal and coat it

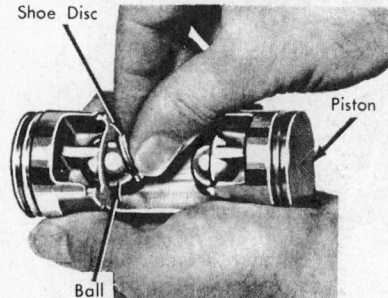

Installing shoe disc (© G.M. Corp.)

with 525 viscosity oil. Assemble it to the suction crossover cover.

12. Start one side of the seal and cover into the "dove-tail" slot in the cylinder.

13. Place a suction crossover cover seal installer between the seal on the opposite side in the "dove-tail" slot to act like a shoe horn.

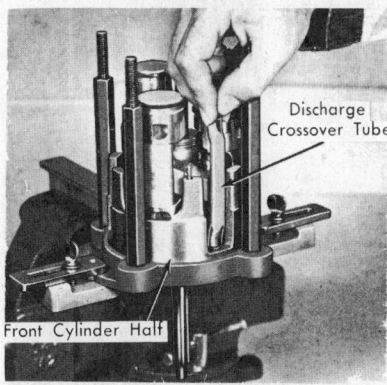

Installing discharge cross-over tube (© G.M. Corp.)

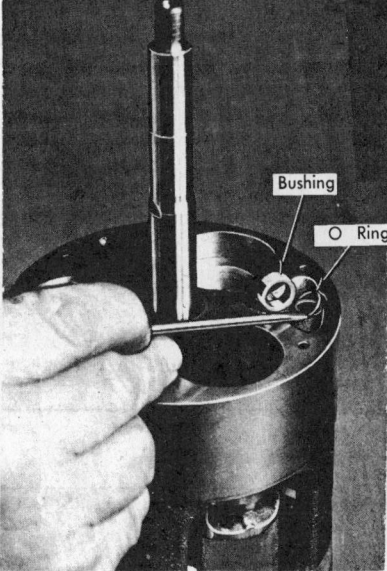

Installing O-ring on cross-over tube (© G.M. Corp.)

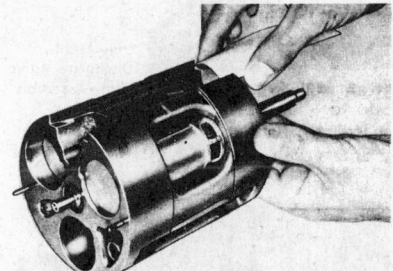

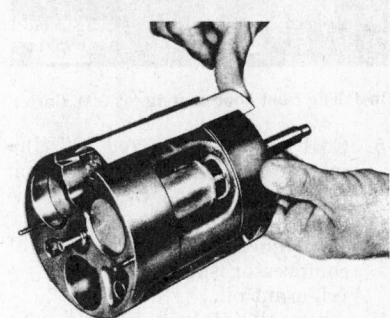

Installing suction crossover cover (© G.M. Corp.)

14. Center the cover and the seal with the ends of the cylinder face.

15. Press down on the cover and snap it into place.

16. Remove the installer by inserting a screwdriver in the hole at one end and lifting.

17. Inspect the cover and seal for proper seating and damage.

18. If pins are damaged, remove with pliers, and replace with new pins by gently tapping.

19. Assemble both service replacement discharge tube "O" rings and bushings onto the cylinder assembly using a small screw driver.

Assembly of Front Suction Valve Reed Disc and Plate, and Front Head, and Installation of Cylinder Unit

1. Place the suction valve reed disc

on the front of the cylinder, aligning it with the dowel pins. suction port and discharge port.

2. Place the front discharge plate on the front head, aligning it with the dowel pins.

3. Lubricate the Teflon surfaces on the front head with Frigidaire 525 viscosity oil.

4. Mark the location of the dowel pin holes on the side of the front head. Place the head on the cylinder assembly, with each hole over the corresponding dowel pins, and tap it lightly with a mallet to seal.

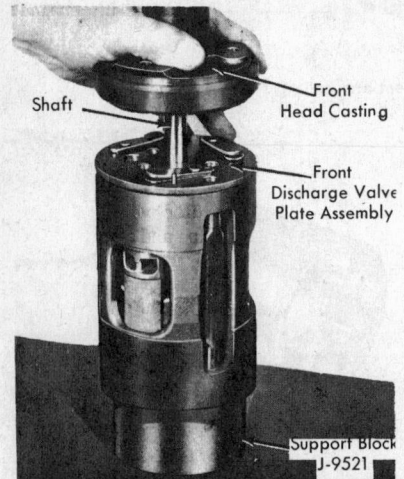

Installing front hood casting (© G.M. Corp.)

5. Coat new "O"-ring and "O" ring groove with clean refrigerant oil and assemble the O-ring in the groove.

6. Coat inner machined surfaces of compressor shell with clean refrigerant oil.

7. Align oil intake tube hole with oil sump location and slide shell down over cylinder mechanism which is being held by the support block.
NOTE: In order to facilitate shell installation and reduce the possi-

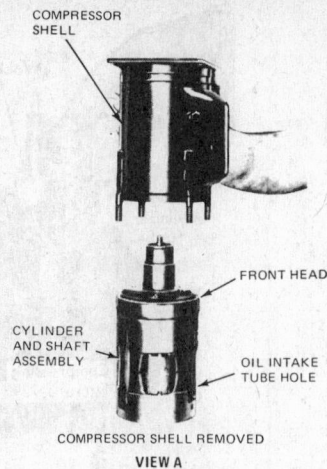

VIEW A

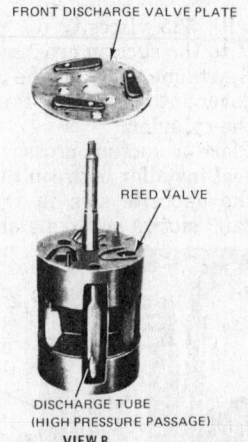

VIEW B

Front discharge valve plate and reed valve (© G.M. Corp.)

bility of O-ring damage, raise the front head slightly.

8. Carefully invert the compressor shell and mechanism. With support bracket in vise, mount the front compressor flange on the bracket.

9. Coat oil pick-up tube hole and new O-ring with clean refrigerant oil. Rotate the compressor

mechanism so that the hole in the oil sump baffle is lined up and install the pick-up tube. Be sure the intake tube and O-ring are properly seated.

10. Insert dowel pins into rear cylinder.

11. The discharge crossover tube rear O-ring and spacer may now be installed.

12. Installed rear suction reed valve, rear discharge valve plate, oil pump gears, rear head and head nuts.

13. Install a new clutch shaft seal assembly. (See section on Clutch Shaft Seal Assembly.)

14. Install coil housing assembly, pulley and bearing and hub and drive plate assembly.

15. Leak test compressor. (See ternal and External.")

Assembly of the Rear Suction Valve Reed Disc, Rear Discharge Valve Plate, Oil Pump and Rear Head

1. Install the shell and cylinder assembly in a holding fixture, rear end up.

2. Orient the cylinder assembly and front head so that the hole in the cylinder assembly is aligned with the reservoir hole in the shell, and install the oil inlet tube and "O" ring.

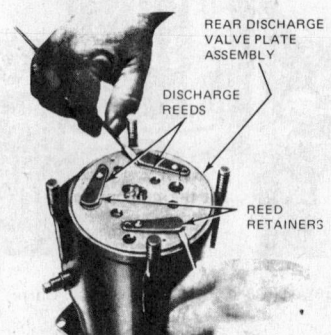

Removing rear discharge valve plate (© G.M. Corp.)

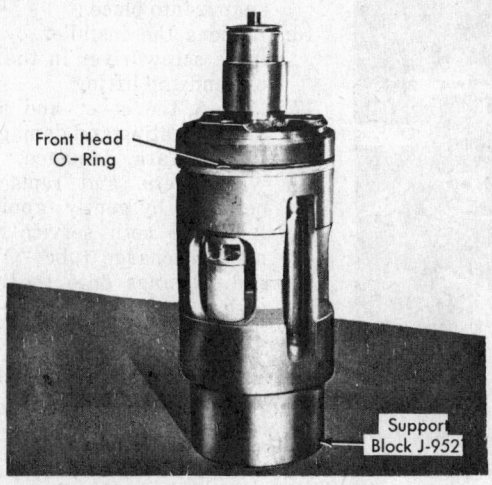

Front hood O-ring installed (© G.M. Corp.)

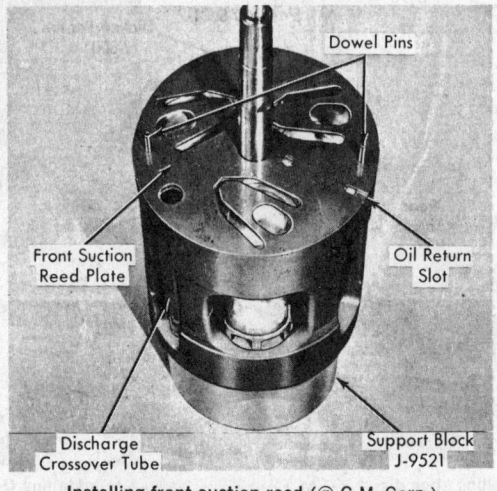

Installing front suction reed (© G.M. Corp.)

3. Place the suction reed valve disc on the rear of the cylinder asembly, aligning it with the dowel pins, suction port and discharge port of the cylinder assembly.

4. Place the rear discharge valve plate on the rear of the cylinder assembly, aligning it with the dowel pins of the cylinder assembly.

5. Put the inner and outer oil pump rotors back together in their previous arrangement, then locate it on the rear head.

6. Place a new shell-to-head "O" ring in the shell, first thoroughly lubricating it with 525 viscosity refrigerator oil.

7. Coat the Teflon surfaces of the rear head with 525 viscosity refrigerator oil.

8. Mark the locations of the dowel pinholes on the side of the rear head.

9. Place the rear head on the compressor, with the holes oriented to the dowel pins.

10. Place new nuts on the threaded shell studs and torque them to 20 foot pounds.

11. Insert new, lubricated, suction and discharge "O" rings into the suction and discharge ports of the rear head.

12. Proceed with assembly of the clutch drive plate and shaft seal.

Clutch Shaft Seal Assembly

1. Wet a new seal seat "O" ring with 525 viscosity refrigerator oil. Install the front head. Make sure that it is not placed in the retaining ring groove.

2. Wet the seal seat with 525 refrigerator oil and place it on a seal remover and installer and

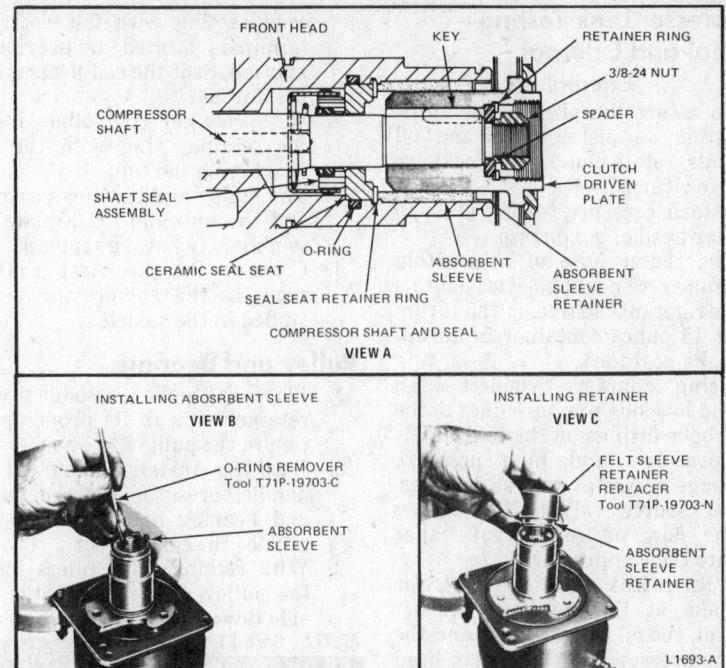

Compressor shaft and seal, absorbent sleeve and retainer (© G.M. Corp.)

insert it in the hub of the front head.

3. Press the installer downward and turn it counterclockwise to release the seal.

5. With retainer ring pliers, set the seal seat retainer ring, flat side down, in the hub of the front head and engage the retainer ring groove.

6. Pressurize the suction side of the compressor with refrigerant, using a compressor leak test fixture.

7. Install, temporarily, the shaft nut.

8. Rotate the compressor shaft several times.

9. Leak test and correct any leaks.

10. Remove and discard the shaft nut.

NOTE: Before proceeding further with assebly, leak test the compressor cage.

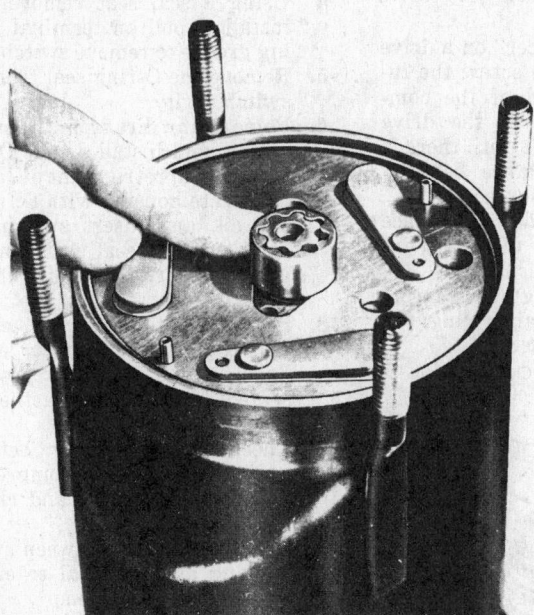

Proper oil pump gear positioning (© G.M. Corp.)

Installing shell over internal mechanism (© G.M. Corp.)

Compressor Leak Testing—Internal and External

External

1. To assure the presence of lubrication on piston rings and oil seals, rotate clutch hub clockwise a few turns.
2. Attach pressure testing plate to rear head of compressor.
3. The center hose of a manifold gauge set should be attached to a refrigerant source. (The drum or 15 ounce container in an upright position).
4. Using adapters, connect high and low side pressure lines to the proper fittings on the test plate.
5. Open low and high pressure gauge valves as well as refrigerant source valves. This allows the flow of refrigerant vapor into the compressor.
6. With a leak detector, check for leaks at the compressor shaft seal, the oil drain fitting, and the compressor front and rear head seals. Shut off manifold gauge valves after checking.
7. If any external leaks have been found, repair them and repeat steps one through six before going on to steps eight through twelve on internal leaks.
8. Remove manifold gauge hoses from the test plate.

Internal

9. The low pressure manifold gauge hose should now be connected to the high pressure fitting on the pressure test plate.
10. In order to allow the refrigerant vapor flow into the compressor, open the low pressure control valve.
11. Note readings on the pressure gauge; close the valve. In the event there is a guage reading drop down to ten pounds within 30 seconds or less, an internal leak is indicated. Check the following points.
    a. Raised section of cylinder face.
    b. Reed valves.
    c. Cross-over tube.
    d. Sealing surfaces on front head, or Teflon seals on rear head.
12. If an internal leak has been detected, correct problem and repeat steps one through eleven. Proceed to step thirteen if corrective procedures were successful or if no leak was originally found.
13. Remove hoses from the test plate.
14. Leave the test plate on compressor until the compressor has been installed on car.

Clutch Coil and Housing

1. Place the clutch coil on the front head casting with the electrical terminals located as previously marked. Seat the coil properly on the dowels.
2. With retainer ring pliers insert the retainer ring with the flat side facing the coil.
3. The pulley and bearing assembly and the hub and drive plate assembly may now be replaced.
4. Connect the electrical connections if the compressor is installed in the vehicle.

Pulley and Bearing

1. Place the pulley assembly wire retainer ring in its proper position in the pulley.
2. With an installer mounted on handle, press or tap the pulley and bearing assembly onto the neck of the compressor.
3. With retainer ring pliers insert the pulley retainer with the flat side toward the bearing.

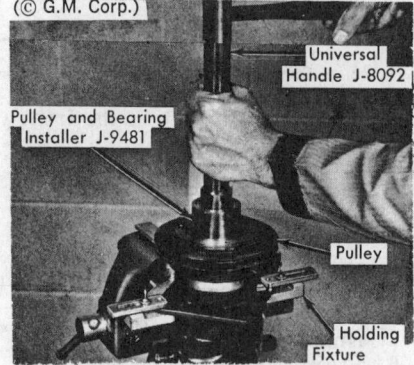
(© G.M. Corp.)

Universal Handle J-8092
Pulley and Bearing Installer J-9481
Pulley
Holding Fixture

Installing pulley and bearing assembly

Clutch Drive Plate

1. Install the woodruff key in the hub of the clutch drive plate so that it projects about 3/16".
2. Align the key with the keyway in the shaft and place the drive plate on the shaft.
3. Screw a "free spacer" on a drive plate installer and screw the installer on the end of the compressor shaft. Force the drive plate onto the shaft until there is about 3/32" clearance between the two frictional surfaces.
4. Take off the installer and place the spacer around the shaft, inside the drive plate hub.
5. With retaining ring pliers, insert the drive plate retainer ring, flat side facing the spacer.

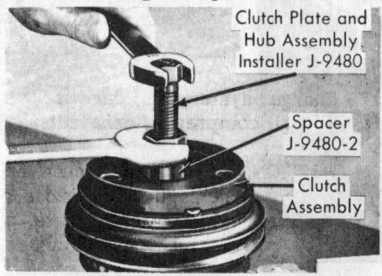

Clutch Plate and Hub Assembly Installer J-9480
Spacer J-9480-2
Clutch Assembly

Installing clutch plate & hub assembly
(© G.M. Corp.)

SERVICE PART NO.	IDENTIFICATION NO. STAMPED SHOE DISC
6557000	0 ("ZERO" SHOE DISC)
6556175	17½
6556180	18
6556185	18½
6556190	19
6556195	19½
6556200	20
6556205	20½
6556210	21
6556215	21½
6556220	22

13-78A

Shoe disc table (© G.M. Corp.)

SERVICE PART NO.	IDENT. NO. ON RACE	THICKNESS
6556000	0	.0920
6556050	5	.0965
6556055	5½	.0970
6556060	6	.0975
6556065	6½	.0980
6556070	7	.0985
6556075	7½	.0990
6556080	8	.0995
6556085	8½	.1000
6556090	9	.1005
6556095	9½	.1010
6556100	10	.1015
6556105	10½	.1020
6556110	11	.1025
6556115	11½	.1030
6556120	12	.1035

13-81A

Thrust race table (© G.M. Corp.)

6. Put on a new shaft nut, using a 9/16" thin wall socket. Torque to 15 foot pounds. The air gap should now be approx. 1/32 (.0313)—1/16 (.0625).

Superheat Switch

1. Disconnect thermal limiter, and discharge system.
2. Remove superheat switch electric connector.
3. Using snap ring pliers, remove the switch retaining ring.
4. Using a seal seat remover and installer, pull on terminal housing groove to remove switch.
5. Remove the O-ring seal from the switch cavity.
6. Remove any dirt from the switch cavity, and install a new O-ring coated with refrigerant oil.
7. Lubricate housing with refrigerant oil, and insert switch into cavity with a seal seat remover and installer. Make sure it bottoms.
8. Reinstall retaining ring, ensuring that sides contact switch housing, and that it is properly seated in the groove in the cavity.
9. Check for continuity between switch housing and compressor.
10. Evacuate, leak test, and charge the system.
11. Check continuity between switch housing and terminal to ensure that contacts are open.
12. Reconnect superheat switch connector.

Dodge

Compressor Oil Level Check

Factory installed compressors originally contain 10 to 11 ounces of refrigerant oil. When the system is in operation some of the oil will be directed to the evaporator, condenser, and drier and remain there. As a result, the amount of oil in the compressor will be less than the initial 10 to 11 ounces.

Chrysler recommends an oil level check whenever the system has lost its charge. To perform an oil level check proceed as follows:

1. Operate system for 15 minutes with hood raised, windows open, and blower switch on high.
2. Shut air conditioner off, without changing any of the above conditions.
3. Remove refrigerant from entire system.
4. Wait ten minutes.
5. Insert dipstick into crankcase filler hole.
6. Measure height on stick and refer to chart below.
7. If reading is below minimum level on dipstick, add clean, dry oil until the minimum reading is reached.
8. If reading is above maximum level, remove oil until maximum reading is reached.
9. Evacuate and recharge system.

Oil Level Reading

Thru 1973, 6 cyl. .... 1⅝"-2⅛"
Thru 1973, 8 cyl. .... 1⅝"-2⅜"
1974-78, All .... 2⅜"

Compressor Removal and Installation

1. Purge the system of refrigerant.
2. Measure and record the compressor oil level.
3. Remove the suction and discharge lines at the compressor fittings.
4. Remove the magnetic clutch wire.
5. Remove the drive belts.
6. Remove the compressor mounting bolts and lift out the compressor.
7. Set the compressor in position.
8. Install the mounting bolts.
9. Install the drive belts.
10. Connect the magnetic clutch wire.
11. Connect the suction and discharge lines to the compressor fittings.
12. Evaculate, leak test and charge the system.

Clutch Overhaul

NOTE: Compressor clutch removal does not require system discharge or compressor removal.

1. Loosen and remove the belts. Disconnect the clutch field wire at the connector.
2. Take out the locking bolt and the washer from the crankshaft at the front center of the clutch.
3. Install a 5/8"-11 x 2 1/2" cap screw into the threaded portion of the hub assembly.
4. Support the clutch with one hand and tighten the cap screw, forcing off the clutch.
5. Take out the three hex head screws holding the clutch field assembly to the compressor and lift off the assembly.
6. Take off the snap ring from the drive hub.
7. a. (Warner)
   Install a drive hub puller so that the three pins of the tool are in the hub and shoe assembly. Tighten the tool bolt until the drive hub is forced from the bearing.

   b. (Electrolock)
   Install a 5/8"-11 x 2-1/2" bolt into the rear of the hub. Hold the pulley and drive the hub from the pulley with a soft hammer.
8. Take off the bearing snap ring from the pulley.
9. Set the pulley assembly in an arbor press, pulley side down and with the bearing hub centered on a tool. Install a tool on the inner race of the bearing and force the bearing from the pulley assembly.
10. Set the pulley assembly pulley side up, on an arbor press and insert a new bearing into the bore. Install a tool against the bearing and press the bearing into position.
11. Install the pulley assembly, pulley side down, on the tool.
12. Start the hub drive into the inner bearing race. Press the hub into position with an arbor press.
13. Install the bearing and hub snap rings.

Compressor Clutch Current Draw Test

The coil may be tested for a short or open circuit if electrical problems are suspected. To test, hook an ammeter in series with a good 12 volt battery and the clutch coil lead. The current draw at 68° F. should be as follows:

Copper wire   1971-74:
               2.7-3.3 amps
             1975-78:
               3.4-4.2 amps
Aluminum wire 1971-74:
               4.0-4.6 amps
             1975-78:
               4.2-5.1 amps

Aluminum coil type clutches are marked "AL".

Condenser Removal and Installation

Typical

1. Purge the system of refrigerant.
2. Disconnect the inlet and outlet connections at the condenser.
3. Remove the bolts retaining the mounting brackets to the radiator assembly.
4. Lift out the condenser.
5. Set the condenser in place.
6. Install the retaining bolts.
7. Connect the inlet and outlet connections to the condenser.
8. Evacuate, charge, and leak test the system.

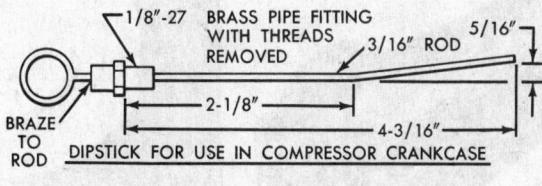

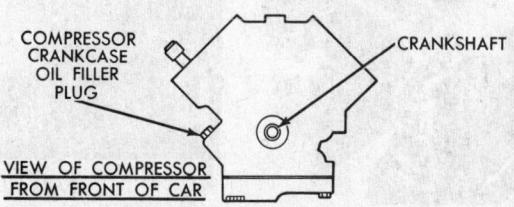

Compressor oil dip stick (© Chrysler Corp.)

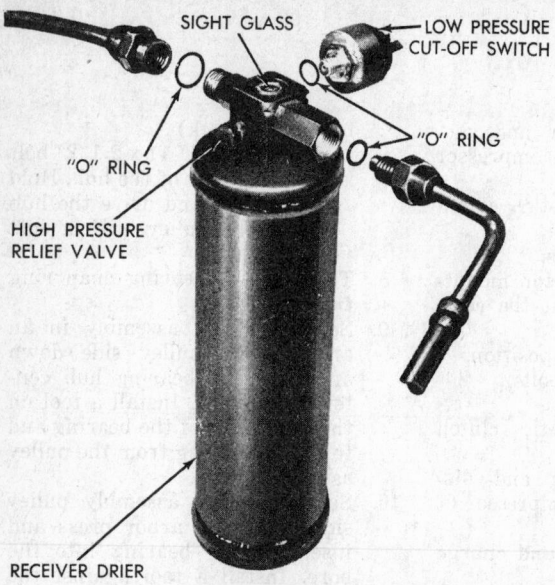

SIGHT GLASS

LOW PRESSURE
CUT-OFF SWITCH

"O" RING

"O" RING

HIGH PRESSURE
RELIEF VALVE

RECEIVER DRIER

Receiver dehydrator (© Chrysler Corp.)

DEFROST DOOR

HEATER CORE

DEFROST
ACTUATOR

REAR
HOUSING

EVAPORATOR
CORE

FRONT HOUSING

Separating unit

Receiver-Drier Removal and Installation

Typical

CAUTION: Moisture hungry receiver driers can become quickly saturated when open. Therefore it is advisable to have all tools needed in this operation close at hand. It is also advisable to do the job as quickly as possible.

1. Purge refrigerant from the system.
2. Disconnect receiver-drier fittings and lift unit out.
3. Uncap new receiver-drier.
4. Coat new "O" rings and fittings with clean refrigerant oil.
5. Position drier and thread line connectors onto drier. (DO NOT overtighten as "O" rings may be damaged.)
6. Evacuate, charge, and leak test the system.

Evaporator Removal and Installation

Typical

Before beginning this procedure, purge refrigerant from the system.

1. Disconnect negative battery cable.
2. Detach refrigerant lines from evaporator core fittings on fire wall, using two wrenches to avoid damaging these components. To avoid contamination of system, cap all refrigerant openings.
3. Remove ash tray and glove box.
4. Remove left and right A/C ducts.

EQUALIZER TUBE

CAPILLARY
SENSING TUBE

EXPANSION
VALVE

SUCTION LINE

Expansion valve (© Chrysler Corp.)

5. Remove four screws retaining distribution duct. Remove duct.
6. Remove two screws retaining center outlet and duct. Remove these components.
7. Disconnect blower motor lead and wiring harness from resistor.
8. Detach rear housing vacuum lines.
9. From the engine side, remove drain tubes.
10. There are 22 rear housing retaining screws. Remove these

screws. Two plugs cover access to the 2 remaining retainer screws. Remove these plugs and, with a long extension, remove the 2 remaining screws. Separate the unit by holding the defroster door in the heat position. This enables the unit to clear the front housing.

11. After removing the 2 engine side screws, remove the 2 screws from each end of the core and carefully remove the core.
12. Align evaporator core in housing, taking care to properly position rubber insulators and seals. Install 2 screws on each end of the core.
13. Engine side screws may now be installed.
14. Holding the defroster door in the heat position, so it will clear the front housing, install rear housing, making sure shaft lines up in hole.
15. Attach vacuum lines.
16. Attach resistor wiring harness.
17. Position and install distribution

EVAPORATOR
CORE

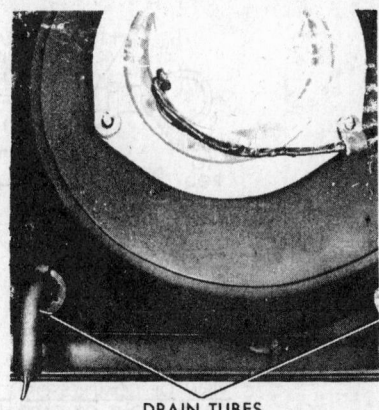

DRAIN TUBES

Evaporator core removal (© Chrysler Corp.)

18. Position and install left and right A/C ducts.
19. Install ash tray and glove box.
20. Coat new "O" rings and connections with clean refrigerant oil and attach lines to evaporator core. (Use two wrenches to avoid damaging the lines or fittings.)
21. Position and install drain tubes.
22. Connect negative battery cable.
23. Evacuate, charge, and leak test the system.
24. Inspect the sealing compound and correct as necessary.

Drain Tubes

Water discharge on vehicle floor may be caused by two conditions: Condensation drainage problems and freeze-up problems. In the event either occurs, check drain tubes to determine whether or not they are open. The drain tubes on truck air conditioning units are routed so that they exit through the front housing into the engine compartment.

Blower Motor Removal and Installation

Typical, Through 1971— Exc. Vans

1. Disconnect the negative battery cable.
2. Remove the glove box and dis-connect the wires from the blower motor resistor and vacuum hoses from the fresh air door actuator.
3. Remove the support bracket attached to the stud on the face of the blower motor housing and back of the instrument panel.
4. Remove the six retaining screws and separate the blower housing from the evaporator housing.
5. To gain access to the two remaining support bracket bolts the air intake grille must be removed. Remove the windshield wiper arms, seven grille retaining screws and lift the grille off the top of the plenum chamber.
6. Remove the two support bracket bolts and lower the blower motor and housing out from under the instrument panel.
7. Place the unit on a bench and disconnect the blower motor air tube and ground wire. Remove the three screws from the outside diameter of the mounting plate and separate the motor from the housing.
8. To install, reverse the removal procedure.

Typical, 1972-78 Exc. Vans

1. Disconnect the negative battery cable.
2. Disconnect the blower motor lead from the resistor, and the ground wire from the fire wall.
3. Remove the three nuts retaining the blower motor mounting plate to the housing, separate the mounting plate and blower motor assembly from the housing.
4. Note the position of the blower wheel to the shaft and remove the wheel from the shaft.
5. Remove the two retaining nuts and separate the motor from the mounting plate.
6. To install, reverse the removal procedure.

Vans

1. Remove the radiator.
2. Remove the cooler tube from the motor.
3. Disconnect the lead wire and ground wire.
4. Remove the blower motor from the housing by removing the three retaining nuts and washers from the mounting plate.
5. To install, reverse the removal procedure.

Clutch

Removal

NOTE: Compressor clutch removal does not require system discharge or compressor removal.

1. Loosen and remove the belts. Disconnect the clutch field wire at the connector.
2. Take out the locking bolt and the washer from the crankshaft at the front center of the clutch.
3. Install a 5/8"-11 x 2 1/2" cap

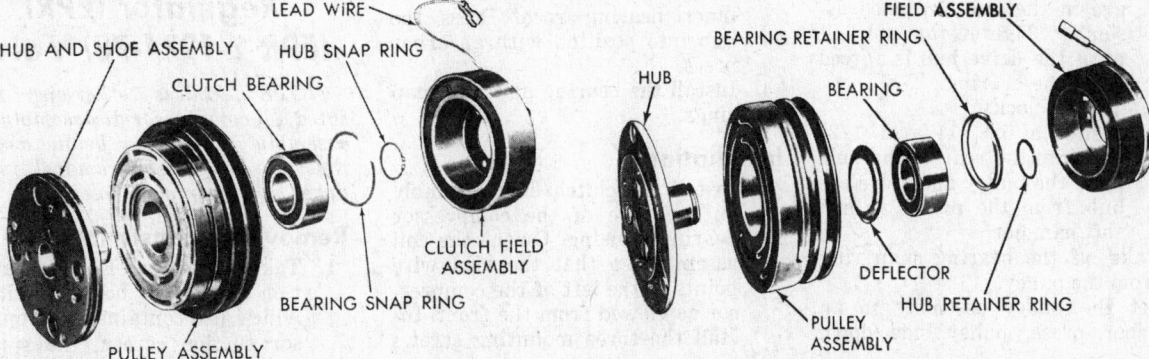

Warner clutch plate type (© Chrysler Corp.)   Pitts Electro-Loc clutch (© Chrysler Corp.)

SINGLE UNIT																																			
INLET AIR WET BULB TEMPERATURE																																			
	53	54	55	56	57	58	59	60	61	62	63	64	65	66	67	68	69	70	71	72	73	74	75	76	77	78	79	80	81	82	83	84	85	86	87	88
MAXIMUM	46	46	46	46	46	46	46	46	46	46	47	47	47	48	49	49	49	50	51	52	53	54	55	55	56	57	58	59	59	60	61	62	63	63	64	65
MINIMUM	39	39	39	39	39	39	40	40	40	40	41	41	41	42	43	43	43	44	45	47	48	49	50	50	51	52	53	54	54	55	56	57	58	58	59	60
DISCHARGE AIR DRY BULB TEMPERATURE																																			
INLET AIR DRY BULB TEMPERATURE MUST BE BETWEEN 75° AND 110°F.																																			

Performance temperature chart (© G.M. Corp.)

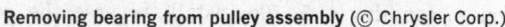

Removing bearing from pulley assembly (© Chrysler Corp.)

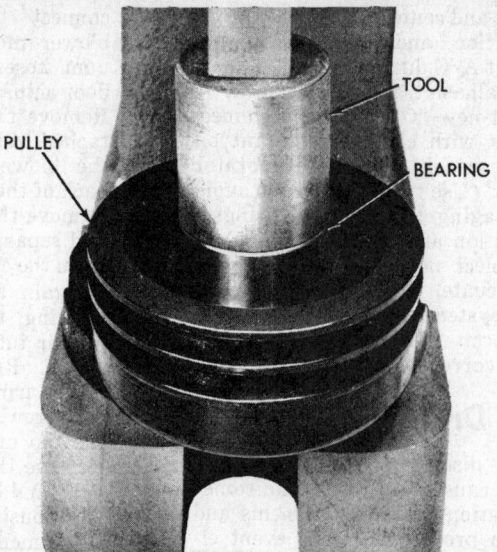

Installing a new bearing in the pulley assembly—typical (© Chrysler Corp.)

screw into the threaded portion of the hub assembly.

4. Support the clutch with one hand and tighten the cap screw, forcing off the clutch.
5. Take out the three hex head screws holding the clutch field assembly to the compressor and lift off the assembly.

Disassembly

1. Remove the snap ring from the drive hub.
2. a. (Warner)
   Install a drive hub puller so that the three pins of the tool are in the hub and shoe assembly. Tighten the tool bolt until the drive hub is forced from the bearing.
   b. (Electrolock)
   Install a 5/8"-11 x 2 1/2" bolt into the rear of the hub. Hold the pulley and drive the hub from the pulley with a soft hammer.
3. Take off the bearing snap ring from the pulley.
4. Set the pulley assembly in an arbor press, pulley side down

and with the bearing hub centered on a tool. Install a tool on the inner race of the bearing and force the bearing from the pulley assembly.

Assembly

1. Set the pulley assembly, pulley side up, on an arbor press and insert a new bearing into the bore. Install a tool against the bearing and press the bearing into position.
2. Install the pulley assembly, pulley side down, on the tool.
3. Start the hub drive into the inner bearing race. Press the hub into position with an arbor press.
4. Install the bearing and hub snap rings.

Installation

1. Install the clutch field assembly on the base of the compressor bearing housing. Orient the coil assembly so that the lead wire points to the left of the compressor as viewed from the front. Install the three mounting screws

and torque to 17 inch pounds.
2. Install the woodruff key in the crankshaft.
3. Install the clutch assembly on the crankshaft.
4. Install the washer and a new self-locking bolt. Prevent the clutch from turning with a spanner wrench inserted in the holes. Torque to 20 foot-pounds.
5. Connect the field lead wire. Reinstall and tighten belts to the specified tension.

Evaporator Pressure Regulator (EPR) (EPR 2-1974-78) Valve

NOTE: Unless otherwise indicated, all compressor disassembly and assembly procedures below assume that all compressor removal procedures have been performed.

Removal and Installation

1. Take out the two EPR valve suction line fitting bolts, the fitting which also contains the compressor suction screen, the spring, and the gasket.
2. Take out the EPR valve and "O" ring from the compressor, by rotating the valve slightly counterclockwise.

CAUTION: Do not handle the EPR valve more than necessary. Valve should be kept lint free and clean and stored in a plastic bag until ready for use.

3. Lubricate a new "O" ring with refrigerant oil, and install on the valve.
4. Using an appropriate tool rotate the valve counterclockwise and install it in the compressor.
5. Set the compressor suction screen in the EPR valve suction line fitting.

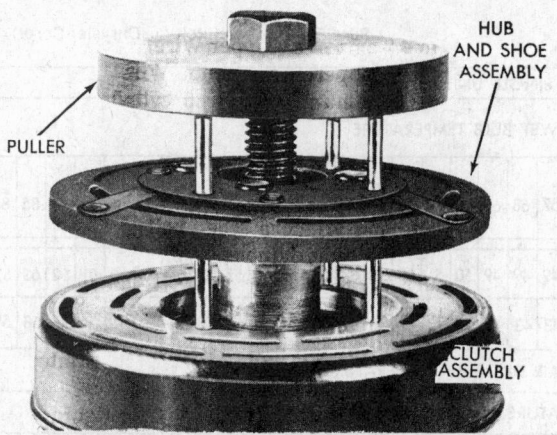

Removing hub and shoe assembly (© Chrysler Corp.)

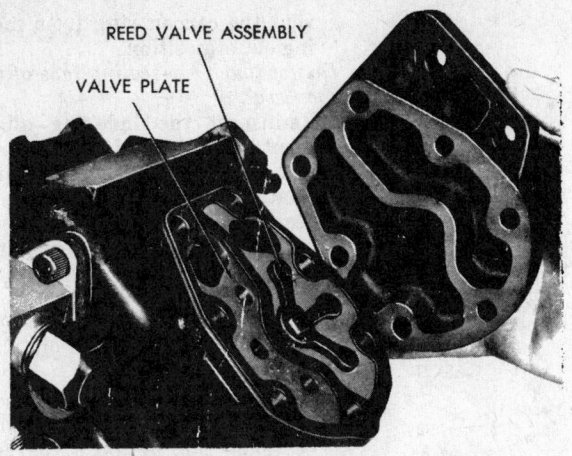

Valve plate—installed position (© Chrysler Corp.)

Installing valve plate and cylinder head (© Chrysler Corp.)

6. Install the suction line fitting gasket, spring and fitting.
7. Install the attaching bolts and torque to 8-14 foot pounds.
8. Recharge system.

Cylinder Head and Valve Plate

Removal

NOTE: This procedure does not require compressor removal or draining of oil.
1. Take out the cylinder head bolts.
2. Lift off the head and valve plate assembly.
3. Separate the head and plate by tapping lightly on the lip with a soft hammer.
4. Inspect valves for damage, and, if damaged, check cylinder bores to see if damage has occurred there. A damaged cylinder head or valve plate must be replaced. A lightly scuffed cylinder bore may be smoothed with crocus cloth. Clean all surfaces with mineral spirits.

Installation

1. Position the reed valve assembly.
2. Install the valve plate gasket, valve plate, cylinder head gasket and cylinder head with the holes over the corresponding studs.

3. Insert the attaching bolts. Tighten alternately.

Piston and Connecting Rod

Removal

1. Take out the sump attaching bolts.
2. Tap lightly with a hammer to separate the sump from the case. (Be careful not to distort the oil pressure relief spring.)
3. Remove the oil relief spring and rubber ball from the crankcase.
4. Lift out the cylinder heads and valve plates.
5. Mark all parts for position.
6. Take off the rod caps.
7. Take out the piston and rod assembly.
8. Inspect pistons and rings for scoring, rod bearings for pitting and chipping, and replace all damaged parts.

Installation

1. With cap removed, install the piston in the bore, using a piston ring compressor.
2. Set the bearing caps in their original positions and install the bolts. Torque to 50-60 inch pounds.
3. Install the valve plates and cylinder heads.

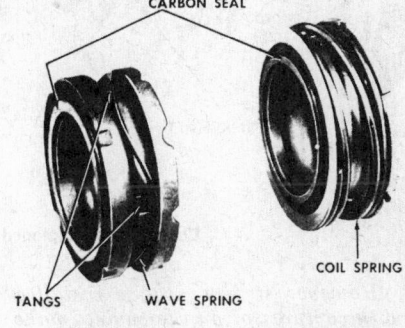

Gas seal identification (© Chrysler Corp.)

4. Set the compressor upside down. Install the pilot studs, gasket, oil pressure relief ball and spring.
5. Set the sump over the pilot studs. Make sure the oil pressure relief spring compresses uniformly.
6. Tighten the sump bolts finger tight. Torque to 14-20 foot pounds.

Crankshaft Bearing Housing and Gas Seal

Removal

NOTE: This procedure does not require compressor removing or draining of oil.

Removing EPR valve (© Chrysler Corp.)

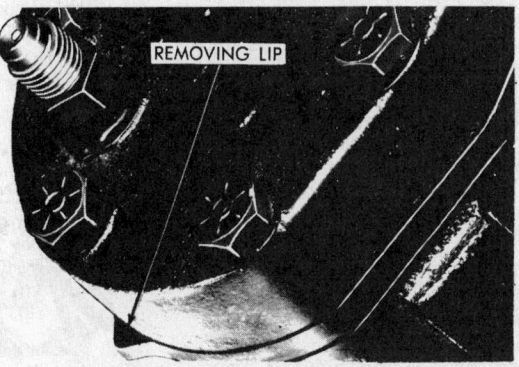

Valve plate and head removing lip (© Chrysler Corp.)

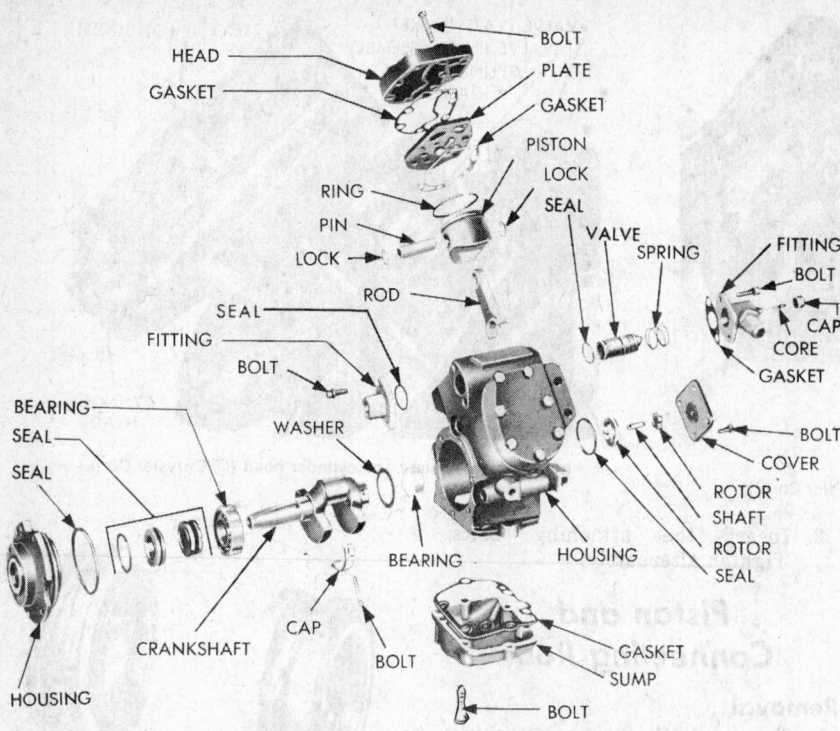

BOLT
HEAD
GASKET
PLATE
GASKET
PISTON
LOCK
RING
SEAL
PIN
VALVE
LOCK
SPRING
FITTING
BOLT
SEAL
ROD
CAP
FITTING
CORE
BOLT
GASKET
BEARING
SEAL
WASHER
BOLT
SEAL
COVER
BEARING
ROTOR
SHAFT
HOUSING
ROTOR
SEAL
CRANKSHAFT
CAP
BOLT
GASKET
SUMP
HOUSING
BOLT

Compressor—exploded view (© Chrysler Corp.)

However if new seal is installed with compressor on the engine, make sure the carbon ring stays in its housing. Generous lubrication of the rotating seal assembly before installing it on the compressor shaft will prevent loss of carbon ring. It is necessary to seal, in order to prevent oil loss, the suction opening located on the back side of the compressor.

Crankshaft bearing housing—removal (© Chrysler Corp.)

NOTE: Approximately 20,000 air conditioner compressors were built with front bearing housing that incorporate a flat section seal in place of the original square seal. This was accomplished to develop field experience on air conditioner compressors with this new seal design.

When the compressor is removed for repair, it is recommended that the seal be checked for type. If the seal is a flat type, a flat section seal must be used.

1. Remove clutch, coil and drive key.
2. Take out the crankshaft bearing housing seal bolts.
3. Pry the bearing housing from the crankcase with two screw-

drivers inserted in the slots provided.
4. Extract bearing housing oil seal.
5. From the bearing housing, take out the "O" ring and the gas seal seat plate. Discard as these components must be replaced.
6. Thoroughly cleanse the front bearing housing with mineral spirits.

Installation

1. Soak new seal seat assembly in clean refrigerant oil. With the smooth (micro finish) side up, install the seal seat assembly in the bearing housing.
2. Position a 1 3/8 in. innerdiameter sleeve over seal seat assembly and lightly tap until seal seat is fully seated in the housing.
3. On the cartridge-type assembly, make sure the tangs of the carbon seal, scallops and retainer tangs index in the mating part slots.
4. With carbon seal up, soak seal assembly in clean refrigerant oil.
5. Hold the seal firmly by the outside edges. At the same time pre-

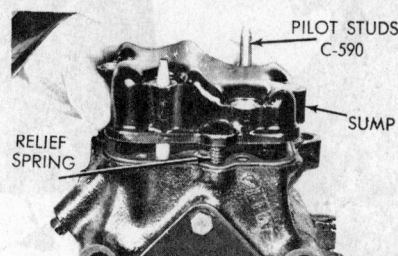

PILOT STUDS
C-590

RELIEF
SPRING

SUMP

Installing compressor sump (© Chrysler Corp.)

vent the carbon ring from moving out of position.
Do not touch the sealing face of the carbon seal.
6. Re-inspect the indexing alignment when the seal stop bottoms against the crankshaft bearing.
7. Coat new bearing housing oil seal with clean refrigerant oil and install. (Be sure seal is not rolled over and is evenly stretched into position.)
8. Using a lint-free cloth wipe the seal seat clean and re-oil with clean refrigerant oil.
9. Carefully install the seal seat and bearing housing. Care must be taken not to touch the seal seat in the bearing housing with the "nose" of the crankshaft.
10. Install 5, 1/4 x 20 screws and draw bearing housing squarely into position. Tighten these screws 1/2 turn at a time per screw so the bearing housing will not jam the ball bearing outer race. When all screws are snugged, tighten them to 100 to 150 inch pounds.
11. Install drive key.
12. Install clutch. Turn crankshaft. A maximum of 10 inch pounds should turn the crankshaft. If shaft won't turn or is overly tight, remove clutch and loosen housing screws til shaft is freer. Repeat tightening procedure outlined in step #10.
13. Check compressor oil level.
14. Re-install clutch. Tighten clutch center bolt to 20-ft.-lbs torque. Install belts and tighten to specifications.
15. Evacuate, charge, and leak test system.

Oil Pump

Removal

1. Take off the oil pump cover plate and oil seal.
2. Take out the drive shaft and rotors.

Installation

1. Orient the pump drive shaft such

Measuring crankshaft axial movement (© Chrysler Corp.)

that the tang end engages in the crankshaft slot.

2. Set the inner rotor on the drive shaft, engaging the drive.
3. Set in the outer rotor and orient it so that it will slide forward over the inner rotor cams. Rotate to check for binding.
4. Install the oil pump cover plate and oil seal.
5. Install the bolts and torque to 100-150 in lb.

Crankshaft and Ball Bearing

Removal

1. Remove the cylinder heads, valve plates, pistons, connecting rods, crankshaft bearing housing and gas seal.
2. Take out the crankshaft and thrust washer from the crankcase.
3. Support the case in an arbor press and force out the bearing.
4. Inspect bearing for cleaniness

COMPRESSOR TIGHTENING TORQUE

NAME	TORQUE
Bolt—Cylinder Cover (Nameplate)	18-24 Ft. lbs.
Bolt—Cylinder Cover	20-26 Ft. lbs.
Bolt—Oil Pump Cover	100-150 In. lbs.
Bolt—Oil Sump	14-20 Ft. lbs.
Bolt—Bearing Housing	100-150 In. lbs.
Screw—Connecting Rod	50-60 In. lbs.
Bolt—Suction Fitting	8-14 Ft. lbs.
Bolt—Discharge Fitting	8-14 Ft. lbs.
Cap—Charging and Gauge Valve	100-130 In. lbs.

and, if necessary, clean with mineral spirits. Coat with refrigerant oil before replacing.

Installation

1. Using a sleeve which bears on the inner race only, press the crankshaft ball bearing on the crankshaft.
2. Make sure the thrust washer is on the rear bearing journal, then set the crankshaft in the crankcase.

3. Orient the crankshaft so that the oil pump shaft is engaged in the crankshaft slot.
4. Install a new gas seal and crankshaft bearing housing. Check to assure free axial movement.
5. Install the pistons and connecting rods.
6. Turn the crankshaft to check for free motion.
7. Install the oil pump, valve plates, and cylinder heads. Use new gaskets.

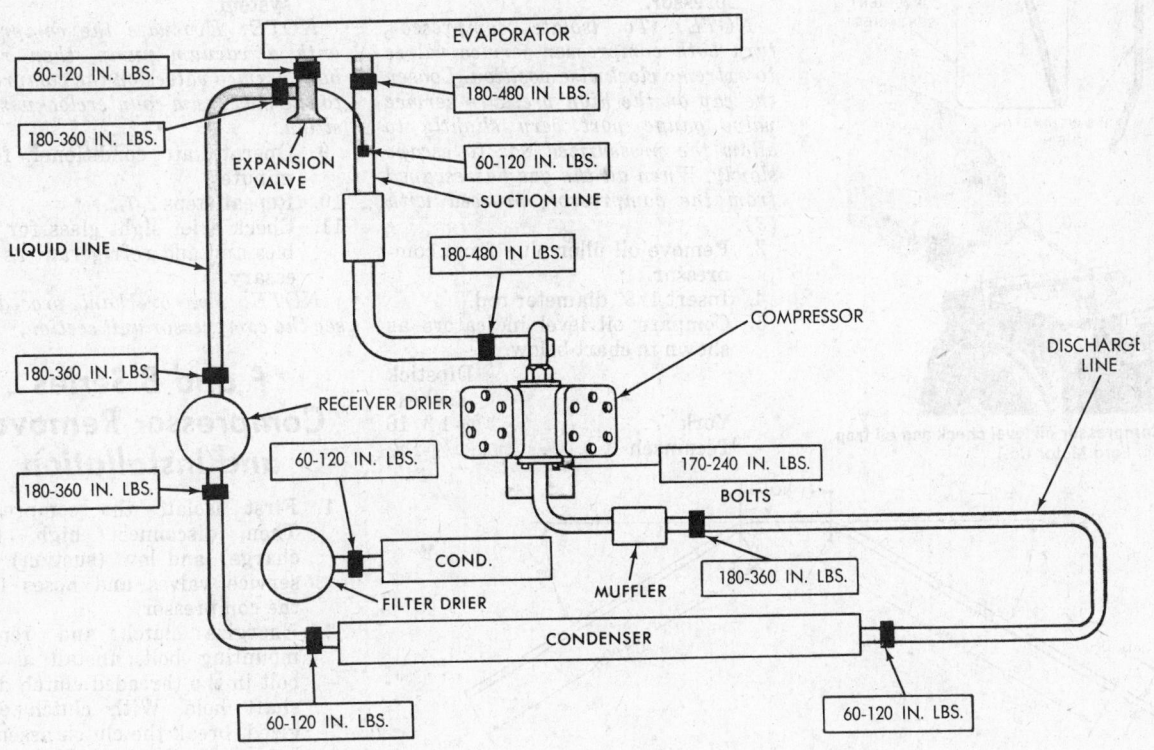

Refrigerant Line Torque—Specifications

Compressor Oil Level Check

Factory installed compressors originally contain 10 to 11 ounces of refrigerant oil.

When the air conditioning system is operating properly, these units do not require periodic oil level checks. The compressor oil level should be inspected, however, if any component of the air conditioner is removed or replaced. Also, if the refrigerant has been discharged too quickly, some oil may escape and the same amount of new oil should be added. Ford recommends Texaco Capella E refrigerant compressor oil or its equivalent.

To determine proper oil capacity follow this procedure:

1. With outside air temperature at

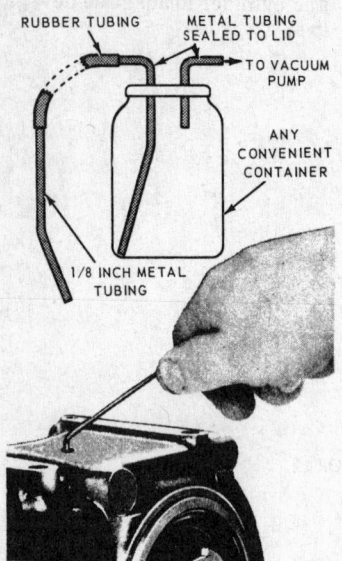

Compressor oil level check and oil trap (© Ford Motor Co.)

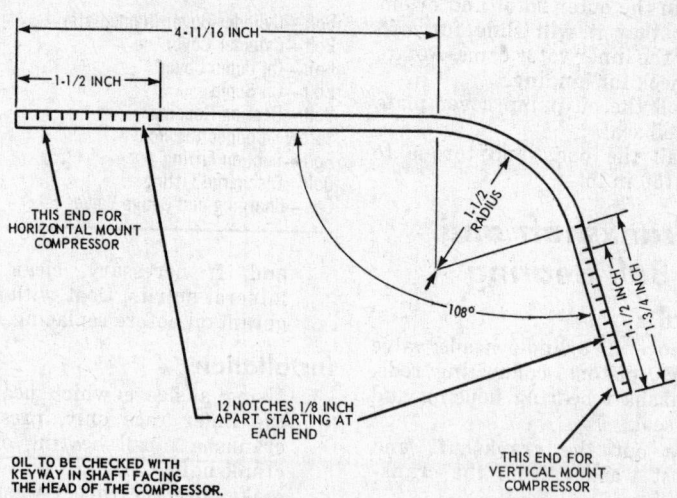

York compressor oil dip stick (© Ford Motor Co.)

least 60 degrees F., run air conditioner with engine at idle speed for 10 minutes.

2. Shut engine off and isolate compressor.

NOTE: To isolate compressor, turn both compressor service valves to extreme clockwise position. Loosen the cap on the high pressure service valve gauge port very slightly to allow the pressurized gas to escape slowly. When all the gas has escaped from the compressor, proceed with (3).

3. Remove oil filler plug from compressor.
4. Insert 1/8" diameter rod.
5. Compare oil level indicators as shown in chart below.

	Dipstick Inches
York	7/8-1 3/16
Tecumseh	7/8-1 5/8

6. Draw out excess or add new oil as indicated by dipstick reading.
7. Replace plug.
8. Connect compressor back into system.

NOTE: Evacuate the compressor with a vacuum pump, then rotate both service valves on the compressor to the maximum counterclockwise position.

9. Operate air conditioner for 5 minutes.
10. Repeat steps 2-7.
11. Check drier sight glass for bubbles and add refrigerant as necessary.

NOTE: For overhaul procedures see the compressor unit section.

F and B Series Compressor Removal and Installation

1. First isolate the compressor. Then disconnect high (discharge) and low (suction) side service valves and hoses from the compressor.
2. Energize clutch and remove mounting bolt. Install a 5/8" bolt in the threaded clutch drive shaft hole. With clutch energized, break the clutch assembly loose from the shaft by tightening the bolt.
3. Disconnect the clutch wire at the bullet connector.
4. Loosen and remove the drive belt. Remove the clutch assembly.

NOTE: When replacing the compressor with a new unit, check the oil level in the old compressor. Oil should now be removed from the new compressor so that it contains the same

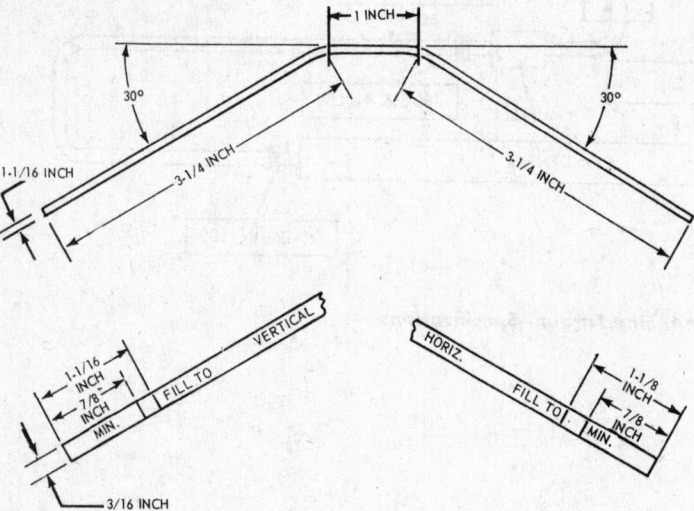

Tecumseh compressor oil dip stick (© Ford Motor Co.)

amount as the old one. This procedure guards against excessive amounts of oil within the system.

5. Install clutch on new compressor and install mounting bolt and washer finger tight. Position compressor on mounting bracket and install the mounting bolts.
6. Connect the clutch wire and energize the clutch. Torque clutch mounting bolt to specifications. Torque compressor mounting bolts to specifications.
7. Install the drive belt and tighten to specifications.
8. Remove shipping plugs from new compressor. Install service valves, using new seals which should be coated with clean refrigerant oil. Tighten to specifications. Do not overtighten. Leak test and evacuate the compressor before connecting it back into the system. Check oil level.
9. Check pressures and charge if necessary.

Evaporator Core Removal and Installation

Through 1972

1. Remove heater-air conditioner.
2. Remove unit's wiring harness.
3. Remove right side mounting bracket from evaporator core.
4. Two housing cover retaining screws are located within the housing glove box. Remove these screws.
5. Remove blower resistor from housing.
6. Remove evaporator core air deflector. Four screws hold it to the air inlet end of the housing.
7. Remove the housing cover by removing the 20 screws retaining it.
8. Now take out the six evaporator

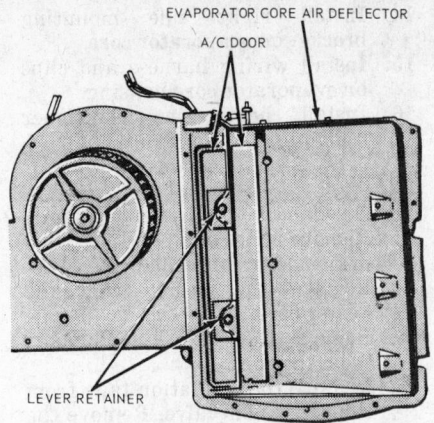

Evaporator core air deflector—Series F & B (© Ford Motor Co.)

core retaining screws and lift out the evaporator core and hoses.
9. Disconnect expansion valve and hoses.
10. Connect expansion valve and hoses and install the expansion valve insulation.
11. Position core in the housing.
12. Align the housing cover on the housing. Install the 20 cover screws. Install evaporator core air deflector.
13. Re-install blower resistor.
14. Install the two housing cover screws located within the housing glove box.
15. The right side mounting bracket may now be installed.
16. Install the wiring harness and wiring harness clips on the housing.
17. Install the heater-air conditioner.

1973-78

1. Disconnect the battery ground.
2. Purge the system.
3. Detach the A/C hose support from the cowl.
4. Remove the insulation tape from

the expansion valve. Then, remove the cover plate and seal from the evaporator housing at the expansion valve.
5. Remove the glove box liner and the right A/C duct. Remove the duct.
6. Disconnect the fresh air-recirc. door vacuum motor hose.
7. Remove the evaporator rear housing from dash panel and inlet boot. Leave front housing in place.
8. Remove the icing switch and mounting plate. Push the wire grommet out of its mounting plate.
9. Remove plenum from evaporator front housing and lower plenum.
10. Remove the expansion valve from the evaporator core.
11. Remove the four attaching screws and remove core from housing.
12. Position core to housing and install and tighten screws.
13. Connect the expansion valve to the evaporator core.
14. Install the icing switch and plenum. Connect the blower and icing switch wires.
15. Install the evaporator rear housing and connect the fresh air inlet boot to the housing.
16. Connect the vacuum hose to the fresh air door vacuum motor.
17. Install the right A/C duct and glove box liner.
18. Install the seal and cover plate on the evaporator housing at the expansion valve.
19. Install the A/C hose support on the cowl.
20. Leak test, evacuate and charge the system.
21. Install the insulating tape on the expansion valve.
22. Connect the battery.

Expansion Valve Removal and Installation

Through 1972

1. Discharge the system.
2. Remove air cleaner.
3. Remove refrigerant hose seal and retainer from dash panel.
4. Remove heater-air conditioner to dash attaching bolts.
5. Remove glove box liner.
6. Remove thermostatic switch attaching screws and move the switch aside.
7. Remove the 2 nuts and bolts which secure the heater-air conditioner to the lower edge of the instrument panel.
8. Remove expansion valve insulation.
9. Disconnect expansion valve refrigerant lines.

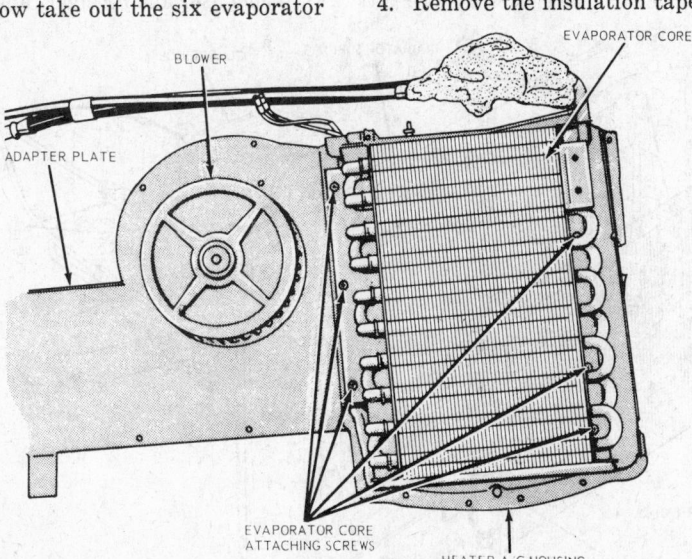

Evaporator core installation (© Ford Motor Co.)

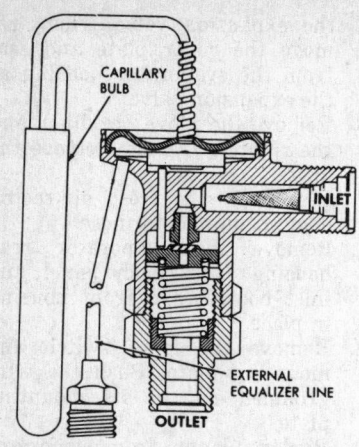

CAPILLARY BULB

INLET

EXTERNAL EQUALIZER LINE

OUTLET

Externally equalized thermostatic expansion valve (© Ford Motor Co.)

1973-78

1. Remove the carburetor air cleaner.
2. Discharge the system.
3. Remove the A/C hose support from the cowl.
4. Remove the insulation tape from the expansion valve and remove the valve from the evaporator core tubes and refrigerant lines.
5. Connect the expansion valve to the evaporator core tubes and the refrigerant lines.
6. Evacuate, charge and leak test the system.
7. Install the insulation tape on the expansion valve.
8. Install the air cleaner.

Blower Motor Removal and Installation

Through 1972

1. Remove heater-air conditioner assembly.
2. Remove wiring harness from housing.
3. Remove right side mounting bracket from evaporator core.
4. Inside the glove box there are two housing cover retaining screws, remove them.
5. Remove blower motor resistor.
6. Remove evaporator core air deflector which is retained by four screws to the air inlet end of the housing.
7. Remove cover from housing by removing the 20 retaining screws.
8. Remove blower and adaptor plate screws and remove blower motor assembly.
9. Separate motor and blower from adaptor plate.
10. Re-install blower and motor to adaptor plate and re-install this unit in the housing.
11. Re-install housing cover. Install evaporator core air deflector.
12. Install blower motor resistor.
13. Install housing cover retainer screws inside glove box.

14. Install right side mounting bracket on evaporator core.
15. Install wiring harness and clips on evaporator core housing.
16. Install heater-air conditioner unit.

1973-78

1. Disconnect the battery ground.
2. Remove the air cleaner and drain the engine coolant.
3. Remove the hoses from the heater core.
4. Remove the A/C hose support from the cowl.
5. Remove the insulation tape from the expansion valve. Remove the cover plate and seal from the evaporator housing at the expansion valve.
6. Remove the glove box liner and right A/C duct. Remove the duct.
7. Disconnect the fresh air door vacuum motor hose.
8. Remove the evaporator rear housing leaving the front housing in place.
9. Remove the icing switch and mounting plate. Remove wire grommet from mounting plate.
10. Remove the plenum from the evaporator front housing.
11. Remove attaching screws and pull evaporator core from housing.
12. Remove heater core from evaporator housing.
13. Unsnap temperature door from arm, remove the arm support and remove the blower.
14. Install the blower in the housing.
15. Install the support and door in the housing.
16. Install the heater and evaporator cores in the housing.
17. Install the plenum and icing switch. Connect the blower and icing switch wires.
18. Install the evaporator rear hous-

ing and connect the fresh air inlet boot to the housing.
19. Connect the vacuum hose to the fresh air door vacuum motor.
20. Install the right A/C duct and glove box liner.
21. Install the seal and cover plate on the evaporator housing at the expansion valve.
22. Install the insulation tape on the expansion valve.
23. Install the A/C hose support on the cowl.
24. Connect the vacuum hoses to the intake manifold and the heater hoses to the heater core.
25. Fill the cooling system and install the air cleaner.
26. Connect the battery.

Condenser Removal and Installation

Through 1972

1. Discharge refrigerant system.
2. The grill upper support should now be removed from the vehicle.
3. The refrigerant lines to the condenser and receiver-dehydrator may now be disconnected.
4. Remove the condenser and receiver-dehydrator attaching screws. Remove condenser.
5. Align the condenser in its mounting postion and install the 5 mounting screws.
6. Re-connect refrigerant lines.
7. Re-install grill upper support.
8. Evacuate, charge, and leak test the system.

1973-78

1. Discharge the system.
2. Remove the hood latch support.
3. Disconnect the refrigerant lines from the receiver and condenser.
4. Remove the bolts and lift out the condenser and receiver tank.

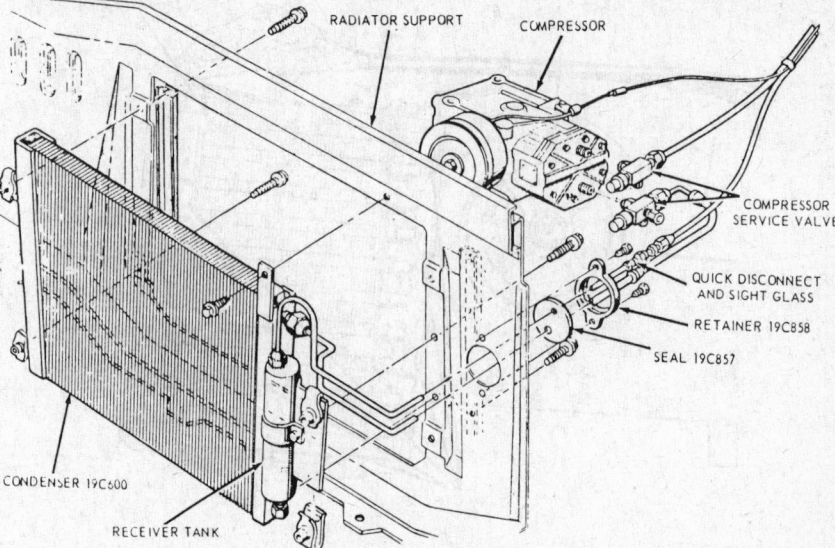

RADIATOR SUPPORT

COMPRESSOR

COMPRESSOR SERVICE VALVE

QUICK DISCONNECT AND SIGHT GLASS

RETAINER 19C858

SEAL 19C857

CONDENSER 19C600

RECEIVER TANK

Condenser installation—Series F & B (© Ford Motor Co.)

5. Separate the receiver tank from the condenser.
6. Assemble the condenser and receiver.
7. Install the condenser on the radiator support and connect the refrigerant lines.
8. Install the hood latch support.
9. Evacuate charge, and leak test the system.

Icing Switch

Typical

1. Disconnect the battery ground cable.
2. Remove the glove box liner and the right A/C duct. Pull the duct from the register and release the clip.
3. Disconnect the fresh air door vacuum motor hose.
4. Remove the evaporator rear housing from the dash panel and boot. Leave the front housing in place.
5. Disconnect the icing switch wires, remove the sensing tube from the core and remove the switch.
6. Secure switch and insert sensing tube four inches into the core.
7. Install evaporator rear housing and connect the fresh air inlet boot to the housing.
8. Connect the vacuum hose to the fresh air door vacuum motor.
9. Install the glove box liner and the right A/C duct.
10. Connect the ground cable to the battery.

E Series
Compressor Removal
and Installation

1. Remove engine cover and carburetor air cleaner.
2. Partially drain cooling system if the vehicle has an auxiliary heater.
3. Detach oil filler tube.
4. Isolate compressor; detach service valves and hoses from compressor; energize clutch and remove mounting bolt.
5. With clutch still energized and drive blets still in place, install a 5/8-11" bolt in the clutch drive shaft hole. Loosen the clutch from the shaft, by tightening the bolt. Disconnect the clutch wire and lift the clutch assembly. Remove field coil.
6. Disconnect heater hoses at the water valve and the T connector if the vehicle has an auxilliary heater.
7. Remove hose retaining bracket, and position hoses aside, on models equipped with a V-8.
8. Remove the bracket that con-

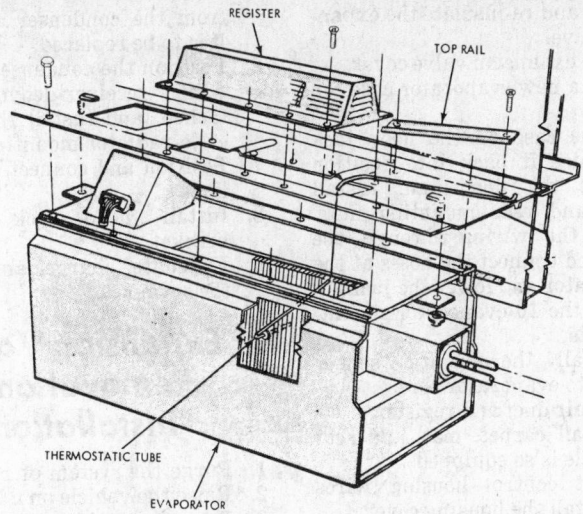

Evaporator—Econoline (© Ford Motor Co.)

nects the compressor to the engine.
9. Remove compressor mounting bolts.
   a. V-8 Models have six mounting bolts. Four bolts are removed from under the vehicle and two are removed from the front of the vehicle.
   b. Six cylinder models have four mounting bolts. Two must be removed from beneath the vehicle. The three front bracket mounting bolts are reached by moving the compressor outboard.
10. Lift out the compressor.
11. Install field coil and connect the wire.
12. Re-install compressor.
13. Re-install compressor-to-engine bracket.
14. On V-8 models, install hose retaining bracket and hoses on engine.
15. If heater hoses were disconneced, re-connect them.
16. Install clutch assembly and mounting bolt. Energize the clutch and torque the clutch mounting bolt to factory specifications.
17. Install proper drive belt and tighten to recommended tension specifications.
18. Align and attach oil filler tube.
19. Using new gaskets, connect the service valves to the compressor. Tighten to proper specifications.
20. Leak test the compressor, evacuate the compressor, and charge the system as necessary.
21. Operate system for 15 minutes. Isolate the compressor (service valves front seated) and check oil level in the compressor.
22. Refill radiator with coolant. (A —15 degree F. permanent antifreeze level is advisable.)
23. Install air cleaner.
24. Install engine cover.

Evaporator Core
Removal and
Installation

Through 1975

1. Remove engine cover.
2. If vehicle has a back seat remove it and pull the mat or carpet out of the way.
3. The control housing cover should now be removed, the wiring harness disconnected and tied out of the way.
4. Remove register from floor.
5. Remove sensor tube from evaporator core.
6. Place a floor jack under the vehicle and align it under the evaporator case. Raise it until it just contacts the case. Detach hoses at the evaporator.
7. Working inside the vehicle, remove the rear air duct by removing the retaining screws.
8. Remove the 10 evaporator mounting bolts.
9. Remove front and rear evaporator retaining nuts (these must be reached from underneath the vehicle). Lower the jack while pulling the wire through the opening in the floor pan. Remove the evaporator from beneath the vehicle.
10. Remove expansion valve cover. Remove the cover seal.
11. Take the gasket off of the top of the case.
12. Take screws out of top rail, which positions the core to the case, and remove the rail.
13. Remove expansion valve insulation and remove expansion valve.
14. Remove 4 core-to-case mounting screws and lift core from case.
15. Install core in case using 4 retaining screws. Install core-positioning top rail.

16. Install and re-insulate the expansion valve.
17. Attach expansion valve cover.
18. Install a new evaporator case top gasket.
19. Put the case on the floor jack and raise it back into position against the floor pan. Install front and rear mounting nuts. Route the wiring through the floor and connect the hoses at the evaporator and lower the jack.
20. Install the 10 evaporator mounting bolts.
21. Re-install thermostatic sensor tube into evaporator core.
22. Install air duct and register.
23. Re-install carpet, mat, and seat if vehicle is so equipped.
24. Connect control housing wires and install the housing cover.
25. Install engine cover.

1976-78

1. Disconnect electrical leads from blower resistor.
2. Disconnect vacuum line from fresh air-recirc. vacuum motor.
3. Remove A/C blower cover.
4. Remove push nut and washer from fresh air-recirc. door shaft.
5. Remove control cable from bracket.
6. Remove T/C blower housing (9 screws).
7. Partially drain cooling system.
8. Remove heater hoses from core.
9. Discharge refrigerant.
10. Remove the battery.
11. Disconnect the suction line and liquid line from the evaporator core and expansion valve.
12. Remove the evaporator drain hose.
13. Remove the evaporator assembly (5 screws)
14. Remove the blend door and housing.
15. Remove the insulation from the expansion valve.
16. Remove the evaporator tube retainer and seal and remove the core.
17. Installation is the reverse of the above. Use new O-rings coated with clean refrigerant oil. Evacuate, charge and leak test the system.

Condenser Removal and Installation

1. Remove hood lock assembly mounting bracket.
2. Purge the system of refrigerant.
3. Detach both refrigerant lines at the receiver tank and condenser. Move the hoses aside and out of the way.
4. Remove the four condenser-to-radiator retaining bolts and remove the condenser and receiver.
5. Remove the receiver-dehydrator

from the condenser assembly if it is to be replaced.
6. Position the condenser assembly, with receiver-dehydrator attached, and install the condenser-to-radiator mounting bolts.
7. Position and connect refrigerant lines.
8. Install hood lock mounting bracket.
9. Evacuate, charge, and leak test the system.

Expansion Valve Removal and Installation

1. Purge the system of refrigerant.
2. Raise the vehicle on a lift.
3. Partially drain coolant system if vehicle is equipped with an auxiliary heater. Remove heater hoses.
4. Remove auxiliary heater-to-floor retaining nuts and push the heater unit up and aside.
5. Detach two refrigerant hoses at the expansion valve.
6. Detach expansion valve cover and seal from evaporator housing.
7. Remove expansion valve insulation.
8. Remove expansion valve.
9. Attach expansion valve to evaporator.
10. Re-insulate expansion valve.
11. Position and install expansion valve seal and cover on the evaporator.
12. Connect the refrigerant lines to the expansion valve.
13. Re-install auxiliary heater and hoses if vehicle is so equipped.
14. Bring the vehicle down and add coolant as required.
15. Evacuate, charge, and leak test the system.

Blower Motor Removal and Installation

1. On 1971-75 models, evaporator assembly must be removed. For

1976-78 models, follow steps 2 & 3. (See "Evaporator Core" removal part of this section.)
2. Now the blower motor wires may be disconnected.
3. Remove blower motor retaining screws and remove the blower motor from the case.
4. Install blower motor in the cage and re-install the retaining screws.
5. The blower motor wires may now be connected.
6. Install evaporator assembly. (See "Evaporator Core" installation part of this section.)

Thermostatic Switch Removal and Installation

1. Fold carpet or mat aside and remove the evaporator register.
2. Extract sensor tube from evaporator core.
3. Take off control housing cover and remove switch lever-to-switch Allen head retaining screw.
4. Disconnect thermostatic switch wires and remove the switch retaining screw.
5. Remove the thermostatic switch and tube assembly.
6. Align thermostatic tube and switch assembly, routing the sensor tube back into the evaporator core. Install switch and connect wires.
7. Install switch lever and control housing cover.
8. Install the floor register and reposition the mat or carpet.
9. Check system operation.

Auxiliary Evaporator Removal and Installation

Through 1975

Auxiliary evaporator location on bus type models is under the first bench seat. The auxiliary evaporator,

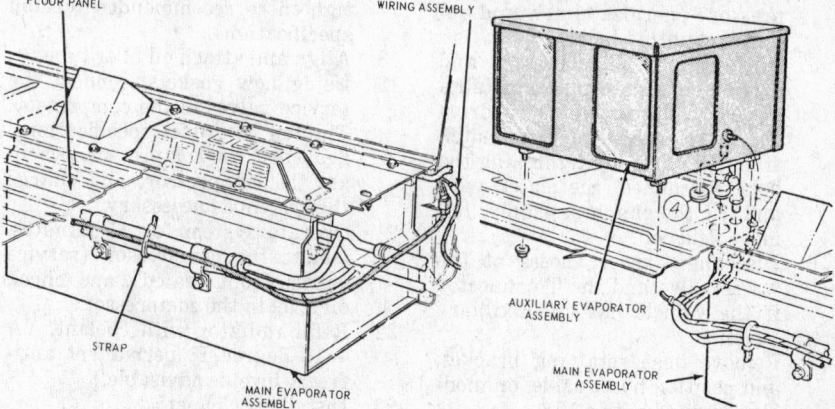

FLOOR PANEL WIRING ASSEMBLY

STRAP

MAIN EVAPORATOR ASSEMBLY

AUXILIARY EVAPORATOR ASSEMBLY

MAIN EVAPORATOR ASSEMBLY

Auxiliary evaporator assembly—Econoline (© Ford Motor Co.)

on van models, is located just forward of the left rear wheel housing.

The same control levers operate both the auxiliary and the main evaporator assemblies.

1. Remove the bench seat and purge refrigerant from the system.
2. Raise the vehicle on a lift. (If a jack is used, to raise the vehicle, install safety stands.)
3. Working under the vehicle, detach the wires harness and the suction and liquid lines.
4. Remove evaporator-to-floor pan retaining nuts.
5. Lower vehicle and remove the evaporator assembly.
6. For installation reverse the above procedure.
7. Evacuate, charge and leak test the system.

1976-78

1. Remove the first bench seat.
2. Remove the auxiliary heater A/C unit cover assembly.
3. Discharge the system.
4. Disconnect the refrigerant lines from the core and expansion valve.
5. Remove the core and seal assembly with the expansion valve.
6. Install in reverse of the above. Evacuate, charge and leak test the system.

York Compressor
Warner Clutch
Removal and Installation

NOTE: On some models it is necessary to loosen the fan shroud so clutch assembly can be removed.

1. Energize the clutch. (Ignition key on accessory and air conditioning controls set in the ON position.)
2. Remove the clutch mounting bolt.
3. Install a 5/8-11 bolt in the threaded clutch drive shaft hole. Tightening this bolt will disengage clutch assembly from compressor crankshaft.
4. Disconnect the magnetic clutch wire.

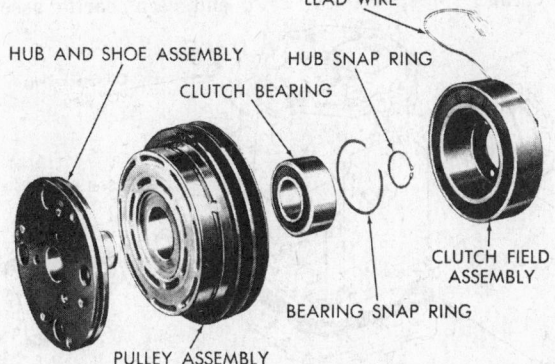

HUB AND SHOE ASSEMBLY HUB SNAP RING
CLUTCH BEARING
CLUTCH FIELD ASSEMBLY
BEARING SNAP RING
PULLEY ASSEMBLY

Warner magnetic clutch (© Ford Motor Co.)

5. Loosen and remove compressor belt.
6. Lift off clutch assembly.
7. Install the woodruff key on the compressor shaft.
8. Align the clutch assembly with the key when installing on the shaft.
9. Energize the clutch.
10. Install the attaching bolt and torque to 18-22 foot pounds.

Electrolock Clutch
Removal and Installation

1. Disconnect the hot wire lead.
2. Remove the pulley assembly as follows:
   a. Remove the 5/16" Nylock capscrew.
   b. Screw a 5/8" N/C (Coarse thread) capscrew into the hub and force the assembly free.
3. Remove the stationary field as follows:
   a. Remove four 1/4" machine screws.
   b. Slide the field off.
4. Place the stationary field in the compressor.
5. Install four, 1/4" machine screws from the field to the compressor hoses. Torque to 8 foot pounds.

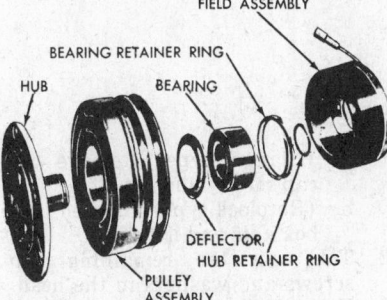

FIELD ASSEMBLY
BEARING RETAINER RING
HUB   BEARING
DEFLECTOR
HUB RETAINER RING
PULLEY ASSEMBLY

Electroloc magnetic clutch (© Ford Motor Co.)

6. Wipe the compressor shaft clean.
7. Align the pulley and hub assembly on the compressor shaft so that the woodruff key is fully seated in the shaft key slot and the top of the key is parallel to the shaft taper.

8. Slide the pulley and hub assembly onto the compressor shaft.
9. Fasten the pulley and hub assembly to the compressor shaft using a 5/16" Nylock cap screw and a large diameter washer. Tighten to 15-22 ft. lbs.
10. Rotate the assembly to check installation.
11. Connect the lead wire to the electrical circuit.

Clutch Bearing Replacement
Removal and Installation

1. Remove the clutch assembly from the compressor shaft.
2. Turn the clutch assembly face down, and remove the external bearing retainer from the rear of the drive plate shaft.
3. Support the clutch, still face downward, on the outer edges, so as to allow clearance for the removal of the drive plate.
4. Insert a 5/8-11 bolt into the center to the drive plate shaft, and hand tighten it. Gradually apply just sufficient pressure on the bolt to free the shaft from the bearing. Remove the drive plate assembly.
5. Carefully inspect the drive face plate for warping or breakage. If damage is evident, the entire clutch assembly must be replaced.
6. Support the pulley face up in a manner that will allow the bearing to be driven out without interference.
7. Secure a plug or other smooth instrument that will allow force to be applied to the inner race of the bearing without any contact with the metal grease catcher. Gently force the bearing out of the pulley.
8. Inspect the new bearing for cleanliness, and clean if necessary.
9. Turn the pulley face down, and support it on its front surface. Do not use a vise to support the pulley by the edges, as this may result in bending.
10. Press the new bearing into the pulley, using force only on the outer race of the bearing. Force cannot be applied to the inner race, as the bearing is not designed for this type of stress. Replace internal bearing retainer.
11. Turn the pulley face up, and support it by the inner race of the bearing. Screw a 5/8-11 bolt into the front of the drive plate, and apply pressure to the bolt to force the drive plate shaft into the inner race of the bearing.
12. Replace the external bearing retainer.

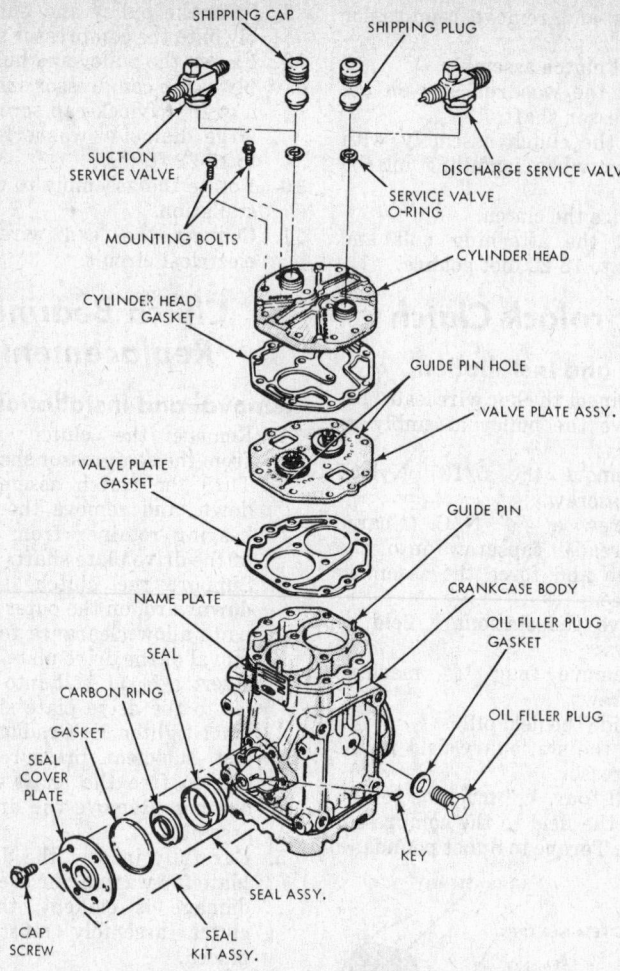

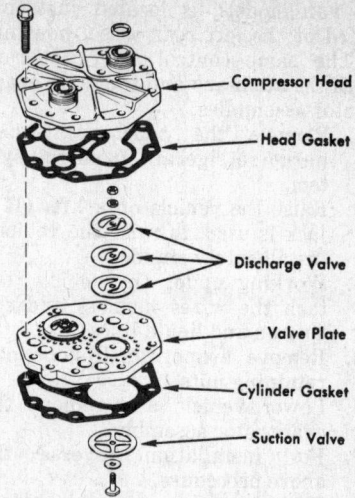

Head and valve plate assembly sequence
(© Ford Motor Co.)

York compressor disassembled
(© Ford Motor Co.)

6. Remove the discharge tube and suction screen assembly from the service valve ports by forcing their entire length up through the head and out the top. Inspect cylinder bores, and replace compressor if they have been scored by valve fragments.

Piston and Connecting Rod

1. Remove the compressor from the engine compartment as described above.
2. Remove the base plate, head and valve plate assemblies (see above).
3. Mark the rod, cap and crankshaft throw with a center punch as a guide for reinstalling.
4. Coat all bearings surfaces with refrigerant oil.
5. Take out the connecting rod bolts.
6. Remove the cap screws.
7. Force the piston upward.
8. Tap out the wristpins.
9. Remove the pistons.

Crankshaft and Seal End Main Bearing

1. Remove the compressor from the engine compartment as described above.
2. Strip down the compressor to crankcase, crankshaft and seal end main bearing assembly.

13. Test the rotation of the pulley on the drive plate shaft. The pulley should rotate freely and without looseness or play.
14. Reinstall the clutch onto the compressor shaft.

Compressor Disassembly

Seal Assembly

1. Remove the clutch and the key from the compressor shaft. Isolate the compressor.
2. Take out the capscrews from the seal plate.
3. Pry the seal plate loose. While doing this, hold a hand under the seal housing to catch the carbon ring, if it should fall.
4. Take the seal assembly from the shaft by prying behind the drive ring (the farthest back on the shaft of the assembly). Do not scratch or burr parts.
5. Remove all old gasket material. Make sure the shaft, seal plate, and gasket surfaces are clean.

Head and Valve Plate

1. Isolate the compressor. Remove the service valves.

a. (Flanged-type) Remove the cap screws and lift off.
b. (Rotolock-type) Loosen the hex nuts and lift off.
2. Take out the remaining cap screws and washers in the head.
3. Take off the valve plate and head by tapping or prying under the ears extending from the valve plate.
4. Separate the head and valve plate by holding the head and tapping the valve plate ears with a soft hammer.
5. Carefully remove all gasket material from parts without nicking or scratching.

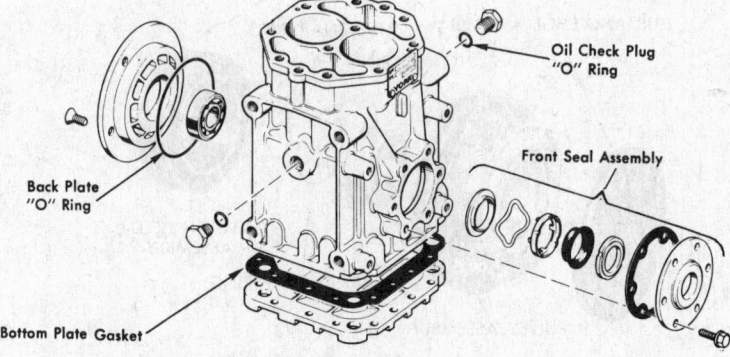

Compressor seal components and gaskets (© Ford Motor Co.)

3. Wash the compressor assembly of all grease and oil in solvent. Dry thoroughly.

4. Heat the assembly *in an oven* to 300 degrees F. Heat *uniformly* to avoid cracking.

5. Apply pressure (at 300 degrees) to the crankshaft and ball bearing assembly. These will push out easily at this temperature.

6. Set the crankshaft, vertically, flywheel end up, in a vise. Grip the shaft at the unmachined center throw.

7. Pry under the race of the bearing with two large screwdrivers and force the bearing upward from the shaft.

Compressor Assembly

Crankshaft and Bearing

1. Be sure that the shaft, the bearing, and the crankcase recess are clean and free from burrs.

2. Heat the crankcase to 300 degrees F, *uniformly* in an oven.

3. Place the bearing in the crankcase recess through the opening in the bottom. Apply force, if needed to the outer race to ensure seating in the bottom of the recess.

4. Let the crankcase cool. Insert the flywheel end of the crankshaft into the inner race of the ball bearing, entering through the rear bearing cover plate opening.

5. Position the crankcase so that it is completely supported on the inner race of the ball bearing; then push the shaft until the cheek of the shaft contacts the inner bearing race.

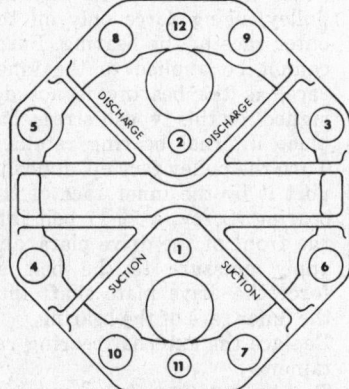

Compressor head capscrew tightening sequence (© Ford Motor Co.)

Piston and Connecting Rod

1. Coat all bearing surfaces with refrigerant oil.

2. Attach connecting rod to the piston.

3. Insert the assembly into the cylinder, so oriented that the wrist-pin roll pin is toward the center of the compressor.

4. Mate the rod, cap, and crankshaft throw. If reusing a connecting rod, realign with the center punch marks made during disassembly. If a new rod is used, line up the vertical match marks cast into the cap and into the rod proper.

5. Insert the connecting rod cap screws and torque to 90-100 inch pounds.

6. Replace the base plate. Replace valve plate assembly, as below.

7. Replace compressor as described above.

Head and Valve Plate

1. Apply a thin film of refrigerant oil to the area of the crankcase to be covered by the cylinder gasket.

2. Place the cylinder gasket on the cylinder, with the dowel pins of the crankcase in the corresponding holes in the gasket.

3. Apply a thin film of refrigeration oil to the top and bottom valve plate areas to be covered by gaskets.

4. Set the valve plate on the cylinder gasket so that the discharge valve assemblies (the smaller diameter assemblies) are facing up, and the dowel pins are in the corresponding holes in the valve plate.

5. Position the head gasket on the valve plate so that the dowel pins are in the corresponding holes in the gasket.

6. Apply a thin film of refrigeration oil to the machined mating surface of the cylinder head.

7. Position the cylinder head on the head gasket with the dowel pins in the corresponding holes in the head.

8. Insert the plain discharge tube into the service valve port marked with "D" and "Disch".

9. Insert the suction screen into the service port marked with an "S" and "Suction". Tap lightly into place until flush.

10. Flange Type Service Valves.
   a. Apply a thin film of refrigeration oil to the valve flanges.
   b. Position a service valve gasket on the cylinder head service valve flanges.
   c. Position the service valve on the proper service valve ports.
   d. Insert the four longer cap screws through the service valve mounting pads, the head, the valve plate and into the crankcase.
   e. Insert the remaining head cap screws with their washers and run in until the heads make contact.
   f. Torque all cap screws to 14-18 foot pounds. Tighten service valve screws first. See illustration.
   g. Retorque all cap screws to 14-18 foot pounds after ½ hour. Replace compressor as described above.

11. Rotolock Service Valves.
   a. Insert the head cap screws and washers and run in until the heads make contact.
   b. Torque the cap screws to 14-18 foot pounds.
   c. Wet the two service valve gaskets with refrigerant oil and place them on the Rotolock head valve bosses.
   d. Screw on the two Rotolock valves on the appropriate bosses.
   e. Retorque all cap screws to 14-18 foot pounds after ½ hour. Replace compressor as described above.

Seal Assembly

1. Remove the old gasket.

2. Clean and lubricate crankshaft surface and new seal with clean refrigerant oil.

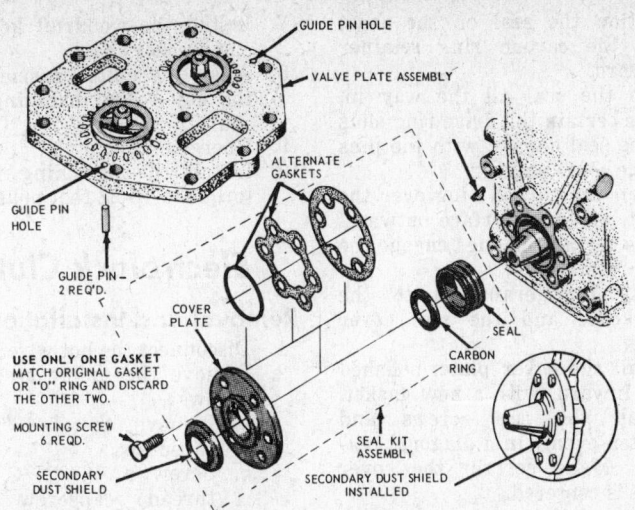

York compressor valve plate and crankshaft seal installation (© Ford Motor Co.)

Labels: GUIDE PIN HOLE, VALVE PLATE ASSEMBLY, ALTERNATE GASKETS, GUIDE PIN HOLE, GUIDE PIN 2 REQ'D., COVER PLATE, SEAL, CARBON RING, USE ONLY ONE GASKET — MATCH ORIGINAL GASKET OR "O" RING AND DISCARD THE OTHER TWO., MOUNTING SCREW 6 REQD., SEAL KIT ASSEMBLY, SECONDARY DUST SHIELD, SECONDARY DUST SHIELD INSTALLED

3. Position the seal on the shaft with the carbon ring retainer outward.
4. Push the seal all the way in, make certain the drive ring slots in the seal engage with the pins on the shaft bearing.
5. Install the carbon ring over the shaft, polished surface outward. Make certain the lugs engage the ring retainer.
6. Apply refrigerant oil to the crankcase and the seal cover plate.
7. Install the cover plate, polished side inward, with a new gasket. Install the cap screws and tighten evenly in a diagonal pattern. Make certain the cover plate is centered.
8. Replace compressor and clutch as described above.

Tecumseh Compressor
Warner Clutch

Removal and Installation

NOTE: On some models it is necessary to loosen the fan shroud so clutch assembly can be removed.

1. Energize the clutch. (Ignition key on accessory and air conditioning controls set in the ON position.)
2. Remove the clutch mounting bolt.
3. Install a ⅝-11 bolt in the threaded clutch drive shaft hole. Tightening this bolt will disengage clutch assembly from compressor crankshaft.
4. Disconnect the magnetic clutch wire.
5. Loosen and remove compressor belt.
6. Lift off clutch assembly.

7. Install the woodruff key on the compressor shaft.
8. Align the clutch assembly with the key when installing on the shaft.
9. Energize the clutch.
10. Install the attaching bolt and torque to 18-22 foot pounds.

Electrolock Clutch

Removal and Installation

1. Disconnect the hot wire lead.
2. Remove the pulley assembly as follows:
   a. Remove the 5/16" Nylock capscrew.
   b. Screw a ⅝" N/C (Coarse thread) capscrew into the hub and force the assembly free.
3. Remove the stationary field as follows:
   a. Remove four ¼" machine screws.
   b. Slide the field off.
4. Place the stationary field in the compressor.
5. Install four, ¼" machine screws from the field to the compressor hoses. Torque to 8 foot pounds.
6. Wipe the compressor shaft clean.
7. Align the pulley and hub assembly on the compressor shaft so that the woodruff key slot and the top of the key is parallel to the shaft taper.
8. Slide the pulley and hub assembly onto the compressor shaft.
9. Fasten the pulley and hub assembly to the compressor shaft using a 5/16" Nylock cap screw and a large diameter washer. Tighten to 15-22 ft lbs.
10. Rotate the assembly to check installation.

11. Connect the lead wire to the electrical circuit.

Clutch Bearing Replacement

1. Remove the clutch assembly from the compressor shaft.
2. Turn the clutch assembly face down, and remove the external bearing retainer from the rear of the drive plate shaft.
3. Support the clutch, still face downward, on the outer edges, so as to allow clearance for the removal of the drive plate.
4. Insert a ⅝-11 bolt into the center to the drive plate shaft, and hand tighten it. Gradually apply just sufficient pressure on the bolt to free the shaft from the bearing. Remove the drive plate assembly.
5. Carefully inspect the drive face plate for warping or breakage. If damage is evident, the entire clutch assembly must be replaced.
6. Support the pulley face up in a manner that will allow the bearing to be driven out without interference.
7. Secure a plug or other smooth instrument that will allow force to be applied to the inner race of the bearing without any contact with the metal grease catcher. Gently force the bearing out of the pulley.
8. Inspect the new bearing for cleanliness, and clean if necessary.
9. Turn the pulley face down, and support it on its front surface. Do not use a vise to support the pulley by the edges, as this may result in bending.
10. Press the new bearing into the pulley, using force only on the outer race of the bearing. Force cannot be applied to the inner race, as the bearing is not designed for this type of stress. Replace internal bearing retainer.
11. Turn the pulley face up, and support it by the inner race of the bearing. Screw a ⅝-11 bolt into the front of the drive plate, and apply pressure to the bolt to force the drive plate shaft into the inner race of the bearing.
12. Replace the external bearing retainer.
13. Test the rotation of the pulley on the drive plate shaft. The pulley should rotate freely and without looseness or play.
14. Reinstall the clutch onto the compressor shaft.

Compressor Disassembly

Seal Assembly

1. Isolate the compressor, and re-

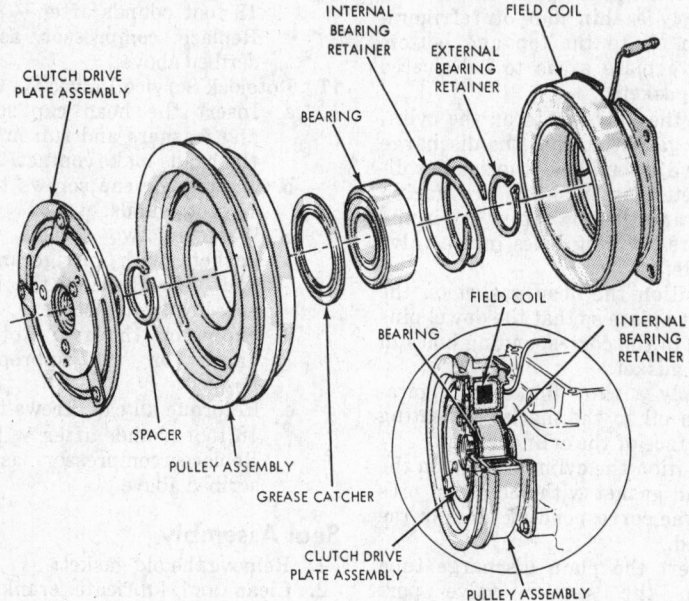

A/C clutch assembly (© Ford Motor Co.)

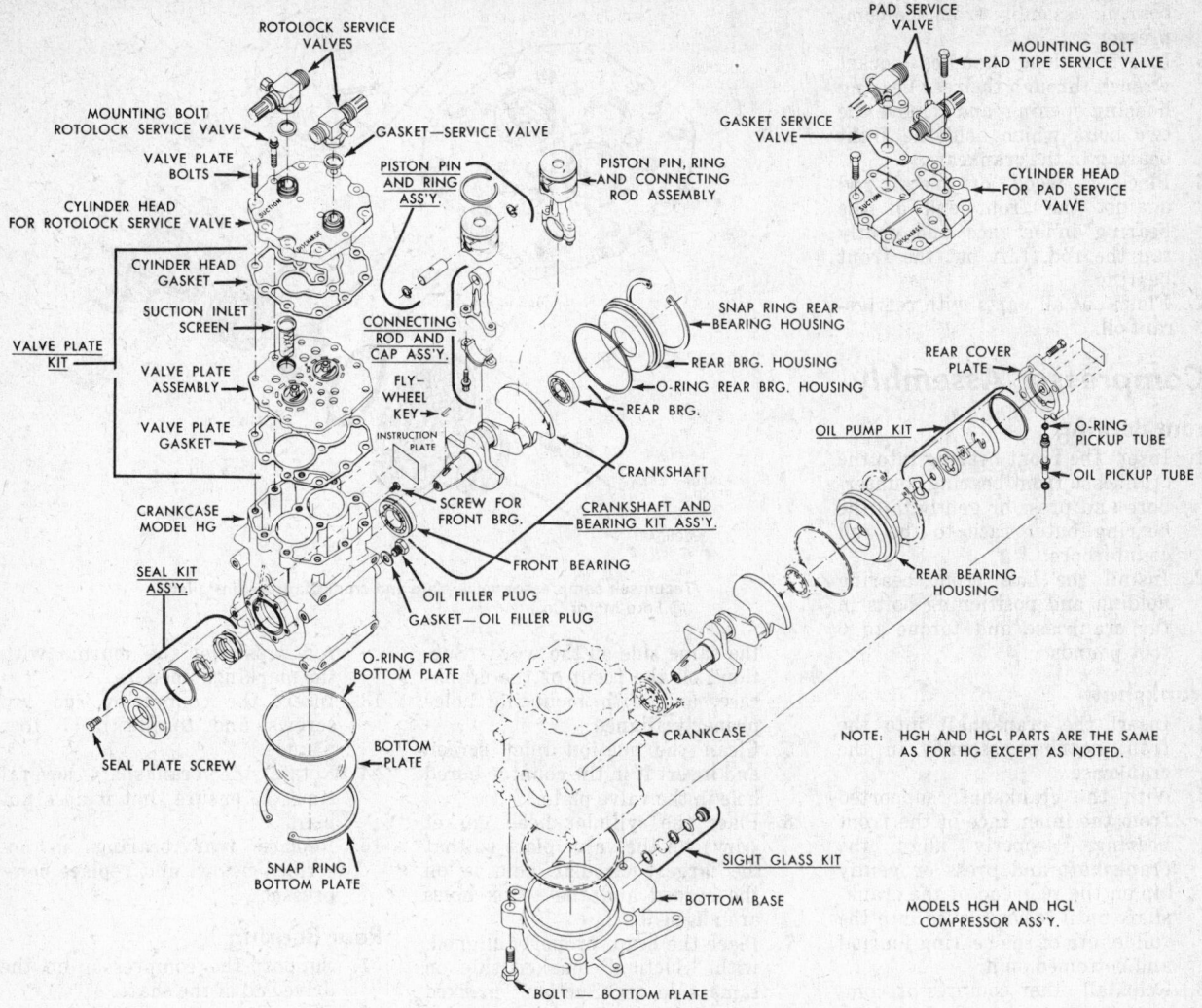

Exploded view—Tecumseh compressor—typical (© Ford Motor Co.)

Labels on the diagram:

ROTOLOCK SERVICE VALVES

PAD SERVICE VALVE

MOUNTING BOLT PAD TYPE SERVICE VALVE

MOUNTING BOLT ROTOLOCK SERVICE VALVE

GASKET—SERVICE VALVE

PISTON PIN AND RING ASS'Y.

PISTON PIN, RING AND CONNECTING ROD ASSEMBLY

GASKET SERVICE VALVE

VALVE PLATE BOLTS

CYLINDER HEAD FOR ROTOLOCK SERVICE VALVE

CYLINDER HEAD FOR PAD SERVICE VALVE

CYINDER HEAD GASKET

SUCTION INLET SCREEN

CONNECTING ROD AND CAP ASS'Y.

SNAP RING REAR BEARING HOUSING

REAR BRG. HOUSING

O-RING REAR BRG. HOUSING

REAR BRG.

REAR COVER PLATE

VALVE PLATE KIT

VALVE PLATE ASSEMBLY

FLY-WHEEL KEY

INSTRUCTION PLATE

OIL PUMP KIT

O-RING PICKUP TUBE

OIL PICKUP TUBE

VALVE PLATE GASKET

CRANKSHAFT

SCREW FOR FRONT BRG.

CRANKSHAFT AND BEARING KIT ASS'Y.

REAR BEARING HOUSING

CRANKCASE MODEL HG

FRONT BEARING

SEAL KIT ASS'Y.

OIL FILLER PLUG

GASKET—OIL FILLER PLUG

O-RING FOR BOTTOM PLATE

NOTE: HGH AND HGL PARTS ARE THE SAME AS FOR HG EXCEPT AS NOTED.

SEAL PLATE SCREW

BOTTOM PLATE

CRANKCASE

SNAP RING BOTTOM PLATE

SIGHT GLASS KIT

BOTTOM BASE

MODELS HGH AND HGL COMPRESSOR ASS'Y.

BOLT — BOTTOM PLATE

---

move the drive belt and clutch.

2. Clean the seal plate and adjoining surfaces.
3. Take out the six bolts in the seal plate.
4. Gently pry the seal plate loose.
5. Pry behind the drive ring and take off the carbon nose and spring assembly.
6. With long nose pliers, pull on the edge of the grommet and remove the rubber seal around the shaft.

Valve Plate

1. Isolate the compressor. Remove the service valves.
2. Remove all bolts from the cylinder head.
3. Tap upward lightly with a fiber hammer on the overhanging edge of the valve plate.
4. Lift off the valve plate and cylinder head assembly.
5. Clean the surface of the cylinder head and cylinder face of all gasket material and dirt.

Rear Bearing

1. Remove the compressor as described above.

2. Set the compressor in an upright position.
3. Clean the rear and base cover plate and adjoining surfaces.
4. Drain the compressor or oil.
5. Remove the rear bearing housing snap ring, using snap ring pliers.
6. Using a slide hammer on the end plate remove the cover plate.
7. Take off and discard the rectangular-sectioned rear bearing cover plate "O" ring.
8. Insert two offset screwdrivers diametrically apart under the inner edge of the bearing outer race, pry evenly on both sides, and force off the rear bearing.
9. Turn the compressor upside down.
10. With snap ring pliers, remove the base plate snap ring.
11. Insert the end of the screwdriver into the spot-welded bracket in the plate and lift the plate out.
12. Take off the rectangular-sectioned base plate sealing "O" ring and discard it.
13. Thoroughly clean the compressor with solvent and dry it thoroughly.

Piston and Connecting Rod

1. Remove the compressor as described above.
2. Remove the seal assembly, cylinder head and valve plate, and rear bearing assembly.
3. Remove the four connecting rod cap screws and lift off the two connecting rod caps. Identify the removed caps so that in reassembly they can be used on their original connecting rods and in the same position.
4. Push the connecting rods, pistons and rings up and out of the crankcase.

Front Bearing and Crankshaft

1. Remove the compressor as described above.
2. Remove the seal assembly, cylinder head and valve plate assembly, pistons and connecting rods.
3. Supporting the crankcase from the rear bearing end, gently tap the front bearing end of the crankshaft toward the compressor until the crankshaft has moved out of the front bearing.
4. Lift out the crankshaft and rear

bearing assembly from the compressor.

5. Insert a long handled socket wrench through the rear bearing housing opening, and remove the two bolts which hold the front bearing in the crankcase.

6. Place a wood or metal rod against the front end of the bearing inner race and gently tap the rod. Lift out the front bearing.

7. Flush out all parts with refrigerant oil.

Compressor Assembly

Front Bearing

1. Insert the front bearing into the crankcase front bearing counterbore and press or gently tap the bearing outer race to the full counterbore.

2. Install the two front bearing holding and positioning bolts in the crankcase and torque to 6 foot pounds.

Crankshaft

1. Insert the crankshaft into the front bearing assembly in the crankcase.

2. With the crankshaft supported from the inner race of the front bearing, properly align the crankshaft and press or gently tap on the rear end of the crankshaft until it has moved into the full length of the bearing journal and bottomed on it.

3. Reinstall other compressor components as described below. Reinstall compressor.

Connecting Rod

1. Properly locate the connecting rods in the pistons and push the piston pins in until they rest against the installed snap rings.

2. Install the snap rings, using application pliers.

3. Replace piston, valve plate, and rear bearing as described below. Reinstall compressor.

Piston and Valve Plate

1. Place the piston, ring, and connecting rod assembly, without the connecting rod cap, in the cylinder bore, oriented so that the connecting rod bearing lines up with the crankshaft journal.

2. Using a piston ring compressor tool, compress the ring against the piston and complete insertion of the piston pin, ring and seal assembly into the cylinder bore.

3. Place the valve gasket (dry) on the cylinder face, properly located.

4. Locate the valve plate assembly on the valve plate gasket so that the letter "S" is visible and on

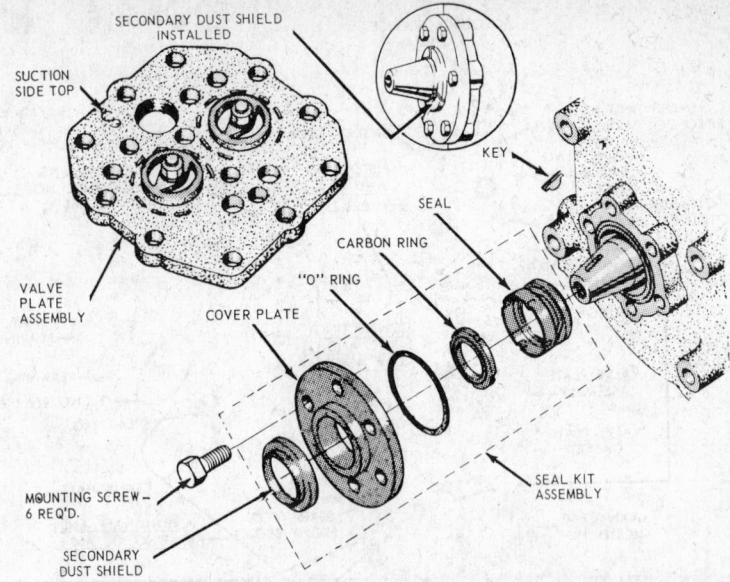

Tecumseh compressor valve plate and crankshaft seal installation (© Ford Motor Co.)

the same side as the word "Suction" on the front of the crankcase, and with mounting holes properly aligned.

5. Clean the suction inlet screen and insert it in the counter-bored hole in the valve plate.

6. Place the cylinder head gasket (dry) on the valve plate so that the largest circular hole is on the screen and the other holes are aligned.

7. Place the head, properly aligned, with "Suction" marked side on same side as "Suction" marked side of the crankcase.

8. Insert the head bolts and tighten finger-tight.

9. Torque the bolts in sequence to 22-24 foot pounds.

10. Turn the compressor over to rest on its head.

11. Place the connecting rod bearings around the crankshaft journal.

12. Properly align the connecting

rod caps over the journal with the markings lined up.

13. Insert the connecting rod cap screws and torque to 7 foot pounds.

14. Rotate the crankshaft several times to ensure that it does not bind.

15. Replace rear bearing as described below, and replace compressor.

Rear Bearing

1. Support the compressor on the drive end of the shaft.

2. Place the rear bearing over the end of the shaft.

3. Press on the inner race of the bearing until the inner race rests against the shaft shoulder.

4. Place a rectangular-sectioned "O" ring in counter-bored hole for rear bearing cover plate, seating it fully.

5. Align the rear bearing housing on top of the bearing, and, using

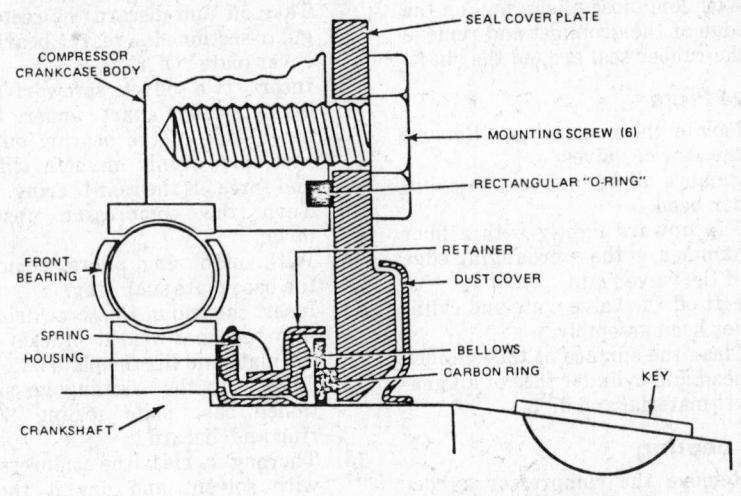

Compressor oil seal sectional view (© Ford Motor Co.)

a press, force it over the bearing and fully into the crankcase counter-bore.

6. Insert the snap ring into the crankcase groove, using snap ring pliers.
7. Turn the compressor until it rests on its rear head.
8. Insert a rectangular-sectioned "O" ring into the crankcase base groove.
9. Place the baseplate into the base counter-bore, with the bracket on the outside.
10. Insert the snap ring into the compressor base plate counterbore on top of the base plate, using snap ring pliers.
11. Press or hammer the base until the snap ring snaps into place.
12. Replace compressor as described above.

Seal Assembly

1. Remove the carbon shaft seal washer from the bellows seal assembly.
2. Coat the exposed surface of the crankshaft with refrigerant oil, and slip the bellows of the seal assembly and shaft washer in refrigerant oil.
3. Place the bellows seal assembly on the shaft, with the end for holding the shaft seal going on last. Push the bellows seal assembly past the shaft taper.
4. Assemble the shaft seal washer in the bellows seal assembly. Assemble the seal washer so that the raised rim is away from the bellows seal assembly and that the notches in the washers line up with the nibs in the bellows seal assembly. Wet the exposed surface of the shaft seal washer with clean refrigerant oil.
5. Place a rectangular-sectioned "O" ring in the crankcase mating surface.

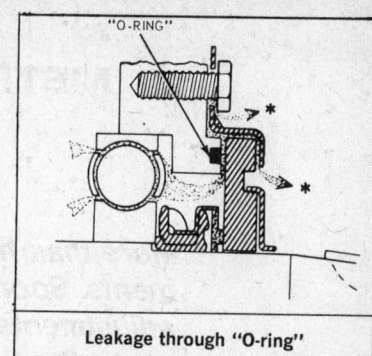

Leakage through "O-ring"

6. Set the front seal plate over the shaft, properly aligned.
7. Push the plate against the crankcase, with hands on both sides.
8. Insert the first two bolts diametrically opposed, preventing the seal plate from popping up.
9. Insert the six cap screws in circular sequence, and torque to 6-10 foot pounds.

CURRENT DRAW LIMITS @ 12.8 VOLTS

Vehicle	A/C Blower Motor	Condenser Fan Motor	Compressor Clutch
Light Truck	18.0 Max. Amps.	—	3.75 Max. Amps.
L-Series	Std. 16.0 Max. Amps.	6.0 Max. Amps.	3.75 Max. Amps.
	High Performance 20.00 Max. Amps.		
C-Series	5 to 5-1/2 Amps.	6.0 Max. Amps.	3.75 Max. Amps.
W-Series	5 to 5-1/2 Amps.	6.0 Max. Amps.	3.75 Max. Amps.
Econoline	19.8 Max. Amps.	—	3.75 Max. Amps.

REFRIGERANT CAPACITIES (REFRIGERANT — 12)

Vehicle	Capacity (Pounds)
Light Truck	2
L-Series	3-1/2
C-Series	2-1/2
W-Series	3-3/8
Econoline	3-1/2 Main
Econoline	4-1/4 Auxiliary

AIR CONDITIONING COMPRESSOR

Torque Limits (Ft-Lbs)		
Description	Tecumseh	York
Cylinder Head	20-24	15-23
Front Seal Plate	54-78 In-Lb.	7-13
Service Valve (Tube-O)	20+10	20+10
Mounting Bolt	20-30	20-30
Oil Filler Plug	18-22	4-11
Clutch Mounting	20-30	20-30
Base Plate	—	14-22
Back Plate	—	9-17
Clutch Brush or Coil	6-12	6-12
Quick Disconnect	10-15	10-15

COMPRESSOR OIL CAPACITIES ①

Description	Vertical	Horizontal
Tecumseh③	7/8 Inch Min.	7/8 Inch Min.
(11 Fluid Ounces)	1-3/8 Inch Min.	1-5/8 Inch. Max.
York 3³	7/8 Min.	13/16 Inch Min.
(10 Fluid Ounces)	1-1/8 Inch. Max.	1-3/16 Inch Max.
Driven Belt Tension (Between Fan Pulley and Air Conditioner Compressor):		
New	140 lbs.	
Used②	110 lbs.	
Minimum②	90 lbs.	
Belt Tension Tool	T63L-8620-A	
Compressor Clutch Run-Out	1/32 Inch. Max.	

① Use Specification ESA-M2C31-A, Ford Part No. C9AZ-19577-A or equivalent.
② Belt Operated for a Minimum of 10 Minutes is Considered a Used Belt.
③ Do not add oil if dip stick indicates proper level of oil between minimum and maximum. If dip stick is below minimum level, add oil up to minimum oil level only.

METRIC CONVERSION BASICS

More than half the cars built today use metric measurements. Soon automotive service procedures will involve adjustments to a metric specification. It's time to understand the SI metric system and how it works.

A HALF-INCH wrench is too small and a nine-sixteenths wrench is too big. The bold head doesn't look abused. Yet it's obvious that something is wrong. What is it? You've just become involved in international affairs! The bolt is metric and your wrenches aren't. And this mismatch of need and capability is going to become increasingly common in professional automotive service shops over the next few years.

All the rest of the world is either using the SI metric system or is in the process of converting to it. While America is not officially metric yet, it's a question we will face in the future. And like most questions, there are pros and cons.

Experts are on record as saying we must convert our inches-feet system to the metric system with all possible haste. Other experts say let's convert, but let's not be hasty. Even such a top-level authority on the subject as John T. Benedict, Manager of Technical Information, Engineering & Research Office, Chrysler Corp., agrees we should change to the metric system. But he votes for a change rate that could take as long as 50 years for the country to become predominantly metric.

The prime mover behind all the noise about metric numbers is the metric activity that's going on all over the world. The rest of the world combined produces more automotive vehicles per year than does the United States. That means more than half the cars built in the world today use metric measurements.

Also, we are no longer the hands-down leader when it comes to producing all kinds of manufactured goods. The European Common Market alone is gigantic; South American and Asian countries are developing their industrial capabilities at impressive rates. If we want to do business with the rest of the world, and world trade is a two-way street, our nuts had better fit their bolts. And our tools must fit both.

Will We Convert?

The question is really not one of whether we will give up our customary English system in favor of the metric system; it is more a question of when this change will occur.

Roy Trowbridge of General Motors supports Chrysler's Benedict in that a slow changeover would be best for the automotive industry—and probably the entire country. Trowbridge points up the fact that when a U.S. car maker sets up a production plant in a foreign country, the whole facility is based on the measurements of that country. Trowbridge calls this practical conversion, because it would be simply out of step to do it any other way. This makes sense, especially since our units of measurements are old fashioned, to say the least.

Ford is having much metric success with its Pinto. Its major components (engine and transmission) originally were built in Europe. You might call Pinto a hybrid because not all measurements on the vehicle are metric. In the case of Dodge's Colt since this car was made in Japan, metrics were used as a matter of course.

Now Ford's Lima, Ohio, assembly plant turns out Pinto engines which feature 100% metric specifications including nuts and bolts. In view of this, and Chevrolet's decision to build its new Chevette small car possibly with mixed SAE and metric dimensions and fasteners, the question isn't: Shall we convert to metrics? We already have. Then too, can we ignore the millions of imported cars that are in daily service in this country?

The metric system is here to stay in the worst possible way, at least for the professional automobile service technician. His tool box must start—if it hasn't already—

THE LANGUAGE OF METRICS

Quantity	Metric Expression	Application
Axle, Mass (weight) distribution per	kilograms per axle	Vehicle ratings, road regulations
Floor Loading	kilograms per square meter	Mass of material which can be safely stored on an existing structure or slab
Fuel Consumption	liter per 100 kilometers	Vehicle performance testing
Fuel Consumption specific	gram per megajoule	Engine testing
Oil Consumption	liters per 1000 kilometers grams per hour	Vehicle testing Engine testing
Oil Consumption specific	grams per megajoule	Engine testing
Oil Economy	kilometers per liter	Vehicle performance testing
Power ratio	kilowatts per metric ton	Vehicle specifications
	kilowgrams per kilowatt	Engine performance
Tire revolutions	revolutions per kilometer	Tire data
Torque	newton-meters	Engine torque, fastener tightening torque, steering torque, general testing
Tractive effort	newtons per kilogram	Vehicle specifications

The metric expressions shown above should be used as language and not be expressed in numbers or signs. Most of them relate to automotive usage, several of them to future automotive metric service expressions.

METRIC CONVERSION BASICS

harboring metric size wrenches as well as the inch and fractional units now most common.

Most mechanics on the line can flash a glance at a capscrew and tell if it is 7/16 in. or ½ in. Show them something close to it in a metric size however, and they will guess anything from a number 6 to a number 12.

Give a tune-up man a contact point gap setting in millimeters, and he will probably do the blank stare routine. Tell the front end man the wheel-base in centimeters and you'll likely draw another blank.

Tell a transmission man that all the nuts, bolts and capscrews used on the box are metric, and once again we find ourselves in never-never land.

What's the answer? We must learn to think metric, as well as know the system itself.

How It Works

The metric system is based on multiples of 10, like our money (100 cents = $1; or 10 pennies = 1 dime, 10 dimes = $1). The foundation of the metric system is the meter (which is a little more than 39 inches in length).

This meter is divided into 1000 parts, each part is called a millimeter. Do not confuse the millimeter with a millionth of a meter just because it sounds like a million. One millimeter is 1/1000th of a meter.

Now if we take 10 millimeters in a row, they equal 1 centimeter, and 100 centimeters equal 1 meter. To repeat:

- 10 millimeters = 1 centimeter
- 100 centimeters = 1 meter
- 1000 millimeters = 1 meter

Since one meter is actually 39.37 inches, a millimeter (mm) is .03937 inches. For all but the most precise work this is rounded off to .0394 in. And there are 25.4 millimeters in one inch.

Four Main Areas

The things that are measured in the customary English system (the one we use) and the metric system—or SI—(the one all other countries are or will be using) can be lumped into four main groupings:

- Length
- Area
- Volume
- Mass (Weight)

When we describe a vehicle as being 4.557 meters in length, powered by a 2000 cubic centimeter engine, with a mass weight of 1210 kilograms and with a front seat area of .83 square meter, we are using all four groups of SI measurements.

To describe the same vehicle in the English system, we would say it is 179.4 inches in length, powered by a 122 cubic inch displacement engine, weighing 2679 pounds, and has a front seat area of 1 square yard.

38" or 95 cm

23" or 57.5 cm

36" or 90 cm

5'5" or 162.5 cm

Metric Tables

METRIC CONVERSION BASICS

BASIC UNITS IN METRIC MEASUREMENT

Quantity	Name of Unit	Symbol
length	meter	m
mass (weight)	kilogram	kg
time	second	s
electric current	ampere	A
electric potential	volt	V
electric energy	kilowatt-hour	kW·h
road speed	kilometers per hr.	km/h
temperature	celsius	°C
torque	newton meters	Nm
luminous intensity	candela	cd

The quantity expressions shown above are some of the most common now used. Not all of these, however, are approved metric standards that are used in the SI measurement method. These are ones that will most likely be encountered in the future in many areas of the automotive service world. Remember, wherever you see a period (·) placed midway between the height of figures, it means the same as today's multiplication (×) sign. The decimal point (.) will continue to be positioned between figures at the bottom of a line.

As you can see, there is more to the metric system than picturing a meter being divided into 100 or 1000 equal parts. This is simply a measurement of length. One must know area, volume and weight (in SI terms this is now called mass), as well.

For most practical purposes, you won't be using any more terms or units of measurement that you now use. The most commonly used units of measure are:

to measure length   meter
       millimeter

METRIC FRACTIONS EXPRESSED AS PREFIXES

Prefix	SI Symbol	Multiplication Factors	Example of use
mega	M	$1\,000\,000 = 10^6$	**mega**watt—1 million watts
kilo	k	$1\,000 = 10^3$	**kilo**meter—1000 meters
hecto	h	$100 = 10^2$	**hecto**liter—100 liters
deka	da	$10 = 10^1$	**deka**gram—10 grams
deci	d	$0.1 = 10^{-1}$	**deci**liter—1/10th of a liter
centi	c	$0.01 = 10^{-2}$	**centi**meter—1/100th of a meter
milli	m	$0.001 = 10^{-3}$	**milli**liter—1/1000th of a liter
micro	μ	$0.000\,001 = 10^{-6}$	**micro**second—1 millionth of a second

Prefixes are used to express fractions of any given quantity or measurement in the SI method. Some of these terms many mechanics will be familiar with today—because they've been using them for a long time. For instance, imported cars measure their engine displacement in cubic centimeters—therefore the centimeter is a commonly used term right now, one you are probably quite familiar with. Power station electrical output is stated as so many megawatts, which when converted tells us how many millions of watts of electrical energy can be generated by a power plant.

Metric fasteners are frequently found on current automotive assemblies. The code carried on the fastener heads indicates: letter, the identity of the manufacturer; numbers, the ISO grade (tensile strength) of that fastener. The table on page 1497 shows the amount of torque recommended for securing metric fasteners according to the head code. For instance, a 14 mm 10K bolt can be torqued to approximately 117 ft./lbs. The torque values given in the four inside columns under Bolt Grade, are for lightly oiled bolts. Bolts that thread into aluminum usually require much less torque.

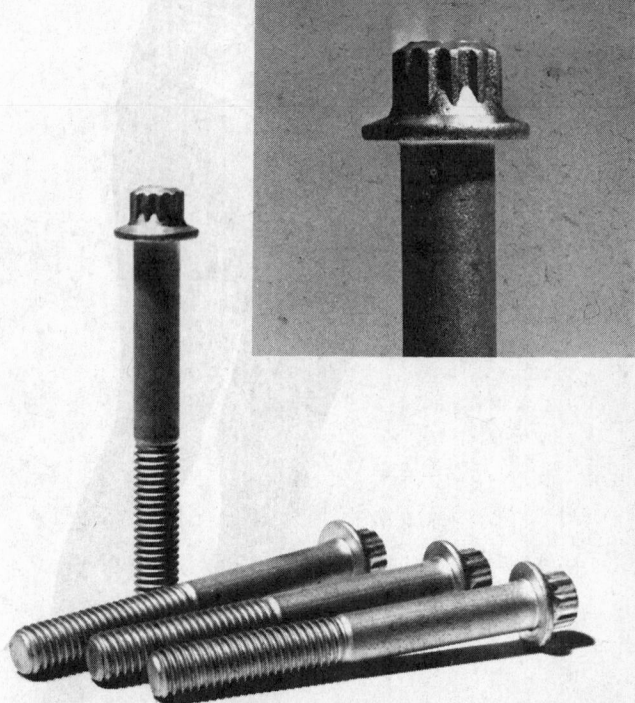

A new type of metric fastener may make its appearance soon. It is the OMFS (Optimum Metric Fastener System) design and it will feature a 12-spline head. These units may call for some new metric handtools in the mechanic's tool box, but most current 12-point metric hand tools should fit the new head, despite its radical new design. Principal advantage of the 12-point head design is to provide more tool contact surfaces during torquing on either applying or removing the fastener, over the current 6-point head design of bolts now in use.

Photos & fasteners from Standard Pressed Steel

METRIC CONVERSION BASICS

to measure weight (mass)	kilogram	
	gram	
to measure volume	liter	
	cubic	
	centimeter	
to measure area	square meter	
	square	
	milimeter	

There are other units, of course, but you're about as likely to see them in a professional manual as you are to see *rods* or *drams* in the manuals you use now.

In normal professional automotive service, we will be dealing often with such prefixes as milli (m) numerical value 0.-001; centi (c) numberical value 0.01; and kilo (k) numberical value 1000. For example, if the distance between town A and town B is 10,000 meters, it is expressed more simply as 10 kilometers.

If you scoffed at the earlier comment that it might be well to take upwards of 50 years to convert our English system to the metric system, maybe now you might be less critical and more sympathetic. As professional mechanics we can get by if we keep our eyes open, and our tool box well stocked. Then if we see a bolt head with metric symbols we simply grab a metric wrench or socket and go to work.

And we will learn metric sizes like we learned English sizes: work with them long enough and you will grab the right wrench at the right time. But will we do so well with area, weight (mass) and volume? Everyone else in the world seems to do it with no particular problem. So there's no reason why you shouldn't grasp it either.

Practical Applications

The most common measure of volume affecting automotive service dealers is the liter. Every time an imported car has its oil drained, cooling system serviced or gearbox oil changed, the refill quantity will likely be specified in liters. If the book says 4.5 liters of oil are needed when the filter is changed, how much is that? A quart is a bit less than a liter. How close would four quarts come to filling the crankcase? Five quarts? A quart equals .946l (liter). Four quarts adds to 3.784l—not enough to do the job. One more quart brings the total to 4.73l. That's less than half a pint over full, and close enough for most jobs.

Suppose the front wheel toe-in spec calls for 3 millimeters. Where should the gauge be set if it's calibrated in fractions of an inch? Since 1 mm equals approximately .039 in. the spec equals .117 in. An eighth of an inch has the decimal equiva-

lent of .125 in. Few could argue that ⅛ in. wouldn't be a reasonable setting.

The number that throws many profes-

sional mechanics for a complete loss is torque spec. Torque is stated as a new unit with a new name—newton meters, ab-

TERMS IN COMMON USE

QUANTITY	WORDS	SYMBOL	APPLICATION
ACCELERATION (linear)	meter per second squared	m/s²	General use
AERODYNAMIC resistance (drag)	newton	N	Drag at stated km/h
AREA	(hectare)	(ha)	Land area (equals 10,000m²)
	square meter	m²	Cargo platforms, vehicle frontal area, glass, fabrics.
	(square centimeter)	(cm²)	
	(square millimeter)	(mm²)	Brake & clutch linings, radiator area, small areas.
CAM PROFILE lift acceleration	(millimeter per degree squared)	(mm/deg²)	General use
CAM PROFILE lift rate	(millimeter per degree)	(mm/deg)	General use
DAMPING	newton-second per meter	N·s/m	Shock absorber data
DENSITY	kilogram per cubic meter	kg/m³	General use
ELECTRICAL CAPACITANCE	farad	F	General use
ELECTRICAL RESISTANCE	ohm	Ω	General use
ELECTRIC CURRENT	ampere	A	General use
ELECTRIC POTENTIAL	volt	V	General use
ENERGY	joule	j	Mechanical, nuclear, electrical, chemical.
	(kilowatt-hour)	(kW·h)	
FORCE	newton	N	Testing, calculations. Clutch, brake, springs.
FREQUENCY	hertz	Hz	Suspension frequency, electronics, acoustics.
	(kilohertz)	(kHz)	
	(megahertz)	(MHz)	
HEAT	joule	J	General use
HEAT TRANSFER, coefficient of	watt per square meter-kelvin	W/(m²·K)	Cooling system radiator capacity.
LENGTH	(kilometer)	(km)	Odometers, maps, etc.
	meter	m	Braking distance, turning radius, building dimensions.
	(milimeter)	(mm)	Product drawings
LUMINOUS FLUX	lumen	lm	Metric unit used in place of footcandle.
LUMINOUS INTENSITY	candela	cd	Bulbs
MASS	metric ton or	(t)	Freight mass (weights)
	(megagram)	(Mg)	
	kilogram	kg	Base SI unit; preferred for calculations; general use.
	(gram)	(g)	Small masses
	(milligram)	(mg)	Minute masses
MASS PER UNIT AREA	kilogram per square meter	kg/m²	Leather, cloth and surface coating.

Metric Tables

METRIC CONVERSION BASICS

breviated Nm.

One foot pound (ft/lb) equals 1.356 Nm. If the specs call for a cylinder head to be torqued to 50 Nm, the correct torque wrench setting in ft/lb is 37. An alternate metric term, kilogram/meters (Kg/m) are not to be used with SI units. They are part of the old gravimetric system.

Look at it this way:

A foot pound is bigger than a newton meter. Converting Nm to ft/lb will always produce a smaller number. As seen above, 50 Nm = 37 ft/lb.

(Non-SI terms in parenthesis)

QUANTITY	WORDS	SYMBOL	APPLICATION
MOMENT OF INERTIA	kilogram-meter squared	kg·m²	General use
MOMENT OF MASS	(kilogram-millimeter)	(kg·mm)	Static balance
MOMENTUM	kilogram-meter per second or	kg·m/s	General use
	newton-second	N·s	
POWER	watt	W	General use; light bulbs, alternate output, engine performance, etc.
	(kilowatt)	(kW)	Rational power.
PRESSURE	(kilopascal)	(kPa)	Tire, coolant, lubricating oil, fuel pump, engine compression, BMEP, manifold.
	(megapascal)	(MPa)	Hydraulic and air brakes, high pressure, hydraulic and air lines, grease guns.
ROAD/VEHICLE FORCE	newton	N	Vehicle dynamics, road and bridge design.
SPRING RATE	(kilonewton per meter)	(kN/m)	Linear stiffness
	(newton-meter per degree)	(N·m/deg)	Torsional stiffness
TEMPERATURE, CELSIUS	(degree Celsius)	(°C)	General use
THERMAL CONDUCTIVITY	watt per meter-kelvin	W/(m·K)	General use
TIME	second	s	The only proper SI unit
	(minute)	(min)	Alternate
	(hour)	(h)	Alternate
	(day)	(d)	Alternate
VEHICLES, Gross Vehicle Mass	kilogram	kg	Specifications, engineering computations
VELOCITY	meter per second	m/s	General use
	(kilometer per hour)	(km/h)	(road speed)
VELOCITY (rotational)	(revolutions per second)	(rev/s)	General use
	(revolutions per minute)	(rev/m)	General use
VELOCITY (volume)	cubic meter per second	m³/s	General use
	(liter per second)	(l/s)	Gas and liquid
VISCOSITY (dynamic)	pascal-second	Pa·s	Fluids
VISCOSITY (kinematic)	square meter per second	m²/s	Fluids
VOLUME	cubic meter	m³	Freight volumes, car luggage capacity.
	(litre)	(l)	Engine displacement
	(cubic decimeter)	(dm³)	General use
	(cubic centimeter)	(cm³)	General use
	(cubic millimeters)	(mm³)	General use

New Abbreviations, Signs

Some of the common abbreviated expressions used up to now in Europe, and frequently here, will change under the proposed standard terminology of the SI method. For instance, road speed stated here as mph (miles per hour) has been expressed in metric standard countries up to now as kph (kilometers per hour). Under the SI method, road speed will be expressed as kilometers per hour, but the standard abbreviation will be km/h.

Another such abbreviation we are accustomed to seeing in metric terms is cc to express engine displacement as being so-many cubic centimeters. The SI method calls for the use of cm³ to express cubic centimeters in its abbreviated form.

In the past when we want to show that two numbers must be multiplied, we have used × as the sign. Now, in SI style, multiplication signs will appear thus: 3·5 = 15. A period sign between figures now means these figures should be multiplied. No doubt there will be other such new ways of handling old indications with the SI system. And we'll have to learn these as we all progress through the SI change-over.

As you begin to become familiar with SI metrics, you may discover that it is not at all important to try relating what a metric measure is compared with the customary English now so well-known. The fact that a part may be 25 millimeters (mm) long really has no relationship to the fact that it is about an inch long. The same goes for 1 liter of antifreeze. It makes no difference that it is not quite a quart. It is expected that during the transition to SI metrics most problems will arise in this area of trying to relate the metric dimension to a known customary dimension. Better to concentrate on using the right tool to achieve a required metric dimension. The only significant difference is that the numeric quantities you'll be dealing with in SI will be somewhat larger than you've been used to in the English system.

Until professional automotive mechanics become fully adjusted to the new language, a lot of care will be needed. And reference charts will be a must. After living with SI for a while, we'll bet you'll find it will become second nature.

METRIC CONVERSION BASICS

Test Your Metric Know-How

No prizes or vacation trips if you get these right. But you will know if you understand what metrics are all about and how they'll relate to your activities as a professional automotive service mechanic.

1 What would you say the major advantage of the metric system is?
- A) It's based on the relationship of the Earth's tides to the Moon
- B) It's based on simple arithmetical multiples of 10s
- C) It has fewer numbers and signs to work with
- D) It has larger numbers to work with
- E) It's smaller number units are easier to understand
- F) It's easier to teach to school-age children

2 Metric fasteners are sized by:
- A) Liters
- D) Centimeters
- B) Hectares
- E) Inches
- C) Millimeters
- F) None of these

3 When working on a Ford Pinto you may find metric fasteners, where they appear, to be color coded. What color will they be?
- A) Silver
- D) Blue
- B) Gray
- E) Green
- C) Red
- F) Black

4 Under the metric system you would express cylinder bore and stroke dimensions in units of:
- A) Dekameters
- D) Centiliters
- B) Amperes
- E) Quarts & ounces
- C) Millimeters
- F) Liters

5 A millimeter is a metric unit of:
- A) Area
- D) Volume
- B) Mass (weight)
- E) Length
- C) Acreage
- F) Width

6 If you were to divide a meter into 1000 parts, each part would be:
- A) a Centimeter
- D) a Millimeter
- B) a Deciliter
- E) a Kilometer
- C) a Decimeter
- F) None of these

7 We now use Fahrenheit as our temperature scale. In the metric system how would temperature be measured?
- A) Calculus
- D) Centripetal
- B) Celsius
- E) Joule
- C) Centimetric
- F) Centigrade

8 If two locations are 20,000 meters apart, how far apart are they when expressed in a smaller number?
- A) 200 millimeters
- D) 20.8 miles
- B) 200 yards
- E) 200,000 inches
- C) 2000 decimeters
- F) 20 kilometers

9 The metric measurement system is known as:
- A) ISO System
- D) The Metric System
- B) SI Units
- E) OFMS System
- C) NATCB System
- F) Conversion System

10 We now state engine displacement as cubic inches (CID). Under the metric standard how would engine displacement be stated?
- A) Cubic liters
- D) Cubic centimeters
- B) Square meters
- E) Centimeters
- C) Liters
- F) Millimeters

11 Torque factors are presently stated as either ft/lbs. or ft/ins. How would torque factors be expressed in metrics?
- A) Mass meters
- D) Foot meters
- B) Milliamperes
- E) Inch centimeters
- C) Newton meters
- F) Kilogram meters

12 One liter of pure water at a temperature of 4° Centigrade is 1000 cubic centimeters in volume. What measurement that we now use is this closest to?
- A) Pint
- D) 2 Quarts
- B) Gallon
- E) Quart
- C) Ounce
- F) 6 Pints

Answers 7B; 8F; 9B; 10D; 11C; 12E
1B; 2C; 3D; 4C; 5E; 6D;

ENGLISH TO METRIC CONVERSION: LIQUID CAPACITY

Liquid or fluid capacity is presently expressed as pints, quarts or gallons, or a combination of all of these. In the metric system the liter (l) will become the basic unit. Fractions of a liter would be expressed as deciliters, centiliters, or most frequently (and commonly) as milliliters.

To convert pints (pts.) to liters (l): multiply the number of pints by .47
To convert liters (l) to pints (pts.): multiply the number of liters by 2.1
To convert quarts (qts.) to liters (l): multiply the number of quarts by .95

To convert liters (l) to quarts (qts.): multiply the number of liters by 1.06
To convert gallons (gals.) to liters (l): multiply the number of gallons by 3.8
To convert liters (l) to gallons (gals.): multiply the number of liters by .26

gals	liters	qts	liters	pts	liters
0.1	0.38	0.1	0.10	0.1	0.05
0.2	0.76	0.2	0.19	0.2	0.10
0.3	1.1	0.3	0.28	0.3	0.14
0.4	1.5	0.4	0.38	0.4	0.19
0.5	1.9	0.5	0.47	0.5	0.24
0.6	2.3	0.6	0.57	0.6	0.28
0.7	2.6	0.7	0.66	0.7	0.33
0.8	3.0	0.8	0.76	0.8	0.38
0.9	3.4	0.9	0.85	0.9	0.43
1	3.8	1	1.0	1	0.5
2	7.6	2	1.9	2	1.0
3	11.4	3	2.8	3	1.4
4	15.1	4	3.8	4	1.9
5	18.9	5	4.7	5	2.4
6	22.7	6	5.7	6	2.8
7	26.5	7	6.6	7	3.3
8	30.3	8	7.6	8	3.8
9	34.1	9	8.5	9	4.3
10	37.8	10	9.5	10	4.7
11	41.6	11	10.4	11	5.2
12	45.4	12	11.4	12	5.7
13	49.2	13	12.3	13	6.2
14	53.0	14	13.2	14	6.6
15	56.8	15	14.2	15	7.1
16	60.6	16	15.1	16	7.6
17	64.3	17	16.1	17	8.0
18	68.1	18	17.0	18	8.5
19	71.9	19	18.0	19	9.0
20	75.7	20	18.9	20	9.5
21	79.5	21	19.9	21	9.9
22	83.2	22	20.8	22	10.4
23	87.0	23	21.8	23	10.9
24	90.8	24	22.7	24	11.4
25	94.6	25	23.6	25	11.8
26	98.4	26	24.6	26	12.3
27	102.2	27	25.5	27	12.8
28	106.0	28	26.5	28	13.2
29	110.0	29	27.4	29	13.7
30	113.5	30	28.4	30	14.2

ENGLISH TO METRIC CONVERSION: TEMPERATURE

To convert Fahrenheit (°F) to Celsius (°C): take number of °F and subtract 32; multiply result by 5; divide result by 9

To convert Celsius (°C) to Fahrenheit (°F): take number of °C and multiply by 9; divide result by 5; add 32 to total

Fahrenheit (F)		Celsius (C)		Fahrenheit (F)		Celsius (C)		Fahrenheit (F)		Celsius (C)	
°F	°C	°C	°F	°F	°C	°C	°F	°F	°C	°C	°F
−40	−40	−38	−36.4	80	26.7	18	64.4	215	101.7	80	176
−35	−37.2	−36	−32.8	85	29.4	20	68	220	104.4	85	185
−30	−34.4	−34	−29.2	90	32.2	22	71.6	225	107.2	90	194
−25	−31.7	−32	−25.6	95	35.0	24	75.2	230	110.0	95	202
−20	−28.9	−30	−22	100	37.8	26	78.8	235	112.8	100	212
−15	−26.1	−28	−18.4	105	40.6	28	82.4	240	115.6	105	221
−10	−23.3	−26	−14.8	110	43.3	30	86	245	118.3	110	230
−5	−20.6	−24	−11.2	115	46.1	32	89.6	250	121.1	115	239
0	−17.8	−22	−7.6	120	48.9	34	93.2	255	123.9	120	248
1	−17.2	−20	−4	125	51.7	36	96.8	260	126.6	125	257
2	−16.7	−18	−0.4	130	54.4	38	100.4	265	129.4	130	266
3	−16.1	−16	3.2	135	57.2	40	104	270	132.2	135	275
4	−15.6	−14	6.8	140	60.0	42	107.6	275	135.0	140	284
5	−15.0	−12	10.4	145	62.8	44	112.2	280	137.8	145	293
10	−12.2	−10	14	150	65.6	46	114.8	285	140.6	150	302
15	−9.4	−8	17.6	155	68.3	48	118.4	290	143.3	155	311
20	−6.7	−6	21.2	160	71.1	50	122	295	146.1	160	320
25	−3.9	−4	24.8	165	73.9	52	125.6	300	148.9	165	329
30	−1.1	−2	28.4	170	76.7	54	129.2	305	151.7	170	338
35	1.7	0	32	175	79.4	56	132.8	310	154.4	175	347
40	4.4	2	35.6	180	82.2	58	136.4	315	157.2	180	356
45	7.2	4	39.2	185	85.0	60	140	320	160.0	185	365
50	10.0	6	42.8	190	87.8	62	143.6	325	162.8	190	374
55	12.8	8	46.4	195	90.6	64	147.2	330	165.6	195	383
60	15.6	10	50	200	93.3	66	150.8	335	168.3	200	392
65	18.3	12	53.6	205	96.1	68	154.4	340	171.1	205	401
70	21.1	14	57.2	210	98.9	70	158	345	173.9	210	410
75	23.9	16	60.8	212	100.0	75	167	350	176.7	215	414

ENGLISH TO METRIC CONVERSION: WEIGHT (MASS)

Current weight measurement is expressed in pounds and ounces (lbs. & ozs.). The metric unit of weight (or mass) is the kilogram (kg). Even although this table does not show conversion of weights (masses) larger than 15 lbs, it is easy to calculate larger units by following the data immediately below.

To convert ounces (oz.) to grams (g): multiply th number of ozs. by 28
To convert grams (g) to ounces (oz.): multiply the number of grams by .035

To convert pounds (lbs.) to kilograms (kg): multiply the number of lbs. by .45
To convert kilograms (kg) to pounds (lbs.): multiply the number of kilograms by 2.2

lbs	kg	lbs	kg	oz	kg	oz	kg
0.1	0.04	0.9	0.41	0.1	0.003	0.9	0.024
0.2	0.09	1	0.4	0.2	0.005	1	0.03
0.3	0.14	2	0.9	0.3	0.008	2	0.06
0.4	0.18	3	1.4	0.4	0.011	3	0.08
0.5	0.23	4	1.8	0.5	0.014	4	0.11
0.6	0.27	5	2.3	0.6	0.017	5	0.14
0.7	0.32	10	4.5	0.7	0.020	10	0.28
0.8	0.36	15	6.8	0.8	0.023	15	0.42

ENGLISH TO METRIC CONVERSION: TORQUE   FT./LBS.

Torque is now expressed as either foot-pounds (ft./lbs.) or inch-pounds (in./lbs.). The metric measurement unit for torque is the Newton-meter (Nm). This unit—the Nm—will be used for all SI metric torque references, both the present ft./lbs. and in./lbs.

ft lbs	N-m	ft lbs	N-m	ft lbs	N-m	ft lbs	N-m
0.1	0.1	33	44.7	74	100.3	115	155.9
0.2	0.3	34	46.1	75	101.7	116	157.3
0.3	0.4	35	47.4	76	103.0	117	158.6
0.4	0.5	36	48.8	77	104.4	118	160.0
0.5	0.7	37	50.7	78	105.8	119	161.3
0.6	0.8	38	51.5	79	107.1	120	162.7
0.7	1.0	39	52.9	80	108.5	121	164.0
0.8	1.1	40	54.2	81	109.8	122	165.4
0.9	1.2	41	55.6	82	111.2	123	166.8
1	1.3	42	56.9	83	112.5	124	168.1
2	2.7	43	58.3	84	113.9	125	169.5
3	4.1	44	59.7	85	115.2	126	170.8
4	5.4	45	61.0	86	116.6	127	172.2
5	6.8	46	62.4	87	118.0	128	173.5
6	8.1	47	63.7	88	119.3	129	174.9
7	9.5	48	65.1	89	120.7	130	176.2
8	10.8	49	66.4	90	122.0	131	177.6
9	12.2	50	67.8	91	123.4	132	179.0
10	13.6	51	69.2	92	124.7	133	180.3
11	14.9	52	70.5	93	126.1	134	181.7
12	16.3	53	71.9	94	127.4	135	183.0
13	17.6	54	73.2	95	128.8	136	184.4
14	18.9	55	74.6	96	130.2	137	185.7
15	20.3	56	75.9	97	131.5	138	187.1
16	21.7	57	77.3	98	132.9	139	188.5
17	23.0	58	78.6	99	134.2	140	189.8
18	24.4	59	80.0	100	135.6	141	191.2
19	25.8	60	81.4	101	136.9	142	192.5
20	27.1	61	82.7	102	138.3	143	193.9
21	28.5	62	84.1	103	139.6	144	195.2
22	29.8	63	85.4	104	141.0	145	196.6
23	31.2	64	86.8	105	142.4	146	198.0
24	32.5	65	88.1	106	143.7	147	199.3
25	33.9	66	89.5	107	145.1	148	200.7
26	35.2	67	90.8	108	146.4	149	202.0
27	36.6	68	92.2	109	147.8	150	203.4
28	38.0	69	93.6	110	149.1	151	204.7
29	39.3	70	94.9	111	150.5	152	206.1
30	40.7	71	96.3	112	151.8	153	207.4
31	42.0	72	97.6	113	153.2	154	208.8
32	43.4	73	99.0	114	154.6	155	210.2

ENGLISH TO METRIC CONVERSION: TORQUE   IN./LBS.

To convert foot-pounds (ft./lbs.) to Newton-meters: multiply the number of ft./lbs. by 1.3

To convert inch-pounds (in./lbs.) to Newton-meters: multiply the number of in./lbs. by .11

in lbs	N-m	in lbs	N-m	in lbs	N-m	in lbs	N-m	in lbs	N-m
0.1	0.01	1	0.11	10	1.13	19	2.15	28	3.16
0.2	0.02	2	0.23	11	1.24	20	2.26	29	3.28
0.3	0.03	3	0.34	12	1.36	21	2.37	30	3.39
0.4	0.04	4	0.45	13	1.47	22	2.49	31	3.50
0.5	0.06	5	0.56	14	1.58	23	2.60	32	3.62
0.6	0.07	6	0.68	15	1.70	24	2.71	33	3.73
0.7	0.08	7	0.78	16	1.81	25	2.82	34	3.84
0.8	0.09	8	0.90	17	1.92	26	2.94	35	3.95
0.9	0.10	9	1.02	18	2.03	27	3.05	36	4.07

ENGLISH TO METRIC CONVERSION: LENGTH

To convert inches (ins.) to millimeters (mm): multiply number of inches by 25.4

To convert millimeters (mm) to inches (ins.): multiply number of millimeters by .04

Inches		Decimals	Milli-meters	Inches to millimeters		Inches		Decimals	Milli-meters	Inches to millimeters	
				inches	mm					inches	mm
	1/64	0.051625	0.3969	0.0001	0.00254		33/64	0.515625	13.0969	0.6	15.24
1/32		0.03125	0.7937	0.0002	0.00508	17/32		0.53125	13.4937	0.7	17.78
	3/64	0.046875	1.1906	0.0003	0.00762		35/64	0.546875	13.8906	0.8	20.32
1/16		0.0625	1.5875	0.0004	0.01016	9/16		0.5625	14.2875	0.9	22.86
	5/64	0.078125	1.9844	0.0005	0.01270		37/64	0.578125	14.6844	1	25.4
3/32		0.09375	2.3812	0.0006	0.01524	19/32		0.59375	15.0812	2	50.8
	7/64	0.109375	2.7781	0.0007	0.01778		39/64	0.609375	15.4781	3	76.2
1/8		0.125	3.1750	0.0008	0.02032	5/8		0.625	15.8750	4	101.6
	9/64	0.140625	3.5719	0.0009	0.02286		41/64	0.640625	16.2719	5	127.0
5/32		0.15625	3.9687	0.001	0.0254	21/32		0.65625	16.6687	6	152.4
	11/64	0.171875	4.3656	0.002	0.0508		43/64	0.671875	17.0656	7	177.8
3/16		0.1875	4.7625	0.003	0.0762	11/16		0.6875	17.4625	8	203.2
	13/64	0.203125	5.1594	0.004	0.1016		45/64	0.703125	17.8594	9	228.6
7/32		0.21875	5.5562	0.005	0.1270	23/32		0.71875	18.2562	10	254.0
	15/64	0.234375	5.9531	0.006	0.1524		47/64	0.734375	18.6531	11	279.4
1/4		0.25	6.3500	0.007	0.1778	3/4		0.75	19.0500	12	304.8
	17/64	0.265625	6.7469	0.008	0.2032		49/64	0.765625	19.4469	13	330.2
9/32		0.28125	7.1437	0.009	0.2286	25/32		0.78125	19.8437	14	355.6
	19/64	0.296875	7.5406	0.01	0.254		51/64	0.796875	20.2406	15	381.0
5/16		0.3125	7.9375	0.02	0.508	13/16		0.8125	20.6375	16	406.4
	21/64	0.328125	8.3344	0.03	0.762		53/64	0.828125	21.0344	17	431.8
11/32		0.34375	8.7312	0.04	1.016	27/32		0.84375	21.4312	18	457.2
	23/64	0.359375	9.1281	0.05	1.270		55/64	0.859375	21.8281	19	482.6
3/8		0.375	9.5250	0.06	1.524	7/8		0.875	22.2250	20	508.0
	25/64	0.390625	9.9219	0.07	1.778		57/64	0.890625	22.6219	21	533.4
13/32		0.40625	10.3187	0.08	2.032	29/32		0.90625	23.0187	22	558.8
	27/64	0.421875	10.7156	0.09	2.286		59/64	0.921875	23.4156	23	584.2
7/16		0.4375	11.1125	0.1	2.54	15/16		0.9375	23.8125	24	609.6
	29/64	0.453125	11.5094	0.2	5.08		61/64	0.953125	24.2094	25	635.0
15/32		0.46875	11.9062	0.3	7.62	31/32		0.96875	24.6062	26	660.4
	31/64	0.484375	12.3031	0.4	10.16		63/64	0.984375	25.0031	27	690.6
1/2		0.5	12.7000	0.5	12.70						

ENGLISH TO METRIC CONVERSION: FORCE

Force is presently measured in pounds (lbs.). This type of measurement is used to measure spring pressure, specifically how many pounds it takes to compress a spring. Our present force unit (the pound) will be replaced in SI metric measurements by the Newton (N). This term will eventually see use in specifications for electric motor brush spring pressures, valve spring pressures, etc.

To convert pounds (lbs.) to Newton (N): multiply the number of lbs. by 4.45

lbs	N	lbs	N	lbs	N	oz	N
0.01	0.04	21	93.4	59	262.4	1	0.3
0.02	0.09	22	97.9	60	266.9	2	0.6
0.03	0.13	23	102.3	61	271.3	3	0.8
0.04	0.18	24	106.8	62	275.8	4	1.1
0.05	0.22	25	111.2	63	280.2	5	1.4
0.06	0.27	26	115.6	64	284.6	6	1.7
0.07	0.31	27	120.1	65	289.1	7	2.0
0.08	0.36	28	124.6	66	293.6	8	2.2
0.09	0.40	29	129.0	67	298.0	9	2.5
0.1	0.4	30	133.4	68	302.5	10	2.8
0.2	0.9	31	137.9	69	306.9	11	3.1
0.3	1.3	32	142.3	70	311.4	12	3.3
0.4	1.8	33	146.8	71	315.8	13	3.6
0.5	2.2	34	151.2	72	320.3	14	3.9
0.6	2.7	35	155.7	73	324.7	15	4.2
0.7	3.1	36	160.1	74	329.2	16	4.4
0.8	3.6	37	164.6	75	333.6	17	4.7
0.9	4.0	38	169.0	76	338.1	18	5.0
1	4.4	39	173.5	77	342.5	19	5.3
2	8.9	40	177.9	78	347.0	20	5.6
3	13.4	41	182.4	79	351.4	21	5.8
4	17.8	42	186.8	80	355.9	22	6.1
5	22.2	43	191.3	81	360.3	23	6.4
6	26.7	44	195.7	82	364.8	24	6.7
7	31.1	45	200.2	83	369.2	25	7.0
8	35.6	46	204.6	84	373.6	26	7.2
9	40.0	47	209.1	85	378.1	27	7.5
10	44.5	48	213.5	86	382.6	28	7.8
11	48.9	49	218.0	87	387.0	29	8.1
12	53.4	50	224.4	88	391.4	30	8.3
13	57.8	51	226.9	89	395.9	31	8.6
14	62.3	52	231.3	90	400.3	32	8.9
15	66.7	53	235.8	91	404.8	33	9.2
16	71.2	54	240.2	92	409.2	34	9.4
17	75.6	55	244.6	93	413.7	35	9.7
18	80.1	56	249.1	94	418.1	36	10.0
19	84.5	57	253.6	95	422.6	37	10.3
20	89.0	58	258.0	96	427.0	38	10.6